BIOLOGY

TENTH EDITION

ELDRA P. SOLOMON
former affiliations:
University of South Florida, Tampa
Hillsborough Community College

CHARLES E. MARTIN
Professor Emeritus, *Rutgers University*

DIANA W. MARTIN
Professor Emeritus, *Rutgers University*

LINDA R. BERG
former affiliations:
University of Maryland, College Park
St. Petersburg College

CENGAGE
Learning·

Australia • Brazil • Mexico • Singapore • United Kingdom • United States

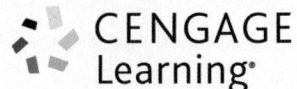
CENGAGE
Learning®

Biology, Tenth Edition
Eldra P. Solomon, Charles E. Martin,
Diana W. Martin, Linda R. Berg

Senior Product Team Manager: Yolanda Cossio

Content Developer: Suzannah Alexander

Content Coordinator: Kellie Petruzzelli

Product Assistant: Victor Luu

Media Developer: Lauren Oliveira

Content Project Manager: H. P. Humphrey

Senior Art Director: John Walker

Manufacturing Planner: Karen Hunt

Rights Acquisitions Specialist: Don Schlotman

Production Service: Whitney Thompson, Lachina

Photo Researcher: Jill Reichenbach, Q2A Bill
Smith

Text Researcher: Jill Krupnik, Q2A Bill Smith

Copy Editor: Kathleen Lafferty

Illustrators: Lachina, Precision Graphics,
Dragonfly Media Group

Text Designer: Jeanne Calabrese

Cover Designer: John Walker

Compositor: Lachina

Cover Image: Tube anemone (*Cerianthus* sp.) showing green fluorescence. The tube anemone is a solitary anthozoan that lives in a long tube that it constructs from sand and mucus. When attacked, this animal can retract rapidly into its tube. Tube anemones feed on plankton, small fish, and crustaceans. They capture and paralyze prey using their tentacles, which are equipped with stinging cells. Photographed in Alor, Indonesia. © altrendo nature/Getty Images

For product information and technology assistance, contact us at
Cengage Learning Customer & Sales Support, 1-800-354-9706.

For permission to use material from this text or product,
submit all requests online at **www.cengage.com/permissions.**
Further permissions questions can be e-mailed to
permissionrequest@cengage.com.

Library of Congress Control Number: 2013952376

College Edition:
ISBN-13: 978-1-285-42358-6
ISBN-10: 1-285-42358-5

Cengage Learning
200 First Stamford Place, 4th Floor
Stamford, CT 06902
USA

Cengage Learning is a leading provider of customized learning solutions with office locations around the globe, including Singapore, the United Kingdom, Australia, Mexico, Brazil, and Japan. Locate your local office at **www.cengage.com/global.**

Cengage Learning products are represented in Canada by Nelson Education, Ltd.

To learn more about Cengage Learning Solutions, visit **www.cengage.com.**

Purchase any of our products at your local college store or at our preferred online store **www.cengagebrain.com.**

Printed in Canada
4 5 17 16 15

To our families, friends, and colleagues who gave freely of their love, support, knowledge, and time as we prepared this tenth edition of *Biology,* and in appreciation of all who teach and learn.

Especially to
My mother, Freda M. Brod, and to Kathleen, Mical, Karla, Amy, Belicia, and Neal

Professors Emeritus A. Gib DeBusk and Guy A. Thompson Jr.

Alan and Jennifer

About the Authors

Eldra P. Solomon has written several leading college textbooks in biology and in human anatomy and physiology. Her books have been translated into more than ten languages. She earned an M.S. from the University of Florida and an M.A. and Ph.D. from the University of South Florida. Dr. Solomon taught biology and nursing students for more than 20 years.

In addition to being a biologist and science author, Dr. Solomon is a biopsychologist with a special interest in the neurophysiology of traumatic experience. Her research has focused on the neurological, endocrine, and psychological effects of trauma, including complex posttraumatic stress disorder and development of maladaptive coping strategies.

Dr. Solomon has presented her research at numerous national and international conferences, and her work has been published in leading professional journals. She has been profiled more than 30 times in leading publications, including *Who's Who in America, Who's Who in Science and Engineering, Who's Who in Medicine and Healthcare, Who's Who in American Education, Who's Who of American Women,* and *Who's Who in the World.*

Charles E. Martin is professor emeritus of cell biology and neuroscience at Rutgers University. He received his Ph.D. in genetics from Florida State University and engaged in postdoctoral research in genetics and membrane biology at the University of Texas at Austin. He has taught general biology as well as undergraduate and graduate level courses in genetics and molecular cell biology throughout his career at Rutgers. An award-winning teacher for more than 30 years, in 2011 Dr. Martin was named Professor of the Year by the Molecular Biosciences Graduate Student Association.

His research on gene regulation of membrane protein enzyme systems in yeast and other fungi illustrates the interdisciplinary nature of the life sciences. He is most proud of the many generations of undergraduate, graduate, and postdoctoral students who contributed to this research and have gone on to productive careers. He continues to be committed to teaching and is grateful for the opportunities to pursue a teaching and research career in what continues to be the most exciting era of the biological sciences.

Diana W. Martin is professor emeritus and former director of general biology in the Division of Life Sciences at Rutgers University. Dr. Martin received an M.S. from Florida State University, where she studied the chromosomes of related plant species to understand their evolutionary relationships. She earned a Ph.D. from the University of Texas at Austin, where she studied the genetics of the fruit fly, *Drosophila melanogaster,* and then conducted postdoctoral research at Princeton University.

Dr. Martin taught general biology and other courses at Rutgers for more than 30 years and has been involved in writing textbooks since 1988. She is immensely grateful that her decision to study biology in college has led to a career that allows her many ways to share her excitement about all aspects of biology.

Linda R. Berg is an award-winning teacher and textbook author. She received a B.S. in science education, an M.S. in botany, and a Ph.D. in plant physiology from the University of Maryland. Her research focused on the evolutionary implications of steroid biosynthetic pathways in various organisms.

Dr. Berg taught at the University of Maryland at College Park for 17 years and at St. Petersburg College in Florida for 8 years. During her career, she taught introductory courses in biology, botany, and environmental science to thousands of students. At the University of Maryland, she received numerous teaching and service awards. Dr. Berg is also the recipient of many national and regional awards, including the National Science Teachers Association Award for Innovations in College Science Teaching, the Nation's Capital Area Disabled Student Services Award, and the Washington Academy of Sciences Award in University Science Teaching.

During her career as a professional science writer, Dr. Berg has authored or coauthored several leading college science textbooks. Her writing reflects her teaching style and love of science.

Brief Contents

Table of Contents

Preface

This tenth edition of Solomon, Martin, Martin, and Berg's *Biology* conveys our vision of the dynamic science of biology and how it affects every aspect of our lives, from our own health and behavior to the challenging global environmental issues that confront us. New discoveries in the biological sciences continue to increase our understanding of both the unity and diversity of life's processes and adaptations. With this understanding, we become ever more aware of our interdependence with the vast diversity of organisms with which we share planet Earth.

BIOLOGY: THE STUDENT-FRIENDLY BIOLOGY BOOK

We want beginning students to experience learning biology as an exciting journey of discovery. In the tenth edition of *Biology,* we explore Earth's diverse organisms, their remarkable adaptations to the environment, and their evolutionary and ecological relationships. We present the workings of science and the contributions of scientists whose discoveries not only expand our knowledge of biology but also help shape and protect the future of our planet. *Biology* provides insight into what science is, how scientists work, what scientists have contributed, and how scientific knowledge affects daily life.

Since the first edition of *Biology,* we have worked very hard to present the principles of biology in an integrated way that is accurate, interesting, and conceptually accessible to students. In this tenth edition of *Biology,* we continue this tradition. We also continue to present biology in an inquiry-based framework. Some professors interpret inquiry as a learning method that takes place in the laboratory as students perform experiments. Laboratory research is certainly an integral part of inquiry-based learning, but inquiry is also a way of learning in which the student actively pursues knowledge outside the laboratory. In *Biology* we have always presented the history of scientific advances, including scientific debates, to help students understand that science is a process—that is, a field of investigative inquiry—as well as a body of knowledge, the product of inquiry. In the tenth edition of *Biology,* we make a concerted effort to further integrate inquiry-based learning into the textbook with the introduction of new features and the expansion of several others (discussed in the following sections).

Throughout the text we stimulate interest by relating concepts to experiences within the student's frame of reference. By helping students make such connections, we facilitate their mastery of general concepts. We hope the combined effect of an engaging writing style and interesting features will motivate and excite students in their study of biology.

THE SOLOMON/MARTIN/MARTIN/BERG LEARNING SYSTEM

In the tenth edition, we have continued to refine our highly successful *Learning System.* This system provides the student with the learning strategies needed to integrate biological concepts and demonstrate mastery of these concepts. Learning biology is challenging because the subject of biology is filled with so many new terms and so many facts that must be integrated into the framework of general biological principles. To help students focus on important principles and concepts, we provide *Learning Outcomes* for the course and *Learning Objectives* for each major section of every chapter. At the end of each section, we provide *Checkpoint* questions based on the *Learning Objectives* so that students can assess their level of understanding of the material presented in the section. At the end of each chapter, we include a *Summary: Focus on Learning Objectives* that is organized around the *Learning Objectives* and emphasizes key terms in context. The *Summary* is followed by *Test Your Understanding,* a set of questions organized according to Bloom's taxonomy. Questions include *Know and Comprehend* multiple-choice exercises as well as a variety of questions that encourage the student to *Apply and Analyze* and *Evaluate and Synthesize* the topics in the chapter.

Students are directed to **www.cengagebrain.com,** a powerful online tool that offers access to course materials such as *Aplia for Biology* and other companion resources. See the Resources for Students section of the Preface for details.

Pedagogical Features

Our *Learning System* includes numerous learning strategies that help students increase their success:

- *NEW* An updated and expanded *art program* reinforces concepts discussed in the text and presents complex processes in clear steps. This edition expands the number of *Key Experiment* figures, which encourage students to evaluate investigative approaches that scientists have taken. *Key Experiment* figures emphasize the scientific process in both classic and modern research; Figure 4-12 is a new example. Also included in this edition are newly designed *Key Point* figures, in which important concepts are stated in process diagrams of complex topics; new examples include Figures 4-11 and 4-15. Many of the *Key Point* figures have numbered parts that show sequences of events in biological processes or life cycles.

- Numerous photographs, both alone and combined with line art, help students grasp concepts by connecting the "real" to the "ideal." The line art uses features such as *orientation icons* to help students put the detailed figures into the broader context. We use symbols and colors consistently throughout the book to help students connect concepts. For example, the same four colors and shapes are used throughout the book to identify guanine, cytosine, adenine, and thymine. Similarly, the same colors are used consistently in illustrations and tables to indicate specific clades of organisms. *Research Method* figures describe why biologists use a particular method and explain how the method is executed. New examples include Figures 4-7 and 15-7.

- *NEW* Many questions have been added, and several types of questions carry special designations: *Predict; Connect; Visualize; Evolution Link; Interpret Data;* or *Science, Technology, and Society.* These questions emphasize that learning is enhanced by many diverse approaches.

- *Inquiring About* boxes explore issues of special relevance to students, such as the effects of smoking, how traumatic experiences affect the body, and breast cancer. These boxes also provide a forum for discussing some interesting topics in more detail, such as the smallest ancient humans, ancient plants and coal formation, hydrothermal vent communities, declining amphibian populations, and stratospheric ozone depletion.

- A list of *Key Concepts* at the beginning of each chapter provides a chapter overview and helps the student focus on important principles discussed in the chapter.

- *Learning Objectives* at the beginning of each major section in the chapter indicate, in behavioral terms, what the student must do to demonstrate mastery of the material in that section.

- Each major section of the chapter is followed by a series of *Checkpoint* questions that assess comprehension by asking the student to describe, explain, compare, contrast, or illustrate important concepts. The *Checkpoint* questions are based on the section *Learning Objectives.*

- *Concept Statement Subheads* introduce sections, previewing and summarizing the key idea or ideas to be discussed in that section.

- *Sequence Summaries* within the text simplify and summarize information presented in paragraph form. For example, paragraphs describing blood circulation through the body or the steps by which cells take in certain materials

are followed by a *Sequence Summary* listing the sequence of structures or steps.

- Numerous *tables,* many illustrated, help the student organize and summarize material presented in the text. Many tables are color-coded.

- A *Summary: Focus on Learning Objectives* at the end of each chapter is organized around the chapter *Learning Objectives.* This summary provides a review of the material, and because selected key terms are boldfaced in the summary, students learn vocabulary words within the context of related concepts.

- *NEW Test Your Understanding* end-of-chapter questions are now organized according to Bloom's taxonomy, providing students with the opportunity to evaluate their understanding of the material in the chapter. *Know and Comprehend* multiple-choice questions reinforce important terms and concepts. *Apply and Analyze* questions challenge students to integrate their knowledge. Higher-level *Evaluate and Synthesize* questions encourage students to apply the concepts just learned to new situations or to make connections among important concepts. Each chapter has one or more *Evolution Link* questions, and many chapters contain one or more *Interpret Data* questions that require students to actively interpret experimental data presented in the chapter. Also included are *Predict, Connect, Visualize,* and *Science, Technology, and Society* questions. Answers to the *Test Your Understanding* questions are provided in Appendix E.

- The *Glossary* at the end of the book, the most comprehensive glossary found in any biology text, provides precise definitions of terms. The *Glossary* is especially useful because it is extensively cross-referenced and includes pronunciations for many terms. The vertical green bar along the margin facilitates rapid access to the *Glossary.* The companion website also includes glossary flash cards with pronunciations.

Course Learning Outcomes

At the end of a successful study of introductory biology, the student can demonstrate mastery of biological concepts by responding accurately to the following *Course Learning Outcomes:*

- Design an experiment to test a given hypothesis, using the procedure and terminology of the scientific method.

- Cite the cell theory and relate the structure of organelles to their functions in both prokaryotic and eukaryotic cells.

- Describe the mechanisms of evolution, explain why evolution is the principal unifying concept in biology, and discuss natural selection as the primary agent of evolutionary change.

- Explain the role of genetic information in all species and discuss applications of genetics that affect society.

- Describe several mechanisms by which cells and organisms transfer information, including the use of nucleic acids in genetic transmission of information, signal transduction, chemical signals (such as hormones and pheromones), electrical signals (such as neural transmission), sounds, and visual displays.

- Provide examples (at various levels of complexity) of interactions among biological systems that illustrate the interdependence of these systems.

- Explain how any given structure is related to its function.

- Argue for or against the classification of organisms in three domains and several kingdoms or supergroups, characterizing each of these clades; based on your knowledge of genetics and evolution, give specific examples of the unity and diversity of organisms in different domains and supergroups.

- Compare the structural adaptations, life processes, and life cycles of a prokaryote, protist, fungus, plant, and animal.

- Define *homeostasis* and give examples of regulatory mechanisms, including feedback systems.

- Trace the flow of matter and energy through a photosynthetic cell and a nonphotosynthetic cell and through the biosphere, comparing the roles of producers, consumers, and decomposers.

- Describe the study of ecology at the levels of an individual organism, a population, a community, and an ecosystem.

WHAT'S NEW: AN OVERVIEW OF *BIOLOGY*, TENTH EDITION

Five themes are interwoven throughout *Biology*: the evolution of life, the transmission of biological information, the flow of energy through living systems, interactions among biological systems, and the inter-relationship of structure and function. As we introduce the concepts of modern biology, we explain how these themes are connected and how life depends on them.

Educators present the major topics of an introductory biology course in a variety of orders. For this reason, we carefully designed the eight parts of this book so that they do not depend heavily on preceding chapters and parts. This flexible organization means that an instructor can present the 57 chapters in any number of sequences with pedagogical success. Chapter 1, which introduces the student to the major principles of biology, provides a comprehensive springboard for future discussions, whether the professor prefers a "top-down" or "bottom-up" approach.

In this edition as in previous editions, we examined every line of every chapter for accuracy and currency, and we made a careful attempt to update every topic and verify all new material. Our efforts have been enhanced by an updated art program with many new illustrations. The following brief summary provides a general overview of the organization of *Biology* and some changes made to the tenth edition.

Part 1 The Organization of Life

The six chapters that make up Part 1 provide basic principles of biology and the concepts of chemistry and cell biology that lay the foundation upon which the remaining parts of the book build. We begin Chapter 1 with a discussion of the promise and challenges of stem cell research. We then introduce the main themes of the book: evolution, information transfer, energy transfer, interactions in biological systems, and the inter-relationship of structure and function. Chapter 1 examines several fundamental concepts in biology and the nature of the scientific process, including a discussion of systems biology. Chapters 2 and 3, which focus on the molecular level of organization, establish the foundations in chemistry necessary for understanding biological processes. Chapters 4, 5, and 6 focus on the cellular level of organization, including cell structure and function, cell membranes, and cell signaling. We have revised these chapters to place greater emphasis on the interdisciplinary nature of cell research and have expanded coverage of transport between the nucleus and cytoplasm as well as the routing of proteins through the endomembrane system.

Part 2 Energy Transfer Through Living Systems

Because all living cells need energy for life processes, the flow of energy through living systems—that is, capturing energy and converting it to usable forms—is a basic theme of *Biology*. Chapter 7 examines how cells capture, transfer, store, and use energy. Chapters 8 and 9 discuss the metabolic adaptations by which organisms obtain and use energy through cellular respiration and photosynthesis.

Part 3 The Continuity of Life: Genetics

We have updated and expanded the eight chapters of Part 3 for the tenth edition. We begin this unit by discussing mitosis and meiosis in Chapter 10. Chapter 11 builds on this foundation as it considers Mendelian genetics and related patterns of inheritance. We then turn our attention to the structure and replication of DNA in Chapter 12. The discussion of RNA and protein synthesis in Chapter 13 includes new insights into how the small percentage of DNA that codes for polypeptides relates to the much larger percentage of the genome that is expressed. We introduce new information derived from the ENCODE project establishing that much of the genome encodes different classes of non-protein-coding RNAs, including microRNAs and long noncoding RNAs. The newly discovered regulatory functions of these RNAs are further explored in Chapter 14, which also includes new information on eukaryotic promoters, enhancers, and silencers as well as on epigenetic inheritance. In Chapter 15 we focus on DNA technology and genomics, including an expanded discussion of rapid DNA sequencing, as well as the importance of gene databases as tools for understanding gene regulation, gene functions, and molecular evolution. These chapters build the necessary foundation for exploring human genetics and the human genome in Chapter 16, which includes new sections on genomic imprinting and on genome-wide association studies. In Chapter 17 we introduce the role of genes in development, emphasizing studies on specific model organisms that have led to spectacular advances in this field; changes include new material on induced pluripotent stem cells as well as a comprehensive view of cancer and its relationship to cell signaling that has developed through the application of genome-wide association studies and whole genome sequencing.

Part 4 The Continuity of Life: Evolution

Although we explore evolution as the cornerstone of biology throughout the book, Part 4 discusses evolutionary concepts in depth. We provide the history behind the discovery of the scientific theory of evolution, the mechanisms by which it occurs, and the methods by which it is studied and tested. Chapter 18 introduces the Darwinian concept of evolution and presents several kinds of evidence that support the scientific theory of evolution. In Chapter 19 we examine evolution at the population level. Chapter 20 describes the evolution of new species and discusses aspects of macroevolution. Chapter 21 summarizes the evolutionary history of life on Earth. In Chapter 22 we recount the evolution of primates, including humans. New molecular and fossil findings, including those relating to recently discovered human relatives such as the Denisovans (a sister species to the Neandertals) and *Australopithecus sediba*, are explored.

Part 5 The Diversity of Life

Emphasizing the cladistic approach, we use an evolutionary framework to discuss each group of organisms. We present current hypotheses of how groups of organisms are related. Chapter 23 has been updated to reflect the effect of recent research on systematics. In this chapter we discuss *why* organisms are classified and provide insight into the scientific process of deciding *how* they are classified. New advances have enabled us to further clarify the connection between evolutionary history and systematics in the tenth edition. Chapter 24 focuses entirely on viruses and subviral agents. Information has been updated and expanded on giant viruses, viral origins, evolutionary

importance of viruses, and recent research on viruses. Chapter 25 is devoted to the prokaryotes, both bacteria and archaea. Information about the evolution, structure, ecology, and phylogeny of archaea has been expanded. Implications of research on the human microbiome are discussed and discussion of antibiotic resistance has been expanded. Chapter 26 describes the protists in the context of five "supergroups" of eukaryotes. Chapters 27 and 28 present the members of the plant kingdom. Chapter 27 considers the evolution of land plants and the evolution of seedless vascular plants. Discussion of the origin and early evolution of angiosperms is included in Chapter 28. Chapter 29 describes the fungi. In Chapters 30 through 32, we discuss the diversity of animals. We have updated the discussions of phylogenetic relationships to reflect recent research.

Part 6 Structure and Life Processes in Plants

Part 6 introduces students to the fascinating plant world. Here we stress relationships between structure and function in plant cells, tissues, organs, and individual organisms. In Chapter 33 we consider plant structure, growth, and differentiation in the context of cell division, cell expansion, cell differentiation, tissue culture, morphogenesis, pattern formation, positional information, and *Arabidopsis* mutants. Chapters 34 through 36 discuss the structural and physiological adaptations of leaves, stems, and roots; these chapters include special consideration of plant transport systems. Chapter 37 describes reproduction in flowering plants, including asexual reproduction, flowers, fruits, and seeds. Chapter 38 focuses on growth responses and regulation of growth, including the latest findings generated by the continuing explosion of knowledge in plant biology, particularly at the molecular level.

Part 7 Structure and Life Processes in Animals

In Part 7 we provide a strong emphasis on comparative animal physiology, showing the structural, functional, and behavioral adaptations that help animals meet environmental challenges. We use a comparative approach to examine how various animal groups have solved both similar and diverse problems. In Chapter 39 we discuss the basic tissues and organ systems of the animal body, homeostasis, and the ways that animals regulate their body temperature. Chapter 40 focuses on different types of body coverings, skeletons, and muscles, and discusses how they function. In Chapters 41 through 43, we discuss neural signaling, neural regulation, and sensory reception. In Chapters 44 through 51, we compare how different animal groups carry on life processes, such as internal transport, internal defense, gas exchange, digestion, reproduction, and development. Each chapter in this part considers the human adaptations for the life processes being discussed. Part 7 ends with a discussion of behavioral adaptations in Chapter 52. Reflecting recent research findings, we have updated or added new material on many topics, including neurotransmitters, cardiovascular disease, evolution of immunity in invertebrates, chronic inflammation, HIV, nutrition, regulation of appetite and energy metabolism, endocrine function, ovarian stem cells, contraception, sexually transmitted infections, and social learning and transmission of culture in vertebrates. The art program has been updated and improved, and new photographs and photomicrographs have been added.

Part 8 The Interactions of Life: Ecology

Part 8 focuses on the dynamics of populations, communities, and ecosystems and on the application of ecological principles to disciplines such as conservation biology. Chapters 53 through 56 give the student an understanding of the ecology of populations, communities, ecosystems, and the biosphere, and Chapter 57 focuses on global environmental issues. Among the many new and updated topics discussed in this unit are Antarctic tundra; the role of archaea in the carbon cycle, nitrogen cycle, and climate change; the Cross River gorilla (*Gorilla gorilla diehli*) as an example of a critically endangered species; new research on global climate change; updated information on stratospheric ozone depletion; and the effect of humans on the biosphere.

A COMPREHENSIVE PACKAGE FOR LEARNING AND TEACHING

A carefully designed supplement package is available to further facilitate learning. In addition to the usual print resources, we are pleased to present student multimedia tools that have been developed in conjunction with the text.

Resources for Students

MindTap, a fully online, highly personalized learning experience built on Cengage Learning content. MindTap combines student learning tools—readings, multimedia, activities, and assessments—into a singular Learning Path that guides students through their course. Instructors personalize the experience by customizing authoritative Cengage Learning content and learning tools, including the ability to add their own content in the Learning Path via apps that integrate into the MindTap framework seamlessly with Learning Management Systems.

MindTap for Biology is easy to use and saves instructors time by allowing them to:

- Seamlessly deliver appropriate content and technology assets from a number of providers to students, as they need them.

- Break course content down into movable objects to promote personalization, encourage interactivity, and ensure student engagement.

- Customize the course—from tools to text—and make adjustments "on the fly," making it possible to intertwine breaking news into their lessons and incorporate today's teachable moments.

- Bring interactivity into learning through the integration of multimedia assets.

- Track students' use, activities, and comprehension in real time, which provides opportunities for early intervention to influence progress and outcomes. Grades are visible and archived so that students and instructors always have access to current standings in the class.

Aplia offers a way to stay on top of coursework with regularly scheduled homework assignments. Interactive tools and additional content are provided to further increase engagement and understanding. Students, ask your instructor about Aplia!

Study Guide to accompany *Biology,* Tenth Edition, by Jennifer Aline Metzler of Ball State University and Robert Yost of Indiana University and Purdue University, Indianapolis. Updated for this edition, the study guide provides the student with many opportunities to review chapter concepts. Multiple-choice study questions, coloring-book exercises, vocabulary-building exercises, and many other types of active-learning tools are provided to suit different cognitive learning styles.

A Problem-based Guide to Basic Genetics by Donald Cronkite of Hope College. This brief guide provides students with a systematic approach to solving genetics problems along with numerous solved problems and practice problems.

Spanish Glossary. This Spanish glossary of biology terms is available to Spanish-speaking students.

Audio Study Tools. This tenth edition of *Biology* is accompanied by useful study tools, which contain valuable information such as reviews of important concepts, key terms, questions, and study tips. Students can download the audio study tools.

Virtual Biology Laboratory 4.0. Now with an upgraded user interface, these 14 online laboratory experiments allow students to "do" science by acquiring data, performing simulated experiments, and using data to explain biological concepts. Assigned activities automatically flow to the instructor's grade book. Self-designed activities ask students to plan their procedures around an experimental question and write up their results.

Additional Resources for Instructors

The instructors' examination copy for this edition lists a comprehensive package of print and multimedia supplements, including online resources, available to qualified adopters. Please ask your local sales representative for details.

Instructor Companion Site. Everything you need for your course in one place! This collection of book-specific lecture and class tools is available online via **www.cengage.com/login.** Access and download PowerPoint presentations, images, instructor's manual, videos, and more.

Cengage Learning Testing Powered by Cognero. A flexible, online system that allows you to import, edit, and manipulate test bank content from the test bank or elsewhere, including your own favorite test questions; create multiple test versions in an instant; and deliver tests from your LMS, your classroom, or wherever you want.

Aplia is a Cengage Learning online homework system dedicated to improving learning by increasing student effort and engagement. Aplia makes it easy for instructors to assign frequent online homework assignments. Aplia provides students with prompt and detailed feedback to help them learn as they work through the questions, and features interactive tutorials to fully engage them in learning course concepts. Automatic grading and powerful assessment tools give instructors real-time reports of student progress, participation, and performance, while Aplia's easy-to-use course management features let instructors flexibly administer course announcements and materials online. With Aplia, students will show up to class fully engaged and prepared, and instructors will have more time to do what they do best . . . teach.

Brooks/Cole Video Library (Featuring BBC Motion Gallery Video Clips). The Brooks/Cole Video Library contains many high-quality videos that can be used alongside the text. A wide range of video topics offer professors a great tool to engage students and help them connect the material to their lives outside of the classroom. Available on the Instructor Companion Site.

ACKNOWLEDGMENTS

The development and production of the tenth edition of *Biology* required extensive interaction and cooperation among the authors and many individuals in our family, social, and professional environments. We thank our editors, colleagues, students, family, and friends for their help and support. Preparing a book of this complexity is challenging and requires a cohesive, talented, and hardworking professional team. We appreciate the contributions of everyone on the editorial and production staff at Brooks/Cole–Cengage Learning who worked on this tenth edition of *Biology*. We thank our senior product team manager, Yolanda Cossio, for her commitment to *Biology* and for working closely with us throughout the entire process of development and production. We greatly appreciate the help of Suzannah Alexander, our very talented content developer, who was a critical part of our team. Suzannah expertly coordinated many aspects of this challenging project, including the complex new art rendered for this edition. She made herself available to advise and help us whenever we needed her, including late at night and during weekends.

We thank Tom Ziolkowski, our market development manager, and Nicole Hamm, our brand manager, whose expertise ensured that you would know about our new edition.

We appreciate the help of content project manager Hal Humphrey, who expertly guided overall production of the project.

We are grateful to product assistant Victor Luu for quickly providing us with resources whenever we needed them.

We thank creative director Rob Hugel, senior art director and cover designer John Walker, and text designer Jeanne Calabrese.

We appreciate the work of Lauren Oliveira, media developer, who coordinated the many high-tech components of the computerized aspects of our *Learning System*. We thank content coordinator Kellie Petruzzelli for coordinating the print supplements.

We are grateful to our production editor, Whitney Thompson of Lachina Publishing Services, for coordinating the many editors involved in the preparation of this edition and bringing together the thousands of complex pieces of the project that together produced *Biology,* Tenth Edition. We value the careful work of our copy editor, Kathleen Lafferty of Roaring Mountain Editorial Services, who helped us maintain consistency and improve the manuscript. We thank the artists at Lachina Publishing Services, Precision Graphics, and Dragonfly Media Group for greatly improving the art program for this book. We appreciate the efforts of photo research manager Jill Reichenbach of Bill Smith Group in helping us find excellent images. We appreciate the help, patience, and hard work of our production team. Our schedule for this project was very demanding. At times, it seemed like the team worked around the clock. When we sent e-mails late at night or during weekends, we often received immediate responses.

These dedicated professionals and many others on the Brooks/Cole team provided the skill, attention, patience, and good humor needed to produce *Biology,* Tenth Edition. We thank them for their help and support throughout this project.

We appreciate the help of obstetrician/gynecologist Dr. Amy Solomon for her input regarding the most recent information on pregnancy, childbirth, contraception, and sexually transmitted infections. We are grateful to Mical Solomon for his computer help. We thank Dr. David Axelrod for insightful discussions on the genetics and biology of cancer.

We thank our families and friends for their understanding, support, and encouragement as we struggled through many revisions and intense deadlines. We especially thank Dr. Kathleen M. Heide, Freda Brod, Alan Berg, Jennifer and Pat Roath, and Margaret Martin for their support and input.

Our colleagues and students who have used our book have provided valuable input by sharing their responses to past editions of *Biology*. We thank them and ask again for their comments and suggestions as they use this new edition. We can be reached via our website at **www.cengagebrain.com** or through our editors at Brooks/Cole, a division of Cengage Learning.

We greatly appreciate and want to acknowledge the participation and help of our contributors:

Peter K. Ducey
Professor and Department Chair
Biological Sciences Department
SUNY Cortland

Lois A. Ball
Meteorologist, Biologist, and Science Educator
University of South Florida

We express our thanks to the many biologists who have read the manuscript during various stages of its development and provided us with valuable suggestions for improving it. Tenth edition reviewers include the following:

Frank K. Ammer, Frostburg State University

Adébiyi Banjoko, Maricopa Community Colleges

Melissa Bartlett, Mohawk Valley Community College

Richard W. Cheney Jr., Christopher Newport University

Kendra Spence Cheruvelil, Michigan State University

Peter Ducey, SUNY Cortland

Cori Fata-Hartley, Michigan State University

Eric Green, Salt Lake Community College

Chris Haynes, Shelton State Community College

Jay Y. S. Hodgson, Armstrong Atlantic State University, Savannah, Georgia

Andrew J. Kreuz, Stevenson University

Gustave K. N. Mbuy, West Chester University of Pennsylvania

Jennifer A. Metzler, Ball State University

Jacalyn Newman, University of Pittsburgh

Ed Perry, Faulkner State Community College

Lori Rose, Hill College

Bruce Stallsmith, University of Alabama in Huntsville

Matt Williford, Faulkner State Community College

Robert Yost, Indiana University and Purdue University Indianapolis

We would also like to thank the hundreds of reviewers of previous editions, both professors and students, who are too numerous to mention. They asked thoughtful questions, provided new perspectives, offered alternative wordings to clarify difficult passages, and informed us of possible errors. We are truly indebted to their excellent feedback. Their suggestions have helped us improve each edition of *Biology*.

To the Student

We have learned a great deal from tens of thousands of students who have taken on the challenge of learning biology. Although they have varied in their life goals and academic preparation, most have found that they needed to modify their approach to learning to be successful.

You already know that memorization and cramming are unsuccessful, and you probably also know that many students fall back on these methods as default strategies. So, what really works?

Use the Wealth of Learning Aids That Accompany *Biology*

The *Learning System* we use in this book is described in the Preface. Using the strategies of the *Learning System* will help you master the language and concepts of biology. You will also want to use the many online tools available to *Biology* students. These tools, described in the Resources for Students section of the Preface, include *Aplia for Biology* and *MindTap* available at **www.cengagebrain.com.** In addition to these learning strategies, you can make the task of learning biology easier by using approaches that have been successful for a broad range of our students over the years.

Be Open to Many Learning Styles

There is a popular belief that each person has an innate "learning style" that is most successful for them. In fact, there is very little scientific evidence to support this view. What works will depend on the nature of the material being learned, and in most cases a mix of activities and a variety of sensory inputs will be most effective. *Biology* includes many kinds of questions to encourage you to think and learn in different ways. Make learning a part of your life as you think, listen, draw, write, argue, describe, speak, observe, explain, and experiment.

Know Your Professor's Expectations

Determine what your professor wants you to know and how your learning will be assessed. Some professors test almost exclusively on material covered in lecture. Others rely on their students' learning most of, or even all, the content assigned in chapters. Find out what your professor's requirements are because the way you study will vary accordingly.

If lectures are the main source of examination questions, make your lecture notes as complete and organized as possible. Before going to class, skim over the chapter, identifying key terms and examining the main figures, so that you can take effective lecture notes. Spend no more than 1 hour on this. Within 24 hours after class, type (or rewrite) your notes. Before typing them, however, read the notes and make marginal notes about anything that is not clear. Then read the corresponding material in your text. Do not copy the information; instead, process it and write out an explanation in your own words. Read the entire chapter, including parts that are not covered in lecture. This extra information will give you breadth of understanding and will help you grasp key concepts. In addition, you should make an effort to employ as many of the techniques described in the next paragraphs as possible.

If the assigned readings in the text are going to be tested, you must use your text intensively. After reading the chapter introduction, read the list of *Learning Objectives* for the first section. These objectives are written in behavioral terms; that is, they ask you to "do" something to demonstrate mastery. The objectives give you a concrete set of goals for each section of the chapter. At the end of each section, you will find *Checkpoint* questions keyed to the *Learning Objectives.* Carefully examine each figure, making certain that you understand what it is illustrating. Answer the question at the end of each *Key Point* figure and at the end of each *Key Experiment.*

Read each chapter section actively. Highlighting and underlining are not always active learning techniques; sometimes they postpone learning. ("This part is important; I'll learn it later.") An active learner always has questions in mind and is constantly making connections. For example, there are many processes that must be understood in biology. Don't try to blindly memorize them; instead, think about causes and effects so that every process becomes a story. Eventually, you'll see that many processes are connected by common elements.

To master the material, you will probably have to read each chapter more than once. Each time will be much easier than the previous time because you'll be reinforcing concepts that you have already partially learned.

Write a chapter outline and flesh out your outline by adding important concepts and boldface terms with definitions in your own words (not copied from the book or cut and pasted). Use this outline when preparing for the exam.

Now it is time to test yourself. Answer the *Test Your Understanding* questions (*Know and Comprehend, Apply and Analyze,* and *Evaluate and Synthesize*) at the end of the chapter. You will sharpen your thinking if you take the time to type or write out your answers. The answers are in Appendix E, but do not be too quick to check them. Think about them and discuss them with your fellow students if possible. Consider each question as a kind of springboard that leads to other questions. Finally, review the *Learning Objectives* in the *Summary* and try to answer them before reading the summary provided.

Learn the Vocabulary

One stumbling block for many students is learning the many terms that make up the language of biology. In fact, it would be much more difficult to learn and communicate if we did not have this terminology because words are really tools for thinking. Learning terminology generally becomes easier if you realize that most biological terms are modular. They consist of mostly Latin and Greek roots; once you learn many of these roots, you will have a good idea of the meaning of a new word even before it is defined. For this reason, we have included Appendix C, Understanding Biological Terms. To be sure that you understand the precise definition of a term, use the Index and the Glossary. The more you use biological terms in speech and writing, the more comfortable you will be with the language of biology.

Develop a Framework for Your Learning

Always aim to get the big picture before adding details. When attempting to learn a complex process, a struggling student will typically begin with the first part, try to learn all the details, and then give up. Instead, begin by making sure that you have a basic understanding of what is happening in the overall process. To encourage you in this way of thinking, we have modeled this approach in *Biology*. As just one example out of many, glycolysis is a multistep process covered in Chapter 8. Before presenting all the details, we provide an overview figure that emphasizes what the process accomplishes.

Form a Study Group

Active learning is facilitated if you do some of your studying collaboratively in a small group. In a study group, the roles of teacher and learner can be interchanged: a good way to learn material is to teach, through a process that cognitive scientists describe as *elaborative rehearsal* (not to be confused with memorization). A study group has other advantages: it can make learning more fun, lets you meet challenges in a nonthreatening environment, and can provide some emotional support. When combined with individual study of text and lecture notes, study groups can be effective learning tools.

Eldra P. Solomon
Charles E. Martin
Diana W. Martin
Linda R. Berg

A View of Life

This is an exciting time to study **biology**, the science of life. Biologists are making remarkable new discoveries that affect every aspect of our lives, including our health, food, safety, relationships with humans and other organisms, and the environment of our planet. New knowledge provides new insights into the human species and the millions of other organisms with which we share planet Earth. Biology affects our personal, governmental, and societal decisions.

One of the most exciting areas of current research is stem cell biology. **Stem cells** are unspecialized cells that have the capacity to divide, giving rise to more stem cells *and* to one or more specialized types of cell. For example, stem cells in the bone marrow differentiate to produce the various types of blood cells. Stem cells also allow the body to repair injury as well as to recover from normal wear and tear. For example, stem cells in the skin continuously divide, and some differentiate to replace skin cells that are constantly worn off from the body's surface.

Basic research in stem cell biology has helped scientists understand how unspecialized cells differentiate to become specific types of cells such as skin cells, white blood cells, or cells lining the intestine. Combined with technological advances, stem cell biology has led to exciting new advances and possibilities in such diverse fields as clinical medicine and ecology. For example, patients with leukemia and certain other cancers are often treated with radiation that destroys blood-producing stem cells in the bone marrow. Thousands of lives are saved each year using procedures in which stem cells are transplanted into the patient's bone marrow.

Researchers are developing methods for using stem cells to treat infertility and to repair spinal cord injury. In the future, stem cells may be used to cure genetic diseases and to treat diseases such as arthritis, Alzheimer's disease, Parkinson's disease, multiple sclerosis, and macular degeneration. Tissue may someday be cultured from a patient to replace organs that are diseased. For example, a diabetic patient may be given a new pancreas.

Biologists continue to discover new types of stem cells within the bodies of plants, humans, and research animals. The most versatile stem cells, called **pluripotent** stem cells, can give rise to all the tissues of the body. Biologists have discovered how to

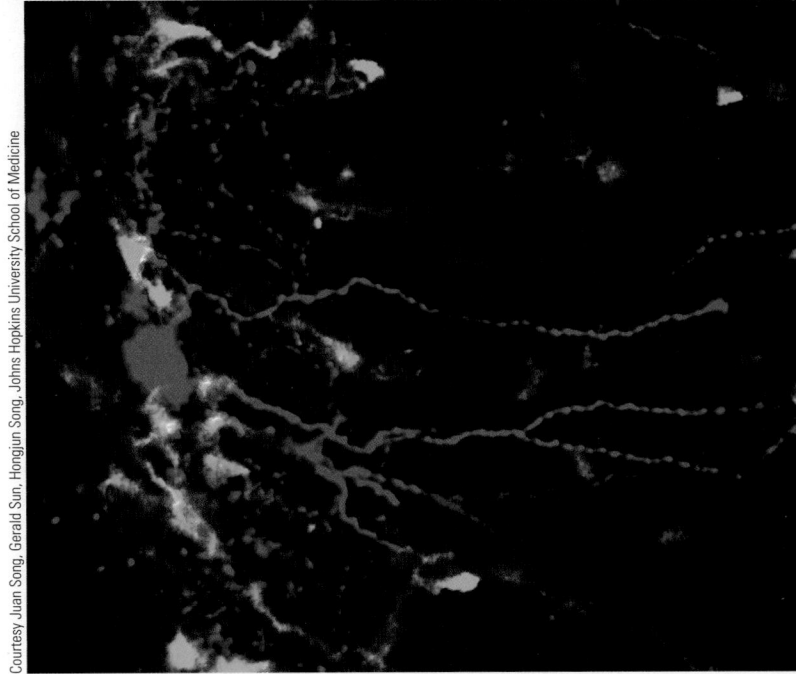

Courtesy Juan Song, Gerald Sun, Hongjun Song, Johns Hopkins University School of Medicine

Neural stem cells in the brain. Neural stem cells (*green*) in the hippocampus gather around a neuron (*purple*). Neural stem cells appear to receive and respond to signals transmitted from one neuron to another.

KEY CONCEPTS

1.1 Basic themes of biology include evolution, interactions of biological systems, inter-relationships of structure and function, information transfer, and energy transfer.

1.2 Characteristics of life include cellular structure, growth and development, self-regulated metabolism, response to stimuli, and reproduction.

1.3 Biological organization is hierarchical and includes chemical, cell, tissue, organ, organ system, and organism levels; ecological organization includes population, community, ecosystem, and biosphere levels.

1.4 Information transfer includes DNA transfer of information from one generation to the next, chemical and electrical signals within and among the cells of every organism, and sensory receptors and response systems that allow organisms to communicate with one another and interact with their environment.

1.5 Individual organisms and entire ecosystems depend on a continuous input of energy. Energy is transferred within cells and from one organism to another.

1.6 Evolution is the process by which populations of organisms change over time, adapting to changes in their environment; the tree of life includes three major branches, or domains.

1.7 Biologists ask questions, develop hypotheses, make predictions, and collect data by careful observation and experiment; based on their results, they come to conclusions and then share their work with other scientists and with the public.

induce pluripotent stem cells by reprogramming the genome of certain adult cells. These *induced pluripotent stem cells (iPSCs)* are similar in many ways to embryonic stem cells. An important advantage of iPSCs is that they give rise to tissues that are genetically identical to those of the patient.

Stem cells may also be used in the future to save endangered species. Researchers have already produced iPSCs from the tissues of an adult snow leopard, a jaguar, a Bengal tiger, and a serval (a medium-sized, slender cat, native to Africa). In the future they hope to clone the iPSCs and to produce eggs and sperm from them.

Recently, the journal *Nature* reported that researchers at the Johns Hopkins University School of Medicine have discovered that neural stem cells in the brain "listen in" on the chemical signals that neurons use to communicate with one another (see photograph). When necessary, the stem cells differentiate into neurons or glial cells, or signal the brain to produce new cells.

The 2012 Nobel Prize in Physiology or Medicine was awarded to John B. Gurdon and Shinya Yamanaka for their contributions to the development of stem cell research. In the 1960s, Gurdon transplanted differentiated cell nuclei taken from tadpoles into frog egg cells. He found that a few of these eggs developed into tadpoles (see Fig. 17-3). More than forty years later Yamanaka and his colleagues identified a combination of four genes from embryonic stem cells that could reprogram certain mature cells to become pluripotent stem cells.

Stem cell research is just one of hundreds of exciting areas of biological research that bring together science, technology, and society. Whatever your college major or career goals, knowledge of biological concepts is a vital tool for understanding our world and for meeting many of the personal, societal, and global challenges that confront us. Among these challenges are the expanding human population, decreasing biological diversity, diminishing natural resources, global climate change, and prevention and cure of diseases, such as heart disease, cancer, diabetes, and Alzheimer's disease. Meeting these challenges will require the combined efforts of biologists and other scientists, health professionals, educators, politicians, and biologically informed citizens.

This book is a starting point for your exploration of biology. It will provide you with the basic knowledge and the tools to become a part of this fascinating science as well as a more informed member of society.

1.1 MAJOR THEMES OF BIOLOGY

LEARNING OBJECTIVE

1 Describe five basic themes of biology.

In this first chapter we introduce five major themes of biology. These themes are interconnected with one another and with almost every concept that we discuss in this book.

1. **Biological systems interact.** Every organism is a biological system made up of millions of other biological systems. Each of its cells is a biological system, as is each organ (e.g., heart and liver) and body system (e.g., cardiovascular system and digestive system). Each of the multitude of microorganisms (e.g., bacteria) that inhabit an organism is also a biological system. Making this concept even more interesting, an organism cannot survive on its own. Every organism is a biological system that is interdependent with many other biological systems. Clearly, scientists can study biological systems and their interactions at many different levels.

2. **Structure and function are inter-related in all biological systems.** The structure of neurons that function to transmit information is very different from the structure of red blood cells, which function to transport oxygen. Similarly, on the level of organisms, the canine teeth of carnivorous mammals are adapted for stabbing their prey and ripping flesh. In contrast, horses and other herbivorous mammals have teeth adapted for cutting off bits of vegetation and grinding plant material. In each case, structure and function are inter-related.

3. **Information must be transmitted within organisms and among organisms.** Each organism must be able to receive information from the surrounding environment. The survival and function of every cell and every organism depend on the orderly transmission of information. As we will learn, evolution depends on the transmission of genetic information from one generation to another.

4. **Life depends on a continuous input of energy from the sun because every activity of a living cell or organism requires energy.** Energy from the sun flows through individual organisms and through ecosystems. Within living cells energy is continuously transferred from one chemical compound to another.

5. **Evolution is the process by which populations of organisms change over time.** Scientists have accumulated a wealth of evidence showing that the diverse life-forms on this planet are related and that populations have *evolved*— that is, have changed over time—from earlier forms of life. The process of *evolution* is the framework for the science of biology and is a major theme of this book.

The interaction of biological systems, the inter-relationship of structure and function, information transfer, energy transfer, and the process of evolution are forces that give life its unique characteristics. You will find reference to one or more of these unifying themes in every chapter of *Biology*. We begin our study of biology by developing a more precise understanding of the fundamental characteristics of living systems and of the levels of biological organization. We then take a closer look at some of the major themes of biology. We end Chapter 1 with a discussion of the process of science.

- *Why are information transmission, energy transfer, and evolution considered basic to life?*
- **CONNECT** *What are some ways in which an organism is dependent on other biological systems?*

1.2 CHARACTERISTICS OF LIFE

LEARNING OBJECTIVE

2   Distinguish between living systems and nonliving things by describing the features that characterize living organisms.

We easily recognize that a pine tree, a butterfly, and a horse are living systems, whereas a rock is not. Despite their diversity, the organisms that inhabit our planet share a common set of characteristics that distinguish them from nonliving things. These features include a precise kind of organization, growth and development, self-regulated metabolism, the ability to respond to stimuli, reproduction, and adaptation to environmental change.

Organisms are composed of cells

Although they vary greatly in size and appearance, all organisms consist of basic units called **cells.** New cells are formed only by the division of previously existing cells. As will be discussed in Chapter 4, these concepts are expressed in the **cell theory,** another fundamental unifying concept of biology.

Some of the simplest life-forms, such as protozoa, are *unicellular* organisms, meaning that each consists of a single cell (**FIG. 1-1a**). In contrast, the body of a maple tree or a buffalo is made of billions of cells (**FIG. 1-1b**). In such complex *multicellular* organisms, life processes depend on the coordinated functions of component cells that are organized to form tissues, organs, and organ systems.

Every cell is enveloped by a protective **plasma membrane** that separates it from the surrounding external environment. The plasma membrane regulates passage of materials between the cell and its environment. Cells have specialized molecules that contain genetic instructions and transmit genetic information. In most cells, the genetic instructions are encoded in deoxyribonucleic acid, more simply known as **DNA.** Cells typically have internal structures called **organelles** that are specialized to perform specific functions.

There are two fundamentally different types of cells: prokaryotic and eukaryotic. *Prokaryotic cells* are exclusive to bacteria and to microscopic organisms called *archaea.* Prokaryotic cells do not have a nucleus or other membrane-enclosed organelles. All other organisms are characterized by their *eukaryotic cells.* These cells typically contain a variety of organelles enclosed by membranes, including a **nucleus,** which houses DNA.

Organisms grow and develop

Biological growth involves an increase in the size of individual cells of an organism, in the number of cells, or in both. Growth

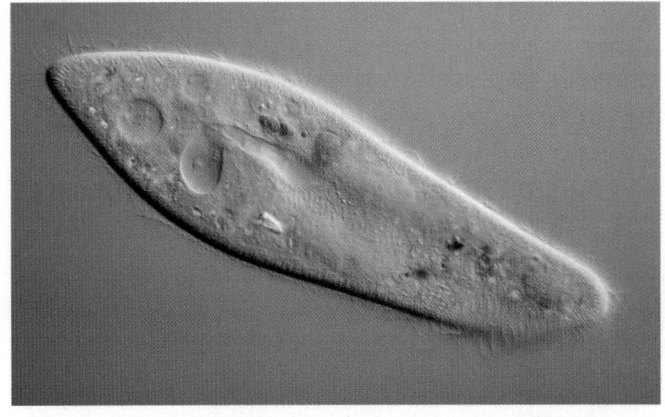

250 μm

(a) Unicellular organisms consist of one cell that performs all the functions essential to life. Ciliates, such as this *Paramecium*, move about by beating their hairlike cilia.

(b) Multicellular organisms, such as this African buffalo (*Syncerus caffer*) and the plants on which it grazes, may consist of billions of cells specialized to perform specific functions.

Figure 1-1 Unicellular and multicellular life-forms

may be uniform in the various parts of an organism, or it may be greater in some parts than in others, causing the body proportions to change as growth occurs. Some organisms—most trees, for example—continue to grow throughout their lives. Many animals have a defined growth period that terminates when a characteristic adult size is reached. An intriguing aspect of the growth process is that each part of the organism typically continues to function as it grows.

Organisms develop as well as grow. **Development** includes all the changes that take place during an organism's life. The structures and body form that develop are exquisitely adapted to the functions the organism must perform. Like many other organisms, every human begins life as a fertilized egg that then grows and develops.

Organisms regulate their metabolic processes

Within all organisms, chemical reactions and energy transformations occur that are essential to nutrition, the growth and repair of cells, and the conversion of energy into usable forms. The sum of all the chemical activities of the organism is its **metabolism.**

Metabolic processes occur continuously in every organism, and they must be carefully regulated to maintain **homeostasis,** an appropriate, balanced internal environment. The term *homeostasis* also refers to the automatic tendency of the organism to maintain a steady state. When a particular substance is required, cell processes that produce it must be turned on. When enough of a cell product has been made, its manufacture must be decreased or turned off. These *homeostatic mechanisms* are self-regulating control systems that are remarkably sensitive and efficient.

The regulation of glucose (a simple sugar) concentration in the blood of complex animals is a good example of a homeostatic mechanism. Your cells require a constant supply of glucose molecules, which they break down to obtain energy. The circulatory system delivers glucose and other nutrients to all the cells. When the concentration of glucose in the blood rises above normal limits, glucose is stored in the liver and in muscle cells. When you have not eaten for a few hours, the glucose concentration begins to fall. Your body mobilizes stored glucose. If necessary, the body converts other stored nutrients to glucose, bringing the glucose concentration in the blood back to normal levels. When the glucose concentration decreases, you also feel hungry and can restore nutrients by eating.

Organisms respond to stimuli

All forms of life respond to **stimuli,** physical or chemical changes in their internal or external environment. Stimuli that evoke a response in most organisms are changes in the color, intensity, or direction of light; changes in temperature, pressure, or sound; and changes in the chemical composition of the surrounding soil, air, or water. Responding to stimuli involves movement, although not always locomotion (moving from one place to another).

In simple organisms, the entire individual may be sensitive to stimuli. Certain unicellular organisms, for example, respond to bright light by retreating. In some organisms, locomotion is achieved by the slow oozing of the cell, the process of *amoeboid movement.* Other organisms move by beating tiny, hairlike extensions of the cell called **cilia** or longer structures known as **flagella** (FIG. 1-2). Some bacteria move by rotating their flagella.

Most animals move very obviously. They wiggle, crawl, swim, run, or fly by contracting muscles. Sponges, corals, and oysters have free-swimming larval stages, but most are **sessile** as adults, meaning that they do not move from place to place. In fact, they may remain firmly attached to a surface, such as the sea bottom or a rock. Many sessile organisms have cilia or flagella that beat rhythmically, bringing them food and oxygen in

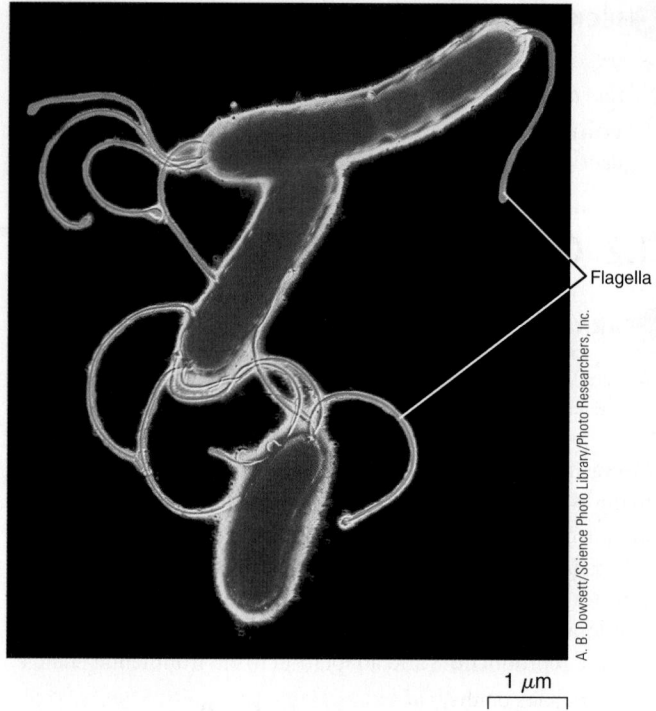

Flagella

1 μm

A. B. Dowsett/Science Photo Library/Photo Researchers, Inc.

Figure 1-2 Biological movement
These bacteria (*Helicobacter pylori*), equipped with flagella for locomotion, have been linked to stomach ulcers. The photograph was taken using a scanning electron microscope. The bacteria are not really red and blue. Their color has been artificially enhanced.

the surrounding water. Complex animals, such as grasshoppers, lizards, and humans, have highly specialized cells that respond to specific types of stimuli. For example, cells in the retina of the vertebrate eye respond to light.

Although their responses may not be as obvious as those of animals, plants do respond to light, gravity, water, touch, and other stimuli. For example, plants orient their leaves to the sun and grow toward light. Many plant responses involve different growth rates of various parts of the plant body. A few plants, such as the Venus flytrap of the Carolina swamps, are very sensitive to touch and catch insects (FIG. 1-3). Their leaves are hinged along the midrib, and they have a scent that attracts insects. Trigger hairs on the leaf surface detect the arrival of an insect and stimulate the leaf to fold. When the edges come together, they interlock, preventing the insect's escape. The leaf then secretes enzymes that kill and digest the insect. The Venus flytrap usually grows in nitrogen-deficient soil. The plant obtains part of the nitrogen required for its growth from the insects it "eats."

Organisms reproduce

At one time, people thought worms arose spontaneously from horsehair in a water trough, maggots from decaying meat, and frogs from the mud of the Nile. Thanks to the work of a great many scientists, beginning with pioneering studies by Italian physician Francesco Redi in the 17th century and French chemist Louis Pasteur in the 19th century, we know that organisms arise only from previously existing organisms.

(a) When hairs on the leaf surface of the Venus flytrap (*Dionaea muscipula*) detect the touch of an insect, the leaf responds by folding.

(b) The edges of the leaf come together and interlock, preventing the fly's escape. The leaf then secretes enzymes that kill and digest the insect.

Figure 1-3 Plants respond to stimuli

Simple organisms, such as amoebas, perpetuate themselves by **asexual reproduction** (FIG. 1-4a). When an amoeba has grown to a certain size, it reproduces by splitting in half to form two new amoebas. Before an amoeba divides, its hereditary material (set of *genes*) is duplicated, and one complete set is distributed to each new cell. Except for size, each new amoeba is similar to the parent cell. The only way that variation occurs among asexually reproducing organisms is by genetic *mutation*, a permanent change in the genes.

In most plants and animals, **sexual reproduction** is carried out by the fusion of an egg and a sperm cell to form a fertilized egg (FIG. 1-4b). The new organism develops from the fertilized egg. Offspring produced by sexual reproduction are the product of the interaction of various genes contributed by the mother and the father. This genetic variation is important in the vital processes of evolution and adaptation.

Populations evolve and become adapted to the environment

The ability of a population to evolve over many generations and adapt to its environment equips it to survive in a changing world. **Adaptations** are inherited characteristics that enhance an organism's ability to survive in a particular environment. The long, flexible tongue of the frog is an adaptation for catching insects. The feathers and light-weight bones of birds are adaptations for flying, and their thick fur coats allow polar bears to survive in frigid temperatures. Adaptations may be structural, physiological, biochemical, behavioral, or a combination of all four (FIG. 1-5). Every biologically successful organism is a complex collection of coordinated adaptations produced through evolutionary processes.

CHECKPOINT 1.2

- *What characteristics distinguish a living organism from a rock?*
- PREDICT *What would be the consequences to an organism if its homeostatic mechanisms failed? Explain your answer.*

Figure 1-4 Asexual and sexual reproduction

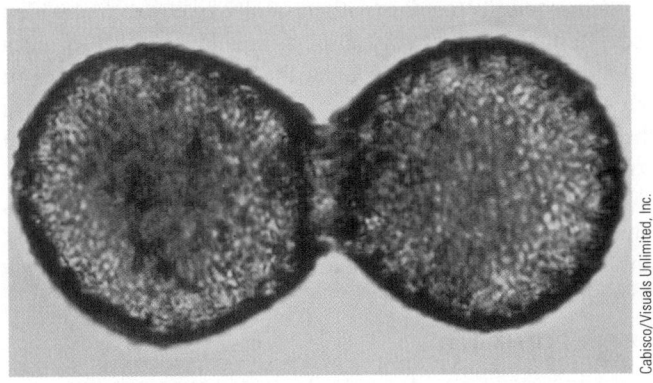

100 μm

(a) **Asexual reproduction.** One individual gives rise to two or more offspring that are similar to the parent. *Difflugia*, a unicellular amoeba, is shown dividing to form two amoebas.

(b) **Sexual reproduction.** Typically, each of two parents contributes a gamete (sperm or egg). Gametes fuse to produce the offspring, which has a combination of the traits of both parents. A pair of tropical flies is shown mating.

Figure 1-5 Adaptations

These Burchell's zebras (*Equus burchelli*), photographed in Tanzania, are behaviorally adapted to position themselves to watch for lions and other predators. Stripes are thought to be an adaptation for visual protection against predators. They serve as camouflage or to break up form when spotted from a distance. The zebra stomach is adapted for feeding on coarse grass passed over by other grazers, an adaptation that helps the animal survive when food is scarce.

1.3 LEVELS OF BIOLOGICAL ORGANIZATION

LEARNING OBJECTIVE

3 Construct a hierarchy of biological organization, including levels characteristic of individual organisms and levels characteristic of ecological systems.

Whether we study a single organism or the world of life as a whole, we can identify a hierarchy of biological organization (FIG. 1-6). At every level, structure and function are precisely coordinated. One way to study a particular level is by looking at its components. Biologists can gain insights about cells by studying atoms and molecules. Learning about a structure by studying its parts is called **reductionism.** However, the whole is more than the sum of its parts. Each level has **emergent properties,** characteristics not found at lower levels. For example, populations of organisms have emergent properties such as population density, age structure, and birth and death rates. The *individuals* that make up a population do not have these characteristics. Consider also the human brain. The brain is composed of billions of neurons (nerve cells). However, we could study every one of these individual neurons and have no clue about the functional capacities of the brain. Only when the neurons interact are the emergent properties, such as the capacity for thought, judgment, and motor coordination, evident.

Organisms have several levels of organization

The chemical level, the most basic level of organization, includes atoms and molecules. An **atom** is the smallest unit of a chemical element that retains the characteristic properties of that element.

For example, an atom of iron is the smallest possible amount of iron. Atoms combine chemically to form **molecules.** Two atoms of hydrogen combine with one atom of oxygen to form a single molecule of water. Although composed of two types of atoms that are gases under conditions found on Earth, water can exist as a gas, liquid, or solid. The properties of water are very different from those of its hydrogen and oxygen components, an example of emergent properties.

At the cellular level, many types of atoms and molecules associate with one another to form *cells*. However, a cell is much more than a heap of atoms and molecules. Its emergent properties make it the basic structural and functional unit of life, the simplest component of living matter that can carry on all the activities necessary for life.

During the evolution of multicellular organisms, cells associated to form **tissues.** For example, most animals have muscle tissue and nervous tissue. Plants have epidermis, a tissue that serves as a protective covering, and vascular tissues that move materials throughout the plant body. In most complex organisms, tissues organize into functional structures called **organs**, such as the heart and stomach in animals and roots and leaves in plants. In animals, each major group of biological functions is performed by a coordinated group of tissues and organs called an **organ system.** The circulatory and digestive systems are examples of organ systems. Functioning together with great precision, organ systems make up a complex, multicellular **organism.** Again, emergent properties are evident. An organism is much more than its component organ systems.

Several levels of ecological organization can be identified

Organisms interact to form still more complex levels of biological organization. All the members of one species living in the same geographic area at the same time make up a **population.** The populations of various types of organisms that inhabit a particular area and interact with one another form a **community.** A community can consist of hundreds of different types of organisms.

A community together with its nonliving environment is an **ecosystem.** An ecosystem can be as small as a pond (or even a puddle) or as vast as the Great Plains of North America or the Arctic tundra. All Earth's ecosystems together are known as the **biosphere.** The biosphere includes all systems of Earth that are inhabited by living organisms: the atmosphere, the hydrosphere (water in any form), and the lithosphere (Earth's crust). The study of how organisms relate to one another and to their physical environment is called **ecology** (derived from the Greek *oikos*, meaning "house").

CHECKPOINT 1.3

- *What are the levels of organization within an organism?*
- PREDICT *At which level do you think more biological systems would be interacting: organism, population, or ecosystem? Justify your answer.*

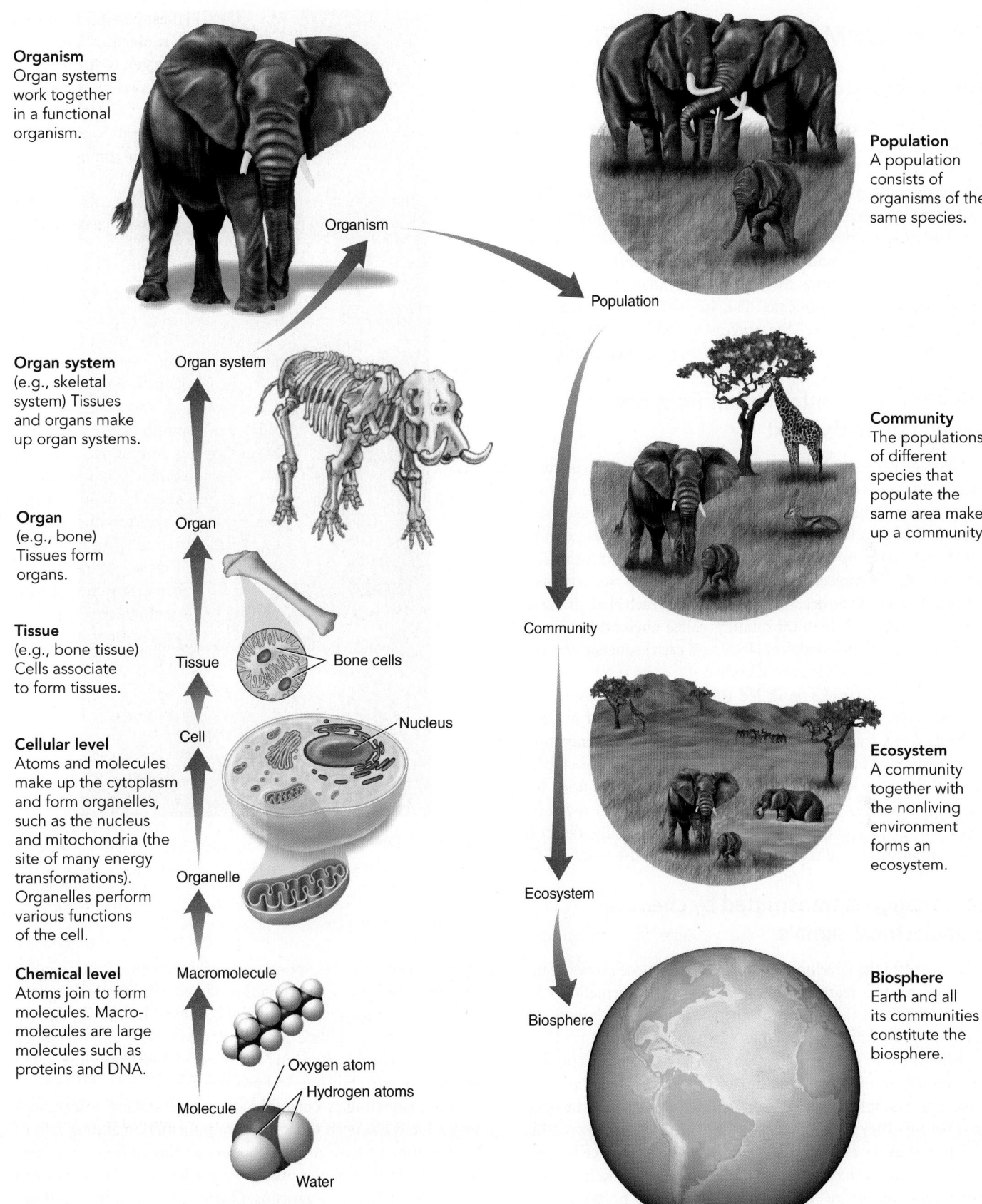

Organism
Organ systems work together in a functional organism.

Organism

Population
A population consists of organisms of the same species.

Population

Organ system
(e.g., skeletal system) Tissues and organs make up organ systems.

Organ system

Organ
(e.g., bone) Tissues form organs.

Organ

Community
The populations of different species that populate the same area make up a community.

Community

Tissue
(e.g., bone tissue) Cells associate to form tissues.

Tissue

Bone cells

Cellular level
Atoms and molecules make up the cytoplasm and form organelles, such as the nucleus and mitochondria (the site of many energy transformations). Organelles perform various functions of the cell.

Cell

Nucleus

Organelle

Ecosystem
A community together with the nonliving environment forms an ecosystem.

Ecosystem

Chemical level
Atoms join to form molecules. Macromolecules are large molecules such as proteins and DNA.

Macromolecule

Oxygen atom

Hydrogen atoms

Molecule

Water

Biosphere

Biosphere
Earth and all its communities constitute the biosphere.

Figure 1-6 *Animation* **The hierarchy of biological organization**
© Cengage Learning

1.4 INFORMATION TRANSFER

LEARNING OBJECTIVE

4 Summarize the importance of information transfer within and between living systems, giving specific examples.

Biological systems receive and respond to information. They also store information. An organism inherits the information it needs to grow, develop, carry on self-regulated metabolism, respond to stimuli, and reproduce. Each organism must also have precise instructions for making the molecules necessary for its cells to communicate. The information an organism requires to carry on these life processes is coded and transmitted in the form of chemical substances and electrical impulses.

DNA transmits information from one generation to the next

Humans give birth only to human babies, not to giraffes or rosebushes. In organisms that reproduce sexually, each offspring is a combination of the traits of its parents. In 1953, James Watson and Francis Crick worked out the structure of DNA, the large molecule that makes up the **genes,** units of hereditary information (**FIG. 1-7**). A DNA molecule consists of two chains of atoms twisted into a helix. As will be described in Chapter 3, each chain is made up of a sequence of chemical subunits called **nucleotides.** There are four types of nucleotides in DNA, and each sequence of three nucleotides is part of the genetic code.

Watson and Crick's work led to the understanding of the genetic code. The information coded in sequences of nucleotides in DNA transmits genetic information from generation to generation. The code works somewhat like an alphabet. The nucleotides can "spell" an amazing variety of instructions for making organisms as diverse as bacteria, frogs, and redwood trees. The genetic code is universal—that is, virtually identical in all organisms—and is a dramatic example of the unity of life.

Information is transmitted by chemical and electrical signals

Genes control the development and functioning of every organism. As you will learn in later chapters, the information carried by the DNA that makes up the genes has many functions, including providing the "recipes" for making all the proteins required by the organism. **Proteins** are large molecules important in determining the structure and function of cells and tissues. For example, brain cells differ from muscle cells in large part because they have different types of proteins. Some proteins are important in communication within and among cells. Certain proteins on the surface of a cell serve as markers so that other cells "recognize" them. Other cell-surface proteins serve as receptors that combine with chemical messengers.

Cells use proteins and many other types of molecules to communicate with one another. In a multicellular organism, cells produce chemical compounds, such as **hormones,** that

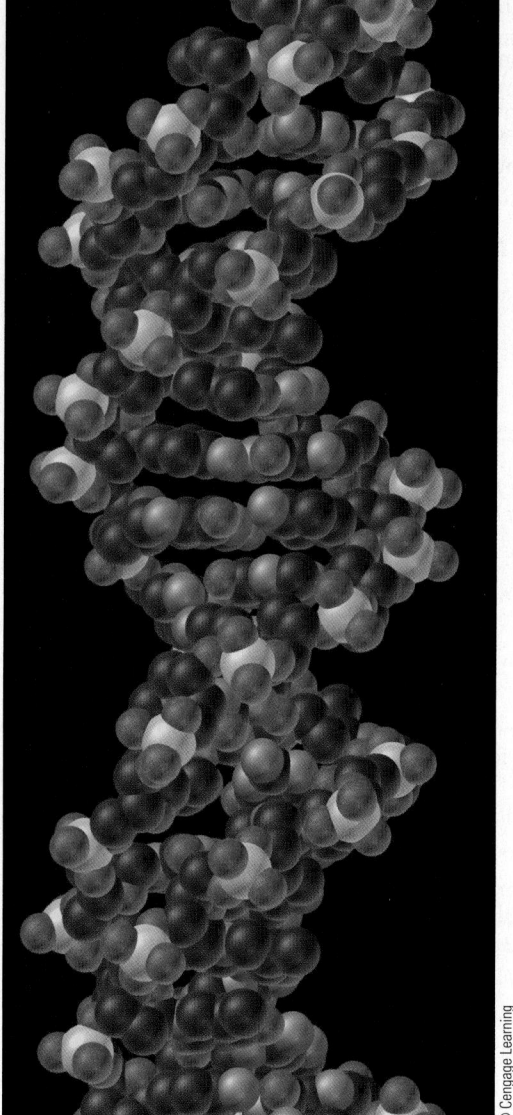

Figure 1-7 DNA

DNA is the hereditary material that transmits information from one generation to the next. As shown in this model, DNA is a macromolecule that consists of two chains of atoms twisted into a helix. Each chain consists of subunits called nucleotides. The sequence of nucleotides makes up the genetic code.

signal other cells. Hormones and other chemical messengers can signal cells in distant organs to secrete a particular required substance or change some metabolic activity. In this way chemical signals help regulate growth, development, and metabolic processes. The mechanisms involved in **cell signaling** often involve complex biochemical processes.

Cell signaling is currently an area of intense research. A major focus has been the transfer of information among cells of the immune system. A better understanding of how cells communicate promises new insights into how the body protects itself against disease organisms. Learning to manipulate cell signaling may lead to new methods of delivering drugs into cells and new treatments for cancer and other diseases.

Many organisms use electrical signals to transmit information. Most animals have nervous systems that transmit information by way of both electrical impulses and chemical compounds known as **neurotransmitters**. Information transmitted from one part of the body to another is important in regulating life processes. In complex animals, the nervous system gives the animal information about its outside environment by transmitting signals from sensory receptors such as the eyes and ears to the brain.

Organisms also communicate information to one another

Organisms communicate information to other organisms by releasing chemicals, sounds, and visual displays. Typically, organisms use a combination of several types of communication signals. A dog may signal aggression by growling, using a particular facial expression, and laying its ears back. Many animals perform complex courtship rituals in which they display parts of their bodies, often elaborately decorated, to attract a mate.

Seaweed algae compete with coral for light and space. Marine biologists studying endangered coral reefs have discovered that certain seaweed algae secrete chemical compounds that kill coral. Researchers have reported that some coral can fight back. When they come into contact with toxic seaweed, the coral release chemical compounds that signal certain species of goby fish. In response to this chemical signal, the fish eat the seaweed. This action helps preserve their coral reef habitat.

CHECKPOINT 1.4

- *What is the function of DNA?*
- *How does a nervous system transmit information?*

1.5 THE ENERGY OF LIFE

LEARNING OBJECTIVE

5   Summarize the flow of energy through ecosystems and contrast the roles of producers, consumers, and decomposers.

The sun provides most of the energy that powers life on Earth. All life processes, including thousands of chemical transactions that maintain life's organization, require a continuous input of energy. Organisms can neither create energy nor use it with complete efficiency. During every energy transaction, some energy is converted to heat and dispersed into the environment. Energy flows through individual organisms and through ecosystems.

A self-sufficient ecosystem consists of a physical environment inhabited by three types of organisms: producers, consumers, and decomposers. These organisms depend on one another and on the environment for nutrients, energy, oxygen, and carbon dioxide. Plants, algae, and certain bacteria are

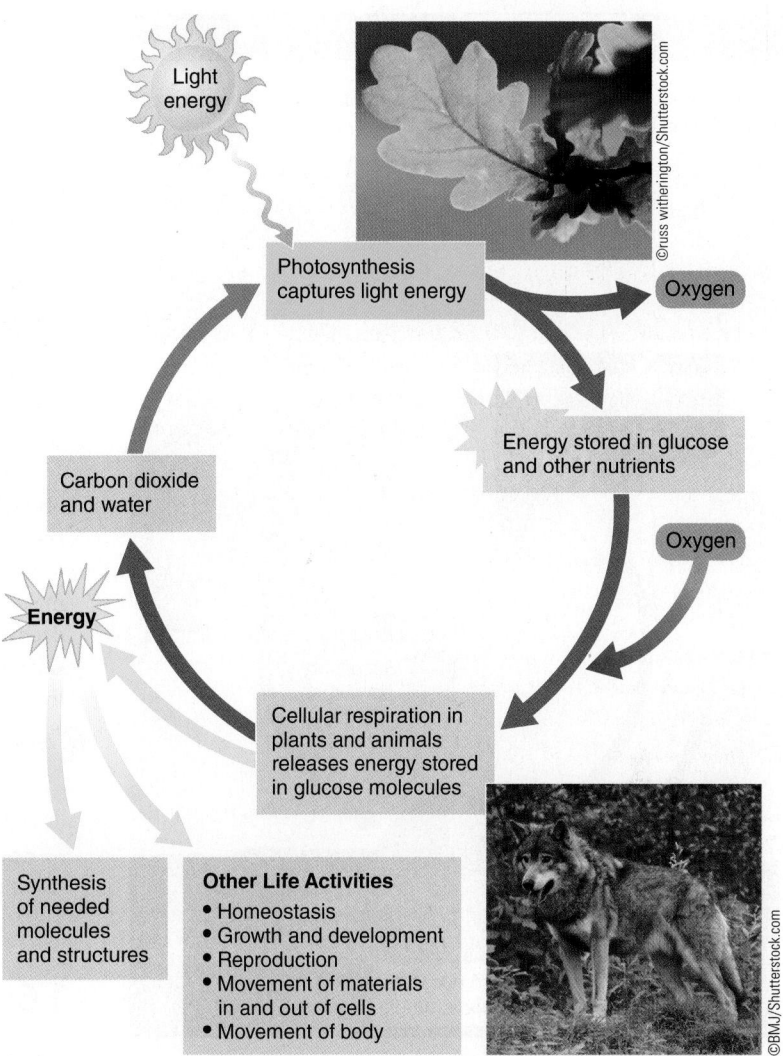

Figure 1-8 *Animation* **Energy flow within and among organisms**

Algae and certain plant cells carry on photosynthesis, a process that uses light energy to produce glucose from carbon dioxide and water. Energy is stored in the chemical bonds of glucose and other nutrients produced from glucose. Through the process of cellular respiration, cells of all organisms, including algae and plant cells, then break down glucose and other nutrients. The energy released can be used to produce needed molecules and to fuel other life activities.

© Cengage Learning

producers, or **autotrophs,** organisms that produce their own food from simple raw materials. Most of these organisms carry on **photosynthesis,** the process during which autotrophs use carbon dioxide, water, and light energy to synthesize complex molecules such as glucose and other sugars (**FIG. 1-8**):

carbon dioxide + water + light energy $\longrightarrow$ glucose + oxygen

The light energy is transformed into chemical energy, which is stored within the chemical bonds of the glucose and other food molecules produced. Oxygen, which is required by the cells of most organisms including plant cells, is produced as a byproduct of photosynthesis.

Recall that all the energy transformations and chemical processes that occur within an organism are referred to as its *metabolism.* Energy is necessary to carry on the metabolic

Energy flows from the sun to producers and then to consumers and decomposers.

Figure 1-9 *Animation* **Energy flow through the biosphere**

Continuous energy input from the sun operates the biosphere. During photosynthesis, producers use the energy from sunlight to make complex molecules from carbon dioxide and water. Primary consumers, such as the caterpillar shown here, obtain energy, nutrients, and other required materials when they eat producers. Secondary consumers, such as the bird, obtain energy, nutrients, and other required materials when they eat primary consumers that have eaten producers. Decomposers obtain their energy and nutrients by breaking down wastes and dead organic material. During every energy transaction, some energy is lost to biological systems, dispersing into the environment as heat.

PREDICT How does air pollution caused by human activity affect the balance of energy flow through the biosphere?

activities essential for growth, repair, and maintenance. Each cell of an organism requires nutrients that contain energy. During **cellular respiration,** cells capture energy stored in glucose and other nutrient molecules through a series of carefully regulated chemical reactions. We can summarize these reactions as follows:

$$\text{glucose} + \text{oxygen} \longrightarrow \text{carbon dioxide} + \text{water} + \text{energy}$$

When chemical bonds are broken during cellular respiration, their stored energy is made available for life processes. Cells use this energy to do work, including the synthesis of required materials, such as new cell components. Virtually all cells carry on cellular respiration.

Animals are **consumers,** or **heterotrophs**—that is, they are organisms that depend on producers for food, energy, and oxygen (**FIG. 1-9**). **Primary consumers** eat producers. **Secondary consumers** eat primary consumers. Consumers obtain energy by breaking down sugars and other nutrients originally produced during photosynthesis. Consumers also contribute to the balance of the ecosystem. For example, consumers produce carbon dioxide required by producers. (Note that producers also carry on cellular respiration.) The metabolism of consumers and producers helps maintain the life-sustaining mixture of gases in the atmosphere.

Most bacteria and fungi are **decomposers,** heterotrophs that obtain nutrients by breaking down nonliving organic material such as wastes, dead leaves and branches, and the bodies of dead organisms. In their process of obtaining energy, decomposers make the components of these materials available for reuse. If decomposers did not exist, nutrients would remain locked up in wastes and dead bodies, and the supply of elements required by living systems would soon be exhausted.

CHECKPOINT 1.5

- **PREDICT** *What components do you think a forest ecosystem might include?*

- **CONNECT** *In what ways do consumers depend on producers? on decomposers? Include energy considerations in your answer.*

1.6 EVOLUTION: THE BASIC UNIFYING CONCEPT OF BIOLOGY

LEARNING OBJECTIVES

6   Demonstrate the binomial system of nomenclature by using specific examples and classify an organism (such as a human) in its domain, kingdom, phylum, class, order, family, genus, and species.

7   Identify the three domains and the kingdoms of living organisms, and give examples of organisms assigned to each group.

8   Give a brief overview of the scientific theory of evolution and explain why it is the principal unifying concept in biology.

9   Apply the concept of natural selection to any given adaptation and suggest a logical explanation of how the adaptation may have evolved.

Evolution is the process by which populations of organisms change over time. The scientific theory of evolution has become the most important unifying concept of biology. As we will discuss, evolution involves passing genes for new traits from one generation to another, leading to differences in populations. The evolutionary perspective is important in every specialized field within biology. Biologists try to understand the structure, function, and behavior of organisms and their interactions with one another by considering them in light of the long, continuing process of evolution. Although we discuss evolution in depth in Chapters 18 through 22, we present a brief overview here to give you the background necessary to understand other aspects of biology. First, we examine how biologists organize the millions of organisms that have evolved, and then we summarize some of the mechanisms that drive evolution.

Biologists use a binomial system for naming organisms

Biologists have identified about 1.9 million kinds (species) of *extant* (currently living) organisms and estimate that several million more remain to be discovered. To study life, we need a system for organizing, naming, and classifying its myriad forms. **Systematics** is the field of biology that studies the diversity of organisms and their evolutionary relationships. **Taxonomy,** a subspecialty of systematics, is the science of naming and classifying organisms. In the 18th century, Carolus Linnaeus, a Swedish botanist, developed a hierarchical system of naming and classifying organisms. Biologists still use this system today, with some modification.

The **species** is a group of organisms with similar structure, function, and behavior. A species consists of one or more populations whose members are capable of breeding with one another; in nature, they do not breed with members of other species. Members of a population contribute to a common **gene pool** (all the genes present in the population) and share a common ancestry. Closely related species are grouped in the next broader category of classification, the **genus** (pl., *genera*).

The Linnaean system of naming species is known as the **binomial system of nomenclature** because each species is assigned a two-part name. The first part of the name is the genus, and the second part, the **specific epithet**, designates a particular species belonging to that genus. The specific epithet is often a descriptive word expressing some quality of the organism. It is always used together with the full or abbreviated generic name preceding it. The generic name is always capitalized; the specific epithet is generally not capitalized. Both names are always italicized or underlined. For example, the domestic dog, *Canis familiaris* (abbreviated *C. familiaris*), and the timber wolf, *Canis lupus* (*C. lupus*), belong to the same genus. The domestic cat, *Felis catus*, belongs to a different genus. The scientific name of the American white oak is *Quercus alba*, whereas the name of the European white oak is *Quercus robur*. Another tree, the white willow, *Salix alba,* belongs to a different genus. The scientific name for our own species is *Homo sapiens* ("wise man").

Taxonomic classification is hierarchical

Just as closely related species may be grouped in a common genus, related genera can be grouped in a more inclusive group, a **family.** Families are grouped into **orders,** orders into **classes,** and classes into **phyla** (sing., *phylum*). Phyla can be assigned to **kingdoms,** and kingdoms are grouped in **domains.** Each formal grouping at any given level is a **taxon** (pl., *taxa*). Note that each taxon is more inclusive than the taxon below it. Together they form a hierarchy ranging from species to domain TABLE 1-1 and FIG. 1-10.

Consider a specific example. The family Canidae, which includes all doglike carnivores (animals that eat mainly meat), consists of 12 genera and about 35 living species. Family Canidae, along with family Ursidae (bears), family Felidae (catlike animals), and several other families that eat mainly meat, are all placed in order Carnivora. Order Carnivora, order Primates (to which chimpanzees and humans belong), and several other orders belong to class Mammalia (mammals). Class Mammalia is grouped with several other classes that include fishes, amphibians, reptiles, and birds in subphylum Vertebrata. The vertebrates belong to phylum Chordata, which is part of kingdom Animalia. Animals are assigned to domain Eukarya.

Systematists classify organisms in three domains

Systematics itself has evolved as scientists have developed new techniques for inferring common ancestry among groups of organisms. Biologists seek to classify organisms based on evolutionary relationships. These relationships are based on shared characteristics that distinguish a particular group. A group of organisms with a common ancestor is a **clade.** Systematists have developed family trees showing proposed evolutionary relationships among organisms. These relationships are based on the patterns of traits shared by organisms and on fossil evidence. Shared characteristics include structural, developmental, behavioral, and molecular similarities. FIGURE 1-11 is a **cladogram,** a branching diagram, that depicts the three domains and several kingdoms of Domain Eukarya. As researchers report new findings, the classification of organisms changes and the branches of cladograms are redrawn.

Although the "tree of life" is a work in progress, most biologists now assign organisms to three domains and to several kingdoms or clades. The late microbiologist Carl Woese (pronounced "woes") was a pioneer in developing molecular approaches to systematics. Woese and his colleagues selected a molecule known as small subunit ribosomal RNA (rRNA) that functions in the process of manufacturing proteins in all organisms. Because its molecular structure differs somewhat

TABLE 1-1	Classification of the Cat, Human, and White Oak Tree		
CATEGORY	CAT	HUMAN	WHITE OAK
Domain	Eukarya	Eukarya	Eukarya
Kingdom	Animalia	Animalia	Plantae
Phylum	Chordata	Chordata	Anthophyta
Subphylum	Vertebrata	Vertebrata	None
Class	Mammalia	Mammalia	Eudicotyledones
Order	Carnivora	Primates	Fagales
Family	Felidae	Hominidae	Fagaceae
Genus	*Felis*	*Homo*	*Quercus*
Species	*Felis catus*	*Homo sapiens*	*Quercus alba*

© Cengage Learning

In the traditional system of classification, biologists classify organisms in a hierarchy of taxonomic categories from species to domain; each category is more general and more inclusive than the one below it.

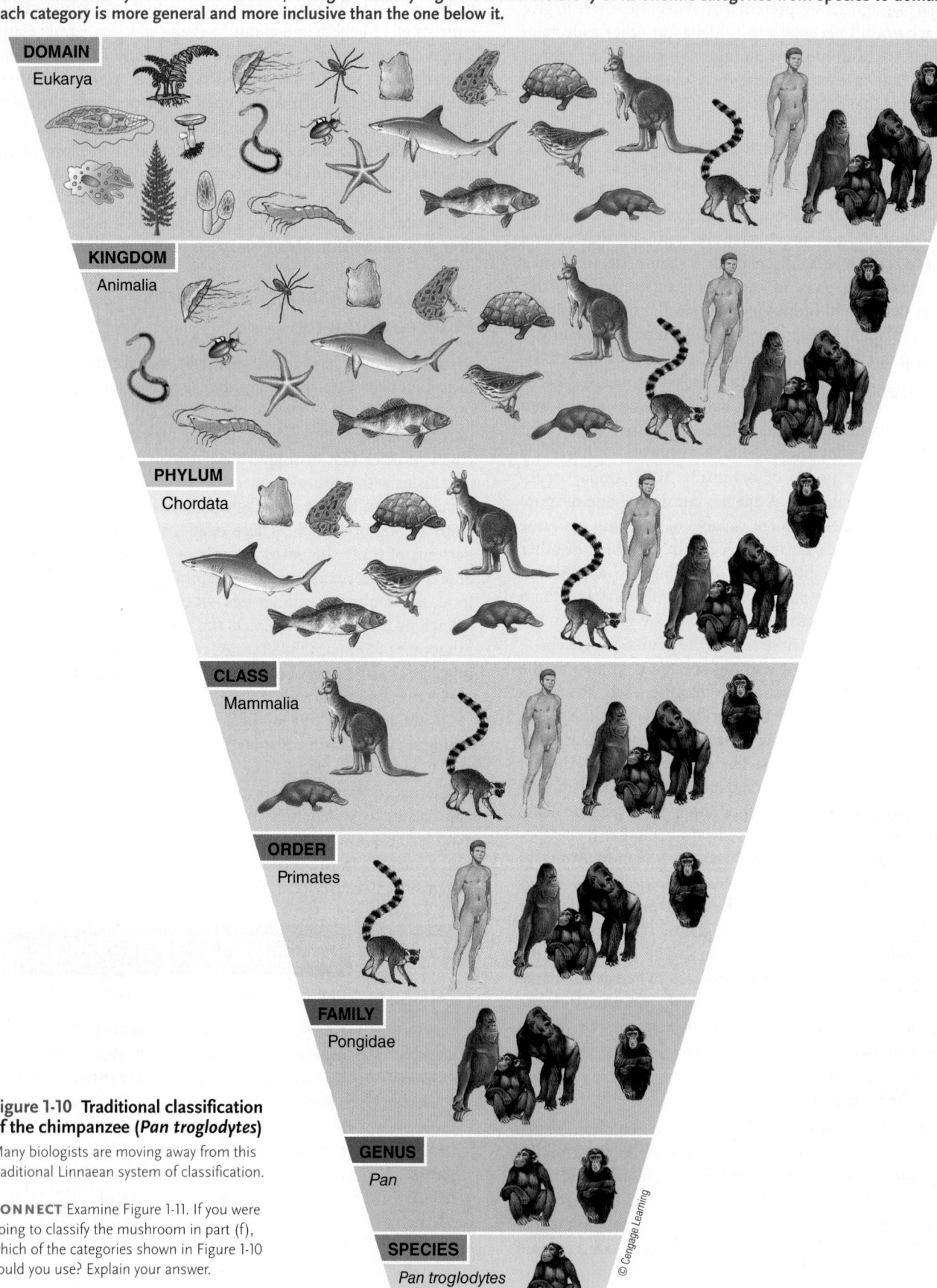

DOMAIN
Eukarya

KINGDOM
Animalia

PHYLUM
Chordata

CLASS
Mammalia

ORDER
Primates

FAMILY
Pongidae

GENUS
Pan

SPECIES
Pan troglodytes

Figure 1-10 Traditional classification of the chimpanzee (*Pan troglodytes*)

Many biologists are moving away from this traditional Linnaean system of classification.

CONNECT Examine Figure 1-11. If you were going to classify the mushroom in part (f), which of the categories shown in Figure 1-10 could you use? Explain your answer.

© Cengage Learning

This cladogram illustrates the evolutionary relationships among the three domains and among major groups of organisms that belong to these domains.

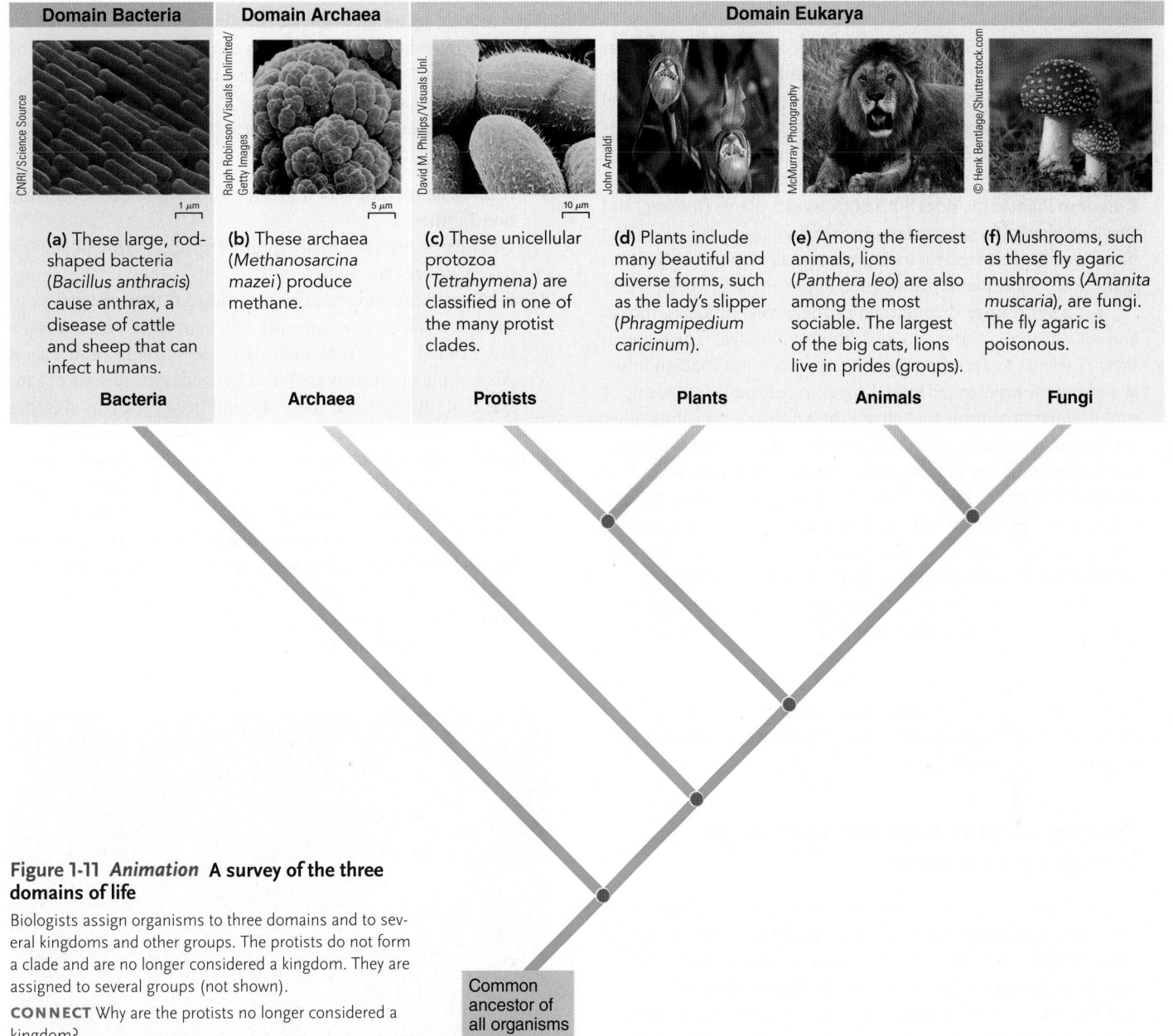

Domain Bacteria	Domain Archaea	Domain Eukarya

(a) These large, rod-shaped bacteria (*Bacillus anthracis*) cause anthrax, a disease of cattle and sheep that can infect humans.

(b) These archaea (*Methanosarcina mazei*) produce methane.

(c) These unicellular protozoa (*Tetrahymena*) are classified in one of the many protist clades.

(d) Plants include many beautiful and diverse forms, such as the lady's slipper (*Phragmipedium caricinum*).

(e) Among the fiercest animals, lions (*Panthera leo*) are also among the most sociable. The largest of the big cats, lions live in prides (groups).

(f) Mushrooms, such as these fly agaric mushrooms (*Amanita muscaria*), are fungi. The fly agaric is poisonous.

Bacteria   **Archaea**   **Protists**   **Plants**   **Animals**   **Fungi**

Common ancestor of all organisms

Figure 1-11 *Animation* **A survey of the three domains of life**

Biologists assign organisms to three domains and to several kingdoms and other groups. The protists do not form a clade and are no longer considered a kingdom. They are assigned to several groups (not shown).

CONNECT Why are the protists no longer considered a kingdom?

in various organisms, Woese hypothesized that the molecular composition of rRNA in closely related organisms would be more similar than in distantly related organisms.

Bacteria have long been recognized as unicellular prokaryotic cells; they are referred to as **prokaryotes.** Woese's findings showed that there are two distinct groups of prokaryotes. He established the domain level of taxonomy and assigned the prokaryotes to two domains: **Bacteria** and **Archaea** (ar´-key-ah). The **eukaryotes,** organisms with eukaryotic cells, are classified in domain **Eukarya.** Woese's work became widely accepted in the mid-1990s.

In the classification system used in this book, each organism is assigned to a domain and to a kingdom or "supergroup." Two kingdoms correspond to the prokaryotic domains: kingdom Archaea corresponds to domain Archaea, and kingdom Bacteria corresponds to domain Bacteria. The remaining kingdoms and groups are assigned to domain Eukarya.

Protists (e.g., algae, slime molds, amoebas, and ciliates) are unicellular, colonial, or simple multicellular organisms that have a eukaryotic cell organization. The word *protist*, from the Greek for "the very first," reflects the idea that protists were the first eukaryotes to evolve. Protists are primarily aquatic organisms with diverse body forms, types of reproduction, modes of nutrition, and lifestyles. Some protists are adapted to carry out photosynthesis. Based mainly on molecular data that have clarified many evolutionary relationships among eukaryotes, the protists are no longer considered a kingdom. As we will learn in Chapter 26, several clades of protists have been identified.

Members of kingdom **Plantae** are complex multicellular organisms adapted to carry out photosynthesis. Among characteristic plant features are the *cuticle* (a waxy covering over aerial parts that reduces water loss) and *stomata* (tiny openings in stems and leaves for gas exchange); many plants have multicellular *gametangia* (organs that protect developing reproductive cells). Kingdom Plantae includes both nonvascular plants (mosses) and vascular plants (ferns, conifers, and flowering plants), those that have tissues specialized for transporting materials throughout the plant body. Most plants are adapted to terrestrial environments.

Kingdom **Fungi** is composed of the yeasts, mildews, molds, and mushrooms. Fungi do not photosynthesize. They obtain their nutrients by secreting digestive enzymes into food and then absorbing the predigested food. Kingdom **Animalia** is made up of multicellular organisms that obtain their nutrition by eating other organisms. Most animals exhibit considerable cell and tissue specialization and body organization. These characters have evolved along with complex sense organs, nervous systems, and muscular systems. Most animals reproduce sexually; they have large, nonmotile (do not move from place to place) eggs and small sperm with flagella that propel them in their journey to find the egg.

We have provided an introduction here to the groups of organisms that make up the tree of life. We will refer to them throughout this book as we consider the many kinds of challenges organisms face and the various adaptations that have evolved in response to them. We discuss the diversity of life in more detail in Chapters 23 through 32, and we summarize classification in Appendix B.

Species adapt in response to changes in their environment

Every organism is the product of numerous interactions between environmental conditions and the genes inherited from its ancestors. If all individuals of a species were exactly alike, any change in the environment might be disastrous to all, and the species would become extinct. Adaptations to changes in the environment occur as a result of evolutionary processes that take place over time and involve many generations.

Natural selection is an important mechanism by which evolution proceeds

Although philosophers and naturalists discussed the concept of evolution for centuries, Charles Darwin and Alfred Wallace first brought a scientific theory of evolution to general attention and suggested a plausible mechanism, **natural selection,** to explain it. In his book *On the Origin of Species by Natural Selection*, published in 1859, Darwin synthesized many new findings in geology and biology. He presented a wealth of evidence supporting his hypothesis that present forms of life descended, with modifications, from previously existing forms.

Darwin's scientific theory of evolution has helped shape the biological sciences to the present day. His work generated a great wave of scientific observation and research that has provided much additional evidence that evolution is responsible for the great diversity of organisms on our planet. Even today, the details of evolutionary processes are a major focus of investigation and discussion.

Darwin based his concept of natural selection on the following four observations:

1. Individual members of a species show some variation from one another.
2. Organisms produce many more offspring than will survive to reproduce (FIG. 1-12).
3. Because more individuals are produced than the environment can support, organisms must compete for necessary, but limited, resources such as food, sunlight, and space. Also, some organisms are killed by predators, disease organisms, or unfavorable natural conditions, such as weather changes. Which organisms are more likely to survive?
4. Individuals with characteristics that enable them to obtain and use resources, escape predators, resist disease organisms, and withstand changes in the environment are more likely to survive to reproductive maturity than those without these characteristics. The survivors that reproduce pass their adaptations for survival on to their offspring. Thus, the best-adapted individuals of a population produce, on average, more offspring than do other individuals. Because of this *differential reproduction*, a greater proportion of the population

Figure 1-12 Egg masses of the wood frog (*Rana sylvatica*)

Many more eggs are produced than can develop into adult frogs. Random events are largely responsible for determining which of these developing frogs will hatch, reach adulthood, and reproduce. However, certain traits of each organism also contribute to the probability for success in its environment. Not all organisms are as prolific as the frog, but the generalization that more organisms are produced than survive is true throughout the living world.

(a) The bill of this 'Akiapola'au male (*Hemignathus munroi*) is adapted for extracting insect larvae from bark. The lower mandible (jaw) is used to peck at and pull off bark, whereas the maxilla (upper jaw) and tongue remove the prey.

(b) 'I'iwi (*Vestiaria cocciniea*) in 'ohi'a blossoms. The bill is adapted for feeding on nectar in tubular flowers.

(c) Palila (*Loxiodes bailleui*) in mamane tree. This finch-billed honeycreeper feeds on immature seeds in pods of the mamane tree. It also eats insects, berries, and young leaves.

Figure 1-13 Adaptation and diversification in Hawaiian honeycreepers
All three species shown here are endangered, mainly because their habitats have been destroyed by humans or species introduced by humans.

becomes adapted to the prevailing environmental conditions and challenges. The environment *selects* the best-adapted organisms for survival. Note that *adaptation involves changes in populations rather than in individual organisms.*

Darwin did not know about DNA or understand the mechanisms of inheritance. Scientists now understand that most variations among individuals are a result of different varieties of genes that code for each characteristic. The ultimate source of these variations is random **mutations,** chemical or physical changes in DNA that persist and can be inherited. *Mutations modify genes and by this process provide the raw material for evolution.*

Populations evolve as a result of selective pressures from changes in their environment

All the genes present in a population make up its **gene pool.** By virtue of its gene pool, a population is a reservoir of variation. Natural selection acts on individuals within a population. Selection favors individuals with genes specifying traits that allow them to respond effectively to pressures exerted by the environment. These organisms are most likely to survive and produce offspring. As successful organisms pass on their genetic recipe for survival, their traits become more widely distributed in the population. Over time, as populations continue to change (and as the environment itself changes, bringing different selective pressures), the members of the population become better adapted to their environment and less like their ancestors.

As a population adapts to environmental pressures and exploits new opportunities for finding food, maintaining safety, and avoiding predators, the population diversifies and new species may evolve. The Hawaiian honeycreepers, a group of related birds, are a good example. When honeycreeper ancestors first reached Hawaii, few other birds were present, so there was little competition. Genetic variation among honeycreepers allowed some to move into different food zones, and over time, species with various types of bills evolved (**FIG. 1-13**; see also Chapter 20 and Fig. 20-18). Some honeycreepers now have long, curved

bills, adapted for feeding on nectar from tubular flowers. Others have short, thick bills for foraging for insects, and still others have adapted for eating seeds.

CHECKPOINT 1.6

- *The scientific name for the African rock python is* Python sebae. *Which name indicates its genus?*
- PREDICT *Why might biologists modify the tree showing the major forms of life?*
- CONNECT *How might you explain the sharp claws and teeth of tigers in terms of natural selection?*

1.7 THE PROCESS OF SCIENCE

LEARNING OBJECTIVES

10 Design a study to test a given hypothesis, using the procedure and terminology of the scientific method.

11 Compare the reductionist and systems approaches to biological research.

Biology is a science. The word *science* comes from a Latin word meaning "to know." Science is a way of thinking and a method of investigating the natural world in a systematic manner. We test ideas, and based on our findings, we modify or reject these ideas. The *process of science* is investigative, dynamic, and often controversial. The observations made, the range of questions asked, and the design of experiments depend on the creativity of the individual scientist. Science is influenced by cultural, social, historical, and technological contexts.

The **scientific method** involves a series of ordered steps. Using the scientific method, scientists make careful observations, ask critical questions, and develop *hypotheses*, which are tentative explanations. Using their hypotheses, scientists make predictions that can be tested by making further observations or by performing experiments. They gather *data,* information that they can analyze, often using computers and sophisticated statistical methods. They interpret the results of their experiments and draw conclusions from them. As we will discuss, scientists develop many hypotheses that cannot be tested by using all of the steps of the scientific method in a rigid way. Scientists use the scientific method as a generalized framework or guide.

Biologists explore every imaginable aspect of life from the structure of viruses and bacteria to the interactions of the communities of our biosphere. Some biologists work mainly in laboratories, and others do their work in the field (**FIG. 1-14**). Perhaps you will decide to become a research biologist and help unravel the complexities of the human brain, discover new hormones that stimulate plants to flower, identify new species of animals or bacteria, or develop new stem cell strategies to treat cancer, AIDS, or heart disease. Applications of basic biological research have provided the technology to transplant kidneys, livers, and hearts; manipulate genes; treat many diseases; and increase world food production. Biology has been a powerful force in providing the quality of life that most of us enjoy. You may choose to enter an applied field of biology, such as environmental science, dentistry, medicine, pharmacology, or veterinary medicine.

Science requires systematic thought processes

Science is systematic. Scientists organize and often quantify knowledge, making it readily accessible to all who wish to build on its foundation. In this way, science is both a personal and a social endeavor. Science is not mysterious. Anyone who understands its rules and procedures can take on its challenges. What distinguishes science is its insistence on rigorous methods to examine a problem. Science seeks to give precise knowledge about the natural world; the supernatural is not accessible to

Figure 1-14 Biologist at work
This biologist studying the rainforest canopy in Costa Rica is part of an international effort to study and preserve tropical rain forests. Researchers study the interactions of organisms and the effects of human activities on the rain forests.

scientific methods of inquiry. Science is not a replacement for philosophy, religion, or art. Being a scientist does not prevent one from participating in other fields of human endeavor, just as being an artist does not prevent one from practicing science.

Deductive reasoning begins with general principles
Scientists use two types of systematic thought processes: deduction and induction. With **deductive reasoning**, we begin with supplied information, called *premises*, and draw conclusions on the basis of that information. Deduction proceeds from general principles to specific conclusions. For example, if you accept the premise that all birds have wings and the second premise that sparrows are birds, you can conclude deductively that sparrows have wings. Deduction helps us discover relationships among known facts. A *fact* is information or knowledge based on evidence.

Inductive reasoning begins with specific observations
Inductive reasoning is the opposite of deduction. We begin with specific observations and draw a conclusion or discover a general principle. For example, you know that sparrows have wings, can fly, and are birds. You also know that robins, eagles, pigeons, and hawks have wings, can fly, and are birds. Considering these facts, you might *induce* that all birds have wings and fly. In this way, you can use the inductive method to organize raw data into manageable categories by answering this question: What do all these facts have in common?

A weakness of inductive reasoning is that conclusions generalize the facts to all possible examples. When we formulate the general principle, we go from many observed examples to all possible examples. This is known as an *inductive leap*. Without it, we could not arrive at generalizations. However, we must be sensitive to exceptions and to the possibility that the conclusion is not valid. For example, the kiwi bird of New Zealand does *not* have functional wings (**FIG. 1-15**). We can never conclusively prove a universal generalization. The generalizations in inductive conclusions come from the creative insight of the human mind, and creativity, however admirable, is not infallible.

Scientists make careful observations and ask critical questions

In 1928, British biologist Alexander Fleming observed that a blue mold had invaded one of his bacterial cultures. He almost discarded it, but then he noticed that the area contaminated by the mold was surrounded by a zone where bacterial colonies did not grow well. The bacteria were disease organisms of the genus *Staphylococcus*, which can cause boils and skin infections. Anything that could kill them was interesting! Fleming saved the mold, a variety of *Penicillium* (blue bread mold), and isolated the antibiotic penicillin from it. However, he had difficulty culturing it.

Even though Fleming recognized the potential practical benefit of penicillin, he did not develop the chemical techniques needed to purify it, and more than ten years passed before the drug was put to significant use. In 1939, Sir Howard Florey and Ernst Boris Chain developed chemical procedures to extract and produce the active agent penicillin from the mold. Florey took

Figure 1-15 Is this animal a bird?

The kiwi bird of New Zealand is about the size of a chicken. Its tiny 2-inch wings cannot be used for flight. The survivor of an ancient order of birds, the kiwi has bristly, hairlike feathers and other characteristics that qualify it as a bird.

the process to laboratories in the United States, and penicillin was first produced to treat wounded soldiers in World War II. In recognition of their work, Fleming, Florey, and Chain shared the 1945 Nobel Prize in Physiology or Medicine.

Chance often plays a role in scientific discovery

Fleming did not set out to discover penicillin. He benefited from the chance growth of a mold in one of his culture dishes. However, we may wonder how many times the same type of mold grew on the cultures of other biologists who failed to make the connection and simply threw away their contaminated cultures. Fleming benefited from chance, but his mind was prepared to make observations and formulate critical questions, and his pen was prepared to publish them. Significant discoveries are usually made by those who are in the habit of looking critically at nature and recognizing a phenomenon or problem. Of course, the technology necessary for investigating the problem must also be available.

A hypothesis is a testable statement

Scientists make careful observations, ask critical questions, and develop hypotheses. A **hypothesis** is a tentative explanation for observations or phenomena. A hypothesis is an abstract idea, but based on their hypotheses, scientists can make predictions that can be tested. For example, we might predict that biology students who study for ten hours will do better on an exam than students who study for one hour. As used here, a *prediction* is a deductive, logical consequence of a hypothesis. It does not have to be a future event. Hypotheses are typically stated in the form of predictions.

In the early stages of an investigation, a scientist typically thinks of many possible hypotheses. A good hypothesis exhibits the following characteristics. (1) It is reasonably consistent with well-established facts. (2) It generates predictions that can be tested (whether the results are positive or negative). (3) Test results should be repeatable by independent observers. (4) It is

falsifiable, which means that it can be proven false, as we will discuss in the next section.

After generating hypotheses, the scientist decides which, if any, could and should be subjected to experimental test. Why not test them all? Time and money are important considerations in conducting research. Scientists must establish priority among the hypotheses to decide which to test first.

A falsifiable hypothesis can be tested In science, a well-stated hypothesis can be tested. If no evidence is found to support it, the hypothesis is rejected. The hypothesis can be shown to be false. Even results that do not support the hypothesis may be valuable and may lead to new hypotheses. If the results do support a hypothesis, a scientist may use them to generate related hypotheses.

A hypothesis is not true just because some of its predictions (the ones people happen to have thought of or have thus far been able to test) have been shown to be true. After all, they could be true by coincidence. In fact, a hypothesis can be supported by data, but it cannot really be *proven* true.

An **unfalsifiable hypothesis** cannot be proven false; in fact, it cannot be scientifically investigated. Belief in an unfalsifiable hypothesis, such as the existence of invisible and undetectable elves, must be rationalized on grounds other than scientific ones.

Models are important in developing and testing hypotheses Hypotheses have many potential sources, including direct observations or even computer simulations. Increasingly in biology, hypotheses may be derived from *models* that scientists have developed to provide a comprehensive explanation for a large number of observations. Examples of such testable models include the model of the structure of DNA and the model of the structure of the plasma membrane (discussed in Chapter 5).

The best design for an experiment can sometimes be established by performing computer simulations. Virtual testing and evaluation are undertaken before the experiment is performed in the laboratory or field. Modeling and computer simulation save time and money.

Many hypotheses can be tested by experiment Many hypotheses can be tested by controlled experiments. Early biologists observed that the nucleus was the most prominent part of the cell, and they hypothesized that cells would be adversely affected if they lost their nuclei. Biologists predicted that if the nucleus were removed from the cell, the cell would die. They then experimented by surgically removing the nucleus of a unicellular amoeba. The amoeba continued to live and move, but it did not grow, and after a few days it died. These results suggested that the nucleus is necessary for the metabolic processes that provide for growth and cell reproduction.

But, the investigators asked, what if the operation itself, not the loss of the nucleus, caused the amoeba to die? They performed a controlled experiment, subjecting two groups of amoebas to the same operative trauma (**FIG. 1-16**). Ideally, an experimental group differs from a control group only with respect to the variable being studied. In the **control group,** the researcher inserted a microloop into each amoeba and pushed it around inside the cell to simulate removal of the nucleus; then the instrument was withdrawn, leaving the nucleus inside. In the **experimental group,** the nucleus was removed; in the control group, it was not.

Amoebas treated with such a sham operation recovered and subsequently grew and divided. This experiment showed that the removal of the nucleus, not simply the operation, caused the death of the amoebas. The conclusion is that amoebas cannot live without their nuclei. The results supported the hypothesis that if cells lose their nuclei, they are adversely affected. We can conclude that the nucleus is essential for the survival of the amoeba.

Researchers must avoid bias

In scientific studies, researchers must do their best to avoid bias or preconceived notions of what should happen. For example, to prevent bias, most medical experiments are carried out in a double-blind fashion. When a drug is tested, one group of patients receives the new medication, and a control group of matched patients receives a placebo (a harmless starch pill similar in size, shape, color, and taste to the pill being tested). This method is called a *double-blind study* because neither the patient nor the physician knows who is getting the experimental drug and who is getting the placebo. The pills or treatments are coded in some way, and the code is broken only after the experiment is over and the results are recorded. Not all experiments can be so neatly designed; for example, it is often difficult to establish appropriate controls.

Scientists interpret the results of experiments and make conclusions

Scientists gather data in an experiment, interpret their results, and then draw conclusions from them. In the amoeba experiment described earlier, investigators concluded that the data supported the hypothesis that the nucleus is essential for the survival of the cell. When the results do support a hypothesis, scientists may use them to generate related hypotheses. Even results that do not support the hypothesis may be valuable and may lead to new hypotheses.

Let us discuss another experiment. Research teams studying chimpanzee populations in Africa have reported that

chimpanzees appear to learn specific ways to use tools from one another. Behavior that is learned from others in a population and passed to future generations is what we call "culture." In the past, most biologists have thought that only humans had culture. It has been difficult to test this type of learning in the field, and the idea has been controversial.

Biologists have asked critical questions about whether chimpanzees learned how to use tools by observing one another. Investigators at Yerkes National Primate Research Center in Atlanta developed a hypothesis that chimpanzees can learn particular ways to use tools by observing other chimps. They

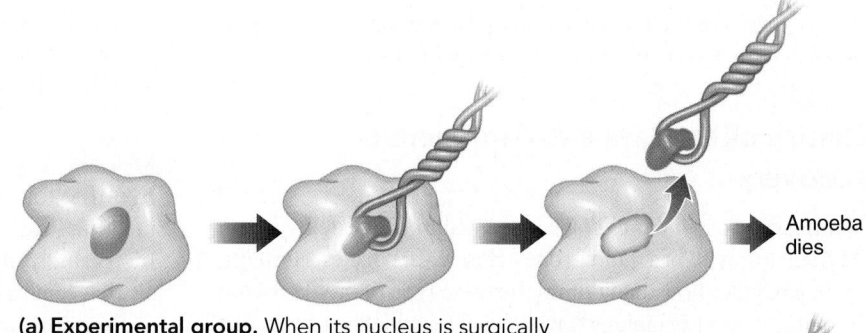
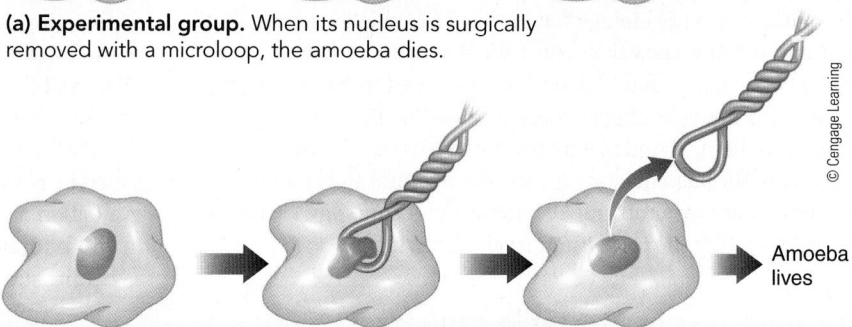

Can chimpanzees learn how to use tools by observing one another?

DEVELOP HYPOTHESIS: Chimpanzees can learn particular ways to use tools by observing other chimps.

PERFORM EXPERIMENTS: One female in each of two groups of 16 chimps was educated in a specific way to use a stick to obtain food. The two educated chimps were then returned to their respective groups. Chimpanzees in a control group were not taught how to use a stick.

David Bygott

RESULTS: Chimpanzees in each experimental group observed the use of the stick by the educated chimp, and a majority began to use the stick in the same way. When tested two months later, many of the chimps in each group continued to use the stick. *The results presented here have been simplified and are based on the number of chimps observed to use the learned method at least ten times.* All but one chimp in each group learned the technology, but a few used it only a few times. Some chimps taught themselves the alternative method and used that alternative. However, most conformed to the group's use of the method that the investigator taught to the educated chimp.

CONCLUSION: Chimpanzees learn specific ways to use tools by observing other chimps. The hypothesis was supported.

SOURCE: Whiten, A., Horner, V., and de Waal, F.B.M. "Conformity to Cultural Norms of Tool Use in Chimpanzees," *Nature*, Vol. 437, Sept. 29, 2005.

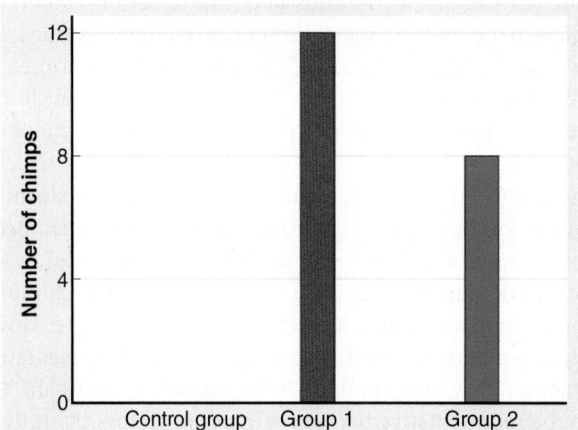

(a) Number of chimpanzees who successfully employed specific method of tool use.

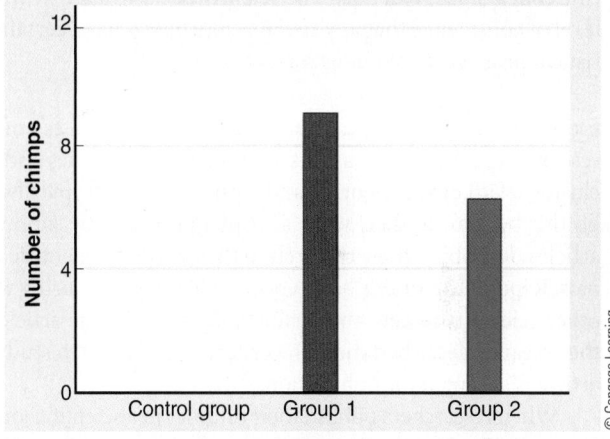

© Cengage Learning

(b) Number of chimpanzees who successfully employed learned method of tool use two months later.

Figure 1-17 An experiment testing learning in chimpanzee populations

In the photo, wild chimpanzees are shown observing a member of their group using a tool.

PREDICT If you repeated this experiment, how do you think your results would compare with those shown here?

predicted that if they taught one chimp to use a stick to obtain food from a dispenser, other chimps would learn the technique from the educated one (**FIG. 1-17**).

These researchers divided chimpanzees into two experimental groups with 16 in each group. Then they taught a high-ranking female in each group to use a stick to obtain food from an apparatus. The two chimps were taught different methods.

One chimp was taught to poke the stick inside the device to free the food. The other was taught to use the stick to lift a hook that removed a blockage, allowing the food to roll forward out of the device.

A third group served as a control group. The chimps in the control group were given access to the sticks and the apparatus with the food inside, but none were taught how to use the sticks.

All the control-group chimps manipulated the apparatus with the stick, but none succeeded in releasing food.

When the chimps were returned to their groups, other chimps observed how the educated chimps used the stick, and a large majority began to use sticks in the same way. The chimps in each experimental group (Group 1 and Group 2 in Fig. 1-17) learned the specific style of using the stick that their educated chimp had been taught. Most used the stick to obtain food at least ten times. Two months later, the apparatus was reintroduced to the chimps. Again, most of the chimps used the learned technique for obtaining food. The results of the experiment supported the hypothesis. The researchers concluded that chimpanzees are capable of culturally transmitting learned technology.

Sampling error can lead to inaccurate conclusions One reason for inaccurate conclusions is *sampling error*. Because not *all* cases of what is being studied can be observed or tested (scientists cannot study every amoeba or every chimpanzee population), scientists must be content with a sample. How can scientists know whether that sample is truly representative of whatever they are studying? If the sample is too small, it may not be representative because of random factors. A study with only two, or even nine, amoebas may not yield reliable data that can be generalized to other amoebas. If researchers test a large number of subjects, they are more likely to draw accurate scientific conclusions (**FIG. 1-18**). The scientist seeks to state with some level of confidence that any specific conclusion has a certain statistical probability of being correct.

Experiments must be repeatable After scientists conduct research and draw conclusions, they share their studies and conclusions with other scientists and with the public. Typically, they do this by sharing their work at conferences and by submitting articles describing their research to the editors of scientific journals. Reputable journals have a peer review process during which other scientists review and evaluate the study. If the article and the research described meet the criteria for a research study, the article is accepted for publication.

When researchers publish their findings in a scientific journal, they typically describe their methods and procedures in sufficient detail so that other scientists can repeat the experiments. When the findings are replicated, the conclusions are, of course, strengthened.

A scientific theory is supported by tested hypotheses

Nonscientists often use the word *theory* incorrectly to refer to a hypothesis or even to some untestable idea they wish to promote. A **scientific theory** is actually an integrated explanation of some aspect of the natural world that is based on a number of hypotheses, each supported by consistent results from many observations or experiments. A scientific theory relates data that previously appeared unrelated. A good scientific theory grows, building on additional facts as they become known. It predicts new facts and suggests new relationships among phenomena. It may even suggest practical applications.

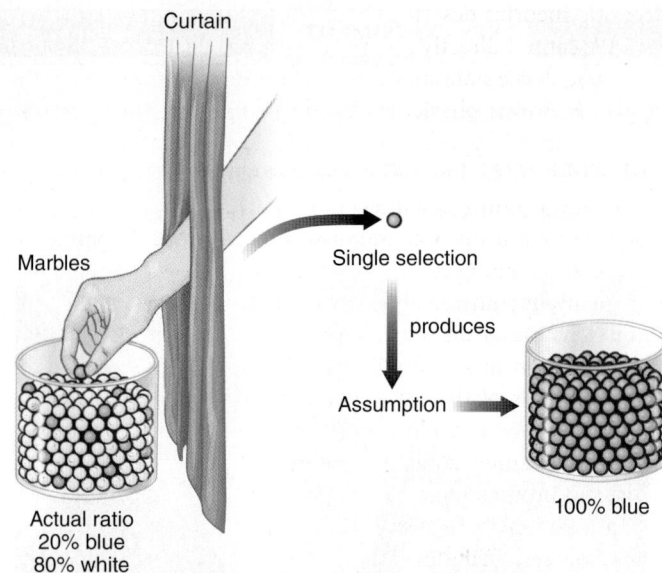

(a) Taking a single selection can result in sampling error. If the only marble selected is blue, we might assume all the marbles are blue.

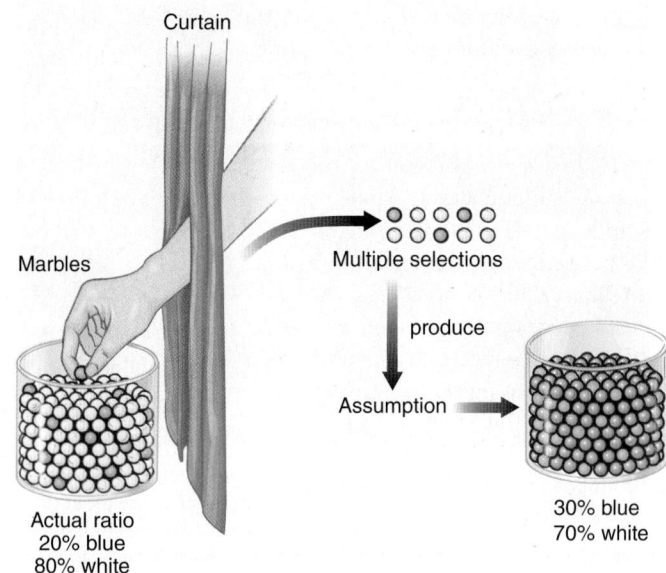

(b) The greater the number of selections we take of an unknown, the more likely we can make valid assumptions about it.

Figure 1-18 *Animation* **Statistical probability**
© Cengage Learning

A scientific theory, by showing the relationships among classes of facts, simplifies and clarifies our understanding of the natural world. As Albert Einstein wrote, "In the whole history of science from Greek philosophy to modern physics, there have been constant attempts to reduce the apparent complexity of natural phenomena to simple, fundamental ideas and relations." Developing scientific theories is indeed a major goal of science.

Many hypotheses cannot be tested by direct experiment

Some well-accepted scientific theories do not lend themselves to hypothesis testing by ordinary experiments. Often, these scientific theories describe events that occurred in the distant past. We cannot directly observe the origin of the universe from a very hot, dense state about 13.7 billion years ago (the Big Bang theory). However, physicists and cosmologists have been able to formulate many hypotheses related to the Big Bang and to test many of the predictions derived from these hypotheses.

Similarly, humans did not observe the evolution of major groups of organisms because that process took place over millions of years and occurred before humans had evolved. However, many hypotheses about evolution have been posed, and predictions based on them have been tested. For example, if complex organisms evolved from simple life-forms, we would find the fossils of the simplest organisms in the oldest strata (rock layers). As we explore more recent strata, we would expect to find increasingly complex organisms. Indeed, scientists have found this progression of simple to complex fossils.

In addition to fossils, evidence for evolution comes from many sources, including physical and molecular similarities between organisms. Evidence also comes from recent and current studies of evolution in action. Many aspects of ongoing evolution can, in fact, be studied in the laboratory or in the field. The evidence for evolution is so compelling that virtually all scientists today accept evolutionary theory as an integral part of biology.

Paradigm shifts accommodate new discoveries

A *paradigm* is a set of assumptions or concepts that constitute a way of thinking about reality. For example, from the time of Aristotle to the mid-19th century, biologists thought that organisms were either plants (kingdom Plantae) or animals (kingdom Animalia). This concept was deeply entrenched. However, with the development of microscopes, investigators discovered tiny life-forms—bacteria and protists—that were neither plant nor animal. Biologists had to make a *paradigm shift*—that is, they changed their view of reality—to accommodate this new knowledge. They assigned these newly discovered organisms to new kingdoms. In a more recent paradigm shift, biologists have revised their idea that we are born with all the brain cells we will ever have. We now understand that certain areas of the brain continue to produce new neurons throughout life.

Systems biology integrates different levels of information

In the *reductionist* approach to biology, researchers study the simplest components of biological processes. Their goal is to synthesize their knowledge of many small parts to understand the whole. Reductionism has been (and continues to be) important in biological research. However, as biologists and their tools have become increasingly sophisticated, huge amounts of data have been generated, bringing the science of biology to a different level.

Systems biology is a field of biology that builds on information provided by the reductionist approach and develops large data sets, typically analyzed by computers. Systems biology is also referred to as *integrative biology*. Reductionism and systems biology are complementary approaches. Using reductionism, biologists have discovered basic information about components, such as molecules, genes, cells, and organs. Systems biologists, who focus on systems as a whole rather than on individual components, need this basic knowledge to study, for example, the *interactions* among various parts and levels of an organism.

Systems biologists integrate data from various levels of complexity with the goal of understanding the big picture—how biological systems function. For example, systems biologists are developing models of different aspects of cell function. Normal cell function depends on the precise actions of hundreds of proteins that relay signals received from other cells. Proteins also relay signals from one part of the cell to another. Researchers are producing detailed maps of the molecular pathways that maintain cell function (FIG. 1-19).

One group of researchers has developed a model consisting of nearly 8000 chemical signals involved in a molecular network that leads to programmed cell death. By understanding cell communication, the interactions of genes and proteins in metabolic pathways, and physiological processes, systems biologists hope to eventually develop a model of the whole organism. Systems biology is increasingly used to study disease processes. For example, the interactions between the pathogen and the host cell can be mapped.

The development of systems biology was fueled by the huge amount of data generated by the Human Genome Project. Researchers working on this and related projects have identified the DNA sequences that make up the *human genome*, the complete set of human genetic material. Computer software developed for the Human Genome Project can analyze very

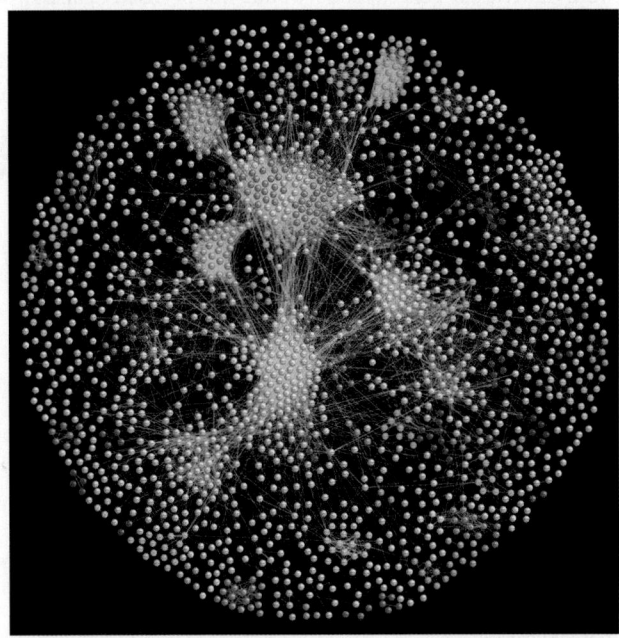

Figure 1-19 Map illustrating interactions among thousands of proteins in a cell of the fruit fly (*Drosophila melanogaster*)

Researchers gain insight into the cell as an integrated system by studying the interaction of its proteins. Because humans and many other animals share a common ancestor with fruit flies, similar cellular mechanisms operate in all of them. Understanding protein interactions and functions helps scientists understand cellular changes in health and disease. Each dot represents one protein.

Source: Guruharsha , KG, et al., 2011 Cell, Elsevier Limited; Image provided by Dr. Guruharsha, Artavanis-Tsakonas group, Harvard Medical School. Cell. 2011 Oct 28; 147(3):690–703.

large data sets. These programs are being used to integrate data about protein interactions and many other aspects of molecular biology. Systems biologists view biology in terms of information systems. Increasingly, they depend on mathematics, statistics, and engineering principles.

Science has ethical dimensions

Scientific investigation depends on a commitment to practical ideals, such as truthfulness and the obligation to communicate results. Honesty is particularly important in science. Consider the great (although temporary) damage done whenever an unprincipled or even desperate researcher, whose career may depend on the publication of a research study, knowingly disseminates false data. Until the deception is uncovered, other researchers may devote thousands of dollars and hours of precious professional labor to futile lines of research inspired by erroneous reports. Deception can also be dangerous, especially in medical research. Fortunately, science tends to correct itself through consistent use of the scientific process. Sooner or later, someone's experimental results are sure to cast doubt on false data.

Research in such areas as stem cells and the human genome brings with it many ethical concerns and responsibilities. For example, how do people safeguard the privacy of genetic information? Suppose that you have a family history of breast cancer and learn from genetic testing that you have one of the BRCA mutations. These mutations increase risk for developing breast cancer and certain other cancers. How can you be certain that knowledge of your individual genetic code would not be used against you when you seek employment or by insurance companies?

Scientists must be ethically responsible and must help educate people about their work, including its benefits relative to its risks. It is significant that at the very beginning of the Human Genome Project, part of its budget was allocated for research on the ethical, legal, and social implications of its findings.

Science, technology, and society interact

Science and technology continuously interact. As scientists doing basic research report new findings, engineers and other inventors develop new products. Many of those products contribute to our quality of life. New technology also provides scientists with more powerful tools for their research and increases the potential for new discoveries. For example, a few years ago determining the genome of any eukaryote required several rooms filled with machines that could sequence the genes. This endeavor also cost millions of dollars. New technologies have revolutionized gene sequencing, allowing complex genomes to be determined quickly, with less equipment, and at far less expense. Plans are under way to determine the genomes of about 10,000 vertebrate species.

Science and technology continue to change society, and these changes present new challenges. In addition to being ethical about their own work, scientists face many societal and political issues surrounding areas such as genetic research, stem cell research, cloning, climate change, and human and animal experimentation. Scientists, and the larger society, will need to determine whether the potential benefits of any research outweigh its ethical risks.

CHECKPOINT 1.7

- *What are the characteristics of a good hypothesis?*
- *Describe a "controlled" experiment.*
- CONNECT *In what ways does systems biology depend on reductionism?*

SUMMARY: FOCUS ON LEARNING OBJECTIVES

1.1 Major Themes of Biology *(page 2)*

1 Describe five basic themes of biology.

- (1) Every organism is a biological system made up of many other biological systems, and every organism is interdependent with many other biological systems. (2) Structure and function are inter-related in all biological systems. (3) Information must be transferred within organisms and among organisms, and organisms must be able to receive information from the nonliving environment. (4) All life processes require a continuous input of energy. (5) Evolution results in populations changing over time.

1.2 Characteristics of Life *(page 3)*

2 Distinguish between living and nonliving things by describing the features that characterize living organisms.

- Every living organism is composed of one or more **cells.** Living things grow by increasing the size and/or number of their cells.
- **Metabolism** includes all the chemical activities that take place in the organism, including the chemical reactions essential to nutrition, growth and repair, and conversion of energy to usable forms. **Homeostasis** refers to the appropriate, balanced internal environment, and the organized tendency of the organism to maintain such a steady state.

- Organisms respond to **stimuli,** physical or chemical changes in their external or internal environment. Responses typically involve movement.
- In **asexual reproduction,** offspring are typically identical to the single parent, except for size. In most plants and animals **sexual reproduction** involves the fusion of an egg and sperm. Genes are typically contributed by two parents, and there is variation in the offspring.
- As populations evolve, they become adapted to their environment. **Adaptations** are traits that increase an organism's ability to survive in its environment.

1.3 Levels of Biological Organization *(page 6)*

3 Construct a hierarchy of biological organization, including levels characteristic of individual organisms and levels characteristic of ecological systems.

- Biological organization is hierarchical. In a complex organism, cells associate to form **tissues** (e.g., muscle and connective tissues) that carry out specific functions. In most multicellular organisms tissues organize to form functional structures called **organs** (e.g., heart or brain). An organized group of tissues and organs form an **organ system** (e.g., the nervous system). Functioning together, organ systems make up a complex, multicellular **organism.**

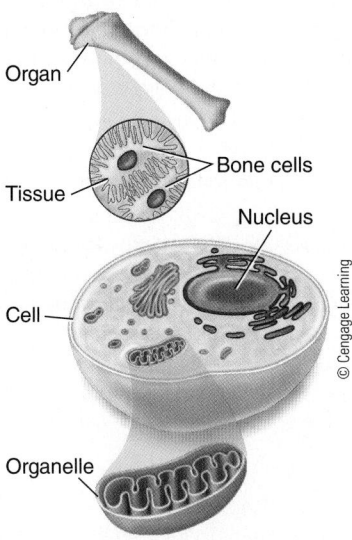

Organ

Tissue

Bone cells

Nucleus

Cell

Organelle

© Cengage Learning

- The basic unit of ecological organization is the **population.** Various populations form **communities,** a community and its physical environment are an **ecosystem,** and all Earth's ecosystems together make up the **biosphere.**

1.4 Information Transfer *(page 8)*

4 Summarize the importance of information transfer within and between living systems, giving specific examples.

- Organisms transmit information chemically, electrically, and behaviorally.
- DNA, which makes up the **genes,** is the hereditary material. Information encoded in DNA is transmitted from one generation to the next. DNA contains the instructions for the development of an organism and for carrying out life processes. Among its many functions, DNA codes for **proteins,** which are important in determining the structure and function of cells and tissues.
- **Hormones,** chemical messengers that transmit messages from one part of an organism to another, are important in **cell signaling.**
- Many organisms use electrical signals to transmit information; most animals have nervous systems that transmit electrical impulses and release **neurotransmitters.**

1.5 The Energy of Life *(page 9)*

5 Summarize the flow of energy through ecosystems and contrast the roles of producers, consumers, and decomposers.

- Activities of living cells require energy. Life depends on continuous energy input from the sun. During **photosynthesis,** plants, algae, and certain bacteria use the energy of sunlight to synthesize complex molecules from carbon dioxide and water.
- Virtually all cells carry on **cellular respiration,** a biochemical process in which they capture the energy stored in nutrients by producers. Some of that energy is then used to synthesize required materials or to carry out other cell activities.
- A self-sufficient ecosystem includes **producers,** or **autotrophs,** which make their own food; **primary consumers,** which eat producers, and typically **secondary consumers** that eat primary consumers; and **decomposers,** which obtain energy by breaking down wastes and dead organisms. Consumers and decomposers are **heterotrophs,** organisms that depend on producers as an energy source and for food and oxygen.

1.6 Evolution: The Basic Unifying Concept of Biology *(page 10)*

6 Demonstrate the binomial system of nomenclature by using specific examples and classify an organism (such as a human) in its domain, kingdom, phylum, class, order, family, genus, and species.

- Millions of species have evolved. A **species** is a group of organisms with similar structure, function, and behavior that, in nature, breed only with one another. Members of a species contribute to a common **gene pool** and share a common ancestry.
- Biologists use a **binomial system of nomenclature** in which the name of each species includes a **genus** name and a **specific epithet.** Traditional taxonomic classification is hierarchical; it includes species, genus, **family, order, class, phylum, kingdom,** and **domain.** Each grouping is referred to as a **taxon.** A group of organisms with a common ancestor is a **clade.**

7 Identify the three domains and the kingdoms of living organisms, and give examples of organisms assigned to each group.

- Bacteria and archaea are **prokaryotes;** all other organisms are **eukaryotes.** Prokaryotes make up two of the three domains.
- Organisms are classified in three domains: **Archaea, Bacteria, and Eukarya** and several kingdoms or clades: **Archaea, Bacteria, Fungi** (e.g., molds and yeasts), **Plantae,** and **Animalia. Protists** (e.g., algae, water molds, slime molds, and amoebas) are now assigned to several clades.

8 Give a brief overview of the scientific theory of evolution and explain why it is the principal unifying concept in biology.

- **Evolution** is the process by which populations change over time in response to changes in the environment. The scientific theory of evolution explains how millions of species came to be and helps us understand the structure, function, behavior, and relationships of organisms.
- **Natural selection,** the major mechanism by which evolution proceeds, favors individuals with traits that enable them to cope with environmental changes. Charles Darwin based his theory of natural selection on his observations that individuals of a species vary, organisms produce more offspring than survive to reproduce, organisms must compete for limited resources, and individuals that are best adapted to their environment are more likely to survive and reproduce, thereby passing on their hereditary information. Their traits become more widely distributed in the population.
- The source of variation in a population is random **mutation.**

9 Apply the concept of natural selection to any given adaptation and suggest a logical explanation of how the adaptation may have evolved.

- When the ancestors of Hawaiian honeycreepers first reached Hawaii, few other birds were present, so there was little competition for food. Through many generations, honeycreepers with longer, more curved bills became adapted for feeding on nectar from tubular flowers. Perhaps those with the longest, most curved bills were best able to survive in this food zone and lived to transmit their genes to their offspring. Those with shorter, thicker bills were more successful foraging for insects and passed their genes to new generations of offspring. Eventually, different species evolved that were adapted to specific food zones.

1.7 The Process of Science *(page 15)*

10 Design a study to test a given hypothesis, using the procedure and terminology of the scientific method.

- The *process of science* is a dynamic approach to investigation. The **scientific method** is a general framework that scientists use in their work; it includes observing, recognizing a problem or stating a critical question, developing a hypothesis, making a prediction that can be tested, making further observations, performing experiments, interpreting results, and drawing conclusions that support or falsify the hypothesis.

- Deductive reasoning and inductive reasoning are two categories of systematic thought used in the scientific method. **Deductive reasoning** proceeds from general principles to specific conclusions and helps us discover relationships among known facts. **Inductive reasoning** begins with specific observations and draws conclusions from them. Inductive reasoning helps us discover general principles.

- A **hypothesis** is a tentative explanation for observations or phenomena. A hypothesis can be tested. If no evidence is found to support it, the hypothesis is rejected.

- A well-designed scientific experiment typically includes both a **control group** and an **experimental group,** and must be as free as possible from bias. The control group should be as closely matched to the experimental group as possible. Ideally, the experimental group differs from the control group only with respect to the variable being studied.

- A **scientific theory** is an integrated explanation of some aspect of the natural world that is based on a number of hypotheses, each supported by consistent results from many observations or experiments.

11 Compare the reductionist and systems approaches to biological research.

- Using reductionism, researchers study the simplest components of biological processes such as molecules or cells. **Systems biology** uses knowledge provided by reductionism. Systems biologists integrate data from various levels of complexity with the goal of understanding how biological systems function.

TEST YOUR UNDERSTANDING

Know and Comprehend

1. Cells (a) are not found among the bacteria (b) always have nuclei (c) are the building blocks of living organisms (d) are made up of tissues (e) a and b are true

2. DNA (a) is produced during cellular respiration (b) functions mainly to transmit information from one species to another (c) cannot be changed (d) is a good example of a biological system (e) makes up the genes

3. Cellular respiration (a) is a process whereby sunlight is used to synthesize cell components with the release of energy (b) occurs in heterotrophs only (c) is carried on by both autotrophs and heterotrophs (d) causes chemical changes in DNA (e) occurs in response to environmental changes

4. Fungi are assigned to domain (a) Protista (b) Archaea (c) Bacteria (d) Eukarya (e) Plantae

5. The scientific name for corn is *Zea mays*. *Zea* is the (a) specific epithet (b) genus (c) class (d) kingdom (e) phylum

6. Darwin suggested that evolution takes place by (a) mutation (b) changes in the individuals of a species (c) natural selection (d) interaction of hormones during competition for resources (e) homeostatic responses to each change in the environment

7. Ideally, an experimental group differs from a control group (a) only with respect to the hypothesis being tested (b) because its subjects are more reliable (c) in that it is less subject to bias (d) in that it is less vulnerable to sampling error (e) only with respect to the variable being studied

Apply and Analyze

8. Which of the following is a correct sequence of levels of biological organization? 1. organ system 2. chemical 3. tissue 4. organ 5. cell (a) 2, 3, 5, 4, 1 (b) 5, 3, 4, 1, 2 (c) 2, 5, 3, 1, 4 (d) 2, 5, 3, 4, 1 (e) 5, 2, 3, 4, 1

9. **VISUALIZE** Draw a simple cladogram illustrating the relationships among the following: Common ancestor of all organisms, Domain Eukarya, Domain Bacteria, Domain Archaea. To which domain do the organisms informally known as protists belong? To which domain do you belong? Refer to Figure 1-11 to check your answer.

10. **PREDICT** What would happen if a homeostatic mechanism failed? Give an example using a homeostatic mechanism at work in your body (other than the regulation of glucose cited in the chapter).

11. What are some characteristics of a good hypothesis? Give an example.

12. **PREDICT** Make a prediction and devise a suitably controlled experiment to test each of the following hypotheses: (a) A type of mold found in your garden produces an effective antibiotic. (b) The growth rate of a bean seedling is affected by temperature. (c) Estrogen alleviates symptoms of Alzheimer's disease in elderly women.

13. Contrast the reductionist approach with systems biology. How are the two approaches complementary? Which approach is more likely to consider emergent properties?

Evaluate and Synthesize

14. **INTERPRET DATA** Compare the two graphs in Figure 1-17. What information does the second graph illustrate? What possible explanation can you give for the differences shown in the two graphs?

15. **EVOLUTION LINK** In what ways does evolution depend on transfer of information? In what ways does transfer of information depend on evolution?

16. **EVOLUTION LINK** How might an understanding of evolutionary processes help a biologist doing research in (a) the development of a new antibiotic to replace one to which bacteria have become resistant? (b) conservation of a specific plant in a rain forest?

17. **SCIENCE, TECHNOLOGY, AND SOCIETY** In the future, gene technology may make it possible for parents to produce children with athletic ability, artistic talent, or high IQ. Do you have any ethical concerns about these possibilities? If so, where and how would you draw the line?

aplia To access course materials, such as Aplia and other companion resources, please visit **www.cengagebrain.com.**

Atoms and Molecules: The Chemical Basis of Life

2

Chemical elements forming in stars. These clusters of young stars photographed by the Hubble Space Telescope are about 20,000 light years away in a giant nebula (NGC 3603) in our Milky Way galaxy.

The only elements formed at the beginning of our universe were hydrogen and helium. Since the "Big Bang," all the naturally occurring heavier elements have formed through processes occurring in stars, such as those depicted in the photograph. In this chapter we consider how the properties of the elements serve as the fundamental underpinnings for the ways structure and function work together in organisms and how these properties enable organisms to store and use both information and energy. Therefore, knowledge of chemistry is essential for understanding organisms and how they function. The chemical similarities among all organisms on Earth provide strong evidence for the evolution of all organisms from a common ancestor and explain why much of what biologists learn from studying bacteria or rats in laboratories can be applied to other organisms, including humans. Furthermore, the basic chemical and physical principles governing organisms are not unique to living things, for they apply to nonliving systems as well.

The success of the Human Genome Project and related studies has relied heavily on biochemistry and *molecular biology,* the chemistry and physics of the molecules that constitute living things. A biochemist may investigate the precise interactions among a cell's atoms and molecules that maintain the energy flow essential to life, and a molecular biologist may study how proteins interact with deoxyribonucleic acid (DNA) in ways that control the expression of certain genes. However, an understanding of chemistry is essential to *all* biologists. An evolutionary biologist may study evolutionary relationships by comparing the DNA of different types of organisms. An ecologist may study how energy is transferred among the organisms living in an estuary or may monitor the biological effects of changes in the salinity of the water. A botanist may study unique compounds produced by plants and may even be a "chemical prospector," seeking new sources of medicinal agents.

In this chapter we lay a foundation for understanding how the structure of atoms determines the way they form chemical bonds to produce complex compounds. Most of our discussion focuses on small, simple substances known as **inorganic compounds.** Among the biologically important groups of inorganic compounds are water, many simple acids and bases, and simple salts. We pay particular attention to water, the most abundant

KEY CONCEPTS

2.1 Carbon, hydrogen, oxygen, and nitrogen are the most abundant elements in living things.

2.2 The chemical properties of an atom are determined by its highest-energy electrons, known as valence electrons.

2.3 A molecule consists of atoms joined by covalent bonds. Other important chemical bonds include ionic bonds. Hydrogen bonds and van der Waals interactions are weak attractions.

2.4 The energy of an electron is transferred in a redox reaction.

2.5 Water molecules are polar, having regions of partial positive charge and partial negative charge that permit them to form hydrogen bonds with one another and with other charged substances.

2.6 Acids are hydrogen ion donors; bases are hydrogen ion acceptors. The pH scale is a convenient measure of the hydrogen ion concentration of a solution.

substance in organisms and on Earth's surface, and we examine how its unique properties affect living things as well as their nonliving environment. In Chapter 3 we extend our discussion to **organic compounds,** carbon-containing compounds that are generally large and complex. In all but the simplest organic compounds, two or more carbon atoms are bonded to each other to form the backbone, or skeleton, of the molecule.

2.1 ELEMENTS AND ATOMS

Elements are substances that cannot be broken down into simpler substances by ordinary chemical reactions. Each element has a **chemical symbol:** usually, it is the first letter or first and second letters of the English or Latin name of the element. For example, O is the symbol for oxygen, C for carbon, H for hydrogen, N for nitrogen, and Na for sodium (from the Latin word *natrium*). Just four elements—oxygen, carbon, hydrogen, and nitrogen—are responsible for more than 96% of the mass of most organisms. Others, such as calcium, phosphorus, potassium, and magnesium, are also consistently present but in smaller quantities. Some elements, such as iodine and copper, are known as *trace elements* because they are required only in minute amounts. TABLE 2-1 lists the elements that make up organisms and briefly explains the importance of each in typical plants and animals.

An **atom** is defined as the smallest portion of an element that retains its chemical properties. Atoms are much too small to be visible under a light microscope, but by sophisticated techniques researchers have been able to photograph the positions of some large atoms in molecules.

The components of atoms are tiny particles of **matter** (anything that has mass and takes up space) known as *subatomic particles*. Physicists have discovered a number of subatomic particles, but for our purposes we need consider only three: electrons, protons, and neutrons. An **electron** is a particle that carries a unit of negative electric charge, a **proton** carries a unit of positive charge, and a **neutron** is an uncharged particle. In an electrically neutral atom, the number of electrons is equal to the number of protons.

Clustered together, protons and neutrons compose the atomic **nucleus.** Electrons, however, have no fixed locations and move rapidly through the mostly empty space surrounding the atomic nucleus.

An atom is uniquely identified by its number of protons

Every element has a fixed number of protons in the atomic nucleus, known as the **atomic number.** It is written as a subscript to the left of the chemical symbol. Thus, $_1H$ indicates that the hydrogen nucleus contains 1 proton, and $_8O$ means that the oxygen nucleus contains 8 protons. The atomic number determines an atom's identity and defines the element.

The **periodic table** is a chart of the elements arranged in order by atomic number (FIG. 2-1 and Appendix A). The periodic table is useful because it lets us simultaneously correlate many of the relationships among the various elements.

Figure 2-1 includes representations of the **electron configurations** of several elements important in organisms. These *Bohr models*, which show the electrons arranged in a series of concentric circles around the nucleus, are convenient to use, but inaccurate. The space outside the nucleus is actually extremely large compared with the nucleus, and as you will see, electrons do not actually circle the nucleus in fixed concentric pathways.

Protons plus neutrons determine atomic mass

The mass of a subatomic particle is exceedingly small, much too small to be conveniently expressed in grams or even

TABLE 2-1	Functions of Elements in Organisms
ELEMENT* (CHEMICAL SYMBOL)	**FUNCTIONS**
O OXYGEN	Required for cellular respiration; present in most organic compounds; component of water
C CARBON	Forms backbone of organic molecules; each carbon atom can form four bonds with other atoms
H HYDROGEN	Present in most organic compounds; component of water; hydrogen ion (H^+) is involved in some energy transfers
N NITROGEN	Component of proteins and nucleic acids; component of chlorophyll in plants
Ca CALCIUM	Structural component of bones and teeth; calcium ion (Ca^{2+}) is important in muscle contraction, conduction of nerve impulses, and blood clotting; associated with plant cell wall
P PHOSPHORUS	Component of nucleic acids and of phospholipids in membranes; important in energy transfer reactions; structural component of bone
K POTASSIUM	Potassium ion (K^+) is a principal positive ion (cation) in interstitial (tissue) fluid of animals; important in nerve function; affects muscle contraction; controls opening of stomata in plants
S SULFUR	Component of most proteins
Na SODIUM	Sodium ion (Na^+) is a principal positive ion (cation) in interstitial (tissue) fluid of animals; important in fluid balance; essential for conduction of nerve impulses; important in photosynthesis in plants
Mg MAGNESIUM	Needed in blood and other tissues of animals; activates many enzymes; component of chlorophyll in plants
Cl CHLORINE	Chloride ion (Cl^-) is principal negative ion (anion) in interstitial (tissue) fluid of animals; important in water balance; essential for photosynthesis
Fe IRON	Component of hemoglobin in animals; activates certain enzymes

*Other elements found in very small (trace) amounts in animals, plants, or both include iodine (I), manganese (Mn), copper (Cu), zinc (Zn), cobalt (Co), fluorine (F), molybdenum (Mo), selenium (Se), boron (B), silicon (Si), and a few others.

© Cengage Learning

The periodic table provides information about the elements: their compositions, structures, and chemical behavior.

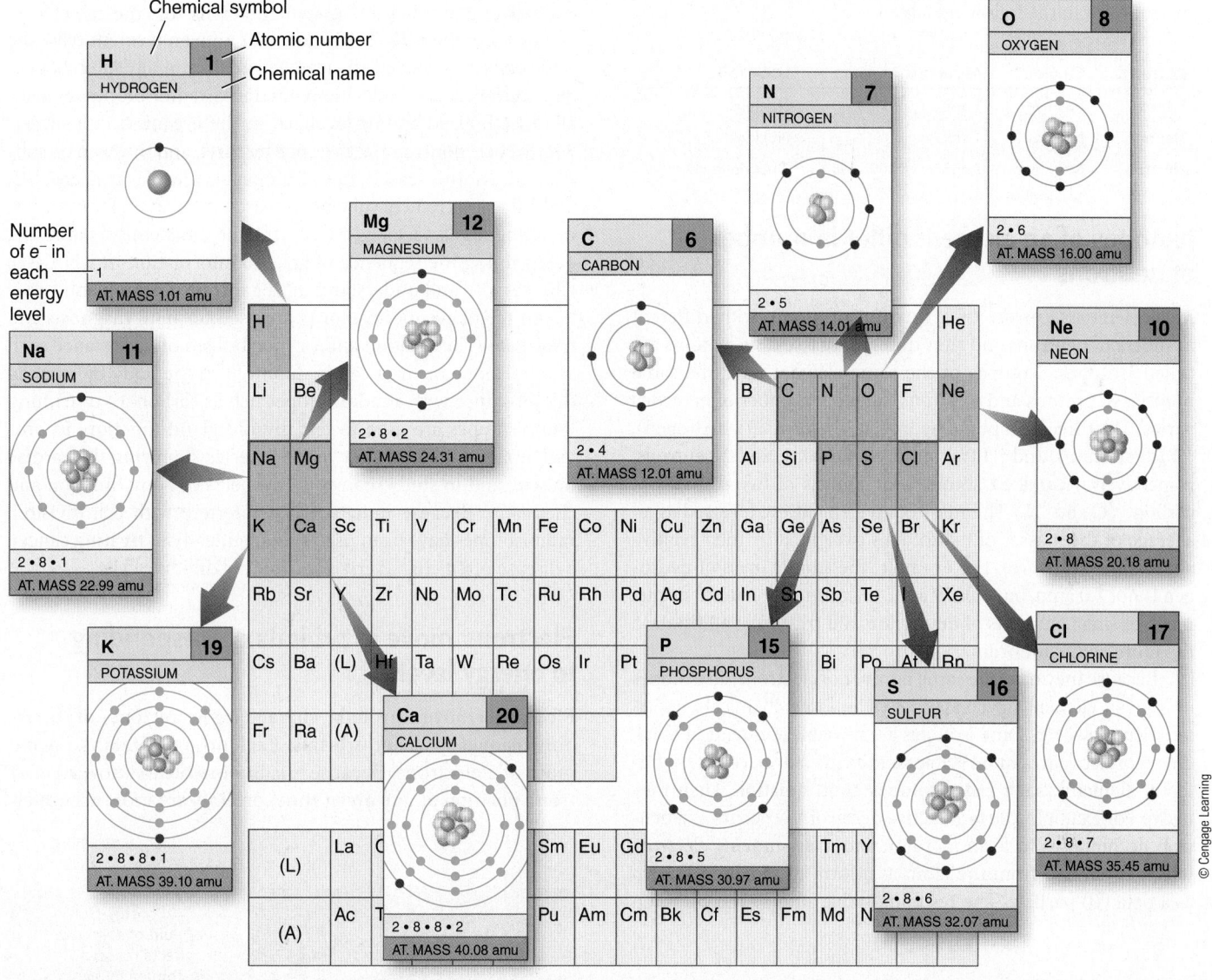

Figure 2-1 The periodic table

Note the Bohr models depicting the electron configuration of atoms of some biologically important elements, plus neon, which is unreactive because its valence shell is full (discussed later in this chapter). Although the Bohr model does not depict electron configurations accurately, it is commonly used because of its simplicity and convenience. A complete periodic table is given in Appendix A.

VISUALIZE Sketch a simple Bohr model for the element fluorine (F), which has an atomic number of 9.

micrograms.[1] Such masses are expressed in terms of the **atomic mass unit (amu),** also called the **dalton** in honor of John Dalton, the English chemist who formulated an atomic theory in the early 1800s. One amu is equal to the approximate mass of a single proton or a single neutron (1.7×10^{-24} g). Protons and neutrons make up almost all the mass of an atom. The mass of a single electron is only about 1/1800 the mass of a proton or neutron.

The **atomic mass** of an atom is a number that indicates approximately how much matter that atom contains compared with another atom. This value is determined by adding the number of protons to the number of neutrons and expressing the result in atomic mass units or daltons.[2] The mass of the electrons is ignored because it is so small. The atomic mass number is indicated by a superscript to the left of the chemical symbol.

[1] Tables of commonly used units of scientific measurement are printed inside the back cover of this text.

[2] Unlike weight, mass is independent of the force of gravity. For convenience, however, we consider mass and weight equivalent. Atomic weight has the same numerical value as atomic mass, but it has no units.

The common form of the oxygen atom, with 8 protons and 8 neutrons in its nucleus, has an atomic number of 8 and a mass of 16 amu. It is indicated by the symbol $^{16}_{8}O$.

The characteristics of protons, electrons, and neutrons are summarized in the following table:

Particle	Charge	Approximate Mass	Location
Proton	Positive	1 amu	Nucleus
Neutron	Neutral	1 amu	Nucleus
Electron	Negative	Approx. 1/1800 amu	Outside nucleus

Isotopes of an element differ in number of neutrons

Most elements consist of a mixture of atoms with different numbers of neutrons and thus different masses. Such atoms are called **isotopes.** Isotopes of the same element have the same number of protons and electrons; only the number of neutrons varies. The three isotopes of hydrogen, $^{1}_{1}H$ (ordinary hydrogen), $^{2}_{1}H$ (deuterium), and $^{3}_{1}H$ (tritium), contain 0, 1, and 2 neutrons, respectively. **FIGURE 2-2** shows Bohr models of two isotopes of carbon, $^{12}_{6}C$ and $^{14}_{6}C$. The mass of an element is expressed as an average of the masses of its isotopes (weighted by their relative abundance in nature). For example, the atomic mass of hydrogen is not 1.0 amu, but 1.0079 amu, reflecting the natural occurrence of small amounts of deuterium and tritium in addition to the more abundant ordinary hydrogen.

Because they have the same number of electrons, all isotopes of a given element have essentially the same chemical characteristics. However, some isotopes are unstable and tend to break down, or decay, to a more stable isotope (usually becoming a different element); such **radioisotopes** emit radiation when they decay. For example, the radioactive decay of $^{14}_{6}C$ occurs as a neutron decomposes to form a proton and a fast-moving electron, which is emitted from the atom as a form of radiation known as a beta (β) particle. The resulting stable atom is the common

form of nitrogen, $^{14}_{7}N$. Using sophisticated instruments, scientists can detect and measure β particles and other types of radiation. Radioactive decay can also be detected by a method known as **autoradiography**, in which radiation causes the appearance of dark silver grains in photographic film (**FIG. 2-3**).

Because the different isotopes of a given element have the same chemical characteristics, they are essentially interchangeable in molecules. Molecules containing radioisotopes are usually metabolized and/or localized in the organism in a similar way to their non-radioactive counterparts, and they can be substituted. For this reason, radioisotopes such as ^{3}H (tritium), ^{14}C, and ^{32}P are extremely valuable research tools used, for example, in dating fossils (see Fig. 18-10), tracing biochemical pathways, determining the sequence of genetic information in DNA (see Chapter 15), and understanding sugar transport in plants.

In medicine, radioisotopes are used for both diagnosis and treatment. The location and/or metabolism of a substance such as a hormone or drug can be followed in the body by labeling the substance with a radioisotope such as carbon-14 or tritium. Radioisotopes are used to test thyroid gland function, to provide images of blood flow in the arteries supplying the cardiac muscle, and to study many other aspects of body function and chemistry. Because radiation can interfere with cell division, radioisotopes have been used therapeutically in treating cancer, a disease often characterized by rapidly dividing cells.

Electrons move in orbitals corresponding to energy levels

Electrons move through characteristic regions of three-dimensional space, or **orbitals.** Each orbital contains a maximum of 2 electrons. Because it is impossible to know an electron's position at any given time, orbitals are most accurately

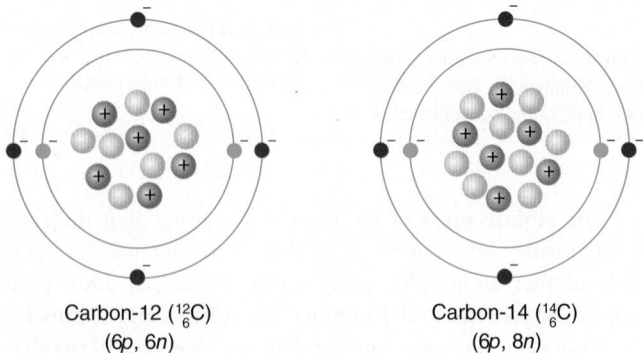

Carbon-12 ($^{12}_{6}C$)
(6*p*, 6*n*)

Carbon-14 ($^{14}_{6}C$)
(6*p*, 8*n*)

Figure 2-2 Isotopes

Carbon-12 ($^{12}_{6}C$) is the most common isotope of carbon. Its nucleus contains 6 protons and 6 neutrons, so its atomic mass is 12. Carbon-14 ($^{14}_{6}C$) is a rare radioactive carbon isotope. It contains 8 neutrons, so its atomic mass is 14.

© Cengage Learning

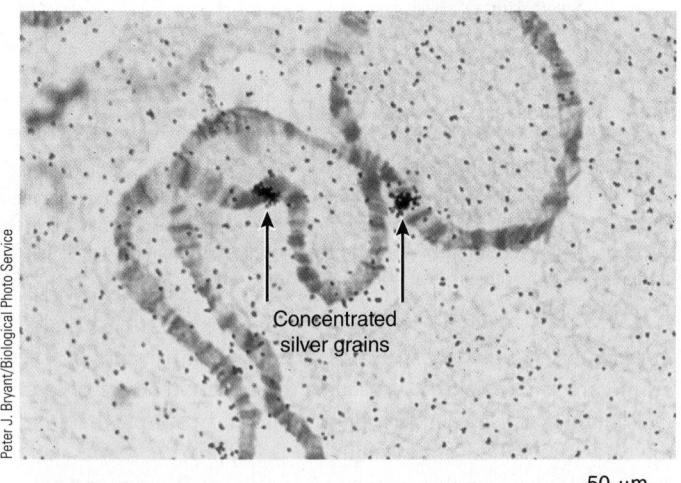

Concentrated silver grains

50 μm

Figure 2-3 Autoradiography

The chromosomes of the fruit fly, *Drosophila melanogaster*, shown in this light micrograph have been covered with photographic film in which silver grains (*dark spots*) are produced when tritium (^{3}H) that has been incorporated into DNA undergoes radioactive decay. The concentrations of silver grains (*arrows*) mark the locations of specific DNA molecules.

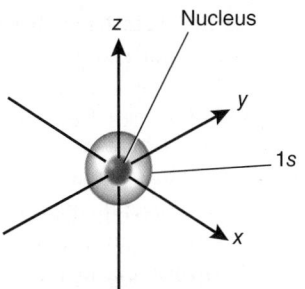

(a) The first principal energy level contains a maximum of 2 electrons, occupying a single spherical orbital (designated 1s). The electrons depicted in the diagram could be present anywhere in the blue area.

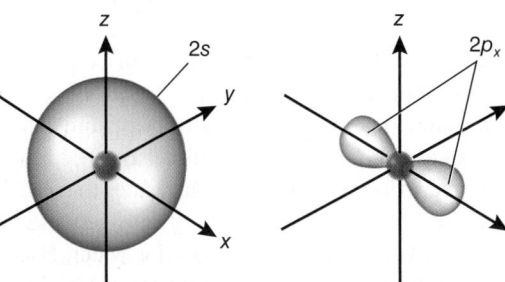
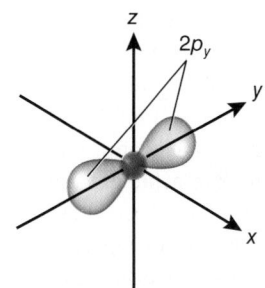
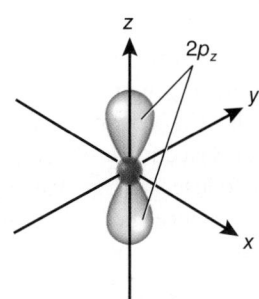

(b) The second principal energy level includes four orbitals, each with a maximum of 2 electrons: one spherical (2s) and three dumbbell-shaped (2p) orbitals at right angles to one another.

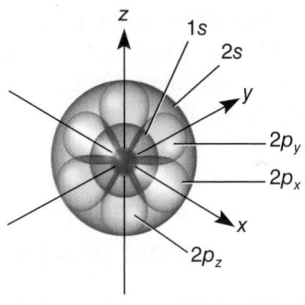

(c) Orbitals of the first and second principal energy levels of a neon atom are shown superimposed. Note that the single 2s orbital plus three 2p orbitals make up neon's full valence shell of 8 electrons. Compare this more realistic view of the atomic orbitals with the Bohr model of a neon atom at right.

(d) Neon atom (Bohr model)

Figure 2-4 *Animation* **Atomic orbitals**
Electrons occupy orbitals corresponding to energy levels. Each orbital is represented as an "electron cloud." The arrows labeled *x*, *y*, and *z* establish the imaginary axes of the atom.
© Cengage Learning

depicted as "electron clouds," shaded areas whose density is proportional to the probability that an electron is present there at any given instant. The energy of an electron depends on the orbital it occupies. Electrons in orbitals with similar energies, said to be at the same *principal energy level,* make up an **electron shell** (FIG. 2-4).

In general, electrons in a shell with a greater average distance from the nucleus have greater energy than those in a shell close to the nucleus. The reason is that energy is required to move a negatively charged electron farther away from the positively charged nucleus. The most energetic electrons, known as valence electrons, are said to occupy the *valence shell.* The valence shell is represented as the outermost concentric ring in a Bohr model. As you will see in the next sections, it is these valence electrons that play a key role in chemical reactions.

An electron can move to an orbital farther from the nucleus by receiving more energy, or it can give up energy and sink to a lower energy level in an orbital nearer the nucleus. Changes in electron energy levels are important in energy conversions in organisms. For example, during photosynthesis, light energy absorbed by chlorophyll molecules causes electrons to move to a higher energy level (see Fig. 9-3).

CHECKPOINT 2.1

- *Do all atoms of an element have the same atomic number? the same atomic mass?*

- *What is a radioisotope? What are some ways radioisotopes are used in biological research?*

- *How do electrons in different orbitals of the same electron shell compare with respect to their energy?*

2.2 CHEMICAL REACTIONS

The chemical behavior of an atom is determined primarily by the number and arrangement of its **valence electrons.** The valence shell of hydrogen or helium is full (stable) when it contains 2 electrons. The valence shell of any other atom is full when it contains 8 electrons. When the valence shell is not full, the atom tends to lose, gain, or share electrons to achieve a full outer shell. The valence shells of all isotopes of an element are identical; for this reason, they have similar chemical properties and can substitute for one another in chemical reactions (e.g., tritium can substitute for ordinary hydrogen).

Elements in the same vertical column (belonging to the same *group*) of the periodic table have similar chemical properties because their valence shells have similar tendencies to lose, gain, or share electrons. For example, chlorine and bromine, included in a group commonly known as the *halogens,* are highly reactive. Because their valence shells have 7 electrons, they tend to gain an electron in chemical reactions. By contrast, hydrogen, sodium, and potassium each have a single valence electron, which they tend to give up or share with another atom. Helium (He) and neon (Ne) belong to a group referred to as the *noble gases.* They are quite unreactive because their valence shells are full. Note in Figure 2-1 the incomplete valence shells of some of the elements important in organisms, including carbon, hydrogen, oxygen, and nitrogen, and compare them with the full valence shell of neon in Figure 2-4d.

Atoms form compounds and molecules

Two or more atoms may combine chemically. When atoms of *different* elements combine, the result is a chemical compound. A **chemical compound** consists of atoms of two or more different elements combined in a fixed ratio. For example, water is a chemical compound composed of hydrogen and oxygen in a ratio of 2:1. Common table salt, sodium chloride, is a chemical compound made up of sodium and chlorine in a 1:1 ratio.

Two or more atoms may become joined very strongly to form a stable particle called a **molecule.** For example, when two atoms of oxygen combine chemically, a molecule of oxygen is formed. Water is a molecular compound, with each molecule consisting of two atoms of hydrogen and one of oxygen. However, as you will see, not all compounds are made up of molecules. Sodium chloride (common table salt) is an example of a compound that is not molecular.

Simplest, molecular, and structural chemical formulas give different information

A **chemical formula** is a shorthand expression that describes the chemical composition of a substance. Chemical symbols indicate the types of atoms present, and subscript numbers indicate the ratios among the atoms. There are several types of chemical formulas, each providing specific kinds of information.

In a **simplest formula** (also known as an *empirical formula*), the subscripts give the smallest whole-number ratios for the atoms present in a compound. For example, the simplest formula for hydrazine is NH_2, indicating a 1:2 ratio of nitrogen to hydrogen. (Note that when a single atom of a type is present, the subscript number 1 is never written.)

In a **molecular formula,** the subscripts indicate the actual numbers of each type of atom per molecule. The molecular formula for hydrazine is N_2H_4, which indicates that each molecule of hydrazine consists of two atoms of nitrogen and four atoms of hydrogen. The molecular formula for water, H_2O, indicates that each molecule consists of two atoms of hydrogen and one atom of oxygen.

A **structural formula** shows not only the types and numbers of atoms in a molecule but also their arrangement. For example, the structural formula for water is H—O—H. As you will learn in Chapter 3, it is common for complex organic molecules with different structural formulas to share the same molecular formula.

One mole of any substance contains the same number of units

The molecular mass of a compound is the sum of the atomic masses of the component atoms of a single molecule; thus, the molecular mass of water, H_2O, is (hydrogen: 2×1 amu) + (oxygen: 1×16 amu), or 18 amu. (Because of the presence of isotopes, atomic mass values are not whole numbers, but for easy calculation each atomic mass value has been rounded to a whole number.) Similarly, the molecular mass of glucose ($C_6H_{12}O_6$), a simple sugar that is a key compound in cell metabolism, is (carbon: 6×12 amu) + (hydrogen: 12×1 amu) + (oxygen: 6×16 amu), or 180 amu.

The amount of an element or compound whose mass in grams is equivalent to its atomic or molecular mass is 1 **mole (mol)**. Thus, 1 mol of water is 18 grams (g), and 1 mol of glucose has a mass of 180 g. The mole is an extremely useful concept because it lets us make meaningful comparisons between atoms and molecules of very different mass. The reason is that *1 mol of any substance always has exactly the same number of units*, whether those units are small atoms or large molecules. The very large number of units in a mole, 6.02×10^{23}, is known as **Avogadro's number,** named for the Italian scientist Amedeo Avogadro, whose pioneering work inspired later scientists to calculate it. Thus, 1 mol (180 g) of glucose contains 6.02×10^{23} molecules, as does 1 mol (2 g) of molecular hydrogen (H_2). Although it is impossible to count atoms and molecules individually, a scientist can calculate them simply by weighing a sample. Molecular biologists usually deal with smaller values, either millimoles (mmol, one-thousandth of a mole) or micromoles (μmol, one-millionth of a mole).

The mole concept also lets us make useful comparisons among solutions. A 1-molar solution, represented by 1 M, contains 1 mol of that substance dissolved in a total volume of 1 liter (L). For example, we can compare 1 L of a 1 M solution of glucose with 1 L of a 1 M solution of sucrose (table sugar, a larger molecule). They differ in the mass of the dissolved sugar (180 g and 340 g, respectively), but they each contain 6.02×10^{23} sugar molecules.

Chemical equations describe chemical reactions

During any moment in the life of an organism—a bacterial cell, a mushroom, or a butterfly—many complex chemical reactions are taking place. Chemical reactions, such as the reaction between glucose and oxygen, can be described by means of chemical equations:

$$C_6H_{12}O_6 + 6\,O_2 \longrightarrow 6\,CO_2 + 6\,H_2O + \text{energy}$$
Glucose   Oxygen   Carbon dioxide   Water

In a chemical equation, the **reactants**—the substances that participate in the reaction—are generally written on the left side, and the **products**—the substances formed by the reaction—are written on the right side. The arrow means "yields" and indicates the direction in which the reaction proceeds.

Chemical compounds react with one another in quantitatively precise ways. The numbers preceding the chemical symbols or formulas (known as *coefficients*) indicate the relative number of atoms or molecules reacting. For example, 1 mol of glucose burned in a fire or metabolized in a cell reacts with 6 mol of oxygen to form 6 mol of carbon dioxide and 6 mol of water.

Many reactions can proceed simultaneously in the reverse direction (to the left) and the forward direction (to the right). At **dynamic equilibrium,** the rates of the forward and reverse reactions are equal (see Chapter 7). Reversible reactions are indicated by double arrows:

$$CO_2 + H_2O \rightleftharpoons H_2CO_3$$
Carbon dioxide   Water   Carbonic acid

In this example the arrows are drawn in different lengths to indicate that when the reaction reaches equilibrium, there will be more reactants (CO_2 and H_2O) than product (H_2CO_3).

CHECKPOINT 2.2

- *What enables a radioisotope to substitute for an ordinary (non-radioactive) atom of the same element in a molecule?*
- *Which kind of chemical formula provides the most information?*
- PREDICT *How many particles would be included in 1 g of hydrogen atoms? in 2 g of hydrogen molecules?*

2.3 CHEMICAL BONDS

LEARNING OBJECTIVE

7 Distinguish among covalent bonds, ionic bonds, hydrogen bonds, and van der Waals interactions. Compare them in terms of the mechanisms by which they form and their relative strengths.

Atoms can be held together by forces of attraction called **chemical bonds.** Each bond represents a certain amount of chemical energy. **Bond energy** is the energy necessary to break a chemical bond. The valence electrons dictate how many bonds an atom can form. The two principal types of strong chemical bonds are covalent bonds and ionic bonds.

In covalent bonds electrons are shared

Covalent bonds involve the sharing of electrons between atoms in a way that results in each atom having a filled valence shell. A molecule consists of atoms joined by covalent bonds. A simple example of a covalent bond is the joining of two hydrogen atoms in a molecule of hydrogen gas, H_2. Each atom of hydrogen has 1 electron, but 2 electrons are required to complete its valence shell. The hydrogen atoms have equal capacities to attract electrons, so neither donates an electron to the other. Instead, the two hydrogen atoms share their single electrons so that the 2 electrons are attracted simultaneously to the 2 protons in the two hydrogen nuclei. The 2 electrons whirl around both atomic nuclei, thus forming the covalent bond that joins the two atoms. Similarly, unlike atoms can also be linked by covalent bonds to form molecules; the resulting compound is a **covalent compound.**

A simple way of representing the electrons in the valence shell of an atom is to use dots placed around the chemical symbol of the element. Such a representation is called the *Lewis structure* of the atom, named for G. N. Lewis, the American chemist who developed this type of notation. In a water molecule, two hydrogen atoms are covalently bonded to an oxygen atom:

$$\text{H} \cdot + \text{H} \cdot + \cdot \ddot{\text{O}} \cdot \longrightarrow \text{H} \! : \! \ddot{\text{O}} \! : \! \text{H}$$

Oxygen has 6 valence electrons; by sharing electrons with two hydrogen atoms, it completes its valence shell of 8. At the same time, each hydrogen atom obtains a complete valence shell of 2. (Note that in the structural formula H—O—H, each pair of shared electrons constitutes a covalent bond, represented by a solid line. Unshared electrons are usually omitted in a structural formula.)

The carbon atom has 4 electrons in its valence shell, all of which are available for covalent bonding:

$$\cdot \dot{\underset{\cdot}{\text{C}}} \cdot$$

When one carbon and four hydrogen atoms share electrons, a molecule of the covalent compound methane, CH_4, is formed:

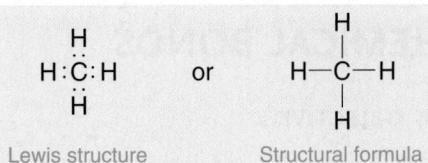

Lewis structure   Structural formula

The nitrogen atom has 5 electrons in its valence shell. Recall that each orbital can hold a maximum of 2 electrons. Usually, 2 electrons occupy one orbital, leaving 3 electrons available for sharing with other atoms:

When a nitrogen atom shares electrons with three hydrogen atoms, a molecule of the covalent compound ammonia, NH_3, is formed:

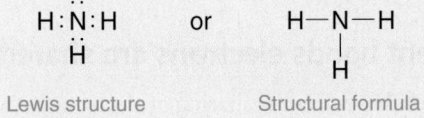

Lewis structure   Structural formula

When one pair of electrons is shared between two atoms, the covalent bond is called a **single covalent bond** (FIG. 2-5a). Two hydrogen atoms share a single pair of electrons. Two oxygen atoms may achieve stability by forming covalent bonds with each other. Each oxygen atom has 6 electrons in its outer shell. To become stable, the two atoms share two pairs of electrons,

forming molecular oxygen (FIG. 2-5b). When two pairs of electrons are shared in this way, the covalent bond is called a *double covalent bond,* which is represented by two parallel, solid lines. Similarly, a *triple covalent bond* is formed when three pairs of electrons are shared between two atoms (represented by three parallel, solid lines).

The number of covalent bonds usually formed by the atoms in biologically important molecules is summarized as follows:

ATOM	SYMBOL	COVALENT BONDS
Hydrogen	H	1
Oxygen	O	2
Carbon	C	4
Nitrogen	N	3
Phosphorus	P	5
Sulfur	S	2

The function of a molecule is related to its shape

In addition to being composed of atoms with certain properties, each kind of molecule has a characteristic size and a general overall shape. Although the shape of a molecule may change (within certain limits), the functions of molecules in living cells are dictated largely by their geometric shapes. A molecule

KEY POINT

Covalent bonds form when atoms share electrons.

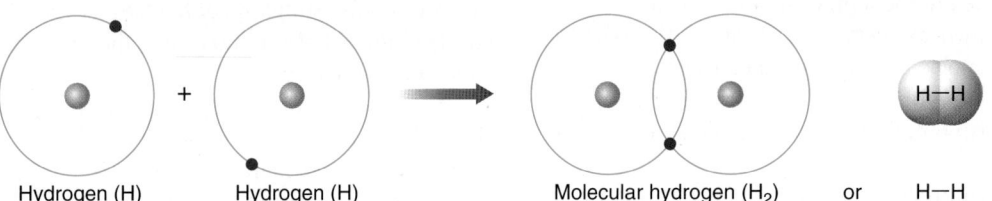

    Hydrogen (H)   Hydrogen (H)   Molecular hydrogen (H_2)   or   H—H

(a) Single covalent bond formation. Two hydrogen atoms achieve stability by sharing a pair of electrons, thereby forming a molecule of hydrogen. In the structural formula on the right, the straight line between the hydrogen atoms represents a single covalent bond.

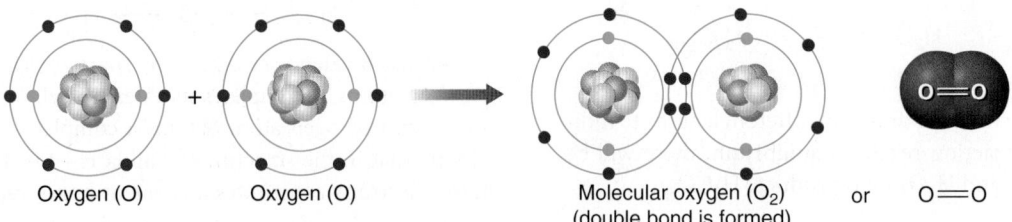

    Oxygen (O)   Oxygen (O)   Molecular oxygen (O_2)   or   O==O
                                      (double bond is formed)

(b) Double covalent bond formation. In molecular oxygen, two oxygen atoms share two pairs of electrons, forming a double covalent bond. The parallel straight lines in the structural formula represent a double covalent bond.

Figure 2-5 *Animation* **Electron sharing in covalent compounds**

VISUALIZE Molecular nitrogen (N_2) is formed when two nitrogen atoms are joined by a triple covalent bond. Use the information in Figure 2-1 to help you sketch a simple Bohr model of N_2.

© Cengage Learning

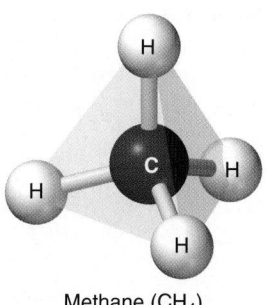

Methane (CH₄)

Figure 2-6 Orbital hybridization in methane

The four hydrogens are located at the corners of a tetrahedron because of hybridization of the valence shell orbitals of carbon.

© Cengage Learning

that consists of two atoms is linear. Molecules composed of more than two atoms may have more complicated shapes. The geometric shape of a molecule provides the optimal distance between the atoms to counteract the repulsion of electron pairs.

When an atom forms covalent bonds with other atoms, the orbitals in the valence shell may become rearranged in a process known as **orbital hybridization,** thereby affecting the shape of the resulting molecule. For example, when four hydrogen atoms combine with one carbon atom to form a molecule of methane (CH_4), the hybridized valence shell orbitals of the carbon form a geometric structure known as a *tetrahedron*, with one hydrogen atom present at each of its four corners (**FIG. 2-6**; see also Fig. 3-2b).

We explore the importance of molecular shape in more detail in Chapter 3 and in our discussion of the properties of water in this chapter.

Covalent bonds can be nonpolar or polar

Atoms of different elements vary in their affinity for electrons. **Electronegativity** is a measure of an atom's attraction for shared electrons in chemical bonds. Very electronegative atoms such as oxygen, nitrogen, fluorine, and chlorine are sometimes called "electron greedy." When covalently bonded atoms have similar electronegativities, the electrons are shared equally and the bond is described as a **nonpolar covalent bond.** The covalent bond of the hydrogen molecule is nonpolar, as are the covalent bonds of molecular oxygen and methane.

In a covalent bond between two different elements, such as oxygen and hydrogen, the electronegativities of the atoms may be different. If so, electrons are pulled closer to the atomic nucleus of the element with the greater electron affinity (in this case, oxygen). A covalent bond between atoms that differ in electronegativity is called a **polar covalent bond.** Such a bond has two dissimilar ends (or poles), one with a partial positive charge and the other with a partial negative charge. Each of the two covalent bonds in water is polar because there is a partial positive charge at the hydrogen end of the bond and a partial negative charge at the oxygen end, where the "shared" electrons are more likely to be.

Covalent bonds differ in their degree of polarity, ranging from those in which the electrons are equally shared (as in the nonpolar hydrogen molecule) to those in which the electrons are much closer to one atom than to the other (as in water). Oxygen is quite electronegative and forms polar covalent bonds with carbon, hydrogen, and many other atoms. Nitrogen is also strongly electronegative, although less so than oxygen.

A molecule with one or more polar covalent bonds can be polar even though it is electrically neutral as a whole. The reason is that a **polar molecule** has one end with a partial positive charge and another end with a partial negative charge. One example is water (**FIG. 2-7**). The polar bonds between the hydrogens and the oxygen are arranged in a V shape, rather than linearly. The oxygen end constitutes the negative pole of the molecule, and the end with the two hydrogens is the positive pole.

Ionic bonds form between cations and anions

Some atoms or groups of atoms are not electrically neutral. A particle with 1 or more units of electric charge is called an ion. An atom becomes an ion if it gains or loses 1 or more electrons. An atom with 1, 2, or 3 electrons in its valence shell tends to lose electrons to other atoms. Such an atom then becomes positively charged because its nucleus contains more protons than the number of electrons orbiting around the nucleus. These positively charged ions are called **cations.** Atoms with 5, 6, or 7 valence electrons tend to gain electrons from other atoms and become negatively charged **anions.**

The properties of ions are quite different from those of the electrically neutral atoms from which they were derived. For example, although chlorine gas is a poison, chloride ions (Cl^-)

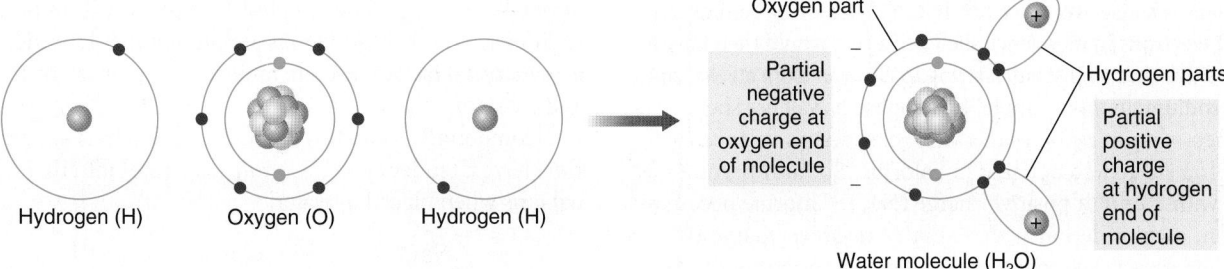

Hydrogen (H)   Oxygen (O)   Hydrogen (H)

Oxygen part

Partial negative charge at oxygen end of molecule

Hydrogen parts

Partial positive charge at hydrogen end of molecule

Water molecule (H₂O)

Figure 2-7 *Animation* Water, a polar molecule

Note that the electrons tend to stay closer to the nucleus of the oxygen atom than to the hydrogen nuclei. The result is a partial negative charge on the oxygen portion of the molecule and a partial positive charge at the hydrogen end. Although the water molecule as a whole is electrically neutral, it is a polar covalent compound.

© Cengage Learning

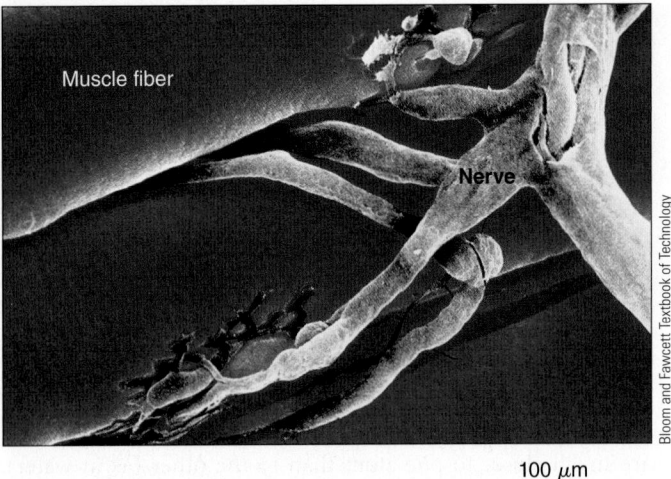

Figure 2-8 Ions and biological processes
Sodium, potassium, and chloride ions are essential for this nerve cell to stimulate these muscle fibers, initiating a muscle contraction. Calcium ions in the muscle cell are required for muscle contraction.

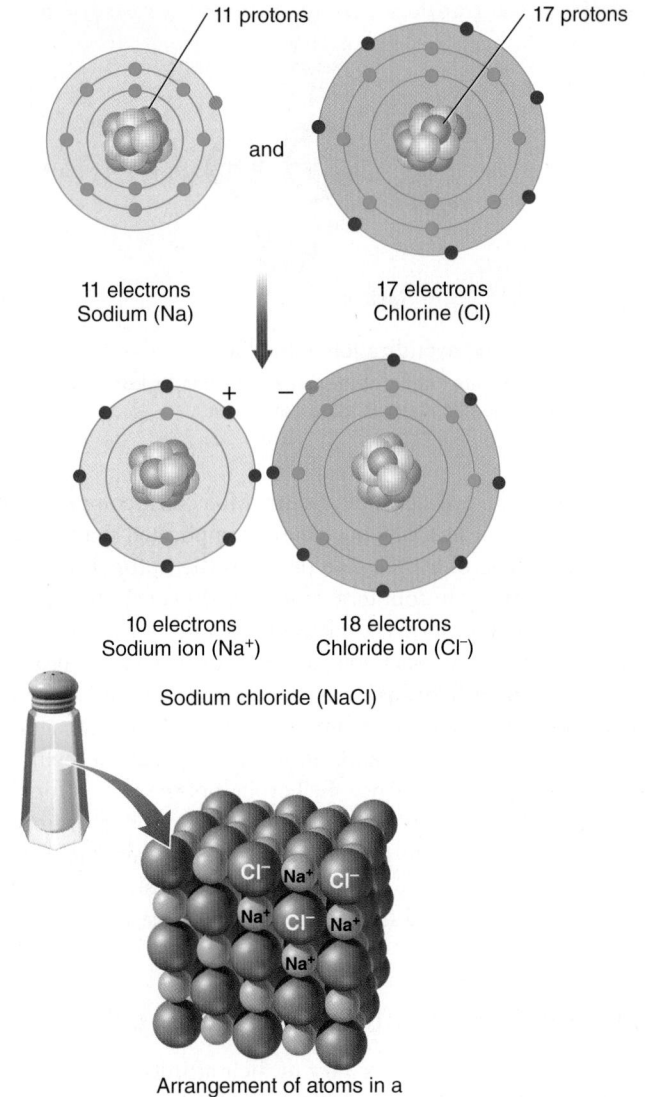

Figure 2-9 *Animation* **Ionic bonding**
Sodium becomes a positively charged ion when it donates its single valence electron to chlorine, which has 7 valence electrons. With this additional electron, chlorine completes its valence shell and becomes a negatively charged chloride ion. These sodium and chloride ions are attracted to one another by their unlike electric charges, forming the ionic compound sodium chloride.
© Cengage Learning

are essential to life (see Table 2-1). Because their electric charges provide a basis for many interactions, cations and anions are involved in energy transformations within the cell, the transmission of nerve impulses, muscle contraction, and many other biological processes (FIG. 2-8).

A group of covalently bonded atoms can also become an ion (*polyatomic ion*). Unlike a single atom, a group of atoms can lose or gain protons (derived from hydrogen atoms) as well as electrons. Therefore, a group of atoms can become a cation if it loses 1 or more electrons or gains 1 or more protons. A group of atoms becomes an anion if it gains 1 or more electrons or loses 1 or more protons.

An **ionic bond** forms as a consequence of the attraction between the positive charge of a cation and the negative charge of an anion. An **ionic compound** is a substance consisting of anions and cations bonded by their opposite charges.

A good example of how ionic bonds are formed is the attraction between sodium ions and chloride ions. A sodium atom has 1 electron in its valence shell. It cannot fill its valence shell by obtaining 7 electrons from other atoms because it would then have a large unbalanced negative charge. Instead, it gives up its single valence electron to a very electronegative atom, such as chlorine, which acts as an electron acceptor (FIG. 2-9). Chlorine cannot give up the 7 electrons in its valence shell because it would then have a large positive charge. Instead, it strips an electron from an electron donor (sodium, in this example) to complete its valence shell.

When sodium reacts with chlorine, sodium's valence electron is transferred completely to chlorine. Sodium becomes a cation, with 1 unit of positive charge (Na^+). Chlorine becomes an anion, a chloride ion with 1 unit of negative charge (Cl^-). These ions attract each other as a result of their opposite charges. This electrical attraction in ionic bonds holds them together to form NaCl, sodium chloride, or common table salt.

The term *molecule* does not adequately explain the properties of ionic compounds such as NaCl. When NaCl is in its solid crystal state, each ion is actually surrounded by six ions

of opposite charge. The simplest formula, NaCl, indicates that sodium ions and chloride ions are present in a 1:1 ratio, but the actual crystal has no discrete molecules composed of 1 Na^+ ion and 1 Cl^- ion.

Compounds joined by ionic bonds, such as sodium chloride, have a tendency to *dissociate* (separate) into their individual ions when placed in water:

$$NaCl \xrightarrow{\text{in } H_2O} Na^+ + Cl^-$$
Sodium chloride   Sodium ion   Chloride ion

In the solid form of an ionic compound (that is in the absence of water), ionic bonds are very strong. Water, however, is an excellent **solvent;** as a liquid it is capable of dissolving many substances, particularly those that are polar or ionic, because of

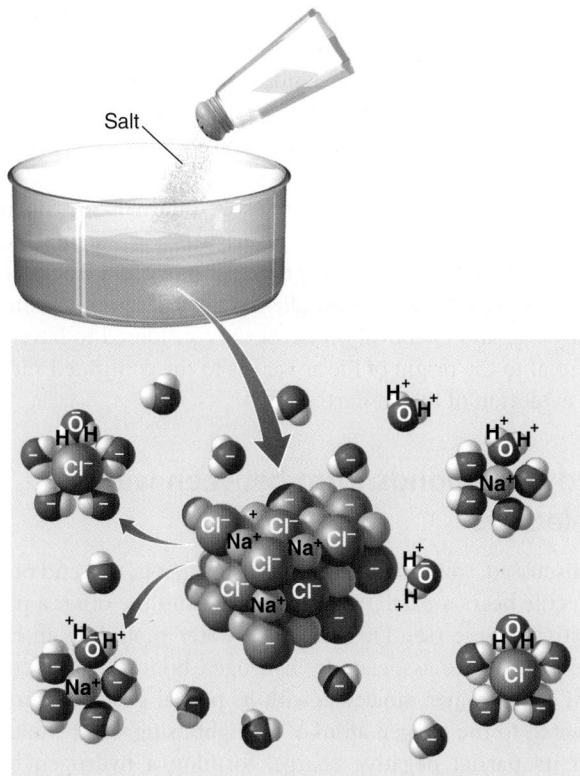

Figure 2-10 *Animation* **Hydration of an ionic compound**
When the crystal of NaCl is added to water, the sodium and chloride ions are pulled apart. When the NaCl is dissolved, each Na⁺ and each Cl⁻ is surrounded by water molecules electrically attracted to it.
© Cengage Learning

the polarity of water molecules. The localized partial positive charge (on the hydrogen atoms) and partial negative charge (on the oxygen atom) on each water molecule attract and surround the anions and cations, respectively, on the surface of an ionic solid. As a result, the solid dissolves. A dissolved substance is referred to as a **solute.** In solution each cation and anion of the ionic compound is surrounded by oppositely charged ends of the water molecules. This process is known as **hydration** (**FIG. 2-10**). Hydrated ions still interact with one another to some extent, but the transient ionic bonds formed are much weaker than those in a solid crystal.

Hydrogen bonds are weak attractions

Another type of bond important in organisms is the **hydrogen bond.** When hydrogen combines with oxygen (or with another relatively electronegative atom such as nitrogen), it acquires a partial positive charge because its electron spends more time closer to the electronegative atom. Hydrogen bonds tend to form between an atom with a partial negative charge and a hydrogen atom that is covalently bonded to oxygen or nitrogen (**FIG. 2-11**). The atoms involved may be in two parts of the same large molecule or in two different molecules. Water molecules interact with one another extensively through hydrogen bond formation.

Hydrogen bonds are readily formed and broken. Although individually relatively weak, hydrogen bonds are collectively strong when present in large numbers. Furthermore, they have

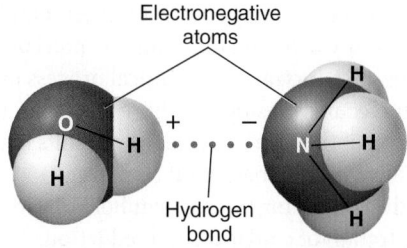

Figure 2-11 *Animation* **Hydrogen bonding**
A hydrogen bond (*dotted line*) can form between two molecules with regions of unlike partial charge. Here, the nitrogen atom of an ammonia molecule is joined by a hydrogen bond to a hydrogen atom of a water molecule.
© Cengage Learning

a specific length and orientation. As you will see in Chapter 3, these features are very important in determining the 3-D structure of large molecules such as DNA and proteins.

van der Waals interactions are weak forces

Even electrically neutral, nonpolar molecules can develop transient regions of weak positive charge and weak negative charge. These slight charges develop as a consequence of the electrons always being in constant motion. A region with a temporary excess of electrons will have a weak negative charge, whereas one with an electron deficit will have a weak positive charge. Adjacent molecules may interact in regions of slight opposite charge. These attractive forces, called **van der Waals interactions,** operate over very short distances and are weaker and less specific than the other types of interactions we have considered. They are most important when they occur in large numbers and when the shapes of the molecules permit close contact between the atoms. Although a single interaction is very weak, the binding force of a large number of these interactions working together can be significant.

CHECKPOINT 2.3

- *Are all compounds composed of molecules? Explain.*
- *What are the ways an atom or a molecule can become an anion or a cation?*
- *How do ionic and covalent bonds differ?*
- **CONNECT** *Under what circumstances can weak forces such as hydrogen bonds and van der Waals interactions play significant roles in biological systems?*

2.4 REDOX REACTIONS

LEARNING OBJECTIVE

8 Distinguish between the terms *oxidation* and *reduction*, and relate these processes to the transfer of energy.

Many energy conversions that go on in a cell involve reactions in which an electron transfers from one substance to another. The reason is that the transfer of an electron also involves the transfer of the energy of that electron. Such an electron transfer is known as an oxidation–reduction reaction, or **redox reaction.**

Oxidation and reduction always occur together. **Oxidation** is a chemical process in which an atom, ion, or molecule *loses* one or more electrons. **Reduction** is a chemical process in which an atom, ion, or molecule *gains* one or more electrons. (The term refers to the fact that the gain of an electron results in the reduction of any positive charge that might be present.)

Rusting—the combining of iron (symbol Fe) with oxygen—is a simple illustration of oxidation and reduction:

$$4\ Fe + 3\ O_2 \longrightarrow 2\ Fe_2O_3$$

Iron (III) oxide

In rusting, each iron atom becomes oxidized as it loses 3 electrons.

$$4\ Fe \longrightarrow 4\ Fe^{3+} + 12e^-$$

The e^- represents an electron, and the + superscript in Fe^{3+} represents an electron deficit. (When an atom loses an electron, it acquires 1 unit of positive charge from the excess of 1 proton. In this example each iron atom loses 3 electrons and acquires 3 units of positive charge.) Recall that the oxygen atom is very electronegative, able to remove electrons from other atoms. In this reaction, oxygen becomes reduced when it accepts electrons from the iron:

$$3\ O_2 + 12e^- \longrightarrow 6\ O^{2-}$$

Redox reactions occur simultaneously because one substance must accept the electrons that are removed from the other. In a redox reaction, one component, the *oxidizing agent*, accepts 1 or more electrons and becomes reduced. Oxidizing agents other than oxygen are known, but oxygen is such a common one that its name was given to the process. Another reaction component, the *reducing agent*, gives up 1 or more electrons and becomes oxidized. In our example, there was a complete transfer of electrons from iron (the reducing agent) to oxygen (the oxidizing agent). Similarly, Figure 2-9 shows that an electron was transferred from sodium (the reducing agent) to chlorine (the oxidizing agent).

Electrons are not easily removed from covalent compounds unless an entire atom is removed. In cells, oxidation often involves the removal of a hydrogen atom (an electron plus a proton that "goes along for the ride") from a covalent compound; reduction often involves the addition of the equivalent of a hydrogen atom (see Chapter 7).

CHECKPOINT 2.4

- *In what form is energy transferred in a redox reaction?*

2.5 WATER

LEARNING OBJECTIVE

9 Explain how hydrogen bonds between adjacent water molecules govern many of the properties of water.

A large part of the mass of most organisms is water. In human tissues the percentage of water ranges from 20% in bones to 85% in brain cells; about 70% of a human's total body weight is water.

As much as 95% of a jellyfish and certain plants is water. Water is the source, through photosynthesis, of the oxygen in the air we breathe, and its hydrogen atoms become incorporated into many organic compounds. Water is also the solvent for most biological reactions and a reactant or product in many chemical reactions.

Water is important not only as an internal constituent of organisms but also as one of the principal environmental factors affecting them (**FIG. 2-12**). Many organisms live in the ocean or in freshwater rivers, lakes, or puddles. Water's unique combination of physical and chemical properties is considered to have been essential to the origin of life as well as to the continued survival and evolution of life on Earth.

Hydrogen bonds form between water molecules

As discussed, water molecules are polar; that is, one end of each molecule bears a partial positive charge and the other a partial negative charge (see Fig. 2-7). The water molecules in liquid water and in ice associate by hydrogen bonds. The hydrogen atom of one water molecule, with its partial positive charge, is attracted to the oxygen atom of a neighboring water molecule, with its partial negative charge, forming a hydrogen bond. An oxygen atom in a water molecule has two regions of partial negative charge, and each of the two hydrogen atoms has a partial positive charge. Each water molecule can therefore form hydrogen bonds with a maximum of four neighboring water molecules (**FIG. 2-13**).

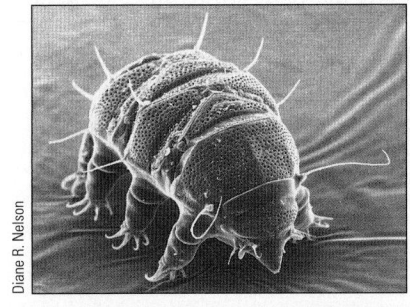

(a) Commonly known as "water bears," tardigrades, such as these members of the genus *Echiniscus*, are small animals (less than 1.2 mm long) that normally live in moist habitats, such as thin films of water on mosses.

100 μm

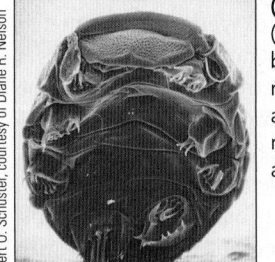

(b) When subjected to desiccation (dried out), tardigrades assume a barrel-shaped form known as a *tun*, remaining in this state, motionless but alive, for as long as 100 years. When rehydrated, they assume their normal appearance and activities.

10 μm

Figure 2-12 The effects of water on an organism

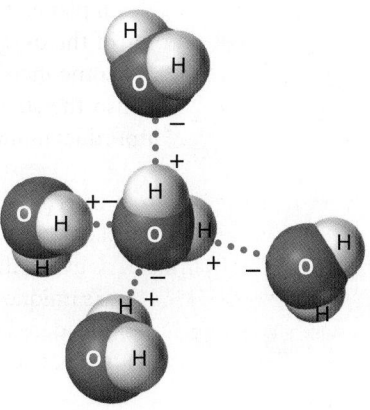

Figure 2-13 Hydrogen bonding of water molecules
Each water molecule can form hydrogen bonds (*dotted lines*) with as many as four neighboring water molecules.
© Cengage Learning

Water molecules have a strong tendency to stick to one another, a property known as **cohesion,** due to the hydrogen bonds among the molecules. Because of the cohesive nature of water molecules, any force exerted on part of a column of water is transmitted to the column as a whole. The major mechanism of water movement in plants (see Chapter 35) depends on the cohesive nature of water. Water molecules also display **adhesion,** the ability to stick to many other kinds of substances, most notably those with charged groups of atoms or molecules on their surfaces. These adhesive forces explain how water makes things wet.

A combination of adhesive and cohesive forces accounts for **capillary action,** which is the tendency of water to move in narrow tubes, even against the force of gravity (FIG. 2-14). For example, water moves through the microscopic spaces between soil particles to the roots of plants by capillary action.

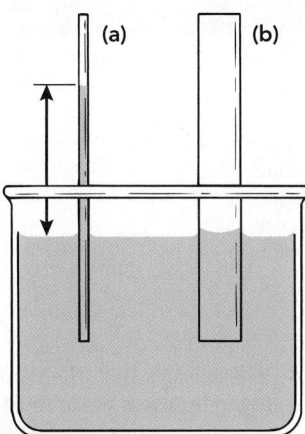

Figure 2-14 Capillary action
(a) In a narrow tube, there is adhesion between the water molecules and the glass wall of the tube. Other water molecules inside the tube are then "pulled along" because of cohesion, which is due to hydrogen bonds between the water molecules. **(b)** In the wider tube, a smaller percentage of the water molecules line the glass wall. As a result, the adhesion is not strong enough to overcome the cohesion of the water molecules beneath the surface level of the container, and water in the tube rises only slightly.
© Cengage Learning

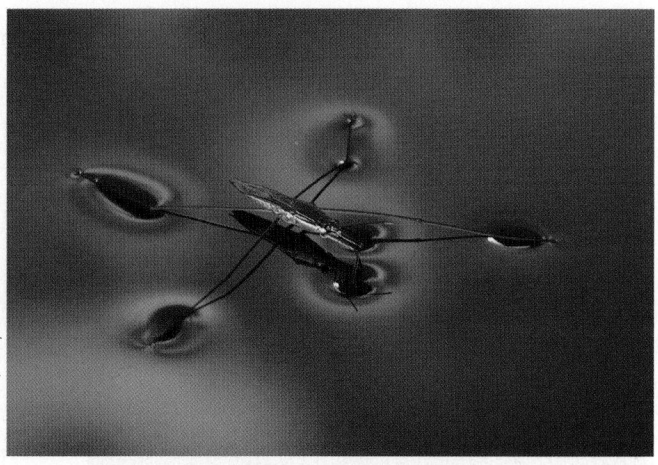

Figure 2-15 Surface tension of water
Hydrogen bonding between water molecules is responsible for the surface tension of water, which causes a dimpled appearance of the surface as this water strider (genus, *Gerris*) walks across it. Fine hairs at the ends of the legs of these insects create highly water-repellent "cushions" of air.

Water has a high degree of **surface tension** because of the cohesion of its molecules, which have a much greater attraction for one another than for molecules in the air. Thus, water molecules at the surface crowd together, producing a strong layer as they are pulled downward by the attraction of other water molecules beneath them (FIG. 2-15).

Water molecules interact with hydrophilic substances by hydrogen bonding

Because its molecules are polar, water is an excellent solvent, a liquid capable of dissolving many kinds of substances, especially polar and ionic compounds. Earlier we discussed how polar water molecules pull the ions of ionic compounds apart so that they dissociate (see Fig. 2-10). Because of its solvent properties and the tendency of the atoms in certain compounds to form ions in solution, water plays an important role in facilitating chemical reactions.

Substances that interact readily with water are **hydrophilic** ("water-loving"). Examples include table sugar (sucrose, a polar compound) and table salt (NaCl, an ionic compound), which dissolve readily in water. Not all substances in organisms are hydrophilic, however. Many **hydrophobic** ("water-fearing") substances found in living things are especially important because of their ability to form associations or structures that are not disrupted. Hydrophobic interactions occur between groups of nonpolar molecules. Such molecules are insoluble in water and tend to cluster together. This tendency is not due to formation of bonds between the nonpolar molecules but rather to the hydrogen-bonded water molecules excluding them and in a sense "driving them together." **Hydrophobic interactions** explain why oil tends to form globules when added to water. Examples of hydrophobic substances include fatty acids and cholesterol, discussed in Chapter 3.

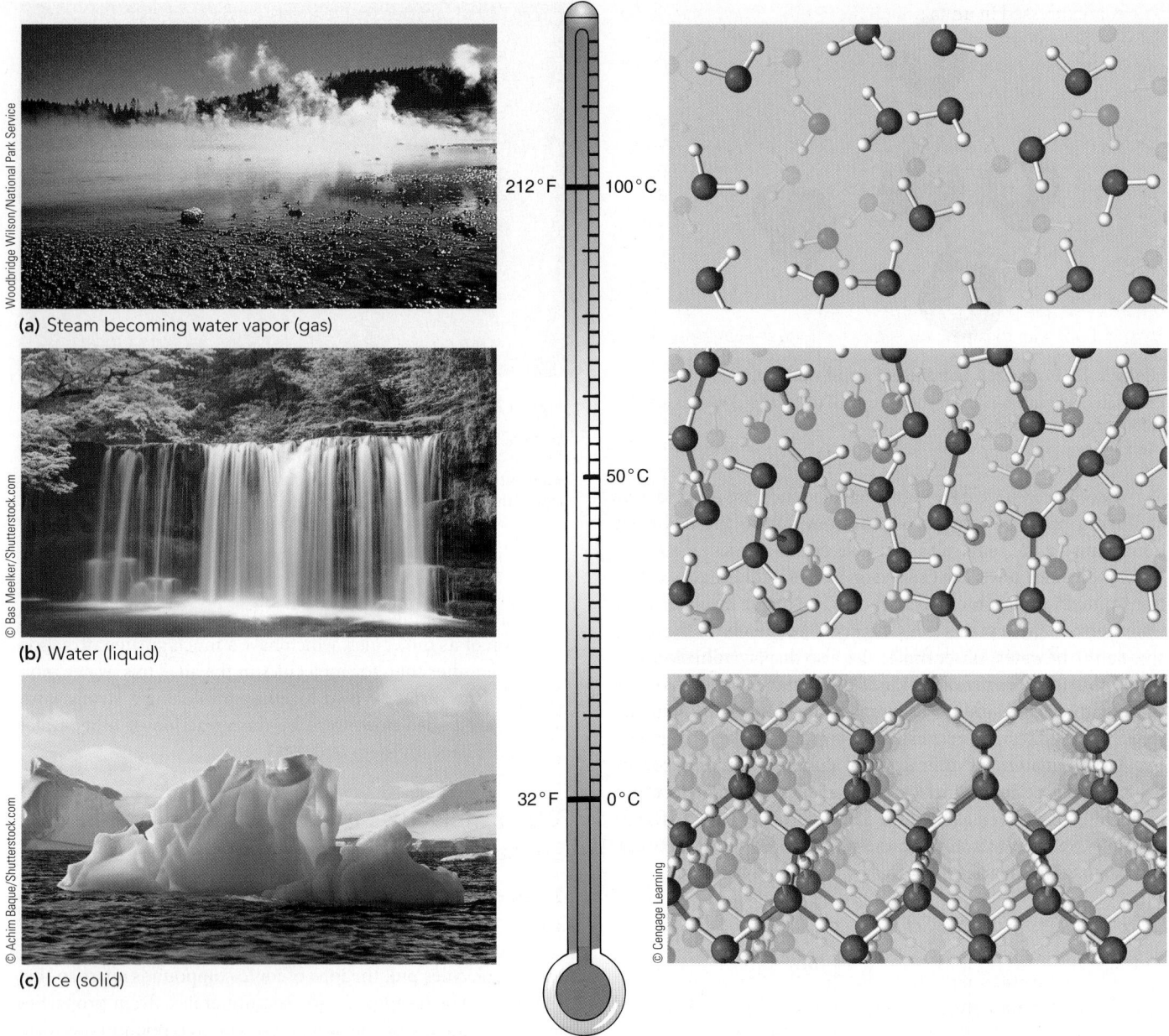

(a) Steam becoming water vapor (gas)

(b) Water (liquid)

(c) Ice (solid)

212°F 100°C

50°C

32°F 0°C

Figure 2-16 Three forms of water

(a) When water boils, as in this hot spring at Yellowstone National Park, many hydrogen bonds are broken, causing steam, which consists of minuscule water droplets, to form. If most of the remaining hydrogen bonds break, the molecules move more freely as water vapor (a gas). (b) Water molecules in a liquid state continually form, break, and re-form hydrogen bonds with one another. (c) In ice, each water molecule participates in four hydrogen bonds with adjacent molecules, resulting in a regular, evenly distanced crystalline lattice structure.

Water helps maintain a stable temperature

Hydrogen bonding explains the way water responds to changes in temperature. Water exists in three forms, which differ in their degree of hydrogen bonding: gas (vapor); liquid; and ice, a crystalline solid (FIG. 2-16). Hydrogen bonds are formed or broken as water changes from one state to another.

Raising the temperature of a substance involves adding heat energy to make its molecules move faster, that is, to increase the energy of motion—**kinetic energy**—of the molecules (see Chapter 7). The term **heat** refers to the *total* amount of kinetic energy in a sample of a substance; **temperature** is a measure of the *average*

kinetic energy of the particles. For the molecules to move more freely, some of the hydrogen bonds of water must be broken.

Much of the energy added to the system is used in breaking the hydrogen bonds, and only a portion of the heat energy is available to speed the movement of the water molecules, thereby increasing the temperature of the water. Conversely, when liquid water changes to ice, additional hydrogen bonds must be formed, making the molecules less free to move and liberating a great deal of heat into the environment.

Heat of vaporization, the amount of heat energy required to change 1 g of a substance from the liquid phase to the vapor phase, is expressed in units called *calories*. A **calorie (cal)** is the amount of heat energy (equivalent to 4.184 joules [J]) required to raise the temperature of 1 g of water 1 degree Celsius (C). Water has a high heat of vaporization—540 cal—because its molecules are held together by hydrogen bonds. The heat of vaporization of most other common liquid substances is much less. As a sample of water is heated, some molecules are moving much faster than others (they have more heat). These faster-moving molecules are more likely to escape the liquid phase and enter the vapor phase (see Fig. 2-16a). When they do, they take their heat with them, lowering the temperature of the sample in a process called *evaporative cooling*. For this reason, the human body can dissipate excess heat as sweat evaporates from the skin, and a leaf can keep cool in the bright sunlight as water evaporates from its surface.

Hydrogen bonding is also responsible for water's high **specific heat;** that is, the amount of energy required to raise the temperature of water is quite large. The specific heat of water is 1 cal/g of water per degree Celsius. Most other common substances, such as metals, glass, and ethyl alcohol, have much lower specific heat values. The specific heat of ethyl alcohol, for example, is 0.59 cal/g/1°C (2.46 J/g/1°C).

Because so much heat input is required to raise the temperature of water (and so much heat is lost when the temperature is lowered), the ocean and other large bodies of water have relatively constant temperatures. Thus, many organisms living in the ocean are provided with a relatively constant environmental temperature. The properties of water are crucial in stabilizing temperatures on Earth's surface. Although surface water is only a thin film relative to Earth's volume, the quantity is enormous compared to the exposed landmass. This relatively large mass of water resists both the warming effect of heat and the cooling effect of low temperatures.

Hydrogen bonding causes ice to have unique properties with important environmental consequences. Liquid water expands as it freezes because the hydrogen bonds joining the water molecules in the crystalline lattice keep the molecules far enough apart to give ice a density about 10% less than the density of liquid water (see Fig. 2-16c). When ice has been heated enough to raise its temperature above 0°C (32°F), the hydrogen bonds are broken, freeing the molecules to slip closer together. The density of water is greatest at 4°C. Above that temperature water begins to expand again as the speed of its molecules increases. As a result, ice floats on the denser cold water.

This unusual property of water has been important to the evolution of life. If ice had a greater density than water, the ice would sink; eventually, all ponds, lakes, and even the ocean would freeze solid from the bottom to the surface, making life impossible. When a deep body of water cools, it becomes covered with floating ice. The ice insulates the liquid water below it, retarding freezing and permitting organisms to survive below the icy surface.

The high water content of organisms helps them maintain relatively constant internal temperatures. Such minimizing of temperature fluctuations is important because biological reactions can take place only within a relatively narrow temperature range.

○HECKPOINT 2.5

- *What properties of a water molecule enable it to participate in the formation of hydrogen bonds?*
- CONNECT *What are some properties of water that result from hydrogen bonding? How do these properties contribute to the role of water as an essential component of organisms?*
- CONNECT *How can weak forces, such as hydrogen bonds, have significant effects in organisms?*

2.6 ACIDS, BASES, AND SALTS

●EARNING OBJECTIVES

10 Contrast acids and bases, and discuss their properties.

11 Convert the hydrogen ion concentration (moles per liter) of a solution to a pH value and describe how buffers help minimize changes in pH.

12 Describe the composition of a salt and explain the ways in which salts are important in organisms.

Water molecules have a slight tendency to ionize, that is, to dissociate into hydrogen ions (H^+) and hydroxide ions (OH^-). The H^+ immediately combines with a negatively charged region of a water molecule, forming a hydronium ion (H_3O^+). However, by convention, H^+, rather than the more accurate H_3O^+, is used. In pure water, a small number of water molecules ionize. This slight tendency of water to dissociate is reversible because hydrogen ions and hydroxide ions reunite to form water.

$$HOH \rightleftharpoons H^+ + OH^-$$

Because each water molecule splits into one hydrogen ion and one hydroxide ion, the concentrations of hydrogen ions and hydroxide ions in pure water are exactly equal (0.0000001 or 10^{-7} mol/L for each ion). Such a solution is said to be neutral, that is, neither acidic nor basic (alkaline).

An acid is a substance that dissociates in solution to yield hydrogen ions (H^+) and anions.

$$Acid \longrightarrow H^+ + anion$$

An **acid** is a proton *donor*. (Recall that a hydrogen ion, or H^+, is nothing more than a proton.) Hydrochloric acid (HCl) is a common inorganic acid.

A **base** is defined as a proton *acceptor*. Most bases are substances that dissociate to yield a hydroxide ion (OH^-) and a cation when dissolved in water.

$$NaOH \longrightarrow Na^+ + OH^-$$

A hydroxide ion can act as a base by accepting a proton (H^+) to form water.

$$OH^- + H^+ \longrightarrow H_2O$$

Sodium hydroxide (NaOH) is a common inorganic base. Some bases do not dissociate to yield hydroxide ions directly. For example, ammonia (NH_3) acts as a base by accepting a proton from water, producing an ammonium ion (NH_4^+) and releasing a hydroxide ion from water.

$$NH_3 + H_2O \longrightarrow NH_4^+ + OH^-$$

pH is a convenient measure of acidity

The degree of a solution's acidity is generally expressed in terms of **pH**, defined as the negative logarithm (base 10) of the hydrogen ion concentration (expressed in moles per liter):

$$pH = -\log_{10}[H^+]$$

The brackets refer to concentration; therefore, $[H^+]$ means "the concentration of hydrogen ions," which is expressed in moles per liter because we are interested in the *number* of hydrogen ions per liter. Because the range of possible pH values is broad, a logarithmic scale (with a 10-fold difference between successive units) is more convenient than a linear scale.

Hydrogen ion concentrations are nearly always less than 1 mol/L. One gram of hydrogen ions dissolved in 1 L of water (a 1 *M* solution) may not sound impressive, but such a solution would be extremely acidic. The logarithm of a number less than 1 is a negative number; thus, the *negative* logarithm corresponds to a *positive* pH value. (Solutions with pH values less than zero can be produced but do not occur under biological conditions.)

Whole-number pH values are easy to calculate. For instance, consider our example of pure water, which has a hydrogen ion concentration of 0.0000001 (10^{-7}) mol/L. The logarithm is –7. The negative logarithm is 7; therefore, the pH is 7. **TABLE 2-2** shows how to calculate pH values from hydrogen ion concentrations and how to do the reverse. For comparison, the table also includes the hydroxide ion concentrations, which can be calculated because the product of the hydrogen ion concentration and the hydroxide ion concentration is 1×10^{-14}:

$$[H^+][OH^-] = 1 \times 10^{-14}$$

Pure water is an example of a **neutral solution;** with a pH of 7, it has equal concentrations of hydrogen ions and hydroxide ions (the concentration of each is 10^{-7} mol/L). An **acidic solution** has a hydrogen ion concentration that is higher than its hydroxide ion concentration and has a pH value of less than 7. For example, the hydrogen ion concentration of a solution with pH 1 is ten times that of a solution with pH 2. A **basic solution** has a hydrogen ion concentration that is lower than its hydroxide ion concentration and has a pH greater than 7.

The pH values of some common substances are shown in **FIG. 2-17**. Although some very acidic compartments exist within cells (see Chapter 4), most of the interior of an animal or plant cell is neither strongly acidic nor strongly basic; rather, it is an

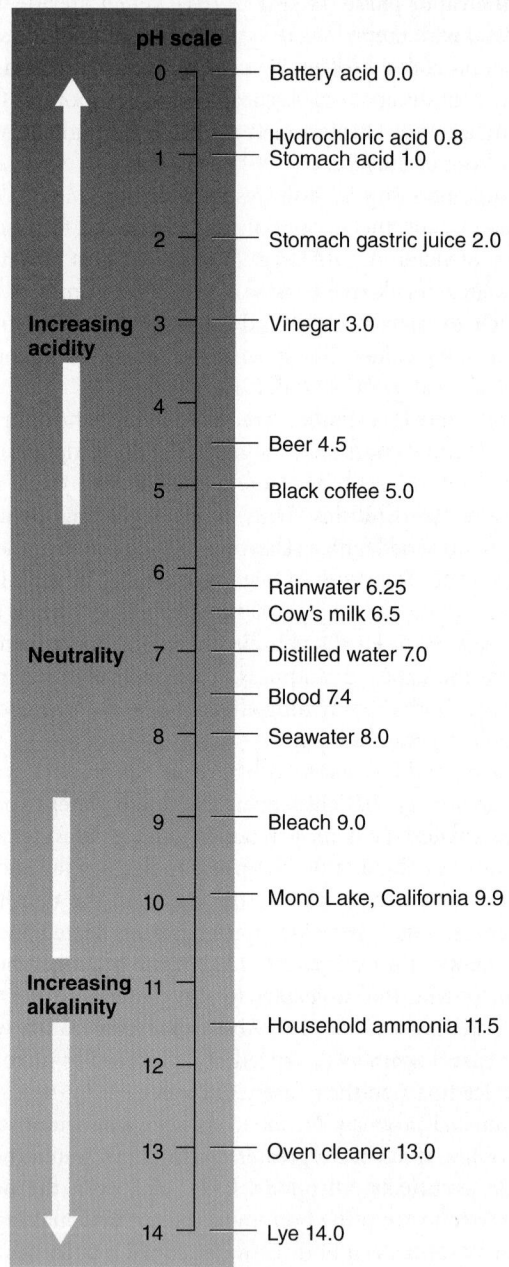

Figure 2-17 pH values of some common solutions

A neutral solution (pH 7) has equal concentrations of H^+ and OH^-. Acidic solutions, which have a higher concentration of H^+ than OH^-, have pH values less than 7; pH values greater than 7 characterize basic solutions, which have an excess of OH^-.

© Cengage Learning

TABLE 2-2	Calculating pH Values and Hydroxide Ion Concentrations from Hydrogen Ion Concentrations			
SUBSTANCE	$[H^+]^*$	LOG $[H^+]$	pH	$[OH^-]^\dagger$
Gastric juice	0.01, 10^{-2}	−2	2	10^{-12}
Pure water, neutral solution	0.0000001, 10^{-7}	−7	7	10^{-7}
Household ammonia	0.00000000001, 10^{-11}	−11	11	10^{-3}

$^*[H^+]$ = hydrogen ion concentration (mol/L)

$^\dagger[OH^-]$ = hydroxide ion concentration (mol/L)

© Cengage Learning

essentially neutral mixture of acidic and basic substances. Certain bacteria are adapted to life in extremely acidic environments (discussed in Chapter 25), but a substantial change in pH is incompatible with life for most cells. The pH of most types of plant and animal cells (and their environment) ordinarily ranges from around 7.2 to 7.4.

Buffers minimize pH change

Many mechanisms operate to maintain appropriate pH values in organisms. For example, the pH of human blood is about 7.4 and must be maintained within very narrow limits. If the blood becomes too acidic (e.g., as a result of respiratory disease), coma and death may result. Excessive alkalinity can result in overexcitability of the nervous system and even convulsions. Organisms contain many natural buffers. A **buffer** is a substance or combination of substances that resists changes in pH when an acid or base is added. A buffering system includes a weak acid or a weak base. A weak acid or weak base does not ionize completely. At any given instant, only a fraction of the molecules are ionized; most are not dissociated.

One of the most common buffering systems functions in the blood of vertebrates (see Chapter 46). Carbon dioxide, produced as a waste product of cell metabolism, enters the blood, the main constituent of which is water. The carbon dioxide reacts with the water to form carbonic acid, a weak acid that dissociates to yield a hydrogen ion and a bicarbonate ion. The following expression describes the buffering system:

$$CO_2 + H_2O \rightleftharpoons H_2CO_3 \rightleftharpoons H^+ + HCO_3^-$$

Carbon dioxide   Water   Carbonic acid   Bicarbonate ion

As the double arrows indicate, all the reactions are reversible. Because carbonic acid is a weak acid, undissociated molecules are always present, as are all the other components of the system. The expression describes the system when it is at dynamic equilibrium, that is, when the rates of the forward and reverse reactions are equal and the relative concentrations of the components are not changing. A system at dynamic equilibrium tends to stay at equilibrium unless a stress is placed on it, which causes it to shift to reduce the stress until it attains a new dynamic equilibrium. A change in the concentration of any component is one such stress. Therefore, the system can be "shifted to the right" by adding reactants or removing products. Conversely, the system can be "shifted to the left" by adding products or removing reactants.

Hydrogen ions are the important products to consider in this system. The addition of excess hydrogen ions temporarily shifts the system to the left as they combine with the bicarbonate ions to form carbonic acid. Eventually, a new dynamic equilibrium is established. At this point the hydrogen ion concentration is similar to the original concentration, and the product of the hydrogen ion and hydroxide ion concentrations is restored to the equilibrium value of 1×10^{-14}.

If hydroxide ions are added, they combine with the hydrogen ions to form water, effectively removing a product and thus shifting the system to the right. As this process occurs, more carbonic acid ionizes, effectively replacing the hydrogen ions that were removed.

Organisms contain many weak acids and weak bases, which allows them to maintain an essential reserve of buffering capacity and helps them avoid pH extremes.

An acid and a base react to form a salt

When an acid and a base are mixed in water, the H^+ of the acid unites with the OH^- of the base to form a molecule of water. The remainder of the acid (an anion) combines with the remainder of the base (a cation) to form a salt. For example, hydrochloric acid reacts with sodium hydroxide to form water and sodium chloride:

$$HCl + NaOH \longrightarrow H_2O + NaCl$$

A **salt** is a compound in which the hydrogen ion of an acid is replaced by some other cation. Sodium chloride, NaCl, is a salt in which the hydrogen ion of HCl has been replaced by the cation Na^+.

When a salt, an acid, or a base is dissolved in water, its dissociated ions can conduct an electric current; these substances are called **electrolytes.** Sugars, alcohols, and many other substances do not form ions when dissolved in water; they do not conduct an electric current and are referred to as *non-electrolytes.*

Cells and extracellular fluids (such as blood) of animals and plants contain a variety of dissolved salts that are the source of the many important mineral ions essential for fluid balance and acid–base balance. Nitrate and ammonium ions from the soil are the important nitrogen sources for plants. In animals, nerve and muscle function, blood clotting, bone formation, and many other aspects of body function depend on ions. Sodium, potassium, calcium, and magnesium are the chief cations present; chloride, bicarbonate, phosphate, and sulfate are important anions. The concentrations and relative amounts of the various cations and anions are kept remarkably constant. Any marked change results in impaired cell functions and may lead to death.

CHECKPOINT 2.6

- A solution has a hydrogen ion concentration of 0.01 mol/L. What is its pH? What is its hydroxide ion concentration? Is it acidic, basic, or neutral? How does the hydrogen ion concentration of this solution differ from one with a pH of 1?

- PREDICT Consider a reversible reaction that is at dynamic equilibrium. What would be the consequences of adding a reactant? adding a product? removing a reactant? removing a product?

- CONNECT What important role do buffers play in organisms? What prevents a strong acid or strong base from working as a buffer?

- What are the features of acids, bases, and salts that cause scientists to refer to them as electrolytes?

2.1 Elements and Atoms (*page 26*)

1 Name the principal chemical elements in living things and provide an important function of each.

- An **element** is a substance that cannot be decomposed into simpler substances by normal chemical reactions. About 96% of an organism's mass consists of carbon, the backbone of organic molecules; hydrogen and oxygen, the components of water; and nitrogen, a component of proteins and nucleic acids.

2 Compare the physical properties (mass and charge) and locations of electrons, protons, and neutrons. Distinguish between the atomic number and the mass number of an atom.

- Each **atom** is composed of a nucleus containing positively charged **protons** and uncharged **neutrons**. Negatively charged **electrons** encircle the nucleus.
- An atom is identified as belonging to a particular element by its number of protons (**atomic number**). The **atomic mass** of an atom is equal to the sum of its protons and neutrons.
- A single proton or a single neutron each has a mass equivalent to one **atomic mass unit (amu).** The mass of a single electron is only about 1/1800 amu.

3 Define the terms *orbital* and *electron shell*. Relate electron shells to principal energy levels.

- In the space outside the nucleus, electrons move rapidly in electron **orbitals.** An **electron shell** consists of electrons in orbitals at the same *principal energy level.* Electrons in a shell distant from the nucleus have greater energy than those in a shell closer to the nucleus.

2.2 Chemical Reactions (*page 30*)

4 Explain how the number of valence electrons of an atom is related to its chemical properties.

- The chemical properties of an atom are determined chiefly by the number and arrangement of its most energetic electrons, known as **valence electrons.** The valence shell of most atoms is full when it contains 8 electrons; that of hydrogen or helium is full when it contains 2. An atom tends to lose, gain, or share electrons to fill its valence shell.

5 Distinguish among simplest, molecular, and structural chemical formulas.

- Different atoms are joined by chemical bonds to form **chemical compounds.** A **chemical formula** gives the types and relative numbers of atoms in a substance.
- A **simplest formula** gives the smallest whole-number ratio of the component atoms. A **molecular formula** gives the actual numbers of each type of atom in a molecule. A **structural formula** shows the arrangement of the atoms in a molecule.

6 Explain why the mole concept is so useful to chemists.

- One **mole** (the atomic or molecular mass in grams) of any substance contains 6.02×10^{23} atoms, molecules, or ions, enabling scientists to "count" particles by weighing a sample. This number is known as **Avogadro's number.**

2.3 Chemical Bonds (*page 31*)

7 Distinguish among covalent bonds, ionic bonds, hydrogen bonds, and van der Waals interactions. Compare them in terms of the mechanisms by which they form and their relative strengths.

- **Covalent bonds** are strong, stable bonds formed when atoms share valence electrons, forming molecules. When covalent bonds are formed, the orbitals of the valence electrons may become rearranged in a process known as **orbital hybridization. Nonpolar covalent bonds** are formed if the electrons are shared equally between the two atoms. **Polar covalent bonds** are formed if one atom is more **electronegative** (has a greater affinity for electrons) than the other.
- An **ionic bond** is formed between a positively charged **cation** and a negatively charged **anion.** Ionic bonds are strong in the absence of water but relatively weak in aqueous solution.
- **Hydrogen bonds** are relatively weak bonds formed when a hydrogen atom with a partial positive charge is attracted to an atom (usually oxygen or nitrogen) with a partial negative charge already bonded to another molecule or in another part of the same molecule.
- **van der Waals interactions** are weak forces based on fluctuating electric charges.

2.4 Redox Reactions (*page 35*)

8 Distinguish between the terms *oxidation* and *reduction,* and relate these processes to the transfer of energy.

- **Oxidation** and **reduction** reactions (**redox reactions**) are chemical processes in which electrons (and their energy) are transferred from a reducing agent to an oxidizing agent. In oxidation, an atom, ion, or molecule loses electrons (and their energy). In reduction, an atom, ion, or molecule gains electrons (and their energy).

2.5 Water (*page 36*)

9 Explain how hydrogen bonds between adjacent water molecules govern many of the properties of water.

- Water is a **polar molecule** because one end has a partial positive charge and the other has a partial negative charge. Because its molecules are polar, water is an excellent **solvent** for ionic or polar **solutes.**
- Water molecules exhibit the property of **cohesion** because they form hydrogen bonds with one another; they also exhibit **adhesion** through hydrogen bonding to substances with ionic or polar regions.
- Water has a high **heat of vaporization.** Hydrogen bonds must be broken for molecules to enter the vapor phase. These molecules carry a great deal of heat, which accounts for *evaporative cooling.*
- Because hydrogen bonds must be broken to raise its temperature, water has a high **specific heat,** which helps organisms maintain a relatively constant internal temperature; this property also helps keep the ocean and other large bodies of water at a constant temperature.
- The hydrogen bonds between water molecules in ice cause it to be less dense than liquid water. Because ice floats, the aquatic environment is less extreme than it would be if ice sank to the bottom.

2.6 Acids, Bases, and Salts (*page 39*)

10 Contrast acids and bases, and discuss their properties.

- **Acids** are proton (hydrogen ion, H^+) donors; **bases** are proton acceptors. An acid dissociates in solution to yield H^+ and an anion. Many bases dissociate in solution to yield hydroxide ions (OH^-), which then accept protons to form water.

11 Convert the hydrogen ion concentration (moles per liter) of a solution to a pH value and describe how buffers help minimize changes in pH.

- **pH** is the negative log of the hydrogen ion concentration of a solution (expressed in moles per liter). A **neutral solution** with equal concentrations of H^+ and OH^- (10^{-7} mol/L) has a pH of 7, an **acidic solution** has a pH less than 7, and a **basic solution** has a pH greater than 7.

- A buffering system is based on a weak acid or a weak base. A **buffer** resists changes in the pH of a solution when acids or bases are added.

12 Describe the composition of a salt and explain the ways in which salts are important in organisms.

- A **salt** is a compound in which the hydrogen atom of an acid is replaced by some other cation. Salts provide the many mineral ions essential for life functions.

Know and Comprehend

1. Which of the following elements is *mismatched* with its properties or function? (a) carbon—forms the backbone of organic compounds (b) nitrogen—component of proteins (c) hydrogen—very electronegative (d) oxygen—can participate in hydrogen bonding (e) all of the above are correctly matched

2. $^{32}_{15}P$, a radioactive form of phosphorus, has (a) an atomic number of 32 (b) an atomic mass of 15 (c) an atomic mass of 47 (d) 32 electrons (e) 17 neutrons

3. Which of the following facts allows you to determine that atom A and atom B are isotopes of the same element? (a) they each have 6 protons (b) they each have 4 neutrons (c) the sum of the electrons and neutrons in each is 14 (d) they each have 4 valence electrons (e) they each have an atomic mass of 14

4. ^1_1H and ^3_1H have (a) different chemical properties because they have different atomic numbers (b) the same chemical properties because they have the same number of valence electrons (c) different chemical properties because they differ in their number of protons and electrons (d) the same chemical properties because they have the same atomic mass (e) the same chemical properties because they have the same number of protons, electrons, and neutrons

5. The orbitals composing an atom's valence electron shell (a) are arranged as concentric spheres (b) contain the atom's least energetic electrons (c) may change shape when covalent bonds are formed (d) never contain more than 1 electron each (e) more than one of the preceding is correct

6. Which of the following bonds and properties are correctly matched? (a) ionic bonds; are strong only if the participating ions are hydrated (b) hydrogen bonds; are responsible for bonding oxygen and hydrogen to form a single water molecule (c) polar covalent bonds; can occur between two atoms of the same element (d) covalent bonds; may be single, double, or triple (e) hydrogen bonds; are stronger than covalent bonds

7. In a redox reaction, (a) energy is transferred from a reducing agent to an oxidizing agent (b) a reducing agent becomes oxidized as it accepts an electron (c) an oxidizing agent accepts a proton (d) a reducing agent donates a proton (e) the electrons in an atom move from its valence shell to a shell closer to its nucleus

8. Water has a high specific heat because (a) hydrogen bonds must be broken to raise its temperature (b) hydrogen bonds must be formed to raise its temperature (c) it is a poor insulator (d) it has low density considering the size of the molecule (e) it can ionize

9. A solution at pH 7 is considered neutral because (a) its hydrogen ion concentration is 0 mol/L (b) its hydroxide ion concentration is 0 mol/L (c) the product of its hydrogen ion concentration and its hydroxide ion concentration is 0 mol/L (d) its hydrogen ion concentration is equal to its hydroxide ion concentration (e) it is nonpolar

10. A solution with a pH of 2 has a hydrogen ion concentration that is ___ the hydrogen ion concentration of a solution with a pH of 4. (a) 1/2 (b) 1/100 (c) 2 times (d) 10 times (e) 100 times

11. Which of the following cannot function as a buffer? (a) phosphoric acid, a weak acid (b) sodium hydroxide, a strong base (c) sodium chloride, a salt that ionizes completely (d) a and c (e) b and c

12. Which of the following statements is true? (a) the number of individual particles (atoms, ions, or molecules) contained in one mole varies depending on the substance (b) Avogadro's number is the number of particles contained in one mole of a substance (c) Avogadro's number is 10^{23} particles (d) one mole of ^{12}C has a mass of 12 g (e) b and d

Apply and Analyze

13. **VISUALIZE** Sketch simple Bohr diagrams for Elements A, B, and C. Element A has 2 electrons in its valence shell (which is complete when it contains 8 electrons). Would you expect element A to share, donate, or accept electrons? What would you expect of element B, which has 4 valence electrons, and element C, which has 7?

14. Consider the following reaction (in water):

$$HCl \longrightarrow H^+ + Cl^-$$

Name the reactant(s) and product(s). Does the expression indicate that the reaction is reversible? Could HCl be used as a buffer?

15. **INTERPRET DATA** Could you safely immerse your hand in water that contains 0.000,000,000,001 g of H^+ per liter?

Evaluate and Synthesize

16. **PREDICT** A hydrogen bond formed between two water molecules is only about 1/20 as strong as a covalent bond between hydrogen and oxygen. In what ways would the physical properties of water be different if these hydrogen bonds were stronger (e.g., 1/10 the strength of covalent bonds)? What if hydrogen bonds were weaker (e.g., 1/100 the strength of covalent bonds)?

17. **EVOLUTION LINK** Scientists have proposed various initiatives to detect water vapor, as well as oxygen and carbon dioxide, in the atmospheres of distant planets. Which of these *biosignatures* (chemical markers that are evidence for life) would you consider the most fundamental indicator that life could have evolved on these planets? Explain your reasoning.

aplia To access course materials, such as Aplia and other companion resources, please visit **www.cengagebrain.com**.

3 The Chemistry of Life: Organic Compounds

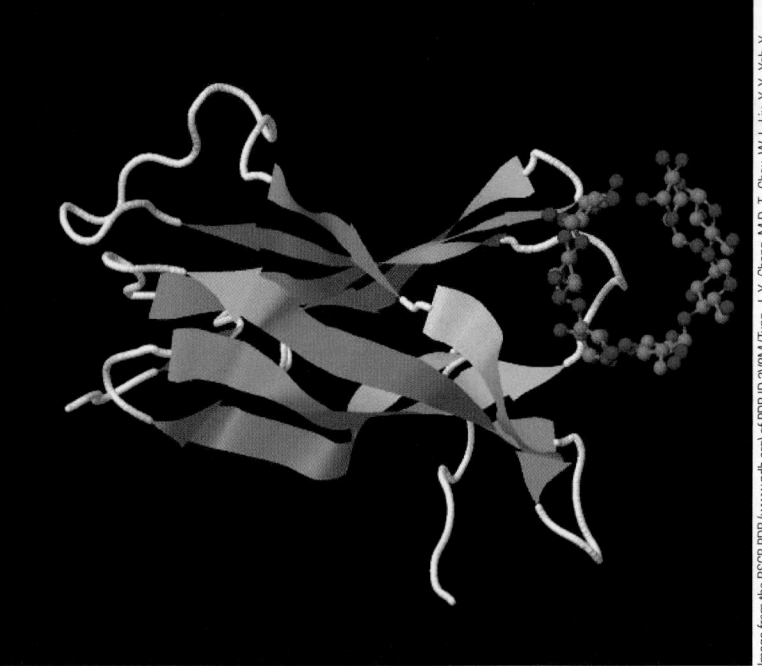

Image from the RSCB PDB (www.pdb.org) of PDB ID 2V8M (Tung, J.-Y., Chang, M.D.-T., Chou, W.-I., Liu, Y.-Y., Yeh, Y., Chang, F., Lin, S., Qui, Z., Sun, Y.-J. (2008) Crystal structures of the starch-binding domain from *Rhizopus oryzae* glucoamylase reveal a polysaccharide-binding path. Biochem. J. 416:27).

A computer-generated model of the starch-binding domain (*yellow* and *white*) of glycamylase, a fungal protein enzyme. The *red* and *gray* ring-shaped structure is β-cyclodextran, a polymer of glucose made by the partial digestion of starch.

Both inorganic and organic forms of carbon occur widely in nature. Organic compounds are those in which carbon atoms are covalently bonded to one another to form the backbone of the molecule. Some very simple carbon compounds are considered inorganic if the carbon is not bonded to another carbon or to hydrogen. The carbon dioxide we exhale as a waste product from the breakdown of organic molecules to obtain energy is an example of an inorganic carbon compound. Organic compounds are so named because at one time it was thought that they could be produced only by living (organic) organisms. In 1828, German chemist Friedrich Wühler synthesized urea, a metabolic waste product. Since that time, scientists have learned to synthesize many organic molecules and have discovered organic compounds not found in any organism.

Organic compounds are extraordinarily diverse, and millions have been identified. There are many reasons for this diversity. Organic compounds can be produced in a wide variety of three-dimensional (3-D) shapes. Furthermore, the carbon atom forms bonds with a greater number of different elements than does any other type of atom. The addition of chemical groups containing atoms of other elements—especially oxygen, nitrogen, phosphorus, and sulfur—can profoundly change the properties of an organic molecule. Extremely large *macromolecules*, which cells construct from simpler modular subunits, are an important source of diversity. For example, protein molecules are built from smaller compounds called *amino acids*. Proteins have many different types of 3-D structures that depend on the composition and arrangements of their amino acid subunits. The function of an individual protein is determined by its structure.

As you study this chapter, you will develop an understanding of the major groups of organic compounds found in organisms, including carbohydrates, lipids, proteins, and nucleic acids (DNA and RNA). Why are these compounds of central importance to all living things? The answer to this question will become more obvious as you study subsequent chapters, in which we explore the evidence that all living things evolved from a common ancestor. Evolution provides a powerful explanation for the similarities of the molecules that make up the structures of cells and tissues, participate in and regulate metabolic reactions, transmit information, and provide energy for life processes.

KEY CONCEPTS

3.1 Carbon atoms join with one another or other atoms to form large molecules with a wide variety of shapes. Hydrocarbons are non-polar, hydrophobic molecules; their properties can be altered by adding functional groups.

3.2 Carbohydrates are composed of sugar subunits (monosaccharides), which can be joined to form disaccharides, storage polysaccharides, and structural polysaccharides.

3.3 Lipids store energy (triacylglycerols) and are the main structural components of cell membranes (phospholipids).

3.4 Proteins have multiple levels of structure and are composed of amino acid subunits joined by peptide bonds. The amino acid sequences of these large molecules produce diverse structures with many different types of functions.

3.5 Nucleic acids (DNA and RNA) are informational molecules composed of long chains of nucleotide subunits. ATP and some other nucleotides have a central role in energy metabolism.

3.6 Biological molecules have recognizable attributes that aid in their identification.

3.1 CARBON ATOMS AND ORGANIC MOLECULES

LEARNING OBJECTIVES

1 Describe the properties of carbon that make it the central component of organic compounds.
2 Define the term *isomer* and distinguish among the three principal isomer types.
3 Identify the major functional groups present in organic compounds and describe their properties.
4 Explain the relationship between polymers and macromolecules.

Carbon has unique properties that allow the formation of the carbon backbones of the large, complex molecules essential to life (FIG. 3-1). Because a carbon atom has 4 valence electrons, it can complete its valence shell by forming a total of 4 covalent bonds (see Fig. 2-2). Each bond can link it to another carbon atom or to an atom of a different element. Carbon is particularly well suited to serve as the backbone of a large molecule because carbon-to-carbon bonds are strong and not easily broken. However, they are not so strong that it would be impossible for cells to break them. Carbon-to-carbon bonds are not limited to single bonds (based on sharing one electron pair).

Two carbon atoms can share two electron pairs with each other, forming double bonds:

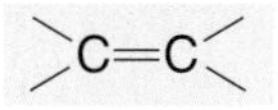

$$\text{>C=C<}$$

In some compounds, triple carbon-to-carbon bonds are formed:

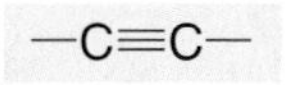

$$-\text{C}\equiv\text{C}-$$

As shown in Figure 3-1, **hydrocarbons,** organic compounds consisting only of carbon and hydrogen, can exist as unbranched or branched chains, or as rings. Rings and chains are joined in some compounds.

The molecules in the cell are analogous to the components of a machine. Each component has a shape that allows it to fill certain roles and to interact with other components (often with a complementary shape). Similarly, the shape of a molecule is important in determining its biological properties and function. Carbon atoms can link to one another and to other atoms to produce a wide variety of 3-D molecular shapes because the four covalent bonds of carbon do not form in a single plane. Instead, as discussed in Chapter 2, the valence electron orbitals become elongated and project from the carbon atom toward the corners of a tetrahedron (FIG. 3-2). The structure is highly symmetrical, with an angle of about 109.5 degrees between any two of these bonds. Keep in mind that for simplicity, many of the figures in this book are drawn as two-dimensional (2-D) graphic representations of 3-D molecules. Even the simplest hydrocarbon

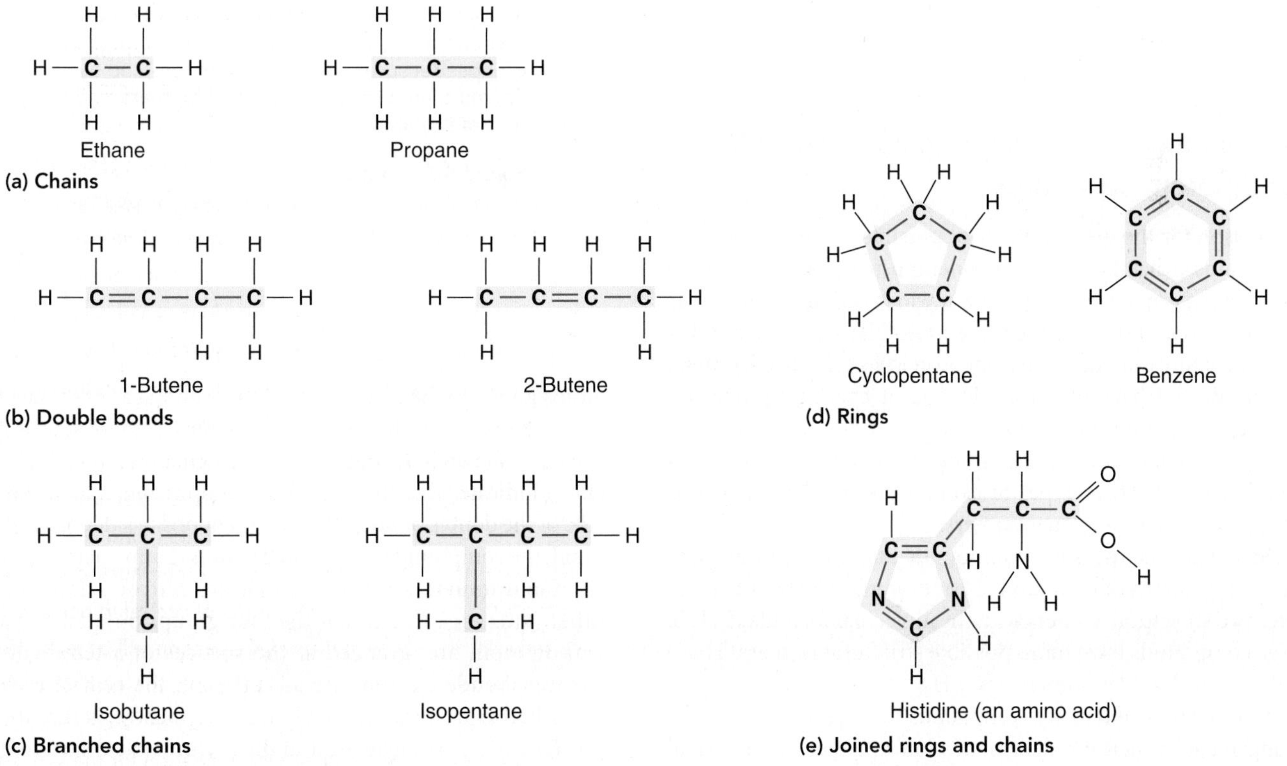

Figure 3-1 Diversity of organic molecules
Note that each carbon atom forms four covalent bonds, producing a wide variety of molecular shapes and sizes.
© Cengage Learning

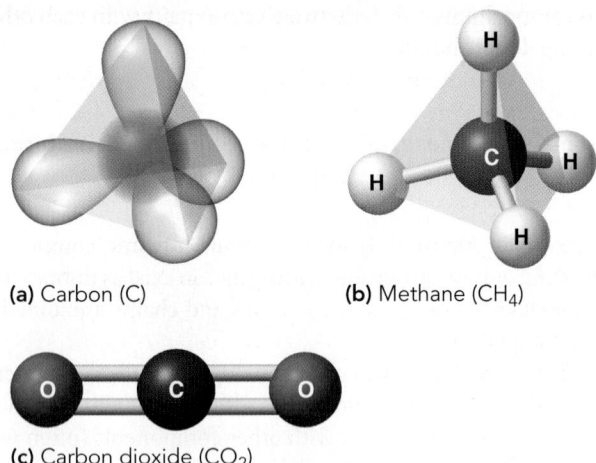

(a) Carbon (C)   **(b)** Methane (CH_4)

(c) Carbon dioxide (CO_2)

Figure 3-2 Carbon bonding

(a) The 3-D arrangement of the electron orbitals of a carbon atom is responsible for **(b)** the tetrahedral architecture of methane. **(c)** In carbon dioxide, oxygen atoms are joined linearly to a central carbon by polar double bonds.

© Cengage Learning

Ethanol (C_2H_6O)   Dimethyl ether (C_2H_6O)

(a) Structural isomers. Structural isomers differ in the covalent arrangement of their atoms.

trans-2-butene   _cis_-2-butene

(b) Geometric isomers. Geometric, or _cis–trans_, isomers have identical covalent bonds but differ in the order in which groups are arranged in space.

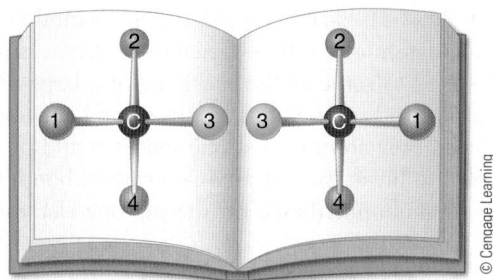

(c) Enantiomers. Enantiomers are isomers that are mirror images of each other. The central carbon is asymmetrical because it is bonded to four different groups. Because of their 3-D structure, the two figures cannot be superimposed no matter how they are rotated.

Figure 3-3 Isomers

Isomers have the same molecular formula, but their atoms are arranged differently.

chains, such as those in Figure 3-1, are not actually straight but have a 3-D zigzag structure.

Generally, there is freedom of rotation around each carbon-to-carbon single bond. This property permits organic molecules to be flexible and to assume a variety of shapes, depending on the extent to which each single bond is rotated. Double and triple bonds do not allow rotation, so regions of a molecule with such bonds tend to be inflexible.

Isomers have the same molecular formula but different structures

One reason for the great number of possible carbon-containing compounds is that the same components usually can link in more than one pattern, generating an even wider variety of molecular shapes. Compounds with the same molecular formulas but different structures and thus different properties are called **isomers.** Isomers do not have identical physical or chemical properties and may have different common names. Cells can distinguish between isomers. Usually, one isomer is biologically active and the other is not. Three types of isomers are structural isomers, geometric isomers, and enantiomers.

Structural isomers are compounds that differ in the covalent arrangements of their atoms. For example, **FIGURE 3-3a** illustrates two structural isomers with the molecular formula C_2H_6O. Large compounds have more possible structural isomers. There can be up to 366,319 isomers of $C_{20}H_{42}$.

Geometric isomers are compounds that are identical in the arrangement of their covalent bonds but different in the spatial arrangement of atoms or groups of atoms. Geometric isomers are present in some compounds with carbon-to-carbon double bonds. Because double bonds are not flexible, as single bonds are,

atoms joined to the carbons of a double bond cannot rotate freely about the axis of the bonds. These _cis–trans_ isomers may be represented as shown in **FIGURE 3-3b**. The designation _cis_ (Latin, "on this side") indicates that the two larger components are on the same side of the double bond. If they are on opposite sides of the double bond, the compound is designated a _trans_ (Latin, "across") isomer.

Enantiomers are isomers that are mirror images of each other (**FIG. 3-3c**). Recall that the four groups bonded to a single carbon atom are arranged at the vertices of a tetrahedron. If the four bonded groups are all different, the central carbon is described as asymmetrical. Figure 3-3c illustrates that the four groups can be arranged around the asymmetrical carbon in two different ways that are mirror images of each other. The two molecules are enantiomers if they cannot be superimposed on each other no matter how they are rotated in space. Although enantiomers have similar chemical properties and most of their physical properties are identical, cells recognize the difference in shape, and usually only one form is found in organisms.

Functional groups change the properties of organic molecules

The existence of isomers is not the only source of variety among organic molecules. The addition of various combinations of atoms generates a vast array of molecules with different properties. Because covalent bonds between hydrogen and carbon are nonpolar, **hydrocarbons** lack distinct charged regions. For this reason, hydrocarbons are insoluble in water and tend to cluster together through **hydrophobic interactions.** "Water fearing," the literal meaning of the term *hydrophobic*, is somewhat misleading. Hydrocarbons interact with water, but much more weakly than the water molecules cohere to one another through hydrogen bonding. Hydrocarbons interact weakly with one another, but the main reason for hydrophobic interactions is that they are driven together in a sense, having been excluded by the hydrogen-bonded water molecules.

However, the characteristics of an organic molecule can be changed dramatically by replacing one or more of the hydrogens with a **functional group,** a group of atoms that help determine the types of chemical reactions and associations in which the compound participates. Most functional groups readily form associations, such as ionic and hydrogen bonds, with other molecules. Polar and ionic functional groups are **hydrophilic** because they associate strongly with polar water molecules.

The properties of the major classes of biologically important organic compounds—carbohydrates, lipids, proteins, and nucleic acids—are largely a consequence of the types and arrangement of functional groups they contain. When we know what kinds of functional groups are present in an organic compound, we can predict its chemical behavior. Note that the symbol R is used to represent the *remainder* of the molecule of which each functional group is a part. For example, the **methyl group,** a common nonpolar hydrocarbon group, is abbreviated $R—CH_3$. As you read the rest of this section, refer to **TABLE 3-1** for the structural formulas of other important functional groups, as well as for additional information.

The **hydroxyl group** (abbreviated $R—OH$) is polar because of the presence of a strongly electronegative oxygen atom. Do not confuse it with the hydroxide ion, OH^-, discussed in Chapter 2. If a hydroxyl group replaces one hydrogen of a hydrocarbon, the resulting molecule can have significantly altered properties. For example, ethane (see Fig. 3-1a) is a hydrocarbon that is a gas at room temperature. If a hydroxyl group replaces a hydrogen atom, the resulting molecule is ethyl alcohol, or ethanol, which is found in alcoholic beverages (see Fig. 3-3a). Ethanol is somewhat cohesive because the polar hydroxyl groups of adjacent molecules interact; it is therefore liquid at room temperature. Unlike ethane, ethyl alcohol dissolves in water because the polar hydroxyl groups interact with the polar water molecules.

The **carbonyl group** consists of a carbon atom that has a double covalent bond with an oxygen atom. This double bond is polar because of the electronegativity of the oxygen; thus, the carbonyl group is hydrophilic. The position of the carbonyl group in the molecule determines the class to which the molecule belongs. An **aldehyde** has a carbonyl group positioned at the end of the carbon skeleton (abbreviated $R—CHO$); a **ketone** has an internal carbonyl group (abbreviated $R—CO—R$).

The **carboxyl group** (abbreviated $R—COOH$) in its non-ionized form consists of a carbon atom joined by a double covalent bond to an oxygen atom as well as by a single covalent bond to another oxygen, which is in turn bonded to a hydrogen atom. Two electronegative oxygen atoms in such close proximity establish an extremely polarized condition, which can cause the hydrogen atom to be stripped of its electron and released as a hydrogen ion (H^+). The resulting ionized carboxyl group has 1 unit of negative charge ($R—COO^-$):

Carboxyl groups are weakly acidic; only a fraction of the molecules ionize in this way. This group therefore exists in one of two hydrophilic states: ionic or polar. Carboxyl groups are essential constituents of amino acids.

An **amino group** (abbreviated $R—NH_2$) in its non-ionized form includes a nitrogen atom covalently bonded to two hydrogen atoms. Amino groups are weakly basic because they are able to accept a hydrogen ion (proton). The resulting ionized amino group has 1 unit of positive charge ($R—NH_3^+$). Amino groups are components of amino acids and of nucleic acids.

A **phosphate group** (abbreviated $R—PO_4H_2$) is weakly acidic. The attraction of electrons by the oxygen atoms can result in the release of one or two hydrogen ions, producing ionized forms with 1 or 2 units of negative charge. Phosphates are constituents of nucleic acids and certain lipids.

The **sulfhydryl group** (abbreviated $R—SH$), consisting of an atom of sulfur covalently bonded to a hydrogen atom, is found in molecules called *thiols*. As you will see, amino acids that contain a sulfhydryl group can make important contributions to the structure of proteins.

Many biological molecules are polymers

Many biological molecules such as proteins and nucleic acids are very large, consisting of thousands of atoms. Such giant molecules are known as **macromolecules.** Most macromolecules are **polymers,** produced by linking small organic compounds called **monomers** (**FIG. 3-4**). Just as all the words in this book have been written by arranging the 26 letters of the alphabet in various combinations, monomers can be grouped to form an almost infinite variety of larger molecules. The thousands of different complex organic compounds present in organisms are constructed from about 40 small, simple monomers. For example, the 20 monomers called *amino acids* can be linked end to end in countless ways to form the polymers known as *proteins.*

TABLE 3-1 | Some Biologically Important Functional Groups

FUNCTIONAL GROUP AND DESCRIPTION	STRUCTURAL FORMULA		CLASS OF COMPOUND CHARACTERIZED BY GROUP
HYDROXYL Polar because electronegative oxygen attracts covalent electrons	R—**OH**		Alcohols Example, ethanol
CARBONYL **Aldehydes:** Carbonyl group carbon is bonded to at least one H atom; polar because electronegative oxygen attracts covalent electrons	R—**C**—H (with double-bond O above)		Aldehydes Example, formaldehyde
Ketones: Carbonyl group carbon is bonded to two other carbons; polar because electronegative oxygen attracts covalent electrons	R—**C**—R (with double-bond O above)		Ketones Example, acetone
CARBOXYL Weakly acidic; can release an H^+	R—**C**—**OH** Non-ionized	R—**C**—**O$^-$** + H^+ Ionized	Carboxylic acids (organic acids) Example, amino acid
AMINO Weakly basic; can accept an H^+	R—**N**(H)(H) Non-ionized	R—**N$^+$**(H)(H)(H) Ionized	Amines Example, amino acid
PHOSPHATE Weakly acidic; one or two H^+ can be released	R—**O**—**P**—**OH** (with O above and OH below) Non-ionized	R—**O**—**P**—**O$^-$** (with O above and O$^-$ below) Ionized	Organic phosphates Example, phosphate ester (as found in ATP)
SULFHYDRYL Helps stabilize internal structure of proteins	R—**SH**		Thiols Example, cysteine

Polymers can be degraded to their component monomers by **hydrolysis reactions** ("to break with water"). In a reaction regulated by a specific enzyme (biological catalyst), a hydrogen from a water molecule attaches to one monomer, and a hydroxyl from water attaches to the adjacent monomer (**FIG. 3-5**).

Monomers become covalently linked by **condensation reactions.** Because the *equivalent* of a molecule of water is removed during the reactions that combine monomers, the term *dehydration synthesis* is sometimes used to describe condensation (see Fig. 3-5). However, in biological systems the synthesis

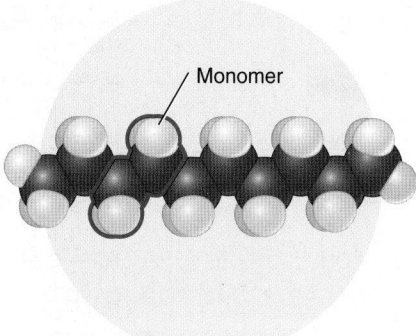

Figure 3-4 A simple polymer

This small polymer of polyethylene is formed by linking two-carbon ethylene (C_2H_4) monomers. One such monomer is outlined in *red*. The structure is represented by a space-filling model, which accurately depicts the 3-D shape of the molecule.

© Cengage Learning

of a polymer is not simply the reverse of hydrolysis, even though the net effect is the opposite of hydrolysis. Synthetic processes such as condensation require energy and are regulated by different enzymes.

In the following sections we examine carbohydrates, lipids, proteins, and nucleic acids. Our discussion begins with the smaller, simpler forms of these compounds and extends to the linking of these monomers to form macromolecules.

CHECKPOINT 3.1

- *What are some of the ways that the features of carbon-to-carbon bonds influence the stability and 3-D structure of organic molecules?*

- VISUALIZE *Draw pairs of simple sketches comparing two (1) structural isomers, (2) geometric isomers, and (3) enantiomers. Why are these differences biologically important?*

- VISUALIZE *Sketch the following functional groups: methyl, amino, carbonyl, hydroxyl, carboxyl, and phosphate. Include both non-ionized and ionized forms for acidic and basic groups.*

- CONNECT *How is the designation that a group is nonpolar, polar, acidic, or basic related to its hydrophilic or hydrophobic properties?*

- VISUALIZE *Draw simple sketches illustrating how the equivalent of a water molecule participates in the reactions of monomers in condensation and hydrolysis.*

3.2 CARBOHYDRATES

LEARNING OBJECTIVE

5 Distinguish among monosaccharides, disaccharides, and polysaccharides; compare storage polysaccharides with structural polysaccharides.

Sugars, starches, and cellulose are **carbohydrates.** Sugars and starches serve as energy sources for cells; cellulose is the main structural component of the walls that surround plant cells. Carbohydrates contain carbon, hydrogen, and oxygen atoms in a ratio of approximately one carbon to two hydrogens to one oxygen ($CH_2O)_n$. The term *carbohydrate*, meaning "hydrate (water) of carbon," reflects the 2:1 ratio of hydrogen to oxygen, the same ratio found in water (H_2O). Carbohydrates contain one sugar unit (monosaccharides), two sugar units (disaccharides), or many sugar units (polysaccharides).

Monosaccharides are simple sugars

Monosaccharides typically contain from three to seven carbon atoms. In a monosaccharide a hydroxyl group is bonded to each carbon except one; that carbon is double-bonded to an oxygen atom, forming a carbonyl group. If the carbonyl group is at the end of the chain, the monosaccharide is an aldehyde; if the carbonyl group is at any other position, the monosaccharide is a ketone. (By convention, the numbering of the carbon skeleton of a sugar begins with the carbon at or nearest the carbonyl end of the open chain.) The large number of polar hydroxyl groups, plus the carbonyl group, gives a monosaccharide hydrophilic properties. FIGURE 3-6 shows simplified, 2-D representations of some common monosaccharides. The simplest carbohydrates are the three-carbon sugars (trioses): glyceraldehyde and dihydroxyacetone. Ribose and deoxyribose are common pentoses, sugars that contain five carbons; they are components of nucleic acids (DNA, RNA, and related compounds). Glucose, fructose, galactose, and other six-carbon sugars are called hexoses. (Note that the names of carbohydrates typically end in -*ose*.)

Glucose ($C_6H_{12}O_6$), the most abundant monosaccharide, is used as an energy source in most organisms. During cellular respiration (see Chapter 8), cells oxidize glucose molecules, converting the stored energy to a form that can be readily used for cell work. Glucose is also used in the synthesis of other types of compounds such as amino acids and fatty acids. Glucose is

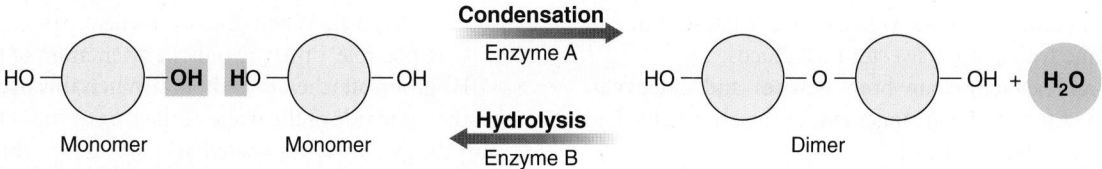

Figure 3-5 Condensation and hydrolysis reactions

Joining two monomers yields a dimer; incorporating additional monomers produces a polymer. Note that condensation and hydrolysis reactions are catalyzed by different enzymes.

© Cengage Learning

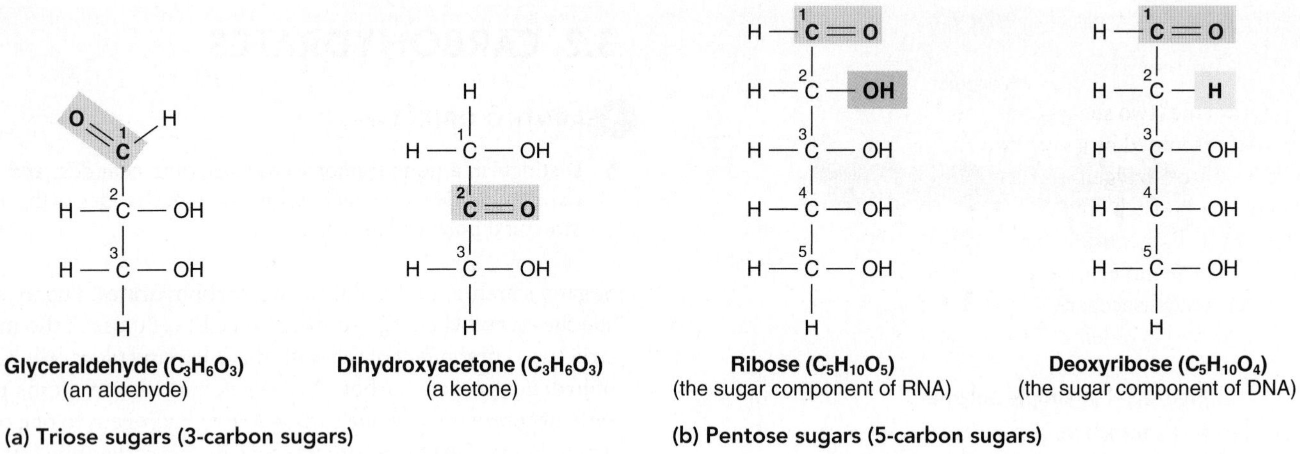

Glyceraldehyde ($C_3H_6O_3$)
(an aldehyde)

Dihydroxyacetone ($C_3H_6O_3$)
(a ketone)

(a) Triose sugars (3-carbon sugars)

Ribose ($C_5H_{10}O_5$)
(the sugar component of RNA)

Deoxyribose ($C_5H_{10}O_4$)
(the sugar component of DNA)

(b) Pentose sugars (5-carbon sugars)

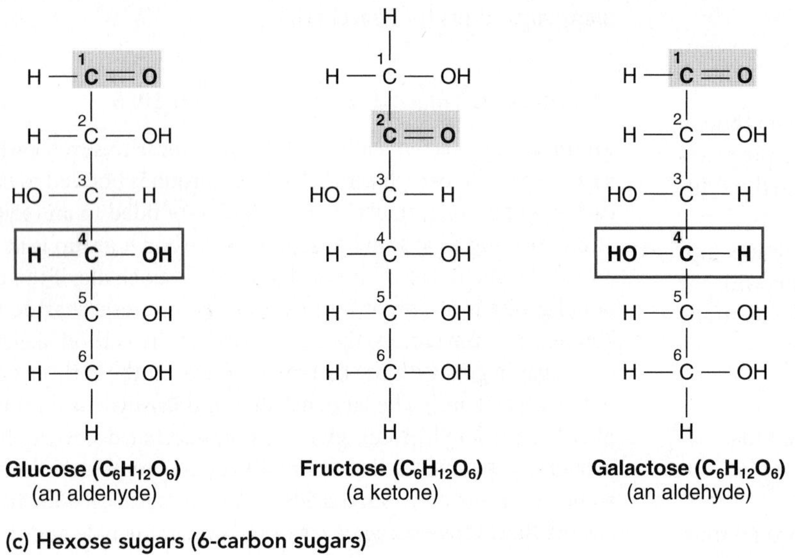

Glucose ($C_6H_{12}O_6$)
(an aldehyde)

Fructose ($C_6H_{12}O_6$)
(a ketone)

Galactose ($C_6H_{12}O_6$)
(an aldehyde)

(c) Hexose sugars (6-carbon sugars)

Figure 3-6 Monosaccharides

Shown are 2-D chain structures of **(a)** three-carbon trioses, **(b)** five-carbon pentoses, and **(c)** six-carbon hexoses. Although it is convenient to show monosaccharides in this form, the pentoses and hexoses are more accurately depicted as ring structures, as in Figure 3-7. The carbonyl group (*gray screen*) is terminal in aldehyde sugars and located in an internal position in ketones. Deoxyribose differs from ribose because deoxyribose has one less oxygen; a hydrogen (*yellow screen*) instead of a hydroxyl group (*blue screen*) is attached to carbon 2. Glucose and galactose differ in the arrangement of the hydroxyl group and hydrogen attached to carbon 4 (*red box*).

© Cengage Learning

so important in metabolism that mechanisms have evolved to maintain its concentration at relatively constant levels in the blood of humans and other complex animals (see Chapter 49).

Glucose and fructose are structural isomers: they have identical molecular formulas, but their atoms are arranged differently. In fructose (a ketone) the double-bonded oxygen is linked to a carbon within the chain rather than to a terminal carbon as in glucose (an aldehyde). Because of these differences, the two sugars have different properties. For example, fructose, found in honey and some fruits, tastes sweeter than glucose.

Glucose and galactose are both hexoses and aldehydes. However, they differ in the arrangement of the atoms attached to asymmetrical carbon atom 4.

The linear formulas in Figure 3-6 give a clear but somewhat unrealistic picture of the structures of some common monosaccharides. As we have mentioned, molecules are not 2-D; in fact, the properties of each compound depend largely on its 3-D structure. Thus, 3-D formulas are helpful in understanding the relationship between molecular structure and biological function.

Molecules of glucose and other pentoses and hexoses in solution are actually rings rather than extended straight carbon chains. Glucose in solution (as in the cell) typically exists as a ring of five carbons and one oxygen. It assumes this configuration when its atoms undergo a rearrangement, permitting a covalent bond to connect carbon 1 to the oxygen attached to carbon 5 (FIG. 3-7). When glucose forms a ring, two isomeric forms are possible, differing only in orientation of the hydroxyl (—OH) group attached to carbon 1. When this hydroxyl group is on the same side of the plane of the ring as the —CH₂OH side group, the glucose is designated beta glucose (β-glucose). When it is on the side (with respect to the plane of the ring) opposite the —CH₂OH side group, the compound is designated alpha glucose (α-glucose). Although the differences between these isomers may seem small, they have important consequences when the rings join to form polymers.

Disaccharides consist of two monosaccharide units

A **disaccharide** (two sugars) contains two monosaccharide rings joined by a glycosidic linkage, consisting of a central oxygen covalently bonded to two carbons, one in each ring (FIG. 3-8). The **glycosidic linkage** of a disaccharide generally forms between carbon 1 of one molecule and carbon 4 of the other molecule. The disaccharide maltose (malt sugar) consists of two covalently linked α-glucose units. Sucrose, common table sugar, consists of a glucose unit combined with a fructose unit. Lactose (the sugar present in milk) consists of one molecule of glucose and one of galactose.

As shown in Figure 3-8, a disaccharide can be hydrolyzed, that is, split by the addition of water, into two monosaccharide units. During digestion, maltose is hydrolyzed to form two molecules of glucose:

maltose + water ⟶ glucose + glucose

Similarly, sucrose is hydrolyzed to form glucose and fructose:

sucrose + water ⟶ glucose + fructose

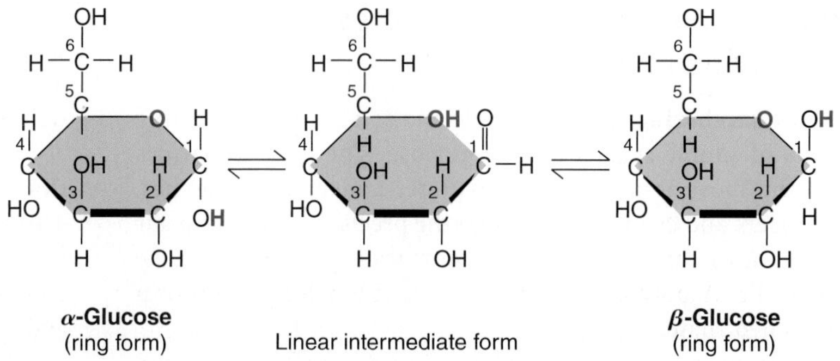

α-Glucose (ring form) — **Linear intermediate form** — **β-Glucose** (ring form)

(a) When dissolved in water, glucose undergoes a rearrangement of its atoms, forming one of two possible ring structures: α-glucose or β-glucose. Although the drawing does not show the complete 3-D structure, the thick, tapered bonds in the lower portion of each ring represent the part of the molecule that would project out of the page toward you.

α-Glucose — **β-Glucose**

(b) The essential differences between α-glucose and β-glucose are more readily apparent in these simplified structures. By convention, a carbon atom is assumed to be present at each angle in the ring unless another atom is shown. Most hydrogen atoms have been omitted.

Figure 3-7 α and β forms of glucose
© Cengage Learning

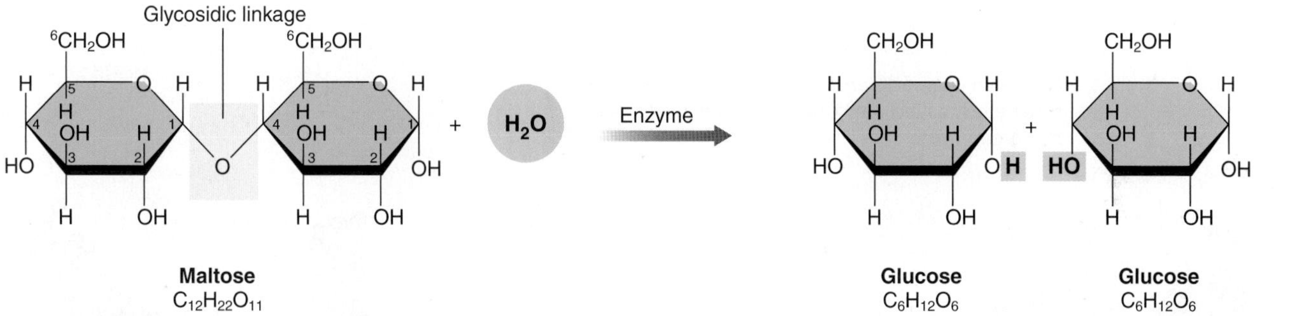

Maltose $C_{12}H_{22}O_{11}$ — **Glucose** $C_6H_{12}O_6$ — **Glucose** $C_6H_{12}O_6$

(a) Maltose may be broken down (as during digestion) to form two molecules of glucose. The glycosidic linkage is broken in a hydrolysis reaction, which requires the addition of water.

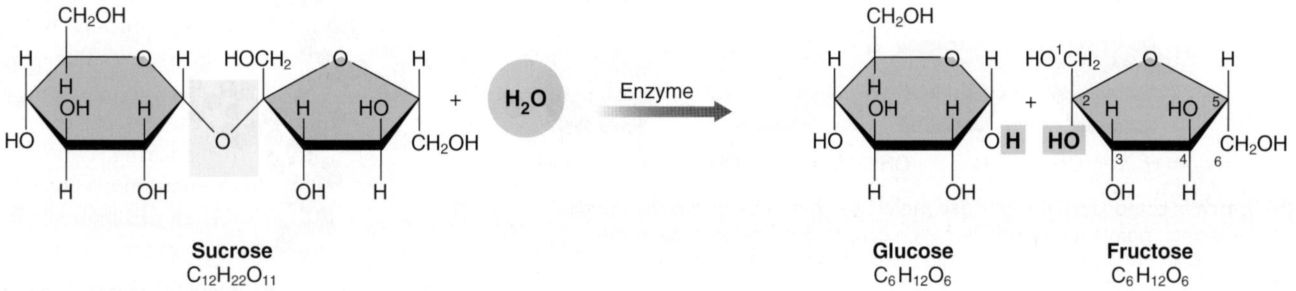

Sucrose $C_{12}H_{22}O_{11}$ — **Glucose** $C_6H_{12}O_6$ — **Fructose** $C_6H_{12}O_6$

(b) Sucrose can be hydrolyzed to yield a molecule of glucose and a molecule of fructose.

Figure 3-8 Hydrolysis of disaccharides
Note that an enzyme is needed to promote these reactions.
© Cengage Learning

Polysaccharides can store energy or provide structure

A **polysaccharide** is a macromolecule consisting of repeating units of simple sugars, usually glucose. The polysaccharides are the most abundant carbohydrates and include starches, glycogen, and cellulose. Although the precise number of sugar units varies, thousands of units are typically present in a single molecule. A polysaccharide may be a single long chain or a branched chain.

Because they are composed of different isomers and because the units may be arranged differently, polysaccharides vary in their properties. Those that can be easily broken down to their subunits are well suited for energy storage, whereas the macromolecular 3-D architecture of others makes them particularly well suited to form stable structures.

Starch, the typical form of carbohydrate used for energy storage in plants, is a polymer consisting of α-glucose subunits. These monomers are joined by α 1—4 linkages, which means that carbon 1 of one glucose is linked to carbon 4 of the next glucose in the chain (**FIG. 3-9**). Starch occurs in two forms: amylose and amylopectin. Amylose, the simpler form, is unbranched. Amylopectin, the more common form, usually consists of about 1000 glucose units in a branched chain.

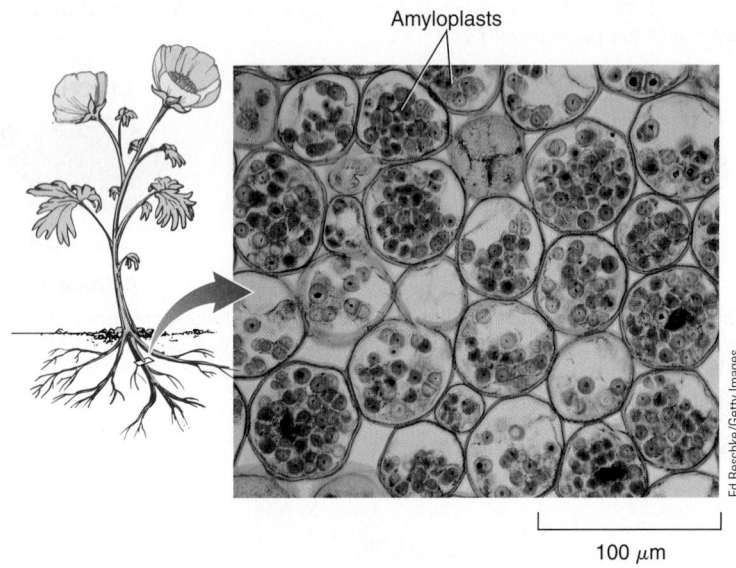

Amyloplasts

100 μm

(a) Starch (stained *purple*) is stored in specialized organelles, called *amyloplasts*, in these cells of a buttercup root.

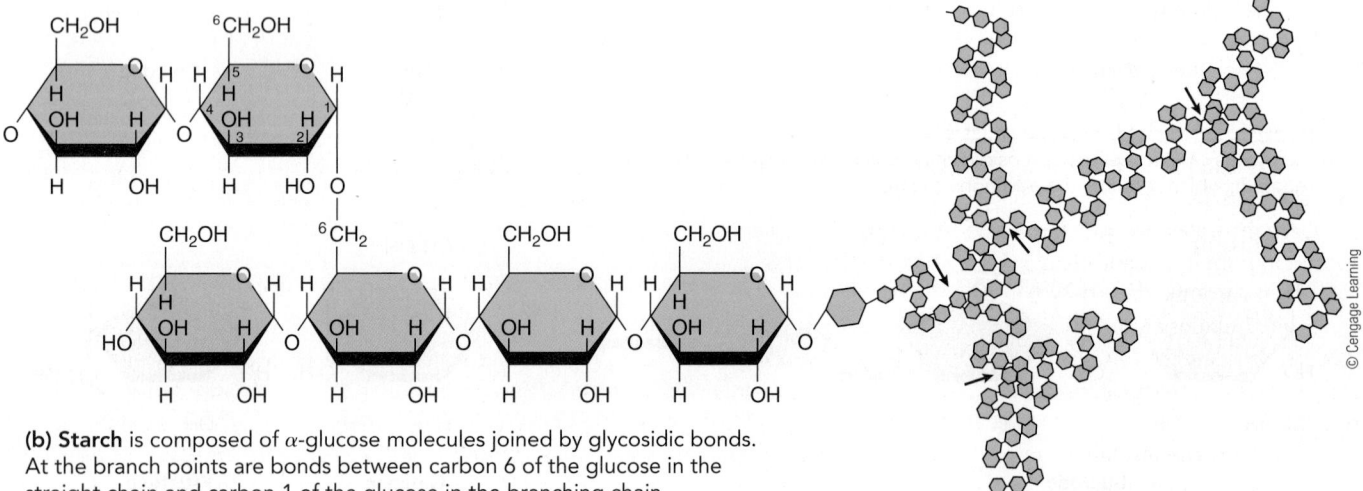

(b) Starch is composed of α-glucose molecules joined by glycosidic bonds. At the branch points are bonds between carbon 6 of the glucose in the straight chain and carbon 1 of the glucose in the branching chain.

(c) Starch consists of highly branched chains; the arrows indicate the branch points. Each chain is actually a coil or helix, stabilized by hydrogen bonds between the hydroxyl groups of the glucose subunits.

Figure 3-9 **Starch, a storage polysaccharide**

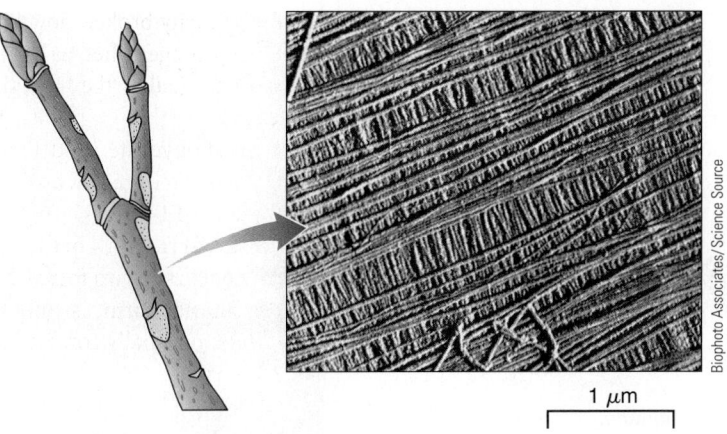

Figure 3-10 *Animation* **Cellulose, a structural polysaccharide**

(a) Cellulose fibers from a cell wall. The fibers shown in this electron micrograph consist of bundles of cellulose molecules that interact through hydrogen bonds.

(b) The cellulose molecule is an unbranched polysaccharide. It consists of about 10,000 β-glucose units joined by glycosidic bonds.

Plant cells store starch mainly as granules within specialized organelles called **amyloplasts** (see Fig. 3-9a); some cells, such as those of potatoes, are very rich in amyloplasts. When energy is needed for cell work, the plant hydrolyzes the starch, releasing the glucose subunits. Virtually all organisms, including humans and other animals, have enzymes that can break α 1—4 linkages.

Glycogen (sometimes referred to as *animal starch*) is the form in which glucose subunits, joined by α 1—4 linkages, are stored as an energy source in animal tissues. Glycogen is similar in structure to plant starch but more extensively branched and more water soluble. In vertebrates, glycogen is stored mainly in liver and muscle cells.

Carbohydrates are the most abundant group of organic compounds on Earth, and cellulose is the most abundant carbohydrate; it accounts for 50% or more of all the carbon in plants (FIG. 3-10). Cellulose is a structural carbohydrate. Wood is about half cellulose, and cotton is at least 90% cellulose. Plant cells are surrounded by strong supporting cell walls consisting mainly of cellulose.

Cellulose is an insoluble polysaccharide composed of many joined glucose molecules. The bonds joining these sugar units are different from those in starch. Recall that starch is composed of α-glucose subunits joined by α 1—4 glycosidic linkages. Cellulose contains β-glucose monomers joined by β 1—4 linkages. These bonds cannot be split by the enzymes that hydrolyze the α linkages in starch. Because humans, like other animals, lack enzymes that digest cellulose, we cannot use it as a nutrient. The

cellulose found in whole grains and vegetables remains fibrous and provides bulk that helps keep our digestive tract functioning properly.

Some microorganisms digest cellulose to glucose. In fact, cellulose-digesting bacteria live in the digestive systems of cows and sheep, enabling these grass-eating animals to obtain nourishment from cellulose. Similarly, the digestive systems of termites contain microorganisms that digest cellulose (see Fig. 26-4b).

Cellulose molecules are well suited for a structural role. The β-glucose subunits are joined in a way that allows extensive hydrogen bonding among different cellulose molecules, and they aggregate in long bundles of fibers (see Fig. 3-10a).

Some modified and complex carbohydrates have special roles

Many derivatives of monosaccharides are important biological molecules. Some form important structural components. The amino sugars galactosamine and glucosamine are compounds in which a hydroxyl group (—OH) is replaced by an amino group (—NH_2). Galactosamine is present in cartilage, a constituent of the skeletal system of vertebrates.

N-acetyl glucosamine (NAG) subunits, joined by glycosidic bonds, compose **chitin,** a main component of the cell walls of fungi and of the external skeletons of insects, crayfish, and other arthropods (FIG. 3-11). Chitin forms very tough structures because, as in cellulose, its molecules interact through multiple

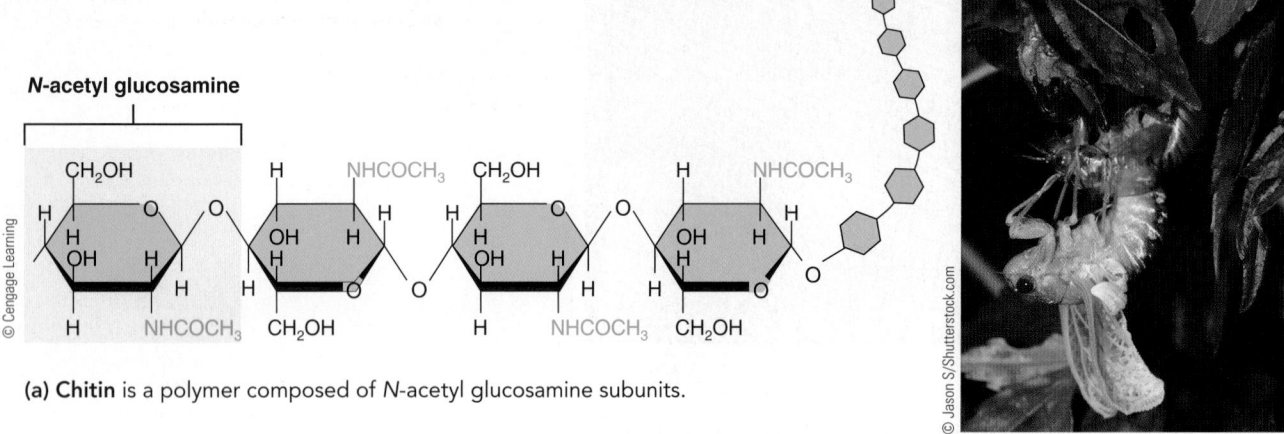

(a) **Chitin** is a polymer composed of *N*-acetyl glucosamine subunits.

(b) Chitin is an important component of the exoskeleton (outer covering) this cicada is shedding.

Figure 3-11 Chitin, a structural polysaccharide

hydrogen bonds. Some chitinous structures, such as the shell of a lobster, are further hardened by the addition of calcium carbonate ($CaCO_3$), an inorganic form of carbon.

Carbohydrates may also combine with proteins to form **glycoproteins,** compounds present on the outer surface of cells other than bacteria. Some of these carbohydrate chains allow cells to adhere to one another, whereas others provide protection. Most proteins secreted by cells are glycoproteins. They include the major components of mucus, a protective material secreted by the mucous membranes of the respiratory and digestive systems. Carbohydrates combine with lipids to form **glycolipids,** compounds on the surfaces of animal cells that allow cells to recognize and interact with one another.

CHECKPOINT 3.2

- VISUALIZE *Draw simple sketches comparing hydrogen bonding in storage polysaccharides, such as starch and glycogen, with structural polysaccharides, such as cellulose and chitin.*

3.3 LIPIDS

LEARNING OBJECTIVE

6 Distinguish among fats, phospholipids, and steroids, and describe the composition, characteristics, and biological functions of each.

Unlike carbohydrates, which are defined by their structure, **lipids** are a heterogeneous group of compounds that are categorized by being soluble in nonpolar solvents (such as ether and chloroform) and relatively insoluble in water. Lipid molecules have these properties because they consist mainly of carbon and hydrogen, with few oxygen-containing functional groups.

Hydrophilic functional groups typically contain oxygen atoms; therefore, lipids, which have little oxygen, tend to be hydrophobic. Among the biologically important groups of lipids are fats, phospholipids, carotenoids (orange and yellow plant pigments), steroids, and waxes. Some lipids are used for energy storage, others serve as structural components of cell membranes, and some are important hormones.

Triacylglycerol is formed from glycerol and three fatty acids

The most abundant lipids in living organisms are **triacylglycerols.** These compounds, commonly known as *fats,* are an economical form of reserve fuel storage because, when metabolized, they yield more than twice as much energy per gram as do carbohydrates. Carbohydrates and proteins can be transformed by enzymes into fats and stored within the cells of adipose (fat) tissue of animals and in some seeds and fruits of plants.

A triacylglycerol molecule (also known as a *triglyceride*) consists of glycerol joined to three fatty acids (FIG. 3-12). **Glycerol** is a three-carbon alcohol that contains three hydroxyl (—OH) groups, and a **fatty acid** is a long, unbranched hydrocarbon chain with a carboxyl group (—COOH) at one end. A triacylglycerol molecule is formed by a series of three condensation reactions. In each reaction, the equivalent of a water molecule is removed as one of the glycerol's hydroxyl groups reacts with the carboxyl group of a fatty acid, resulting in the formation of a covalent linkage known as an **ester linkage** (see Fig. 3-12b). The first reaction yields a **monoacylglycerol** (*monoglyceride*); the second, a **diacylglycerol** (*diglyceride*); and the third, a triacylglycerol. During digestion, triacylglycerols are hydrolyzed to produce fatty acids and glycerol (see Chapter 47). Diacylglycerol is an important molecule for sending signals within the cell (see Chapters 6 and 49).

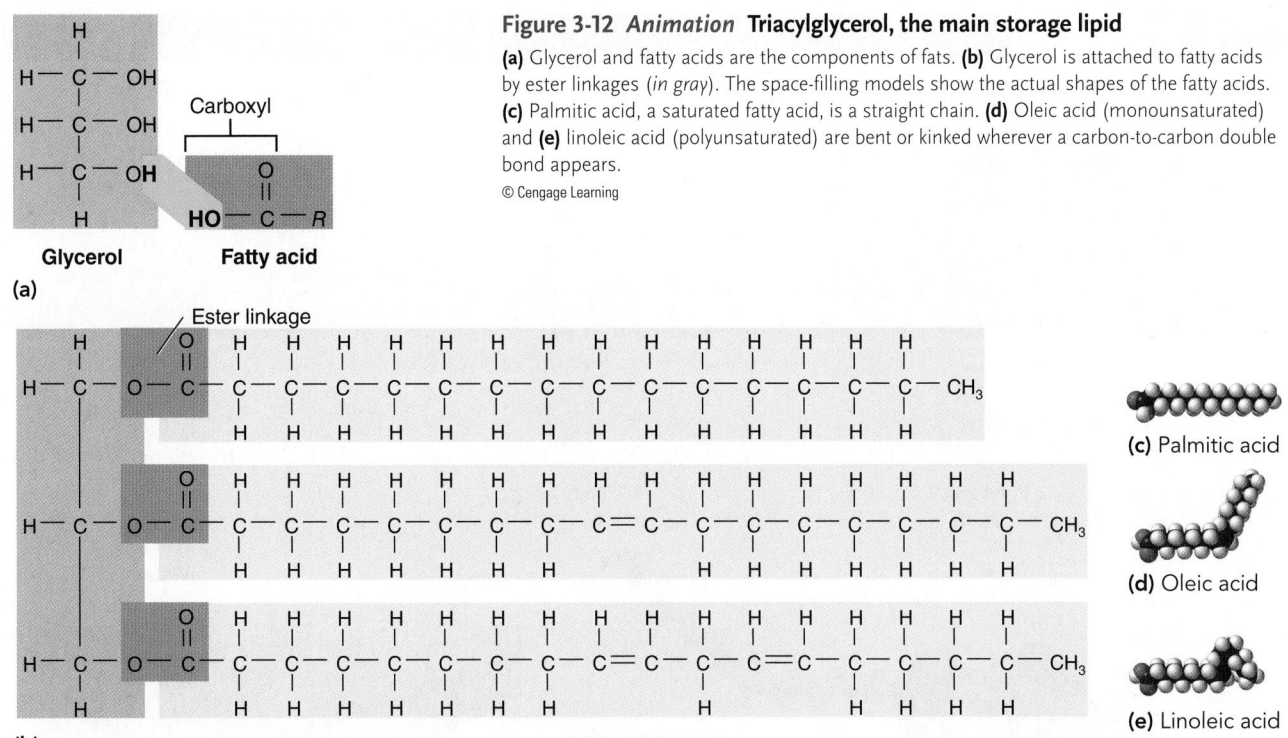

Figure 3-12 *Animation* **Triacylglycerol, the main storage lipid**

(a) Glycerol and fatty acids are the components of fats. **(b)** Glycerol is attached to fatty acids by ester linkages (*in gray*). The space-filling models show the actual shapes of the fatty acids. **(c)** Palmitic acid, a saturated fatty acid, is a straight chain. **(d)** Oleic acid (monounsaturated) and **(e)** linoleic acid (polyunsaturated) are bent or kinked wherever a carbon-to-carbon double bond appears.

© Cengage Learning

(c) Palmitic acid

(d) Oleic acid

(e) Linoleic acid

A triacylglycerol

Saturated and unsaturated fatty acids differ in physical properties

About 30 different fatty acids are commonly found in lipids, and they typically have an even number of carbon atoms. For example, butyric acid, present in rancid butter, has four carbon atoms. Oleic acid, with 18 carbons, is the most widely distributed fatty acid in nature and is found in most animal and plant fats.

Saturated fatty acids contain the maximum possible number of hydrogen atoms. Palmitic acid, a 16-carbon fatty acid, is a common saturated fatty acid (see Fig. 3-12c). Fats high in saturated fatty acids, such as animal fat and solid vegetable shortening, tend to be solid at room temperature. The reason is that even electrically neutral, nonpolar fatty acyl chains can develop transient regions of weak positive charge and weak negative charge. This situation occurs because the constant motion of their electrons causes some regions to have a temporary excess of electrons, whereas others have a temporary electron deficit. These slight opposite charges result in van der Waals interactions between adjacent molecules (see Chapter 2). Although van der Waals interactions are weakly attractive, they are strong when many occur along hydrocarbon chains that can closely align together. These van der Waals interactions tend to make a substance more solid by limiting the motion of its molecules.

Unsaturated fatty acids include one or more adjacent pairs of carbon atoms joined by a double bond. Therefore, they are not fully saturated with hydrogen. Fatty acids with one double bond are **monounsaturated fatty acids,** whereas those with more than one double bond are **polyunsaturated fatty acids.** Oleic acid is a monounsaturated fatty acid, and linoleic acid

is a common polyunsaturated fatty acid (see Figs. 3-12d and e). Fats containing a high proportion of monounsaturated or polyunsaturated fatty acids tend to be liquid at room temperature. The reason is that each double bond produces a bend in the hydrocarbon chain that prevents it from aligning closely with an adjacent chain. This limits the number of van der Waals interactions between chains, permitting freer molecular motion.

Food manufacturers commonly hydrogenate or partially hydrogenate cooking oils to make margarine and other foodstuffs, converting unsaturated fatty acids to saturated fatty acids and making the fat more solid at room temperature. This process makes the fat less healthful because saturated fatty acids in the diet are known to increase the risk of cardiovascular disease (see Chapter 44). The hydrogenation process has yet another effect. Note that in the naturally occurring unsaturated fatty acids oleic acid and linoleic acid shown in Figure 3-12, the two hydrogens flanking each double bond are on the same side of the hydrocarbon chain (the *cis* configuration). When fatty acids are artificially hydrogenated, the double bonds can become rearranged, resulting in a *trans* configuration, analogous to the arrangement shown in Figure 3-3b. *Trans* fatty acids are technically unsaturated, but they mimic many of the properties of saturated fatty acids. Because the *trans* configuration does not produce a bend at the site of the double bond, *trans* fatty acids are more solid at room temperature than *cis* fatty acids; like saturated fatty acids, they increase the risk of cardiovascular disease.

At least two unsaturated fatty acids (linoleic acid and arachidonic acid) are essential nutrients that must be obtained from food because the human body cannot synthesize them. However,

A lipid bilayer forms when phospholipids interact with water.

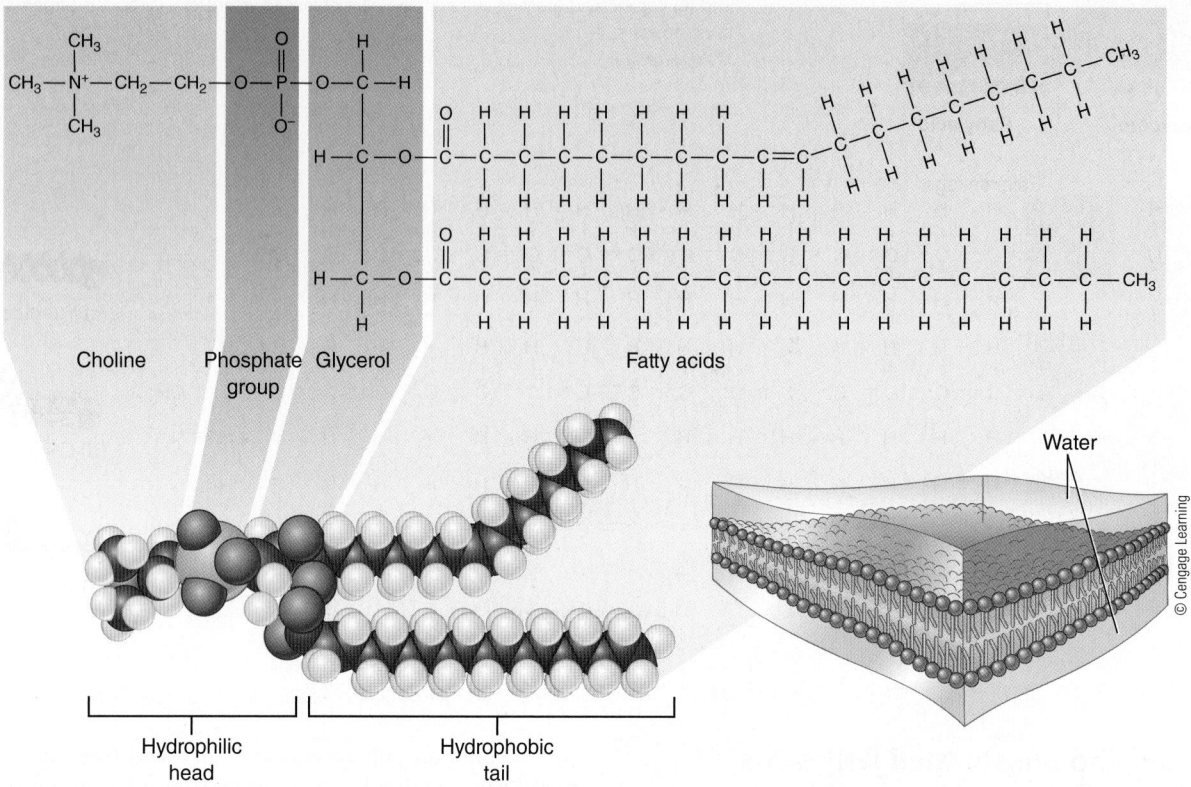

Choline Phosphate Glycerol Fatty acids
group

Water

Hydrophilic head   Hydrophobic tail

© Cengage Learning

(a) Phospholipid (lecithin). A phospholipid consists of a hydrophobic tail, made up of two fatty acids, and a hydrophilic head, which includes a glycerol bonded to a phosphate group, which is in turn bonded to an organic group that can vary. Choline is the organic group in lecithin (or phosphatidylcholine), the molecule shown. The fatty acid at the top of the figure is monounsaturated; it contains one double bond that produces a characteristic bend in the chain.

(b) Phospholipid bilayer. Phospholipids form lipid bilayers in which the hydrophilic heads interact with water and the hydrophobic tails are in the bilayer interior.

Figure 3-13 *Animation* **A phospholipid and a phospholipid bilayer**

PREDICT Examine Figure 3-12. Would you expect the structure of triacylglycerols to enable them to interact with water to form a bilayer? Support your answer.

the amounts required are small, and deficiencies are rarely seen. There is no dietary requirement for saturated fatty acids.

Phospholipids are components of cell membranes

Phospholipids belong to a group of lipids called **amphipathic lipids,** in which one end of each molecule is hydrophilic and the other end is hydrophobic (**FIG. 3-13**). The two ends of a phospholipid differ both physically and chemically. A **phospholipid** consists of a glycerol molecule attached at one end to two fatty acids and at the other end to a phosphate group linked to an organic compound such as choline. The organic compound usually contains nitrogen. (Note that phosphorus and nitrogen are absent in triacylglycerols, as shown in Fig. 3-12b.) The fatty acid portion of the molecule (containing the two hydrocarbon "tails") is hydrophobic and not soluble in water. However, the

portion composed of glycerol, phosphate, and the organic base (the "head" of the molecule) is ionized and readily water soluble. The amphipathic properties of phospholipids cause them to form lipid bilayers in aqueous (watery) solution (see Fig. 3-13b). Thus, they are uniquely suited as the fundamental components of cell membranes (discussed in Chapter 5).

Carotenoids and many other pigments are derived from isoprene units

The orange and yellow plant pigments called **carotenoids** are classified with the lipids because they are insoluble in water and have an oily consistency. These pigments, found in the cells of plants, play a role in photosynthesis. Carotenoid molecules, such as β-carotene, and many other important pigments consist of five-carbon hydrocarbon monomers known as *isoprene units* (**FIG. 3-14**).

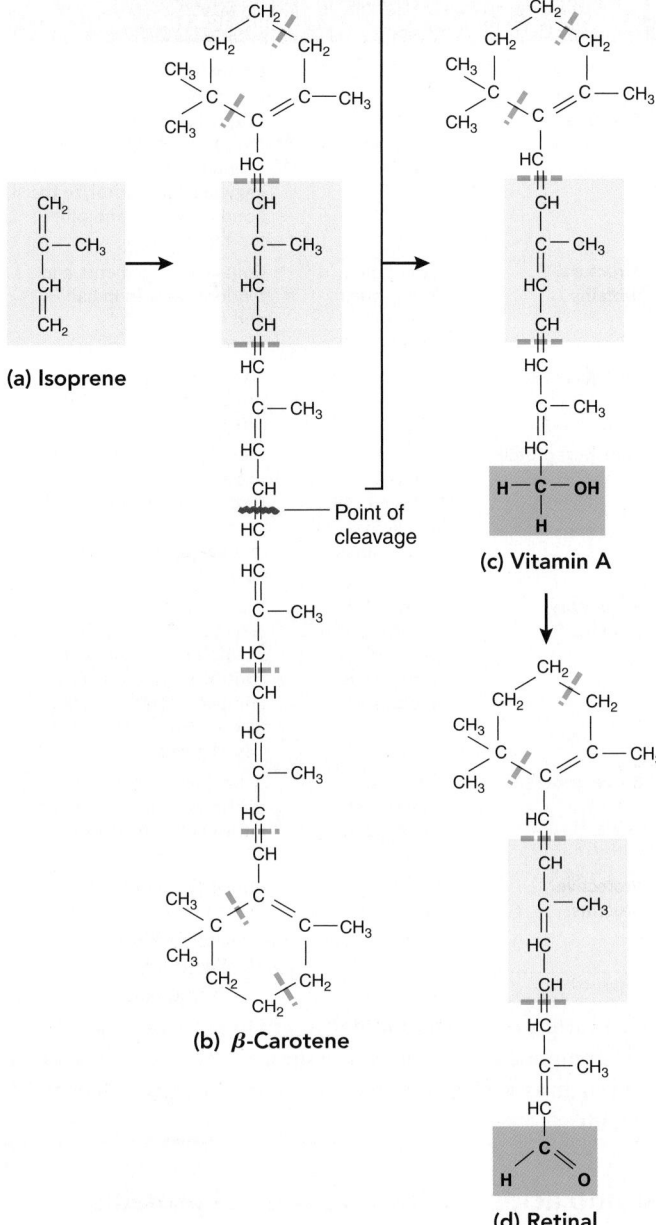

Figure 3-14 Isoprene-derived compounds

(a) An isoprene subunit. **(b)** β-carotene, with dashed lines indicating the boundaries of the individual isoprene units within. The wavy line is the point at which most animals cleave the molecule to yield two molecules of **(c)** vitamin A. Vitamin A is converted to the visual pigment **(d)** retinal.

© Cengage Learning

(a) **Isoprene**

(b) **β-Carotene**

Point of cleavage

(c) **Vitamin A**

(d) **Retinal**

Most animals convert carotenoids to vitamin A, which can then be converted to the visual pigment **retinal.** Three groups of animals—the mollusks, insects, and vertebrates—have eyes that use retinal in the process of light reception.

Notice that carotenoids, vitamin A, and retinal all have a pattern of double bonds alternating with single bonds. The electrons that make up these bonds can move about relatively easily when light strikes the molecule. Such molecules are *pigments;* they tend to be highly colored because the mobile electrons cause them to strongly absorb light of certain wavelengths and reflect light of other wavelengths.

(a) **Cholesterol** is an essential component of animal cell membranes.

Indicates double bond

(b) **Cortisol** is a steroid hormone secreted by the adrenal glands.

Figure 3-15 Steroids

Four attached rings—three six-carbon rings and one with five carbons—make up the fundamental structure of a steroid. Note that some carbons are shared by two rings. In these simplified structures, a carbon atom is present at each angle of a ring; the hydrogen atoms attached directly to the carbon atoms have not been drawn. Steroids are mainly distinguished by their attached functional groups.

© Cengage Learning

Steroids contain four rings of carbon atoms

A **steroid** consists of carbon atoms arranged in four attached rings; three of the rings contain six carbon atoms, and the fourth contains five (**FIG. 3-15**). The length and structure of the side chains that extend from these rings distinguish one steroid from another. Like carotenoids, steroids are synthesized from isoprene units.

Among the steroids of biological importance are cholesterol, bile salts, reproductive hormones, and cortisol as well as other hormones secreted by the adrenal cortex. Cholesterol is an essential structural component of animal cell membranes, but when excess cholesterol in blood forms plaques on artery walls, the risk of cardiovascular disease increases (see Chapter 44).

Plant cell membranes contain molecules similar to cholesterol. Interestingly, some of these plant steroids block the intestine's absorption of cholesterol. Bile salts emulsify fats in the intestine so that they can be enzymatically hydrolyzed. Steroid hormones regulate certain aspects of metabolism in a variety of animals and plants.

Some chemical mediators are lipids

Animal cells secrete chemicals to communicate with one another or to regulate their own activities. Some chemical mediators are produced by the modification of fatty acids that have been removed from membrane phospholipids. These mediators include *prostaglandins*, which have varied roles, including promoting inflammation and smooth muscle contraction. Certain hormones, such as the juvenile hormone of insects, are also fatty acid derivatives (discussed in Chapter 49).

CHECKPOINT 3.3

- *How do the shapes of saturated, unsaturated, and trans fatty acids cause them to differ in their properties?*
- *Explain why the structure of phospholipids enables them to form lipid bilayers in aqueous conditions, whereas triacylglycerols and diacylglycerols do not.*

3.4 PROTEINS

LEARNING OBJECTIVES

7   Give an overall description of the structure and functions of proteins.

8   Describe the features that are shared by all amino acids and explain how amino acids are grouped into classes based on the characteristics of their side chains.

9   Distinguish among the four levels of organization of protein molecules.

Proteins, macromolecules composed of amino acids, are the most versatile cell components. Each type of protein consists of a unique set of amino acids linked in a chain that can be from several amino acids to hundreds of amino acids in length. Each cell type contains a characteristic set of thousands of different types of proteins that largely determine what the cell looks like and how it functions. Muscle cells, for example, contain large amounts of the proteins myosin and actin, which are responsible for their ability to contract. The most abundant protein in red blood cells is hemoglobin, which is responsible for the specialized function of oxygen transport.

In this section we will see that the sequence of the amino acids in a protein chain determines its shape, which in turn determines its function. As will be discussed in Chapter 16, scientists have succeeded in sequencing all the genetic information in a human cell as well as in many other organisms. Some of this information is used to specify the amino acid sequences of cellular proteins, and biologists are now using this information to understand the structures, interactions, and functions of these multifaceted macromolecules that are of central importance in the chemistry of life.

TABLE 3-2 shows many of the different types of functions performed by proteins. Proteins are involved in virtually all aspects of metabolism because most **enzymes**, molecules that accelerate the thousands of different chemical reactions that take place in an organism, are proteins. Proteins are assembled

TABLE 3-2	Major Classes of Proteins and Their Functions	
PROTEIN CLASS	**FUNCTIONS**	**EXAMPLES**
Enzymes	Catalyze specific chemical reactions	Enzymes, thousand of different proteins that speed chemical reactions (e.g., enzymes that hydrolyze the bonds in starch and other food molecules)
Structural proteins	Strengthen and protect cells and tissues	Collagen in ligaments and tendons; keratin in hair and skin tissue
Storage proteins	Store nutrients	Ovalbumin: egg white protein; zein: corn seed protein
Transport proteins	Move substances between cells and across cell membranes	Hemoglobin: O_2 transport in blood; Proteins in cell membranes: transport of glucose, ions, and amino acids
Regulatory proteins	Control the activities of proteins, genes, cells and tissues	Protein kinases: control activities of other proteins; hormones (e.g., insulin): control activities of cells and tissues; transcription factors: control the activities of genes
Motile proteins	Generate movement in cells and tissues	Actin/myosin: muscle contraction and movement of intracellular structures
Protective proteins	Defend against foreign invaders	Antibodies: bind to specific foreign proteins; defensins: inhibit growth of microbial invaders

© Cengage Learning

into a variety of shapes, allowing them to serve as major structural components of cells and tissues. For this reason, growth and repair as well as maintenance of the organism depend of proteins.

Amino acids are the subunits of proteins

Amino acids, the constituents of proteins, have an amino group ($-NH_2$) and a carboxyl group ($-COOH$) bonded to the same asymmetrical carbon atom, known as the **alpha carbon**. Amino acids in solution at neutral pH are mainly *dipolar ions;* that is, they possess a positive charge at one end and a negative charge at the opposite end. This is generally how amino acids exist at cell pH. Each carboxyl group ($-COOH$) donates a proton and becomes ionized ($-COO^-$), whereas each amino group ($-NH_2$) becomes ionized as it accepts a proton and becomes $-NH_3^+$ (**FIG. 3-16**). Because of the ability of their amino and carboxyl groups to accept and release protons, amino acids in solution resist changes in acidity and alkalinity and are therefore important biological buffers.

Twenty amino acids are commonly found in proteins, each identified by the variable side chain (R group) bonded to the α carbon (**FIG. 3-17**). Glycine, the simplest amino acid, has a hydrogen atom as its R group; alanine has a methyl ($-CH_3$) group. The amino acids are grouped in Figure 3-17 by the properties of their

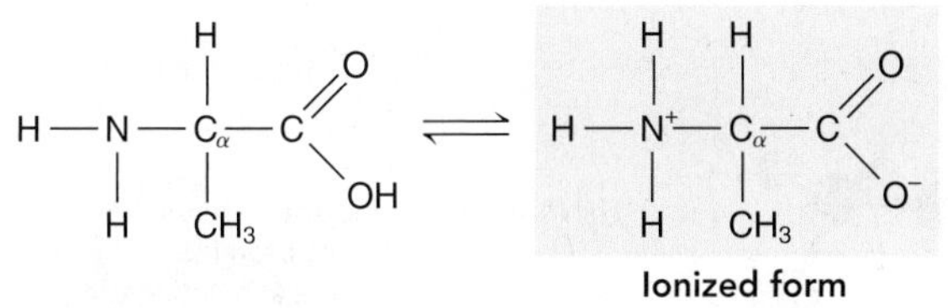

Figure 3-16 An amino acid at pH 7

In living cells, amino acids exist mainly in their ionized form, as dipolar ions.

© Cengage Learning

side chains. These broad groupings actually include amino acids with a fairly wide range of properties. Amino acids classified as having *nonpolar* side chains tend to have hydrophobic properties, whereas those classified as *polar* are more hydrophilic. An acidic amino acid has a side chain that contains a carboxyl group. At cell pH the carboxyl group is dissociated, giving the *R* group a negative charge. A basic amino acid becomes positively charged when an amino group in its side chain accepts a hydrogen ion. Acidic and basic side chains are ionic at cell pH and therefore hydrophilic.

Some proteins have unusual amino acids in addition to the 20 common ones. These rare amino acids are produced by the modification of common amino acids after they have become part of a protein. For example, after they have been incorporated into collagen, lysine and proline may be converted to hydroxylysine and hydroxyproline. These amino acids can form cross links between the peptide chains that make up collagen. Such cross links produce the firmness and great strength of the collagen molecule, which is a major component of cartilage, bone, and other connective tissues.

With some exceptions, prokaryotes and plants synthesize all their needed amino acids from simpler substances. If the proper raw materials are available, the cells of animals can manufacture some, but not all, of the biologically significant amino acids. **Essential amino acids** are those an animal cannot synthesize in amounts sufficient to meet its needs and must obtain from the diet. Animals differ in their biosynthetic capacities; what is an essential amino acid for one species may not be for another.

The essential amino acids for humans are isoleucine, leucine, lysine, methionine, phenylalanine, threonine, tryptophan, valine, and histidine. Arginine is added to the list for children because they do not synthesize enough to support growth.

Peptide bonds join amino acids

Amino acids combine chemically with one another by a condensation reaction that bonds the carboxyl carbon of one molecule to the amino nitrogen of another (FIG. 3-18). The covalent carbon-to-nitrogen bond linking two amino acids is a **peptide bond.** When two amino acids combine, a **dipeptide** is formed; a longer chain of amino acids is a **polypeptide.** A protein consists of one or more polypeptide chains. Each polypeptide has a free amino group at one end and a free carboxyl group (belonging to the last amino acid added to the chain) at the opposite end. The other amino and carboxyl groups of the amino acid monomers (except those in side chains) are part of the peptide bonds. The process by which polypeptides are synthesized is discussed in Chapter 13.

A polypeptide may contain hundreds of amino acids joined in a specific linear order. The flexible backbone of the polypeptide chain includes the repeating sequence

These backbones consist of the covalently linked amino nitrogen, α-carbon, and carboxyl group carbon atoms and all other atoms *except those in the R groups.* The two bonds that link the α carbon atoms to the amino and carboxyl groups along with the peptide bonds (shown in blue) form the backbone, whereas the *R* groups of the amino acids extend from the α-carbon atoms.

An almost infinite variety of protein molecules is possible, differing from one another in the number, types, and sequences of amino acids they contain. The 20 types of amino acids found in proteins may be thought of as letters of a protein alphabet; each protein is a very long sentence made up of amino acid letters.

Proteins have four levels of organization

The polypeptide chains making up a protein are twisted or folded to form a macromolecule with a specific *conformation*, or 3-D shape. Some polypeptide chains form long fibers, whereas *globular proteins* are tightly folded into compact, roughly spherical shapes. There is a close relationship between a protein's conformation and its function. For example, a typical enzyme is a globular protein with a unique shape that allows it to catalyze a specific chemical reaction. Similarly, the shape of a protein hormone enables it to combine with receptors on its target cell (the cell on which the hormone acts). Scientists recognize four main levels of protein organization: primary, secondary, tertiary, and quaternary.

Primary structure is the amino acid sequence The sequence of amino acids, joined by peptide bonds, is the **primary structure** of a polypeptide chain. As discussed in Chapter 13, this sequence is specified by the instructions in a gene. The primary structures of thousands of proteins are known. For example, glucagon, a hormone secreted by the pancreas, is a small polypeptide, consisting of only 29 amino acid units (FIG. 3-19).

Primary structure is always represented in a simple, linear, "beads-on-a-string" form. However, the overall conformation of a protein is far more complex, involving interactions among the various amino acids that make up the primary structure of the molecule. Therefore, the higher orders of structure—secondary, tertiary, and quaternary—that produce the 3-D shape of the molecule ultimately derive from the specific amino acid sequence (the primary structure).

Secondary structure results from hydrogen bonding involving the backbone Some regions of a polypeptide exhibit **secondary structure,** which is highly regular. The two most common types of secondary structure are the α-helix and the β-pleated sheet; the designations α and β refer simply to the order in which these two types of secondary structure were discovered. An **α-helix** is a region where a polypeptide chain forms a uniform helical coil (FIG. 3-20a). The helical structure

(a) Nonpolar (hydrophobic)

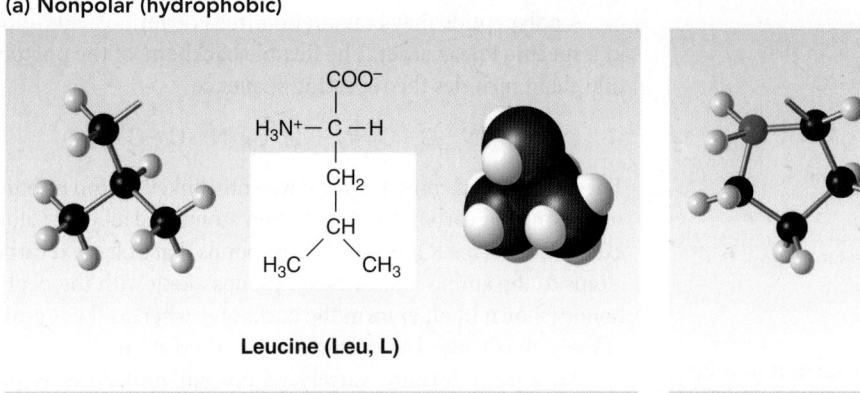

Leucine (Leu, L)

Proline (Pro, P)

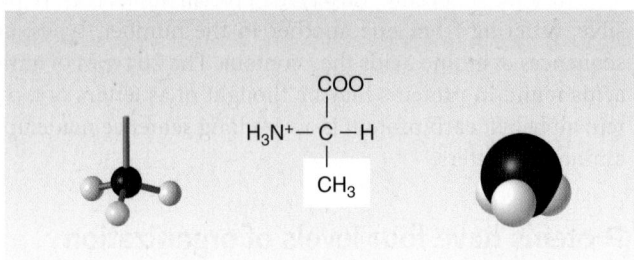

Alanine (Ala, A)

Valine (Val, V)

(b) Polar, uncharged

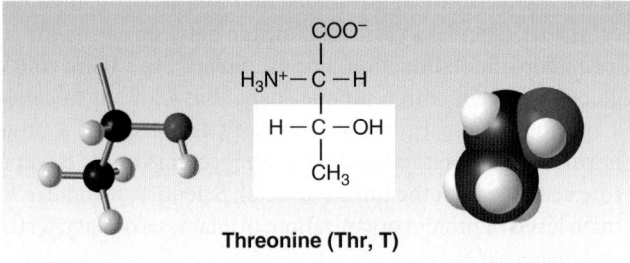

Threonine (Thr, T)

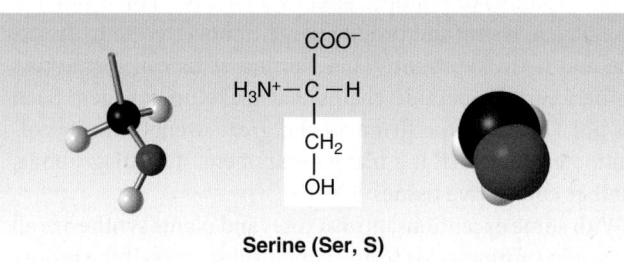

Serine (Ser, S)

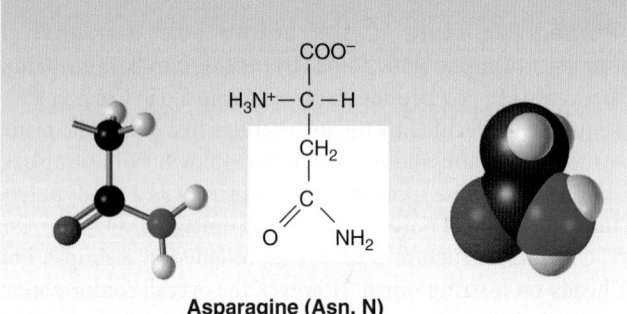

Asparagine (Asn, N)

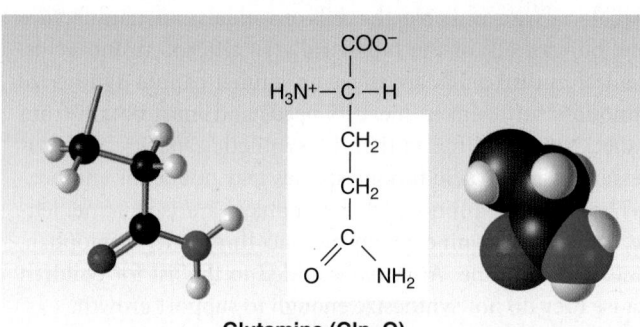

Glutamine (Gln, Q)

(c) Acidic

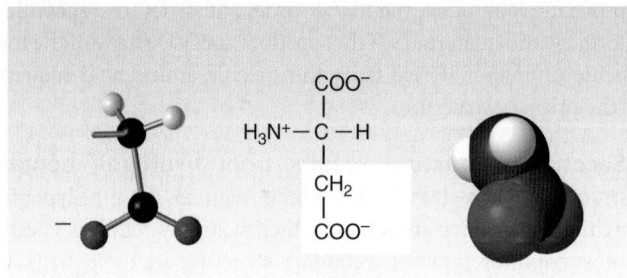

Aspartic acid (Asp, D)

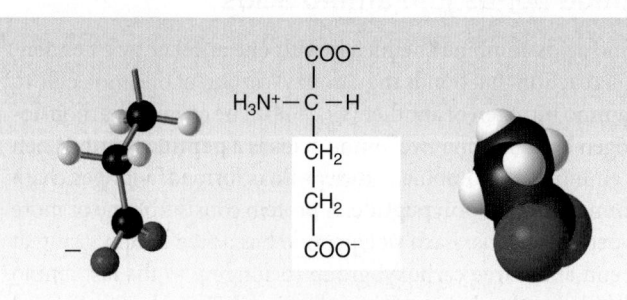

Glutamic acid (Glu, E)

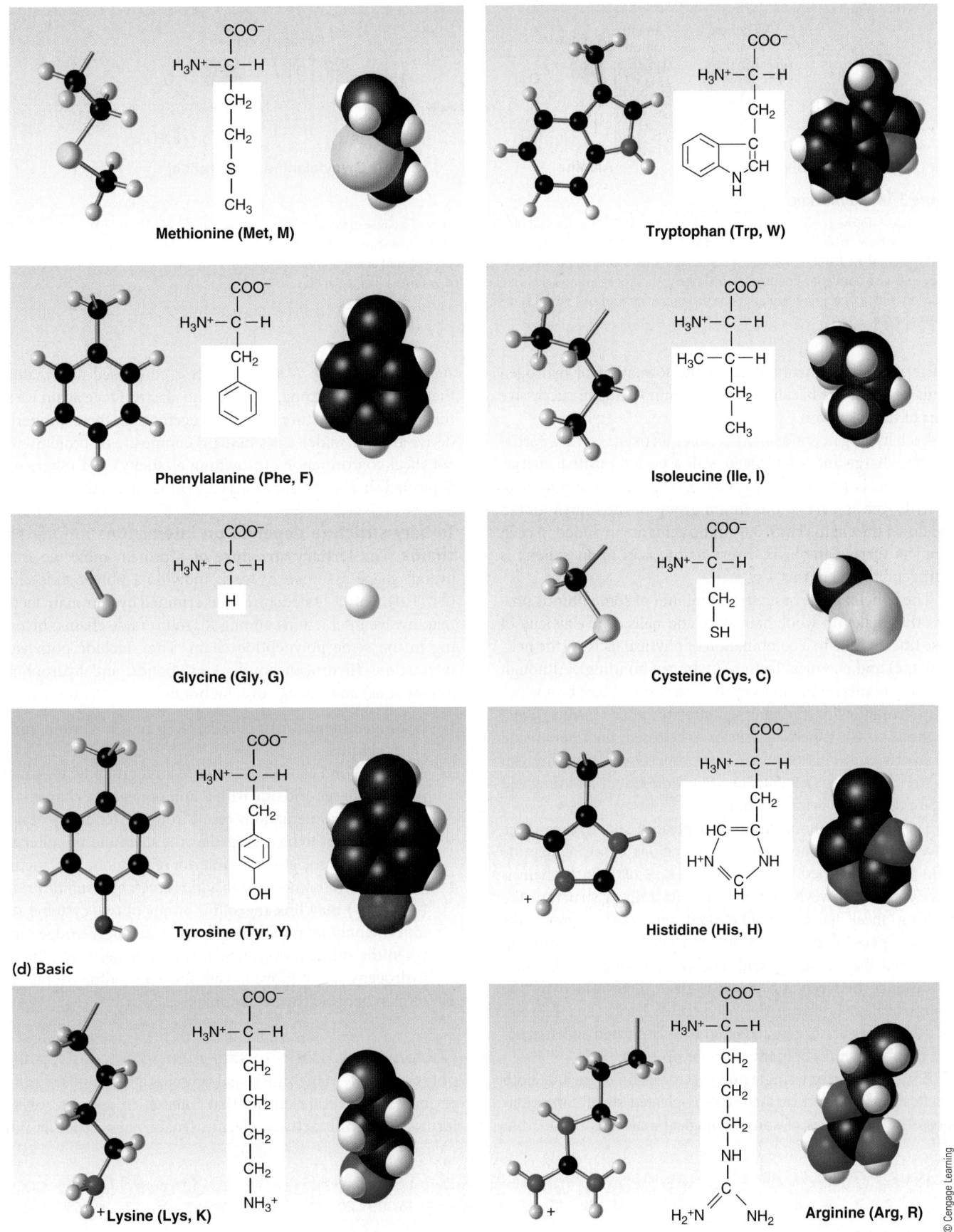

Methionine (Met, M)

Tryptophan (Trp, W)

Phenylalanine (Phe, F)

Isoleucine (Ile, I)

Glycine (Gly, G)

Cysteine (Cys, C)

Tyrosine (Tyr, Y)

Histidine (His, H)

(d) Basic

+Lysine (Lys, K)

Arginine (Arg, R)

© Cengage Learning

Figure 3-17 The 20 common amino acids

(a) Nonpolar amino acids (*yellow background*) have side chains that are relatively hydrophobic, whereas **(b)** polar amino acids (*green background*) have relatively hydrophilic side chains. Carboxyl groups and amino groups are electrically charged at cell pH; therefore, **(c)** acidic (*red background*) and **(d)** basic (*blue background*) amino acids are hydrophilic. The standard three-letter and one-letter abbreviations appear beside the amino acid names.

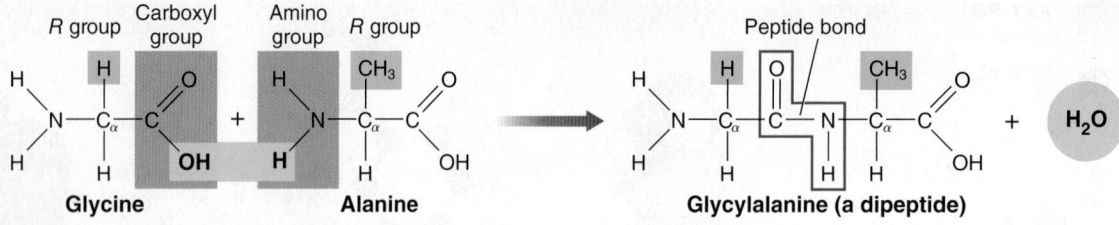

Figure 3-18 *Animation* **Peptide bonds**

A dipeptide is formed by a condensation reaction, that is, by the removal of the equivalent of a water molecule from the carboxyl group of one amino acid and the amino group of another amino acid. The resulting peptide bond is a covalent, carbon-to-nitrogen bond. Note that the carbon is also part of a carbonyl group and that the nitrogen is also covalently bonded to a hydrogen. Additional amino acids can be added to form a long polypeptide chain with a free amino group at one end and a free carboxyl group at the other.

© Cengage Learning

is determined and maintained by the formation of hydrogen bonds between the backbones of the amino acids in successive turns of the spiral coil.

Each hydrogen bond forms between an oxygen with a partial negative charge and a hydrogen with a partial positive charge. The oxygen is part of the carboxyl group of one amino acid; the hydrogen is part of the amino group of the fourth amino acid down the chain. Thus, 3.6 amino acids are included in each complete turn of the helix. Every amino acid in an α-helix is hydrogen bonded in this way.

The α-helix is the basic structural unit of some fibrous proteins that make up wool, hair, skin, and nails. The elasticity of these fibers is due to a combination of physical factors (the helical shape) and chemical factors (hydrogen bonding). Although hydrogen bonds maintain the helical structure, these bonds can be broken, allowing the fibers to stretch under tension (like a telephone cord). When the tension is released, the fibers recoil and hydrogen bonds re-form. This property explains why you can stretch the hairs on your head to some extent and they will snap back to their original length.

The hydrogen bonding in a **β-pleated sheet** takes place between the backbones of different regions of a polypeptide chain that has turned back on itself (FIG. 3-20b). Each chain is fully extended; however, because each has a zigzag structure, the resulting "sheet" has an overall pleated conformation (much like a sheet of paper that has been folded to make a fan). Although the pleated sheet is strong and flexible, it is not elastic because the distance between the pleats is fixed, determined by the strong covalent bonds of the polypeptide backbones. Fibroin, the protein of silk, is characterized by a β-pleated sheet structure, as are the cores of many globular proteins.

It is common for a single polypeptide chain to include both α-helical regions and regions with β-pleated sheet conformations. The properties of some biological materials result from

such combinations. A spider's web is composed of a material that is extremely strong, flexible, and elastic. Once again we see function and structure working together, as these properties derive from a spider silk's being a composite of proteins with α-helical conformations (providing elasticity) and others with β-pleated sheet conformations (providing strength).

Tertiary structure depends on interactions among side chains The **tertiary structure** of a protein molecule is the overall shape assumed by each individual polypeptide chain (FIG. 3-21). This 3-D structure is determined by four main factors that involve interactions among *R* groups (side chains) belonging to the same polypeptide chain. They include both weak interactions (hydrogen bonds, ionic bonds, and hydrophobic interactions) and strong covalent bonds.

1. Hydrogen bonds form between *R* groups of certain amino acid subunits.
2. An ionic bond can occur between an *R* group with a unit of positive charge and one with a unit of negative charge.
3. Hydrophobic interactions result from the tendency of nonpolar *R* groups to be excluded by the surrounding water and therefore to associate in the interior of the globular structure.
4. Covalent bonds known as *disulfide bonds* or *disulfide bridges* (—S—S—) may link the sulfur atoms of two cysteine subunits belonging to the same chain. A disulfide bridge forms when the sulfhydryl groups of two cysteines react; the two hydrogens are removed, and the two sulfur atoms that remain become covalently linked.

Quaternary structure results from interactions among polypeptides Many functional proteins are composed of two or more polypeptide chains that interact in specific ways to form a biologically active molecule. **Quaternary structure** is the

Figure 3-19 Primary structure of a polypeptide

Glucagon is a very small polypeptide made up of 29 amino acids. The linear sequence of amino acids is indicated by ovals containing their abbreviated names (see Fig. 3-17).

© Cengage Learning

Secondary structure is highly regular.

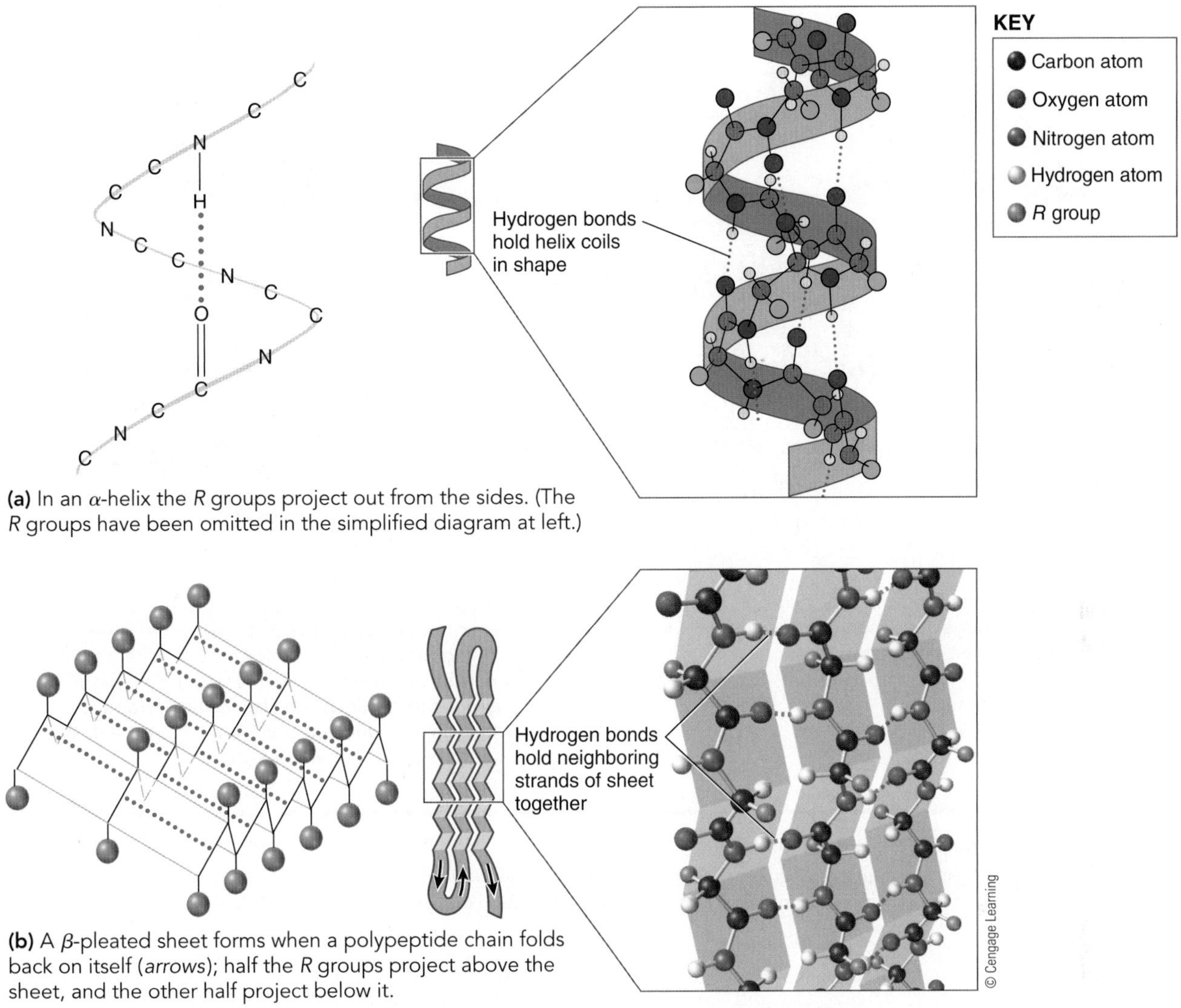

KEY
- Carbon atom
- Oxygen atom
- Nitrogen atom
- Hydrogen atom
- *R* group

(a) In an α-helix the *R* groups project out from the sides. (The *R* groups have been omitted in the simplified diagram at left.)

Hydrogen bonds hold helix coils in shape

Hydrogen bonds hold neighboring strands of sheet together

(b) A β-pleated sheet forms when a polypeptide chain folds back on itself (*arrows*); half the *R* groups project above the sheet, and the other half project below it.

© Cengage Learning

Figure 3-20 Secondary structure of a protein

Hydrogen bonding of the backbone can produce two types of secondary structure: the α-helix and the β-pleated sheet.

PREDICT How might secondary structure be affected if some amino acids in the polypeptide chain were to include an additional carbon inserted between the α carbon and the peptide bond carbon?

resulting 3-D structure. The same types of interactions that produce secondary and tertiary structure can also occur between the polypeptide chains to contribute to quaternary structure; they include hydrogen bonding, ionic bonding, hydrophobic interactions, and disulfide bridges.

A functional antibody molecule, for example, consists of four polypeptide chains joined by disulfide bridges (**FIG. 3-22**). Disulfide bridges are a common feature of antibodies and other proteins secreted from cells. These strong bonds can link regions of the same polypeptide chain as well as form links between different polypeptide chains in proteins with quaternary structure.

Hemoglobin, the protein in red blood cells responsible for oxygen transport, is an example of a globular protein with a quaternary structure (**FIG. 3-23a**). Hemoglobin consists of 574

amino acids arranged in four polypeptide chains: two identical chains called *alpha chains* and two identical chains called *beta chains*.

Collagen, mentioned previously, has a fibrous type of quaternary structure that allows it to function as the major strengthener of animal tissues. It consists of three polypeptide chains wound about one another and bound by cross links between their amino acids (**FIG. 3-23b**).

The amino acid sequence of a protein determines its conformation

In 1972, U.S. researcher Christian B. Anfinsen was awarded the Nobel Prize in Chemistry for his studies on protein folding,

Tertiary structure depends on side chain interactions.

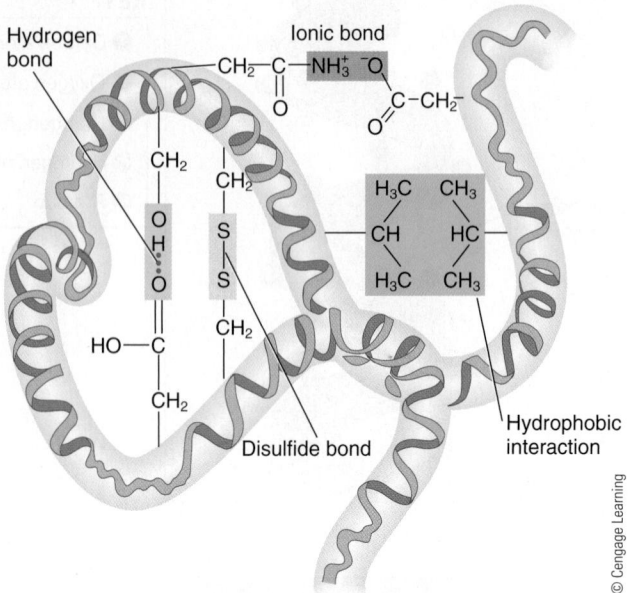

(a) Hydrogen bonds, ionic bonds, hydrophobic interactions, and disulfide bridges between *R* groups hold the parts of the molecule in the designated shape.

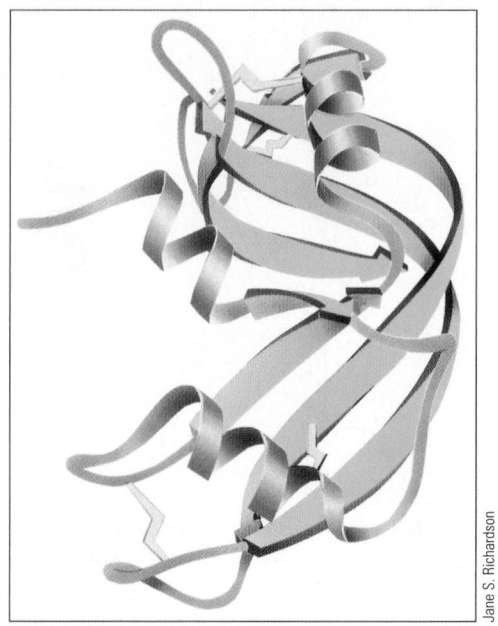

(b) In this drawing, α-helical regions are represented by *purple coils*, β-pleated sheets by *broad green ribbons*, and connecting regions by *narrow tan ribbons*. The interactions among *R* groups that stabilize the bends and foldbacks that give the molecule its overall conformation (tertiary structure) are represented in *yellow*. This protein is bovine ribonuclease a.

Figure 3-21 *Animation* **Tertiary structure of a protein**

PREDICT Examine Figure 3-17. Would you expect the *R* groups of the following pairs of amino acids to interact? If so, by what mechanism: leucine and valine, alanine and serine, lysine and arginine, or glutamic acid and lysine?

which demonstrated that, at least under defined experimental conditions in vitro (outside a living cell), a polypeptide can spontaneously undergo folding processes that yield its normal, functional conformation. Since Anfinsen's pioneering work, many researchers studying various proteins and using a variety of highly sophisticated approaches have amassed evidence supporting the widely held conclusion that amino acid sequence is the ultimate determinant of protein conformation.

However, because conditions in vivo (in the cell) are quite different from defined laboratory conditions, proteins do not always fold spontaneously into their correct shape. On the contrary, scientists have learned that intracellular proteins known as **molecular chaperones** assist the folding of other protein molecules. Molecular chaperones are thought to make the folding process more orderly and efficient, and to prevent partially folded proteins from becoming inappropriately aggregated. However, there is no evidence that molecular chaperones actually dictate the folding pattern. For this reason, the existence of chaperones is not an argument against the idea that amino acid sequence determines conformation.

Protein conformation determines function The overall structure of a protein helps determine its biological activity. A single protein may have more than one distinct structural region, called a **domain.** Each domain is a part of the amino acid sequence that can be independently folded and function free from the other parts of the protein. Many proteins are modular, consisting of two or more domains that are connected by less compact regions of the polypeptide chain.

Each domain may have a different function. Because a domain is a stable structure that has a specific function, one type of domain might be found in a number of different proteins that are generated through the process of molecular evolution. Researchers are now able to use genetic engineering methods to make recombinant proteins with new functions by linking together domains from different proteins.

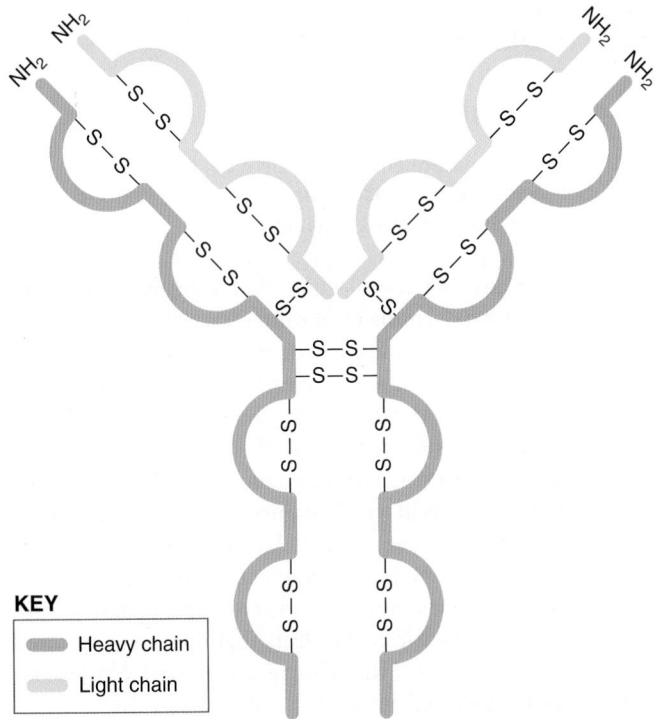

KEY

■ Heavy chain

■ Light chain

Figure 3-22 Disulfide bonds contribute to the structure of an antibody molecule

Antibody molecules consist of two light-chain and two heavy-chain polypeptides linked together by disulfide bonds. Each of the four chains also contains internal disulfide linkages that contribute to multiple structural and functional domains within each polypeptide.

© Cengage Learning

Small changes in the primary structure of a protein can alter its shape and affect its function The biological activity of a protein can be disrupted by a change in amino acid sequence that results in a change in conformation. For example, the genetic disease known as *sickle cell anemia* is due to a mutation that causes the substitution of the amino acid valine for glutamic acid at position 6 (the sixth amino acid from the amino end) in the beta chain of hemoglobin. The substitution of valine (which has a nonpolar side chain) for glutamic acid (which has a charged side chain) makes the hemoglobin less soluble and more likely to form crystal-like structures. This alteration of the hemoglobin affects the red blood cells, changing them to the crescent or sickle shapes that characterize this disease (see Fig. 16-9).

The biological activity of a protein may also be affected by agents or conditions that change its 3-D structure. When a protein is heated, subjected to significant pH changes, or treated with certain chemicals, its structure becomes disordered and the coiled peptide chains unfold, yielding a more random conformation. This unfolding, which is mainly due to the disruption of hydrogen bonds and ionic bonds, is typically accompanied by a loss of

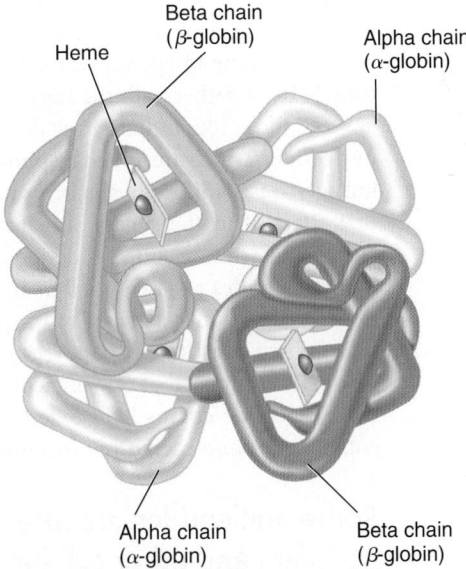

(a) Hemoglobin, a globular protein, consists of four polypeptide chains, each joined to an iron-containing molecule, a heme.

(b) Collagen, a fibrous protein, is a triple helix consisting of three long polypeptide chains.

Figure 3-23 Quaternary structure of a protein

Proteins with two or more polypeptide chains have quaternary structure.

© Cengage Learning

normal function. Such changes in shape and the accompanying loss of biological activity are termed *denaturation* of the protein.

For example, a denatured enzyme would lose its ability to catalyze a chemical reaction. An everyday example of denaturation occurs when we fry an egg. The consistency of the egg white protein, known as *albumin*, changes to a solid. Denaturation generally cannot be reversed (you cannot "unfry" an egg). Under certain conditions, however, some proteins have been denatured and have returned to their original shape and biological activity when normal environmental conditions were restored.

Protein conformation is studied through a variety of methods The architecture of a protein can be ascertained directly through various types of analysis, such as the X-ray diffraction studies discussed in Chapter 12 or nuclear magnetic resonance (NMR) methods. These methods provide the precise mathematical coordinates of individual atoms within the protein molecule that can be visualized by computer graphics as 3-D structures (see Fig. 3-23). Researchers can now use these computer-based models to simulate how molecules such as drugs (or another protein) might specifically bind to a protein to alter its function.

Today a protein's primary structure can be determined rapidly through the application of genetic engineering techniques (discussed in Chapter 15) or by the use of sophisticated technology such as mass spectrometry. Researchers use these amino acid sequence data to predict a protein's higher levels of structure. As we have seen, side chains interact in relatively predictable ways, such as through ionic and hydrogen bonds. In addition, regions with certain types of side chains are more likely to form α-helices or β-pleated sheets. Computer programs make such predictions that can be verified by X-ray diffraction or NMR methods.

Researchers also use computers to search databases to find polypeptides with similar sequences. If the conformations of any of those polypeptides or portions have already been determined directly by X-ray diffraction or other techniques, this information can be extrapolated to make similar correlations between amino acid sequence and 3-D structure for the protein under investigation.

Misfolded proteins are implicated in human diseases Studies on the mechanisms of protein folding, and on the relationship between the activity of a protein and its conformation, are increasingly of medical importance. For example, as discussed in Chapter 24, mad cow disease and related diseases in humans and other animals are caused by misfolded proteins called *prions*. Other serious diseases in which misfolded proteins appear to play an important role include Alzheimer's disease (see Chapter 41) and Huntington's disease (see Chapter 16).

CHECKPOINT 3.4

- **VISUALIZE** *Draw the structural formula of a simple amino acid. What is the importance of the carboxyl group, amino group, and R group?*
- *Explain how the primary structure of a polypeptide influences its secondary and tertiary structures.*

3.5 NUCLEIC ACIDS

LEARNING OBJECTIVE

10 Describe the components of a nucleotide. Name some nucleic acids and nucleotides, and discuss the importance of these compounds in living organisms.

Nucleic acids are macromolecules that transmit hereditary information and determine what proteins a cell manufactures. Two classes of nucleic acids are found in cells: deoxyribonucleic acid and ribonucleic acid. **Deoxyribonucleic acid (DNA)** makes up the genes, the hereditary material of the cell, and contains instructions for making all the proteins as well as all the RNA the organism needs. **Ribonucleic acid (RNA)** participates in the process in which amino acids are linked to form polypeptides. Some types of RNA, known as **ribozymes** (see Chapter 13 and Fig. 21-5) can even act as specific biological catalysts. Like proteins, nucleic acids are large, complex molecules. The name *nucleic acid* reflects that they are acidic and were first identified, by Swiss biochemist Friedrich Miescher in 1870, in the nuclei of pus cells.

Nucleic acids are polymers of **nucleotides,** molecular units that consist of (1) a five-carbon sugar, either **deoxyribose** (in DNA) or **ribose** (in RNA); (2) one or more phosphate groups, which make the molecule acidic; and (3) a nitrogenous base, a ring compound that contains nitrogen. The nitrogenous base may be either a double-ring **purine** or a single-ring **pyrimidine** (FIG. 3-24). DNA commonly contains the purines **adenine (A)** and **guanine (G),** the pyrimidines **cytosine (C)** and **thymine (T),** the sugar deoxyribose, and phosphate. RNA contains the purines adenine and guanine, and the pyrimidines cytosine and **uracil (U),** together with the sugar ribose, and phosphate.

The molecules of nucleic acids are made of linear chains of nucleotides, which are joined by **phosphodiester linkages**. Each linkage consists of a phosphate group and the covalent bonds that attach it to the sugars of adjacent nucleotides (FIG. 3-25). Note that each nucleotide is defined by its particular base and that nucleotides can be joined in any sequence. A nucleic acid molecule is uniquely defined by its specific sequence of nucleotides, which constitutes a code (see Chapter 13). Whereas RNA is usually composed of one nucleotide chain, DNA consists of two nucleotide chains held together by hydrogen bonds and entwined around each other in a double helix (see Fig. 1-7).

Some nucleotides are important in energy transfers and other cell functions

In addition to their importance as subunits of DNA and RNA, nucleotides perform other vital functions in living cells. **Adenosine triphosphate (ATP),** composed of adenine, ribose, and three phosphates (see Fig. 7-5), is of major importance as the primary energy currency of all cells (see Chapter 7). The two terminal phosphate groups are joined to the nucleotide by covalent bonds. These bonds are traditionally indicated by wavy lines, which indicate that ATP can transfer a phosphate group to

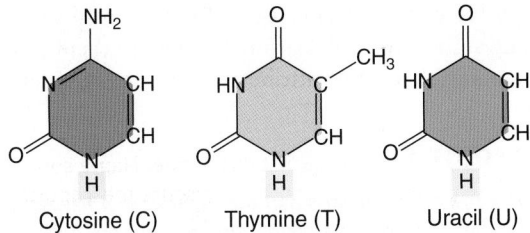

(a) Pyrimidines. The three major pyrimidine bases found in nucleotides are cytosine, thymine (in DNA only), and uracil (in RNA only).

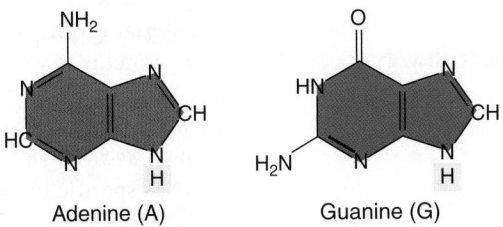

(b) Purines. The two major purine bases found in nucleotides are adenine and guanine.

Figure 3-24 *Animation* Components of nucleotides
The hydrogens indicated by the *yellow screens* are removed when the base is attached to a sugar.
© Cengage Learning

another molecule, making that molecule more reactive. In this way ATP is able to donate some of its chemical energy. Most of the readily available chemical energy of the cell is associated with the phosphate groups of ATP. Like ATP, **guanosine triphosphate (GTP),** a nucleotide that contains the base guanine, can transfer energy by transferring a phosphate group and also has a role in cell signaling (see Chapter 6).

A nucleotide may be converted to an alternative form with specific cell functions. ATP, for example, is converted to **cyclic adenosine monophosphate (cyclic AMP, or cAMP)** by the enzyme adenylyl cyclase (FIG. 3-26). Cyclic AMP regulates certain cell functions, such as cell signaling, and is important in the mechanism by which some hormones act. A related molecule, **cyclic guanosine monophosphate (cGMP),** also plays a role in certain cell signaling processes.

Cells contain several dinucleotides, which are of great importance in metabolic processes. For example, as discussed in Chapter 7, **nicotinamide adenine dinucleotide** has a primary role in oxidation and reduction reactions in cells. It can exist in an oxidized form (NAD^+) that is converted to a reduced form (**NADH**) when it accepts electrons (in association with hydrogen; see Fig. 7-7). These electrons, along with their energy, are transferred to other molecules.

CHECKPOINT 3.5

- **VISUALIZE** *Sketch a pyrimidine nucleotide subunit that would be found only in RNA. Circle and label the three components that make up this type of nucleotide. Explain what changes in the functional groups of this subunit would have to occur for it to be found in a DNA molecule.*

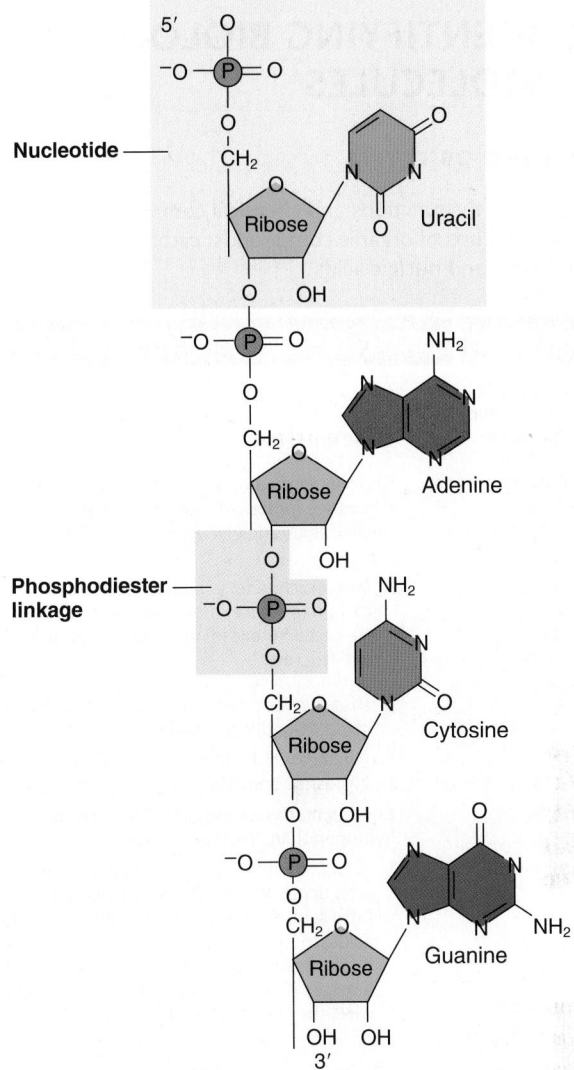

Figure 3-25 RNA, a nucleic acid
Nucleotides, each with a specific base, are joined by phosphodiester linkages.
© Cengage Learning

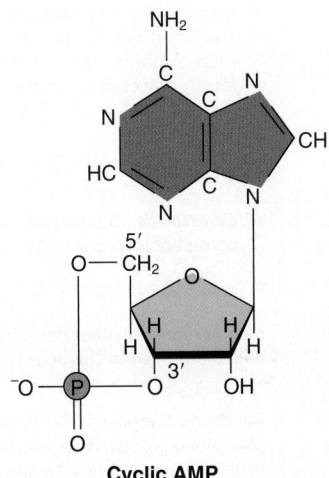

Cyclic AMP

Figure 3-26 Cyclic adenosine monophosphate (cAMP)
The single phosphate is part of a ring connecting two regions of the ribose.
© Cengage Learning

3.6 IDENTIFYING BIOLOGICAL MOLECULES

Although the classes of biological molecules may seem overwhelming at first, you will learn to distinguish them readily by understanding their chief attributes. They are summarized in TABLE 3-3.

CHECKPOINT 3.6

- *How can you distinguish a pentose sugar from a hexose sugar? a disaccharide from a sterol? an amino acid from a monosaccharide? a phospholipid from a triacylglycerol? a protein from a polysaccharide? a nucleic acid from a protein?*

TABLE 3-3	Classes of Biologically Important Organic Compounds		
CLASS AND COMPONENT ELEMENTS	**DESCRIPTION**	**HOW TO RECOGNIZE**	**PRINCIPAL FUNCTION IN LIVING SYSTEMS**
CARBOHYDRATES C, H, O	Contain approximately 1 C:2 H:1 O (but make allowance for loss of oxygen when sugar units as nucleic acids and glycoproteins are linked)	Count the carbons, hydrogens, and oxygens.	Cell fuel; energy storage; structural component of plant cell walls; component of other compounds such as nucleic acids and glycoproteins
	1. *Monosaccharides* (simple sugars). Mainly five-carbon (pentose) molecules such as ribose or six-carbon (hexose) molecules such as glucose and fructose	Look for the ring shapes:	Cell fuel; components of other compounds
	2. *Disaccharides.* Two sugar units linked by a glycosidic bond, e.g., maltose, sucrose	Count sugar units.	Components of other compounds; form of sugar transported in plants
	3. *Polysaccharides.* Many sugar units linked by glycosidic bonds, e.g., glycogen, cellulose	Count sugar units.	Energy storage; structural components of plant cell walls
LIPIDS C, H, O (sometimes N, P)	Contain much less oxygen relative to carbon and hydrogen than do carbohydrates		Energy storage; cellular fuel, components of cells; thermal insulation
	1. *Fats.* Combination of glycerol with one to three fatty acids. Monoacylglycerol contains one fatty acid; diacylglycerol contains two fatty acids; triacylglycerol contains three fatty acids. If fatty acids contain double carbon-to-carbon linkages (C═C), they are unsaturated; otherwise, they are saturated.	Look for glycerol at one end of molecule:	Cell fuel; energy storage
	2. *Phospholipids.* Composed of glycerol attached to one or two fatty acids and to an organic base containing phosphorus	Look for glycerol and side chain containing phosphorus and nitrogen.	Components of cell membranes
	3. *Steroids.* Complex molecules containing carbon atoms arranged in four attached rings (Three rings contain six carbon atoms each, and the fourth ring contains five.)	Look for four attached rings:	Some are hormones; others include cholesterol, bile salts, vitamin D; components of cell membranes
	4. *Carotenoids.* Orange and yellow pigments; consist of isoprene units	Look for isoprene units.	Converted to retinal (important in photoreception) and vitamin A
PROTEINS C, H, O, N (usually S)	One or more polypeptides (chains of amino acids) coiled or folded in characteristic shapes	Look for amino acid units joined by C—N bonds.	Serve as enzymes; structural components; muscle proteins; hemoglobin
NUCLEIC ACIDS C, H, O, N, P	Backbone composed of alternating pentose and phosphate groups, from which nitrogenous bases project. DNA contains the sugar deoxyribose and the bases guanine, cytosine, adenine, and thymine. RNA contains the sugar ribose and the bases guanine, cytosine, adenine, and uracil. Each molecular subunit, called a nucleotide, consists of a pentose, a phosphate, and a nitrogenous base.	Look for a pentose–phosphate backbone. DNA forms a double helix.	Storage, transmission, and expression of genetic information; some important in energy transfers, cell signaling, and other aspects of metabolism.

3.1 Carbon Atoms and Organic Molecules (page 45)

1 Describe the properties of carbon that make it the central component of organic compounds.

- Each carbon atom forms four covalent bonds with up to four other atoms; these bonds are single, double, or triple bonds. Carbon atoms form straight or branched chains, or join into rings. Carbon forms covalent bonds with a greater number of different elements than does any other type of atom.

2 Define the term *isomer* and distinguish among the three principal isomer types.

- **Isomers** are compounds with the same molecular formula but different structures.
- **Structural isomers** differ in the covalent arrangements of their atoms. **Geometric isomers,** or *cis–trans isomers,* differ in the spatial arrangements of their atoms. **Enantiomers** are isomers that are mirror images of each other. Cells can distinguish between these configurations.

3 Identify the major functional groups present in organic compounds and describe their properties.

- **Hydrocarbons,** organic compounds consisting of only carbon and hydrogen, are nonpolar and **hydrophobic.** The **methyl group** is a hydrocarbon group.
- Polar and ionic functional groups interact with one another and are **hydrophilic.** Partial charges on atoms at opposite ends of a bond are responsible for the polar property of a functional group. **Hydroxyl** and **carbonyl groups** are polar.
- **Carboxyl** and **phosphate groups** are acidic, becoming negatively charged when they release hydrogen ions. The **amino group** is basic, becoming positively charged when it accepts a hydrogen ion.

4 Explain the relationship between polymers and macromolecules.

- Long chains of **monomers** (similar organic compounds) linked through **condensation reactions** are called **polymers.** Large polymers such as polysaccharides, proteins, and DNA are referred to as **macromolecules.** They can be broken down by **hydrolysis reactions.**

3.2 Carbohydrates (page 49)

5 Distinguish among monosaccharides, disaccharides, and polysaccharides; compare storage polysaccharides with structural polysaccharides.

- **Carbohydrates** contain carbon, hydrogen, and oxygen in a ratio of approximately one carbon to two hydrogens to one oxygen. **Monosaccharides** are simple sugars such as glucose, fructose, and ribose. Two monosaccharides join by a **glycosidic linkage** to form a **disaccharide** such as maltose or sucrose.

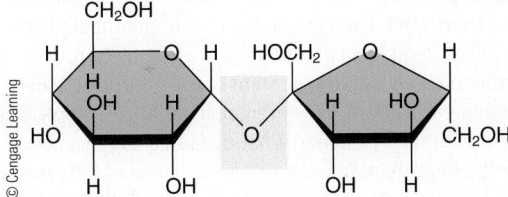

- Most carbohydrates are **polysaccharides,** long chains of repeating units of a simple sugar. Carbohydrates are typically stored in plants as the polysaccharide **starch** and in animals as the polysaccharide **glycogen.** The cell walls of plants are composed mainly of the structural polysaccharide **cellulose.**

3.3 Lipids (page 54)

6 Distinguish among fats, phospholipids, and steroids, and describe the composition, characteristics, and biological functions of each.

- **Lipids** are composed mainly of hydrocarbon-containing regions, with few oxygen-containing (polar or ionic) functional groups. Lipids have a greasy or oily consistency and are relatively insoluble in water.
- **Triacylglycerol,** the main storage form of fat in organisms, consists of a molecule of **glycerol** combined with three **fatty acids. Monoacylglycerols** and **diacylglycerols** contain one and two fatty acids, respectively. A fatty acid can be either **saturated** with hydrogen or **unsaturated.**

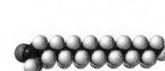

- **Phospholipids** are structural components of cell membranes. A phospholipid consists of a glycerol molecule attached at one end to two fatty acids and at the other end to a phosphate group linked to an organic compound such as choline.
- **Steroid** molecules contain carbon atoms arranged in four attached rings. Cholesterol, bile salts, and certain hormones are important steroids.

3.4 Proteins (page 58)

7 Give an overall description of the structure and functions of proteins.

- **Proteins** are complex macromolecules made of simpler subunits, called **amino acids,** joined by **peptide bonds.** Two amino acids combine to form a **dipeptide.** A longer chain of amino acids is a **polypeptide.** Proteins are the most versatile class of biological molecules, serving a variety of functions, such as **enzymes,** structural components, and cell regulators.
- Proteins are composed of various linear sequences of 20 different amino acids.

8 Describe the features that are shared by all amino acids and explain how amino acids are grouped into classes based on the characteristics of their side chains.

$$H_3N^+ - \overset{\overset{\textstyle COO^-}{\textstyle |}}{\underset{\underset{\textstyle CH_3}{\textstyle |}}{C}} - H$$

- All amino acids contain an amino group and a carboxyl group. Amino acids vary in their side chains, which dictate their chemical properties: nonpolar, polar, acidic, or basic. Amino acids generally exist as dipolar ions at cell pH and serve as important biological buffers.

9 Distinguish among the four levels of organization of protein molecules.

- **Primary structure** is the linear sequence of amino acids in the polypeptide chain.
- **Secondary structure** is a regular conformation, such as an **α-helix** or a **β-pleated sheet;** it is due to hydrogen bonding between elements of the backbones of the amino acids.
- **Tertiary structure** is the overall shape of the polypeptide chains, as dictated by chemical properties and interactions of the side chains of specific amino acids. Hydrogen bonds, ionic bonds, hydrophobic interactions, and disulfide bridges contribute to tertiary structure.
- **Quaternary structure** is determined by the association of two or more polypeptide chains.

3.5 Nucleic Acids *(page 66)*

10 Describe the components of a nucleotide. Name some nucleic acids and nucleotides, and discuss the importance of these compounds in living organisms.

- The nucleic acids **DNA** and **RNA,** composed of long chains of nucleotide subunits, store and transfer information that specifies the sequence of amino acids in proteins and ultimately the structure and function of the organism.

- **Nucleotides** are composed of a two-ring **purine** or one-ring **pyrimidine** nitrogenous base, a five-carbon sugar (**ribose** or **deoxyribose**), and one or more phosphate groups.

- **ATP (adenosine triphosphate)** is a nucleotide of special significance in energy metabolism. NAD$^+$ is also involved in energy metabolism through its role as an electron (hydrogen) acceptor in biological oxidation and reduction reactions.

3.6 Identifying Biological Molecules *(page 68)*

11 Compare the functions and chemical compositions of the major groups of organic compounds: carbohydrates, lipids, proteins, and nucleic acids.

- Review Table 3-3.

TEST YOUR UNDERSTANDING

Know and Comprehend

1. Carbon is particularly well suited to be the backbone of organic molecules because (a) it can form both covalent bonds and ionic bonds (b) its covalent bonds are very irregularly arranged in three-dimensional space (c) its covalent bonds are the strongest chemical bonds known (d) it can bond to atoms of a large number of other elements (e) all the bonds it forms are polar

2. **VISUALIZE** The structures depicted are (a) enantiomers (b) different views of the same molecule (c) geometric (*cis–trans*) isomers (d) both geometric isomers and enantiomers (e) structural isomers

3. Which of the following is a nonpolar molecule? (a) water, H_2O (b) ammonia, NH_3 (c) methane, CH_4 (d) ethane, C_2H_6 (e) more than one of the preceding is/are correct.

4. The synthetic process by which monomers are covalently linked is (a) hydrolysis (b) isomerization (c) condensation (d) glycosidic linkage (e) ester linkage

5. A monosaccharide designated as an aldehyde sugar contains (a) a terminal carboxyl group (b) an internal carboxyl group (c) a terminal carbonyl group (d) an internal carbonyl group (e) a terminal carboxyl group and an internal carbonyl group

6. Structural polysaccharides typically (a) have extensive hydrogen bonding between adjacent molecules (b) are much more hydrophilic than storage polysaccharides (c) have much stronger covalent bonds than do storage polysaccharides (d) consist of alternating α-glucose and β-glucose subunits (e) form helical structures in the cell

7. Saturated fatty acids are so named because they are saturated with (a) hydrogen (b) water (c) hydroxyl groups (d) glycerol (e) double bonds

8. Fatty acids in phospholipids and triacylglycerols interact with one another by (a) disulfide bridges (b) van der Waals interactions (c) covalent bonds (d) hydrogen bonds (e) fatty acids do not interact with one another

9. Which of the following levels of protein structure may be affected by hydrogen bonding? (a) primary and secondary (b) primary and tertiary (c) secondary, tertiary, and quaternary (d) primary, secondary, and tertiary (e) primary, secondary, tertiary, and quaternary

10. Which of the following associations between *R* groups are the strongest? (a) hydrophobic interactions (b) hydrogen bonds (c) ionic bonds (d) peptide bonds (e) disulfide bridges

11. Each phosphodiester linkage in DNA or RNA includes a phosphate joined by covalent bonds to (a) two bases (b) two sugars (c) two additional phosphates (d) a sugar, a base, and a phosphate (e) a sugar and a base

Apply and Analyze

12. **PREDICT** Do any of the amino acid side groups shown below have the potential to form an ionic bond with any of the other side groups shown? If so, which pair(s) could form such an association?
    (a) —CH_3
    (b) —CH_2—COO^-
    (c) —CH_2—CH_2—NH_3^+
    (d) —CH_2—CH_2—COO^-
    (e) —CH_2—OH

Evaluate and Synthesize

13. **PREDICT** Like oxygen, sulfur forms two covalent bonds. However, sulfur is far less electronegative. In fact, it is approximately as electronegative as carbon. How would the properties of the various classes of biological molecules be altered if you were to replace all the oxygen atoms with sulfur atoms?

14. Hydrogen bonds and van der Waals interactions are much weaker than covalent bonds, yet they are vital to organisms. Explain, providing some specific examples.

15. **EVOLUTION LINK** In what ways are all species alike biochemically? Identify some ways in which species may differ from one another biochemically. What do these similarities and differences suggest about the history of life on Earth?

16. **EVOLUTION LINK** The total number of possible amino acid sequences in a polypeptide chain is staggering. Given that there are 20 amino acids, potentially there could be 20^{100} different amino acid sequences just for polypeptides only 100 amino acids in length. However, the actual number of different polypeptides occurring in organisms is only a tiny fraction of this potential. What insight does this finding provide into the evolutionary process?

17. **EVOLUTION LINK** Each amino acid could potentially exist as one of two possible enantiomers, known as the D-form and the L-form (based on the arrangement of the groups attached to the asymmetric α carbon). However, in all organisms, only L-amino acids are found in proteins. What does this suggest about the evolution of proteins?

aplia To access course materials, such as Aplia and other companion resources, please visit **www.cengagebrain.com.**

Organization of the Cell

The cell is the smallest unit that can carry out all activities we associate with life. When provided with essential nutrients and an appropriate environment, some cells can be kept alive and growing in the laboratory for many years. By contrast, no isolated part of a cell is capable of sustained survival. As you read this chapter, recall the discussion of systems biology in Chapter 1. Even as we describe individual components of cells, each of which is a system in itself, we discuss how these structures work together to generate ever more complex interacting biological systems within the cell. The cell itself is a highly intricate biological system, and groups of cells make up tissues, organs, and organisms that are biological systems of ever increasing complexity.

Most prokaryotes and many protists and fungi consist of a single cell. In contrast, most plants and animals are composed of millions of cells. Cells are the building blocks of complex multicellular organisms. Although they are basically similar, cells are also extraordinarily diverse and versatile. They are modified in a variety of ways to carry out specialized functions.

Cell biology is an interdisciplinary science. It draws on increasingly sophisticated tools from a wide array of scientific fields in the quest to better understand cellular structure and function. For example, investigation of the cytoskeleton (cell skeleton), currently an active and exciting area of research, has been greatly enhanced by advances in microscopy, biochemistry, molecular biology, genetics, and computational biology. The photomicrograph illustrates the extensive distribution of cytoskeletal fibers known as microfilaments (composed of the protein actin) and intermediate filaments (composed of the protein keratin) in cells. Biochemical studies of microfilaments, combined with computer-generated 3-D images of their protein subunits, have led to deep insights into how cells rapidly assemble and disassemble these dynamic and complex structures. Further understanding of microfilament functions comes from molecular genetic studies of cells that contain mutations in genes that regulate microfilament assembly. Through this combination of approaches, scientists are developing a comprehensive picture of how the cytoskeletal system maintains cell shape and functions in cell movement.

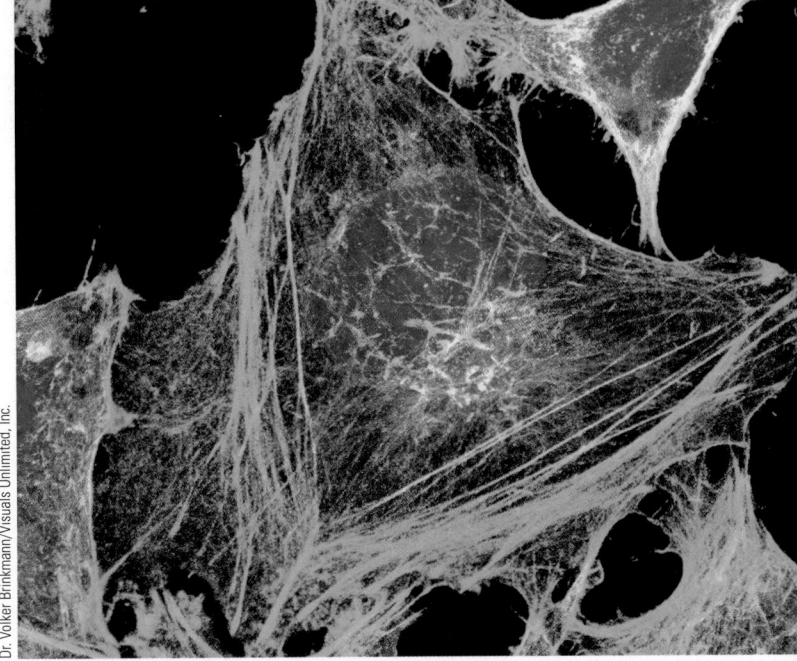

Dr. Volker Brinkmann/Visuals Unlimited, Inc.

The cytoskeleton. The cell shown here was stained with fluorescent molecules that bind to DNA (*blue*), microfilaments (*green*), and intermediate filaments (*red*). This type of microscopy, known as confocal fluorescence microscopy, shows the extensive distribution of microfilaments in this cell.

KEY CONCEPTS

4.1 The cell is the basic unit of life, its organization is critical to its ability to carry out all life activities, and its size and shape are adapted for its functions.

4.2 Biologists connect cellular structures to their functions with an interdisciplinary approach that includes microscopy, biochemistry, genetics, and computational methods.

4.3 Unlike prokaryotic cells, eukaryotic cells have internal membranes that divide the cell into compartments, allowing cells to conduct specialized activities within separate, small spaces.

4.4 In eukaryotic cells, genetic information coded in DNA is located in the nucleus, which is typically the most prominent organelle in the cell.

4.5 Among the many organelles in the cytoplasm are ribosomes, which synthesize proteins; endoplasmic reticulum and Golgi complexes, which process proteins; and mitochondria and chloroplasts, which convert energy from one form to another.

4.6 The cytoskeleton is a dynamic internal framework that functions in various types of cell movement.

4.7 Most eukaryotic cells are surrounded by a cell coat; in addition, many animal cells are surrounded by an extracellular matrix; cells of most bacteria, archaea, fungi, and plants are surrounded by a cell wall.

4.1 THE CELL: BASIC UNIT OF LIFE

LEARNING OBJECTIVES

1 Connect the cell theory to the evolution of life.
2 Relate the organizational similarities of all cells to the need to conduct essential life functions.
3 Explain the functional significance of cell size and cell shape.

Cells, the building blocks of organisms, are dramatic examples of the underlying unity of all living things.

The cell theory is a unifying concept in biology

From their own microscopic observations and those of other scientists, botanist Matthias Schleiden in 1838 and zoologist Theodor Schwann in 1839, reasoned that all plants and animals consist of cells. Later, Rudolf Virchow, another German scientist, observed cells dividing and giving rise to daughter cells. In 1855, Virchow proposed that new cells form only by the division of previously existing cells. The work of Schleiden, Schwann, and Virchow contributed greatly to the development of the **cell theory**, the unifying concept that (1) cells are the basic living units of organization and function in all organisms and (2) all cells come from other cells.

Around 1880, another German biologist, August Weismann, added an important corollary to Virchow's concept by pointing out that the ancestry of all the cells alive today can be traced back to ancient times. Evidence that all living cells have a common origin is provided by the basic similarities in their structures and in the molecules of which they are made. When we examine a variety of diverse organisms, ranging from simple bacteria to the most complex plants and animals, we find striking similarities at the cellular level. Careful studies of shared cell characteristics help us trace the evolutionary history of various organisms and furnish powerful evidence that all organisms alive today had a common origin.

The organization and basic functions of all cells are similar

The organization of cells and their small size are critical properties that allow them to maintain an appropriate internal environment necessary for their biochemical systems to function. For the cell to maintain its interior composition, its contents must be separated from the external environment. The **plasma membrane** is a structurally distinctive surface membrane that surrounds all cells. By making the interior of the cell an enclosed compartment, the plasma membrane allows the chemical composition of the cell to be different from that outside the cell. The plasma membrane has unique properties that enable it to serve as a selective barrier between the cell contents and the exterior, permitting the cell to exchange materials with the environment and accumulate substances needed to drive its biochemical reactions.

Cells have internal structures, called **organelles**, that are specialized to carry out cell activities. Most of the organelles of eukaryotic cells (cells that possess a nucleus) consist of one or more membrane-enclosed compartments capable of regulating their own internal environments for specialized functions such as converting energy to usable forms, processing nutrients, and recycling damaged or unneeded structures. Genetic information stored in DNA molecules of all cells, for example, is contained, duplicated, and transcribed in the nuclear compartment of eukaryotic cells (see Chapter 15).

Cell membranes also serve as organizing surfaces for interacting proteins that function in certain types of biochemical reactions. These stepwise reactions are more efficient and more rapid when their components are arranged to minimize the distance reactants must travel. As you will see in this chapter and in Chapter 8, the inner membrane of the eukaryotic mitochondrial compartment (as well as the plasma membrane of prokaryotes) contains tightly packed protein complexes that rapidly exchange electrons and protons, converting energy from food stores into energy-rich ATP that is used in hundreds of different biochemical reactions that occur every moment in a living cell. These chemical reactions that convert energy from one form to another are essentially the same in all cells, from the reactions in bacteria to those of large, multicellular plants and animals. Such fundamental similarities are strong evidence of their evolutionary relationships.

Cell size is limited

Although their sizes vary over a wide range (**FIG. 4-1**), most cells are microscopic and must be measured by very small units. The basic unit of linear measurement in the metric system (see inside back cover) is the meter (m), which is just a little longer than 1 yard. One millimeter (mm) is 1/1000 of 1 meter and is about as long as the bar enclosed in parentheses (-). The micrometer (μm) is the most convenient unit for measuring cells. A bar 1 μm long is 1/1,000,000 (1 millionth) of a meter, or 1/1000 of a millimeter, which is far too short to be seen with the unaided eye. Most of us have difficulty thinking about units that are too small to see, but it is helpful to remember that a micrometer has the same relationship to a millimeter that a millimeter has to a meter (1/1000).

As small as it is, the micrometer is actually too large to measure most cell components. For this purpose biologists use the nanometer (nm), which is 1/1,000,000,000 (1 billionth) of a meter, or 1/1000 of a micrometer. To mentally move down to the world of the nanometer, recall that a millimeter is 1/1000 of a meter, a micrometer is 1/1000 of a millimeter, and a nanometer is 1/1000 of a micrometer.

A few specialized algae and animal cells are large enough to be seen with the naked eye. A human egg cell, for example, is about 130 μm in diameter, or approximately the size of the period at the end of this sentence. The largest cells are birds' eggs, but they are not typical cells because they include large amounts of food reserves: the yolk and the egg white. The functioning part of the cell is a small mass on the surface of the yolk.

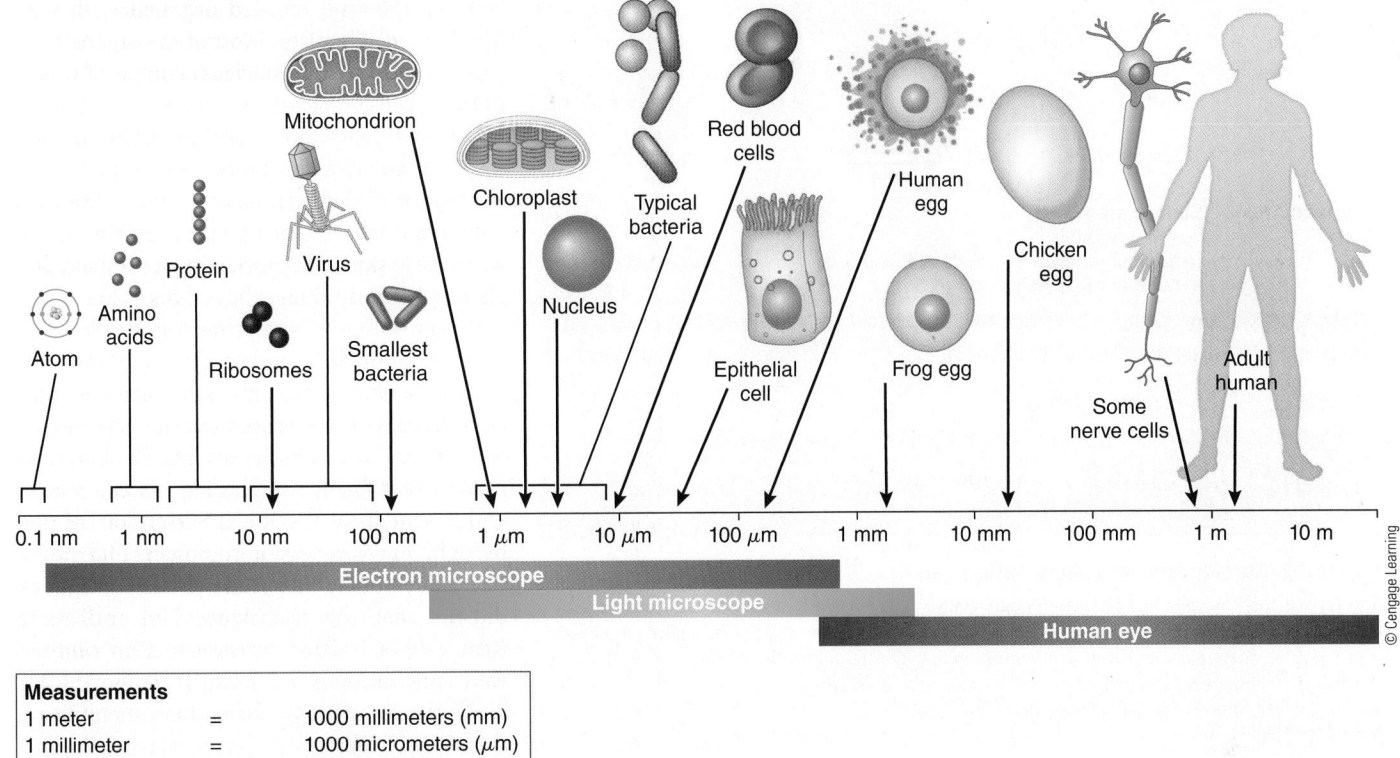

Measurements		
1 meter	=	1000 millimeters (mm)
1 millimeter	=	1000 micrometers (μm)
1 micrometer	=	1000 nanometers (nm)

Figure 4-1 Biological size and cell diversity

We can compare relative size from the chemical level to the level of an entire organism by using a logarithmic scale (multiples of 10). The prokaryotic cells of most bacteria range in size from 1 to 10 μm long. Most eukaryotic cells are between 10 and 30 μm in diameter. Mitochondria are about the size of small bacteria, whereas chloroplasts are usually larger, about 5 μm long. Ova (egg cells) are among the largest cells. Although microscopic, some nerve cells are very long. The cells shown here are not drawn to scale.

Why are most cells so small? If you consider what a cell must do to maintain its functions and to grow, it may be easier to understand the reasons for its small size. A cell must take in food and other materials and must rid itself of waste products generated by metabolic reactions. Everything that enters or leaves a cell must pass through its plasma membrane. The plasma membrane contains specialized "pumps" and channels with "gates" that selectively regulate the passage of materials into and out of the cell. The plasma membrane must be large enough relative to the cell volume to keep up with the demands of regulating the passage of materials. Thus, a critical factor in determining cell size is the ratio of its surface area (the plasma membrane) to its volume (FIG. 4-2).

As a cell becomes larger, its volume increases at a greater rate than its surface area (its plasma membrane), which effectively places an upper limit on cell size. Above some critical size, the number of molecules required by the cell could not be transported into the cell fast enough to sustain its needs. In addition, the cell would not be able to regulate its concentration of various ions or efficiently export its wastes.

Of course, not all cells are spherical or cuboid. Because of their shapes, some very large cells have relatively favorable ratios of surface area to volume. In fact, some variations in cell shape represent a strategy for increasing the ratio of surface area to volume. For example, many large plant cells are long and thin, which increases their surface area–to-volume ratio. Some cells, such as epithelial cells lining the small intestine, have fingerlike projections of the plasma membrane, called **microvilli**, that significantly increase the surface area for absorbing nutrients and other materials (see Fig. 47-10).

Another reason for the small size of cells is that, once inside, molecules must be transported to the locations where they are converted into other forms. Because cells are small, the distances molecules travel within them are relatively short. Thus, molecules are rapidly available for cell activities.

Cell size and shape are adapted to function

The sizes and shapes of cells are adapted to the particular functions they perform. Some cells, such as amoebas and white blood cells, change their shape as they move about. Sperm cells have long, whiplike tails, called *flagella*, for locomotion. Nerve cells have long, thin extensions that enable them to transmit messages over great distances. The extensions of some nerve cells in the human body may be as long as 1 meter! Certain epithelial cells are almost rectangular and are stacked much like building blocks to form sheetlike tissues. (Epithelial tissue covers the body and lines body cavities.)

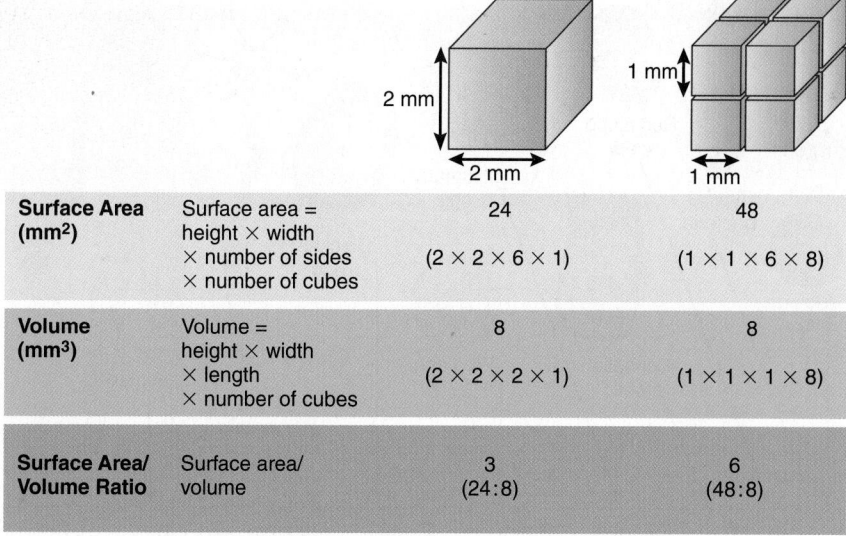

Surface Area (mm²)	Surface area = height × width × number of sides × number of cubes	24 (2 × 2 × 6 × 1)	48 (1 × 1 × 6 × 8)
Volume (mm³)	Volume = height × width × length × number of cubes	8 (2 × 2 × 2 × 1)	8 (1 × 1 × 1 × 8)
Surface Area/ Volume Ratio	Surface area/ volume	3 (24:8)	6 (48:8)

Figure 4-2 Surface area–to-volume ratio

The surface area of a cell must be large enough relative to its volume to allow adequate exchange of materials with the environment. Although their volumes are the same, eight small cells have a much greater surface area (plasma membrane) in relation to their total volume than one large cell does. In the example shown, the ratio of the total surface area to total volume of eight 1 mm cubes is double the surface area–to-volume ratio of the single large cube.
© Cengage Learning

CHECKPOINT 4.1

- **CONNECT** *How does the cell theory contribute to our understanding of the evolution of life?*
- **PREDICT** *Could a cell function if it were not enclosed by a selective barrier (i.e., a plasma membrane)?*
- *What molecule is used for information storage in all cells?*
- *What convenient form of chemical energy is used by all cells?*
- *Why is the relationship between surface area and volume of a cell important in determining cell size limits?*

4.2 METHODS FOR STUDYING CELLS

LEARNING OBJECTIVE

4 Compare methods that biologists use to study cells and point out the ways in which many of these approaches are complementary.

One of the most important tools biologists use for studying cell structures is the microscope. Using a microscope he had made, Robert Hooke, an English scientist, first described cells in 1665 in his book *Micrographia*. Hooke examined a piece of cork and then drew and described what he saw. Hooke chose the term *cell* because the tissue reminded him of the small rooms monks lived in. Interestingly, what Hooke saw were not actually living

cells but the walls of dead cork cells (**FIG. 4-3a**). Much later, scientists recognized that the interior enclosed by the walls is the important part of living cells.

A few years after Hooke's discovery and inspired by Hooke's work, Dutch naturalist Antonie van Leeuwenhoek viewed living cells with small lenses that he made. Leeuwenhoek was highly skilled at fabricating lenses and was able to magnify images more than 200 times. Among his important discoveries were bacteria, protists, blood cells, and sperm cells. Leeuwenhoek was among the first scientists to report cells in animals. He was a merchant and was not formally trained as a scientist, but his skill, curiosity, and diligence in sharing his discoveries with scientists at the Royal Society of London brought an awareness of microscopic life to the scientific world. Unfortunately, Leeuwenhoek did not share his techniques. Not until more than 170 years later, in the late 19th century, were microscopes sufficiently developed for biologists to seriously focus their attention on the study of cells.

Light microscopes are used to study stained or living cells

The **light microscope (LM),** the type used by most students, consists of a tube with glass lenses at each end. Because it contains several lenses, the modern light microscope is referred to as a *compound microscope*. Visible light passes through the specimen being observed and through the lenses. Light is refracted (bent) by the lenses, magnifying the image. Images obtained with light microscopes are referred to as light micrographs, or LMs.

Two features of a microscope determine how clearly a small object can be viewed: magnification and resolving power. **Magnification** is the ratio of the size of the image seen with the microscope to the actual size of the object. The best light microscopes usually magnify an object no more than 2000 times. **Resolution**, or **resolving power**, is the capacity to distinguish fine detail in an image; it is defined as the minimum distance between two points at which they can both be seen separately rather than as a single, blurred point. Resolving power depends on the quality of the lenses and the wavelength of the illuminating light. As the wavelength decreases, the resolution increases.

The visible light used by light microscopes has wavelengths ranging from about 400 nm (violet) to 700 nm (red); this limits the resolution of the light microscope to details no smaller than the diameter of a small bacterial cell (about 0.2 μm or 200 nm). By the early 20th century, refined versions of the light microscope became available.

The interior of many cells is transparent, and it is difficult to discern specific structures. Organic chemists contributed greatly

WHY IS IT USED? Cells are too small to be studied with the naked eye. Biologists use microscopes to view cells and structures inside cells. Many types of microscopes have been developed. Panels b–f are photomicrographs of *Paramecium*, a ciliated protist, made using several kinds of light microscopes.

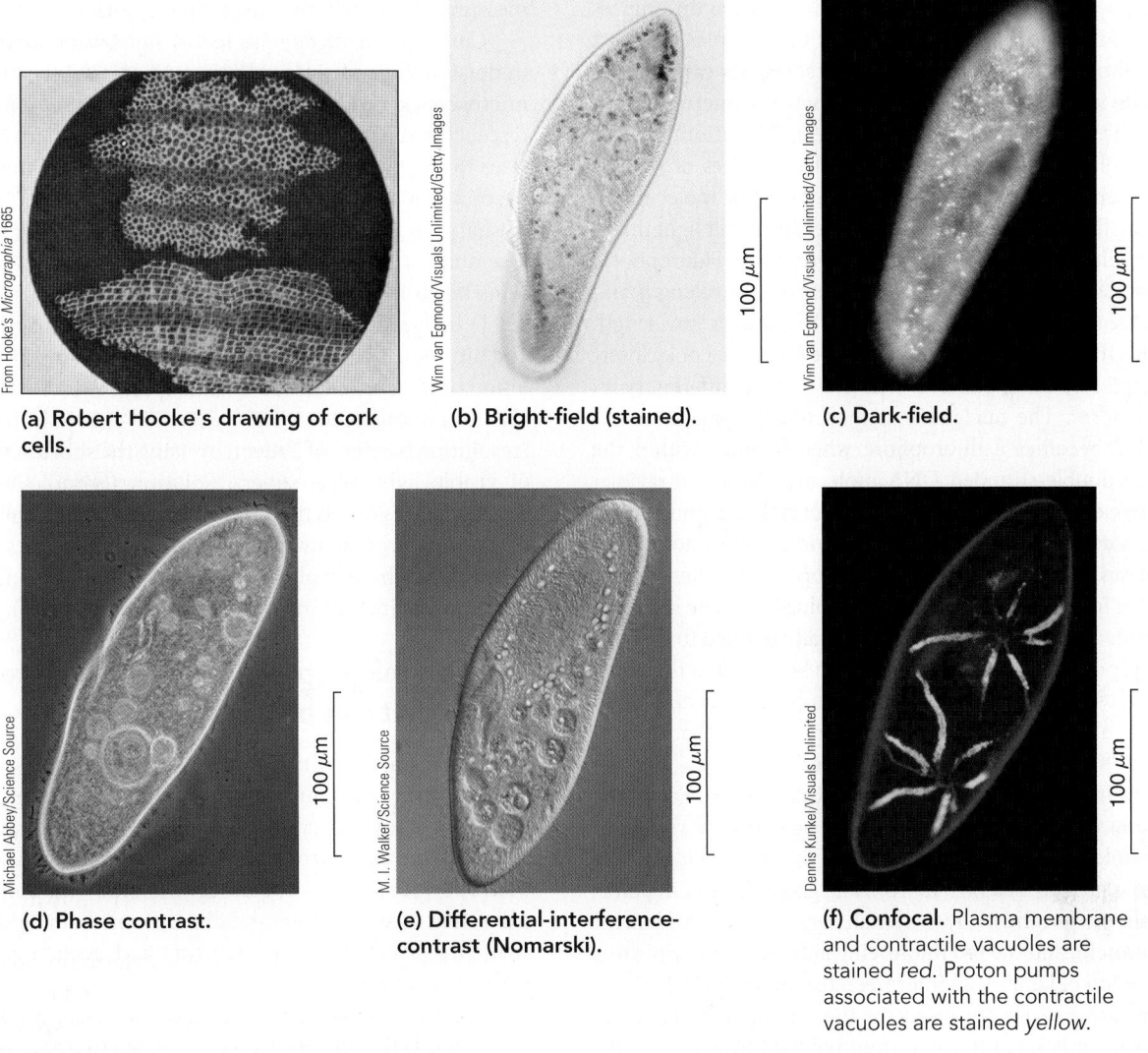

(a) **Robert Hooke's drawing of cork cells.**

(b) **Bright-field (stained).**

(c) **Dark-field.**

(d) **Phase contrast.**

(e) **Differential-interference-contrast (Nomarski).**

(f) **Confocal.** Plasma membrane and contractile vacuoles are stained *red*. Proton pumps associated with the contractile vacuoles are stained *yellow*.

HOW IS IT DONE? Using a microscope that he constructed, Robert Hooke looked at a thin slice of cork and drew what he saw. Biologists now view cells in more detail using the more sophisticated microscopes and techniques that have been developed. In the light microscope, a beam of light passes through the specimen being observed and through the lenses. The lenses refract the light, which magnifies the image. Bright-field microscopy can be enhanced by staining. The phase contrast and differential-interference-contrast microscopes enhance detail by increasing the differences in optical density in different regions of the cells. Confocal microscopes use a laser to illuminate a thin slice of the cell within the focal plane.

Figure 4-3 *Animation* **Using light microscopy**

to light microscopy by developing biological stains that enhance contrast in the microscopic image. Staining has enabled biologists to discover the many different internal cell structures, the organelles. Unfortunately, most methods used to prepare and stain cells for observation also kill them in the process.

Light microscopes with special optical systems now permit biologists to study living cells. In *bright-field microscopy*, an image is formed by transmitting light through a cell (or other specimen) (FIG. 4-3b). Because there is little contrast, the details of cell structure are not visible. In *dark-field microscopy*, rays of light are directed from the side, and only light scattered by the specimen enters the lenses. The cell is seen as a bright image against a dark background (FIG. 4-3c). The specimen does not need to be stained.

Phase contrast microscopy and Nomarski differential-interference-contrast microscopy take advantage of variations in density within the cell (FIG. 4-3d and e). These differences in density affect how various regions of the cytoplasm refract (bend) light. Using these microscopes, scientists can observe living cells in action and can view numerous internal structures that are constantly changing shape and location.

Cell biologists today widely use different types of fluorescence microscopes to detect the locations of specific molecules in cells. In the fluorescence microscope, filters transmit light that is emitted by fluorescent molecules, or fluorophores. Fluorophores are molecules that absorb light energy of one wavelength and then release some of that energy as light of a longer wavelength (like paints that glow under black light, see Fig. 9-3). Look closely at the chapter-opening photo, which shows three different types of fluorophores. The nucleus is stained with an organic compound that becomes a fluorophore when it binds within the grooves of double-stranded DNA molecules. When ultraviolet light is passed through the specimen, each fluorophore molecule absorbs the energy from an ultraviolet light photon and then releases part of that energy in the form of another photon that has the longer wavelength of visible blue light. The red color comes from a fluorophore that is chemically bonded to an antibody that specifically binds to the protein keratin, and the green color is derived from a fluorophore that is chemically bonded to phalloidin, a molecule isolated from a mushroom that specifically binds to microfilaments. **Antibodies** are proteins derived from the immune system (discussed in Chapter 45). Each type of antibody molecule can bind to only one specific region of another molecule (such as a small patch of amino acids on the surface of a protein; see Fig. 45-10). The green fluorescing molecule in the photo binds only to a small region of a microfilament subunit protein, and the red fluorescing antibody binds only to a specific region of keratin, an intermediate filament protein. Note that there are also yellow colors in the photograph. They are caused by the mixing of the green and red light from the microfilament and intermediate filament proteins that are very close to one another. This method (using highly sensitive microscopy) is commonly used to study the corresponding localizations (and possible interactions) of different proteins in cells.

Fluorescence microscopy using antibody labels must be done with nonliving cells that have been "fixed" by chemicals that form cross links between cellular proteins and other macromolecules to preserve them in their normal locations. Biologists today also employ types of fluorescent molecules to study the internal dynamics of living cells (in vivo). Certain types of fluorophores have been developed that will naturally diffuse into cells without killing them. Available today are wide arrays of compounds used to detect changes in intracellular pH, ion concentrations within intracellular compartments, and electrical charge differences across membranes. This is done by measuring shifts in the wavelength or the intensity of the emitted fluorescence. Biologists also use genetic engineering methods (see Chapter 15) to link the gene coding for a protein under study to part of another gene that encodes a green fluorescent protein (GFP) derived from a species of jellyfish. When the protein encoded by the modified gene is synthesized by the cell, it contains the GFP amino acid sequence as a "tag." The GFP tag functions as a fluorophore, allowing the intracellular movement of these "tagged" proteins to be tracked and measured in live cells by sensitive photodetectors.

Confocal microscopy has led to significant advances in our understanding of intracellular structural dynamics. These microscopes produce a sharper image than standard fluorescence microscopy (FIG. 4-3f). A confocal microscope uses a laser to excite fluorophores in just a thin "slice" through a cell, enabling an investigator to visualize objects in a single plane of sharp focus. In the chapter-opening photo, a computer has assembled a series of stacked images of a series of optical sections taken from the bottom to the top of the cell to construct a 3-D image. The use of powerful computer-imaging methods and ultrasensitive photodetectors has greatly improved the resolution of structures labeled by fluorescent dyes. Recent developments in fluorescence imaging have led to breakthroughs in the "resolution barrier" of 200 nm by using the shortest wavelengths of visible light. New super-resolution technologies can now resolve images of less than 70 nm derived from single molecules of fluorophores in living cells. For example, these advanced technologies have enabled researchers to monitor the intracellular movement within cells of brain tissue (FIG 4-4).

Electron microscopes provide a high-resolution image that can be greatly magnified

Even with improved microscopes and techniques for staining cells, ordinary light microscopes can distinguish only the gross details of many cell parts (FIG. 4-5a). With the development of the **electron microscope (EM),** which came into wide use in the 1950s, researchers could begin to study the fine details, or **ultrastructure,** of cells at the dimensions that could define the structure of cellular compartments and structures associated with cell membranes.

Because electrons have very short wavelengths on the order of about 0.1–0.2 nm, electron microscopes have resolving powers of just less than 1 nm. This high degree of resolution permits magnifications of more than 1 million times, compared with typical magnifications of no more than 1500 to 2000 times in light microscopy.

The image formed by the electron microscope is not directly visible. The electron beam itself consists of energized electrons, which, because of their negative charge, can be focused by electromagnets just as images are focused by glass lenses in a light microscope (FIG. 4-5b). Two main types of electron microscopes are the **transmission electron microscope (TEM)** and the **scanning electron microscope (SEM).** The acronyms TEM and SEM also identify that a micrograph was prepared using a transmission or scanning EM. Electron micrographs are black and white. They are often artificially colorized to highlight various structures.

In transmission electron microscopy, the specimen is embedded in plastic and then cut into extraordinarily thin sections

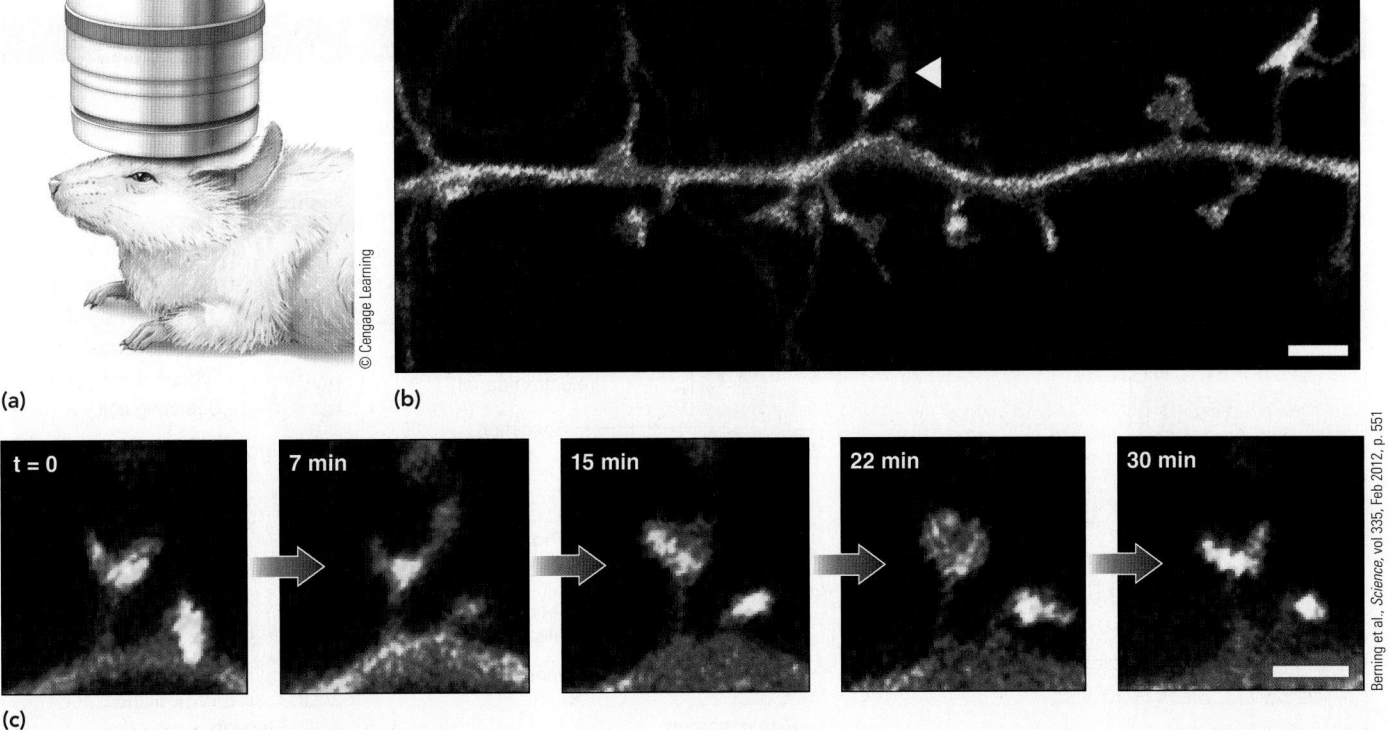

Berning et al., *Science*, vol 335, Feb 2012, p. 551

(a) **(b)**

t = 0   7 min   15 min   22 min   30 min

(c)

Figure 4-4 Super-resolution light microscopy

This series of super-resolution micrographs tracks the rapid changes in nerve cell connections in a living mouse. **(a)** The surface layers of the brain of an anesthetized mouse fitted with an optical coverglass in the top of its skull are viewed through the microscope. The genetically engineered mouse produces a modified form of green fluorescent protein only in its nerve cells. **(b)** An extension of a single brain cell shows its "spine" projections that connect to neighboring nerve cells. **(c)** Video frames spaced 7–8 minutes apart show the changes in shape that occur in a single "spine" over a period of 30 minutes. The white scale bars in the photos represent 1 μm.

(50 to 100 nm thick) with a glass or diamond knife. A section is then placed on a small metal grid. The electron beam passes through the specimen and then falls onto a photographic plate or a fluorescent screen. When you look at TEMs in this chapter (and elsewhere), keep in mind that each represents only a thin cross section of a cell.

Researchers can also detect certain specific molecules in electron microscope images by using antibody molecules to which very tiny gold particles are bound. The dense gold particles block the electron beam and identify the location of the proteins recognized by the antibodies as precise black spots on the electron micrograph.

In the scanning electron microscope, the electron beam does not pass through the specimen. Instead, the specimen is coated with a thin film of gold or some other metal. When the electron beam strikes various points on the surface of the specimen, secondary electrons are emitted whose intensity varies with the contour of the surface. The recorded emission patterns of the secondary electrons give a 3-D picture of the surface (**FIG. 4-5c**). The SEM provides information about the shape and external features of the specimen that cannot be obtained with the TEM.

Note that the LM, TEM, and SEM are focused by similar principles. A beam of light or an electron beam is directed by the condenser lens onto the specimen, and it is magnified by the objective lens and the eyepiece in the light microscope or by the objective lens and the projector lens in the TEM. The TEM image is focused onto a fluorescent screen, and the SEM image is viewed on a type of television screen. Lenses in electron microscopes are actually electromagnets that bend the beam of electrons.

Biologists use biochemical and genetic methods to connect cell structures with their functions

The EM and light microscopes are powerful tools for studying cell structure, but they have limitations. The methods used to prepare cells for electron microscopy kill them and may alter their structure. Furthermore, electron microscopy provides few clues about the functions of organelles and other cell components. To determine what organelles actually do, researchers use a variety of biochemical techniques.

Cell fractionation is a technique for separating (fractionating) different parts of cells so that they can be studied by physical and chemical methods. Generally, cells are broken apart in a blender. The resulting mixture, called the *cell homogenate,* is subjected to centrifugal force by spinning in a **centrifuge** (**FIG. 4-6a**). **Differential centrifugation** involves the separation of cell components through a series of centrifugation stages run at increasingly higher speeds. This allows various cell components to be separated on the basis of their different sizes and densities (**FIG. 4-6b**). At each step centrifugal force separates the extract into two fractions: a pellet and a supernatant. The *pellet* that forms at the bottom of the tube contains heavier materials packed together. (In the first low-speed step, this pellet is typically

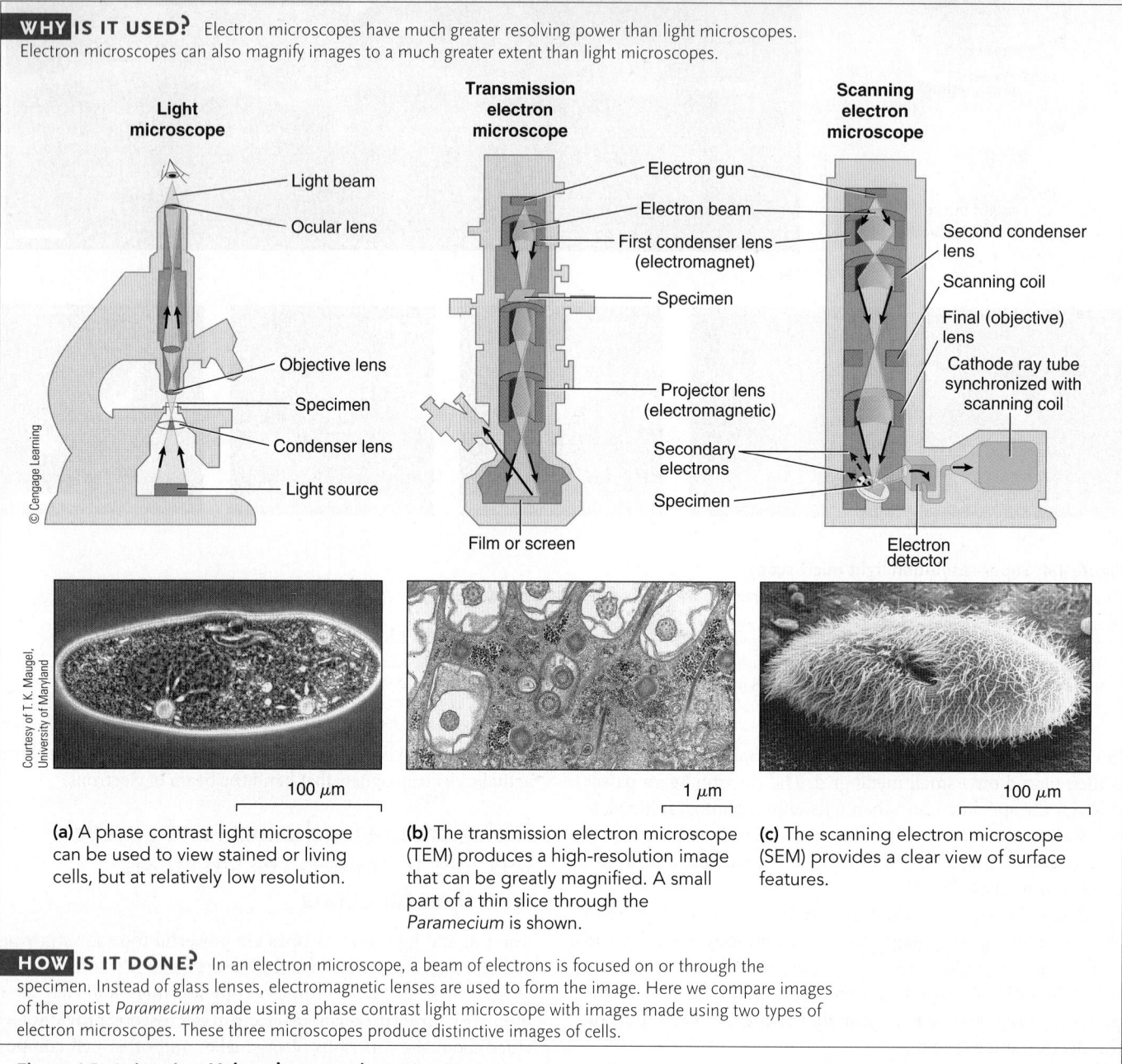

WHY IS IT USED? Electron microscopes have much greater resolving power than light microscopes. Electron microscopes can also magnify images to a much greater extent than light microscopes.

Light microscope

- Light beam
- Ocular lens
- Objective lens
- Specimen
- Condenser lens
- Light source

Transmission electron microscope

- Electron gun
- Electron beam
- First condenser lens (electromagnet)
- Specimen
- Projector lens (electromagnetic)
- Film or screen

Scanning electron microscope

- Second condenser lens
- Scanning coil
- Final (objective) lens
- Cathode ray tube synchronized with scanning coil
- Secondary electrons
- Specimen
- Electron detector

100 μm

(a) A phase contrast light microscope can be used to view stained or living cells, but at relatively low resolution.

1 μm

(b) The transmission electron microscope (TEM) produces a high-resolution image that can be greatly magnified. A small part of a thin slice through the *Paramecium* is shown.

100 μm

(c) The scanning electron microscope (SEM) provides a clear view of surface features.

HOW IS IT DONE? In an electron microscope, a beam of electrons is focused on or through the specimen. Instead of glass lenses, electromagnetic lenses are used to form the image. Here we compare images of the protist *Paramecium* made using a phase contrast light microscope with images made using two types of electron microscopes. These three microscopes produce distinctive images of cells.

Figure 4-5 *Animation* Using electron microscopes

composed of nuclei.) The *supernatant,* the liquid above the pellet, contains lighter organelles, dissolved molecules, and ions.

After the pellet is removed, the supernatant is centrifuged again at a higher speed to obtain a pellet that contains the next-heaviest cell components, for example, mitochondria and chloroplasts. To separate smaller, less dense components, the supernatant is then centrifuged in the powerful ultracentrifuge, which can spin at speeds exceeding 100,000 revolutions per minute (rpm), generating a centrifugal force of 500,000 × G (1 G is equal to the force of gravity).

Pellets can be resuspended and their components further purified by **density gradient centrifugation**. In this procedure, the ultracentrifuge tube is filled with a series of solutions of decreasing density. For example, sucrose solutions can be used. The concentration of sucrose is highest at the bottom of the tube and decreases gradually so that it is lowest at the top. The resuspended pellet is placed in a layer on top of the density gradient. Because the densities of organelles differ, each migrates during centrifugation to a position in the sucrose gradient that corresponds to its own density (FIG. 4-6c). These purified organelles can then be studied to determine what kinds of proteins and other molecules they might contain, or what types of biochemical reactions take place within them.

Antibodies are also widely used in laboratories to detect specific proteins in subcellular fractions, measure their intracellular

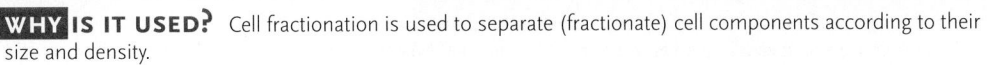

WHY IS IT USED? Cell fractionation is used to separate (fractionate) cell components according to their size and density.

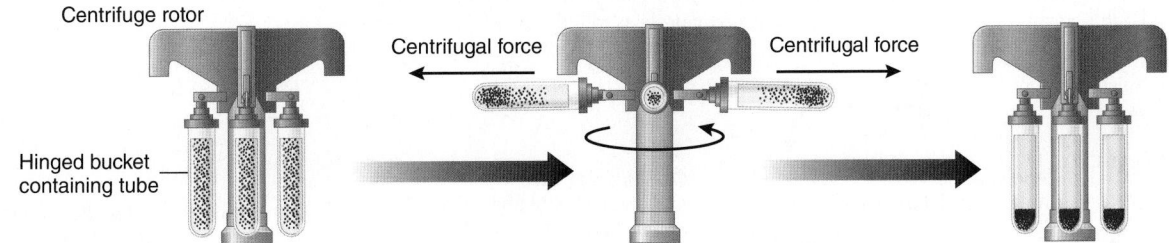

(a) Centrifugation. Due to centrifugal force, large or very dense particles move toward the bottom of a tube and form a pellet.

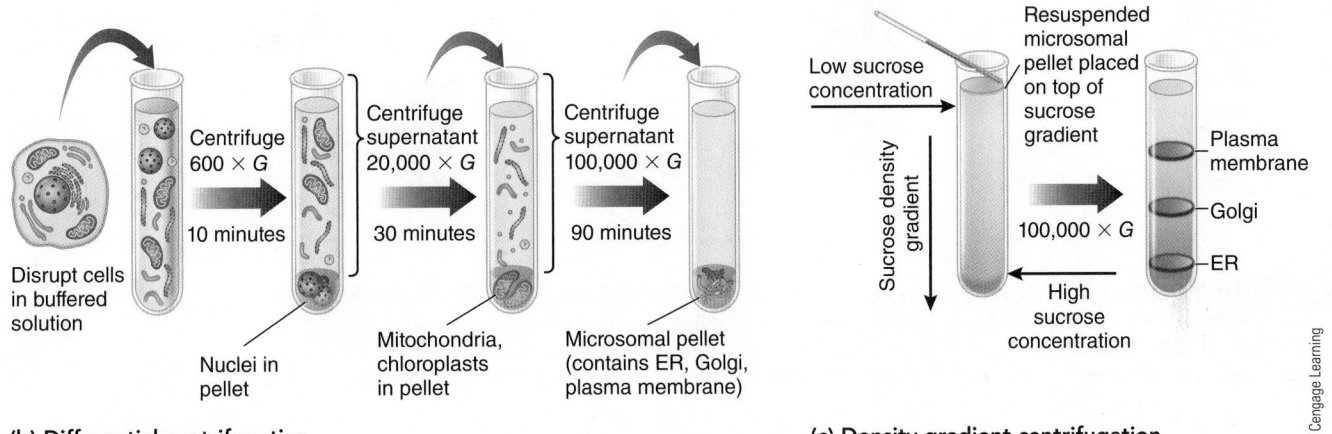

(b) Differential centrifugation.

(c) Density gradient centrifugation.

© Cengage Learning

HOW IS IT DONE? Cells are broken up in a blender. The cell homogenate (the resulting mixture) is then spun in a centrifuge. As a result of centrifugal force, the heaviest cell components, the nuclei, form a pellet at the bottom of the tube. The supernatant (the liquid above the pellet) can then be spun at a higher speed. The next heaviest component, the mitochondria and chloroplasts, form a pellet, and the supernatant can be spun at a higher speed. This process can be repeated several times. The pellet can be further purified by density gradient centrifugation (see text for further explanation).

Figure 4-6 Cell fractionation

levels, and study how they interact with other proteins. One application involves using antibodies that are coated onto small polymer beads to identify proteins that might bind to and function together with a protein of interest (**FIG. 4-7**). For example, an antibody that is specific for a microtubule subunit protein might be used to find other proteins that bind to microtubules and regulate their activity. The antibody-coated beads would be added to a cell extract (or a purified cell fraction) and then washed extensively to remove all substances in the extract that did not bind to the antibody-bound microtubule subunit protein. Proteins that remain attached to the antibody-bound subunit molecule can then be released from the antibody and analyzed to determine their identities.

Cell biologists also use genetic methods together with microscopy or biochemical methods to connect cellular proteins with their functions. When a protein has been identified as a critical component in a cell structure, researchers can use genetic engineering methods to alter or delete the gene that encodes that protein, effectively "turning off" its activity. By observing the differences between cells that contain the genetically altered protein with normal cells, researchers can gain insights into its function and how it interacts with other cellular proteins.

CHECKPOINT 4.2

- **CONNECT** *How does resolution limit the effective magnification that can be achieved in microscopy?*
- *What are the advantages of using many varied methods to connect structure and function in cells? Give some examples.*

WHY IS IT USED? An antibody is a protein derived from the immune system that can bind with high specificity to a small region of another protein. Antibodies are widely used to identify the location of a specific protein in a cell by immunofluorescence microscopy. Immunoprecipitation methods employ antibodies to isolate and purify a specific protein from a complex mixture of proteins found in a cellular fraction.

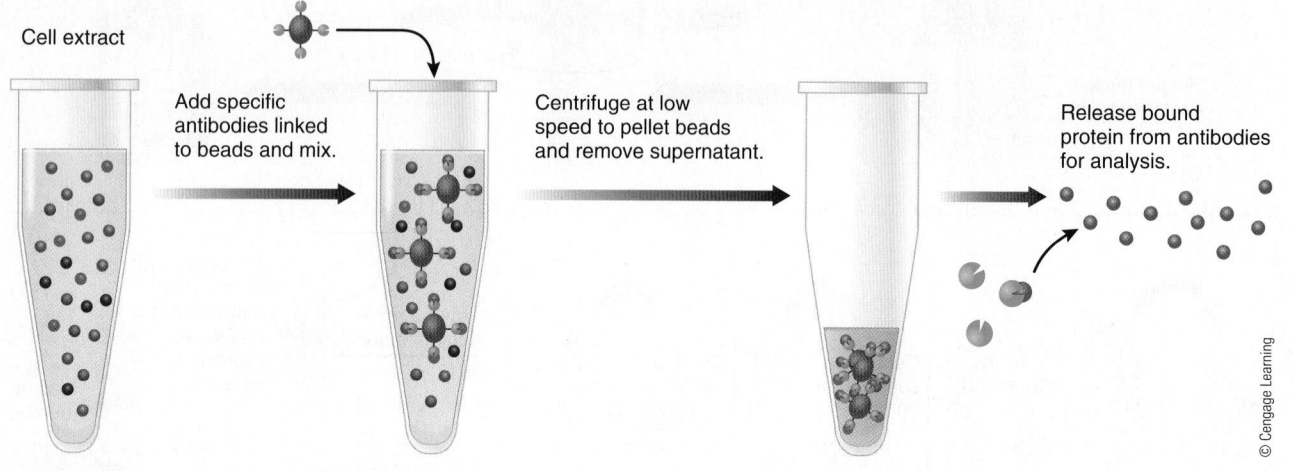

HOW IS IT DONE? Disrupt and fractionate the cells. Suspend a soluble fraction in a buffered solution. Add an antibody that is specific for the protein of interest covalently linked to small polymer beads. Allow the protein to bind to the bead-linked antibodies, and then pellet the beads with low-speed centrifugation (less than 1000 × G). After washing the pellet to remove unbound proteins, separate the antibody from its bound protein and subject the released protein to biochemical analysis.

Figure 4-7 Immunoprecipitation

4.3 PROKARYOTIC AND EUKARYOTIC CELLS

LEARNING OBJECTIVES

5 Compare and contrast the general characteristics of prokaryotic and eukaryotic cells, and contrast plant and animal cells.

6 Describe three functions of cell membranes.

Recall from Chapter 1 that two basic types of cells are known: **prokaryotic cells** and **eukaryotic cells**. Bacteria and archaea are prokaryotic cells. All other known organisms consist of one or more eukaryotic cells.

Organelles of prokaryotic cells are not surrounded by membranes

Prokaryotic cells are typically smaller than eukaryotic cells. In fact, the average prokaryotic cell is only about 1/10 the diameter of the average eukaryotic cell. In prokaryotic cells, the DNA is typically located in a limited region of the cell called a **nuclear area,** or **nucleoid.** Unlike the nucleus of eukaryotic cells, the nuclear area is not enclosed by a membrane (**FIG. 4-8**). The term *prokaryotic,* meaning "before the nucleus," refers to this major difference between prokaryotic and eukaryotic cells. Other types of internal membrane–enclosed organelles are also absent in prokaryotic cells.

Like eukaryotic cells, prokaryotic cells have a plasma membrane that surrounds the cell. The plasma membrane confines the contents of the cell to an internal compartment. In some prokaryotic cells, the plasma membrane may be folded inward to form a complex of membranes along which many of the cell's metabolic reactions take place. Most prokaryotic cells have **cell walls**, which are extracellular structures that enclose the entire cell, including the plasma membrane.

Many prokaryotes have **flagella** (sing., *flagellum*), long fibers that project from the surface of the cell. Prokaryotic flagella, which operate like propellers, are important in locomotion. Their structure is different from that of flagella found in eukaryotic cells. Some prokaryotes also have hairlike projections called *fimbriae,* which are used to adhere to one another or to attach to cell surfaces of other organisms.

The dense internal material of the bacterial cell contains **ribosomes**, small complexes of ribonucleic acid (RNA) and protein that synthesize polypeptides. The ribosomes of prokaryotic cells are smaller than those found in eukaryotic cells. Prokaryotic cells also contain storage granules that hold glycogen, lipid, or phosphate compounds. This chapter focuses primarily on eukaryotic cells. Prokaryotes are discussed in more detail in Chapter 25.

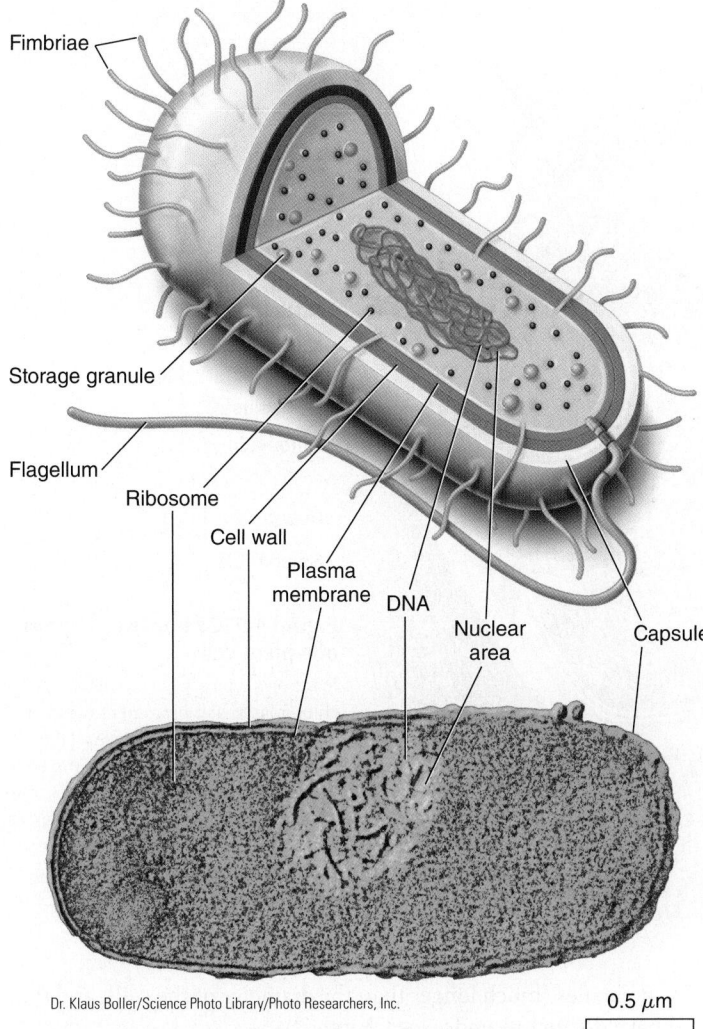

Fimbriae

Storage granule

Flagellum

Ribosome

Cell wall

Plasma membrane

DNA

Nuclear area

Capsule

Dr. Klaus Boller/Science Photo Library/Photo Researchers, Inc.

0.5 µm

Figure 4-8 *Animation* **Structure of a prokaryotic cell**

This colorized TEM shows a thin lengthwise slice through an *Escherichia coli* bacterium. Note the prominent nuclear area containing the genetic material (DNA). *E. coli* is a normal inhabitant of the human intestine, but under certain conditions some strains can cause infections.

© Cengage Learning

Membranes divide the eukaryotic cell into compartments

Eukaryotic cells are characterized by highly organized membrane-enclosed organelles, including a prominent *nucleus,* which contains DNA, the hereditary material. The term *eukaryotic* means "true nucleus." Early biologists thought that cells consisted of a homogeneous jelly, which they called *protoplasm.* With the electron microscope and other modern research tools, perception of the environment within the cell has been greatly expanded. We now know that the cell is highly organized and complex (FIGS. 4-9 and 4-10). The eukaryotic cell has its own control center, internal transportation system, power plants, factories for making needed materials, packaging plants, and even a "self-destruct" system.

Biologists refer to the part of the cell outside the nucleus as **cytoplasm** and the part of the cell within the nucleus as **nucleoplasm**. Various organelles are suspended within the fluid component of the cytoplasm, which is called the

cytosol. Eukaryotic proteins are synthesized in the cytoplasmic compartment by ribosomes. These large macromolecular complexes can function as components of the cytosol when they produce soluble proteins. Alternatively, they can be firmly bound to the cytosolic surfaces of membranes, where they form proteins that are either attached to membranes or enclosed in compartments bounded by membranes. The term *cytoplasm* includes both the cytosol and all the organelles other than the nucleus.

The many specialized organelles of eukaryotic cells solve some of the problems associated with large size, so eukaryotic cells can be larger than prokaryotic cells. Eukaryotic cells also differ from prokaryotic cells in having a supporting framework, or cytoskeleton, important in maintaining shape and transporting materials within the cell.

Some organelles are present only in specific cells. For example, *chloroplasts,* structures that trap sunlight for energy conversion, are only in cells that carry on photosynthesis, such as certain plant or algal cells. Cells of fungi and plants are surrounded by a *cell wall* external to the plasma membrane. Plant cells also contain a large, membrane-enclosed *vacuole.* We discuss these and other differences among major types of cells throughout this chapter.

The unique properties of biological membranes allow eukaryotic cells to carry on many diverse functions

Cell membranes have unique properties that enable membranous organelles to carry out a wide variety of functions. For example, cell membranes never have free ends. As a result, a membranous organelle always contains at least one enclosed internal space or compartment. These membrane-enclosed compartments allow certain cell activities to be localized within specific regions of the cell. Reactants confined to only a small part of the total cell volume are far more likely to come in contact, dramatically increasing the rate of the reaction. Membrane-enclosed compartments also keep certain reactive compounds away from other parts of the cell that they might adversely affect, allowing many different activities to go on simultaneously.

Membranes serve as important work surfaces. For example, many chemical reactions in cells are carried out by enzymes that are bound to membranes. Because the enzymes that carry out successive steps of a series of reactions are organized close together on a membrane surface, certain series of chemical reactions occur more rapidly.

Membranes allow cells to store energy. The membrane serves as a barrier that is somewhat analogous to a dam on a river. As we will discuss in Chapter 5, there is both an electric charge difference and a concentration difference of ions on the two sides of certain cell membranes. These differences constitute an *electrochemical gradient.* Such gradients store energy and so have *potential energy* (discussed in Chapter 7). As particles of a substance move across the membrane from the side of higher concentration to the side of lower concentration, the cell can convert some of this potential energy to the chemical energy of

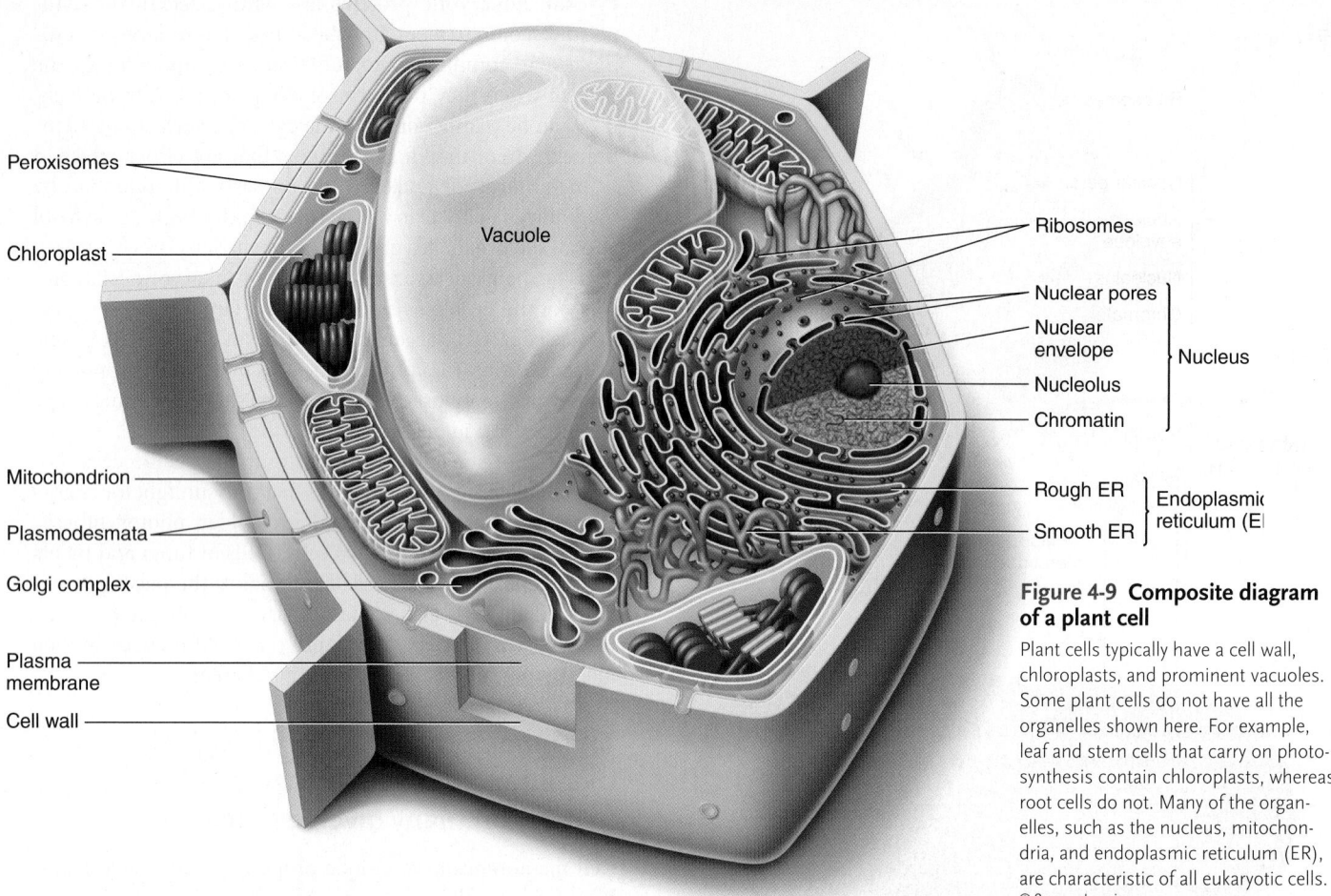

Peroxisomes

Chloroplast

Vacuole

Mitochondrion

Plasmodesmata

Golgi complex

Plasma membrane

Cell wall

Ribosomes

Nuclear pores

Nuclear envelope

Nucleolus

Chromatin

Nucleus

Rough ER

Smooth ER

Endoplasmic reticulum (ER)

Figure 4-9 Composite diagram of a plant cell

Plant cells typically have a cell wall, chloroplasts, and prominent vacuoles. Some plant cells do not have all the organelles shown here. For example, leaf and stem cells that carry on photosynthesis contain chloroplasts, whereas root cells do not. Many of the organelles, such as the nucleus, mitochondria, and endoplasmic reticulum (ER), are characteristic of all eukaryotic cells.
© Cengage Learning

ATP molecules. This process of energy conversion (discussed in Chapters 7, 8, and 9) is a basic mechanism that cells use to capture and convert the energy necessary to sustain life.

CHECKPOINT 4.3

- *What features do prokaryotic and eukaryotic cells have in common?*
- *What are three ways that a plant cell might differ from an animal cell?*
- *In what ways do membrane-enclosed organelles facilitate cell metabolism?*

4.4 THE CELL NUCLEUS

LEARNING OBJECTIVE

7   Relate the structure of the nucleus to its function as the control center of the eukaryotic cell.

The **nucleus** is typically the most prominent organelle in the cell. It is usually spherical or oval in shape and averages 5 μm in diameter. Most of the cell's DNA is located inside the nucleus. Unlike prokaryotic cells, whose DNA is in the form of circular molecules, eukaryotic DNA molecules are very long, linear

molecules (much longer than the diameter of the cell), and they have distinctive ends (see Chapter 10).

The **nuclear envelope** consists of two concentric membranes that separate the nuclear contents from the surrounding cytoplasm (FIG. 4-11). These membranes are separated by about 20 to 40 nm. At intervals the membranes come together to form **nuclear pores,** which are the largest and most complex assembly of proteins in most eukaryotic cells. Each nuclear pore consists of 500–1000 molecules made up of many copies of about 30 different proteins. A typical vertebrate cell that is not growing or dividing will usually have several thousand pores distributed across its nuclear surface. These complexes regulate the passage of large materials (including large proteins as well as macromolecular complexes such as ribosomes) across the nuclear membranes; however, ions and biological molecules (including small proteins) can pass freely through the nuclear pores. Each protein from the cytoplasm that must be actively transported through the pore contains a *nuclear localization signal (NLS)* as part of its amino acid sequence. Special proteins called *importins* bind with the NLS sequence, forming a "cargo complex" that can then be captured by the nuclear pore machinery and transported through the pore into the nucleus. These sequences are not only found on soluble proteins synthesized by ribosomes, but also on proteins that are embedded in cytoplasmic membranes and must then be moved to the inner nuclear membrane (FIG. 4-12).

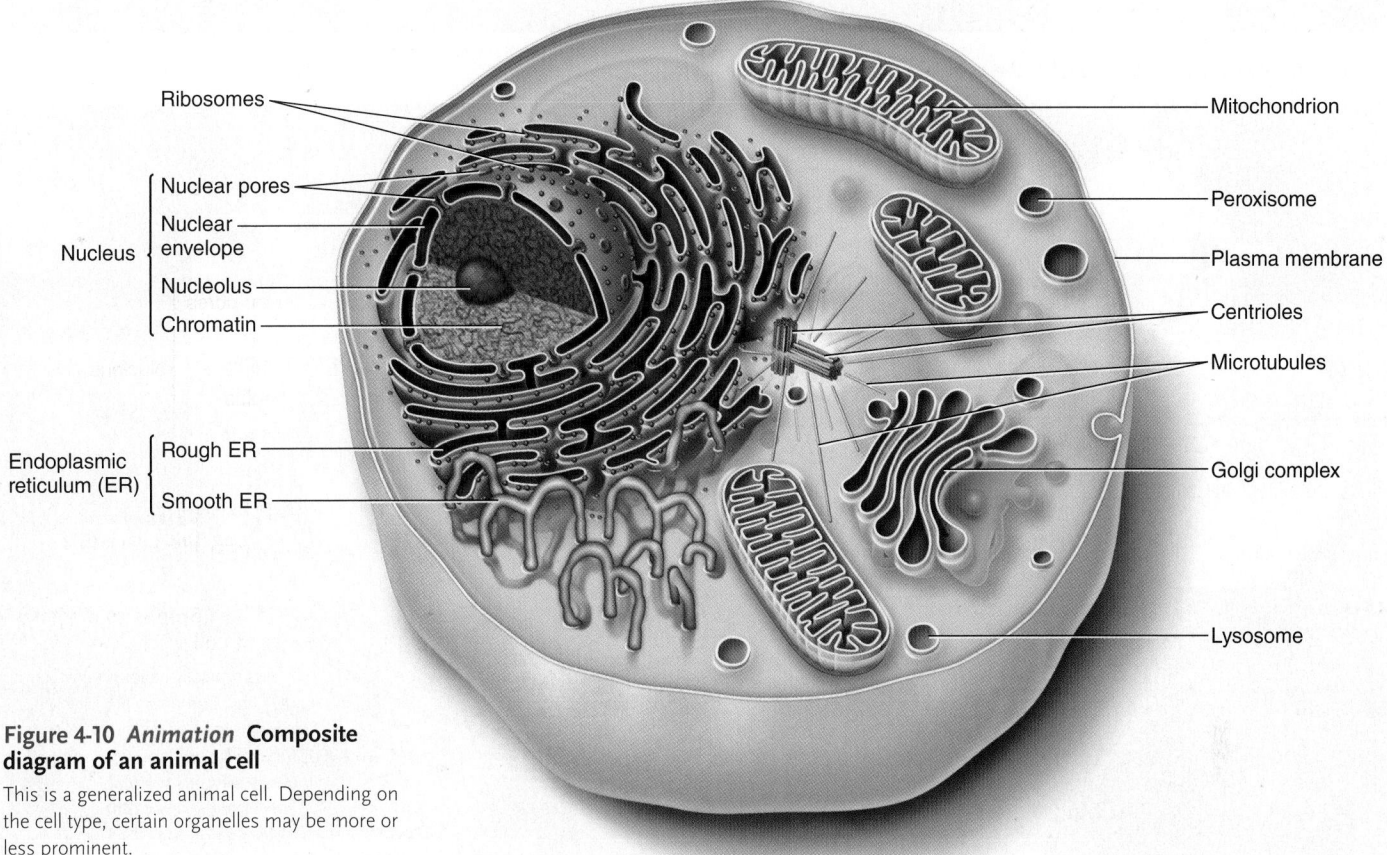

Ribosomes

Nuclear pores

Nuclear envelope

Nucleus

Nucleolus

Chromatin

Endoplasmic reticulum (ER)

Rough ER

Smooth ER

Mitochondrion

Peroxisome

Plasma membrane

Centrioles

Microtubules

Golgi complex

Lysosome

Figure 4-10 *Animation* **Composite diagram of an animal cell**

This is a generalized animal cell. Depending on the cell type, certain organelles may be more or less prominent.
© Cengage Learning

The transport mechanism through nuclear pores can be very fast. Fluorescent-labeled proteins containing an NLS sequence have been observed moving through the nuclear pores at rates approaching 2000 molecules/second, whereas deletion of the NLS sequence from the same protein has been shown to effectively block its entry into the nucleus.

A fibrous network of protein filaments, called the nuclear lamins, forms the nuclear lamina, an inner lining for the nuclear envelope. The *nuclear lamina* supports the inner nuclear membrane and helps organize the nuclear contents. It also plays a role in DNA duplication and in regulating the cycle of growth and division (the *cell cycle*; see Chapter 10). Mutations in genes encoding proteins that make up the nuclear lamina are associated with several human genetic diseases, including some muscular dystrophies and premature aging (progeria).

When a cell divides, the information stored in DNA must be duplicated exactly through a process called replication (discussed in Chapter 12). Each copy is then passed intact to one of the two daughter cells. DNA molecules include sequences of nucleotides called **genes,** which contain the chemically coded instructions for producing the proteins and specialized RNA molecules needed by the cell. The nucleus controls protein synthesis by transcribing its information from DNA into **messenger RNA (mRNA)** molecules, which are then moved into the cytoplasm, where proteins are manufactured by ribosomes.

DNA in the nucleus is associated with RNA and certain proteins, forming a complex known as **chromatin** (see Chapter 10). In the light microscope this complex appears as a network of granules and strands in nondividing cells. Although chromatin appears disorganized, it is not. Because DNA molecules are extremely long and thin, each molecule is packed inside the nucleus in a regular fashion as part of a structure called a **chromosome.** In dividing cells, the chromosomes become visible as distinct threadlike structures. If the DNA molecules in the 46 chromosomes of one human cell could be stretched end to end, they would extend for 2 meters!

Most nuclei have one or more compact structures called **nucleoli** (sing., *nucleolus*). A nucleolus, which is *not* enclosed by a membrane, usually stains differently than the surrounding chromatin. Each nucleolus contains a *nucleolar organizer,* made up of overlapping regions of DNA from several chromosomes. This DNA contains instructions for making the type of RNA found in ribosomes. This **ribosomal RNA (rRNA)** is synthesized in the nucleolus.

Although it is tempting to think of the nucleus as a relatively quiet region that is reserved for information storage (much like a library), it is a center of constant activity, and there is a very high volume of traffic in both directions through the nuclear pores.

Large amounts of energy are consumed in the nucleus in the form of ATP and other nucleoside triphosphate molecules that are used to synthesize mRNA and rRNA molecules. Energy is also used to transport the large and small materials that are constantly moving in and out of the nucleus through the nuclear pores. This movement can be illustrated by considering the assembly of ribosomes, organelles that function in the cytoplasm but are assembled in the nucleolus.

The nucleus contains DNA and is the control center of the cell.

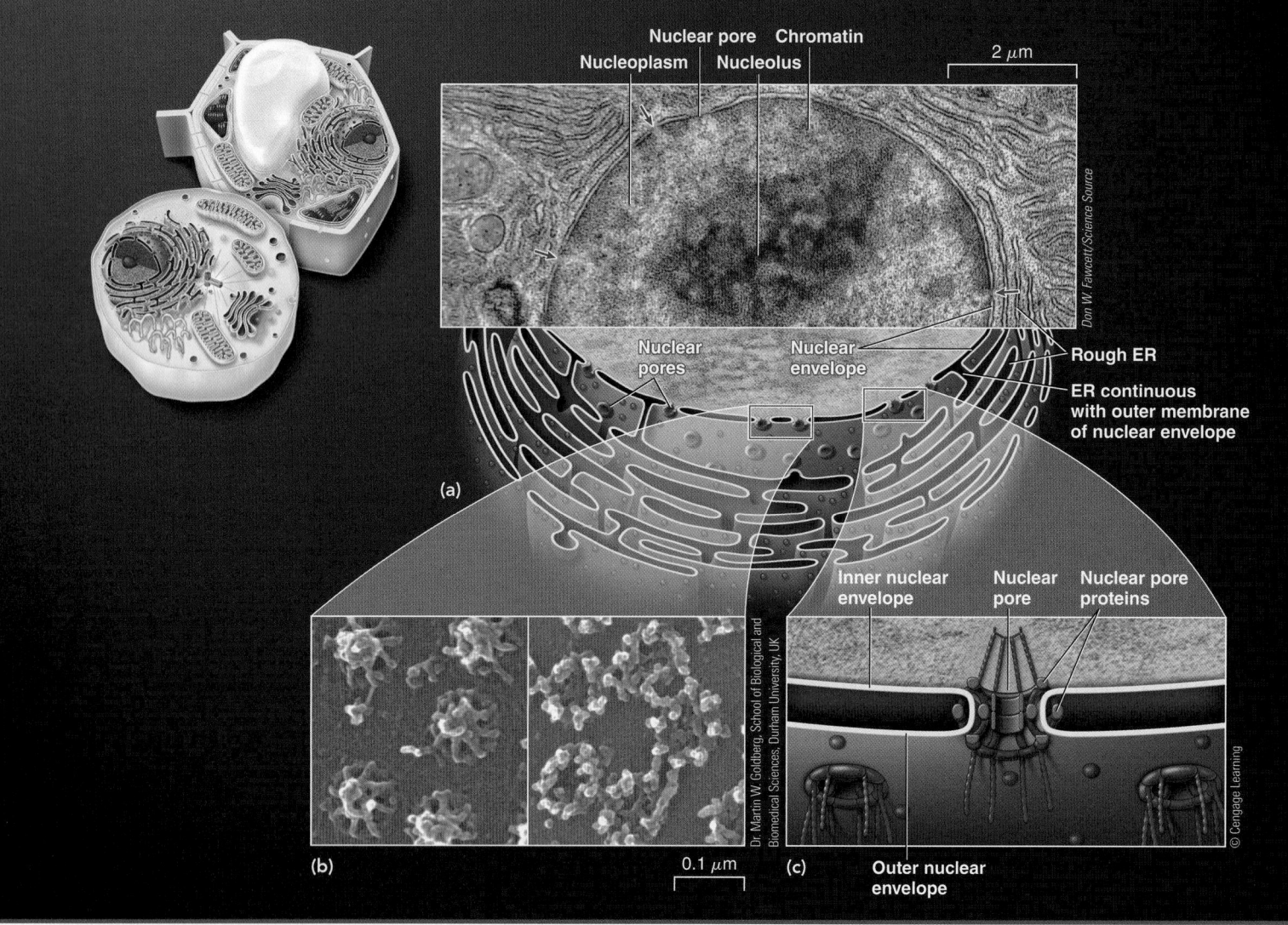

Figure 4-11 *Animation* **The cell nucleus**

(a) The TEM and interpretive drawing show that the nuclear envelope is composed of two concentric membranes connected through nuclear pores that perforate both membranes. The outer membrane of the nuclear envelope is continuous with the membrane of the ER (endoplasmic reticulum) and is usually covered with attached ribosomes. The nucleolus is not enclosed by a membrane. (b) SEM images of the nuclear pores show the inner nucleoplasmic face (*left*), which forms a basket-like structure, and the outer (cytoplasmic) face (*right*), which consists of eight sub-units with fibers that extend into the cytoplasm. (c) The nuclear pores, which are made up of many copies of more than 30 different proteins, form channels between the nucleoplasm and the cytoplasm.

PREDICT In which direction would you expect ribosomes to pass through the nuclear pores?

Ribosomes manufacture proteins in the cytoplasm

Ribosomes are organelles that can be found free in the cytosol or attached to certain cytoplasmic membranes. These small particles contain the enzyme activities necessary to form peptide bonds, which join amino acids to produce proteins (see Chapter 3). Each ribosome has two main components: a large subunit and a small subunit. When the two ribosome subunits join in a complex with an mRNA molecule, they function as manufacturing plants that assemble polypeptides. Cells that actively produce large amounts of proteins may have millions of ribosomes, and the cell can change the number of ribosomes present to meet its metabolic needs. We will discuss much more about ribosomes in Chapter 13.

Ribosomes are molecular machines made up of both proteins and RNA molecules. The two subunits of each eukaryotic ribosome actually consist of more than 80 different proteins and three different rRNA molecules. The mRNAs that encode the ribosomal proteins are first copied from their respective genes in the nucleus. They must then be transported through

Do large inner nuclear membrane proteins need the nuclear pore machinery to move from the rough ER to their location on the inside of the nucleus?

HYPOTHESIS: Large inner nuclear membrane proteins are transported from their cytoplasmic sites of synthesis to their inner membrane locations by the nuclear pore machinery.

EXPERIMENT: Previous experiments had shown that soluble proteins move into the nucleus only after they are bound to specialized "importin" proteins to form a "cargo complex" that can then be captured and transported through the nuclear pore. Megan King, Patrick Lusk, and Günter Blobel of Rockefeller University wanted to know if large proteins incorporated in the inner nuclear membrane are transported from their site of synthesis in the cytoplasm to the inner nuclear membrane by the same mechanism. They engineered a gene that encodes a GFP-tagged form of a large inner nuclear membrane protein and placed it into mutant yeast cells that contain a mutation in an importin protein. This mutant form

of importin is active in cells grown at 25°C, but becomes unfolded and inactive in cells exposed to 34°C. The researchers then used confocal and immuno-gold-transmission electron microscopy to compare the location of this protein in cells grown at the different temperatures.

RESULTS AND CONCLUSION: At 25°C, the GFP-tagged protein and the antibody-linked gold particles are concentrated at the inner nuclear membrane. In the cells grown at 34°C, with defective importin proteins, the GFP-tagged protein accumulates at the outer nuclear membrane and on cytoplasmic ER membranes. Thus, the nuclear pore machinery is required for the membrane protein to move to its correct cellular location within the nucleus.

SOURCE: King, M.C., Lusk, C.P., and Blobel, G. "Karyopherin-Mediated Import of Integral Inner Nuclear Membrane Proteins." *Nature*, Vol. 442, 2006, 1003–1007.

25°C
(importin proteins active)

34°C
(importin proteins inactive)

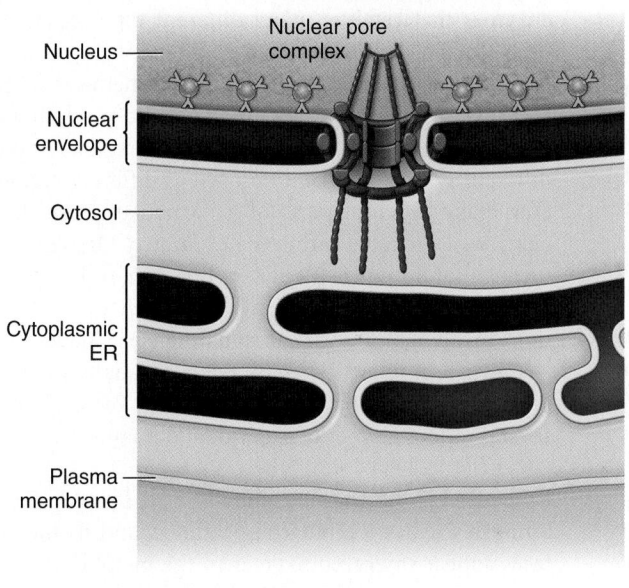
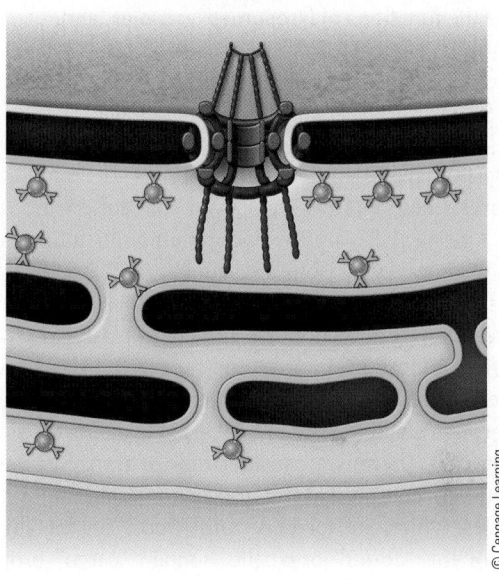

© Cengage Learning

KEY

	GFP-labeled protein fluorescence
	Antibody-linked gold particle

Figure 4-12 Translocation of membrane-bound proteins through the nuclear pore

PREDICT Soluble proteins that are actively transported through the nuclear pore contain a short series of amino acids called a nuclear localization signal that binds to importin proteins. The researchers identified a similar, but not identical, sequence on the inner nuclear membrane protein that they described in this experiment. If that sequence were a true importin binding site, what would you predict would happen if you were to mix importin-coated beads with an extract of inner nuclear membrane proteins?

the nuclear pores into the cytoplasm, where they are translated by ribosomes to synthesize their corresponding protein. Then, these newly formed ribosomal proteins must be imported back into the nucleus. Many of these proteins include an NLS sequence that is recognized by the nuclear import machinery. It is then packaged as cargo bound to carrier proteins that

allow it to be translocated through the pore. In the nucleus, these ribosomal proteins are assembled with rRNAs into ribosomal subunits. Once assembled, the newly formed ribosome subunit is then transported from the nucleolar region to the inside of the nuclear envelope, where it exits the nucleus through a nuclear pore.

- **CONNECT** *How does the structure of the nucleus affect how information stored in DNA is used by the cell?*
- *In what ways is transport through the nuclear pores selective?*
- *How does the movement of small molecules and large macromolecular structures between the nucleoplasmic and the cytoplasmic compartments differ?*

4.5 MEMBRANOUS ORGANELLES IN THE CYTOPLASM

LEARNING OBJECTIVES

8 Distinguish between smooth and rough endoplasmic reticulum in terms of both structure and function.

9 Trace the path of proteins synthesized in the rough endoplasmic reticulum as they are processed, modified, and sorted by the Golgi complex and then transported to specific destinations.

10 Compare the functions of lysosomes, vacuoles, and peroxisomes.

11 Contrast the functions of mitochondria and chloroplasts, and discuss ATP synthesis by each of these organelles.

Cell biologists have identified many types of membrane-enclosed organelles in the cytoplasm of eukaryotic cells (**TABLE 4-1**). Among these are the endoplasmic reticulum, Golgi complex, lysosomes, peroxisomes, vacuoles, mitochondria, and chloroplasts. Most of the cytoplasmic organelles in a typical cell—with the exception of mitochondria and chloroplasts—are components of the **endomembrane system** (see Fig. 4-15), a network of organelles (including the plasma membrane) that exchange materials through small membrane-enclosed **transport vesicles.**

In living cells, there is a constant flow of transport vesicle "traffic" between components of the endomembrane system. These vesicles, which are formed from "buds" on the surface membrane of the "donor" organelle, contain "cargo" proteins derived from the organelle's internal compartment or **lumen.** Each vesicle has proteins embedded in its membrane that are specific routing signals for its destination organelle. The vesicle is then transported on cytoskeletal "tracks" by molecular motors (which we will discuss later in this chapter) to the surface of its "target" organelle membrane. On contact, the transport vesicle fuses with the target membrane, releasing its contents into the compartment of the acceptor organelle. This fusion of the two membranes results from interactions between targeting proteins in the vesicle and receptor proteins on the acceptor membrane. As the two membranes fuse, the surface area of the acceptor membrane is expanded by the integration of the vesicle membrane.

The endoplasmic reticulum is a multifunctional network of membranes

One of the most prominent features in the electron micrographs in Figure 4-11 is a maze of parallel internal membranes that encircle the nucleus and extend into many regions of the cytoplasm. This complex of membranes, the **endoplasmic reticulum (ER),** forms a network that makes up a significant part of the total volume of the cytoplasm in many cells. A higher magnification TEM of the ER is shown in **FIGURE 4-13**. Remember that a TEM represents only a thin cross section of the cell, so there is a tendency to interpret the ER as a series of tubes. In fact, ER membranes consist of a number of different domains that have distinct structures and functions. Many are a series of tightly packed and flattened, saclike structures that form interconnected compartments within the cytoplasm; others are highly curved and tubular; and other parts assume shapes that come in contact with most every other organelle in the cell, including the plasma membrane. Although ER domains have different functions, their membranes are connected, and they enclose a continuous internal compartment called the *ER lumen.* The membranes of other organelles in the endomembrane system are not directly connected to the ER; they form distinct and separate compartments within the cytoplasm.

Each functional domain of the ER membrane and its corresponding region of the ER lumen contains a unique set of enzymes that catalyze many different types of chemical reactions. In some cases, the membranes serve as a framework for enzymes that carry out sequential biochemical reactions. The two surfaces of the membrane (cytosolic and luminal) contain different sets of enzymes and represent regions of the cell with different synthetic capabilities, just as different regions of a factory make different parts of a particular product. Still other enzymes are located within the ER lumen. One common feature throughout the entire ER lumen, however, is that it is the storage site of high levels of calcium, which is used for a number of functions involving intracellular signaling (see Chapter 6).

Two prominent types of ER can usually be easily distinguished in TEMs: *rough ER* and *smooth ER*, each of which has several different types of functional domains.

Smooth ER synthesizes lipids and breaks down toxins

Smooth ER has a tubular appearance, and its membrane surfaces appear smooth. Enzymes in the membranes on the cytosolic surface of the smooth ER catalyze the synthesis of many lipids and carbohydrates. The smooth ER is the primary site for the synthesis of phospholipids and cholesterol needed to make cell membranes. In some cells, such as those found in fat tissue, a region of the smooth ER will have enzymes that form lipid droplets, which store energy reserves in the form of triacylglycerols. Other domains of the smooth ER synthesize steroid hormones, including reproductive hormones, from cholesterol. In liver cells, smooth ER is also important in enzymatically breaking down stored glycogen. (The liver helps regulate the concentration of glucose in the blood. See Chapter 49.)

Whereas smooth ER may be a minor membrane component in some cells, extensive amounts of smooth ER are present in others. For example, large amounts of smooth ER are present in human liver cells where, in addition to lipid biosynthesis, it serves as a major detoxification site. Enzymes located along the smooth ER membrane of liver cells break down toxic chemicals such as carcinogens (cancer-causing agents) and many drugs, including alcohol, amphetamines, and barbiturates. The cell

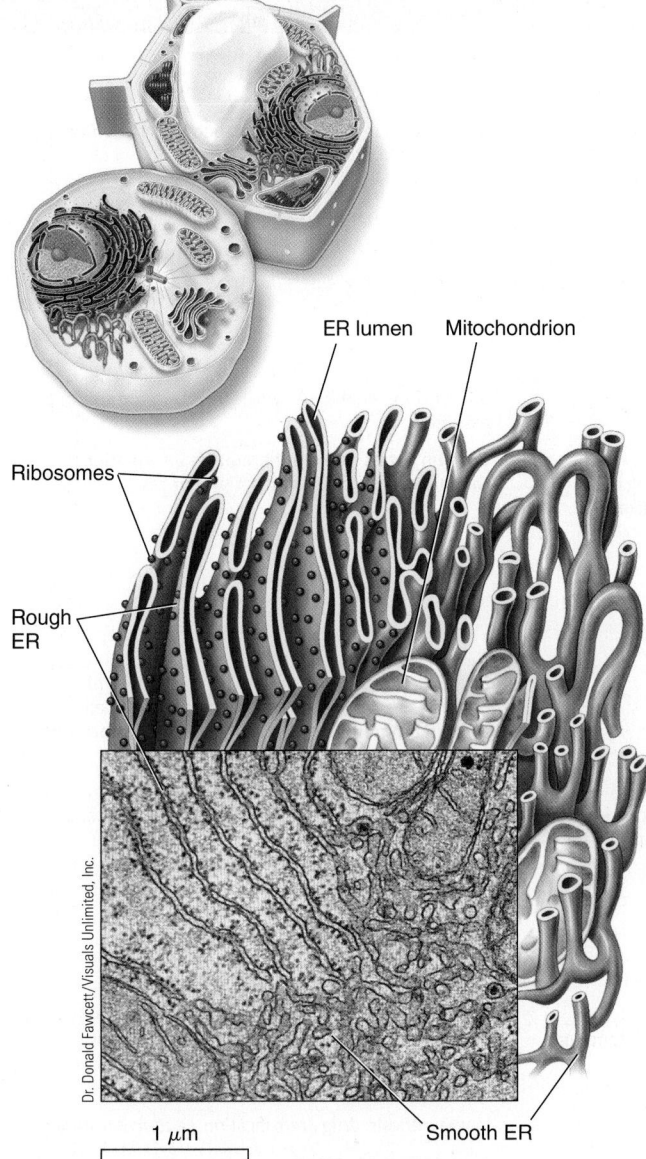

ER lumen   Mitochondrion

Ribosomes

Rough
ER

Dr. Donald Fawcett/Visuals Unlimited, Inc.

1 μm

Smooth ER

Figure 4-13 Endoplasmic reticulum (ER)
The TEM shows both rough and smooth ER in a liver cell.
© Cengage Learning

then converts these compounds to water-soluble products that it excretes. Interestingly, alcohol and many other drugs stimulate liver cells to produce additional smooth ER, increasing the rate that these cells can detoxify the drugs. This mechanism is of limited value, however, because alcohol abuse causes liver inflammation that can lead to cirrhosis and eventual liver failure.

Rough ER is involved in the synthesis of secreted and membrane proteins The outer surface of the **rough ER** is studded with ribosomes that appear as dark granules. Notice in Figure 4-13 that the lumen side of the rough ER appears bare, whereas the outer surface (the cytosolic side) looks rough. The ribosomes attached to the rough ER are known as *bound ribosomes* to differentiate them from *free ribosomes*, which are suspended in the cytosol.

The nuclear envelope is one functional domain of the rough ER. Figure 4-11 shows that the outer and inner nuclear membranes are actually continuous with the rough ER. Ribosomes are typically present on the outer (cytosolic) surface of the nuclear envelope, but are not found on the inner (nucleoplasmic) surface.

The rough ER plays a central role in the synthesis and assembly of proteins, including those that are exported from the cell (such as digestive enzymes), proteins destined for the compartments of other organelles, and proteins that function as an integral part of a membrane. All are synthesized by ribosomes bound to the ER membrane. When an mRNA encoding one of these types of proteins enters the cytosol, it first associates with a ribosome. The ribosome is then transported to the ER membrane cytosolic surface, where it forms a tight seal with the membrane. As the polypeptide is synthesized, it passes through a tunnel within the ribosome and into an ER pore. Polypeptides that are destined for export from the cell or to the compartment of another organelle pass directly into the ER lumen. Proteins that are destined to become embedded in the ER membrane will only partially pass through the ER pore before being integrated into the membrane lipid bilayer (see Chapter 5).

In the ER lumen, proteins are assembled and may be modified by enzymes that add carbohydrates or lipids to them. Other enzymes, called **molecular chaperones**, in the ER lumen catalyze the efficient folding of proteins into proper conformations. Proteins that are not processed correctly—for example, proteins that are misfolded—are transported back to the cytosol. There they are degraded by **proteasomes**, protein complexes that direct the destruction of defective proteins. Properly processed proteins that are to be routed to other compartments within the cell, however, are packaged into the small transport vesicles that bud off the ER membrane. They are then transported to and fused with the membrane of the appropriate target organelle.

The ER is the primary site of membrane assembly for components of the endomembrane system

Most cell membranes are first assembled in the endoplasmic reticulum. As we noted in Chapter 3 (and discuss in more detail in Chapter 5), the core of biological membranes consists of a phospholipid bilayer. These phospholipids are synthesized on the cytosolic surface of the smooth ER and then integrated into the membrane bilayer, causing the surfaces of the membrane to expand. Proteins that are to be embedded into newly formed membrane are synthesized by ribosomes attached to the cytosolic surface of the rough ER membrane. By contrast, membranes in other organelles within the endomembrane system grow when transport vesicles from the ER fuse with them.

The Golgi complex processes, sorts, and routes proteins from the ER to different parts of the endomembrane system

The **Golgi complex** (also known as the *Golgi body* or *Golgi apparatus*) was first described in 1898 by Italian microscopist

TABLE 4-1	Eukaryotic Cell Structures and Their Functions	

STRUCTURE	DESCRIPTION	FUNCTION
CELL NUCLEUS		
Nucleus 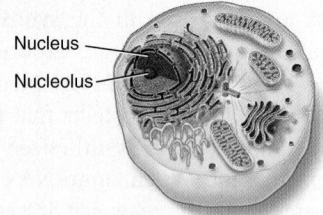 Nucleus Nucleolus	Large structure surrounded by double membrane; contains nucleolus and chromosomes	Information in DNA is transcribed in RNA synthesis; specifies cell proteins
Nucleolus	Granular body within nucleus; consists of RNA and protein	Site of ribosomal RNA synthesis; ribosome subunit assembly
Chromosomes 	Composed of chromatin, a complex of DNA and protein; condense during cell division, becoming visible as rodlike structures	Contain genes (units of hereditary information) that govern structure and activity of cell
CYTOPLASMIC ORGANELLES		
Plasma membrane 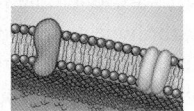	Membrane boundary of cell	Encloses cell contents; regulates movement of materials in and out of cell; helps maintain cell shape; communicates with other cells (also present in prokaryotes)
Ribosomes 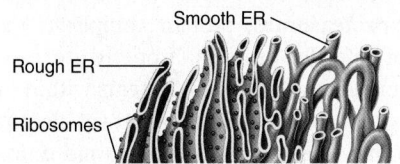 Smooth ER Rough ER Ribosomes	Granules composed of RNA and protein; some attached to ER, some free in cytosol	Synthesize polypeptides in both prokaryotes and eukaryotes
Endoplasmic reticulum (ER)	Network of internal membranes extending through cytoplasm	Synthesizes lipids and modifies many proteins; origin of intracellular transport vesicles that carry proteins
Smooth	Lacks ribosomes on outer surface	Lipid synthesis; drug detoxification; calcium ion storage
Rough	Ribosomes stud outer surface	Manufactures proteins
Golgi complex 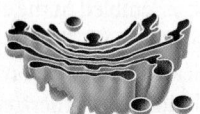	Stacks of flattened membrane sacs	Modifies proteins; packages secreted proteins; sorts other proteins to vacuoles and other organelles
Lysosomes 	Membranous sacs (in animals)	Contain enzymes that break down ingested materials; break down damaged or unneeded organelles and proteins
Vacuoles 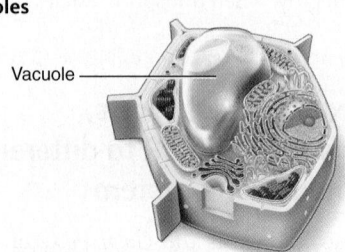 Vacuole	Membranous sacs (mostly in plants, fungi, algae)	Store materials, wastes, water; maintain hydrostatic pressure

TABLE 4-1 — Eukaryotic Cell Structures and Their Functions

STRUCTURE		DESCRIPTION	FUNCTION
Peroxisomes		Membranous sacs containing a variety of enzymes	Site of many diverse metabolic reactions; e.g., break down fatty acids
Mitochondria		Sacs consisting of two membranes; inner membrane is folded to form cristae and encloses matrix	Site of most reactions of cellular respiration; transformation of energy originating from glucose or lipids into ATP energy
Plastids (e.g., chloroplasts)		Double-membrane structure enclosing internal thylakoid membrane; chloroplasts contain chlorophyll in thylakoid membrane	Chloroplasts are site of photosynthesis; chlorophyll captures light energy; ATP and other energy-rich compounds are produced and then used to convert CO_2 to carbohydrate
CYTOSKELETON			
Microtubules		Hollow tubes made of subunits of tubulin protein	Provide structural support; have role in cell and organelle movement and cell division; components of cilia, flagella, centrioles, basal bodies
Microfilaments		Solid, rodlike structures consisting of actin protein	Provide structural support; play role in cell and organelle movement and cell division
Intermediate filaments		Tough fibers made of protein	Help strengthen cytoskeleton; stabilize cell shape
Centrioles		Pair of hollow cylinders located near nucleus; each centriole consists of nine microtubule triplets (9 × 3 structure)	Mitotic spindle forms between centrioles during animal cell division; may anchor and organize microtubule formation in animal cells; absent in most plant cells
Cilia		Relatively short projections extending from surface of cell; covered by plasma membrane; made of two central and nine pairs of peripheral microtubules (9 + 2 structure)	Movement of some unicellular organisms; used to move materials on surface of some tissues; important in cell signaling
Flagella		Long projections made of two central and nine pairs of peripheral microtubules (9 + 2 structure); extend from surface of cell; covered by plasma membrane	Cell locomotion by sperm cells and some unicellular organisms

© Cengage Learning

Camillo Golgi, who found a way to specifically stain this organelle. This membrane system modifies and sorts proteins that it receives from the ER and then packages them into transport vesicles destined to different components of the endomembrane system. Researchers have studied the function of the ER, Golgi complex, and other interacting organelles by radioactively labeling newly manufactured molecules and observing the timing of their movement through the cell. The radiolabeled proteins are first detected in the lumen of the rough ER. They then pass to the Golgi complex and then to some final destination such as the plasma membrane or lysosomes.

In many cells, the Golgi complex consists of stacks of flattened membranous sacs called **cisternae** (sing., *cisterna*). Each cisterna has an internal space, or lumen. The Golgi complex contains a number of separate compartments as well as some that are interconnected.

Each Golgi stack has three areas, referred to as the *cis face*, the *trans face*, and a *medial region* in between. Typically, the *cis* face (the entry surface) is located nearest the nucleus and receives materials from transport vesicles bringing molecules from the ER. The *trans* face (the exit surface) is closest to the plasma membrane. It packages molecules in vesicles and transports them out of the Golgi.

In a cross-sectional view like that in the TEM in FIGURE 4-14, many ends of the sheetlike layers of Golgi membranes are distended because they are filled with cell products, an arrangement characteristic of well-developed Golgi complexes in many cells. In some animal cells, the Golgi complex lies at one side of the nucleus; other animal cells and plant cells have many Golgi complexes dispersed throughout the cell.

Cells that secrete large amounts of *glycoproteins* have many Golgi stacks. (Recall from Chapter 3 that a glycoprotein is a

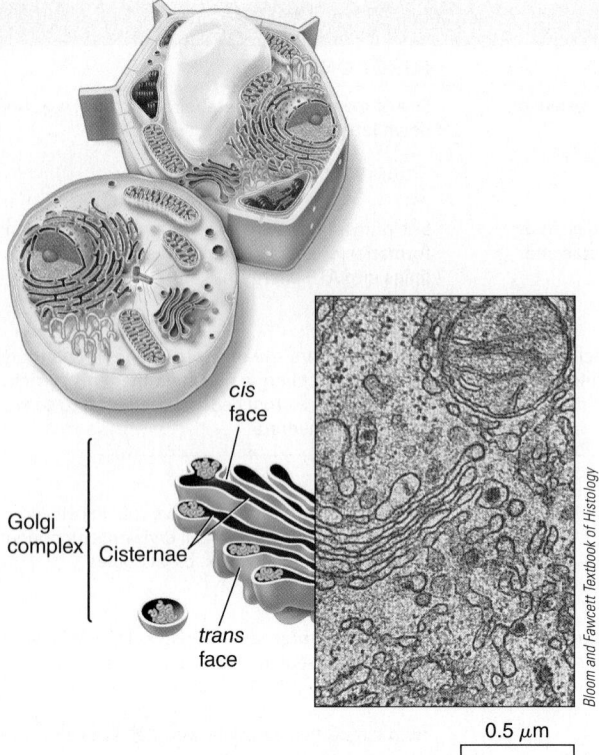

cis face

Golgi complex

Cisternae

trans face

0.5 μm

Bloom and Fawcett Textbook of Histology

Figure 4-14 TEM and an interpretive drawing of the Golgi complex
© Cengage Learning

protein with a covalently attached carbohydrate.) Golgi complexes of plant cells produce extracellular polysaccharides that are used as components of the cell wall. In animal cells, the Golgi complex manufactures lysosomes and complex glycoproteins that are secreted to form the extracellular matrix as well as mucus that coats epithelial cells.

Secretory proteins that are assembled in the rough ER lumen are first transported to the *cis* face of the Golgi complex as cargo in the small transport vesicles formed from the ER membrane. These proteins then progress through the Golgi stack, undergoing successive modifications until they reach the *trans* face of the Golgi, where they are packaged into vesicles and shipped to their destination membranes. Vesicles formed from the *cis* Golgi have also been found that move back to the ER as a way of recycling proteins involved in ER packaging and vesicle formation.

What are the mechanisms that move these proteins through the different components of the Golgi complex? One hypothesis holds that the glycoproteins are enclosed in new vesicles that shuttle them from one compartment to another within the Golgi complex. A competing hypothesis postulates that the cisternae themselves may move from *cis* to *trans* positions. In the latter case, new *cis* Golgi compartments would be continuously formed by the fusion of vesicles originating from the ER. As each new *cis* compartment is formed, the previous one moves outward in the *trans* direction until it becomes the *trans* compartment. Both hypotheses may be correct: glycoproteins may be transported by both methods.

Regardless of how proteins are moved through the Golgi complex, while there they are modified in different ways, resulting in the formation of complex biological molecules. For example, the carbohydrate part of a glycoprotein (first added to

proteins in the rough ER) may be modified. In some cases, the carbohydrate component may be a "sorting signal," a cellular zip code that tags the protein, routing it to a specific organelle.

Glycoproteins are packaged in transport vesicles in the *trans* face. These vesicles pinch off from the Golgi membrane and transport their contents to a specific destination. Vesicles transporting products for export from the cell fuse with the plasma membrane. The vesicle membrane becomes part of the plasma membrane, and the glycoproteins are secreted from the cell. Other vesicles may store glycoproteins for secretion at a later time, and still others are routed to various organelles of the endomembrane system such as the lysosomes. **FIGURE 14-15** illustrates the multiple paths followed by proteins that are processed through the Golgi complex. In summary, here is a typical sequence followed by a glycoprotein destined for secretion from the cell.

polypeptides synthesized on ribosomes → protein assembled and carbohydrate component added in lumen of ER → transport vesicles move glycoprotein to Golgi (*cis* face) → glycoprotein further modified in Golgi → in *trans* face, glycoproteins packaged in transport vesicles → glycoproteins transported to plasma membrane → contents released from cell

Lysosomes are compartments for digestion

Lysosomes are small sacs of digestive enzymes dispersed in the cytoplasm of most animal cells (**FIG. 4-16**). Researchers have identified about 40 different digestive enzymes in lysosomes.

Because lysosomal enzymes are active under rather acidic conditions, the lysosome maintains a pH of about 5 in its interior. The powerful enzymes and low pH that the lysosome maintains provide an excellent example of the importance of separating functions within the cell into different compartments. Under most normal conditions, the lysosome membrane confines its enzymes and their actions. However, some forms of tissue damage are related to "leaky" lysosomes.

Primary lysosomes are formed by budding from the Golgi complex. Their hydrolytic enzymes are synthesized in the rough ER. As these enzymes pass through the lumen of the ER, sugars attach to each molecule, identifying it as bound for a lysosome. This signal permits the Golgi complex to sort the enzyme to the lysosomes rather than to export it from the cell.

Lysosomes act by fusing their membranes with vesicles that contain material to be digested. For example, bacteria (or cellular debris) that are engulfed by scavenger cells are captured from the exterior by endocytotic vesicles (see Chapter 5) that form from the plasma membrane (see Figs. 4-16, 5-20, and 5-22). One or more primary lysosomes fuse with a vesicle containing the ingested material, forming a larger vesicle called a *secondary lysosome*. Powerful enzymes in the secondary lysosome come in contact with the ingested molecules and degrade them into their components. Lysosomes can also break down damaged organelles (such as mitochondria) that have been captured in vesicles, allowing their components to be recycled or used as an energy source.

In certain genetic diseases of humans, known as *lysosomal storage diseases*, one of the digestive enzymes normally present in lysosomes is absent. Its substrate (a substance the enzyme would normally break down) accumulates in the cell, which eventually

Proteins are routed through the endomembrane system by membrane-bound targeting signals.

1. Membrane-bound ribosomes insert proteins into the rough ER lumen.

2. Sugars are added in the ER lumen, forming glycoproteins.

3. Transport vesicles containing *cis* Golgi targeting signals on their surface dock with receptor molecules on the *cis* Golgi surface. Fusion of the vesicle with the Golgi membrane releases glycoproteins into a Golgi cisterna.

4. Glycoprotein sugars on proteins modified further in Golgi.

5. Glycoproteins are packaged on Golgi *trans* face into transport vesicles with targeting signals for the plasma membrane (secretory vesicles) or for lysosomes.

6. Secretory vesicles dock with targeting signals on the plasma membrane, triggering membrane fusion and release of contents from cell. Proteins and lipids from the secretory transport vesicle membrane become part of plasma membrane.

7. Endocytic vesicles containing lysosome targeting signals bud from plasma membrane.

8. Lysosomes released from *trans* Golgi fuse with endocytic vesicles.

9. Lysosomes released from *trans* Golgi may also fuse with vesicles containing damaged organelles, degrading and recycling contents.

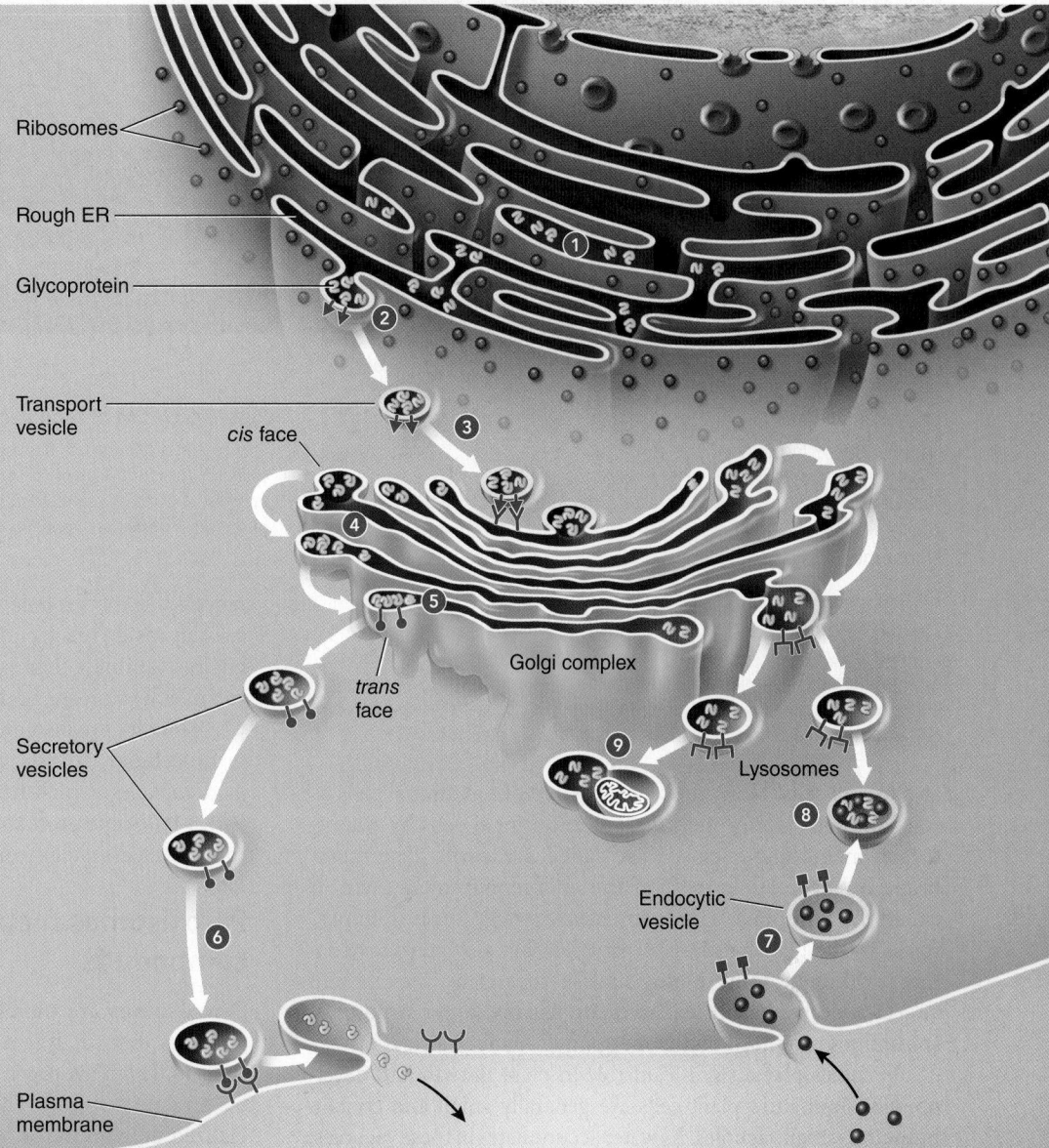

Figure 4-15 *Animation* **The endomembrane system**

This diagram illustrates the sorting and passage of glycoproteins destined for secretion or for packaging in lysosomes. Proteins synthesized by membrane-bound ribosomes are inserted into the ER lumen where sugars are added, forming glycoproteins. They are packaged into vesicles and transported to the *cis* Golgi membrane surface. Targeting signals on the vesicle membrane surface dock with receptor proteins on the Golgi membrane. Following membrane fusion, the vesicle cargo glycoproteins are released into the Golgi complex, where their sugars are further modified. At the Golgi *trans* face, the glycoproteins are sorted into vesicles containing targeting signals either for the plasma membrane or for lysosomes. Vesicles containing secretory proteins dock with appropriate targeting signals on the plasma membrane, triggering membrane fusion and release of the proteins to the exterior. Vesicles containing lysosomal proteins are packaged in lysosomal vesicles, which can fuse with either endocytic vesicles or membranes that enclose damaged organelles.

PREDICT What would happen to proteins destined for secretion if the targeting signal on their vesicles derived from the *trans* Golgi were missing?
© Cengage Learning

interferes with cell activities. An example is Tay-Sachs disease, an inherited disease in which a normal lipid cannot be broken down (discussed in Chapter 16). The accumulation of this lipid in brain cells causes severe intellectual disability, blindness, and death before age 4.

Vacuoles are large, fluid-filled sacs with a variety of functions

Although lysosomes have been identified in almost all kinds of animal cells, biologists have not identified them in plant

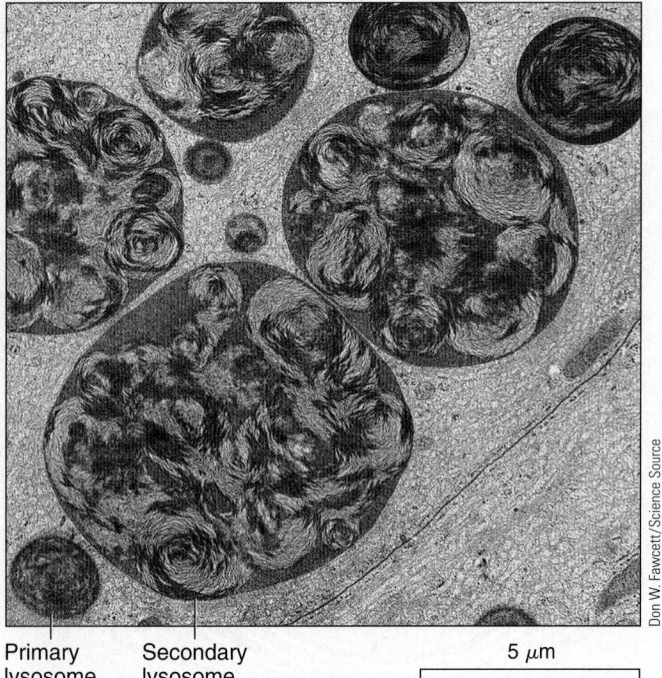

Primary lysosome   Secondary lysosome   5 μm

Don W. Fawcett/Science Source

Figure 4-16 Lysosomes

The dark vesicles in this TEM are lysosomes, compartments that separate powerful digestive enzymes from the rest of the cell. Primary lysosomes bud off from the Golgi complex. After a lysosome takes in material to be digested, it is known as a secondary lysosome. The large vesicles shown here are secondary lysosomes containing various materials being digested.

and fungal cells. Many functions carried out in animal cells by lysosomes are performed in plant and fungal cells by a large, single, membrane-enclosed sac called a **vacuole**. The membrane of the vacuole, part of the endomembrane system, is called the **tonoplast**. The term *vacuole*, which means "empty," refers to the fact that these organelles have no internal structure. Although some biologists use the terms *vacuole* and *vesicle* interchangeably, vacuoles are usually larger structures, sometimes produced by the merging of many vesicles.

Vacuoles play a significant role in plant growth and development. Immature plant cells are generally small and contain numerous small vacuoles. As water accumulates in these vacuoles, they tend to coalesce, forming a large central vacuole. Because the vacuole contains a high concentration of solutes (dissolved materials), it takes in water and pushes outward on the cell wall. This hydrostatic pressure, called *turgor pressure,* provides much of the mechanical strength of plant cells. A plant cell increases in size mainly by adding water to this central vacuole. As much as 80% of the volume of a plant cell may be occupied by a large central vacuole containing water, stored food, salts, pigments, and metabolic wastes (see Fig. 4-9). The vacuole is important in maintaining the intracellular environment. For example, it helps maintain appropriate pH by taking in excess hydrogen ions. The vacuole may serve as a storage compartment for inorganic compounds. In seeds, vacuoles store molecules such as proteins.

Plants lack organ systems for disposing of toxic metabolic waste products. Plant vacuoles are like lysosomes in that they contain hydrolytic enzymes and break down wastes as well as unneeded organelles and other cell components. Wastes may be

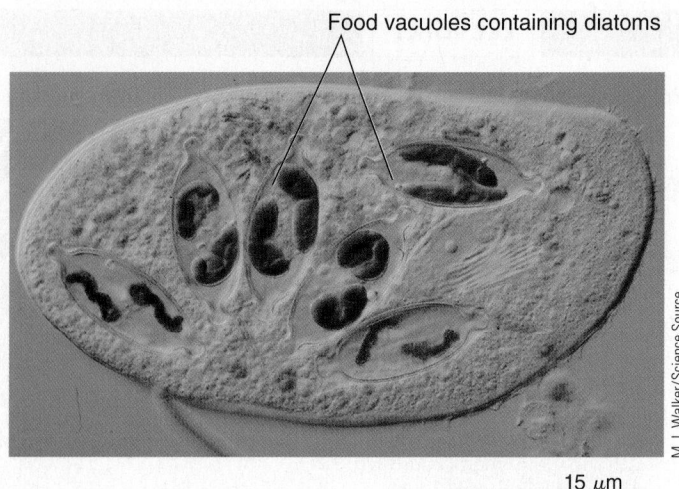

Food vacuoles containing diatoms

15 μm

M. I. Walker/Science Source

Figure 4-17 LM of food vacuoles

This protist, *Chilodonella*, has ingested many small, photosynthetic protists called diatoms (*dark areas*) that have been enclosed in food vacuoles. From the number of diatoms scattered about its cell, one might conclude that *Chilodonella* has a rather voracious appetite.

recycled in the vacuole, or they may aggregate and form small crystals inside the vacuole. Compounds that are noxious to herbivores (animals that eat plants) may be stored in some plant vacuoles as a means of defense.

Vacuoles are also present in many types of animal cells and in unicellular protists such as protozoa. Most protozoa have *food vacuoles,* which fuse with lysosomes that digest the food (FIG. 4-17). Some protozoa also have **contractile vacuoles**, which remove excess water from the cell (discussed in Chapter 26).

Peroxisomes metabolize small organic compounds

Peroxisomes are membrane-enclosed organelles containing enzymes that catalyze a diverse assortment of metabolic reactions in which hydrogen is transferred from various compounds to oxygen (FIG. 4-18). They are formed by budding from a specialized domain in the smooth ER. Peroxisomes get their name from the fact that during these oxidation reactions, they produce hydrogen peroxide (H_2O_2). Hydrogen peroxide detoxifies certain compounds, but were this compound to escape from the peroxisome, it would damage other membranes in the cell. Peroxisomes contain the enzyme *catalase*, which rapidly splits excess hydrogen peroxide to water and oxygen, rendering it harmless.

Peroxisomes are found in large numbers in cells that synthesize, store, or degrade lipids. One of their main functions is to break down fatty acid molecules. Peroxisomes synthesize certain phospholipids that are components of the insulating covering of nerve cells. In fact, certain neurological disorders occur when peroxisomes do not perform this function. Mutations that cause synthesis of abnormal membrane lipids by peroxisomes are linked to some forms of intellectual disability.

When yeast cells are grown in an alcohol-rich medium, they manufacture large peroxisomes containing an enzyme that degrades the alcohol. Peroxisomes in human liver and kidney

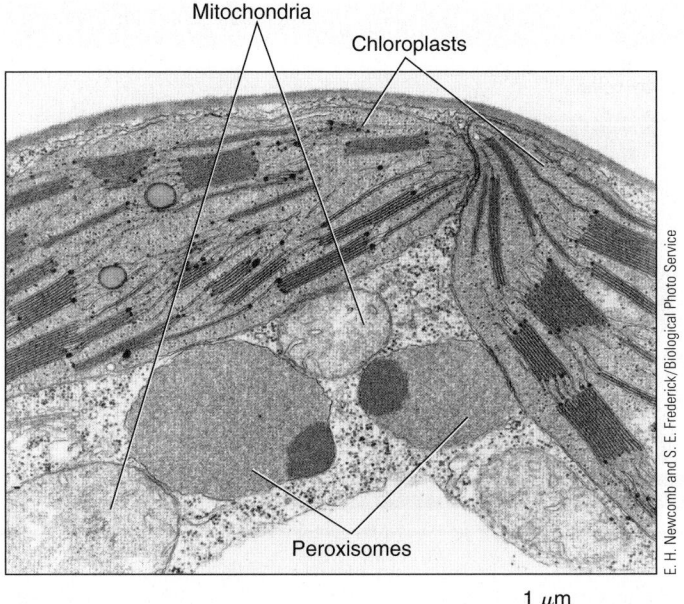

Mitochondria

Chloroplasts

Peroxisomes

1 µm

E. H. Newcomb and S. E. Frederick/Biological Photo Service

Figure 4-18 Peroxisomes

In this TEM of a tobacco (*Nicotiana tabacum*) leaf cell, peroxisomes are in close association with chloroplasts and mitochondria. These organelles may cooperate in carrying out some metabolic activities.

cells detoxify certain toxic compounds, including ethanol, the alcohol in alcoholic beverages.

In plant seeds, specialized peroxisomes, called *glyoxysomes,* contain enzymes that convert stored fats to sugars. The sugars are used by the young plant as an energy source and as a component for synthesizing other compounds. Animal cells lack glyoxysomes and cannot convert fatty acids into sugars.

Mitochondria and chloroplasts are energy-converting organelles

The energy a cell obtains from its environment is usually in the form of chemical energy in food molecules (such as glucose) or in the form of light energy. These types of energy must be converted to forms that cells can use more conveniently. Some energy conversions occur in the cytosol, but other types take place in **mitochondria** and **chloroplasts**, organelles specialized to facilitate the conversion of energy from one form to another.

Chemical energy is most commonly stored in ATP. Recall from Chapter 3 that the chemical energy of ATP can be used to drive a variety of chemical reactions in the cell. FIGURE 4-19 summarizes the main activities relating to energy conversion that take place in mitochondria, found in almost all eukaryotic cells, and in chloroplasts, found only in algae and certain plant cells.

Mitochondria and chloroplasts grow and reproduce themselves. They contain small amounts of DNA that code for a small number of the proteins found in these organelles. These proteins are synthesized by mitochondrial or chloroplast ribosomes. Interestingly, these ribosomes are similar to the ribosomes of prokaryotes. The existence of a separate set of ribosomes and DNA molecules in mitochondria and chloroplasts and their similarity in size to many bacteria provide support for **serial endosymbiosis** (discussed in Chapters 21 and 26; see Figs. 21-8

and 26-2). According to this hypothesis, mitochondria and chloroplasts evolved from prokaryotes that took up residence inside larger eukaryotic cells. Eventually, these symbiotic prokaryotes lost the ability to function as autonomous organisms.

Mitochondria make ATP through aerobic respiration

Virtually all eukaryotic cells (plant, animal, fungal, and protist) contain complex organelles called **mitochondria** (sing., *mitochondrion*). These organelles are the site of **aerobic respiration**, an oxygen-requiring process that includes most of the reactions that convert the chemical energy present in certain foods to ATP (discussed in Chapter 8). During aerobic respiration, carbon, hydrogen, and oxygen atoms are removed from food molecules, such as glucose, and converted to carbon dioxide and water.

Mitochondria are most numerous in cells that are very active and therefore have high energy requirements. More than 1000 mitochondria have been counted in a single liver cell! These organelles vary in size, ranging from 2 to 8 µm in length, and change size and shape rapidly. Mitochondria usually give rise to other mitochondria by growth and subsequent division.

Each mitochondrion is enclosed by a double membrane, which forms two *different* compartments within the organelle: the intermembrane space and the matrix (FIG. 4-20; we will provide more detailed descriptions of mitochondrial structure in Chapter 8). The **intermembrane space** is the compartment formed between the outer and inner mitochondrial membranes. The **matrix,** the compartment enclosed by the inner mitochondrial membrane, contains enzymes that break down food molecules and convert their energy to other forms of chemical energy.

The *outer mitochondrial membrane* is smooth and allows many small molecules to pass through it. By contrast, the *inner mitochondrial membrane* has numerous folds and strictly regulates the types of molecules that can move across it. The folds, called **cristae** (sing., *crista*), extend into the matrix. Cristae greatly increase the surface area of the inner mitochondrial membrane, providing a surface for the chemical reactions that transform the chemical energy in food molecules into the energy of ATP. The membrane contains the enzymes and other proteins needed for these reactions.

Mitochondria play an important role in programmed cell death, or **apoptosis.** Unlike **necrosis,** which is uncontrolled cell death that causes inflammation and damages other cells, apoptosis is a normal part of development and maintenance. For example, during the metamorphosis of a tadpole to a frog, the cells of the tadpole tail must die. The hand of a human embryo is webbed until apoptosis destroys the tissue between the fingers. Cell death also occurs in the adult. For example, cells that are no longer functional because they have aged or become damaged are destroyed by apoptotic mechanisms and replaced by new cells.

Mitochondria initiate cell death in several different ways. For example, they can interfere with energy metabolism or activate enzymes that mediate cell destruction. When a mitochondrion is injured, large pores open in its membrane, and cytochrome *c*, a protein important in energy conversions, is released into the cytoplasm. Cytochrome *c* triggers apoptosis by activating

Mitochondria and chloroplasts convert energy into forms that can be used by cells.

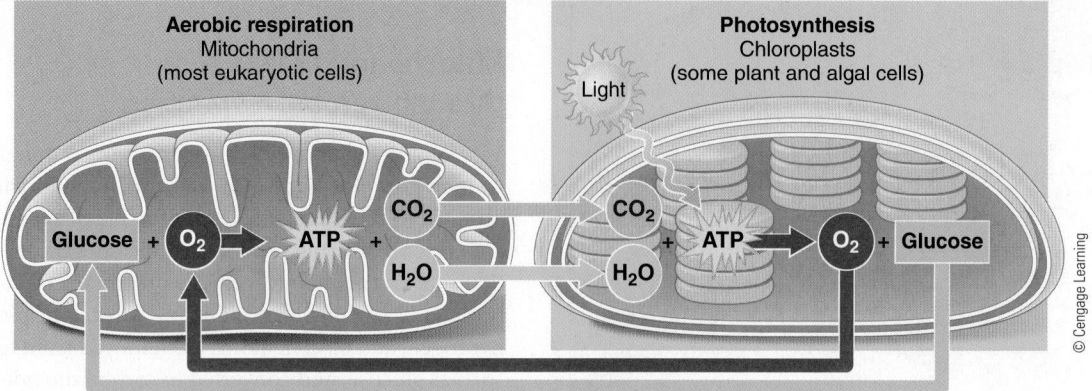

Figure 4-19 Aerobic respiration and photosynthesis

Aerobic respiration takes place in the mitochondria of virtually all eukaryotic cells. In this process, some of the chemical energy in glucose is transferred to ATP. Photosynthesis, which is carried out in chloroplasts in some plant and algal cells, converts light energy to ATP and to other forms of chemical energy. This energy is used to synthesize glucose from carbon dioxide and water.

PREDICT What do you think would happen in the leaf cells of a plant placed in the dark (but supplied with O_2, CO_2, and H_2O) for several days?

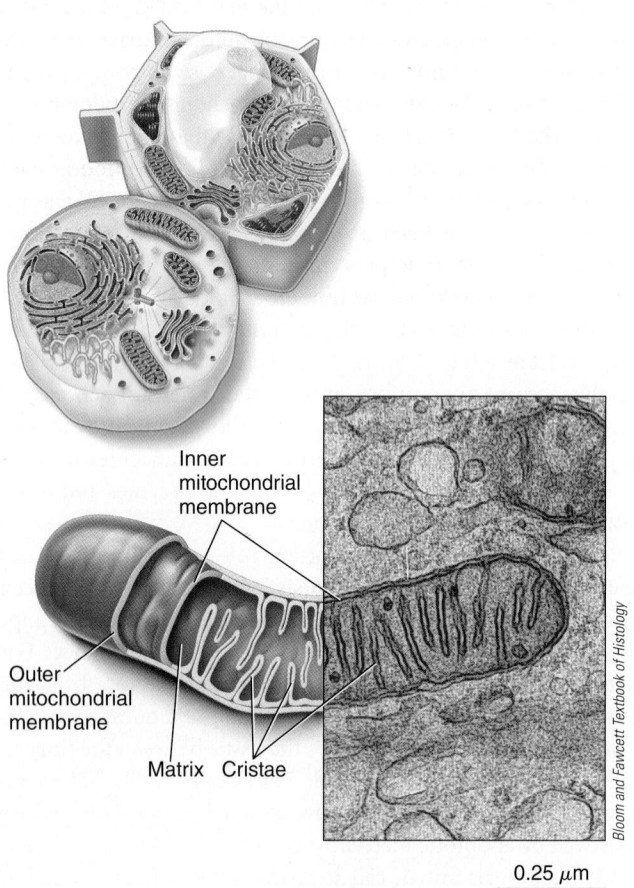

Inner mitochondrial membrane

Outer mitochondrial membrane

Matrix   Cristae

0.25 μm

Bloom and Fawcett Textbook of Histology

Figure 4-20 *Animation* Mitochondria

Aerobic respiration takes place within mitochondria. Cristae are evident in the TEM as well as in the drawing. The drawing shows the relationship between the inner and outer mitochondrial membranes.
© Cengage Learning

enzymes known as **caspases**, which cut up vital compounds in the cell.

Inappropriate inhibition of apoptosis may contribute to a variety of diseases, including cancer. On the other hand, too much apoptosis may deplete needed cells and lead to death of brain cells associated with Alzheimer's disease, Parkinson's disease, and stroke. Mutations that promote apoptosis may be an important mechanism in mammalian aging. Pharmaceutical companies are developing drugs that block apoptosis. However, cell dynamics are extremely complex, and blocking apoptosis could lead to a worse fate, including cancer.

Each mitochondrion in a mammalian cell has 5 to 10 identical, circular molecules of DNA, accounting for up to 1% of the total DNA in the cell. Mitochondrial DNA mutates far more frequently than nuclear DNA. Mutations in mitochondrial DNA have been associated with certain genetic diseases, including a form of young adult blindness, and certain types of progressive muscle degeneration. Mitochondria also affect health and aging by leaking electrons. These electrons form **free radicals**, which are toxic, highly reactive compounds with unpaired electrons. The electrons bond with other compounds in the cell, interfering with normal function.

Chloroplasts convert light energy to chemical energy through photosynthesis

Certain plant and algal cells carry on **photosynthesis**, a set of reactions during which light energy is transformed into the chemical energy of glucose and other carbohydrates. **Chloroplasts** are organelles that contain **chlorophyll**, a green pigment that traps light energy for photosynthesis. Chloroplasts also contain a variety of light-absorbing yellow and orange pigments known as **carotenoids** (see Chapter 3). A unicellular alga may have only

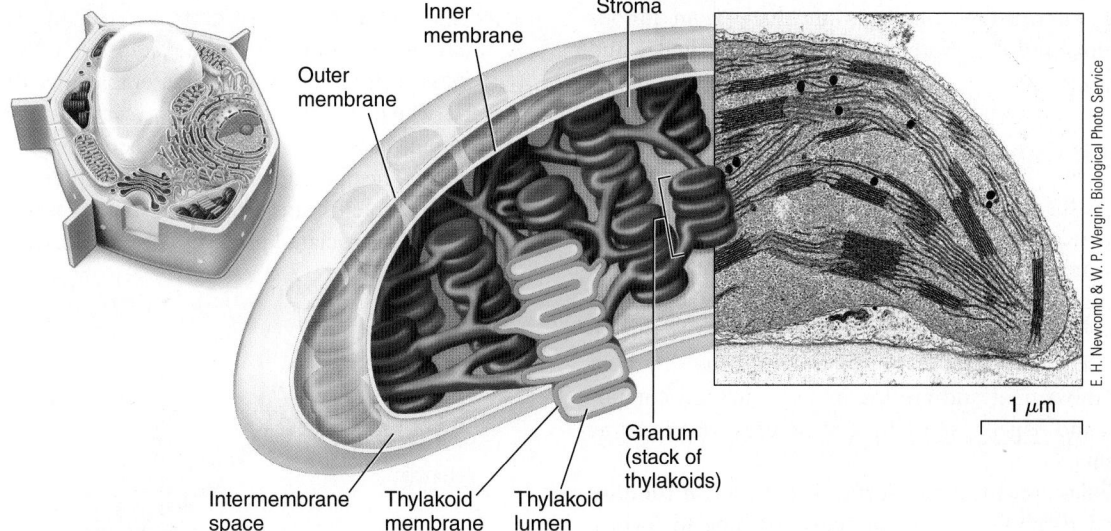

Figure 4-21 *Animation* **A chloroplast, the organelle of photosynthesis**

This TEM shows part of a chloroplast from a corn leaf cell. Chlorophyll and other photosynthetic pigments are in the thylakoid membranes. One granum is cut open to show the thylakoid lumen. The inner chloroplast membrane may or may not be continuous with the thylakoid membrane (*as shown*).
© Cengage Learning

a single large chloroplast, whereas a leaf cell may have 20 to 100. Chloroplasts tend to be somewhat larger than mitochondria, with lengths usually ranging from about 5 to 10 μm or longer.

Chloroplasts are typically disc-shaped structures and, like mitochondria, have a system of folded membranes (**FIG. 4-21**; we will provide a more detailed description of chloroplast structure in Chapter 9). Two membranes enclose the chloroplast and separate it from the cytosol. The inner membrane encloses a fluid-filled space called the **stroma,** which contains enzymes that produce carbohydrates from carbon dioxide and water, using energy trapped from sunlight. A system of internal membranes, which consists of an interconnected set of flat, disclike sacs called **thylakoids**, is suspended in the stroma. The thylakoids are arranged in stacks called **grana** (sing., *granum*).

The thylakoid membrane encloses the innermost compartments within the chloroplast, the **thylakoid lumen.** Chlorophyll is present in the thylakoid membrane, which, like the inner mitochondrial membranes, is involved in the formation of ATP. Energy absorbed from sunlight by the chlorophyll molecules excites electrons; the energy in these excited electrons is then used to produce ATP and other molecules that transfer chemical energy.

Chloroplasts belong to a group of organelles, known as **plastids**, that produce and store food materials in cells of plants and algae. All plastids develop from **proplastids**, precursor organelles found in less specialized plant cells, particularly in growing, undeveloped tissues. Depending on the specific functions a cell will eventually have, its proplastids can develop into a variety of specialized mature plastids. They are extremely versatile organelles; in fact, under certain conditions even mature plastids can convert from one form to another.

Chloroplasts are produced when proplastids are stimulated by exposure to light. **Chromoplasts** contain pigments that give certain flowers and fruits their characteristic colors, and these colors attract animals that serve as pollinators or as seed dispersers. **Leukoplasts** are unpigmented plastids; they include

amyloplasts (see Fig. 3-9), which store starch in the cells of many seeds, roots, and tubers (such as white potatoes).

CHECKPOINT 4.5

- *How do the structure and function of rough ER differ from those of smooth ER?*
- *How does the structure of the Golgi complex allow it to carry out its functions?*
- **CONNECT** *Outline the sequence of events that must take place for a protein to be manufactured and then secreted from the cell.*
- **CONNECT** *In what ways are chloroplasts like mitochondria? How are they different?*

4.6 THE CYTOSKELETON

LEARNING OBJECTIVES

12 Contrast the structure and functions of the cytoskeleton with the structure and functions of the membranous components of the cell.

13 Relate the structure of cilia and flagella to their functions.

Scientists watching cells growing in the laboratory see that they frequently change shape and that many types of cells move about. The **cytoskeleton,** a dense network of protein fibers, gives cells mechanical strength, shape, and their ability to move (**FIG. 4-22**). The cytoskeleton also functions in cell division and in the transport of materials within the cell.

The cytoskeleton is highly dynamic and constantly changing. Its framework is made of three major types of protein filaments: microtubules, microfilaments, and intermediate filaments. Both microfilaments and microtubules are formed from beadlike, globular protein subunits, which can be rapidly assembled and disassembled. Intermediate filaments are made from fibrous protein subunits and are more stable than microtubules and microfilaments.

Microtubules are hollow cylinders

Microtubules, the thickest filaments of the cytoskeleton, are rigid, hollow rods about 25 nm in outside diameter and up to several micrometers in length. In addition to playing a structural role in the cytoskeleton, these extremely adaptable structures are involved in the movement of chromosomes during cell division. They serve as tracks for several other kinds of intracellular movement and are the major structural components of cilia and flagella, specialized structures used in some cell movements.

Microtubules consist of two forms of the protein **tubulin:** α-tubulin and β-tubulin. These proteins combine to form a dimer. (Recall from Chapter 3 that a dimer forms from the association of two simpler units, referred to as monomers.) A microtubule elongates by the addition of tubulin dimers (FIG. 4-23). Microtubules are disassembled by the removal of dimers, which are recycled to form new microtubules. Each microtubule has polarity, and its two ends are referred to as *plus* and *minus*. The plus end elongates more rapidly.

Other proteins are important in microtubule function. **Microtubule-associated proteins (MAPs)** are classified into two

---

KEY POINT

The cytoskeleton consists of networks of several types of fibers that support the cell and are important in cell movement.

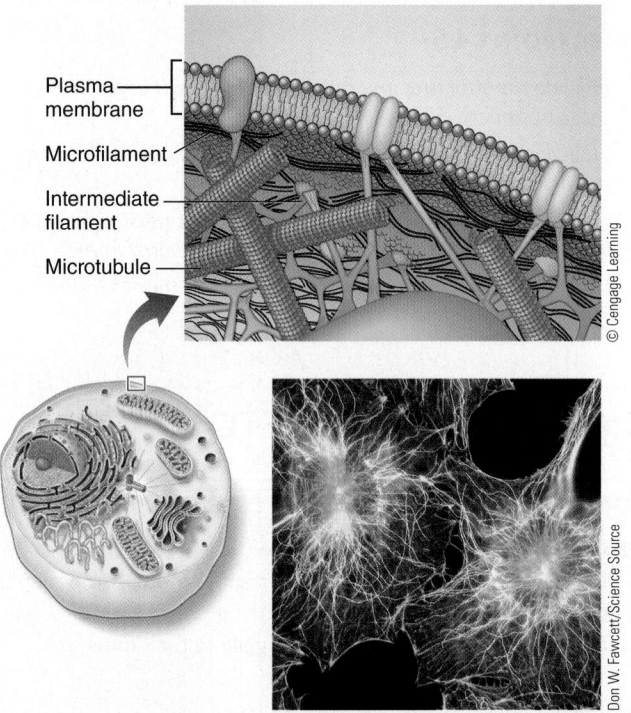

Figure 4-22 The cytoskeleton

The cytoskeleton of eukaryotic cells consists of networks of several types of fibers, including microtubules, microfilaments, and intermediate filaments. The cytoskeleton contributes to the shape of the cell, anchors organelles, and sometimes rapidly changes shape during cell locomotion. This fluorescent LM shows the cytoskeleton of two fibroblast cells (microtubules, *yellow;* microfilaments, *blue;* nuclei, *green*).

PREDICT What might happen in a cell in which all the cytoskeletal fibers associated with the plasma membrane were removed?

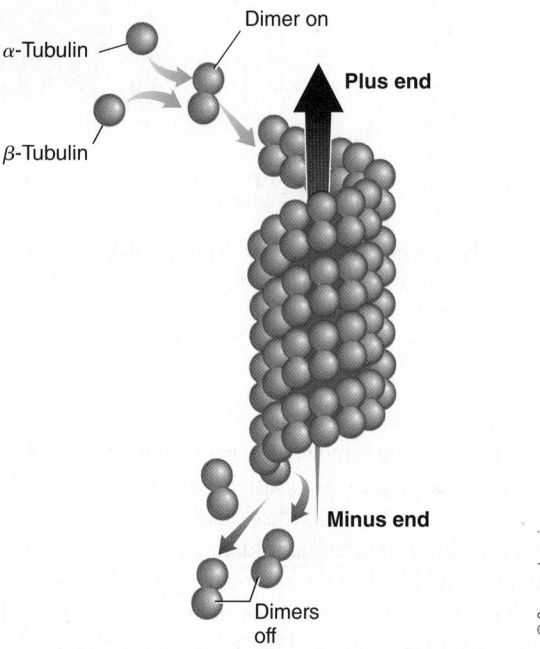

(a) Microtubules are manufactured in the cell by adding dimers of α-tubulin and β-tubulin to an end of the hollow cylinder. Notice that the cylinder has polarity. The end shown at the top of the figure is the fast-growing, or plus, end; the opposite end is the minus end. Each turn of the spiral requires 13 dimers.

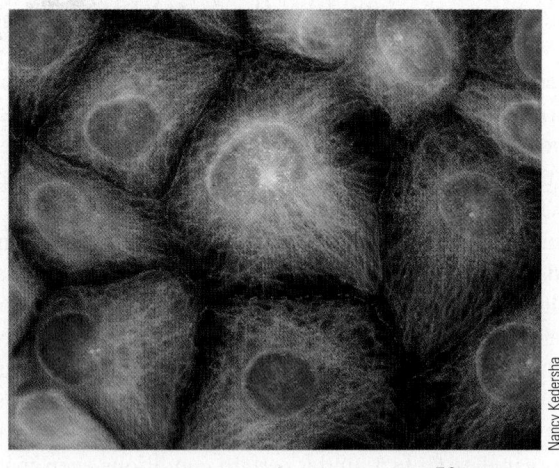

50 μm

(b) Fluorescent LM showing microtubules in *green.* A microtubule-organizing center (*pink dot*) is visible beside or over most of the cell nuclei (*blue*).

Figure 4-23 Organization of microtubules

groups: structural MAPs and motor MAPs. *Structural MAPs* may help regulate microtubule assembly, and they cross-link microtubules to other cytoskeletal polymers. *Motor MAPs* use ATP energy to produce movement.

What are the mechanisms by which organelles and other materials move within the cell? Nerve cells typically have long extensions called axons that transmit signals to other nerve cells, muscle cells, or cells that produce hormones. Because of the axon's length and accessibility and because other cells use similar transport mechanisms, researchers have used the axon as a model for studying the transport of organelles within the cell. Many organelles, including mitochondria, transport vesicles, and secretory vesicles, are attached by molecular motor proteins to microtubules that serve as tracks to different cellular locations.

One motor protein, **kinesin**, moves organelles toward the plus end of a microtubule (**FIG. 4-24**). **Dynein,** another motor protein, transports organelles in the opposite direction, toward the minus end. This dynein movement is referred to as *retrograde transport*. A protein complex called *dynactin* is also required for retrograde transport. Dynactin is an adapter protein that links dynein to the microtubule and the organelle. At times, for example, during peroxisome transport, kinesins and dyneins may work together.

Centrosomes and centrioles function in cell division

For microtubules to act as a structural framework or participate in cell movement, they must be anchored to other parts of the cell. In nondividing cells, the minus ends of microtubules appear to be anchored in regions called **microtubule-organizing centers** (**MTOCs;** see Fig. 4-23b). In animal cells, the main MTOC is the cell center, or **centrosome**.

In many cells, including almost all animal cells, the centrosome contains two structures called **centrioles** (**FIG. 4-25**). These structures are oriented within the centrosome at right angles to each other. They are known as *9 × 3 structures* because they consist of nine sets of three attached microtubules arranged to form a hollow cylinder. The centrioles are duplicated before cell division and may play a role in some types of microtubule assembly. Most plant cells and fungal cells have an MTOC but lack centrioles, which suggests that centrioles are not essential to most microtubule assembly processes and that alternative assembly mechanisms are present.

The ability of microtubules to assemble and disassemble rapidly is seen during cell division, when much of the cytoskeleton disassembles (discussed in Chapter 10). At that time, tubulin subunits organize into a structure called the **mitotic spindle**, which serves as a framework for the orderly distribution of chromosomes during cell division.

Cilia and flagella are composed of microtubules

Thin, movable structures, important in cell movement, project from surfaces of many cells. If a cell has one, or only a few, of these appendages and if they are long (typically about 200 μm)

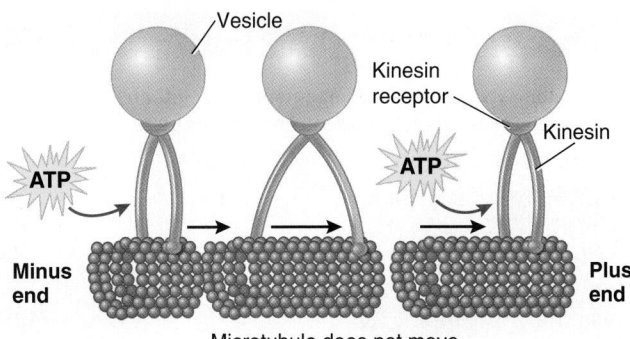

Figure 4-24 *Animation* **A model of a kinesin motor**

A kinesin molecule attaches to a specific receptor on the vesicle. Energy from ATP powers the kinesin molecule so that it changes its conformation and "walks" along the microtubule, carrying the vesicle along. (Size relationships are exaggerated for clarity.)
© Cengage Learning

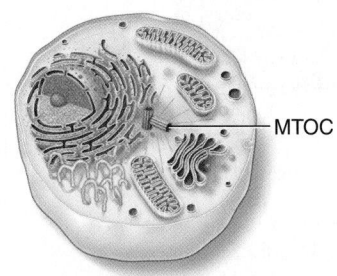

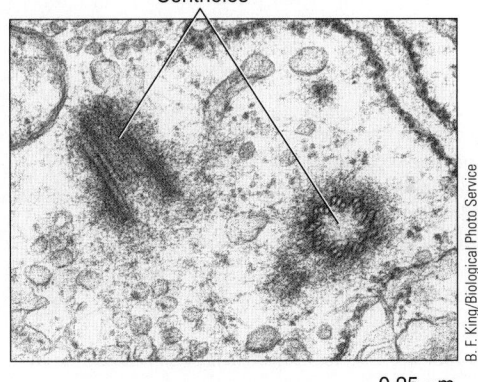

(a) In the TEM, the centrioles are positioned at right angles to each other, near the nucleus of a nondividing animal cell.

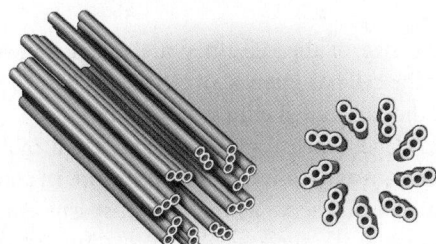

(b) Note the 9 × 3 arrangement of microtubules. The centriole on the right has been cut transversely.

Figure 4-25 Centrioles
© Cengage Learning

A cilium consists of a 9 + 2 arrangement of microtubules surrounded by the plasma membrane; dynein proteins move the microtubules by forming and breaking cross bridges on adjacent pairs of microtubules.

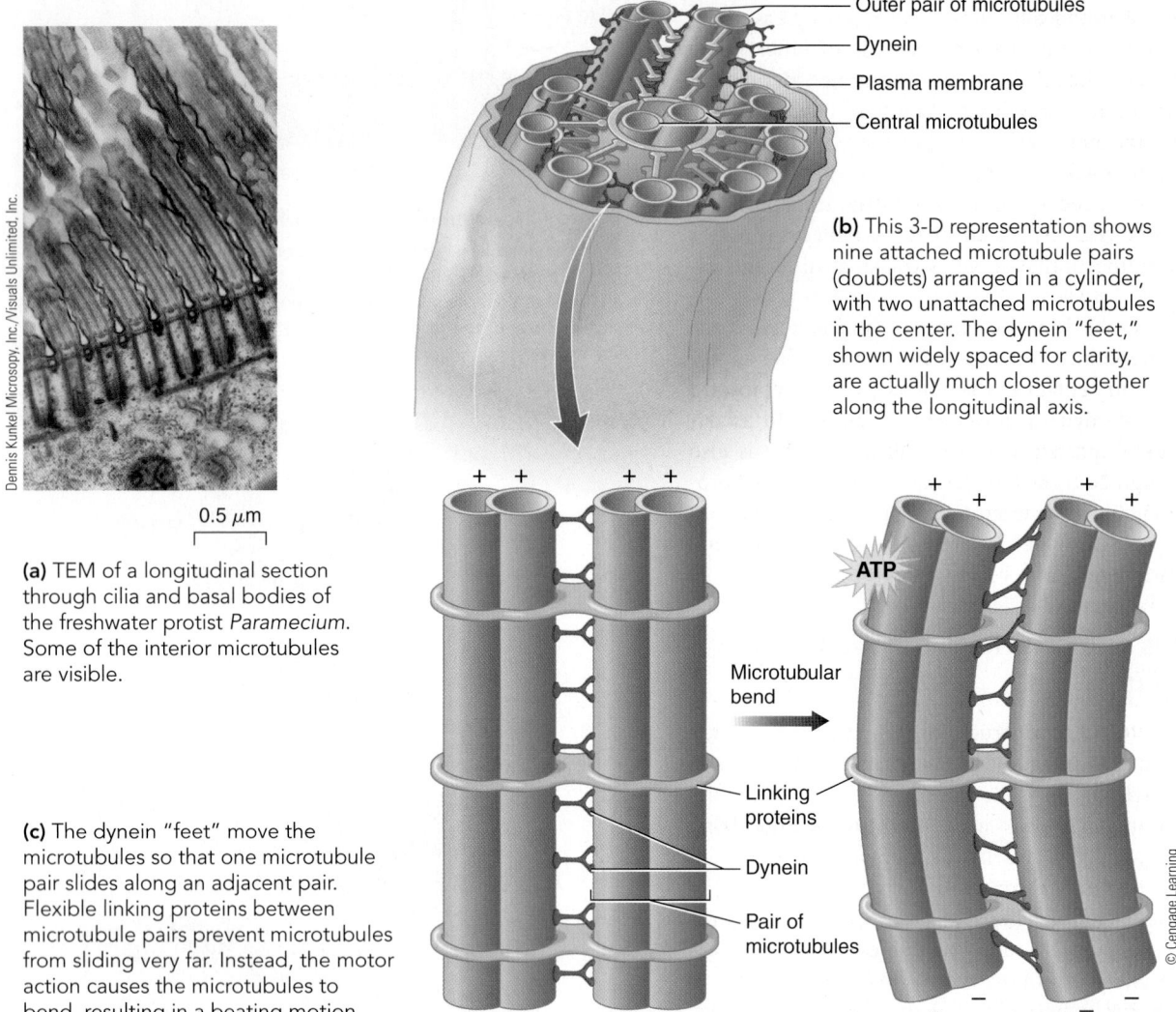

Outer pair of microtubules

Dynein

Plasma membrane

Central microtubules

(b) This 3-D representation shows nine attached microtubule pairs (doublets) arranged in a cylinder, with two unattached microtubules in the center. The dynein "feet," shown widely spaced for clarity, are actually much closer together along the longitudinal axis.

Dennis Kunkel Microsopy, Inc./Visuals Unlimited, Inc.

0.5 μm

(a) TEM of a longitudinal section through cilia and basal bodies of the freshwater protist *Paramecium.* Some of the interior microtubules are visible.

ATP

Microtubular bend

(c) The dynein "feet" move the microtubules so that one microtubule pair slides along an adjacent pair. Flexible linking proteins between microtubule pairs prevent microtubules from sliding very far. Instead, the motor action causes the microtubules to bend, resulting in a beating motion.

Linking proteins

Dynein

Pair of microtubules

© Cengage Learning

Figure 4-26 Structure and movement of cilia

PREDICT How might the action of cilia be affected if the linking proteins were inflexible? if they were absent?

relative to the size of the cell, they are called **flagella** (sing., *flagellum*). If the cell has many short (typically 2 to 10 μm long) appendages, they are called **cilia** (sing., *cilium*).

Cilia and flagella help unicellular and small multicellular organisms move through a watery environment. In animals and certain plants, flagella serve as the tails of sperm cells. In animals, cilia occur on the surfaces of cells that line internal ducts of the body (such as respiratory passageways). Cells use cilia to move liquids and particles across the cell surface. Investigators have shown that cilia also serve as the cell's antenna and play important roles in cell signaling.

Eukaryotic cilia and flagella are structurally alike (but different from bacterial flagella). Each consists of a slender, cylindrical stalk covered by an extension of the plasma membrane. The core of the stalk contains a group of microtubules arranged so that there are nine attached pairs of microtubules around the circumference and two unpaired microtubules in the center (**FIG. 4-26**). This *9 + 2 arrangement* of microtubules is characteristic of virtually all eukaryotic cilia and flagella.

The microtubules in cilia and flagella move by sliding in pairs past each other. The sliding force is generated by dynein proteins, powered by ATP. The dynein "feet" move the microtubule pairs by forming and breaking cross bridges on adjacent pairs of microtubules. Each microtubule pair "walks" along its neighbor. Flexible linking proteins between microtubule pairs prevent microtubules from sliding very far. As a result, the motor action causes the microtubules to bend, resulting in a beating motion (see Fig. 4-26c). Cilia typically move like the arms of a swimmer, alternating a power stroke in one direction with a recovery stroke in the opposite direction. They exert a force that is parallel to the cell surface. In contrast, a flagellum moves like a whip, exerting a force perpendicular to the cell surface.

Each cilium or flagellum is anchored in the cell by a **basal body,** which has nine sets of three attached microtubules in a cylindrical array (9×3 *structure*). The basal body appears to be the organizing structure for the cilium or flagellum when it first begins to form. However, experiments have shown that as growth proceeds, the tubulin subunits are added much faster to the tips of the microtubules than to the base.

Basal bodies and centrioles may be functionally related as well as structurally similar. In fact, centrioles are typically found in the cells of eukaryotic organisms that produce flagellated or ciliated cells; they include animals, certain protists, a few fungi, and a few plants. Both basal bodies and centrioles replicate themselves.

Almost every vertebrate cell has a **primary cilium,** a single cilium on the cell surface that serves as a cellular antenna. The primary cilium has receptors on its surface that bind with specific molecules outside the cell or on surfaces of other cells. Primary cilia play important roles in many signaling pathways that regulate growth and specialization of cells during embryonic development. They also help maintain healthy tissues. Malfunction of primary cilia has been associated with several human disorders, including developmental defects, degeneration of the retina, and polycystic kidney disease.

Microfilaments consist of intertwined strings of actin

Microfilaments, also called *actin filaments*, are flexible, solid fibers about 7 nm in diameter. Each microfilament consists of two intertwined polymer chains of beadlike **actin** molecules (**FIG. 4-27**). Microfilaments are linked with one another and with other proteins by linker proteins. They form bundles of fibers that provide mechanical support for various cell structures.

In many cells, a network of microfilaments is visible just inside the plasma membrane, a region called the *cell cortex.* Microfilaments give the cell cortex a gel-like consistency compared to the more fluid state of the cytosol deeper inside the cell. The microfilaments in the cell cortex help determine the shape of the cell and are important in its movement.

Microfilaments themselves cannot contract, but they can generate movement in two ways: by rapidly assembling and disassembling or by functioning with molecular motors. Muscle cells have two types of specialized filaments: one composed mainly of the protein **myosin** and another composed mainly of the protein actin. ATP bound to myosin provides energy for

(a) A microfilament consists of two intertwined strings of beadlike actin molecules.

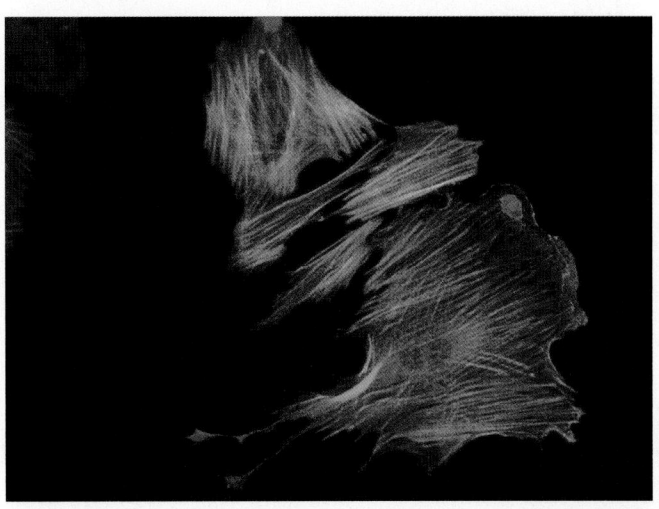

100 μm

(b) Many bundles of microfilaments (*green*) are evident in this fluorescent LM of fibroblasts, cells found in connective tissue.

Figure 4-27 Microfilaments
© Cengage Learning

muscle contraction. When ATP is hydrolyzed to ADP, myosin binds to actin and causes the microfilament to slide. When thousands of filaments slide in this way, the muscle cell shortens. Thus, ATP, actin, and myosin generate the forces that contract muscles (discussed in Chapter 40).

In nonmuscle cells, actin also associates with myosin, forming contractile structures involved in various cell movements. For example, in animal cell division, contraction of a ring of actin associated with myosin constricts the cell, forming two daughter cells (discussed in Chapter 10).

Some cells change their shape quickly in response to changes in the outside environment. Amoebas, human white blood cells, and cancer cells are among the many cell types that can creep along a surface, a process that includes changes in shape. Such responses depend on external signals that affect microfilament, as well as microtubule, assembly. Actin filaments push the plasma membrane outward, forming cytoplasm-filled bulges called **pseudopodia** ("false feet"). The pseudopodia adhere to the surface. Contractions of microfilaments at the opposite end of the cell force the cytoplasm forward in the direction of locomotion. Microtubules, myosin, and other proteins also appear necessary for cell creeping.

As mentioned earlier in the chapter, some types of cells have microvilli, projections of the plasma membrane that increase the surface area of the cell for transporting materials across the plasma membrane. Composed of bundles of microfilaments, microvilli extend and retract as the microfilaments assemble and disassemble.

Intermediate filaments help stabilize cell shape

Intermediate filaments are tough, flexible fibers about 10 nm in diameter (FIG. 4-28). They provide mechanical strength and help stabilize cell shape. These filaments are abundant in regions of a cell that may be subject to mechanical stress applied from outside the cell. Intermediate filaments prevent the cell from stretching excessively in response to outside forces. Certain proteins cross-link intermediate filaments with other types of filaments and mediate interactions between them.

All eukaryotic cells have microtubules and microfilaments, but only some animal groups, including vertebrates, are known to have intermediate filaments. Even when present, intermediate filaments vary widely in protein composition and size among different cell types and different organisms. Examples of intermediate filaments are the keratins found in the epithelial cells of vertebrate skin and neurofilaments found in vertebrate nerve cells.

Certain mutations in genes coding for intermediate filaments weaken the cell and are associated with several diseases. For example, in the neurodegenerative disease amyotrophic lateral sclerosis (ALS, or Lou Gehrig's disease), abnormal neurofilaments have been identified in nerve cells that control muscles. This condition interferes with normal transport of materials in the nerve cells and leads to degeneration of the cells. The resulting loss of muscle function is typically fatal.

CHECKPOINT 4.6

- *In what ways do the functions of the cytoskeleton differ from those of the endomembrane system?*
- *How are microfilaments and microtubules similar? How are they different?*
- *What roles do microtubules play in movement by cilia and flagella?*

4.7 CELL COVERINGS

LEARNING OBJECTIVE

14 Compare the roles of the glycocalyx, extracellular matrix, and cell wall.

Many cells are surrounded by a **glycocalyx,** or *cell coat,* formed by polysaccharide side chains of proteins and lipids that are part of the plasma membrane. The glycocalyx protects the cell and may help keep other cells at a distance. Certain molecules of the glycocalyx enable cells to recognize one another, to make contact, and in some cases to form adhesive or communicating

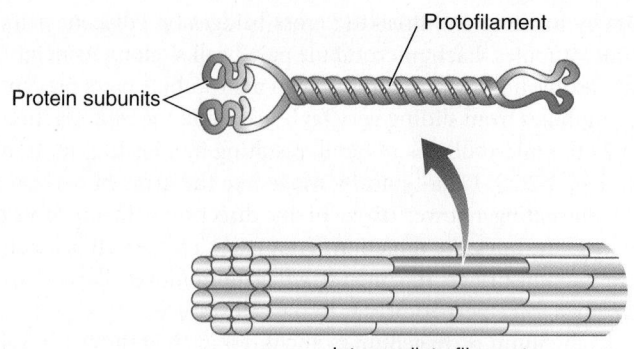

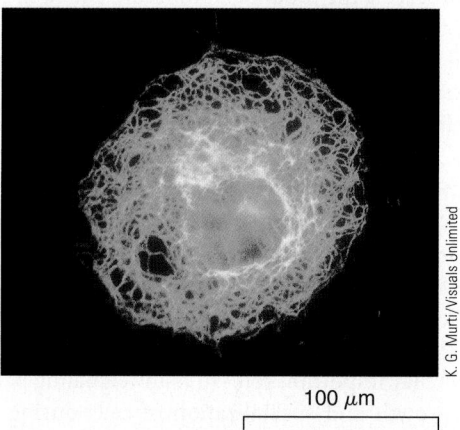

(a) Intermediate filaments are flexible rods about 10 nm in diameter. Each intermediate filament consists of components, called protofilaments, composed of coiled protein subunits.

K. G. Murti/Visuals Unlimited

100 μm

(b) Intermediate filaments are stained *green* in this human cell isolated from a tissue culture.

Figure 4-28 *Animation* **Intermediate filaments**
© Cengage Learning

associations (discussed in Chapter 5). Other molecules of the cell coat contribute to the mechanical strength of multicellular tissues.

Many animal cells are also surrounded by an **extracellular matrix (ECM),** which they secrete. The ECM consists of a gel of carbohydrates and fibrous proteins (FIG. 4-29). The main structural protein in the ECM is *collagen*, a protein that forms very tough fibers (see Fig. 3-23b). Certain glycoproteins of the ECM, called **fibronectins**, organize the matrix and help cells attach to it. Fibronectins bind to protein receptors that extend from the plasma membrane.

Integrins are receptor proteins in the plasma membrane. They maintain adhesion between the ECM and the intermediate filaments and microfilaments inside the cell. These proteins activate many cell signaling pathways that communicate information from the ECM, and they control signals inside the cell that regulate the differentiation and survival of the cell. When cells are not appropriately anchored, apoptosis is initiated. Integrins are also important in organizing the cytoskeleton so that cells assume a definite shape.

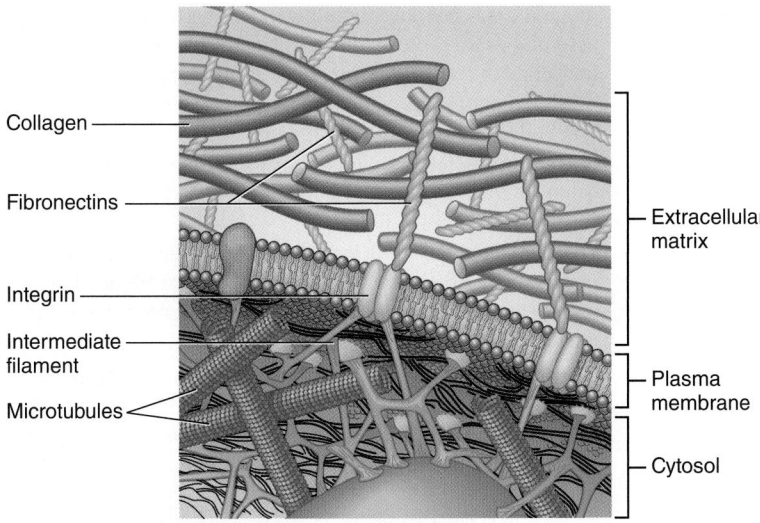

Collagen

Fibronectins

Integrin

Intermediate filament

Microtubules

Extracellular matrix

Plasma membrane

Cytosol

Figure 4-29 The extracellular matrix (ECM)

Fibronectins, glycoproteins of the ECM, bind to integrins and other receptors in the plasma membrane.
© Cengage Learning

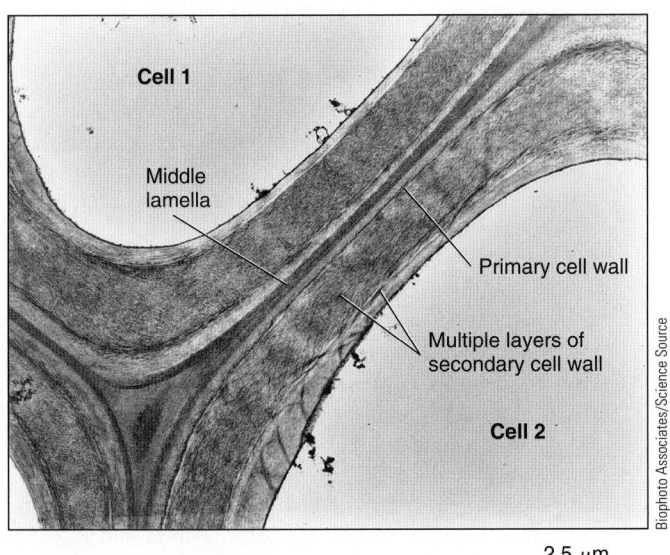

Cell 1

Middle lamella

Primary cell wall

Multiple layers of secondary cell wall

Cell 2

2.5 μm

Biophoto Associates/Science Source

The cells of most bacteria, archaea, fungi, and plant cells are surrounded by a cell wall. Plant cells have thick cell walls that contain tiny fibers composed of the polysaccharide **cellulose** (see Fig. 3-10). Other polysaccharides in the plant cell wall form cross links between the bundles of cellulose fibers. Cell walls provide structural support, protect plant cells from disease-causing organisms, and prevent them from bursting due to increased hydrostatic pressure.

A growing plant cell secretes a thin, flexible *primary cell wall*. As the cell grows, the primary cell wall increases in size (**FIG. 4-30**). After the cell stops growing, either new wall material is secreted that thickens and solidifies the primary wall, or multiple layers of a *secondary cell wall* with a different chemical composition are formed between the primary wall and the plasma membrane. Wood is made mainly of secondary cell walls. The **middle lamella**, a layer of gluelike polysaccharides called *pectins*, lies between the primary cell walls of adjacent cells. The middle lamella causes the cells to adhere tightly to one another. (For more information on plant cell walls, see the discussion of the ground tissue system in Chapter 33.)

CHECKPOINT 4.7

- *What are the functions of the glycocalyx?*
- *How do the functions of fibronectins and integrins differ?*

Figure 4-30 *Animation* Plant cell walls

The cell walls of two adjacent plant cells are labeled in this TEM. The cells are cemented together by the middle lamella, a layer of gluelike polysaccharides called pectins. A growing plant cell first secretes a thin primary wall that is flexible and can stretch as the cell grows. The thicker layers of the secondary wall are secreted inside the primary wall after the cell stops elongating.

SUMMARY: FOCUS ON LEARNING OBJECTIVES

4.1 The Cell: Basic Unit of Life *(page 72)*

1 Connect the cell theory to the evolution of life.

- The **cell theory** holds that (1) cells are the basic living units of organization and function in all organisms and (2) all cells come from other cells. It explains that the ancestry of all the cells alive today can be traced back to ancient times. Evidence that all living cells have evolved from a common ancestor is supported by the basic similarities in their structures and in their molecular composition.

2 Relate the organizational similarities of all cells to the need to conduct essential life functions.

- Every cell is surrounded by a **plasma membrane** that separates it from its external environment. The plasma membrane allows the cell to maintain internal conditions that may be very

different from those of the outer environment. The plasma membrane also allows the cell to selectively exchange materials with its outer environment. Cells have many **organelles**, internal structures that carry out specific functions.

- All cells have similar mechanisms for information transfer and chemical reactions that convert energy from one form to another.

3 Explain the functional significance of cell size and cell shape.

- A critical factor in determining cell size is the ratio of the plasma membrane (surface area) to the cell's volume. The plasma membrane must be large enough relative to the cell volume to regulate the passage of materials into and out of the cell. For this reason, most cells are microscopic.

- The size and shape of a cell are largely dictated by the functions it must perform.

4.2 Methods for Studying Cells (page 74)

4 Compare methods that biologists use to study cells and point out the ways in which many of these approaches are complementary.

- Biologists use **light microscopes, electron microscopes,** and a variety of chemical methods, including the binding of specific antibodies, to study cells and learn about cell structure. The electron microscope has superior **resolving power,** enabling investigators to see details of cell structures not observable with conventional microscopes. Fluorescence microscopy can be used to track the locations and movements of specific tagged molecules within cells.
- Cell biologists use the technique of **cell fractionation** for purifying organelles as well as genetic methods to gain information about the function of cell structures.

4.3 Prokaryotic and Eukaryotic Cells (page 80)

5 Compare and contrast the general characteristics of prokaryotic and eukaryotic cells, and contrast plant and animal cells.

- **Prokaryotic cells** are enclosed by a plasma membrane but have little or no internal membrane organization. They have a **nuclear area** rather than a membrane-enclosed nucleus. Prokaryotic cells typically have a **cell wall** and **ribosomes,** and may have propeller-like **flagella.**
- **Eukaryotic cells** have a membrane-enclosed nucleus, and their **cytoplasm** contains a variety of organelles; the fluid component of the cytoplasm is the **cytosol.**
- Plant cells differ from animal cells in that plant cells have rigid cell walls, plastids, and large vacuoles, which are important in plant growth and development.

6 Describe three functions of cell membranes.

- Membranes divide the eukaryotic cell into compartments, allowing it to conduct specialized activities within small areas of the cytoplasm, concentrate reactants, and organize metabolic reactions. Small membrane-enclosed sacs, called **vesicles,** transport materials between compartments.
- Membranes are important in energy storage and conversion.
- Membranes serve as work surfaces for certain chemical reactions.

4.4 The Cell Nucleus (page 82)

7 Relate the structure of the nucleus to its function as the control center of the eukaryotic cell.

- The **nucleus** contains genetic information coded in DNA. The nucleus is bounded by a **nuclear envelope,** consisting of a double membrane perforated with **nuclear pores** that communicate with the cytoplasm.
- DNA in the nucleus associates with protein to form **chromatin,** which is organized into **chromosomes.** During cell division, the chromosomes condense and become visible as threadlike structures.
- DNA transcribes its information in messenger RNA (mRNA) molecules, which enter the cytoplasm to provide information for protein synthesis by ribosomes.
- The **nucleolus** is a region in the nucleus that is the site of ribosomal RNA (rRNA) synthesis and ribosome assembly.

4.5 Membranous Organelles in the Cytoplasm (page 86)

8 Distinguish between smooth and rough endoplasmic reticulum in terms of both structure and function.

- The **endoplasmic reticulum (ER)** is a network of folded internal membranes in the cytosol. **Smooth ER** is the site of lipid synthesis, calcium ion storage, and detoxifying enzymes.
- **Rough ER** is studded along its outer surface with **ribosomes** that manufacture polypeptides. Polypeptides synthesized on

rough ER may be moved into the ER lumen, where they are assembled into proteins and modified by the addition of a carbohydrate or lipid. These proteins may then be transferred to other compartments within the cell by small transport vesicles that bud off from the ER membrane.

9 Trace the path of proteins synthesized in the rough endoplasmic reticulum as they are processed, modified, and sorted by the Golgi complex and then transported to specific destinations.

- The **Golgi complex** consists of stacks of flattened membranous sacs called **cisternae** that process, sort, and modify proteins synthesized on the rough ER. The Golgi complex also manufactures lysosomes.
- Glycoproteins are transported from the ER to the *cis* face of the Golgi complex by transport vesicles, which are formed by membrane budding. The Golgi complex modifies carbohydrates and lipids that were added to proteins by the ER and packages them in vesicles.
- Glycoproteins exit the Golgi through vesicles that are formed at its *trans* face. The Golgi routes some proteins to the plasma membrane for export from the cell. Others are transported to lysosomes or other organelles within the cytoplasm.

10 Compare the functions of lysosomes, vacuoles, and peroxisomes.

- **Lysosomes** contain enzymes that break down worn-out cell structures, bacteria, and debris taken into cells.
- **Vacuoles** store materials, water, and wastes. They maintain hydrostatic pressure in plant cells.
- **Peroxisomes** are important in lipid metabolism and detoxify harmful compounds such as ethanol. They produce hydrogen peroxide, but contain the enzyme catalase, which degrades this toxic compound.

11 Contrast the functions of mitochondria and chloroplasts, and discuss ATP synthesis by each of these organelles.

- **Mitochondria,** organelles enclosed by a double membrane, are the sites of aerobic respiration. The inner membrane is folded, forming **cristae** that increase its surface area.
- The cristae and the compartment enclosed by the inner membrane, the **matrix,** contain enzymes for the reactions of **aerobic respiration.** During aerobic respiration, nutrients are broken down in the presence of oxygen. Energy captured from nutrients is packaged in ATP, and carbon dioxide and water are produced as byproducts.
- **Plastids** are organelles that produce and store food in the cells of plants and algae.
- **Chloroplasts** are plastids that carry out **photosynthesis.**
- The inner membrane of the chloroplast encloses a fluid-filled space, the **stroma.**
- **Grana,** stacks of interconnected disclike membranous sacs called **thylakoids,** are suspended in the stroma.
- During photosynthesis, **chlorophyll,** the green pigment found in the thylakoid membranes, traps light energy. This energy is converted to chemical energy in ATP and used to synthesize carbohydrates from carbon dioxide and water.

4.6 The Cytoskeleton (page 95)

12 Contrast the structure and functions of the cytoskeleton with the structure and functions of the membranous components of the cell.

- The **cytoskeleton** is a dynamic internal protein fiber framework that includes microtubules, microfilaments, and intermediate filaments. The cytoskeleton provides structural support and functions in various types of cell movement, including transport of materials in the cell.

- **Microtubules** are hollow cylinders assembled from subunits of the protein **tubulin.** In cells that are not dividing, the minus ends of microtubules are anchored in **microtubule-organizing centers (MTOCs).** The main MTOC of animal cells is the **centrosome,** which usually contains two centrioles. Each centriole has a 9 × 3 arrangement of microtubules.
- **Microfilaments,** or **actin** filaments, formed from subunits of the protein actin, are important in cell movement.
- **Intermediate filaments** strengthen the cytoskeleton and stabilize cell shape.

13 Relate the structure of cilia and flagella to their functions.
- **Cilia** and **flagella** are thin, movable structures that project from the cell surface and function in movement. Each consists of a 9 + 2 arrangement of microtubules, and each is anchored in the cell by a basal body that has a 9 × 3 organization of microtubules. Cilia are short, and flagella are long.

4.7 Cell Coverings *(page 100)*

14 Compare the roles of the glycocalyx, extracellular matrix, and cell wall.
- Most cells are surrounded by a **glycocalyx,** or *cell coat,* formed by polysaccharides extending from the plasma membrane.
- Many animal cells are also surrounded by an **extracellular matrix (ECM)** consisting of carbohydrates and protein. **Fibronectins** are glycoproteins of the ECM that bind to **integrins,** receptor proteins in the plasma membrane.
- Cells of most bacteria, archaea, fungi, and plant cells are surrounded by a cell wall made mainly of carbohydrates. Plant cells secrete **cellulose** and other polysaccharides that form rigid cell walls.

TEST YOUR UNDERSTANDING

Know and Comprehend

1. Which of the following is the most fundamental feature that enables the cell to function as a distinct entity, separate from its environment? (a) nucleus (b) ribosomes (c) nucleic acids (d) plasma membrane (e) protein
2. Which of the following is/are too small to have been discovered before the invention of the electron microscope? (a) chloroplasts (b) mitochondria (c) nucleus (d) ribosomes (e) central vacuole
3. Which of the following would *not* be found in prokaryotic cells? (a) cell wall (b) ribosomes (c) plasma membrane (d) nucleus (e) propeller-like flagellum
4. Which of the following would you expect to find passing through a nuclear pore? (a) ribosome subunit (b) ribosomal protein (c) messenger RNA (d) water (e) all of the preceding
5. Select the sequence that most accurately describes glycoprotein processing in the eukaryotic cell. 1. *cis* face of Golgi 2. smooth ER 3. rough ER 4. *trans* face of Golgi 5. transport vesicle (a) 1, 2, 3, 4, 5 (b) 1, 4, 5, 3, 2 (c) 2, 3, 4, 5, 1 (d) 3, 5, 1, 4 (e) 2, 5, 1, 4
6. To carry out its life functions, every plant cell requires (a) chloroplasts and ribosomes (b) mitochondria and ribosomes (c) mitochondria, chloroplasts, and lysosomes (d) mitochondria and chloroplasts (e) chloroplasts and a vacuole
7. Microtubules (a) have constant diameters, but vary in length (b) vary in diameter, but are constant in length (c) vary in both length and diameter (d) are constant in both length and diameter
8. All of the following are true of integrins *except* (a) they are receptor proteins (b) they help organize the cytoskeleton (c) they are part of the ECM (d) they are important in cell signaling (e) they are located in the plasma membrane

Apply and Analyze

9. **VISUALIZE** Illustrate the relationships among the membranes and compartments of (a) mitochondria and (b) chloroplasts by sketching simple concentric circles to represent the membranes. Choose the appropriate labels for each sketch from the following list: inner membrane, outer membrane, thylakoid membrane, stroma, thylakoid lumen, matrix.

10. Three observers independently measured the same cells and reported their results in the table:

Observer	Average Diameter of 10 cells
1	0.02 μm
2	0.002 mm
3	2×10^4 nm

Which, if any, of their measurements agree? Which observer is most likely to be correct if the cells are prokaryotic cells? if they are eukaryotic cells?

Evaluate and Synthesize

11. Why does a eukaryotic cell need both membranous organelles and fibrous cytoskeletal components? Justify your answer.
12. **INTERPRET DATA** An investigator has isolated two mutant strains of yeast that fail to secrete a short polypeptide. She fractionated the cells and compared the concentrations of this polypeptide in the different cellular fractions with those found in normal cells. What point in the secretory pathway would you expect would be defective for mutant a and mutant b?

Yeast Strain	Secretory Vesicles	Golgi Complex	ER
Normal	90	5	5
Mutant a	4	2	94
Mutant b	2	85	13

13. **EVOLUTION LINK** What types of similarities in cell structure and function tell biologists about the common origin of organisms? Explain.
14. **SCIENCE, TECHNOLOGY, AND SOCIETY** Students majoring in one of the life sciences also take courses in chemistry, physics, and mathematics. Based on what you've learned in this chapter, do you think that these requirements are justified?

aplia   To access course materials, such as Aplia and other companion resources, please visit **www.cengagebrain.com.**

5 Biological Membranes

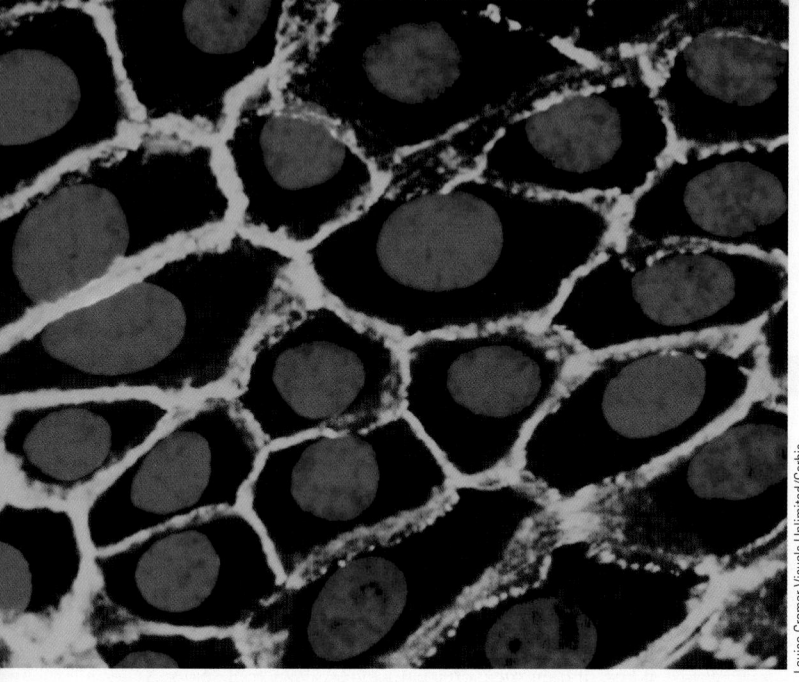

Louise Cramer Visuals Unlimited/Corbis

Membrane proteins. Cadherins, proteins that span the plasma membrane and participate in adhering junctions with adjacent cells, are stained *green* in this fluorescence LM. Nuclei are stained *purple*.

KEY CONCEPTS

5.1 Biological membranes are composed of fluid phospholipid bilayers that form closed cellular compartments. Proteins embedded in the lipid core of the membrane maintain internal cellular environments by selectively controlling the movement of molecules across the lipid bilayers.

5.2 Membrane proteins have many functions, including transport of materials, enzymatic activity, transfer of information, and recognition of other cells.

5.3 Transport proteins move ions and small polar molecules through cell membranes.

5.4 Many small molecules, including water, move through biological membranes by diffusion, the net movement of particles from a region of higher to a region of lower concentration.

5.5 Cells must internalize many substances at higher concentrations than they are outside the cell. These substances must be actively transported against a concentration gradient, requiring a direct expenditure of energy.

5.6 Cells eject products by exocytosis and import materials by endocytosis; both processes require cells to expend energy.

5.7 Cells in close contact may form specialized junctions with one another between their plasma membranes.

The evolution of biological membranes that separate the cell from its external environment was an essential step in the origin of life. Later, these membranes made the evolution of complex cells possible. The extensive internal membranes of eukaryotic cells form multiple compartments with unique chemical environments that allow highly specialized activities to take place.

Membrane proteins are critical molecules in cellular activities. Some proteins associated with the plasma membrane transport materials, whereas others transmit information or serve as enzymes. Still others, known as *cell adhesion molecules,* are important in connecting cells to one another to form tissues.

In Chapter 4 we discussed a variety of cell organelles and how they interact to perform cell activities. In this chapter we focus on the structure and functions of the plasma membrane that surrounds the cell and on the membranes that surround many organelles. We first consider what is known about the composition and structure of biological membranes. Then we provide an overview of the many vital functions of cell membranes, including transport of materials and information transfer. We discuss how cells transport various materials, from ions to complex molecules and even bacteria, across membranes. Finally, we examine specialized structures that allow membranes of different cells to interact. Although much of our discussion centers on the structure and functions of plasma membranes, many of the concepts apply to other cell membranes.

5.1 THE STRUCTURE OF BIOLOGICAL MEMBRANES

LEARNING OBJECTIVES

1 Evaluate the importance of membranes to cells, emphasizing their various functions.
2 Describe the fluid mosaic model of cell membrane structure.
3 Relate properties of the lipid bilayer to properties and functions of cell membranes.
4 Describe the ways that membrane proteins associate with the lipid bilayer.

To carry out the many chemical reactions necessary to sustain life, the cell must maintain an appropriate internal environment. Every cell is surrounded by a **plasma membrane** that physically separates it from the outside world and defines it as a distinct entity. By regulating passage of materials into and out of the cell, the plasma membrane helps maintain a life-supporting internal environment. As we discussed in Chapter 4, eukaryotic cells are characterized by numerous organelles that are surrounded by membranes. Some of these organelles—including the nuclear envelope, endoplasmic reticulum (ER), Golgi complex, lysosomes, vesicles, and vacuoles—form the endomembrane system, which extends throughout the cell.

Biological membranes are complex, dynamic structures made of lipid and protein molecules that are in constant motion. The properties of membranes allow them to perform many vital functions. They regulate the passage of materials, divide the cell into compartments, serve as surfaces for chemical reactions, adhere to and communicate with other cells, and transmit signals between the environment and the interior of the cell. Membranes are also an essential part of energy transfer and storage systems. How do the properties of cell membranes enable the cell to carry on such varied functions?

Long before the development of the electron microscope, scientists knew that membranes consist of both lipids and proteins. Work by researchers in the 1920s and 1930s had provided clues that the core of cell membranes consists of lipids, mostly phospholipids (see Chapter 3). Mammalian red blood cells have only a plasma membrane and no internal membrane compartments. By comparing their surface area with the total number of lipid molecules per cell, early investigators had calculated that the membrane is no more than two phospholipid molecules thick. Studies by early researchers also showed that when purified membrane lipids were dispersed in water they could form flexible, self-sealing bilayers that would spontaneously round up to form closed compartments or vesicles.

Phospholipids form bilayers in water

Phospholipid molecules have unique attributes that allow them to form bilayered structures, and they are primarily responsible for the physical properties of biological membranes. Recall from Chapter 3 that phospholipids are **amphipathic molecules,** consisting of two *hydrophobic* ("water-fearing") fatty acid chains linked to two of the three carbons of a glycerol molecule (see Fig. 3-13). Bonded to the third carbon of the glycerol is a negatively charged, *hydrophilic* ("water-loving") phosphate group, which in turn is linked to a polar, hydrophilic organic group. Molecules of this type have distinct hydrophobic and hydrophilic regions, and all lipids that make up the core of biological membranes have amphipathic characteristics that cause them to interact with water in predictable ways (**FIG. 5-1**).

Because one end of each phospholipid associates freely with water and the opposite end does not, the most stable orientation for them to assume in water results in the formation of a bilayer structure (see Fig. 5-1a). This arrangement allows the hydrophilic heads of the phospholipids to be in contact with the aqueous medium, while their oily tails, the hydrophobic fatty

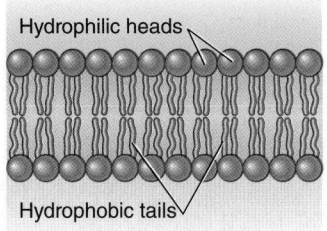
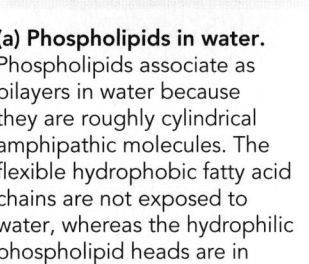
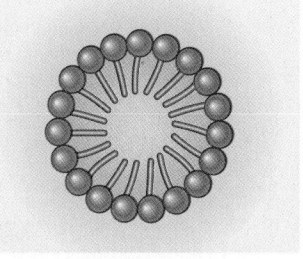

(a) Phospholipids in water. Phospholipids associate as bilayers in water because they are roughly cylindrical amphipathic molecules. The flexible hydrophobic fatty acid chains are not exposed to water, whereas the hydrophilic phospholipid heads are in contact with water.

(b) Detergent in water. Detergent molecules are roughly cone-shaped amphipathic molecules that associate in water as spherical structures.

Figure 5-1 *Animation* **Properties of lipids in water**

acid chains, are buried in the interior of the structure away from the water molecules. This hydrophobic core of the bilayer is the barrier that prevents many types of small, hydrophilic molecules (including ions, amino acids, and organic metabolites) from passing from one side of a membrane to the other, and it allows a cell to maintain different chemical environments within each membrane compartment.

Physical studies on artificial vesicles formed from isolated membrane lipids also revealed another important property associated with the hydrophobic core: at normal biological temperatures, the fatty acid chains in the core are in constant motion as parts of hydrocarbon chains rotate around the C—C bonds. This constant hydrocarbon chain motion gives the bilayer the property of a *liquid crystal,* or a 2-dimensional fluid. The bilayers are crystal-like in that the lipids form an ordered array, with the head groups on the outer surfaces of the bilayer and the fatty acid chains on the inside. Thus, the molecules are free to rotate and can move laterally within their single layer, or *leaflet* of the bilayer. Under normal conditions a single phospholipid exchanges places with its neighbor in about 10^{-7} seconds and can travel laterally across the surface of the cell in seconds. By contrast, the "flip-flop" movement of a phospholipid molecule from one leaflet of an artificial bilayer to the other side is a rare event (on the order of days) because it requires the large, hydrophilic head group to pass through the hydrophobic core of the membrane (**FIG. 5-2**). In cells, this trans-bilayer movement is facilitated by special membrane proteins commonly referred to as "flippases."

Note that amphipathic properties alone do not predict the ability of lipids to associate as a bilayer. Shape is also important. Phospholipids tend to have uniform widths; their roughly cylindrical shapes, together with their amphipathic properties, are responsible for bilayer formation. By contrast, many common detergents are amphipathic molecules, each containing a single hydrocarbon chain (like a fatty acid) at one end and a hydrophilic region at the other. These molecules are roughly cone-shaped, with the hydrophilic end forming the broad base and the hydrocarbon tail leading to the point. Because of their

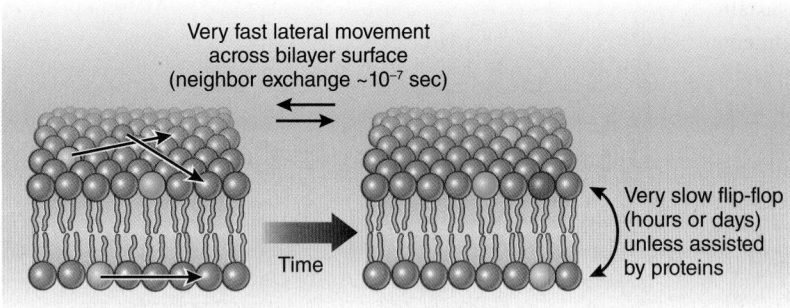

Figure 5-2 Membrane fluidity

The ordered arrangement of phospholipid molecules makes the cell membrane a liquid crystal. The hydrocarbon chains are in constant motion, allowing each molecule to rapidly move laterally on the same side of the bilayer. Flip-flop from one side of the bilayer to the other is a rare event, and in cells it is facilitated by special membrane proteins.
© Cengage Learning

shapes, these molecules do not associate as bilayers but instead tend to form spherical structures in water (see Fig. 5-1b). Detergents can "solubilize" oil because the oil molecules associate with the hydrophobic interiors of the spheres.

The fluid mosaic model explains membrane structure

With the development of the electron microscope in the 1950s, cell biologists were able to see the plasma membrane for the first time. One of their most striking observations was how uniform and thin the membranes appeared to be. The electron microscope revealed a three-layered structure, something like a railroad track, suggesting that the plasma membrane was a uniform structure no more than 10 nm thick (**FIG. 5-3**). This finding was surprising because membrane proteins were found to be far from uniform, and they vary widely in composition and size.

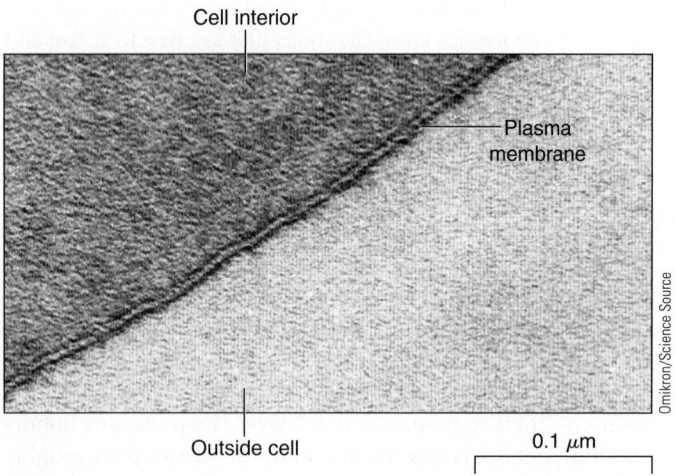

Figure 5-3 TEM of the plasma membrane of a mammalian red blood cell

The plasma membrane separates the cytosol (*darker region*) from the external environment (*lighter region*). The hydrophilic heads of the phospholipids are the parallel dark lines, and the hydrophobic tails are the light zone between them.

Some of these membrane-bound proteins are quite large, with diameters much greater than 10 nm.

Studies of many membrane proteins showed that one region (or domain) of the molecule could always be found on one side of the bilayer, whereas another part of the protein might be located on the opposite side. These studies indicated that many membrane proteins extend completely through the lipid bilayer and that membranes contain many types of proteins of different shapes and sizes.

In 1972, S. Jonathan Singer and Garth Nicolson of the University of California at San Diego proposed a model of membrane structure that represented a synthesis of the known properties of biological membranes. According to their **fluid mosaic model,** a cell membrane consists of a fluid bilayer of phospholipid molecules in which the proteins are embedded or otherwise associated, much like the tiles in a mosaic picture. This mosaic pattern is not static, however, because the positions of many of the proteins are constantly changing as they move about like icebergs in a fluid sea of phospholipids. **FIGURE 5-4** depicts the plasma membrane of a eukaryotic cell according to the fluid mosaic model; prokaryotic plasma membranes are discussed in Chapter 25.

Biological membranes are two-dimensional fluids

The fluid qualities of lipid bilayers also allow molecules embedded in them to move along the plane of the membrane (as long as they are not anchored in some way), as David Frye and Michael Edidin elegantly demonstrated in 1970. They conducted experiments in which they followed the movement of membrane proteins on the surface of two cells that had been joined (**FIG. 5-5**). When the plasma membranes of a mouse cell and a human cell are fused, within minutes at least some of the membrane proteins from each cell migrate and become randomly distributed over the single, continuous plasma membrane that surrounds the joined cells. Frye and Edidin showed that the fluidity of the lipids in the membrane allows many of the proteins to move, producing an ever-changing configuration. Not all plasma membrane proteins exhibit this behavior, however. Membrane proteins that are physically anchored to the cytoskeleton (see Figs. 4-29 and 5-4) or the extracellular matrix tend to remain in fixed positions on the cell surface.

For a membrane to function properly, its hydrophobic core must be in an optimal fluid state. At normal growing temperatures, cell membranes are fluids with a consistency similar to a vegetable oil at room temperature. If cells are rapidly cooled to lower than normal temperatures, however, membrane functions, such as the transport of certain substances, are inhibited or cease because the motion of the fatty acid chains is slowed and the lipid bilayer becomes too rigid.

Certain properties of membrane lipids have significant effects on the fluidity of the bilayer. Recall from Chapter 3 that fats

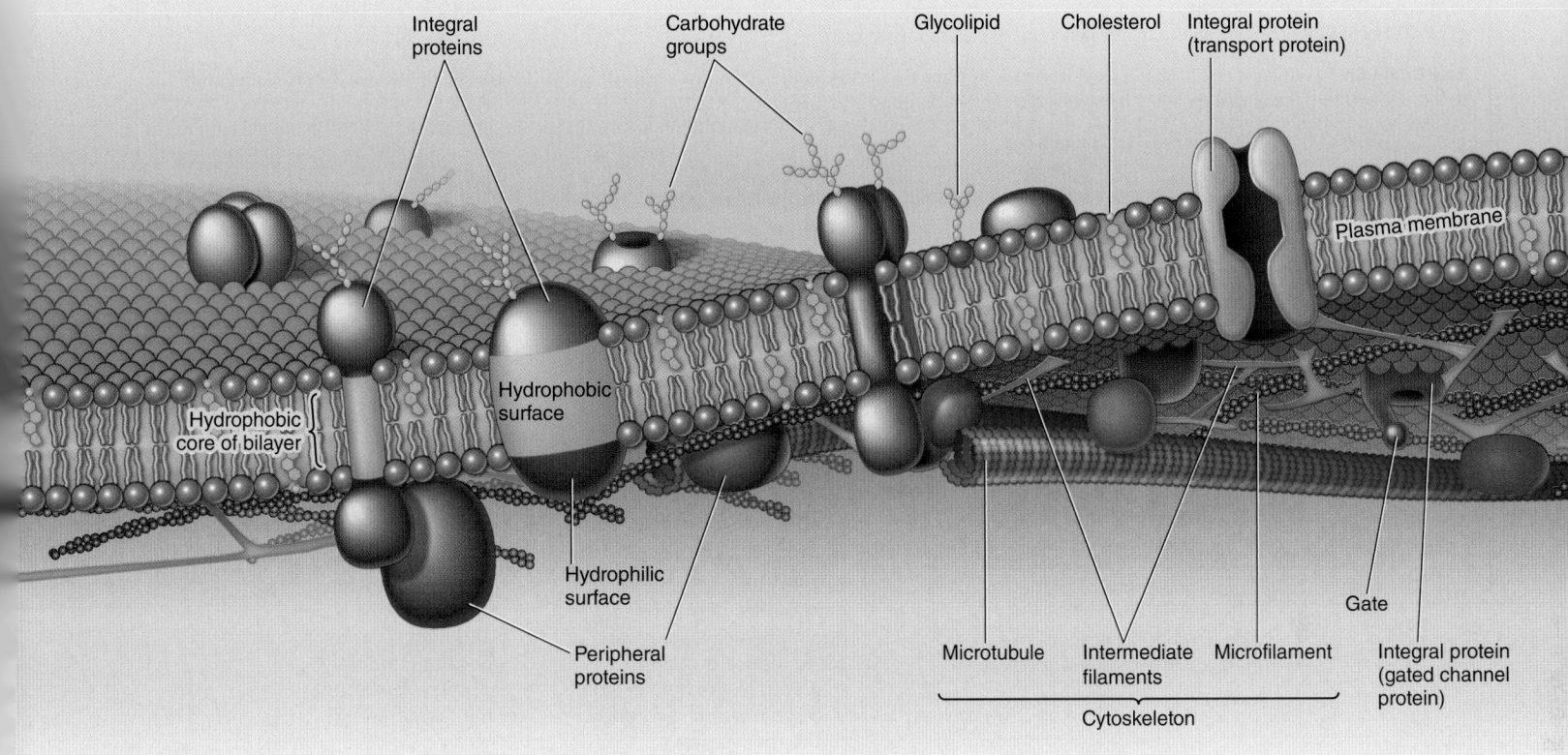

Figure 5-4 Detailed structure of the plasma membrane

Cell membranes are composed of a fluid bilayer of phospholipids in which proteins move about like icebergs in a sea. Although the lipid bilayer consists mainly of phospholipids, other lipids, such as cholesterol and glycolipids, are present. Peripheral proteins are loosely associated with the bilayer, whereas integral proteins are tightly bound. The integral proteins shown here are transmembrane proteins that extend through the bilayer. They have hydrophilic regions on both sides of the bilayer connected by a hydrophobic membrane-spanning region. Some types of integral membrane proteins are anchored in the membrane by proteins that attach them to the cytoskeleton or the extracellular matrix. Glycolipids (carbohydrates attached to lipids) and glycoproteins (carbohydrates attached to proteins) are exposed on the extracellular surface; they play roles in cell recognition and adhesion to other cells.
© Cengage Learning

containing large amounts of saturated fatty acids tend to be solid at room temperature due to the effects of many van der Waals interactions that occur between closely aligned fatty acid chains. Fats or oils with mono or polyunsaturated fatty acids, however, tend to be liquid at room temperature due to the presence of double bonds that produce kinks in their fatty acid chains. Such kinks prevent the close packing necessary for these interactions to occur.

Many organisms have regulatory mechanisms for maintaining cell membranes in an optimally fluid state. Some organisms compensate for environmental temperature differences by altering the fatty acid content of their membrane lipids. Arctic fish, for example, tend to have higher levels of unsaturated fatty acids in their membrane lipids than tropical species in order to maintain optimal membrane fluidity at the lower

ocean temperatures. Other organisms, such as plants and many microorganisms, will adjust the ratios of saturated and unsaturated fatty acids in their membrane lipids to optimize fluidity in response to changes in environmental temperatures that occur during the growing season.

Some membrane lipids stabilize membrane fluidity within certain limits. One such "fluidity buffer" is cholesterol, a steroid found in animal cell membranes. A cholesterol molecule is largely hydrophobic but is slightly amphipathic because of the presence of a single hydroxyl group (see Fig. 3-15a). In a phospholipid bilayer, this hydroxyl group associates with the hydrophilic heads of the phospholipids; the hydrophobic remainder of the cholesterol molecule fits between the fatty acid hydrocarbon chains (see Fig. 5-4).

Do proteins embedded in a biological membrane move about?

HYPOTHESIS: Proteins are able to move laterally in the plasma membrane.

EXPERIMENT: Larry Frye and Michael Edidin labeled membrane proteins of mouse and human cells using different-colored fluorescent dyes to distinguish between them. They then fused mouse and human cells to produce hybrid cells.

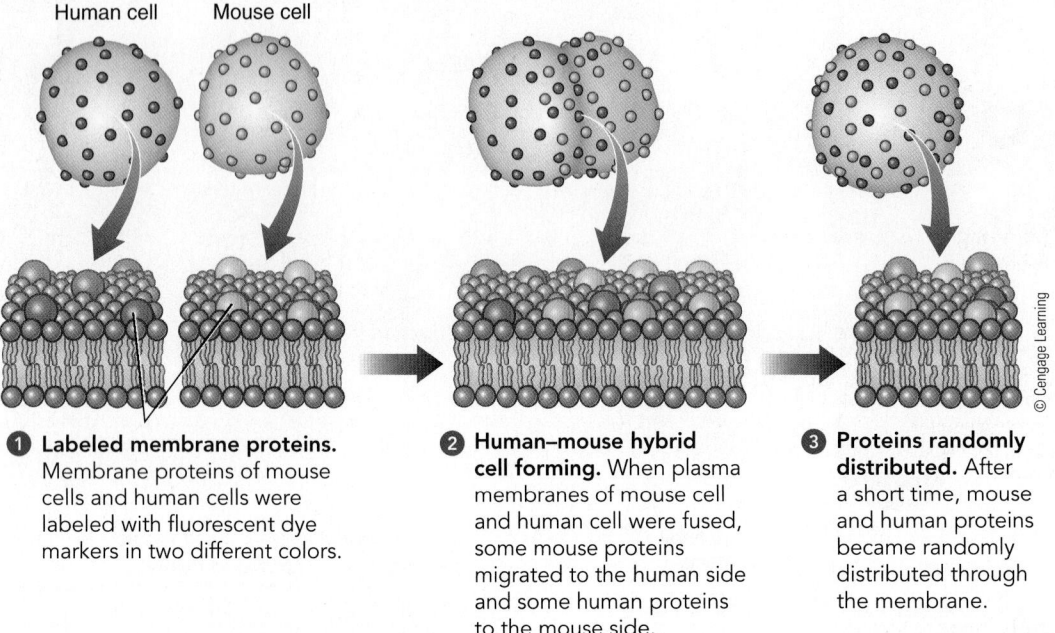

1. **Labeled membrane proteins.** Membrane proteins of mouse cells and human cells were labeled with fluorescent dye markers in two different colors.

2. **Human–mouse hybrid cell forming.** When plasma membranes of mouse cell and human cell were fused, some mouse proteins migrated to the human side and some human proteins to the mouse side.

3. **Proteins randomly distributed.** After a short time, mouse and human proteins became randomly distributed through the membrane.

RESULTS AND CONCLUSION: After a brief period of incubation, mouse and human cells intermixed over the surface of the hybrid cells. After about 40 minutes, the proteins of each species had become randomly distributed through the entire hybrid plasma membrane. This experiment demonstrated that proteins in the plasma membrane do move.

Figure 5-5 *Animation* **Frye and Edidin's experiment**

PREDICT What would be the expected distribution of fluorescence in the fused cell if a membrane protein linked to a microtubule on its cytoplasmic side were the protein that was labeled with the green fluorescent antibody?

SOURCE: Frye, L.D., and M. Edidin. "The Rapid Intermixing of Cell Surface Antigens after Formation of Mouse-Human Heterokaryons." *Journal of Cell Science,* Vol. 7, 319–335, 1970.

At low temperatures cholesterol molecules act as "spacers" between the hydrocarbon chains, restricting van der Waals interactions that would promote solidifying. Cholesterol also helps prevent the membrane from becoming weakened or unstable at higher temperatures. The reason is that the cholesterol molecules interact strongly with the portions of the hydrocarbon chains closest to the phospholipid head. This interaction restricts motion in these regions. Plant cells have steroids other than cholesterol that carry out similar functions.

Biological membranes fuse and form closed vesicles

Lipid bilayers, particularly those in the liquid-crystalline state, have additional important physical properties. Bilayers tend to resist forming free ends; as a result, they are self-sealing and under most conditions spontaneously round up to form closed vesicles. Lipid bilayers are also flexible, allowing cell membranes to change shape without breaking, and under appropriate conditions lipid bilayers fuse with other bilayers.

Membrane fusion is an important cell process. When a vesicle fuses with another membrane, both membrane bilayers and their compartments become continuous. Various transport vesicles form from, and also merge with, membranes of the ER and Golgi complex, facilitating the transfer of materials from one compartment to another. A vesicle fuses with the plasma membrane when a product is secreted from the cell.

Membrane proteins include integral and peripheral proteins

The two major classes of membrane proteins, integral proteins and peripheral proteins, are defined by how tightly they are associated with the lipid bilayer (**FIG. 5-6**). **Integral membrane**

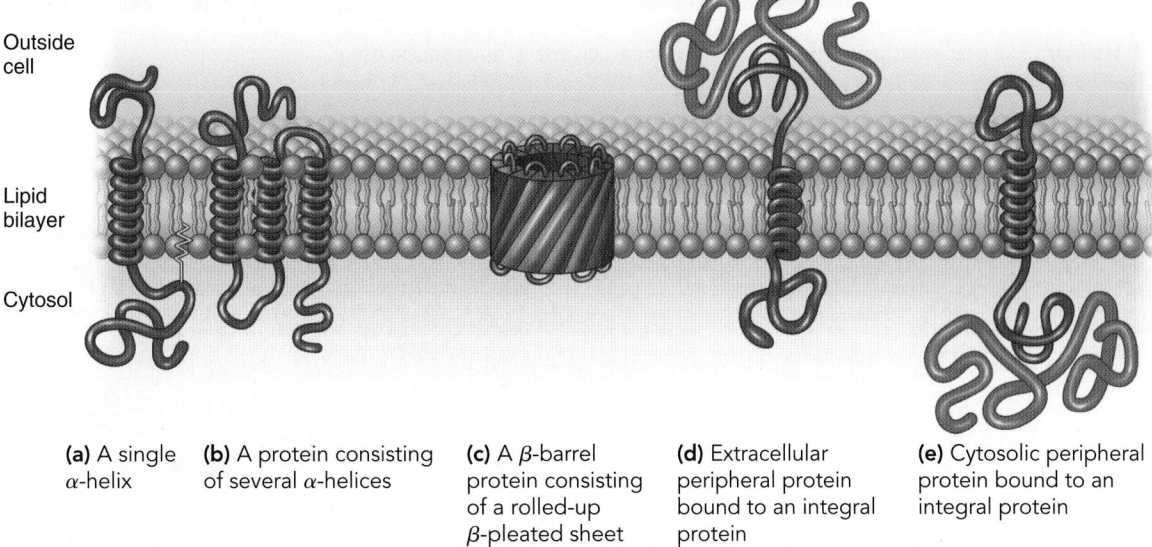

Outside
cell

Lipid
bilayer

Cytosol

(a) A single α-helix

(b) A protein consisting of several α-helices

(c) A β-barrel protein consisting of a rolled-up β-pleated sheet

(d) Extracellular peripheral protein bound to an integral protein

(e) Cytosolic peripheral protein bound to an integral protein

Figure 5-6 *Animation* **Membrane proteins**

Three transmembrane proteins are shown in **(a)**, **(b)**, and **(c)**. Water molecules and ions pass through specific types of protein pores in the membrane formed between groups of α-helices or through β-barrel structures formed as in **(c)**. Peripheral proteins are bound to integral proteins by noncovalent interactions **(d, e)**.
© Cengage Learning

proteins are firmly bound to the membrane. Cell biologists usually can release them only by disrupting the bilayer with detergents. These proteins are amphipathic. Their hydrophilic regions extend out of the cell or into the cytoplasm, whereas their hydrophobic regions interact with the fatty acid tails of the membrane phospholipids.

Some integral proteins do not extend all the way through the membrane. Many others, called **transmembrane proteins**, extend completely through the membrane. Some span the membrane only once, whereas others wind back and forth as many as 24 times. The most common kind of transmembrane region of a membrane protein is an α-helix (see Chapter 3) with hydrophobic amino acid side chains that project out from the helix into the hydrophobic region of the lipid bilayer. Some proteins span the membrane in the form of rolled-up β-pleated sheets. These protein formations are barrel-shaped and form pores through which water and other substances can pass.

Peripheral membrane proteins are not embedded in the lipid bilayer. They are located on the inner or outer surface of the plasma membrane, usually bound to exposed regions of integral proteins by noncovalent interactions. Peripheral proteins can be easily removed from the membrane without disrupting the structure of the bilayer.

One of the most remarkable demonstrations that proteins are actually embedded in the lipid bilayer comes from *freeze–fracture* electron microscopy, a technique that splits the membrane into halves. The researcher can literally see the two halves of the membrane from "inside out." When cell biologists examine membranes in this way, they observe numerous particles on the fracture faces (**FIG. 5-7**). The particles are clearly integral membrane proteins because researchers never see them in freeze–fractured artificial lipid bilayers. These findings profoundly influenced the development of the fluid mosaic model.

Proteins are oriented asymmetrically across the bilayer

Figure 5-4 shows that membrane protein molecules are *asymmetrically oriented*. A specific external domain of an intrinsic membrane protein is always found on the same side of the membrane. For example, domains with attached carbohydrate groups of plasma membrane proteins are only found on the external side of the membrane. This asymmetry is produced by the highly specific way in which each protein is inserted into the bilayer. Integral membrane proteins that will be associated with the cell's outer surface are manufactured like proteins destined to be exported from the cell.

As discussed in Chapter 4, proteins destined to be exported through the cell's outer surface are initially manufactured by ribosomes on the rough ER. They pass through the ER membrane into the ER lumen, where sugars are added, making them **glycoproteins.** For integral membrane proteins, only a part of each protein passes through the ER membrane, so each completed protein has some regions that are located in the ER lumen and other regions that remain in the cytosol. Because the enzymes that attach the sugars to certain amino acids of the

WHY IS IT USED? The freeze–fracture method is used to split the lipid bilayer apart so that its components can be analyzed.

HOW IS IT DONE?

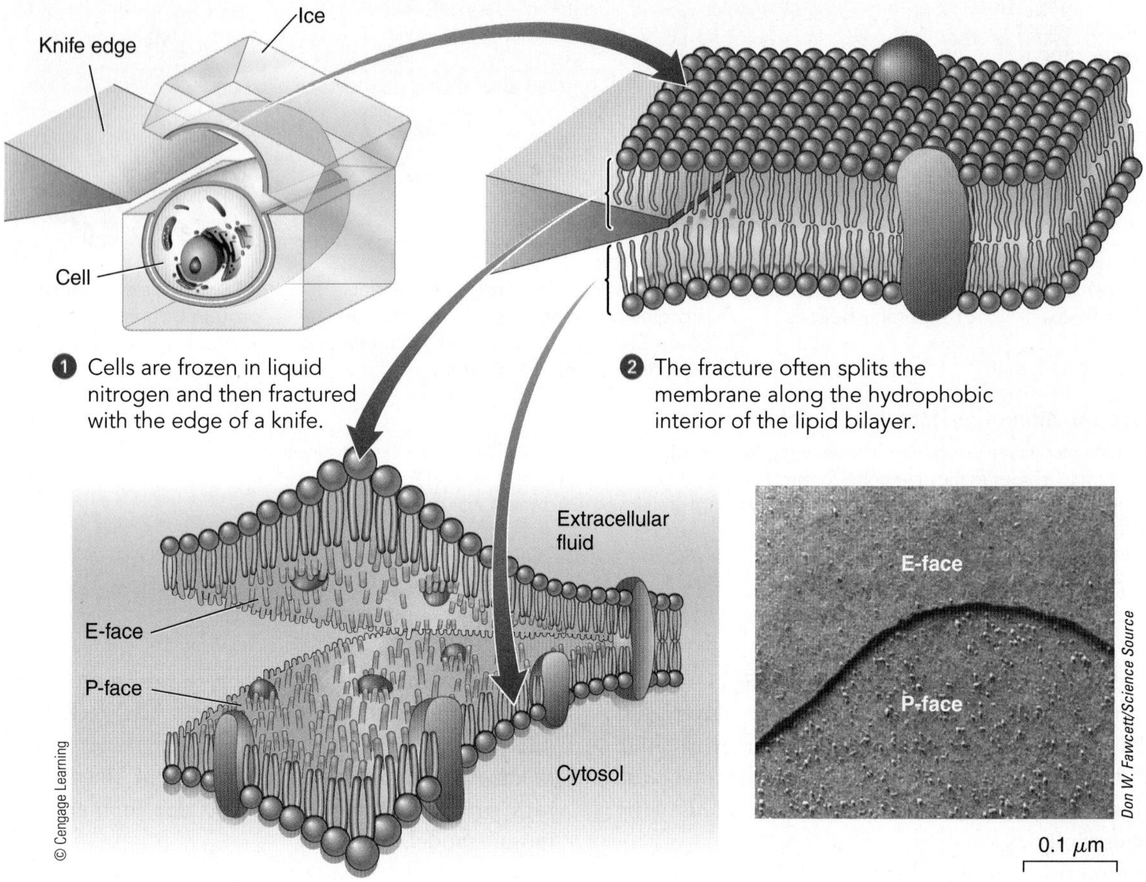

Knife edge
Ice
Cell

1 Cells are frozen in liquid nitrogen and then fractured with the edge of a knife.

2 The fracture often splits the membrane along the hydrophobic interior of the lipid bilayer.

Extracellular fluid
E-face
P-face
Cytosol

© Cengage Learning

E-face
P-face

Don W. Fawcett/Science Source

0.1 μm

3 Two complementary fracture faces result. The inner half-membrane presents the P-face (or protoplasmic face), and the outer half-membrane presents the E-face (or external face). Integral proteins, including transmembrane proteins, are inserted through the lipid bilayer.

4 An electron microscope is used to view the interior surfaces (faces) of the two layers. In this TEM the particles (which appear as bumps) represent large integral proteins.

Figure 5-7 Freeze–fracture method

protein are found only in the ER lumen, carbohydrates can be added only to the parts of the membrane proteins that protrude into that compartment.

In **FIGURE 5-8** follow from top to bottom the vesicle budding and membrane fusion events that are part of the transport process. You can see that the same region of the protein that protruded into the ER lumen is also transported to the lumen of the Golgi complex. There additional enzymes further modify the carbohydrate chains. Within the Golgi complex, the glycoprotein is sorted and directed to the plasma membrane. The modified region of the protein remains inside the membrane compartment of a transport vesicle as it buds from the Golgi

complex. Note that when the transport vesicle fuses with the plasma membrane, the inside layer of the transport vesicle becomes the outside layer of the plasma membrane. The carbohydrate chain extends to the exterior of the cell surface.

In summary, here is the sequence:

sugars added to protein in ER lumen → glycoprotein transported to Golgi complex, where it is further modified → glycoprotein transported to plasma membrane → transport vesicle fuses with plasma membrane → inside layer of transport vesicle becomes outer layer of plasma membrane

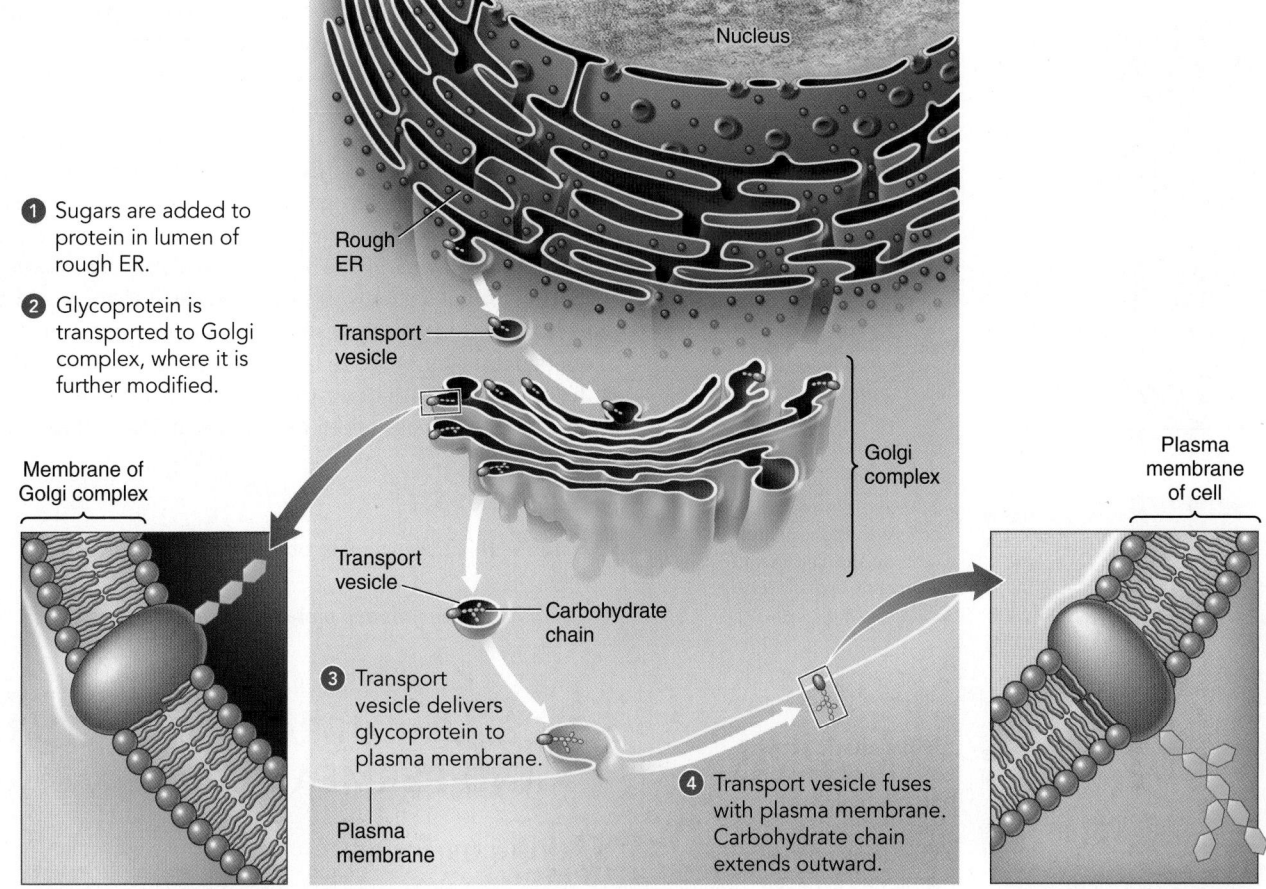

1 Sugars are added to protein in lumen of rough ER.

2 Glycoprotein is transported to Golgi complex, where it is further modified.

Membrane of Golgi complex

Nucleus

Rough ER

Transport vesicle

Golgi complex

Transport vesicle

Carbohydrate chain

3 Transport vesicle delivers glycoprotein to plasma membrane.

Plasma membrane

4 Transport vesicle fuses with plasma membrane. Carbohydrate chain extends outward.

Plasma membrane of cell

Figure 5-8 Synthesis and orientation of a membrane protein

The orientation of a protein in the plasma membrane is determined by the pathway of its synthesis and transport. The surface of the rough ER membrane that faces the lumen of the rough ER also faces the lumen of the Golgi complex and membrane transport vesicles. However, when a vesicle fuses with the plasma membrane, the inner surface of the vesicle becomes the extracellular surface of the plasma membrane.
© Cengage Learning

CHECKPOINT 5.1

- *What molecules are responsible for the physical properties of a cell membrane?*
- VISUALIZE *Draw a simple sketch illustrating how a transmembrane protein might be positioned in a lipid bilayer.*
- *What is the pathway used by cells to place carbohydrates on plasma membrane proteins?*

5.2 OVERVIEW OF MEMBRANE PROTEIN FUNCTIONS

LEARNING OBJECTIVE

5 Summarize the functions of membrane proteins.

Why does the plasma membrane require so many different proteins? This diversity reflects the multitude of activities that take place in or on the membrane. Proteins associated with the membrane are essential for most of these activities. Generally, plasma membrane proteins fall into several broad functional categories, as shown in FIGURE 5-9. Some membrane proteins anchor the cell to its substrate. For example, *integrins,* proteins bound to microfilaments inside the cell, attach the cell to the extracellular matrix (Fig. 5-9a). Integrins also serve as receptors, or docking sites, for proteins of the extracellular matrix (see Fig. 4-29).

Many membrane proteins are involved in the transport of molecules across the membrane. Some form channels that selectively allow the passage of specific ions or molecules (Fig. 5-9b). Other proteins are pumps that use ATP, or other energy sources, to actively transport solutes across the membrane (Fig. 5-9c).

Certain membrane proteins are enzymes that catalyze reactions near the cell surface (Fig. 5-9d). In mitochondrial or chloroplast membranes, enzymes that catalyze a series of reactions in cellular respiration or photosynthesis may be organized together in sequence to allow the organelle to efficiently regulate those reactions.

Some membrane proteins are receptors that receive information from other cells in the form of chemical or electrical signals. Most vertebrate cells have receptors for hormones released by endocrine glands. Information may be transmitted from

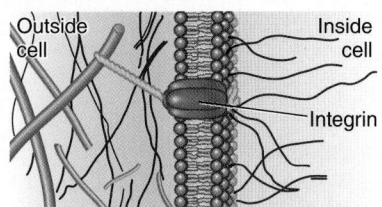

(a) Anchoring. Some membrane proteins, such as integrins, anchor the cell to the extracellular matrix; they also connect to microfilaments within the cell.

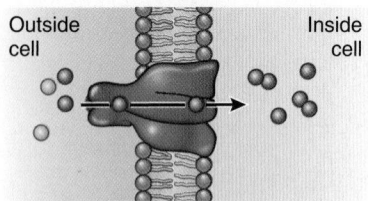

(b) Passive transport. Certain proteins form channels for selective passage of ions or molecules.

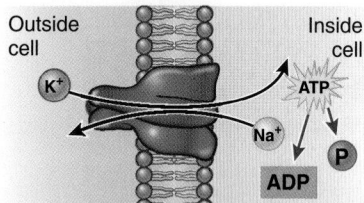

(c) Active transport. Some transport proteins pump solutes across the membrane, which requires a direct input of energy.

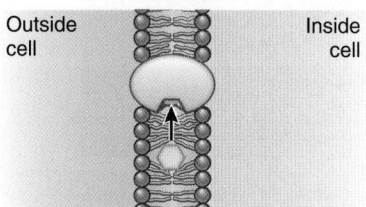

(d) Enzymatic activity. Many membrane-bound enzymes catalyze reactions that take place within or along the membrane surface.

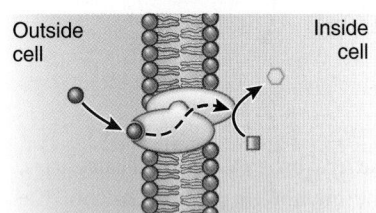

(e) Signal transduction. Some receptors bind with signal molecules such as hormones and transmit information into the cell.

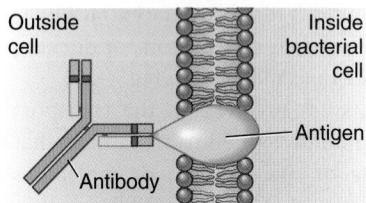

(f) Cell recognition. Some glycoproteins function as identification tags. For example, bacterial cells have surface proteins, or antigens, that human cells recognize as foreign.

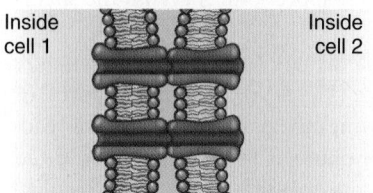

(g) Intercellular junction. Cell adhesion proteins attach membranes of adjacent cells.

Figure 5-9 Some functions of membrane proteins

Cell proteins perform many functions, including transporting materials, serving as enzymes for chemical reactions, and transmitting information.
© Cengage Learning

proteins in the plasma membrane to the cell interior by *signal transduction* (discussed in Chapter 6; Fig. 5-9e).

Some membrane proteins serve as identification tags that other cells recognize. For example, certain cells recognize the surface proteins, or *antigens,* of bacterial cells as foreign. Antigens stimulate immune defenses that destroy the bacteria (Fig. 5-9f).

When certain cells recognize one another, they connect to form tissues. Some membrane proteins form junctions between adjacent cells (Fig. 5-9g). These proteins may also serve as anchoring points for networks of cytoskeletal elements. In the remaining sections of this chapter, we will discuss the functions of cell membrane proteins in transporting material into and out of the cell, and we will discuss junctions between cells. We will discuss other functions of cell membranes in many of the chapters that follow.

CHECKPOINT 5.2

- *How do proteins function in transporting materials into the cell?*
- *What roles do membrane proteins play in cell interactions?*

5.3 CELL MEMBRANE STRUCTURE AND PERMEABILITY

LEARNING OBJECTIVE

6 Describe the importance of selectively permeable membranes and compare the functions of carrier proteins and channel proteins.

A membrane is *permeable* to a given substance if it allows that substance to pass through and impermeable if it does not. The fluid mosaic structure of biological membranes allows them to function as **selectively permeable membranes;** that is, they let some, but not all, substances pass through them. In response to varying environmental conditions or cell needs, a membrane may be a barrier to a particular substance at one time and actively promote its passage at another time. By regulating chemical traffic across its plasma membrane, a cell controls its volume and its internal ionic and molecular composition. This regulation allows the molecular composition of the cell to be quite different from that of its external environment.

Biological membranes present a barrier to polar molecules

In general, biological membranes are most permeable to small nonpolar (hydrophobic) molecules. Such molecules can pass through the hydrophobic lipid bilayer. Gases such as oxygen and carbon dioxide are small, nonpolar molecules that cross the lipid bilayer rapidly. Although they are polar, water molecules are small enough to pass through gaps that occur as a fatty acid chain momentarily moves out of the way. As a result, water molecules slowly cross the lipid bilayer.

The lipid bilayer of the plasma membrane is relatively impermeable to charged ions of any size, so ions and most large polar molecules pass through the bilayer slowly. Ions are important in cell signaling and many other physiological processes. For example, many cell processes, such as muscle contraction, depend on changes in the cytoplasmic concentration of calcium ions. Glucose, amino acids, and most other compounds required in metabolism are polar molecules that also pass through the lipid bilayer slowly. This property is advantageous to cells because the impermeability of the plasma membrane prevents them from diffusing out. How then do cells obtain the ions and polar molecules they require?

Transport proteins transfer molecules across membranes

Systems of *transport proteins* that move ions, amino acids, sugars, and other needed polar molecules through membranes apparently evolved very early in the origin of cells. These transmembrane proteins have been found in all biological membranes. Two main types of membrane transport proteins are carrier proteins and channel proteins. Each type of transport protein transports a specific type of ion or molecule or a group of related substances.

Carrier proteins, also called *transporters,* bind specific ions or molecules and undergo changes in shape, resulting in movement of the solute across the membrane. Transfer of solutes by carrier proteins located within the membrane is called **carrier-mediated transport.** As we will discuss, the two forms of carrier-mediated transport—facilitated diffusion and carrier-mediated active transport—differ in their capabilities and energy sources.

ABC transporters make up a large, important group of carrier proteins. The acronym *ABC* stands for ATP-binding cassette. Found in the cell membranes of all species, ABC transporters use energy donated by ATP to transport certain ions, sugars, and polypeptides across cell membranes. Scientists have identified about 48 ABC transporters in human cells. Mutations in the genes encoding these proteins cause or contribute to many human disorders, including cystic fibrosis and certain neurological diseases. ABC transporters transport hydrophobic drugs out of the cell. This response can be a problem clinically because certain transporters remove antibiotics, antifungal drugs, and anticancer drugs.

Channel proteins form tunnels, called pores, through the membrane. Many of these channels are *gated,* which means that they can be opened and closed. Cells regulate the passage of materials through the channels by opening and closing the gates in response to electrical changes, chemical stimuli, or mechanical stimuli. Water and specific types of ions are transported through channels. There are numerous ion channels in every membrane of every cell.

Porins are transmembrane channel proteins that allow various solutes or water to pass through membranes. These channel proteins are rolled-up, barrel-shaped β-pleated sheets that form pores. Researchers Peter Agre of the Johns Hopkins School of Medicine in Baltimore, Maryland, and Roderick MacKinnon of the Howard Hughes Medical Institute at Rockefeller University in New York City shared the 2003 Nobel Prize in Chemistry for their work on transport proteins. Agre identified transmembrane proteins called **aquaporins** that function as gated water channels.

Aquaporins facilitate the rapid transport of water through the plasma membrane. About a billion water molecules per second can pass through an aquaporin! These channels are very selective and do not permit passage of ions and other small molecules. In some cells, such as those lining the kidney tubules of mammals, aquaporins respond to specific signals from hormones. Aquaporins help prevent dehydration by returning water from the kidney tubules into the blood.

CHECKPOINT 5.3

- *What types of molecules pass easily through the plasma membrane?*
- *What are the two main types of transport proteins? What are their functions?*
- *What are aquaporins? What is their function?*

5.4 PASSIVE TRANSPORT

LEARNING OBJECTIVES

7 Contrast simple diffusion with facilitated diffusion.

8 Define *osmosis* and solve simple problems involving osmosis; for example, predict whether cells will swell or shrink under various osmotic conditions.

Passive transport does not require the cell to expend metabolic energy. Many ions and small molecules move through membranes by *diffusion.* Two types of diffusion are simple diffusion and facilitated diffusion.

Diffusion occurs down a concentration gradient

Some substances pass into or out of cells and move about within cells by **diffusion,** a physical process based on random motion. All atoms and molecules possess kinetic energy, or energy of motion, at temperatures above absolute zero (0 K, $-273°C$, or $-459.4°F$). Matter may exist as a solid, liquid, or gas, depending on the freedom of movement of its constituent particles (atoms, ions, or molecules). The particles of a solid are closely packed, and the forces of attraction between them let them vibrate but not move around. In a liquid the particles are farther apart; the intermolecular attractions are weaker, and the particles move about with considerable freedom. In a gas the particles are so far apart that intermolecular forces are negligible; molecular movement is restricted only by the walls of the container that encloses the gas. Atoms and molecules in liquids and gases move in a kind of "random walk," changing directions as they collide.

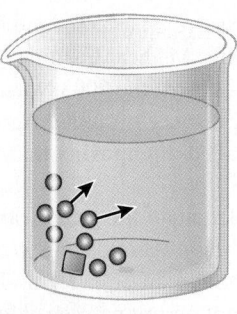 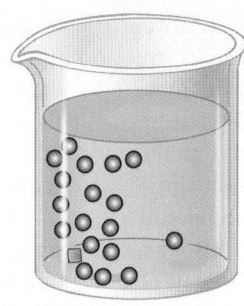 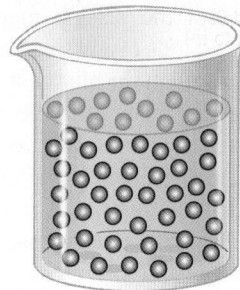

① When lump of sugar is dropped into beaker of pure water, sugar molecules begin to dissolve and diffuse through water.

② Sugar molecules continue to dissolve and diffuse through water.

③ Eventually, sugar molecules become distributed randomly throughout water.

Figure 5-10 Diffusion
© Cengage Learning

Although the movement of individual particles is undirected and unpredictable, we can nevertheless make predictions about the behavior of groups of particles. If the particles are not evenly distributed, at least two regions exist: one with a higher concentration of particles and the other with a lower concentration. Such a difference in the concentration of a substance from one place to another establishes a **concentration gradient.**

In diffusion the random motion of particles results in their net movement "down" their own concentration gradient, from the region of higher concentration to the one of lower concentration. This motion does not mean that individual particles are prohibited from moving "against" the gradient. However, because there are initially more particles in the region of high concentration, it logically follows that more particles move randomly from there into the low-concentration region than the reverse (**FIG. 5-10**).

Thus, if a membrane is permeable to a substance, there is net movement from the side of the membrane where it is more highly concentrated to the side where it is less concentrated. Such a gradient across the membrane is a form of stored energy. Stored energy is *potential energy,* which is the capacity to do work as a result of position or state. The stored energy of the concentration gradient is released when ions or molecules move from a region of high concentration to one of low concentration. For this reason, movement down a concentration gradient is spontaneous. (Forms of energy and spontaneous processes are discussed in greater detail in Chapter 7.)

Diffusion occurs rapidly over very short distances. The rate of diffusion is determined by the movement of the particles, which in turn is a function of their size and shape, their electric charges, and the temperature. As the temperature rises, particles move faster, and the rate of diffusion increases.

Particles of different substances in a mixture diffuse independently of one another. Diffusion moves solutes toward a state of equilibrium. If particles are not added to or removed from the system, a state of **dynamic equilibrium** is reached. In this condition, the particles are uniformly distributed, and there is no net change in the system. Particles continue to move back and forth across the membrane, but they move at equal rates and in both directions.

In organisms equilibrium is rarely attained. For example, human cells continually produce carbon dioxide as sugars and other molecules are metabolized during aerobic respiration. Carbon dioxide readily diffuses across the plasma membrane but then is rapidly removed by the blood. This limits the opportunity for the molecules to re-enter the cell, so a sharp concentration gradient of carbon dioxide molecules always exists across the plasma membrane.

In **simple diffusion** through a biological membrane, small, nonpolar (uncharged) solute molecules move directly through the membrane down their concentration gradient. Oxygen and carbon dioxide can rapidly diffuse through the membrane. The rate of simple diffusion is directly related to the concentration of the solute; the more concentrated the solute, the more rapid the diffusion.

Osmosis is diffusion of water across a selectively permeable membrane

Osmosis is a special kind of diffusion that involves the net movement of water (the principal *solvent* in biological systems) through a selectively permeable membrane from a region of higher concentration of water to a region of lower concentration. Water molecules pass freely in both directions, but as in all types of diffusion, *net* movement is from the region where the water molecules are more concentrated to the region where they are less concentrated. Most *solute molecules* (such as sugar and salt) cannot diffuse freely through the selectively permeable membranes of the cell.

The principles involved in osmosis can be illustrated using an apparatus called a U-tube (**FIG. 5-11**). The U-tube is divided into two sections by a selectively permeable membrane that allows solvent (water) molecules to pass freely but excludes solute molecules. A water/solute solution is placed on one side, and pure water is placed on the other. The side containing the solute has a lower *effective water concentration* than the pure water side does. The reason is that the solute particles, which are charged (ionic) or polar, interact with the partial electric charges on the polar water molecules. Many of the water molecules are thus "bound up" and no longer free to diffuse across the membrane.

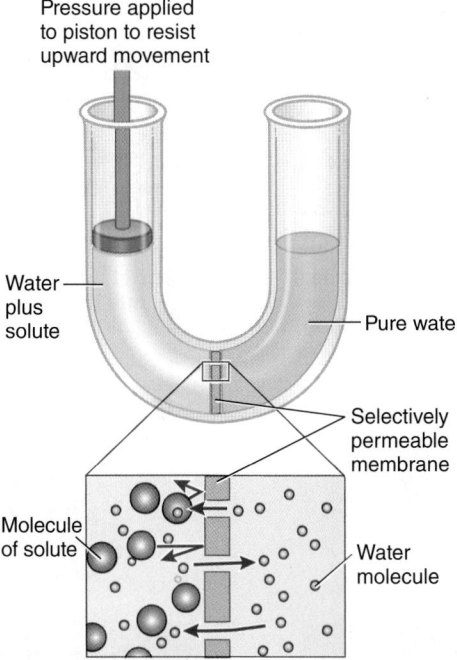

Figure 5-11 *Animation* **Osmosis**

The U-tube contains pure water on the right and water plus a solute on the left, separated by a selectively permeable membrane. Water molecules cross the membrane in both directions (*blue arrows*). Solute molecules cannot cross (*red arrows*). The fluid level would normally rise on the left and fall on the right because net movement of water would be to the left. However, the piston prevents the water from rising. The force that must be exerted by the piston to prevent the rise in fluid level is equal to the osmotic pressure of the solution.
© Cengage Learning

Because of the difference in effective water concentration, there is net movement of water molecules from the pure water side (with a high effective concentration of water) to the water/solute side (with a lower effective concentration of water). As a result, the fluid level drops on the pure water side and rises on the water/solute side. Because the solute molecules do not diffuse across the membrane, equilibrium is never attained. Net movement of water continues, and the fluid level rises on the side containing the solute. The weight of the rising column of fluid eventually exerts enough pressure to stop further changes in fluid levels, although water molecules continue to pass through the selectively permeable membrane in both directions.

We define the **osmotic pressure** of a solution as the pressure that must be exerted on the side of a selectively permeable membrane containing the higher concentration of solute to prevent the diffusion of water (by osmosis) from the side containing the lower solute concentration. In the U-tube example, you could measure the osmotic pressure by inserting a piston on the water/solute side of the tube and measuring how much pressure must be exerted by the piston to prevent the rise of fluid on that side of the tube. A solution with a high solute concentration has a low effective water

concentration and a high osmotic pressure; conversely, a solution with a low solute concentration has a high effective water concentration and a low osmotic pressure.

Two solutions may be isotonic Salts, sugars, and other substances are dissolved in the fluid compartment of every cell. These solutes give the cytosol a specific osmotic pressure. **TABLE 5-1** summarizes the movement of water into and out of a solution (or cell) depending on relative solute concentrations.

When a cell is placed in a fluid with exactly the same osmotic pressure, no net movement of water molecules occurs, either into or out of the cell. The cell neither swells nor shrinks. Such a fluid is of equal solute concentration, or **isotonic,** to the fluid within the cell. Normally, your blood plasma (the fluid component of blood) and all your other body fluids are isotonic to your cells; they contain a concentration of water equal to that in the cells. A solution of 0.9% sodium chloride (sometimes called *physiological saline*) is isotonic to the cells of humans and other mammals. Human red blood cells placed in 0.9% sodium chloride neither shrink nor swell (**FIG. 5-12a**).

One solution may be hypertonic and the other hypotonic If the surrounding fluid has a concentration of dissolved substances greater than the concentration within the cell, the fluid has a higher osmotic pressure than the cell and is said to be **hypertonic** to the cell. Because a hypertonic solution has a lower effective water concentration, a cell placed in such a solution shrinks as it loses water by osmosis. Human red blood cells placed in a solution of 1.3% sodium chloride shrivel (**FIG. 5-12b**).

If the surrounding fluid contains a lower concentration of dissolved materials than does the cell, the fluid has a lower osmotic pressure and is said to be **hypotonic** to the cell; water then enters the cell and causes it to swell. Red blood cells placed in a solution of 0.6% sodium chloride gain water, swell (**FIG. 5-12c**), and may eventually burst. Many cells that normally live in hypotonic environments have adaptations to prevent excessive water accumulation. For example, *Paramecium* and certain other ciliates (members of the supergroup Chromalveolates) have contractile vacuoles that expel excess water (see Fig. 26-8).

Turgor pressure is the internal hydrostatic pressure usually present in walled cells The cells of most prokaryotes, algae, plants, and fungi have relatively rigid cell walls. These cells can withstand, without bursting, an external medium that is very dilute, containing only a very low concentration of solutes. Because of the substances dissolved in the cytoplasm, the cells are hypertonic to the outside medium (conversely, the

TABLE 5-1	Osmotic Terminology		
SOLUTE CONCENTRATION IN SOLUTION A	SOLUTE CONCENTRATION IN SOLUTION B	TONICITY	DIRECTION OF NET MOVEMENT OF WATER
Greater	Less	A hypertonic to B; B hypotonic to A	B to A
Less	Greater	B hypertonic to A; A hypotonic to B	A to B
Equal	Equal	A and B are isotonic to each other	No net movement

© Cengage Learning

outside medium is hypotonic to the cytoplasm).

Water moves into the cells by osmosis, filling their central vacuoles and distending the cells. The cells swell, building up **turgor pressure** against the rigid cell walls (FIG. 5-13a). The cell walls stretch only slightly, and a steady state is reached when their resistance to stretching prevents any further increase in cell size and thereby halts the *net movement* of water molecules into the cells. (Of course, molecules continue to move back and forth across the plasma membrane.) Turgor pressure in the cells is an important factor in supporting the body of nonwoody plants.

If a cell that has a cell wall is placed in a hypertonic medium, the cell loses water to its surroundings. Its contents shrink, and the plasma membrane separates from the cell wall, a process known as **plasmolysis** (FIGS. 5-13b and 5-13c). Plasmolysis occurs in plants when the soil or water around them contains high concentrations of salts or fertilizers. It also explains why lettuce becomes limp in a salty salad dressing and why a picked flower wilts from lack of water.

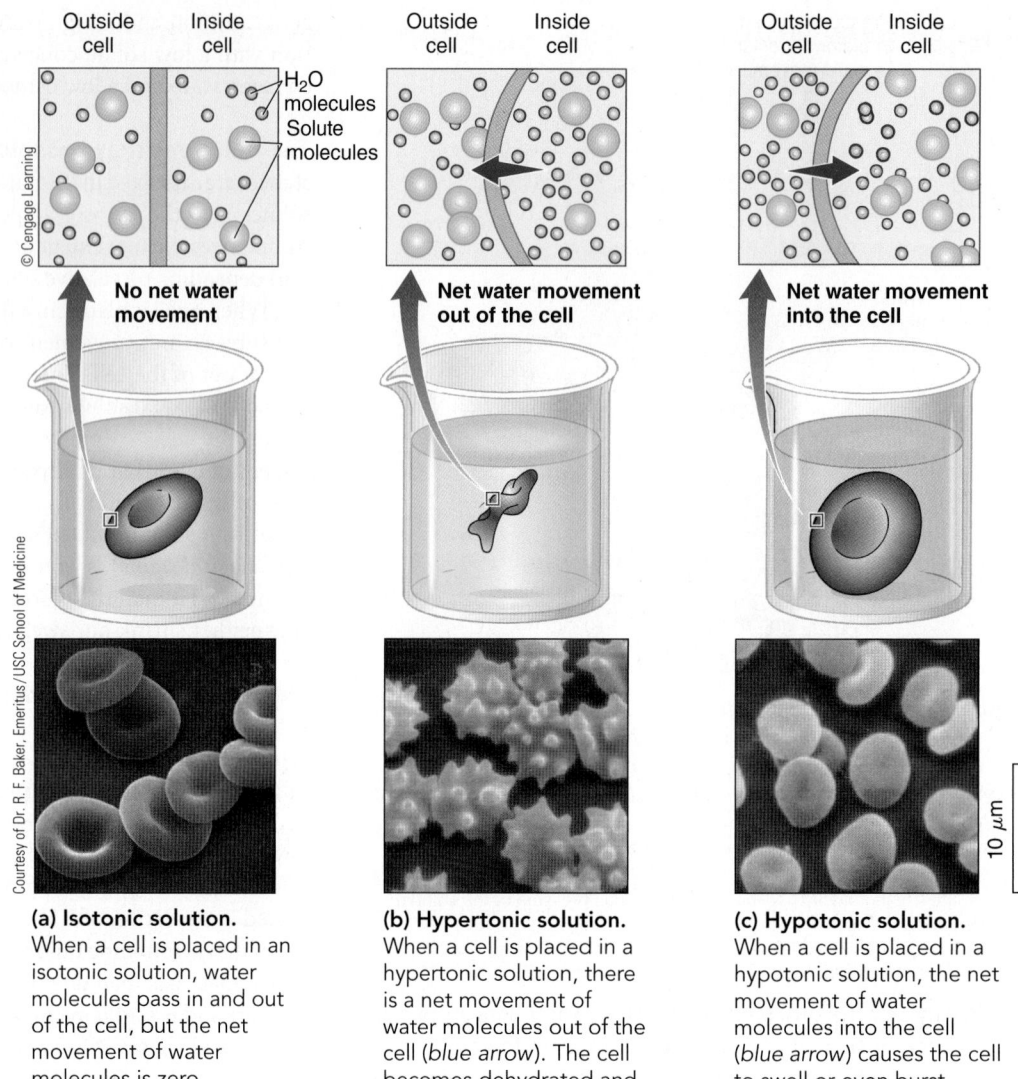

(a) Isotonic solution.
When a cell is placed in an isotonic solution, water molecules pass in and out of the cell, but the net movement of water molecules is zero.

(b) Hypertonic solution.
When a cell is placed in a hypertonic solution, there is a net movement of water molecules out of the cell (*blue arrow*). The cell becomes dehydrated and shrunken.

(c) Hypotonic solution.
When a cell is placed in a hypotonic solution, the net movement of water molecules into the cell (*blue arrow*) causes the cell to swell or even burst.

Figure 5-12 *Animation* **The responses of animal cells to osmotic pressure differences**

Facilitated diffusion occurs down a concentration gradient

In all processes in which substances move across membranes by diffusion, the net transfer of those molecules from one side to the other occurs as a result of a concentration gradient. We have seen that small, uncharged (nonpolar) solute molecules, such as oxygen and carbon dioxide, move directly through the membrane down their concentration gradient by simple diffusion. In **facilitated diffusion,** a specific transport protein makes the membrane permeable to a particular solute, such as a specific ion or polar molecule. A specific solute can be transported from inside the cell to the outside or from the outside to the inside, but net movement is always from a region of higher solute concentration to a region of lower concentration. Channel proteins and carrier proteins facilitate diffusion by different mechanisms.

Channel proteins form hydrophilic channels through membranes Most channel proteins form narrow channels that

transport specific ions down their gradients (FIG. 5-14). (As we will discuss, because ions are charged particles, these gradients are electrochemical gradients.) These *ion channels* are referred to as *gated channels* because they can open and close. As many as 100 million ions per second can pass through an open ion channel! Channels can facilitate transport only down a concentration gradient. They cannot actively transport ions from a region of lower concentration to a region of higher concentration.

Carrier proteins undergo a change in shape Transport of solutes through carrier proteins is slower than through channel proteins. The carrier protein binds with one or more solute molecules on one side of the membrane. The protein then undergoes a conformational change (change in shape) that moves the solute to the other side of the membrane.

As an example of facilitated diffusion by a carrier protein, let us consider glucose transport. A carrier protein known as *glucose transporter 1,* or *GLUT 1,* transports glucose into red blood cells (FIG. 5-15). Each red blood cell has approximately

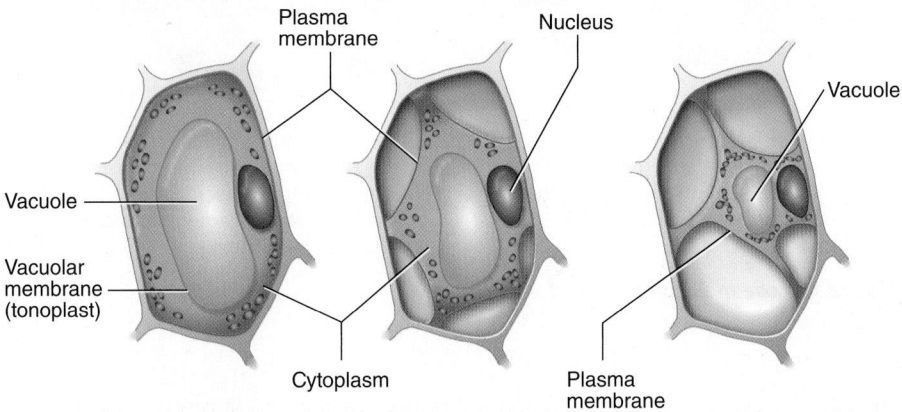

(a) In hypotonic surroundings, the vacuole of a plant cell fills with water, but the rigid cell walls prevent the cell from expanding. The cells of this healthy plant are turgid.

(b) When the plant is exposed to a hypertonic solution, its cells become plasmolyzed as they lose water.

(c) The plant wilts and eventually dies.

Figure 5-13 *Animation* **Turgor pressure and plasmolysis**
© Cengage Learning

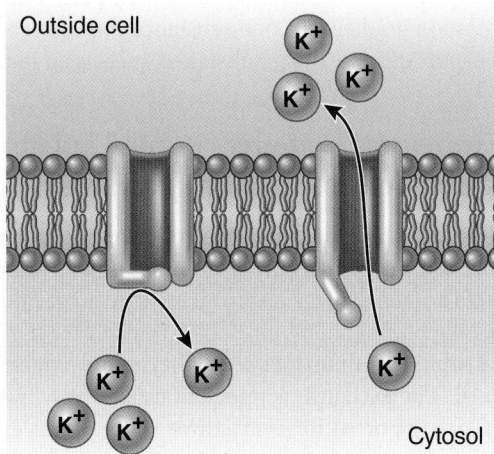

Figure 5-14 Facilitated diffusion of potassium ions
In response to an electrical stimulus, the gate of the potassium ion channel opens, allowing potassium to diffuse out of the cell.
© Cengage Learning

200,000 GLUT 1 transporters in its plasma membrane. The concentration of glucose is higher in the blood plasma than in red blood cells, so glucose diffuses down its concentration gradient into these blood cells. The GLUT 1 transporter facilitates glucose diffusion, allowing glucose to enter the cell about 50,000 times as rapidly as it could by simple diffusion.

Red blood cells keep the internal concentration of glucose low by immediately adding a phosphate group to entering glucose molecules, converting them to highly charged glucose phosphates that cannot pass back through the membrane. Because glucose phosphate is a different molecule, it does not contribute to the glucose concentration gradient. Thus, a steep concentration gradient for glucose is continually maintained, and glucose rapidly diffuses into the cell, only to be immediately changed to the phosphorylated form. Facilitated diffusion is powered by the concentration gradient.

Researchers have studied facilitated diffusion of glucose using *liposomes,* artificial vesicles enclosed by phospholipid bilayers. The phospholipid membrane of a liposome does not allow the passage of glucose unless a glucose transporter has been incorporated into the liposome membrane. Glucose transporters and similar carrier proteins temporarily bind to the molecules they transport. This mechanism appears to be similar to the way an enzyme binds with its substrate, the molecule on which it acts (discussed in Chapter 7). In addition, as in enzyme action, binding apparently changes the shape of the carrier protein. This change allows the glucose molecule to be released on the inside of the cell. According to this model, when the glucose is released into the cytoplasm, the carrier protein reverts to its original shape and is available to bind another glucose molecule on the outside of the cell.

Because carrier proteins bind to their substrate and undergo changes in shape, they operate at a defined maximum rate, which ranges from about 100 to 10,000 ions (or molecules) per second. Unlike channel proteins, carrier proteins become saturated when there is a high concentration of the molecule being transported because each cell has a finite number of each type of carrier protein. When the concentration of solute molecules to be transported reaches a certain level, all the carrier proteins are working at their maximum rate.

It is a common misconception that diffusion, whether simple or facilitated, is somehow "free of cost" and that only active transport mechanisms require energy. Because diffusion always

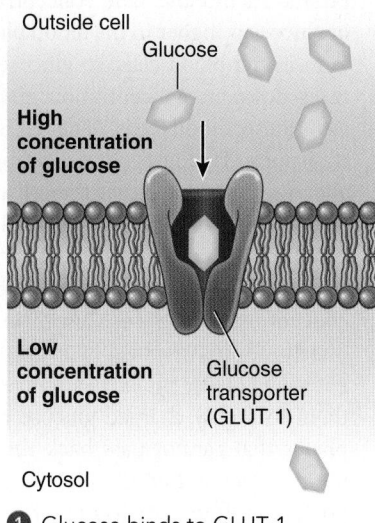
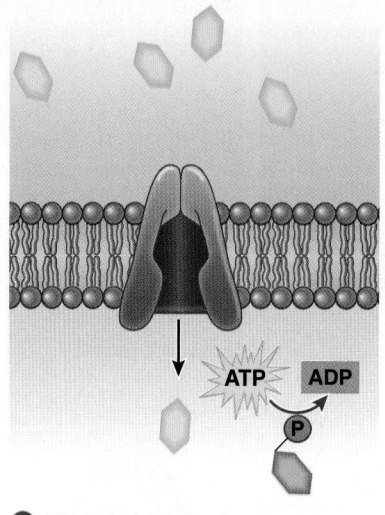
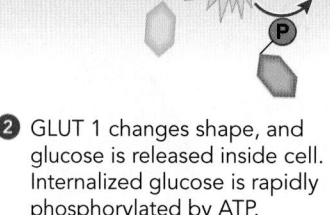
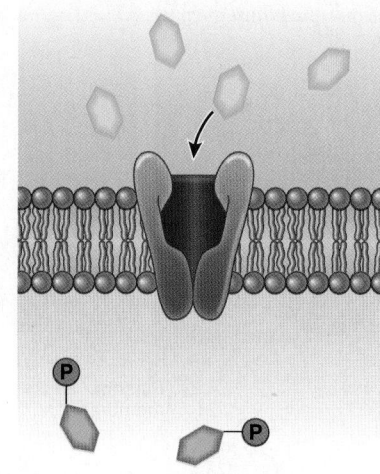

Outside cell

Glucose

High concentration of glucose

Low concentration of glucose

Glucose transporter (GLUT 1)

Cytosol

1 Glucose binds to GLUT 1.

ATP   ADP

2 GLUT 1 changes shape, and glucose is released inside cell. Internalized glucose is rapidly phosphorylated by ATP.

3 GLUT 1 returns to its original shape. Phosphorylation of internalized glucose molecules maintains a large concentration gradient across the plasma membrane.

Figure 5-15 *Animation* **Facilitated diffusion of glucose molecules**
Facilitated diffusion requires the potential energy of a concentration gradient.
© Cengage Learning

involves the net movement of a substance down its concentration gradient, we say that the concentration gradient "powers" the process. However, energy is required to do the work of establishing and maintaining the gradient. In our example of facilitated diffusion of glucose, the cell maintains a steep concentration gradient (high outside, low inside) by phosphorylating the glucose molecules once they enter the cell. One ATP molecule is spent for every glucose molecule phosphorylated, and there are additional costs, such as the energy required to make the enzymes that carry out the reaction.

CHECKPOINT 5.4

- **PREDICT** *What would happen if a plant cell were placed in an isotonic solution? a hypertonic environment? a hypotonic environment? How would you modify your predictions for an animal cell?*

- *What is the immediate source of energy for simple diffusion? for facilitated diffusion?*

- *In what direction is there a net movement of particles along their concentration gradient? Would your answers be different for facilitated diffusion compared with simple diffusion?*

5.5 ACTIVE TRANSPORT

LEARNING OBJECTIVE

9   Describe active transport, including cotransport.

Although adequate amounts of a few substances move across cell membranes by diffusion, cells must actively transport many solutes *against* a concentration gradient. The reason is that cells

require many substances in higher concentrations than their concentration outside the cell.

Both diffusion and active transport require energy. The energy for diffusion is provided by a concentration gradient for the substance being transported. Active transport requires the cell to expend metabolic energy directly to power the process.

An **active transport** system can pump materials from a region of low concentration to a region of high concentration. The energy stored in the concentration gradient not only is unavailable to the system but actually works against it. For this reason, the cell needs some other source of energy. In many cases, cells use ATP energy directly. However, active transport may be coupled to ATP indirectly. In indirect active transport, a concentration gradient for one substance provides energy for the cotransport of some other substance, such as an ion.

Active transport systems "pump" substances against their concentration gradients

One of the most striking examples of an active transport mechanism is the **sodium–potassium pump** found in all animal cells (**FIG. 5-16**). The pump is an ABC transporter, a specific carrier protein in the plasma membrane. This transporter uses energy from ATP to pump sodium ions out of the cell and potassium ions into the cell. The exchange is unequal: usually only two potassium ions are imported for every three sodium ions exported. Because these particular concentration gradients involve ions, an electrical potential (separation of electric charges) is generated across the membrane; that is, the membrane is *polarized*.

Both sodium and potassium ions are positively charged, but because there are fewer potassium ions inside relative to the

The sodium–potassium pump is a carrier protein that maintains an electrochemical gradient across the plasma membrane.

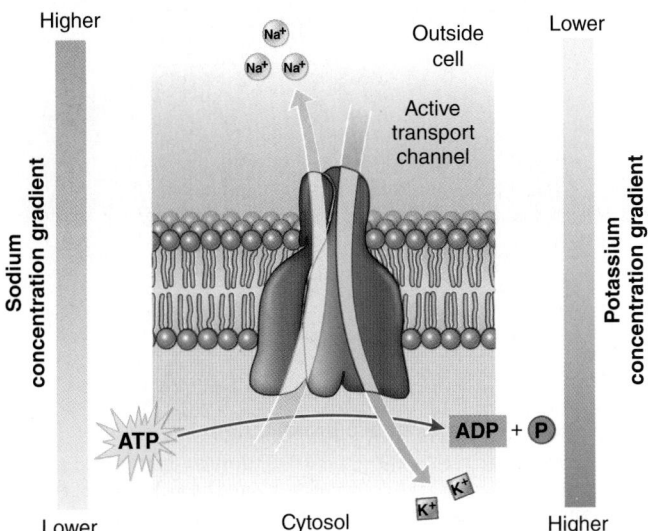

Figure 5-16 A model for the pumping cycle of the sodium–potassium pump

PREDICT What would be the effect on the membrane's electro-chemical gradient if the Na⁺/K⁺ pump moved an equal number of Na⁺ and K⁺ ions in each pump cycle? Would this produce an electrical gradient? Would this produce an ion (or chemical) gradient?
© Cengage Learning

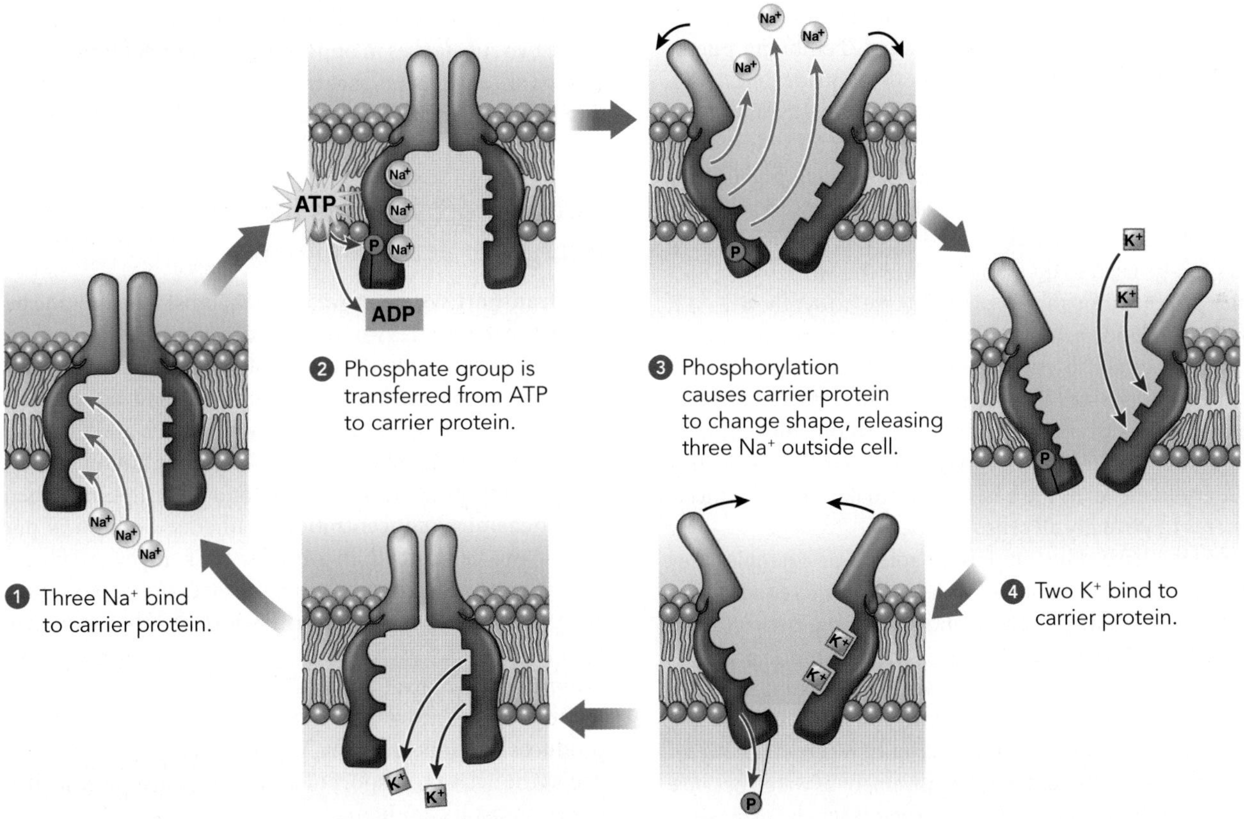

(a) The sodium–potassium pump is a carrier protein that requires energy from ATP. In each complete pumping cycle, the energy of one molecule of ATP is used to export three sodium ions (Na⁺) and import two potassium ions (K⁺).

❶ Three Na⁺ bind to carrier protein.

❷ Phosphate group is transferred from ATP to carrier protein.

❸ Phosphorylation causes carrier protein to change shape, releasing three Na⁺ outside cell.

❹ Two K⁺ bind to carrier protein.

❺ Phosphate is released.

❻ Phosphate release causes carrier protein to return to its original shape. Two K⁺ ions are released inside cell.

(b) Follow the steps illustrating a model of active transport by the sodium–potassium pump.

sodium ions outside, the inside of the cell is negatively charged relative to the outside. The unequal distribution of ions establishes an *electrical gradient* that drives ions across the plasma membrane. Sodium–potassium pumps help maintain a separation of charges across the plasma membrane. This separation is called a **membrane potential.** Because there is both an electric charge difference and a concentration difference on the two sides of the membrane, the gradient is called an **electrochemical gradient.** Such gradients store energy that is used to drive other transport systems. So important is the electrochemical gradient produced by these pumps that some cells (such as nerve cells) expend more than 25% of their total available energy just to power this one transport system.

Sodium–potassium pumps (as well as all other ATP-driven pumps) are transmembrane proteins that extend entirely through the membrane. By undergoing a series of conformational changes, the pumps exchange sodium for potassium across the plasma membrane. Unlike what occurs in facilitated diffusion, at least one of the conformational changes in the pump cycle requires energy, which is provided by ATP. The shape of the pump protein changes as a phosphate group from ATP first binds to it and is subsequently removed later in the pump cycle.

The use of electrochemical membrane potentials for energy storage is not confined to the plasma membranes of animal cells. Cells of bacteria, fungi, and plants use carrier proteins, known as proton pumps, to actively transport hydrogen ions (which are protons) out of the cell. These ATP-driven membrane pumps transfer protons from the cytosol to the outside (FIG. 5-17). Removal of positively charged protons from the cytoplasm of these cells results in a large difference in the concentration of protons between the outside and inside of the cell. The outside of the cells is positively charged relative to the inside of the plasma membrane. The energy stored in these electrochemical gradients can be used to do many kinds of cell work.

Other proton pumps are used in "reverse" to synthesize ATP. Bacteria, mitochondria, and chloroplasts use energy from food or sunlight to establish proton concentration gradients (discussed in Chapters 8 and 9). When the protons diffuse through the proton carriers from a region of high proton concentration to one of low concentration, ATP is synthesized. These electrochemical gradients form the basis for the major energy conversion systems in virtually all cells.

Ion pumps have other important roles. For example, they are instrumental in the ability of an animal cell to equalize the osmotic pressures of its cytoplasm and its external environment. If an animal cell does not control its internal osmotic pressure, its contents become hypertonic relative to the exterior. Water enters by osmosis, causing the cell to swell and possibly burst (see Fig. 5-12c). By controlling the ion distribution across the membrane, the cell indirectly controls the movement of water, because when ions are pumped out of the cell, water leaves by osmosis.

Carrier proteins can transport one or two solutes

You may have noticed that some carrier proteins, such as proton pumps, transport one type of substance in one direction. These

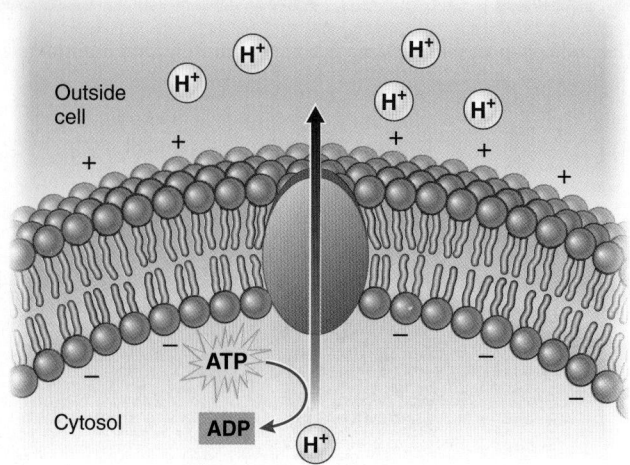

Figure 5-17 A model of a proton pump

Proton pumps use the energy of ATP to transport protons (hydrogen ions) across membranes. The energy of the electrochemical gradient established can then be used for other processes.

© Cengage Learning

carrier proteins are called **uniporters.** Other carrier proteins, **symporters,** move two types of substances in one direction. For example, a specific carrier protein transports both sodium and glucose into the cell. Still other carrier proteins, **antiporters,** move two substances in opposite directions. Sodium–potassium pumps are antiporters that transport sodium ions out of the cell and potassium ions into the cell. Both symporters and antiporters cotransport solutes.

Cotransport systems indirectly provide energy for active transport

A **cotransport** system moves solutes across a membrane by *indirect active transport.* Two solutes are transported at the same time. The movement of one solute down its concentration gradient provides energy for transport of some other solute up its concentration gradient. However, an energy source such as ATP is required to power the pump that produces the concentration gradient.

Sodium–potassium pumps (and other pumps) generate electrochemical concentration gradients. Sodium is pumped out of the cell and then diffuses back in by moving down its concentration gradient. This process generates sufficient energy to power the active transport of other essential substances. In these systems, a carrier protein *cotransports* a solute *against* its concentration gradient while sodium, potassium, or hydrogen ions move *down* their gradient. Energy from ATP produces the ion gradient. Then the energy of this gradient drives the active transport of a required substance, such as glucose, against its gradient.

We have seen how glucose can be moved into the cell by facilitated diffusion. Glucose can also be cotransported into the cell. The sodium concentration inside the cell is kept low by the ATP-requiring sodium–potassium pumps that actively transport sodium ions out of the cell. In glucose cotransport, a carrier protein transports both sodium and glucose (FIG. 5-18).

A carrier protein transports sodium ions down their concentration gradient and uses that energy to cotransport glucose molecules against their concentration gradient.

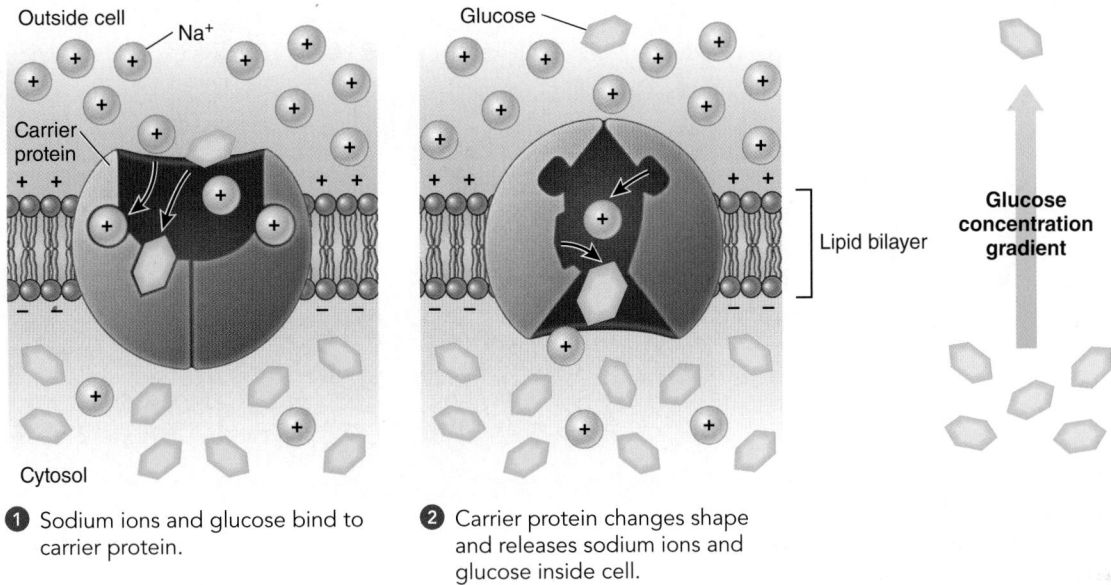

① Sodium ions and glucose bind to carrier protein.

② Carrier protein changes shape and releases sodium ions and glucose inside cell.

Figure 5-18 A model for the cotransport of glucose and sodium ions

This carrier protein is a symporter.

PREDICT What would be the effect on glucose transport through this system if the cell in the figure were treated with a drug that inhibited the Na^+/K^+ pump?

© Cengage Learning

As sodium moves into the cell along its concentration gradient, the carrier protein captures the energy released and uses it to transport glucose into the cell. Thus, this indirect active transport system for glucose is "driven" by the cotransport of sodium.

CHECKPOINT 5.5

- *What is the immediate energy source for active transport?*
- *What is the immediate energy source for cotransport?*

5.6 EXOCYTOSIS AND ENDOCYTOSIS

LEARNING OBJECTIVE

10 Compare exocytotic and endocytotic transport mechanisms.

Individual molecules and ions pass through the plasma membrane by simple and facilitated diffusion and by carrier-mediated active transport. Some larger materials, such as large molecules, particles of food, and even small cells, are also moved into or out of cells. They are transported by exocytosis and endocytosis. Like active transport, these processes require cells to expend energy directly.

In exocytosis, vesicles export large molecules

In **exocytosis,** a cell ejects waste products, or specific products of secretion such as hormones, by the fusion of a vesicle with the plasma membrane (**FIG. 5-19**). As the contents of the vesicle are released from the cell, the membrane of the secretory vesicle is incorporated into the plasma membrane. This process is the primary mechanism by which plasma membranes grow larger.

In endocytosis, the cell imports materials

In **endocytosis,** materials are taken into the cell. Several types of endocytotic mechanisms operate in biological systems, including phagocytosis, pinocytosis, and receptor-mediated endocytosis.

In **phagocytosis** (literally, "cell eating"), the cell ingests large solid particles such as food or bacteria (**FIG. 5-20**). Certain protists ingest food by phagocytosis. Some types of vertebrate cells, including certain white blood cells, ingest bacteria and other particles by phagocytosis. During ingestion, folds of the plasma membrane enclose the cell or particle. When the membrane has encircled the particle, the membrane fuses at the point of contact, forming a vacuole. The vacuole may then fuse with lysosomes, which degrade the ingested material.

In **pinocytosis** ("cell drinking"), the cell takes in dissolved materials (**FIG. 5-21**). Tiny droplets of fluid are trapped by folds in the plasma membrane, which pinch off into the cytosol as tiny

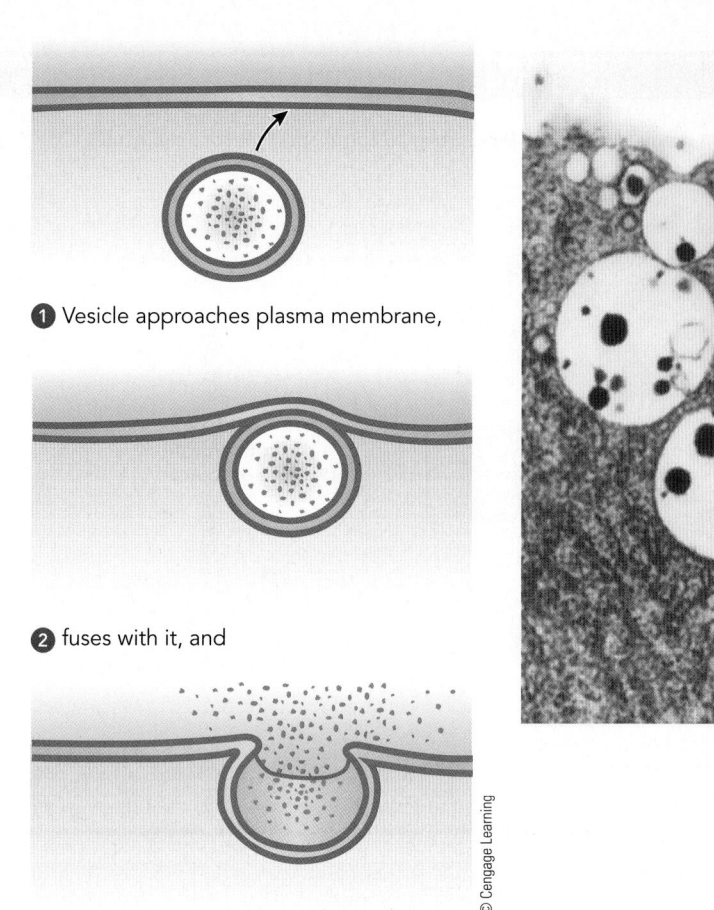

1 Vesicle approaches plasma membrane,

2 fuses with it, and

3 releases its contents outside cell.

© Cengage Learning

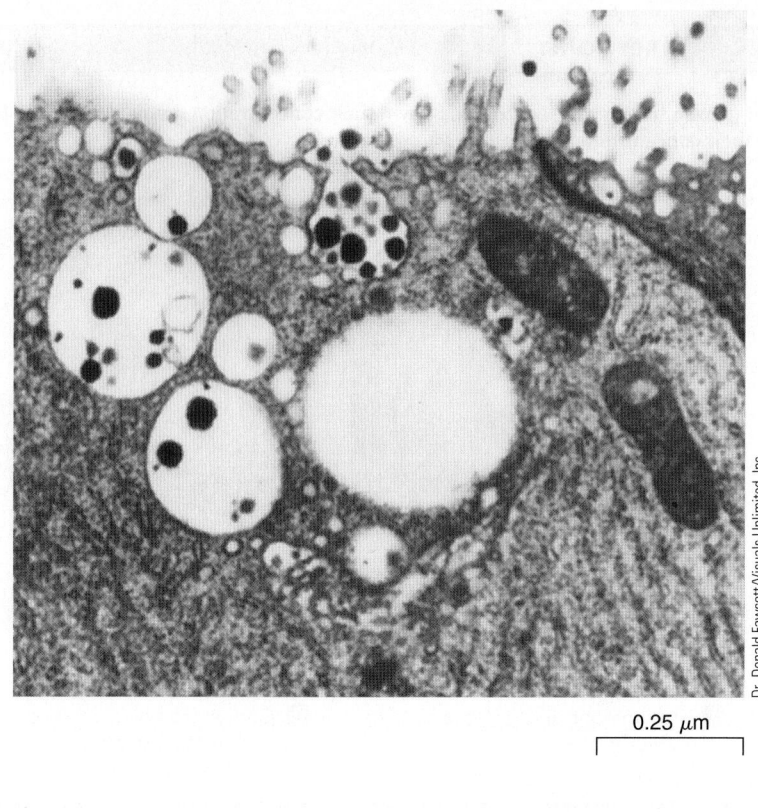

Dr. Donald Fawcett/Visuals Unlimited, Inc.

0.25 µm

Figure 5-19 Exocytosis

The TEM shows exocytosis of the protein components of milk by a mammary gland cell.

vesicles. As the liquid contents of these vesicles are slowly transferred into the cytosol, the vesicles become progressively smaller.

In a third type of endocytosis, **receptor-mediated endocytosis,** specific molecules combine with receptor proteins in the plasma membrane. Receptor-mediated endocytosis is the main mechanism by which eukaryotic cells take in macromolecules. As an example, let us look at how mammalian cells take up cholesterol from the blood. Cells use cholesterol as a component of cell membranes and as a precursor of steroid hormones. Cholesterol is transported in the blood as part of particles called *low-density lipoproteins (LDLs;* popularly known as "bad cholesterol").

Much of the receptor-mediated endocytosis pathway was detailed through studies at the University of Texas Health Science Center by Michael Brown and Joseph Goldstein on the LDL receptor. These researchers were awarded the Nobel Prize in Physiology or Medicine in 1985 for their pioneering work. Their findings have important medical implications because cholesterol that remains in the blood instead of entering the cells can be deposited in the artery walls, which increases the risk of cardiovascular disease.

When it needs cholesterol, the cell makes LDL receptors. The receptors are concentrated in *coated pits,* depressed regions on the cytoplasmic surface of the plasma membrane. Each pit is coated by a layer of a protein, called *clathrin,* found just below the plasma membrane. A molecule that binds specifically to a receptor is called a **ligand.** In this case, LDL is the ligand. After the LDL binds with a receptor, the coated pit forms a *coated vesicle* by endocytosis.

FIGURE 5-22 shows the uptake of an LDL particle. Seconds after the vesicle moves into the cytoplasm, the coating dissociates from it, leaving an uncoated vesicle. The vesicles fuse with small compartments called *endosomes.* The LDL and LDL receptors separate, and the receptors are transported to the plasma membrane, where they are recycled. LDL is transferred to a lysosome, where it is broken down. Cholesterol is released into the cytosol for use by the cell. A simplified summary of receptor-mediated endocytosis follows:

ligand molecules bind to receptors in coated pits of plasma membrane → coated vesicle forms by endocytosis → coating detaches from vesicle → uncoated vesicle fuses with endosome → ligands separate from receptors, which are recycled; endosome fuses with primary lysosome, forming secondary lysosome → contents of secondary lysosome are digested and released into the cytosol

The recycling of LDL receptors to the plasma membrane through vesicles causes a problem common to all cells that use endocytotic and exocytotic mechanisms: the plasma membrane changes size as the vesicles bud off from it or fuse with it. A type of phagocytic cell known as a *macrophage,* for example, ingests the equivalent of its entire plasma membrane in about 30 minutes, requiring an equivalent amount of recycling or new membrane synthesis for the cell to maintain its surface area. On the other hand, cells that are constantly

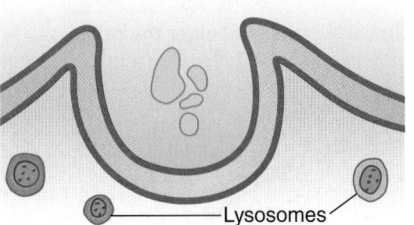

① Folds of plasma membrane surround particle to be ingested, forming small vacuole around it.

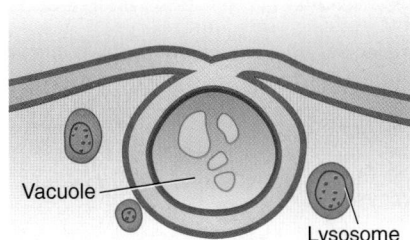

② Vacuole then pinches off inside cell.

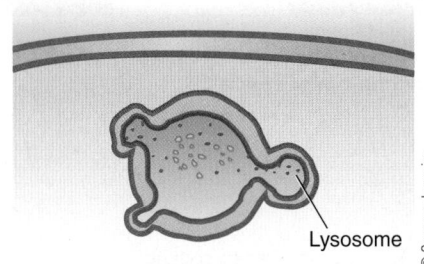

③ Lysosomes fuse with vacuole and pour potent hydrolytic enzymes onto ingested material.

© Cengage Learning

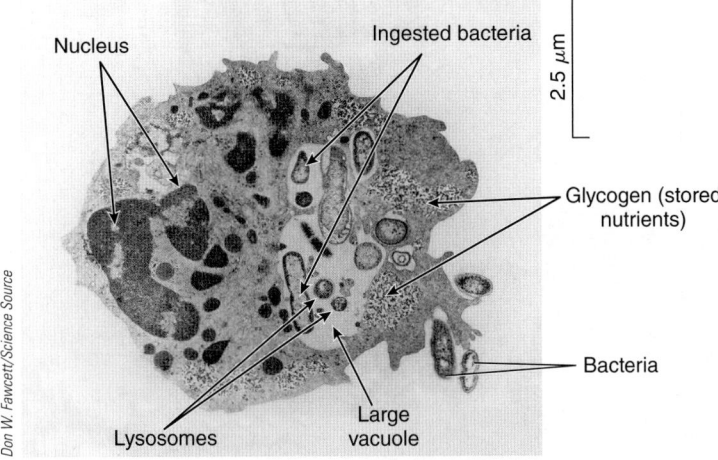

Don W. Fawcett/Science Source

Figure 5-20 *Animation* **Phagocytosis**

In this type of endocytosis, a cell ingests relatively large solid particles. The white blood cell (a neutrophil) shown in the TEM is phagocytizing bacteria. The vacuoles contain bacteria that have already been ingested. Lysosomes contain digestive enzymes that break down the ingested material. Other bacteria are visible outside the cell.

involved in secretion must return an equivalent amount of membrane to the interior of the cell for each vesicle that fuses with the plasma membrane; if not, the cell surface would continue to expand even though the growth of the cell itself may be arrested.

CHECKPOINT 5.6

- *In what ways are exocytosis and endocytosis similar?*
- *How are the processes of phagocytosis and pinocytosis different?*
- *What is the sequence of events in receptor-mediated endocytosis?*

5.7 CELL JUNCTIONS

LEARNING OBJECTIVE

11 Compare the structures and functions of anchoring junctions, tight junctions, gap junctions, and plasmodesmata.

Cells in close contact with one another typically develop specialized intercellular junctions. These structures may allow neighboring cells to form strong connections with one another, prevent the passage of materials, or establish rapid communication between adjacent cells. Several types of junctions connect animal cells, including anchoring junctions, tight junctions, and gap junctions. Plant cells are connected by plasmodesmata.

Anchoring junctions connect cells of an epithelial sheet

Adjacent epithelial cells, such as those found in the outer layer of the mammalian skin, are so tightly bound to each other by *anchoring junctions* that strong mechanical forces are required to separate them. These junctions do not prevent the passage of materials between adjacent cells. Two common types of anchoring junctions are desmosomes and adhering junctions.

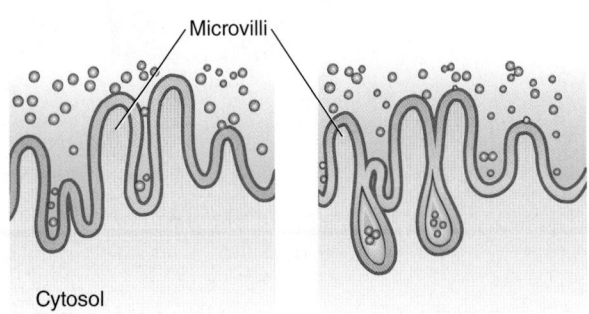

① Tiny droplets of fluid are trapped by folds of plasma membrane.

② These pinch off into cytosol as small fluid-filled vesicles.

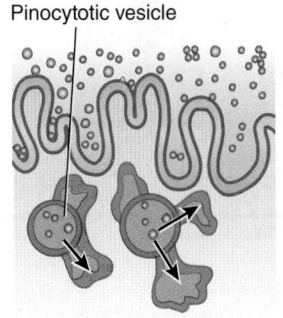

③ Contents of these vesicles are then slowly transferred to cytosol.

Figure 5-21 **Pinocytosis, or "cell drinking"**
© Cengage Learning

In receptor-mediated endocytosis, specific macromolecules bind to receptor proteins, accumulate in coated pits, and enter the cell in clathrin-coated vesicles.

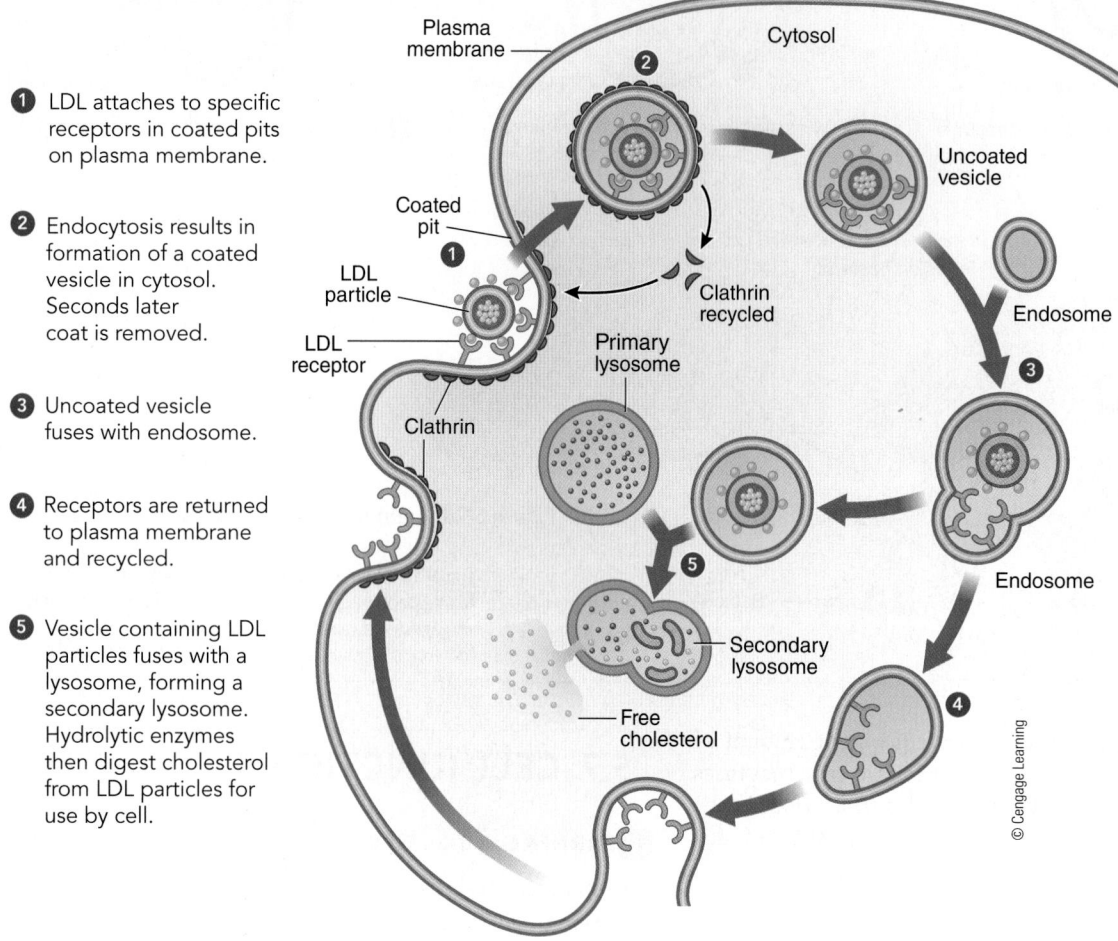

❶ LDL attaches to specific receptors in coated pits on plasma membrane.

❷ Endocytosis results in formation of a coated vesicle in cytosol. Seconds later coat is removed.

❸ Uncoated vesicle fuses with endosome.

❹ Receptors are returned to plasma membrane and recycled.

❺ Vesicle containing LDL particles fuses with a lysosome, forming a secondary lysosome. Hydrolytic enzymes then digest cholesterol from LDL particles for use by cell.

(a) Uptake of low-density lipoprotein (LDL) particles, which transport cholesterol in the blood.

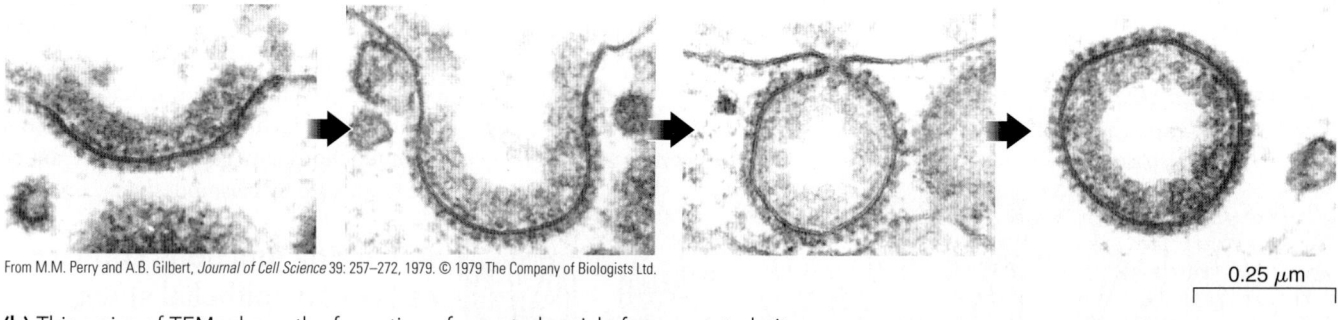

From M.M. Perry and A.B. Gilbert, *Journal of Cell Science* 39: 257–272, 1979. © 1979 The Company of Biologists Ltd.

(b) This series of TEMs shows the formation of a coated vesicle from a coated pit.

Figure 5-22 Receptor-mediated endocytosis

PREDICT In some humans, the membrane-spanning region of the LDL receptor protein is missing. How would that affect the ability of the cell to take up needed cholesterol? What would be the effect on cholesterol levels in the bloodstream?

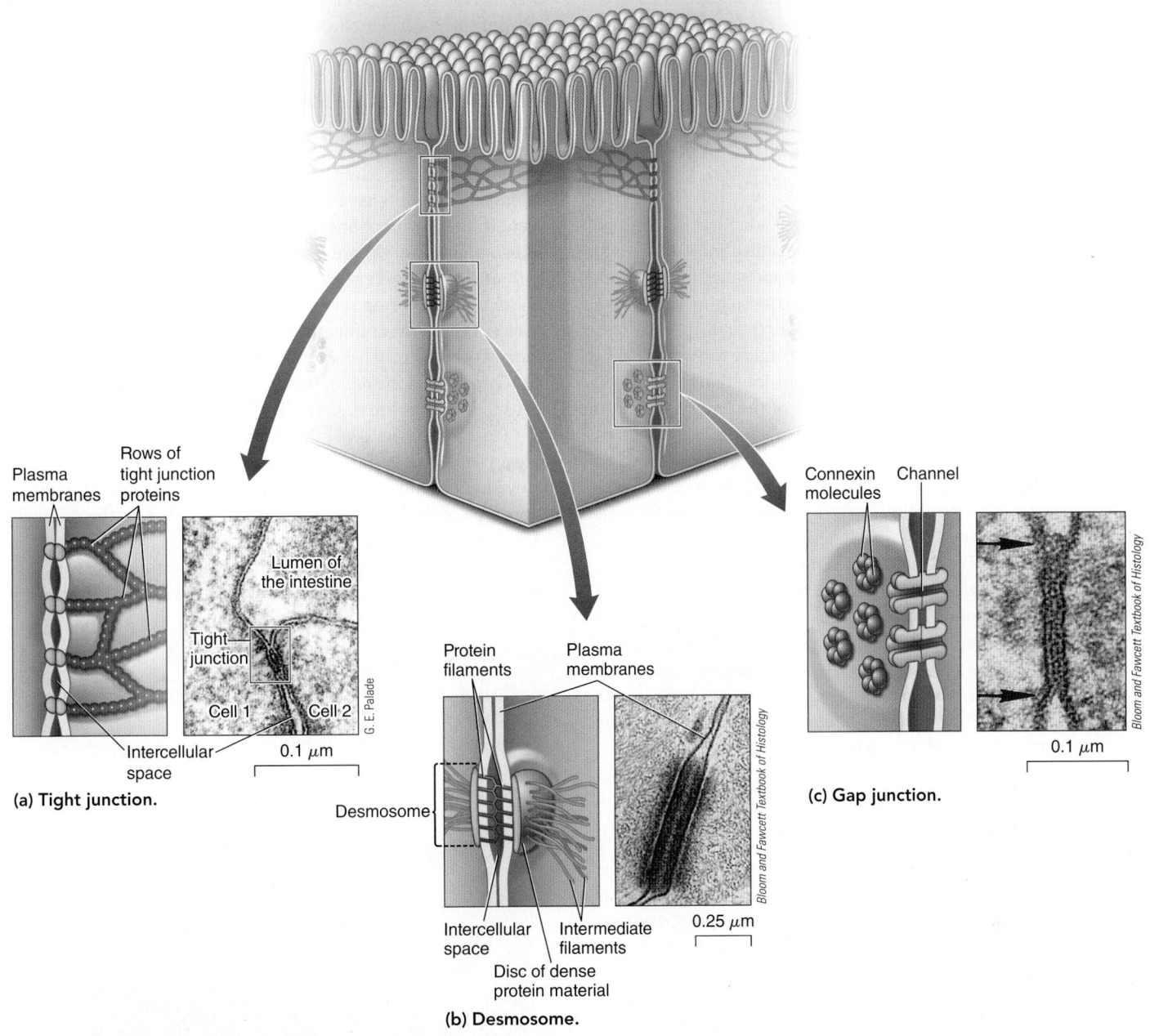

Plasma membranes
Rows of tight junction proteins

Lumen of the intestine

Tight junction

Cell 1   Cell 2

G. E. Palade

0.1 μm

Intercellular space

(a) Tight junction.

Protein filaments
Plasma membranes

Desmosome

Intercellular space
Intermediate filaments
Disc of dense protein material

Bloom and Fawcett Textbook of Histology

0.25 μm

(b) Desmosome.

Connexin molecules   **Channel**

Bloom and Fawcett Textbook of Histology

0.1 μm

(c) Gap junction.

Figure 5-23 *Animation* **Cellular junctions are protein complexes that join sections of membranes between adjacent cells.**

Not all three types of junctions depicted in this composite drawing would be present between the same two cells. **(a) Tight junctions prevent the passage of materials through spaces between cells. (b) Desmosomes anchor cells together in strong sheets.** The dense structure in the TEM is a desmosome. Each desmosome consists of a pair of buttonlike discs associated with the plasma membranes of adjacent cells plus the intercellular protein filaments that connect them. Intermediate filaments in the cells are attached to the discs and to other desmosomes that connect other adjoining cells. **(c) Gap junctions allow the transfer of small molecules and ions between adjacent cells.** The model of a gap junction shown above is based on electron microscopic and X-ray diffraction data.

Desmosomes are points of attachment between cells (**FIG. 5-23b**). They hold cells together at one point as a rivet or a spot weld does. Desmosomes allow cells to form strong sheets, and substances still pass freely through the spaces between the plasma membranes. Each desmosome is made up of regions of dense material associated with the cytosolic sides of the two plasma membranes, plus protein filaments that cross the narrow intercellular space between them. Desmosomes are anchored to systems of intermediate filaments inside the cells. Thus, the intermediate filament networks of adjacent cells are connected. As a result, mechanical stresses are distributed throughout the tissue.

Adhering junctions cement cells together. **Cadherins,** transmembrane proteins that are components of these junctions, form a continuous adhesion belt around each cell, binding the cell to neighboring cells. These junctions connect to microfilaments of the cytoskeleton. The cadherins of adhering junctions are a potential path for signals from the outside environment to be transmitted to the cytoplasm.

Tight junctions seal off intercellular spaces between some animal cells

Tight junctions are literally areas of tight connections between the membranes of adjacent cells. These connections are so tight that no space remains between the cells and substances cannot leak between them. TEMs of tight junctions show that in the region of the junction the plasma membranes of the two cells are held together by proteins in actual contact with each other. However, as shown in **FIGURE 5-23a**, tight junctions are located intermittently. The plasma membranes of the two cells are not fused over their entire surface.

Cells connected by tight junctions seal off body cavities. For example, tight junctions between cells lining the intestine prevent substances in the intestine from passing between the cells and directly entering the blood. The sheet of cells thus acts as a selective barrier. Food substances must be transported across the plasma membranes and *through* the intestinal cells before they enter the blood. This arrangement helps prevent toxins and other unwanted materials from entering the blood and also prevents nutrients from leaking out of the intestine. Tight junctions are also present between the cells that line capillaries in the brain. They form the *blood–brain barrier*, which prevents many substances in the blood from passing into the brain.

Gap junctions allow the transfer of small molecules and ions

A **gap junction** is like a desmosome in that it bridges the space between cells; however, the space it spans is somewhat narrower (**FIG. 5-23c**). Gap junctions also differ in that they are communicating junctions. They not only connect the plasma membranes but also contain channels connecting the cytoplasm of adjacent cells.

Gap junctions are composed of *connexin,* an integral membrane protein. Groups of six connexin molecules cluster to form a cylinder that spans the plasma membrane. The connexin cylinders on adjacent cells become tightly joined. The two cylinders form a channel about 1.5 nm in diameter. Small inorganic particles (such as ions) and some regulatory molecules (such as cyclic AMP, which is illustrated in Figure 3-26) pass through the channels, but larger molecules are excluded. When a marker substance is injected into one of a group of cells connected by gap junctions, the marker passes rapidly into the adjacent cells but does not enter the space between the cells.

Gap junctions provide for rapid chemical and electrical communication between cells. Cells control the passage of materials through gap junctions by opening and closing the channels. Cells in the pancreas, for example, are linked by gap junctions in such a way that if one of a group of cells is stimulated to secrete insulin, the signal is passed through the junctions to the other cells in the cluster. This mechanism ensures a coordinated response to the initial signal. Gap junctions allow

some nerve cells to be electrically coupled. Cardiac muscle cells are linked by gap junctions that permit the flow of ions necessary to synchronize contractions of the heart.

Plasmodesmata allow certain molecules and ions to move between plant cells

Because plant cells have walls, they do not need desmosomes for strength. Plant cells have connections that are functionally equivalent to the gap junctions of some animal cells. **Plasmodesmata** (sing., *plasmodesma*) are channels 20 to 40 nm wide that pass through the cell walls of adjacent plant cells, connecting the cytoplasm of neighboring cells (**FIG. 5-24**). The plasma membranes of adjacent cells are continuous with one another through the plasmodesmata. Most plasmodesmata contain a narrow cylindrical structure, called the *desmotubule,* which runs through the channel and connects the smooth ER of the two adjacent cells.

Plasmodesmata generally allow molecules and ions, but not organelles, to pass through the openings from cell to cell. The movement of ions through the plasmodesmata allows for a very slow type of electrical signaling in plants. Whereas the channels of gap junctions have a fixed diameter, plant cells can dilate the plasmodesmata channels. Certain proteins and RNA can pass through plasmodesmata. Some plant viruses spread infection by passing through these junctions.

CHECKPOINT 5.7

- *How are desmosomes and tight junctions functionally similar? How do they differ?*
- **CONNECT** *What is the justification for considering gap junctions and plasmodesmata to be functionally similar? How do they differ structurally?*

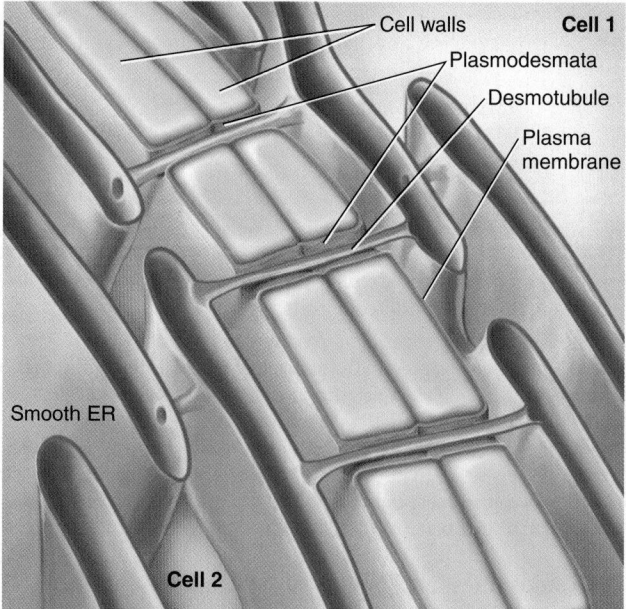

Figure 5-24 Plasmodesmata

Most plant cells have plasmodesmata that connect the cytoplasm of adjacent cells. Cytoplasmic channels through the cell walls of adjacent plant cells allow passage of water, ions, and small molecules. The channels are lined with the fused plasma membranes of the two adjacent cells.

© Cengage Learning

5.1 The Structure of Biological Membranes (page 104)

1 Evaluate the importance of membranes to cells, emphasizing their various functions.

- The **plasma membrane** physically separates the interior of the cell from the extracellular environment, receives information about changes in the environment, regulates the passage of materials into and out of the cell, and communicates with other cells.
- Biological membranes form compartments within eukaryotic cells that allow a variety of separate functions. Membranes participate in and serve as surfaces for biochemical reactions.

2 Describe the fluid mosaic model of cell membrane structure.

- According to the **fluid mosaic model,** membranes consist of a fluid phospholipid bilayer in which a variety of proteins are embedded. The phospholipid molecules are **amphipathic:** they have hydrophobic and hydrophilic regions. The hydrophilic heads of the phospholipids are at the two surfaces of the bilayer, and their hydrophobic fatty acid chains are in the interior.

3 Relate properties of the lipid bilayer to properties and functions of cell membranes.

- In almost all biological membranes, the lipids of the bilayer are in a fluid or liquid-crystalline state, which allows the lipid molecules to move rapidly in the plane of the membrane. Proteins also move within the membrane.
- Lipid bilayers are flexible and self-sealing and can fuse with other membranes. These properties allow the cell to transport materials from one region of the cell to another; materials are transported in vesicles that bud from one cell membrane and then fuse with some other membrane.

4 Describe the ways that membrane proteins associate with the lipid bilayer.

- **Integral membrane proteins** are embedded in the bilayer with their hydrophilic surfaces exposed to the aqueous environment and their hydrophobic surfaces in contact with the hydrophobic interior of the bilayer. **Transmembrane proteins** are integral proteins that extend completely through the membrane.
- **Peripheral membrane proteins** are associated with the surface of the bilayer, usually bound to exposed regions of integral proteins, and are easily removed without disrupting the structure of the membrane.

5.2 Overview of Membrane Protein Functions (page 111)

5 Summarize the functions of membrane proteins.

- Membrane proteins anchor cells, transport materials, act as enzymes or receptors, recognize cells and communicate with them, and structurally link cells.

5.3 Cell Membrane Structure and Permeability (page 112)

6 Describe the importance of selectively permeable membranes and compare the functions of carrier proteins and channel proteins.

- Biological membranes are **selectively permeable membranes:** they allow the passage of some substances but not others. By regulating passage of molecules that enter and leave the cell and its compartments, the cell controls its volume and the internal composition of ions and molecules.
- Membrane *transport proteins* facilitate the passage of certain ions and molecules through biological membranes. *Carrier proteins* are transport proteins that undergo a series of conformational changes as they bind and transport a specific solute. **ABC transporters** are carrier proteins that use energy from ATP to transport solutes.

- *Channel proteins* are transport proteins that form passageways through which water and certain ions travel through the membrane. *Porins* are channel proteins that form relatively large pores through the membrane for passage of water and certain solutes.

5.4 Passive Transport (page 113)

7 Contrast simple diffusion with facilitated diffusion.

- **Diffusion** is the net movement of a substance down its **concentration gradient** from a region of greater concentration to one of lower concentration. Diffusion and osmosis are physical processes that do not require the cell to directly expend metabolic energy.
- In **simple diffusion** through a biological membrane, solute molecules or ions move directly through the membrane down their concentration gradient. **Facilitated diffusion** uses specific transport proteins to move solutes across a membrane. As in simple diffusion, net movement is always from a region of higher to a region of lower solute concentration. Facilitated diffusion cannot work against a concentration gradient.

8 Define *osmosis* and solve simple problems involving osmosis; for example, predict whether cells will swell or shrink under various osmotic conditions.

- **Osmosis** is a kind of diffusion in which molecules of water pass through a selectively permeable membrane from a region where water has a higher effective concentration to a region where its effective concentration is lower.
- The concentration of dissolved substances (solutes) in a solution determines its **osmotic pressure.** Cells regulate their internal osmotic pressures to prevent shrinking or bursting.
- An **isotonic** solution has an equal solute concentration compared with that of another fluid, for example, the fluid within the cell.
- When placed in a **hypertonic** solution, one that has a greater solute concentration than that of the cell, a cell loses water to its surroundings; plant cells undergo **plasmolysis,** a process in which the plasma membrane separates from the cell wall.
- When cells are placed in a **hypotonic** solution, one that has a lower solute concentration than the solute concentration of the cell, water enters the cells and causes them to swell.
- Plant cells withstand high internal hydrostatic pressure because their cell walls prevent them from expanding and bursting. Water moves into plant cells by osmosis and fills the central vacuoles. The cells swell, building up **turgor pressure** against the supportive cell walls.

5.5 Active Transport (page 118)

9 Describe active transport, including cotransport.

- In **active transport,** the cell expends metabolic energy to move ions or molecules across a membrane against a concentration gradient. For example, the **sodium–potassium pump** uses ATP to pump sodium ions out of the cell and potassium ions into the cell.
- In **cotransport,** also called *indirect active transport,* two solutes are transported at the same time. An ATP-powered pump maintains a concentration gradient. Then a carrier protein cotransports two solutes. It transports one solute down its concentration gradient and uses the energy released to move another solute against its concentration gradient.

5.6 Exocytosis and Endocytosis *(page 121)*

10 Compare exocytotic and endocytotic transport mechanisms.

- The cell expends metabolic energy to carry on exocytosis and endocytosis. In **exocytosis,** the cell ejects waste products or secretes substances such as mucus by fusion of vesicles with the plasma membrane. This process increases the surface area of the plasma membrane.

- In **endocytosis,** materials such as food particles are moved into the cell. A portion of the plasma membrane envelops the material, enclosing it in a vesicle or vacuole that is then released inside the cell. This process decreases the surface area of the plasma membrane.

- Three types of endocytosis are phagocytosis, pinocytosis, and receptor-mediated endocytosis.

- In **phagocytosis,** the plasma membrane encloses a large particle such as a bacterium, forms a vacuole around it, and moves it into the cell. In **pinocytosis,** the cell takes in dissolved materials by forming tiny vesicles around droplets of fluid trapped by folds of the plasma membrane.

- In **receptor-mediated endocytosis,** specific receptors in coated pits along the plasma membrane bind **ligand** molecules. These pits, coated by the protein clathrin, form coated vesicles

by endocytosis. The vesicles fuse with lysosomes, and their contents are digested and released into the cytosol.

5.7 Cell Junctions *(page 123)*

11 Compare the structures and functions of anchoring junctions, tight junctions, gap junctions, and plasmodesmata.

- Cells in close contact with one another may form intercellular junctions. *Anchoring junctions* include desmosomes and adhering junctions; they are found between cells that form a sheet of tissue. **Desmosomes** spot-weld adjacent animal cells together. **Adhering junctions** are formed by **cadherins,** transmembrane proteins that cement cells together.

- **Tight junctions** seal membranes of adjacent animal cells together, preventing substances from moving through the spaces between the cells.

- **Gap junctions,** composed of the protein connexin, form channels that allow communication between the cytoplasm of adjacent animal cells.

- **Plasmodesmata** are channels connecting adjacent plant cells. Openings in the cell walls allow the plasma membranes and cytosol to be continuous; certain molecules and ions can pass from cell to cell.

TEST YOUR UNDERSTANDING

Know and Comprehend

1. Transmembrane proteins (a) are peripheral proteins (b) are receptor proteins (c) extend completely through the membrane (d) extend along the surface of the membrane (e) are secreted from the cell

2. Which of the following is *not* a function of the plasma membrane? (a) transports materials (b) helps structurally link cells (c) has receptors that relay signals (d) anchors the cell to the extracellular matrix (e) manufactures proteins

3. ABC transporters (a) use the energy of ATP hydrolysis to transport certain ions and sugars (b) are important in facilitated diffusion of certain ions (c) are a small group of channel proteins (d) are found mainly in plant cell membranes (e) permit passive diffusion through their channels

4. When plant cells are in a hypotonic medium, they (a) undergo plasmolysis (b) build up turgor pressure (c) wilt (d) decrease pinocytosis (e) lose water to the environment

5. Which of the following processes requires the cell to expend metabolic energy directly (e.g., from ATP)? (a) osmosis (b) facilitated diffusion (c) all forms of carrier-mediated transport (d) active transport (e) simple diffusion

6. Electrochemical gradients (a) power simple diffusion (b) are established by pinocytosis (c) are necessary for transport by aquaporins (d) are established by concentration gradients (e) are a result of both an electric charge difference and a concentration difference between the two sides of the membrane

7. In cotransport (indirect active transport) (a) a uniporter moves a solute across a membrane against its concentration gradient (b) the movement of one solute down its concentration gradient provides energy for transport of some other solute up its concentration gradient (c) a channel protein moves ions by facilitated diffusion (d) osmosis powers the movement of ions against their concentration gradient (e) sodium is directly transported in one direction, and potassium is indirectly transported in the same direction

8. Junctions that permit the transfer of water, ions, and molecules between adjacent plant cells are (a) tight junctions (b) adhering junctions (c) desmosomes (d) gap junctions (e) plasmodesmata

Apply and Analyze

9. **PREDICT** A laboratory technician accidentally places red blood cells in a hypertonic solution. What happens? (a) They undergo plasmolysis (b) They build up turgor pressure (c) They swell (d) They pump solutes out (e) They become dehydrated and shrunken

Evaluate and Synthesize

10. **INTERPRET DATA** GLUT 4 is a glucose transporter that functions in adipose (fat) cell plasma membranes. An analysis of adipose cells exposed to insulin showed that a single cell could import glucose at a maximum rate of about 1×10^8 molecules/second. Under the same conditions, unstimulated cells could only transport a maximum of about 1×10^7 molecules/second. What does this finding tell you about the relative number of GLUT 4 transporters functioning in the plasma membranes of stimulated versus unstimulated cells?

11. **CONNECT** Most adjacent plant cells are connected by plasmodesmata, whereas only certain adjacent animal cells are connected through gap junctions. What might account for this difference?

12. **EVOLUTION LINK** Explain to your roommate why the evolution of biological membranes was an essential step in the origin of life. Give arguments supporting (or challenging) this hypothesis.

13. **EVOLUTION LINK** Transport proteins have been found in all biological membranes. What hypothesis could you make regarding whether these molecules evolved early or later in the history of cells? Argue in support of your hypothesis.

To access course materials, such as Aplia and other companion resources, please visit **www.cengagebrain.com**.

Cell Communication

6

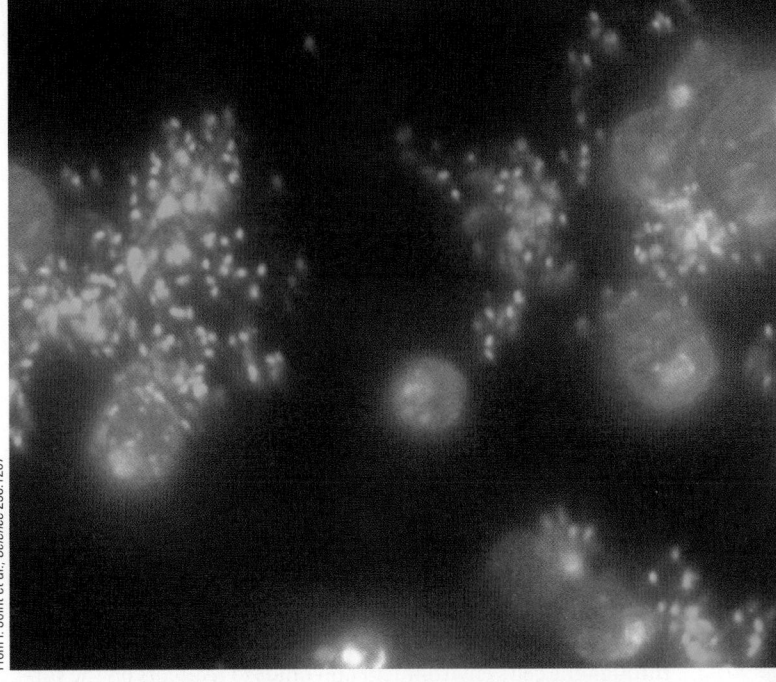

From I. Joint et al., *Science* 298:1207

Cell-to-cell communication across the prokaryote–eukaryote boundary.
Spores of the green seaweed *Enteromorpha* (Domain Eukarya) attach
to biofilm-forming bacteria (Domain Bacteria) in response to chemical
compounds released by the bacteria. The bacteria (*blue*) were stained
and visualized with blue light. The spores appear *red* because of the
fluorescence of chlorophyll within them.

To maintain homeostasis, the cells of a multicellular organism must continuously communicate with one another. Many different mechanisms for transmitting information between cells, tissues, and organs have evolved, including electrical signaling and many types of chemical signaling. Organisms also communicate with other members of their species by secreting chemical signals. For example, bacteria release chemical signals that diffuse among nearby bacteria. As the population of bacteria increases, the concentration of the chemical signal increases. Through a process known as *quorum sensing,* the bacteria sense when a certain critical concentration of a signal molecule is reached. The bacteria respond by activating a specific biological process. For example, they may form a **biofilm,** a community of microorganisms attached to a solid surface. Forming a biofilm requires the coordinated activity of numerous bacteria.

Over billions of years, elaborate systems of cell signaling have evolved. Organisms of different species, and even different kingdoms and domains, communicate with one another. For example, the chemical signals released by bacteria can be intercepted by other organisms. During its life cycle, the green seaweed *Enteromorpha* produces spores that move about and temporarily attach to a surface. The spores sense a chemical signal released by bacteria that form biofilms. In response to the chemical signal, the spores move toward the biofilm and attach to individual bacteria that are part of the biofilm surface (see photograph).

Thousands of chemical reactions are involved in responding to signal molecules and regulating the communication among molecules necessary to maintain homeostasis. Many researchers are working on the molecular level to understand how proteins receive messages and relay signals. They are learning how proteins act as molecular switches, activating and deactivating molecules in complex signaling pathways.

Some cell biologists are using a systems biology approach to understand the intricate, dynamic interactions involved in cell communication. Biologists, biochemists, physicists, and scientists from many other disciplines are working together to understand how elaborate signaling systems within the cell interact to regulate cell functions. They are studying how information is transferred between cells at the tissue and organ levels, and

KEY CONCEPTS

6.1 Cells communicate by signaling one another, a complex process that involves production of signaling molecules, reception of the signal, signal transduction, and a response.

6.2 Cells signal one another using chemical compounds such as neurotransmitters, hormones, and other signaling molecules.

6.3 A signaling molecule binds to a receptor molecule on the cell surface or inside the target cell.

6.4 In signal transduction, a cell converts an extracellular signal into an amplified intracellular signal that triggers some change in the cell (the response).

6.5 Cells respond to signals in many different ways: by opening or closing ion channels, altering enzyme activities that activate or inhibit specific genes, modifying metabolic activity, or triggering changes in cellular structure or functions such as growth and motility.

6.6 Similarities in cell communication among diverse organisms suggest that the molecules and mechanisms used in information transfer evolved long ago.

throughout the organism. Faulty signaling in cells and between cells can cause a variety of diseases, including cancer and diabetes.

In Chapter 1 we introduced five basic themes of biology. One theme, transmission of information, is the main focus of this chapter. In this chapter we discuss how cells send and receive signals. We consider how information crosses the plasma membrane and is transmitted through signaling systems. We describe some of the responses that cells make. Finally, we discuss the evolution of cell communication. Increased understanding of the mechanisms of cell communication may suggest new strategies for preventing and treating diseases.

6.1 CELL COMMUNICATION: AN OVERVIEW

LEARNING OBJECTIVE

1   Describe the four main processes essential for cells to communicate.

When food is scarce, the amoeba-like cellular slime mold *Dictyostelium* secretes the compound **cyclic adenosine monophosphate (cAMP).** This chemical compound diffuses through the cell's environment and binds to *receptors* on the surfaces of nearby cells. The activated receptors send signals into the cells that result in movement toward the cAMP. Hundreds of slime molds come together and form a multicellular slug-shaped colony (**FIG. 6-1**; also see Fig. 26-20). (When conditions are favorable, the cells of the slug form a stalked fruiting body with spores at the top; when released, each spore gives rise to an amoeba-like cell.)

Even though they do not move from one place to another, plants also communicate with one another. For example, diseased maple trees send airborne chemical signals that are received by uninfected trees nearby. Cells of the uninfected trees respond by increasing their chemical defenses so that they are more resistant to the disease-causing organisms. Plants also send signals to insects. When tobacco plants are attacked by herbivorous insects, the plants release volatile chemicals. In response to these signals, the insects lay fewer eggs, thus reducing the number of insects feeding on the plants. Predator insects that eat the eggs of the herbivorous insects respond to the plant signals by eating more of the herbivorous insect eggs. Thus, natural selection has resulted in plant signals that herbivorous insects detect and avoid and that carnivorous insects detect and approach. This system of information transfer helps protect plants from herbivorous insects.

To survive, organisms must receive signals from the outside environment and effectively respond to them. To grow, develop, and function, the cells of a multicellular organism must also communicate with one another. In plants and animals, *hormones* and other regulatory molecules serve as important chemical signals between various cells and organs. In animals, neurons (nerve cells) transmit information electrically and chemically.

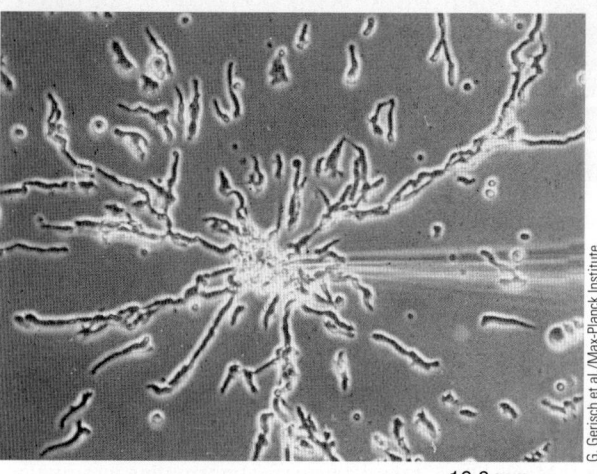

16.6 mm

Figure 6-1 Cell signaling in cellular slime molds
When food is scarce, the amoeba-like cellular slime mold *Dictyostelium* secretes the chemical compound cyclic AMP (cAMP). The slime molds respond to this chemical signal by aggregating. Converging streams of hundreds of individuals come together and form a multicellular colony.

G. Gerisch et al./Max-Planck Institute

The term **cell signaling** refers to the mechanisms by which cells communicate with one another. If the cells are physically close to one another, a signaling molecule on one cell may combine with a *receptor* (a macromolecule that binds with signaling molecules) on another cell. Most commonly, cells communicate by sending chemical signals over some distance. As we will discuss, cell signaling must be precisely regulated. Cell signaling involves a sequence of four main processes (**FIG. 6-2**). We summarize them here and discuss them in more detail in the following sections of this chapter.

1.   **Signal transmission.** In chemical signaling, a cell must synthesize and release signaling molecules. For example, specialized cells in the vertebrate pancreas secrete the hormone insulin. If the **target cells,** the cells that can respond to the signal, are not in close proximity, the signal must be transported to them. The circulatory system transports insulin to target cells throughout the body. Next, target cells must receive, relay, and respond to the information signaled.

2.   **Reception.** Reception is the process of receiving an incoming signal. **Receptors** are large proteins or glycoproteins that bind with signaling molecules. Many types of signaling molecules do not actually enter the target cell. They bind to specific receptors on the surface of the target cell. Insulin, for example, binds to insulin receptors, which are transmembrane proteins.

3.   **Signal transduction** is the process by which a cell converts an extracellular signal into an intracellular signal (or signals) that causes a response. Signal transduction typically involves a chain of molecules that relay information. When insulin binds to an insulin receptor, the signal is relayed through several different signaling pathways.

4.   **Response.** The final molecule in the signaling pathway converts the signal into a response that alters some cell process.

For example, as we will discuss in Chapter 49, insulin stimulates cells to take up glucose from the blood. This response lowers the concentration of glucose in the blood. In some cell types, insulin is also involved in the regulation of fat and protein metabolism. Many signaling molecules stimulate ion channels in the plasma membrane to open or close. Still other signaling molecules activate or inhibit specific genes in the nucleus. These responses can result in changes in cell division and other aspects of cell development. After a signaling molecule has done its job, its action must be stopped, and as part of the signaling process, certain mechanisms operate to terminate the signal.

CHECKPOINT 6.1

- *What is the sequence of events that takes place in cell signaling?*
- **CONNECT** *What key role does signal transduction play in the signaling process?*

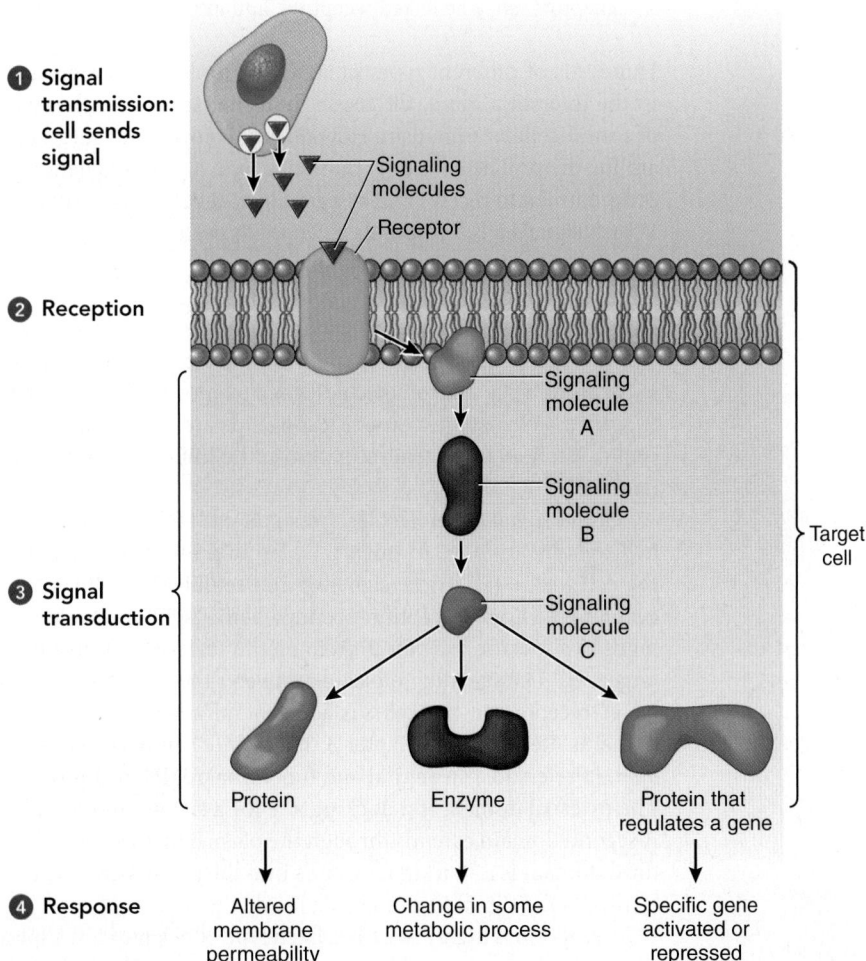

1 Signal transmission: cell sends signal

Signaling molecules

Receptor

2 Reception

Signaling molecule A

Signaling molecule B

Target cell

3 Signal transduction

Signaling molecule C

Protein

Enzyme

Protein that regulates a gene

4 Response

Altered membrane permeability

Change in some metabolic process

Specific gene activated or repressed

Figure 6-2 *Animation* **Overview of cell signaling**

When a signaling molecule binds to a receptor molecule on a target cell, the receptor activates a signal transduction pathway, leading to some response in the cell.

PREDICT What would happen if the cell were exposed to a drug that blocked the activity of signaling molecule A?

© Cengage Learning

6.2 SENDING SIGNALS

LEARNING OBJECTIVE

2 Compare three types of signaling molecules: neurotransmitters, hormones, and local regulators.

Cells communicate in several ways, including directly through cell junctions, by way of electrical signals, temporary cell-to-cell contact, and chemical signals. Recall from Chapter 5 that gap junctions in animal cells allow rapid chemical and electrical communication between adjacent cells. For example, the gap junctions between cardiac muscle cells of the heart wall allow the rapid flow of ions necessary for synchronized contraction. Plasmodesmata between adjacent plant cells also allow signal molecules to pass quickly from one cell to another.

Cells that are not directly connected also communicate with one another. In animals, some neurons communicate with electrical signals. Most neurons, however, signal one another by releasing chemical compounds called **neurotransmitters** (**FIGS. 6-3a** and 41-11). Neurotransmitter molecules diffuse across *synapses,* tiny gaps between neurons. More than 60 different neurotransmitters have been identified, including acetylcholine, norepinephrine, dopamine, serotonin, and several amino acids and peptides.

Cells synthesize many different types of chemical signals and deliver them in various ways. In animals certain cells in the immune system produce specific chemical compounds that are displayed on the cell surface. These cells recognize the chemical signals and communicate with one another by making direct contact (**FIGS. 6-3b**, 45-6, and 45-7).

Specialized cells in plants and animals produce signaling molecules called **hormones.** Hormones may be synthesized by neighboring cells or by specialized organs or tissues some distance from the target cells. In animals many hormones are produced by **endocrine glands.** These glands have no ducts; they secrete their hormones into the surrounding interstitial fluid. Typically, hormones diffuse into capillaries and are transported by the blood to target cells (**FIG. 6-3c**).

Some cells produce local regulators that signal cells in close proximity. A **local regulator** is a signaling molecule that diffuses through the **interstitial fluid,** the fluid surrounding the cells, and acts on nearby cells. This process is called **paracrine regulation** (**FIG. 6-3d**). Some local regulators are considered hormones.

Local regulators include local chemical mediators such as growth factors, histamine, nitric oxide, and prostaglandins.

Histamine is a local regulator that is stored in certain cells of the immune system and is released in response to allergic reactions, injury, or infection. Histamine causes blood vessels to dilate and capillaries to become more permeable. **Nitric oxide (NO),** another local regulator, is a gas produced by many types of cells, including plant and animal cells. Nitric oxide released by cells lining blood vessels relaxes smooth muscle in the blood vessel walls. As a result, the blood vessels dilate, decreasing blood pressure.

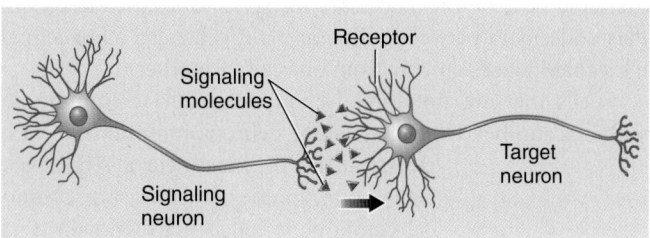

(a) Neurons transmit signals across synapses. (Distances across the synapses are exaggerated for clarity.)

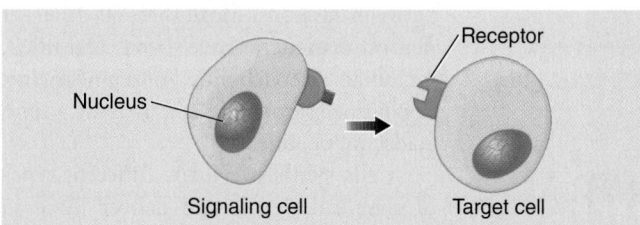

(b) Some cells signal one another by making direct contact.

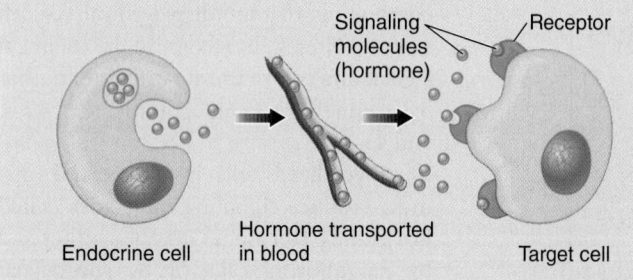

(c) Many hormones are transported by the blood to target cells.

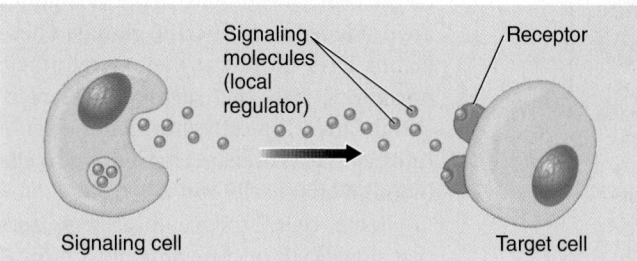

(d) In paracrine regulation a local regulator diffuses to target cells.

Figure 6-3 Some types of cell signaling
Different types of cells may communicate in different ways.
© Cengage Learning

Prostaglandins are local hormones that are paracrine regulators. Prostaglandins modify cAMP levels and interact with other signaling molecules to regulate metabolic activities. For example, some prostaglandins stimulate smooth muscle to contract.

CHECKPOINT 6.2

- *What are neurotransmitters?*
- *How are animal hormones typically transported to target cells?*
- **CONNECT** *How is paracrine regulation similar to endocrine regulation? How does it differ?*

6.3 RECEPTION

LEARNING OBJECTIVES

3 Identify mechanisms that make reception a highly specific process.
4 Briefly compare ion channel–linked receptors, G protein–linked receptors, enzyme-linked receptors, and intracellular receptors.

Hundreds of different types of signaling molecules are present in the interstitial fluid, the tissue fluid that surrounds the cells of a multicellular organism. How do cells know which messages are for them? The answer is that each type of cell is genetically programmed to receive and respond to specific types of signals. Which signals a cell responds to depends on the specific receptors it is programmed to synthesize.

A signaling molecule, such as insulin, that binds to a specific receptor acts as a **ligand.** A ligand is a molecule, other than an enzyme, that binds specifically to a macromolecule (usually a protein), forming a macromolecule-ligand complex. The complex triggers a biological response. Most ligands are hydrophilic molecules that bind to protein receptors on the surface of target cells (**FIG. 6-4a**).

Some signaling molecules are small enough or sufficiently hydrophobic to move through the plasma membrane and enter the cell (**FIG. 6-4b**). These signaling molecules bind with intracellular receptors. Reception occurs when a signaling molecule binds to a specific receptor protein on the surface of, or inside, a target cell. The signaling molecule activates the receptor.

A receptor on the cell surface generally has at least three domains. Recall from Chapter 3 that in biochemistry, the term *domain* refers to a structural and functional region of a protein. The external domain is a docking site for a signaling molecule. A second domain extends through the plasma membrane, and a third domain is a "tail" that extends into the cytoplasm. The tail transmits the signal to a molecule inside the cell.

Reception is highly selective. Each type of receptor has a specific shape. The receptor binding site is somewhat like a lock, and signaling molecules are like different keys. Only the signaling molecule that fits the specific receptor can influence the metabolic machinery of the cell. Receptors are important in determining the specificity of cell communication.

Different types of cells can produce different types of receptors. Any one cell makes many different receptors. Furthermore, a cell may synthesize different kinds of receptors at different stages in its life cycle or in response to different conditions. Another consideration is that the same signal can have different meanings for various target cells.

Some receptors are specialized to respond to signals other than chemical signals. For example, in the vertebrate eye, a receptor called **rhodopsin** is activated by light. Rhodopsin is part of a signal transduction pathway that leads to vision in dim light. Plants and some algae have **phytochromes,** a family of blue-green pigment proteins that are activated by red light. Activation can lead to changes such as flowering. Plants, some algae, and at least some animals have **cryptochromes,** pigments that absorb blue light. Cryptochromes play a role in biological rhythms.

Cells regulate reception

An important mechanism that cells use to regulate reception is increasing or decreasing the number of each type of receptor. Depending on the needs of the cell, receptors are synthesized or degraded. For example, when the concentration of the hormone insulin is too high for an extended period, cells decrease the number of their insulin receptors. This process is called **receptor down-regulation.** In the case of insulin, receptor down-regulation suppresses the sensitivity of target cells to the hormone. Insulin stimulates cells to take in glucose by facilitated diffusion, so receptor down-regulation decreases the ability of cells to take in glucose. Receptor down-regulation often involves transporting receptors to lysosomes, where they are destroyed.

Receptor up-regulation occurs in response to low hormone concentrations. In this process, a greater number of receptors are synthesized, and their increased numbers on the plasma membrane make it more likely that the signal will be received by a receptor on the cell. Receptor up-regulation thus *amplifies* the signaling molecule's effect on the cell. Receptor up-regulation and down-regulation are controlled in part by signals to genes that code for the receptors.

Three types of receptors occur on the cell surface

Three main types of receptors on the cell surface are ion channel–linked receptors, G protein–linked receptors, and enzyme-linked receptors.

Ion channel–linked receptors convert chemical signals into electrical signals
Ion channel–linked receptors are found in the plasma membrane. These receptors, which have been extensively studied in neurons and muscle cells, convert chemical signals into electrical signals (**FIG. 6-5a**). The receptor itself serves as a channel.

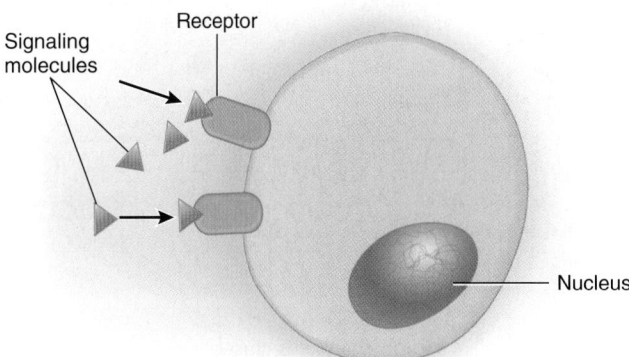

(a) Hydrophilic (water soluble) signaling molecules cannot pass through the plasma membrane. They bind to receptors in the plasma membrane.

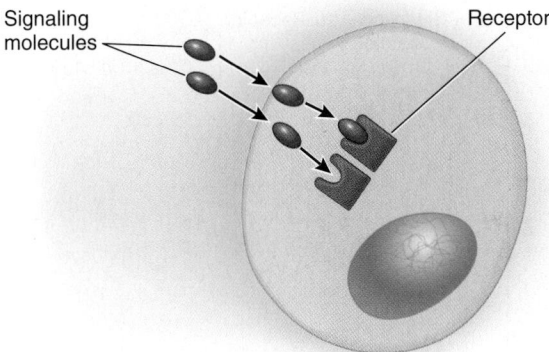

(b) Hydrophobic (lipid soluble) signaling molecules cross the plasma membrane and bind with receptors inside the cell (cytosol or nucleus).

Figure 6-4 Cell-surface and intracellular receptors
© Cengage Learning

Ion channel–linked receptors are also called **ligand-gated channels,** which means that the ion channel opens or closes in response to the binding of the signaling molecule (ligand). A part of the receptor (protein) that makes up the channel forms the gate. The receptor responds to certain signals by changing its shape, opening or closing the gate. Typically, the gate of an ion channel remains closed until a ligand binds to the receptor.

For example, **acetylcholine** is a neurotransmitter that binds to an acetylcholine receptor. This receptor is a ligand-gated sodium ion channel that is important in muscle contraction. When acetylcholine binds to the receptor, the channel opens, allowing sodium ions to enter the cell. The influx of sodium ions decreases the electric charge difference across the membrane (depolarization), which can lead to muscle contraction. After a brief time, the ligand dissociates from the receptor and the gate

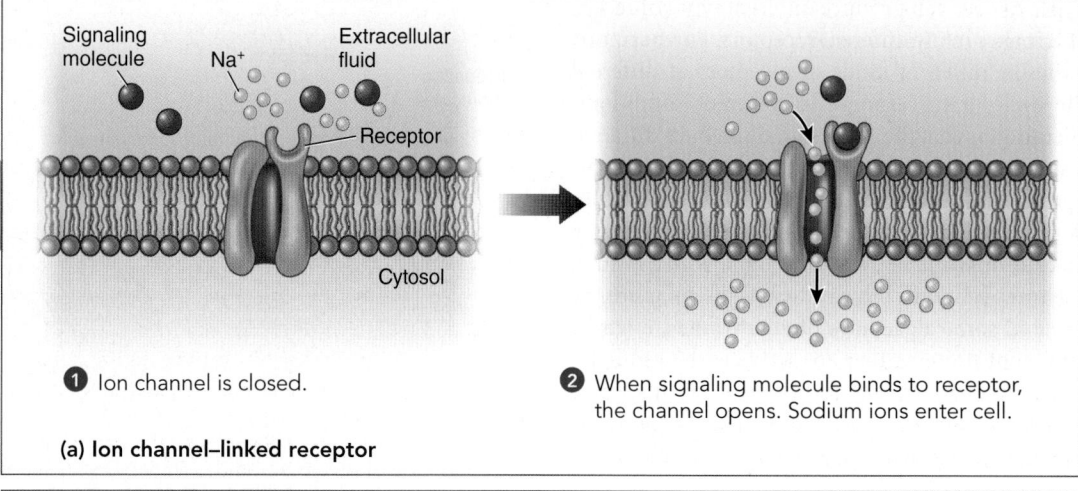

① Ion channel is closed.

② When signaling molecule binds to receptor, the channel opens. Sodium ions enter cell.

(a) Ion channel–linked receptor

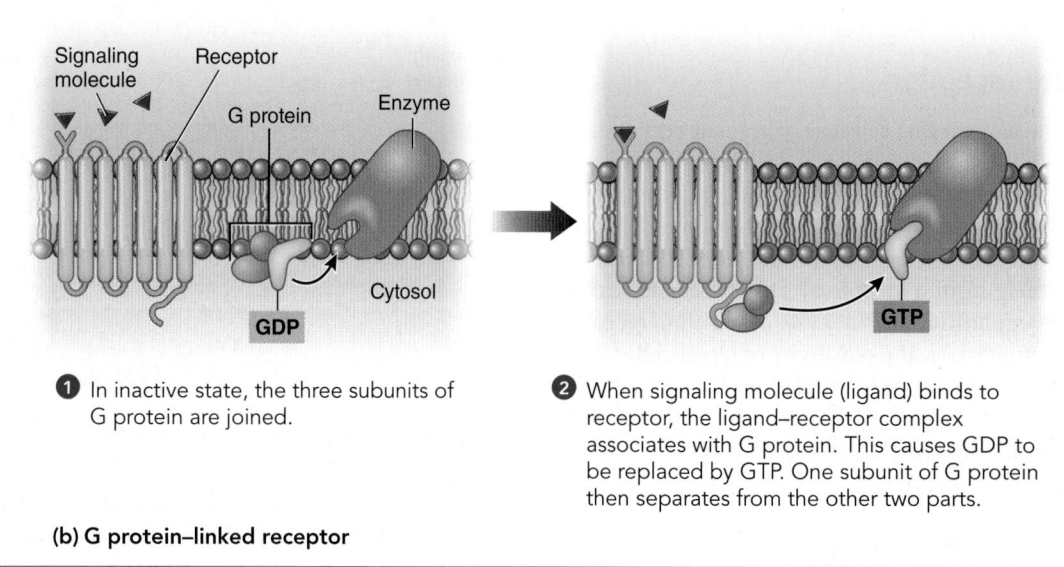

① In inactive state, the three subunits of G protein are joined.

② When signaling molecule (ligand) binds to receptor, the ligand–receptor complex associates with G protein. This causes GDP to be replaced by GTP. One subunit of G protein then separates from the other two parts.

(b) G protein–linked receptor

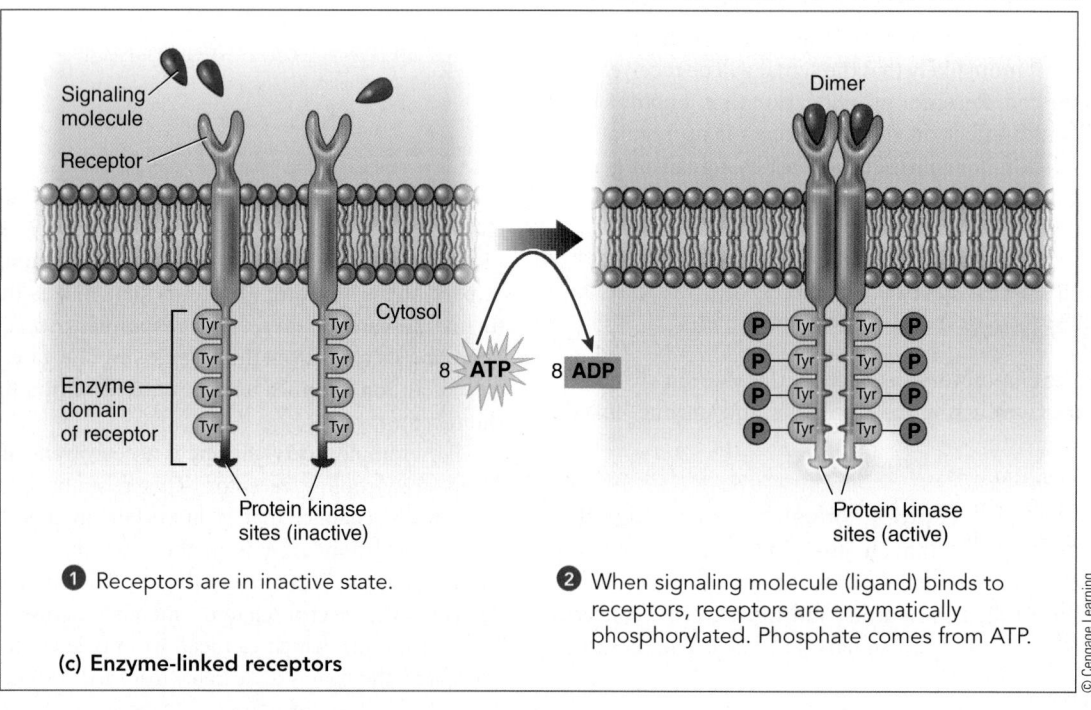

① Receptors are in inactive state.

② When signaling molecule (ligand) binds to receptors, receptors are enzymatically phosphorylated. Phosphate comes from ATP.

(c) Enzyme-linked receptors

Figure 6-5 **Three types of cell-surface receptors**

© Cengage Learning

closes the channel. As we will discuss in Chapter 41, some ion channels, called voltage-activated channels, are regulated by electrical signals.

G protein–linked receptors link signaling molecules to signal transduction pathways

G protein–linked receptors (also called G protein–coupled receptors) are a large family of transmembrane proteins that loop back and forth through the plasma membrane seven times (**FIG. 6-5b**). The receptor consists of seven transmembrane helices connected by loops that extend into the cytosol or outside the cell. G protein–linked receptors couple certain signaling molecules to various signal transduction pathways inside the cell. The outer part of the receptor has a binding site for a signaling molecule, and the part of the receptor that extends into the cytosol has a binding site for a specific **G protein.**

G proteins bind *guanine nucleotides.* When a signaling molecule binds with a G protein–linked receptor, the receptor changes shape. This change allows the G protein to associate with the receptor.

G protein–linked receptors are found in all eukaryotes. These receptors bind with hundreds of different signaling molecules. Some of the many critical processes that depend on G protein–linked receptors are vision, sense of smell, regulation of mood and behavior, and regulation of the immune system. As you might imagine, understanding how G protein–linked receptors work is medically important. About 60% of prescription medications currently in use act on these receptors. More than 900 G protein–linked receptors have been identified in mammals, and more than 400 of these receptors are potential targets for pharmaceutical interventions.

Enzyme-linked receptors function directly as enzymes or are linked to enzymes

Enzyme-linked receptors are transmembrane proteins with a binding site for a signaling molecule outside the cell and an enzyme component inside the cell. Recall from Chapter 3 that *enzymes* catalyze specific chemical reactions; most enzymes are proteins.

Proteins called *tyrosine kinases* make up a major group of enzyme-linked receptors found on the plasma membrane. The domain of the protein that extends into the cytosol is a **tyrosine kinase** enzyme that transfers the terminal phosphate group from ATP to the amino acid *tyrosine* that is part of a protein. This process is called **phosphorylation.** Tyrosine kinase receptors bind certain hormones that regulate many cellular processes and are important in development. They include insulin and *growth factors,* such as nerve growth factor.

Before the tyrosine kinase can phosphorylate a signaling protein in the cell, it must first be activated. When signal molecules bind to two tyrosine kinase receptors, the receptor proteins move closer together in the plasma membrane and pair, forming a *dimer.* A conformational change (change in shape) takes place, which allows the tyrosine kinase part of each receptor to add a phosphate from an ATP molecule to certain tyrosines on the other member of the dimer (**FIG. 6-5c**). Once activated, tyrosine kinase enzymes can phosphorylate signaling proteins inside the cell (discussed in the next section).

Many plant cell-surface receptors are enzyme-linked receptors. **Brassinosteroids (BRs),** a group of plant steroid hormones, regulate many plant processes, including cell division, cell elongation, and flower development (discussed in Chapter 38). Unlike animal steroid hormones, which typically bind with intracellular receptors, BRs bind with a protein kinase receptor in the plasma membrane. In plant protein kinase receptors, serine and threonine, rather than tyrosine, appear to be the amino acids that are phosphorylated.

The gas **ethylene** is a plant hormone that regulates a variety of processes, including seed germination and ripening of fruit. Ethylene is also important in plant responses to stressors. The ethylene receptor has two components. Each component has a domain that is an enzyme, a histidine kinase, that extends into the cell. Receptors with histidine kinase domains are also present in bacterial and yeast cells.

Some receptors are located inside the cell

Certain receptors are found in the cytosol or in the nucleus. Most of these intracellular receptors are **transcription factors,** proteins that regulate the expression of specific genes. The signaling molecules that bind with intracellular receptors are small, hydrophobic molecules that can diffuse across the membranes of target cells (see Fig. 6-4b).

In animal cells most steroid hormones, such as the molting hormone *ecdysone* in insects and *cortisol* in vertebrates, enter target cells and combine with receptor molecules in the cytosol. Vitamins A and D and nitric oxide also bind with intracellular receptors. After binding, the ligand–receptor complex moves into the nucleus. *Thyroid hormones* (which are not steroids) bind to receptors already bound to DNA inside the nucleus.

CHECKPOINT 6.3

- *How does a receptor "know" which signaling molecules to bind?*
- **PREDICT** *Under what conditions might receptor up-regulation occur? receptor down-regulation?*
- **CONNECT** *What do the three main types of cell-surface receptors have in common? How do they differ?*
- **CONNECT** *What is the function of most intracellular receptors?*

6.4 SIGNAL TRANSDUCTION

LEARNING OBJECTIVES

5 Compare the actions of the main types of receptors in signal transduction.
6 Trace the sequence of events in signal transduction for each of the following second messengers: cyclic AMP, inositol trisphosphate (IP$_3$), diacylglycerol (DAG), and calcium ions.

Many regulatory molecules transmit information to the cell's interior without physically crossing the plasma membrane. Instead, they activate membrane proteins, which then *transduce* the signal. The first component in a signal transduction pathway is typically the receptor, which may be a transmembrane protein with a domain exposed on the extracellular surface. Each type of receptor activates a different signal transduction pathway.

In a typical signaling pathway, a signaling molecule binds with a cell-surface receptor and activates it by changing the shape of the receptor tail that extends into the cytoplasm. The signal may then be relayed through a sequence of proteins that are intracellular signaling molecules. Typically, the proteins in

this chain are protein kinases that relay and amplify the original signal by adding phosphates to the next protein molecule in the pathway. This chain of signaling molecules in the cell that relays and intensifies the signal is called a *signaling pathway* or *signaling cascade* (FIG. 6-6).

More than 3000 signaling proteins have been identified in the cells of mammals alone. Malfunctions in signal transduction pathways have been linked to major diseases, including cancer, heart disease, diabetes, and autoimmune diseases.

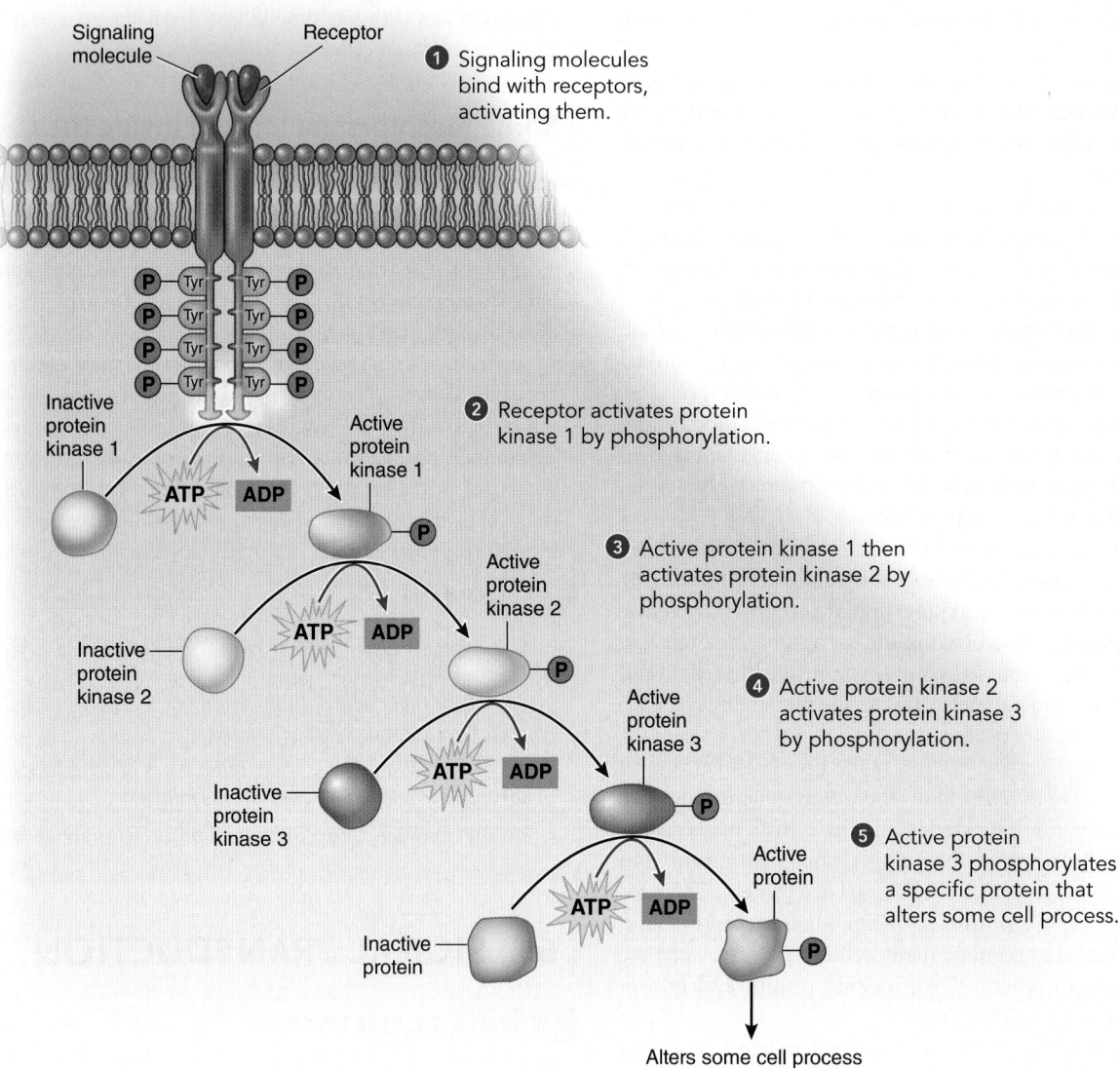

Figure 6-6 A phosphorylation cascade

Many signaling pathways are protein kinase cascades. When the receptor is activated, each protein kinase activates the next protein kinase in the pathway by phosphorylation of one or more of its threonine, serine, or tyrosine residues. Addition of a phosphate group typically changes the shape of the molecule. Activation of the last protein in the chain changes some cell process or turns on (or turns off) specific genes. (*P* represents phosphate.) Note that the number of protein kinases varies from pathway to pathway.
© Cengage Learning

Signaling molecules can act as molecular switches

Each component in a signaling pathway acts as a molecular "switch," which can be in an active ("on") state or an inactive ("off") state. When an intracellular signaling molecule receives a signal, it is activated (turned on). Another process then occurs that inactivates (turns off) the signaling molecule. To transmit a new signal, every activated molecule in a signaling pathway must be inactivated. Molecular switches are typically regulated by the addition or removal of phosphate groups.

Each time a signal molecule binds to a receptor, a signaling pathway is turned on. As the signal is transmitted to each protein kinase in the chain, the protein is phosphorylated. Recall that in phosphorylation, an enzyme transfers phosphate groups from one molecule to another. As phosphate is transferred, the signal passes from protein kinase 1 to protein kinase 2 to protein kinase 3 and on down the chain (see Fig. 6-6). Each protein kinase enzyme in the pathway will specifically phosphorylate one or more tyrosine, serine, or threonine residues on its target protein. Phosphorylation typically activates a protein kinase, although in some cases it inhibits protein kinase activity. A signaling pathway in which a series of protein kinase molecules are phosphorylated is referred to as a *protein kinase cascade*. The last protein kinase in the cascade activates the target protein by phosphorylation. The target protein alters some process in the cell.

After the signal passes from protein kinase 1 to protein kinase 2, protein kinase 1 is usually inactivated so that it can be available to transmit a new signal. A **phosphatase** is an enzyme that catalyzes the removal of a phosphate group by hydrolysis in a process called *dephosphorylation*. Protein phosphatases help regulate protein kinase cascades. Just as a cell contains hundreds of different protein kinases, it also contains many types of protein phosphatases. Rapid removal of phosphate groups by protein phosphatases is an important regulatory mechanism for protein kinase cascades. Such regulation ensures that these pathways operate only in response to the binding of a signal molecule to a receptor.

Ion channel–linked receptors open or close channels

The gates of many ion channels remain closed until a ligand binds to the receptor. For example, when the neurotransmitter acetylcholine binds to an acetylcholine receptor (which is an ion channel–linked receptor), the channel opens, allowing sodium ions to enter the cell (see Fig. 6-5a). Depending on the type of cell, the influx of sodium ions can result in transmission of a neural impulse or in muscle contraction.

Gamma-aminobutyric acid (GABA) is a neurotransmitter that binds to GABA receptors. One class of GABA receptors consists of ligand-gated chloride ion channels. When GABA binds to the receptor, the channel opens. Chloride ions, which are negatively charged, rush out of or into the neuron, depending on the conditions (electrochemical gradient) in the cell. Typically, chloride ions enter the cell, which inhibits transmission of neural impulses. Thus, GABA *inhibits* neural signaling. Barbiturates and benzodiazepine drugs such as Valium bind to GABA receptors. When that occurs, lower amounts of GABA are required to open the chloride channels and inhibit neural impulses. This action results in a tranquilizing effect. (Not all GABA receptors are *themselves* ion channels; some are G protein–linked receptors or enzyme-linked receptors that trigger a series of reactions resulting in activation of other proteins that serve as ion channels.)

G protein–linked receptors initiate signal transduction

As discussed in the last section, G protein–linked receptors activate G proteins, a group of regulatory proteins important in many signal transduction pathways. These proteins are found in fungi (yeasts), protists, plants, and animals. They are involved in the action of some plant and many animal hormones. Some G proteins regulate channels in the plasma membrane, allowing ions to enter or exit the cell. Other G proteins are involved in the perception of sight and smell. In 1994, Alfred G. Gilman of the University of Texas and Martin Rodbell of the National Institute of Environmental Health Sciences were awarded the Nobel Prize in Physiology or Medicine for their discovery of G proteins and their role in signal transduction. In 2012, the Nobel Prize in Chemistry was awarded to Robert Lefkowitz at the Howard Hughes Medical Institute of Duke University and Brian Kobilka of Stanford University for their work on the structure and function of G protein–linked receptors.

In its inactive state, the G protein consists of three subunits that are joined (see Fig. 6-5b). One subunit is linked to a molecule of **guanosine diphosphate (GDP),** a molecule similar to ADP but containing the base guanine instead of adenine. When a signaling molecule binds with the receptor, the GDP is released and is replaced by **guanosine triphosphate (GTP),** a nucleotide that, like ATP, functions in energy transactions.

Binding of GTP to the subunit of the G protein alters its shape, allowing it to separate from the other two subunits and bind with its protein target. The G protein subunit linked to the GTP, however, is also a GTPase, an enzyme that catalyzes the hydrolysis of GTP to GDP. This action, a process that releases energy, deactivates the G protein. In its inactive state, the G protein subunit rejoins the other two subunits.

When activated, a G protein initiates signal transduction by binding with a specific protein in the cell. In some cases, G proteins *directly* activate enzymes that catalyze changes in certain proteins. These changes lead to alterations in cell function. More commonly, the G protein relays the information from the signaling molecule, now referred to as the **first messenger,** to a *second messenger*.

Second messengers are intracellular signaling agents

Second messengers are ions or small molecules that relay signals inside the cell. When receptors are activated, second messengers

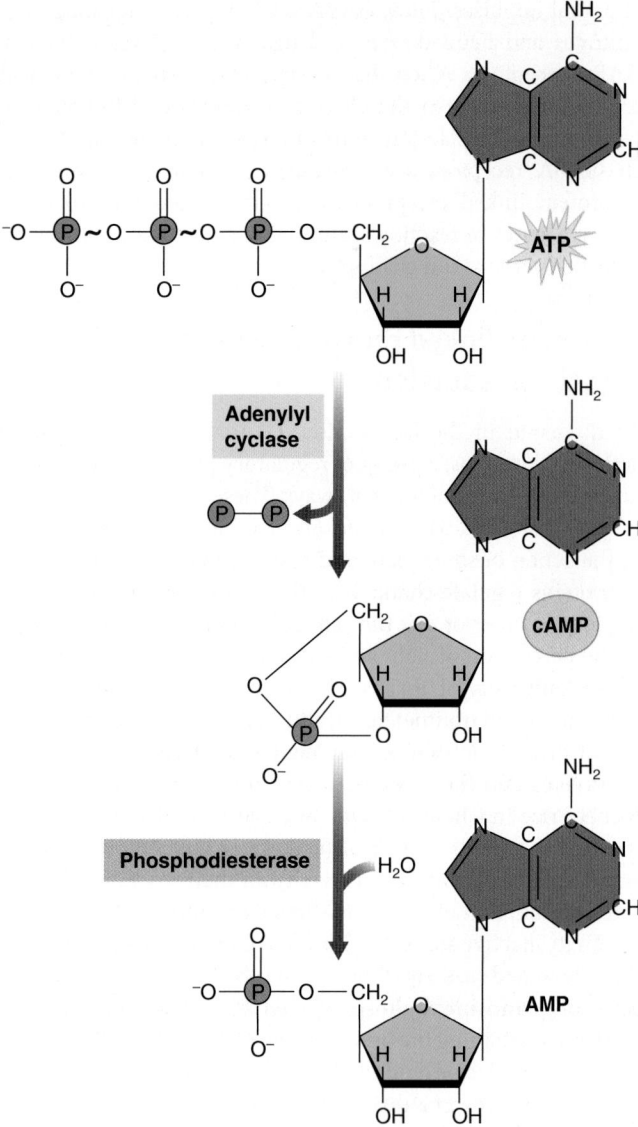

Figure 6-7 Synthesis and inactivation of cyclic AMP

Cyclic AMP (cAMP) is a second messenger produced from ATP that relays a signal from the plasma membrane to the cytosol, affecting many cellular processes. The enzyme adenylyl cyclase catalyzes the reaction. Cyclic AMP is inactivated by the enzyme phosphodiesterase, which converts it to adenosine monophosphate (AMP).

© Cengage Learning

are produced in large quantities. Second messengers rapidly diffuse through the cell (or membrane), relaying the signal. Thus, second messengers amplify the signal.

Second messengers pass the signal along to other signaling proteins or target proteins. The signal typically passes through a chain of proteins and other molecules. The last molecule in the sequence activates the final response. Second messengers are not enzymes, but some regulate specific enzymes, such as protein kinases. Others bind to ion channels, opening or closing them.

Cyclic AMP is a second messenger Most G proteins shuttle a signal between the receptor and a second messenger. In many signaling cascades in prokaryotic and animal cells, the second messenger is cyclic AMP (cAMP) (**FIG. 6-7**). Researcher Earl Sutherland identified cAMP as a second messenger in the 1960s

and was awarded the 1971 Nobel Prize in Physiology or Medicine for his pioneering work.

When the G protein undergoes a conformational change (change in shape), it binds with and activates **adenylyl cyclase,** an enzyme on the cytoplasmic side of the plasma membrane. The type of G protein that activates adenylyl cyclase is known as a stimulatory G protein, or G_s. (Some G proteins, denoted as G_i, inhibit enzymes.) Note that the G protein couples the ligand–receptor complex to adenylyl cyclase action (**FIG. 6-8**).

When activated, adenylyl cyclase catalyzes the production of cAMP from ATP. By coupling the signaling molecule–receptor complex to an enzyme that rapidly generates multiple cAMP molecules, G proteins rapidly amplify the effects of the original signaling molecule. The pathway is regulated in part by **phosphodiesterase,** an enzyme that converts cAMP to adenosine monophosphate (AMP) (see Fig. 6-7). This action is an off switch that rapidly inactivates cAMP when the receptor becomes inactive.

Figure 6-8 illustrates the sequence of events in a signaling pathway involving G protein and cyclic AMP that activates a group of protein kinase enzymes, referred to as *protein kinase A.* Recall that protein kinases add phosphate groups to target proteins. When a protein is phosphorylated, its function is altered and it triggers a chain of reactions leading to some response in the cell, such as a metabolic change.

The **substrates** (the substances on which an enzyme acts) for protein kinases are different in various cell types. Consequently, the effect of the enzyme varies depending on the substrate. For example, in skeletal muscle cells, protein kinase A activates enzymes that break down glycogen to glucose, providing the muscle cells with energy. In certain neurons in the brain, the same enzyme activates the reward system by regulating action of the neurotransmitter dopamine.

We can summarize the sequence of events during signal transduction involving a G protein and cyclic AMP, beginning with the binding of the signaling molecule to the receptor and leading to a change in some cell function:

> signaling molecule (first messenger) binds to G protein–linked receptor → activates G protein → activates adenylyl cyclase → catalyzes the formation of cAMP (second messenger) → activates protein kinase → phosphorylates proteins → response in cell

Some G proteins use phospholipids as second messengers Certain signaling molecule–receptor complexes activate a G protein that then activates the membrane-bound enzyme *phospholipase C* (**FIG. 6-9**). This enzyme splits a membrane phospholipid, PIP_2 (phosphotidylinositol-4,5-bisphosphate), into two products: **inositol trisphosphate (IP₃)** and **diacylglycerol (DAG).** Both act as second messengers.

DAG remains in the plasma membrane, where in combination with calcium ions it activates *protein kinase C* enzymes. Depending on the type of cell and the specific protein kinase C, the response of the cell can include growth, a change in cell pH, or regulation of certain ion channels. Protein kinase C stimulates contraction of smooth muscle in the digestive system and in other organs of the body.

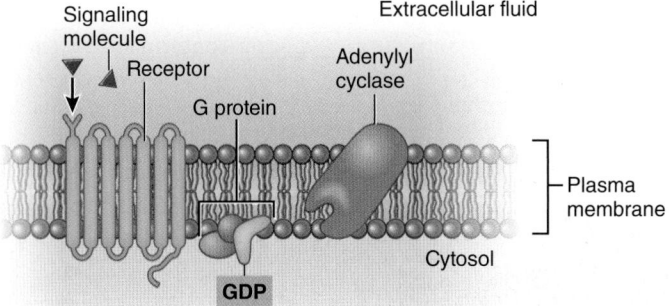

1 Signaling molecule binds with G protein–linked receptor in plasma membrane.

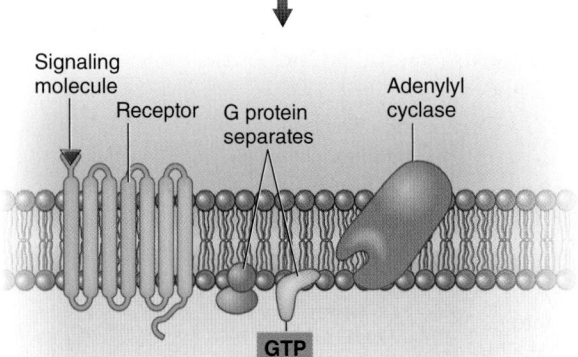

2 Signaling molecule–receptor complex activates G protein. GDP is replaced by GTP.

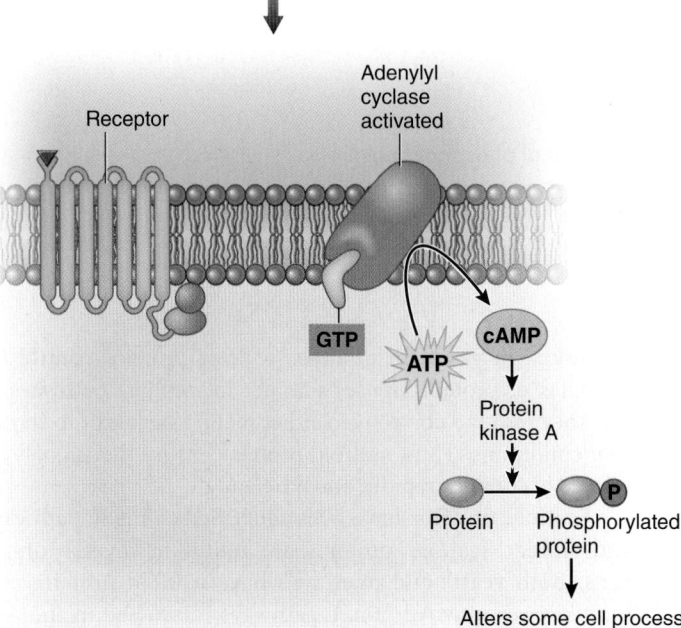

3 G protein activates adenylyl cyclase, which then catalyzes synthesis of cAMP. cAMP activates protein kinase A, which then phosphorylates specific proteins, leading to some response in cell.

Figure 6-8 Signal transduction involving a G protein and cyclic AMP

When a signaling molecule binds to a G protein–linked receptor, a G protein is activated. In this pathway the second messenger, cyclic AMP, is produced and activates a protein kinase.
© Cengage Learning

IP_3 is a member of a family of inositol phosphate messengers, some of which can donate phosphate groups to proteins. IP_3 binds to calcium channels in the endoplasmic reticulum (ER), causing them to open and release calcium ions into the cytosol. We can summarize this sequence of events as follows:

signaling molecule binds to G protein–linked receptor → activates G protein → activates phospholipase → splits PIP_2 → inositol trisphosphate (IP_3) + diacylglycerol (DAG)

DAG → activates protein kinase enzymes → phosphorylates proteins → some response in the cell

IP_3 → binds to calcium channels in ER → calcium ions released into the cytosol → some response in the cell

Calcium ions are important messengers Calcium ions (Ca^{2+}) have important functions in many cell processes, including microtubule disassembly, muscle contraction, blood clotting, secretion, and activation of certain cells in the immune system. Calcium ions are critical in neural signaling, including the pathways involved in learning. These ions are also essential in fertilization of an egg and in the initiation of development.

Ion pumps in the plasma membrane normally maintain a low calcium ion concentration in the cytosol compared to its concentration in the extracellular fluid. Calcium ions are also stored in the endoplasmic reticulum. When Ca^{2+} gates open in the plasma membrane or endoplasmic reticulum, the Ca^{2+} concentration rises in the cytosol.

Calcium ions can act alone, but typically they exert their effects by binding to certain proteins. **Calmodulin,** found in all eukaryotic cells, is an important Ca^{2+} binding protein. When 4 Ca^{2+} bind to a calmodulin molecule, the molecule changes shape and can then activate certain enzymes. Calmodulin combines with a number of different enzymes, including protein kinases and protein phosphatases, and alters their activity.

Many activated intracellular receptors are transcription factors

Some hydrophobic signaling molecules diffuse across the membranes of target cells and bind with intracellular receptors in the cytosol or in the nucleus. For example, cortisol receptors are located in the cytosol. (Cortisol is a steroid hormone produced in the adrenal glands; its structure is shown in Figure 3-15b.) Thyroid hormones pass into the nucleus and bind with receptors that are bound to DNA in the nucleus. Many intracellular receptors are transcription factors that regulate the expression of specific genes. When a signaling molecule binds to a receptor, the receptor is activated. The ligand–receptor complex binds to a specific region of DNA and activates or represses specific genes (**FIG. 6-10**).

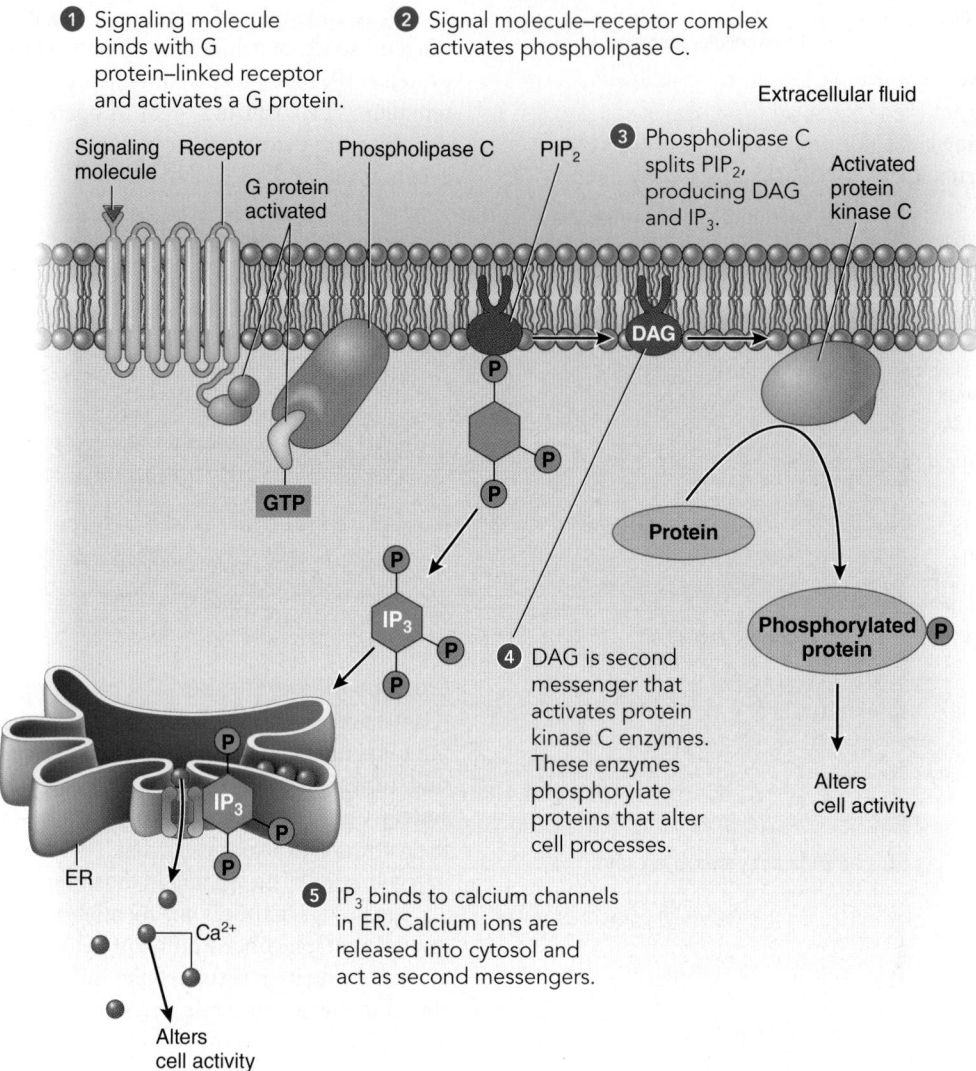

Figure 6-9 Signal transduction involving a G protein activation of phospholipase C

Activated phospholipase C splits PIP$_2$, producing two second messengers: IP$_3$ and DAG. IP$_3$ binds to calcium channels in the endoplasmic reticulum (ER); calcium ions are released into the cytosol and act as second messengers. DAG activates protein kinase C, a family of enzymes that activates signaling pathways by phosphorylating proteins; calcium ions are needed for the activation of protein kinase C.
© Cengage Learning

Gene activation can take place quickly, within about 30 minutes. Messenger RNA is produced and carries the code for synthesis of a particular protein into the cytoplasm. In combination with ribosomes, messenger RNA manufactures specific proteins that can alter cell activity.

Scaffold proteins increase efficiency

Signal transduction is a rapid, precise process. Enzymes must be organized so that they are available as needed for signaling pathways. **Scaffold proteins** organize groups of intracellular signaling molecules into signaling complexes (FIG. 6-11). Many kinases, phosphatases, and other signal transduction proteins are components of multiple signal transduction pathways. Scaffold proteins position enzymes close to the proteins they regulate, increasing the probability that they will react with one another in the correct pathway appropriate for the cell's physiological condition. In yeast, for example, scaffold proteins control the activities of protein kinases that act in multiple pathways so that they respond correctly to either mating signals or to starvation conditions. Thus, scaffold proteins ensure that signals are relayed accurately, rapidly, and efficiently.

Scaffold proteins have been identified in many pathways, and similar scaffold proteins are found in diverse organisms. Both yeasts and mammals have scaffold proteins that bind kinases in MAP kinase pathways (discussed in the next section).

Signals can be transmitted in more than one direction

Integrins, transmembrane proteins that connect the cell to the extracellular matrix, transduce signals in two directions. They transmit signals from outside the cell to the cell interior and also transmit information about the cell interior to the extracellular matrix.

Cell biologists have demonstrated that when certain signaling molecules bind to integrins in the plasma membrane, specific signal transduction pathways are activated. Interestingly, growth factors and certain molecules of the extracellular matrix may modulate one another's messages. Integrins also respond to information received from inside the cell. This *inside-out signaling* affects how selective integrins are with respect to the molecules to which they bind and how strongly they bind to them.

CHECKPOINT 6.4

- *How is an extracellular signal converted to an intracellular signal in signal transduction? Give a specific example.*
- CONNECT *What feature do all second messengers have in common?*
- PREDICT *How might a signaling pathway be affected if a scaffold protein specific to that pathway were absent?*

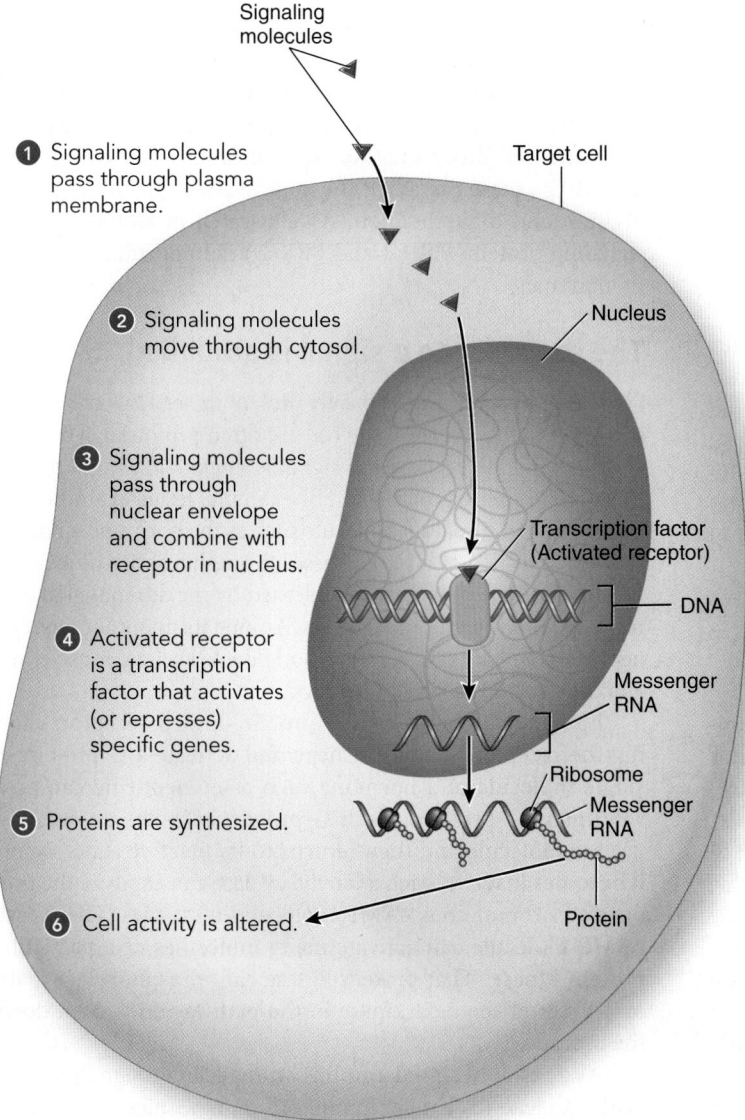

Figure 6-10 Intracellular receptors

Many intracellular receptors are transcription factors; when activated, they activate or repress specific genes.

© Cengage Learning

6.5 RESPONSES TO SIGNALS

LEARNING OBJECTIVES

7 Describe three types of responses that cells make to signaling molecules.

8 Contrast signal amplification with signal termination.

As we have learned, signaling molecules activate signal transduction pathways that bring about specific responses in the cell. Most of these responses fall into three categories: ion channels open or close; enzyme activity is altered, leading to metabolic changes and other effects; and specific gene activity may be turned on or off. Various mechanisms and pathways interact to produce specific actions that are responsible for the structure and function of the cell. They are responsible for metabolic activity, movement, growth, cell division, development, and further information transfer.

In animals neurons release neurotransmitters that excite or inhibit other neurons or muscle cells by affecting ion channels. For example, when acetylcholine binds with a receptor on a target neuron, an ion channel opens and allows passage of sodium and potassium ions. The resulting change in ion permeability can activate the neuron so that it transmits a neural impulse. Serotonin and some other neurotransmitters work indirectly through G proteins and cyclic AMP. In this chain of events, cAMP activates a kinase that phosphorylates a protein, which then closes potassium ion channels. This action leads to transmission of a neural impulse. Some G proteins directly open or close ion channels.

Some receptors directly affect enzyme activity, whereas others initiate signal transduction pathways in which enzymes are altered by components of the pathway. When bacteria infect the body, they release certain peptides. Neutrophils, a type of white blood cell, have cell-surface receptors that detect these peptides. When the bacterial peptides bind to a neutrophil's receptors, enzymes are activated that lead to assembly of microfilaments (actin filaments) and microtubules. Contractions of microfilaments at the far end of the neutrophil force the cytoplasm forward. This action allows the neutrophil to move toward the invading bacteria and destroy them. Microtubules and a variety of proteins, including the contractile protein myosin, appear necessary for this movement.

Some signaling molecules affect gene activity. For example, some signaling molecules activate genes that lead to the manufacture of proteins needed for growth and cell division. In both plants and animals, steroid hormones regulate development by causing changes in the

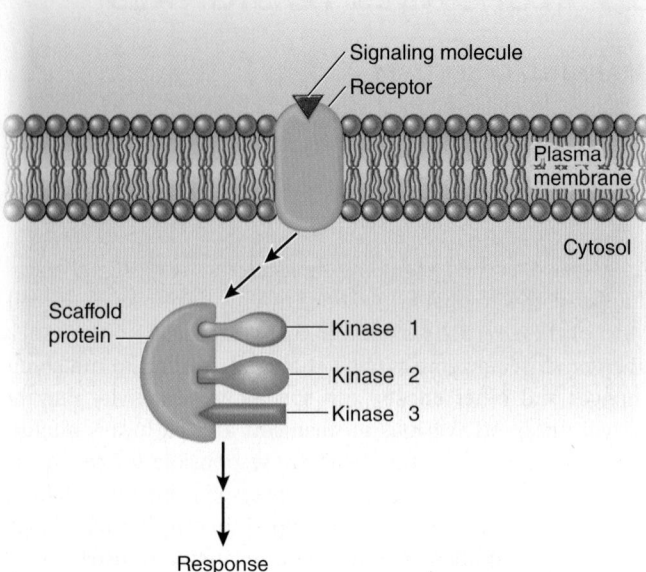

Figure 6-11 Scaffold proteins

These proteins direct signaling traffic by organizing groups of signaling molecules into a complex to make signal transduction faster, more precise, and more efficient. Scaffold proteins are involved in regulating the activity of protein kinases that are used in multiple signal transduction pathways, ensuring that the correct pathway is used for a given physiological condition.
© Cengage Learning

expression of specific genes. In animal cells some steroid hormones bind to nuclear receptors and directly regulate expression of specific target genes. In plant cells steroid hormones bind to receptors on the cell surface. The signal is then transmitted through a chain of molecules, eventually leading to changes in gene expression. Plant hormones will be discussed in greater detail in Chapter 38, and animal hormones are the focus of Chapter 49.

Ras pathways involve tyrosine kinase receptors and G proteins

Some tyrosine kinase receptors activate G proteins. For example, when growth factors bind to tyrosine kinase receptors, **Ras proteins** are activated. Ras proteins are a group of small G proteins that were originally isolated from *Ra*t *s*arcoma cancer cells. Like other G proteins, Ras proteins are active when bound to GTP. Ras proteins are molecular switches that regulate signaling networks inside the cell. When activated, Ras triggers a cascade of reactions called the Ras pathway. In this pathway a tyrosine in specific kinase proteins is phosphorylated, leading to critical cell responses.

Ras pathways are important in gene expression, cell division, cell movement, cell differentiation, cell adhesion, embryonic development, and apoptosis. For example, to initiate DNA synthesis, fibroblasts (a type of connective tissue cell) require the presence of two growth factors, epidermal growth factor and platelet-derived growth factor. In one study investigators injected fibroblasts with antibodies that inactivate Ras proteins by binding to them. The fibroblasts no longer synthesized DNA

in response to growth factors. Data from this and similar experiments led to the conclusion that Ras proteins are important in signal transduction involving growth factors.

Ras genes code for Ras proteins. Certain mutations in *Ras* genes result in mutant Ras proteins that bind GTP but cannot hydrolyze it. The mutant Ras proteins are stuck in the "on" state, resulting in unregulated cell division. This condition is associated with several types of human cancer. In fact, mutations in *Ras* genes have been identified in about one-third of all human cancers. Drugs that inhibit specific tyrosine kinase receptors and Ras pathways are being developed to treat cancer.

One Ras pathway that has been extensively studied is the *MAP kinase pathway,* also known as the *ERK pathway; MAP* is an acronym for "mitogen-activated protein" (mitogens induce mitosis, the nuclear division associated with eukaryotic cell division). *ERK* is an acronym for "extracellular signal-regulated kinases." Several distinct groups of MAP kinases have been described. The pathway illustrated in **FIGURE 6-12** shows three main MAP protein kinases: Raf, Mek, and ERK. Proteins in the MAP pathway phosphorylate a nuclear protein that combines with other proteins to form a transcription factor. When specific genes are activated, proteins needed for cell growth, cell division, and cell differentiation (specialization) are synthesized. The MAP kinase cascade is the main signaling pathway for cell division and differentiation. As illustrated in **FIGURE 6-13**, the signaling proteins ERK 1 and ERK 2 are important in fertility in mammals.

The response to a signal is amplified

Signaling molecules are typically present in very low concentrations, yet their effects on the cell are often profound. This situation is possible because the signal is *amplified,* as it is relayed through a signaling pathway. For example, let us examine how the action of a signaling molecule such as the hormone epinephrine is magnified as a signal passes through a series of proteins inside the cell. Epinephrine is released by the adrenal glands in response to danger or other stress. Among its many actions, epinephrine increases heart rate, blood flow to skeletal muscle, and glucose concentration in the blood.

Epinephrine binds to a G protein–linked receptor, causing the receptor to change shape and activate a G protein. A single molecule of a hormone such as epinephrine can activate many G proteins. Each G protein activates an adenylyl cyclase molecule and then returns to its inactive state. Before it becomes inactive, each adenylyl cyclase can catalyze the production of numerous cAMP molecules (**FIG. 6-14**). Then, each cAMP molecule can activate many molecules of a particular protein kinase. That protein kinase can phosphorylate many molecules of the next kinase in the pathway and so on down the cascade.

As a result of **signal amplification,** a single signaling molecule can lead to changes in millions of molecules at the end of a signaling cascade. The response is much greater than would be possible if each signaling molecule acted alone. This process of magnifying the strength of a signaling molecule explains how just a few signaling molecules can lead to major responses in the cell.

When growth factors bind to their receptors, they activate the G protein Ras; Ras activates a MAP-kinase signaling pathway, leading to activation (or repression) of specific genes and ultimately to protein activity that affects some cell process.

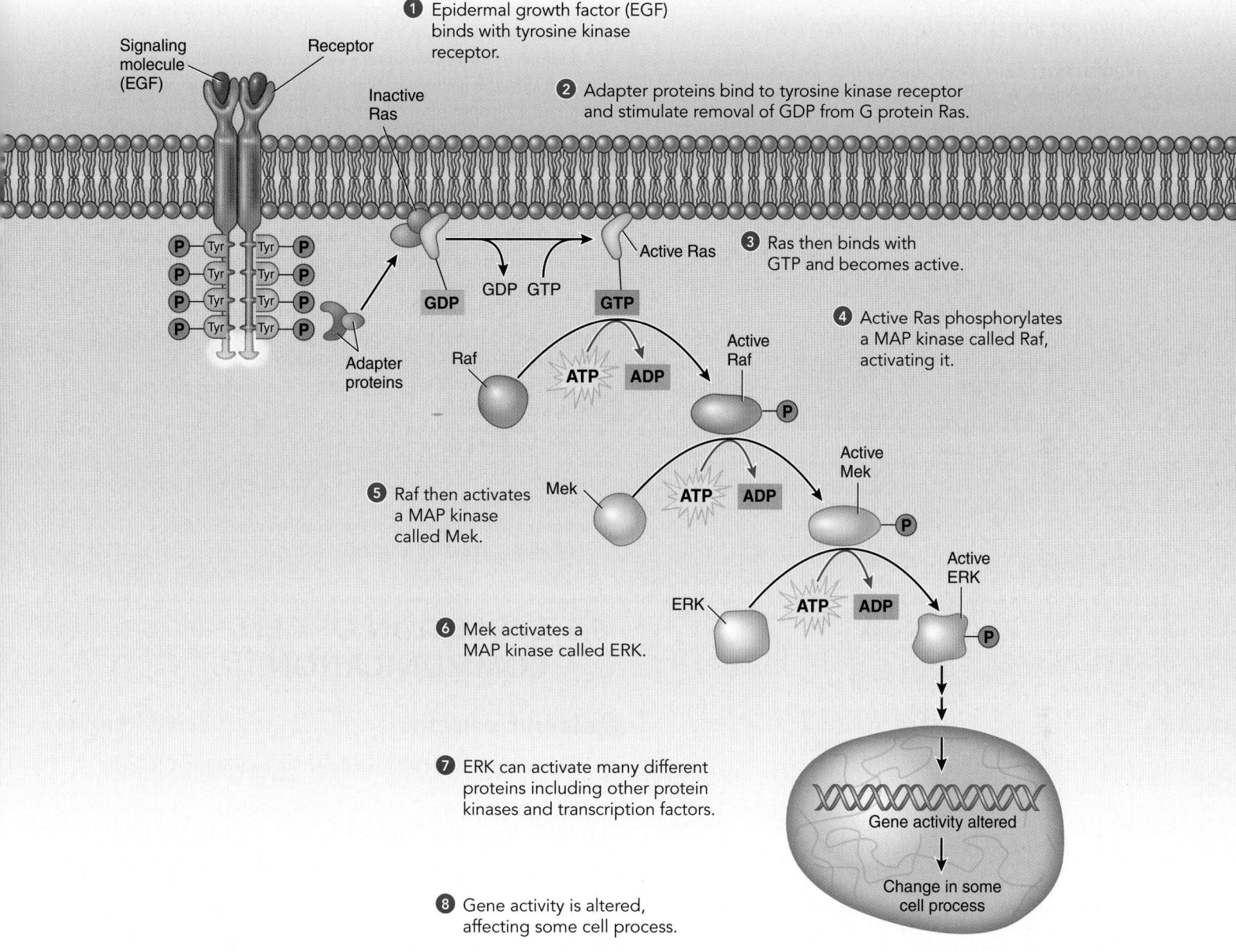

Figure 6-12 A highly simplified Ras/MAP kinase signaling pathway

In the pathway illustrated here, epidermal growth factor (EGF) binds with a tyrosine kinase receptor, leading to activation of the small G protein Ras. Then, Ras activates a MAP-kinase signaling pathway. A series of MAP kinases in the pathway are activated by phosphorylation. The final MAP kinase can regulate several transcription factors, leading to changes in gene expression that affect cell processes.

PREDICT In many cancers the Ras protein is mutated so that it is always active, promoting continuous cell growth and division. Starting with the EGF receptor, what proteins in the signaling pathway might be good targets for an inhibitory drug that would prevent the mutant Ras protein from promoting cancerous growth?

© Cengage Learning

Signals must be terminated

Once a signal has done its job, it must be terminated. *Signal termination* returns the receptor and each of the components of the signal transduction pathway to their inactive states. This action allows the magnitude of the response to reflect the strength of the signal. Molecules in the system must also be ready to respond to new signals.

We have seen that after a G protein is activated, a subunit of the G protein, a GTPase, catalyzes the hydrolysis of GTP to GDP. This action inactivates the G protein. In the cyclic AMP pathway, any increase in cAMP concentration is temporary. Cyclic AMP

Do the signaling molecules ERK 1 and ERK 2 have key functions in signaling pathways leading to maturation of oocytes (eggs) and ovulation (release of a mature egg from the ovary) in mammals?

HYPOTHESIS: The signaling molecules ERK 1 and ERK 2 are key target molecules of the signal sent by the reproductive hormone, luteinizing hormone (LH), and thus are important in oocyte maturation and ovulation in mammals.

EXPERIMENT: The researchers produced a line of mice that were deficient in both ERK 1 and ERK 2. They performed biochemical analyses on the signaling pathway activated by LH. They also examined the ovaries of the mice.

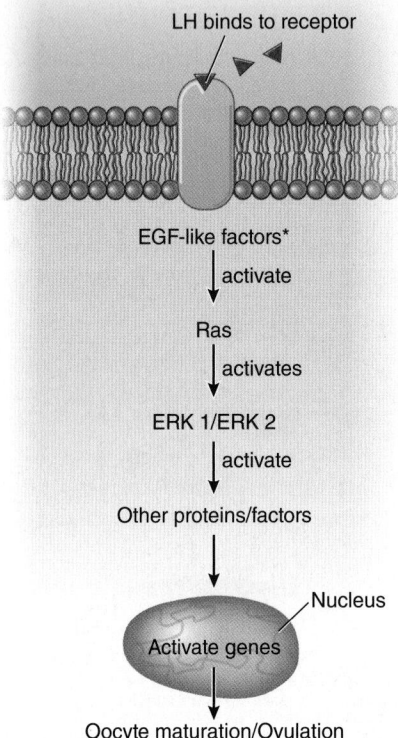

RESULTS AND CONCLUSION: In mice deficient in both ERK 1 and ERK 2, the signaling pathway from LH was disrupted. As a result, genes regulating oocyte maturation and ovulation were not activated. Oocytes failed to mature and ovulate, and the mice also exhibited reproductive disorders associated with changes in levels of reproductive hormones.

This experiment demonstrates that ERK 1 and ERK 2 are key molecules in a signaling pathway that regulates reproduction in mice. These results may lead to greater understanding of certain ovarian disorders that result in infertility in humans.

*EGF = epidermal growth factor

SOURCE: Fan, H.-Y. et al., *Science*, Vol. 324, 938–941, 2009.

Figure 6-13 Identifying key molecules in a signaling pathway that regulates reproduction in mammals

PREDICT What do you think would happen if an altered receptor were to become activated in the absence of LH?

© Cengage Learning

is rapidly inactivated by a phosphodiesterase, which converts it to adenosine monophosphate (AMP). Thus, the concentration of cAMP depends on the activity of both adenylyl cyclase, which produces it, and of phosphodiesterase, which breaks it down (see Fig. 6-8). Recall also that in many signaling pathways, each protein kinase activates the next protein kinase in the chain by phosphorylating it; a phosphatase then inactivates it by removing the phosphate group.

Failure to terminate signals can lead to dire consequences. For example, the bacterium that causes cholera is ingested when people drink contaminated water. Cholera is prevalent in areas where water is contaminated with human feces. The cholera bacterium releases a toxin that activates G proteins in the epithelial cells lining the intestine. The toxin chemically changes the G protein so that it no longer switches off. As a result, the G protein continues to stimulate adenylyl cyclase to make cAMP. The cells lining the intestine malfunction, allowing a large flow of chloride ions into the intestine. Water and other ions follow, leading to the severe, watery diarrhea that characterizes cholera. The disease is treated by replacing the lost fluid. If untreated, this G protein malfunction can cause death.

CHECKPOINT 6.5

- *What are some cell responses to signals?*
- **VISUALIZE** *Draw a simple sketch illustrating signal amplification.*
- **PREDICT** *What are some of the potential consequences of failure of signal termination?*

6.6 EVOLUTION OF CELL COMMUNICATION

LEARNING OBJECTIVE

9 Cite evidence supporting a long evolutionary history for cell signaling molecules.

In this chapter we have examined how the cells of a multicellular organism signal one another and have described a few of the many signal transduction pathways within cells. We have described quorum sensing and other examples of communication between members of a species. We have also discussed communication among members of different species, such as signaling between plants and insects. In our discussion we have noted many similarities in the types of signals used and in the molecules that cells use to relay signals from the cell surface to the molecules that carry out a specific response. Some signal transduction pathways found in organisms as diverse as yeasts and animals are quite similar.

G proteins, protein kinases, and phosphatases have been highly conserved and are part of signaling pathways in most organisms. Certain disease-causing bacteria have signal transduction pathways similar to those found in eukaryotes. Bacteria may use some of these signal mechanisms to interfere with normal function in the eukaryotic cells they infect. Such similarities in many species suggest evolutionary relationships.

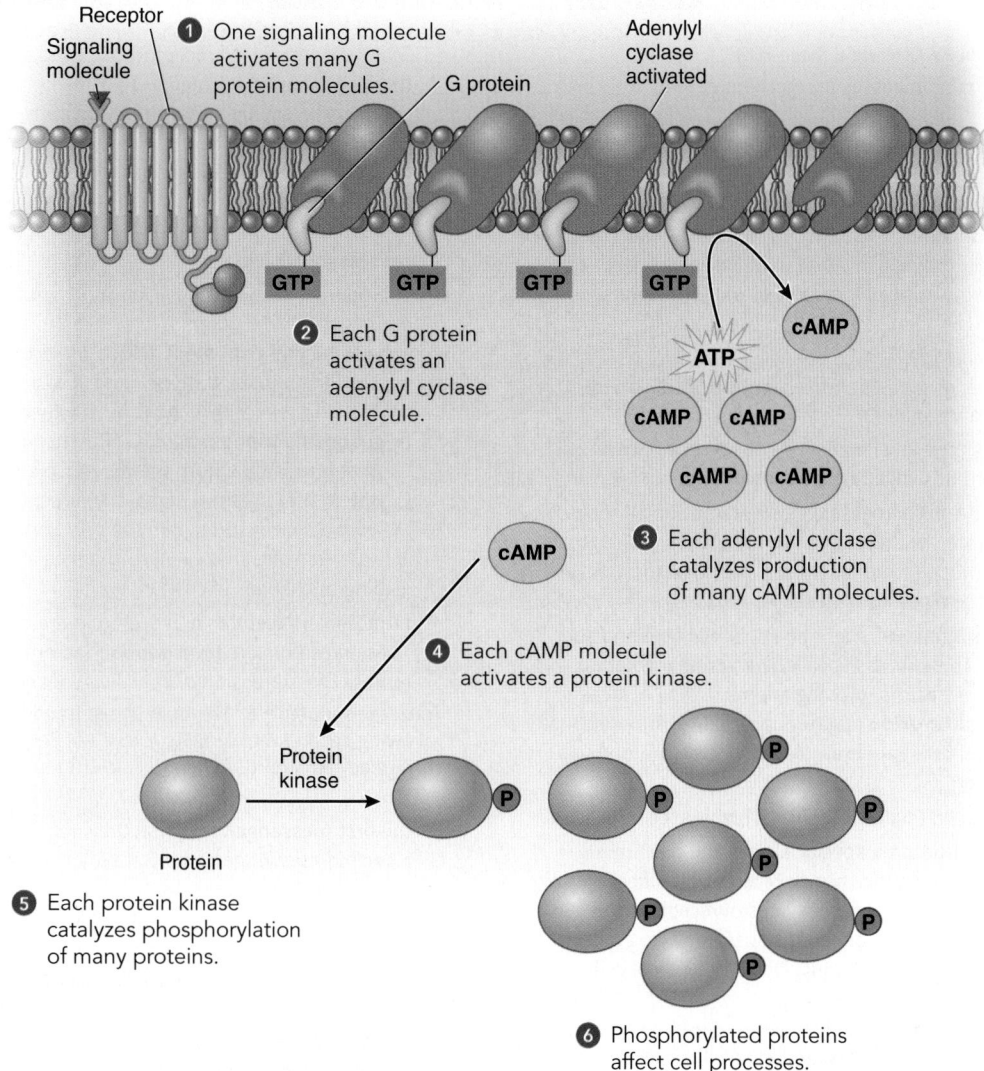

1 One signaling molecule activates many G protein molecules.

Receptor

Signaling molecule

G protein

Adenylyl cyclase activated

GTP GTP GTP GTP

2 Each G protein activates an adenylyl cyclase molecule.

ATP

cAMP

cAMP cAMP

cAMP cAMP

3 Each adenylyl cyclase catalyzes production of many cAMP molecules.

cAMP

4 Each cAMP molecule activates a protein kinase.

Protein kinase

Protein

P P P P P P P

5 Each protein kinase catalyzes phosphorylation of many proteins.

6 Phosphorylated proteins affect cell processes.

Figure 6-14 Signal amplification

The signal is amplified at each step in the pathway so that one activated receptor can give rise to thousands of final products (proteins at the end of the pathway). The response is far greater than what you might expect from a single receptor.

© Cengage Learning

Similarities in cell signaling suggest that the molecules and mechanisms used in cell communication are very old. The evidence suggests that cell communication first evolved in prokaryotes and continued to change over time as new types of organisms evolved. However, some cell signaling molecules have not changed very much over time, which suggests that the importance of these pathways to cell survival has restricted any evolutionary changes that might have made them less effective. Thus, these pathways have weathered the demands of natural selection through millions of years of evolution.

Choanoflagellates, unicellular protists (see Chapter 26) that are thought to be the most recent ancestors of animals, are used as models for studying the early evolution of animals.

Choanoflagellates have many of the same proteins found in animals. These tiny organisms have protein kinases similar to those of animals, and a G protein–linked receptor has been identified. These findings indicate that important proteins necessary for cell communication in animals had evolved long before the evolution of animals. As we will discuss in later chapters, similarities and differences in basic molecules, such as G proteins, can be used to trace evolutionary pathways.

CHECKPOINT 6.6

- **CONNECT** *Choanoflagellates and animals have similar protein kinases. What does that suggest about their cell signaling mechanisms?*

6.1 Cell Communication: An Overview *(page 130)*

1 Describe the four main processes essential for cells to communicate.

- Cells communicate by **cell signaling,** which consists of four main processes: (1) synthesis, release, and transport of signaling molecules; (2) **reception** of information by **target cells;** (3) **signal transduction,** the process by which a receptor converts an extracellular signal into an intracellular signal and relays the signal, leading to a cellular response; and (4) **response** by the cell, for example, some metabolic process may be altered.

6.2 Sending Signals *(page 131)*

2 Compare three types of signaling molecules: neurotransmitters, hormones, and local regulators.

- Most neurons (nerve cells) signal one another by releasing chemical compounds called **neurotransmitters.**

- **Hormones** are chemical messengers in plants and animals. In animals they are secreted by **endocrine glands,** glands that have no ducts. Most hormones diffuse into capillaries and are transported by the blood to target cells.

- **Local regulators** diffuse through the interstitial fluid and act on nearby cells. This process is called **paracrine regulation.** Histamine, growth factors, **prostaglandins** (a type of local hormone), and **nitric oxide** (a gaseous signaling molecule that passes into target cells) are examples of local regulators.

6.3 Reception *(page 132)*

3 Identify mechanisms that make reception a highly specific process.

- Each type of receptor has a specific shape, and only the signaling molecule that fits the specific receptor can affect the cell. A cell can have many different types of receptors and can make different receptors at different stages in its life cycle or in response to different conditions. Different types of cells can have different types of receptors.

4 Briefly compare ion channel–linked receptors, G protein–linked receptors, enzyme-linked receptors, and intracellular receptors.

- When a signaling molecule binds to an **ion channel–linked receptor,** the ion channel opens or, in some cases, closes.

- **G protein–linked receptors** are transmembrane proteins that extend into the cytosol or outside the cell. These receptors couple specific signaling molecules to signal transduction pathways inside the cell. The tail of the receptor that extends into the cytosol has a binding site for a specific **G protein,** a regulatory protein that binds to GTP.

- **Enzyme-linked receptors** are transmembrane proteins with a binding site for a signaling molecule outside the cell and a binding site for an enzyme inside the cell. Many enzyme-linked receptors are *tyrosine kinases* in which the enzyme is part of the receptor.

- *Intracellular receptors* are located in the cytosol or in the nucleus. Their ligands are small, hydrophobic molecules that diffuse across the plasma membrane.

6.4 Signal Transduction *(page 135)*

5 Compare the actions of the main types of receptors in signal transduction.

- Ion channel–linked receptors convert chemical signals into electrical signals. The gates of many ion channels remain closed until ligands bind to them.

- Many enzyme-linked receptors are tyrosine kinases, enzymes that phosphorylate proteins. Tyrosine kinase receptors activate several different signal transduction pathways. In a *protein kinase cascade,* each molecule in the signaling pathway is phosphorylated by the preceding protein kinase in the chain. The last protein kinase in the cascade activates the target protein by phosphorylation. The target protein alters some process in the cell.

- G protein–linked receptors activate G proteins. A G protein consists of three subunits. It is linked to a molecule of **guanosine diphosphate (GDP),** a molecule similar to ADP but containing the base guanine instead of adenine. When a ligand binds with the receptor, the GDP is released and is replaced by **guanosine triphosphate (GTP).** Then one subunit of the G protein separates from the other two subunits. The activated G protein may initiate a signal transduction pathway by binding with a specific protein in the cell. Some G proteins directly activate enzymes that catalyze changes in certain proteins, leading to changes in cell function.

- Intracellular receptors are located in the cytosol or nucleus. These receptors are **transcription factors** that activate or repress the expression of specific genes.

6 Trace the sequence of events in signal transduction for each of the following second messengers: cyclic AMP, inositol trisphosphate (IP$_3$), diacylglycerol (DAG), and calcium ions.

- In many signaling systems, the signaling molecule serves as the **first messenger.** Information is relayed by the G protein to a **second messenger,** an intracellular signaling molecule.

- When certain G proteins undergo a conformational change, they bind with and activate **adenylyl cyclase,** an enzyme on the cytoplasmic side of the plasma membrane. Adenylyl cyclase catalyzes the formation of **cyclic AMP (cAMP)** from ATP. Cyclic AMP is a second messenger that activates certain **protein kinase** enzymes that phosphorylate certain proteins. The phosphorylated protein triggers a chain of reactions that lead to some response in the cell.

- Certain G proteins activate the membrane-bound enzyme phospholipase C. This enzyme splits a phospholipid, PIP$_2$ (phosphotidylinositol-4,5-bisphosphate), into two products, **inositol trisphosphate (IP$_3$)** and **diacylglycerol (DAG).** IP$_3$ is a second messenger that can donate phosphate groups to proteins. IP$_3$ binds to calcium channels in the endoplasmic reticulum, which causes the channels to open and release calcium ions into the cytosol. DAG is a second messenger that activates certain protein kinase enzymes. These enzymes phosphorylate a variety of target proteins.

- Calcium ions can also act as second messengers. They typically combine with the protein **calmodulin,** which then affects the activity of protein kinases and protein phosphatases.

6.5 Responses to Signals *(page 141)*

7 Describe three types of responses that cells make to signaling molecules.

- In response to signaling molecules, ion channels open or close, enzyme activity changes, leading to metabolic changes and other effects, and specific genes are activated or repressed. These responses can affect cell shape, cell growth, cell division, cell differentiation, and metabolism.

8 Contrast signal amplification with signal termination.

- **Signal amplification** is the process of enhancing the cell's response to a signal as the signal is relayed through a signal transduction pathway. Before it becomes inactive, each enzyme can catalyze the production of numerous product molecules.

- *Signal termination* is the process of inactivating the receptor and each component of the signal transduction pathway once they have done their jobs. Signal termination allows molecules in the system to respond to new signals.

6.6 Evolution of Cell Communication *(page 144)*

9 Cite evidence supporting a long evolutionary history for cell signaling molecules.

- Molecules important in cell signaling first evolved in prokaryotes. G proteins, protein kinases, and phosphatases have been highly conserved and are part of most signaling pathways.

TEST YOUR UNDERSTANDING

Know and Comprehend

1. During signal transduction (a) the cell converts an extracellular signal into an intracellular signal that leads to a change in some cell process (b) a signaling molecule directly activates or represses several genes (c) each enzyme catalyzes production of one molecule of product (d) enzymes in the signal cascade remain active until the last component of the pathway alters a cellular process (e) the signal is terminated by cyclic AMP

2. When a signaling molecule binds with a receptor, (a) G proteins are inactivated (b) a third messenger is activated (c) cell signaling is terminated (d) cAMP is produced by the receptor (e) the receptor becomes activated

3. G protein–linked receptors (a) inactivate G proteins (b) activate first messengers (c) consist of 18 transmembrane alpha helices (d) have a tail that extends into the cytosol with a binding site for a G protein (e) are located in the cytoplasm or nucleus

4. An enzyme-linked receptor (a) is a cytoplasmic protein (b) would not be found on plant cell surfaces (c) forms a dimer with another enzyme-linked receptor when a ligand binds to it (d) is typically an adenylyl cyclase molecule (e) typically activates ion channels

5. G proteins (a) relay a message from an activated receptor to an enzyme that activates a second messenger (b) are GTP molecules (c) terminate cell signaling (d) directly activate protein kinases (e) function as first messengers

6. Calcium ions (a) can act as second messengers (b) split calmodulin (c) are kept at higher concentration in the cytosol than in the extracellular fluid (d) are produced in the ER by protein kinases and protein phosphatases (e) typically terminate signaling cascades

7. When growth hormone binds to an enzyme-linked receptor, (a) G proteins are amplified into a cascade of molecules (b) the enzyme portion of the receptor becomes dephosphorylated (c) the receptor becomes activated and phosphorylates signaling proteins in the cell (d) an ion channel is opened (e) an immediate signal is sent into the nucleus, and specific genes are activated or inhibited

8. Scaffold proteins (a) release kinases and phosphatases into the extracellular fluid (b) bind G proteins to cell membranes (c) increase accuracy but slow signaling cascades (d) organize groups of intracellular signaling molecules into signaling complexes (e) are transcription factors found mainly in plant cells

9. Each adenylyl cyclase molecule produces many cAMP molecules in an example of (a) receptor up-regulation (b) receptor down-regulation (c) signal amplification (d) scaffolding (e) similarities produced by evolution

Apply and Analyze

10. In response to a hormone secreted by a cell of the opposite mating type, a yeast cell undergoes a complex series of physiological changes involving the activity of about 200 genes and cytoplasmic proteins. They include blocking DNA synthesis, growing toward the mating partner, fusion of the plasma membranes of the two cells, and fusion of their nuclear membranes. Explain how all these events can be controlled through a complex signaling cascade that is triggered by the binding of the hormone to a G protein–linked receptor.

11. More than five hundred genes have been identified in the human genome that code for protein kinases. What does such identification imply regarding the role of protein kinases in cellular functions? Explain your answer.

12. Which is the correct sequence? 1. protein kinase activated 2. adenylyl cyclase activated 3. cAMP produced 4. proteins phosphorylated 5. G protein activated (a) 1, 2, 3, 5, 4 (b) 5, 3, 2, 1, 4 (c) 5, 2, 3, 4, 1 (d) 5, 2, 3, 1, 4 (e) 2, 3, 1, 4, 5

13. Which is the correct sequence? 1. phospholipase activated 2. G protein activated 3. PIP_2 split 4. proteins phosphorylated 5. DAG produced (a) 1, 2, 5, 3, 4 (b) 2, 1, 3, 5, 4 (c) 4, 2, 3, 1, 5 (d) 5, 2, 3, 1, 4 (e) 2, 3, 5, 4, 1

Evaluate and Synthesize

14. **EVOLUTION LINK** Cell signaling in plant and animal cells is similar in some ways and different in others. Offer one or more hypotheses for these similarities and differences, and cite specific examples.

15. **EVOLUTION LINK** Some of the same G protein–linked receptors and signal transduction pathways found in plants and animals have been identified in fungi and algae. What does that suggest about the evolution of these molecules? What does it suggest about these molecules and pathways?

16. **SCIENCE, TECHNOLOGY, AND SOCIETY** Mutant tyrosine kinase signaling proteins are implicated in many types of human cancer. Hundreds of millions of dollars are required for the basic research and development of each new drug; consequently, many of these drugs are very expensive when used for cancer treatments. In 2012, eight tyrosine kinase inhibitors were approved for cancer treatments by the U.S. Food and Drug Administration, and more than one hundred potential inhibitors were undergoing clinical trials for approval. What do you think would be some of the difficulties of finding these drugs given that similar kinases are active in normal cells? Do you think new medications of this type should be developed through government-sponsored research? Why or why not? If not, what alternatives do you propose?

aplia To access course materials, such as Aplia and other companion resources, please visit **www.cengagebrain.com**.

Energy and Metabolism

Giant panda (*Ailuropoda melanoleuca*). The chemical energy produced by photosynthesis and stored in bamboo leaves transfers to the panda as it eats.

Keren Su/Corbis

A ll living things require energy to carry out life processes. It may seem obvious that cells need energy to grow and reproduce, but even nongrowing cells need energy simply to maintain themselves. Cells obtain energy in many forms, but that energy can seldom be used directly to power cell processes. For this reason, cells have mechanisms that convert energy from one form to another. The ordered systems of the cell provide the information that makes these energy transformations possible. Because most components of these energy conversion systems evolved very early in the history of life, many aspects of energy metabolism tend to be similar in a wide range of organisms.

The sun is the ultimate source of almost all the energy that powers life; this *radiant energy* flows from the sun as electromagnetic waves. Plants and other photosynthetic organisms capture about 0.02% of the sun's energy that reaches Earth. As discussed in Chapter 9, photosynthetic organisms convert radiant energy to *chemical energy* in the bonds of organic molecules. This chemical energy becomes available to plants, animals such as the giant panda shown in the photograph, and other organisms through the process of cellular respiration. In cellular respiration, discussed in Chapter 8, organic molecules are broken apart, and their energy is converted to more immediately usable forms.

This chapter focuses on some of the basic principles that govern how cells capture, transfer, store, and use energy. We discuss the functions of adenosine triphosphate (ATP) and other molecules used in energy conversions, including those that transfer electrons in oxidation–reduction (redox) reactions. We also pay particular attention to the essential role of enzymes in cell energy dynamics. The flow of energy in ecosystems is discussed in Chapter 55.

KEY CONCEPTS

7.1 Energy, the capacity to do work, can be kinetic energy (energy of motion) or potential energy (energy due to position or state).

7.2 Energy cannot be created or destroyed (the first law of thermodynamics), but the total amount of energy available to do work in a closed system decreases over time (the second law of thermodynamics). Organisms do not violate the laws of thermodynamics because, as open systems, they use energy obtained from their surroundings to do work.

7.3 In cells energy-releasing (exergonic) processes drive energy-requiring (endergonic) processes.

7.4 ATP plays a central role in cell energy metabolism by linking exergonic and endergonic reactions. ATP transfers energy by transferring a phosphate group.

7.5 The transfer of electrons in redox reactions is another way that cells transfer energy.

7.6 As biological catalysts, enzymes increase the rate of specific chemical reactions. The activity of an enzyme is influenced by temperature, pH, the presence of cofactors, and inhibitors and activators.

7.1 BIOLOGICAL WORK

LEARNING OBJECTIVES

1 Define *energy*, emphasizing how it is related to work and to heat.
2 Use examples to contrast potential energy and kinetic energy.

Energy, one of the most important concepts in biology, can be understood in the context of **matter,** which is anything that has mass and takes up space. **Energy** is defined as the capacity to do work, which is any change in the state or motion of matter. Technically, mass is a form of energy, which is the basis behind the energy generated by the sun and other stars. More than 4 billion kilograms of matter per second are converted into energy in the sun.

Biologists generally express energy in units of work, or **kilojoules (kJ).** It can also be expressed in units of *heat energy*—**kilocalories (kcal)**—thermal energy that flows from an object with a higher temperature to an object with a lower temperature. One kilocalorie is equal to 4.184 kJ. Heat energy cannot do cell work because a cell is too small to have regions that differ in temperature. For that reason, the unit most biologists prefer today is the kilojoule. However, we will use both units because references to the kilocalorie are common in the scientific literature.

Organisms carry out conversions between potential energy and kinetic energy

When an archer draws a bow, **kinetic energy,** the energy of motion, is used and work is performed (FIG. 7-1). The resulting tension in the bow and string represents stored, or potential, energy. **Potential energy** is the capacity to do work as a result of position or state. When the string is released, this potential energy is converted to kinetic energy in the motion of the bow, which propels the arrow.

Most actions of an organism involve a series of energy transformations that occur as kinetic energy is converted to potential energy or as potential energy is converted to kinetic energy. *Chemical energy,* potential energy stored in chemical bonds, is of particular importance to organisms. In our example the chemical energy of food molecules is converted to kinetic energy in the muscle cells of the archer. The contraction of the archer's muscles, like many of the activities performed by an organism, is an example of *mechanical energy,* which performs work by moving matter.

CHECKPOINT 7.1

- **VISUALIZE** *Draw simple sketches showing (1) a person stretching a spring, (2) the stretched spring, and (3) the release of tension. Label work that is done, potential energy, and kinetic energy.*

7.2 THE LAWS OF THERMODYNAMICS

LEARNING OBJECTIVE

3   State the first and second laws of thermodynamics, and discuss the implications of these laws as they relate to organisms.

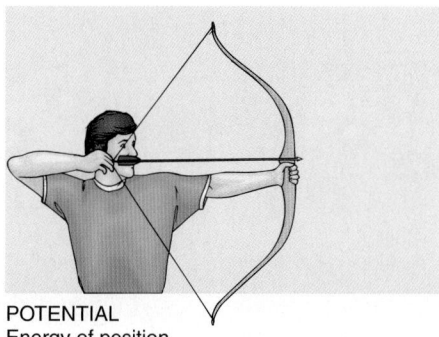

POTENTIAL
Energy of position

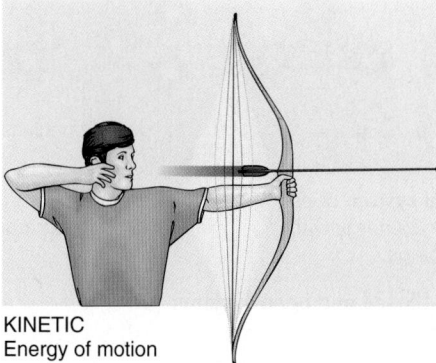

KINETIC
Energy of motion

Figure 7-1 Potential versus kinetic energy
The potential chemical energy released by cellular respiration is converted to kinetic energy in the muscles, which do the work of drawing the bow. The potential energy stored in the drawn bow is transformed into kinetic energy as the bowstring pushes the arrow toward its target.
© Cengage Learning

Thermodynamics, the study of energy and its transformations, governs all activities of the universe, from the life and death of cells to the life and death of stars. When considering thermodynamics, scientists use the term *system* to refer to an object that they are studying, whether a cell, an organism, or planet Earth. The rest of the universe other than the system being studied constitutes the *surroundings*. A **closed system** does not exchange energy with its surroundings, whereas an **open system** can exchange energy with its surroundings (FIG. 7-2). Biological systems are open systems. Two laws about energy—the first and second laws of thermodynamics—apply to all things in the universe.

The total energy in the universe does not change

According to the **first law of thermodynamics,** energy cannot be created or destroyed, although it can be transferred or converted from one form to another, including conversions between matter and energy. As far as we know, the total mass-energy present in the universe when it formed, almost 14 billion years ago, equals the amount of energy present in the universe today. It is all the energy that can ever be present in the universe. Similarly, the energy of any system plus its surroundings is constant.

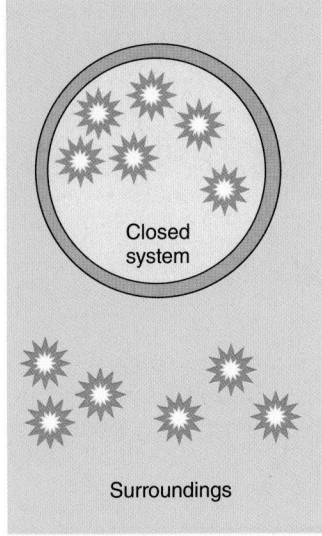
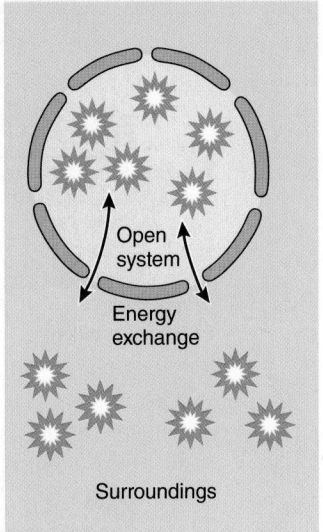

(a) A closed system does not exchange energy with its surroundings.

(b) An open system exchanges energy with its surroundings.

Figure 7-2 Closed and open systems
© Cengage Learning

A system may absorb energy from its surroundings or may give up some energy to its surroundings, but the total energy content of that system plus its surroundings is always the same.

As specified by the first law of thermodynamics, organisms cannot create the energy they require to live. Instead, they must capture energy from the environment and transform it to a form that can be used for biological work.

The entropy of the universe is increasing

The **second law of thermodynamics** states that when energy is converted from one form to another, some usable energy—that is, energy available to do work—is converted into heat that disperses into the surroundings (see Figure 55-1 for an illustration of energy flow through an ecosystem). As you learned in Chapter 2, **heat** is the kinetic energy of randomly moving particles. Unlike *heat energy,* which flows from an object with a higher temperature to one with a lower temperature, this random motion cannot perform work. As a result, the amount of usable energy available to do work in the universe decreases over time.

It is important to understand that the second law of thermodynamics is consistent with the first law; that is, the total amount of energy in the universe is *not* decreasing with time. However, the total amount of energy in the universe that is available to do work is decreasing over time.

Less-usable energy is more diffuse, or disorganized. **Entropy (S)** is a measure of this disorder, or randomness; organized, usable energy has a low entropy, whereas disorganized energy, such as heat, has a high entropy.

Entropy is continuously increasing in the universe in all natural processes. Maybe at some time, billions of years from now, all energy will exist as heat uniformly distributed throughout the universe. If that happens, the universe will cease to operate because no work will be possible. Everything will be at the same temperature, so there will be no way to convert the thermal energy of the universe into usable mechanical energy.

As a consequence of the second law of thermodynamics, no process requiring an energy conversion is ever 100% efficient because much of the energy is dispersed as heat, increasing entropy. For example, an automobile engine, which converts the chemical energy of gasoline to mechanical energy, is between 20% and 30% efficient. Thus, only 20% to 30% of the original energy stored in the chemical bonds of the gasoline molecules is actually transformed into mechanical energy; the other 70% to 80% dissipates as waste heat. Energy use in your cells is about 40% efficient, with the remaining energy given to the surroundings as heat.

Organisms have a high degree of organization, and at first glance they may appear to refute the second law of thermodynamics. As organisms grow and develop, they maintain a high level of order and do not appear to become more disorganized. However, organisms are open systems; they maintain their degree of order over time only with the constant input of energy from their surroundings. That is why plants must photosynthesize and animals must eat. Although the order within organisms may tend to increase temporarily, the total entropy of the universe (organisms plus surroundings) always increases over time.

CHECKPOINT 7.2

- *What is the first law of thermodynamics? the second law?*
- **CONNECT** *Life is sometimes described as a constant struggle against the second law of thermodynamics. How do organisms succeed in this struggle without violating the second law?*

7.3 ENERGY AND METABOLISM

LEARNING OBJECTIVES

4   Discuss how changes in free energy in a reaction are related to changes in entropy and enthalpy.

5   Distinguish between exergonic and endergonic reactions, and give examples of how they may be coupled.

6   Compare the energy dynamics of a reaction at equilibrium with the dynamics of a reaction not at equilibrium.

The chemical reactions that enable an organism to carry on its activities—to grow, move, maintain and repair itself; reproduce; and respond to stimuli—together make up its metabolism. Recall from Chapter 1 that **metabolism** is the sum of all the chemical activities taking place in an organism. An organism's metabolism consists of many intersecting series of chemical reactions, or pathways. Two main types of metabolism are anabolism and catabolism. **Anabolism** includes the various pathways in which complex molecules are synthesized from simpler substances, such as in the linking of amino acids to form proteins. **Catabolism** includes the pathways in which larger molecules are broken down into smaller ones, such as in the degradation of starch to form monosaccharides.

As you will see, these changes involve not only alterations in the arrangement of atoms but also various energy transformations. Catabolism and anabolism are complementary processes; catabolic pathways involve an overall release of energy, some of which powers anabolic pathways, which have an overall energy requirement. In the following sections, we discuss how to predict whether a particular chemical reaction requires energy or releases it.

Enthalpy is the total potential energy of a system

In the course of any chemical reaction, including the metabolic reactions of a cell, chemical bonds break and new and different bonds may form. Every specific type of chemical bond has a certain amount of *bond energy,* defined as the energy required to break that bond. The total bond energy is essentially equivalent to the total potential energy of the system, a quantity known as **enthalpy (*H*).**

Free energy is available to do cell work

Entropy and enthalpy are related by a third type of energy, termed **free energy (*G*),** which is the amount of energy available to do work under the conditions of a biochemical reaction. (*G*, also known as "Gibbs free energy," is named for J.W. Gibbs, a Yale professor who was one of the founders of the science of thermodynamics.) Free energy, the only kind of energy that can do cell work, is the aspect of thermodynamics of greatest interest to a biologist. Enthalpy, free energy, and entropy are related by the equation

$$H = G + TS$$

in which *H* is enthalpy; *G* is free energy; *T* is the absolute temperature of the system, expressed in Kelvin units; and *S* is entropy. Disregarding temperature for the moment, enthalpy (the total energy of a system) is equal to free energy (the usable energy) plus entropy (the unusable energy).

A rearrangement of the equation shows that as entropy increases, the amount of free energy decreases:

$$G = H - TS$$

If we assume that entropy is zero, the free energy is simply equal to the total potential energy (enthalpy); an increase in entropy reduces the amount of free energy.

What is the significance of the temperature (*T*)? Remember that as the temperature increases, there is an increase in random molecular motion, which contributes to disorder and multiplies the effect of the entropy term.

Chemical reactions involve changes in free energy

Biologists analyze the role of energy in the many biochemical reactions of metabolism. Although the total free energy of a system (*G*) cannot be effectively measured, the equation $G = H - TS$ can be extended to predict whether a particular chemical reaction will release

energy or require an input of energy. The reason is that *changes* in free energy can be measured. Scientists use the Greek capital letter delta (Δ) to denote any change that occurs in the system between its initial state before the reaction and its final state after the reaction. To express what happens with respect to energy in a chemical reaction, the equation becomes

$$\Delta G = \Delta H - T\Delta S$$

Notice that the temperature does not change; it is held constant during the reaction. Thus, the change in free energy (Δ*G*) during the reaction is equal to the change in enthalpy (Δ*H*) minus the product of the absolute temperature (*T*) in Kelvin units multiplied by the change in entropy (Δ*S*). Scientists express Δ*G* and Δ*H* in kilojoules or kilocalories per mole; they express Δ*S* in kilojoules or kilocalories per Kelvin unit.

Free energy decreases during an exergonic reaction

An **exergonic reaction** releases energy and is said to be a spontaneous or a "downhill" reaction, from higher to lower free energy (FIG. 7-3a). Because the total free energy in its final state is less than the total free energy in its initial state, Δ*G* is a negative number for exergonic reactions.

The term *spontaneous* may give the false impression that such reactions are always instantaneous. In fact, spontaneous reactions do not necessarily occur readily; some are extremely slow. The reason is that energy, known as *activation energy,* is required to initiate every reaction, even a spontaneous one. We discuss activation energy later in the chapter.

Free energy increases during an endergonic reaction

An **endergonic reaction** is a reaction in which there is a gain of free energy (FIG. 7-3b). Because the free energy of the products is greater than the free energy of the reactants, Δ*G* has a positive value. Such a reaction cannot take place in isolation. Instead, it must occur in such a way that energy can be supplied from the

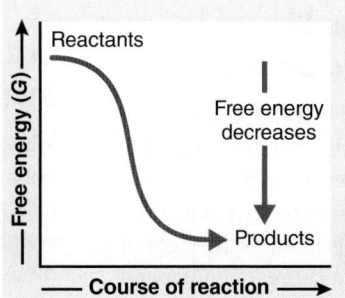
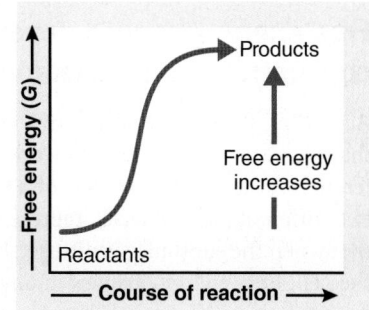

(a) In an exergonic reaction, there is a net loss of free energy. The products have less free energy than was present in the reactants, and the reaction proceeds spontaneously.

(b) In an endergonic reaction, there is a net gain of free energy. The products have more free energy than was present in the reactants.

© Cengage Learning

Figure 7-3 *Animation* **Exergonic and endergonic reactions**

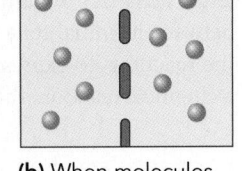

Concentration gradient

(a) A concentration gradient is a form of potential energy.

Exergonic process (occurs spontaneously)

(b) When molecules are evenly distributed, they have high entropy.

Figure 7-4 Entropy and diffusion
The tendency of entropy to increase can be used to produce work, in this case, diffusion.
© Cengage Learning

surroundings. Of course, many energy-requiring reactions take place in cells, and as you will see, metabolic mechanisms have evolved that supply the energy to "drive" these nonspontaneous cell reactions in a particular direction.

Diffusion is an exergonic process

In Chapter 5, you saw that randomly moving particles diffuse down their own concentration gradient (FIG. 7-4). Although the movements of the individual particles are random, net movement of the group of particles seems to be directional. What provides energy for this apparently directed process? A **concentration gradient,** with a region of higher concentration and another region of lower concentration, is an orderly state. A cell must expend energy to produce a concentration gradient. Because work is done to produce this order, a concentration gradient is a form of potential energy. As the particles move about randomly, the gradient becomes degraded. Thus, free energy decreases as entropy increases.

In cellular respiration and photosynthesis, the potential energy stored in a concentration gradient of hydrogen ions (H$^+$) is transformed into chemical energy in adenosine triphosphate (ATP) as the hydrogen ions pass through a membrane down their concentration gradient. This important concept, known as *chemiosmosis,* is discussed in detail in Chapters 8 and 9.

Free-energy changes depend on the concentrations of reactants and products

According to the second law of thermodynamics, any process that increases entropy can do work. As we have discussed, differences in the concentration of a substance, such as between two different parts of a cell, represent a more orderly state than that when the substance is diffused homogeneously throughout the cell. Free-energy changes in any chemical reaction depend mainly on the difference in bond energies (enthalpy, *H*) between reactants and products. Free energy also depends on *concentrations* of both reactants and products.

In most biochemical reactions, there is little intrinsic free-energy difference between reactants and products. Such reactions are reversible, indicated by drawing double arrows:

$$A \rightleftharpoons B$$

At the beginning of a reaction, only the reactant molecules (A) may be present. As the reaction proceeds, the concentration of the reactant molecules decreases and the concentration of the product molecules (B) increases. As the concentration of the product molecules increases, they may have enough free energy to initiate the reverse reaction. The reaction thus proceeds in both directions simultaneously; if undisturbed, it eventually reaches a state of **dynamic equilibrium,** in which the rate of the reverse reaction equals the rate of the forward reaction. At equilibrium there is no net change in the system; a reverse reaction balances every forward reaction.

At a given temperature and pressure, each reaction has its own characteristic equilibrium. For any given reaction, chemists can perform experiments and calculations to determine the relative concentrations of reactants and products present at equilibrium. If the reactants have much greater intrinsic free energy than the products, the reaction goes almost to completion; that is, it reaches equilibrium at a point at which most of the reactants have been converted to products. Reactions in which the reactants have much less intrinsic free energy than the products reach equilibrium at a point where very few of the reactant molecules have been converted to products.

If you increase the initial concentration of A, the reaction will "shift to the right," and more A will be converted to B. A similar effect can be obtained if B is removed from the reaction mixture. The reaction always shifts in the direction that reestablishes equilibrium so that the proportions of reactants and products characteristic of that reaction at equilibrium are restored. The opposite effect occurs if the concentration of B increases or if A is removed; here the system "shifts to the left." The actual free-energy change that occurs during a reaction is defined mathematically to include these effects, which stem from the relative initial concentrations of reactants and products. Cells use energy to manipulate the relative concentrations of reactants and products of almost every reaction. Cell reactions are virtually never at equilibrium. By displacing their reactions far from equilibrium, cells supply energy to endergonic reactions and direct their metabolism according to their needs.

Cells drive endergonic reactions by coupling them to exergonic reactions

Many metabolic reactions, such as protein synthesis, are anabolic and endergonic. Because an endergonic reaction cannot take place without an input of energy, endergonic reactions are coupled to exergonic reactions. In **coupled reactions,** the thermodynamically favorable exergonic reaction provides the energy required to drive the thermodynamically unfavorable endergonic reaction. The endergonic reaction proceeds only if it absorbs free energy released by the exergonic reaction to which it is coupled.

Consider the free-energy change, ΔG, in the reaction

$$(1)\ A \longrightarrow B \qquad \Delta G = +20.9\ kJ/mol\ (+5\ kcal/mol)$$

Because ΔG has a positive value, you know that the product of this reaction has more free energy than the reactant. It is an endergonic reaction. It is not spontaneous and does not take place without an energy source.

By contrast, consider the reaction

$$(2)\ C \longrightarrow D \qquad \Delta G = -33.5\ \text{kJ/mol}\ (-8\ \text{kcal/mol})$$

The negative value of ΔG tells you that the free energy of the reactant is greater than the free energy of the product. This exergonic reaction proceeds spontaneously.

You can add reactions 1 and 2 as follows:

$(1)\ A \longrightarrow B$	$\Delta G = +20.9\ \text{kJ/mol}\ (+5\ \text{kcal/mol})$
$(2)\ C \longrightarrow D$	$\Delta G = -33.5\ \text{kJ/mol}\ (-8\ \text{kcal/mol})$
Overall	$\Delta G = -12.6\ \text{kJ/mol}\ (-3\ \text{kcal/mol})$

Because thermodynamics considers the overall changes in these two reactions, which show a net negative value of ΔG, the two reactions taken together are exergonic.

That scientists can write reactions this way is a useful bookkeeping device, but it does not mean that an exergonic reaction mysteriously transfers energy to an endergonic "bystander" reaction. However, these reactions are coupled if their pathways are altered so a common intermediate links them. Reactions 1 and 2 might be coupled by an intermediate (I) in the following way:

$(3)\ A + C \longrightarrow I$	$\Delta G = -8.4\ \text{kJ/mol}\ (-2\ \text{kcal/mol})$
$(4)\ I \longrightarrow B + D$	$\Delta G = -4.2\ \text{kJ/mol}\ (-1\ \text{kcal/mol})$
Overall	$\Delta G = -12.6\ \text{kJ/mol}\ (-3\ \text{kcal/mol})$

Note that reactions 3 and 4 are sequential. Thus, the reaction pathways have changed, but overall the reactants (A and C) and products (B and D) are the same, and the free-energy change is the same.

Generally, for each endergonic reaction occurring in a living cell there is a coupled exergonic reaction to drive it. Often the exergonic reaction involves the breakdown of ATP. Now let's examine specific examples of the role of ATP in energy coupling.

CHECKPOINT 7.3

- **CONNECT** *Consider the free-energy change in a reaction in which enthalpy decreases and entropy increases. Is ΔG zero, or does it have a positive value or a negative value? Is the reaction endergonic or exergonic?*

- **PREDICT** *Reaction 1 is at equilibrium; reaction 2 is not. Can either reaction do work? If so, which one?*

7.4 ATP, THE ENERGY CURRENCY OF THE CELL

LEARNING OBJECTIVE

7   Explain how the chemical structure of ATP allows it to transfer a phosphate group and discuss the central role of ATP in the overall energy metabolism of the cell.

In all living cells, energy is temporarily packaged within a remarkable chemical compound called **adenosine triphosphate (ATP),** which holds readily available energy for very short periods. We may think of ATP as the energy currency of the cell. When you work to earn money, you might say that your energy is symbolically stored in the money you earn. The energy the cell requires for immediate use is temporarily stored in ATP, which is like cash. When you earn extra money, you may deposit some in the bank; similarly, a cell may deposit energy in the chemical bonds of lipids, starch, or glycogen. Moreover, just as you dare not make less money than you spend, the cell must avoid energy bankruptcy, which would mean its death. Finally, just as you probably do not keep money you earn very long, the cell continuously spends its ATP, which must be replaced immediately.

ATP is a nucleotide consisting of three main parts: adenine, a nitrogen-containing organic base; ribose, a five-carbon sugar; and three phosphate groups, identifiable as phosphorus atoms surrounded by oxygen atoms (**FIG. 7-5**). Notice that the phosphate groups are bonded to the end of the molecule in a series,

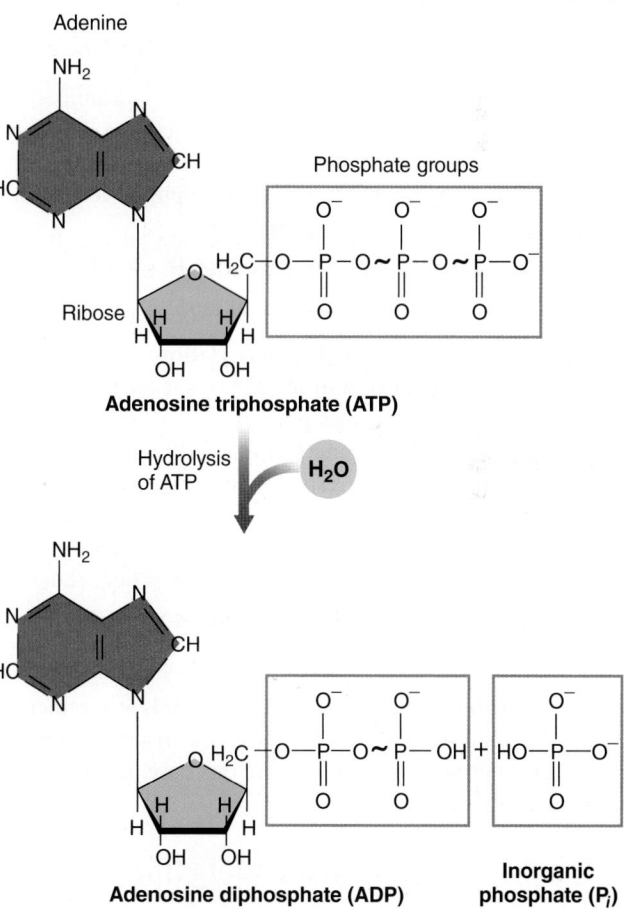

Figure 7-5 ATP and ADP

ATP, the energy currency of all living things, consists of adenine, ribose, and three phosphate groups. The hydrolysis of ATP, an exergonic reaction, yields ADP and inorganic phosphate. (The black wavy lines indicate unstable bonds. These bonds allow the phosphates to be transferred to other molecules, making them more reactive.)
© Cengage Learning

rather like three cars behind a locomotive, and, like the cars of a train, they can be attached and detached.

ATP donates energy through the transfer of a phosphate group

When the terminal phosphate is removed from ATP, the remaining molecule is adenosine diphosphate (ADP) (see Fig. 7-5). If the phosphate group is not transferred to another molecule, it is released as inorganic phosphate (P_i). This exergonic reaction has a relatively large negative value of ΔG. (Calculations of the free energy of ATP hydrolysis vary somewhat, but range between about -28 and -37 kJ/mol, or -6.8 to -8.7 kcal/mol.)

$$(5)\ \text{ATP} + H_2O \longrightarrow \text{ADP} + P_i$$
$$\Delta G = -32\ \text{kJ/mol}\ (-7.6\ \text{kcal/mol})$$

Reaction 5 can be coupled to endergonic reactions in cells. Consider the following endergonic reaction, in which two monosaccharides, glucose and fructose, form the disaccharide sucrose:

$$(6)\ \text{glucose} + \text{fructose} \longrightarrow \text{sucrose} + H_2O$$
$$\Delta G = +27\ \text{kJ/mol}\ (+6.5\ \text{kcal/mol})$$

With a free-energy change of -32 kJ/mol (-7.6 kcal/mol), the hydrolysis of ATP in reaction 5 can drive reaction 6, but only if the reactions are coupled through a common intermediate.

The following series of reactions is a simplified version of an alternative pathway that some bacteria use:

$$(7)\ \text{glucose} + \text{ATP} \longrightarrow \text{glucose-P} + \text{ADP}$$
$$(8)\ \text{glucose-P} + \text{fructose} \longrightarrow \text{sucrose} + P_i$$

Recall from Chapter 6 that **phosphorylation** is a reaction in which a phosphate group is transferred to some other compound. In reaction 7 glucose becomes phosphorylated to form glucose phosphate (glucose-P), the intermediate that links the two reactions. Glucose-P, which corresponds to I in reactions 3 and 4, reacts exergonically with fructose to form sucrose. For energy coupling to work in this way, reactions 7 and 8 must occur in sequence. It is convenient to summarize the reactions thus:

$$(9)\ \text{glucose} + \text{fructose} + \text{ATP} \longrightarrow \text{sucrose} + \text{ADP} + P_i$$
$$\Delta G = -5\ \text{kJ/mol}\ (-1.2\ \text{kcal/mol})$$

When you encounter an equation written in this way, remember that it is actually a summary of a series of reactions and that transitory intermediate products (in this case, glucose-P) are sometimes not shown.

ATP links exergonic and endergonic reactions

We have just discussed how the transfer of a phosphate group from ATP to some other compound is coupled to endergonic reactions in the cell. Conversely, adding a phosphate group to adenosine monophosphate, or AMP (forming ADP), or to ADP (forming ATP) requires coupling to exergonic reactions in the cell.

$$\text{AMP} + P_i + \text{energy} \longrightarrow \text{ADP}$$
$$\text{ADP} + P_i + \text{energy} \longrightarrow \text{ATP}$$

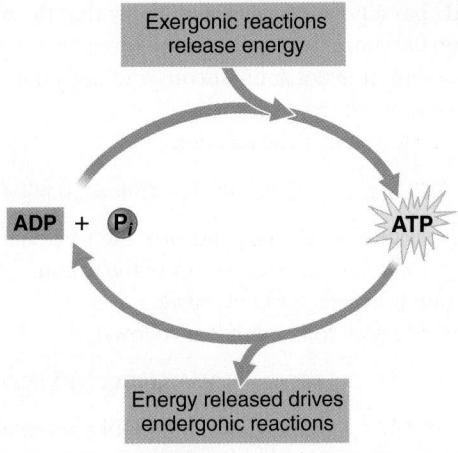

Figure 7-6 ATP links exergonic and endergonic reactions
Exergonic reactions in catabolic pathways (*top*) supply energy to drive the endergonic formation of ATP from ADP. Conversely, the exergonic hydrolysis of ATP supplies energy to endergonic reactions in anabolic pathways (*bottom*).
© Cengage Learning

Thus, ATP occupies an intermediate position in the metabolism of the cell and is an important link between exergonic reactions, which are generally components of *catabolic pathways,* and endergonic reactions, which are generally part of *anabolic pathways* (**FIG. 7-6**).

The cell maintains a very high ratio of ATP to ADP

The cell maintains a ratio of ATP to ADP far from the equilibrium point. ATP is constantly formed from ADP and inorganic phosphate as nutrients break down in cellular respiration or as photosynthesis traps the radiant energy of sunlight. At any time, a typical cell contains more than 10 ATP molecules for each ADP molecule. Because the cell maintains the ATP concentration at such a high level (relative to the concentration of ADP), its hydrolysis reaction is even more strongly exergonic and more able to drive the endergonic reactions to which it is coupled.

Although the cell maintains a high ratio of ATP to ADP, the cell cannot store large quantities of ATP. The concentration of ATP is always very low, less than 1 mmol/L. In fact, studies suggest that a bacterial cell has no more than a 1-second supply of ATP. Thus, it uses ATP molecules almost as quickly as they are produced. A healthy adult human at rest uses about 45 kg (100 lb) of ATP each day, but the amount present in the body at any given moment is less than 1 g (0.035 oz). Each second in every cell, an estimated 10 million molecules of ATP are made from ADP and phosphate, and an equal number of ATPs transfer their phosphate groups, along with their energy, to whatever chemical reactions need them.

CHECKPOINT 7.4

- *The conversion of substance X to substance Y is an endergonic reaction, but it can be driven if coupled to the hydrolysis of ATP. Write simple equations illustrating how that can occur.*

- **PREDICT** *What would likely happen in a cell if the ratio of ATP to ADP were to become 1:1?*

7.5 ENERGY TRANSFER IN REDOX REACTIONS

8   Relate the transfer of electrons (or hydrogen atoms) to the transfer of energy.

You have seen that cells transfer energy through the transfer of a phosphate group from ATP. Energy is also transferred through the transfer of electrons. As discussed in Chapter 2, oxidation is the chemical process in which a substance loses electrons, whereas reduction is the complementary process in which a substance gains electrons. Because electrons released during an oxidation reaction cannot exist in the free state in living cells, every oxidation reaction must be accompanied by a reduction reaction in which the electrons are accepted by another atom, ion, or molecule. Oxidation and reduction reactions are often called redox reactions because they occur simultaneously. The substance that becomes oxidized gives up energy as it releases electrons, and the substance that becomes reduced receives energy as it gains electrons.

Redox reactions often occur in a series as electrons are transferred from one molecule to another. These electron transfers, which are equivalent to energy transfers, are an essential part of cellular respiration, photosynthesis, and many other chemical processes. Redox reactions, for example, release the energy stored in food molecules so that ATP can be synthesized using that energy.

Most electron carriers transfer hydrogen atoms

Generally, it is not easy to remove one or more electrons from a covalent compound; it is much easier to remove a whole atom. For this reason, redox reactions in cells usually involve the transfer of a hydrogen atom rather than just an electron. A hydrogen atom contains an electron, plus a proton that does not participate in the oxidation–reduction reaction.

When an electron, either singly or as part of a hydrogen atom, is removed from an organic compound, it takes with it some of the energy stored in the chemical bond of which it was a part. That electron, along with its energy, is transferred to an acceptor molecule. An electron progressively loses free energy as it is transferred from one acceptor to another.

One of the most common acceptor molecules in cellular processes is **nicotinamide adenine dinucleotide (NAD$^+$)**. When NAD$^+$ becomes reduced, it temporarily stores large amounts of free energy. Here is a generalized equation showing the transfer of hydrogen from a compound, which we call X, to NAD$^+$:

$$XH_2 + NAD^+ \longrightarrow X + NADH + H^+$$
$$\text{Oxidized} \qquad\qquad \text{Reduced}$$

Note that the NAD$^+$ becomes reduced when it combines with hydrogen. NAD$^+$ is an ion with a net charge of +1. When 2 electrons and 1 proton are added, the charge is neutralized and the reduced form of the compound, **NADH,** is produced (**FIG. 7-7**). (Although the correct way to write the reduced form of NAD$^+$ is NADH + H$^+$, for simplicity we present the reduced form as NADH in this book.) Some energy stored in the bonds holding the hydrogen atoms to molecule X has been transferred by this

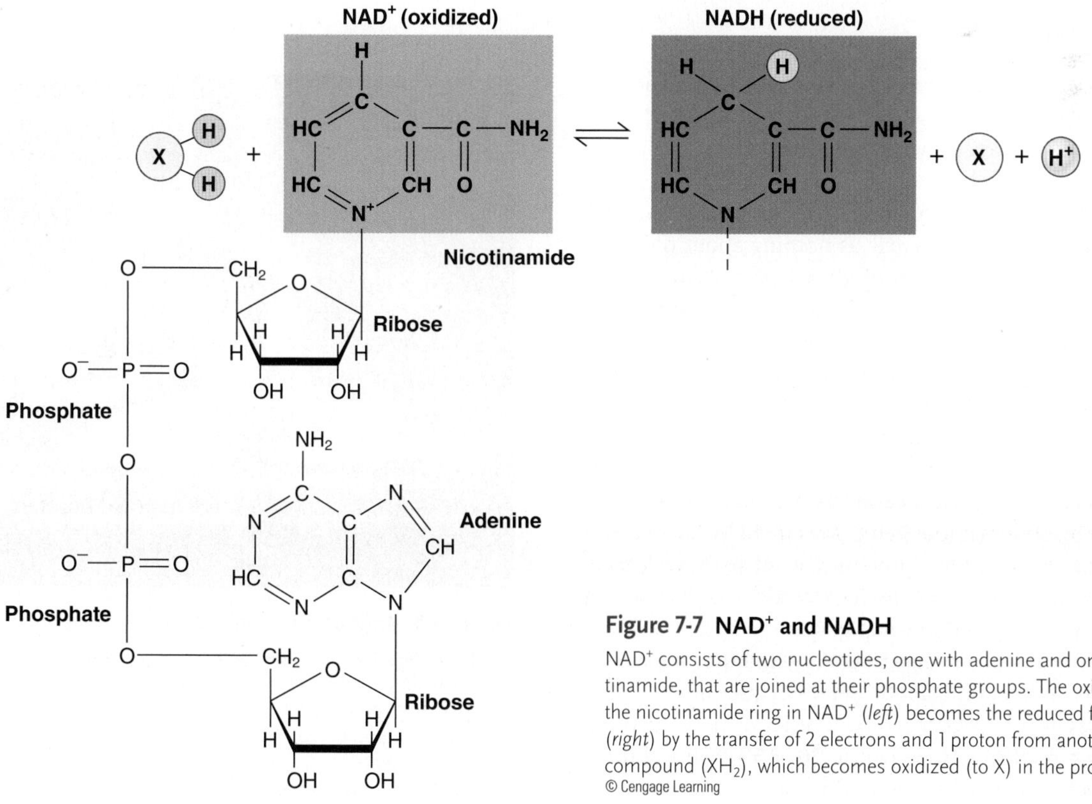

Figure 7-7 NAD$^+$ and NADH

NAD$^+$ consists of two nucleotides, one with adenine and one with nicotinamide, that are joined at their phosphate groups. The oxidized form of the nicotinamide ring in NAD$^+$ (*left*) becomes the reduced form in NADH (*right*) by the transfer of 2 electrons and 1 proton from another organic compound (XH$_2$), which becomes oxidized (to X) in the process.
© Cengage Learning

redox reaction and is temporarily held by NADH. When NADH transfers the electrons to some other molecule, some of their energy is transferred. This energy is usually then transferred through a series of reactions that ultimately result in the formation of ATP (discussed in Chapter 8).

Nicotinamide adenine dinucleotide phosphate ($NADP^+$) is a hydrogen acceptor that is chemically similar to NAD^+ but has an extra phosphate group. Unlike NADH, the reduced form of $NADP^+$, abbreviated **NADPH,** is not involved in ATP synthesis. Instead, the electrons of NADPH are used more directly to provide energy for certain reactions, including certain essential reactions of photosynthesis (discussed in Chapter 9).

Other important hydrogen acceptors or electron acceptors are FAD and the cytochromes. Flavin adenine dinucleotide (FAD) is a nucleotide that accepts hydrogen atoms and their electrons; its reduced form is $FADH_2$. The cytochromes are proteins that contain iron; the iron component accepts electrons from hydrogen atoms and then transfers these electrons to some other compound. Like NAD^+ and $NADP^+$, FAD and the cytochromes are electron transfer agents. Each exists in a *reduced state,* in which it has more free energy, or in an *oxidized state,* in which it has less. Each is an essential component of many redox reaction sequences in cells.

CHECKPOINT 7.5

- **PREDICT** *Which has the most energy, the oxidized form of a substance or its reduced form? What is responsible for the difference?*

7.6 ENZYMES

LEARNING OBJECTIVES

9 Explain how an enzyme lowers the required energy of activation for a reaction.

10 Describe specific ways enzymes are regulated.

The principles of thermodynamics help us predict whether a reaction can occur, but they tell us nothing about the speed of the reaction. The breakdown of glucose, for example, is an exergonic reaction, yet a glucose solution stays unchanged virtually indefinitely in a bottle if it is kept free of bacteria and molds and not subjected to high temperatures or strong acids or bases. Cells cannot wait for centuries for glucose to break down, nor can they use extreme conditions to cleave glucose molecules. Cells regulate the rates of chemical reactions with **enzymes,** which are biological **catalysts** that increase the speed of a chemical reaction without being consumed by the reaction. Although most enzymes are proteins, scientists have learned that some types of RNA molecules have catalytic activity as well (catalytic RNA is discussed in Chapter 13).

Cells require a steady release of energy, and they must regulate that release to meet metabolic energy requirements. Metabolic processes generally proceed by a series of steps such that a molecule may go through as many as 20 or 30 chemical transformations before it reaches some final state. Even then, the seemingly completed molecule may enter yet another chemical pathway and become totally transformed or consumed to release energy. The changing needs of the cell require a system of flexible metabolic control. The key directors of this control system are enzymes.

The catalytic ability of some enzymes is truly impressive. For example, hydrogen peroxide (H_2O_2) breaks down extremely slowly if the reaction is uncatalyzed, but a single molecule of the enzyme *catalase* brings about the decomposition of 40 million molecules of hydrogen peroxide per second! Catalase has the highest catalytic rate known for any enzyme. It protects cells by destroying hydrogen peroxide, a poisonous substance produced as a byproduct of some cell reactions. The bombardier beetle uses the enzyme catalase as a defense mechanism (**FIG. 7-8**).

All reactions have a required energy of activation

All reactions, whether exergonic or endergonic, have an energy barrier known as the **energy of activation** (E_A), or **activation energy,** which is the energy required to break the existing bonds and begin the reaction. In a population of molecules of any kind, some have a relatively high kinetic energy, whereas others have a lower energy content. Only molecules with a relatively high kinetic energy are likely to react to form the product.

Even a strongly exergonic reaction, one that releases a substantial quantity of energy as it proceeds, may be prevented from proceeding by the activation energy required to begin the reaction. For example, molecular hydrogen and molecular oxygen can react explosively to form water:

$$2\,H_2 + O_2 \longrightarrow 2\,H_2O$$

Dr. Thomas Eisner/Visuals Unlimited, Inc.

Figure 7-8 *Animation* **Catalase as a defense mechanism**

When threatened, a bombardier beetle (*Stenaptinus insignis*) uses the enzyme catalase to decompose hydrogen peroxide. The oxygen gas formed in the decomposition ejects water and other chemicals with explosive force. Because the reaction releases a great deal of heat, the water comes out as steam. (A wire attached by a drop of adhesive to the beetle's back immobilizes it. The researcher prodded its leg with the dissecting needle on the left to trigger the ejection.)

This reaction is spontaneous, yet hydrogen and oxygen can be safely mixed as long as all sparks are kept away because the required activation energy for this particular reaction is relatively high. A tiny spark provides the activation energy that allows a few molecules to react. Their reaction liberates so much heat that the rest react, producing an explosion. Such an explosion occurred on the space shuttle *Challenger* on January 28, 1986 (**FIG. 7-9**). The failure of a rubber O-ring to seal properly caused the liquid hydrogen in the tank attached to the shuttle to leak and start burning. When the hydrogen tank ruptured a few seconds later, the resulting force burst the nearby oxygen tank as well, mixing hydrogen and oxygen and igniting a huge explosion.

An enzyme lowers a reaction's activation energy

Like all catalysts, enzymes affect the rate of a reaction by lowering the activation energy (E_A) necessary to initiate a chemical reaction (**FIG. 7-10**). If molecules need less energy to react because the activation barrier is lowered, a larger fraction of the reactant molecules reacts at any one time. As a result, the reaction proceeds more quickly.

Although an enzyme lowers the activation energy requirement for a reaction, it has no effect on the overall free-energy change; that is, an enzyme can promote only a chemical reaction

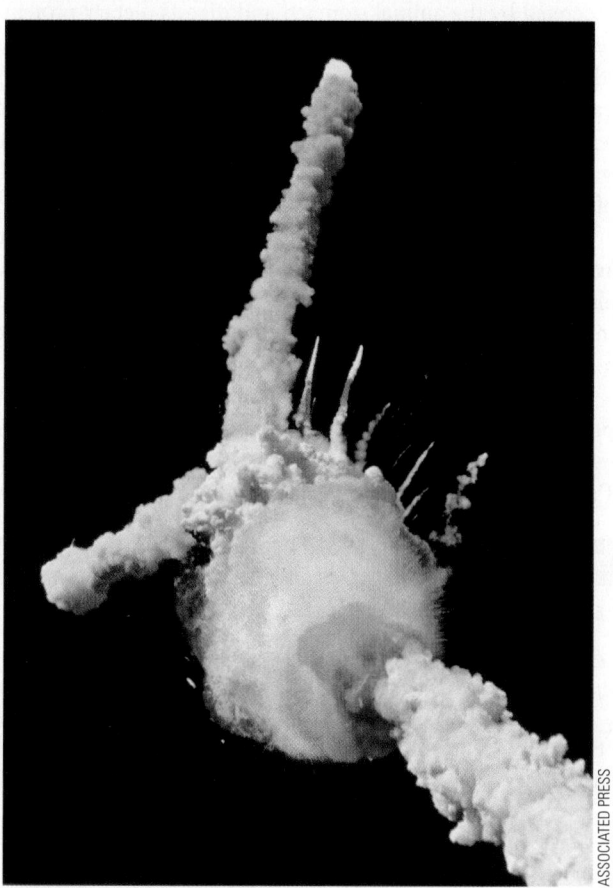

Figure 7-9 The space shuttle *Challenger* explosion

This disaster resulted from an explosive exergonic reaction between hydrogen and oxygen. All seven crew members died in the accident on January 28, 1986.

An enzyme lowers the activation energy of a reaction but does not alter the free-energy change.

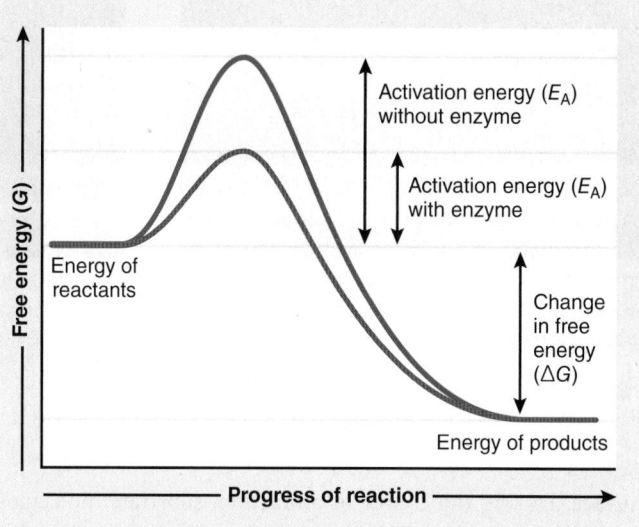

Figure 7-10 *Animation* **Activation energy and enzymes**

An enzyme speeds up a reaction by lowering its activation energy (E_A). In the presence of an enzyme, reacting molecules require less kinetic energy to complete a reaction.

PREDICT Could this reaction have proceeded if ΔG had a positive value?

© Cengage Learning

that could proceed without it. If the reaction goes to equilibrium, no catalyst can cause it to proceed in a thermodynamically unfavorable direction or can influence the final concentrations of reactants and products. Enzymes simply speed up reaction rates.

An enzyme works by forming an enzyme–substrate complex

An uncatalyzed reaction depends on random collisions among reactants. Because of its ordered structure, an enzyme reduces this reliance on random events. It controls the reaction by forming an unstable intermediate complex with the **substrate,** the substance on which it acts. When the **enzyme–substrate complex,** or *ES complex,* breaks up, the product is released; the original enzyme molecule is regenerated and is free to form a new ES complex:

enzyme + substrate(s) ⟶ ES complex

ES complex ⟶ enzyme + product(s)

The enzyme itself is not permanently altered or consumed by the reaction and can be reused.

As shown in **FIGURE 7-11a,** every enzyme contains one or more **active sites,** regions to which the substrate binds, to form the ES complex. The active sites of some enzymes are grooves or cavities in the enzyme molecule, formed by amino acid side chains. The active sites of most enzymes are located close to the

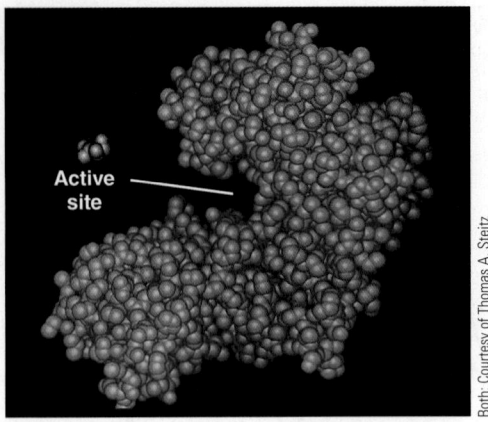

Active site

(a) Prior to forming an ES complex, the enzyme's active site is the furrow where the substrate will bind.

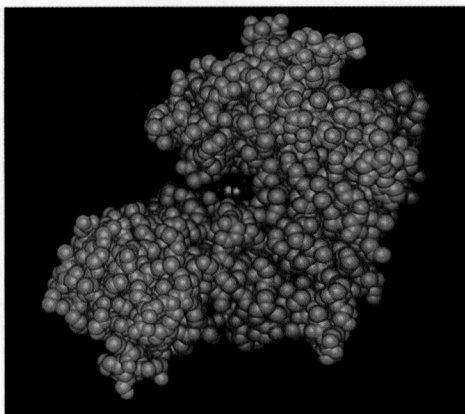

(b) The binding of the substrate to the active site induces a change in the conformation of the active site.

Both: Courtesy of Thomas A. Steitz

Figure 7-11 *Animation* **An enzyme–substrate complex**

This computer graphic model shows the enzyme hexokinase (*blue*) and its substrate, glucose (*red*).

surface. During the course of a reaction, substrate molecules occupying these sites are brought close together and react with one another.

The shape of the enzyme does not seem exactly complementary to that of the substrate. The binding of the substrate to the enzyme molecule causes a change, known as **induced fit,** in the shape of the enzyme (FIG. 7-11b). Usually, the shape of the substrate also changes slightly, in a way that may distort its chemical bonds. The proximity and orientation of the reactants, together with strains in their chemical bonds, facilitate the breakage of old bonds and the formation of new ones. Thus, the substrate is changed into a product, which diffuses away from the enzyme. The enzyme is then free to catalyze the reaction of more substrate molecules to form more product molecules.

Enzymes are specific

Enzymes catalyze virtually every chemical reaction that takes place in an organism. Because the shape of the active site is closely related to the shape of the substrate, most enzymes are highly specific. Most catalyze only a few closely related chemical reactions or, in many cases, only one particular reaction. The enzyme urease, which decomposes urea to ammonia and carbon dioxide, attacks no other substrate. The enzyme sucrase splits only sucrose; it does not act on other disaccharides, such as maltose or lactose. A few enzymes are specific only to the extent that they require the substrate to have a certain kind of chemical bond. For example, lipase, secreted by the pancreas, splits the ester linkages connecting the glycerol and fatty acids of a wide variety of fats.

Scientists usually name enzymes by adding the suffix *-ase* to the name of the substrate. The enzyme sucrase, for example, splits sucrose into glucose and fructose. A few enzymes retain traditional names that do not end in *-ase;* some of these end in *-zyme.* For example, lysozyme (from the Greek *lysis,* "a loosening") is an enzyme found in tears and saliva; it breaks down bacterial cell walls. Other examples of enzymes with traditional names are pepsin and trypsin, which break peptide bonds in proteins.

Scientists classify enzymes that catalyze similar reactions into groups, although each particular enzyme in the group may catalyze only one specific reaction. TABLE 7-1 describes the six classes of enzymes that biologists recognize. Each class is divided into many subclasses. For example, sucrase, mentioned earlier, is called a glycosidase because it cleaves a glycosidic linkage. Glycosidases are a subclass of the hydrolases (see Figure 3-8b for the hydrolysis of sucrose). Phosphatases, enzymes that remove phosphate groups by hydrolysis, are also hydrolases. Kinases, enzymes that transfer phosphate groups to substrates, are transferases.

Many enzymes require cofactors

Some enzymes consist only of a protein. The enzyme pepsin, which is secreted by the animal stomach and digests dietary protein by breaking certain peptide bonds, is exclusively a protein molecule. Other enzymes have two components: a protein called the *apoenzyme* and an additional chemical component called a **cofactor.** Neither the apoenzyme nor the cofactor alone has catalytic activity; only when the two are combined does the enzyme function. A cofactor may be inorganic, or it may be an organic molecule.

Some enzymes require a specific metal ion as a cofactor. Two very common inorganic cofactors are magnesium ions and calcium ions. Most of the trace elements, such as iron, copper, zinc, and manganese—all of which organisms require in very small amounts—function as cofactors.

An organic, nonpolypeptide compound that binds to the apoenzyme and serves as a cofactor is called a **coenzyme.** Most coenzymes are carrier molecules that transfer electrons or part of a substrate from one molecule to another. We have already

TABLE 7-1	Important Classes of Enzymes
ENZYME CLASS	**FUNCTION**
Oxidoreductases	Catalyze oxidation–reduction reactions
Transferases	Catalyze the transfer of a functional group from a donor molecule to an acceptor molecule
Hydrolases	Catalyze hydrolysis reactions
Isomerases	Catalyze conversion of a molecule from one isomeric form to another
Ligases	Catalyze certain reactions in which two molecules become joined in a process coupled to the hydrolysis of ATP
Lyases	Catalyze certain reactions in which double bonds form or break

© Cengage Learning

introduced some examples of coenzymes in this chapter. NADH, NADPH, and $FADH_2$ are coenzymes; they transfer electrons.

ATP functions as a coenzyme; it is responsible for transferring phosphate groups. Yet another coenzyme, coenzyme A, is involved in the transfer of groups derived from organic acids. Most vitamins, which are organic compounds that an organism requires in small amounts but cannot synthesize itself, are coenzymes or components of coenzymes (see descriptions of vitamins in Table 47-3).

Enzymes are most effective at optimal conditions

Enzymes generally work best under certain narrowly defined conditions, such as appropriate temperature, pH (FIG. 7-12), and ion concentration. Any departure from optimal conditions adversely affects enzyme activity.

Each enzyme has an optimal temperature Most enzymes have an optimal temperature, at which the rate of reaction is fastest. For human enzymes, the temperature optima are near the human body temperature (35°C to 40°C). Enzymatic reactions occur slowly or not at all at low temperatures. As the temperature increases, molecular motion increases, resulting in more molecular collisions. The rates of most enzyme-controlled reactions therefore increase as the temperature increases, within limits (see Fig. 7-12a). High temperatures rapidly denature most enzymes. The molecular conformation (3-D shape) of the protein becomes altered as the hydrogen bonds responsible for its secondary, tertiary, and quaternary structures are broken. Because this inactivation is usually not reversible, activity is not regained when the enzyme is cooled.

Most organisms are killed by even a short exposure to high temperature; their enzymes are denatured, and they are unable to continue metabolism. There are a few stunning exceptions to this rule. Certain species of archaea (see Chapter 1 for a description of the archaea), and also some bacteria, can survive in the waters of hot springs, such as those in Yellowstone National Park, where the temperature is almost 100°C; these organisms are responsible for the brilliant colors in the terraces of the hot springs (FIG. 7-13). Still other archaea live at temperatures not much above that of boiling water, near deep-sea vents, where the extreme pressure keeps water in its liquid state (see Chapter 25 for a discussion of archaea that live in extreme habitats; see also Chapter 55).

Each enzyme has an optimal pH Most enzymes are active only over a narrow pH range and have an optimal pH, at which the rate of reaction is fastest. The optimal pH for most human enzymes is between 6 and 8. Recall from Chapter 2 that *buffers* minimize pH changes in cells so that the pH is maintained within a narrow limit. Pepsin, a protein-digesting enzyme secreted by cells lining the stomach, is an exception; it works only in a very acidic medium, optimally at pH 2 (see Fig. 7-12b). In contrast, trypsin, a protein-splitting enzyme secreted by the pancreas, functions best under the slightly basic conditions found in the small intestine.

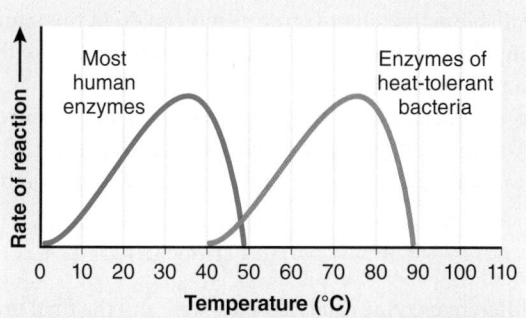

(a) Generalized curves for the effect of temperature on enzyme activity. As temperature increases, enzyme activity increases until it reaches an optimal temperature. Enzyme activity abruptly falls after it exceeds the optimal temperature because the enzyme, being a protein, denatures.

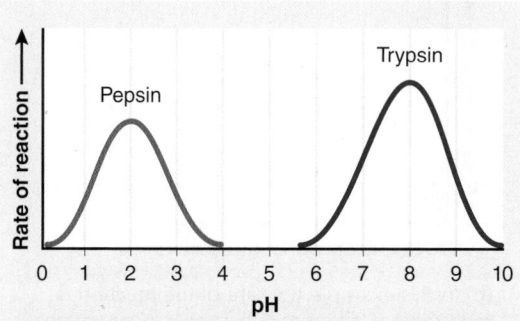

(b) Enzyme activity is very sensitive to pH. Pepsin is a protein-digesting enzyme in the very acidic stomach juice. Trypsin, secreted by the pancreas into the slightly basic small intestine, digests polypeptides.

Figure 7-12 The effects of temperature and pH on enzyme activity

Substrate and enzyme concentrations are held constant in the reactions illustrated.
© Cengage Learning

Figure 7-13 Grand Prismatic Spring in Yellowstone National Park

The world's third-largest spring, about 61 m (200 ft) in diameter, the Grand Prismatic Spring teems with heat-tolerant archaea. The rings around the perimeter, where the water is slightly cooler, get their distinctive colors from the various kinds of archaea living there.

© Rawin Cheasagul/Shutterstock.com

The activity of an enzyme may be markedly changed by any alteration in pH, which in turn alters electric charges on the enzyme. Changes in charge affect the ionic bonds that contribute to tertiary and quaternary structure, thereby changing the protein's conformation and activity. Many enzymes become inactive, and usually irreversibly denatured, when the medium is made very acidic or very basic.

Enzymes are organized into teams in metabolic pathways

Enzymes play an essential role in reaction coupling because they usually work in sequence, with the product of one enzyme-controlled reaction serving as the substrate for the next. You can picture the inside of a cell as a factory with many different assembly (and disassembly) lines operating simultaneously. An assembly line consists of a number of enzymes. Each enzyme carries out one step, such as changing molecule A into molecule B. Then molecule B is passed along to the next enzyme, which converts it into molecule C, and so on. Such a series of reactions is called a **metabolic pathway.**

$$\text{A} \xrightarrow{\text{Enzyme 1}} \text{B} \xrightarrow{\text{Enzyme 2}} \text{C}$$

Each of these reactions is reversible, even though an enzyme catalyzes it. An enzyme does not itself determine the direction of the reaction it catalyzes. However, the overall reaction sequence is portrayed as proceeding from left to right. Recall that if there is little intrinsic free-energy difference between the reactants and products for a particular reaction, the direction of the reaction is determined mainly by the relative concentrations of reactants and products.

In metabolic pathways, both intermediate and final products are often removed and converted to other chemical compounds. Such removal drives the sequence of reactions in a particular direction. Let us assume that reactant A is continually supplied and that its concentration remains constant. Enzyme 1 converts reactant A to product B. The concentration of B is always lower than the concentration of A because B is removed as it is converted to C in the reaction catalyzed by enzyme 2. If C is removed as quickly as it is formed (perhaps by leaving the cell), the entire reaction pathway is "pulled" toward C.

In some cases, the enzymes of a metabolic pathway bind to one another to form a multienzyme complex that efficiently transfers intermediates in the pathway from one active site to another. An example of one such multienzyme complex, pyruvate dehydrogenase, is discussed in Chapter 8.

The cell regulates enzymatic activity

Enzymes regulate the chemistry of the cell, but what controls the enzymes? One regulatory mechanism involves controlling the amount of enzyme produced. A specific gene directs the synthesis of each type of enzyme. The gene, in turn, may be switched on by a signal from a hormone or by some other signal molecule. When the gene is switched on, the enzyme is synthesized. The total amount of enzyme present then influences the overall cell reaction rate.

If the pH and temperature are kept constant (as they are in most cells), the rate of the reaction can be affected by the substrate concentration or by the enzyme concentration. If an excess of substrate is present, the enzyme concentration is the rate-limiting factor. The initial rate of the reaction is then directly proportional to the enzyme concentration (**FIG. 7-14a**).

If the enzyme concentration is kept constant, the rate of an enzymatic reaction is proportional to the concentration of substrate present. Substrate concentration is the rate-limiting factor at lower concentrations; the rate of the reaction is therefore directly proportional to the substrate concentration. However, at higher substrate concentrations, the enzyme molecules become saturated with substrate; that is, substrate molecules are bound to all available active sites of enzyme molecules. In this situation, increasing the substrate concentration does not increase the net reaction rate (**FIG. 7-14b**).

The product of one enzymatic reaction may control the activity of another enzyme, especially in a sequence of enzymatic reactions. For example, consider the metabolic pathway

$$\text{A} \xrightarrow{\text{Enzyme 1}} \text{B} \xrightarrow{\text{Enzyme 2}} \text{C} \xrightarrow{\text{Enzyme 3}} \text{D} \xrightarrow{\text{Enzyme 4}} \text{E}$$

A different enzyme catalyzes each step, and the final product E may inhibit the activity of enzyme 1. When the concentration of E is low, the sequence of reactions proceeds rapidly. However, an increasing concentration of E serves as a signal for enzyme 1

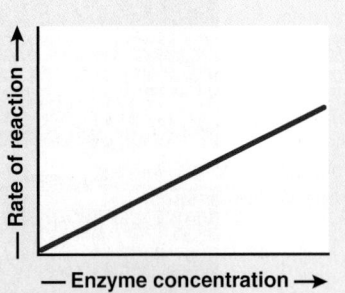

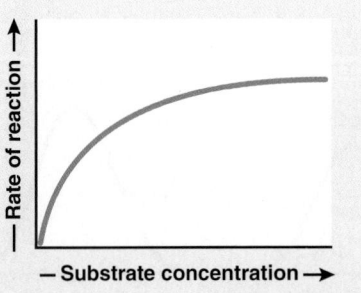

(a) In this example the rate of reaction is measured at different enzyme concentrations, with an excess of substrate present. (Temperature and pH are constant.) The rate of the reaction is directly proportional to the enzyme concentration.

(b) In this example the rate of the reaction is measured at different substrate concentrations, and enzyme concentration, temperature, and pH are constant. If the substrate concentration is relatively low, the reaction rate is directly proportional to substrate concentration. However, higher substrate concentrations do not increase the reaction rate because the enzymes become saturated with substrate.

Figure 7-14 The effects of enzyme concentration and substrate concentration on the rate of a reaction
© Cengage Learning

to slow down and eventually to stop functioning. Inhibition of enzyme 1 stops the entire reaction sequence. This type of enzyme regulation, in which the formation of a product inhibits an earlier reaction in the sequence, is called **feedback inhibition** (FIG. 7-15).

Another method of enzymatic control focuses on the activation of enzyme molecules. In their inactive form, the active sites of the enzyme are inappropriately shaped, so the substrates do not fit. Among the factors that influence the shape of the enzyme are pH, the concentration of certain ions, and the addition of phosphate groups to certain amino acids in the enzyme.

Some enzymes have a receptor site, called an **allosteric site,** on some region of the enzyme molecule other than the active site. (The word *allosteric* means "another space.") When a substance binds to an enzyme's allosteric site, the conformation of the enzyme's active site changes, thereby modifying the enzyme's activity. Substances that affect enzyme activity by binding to allosteric sites are called **allosteric regulators.** Some allosteric regulators are allosteric inhibitors that keep the enzyme in its inactive shape. Conversely, the activities of allosteric activators result in an enzyme with a functional active site.

The enzyme *cyclic AMP–dependent protein kinase* is an allosteric enzyme regulated by a protein that binds reversibly to the allosteric site and inactivates the enzyme. Protein kinase is in this inactive form most of the time (FIG. 7-16). When protein kinase activity is needed, the compound cyclic AMP (cAMP; see Figure 3-26 for the structure) contacts the enzyme–inhibitor complex and removes the inhibitory protein, thereby activating the protein kinase. Activation of protein kinases by cAMP is an important aspect of the mechanism of cell signaling, including the action of certain hormones (see Chapters 6 and 49 for discussions of cell signaling).

Enzymes are inhibited by certain chemical agents

Most enzymes are inhibited or even destroyed by certain chemical agents. Enzyme inhibition may be reversible or irreversible.

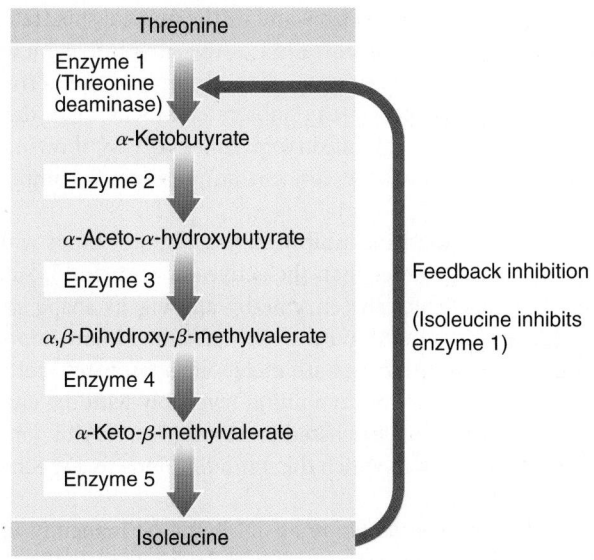

Figure 7-15 Feedback inhibition

Bacteria synthesize the amino acid isoleucine from the amino acid threonine. The isoleucine pathway involves five steps, each catalyzed by a different enzyme. When enough isoleucine accumulates in the cell, the isoleucine inhibits threonine deaminase, the enzyme that catalyzes the first step in this pathway.
© Cengage Learning

Reversible inhibition occurs when an inhibitor forms weak chemical bonds with the enzyme. Reversible inhibition can be competitive or noncompetitive.

In **competitive inhibition,** the inhibitor competes with the normal substrate for binding to the active site of the enzyme (FIG. 7-17a). Usually, a competitive inhibitor is structurally similar to the normal substrate and fits into the active site and combines with the enzyme. However, it is not similar enough to substitute fully for the normal substrate in the chemical reaction, and the enzyme cannot convert it to product molecules. A competitive inhibitor occupies the active site only temporarily and does not permanently damage the enzyme. In competitive inhibition an active site is occupied by

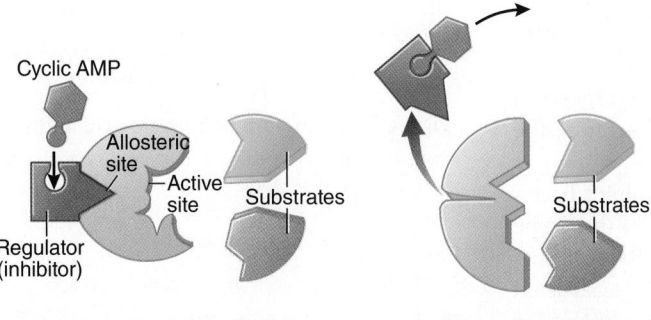

(a) Inactive form of the enzyme. The enzyme protein kinase is inhibited by a regulatory protein that binds reversibly to its allosteric site. When the enzyme is in this inactive form, the shape of the active site is modified so that the substrate cannot combine with it.

(b) Active form of the enzyme. Cyclic AMP removes the allosteric inhibitor and activates the enzyme.

(c) Enzyme–substrate complex. The substrate can then combine with the active site.

Figure 7-16 *Animation* **An allosteric enzyme**
© Cengage Learning

the inhibitor part of the time and by the normal substrate part of the time. If the concentration of the substrate is increased relative to the concentration of the inhibitor, the active site is usually occupied by the substrate. Biochemists demonstrate competitive inhibition experimentally by showing that increasing the substrate concentration reverses competitive inhibition.

In **noncompetitive inhibition** the inhibitor binds with the enzyme at a site other than the active site (FIG. 7-17b). Such an inhibitor inactivates the enzyme by altering its shape so that the active site cannot bind with the substrate. Many important noncompetitive inhibitors are metabolic substances that regulate enzyme activity by combining reversibly with the enzyme. Allosteric inhibition, discussed previously, is a type of noncompetitive inhibition in which the inhibitor binds to a special site, the allosteric site.

In *irreversible inhibition* an inhibitor permanently inactivates or destroys an enzyme when the inhibitor combines with one of the enzyme's functional groups, either at the active site or elsewhere. Many poisons are irreversible enzyme inhibitors. For example, heavy metals such as mercury and lead bind irreversibly to and denature many proteins, including enzymes. Certain nerve gases poison the enzyme acetylcholinesterase, which is important for the functioning of nerves and muscles. Cytochrome oxidase, one of the enzymes that transports electrons in cellular respiration, is especially sensitive to cyanide. Death results from cyanide poisoning because cytochrome oxidase is irreversibly inhibited and no longer transfers electrons from its substrate to oxygen.

Some drugs are enzyme inhibitors

Physicians treat many bacterial infections with drugs that directly or indirectly inhibit bacterial enzyme activity. For example, sulfa drugs have a chemical structure similar to that of the nutrient *para-aminobenzoic acid (PABA)* (FIG. 7-18). When PABA is available, microorganisms can synthesize the vitamin *folic acid*, which is necessary for growth. Humans do not synthesize folic acid from PABA. For this reason, sulfa drugs selectively affect bacteria. When a sulfa drug is present, the drug competes with PABA for the active site of the bacterial enzyme. When bacteria use the sulfa drug instead of PABA, they synthesize a compound that cannot be used to make folic acid. Therefore, the bacterial cells are unable to grow.

Penicillin and related antibiotics irreversibly inhibit a bacterial enzyme called *transpeptidase.* This enzyme establishes some of the chemical linkages in the bacterial cell wall. Bacteria susceptible to these antibiotics cannot produce properly constructed cell walls and are prevented from multiplying effectively. Human cells do not have cell walls and therefore do not use this enzyme. Thus, except for individuals allergic to it, penicillin is harmless to humans. Unfortunately, during the years since it was introduced, resistance to penicillin has evolved in many bacterial strains. The resistant bacteria fight back with an enzyme of their own, penicillinase, which breaks down the penicillin and renders it ineffective. Because bacteria evolve at such a rapid rate, drug resistance is a growing problem in medical practice.

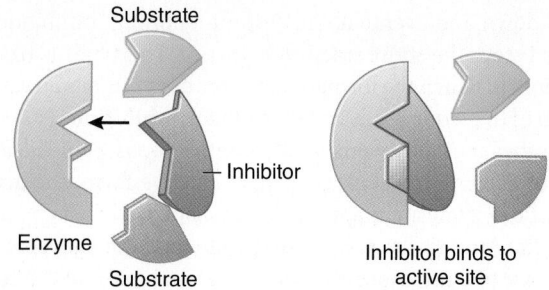

(a) **Competitive inhibition.** The inhibitor competes with the normal substrate for the active site of the enzyme. A competitive inhibitor occupies the active site only temporarily.

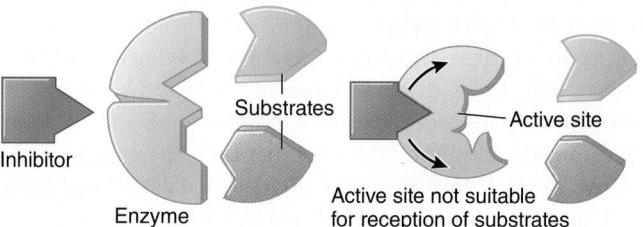

(b) **Noncompetitive inhibition.** The inhibitor binds with the enzyme at a site other than the active site, altering the shape of the enzyme and thereby inactivating it.

Figure 7-17 *Animation* **Competitive and noncompetitive inhibition**
© Cengage Learning

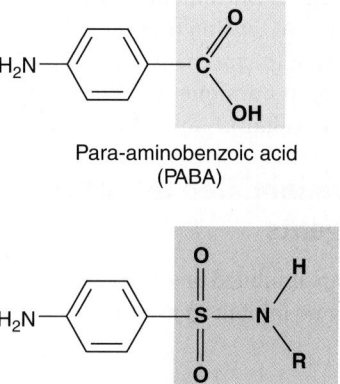

Para-aminobenzoic acid
(PABA)

Generic sulfonamide
(Sulfa drug)

Figure 7-18 **Para-aminobenzoic acid and sulfonamides**
Sulfa drugs inhibit an enzyme in bacteria necessary for the synthesis of folic acid, an important vitamin required for growth. (Note the unusual structure of the sulfonamide molecule; sulfur, which commonly forms two covalent bonds, forms six instead.)
© Cengage Learning

CHECKPOINT 7.6

- **CONNECT** *What effect does an enzyme have on the required activation energy of a reaction?*
- *How does the function of the active site of an enzyme differ from that of an allosteric site?*
- **CONNECT** *How are temperature and pH optima of an enzyme related to its structure and function?*
- *Is allosteric inhibition competitive or noncompetitive?*

7.1 Biological Work (page 148)

1 Define *energy*, emphasizing how it is related to work and to heat.

- **Energy** is the capacity to do work (expressed in **kilojoules, kJ**). Energy can be conveniently measured as *heat energy*, thermal energy that flows from an object with a higher temperature to an object with a lower temperature; the unit of heat energy is the **kilocalorie (kcal),** which is equal to 4.184 kJ. Heat energy cannot do cell work.

2 Use examples to contrast potential energy and kinetic energy.

- **Potential energy** is stored energy; **kinetic energy** is energy of motion.
- All forms of energy are interconvertible. For example, photosynthetic organisms capture radiant energy and convert some of it to *chemical energy*, a form of potential energy that powers many life processes, such as muscle contraction.

7.2 The Laws of Thermodynamics (page 149)

3 State the first and second laws of thermodynamics, and discuss the implications of these laws as they relate to organisms.

- A **closed system** does not exchange energy with its surroundings. Organisms are **open systems** that do exchange energy with their surroundings.
- The **first law of thermodynamics** states that energy cannot be created or destroyed but can be transferred and changed in form. The first law explains why organisms cannot produce energy; but as open systems, they continuously capture it from the surroundings.
- The **second law of thermodynamics** states that disorder (entropy) in the universe, a closed system, is continuously increasing. No energy transfer is 100% efficient; some energy is dissipated as **heat,** random motion that contributes to **entropy (S),** or disorder. As open systems, organisms maintain their ordered states at the expense of their surroundings.

7.3 Energy and Metabolism (page 150)

4 Discuss how changes in free energy in a reaction are related to changes in entropy and enthalpy.

- As entropy increases, the amount of **free energy** decreases, as shown in the equation $G = H - TS$, in which G is the free energy, H is the **enthalpy** (total potential energy of the system), T is the absolute temperature (expressed in Kelvin units), and S is entropy.
- The equation $\Delta G = \Delta H - T\Delta S$ indicates that the change in free energy (ΔG) during a chemical reaction is equal to the change in enthalpy (ΔH) minus the product of the absolute temperature (T) multiplied by the change in entropy (ΔS).

5 Distinguish between exergonic and endergonic reactions, and give examples of how they may be coupled.

- An **exergonic reaction** has a negative value of ΔG; that is, free energy decreases. Such a reaction is spontaneous; it releases free energy that can perform work.
- Free energy increases in an **endergonic reaction.** Such a reaction has a positive value of ΔG and is nonspontaneous. In a **coupled reaction,** the input of free energy required to drive an endergonic reaction is supplied by an exergonic reaction.

6 Compare the energy dynamics of a reaction at equilibrium with the dynamics of a reaction not at equilibrium.

- When a chemical reaction is in a state of **dynamic equilibrium,** the rate of change in one direction is exactly the same as the rate of change in the opposite direction; the system can do no work because the free-energy difference between the reactants and products is zero.
- When the concentration of reactant molecules is increased, the reaction shifts to the right and more product molecules are formed until equilibrium is re-established.

7.4 ATP, the Energy Currency of the Cell (page 153)

7 Explain how the chemical structure of ATP allows it to transfer a phosphate group and discuss the central role of ATP in the overall energy metabolism of the cell.

- **Adenosine triphosphate (ATP)** is the immediate energy currency of the cell. It donates energy by means of its terminal phosphate group, which is easily transferred to an acceptor molecule. ATP is formed by the **phosphorylation** of **adenosine diphosphate (ADP),** an endergonic process that requires an input of energy.
- ATP is the common link between exergonic and endergonic reactions and between **catabolism** (degradation of large complex molecules into smaller, simpler molecules) and **anabolism** (synthesis of complex molecules from simpler molecules).

7.5 Energy Transfer in Redox Reactions (page 155)

8 Relate the transfer of electrons (or hydrogen atoms) to the transfer of energy.

- Energy is transferred in **oxidation–reduction (redox) reactions.** A substance becomes oxidized as it gives up one or more electrons to another substance, which becomes reduced in the process. Electrons are commonly transferred as part of hydrogen atoms.
- NAD^+ and $NADP^+$ accept electrons as part of hydrogen atoms and become reduced to form **NADH** and **NADPH,** respectively. These electrons (along with some of their energy) can be transferred to other acceptors.

7.6 Enzymes (page 156)

9 Explain how an enzyme lowers the required energy of activation for a reaction.

- An **enzyme** is a biological **catalyst;** it greatly increases the speed of a chemical reaction without being consumed.
- An enzyme works by lowering the **activation energy (E_A),** the energy necessary to get a reaction going. The **active site** of an enzyme is a 3-D region where **substrates** come into close contact and thereby react more readily. When a substrate binds to an active site, an **enzyme–substrate complex** forms in which the shapes of the enzyme and substrate change slightly. This **induced fit** facilitates the breaking of bonds and formation of new ones.

10 Describe specific ways enzymes are regulated.

- Enzymes work best at specific temperature and pH conditions.
- Some enzymes can only function if an additional chemical component, called a **cofactor,** is present. A cofactor may be inorganic (e.g., a certain metal ion), or it may be a non-polypeptide organic molecule referred to as a **coenzyme.**
- A cell can regulate enzymatic activity by controlling the amount of enzyme produced and by regulating metabolic conditions that influence the shape of the enzyme.
- Some enzymes have **allosteric sites,** noncatalytic sites to which an **allosteric regulator** binds, changing the enzyme's activity. Some allosteric enzymes are subject to **feedback**

inhibition, in which the formation of an end product inhibits an earlier reaction in the metabolic pathway.

- *Reversible inhibition* occurs when an inhibitor forms weak chemical bonds with the enzyme. Reversible inhibition may be **competitive,** in which the inhibitor competes with the substrate for the active site, or **noncompetitive,** in which the inhibitor binds with the enzyme at a site other than the active site. *Irreversible inhibition* occurs when an inhibitor combines with an enzyme and permanently inactivates it.

TEST YOUR UNDERSTANDING

Know and Comprehend

1. Which of the following can do work in a cell? (a) entropy (b) heat (c) heat energy (d) all the preceding (e) none of the preceding

2. In a chemical reaction occurring in a cell, free energy is equivalent to (a) heat energy (b) heat (c) disorder (d) potential energy (e) more than one of the preceding options are true

3. Cells are able to function because they (a) are closed systems (b) have mechanisms that transform energy from the environment into useful forms (c) can use enzymes to convert endergonic reactions into spontaneous reactions (d) all the preceding

4. Diffusion is an (a) endergonic process because free energy increases (b) endergonic process because free energy decreases (c) exergonic process because entropy increases (d) exergonic process because entropy decreases (e) more than one of the preceding options are true

5. A spontaneous reaction is one in which the change in free energy (ΔG) has a _____ value. (a) positive (b) negative (c) positive or negative (d) none of the preceding (ΔG has no measurable value)

6. Healthy living cells maintain (a) ATP and ADP at equilibrium (b) equal concentrations of ATP and ADP (c) an ATP/ADP ratio of at least 10:1 (d) an ATP/ADP ratio of no more than 1:10 (e) most of the cell's stored energy in the form of ATP

7. The required energy of activation of a reaction (a) is fixed and cannot be altered (b) can be lowered by a specific enzyme (c) can be raised by a specific enzyme (d) b or c, depending on the enzyme (e) none of the preceding

8. "Induced fit" means that when a substrate binds to an enzyme's active site, (a) it fits perfectly, like a key in a lock (b) the substrate and enzyme undergo conformational changes (c) a site other than the active site undergoes a conformational change (d) the substrate and the enzyme become irreversibly bound to each other (e) c and d

9. The function of a biochemical pathway is to (a) supply energy to reactions (b) drive a sequence of reactions in a particular direction (c) maintain chemical equilibrium (d) make energy available to endergonic reactions (e) any of the preceding, depending on the pathway

Apply and Analyze

10. **PREDICT** Which of the following reactions could be coupled to an endergonic reaction with ΔG = +3.56 kJ/mol?
    - (a) A $\longrightarrow$ B, ΔG = +6.08 kJ/mol
    - (b) C $\longrightarrow$ D, ΔG = +3.56 kJ/mol
    - (c) E $\longrightarrow$ F, ΔG = 0 kJ/mol
    - (d) G $\longrightarrow$ H, ΔG = −1.22 kJ/mol
    - (e) I $\longrightarrow$ J, ΔG = −5.91 kJ/mol

Evaluate and Synthesize

11. **PREDICT** In the following reaction series, which enzyme(s) is/are most likely to have an allosteric site to which the end product E binds? (a) enzyme 1 (b) enzyme 2 (c) enzyme 3 (d) enzyme 4 (e) enzymes 3 and 4

Enzyme 1   Enzyme 2   Enzyme 3   Enzyme 4
A $\longrightarrow$ B $\longrightarrow$ C $\longrightarrow$ D $\longrightarrow$ E

12. **EVOLUTION LINK** All organisms use ATP/ADP as central links between exergonic and endergonic reactions. What does that suggest about the evolution of energy metabolism?

13. **EVOLUTION LINK** Some have argued that "evolution is impossible because the second law of thermodynamics states that entropy always increases; therefore natural processes cannot give rise to greater complexity." In what ways is this statement a misunderstanding of the laws of thermodynamics?

14. **INTERPRET DATA** Does the figure illustrate an exergonic reaction or an endergonic reaction? How do you know?

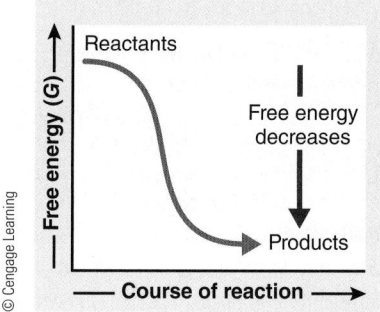

15. **INTERPRET DATA** Reactions 1 and 2 happen to have the same standard free-energy change: ΔG = −41.8 kJ/mol (−10 kcal/mol). Reaction 1 is at equilibrium, but reaction 2 is far from equilibrium. Is either reaction capable of performing work? If so, which one?

16. **INTERPRET DATA** You are performing an experiment in which you are measuring the rate at which succinate is converted to fumarate by the enzyme succinic dehydrogenase. You decide to add a little malonate to make things interesting. You observe that the reaction rate slows markedly and hypothesize that malonate is inhibiting the reaction. Design an experiment that will help you decide whether malonate is acting as a competitive inhibitor or a noncompetitive inhibitor.

aplia To access course materials, such as Aplia and other companion resources, please visit **www.cengagebrain.com.**

How Cells Make ATP: Energy-Releasing Pathways

8

Digital Vision/Getty Images

Grizzly bear (*Ursus arctos*). This grizzly, shown attempting to eat a jumping salmon, may also eat fruit, nuts, roots, insects, and small vertebrates such as mice and ground squirrels.

Cells are tiny factories that use the information inherent in their orderly systems to enable them to process materials on the molecular level, through thousands of metabolic reactions. Cells exist in a dynamic state and are continuously building up and breaking down the many different cell constituents. As you learned in Chapter 7, metabolism has two complementary components: **catabolism,** which releases energy by splitting complex molecules into smaller components; and **anabolism,** the synthesis of complex molecules from simpler building blocks. Anabolic reactions produce proteins, nucleic acids, lipids, polysaccharides, and other molecules that help maintain the cell or the organism. Most anabolic reactions are endergonic and require ATP or some other energy source to drive them.

Every organism must extract energy from food molecules that it either manufactures by photosynthesis or obtains from the environment. Grizzly bears, such as the one in the photograph, obtain organic molecules from their varied plant and animal diets. How do they obtain energy from these organic molecules? First, the food molecules are broken down by digestion into simpler components that are absorbed into the blood and transported to all the cells. The catabolic processes that convert the energy in the chemical bonds of nutrients to chemical energy stored in ATP subsequently occur inside cells, usually through a process known as **cellular respiration.**

Cellular respiration may be either aerobic or anaerobic. *Aerobic respiration* requires oxygen, whereas *anaerobic pathways,* which include anaerobic respiration and fermentation, do not require oxygen. In the process of *organismal respiration* (discussed in Chapter 46), your lungs provide a steady supply of oxygen that enables your cells to capture energy through aerobic respiration, which is by far the most common pathway and the main subject of this chapter. All three pathways—aerobic respiration, anaerobic respiration, and fermentation—are exergonic and release free energy that can be captured by the cell.

KEY CONCEPTS

8.1 Aerobic respiration is an exergonic redox process in which glucose becomes oxidized, oxygen becomes reduced, and energy is captured to make ATP.

8.2 Aerobic respiration consists of four stages: glycolysis, formation of acetyl coenzyme A, the citric acid cycle, and the electron transport chain and chemiosmosis.

8.3 Nutrients other than glucose, including many carbohydrates, lipids, and amino acids, can be oxidized by aerobic respiration.

8.4 Anaerobic respiration and fermentation are ATP-yielding redox processes in which glucose becomes oxidized, but oxygen does not become reduced. Instead, these processes involve the reduction of inorganic substances (in anaerobic respiration) or organic substances (in fermentation).

8.1 REDOX REACTIONS

LEARNING OBJECTIVE

1 Write a summary reaction for aerobic respiration that shows which reactant becomes oxidized and which becomes reduced.

Most eukaryotes and prokaryotes carry out **aerobic respiration,** a form of cellular respiration requiring molecular oxygen (O_2). During aerobic respiration, nutrients are catabolized to carbon dioxide and water. Most cells use aerobic respiration to obtain energy from glucose, which enters the cell through a specific transport protein in the plasma membrane (see discussion of facilitated diffusion in Chapter 5). The overall reaction pathway for the aerobic respiration of glucose is summarized as follows:

$$C_6H_{12}O_6 + 6\ O_2 + 6\ H_2O \longrightarrow$$
$$6\ CO_2 + 12\ H_2O + energy\ (in\ the\ chemical\ bonds\ of\ ATP)$$

Note that water is shown on both sides of the equation because it is a reactant in some reactions and a product in others. For purposes of discussion, the equation for aerobic respiration can be simplified to indicate that there is a net yield of water:

$$\overset{\overbrace{\qquad Oxidation \qquad}}{C_6H_{12}O_6} + 6\ \underset{\underbrace{\qquad Reduction \qquad}}{O_2} \rightarrow 6\ CO_2 + 6\ H_2O + energy\ (in\ the\ chemical\ bonds\ of\ ATP)$$

If we analyze this summary reaction, it appears that CO_2 is produced by the removal of hydrogen atoms from glucose. Conversely, water seems to be formed as oxygen accepts the hydrogen atoms. Because the transfer of hydrogen atoms is equivalent to the transfer of electrons, this process is a **redox reaction** in which glucose becomes *oxidized* and oxygen becomes *reduced* (see discussion of redox reactions in Chapters 2 and 7).

The products of the reaction would be the same if the glucose were simply placed in a test tube and burned in the presence of oxygen. However, if a cell were to burn glucose, its energy would be released all at once as heat, which not only would be unavailable to the cell but also would actually destroy it. For this reason, cells do not transfer hydrogen atoms directly from glucose to oxygen. Aerobic respiration includes a series of redox reactions in which electrons associated with the hydrogen atoms in glucose are transferred to oxygen in a series of steps (**FIG. 8-1**). During this process, the free energy of the electrons is coupled to ATP synthesis.

CHECKPOINT 8.1

- *Does glucose become oxidized or reduced in aerobic respiration?*
- **CONNECT** *What is the specific role of oxygen in most cells?*

8.2 THE FOUR STAGES OF AEROBIC RESPIRATION

LEARNING OBJECTIVES

2 List and give a brief overview of the four stages of aerobic respiration.
3 Indicate where each stage of aerobic respiration takes place in a eukaryotic cell.
4 Add up the energy captured (as ATP, NADH, and $FADH_2$) in each stage of aerobic respiration.
5 Define *chemiosmosis* and explain how a gradient of protons is established across the inner mitochondrial membrane.
6 Describe the process by which the proton gradient drives ATP synthesis in chemiosmosis.

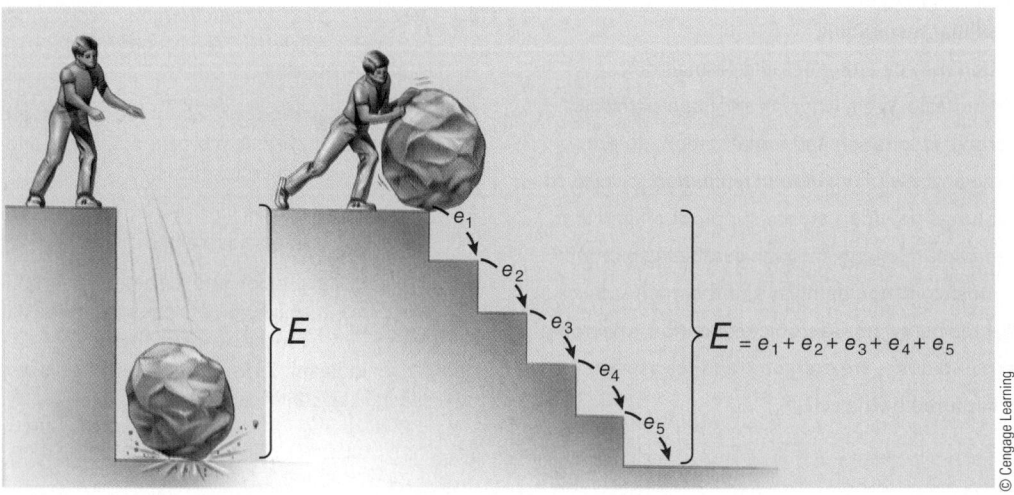

Figure 8-1 *Animation* **Changes in free energy**
The release of energy from a glucose molecule is analogous to the liberation of energy by a falling object. The total energy released (E) is the same whether it occurs all at once or in a series of steps.

The chemical reactions of the aerobic respiration of glucose are grouped into four stages (FIG. 8-2 and TABLE 8-1; see also the summary equations at the end of this chapter). In eukaryotes the first stage (glycolysis) takes place in the cytosol, and the remaining stages take place inside mitochondria. Most bacteria and archaea also carry out these processes, but because prokaryotic cells lack mitochondria, the reactions of aerobic respiration occur in the cytosol and in association with the plasma membrane.

1. *Glycolysis.* A six-carbon glucose molecule is converted to two three-carbon molecules of pyruvate.[1] Some of the energy of glucose is captured with the formation of two kinds of energy carriers, ATP and NADH.[2] See Chapter 7 to review how ATP transfers energy by transferring a phosphate group (see Figs. 7-5 and 7-6). NADH is a reduced molecule that transfers energy by transferring electrons (see Fig. 7-7).

2. *Formation of acetyl coenzyme A.* Each pyruvate enters a mitochondrion and is oxidized to a two-carbon group (an acetyl group) that combines with coenzyme A, forming acetyl coenzyme A. NADH is produced, and carbon dioxide is released as a waste product.

3. *The citric acid cycle.* The acetyl group of acetyl coenzyme A combines with a four-carbon molecule (oxaloacetate) to form a six-carbon molecule (citrate). In the course of the cycle, citrate is recycled to oxaloacetate, and carbon dioxide is released as a waste product. Energy is captured as ATP and the reduced, high-energy compounds NADH and $FADH_2$ (see Chapter 7 to review $FADH_2$).

4. *Electron transport and chemiosmosis.* The electrons removed from glucose during the preceding stages are transferred from NADH and $FADH_2$ to a chain of electron acceptor

---

[1] Pyruvate and many other compounds in cellular respiration exist as anions at the pH found in the cell. They sometimes associate with H^+ to form acids. For example, pyruvate forms pyruvic acid. In some textbooks these compounds are presented in the acid form.

[2] Although the correct way to write the reduced form of NAD^+ is NADH + H^+, for simplicity we present the reduced form as NADH throughout this book.

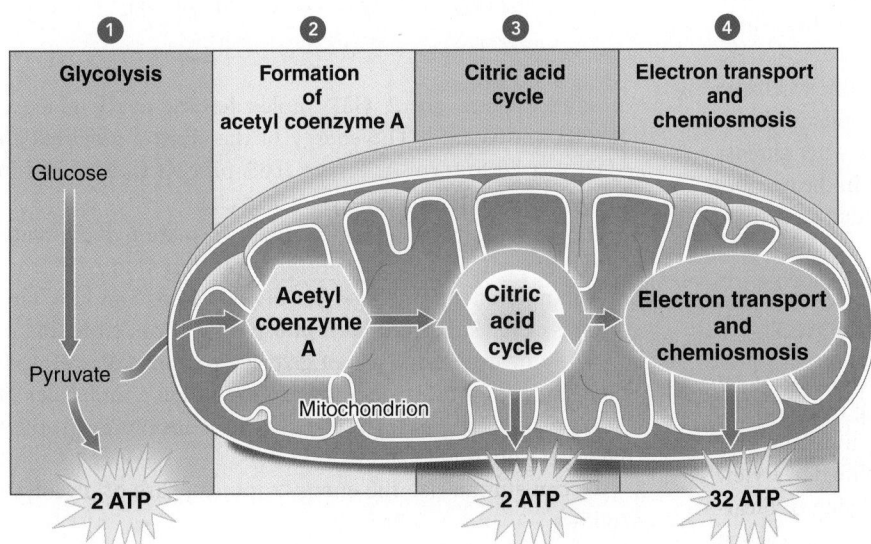

Figure 8-2 *Animation* **The four stages of aerobic respiration**

The stages of aerobic respiration occur in specific locations. Glycolysis, the first stage, occurs in the cytosol. Pyruvate, the product of glycolysis, enters a mitochondrion, where cellular respiration continues with the formation of acetyl CoA, the citric acid cycle, and electron transport and chemiosmosis. Most ATP is synthesized by chemiosmosis.
© Cengage Learning

TABLE 8-1	Summary of Aerobic Respiration		
STAGE	SUMMARY	SOME STARTING MATERIALS	SOME END PRODUCTS
1. Glycolysis (in cytosol)	Series of reactions in which glucose is degraded to pyruvate; net profit of 2 ATPs; electrons are transferred to carriers; can proceed anaerobically	Glucose, ATP, NAD^+, ADP, P_i	Pyruvate, ATP, NADH
2. Formation of acetyl CoA (in mitochondria)	Pyruvate is degraded and combined with coenzyme A to form acetyl CoA; electrons are transferred to carriers; CO_2 is released	Pyruvate, coenzyme A, NAD^+	Acetyl CoA, CO_2, NADH
3. Citric acid cycle (in mitochondria)	Series of reactions in which the acetyl portion of acetyl CoA is degraded to CO_2; electrons are transferred to carriers; ATP is synthesized	Acetyl CoA, H_2O, NAD^+, FAD, ADP, P_i	CO_2, NADH, $FADH_2$, ATP
4. Electron transport and chemiosmosis (in mitochondria)	Chain of several electron transport molecules; electrons are passed along chain; released energy is used to form a proton gradient; ATP is synthesized as protons diffuse down the gradient; oxygen is final electron acceptor	NADH, $FADH_2$, O_2, ADP, P_i	ATP, H_2O, NAD^+, FAD

© Cengage Learning

compounds. These electrons are ultimately passed to the final electron acceptor, oxygen, and water is formed. As the electrons are passed from one electron acceptor to another, some of their energy is used to transport hydrogen ions (protons) across the inner mitochondrial membrane, forming a proton gradient. In a process known as *chemiosmosis* (described later), the energy of this proton gradient is used to produce ATP.

Most reactions involved in aerobic respiration are one of three types: dehydrogenations, decarboxylations, and those we informally categorize as preparation reactions. **Dehydrogenations** are reactions in which two hydrogen atoms (actually, 2 electrons plus 1 or 2 protons) are removed from the substrate and transferred to NAD^+ or FAD. **Decarboxylations** are reactions in which part of a carboxyl group (—COOH) is removed from the substrate as a molecule of CO_2. The carbon dioxide you exhale with each breath is derived from decarboxylations that occur in your cells. The rest of the reactions are preparation reactions in which molecules undergo rearrangements and other changes so that they can undergo further dehydrogenations or decarboxylations. As you examine the individual reactions of aerobic respiration, you will encounter these three basic types.

In following the reactions of aerobic respiration, it helps to do some bookkeeping as you go along. Because glucose is the starting material, it is useful to express changes on a per glucose basis. We will pay particular attention to changes in the number of carbon atoms per molecule and to steps in which some type of energy transfer takes place.

In glycolysis, glucose yields two pyruvates

The word **glycolysis** comes from Greek words meaning "sugar splitting," which refers to the sugar glucose being metabolized. Glycolysis does not require oxygen and proceeds under aerobic or anaerobic conditions. **FIGURE 8-3** shows a simplified overview of glycolysis, in which a glucose molecule consisting of 6 carbons is converted to 2 molecules of **pyruvate,** a three-carbon molecule. Some of the energy in the glucose is captured; there is a net yield of 2 ATP molecules and 2 NADH molecules. The reactions of glycolysis take place in the cytosol, where the necessary reactants, such as ADP, NAD^+, and inorganic phosphate, float freely and are used as needed.

The glycolysis pathway consists of a series of reactions, each of which is catalyzed by a specific enzyme (**FIG. 8-4,** pages 172–173). Glycolysis is divided into two major phases: the first includes endergonic reactions that require ATP, and the second includes exergonic reactions that yield ATP and NADH.

The first phase of glycolysis requires an investment of ATP The first phase of glycolysis is sometimes called the "energy investment phase" (see Fig. 8-4, steps l to 5). Glucose is a relatively stable molecule and is not easily broken down. In two separate **phosphorylation** reactions, a phosphate group is transferred from ATP to the sugar. The resulting phosphorylated sugar (fructose-1,6-bisphosphate) is less stable and is broken enzymatically into 2 three-carbon molecules, dihydroxyacetone phosphate and glyceraldehyde-3-phosphate (G3P). The dihydroxyacetone phosphate is enzymatically converted to G3P, so the products at this point in glycolysis are 2 molecules of G3P per glucose. We can summarize this portion of glycolysis as follows:

$$\text{glucose} + 2\,\text{ATP} \longrightarrow 2\,\text{G3P} + 2\,\text{ADP}$$

Six-carbon compound   Three-carbon compound

The second phase of glycolysis yields NADH and ATP The second phase of glycolysis is sometimes called the "energy capture phase" (see Fig. 8-4, steps 6 to 10). Each G3P is converted to pyruvate. In the first step of this process, each G3P is oxidized by the removal of 2 electrons (as part of 2 hydrogen atoms). These immediately combine with the hydrogen carrier molecule, NAD^+:

$$NAD^+ + 2\,H \longrightarrow NADH + H^+$$

Oxidized   (From G3P)   Reduced

Because there are 2 G3P molecules for every glucose, 2 NADH are formed. The energy of the electrons carried by NADH is used to form ATP later. This process is discussed in conjunction with the electron transport chain.

In two of the reactions leading to the formation of pyruvate, ATP forms when a phosphate group is transferred to ADP from a phosphorylated intermediate (see Fig. 8-4, steps 7 and l0). This process is called **substrate-level phosphorylation.** Note that in the energy investment phase of glycolysis 2 molecules of ATP are consumed, but in the energy capture phase 4 molecules of ATP are produced. Thus, glycolysis yields a net energy profit of *2 ATPs* per glucose.

We can summarize the energy capture phase of glycolysis as follows:

$$2\,\text{G3P} + 2\,NAD^+ + 4\,\text{ADP} \longrightarrow$$
$$2\,\text{pyruvate} + 2\,\text{NADH} + 4\,\text{ATP}$$

Pyruvate is converted to acetyl CoA

In eukaryotes the pyruvate molecules formed in glycolysis enter the mitochondria, where they are converted to **acetyl coenzyme A (acetyl CoA).** These reactions occur in the cytosol of aerobic prokaryotes. In this series of reactions, pyruvate undergoes a process known as *oxidative decarboxylation.* First, a carboxyl group is removed as carbon dioxide, which diffuses out of the cell (**FIG. 8-5**). Then the remaining two-carbon fragment becomes oxidized, and NAD^+ accepts the electrons removed during the oxidation. Finally, the oxidized two-carbon fragment, an acetyl group, becomes attached to **coenzyme A,** yielding acetyl

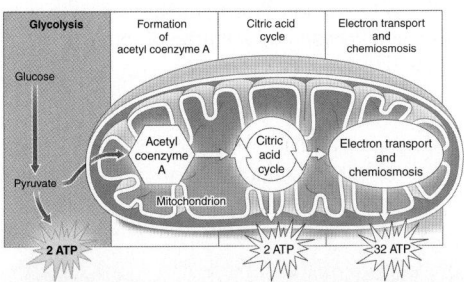

GLYCOLYSIS

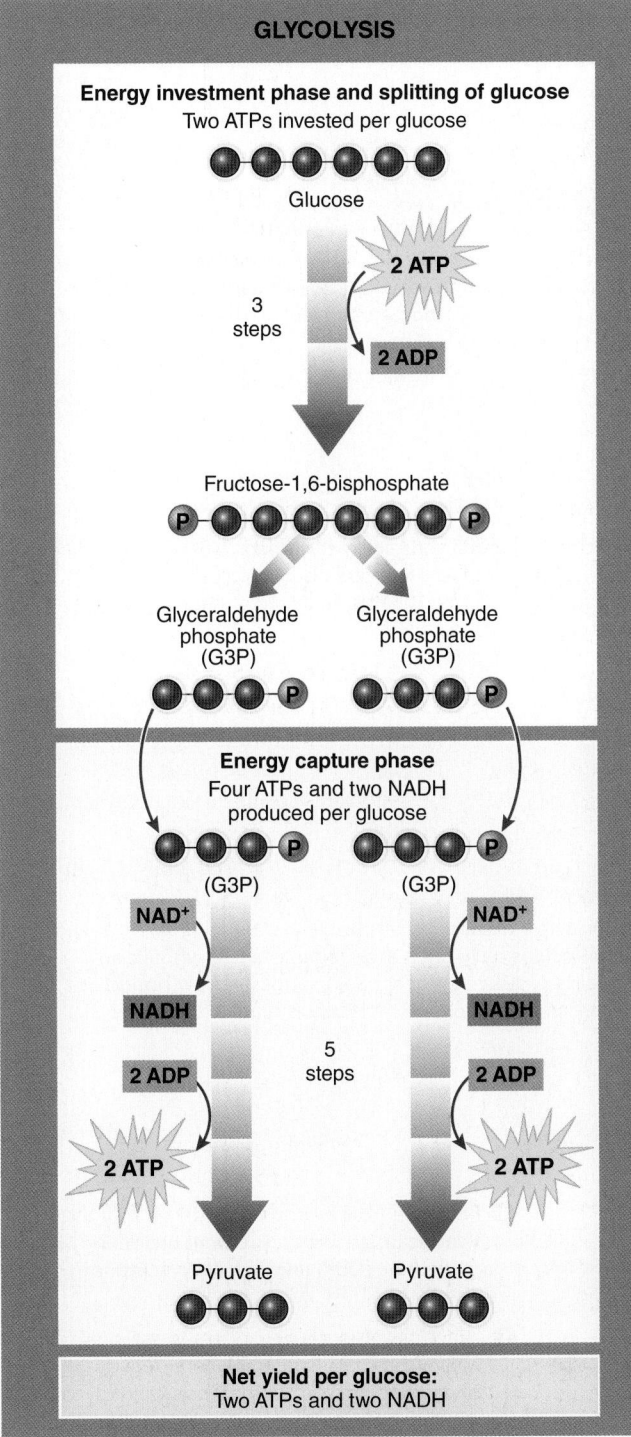

Energy investment phase and splitting of glucose
Two ATPs invested per glucose

Glucose

2 ATP

3 steps

2 ADP

Fructose-1,6-bisphosphate

Glyceraldehyde phosphate (G3P) Glyceraldehyde phosphate (G3P)

Energy capture phase
Four ATPs and two NADH produced per glucose

(G3P) (G3P)

NAD⁺ NAD⁺

NADH NADH

2 ADP 2 ADP

5 steps

2 ATP 2 ATP

Pyruvate Pyruvate

Net yield per glucose:
Two ATPs and two NADH

CoA. *Pyruvate dehydrogenase,* the enzyme that catalyzes these reactions, is an enormous multienzyme complex consisting of 72 polypeptide chains! Recall from Chapter 7 that coenzyme A transfers groups derived from organic acids. In this case, coenzyme A transfers an acetyl group, which is related to acetic acid. Coenzyme A is manufactured in the cell from one of the B vitamins, pantothenic acid.

The overall reaction for the formation of acetyl coenzyme A is

$$2 \text{ pyruvate} + 2 \text{ NAD}^+ + 2 \text{ CoA} \longrightarrow 2 \text{ acetyl CoA} + 2 \text{ NADH} + 2 \text{ CO}_2$$

Note that the original glucose molecule has now been partially oxidized, yielding 2 acetyl groups and 2 CO_2 molecules. The electrons removed have reduced NAD^+ to NADH. At this point in aerobic respiration, 4 NADH molecules have been formed as a result of the catabolism of a single glucose molecule: 2 during glycolysis and 2 during the formation of acetyl CoA from pyruvate. Keep in mind that these NADH molecules will be used later (during electron transport) to form additional ATP molecules.

The citric acid cycle oxidizes acetyl groups derived from acetyl CoA

The **citric acid cycle** is also known as the **tricarboxylic acid (TCA) cycle** and the **Krebs cycle,** after Hans Krebs, a German biochemist who assembled the accumulated contributions of many scientists and worked out the details of the cycle in the 1930s. He received a Nobel Prize in Physiology or Medicine in 1953 for this contribution. A simplified overview of the citric acid cycle, which takes place in the matrix of the mitochondria, is given in **FIGURE 8-6** on page 174. The 8 steps of the citric acid cycle are shown in **FIGURE 8-7** on page 175. A specific enzyme catalyzes each reaction.

The first reaction of the cycle occurs when acetyl CoA transfers its two-carbon acetyl group to the four-carbon acceptor compound **oxaloacetate,** forming **citrate,** a six-carbon compound.

$$\underset{\substack{\text{Four-carbon}\\\text{compound}}}{\text{oxaloacetate}} + \underset{\substack{\text{Two-carbon}\\\text{compound}}}{\text{acetyl CoA}} \longrightarrow \underset{\substack{\text{Six-carbon}\\\text{compound}}}{\text{citrate}} + \text{CoA}$$

The citrate then goes through a series of chemical transformations, losing first one and then a second carboxyl group as CO_2. One ATP is formed (per acetyl group) by substrate-level

Figure 8-3 *Animation* An overview of glycolysis

Glycolysis includes both energy investment and energy capture. The black spheres represent carbon atoms. The energy investment phase of glycolysis leads to the splitting of sugar; ATP and NADH are produced during the energy capture phase. During glycolysis, each glucose molecule is converted to 2 pyruvates, with a net yield of 2 ATP molecules and 2 NADH molecules.

PREDICT Which do you think has more energy value to the cell, 1 molecule of glucose or 2 molecules of G3P? 2 molecules of G3P or 2 molecules of pyruvate?
© Cengage Learning

Figure 8-4
Animation A
detailed look at
glycolysis

A specific enzyme
catalyzes each of the
reactions in glycolysis.
Note the net yield of
2 ATP molecules and
2 NADH molecules.
(The black wavy lines
indicate bonds that
permit the phosphates
to be readily transferred
to other molecules, in
this case, ADP.)
© Cengage Learning

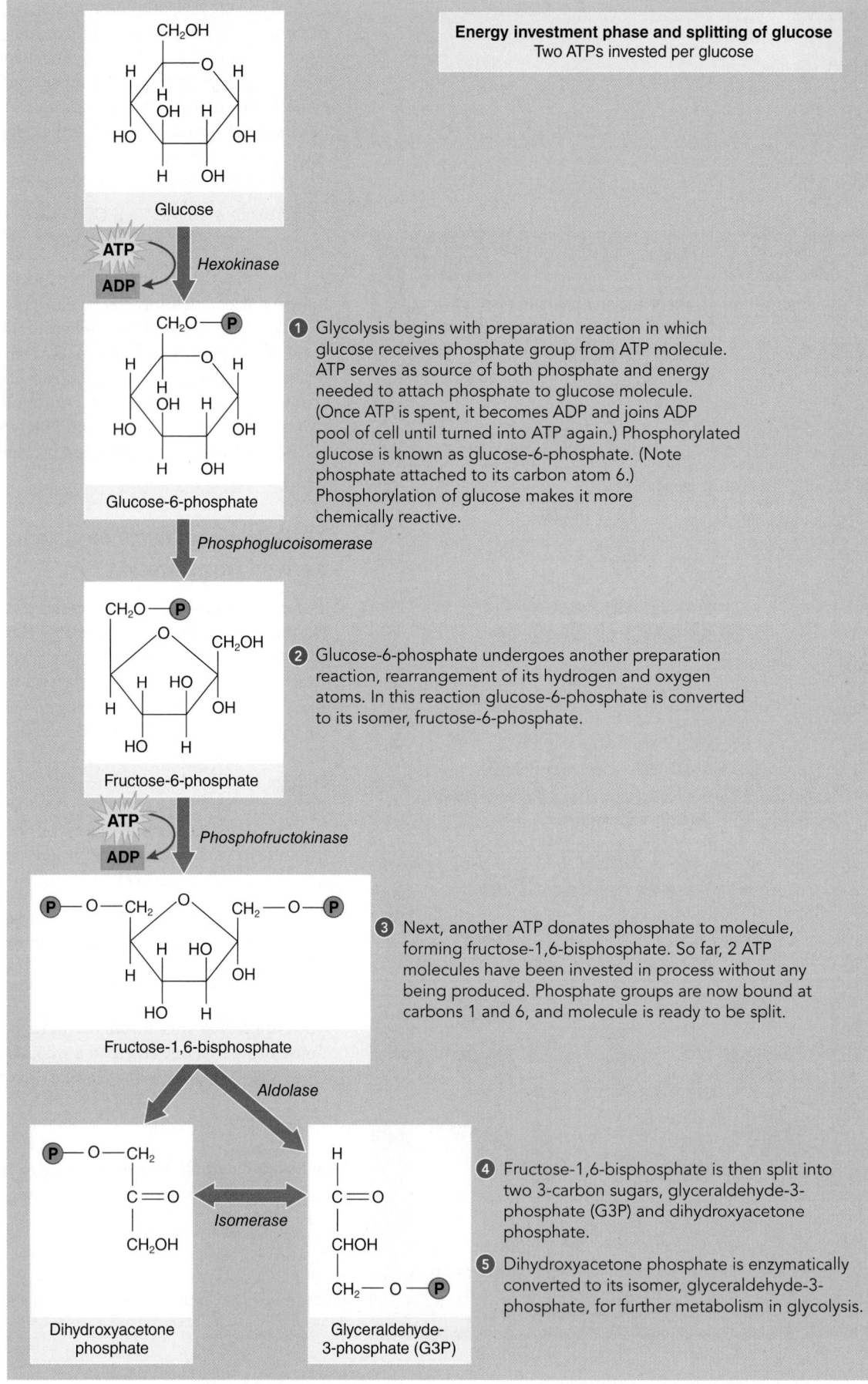

Energy investment phase and splitting of glucose
Two ATPs invested per glucose

Glucose

Hexokinase

Glucose-6-phosphate

1 Glycolysis begins with preparation reaction in which glucose receives phosphate group from ATP molecule. ATP serves as source of both phosphate and energy needed to attach phosphate to glucose molecule. (Once ATP is spent, it becomes ADP and joins ADP pool of cell until turned into ATP again.) Phosphorylated glucose is known as glucose-6-phosphate. (Note phosphate attached to its carbon atom 6.) Phosphorylation of glucose makes it more chemically reactive.

Phosphoglucoisomerase

Fructose-6-phosphate

2 Glucose-6-phosphate undergoes another preparation reaction, rearrangement of its hydrogen and oxygen atoms. In this reaction glucose-6-phosphate is converted to its isomer, fructose-6-phosphate.

Phosphofructokinase

Fructose-1,6-bisphosphate

3 Next, another ATP donates phosphate to molecule, forming fructose-1,6-bisphosphate. So far, 2 ATP molecules have been invested in process without any being produced. Phosphate groups are now bound at carbons 1 and 6, and molecule is ready to be split.

Aldolase

Isomerase

Dihydroxyacetone phosphate

Glyceraldehyde-3-phosphate (G3P)

4 Fructose-1,6-bisphosphate is then split into two 3-carbon sugars, glyceraldehyde-3-phosphate (G3P) and dihydroxyacetone phosphate.

5 Dihydroxyacetone phosphate is enzymatically converted to its isomer, glyceraldehyde-3-phosphate, for further metabolism in glycolysis.

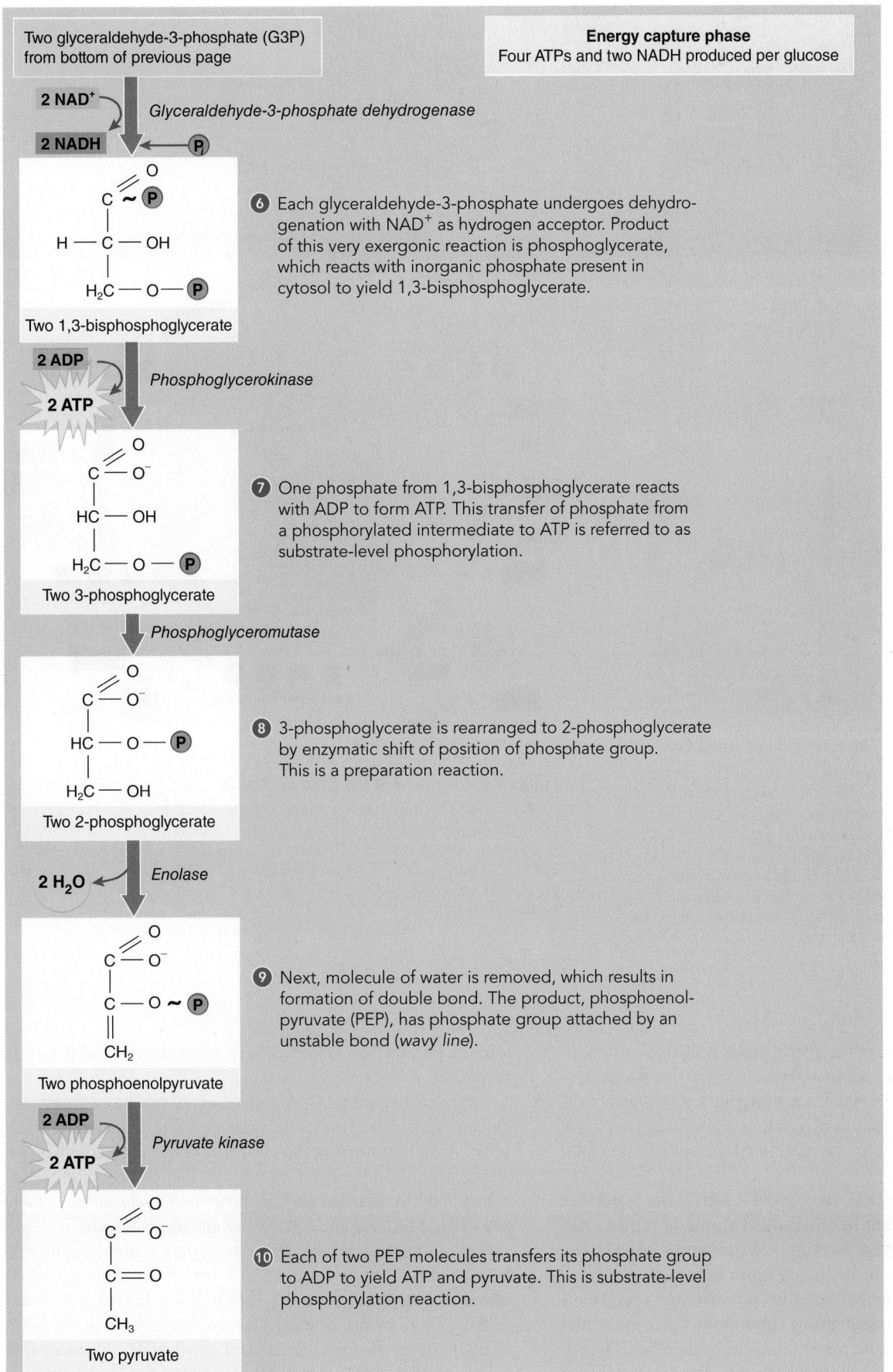

Two glyceraldehyde-3-phosphate (G3P) from bottom of previous page

2 NAD⁺

Glyceraldehyde-3-phosphate dehydrogenase

2 NADH ← Pᵢ

Two 1,3-bisphosphoglycerate

6 Each glyceraldehyde-3-phosphate undergoes dehydrogenation with NAD⁺ as hydrogen acceptor. Product of this very exergonic reaction is phosphoglycerate, which reacts with inorganic phosphate present in cytosol to yield 1,3-bisphosphoglycerate.

2 ADP

Phosphoglycerokinase

2 ATP

Two 3-phosphoglycerate

7 One phosphate from 1,3-bisphosphoglycerate reacts with ADP to form ATP. This transfer of phosphate from a phosphorylated intermediate to ATP is referred to as substrate-level phosphorylation.

Phosphoglyceromutase

Two 2-phosphoglycerate

8 3-phosphoglycerate is rearranged to 2-phosphoglycerate by enzymatic shift of position of phosphate group. This is a preparation reaction.

2 H₂O

Enolase

Two phosphoenolpyruvate

9 Next, molecule of water is removed, which results in formation of double bond. The product, phosphoenolpyruvate (PEP), has phosphate group attached by an unstable bond (*wavy line*).

2 ADP

Pyruvate kinase

2 ATP

Two pyruvate

10 Each of two PEP molecules transfers its phosphate group to ADP to yield ATP and pyruvate. This is substrate-level phosphorylation reaction.

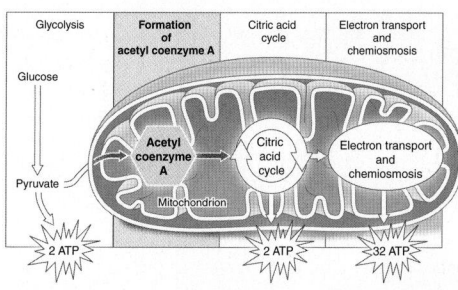

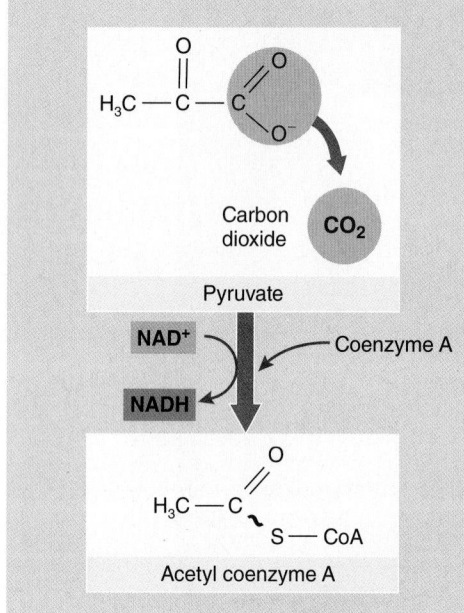

Figure 8-5 *Animation* **The formation of acetyl CoA**

This series of reactions is catalyzed by the multienzyme complex pyruvate dehydrogenase. Pyruvate, a three-carbon molecule that is the end product of glycolysis, enters the mitochondrion and undergoes oxidative decarboxylation. First, the carboxyl group is split off as carbon dioxide. Then, the remaining two-carbon fragment is oxidized, and its electrons are transferred to NAD⁺. Finally, the oxidized two-carbon group, an acetyl group, is attached to coenzyme A. CoA has a sulfur atom that forms a bond, shown as a black wavy line, with the acetyl group. When this bond is broken, the acetyl group can be readily transferred to another molecule.
© Cengage Learning

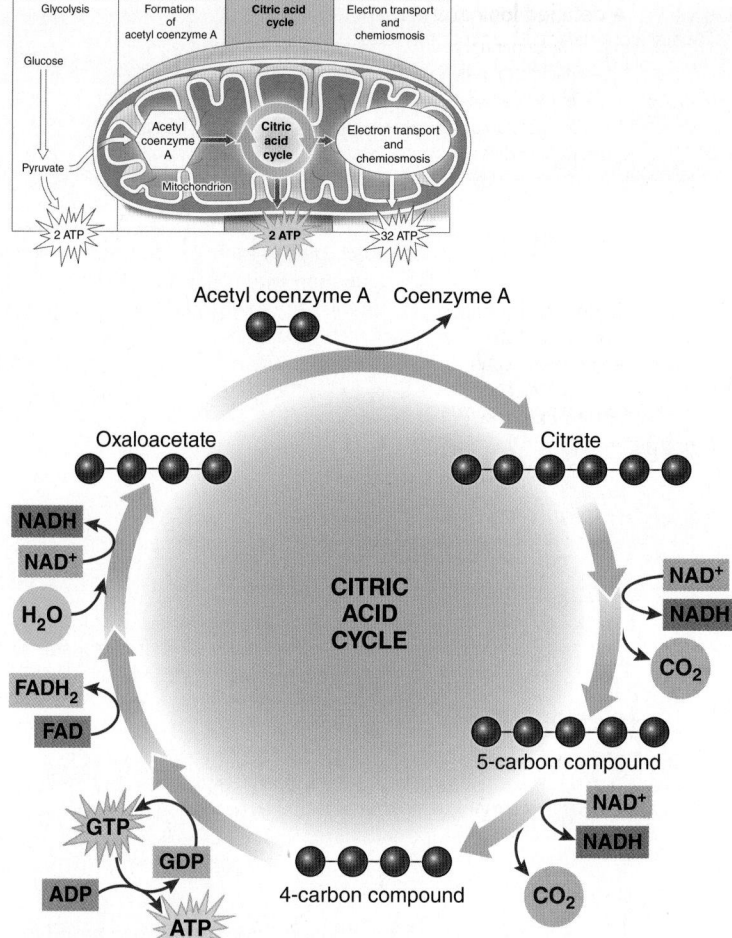

Figure 8-6 Overview of the citric acid cycle

For every glucose, 2 acetyl groups enter the citric acid cycle (*top*). Each two-carbon acetyl group combines with a four-carbon compound, oxaloacetate, to form the six-carbon compound citrate. Two CO_2 molecules are removed, and energy is captured as 1 ATP, 3 NADH, and 1 $FADH_2$ per acetyl group (or 2 ATPs, 6 NADH, and 2 $FADH_2$ per glucose molecule).
© Cengage Learning

phosphorylation. Most of the energy made available by the oxidative steps of the cycle is transferred as energy-rich electrons to NAD⁺, forming NADH. For each acetyl group that enters the citric acid cycle, 3 molecules of NADH are produced (steps 3, 4, and 8). Electrons are also transferred to the electron acceptor FAD, forming $FADH_2$.

In the course of the citric acid cycle, 2 molecules of CO_2 and the equivalent of 8 hydrogen atoms (8 protons and 8 electrons) are removed, forming 3 NADH and 1 $FADH_2$. You may wonder why more hydrogen equivalents are generated by these reactions than entered the cycle with the acetyl CoA molecule. These hydrogen atoms come from water molecules that are added during the reactions of the cycle. The CO_2 produced accounts for the 2 carbon atoms of the acetyl group that entered the citric acid cycle. At the end of each cycle, the four-carbon oxaloacetate has been regenerated, and the cycle continues.

Because two acetyl CoA molecules are produced from each glucose molecule, two cycles are required per glucose molecule. After two turns of the cycle, the original glucose has lost all its carbons and may be regarded as having been completely consumed. To summarize, the citric acid cycle yields 4 CO_2, 6 NADH, 2 $FADH_2$, and 2 ATPs per glucose molecule.

At this point in aerobic respiration, only 4 molecules of ATP have been formed per glucose by substrate-level phosphorylation: 2 during glycolysis and 2 during the citric acid cycle. Most of the energy of the original glucose molecule is in the form of high-energy electrons in NADH and $FADH_2$. Their energy will be used to synthesize additional ATP through the electron transport chain and chemiosmosis.

Figure 8-7 A detailed look at the citric acid cycle

Begin with step 1, in the upper right corner, where an acetyl group donated by acetyl coenzyme A attaches to oxaloacetate. Follow the steps in the citric acid cycle to see that the entry of a two-carbon acetyl group is balanced by the release of 2 molecules of CO_2. Electrons are transferred to NAD^+ or FAD, yielding NADH and $FADH_2$, respectively, and ATP is formed by substrate-level phosphorylation.
© Cengage Learning

1 Unstable bond attaching acetyl group to coenzyme A breaks. 2-carbon acetyl group becomes attached to 4-carbon oxaloacetate molecule, forming citrate, a 6-carbon molecule with three carboxyl groups. Coenzyme A is free to combine with another 2-carbon group and repeat process.

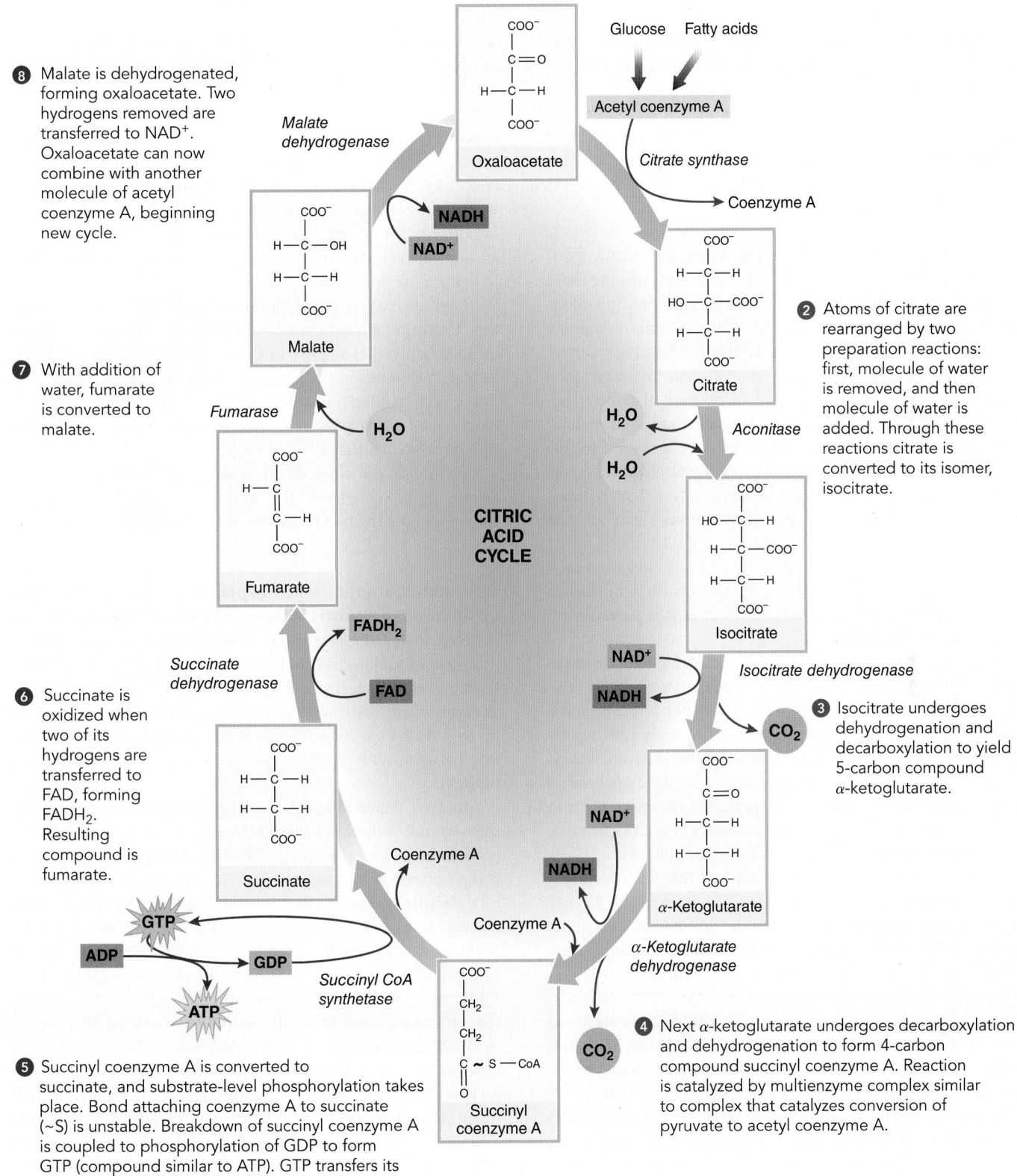

8 Malate is dehydrogenated, forming oxaloacetate. Two hydrogens removed are transferred to NAD^+. Oxaloacetate can now combine with another molecule of acetyl coenzyme A, beginning new cycle.

7 With addition of water, fumarate is converted to malate.

6 Succinate is oxidized when two of its hydrogens are transferred to FAD, forming $FADH_2$. Resulting compound is fumarate.

5 Succinyl coenzyme A is converted to succinate, and substrate-level phosphorylation takes place. Bond attaching coenzyme A to succinate (~S) is unstable. Breakdown of succinyl coenzyme A is coupled to phosphorylation of GDP to form GTP (compound similar to ATP). GTP transfers its phosphate to ADP, yielding ATP.

2 Atoms of citrate are rearranged by two preparation reactions: first, molecule of water is removed, and then molecule of water is added. Through these reactions citrate is converted to its isomer, isocitrate.

3 Isocitrate undergoes dehydrogenation and decarboxylation to yield 5-carbon compound α-ketoglutarate.

4 Next α-ketoglutarate undergoes decarboxylation and dehydrogenation to form 4-carbon compound succinyl coenzyme A. Reaction is catalyzed by multienzyme complex similar to complex that catalyzes conversion of pyruvate to acetyl coenzyme A.

The electron transport chain is coupled to ATP synthesis

Let us consider the fate of all the electrons removed from a molecule of glucose during glycolysis, acetyl CoA formation, and the citric acid cycle. Recall that these electrons were transferred as part of hydrogen atoms to the acceptors NAD^+ and FAD, forming NADH and $FADH_2$. These reduced compounds now enter the **electron transport chain,** where the high-energy electrons of their hydrogen atoms are shuttled from one acceptor to another. As the electrons are passed along in a series of exergonic redox reactions, some of their energy is used to drive the synthesis of ATP, which is an endergonic process. Because ATP synthesis (by phosphorylation of ADP) is coupled to the redox reactions in the electron transport chain, the entire process is known as **oxidative phosphorylation.**

The electron transport chain transfers electrons from NADH and $FADH_2$ to oxygen The electron transport chain is a series of electron carriers embedded in the inner mitochondrial membrane of eukaryotes and in the plasma membrane of aerobic prokaryotes. Like NADH and $FADH_2$, each carrier exists in an oxidized form or a reduced form. Electrons pass down the electron transport chain in a series of redox reactions that works much like a bucket brigade, the old-time chain of people who passed buckets of water from a stream to one another to a building that was on fire. In the electron transport chain, each acceptor molecule becomes alternately reduced as it accepts electrons and oxidized as it gives them up. The electrons entering the electron transport chain have a relatively high energy content. They lose some of their energy at each step as they pass along the chain of electron carriers (just as some of the water spills out of the bucket as it is passed from one person to another).

Members of the electron transport chain include the flavoprotein *flavin mononucleotide (FMN)*, the lipid *ubiquinone* (also called *coenzyme Q or CoQ*), several *iron–sulfur proteins,* and a group of closely related iron-containing proteins called *cytochromes* (**FIG. 8-8**). Each electron carrier has a different mechanism for accepting and passing electrons. As cytochromes accept and donate electrons, for example, the charge on the iron atom, which is the electron carrier portion of the cytochromes, alternates between Fe^{2+} (reduced) and Fe^{3+} (oxidized).

Scientists have extracted and purified the electron transport chain from the inner mitochondrial membrane as four large, distinct protein complexes, or groups, of acceptors. *Complex I (NADH–ubiquinone oxidoreductase)* accepts electrons from NADH molecules that were produced during glycolysis, the formation of acetyl CoA, and the citric acid cycle. *Complex II (succinate–ubiquinone reductase)* accepts electrons from $FADH_2$ molecules that were produced during the citric acid cycle. Complexes I and II both produce the same product, reduced ubiquinone, which is the substrate of *complex III (ubiquinone–cytochrome c oxidoreductase).* That is, complex III accepts electrons from reduced ubiquinone and passes them on to cytochrome *c. Complex IV (cytochrome c oxidase)* accepts electrons from cytochrome *c* and uses these electrons to reduce molecular oxygen, forming water in the process. The electrons simultaneously unite with protons from the surrounding medium to form hydrogen, and the chemical reaction between hydrogen and oxygen produces water.

Because oxygen is the final electron acceptor in the electron transport chain, organisms that respire aerobically require oxygen. What happens when cells that are strict aerobes are deprived of oxygen? The last cytochrome in the chain retains its electrons when no oxygen is available to accept them. When that occurs, each acceptor molecule in the chain retains its electrons (each remains in its reduced state), and the entire chain is blocked all the way back to NADH. Because oxidative phosphorylation is coupled to electron transport, no additional ATP is produced by way of the electron transport chain. Most cells of multicellular organisms cannot live long without oxygen because the small amount of ATP they produce by glycolysis alone is insufficient to sustain life processes.

Lack of oxygen is not the only factor that interferes with the electron transport chain. Some poisons, including cyanide, inhibit the normal activity of the cytochromes. Cyanide binds tightly to the iron in the last cytochrome in the electron transport chain, making it unable to transport electrons to oxygen. It blocks the further passage of electrons through the chain, and ATP production ceases.

Although the flow of electrons in electron transport is usually tightly coupled to the production of ATP, some organisms uncouple the two processes to produce heat (see *Inquiring About: Electron Transport and Heat*).

The chemiosmotic model explains the coupling of ATP synthesis to electron transport in aerobic respiration For decades, scientists were aware that oxidative phosphorylation occurs in mitochondria, and many experiments had shown that the transfer of 2 electrons from each NADH to oxygen (via the electron transport chain) usually results in the production of up to 3 ATP molecules. However, for a long time, the connection between ATP synthesis and electron transport remained a mystery.

In 1961, Peter Mitchell, a British biochemist, proposed the *chemiosmotic model,* which was based on his experiments and on theoretical considerations. One type of experiment involved using bacteria as a model system (**FIG. 8-9**). Because the respiratory electron transport chain is located in the plasma membrane of an aerobic bacterial cell, the bacterial plasma membrane can be considered comparable to the inner mitochondrial membrane. Mitchell demonstrated that if bacterial cells were placed in an acidic environment (i.e., an environment with a high hydrogen ion, or proton, concentration), the cells synthesized ATP even if electron transport was not taking place. On the basis of these and other experiments, Mitchell proposed that electron transport and ATP synthesis are coupled by means of a proton gradient across the inner mitochondrial membrane in eukaryotes (or across the plasma membrane in bacteria). His model was so radical that it

Electron carriers in the mitochondrial inner membrane transfer electrons from NADH and FADH$_2$ to oxygen.

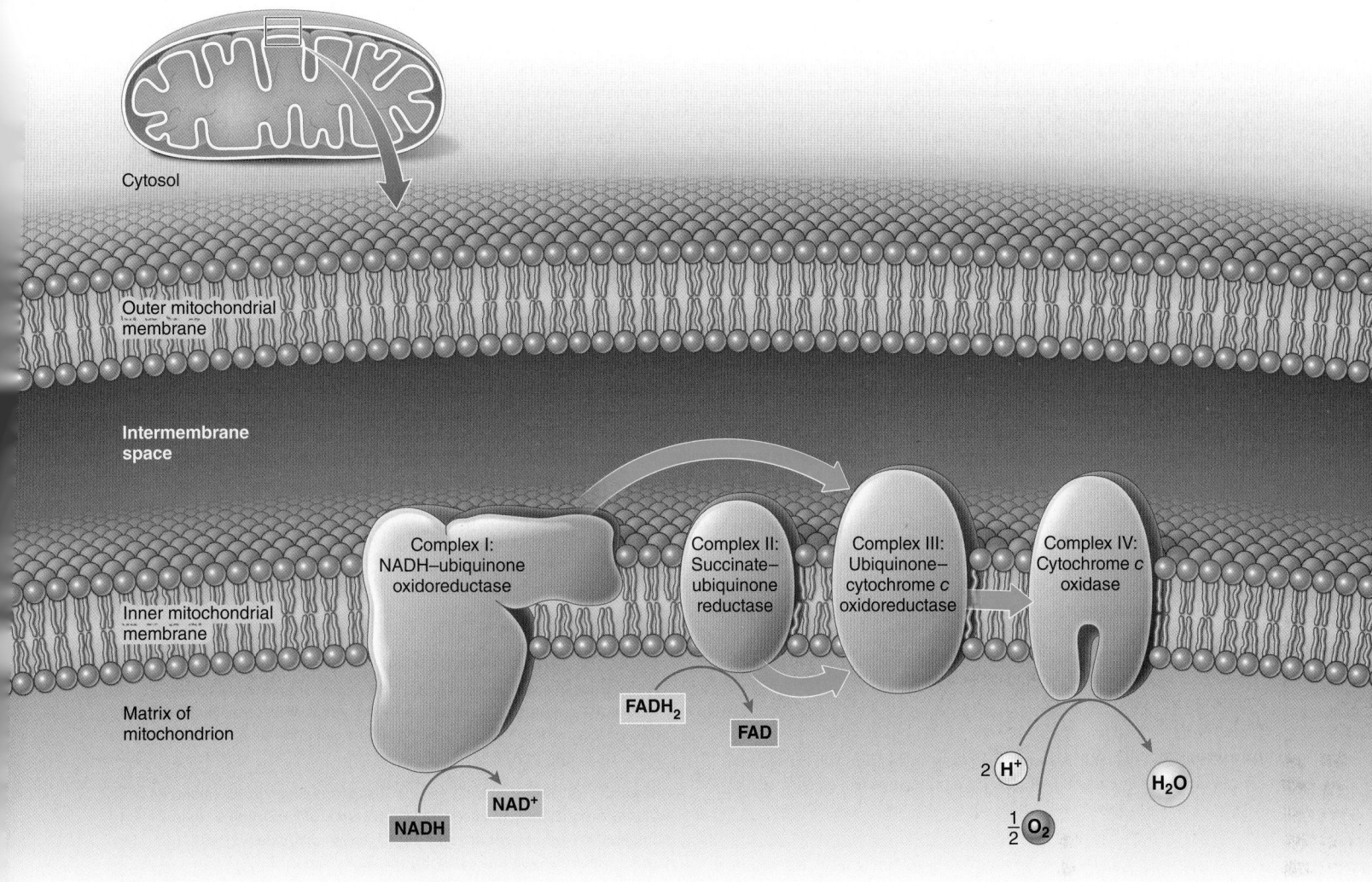

Figure 8-8 *Animation* **An overview of the electron transport chain**

Electrons fall to successively lower energy levels as they are passed along the four complexes of the electron transport chain located in the inner mitochondrial membrane. (The *blue arrows* indicate the pathway of electrons.) The carriers within each complex become alternately reduced and oxidized as they accept and donate electrons. The terminal acceptor is oxygen; one of the 2 atoms of an oxygen molecule (written as $\frac{1}{2}$ O$_2$) accepts 2 electrons, which are added to 2 protons from the surrounding medium to produce water.

PREDICT How would the flow of electrons be affected if complex III were missing?
© Cengage Learning

was not immediately accepted, but by 1978, so much evidence had accumulated in support of **chemiosmosis** that Mitchell was awarded the Nobel Prize in Chemistry that year.

The electron transport chain establishes the proton gradient; some of the energy released as electrons pass down the electron transport chain is used to move protons (H$^+$) across a membrane. In eukaryotes the protons are moved across the inner mitochondrial membrane into the intermembrane space (FIG. 8-10). Hence, the inner mitochondrial membrane separates a space with a higher concentration of protons (the intermembrane space) from a space with a lower concentration of protons (the mitochondrial matrix).

Protons are moved across the inner mitochondrial membrane by three of the four electron transport complexes (complexes I,

Electron Transport and Heat

What is the source of our body heat? Essentially, it is a byproduct of various exergonic reactions, especially those involving the electron transport chains in our mitochondria. Some cold-adapted animals, hibernating animals, and newborn animals produce unusually large amounts of heat by uncoupling electron transport from ATP production. These animals have adipose tissue (tissue in which fat is stored) that is brown. The brown color comes from the large number of mitochondria found in the brown adipose tissue cells. The inner mitochondrial membranes of these mitochondria contain an uncoupling protein that produces a passive proton channel through which protons flow into the mitochondrial matrix. As a consequence, most of the energy of glucose is converted to heat rather than to chemical energy in ATP.

Certain plants, which are not generally considered "warm" organisms, also have the ability to produce large amounts of heat. Skunk cabbage (*Symplocarpus foetidus*), for example, lives in North American swamps and wet woodlands and generally flowers during February and March when the ground is still covered with snow (see figure). Its uncoupled mitochondria generate large amounts of heat, enabling the plant to melt the snow and attract insect pollinators by vaporizing certain odiferous molecules into the surrounding air. The flower temperature of skunk cabbage is 15° to 22°C (59° to 72°F) when the air surrounding it is −15° to 10°C (5° to 50°F). Skunk cabbage flowers maintain this temperature for two weeks or more. Other plants, such as split leaf philodendron (*Philodendron selloum*) and sacred lotus (*Nelumbo nucifera*), also generate heat when they bloom and maintain their temperatures within precise limits.

Some plants generate as much or more heat per gram of tissue than animals in flight, which have long been considered the greatest heat producers in the living world. The European plant lords-and-ladies (*Arum maculatum*), for example, produces 0.4 J (0.1 cal) of heat per second per gram of tissue, whereas a hummingbird in flight produces 0.24 J (0.06 cal) per second per gram of tissue.

Ed Reschke/Getty Images

Skunk cabbage (*Symplocarpus foetidus*)

This plant not only produces a significant amount of heat when it flowers but also regulates its temperature within a specific range.

KEY EXPERIMENT

What is the mechanism of oxidative phosphorylation?

HYPOTHESIS: Peter Mitchell proposed that the cell uses energy released during electron transport to create a proton gradient across a membrane. The potential energy inherent in that gradient then drives the synthesis of ATP.

EXPERIMENT: Aerobic bacteria were placed in an acid (high H^+ concentration) environment, thus creating a proton gradient across the plasma membrane. It was done under conditions in which no electron transport was occurring.

RESULTS AND CONCLUSION: The bacteria synthesized ATP in the absence of aerobic respiration. These results supported Mitchell's view that a proton gradient across a membrane is an essential link in the conversion of the electrical energy of the electron transport chain to chemical energy in ATP.

SOURCE: Mitchell, P. "David Keilin's Respiratory Chain Concept and Its Chemiosmotic Consequences." *Nobel Lectures, Chemistry 1971–1980.* Singapore: World Scientific, 1993. http://www.nobelprize.org/nobel_prizes/chemistry/laureates/1978/mitchell-lecture.html

Figure 8-9 Evidence for chemiosmosis

PREDICT What do you think would happen if the cells were placed in a basic (low H^+ concentration) environment?

© Cengage Learning

Bacterial cytoplasm (low acid)

ATP Synthesized

Plasma membrane

Acidic environment

III, and IV) (**FIG. 8-11a**). Like water behind a dam, the resulting proton gradient is a form of potential energy that can be harnessed to provide the energy for ATP synthesis.

Diffusion of protons from the intermembrane space, where they are highly concentrated, through the inner mitochondrial membrane to the matrix of the mitochondrion is limited to specific channels formed by a fifth enzyme complex, **ATP synthase,** a transmembrane protein. Portions of these complexes project from the inner surface of the membrane (the surface that faces the matrix) and are visible by electron microscopy (**FIG. 8-11b**). Diffusion of the protons down their gradient, through the ATP synthase complex, is exergonic because the entropy of the system increases. This exergonic process provides the energy for ATP production, although the exact mechanism by which ATP synthase catalyzes the phosphorylation of ADP is still not completely understood.

In 1997, Paul Boyer of the University of California at Los Angeles and John Walker of the Medical Research Council Laboratory of Molecular Biology, Cambridge, England, shared the Nobel Prize in Chemistry for the discovery that ATP synthase functions in an unusual way. Experimental evidence strongly suggests that ATP synthase acts like a highly efficient molecular motor. During the production of ATP from ADP and inorganic phosphate, a central structure of ATP synthase rotates, possibly in response to the force of protons moving through the enzyme complex. The rotation apparently alters the conformation of the catalytic subunits in a way that drives ATP synthesis.

Chemiosmosis is a fundamental mechanism of energy coupling in cells; it allows exergonic redox reactions to drive the endergonic reaction in which ATP is produced by phosphorylating ADP. In photosynthesis (discussed in Chapter 9), ATP is produced by a comparable process.

Aerobic respiration of one glucose yields a maximum of 36 to 38 ATPs

Let us now review where biologically useful energy is captured in aerobic respiration and calculate the total energy yield from the complete oxidation of glucose. **FIGURE 8-12** summarizes the arithmetic involved.

1. In glycolysis glucose is activated by the addition of phosphates from 2 ATP molecules and converted ultimately to 2 pyruvates + 2 NADH + 4 ATPs, yielding a net profit of 2 ATPs.
2. The 2 pyruvates are metabolized to 2 acetyl CoA + 2 CO_2 + 2 NADH.
3. In the citric acid cycle, the 2 acetyl groups from CoA are metabolized to 4 CO_2 + 6 NADH + 2 $FADH_2$ + 2 ATPs.

Because the oxidation of NADH in the electron transport chain yields up to 3 ATPs per molecule, the total of 10 NADH molecules can yield up to 30 ATPs. The 2 NADH molecules from glycolysis, however, yield either 2 or 3 ATPs each. The reason is that certain types of eukaryotic cells must expend energy to shuttle the electrons from NADH produced by glycolysis across the mitochondrial membrane (to be discussed shortly). Prokaryotic cells lack mitochondria; hence, they have no need to shuttle electrons. For this reason, bacteria are able to generate 3 ATPs for every NADH, even those produced during glycolysis. Thus, the maximum number of ATPs formed using the energy from NADH is 28 to 30, depending on the cell.

The oxidation of $FADH_2$ yields 2 ATPs per molecule (recall that electrons from $FADH_2$ enter the electron transport chain at a different location than those from NADH), so the 2 $FADH_2$ molecules produced in the citric acid cycle yield 4 ATPs.

4. Summing all the ATPs (2 from glycolysis, 2 from the citric acid cycle, and 32 to 34 from electron transport and chemiosmosis), you can see that the complete aerobic metabolism of one molecule of glucose yields a maximum of 36 to 38 ATPs. Most ATP is generated by oxidative phosphorylation, which involves the electron transport chain and chemiosmosis. Only 4 ATPs are formed by substrate-level phosphorylation in glycolysis and the citric acid cycle.

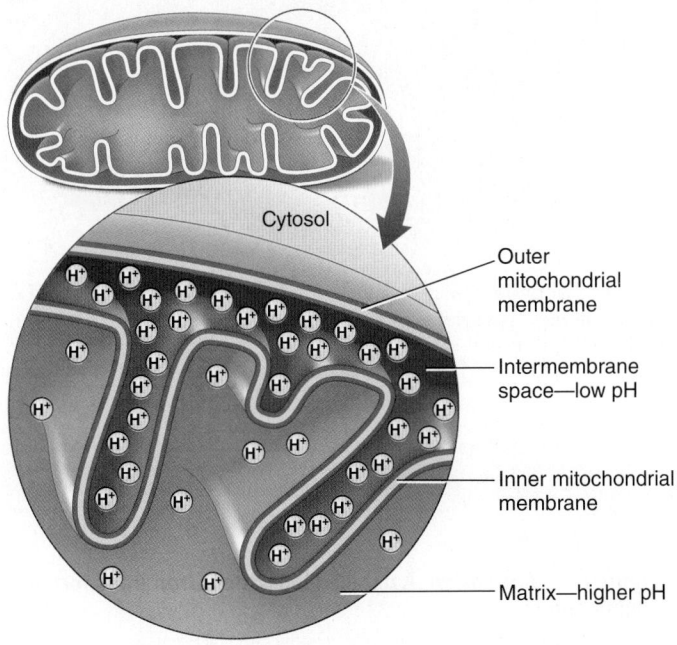

Cytosol

Outer mitochondrial membrane

Intermembrane space—low pH

Inner mitochondrial membrane

Matrix—higher pH

Figure 8-10 The accumulation of protons (H⁺) within the intermembrane space

As electrons move down the electron transport chain, the electron transport complexes move protons (H⁺) from the matrix to the intermembrane space, creating a proton gradient. The high concentration of H⁺ in the intermembrane space lowers the pH.
© Cengage Learning

The electron transport chain forms a concentration gradient for H⁺, which diffuses through ATP synthase complexes, producing ATP.

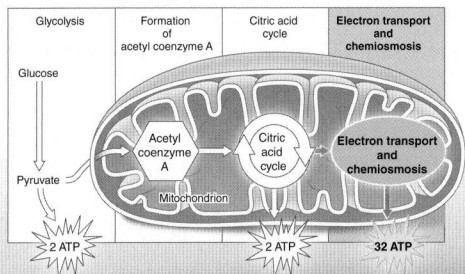

(a) The electron transport chain in the inner mitochondrial membrane includes three proton pumps that are located in three of the four electron transport complexes. (The *blue arrows* indicate the pathway of electrons; and the *white arrows*, the pathway of protons.) The energy released during electron transport is used to transport protons (H⁺) from the mitochondrial matrix to the intermembrane space, where a high concentration of protons accumulates. The protons cannot diffuse back into the matrix except through special channels in ATP synthase in the inner membrane. The flow of the protons through ATP synthase provides the energy for generating ATP from ADP and inorganic phosphate (P_i). In the process, the inner part of ATP synthase rotates (*thick purple arrows*) like a motor.

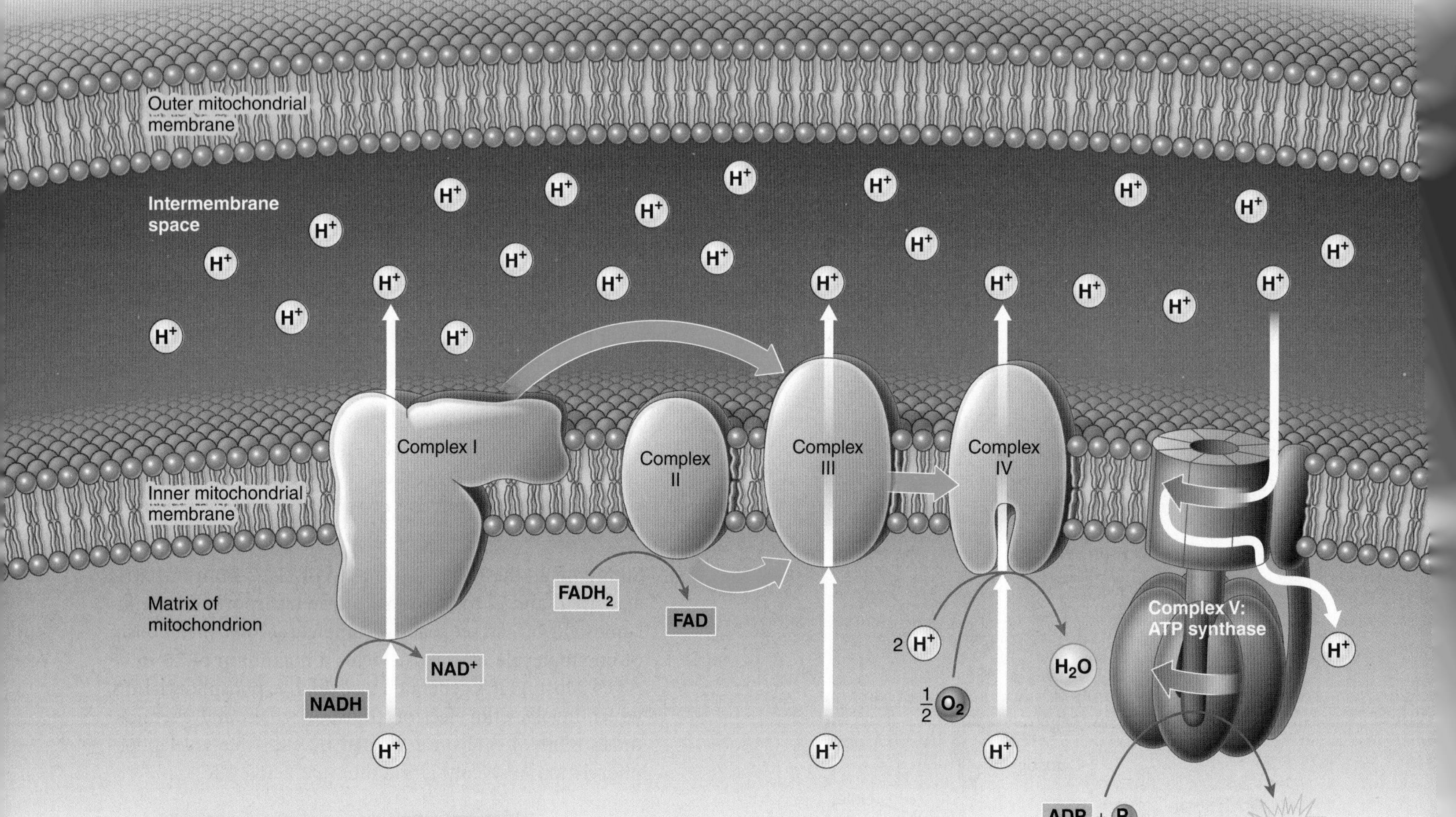

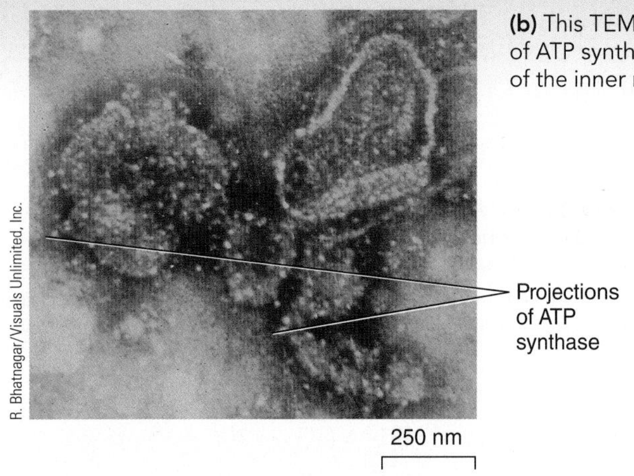

Cytosol

Outer mitochondrial membrane

Intermembrane space

Inner mitochondrial membrane

Matrix of mitochondrion

Complex I

Complex II

Complex III

Complex IV

FADH₂

FAD

NADH

NAD⁺

2 H⁺

½ O₂

H₂O

Complex V: ATP synthase

ADP + P$_i$

ATP

H⁺

(b) This TEM shows hundreds of projections of ATP synthase complexes along the surface of the inner mitochondrial membrane.

Projections of ATP synthase

250 nm

Figure 8-11 *Animation* **A detailed look at electron transport and chemiosmosis**

PREDICT What would be the immediate effect on electron flow if the ATP synthase complexes were removed?

© Cengage Learning

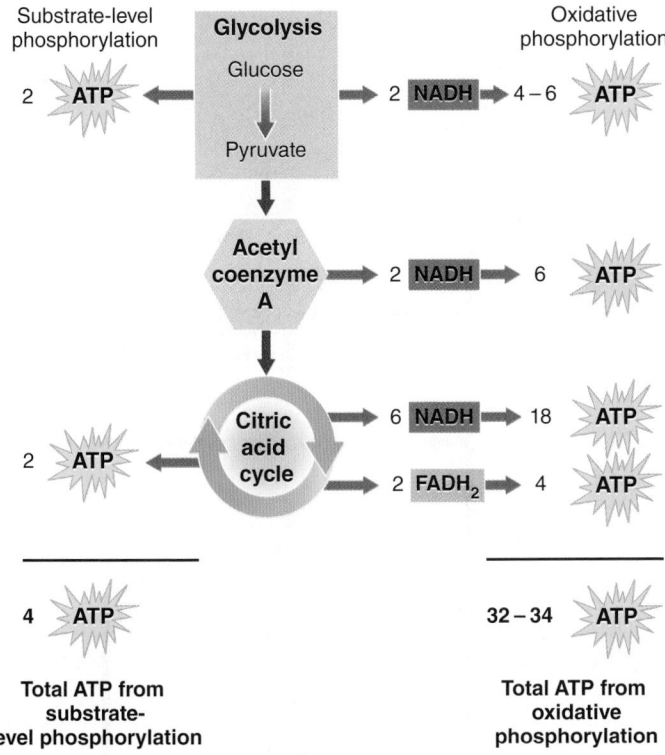

Figure 8-12 Energy yield from the complete oxidation of glucose by aerobic respiration

Most ATP is produced by electron transport and chemiosmosis (oxidative phosphorylation).

© Cengage Learning

We can analyze the efficiency of the overall process of aerobic respiration by comparing the free energy captured as ATP to the total free energy in a glucose molecule. Recall from Chapter 6 that although heat energy cannot power biological reactions, it is convenient to measure energy as heat. This is done through the use of a calorimeter, an instrument that measures the heat of a reaction. A sample is placed in a compartment surrounded by a chamber of water. As the sample burns (becomes oxidized), the temperature of the water rises, providing a measure of the heat released during the reaction.

When 1 mol of glucose is burned in a calorimeter, some 686 kcal (2870 kJ) are released as heat. The free energy temporarily held in the phosphate bonds of ATP is about 7.6 kcal (31.8 kJ) per mole. When 36 to 38 ATPs are generated during the aerobic respiration of glucose, the free energy trapped in ATP amounts to 7.6 kcal/mol × 36, or about 274 kcal (1146 kJ) per mole. Thus, the efficiency of aerobic respiration is 274/686, or about 40%. (By comparison, a steam power plant has an efficiency of 35% to 36% in converting its fuel energy into electricity.) The remainder of the energy in the glucose is released as heat.

Mitochondrial shuttle systems harvest the electrons of NADH produced in the cytosol The inner mitochondrial membrane is not permeable to NADH, which is a large molecule. Therefore, the NADH molecules produced in the cytosol during glycolysis cannot diffuse into the mitochondria to transfer their electrons to the electron transport chain. Unlike ATP

and ADP, NADH does not have a carrier protein to transport it across the membrane. Instead, several systems have evolved to transfer just the *electrons* of NADH, not the NADH molecules themselves, into the mitochondria.

In liver, kidney, and heart cells, a special shuttle system transfers the electrons from NADH through the inner mitochondrial membrane to an NAD^+ molecule in the matrix. These electrons are transferred to the electron transport chain in the inner mitochondrial membrane, and up to 3 molecules of ATP are produced per pair of electrons.

In skeletal muscle, brain, and some other types of cells, another type of shuttle operates. Because this shuttle requires more energy than the shuttle in liver, kidney, and heart cells, the electrons are at a lower energy level when they enter the electron transport chain. They are accepted by ubiquinone rather than by NAD^+ and so generate a maximum of 2 ATP molecules per pair of electrons. For this reason, the number of ATPs produced by aerobic respiration of 1 molecule of glucose in skeletal muscle cells is 36 rather than 38.

Cells regulate aerobic respiration

Aerobic respiration requires a steady input of fuel molecules and oxygen. Under normal conditions these materials are adequately provided and do not affect the rate of respiration. Instead, the rate of aerobic respiration is regulated by how much ADP and phosphate are available, with ATP synthesis continuing until most of the ADP has been converted to ATP. At this point oxidative phosphorylation slows considerably, which in turn slows down the citric acid cycle.

Glycolysis is partly controlled by feedback regulation (see Figure 7-15 for an illustration of feedback regulation) exerted on the enzyme phosphofructokinase, which catalyzes an early reaction of glycolysis (see Fig. 8-4). The active site of phosphofructokinase binds ATP and fructose-6-phosphate. However, the enzyme has two allosteric sites: an inhibitor site to which ATP binds when present at very high levels and an activator site to which AMP (adenosine monophosphate, a molecule formed when two phosphates are removed from ATP) binds. Therefore, this enzyme is inactivated when ATP levels are high and activated when they are low. Respiration proceeds when the enzyme becomes activated, thus generating more ATP.

CHECKPOINT 8.2

- *How much ATP is made available to the cell from a single glucose molecule by the operation of (1) glycolysis, (2) the formation of acetyl CoA, (3) the citric acid cycle, and (4) the electron transport chain and chemiosmosis? Where does each of these processes take place in a eukaryotic cell?*

- **CONNECT** *What essential role does each of the following play in chemiosmotic ATP synthesis: (1) electron transport chain, (2) proton gradient, and (3) ATP synthase complex?*

- **CONNECT** *What are the roles of NAD^+, FAD, and oxygen in aerobic respiration?*

- **PREDICT** *What do you expect would happen to the rate of oxidative phosphorylation if most of the ADP in the cell were to become converted to ATP?*

8.3 ENERGY YIELD OF NUTRIENTS OTHER THAN GLUCOSE

LEARNING OBJECTIVE

7 Summarize how the products of protein and lipid catabolism enter the same metabolic pathway that oxidizes glucose.

Many organisms, including humans, depend on nutrients other than glucose as a source of energy. In fact, you usually obtain more of your energy by oxidizing fatty acids than by oxidizing glucose. Amino acids derived from protein digestion are also used as fuel molecules. Such nutrients are transformed into one of the metabolic intermediates that are fed into glycolysis or the citric acid cycle (FIG. 8-13).

Amino acids are metabolized by reactions in which the amino group ($—NH_2$) is first removed, a process called **deamination.** In mammals and some other animals, the amino group is converted to urea (see Figure 48-1 for the biochemical pathway) and excreted, but the carbon chain is metabolized and eventually is used as a reactant in one of the steps of aerobic respiration. The amino acid alanine, for example, undergoes deamination to become pyruvate, the amino acid glutamate is converted to α-ketoglutarate, and the amino acid aspartate yields oxaloacetate. Pyruvate enters aerobic respiration as the end product of glycolysis, and α-ketoglutarate and oxaloacetate both enter aerobic respiration as intermediates in the citric acid cycle. Ultimately, the carbon chains of all the amino acids are metabolized in this way.

Each gram of lipid in the diet contains 9 kcal (38 kJ), more than twice as much energy as 1 g of glucose or amino acids, which have about 4 kcal (17 kJ) per gram. Lipids are rich in energy because they are highly reduced; that is, they have many hydrogen atoms and few oxygen atoms. When completely oxidized in aerobic respiration, a molecule of a six-carbon fatty acid generates up to 44 ATPs (compared with 36 to 38 ATPs for a molecule of glucose, which also has 6 carbons).

Both the glycerol and fatty acid components of a triacylglycerol (see Figure 3-12 for structures) are used as fuel; phosphate is added to glycerol, converting it to G3P or another compound that enters glycolysis. Fatty acids are oxidized and split enzymatically into two-carbon acetyl groups that are bound to coenzyme A; that is, fatty acids are converted to acetyl CoA. This process, which occurs in the mitochondrial matrix, is called **β-oxidation** (beta-oxidation). Acetyl groups from CoA molecules formed by β-oxidation enter the citric acid cycle.

CHECKPOINT 8.3

- CONNECT *How can a person obtain energy from a low-carbohydrate diet?*
- *What process must occur before amino acids enter the aerobic respiratory pathway?*
- VISUALIZE *Draw a simple diagram indicating where fatty acids enter the aerobic respiratory pathway.*

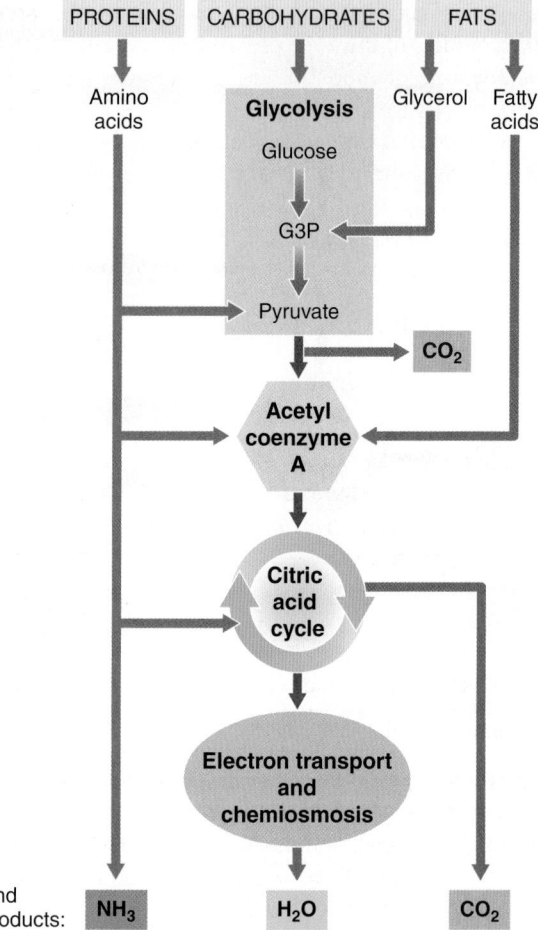

Figure 8-13 *Animation* **Energy from proteins, carbohydrates, and fats**

Products of the catabolism of proteins, carbohydrates, and fats enter glycolysis or the citric acid cycle at various points. This diagram is greatly simplified and illustrates only a few of the principal catabolic pathways.
© Cengage Learning

8.4 ANAEROBIC RESPIRATION AND FERMENTATION

LEARNING OBJECTIVE

8 Compare and contrast anaerobic respiration and fermentation. Include the mechanism of ATP formation, the final electron acceptor, and the end products.

Anaerobic respiration, which does not use oxygen as the final electron acceptor, is performed by some prokaryotes that live in anaerobic environments, such as waterlogged soil, stagnant ponds, and animal intestines. As in aerobic respiration, electrons are transferred in anaerobic respiration from glucose to NADH; they then pass down an electron transport chain that is coupled to ATP synthesis by chemiosmosis. However, an inorganic substance such as nitrate (NO_3^-) or sulfate (SO_4^{2-})

replaces molecular oxygen as the terminal electron acceptor. The end products of this type of anaerobic respiration are carbon dioxide, one or more reduced inorganic substances, and ATP. The following equation summarizes one representative type of anaerobic respiration, which is part of the biogeochemical cycle known as the **nitrogen cycle** (discussed in Chapter 55).

$$C_6H_{12}O_6 + 12\,KNO_3 \longrightarrow$$

Potassium nitrate

$$6\,CO_2 + 6\,H_2O + 12\,KNO_2 + energy$$

Potassium nitrite **(in the chemical bonds of ATP)**

Certain other bacteria, as well as some fungi, regularly use **fermentation,** an anaerobic pathway that does not involve an electron transport chain. During fermentation only 2 ATPs are formed per glucose (by substrate-level phosphorylation during glycolysis). One might expect that a cell that obtains energy from glycolysis would produce pyruvate, the end product of glycolysis. However, that cannot happen because every cell has a limited supply of NAD^+, and NAD^+ is required for glycolysis to continue. If virtually all NAD^+ becomes reduced to NADH during glycolysis, glycolysis stops and no more ATP is produced.

In fermentation, NADH molecules transfer their hydrogen atoms to organic molecules, thus regenerating the NAD^+ needed to keep glycolysis going. The resulting relatively reduced organic molecules (commonly, alcohol or lactate) tend to be toxic to the cells and are essentially waste products.

TABLE 8-2 compares aerobic respiration, anaerobic respiration, and fermentation.

Alcohol fermentation and lactate fermentation are inefficient

Yeasts are **facultative anaerobes** that carry out aerobic respiration when oxygen is available but switch to *alcohol fermentation* when deprived of oxygen (FIG. 8-14a). These eukaryotic, unicellular fungi have enzymes that decarboxylate pyruvate, releasing carbon dioxide and forming a two-carbon compound called *acetaldehyde.* NADH produced during glycolysis transfers hydrogen atoms to acetaldehyde, reducing it to *ethyl alcohol* (FIG. 8-14b). Alcohol fermentation is the basis for the production of beer, wine, and other alcoholic beverages. Yeast cells are also used in baking to produce the carbon dioxide that causes dough to rise; the alcohol evaporates during baking.

Certain fungi and bacteria perform *lactate (lactic acid) fermentation*. In this alternative pathway, NADH produced during glycolysis transfers hydrogen atoms to pyruvate, reducing it to *lactate* (FIG. 8-14c). The ability of some bacteria to produce lactate is exploited by humans, who use these bacteria to make yogurt and ferment cabbage for sauerkraut.

Vertebrate muscle cells also produce lactate. Exercise can cause fatigue and muscle cramps possibly due to insufficient oxygen, the depletion of fuel molecules, and the accumulation of lactate during strenuous activity. This buildup of lactate occurs because muscle cells shift briefly to lactate fermentation if the amount of oxygen delivered to muscle cells is insufficient to support aerobic respiration. The shift is only temporary, however, and oxygen is required for sustained work. About 80% of the lactate is eventually exported to the liver, where it is used to regenerate more glucose for the muscle cells. The remaining 20% of the lactate is metabolized in muscle cells in the presence of oxygen. For this reason, you continue to breathe heavily after you have stopped exercising: the additional oxygen is needed to oxidize lactate, thereby restoring the muscle cells to their normal state.

Although humans use lactate fermentation to produce ATP for only a few minutes, a few animals can live without oxygen for much longer periods. The red-eared slider, a freshwater turtle, remains underwater for as long as two weeks. During this time, it is relatively inactive and therefore does not expend a great deal of energy. It relies on lactate fermentation for ATP production.

Both alcohol fermentation and lactate fermentation are highly inefficient because the fuel is only partially oxidized. Alcohol, the end product of fermentation by yeast cells, can be burned and is even used as automobile fuel; obviously, it

TABLE 8-2	A Comparison of Aerobic Respiration, Anaerobic Respiration, and Fermentation		
	AEROBIC RESPIRATION	**ANAEROBIC RESPIRATION**	**FERMENTATION**
Immediate fate of electrons in NADH	Transferred to electron transport chain	Transferred to electron transport chain	Transferred to organic molecule
Terminal electron acceptor of electron transport chain	O_2	Inorganic substances such as NO_3^- or SO_4^{2-}	No electron transport chain
Reduced product(s) formed	Water	Relatively reduced inorganic substances	Relatively reduced organic compounds (commonly, alcohol or lactate)
Mechanism of ATP synthesis	Oxidative phosphorylation/chemiosmosis; also substrate-level phosphorylation	Oxidative phosphorylation/chemiosmosis; also substrate-level phosphorylation	Substrate-level phosphorylation only (during glycolysis)

Fermentation regenerates NAD⁺ needed for glycolysis.

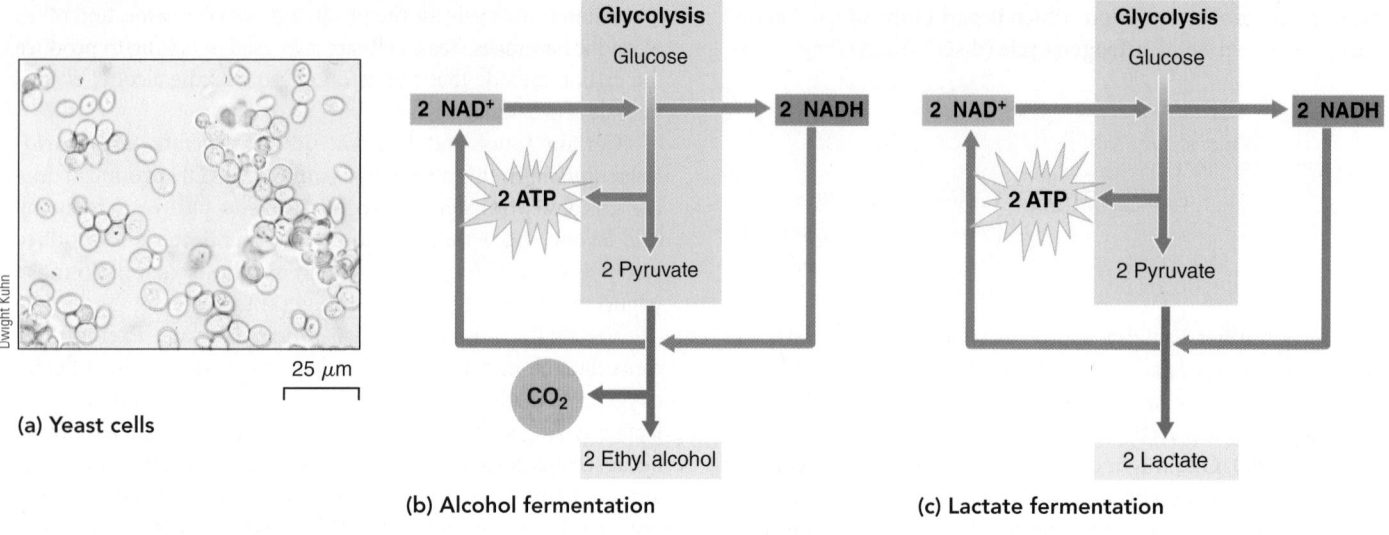

(a) Yeast cells

(b) Alcohol fermentation

(c) Lactate fermentation

Figure 8-14 *Animation* **Fermentation**

(a) Light micrograph of live brewer's yeast (*Saccharomyces cerevisiae*). Yeast cells have mitochondria and carry on aerobic respiration when O_2 is present. In the absence of O_2, yeasts carry on alcohol fermentation. **(b, c)** Glycolysis is the first part of fermentation pathways. In alcohol fermentation **(b)**, CO_2 is split off, and the two-carbon compound ethyl alcohol is the end product. In lactate fermentation **(c)**, the final product is the three-carbon compound lactate. In both alcohol and lactate fermentation, there is a net gain of only 2 ATPs per molecule of glucose. Note that the NAD⁺ used during glycolysis is regenerated during both alcohol fermentation and lactate fermentation.

CONNECT Which is more efficient in terms of yield of ATP per glucose, anaerobic respiration or fermentation?
© Cengage Learning

contains a great deal of energy that the yeast cells cannot extract using anaerobic methods. Lactate, a three-carbon compound, contains even more energy than the two-carbon alcohol. In contrast, all available energy is removed during aerobic respiration because the fuel molecules become completely oxidized to CO_2. A net profit of only 2 ATPs is produced by the fermentation of 1 molecule of glucose, compared with up to 36 to 38 ATPs when oxygen is available.

The inefficiency of fermentation necessitates a large supply of fuel. To perform the same amount of work, a cell engaged in fermentation must consume up to 20 times as much glucose or other carbohydrate per second as a cell using aerobic respiration. For this reason, your skeletal muscle cells store large

quantities of glucose in the form of glycogen, which enables them to metabolize anaerobically for short periods.

CHECKPOINT 8.4

- **CONNECT** *What is the fate of hydrogen atoms removed from glucose during glycolysis when oxygen is present in muscle cells? How does it compare with the fate of hydrogen atoms removed from glucose when the amount of available oxygen is insufficient to support aerobic respiration?*

- *What accounts for the ATP yield of fermentation being only a tiny fraction of the yield from aerobic respiration?*

- *Is chemiosmosis involved in fermentation? in anaerobic respiration?*

SUMMARY: FOCUS ON LEARNING OBJECTIVES

8.1 Redox Reactions *(page 166)*

1 Write a summary reaction for aerobic respiration that shows which reactant becomes oxidized and which becomes reduced.

$$C_6H_{12}O_6 + 6\,O_2 \longrightarrow 6\,CO_2 + 6\,H_2O + \text{energy}$$

Oxidation / Reduction

- **Aerobic respiration** is a catabolic process in which a fuel molecule such as glucose is broken down to form carbon dioxide and water. It includes **redox reactions** that result in the transfer of electrons from glucose (which becomes *oxidized*) to oxygen (which becomes *reduced*).

- Energy released during aerobic respiration is used to produce up to 36 to 38 ATPs per molecule of glucose.

8.2 The Four Stages of Aerobic Respiration *(page 166)*

2 List and give a brief overview of the four stages of aerobic respiration.

- The chemical reactions of aerobic respiration occur in four stages: glycolysis, formation of acetyl CoA, the citric acid cycle, and the electron transport chain and chemiosmosis.
- During **glycolysis** a molecule of glucose is degraded to 2 molecules of **pyruvate. Substrate-level phosphorylation** produces 2 ATP molecules during glycolysis, and 4 hydrogen atoms are removed and used to produce 2 NADH.
- During the formation of **acetyl CoA,** the 2 pyruvate molecules each lose a molecule of carbon dioxide, and the remaining acetyl groups each combine with **coenzyme A,** producing 2 molecules of acetyl CoA; 1 NADH is produced per pyruvate.
- Each acetyl group from acetyl CoA enters the **citric acid cycle** by combining with a four-carbon compound, **oxaloacetate,** to form **citrate,** a six-carbon compound. Two acetyl CoA molecules enter the cycle for every glucose molecule. For every 2 carbons that enter the cycle as part of an acetyl CoA molecule, 2 leave as carbon dioxide. For every acetyl CoA, hydrogen atoms are transferred to 3 NAD$^+$ and 1 FAD; only 1 ATP is produced by substrate-level phosphorylation.
- Hydrogen atoms (or their electrons) removed from fuel molecules are transferred from one electron acceptor to another down an **electron transport chain** located in the mitochondrial inner membrane; ultimately, these electrons reduce molecular oxygen, forming water. In **oxidative phosphorylation** the redox reactions in the electron transport chain are coupled to synthesis of ATP through the mechanism of **chemiosmosis.**

3 Indicate where each stage of aerobic respiration takes place in a eukaryotic cell.

- Glycolysis occurs in the cytosol, and the remaining stages of aerobic respiration take place in the mitochondria.

4 Add up the energy captured (as ATP, NADH, and FADH$_2$) in each stage of aerobic respiration.

- In glycolysis each glucose molecule produces 2 NADH and 2 ATPs (net). The conversion of 2 pyruvates to acetyl CoA results in the formation of 2 NADH. In the citric acid cycle, the 2 acetyl CoA molecules are metabolized to form 6 NADH, 2 FADH$_2$, and 2 ATPs. To summarize, we have 4 ATPs, 10 NADH, and 2 FADH$_2$.
- When electrons donated by the 10 NADH and 2 FADH$_2$ pass through the electron transport chain, 32 to 34 ATPs are produced by chemiosmosis. Therefore, each glucose molecule yields a total of up to 36 to 38 ATPs.

5 Define *chemiosmosis* and explain how a gradient of protons is established across the inner mitochondrial membrane.

- In chemiosmosis some of the energy of the electrons in the electron transport chain is used to pump protons across the inner mitochondrial membrane into the intermembrane space. This pumping establishes a proton gradient across the inner mitochondrial membrane. Protons (H$^+$) accumulate within the intermembrane space, lowering the pH.

6 Describe the process by which the proton gradient drives ATP synthesis in chemiosmosis.

- The diffusion of protons through channels formed by the enzyme **ATP synthase,** which extends through the inner mitochondrial membrane from the intermembrane space to the mitochondrial matrix, provides the energy to synthesize ATP.

8.3 Energy Yield of Nutrients Other Than Glucose *(page 180)*

7 Summarize how the products of protein and lipid catabolism enter the same metabolic pathway that oxidizes glucose.

- Amino acids undergo **deamination,** and their carbon skeletons are converted to metabolic intermediates of aerobic respiration.
- Both the glycerol and fatty acid components of lipids are oxidized as fuel. Fatty acids are converted to acetyl CoA molecules by the process of β**-oxidation.**

8.4 Anaerobic Respiration and Fermentation *(page 180)*

8 Compare and contrast anaerobic respiration and fermentation. Include the mechanism of ATP formation, the final electron acceptor, and the end products.

- In **anaerobic respiration** electrons are transferred from fuel molecules to an electron transport chain that is coupled to ATP synthesis by chemiosmosis; the final electron acceptor is an inorganic substance such as nitrate or sulfate, not molecular oxygen.
- **Fermentation** is an anaerobic process that does not use an electron transport chain. There is a net gain of only 2 ATPs per glucose; they are produced by substrate-level phosphorylation during glycolysis. To maintain the supply of NAD$^+$ essential for glycolysis, hydrogen atoms are transferred from NADH to an organic compound derived from the initial nutrient.
- Yeast cells carry out *alcohol fermentation,* in which ethyl alcohol and carbon dioxide are the final waste products.
- Certain fungi, prokaryotes, and animal cells carry out *lactate (lactic acid) fermentation,* in which hydrogen atoms are added to pyruvate to form lactate, a waste product.

Summary Reactions for Aerobic Respiration

Summary reaction for the complete oxidation of glucose:

$$C_6H_{12}O_6 + 6\ O_2 + 6\ H_2O \longrightarrow$$
$$6\ CO_2 + 12\ H_2O + \text{energy (36 to 38 ATP)}$$

Summary reaction for glycolysis:

$$C_6H_{12}O_6 + 2\ ATP + 2\ ADP + 2\ P_i + 2\ NAD^+ \longrightarrow$$
$$2\ \text{pyruvate} + 4\ ATP + 2\ NADH + H_2O$$

Summary reaction for the conversion of pyruvate to acetyl CoA:

$$2\ \text{pyruvate} + 2\ \text{coenzyme A} + 2\ NAD^+ \longrightarrow$$
$$2\ \text{acetyl CoA} + 2\ CO_2 + 2\ NADH$$

Summary reaction for the citric acid cycle:

$$2\ \text{acetyl CoA} + 6\ NAD^+ + 2\ FAD + 2\ ADP + 2\ P_i + 2\ H_2O$$
$$\longrightarrow 4\ CO_2 + 6\ NADH + 2\ FADH_2 + 2\ ATP + 2\ CoA$$

Summary reactions for the processing of the hydrogen atoms of NADH and FADH$_2$ in the electron transport chain:

$$NADH + 3\ ADP + 3\ P_i + \tfrac{1}{2}\ O_2 \longrightarrow NAD^+ + 3\ ATP + H_2O$$

$$FADH_2 + 2\ ADP + 2\ P_i + \tfrac{1}{2}\ O_2 \longrightarrow FAD^+ + 2\ ATP + H_2O$$

Summary Reactions for Fermentation

Summary reaction for lactate fermentation:

$$C_6H_{12}O_6 \longrightarrow 2\ \text{lactate} + \text{energy (2 ATP)}$$

Summary reaction for alcohol fermentation:

$$C_6H_{12}O_6 \longrightarrow 2\ CO_2 + 2\ \text{ethyl alcohol} + \text{energy (2 ATP)}$$

Know and Comprehend

1. A chemical process during which a substance gains electrons and energy is called (a) oxidation (b) oxidative phosphorylation (c) deamination (d) reduction (e) dehydrogenation

2. The reactions of ___ take place within the cytosol of eukaryotic cells. (a) glycolysis (b) oxidation of pyruvate (c) the citric acid cycle (d) chemiosmosis (e) the electron transport chain

3. Before pyruvate enters the citric acid cycle, it is decarboxylated, oxidized, and combined with coenzyme A, forming acetyl CoA, carbon dioxide, and one molecule of (a) NADH (b) $FADH_2$ (c) ATP (d) ADP (e) $C_6H_{12}O_6$

4. In the first step of the citric acid cycle, an acetyl group from acetyl CoA reacts with oxaloacetate to form (a) pyruvate (b) citrate (c) NADH (d) ATP (e) CO_2

5. Which of the following is the major source of electrons that flow through the mitochondrial electron transport chain? (a) H_2O (b) ATP (c) NADH (d) ATP synthase (e) coenzyme A

6. The "aerobic" part of aerobic cellular respiration occurs during (a) glycolysis (b) the conversion of pyruvate to acetyl CoA (c) the citric acid cycle (d) electron transport (e) all the preceding are aerobic processes

7. Substrate-level phosphorylation (a) occurs through a chemiosmotic mechanism (b) accounts for most of the ATP formed during aerobic cellular respiration (c) occurs during the conversion of pyruvate to acetyl CoA (d) occurs during glycolysis and the citric acid cycle (e) requires high energy electrons from NADH

8. A net profit of only 2 ATPs can be produced anaerobically from the ___ of one molecule of glucose, compared with a maximum of 38 ATPs produced in ___. (a) fermentation; anaerobic respiration (b) aerobic respiration; fermentation (c) aerobic respiration; anaerobic respiration (d) dehydrogenation; decarboxylation (e) fermentation; aerobic respiration

9. When deprived of oxygen, yeast cells obtain energy by fermentation, producing carbon dioxide, ATP, and (a) acetyl CoA (b) ethyl alcohol (c) lactate (d) pyruvate (e) citrate

Apply and Analyze

10. Which of the following is a correct ranking of molecules with respect to their energy value in glycolysis (*note:* > means "greater than")? (a) two pyruvates > one glucose (b) one glucose > one fructose-1,6-bisphosphate (c) two glyceraldehyde-3-phosphates (G3P) > one glucose (d) two pyruvates > one fructose-1,6-bisphosphate (e) two pyruvates > two glyceraldehyde-3-phosphates (G3P)

11. Which of the following is a correct ranking of molecules, according to their energy value in oxidative phosphorylation (*note:* > means "greater than")? (a) ATP > NADH (b) NAD^+ > NADH (c) FAD > $FADH_2$ (d) NADH > ATP

Evaluate and Synthesize

12. **CONNECT** Explain why the proton gradient formed during chemiosmosis represents a state of low entropy. (You may wish to refer to the discussion of entropy in Chapter 7.)

13. **CONNECT** How are the endergonic reactions of the first phase of glycolysis coupled to the hydrolysis of ATP, which is exergonic? How are the exergonic reactions of the second phase of glycolysis coupled to the endergonic synthesis of ATP and NADH?

14. **PREDICT** Could the inner mitochondrial membrane carry out its functions in the coupling of electron transport and ATP synthesis if its lipid bilayer were readily permeable to hydrogen ions (protons)?

15. **VISUALIZE** Draw a simple sketch illustrating an inner mitochondrial membrane that is actively involved in chemiosmosis and label the two compartments it separates. Add the ATP synthase complex, indicate the proton gradient, and specify in which compartment ATP is synthesized.

16. **CONNECT** When you lose weight, where does it go?

17. **EVOLUTION LINK** The reactions of glycolysis are identical in *all* organisms—prokaryotes, protists, fungi, plants, and animals—that obtain energy from glucose catabolism. What does this universality suggest about the evolution of glycolysis?

18. **EVOLUTION LINK** Molecular oxygen is so reactive that it would not exist in Earth's atmosphere today if it were not constantly replenished by organisms that release oxygen as a waste product of photosynthesis. What does that fact suggest about the evolution of aerobic respiration and oxygen-releasing photosynthetic processes?

To access course materials, such as Aplia and other companion resources, please visit **www.cengagebrain.com.**

Photosynthesis: Capturing Light Energy

9

Look at all the living things that surround you: the trees, your pet goldfish, your own body. Most of that biomass is made up of carbon-based biological molecules. What is the ultimate source of all that carbon? Surprising to some, the source is carbon dioxide from the air. Your cells cannot take carbon dioxide from the air and incorporate it into organic molecules, but some plant cells can. They do so through **photosynthesis,** the sequence of events in which the orderly systems in these cells provide the information needed to convert light energy into the stored chemical energy of organic molecules. Photosynthesis is the first step in the flow of energy through most of the living world, capturing the vast majority of the energy that living organisms use. Photosynthesis not only sustains plants (see photograph) and other photosynthetic organisms such as algae and photosynthetic prokaryotes but also indirectly supports most non-photosynthetic organisms such as animals, fungi, protozoa, and most prokaryotes.

Each year photosynthetic organisms convert CO_2 into billions of tons of organic molecules. These molecules have two important roles in both photosynthetic and non-photosynthetic organisms: they are both the building blocks of cells and, as we saw in Chapter 8, a source of chemical energy that fuels the metabolic reactions that sustain almost all life. Photosynthesis also releases O_2, which is essential to aerobic cellular respiration, the process by which plants, animals, and most other organisms convert this chemical energy to ATP to power cellular processes.

In this chapter we first examine how light energy is used in the synthesis of ATP and other molecules that temporarily hold chemical energy but are unstable and cannot be stockpiled in the cell. We then see how their energy powers the anabolic pathway by which a photosynthetic cell synthesizes stable organic molecules from the simple inorganic compounds CO_2 and water. Finally, we explore the role of photosynthesis in plants and in Earth's environment.

©vovan/Shutterstock.com

Photosynthesis. These trees use light energy to power the processes that incorporate CO_2 into organic molecules.

KEY CONCEPTS

9.1 Light energy powers photosynthesis, which is essential to plants and most life on Earth.

9.2 Photosynthesis occurs in chloroplasts and requires the pigment chlorophyll.

9.3 Photosynthesis is a redox process.

9.4 Light-dependent reactions convert light energy to the chemical energy of NADPH and ATP.

9.5 Carbon fixation reactions incorporate CO_2 into organic molecules.

9.6 Most photosynthetic organisms are photoautotrophs.

9.7 Photosynthesis is important to plants and also other organisms.

9.1 LIGHT AND PHOTOSYNTHESIS

LEARNING OBJECTIVE

1 Describe the physical properties of light and explain the relationship between a wavelength of light and its energy.

Because most life on this planet depends on light, either directly or indirectly, it is important to understand the nature of light and its essential role in photosynthesis. Visible light represents a very small portion of a vast, continuous range of radiation called the *electromagnetic spectrum* (FIG. 9-1). All radiation in this spectrum travels as waves. A **wavelength** is the distance from one wave peak to the next.

At one end of the electromagnetic spectrum are gamma rays, which have very short wavelengths measured in fractions of nanometers, or nm (1 nanometer equals 10^{-9} m, one-billionth of a meter). At the other end of the spectrum are radio waves, with wavelengths so long they can be measured in kilometers. The portion of the electromagnetic spectrum from 380 to 760 nm is called the *visible spectrum* because we humans can see it. The visible spectrum includes all the colors of the rainbow (FIG. 9-2); violet has the shortest wavelength, and red has the longest.

Light is composed of small particles, or packets, of energy called **photons.** The energy of a photon is inversely proportional to its wavelength: shorter-wavelength light has more energy per photon than longer-wavelength light.

Why does photosynthesis depend on light detectable by the human eye (visible light) rather than on some other wavelength(s) of radiation? We know that radiation within the visible-light portion of the spectrum excites certain types of biological molecules, moving electrons into higher energy levels. Radiation with wavelengths longer than those of visible light does not have enough energy to excite these biological molecules. Radiation with wavelengths shorter than those of visible light is so energetic that it disrupts the bonds of many biological molecules. Thus, visible light has just the right amount of energy to cause the kinds of reversible changes in the molecules that are useful in photosynthesis.

When a molecule absorbs a photon of light energy, one of its electrons becomes energized, which means that the electron shifts from a lower-energy atomic orbital to a high-energy orbital that is more distant from the atomic nucleus. One of two things then happens to the energized electron, depending on the atom and its surroundings (FIG. 9-3). The atom may return to its **ground state,** which is the condition in which all its electrons are in their normal, lowest-energy levels. When an electron returns to its ground state, its energy dissipates as heat, and/or as an emission of light of a longer wavelength than the absorbed light; this emission of light is called **fluorescence.** Alternatively, the energized electron may leave the atom and be accepted by an electron acceptor molecule, which becomes reduced in the process; this is what occurs in photosynthesis.

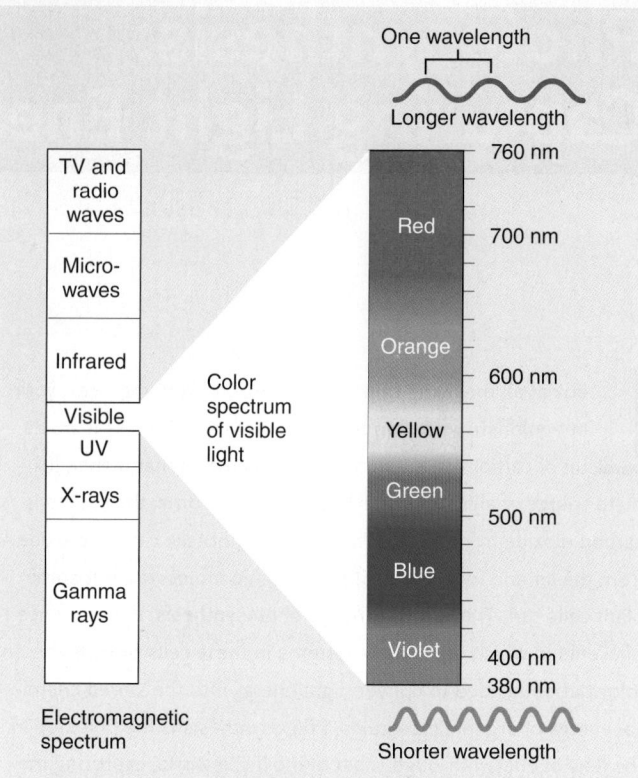

Figure 9-1 The electromagnetic spectrum

Waves in the electromagnetic spectrum have similar properties but different wavelengths. Radio waves are the longest (and least energetic) waves, with wavelengths as long as 20 km. Gamma rays are the shortest (and most energetic) waves. Visible light represents a small fraction of the electromagnetic spectrum and consists of a mixture of wavelengths ranging from about 380 to 760 nm. The energy from visible light is used in photosynthesis.
© Cengage Learning

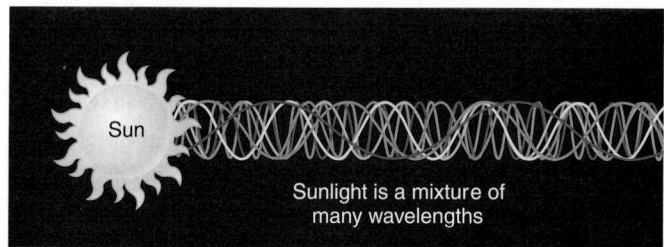

Figure 9-2 Visible radiation emitted from the sun

Electromagnetic radiation from the sun includes ultraviolet radiation and visible light of varying colors and wavelengths.
© Cengage Learning

Now that you understand some of the properties of light, let us consider the organelles that use light for photosynthesis.

CHECKPOINT 9.1

- *Which color of light has the longer wavelength, violet or red?*
- **CONNECT** *Which color of light has the higher energy per photon, violet or red?*

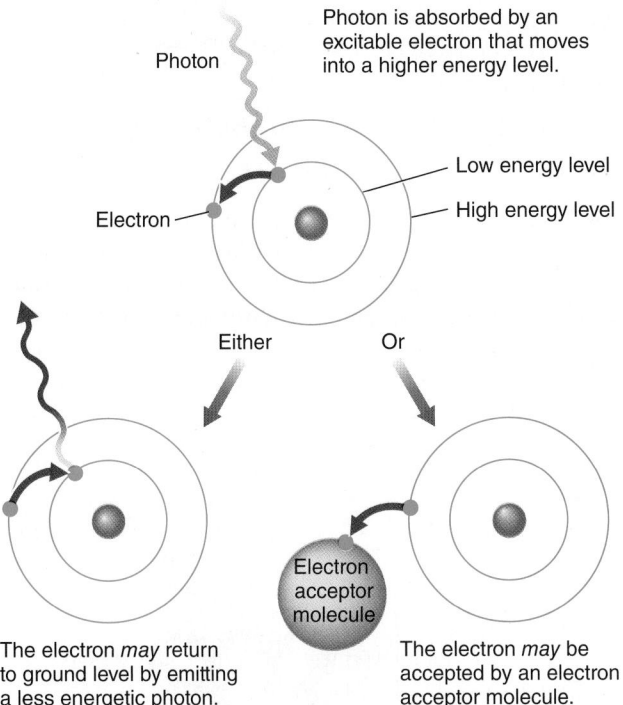

Photon

Photon is absorbed by an excitable electron that moves into a higher energy level.

Electron

Low energy level

High energy level

Either   Or

Electron acceptor molecule

The electron *may* return to ground level by emitting a less energetic photon.

The electron *may* be accepted by an electron acceptor molecule.

Figure 9-3 Interactions between light and atoms or molecules

(*Top*) When a photon of light energy strikes an atom or a molecule of which the atom is a part, the energy of the photon may push an electron to an orbital farther from the nucleus (i.e., into a higher energy level). (*Lower left*) If the electron returns to the lower, more stable energy level, the energy may be released as a less energetic, longer-wavelength photon, known as fluorescence (*shown*), or as heat. (*Lower right*) If the appropriate electron acceptors are available, the electron may leave the atom. During photosynthesis, an electron acceptor captures the energetic electron and passes it to a chain of acceptors.

© Cengage Learning

9.2 CHLOROPLASTS

LEARNING OBJECTIVES

2   Diagram the internal structure of a chloroplast and explain how its components interact and facilitate the process of photosynthesis.

3   Describe what happens to an electron in a biological molecule such as chlorophyll when a photon of light energy is absorbed.

If you examine a section of leaf tissue in a microscope, you see that the green pigment, chlorophyll, is not uniformly distributed in the cell but is confined to organelles called chloroplasts. In plants, **chloroplasts** lie mainly inside the leaf in the cells of the **mesophyll,** a layer that includes many air spaces and a very high concentration of water vapor (**FIG. 9-4a**). The interior of the leaf exchanges gases with the outside through microscopic pores, called **stomata** (sing., *stoma*). Each mesophyll cell has 20 to 100 chloroplasts.

The chloroplast, like the mitochondrion, is enclosed by outer and inner membranes (**FIG. 9-4b**). The inner membrane encloses a fluid-filled region called the **stroma,** which contains most of

the enzymes required to produce carbohydrate molecules. Suspended in the stroma is a third system of membranes that forms an interconnected set of flat, disclike sacs called **thylakoids.**

The thylakoid membrane encloses a fluid-filled interior space, the **thylakoid lumen.** In some regions of the chloroplast, thylakoid sacs are arranged in stacks called **grana** (sing., *granum*). Each granum looks something like a stack of coins, with each "coin" being a thylakoid. Some thylakoid membranes extend from one granum to another. Thylakoid membranes, like the inner mitochondrial membrane (see Chapter 8), are involved in ATP synthesis. (Photosynthetic prokaryotes have no chloroplasts, but thylakoid membranes are often arranged around the periphery of the cell as infoldings of the plasma membrane.)

Chlorophyll is found in the thylakoid membrane

Thylakoid membranes contain several kinds of *pigments,* which are substances that absorb visible light. Different pigments absorb light of different wavelengths. **Chlorophyll,** the main pigment of photosynthesis, absorbs light primarily in the blue and red regions of the visible spectrum. Green light is not appreciably absorbed by chlorophyll. Plants usually appear green because some of the green light that strikes them is scattered or reflected.

A chlorophyll molecule has two main parts, a complex ring and a long side chain (**FIG. 9-5**). The ring structure, called a *porphyrin ring,* is made up of joined smaller rings composed of carbon and nitrogen atoms; the porphyrin ring absorbs light energy. The porphyrin ring of chlorophyll is strikingly similar to the heme portion of the red pigment hemoglobin in red blood cells. However, unlike heme, which contains an atom of iron in the center of the ring, chlorophyll contains an atom of magnesium in that position. The chlorophyll molecule also contains a long, hydrocarbon side chain that makes the molecule extremely nonpolar and anchors the chlorophyll in the membrane.

All chlorophyll molecules in the thylakoid membrane are associated with specific *chlorophyll-binding proteins;* biologists have identified about 15 different kinds. Each thylakoid membrane is filled with precisely oriented chlorophyll molecules and chlorophyll-binding proteins that facilitate the transfer of energy from one molecule to another.

There are several kinds of chlorophyll. The most important is **chlorophyll *a*,** the pigment that initiates the light-dependent reactions of photosynthesis. **Chlorophyll *b*** is an accessory pigment that also participates in photosynthesis. It differs from chlorophyll *a* only in a functional group on the porphyrin ring: the methyl group ($-CH_3$) in chlorophyll *a* is replaced in chlorophyll *b* by a terminal carbonyl group ($-CHO$). This difference shifts the wavelengths of light absorbed and reflected by chlorophyll *b*, making it appear yellow-green, whereas chlorophyll *a* appears bright green.

Chloroplasts have other accessory photosynthetic pigments, such as **carotenoids,** which are yellow and orange (see Fig. 3-14). Carotenoids absorb different wavelengths of light than chlorophyll, thereby expanding the spectrum of light that provides energy for photosynthesis. Chlorophyll may be excited by light directly by energy passed to it from the light source or indirectly by energy passed to it from accessory pigments that have become excited by light. When a carotenoid molecule is excited, its energy can be transferred to chlorophyll *a*.

In addition, carotenoids are antioxidants that inactivate highly reactive forms of oxygen generated in the chloroplasts.

Chlorophyll is the main photosynthetic pigment

As you have seen, the thylakoid membrane contains more than one kind of pigment. An instrument called a *spectrophotometer* measures the relative abilities of different pigments to absorb different wavelengths of light. The **absorption spectrum** of a pigment is a plot of its absorption of light of different wavelengths. FIGURE 9-6a shows the absorption spectra for chlorophylls *a* and *b*.

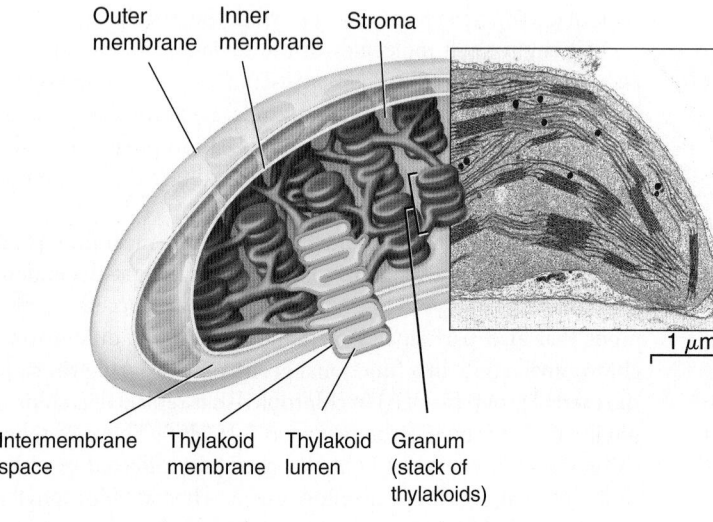

(a) This leaf cross section reveals that the mesophyll is the photosynthetic tissue. CO_2 enters the leaf through tiny pores or stomata, and H_2O is carried to the mesophyll in veins.

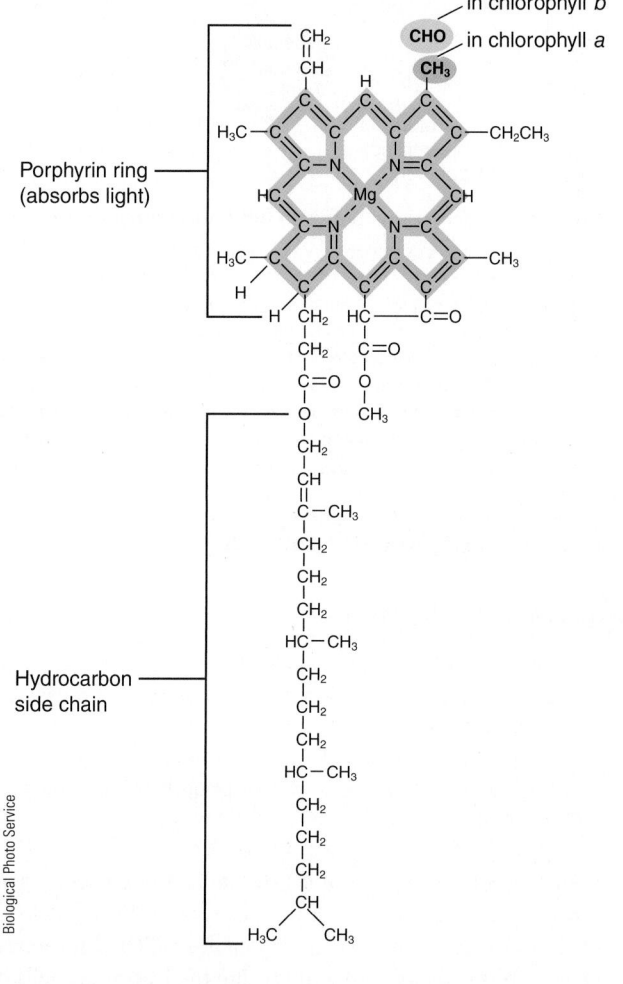

(b) In the chloroplast, pigments necessary for the light-capturing reactions of photosynthesis are part of thylakoid membranes, whereas the enzymes for the synthesis of carbohydrate molecules are in the stroma.

Figure 9-4 *Animation* **The site of photosynthesis**
© Cengage Learning

E. H. Newcomb and W. P. Wergin, Biological Photo Service

Figure 9-5 The structure of chlorophyll

Chlorophyll consists of a porphyrin ring and a hydrocarbon side chain. The porphyrin ring, with a magnesium atom in its center, contains a system of alternating double and single bonds; these bonds are commonly found in molecules that strongly absorb certain wavelengths of visible light and reflect others (chlorophyll reflects green). Note that at the top right corner of the diagram, the methyl group (—CH$_3$) distinguishes chlorophyll *a* from chlorophyll *b*, which has a carbonyl group (—CHO) in this position. The hydrophobic hydrocarbon side chain anchors chlorophyll to the thylakoid membrane.
© Cengage Learning

An **action spectrum** of photosynthesis is a graph of the relative effectiveness of different wavelengths of light. To obtain an action spectrum, scientists measure the rate of photosynthesis at each wavelength for leaf cells or tissues exposed to monochromatic light (light of one wavelength) (**FIG. 9-6b**).

In a classic biology experiment, German biologist T. W. Engelmann obtained the first action spectrum in 1882. Engelmann's experiment, described in **FIGURE 9-7**, took advantage of the shape of the chloroplast in a species of a filamentous green alga. Engelmann exposed these cells, each of which contained a long chloroplast that filled most of the cell, to a color spectrum produced by passing light through a prism. He hypothesized that if chlorophyll were indeed responsible for photosynthesis, the process would take place most rapidly in the areas where the chloroplast was illuminated by the colors most strongly absorbed by chlorophyll.

Yet how could photosynthesis be measured in those technologically unsophisticated days? Engelmann knew that photosynthesis produces oxygen and that certain motile bacteria are

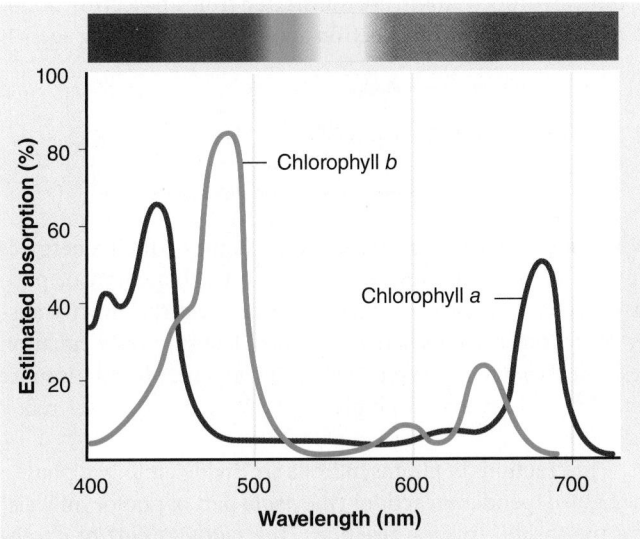

(a) Chlorophylls *a* and *b* absorb light mainly in the blue (422 to 492 nm) and red (647 to 760 nm) regions.

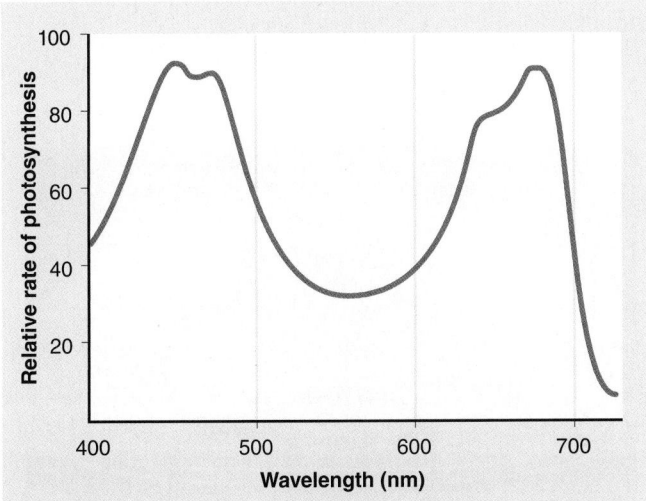

(b) The action spectrum of photosynthesis indicates the effectiveness of various wavelengths of light in powering photosynthesis. Many plant species have action spectra for photosynthesis that resemble the generalized action spectrum shown here.

Figure 9-6 A comparison of the absorption spectra for chlorophylls *a* and *b* with the action spectrum for photosynthesis
© Cengage Learning

Is a pigment in the chloroplast responsible for photosynthesis?

HYPOTHESIS: Engelmann hypothesized that chlorophyll was the main photosynthetic pigment. Accordingly, he predicted that he would observe differences in the amount of photosynthesis, as measured by the amount of oxygen produced, depending on the wavelengths of light used and that these wavelengths would be consistent with the known absorption spectrum of chlorophyll.

EXPERIMENT: Engelmann used the filamentous alga *Cladophora*, in which each cell has a long chloroplast, in his experiments. He used a prism to expose the cells to light that had been separated into various wavelengths. He estimated the formation of oxygen (which he knew was a product of photosynthesis) by exploiting that certain aerobic bacteria would be attracted to the oxygen. As a control (*not shown*), he also exposed the bacteria to the spectrum of light in the absence of the algal cells.

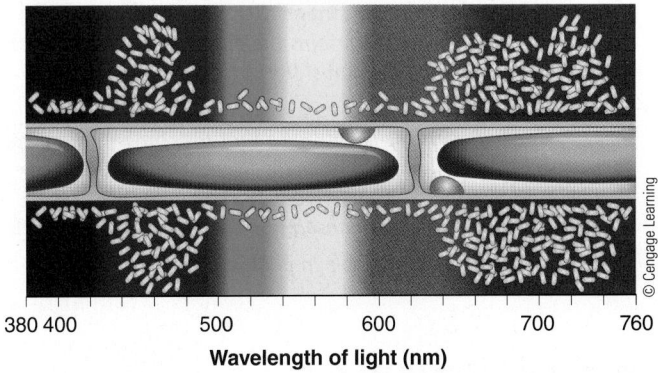

RESULTS AND CONCLUSION: Although the bacteria alone (*control*) showed no preference for any particular wavelength, large numbers were attracted to the photosynthesizing cells in red or blue light, wavelengths that are strongly absorbed by chlorophyll (see Fig. 9-6). Thus, Engelmann concluded that chlorophyll is responsible for photosynthesis.

SOURCE: Englemann's work is reviewed by R.P. Hangarter and H. Gest, "Pictorial Demonstrations of Photosynthesis," in J. Govindjee, T. Beatty, H. Gest, and J. F. Allen (eds.), *Discoveries in Photosynthesis* (The Netherlands: Springer, 2005).

Figure 9-7 *Animation* The first action spectrum of photosynthesis

PREDICT What would have been the distribution of bacteria if the main photosynthetic pigment had turned out to be one with maximal absorption in the yellow and green parts of the spectrum, and very low absorption in other wavelengths?

attracted to areas of high oxygen concentration. He determined the action spectrum of photosynthesis by observing that the bacteria swam toward the parts of the algal filaments in the blue and red regions of the spectrum. How did Engelmann know bacteria were not simply attracted to blue or red light? Engelmann exposed bacteria to the spectrum of visible light in the absence of algal cells as a control. The bacteria showed no preference for any particular wavelength of light. Because the action spectrum of photosynthesis closely matched the absorption spectrum of chlorophyll, Engelmann concluded that chlorophyll in the chloroplasts (and not another compound in another organelle) is responsible for photosynthesis. Numerous studies using sophisticated instruments have since confirmed Engelmann's conclusions.

If you examine Figure 9-6 closely, you will observe that the action spectrum of photosynthesis does not parallel the absorption spectrum of chlorophyll exactly. This difference occurs because accessory pigments, such as carotenoids, transfer some of the energy of excitation produced by green light to chlorophyll molecules. The presence of these accessory photosynthetic pigments can be demonstrated by chemical analysis of almost any leaf, although it is obvious in temperate climates when leaves change color in the fall. Toward the end of the growing season, chlorophyll breaks down (and its magnesium is stored in the permanent tissues of the tree), leaving orange and yellow accessory pigments in the leaves.

◉HECKPOINT 9.2

- **VISUALIZE** Draw a simple straight line to represent the chloroplast membrane that contains the photosynthetic pigments. What color should it be? Label the two compartments it separates.

- **CONNECT** What is the significance that the combined absorption spectra of chlorophylls a and b roughly match the action spectrum of photosynthesis? Would photosynthesis be more efficient if their individual absorption spectra coincided exactly?

- Does fluorescence play a role in photosynthesis?

9.3 OVERVIEW OF PHOTOSYNTHESIS

◉EARNING OBJECTIVES

4   Describe photosynthesis as a redox process.
5   Distinguish between the light-dependent reactions and carbon fixation reactions of photosynthesis.

During photosynthesis a cell uses light energy captured by chlorophyll to power the synthesis of carbohydrates. The overall reaction of photosynthesis can be summarized as

$$6\,CO_2 + 12\,H_2O \xrightarrow[\text{Chlorophyll}]{\text{Light energy}} C_6H_{12}O_6 + 6\,O_2 + 6\,H_2O$$

Carbon dioxide   Water   Glucose   Oxygen   Water

The equation is most accurately written in the form just given, with H_2O on both sides, because water is a reactant in some reactions and a product in others. Furthermore, all the oxygen produced comes from water, so 12 water molecules are required to produce 12 oxygen atoms. However, because there is no net yield of H_2O, we can simplify the summary equation of photosynthesis for purposes of discussion:

$$6\,CO_2 + 6\,H_2O \xrightarrow[\text{Chlorophyll}]{\text{Light}} C_6H_{12}O_6 + 6\,O_2$$

When you analyze this process, it appears that hydrogen atoms are transferred from H_2O to CO_2 to form carbohydrate, so you can recognize it as a redox reaction. Recall from Chapter 7 that in a redox reaction one or more electrons, usually as part of one or more hydrogen atoms, are transferred from an electron donor (a reducing agent) to an electron acceptor (an oxidizing agent).

$$\underbrace{6\,CO_2 + 6H_2O \xrightarrow[\text{Chlorophyll}]{\text{Light}} C_6H_{12}O_6}_{\text{Oxidation}} {}^{\text{Reduction}}\; + \; 6\,O_2$$

When the electrons are transferred, some of their energy is transferred as well. However, the summary equation of photosynthesis is somewhat misleading because no direct transfer of hydrogen atoms actually occurs. The summary equation describes *what* happens but not *how* it happens. The *how* is more complex and involves multiple steps, many of which are redox reactions.

The reactions of photosynthesis are divided into two phases: the light-dependent reactions (the *photo* part of photosynthesis) and the carbon fixation reactions (the *synthesis* part of photosynthesis). Each set of reactions occurs in a different part of the chloroplast: the light-dependent reactions in association with the thylakoids and the carbon fixation reactions in the stroma (FIG. 9-8).

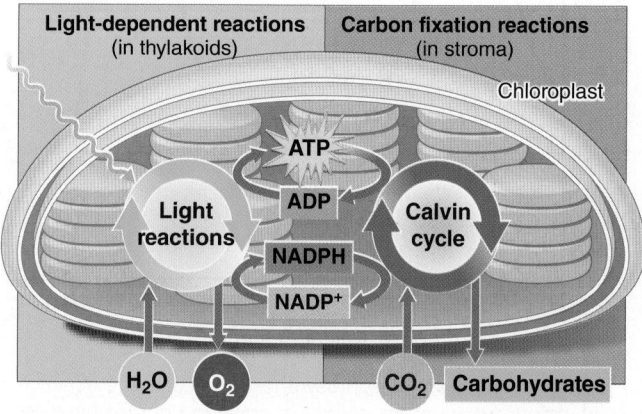

Figure 9-8 *Animation* **An overview of photosynthesis**

The light-dependent reactions in the thylakoids capture energy as ATP and NADPH, which power the carbon fixation reactions in the stroma.
© Cengage Learning

ATP and NADPH are the products of the light-dependent reactions: An overview

Light energy is converted to chemical energy in the **light-dependent reactions,** which are associated with the thylakoids. The light-dependent reactions begin as chlorophyll captures light energy, which causes one of its electrons to move to a higher energy state. The energized electron is transferred to an acceptor molecule and is replaced by an electron from H_2O. When that happens, H_2O is split and molecular oxygen is released (**FIG. 9-9**). Some energy of the energized electrons is used to phosphorylate **adenosine diphosphate (ADP),** forming **adenosine triphosphate (ATP).** In addition, the coenzyme **nicotinamide adenine dinucleotide phosphate (NADP$^+$)** becomes reduced, forming **NADPH.**[1] NADPH is an electron carrier similar to NADH, differing by the addition of a phosphate group. Unlike NADH, which is generally associated with catabolic pathways like aerobic cellular respiration, NADPH has the ability to provide high-energy electrons to power certain reactions in anabolic pathways, such as the carbon fixation reactions of photosynthesis. Thus, the products of the light-dependent reactions, ATP and NADPH, are both needed in the energy-requiring carbon fixation reactions.

Carbohydrates are produced during the carbon fixation reactions: An overview

The ATP and NADPH molecules produced during the light-dependent phase are suited for transferring chemical energy but not for long-term energy storage. For this reason, some of their energy is transferred to chemical bonds in carbohydrates, which can be produced in large quantities and stored for future use. Known as **carbon fixation,** these reactions "fix" carbon atoms from CO_2 to existing skeletons of organic molecules. Because the

Figure 9-9 Oxygen produced by photosynthesis
On sunny days, the oxygen released by aquatic plants is sometimes visible as bubbles in the water. This plant (*Elodea*) is actively carrying on photosynthesis.

carbon fixation reactions have no direct requirement for light, they were previously referred to as the "dark" reactions. However, they do not require darkness; in fact, many of the enzymes involved in carbon fixation are much more active in the light than in the dark. Furthermore, carbon fixation reactions depend on the products of the light-dependent reactions. Carbon fixation reactions take place in the stroma of the chloroplast.

Now that we have presented an overview of photosynthesis, let us examine the entire process more closely.

CHECKPOINT 9.3

- **CONNECT** *Which is more oxidized, oxygen that is part of a water molecule or molecular oxygen?*
- **CONNECT** *In what ways do the carbon fixation reactions depend on the light-dependent reactions?*

9.4 THE LIGHT-DEPENDENT REACTIONS

LEARNING OBJECTIVES

6 Describe the flow of electrons through photosystems I and II in the noncyclic electron transport pathway and the products produced. Contrast this flow with cyclic electron transport.

7 Explain how a proton (H$^+$) gradient is established across the thylakoid membrane and how this gradient functions in ATP synthesis.

In the light-dependent reactions, the radiant energy from sunlight phosphorylates ADP, producing ATP, and reduces NADP$^+$, forming NADPH. The light energy that chlorophyll captures is temporarily stored in these two compounds. The light-dependent reactions are summarized as follows:

$$12\ H_2O + 12\ NADP^+ + 18\ ADP + 18\ P_i \xrightarrow[\text{Chlorophyll}]{\text{Light}} 6\ O_2 + 12\ NADPH + 18\ ATP$$

Photosystems I and II each consist of a reaction center and multiple antenna complexes

The light-dependent reactions of photosynthesis begin when chlorophyll *a* and/or accessory pigments absorb light. According to the currently accepted model, chlorophylls *a* and *b* and accessory pigment molecules are organized with pigment-binding proteins in the thylakoid membrane into units called **antenna complexes.** The pigments and associated proteins are arranged as highly ordered groups that include about 250 chlorophyll molecules associated with specific enzymes and other proteins. Each antenna complex absorbs light energy and transfers it to its **reaction center,** which consists of chlorophyll molecules and

---

[1] Although the correct way to write the reduced form of NADP$^+$ is NADPH + H$^+$, for simplicity's sake we present the reduced form as NADPH throughout this book.

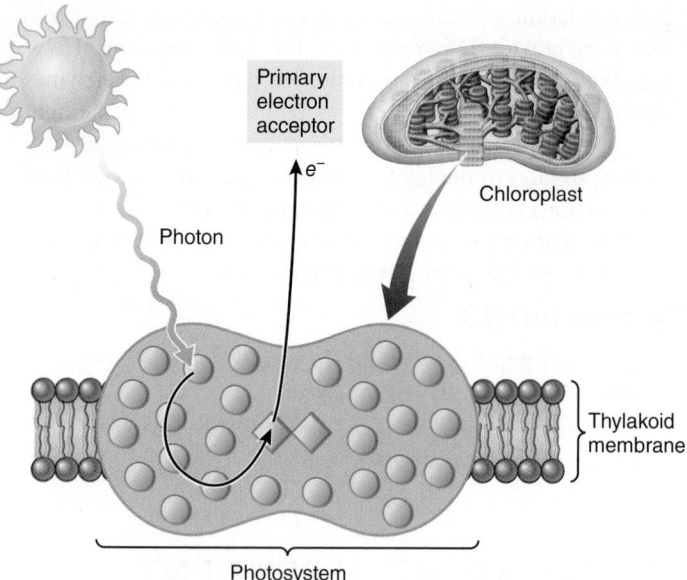

Figure 9-10 *Animation* **Schematic view of a photosystem**

Chlorophyll molecules (*green circles*) and accessory pigments (*not shown*) are arranged in light-harvesting arrays, or antenna complexes. A portion of one such complex within a photosystem is depicted. Each complex consists of several hundred pigment molecules, in association with special proteins (*not shown*). These proteins hold the pigments in a highly ordered spatial array such that when a molecule in an antenna complex absorbs a photon, energy derived from that photon is readily passed from one pigment molecule to another (*black arrow*). When this energy reaches one of the two chlorophyll molecules in the reaction center (*green diamonds*), an electron becomes energized and is accepted by a primary electron acceptor.

© Cengage Learning

proteins, including electron transfer components, that participate directly in photosynthesis (**FIG. 9-10**). Energy derived from light is converted to chemical energy in the reaction centers by a series of electron transfer reactions.

Two types of photosynthetic units, designated photosystem I and photosystem II, are involved in photosynthesis. Their reaction centers are distinguishable because they are associated with proteins in a way that causes a slight shift in their absorption spectra. Ordinary chlorophyll *a* has a strong absorption peak at about 660 nm. In contrast, the reaction center of **photosystem I** consists of a pair of chlorophyll *a* molecules with an absorption peak at 700 nm and is referred to as **P700.** The reaction center of **photosystem II** is made up of a pair of chlorophyll *a* molecules with an absorption peak of about 680 nm and is referred to as **P680.**

When a pigment molecule absorbs light energy, that energy is passed, through a process known as *resonance,* directly from one pigment molecule to another within the antenna complex until it reaches the reaction center. When the energy reaches a molecule of P700 (in a photosystem I reaction center) or P680 (in a photosystem II reaction center), an electron is then raised to a higher energy level. As we explain in the next section, this energized electron can be donated to an electron acceptor that becomes reduced in the process.

Noncyclic electron transport produces ATP and NADPH

Let us begin our discussion of **noncyclic electron transport** with the events associated with photosystem I (**FIG. 9-11**). A pigment molecule in an antenna complex associated with photosystem I absorbs a photon of light. The absorbed energy is transferred from one pigment molecule to another until it reaches the reaction center, where it excites an electron in a molecule of P700. This energized electron is transferred to a primary electron acceptor, a special molecule of chlorophyll *a,* which is the first of several electron acceptors in a series. The energized electron is passed along an **electron transport chain** from one electron acceptor to another, until it is passed to *ferredoxin,* an iron-containing protein. Ferredoxin transfers the electron to $NADP^+$ in the presence of the enzyme *ferredoxin–NADP$^+$ reductase.*

Although single electrons pass down the electron transport chain, two are needed to reduce $NADP^+$. When $NADP^+$ accepts two electrons, they unite with a proton (H^+); thus, the reduced form of $NADP^+$ is NADPH, which is released into the stroma. P700 becomes positively charged when it gives up an electron to the primary electron acceptor; the missing electron is replaced by one donated by photosystem II.

As in photosystem I, photosystem II becomes activated when a pigment molecule in an antenna complex absorbs a photon of light energy. The energy is transferred to the reaction center, where it causes an electron in a molecule of P680 to move to a higher energy level. This energized electron is accepted by a primary electron acceptor (a highly modified chlorophyll molecule known as *pheophytin*) and then passes along an electron transport chain until it is donated to P700 in photosystem I.

How is the electron that has been donated to the electron transport chain replaced? This process occurs through **photolysis** (light splitting) of water, which not only yields electrons but also is the source of almost all the oxygen in Earth's atmosphere. A molecule of P680 that has given up an energized electron to the primary electron acceptor is positively charged ($P680^+$). $P680^+$ is an oxidizing agent so strong that it pulls electrons away from an oxygen atom that is part of an H_2O molecule. In a reaction catalyzed by a unique, manganese-containing enzyme, water is broken into its components: two electrons, two protons, and oxygen. Each electron is donated to a P680 molecule, which then loses its positive charge; the protons are released into the thylakoid lumen. Because oxygen does not exist in atomic form, the oxygen produced by splitting one H_2O molecule is written $\frac{1}{2} O_2$. Two water molecules must be split to yield one oxygen molecule. The photolysis of water is a remarkable reaction, but its name is somewhat misleading because it implies that water is broken by light. Actually, light splits water indirectly by causing P680 to become oxidized.

Noncyclic electron transport is a continuous linear process In the presence of light, there is a continuous, one-way flow of electrons from the ultimate electron source, H_2O, to the terminal electron acceptor, $NADP^+$. Water undergoes enzymatically catalyzed photolysis to replace energized electrons donated to the electron transport chain by molecules of P680

Noncyclic electron transport converts light energy to chemical energy in ATP and NADPH.

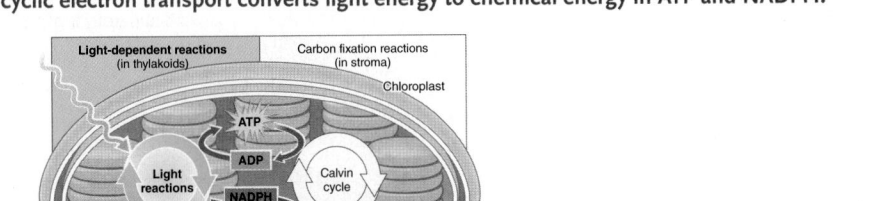

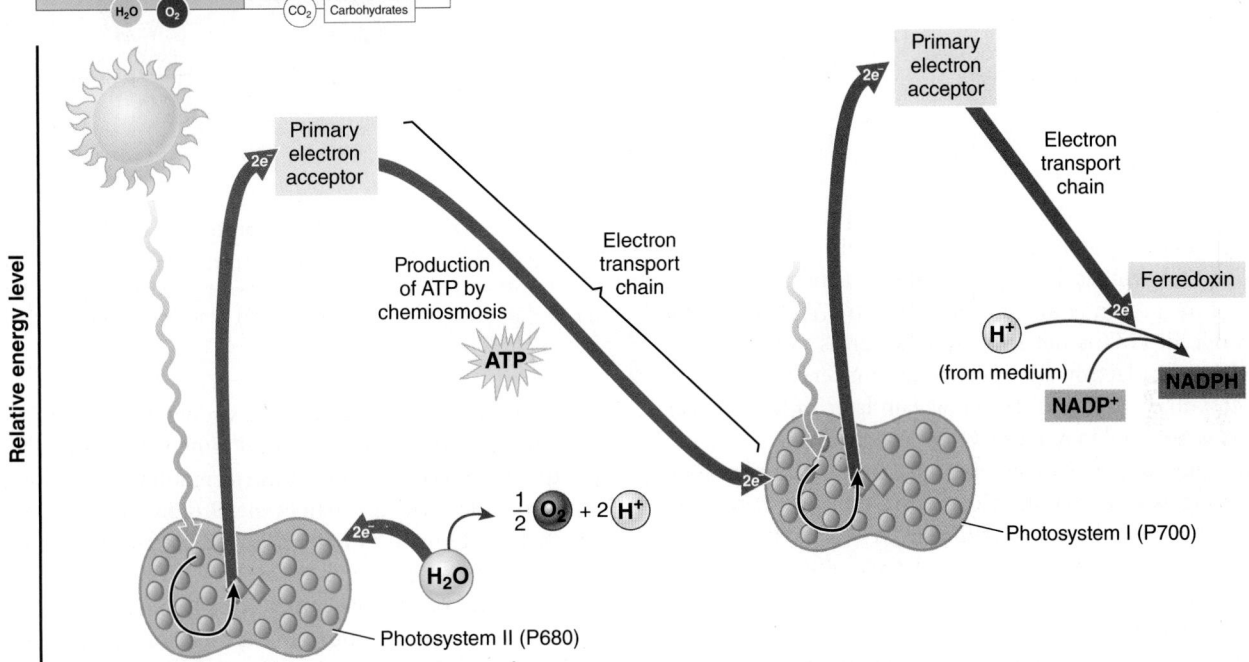

① Electrons are supplied to system from the splitting of H_2O by photosystem II, with release of O_2 as byproduct. When photosystem II is activated by absorbing photons, electrons are passed along the electron transport chain and are eventually donated to photosystem I.

② Electrons in photosystem I are "re-energized" by absorption of additional light energy and are passed to $NADP^+$, forming NADPH.

Figure 9-11 *Animation* **Noncyclic electron transport**

In noncyclic electron transport, the formation of ATP is coupled to one-way flow of energized electrons (*orange arrows*) from H_2O (*lower left*) to $NADP^+$ (*far right*). Single electrons actually pass down the electron transport chain; two are shown in this figure because two electrons are required to form one molecule of NADPH.

PREDICT What would happen if photosystem I were missing?

© Cengage Learning

in photosystem II. These electrons travel down the electron transport chain that connects photosystem II with photosystem I. Thus, they provide a continuous supply of replacements for energized electrons that have been given up by P700.

As electrons are transferred along the electron transport chain that connects photosystem II with photosystem I, they lose energy. Some of the energy released is used to pump protons across the thylakoid membrane, from the stroma to the thylakoid lumen, producing a proton gradient. The energy of this proton gradient is harnessed to produce ATP from ADP by *chemiosmosis*, which we discuss later in this chapter. ATP and NADPH, the products of the light-dependent reactions, are

released into the stroma, where both are required by the carbon fixation reactions.

Cyclic electron transport produces ATP but no NADPH

Only photosystem I is involved in **cyclic electron transport,** the simplest light-dependent reaction. The pathway is cyclic because energized electrons that originate from P700 at the reaction center eventually return to P700. In the presence of light, electrons flow continuously through an electron transport chain within the thylakoid membrane. As they pass from one

TABLE 9-1	A Comparison of Noncyclic and Cyclic Electron Transport	
	NONCYCLIC ELECTRON TRANSPORT	**CYCLIC ELECTRON TRANSPORT**
Electron source	H_2O	None—electrons cycle through the system
Oxygen released?	Yes (from H_2O)	No
Terminal electron acceptor	$NADP^+$	None—electrons cycle through the system
Form in which energy is temporarily captured	ATP (by chemiosmosis); NADPH	ATP (by chemiosmosis)
Photosystem(s) required	PS I (P700) and PS II (P680)	PS I (P700) only

© Cengage Learning

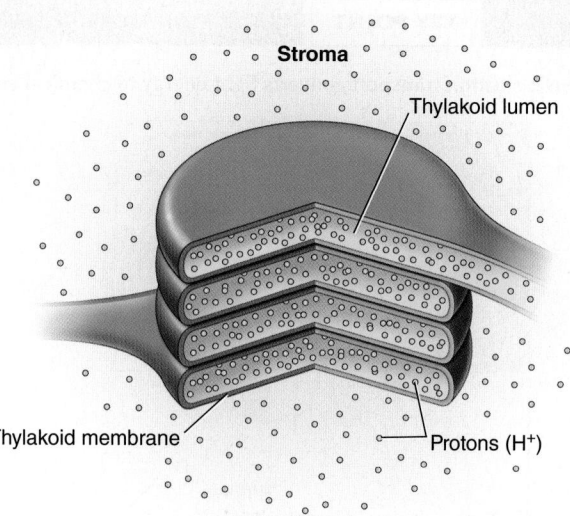

Figure 9-12 The accumulation of protons in the thylakoid lumen
As electrons move down the electron transport chain, protons (H^+) move from the stroma to the thylakoid lumen, creating a proton gradient. The greater concentration of H^+ in the thylakoid lumen lowers the pH.
© Cengage Learning

acceptor to another, the electrons lose energy, some of which is used to pump protons across the thylakoid membrane. An enzyme (ATP synthase) in the thylakoid membrane uses the energy of the proton gradient to manufacture ATP. NADPH is not produced, H_2O is not split, and oxygen is not generated. By itself, cyclic electron transport could not serve as the basis of photosynthesis because, as we explain later in this chapter, NADPH is required to reduce CO_2 to carbohydrate.

The significance of cyclic electron transport to photosynthesis in plants is unclear. Cyclic electron transport may occur in plant cells when there is too little $NADP^+$ to accept electrons from ferredoxin. There is evidence that cyclic electron flow may help maintain the optimal ratio of ATP to NADPH required for carbon fixation as well as provide extra ATP to power other ATP-requiring processes in chloroplasts. Biologists generally agree that ancient bacteria used this process to produce ATP from light energy. A reaction pathway analogous to cyclic electron transport in plants is present in some modern photosynthetic prokaryotes. Noncyclic and cyclic electron transport are compared in TABLE 9-1.

ATP synthesis occurs by chemiosmosis

Each member of the electron transport chain that links photosystem II to photosystem I can exist in an oxidized (lower-energy) form and a reduced (higher-energy) form. The electron accepted from P680 by the primary electron acceptor is highly energized; it is passed from one carrier to the next in a series of exergonic redox reactions, losing some of its energy at each step. Some of the energy given up by the electron is not lost by the system, however; it is used to provide energy for ATP synthesis. Because the synthesis of ATP (i.e., the phosphorylation of ADP) is coupled to the transport of electrons that have been energized by photons of light, the process is called **photophosphorylation.**

The chemiosmotic model explains the coupling of ATP synthesis and electron transport As discussed earlier, the pigments and electron acceptors of the light-dependent reactions are embedded in the thylakoid membrane. Energy released from electrons traveling through the chain of acceptors

is used to pump protons from the stroma, across the thylakoid membrane, and into the thylakoid lumen (FIG. 9-12). Thus, the pumping of protons results in the formation of a proton gradient across the thylakoid membrane. Protons also accumulate in the thylakoid lumen as water is split during noncyclic electron transport. Because protons are actually hydrogen ions (H^+), the accumulation of protons causes the pH of the thylakoid interior to fall to a pH of about 5 in the thylakoid lumen, compared with a pH of about 8 in the stroma. This difference of about 3 pH units across the thylakoid membrane means that there is an approximately thousand-fold difference in hydrogen ion concentration.

The proton gradient has a great deal of free energy because of its state of low entropy. How does the chloroplast convert that energy to a more useful form? According to the general principles of diffusion, the concentrated protons inside the thylakoid might be expected to diffuse out readily. However, they are prevented from doing so because the thylakoid membrane is impermeable to H^+ except through certain channels formed by the enzyme **ATP synthase.** This enzyme, a transmembrane protein also found in mitochondria, forms complexes so large they can be seen in electron micrographs (see Fig. 8-11b). ATP synthase complexes project into the stroma. As the protons diffuse through an ATP synthase complex, free energy decreases as a consequence of an increase in entropy. Each ATP synthase complex couples this exergonic process of diffusion down a concentration gradient to the endergonic process of phosphorylation of ADP to form ATP, which is released into the stroma (FIG. 9-13). The movement of protons through ATP synthase is thought to induce changes in the conformation of the enzyme that are necessary for the synthesis of ATP. It is estimated that for every four protons that move through ATP synthase, one ATP molecule is synthesized.

Electron carriers associated with the thylakoid membrane transfer energized electrons from water to NADP$^+$, forming NADPH. ATP is generated by chemiosmosis.

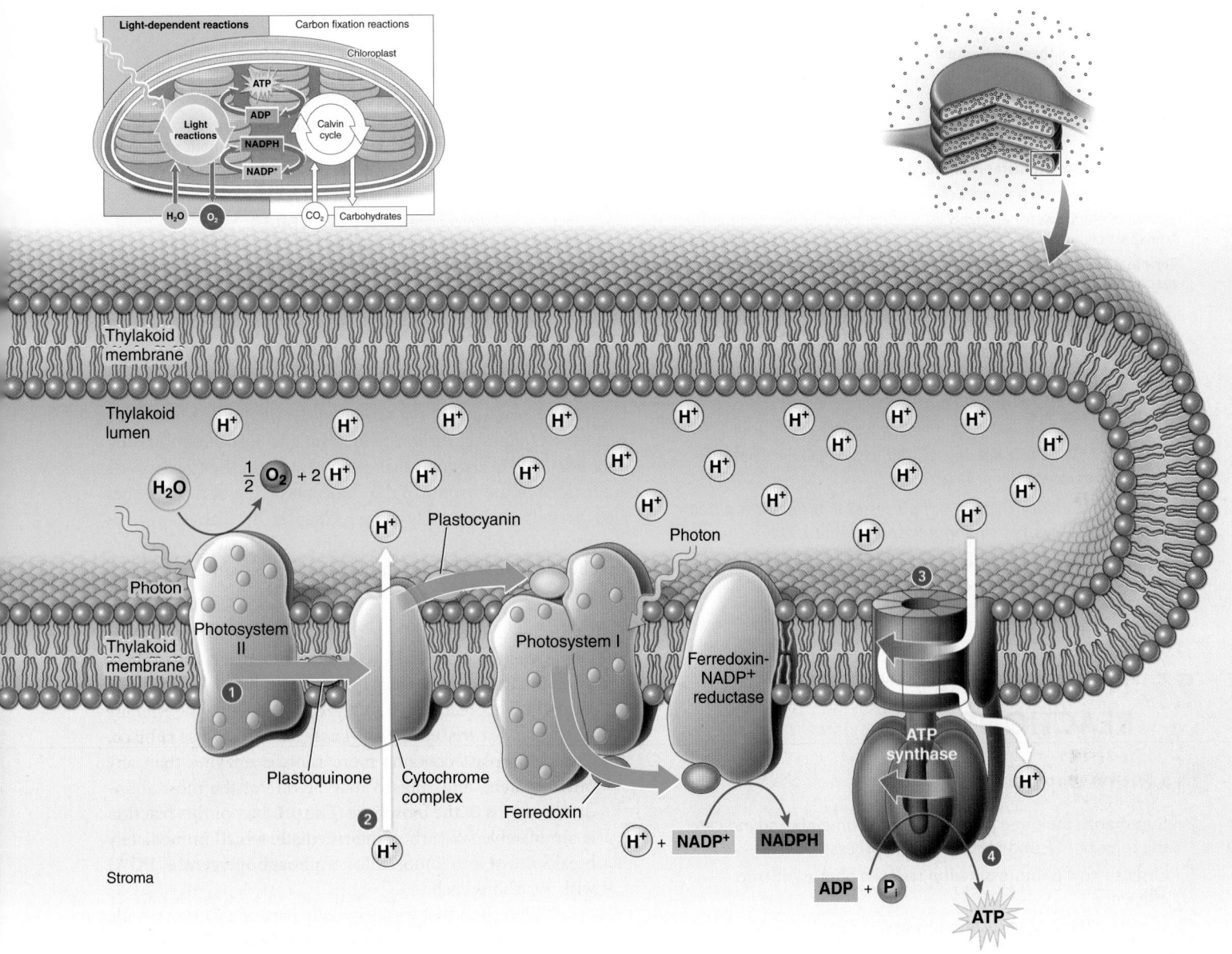

① Blue arrows indicate pathway of electrons along electron transport chain in thylakoid membrane. Electron carriers within membrane become alternately reduced and oxidized as they accept and donate electrons.

② Energy released during electron transport is used to transport H$^+$ from the stroma to the thylakoid lumen, where a high concentration of H$^+$ accumulates.

③ H$^+$ are prevented from diffusing back into stroma except through special channels in ATP synthase in the thylakoid membrane.

④ H$^+$ flows through ATP synthase, generating ATP.

Figure 9-13 *Animation* **A detailed look at electron transport and chemiosmosis**

PREDICT What would be the result if the cytochrome complex were able to transport electrons to photosystem I but an inhibitor prevented it from pumping protons across the thylakoid membrane?
© Cengage Learning

The mechanism by which the phosphorylation of ADP is coupled to diffusion down a proton gradient is called **chemiosmosis.** As the essential connection between the electron transport chain and the phosphorylation of ADP, chemiosmosis is a basic mechanism of energy coupling in cells. You may recall from Chapter 8 that chemiosmosis also occurs in aerobic respiration (see TABLE 9-2).

TABLE 9-2 A Comparison of Photosynthesis and Aerobic Respiration

	PHOTOSYNTHESIS	AEROBIC RESPIRATION
Type of metabolic reaction	Anabolism	Catabolism
Raw materials	CO_2, H_2O	$C_6H_{12}O_6$, O_2
End products	$C_6H_{12}O_6$, O_2	CO_2, H_2O
Which cells have these processes?	Cells that contain chlorophyll (certain cells of plants, algae, and some prokaryotes)	Every actively metabolizing cell has aerobic respiration or some other energy-releasing pathway
Sites involved (in eukaryotic cells)	Chloroplasts	Cytosol (glycolysis); mitochondria
ATP production	By photophosphorylation (a chemiosmotic process)	By substrate-level phosphorylation and by oxidative phosphorylation (a chemiosmotic process)
Principal electron transfer compound	$NADP^+$ is reduced to form NADPH*	NAD^+ is reduced to form NADH*
Location of electron transport chain	Thylakoid membrane	Mitochondrial inner membrane (cristae)
Source of electrons for electron transport chain	In noncyclic electron transport: H_2O (undergoes photolysis to yield electrons, protons, and oxygen)	Immediate source: NADH, $FADH_2$ Ultimate source: glucose or other carbohydrate
Terminal electron acceptor for electron transport chain	In noncyclic electron transport: $NADP^+$ (becomes reduced to form NADPH)	O_2 (becomes reduced to form H_2O)

*NADPH and NADH are very similar hydrogen (i.e., electron) carriers, differing only in a single phosphate group. However, NADPH generally works with enzymes in anabolic pathways, such as photosynthesis. NADH is associated with catabolic pathways, such as cellular respiration.

© Cengage Learning

CHECKPOINT 9.4

- **CONNECT** *What role does molecular oxygen play in photosynthesis? in aerobic cellular respiration?*
- **CONNECT** *Describe the series of processes that link light absorption with ATP synthesis (photophosphorylation).*
- **PREDICT** *Can cyclic electron transport alone support photosynthesis? Explain your answer.*

9.5 THE CARBON FIXATION REACTIONS

LEARNING OBJECTIVES

8 Summarize the three phases of the Calvin cycle and indicate the roles of ATP and NADPH in the process.

9 Discuss how photorespiration reduces photosynthetic efficiency.

10 Compare the C_4 and CAM pathways.

In carbon fixation, the energy of ATP and NADPH is used in the formation of organic molecules from CO_2. The carbon fixation reactions may be summarized as follows:

$$12 \text{ NADPH} + 18 \text{ ATP} + 6 \text{ CO}_2 \longrightarrow$$
$$C_6H_{12}O_6 + 12 \text{ NADP}^+ + 18 \text{ ADP} + 18 \text{ P}_i + 6 \text{ H}_2O$$

Most plants use the Calvin cycle to fix carbon

Carbon fixation occurs in the stroma through a sequence of 13 reactions known as the **Calvin cycle.** During the 1950s, University of California researchers Melvin Calvin, Andrew Benson, and others elucidated the details of this cycle. Calvin was awarded a Nobel Prize in Chemistry in 1961.

The 13 reactions of the Calvin cycle are divided into three phases: CO_2 uptake, carbon reduction, and RuBP regeneration (**FIG. 9-14**). All 13 enzymes that catalyze steps in the Calvin cycle are located in the stroma of the chloroplast. Ten of the enzymes also participate in glycolysis (see Chapter 8). These enzymes catalyze reversible reactions, degrading carbohydrate molecules in cellular respiration and synthesizing carbohydrate molecules in photosynthesis.

1. *CO_2 uptake.* The first phase of the Calvin cycle consists of a single reaction in which a molecule of CO_2 reacts with a phosphorylated five-carbon compound, **ribulose bisphosphate (RuBP).** This reaction is catalyzed by the enzyme *ribulose bisphosphate carboxylase/oxygenase,* also known as **rubisco.** The chloroplast contains more rubisco enzyme than any other protein, and rubisco may be one of the most abundant proteins in the biosphere. The product of this reaction is an unstable six-carbon intermediate, which immediately breaks down into 2 molecules of **phosphoglycerate (PGA)** with 3 carbons each.

   The carbon that was originally part of a CO_2 molecule is now part of a carbon skeleton; the carbon has been "fixed." The Calvin cycle is also known as the **C_3 pathway** because the product of the initial carbon fixation reaction is a three-carbon compound. Plants that initially fix carbon in this way are called **C_3 plants.**

2. *Carbon reduction.* The second phase of the Calvin cycle consists of two steps in which the energy and reducing power from ATP and NADPH (both produced in the light-dependent reactions) are used to convert the PGA molecules to **glyceraldehyde-3-phosphate (G3P).** As shown in Figure 9-14, for every 6 carbons that enter the cycle as CO_2, 6 carbons can leave the system as 2 molecules of G3P, to be used in carbohydrate synthesis. Each of these three-carbon molecules of G3P is essentially half a hexose (six-carbon sugar) molecule. (In fact, you may recall that G3P is a key intermediate in the splitting of sugar in glycolysis; see Figs. 8-3 and 8-4.)

ATP and NADPH provide the energy that drives carbon fixation in the Calvin cycle.

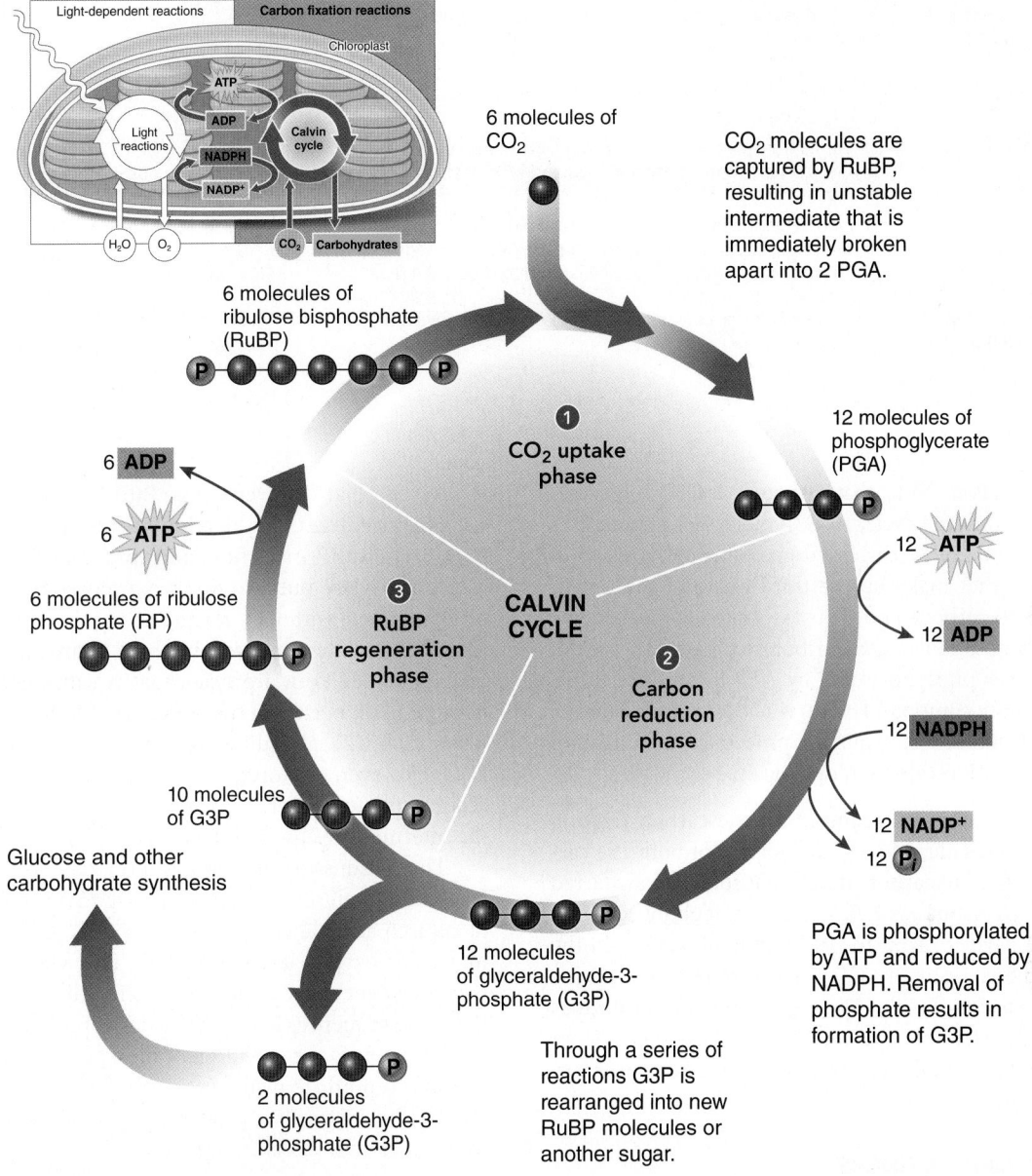

Figure 9-14 *Animation* **A detailed look at the Calvin cycle**

❶ This diagram, in which carbon atoms are black spheres, shows that 6 molecules of CO₂ must be "fixed" (incorporated into pre-existing carbon skeletons) in the CO₂ uptake phase to produce one molecule of a six-carbon sugar such as glucose. ❷ Glyceraldehyde-3-phosphate (G3P) is formed in the carbon reduction phase. For every glucose formed, 2 G3P molecules "leave" the cycle. ❸ Ribulose bisphosphate (RuBP) is regenerated, and a new cycle

can begin. Although these reactions do not require light directly, the energy that drives the Calvin cycle comes from ATP and NADPH, which are the products of the light-dependent reactions.

PREDICT What would happen if the 2:10 ratio of G3P molecules leaving the cycle to those remaining to regenerate RuBP were to become a 4:8 ratio?
© Cengage Learning

The reaction of 2 molecules of G3P is exergonic and leads to the formation of glucose or fructose. In some plants glucose and fructose are then joined to produce sucrose (common table sugar). (Sucrose can be harvested from sugarcane, sugar beets, and maple sap.) The plant cell also uses glucose to produce starch or cellulose.

| | TABLE 9-3 | Summary of Photosynthesis |

REACTION SERIES	SUMMARY OF PROCESS	NEEDED MATERIALS	END PRODUCTS
Light-dependent reactions (take place in thylakoid membranes)	Energy from sunlight used to split water, manufacture ATP, and reduce $NADP^+$		
Photochemical reactions	Chlorophyll-activated; reaction center gives up photoexcited electron to electron acceptor	Light energy; pigments (chlorophyll)	Electrons
Electron transport	Electrons transported along chain of electron acceptors in thylakoid membranes; electrons reduce $NADP^+$; splitting of water provides some H^+ that accumulates inside thylakoid space	Electrons, $NADP^+$, H_2O, electron acceptors	NADPH, O_2
Chemiosmosis	H^+ permitted to diffuse across the thylakoid membrane down their gradient; they cross the membrane through special channels in ATP synthase complex; energy released is used to produce ATP	Proton gradient, $ADP + P_i$, ATP synthase	ATP
Carbon fixation reactions (take place in stroma)	Carbon fixation: carbon dioxide used to make carbohydrate	Ribulose bisphosphate, CO_2, ATP, NADPH, necessary enzymes	Carbohydrates, $ADP + P_i$, $NADP^+$

© Cengage Learning

3. *RuBP regeneration.* Note that although 2 G3P molecules are removed from the cycle, 10 G3P molecules remain; this represents a total of 30 carbon atoms. Through a series of ten reactions that make up the third phase of the Calvin cycle, these 30 carbons and their associated atoms become rearranged into 6 molecules of ribulose phosphate, each of which becomes phosphorylated by ATP to produce RuBP, the five-carbon compound with which the cycle started. These RuBP molecules begin the process of CO_2 fixation and eventual G3P production once again.

In summary, the inputs required for the carbon fixation reactions are 6 molecules of CO_2 (the source of both the carbons and the oxygens in carbohydrate), phosphates transferred from ATP, and electrons (as hydrogen) provided by NADPH (but ultimately derived from the photolysis of water). In the end, the 6 carbons from the CO_2 are accounted for by the harvest of a hexose molecule. The remaining G3P molecules are used to synthesize the RuBP molecules with which more CO_2 molecules may combine. **TABLE 9-3** provides a summary of photosynthesis.

Photorespiration reduces photosynthetic efficiency

Many C_3 plants, including certain agriculturally important crops such as soybeans, wheat, and potatoes, do not yield as much carbohydrate from photosynthesis as might be expected, especially during periods of very hot temperature in summer. This phenomenon is a consequence of trade-offs between the plant's need for CO_2 and its need to prevent water loss. Recall that most photosynthesis occurs in mesophyll cells inside the leaf and that the entry and exit of gases from the interior of the leaf are regulated by stomata, tiny pores concentrated on the underside of the leaf (see Fig. 9-4a). On hot, dry days, plants close their stomata to conserve water. Once the stomata close, photosynthesis rapidly uses up the CO_2 remaining in the leaf and produces O_2, which accumulates in the chloroplasts.

Recall that the enzyme RuBP carboxylase/oxygenase (rubisco) catalyzes CO_2 fixation in the Calvin cycle by attaching CO_2 to RuBP. As its full name implies, rubisco acts not only as a carboxylase but also as an oxygenase because high levels of O_2 compete with CO_2 for the active site of rubisco. Some of the intermediates involved in the Calvin cycle are degraded to CO_2 and H_2O in a process that is called **photorespiration** because (1) it occurs in the presence of light, and as in aerobic respiration, (2) it requires oxygen and (3) produces CO_2 and H_2O. However, photorespiration does not produce ATP, and it reduces photosynthetic efficiency because it removes some of the intermediates used in the Calvin cycle.

The reasons for photorespiration are incompletely understood, although scientists hypothesize that it reflects the origin of rubisco at an ancient time when CO_2 levels were high and molecular oxygen levels were low. This view is supported by recent evidence that some amino acid sequences in rubisco are similar to sequences in certain bacterial proteins that apparently evolved prior to the evolution of the Calvin cycle. Genetic engineering to produce plants with rubisco that has a much lower affinity for oxygen is a promising area of research to improve yields of certain valuable crop plants.

The initial carbon fixation step differs in C_4 plants and in CAM plants

Photorespiration is not the only problem faced by plants engaged in photosynthesis. Because CO_2 is not a very abundant gas (composing only 0.04% of the atmosphere), it is not easy for plants to obtain the CO_2 they need. As you have learned, when conditions are hot and dry, the stomata close to reduce the loss of water vapor, greatly diminishing the supply of CO_2. Ironically, CO_2 is potentially less available at the very times when maximum sunlight is available to power the light-dependent reactions.

Many plant species living in hot, dry environments have adaptations that facilitate carbon fixation. **C_4 plants** first fix CO_2 into a four-carbon compound, **oxaloacetate. CAM plants** initially fix carbon at night through the formation of oxaloacetate. These special pathways found in C_4 and CAM plants precede the Calvin cycle (C_3 pathway); they do not replace it.

The C_4 pathway efficiently fixes CO_2 at low concentrations

The **C_4 pathway,** in which CO_2 is fixed through the formation of oxaloacetate, occurs not only before the C_3 pathway but also in different cells. Leaf anatomy is usually distinctive in C_4 plants. The photosynthetic mesophyll cells are closely associated with prominent, chloroplast-containing **bundle sheath** cells, which tightly encircle the veins of the leaf (FIG. 9-15). The C_4 pathway occurs in the mesophyll cells, whereas the Calvin cycle takes place within the bundle sheath cells.

The key component of the C_4 pathway is a remarkable enzyme that has an extremely high affinity for CO_2, binding it effectively even at unusually low concentrations. This enzyme, **PEP carboxylase,** catalyzes the reaction by which CO_2 reacts with the three-carbon compound *phosphoenolpyruvate (PEP)*, forming oxaloacetate (FIG. 9-16).

In a step that requires NADPH, oxaloacetate is converted to some other four-carbon compound, usually malate. The malate then passes to chloroplasts within bundle sheath cells, where a different enzyme catalyzes the decarboxylation of malate to yield pyruvate (which has three carbons) and CO_2. NADPH is formed, replacing the one used earlier.

$$\text{malate} + \text{NADP}^+ \longrightarrow \text{pyruvate} + CO_2 + \text{NADPH}$$

The CO_2 released in the bundle sheath cell combines with ribulose bisphosphate in a reaction catalyzed by rubisco and goes through the Calvin cycle in the usual manner. The pyruvate formed in the decarboxylation reaction returns to the mesophyll cell, where it reacts with ATP to regenerate phosphoenolpyruvate.

Because the C_4 pathway captures CO_2 and provides it to the bundle sheath cells so efficiently, CO_2 concentration within the bundle sheath cells is about 10 to 60 times as great as its concentration in the mesophyll cells of plants having only the C_3 pathway. Photorespiration is negligible in C_4 plants such as crabgrass because the concentration of CO_2 in bundle sheath cells (where rubisco is present) is always high.

The combined C_3–C_4 pathway involves the expenditure of 30 ATPs per hexose rather than the 18 ATPs used by the C_3 pathway alone. The extra energy expense required to regenerate PEP from pyruvate is worthwhile at high light intensities because it ensures a high concentration of CO_2 in the bundle sheath cells and permits them to carry on photosynthesis at a rapid rate. At lower light intensities and temperatures, C_3 plants are favored. For example, winter rye, a C_3 plant, grows lavishly in cool weather, when crabgrass cannot because it requires more energy to fix CO_2.

CAM plants fix CO_2 at night

Plants living in dry, or *xeric,* conditions have a number of structural adaptations that enable them to survive. Many xeric plants have physiological adaptations as well, including a special carbon fixation pathway, the **crassulacean acid metabolism (CAM) pathway.** The name comes from the stonecrop plant family (the Crassulaceae), which uses the CAM pathway, although the pathway has evolved independently in some members of more than 25 other plant

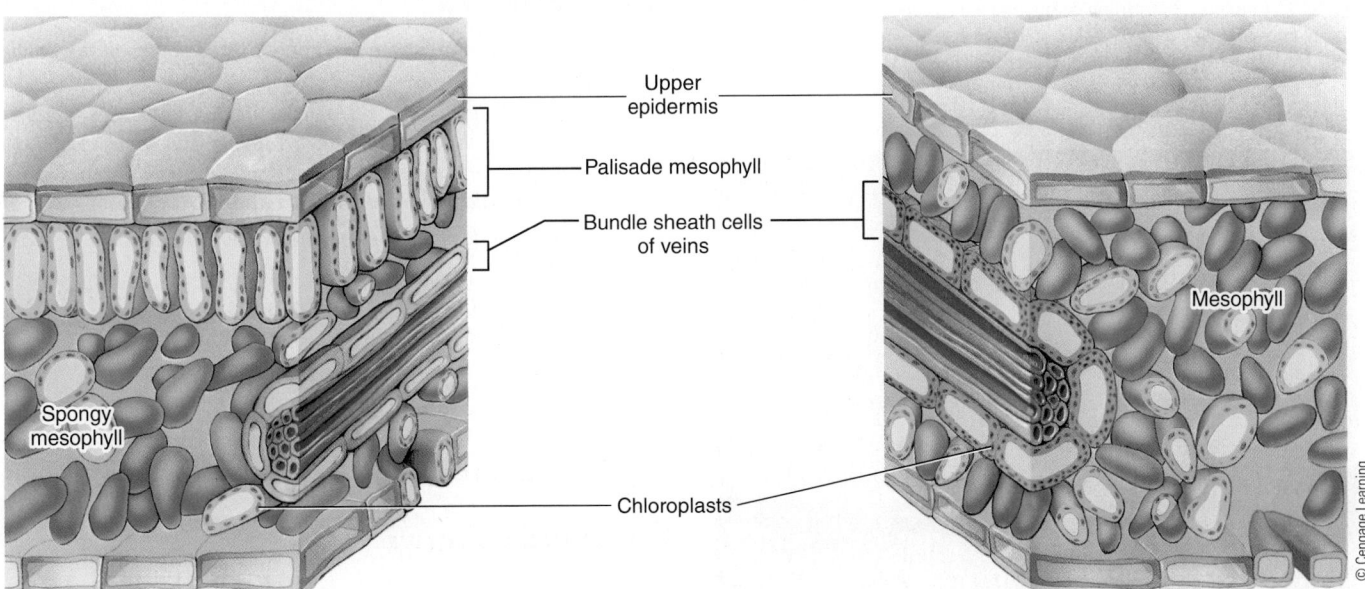

(a) In C_3 plants, the Calvin cycle takes place in the mesophyll cells and the bundle sheath cells are nonphotosynthetic.

(b) In C_4 plants, reactions that fix CO_2 into four-carbon compounds take place in the mesophyll cells. The four-carbon compounds are transferred from the mesophyll cells to the photosynthetic bundle sheath cells, where the Calvin cycle takes place.

Figure 9-15 **C_3 and C_4 plant structure compared**

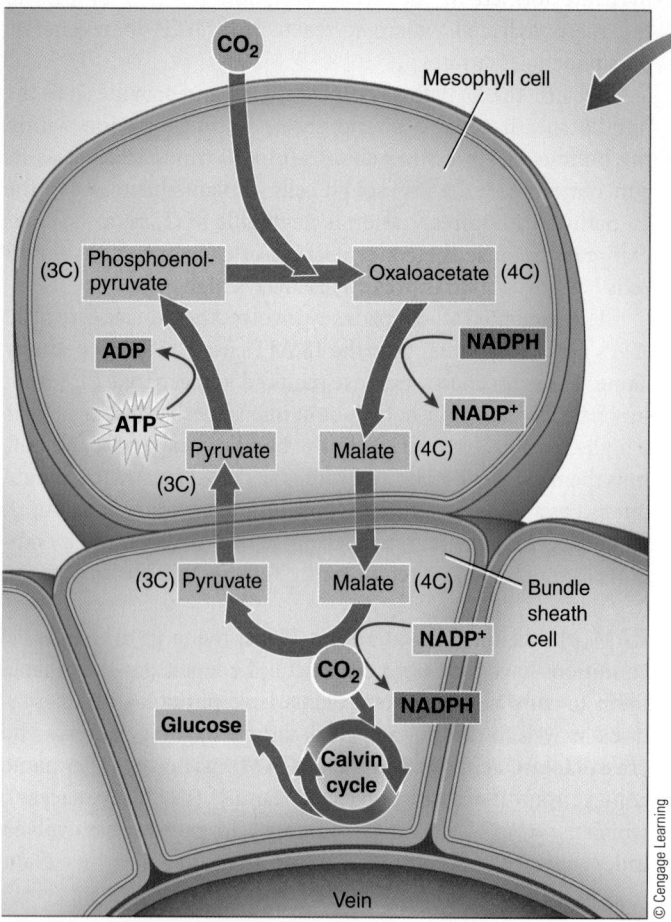

Figure 9-16 *Animation* **Summary of the C₄ pathway**

CO_2 combines with phosphoenolpyruvate (PEP) in the chloroplasts of mesophyll cells, forming a four-carbon compound that is converted to malate. Malate goes to the chloroplasts of bundle sheath cells, where it is decarboxylated. The CO_2 released in the bundle sheath cell is used to make carbohydrate by way of the Calvin cycle.

Figure 9-17 *Animation* **A typical CAM plant**

Prickly pear cactus (*Opuntia*) is a CAM plant. The more than 200 species of *Opuntia* living today originated in various xeric habitats in North and South America.

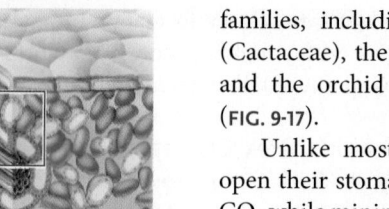

families, including the cactus family (Cactaceae), the lily family (Liliaceae), and the orchid family (Orchidaceae) (FIG. 9-17).

Unlike most plants, CAM plants open their stomata at night, admitting CO_2 while minimizing water loss. They use the enzyme PEP carboxylase to fix CO_2, forming oxaloacetate, which is converted to malate and stored in cell vacuoles. During the day, when stomata are closed and gas exchange cannot occur between the plant and the atmosphere, CO_2 is removed from malate by a decarboxylation reaction. Now the CO_2 is available within the leaf tissue to be fixed into sugar by the Calvin cycle (C₃ pathway).

The CAM pathway is very similar to the C₄ pathway but with important differences. C₄ plants initially fix CO_2 into four-carbon organic acids in mesophyll cells. The acids are later decarboxylated to produce CO_2, which is fixed by the C₃ pathway in the bundle sheath cells. In other words, the C₄ and C₃ pathways occur in *different locations* within the leaf of a C₄ plant. In CAM plants the initial fixation of CO_2 occurs at night. Decarboxylation of malate and subsequent production of sugar from CO_2 by the normal C₃ photosynthetic pathway occur during the day. In other words, the CAM and C₃ pathways occur at *different times* within the same cell of a CAM plant.

Although it does not promote rapid growth the way that the C₄ pathway does, the CAM pathway is a very successful adaptation to xeric conditions. CAM plants can exchange gases for photosynthesis and reduce water loss significantly. Plants with CAM photosynthesis survive in deserts where neither C₃ nor C₄ plants can.

CHECKPOINT 9.5

- Describe what happens in each of the three phases of the Calvin cycle.
- CONNECT A decrease in entropy occurs during the CO_2 uptake phase of the Calvin cycle, as freely moving CO_2 molecules become fixed into a carbon skeleton. How can this occur without direct participation by NADPH and/or ATP?
- In what ways does photorespiration differ from aerobic cellular respiration?
- CONNECT Do C₃, C₄, and CAM plants all have rubisco? PEP carboxylase?

9.6 METABOLIC DIVERSITY

LEARNING OBJECTIVE

11 Contrast photoautotrophs and chemoheterotrophs with respect to their energy and carbon sources.

Land plants, algae, and certain prokaryotes are known as **photoautotrophs.** They are **phototrophs** because they use light energy to make ATP and NADPH, which temporarily hold chemical energy but are unstable and cannot be stockpiled in the cell. They are **autotrophs** (from the Greek *auto*, which means "self," and *trophos*, which means "nourishing") that synthesize complex organic compounds from simpler, inorganic raw materials. The chemical energy of ATP and NADPH then drives carbon fixation, the anabolic pathway in which stable organic molecules are synthesized from CO_2 and water. These organic compounds are used not only as starting materials to synthesize all the other organic compounds the photosynthetic organism needs (such as complex carbohydrates, amino acids, and lipids) but also for energy storage. Glucose and other carbohydrates produced during photosynthesis are relatively reduced compounds that can be subsequently oxidized by aerobic respiration or by some other catabolic pathway (see Chapter 8).

In contrast, animals, fungi, and most prokaryotes are known as **chemoheterotrophs.** They are **chemotrophs** because they obtain energy from chemicals, typically by redox reactions (see Chapters 7 and 8). They are **heterotrophs** (from the Greek *heter*, which means "other," and *trophos*, which means "nourishing") because they cannot fix carbon; they use organic molecules produced by *other* organisms as the building blocks from which they synthesize the carbon compounds they need.

We are so familiar with plants as photoautotrophs and animals such as ourselves as chemoheterotrophs that we tend to think that all organisms should fit into these two "mainstream" categories. Two other types of nutrition are found in certain prokaryotes. A few bacteria, known as nonsulfur purple bacteria, are **photoheterotrophs,** able to use light energy but unable to carry out carbon fixation, so they must obtain carbon from organic compounds. Some other prokaryotes are **chemoautotrophs,** which obtain their energy from the oxidation of reduced inorganic molecules such as hydrogen sulfide (H_2S), nitrite (NO_2^+), or ammonia (NH_3). Some of this captured energy is subsequently used to carry out carbon fixation.

CHECKPOINT 9.6

- CONNECT *How does a green plant obtain energy? carbon? How does your body obtain these things?*

9.7 PHOTOSYNTHESIS IN PLANTS AND IN THE ENVIRONMENT

LEARNING OBJECTIVE

12 State the importance of photosynthesis both in a plant and to other organisms.

Although we characterize plants as photoautotrophs, not all plant cells carry out photosynthesis, and even cells with chloroplasts also possess mitochondria and carry out aerobic respiration. In fact, respiration using the organic molecules the plant has made for itself is the direct source of ATP needed for most plant metabolism.

Several mechanisms regulate the relative activities of photosynthesis and aerobic respiration in plants. Although the enzymes of the Calvin cycle do not require light to function, they are actually regulated by light. As a consequence of the light-requiring reactions, the stroma becomes more basic (approximately pH 8), activating rubisco and other Calvin cycle enzymes. In contrast, light tends to inhibit the enzymes of glycolysis in the cytosol. Hence, photosynthesis, not aerobic respiration, is favored in the light. When light is very dim, at a point known as the *light compensation point,* photosynthesis still occurs, but it is not evident because the rate of CO_2 fixation by photosynthesis is equal to the rate of CO_2 release through aerobic respiration. On the other hand, when light is very bright, photorespiration can significantly diminish photosynthetic yields.

As we have seen in this chapter, the reactions of the Calvin cycle provide a net yield of the three-carbon phosphorylated sugar, G3P. What are the various fates of G3P in the plant? Consider a leaf cell actively conducting photosynthesis. A series of enzymes may convert some of the G3P to glucose and then to starch. This starch is stored in starch granules that form inside chloroplasts. It has been recently shown that when this starch is broken down, the disaccharide maltose is typically formed (see Fig. 3-8a). Maltose is transported out of the chloroplast and then cleaved in the cytosol, providing glucose for aerobic respiration. Not all G3P ends up as carbohydrate; some is ultimately converted to amino acids, fatty acids, and other organic molecules needed by the photosynthetic cell.

Some of the G3P is exported to the cytosol, where enzymes convert it to the disaccharide sucrose (see Fig. 3-8b). Sucrose is then actively transported out of the cell, moves through the vascular system of the plant (see Chapter 35 for a discussion of plant transport), and is actively transported into the various cells. Sucrose can be broken down into glucose and fructose, which are used in aerobic respiration or as starting points for the synthesis of the various organic molecules the cells need, such as amino acids, lipids, and carbohydrates. Important carbohydrates include cellulose for cell walls (see Fig. 3-10) and starch, particularly in starch-storing structures such as roots (see Fig. 3-9a) and developing seeds and tubers (such as potatoes).

The benefits of photosynthesis in the environment are staggering. Of course, by fixing carbon, photoautotrophs are the ultimate source of virtually all organic molecules used as energy and carbon sources by chemoheterotrophs such as ourselves (for an exception, see *Inquiring About: Life without the Sun* in Chapter 55). In carrying out carbon fixation, photoautotrophs remove CO_2 from the atmosphere, thereby slowing climate change (see Chapter 57). Also of prime importance is that photolysis of water by photosystem II releases the O_2 that all aerobic organisms require for aerobic respiration. Molecular oxygen is so reactive that it could not be maintained in the atmosphere if it were not constantly replenished in this way.

As discussed in Chapter 21, the evolution of oxygen-producing photosynthesis was a critical event in the history of life on Earth. It not only permitted the evolution of aerobic organisms, but it also made terrestrial life possible because in the stratosphere O_2 is converted to ozone (O_3), which shields the planet from damaging ultraviolet light.

CHECKPOINT 9.7

- **CONNECT** *How does a root cell obtain energy? organic molecules?*
- **CONNECT** *What is the source of molecular oxygen in Earth's atmosphere?*

SUMMARY: FOCUS ON LEARNING OBJECTIVES

9.1 Light and Photosynthesis *(page 186)*

1 Describe the physical properties of light and explain the relationship between a wavelength of light and its energy.

- Light consists of particles called **photons** that move as waves.
- Photons with shorter **wavelengths** have more energy than those with longer wavelengths.

9.2 Chloroplasts *(page 187)*

2 Diagram the internal structure of a chloroplast and explain how its components interact and facilitate the process of photosynthesis.

- In plants **photosynthesis** occurs in chloroplasts, which are located mainly within **mesophyll** cells inside the leaf.
- **Chloroplasts** are organelles enclosed by a double membrane; the inner membrane encloses the **stroma** in which membranous, saclike **thylakoids** are suspended. Thylakoids enclose the **thylakoid lumen.** Thylakoids arranged in stacks are called **grana.**
- **Chlorophyll *a*, chlorophyll *b*, carotenoids,** and other photosynthetic pigments are components of the thylakoid membranes of chloroplasts.

3 Describe what happens to an electron in a biological molecule such as chlorophyll when a photon of light energy is absorbed.

- Photons excite biological molecules such as chlorophyll and other photosynthetic pigments, causing one or more electrons to become energized. These energized electrons may be accepted by electron acceptor compounds.
- The combined **absorption spectra** of chlorophylls *a* and *b* are similar to the **action spectrum** for photosynthesis.

9.3 Overview of Photosynthesis *(page 190)*

4 Describe photosynthesis as a redox process.

- During photosynthesis, light energy is captured and converted to the chemical energy of carbohydrates; hydrogens from water are used to reduce carbon, and oxygen derived from water becomes oxidized, forming molecular oxygen.

5 Distinguish between the light-dependent reactions and carbon fixation reactions of photosynthesis.

- In the **light-dependent reactions,** electrons energized by light are used to generate **ATP** and **NADPH;** these compounds provide energy for the formation of carbohydrates during the **carbon fixation reactions.**

9.4 The Light-Dependent Reactions *(page 191)*

6 Describe the flow of electrons through photosystems I and II in the noncyclic electron transport pathway and the products produced. Contrast this flow with cyclic electron transport.

- **Photosystems I** and **II** are the two types of photosynthetic units involved in photosynthesis. Each photosystem includes chlorophyll molecules and accessory pigments organized with pigment-binding proteins into **antenna complexes.**
- Only a special pair of chlorophyll *a* molecules in the **reaction center** of an antenna complex give up energized electrons to a nearby electron acceptor. **P700** is in the reaction center for photosystem I; **P680** is in the reaction center for photosystem II.

- During the noncyclic light-dependent reactions, known as **noncyclic electron transport,** ATP and NADPH are formed.
- Electrons in photosystem I are energized by the absorption of light and passed through an **electron transport chain** to $NADP^+$, forming NADPH. Electrons given up by P700 in photosystem I are replaced by electrons from P680 in photosystem II.
- A series of redox reactions takes place as energized electrons are passed along the electron transport chain from photosystem II to photosystem I. Electrons given up by P680 in photosystem II are replaced by electrons made available by the **photolysis** of H_2O; oxygen is released in the process.
- During **cyclic electron transport,** electrons from photosystem I are eventually returned to photosystem I. ATP is produced by **chemiosmosis,** but no NADPH or oxygen is generated.

7 Explain how a proton (H^+) gradient is established across the thylakoid membrane and how this gradient functions in ATP synthesis.

- **Photophosphorylation** is the synthesis of ATP coupled to the transport of electrons energized by photons of light. Some of the energy of the electrons is used to pump protons across the thylakoid membrane, providing the energy to generate ATP by chemiosmosis.
- As protons diffuse through **ATP synthase,** an enzyme complex in the thylakoid membrane, ADP is phosphorylated to form ATP.

9.5 The Carbon Fixation Reactions *(page 196)*

8 Summarize the three phases of the Calvin cycle and indicate the roles of ATP and NADPH in the process.

- The carbon fixation reactions proceed by way of the **Calvin cycle,** also known as the **C_3 pathway.**
- In the CO_2 uptake phase of the Calvin cycle, CO_2 is combined with **ribulose bisphosphate (RuBP),** a five-carbon sugar, by the enzyme ribulose bisphosphate carboxylase/oxygenase, commonly known as **rubisco,** forming the three-carbon molecule **phosphoglycerate (PGA).**
- In the carbon reduction phase of the Calvin cycle, the energy and reducing power of ATP and NADPH are used to convert PGA molecules to **glyceraldehyde-3-phosphate (G3P).** For every 6 CO_2 molecules fixed, 12 molecules of G3P are produced, and 2 molecules of G3P leave the cycle to produce the equivalent of 1 molecule of glucose.
- In the RuBP regeneration phase of the Calvin cycle, the remaining G3P molecules are modified to regenerate RuBP.

9 Discuss how photorespiration reduces photosynthetic efficiency.

- In **photorespiration** C_3 plants consume oxygen and generate CO_2 by degrading Calvin cycle intermediates but do not produce ATP. Photorespiration is significant on bright, hot, dry days when plants close their stomata, conserving water but preventing the passage of CO_2 into the leaf.

10 Compare the C_4 and CAM pathways.

- In the **C_4 pathway,** the enzyme **PEP carboxylase** binds CO_2 effectively, even when CO_2 is at a low concentration. C_4

reactions take place within mesophyll cells. The CO_2 is fixed in oxaloacetate, which is then converted to malate. The malate moves into a **bundle sheath** cell, and CO_2 is removed from it. The released CO_2 then enters the Calvin cycle.

- The **crassulacean acid metabolism (CAM)** pathway is similar to the C_4 pathway. PEP carboxylase fixes carbon at night in the mesophyll cells, and the Calvin cycle occurs during the day in the same cells.

9.6 Metabolic Diversity (page 200)

11 Contrast photoautotrophs and chemoheterotrophs with respect to their energy and carbon sources.

- **Photoautotrophs** use light as an energy source and are able to incorporate atmospheric CO_2 into pre-existing carbon skeletons.
- **Chemoheterotrophs** obtain energy by oxidizing chemicals and obtain carbon as organic molecules from other organisms.

9.7 Photosynthesis in Plants and in the Environment (page 201)

12 State the importance of photosynthesis both in a plant and to other organisms.

- Photosynthesis is the ultimate source of all chemical energy and organic molecules available to photoautotrophs, such as plants, and to virtually all other organisms as well. It also constantly replenishes the supply of oxygen in the atmosphere, vital to all aerobic organisms.

Summary Reactions for Photosynthesis

The light-dependent reactions (noncyclic electron transport):

$$12\ H_2O + 12\ NADP^+ + 18\ ADP + 18\ P_i \xrightarrow[\text{Chlorophyll}]{\text{Light}}$$
$$6\ O_2 + 12\ NADPH + 18\ ATP$$

The carbon fixation reactions (Calvin cycle):

$$12\ NADPH + 18\ ATP + 6\ CO_2 \longrightarrow$$
$$C_6H_{12}O_6 + 12\ NADP^+ + 18\ ADP + 18\ P_i + 6\ H_2O$$

By canceling the common items on opposite sides of the arrows in these two coupled equations, we obtain the simplified overall equation for photosynthesis:

$$6\ CO_2 + 12\ H_2O \xrightarrow[\text{Chlorophyll}]{\text{Light energy}} C_6H_{12}O_6 + 6\ O_2 + 6\ H_2O$$

Carbon dioxide   Water   Glucose   Oxygen   Water

TEST YOUR UNDERSTANDING

Know and Comprehend

1. Where is chlorophyll located in the chloroplast? (a) thylakoid membranes (b) stroma (c) matrix (d) thylakoid lumen (e) between the inner and outer membranes

2. In photolysis some of the energy captured by chlorophyll is used to split (a) CO_2 (b) ATP (c) NADPH (d) H_2O (e) both b and c

3. In plants, the final electron acceptor in noncyclic electron transport is (a) $NADP^+$ (b) CO_2 (c) H_2O (d) O_2 (e) G3P

4. In ___, electrons that have been energized by light contribute their energy to add phosphate to ADP, producing ATP. (a) crassulacean acid metabolism (b) the Calvin cycle (c) photorespiration (d) C_4 pathways (e) photophosphorylation

5. The Calvin cycle begins when CO_2 reacts with (a) phosphoenolpyruvate (b) glyceraldehyde-3-phosphate (c) ribulose bisphosphate (d) oxaloacetate (e) phosphoglycerate

6. The enzyme directly responsible for almost all carbon fixation on Earth is (a) rubisco (b) PEP carboxylase (c) ATP synthase (d) phosphofructokinase (e) maltase

7. In C_4 plants C_4 and C_3 pathways occur at different ___, whereas in CAM plants CAM and C_3 pathways occur at different ___. (a) times of day; locations within the leaf (b) seasons; locations (c) locations; times of day (d) locations; seasons (e) times of day; seasons

8. An organism characterized as a photoautotroph obtains energy from ___ and carbon from ___. (a) light; organic molecules (b) light; CO_2 (c) organic molecules; organic molecules (d) organic molecules; CO_2 (e) O_2; CO_2

Apply and Analyze

9. **VISUALIZE** Draw a simple sketch illustrating a thylakoid membrane that is actively involved in chemiosmosis and label the two compartments it separates. Add the ATP synthase complex, indicate the proton gradient, and specify in which compartment ATP is synthesized.

10. **CONNECT** Compare the sketch you drew for question 9 (above) with the one you drew for Chapter 8, Question 15. In each sketch, are protons being pumped into the innermost compartment or out of it? Does it matter?

11. **CONNECT** Must all autotrophs use light energy? Explain.

12. **CONNECT** Only some plant cells have chloroplasts, but all actively metabolizing plant cells have mitochondria. Explain.

13. **CONNECT** High-energy electrons from glucose are used to drive ATP synthesis in aerobic respiration. How did these electrons become so energetic?

Evaluate and Synthesize

14. **PREDICT** What would life be like for photoautotrophs if there were no chemoheterotrophs? for chemoheterotrophs if there were no photoautotrophs?

15. **EVOLUTION LINK** Propose an explanation for bacteria, chloroplasts, and mitochondria all having ATP synthase complexes.

16. **INTERPRET DATA** The figure depicts the absorption spectrum of a plant pigment. What colors or wavelengths does it absorb? What is the color of this pigment?

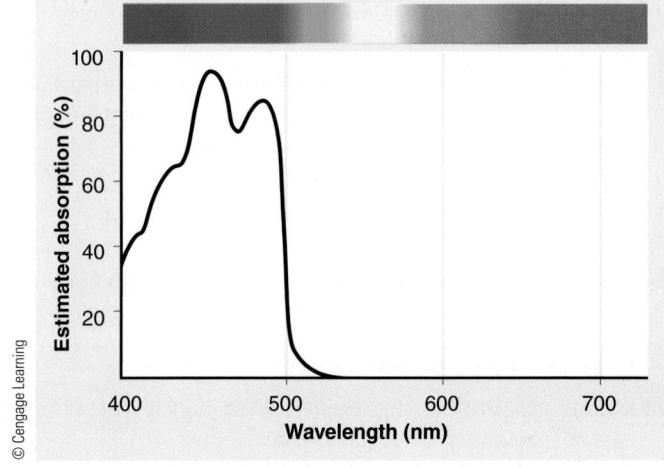

© Cengage Learning

17. **SCIENCE, TECHNOLOGY, AND SOCIETY** What strategies may be employed in the future to increase world food supply? Base your answer on your knowledge of photosynthesis and related processes.

aplia To access course materials, such as Aplia and other companion resources, please visit **www.cengagebrain.com**.

10 Chromosomes, Mitosis, and Meiosis

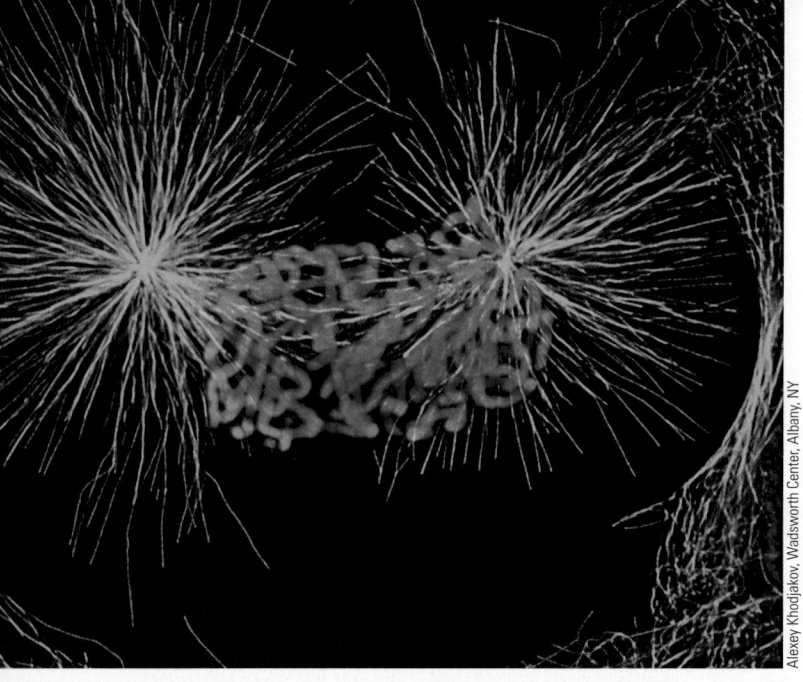

Fluorescence LM of a cultured newt lung cell in mitosis (early prometaphase). The nuclear envelope has broken down, and the microtubules of the mitotic spindle (*green*) now interact with the chromosomes (*blue*).

KEY CONCEPTS

10.1 In eukaryotic cells, DNA is wound around specific proteins to form chromatin, which in turn is folded and packaged to make individual chromosomes.

10.2 In mitosis, duplicated chromosomes separate (split apart) and are evenly distributed into two daughter nuclei. Mitosis is an important part of the cell cycle, which consists of the successive stages through which a cell passes.

10.3 An internal genetic program interacts with external signals to regulate the cell cycle.

10.4 Meiosis, which reduces the number of chromosome sets from diploid to haploid, is necessary to maintain the normal chromosome number when two cells join during sexual reproduction. Meiosis helps increase genetic variation among offspring.

10.5 Meiosis and gamete production precede fertilization in the life cycles of sexually reproducing organisms.

P re-existing cells divide to form new cells. This remarkable process enables an organism to grow, repair damaged parts, and reproduce. Cells serve as the essential link between generations. Even the simplest cell contains a large amount of precisely coded genetic information in the form of deoxyribonucleic acid (DNA). When a cell divides, the information contained in the DNA must be faithfully replicated and the copies then transmitted to each daughter cell through a precisely choreographed series of steps (see photograph).

DNA is a very long, thin molecule that could easily become tangled and broken, and a eukaryotic cell's nucleus contains a huge amount of DNA. In this chapter we consider how eukaryotes accommodate the genetic material by packaging each DNA molecule with proteins to form a structure called a *chromosome*, each of which contains large amounts of genetic information.

We then consider *mitosis*, the highly regimented process that ensures a parent cell transmits one copy of every chromosome to each of its two daughter cells. In this way, the chromosome number is preserved through successive mitotic divisions. Most *somatic cells* (body cells) of eukaryotes divide by mitosis. Mitosis is an active area of biological research, and for good reason: errors in mitosis can result in a host of disorders and diseases such as cancer, a disease condition in which cells divide at an inappropriate rate and become invasive. Thus, a clearer understanding of mitosis has the potential to improve our treatment of many diseases.

Finally, we discuss *meiosis*, a process that reduces the chromosome number by half. Sexual life cycles in eukaryotes require meiosis. Sexual reproduction involves the fusion of two sex cells, or *gametes*, to form a fertilized egg called a *zygote*. Meiosis makes it possible for each gamete to contain only half the number of chromosomes in the parent cell, thereby preventing the zygotes from having twice as many chromosomes as the parents.

10.1 EUKARYOTIC CHROMOSOMES

LEARNING OBJECTIVES

LEARNING OBJECTIVES

1 Discuss the significance of chromosomes in terms of their information content.

2 Explain how DNA is packed into chromosomes in eukaryotic cells.

The major carriers of genetic information in eukaryotes are the **chromosomes,** which lie within the cell nucleus. Although *chromosome* means "colored body," chromosomes are virtually colorless; the term refers to their ability to be stained by certain dyes. In the 1880s, light microscopes had been improved to the point that scientists such as German biologist Walther Fleming began to observe chromosomes during cell division. In 1903, American biologist Walter Sutton and German biologist Theodor Boveri noted independently that chromosomes were the physical carriers of genes, the genetic factors Gregor Mendel discovered in the 19th century (discussed in Chapter 11).

Chromosomes are made of **chromatin,** a material consisting of DNA and associated proteins. When a cell is not dividing, the chromosomes are present but in an extended, partially unraveled form. Chromatin consists of long, thin threads that are somewhat aggregated, giving them a granular appearance when viewed with the electron microscope (see Fig. 4-11a). During cell division, the chromatin fibers condense and the chromosomes become visible as distinct structures (**FIG. 10-1**).

DNA is organized into informational units called genes

An organism may have thousands of *genes.* For example, humans have more than 20,000 genes that code for proteins. The concept of the gene has changed considerably since the science of genetics began, but our traditional definitions have always centered on the gene as an informational unit. By providing information needed to carry out one or more specific cell functions, a gene affects some characteristic of the organism. For example, genes govern eye color in humans, wing length in flies, and seed color in peas. As you will learn in later chapters, these concepts are being expanded as scientists study the many ways information stored in DNA controls the workings of the cell.

DNA is packaged in a highly organized way in chromosomes

Prokaryotic and eukaryotic cells differ markedly in their DNA content as well as in the organization of DNA molecules. The bacterium *Escherichia coli* normally contains about 4×10^6 base pairs (almost 1.35 mm) of DNA in its single, circular DNA molecule. In fact, the total length of its DNA is about 1000 times as

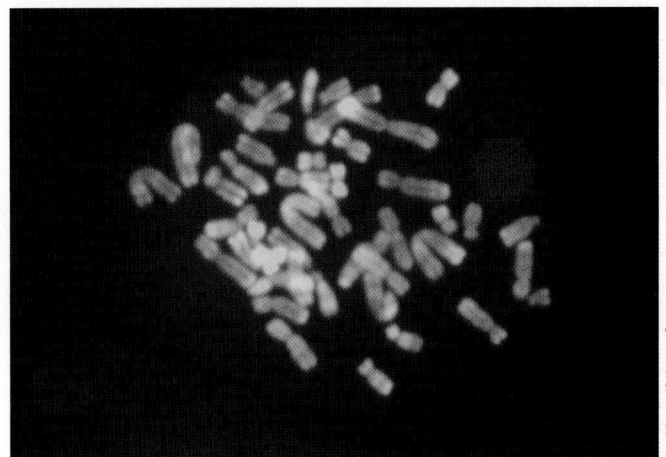

Figure 10-1 Chromosomes
Human chromosomes from an unidentified cell are shown in this fluorescence LM.

Jan Hinsch/Science Source

10 μm

long as the length of the cell itself. Therefore, the DNA molecule is, with the help of proteins, twisted and folded compactly to fit inside the bacterial cell (see Fig. 25-2).

A typical eukaryotic cell contains much more DNA than a bacterium does, and it is organized in the nucleus as multiple chromosomes that vary widely in size and number among different species. Although a human nucleus is about the size of a large bacterial cell, it contains more than 1000 times the amount of DNA found in *E. coli.* The DNA content of a human sperm cell is about 3×10^9 base pairs; stretched end to end, it would measure almost 1 m long. Remarkably, this DNA fits into a nucleus with a diameter of only 10 μm.

How does a eukaryotic cell pack its DNA into the chromosomes? Chromosome packaging is facilitated by certain proteins known as **histones.**[1] Histones have a positive charge because they have a high proportion of amino acids with basic side chains (see Chapter 3). The positively charged histones associate with DNA, which has a negative charge because of its phosphate groups, to form structures called **nucleosomes.** The fundamental unit of each nucleosome consists of a beadlike structure with 146 base pairs of DNA wrapped around a disc-shaped core of eight histone molecules (two each of four different histone types) (**FIG. 10-2**). Although the nucleosome was originally defined as a bead plus a DNA segment that links it to an adjacent bead, today the term more commonly refers only to the bead itself (i.e., the eight histones and the DNA wrapped around them).

Nucleosomes function like tiny spools, preventing DNA from becoming tangled. You can see the importance of this role in **FIGURE 10-3**, which illustrates the enormous length of DNA that unravels from a mouse chromosome after researchers have removed the histones. The role of histones is more than simply

---

[1] A few types of eukaryotic cells lack histones. Conversely, histones occur in one group of prokaryotes, the archaea (see Chapter 25).

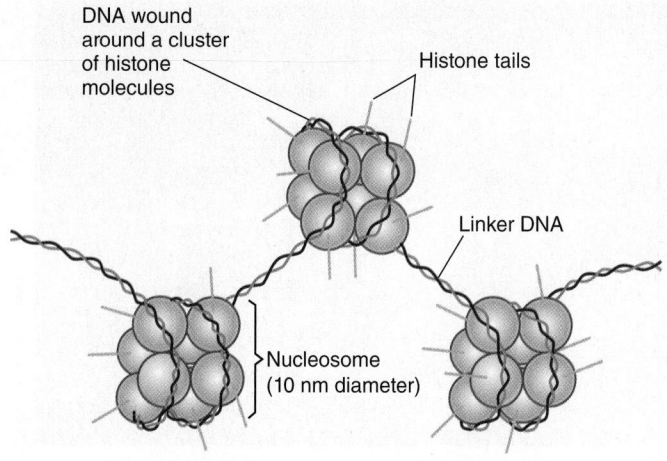

DNA wound around a cluster of histone molecules

Histone tails

Linker DNA

Nucleosome (10 nm diameter)

(a) A model for the structure of a nucleosome. Each nucleosome bead contains a set of eight histone molecules, forming a protein core around which the double-stranded DNA winds. The DNA surrounding the histones consists of 146 nucleotide pairs; another segment of DNA, about 60 nucleotide pairs long, links nucleosome beads.

Figure 10-2 Nucleosomes
© Cengage Learning

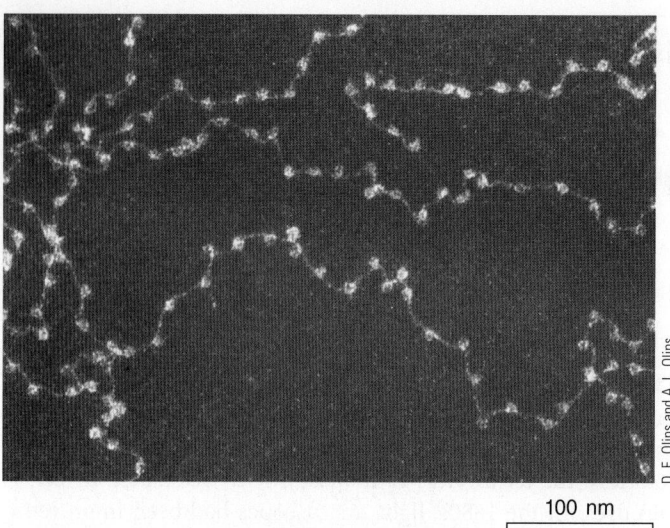

100 nm

(b) TEM of nucleosomes from the nucleus of a chicken cell. Normally, nucleosomes are packed more closely together, but the preparation procedure has spread them apart, revealing the DNA linkers.

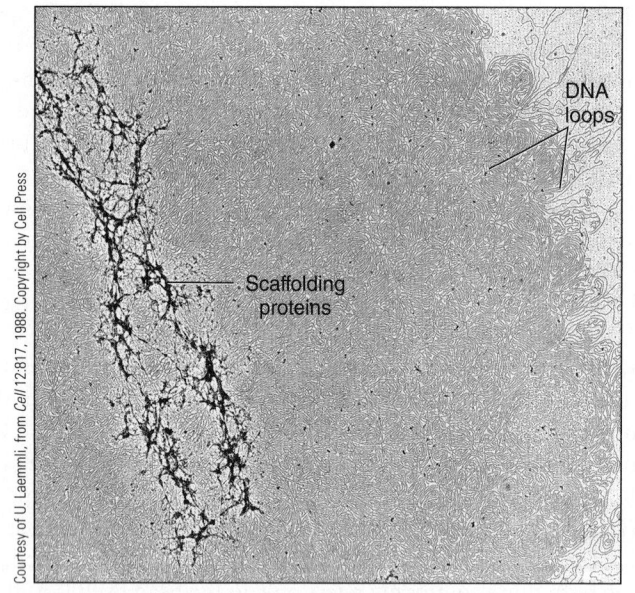

DNA loops

Scaffolding proteins

2 μm

Figure 10-3 TEM of a mouse chromosome depleted of histones
Notice how densely packed the DNA strands are, even though they have been released from the histone proteins that organize them into tightly coiled structures.

structural because their arrangement also affects the activity of the DNA with which they are associated. Histones are increasingly viewed as an important part of the regulation of gene expression, that is, whether genes are turned off or on. We discuss gene regulation by histones in Chapter 14.

The wrapping of DNA into nucleosomes represents the first level of chromosome structure. **FIGURE 10-4** shows the higher-order structures of chromatin leading to the formation of a condensed chromosome. The nucleosomes themselves are 10 nm in diameter. The packed nucleosome state occurs when a fifth type of histone, known as *histone H1*, associates with the linker DNA, packing adjacent nucleosomes together to form a compacted 30 nm chromatin fiber. In extended chromatin these fibers form large, coiled loops held together by **scaffolding proteins,** nonhistone proteins that help maintain chromosome structure. The loops then interact to form the condensed chromatin found in a chromosome. Cell biologists have identified a group of proteins, collectively called **condensin,** required for chromosome compaction. Condensin binds to DNA and wraps it into coiled loops that are compacted into a chromosome.

Chromosome number and informational content differ among species

Every individual of a given species has a characteristic number of chromosomes in the nuclei of its somatic (body) cells. However, it is not the number of chromosomes that makes each species unique but the information the genes specify. Most human somatic cells have exactly 46 chromosomes, but humans are not humans merely because we have 46 chromosomes. Some other species—the olive tree, for example—also have 46. Some

When a cell prepares to divide, its chromosomes become thicker and shorter as their long chromatin fibers are compacted.

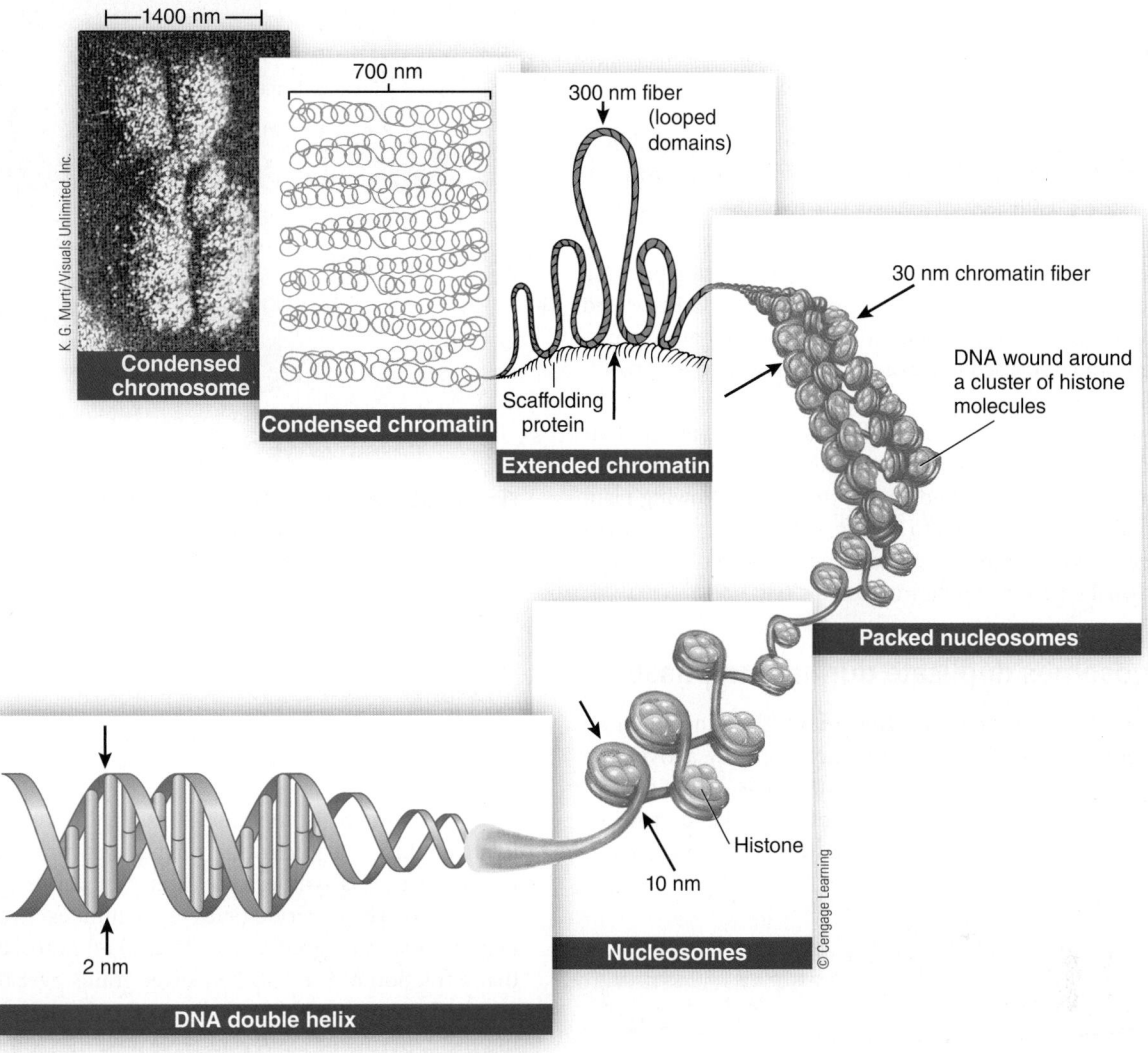

Figure 10-4 *Animation* **Organization of a eukaryotic chromosome**

This diagram shows how DNA is packaged into highly condensed metaphase chromosomes. First, DNA is wrapped around histone proteins to form nucleosomes. Then, the nucleosomes are compacted into chromatin fibers, which are coiled into looped domains. The looped domains are compacted, ultimately forming chromosomes.

VISUALIZE For which of the states of chromatin illustrated is histone H1 most responsible?

humans have an abnormal chromosome composition with more or fewer than 46 (see Fig. 16-5).

Other species have different chromosome numbers. A certain species of roundworm has only 2 chromosomes in each cell, whereas some crabs have as many as 200, and some ferns have more than 1000. Most animal and plant species have between 8 and 50 chromosomes per somatic cell. Quantities above and below these numbers are uncommon. The number of chromosomes a species has does not indicate the species' complexity or its status within a particular domain or kingdom.

CHECKPOINT 10.1

- *What are the informational units on chromosomes called? Of what do these informational units consist?*
- **CONNECT** *How is the large discrepancy between DNA length and nucleus size addressed in eukaryotic cells?*

10.2 THE CELL CYCLE AND MITOSIS

LEARNING OBJECTIVES

3 Identify the stages in the eukaryotic cell cycle and describe their principal events.

4 Describe the structure of a duplicated chromosome, including the sister chromatids, centromeres, and kinetochores.

5 Explain the significance of mitosis and describe the process.

When cells reach a certain size, they usually either stop growing or divide. Not all cells divide; some, such as skeletal muscle and red blood cells, do not normally divide once they are mature. Other cells undergo a sequence of activities required for growth and cell division.

The stages through which a cell passes from one cell division to the next are collectively referred to as the **cell cycle.** Timing of the cell cycle varies widely, but in actively growing plant and animal cells, it is about 8 to 20 hours. The cell cycle consists of two main phases, interphase and M phase, both of which can be distinguished under a light microscope (**FIG. 10-5**).

Chromosomes duplicate during interphase

Most of a cell's life is spent in **interphase,** the time when no cell division is occurring. A cell is metabolically active during interphase, synthesizing needed materials (proteins, lipids, and other biologically important molecules) and growing. Here is the sequence of interphase and M phase in the eukaryotic cell cycle:

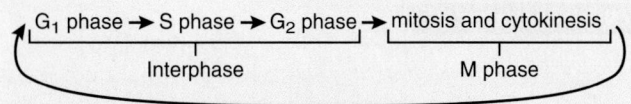

The time between the end of mitosis and the beginning of the S phase is termed the **G_1 phase** (G stands for *gap*, an interval during which no DNA synthesis occurs). Growth and normal metabolism take place during the G_1 phase, which is typically the longest phase. Cells that are not dividing usually become arrested in this part of the cell cycle and are said to be in a state called G_0. Toward the end of G_1, the enzymes required for DNA synthesis become more active. Synthesis of these enzymes, along with proteins needed to initiate cell division (discussed later in this chapter), enable the cell to enter the S phase.

During the **synthesis phase,** or **S phase,** DNA replicates and histone proteins are synthesized so that the cell can make duplicate copies of its chromosomes. How did researchers identify the S phase of the cell cycle? In the early 1950s, researchers demonstrated that cells preparing to divide duplicate their chromosomes at a relatively restricted time interval during interphase and not during early mitosis, as previously hypothesized. These

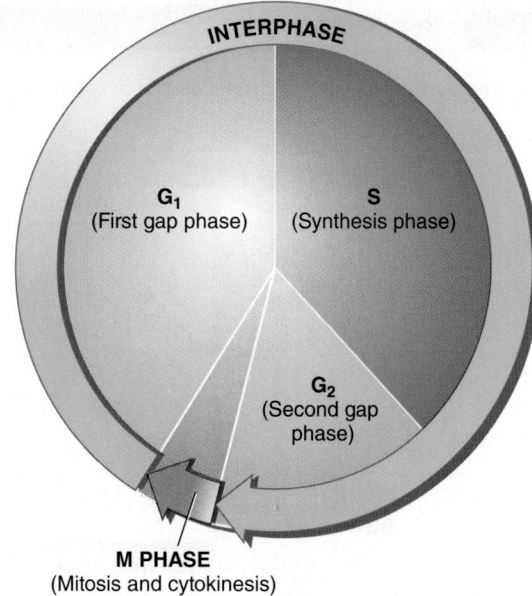

Figure 10-5 *Animation* **The eukaryotic cell cycle**

The cell cycle, a successive series of events in the life of a cell, includes interphase (G_1, S, and G_2) and M phase (mitosis and cytokinesis). Proportionate amounts of time spent at each stage vary among species, cell types, and growth conditions. If the cell cycle were a period of 12 hours, G_1 would be about 5 hours, S would be 4.5 hours, G_2 would be 2 hours, and M phase would be 30 minutes.
© Cengage Learning

investigators used isotopes, such as ^{3}H, to synthesize radioactive thymidine, a nucleotide that is incorporated specifically into DNA as it is synthesized. After radioactive thymidine was supplied for a brief period (such as 30 minutes) to actively growing cells, *autoradiography* (see Fig. 2-3) on exposed film showed that a fraction of the cells had silver grains over their chromosomes. The nuclei of these cells were radioactive because during the experiment DNA had replicated. DNA replication was not occurring in the cells that did not have radioactively labeled chromosomes. Researchers therefore inferred that the proportion of labeled cells out of the total number of cells provides a rough estimate of the length of the S phase relative to the rest of the cell cycle.

After it completes the S phase, the cell enters a second gap phase, the **G_2 phase.** At this time, increased protein synthesis occurs as the final steps in the cell's preparation for division take place. For many cells, the G_2 phase is short relative to the G_1 and S phases.

M phase involves two main processes, mitosis and cytokinesis. **Mitosis,** the nuclear division that produces two nuclei containing chromosomes identical to the parental nucleus, begins at the end of the G_2 phase. Cytokinesis, which generally begins before mitosis is complete, is the division of the cell cytoplasm to form two cells.

Mitosis is a continuous process, but for descriptive purposes, it is divided into five stages:

prophase → prometaphase → metaphase→ anaphase → telophase

Study **FIGURE 10-6** while you read the following descriptions of these stages as they occur in a typical plant or animal cell.

During prophase, duplicated chromosomes become visible with the microscope

The first stage of mitosis, **prophase,** begins with chromosome compaction, when the long chromatin fibers that make up the chromosomes begin a coiling process that makes them shorter and thicker. The compacted chromatin can then be distributed to the daughter cells with less likelihood of tangling.

When stained with certain dyes and viewed through the light microscope, chromosomes become visible as darkly stained bodies as prophase progresses. It is now apparent that each chromosome was duplicated during the preceding S phase and consists of a pair of **sister chromatids,** which contain identical, double-stranded DNA sequences. Each chromatid includes a constricted region called the **centromere.** Sister chromatids are tightly associated in the vicinity of their centromeres (**FIG. 10-7**). The chemical basis for this close association at the centromeres consists of precise DNA sequences that are tightly bound to specific proteins.

For example, sister chromatids are physically linked by a ring-shaped protein complex called **cohesin.** Cohesins extend along the length of the sister chromatid arms and are particularly concentrated at the centromere (**FIG. 10-8**). These cohesins, which hold the duplicated chromosomes together from their synthesis in S phase onward, help ensure accurate chromosome separation during mitosis.

Attached to each centromere is a **kinetochore,** a multiprotein complex to which **microtubules** can bind. These microtubules function in chromosome distribution during mitosis, in which one copy of each chromosome is delivered to each daughter cell.

A dividing cell can be described as a globe, with an equator that determines the midplane (equatorial plane) and two opposite poles. This terminology is used for all cells regardless of their actual shape. Microtubules radiate from each pole, and some of these protein fibers elongate toward the chromosomes, forming the **mitotic spindle,** a structure that separates the duplicated chromosomes during anaphase (**FIG. 10-9**). The *minus* ends of these microtubules are at the poles, and the *plus* ends extend to the cell's midplane. It might be helpful to review Figure 4-23a, which shows the organization of microtubules as linear polymers of the protein *tubulin*. The organization and function of the spindle require the presence of motor proteins and a variety of signaling molecules.

Animal cells differ from plant cells in the details of mitotic spindle formation. In both types of dividing cells, each pole contains a region, the **microtubule-organizing center,** from which extend the microtubules that form the mitotic spindle. The electron microscope shows that microtubule-organizing centers in certain plant cells consist of fibrils with little or no discernible structure.

In contrast, animal cells have a pair of **centrioles** in the middle of each microtubule-organizing center (see Fig. 4-25). The centrioles are surrounded by fibrils that make up the **pericentriolar material.** The spindle microtubules terminate in the pericentriolar material, but they do not actually touch the centrioles. Although cell biologists once thought spindle formation in animal cells required centrioles, their involvement is probably coincidental. Current evidence suggests that centrioles organize the pericentriolar material and ensure its duplication when the centrioles duplicate.

Each of the two centrioles is duplicated during the S phase of interphase, yielding two centriole pairs. Late in prophase, microtubules radiate from the pericentriolar material surrounding the centrioles; these clusters of microtubules are called **asters.** The two asters migrate to opposite sides of the nucleus, establishing the two poles of the mitotic spindle.

Prometaphase begins when the nuclear envelope breaks down

During **prometaphase,** the nuclear envelope fragments so that the spindle microtubules come into contact with the chromosomes; units of the disassembled nuclear envelope are sequestered in vesicles to be used later, to assemble nuclear envelopes for the daughter cells. The nucleolus shrinks and usually disappears, and the mitotic spindle is completely assembled. At the start of prometaphase, the duplicated chromosomes are scattered throughout the nuclear region (see the chapter-opening photograph, which shows early prometaphase). The spindle microtubules grow and shrink as they move toward the center of the cell in a "search and capture" process. Their random, dynamic movements give the appearance that they are "searching" for the chromosomes. If a microtubule comes near a centromere, one of the kinetochores of a duplicated chromosome "captures" it. As the now-tethered chromosome continues moving toward the cell's midplane, the unattached kinetochore of its sister chromatid becomes connected to a spindle microtubule from the cell's other pole.

During the chromosomes' movements toward the cell's midplane, long microtubules are shortened by the removal of tubulin subunits, and short microtubules are lengthened by the addition of tubulin subunits. Evidence indicates that shortening and lengthening occur at the kinetochore end (the plus end) of the microtubule, not at the spindle pole end (the minus end). This shortening or lengthening occurs while the spindle microtubule remains firmly tethered to the kinetochore. Motor proteins located at the kinetochores may be involved in this tethering of the spindle microtubule. These motor proteins may work in a similar fashion to the kinesin motor shown in Figure 4-24.

To summarize the events of prometaphase, sister chromatids of each duplicated chromosome become attached at their kinetochores to spindle microtubules extending from opposite

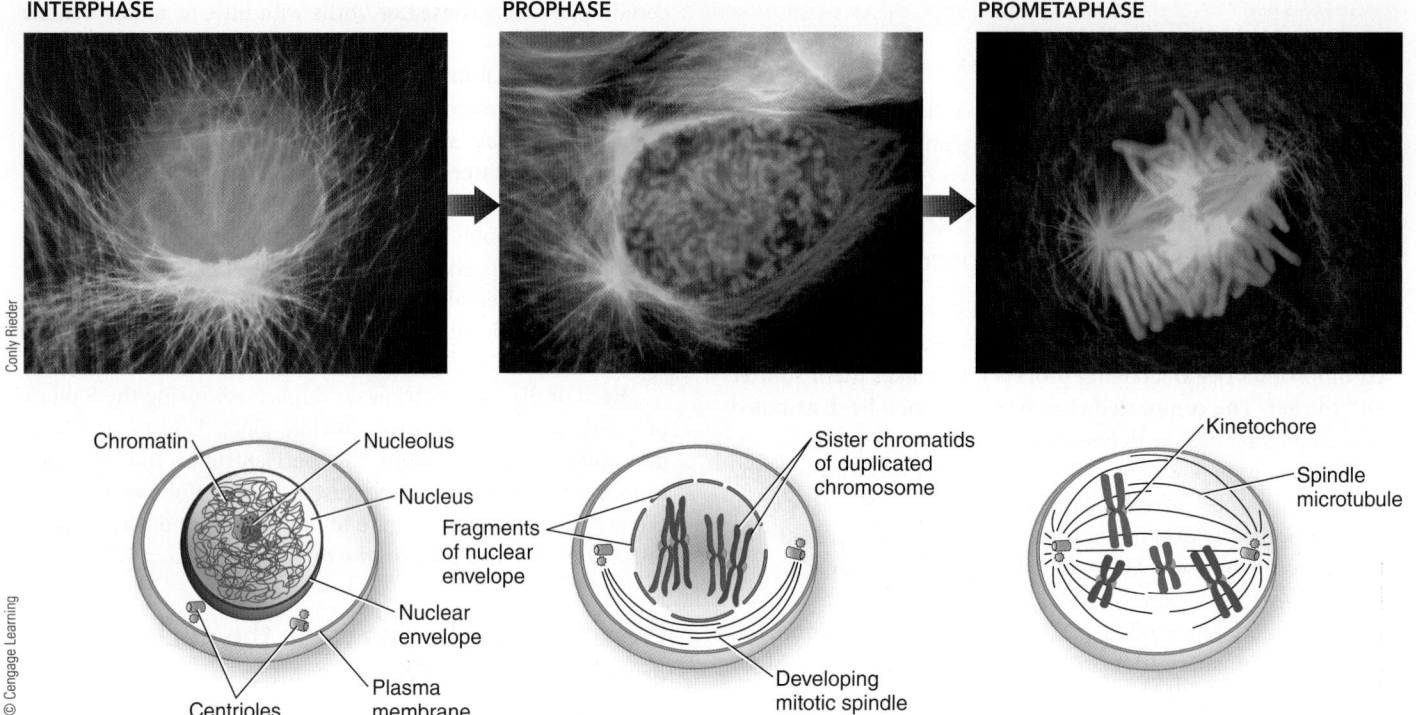

INTERPHASE **PROPHASE** **PROMETAPHASE**

Chromatin — Nucleolus
— Nucleus
Fragments — Sister chromatids
of nuclear of duplicated
envelope chromosome
Nuclear
envelope
Centrioles — Plasma
membrane
— Developing
mitotic spindle

Kinetochore
— Spindle
microtubule

(a) Cell carries out normal life activities. Chromosomes become duplicated.

(b) Long fibers of chromatin condense as compact mitotic chromosomes, each consisting of two chromatids attached at their centromeres. Cytoskeleton is disassembled, and mitotic spindle forms between centrioles, which have moved to poles of cell. Nuclear envelope begins to disassemble.

(c) Spindle microtubules attach to kinetochores of chromosomes. Chromosomes begin to move toward cell's midplane.

Figure 10-6 *Animation* **Interphase and the stages of mitosis**

The fluorescence LMs are of cultured lung cells of a newt, *Taricha granulosa*. (Chromosomes and chromatin, *blue*; microtubules, *yellow/green*.) The drawings depict generalized animal cells with a diploid chromosome number of 4; the sizes of the nuclei and chromosomes are exaggerated to show the structures more clearly. Cells of most plants lack centrioles.

poles of the cell, and the chromosomes begin to move toward the cell's midplane. As the cell transitions from prometaphase to metaphase, cohesins dissociate from the sister chromatid arms, freeing them from one another, although some cohesins remain in the centromere regions.

Duplicated chromosomes line up on the midplane during metaphase

During **metaphase**, all the cell's chromosomes align at the cell's midplane, or **metaphase plate**. As already mentioned, one of the two sister chromatids of each chromosome is attached by its kinetochore to microtubules from one pole, and its sister chromatid is attached by its kinetochore to microtubules from the opposite pole.

The mitotic spindle has three types of microtubules: polar microtubules, kinetochore microtubules, and astral microtubules (see **FIG. 10-9**). *Polar microtubules*, also known as *nonkinetochore microtubules*, extend from each pole to the equatorial region, where they overlap and interact with nonkinetochore microtubules from the opposite pole. *Kinetochore microtubules* extend from each pole and attach to chromosomes at their kinetochores. *Astral microtubules* are the short microtubules that form asters at each pole.

Each chromatid is completely condensed and appears distinct during metaphase. Because individual chromosomes are more obvious at metaphase than at any other time, the **karyotype,** or chromosome composition, is usually checked at this stage for chromosome abnormalities (see Chapter 16). As the mitotic cell transitions from metaphase to anaphase, the remaining cohesin proteins joining the sister chromatids at their centromere dissociate.

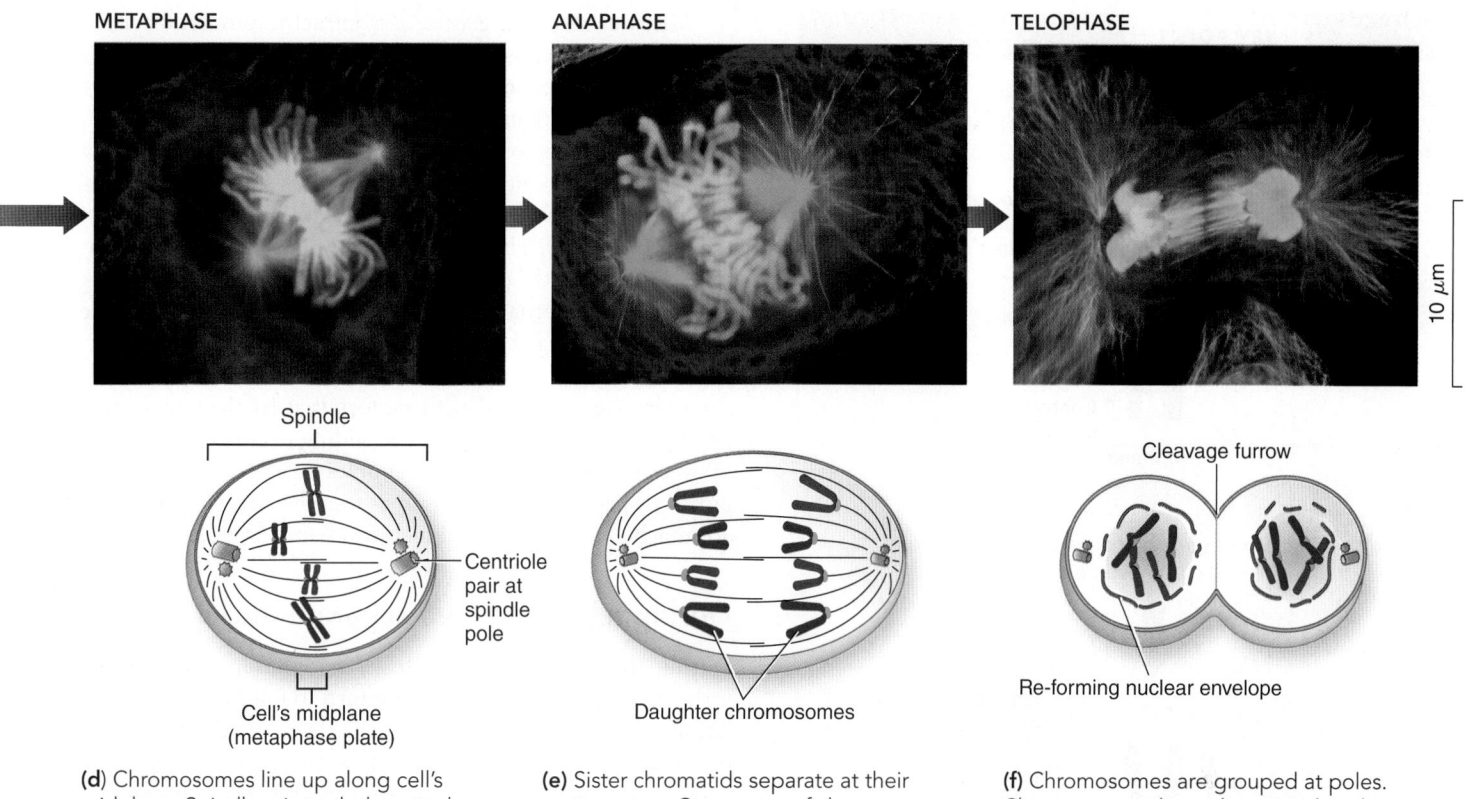

METAPHASE

ANAPHASE

TELOPHASE

10 μm

Spindle

Centriole pair at spindle pole

Cell's midplane (metaphase plate)

Cleavage furrow

Daughter chromosomes

Re-forming nuclear envelope

(d) Chromosomes line up along cell's midplane. Spindle microtubules attach each chromosome to both poles.

(e) Sister chromatids separate at their centromeres. One group of chromosomes moves toward each pole of cell. Spindle poles move farther apart.

(f) Chromosomes are grouped at poles. Chromosomes decondense, and nuclear envelopes begin to form. Cytokinesis produces two daughter cells.

Figure 10-6 *Continued*

During anaphase, chromosomes move toward the poles

Anaphase begins as the sister chromatids separate. Once the chromatids are no longer attached to their duplicates, each chromatid is called a *chromosome*. The now-separate chromosomes move to opposite poles, using the spindle microtubules as tracks. The kinetochores, still attached to kinetochore microtubules, lead the way, with the chromosome arms trailing behind. Anaphase ends when all the chromosomes have reached the poles.

Cell biologists are making significant progress in understanding the overall mechanism of chromosome movement in anaphase. Chromosome movements are studied in several ways. The number of microtubules at a particular stage or after certain treatments is determined by carefully analyzing electron micrographs. Researchers also physically perturb living cells that are

Figure 10-7 Sister chromatids and centromeres

The sister chromatids, each consisting of tightly coiled chromatin fibers, are tightly associated at their centromere regions, indicated by the brackets. Associated with each centromere is a kinetochore, which serves as a microtubule attachment site. Kinetochores and microtubules are not visible in this TEM of a metaphase chromosome.

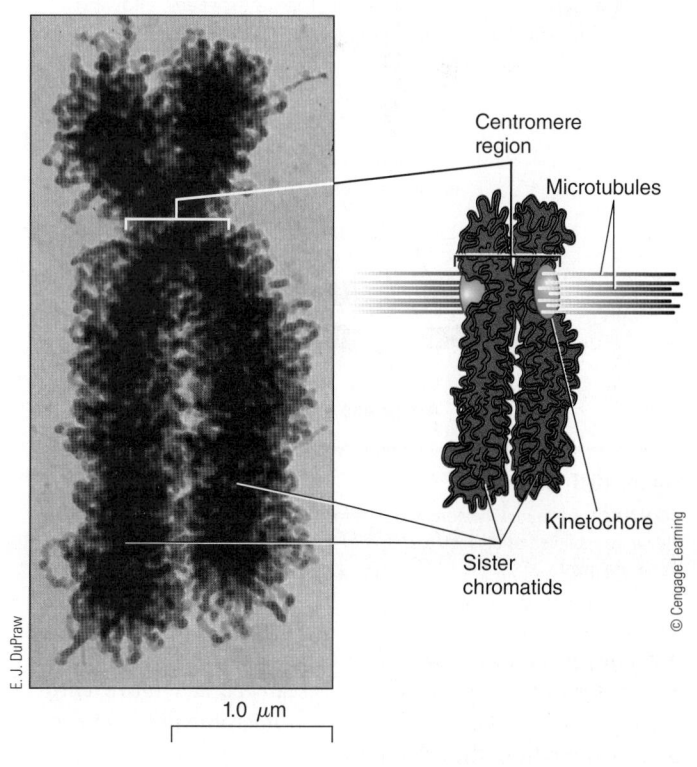

Centromere region

Microtubules

Kinetochore

Sister chromatids

E. J. DuPraw

1.0 μm

© Cengage Learning

When chromosomes duplicate, sister chromatids are initially linked to one another by protein complexes called cohesins. Cohesin linkages are particularly concentrated in the vicinity of the centromere.

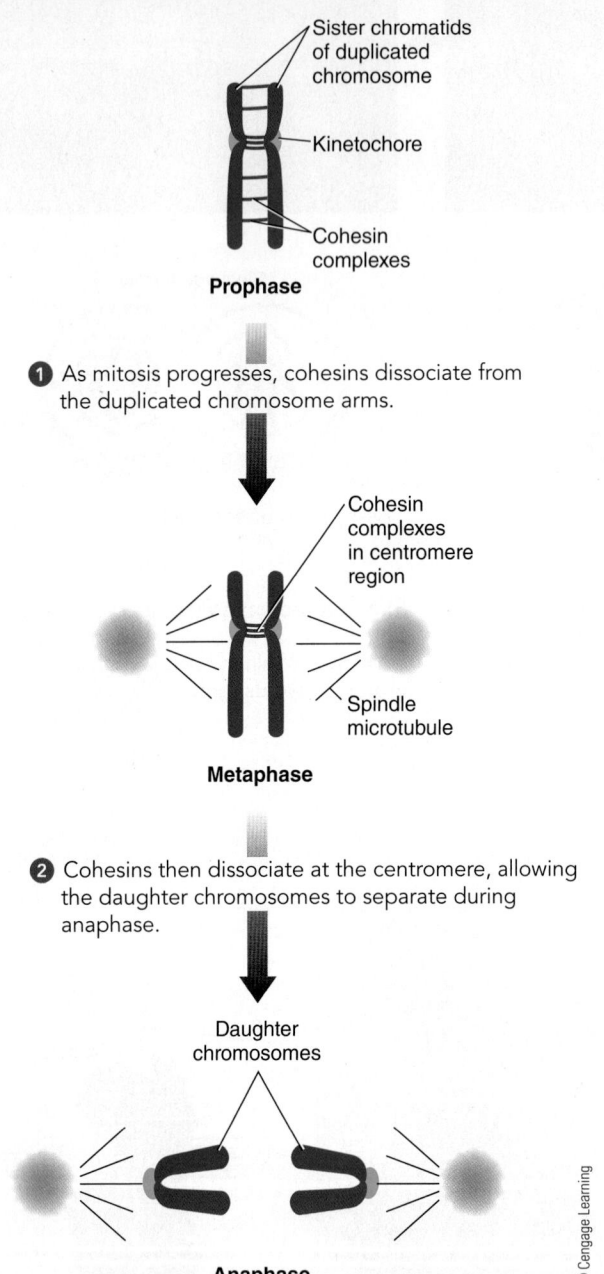

Sister chromatids of duplicated chromosome

Kinetochore

Cohesin complexes

Prophase

1 As mitosis progresses, cohesins dissociate from the duplicated chromosome arms.

Cohesin complexes in centromere region

Spindle microtubule

Metaphase

2 Cohesins then dissociate at the centromere, allowing the daughter chromosomes to separate during anaphase.

Daughter chromosomes

Anaphase

© Cengage Learning

Figure 10-8 Cohesins

CONNECT As you learn more about mitosis, consider what would happen if the cohesins failed to dissociate or if they were to dissociate prematurely, in late prophase.

dividing, using microlaser beams or mechanical devices known as *micromanipulators*. A skilled researcher can move chromosomes, break their connections to microtubules, and even remove them from the cell entirely.

Microtubules lack elastic or contractile properties. Then how do the chromosomes move apart? Are they pushed or pulled, or do other forces operate? Microtubules are dynamic structures, with *tubulin* subunits constantly being removed from their ends and others being added. Evidence indicates that during anaphase, kinetochore microtubules shorten, or *depolymerize*, at their *plus* ends, that is, closest toward the midplane of the cell (FIG. 10-10). This shortening mechanism pulls the chromosomes toward the poles.

A second mechanism also plays a role in chromosome separation. During anaphase, the spindle as a whole elongates, at least partly because polar microtubules originating at opposite poles are associated with motors that let them slide past one another at the midplane. The sliding decreases the degree of overlap, thereby "pushing" the poles apart. This mechanism indirectly causes the chromosomes to move apart because they are attached to the poles by kinetochore microtubules.

During telophase, two separate nuclei form

During the final stage of mitosis, **telophase,** chromosomes arrive at the poles, and there is a return to interphase-like conditions. The chromosomes decondense by partially uncoiling. A new nuclear envelope forms around each set of chromosomes, made at least in part from small vesicles and other components derived from the old nuclear envelope. The spindle microtubules disappear, and the nucleoli reorganize.

Cytokinesis forms two separate daughter cells

Cytokinesis, the division of the cytoplasm to yield two daughter cells, is the last step in M phase and usually overlaps mitosis, generally beginning during telophase. Cytokinesis of an animal or fungal cell (e.g., yeast) begins as an *actomyosin contractile ring* is assembled and attached to the plasma membrane. The contractile ring encircles the cell in the equatorial region, at right angles to the spindle (FIG. 10-11a). The contractile ring consists of an association between actin and myosin filaments; it is thought that the motor activity of myosin moves actin filaments to cause the constriction, similar to the way actin and myosin cause muscle contraction (see Fig. 40-11). The ring contracts, producing a **cleavage furrow** that gradually deepens and eventually separates the cytoplasm into two daughter cells, each with a complete nucleus. The contractile ring then disassembles.

In plant cells cytokinesis occurs by forming a **cell plate** (FIG. 10-11b), a partition constructed in the equatorial region of the spindle that grows laterally toward the cell wall. The cell plate forms as a line of vesicles originating in the *Golgi complex*. The vesicles contain materials to construct both a primary cell wall for each daughter cell and a middle lamella that cements the primary cell walls together. The vesicle membranes fuse to become the plasma membrane of each daughter cell.

Multinucleated cells form if mitosis is not followed by cytokinesis; this is a normal condition for certain cell types. For example, the body of plasmodial slime molds consists of a multinucleate mass of cytoplasm (see Fig. 26-19a).

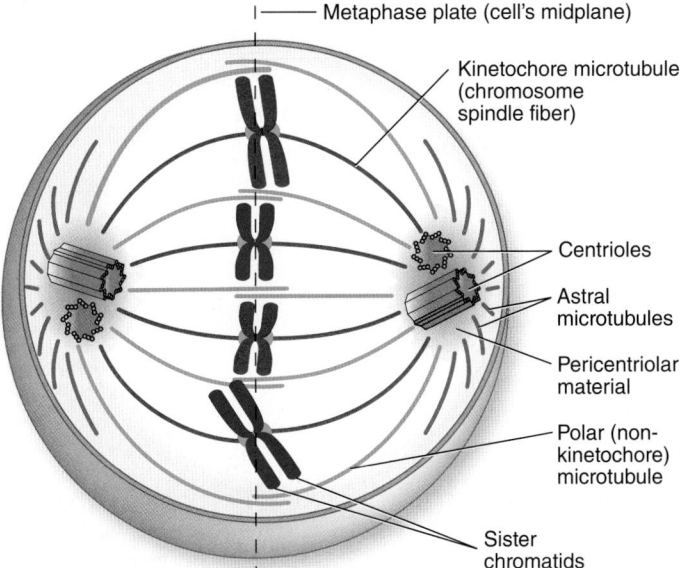

Metaphase plate (cell's midplane)

Kinetochore microtubule (chromosome spindle fiber)

Centrioles

Astral microtubules

Pericentriolar material

Polar (non-kinetochore) microtubule

Sister chromatids

Figure 10-9 The mitotic spindle

One end of each microtubule of this animal cell is associated with one of the poles. Astral microtubules (*green*) radiate in all directions, forming the aster. Kinetochore microtubules (*red*) connect the kinetochores to the poles, and polar (nonkinetochore) microtubules (*blue*) overlap at the midplane.
© Cengage Learning

Mitosis produces two cells genetically identical to the parent cell

The remarkable regularity of the process of cell division ensures that each daughter nucleus receives exactly the same number and kinds of chromosomes that the parent cell had. Thus, with a few exceptions, every cell of a multicellular organism has the same genetic makeup. If a cell receives more or fewer than the characteristic number of chromosomes through some malfunction of the cell division process, the resulting cell may show marked abnormalities and often cannot survive.

Mitosis provides for the orderly distribution of chromosomes (and of centrioles, if present), but what about the various cytoplasmic organelles? For example, all eukaryotic cells, including plant cells, require mitochondria. Likewise, photosynthetic plant cells cannot carry out photosynthesis without chloroplasts. These organelles contain their own DNA and appear to form by the division of previously existing mitochondria or plastids or their precursors. This nonmitotic division process is similar to prokaryotic cell division (discussed in the next section) and generally occurs during interphase. Because many copies of each organelle are present in each cell, organelles are apportioned with the cytoplasm that each daughter cell receives during cytokinesis.

Lacking nuclei, prokaryotes divide by binary fission

Bacteria and archaea contain much less DNA than do most eukaryotic cells, but precise distribution of the genetic material into two daughter cells is still a formidable process.

Do spindle microtubules move chromosomes by a shortening (i.e., depolymerization) of microtubules at the spindle poles or at the kinetochore ends?

HYPOTHESIS: Spindle microtubules move chromosomes toward the spindle poles by a mechanism in which the spindle microtubules are shortened at their kinetochore ends.

EXPERIMENT: Microtubules in pig kidney cells in early anaphase were labeled with a fluorescent dye that specifically attaches to microtubules.

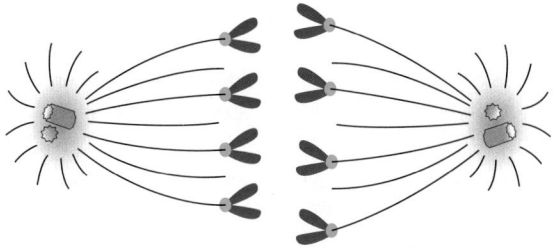

The researchers marked the microtubules by using a laser microbeam to bleach the dye while keeping the microtubules intact.

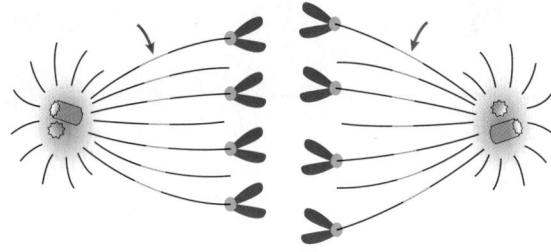

RESULTS AND CONCLUSION: The chromosomes moved toward the bleached areas of the spindle microtubules, indicating a shortening of the microtubules on the kinetochore side. The microtubules on the polar ends did not shorten.

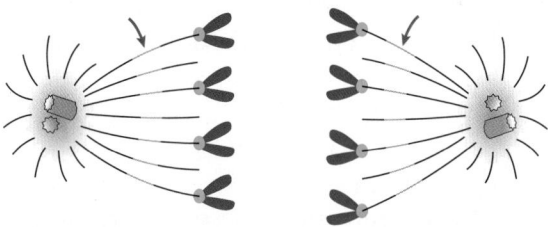

Chromosomes moved poleward because they remained anchored to the kinetochore microtubules as tubulin subunits were removed at the kinetochore ends of the microtubules.

SOURCE: Gorbsky, G.J., P.J. Sammak, and G.G. Borisy (1987) "Chromosomes move poleward in anaphase along stationary microtubules that coordinately disassemble from their kinetochore ends." *Journal of Cell Biology* 104: 9–18.

Figure 10-10 *Animation* **Using laser photobleaching to determine how chromosomes are transported toward the spindle poles during anaphase**

PREDICT How would the spindle microtubules have looked during anaphase if they had disassembled at their polar ends instead of their kinetochore ends?

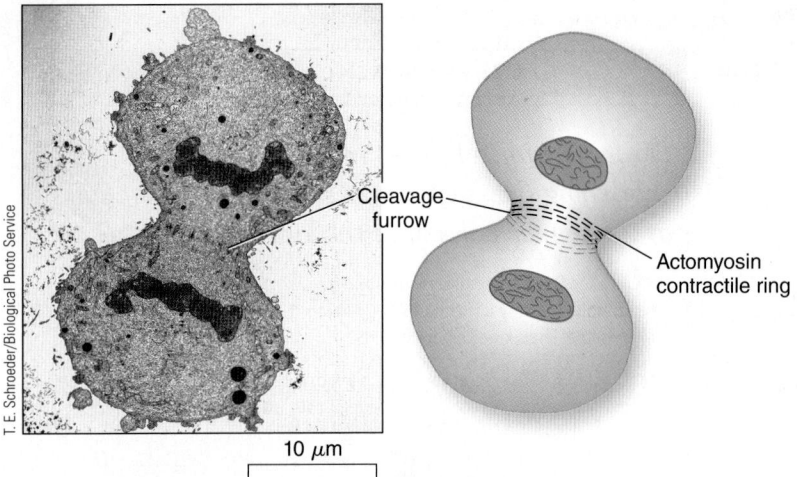

(a) TEM of the equatorial region of a cultured animal cell undergoing cytokinesis. Note the cleavage furrow. Dividing fungal cells also have a contractile ring that causes cytokinesis.

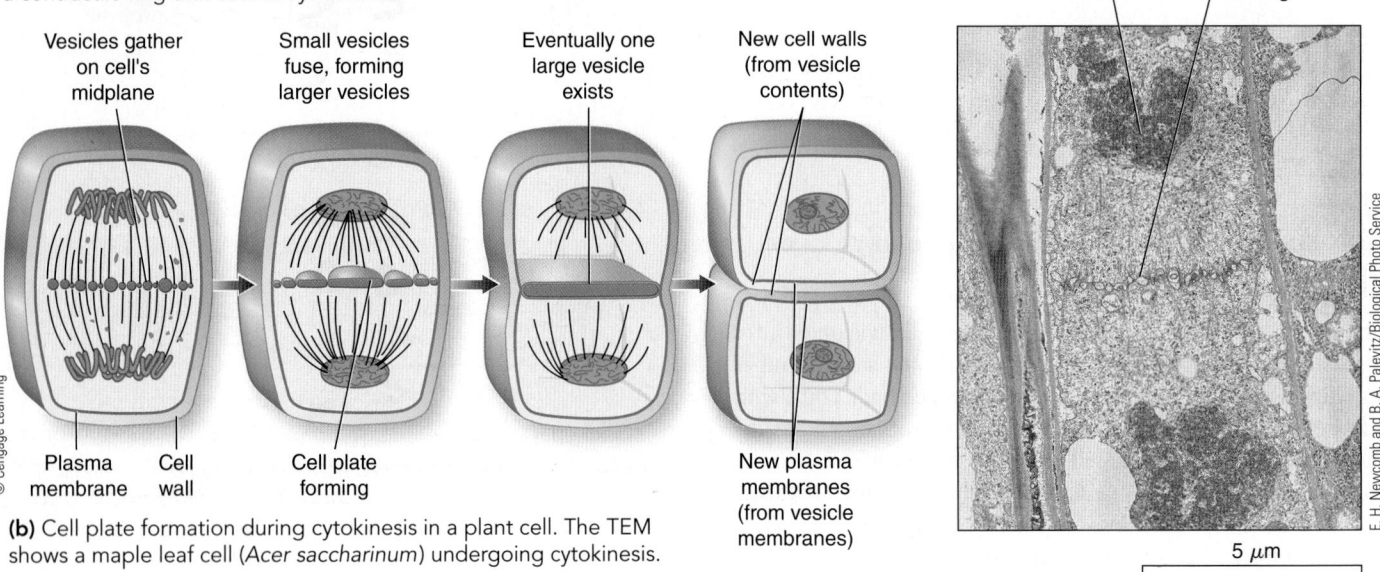

Vesicles gather on cell's midplane

Small vesicles fuse, forming larger vesicles

Eventually one large vesicle exists

New cell walls (from vesicle contents)

Plasma membrane

Cell wall

Cell plate forming

New plasma membranes (from vesicle membranes)

Nucleus

Cell plate forming

(b) Cell plate formation during cytokinesis in a plant cell. The TEM shows a maple leaf cell (*Acer saccharinum*) undergoing cytokinesis.

Figure 10-11 *Animation* **Cytokinesis in animal and plant cells**
The nuclei in both TEMs are in telophase. The drawings show 3-D relationships.

Prokaryotic DNA usually consists of a single, circular chromosome that is packaged with associated proteins. Although the distribution of genetic material in dividing prokaryotic cells is a simpler process than mitosis, it nevertheless is very precise, to ensure that the daughter cells are genetically identical to the parent cell.

Prokaryotes reproduce asexually, generally by **binary fission,** a process in which one cell divides into two daughter cells (**FIG. 10-12**). The circular DNA molecule replicates, resulting in two identical chromosomes. DNA replication begins at a single site on the bacterial chromosome, called the *origin of replication*. DNA synthesis proceeds from that point in both directions until they eventually meet (see Fig. 12-16).

Following replication, the daughter chromosomes separate and move to opposite ends of the elongating cell. Cytokinesis between the daughter chromosomes is controlled by the **Z ring,** a protein scaffold that holds about ten different proteins around the cell's midsection. There the plasma membrane grows inward between the two DNA copies, dividing the cell's cytoplasm in half, and a new transverse cell wall is synthesized between the two cells. (Bacterial reproduction is described further in Chapter 25.)

CHECKPOINT 10.2

- *What are the stages of the cell cycle? During which stage does DNA replicate?*
- *What are the stages of mitosis, and what happens in each stage?*

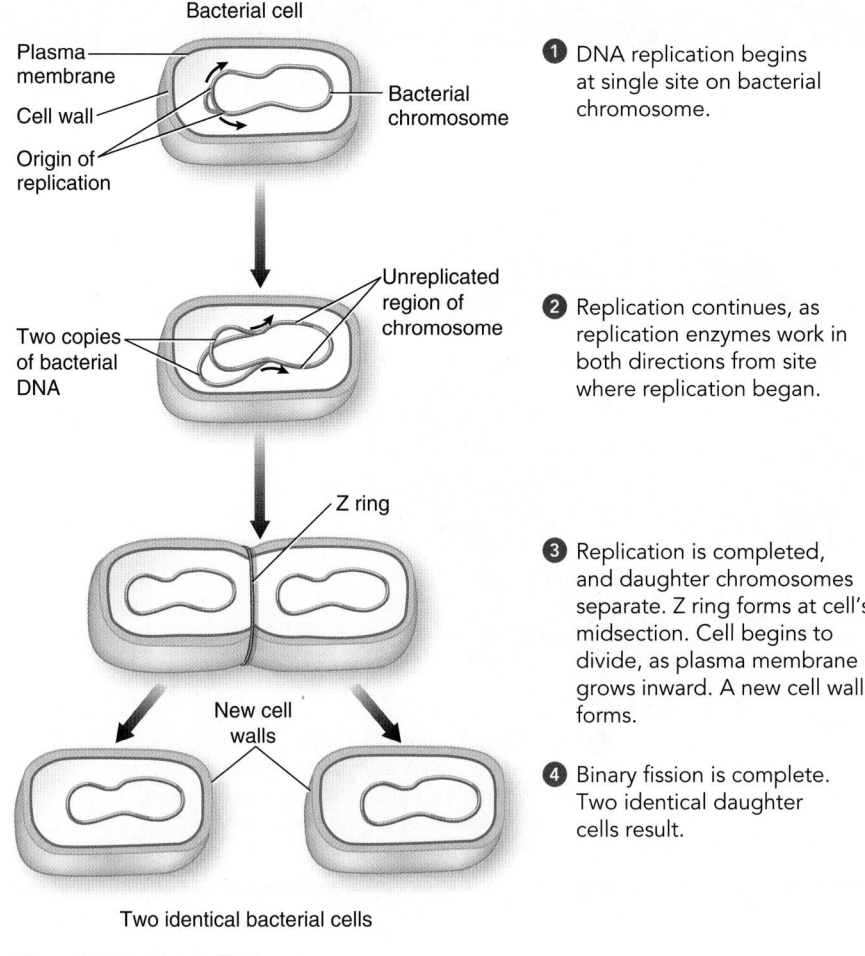

Bacterial cell

Plasma membrane

Cell wall

Origin of replication

Bacterial chromosome

Unreplicated region of chromosome

Two copies of bacterial DNA

Z ring

New cell walls

Two identical bacterial cells

1 DNA replication begins at single site on bacterial chromosome.

2 Replication continues, as replication enzymes work in both directions from site where replication began.

3 Replication is completed, and daughter chromosomes separate. Z ring forms at cell's midsection. Cell begins to divide, as plasma membrane grows inward. A new cell wall forms.

4 Binary fission is complete. Two identical daughter cells result.

Figure 10-12 Binary fission

Binary fission is a precisely orchestrated sequence of events that ensures each bacterial daughter cell has identical genetic material. The bacterial chromosome is much longer than depicted here and is tethered to the plasma membrane at one spot (not shown).
© Cengage Learning

10.3 REGULATION OF THE CELL CYCLE

⬤ LEARNING OBJECTIVE

6 Explain some ways in which the cell cycle is controlled.

When conditions are optimal, some prokaryotic cells can divide every 20 minutes. The generation times of eukaryotic cells are generally much longer, although the frequency of cell division varies widely among different species and among different tissues of the same species. Some skeletal muscle cells usually stop dividing after the first few months of life, whereas blood-forming cells, digestive tract cells, and skin cells divide frequently throughout the life of the organism. Under optimal conditions of nutrition, temperature, and pH, the length of the eukaryotic cell cycle is constant for any given cell type. Under less-favorable conditions, however, the generation time may be longer.

Certain regulatory molecules that control the cell cycle are common to all eukaryotes. Genetically programmed in the cell's nucleus, these regulatory molecules are components of the *cell cycle control system* found in organisms as diverse as yeast, clams, frogs, humans, and plants. Regulatory molecules trigger a specific sequence of events during the cell cycle.

Because the cell cycle consists of hundreds of sequential events that proceed in an orderly manner, a failure to carefully control these events can have disastrous consequences. Control mechanisms in the genetic program, called **cell-cycle checkpoints**, temporarily block key events from being initiated during the cell cycle. Cell-cycle checkpoints ensure that all the events of a particular stage have been completed before the next stage begins (**FIG. 10-13**). The checkpoints are inactivated after they have done their job so the cell cycle can proceed.

Genes that encode molecules involved in checkpoints are critically important to the cell cycle. If a checkpoint gene is defective, it can result in cancer or other serious diseases. Consider what might happen if the metaphase-anaphase checkpoint molecules were nonfunctional. In this case, anaphase might be initiated too early, before all chromosomes were properly attached to spindle fibers. The resulting daughter cells might have too few or too many chromosomes. An abnormal number of chromosomes is associated with Down syndrome as well as many cancers.

FIGURE 10-14 shows some of the key molecules involved in regulating the cell cycle. Among them are **protein kinases**, enzymes that activate or inactivate other proteins by *phosphorylating* (adding phosphate groups to) them. The protein kinases involved in controlling the cell cycle are **cyclin-dependent kinases (Cdks)**. The activity of various Cdks increases and then decreases as the cell moves through the cell cycle. Cdks are active only when they bind tightly to regulatory proteins called **cyclins.** The cyclins are so named because their levels fluctuate predictably during the cell cycle (i.e., they "cycle," or are alternately synthesized and degraded as part of the cell cycle).

Three scientists who began their research in the 1970s and 1980s on the roles of protein kinases and cyclins in the cell cycle (Leland Hartwell from the United States, and Paul Nurse and Tim Hunt from Great Britain) were awarded the Nobel Prize in Physiology or Medicine in 2001. Their discoveries were cited as important not only in working out the details of the fundamental cell process of mitosis but also in understanding why cancer cells divide when they should not. For example, cyclin levels are often higher than normal in human cancer cells.

When a specific Cdk associates with a specific cyclin, it forms a **cyclin–Cdk complex**. Cyclin–Cdk complexes phosphorylate

When a cell has not completed the steps leading to a cell-cycle checkpoint, the checkpoint is active and halts the cell cycle. When the necessary steps are completed, the checkpoint is inactivated, and the cell cycle proceeds.

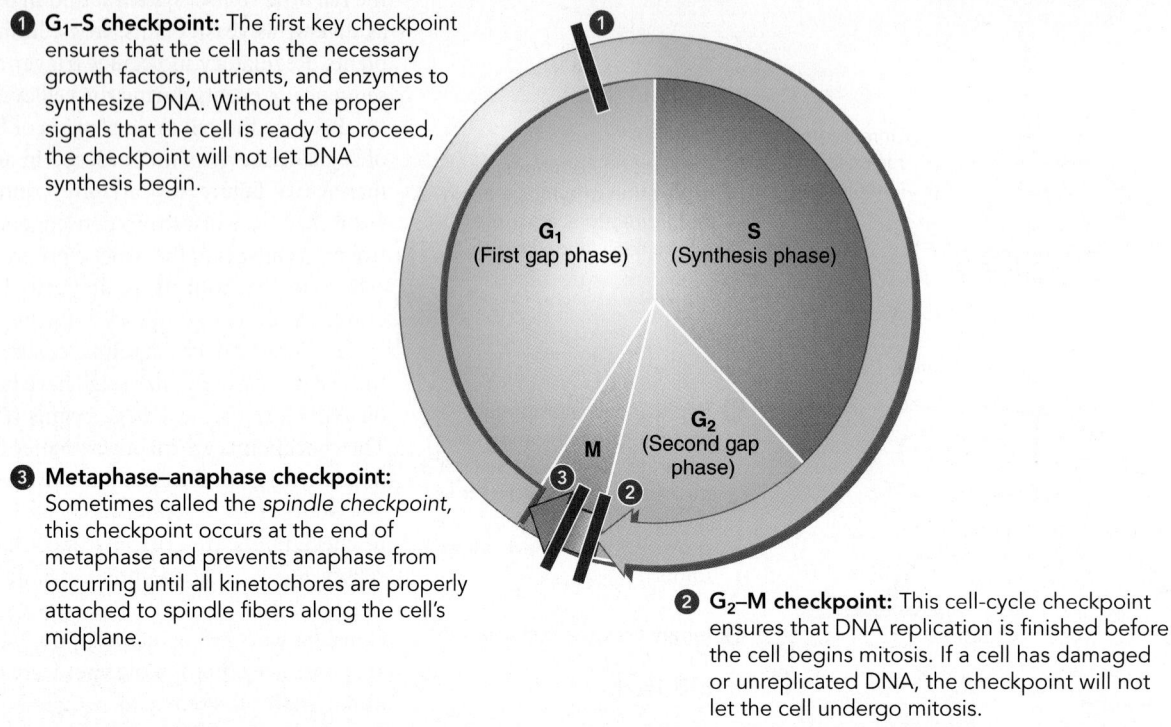

1 **G₁–S checkpoint:** The first key checkpoint ensures that the cell has the necessary growth factors, nutrients, and enzymes to synthesize DNA. Without the proper signals that the cell is ready to proceed, the checkpoint will not let DNA synthesis begin.

3 **Metaphase–anaphase checkpoint:** Sometimes called the *spindle checkpoint,* this checkpoint occurs at the end of metaphase and prevents anaphase from occurring until all kinetochores are properly attached to spindle fibers along the cell's midplane.

2 **G₂–M checkpoint:** This cell-cycle checkpoint ensures that DNA replication is finished before the cell begins mitosis. If a cell has damaged or unreplicated DNA, the checkpoint will not let the cell undergo mitosis.

Figure 10-13 Key checkpoints in the cell cycle

The cell cycle consists of hundreds of sequential events. The *red bars* show three important checkpoints that determine that previous steps are completed so that the next steps may proceed. Each checkpoint is inactivated after it has performed its function, allowing the cell cycle to continue.

VISUALIZE Consider a cell with two chromosomes. Sketch what it might look like if it were arrested at the metaphase–anaphase checkpoint.
© Cengage Learning

enzymes and other proteins. Some of these proteins become activated when they are phosphorylated, and others become inactivated. For example, phosphorylation of the protein p27, known to be a major inhibitor of cell division, is thought to initiate degradation of the protein. As various enzymes are activated or inactivated by phosphorylation, the activities of the cell change. Thus, a decrease in a cell's level of p27 causes a nondividing cell to resume division.

Eukaryotic cells form four major cyclin–Cdk complexes: G₁-Cdk, G₁/S-Cdk, S-Cdk, and M-Cdk. Each cyclin–Cdk complex phosphorylates a different group of proteins. G₁-Cdk prepares the cell to pass from the G₁ phase to the S phase, and then G₁/S-Cdk commits the cell to undergo DNA replication. S-Cdk initiates DNA replication. M-Cdk promotes the events of mitosis, including chromosome condensation, nuclear envelope breakdown, and mitotic spindle formation.

M-Cdk also activates another enzyme complex, the **anaphase-promoting complex (APC),** toward the end of metaphase. APC initiates anaphase by allowing degradation of the cohesins and other proteins that hold the sister chromatids together during metaphase. As a result, the sister chromatids separate as two daughter chromosomes. At this point, cyclin is

degraded to negligible levels and M-Cdk activity drops, allowing the mitotic spindle to disassemble and the cell to exit mitosis.

Certain drugs can stop the cell cycle at a specific checkpoint. Some of them prevent DNA synthesis, whereas others inhibit the synthesis of proteins that control the cycle or inhibit the synthesis of structural proteins that contribute to the mitotic spindle. Because one of the distinguishing features of most cancer cells is their high rate of cell division relative to that of most normal somatic cells, cancer cells are greatly affected by these drugs. Many side effects of certain anticancer drugs (such as nausea and hair loss) are due to the drugs' effects on normal cells that divide rapidly in the digestive system and hair follicles.

In plant cells certain hormones stimulate mitosis. They include the **cytokinins,** a group of plant hormones that promote mitosis both in normal growth and in wound healing (see Chapter 38). Similarly, animal hormones, such as certain steroids, stimulate growth and mitosis (see Chapter 49).

Protein **growth factors,** which are active at extremely low concentrations, stimulate mitosis in some animal cells. Of the approximately 50 protein growth factors known, some act only on specific types of cells, whereas others act on a broad range of cell types. For example, the effects of the growth factor

Cyclin-dependent kinases (Cdks) control the phosphorylation of other proteins, thereby regulating the transitions between phases of the cell cycle.

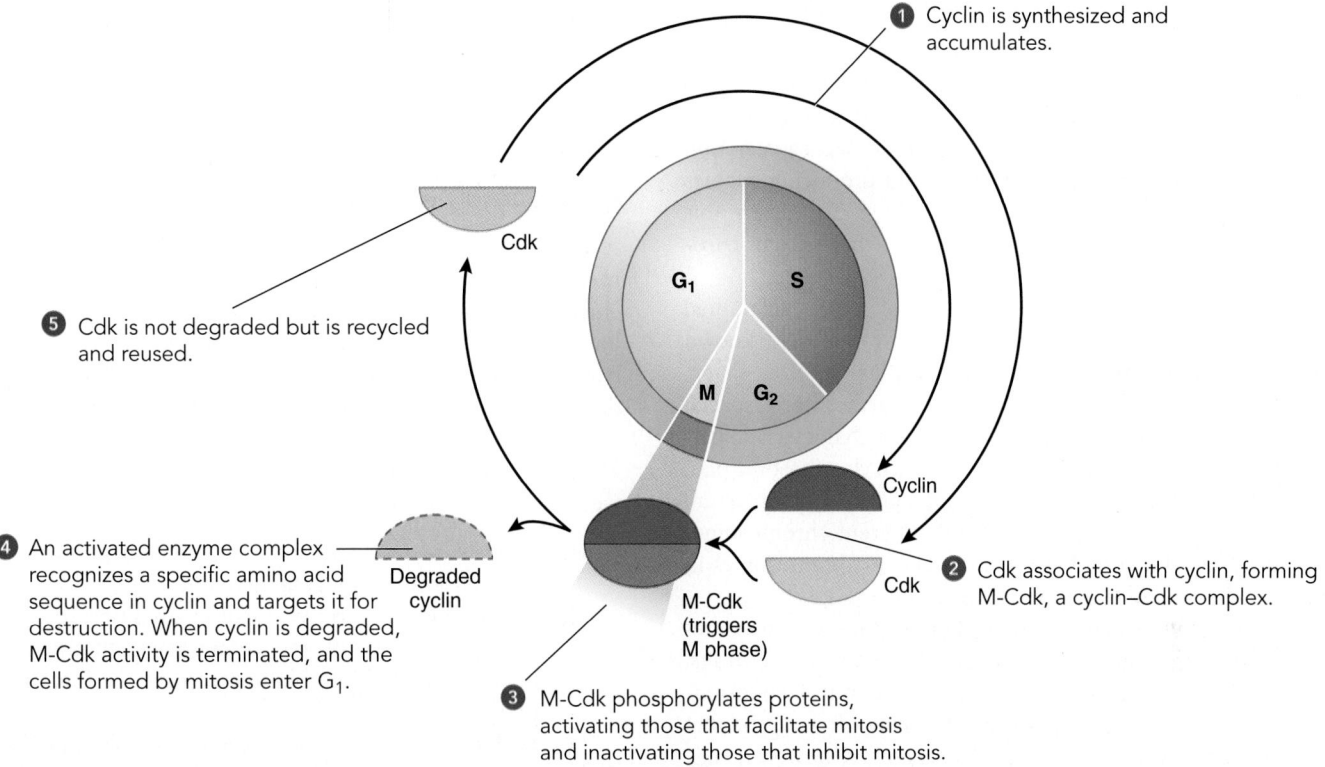

❶ Cyclin is synthesized and accumulates.

❺ Cdk is not degraded but is recycled and reused.

❹ An activated enzyme complex recognizes a specific amino acid sequence in cyclin and targets it for destruction. When cyclin is degraded, M-Cdk activity is terminated, and the cells formed by mitosis enter G_1.

Degraded cyclin

M-Cdk (triggers M phase)

Cyclin

Cdk

❷ Cdk associates with cyclin, forming M-Cdk, a cyclin–Cdk complex.

❸ M-Cdk phosphorylates proteins, activating those that facilitate mitosis and inactivating those that inhibit mitosis.

Figure 10-14 Molecular control of the cell cycle

This diagram is a simplified view of the control system that triggers the cell to move from G_2 to M phase.

PREDICT What might happen if cyclin were not degraded in step 4?
© Cengage Learning

erythropoietin are limited to cells that will develop into red blood cells, but *epidermal growth factor* stimulates many cell types to divide. Many types of cancer cells divide in the absence of growth factors.

CHECKPOINT 10.3

- *What are cell-cycle checkpoints?*
- *What are two molecular controls that trigger the onset of different stages of the cell cycle?*

10.4 SEXUAL REPRODUCTION AND MEIOSIS

LEARNING OBJECTIVES

7 Differentiate between asexual and sexual reproduction.

8 Distinguish between haploid and diploid cells, and define *homologous chromosomes*.

9 Explain the significance of meiosis and describe the process.

10 Contrast mitosis and meiosis, emphasizing the different outcomes.

Although the details of the reproductive process vary greatly among different kinds of eukaryotes, biologists distinguish two basic types of reproduction: asexual and sexual. In **asexual reproduction** a single parent splits, buds, or fragments to produce two or more individuals. In most kinds of eukaryotic asexual reproduction, all the cells are the result of mitotic divisions, so their genes and inherited traits are like those of the parent. Such a group of genetically identical organisms is called a **clone.** In asexual reproduction organisms that are well adapted to their environment produce new generations of similarly adapted organisms. Asexual reproduction occurs rapidly and efficiently, partly because the organism does not need to expend time and energy finding a mate.

In contrast, **sexual reproduction** involves the union of two sex cells, or **gametes,** to form a single cell called a **zygote.** Usually, two different parents contribute the gametes, but in some cases a single parent furnishes both gametes. In the case of animals and plants, the egg and sperm cells are the gametes, and the fertilized egg is the zygote.

Sexual reproduction results in genetic variation among the offspring. (*How* this genetic variation arises is discussed later in this chapter and in Chapter 11.) Because the offspring produced by sexual reproduction are not genetically identical to their parents or to each other, some offspring may be able to survive environmental changes better than either parent does. However, one disadvantage of sexual reproduction is that some offspring with a different combination of traits may be less likely to survive than their parents.

There is a potential problem in eukaryotic sexual reproduction: if each gamete had the same number of chromosomes as the parent cell that produced it, the zygote would have twice as many chromosomes. This doubling would occur generation after generation. How do organisms avoid producing zygotes with ever-increasing chromosome numbers? To answer this question, we need more information about the types of chromosomes found in cells.

Each chromosome in a somatic cell of a plant or animal normally has a partner chromosome. The two partners, known as **homologous chromosomes,** are similar in size, shape, and the position of their centromeres. Furthermore, special chromosome-staining procedures produce a characteristic pattern of bands evident in the members of each chromosome pair. In most species chromosomes vary enough in their structure that cell biologists can distinguish the different chromosomes and match up the homologous pairs. The 46 chromosomes in human cells constitute 23 homologous pairs.

The most important feature of homologous chromosomes is that they carry information about the same genetic traits, although this information is not necessarily identical. For example, each member of a homologous pair may carry a gene that specifies hemoglobin structure. One member, however, may have the information for the normal hemoglobin β chain (see Fig. 3-23a), whereas the other may specify the abnormal form of hemoglobin associated with sickle cell anemia (see Chapter 16). Homologous chromosomes can therefore be contrasted with the two members of a pair of sister chromatids, which are precisely identical to each other.

A set of chromosomes has one of each kind of chromosome; in other words, it contains one member of each homologous pair. If a cell or nucleus contains two sets of chromosomes, it is said to have a **diploid** chromosome number. If it has only a single set of chromosomes, it has the **haploid** number.

In humans the diploid chromosome number is 46 and the haploid number is 23. When a sperm and egg fuse at fertilization, each gamete is haploid, contributing one set of chromosomes; the diploid number is thereby restored in the fertilized egg (zygote). When the zygote divides by mitosis to form the first two cells of the embryo, each daughter cell receives the diploid number of chromosomes, and subsequent mitotic divisions repeat this. Thus, somatic cells are diploid.

An individual whose cells have three or more sets of chromosomes is **polyploid.** Polyploidy is relatively rare among animals but common among plants (see Chapter 20). In fact, polyploidy has been an important mechanism of plant evolution. As many as 80% of all flowering plants are polyploid. Polyploid plants are often larger and hardier than diploid members of the same group. Many commercially important plants, such as wheat and cotton, are polyploid.

The chromosome number found in the gametes of a particular species is represented as n, and the zygotic chromosome number is represented as $2n$. If the organism is not polyploid, the haploid chromosome number is equal to n and the diploid number is equal to $2n$; thus, in humans, $n = 23$ and $2n = 46$. For simplicity, in the rest of this chapter, the organisms used as examples are not polyploid. We use diploid and $2n$ interchangeably, and haploid and n interchangeably, although the terms are not strictly synonymous.

Meiosis produces haploid cells with unique gene combinations

A cell division that reduces chromosome number is called **meiosis.** The term *meiosis* means "to make smaller," and the chromosome number is reduced by one-half. In meiosis a diploid cell undergoes two cell divisions, potentially yielding four haploid cells. It is important to note that the haploid cells produced by this process do not contain just any combination of chromosomes, but one member of each homologous pair.

The events of meiosis are similar to the events of mitosis, with four important differences:

1. Meiosis involves two successive nuclear and cytoplasmic divisions, producing up to four cells.
2. Despite two successive nuclear divisions, the DNA and other chromosome components duplicate only once, during the interphase preceding the first meiotic division.
3. Each of the four cells produced by meiosis contains the haploid chromosome number, that is, only one chromosome set containing only one representative of each homologous pair.
4. During meiosis, each homologous chromosome pair is shuffled, so the resulting haploid cells each have a virtually unique combination of genes.

Meiosis typically consists of two nuclear and cytoplasmic divisions, designated the *first* and *second meiotic divisions,* or simply **meiosis I** and **meiosis II** (FIG. 10-15). Each includes prophase, metaphase, anaphase, and telophase stages. During meiosis I, the partner homologous chromosomes physically pair with each other and subsequently separate and move into different nuclei. In meiosis II, the sister chromatids that make up each duplicated chromosome separate from each other and are distributed to two different nuclei. The following discussion describes meiosis in an organism with a diploid chromosome number of 4. Refer to FIG. 10-16 as you read.

Prophase I includes synapsis and crossing-over

As occurs during mitosis, the chromosomes duplicate in the S phase of interphase, before the complex movements of meiosis actually begin. Each duplicated chromosome consists of two chromatids, which are linked by cohesins. During **prophase I,** while the chromatids are still elongated and thin, the homologous chromosomes come to lie lengthwise side by side. This process is called **synapsis,** which means "fastening together." For example, in an animal cell with a diploid number of four, synapsis results in two homologous pairs.

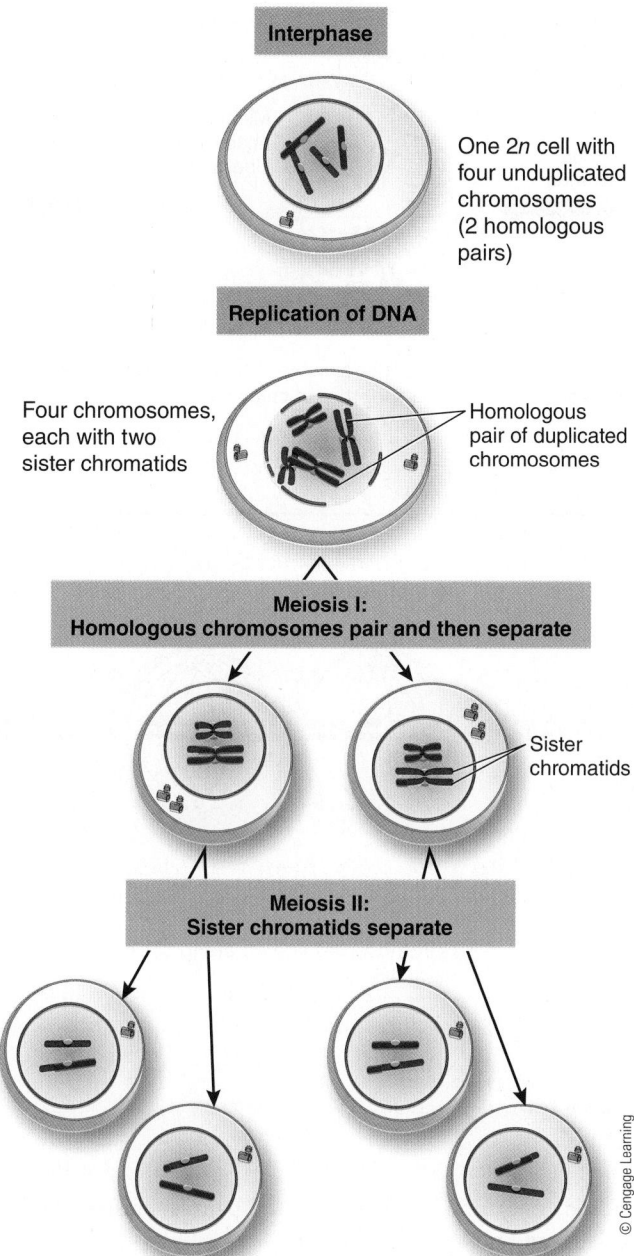

Interphase

One 2*n* cell with four unduplicated chromosomes (2 homologous pairs)

Replication of DNA

Four chromosomes, each with two sister chromatids

Homologous pair of duplicated chromosomes

Meiosis I:
Homologous chromosomes pair and then separate

Sister chromatids

Meiosis II:
Sister chromatids separate

© Cengage Learning

Four haploid (*n*) cells, each with two unduplicated chromosomes

Figure 10-15 Overview of meiosis

This figure begins with a diploid cell with four unduplicated chromosomes. The chromosomes derived from one parent are shown in *blue*, and those from the other parent are *red*. Homologous pairs are similar in size and shape.

One member of each homologous pair is called the *maternal homologue* because it was originally inherited from the female parent; the other member of a homologous pair is the *paternal homologue* because it was inherited from the male parent. Because each chromosome duplicated during interphase and now consists of two chromatids, synapsis results in the association of four chromatids. The resulting association is a **tetrad.** The number of tetrads per prophase I cell is equal to the haploid chromosome number. In an animal cell with a diploid number of 4, there are 2 tetrads (and a total of 8 chromatids); in

a human cell at prophase I, there are 23 tetrads (and a total of 92 chromatids).

Homologous chromosomes become closely associated during synapsis. Electron microscopic observations reveal that a characteristic structure, the **synaptonemal complex,** forms along the entire length of the synapsed homologues (**FIG. 10-17**). This proteinaceous structure holds the synapsed homologues together and is thought to play a role in chromosome **crossing-over,** a process in which enzymes break and rejoin DNA molecules, allowing paired homologous chromosomes to exchange genetic material. Crossing-over produces new combinations of genes. The **genetic recombination** from crossing-over greatly enhances the genetic variation—that is, new combinations of traits—among sexually produced offspring. Some biologists think that recombination is the main reason for sexual reproduction in eukaryotes.

In addition to the unique processes of synapsis and crossing-over, events similar to those in mitotic prophase also occur during prophase I. A spindle forms, consisting of microtubules and other components. In animal cells one pair of centrioles moves to each pole, and astral microtubules form. The nuclear envelope disappears in late prophase I, and in cells with large and distinct chromosomes, the structure of the tetrads can be seen clearly with the microscope.

The sister chromatids remain closely aligned along their lengths. However, the centromeres (and kinetochores) of the homologous chromosomes become separated from one another. In late prophase I, the homologous chromosomes are held together only at specific regions, called **chiasmata** (sing., *chiasma*). Each chiasma originates at a crossing-over site, that is, a site at which homologous chromatids exchanged genetic material and rejoined, producing an X-shaped configuration (**FIG. 10-18**). At the chiasmata, cohesins hold homologous chromosomes together after the synaptonemal complex has been disassembled. Later, the cohesins dissociate from the chiasmata, freeing the homologous chromosome arms from one another. The consequences of crossing-over and genetic recombination are discussed in Chapter 11 (e.g., see Fig. 11-12).

During meiosis I, homologous chromosomes separate

Metaphase I occurs when the tetrads align on the midplane. Both sister kinetochores of one duplicated chromosome are attached by spindle fibers to the same pole, and both sister kinetochores of the other duplicated homologous chromosome are attached to the opposite pole. (By contrast, sister kinetochores of each duplicated chromosome are attached to opposite poles in mitosis.)

During **anaphase I,** the paired homologous chromosomes separate, or disjoin, and move toward opposite poles. Each pole receives a random combination of maternal and paternal chromosomes, but only one member of each homologous pair is present at each pole. The sister chromatids remain united at their centromere regions. Again, this process differs from mitotic anaphase, in which the sister chromatids separate and move to opposite poles.

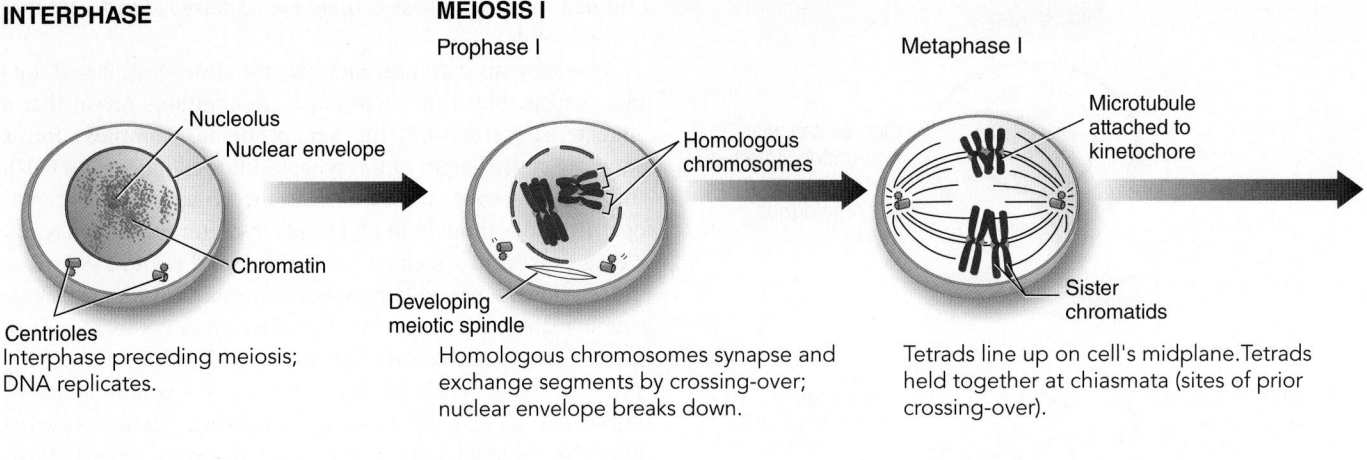

INTERPHASE

MEIOSIS I

Prophase I

Metaphase I

Nucleolus

Nuclear envelope

Chromatin

Centrioles
Interphase preceding meiosis;
DNA replicates.

Homologous
chromosomes

Developing
meiotic spindle

Homologous chromosomes synapse and
exchange segments by crossing-over;
nuclear envelope breaks down.

Microtubule
attached to
kinetochore

Sister
chromatids

Tetrads line up on cell's midplane. Tetrads
held together at chiasmata (sites of prior
crossing-over).

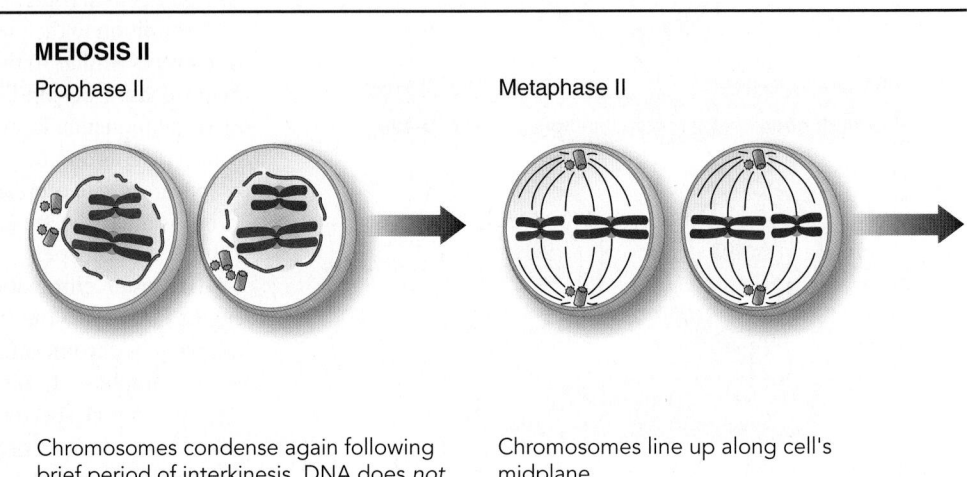

MEIOSIS II

Prophase II

Metaphase II

Chromosomes condense again following
brief period of interkinesis. DNA does *not*
replicate again.

Chromosomes line up along cell's
midplane.

Figure 10-16 *Animation* **Interphase and the stages of meiosis**

Meiosis consists of two nuclear divisions, meiosis I (*top row*) and meiosis II (*bottom row*). The drawings depict generalized animal cells with a diploid chromosome number of 4; the sizes of the nuclei and chromosomes are exaggerated to show the structures more clearly. Cells of most plants lack centrioles.

During **telophase I,** the chromatids generally decondense somewhat, the nuclear envelope may reorganize, and cytokinesis may take place. Each telophase I nucleus contains the haploid number of chromosomes, but each chromosome is a duplicated chromosome (it consists of a pair of chromatids). In our example, 2 duplicated chromosomes lie at each pole, for a total of 4 chromatids; humans have 23 duplicated chromosomes (46 chromatids) at each pole.

An interphase-like stage called **interkinesis** usually follows. Interkinesis is not a true interphase: there is no S phase and therefore no DNA replication. Interkinesis is brief in most organisms and absent in some.

Chromatids separate in meiosis II

Because the chromosomes usually remain partially condensed between divisions, the prophase of the second meiotic division is brief. **Prophase II** is similar to mitotic prophase in many respects. There is no pairing of homologous chromosomes

(indeed, only one member of each pair is present in each nucleus) and no crossing-over.

During **metaphase II,** the chromosomes line up on the midplanes of their cells. You can easily distinguish the first and second metaphases in diagrams; at metaphase I the chromatids are arranged in bundles of four (tetrads), and at metaphase II they are in groups of two (as in mitotic metaphase). This difference is not always so obvious in living cells.

During **anaphase II,** the chromatids, attached to spindle fibers at their kinetochores, separate and move to opposite poles, just as they would at mitotic anaphase. As in mitosis, each former chromatid is now referred to as a *chromosome*. Thus, at **telophase II** there is one representative for each homologous pair at each pole. Each is an unduplicated (single) chromosome. Nuclear envelopes re-form, the chromosomes gradually elongate to form chromatin fibers, and cytokinesis occurs.

The two successive divisions of meiosis yield four haploid nuclei, each containing *one* of each kind of chromosome. Each resulting haploid cell has a different combination of genes.

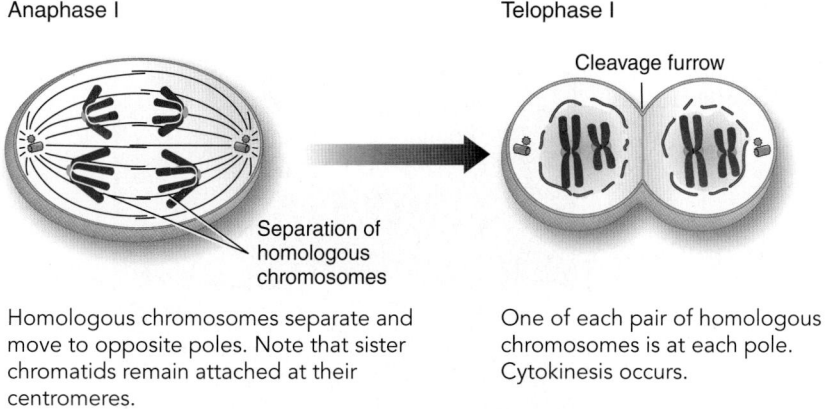

Anaphase I

Telophase I

Cleavage furrow

Separation of homologous chromosomes

Homologous chromosomes separate and move to opposite poles. Note that sister chromatids remain attached at their centromeres.

One of each pair of homologous chromosomes is at each pole. Cytokinesis occurs.

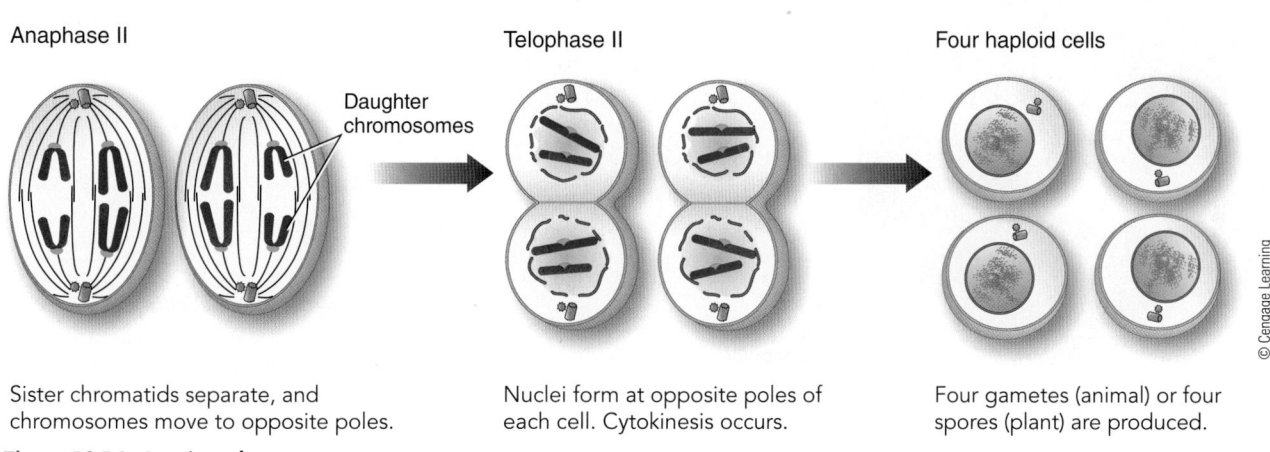

Anaphase II

Telophase II

Four haploid cells

Daughter chromosomes

© Cengage Learning

Sister chromatids separate, and chromosomes move to opposite poles.

Nuclei form at opposite poles of each cell. Cytokinesis occurs.

Four gametes (animal) or four spores (plant) are produced.

Figure 10-16 *Continued*

This genetic variation has two sources: (1) DNA segments are exchanged between maternal and paternal homologues during crossing-over; and (2) during meiosis, the maternal and paternal chromosomes of homologous pairs separate independently, and the chromosomes are "shuffled" so that each member of a pair becomes randomly distributed to one of the poles at anaphase I.

Mitosis and meiosis lead to contrasting outcomes

Although mitosis and meiosis share many similar features, specific distinctions between these processes result in the formation of different types of cells. Mitosis is a single nuclear division in which sister chromatids separate from each other and are distributed to the two daughter cells, which are genetically identical to each other and to the original cell. A diploid cell that undergoes mitosis produces two diploid cells. Similarly, a haploid cell that undergoes mitosis produces

two haploid cells. (Some eukaryotic organisms—e.g., certain yeasts—are haploid, as are plants at certain stages of their life cycles.) Homologous chromosomes do not associate physically at any time in mitosis.

In meiosis, a diploid cell undergoes two successive nuclear divisions, meiosis I and meiosis II. In prophase I of meiosis, the homologous chromosomes undergo synapsis to form tetrads. Homologous chromosomes separate during meiosis I, and sister chromatids separate during meiosis II. Meiosis ends with the formation of four genetically different, haploid daughter cells. The fates of these cells depend on the type of life cycle; in animals they differentiate as gametes, whereas in plants they become spores.

CHECKPOINT 10.4

- *Are homologous chromosome pairs present in a diploid cell? Are they present in a haploid cell?*
- CONNECT *How does the outcome of meiosis differ from the outcome of mitosis?*
- *Can haploid cells divide by mitosis? by meiosis?*

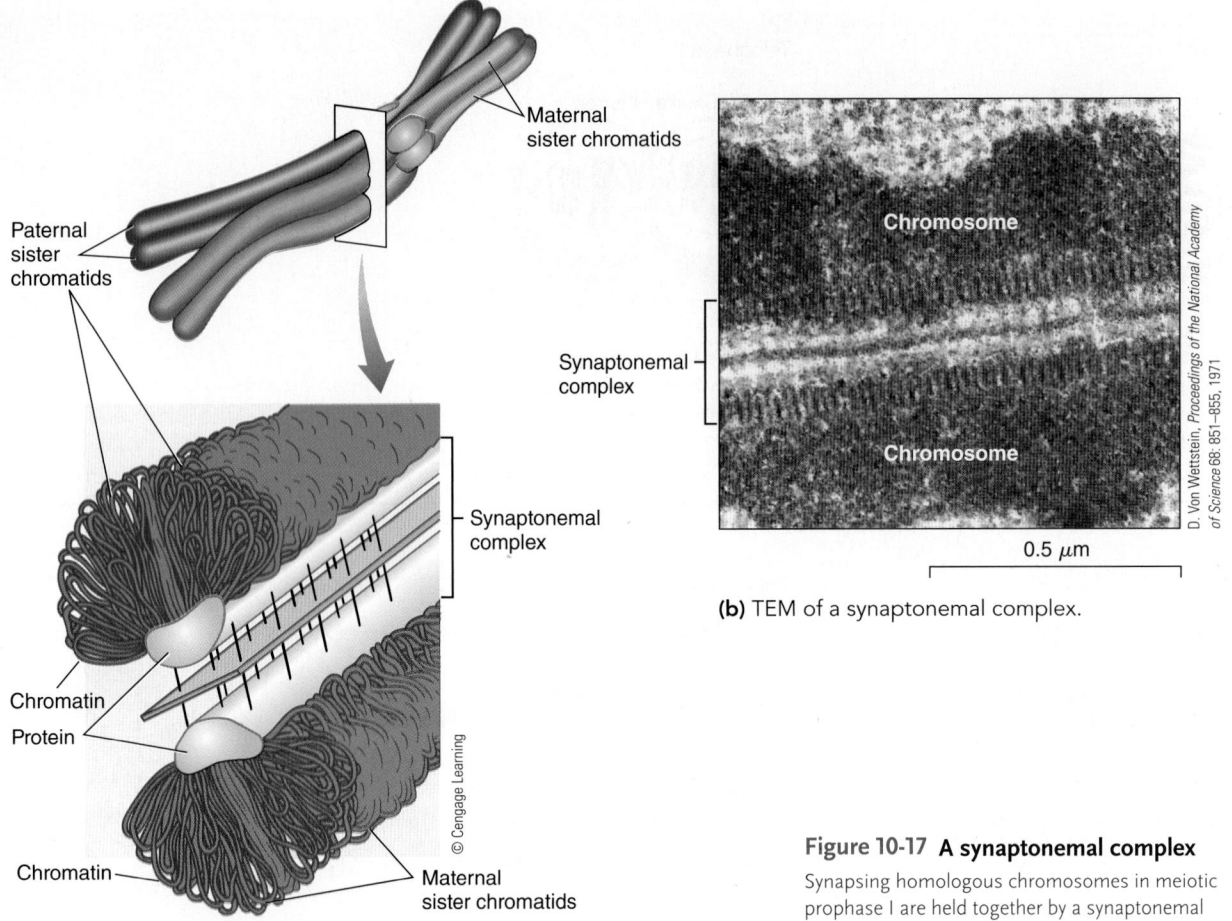

Paternal sister chromatids

Maternal sister chromatids

Synaptonemal complex

Chromatin
Protein

Chromatin

Maternal sister chromatids

(a) A 3-D model of a tetrad with a complete synaptonemal complex.

Chromosome

Synaptonemal complex

Chromosome

0.5 μm

D. Von Wettstein, Proceedings of the National Academy of Science 68: 851–855, 1971

(b) TEM of a synaptonemal complex.

Figure 10-17 A synaptonemal complex

Synapsing homologous chromosomes in meiotic prophase I are held together by a synaptonemal complex, composed mainly of protein.

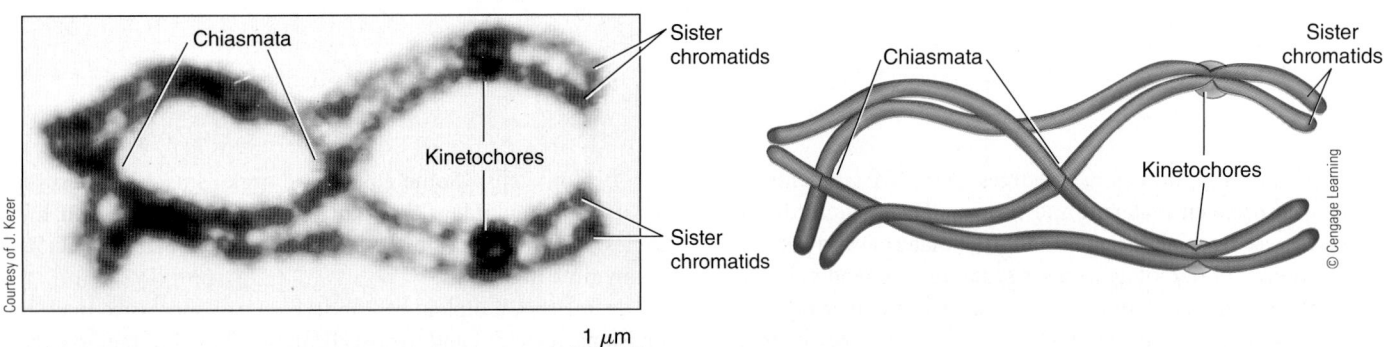

Chiasmata

Sister chromatids

Kinetochores

Sister chromatids

Courtesy of J. Kezer

1 μm

(a) LM of a tetrad during late prophase I of a male meiotic cell (spermatocyte) from a salamander.

Chiasmata

Sister chromatids

Kinetochores

(b) A drawing showing the structure of the tetrad. The paternal chromatids are *blue*, and the maternal chromatids are *red*.

Figure 10-18 *Animation* A meiotic tetrad with two chiasmata

The two chiasmata are the result of separate crossing-over events.

10.5 SEXUAL LIFE CYCLES

LEARNING OBJECTIVE

11 Compare the roles of mitosis and meiosis in various generalized life cycles.

Because sexual reproduction is characterized by the fusion of two haploid sex cells to form a diploid zygote, it follows that in a sexual life cycle, meiosis must occur before gametes can form. The timing of meiosis in the life cycle varies among species.

In animals and a few other organisms, meiosis leads directly to gamete production (**FIG. 10-19a**). An organism's somatic cells multiply by mitosis and are diploid; the only haploid cells

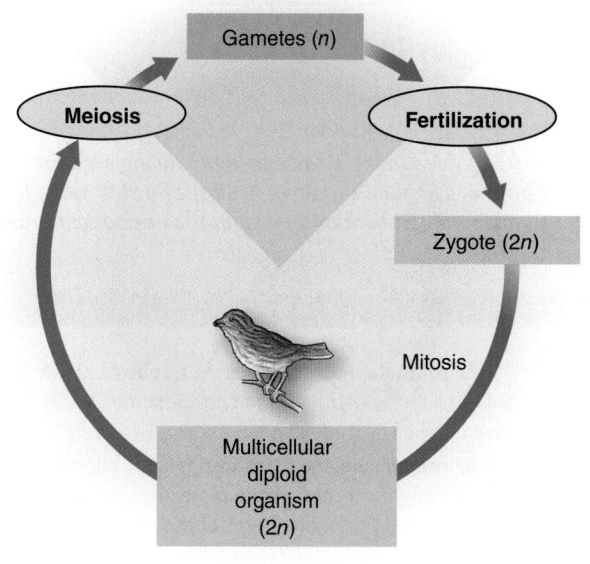

(a) Animals

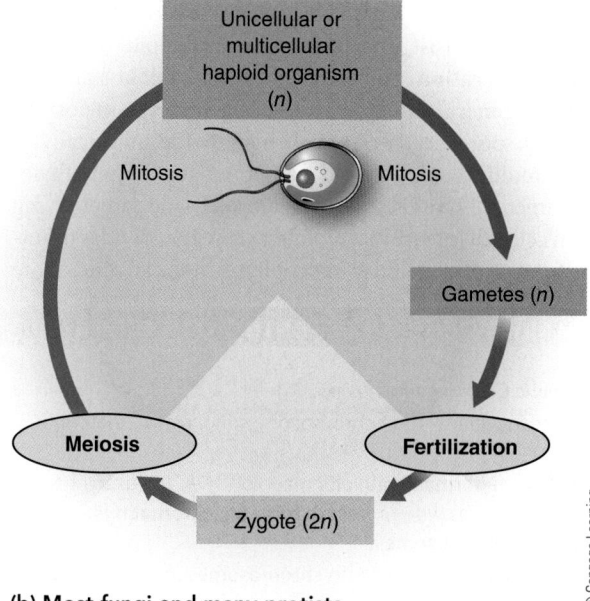

(b) Most fungi and many protists

© Cengage Learning

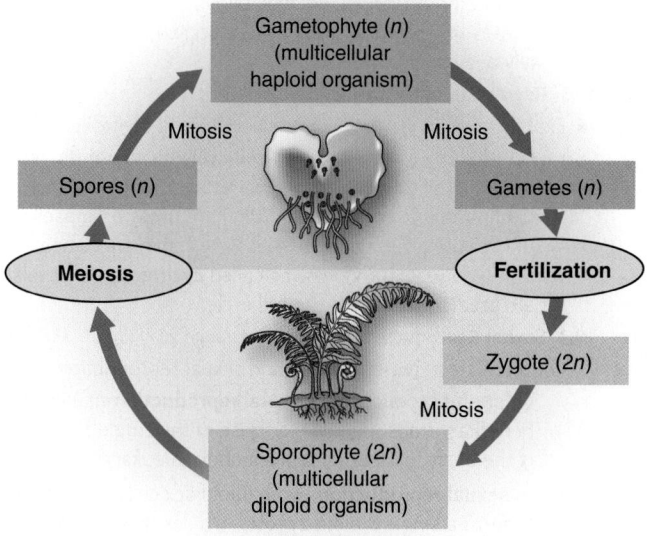

(c) Plants, some algae, and some fungi

Figure 10-19 *Animation* **Representative life cycles**

Each species has a characteristic number of chromosomes that does not change because the doubling of chromosomes that occurs during fertilization is compensated for by the reduction in chromosome number that occurs during meiosis. The color code and design here is used throughout the rest of the book. For example, in all life cycles the haploid (*n*) generation is shown in *purple*, and the diploid (2*n*) generation is *gold*. The processes of meiosis and fertilization always link the haploid and diploid generations.

produced are the gametes. Gametes develop when **germ line cells,** which give rise to the next generation, undergo meiosis.

The formation of gametes is known as **gametogenesis.** Male gametogenesis, called **spermatogenesis,** forms four haploid sperm cells for each cell that enters meiosis. (See Chapter 50 and Fig. 50-5 for a detailed description of spermatogenesis.)

In contrast, female gametogenesis, called **oogenesis,** forms a single egg cell, or *ovum,* for every cell that enters meiosis. In this process, most of the cytoplasm goes to only one of the two cells produced during each meiotic division. At the end of meiosis I, one nucleus is retained and the other, called the first *polar body,* often degenerates. Similarly, at the end of meiosis II, one nucleus becomes another polar body and the other nucleus

survives. In this way, one haploid nucleus receives most of the accumulated cytoplasm and nutrients from the original meiotic cell. (See Chapter 50 and Fig. 50-11 for a detailed description of oogenesis.)

Although meiosis occurs at some point in a sexual life cycle, it does not always *immediately* precede gamete formation. Many eukaryotes, including most fungi and many protists, remain haploid (their cells dividing mitotically) throughout most of their life cycles, with individuals being unicellular or multicellular. Two haploid gametes (produced by mitosis) fuse to form a diploid zygote that undergoes meiosis to restore the haploid state (**FIG. 10-19b**). Examples of these types of life cycles are found in Figures 26-16 and 29-9.

Plants and some algae and fungi have complicated life cycles (FIG. 10-19c). These life cycles, characterized by an **alternation of generations,** consist of a multicellular diploid stage, the **sporophyte generation,** and a multicellular haploid stage, the **gametophyte generation.** Sporophyte cells undergo meiosis to form haploid spores, each of which may divide mitotically to produce a multicellular haploid gametophyte. Gametophytes produce gametes by mitosis. The female and male gametes (egg and sperm cells) fuse to form a diploid zygote that divides mitotically to form a multicellular sporophyte. You can find more detailed descriptions of alternation of generations in plants in Chapters 27 and 28.

CHECKPOINT 10.5

- **CONNECT** *In animals, which cells are produced by mitosis? by meiosis? Which cells are produced by these processes in plants?*
- **CONNECT** *Assuming a sexually reproducing organism such as a protist is haploid throughout most of its life cycle, how do meiosis and fertilization maintain the normal chromosome number?*

SUMMARY: FOCUS ON LEARNING OBJECTIVES

10.1 Eukaryotic Chromosomes *(page 205)*

1 Discuss the significance of chromosomes in terms of their information content.
- Genes are informational units made of DNA. In eukaryotes, DNA associates with protein to form the **chromatin** fibers that make up **chromosomes.**

2 Explain how DNA is packed into chromosomes in eukaryotic cells.
- The organization of eukaryotic DNA into chromosomes allows the DNA to be accurately replicated and sorted into daughter cells without tangling. In eukaryotic cells DNA is associated with **histones** (basic proteins) to form **nucleosomes,** each of which consists of a histone bead with DNA wrapped around it. Nucleosomes are organized into large, coiled loops held together by nonhistone **scaffolding proteins.**

10.2 The Cell Cycle and Mitosis *(page 208)*

3 Identify the stages in the eukaryotic cell cycle and describe their principal events.
- The eukaryotic **cell cycle** is the period from the beginning of one division to the beginning of the next. The cell cycle consists of interphase and M phase.
- **Interphase** consists of the first gap phase (G_1), the synthesis phase (S), and the second gap phase (G_2). During the **G_1 phase,** the cell grows and prepares for the S phase. During the **S phase,** DNA and the chromosome proteins are synthesized, and chromosome duplication occurs. During the **G_2 phase,** protein synthesis increases in preparation for cell division.
- **M phase** consists of **mitosis,** the nuclear division that produces two nuclei identical to the parental nucleus, and **cytokinesis,** the division of the cytoplasm to yield two daughter cells.

4 Describe the structure of a duplicated chromosome, including the sister chromatids, centromeres, and kinetochores.
- A duplicated chromosome consists of a pair of **sister chromatids,** which contain identical DNA sequences. Each chromatid includes a constricted region called a **centromere.** Sister chromatids are tightly associated in the region of their centromeres. Attached to each centromere is a **kinetochore,** a protein structure to which microtubules can bind.

5 Explain the significance of mitosis and describe the process.
- Mitosis assures that the chromosome number is preserved when one eukaryotic cell divides to form two. In mitosis, identical chromosomes are distributed to each pole of the cell, and a nuclear envelope forms around each set.
- During **prophase,** the structure of the duplicated chromosomes becomes apparent as the chromatin condenses; each is composed of a pair of identical sister chromatids. The nuclear envelope begins to disassemble, and the **mitotic spindle** begins to form.

- During **prometaphase,** spindle microtubules attach to kinetochores of chromosomes, and chromosomes begin to move toward the cell's midplane.
- During **metaphase,** the chromosomes are aligned on the cell's midplane, or **metaphase plate;** the mitotic spindle is complete, and the kinetochores of the sister chromatids are attached by microtubules to opposite poles of the cell.
- During **anaphase,** the sister chromatids separate and move to opposite poles. Each former chromatid is now a chromosome.
- During **telophase,** a nuclear envelope re-forms around each set of chromosomes, nucleoli become apparent, the chromosomes uncoil, and the spindle disappears. Cytokinesis generally begins in telophase.

10.3 Regulation of the Cell Cycle *(page 215)*

6 Explain some ways in which the cell cycle is controlled.
- Control mechanisms, called **cell-cycle checkpoints,** temporarily block key events from being initiated during the cell cycle. **Cyclin-dependent kinases (Cdks)** are **protein kinases** involved in regulating the cell cycle. Cdks are active only when they bind tightly to regulatory proteins called **cyclins.** Cyclin levels fluctuate predictably during the cell cycle.

10.4 Sexual Reproduction and Meiosis *(page 217)*

7 Differentiate between asexual and sexual reproduction.
- Offspring produced by **asexual reproduction** usually have hereditary traits identical to those of the single parent. Mitosis is the basis for asexual reproduction in eukaryotic organisms.
- In **sexual reproduction** two haploid sex cells, or **gametes,** fuse to form a single diploid **zygote.** In a sexual life cycle, meiosis must occur before gametes can be produced.

8 Distinguish between haploid and diploid cells, and define *homologous chromosomes.*
- A **diploid** cell has a characteristic number of chromosome pairs per cell. The members of each pair, called **homologous chromosomes,** are similar in length, shape, and other features and carry genes affecting the same kinds of attributes of the organism.
- A **haploid** cell contains only one member of each homologous chromosome pair.

9 Explain the significance of meiosis and describe the process.
- A diploid cell undergoing **meiosis** completes two successive cell divisions, yielding four haploid cells. Sexual life cycles in eukaryotes require meiosis, which makes it possible for each gamete to contain only half the number of chromosomes in the parent cell.
- **Meiosis I** begins with **prophase I,** in which the members of a homologous pair of chromosomes physically join by the process of **synapsis. Crossing-over** is a process of **genetic recombination** during which homologous (nonsister) chromatids exchange segments of DNA strands.

- At **metaphase I**, **tetrads**—each consisting of a pair of homologous chromosomes held together by one or more **chiasmata**—line up on the metaphase plate. The members of each pair of homologous chromosomes separate during meiotic **anaphase I** and are distributed to different nuclei. Each nucleus contains the haploid number of chromosomes; each chromosome consists of two chromatids.

- During **meiosis II**, the two chromatids of each chromosome separate, and one is distributed to each daughter cell. Each former chromatid is now a chromosome.

10 Contrast mitosis and meiosis, emphasizing the different outcomes.

- Mitosis involves a single nuclear division in which the two daughter cells formed are genetically identical to each other and to the original cell. Synapsis of homologous chromosomes does not occur during mitosis.

- Meiosis involves two successive nuclear divisions and forms four haploid cells. Synapsis of homologous chromosomes occurs during prophase I of meiosis.

10.5 Sexual Life Cycles (page 222)

11 Compare the roles of mitosis and meiosis in various generalized life cycles.

- Somatic cells of animals are diploid and are produced by mitosis. The only haploid cells are gametes, produced by **gametogenesis,** which in animals occurs by meiosis.

- Most fungi and many protists are haploid and are produced by mitosis. The only diploid stage is the zygote, which undergoes meiosis to restore the haploid state.

- The life cycle of plants, some algae, and some fungi includes an **alternation of generations.** The multicellular diploid **sporophyte generation** forms haploid spores by meiosis. Each spore divides mitotically to form a multicellular haploid **gametophyte generation,** which produces gametes by mitosis. Two haploid gametes then fuse to form a diploid zygote, which divides mitotically to produce a new sporophyte generation.

TEST YOUR UNDERSTANDING

Know and Comprehend

1. A nucleosome consists of (a) DNA and scaffolding proteins (b) scaffolding proteins and histones (c) DNA and histones (d) DNA, histones, and scaffolding proteins (e) histones only

2. At which of the following stages do human skin cell nuclei have the same DNA content? (a) early mitotic prophase and late mitotic telophase (b) G_1 and G_2 (c) G_1 and early mitotic prophase (d) G_1 and late mitotic telophase (e) G_2 and late mitotic telophase

3. In a cell at ___, each chromosome consists of a pair of attached chromatids. (a) mitotic prophase (b) meiotic prophase II (c) meiotic prophase I (d) meiotic anaphase I (e) all the preceding

4. The molecular tether that links sister chromatids of a duplicated chromosome to each other is (a) condensin (b) actin (c) myosin (d) cohesin (e) actomyosin

5. In an animal cell at mitotic metaphase, you would expect to find (a) two pairs of centrioles located on the metaphase plate (b) a pair of centrioles inside the nucleus (c) a pair of centrioles within each microtubule-organizing center (d) a centriole within each centromere (e) no centrioles

6. A diploid nucleus at early mitotic prophase has ___ set(s) of chromosomes; a diploid nucleus at mitotic telophase has ___ set(s) of chromosomes. (a) 1; 1 (b) 1; 2 (c) 2; 2 (d) 2; 1 (e) not enough information has been given

7. You would expect to find a synaptonemal complex in a cell at (a) mitotic prophase (b) meiotic prophase I (c) meiotic prophase II (d) meiotic anaphase I (e) meiotic anaphase II

8. A chiasma links a pair of (a) homologous chromosomes at prophase II (b) homologous chromosomes at late prophase I (c) sister chromatids at metaphase II (d) sister chromatids at mitotic metaphase (e) sister chromatids at metaphase I

Apply and Analyze

9. **VISUALIZE** Sketch a mitotic prophase chromosome and label sister chromatids, sister centromeres, and sister kinetochores.

10. **CONNECT** Does the DNA content of the cell change from the beginning of interphase to the end of interphase? Does the number of chromosomes change? Explain.

11. Fill out the following table for a cell with 5 chromosomes undergoing mitosis. Is this cell haploid or diploid? How do you know?

	Number of Duplicated Chromosomes	Number of Unduplicated Chromosomes	Number of Kinetochores
Prophase (number per cell)			
Metaphase (number per cell)			
Telophase (number per nucleus)			

12. Fill out the following table for a diploid cell with 14 chromosomes undergoing meiosis.

	Number of Duplicated Chromosomes	Number of Unduplicated Chromosomes	Number of Tetrads
Beginning of prophase I (number per cell)			
End of prophase I (number per cell)			
End of telophase II (number per nucleus)			

In questions 13 and 14, decide whether each is an example of sexual or asexual reproduction, and state why.

13. A diploid queen honeybee produces haploid eggs by meiosis. Some of these eggs are never fertilized and develop into haploid male honeybees (drones).

14. Seeds develop after a flower has been pollinated with pollen from the same plant.

Evaluate and Synthesize

15. **EVOLUTION LINK** How does mitosis provide evidence for relationships among eukaryotes as diverse as mammals and seaweeds?

16. **EVOLUTION LINK** Some organisms—for example, certain fungi—reproduce asexually when the environment is favorable and sexually when the environment becomes unfavorable. What might be the evolutionary advantage of sexual reproduction with the associated process of meiosis during unfavorable conditions?

aplia To access course materials, such as Aplia and other companion resources, please visit **www.cengagebrain.com.**

Gregor Mendel. This painting shows Mendel with his pea plants in the monastery garden at Brünn, Austria (now Brüno, Czech Republic).

© Pictorial Press Ltd/Alamy

KEY CONCEPTS

11.1 The experiments of Gregor Mendel, a pioneer in the field of genetics, revealed the basic principles of inheritance. In Mendel's principle of segregation, members of a gene pair segregate (separate) from one another prior to gamete formation. In the principle of independent assortment, members of different gene pairs assort independently (randomly) into gametes.

11.2 You can use probability to predict Mendelian inheritance: the product rule shows how to combine the probabilities of independent events, and the sum rule shows how to combine the probabilities of mutually exclusive events.

11.3 Chromosome behavior during meiosis helps explain Mendel's principles of inheritance.

11.4 Distinctive inheritance patterns (i.e., "extensions" to Mendel's principles) characterize some traits.

Human characters such as eye color and hair color, along with a multitude of other characteristics, are passed on from one generation to another. **Heredity,** the transmission of genetic information from parent to offspring, generally follows predictable patterns in organisms as diverse as humans, penguins, baker's yeast, and sunflowers. **Genetics,** the science of heredity, studies both genetic similarities and **genetic variation,** the differences between parents and offspring or among individuals of a population.

The study of inheritance as a modern branch of science began in the mid-19th century with the work of Gregor Mendel (1822–1884), a monk who bred pea plants (see picture). Mendel was the first scientist to effectively apply quantitative methods to the study of inheritance. He did not merely describe his observations; he planned his experiments carefully, recorded the data, and analyzed the results mathematically. Although unappreciated during his lifetime, his work was rediscovered in 1900.

During the decades following the rediscovery of Mendel's findings, geneticists extended Mendel's principles by correlating the transmission of genetic information from generation to generation with the behavior of chromosomes during *meiosis.* By studying a variety of organisms, geneticists verified Mendel's findings and added a growing list of so-called exceptions to his principles.

Geneticists study not only the transmission of genes but also the expression of genetic information. As you will see in this chapter and those that follow, understanding the relationships between an organism's genetic information and its characteristics has become increasingly sophisticated as biologists have learned more about the orderly transmission of information in cells from one generation to another.

11.1 MENDEL'S PRINCIPLES OF INHERITANCE

LEARNING OBJECTIVES

1 Define the terms *phenotype, genotype, locus, allele, dominant allele, recessive allele, homozygous,* and *heterozygous*.
2 Describe Mendel's principles of segregation and independent assortment.
3 Distinguish among monohybrid, dihybrid, and test crosses.
4 Explain Mendel's principles of segregation and independent assortment, given what scientists now know about genes and chromosomes.

Gregor Mendel was not the first plant breeder. At the time he began his work, breeders had long recognized the existence of **hybrid** plants and animals, the offspring of two genetically dissimilar parents. When Mendel began his breeding experiments in 1856, two main concepts about inheritance were widely accepted. First, all hybrid plants that are the offspring of genetically pure, or **true-breeding,** parents are similar in appearance. Second, when these hybrids mate with each other, they do not breed true; their offspring show a mixture of traits. Some look like their parents, and some have features like those of their grandparents.

Mendel's genius lay in his ability to recognize a pattern in the way the parental traits reappear in the offspring of hybrids. Before Mendel, no one had categorized and counted the offspring and analyzed these regular patterns over several generations to the extent he did. Just as geneticists do today, Mendel chose the organism for his experiments very carefully. The garden pea, *Pisum sativum*, had several advantages. Pea plants are easy to grow, and many varieties were commercially available. Another advantage of pea plants is that controlled pollinations are relatively easy to conduct. Pea flowers have both male and female parts and naturally self-pollinate. However, the anthers (the male parts of the flower that produce pollen) can be removed to prevent self-fertilization (**FIG. 11-1**). Pollen from

RESEARCH METHOD

WHY IS IT USED? Garden peas normally self-fertilize during reproduction; that is, the male and female gametes are from the same flower. Because pea petals completely enclose the reproductive parts, there is little chance of natural cross-pollination between separate flowers. Cross-pollination enables the researcher to study various patterns of inheritance in peas.

HOW IS IT DONE?

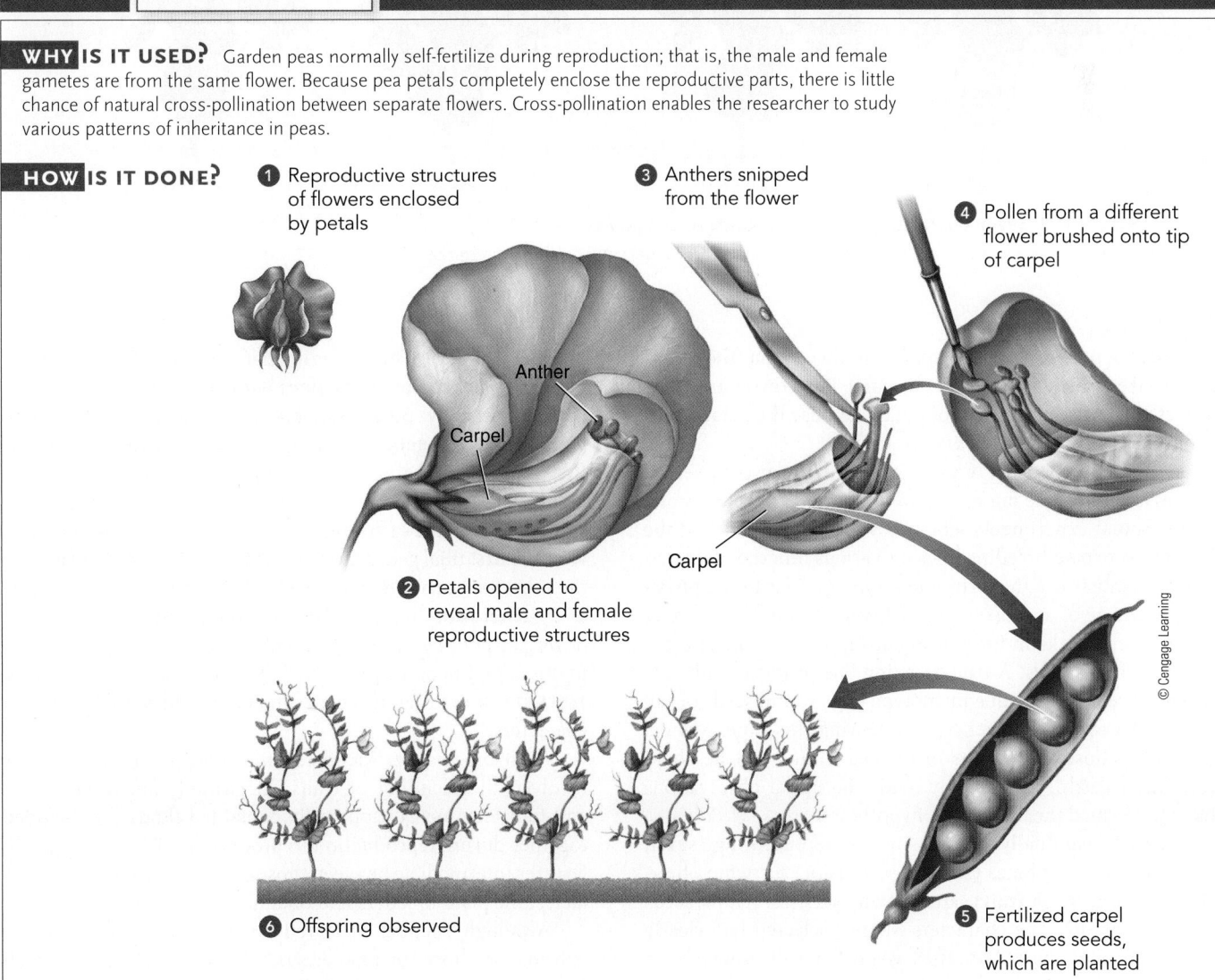

1 Reproductive structures of flowers enclosed by petals

Anther

Carpel

2 Petals opened to reveal male and female reproductive structures

3 Anthers snipped from the flower

4 Pollen from a different flower brushed onto tip of carpel

Carpel

5 Fertilized carpel produces seeds, which are planted

6 Offspring observed

© Cengage Learning

Figure 11-1 *Animation* **How garden peas are cross-pollinated**

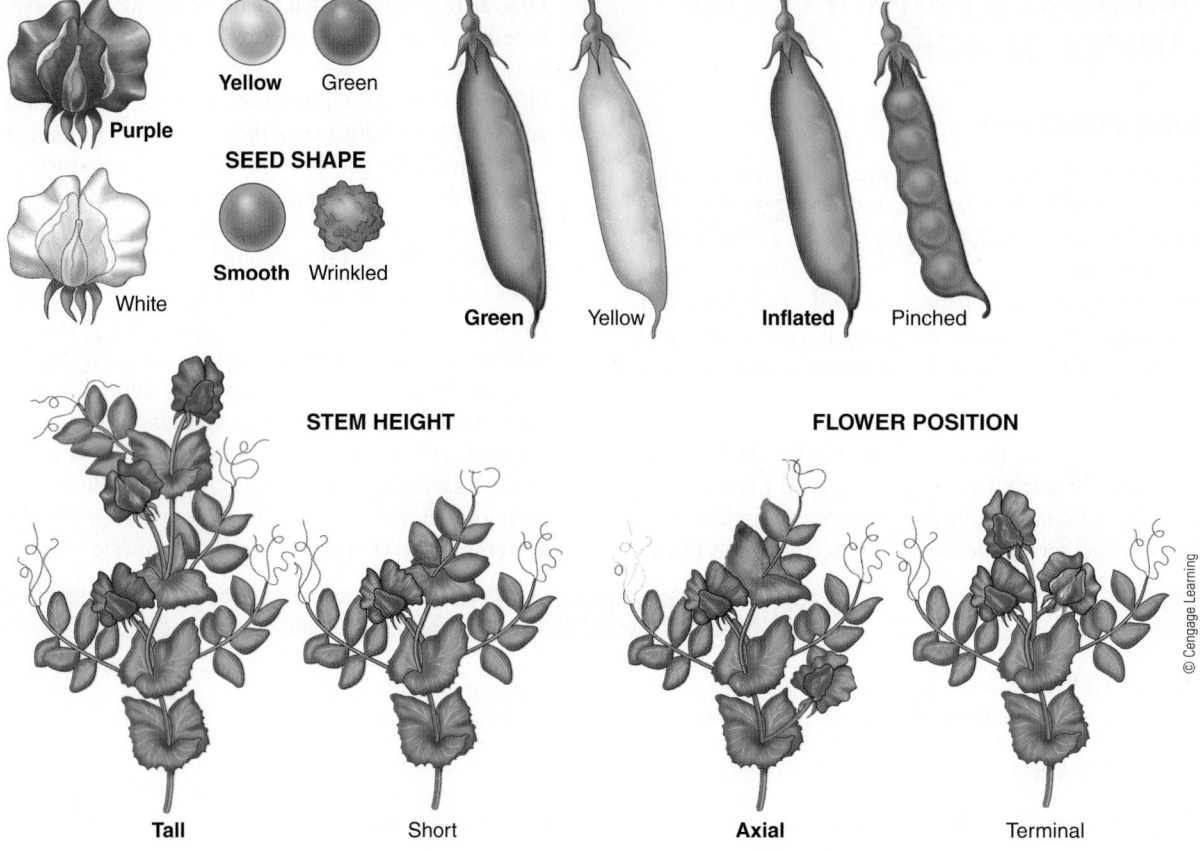

FLOWER COLOR — Purple / White

SEED COLOR — Yellow / Green

SEED SHAPE — Smooth / Wrinkled

POD COLOR — Green / Yellow

POD SHAPE — Inflated / Pinched

STEM HEIGHT — Tall / Short

FLOWER POSITION — Axial / Terminal

© Cengage Learning

Figure 11-2 Seven characters in Mendel's study of pea plants
Each character had two clearly distinguishable phenotypes; the dominant phenotype is boldface.

a different source can then be applied to the stigma (the receptive surface of the carpel, or female part). Pea flowers are easily protected from other sources of pollen because the petals completely enclose the reproductive structures.

Mendel obtained his original pea seeds from commercial sources and did some important preliminary work before starting his actual experiments. For two years he verified that the varieties were true-breeding lines for various inherited features. Today, scientists use the term **phenotype** to refer to the physical appearance of an organism and the term **genotype** to refer to the genetic makeup for that organism, which is most often expressed in symbols. A true-breeding line produces only offspring expressing the same phenotype (e.g., round seeds or tall plants) generation after generation. During this time, Mendel apparently chose those traits of his pea strains that he could study most easily. He probably made the initial observations that later formed the basis of his hypotheses.

Mendel eventually chose strains representing seven **characters**, the attributes (such as seed color) for which heritable differences, or **traits,** are known (such as yellow seeds and green seeds). The characters Mendel selected had clearly contrasting phenotypes (**FIG. 11-2**). Mendel's results were easy to analyze because he chose easily distinguishable phenotypes and limited the genetic variation studied in each experiment.

Mendel began his experiments by crossing plants from two different true-breeding lines with contrasting phenotypes; these genetically pure individuals constituted the **parental generation,** or **P generation.** In every case, the members of the first generation of offspring all looked alike and resembled one of the two parents. For example, when he crossed tall plants with short plants, all the offspring were tall (**FIG. 11-3**). These offspring were the first filial generation, or the **F$_1$ generation** (*filial* is from the Latin for "sons and daughters"). The second filial generation, or **F$_2$ generation,** resulted from a cross between F$_1$ individuals or by self-pollination of F$_1$ individuals. Mendel's F$_2$ generation in this experiment included 787 tall plants and 277 short plants. **TABLE 11-1** shows Mendel's experimental results for all seven pea characters.

Most breeders in Mendel's time thought that inheritance involved the blending of traits. In *blending inheritance* male and female gametes supposedly contained fluids that blended together during reproduction to produce hybrid offspring with features intermediate between those of the mother and father. In fact, some plant breeders had obtained such hybrids.

Although Mendel observed some intermediate types of hybrids, he chose for further study those F$_1$ hybrids in which "hereditary factors" (as he called them) from one of the parents apparently masked the expression of those factors from the

When the F₁ generation of tall pea plants is self-pollinated, what phenotypes appear in the F₂ generation?

HYPOTHESIS: Although only the "factor" (gene) for tall height is expressed in the F₁ generation, Mendel hypothesized that the factor for short height is not lost. He predicted that the short phenotype would reappear in the F₂ generation.

EXPERIMENT: Mendel crossed true-breeding tall pea plants with true-breeding short pea plants, yielding only tall offspring in the F₁ generation. He then allowed these F₁ individuals to self-pollinate to yield the F₂ generation.

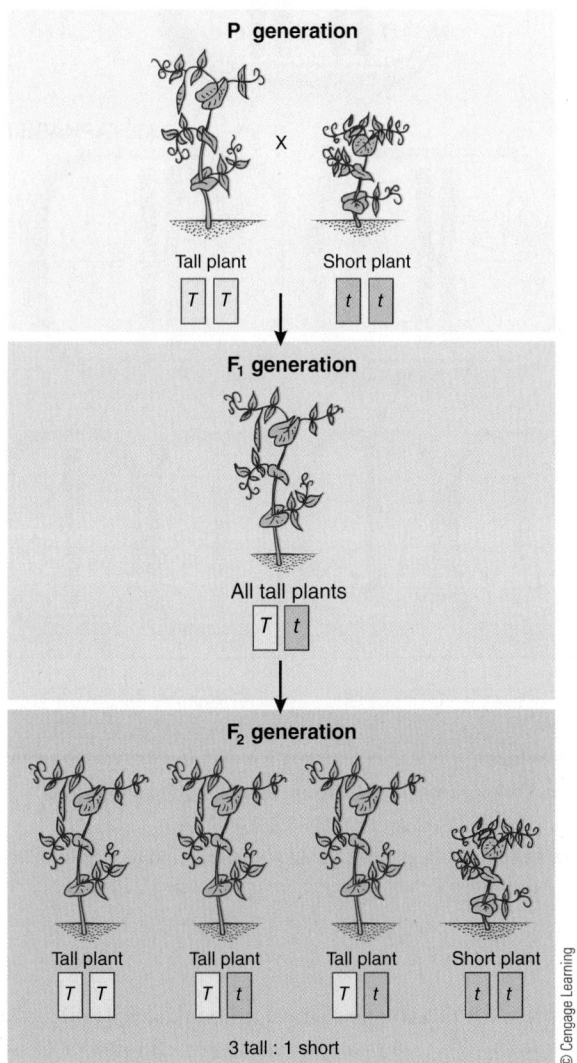

RESULTS AND CONCLUSION: The F₂ generation included 787 tall and 277 short plants, a ratio of about 3:1. Thus, Mendelian traits pass to successive generations in fixed ratios.

SOURCE: An extensively annotated English translation of Mendel's original 1866 paper is included in *Gregor Mendel's Experiments on Plant Hybrids: A Guided Study* (1993) Corcos, A.F. and F.V. Monaghan, Rutgers University Press.

Figure 11-3 One of Mendel's pea crosses

PREDICT Would you expect to obtain predictable results if you were to cross an F₂ generation tall plant with an F₂ generation short plant? Explain your answer.

CHARACTERS AND TRAITS IN PEAS			F₂ GENERATION	
CHARACTER	**DOMINANT TRAIT** ×	**RECESSIVE TRAIT**	**DOMINANT: RECESSIVE**	**RATIO**
Stem height	Tall ×	Short	787:277	2.84:1
Flower color	Purple ×	White	705:224	3.15:1
Flower position	Axial ×	Terminal	651:207	3.14:1
Pod color	Green ×	Yellow	428:152	2.82:1
Pod shape	Inflated ×	Pinched	882:299	2.95:1
Seed color	Yellow ×	Green	6022:2001	3.01:1
Seed shape	Smooth ×	Wrinkled	5474:1850	2.96:1

TABLE 11-1 Mendel's Experimental Results for Seven Characters

© Cengage Learning

other parent. Other breeders had also observed these types of hybrids, but they had not explained them. Using modern terms, we say that the factor expressed in the F₁ generation (tallness, in our example) is **dominant;** the one hidden in the F₁ (shortness) is **recessive.** Dominant traits mask recessive ones when both are present in the same individual. Although scientists know today that dominance is not always observed, the realization that dominance can occur was not consistent with the notion of blending inheritance.

Mendel's results were inconsistent with the hypothesis of blending inheritance in a more compelling way. Once two fluids have blended, it is very difficult to imagine how they can separate. However, in the preceding example, in the F₁ generation the hereditary factor(s) that controlled shortness clearly was not lost or blended inseparably with the hereditary factor(s) that controlled tallness because shortness reappeared in the F₂ generation. Mendel was very comfortable with the theoretical side of biology because he was also a student of physics and mathematics. He therefore proposed that each kind of inherited feature of an organism is controlled by two factors that behave like discrete particles and are present in every individual.

To Mendel these hereditary factors were abstractions: he knew nothing about chromosomes and DNA. These factors are essentially what scientists today call **genes,** units of heredity that affect an organism's traits. At the molecular level, a gene has been thought of as a DNA sequence that contains the information to make an RNA or protein product with a specific function. However, as you will see in Chapter 13, this definition of a gene is being reconsidered as more information accumulates about the many functions of DNA in the cell.

Mendel's experiments led to his discovery and explanation of the major principles of heredity, which we now know as the principles of segregation and independent assortment (see **TABLE 11-2** for a summary of Mendel's model of inheritance). We discuss the first principle next and the second later in the chapter.

Alleles separate before gametes are formed: the principle of segregation

Alternative forms of a gene are called **alleles.** In the example in Figure 11-3, each F₁ generation tall plant had two different

TABLE 11-2	Mendel's Model of Inheritance

1. Alternative forms of a "factor" (what we now call a *gene*) account for variations in inherited traits.

Although Mendel observed only two forms (what we now call *alleles*) for each factor he studied, we now know that many genes have more than two alleles.

2. Inherited traits pass from parents to offspring as unmodified factors.

Mendel did not observe offspring of intermediate appearance, as a hypothesis of blending inheritance would have predicted. Exceptions to this concept are known today.

3. Each individual has two sets of factors, one of each pair inherited from the mother and one from the father.

It does not matter which parent contributes which set of factors.

4. The paired factors separate prior to the formation of reproductive cells (the principle of segregation).

Because of *meiosis*, which was discovered after Mendel's time, each parent passes one set of factors to each offspring.

5. Factors may be expressed or hidden in a given generation, but they are never lost.

For example, factors not expressed in the F_1 generation reappear in some F_2 individuals.

6. Each factor is passed to the next generation independently from all other factors (the principle of independent assortment).

Research since Mendel's time has revealed that exceptions to his model are very common.

© Cengage Learning

Mendel's principle of segregation is related to the events of meiosis: the separation of homologous chromosomes during meiosis results in the segregation of alleles.

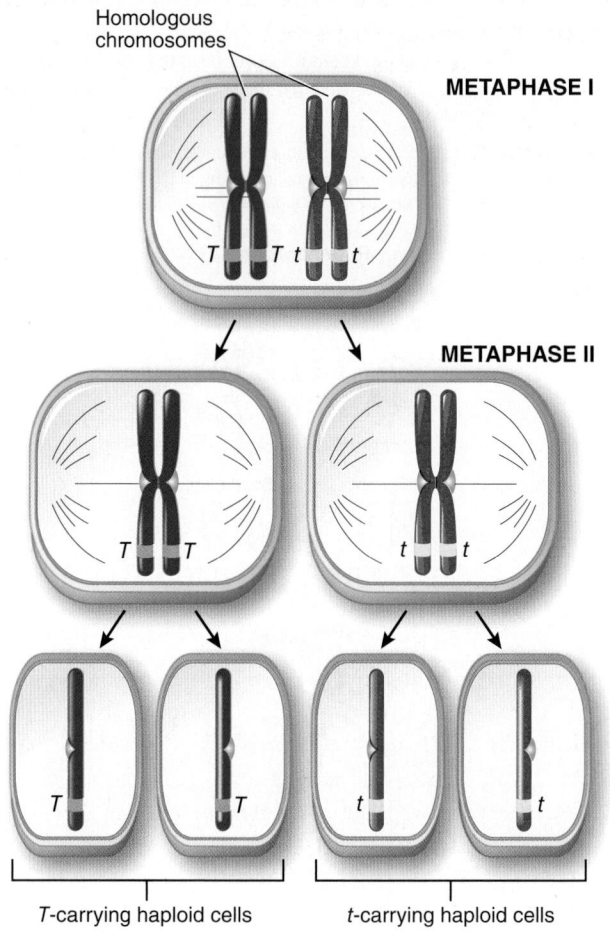

Figure 11-4 *Animation* Chromosomes and segregation

Note that half the haploid cells carry *T* and half carry *t*.

CONNECT What kinds of cells would the haploid cells be if meiosis were occurring in a plant? in an animal?

© Cengage Learning

alleles that control plant height: a **dominant allele** for tallness (which we designate *T*) and a **recessive allele** for shortness (designated *t*). Because the tall allele was dominant, these plants were tall. To explain his experimental results, Mendel proposed an idea now known as the **principle of segregation.** Using modern terminology, the principle of segregation states that before sexual reproduction occurs, the two alleles carried by an individual parent must become separated (i.e., segregated).

Recall that during meiosis, homologous chromosomes, and therefore the alleles that reside on them, separate. (You may want to review meiosis in Figure 10-16.) As a result, each sex cell (egg or sperm) formed contains only one allele of each pair. (Later, at the time of fertilization, each haploid gamete contributes one chromosome from each homologous pair and therefore one allele for each gene pair.) An essential feature of meiosis is that the alleles remain intact (one does not mix with or eliminate the other); thus, recessive alleles are not lost and can reappear in the F_2 generation. In our example, before the F_1 plants formed gametes, the allele for tallness segregated from the allele for shortness; so, half the gametes contained a *T* allele, and the other half, a *t* allele (FIG. 11-4).

The random process of fertilization led to three possible combinations of alleles in the F_2 offspring: one-fourth with two tallness alleles (*TT*), one-fourth with two shortness alleles (*tt*), and one-half with one allele for tallness and one for shortness (*Tt*). Because both *TT* and *Tt* plants are tall, Mendel expected approximately three-fourths (787 of the 1064 plants he obtained)

to express the phenotype of the dominant allele (tall) and about one-fourth (277/1064) to express the phenotype of the recessive allele (short). We will explain the mathematical reasoning behind these predictions later in the chapter.

Alleles occupy corresponding loci on homologous chromosomes

Today scientists know that each unduplicated chromosome consists of one long, linear DNA molecule and that each gene is actually a segment of that DNA molecule. We also know that homologous chromosomes not only are similar in size and shape but also usually have the same genes (often with different alleles) located in corresponding positions. The term **locus** (pl., *loci*) originally designated the location of a particular gene on the chromosome (**FIG. 11-5**). We are actually referring to a segment of the DNA that has the information for controlling some aspect of the structure or function of the organism. One locus may govern seed color, another seed shape, another shape of the pods, and so on. Traditional genetic methods can infer the existence of a particular locus only if at least two allelic variants of that locus, producing contrasting phenotypes (e.g., yellow peas versus green peas), are available for study. In the simplest cases, an individual can express one (yellow) or the other (green), but not both.

Alleles are therefore genes that govern variations of the same character (yellow versus green seed color) and occupy corresponding loci on homologous chromosomes. Geneticists

assign each allele of a locus a single letter or group of letters as its symbol.[1]

Although geneticists often use more complicated forms of notation, it is customary when working simple genetics problems to indicate a dominant allele with a capital letter. A recessive allele with the same letter is indicated in lowercase.

Remember that the term *locus* designates not only a position on a chromosome but also a type of gene controlling a particular character; thus, Y (yellow) and y (green) represent a specific pair of alleles of a locus involved in determining seed color in peas. Although you may initially be uncomfortable with geneticists sometimes using the term *gene* to specify a locus and at other times to specify one of the alleles of that locus, the meaning is usually clear from the context.

A monohybrid cross involves individuals with different alleles of a given locus

The basic principles of genetics and the use of genetics terms are best illustrated by examples. In the simplest case, a **monohybrid cross,** the inheritance of two different alleles of a single locus, is studied. **FIGURE 11-6** illustrates a monohybrid cross featuring a locus that governs coat color in guinea pigs. The female comes from a true-breeding line of black guinea pigs. We say that she is **homozygous** for black because the two alleles she carries for this locus are identical. The brown male is also from a true-breeding line and is homozygous for brown. What color would you expect the F_1 offspring to be? Actually, it is impossible to make such a prediction without more information.

---

[1] Early geneticists developed their own symbols to represent genes and alleles. Later, groups of scientists met and decided on specific symbols for a given research organism, such as the fruit fly, but each research group had its own rules for assigning symbols. Universally accepted rules for assigning symbols for genes and alleles still do not exist.

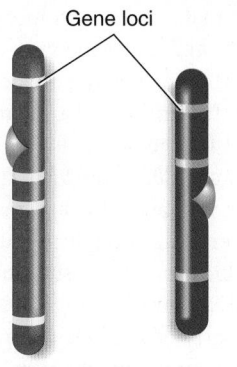

A gamete has one set of chromosomes, the *n* number. It carries *one* chromosome of *each* homologous pair. A given gamete can have only *one* gene of any particular pair of alleles.

When the gametes fuse, the resulting zygote is diploid (2*n*) and has homologous pairs of chromosomes. For purposes of illustration, these chromosomes are shown physically paired.

(a) One member of each pair of homologous chromosomes is of maternal origin (*red*), and the other is paternal (*blue*).

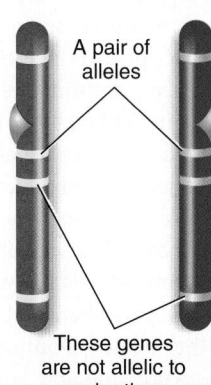

Gene loci

A pair of alleles

These genes are not allelic to each other

(b) These chromosomes are nonhomologous. Each chromosome is made up of hundreds or thousands of genes. A locus is the specific place on a chromosome where a gene is located.

(c) These chromosomes are homologous. Alleles are members of a gene pair that occupy corresponding loci on homologous chromosomes.

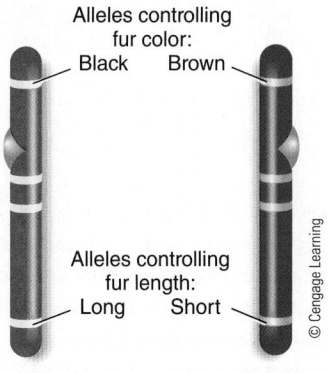

Alleles controlling fur color:
Black   Brown

Alleles controlling fur length:
Long   Short

(d) Alleles govern the same character but do not necessarily contain the same information.

Figure 11-5 *Animation* **Gene loci and their alleles**

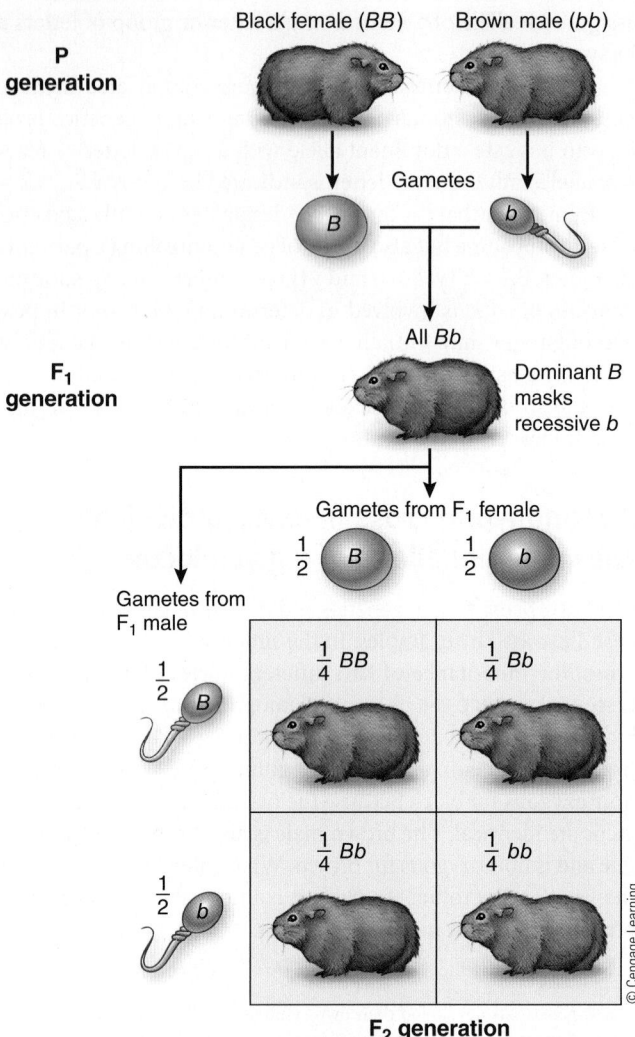

P generation

Black female (*BB*)   Brown male (*bb*)

Gametes

B   *b*

All *Bb*

F₁ generation

Dominant *B* masks recessive *b*

Gametes from F₁ female

$\frac{1}{2}$ *B* $\frac{1}{2}$ *b*

Gametes from F₁ male

$\frac{1}{2}$ *B*

$\frac{1}{4}$ *BB* $\frac{1}{4}$ *Bb*

$\frac{1}{2}$ *b*

$\frac{1}{4}$ *Bb* $\frac{1}{4}$ *bb*

© Cengage Learning

F₂ generation

Figure 11-6 *Animation* **A monohybrid cross in guinea pigs**
In this example a homozygous black guinea pig is mated with a homozygous brown guinea pig. The F₁ generation includes only black individuals. However, the mating of two of these F₁ offspring yields F₂ generation offspring in the expected ratio of 3 black to 1 brown, indicating that the F₁ individuals are heterozygous.

PREDICT What results would you expect if the P generation female were homozygous brown and the male homozygous black?

In this particular case, the F₁ offspring are black, but they are *heterozygous*, meaning that they carry two different alleles for this locus. The brown allele determines coat color only in a homozygous brown individual; it is a recessive allele. The black allele influences coat color in both homozygous black and heterozygous individuals; it is a dominant allele. On the basis of this information, we can use symbols to designate the dominant black allele *B* and the recessive brown allele *b*.

During meiosis in the female parent (*BB*), the two *B* alleles separate according to Mendel's principle of segregation, so each egg has only one *B* allele. In the male (*bb*) the two *b* alleles separate, so each sperm has only one *b* allele. The fertilization of each *B* egg by a *b* sperm results in heterozygous F₁ offspring, each *Bb*; that is, each individual has one allele for brown coat and one for black coat. Because it is the only possible combination of alleles

present in the eggs and sperm, all the F₁ offspring are *Bb*. Also, note that it does not make any difference whether the offspring receive the dominant allele from the mother or father guinea pig.

A Punnett square predicts the ratios of the various offspring of a cross During meiosis in heterozygous black guinea pigs (*Bb*), the chromosome containing the *B* allele becomes separated from its homologue (the chromosome containing the *b* allele), so each normal sperm or egg contains *B* or *b*, but never both. Heterozygous *Bb* individuals form gametes containing *B* alleles and gametes containing *b* alleles in equal numbers. Because no special attraction or repulsion occurs between an egg and a sperm containing the same allele, fertilization is a random process.

As you can see in Figure 11-6, the possible combinations of eggs and sperm at fertilization can be represented in the form of a grid, or **Punnett square,** devised by early English geneticist Sir Reginald Punnett. The types of gametes (and their expected frequencies) from one parent are listed across the top, and those from the other parent are listed along the left side. The squares are then filled in with the resulting F₂ combinations. Three-fourths of all F₂ offspring have the genetic constitution *BB* or *Bb* and are phenotypically black; one-fourth have the genetic constitution *bb* and are phenotypically brown. The genetic mechanism that governs the approximate 3:1 F₂ ratios obtained by Mendel in his pea-breeding experiments is again evident. These ratios are called *monohybrid F₂ phenotypic ratios*.

The phenotype of an individual does not always reveal its genotype As mentioned earlier, an organism's phenotype is its appearance with respect to a certain inherited trait. However, because some alleles may be dominant and others recessive, we cannot always determine, simply by examining its phenotype, which alleles are carried by an organism. In the cross we have been considering, the parents are from true-breeding lines, so the genotype of the female parent is homozygous dominant, *BB*, and her phenotype is black. The genotype of the male parent is homozygous recessive, *bb*, and his phenotype is brown. The genotype of all the F₁ offspring is heterozygous, *Bb*, and their phenotype is black. To prevent confusion, we always indicate the genotype of a heterozygous individual by writing the symbol for the dominant allele first and the recessive allele second (always *Bb*, never *bB*). The expression of dominance partly explains why an individual may resemble one parent more than the other, even though the two parents contribute equally to their offspring's genetic constitution.

Dominance is not predictable; that is, dominance is not an intrinsic part of an allele but is the property of an allele relative to other alleles. Dominance is produced by the mechanism of gene expression and can be determined only by experiment. In one species of animal, black coat may be dominant to brown; in another species, brown may be dominant to black. In a population, the dominant phenotype is not necessarily more common than the recessive phenotype.

A test cross can detect heterozygosity Guinea pigs with the genotypes *BB* and *Bb* are alike phenotypically; they both have black coats. How, then, can we know the genotype of a black guinea pig?

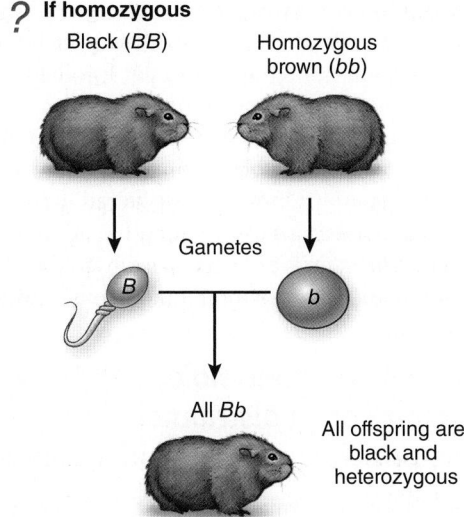

? If homozygous

Black (*BB*)   Homozygous brown (*bb*)

Gametes

B   *b*

All *Bb*

All offspring are black and heterozygous

(a) If a black guinea pig is mated with a brown guinea pig and all the offspring are black, the black parent probably has a homozygous genotype.

Figure 11-7 *Animation* **A test cross in guinea pigs**
In this illustration, a test cross is used to determine the genotype of a black guinea pig.

PREDICT If this cross were to produce a single offspring, what might you conclude if it happened to be black? brown?

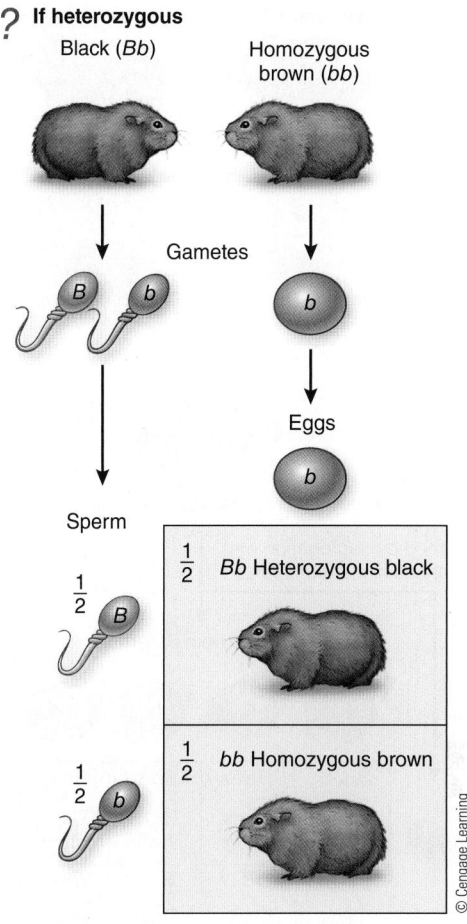

? If heterozygous

Black (*Bb*)   Homozygous brown (*bb*)

Gametes

B   *b*   *b*

Eggs

b

Sperm

$\frac{1}{2}$ *B*

$\frac{1}{2}$ *b*

$\frac{1}{2}$ *Bb* Heterozygous black

$\frac{1}{2}$ *bb* Homozygous brown

© Cengage Learning

(b) If any of the offspring is brown, the black guinea pig must be heterozygous. The expected phenotypic ratio is 1 black to 1 brown.

Geneticists find out by performing an experimental cross known as a **test cross,** in which an individual of unknown genotype is crossed with a recessive individual (**FIG. 11-7**). In a test cross, the alleles carried by the gametes from the parent of unknown genotype are never "hidden" in the offspring by dominant alleles contributed by the other parent. Therefore, you can deduce the genotypes of all offspring directly from their phenotypes.

Mendel conducted several test crosses; for example, he bred F_1 (tall) pea plants with homozygous recessive (*tt*) short ones. He reasoned that the F_1 individuals were heterozygous (*Tt*) and would be expected to produce equal numbers of *T* and *t* gametes. Because the homozygous short parents (*tt*) were expected to produce only *t* gametes, Mendel hypothesized that he would obtain equal numbers of tall (*Tt*) and short (*tt*) offspring. His results agreed with his hypothesis, providing additional evidence for the hypothesis that there is 1:1 segregation of the alleles of a heterozygous parent. Thus, Mendel's principle of segregation not only explained the known facts, such as the 3:1 monohybrid F_2 phenotypic ratio, but also let him successfully anticipate the results of other experiments, in this case, the 1:1 test-cross phenotypic ratio.

A dihybrid cross involves individuals that have different alleles at two loci

Monohybrid crosses involve a pair of alleles of a single locus. Mendel also analyzed crosses involving alleles of two or more loci. A mating between individuals with different alleles at two loci is called a **dihybrid cross.** Consider the case of two pairs of alleles carried on nonhomologous chromosomes (i.e., one pair of alleles is in one pair of homologous chromosomes, and the other pair of alleles is in a different pair of homologous chromosomes). Each pair of alleles is inherited independently; that is, each pair segregates during meiosis independently of the other.

An example of a dihybrid cross carried through the F_2 generation is shown in **FIGURE 11-8**. In this example black is dominant to brown, and short hair is dominant to long hair. When a homozygous, black, short-haired guinea pig (*BBSS*) and a homozygous, brown, long-haired guinea pig (*bbss*) are mated, the *BBSS* animal produces gametes that are all *BS*, and the *bbss* individual produces gametes that are all *bs*. Each gamete contains one allele for each of the two loci. The union of the *BS* and *bs* gametes yields only individuals with

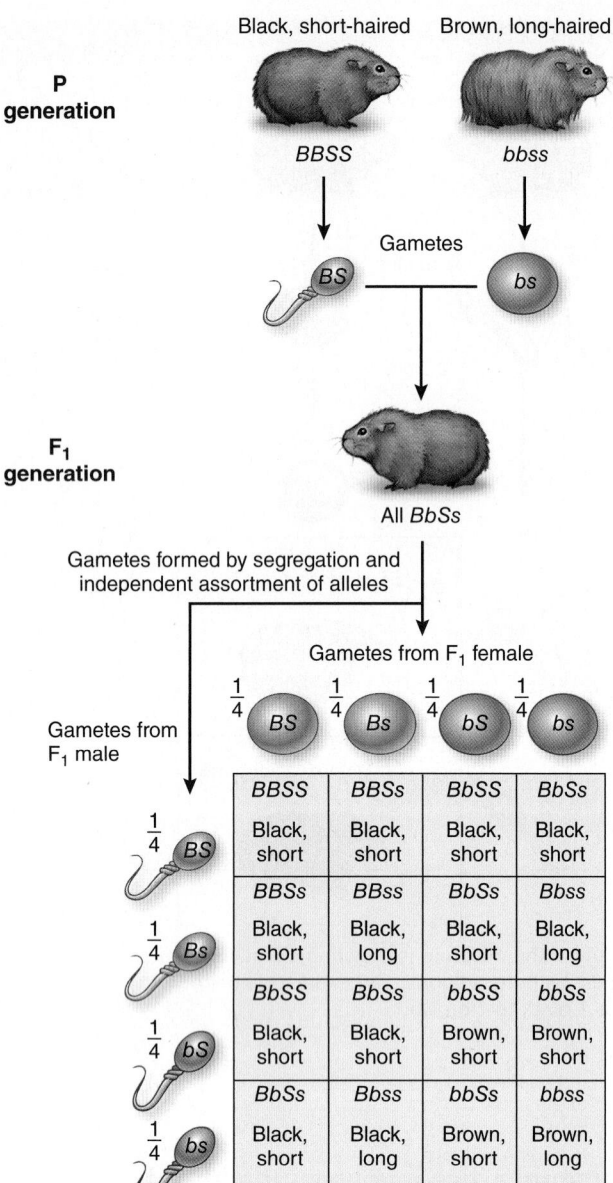

Each F_1 guinea pig produces four kinds of gametes with equal probability: *BS, Bs, bS,* and *bs.* Hence, the Punnett square has 16 (i.e., 4^2) squares representing the F_2 offspring, some of which are genotypically or phenotypically alike. There are 9 chances in 16 of obtaining a black, short-haired individual; 3 chances in 16 of obtaining a black, long-haired individual; 3 chances in 16 of obtaining a brown, short-haired individual; and 1 chance in 16 of obtaining a brown, long-haired individual. This 9:3:3:1 phenotypic ratio is expected in a dihybrid F_2 if the hair color and hair length loci are on nonhomologous chromosomes.

Alleles on nonhomologous chromosomes are randomly distributed into gametes: the principle of independent assortment

On the basis of results similar to the guinea pig example, Mendel formulated the principle of inheritance, now known as Mendel's **principle of independent assortment**, which states that members of any gene pair segregate from each other independently of the members of the other gene pairs. This mechanism occurs in a regular way to ensure that each gamete contains one allele for each locus, but the alleles of different loci are assorted at random with respect to each other in the gametes. The independent assortment of these alleles can result in **genetic recombination** (or simply **recombination**), the process of assorting and passing alleles to offspring in new combinations that are different from those in the parents.

Today we recognize that independent assortment is related to the events of meiosis. It occurs because two pairs of homologous chromosomes can be arranged in two different ways at metaphase I of meiosis (**FIG. 11-9**). These arrangements occur randomly, with approximately half the meiotic cells oriented one way and half oriented the opposite way. The orientation of the homologous chromosomes on the metaphase plate then determines the way they subsequently separate and disperse into the haploid cells. (As you will soon see, however, independent assortment does not always occur.)

Recognition of Mendel's work came during the early 20th century

Mendel reported his findings at a meeting of the Brünn Society for the Study of Natural Science (in what is now the Czech Republic), and he published his results in the society's report in 1866. At that time biology was largely a descriptive science, and biologists had little interest in applying quantitative and experimental methods such as those Mendel had used. Other biologists of the time did not appreciate the importance of his results nor his interpretations of those results. For 34 years, his findings were largely neglected.

In 1900, Hugo DeVries in Holland, Carl Correns in Germany, and Erich von Tschermak in Austria recognized Mendel's principles in their own experiments; they later discovered Mendel's paper and found that it explained their own research observations. By this time biologists had a much greater appreciation of the value of quantitative experimental methods.

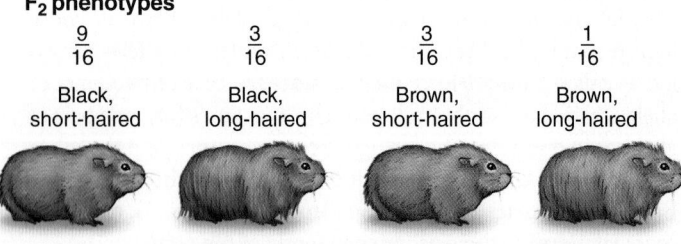

Figure 11-8 *Animation* **A dihybrid cross in guinea pigs**

When a black, short-haired guinea pig is crossed with a brown, long-haired one, all the offspring are black and have short hair. However, when two members of the F_1 generation are crossed, the ratio of phenotypes is 9:3:3:1.

PREDICT Would these results be different if the P generation genotypes were *BBss* and *bbSS*? Explain your answer.

the genotype *BbSs.* All these F_1 offspring are heterozygous for hair color and for hair length, and all are phenotypically black and short-haired.

Mendel's principle of independent assortment—that factors for different characteristics separate independently of one another prior to gamete formation—is a direct consequence of the events of meiosis.

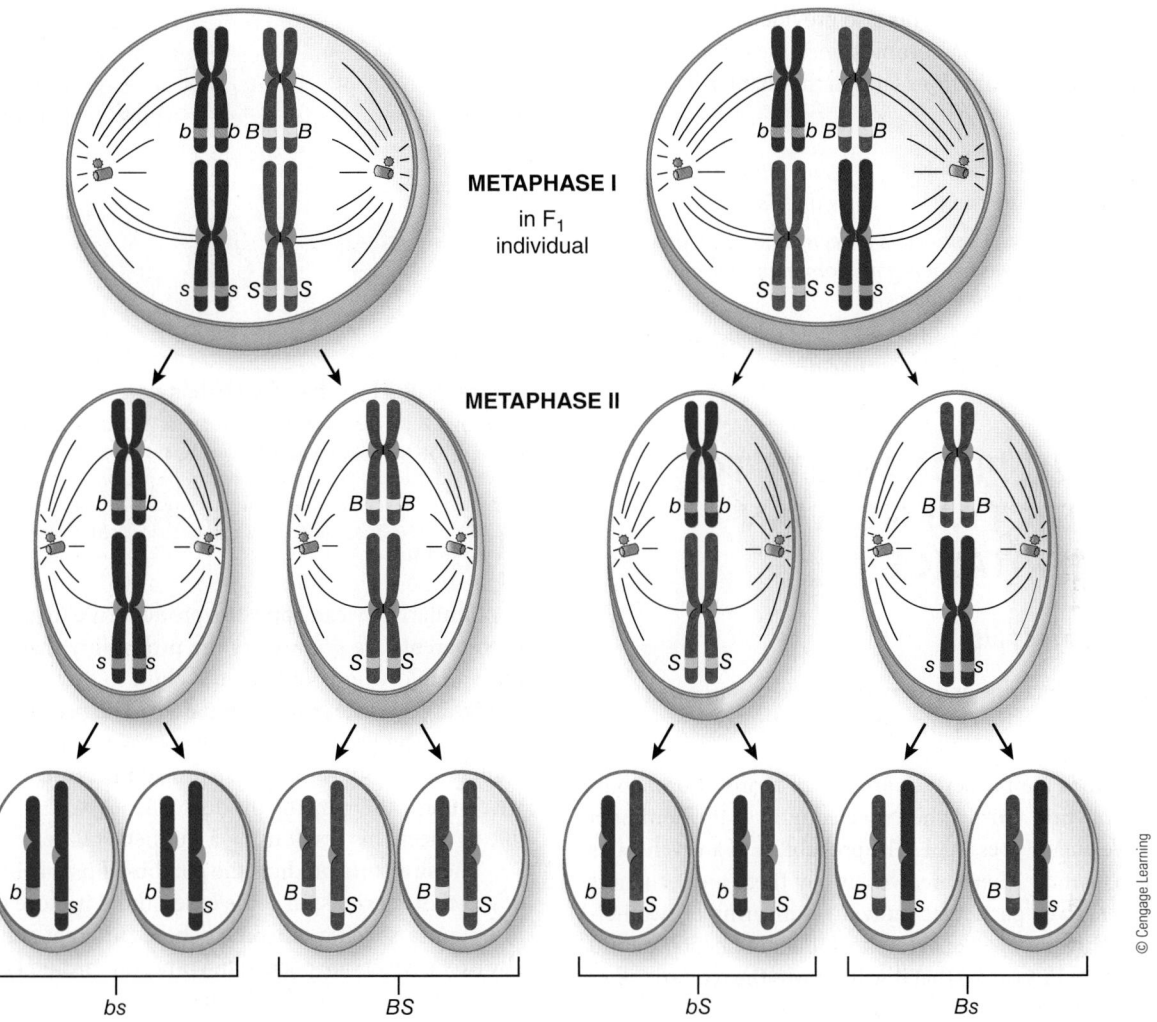

METAPHASE I in F₁ individual

METAPHASE II

bs   BS   bS   Bs

© Cengage Learning

Figure 11-9 *Animation* **Meiosis and independent assortment**

Two different pairs of homologous chromosomes can line up two different ways at metaphase I in an F₁ individual and be subsequently distributed. A cell with the orientation shown at the left produces half *BS* and half *bs* gametes. Conversely, the cell at the right produces half *Bs* and half *bS* gametes. Because approximately half of the meiotic cells at metaphase I are of each type, the ratio of the four possible types of gametes is 1:1:1:1.

CONNECT In Figure 11-8 the genotypes of the P generation were *BBSS* and *bbss*. Would this diagram be drawn differently if the P generation genotypes were *BBss* and *bbSS*? Explain your answer.

Correns gave credit to Mendel by naming the basic laws of inheritance after him.

Although gametes and fertilization were known at the time Mendel carried out his research, mitosis and meiosis had not yet been discovered. It is truly remarkable that Mendel formulated his ideas mainly on the basis of mathematical abstractions. Today his principles are much easier to understand because we relate the transmission of genes to the behavior of chromosomes.

The details of mitosis and meiosis were described during the late 19th century, and in 1902, American biologist

Walter Sutton and German biologist Theodor Boveri independently pointed out the connection between Mendel's segregation of alleles and the separation of homologous chromosomes during meiosis. This connection developed into the **chromosome theory of inheritance,** or the *Sutton–Boveri theory*, which stated that inheritance can be explained by assuming that genes are linearly arranged in specific locations along the chromosomes.

The chromosome theory of inheritance was initially controversial because at that time there was no direct evidence that genes are found on chromosomes. However, new research provided the findings necessary for wider acceptance and extension of these ideas and their implications. For example, the work of American geneticist Thomas Hunt Morgan in 1910 provided evidence for the location of a particular gene (white eye color) on a specific chromosome (the X chromosome) in fruit flies. Morgan and his graduate students also provided insight into the way genes are organized on chromosomes; we discuss some of Morgan's research contributions later in the chapter.

CHECKPOINT 11.1

- *What is the maximum number of different alleles for a particular locus that can be present in a single diploid individual?*
- **CONNECT** *Can Mendel's principle of segregation be illustrated by a cross between two homozygous dominant individuals? two homozygous recessive individuals?*
- **CONNECT** *Can a monohybrid cross be used to illustrate Mendel's principle of independent assortment?*

11.2 USING PROBABILITY TO PREDICT MENDELIAN INHERITANCE

LEARNING OBJECTIVE

5 Apply the product rule and sum rule appropriately when predicting the outcomes of genetic crosses.

All genetic ratios are properly expressed in terms of probabilities. In monohybrid crosses the expected ratio of the dominant and recessive phenotypes is 3:1. The probability of an event is its expected frequency. Therefore, we can say there are 3 chances in 4 (or ¾) that any particular individual offspring of two heterozygous individuals will express the dominant phenotype and there is 1 chance in 4 (or ¼) that it will express the recessive phenotype. Although we sometimes speak in terms of percentages, probabilities are calculated as fractions (such as ¾) or decimal fractions (such as 0.75). If an event is certain to occur, its probability is 1; if it is certain *not* to occur, its probability is 0. A probability can be 0, 1, or some number between 0 and 1.

The Punnett square lets you combine two or more probabilities. When you use a Punnett square, you are following two important statistical principles known as the product rule and the sum rule. The **product rule** predicts the combined probabilities of independent events. Events are *independent* if the occurrence of one does not affect the probability that the other will occur. For example, the probability of obtaining heads on the first toss of a coin is ½; the probability of obtaining heads on the second toss (an independent event) is also ½. If two or more events are independent of each other, the probability of both occurring is the product of their individual probabilities. If this concept seems strange, keep in mind that when we multiply two numbers that are less than 1, the product is a smaller number. Therefore, the probability of obtaining heads two times in a row is ½ × ½ = ¼, or 1 chance in 4 (**FIG. 11-10**).

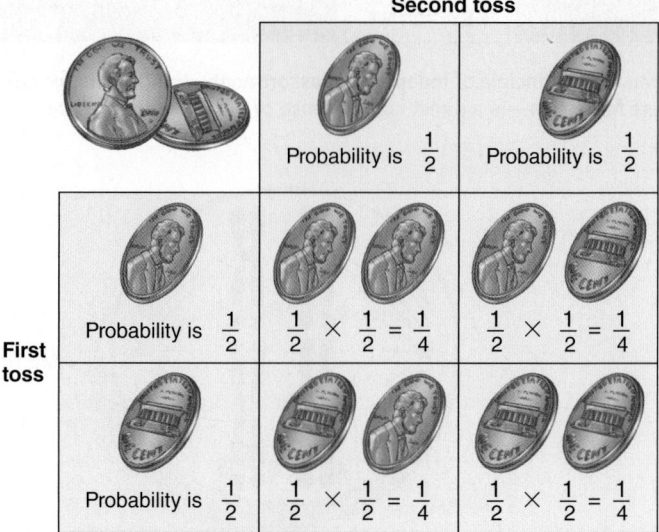

Figure 11-10 The rules of probability

For each coin toss, the probability of heads is ½, and the probability of tails is also ½. Because the outcome of the first toss is independent of the outcome of the second, the combined probabilities of the outcomes of successive tosses are calculated by multiplying their individual probabilities (according to the product rule: ½ x ½ = ¼). These same rules of probability predict genetic events.
© Cengage Learning

Similarly, we can apply the product rule to genetic events. If both parents are *Bb*, what is the probability that they will produce a child who is *bb*? For the child to be *bb*, he or she must receive a *b* gamete from each parent. The probability of a *b* egg is ½, and the probability of a *b* sperm is also ½. Like the outcomes of the coin tosses, these probabilities are independent, so we combine them by the product rule (½ × ½ = ¼). You might like to check this result using a Punnett square.

The **sum rule** predicts the combined probabilities of *mutually exclusive* events. In some cases, there is more than one way to obtain a specific outcome. These different ways are mutually exclusive; if one occurs, the other(s) cannot. For example, if both parents are *Bb*, what is the probability that their first child will also have the *Bb* genotype? There are two different ways these parents can have a *Bb* child: either a *B* egg combines with a *b* sperm (probability ¼) or a *b* egg combines with a *B* sperm (probability ¼).

Naturally, if there is more than one way to get a result, the chances of its being obtained improve; we therefore combine the probabilities of mutually exclusive events by summing (adding) their individual probabilities. The probability of obtaining a *Bb* child in our example is therefore ¼ + ¼ = ½. (Because there is only one way these heterozygous parents can produce a homozygous recessive child, *bb*, that probability is only ¼. The probability of a homozygous dominant child, *BB*, is likewise ¼.)

The rules of probability can be applied to a variety of calculations

The rules of probability have wide applications. For example, what are the probabilities that a family with two (and only two) children will have two girls, two boys, or one girl and one boy? For

How do you solve simple Mendelian genetics problems? Such problems can be fun and easy to work if you follow certain conventions and are methodical in your approach.

1. Always use standard designations for the generations. The generation in which a particular genetic experiment is begun is called the P, or parental, generation. Offspring of this generation are called the F_1, or first filial, generation. The offspring resulting when two F_1 individuals are bred constitute the F_2, or second filial, generation.

2. Write down a key for the symbols you are using for the allelic variants of each locus. Use an uppercase letter to designate a dominant allele and a lowercase letter to designate a recessive allele. Use the same letter of the alphabet to designate both alleles of a particular locus. If you are not told which allele is dominant and which is recessive, the phenotype of the F_1 generation is a good clue.

3. Determine the genotypes of the parents of each cross by using the following types of evidence:
   - Are they from true-breeding lines? If so, they should be homozygous.
   - Can their genotypes be reliably deduced from their phenotypes? Usually they can, if they express the recessive phenotype.
   - Do the phenotypes of their offspring provide any information? An example is included in part 8.

4. Indicate the possible kinds of gametes formed by each of the parents. It is helpful to draw a circle around the symbols for each kind of gamete.

- If it is a monohybrid cross, we apply the principle of segregation; that is, a heterozygote Aa forms two kinds of gametes, A and a. A homozygote, such as aa, forms only one kind of gamete, a.

- If it is a dihybrid cross, we apply the principles of segregation and independent assortment. For example, an individual heterozygous for two loci would have the genotype AaBb. Allele A segregates from a, and B segregates from b. The assortment of A and a into gametes is independent of the assortment of B and b. A is equally likely to end up in a gamete with B or b. The same is true for a. Thus, an individual with the genotype AaBb produces four kinds of gametes in equal amounts: AB, Ab, aB, and ab.

5. Set up a Punnett square, placing the possible types of gametes from one parent down the left side and the possible types from the other parent across the top.

6. Fill in the Punnett square. Avoid confusion by consistently placing the dominant allele first and the recessive allele second in heterozygotes (Aa, never aA). If it is a dihybrid cross, always write the two alleles of one locus first and the two alleles of the other locus second. It does not matter which locus you choose to write first, but once you have decided on the order, it is crucial to maintain it consistently. So, if the individual is heterozygous for both loci, you will always use the form AaBb. Writing this particular genotype as aBbA, or even as BbAa, would cause confusion.

7. If you do not need to know the frequencies of all the expected genotypes and phenotypes, you can use the rules of probability as a shortcut instead of making a Punnett square. For example, if both parents are AaBb, what is the probability of an AABB offspring? To be AA, the offspring must receive an A gamete from each parent. The probability that a given gamete is A is ½, and each gamete represents an independent event, so combine their probabilities by multiplying ($\frac{1}{2} \times \frac{1}{2} = \frac{1}{4}$). The probability of BB is calculated similarly and is also ¼. The probability of AA is independent of the probability of BB, so again, use the product rule to obtain their combined probabilities ($\frac{1}{4} \times \frac{1}{4} = \frac{1}{16}$).

8. Very often the genotypes of the parents can be deduced from the phenotypes of their offspring. In peas, for example, the allele for yellow seeds (Y) is dominant to the allele for green seeds (y). Suppose plant breeders cross two plants with yellow seeds, but you do not know whether the yellow-seeded plants are homozygous or heterozygous. You can designate the cross as: Y_ x Y_. The offspring of the cross are allowed to germinate and grow, and their seeds are examined. Of the offspring, 74 produce yellow seeds, and 26 produce green seeds. Knowing that green-seeded plants are recessive, yy, the answer is obvious: Each parent contributed a y allele to the green-seeded offspring, so the parents' genotypes must have been Yy x Yy.

---

purposes of discussion, let's assume male and female births are equally probable. The probability of having a girl first is ½, and the probability of having a girl second is also ½. They are independent events, so we combine their probabilities by multiplying: $\frac{1}{2} \times \frac{1}{2} = \frac{1}{4}$. Similarly, the probability of having two boys is ¼.

In families with both a girl and a boy, the girl can be born first or the boy can be born first. The probability that a girl will be born first is ½, and the probability that a boy will be born second is also ½. We use the product rule to combine the probabilities of these two independent events: $\frac{1}{2} \times \frac{1}{2} = \frac{1}{4}$. Similarly, the probability that a boy will be born first and a girl second is ¼. These two kinds of families represent mutually exclusive outcomes, that is, two different ways of obtaining a family with one boy and one girl. Having two different ways of obtaining the desired result improves our chances, so we use the sum rule to combine the probabilities: $\frac{1}{4} + \frac{1}{4} = \frac{1}{2}$.

In working with probabilities, keep in mind a point that many gamblers forget: Chance has no memory. If events are truly independent, past events have no influence on the probability of the occurrence of future events. Probability only has predictive value in the long run, over many trials. (Recall that Mendel counted hundreds of offspring for each cross, which was one of the reasons for his success.) When working probability problems, common sense is more important than blindly memorizing rules. Examine your results to see whether they appear reasonable; if they do not, re-evaluate your assumptions. (See *Inquiring About: Solving Genetics Problems* for procedures to solve genetics problems, including when to use the rules of probability.)

CHECKPOINT 11.2

- **PREDICT** *Use the rules of probability to answer the following question: in a cross between homozygous pea plants with yellow, round seeds (YYRR) and homozygous pea plants with green, wrinkled seeds (yyrr), what is the probability of an F_2 plant having yellow, round seeds?*

- **CONNECT** *In answering the previous question, did you use the product rule, the sum rule, or both?*

11.3 INHERITANCE AND CHROMOSOMES

LEARNING OBJECTIVES

6 Define *linkage* and relate it to specific events in meiosis.
7 Show how data from a two-point test cross can be used to distinguish between independent assortment and linkage.
8 Discuss the genetic determination of sex and the inheritance of X-linked genes in mammals.

It is a measure of Mendel's genius that he worked out the principles of segregation and independent assortment without knowing anything about meiosis or the chromosome theory of inheritance. The chromosome theory of inheritance also helps explain certain apparent exceptions to Mendelian inheritance. One of these so-called exceptions involves linked genes.

Linked genes do not assort independently

Beginning around 1910, the research of geneticist Thomas Hunt Morgan and his graduate students extended the concept of the chromosome theory of inheritance. Morgan's research organism was the fruit fly (*Drosophila melanogaster*). Just as the garden pea was an excellent model research organism for Mendel's studies, the fruit fly was perfect for extending general knowledge about inheritance. Fruit flies have a short life cycle—just 14 days—and their small size means that thousands can be kept in a research lab. The large number of individuals increases the chance of identifying mutants. Also, fruit flies have just four pairs of chromosomes, one of which is a pair of sex chromosomes.

By carefully analyzing the results of crosses involving fruit flies, Morgan and his students demonstrated that genes are arranged in a linear order on each chromosome. Morgan also showed that independent assortment does not apply if the two loci lie close together in the same pair of homologous chromosomes. In fruit flies there is a locus controlling wing shape: the dominant allele *V* for normal wings and the recessive allele *v* for abnormally short, or vestigial, wings. Another locus controls body color: the dominant allele *B* for gray body and the recessive allele *b* for black body. If a homozygous *BBVV* fly is crossed with a homozygous *bbvv* fly, the F$_1$ flies all have gray bodies and normal wings, and their genotype is *BbVv*.

Because these loci happen to lie close to each other in the *same pair* of homologous chromosomes, their alleles do not assort independently; instead, they are **linked genes** that tend to be inherited together. **Linkage** is the tendency for a group of genes on the same chromosome to be inherited together in successive generations. You can readily observe linkage in the results of a test cross in which heterozygous F$_1$ flies (*BbVv*) are mated with homozygous recessive (*bbvv*) flies (**FIG. 11-11**). Because heterozygous individuals are mated to homozygous recessive individuals, this test cross is similar to the test cross described earlier. However, it is called a **two-point test cross** because alleles of two loci are involved.

If the loci governing these traits were unlinked—that is, on different chromosomes—the heterozygous parent in a test cross would produce four kinds of gametes (*BV*, *Bv*, *bV*, and *bv*) in equal numbers. This independent assortment would produce offspring with new gene combinations not present in the parental generation. Recall that any process that leads to new allele combinations is called genetic recombination. In our example, gametes *Bv* and *bV* are *recombinant types*. The other two kinds of gametes, *BV* and *bv*, are *parental types* because they are identical to the gametes produced by the P generation.

Of course, the homozygous recessive parent produces only one kind of gamete, *bv*. Thus, if independent assortment were to occur in the F$_1$ flies, approximately 25% of the test-cross offspring would be gray-bodied and normal-winged (*BbVv*), 25% black-bodied and normal-winged (*bbVv*), 25% gray-bodied and vestigial-winged (*Bbvv*), and 25% black-bodied and vestigial-winged (*bbvv*). Note that the two-point test cross lets us determine the genotypes of the offspring directly from their phenotypes.

By contrast, the alleles of the loci in our example do not undergo independent assortment because they are linked. Alleles at different loci but close to one another on a given chromosome tend to be inherited together; because chromosomes pair and separate during meiosis as units, the alleles at different loci on a given chromosome tend to be inherited as units. If linkage were complete, only parental-type flies would be produced, with approximately 50% having gray bodies and normal wings (*BbVv*) and 50% having black bodies and vestigial wings (*bbvv*).

However, in our example the offspring also include some gray-bodied, vestigial-winged flies and some black-bodied, normal-winged flies. They are recombinant flies, having received a recombinant-type gamete from the heterozygous F$_1$ parent. Each recombinant-type gamete arose by **crossing-over** between these loci in a meiotic cell of a heterozygous female fly. (Fruit flies are unusual in that crossing-over occurs only in females and not in males; it is far more common for crossing-over to occur in both sexes of a species.) When chromosomes pair and undergo synapsis, crossing-over occurs as homologous (nonsister) chromatids exchange segments of chromosome material by a process of breakage and rejoining catalyzed by enzymes (**FIG. 11-12**). Also see the discussion of prophase I in Chapter 10, Section 10.4.

Calculating the frequency of crossing-over reveals the linear order of linked genes on a chromosome

In our fruit fly example (see Fig. 11-11), 217 of the offspring are recombinant types: gray flies with vestigial wings, *Bbvv* (111 of the total); and black flies with normal wings, *bbVv* (106 of the total). The remaining 1051 offspring are parental types. These data can be used to calculate the percentage of crossing-over between the loci (**TABLE 11-3**). You do this calculation by adding the number of individuals in the two recombinant types of offspring, dividing by the *total number of offspring*, and multiplying by 100: $217 \div 1268 = 0.17; 0.17 \times 100 = 17\%$. Thus, the *V* locus and the *B* locus have 17% recombination between them.

How can linkage be recognized in fruit flies?

HYPOTHESIS: Linkage can be recognized when an excess of parental-type offspring and a deficiency of recombinant-type offspring are produced in a two-point test cross.

EXPERIMENT: Fruit flies from pure lines with gray bodies and normal wings (*BBVV*) and black bodies and vestigial wings (*bbvv*) (P generation) are crossed. The heterozygous F₁ flies with gray, normal wings (*BbVv*) are then crossed with flies that have black, vestigial wings (*bbvv*). If the alleles for color and wing shape are not linked (i.e., the alleles assort independently), the offspring will consist of an equal number of each of four phenotypes.

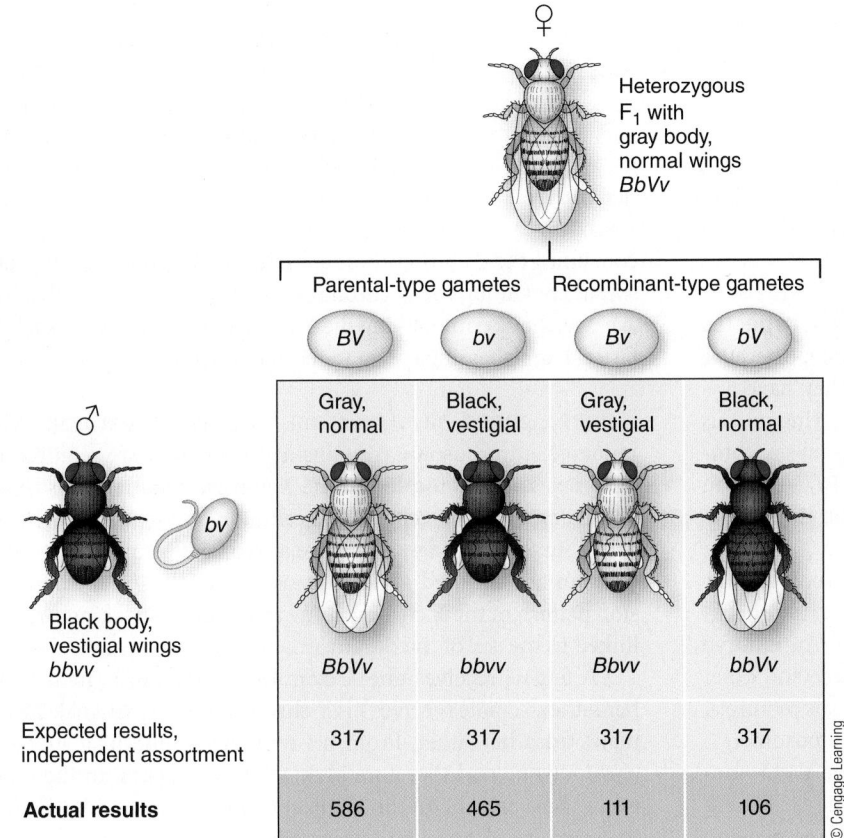

	Parental-type gametes		Recombinant-type gametes	
	BV	*bv*	*Bv*	*bV*
	Gray, normal	Black, vestigial	Gray, vestigial	Black, normal
	BbVv	*bbvv*	*Bbvv*	*bbVv*
Expected results, independent assortment	317	317	317	317
Actual results	586	465	111	106

RESULTS AND CONCLUSION: Of the 1268 offspring (*bottom row*) of an actual cross, a total of 1051 belong to one of the two parental types (83%), and 217 belong to one of the two recombinant types (17%). Thus, loci for wing length and body color are linked on a homologous chromosome pair..

SOURCE: Morgan, T.H. (1914) "No crossing over in the male of Drosophila of genes in the second and third pairs of chromosomes." *Biological Bulletin* 26: 195–204.

Figure 11-11 A two-point test cross to detect linkage in fruit flies

PREDICT What results would you expect if the P generation flies were *BBvv* and *bbVV*? What would be the genotypes and phenotypes of the parental classes? the recombinant classes?

During a single meiotic division, crossing-over may occur at several different points along the length of each homologous chromosome pair. In general, crossing-over is more likely to occur between two loci if they lie far apart on the chromosome and less likely to occur if they lie close together. Because of this rough correlation between the frequency of recombination of two loci and the linear distance between them, a genetic map of the chromosome can be generated by converting the percentage of recombination to **map units.** By convention, 1% recombination between two loci equals a distance of 1 map unit, so the loci in our example are 17 map units apart.

Scientists have determined the frequencies of recombination between specific linked loci in many species. All the experimental results are consistent with the hypothesis that genes are present in a linear order in the chromosomes. **FIGURE 11-13** illustrates the traditional method for determining the linear order of genes in a chromosome.

By putting together the results of numerous crosses, scientists have developed detailed linkage maps for many eukaryotes, including the fruit fly, the mouse, yeast, *Neurospora* (a fungus), and many plants, especially those that are important crops. In addition, researchers have used genetic methods to develop a detailed map for *Escherichia coli*—a bacterium with a single, circular DNA molecule—and many other prokaryotes and viruses. They have made much more sophisticated maps of chromosomes of various species by means of recombinant DNA technology (discussed in Chapter 15). Using these techniques, the *Human Genome Project* has produced detailed maps of human chromosomes (discussed in Chapter 16).

Sex is generally determined by sex chromosomes

In some species, environmental factors exert a large control over an individual's sex. However, genes are the most important sex determinants in most eukaryotic organisms. The major sex-determining genes of mammals, birds, and many insects are carried by **sex chromosomes.** Typically, members of one sex have a pair of similar sex chromosomes and produce gametes that are all identical in sex chromosome constitution. The members of the other sex have two different sex chromosomes and produce two kinds of gametes, each bearing a single kind of sex chromosome.

The cells of female mammals, including humans, contain two **X chromosomes.** In contrast, the males have a single X chromosome and a smaller **Y chromosome** that bears only a few active genes (**FIG. 11-14**). For example, human females have 22 pairs of **autosomes,** which are chromosomes other than the sex chromosomes, plus a pair of X chromosomes; males have 22 pairs of autosomes plus one X chromosome and one Y chromosome. Domestic cats have 19 pairs of autosomes, to which are added a pair of X chromosomes

TABLE 11-3	How to Determine Recombination Frequency from a Two-Point Test Cross			
	TEST-CROSS RESULTS (FROM FIG. 11-11)			
TYPE OF OFFSPRING	PARENTAL TYPE		RECOMBINANT TYPE	
Phenotype	Gray, normal	Black, vestigial	Gray, vestigial	Black, normal
Genotype	*BbVv*	*bbvv*	*Bbvv*	*bbVv*
Number of offspring	586	465	111	106
Calculations of recombination frequency	1. Number of parental-type offspring = 1051 2. Number of recombinant-type offspring = 217 3. Total number of offspring = 1051 + 217 = 1268 4. Recombination frequency = 217/1268 × 100 = 17%			

© Cengage Learning

in females, or an X plus a Y in males. In contrast, sex determination is based on different genetic systems in animals other than mammals (**TABLE 11-4**).

The Y chromosome determines male sex in most species of mammals Do male humans have a male phenotype because they have only one X chromosome, or because they have a Y chromosome? Much of the traditional evidence bearing on this question comes from studies of people with abnormal sex chromosome constitutions (see discussion in Chapter 16). A person with an XXY constitution is a nearly normal male in external appearance, although his testes are underdeveloped (Klinefelter syndrome). A person with one X but no Y chromosome has the overall appearance of a female but has defects such as short stature and undeveloped ovaries (Turner syndrome). An embryo with a Y but no X does not survive. Based on these and other observations, biologists concluded that all individuals require at least one X, and the Y is the male-determining chromosome.

Geneticists have identified several genes on the Y chromosome that are involved in determination of the male sex. The *sex reversal on Y (SRY) gene*, the major male-determining gene on the Y chromosome, acts as a "genetic switch" that causes testes to develop in the fetus. The developing testes then secrete the hormone *testosterone*, which causes other male characteristics to develop. Several other genes on the Y chromosome also play a role in sex determination, as do many genes on the X chromosome, which explains why an XXY individual does not have a completely normal male phenotype. Some genes on the autosomes also affect sex development.

Evidence suggests that the X and Y chromosomes of mammals originated as a homologous pair of autosomes. During the evolution of the X and Y chromosomes, almost all of the original functional genes were retained on the X chromosome and lost from the Y chromosome. Today, about 95% of the Y chromosome is male-specific. Either the Y chromosome retained certain genes that coded for proteins that determine maleness or perhaps mutations occurred in existing genes on the Y chromosome that made it the male-determining chromosome.

Thus, the sex chromosomes are not truly homologous in their present forms because they are not similar in size, shape, or genetic constitution. Nevertheless, the Y chromosome has short, homologous "pairing regions" at its tips that lets it synapse and exchange genetic material with the X chromosome during meiosis.

Half the sperm contain an X chromosome, and half contain a Y chromosome. All normal eggs bear a single X chromosome. Fertilization of an X-bearing egg by an X-bearing sperm results in an XX (female) zygote, or fertilized egg; fertilization by a Y-bearing sperm results in an XY (male) zygote.

You might expect to have equal numbers of X- and Y-bearing sperm and a 1:1 ratio of females to males. However, in humans more males are conceived than females, and more males die before birth. Even at birth the ratio is not 1:1; about 106 boys are born for every 100 girls. The Y-bearing sperm appear to have a selective advantage, possibly because sperm containing the Y chromosome are smaller and have less mass than sperm containing the X chromosome; it is hypothesized that, on average, the sperm carrying the Y chromosome can swim slightly faster than X-bearing sperm to reach the egg.

X-linked genes have unusual inheritance patterns The human X chromosome contains many loci that are required in both sexes. Genes located in the X chromosome, such as those governing color perception and blood clotting, are sometimes called *sex-linked genes*. It is more appropriate, however, to refer to them as **X-linked genes** because they follow the transmission pattern of the X chromosome and, strictly speaking, are not linked to the sex of the organism.

A female receives one X from her mother and one X from her father. A male receives his Y chromosome, which makes him male, from his father. From his mother he inherits a single X chromosome and therefore all his X-linked genes. In the male every allele present on the X chromosome is expressed, whether that allele was dominant or recessive in the female parent.

A male is neither homozygous nor heterozygous for his X-linked alleles; instead, he is always **hemizygous;** that is, he has only one copy of each X-linked gene. The significance of hemizygosity in males, including human males, is that rare, recessive X-linked genes are expressed, making males more likely to be affected by the many genetic disorders associated with the X chromosome. In comparison, only a few genes are located exclusively on the Y chromosome. X and Y chromosomes also have short areas where they carry the same genes, enabling the X and Y to pair during meiosis I.

We will use a simple system of notation for problems involving X linkage, indicating the X chromosome and incorporating specific alleles as superscripts. For example, the symbol X^e signifies a recessive X-linked allele for color blindness and X^E the dominant X-linked allele for normal color vision. The Y chromosome is written without superscripts because it does not carry the locus of interest. Two recessive X-linked alleles must be present in a female for the abnormal phenotype to be expressed (X^eX^e), whereas in the hemizygous male a single recessive allele

The exchange of segments between chromatids of homologous chromosomes is the mechanism of recombination of linked genes.

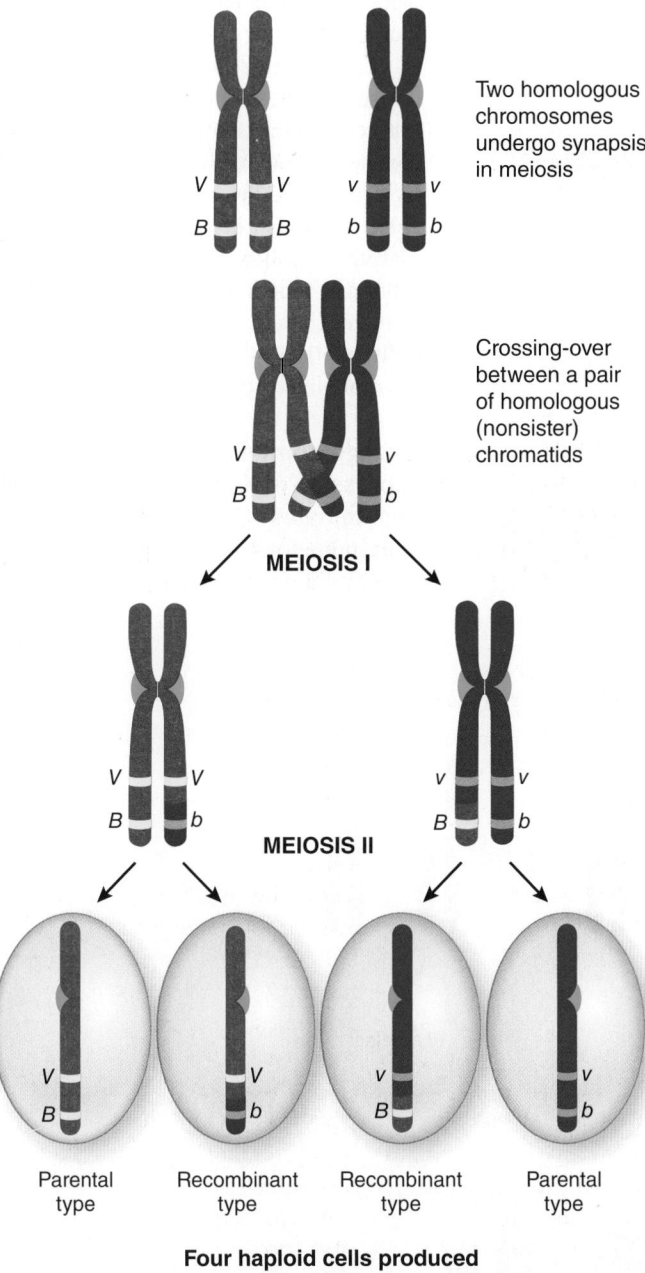

Two homologous chromosomes undergo synapsis in meiosis

Crossing-over between a pair of homologous (nonsister) chromatids

MEIOSIS I

MEIOSIS II

Parental type | Recombinant type | Recombinant type | Parental type

Four haploid cells produced

Figure 11-12 *Animation* Crossing-over

Genes located far apart on a chromosome have a greater probability of being separated by crossing-over than do genes that are closer together.

VISUALIZE Draw a series of simple sketches illustrating what would happen in meiosis if the genotypes of the P generation were *BBvv* and *bbVV*.
© Cengage Learning

(X^eY) causes the abnormal phenotype. As a consequence, these abnormal alleles are much more frequently expressed in male offspring. A heterozygous female may be a *carrier*, an individual who possesses one copy of a mutant recessive allele but does not express it in the phenotype (X^EX^e).

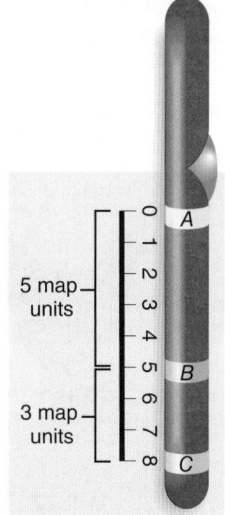
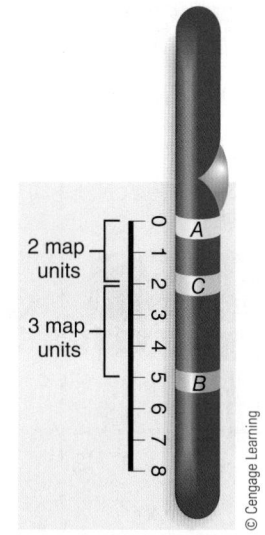

(a) If the recombination between *A* and *C* is 8% (8 map units), then *B* must be in the middle.

(b) If the recombination between *A* and *C* is 2%, then *C* must be in the middle.

Figure 11-13 Gene mapping

Gene order (i.e., which locus lies between the other two) is determined by the percentage of recombination between each of the possible pairs. In this hypothetical example, the percentage of recombination between locus *A* and locus *B* is 5% (corresponding to 5 map units) and between *B* and *C* is 3% (3 map units). There are two alternatives for the linear order of these loci.

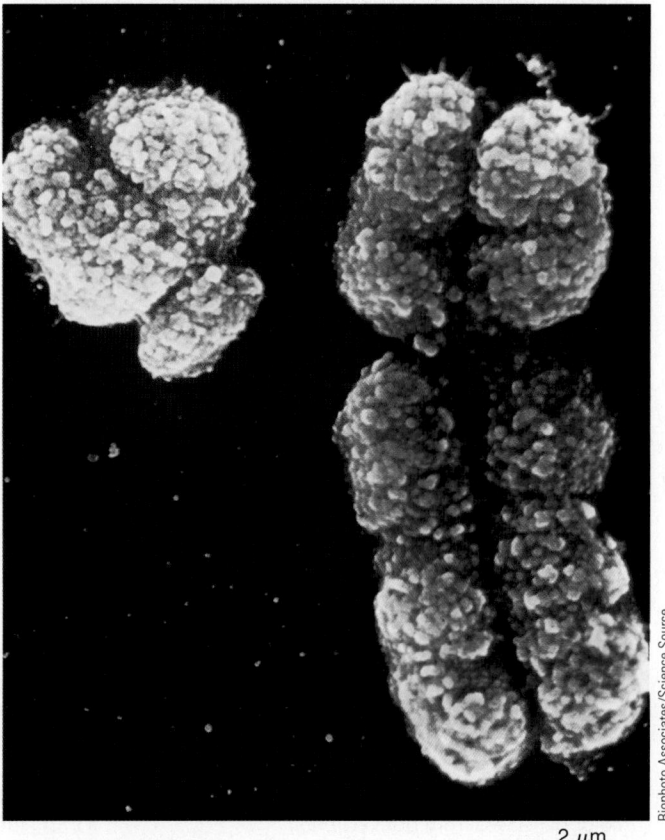

2 μm

Biophoto Associates/Science Source

Figure 11-14 SEM of human Y chromosome (*left*) and X chromosome (*right*)

Each chromosome is in the duplicated state and consists of two identical chromatids.

TABLE 11-4

TABLE 11-4 | **Representative Examples of Sex Determination in Animals**

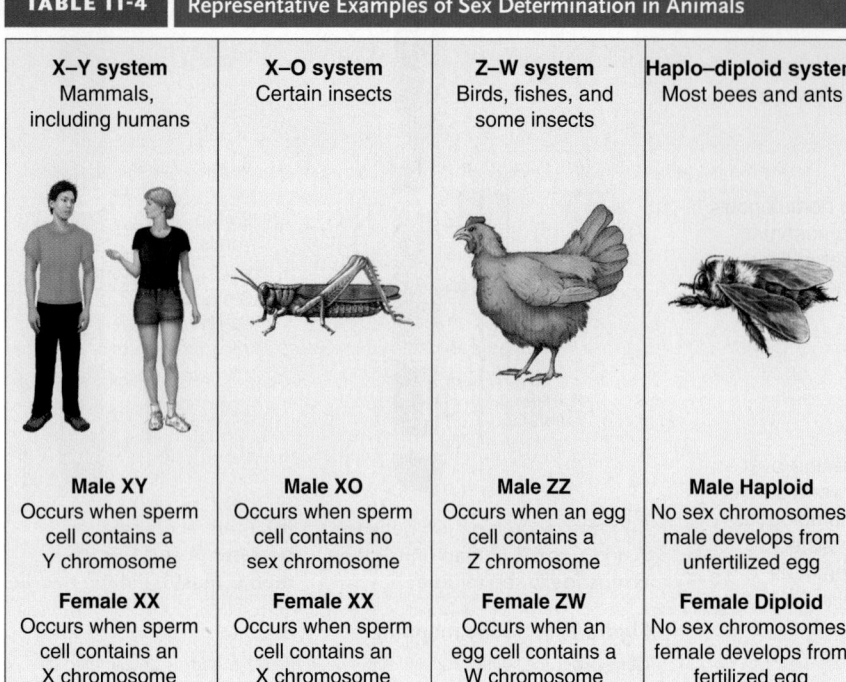

X–Y system Mammals, including humans	X–O system Certain insects	Z–W system Birds, fishes, and some insects	Haplo–diploid system Most bees and ants
Male XY Occurs when sperm cell contains a Y chromosome	**Male XO** Occurs when sperm cell contains no sex chromosome	**Male ZZ** Occurs when an egg cell contains a Z chromosome	**Male Haploid** No sex chromosomes; male develops from unfertilized egg
Female XX Occurs when sperm cell contains an X chromosome	**Female XX** Occurs when sperm cell contains an X chromosome	**Female ZW** Occurs when an egg cell contains a W chromosome	**Female Diploid** No sex chromosomes; female develops from fertilized egg

© Cengage Learning

To be expressed in a female, a recessive X-linked allele must be inherited from both parents. A color-blind female, for example, must have a color-blind father and a mother who is homozygous or heterozygous for a recessive allele for color blindness (**FIG. 11-15**).

The homozygous combination is unusual because the frequency of alleles for color blindness is relatively low. In contrast, a color-blind male need only have a mother who is heterozygous for color blindness; his father can have normal vision. Therefore, X-linked recessive traits are generally much more common in males than in females, which may partially explain why human male embryos are more likely to die.

Dosage compensation equalizes the expression of X-linked genes in males and females The X chromosome contains numerous genes required by both sexes, yet a normal female has two copies

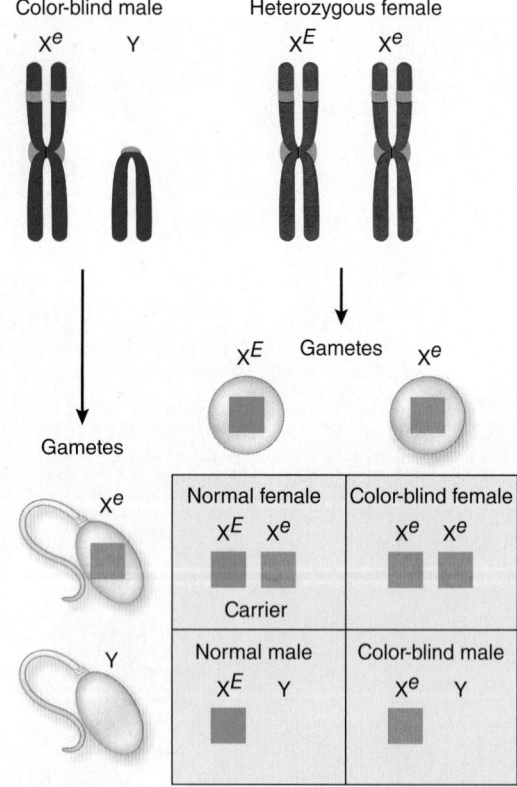

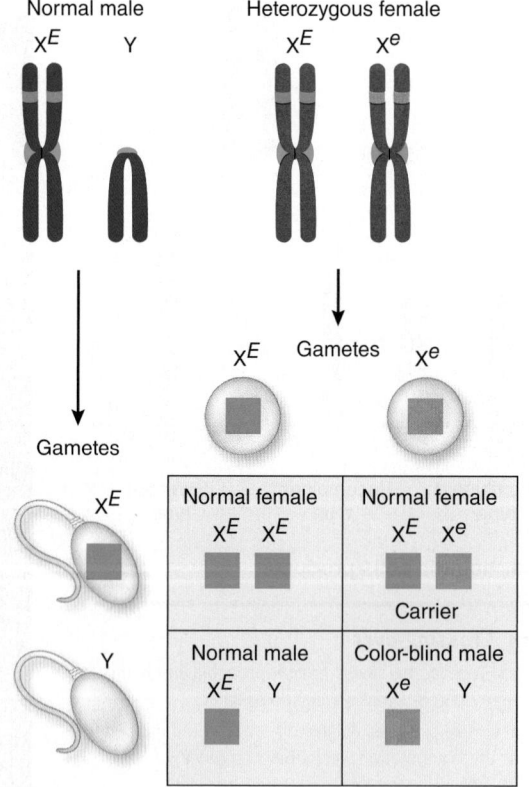

(a) To be color-blind, a female must inherit alleles for color blindness from both parents.

(b) If a normal male mates with a carrier (heterozygous) female, half of their sons would be expected to be color-blind and half of their daughters would be expected to be carriers.

KEY
- Normal color vision
- Color-blind

Figure 11-15 *Animation* **X-linked red–green color blindness**

Note that the Y chromosome does not carry a gene for color vision.
© Cengage Learning

In female mammals one of the two X chromosomes is randomly inactivated during early development, equalizing the level of expression of genes at loci on the X chromosome in males and females.

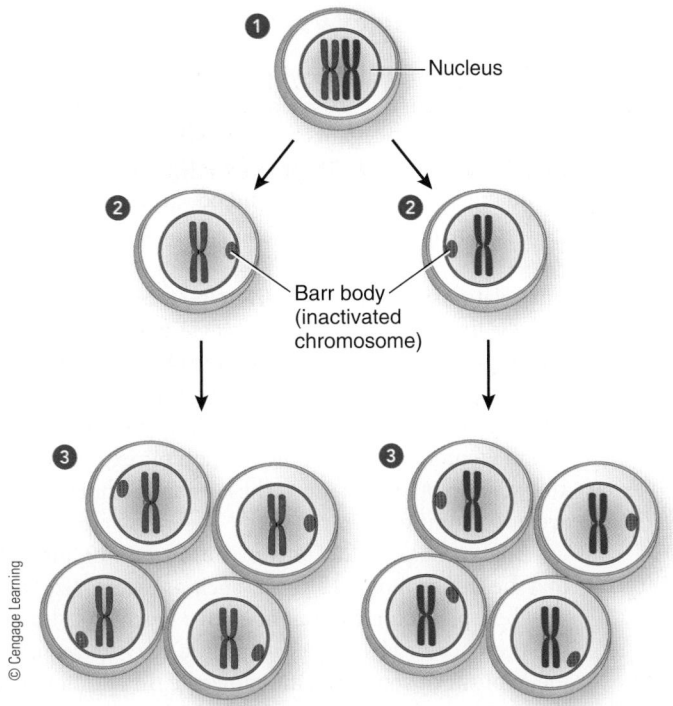

1 Nucleus

Barr body (inactivated chromosome)

2 **2**

3 **3**

(a) Inactivation of one X chromosome in female cells.

© Cengage Learning

1 The zygote and early embryonic cells have two X chromosomes, one from each parent.

2 Random inactivation of one X chromosome occurs early in development. Roughly half the cells inactivate one X chromosome (*left cell*), and the other half inactivate the second X chromosome (*right cell*). The inactive X chromosome is visible as a Barr body near the nuclear envelope.

3 Chromosome inactivation persists through subsequent mitotic divisions, resulting in patches of cells in the adult body.

© Sari ONeal/Shutterstock.com

(b) A calico cat has X-linked genes for both black and yellow (or orange) fur. Because of random X chromosome inactivation, black is expressed in some groups of cells and orange is expressed in others. (The patches of white fur are due to the presence of other genes that affect fur color.)

Figure 11-16 *Animation* **Dosage compensation in female mammals**

PREDICT What is most likely to be the sex of the cat in the photo? Could you make such a prediction if the cat were black or orange?

("doses") for each locus, whereas a normal male has only one. **Dosage compensation** is a mechanism that makes equivalent the two doses in the female and the single dose in the male. Because of dosage compensation, males and females produce the same amounts of proteins coded by X-linked genes. Male fruit flies accomplish this by making their single X chromosome more active. In most fly tissues, the metabolic activity of a single male X chromosome is equal to the combined metabolic activity of the two X chromosomes present in the female.

Our understanding of dosage compensation in humans and other mammals is incomplete, but the process generally involves the random inactivation of one of the two X chromosomes in the female (**FIG. 11-16a**). During interphase, a dark spot of chromatin is visible at the edge of the nucleus of each female mammalian cell when stained and observed with a microscope. This dark spot, a **Barr body,** is a dense, metabolically inactive X chromosome. The other X chromosome resembles the metabolically active autosomes; during interphase, it is a greatly extended thread that is not evident in light microscopy. From this and other evidence, British geneticist Mary Lyon hypothesized in 1961 that in most cells of a female mammal, only one of the two X chromosomes is active; the other is inactive and is condensed as a Barr body. Actually, however, X chromosome inactivation is never complete; as many as 25% of the genes on the inactive chromosome are expressed to some degree.

X chromosome inactivation is a random event in each somatic (body) cell of the female embryo. A female mammal that is heterozygous at an X-linked locus expresses one of the alleles in about half her cells and the other allele in the other half. X chromosome inactivation is sometimes evident in the phenotype. Mice and cats have several X-linked genes for certain coat colors. Females that are heterozygous for such genes may show patches of one coat color in the middle of areas of the other coat color. This phenomenon, called *variegation,* is evident in tortoiseshell and calico cats (**FIG. 11-16b**). Early in development, when relatively few cells are present, X chromosome inactivation occurs randomly in each cell. X inactivation is then maintained during all subsequent divisions of that cell line. When any one of these cells divides by mitosis, the cells of the resulting *clone* (a group of genetically identical cells) all have the same active X chromosome; therefore, a patch of cells that all express the same color develops.

You might wonder why variegation is not always apparent in females heterozygous at X-linked loci. The answer is that although variegation usually occurs, we may need to use special techniques to observe it. For example, color blindness is caused by a defect involving the pigments in the cone cells in the retina of the eye. In at least one type of red–green color blindness, the retina of a heterozygous female actually contains patches of abnormal cones, but the patches of normal cones are enough to provide normal color vision. Variegation can be very hard to observe in cases where cell products become mixed in bodily fluids. For instance, in females heterozygous for the allele that causes hemophilia, only half the cells responsible for producing a specific blood-clotting factor do so, but they produce enough to ensure that the blood clots normally.

CHECKPOINT 11.3

- *What ratio of genotypes to phenotypes is observed in a two-point test cross if genes are unlinked?*
- CONNECT *Does the mechanism of recombination of unlinked genes require crossing-over? Explain your answer.*
- CONNECT *Two loci exhibit 5% recombination between them. How many map units apart are they?*
- *Which chromosome determines the male sex in humans and other mammals?*
- CONNECT *In what way is variegation of X-linked genes related to the mechanism responsible for sex chromosome dosage compensation in humans and other mammals?*

11.4 EXTENSIONS OF MENDELIAN GENETICS

LEARNING OBJECTIVES

9 Explain some of the ways genes may interact to affect the phenotype and discuss how a single gene can affect many features of the organism simultaneously.

10 Distinguish among incomplete dominance, codominance, multiple alleles, epistasis, and polygenic inheritance.

11 Describe *norm of reaction* and give an example.

The relationship between a given locus and the trait it controls may or may not be simple. A single pair of alleles of a locus may regulate the appearance of a single trait (such as tall versus short in garden peas). Alternatively, a pair of alleles may participate in the control of several traits, or alleles of many loci may interact to affect the phenotypic expression of a single character. Not surprisingly, these more complex relationships are quite common.

You can assess the phenotype on one or many levels. It may be a morphological trait, such as shape, size, or color. It may be a physiological trait or even a biochemical trait, such as the presence or absence of a specific enzyme required for the metabolism of some specific molecule. In addition, changes in the environmental conditions under which the organism develops may alter the phenotypic expression of genes.

Dominance is not always complete

Studies of the inheritance of many traits in a wide variety of organisms have shown that one member of a pair of alleles may not be completely dominant over the other. In such instances, it is inaccurate to use the terms *dominant* and *recessive*.

The plants commonly called four o'clocks (*Mirabilis jalapa*) may have red or white flowers. Each color breeds true when these plants are self-pollinated. What flower color might you expect in the offspring of a cross between a red-flowering plant and a white-flowering one? Without knowing which is dominant, you might predict that all would have red flowers or all would have white flowers. German botanist Carl Correns, one of the rediscoverers of Mendel's work, first performed this cross and found that all F_1 offspring have pink flowers!

Does this result in any way indicate that the principles of inheritance that Mendel deduced are wrong? Did the parental traits blend inseparably in the offspring? Quite the contrary, for when two of these pink-flowered plants are crossed, red-flowered, pink-flowered, and white-flowered offspring appear in a ratio of 1:2:1 (**FIG. 11-17**). In this instance as in all other aspects of the scientific process, results that differ from those hypothesized prompt scientists to re-examine and modify their hypotheses to account for the exceptional results. The pink-flowered plants are clearly the heterozygous individuals, and neither the red allele nor the white allele is completely dominant. When the heterozygote has a phenotype intermediate between those of its two parents, the

In incomplete dominance an F₁ heterozygote has a phenotype intermediate between its parents.

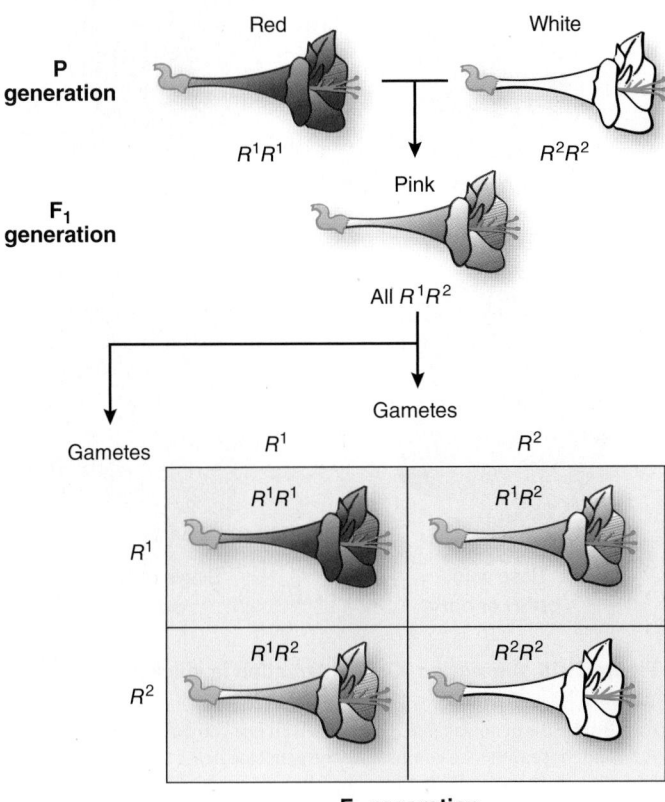

Figure 11-17 *Animation* **Incomplete dominance in four o'clocks**

Two incompletely dominant alleles, R^1 and R^2, are responsible for red, white, and pink flower colors. Red-flowered plants are R^1R^1, white-flowered plants are R^2R^2, and heterozygotes (R^1R^2) are pink. Note that uppercase letters are used for both alleles because neither is recessive to the other.

PREDICT What ratio of offspring would you expect in four o'clocks if you crossed a pink-flowered plant with a white-flowered plant? a pink-flowered plant with a red-flowered plant?
© Cengage Learning

genes show **incomplete dominance.** In these crosses, the genotypic and phenotypic ratios are identical.

Incomplete dominance is not unique to four o'clocks, and additional examples of incomplete dominance are known in both plants and animals. For example, true-breeding white chickens and true-breeding black chickens produce bluish gray offspring, or Andalusian blues, when crossed.

In both cattle and horses, reddish coat color is not completely dominant to white coat color. Heterozygous individuals have a mixture of reddish hairs and white hairs, which is called *roan.* If you saw a white mare nursing a roan foal, what would you guess was the coat color of the foal's father? Because the reddish and white colors are expressed independently (hair by

TABLE 11-5	**ABO Blood Types**		
PHENOTYPE (BLOOD TYPE)	GENOTYPE	SURFACE ANTIGEN ON RBC	SERUM ANTIBODIES TO A OR B ANTIGENS
A	I^AI^A or I^Ai	A	Anti-B
B	I^BI^B or I^Bi	B	Anti-A
AB	I^AI^B	A, B	None
O	ii	None	Anti-A, anti-B

*This table and the discussion of the ABO system have been simplified somewhat. Note that the body produces antibodies against the antigens *lacking* on its own red blood cells (RBCs). Because of their specificity for the corresponding antigens, these antibodies are used in standard tests to determine blood types.

© Cengage Learning

hair) in the roan heterozygote, scientists sometimes refer to it as a case of **codominance.** Strictly speaking, the term *incomplete dominance* refers to instances in which the heterozygote is intermediate in phenotype, and *codominance* refers to instances in which the heterozygote simultaneously expresses the phenotypes of both types of homozygotes.

Humans have four blood types (A, B, AB, and O), collectively called the *ABO blood group.* The human ABO blood group is an excellent example of codominant alleles. Blood types A, B, AB, and O are controlled by three alleles representing a single locus (**TABLE 11-5**). Allele I^A codes for the synthesis of a specific glycoprotein, antigen A, which is expressed on the surface of red blood cells. (Immune responses are discussed in Chapter 45.) *Antigens* are substances capable of stimulating an immune response. Allele I^B leads to the production of antigen B, a different but related glycoprotein. Allele *i* is allelic to I^A and I^B, but it does not code for an antigen.

Individuals with the genotype I^AI^A or I^Ai have blood type A. People with genotype I^BI^B or I^Bi have blood type B. Those with genotype I^AI^B have blood type AB, and those with genotype *ii* have blood type O. These results show that neither allele I^A nor allele I^B is dominant to the other. Both alleles are expressed phenotypically in the heterozygote and are therefore codominant to each other, although each is dominant to allele *i*.

At one time, determining blood types was used to settle cases of disputed parentage. Although blood-type tests can *exclude* someone as a possible parent of a particular child, they can never prove that a certain person is the parent; they only determine that he or she *could* be. Could a man with blood type AB be the father of a child with blood type O? Could a woman with blood type O be the mother of a child with blood type AB? Could a type B child with a type A mother have a type A father or a type O father?[2]

Multiple alleles for a locus may exist in a population

Most of our examples so far have dealt with situations in which each locus was represented by a maximum of two allelic variants

---

[2] The answer to all these questions is no.

(three in the example of blood types). It is true that a single diploid individual has a maximum of two different alleles for a particular locus and that a haploid gamete has only one allele for each locus. However, if you survey a population, you may find more than two alleles for a particular locus, as you saw with the ABO blood group (see Table 11-5). If three or more alleles for a given locus exist within the population, that locus has **multiple alleles.**

Research has shown that many loci have multiple alleles. Some alleles can be identified by the activity of a certain enzyme or by some other biochemical feature but do not produce an obvious phenotype. Others produce a readily recognizable phenotype, and certain patterns of dominance can be discerned when the alleles are combined in various ways. In other series of multiple alleles, certain alleles may be codominant and others incompletely dominant. In such cases, the heterozygotes commonly have phenotypes intermediate between those of their parents.

A single gene may affect multiple aspects of the phenotype

In our examples the relationship between a gene and its phenotype has been direct, precise, and exact, and the loci have controlled the appearance of single traits. However, the relationship of gene to trait may not have such a simple genetic basis. Most genes affect several different characters. The ability of a single gene to have multiple effects is known as **pleiotropy.** Most cases of pleiotropy can be traced to a single fundamental cause. For example, a defective enzyme may affect the functioning of many types of cells. Pleiotropy is evident in many genetic diseases in which a single pair of alleles causes multiple symptoms. For example, people who are homozygous for the recessive allele that causes cystic fibrosis produce abnormally thick mucus in many parts of the body, including the respiratory, digestive, and reproductive systems (cystic fibrosis is discussed in Chapter 16).

Alleles of different loci may interact to produce a phenotype

Several pairs of alleles may interact to affect a single phenotype, or one pair may inhibit or reverse the effect of another pair. One example of gene interaction is illustrated by the inheritance of combs in chickens, where two genes may interact to produce a novel phenotype (FIG. 11-18). The allele for a rose comb, *R*, is dominant to that for a single comb, *r*. A second, unlinked pair of alleles governs the inheritance of a pea comb, *P*, versus a single comb, *p*. A single-comb chicken is homozygous for the recessive allele at both loci (*pprr*). A rose comb chicken is either *ppRR* or *ppRr*, and a pea comb chicken is either *PPrr* or *Pprr*. When an *R* and *P* occur in the same individual, the phenotype is neither pea comb nor rose comb but a completely different type, a walnut comb. The walnut comb phenotype is produced whenever a chicken has one or two *R* alleles, plus one or two *P* alleles (that is, *PPRR, PpRR, PPRr,* or *PpRr*).

Epistasis is a common type of gene interaction in which the presence of certain alleles of one locus can prevent or mask the

© Cengage Learning

Figure 11-18 *Animation* **Gene interaction in chickens**

Two gene pairs govern four chicken comb phenotypes. Chickens with walnut combs have the genotype *P_R_*. Chickens with pea combs have the genotype *P_rr*, and those with rose combs have the genotype *ppR_*. Chickens that are homozygous recessive for both loci, *pprr*, have a single comb. (The blanks represent either dominant or recessive alleles.)

PREDICT What types of combs, and in what ratio, would you expect among the offspring of two heterozygous walnut comb chickens, *PpRr*?

expression of alleles of a different locus and express their own phenotype instead. (The term *epistasis* means "standing on.") Unlike the chicken example, no novel phenotypes are produced in epistasis.

Coat color in Labrador retrievers is an example of epistasis that involves a gene for pigment and a gene for depositing color in the coat (FIG. 11-19). The two alleles for the pigment gene are *B* for black coat and its recessive counterpart, *b*, for brown coat. The gene for depositing color in the coat has two alleles, *E* for the expression of black and brown coats and *e*, which is epistatic and blocks the expression of the *B/b* gene. The epistatic allele is recessive and therefore is expressed as a yellow coat only in the homozygous condition (*ee*), regardless of the combination of *B* and *b* alleles in the genotype.

In polygenic inheritance, the offspring exhibit a continuous variation in phenotypes

Many human characters—such as height, body form, and skin pigmentation—are not inherited through alleles at a single locus. The same holds true for many commercially important

Figure 11-19 *Animation* **Epistasis in Labrador retrievers**

Two gene pairs interact to govern coat color in Labrador retrievers. Black Labs (*left*) have the genotype *B_E_*; chocolate Labs (*middle*) have the genotype *bbE_*; and yellow Labs (*right*) have the genotype *B_ee* or *bbee*. (The blanks represent either dominant or recessive alleles.)

PREDICT Can two black Labs have a yellow puppy?

characters in domestic plants and animals, such as milk and egg production. Alleles at several, perhaps many, loci affect each character. The term **polygenic inheritance** is applied when multiple independent pairs of genes have similar and additive effects on the same character.

As many as 60 loci account for the inheritance of skin pigmentation in humans. To keep things simple in this example, we illustrate the principle of polygenic inheritance in human skin pigmentation with pairs of alleles at three unlinked loci. They can be designated *A* and *a, B* and *b,* and *C* and *c.* The capital letters represent incompletely dominant alleles producing dark skin. The more capital letters, the darker the skin because the alleles affect skin pigmentation in an additive fashion. A person with the darkest skin possible would have the genotype *AABBCC,* and a person with the lightest skin possible would have the genotype *aabbcc.* The F$_1$ offspring of an *aabbcc* person and an *AABBCC* person are all *AaBbCc* and have an intermediate skin color. The F$_2$ offspring of two such triple heterozygotes (*AaBbCc × AaBbCc*) would have skin pigmentation ranging from very dark to very light (**FIG. 11-20**).

Polygenic inheritance is characterized by an F$_1$ generation that is intermediate between the two completely homozygous parents and by an F$_2$ generation that shows wide variation between the two parental types. When the number of individuals in a population is plotted against the amount of skin pigmentation and the points are connected, the result is a bell-shaped curve, a *normal distribution curve*. Most of the F$_2$ generation individuals have one of the intermediate phenotypes; only a few show the extreme phenotypes of the original P generation (i.e., the grandparents). On average, only 1 of 64 is as dark as the very dark grandparent, and only 1 of 64 is as light as the very light grandparent. The alleles *A, B,* and *C* each produce about the same amount of darkening of the skin; hence, the genotypes *AaBbCc, AABbcc, AAbbCc, aaBBcc, aaBBCc, AabbCC,* and *aaBbCC* all produce similar intermediate phenotypes.

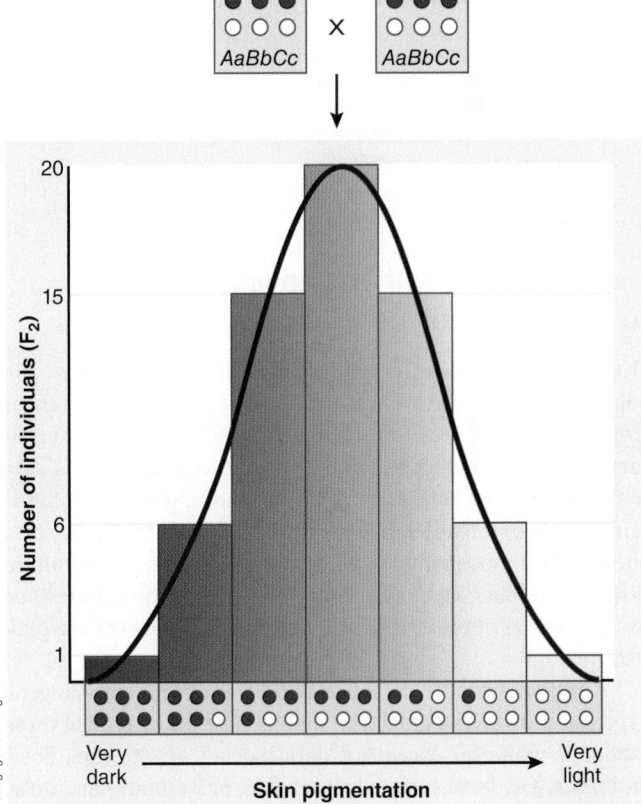

Figure 11-20 Polygenic inheritance in human skin pigmentation

This simplified example assumes that skin pigmentation in humans is governed by alleles of three unlinked loci. The alleles producing dark skin (*A, B,* and *C*) are represented by capital letters, but they are not dominant. Instead, their effects are additive. The number of dark dots, each signifying an allele producing dark skin, is counted to determine the phenotype. A wide range of phenotypes is possible when individuals of intermediate phenotype mate and have offspring (*AaBbCc × AaBbCc*). The expected distribution of phenotypes is consistent with the superimposed normal distribution curve.

PREDICT In this simplified model, what are all the possible genotypes of an individual from the class in the middle of the normal distribution curve?

Figure 11-21 Influence of the environment on hydrangea flower color

These hydrangeas are all the same genotype, but their flowers vary in color depending on availability of aluminum, which in turn is dependent on soil pH. Blue flowers develop in a soil pH of 5.5 or less, purple flowers in a soil pH of 5.6 to 6.4, and pink flowers in a soil pH of 6.5 to 7.0.

(a) The Himalayan genotype at warm temperatures.

(b) The Himalayan genotype at cold temperatures.

Figure 11-22 *Animation* **Influence of the environment on rabbit coat color**

These rabbits have the same genotype for coat color, but have been exposed to different temperatures.
© Cengage Learning

Genes interact with the environment to shape phenotype

Hydrangeas are shrubs grown for their attractive flowers. The color of certain genetically identical hydrangea flowers ranges from blue to purple to pink depending on the level of aluminum in the soil before the flowers begin to develop. It is easy to manipulate the level of aluminum by adding alum or aluminum sulfate to the soil or by lowering the soil pH (in acidic soils, aluminum is more soluble). Under these conditions, hydrangea flowers are blue. In alkaline soils (made by adding limestone to the soil), aluminum is less soluble, and the flowers are pink (FIG. 11-21).

Another example of the effect of environment on gene expression involves Himalayan rabbits. The phenotype of these rabbits is white fur except for dark patches on the ears, nose, and paws. The local surface temperature of a rabbit's ears, nose, and paws is colder in the rabbit's natural environment, and this temperature difference causes the production of dark fur. If you raise rabbits with the Himalayan genotype at a warm temperature (30°C), the rabbits are completely white, with no dark patches on the ears, nose, or paws (FIG. 11-22a). If you raise Himalayan rabbits at a cooler temperature (25°C), however, they develop the characteristic dark patches of fur (FIG. 11-22b). Thus, genes can function differently in different environments.

Now let's examine a human example—height—in the context of genes and the environment. The inheritance of height in humans is polygenic and involves alleles representing ten or

more loci. Because many genes are involved and because height is modified by a variety of environmental conditions, such as diet and general health, the height of most adults ranges from 4 ft 2 in. (1.25 m) to 7 ft 2 in. (2.15 m). The genes that affect height set limits for the phenotype—no human is 12 ft tall, for example—but the environment shapes the phenotype within its genetic limits. The range of phenotypic possibilities that can develop from a single genotype under different environmental conditions is known as the **norm of reaction.** In certain genotypes the norm of reaction is quite limited. In other genotypes, such as those involved in human height, the norm of reaction is quite broad.

Although interactions between genes and the environment influence the phenotypes of many characters, it is difficult to determine the exact contributions of genes and the environment to a given phenotype. In some cases, the environment regulates the activity of certain genes, turning them on in some environmental conditions and off in others. Investigators argued for years about the relative importance of *nature* (genetics/biology) versus *nurture* (environmental influences) regarding such human characters as intelligence, depression, bipolar disorder, and schizophrenia. Studies with identical twins raised in different environments indicate that individuals may inherit genetic potential or vulnerabilities, but environmental conditions may influence the expression of the genotype. Thus, phenotypic expression depends on a combination of nature *and* nurture.

●HECKPOINT 11.4

- **CONNECT** *The human ABO blood groups are considered an example of both dominance and codominance. However, flower color in four o'clocks is considered an example of incomplete dominance. Explain.*

- *What is the difference between multiple alleles and polygenic inheritance?*

- *How do examples of pleiotropy and epistasis differ in terms of the minimum number of loci required?*

- **CONNECT** *How is the concept of norm of reaction related to the lack of yellow hydrangea flowers illustrated in Figure 11-21?*

11.1 Mendel's Principles of Inheritance *(page 227)*

1 Define the terms *phenotype, genotype, locus, allele, dominant allele, recessive allele, homozygous,* and *heterozygous.*

- **Genes** are in chromosomes, and the site a gene occupies in the chromosome is its **locus.** Different forms of a particular gene are **alleles;** they occupy corresponding loci on homologous chromosomes.

- An individual that carries two identical alleles is **homozygous** for that locus. If the two alleles are different, that individual is **heterozygous** for that locus.

- One allele, the **dominant allele**, may mask the expression of the other allele, the **recessive allele**, in a heterozygous individual. For this reason two individuals with the same appearance, or **phenotype,** may differ from each other in their genetic makeup (i.e., their combination of alleles), or **genotype.**

2 Describe Mendel's principles of segregation and independent assortment.

- According to Mendel's **principle of segregation,** during meiosis the alleles for each locus separate, or segregate, from each other. When haploid gametes are formed, each contains only one allele for each locus.

- According to Mendel's **principle of independent assortment,** alleles of different loci are distributed randomly into gametes. The result can be **genetic recombination,** the production of new allele combinations that were not present in the **parental (P) generation.**

3 Distinguish among monohybrid, dihybrid, and test crosses.

- A cross between homozygous parents (P generation) that differ from each other with respect to their alleles at one locus is called a **monohybrid cross;** if they differ at two loci, it is called a **dihybrid cross.** The first generation of offspring, called the F_1 **generation,** is heterozygous; the generation produced by a cross of two F_1 individuals is the F_2 **generation.**

- A test cross between an individual of unknown genotype and a recessive individual helps determine the unknown genotype.

4 Explain Mendel's principles of segregation and independent assortment, given what scientists now know about genes and chromosomes.

- Segregation of alleles is a direct result of homologous chromosomes separating during meiosis.

- Independent assortment occurs because there are two ways in which two pairs of homologous chromosomes can be arranged at metaphase I of meiosis. The orientation of homologous chromosomes on the metaphase plate determines the way chromosomes are distributed into haploid cells.

11.2 Using Probability to Predict Mendelian Inheritance *(page 236)*

5 Apply the product rule and sum rule appropriately when predicting the outcomes of genetic crosses.

- According to the **product rule,** the probability of two independent events occurring together can be calculated by multiplying the probabilities of each event occurring separately.

- According to the **sum rule,** the probability of an outcome that can be obtained in more than one way can be calculated by adding the separate probabilities.

11.3 Inheritance and Chromosomes *(page 238)*

6 Define *linkage* and relate it to specific events in meiosis.

- **Linkage** is the tendency for a group of genes on the same chromosome to be inherited together. Independent assortment does

not apply if two loci are linked close together on the same pair of homologous chromosomes.

- Recombination of linked genes can result from **crossing-over** (breaking and rejoining of homologous chromatids) in meiotic prophase I. (Recall from Section 11.1 that recombination can also result from independent assortment of unlinked genes.)

7 Show how data from a two-point test cross can be used to distinguish between independent assortment and linkage.

- To distinguish between independent assortment of unlinked genes and linked genes, perform a **two-point test cross** between an individual that is heterozygous at both loci and an individual that is homozygous recessive for both.

- Linkage is recognized when an excess of parental-type offspring and a deficiency of recombinant-type offspring are produced in a two-point test cross.

8 Discuss the genetic determination of sex and the inheritance of X-linked genes in mammals.

- The sex of humans and other mammals is determined by the **X and Y chromosomes.** Normal female mammals have two X chromosomes; normal males have one X and one Y. The fertilization of an X-bearing egg by an X-bearing sperm results in a female (XX) zygote. The fertilization of an X-bearing egg by a Y-bearing sperm results in a male (XY) zygote.

- The Y chromosome determines male sex in mammals. The X chromosome contains many important genes unrelated to sex determination that are required by both males and females. A male receives all his **X-linked genes** from his mother. A female receives X-linked genes from both parents.

11.4 Extensions of Mendelian Genetics *(page 244)*

9 Explain some of the ways genes may interact to affect the phenotype and discuss how a single gene can affect many features of the organism simultaneously.

- **Pleiotropy** is the ability of one gene to have several effects on different characters. Most cases of pleiotropy can be traced to a single cause, such as a defective enzyme. Alternatively, alleles of many loci may interact to affect the phenotypic expression of a single character.

10 Distinguish among incomplete dominance, codominance, multiple alleles, epistasis, and polygenic inheritance.

- Dominance does not always apply; some alleles demonstrate **incomplete dominance,** in which the heterozygote is intermediate in phenotype, or **codominance,** in which the heterozygote simultaneously expresses the phenotypes of both homozygotes.

- **Multiple alleles,** three or more alleles that can potentially occupy a particular locus, may exist in a population. A diploid individual has any two of the alleles; a haploid individual or gamete has only one.

- In **epistasis** an allele of one locus can mask the expression of alleles of a different locus.

- In **polygenic inheritance** multiple independent pairs of genes may have similar and additive effects on the phenotype.

11 Describe *norm of reaction* and give an example.

- The range of phenotypic possibilities that can develop from a single genotype under different environmental conditions is known as the **norm of reaction.**

- Many genes are involved in the inheritance of height in humans. Also, height is modified by a variety of environmental conditions, such as diet and general health. The genes that affect height set the norm of reaction—that is, the limits—for the phenotype, and the environment molds the phenotype within its norm of reaction.

Know and Comprehend

1. One of the autosomal loci controlling eye color in fruit flies has two alleles: one for brown eyes and the other for red eyes. Fruit flies from a true-breeding line with brown eyes were crossed with flies from a true-breeding line with red eyes. The F_1 flies had red eyes. What conclusion can be drawn from this experiment? (a) these alleles underwent independent assortment (b) these alleles underwent segregation (c) these genes are X-linked (d) the allele for red eyes is dominant to the allele for brown eyes (e) all the preceding are true

2. The F_1 flies described in question 1 were mated with brown-eyed flies from a true-breeding line. What phenotypes would you expect the offspring to have? (a) all with red eyes (b) all with brown eyes (c) half with red eyes and half with brown eyes (d) red-eyed females and brown-eyed males (e) brown-eyed females and red-eyed males

3. The type of cross described in question 2 is (a) an F_2 cross (b) a dihybrid cross (c) a test cross (d) a two-point test cross (e) none of the preceding

4. Individuals of genotype *AaBb* were crossed with *aabb* individuals. Approximately equal numbers of the following classes of offspring were produced: *AaBb*, *Aabb*, *aaBb*, and *aabb*. These results illustrate Mendel's principle(s) of (a) linkage (b) independent assortment (c) segregation (d) a and c (e) b and c

Apply and Analyze

5. Assume that the ratio of females to males is 1:1. A couple already has two daughters and no sons. If they plan to have a total of six children, what is the probability that they will have four more girls? (a) ¼ (b) ⅛ (c) 1⁄16 (d) 1⁄32 (e) 1⁄64

6. Red–green color blindness is an X-linked recessive disorder in humans. Your friend is the daughter of a color-blind father. Her mother had normal color vision, but her maternal grandfather was color-blind. What is the probability that your friend is color-blind? (a) 1 (b) ½ (c) ¼ (d) ¾ (e) 0

7. When two long-winged flies were mated, the offspring included 77 with long wings and 24 with short wings. Is the short-winged condition dominant or recessive? What are the genotypes of the parents?

8. The long hair of Persian cats is recessive to the short hair of Siamese cats, but the black coat color of Persians is dominant to the brown-and-tan coat color of Siamese. Make up appropriate symbols for the alleles of these two unlinked loci. If a pure black, long-haired Persian is mated to a pure brown-and-tan, short-haired Siamese, what will be the appearance of the F_1 offspring? If two of these F_1 cats are mated, what is the chance that a long-haired, brown-and-tan cat will be produced in the F_2 generation? (Use the shortcut probability method to obtain your answer; then check it with a Punnett square.)

9. Mr. and Mrs. Smith are concerned because their own blood types are A and B, respectively, but their new son, Richard, is blood type O. Could Richard be the child of these parents?

10. A walnut comb rooster is mated to three hens. Hen A, which has a walnut comb, has offspring in the ratio of 3 walnut to 1 rose. Hen B, which has a pea comb, has offspring in the ratio of 3 walnut to 3 pea to 1 rose to 1 single. Hen C, which has a walnut comb, has only walnut comb offspring. What are the genotypes of the rooster and the three hens?

11. Individuals of genotype *AaBb* were mated to individuals of genotype *aabb*. One thousand offspring were counted, with the following results: 474 *Aabb*, 480 *aaBb*, 20 *AaBb*, and 26 *aabb*. What type of cross is it? Are these loci linked? What are the two parental classes and the two recombinant classes of offspring? What is the percentage of recombination between these two loci? How many map units apart are they?

12. Genes *A* and *B* are 6 map units apart, and *A* and *C* are 4 map units apart. Which gene is in the middle if *B* and *C* are 10 map units apart? Which is in the middle if *B* and *C* are 2 map units apart?

Evaluate and Synthesize

13. **VISUALIZE** Sketch a series of diagrams showing each of the following, making sure to end each series with haploid cells:
    (a) How a pair of alleles for a single locus segregate in meiosis
    (b) How the alleles of two unlinked loci assort independently in meiosis
    (c) How the alleles of two linked loci undergo genetic recombination

14. Can you always ascertain an organism's genotype for a particular locus if you know its phenotype? Conversely, if you are given an organism's genotype for a locus, can you always reliably predict its phenotype? Explain.

15. **CONNECT** Compare the mechanisms of genetic recombination in linked and unlinked genes.

16. **EVOLUTION LINK** Darwin's theory of evolution by natural selection is based on four observations about the natural world. One of them is that each individual has a combination of traits that makes it uniquely different. Darwin recognized that much of this variation among individuals must be inherited, but he did not know about Mendel's mechanism of inheritance. Based on what you have learned in this chapter, briefly explain the variation among individuals that Darwin observed.

17. **INTERPRET DATA** Using the graph in Figure 11-20, determine how many offspring were involved in the hypothetical cross studying skin color. What percentage had the lightest skin possible? the darkest skin possible?

aplia. To access course materials, such as Aplia and other companion resources, please visit **www.cengagebrain.com**.

DNA: The Carrier of Genetic Information

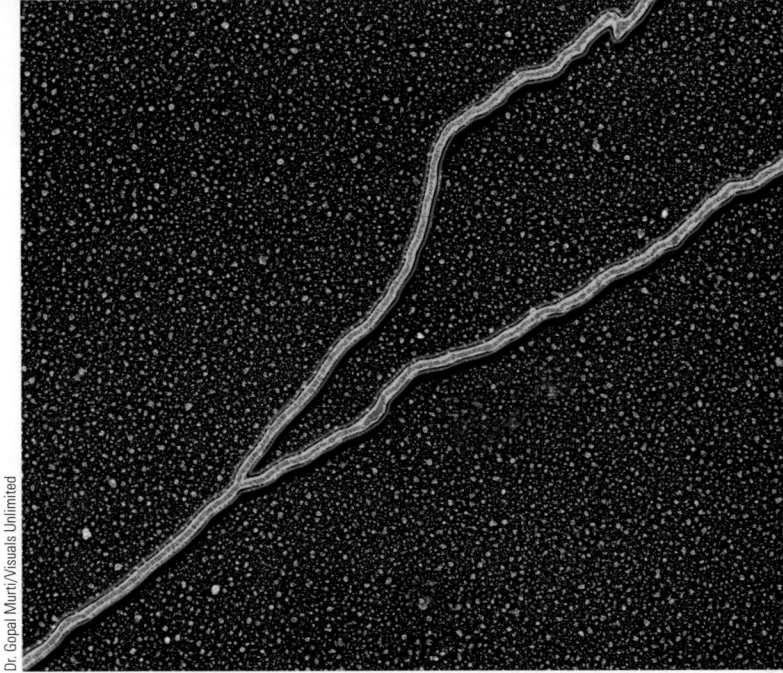

Electron micrograph of DNA replication. During replication, two DNA molecules are synthesized from the original parent molecule. Replication is occurring at the Y-shaped structure, which is called a *replication fork*.

After the rediscovery of Mendel's principles in 1900, geneticists conducted experiments to learn how genes are arranged in chromosomes and how they are transmitted from generation to generation. However, basic questions remained unanswered through most of the first half of the 20th century. What are genes made of? How do genes work? The studies of inheritance patterns described in Chapter 11 did not answer these questions. However, they provided a foundation of knowledge that enabled scientists to make predictions about the molecular (chemical) nature of genes and how genes function.

Scientists generally agreed that genes would have to store information in a form that the cell could retrieve and use, but they also had to account for other properties of genes. For example, experiments on a variety of organisms had demonstrated that genes are usually stable and passed unchanged from generation to generation. However, occasionally a gene converted to a different form; such genetic changes, called *mutations*, were then transmitted unchanged to future generations.

As the science of genetics was developing, biochemists were making a growing effort to correlate the known properties of genes with the nature of various biological molecules. What kind of molecule could store information? How could that information be retrieved and used to direct cell functions? What kind of molecule could be relatively stable but have the capacity to change, resulting in a mutation, under certain conditions?

As they learned more about the central role of proteins in virtually every aspect of cell structure and metabolism, some scientists considered proteins the prime candidates for the genetic material. However, protein did not turn out to be the molecule that governs inheritance. In this chapter we discuss how researchers discovered that **deoxyribonucleic acid (DNA),** a nucleic acid once thought unremarkable, is the molecular basis of inheritance. We explore the unique features of DNA, including its structure and replication (see photograph), that enable it to carry out this role.

KEY CONCEPTS

12.1 In the 1920s, evidence began to accumulate that DNA is the hereditary material.

12.2 The DNA molecule consists of two strands that wrap around each other to form a double helix; the order of its building blocks provides for the storage of genetic information. The DNA building blocks consist of four different nucleotide subunits, designated T, C, A, and G. The pairing of nucleotide subunits occurs based on precise pairing rules: T pairs with A, and C pairs with G.

12.3 DNA replication involves the action of many different proteins. It results in two identical double-stranded DNA molecules and provides the molecular mechanism for passing genetic information from one generation to the next.

12.1 EVIDENCE OF DNA AS THE HEREDITARY MATERIAL

LEARNING OBJECTIVES

1 Summarize the evidence that accumulated during the 1940s and early 1950s demonstrating that DNA is the genetic material.
2 State the questions that these classic experiments addressed: Griffith's transformation experiment, Avery's contribution to Griffith's work, and the Hershey–Chase experiments.

During the 1930s and early 1940s, most geneticists paid little attention to DNA, convinced that the genetic material must be protein. Given the accumulating evidence that genes control production of proteins (discussed in Chapter 13), it certainly seemed likely that genes themselves must also be proteins. Scientists knew proteins consisted of more than 20 different kinds of amino acids in many different combinations, which conferred unique properties on each type of protein. Given their complexity and diversity compared with other molecules, proteins seemed to be the "stuff" of which genes are made.

In contrast, scientists had established that DNA and other nucleic acids were made of only four nucleotides, and what was known about their arrangement made them relatively uninteresting to most researchers. For this reason, several early clues to the role of DNA were not widely noticed.

DNA is the transforming factor in bacteria

One of these clues had its origin in 1928, when Frederick Griffith, a British medical officer, made a curious observation concerning two strains of pneumococcus bacteria (FIG. 12-1). A smooth (S) strain, named for its formation of smooth colonies on a solid growth medium, exhibited **virulence,** the ability to cause

disease and often death, in its host. When living cells of this strain were injected into mice, the animals contracted pneumonia and died. Not surprisingly, the injected animals survived if the cells were first killed with heat. A related rough (R) strain of bacteria, which forms colonies with a rough surface, exhibited **avirulence,** or inability to produce pathogenic effects; mice injected with either living or heat-killed cells of this strain survived. However, when Griffith injected mice with a mixture of *heat-killed* virulent S cells and *live* avirulent R cells, a high proportion of the mice died. Griffith then isolated living S cells from the dead mice.

Can a genetic trait be transmitted from one bacterial strain to another?

HYPOTHESIS: The ability of pneumococcus bacteria to cause disease can be transmitted from the virulent strain (smooth, or S cells) to the avirulent strain (rough, or R cells).

EXPERIMENT: Griffith performed four experiments on mice, using the two strains of pneumococci: (1) injection of mice with live rough cells, (2) injection with live smooth cells, (3) injection with heat-killed smooth cells, and (4) injection with both live rough cells and heat-killed smooth cells.

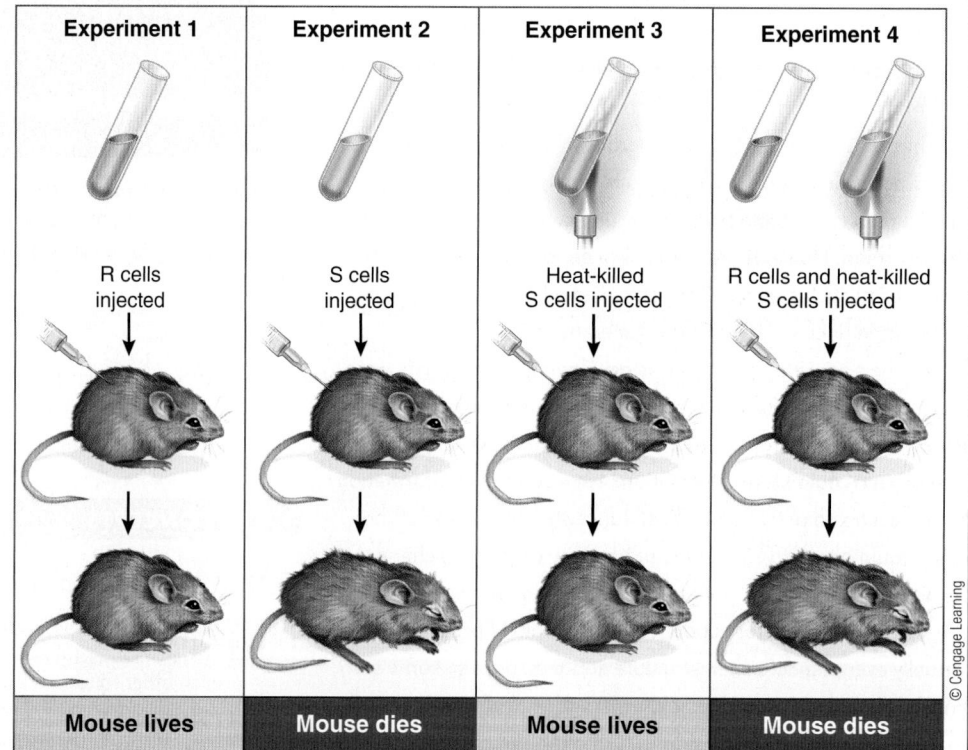

RESULTS AND CONCLUSION: Although neither the rough strain nor the heat-killed smooth strain could kill a mouse, a combination of the two did. Autopsy of the dead mouse showed the presence of living S-strain pneumococci. These results indicated that some substance in the heat-killed S cells had transformed the living R cells into a virulent form.

SOURCE: Griffith, F. (1928) The significance of Pneumococcal types. *Journal of Hygiene* 27(2): 113–159.

Figure 12-1 *Animation* Griffith's transformation experiments

Griffith was trying to develop a vaccine against pneumonia when he serendipitously discovered the phenomenon of transformation.

PREDICT What outcome would be expected for experiment 3 if the heat-killed S cells were treated with enzymes that would specifically (1) digest the cell walls of the heat-killed cells? (2) digest proteins in the heat-killed cells? (3) break down nucleic acids in the heat-killed cells?

Because neither the heat-killed S strain nor the living R strain could be converted to the living virulent form when injected by itself in the control experiments, it seemed that something in the heat-killed cells converted the avirulent cells to the lethal form. This type of permanent genetic change in which the properties of one strain of dead cells are conferred on a different strain of living cells is called **transformation.** Scientists hypothesized that some chemical substance (a *transforming principle,* or *factor*) was transferred from the dead bacteria to the living cells and caused transformation.

In 1944, Oswald T. Avery and his colleagues Colin M. MacLeod and Maclyn McCarty chemically identified Griffith's transforming factor as DNA. They did so through a series of careful experiments in which they lysed (split open) S cells and separated the cell contents into several fractions: lipids, proteins, polysaccharides, and nucleic acids (DNA and RNA). They tested each fraction to see if it could transform living R cells into S cells. The experiments using lipids, polysaccharides, and proteins did not cause transformation. However, when Avery treated living R cells with nucleic acids extracted from S cells, the R cells were transformed into S cells.

Although today scientists consider these results to be the first definitive demonstration that DNA is the genetic material, not all scientists of the time were convinced. Many thought that the findings might apply only to bacteria and might not have any relevance for the genetics of eukaryotes. During the next few years, new evidence accumulated that the haploid nuclei of pollen grains and gametes such as sperm contain only half the amount of DNA found in diploid somatic cells of the same species. (*Somatic* cells are body cells and never become gametes.) Because scientists generally accepted that genes are located on chromosomes, these findings correlating DNA content with chromosome number provided strong circumstantial evidence for DNA's importance in eukaryotic inheritance.

DNA is the genetic material in certain viruses

In 1952, geneticists Alfred Hershey and Martha Chase performed a series of elegant experiments on the reproduction of viruses that infect bacteria, known as **bacteriophages** or **phages** (discussed in Chapter 24). When they planned their experiments, they knew that phages reproduce inside a bacterial cell, eventually causing the cell to break open and release large numbers of new viruses. Because electron microscopic studies had shown that only part of an infecting phage actually enters the cell, they reasoned that the genetic material should be included in that portion (**FIG. 12-2**).

As shown in **FIGURE 12-3**, they labeled the viral protein of one sample of phages with ^{35}S, a radioactive isotope of sulfur, and the viral DNA of a second sample with ^{32}P, a radioactive isotope of phosphorus. Recall from Chapter 3 that proteins contain sulfur as part of the amino acids cysteine and methionine

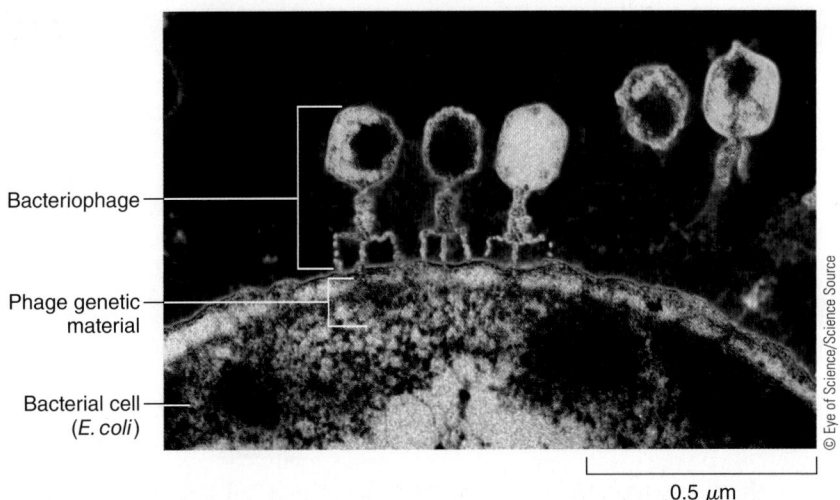

Bacteriophage

Phage genetic material

Bacterial cell (*E. coli*)

0.5 μm

Figure 12-2 Bacteriophages on a bacterial cell

This colorized TEM shows several phages attached to the bacterium *Escherichia coli*. Note the viral genetic material being injected into the bacterium.

and that nucleic acids contain phosphate groups. The phages in each sample attached to bacteria, and the researchers shook the phages off by agitating the sample in a blender. Then they centrifuged the samples.

In the sample in which they had labeled the proteins with ^{35}S, they subsequently found radioactivity in the supernatant, indicating that the protein had not entered the cells. In the sample in which they had labeled the DNA with ^{32}P, they found radioactivity associated with the bacterial cells (in the pellet): DNA had actually entered the cells. Hershey and Chase concluded that phages inject their DNA into bacterial cells, leaving most of their protein on the outside. This finding emphasized the significance of DNA in viral reproduction, and many scientists saw it as an important demonstration of DNA's role as the hereditary material.

CHECKPOINT 12.1

- **CONNECT** *How did Griffith's work provide a foundation for the experiments of Avery and his colleagues pointing to DNA as the essential genetic material?*

- **CONNECT** *How did Hershey and Chase use their knowledge of the chemical composition of proteins and nucleic acids to design their experiments establishing that DNA is the genetic material in bacteriophages?*

12.2 THE STRUCTURE OF DNA

LEARNING OBJECTIVES

3 Explain how nucleotide subunits link to form a single DNA strand.

4 Describe how the two strands of DNA are oriented with respect to each other.

5 State the base-pairing rules for DNA and describe how complementary bases bind to each other.

Is DNA or protein the genetic material in bacterial viruses (phages)?

HYPOTHESIS: DNA is the genetic material in bacterial viruses.

EXPERIMENT: Hershey and Chase produced phage populations with either radioactively labeled DNA or radioactively labeled protein. In both cases, they infected bacteria with the phages and then determined whether DNA or protein was injected into bacterial cells to direct the formation of new viral particles.

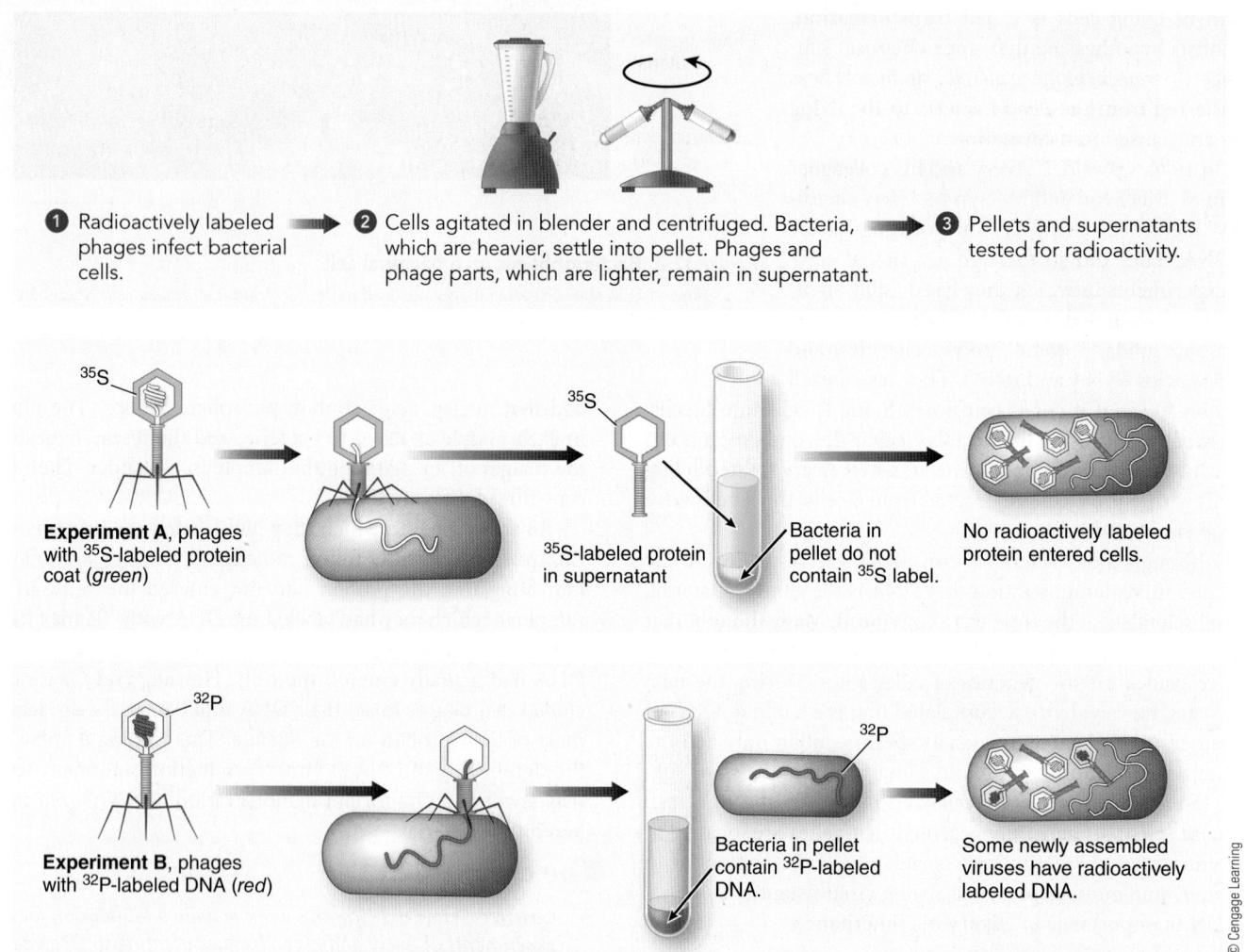

1 Radioactively labeled phages infect bacterial cells. ➡ **2** Cells agitated in blender and centrifuged. Bacteria, which are heavier, settle into pellet. Phages and phage parts, which are lighter, remain in supernatant. ➡ **3** Pellets and supernatants tested for radioactivity.

^{35}S

Experiment A, phages with ^{35}S-labeled protein coat (*green*)

^{35}S

^{35}S-labeled protein in supernatant

Bacteria in pellet do not contain ^{35}S label.

No radioactively labeled protein entered cells.

^{32}P

Experiment B, phages with ^{32}P-labeled DNA (*red*)

^{32}P

Bacteria in pellet contain ^{32}P-labeled DNA.

Some newly assembled viruses have radioactively labeled DNA.

© Cengage Learning

RESULTS AND CONCLUSION: The researchers could separate phage protein coats labeled with the radioactive isotope ^{35}S from infected bacterial cells without affecting viral reproduction. However, they could not separate viral DNA labeled with the radioactive isotope ^{32}P from infected bacterial cells, thus demonstrating that viral DNA enters the bacterial cells and is required for the synthesis of new viral particles. Thus, DNA is the genetic material in phages.

SOURCE: Hershey, A.D. and M. Chase (1952) Independent functions of viral protein and nucleic acid in growth of bacteriophage. *The Journal of General Physiology* 36(1): 39–56.

Figure 12-3 *Animation* **The Hershey–Chase experiments**

CONNECT Explain how this experiment reinforces the earlier findings of Avery, McLeod, and McCarty that genes are composed of nucleic acids.

Scientists did not generally accept DNA as the genetic material until 1953, when American scientist James Watson and British scientist Francis Crick, both working in England, proposed a model for its structure that had extraordinary explanatory power. The story of how the structure of DNA was figured out is one of the most remarkable chapters in the history of modern biology (TABLE 12-1). As you will see in the following discussion, scientists already knew a great deal about DNA's physical and chemical properties when Watson and Crick became interested in the problem; in fact, Watson and Crick did not conduct any experiments or gather any new data. Their all-important contribution was to integrate all the available information into a model that demonstrated how the molecule can both carry information for making proteins and serve as its own **template** (pattern or guide) for its duplication.

TABLE 12-1	A Time Line of Selected Historical DNA Discoveries

DATE	DISCOVERY
1871	**Friedrich Miescher** reports discovery of new substance, *nuclein,* from cell nuclei. Nuclein is now known to be a mixture of DNA, RNA, and proteins.
1928	**Frederick Griffith** finds a substance in heat-killed bacteria that "transforms" living bacteria.
1944	**Oswald Avery, Colin MacLeod,** and **Maclyn McCarty** chemically identify Griffith's transforming principle as DNA.
1949	**Erwin Chargaff** reports relationships among DNA bases that provide a clue to the structure of DNA.
1952	**Alfred Hershey** and **Martha Chase** demonstrate that DNA, not protein, is involved in viral reproduction.
1952	**Rosalind Franklin** produces X-ray diffraction images of DNA.
1953	**James Watson** and **Francis Crick** propose a model of the structure of DNA; this contribution is widely considered the start of a revolution in molecular biology that continues to the present.
1958	**Matthew Meselson** and **Franklin Stahl** demonstrate that DNA replication is semiconservative.
1962	**James Watson, Francis Crick,** and **Maurice Wilkins** are awarded the Nobel Prize in Physiology or Medicine for discoveries about the molecular structure of nucleic acids.*
1969	**Alfred Hershey** is awarded the Nobel Prize in Physiology or Medicine for discovering the replication mechanism and genetic structure of viruses.

*Rosalind Franklin could not share the prize because she was deceased.

© Cengage Learning

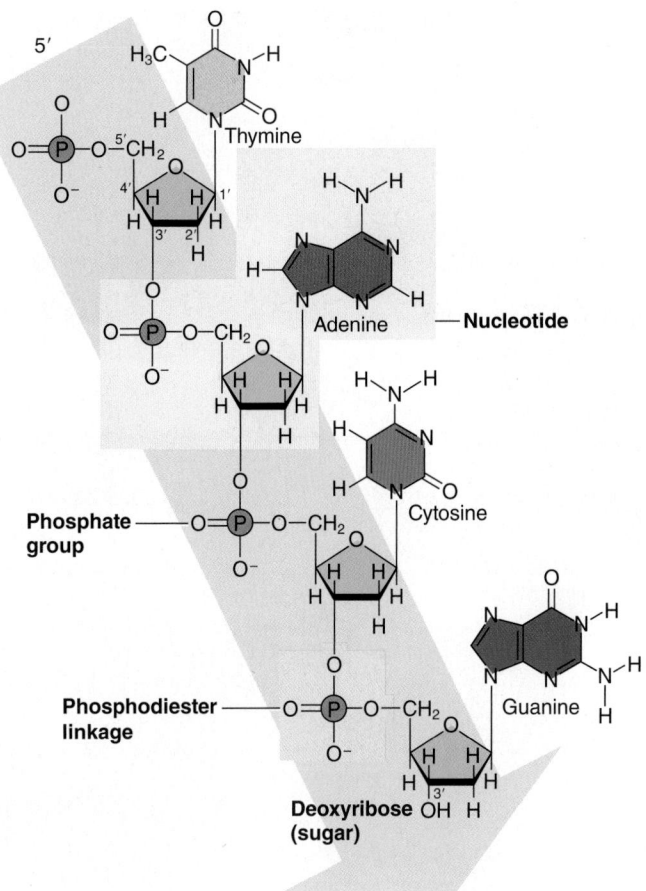

Figure 12-4 *Animation* **The nucleotide subunits of DNA**

A single strand of DNA consists of a backbone (*superimposed on blue screen*) of phosphate groups alternating with the sugar deoxyribose (*green*). Phosphodiester linkages (*pink*) join sugars of adjacent nucleotides. Linked to the 1′ carbon of each sugar is one of four nitrogenous bases (*top to bottom*): thymine, adenine, cytosine, and guanine. (*The nucleotide containing the base adenine is highlighted yellow.*) Note the polarity of the polynucleotide chain, with the 5′ end at the top of the figure and the 3′ end at the bottom.

© Cengage Learning

Nucleotides can be covalently linked in any order to form long polymers

As discussed in Chapter 3, each DNA building block is a **nucleotide** consisting of the pentose sugar **deoxyribose,** a phosphate, and one of four nitrogenous bases (**FIG. 12-4**). It is conventional to number the atoms in a molecule using a system devised by organic chemists. Accordingly, in nucleic acid chemistry the individual carbons in each sugar and each base are numbered. The carbons in a base are designated by numerals, but the carbons in a sugar are distinguished from those in the base by prime symbols, such as 2′. The nitrogenous base is attached to the 1′ carbon of the sugar, and the phosphate is attached to the 5′ carbon. The bases include two **purines, adenine (A)** and **guanine (G),** and two **pyrimidines, thymine (T)** and **cytosine (C).**

The nucleotides are linked by covalent bonds to form an alternating sugar–phosphate backbone. The 3′ carbon of one sugar is bonded to the 5′ phosphate of the adjacent sugar to form a 3′, 5′ **phosphodiester linkage.** The result is a polymer of indefinite length, with the nucleotides linked in any order. Scientists now know that most DNA molecules found in cells are millions of bases long. Figure 12-4 also shows that a single polynucleotide chain is directional. No matter how long the chain may be, one end, the **5′ end,** has a 5′ carbon attached to a phosphate, and the other end, the **3′ end,** has a 3′ carbon attached to a hydroxyl group.

By 1949, Erwin Chargaff and his colleagues at Columbia University had determined the base composition of DNA from several organisms and tissues. They found a simple relationship among the bases that turned out to be an important clue to the structure of DNA. Regardless of the source of the DNA, in

Chargaff's words the "ratios of purines to pyrimidines and also of adenine to thymine and of guanine to cytosine were not far from 1" (**TABLE 12-2**). In other words, in double-stranded DNA molecules, the number of purines equals the number of pyrimidines, the number of adenines equals the number of thymines (A equals T), and the number of guanines equals the number of cytosines (G equals C). These equalities are called **Chargaff's rules.**

TABLE 12-2	Base Compositions in DNA from Selected Organisms

	PERCENTAGE OF DNA BASES				RATIOS	
SOURCE OF DNA	A	T	G	C	A/T	G/C
E. coli	26.1	23.9	24.9	25.1	1.09	0.99
Yeast	31.3	32.9	18.7	17.1	0.95	1.09
Sea urchin sperm	32.5	31.8	17.5	18.2	1.02	0.96
Herring sperm	27.8	27.5	22.2	22.6	1.01	0.98
Human liver	30.3	30.3	19.5	19.9	1.00	0.98
Corn (*Zea mays*)	25.6	25.3	24.5	24.6	1.01	1.00

© Cengage Learning

Figure 12-5 Rosalind Franklin

Franklin was a gifted scientist whose contributions helped elucidate the double-helix structure of DNA.

DNA is made of two polynucleotide chains intertwined to form a double helix

Key information about the structure of DNA came from X-ray diffraction studies on crystals of purified DNA, carried out by British scientist Rosalind Franklin at King's College from 1951 to 1953 (**FIG. 12-5**).

X-ray diffraction, a powerful method for determining the 3-D structure of a molecule, can determine the distances between the atoms of molecules arranged in a regular, repeating crystalline structure (**FIG. 12-6**). X-rays have such short wavelengths that they can be scattered by the electrons surrounding the atoms in a molecule. Atoms with dense electron clouds (such as phosphorus and oxygen) tend to deflect electrons more strongly than atoms with lower atomic numbers. Exposing a crystal to an intense beam of X-rays causes the regular arrangement of its atoms to diffract, or scatter, the X-rays in specific ways. The pattern of diffracted X-rays appears on photographic film as dark spots. Mathematical analysis of the pattern and distances between the spots yields the precise distances between atoms and their orientation within the molecules.

WHY IS IT USED? X-ray diffraction can be used to determine the regular array of atoms in a crystalline sample of, for example, DNA. Because each type of crystal has its own characteristic pattern, the three-dimensional structure of the molecule being studied can be deduced.

HOW IS IT DONE?

1. Researchers direct a narrow beam of X-rays at a single crystal of DNA. X-rays are diffracted (bent) at specific angles based on the density of electrons of different atoms. Important clues about DNA structure are provided by detailed mathematical analysis of measurements of the spots, which are formed by X-rays hitting the photographic plate

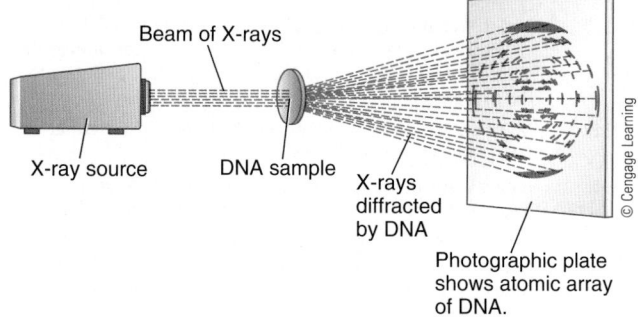

2. X-ray diffraction image of DNA. This diagonal pattern of spots stretching from 11 to 5 and from 1 to 7 (as on a clock face) provides evidence for the helical structure of DNA. The elongated horizontal patterns at the top and bottom indicate that the purine and pyrimidine bases are stacked 0.34 nm apart and are perpendicular to the axis of the DNA molecule.

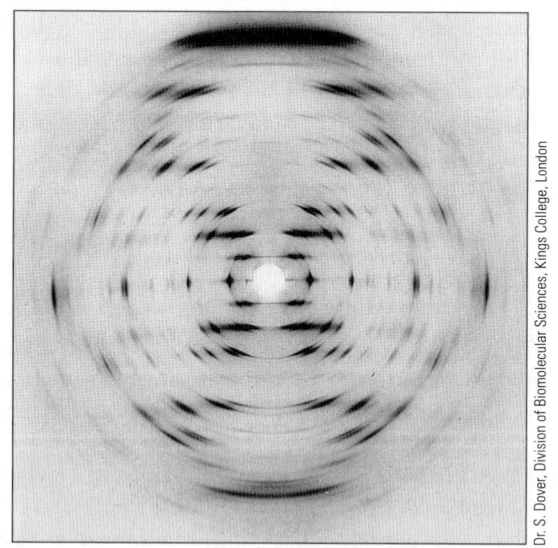

Figure 12-6 How X-ray diffraction works

Franklin had already produced X-ray crystallographic films of DNA patterns when Watson and Crick began to pursue the problem of DNA structure. Her pictures clearly showed that DNA has a type of helical structure, and three major types of regular, repeating patterns in the molecule (with the dimensions 0.34 nm, 3.4 nm, and 2.0 nm) were evident. Franklin and colleague Maurice Wilkins had inferred from these patterns that the nucleotide bases (which are flat molecules) are stacked like the rungs of a ladder. Using this information, Watson and Crick began to build scale models of the DNA components and then fit them together to correlate with the experimental data (FIG. 12-7).

After several trials, they worked out a model that fit the existing data (FIG. 12-8). The nucleotide chains conformed to the dimensions of the X-ray data only if each DNA molecule consisted of *two* polynucleotide chains arranged in a coiled **double helix.** In their model the sugar–phosphate backbones of the two chains form the outside of the helix. The bases belonging to the two chains associate as pairs along the central axis of the helix. The reasons for the repeating patterns of 0.34 nm and 3.4 nm measurements are readily apparent from the model: each pair of bases is exactly 0.34 nm from the adjacent pairs above and below. Because exactly 10 base pairs are present in each full turn of the helix, each turn constitutes 3.4 nm of length. To fit the data, the two chains must run in opposite directions; therefore, each end of the double helix must have an exposed 5′ phosphate on one strand and an exposed 3′ hydroxyl group (—OH) on the other. Because the two strands run in opposite directions, they are **antiparallel** to each other (FIG. 12-9a).

In double-stranded DNA, hydrogen bonds form between A and T and between G and C

Other features of the Watson and Crick model integrated critical information about the chemical composition of DNA with the X-ray diffraction data. The X-ray diffraction studies indicated that the double helix has a precise and constant width, as shown by the 2.0 nm measurements. This finding is actually consistent with Chargaff's rules. As Figure 12-4 shows, each pyrimidine (cytosine or thymine) contains only one ring of atoms, whereas each purine (guanine or adenine) contains two rings. Study of the models made it clear to Watson and Crick that if each rung of the ladder contained one purine and one pyrimidine, the width of the helix at each base pair would be exactly 2.0 nm. By contrast, the combination of two purines (each of which is 1.2 nm wide) would be wider than 2.0 nm and that of two pyrimidines would be narrower, so the diameter would not be constant. Further examination of the model showed that adenine can pair with thymine (and guanine with cytosine) in such a way that hydrogen bonds form between them; the opposite combinations, cytosine with adenine and guanine with thymine, do not lead to favorable hydrogen bonding.

The nature of the hydrogen bonding between adenine and thymine and between guanine and cytosine is shown in FIGURE 12-9b. Two hydrogen bonds form between adenine and thymine, and three form between guanine and cytosine. This concept of specific base pairing neatly explains Chargaff's rules.

Figure 12-7 James Watson and Francis Crick

Watson (*left*) and Crick (*right*) are shown with their model of DNA's double helix.

© A. Barrington Brown/Science Source

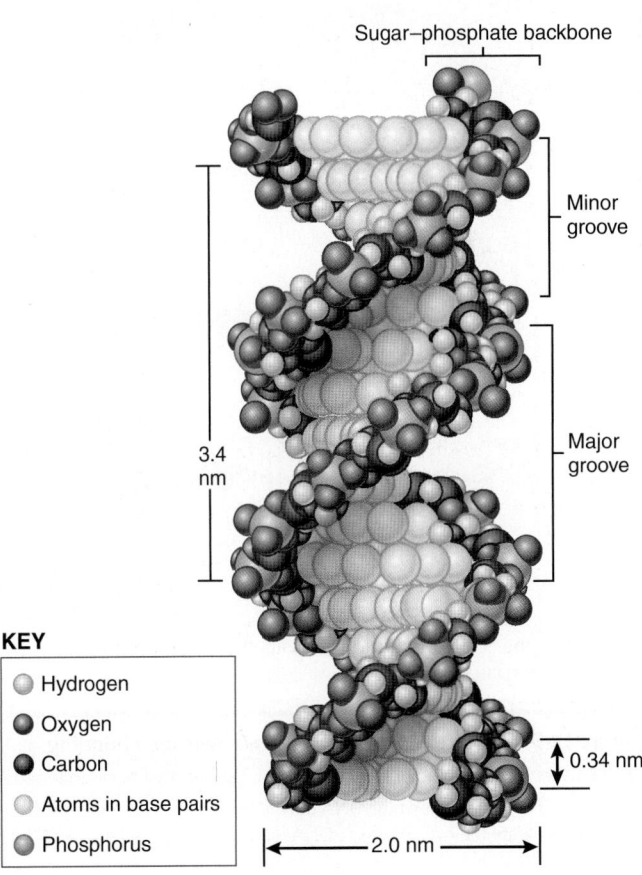

Sugar–phosphate backbone

Minor groove

Major groove

3.4 nm

0.34 nm

2.0 nm

KEY

- ○ Hydrogen
- ● Oxygen
- ● Carbon
- ○ Atoms in base pairs
- ○ Phosphorus

Figure 12-8 A three-dimensional model of the DNA double helix

The measurements match those derived from X-ray diffraction images.
© Cengage Learning

Base pairing and the sequence of bases in DNA provide a foundation for understanding both DNA replication and the inheritance of genetic material.

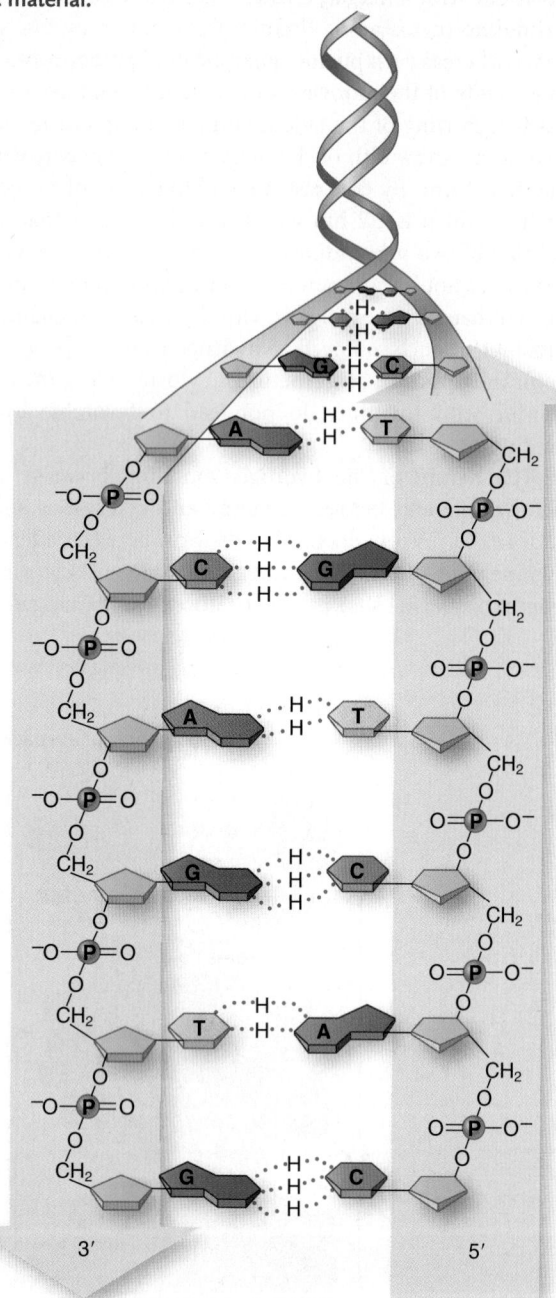

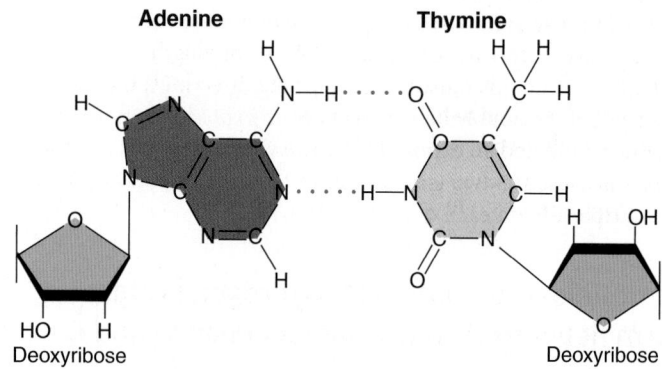

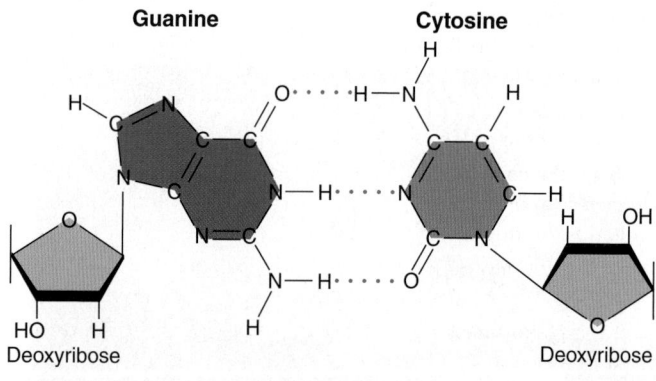

(a) The two sugar–phosphate chains run in opposite directions. This orientation permits the complementary bases to pair.

(b) Hydrogen bonding between base pairs adenine (A) and thymine (T) (*top*) and guanine (G) and cytosine (C) (*bottom*). The A:T pair has two hydrogen bonds; the G:C pair has three.

Figure 12-9 *Animation* **Base pairing and hydrogen bonding**

The two strands of a DNA double helix are hydrogen-bonded between the bases.

PREDICT The two strands of the DNA double helix are held together by the hydrogen bonding between the paired bases. Which would require more energy to separate the two strands, a DNA molecule composed mostly of G:C base pairs or a DNA molecule with mostly A:T base pairs?
© Cengage Learning

The amount of cytosine must equal the amount of guanine because every cytosine in one chain must have a paired guanine in the other chain. Similarly, every adenine in the first chain must have a thymine in the second chain. The sequences of bases in the two chains exhibit **complementary base pairing;** that is, the sequence of nucleotides in one chain dictates the complementary sequence of nucleotides in the other. For example, if one strand has the sequence

$$3'—AGCTAC—5'$$

the other strand has the complementary sequence

$$5'—TCGATG—3'$$

The double-helix model strongly suggested that the sequence of bases in DNA provides for the storage of genetic information and that this sequence ultimately relates to the sequences of amino acids in proteins. Although restrictions limit how the bases on opposite strands pair with each other, the number of possible linear sequences of bases in a strand is virtually unlimited. Because a DNA molecule in a cell can be millions of nucleotides long, it can store enormous amounts of information, usually consisting of hundreds of genes.

CHECKPOINT 12.2

- *What types of subunits make up a single strand of DNA? How are the subunits linked?*
- *What is the structure of double-stranded DNA as determined by Watson and Crick?*
- **CONNECT** *How do Chargaff's rules relate to the structure of DNA?*

12.3 DNA REPLICATION

LEARNING OBJECTIVES

6 Cite evidence from Meselson and Stahl's experiment that enabled scientists to differentiate between semiconservative replication of DNA and alternative models.

7 Summarize how DNA replicates and identify some unique features of the process.

8 Explain the complexities of DNA replication that make the process (a) bidirectional and (b) continuous in one strand and discontinuous in the other.

9 Discuss how enzymes proofread and repair errors in DNA.

10 Define *telomere* and describe the possible connections between telomerase and cell aging and between telomerase and cancer.

Two immediately apparent and distinctive features of the Watson–Crick model made it seem plausible that DNA is the genetic material. We have already mentioned that the sequence of bases in DNA can carry coded information. The model also suggested a way in which the sequence of nucleotides in DNA could be precisely copied, a process known as **DNA replication.**

The connection between DNA replication and the behavior of chromosomes in mitosis was obvious to Watson and Crick. A chromosome becomes duplicated so that it consists of two identical sister chromatids that later separate at anaphase; the genetic material must be precisely duplicated and distributed to the daughter cells. They noted, in a classic and now famous understatement at the end of their first brief paper, "It has not escaped our notice that the specific pairing we have postulated

immediately suggests a possible copying mechanism for the genetic material."

The model suggested that because the nucleotides pair with each other in complementary fashion, each strand of the DNA molecule could serve as a template for synthesizing the opposite strand. It would simply be necessary for the hydrogen bonds between the two strands to break (recall from Chapter 2 that hydrogen bonds are relatively weak) and the two chains to separate. Each strand of the double helix could then pair with new complementary nucleotides to replace its missing partner. The result would be two DNA double helices, each identical to the original one and consisting of one original strand from the parent molecule and one newly synthesized complementary strand. This type of information copying is called **semiconservative replication** (FIG. 12-10a).

Meselson and Stahl verified the mechanism of semiconservative replication

Although the semiconservative replication mechanism suggested by Watson and Crick was (and is) a simple and compelling model, experimental proof was needed to establish that DNA in fact replicates in that manner. Researchers first needed to rule out other possibilities. With **conservative replication,** both parent (or old) strands would remain together, and the two newly synthesized strands would form a second double helix (FIG. 12-10b). As a third hypothesis, the parental and newly synthesized strands might become randomly mixed during the replication process, that is, *dispersive replication* (FIG. 12-10c). To discriminate among semiconservative replication and the other models, investigators had to distinguish between old and newly synthesized strands of DNA.

One technique is to use a heavy isotope of nitrogen, ^{15}N (ordinary nitrogen is ^{14}N), to label the bases of the DNA strands, making them more dense. Using **density gradient centrifugation,** scientists can separate large molecules such as DNA on the basis of differences in their density (see Fig. 4-6). When DNA is mixed with a solution containing cesium chloride (CsCl) and centrifuged at high speed, the solution forms a density gradient in the centrifuge tube, ranging from a region of lowest density at the top to one of highest density at the bottom. During centrifugation, the DNA molecules migrate to the region of the gradient identical to their own density.

In 1958, Matthew Meselson and Franklin Stahl at the California Institute of Technology grew the bacterium *Escherichia coli* on a medium that contained ^{15}N in the form of ammonium chloride (NH_4Cl). The cells used the ^{15}N to synthesize bases, which then became incorporated into DNA (FIG. 12-11). The resulting DNA molecules, which contained heavy nitrogen, were extracted from some of the cells. When the researchers subjected them to density gradient centrifugation, they accumulated in the high-density region of the gradient. The team transferred the rest of the bacteria (which also contained ^{15}N-labeled DNA) to a different growth medium in which the NH_4Cl contained the naturally abundant, lighter ^{14}N isotope and then allowed them to undergo additional cell divisions.

Meselson and Stahl expected the newly synthesized DNA strands to be less dense because they incorporated bases

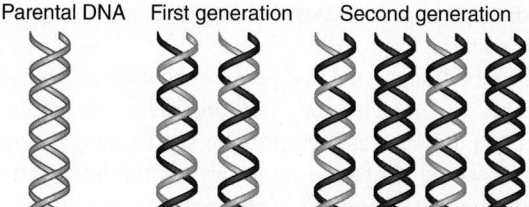

(a) Hypothesis 1: Semiconservative replication

Parental DNA   First generation   Second generation

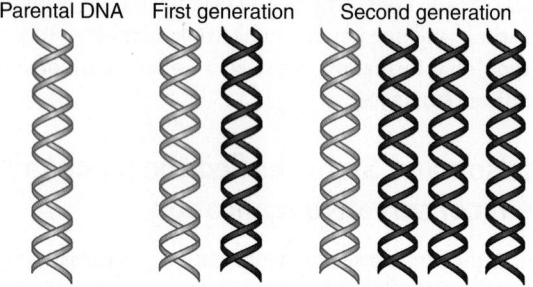

(b) Hypothesis 2: Conservative replication

Parental DNA   First generation   Second generation

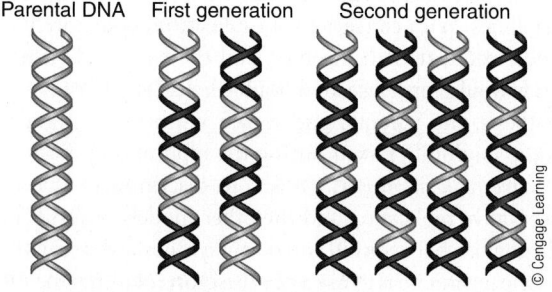

© Cengage Learning

(c) Hypothesis 3: Dispersive replication

Figure 12-10 Alternative models of DNA replication

The hypothesized arrangement of old (*light blue*) and newly synthesized (*dark blue*) DNA strands after one and two generations, according to **(a)** the semiconservative model, **(b)** the conservative model, and **(c)** the dispersive model.

containing the lighter ^{14}N isotope. Indeed, double-stranded DNA from cells isolated after one generation had an intermediate density, indicating that they contained half as many ^{15}N atoms as the "parent" DNA. This finding supported the semiconservative replication model, which predicted that each double helix would contain a previously synthesized strand (heavy in this case) and a newly synthesized strand (light in this case). It was also consistent with the dispersive model, which would also yield one class of molecules, all with intermediate density. It was inconsistent with the conservative model, which predicted two classes of double-stranded molecules: those with two heavy strands and those with two light strands.

After another cycle of cell division in the medium with the lighter ^{14}N isotope, two types of DNA appeared in the density gradient, exactly as predicted by the semiconservative replication model. One consisted of DNA with a density intermediate between ^{15}N-labeled DNA and ^{14}N-labeled DNA, whereas the

What is the mechanism of DNA replication?

HYPOTHESIS: Figure 12-10 depicts three hypotheses of DNA replication and predicts the arrangement of old and newly synthesized DNA strands after one or two generations according to each hypothesis.

EXPERIMENT: Meselson and Stahl grew bacteria (*E. coli*) in heavy nitrogen (^{15}N) growth medium for many generations so that all the DNA strands would be heavy. Then they transferred some of the cells to light nitrogen (^{14}N) medium so that the newly synthesized strands would be light. They isolated DNA from bacterial cells after 20 minutes (one generation) and 40 minutes (two generations), and centrifuged it to separate DNA into bands based on density.

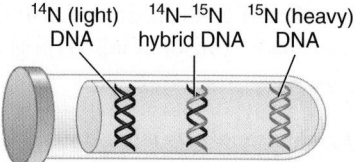

^{14}N (light) DNA ^{14}N–^{15}N hybrid DNA ^{15}N (heavy) DNA

In density gradient centrifugation, the concentration of CsCl is highest at the bottom of the tube and lowest at the top. DNA molecules move to positions where their density equals that of the CsCl solution in which they are centrifuged.

RESULTS AND CONCLUSION: Based on the observed density of the DNA molecules in each generation, Meselson and Stahl concluded that the semiconservative model accurately predicts the mechanism of DNA replication.

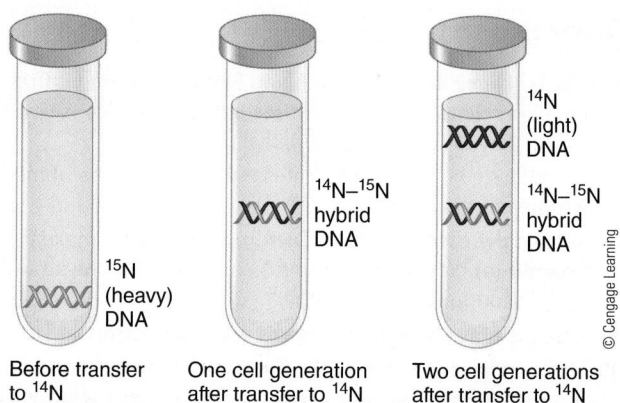

© Cengage Learning

^{15}N (heavy) DNA

^{14}N–^{15}N hybrid DNA

^{14}N (light) DNA

^{14}N–^{15}N hybrid DNA

Before transfer to ^{14}N   One cell generation after transfer to ^{14}N   Two cell generations after transfer to ^{14}N

The location of DNA molecules within the centrifuge tube can be determined by UV optics. DNA solutions absorb strongly at 260 nm.

SOURCE: Meselson, M. and F.W. Stahl (1958). The replication of DNA in *Escherichia coli. The Proceedings of the National Academy of Sciences* (U.S.A.) 44(7): 671–682.

Figure 12-11 The Meselson–Stahl experiment

VISUALIZE Draw the positions of the second-generation DNA molecules you would expect to see in the tube on the right (two generations) if they were replicated by (1) the conservative replication hypothesis and (2) the dispersive replication hypothesis.

other contained only DNA with a density of ^{14}N-labeled DNA. This finding refuted the dispersive model, which predicted that all strands would have an intermediate density.

Semiconservative replication explains the perpetuation of mutations

The recognition that DNA could be copied by a semiconservative mechanism suggested how DNA could explain a third essential characteristic of genetic material, the ability to mutate. It was long known that **mutations,** or genetic changes, could arise in genes and then be transmitted faithfully to succeeding generations. For example, a mutation in the fruit fly (*Drosophila melanogaster*) produces vestigial wings.

When the double-helix model was proposed, it seemed plausible that mutations could represent a change in the sequence of bases in the DNA. You could predict that if DNA is copied by a mechanism involving complementary base pairing, any change in the sequence of bases on one strand would produce a new sequence of complementary bases during the next replication cycle. The new base sequence would then transfer to daughter molecules by the same mechanism used to copy the original genetic material as if no change had occurred.

For the example in **FIGURE 12-12,** an adenine base in one of the DNA strands has been changed to guanine. This change could occur by a rare error in DNA replication or by one of several other known mechanisms. As discussed later, certain enzymes correct errors when they occur, but not all errors are corrected properly. By one estimate, the rate of uncorrected errors that occur during DNA replication is equal to about one nucleotide in a billion. When the DNA molecule containing an error replicates (left side of Figure 12-12), one of the strands gives rise to a molecule exactly like its parent strand; the other (mutated) strand gives rise to a molecule with a new combination of bases that is transmitted generation after generation.

DNA replication requires protein "machinery"

Although semiconservative replication by base pairing appears simple and straightforward, the actual process is highly regulated and requires a "replication machine" containing many types of protein and enzyme molecules. Many essential features of DNA replication are common to all organisms, although prokaryotes and eukaryotes differ somewhat because their DNA is organized differently. In most bacterial cells, such as *E. coli*, all or most of the DNA is in the form of a single, *circular*, double-stranded DNA molecule. In contrast, each unreplicated eukaryotic chromosome contains a single, *linear*, double-stranded molecule associated with at least as much protein (by mass) as DNA.

The process of DNA replication requires a large number of proteins with different functions, many of which are organized as multimolecular "machines." For example, in the unicellular yeast *Saccharomyces cerevisiae*, which is considered a relatively "simple" eukaryote, 88 genes are involved in DNA replication!

DNA strands must be unwound during replication DNA replication begins at specific sites on the DNA molecule, called **origins of replication,** where small sections of the double helix unwind. **DNA helicases** are helix-destabilizing enzymes (several have been identified) that bind to DNA at the origin of replication and break hydrogen bonds, thereby separating the two strands

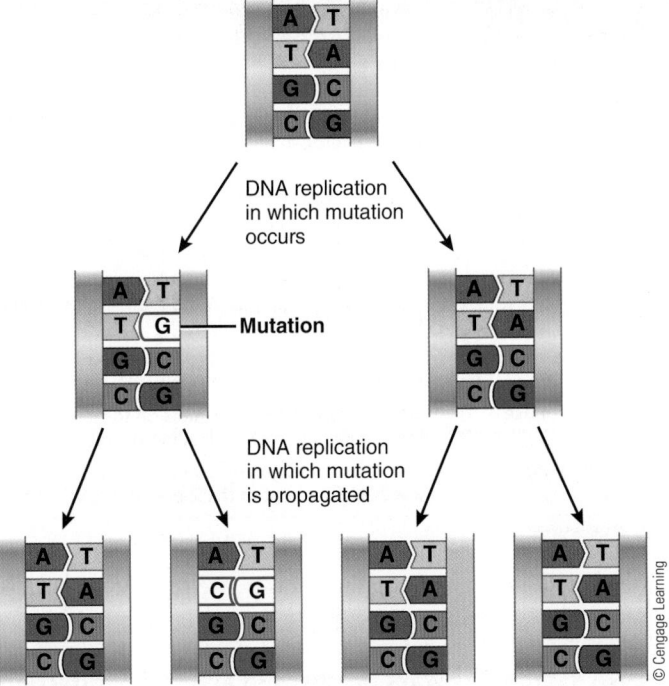

Figure 12-12 The perpetuation of a mutation

The process of DNA replication can stabilize a mutation (*outlined in red*) so that it is transmitted to future generations.

(**FIG. 12-13; TABLE 12-3**). Both DNA strands replicate at the same time at the junction between the separated strands, which is a Y-shaped structure called the **replication fork** (see chapter opener photograph). Helicase travels along the helix, opening the double helix like a zipper during movement of the replication fork.

Once DNA helicases separate the strands, **single-strand binding (SSB) proteins** bind to single DNA strands and stabilize them, which prevents the double helix from re-forming until the strands are replicated. SSB proteins also prevent the hydrolysis of the single-strand regions by other enzymes (*nucleases*; as we discuss later in the chapter, nucleases are involved in DNA repair).

Watson and Crick recognized that in their double-helix model, the two DNA strands wrap around each other like the strands of a rope. If you try to pull the strands apart, the rope must either rotate or twist into tighter coils. You could expect similar results when complementary DNA strands are separated for replication. As the DNA strands unwind to open for replication, torsional strain from supercoiling (excessive twisting) occurs in another part of the DNA molecule. Enzymes called **topoisomerases** produce breaks in the DNA molecules and then rejoin the strands, relieving strain and effectively preventing supercoiling and knot formation during replication. There are two ways that topoisomerases reduce supercoiling. Some topoisomerases produce a temporary break in the polynucleotide backbone of a single strand of DNA, pass that strand through the excessively twisted part, and then reseal the break. Other topoisomerases break both DNA strands, pass some of the helix between the cut ends, and then reseal the break. Regardless of

TABLE 12-3	Proteins Involved in DNA Replication
ENZYME	**FUNCTION**
Helicase	Opens the double helix at replication forks by disrupting the hydrogen bonds that hold the two strands together.
Single-strand binding (SSB) protein	Binds to single strands of DNA and prevents the helix from re-forming before it can be used as a template for replication.
Topoisomerase	Breaks one or both DNA strands, preventing excessive coiling during replication, and then rejoins them in a more relaxed configuration.
DNA polymerase	Links nucleotide subunits to form a new DNA strand complementary to a DNA template.
DNA primase	Synthesizes short RNA primers on the lagging strand. Begins replication of the leading strand.
DNA ligase	Links Okazaki fragments by joining the 3′ end of the new DNA fragment to the 5′ end of the adjoining DNA.

© Cengage Learning

their modes of action, topoisomerases give replicating DNA a more relaxed configuration.

DNA synthesis always proceeds in a 5′ ⟶ 3′ direction

The enzymes that catalyze the linking of successive nucleotide subunits are called **DNA polymerases.** They add nucleotides only to the 3′ end of a growing polynucleotide strand, and this strand must be paired with the DNA template strand (**FIG. 12-14**). Nucleotides with three phosphate groups are substrates for the polymerization reaction. As the nucleotides become linked, two of the phosphates are removed. These reactions are strongly exergonic and do not need additional energy. Because the polynucleotide chain is elongated by the linkage of the 5′ phosphate group of the next nucleotide subunit to the 3′ hydroxyl group of the sugar at the end of the existing strand, the new strand of DNA always grows in the 5′ ⟶ 3′ direction. Some DNA polymerases are very efficient in joining nucleotides to the growing polypeptide chain. *DNA Pol III*, which is one of five DNA polymerases that have been identified in the bacterium *E. coli*, can join 1200 nucleotides per minute.

DNA synthesis requires an RNA primer

As mentioned, DNA polymerases add nucleotides only to the 3′ end of an *existing* polynucleotide strand. Then how is DNA synthesis initiated once the two strands are separated? The answer is that first a short piece of RNA (5 to 14 nucleotides) called an **RNA primer** is synthesized at the point where replication begins (**FIG. 12-15**).

RNA, or **ribonucleic acid** (see Chapters 3 and 13), is a nucleic acid polymer consisting of nucleotide subunits that can associate by complementary base pairing with the single-strand DNA template. The RNA primer is synthesized by **DNA primase**, an enzyme that starts a new strand of RNA opposite a short stretch of the DNA template strand. After a few nucleotides have been added, DNA polymerase displaces the primase and subsequently adds subunits to the 3′ end of the short RNA primer. Specific enzymes later degrade the primer (discussed in the next section), and the space fills in with DNA.

DNA replication is discontinuous in one strand and continuous in the other

We mentioned earlier that the complementary DNA strands are antiparallel. DNA synthesis proceeds only in the 5′ ⟶ 3′ direction, which means that the strand being copied is being read in the 3′ ⟶ 5′ direction. Thus, it may seem necessary to copy one of the strands starting at one end of the double helix and the other strand starting at the opposite end. Some viruses replicate their DNA in this way, but this replication method is not workable in the extremely long DNA molecules in eukaryotic chromosomes.

Instead, as previously mentioned, DNA replication begins at origins of replication, and both strands replicate at the same time at a replication fork (see Fig. 12-15). The position of the replication fork is constantly moving as replication proceeds. Two identical DNA polymerase molecules catalyze replication. One of these adds nucleotides to the 3′ end of the new strand that is always growing *toward* the replication fork. Because this strand is synthesized smoothly and continuously, it is called the **leading strand.**

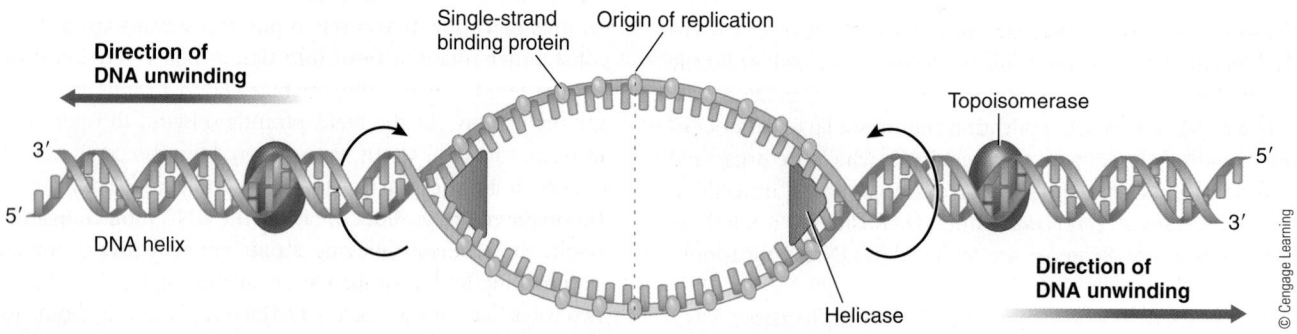

Figure 12-13 DNA strand separation during DNA replication

DNA strands are separated at an origin of replication by ATP-powered DNA helicase enzymes to create a "replication bubble" with a Y-shaped replication fork at each end of the bubble. Single-strand binding proteins bind to the unpaired strands to keep them from re-forming the helix. Unwinding the strands twists and creates strain on the double helix ahead of the replication fork. Topoisomerase enzymes relieve this strain by breaking, swiveling, and rejoining the paired strands.

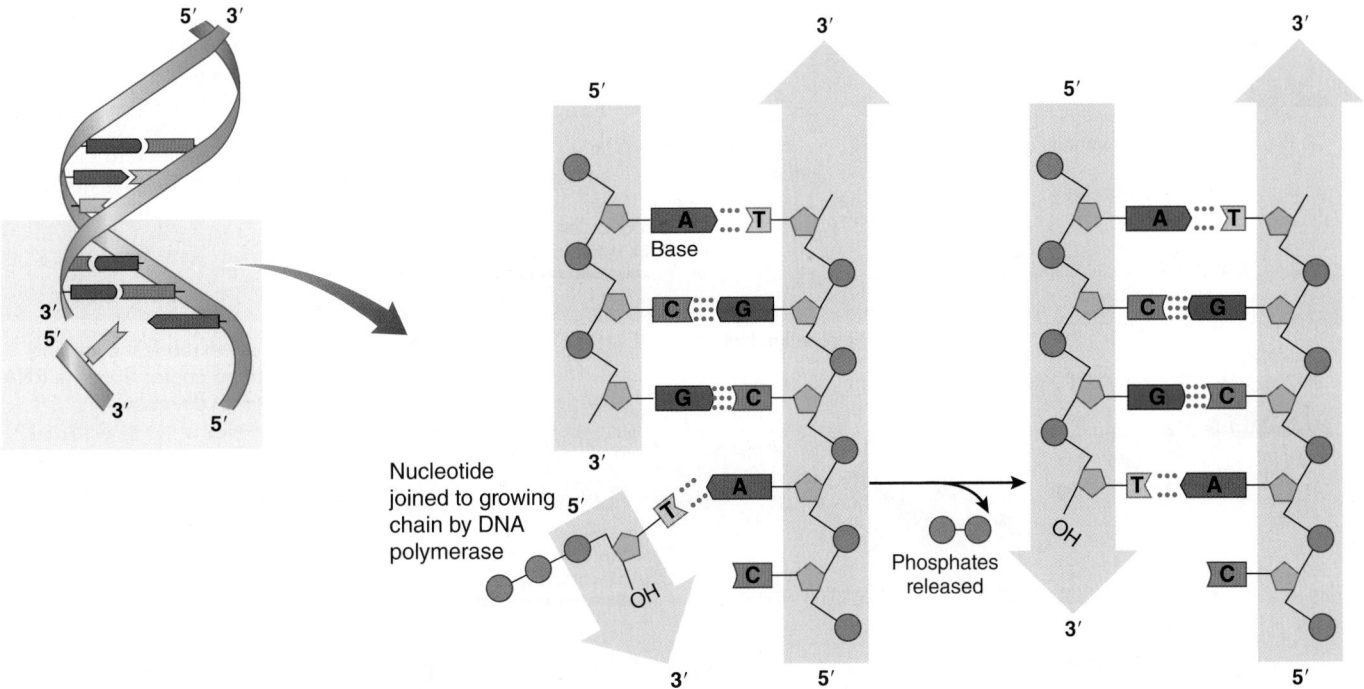

Figure 12-14 *Animation* **A simplified view of DNA replication**

One nucleotide at a time is added to the 3′ end of a growing chain.
© Cengage Learning

A separate DNA polymerase molecule adds nucleotides to the 3′ end of the other new strand. Called the **lagging strand**, it is always growing *away* from the replication fork. Only short pieces can be synthesized, because if the DNA polymerase were to add continuously to the 3′ end of that strand, it would need to move far from the replication fork. These 100- to 2000-nucleotide pieces are called **Okazaki fragments** after their discoverer, Japanese molecular biologist Reiji Okazaki.

DNA primase periodically catalyzes the synthesis of a new RNA primer on the lagging strand as more of its template strand becomes unwound by the helicase. A separate RNA primer initiates each Okazaki fragment, which DNA polymerase then extends toward the 5′ end of the previously synthesized fragment. When the RNA primer of the previously synthesized fragment is reached, DNA polymerase degrades and replaces the primer with DNA. The fragments are then joined by **DNA ligase,** an enzyme that links the 3′ hydroxyl of one Okazaki fragment to the 5′ phosphate of the DNA immediately next to it, forming a phosphodiester linkage. (As you will see later in the chapter, DNA ligase also rejoins broken phosphodiester bonds during DNA repair.)

DNA synthesis is bidirectional When double-stranded DNA separates, two replication forks form, creating a replication bubble. Within this bubble the molecule replicates in both directions from the origin of replication. **FIGURE 12-16a** shows how the lagging and leading strands are arranged at the two replication forks.

Unlike the linear DNA molecules found in eukaryotic cells, DNA in bacteria are in the form of circles, with no free ends.

FIGURE 12-16b shows a replicating plasmid in bacteria. **Plasmids** are small, circular DNA molecules that carry genes separate from those on a bacterial chromosome. Because plasmids are so small compared to the circular bacterial chromosome, they can be clearly photographed during replication. In bacteria each circular DNA molecule usually has only one origin of replication, so the two replication forks proceed around the circle and eventually meet at the other side to complete the formation of two new DNA molecules (**FIG. 12-16c**).

By contrast, a eukaryotic chromosome consists of one long, linear DNA molecule, so having multiple origins of replication speeds the replication process (**FIG. 12-16d** and **e**). (In mammals, between 20,000 and 50,000 origins of replication are involved in the replication of the DNA during a cell cycle.) Synthesis continues at each replication fork until it meets a newly synthesized strand coming from the opposite direction. The result is a chromosome containing two DNA double helices, each corresponding to a chromatid.

Enzymes proofread and repair errors in DNA

DNA replication occurs only once during each cell generation, and it is important that the process be as accurate as possible to avoid harmful, or possibly even lethal, mutations. Although base pairing during DNA replication is very accurate, errors do occur. Mechanisms have evolved that ensure that errors in replication are corrected. During replication, DNA polymerases proofread each newly added nucleotide against its template nucleotide. When an error in base pairing is found, DNA polymerase immediately removes the incorrect nucleotide and

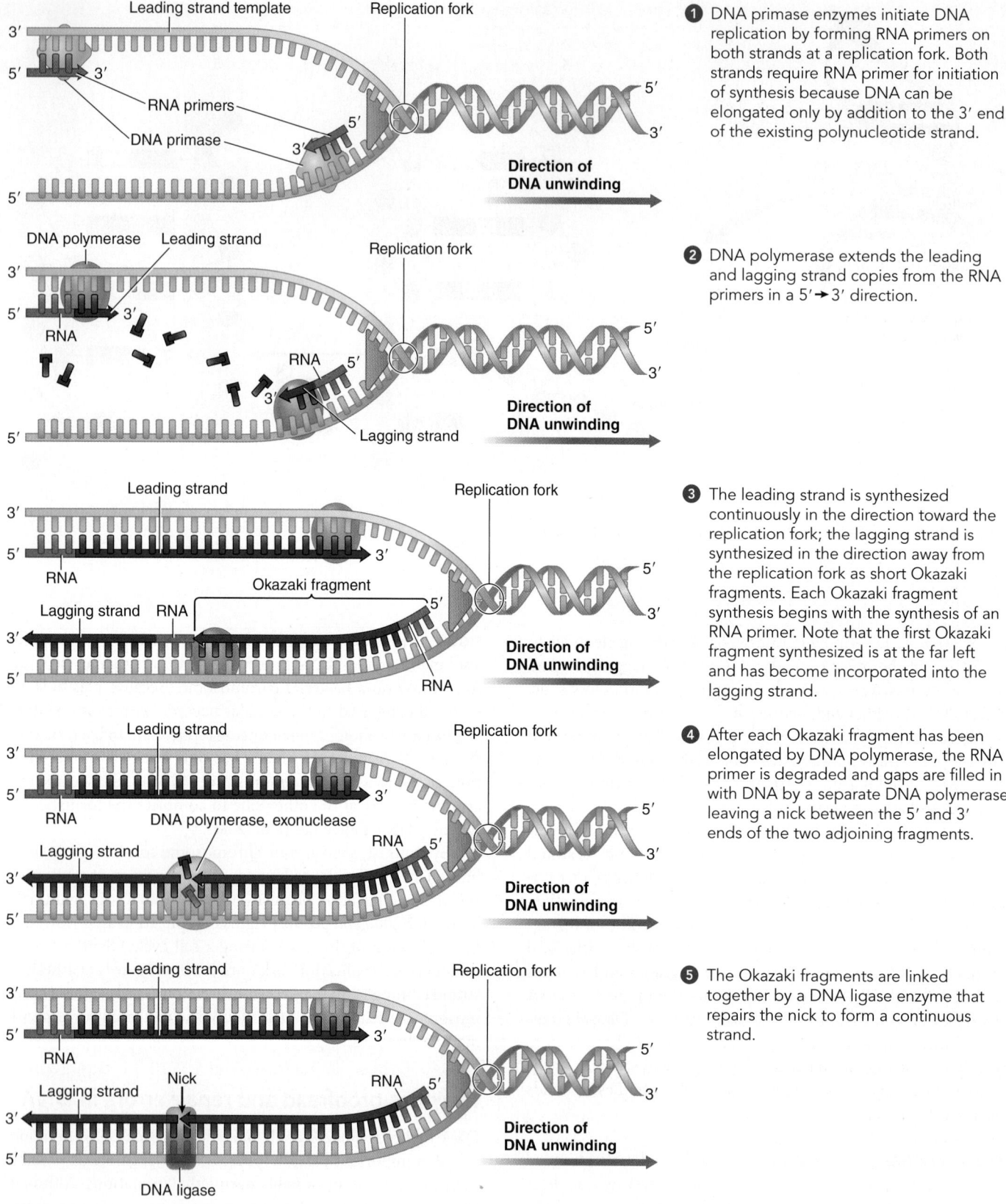

1 DNA primase enzymes initiate DNA replication by forming RNA primers on both strands at a replication fork. Both strands require RNA primer for initiation of synthesis because DNA can be elongated only by addition to the 3' end of the existing polynucleotide strand.

2 DNA polymerase extends the leading and lagging strand copies from the RNA primers in a 5'→3' direction.

3 The leading strand is synthesized continuously in the direction toward the replication fork; the lagging strand is synthesized in the direction away from the replication fork as short Okazaki fragments. Each Okazaki fragment synthesis begins with the synthesis of an RNA primer. Note that the first Okazaki fragment synthesized is at the far left and has become incorporated into the lagging strand.

4 After each Okazaki fragment has been elongated by DNA polymerase, the RNA primer is degraded and gaps are filled in with DNA by a separate DNA polymerase, leaving a nick between the 5' and 3' ends of the two adjoining fragments.

5 The Okazaki fragments are linked together by a DNA ligase enzyme that repairs the nick to form a continuous strand.

Figure 12-15 *Animation* **An overview of DNA replication**
© Cengage Learning

inserts the correct one. A few uncorrected mutations still occur, but they are very infrequent, on the order of one error for every 10^9 or 10^{10} base pairs.

When errors have been left uncorrected by the normal repair activities of DNA polymerase during DNA replication, cells make use of other repair mechanisms (although exactly

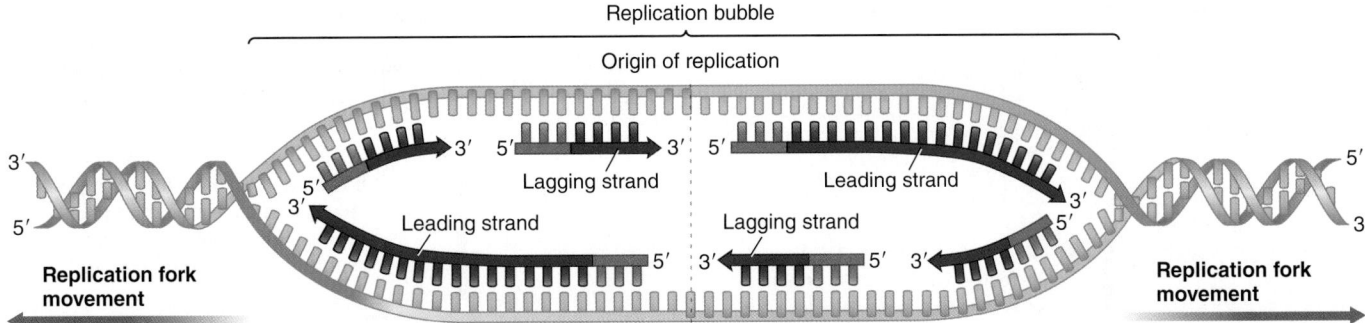

(a) Leading and lagging strand synthesis at the two replication forks of a replication bubble.

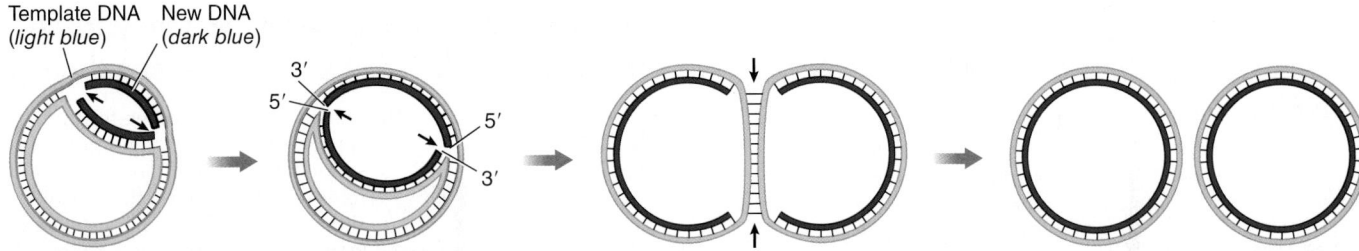

(b) Most bacterial plasmids and chromosomes have only one origin of replication. Because DNA synthesis proceeds from that point in both directions, two replication forks form (*black arrows*), travel around the circle, and eventually meet to form two chromosomes.

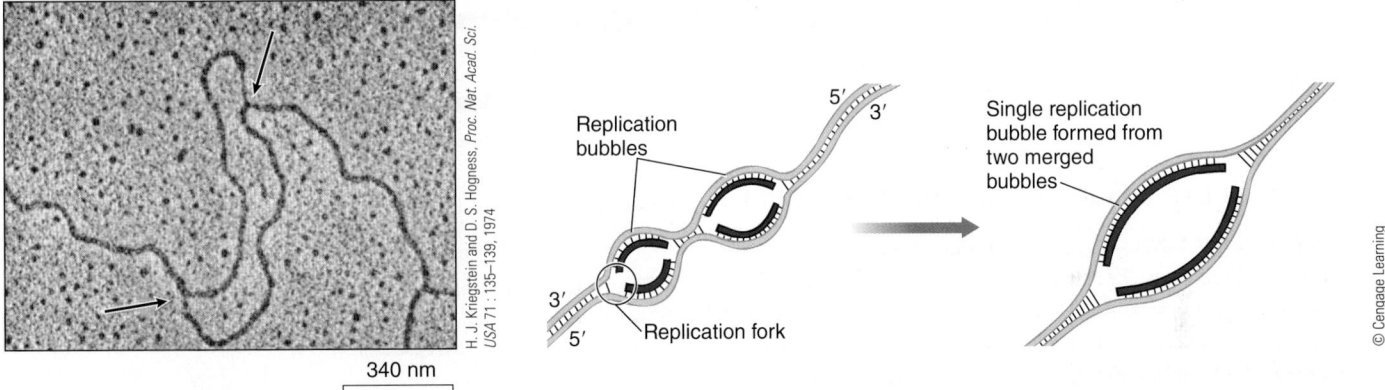

(c) TEM of two replication forks (*black arrows*) in a segment of a eukaryotic chromosome that has partly replicated.

(d) Eukaryotic chromosome DNA contains multiple origins of replication. DNA synthesis proceeds in both directions from each origin until adjacent replication bubbles eventually merge.

Figure 12-16 Bidirectional DNA replication in bacteria and eukaryotes

how DNA repair enzymes identify these rare, often-subtle errors in the vast amount of normal DNA is not well understood). In **mismatch repair** special enzymes recognize the incorrectly paired nucleotides and remove them; DNA polymerases then fill in the missing nucleotides. The observation that individuals with a hereditary defect in a mismatch repair enzyme are likely to develop a type of colon cancer demonstrates the crucial role of mismatch repair enzymes in ensuring the fidelity of DNA replication from one generation to the next.

In Chapter 13 you will discover that some types of radiation and chemicals, both in the environment and within the cells themselves, cause mutations in DNA. These mutations are almost always harmful, and cells usually correct mutations by using one or more DNA repair enzymes. About 100 kinds of repair enzymes in the bacterium *E. coli*, and 130 kinds in human cells have been discovered so far.

One type of DNA repair—**nucleotide excision repair**—is commonly used to repair DNA lesions (deformed DNA) caused by the sun's ultraviolet radiation or by harmful chemicals (**FIG. 12-17**). Three enzymes are involved in nucleotide excision repair: a nuclease to cut out the damaged DNA, a DNA polymerase to add the correct nucleotides, and DNA ligase to close the breaks in the sugar–phosphate backbone. Individuals suffering from the disease *xeroderma pigmentosum* have an inherited defect in a nucleotide excision repair enzyme. Affected individuals develop many skin cancers at an early age because DNA lesions caused by ultraviolet radiation are not repaired.

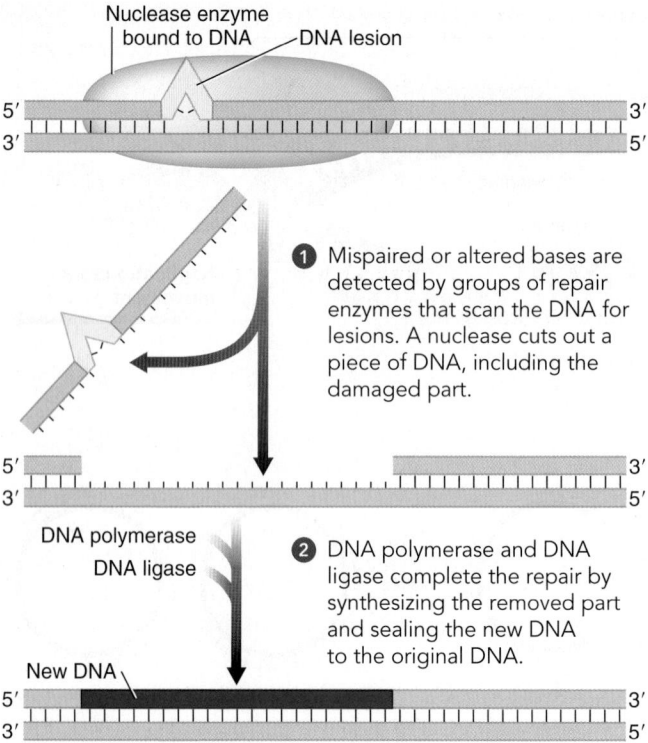

1 Mispaired or altered bases are detected by groups of repair enzymes that scan the DNA for lesions. A nuclease cuts out a piece of DNA, including the damaged part.

2 DNA polymerase and DNA ligase complete the repair by synthesizing the removed part and sealing the new DNA to the original DNA.

Figure 12-17 Nucleotide excision repair of damaged DNA
© Cengage Learning

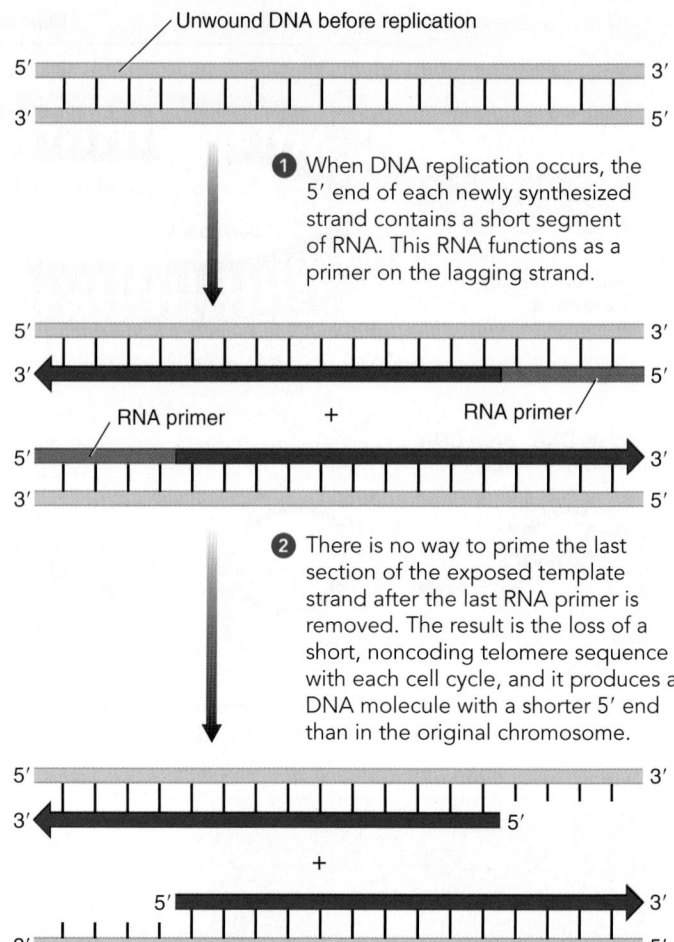

1 When DNA replication occurs, the 5' end of each newly synthesized strand contains a short segment of RNA. This RNA functions as a primer on the lagging strand.

2 There is no way to prime the last section of the exposed template strand after the last RNA primer is removed. The result is the loss of a short, noncoding telomere sequence with each cell cycle, and it produces a DNA molecule with a shorter 5' end than in the original chromosome.

(a) Eukaryotic DNA replication shortens the ends of chromosomes each generation.

Figure 12-18 Telomeres
Repetitive telomere sequences protect the ends of eukaryotic chromosomes as they shorten during DNA replication
© Cengage Learning

Telomeres cap eukaryotic chromosome ends

Unlike bacterial DNA, which is circular, eukaryotic chromosomes have free ends. Because DNA replication is discontinuous in the lagging strand, DNA polymerases do not complete replication of the strand neatly. At the end of the DNA, a small portion is left unreplicated, and a small, single-strand segment of the DNA is lost with each cell cycle (**FIG. 12-18a**).

The important genetic information is retained because chromosomes have protective end caps, or telomeres, that do not contain protein-coding genes. **Telomeres** consist of short, noncoding, guanine-rich DNA sequences that repeat many times (**FIG. 12-18b**). For example, in human sperm and egg cells, the sequence 5'—TTAGGG—3' is repeated more than 1000 times at the ends of each chromosome. Therefore, although a small amount of telomeric DNA fails to replicate each time the cell divides, a cell can divide many times before it starts losing essential genetic information.

Telomerase, a special DNA replication enzyme, can lengthen telomeric DNA by adding back these repetitive nucleotide sequences to the ends of eukaryotic chromosomes. This enzyme—which was discovered in 1984 by 2009 Nobel laureates Elizabeth Blackburn and Carol Greider—is typically active in cells that divide an unlimited number of times, including protozoa and other unicellular eukaryotes, and most types of **cancer cells.** In humans and other mammals, active telomerase is usually present in germ line cells (which give rise to eggs and

sperm) and rapidly dividing cells (such as blood cells, cells lining the intestine, and skin cells), but not in most somatic cells of adult tissues.

When most cells divide for repair or replacement, their chromosome ends shorten. Research evidence suggests that the shortening of telomeres may contribute to *cell aging* and **apoptosis,** which is programmed cell death. The pioneering studies of American biologist Leonard Hayflick in the 1960s showed that when normal somatic cells of the human body are grown in culture, they lose their ability to divide after a limited number of cell divisions. Furthermore, the number of cell divisions is determined by the age of the individual from whom the cells were taken. Cells from a 70-year-old can divide only 20 to 30 times, compared with those from an infant, which can divide 80 to 90 times.

Many investigators have observed correlations between telomerase activity and the ability of cells to undergo unlimited

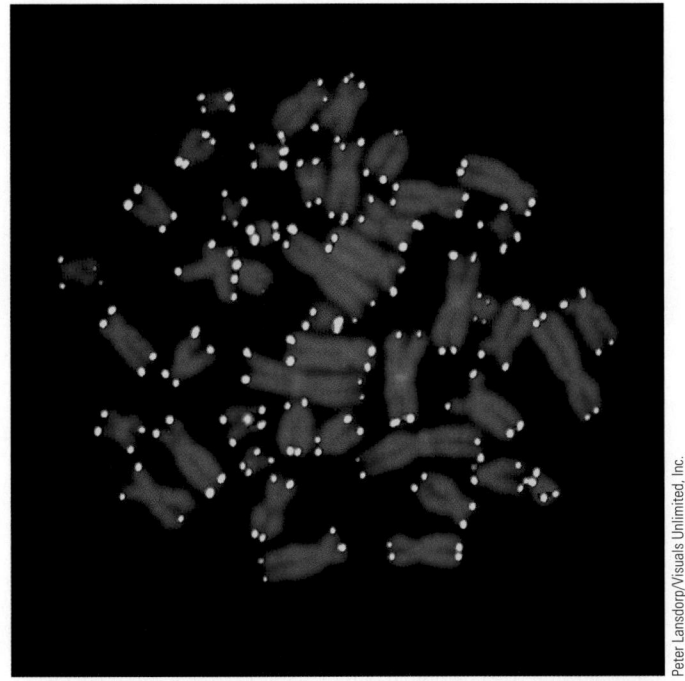

Peter Lansdorp/Visuals Unlimited, Inc.

10 μm

(b) Repetitive telomere sequences cap the ends of linear DNA in eukaryotic chromosomes. LM of duplicated human chromosomes with fluorescent-tagged telomeres (*yellow*).

Figure 12-18 *Continued*

divisions without showing signs of the aging process. However, the evidence of a causal relationship was not found until the early 2000s, when scientists used recombinant DNA technology (see Chapter 15) to infect cultured normal human cells with a virus that carried the genetic information coding for the

catalytic subunit of telomerase. The infected cells not only produced active telomerase, which elongated the telomeres significantly, but also continued to divide long past the point at which cell divisions would normally cease.

Telomeres and telomerase are an active focus of research, for both scientific and practical reasons. The ability to give somatic cells telomerase so that they can divide many times more than they ordinarily would has many potential therapeutic applications, especially if lost or injured cells must be replaced. However, giving such cells the property of unlimited cell division also has potentially serious consequences. For example, if transplanted into the body, cells with active telomerase might behave like cancer cells.

Cancer cells have the ability to divide hundreds of times in culture; in fact, they are virtually immortal. Researchers have demonstrated that in cancer cells of the prostate and pancreas, telomere DNA sequences shorten to a critical point at which time telomerase is reactivated, which may explain their ability to proliferate in a rapid and uncontrolled manner. Most cancer cells, including human cancers of the breast, lung, colon, prostate gland, and pancreas, have active telomerase, which allows them to maintain telomere length and possibly to resist apoptosis. Given that most adult differentiated cells have very low telomerase activity, there are now many types of cancer drugs in development that target the telomerase enzyme as an anticancer therapy.

CHECKPOINT 12.3

- **CONNECT** *How did the ability to distinguish old and newly synthesized DNA strands enable Meselson and Stahl to verify that DNA replication is semiconservative?*

- *What feature of DNA structure causes DNA replication to be continuous for one strand but discontinuous for the other?*

- **CONNECT** *What is the reason that eukaryotic cells require telomerase, but bacterial cells do not?*

SUMMARY: FOCUS ON LEARNING OBJECTIVES

12.1 Evidence of DNA as the Hereditary Material *(page 252)*

1 Summarize the evidence that accumulated during the 1940s and early 1950s demonstrating that DNA is the genetic material.

- Many early geneticists thought genes were made of proteins. They knew proteins were complex and variable, whereas they thought nucleic acids were simple molecules with a limited ability to store information.

- Several lines of evidence supported the idea that **DNA (deoxyribonucleic acid)** is the genetic material. In **transformation** experiments the DNA of one strain of bacteria can endow related bacteria with new genetic characteristics.

- When a bacterial cell becomes infected with a **bacteriophage** (virus), only the DNA from the virus enters the cell; this DNA is sufficient for the virus to reproduce and form new viral particles.

2 State the questions that these classic experiments addressed: Griffith's transformation experiment, Avery's contribution to Griffith's work, and the Hershey–Chase experiments.

- Griffith's transformation experiment addressed this question: Can a genetic trait be transmitted from one bacterial strain to another? (The answer is yes.)

- Avery's experiments addressed this question: What molecule is responsible for bacterial transformation? (The answer is DNA.)

- The Hershey–Chase experiments addressed this question: Is DNA or protein the genetic material in bacterial viruses (phages)? (The answer is DNA.)

12.2 The Structure of DNA *(page 253)*

3 Explain how nucleotide subunits link to form a single DNA strand.

- Watson and Crick's model of the structure of DNA demonstrated how information can be stored in the molecule's structure and how DNA molecules can serve as **templates** for their own replication.

- DNA is a polymer of **nucleotides**. Each nucleotide subunit contains a nitrogenous base, which may be one of the **purines** (**adenine** or **guanine**) or one of the **pyrimidines** (**thymine** or **cytosine**). Each base covalently links to a pentose sugar, **deoxyribose,** which covalently bonds to a phosphate group.

- The backbone of each single DNA chain is formed by alternating sugar and phosphate groups, joined by covalent phosphodiester linkages. Each phosphate group attaches to the 5′ carbon of one deoxyribose and to the 3′ carbon of the neighboring deoxyribose.

4 Describe how the two strands of DNA are oriented with respect to each other.

- Each DNA molecule consists of two polynucleotide chains that associate as a **double helix.** The two chains are **antiparallel** (running in opposite directions); at each end of the DNA molecule, one chain has a phosphate attached to a 5′ deoxyribose carbon, the **5′ end,** and the other has a hydroxyl group attached to a 3′ deoxyribose carbon, the **3′ end.**

5 State the base-pairing rules for DNA and describe how complementary bases bind to each other.

- Hydrogen bonding between specific base pairs holds together the two chains of the helix. Adenine (A) forms two hydrogen bonds with thymine (T), and guanine (G) forms three hydrogen bonds with cytosine (C).

- **Complementary base pairing** between A and T and between G and C is the basis of Chargaff's rules, which state that A equals T and that G equals C.

- Because complementary base pairing holds together the two strands of DNA, it is possible to predict the base sequence of one strand if you know the base sequence of the other strand.

12.3 DNA Replication *(page 259)*

6 Cite evidence from Meselson and Stahl's experiment that enabled scientists to differentiate between semiconservative replication of DNA and alternative models.

- When *E. coli* cells are grown for many generations in a medium containing heavy nitrogen (^{15}N), they incorporate the ^{15}N into their DNA. When researchers transfer cells from a ^{15}N medium to a ^{14}N medium and isolate them after either one or two generations, the density of the DNA in each group is what would be expected if DNA replication were semiconservative. In **semiconservative replication** each daughter double helix consists of an original strand from the parent molecule and a newly synthesized complementary strand.

7 Summarize how DNA replicates and identify some unique features of the process.

- During **DNA replication,** the two strands of the double helix unwind. Each strand serves as a template for forming a new, complementary strand. Replication is initiated as **DNA primase** synthesizes a short **RNA primer. DNA polymerase** then adds new nucleotide subunits to the growing DNA strand.

- Additional enzymes and other proteins are required to unwind and stabilize the separated DNA helix. **DNA helicases** open the double helix, and **topoisomerases** prevent tangling and knotting.

8 Explain the complexities of DNA replication that make the process (a) bidirectional and (b) continuous in one strand and discontinuous in the other.

- DNA replication is bidirectional, starting at the **origin of replication** and proceeding in both directions from that point. A eukaryotic chromosome may have multiple origins of replication and may be replicating at many points along its length at any one time.

- DNA synthesis always proceeds in a 5′ ⟶ 3′ direction, which requires that one DNA strand, the **lagging strand,** be synthesized discontinuously, as short **Okazaki fragments.** DNA primase synthesizes short RNA primers on the lagging strand, and **DNA ligase** links Okazaki fragments of newly synthesized DNA. The opposite strand, the **leading strand,** is synthesized continuously.

9 Discuss how enzymes proofread and repair errors in DNA.

- During replication, DNA polymerases proofread each newly added nucleotide against its template nucleotide. When an error in base pairing is found, DNA polymerase immediately removes the incorrect nucleotide and inserts the correct one.

- In **mismatch repair** enzymes recognize incorrectly paired nucleotides and remove them; DNA polymerases then fill in the missing nucleotides.

- **Nucleotide excision repair** is commonly used to repair DNA lesions caused by the sun's ultraviolet radiation or by harmful chemicals. Three enzymes are involved: a nuclease to cut out the damaged DNA, a DNA polymerase to add the correct nucleotides, and DNA ligase to close the breaks in the sugar–phosphate backbone.

10 Define *telomere* and describe the possible connections between telomerase and cell aging and between telomerase and cancer.

- Eukaryotic chromosome ends, called **telomeres,** are short, noncoding, repetitive DNA sequences. Telomeres shorten slightly with each cell cycle but can be extended by the enzyme **telomerase.**

- The absence of telomerase activity in certain cells may be a cause of *cell aging,* in which cells lose their ability to divide after a limited number of cell divisions.

- Most **cancer cells,** including human cancers of the breast, lung, colon, prostate gland, and pancreas, have telomerase to maintain telomere length and possibly to resist **apoptosis.**

Know and Comprehend

1. When Griffith injected mice with a combination of live rough-strain and heat-killed smooth-strain pneumococci, he discovered that (a) the mice were unharmed (b) the dead mice contained living rough-strain bacteria (c) the dead mice contained living smooth-strain bacteria (d) DNA had been transferred from the smooth-strain bacteria to the mice (e) DNA had been transferred from the rough-strain bacteria to the smooth-strain bacteria

2. Which of the following inspired Avery and his colleagues to perform the experiments demonstrating that the transforming factor in bacteria is DNA? (a) that A is equal to T and that G is equal to C (b) Watson and Crick's model of DNA structure (c) Meselson and Stahl's studies on DNA replication in *E. coli* (d) Griffith's experiments on smooth and rough strains of pneumococci (e) Hershey and Chase's experiments on the reproduction of bacteriophages

3. In the Hershey–Chase experiment with bacteriophages, (a) harmless bacterial cells permanently transformed into virulent cells (b) DNA was shown to be the transforming factor of earlier bacterial transformation experiments (c) the replication of DNA was conclusively shown to be semiconservative (d) viral DNA was shown to enter bacterial cells and cause the production of new viruses within the bacteria (e) viruses injected their proteins, not their DNA, into bacterial cells

4. The two complementary strands of the DNA double helix are held to each other by (a) ionic bonds between deoxyribose molecules (b) ionic bonds between phosphate groups (c) covalent bonds between nucleotide bases (d) covalent bonds between deoxyribose molecules (e) hydrogen bonds between nucleotide bases

5. If a segment of DNA is 5′—CATTAC—3′, the complementary DNA strand is (a) 3′—CATTAC—5′ (b) 3′—GTAATG—5′ (c) 5′—CATTAC—3′ (d) 5′—GTAATG—3′ (e) 5′—CATTAC—5′

6. Each DNA strand has a backbone that consists of alternating (a) purines and pyrimidines (b) nucleotide bases (c) hydrogen bonds and phosphodiester linkages (d) deoxyribose and phosphate (e) phosphate and phosphodiester linkages

7. The experiments in which Meselson and Stahl grew bacteria in heavy nitrogen conclusively demonstrated that DNA (a) is a double helix (b) replicates semiconservatively (c) consists of repeating nucleotide subunits (d) has complementary base pairing (e) is always synthesized in the 5′ ⟶ 3′ direction

8. The statement "DNA replicates by a semiconservative mechanism" means that (a) only one DNA strand is copied (b) first one DNA strand is copied and then the other strand is copied (c) the two strands of a double helix have identical base sequences (d) some portions of a single DNA strand are old and other portions are newly synthesized (e) each double helix consists of one old and one newly synthesized strand

9. Topoisomerases (a) synthesize DNA (b) synthesize RNA primers (c) join Okazaki fragments (d) break and rejoin DNA to reduce torsional strain (e) prevent single DNA strands from joining to form a double helix

10. A lagging strand forms by (a) joining primers (b) joining Okazaki fragments (c) joining leading strands (d) breaking up a leading strand (e) joining primers, Okazaki fragments, and leading strands

11. The immediate source of energy for DNA replication is (a) the hydrolysis of nucleotides with three phosphate groups (b) the oxidation of NADPH (c) the hydrolysis of ATP (d) electron transport (e) the breaking of hydrogen bonds

12. Which of the following statements about eukaryotic chromosomes is *false*? (a) eukaryotic chromosomes have free ends (b) telomeres contain protein-coding genes (c) telomerase lengthens telomeric DNA (d) telomere shortening may contribute to cell aging (e) cells with active telomerase may undergo many cell divisions

Apply and Analyze

13. Many cancer cells are immortal and can be cultured in the laboratory for many years. Many of these cell lines have highly active telomerase activities. Why would that affect the ability of cancer cells to sustain growth and cell division?

14. **VISUALIZE** Construct a diagram of a replication fork. Label the 3′ and 5′ ends of the leading strand, lagging strand, and the two strands of the DNA double helix.

Evaluate and Synthesize

15. Explain to a friend the characteristics a molecule must have to function as genetic material and then explain the important features of the structure of DNA that are consistent with its role as the chemical basis of heredity.

16. **INTERPRET DATA** In the Hershey–Chase experiment, the radioactive label ^{32}P was present inside bacterial cells (i.e., in the pellet), whereas the radioactive label ^{35}S was present outside bacterial cells (in the supernatant). What would the researchers have concluded had the reverse been true, that is, if the radioactive label ^{35}S were inside the cells and the radioactive label ^{32}P were outside the cells?

17. **EVOLUTION LINK** How does DNA being the universal molecule of inheritance in all cells support evolutionary theory?

aplia   To access course materials, such as Aplia and other companion resources, please visit **www.cengagebrain.com.**

13 Gene Expression

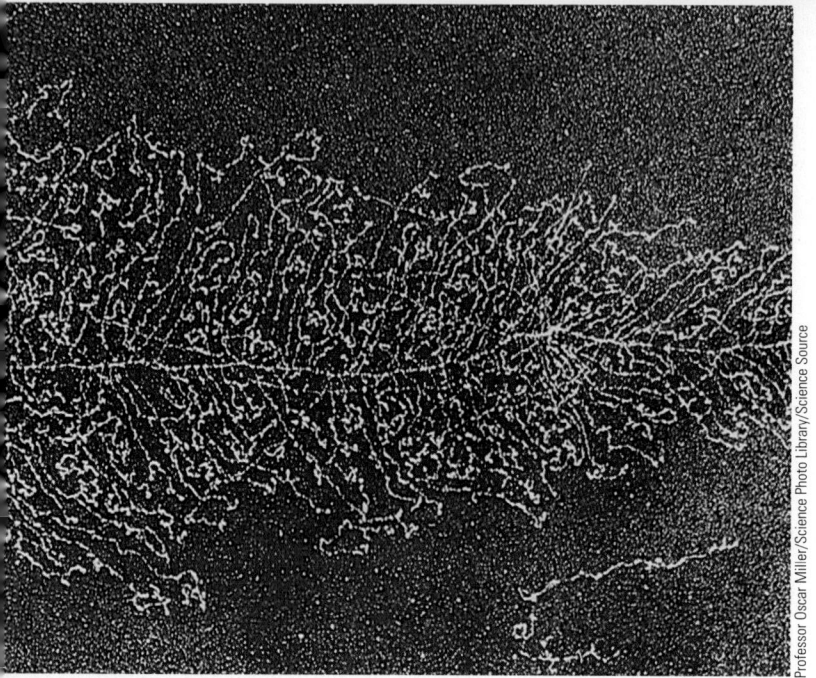

Visualizing transcription. In this TEM, RNA molecules (*lateral strands*) are synthesized as complementary copies of a DNA template (*central axis*).

Professor Oscar Miller/Science Photo Library/Science Source

I n Chapter 12 we discussed how the cell accurately replicates the nucleotide sequence in DNA to pass the genetic material unaltered to the next generation. The basic features of that process are the same in all cells studied to date, from the simplest bacterial cells to intricate human cells. In the half century before that discovery, however, scientists struggled to determine what genes were and how they worked. Early studies suggested that some genes contained information that specified the structure of proteins, but how genes and proteins were connected was unclear. With the discovery of the structure of DNA, it became apparent that these molecules could store the information needed to specify the amino acid sequences for the many different proteins a cell needs to carry out its functions. It would take more than a decade of intense investigation by many scientists before a fundamental understanding was reached of how cells convert DNA information into the amino acid sequences of proteins, a process that we refer to as *gene expression*. It would take many more years of intense work to understand that only a small part of the total DNA in eukaryotic cells actually encodes proteins and that many other parts of the genome contain genes that encode RNA molecules that regulate gene expression.

In this chapter we examine the early evidence that indicated that many genes specify the structure of proteins. We then consider how DNA governs the phenotype of the organism at the molecular level through the process of gene expression. For protein-coding genes, this process involves a series of steps in which the sequence of bases in DNA specifies the makeup of the cell's proteins. The proteins affect the phenotype in some way, ranging from readily observable physical traits to subtle changes detectable only at the biochemical level. The first major step of gene expression is *transcription*, the synthesis of RNA molecules complementary to the DNA (see photograph). The second major step is *translation*, in which the RNA acts as a coded template to direct polypeptide synthesis.

13.1 DISCOVERY OF THE GENE–PROTEIN RELATIONSHIP

LEARNING OBJECTIVES

1 Summarize the early evidence indicating that some genes specify the structure of proteins.
2 Describe how Beadle and Tatum's experiments connected certain genes to enzymes and proteins.

The idea that genes and proteins are connected originated early in the 20th century, shortly after scientists rediscovered Mendel's principles. In the first edition of his book *Inborn Errors of Metabolism* (1908), Archibald Garrod, a physician and biochemist, discussed a rare genetic disease called *alkaptonuria,* which has a simple recessive inheritance pattern. The condition involves the metabolic pathway that breaks down the amino acid tyrosine, ultimately converting it to carbon dioxide and water. An intermediate in this pathway, homogentisic acid, accumulates in the urine of affected people, turning it black when exposed to air (**FIG. 13-1**). Other symptoms of alkaptonuria include the later development of arthritis and, in men, stones in the prostate gland.

In Garrod's time scientists knew about enzymes but did not recognize that they were proteins. Garrod hypothesized that people with alkaptonuria lack the enzyme that normally oxidizes homogentisic acid. Before the second edition of his book was published, in 1923, researchers had found that affected people do indeed lack the enzyme that oxidizes homogentisic acid. Garrod's hypothesis was correct: a mutation in this gene is associated with the absence of a specific enzyme. Shortly thereafter, in 1926, U.S. biochemist James Sumner purified a different enzyme, urease, and showed it to be a protein. It was the first clear identification of an enzyme as a protein. In 1946, Sumner was awarded the Nobel Prize in Chemistry for being the first to crystallize an enzyme.

Beadle and Tatum proposed the one-gene, one-enzyme hypothesis

A major advance in understanding the relationship between genes and enzymes came in the early 1940s with the work of U.S. geneticists George Beadle, Edward Tatum, and their associates, who developed a new approach to the problem. Researchers knew that a series of biosynthetic reactions controlled specific phenotypes, but it was not clear whether the genes themselves were acting as enzymes or whether they controlled the workings of enzymes in some indirect way.

Rather than try to identify the enzymes affected by single genes, Beadle and Tatum decided to look for mutations interfering with known metabolic reactions that produce essential molecules, such as amino acids and vitamins. They chose as an experimental organism a fungus, the pink-orange bread mold *Neurospora,* for several important reasons. Wild-type (an individual with a normal phenotype) *Neurospora* is easy to grow in culture; it manufactures all its essential biological molecules

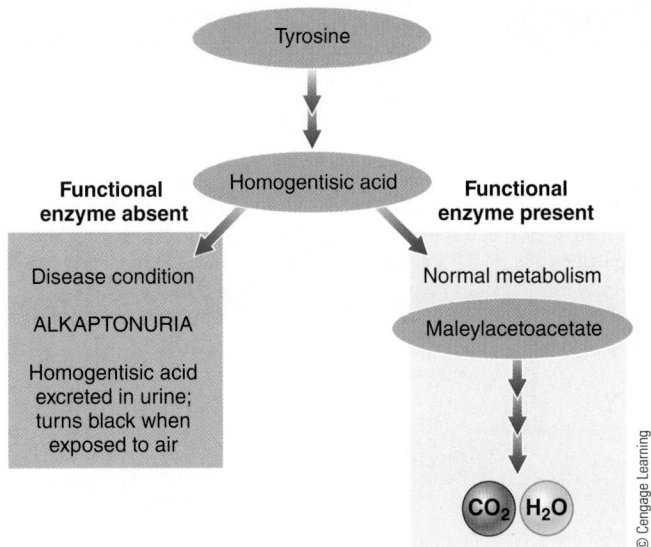

Figure 13-1 An "inborn error of metabolism"
Garrod proposed that alkaptonuria is a genetic disease caused by the absence of the enzyme homogentisic acid oxidase, which normally converts homogentisic acid to maleylacetoacetate. Thus, the acid accumulates in the blood and is excreted in the urine, which turns black on contact with the air.

when grown on a simple growth medium (minimal medium) containing only sugar, salts, and the vitamin biotin. However, a mutant *Neurospora* strain that cannot make a substance such as an amino acid can still grow if that substance is added to the growth medium.

Neurospora is also an ideal experimental organism because it grows primarily as a haploid organism. The haploid condition allows a researcher to immediately identify a recessive mutant allele; there is no homologous chromosome that might carry a dominant allele that would mask its expression. Furthermore, using simple genetic crossing methods it was possible to show that a mutant phenotype could be the result of a single mutation.

Beadle and Tatum exposed thousands of haploid wild-type *Neurospora* asexual spores to X-rays or ultraviolet radiation to produce mutant strains to find those that could grow only on minimal medium that contained a single type of vitamin or amino acid, such as arginine, as a supplement. Further testing of the mutant strain on media containing different combinations of precursors in the arginine biosynthetic pathway identified which compound was required and the actual enzymatic step that was blocked (**FIG. 13-2**).

The work on *Neurospora* revealed that each mutant strain had a mutation in only one gene and that each gene affected only one enzyme-catalyzed chemical reaction. Beadle and Tatum stated this one-to-one correspondence between genes and enzymes as the *one-gene, one-enzyme hypothesis*. In 1958, they received the Nobel Prize in Physiology or Medicine for discovering that genes regulate specific chemical events. The idea that a gene encodes the information to produce a single enzyme held for almost a decade, until additional findings required a modification of this definition.

In the late 1940s, researchers began to understand that genes control not only enzymes but other proteins as well. U.S. chemist

Can individual genes code for specific biosynthetic enzymes?

HYPOTHESIS: Nutrient-requiring mutations induced in *Neurospora*, a type of bread mold, correspond to the absence of functional biosynthetic enzymes.

EXPERIMENT: Beadle, Tatum, and their colleagues isolated a group of mutant *Neurospora* strains that could grow only if the amino acid arginine was added to the minimal medium (MM). Their genetic tests showed that (1) the arginine-requiring phenotype of each mutant strain was the result of a single mutation and (2) the arginine-requiring strains could be grouped into three classes (I, II, and III), with the mutation in each class located in a different gene. It was known that two compounds, ornithine and citrulline, are precursors to arginine in its biosynthetic pathway, as shown below:

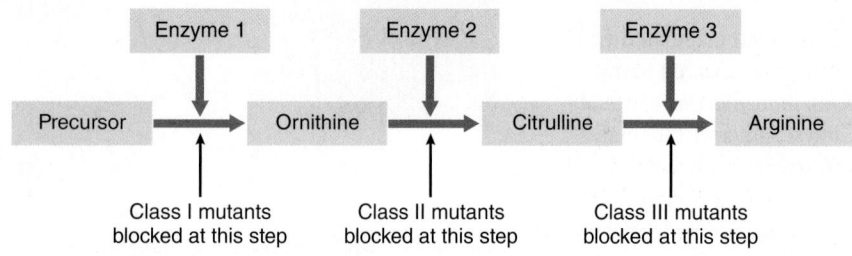

The different mutants were tested on medium containing arginine and the two different intermediates to see in what step of the arginine biosynthetic pathway each class of mutants was blocked.

RESULTS AND CONCLUSION: Mutant type I grows on ornithine, citrulline, and arginine and must therefore be blocked prior to the formation of all three compounds. Mutant class II cannot grow on ornithine, but allows the conversion of citrulline to arginine and must be blocked at a point after the synthesis of ornithine. Mutant class III can only grow on arginine and must be blocked after the synthesis of both ornithine and citrulline. On the basis of many similar experiments, the researchers concluded that each gene controls the production of a single enzyme.

SOURCE: Beadle, G.W. and E.L. Tatum (1941) "Genetic Control of Biochemical Reactions in *Neurospora*. *Proceedings of the National Academy of Sciences*. Vol. 27:499–506

***Neurospora* strains**

	Wild type	Class I	Class II	Class III
Minimal medium				
Minimal medium + ornithine				
Minimal medium + citrulline				
Minimal medium + arginine				

Figure 13-2 The Beadle–Tatum experiments

PREDICT Which of the supplements would support growth of a mutant strain missing both Enzyme 1 and Enzyme 2?

© Cengage Learning

Linus Pauling and his colleagues demonstrated in 1949 that a mutation of a single gene alters the structure of hemoglobin. This particular mutant form of hemoglobin is associated with the genetic disease sickle cell anemia (discussed in Chapter 16). British biochemist Vernon Ingram extended Pauling's research in 1957 when he determined that sickle cell hemoglobin and normal hemoglobin differ by only one amino acid.

Studies by other scientists showed that many proteins are constructed from two or more polypeptide chains, each of which is under the control of a different locus. For example, hemoglobin contains two types of polypeptide chains, the α and β subunits. See the structure of hemoglobin in Figure 3-23a. Sickle cell anemia results from a mutation affecting the β subunits.

Scientists therefore extended the definition of a gene to include that one gene is responsible for one polypeptide chain. Even this definition has proved only partially correct, although as you will see later in this chapter, scientists still define a gene in terms of its product.

Although the elegant work of Beadle and Tatum and others demonstrated that genes are expressed in the form of proteins, the mechanism of gene expression was completely unknown. After Watson and Crick's discovery of the structure of DNA, many scientists worked to understand exactly how gene expression takes place. We begin with an overview of gene expression and then consider the various steps in more detail.

CHECKPOINT 13.1

- *What is the one-gene, one-enzyme hypothesis?*
- **CONNECT** *How did the work of each of the following scientists—Garrod, Beadle and Tatum, and Pauling—contribute to our understanding of the relationship between genes and proteins?*

13.2 INFORMATION FLOW FROM DNA TO PROTEIN: AN OVERVIEW

LEARNING OBJECTIVES

3 Outline the flow of genetic information in cells, from DNA to RNA to polypeptide.
4 Compare the structures of DNA and RNA.
5 Explain why the genetic code is said to be redundant and virtually universal, and discuss how these features may reflect its evolutionary history.

Although the sequence of bases in DNA determines the sequence of amino acids in polypeptides, cells do not use the information in DNA directly. Instead, a related nucleic acid, **ribonucleic acid (RNA)**, links DNA and protein. When a gene that codes for a protein is expressed, first an RNA copy is made of the information in the DNA. It is this RNA copy that provides the information that directs polypeptide synthesis.

Like DNA, RNA is a polymer of nucleotides, but it has some important differences (**FIG. 13-3**). RNA is usually single-stranded,

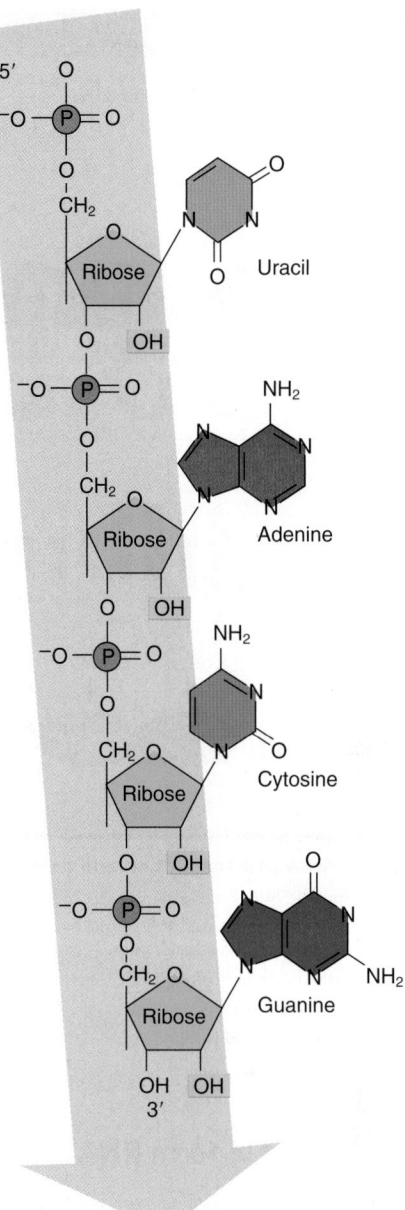

Figure 13-3 Animation The nucleotide structure of RNA

The nucleotide subunits of RNA are joined by 5' ⟶ 3' phosphodiester linkages, like those found in DNA. Adenine, guanine, and cytosine are present, as in DNA, but the base uracil replaces thymine. All four nucleotide types contain the five-carbon sugar ribose, which has a hydroxyl group (*blue*) on its 2' carbon atom.
© Cengage Learning

although internal regions of some RNA molecules may have complementary sequences that allow the strand to fold back and pair to form short, double-stranded segments. As shown in Figure 13-3, the sugar in RNA is **ribose,** which is similar to deoxyribose of DNA but has a hydroxyl group at the 2' position. (Compare ribose with the deoxyribose of DNA, shown in Figure 12-4, which has a hydrogen at the 2' position.) The base **uracil** substitutes for thymine and, like thymine, is a pyrimidine that can form two hydrogen bonds with adenine. Hence, uracil and adenine are a complementary pair.

Protein synthesis requires two major steps: DNA $\xrightarrow{\text{transcription}}$ RNA $\xrightarrow{\text{translation}}$ protein

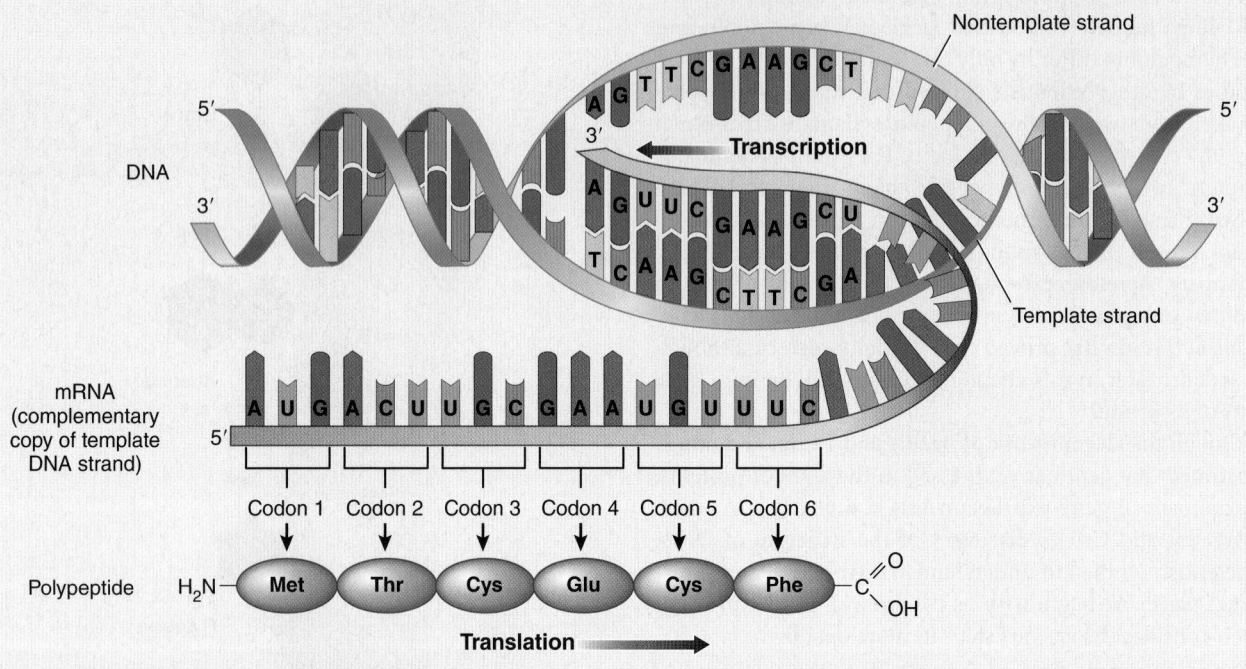

Figure 13-4 An overview of transcription and translation

In transcription messenger RNA is synthesized as a complementary copy of one of the DNA strands, the template strand. Messenger RNA carries genetic information in the form of sets of three bases, or codons, each of which specifies one amino acid. Codons are translated consecutively, thus specifying the linear sequence of amino acids in the polypeptide chain. Translation requires tRNA and ribosomes (*not shown*). The figure depicts transcription and translation in bacteria. In eukaryotes transcription takes place in the nucleus and translation occurs in the cytosol.

VISUALIZE Write out the sequence of bases corresponding to the first six codons on the above mRNA with its corresponding bases on the template strand. Label the ends of each strand with the correct polarity.
© Cengage Learning

DNA is transcribed to form RNA

The process by which RNA is synthesized resembles DNA replication in that the sequence of bases in the RNA strand is determined by complementary base pairing with one of the DNA strands, the *template strand* (FIG. 13-4). Because RNA synthesis takes the information in one kind of nucleic acid (DNA) and copies it as another nucleic acid (RNA), this process is called **transcription** ("copying").

Three main kinds of RNA molecules required for protein synthesis are transcribed: messenger RNA, transfer RNA, and ribosomal RNA. **Messenger RNA (mRNA)** is a single strand of RNA that carries the information for making a protein. Each of the 45 or so kinds of **transfer RNAs (tRNAs)** is a single strand of RNA that folds back on itself to form a specific shape. Each kind of tRNA bonds with only one kind of amino acid and carries it to a *ribosome*. (Because there are more kinds of tRNA molecules than there are amino acids, many amino acids are carried by two or more kinds of tRNA molecules.) **Ribosomal RNA (rRNA),** which is in a globular form, is an important part of the structure of ribosomes and has catalytic functions needed during protein synthesis.

RNA is translated to form a polypeptide

Following transcription, the transcribed information in the mRNA is used to specify the amino acid sequence of a polypeptide (see Fig. 13-4). This process is called **translation** because it involves conversion of the "nucleic acid language" in the mRNA molecule into the "amino acid language" of protein.

In translation of the genetic instructions to form a polypeptide, a sequence of three consecutive bases in mRNA, called a **codon,** specifies one amino acid. For example, one codon that corresponds to the amino acid phenylalanine is 5′—UUC—3′. Because each codon consists of three nucleotides, the code is described as a **triplet code.** The assignments of codons for amino acids and for start and stop signals are collectively named the **genetic code** (FIG. 13-5).

Transfer RNAs are crucial parts of the decoding machinery because they act as "adapters" that connect amino acids and nucleic acids. This mechanism is possible because each tRNA can (1) link with a specific amino acid and (2) recognize the appropriate mRNA codon for that particular amino acid (FIG. 13-6). A particular tRNA can recognize a particular codon because it has a sequence of three bases, called the **anticodon,**

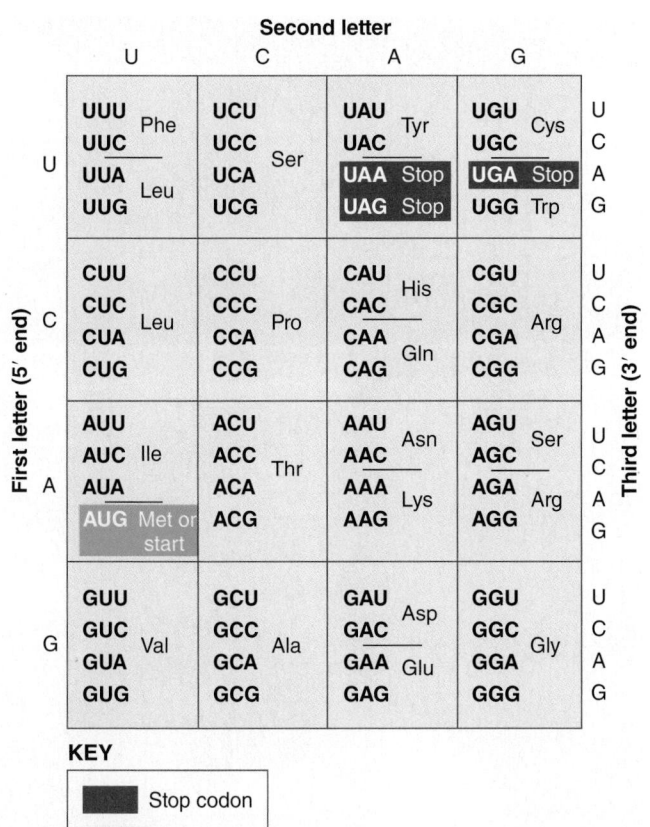

Figure 13-5 The genetic code

The genetic code specifies all possible combinations of the three bases that compose codons in mRNA. Of the 64 possible codons, 61 specify amino acids (see Figure 3-17 for an explanation of abbreviations). The codon AUG specifies the amino acid methionine and also signals the ribosome to initiate translation ("start"). Three codons—UAA, UGA, and UAG—do not specify amino acids; they terminate polypeptide synthesis ("stop").

that hydrogen-bonds with the mRNA codon by complementary base pairing. The exact anticodon that is complementary to the codon for phenylalanine in our example is 3′—AAG—5′.

Translation requires that each tRNA anticodon be hydrogen-bonded to the complementary mRNA codon and that the amino acids carried by the tRNAs be linked in the order specified by the sequence of codons in the mRNA. **Ribosomes,** the site of translation, are organelles composed of two different subunits, each containing protein and rRNA. (The structure and function of ribosomes were introduced in Chapter 4.) Ribosomes attach to the 5′ end of the mRNA and travel along it, allowing the tRNAs to attach sequentially to the codons of mRNA. In this way the amino acids carried by the tRNAs take up the proper position to be joined by *peptide bonds* in the correct sequence to form a polypeptide.

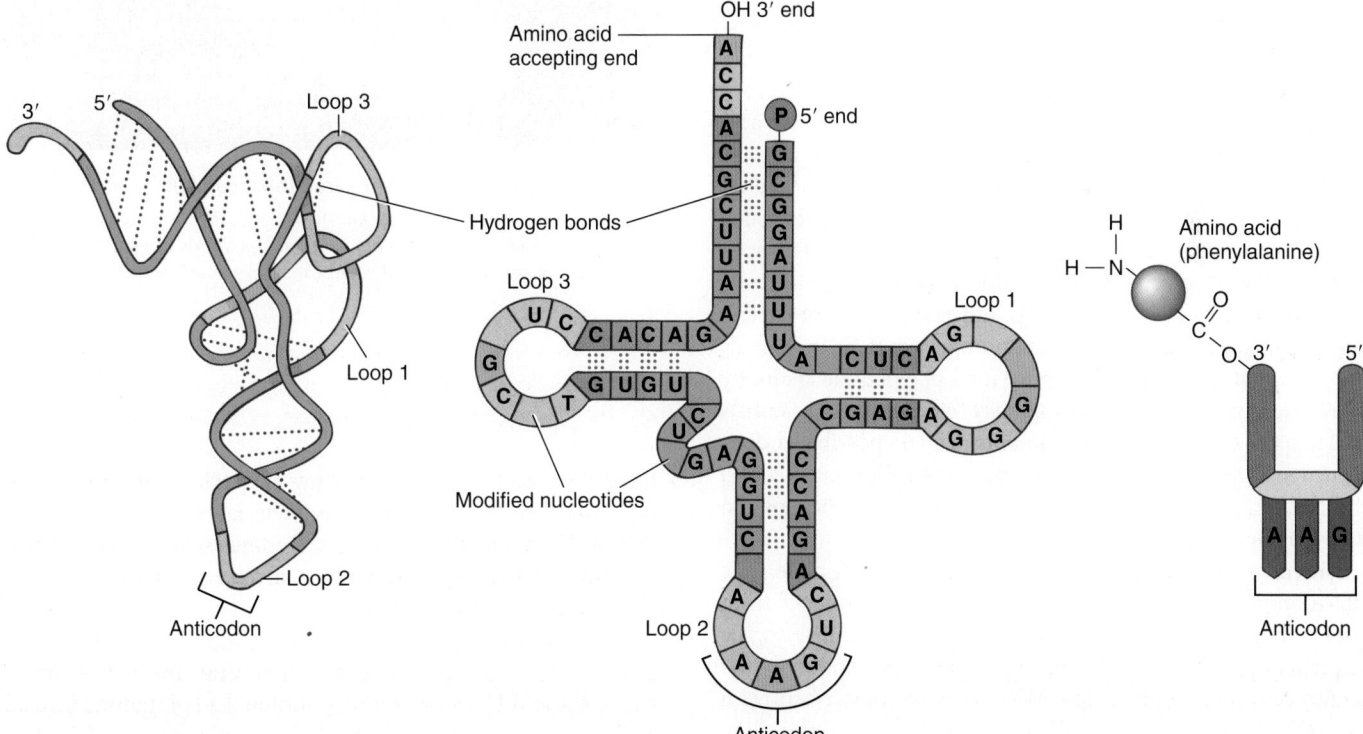

(a) The 3-D shape of a tRNA molecule is determined by hydrogen bonds formed between complementary bases.

(b) One loop contains the anticodon; these unpaired bases pair with a complementary mRNA codon. The amino acid attaches to the terminal nucleotide at the hydroxyl (OH) 3′ end.

(c) This stylized diagram of an aminoacyl–tRNA shows that the amino acid attaches to tRNA by its carboxyl group, leaving its amino group exposed for peptide bond formation.

Figure 13-6 Three representations of a tRNA molecule

Biologists cracked the genetic code in the 1960s

Before the genetic code was deciphered, scientists had become interested in how a genetic code might work. The Watson and Crick model of DNA showed it to be a linear sequence of four different nucleotides. If each nucleotide coded for a single amino acid, the genetic code could specify only 4 amino acids, not the 20 found in the vast variety of proteins in the cell.

Scientists saw that the DNA bases could serve as a four-letter "alphabet" and hypothesized that three-letter combinations of the four bases would make it possible to form a total of 64 "words," more than enough to specify all the naturally occurring amino acids. In 1961, Crick and British scientist Sydney Brenner concluded from experimental evidence that the code used nonoverlapping triplets of bases. They predicted that the code is read, one triplet at a time, from a fixed starting point that establishes the *reading frame* for the genetic message.

U.S. biochemists Marshall Nirenberg and Heinrich Matthaei obtained the first experimental evidence indicating the assignment of triplets to specific amino acids. Using artificial mRNA molecules constructed with known base sequences, they studied protein synthesis in purified *cell-free systems* derived from the bacterium *Escherichia coli* to determine which amino acids would be incorporated into protein. For example, when they added the synthetic mRNA polyuridylic acid (UUUUUUUUU . . .) to a mixture of purified ribosomes, aminoacyl–tRNAs (amino acids linked to their respective tRNAs), and essential cofactors needed to synthesize polypeptide, only phenylalanine was incorporated into the resulting polypeptide chain. The inference was inescapable that the UUU triplet codes for phenylalanine. Similar experiments showed that polyadenylic acid (AAAAAAAAA . . .) codes for a polypeptide of lysine and that polycytidylic acid (CCCCCCCCC . . .) codes for a polypeptide of proline.

By using mixed nucleotide polymers (such as a random polymer of A and C) as artificial messengers, researchers such as H. Gobind Khorana, then at the University of Wisconsin, assigned the other codons to specific amino acids. However, three of the codons—UAA, UGA, and UAG—did not specify any amino acids. These codons are now known to be the signals that terminate the coding sequence for a polypeptide chain. By 1967, the genetic code was completely "cracked," and scientists had identified the coding assignments of all 64 possible codons shown in Figure 13-5. In 1968, Nirenberg and Khorana received the Nobel Prize in Physiology or Medicine for their work in deciphering the genetic code.

Remember that the genetic code is an mRNA code. The tRNA anticodon sequences, as well as the DNA sequence from which the message is transcribed, are complementary to the sequences in Figure 13-5. For example, the mRNA codon for the amino acid methionine is 5′—AUG—3′. It is transcribed from the DNA base sequence 3′—TAC—5′, and the corresponding tRNA anticodon is 3′—UAC—5′.

The genetic code is virtually universal

Perhaps the single most remarkable feature of the code is that it is essentially universal. Over the years, biologists have examined

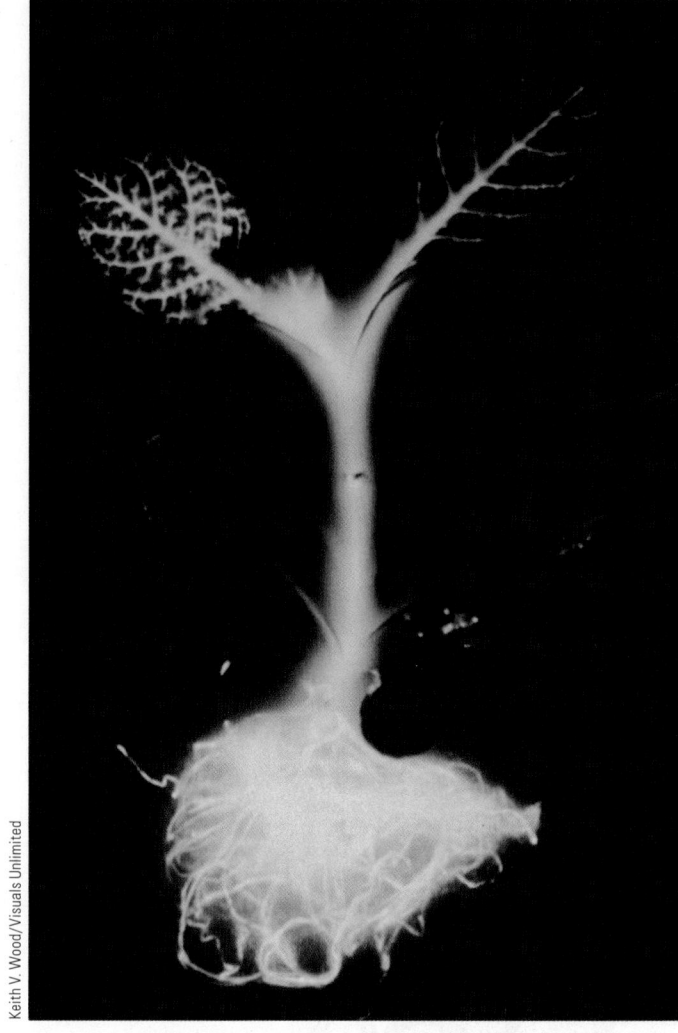

Keith V. Wood/Visuals Unlimited

Figure 13-7 Genetically engineered tobacco plant
This plant glows as it expresses the luciferase gene from a firefly. Luciferase is an enzyme that catalyzes a reaction that produces a flash of light. This classic experiment indicated that animal genes can be expressed in plants.

the genetic code in a diverse array of species and found it the same in organisms as different as bacteria, redwood trees, jellyfish, and humans (**FIG. 13-7**). These findings strongly suggest that the code is an ancient legacy that evolved early in the history of life (discussed in Chapter 21).

Scientists have discovered some minor exceptions to the universality of the genetic code. In several unicellular protozoa, UAA and UGA code for the amino acid glutamine instead of functioning as stop signals. Other exceptions are found in some mitochondria, which contain their own DNA and protein synthesis machinery for some genes. These slight coding differences vary with the organism, but keep in mind that in each case, all the other coding assignments are identical to those of the standard genetic code.

The genetic code is redundant

Given 64 possible codons and only 20 common amino acids, it is not surprising that more than one codon specifies certain amino acids. This redundancy in the genetic code has certain characteristic patterns. The codons CCU, CCC, CCA, and CCG are "synonymous" in that they all code for the amino acid proline. The only difference among the 4 codons involves the nucleotide at the 3′ end of the triplet. Although the code may be read three nucleotides at a time, only the first two nucleotides seem to contain specific information for proline. A similar pattern can be seen for several other amino acids. Only methionine and tryptophan are specified by single codons. Each of the other amino acids is specified by 2 to 6 different codons.

Crick first proposed this apparent breach of the base-pairing rules as the **wobble hypothesis.** He reasoned that the third nucleotide of a tRNA anticodon (which is the 5′ base of the anticodon sequence) may sometimes form hydrogen bonds with more than one kind of third nucleotide (the 3′ base) of a codon. Investigators later demonstrated wobble experimentally by determining the anticodon sequences of tRNA molecules and testing their specificities in artificial systems. Some tRNAs bond exclusively to one codon, but other tRNA molecules pair with as many as three separate codons that differ in their third nucleotide but that specify the same amino acid. Thus, the wobble hypothesis accounts for the possible variation in base pairing between the third base of a codon and the corresponding base in its anticodon. "Wobble" results in several acceptable forms of base pairing between mRNA and tRNA.

CHECKPOINT 13.2

- VISUALIZE *Sketch a simple flow diagram that shows the relationships among the following: RNA, translation, DNA, transcription, and polypeptide.*
- *How are the structures of DNA and RNA similar? How are they different?*

13.3 TRANSCRIPTION

LEARNING OBJECTIVES

6 Compare the processes of transcription and DNA replication, identifying both similarities and differences.
7 Compare bacterial and eukaryotic mRNAs, and explain the functional significance of their structural differences.

Now that we have presented an overview of information flow from DNA to RNA to polypeptide, let us examine the entire process more closely. The first stage in the flow of information from DNA to polypeptide is the transcription of a DNA nucleotide sequence into an RNA nucleotide sequence. Stanford University biologist Roger Kornberg painstakingly worked out the details of transcription, for which he was awarded the 2006 Nobel Prize in Chemistry.

In eukaryotic transcription, most RNA synthesis requires one of three **RNA polymerases,** enzymes present in all cells.

The three RNA polymerases differ in the kinds of RNA synthesis they catalyze. *RNA polymerase I* catalyzes the synthesis of several kinds of rRNA molecules that are components of the ribosome, *RNA polymerase II* catalyzes the production of the protein-coding mRNA, and *RNA polymerase III* catalyzes the synthesis of tRNA and one of the rRNA molecules.

RNA polymerases require DNA as a template and have many similarities to the DNA polymerases discussed in Chapter 12. Like DNA polymerases, RNA polymerases carry out synthesis in the 5′ ⟶ 3′ direction; that is, they begin at the 5′ end of the RNA molecule being synthesized and then continue to add nucleotides at the 3′ end until the molecule is complete (FIG. 13-8). RNA polymerases use nucleotides with three phosphate groups (e.g., ATP and GTP) as substrates for the polymerization reaction. As the nucleotides link to the 3′ end of the RNA, two of the phosphates are removed. These reactions are strongly exergonic and do not need the input of additional energy.

Recall from Chapter 12 that whenever nucleic acid molecules associate by complementary base pairing, the two strands are **antiparallel**. Just as the two paired strands of DNA are antiparallel, the template strand of the DNA and the complementary RNA strand are also antiparallel. Therefore, when transcription takes place, as RNA is synthesized in its 5′ ⟶ 3′ direction, the DNA template is read in its 3′ ⟶ 5′ direction.

Scientists conventionally refer to a sequence of bases in a gene or the mRNA sequence transcribed from it as upstream or downstream from some reference point. *Upstream* is toward the 5′ end of the mRNA sequence or the 3′ end of the template DNA strand. *Downstream* is toward the 3′ end of the RNA or the 5′ end of the template DNA strand.

Upstream		Downstream
5′—A—T—G—A—C—T—3′		nontemplate DNA strand
3′—T—A—C—T—G—A—5′		template DNA strand
Direction of transcription ⟶		
5′—A—U—G—A—C—U—3′		RNA

The synthesis of mRNA includes initiation, elongation, and termination

In both bacteria and eukaryotes, the nucleotide sequence in DNA to which RNA polymerase and associated proteins initially bind is called the **promoter.** Because the promoter is not transcribed, RNA polymerase moves past the promoter to begin transcription of the protein-coding sequence of DNA. Different genes may have slightly different promoter sequences, so the cell can direct which genes are transcribed at any one time. Bacterial promoters are usually about 40 bases long and lie in the DNA just upstream of the point at which transcription will begin. Once RNA polymerase has recognized the correct promoter, it unwinds the DNA double helix and *initiates*

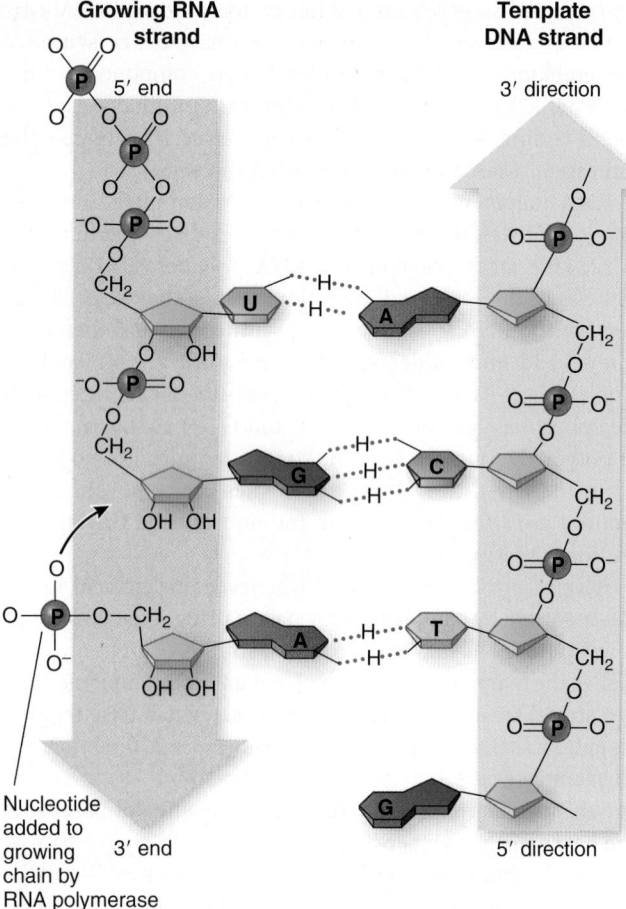

Growing RNA strand

5′ end

3′ end

Nucleotide added to growing chain by RNA polymerase

Template DNA strand

3′ direction

5′ direction

Figure 13-8 A molecular view of transcription

Incoming nucleotides with three phosphates pair with complementary bases on the template DNA strand (*right*). RNA polymerase cleaves two phosphates (*not shown*) from each nucleotide and covalently links the remaining phosphate to the 3′ end of the growing RNA chain. Thus, RNA, like DNA, is synthesized in the 5′ ⟶ 3′ direction.

© Cengage Learning

transcription (**FIG. 13-9**). Unlike DNA synthesis, RNA synthesis does not require a primer. However, transcription requires several proteins in addition to RNA polymerase, which we discuss in Chapter 14.

The first nucleotide at the 5′ end of a new mRNA chain initially retains its triphosphate group (see Fig. 13-8). However, during the *elongation* stage of transcription, as each additional nucleotide is incorporated at the 3′ end of the growing RNA molecule, two of its phosphates are removed in an exergonic reaction that leaves the remaining phosphate to become part of the sugar–phosphate backbone (as in DNA replication). The last nucleotide to be incorporated has an exposed 3′-hydroxyl group.

Elongation of RNA continues until *termination*, when RNA polymerase recognizes a termination sequence consisting of a specific sequence of bases in the DNA template. This signal leads to a separation of RNA polymerase from the template DNA and the newly synthesized RNA.

Termination of transcription occurs by different mechanisms in bacteria and eukaryotes. In bacteria transcription stops at the end of the termination sequence. When RNA polymerase comes to the termination sequence, it releases the DNA template

and the new RNA strand. In eukaryotic cells RNA polymerase adds nucleotides to the mRNA molecule after it passes the termination sequence. The 3′ end of the mRNA becomes separated from RNA polymerase about 10 to 35 nucleotides past the termination sequence.

Only one strand in a protein-coding region of DNA is used as a template. For example, consider a segment of DNA that contains the following DNA base sequence in the template strand

3′—TAACGGTCT—5′

If the complementary DNA strand

5′—ATTGCCAGA—3′

were used as a template, a message would be produced specifying an entirely different (and probably nonfunctional) protein. However, just because one strand is transcribed for a given gene does not mean that the same DNA strand is always the template for all genes throughout the length of a chromosome-sized DNA molecule. Instead, a particular strand may serve as the template strand for some genes and the nontemplate strand for others (**FIG. 13-10**).

Messenger RNA contains base sequences that do not directly code for protein

A completed mRNA molecule contains more than the nucleotide sequence that codes for a protein. **FIGURE 13-11** shows a typical bacterial mRNA. (The unique features of eukaryotic mRNA are discussed in the next section.) In both bacteria and eukaryotes, RNA polymerase starts transcription of a gene upstream of the protein-coding DNA sequence. As a result, the mRNA has a noncoding **leader sequence** at its 5′ end. The leader has recognition sites for ribosome binding, which properly position the ribosomes to translate the message. The **start codon** follows the leader sequence and signals the beginning of the *protein-coding sequence* that contains the actual message for the polypeptide. Unlike coding in eukaryotic cells, in bacterial cells it is common for more than one polypeptide to be encoded by a single mRNA molecule (see Chapter 14). At the end of each coding sequence, a **stop codon** signals the end of the protein. The stop codons—UAA, UGA, and UAG—end both bacterial and eukaryotic messages. They are followed by noncoding 3′ *trailing sequences*, which vary in length.

Eukaryotic mRNA is modified after transcription and before translation

There are differences between bacterial and eukaryotic mRNA. Bacterial mRNAs are used immediately after transcription, without further processing. In eukaryotes the original transcript, called *precursor mRNA* or **pre-mRNA**, is modified in several ways while it is still in the nucleus. These *posttranscriptional modification and processing* activities produce mature mRNA for transport and translation.

Modification of the eukaryotic message begins when the growing RNA transcript is about 20 to 30 nucleotides long. At

The mRNA is synthesized in the 5' to 3' direction as the template strand of the DNA molecule is read in the 3' to 5' direction.

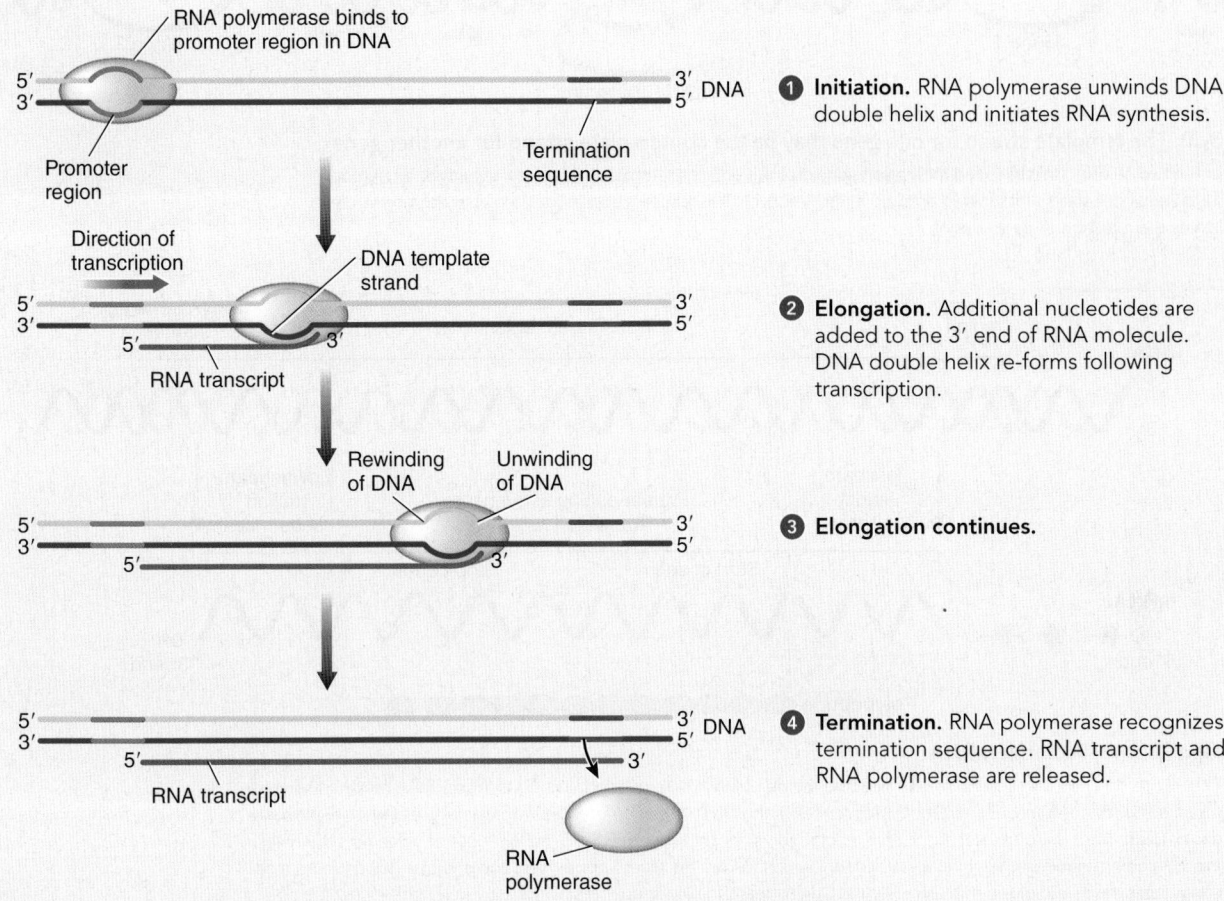

1 Initiation. RNA polymerase unwinds DNA double helix and initiates RNA synthesis.

2 Elongation. Additional nucleotides are added to the 3' end of RNA molecule. DNA double helix re-forms following transcription.

3 Elongation continues.

4 Termination. RNA polymerase recognizes termination sequence. RNA transcript and RNA polymerase are released.

Figure 13-9 *Animation* **An overview of transcription: initiation, elongation, and termination**

CONNECT RNA polymerase is a complex of 10–12 proteins that act together as a molecular machine to initiate and continue transcription. Based on what you have learned about DNA synthesis, name two types of functions you would expect these proteins to perform.
© Cengage Learning

that point, enzymes add a *5' cap* to the 5' end of the mRNA chain (**FIG. 13-12**). The cap is in the form of an unusual nucleotide, 7-methylguanosine, which is linked to the mRNA transcript by three phosphate groups. Eukaryotic ribosomes cannot bind to an uncapped message.

Capping may also protect the mRNA from certain types of degradation and may therefore partially explain why eukaryotic mRNAs are much more stable than bacterial mRNAs. Eukaryotic mRNAs have half-lives ranging from 30 minutes to as long as 24 hours. The average half-life of an mRNA molecule in a mammalian cell is about 10 hours compared with 2 minutes in a bacterial cell.

A second modification of eukaryotic mRNA, known as **polyadenylation,** may occur at the 3' end of the molecule. Near the 3' end of a completed message usually lies a sequence of bases that serves as a signal for adding many adenine-containing nucleotides, known as a **poly-A tail** (for *polyadenylated*).

Enzymes in the nucleus recognize the signal for the poly-A tail and cut the mRNA molecule at that site. This process is followed by the enzymatic addition of a string of 100 to 250 adenine nucleotides to the 3' end. Polyadenylation appears to have several functions: it helps export the mRNA from the nucleus, stabilizes some mRNAs against degradation in the cytosol (the longer the poly-A tail, the longer the molecule persists in the cytosol), and facilitates initiation of translation by helping ribosomes recognize the mRNA.

Both noncoding and coding sequences are transcribed from eukaryotic genes One of the greatest surprises in the history of molecular biology was the finding that most eukaryotic genes have *interrupted coding sequences*: long sequences of bases within the protein-coding sequences of the gene do not code for amino acids in the final polypeptide product. The noncoding regions within the gene are called **introns** (*in*tervening

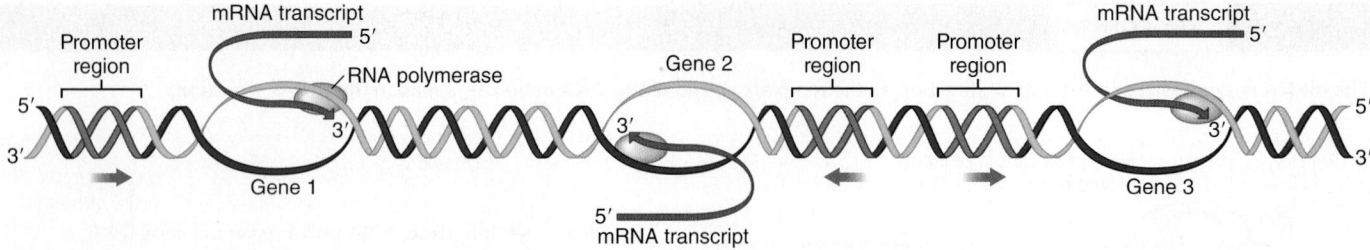

Figure 13-10 The template strand for one gene may be the nontemplate strand for another gene

Only one of the two strands is transcribed for a given gene, but the opposite strand may be transcribed for a neighboring gene. Each transcript starts at its own promoter region (*orange*). The orange arrow associated with each promoter region indicates the direction of transcription.
© Cengage Learning

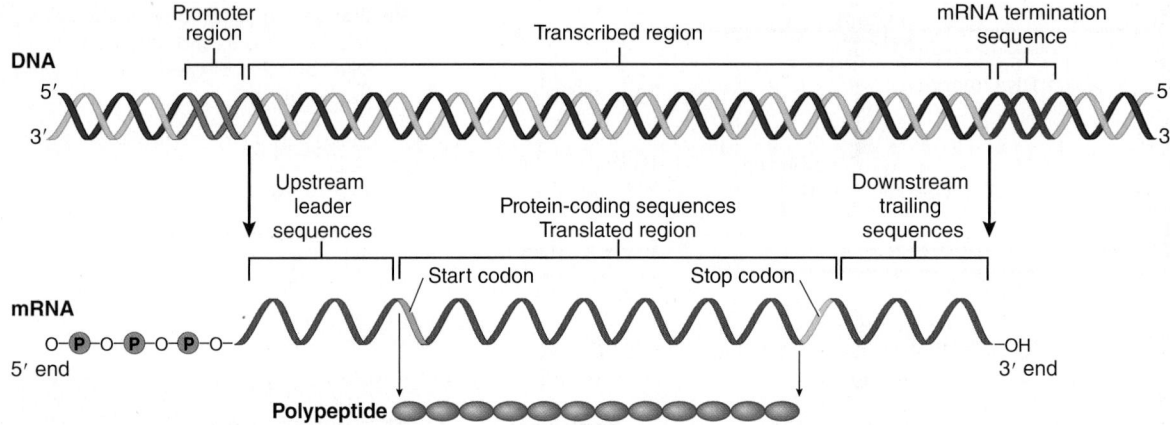

Figure 13-11 Bacterial mRNA

This figure compares a bacterial mRNA with the region of DNA from which it was transcribed. RNA polymerase recognizes, but does not transcribe, promoter sequences in the DNA. Initiation of RNA synthesis occurs five to eight bases downstream from the promoter. Ribosome recognition sites are located in the 5′ mRNA upstream leader sequences. Protein-coding sequences begin at a start codon, which follows the leader sequences, and end at a downstream stop codon near the 3′ end of the mRNA molecule. Downstream trailing sequences, which vary in length, follow the polypeptide-coding sequences.
© Cengage Learning

sequences), as opposed to **exons** (*ex*pressed sequences), which are parts of the protein-coding sequence.

A typical eukaryotic gene may have multiple exons and introns, although the number varies. For example, the human β-globin gene, which produces one component of hemoglobin, contains 2 introns and 3 exons; the human gene for albumin contains 14 introns and 15 exons; and the human gene for titin, the largest protein known, found in muscle cells, contains 233 introns and 234 exons. In many cases, the combined lengths of the intron sequences are much greater than the combined lengths of the exons. For instance, the ovalbumin gene contains about 7700 base pairs, whereas the total of all the exon sequences is only 1859 base pairs.

When a gene that contains introns is transcribed, the entire gene is copied as a large RNA transcript, the pre-mRNA. A pre-mRNA molecule contains both intron and exon sequences. (Note that *intron* and *exon* refer to corresponding nucleotide sequences in both DNA and RNA.) For the pre-mRNA to be made into a functional message, it must be capped and have a poly-A tail added, and the introns must be removed and the exons spliced together to form a continuous protein-coding message.

Splicing itself occurs by several different mechanisms, depending on which type of RNA is involved. In many instances, splicing involves the association of several **small nuclear ribonucleoprotein complexes (snRNPs,** pronounced "snurps") to form a large ribonucleoprotein complex called a **spliceosome.** Spliceosomes, which are similar in size to ribosomes, catalyze the reactions that remove introns from pre-mRNA. In other instances, the RNA within the intron acts as an RNA catalyst, or **ribozyme,** splicing itself without the use of a spliceosome or protein enzymes. Following pre-mRNA processing, the mature mRNA is transported through a nuclear pore into the cytosol to be translated by a ribosome.

Posttranscriptional modification in eukaryotes is summarized as follows:

> pre-mRNA containing introns and exons → 5′ end of pre-mRNA capped with modified nucleotide → poly-A tail added to 3′ end → introns removed and exons spliced together → mature mRNA transported into cytosol → translation at ribosome

After transcription in eukaryotes, pre-mRNA undergoes extensive modification to produce mature, functional mRNA.

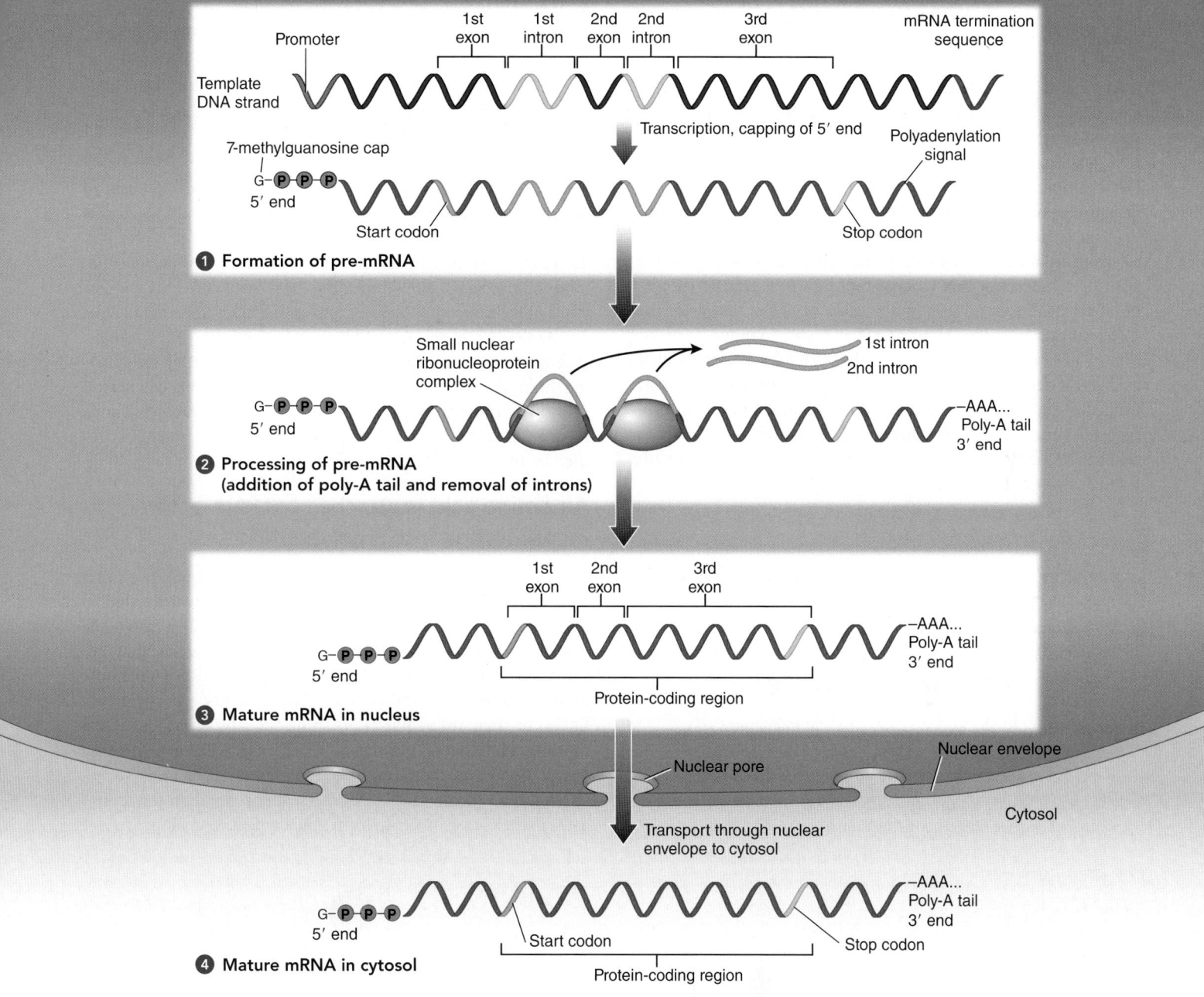

Figure 13-12 *Animation* Posttranscriptional modification of eukaryotic messenger RNA

PREDICT What modifications would take place and what would be the fate of a transcribed pre-mRNA that had a mutation in its polyadenylation signal?

© Cengage Learning

Biologists debate the evolution of eukaryotic gene structure

The reason for the intricate structure of eukaryotic genes and its potential evolutionary advantage is a matter of ongoing discussion. Why do introns occur in most eukaryotic nuclear genes but not in the genes of most bacteria (or of mitochondria and chloroplasts)? How did this remarkable genetic system with interrupted coding sequences ("split genes") evolve, and why has it survived? It seems incredible that as much as 75% of the original transcript of a eukaryotic nuclear gene must be *removed* to make a working message.

Some intron-containing genes can encode more than one protein, depending on which regions are spliced together as exons, a process known as *alternative splicing* (see Fig. 14-15). Intron excision provides one of many ways in which present-day eukaryotes regulate expression of their genes. This opportunity for control may balance the energy cost of maintaining a large load of noncoding DNA.

Another proposal is that many exons are nucleotide sequences that code for **protein domains,** regions of protein tertiary structure that may have important functions. For example, the active site of an enzyme may compose one domain. A different domain may enable that enzyme to bind to a particular cell structure, and yet another may be a site involved in allosteric regulation (see Chapter 7). New proteins with new functions may emerge rapidly when genetic recombination produces novel combinations of exons that code for different proteins. This hypothesis has become known as *evolution by exon shuffling*. It has been supported by examples such as the low-density lipoprotein (LDL) receptor protein, a protein found on the surface of human cells that binds to cholesterol transport molecules (see Chapter 5). The LDL receptor protein has domains that are related to parts of other proteins with totally different functions.

CHECKPOINT 13.3

- **CONNECT** *In what ways are DNA polymerase and RNA polymerase similar? How do they differ?*
- *A certain template DNA strand has the following nucleotide sequence:*

  3′—TACTGCATAATGATT—5′

  What would be the sequence of codons in the mRNA transcribed from this strand? What would be the nucleotide sequence of the complementary nontemplate DNA strand?
- **CONNECT** *What features do mature eukaryotic mRNA molecules have that bacterial mRNAs lack?*

13.4 TRANSLATION

LEARNING OBJECTIVES

8 Identify the features of tRNA that are important in decoding genetic information and converting it into "protein language."
9 Explain how ribosomes function in polypeptide synthesis.
10 Describe the processes of initiation, elongation, and termination in polypeptide synthesis.
11 Describe a polysome in bacterial cells.

Translation adds another level of complexity to the process of information transfer because it involves conversion of the triplet nucleic acid code to the 20–amino acid alphabet of polypeptides. Translation requires the coordinated functioning of more than 100 kinds of macromolecules, including the protein and RNA components of the ribosomes, mRNA, and amino acids linked to tRNAs.

Amino acids are joined by *peptide bonds* to form proteins. (See the formation of a peptide bond in Figure 3-18.) This joining links the amino and carboxyl groups of adjacent amino acids. The translation process ensures both that peptide bonds form and that the amino acids link in the correct sequence specified by the codons in the mRNA.

An amino acid is attached to tRNA before incorporation into a polypeptide

How do the amino acids align in the proper sequence so that they can link? Crick proposed that a molecule was needed to bridge the gap between mRNA and proteins. This molecule was determined to be tRNA. DNA contains genes that are transcribed to form the tRNAs. Each kind of tRNA molecule binds to a specific amino acid. Amino acids are covalently linked to their respective tRNA molecules by enzymes, called **aminoacyl–tRNA synthetases,** which use ATP as an energy source (FIG. 13-13). The resulting complexes, called **aminoacyl–tRNAs,** bind to the mRNA coding sequence to align the amino acids in the correct order to form the polypeptide chain.

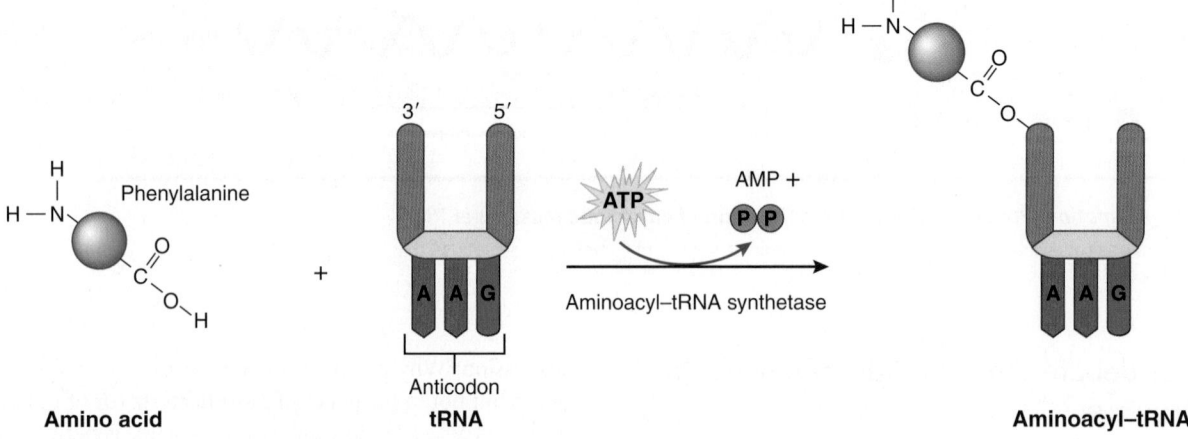

Figure 13-13 Linkage of a specific amino acid to the appropriate tRNA

Amino acids are covalently linked to their respective tRNA molecules by aminoacyl–tRNA synthetases, which use ATP as an energy source.
© Cengage Learning

The tRNAs are polynucleotide chains 70 to 80 nucleotides long, each with several unique base sequences as well as some sequences that are common to all. Although considerably smaller than mRNA or rRNA molecules, tRNA molecules have an intricate structure (see Fig. 13-6). A tRNA molecule has several properties essential to its function: (1) it has an anticodon, a complementary binding sequence for the correct mRNA codon; (2) it is recognized by an aminoacyl–tRNA synthetase that adds the correct amino acid; (3) it has an attachment site for the specific amino acid specified by the anticodon (the site lies about 180 degrees from the anticodon); and (4) it is recognized by ribosomes.

Complementary base pairing within each tRNA molecule causes it to be doubled back and folded. Three or more loops of unpaired nucleotides are formed, one of which contains the anticodon. The amino acid–binding site is at the 3′ end of the molecule. The carboxyl group of the amino acid is covalently bound to the exposed 3′ hydroxyl group of the terminal nucleotide, leaving the amino group of the amino acid free to participate in peptide bond formation. The pattern of folding in tRNAs keeps a constant distance between the anticodon and amino acid, allowing precise positioning of the amino acids during translation.

The components of the translational machinery come together at the ribosomes

The importance of ribosomes and of protein synthesis in cell metabolism is exemplified by a rapidly growing *E. coli* cell, which contains some 15,000 ribosomes, nearly a third of the total mass of the cell. Although bacterial and eukaryotic ribosomes are not identical, ribosomes from all organisms share many fundamental features and basically work in the same way. Ribosomes consist of two subunits, both made up of protein (about 40% by weight) and rRNA (about 60% by weight). Unlike mRNA and tRNA, rRNA has catalytic functions and does not transfer information. The ribosomal proteins do not appear to be catalytic but instead contribute to the overall structure and stability of the ribosome.

Researchers have isolated each ribosomal subunit intact in the laboratory and then separated each into its RNA and protein constituents. In bacteria the smaller of these subunits contains 21 proteins and one rRNA molecule, and the larger subunit contains 34 proteins and two rRNA molecules. Under certain conditions researchers can reassemble each subunit into a functional form by adding each component in its correct order. Using this approach, together with sophisticated electron microscopic and crystallography studies, researchers have determined the 3-D structure of the ribosome (**FIG. 13-14a**), the way it is assembled in the living cell and the functions of its components. The large subunit contains a depression on one surface into which the small subunit fits. During translation, the mRNA fits in a groove between the contact surfaces of the two subunits.

The ribosome has four binding sites, one for mRNA and three for tRNAs. Thus, the ribosome holds the mRNA template, the aminoacyl–tRNA molecules, and the growing polypeptide chain in the correct orientation so that the genetic code can be read and the next peptide bond formed. Transfer RNA molecules attach to three depressions on the ribosome, the A, P, and E binding sites (**FIG. 13-14b**). The *P site*, or peptidyl site, is so named because the tRNA holding the growing polypeptide

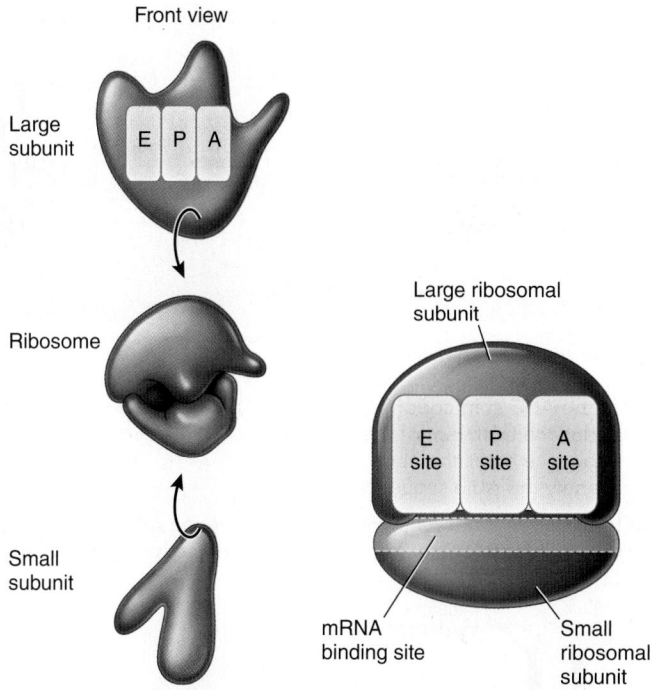

(a) A ribosome consists of two subunits, one large and one small. This model of a ribosome is based on a 3-D reconstruction of electron microscopic images.

(b) The mRNA passes through a groove between the two ribosomal subunits. Each ribosome contains three binding sites for tRNA molecules.

Figure 13-14 *Animation* **Ribosome structure**
© Cengage Learning

chain occupies the P site. The *A site* is named the aminoacyl site because the aminoacyl–tRNA delivering the next amino acid in the sequence binds at this location. The *E site* (for *exit*) is where tRNAs that have delivered their amino acids to the growing polypeptide chain exit the ribosome.

Translation begins with the formation of an initiation complex

The process of protein synthesis has three distinct stages: initiation, repeating cycles of elongation, and termination. Initiation of protein synthesis is essentially the same in all organisms. Here we describe initiation in bacteria and then briefly discuss some differences between bacterial and eukaryotic initiation.

The **initiation** of translation uses proteins called *initiation factors*, which become attached to the small ribosomal subunit. In bacteria three different initiation factors attach to the small subunit, which then binds to the mRNA molecule in the region of the AUG start codon (**FIG. 13-15**). A leader sequence upstream (toward the 5′ end) of the AUG helps the ribosome identify the AUG sequence that signals the beginning of the mRNA code for a protein.

The tRNA that bears the first amino acid of the polypeptide is the *initiator tRNA*. The amino acid methionine is bound to the initiator tRNA; as a result, the first amino acid on new polypeptides is methionine. (Quite often, the methionine is removed later.) In bacteria the initiator methionine is modified by the addition of a one-carbon group derived from formic acid and is designated *f*Met for *N*-formylmethionine. The fMet–initiator tRNA, which has the anticodon 3′–UAC–5′, binds to the AUG start codon, releasing

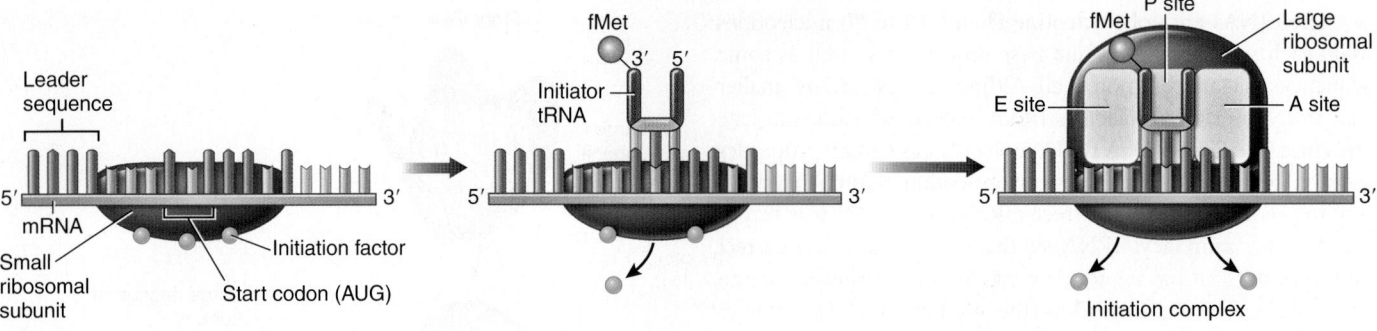

① The small ribosomal subunit, with initiation factors, binds to the mRNA at the AUG start codon. The leader sequence upstream of the AUG sequence helps the ribosome identify the AUG sequence.

② The initiator tRNA binds to the start codon, and one of the initiation factors is released. Energy is provided by the hydrolysis of GTP.

③ When the large ribosomal subunit binds to the small subunit and the remaining initiation factors are released, the initiation complex is complete. Hydrolysis of GTP bound to an initiation factor provides energy for assembly of the complex.

(a) Initiation in bacteria involves a ribosome, an mRNA, an initiator tRNA to which formylated methionine (fMet) is bound, and several protein initiation factors.

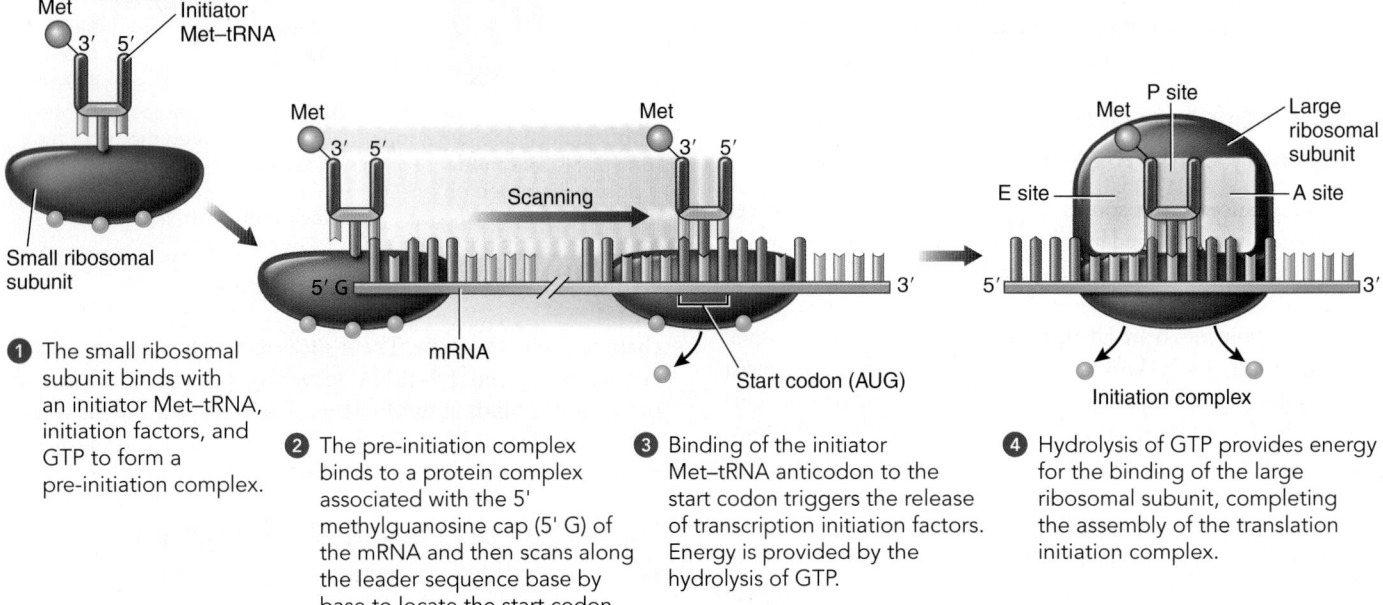

① The small ribosomal subunit binds with an initiator Met–tRNA, initiation factors, and GTP to form a pre-initiation complex.

② The pre-initiation complex binds to a protein complex associated with the 5' methylguanosine cap (5' G) of the mRNA and then scans along the leader sequence base by base to locate the start codon.

③ Binding of the initiator Met–tRNA anticodon to the start codon triggers the release of transcription initiation factors. Energy is provided by the hydrolysis of GTP.

④ Hydrolysis of GTP provides energy for the binding of the large ribosomal subunit, completing the assembly of the translation initiation complex.

(b) Initiation of translation in eukaryotes involves binding of a pre-initiation complex with the small ribosomal subunit to the mRNA cap. The complex then scans along the leader sequence to the start codon, where the assembly of the ribosome is completed.

Figure 13-15 The initiation of translation in bacteria and eukaryotes
© Cengage Learning

one of the initiation factors in the process. The *initiation complex* is complete when the large ribosomal subunit binds to the small subunit and the remaining initiation factors are released.

In eukaryotes initiation of translation differs in several respects. First, the methionine of the initiator tRNA is unmodified. Second, the small subunit, along with about 10 protein initiation factors, forms an initiation complex that binds to the 5' cap of the mRNA. This complex moves down the noncoding leader sequence by an ATP-dependent ratcheting mechanism until it reaches the start codon, which is embedded within a short sequence (such as ACC*AUG*G) that indicates the site (the underlined AUG) of translation initiation. At that point, the entire ribosome is assembled to initiate translation. The ribosome assembly process in both prokaryotes and eukaryotes

requires energy from guanosine triphosphate (GTP), an energy transfer molecule similar to ATP.

During elongation, amino acids are added to the growing polypeptide chain

FIGURE 13-16 outlines the four steps in a cycle of **elongation,** the stage of translation in which amino acids are added one by one to the growing polypeptide chain. Elongation is essentially the same in bacteria and eukaryotes. The appropriate aminoacyl–tRNA recognizes the codon in the A site and binds there by base pairing of its anticodon with the complementary mRNA codon. This binding step requires several proteins called *elongation factors* in addition to energy from GTP.

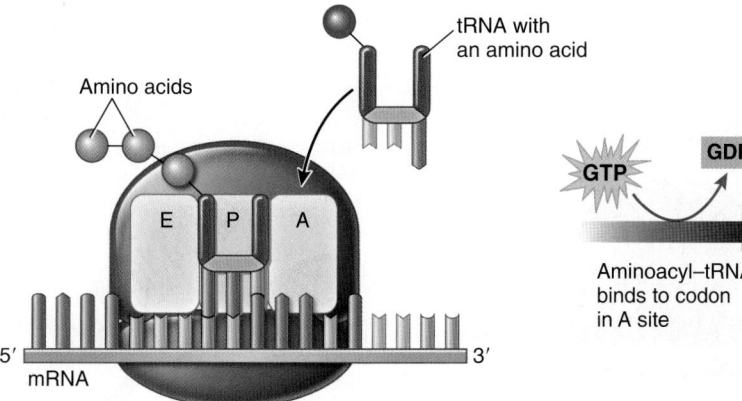

tRNA with an amino acid

Amino acids

E P A

5′ 3′
mRNA

1 The polypeptide chain is covalently bonded to the tRNA that carries the amino acid most recently added to the chain. This tRNA is in the P site of the ribosome.

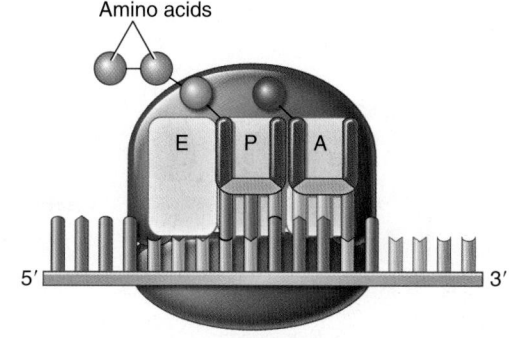

GTP → GDP

Aminoacyl–tRNA binds to codon in A site

Amino acids

E P A

5′ 3′

2 An aminoacyl–tRNA binds to the A site by complementary base pairing between the tRNA's anticodon and the mRNA's codon.

Ribosome ready to accept another aminoacyl–tRNA

Peptide bond formation

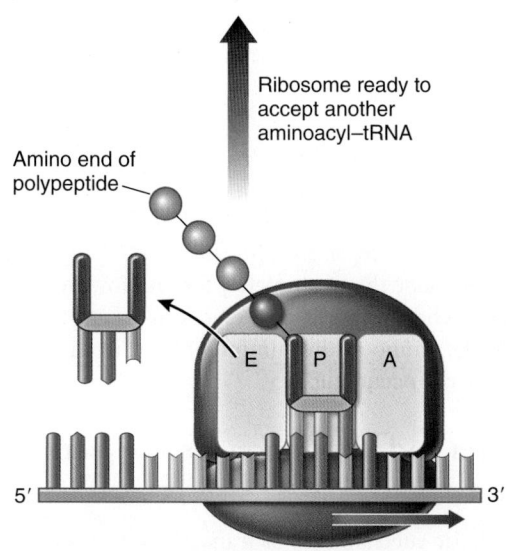

Amino end of polypeptide

E P A

5′ 3′

GDP ← GTP

Translocation toward 3′ end of mRNA

4 In the translocation step, the ribosome moves one codon toward the 3′ end of mRNA. As a result, the growing polypeptide chain is transferred to the P site. The uncharged tRNA in the E site exits the ribosome.

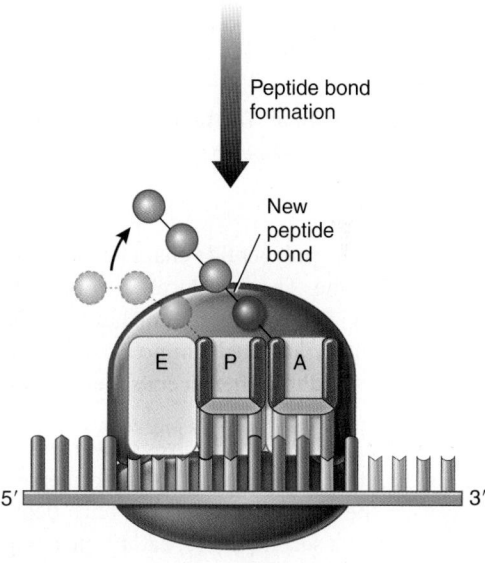

New peptide bond

E P A

5′ 3′

3 The growing polypeptide chain detaches from the tRNA molecule in the P site and becomes attached by a peptide bond to the amino acid linked to the tRNA at the A site.

Figure 13-16 An elongation cycle in translation

This illustration begins after a short chain of amino acids has formed. Each repetition of the elongation process adds one amino acid to the growing polypeptide chain.
© Cengage Learning

The amino group of the amino acid at the A site is now aligned with the carboxyl group of the preceding amino acid attached to the growing polypeptide chain at the P site. A peptide bond forms between the amino group of the new amino acid and the carboxyl group of the preceding amino acid. In this way the polypeptide is released from tRNA at the P site and attaches to the aminoacyl–tRNA at the A site. This reaction is spontaneous (it does not require additional energy) because ATP transferred energy during formation of the aminoacyl–tRNA. Peptide bond formation, however, does require the enzyme **peptidyl transferase**. Remarkably, this enzyme is not a protein but a ribozyme that is an RNA component of the large ribosomal subunit. Two biochemists, Thomas Cech of the University of Colorado and Sidney Altman of Yale University, received a Nobel Prize in Chemistry in 1989 for their independent discovery of ribozymes.

Recall from Chapter 3 that polypeptide chains have direction, or polarity. The amino acid on one end of the polypeptide chain has a free amino group (the amino end), and the amino acid at the other end has a free carboxyl group (the carboxyl end). Protein synthesis always proceeds from the amino end to the carboxyl end of the growing polypeptide chain:

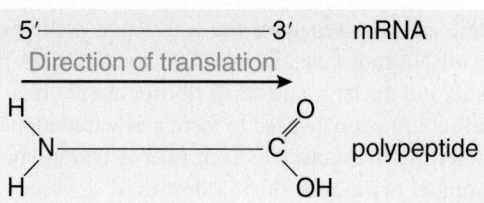

5′——————————3′   mRNA
Direction of translation

H   O
  \   ‖
   N——————C   polypeptide
  /   \
H   OH

In **translocation** the ribosome moves down the mRNA by one codon. As a result, the mRNA codon specifying the next

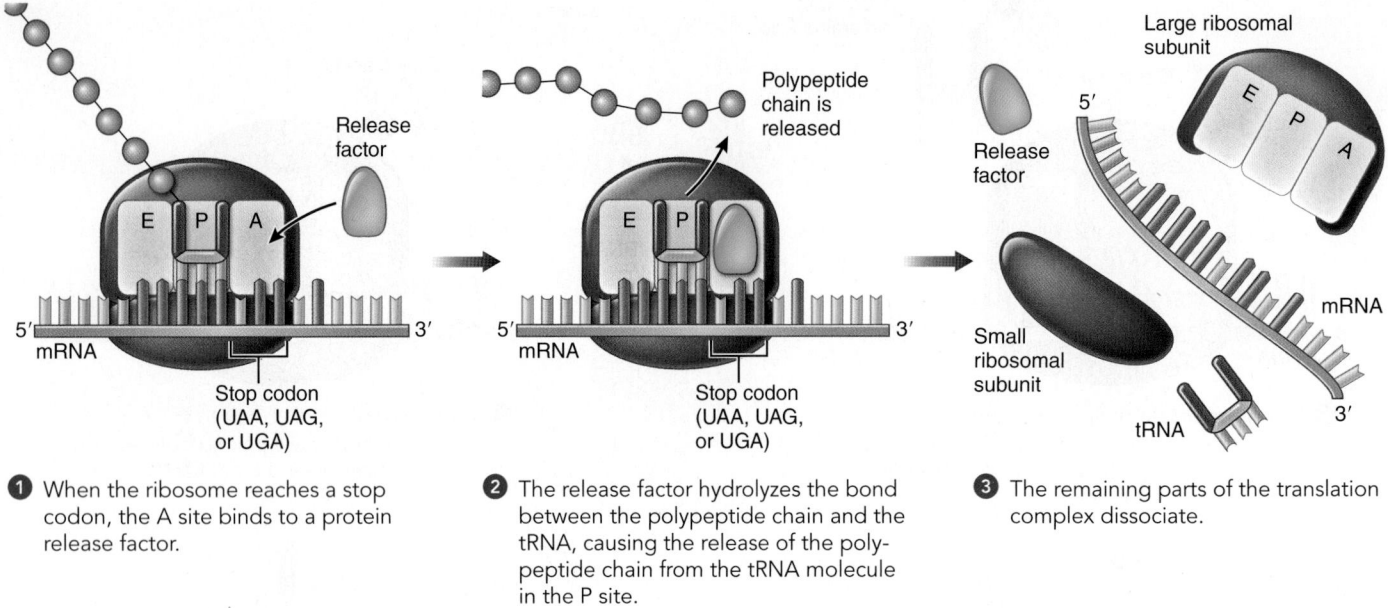

1 When the ribosome reaches a stop codon, the A site binds to a protein release factor.

2 The release factor hydrolyzes the bond between the polypeptide chain and the tRNA, causing the release of the polypeptide chain from the tRNA molecule in the P site.

3 The remaining parts of the translation complex dissociate.

Figure 13-17 *Animation* **The termination of translation**

A stop signal terminates polypeptide synthesis because no tRNA molecules exist with an anticodon complementary to a stop codon.
© Cengage Learning

amino acid in the polypeptide chain becomes positioned in the unoccupied A site. The translocation process requires energy, which again GTP supplies. The uncharged tRNA—that is, the tRNA without an attached amino acid—is moved from the P site to the E site. From there it exits the ribosome and joins the cytosolic pool of tRNAs.

Because translocation involves movement of the ribosome in the 3′ direction along the mRNA molecule, translation always proceeds in the 5′ ⟶ 3′ direction. Each peptide bond forms in a fraction of a second; by repeating the elongation cycle, an average-sized protein of about 360 amino acids is assembled by a bacterium in approximately 18 seconds and by a eukaryotic cell in a little more than 1 minute.

One of three stop codons signals the termination of translation

Termination is the final stage of translation. In termination the synthesis of the polypeptide chain is terminated by a *release factor*, a protein that recognizes the stop codon at the end of the coding sequence. (Because no tRNA molecule binds to a stop codon, the codon is available for binding by the release factor.) When the release factor binds to the A site, the bond between the tRNA in the P site and the last amino acid of the polypeptide chain breaks (**FIG. 13-17**). This hydrolysis reaction frees the newly synthesized polypeptide and also separates the translation complex into its parts: the mRNA molecule, the release factor, the tRNA molecule in the P site, and the large and small ribosomal subunits. The two ribosomal subunits can be used to form a new initiation complex with another mRNA molecule. Each mRNA is translated only a limited number of times before it is destroyed.

Specialized ribosome-associated proteins, called **molecular chaperones,** assist in the folding of the newly synthesized polypeptide chain into its three-dimensional active shape (see

Chapter 3). The sequence of amino acids in the polypeptide chain dictates the shape that it ultimately forms. However, without the molecular chaperone's help, interactions among the various regions of the amino acid chain might prevent the proper folding process from occurring.

Transcription and translation are coupled in bacteria

Although the basic mechanisms of transcription and translation are quite similar in all organisms, significant differences exist between bacteria and eukaryotes (**FIG. 13-18**). Bacteria lack a nucleus, whereas eukaryotic chromosomes are confined to the cell nucleus and protein synthesis takes place in the cytosol. In eukaryotes mRNA must move from the nucleus (the site of transcription) to the cytosol (the site of translation, or polypeptide synthesis). Also, as we discussed earlier in this chapter, extensive mRNA processing occurs in eukaryotic cells before translation.

In *E. coli* and other bacteria, translation begins very soon after transcription, and several ribosomes are attached to the same mRNA (**FIG. 13-19**). The end of the mRNA molecule synthesized first during transcription is also the first to be translated to form a polypeptide. Ribosomes bind to the 5′ end of the growing mRNA and initiate translation long before the message is fully synthesized. As many as 15 ribosomes may bind to a single mRNA. An mRNA molecule that is bound to clusters of ribosomes constitutes a **polyribosome,** or *polysome*. Polyribosomes also occur in eukaryotic cells, but they are assembled in the cytoplasm, not while transcription is still occurring.

Although many polypeptide chains can be actively synthesized on a single mRNA at any one time, the half-life of mRNA molecules in bacterial cells is only about 2 minutes. (Half-life is the time it takes for half the molecules to be degraded.) Usually, degradation of the 5′ end of the bacterial mRNA begins even before

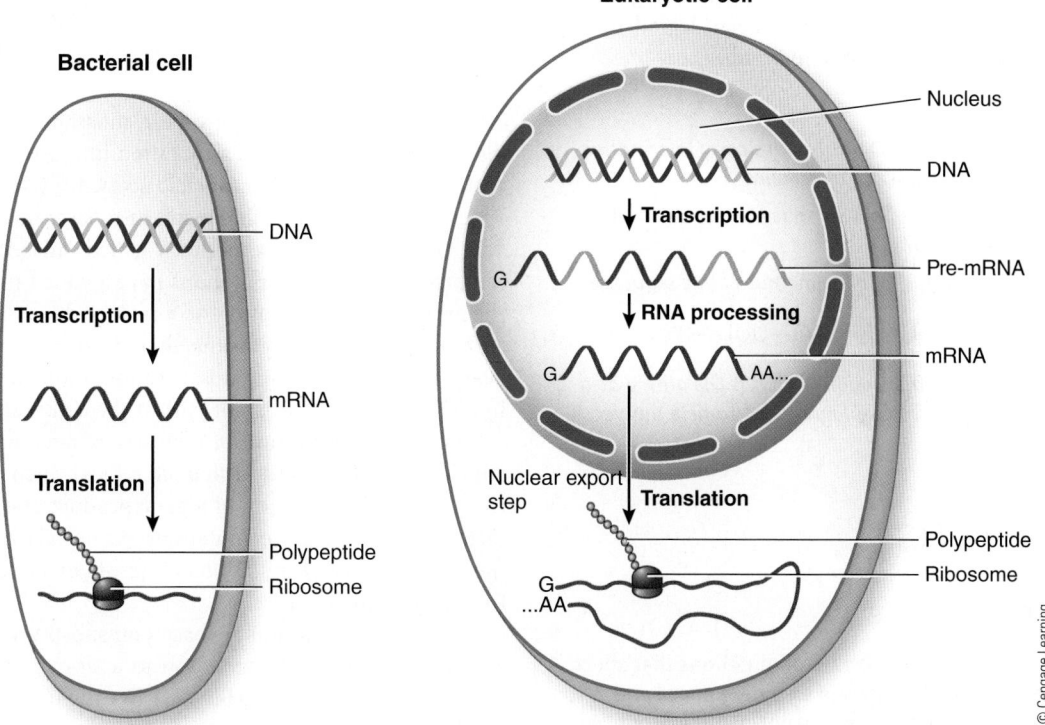

Bacterial cell

DNA

Transcription

mRNA

Translation

Polypeptide

Ribosome

Eukaryotic cell

Nucleus

DNA

Transcription

G — Pre-mRNA

RNA processing

G — AA... mRNA

Nuclear export step

Translation

G ...AA

Polypeptide

Ribosome

(a) In a bacterial cell, mRNA is immediately ready for translation by ribosomes.

(b) In a eukaryotic cell, RNA processing occurs in the nucleus before the mRNA exits the nucleus for translation.

Figure 13-18 Summary: the flow of genetic information in bacteria and eukaryotes

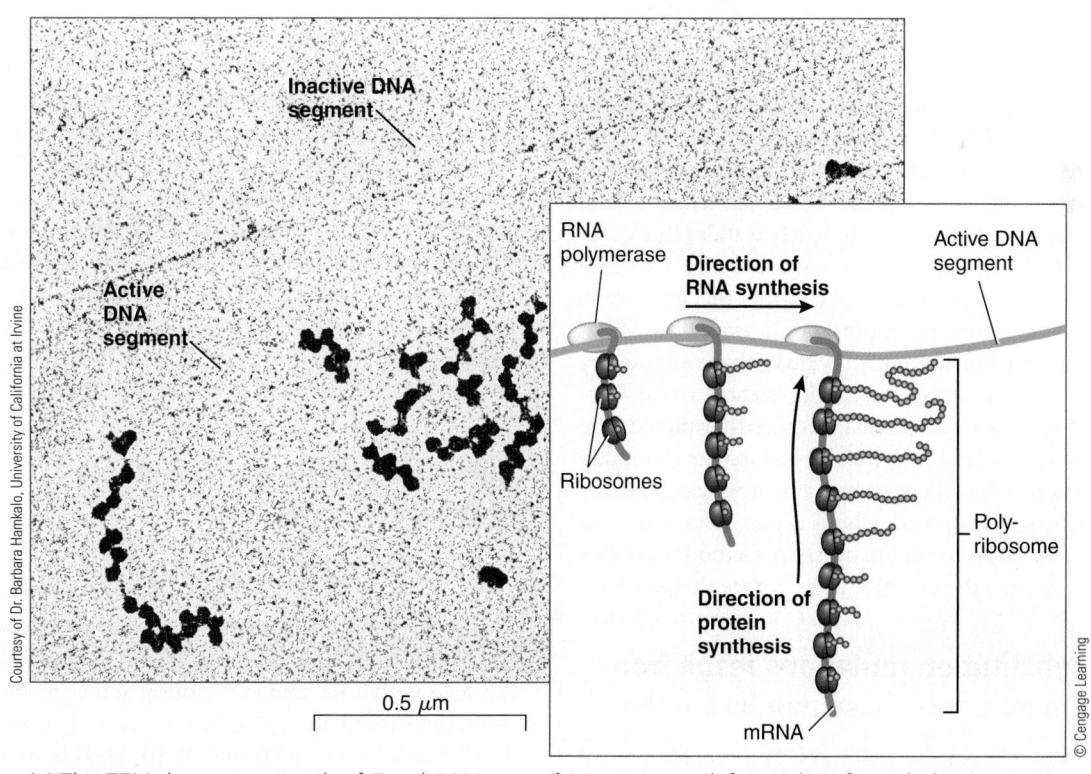

Inactive DNA segment

Active DNA segment

RNA polymerase

Direction of RNA synthesis

Active DNA segment

Ribosomes

Direction of protein synthesis

Poly-ribosome

mRNA

0.5 μm

(a) This TEM shows two strands of *E. coli* DNA, one inactive and the other actively producing mRNA. Protein synthesis begins while the mRNA is being completed.

(b) A sequence (*left to right*) of coupled transcription and translation. Note that several ribosomes translate each mRNA molecule simultaneously.

Figure 13-19 Coupled transcription and translation in bacteria

the first polypeptide is complete. Once the ribosome recognition sequences at the 5′ end of the mRNA have been degraded, no more ribosomes attach and initiate protein synthesis.

CHECKPOINT 13.4

- *What are ribosomes made of? Do ribosomes carry information to specify the amino acid sequence of proteins?*
- *What happens in each stage of polypeptide synthesis: initiation, elongation, and termination?*
- *A certain mRNA strand has the following nucleotide sequence:*

   5′—AUG—ACG—UAU—AAC—UUU—3′

What is the anticodon for each codon? What is the amino acid sequence of the polypeptide? (Use Figure 13-5 to help answer this question.)

13.5 MUTATIONS

LEARNING OBJECTIVE

12 Give examples of the different classes of mutations that affect the base sequence of DNA in protein encoding genes and explain the effects that each has on the polypeptide produced.

One of the first major discoveries about genes was that they undergo **mutations,** changes in the nucleotide sequence of the DNA. However, the overall mutation rate is much higher than the frequency of damage to DNA because all organisms have systems of enzymes that repair certain kinds of DNA damage.

If a DNA sequence has been changed and not corrected, DNA replication copies the altered sequence just as it would copy a normal sequence, making the mutation stable over an indefinite number of generations. In most cases, the mutant allele has no greater tendency than the original allele to mutate again.

Although most uncorrected mutations are either *silent* (and have no discernible effect) or harmful (such as those that cause inherited diseases and cancer), a few are useful. Because they can be passed to the next generation in cell lines that give rise to gametes, mutations are of vital importance in evolution. Mutations provide the variation among individuals that evolutionary forces act on. Mutations are also useful in research (recall Beadle and Tatum's work with mutations) because they provide the diversity of genetic material that enables researchers to modify genes so as to study their inheritance and their molecular functions. Scientists now determine where a particular mutation occurs in a gene by using recombinant DNA methods to isolate the gene and determine its sequence of bases (see Chapter 15).

Base-pair substitution mutations result from the replacement of one base pair by another

Mutations can alter genes in several ways. The simplest type of mutation, called a **base-pair substitution,** involves a change in only one pair of nucleotides. Often these mutations result from errors in base pairing during the replication process. For example, a G:C, C:G, or T:A pair might replace an A:T base pair. Such a mutation may cause the altered DNA to be transcribed as an altered mRNA. The altered mRNA may then be translated into a polypeptide chain with only one amino acid different from the normal sequence.

Silent mutations are base-pair substitutions that have no discernible effect, such as a mutation in a protein-coding gene that does not alter the amino acid sequence (FIGS. 13-20a and b). Base-pair substitutions that result in replacement of one amino acid by another are sometimes called **missense mutations** (FIG. 13-20c). Missense mutations have a wide range of effects. If the amino acid substitution occurs at or near the active site of an enzyme, the activity of the altered protein may decrease or even be destroyed. Some missense mutations involve a change in an amino acid that is not part of the active site. Others may result in the substitution of a closely related amino acid (one with very similar chemical characteristics). Such mutations have no effect on the function of a gene product and may be undetectable. Because these mutations occur relatively frequently, the true number of mutations in an organism is much greater than is actually observed.

Nonsense mutations are base-pair substitutions that convert an amino acid–specifying codon to a stop codon (FIG. 13-20d). A nonsense mutation usually destroys the function of the gene product; in the case of a protein-specifying gene, the part of the polypeptide chain that follows the mutant stop codon is missing.

Frameshift mutations result from the insertion or deletion of base pairs

In **frameshift mutations**, one or two nucleotide pairs are inserted into or deleted from the molecule, altering the reading frame. As a result, codons downstream of the insertion or deletion site specify an entirely new sequence of amino acids. Depending on where the insertion or deletion occurs in the gene, different effects are generated. Frameshift mutations may produce a stop codon within a short distance of the mutation (FIG. 13-20e). This codon terminates the already altered polypeptide chain. In addition to producing a stop codon, frameshift mutations can result in an altered amino acid sequence immediately after the change (FIG. 13-20f). A frameshift mutation in a gene specifying an enzyme usually results in a complete loss of enzyme activity.

Some mutations involve mobile genetic elements

One type of mutation is caused by DNA sequences that "jump" into the middle of a gene. These movable sequences of DNA are known as **mobile genetic elements,** or **transposons.** Mobile genetic elements not only disrupt the functions of some genes but also inactivate some previously active genes. Transposons were discovered in maize (corn) by U.S. geneticist Barbara McClintock in the 1950s (FIG. 13-21). She observed that certain genes appeared to be turned off and on spontaneously. She deduced that the mechanism involved a segment of DNA that moved from one region of a chromosome to another, where it would inactivate a gene. In recognition of her insightful

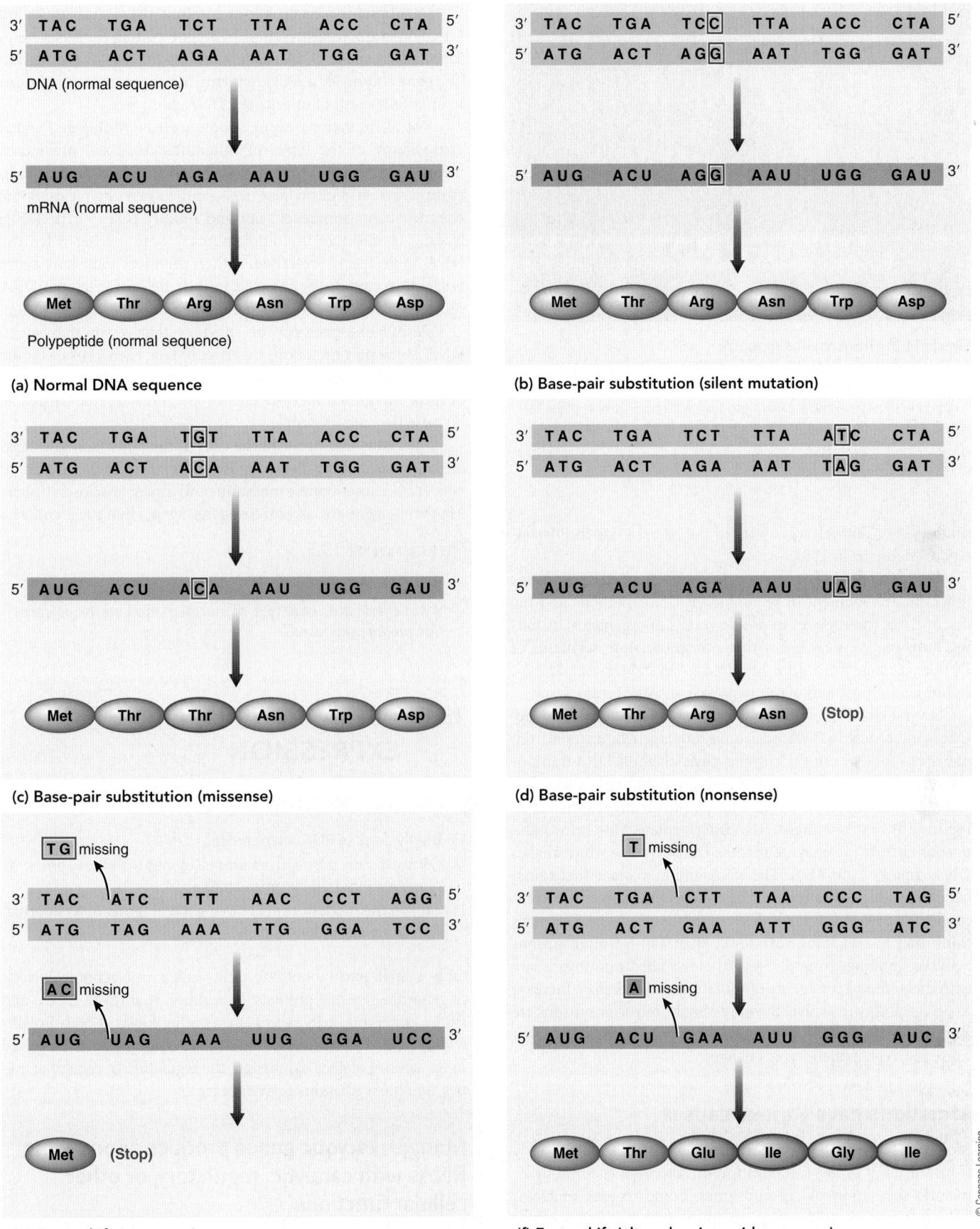

(a) Normal DNA sequence

(b) Base-pair substitution (silent mutation)

(c) Base-pair substitution (missense)

(d) Base-pair substitution (nonsense)

(e) Frameshift (nonsense)

(f) Frameshift (altered amino acid sequence)

Figure 13-20 *Animation* **Types of mutations**

Compare the normal DNA sequence in part **(a)** with the mutations in parts **(b)** to **(f)**.

Figure 13-21 Transposons in maize

The white lines and streaks in these kernels are caused by transposons that have moved from one location to another, turning off pigment production in certain cells. The number of cell divisions during the development of the kernel since the transposon turned off the gene determines how much pigment is produced. Some kernels are colorless (unpigmented) because the pigment-producing genes were turned off very early in development.

findings, McClintock was awarded the Nobel Prize in Physiology or Medicine in 1983.

Biologists did not appreciate the significance of transposons until the development of recombinant DNA methods and the discovery of transposons in a wide variety of organisms, including humans. We now know that transposons are segments of DNA that range from a few hundred to several thousand bases. In humans almost half of our genome consists of transposons.

Several kinds of mobile genetic elements have been identified. One of these, called a *DNA transposon*, moves genetic material from one site to another using a "cut-and-paste" method. DNA transposons are common in bacteria and animals, including humans.

Many mobile genetic elements are **retrotransposons** which replicate by first forming an RNA intermediate. Most retrotransposons encode the enzyme **reverse transcriptase** which makes DNA copies of the RNA. The RNA intermediate is first translated to produce the enzyme. The enzyme then copies the RNA, producing a DNA fragment that can jump into a gene. Scientists think that human cells have fewer than 100 retrotransposons actively jumping around. Nevertheless, retrotransposons are extremely important agents of mutation and therefore increase a species' ability to evolve. By one estimate, retrotransposons are responsible for about 10% of naturally occurring mutations that cause noticeable phenotypic changes.

Mutations have various causes

Most types of mutations occur infrequently but spontaneously from mistakes in DNA replication or defects in the mitotic or meiotic separation of chromosomes. Some regions of DNA, known as *mutational hot spots*, are much more likely than others to undergo mutation. An example is a short stretch of repeated nucleotides, which can cause DNA polymerase to "slip" while reading the template during replication.

Mutations in certain genes increase the overall mutation rate. For example, a mutation in a gene coding for DNA polymerase may make DNA replication less precise, or a mutation in a gene coding for a repair enzyme may allow more mutations to arise as a result of unrepaired DNA damage.

Not all mutations occur spontaneously. **Mutagens,** which cause many of the types of mutations discussed previously, include various types of radiation—such as X-rays, gamma rays, cosmic rays, and ultraviolet rays—and certain chemicals. Some chemical mutagens react with and modify bases in the DNA, leading to mistakes in complementary base pairing when the DNA molecule is replicated. Other chemical mutagens cause nucleotide pairs to be inserted into or deleted from the DNA molecule, changing the normal reading frame during replication.

Despite the presence of enzymes that repair damage to DNA, some new mutations do arise. In fact, each of us has some mutant alleles that neither of our parents had. Although some of these mutations alter the phenotype, most are not noticeable because they are recessive.

Mutations that occur in the cells of the body (somatic cells) are not passed on to offspring. However, these mutations are of concern because somatic mutations and cancer are closely linked. Many mutagens are also **carcinogens,** agents that cause cancer.

CHECKPOINT 13.5

- *What are the main types of mutations?*
- *What effects does each type of mutation have on the polypeptide product produced?*

13.6 VARIATIONS IN GENE EXPRESSION

LEARNING OBJECTIVES

13 Briefly discuss RNA interference.

14 Compare how retroviruses express genes with the expression of genes found on chromosomes.

15 Explain the modern concept of a gene in terms of its information content and function.

Only a small part (about 2%) of the eukaryotic genome consists of genes that encode proteins. In addition to mRNA, rRNA, and tRNA, eukaryotic cells express many other kinds of non-protein-coding RNA that are an essential part of a complex system involved in the control of gene expression, the regulation of cellular activities, and the regulation of development.

Many eukaryotic genes produce "noncoding" RNAs with catalytic, regulatory, or other cellular functions

Small nuclear RNA (snRNA) molecules bind to specific proteins to form a small nuclear ribonucleoprotein complex (snRNP), which in turn combines with other snRNPs to form a spliceosome. (Recall from the discussion earlier in the chapter that a spliceosome catalyzes intron removal.)

Some polypeptides enter the endoplasmic reticulum (ER) as they are being synthesized (see Chapter 4). **Signal-recognition particle RNA (SRP RNA),** in combination with proteins, directs the ribosome–mRNA–polypeptide complex to the rough ER shortly after translation begins. Translation continues after the complex binds to an integral protein in the ER membrane, and the newly synthesized protein is then distributed to its destination within the cell or outside the cell (e.g., as a secretion).

A special group of RNA molecules, known as **small nucleolar RNAs (snoRNAs),** processes pre-rRNA molecules in the nucleolus during ribosome subunit formation. Small nucleolar RNAs bind to complementary regions of pre-rRNA and target sites for cleavage or methylation. A pre-rRNA molecule is larger than the combined size of the three rRNA molecules made from it. In mammalian cells about 48% of the pre-rRNA molecule is removed during the cleavage steps. Pre-rRNA may have methyl groups added to regions that are to be saved during processing.

The discovery that noncoding RNAs have the ability to regulate gene expression was big news in biology. A phenomenon known as **RNA interference (RNAi),** for example, is found in many organisms and is now widely used as a research and medical tool (see Chapters 15 and 17). In RNAi, certain small RNA molecules interfere with the expression of genes or their RNA transcripts. RNA interference involves small interfering RNAs, microRNAs, and other kinds of short RNA molecules.

Small interfering RNAs (siRNAs) are double-stranded molecules that are 20 to 25 nucleotides in length. These RNA molecules help control damage from transposons (discussed in the previous section) and viral infections; they also have been shown to regulate the expression of protein-coding genes.

MicroRNAs (miRNAs) are single-stranded RNA molecules 20 to 25 nucleotides long that inhibit the translation of mRNAs involved in many biological processes, such as growth and development (**FIG. 13-22**). MicroRNAs are transcribed from genes and then shortened before they are combined with proteins to form a complex that inhibits the expression of mRNA molecules with sequences complementary to the miRNA. More than 1600 distinct miRNA genes have been identified in the human genome. miRNAs also have an important role in preventing cancer: cancer cells have measurably lower levels of miRNAs than non-cancer cells. A breakdown in miRNA function has also been implicated in heart disease and possibly Parkinson's disease.

Piwi-associated RNAs (piRNAs) are 26 to 31 nucleotide RNAs that appear to be the most abundant class of small RNAs produced in mammalian cells. One of their primary functions is to suppress retrotransposon activity and other mobile DNA elements in sperm-producing and other germ line cells.

Recently, the ENCODE project, which involves collaborative genome-wide studies from many different laboratories, has called attention to the importance of **long noncoding RNAs (lncRNAs).** These RNAs are molecules larger than 200 bases that are transcribed from genes distributed throughout the genome. Some are found within introns of genes, whereas many others are found in the vast intergenic regions of the genome between protein-coding genes. These molecules appear to act

Millar and Gubler (2005) The Plant Cell 17, 705–721

Figure 13-22 The effect of an miRNA molecule on plant growth

A normal *Arabidopsis* plant (*left*) serves as a control. The small size and other developmental defects exhibited by the plant on the *right* are the result of genetic modifications introduced by researchers that inhibit regulation by miRNA.

as scaffolds that assemble ribonucleoprotein complexes that regulate chromatin structure over long distances in the genome. Some of these RNAs have been shown to act as "global positioning" regulators that control genes that specify the location of a cell within the body. One such gene, called "HOTAIR," has been implicated in certain types of cancer. When HOTAIR is expressed in breast cancer cells, they lose their positional identity and become metastatic, allowing them to migrate and grow in different regions of the body. Elevated levels of lncRNAs such as HOTAIR in body tissues are also predictive of metastasis in colon and liver cancers, suggesting that they might be used to diagnose the severity of these diseases.

The definition of a gene has evolved

At the beginning of this chapter, we traced the development of ideas regarding the nature of the gene. For a time, scientists found it useful to define a *gene* as "a sequence of nucleotides that codes for one polypeptide chain." As you will see in Chapter 14, studies have now shown that in eukaryotic cells, a single gene may produce more than one polypeptide chain by modifications in the way the mRNA is processed. We have also learned from genomic research studies that although DNA coding for polypeptides makes up only about 2% of our genome, about 80% of our genome is expressed. Many of those additional genes are now known to produce the various kinds of noncoding RNA molecules (see **TABLE 13-1**).

It is perhaps most useful to define a gene in terms of its product: a **gene** is a DNA nucleotide sequence that carries the information needed to produce a specific RNA or polypeptide product.

The usual direction of information flow has exceptions

For several decades, a central premise of molecular biology was that genetic information always flows from DNA to RNA to protein. Through his studies of viruses, U.S. biologist Howard Temin discovered an important exception to this rule in 1964. Although viruses are not cellular organisms, they contain a single type of nucleic acid and reproduce in a host cell. Temin

TABLE 13-1 Selected Kinds of RNA in Eukaryotic Cells

KIND OF RNA	FUNCTION
Messenger RNA (mRNA)	Specifies the amino acid sequence of a protein
Transfer RNA (tRNA)	Binds to specific amino acid and serves as adapter molecule when amino acids are incorporated into growing polypeptide chain
Ribosomal RNA (rRNA)	Has both structural and catalytic (ribozyme) roles in ribosome
Small nuclear RNA (snRNA)	Involved in intron removal and regulation of transcription; part of spliceosome particles
Signal-recognition particle RNA (SRP RNA)	Helps direct the ribosome–mRNA–polypeptide complex to the rough ER shortly after translation begins
Small nucleolar RNA (snoRNA)	Processes pre-rRNA in nucleolus during formation of ribosome subunits
Small interfering RNA (siRNA)	Controls transposons and fights viral infections
MicroRNA (miRNA)	Controls expression of genes involved in growth and development by inhibiting the translation of mRNA; sometimes involved in activation of genes; involved in diseases, including prevention of cancer
Piwi-associated RNAs (piRNA)	Block expression of transposons in germ cells; can affect chromatin structure, regulating gene activity
Long noncoding RNA (lncRNA)	Controls expression of genes involved in growth and development; involved in numerous genetic diseases and different types of cancer

was studying unusual, cancer-causing tumor viruses that have RNA, rather than DNA, as their genetic material. He found that infection of a host cell by one of these viruses is blocked by inhibitors of DNA synthesis and also by inhibitors of transcription. These findings suggested that DNA synthesis and transcription are required for RNA tumor viruses to multiply and that there must be a way for information to flow in the "reverse" direction, from RNA to DNA.

Temin proposed that a **DNA provirus** forms as an intermediary in the replication of RNA tumor viruses. This hypothesis required a new kind of enzyme that would synthesize DNA using RNA as a template. In 1970, Temin and U.S. biologist David Baltimore discovered the RNA-directed DNA polymerase, *reverse transcriptase*, is found in all RNA tumor viruses, and they shared the Nobel Prize in Physiology or Medicine in 1975 for their discovery. (Some RNA viruses that do not produce tumors, however, replicate without using a DNA intermediate.) FIGURE 13-23 shows the steps of RNA tumor virus reproduction. Because they reverse the usual direction of information flow, viruses that require reverse transcriptase are called **retroviruses.** The **human immunodeficiency virus, HIV-1,** that causes **AIDS,** the **acquired immunodeficiency syndrome,** is the most widely known retrovirus. Because retrotransposons use reverse transcriptase, some biologists hypothesize that certain retrotransposons may have evolved from retroviruses, or vice versa. As you will see in Chapter 15, the reverse transcriptase enzyme has become an important research tool for molecular biologists.

CHECKPOINT 13.6

- *What is RNA interference?*
- *What is a gene?*
- *How do retroviruses use the enzyme reverse transcriptase?*

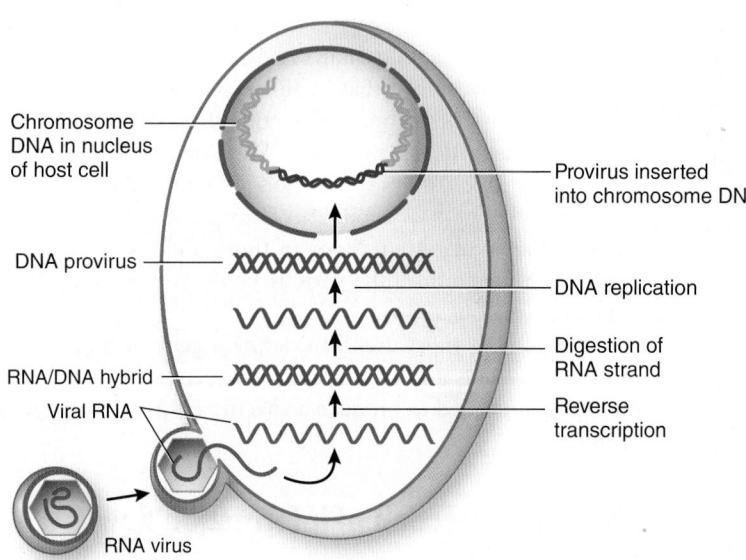

(a) After an RNA tumor virus enters the host cell, the viral enzyme reverse transcriptase synthesizes a DNA strand that is complementary to the viral RNA. Next, the RNA strand is degraded and a complementary DNA strand is synthesized, thus completing the double-stranded DNA provirus, which is then integrated into the host cell's DNA.

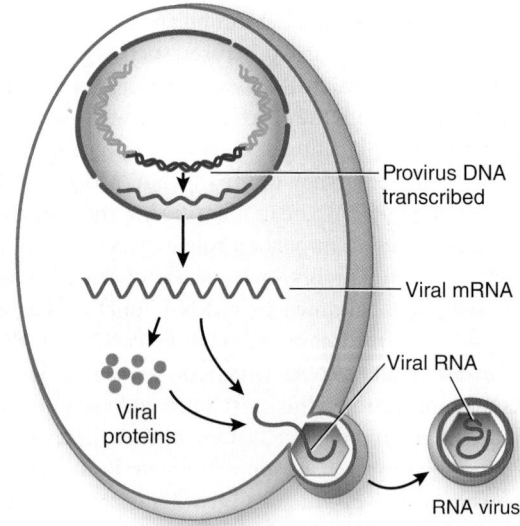

(b) The provirus DNA is transcribed, and the resulting viral mRNA is translated to form viral proteins. Additional viral RNA molecules are produced and then incorporated into mature viral particles enclosed by protein coats.

Figure 13-23 The infection cycle of an RNA tumor virus
© Cengage Learning

13.1 Discovery of the Gene–Protein Relationship *(page 271)*

1 Summarize the early evidence indicating that some genes specify the structure of proteins.

- Garrod's work on inborn errors of metabolism provided evidence that genes specify proteins. Garrod studied a rare genetic disease called *alkaptonuria* and hypothesized that people with alkaptonuria lack the enzyme that normally oxidizes homogentisic acid.

2 Describe how Beadle and Tatum's experiments connected certain genes to enzymes and proteins.

- Beadle and Tatum exposed *Neurospora* spores to X-rays or ultraviolet radiation to induce mutant strains and then identified strains that carried a mutation preventing the fungus from producing a chemical essential for growth. Their work revealed that each mutant strain had a mutation in only one gene and that each gene affected only one enzyme.

13.2 Information Flow from DNA to Protein: An Overview *(page 273)*

3 Outline the flow of genetic information in cells, from DNA to RNA to polypeptide.

- The process by which information encoded in DNA specifies the sequences of amino acids in proteins involves transcription and translation. During **transcription,** an RNA molecule complementary to the template DNA strand is synthesized. **Messenger RNA (mRNA)** molecules contain information that specifies the amino acid sequences of polypeptide chains.

- During **translation,** a polypeptide chain specified by the mRNA is synthesized. Each sequence of three nucleotide bases in the mRNA constitutes a **codon,** which specifies one amino acid in the polypeptide chain, or a start or stop signal. Translation requires tRNAs and cell machinery, including ribosomes.

4 Compare the structures of DNA and RNA.

- Like DNA, RNA is formed from nucleotide subunits. However, in RNA each subunit contains the sugar **ribose,** a base (**uracil,** adenine, guanine, or cytosine), and three phosphates. Like DNA, RNA subunits are covalently joined by 5′–3′ linkages to form an alternating sugar–phosphate backbone.

5 Explain why the genetic code is said to be redundant and virtually universal, and discuss how these features may reflect its evolutionary history.

- The **genetic code** is read from mRNA as a series of codons that specify a single sequence of amino acids. The genetic code is redundant because some amino acids are specified by more than one codon. Only a few minor variations are exceptions to the genetic code found in all organisms, strongly suggesting that all organisms are descended from a common ancestor.

13.3 Transcription *(page 277)*

6 Compare the processes of transcription and DNA replication, identifying both similarities and differences.

- **RNA polymerases,** involved in RNA synthesis, have many similarities to DNA polymerases involved in DNA replication. Both enzymes carry out synthesis in the 5′ ⟶ 3′ direction. Both use nucleotides with three phosphate groups as substrates, removing two of the phosphates as the nucleotides are covalently linked to the 3′ end of the newly synthesized strand.

- Just as the paired strands of DNA are **antiparallel,** the template strand of the DNA and its complementary RNA strand are antiparallel. As a result, the DNA template strand is read in its 3′ ⟶ 5′ direction as RNA is synthesized in its 5′ ⟶ 3′ direction.

- The same base-pairing rules are followed as in DNA replication, except that uracil is substituted for thymine.

7 Compare bacterial and eukaryotic mRNAs, and explain the functional significance of their structural differences.

- Eukaryotic genes and their mRNA molecules are more complicated than those of bacteria. After transcription, a *5′ cap* (a modified guanosine triphosphate) is added to the 5′ end of a eukaryotic mRNA molecule. The molecule also has a **poly-A tail** of adenine-containing nucleotides added at the 3′ end. These modifications may protect eukaryotic mRNA molecules from degradation, giving them a longer lifespan than bacterial mRNA.

- In many eukaryotic genes, the coding regions, called **exons,** are interrupted by noncoding regions, called **introns.** Both exons and introns are transcribed, but the introns are later removed from the original **pre-mRNA** and the exons are spliced together to produce a continuous polypeptide-coding sequence.

13.4 Translation *(page 282)*

8 Identify the features of tRNA that are important in decoding genetic information and converting it into "protein language."

- **Transfer RNAs (tRNAs)** are the "decoding" molecules in the translation process. Each tRNA molecule is specific for only one amino acid. One part of the tRNA molecule contains an **anticodon** that is complementary to a codon of mRNA. Attached to one end of the tRNA molecule is the amino acid specified by the complementary mRNA codon.

9 Explain how ribosomes function in polypeptide synthesis.

- Each **ribosome** is made of a large and a small subunit; each subunit contains **ribosomal RNA (rRNA)** and many proteins. Ribosomes couple the tRNAs to their proper codons on the mRNA, catalyze the formation of peptide bonds between amino acids, and translocate the mRNA so that the next codon can be read.

10 Describe the processes of initiation, elongation, and termination in polypeptide synthesis.

- **Initiation** is the first stage of translation. *Initiation factors* bind to the small ribosomal subunit, which then binds to mRNA in the region of AUG, the **start codon.** The *initiator tRNA* binds to the start codon, followed by binding of the large ribosomal subunit.

- **Elongation** is a cyclic process in which amino acids are added one by one to the growing polypeptide chain. Elongation proceeds in the 5′ ⟶ 3′ direction along the mRNA.

- **Termination,** the final stage of translation, occurs when the ribosome reaches one of three **stop codons.** The A site binds to a *release factor,* which triggers the release of the completed polypeptide chain and dissociation of the translation complex.

11 Describe a polyribosome in bacterial cells.

- Unlike the processes in eukaryotic cells, transcription and translation are coupled in bacterial cells. Bacterial ribosomes bind to the 5′ end of the growing mRNA and initiate translation before the message is fully synthesized. As many as 15 ribosomes may bind to a single mRNA, forming a **polyribosome.**

13.5 Mutations *(page 288)*

12 Give examples of the different classes of mutations that affect the base sequence of DNA in protein encoding genes and explain the effects that each has on the polypeptide produced.

- A **base-pair substitution** is a **mutation** that may alter or destroy the function of a protein if a codon changes so that it specifies a different amino acid (a **missense mutation**) or becomes a stop codon (a **nonsense mutation**).

- The insertion or deletion of one or two base pairs in a gene invariably destroys the function of that protein because it results in a **frameshift mutation,** which changes the codon sequences downstream from the mutation.
- One type of mutation is caused by movable DNA sequences, known as **transposons,** that "jump" into the middle of a gene. Many transposons are **retrotransposons,** which replicate by forming an RNA intermediate; **reverse transcriptase** converts them to their original DNA sequence before they jump into a gene.

13.6 Variations in Gene Expression *(page 290)*

13 Briefly discuss RNA interference.
- **RNA interference (RNAi)** occurs when small RNA molecules, such as **small interfering RNAs (siRNAs)** and **microRNAs**

(miRNAs), interfere with the expression of genes or their RNA transcripts.

14 Compare how retroviruses express genes with the expression of genes found on chromosomes.
- **Retroviruses** are viruses that synthesize DNA from an RNA template. In retroviruses the flow of genetic information is reversed by the enzyme **reverse transcriptase.** HIV-1, the virus that causes AIDS, is a retrovirus.

15 Explain the modern concept of a gene in terms of its information content and function.
- A **gene** is a unit of inheritance. In molecular terms it is a DNA nucleotide sequence that carries information needed to produce a specific RNA or protein product.

TEST YOUR UNDERSTANDING

Know and Comprehend

1. Beadle and Tatum (a) predicted that tRNA molecules would have anticodons (b) discovered the genetic disease alkaptonuria (c) showed that the genetic disease sickle cell anemia is caused by a change in a single amino acid in a hemoglobin polypeptide chain (d) worked out the genetic code (e) studied the relationship between genes and enzymes in *Neurospora*

2. What is the correct order of information flow in bacterial and eukaryotic cells? (a) DNA ⟶ mRNA ⟶ protein (b) protein ⟶ mRNA ⟶ DNA (c) DNA ⟶ protein ⟶ mRNA (d) protein ⟶ DNA ⟶ mRNA (e) mRNA ⟶ protein ⟶ DNA

3. During transcription, how many RNA nucleotide bases would usually be encoded by a sequence of 99 DNA nucleotide bases? (a) 297 (b) 99 (c) 33 (d) 11 (e) answer is impossible to determine with the information given

4. The genetic code is defined as a series of ___ in ___. (a) anticodons; tRNA (b) codons; DNA (c) anticodons; mRNA (d) codons; mRNA (e) codons and anticodons; rRNA

5. RNA differs from DNA in that the base ___ is substituted for ___. (a) adenine; uracil (b) uracil; thymine (c) guanine; uracil (d) cytosine; guanine (e) guanine; adenine

6. RNA grows in the ___ direction as RNA polymerase moves along the template DNA strand in the ___ direction. (a) 5′ ⟶ 3′; 3′ ⟶ 5′ (b) 3′ ⟶ 5′; 3′ ⟶ 5′ (c) 5′ ⟶ 3′; 5′ ⟶ 3′ (d) 3′ ⟶ 3′; 5′ ⟶ 5′ (e) 5′ ⟶ 5′; 3′ ⟶ 3′

7. Which of the following is/are *not* found in a bacterial mRNA molecule? (a) stop codon (b) upstream leader sequences (c) downstream trailing sequences (d) start codon (e) promoter sequences

8. Which of the following is/are typically removed from pre-mRNA during nuclear processing in eukaryotes? (a) upstream leader sequences (b) poly-A tail (c) introns (d) exons (e) all the preceding

9. The role of tRNA is to transport (a) amino acids to the ribosome (b) amino acids to the nucleus (c) initiation factors to the ribosome (d) mRNA to the ribosome (e) release factors to the ribosome

10. Suppose you mix the following components of protein synthesis in a test tube: amino acids from a rabbit, ribosomes from a dog, tRNAs from a mouse, mRNA from a chimpanzee, and necessary enzymes plus an energy source from a giraffe. If protein synthesis occurs, which animal's protein will be made? (a) rabbit (b) dog (c) mouse (d) chimpanzee (e) giraffe

11. The ___ catalyzes the excision of introns from pre-mRNA. (a) ribosome (b) spliceosome (c) RNA polymerase enzyme (d) aminoacyl–tRNA synthetase enzyme (e) reverse transcriptase enzyme

12. A nonsense mutation (a) causes one amino acid to be substituted for another in a polypeptide chain (b) results from the deletion of one or two bases, leading to a shift in the reading frame (c) results from the insertion of one or two bases, leading to a shift in the reading frame (d) results from the insertion of a transposon (e) usually results in the formation of an abnormally short polypeptide chain

Apply and Analyze

13. Compare and contrast the formation of mRNA in bacterial and eukaryotic cells. How do the differences affect the way in which each type of mRNA is translated? Does one system have any obvious advantage in terms of energy cost? Which system offers greater opportunities for control of gene expression?

14. Explain to a friend the experimental strategy that was used to decipher the genetic code.

Evaluate and Synthesize

15. Biologists hypothesize that transposons eventually lose the ability to replicate and therefore remain embedded in DNA without moving around. Based on what you have learned in this chapter, suggest a possible reason for this loss.

16. **VISUALIZE** Sketch a figure to show how reverse transcription is catalyzed by the enzyme reverse transcriptase. Label the nucleic acids involved and produced by the process and draw arrows to indicate the direction of synthesis along the template.

17. **INTERPRET DATA** You are repeating Beadle and Tatum's classic experiment in biology lab. First, you expose *Neurospora* spores to X-rays, establish a culture from an irradiated spore, and then grow it on (1) a minimal medium plus vitamins, (2) a minimal medium plus amino acids, and (3) a minimal medium. Interpret your results if the fungus grows in *all three* media. Interpret your results if the fungus does not grow in *any* of the media.

18. **EVOLUTION LINK** Because introns are present in all eukaryotes and a few prokaryotes, biologists hypothesize that introns evolved very early in the history of life. If that is true, suggest why the majority of prokaryotes living today lost most of their introns during the course of evolution.

To access course materials, such as Aplia and other **aplia** companion resources, please visit **www.cengagebrain.com.**

Gene Regulation

Each type of cell in a multicellular organism has a characteristic shape, carries out very specific activities, and makes a distinct set of proteins. With few exceptions, however, all cells in an organism contain the same genetic information. Why, then, aren't cells identical in structure, molecular composition, and function? Cells differ because gene expression is regulated, and only certain subsets of the total genetic information are expressed in any given cell (see photograph).

What mechanisms control gene expression? Consider a gene coding for a protein that is an enzyme. Expressing that gene involves three basic steps: transcribing the gene to form messenger RNA (mRNA), translating the mRNA into protein, and activating the protein so that it can catalyze a specific reaction in the cell. Thus, gene expression results from a series of processes, each of which is regulated in many ways. The control mechanisms use various signals, some originating within the cell and others coming from other cells or from the environment. These signals interact with DNA, RNA, or protein.

Mechanisms that regulate gene expression include control of the amount of mRNA transcribed, of the rate of translation of mRNA, and of the activity of the protein product. These controls are accomplished in several ways. For example, the rate of transcription and the rate of mRNA degradation both control the amount of available mRNA.

Although bacteria are not multicellular, regulation of gene expression is also essential for their survival. Gene regulation in bacteria often involves controlling the transcription of genes whose products are involved in resource use. In eukaryotes, by contrast, fine-tuning the control systems occurs at *all* levels of gene regulation. Involving all levels of gene regulation is consistent with the greater complexity of eukaryotic cells and the need for developmental controls in multicellular organisms.

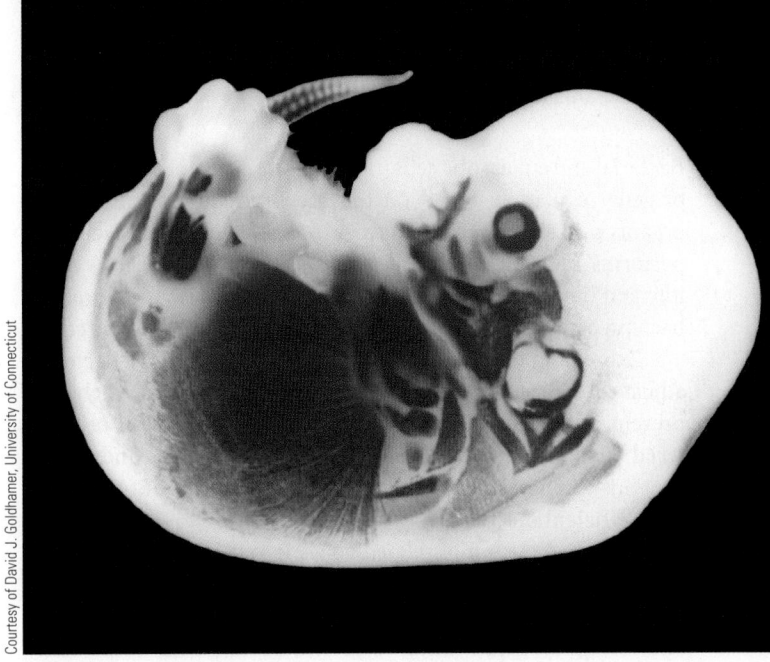

Courtesy of David J. Goldhamer, University of Connecticut

Specific gene expression in a mouse embryo. This 13.5-day-old mouse embryo has been genetically engineered so that cells that transcribe and express the myogenin gene stain blue. The myogenin gene is expressed only in those cells of the embryo that will give rise to muscle tissue. (The gestation period for a mouse is 18 to 20 days.)

KEY CONCEPTS

14.1 Cells regulate which genes will be expressed and when. Gene regulation in bacteria is primarily in response to stimuli in the environment; gene regulation in eukaryotes helps maintain homeostasis and provides for development.

14.2 Gene regulation in bacteria occurs primarily at the level of transcription; regulation of transcription can be either positive or negative.

14.3 Gene regulation in eukaryotes occurs at the levels of transcription, posttranscription, translation, and posttranslation.

14.1 GENE REGULATION IN BACTERIA AND EUKARYOTES: AN OVERVIEW

Bacterial and eukaryotic cells use distinctly different mechanisms of gene regulation, based on the specific requirements of the organism. Bacterial cells can exist independently, and each cell performs all its own essential functions. Because they grow rapidly and have relatively short times between cell division events, bacteria carry fewer chemical components than eukaryotic cells.

Expression of bacterial genes occurs when it helps bacteria adjust to their environment, including any food that might be present in the surroundings. The primary requirement of bacterial gene regulation is the production of enzymes and other proteins when needed, and **transcriptional-level control** is the most efficient mechanism.

The organization of related genes into groups that are rapidly turned on and off as units allows the synthesis of only the gene products needed at any particular time. This type of regulation requires rapid turnover of mRNA molecules to prevent messages from accumulating and continuing to be translated when they are not needed. Bacteria rarely regulate enzyme levels by degrading proteins. Once the synthesis of a protein ends, the previously synthesized protein molecules are diluted so rapidly in subsequent cell divisions that breaking them down is usually not necessary. Only when cells are starved or deprived of essential amino acids do protein-digesting enzymes recycle amino acids by breaking down proteins no longer needed for survival.

Eukaryotic cells have different regulatory requirements. Although transcriptional-level control predominates, control at other levels of gene expression is also very important, especially in multicellular organisms in which groups of cells cooperate with one another in a division of labor. Because a single gene is regulated in different ways in different types of cells, eukaryotic gene regulation is complex.

Eukaryotic cells usually have a long lifespan during which they may need to maintain homeostasis while responding to many different stimuli. New enzymes are not necessarily synthesized each time the cells respond to a stimulus; instead, in many instances preformed enzymes and other proteins are rapidly converted from an inactive to an active state. Some cells have a large store of inactive messenger RNA; for example, the mRNA of an egg cell becomes activated when the egg is fertilized.

Much gene regulation in multicellular organisms is focused on the differential expression of genes in the cells in various tissues. Each type of cell has certain genes that are active and other genes that may never be used. For example, developing red blood cells produce the oxygen transport protein hemoglobin; muscle cells, on the other hand, never produce hemoglobin but instead produce myoglobin, a related protein coded for by a different gene that stores oxygen in muscle tissues. Apparently, the

Figure 14-1 *Animation* **A lean pig with a mutation in the *IGF2* gene**

Years ago, farmers discovered a pig with more muscle and less fat. They selectively bred this pig, and today most commercial pig populations have the mutation that confers this trait. The mutation, which has been identified as a substitution of a guanine nucleotide for an adenine nucleotide, occurs in a non-protein-coding region of the *IGF2* gene.

selective advantages of cell cooperation in multicellular eukaryotes far outweigh the detrimental effects of carrying a load of inactive genes through many cell divisions.

The domesticated pig provides another example of how gene regulation in multicellular organisms affects expression in different tissues. A genetic mutation in pigs makes them develop more muscle tissue (**FIG. 14-1**). The mutation is in a gene, designated **insulin-like growth factor 2 (*IGF2*),** which codes for a protein produced by both muscle and liver tissues. The biologists found that a *base-substitution mutation* (see Chapter 13) in the *IGF2* gene makes the gene three times more active in pig muscles, resulting in leaner meat. However, the mutation does not change expression of the gene in liver cells. Interestingly, this mutation is not in the protein-coding portion of the gene; instead it is a regulatory mutation located in an intron.

CHECKPOINT 14.1

- *What is the most efficient mechanism of gene regulation in bacteria?*
- **CONNECT** *How would you contrast gene regulation in prokaryotes with that in eukaryotes?*

14.2 GENE REGULATION IN BACTERIA

The *Escherichia coli* bacterium is common in the intestines of humans and other mammals. It has 4288 genes that code for proteins, approximately 60% of which have known functions. Some of these genes encode proteins that are always needed (such as enzymes involved in glycolysis). These genes, which are constantly transcribed, are **constitutive genes,** and biologists describe them as *constitutively expressed.* Other proteins are needed only when the bacterium is growing under certain conditions.

For instance, bacteria living in the colon of an adult cow are not normally exposed to the milk sugar lactose, a disaccharide. If those cells ended up in the colon of a calf, however, they would have lactose available as a source of energy. This situation presents a dilemma. Should a bacterial cell invest energy and materials to produce lactose-metabolizing enzymes just in case it ends up in the digestive system of a calf? If *E. coli* cells could not produce those enzymes, however, they might starve in the middle of an abundant food supply. Thus, *E. coli* functions by regulating many enzymes to use available organic molecules efficiently.

Cell metabolic activity is controlled in two ways. One is by regulating the *activity* of certain enzymes (how effectively an enzyme molecule works), and the other is by regulating the *number* of enzyme molecules present in each cell. Some enzymes are regulated in both ways.

An *E. coli* cell growing on glucose needs about 800 different enzymes. Some are present in large numbers, whereas only a few molecules of others are required. For the cell to function properly, the quantity of each enzyme must be efficiently controlled.

Bacteria respond to changing environmental conditions. If lactose is added to a culture of *E. coli* cells, they rapidly synthesize the three enzymes needed to metabolize lactose. Bacteria respond so efficiently because functionally related genes—such as the three genes involved in lactose metabolism—are regulated together in gene complexes called *operons.*

Operons in bacteria facilitate the coordinated control of functionally related genes

French researchers François Jacob and Jacques Monod are credited with the first demonstration, in 1961, of gene regulation at the biochemical level through their elegant studies on the genes that code for the enzymes that metabolize lactose. In 1965, they received the Nobel Prize in Physiology or Medicine for their discoveries relating to genetic control of enzymes.

To use lactose as an energy source, *E. coli* cells first cleave the sugar into the monosaccharides glucose and galactose, using the enzyme β-galactosidase. Another enzyme converts galactose to glucose, and enzymes in the glycolysis pathway further break down the resulting two glucose molecules (see Chapter 8).

E. coli cells growing on glucose produce very little β-galactosidase enzyme. However, each cell grown on lactose as the sole carbon source has several thousand β-galactosidase molecules, accounting for about 3% of the cell's total protein. Amounts of two other enzymes, lactose permease and galactoside transacetylase, also increase when the cells are grown on lactose. The cell needs permease to transport lactose efficiently across the bacterial plasma membrane; without it, only small amounts of lactose enter the cell. The transacetylase may function in a minor aspect of lactose metabolism, although its role is not clear.

Jacob and Monod identified mutant strains of *E. coli* in which a single genetic defect wiped out all three enzymes. This finding, along with other information, led the researchers to conclude that the DNA coding sequences for all three enzymes are linked as a unit on the bacterial DNA and are controlled by a common mechanism. Each enzyme-coding sequence is a *structural gene.* Jacob and Monod coined the term **operon** for a gene complex consisting of a group of structural genes with related functions plus the closely linked DNA sequences responsible for controlling them. The structural genes of the *lac* operon (lactose operon)—*lacZ, lacY,* and *lacA*—code for β-galactosidase, lactose permease, and transacetylase, respectively (**FIG. 14-2**).

Transcription of the *lac* operon begins as RNA polymerase binds to a single **promoter** region *upstream* from the coding sequences. It then proceeds to transcribe the DNA, forming a single mRNA molecule that contains the coding information for all three enzymes. Each enzyme-coding sequence on this mRNA contains its own start and stop codons; thus, the mRNA is translated to form three separate polypeptides. Because all three enzymes are translated from the same mRNA molecule, their synthesis is coordinated by turning a single molecular "switch" on or off.

The switch that controls mRNA synthesis is the **operator,** which is a sequence of bases upstream from the first structural gene in the operon. In the absence of lactose, a **repressor protein** called the *lactose repressor* binds tightly to the operator. RNA polymerase binds to the promoter but is blocked from transcribing the protein-coding genes of the *lac* operon.

The lactose repressor protein is encoded by a *repressor gene,* which in this case is a protein-coding gene located upstream from the promoter site. Unlike the *lac* operon genes, the repressor gene is constitutively expressed and is therefore constantly transcribed; the cell continuously produces small amounts of the repressor protein.

The repressor protein binds specifically to the *lac* operator sequence. When cells grow in the absence of lactose, a repressor molecule nearly always occupies the operator site. When the operator site is briefly free of the repressor, the cell synthesizes a small amount of mRNA. However, the cell synthesizes very few enzyme molecules because *E. coli* mRNA is degraded rapidly (it has a half-life of about 2 to 4 minutes).

Lactose "turns on," or *induces,* the transcription of the *lac* operon because the lactose repressor protein contains a second functional region (an *allosteric site;* see Chapter 7) separate from its DNA-binding site. This allosteric site binds to allolactose, a structural isomer made from lactose. If lactose is in the growth medium, a few molecules enter the cell and are converted to allolactose by the few β-galactosidase molecules present. The binding of a molecule of allolactose to the repressor alters the shape of the protein so that its DNA-binding site no longer recognizes the operator. When the repressor molecules have allolactose bound to them and are therefore inactivated, RNA polymerase actively transcribes the structural genes of the operon.

The *E. coli* cell continues to produce β-galactosidase and the other *lac* operon enzymes until it uses up virtually all the lactose. When the intracellular level of lactose drops, so too does

The structural genes of the *lac* operon are coordinately controlled and transcribed as a single mRNA.

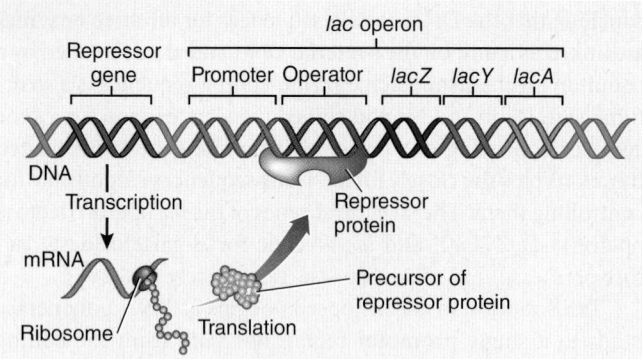

(a) Lactose absent. In the absence of lactose, a repressor protein, encoded by another gene, binds to a region known as the operator, thereby blocking transcription of the structural genes.

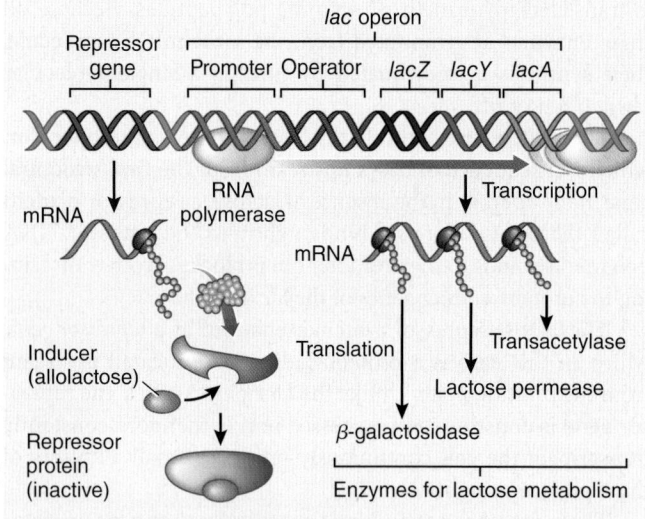

(b) Lactose present. When lactose is present, it is converted to allolactose, which binds to the repressor at an allosteric site, altering the structure of the protein so it no longer binds to the operator. As a result, RNA polymerase is able to transcribe the structural genes.

Figure 14-2 The *lac* operon

PREDICT What would be the effect on *lac* operon activity if the gene encoding the *lac* repressor were deleted from the DNA?

the concentration of allolactose. Therefore, allolactose dissociates from the repressor protein, which then assumes a shape that lets the repressor bind to the operator region and shut down transcription of the operon.

Jacob and Monod isolated genetic mutants to study the *lac* operon How did Jacob and Monod elucidate the functioning of the *lac* operon? Their approach involved the use of mutant strains, which even today are an essential tool of researchers trying to unravel the components of various regulatory systems.

Genetic crosses of mutant strains allow investigators to determine the *map positions* (linear order) of the genes on the DNA and to infer normal gene functions by studying what happens when they are missing or altered. Researchers usually combine this information with the results of direct biochemical studies.

Jacob and Monod divided their mutant strains into two groups, based on whether a particular mutation affected only one enzyme or all three (**FIG. 14-3**). In one group, only one enzyme of the three—β-galactosidase, lactose permease, or galactoside transacetylase—was affected. Subsequent gene-mapping studies showed that these were mutations in structural genes located next to each other in a linear sequence.

Jacob and Monod also studied regulatory mutants in which a single mutation affected the expression of all three enzymes. Some of the regulatory mutants transcribed the structural genes of the *lac* operon at a significant rate, even in the absence of lactose, causing the cell to waste energy in producing unneeded enzymes. One group of these gene mutations had map positions just outside the *lac* operon. Jacob and Monod hypothesized the existence of a repressor gene that codes for a repressor molecule (later found to be a protein). Although the specific defect may vary, the members of this group of mutants produce defective repressors; hence, no binding to the *lac* operator and promoter takes place, and the *lac* operon is constitutively expressed.

In contrast, some regulatory mutants failed to transcribe the *lac* operon even when lactose was present. Researchers eventually found that each of these abnormal genes coded for a repressor with an abnormal binding site that prevented allolactose from binding, although the repressor could still bind to the operator. Once bound to the operator, such a mutant repressor remained bound, keeping the operon "turned off."

The genes responsible for another group of regulatory mutants had map positions within the *lac* operon but did not directly involve any of the three structural genes. Jacob and Monod hypothesized that the members of this group produced normal repressor molecules but had abnormal operator sequences incapable of binding to the repressor.

An inducible gene is not transcribed unless a specific inducer inactivates its repressor Geneticists call the *lac* operon an **inducible operon**. A repressor usually controls an inducible gene or operon by keeping it "turned off." The presence of an **inducer** (in this case, allolactose) inactivates the repressor, permitting the gene or operon to be transcribed. Inducible genes or operons usually code for enzymes that are part of catabolic pathways, which break down molecules to provide both energy and components for anabolic reactions. This type of regulatory system lets the cell save the energy cost of making enzymes when no substrates are available on which they can act.

A repressible gene is transcribed unless a specific repressor–corepressor complex is bound to the DNA Another type of gene regulation system in bacteria is associated mainly with anabolic pathways such as those in which cells synthesize amino acids, nucleotides, and other essential biological molecules from simpler materials. Enzymes coded by repressible genes normally regulate these pathways.

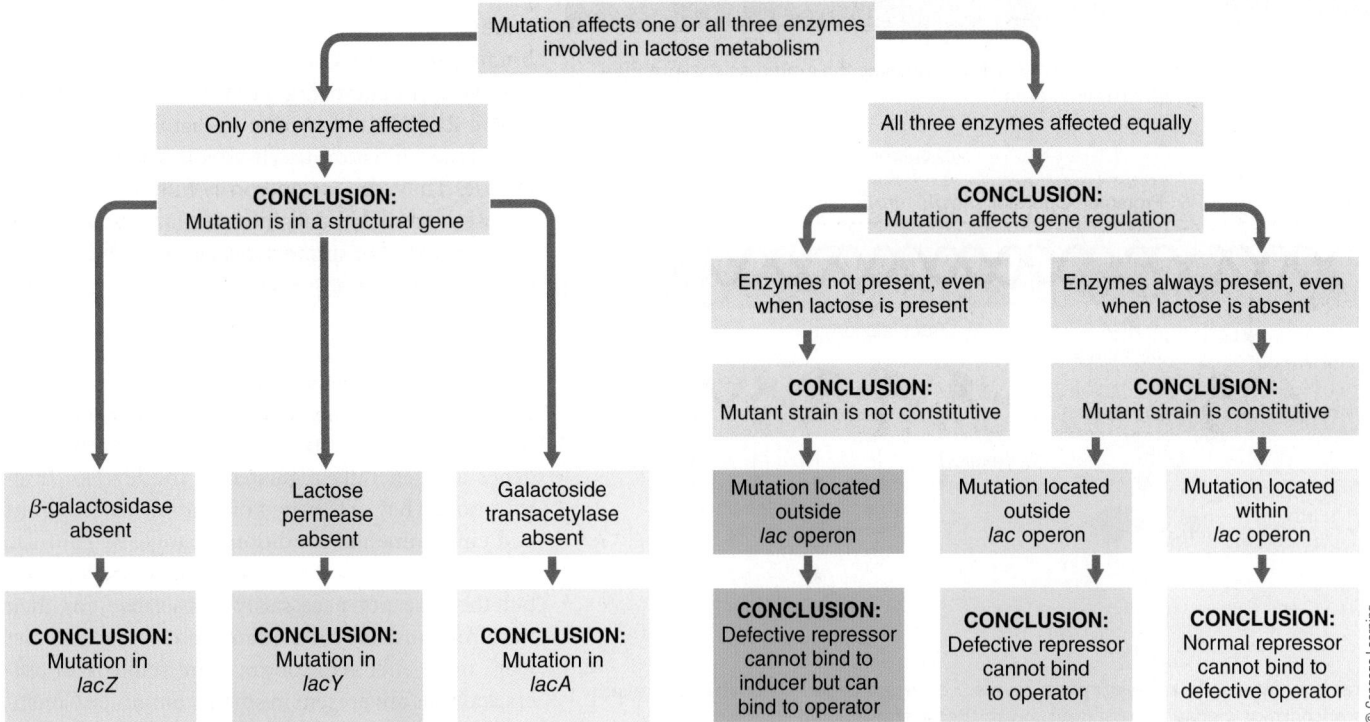

Figure 14-3 Genetic and biochemical characterization of the *lac* operon

Jacob and Monod analyzed the properties of various mutant strains of *E. coli* to deduce the structure and function of the *lac* operon. This diagram shows how Jacob and Monod inferred conclusions based on their experimental results.

Repressible operons and genes are usually "turned on"; they are turned off only under certain conditions. In most cases, the molecular signal for regulating these genes is the end product of the anabolic pathway. When the supply of the end product (such as an amino acid) becomes low, all enzymes in the pathway are actively synthesized. When intracellular levels of the end product rise, enzyme synthesis is repressed. Because the growing cell continuously needs compounds such as amino acids, the most effective mechanism for the cell is to keep the genes that control their production turned on except when a large supply of the amino acid is available. The ability to turn the genes off prevents cells from overproducing amino acids and other molecules that are essential but expensive to make in terms of energy.

The *trp* operon (tryptophan operon) is an example of a repressible system. In both *E. coli* and a related bacterium, *Salmonella*, the *trp* operon consists of five structural genes that code for the enzymes the cell needs to synthesize the amino acid tryptophan; these genes are clustered as a transcriptional unit with a single promoter and a single operator (**FIG. 14-4a**). A distant repressor gene codes for a diffusible repressor protein, which differs from the lactose repressor in that the cell synthesizes it in an inactive form that cannot bind to the operator region of the *trp* operon.

The DNA-binding site of the repressor becomes effective only when tryptophan, its **corepressor,** binds to an allosteric site on the repressor (**FIG. 14-4b**). When intracellular tryptophan levels are low, the repressor protein is inactive and cannot bind

to the operator region of the DNA. The enzymes required for tryptophan synthesis are produced, and the intracellular concentration of tryptophan increases. Some of it binds to the repressor, altering the repressor's shape so that it binds tightly to the operator and switches the operon off, thereby blocking transcription.

Negative regulators inhibit transcription; positive regulators stimulate transcription The features of the *lac* and *trp* operons we have described so far are examples of **negative control,** a regulatory mechanism in which the DNA binding regulatory protein is a repressor that turns off transcription of the gene. **Positive control** is regulation by **activator proteins** that bind to DNA and thereby stimulate the transcription of a gene. The *lac* operon is controlled by both a negative regulator (the lactose repressor) and a positively acting activator protein.

Positive control of the *lac* operon requires the cell to recognize the absence of the sugar glucose, which is the initial substrate in the glycolysis pathway. Lactose, like glucose, undergoes stepwise breakdown to yield energy. However, because glucose is a product of the catabolic hydrolysis of lactose, it is most efficient for *E. coli* cells to use the available supply of glucose first, sparing the cell the considerable energy cost of making additional enzymes such as β-galactosidase (**FIG. 14-5a**).

The *lac* operon has a very inefficient promoter element; that is, it has a low affinity for RNA polymerase, even when the repressor protein is inactivated. However, a DNA sequence adjacent to the promoter site is a binding site for another regulatory

Structural genes coding for enzymes that synthesize the amino acid tryptophan are organized in a repressible operon.

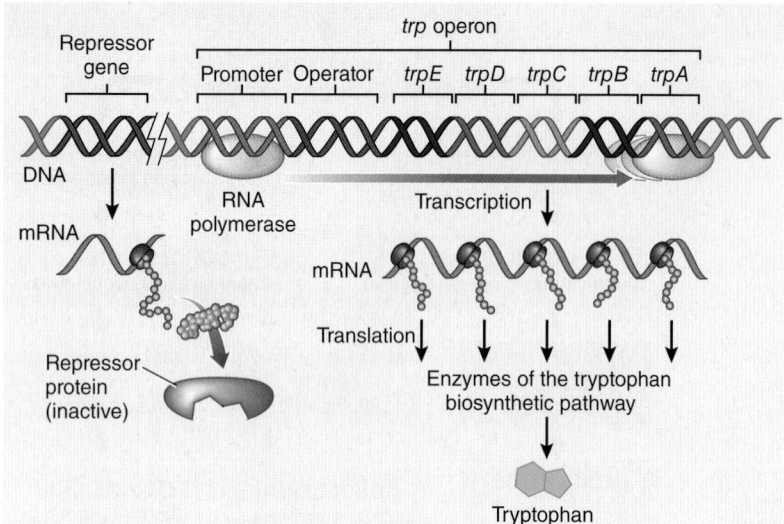

(a) Low intracellular tryptophan levels. Repressor protein is unable to prevent transcription because it cannot bind to the operator.

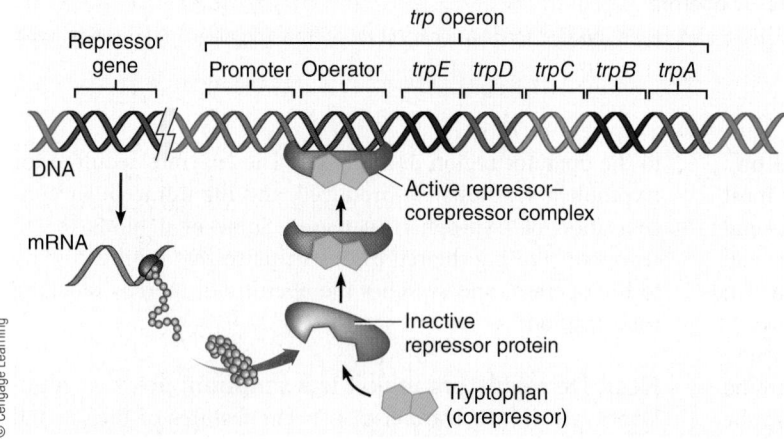

(b) High intracellular tryptophan levels. The amino acid tryptophan binds to an allosteric site on the repressor protein, changing its conformation. The resulting active form of the repressor binds to the operator region, blocking transcription of the operon until tryptophan is again required by the cell.

Figure 14-4 The *trp* operon

PREDICT What would be the effect on transcription of the *trp* operon if the *trp* operator region were mutated or deleted?

protein, the **catabolite activator protein (CAP).** When activated, CAP stimulates transcription of the *lac* operon and several other bacterial operons.

In its active form, CAP is bound by an allosteric site to **cyclic AMP,** or **cAMP,** an alternative form of adenosine monophosphate. Examine Figure 3-26 to see the chemical structure of cAMP. As the cells become depleted of glucose, cAMP levels increase (**FIG. 14-5b**). After the cAMP molecules bind to CAP,

the resulting active complex then binds to the CAP-binding site near the *lac* operon promoter. This binding of active CAP bends DNA's double helix (**FIG. 14-6**), strengthening the affinity of the promoter region for RNA polymerase so that the rate of transcriptional initiation accelerates in the presence of lactose. Thus, the *lac* operon is fully active only if lactose is available and intracellular glucose levels are low. **TABLE 14-1** summarizes negative and positive controls in bacteria.

Constitutive genes are transcribed at different rates Many gene products encoded by *E. coli* DNA are needed only under certain environmental or nutritional conditions. As we have seen, these genes are generally regulated at the level of transcription. They are turned on and off as metabolic and environmental conditions change. By contrast, constitutive genes are continuously transcribed, but they are not necessarily transcribed (or their mRNAs translated) at the same rate. Some enzymes work more effectively or are more stable than others and thus are present in smaller amounts. Constitutive genes that encode proteins required in large amounts generally have greater expression—that is, are transcribed more rapidly—than genes coding for proteins required in smaller amounts. The promoter elements of these genes control their transcriptional rate. Constitutive genes with efficient ("strong") promoters bind RNA polymerase more frequently and consequently transcribe more mRNA molecules than those with inefficient ("weak") promoters.

Genes coding for repressor or activator proteins that regulate metabolic enzymes are usually constitutive and produce their protein products constantly. Because each cell usually needs relatively few molecules of any specific repressor or activator protein, promoters for those genes tend to be relatively weak.

Some posttranscriptional regulation occurs in bacteria

As you have seen, much of the variability in protein levels in *E. coli* is determined by transcriptional-level control. However, regulatory mechanisms after transcription, known as **posttranscriptional controls,** also occur at various levels of gene expression.

Translational controls are posttranscriptional controls that regulate the rate at which an mRNA molecule is translated. Because the lifespan of an mRNA molecule in a bacterial cell is very short, one that is translated rapidly produces more proteins than one that is translated slowly. Some mRNA molecules in *E. coli* are translated as much as 1000 times faster than other mRNAs. Most of the differences appear to be due to the rate at which ribosomes attach to the mRNA and begin translation.

Posttranslational controls act as switches that activate or inactivate one or more existing enzymes, thereby letting the cell respond to changes in the intracellular concentrations of essential molecules, such as amino acids. A common posttranslational control adjusts the rate of synthesis in a metabolic pathway through **feedback inhibition.** An example of feedback inhibition is diagrammed in Figure 7-15. The end product of a metabolic pathway binds to an allosteric site on the first enzyme in the pathway, temporarily inactivating the enzyme. When the first enzyme in the pathway does not function, all the succeeding enzymes are deprived of substrates. Notice that feedback inhibition differs from the repression caused by tryptophan discussed previously. In that case, the end product of the pathway (tryptophan) prevented the formation of *new* enzymes. Feedback inhibition acts as a fine-tuning mechanism that regulates the activity of the *existing* enzymes in a metabolic pathway.

CHECKPOINT 14.2

- *What is the function of each of the parts of the* lac *operon (promoter, operator, and structural genes)?*
- CONNECT *What structural features does the* trp *operon share with the* lac *operon?*
- *Why do scientists define the* trp *operon as repressible and the* lac *operon as inducible?*
- *How is glucose involved in positive control of the* lac *operon?*

14.3 GENE REGULATION IN EUKARYOTIC CELLS

LEARNING OBJECTIVES

6 Discuss the structure of a typical eukaryotic gene and the DNA elements involved in regulating that gene.

7 Give examples of some of the ways eukaryotic DNA-binding proteins bind to DNA.

8 Illustrate how a change in chromosome structure may affect the activity of a gene.

9 Explain how a gene in a multicellular organism may produce different products in different types of cells.

10 Identify some of the types of regulatory controls that operate in eukaryotes after mature mRNA is formed.

Like bacterial cells, eukaryotic cells respond to changes in their environment. In addition, multicellular eukaryotes require regulation modes that let individual cells commit to specialized roles and let groups of cells organize into tissues and organs. These processes are accomplished mainly by gene expression at multiple levels, involving transcriptional, posttranscriptional, translational, and posttranslational controls. In Chapters 12 and 13, we observed that

KEY POINT

The lactose promoter by itself is weak and binds RNA polymerase inefficiently even when the lactose repressor is inactive.

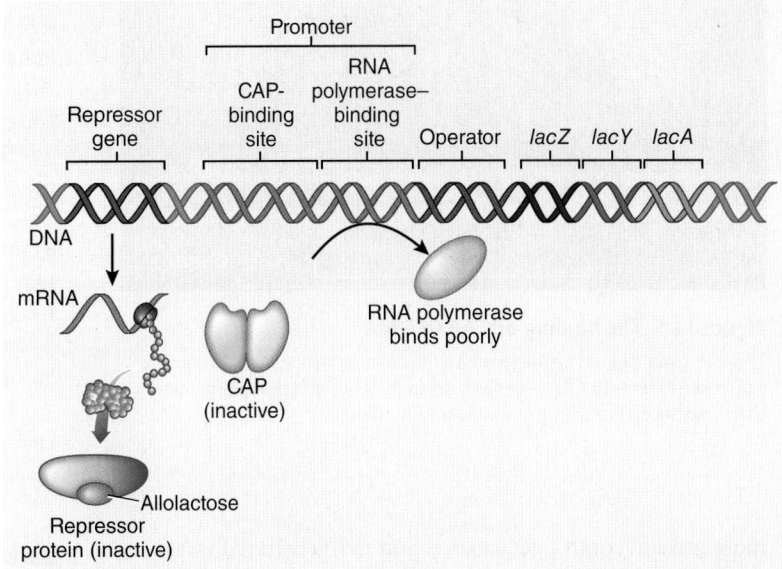

(a) Lactose high, glucose high, cAMP low. When glucose levels are high, cAMP is low. CAP is in an inactive form and cannot stimulate transcription. Transcription occurs at a low level or not at all.

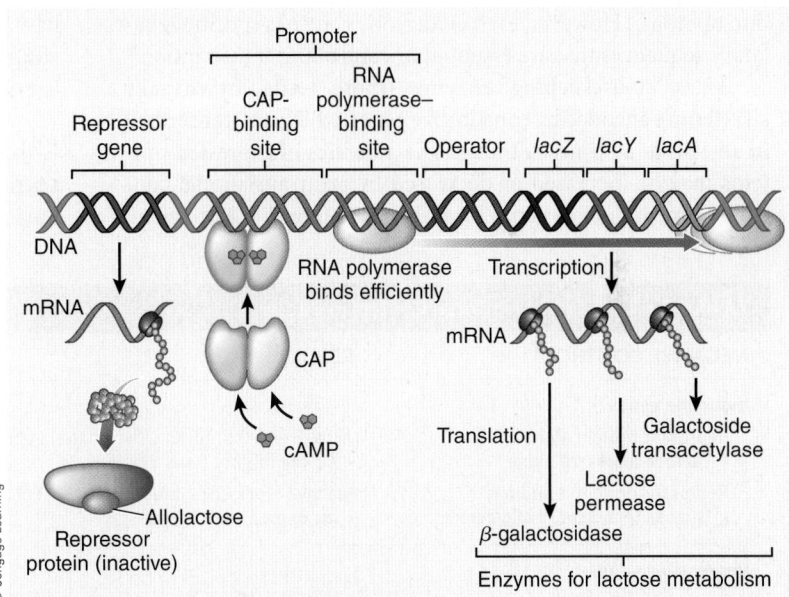

(b) Lactose high, glucose low, cAMP high. When glucose concentrations are low, each CAP polypeptide has cAMP bound to its allosteric site. The active form of CAP binds to the DNA sequence, and transcription becomes activated.

Figure 14-5 Positive control of the *lac* operon

PREDICT What would be the effect on *lac* operon transcription rates in a cell that contains a mutation that activates adenyl cyclase in the absence of glucose?

in eukaryotes all aspects of information transfer—including replication, transcription, and translation—are more complicated than in prokaryotes. Not surprisingly, this complexity offers

© Cengage Learning

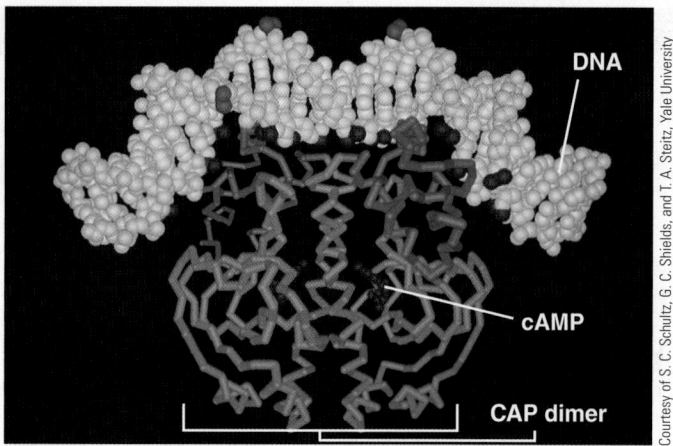

Figure 14-6 The binding of CAP to DNA

This computer-generated image shows the bend formed in the DNA double helix when it binds to CAP, a dimer consisting of two identical polypeptide chains, each of which binds one molecule of cAMP.

Courtesy of S. C. Schultz, G. C. Shields, and T. A. Steitz, Yale University

to environmental threats or stimuli, such as heavy-metal ingestion, viral infection, and heat shock. For example, when a cell is exposed to high temperature, many proteins fail to fold properly. These unfolded proteins elicit a survival response in which *heat-shock genes* are transcribed and heat-shock proteins are synthesized. Although the functions of most heat-shock proteins in eukaryotic cells are unknown, some are **molecular chaperones** that help newly synthesized proteins fold into their proper shape (discussed in Chapters 3 and 13).

Some genes are expressed only during certain periods in the life of the organism; they are controlled by *temporal regulation.* Finally, some genes are under *tissue-specific regulation.* For example, a gene involved in production of a particular enzyme may be regulated by one stimulus (such as a hormone) in muscle tissue, by an entirely different stimulus in pancreatic cells, and by a third stimulus in liver cells. We explore these types of regulation in more detail in Chapter 17.

FIGURE 14-7 summarizes the regulation of gene expression in eukaryotes. You may want to refer to it as you read the following discussion.

more ways to control gene expression in finely tuned ways that can be rapidly reversed as the cell encounters new conditions.

Unlike many bacterial genes, eukaryotic genes are not typically arranged in operon-like clusters. (A notable exception is the nematodes, or roundworms, in which some genes are organized into operons.) However, each eukaryotic gene has specific regulatory sequences that are essential in controlling transcription.

Many "housekeeping" enzymes (those needed by virtually all cells) are encoded by constitutive genes, which are transcribed in all cells at all times, although the activities of expressed proteins may be increased or decreased by posttranslational controls. Some inducible genes have also been found; they respond

Eukaryotic transcription is controlled at many sites and by many regulatory molecules

Like bacterial genes, most genes of multicellular eukaryotes are controlled at the transcriptional level. As you will see, various base sequences in the DNA are important in transcriptional control. In addition, regulatory proteins and the way the DNA is organized in the chromosome affect the rate of transcription.

Chromosome organization affects the expression of some genes A chromosome is not only a bearer of genes. Various arrangements of a chromosome's ordered components

TABLE 14-1	Transcriptional Control in Bacteria*
NEGATIVE CONTROL	**RESULT**
Inducible genes	
Repressor protein alone	Active repressor turns off regulated gene(s)
• Lactose repressor alone	• *lac* operon not transcribed
Repressor protein + inducer	Inactive repressor–inducer complex fails to turn off regulated gene(s)
• Lactose repressor + allolactose	• *lac* operon transcribed
Repressible genes	
Repressor protein alone	Inactive repressor fails to turn off regulated gene(s)
• Tryptophan repressor alone	• *trp* operon transcribed
Repressor protein + corepressor	Active repressor–corepressor complex turns off regulated gene(s)
• Tryptophan repressor + tryptophan	• *trp* operon not transcribed
POSITIVE CONTROL	**RESULT**
Activator protein alone	Activator alone cannot stimulate transcription of regulated gene(s)
• CAP alone	• Transcription of *lac* operon not stimulated
Activator complex	Functional activator complex stimulates transcription of regulated gene(s)
• CAP + cAMP	• Transcription of *lac* operon stimulated

*Note: A general description of each type is followed by a specific example.

© Cengage Learning

increase or decrease expression of the genes it contains. Figure 10-4 shows the organization of a eukaryotic chromosome. In multicellular eukaryotes only a subset of the genes in a cell is active at any one time. The inactivated genes differ among cell types and in many cases are irreversibly dormant.

Some of the inactive genes lie in highly compacted chromatin, visible microscopically as densely stained regions of chromosomes during cell division. These regions of chromatin remain tightly coiled and bound to chromosome proteins throughout the cell cycle; even during interphase, they are visible as darkly stained fibers called **heterochromatin** (FIG. 14-8a). Evidence suggests that most of the heterochromatin DNA is not transcribed. For example, most of the inactive X chromosome of the two X chromosomes in female mammals is heterochromatic and appears as a *Barr body* (see Fig. 11-16).

Active genes are associated with a more loosely packed chromatin structure called **euchromatin** (FIG. 14-8b). The exposure of the DNA in euchromatin lets it interact with *transcription factors* (discussed later in this chapter) and other regulatory proteins.

Chemical changes to histones alter chromatin structure
Cells may have several ways to change chromatin structure from heterochromatin to euchromatin. One way is to chemically modify **histones,** the proteins that associate with DNA to form **nucleosomes.** Each histone molecule has a so-called tail, a string of amino acids that extends from the DNA-wrapped nucleosome (see Fig. 10-2). Methyl groups, acetyl groups, sugars, and even proteins chemically attach to the histone tail and may expose or hide genes, turning them on or off. Nucleosomes of genes that are actively transcribed typically show a methylation pattern associated with the protein known as histone H3. If this histone is part of a nucleosome associated with an active promoter region, lysine 4 of the polypeptide chain is usually trimethylated. However, lysine 36 of this histone is trimethylated if it is part of a nucleosome associated with the transcribed region of the same gene. These chromatin signatures are now widely used to detect the presence of actively transcribed genes.

Such chemical modifications of histones are the focus of active research because they appear to influence gene expression and therefore offer promise of new approaches to treat cancer and other diseases with a genetic component. Biologists have found general correlations between gene activity and certain chemical groups attached to or absent from histone tails. For example, genes that are associated with nucleosomes with histone tails that are flagged with acetyl groups generally tend to be expressed.

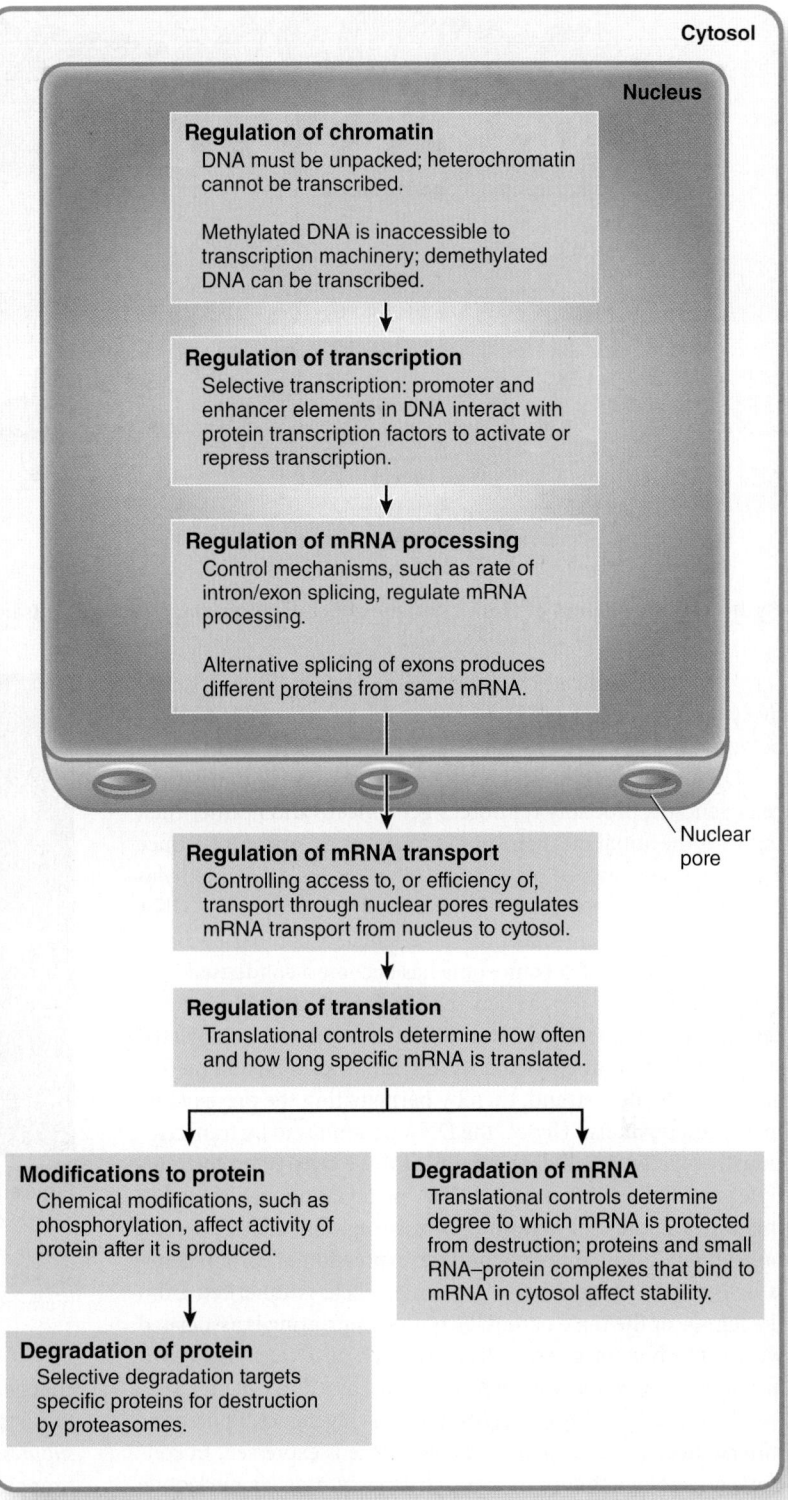

Figure 14-7 Gene regulation in eukaryotic cells
© Cengage Learning

Gene inactivation by DNA methylation is an example of epigenetic inheritance
Inactive genes of vertebrates and some other organisms typically show a pattern of **DNA methylation** in which the DNA has been chemically altered by enzymes that add methyl groups to certain cytosine nucleotides in DNA. (The resulting 5-methylcytosine still pairs with guanine in the usual

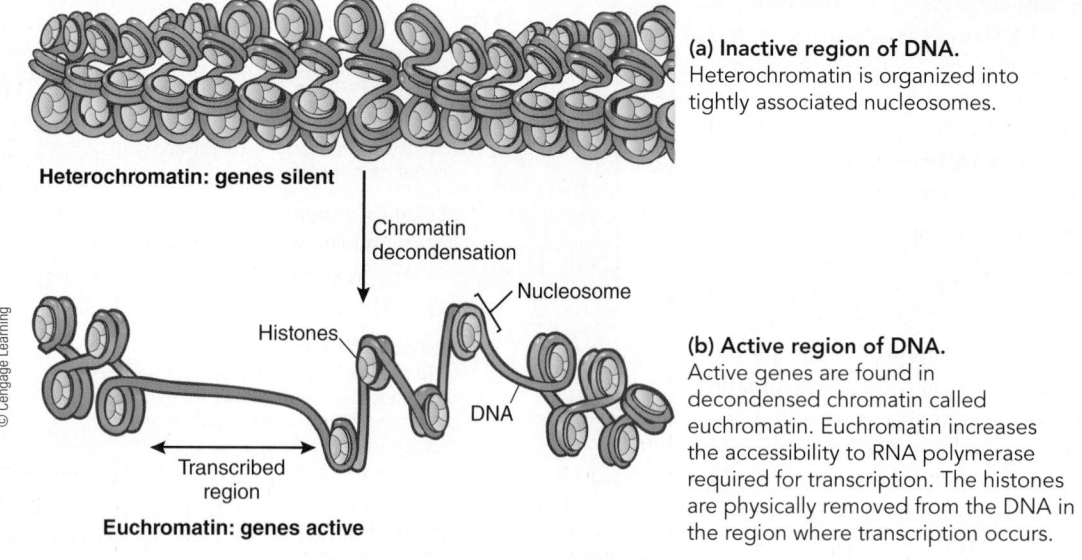

Heterochromatin: genes silent

(a) Inactive region of DNA.
Heterochromatin is organized into tightly associated nucleosomes.

Chromatin decondensation

Nucleosome

Histones

DNA

Transcribed region

Euchromatin: genes active

(b) Active region of DNA.
Active genes are found in decondensed chromatin called euchromatin. Euchromatin increases the accessibility to RNA polymerase required for transcription. The histones are physically removed from the DNA in the region where transcription occurs.

Figure 14-8 Regulation of chromatin: the effect of chromatin structure on transcription

way.) Evidence indicates that certain regulatory proteins selectively bind to methylated DNA and make it inaccessible to RNA polymerase and other proteins involved in transcription.

DNA methylation, which is found in inactive regions of the cell's genome, probably reinforces gene inactivation rather than serves as the initial mechanism to silence genes. Apparently, once a gene has been turned off by some other means, DNA methylation ensures that it will remain inactive. For example, the DNA of the inactive X chromosome of a female mammal becomes methylated after the chromosome has become a condensed Barr body (see Chapter 11). When the DNA replicates in a mitotic cell division, each double strand of DNA has one methylated strand and one unmethylated strand. Enzymes then add methyl groups to the new strand, thereby perpetuating the pre-existing methylation pattern. Hence, the DNA continues to be transcriptionally inactive in all descendants of these cells, often for multiple generations. In mammals DNA methylation maintained in this way accounts for **genomic imprinting** (also called *parental imprinting*) in which the phenotypic expression of certain genes is determined by whether a particular allele is inherited from the female or the male parent. Genomic imprinting is associated with some human genetic diseases, such as *Prader–Willi syndrome* and *Angelman syndrome* (both discussed in Chapter 16).

Genomic imprinting is an example of epigenetics. **Epigenetic inheritance** involves changes in how a gene is expressed. In contrast, *genetic inheritance* involves changes in a gene's nucleotide sequence. Because DNA methylation patterns tend to be passed to successive cell generations during cell division, methylation provides a mechanism for epigenetic inheritance. Evidence continues to accumulate that epigenetic inheritance is an important mechanism of gene regulation over the lifetime of an individual. For example, the methylation of genes increases in many people as they age and decreases in others, perhaps in response to signals from the environment.

New phenotypic traits can arise from epigenetic changes even though the nucleotide sequence of the gene itself has not

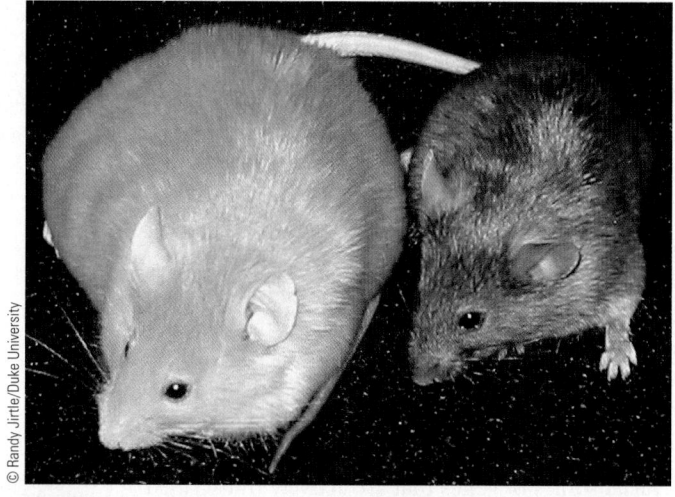

Figure 14-9 Epigenetic variation in the *agouti* gene in mice
The DNA of these mice is almost identical, yet their coat colors and body shapes vary significantly. These mice differ in the presence or absence of epigenetic "marks," such as methyl groups on DNA in the region of a gene known as *agouti*.

changed (**FIG. 14-9**). Epigenetic changes have also been linked to certain human diseases, such as cancer. For example, *tumor suppressor genes* that inhibit cell division (and cancer) are sometimes inactivated epigenetically, resulting in cancer. Chapter 17 discusses cancer and tumor suppressor genes. Thus, some inherited characteristics do not depend exclusively on the sequence of nucleotides in DNA.

Gene amplification increases the number of copies of a gene A single gene cannot always provide enough copies of its mRNA to meet the cell's needs. The requirement for high levels of certain products is met if multiple copies of the genes that encode them are present in the chromosome. Genes of this type, whose products are essential for all cells, may occur as

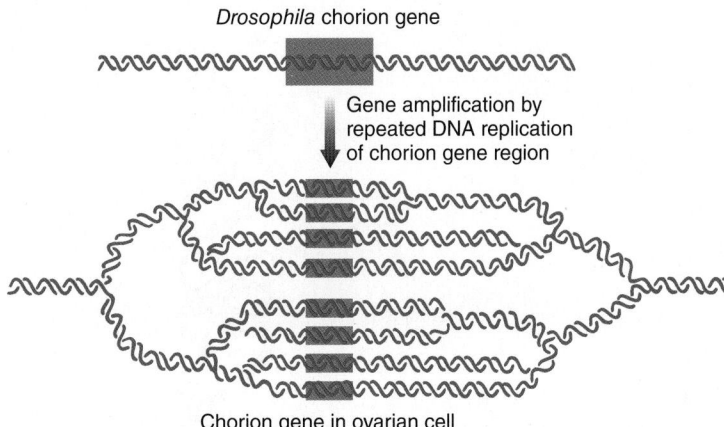

Drosophila chorion gene

Gene amplification by repeated DNA replication of chorion gene region

Chorion gene in ovarian cell

Figure 14-10 Regulation of transcription: gene amplification

In *Drosophila* multiple replications of a small region of the chromosome amplify the chorion (eggshell) protein genes. Replication is initiated at a discrete chromosome origin of replication (*top pink box*) for each copy of the gene that is produced. Replication is randomly terminated, resulting in a series of forked structures in the chromosome.
© Cengage Learning

multiple copies arranged one after another along the chromosome, in *tandemly repeated gene sequences.* Histone genes are usually present as multiple copies of 50 to 500 genes in cells of multicellular organisms. Similarly, multiple copies (150 to 450) of genes for rRNA and tRNA are present in cells.

Other genes, which are required only by a small group of cells, are selectively replicated in those cells in a process called **gene amplification.** For example, the *Drosophila* chorion (eggshell) gene product is a protein made specifically in cells of the female insect's reproductive tract. These cells make massive amounts of the protein that envelops and protects the zygote. Amplifying the chorion protein gene by DNA replication meets the demand for chorion mRNA. In other words, the DNA in that small region of the chromosome is copied many times (FIG. 14-10). In other cells of the insect body, however, the gene exists as a single copy in the chromosome.

Eukaryotic promoters vary in efficiency depending on the presence of functional promoter elements In eukaryotic as well as bacterial cells, the transcription of any gene requires a base pair where transcription begins, known as the *transcription initiation site,* or *start site,* plus a sequence of bases, the *promoter,* to which RNA polymerase binds. The promoter is a relatively short DNA sequence located adjacent to the gene it regulates.

In almost all multicellular eukaryotes, the promoter contains an element called the **TATA box,** which consists of a sequence of DNA nucleotides such as TATAAA. It is located about 25 to 30

base pairs upstream from the transcription initiation site (FIG. 14-11). Studies show that mutations in the TATA box reduce the rate of transcription.

Some eukaryotic genes have promoter elements in addition to the TATA box. Like the TATA box, these promoter elements have a regulatory function and facilitate the expression of the gene. Many eukaryotic promoters have a promoter element of one or more sequences of nucleotide bases located within a short distance (e.g., 80 base pairs) upstream from the RNA polymerase–binding site. Mutations in these elements, like mutations in the TATA box, lower the basic level of transcription.

Enhancers increase the rate of transcription, whereas silencers decrease the rate of transcription Regulated eukaryotic genes commonly have DNA sequences called **enhancers** that help form an active transcription initiation complex. In this way, enhancers increase the rate of RNA synthesis, often by several orders of magnitude.

Enhancers are remarkable in many ways. Although they are present in all cells, a particular enhancer is functional only in certain types of cells. An enhancer regulates a gene on the same DNA molecule from very long distances—up to thousands of base pairs away from the promoter—and is either upstream or downstream from the promoter it regulates. Furthermore, if an enhancer element is experimentally cut out of the DNA and inverted, it still regulates the gene it normally controls.

Many eukaryotic genes are associated with **silencers,** DNA sequences that can decrease transcription. Like enhancers, silencers are often located long distances from the genes they regulate. As you will see, enhancers and silencers work by interacting with DNA-binding proteins that regulate transcription.

Transcription factors are regulatory proteins with several functional domains We previously discussed some DNA-binding proteins that regulate transcription in bacteria, including the lactose repressor, the tryptophan repressor, and the catabolite activator protein (CAP). Researchers have identified many more DNA-binding proteins that regulate transcription in eukaryotes than in bacteria; these eukaryotic proteins are collectively known as **transcription factors.** Researchers have identified more than 2000 transcription factors in humans.

It is useful to compare regulatory proteins in bacteria and eukaryotes. Many regulatory proteins are modular molecules; that is, they have more than one **domain,** a region with its own tertiary structure and function. Each eukaryotic transcription

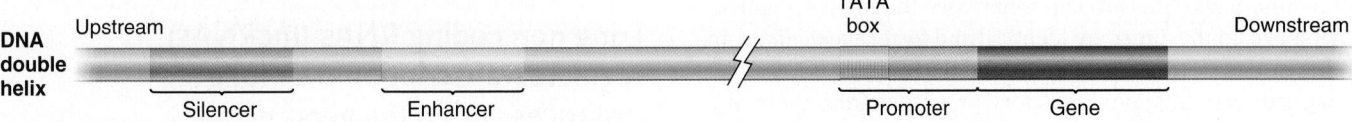

DNA double helix	Upstream					TATA box		Downstream
		Silencer	Enhancer			Promoter	Gene	

Figure 14-11 Regulation of transcription: eukaryotic elements

Eukaryotic regulatory elements include the promoter and various enhancers and silencers. The promoter elements help determine the basic level of transcription that normally occurs, whereas the enhancers (and silencers) have the potential to greatly increase (or decrease) that rate of transcription. (For simplicity, the DNA double helix is represented as a rod.)
© Cengage Learning

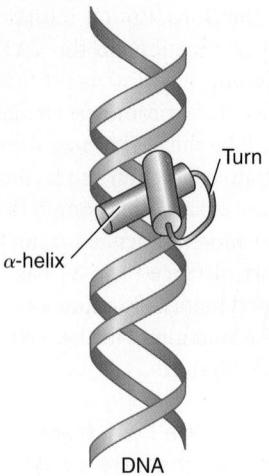

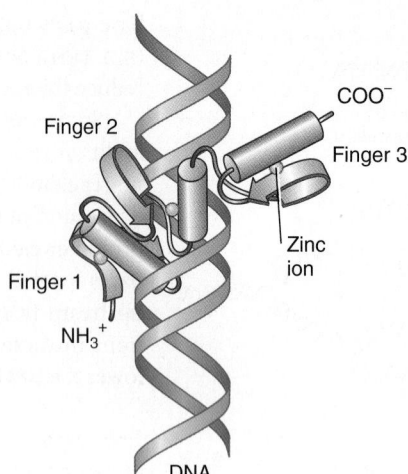

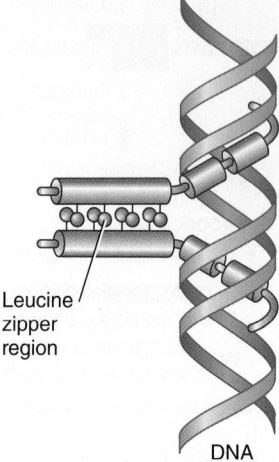

(a) Helix-turn-helix. This portion of a transcription factor (*purple*) contains the helix-turn-helix arrangement. The recognition helix is inserted into the groove of the DNA and is connected to a second helix that helps hold it in place by a sequence of amino acids that form a sharp bend.

(b) Zinc fingers. Regions of certain transcription factors form projections known as "zinc fingers," which insert into the grooves of the DNA and bind to specific base sequences.

(c) Leucine zipper. This leucine zipper protein is a dimer, held together by hydrophobic interactions involving side chains of leucine and other amino acids.

Figure 14-12 Regulation of transcription: regulatory proteins

In these illustrations cylinders represent α-helical regions of regulatory proteins and ribbons represent β-pleated sheets.
© Cengage Learning

factor, like the regulatory proteins of bacteria, has a DNA-binding domain plus at least one other domain that is either an activator or a repressor of transcription for a given gene.

Many transcription factors in eukaryotes (and regulatory proteins in bacteria) have similar DNA-binding domains. One example is the *helix-turn-helix* arrangement, consisting of two α-helical segments. One of them, the recognition helix, inserts into the major groove of the DNA without unwinding the double helix. The other helps hold the first one in place (**FIG. 14-12a**). The "turn" in the helix-turn-helix is a sequence of amino acids that forms a sharp bend in the molecule.

Other transcription factors have DNA-binding domains with multiple "zinc fingers," loops of amino acids held together by zinc ions. Each loop includes an α-helix that fits into a groove of the DNA (**FIG. 14-12b**). Certain amino acid functional groups exposed in each finger have been shown to recognize specific DNA sequences.

Many transcription factors have DNA-binding domains that function only as pairs, or *dimers.* Many of these transcription factors are known as *leucine zipper proteins* because they are held together by the side chains of leucine and other hydrophobic amino acids (**FIG. 14-12c**). In some cases, the two polypeptides that make up the dimer are identical and form a *homodimer.* In other instances, they are different, and the resulting *heterodimer* may have very different regulatory properties from those of a homodimer. For a simple and speculative example, let us assume that three regulatory proteins—A, B, and C—are involved in controlling a particular set of genes. These three proteins might associate as dimers in six different ways: three kinds of homodimers (AA, BB, and CC) and three kinds of heterodimers (AB,

AC, and BC). Such multiple combinations of regulatory proteins greatly increase the number of possible ways that transcription is controlled.

Transcription in eukaryotes requires multiple regulatory proteins that are bound to different parts of the promoter. The *general transcriptional machinery* is a protein complex that binds to the TATA box of the promoter near the transcription initiation site. That complex is required for RNA polymerase to bind, thereby initiating transcription.

Both enhancers and silencers become functional when specific transcription factors bind to them. **FIGURE 14-13** shows interactions involving an enhancer and a transcription factor that acts as an **activator protein.** Each activator protein has at least two functional domains: a DNA recognition site that usually binds to an enhancer and a gene activation site that contacts the target in the general transcriptional machinery. The DNA between the enhancer and promoter elements forms a loop that lets an activator protein bound to an enhancer come in contact with proteins associated with the general transcriptional machinery, increasing the rate of transcription.

Long non-coding RNAs (lncRNAs) regulate transcription over long distances within the genome

In mammals as many as four times as many genes encode **long non-coding RNAs (lncRNAs)** as those that actually encode proteins. These RNAs, which are greater than 200 bases in length, are processed like mRNAs through capping, intron

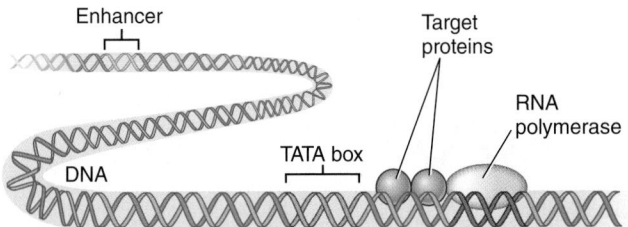

(a) Low rate of transcription. This gene is transcribed at a basic level when the general transcriptional machinery, including RNA polymerase, is bound to the promoter.

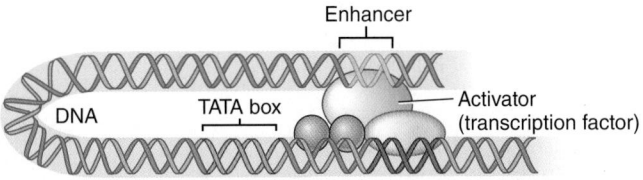

(b) High rate of transcription. A transcription factor that functions as an activator binds to an enhancer. The intervening DNA forms a loop, allowing the transcription factor to contact one or more target proteins in the general transcriptional machinery, thereby increasing the rate of transcription. This diagram is highly simplified, and many more target proteins than the two shown are involved.

Figure 14-13 Regulation of transcription: the stimulation of transcription by an enhancer
© Cengage Learning

splicing, and the addition of poly-A tails. Although more than 20,000 of these RNAs have been identified in humans, it has been difficult to determine their functions. Emerging evidence suggests that many of these RNAs regulate gene expression by a diverse set of mechanisms. Xist, one of the first lncRNAs shown to have a regulatory function, is responsible for the inactivation of X chromosomes in females. This lncRNA, which is required for the silencing of hundreds of genes on the inactive X chromosome in females (see Chapter 11), effectively coats the inactivated X chromosome and is associated with extensive DNA methylation within the silenced chromosome.

Other lncRNAs control gene expression by mechanisms shown in **FIGURE 14-14.** Some lncRNAs act as molecular sponges or "decoys" that bind up transcription factors, preventing them from binding to the promoters of active genes. Others act as molecular scaffolds that assemble regulatory proteins into molecular complexes within the nucleus. They can also act as guides that target chromatin-modifying enzymes to a specific region of DNA or as components of enhancer complexes by anchoring to the enhancer and presenting bound transcription factors to a distant promoter.

The mRNAs of eukaryotes are subject to many types of posttranscriptional control

The half-life of bacterial mRNA is usually minutes long; eukaryotic mRNA, even when it turns over rapidly, is far more stable. Bacterial mRNA is transcribed in a form that is translated immediately. In contrast, eukaryotic mRNA molecules undergo further modification and processing before they are used in protein synthesis. The message is capped, polyadenylated, spliced, and then transported from the nucleus to the cytoplasm to initiate translation (see Fig. 13-12). These events represent potential control points for translation of the message and the production of its encoded protein.

The addition of a poly-A tail to eukaryotic mRNA, for example, appears necessary to initiate translation. Researchers

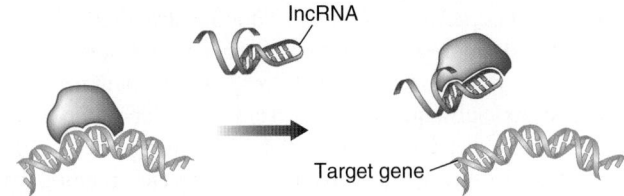

(a) Decoys. lncRNA molecular decoys bind transcription factors to prevent them from activating their target genes.

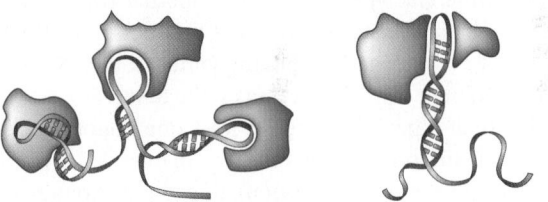

(b) Scaffolds. lncRNA molecular scaffolds bring proteins into complexes that can regulate transcription.

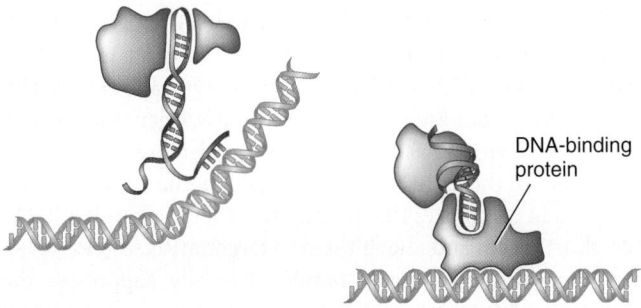

(c) Guides. lncRNAs bind to DNA (*left*) or DNA-binding proteins (*right*) to guide chromatin-modifying enzymes to sites in the genome.

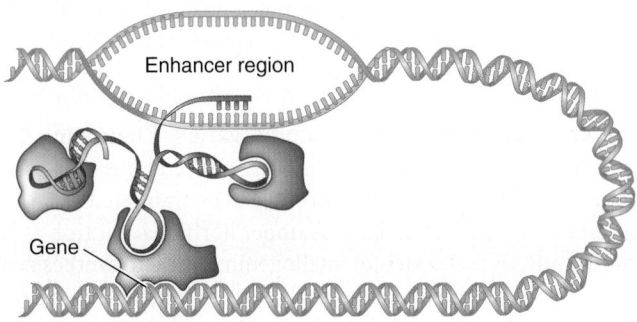

(d) Enhancer complex components. lncRNAs interact with enhancer regions to activate distant genes on the same chromosome.

Figure 14-14 Regulation of transcription by long non-coding RNAs
© Cengage Learning

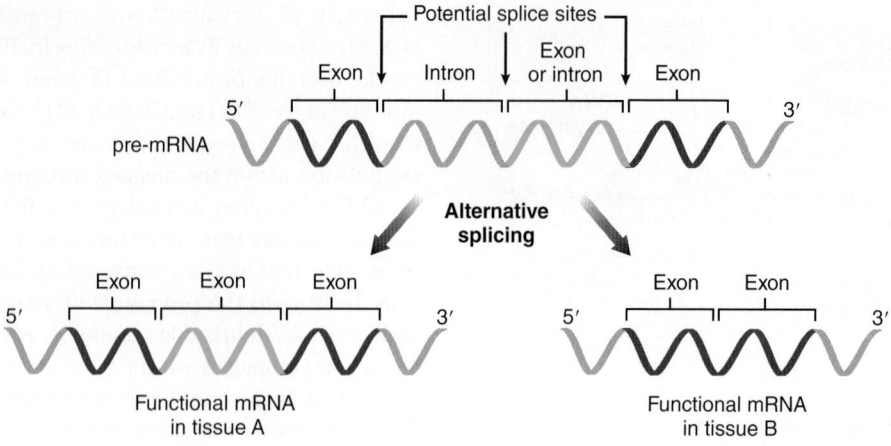

Figure 14-15 Regulation of mRNA processing: alternative splicing

In some cases, a pre-mRNA molecule is processed in more than one way to yield two or more mature mRNAs, each of which encodes a related, but different, protein. In this generalized example, the gene contains a segment that can be an exon in tissue A (*left*) but an intron in tissue B (*right*).
© Cengage Learning

have shown that mRNAs with long poly-A tails are efficiently translated, whereas mRNAs with short poly-A tails are essentially dormant. **Polyadenylation,** which commonly occurs in the nucleus, can also take place in the cytosol. When the short poly-A tail of an mRNA is elongated, the mRNA becomes activated and is then translated.

Some pre-mRNAs are processed in more than one way Investigators have discovered several forms of regulation involving mRNA processing. In some instances, the same gene produces one type of protein in one tissue and a related but somewhat different type of protein in another tissue. This difference is possible because many genes produce pre-mRNA molecules that have **alternative splicing** patterns; that is, they are spliced in different ways depending on the tissue.

Typically, such a gene includes at least one segment that can be either an **intron** or an **exon** (see Chapter 13). As an intron, the sequence is removed, but as an exon, it is retained. Through alternative splicing, the cells in each tissue produce their own version of mRNA corresponding to the particular gene (**FIG. 14-15**). For example, this mechanism produces different forms of troponin, a protein that regulates muscle contraction, in different muscle tissues.

The stability of mRNA molecules varies Controlling the lifespan of a particular kind of mRNA molecule permits control over the number of protein molecules translated from it. In some cases, mRNA stability is under hormonal control, as is true for mRNA that codes for vitellogenin, a protein synthesized in the liver of certain female animals, such as frogs and chickens. Vitellogenin is transported to the oviduct, where it is used in forming egg yolk proteins.

The hormone estradiol regulates vitellogenin synthesis. When estradiol levels are high, the half-life of vitellogenin mRNA

in frog liver is about 500 hours. When researchers deprive cells of estradiol, the half-life of the mRNA drops rapidly, to less than 165 hours. This change quickly lowers cell vitellogenin mRNA levels and decreases synthesis of the vitellogenin protein. In addition to affecting mRNA stability, the hormone seems to control the rate at which the mRNA is synthesized.

Small non-coding RNAs control eukaryotic gene expression in multiple ways In 2006, Andrew Fire and Craig Mello were awarded the Nobel Prize in Physiology or Medicine for their discovery of a phenomenon called **RNA interference (RNAi).** These investigators observed that the injection of just a few molecules per cell of very short double-stranded RNA molecules could suppress the expression of genes that had sequences in their 3′ mRNA regions that were complementary to the small RNAs. Subsequent studies from many laboratories have shown that several major types of small non-protein-coding regulatory RNAs are found in eukaryotic species, ranging from single-celled fungi to plants and animals. These small RNAs (see Table 13-1) are now known to make up one of the largest classes of gene regulators, controlling the expression of least half of all human genes. One type involves **micro RNAs (miRNAs),** which are transcribed from genes found in the nucleus. Other **small interfering RNAs (siRNAs)** are not typically encoded by nuclear genes, but are copied from the RNA genomes of certain types of viruses as part of a cellular defense mechanism. Both types of these siRNAs are activated by the same enzymes and protein complexes that process the double-stranded RNAs into short pieces and attach them to target mRNAs (**FIG. 14-16**). A bound miRNA–protein complex typically suppresses the expression of its target mRNA by either blocking its translation or triggering the degradation of the mRNA molecule.

Small RNAs are also known to act in the nucleus to remodel chromatin structure, inducing the formation of heterochromatin

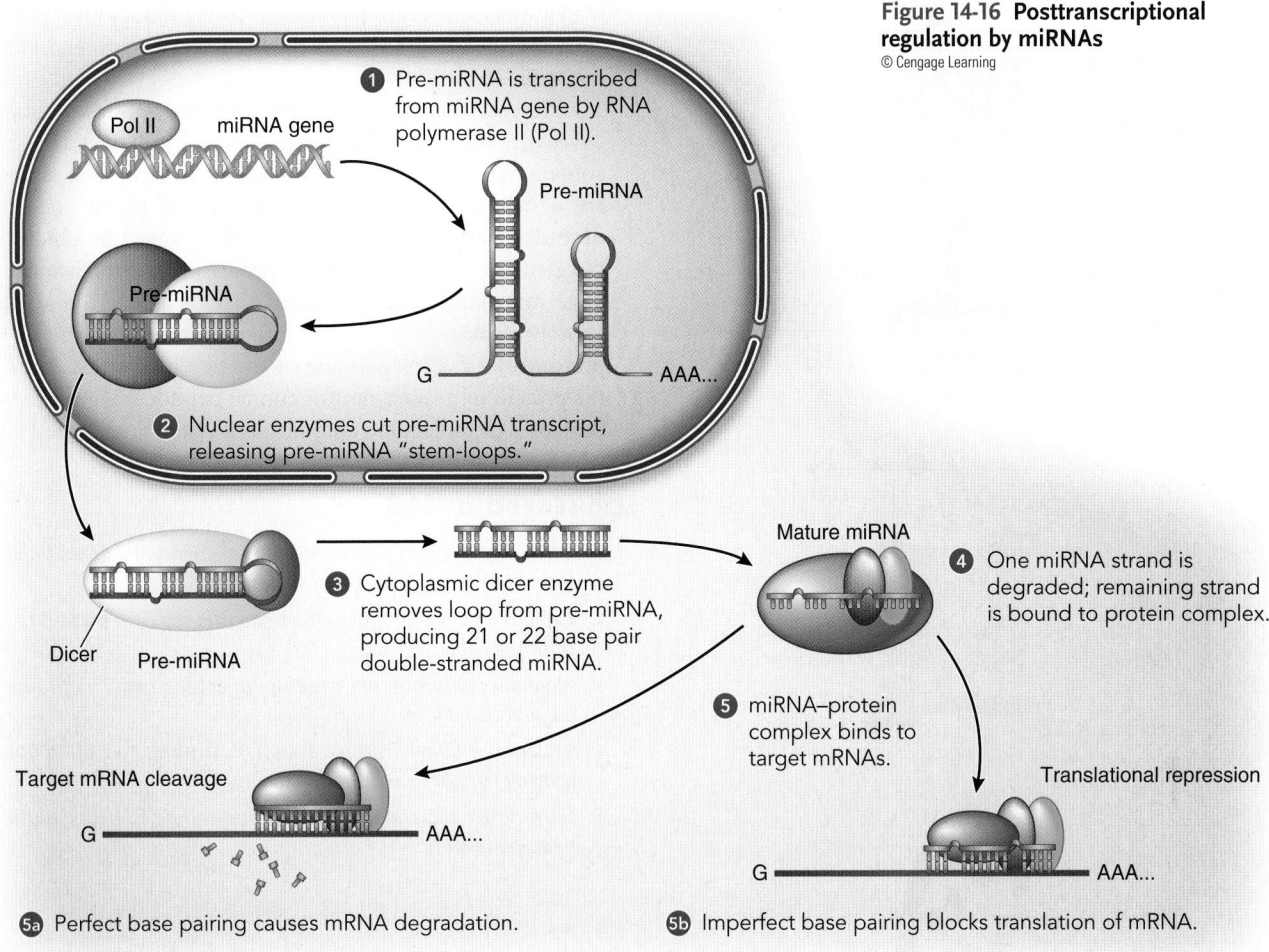

① Pre-miRNA is transcribed from miRNA gene by RNA polymerase II (Pol II).

Pol II   miRNA gene

Pre-miRNA

Pre-miRNA

G   AAA...

② Nuclear enzymes cut pre-miRNA transcript, releasing pre-miRNA "stem-loops."

Dicer   Pre-miRNA

③ Cytoplasmic dicer enzyme removes loop from pre-miRNA, producing 21 or 22 base pair double-stranded miRNA.

Mature miRNA

④ One miRNA strand is degraded; remaining strand is bound to protein complex.

⑤ miRNA–protein complex binds to target mRNAs.

Target mRNA cleavage

Translational repression

G   AAA...

G   AAA...

⑤a Perfect base pairing causes mRNA degradation.

⑤b Imperfect base pairing blocks translation of mRNA.

---

that blocks gene expression over large regions of chromosomes. In plants small RNA molecules have been identified that can either block or activate transcription of specific genes by binding to certain sites on their promoters.

Posttranslational chemical modifications may alter the activity of eukaryotic proteins

Another way to control gene expression is by regulating the activity of the gene product. Many metabolic pathways in eukaryotes, as in bacteria, contain allosteric enzymes regulated through feedback inhibition. In addition, after they are synthesized, many eukaryotic proteins are extensively modified.

In *proteolytic processing* proteins are synthesized as inactive precursors, which are converted to an active form by the removal of a portion of the polypeptide chain. For example, proinsulin contains 86 amino acids. Removing 35 amino acids

yields the hormone insulin, which consists of two polypeptide chains containing 30 and 21 amino acids, respectively, linked by disulfide bridges.

Chemical modification, by adding or removing functional groups, reversibly alters the activity of an enzyme. One common way to modify the activity of an enzyme or other protein is to add or remove phosphate groups. Enzymes that add phosphate groups are called **kinases;** those that remove them are **phosphatases.** For example, the cyclin-dependent kinases discussed in Chapter 10 help control the cell cycle by adding phosphate groups to certain key proteins, causing them to become activated or inactivated. Chemical modifications such as protein phosphorylation also let the cell respond rapidly to certain hormones or to fast-changing environmental or nutritional conditions.

Posttranslational control of gene expression can also involve *protein degradation.* The amino acid found at the amino end of a protein correlates with whether the protein is long-lived or short-lived. Some proteins, for example, have a 50% turnover

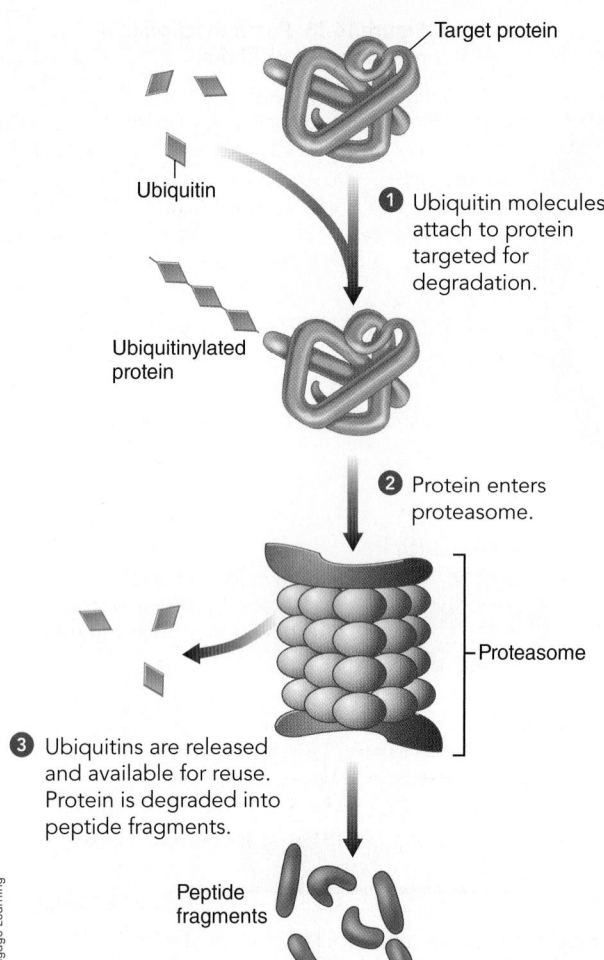

Target protein

Ubiquitin

1 Ubiquitin molecules attach to protein targeted for degradation.

Ubiquitinylated protein

2 Protein enters proteasome.

Proteasome

3 Ubiquitins are released and available for reuse. Protein is degraded into peptide fragments.

Peptide fragments

© Cengage Learning

of synthesis and degradation every three days. Other proteins, such as proteins in the lens of the human eye, may persist for the entire life of the individual.

Proteins that are selectively targeted for destruction are covalently bonded to **ubiquitin,** a small polypeptide tag that contains 76 amino acids. Additional molecules of ubiquitin are added, forming short chains attached to the soon-to-be degraded protein (**FIG. 14-17**). A protein tagged by ubiquitin is targeted for degradation in a proteasome. **Proteasomes** are large macromolecular structures that recognize the ubiquitin tags. **Proteases** (protein-degrading enzymes) associated with proteasomes hydrolyze some of the peptide bonds, degrading the protein into short, nonfunctional peptide fragments. As the protein is degraded, the ubiquitin molecules are released intact, which allows them to be used again.

CHECKPOINT 14.3

- **CONNECT** *Explain how regulation of eukaryotic genes differs from regulation of bacterial genes.*
- *Explain why certain genes in eukaryotic cells are present in multiple copies.*
- *How does chromosome structure affect the activity of some eukaryotic genes?*
- **VISUALIZE** *Draw a simple sketch illustrating how alternative splicing can give rise to different proteins.*
- *Summarize three ways in which non-coding RNAs regulate gene expression.*

Figure 14-17 Degradation of protein: ubiquitin and proteasomes

---

SUMMARY: FOCUS ON LEARNING OBJECTIVES

14.1 Gene Regulation in Bacteria and Eukaryotes: An Overview *(page 296)*

1 Explain why bacterial and eukaryotic cells have different mechanisms of gene regulation.

- Bacterial cells grow rapidly and have a relatively short lifespan. **Transcriptional-level control** is the most common way for the prokaryotic cell to regulate gene expression.
- Eukaryotic cells usually have a relatively long lifespan during which they must maintain homeostasis in response to many different stimuli. Regulatory control at all levels of gene expression allows cells to rapidly and reversibly respond to changing physiological conditions.

14.2 Gene Regulation in Bacteria *(page 296)*

2 Define *operon* and explain the functions of the operator and promoter regions.

- An **operon** is a gene complex consisting of a group of *structural genes* with related functions plus the closely linked DNA sequences responsible for controlling them.
- Each operon has a single **promoter** region upstream from the protein-coding regions; the promoter is where RNA polymerase first binds to DNA before transcription begins.
- The **operator** serves as the regulatory switch for transcriptional-level control of the operon. The binding of a **repressor protein**

to the operator sequence prevents transcription; although RNA polymerase binds to the promoter, it is blocked from transcribing the structural genes. When the repressor is not bound to the operator, transcription proceeds.

3 Distinguish among inducible, repressible, and constitutive genes.

- An **inducible operon,** such as the *lac* operon, is normally turned off. The repressor protein is synthesized in an active form that binds to the operator. If lactose is present, it is converted to allolactose, the **inducer,** which binds to the repressor protein. The altered repressor cannot bind to the operator, and the operon is turned on.
- A **repressible operon,** such as the *trp* operon, is normally turned on. The repressor protein is synthesized in an inactive form that cannot bind to the operator. A metabolite (usually the end product of a metabolic pathway) acts as a **corepressor.** When corepressor levels are high, a corepressor molecule binds to the repressor, which can now bind to the operator and turn off transcription of the operon.
- **Constitutive genes** are neither inducible nor repressible; they are active at all times. The activity of constitutive genes is controlled by how efficiently RNA polymerase binds to their promoter regions.

4 Differentiate between positive and negative control, and show how both types of control operate in regulating the *lac* operon.

- Repressible and inducible operons are under **negative control.** When the repressor protein binds to the operator, transcription of the operon is turned off.

- Some inducible operons are also under **positive control,** in which an activator protein binds to the DNA and stimulates transcription of the gene. **Catabolite activator protein (CAP)** activates the *lac* operon; CAP binds to the promoter region, stimulating transcription by binding RNA polymerase tightly. To bind, CAP requires **cyclic AMP (cAMP),** which increases in the cell as glucose decreases.

5 Describe the types of posttranscriptional control in bacteria.

- Some **posttranscriptional controls** operate in bacteria. A **translational control** is a posttranscriptional control that regulates the rate of translation of a particular mRNA. **Posttranslational controls** include **feedback inhibition** of key enzymes in some metabolic pathways.

14.3 Gene Regulation in Eukaryotic Cells *(page 301)*

6 Discuss the structure of a typical eukaryotic gene and the DNA elements involved in regulating that gene.

- Eukaryotic genes are not normally organized into operons. Regulation of eukaryotic genes occurs at the levels of transcription, mRNA processing, translation, and modifications of the protein product.

- Transcription of a gene requires a *transcription initiation site,* where transcription begins, plus a promoter to which RNA polymerase binds. In almost all multicellular eukaryotes, the promoter contains an element called the **TATA box** that has a regulatory function and facilitates expression of the gene.

- Some eukaryotic genes have **enhancers** and **silencers** located thousands of bases away from the promoter. These regulatory elements increase or decrease the rate of transcription.

7 Give examples of some of the ways eukaryotic DNA-binding proteins bind to DNA.

- Eukaryotic genes are controlled by DNA-binding proteins called **transcription factors.** Many are transcriptional activators; others are transcriptional repressors.

- Each transcription factor has a DNA-binding **domain.** Some transcription factors have a helix-turn-helix arrangement and insert one of the helices into the DNA. Other transcription factors have loops of amino acids held together by zinc ions; each loop includes an α-helix that fits into the DNA. Some transcription factors are *leucine zipper proteins* that associate as dimers that insert into the DNA.

8 Illustrate how a change in chromosome structure may affect the activity of a gene.

- Densely packed regions of chromosomes, called **heterochromatin,** contain inactive genes. Active genes are associated with a loosely packed chromatin structure called **euchromatin.** Methyl groups, acetyl groups, sugars, and proteins may chemically attach to the **histone** tail, a string of amino acids that extends from the DNA-wrapped **nucleosome,** and may expose or hide genes, turning them on or off.

- **DNA methylation** perpetuates gene inactivation. **Epigenetic inheritance** is an important mechanism of gene regulation that involves changes in how a gene is expressed. Because DNA methylation patterns tend to be repeated in successive cell generations, they provide a mechanism for epigenetic inheritance.

- Some genes whose products are required in large amounts exist as multiple copies in the chromosome. In the process of **gene amplification,** some cells selectively amplify genes by DNA replication.

9 Explain how a gene in a multicellular organism may produce different products in different types of cells.

- As a result of **alternative splicing,** a single gene produces different forms of a protein in different tissues depending on how the pre-mRNA is spliced. Typically, such a gene contains a segment that can be either an intron or an exon. As an intron, the sequence is removed, and as an exon, the sequence is retained.

10 Identify some of the types of regulatory controls that operate in eukaryotes after mature mRNA is formed.

- Certain regulatory mechanisms increase the stability of mRNA, allowing more protein molecules to be synthesized before the mRNA is degraded. Sometimes mRNA stability is under hormonal control.

- Posttranslational control of eukaryotic genes occurs by feedback inhibition or by chemical modifications that change the protein's structure. The function of a protein is changed by **kinases** adding phosphate groups or by **phosphatases** removing phosphates.

- Posttranslational control of gene expression also involves *protein degradation.* Proteins targeted for destruction are covalently bonded to **ubiquitin,** which tags it for degradation in a **proteasome,** a large macromolecular structure. **Proteases** associated with proteasomes degrade the protein into short peptide fragments.

TEST YOUR UNDERSTANDING

Know and Comprehend

1. The regulation of most bacterial genes occurs at the level of (a) transcription (b) translation (c) replication (d) posttranslation (e) postreplication

2. The operator of an operon (a) encodes information for the repressor protein (b) is the binding site for the inducer (c) is the binding site for the repressor protein (d) is the binding site for RNA polymerase (e) encodes the information for the CAP

3. Feedback inhibition is an example of control at the level of (a) transcription (b) translation (c) posttranslation (d) replication (e) all the preceding

4. The "zipper" of a leucine zipper protein attaches (a) specific amino acids to specific DNA base pairs (b) two polypeptide chains to each other (c) one DNA region to another DNA region (d) amino acids to zinc atoms (e) RNA polymerase to the operator

5. Inactive genes tend to be found in (a) highly condensed chromatin, known as euchromatin (b) decondensed chromatin, known as euchromatin (c) highly condensed chromatin, known as heterochromatin (d) decondensed chromatin, known as heterochromatin (e) chromatin that is not organized as nucleosomes

6. Which of the following is characteristic of genes and gene regulation in bacteria but *not* in eukaryotes? (a) presence of enhancers (b) capping of mRNAs (c) many chromosomes per cell (d) binding of DNA to regulatory proteins (e) no requirement for exon splicing

7. Which of the following is characteristic of genes and gene regulation in *both* bacteria and eukaryotes? (a) promoters (b) noncoding DNA within coding sequences (c) enhancers (d) operons (e) DNA located in a nucleus

8. Through alternative splicing, eukaryotes (a) reinforce gene inactivation (b) prevent transcription of heterochromatin (c) produce related but different proteins in different tissues (d) amplify genes to meet the requirement of high levels of a gene product (e) bind transcription factors to enhancers to activate transcription

Apply and Analyze

9. A mutation that inactivates the repressor gene of the *lac* operon results in (a) the continuous transcription of the structural genes (b) no transcription of the structural genes (c) the binding of the repressor to the operator (d) no production of RNA polymerase (e) no difference in the rate of transcription

10. Which of the following is an example of positive control? (a) transcription occurs when a repressor binds to an inducer (b) transcription cannot occur when a repressor binds to a corepressor (c) transcription is stimulated when an activator protein binds to DNA (d) a and b (e) a and c

11. A repressible operon codes for the enzymes of the following pathway:

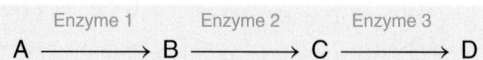

    Which component of the pathway is most likely to be the corepressor for that operon? (a) substance A (b) substance B or C (c) substance D (d) enzyme 1 (e) enzyme 3

Evaluate and Synthesize

12. **PREDICT** Compare the types of bacterial genes associated with inducible operons, those associated with repressible operons, and those that are constitutive. Predict the category into which each of the following would most likely fit: (a) a gene that codes for RNA polymerase, (b) a gene that codes for an enzyme required to break down maltose, and (c) a gene that codes for an enzyme used in the synthesis of adenine.

13. **INTERPRET DATA** Develop a simple hypothesis that would explain the behavior of each of the following types of mutants in *E. coli*.
    Mutant a: The map position of this mutation is in the *trp* operon. The mutant cells are constitutive; that is, they produce all the enzymes coded for by the *trp* operon, even if large amounts of tryptophan are present in the growth medium.
    Mutant b: The map position of this mutation is in the *trp* operon. The mutant cells do not produce any enzymes coded for by the *trp* operon under any conditions.
    Mutant c: The map position of this mutation is some distance from the *trp* operon. The mutant cells are constitutive; that is, they produce all the enzymes coded for by the *trp* operon, even if the growth medium contains large amounts of tryptophan.

14. **INTERPRET DATA** You are studying the rate of transcription of a particular eukaryotic gene. When the DNA located several thousand bases upstream from the gene is removed, the transcription rate of the gene decreases dramatically. How would you interpret these results?

15. The regulatory gene that codes for the tryptophan repressor is not tightly linked to the *trp* operon. Would it be advantageous if it were? Explain your answer.

16. **EVOLUTION LINK** Suggest why evolution resulted in multiple levels of gene regulation in multicellular eukaryotes.

17. **EVOLUTION LINK** Comparing DNA sequences in different species indicates that more DNA segments that do not code for protein have been conserved (unchanged) than protein-coding regions. These non-protein-coding regions are interpreted as gene regulatory elements. Suggest why gene regulatory elements have not undergone many changes during the course of evolution.

aplia To access course materials, such as Aplia and other companion resources, please visit **www.cengagebrain.com.**

DNA Technology and Genomics

Beginning in the mid-1970s, the development of new ways to study DNA led to radically new research approaches, such as **recombinant DNA** technology, in which researchers splice together DNA from different organisms in the laboratory. One goal of this technology is to enable scientists to obtain many copies of a specific DNA segment for the purpose of studying it. Because new methods to analyze DNA are continually emerging, we do not attempt to explore them all here. Instead, we discuss some of the major approaches that have provided a foundation for the technologies commonly used by molecular geneticists.

We then consider how studies of DNA sequences have helped scientists understand the organization of genes and the relationship between genes and their products. In fact, most of our current knowledge of the structure and control of eukaryotic genes, and of the roles of genes in development, comes from applying these methods. DNA sequencing has also revolutionized systematics by clarifying many evolutionary relationships.

This chapter also explores some of the practical applications of DNA technologies. Altering the DNA of an organism to produce new genes with new traits, called **genetic engineering,** is used in many ways, ranging from basic research to the production of strains of bacteria that manufacture useful protein products and to the development of plants and animals that express foreign genes (see photograph).

The development of DNA cloning and related techniques has led to the emergence of *biotechnology*, which is the commercial or industrial use of cells or organisms. Today, biotechnology includes numerous applications in such diverse areas as medicine, foods and agriculture, and forensic science.

Courtesy of GloFish®

Transgenic fish (GloFish®). These aquarium fish glow green, red, or yellow because researchers inserted a foreign gene encoding a fluorescent protein from either jellyfish or sea corals.

KEY CONCEPTS

15.1 Scientists use DNA cloning to isolate and produce many copies of specific DNA sequences.

15.2 Biologists study DNA and gene expression using gel electrophoresis, blotting methods, polymerase chain reaction, automated sequencing, DNA microchip analysis, bioinformatics software, and other tools.

15.3 Genomic studies uncover the structure, function, and evolution of genomes.

15.4 DNA technology and genomics have wide applications, from medical to forensic to agricultural.

15.5 Scientists must assess the risks of each new recombinant organism.

15.1 DNA CLONING

LEARNING OBJECTIVES

1 Explain how a typical restriction enzyme cuts DNA molecules and give examples of the ways in which these enzymes are used in recombinant DNA technology.
2 Distinguish between a genomic DNA library and a complementary DNA (cDNA) library; explain why one would clone the same eukaryotic gene from both a genomic DNA library and a cDNA library.
3 Explain how researchers use a DNA probe.
4 Describe how the polymerase chain reaction amplifies DNA in vitro.

Recombinant DNA technology was not developed quickly. It actually had its roots in the 1940s with genetic studies of bacteria and **bacteriophages** ("bacteria eaters"), the viruses that infect them. (See Figure 12-2, which shows bacteriophages attached to the bacterium *Escherichia coli.*) Molecular biologist Paul Berg was the first investigator to construct a recombinant DNA molecule; in 1980, he shared the Nobel Prize in Chemistry for his contributions to recombinant DNA technology. After decades of basic research and the accumulation of extensive knowledge, the technology has become highly developed, and many scientists now use these methods.

In traditional recombinant DNA technology, scientists use **restriction enzymes** from bacteria to cut DNA molecules only in specific places. Restriction enzymes enable researchers to cut DNA into manageable segments. Each fragment is then incorporated into a suitable **vector** molecule, a carrier capable of transporting the DNA fragment into a cell.

Bacteriophages and DNA molecules called plasmids are two examples of vectors. Bacterial DNA is generally circular; most **plasmids** are separate, much smaller, circular DNA molecules that may be present and replicate inside a bacterial cell, such as *E. coli.* Researchers introduce plasmids into bacterial cells by a method called **transformation,** the uptake of foreign DNA by cells (see Chapter 12). Once a plasmid enters a cell, it is replicated and distributed to daughter cells during cell division. When a *recombinant plasmid*—one that has foreign DNA spliced into it—replicates in this way, many identical copies of the foreign DNA are made, thereby **cloning** the foreign DNA.

Restriction enzymes are "molecular scissors"

Discovering restriction enzymes was a major breakthrough in developing recombinant DNA technology. Today, large numbers of different types of restriction enzymes, each with its own characteristics, are readily available to researchers. For example, a restriction enzyme known as *Hind*III searches a DNA molecule for the restriction site 5′—AAGCTT—3′, and when it finds that site, it cuts the DNA. (A *restriction site* is a DNA sequence containing the cleavage site that is cut by a particular restriction enzyme.) The sequence 5′—GAATTC—3′ is cut by another restriction enzyme known as *Eco*RI. The names of restriction enzymes are generally derived from the names of the bacteria from which they were originally isolated. Thus, *Hind*III and *Eco*RI are derived from *Hemophilus influenzae* and *E. coli,* respectively.

Why do bacteria produce such enzymes? During infection, a bacteriophage injects its DNA into a bacterial cell. The bacterium can defend itself if it has restriction enzymes that can attack the bacteriophage DNA. The bacterial cell protects its own DNA from breakdown by modifying it after replication. An enzyme adds a methyl group to one or more bases in each restriction site so that the restriction enzyme does not recognize and cut the bacterial DNA.

Restriction enzymes enable scientists to cut DNA from chromosomes into shorter fragments in a controlled way. Many of the restriction enzymes used for recombinant DNA studies cut **palindromic** sequences, which means that the base sequence of one strand reads the same as its complement when both are read in the 5′ ⟶ 3′ direction. Thus, in our *Hind*III example, both strands read 5′—AAGCTT—3′, which as a double-stranded molecule is diagrammed as follows:

$$5'—AAGCTT—3'$$
$$3'—TTCGAA—5'$$

By cutting both strands of the DNA but in a staggered fashion, these enzymes produce fragments with identical, complementary, single-stranded ends:

5′—A	AGCTT—3′
3′—TTCGA	A—5′

These ends are called *sticky ends* because they pair by hydrogen bonding with the complementary, single-stranded ends of other DNA molecules that have been cut with the same enzyme (**FIG. 15-1**). Once the sticky ends of two molecules have been

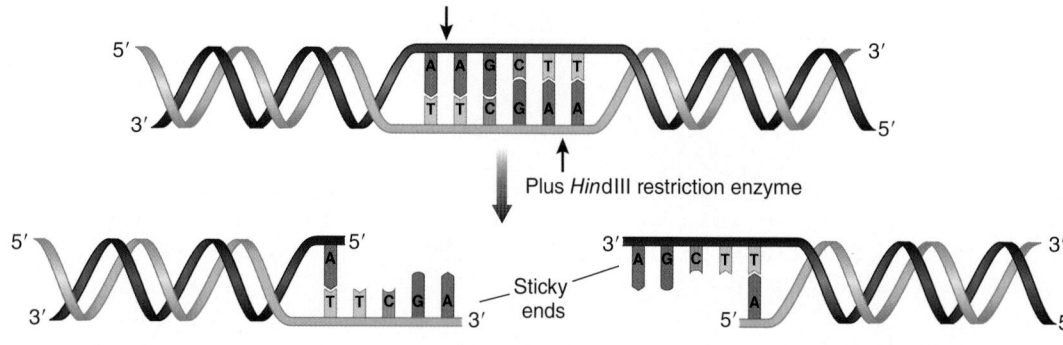

Plus *Hind*III restriction enzyme

Sticky ends

Figure 15-1 *Animation* Cutting DNA with a restriction enzyme

Many restriction enzymes, such as *Hind*III, cut DNA at sequences that are palindromic, producing complementary sticky ends. The small black arrows designate the enzyme's cleavage sites.
© Cengage Learning

joined in this way, they are treated with **DNA ligase,** an enzyme that covalently links the two DNA fragments to form a stable recombinant DNA molecule. (DNA ligase was discussed in the section on DNA replication in Chapter 12.)

Recombinant DNA forms when DNA is spliced into a vector

In recombinant DNA technology, geneticists cut both the DNA to be cloned and plasmid DNA with the same restriction enzyme. The two DNA samples are then mixed under conditions that facilitate hydrogen bonding between the complementary bases of the sticky ends, and the nicks in the resulting recombinant DNA are sealed by DNA ligase (**FIG. 15-2**).

The plasmids now used in recombinant DNA work have been engineered in the laboratory to include features helpful in isolating and analyzing cloned DNA (**FIG. 15-3**). Among them are an origin of replication (see Chapter 12), one or more restriction sites, and genes that confer resistance to particular antibiotics or that let the cells use a specific nutrient. These features allow researchers to select cells transformed by recombinant plasmids and identify cells that contain plasmids with inserted DNA. For

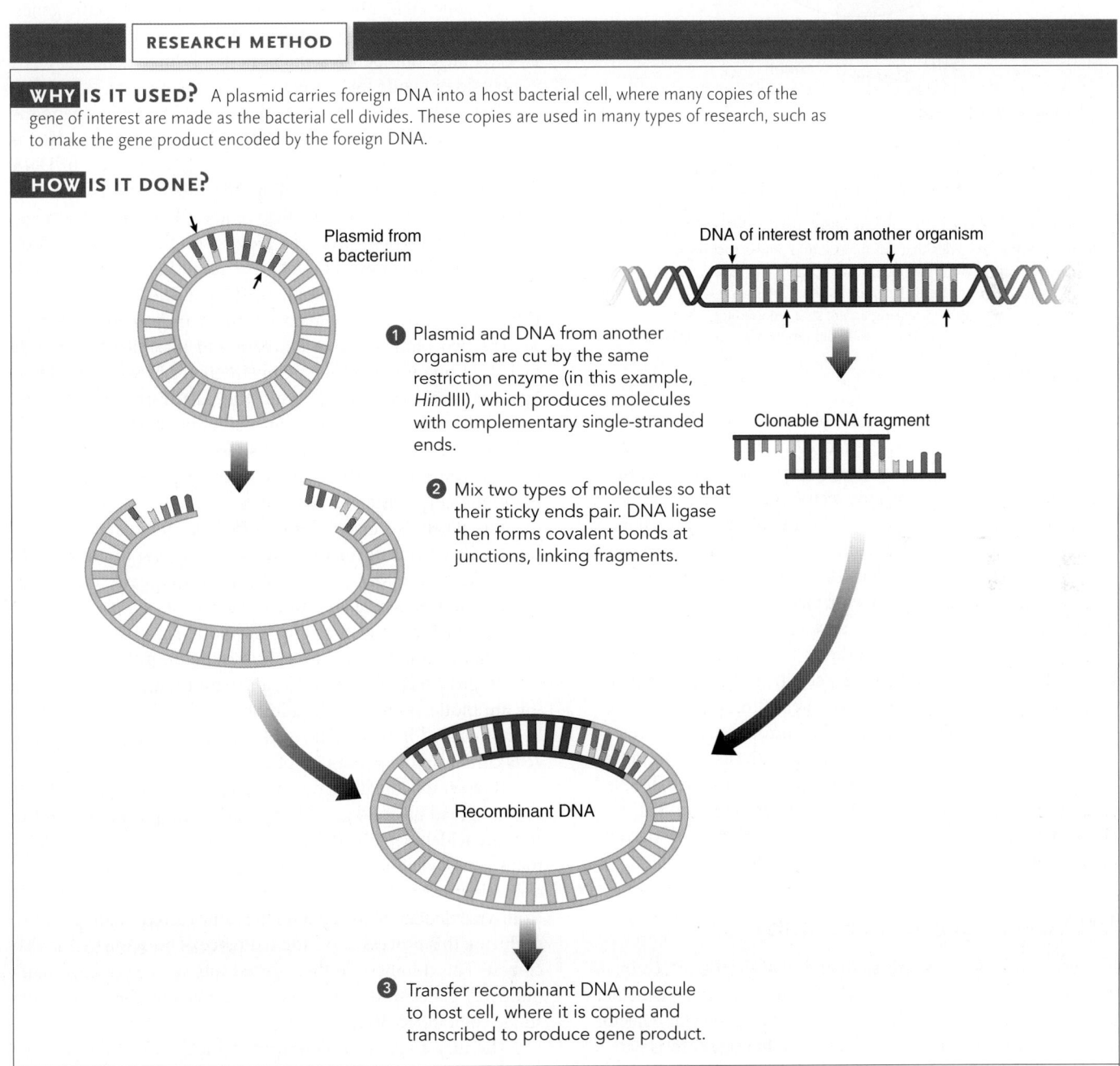

RESEARCH METHOD

WHY IS IT USED? A plasmid carries foreign DNA into a host bacterial cell, where many copies of the gene of interest are made as the bacterial cell divides. These copies are used in many types of research, such as to make the gene product encoded by the foreign DNA.

HOW IS IT DONE?

Plasmid from a bacterium

DNA of interest from another organism

❶ Plasmid and DNA from another organism are cut by the same restriction enzyme (in this example, *Hind*III), which produces molecules with complementary single-stranded ends.

Clonable DNA fragment

❷ Mix two types of molecules so that their sticky ends pair. DNA ligase then forms covalent bonds at junctions, linking fragments.

Recombinant DNA

❸ Transfer recombinant DNA molecule to host cell, where it is copied and transcribed to produce gene product.

Figure 15-2 *Animation* **Splicing foreign DNA (gene of interest) into a cloning vector (in this example, a bacterial plasmid) to make recombinant DNA**
© Cengage Learning

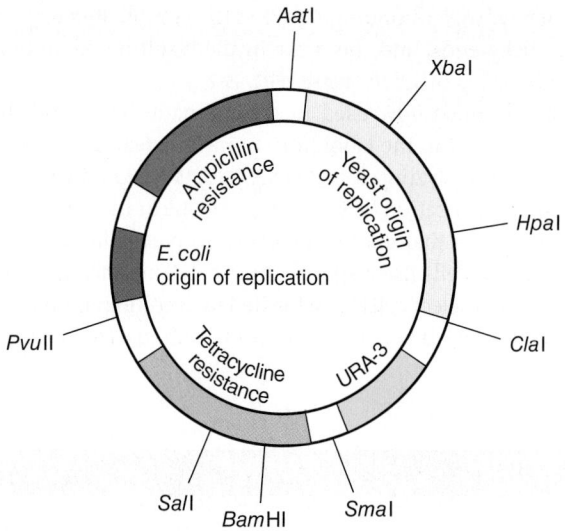

Figure 15-3 Plasmid

This plasmid vector has many useful features. It was constructed from DNA fragments isolated from plasmids, *E. coli* genes, and yeast genes. The two origins of replication, one for *E. coli* and one for yeast, *Saccharomyces cerevisiae*, let it replicate independently in either type of cell. Letters on the outer circle designate sites for restriction enzymes that cut the plasmid only in that position. Resistance genes for the antibiotics ampicillin and tetracycline and the yeast *URA*-3 gene are also shown. The *URA*-3 gene is useful when transforming cells lacking an enzyme required for uracil synthesis. Cells that take up the plasmid grow on a uracil-deficient medium.

© Cengage Learning

example, cells transformed with a plasmid that includes a gene for resistance to the antibiotic tetracycline can grow in a medium that contains tetracycline, whereas untransformed cells cannot.

A limiting property of any vector is the size of the DNA fragment it can effectively carry. The size of a DNA segment is often given in kilobases (kb), with 1 kb equal to 1000 base pairs. Fragments that are smaller than 10 kb are usually inserted into plasmids for use in *E. coli*. However, larger fragments require the use of specially designed vectors such as *bacterial artificial chromosomes* (BACs) that can include up to about 350 kb of extra DNA. This feature made BACs especially useful in the *Human Genome Project*, which is discussed later in this chapter.

Recombinant DNA can also be introduced into cells of eukaryotic organisms. For example, geneticists use modified viruses as vectors in mammal cells. These viruses are disabled so that they do not kill the cells they infect. Instead, the viral DNA molecules as well as any foreign DNA they carry become incorporated into the cell's chromosomes after infection.

DNA can be cloned inside cells

Because a single gene is only a small part of the total DNA in an organism, isolating the piece of DNA containing that particular gene is like finding a needle in a haystack: a powerful detector is needed. Today, many methods enable biologists to isolate a specific nucleotide sequence from an organism. We start by discussing methods in which DNA is cloned inside bacterial cells.

We use the cloning of human DNA as an example, although the procedure is applicable for any organism.

A genomic DNA library contains fragments of all DNA in the genome The total DNA in a cell is called its **genome.** For example, if DNA is extracted from human cells, we refer to it as human genomic DNA. A **genomic DNA library** is a collection of thousands of DNA fragments that represent all the DNA in the genome. Typically, each fragment is inserted into a plasmid, which is then incorporated into a bacterial cell. Thus, a human genomic DNA library is stored in a collection of recombinant bacteria; each cell has a different fragment of human DNA incorporated into its plasmid. Scientists use genomic DNA libraries to isolate and study specific genes. Because protein-coding genes are only about 1.5% of human DNA, a huge number of sequences in a human genomic DNA library do not code for protein.

The basic cloning methods that are used to make genomic DNA libraries are shown in **FIGURE 15-4.** Genomic DNA that is cut with a restriction enzyme generates a population of DNA fragments that vary in size and in the genetic information they carry, but all have identical sticky ends. Geneticists treat plasmid DNA that will be used as a vector with the same restriction enzyme, which converts the circular plasmids into linear molecules with sticky ends complementary to those of the human DNA fragments. Recombinant plasmids are produced by mixing the two kinds of DNA (human and plasmid) to promote hydrogen bonding between complementary bases. Then, DNA ligase is used to covalently bond the paired ends of the plasmid and human DNA, forming recombinant DNA. Unavoidably, nonrecombinant plasmids also form (*not shown in figure*) because some plasmids revert to their original circular shape without incorporating foreign DNA.

The geneticists insert plasmids into antibiotic-sensitive bacterial cells by transformation. Because the ratio of plasmids to cells is kept very low, it is rare for a cell to receive more than one plasmid molecule, and not all cells receive a plasmid. The researchers incubate normally antibiotic-sensitive cells on a nutrient medium that includes antibiotics; the only cells that are able to grow have incorporated a plasmid that contains a gene for antibiotic resistance. In addition, the plasmid has usually been modified in ways that enable researchers to identify those cells containing recombinant plasmids.

To assemble the library, a dilute sample of the bacterial culture is spread on solid growth medium (agar plates) so that the cells are widely separated. Each plasmid-containing cell divides many times, yielding a visible **colony,** which is a clone of genetically identical cells originating from a single cell. All the cells from a particular colony contain the same recombinant plasmid, so during this process a specific sequence of human DNA is also cloned. The colonies are then sorted into a collection of many multiwell plates so that each well contains only the cells from a single colony (**FIG. 15-5**).

The major task is to determine which colony out of thousands contains a cloned fragment of interest. Specific DNA sequences are identified in various ways.

WHY IS IT USED? A genomic library is a collection of all the DNA fragments in an organism's genome. Each DNA fragment is "stored" in a clone of recombinant bacterial cells. Scientists use genomic libraries to isolate and study genes.

HOW IS IT DONE?

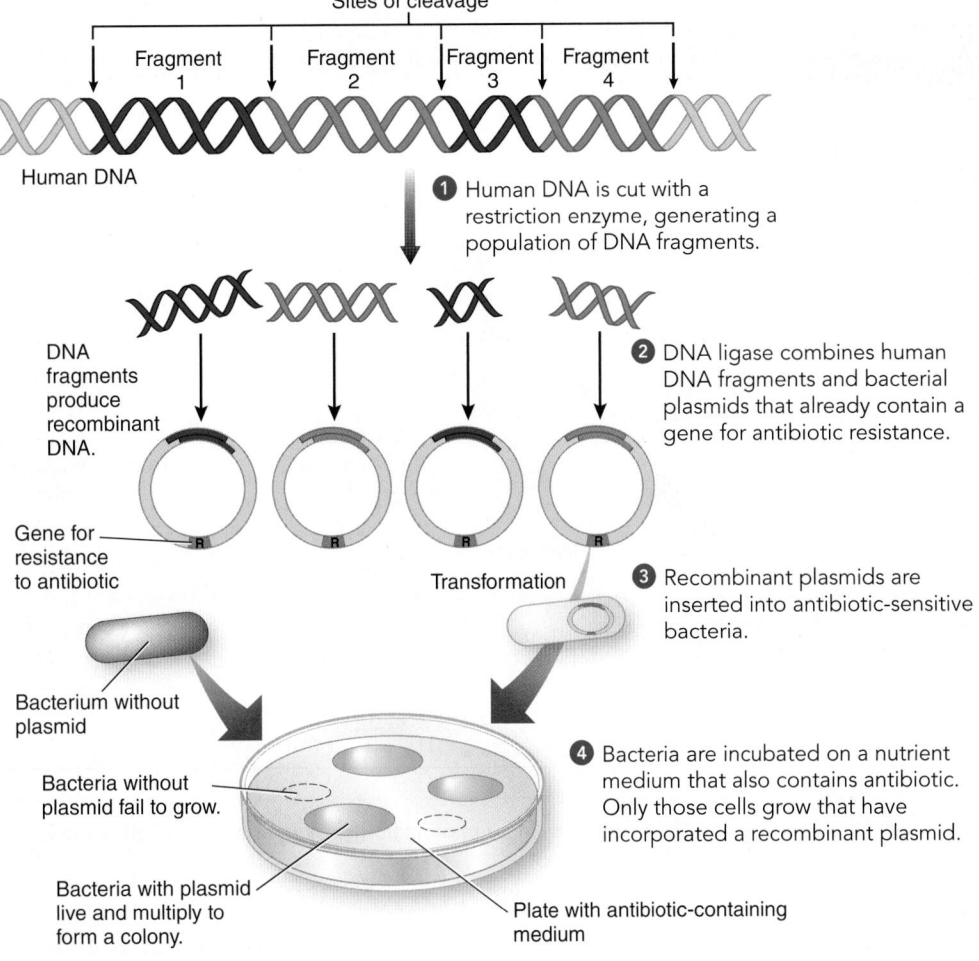

Sites of cleavage

Fragment 1   Fragment 2   Fragment 3   Fragment 4

Human DNA

DNA fragments produce recombinant DNA.

Gene for resistance to antibiotic

Bacterium without plasmid

Transformation

Bacteria without plasmid fail to grow.

Bacteria with plasmid live and multiply to form a colony.

Plate with antibiotic-containing medium

1 Human DNA is cut with a restriction enzyme, generating a population of DNA fragments.

2 DNA ligase combines human DNA fragments and bacterial plasmids that already contain a gene for antibiotic resistance.

3 Recombinant plasmids are inserted into antibiotic-sensitive bacteria.

4 Bacteria are incubated on a nutrient medium that also contains antibiotic. Only those cells grow that have incorporated a recombinant plasmid.

Figure 15-4 Producing a genomic DNA library
© Cengage Learning

A complementary DNA probe detects a specific DNA sequence Suppose that a researcher wishes to screen the thousands of recombinant DNA molecules in bacterial cells to find the specific DNA sequence of interest. One approach to detecting a specific DNA sequence involves using a **DNA probe,** which is a segment of DNA that is homologous to part of the sequence of interest. The DNA probe is usually a radioactively labeled segment of single-stranded DNA that can undergo **hybridization,** i.e. attach by base pairing to complementary base sequences in the target DNA.

To find the specific DNA, biologists transfer the DNA from cells of a genomic library to a nitrocellulose or nylon membrane. They then incubate the membrane with the radioactive probe mixture to let the probes hybridize with any complementary strands of DNA that may be present. Each spot on the membrane containing DNA that is complementary to that particular probe becomes radioactive and is detected by *autoradiography* (see Chapter 2), using X-ray film. Each spot on the film therefore identifies a colony containing a plasmid that includes the DNA of interest. The biologist then returns to the genomic library and

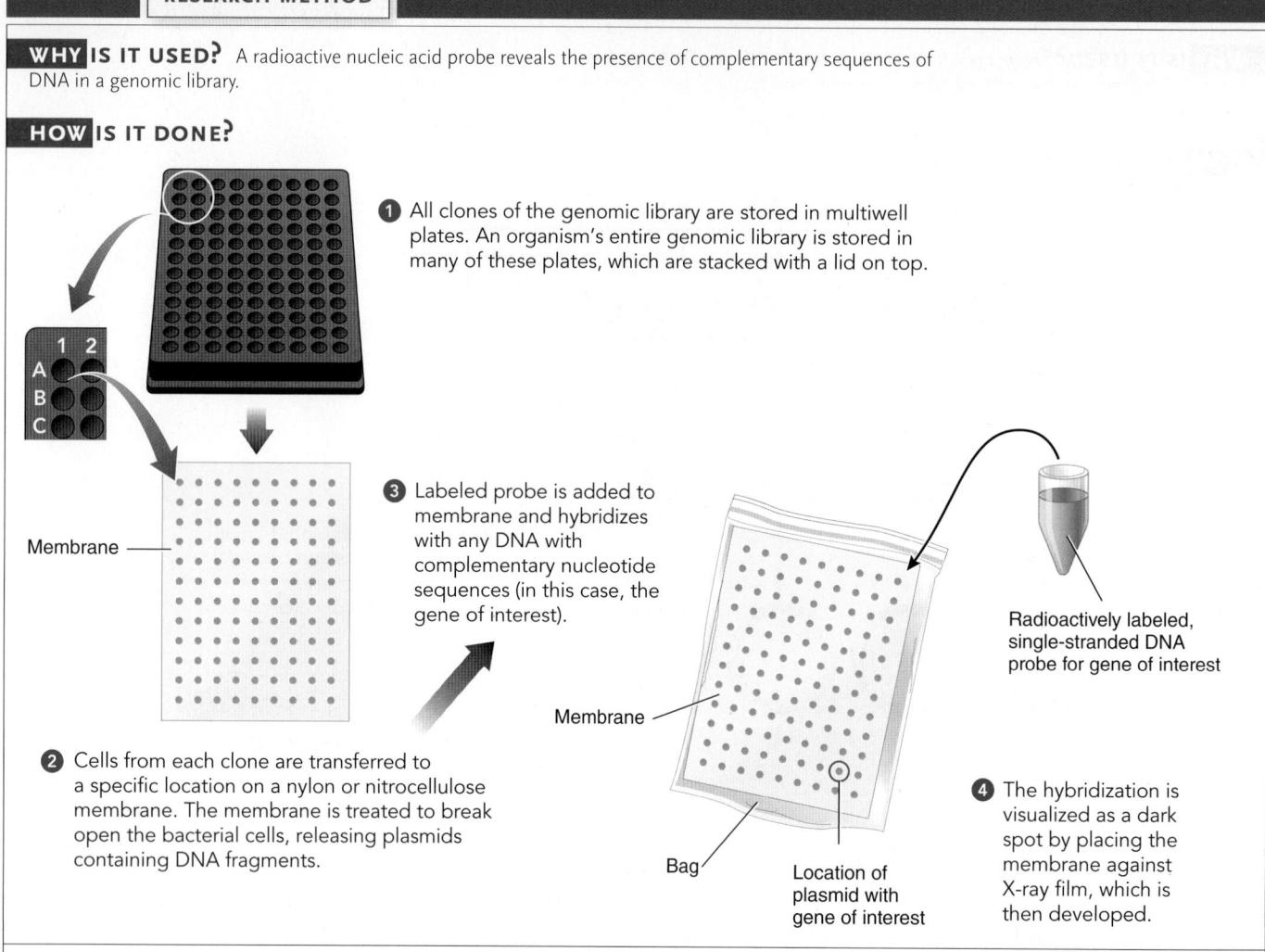

WHY IS IT USED? A radioactive nucleic acid probe reveals the presence of complementary sequences of DNA in a genomic library.

HOW IS IT DONE?

1 All clones of the genomic library are stored in multiwell plates. An organism's entire genomic library is stored in many of these plates, which are stacked with a lid on top.

Membrane

3 Labeled probe is added to membrane and hybridizes with any DNA with complementary nucleotide sequences (in this case, the gene of interest).

Radioactively labeled, single-stranded DNA probe for gene of interest

Membrane

2 Cells from each clone are transferred to a specific location on a nylon or nitrocellulose membrane. The membrane is treated to break open the bacterial cells, releasing plasmids containing DNA fragments.

4 The hybridization is visualized as a dark spot by placing the membrane against X-ray film, which is then developed.

Bag

Location of plasmid with gene of interest

Figure 15-5 *Animation* **Identifying a specific DNA sequence in a genomic library by hybridization with a nucleic acid probe**
© Cengage Learning

removes cells that contain the cloned sequence. The plasmid can then be amplified by taking a small number of cells from one of the wells of the plate, inoculating them in a sterile culture, and allowing them to grow and divide to produce millions of new bacteria, all containing the recombinant DNA of interest.

A cDNA library is complementary to mRNA and does not contain introns

Researchers frequently wish to clone intact genes while avoiding **introns,** which are regions of eukaryotic genes that do not code for proteins (see Chapter 12). Scientists also may wish to clone only genes that are expressed in a particular cell type. In such cases, they construct libraries of DNA copies of mature mRNA from which introns have been removed. The copies, known as **complementary DNA (cDNA)** because they are complementary to mRNA, also lack introns. The researchers first use the enzyme **reverse transcriptase** (see Chapter 13) to synthesize single-stranded cDNA. DNA polymerase is then used to make the cDNA double-stranded (FIG. 15-6). A **cDNA library** is formed

using mRNA from a single cell type as the starting material. The double-stranded cDNA molecules are inserted into plasmid or virus vectors, which then multiply in bacterial cells.

Analyzing cDNA clones lets investigators read the amino acid sequence of the protein from knowledge of the genetic code (see Fig. 13-5). Because the cDNA copy of the mRNA does not contain intron sequences, comparing the DNA base sequences in cDNA and the DNA in genomic DNA libraries reveals the locations of intron and exon coding sequences in the gene.

Cloned cDNA sequences are also useful when geneticists want to produce a eukaryotic protein in bacteria. When they introduce an intron-containing human gene, such as the gene for human growth hormone, into a bacterium, it cannot remove the introns from the transcribed RNA to make a functional mRNA for producing its protein product. If they insert a cDNA clone of the gene into the bacterium, however, its transcript contains an uninterrupted coding region. A functional protein is synthesized if the geneticist inserts the gene downstream of an appropriate bacterial promoter and places the appropriate bacterial translation initiation sequences in it.

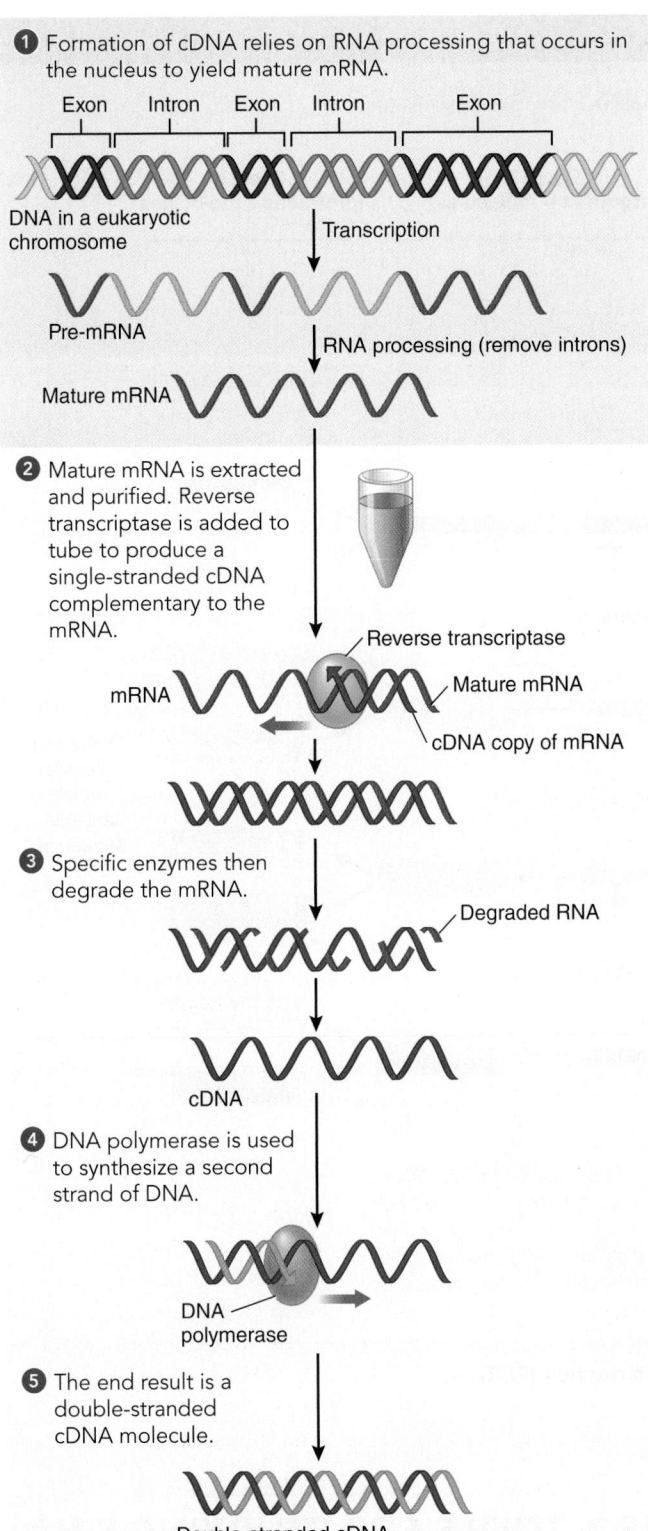

1 Formation of cDNA relies on RNA processing that occurs in the nucleus to yield mature mRNA.

Exon Intron Exon Intron Exon

DNA in a eukaryotic chromosome

↓ Transcription

Pre-mRNA

↓ RNA processing (remove introns)

Mature mRNA

2 Mature mRNA is extracted and purified. Reverse transcriptase is added to tube to produce a single-stranded cDNA complementary to the mRNA.

Reverse transcriptase

mRNA — Mature mRNA

— cDNA copy of mRNA

3 Specific enzymes then degrade the mRNA.

Degraded RNA

cDNA

4 DNA polymerase is used to synthesize a second strand of DNA.

DNA polymerase

5 The end result is a double-stranded cDNA molecule.

Double-stranded cDNA

© Cengage Learning

Figure 15-6 *Animation* **The formation of cDNA**

The polymerase chain reaction amplifies DNA in vitro

The methods just described for amplifying, or making multiple copies of, a specific DNA sequence involve cloning DNA in cells, usually those of bacteria. These processes are time consuming and require an adequate DNA sample as the starting material. The **polymerase chain reaction (PCR),** which biochemist Kary Mullis developed in 1985, lets researchers amplify a tiny sample of DNA *without cloning in a cell.* The DNA is amplified millions of times in a few hours. In 1993, Mullis received the Nobel Prize in Chemistry for his work.

In PCR, DNA polymerase uses nucleotides and primers to replicate a DNA sequence in vitro (outside a living organism), thereby producing two DNA molecules (**FIG. 15-7**). Researchers use automated equipment to heat and cool the reaction mixture repeatedly. The two strands of each DNA molecule are denatured in the heating phases. In the cooling phases, the primers pair with the target sequences, which are then replicated by the DNA polymerase. The first cycle generates 2 double-stranded molecules, the second cycle generates 4, and the third cycle generates 8, so the number of DNA molecules doubles in each cycle. After 20 heating and cooling cycles, this exponential process yields 2^{20}, or more than one million, copies of the target sequence!

Because the reaction can be carried out efficiently only if the DNA polymerase can remain stable through many heating cycles, researchers use a heat-resistant DNA polymerase, known as *Taq* polymerase. The name of this enzyme reflects its source, *Thermus aquaticus,* a bacterium that lives in hot springs in Yellowstone National Park. Because the water in this environment is close to the boiling point, all enzymes in *T. aquaticus* have evolved to be stable at high temperatures. (Recall from Chapter 3 that, when heated, most proteins are irreversibly denatured; they change their overall shape and lose their biological activity.) Similar heat-resistant enzymes have evolved in bacteria or archaea living in deep-sea thermal vents.

The PCR technique has many applications. It makes cloning easier and allows rapid and sophisticated detection of specific nucleic acid sequences for many purposes. It enables researchers to amplify and analyze tiny DNA samples from a variety of sources, ranging from crime scenes to archaeological remains. For example, investigators have used PCR to analyze DNA obtained from the bones of Neanderthals (see Chapter 22) and prehistoric animals. The technique has also been adapted to detect differences in the DNA of human and animal populations as well as to measure mRNA levels of specific genes in diseased and normal tissues.

A limitation of the PCR technique is that it is almost too sensitive. Even a tiny amount of contaminant DNA in a sample may become amplified if it includes a DNA sequence complementary to the primers, potentially leading to an erroneous conclusion. Researchers take appropriate precautions to avoid this and other technical pitfalls.

CHECKPOINT 15.1

- **VISUALIZE** *An arrow marks the position where a particular restriction enzyme cuts this DNA sequence: 5′ G↓GATCC. Write out the sequence and the position of the cut for the complementary strand.*
- **CONNECT** *What do genomic DNA libraries and cDNA libraries have in common? How do they differ?*
- *How are DNA probes used?*

WHY IS IT USED? PCR takes an extremely small sample of DNA and amplifies it to millions of copies in a few hours.

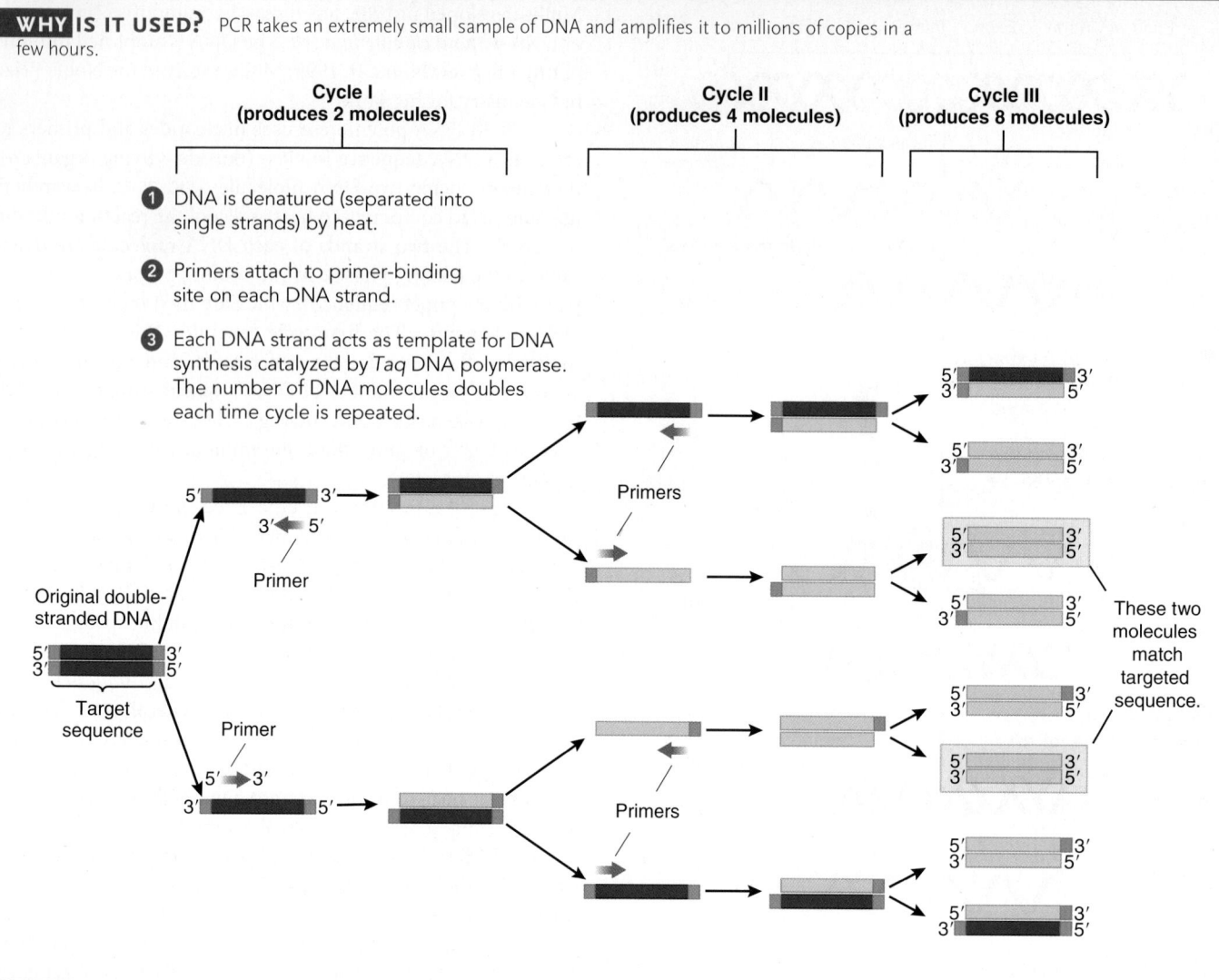

Cycle I
(produces 2 molecules)

Cycle II
(produces 4 molecules)

Cycle III
(produces 8 molecules)

1. DNA is denatured (separated into single strands) by heat.

2. Primers attach to primer-binding site on each DNA strand.

3. Each DNA strand acts as template for DNA synthesis catalyzed by *Taq* DNA polymerase. The number of DNA molecules doubles each time cycle is repeated.

Original double-stranded DNA

Target sequence

Primer

Primer

Primers

Primers

These two molecules match targeted sequence.

HOW IS IT DONE? The initial reaction mixture contains a very small amount of double-stranded DNA, DNA precursors (deoxyribonucleotides), specific nucleic acid primers, and heat-resistant *Taq* polymerase. PCR involves repeated cycles of denaturation, primer attachment, and strand elongation. Starting from one double-stranded DNA molecule, at the end of cycle 3 two of the eight double-stranded molecules will consist of only the target sequence (*highlighted in yellow*). At the end of cycle 21 over one million double-stranded copies of the target will be formed, and after 30 cycles over one billion copies will be produced.

Figure 15-7 *Animation* **Amplification of DNA by the polymerase chain reaction (PCR)**
© Cengage Learning

- **PREDICT** *The size of the human genome is 3×10^9 base pairs of DNA. What is the minimum number of colonies of E. coli that would have to be isolated to clone the entire genome if the plasmid used could only contain a 10 kb (1×10^4 base pair) fragment? What is the minimum number that would be required if a BAC vector that accepts a 350 kb fragment were used? Would it be sufficient for an experimenter to isolate the calculated minimum number of colonies? Explain your answer.*

- *What advantages does the PCR method have over gene cloning in bacterial vectors?*

15.2 TOOLS FOR STUDYING DNA

LEARNING OBJECTIVES

5 Distinguish among DNA, RNA, and protein blotting.

6 Describe how gene databases are used to gain insights into gene structure, expression, and function.

7 Discuss how Northern blotting, RT–PCR, and DNA chips are used to compare levels of gene expression among normal versus diseased tissues.

DNA cloning and PCR methods have had major effects on genetic analysis because they provide enough genetic material to address certain fundamental questions about specific genes and their RNA or protein products. For example, what is the base sequence that encompasses a gene? When—and how—is a gene expressed? Is a given gene identical from one individual to the next? Is it different from one species to the next? Where is the gene found within the genome? And most, important, what is the function of the gene? In this and the following section, we consider various techniques that help address these questions. We then examine some of the practical applications of DNA technology in the final section of the chapter.

Gel electrophoresis is used for separating macromolecules

Mixtures of certain macromolecules—proteins, polypeptides, DNA fragments, or RNA—are separated by **gel electrophoresis,** a method that exploits the charged groups on these molecules that cause them to migrate in an electrical field. FIGURE 15-8 illustrates gel electrophoresis of DNA molecules. Nucleic acids migrate through the gel toward the positive pole of the electric field because they are negatively charged due to their phosphate groups. Because the gel slows down the large molecules more than the small molecules, the DNA is separated by size. Including DNA fragments of known size as standards ensures accurate measurements of the molecular weights of the unknown fragments. The gel is stained after the DNA fragments are separated to show their number and location.

DNA, RNA, and protein blots detect differences in related molecules separated by gel electrophoresis

The development of recombinant DNA methods revolutionized the way scientists could study genes and the ways in which they were expressed. Using homologous probes produced from cloned DNA sequences (and, similarly, highly specific antibodies against proteins), scientists were able to identify and measure specific DNA fragments, RNAs, and proteins from cells and tissues. Many of the findings of these early studies relied on the development of *gel electrophoresis blotting methods*. For example, the DNA **Southern blot** method (named after Edwin Southern of Edinburgh University) allowed investigators to discriminate between the normal and mutant alleles of a gene with a **restriction fragment length polymorphism (RFLP)** (FIG. 15-9).

RFLPs are caused by mutations in the DNA that create (or destroy) a specific restriction enzyme site within a gene. Different lengths of DNA fragments are produced from alleles that differ in their restriction sites when the genomic DNA is cut with that specific enzyme. To identify individuals who might have a mutant allele of the affected gene, a sample of their genomic DNA is digested with that enzyme, and the fragments are separated by gel electrophoresis. After the DNA is transferred from the gel to a nylon membrane (by blotting), it is hybridized with a radioactively labeled probe that is complementary to the gene. The washed membrane is then exposed to an X-ray film. Only the DNA fragments that hybridize to the probe are visible on the developed film, even though each lane of the gel contains pieces of all the DNA in the individual's genome.

The Southern blot procedure has had widespread applications, such as diagnosing certain types of genetic disorders, conducting paternity testing, and analyzing evidence found at crime scenes. RFLP analysis has also been used to help map the exact location of gene mutations, such as the mutation that causes cystic fibrosis. Although the RFLP technique has provided useful information in many areas of biology, it is rapidly being replaced by newer methods, such as automated DNA sequencing (discussed in the next section).

Similar blotting techniques are used to study RNA and proteins. When RNA molecules separated by electrophoresis are transferred to a membrane and detected using a nucleic acid probe, the result is called a **Northern blot** (as a pun on Southern's name being associated with his technique). In the same spirit, the term **Western blot** is applied to a blot consisting of proteins or polypeptides previously separated by gel electrophoresis. (So far, no one has invented a type of blot that could be called an Eastern blot.) In the case of Western blotting, scientists recognize the polypeptides of interest by using labeled antibody molecules that bind to them specifically. For example, Western blotting is used diagnostically to detect the presence of proteins specific to HIV-1, the AIDS virus.

Automated DNA sequencing methods have been developed

One of the first ways DNA is characterized is to determine its nucleotide sequence. From this primary information, investigators can then use a cloned piece of the DNA as a research tool for many different applications.

Almost all **DNA sequencing** today is carried out through automated machines that can sequence huge amounts of DNA quickly and reliably. Initially, DNA sequencing relied on the *chain termination method* (FIG. 15-10) that British biochemist and Nobel laureate Fred Sanger developed in 1974. In 1980, Sanger shared the Nobel Prize in Chemistry for this contribution.

The chain termination method of DNA sequencing is based on replicating a cloned DNA strand in a reaction mixture containing small amounts of modified synthetic nucleotides, known as *dideoxynucleotides*. A dideoxynucleotide lacks a hydroxyl group on its 3′ carbon. When a dideoxynucleotide is incorporated into a new DNA strand, growth of the new strand is terminated because DNA polymerase cannot form a phosphodiester linkage at the modified 3′ carbon atom. Figure 15-10 shows how a small amount of dideoxyATP (ddATP) added to the reaction mixture will produce a series of DNA fragments of different lengths, all ending in dideoxyA. Similarly, a reaction mixture containing small amounts of ddGTP would produce a series of DNA fragments corresponding to the positions of guanine bases on the new strand.

To sequence a DNA fragment, radioactively labeled fragments from four different reaction mixtures, each with a different dideoxynucleotide, are separated by high-resolution gel electrophoresis. The positions of the newly synthesized fragments in the gel can then be determined by autoradiography.

WHY IS IT USED? Gel electrophoresis separates molecules such as nucleic acids based on their size and charge as they migrate through an electrical field.

HOW IS IT DONE?

① A researcher sets up an electrical field in a gel material, consisting of agarose or polyacrylamide, which is poured as a thin slab on a glass or Plexiglas holder. After the gel has solidified, the researcher loads samples containing a mixture of macromolecules (in this case, DNA fragments of different sizes) in wells formed at one end of the gel.

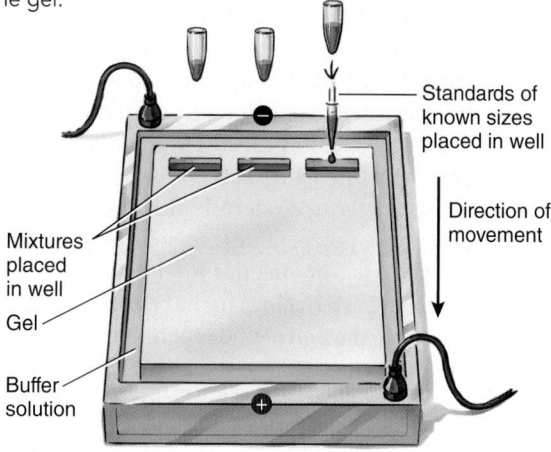

Standards of known sizes placed in well

Direction of movement

Mixtures placed in well

Gel

Buffer solution

② When an electrical current is applied, the negatively charged DNA molecules move from the negative pole to the positive pole. The smallest DNA fragments travel the longest distance. The DNA bands are invisible at this point.

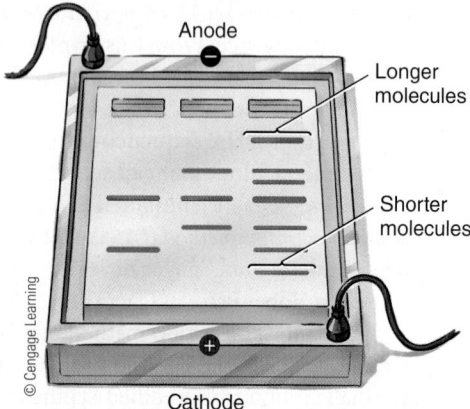

Anode

Longer molecules

Shorter molecules

© Cengage Learning

Cathode

③ This gel contains separated DNA fragments. The gel is stained with ethidium bromide, a dye that binds to DNA and is fluorescent under UV light.

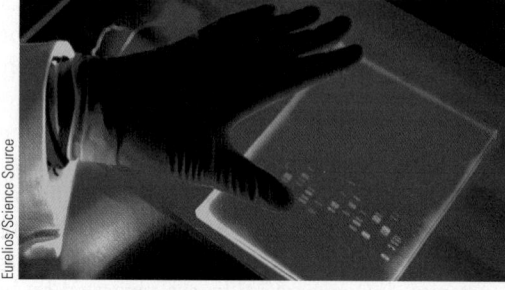

Eurelios/Science Source

Figure 15-8 *Animation* Gel electrophoresis

Because the high-resolution gel can separate DNA fragments that differ in length by only a single nucleotide, the researcher can read off the sequence in the newly synthesized DNA one base at a time.

Today's "next-generation" automatic sequencing machines employ fluorescently labeled nucleotides. Each labeled nucleotide fluoresces at a different wavelength that is read by sensitive photodetectors, dramatically increasing the speed and reducing the cost of DNA sequencing. These machines can decode about 1.5 million bases in a 24-hour period. These advances in rapid sequencing technology have made it possible for researchers to determine the nucleotide sequences of entire genomes in a wide variety of bacteria, archaea, and eukaryotic organisms.

Gene databases are powerful research tools

We are in the middle of an extraordinary explosion of gene sequence data, largely due to automated sequencing methods. Much of this research received its initial impetus from the *Human Genome Project,* which began in 1990. The sequencing of the three billion base pairs of the human genome was essentially completed in 2001. Scientists now rely on massive databases of DNA sequence information that can be accessed and analyzed through the Internet. Published nucleotide sequences from DNA sequencing projects are now collected by the *International Nucleotide Sequence Database Collaboration,* which comprises the *GenBank* at the *United States National Center for Biotechnology Information (NCBI),* the *DNA DataBank of Japan (DDBJ),* and the *European Molecular Biology Laboratory (EMBL).* These sites also provide numerous other databases, including protein sequence data derived from the stored DNA sequences. All the sites collect and share their sequence data on a daily basis.

The genomes of over 1000 organisms from all three domains of life have now been sequenced. Geneticists use powerful **bioinformatics** software and supercomputers provided with these databases to compare newly discovered sequences and analyze gene expression data and protein structures. For example, an investigator can enter a newly determined human gene sequence (or the corresponding amino acid sequence of its encoded protein) into analysis software called *blast* at the NCBI website (**FIG. 15-11**). National Library of Medicine

WHY IS IT USED? The Southern blotting technique allows researchers to compare specific DNA sequences from complex mixtures of DNA fragments. In this example, DNA encoding a specific gene is compared among three individuals using RFLP analysis.

HOW IS IT DONE?

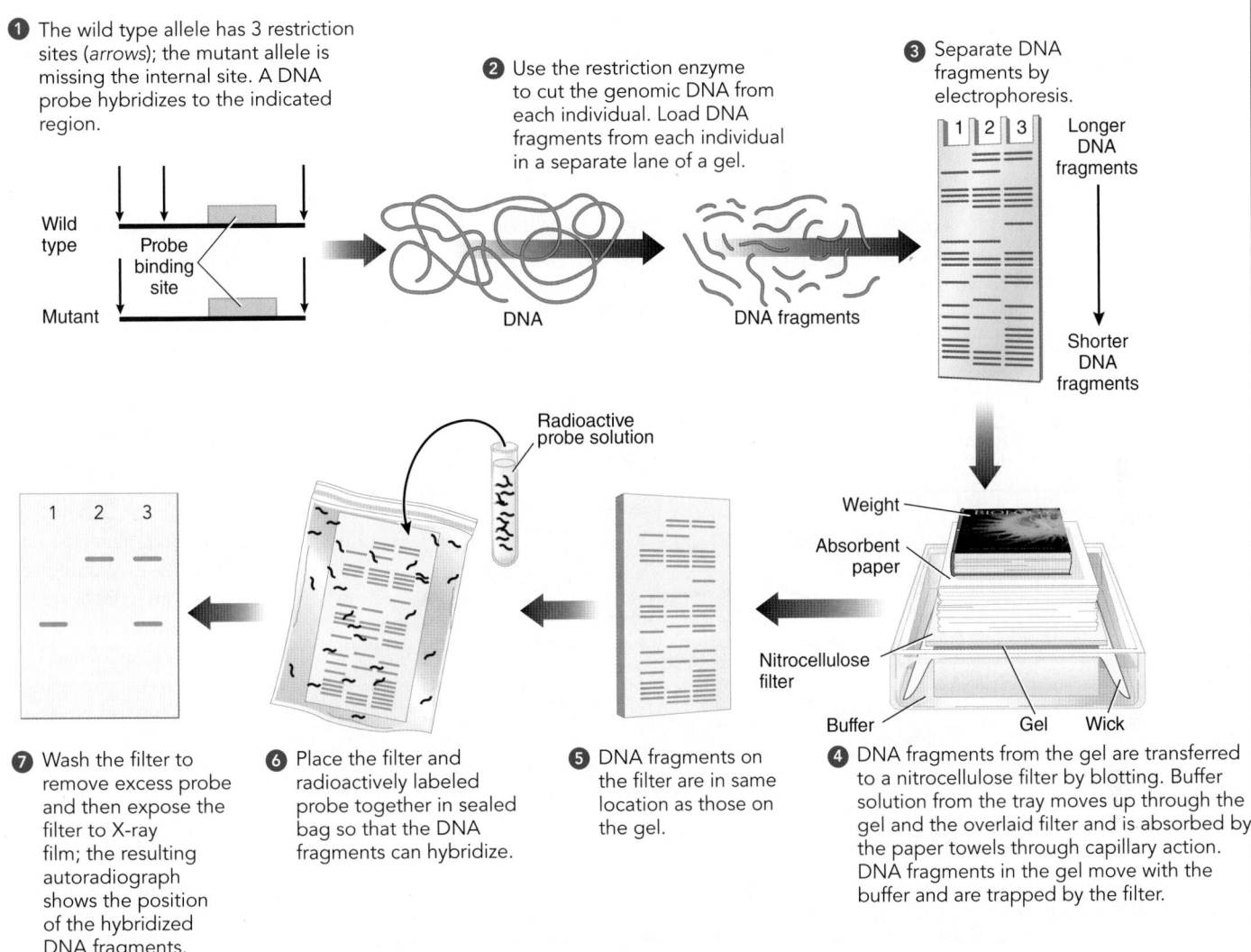

1. The wild type allele has 3 restriction sites (*arrows*); the mutant allele is missing the internal site. A DNA probe hybridizes to the indicated region.

2. Use the restriction enzyme to cut the genomic DNA from each individual. Load DNA fragments from each individual in a separate lane of a gel.

3. Separate DNA fragments by electrophoresis.

4. DNA fragments from the gel are transferred to a nitrocellulose filter by blotting. Buffer solution from the tray moves up through the gel and the overlaid filter and is absorbed by the paper towels through capillary action. DNA fragments in the gel move with the buffer and are trapped by the filter.

5. DNA fragments on the filter are in same location as those on the gel.

6. Place the filter and radioactively labeled probe together in sealed bag so that the DNA fragments can hybridize.

7. Wash the filter to remove excess probe and then expose the filter to X-ray film; the resulting autoradiograph shows the position of the hybridized DNA fragments.

Figure 15-9 The Southern blotting technique

CONNECT Compare Steps 1 and 7. Which of the lanes in Step 7 corresponds to a heterozygous individual? a homozygous mutant? a homozygous wild type individual?

© Cengage Learning

supercomputers then search the entire database (or specified parts of the database) and supply detailed comparisons between closely related gene sequences from other organisms and the newly discovered gene. This information can give important clues about the function of the gene and structural features about its encoded protein. These types of comparative data can also yield valuable information about the evolutionary relationships among genes as well as the variability among gene sequences within a population. DNA analysis of many different species using these methods has added much new information in support of evolution (see Chapter 18).

Reverse transcription of mRNA to cDNA is used to measure gene expression in a number of ways

The ability to detect and measure the levels of an expressed RNA in different cell types can provide important clues to its gene's expression and function. Using the Northern blot procedure, investigators are typically limited to measuring the levels of RNA expression of a single gene (identified by the radioactive probe) from a cellular preparation. Furthermore, a single Northern blot experiment usually requires several days from

WHY IS IT USED? The dideoxy chain termination technique illustrates a method of DNA sequencing.

HOW IS IT DONE?

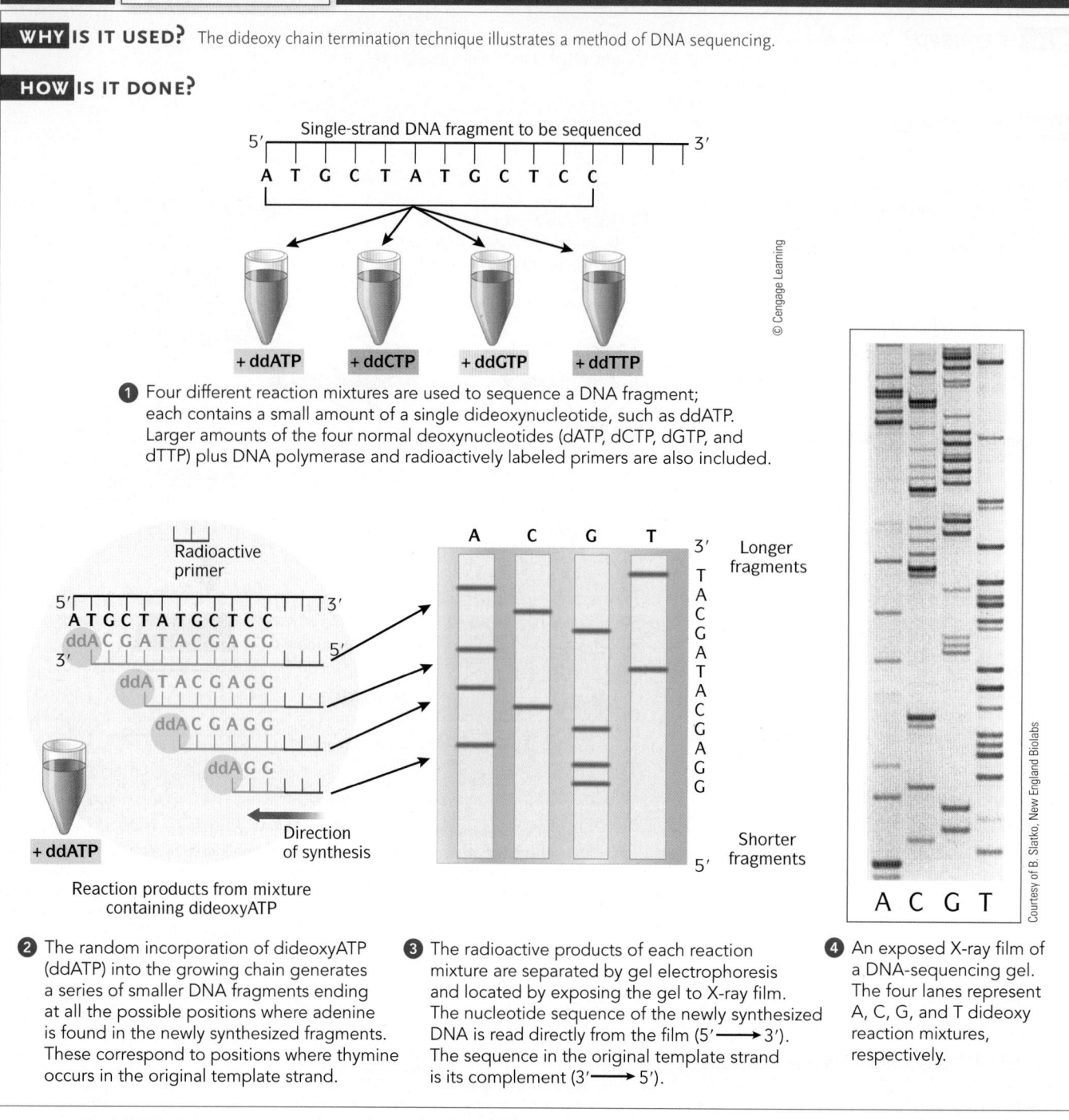

① Four different reaction mixtures are used to sequence a DNA fragment; each contains a small amount of a single dideoxynucleotide, such as ddATP. Larger amounts of the four normal deoxynucleotides (dATP, dCTP, dGTP, and dTTP) plus DNA polymerase and radioactively labeled primers are also included.

② The random incorporation of dideoxyATP (ddATP) into the growing chain generates a series of smaller DNA fragments ending at all the possible positions where adenine is found in the newly synthesized fragments. These correspond to positions where thymine occurs in the original template strand.

③ The radioactive products of each reaction mixture are separated by gel electrophoresis and located by exposing the gel to X-ray film. The nucleotide sequence of the newly synthesized DNA is read directly from the film (5'⟶3'). The sequence in the original template strand is its complement (3'⟶5').

④ An exposed X-ray film of a DNA-sequencing gel. The four lanes represent A, C, G, and T dideoxy reaction mixtures, respectively.

Figure 15-10 The dideoxy chain termination method of DNA sequencing

the time the RNA samples are isolated until the results can be read. The ability to copy an mRNA into a cDNA using reverse transcriptase and then amplify the cDNA by PCR has led to the development of rapid and precise methods for measuring the levels of many different RNA species from a cell type.

Automated RT–PCR machines use special fluorescent DNA primers to measure mRNA levels For RT–PCR analysis, total cellular RNA fractions are first copied to cDNA molecules using reverse transcriptase. Each well of a transparent multiwell plate contains a portion of the cDNA sample, DNA primer molecules for a specific mRNA sequence, and fluorescent probe molecules used to detect the cDNAs. As the cDNA in each well is amplified, the intensity of the emitted fluorescent light in each well is proportional to the number of double-stranded cDNA molecules produced by the PCR reaction. By measuring the number of amplification cycles required to achieve a certain level of fluorescence, the researcher can

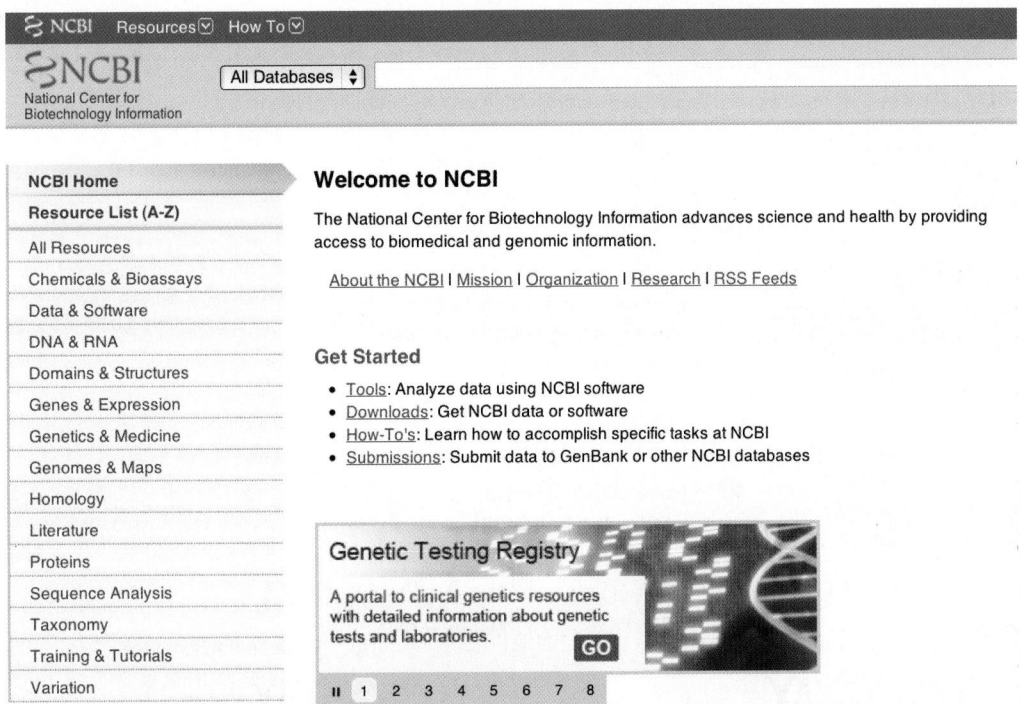

Figure 15-11 National Center for Biotechnology Information (NCBI) tools for DNA analysis
This website gives scientists and the public access to DNA and protein databases in addition to bioinformatics software tools for their analysis.

calculate the original number of the targeted mRNA molecules present in that type of cell.

DNA microarrays use reverse-transcribed mRNAs to study large-scale patterns of gene expression DNA microarrays *(DNA chips)* are used to simultaneously compare the expression of thousands of genes from cells grown under different conditions (FIG. 15-12). The microarrays are glass slides with hundreds of spots, or *microdots,* prepared by mechanical robots. Each microdot contains many copies of a single-stranded cDNA molecule that is a complementary copy of a specific mRNA produced by the organism being studied.

Researchers then isolate mature mRNA molecules from two cell populations, such as liver cells treated with a newly developed drug and control liver cells not treated with the drug. Liver cells are often tested because the liver produces many enzymes that metabolize foreign molecules, including drug molecules. Drugs that the liver cannot metabolize, or metabolizes weakly, may be too toxic to be used.

Researchers use the isolated mRNA molecules to prepare cDNA molecules that are tagged with different-colored fluorescent dyes for each cell population. For example, the treated cells' cDNA molecules might be labeled with red dye and the untreated cells' cDNA molecules with green dye. Researchers add the two cDNA populations to the array, which then hybridize (form base pairs) with complementary DNA spots on the array.

Investigators then scan the array with lasers to measure the levels of red and green fluorescence in each microdot. Computer analysis of the ratio of red to green fluorescence at each spot in the array produces a color-coded readout that researchers can further analyze. For example, a medical researcher might compare the overall pattern of gene activity in cells treated with an experimental drug with that of cells exposed to known drugs and toxins. If the pattern of gene activity for the treated cells matches that of cells treated with a toxin that damages the liver, the drug will probably not go into clinical trials.

DNA microarrays enable researchers to compare the activities of thousands of genes in cell types from various tissue samples and in normal and diseased cells. Because cancer and other diseases exhibit altered patterns of gene expression, DNA microarrays have the potential to identify disease-causing genes or the proteins they code for, which can then be targeted by therapeutic drugs.

For an example of an application, let us consider a specific cancer. Diffuse large B-cell lymphoma is a cancer of white blood cells that has different subtypes. Researchers used microarrays to identify 17 genes that are active in various combinations for each subtype. These studies determined which combination of those 17 genes is active in each subtype of this form of cancer. Armed with the knowledge of which cancer subtype a particular patient has, physicians can then choose the treatment that will probably be most effective for that patient. Examining patterns of gene activity with DNA microarrays also helps medical scientists identify which patients will probably remain free of cancer after treatment and which will probably relapse.

CHECKPOINT 15.2

- CONNECT *Explain how you would compare the expression of a gene in a cancer cell and a normal cell using a Northern Blot, RT–PCR, and a DNA chip.*

- *Explain how you might use GenBank to obtain information about the function of a gene that you cloned and sequenced from a tumor cell.*

WHY IS IT USED? DNA microarray analysis indicates gene expression levels of hundreds or even thousands of specific genes in different cell types. This technique helps scientists understand which genes are active (or inactive) in untreated (control) cells as well as in cells treated with a specific drug.

HOW IS IT DONE?

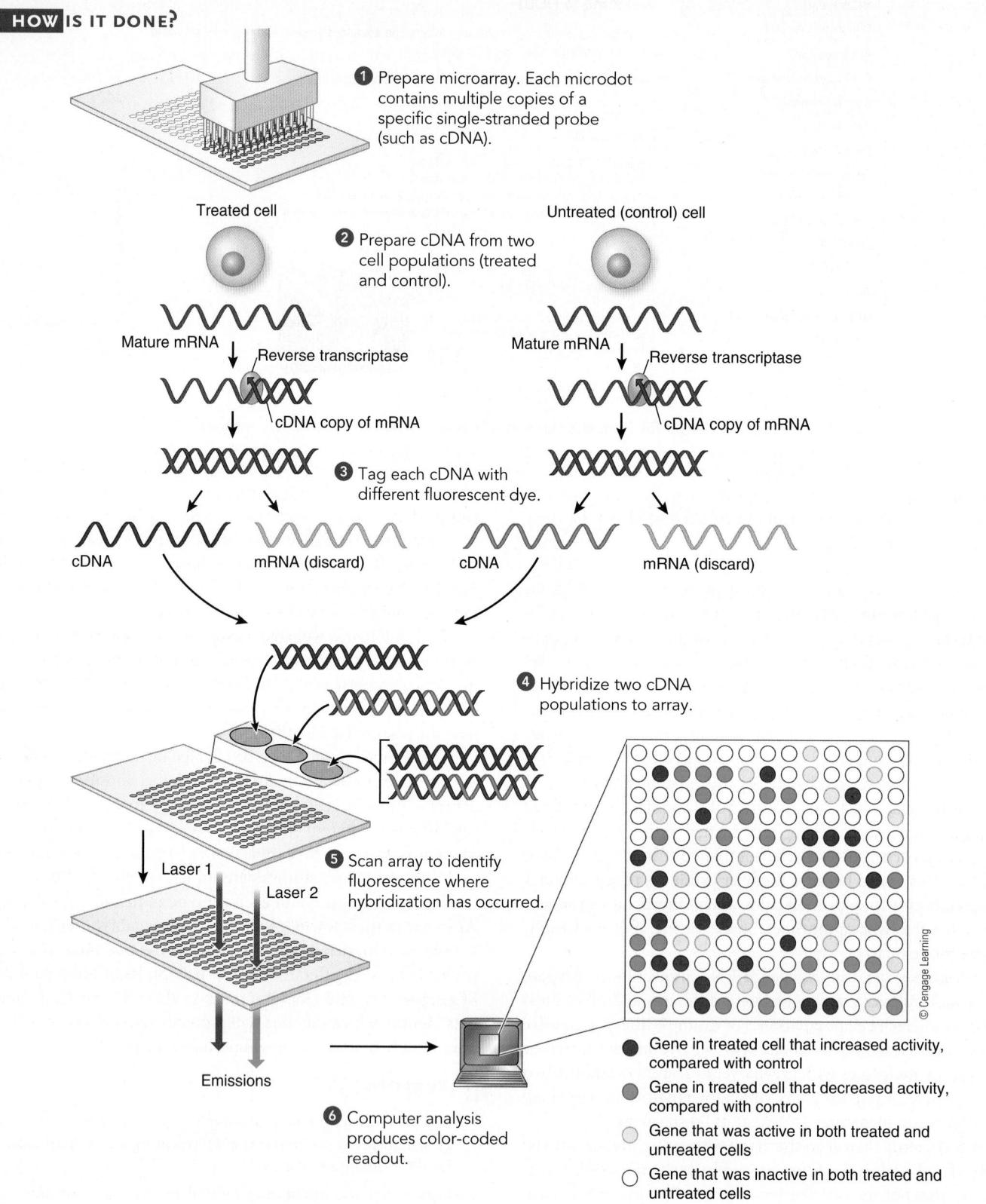

Figure 15-12 **DNA microarray analysis**

15.3 GENOMICS

LEARNING OBJECTIVES

8 Describe how genome-wide association studies have led to new understandings about the structure and function of the genome.

9 Explain how targeted gene silencing and knockout mutations can give insights into the functions of a gene.

Now that we have sequenced the genomes of humans and more than one thousand other species, what do we do with that information? **Genomics** is the emerging field of biology that studies entire genomes to identify all the genes, determine their functions, and investigate how they act together. Through the use of new rapid DNA sequencing and DNA chip technologies, investigators from many laboratories can now work together to conduct *genome-wide association studies* that collect simultaneous sequence, expression, and functional data from entire genomes for analysis.

Collaborative genome-wide association studies have radically changed our view of the human genome

One **genome-wide association study (GWAS)**, the *ENCODE (ENCyclopedia Of DNA Elements) Consortium,* is spearheaded by an international group of scientists who use different types of high-speed DNA sequencing and DNA chip-based methods; their goal is to identify all the functional regions of the human genome. In 2012, the initial results of this study were published in 30 different scientific papers. These papers, which summarized the results of 1640 genome-wide data sets from 147 different cell types, dramatically changed our understanding of the complexity of the human genome. Previous data, including that from the Human Genome Project, had indicated that only 1.5% of the genome consisted of protein-coding genes and that probably no more than 5% of the total genome encoded functional RNAs. These findings led to a long-held idea that the vast majority of the human genome was composed of "gene deserts" filled with "junk DNA." The combined studies of the ENCODE project showed, however, that as much as 75% of the genome contains elements linked to biochemical functions. Most of these encode non-protein-coding RNAs that appear to be involved with the regulation of expression of protein-coding genes.

These findings of these functional, non-protein-coding RNAs have led to important insights concerning cancer and human genetic diseases. For example, previous studies had showed that about 90% of all single base mutations (**single nucleotide polymorphisms,** or **SNPs**) that are associated with cancers and genetic diseases in humans occur in areas of the genome not associated with protein-coding genes. The ENCODE study showed that many of these disease-linked mutations were associated with the non-protein-coding regulatory regions. Now that these connections have been discovered, the goal is to uncover how these non-protein-coding genes act to regulate interacting genetic systems in different cell types.

Comparative genomic databases are tools for uncovering gene functions

To aid in analyzing the human genome, investigators use *comparative genomics* involving sequencing and mapping studies on other organisms. Comparison of the DNA sequences and chromosome organization of related genes from different species is a powerful tool for identifying the elements essential for their functions. If a human gene has an unknown function, researchers can often deduce clues about its role by studying the equivalent gene in another species, such as a mouse or rat.

Mice and rats were obvious choices for DNA sequencing because they have been studied for almost a century and much is known about their biology. They are also sufficiently different from humans that any conserved (i.e., identical) sequences found in both rodents and humans are probably functionally important.

In addition, mice are similar to humans in that both share many physiological traits, including some of the same diseases. To demonstrate the shared similarity between the mouse and human genomes, a consortium of scientists published a study that compared mouse chromosome 16 with the human genome. Only 14 of the 731 known genes on the mouse chromosome had no human counterpart; the remaining 717 genes appeared in one form or another in the human genome.

Investigators have also sequenced the genomes of other model vertebrate organisms, such as the zebrafish and pufferfish. Biologists have used the zebrafish for many years as a model of vertebrate development because its embryonic stages are transparent and therefore convenient to examine. The pufferfish has the smallest known genome of all vertebrates. By comparing the human genome to that of the pufferfish, researchers have identified shared sequences that have not changed in several hundred million years. Because these sequences have resisted the selective forces of evolution, they may be genes or regulatory elements that direct vertebrate development and are essential in all vertebrates.

Genome analysis of commercially important organisms, such as salmon, chickens, pigs, and rice, is also a priority. For example, scientists have published the rice genome, the genome of the parasite that causes malaria, and the genome of the parasite that causes African sleeping sickness.

One of the most significant genome-sequencing efforts from an evolutionary perspective focuses on nonhuman primates, such as the chimpanzee (whose sequence data were published in 2005). Comparing the DNA sequences of humans and our closest living relatives is helping biologists understand the genetic changes that occurred during human evolution, including which genes govern mental and linguistic capabilities.

RNA interference is used to study gene functions

RNA interference (RNAi), first discovered in the mid-1990s, can be used to determine the function of a specific gene quickly. You may recall from Chapter 13 that RNAi is caused by small RNA molecules that interfere with the expression of genes (i.e., "silence" them) or, more precisely, their mature mRNA transcripts. RNAi involves small interfering RNAs, microRNAs, and a few other kinds of short RNA molecules.

After a protein-coding gene is identified, the function of that gene can be studied by using RNAi to inactivate the gene. To do so, biologists synthesize a short stretch of RNA that is complementary to part of the DNA sequence of the gene being examined. The RNA is put into cells, and after the gene is silenced, biologists observe any changes in the phenotype to help determine the function of the missing protein. For example, silencing a gene might cause a defect in certain connections of nerve cells in the brain, which would indicate that the gene has a role in the development of the nervous system.

RNAi silencing is now used in large-scale *functional genomics* studies to systematically examine gene functions in organisms such as the nematode worm *C. elegans;* plants such as wheat, *Arabidopsis,* and maize; and human cells grown in culture. In these studies, RNAi libraries are constructed based on gene sequence data. In the case of *C. elegans,* an RNAi-producing gene is cloned into the *E. coli* bacteria. As the bacteria are eaten by the worms, the RNA is delivered to its intestinal tract, where it is absorbed. These methods have identified numerous new genes involved in gametogenesis and development.

Targeted gene "knockouts" reveal gene functions in mice Another research tool that reveals gene function is *gene targeting,* a procedure in which the researcher chooses and "knocks out," or inactivates, a single gene in an organism. **Knockout mice** have been particularly useful in studying the roles of orthologous genes with unknown functions in mammals, including humans. **Orthologous genes** are homologous genes that are present in different species because they were inherited from a common ancestor. The roles of the inactivated gene are determined by observing the phenotype of the mice bearing the knockout gene. If the gene codes for a protein, for example, studying individuals who lack that protein helps the researcher identify the function of the protein. Because 98% to 99% of the loci of mice are orthologous to human loci, information about knockout genes in mice also provides details about human genes.

Unlike RNA interference, gene targeting is a lengthy procedure. It takes about one year for scientists to develop a new strain of knockout mice through gene targeting. A nonfunctional, or knockout, gene is cloned and introduced into mouse **embryonic stem cells (ES cells)** (also see Chapters 1 and 17). In a tiny fraction of these ES cells, the introduced knockout gene and the gene on the chromosome will exchange DNA segments in a process called *homologous recombination.* In this way, the knockout allele replaces the normal allele in the mouse chromosome.

ES cells, like cancer cells, grow in culture indefinitely. Most important, if placed in a mouse embryo, ES cells divide and produce all the cell types normally found in the mouse. Researchers inject the knockout ES cells into early mouse embryos and let the mice develop to maturity. Because many genes are essential to life, researchers usually employ special knockout vectors that selectively inactivate the gene in only one cell type, such as a liver or nerve cell. Research laboratories around the world have developed thousands of different strains of knockout mice, each displaying its own characteristic phenotype, and the number continues to grow.

Gene targeting in mice is providing answers to basic biological questions relating to the development of embryos, the development of the nervous system, and the normal functioning of the immune system. This technique has great potential for revealing more about various human diseases, especially because many diseases have a genetic component. Geneticists are using gene targeting to study cancer, heart disease, sickle cell anemia, respiratory diseases such as cystic fibrosis, and other disorders.

CHECKPOINT 15.3

* **PREDICT** *Exposing* C. elegans *to an RNAi molecule homologous to a non-protein-coding RNA gene resulted in the worms' inability to respond to touch. What predictions could you make about the function of that gene in development?*

15.4 APPLICATIONS OF DNA TECHNOLOGIES

LEARNING OBJECTIVE

10 Describe at least one important application of recombinant DNA technology in each of the following fields: medicine, DNA fingerprinting, and transgenic organisms.

Recombinant DNA technology has provided a new and unique set of tools for examining fundamental questions about cells as well as new approaches to applied problems in many other fields. In some areas the production of genetically modified proteins and organisms has begun to have considerable effect on our lives. The most striking of these has been in medicine.

DNA technology has revolutionized medicine

In increasing numbers of cases, doctors perform *genetic tests* to determine whether an individual has a particular genetic mutation associated with such disorders as Huntington's disease, hemophilia, cystic fibrosis, Tay-Sachs disease, breast cancer, and sickle cell anemia. **Gene therapy,** the use of specific DNA to treat a genetic disease by correcting the genetic problem, is another application of DNA technology that is currently in its infancy. Because genetic testing and gene therapy focus almost exclusively on humans, these applications of DNA technology are discussed in Chapter 16. We focus our discussion here on the use of DNA technology to produce medical products.

Human *insulin* produced by *E. coli* became one of the first genetically modified proteins approved for use by humans. Before the use of recombinant DNA techniques to generate genetically altered bacteria capable of producing the human hormone, insulin was derived exclusively from other animals. Many diabetic patients become allergic to the insulin from animal sources because its amino acid sequence differs slightly from that of human insulin. The ability to produce the human hormone by recombinant DNA methods has resulted in significant medical benefits to insulin-dependent diabetics.

Genetically modified human *growth hormone (GH)* is available to children who need it to overcome growth deficiencies, specifically pituitary dwarfism (see Chapter 49). In the past, human GH was obtainable only from cadavers. Only small amounts were available, and evidence suggested that some of the preparations from human cadavers were contaminated with infectious agents similar to those that cause bovine spongiform encephalopathy, or mad cow disease (discussed in Chapter 24).

The list of products produced by genetic engineering is continually growing. For example, *tissue plasminogen activator (TPA)*, a protein that prevents or dissolves blood clots, is used to treat stroke caused by a blood clot. If administered shortly after the onset of symptoms, TPA can break up the clot and restore blood flow to the affected tissues. *Tissue growth factor-β (TGF-β)* promotes the growth of blood vessels and skin, and is used in wound and burn healing. Researchers also use TGF-β in tissue engineering, a developing technology to meet the pressing need for human tissues and, eventually, organs for transplantation by growing them from cell cultures.

The U.S. Food and Drug Administration (FDA) has approved tissue-engineered skin grafts for individuals who have been badly burned, and tissue-engineered cartilage is used for joint repairs.

Hemophilia A is treated with *human clotting factor VIII*. Before the development of recombinant DNA techniques, factor VIII was available only from human or animal-derived blood, which posed a risk of transmitting infectious agents, such as HIV-1.

Recombinant DNA technology is increasingly used to produce vaccines that provide safe and effective immunity against infectious diseases in humans and animals. One way to develop a recombinant vaccine is to clone a gene for a surface protein produced by the disease-causing agent, or pathogen; the researcher then introduces the gene into a non-disease-causing vector. The vaccine, when delivered into the human or animal host, stimulates an immune response to the surface-exposed protein. As a result, if the pathogen carrying that specific surface protein is encountered, the immune system targets it for destruction. Human examples of antiviral recombinant vaccines include vaccines for influenza A, hepatitis B, and polio.

Recombinant vaccines are also being developed against certain bacterial diseases and human cancers. For example, in 2006 the FDA approved recombinant vaccines against several types of human papillomavirus (HPV). These vaccines, which are administered to females and males age 9 to 26, protect against the 70% of cervical cancers caused by HPV. They also provide protection against genital warts, anal cancers, and several other cancers.

DNA fingerprinting has numerous applications

The analysis of DNA fragments extracted from an individual, which is unique to that individual, is known as **DNA fingerprinting.** DNA fingerprinting has many applications in humans and other organisms, as in the following examples:

1. Analyzing evidence found at crime scenes (forensic analysis) (FIG. 15-13)
2. Exonerating prisoners wrongfully convicted of a crime
3. Identifying mass disaster victims
4. Proving parentage in dogs for pedigree registration purposes
5. Identifying human cancer cell lines
6 Studying endangered species in conservation biology
7. Tracking tainted foods
8. Studying the genetic ancestry of human populations
9. Clarifying disputed parentage

Historically, DNA fingerprinting relied on RFLPs (discussed earlier). Today, many techniques for DNA fingerprinting rely on PCR amplification, restriction enzyme digestion, and Southern blot hybridization to detect molecular markers. The most useful molecular markers are highly polymorphic within the human population (recall the earlier discussion of *polymorphism*). **Short tandem repeats (STRs)** are molecular markers that are short sequences of repetitive DNA, up to 200 nucleotide bases with a simple repeat pattern such as GTGTGTGTGT or CAGCAGCAGCAG. Because they vary in length from one individual to another, STRs are highly polymorphic in the population. This characteristic makes them useful with a high degree of certainty in identifying individuals.

If enough markers are compared, the odds that two people taken at random from the general population would have identical DNA profiles may be as low as one in several billion. The FBI uses a set of STRs from 13 different markers to establish a unique DNA profile for an individual. These STR loci are amplified by PCR and separated by gel electrophoresis to produce the individual's profile. Such a profile distinguishes that person from every other individual in the United States, except an identical twin.

Recall that DNA is also found in the cell's mitochondria. Whereas nuclear DNA occurs in two copies per cell—one on each of the homologous chromosomes—mitochondrial DNA has as many as 100,000 copies per cell. Therefore, mitochondrial DNA is the molecule of choice for DNA fingerprinting in which biological samples have been damaged, as in identifying exhumed human remains.

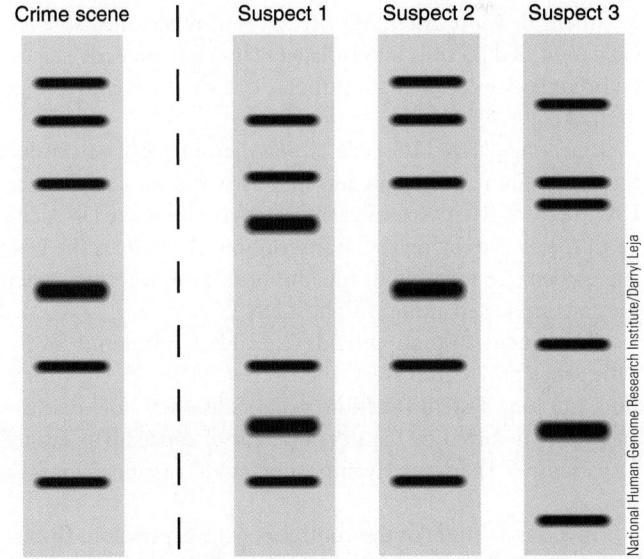

Figure 15-13 *Animation* **DNA fingerprinting**

These DNA fingerprints show DNA from a crime scene along with the DNA profiles of three suspects. Can you pick the suspect whose DNA profile matches blood from the crime scene?

DNA fingerprinting has revolutionized law enforcement. In 1990, the Federal Bureau of Investigation established the Combined DNA Index System (CODIS), which consists of DNA databases from all 50 states and the District of Columbia. A DNA profile of an unknown suspect can be compared with millions of DNA profiles of convicted offenders in the database, often resulting in identification of the suspect. The DNA from the unknown suspect may come from blood, semen, bones, teeth, hair, saliva, urine, or feces left at the crime scene. Tiny amounts of human DNA have even been extracted from cigarette butts, licked envelopes or postage stamps, dandruff, fingerprints, razor blades, chewing gum, wristwatches, earwax, debris from under fingernails, and toothbrushes.

If applied properly, DNA fingerprinting has the power to identify the guilty and exonerate the innocent. Hundreds of convicted individuals have won new trials and have been subsequently released from incarceration based on correlating DNA profiles with physical evidence from the crime scene. One limitation is that the DNA samples are usually small and may have been degraded. Obviously, great care must be taken to prevent contamination of the samples, especially if the PCR technique is used to amplify DNA.

Transgenic organisms have foreign DNA incorporated into their cells

Plants and animals in which foreign genes have been incorporated are called **transgenic organisms.** Researchers use varied approaches to insert foreign genes into plant or animal cells. They often use viruses as vectors, although other methods, such as the direct injection of DNA into cells, have also been applied.

Transgenic animals are valuable in research Transgenic animals are usually produced by injecting the DNA of a particular gene into the nucleus of a fertilized egg cell or of ES cells. The researcher then implants the eggs into the uterus of a female and lets them develop. Alternatively, the researcher injects genetically modified ES cells into isolated *blastocysts,* an early stage in embryonic development, and then implants them into a foster mother.

Injecting DNA into cells is not the only way to produce transgenic animals. Researchers may use viruses as recombinant DNA vectors. RNA viruses called **retroviruses** make DNA copies of themselves by reverse transcription. Sometimes the DNA copies become integrated into the host chromosomes, where they are replicated along with host DNA.

Transgenic animals provide valuable applications over a wide range of research areas, such as regulation of gene expression, immune system function, genetic diseases, viral diseases, and genes involved in the development of cancer. The laboratory mouse is a particularly important model organism for these studies.

In a classic study of the control of gene expression, University of Pennsylvania geneticist Ralph L. Brinster in 1983 produced transgenic mice carrying a gene for rat growth hormone (**FIG. 15-14;** see Fig. 17-18). Brinster and his co-workers wanted to understand the controls that let certain genes be expressed in some tissues and not in others. The pituitary gland of a mouse normally produces small amounts of GH, and these researchers reasoned that other tissues might also be capable of producing the hormone. First, they isolated the GH gene from a library of genomic rat DNA. They combined it with the promoter region of a mouse gene that normally produces metallothionein, a protein that is active in the liver and whose synthesis is stimulated by the presence of toxic amounts of heavy metals such as zinc. The researchers used metallothionein regulatory sequences as a switch to turn the production of rat GH on and off at will. After injecting the modified gene into mouse embryo cells, they implanted the embryos into the uterus of a mouse and let them develop.

Because of the difficulty in manipulating the embryos without damaging them, the gene transplant succeeded in only a small percentage of the animals. When exposed to small amounts of zinc, these transgenic mice produced large amounts of growth hormone because the liver is a much larger organ than the pituitary gland. The mice grew rapidly, and one mouse, which developed from an embryo that had received two copies of the GH gene, grew to more than double the normal size. As might be expected, such mice often transmit their increased growth capability to their offspring.

Transgenic animals can produce genetically modified proteins Certain transgenic animal strains produce milk containing foreign proteins of therapeutic or commercial importance. For example, researchers have introduced into sheep the

Figure 15-14 A transgenic mouse
The mouse on the *right* is normal, whereas the mouse on the *left* is a transgenic animal that expresses rat growth hormone in large amounts.

Figure 15-15 Transgenic "pharm" cows

These cows contain a human gene that codes for lactoferrin, a protein found in human mothers' milk and in secretions such as tears, saliva, bile, and pancreatic fluids. Lactoferrin is one of the immune system's lines of defense against disease-causing organisms. The cows secrete human lactoferrin in their milk.

gene for the human protein tissue plasminogen activator (TPA). Producing transgenic livestock (such as pigs, sheep, cows, and goats) that secrete foreign proteins in their milk is known informally as "pharming," a combination of "pharmaceuticals" and "farming" (**FIG. 15-15**).

In pharming recombinant genes are fused to the regulatory sequences of the milk protein genes, and such genes are therefore activated only in mammary tissues involved in milk production. The advantages of obtaining a protein drug from milk are that it can potentially be produced in large quantities and it can be harvested simply by milking the animal, with the protein then purified from the milk. The introduction of the gene does not harm the animals, and because the offspring of the transgenic animal usually produce the recombinant protein, transgenic strains are established.

Transgenic plants are increasingly important in agriculture People have selectively bred plants for thousands of years. The success of such efforts depends on desirable traits in the variety of plant selected or in closely related wild or domesticated plants whose traits are transferred by crossbreeding. Local varieties or closely related species of cultivated plants often have traits, such as disease resistance, that agriculturalists could advantageously introduce into varieties more useful to modern human needs. If genes are introduced into plants from strains or species with which they do not ordinarily interbreed, the possibilities for improvement increase greatly.

The most widely used vector system for introducing recombinant genes into many types of plant cells is the crown gall bacterium, *Agrobacterium tumefaciens*. This bacterium normally produces plant tumors by introducing a plasmid, called the *Ti plasmid*, into the cells of its host; *Ti* stands for "tumor-inducing." The Ti plasmid induces abnormal growth by forcing the plant cells to produce elevated levels of a plant growth hormone called *cytokinin* (discussed in Chapter 38).

Geneticists "disarm" the Ti plasmid so that it does not induce tumor formation and then use it as a vector to insert genes into plant cells. The cells into which the altered plasmid is introduced are essentially normal except for the inserted genes. Genes placed in the plant genome in this fashion may be transmitted sexually, via seeds, to the next generation, but they can also be propagated asexually, for example, by taking cuttings.

Unfortunately, not all plants take up DNA readily, particularly the cereal grains that are a major food source for humans. One useful approach has been the development of a genetic "shotgun." Researchers coat microscopic gold or tungsten fragments with DNA and then shoot them into plant cells, penetrating the cell walls. Some of the cells retain the DNA, which transforms them. Those cells can then be cultured and used to regenerate an entire plant (see Fig. 17-2). Geneticists have successfully used such an approach to transfer a gene for resistance to a bacterial disease into cultivated rice from one of its wild relatives.

An additional complication of plant genetic modification is that about 120 plant genes lie in the DNA of the chloroplasts; the other 3000 or so genes that chloroplasts require to function are in the nucleus. Chloroplasts are essential in photosynthesis, which is the basis for plant productivity. Because great agricultural potential hinges on improving the productivity of photosynthesis, developing methods for changing the part of the plant's DNA within the chloroplast is desirable. Dozens of labs are currently studying methods of chloroplast engineering. One major emphasis of this work has focused on altering the enzymatic properties of **rubisco,** the key carbon-fixing enzyme of photosynthesis (see Chapter 9), by changing one of its structural and regulatory genes in the chloroplast genome.

The United States is the world's top producer of transgenic crops, also known as *genetically modified (GM) crops*. In 2012, U.S. farmers planted 66.8 million hectares (165 million acres) of GM crops. Globally, in 2011 a total of 160 million hectares of GM crops were planted in 29 countries. These GM crops represented 75% of the worldwide soybean, 32% of the corn, 68% of the cotton, and 26% of the canola crops that year.

Transgenic plants have various applications Agricultural geneticists have developed GM plants that are resistant to insect pests, viral and fungal diseases, heat, cold, herbicides, salty or acidic soil, and drought. For example, genetically modified papaya trees, which are resistant to the virus that causes ring spot, a plant disease, have been grown in Hawaii since 1999. Varieties of corn that have been genetically modified to be resistant to drought have been introduced into crop-growing regions of the United States and Africa (**FIG. 15-16**).

Corn has also been genetically modified to contain the *Bt* gene, a bacterial gene that codes for a protein with insecticidal properties; *Bt* stands for the bacterium's scientific name, *Bacillus*

(a) Note the poor yield in genetically unmodified corn plants used as a control.

(b) Genetically modified corn plants withstood drought better than the unmodified corn.

Figure 15-16 Field test of transgenic corn with resistance to drought

thuringiensis. Bt corn, introduced in the United States in 1996, does not need periodic sprays of chemical insecticides to control the European corn borer, the most damaging insect pest of corn in the United States and Canada.

Like some transgenic animals, certain transgenic plants can potentially be "pharmed" to produce large quantities of medically important proteins. Examples include antibodies against the herpes virus and a patient-specific vaccine for lymphoma, a type of cancer.

Some people are concerned about the health effects of consuming foods derived from GM crops and believe that such foods should be restricted. For example, critics say that some consumers may develop food allergies. Scientists also recognize this concern and routinely screen new GM crops for allergenicity. There is also an ongoing controversy as to whether GM foods should be labeled. The FDA and many scientists think such labeling would be counterproductive because it would increase public anxiety over a technology that they consider as safe as traditional breeding methods. Labeling of GM foods is required by the European Union member countries, India, and some other countries, but it is not required in the United States.

CHECKPOINT 15.4

- *In what way has the production of human insulin by recombinant DNA methods had significant medical advantages for diabetics?*
- **CONNECT** *What are short tandem repeats (STRs), and why are they so useful in DNA fingerprinting?*
- **CONNECT** *Why do gene targeting and mutagenesis screening in mice have potential benefits for humans?*

15.5 DNA TECHNOLOGY HAS RAISED SAFETY CONCERNS

LEARNING OBJECTIVE

11 Describe at least two safety issues associated with recombinant DNA technology and explain how these issues are being addressed.

When recombinant DNA technology was introduced in the early 1970s, many scientists considered the potential misuses at least as significant as the possible benefits. The possibility that an organism with undesirable environmental effects might be accidentally produced was of great concern. Scientists feared that new strains of bacteria or other organisms, with which the world has no previous experience, might be difficult to control. The scientists who developed the recombinant DNA methods insisted on stringent guidelines for making the new technology safe.

Experiments in thousands of university and industry laboratories since then have shown that recombinant DNA manipulations can be carried out safely. Researchers perform experiments that might present unusual risks in facilities designed to hold pathogenic organisms; this level of containment ensures that researchers can work with them safely. So far, no evidence suggests that researchers have accidentally cloned hazardous genes or have released dangerous organisms into the environment. However, malicious *intentional* manipulation of dangerous genes certainly remains a concern.

As the safety of the experiments has been established, scientists have relaxed many of the restrictive guidelines for using recombinant DNA. Stringent restrictions still exist, however, in certain areas of recombinant DNA research where there are known dangers or where questions about possible effects on the environment are still unanswered.

These restrictions are most evident in research that proposes to introduce transgenic organisms into the wild, such as agricultural strains of plants whose seeds or pollen might spread in an uncontrolled manner. A great deal of research now focuses on determining the effects of introducing transgenic organisms into a natural environment. Carefully conducted tests have shown that transgenic organisms are not dangerous to the environment simply because they are transgenic.

It is, however, important to assess the risks of each new recombinant organism. Scientists determine whether the organism has characteristics that might cause it to be environmentally hazardous under certain conditions. For example, if geneticists have modified a transgenic crop plant to resist an herbicide, could that gene be transferred, via pollen or some other route, to that plant's weedy relatives, generating herbicide-resistant "superweeds"? In one test scientists crossed transgenic oilseed rape plants that contained the *Bt* gene with their wild relative, which is a weed. They then crossed the resulting hybrids with the wild relative again and tested its ability to compete with other weeds in a field of wheat. The transgenic weed was a poor competitor and had less effect on wheat production than its wild relatives in a control field. These results, although encouraging for wheat production, must be interpreted with care. Scientists must evaluate each transgenic crop plant individually to see if there is gene flow to wild relatives and, if so, what effects might result.

Other concerns relate to plants modified to produce pesticides, such as the *Bt* toxin. The future of the *Bt* toxin in transgenic crops is not secure because low levels of the insecticide could potentially provide ideal conditions for selection for resistant individuals. It appears that genetic resistance to the *Bt* toxin may evolve in populations of insects feeding on transgenic plants in the same way that chemical resistance to chemical insecticides evolves.

Another concern is that nonpest species could be harmed. For example, in 1999 people paid a great deal of attention to the finding that monarch butterfly larvae raised in the laboratory are harmed if they are fed pollen from *Bt* corn plants. However, studies conducted over the next decade suggest that monarch larvae living in a natural environment do not consume enough pollen to cause damage. Also, evidence indicates that the harm to nonpest species from spraying pesticides is far greater than any harm from transgenic crops.

Environmental concerns about transgenic animals also exist. Several countries are in the process of developing fast-growing transgenic fish, usually by inserting a gene that codes for a growth hormone. Transgenic Atlantic salmon, for example, grow up to six times as fast as nontransgenic salmon grown for human consumption. The transgenic fish do not grow larger than other fish, just faster. The benefits of such genetically enhanced fish include reduced pressure on wild fisheries and less pollution from fish farms. However, if the transgenic fish escaped from the fish farm, what effect would they have on wild relatives? To address this concern, the researchers are developing only nonreproducing female transgenic salmon.

To summarize, DNA technology in agriculture offers many potential benefits, including higher yields by providing disease resistance, more nutritious foods, and the reduced use of chemical pesticides. However, like other kinds of technology, genetic engineering poses some risks, such as the risk that GM plants and animals could pass their foreign genes to wild relatives, causing unknown environmental problems. The science of *risk assessment*, which uses statistical methods to quantify risks so that they can be compared and contrasted, will help society decide whether to ignore, reduce, or eliminate specific risks of molecularly modified organisms.

CHECKPOINT 15.5

- *What are some of the environmental concerns regarding transgenic organisms? What kinds of information does society need to determine if these concerns are valid?*

SUMMARY: FOCUS ON LEARNING OBJECTIVES

15.1 DNA Cloning (page 314)

1 Explain how a typical restriction enzyme cuts DNA molecules and give examples of the ways in which these enzymes are used in recombinant DNA technology.

- **Recombinant DNA** technology isolates and amplifies specific sequences of DNA by incorporating them into **vector** DNA molecules. Researchers then **clone**—propagate and amplify—the resulting recombinant DNA in organisms such as *E. coli*.

- Researchers use **restriction enzymes** to cut DNA into specific fragments. Each type of restriction enzyme recognizes and cuts DNA at a highly specific base sequence. Many restriction enzymes cleave DNA sequences to produce complementary, single-stranded sticky ends.

- Geneticists often construct recombinant DNA molecules by allowing the ends of a DNA fragment and a vector, both cut

with the same restriction enzyme, to associate by complementary base pairing. Then **DNA ligase** covalently links the DNA strands to form a stable recombinant molecule.

2 Distinguish between a genomic DNA library and a complementary DNA (cDNA) library; explain why one would clone the same eukaryotic gene from both a genomic DNA library and a cDNA library.

- A **genomic DNA library** contains cloned copies of thousands of DNA fragments that represent the total DNA of an organism.

- A **cDNA library** contains cloned copies of mature mRNA isolated from eukaryotic cells produced by using **reverse transcriptase**. These copies, known as **complementary DNA (cDNA),** are incorporated into recombinant DNA vectors.

- Genes present in genomic DNA from eukaryotes contain **introns,** regions that do not code for protein. Those genes can be amplified in bacteria, but the protein is not properly expressed. Because introns have been removed from mature mRNA molecules, eukaryotic genes in cDNA libraries can be expressed in bacteria to produce functional protein products.

3 Explain how researchers use a DNA probe.

- Researchers use a radioactive DNA sequence as a **DNA probe** to screen thousands of recombinant DNA molecules in bacterial cells to find the colony that contains the DNA of interest.

4 Describe how the polymerase chain reaction amplifies DNA in vitro.

- The **polymerase chain reaction (PCR)** is a widely used, automated, in vitro technique in which researchers use specific primers to target a particular DNA sequence. The sequence is amplified by repeatedly heating the target DNA to separate the strands and then copying the single-stranded target sequences using a heat-resistant DNA polymerase.
- Using PCR, scientists amplify and analyze tiny DNA samples taken from various sites, from crime scenes to archaeological remains.

15.2 Tools for Studying DNA *(page 320)*

5 Distinguish among DNA, RNA, and protein blotting.

- A **Southern blot** detects DNA fragments by separating them using gel electrophoresis and then transferring them to a nitrocellulose or nylon membrane. A probe is then **hybridized** by complementary base pairing to the DNA bound to the membrane, and the band or bands of DNA are identified by autoradiography or chemical luminescence.
- When RNA molecules that are separated by electrophoresis are transferred to a membrane, the result is a **Northern blot.**
- A **Western blot** consists of proteins or polypeptides previously separated by gel electrophoresis.

6 Describe how gene databases are used to gain insights into gene structure, expression, and function.

- Bioinformatics software can be used to locate and compare the nucleotide sequence of newly cloned DNA sequences with related genes from other organisms. Knowing the function of the gene in a related organism will provide insights into the function of the new gene.
- DNA and protein sequences in databases can be analyzed using bioinformatics software to identify and compare the structures and functions of proteins expressed in related organisms.

7 Discuss how Northern blotting, RT–PCR, and DNA chips are used to compare levels of gene expression among normal versus diseased tissues.

- Northern blotting identifies a specific mRNA species in gel electrophoresis blots of mRNA isolated from different tissues using a radioactive probe.
- RT–PCR uses automated machines to measure PCR-amplified levels of cDNAs from reverse-transcribed mRNAs. Many different mRNAs from different tissues can be measured simultaneously by using fluorescent primers specific for each target mRNA.

- DNA chips can compare mRNA levels over entire genomes. The mRNAs from normal and diseased tissues are reverse transcribed to create different fluorescent color-coded cDNAs from each population. The cDNAs are then annealed cDNAs in microdots that were synthesized using previously sequenced genomic DNA data.

15.3 Genomics *(page 327)*

8 Describe how genome-wide association studies have led to new understandings about the structure and function of the genome.

- Studies such as the ENCODE project have shown that large regions of the genome that are devoid of protein-coding genes contain genes for non-protein-coding RNAs that are involved in the regulation of protein-coding genes. Defects in these regions are associated with genetic markers for many types of genetic and cancer disease states.

9 Explain how targeted gene silencing and knockout mutations can give insights into the functions of a gene.

- These methods prevent the expression and function of a targeted gene, allowing the investigator to observe its effects on the organism. The connection of the silenced or knocked-out gene to a phenotypic defect indicates that the gene plays some essential role in that biochemical or developmental process.

15.4 Applications of DNA Technologies *(page 328)*

10 Describe at least one important application of recombinant DNA technology in each of the following fields: medicine, DNA fingerprinting, and transgenic organisms.

- Genetically altered bacteria produce many important human protein products, including insulin, growth hormone, tissue plasminogen activator (TPA), tissue growth factor-β (TGF-β), and clotting factor VIII.
- **DNA fingerprinting** is the analysis of an individual's DNA. It is based on a variety of **short tandem repeats (STRs)**, molecular markers that are highly polymorphic within the human population. DNA fingerprinting has applications in areas such as law enforcement, issues of disputed parentage, and tracking tainted foods.
- **Transgenic organisms** have foreign DNA incorporated into their genetic material. Transgenic livestock produce foreign proteins in their milk. Transgenic plants have great potential in agriculture.

15.5 DNA Technology Has Raised Safety Concerns *(page 332)*

11 Describe at least two safety issues associated with recombinant DNA technology and explain how these issues are being addressed.

- Some people are concerned about the safety of transgenic organisms. To address these concerns, scientists carry out recombinant DNA technology under specific safety guidelines.
- The introduction of transgenic plants and animals into the natural environment, where they may spread in an uncontrolled manner, is an ongoing concern.

Know and Comprehend

1. A plasmid (a) can be used as a DNA vector (b) is a type of bacteriophage (c) is a type of cDNA (d) is a retrovirus (e) b and c

2. DNA molecules with complementary sticky ends associate by (a) covalent bonds (b) hydrogen bonds (c) ionic bonds (d) disulfide bonds (e) phosphodiester linkages

3. Human DNA and a particular plasmid both have sites that are cut by the restriction enzymes *Hind*III and *Eco*RI. To make recombinant DNA, the scientist should (a) cut the plasmid with *Eco*RI and the human DNA with *Hind*III (b) use *Eco*RI to cut both the plasmid and the human DNA (c) use *Hind*III to cut both the plasmid and the human DNA (d) a or b (e) b or c

4. Which technique rapidly replicates specific DNA fragments without cloning in cells? (a) gel electrophoresis (b) cDNA libraries (c) DNA probe (d) restriction fragment length polymorphism (e) polymerase chain reaction

5. The PCR technique uses (a) heat-resistant DNA polymerase (b) reverse transcriptase (c) DNA ligase (d) restriction enzymes (e) b and c

6. A cDNA clone contains (a) introns (b) exons (c) anticodons (d) a and b (e) b and c

7. The dideoxynucleotides ddATP, ddTTP, ddGTP, and ddCTP are important in DNA sequencing because they (a) cause premature termination of a growing DNA strand (b) are used as primers (c) cause the DNA fragments that contain them to migrate more slowly through a sequencing gel (d) are not affected by high temperatures (e) have more energy than deoxynucleotides

8. Gel electrophoresis separates nucleic acids on the basis of differences in (a) length (molecular weight) (b) charge (c) nucleotide sequence (d) relative proportions of adenine and guanine (e) relative proportions of thymine and cytosine

9. A genomic DNA library (a) represents all the DNA in a specific chromosome (b) is made using reverse transcriptase (c) is stored in a collection of recombinant bacteria (d) is a DNA copy of mature mRNAs (e) allows researchers to amplify a tiny sample of DNA

10. Tissue growth factor-β (a) is a DNA probe for recombinant plasmids (b) is a product of DNA technology used in tissue engineering (c) is necessary to make a cDNA library (d) cannot be synthesized without a heat-resistant DNA polymerase (e) is isolated by the Southern blot technique

11. These highly polymorphic molecular markers are useful in DNA fingerprinting: (a) plasmid vectors (b) cloned DNA sequences (c) palindromic DNA sequences (d) short tandem repeats (e) complementary DNAs

Apply and Analyze

12. What are some of the problems that might arise if you were trying to produce a eukaryotic protein in a bacterium? How might using transgenic plants or animals help solve some of these problems?

13. **CONNECT** Would genetic engineering be possible if we did not know a great deal about the genetics of bacteria? Explain.

Evaluate and Synthesize

14. **INTERPRET DATA** STR analysis was performed on blood from a crime scene as well as from three suspects. The crime scene DNA is marked Evidence. Which suspect's DNA matches the evidence?

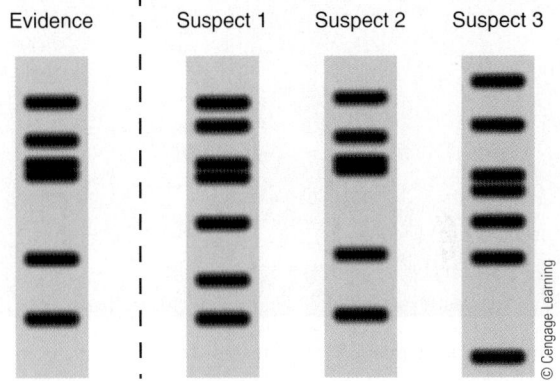

15. **EVOLUTION LINK** DNA technology, such as the production of transgenic animals, is possible only because widely different organisms have essentially identical genetic systems (DNA ⟶ RNA ⟶ protein). What is the evolutionary significance of the universality of genetic systems in organisms as diverse as bacteria and pigs?

16. **SCIENCE, TECHNOLOGY, AND SOCIETY** What are some ways that recent DNA technologies might affect you personally?

aplia To access course materials, such as Aplia and other companion resources, please visit **www.cengagebrain.com.**

16 | Human Genetics and the Human Genome

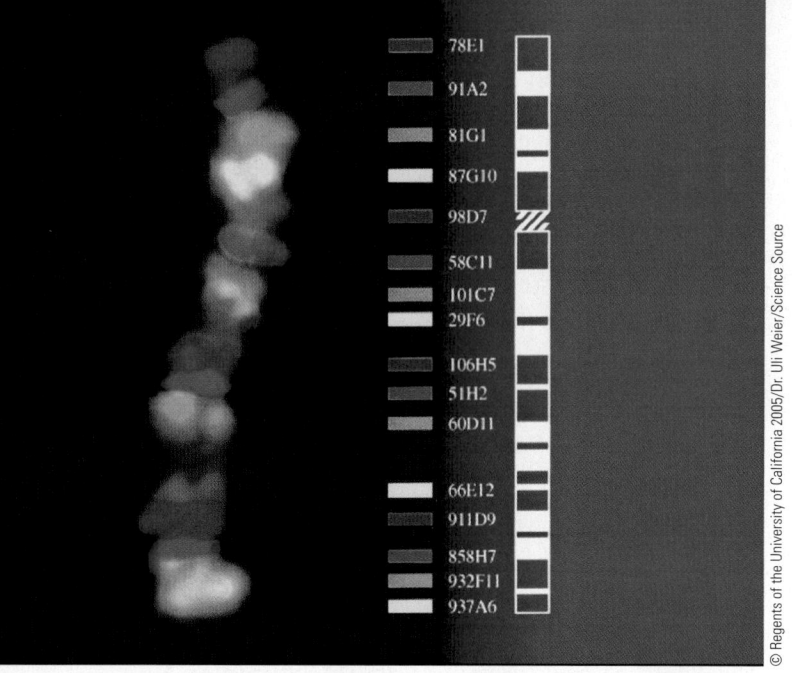

The labels on the chromosome key read (top to bottom): 78E1, 91A2, 81G1, 87G10, 98D7, 58C11, 101C7, 29F6, 106H5, 51H2, 60D11, 66E12, 911D9, 858H7, 932F11, 937A6

Rapid chromosome mapping. This fluorescent LM of human chromosome 10 (*left*) has different-colored fluorescent probes attached to specific DNA sequences. (Loci and key are at *right*.) This technique allows researchers to efficiently determine certain chromosome rearrangements and abnormalities.

KEY CONCEPTS

16.1 Pedigree analysis and karyotype analysis are traditional methods used to study human genetics; investigations of molecular aspects of the human genome, including genome-wide association studies and comparative genomics, are increasingly used today.

16.2 Chromosome abnormalities such as aneuploidies—the presence or absence of a single chromosome—can cause serious medical conditions.

16.3 Mutations in single genes can produce genetic disorders.

16.4 Researchers are developing gene therapy methods that could potentially correct certain genetic disorders.

16.5 Genetic testing and counseling help individuals make reproductive decisions.

16.6 Certain advances in human genetics raise ethical questions.

The principles of genetics apply to all organisms, including humans. However, important differences separate genetic research on humans and genetic research on other organisms. To study inheritance in other species, geneticists may conduct controlled matings between individuals and raise their offspring under carefully controlled conditions. Of course, setting up experimental matings in humans is unethical and illegal.

Despite the inherent difficulties of studying inheritance in humans, **human genetics,** the science of inherited variation in humans, is progressing rapidly (see figure). The medical attention given to human genetic diseases has expanded our knowledge of human genetics. Our most detailed understanding comes from diseases caused by defects in single genes that exhibit a Mendelian inheritance pattern. Valuable insights have come from studies of other organisms that have similar genes, and many human inheritance questions have been explained using model organisms, such as bacteria, yeasts, worms, fruit flies, and mice.

Researchers also conduct population studies of large extended families with a high incidence of a particular type of human genetic disease. These studies use genetic information and technologies derived from the Human Genome Project to uncover genes associated with many different types of *complex genetic diseases.* For these diseases, many different genes interact to cause various conditions, including susceptibility to certain types of cancer, diabetes, kidney disease, obesity, and behavioral disorders such as schizophrenia, depression, and autism.

In this chapter we first examine the methods of human genetics, including new research directions based on DNA technologies discussed in Chapter 15, and then we discuss a variety of human genetic disorders. We explore the use of gene therapy as well as the application of genetic testing, screening, and counseling for families at risk. The chapter concludes with a discussion of ethical issues related to human genetics.

16.1 STUDYING HUMAN GENETICS

LEARNING OBJECTIVES

1 Distinguish between karyotyping and pedigree analysis.
2 Discuss how gene databases and genomic methods are used to study human genetic diseases.
3 Discuss the importance of comparative genomics to the study of human genetics.

Human geneticists use a variety of methods that enable them to help identify genetic defects and to make inferences about a trait's mode of inheritance. We consider three of these methods: the identification of chromosomes by karyotyping, the analysis of family inheritance patterns using pedigrees, and DNA sequencing and mapping of genes through genome projects. Investigators often study human inheritance most effectively by combining these and other approaches.

Human chromosomes are studied by karyotyping

Some well-known human genetic disorders involve changes in the structure or number of chromosomes. The normal number of chromosomes for the human species is 46, made up of 44 *autosomes* (22 pairs) and 2 sex chromosomes (1 pair). A *karyotype* (from Greek, meaning "nucleus") illustrates an individual's chromosome composition, showing both the chromosome number and any large structural defects in the chromosomes.

Human chromosomes are visible only in dividing cells (see Chapter 10), which are difficult to obtain directly from the body. Researchers typically use blood because white blood cells can be induced to divide in a culture medium by treating them with certain chemicals. Other sources of dividing cells include skin and, for prenatal chromosome studies, chorionic villi or fetal cells shed into the amniotic fluid (discussed later in this chapter).

In karyotyping biologists culture dividing human cells and then treat them with the drug *colchicine,* which arrests the cells at mitotic metaphase or late prophase, when the chromosomes are most highly condensed. Next, the researchers immerse the cells in a hypotonic solution; the cells swell, and the chromosomes spread out and are easily observed. The investigators then flatten the cells on microscope slides and stain the chromosomes to reveal the patterns of bands, which are unique for each homologous pair. After the microscopic image has been scanned into a computer, the homologous pairs are electronically matched and placed together (FIG. 16-1).

By convention, geneticists identify chromosomes by length; position of the centromere; banding patterns, which are produced by staining chromosomes with dyes that produce dark and light crossbands of varying widths; and other features such as *satellites*, tiny knobs of chromosome material at the tips of

WHY IS IT USED? Researchers use karyotypes to screen individuals for chromosome abnormalities.

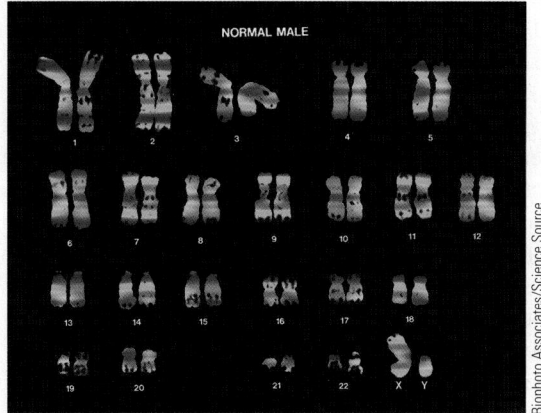

NORMAL MALE

Biophoto Associates/Science Source

HOW IS IT DONE? Karyotypes are commonly prepared from cultures of white blood cells. Researchers place prepared cells on a slide, and the chromosomes spill out of the ruptured nuclei. The light microscope under which the slide is viewed is connected to a computer with image analysis software, which helps the biologist match homologous chromosomes and organize them by size. Before computers were used, researchers cut and pasted chromosomes from photographs.

The chromosomes in the photograph have been "painted" by different-colored fluorescent dyes that hybridize to specific pairs of homologous chromosomes. Painting helps the researcher identify chromosomes. A normal human male karyotype is shown.

Figure 16-1 *Animation* **Karyotyping**

certain chromosomes. With the exception of the sex chromosomes, all chromosomes but chromosome 21, which is smaller than chromosome 22, are numbered and lined up in order of size. The largest human chromosome (chromosome 1) is about five times as long as the smallest one (chromosome 21), but there are only slight size differences among some of the intermediate-sized chromosomes. The X and Y chromosomes of a normal male are homologous only at their tips; normal females have two X chromosomes and no Y chromosome. Differences from the normal karyotype—that is, deviations in chromosome number or structure—are associated with certain disorders such as Down syndrome (discussed later in the chapter).

Another way to distinguish chromosomes in a karyotype is by **fluorescent in situ hybridization (FISH).** A geneticist tags a DNA strand complementary to DNA in a specific chromosome with a fluorescent dye. The DNA in the chromosome is denatured—that is, the two strands are separated—so the tagged strand can bind to it. A different dye is used for each chromosome, "painting" each with a different color (see Fig. 16-1). A chromosome that is multicolored (*not shown*) indicates breakage and fusion of chromosomes, an abnormality associated with certain genetic diseases and many types of cancer.

Family pedigrees help identify certain inherited conditions

Early studies of human genetics usually dealt with readily identified pairs of contrasting traits and their distribution among members of a family. A "family tree" that shows inheritance patterns—the transmission of genetic traits within a family over several generations—is known as a **pedigree.** Pedigree analysis remains widely used, even in today's world of powerful molecular genetic techniques, because it helps molecular geneticists determine the exact inter-relationships of the DNA molecules they analyze from related individuals. Pedigree analysis is also an important tool of genetic counselors and clinicians. However, because human families tend to be small and information about certain family members, particularly deceased relatives, may not be available, pedigree analysis has limitations.

Pedigrees are produced using standardized symbols. Examine **FIGURE 16-2**, which shows a pedigree for **albinism,** a lack of the pigment melanin in the skin, hair, and eyes. Each horizontal row represents a separate generation, with the earliest generation (Roman numeral I) at the top and the most recent generation at the bottom. Within a given generation, the individuals are usually numbered consecutively, from left to right, using Arabic numbers. A horizontal line connects two parents, and a vertical line drops from the parents to their children. For example, individuals II-3 and II-4 are parents of four offspring (III-1, III-2, III-3, and III-4). Note that individuals in a given generation can be genetically unrelated. For example, II-1, II-2, and II-3 are unrelated to II-4 and II-5. Within a group of siblings, the oldest is on the left, and the youngest is on the right.

The allele for albinism cannot be dominant; if it were, at least one of female III-2's parents would have been an albino. This pedigree is explained if albinism is inherited as an autosomal (not carried on a sex chromosome) recessive allele. In such cases, two phenotypically normal parents could produce an albino offspring because they are heterozygotes and could each transmit a recessive allele.

Studying pedigrees enables human geneticists to predict how phenotypic traits that are governed by the genotype at a single locus are inherited. More than 12,000 traits have been described in humans. Pedigree analysis most often identifies three modes of single-locus inheritance: autosomal dominant, autosomal recessive, and X-linked recessive. We define and discuss examples of these inheritance modes later in this chapter.

Human gene databases allow geneticists to map the locations of genes on chromosomes

Although pedigree analysis can provide information about the type of inheritance associated with a single gene, it does not provide specific information about where it is located or how that form of the gene causes the disease. Furthermore, most genetic diseases involve many genes that produce complex patterns of inheritance. How do you identify a specific gene and find its location in an extremely complex genome if you know nothing about it?

(a) An albino Cuna Indian girl plays with her friends. Albinism is common among the Cuna people living on the San Blas Islands of Panama.

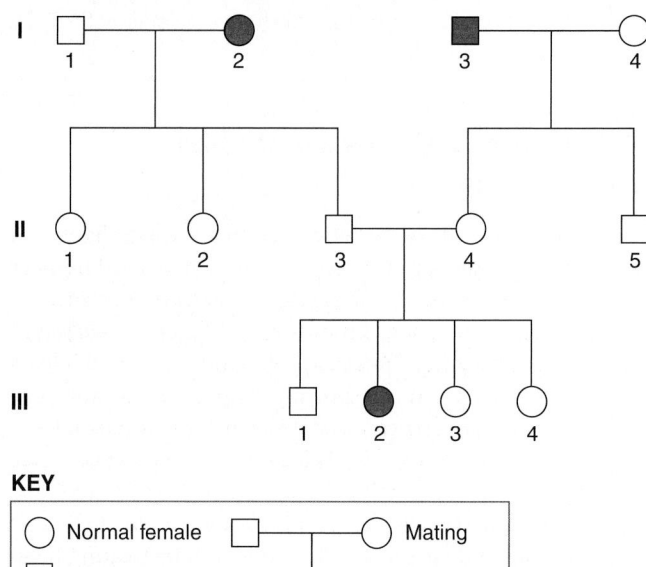

(b) By studying family histories, a researcher can determine the genetic mechanism of an inherited trait. In this example, III-2 represents an albino girl with two phenotypically normal parents, II-3 and II-4.

Figure 16-2 *Animation* **A pedigree for albinism**
© Cengage Learning

Using the database of the **Human Genome Project,** researchers have now generated maps of each chromosome that show the precise location of each gene. These maps allow investigators to locate the region where the gene resides on a specific chromosome and then to analyze genes within that region to determine which one is associated with the disease. Once the disease-causing gene is identified, geneticists can initiate studies to determine its function and how it interacts with other genes associated with the disease.

Single nucleotide polymorphisms are used as genomic landmarks to locate genes linked to genetic disease

Single nucleotide polymorphisms (SNPs, pronounced "snips") are single-letter differences in DNA sequences found every few hundred bases when the genomes of individuals in large populations are compared. In 2012 the **1000 Genomes Project** (see Chapter 15), a group of scientists in eight countries, published data derived from 1092 individuals from 14 different populations, providing a map of 38 million SNPs that span all the human chromosomes. Investigators now use DNA chips to compare the distribution of more than one million SNPs in DNA isolated from individuals affected by a given genetic disease with those who are not affected. By identifying a specific SNP that consistently appears in the affected individuals but that is rare in unaffected persons, researchers can locate a small region of the genome that contains a gene now associated with that disease. FIGURE 16-3 shows the strategy used to locate affected genes using SNP mapping methods. Once the region of an affected gene is identified, investigators use genomic and analytical methods (discussed in Chapter 15) to identify the specific gene that is associated with the disease. For example, researchers would first compare the DNA sequences of genes in the region to determine if one had an unusual mutation that might be the cause of the genetic defect.

Genome-wide association studies demonstrate the genetic complexity of major human diseases

In humans only about 1.5% of the genome specifies the amino acid sequences of polypeptides. However, reports from the **ENCODE Project** have now shown that at least 80% of the human genome is involved in biological functions. These DNA sequences include non-protein-coding regulatory elements and non-protein-coding genes that produce RNAs with unknown functions. Many times, RNA encoding genes are found in introns as well as in the regions between genes that were previously thought to be "gene deserts" or "junk DNA" with no biological functions.

Researchers are now actively engaged in **genome-wide association studies (GWAS)** to compare the genomes of individuals with a particular disease to individuals without that disease. In one study, for example, researchers used SNP analysis to survey the genomes of more than 14,000 British people to compare genetic patterns in healthy people with those in people with one of seven common diseases. The phenotypes of these diseases exhibit complex inheritance patterns, with many different genes contributing to susceptibility. In this particular study, scientists identified 24 SNP variations associated with bipolar disorder, 1 with coronary artery disease, 9 with Crohn's disease, 3 with rheumatoid arthritis, 7 with type 1 diabetes, and 3 with type 2 diabetes.

Many genetic variations associated with complex human diseases are found in non-protein-coding genes

The results of many genome-wide studies conducted since 2005 have shown that almost 90% of SNP variants associated with disease-linked regions are found outside of protein-coding genes. Many of them appear to be associated with enhancer regions or genes that encode regulatory RNAs. For example, one gene associated with one of the highest risks for schizophrenia encodes a microRNA (miRNA; see Chapter 13). Thirty percent of individuals who have a deleted form of this gene develop schizophrenia during adolescence or early childhood. Recent studies in mice and humans have shown that this miRNA regulates the expression of an enzyme in nerve cells involved in memory formation.

How do investigators determine the functions of disease-associated genes?

Determining the genomic location of a disease-associated gene is simply the first step toward developing a treatment for the disease. To develop effective treatment strategies, scientists must first identify the specific gene that causes the disease. Then, they must determine its function, how it interacts with other genes, and how its expression is regulated in different tissues. Many of these studies involve the use of comparative genomics with other organisms to identify disease genes that might have highly conserved and identical functions.

Comparative genomics has revealed DNA identical in both mouse and human genomes

A comparison between the mouse and human genomes has shown that about five hundred DNA segments longer than two hundred base pairs are 100% conserved (i.e., identical). This remarkable degree of conservation has interesting evolutionary implications: it means that these segments have not mutated in the past 75 million years, since mice and humans shared a common ancestor. During this time, other DNA sections underwent considerable mutation and selection, allowing mice and humans to diverge to their present states. Although the functions of these unchanged elements are not yet determined, they clearly have a vital role. (If they were not essential in their present form, they would have undergone

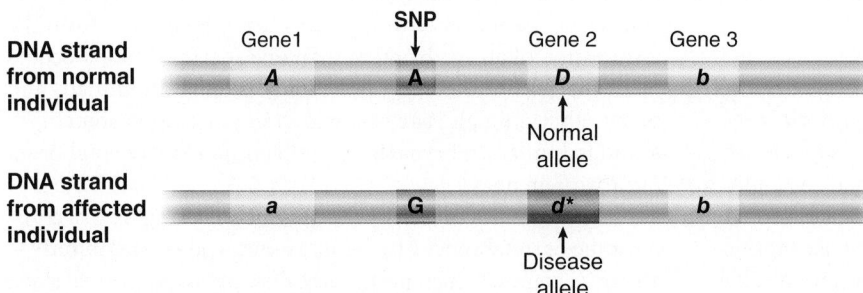

Figure 16-3 Single nucleotide polymorphisms (SNPs) as landmarks for locating disease-causing genes

Microarray analysis comparing the DNA from many affected and normal individuals can identify single nucleotide polymorphisms that are closely linked to genes involved in genetic disorders. In this example the SNP polymorphism consists of an A (adenine) base near the normal gene, and a G (guanine) base in that position in the DNA of an affected individual. SNP markers are usually not directly involved in the genetic defect. Researchers use bioinformatics and molecular genetic methods to identify which gene in the region of a SNP is associated with the disease and to investigate its function.

© Cengage Learning

mutation and selection.) These strong evolutionary connections allow investigators to use the mouse as a powerful research tool to detect and investigate the functions of human disease-causing genes that are common between the two species.

Researchers use mouse models to study human genetic diseases Many questions relating to human genetic diseases are difficult to answer because of the ethical issues involved in using humans as test subjects. However, research on any disease is greatly facilitated if an animal model is used for experimentation. A good example is *cystic fibrosis,* a genetic disease caused by a single gene mutation inherited as a recessive allele. Researchers used *gene targeting* to produce strains of mice that were either homozygous or heterozygous for cystic fibrosis by replacing the normal mouse genes with ones containing the same types of mutations found in human cystic fibrosis patients.

The allele that causes cystic fibrosis is a mutant form of a locus involved in controlling the body's water and electrolyte balance. Geneticists have cloned this gene and found that it codes for a protein, the *CFTR protein,* that serves as a chloride ion channel in the plasma membrane. (*CFTR* stands for *cystic fibrosis transmembrane conductance regulator.*) This ion channel transports chloride ions out of the cells lining the digestive tract and the respiratory system. When the chloride ions leave the cells, water follows by osmosis. Thus, the normal secretions of these cells are relatively watery. Because the cells of individuals with cystic fibrosis lack normal chloride ion channels, the individuals' secretions have a low water content, and their sweat is very salty. Cells of heterozygous individuals have only half the usual number of functional CFTR ion channels, but they are enough to maintain the normal fluidity of their secretions.

Using these mouse models, researchers have focused on understanding the way in which the CFTR channel is activated or inactivated. This information is used to design drugs that enhance chloride transport through the CFTR channels. Such drugs have the potential to treat cystic fibrosis in humans by activating the mutant channels.

Sequencing of individual genomes is an emerging strategy for diagnosing and treating rare genetic diseases The completion of the Human Genome Project required the efforts of thousands of scientists over 13 years and cost more than $3 billion. Remarkably, the technological advances gained from this project have dramatically lowered the cost of genomic sequencing to the point that it is now becoming economically feasible to rapidly sequence an individual's genome to diagnose a genetic disease. One recent case involved twins who were born with a severe neuromuscular disorder. Some characteristics of their condition resembled a rare genetic disease caused by mutations in the genes that are involved in the synthesis of the neurotransmitter dopamine, which is used to transmit signals between nerves and muscle cells (see Chapter 41). Providing the patients dopamine, however, did not relieve the symptoms. When the twins were age 14, the genomes of both were sequenced. Analysis of the genomes found a mutation in a gene known as *SPR,* which is required for the synthesis of a cofactor used by enzymes that synthesize both dopamine and a second neurotransmitter, serotonin. (A subsequent analysis of the parent's DNA revealed that both the mother and father were heterozygous for different mutations within the same gene.) When the twins were provided with both dopamine and serotonin, their symptoms were improved to the point that both could participate in athletic programs.

CHECKPOINT 16.1

- What kinds of information can a human karyotype provide?
- CONNECT *How does pedigree analysis complement other methods for studying human genetics?*
- Describe two ways in which genome database information is used to study human disease genes.
- CONNECT *How does the evolutionary relationship between mice and humans overcome some of the difficulties in studying human inheritance?*

16.2 ABNORMALITIES IN CHROMOSOME NUMBER AND STRUCTURE

LEARNING OBJECTIVES

4 Explain how nondisjunction in meiosis is responsible for chromosome abnormalities such as Down syndrome, Klinefelter syndrome, and Turner syndrome.
5 Distinguish among the following structural abnormalities in chromosomes: translocations, deletions, and fragile sites.
6 Explain how genomic imprinting influences inheritance patterns.

Polyploidy, the presence of multiple sets of chromosomes, is common in plants but rare in animals. It may arise from the failure of chromosomes to separate during meiosis or from the fertilization of an egg by more than one sperm. When it occurs in all the cells of the body, polyploidy is lethal in humans and many other animals. For example, *triploidy* ($3n$) is sometimes found in human embryos that have been spontaneously aborted in early pregnancy.

Abnormalities caused by the presence of a single extra chromosome or the absence of a chromosome—called **aneuploidy**—are more common than polyploidy. **Disomy** is the normal state: two of each kind of chromosome in a cell or individual. In **trisomy** a cell or individual has two copies of each chromosome except for one, which has three copies; a trisomic chromosome number is designated $2n + 1$. In **monosomy** an individual lacks one member of a pair of chromosomes; a monosomic chromosome number is designated $2n - 1$. **TABLE 16-1** summarizes some disorders that aneuploidies produce.

Aneuploidies generally arise as a result of an abnormal meiotic (or, rarely, mitotic) division in which chromosomes fail to separate at anaphase. This phenomenon, called **nondisjunction,** can occur with the autosomes or with the sex chromosomes.

TABLE 16-1	Chromosome Abnormalities: Disorders Produced by Aneuploidies	
KARYOTYPE	**COMMON NAME**	**CLINICAL DESCRIPTION**
Trisomy 13	Patau syndrome	Multiple defects, with death typically by age 3 months.
Trisomy 18	Edwards syndrome	Ear deformities, heart defects, spasticity, and other damage; death typically by age 1 year, but some survive much longer.
Trisomy 21	Down syndrome	Overall frequency is about 1 in 1000 live births. Most conceptions involving true trisomy occur in older (age 35+) mothers, but translocation resulting in the equivalent of trisomy is not related to age. Trisomy 21 is characterized by a fold of skin above the eye, varying degrees of intellectual disability, short stature, protruding furrowed tongue, transverse palmar crease, cardiac deformities, and increased risk of leukemia and Alzheimer's disease.
X0	Turner syndrome	Short stature, webbed neck, sometimes slight intellectual disability; ovaries degenerate in late embryonic life, leading to rudimentary sexual characteristics; gender is female; no Barr bodies.
XXY	Klinefelter syndrome	Male with slowly degenerating testes, enlarged breasts; one Barr body per cell.
XYY	XYY karyotype	Many males have no unusual symptoms; others are unusually tall, with heavy acne, and some tendency to mild intellectual disability.
XXX	Triplo-X	Despite three X chromosomes, usually fertile females with normal intelligence; two Barr bodies per cell.

© Cengage Learning

In meiosis chromosome nondisjunction may occur during the first or the second meiotic division (or both). For example, two X chromosomes that fail to separate at either the first or the second meiotic division may both enter the egg nucleus. Alternatively, the two joined X chromosomes may go into a *polar body*, leaving the egg with no X chromosome. (Recall from Chapter 10 that a polar body is a nonfunctional haploid cell produced during oogenesis; also see Fig. 50-11).

Nondisjunction of the XY pair during the first meiotic division in the male may lead to the formation of a sperm with both X and Y chromosomes or a sperm with neither an X nor a Y chromosome (FIG. 16-4). Similarly, nondisjunction at the second meiotic division can produce sperm with two Xs or two Ys. When an abnormal gamete unites with a normal one, the resulting zygote has a chromosome abnormality that will be present in every cell of the body.

Meiotic nondisjunction results in an abnormal chromosome number at the zygote stage of development, so all cells in the individual have an abnormal chromosome number. In contrast, nondisjunction during a mitotic division occurs sometime later in development and leads to the establishment of a clone of abnormal cells in an otherwise normal individual. Such a mixture of cells with different chromosome numbers may or may not affect somatic (body), or germ line (reproductive) tissues.

Different numbers of chromosomes are common in many cancer cells, particularly those in solid tumors. Whether aneuploidy is a cause or a consequence of cancer, however, is not yet clear.

Recognizable chromosome abnormalities are seen in less than 1% of all live births, but substantial evidence suggests that the rate at conception is much higher. At least 17% of pregnancies recognized at eight weeks will end in spontaneous abortion (miscarriage). Approximately half of these spontaneously aborted embryos have major chromosome abnormalities, including autosomal trisomies (such as trisomy 21), triploidy, tetraploidy, and Turner syndrome (X0), in which the 0 refers to the absence of a second sex chromosome. Autosomal monosomies are exceedingly rare, possibly because they induce a spontaneous abortion very early in the pregnancy, before a woman is even aware that she is pregnant. Some investigators give surprisingly high estimates (50% or more) for the loss rate of very early embryos. Chromosome abnormalities probably induce many of these spontaneous abortions.

Down syndrome is usually caused by trisomy 21

Down syndrome is one of the most common chromosome abnormalities in humans. (The term *syndrome* refers to a set of symptoms that usually occur together in a particular disorder.) It was named after J. Langdon Down, the British physician who in 1866 first described the condition. Affected individuals have abnormalities of the face, eyelids, tongue, hands, and other parts of the body and are intellectually disabled (FIG. 16-5a). They are also unusually susceptible to certain diseases, such as leukemia and Alzheimer's disease.

Genetic studies have revealed that most people with Down syndrome have 47 chromosomes because of *autosomal trisomy*: this condition is known as **trisomy 21** (FIG. 16-5b). Nondisjunction during meiosis is responsible for the presence of the extra chromosome. Although no genetic information is missing in these individuals, the extra copies of chromosome 21 genes bring about some type of genetic imbalance that causes abnormal physical and mental development. Down syndrome is quite variable in expression, with some individuals far more severely affected than others.

Researchers using DNA technologies have pinpointed genes on chromosome 21 that affect mental development as well as possible *oncogenes* (cancer-causing genes) and genes that may be involved in Alzheimer's disease. (Like many human conditions, cancer and Alzheimer's disease have both genetic and environmental components.) A study published in 2013 reported that one cause of Down syndrome appears to be the result of the overexpression of a chromosome 21 gene that encodes the micro RNA, miR155. This miRNA negatively regulates the expression of a protein involved in the endomembrane traffic of receptor proteins in the brain. The resulting underexpression of this gene in a mouse model causes severe defects in neuron functions.

Aneuploidy may occur by meiotic nondisjunction, the abnormal segregation of chromosomes during meiosis.

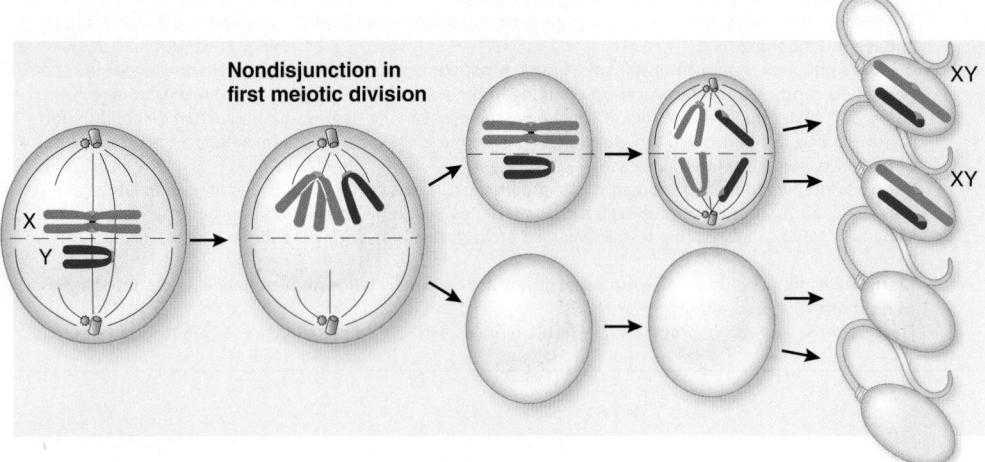

(a) First meiotic division nondisjunction results in two XY sperm and two sperm with neither an X nor a Y.

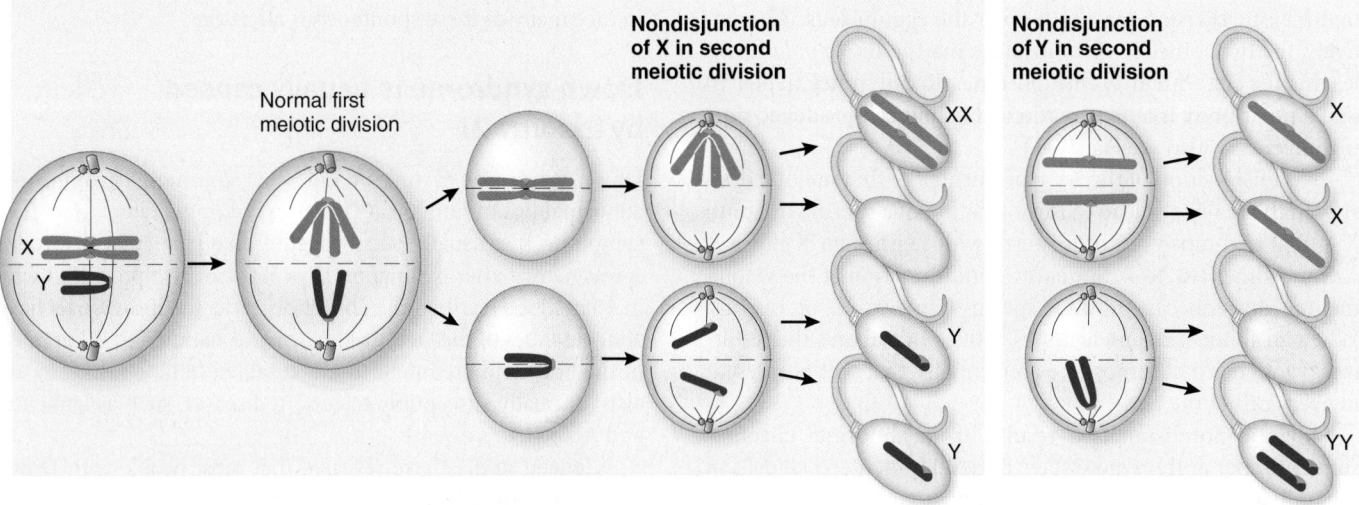

(b) Second meiotic division nondisjunction of the X chromosome results in one sperm with two X chromosomes, two with one Y each, and one with no sex chromosomes. Nondisjunction of the Y chromosome results in one sperm with two Y chromosomes, two with one X each, and one with no sex chromosome (*box on right*).

Figure 16-4 *Animation* **Meiotic nondisjunction**

In these examples of nondisjunction of the sex chromosomes in the human male, only the X (*purple*) and Y (*blue*) chromosomes are shown. (The positions of the metaphase chromosomes have been modified to save space.)

PREDICT Which of the sperm illustrated here could give rise to an offspring with Klinefelter syndrome? Turner syndrome? XYY karyotype?

© Cengage Learning

Down syndrome occurs in all ethnic groups in about 1 out of 1000 live births in the U.S. Its incidence increases markedly with increasing maternal age. The occurrence of Down syndrome is not affected by the father's age (although other disorders are, including schizophrenia and achondroplasia, the most common form of dwarfism). Down syndrome is 68 times as likely in the offspring of mothers of age 45 than in the offspring of mothers age 20. However, most babies with Down syndrome in the United States are born to mothers younger than 35, in part because these women greatly outnumber older mothers and in part because about 90% of older women who undergo prenatal testing terminate the pregnancy if Down syndrome is diagnosed.

The relationship between increased incidence in Down syndrome and maternal age has been studied for decades, but there is no explanation. Scientists have proposed several hypotheses to explain the maternal age effect, but none is supported unequivocally. One explanation is that older women have held

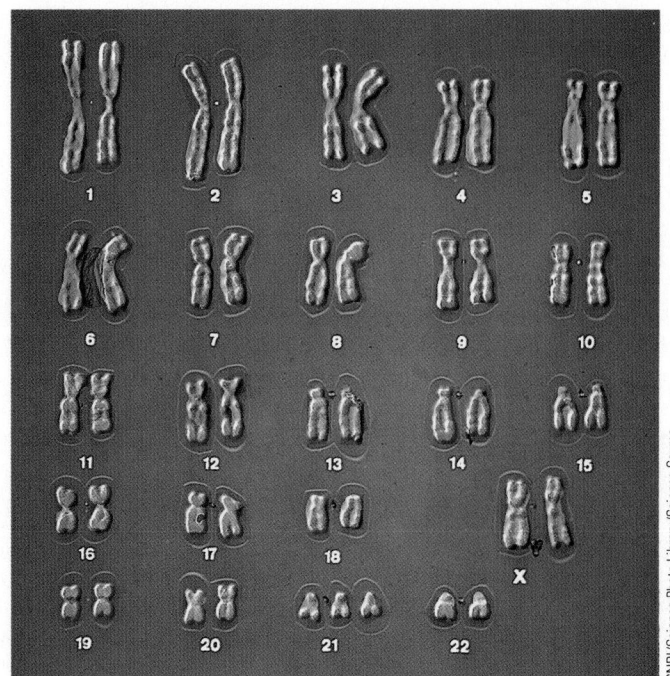

(a) This girl with Down syndrome is working on a math problem with her teacher. Some individuals with Down syndrome learn to read and write.

(b) Note the presence of an extra chromosome 21 in this colorized karyotype of a female with Down syndrome.

Figure 16-5 Down syndrome

eggs in suspended meiosis too long, leading to a deterioration of the meiotic spindle apparatus. (A woman is born with all the oocytes she will ever have; these oocytes remain in prophase I of meiosis until ovulation.) Another possibility is that an aging uterus is less likely to reject an abnormal fetus than a younger uterus.

Most sex chromosome aneuploidies are less severe than autosomal aneuploidies

Sex chromosome aneuploidies are tolerated relatively well (see Table 16-1) at least in part because of the mechanism of **dosage compensation:** mammalian cells compensate for extra X chromosome material by rendering all but one X chromosome inactive. The inactive X is seen as a **Barr body,** a region of darkly stained, condensed chromatin next to the nuclear envelope of an interphase nucleus (see Fig. 11-16). Investigators have used the presence of the Barr body in the cells of normal females as an initial screen to determine whether an individual is genetically female or male. However, as you will see shortly in the context of sex chromosome aneuploidies, the Barr body test has limitations.

Individuals with **Klinefelter syndrome** are males with 47 chromosomes, including two Xs and one Y. They have small testes, produce few or no sperm, and are therefore sterile. The hypothesis that the Y chromosome is the major determinant of the male phenotype has been substantiated by the identification of at least one gene on the Y chromosome that acts as a genetic switch, directing male development. Males with Klinefelter syndrome tend to be unusually tall and have female-like breast development. About half show some intellectual disability, but many live relatively normal lives. However, each of their cells has one Barr body. On the basis of such a test, they would be erroneously classified as females. About 1 in 1000 live-born males has Klinefelter syndrome.

The sex chromosome composition for **Turner syndrome,** in which an individual has only one sex chromosome, an X chromosome, is designated X0. Because they lack the male-determining effect of the Y chromosome, individuals with Turner syndrome develop as females. However, both their internal and their external genital structures are underdeveloped, and they are sterile. Apparently, a second X chromosome is necessary for normal development of ovaries in a female embryo. Examination of the cells of these individuals reveals no Barr bodies because there is no extra X chromosome to be inactivated. Using the standards of the Barr body test, such an individual would be classified erroneously as a male. About 1 in 2500 live-born females has Turner syndrome.

People with an X chromosome plus two Y chromosomes are phenotypically males, and they are fertile. Other characteristics of these individuals (tall, often with severe acne) hardly qualify as a syndrome; hence, the term **XYY karyotype** is used. Some years ago, several widely publicized studies suggested that males with this condition are more likely to display criminal tendencies and thus to be imprisoned than those without this condition. However, these studies were flawed because they were based on small numbers of XYY males, without adequate or well-matched control studies of XY males. The prevailing

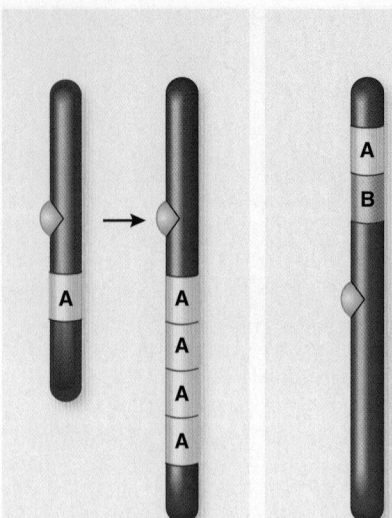

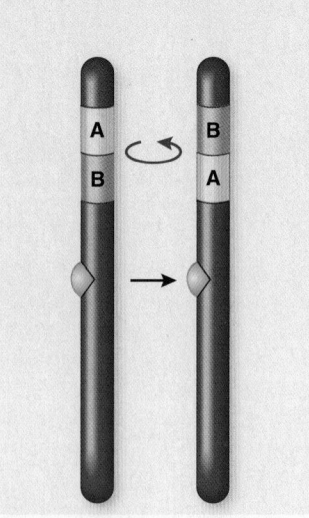

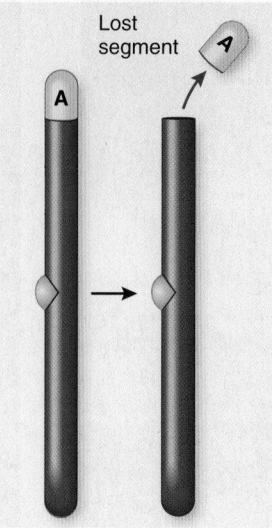

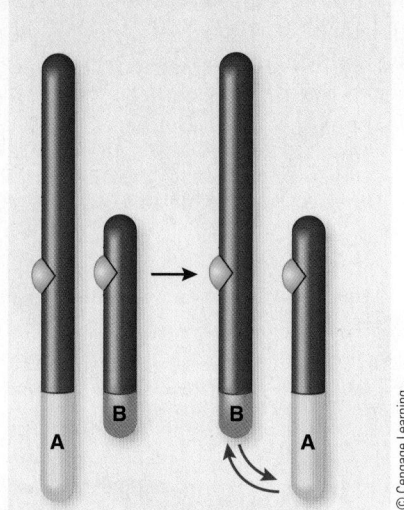

(a) A **duplication** is a repeated segment of a chromosome. In this example, segment A is repeated.

(b) An **inversion** is a chromosome segment with a reversed orientation. An inversion does not change the amount of genetic material in the chromosome, only its arrangement.

(c) A **deletion** is the loss of a chromosome segment. A deletion can occur at the tip (*shown*) or within the chromosome.

(d) A **reciprocal translocation** occurs when two nonhomologous chromosomes exchange segments.

Figure 16-6 *Animation* **Common abnormalities of chromosome structure**

opinion in medical genetics today is that there are many undiagnosed XYY males in the general population who do not have overly aggressive or criminal behaviors and are unlikely to be incarcerated.

Abnormalities in chromosome structure cause certain disorders

Chromosome abnormalities are caused not only by changes in chromosome number but also by distinct changes in the structure of one or more chromosomes. The breakage and rejoining of chromosome parts result in four structural changes within or between chromosomes: duplications, inversions, deletions, and translocations (**FIG. 16-6**). Breaks in chromosomes are the result of errors in replication or recombination. In a **duplication** a segment of the chromosome is repeated one or more times; these repeats often appear in tandem with one another. The orientation of a chromosome segment is reversed in an **inversion.** In a **deletion** breakage causes loss of part of a chromosome along with the genes on that segment. A deletion can occur at the end of a chromosome or on an internal part of the chromosome. In some cases of **translocation,** a chromosome fragment breaks off and attaches to a nonhomologous chromosome. In a *reciprocal translocation,* two nonhomologous chromosomes exchange segments.

Here we consider three simple examples of chromosomal structural changes that result in abnormal phenotypes: translocations, deletions, and *fragile sites,* which are chromosome sites that are susceptible to breakage.

Translocation is the attachment of part of one chromosome to another The consequences of translocations vary

considerably. They include deletions, in which some genes are missing, and duplications, in which extra copies of certain genes are present. In about 4% of individuals with Down syndrome, only 46 chromosomes are present, but one is abnormal. The large arm of chromosome 21 has been translocated to the large arm of another chromosome, usually chromosome 14. Individuals with *translocation Down syndrome* have one chromosome 14, one combined 14/21 chromosome, and two normal copies of chromosome 21. All or part of the genetic material from chromosome 21 is thus present in triplicate. When studying the karyotypes of the parents in such cases, geneticists usually find that either the mother or the father has only 45 chromosomes, although she or he is generally phenotypically normal. The parent with 45 chromosomes has one chromosome 14, one combined 14/21 chromosome, and one chromosome 21; although the karyotype is abnormal, there is no extra genetic material. In contrast to trisomy 21, translocation Down syndrome can run in families, and its incidence is not related to maternal age.

A deletion is the loss of part of a chromosome Sometimes chromosomes break and fail to rejoin. Such breaks result in deletions of as little as a few base pairs to as much as an entire chromosome arm. As you might expect, large deletions are generally lethal, whereas small deletions may have no effect or may cause recognizable human disorders.

One deletion disorder (1 in 20,000–50,000 live births) is **cri du chat syndrome,** in which part of the short arm of chromosome 5 is deleted. As in most deletions, the exact point of breakage in chromosome 5 varies from one individual to another; some cases of cri du chat involve a small loss, whereas others involve a more substantial deletion of base pairs. Infants born with cri du chat syndrome typically have a small head with altered features described as a "moon face" and a distinctive cry that sounds like a kitten mewing. (The name literally means "cry of the cat" in French.) Affected individuals usually survive beyond childhood but exhibit severe intellectual disability.

Fragile sites are weak points at specific locations in chromatids A **fragile site** is a place where part of a chromatid appears to be attached to the rest of the chromosome by a thin thread of DNA. Fragile sites occur at corresponding locations on both chromatids of a chromosome. They have been identified on the X chromosome as well as on certain autosomes. The location of a fragile site is exactly the same in each one of an individual's cells as well as in cells of other family members. Scientists report growing evidence that cancer cells may have breaks at these fragile sites. Whether cancer destabilizes the fragile sites, leading to breakage, or the fragile sites themselves contain genes that contribute to cancer is unknown at this time.

In **fragile X syndrome,** also known as *Martin–Bell syndrome,* a fragile site occurs near the tip of the X chromosome, where the fragile X gene contains a nucleotide triplet CGG that repeats from 200 to more than 1000 times (FIG. 16-7). In a normal chromosome, CGG repeats up to 50 times.

Fragile X syndrome is the most common cause of inherited intellectual disability. The effects of fragile X syndrome, which are more pronounced in males than in females, range from mild learning and attention deficit disorders to severe intellectual disability and hyperactivity. According to the National Fragile X Foundation, about 80% of boys and 35% of girls with fragile X syndrome are at least mildly intellectually disabled. Females with fragile X syndrome are usually heterozygous (because their other X chromosome is normal) and are therefore more likely to have normal intelligence.

The discovery of the fragile X gene in 1991 and the development of the first fragile X mouse model in 1994 have provided researchers with ways to develop and test potential treatments, including gene therapy. At the microscopic level, the nerve cells of individuals with fragile X syndrome have malformed dendrites (the part of the nerve cell that receives nerve impulses from other nerve cells). At the molecular level, the triplet repeats associated with fragile X syndrome disrupt the functioning of a gene that codes for a certain protein, designated *fragile X mental retardation protein (FMRP).* In normal cells FMRP binds to many different mRNA molecules in nerve cells and regulates their translation. In cells of individuals with fragile X syndrome, the mutated allele does not produce functional FMRP, resulting in excess translation of its mRNA targets.

Genomic imprinting may determine whether inheritance is from the male or female parent

Some traits are the result of **genomic** or **parental imprinting,** in which the expression of a gene in a given tissue or developmental

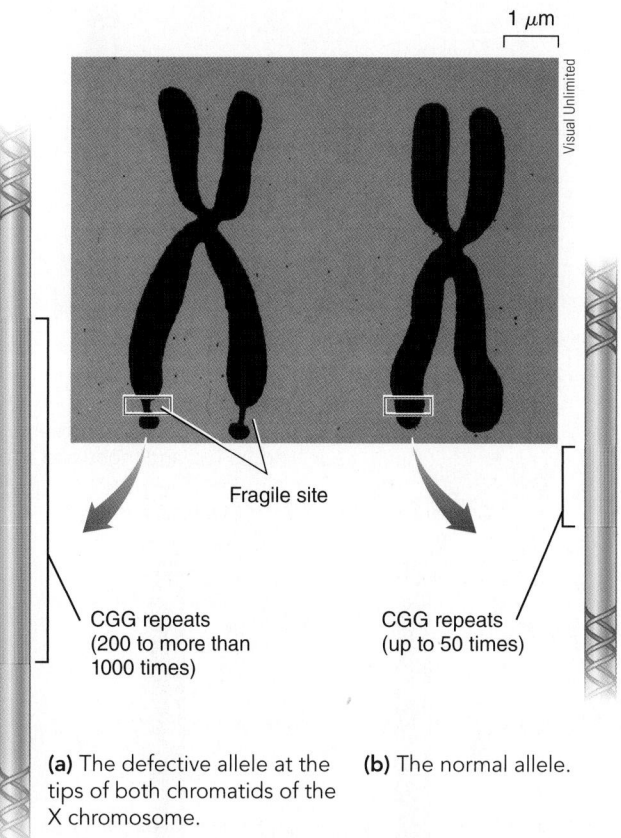

1 μm

Fragile site

CGG repeats (200 to more than 1000 times)

CGG repeats (up to 50 times)

(a) The defective allele at the tips of both chromatids of the X chromosome.

(b) The normal allele.

Figure 16-7 Fragile X syndrome

This colorized SEM shows an X chromosome with a fragile site and a normal X chromosome.

© Cengage Learning

stage is based on its parental origin; that is, it is based on whether the individual inherits the gene from the male or the female parent. For some imprinted genes, the paternally inherited allele is always repressed (not expressed); for other imprinted genes, the maternally inherited allele is always repressed. Thus, maternal and paternal genomes have different imprints that result in differential gene expression in the embryo. As discussed in Chapter 14, *epigenetic inheritance* refers to changes in how a gene is expressed without any change in the coding of the DNA bases. Epigenetic inheritance may cause genomic imprinting.

Two rare genetic disorders provide a fascinating demonstration of genomic imprinting. In *Prader-Willi syndrome (PWS),* individuals become compulsive overeaters and obese; they are also short in stature and have mild to moderate intellectual disabilities. In *Angelman syndrome (AS),* affected individuals are hyperactive, intellectually disabled and unable to speak, and they suffer from seizures. A small *deletion* of several loci from the same region of chromosome 15 causes both PWS and AS (FIG. 16-8a). One of these deleted loci is responsible for PWS, another for AS.

Pedigree analysis has shown that when the person inherits the deletion from the father, PWS occurs, whereas when the person inherits the deletion from the mother, AS occurs. This inheritance pattern suggests that the normal PWS gene is expressed only in the paternal chromosome and that the normal

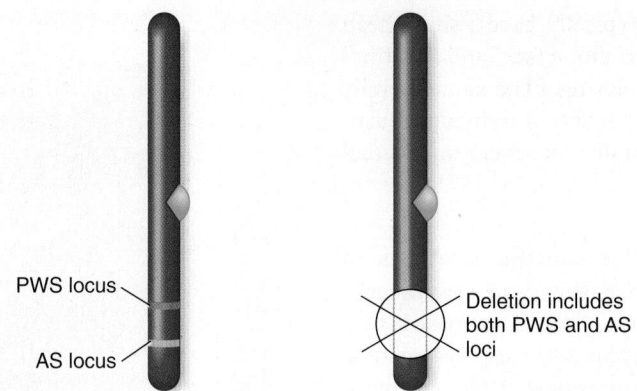

(a) Homologous chromosome 15, showing (*left*) close proximity of the PWS and AS loci and (*right*) site of deletion that includes both PWS and AS loci.

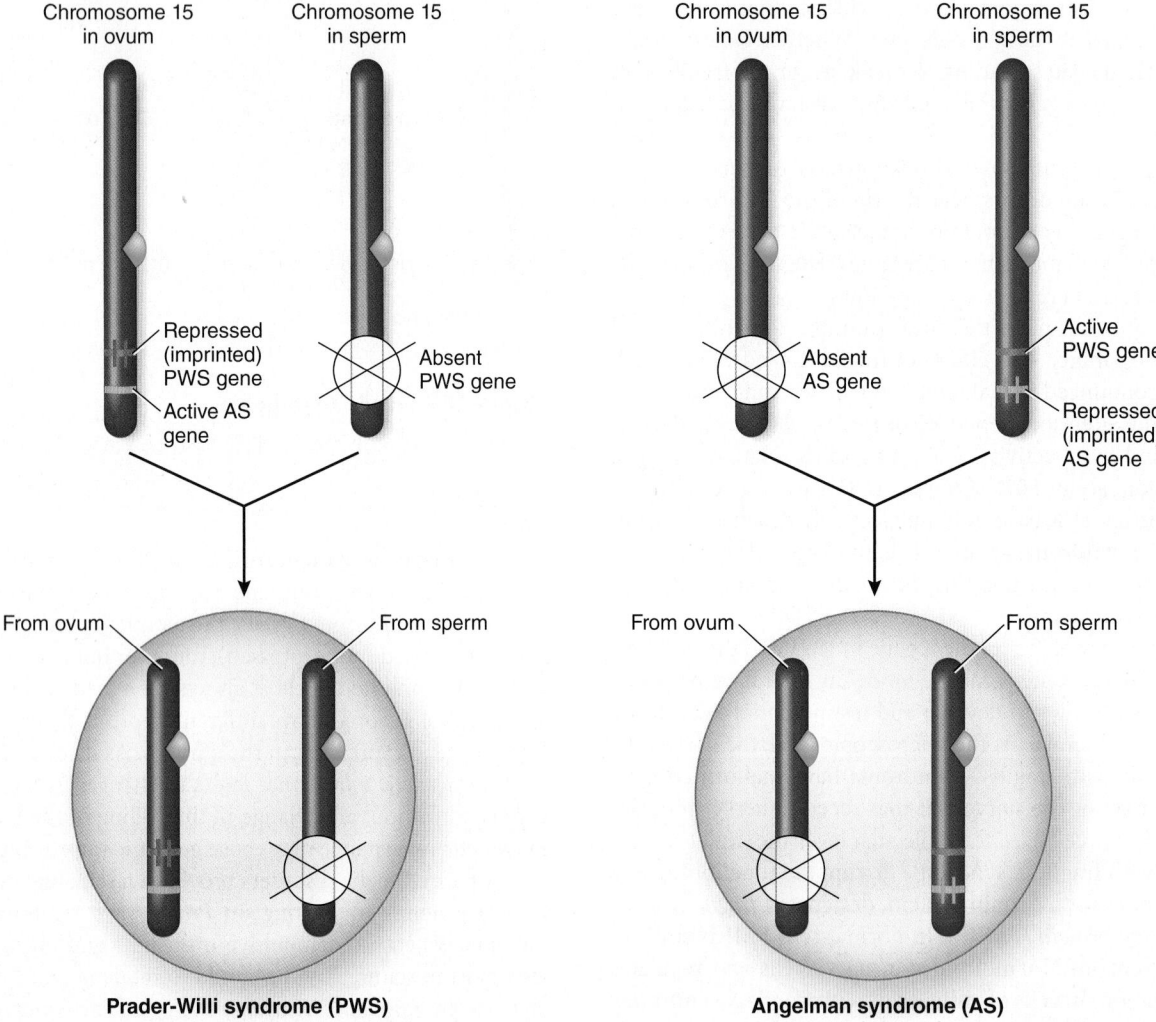

Prader-Willi syndrome (PWS)

Angelman syndrome (AS)

(b) This individual has Prader-Willi syndrome because the sperm contributed a chromosome 15 in which the PWS gene had been deleted. The chromosome with the repressed (imprinted) PWS gene from the egg is unable to compensate for the deletion.

(c) This individual has Angelman syndrome because the egg contributed a chromosome 15 in which the AS gene had been deleted. The chromosome with the repressed (imprinted) AS gene from the sperm is unable to compensate for the deletion.

Figure 16-8 Genomic imprinting and development of Prader-Willi syndrome (PWS) and Angelman syndrome (AS)

The PWS gene is active in the male gamete and repressed (imprinted) in the female gamete. Conversely, the AS gene is active in the female gamete and repressed (imprinted) in the male gamete.

© Cengage Learning

AS gene is expressed only in the maternal chromosome. PWS occurs because the repressed (imprinted) PWS gene from the mother cannot make up for the absent PWS gene in the paternal chromosome (FIG. 16-8b). Similarly, AS occurs because the repressed (imprinted) AS gene from the father cannot make up for the absent AS gene in the maternal chromosome (FIG. 16-8c).

CHECKPOINT 16.2

- VISUALIZE *Draw a simple sketch illustrating how nondisjunction in meiosis can lead to Down syndrome.*
- *What is the chromosome abnormality in cri du chat syndrome?*
- *What is the chromosome abnormality in fragile X syndrome?*
- CONNECT *What types of chemical modifications occur in DNA or histones that result in genomic imprinting? (See Chapter 14.)*

16.3 GENETIC DISEASES CAUSED BY SINGLE-GENE MUTATIONS

LEARNING OBJECTIVE

7 State whether each of the following genetic defects is inherited as an autosomal recessive, autosomal dominant, or X-linked recessive trait: phenylketonuria (PKU), sickle cell anemia, cystic fibrosis, Tay-Sachs disease, Huntington's disease, and hemophilia A.

You have seen that several human disorders involve chromosome abnormalities. Hundreds of human disorders, however, involve enzyme defects caused by mutations of single genes. Alkaptonuria (discussed in Chapter 13) and phenylketonuria (discussed below) are examples of these disorders, each of which is sometimes referred to as an **inborn error of metabolism,** a metabolic disorder caused by the mutation of a gene that codes for an enzyme needed in a biochemical pathway. Both PKU and alkaptonuria involve blocks in the metabolism of specific amino acids.

Many genetic diseases are inherited as autosomal recessive traits

Many human genetic diseases have a simple autosomal recessive inheritance pattern and therefore are expressed only in the homozygous state. Why are these traits recessive? Most recessive mutations result in a mutant allele that encodes a product that no longer works (either there is not enough gene product or it is a defective gene product). In the heterozygous state, there is one functional copy of the gene and one mutated, nonfunctional copy. The normal copy of the gene generally produces enough protein to meet the cell's needs. In homozygous recessive individuals, *both* alleles of the gene are nonfunctional, and the cell's needs are not met. As a result, the person shows symptoms of disease.

Phenylketonuria results from an enzyme deficiency
Phenylketonuria (PKU), which is most common in individuals of western European descent, is an autosomal recessive disease caused by a defect of amino acid metabolism. It affects about 1 in 10,000-15,000 live births in North America. Homozygous recessive individuals lack an enzyme that converts the amino acid phenylalanine to another amino acid, tyrosine. These individuals accumulate high levels of phenylalanine, phenylpyruvic acid, and similar compounds.

The accumulating phenylalanine is converted to phenylketones, which damage the central nervous system, including the brain, in children. The ultimate result in untreated cases is severe intellectual disability. An infant with PKU is usually healthy at birth because its mother, who is heterozygous, breaks down excess phenylalanine for both herself and her fetus. However, during infancy and early childhood, the accumulation of toxic products eventually causes irreversible damage to the central nervous system.

Beginning in the 1950s, infants with PKU have been identified early and placed on a low-phenylalanine diet, dramatically alleviating their symptoms. The diet is difficult to adhere to because it contains no meat, fish, dairy products, breads, or nuts. Also, individuals with PKU should not consume the sugar substitute aspartame, found in many diet drinks and foods, because it contains phenylalanine. Biochemical tests for PKU have been developed, and screening of newborns through a simple blood test is required in the United States. Because of these screening programs and the availability of effective treatment, thousands of PKU-diagnosed children have not developed severe intellectual disability. Most must continue the diet through at least adolescence. Doctors now recommend that patients stay on the diet throughout life because some adults who have discontinued the low-phenylalanine diet experience certain mental problems, such as difficulty in concentration and short-term memory loss.

Ironically, the success of PKU treatment in childhood presents a new challenge today. If a homozygous female who has discontinued the special diet becomes pregnant, the high phenylalanine levels in her blood can damage the brain of the fetus she is carrying, even though that fetus is heterozygous. Therefore, she must resume the diet, preferably before becoming pregnant. This procedure is usually (although not always) successful in preventing the effects of *maternal PKU*. It is especially important for women with PKU to be aware of maternal PKU and to obtain appropriate counseling and medical treatment during pregnancy.

Sickle cell anemia results from a hemoglobin defect
Sickle cell anemia is inherited as an autosomal recessive trait. The disease is most common in people of African descent (approximately 1 in 500 African Americans), and about 1 in 12 African Americans is heterozygous. Under low oxygen conditions, the red blood cells of an individual with sickle cell anemia are shaped like sickles, or half-moons, whereas normal red blood cells are biconcave discs.

The mutation that causes sickle cell anemia was first identified in 1957. The sickled cells contain abnormal hemoglobin molecules, which have the amino acid valine instead of glutamic acid at position 6 (the sixth amino acid from the amino terminal end) in the β-globin chain (see Fig. 3-23a). The substitution of valine for glutamic acid makes the hemoglobin molecules stick to one another to form fiber-like structures that change the shape of some of the red blood cells. This sickling occurs in the veins after the oxygen has been released from the hemoglobin. The blood cells' abnormal sickled shape slows blood flow and blocks small blood vessels (FIG. 16-9), resulting in tissue damage from lack of oxygen and essential nutrients, and in episodes of pain. Because sickled red blood cells also have a shorter lifespan than normal red blood cells, many affected individuals have severe anemia.

Treatments for sickle cell anemia include pain-relief measures, transfusions, and more recently, medicines such as hydroxyurea, which activates the gene for the production of normal fetal hemoglobin (this gene is generally not expressed after birth). The presence of normal fetal hemoglobin in the red blood cells dilutes the sickle cell hemoglobin, thereby minimizing the painful episodes and reducing the need for blood transfusions. The long-term effects of hydroxyurea are not known at this time, but there are concerns that it may induce tumor formation.

Ongoing research is directed toward providing gene therapy for sickle cell anemia using a mouse model for the disease. The first gene therapy treatments in mice used a mouse retrovirus as a *vector,* a carrier that transfers the genetic information. However, the retrovirus did not effectively transport the normal gene for hemoglobin into the bone marrow, where stem cells produce new blood cells. Current research involves the use of other viral vectors to engineer stem cells isolated from the individual to produce fetal hemoglobin in the adult animal.

The reason that the sickle cell allele occurs at a higher frequency in parts of Africa and Asia than in other parts of the world is well established. Individuals who are heterozygous (Hb^AHb^S) and carry alleles for both normal hemoglobin (Hb^A) and sickle cell hemoglobin (Hb^S) are more resistant than others to the malarial parasite, *Plasmodium falciparum,* which causes a severe and often fatal form of malaria. The malarial parasite, which spends part of its life cycle inside red blood cells, does not thrive when sickle cell hemoglobin is present. (An individual heterozygous for sickle cell anemia produces both normal and sickle cell hemoglobin.) Areas in Africa where *falciparum* malaria occurs correlate well with areas in which the frequency of the sickle cell allele is more common in the human population. Thus, Hb^AHb^S individuals, possessing one copy of the mutant sickle cell allele, have a selective advantage over both types of homozygous individuals, Hb^AHb^A (who may die of malaria) and H^bSHb^S (who may die of sickle cell anemia). This phenomenon, known as *heterozygote advantage,* is discussed further in Chapter 19 (see Fig. 19-7).

Cystic fibrosis results from defective ion transport **Cystic fibrosis** is the most common autosomal recessive disorder in children of European descent (1 in 2500 births). About 1 in 25 individuals in the United States is a heterozygous carrier of the mutant cystic fibrosis allele. Abnormal secretions characterize this disorder. The most severe effect is on the respiratory system, where abnormally viscous mucus clogs the airways. The cilia that line the bronchi cannot easily remove the mucus, and it becomes a growth medium for dangerous bacteria. These bacteria or their toxins attack the surrounding tissues, leading to recurring pneumonia and other complications. The heavy mucus also occurs elsewhere in the body, causing digestive difficulties and other effects.

As discussed earlier, the gene responsible for cystic fibrosis codes for CFTR, the protein that regulates the transport of chloride ions across cell membranes. The defective protein, found in plasma membranes of epithelial cells lining the passageways of the lungs, intestines, pancreas, liver, sweat glands, and reproductive organs, results in the production of an unusually thick mucus that eventually leads to tissue damage. Although many forms of cystic fibrosis exist that vary somewhat in the severity of symptoms, the disease is almost always serious.

Antibiotics are used to control bacterial infections, and daily physical therapy is required to clear mucus from the respiratory system (FIG. 16-10). Treatment with *Dornase Alpha (DNase),* an enzyme produced by recombinant DNA technology, helps break down the mucus. Without treatment, death would occur in infancy. With treatment, the average life expectancy for individuals with cystic fibrosis is currently about 38 years. Because of the serious limitations of available treatments, gene therapy for cystic fibrosis is under development.

The most severe mutant allele for cystic fibrosis predominates in northern Europe, and another, somewhat less serious, mutant allele is more prevalent in southern Europe. Presumably, these mutant alleles are independent mutations that have been maintained by natural selection. Some experimental evidence supports the hypothesis that heterozygous individuals are less likely than other individuals to die from infectious diseases that

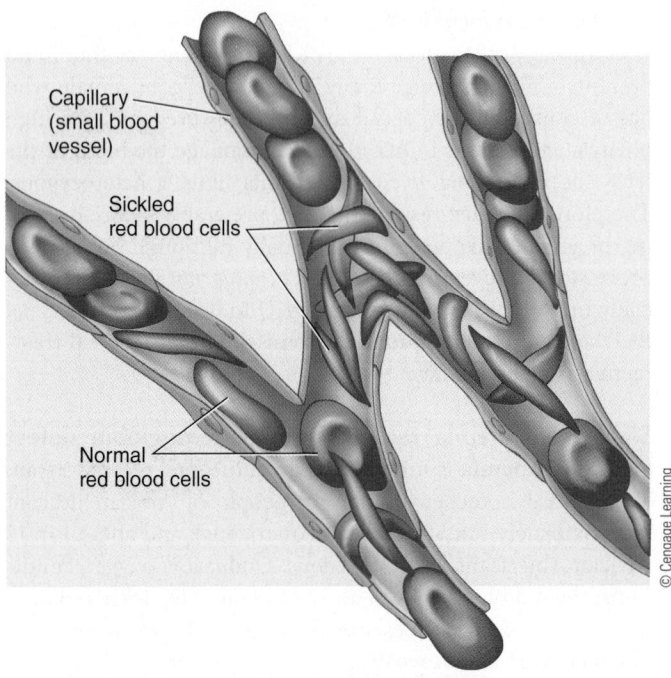

© Cengage Learning

Figure 16-9 *Animation* **Sickle cell anemia**
Sickled red blood cells do not pass through small blood vessels as easily as unsickled red blood cells do. The sickled cells can cause blockages that prevent oxygen from being delivered to tissues.

Labels in figure: Capillary (small blood vessel); Sickled red blood cells; Normal red blood cells

Figure 16-10 Treating cystic fibrosis

This brother and sister are being treated for cystic fibrosis with nebulizer therapy, which loosens the mucus. The boy is also wearing a vest that creates chest percussion or gentle pounding on the chest, to clear mucus from clogged airways in the lungs.

cause severe diarrhea, such as cholera, another possible example of heterozygote advantage.

Tay-Sachs disease results from abnormal lipid metabolism in the brain Tay-Sachs disease is an autosomal recessive disease that affects the central nervous system and results in blindness and severe intellectual disability. The symptoms begin within the first year of life and result in death before the age of five years. Because of the absence of an enzyme, a normal membrane lipid in brain cells fails to break down properly and accumulates in intracellular organelles called *lysosomes* (discussed in Chapter 4). The lysosomes swell and cause nerve cells to malfunction. Although research is ongoing, no effective treatment for Tay-Sachs disease is available at this time.

The abnormal allele is especially common in the United States among Jewish individuals whose ancestors came from eastern and central Europe (Ashkenazi Jews). In contrast, Jewish individuals whose ancestors came from the Mediterranean region (Sephardic Jews) have a very low frequency of the allele.

Some genetic diseases are inherited as autosomal dominant traits

Huntington's disease (HD), named after George Huntington, the U.S. physician who first described it in 1872, is caused by a rare autosomal dominant allele that affects the central nervous system. The disease causes severe mental and physical deterioration, uncontrollable muscle spasms, and personality changes, and death ultimately results. No effective treatment has been found. Every child of an affected individual has a 50% chance

of also being affected (and, if affected, of passing the abnormal allele to his or her offspring). Ordinarily, we would expect a dominant allele with such devastating effects to occur only as a new mutation and not to be transmitted to future generations. Because HD symptoms typically do not appear until relatively late in life (most people do not develop the disease until they are in their 40s), affected individuals are likely to produce children before the disease develops (**FIG. 16-11**). In North America HD occurs in 1 in 20,000 live births.

The gene responsible for HD is located on the short end of chromosome 4. The mutation is a nucleotide triplet (CAG) that is repeated many times; the normal allele repeats CAG from 6 to 35 times, whereas the mutant allele repeats CAG from 40 to more than 150 times. Because CAG codes for the amino acid glutamine, the resulting protein, called "huntingtin," has a long strand of glutamines. The number of nucleotide triplet repeats seems to be important in determining the age of onset and the severity of the disease; larger numbers of repeats correlate with an earlier age of onset and greater severity.

Much research now focuses on how the mutation is linked to neurodegeneration in the brain. A mouse model of HD is providing valuable clues about the development of the disease. Using this model, researchers have demonstrated that the defective version of huntingtin binds to enzymes called *acetyltransferases* in brain cells, blocking their action. Acetyltransferases are involved in turning genes on for expression, so in the brain cells of HD individuals much of normal transcription cannot occur. Once neurologists better understand HD's mechanism of action on nerve cells, it may be possible to develop effective treatments to slow the progression of the disease.

Cloning of the HD allele became the basis for tests that allow those at risk to learn presymptomatically if they carry the allele. The decision to be tested for any genetic disease is

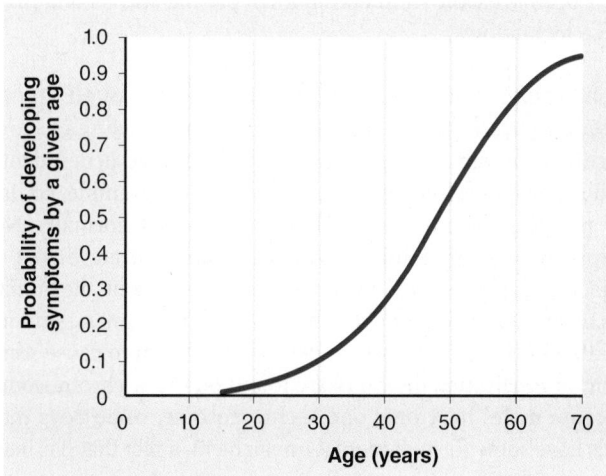

Figure 16-11 Age of onset of Huntington's disease

This graph shows the cumulative probability that an individual carrying a Huntington's disease allele will have developed symptoms at a given age. (Adapted from Harper, P.S., *Genetic Counseling*, 5th ed., Butterworth-Heinemann, Oxford, 1998.)

understandably a highly personal one. The information is, of course, invaluable for those who must decide whether or not to have children. However, someone who tests positive for the HD allele must then live with the virtual certainty of eventually developing this devastating and incurable disease. Researchers hope that information from affected individuals who choose to be identified before the onset of symptoms may ultimately contribute to the development of effective treatments.

Some genetic diseases are inherited as X-linked recessive traits

Hemophilia was once referred to as a disease of royalty because of its high incidence among male descendants of England's Queen Victoria, but it is also found in many nonroyal pedigrees. There are two primary forms of hemophilia. Both are caused by the absence of a blood-clotting protein encoded by genes on the X chromosome. Hemophilia A is the most common form of the disease involving clotting factor VIII. Hemophilia B (in Queen Victoria's lineage) is associated with a clotting factor IX deficiency. Both forms are characterized by severe internal bleeding in the head, joints, and other areas from even a slight wound. The mode of inheritance is X-linked recessive. Thus, affected individuals are almost exclusively male, having inherited the abnormal allele on the X chromosome from their heterozygous carrier mothers. (For a female to be affected by an X-linked trait, she would have to inherit the defective allele from both parents, whereas an affected male need only inherit one defective allele from his mother.)

Treatments for both types of hemophilia consist of blood transfusions and the administration of appropriate clotting factor (the missing gene product) by injection. Unfortunately, these treatments are costly. During the 1980s, many clotting factor preparations made from human plasma were unknowingly contaminated with HIV, and many men with hemophilia subsequently died from AIDS. Since 1992, virus-free clotting factors have been available from both human plasma and recombinant DNA technology.

Geneticists are beginning to unravel X-linked genes affecting intelligence The X chromosome contains a disproportionate number of the more than two hundred genes identified so far that affect cognitive abilities by, for example, coding for proteins required for the brain to function normally. Not surprisingly, many kinds of mental impairment are linked to defects in genes on the X chromosome. By one count, the human X chromosome contains less than 4% of the human genome, yet 10% of the genes in which defects are known to cause some form of intellectual disability are found on the X chromosome. Because males have only one X chromosome, more boys than girls have some form of mental impairment, a fact that has been observed for more than a century.

CHECKPOINT 16.3

- *Which of the following genetic diseases is/are inherited as an autosomal recessive trait: phenylketonuria, Huntington's disease, Tay-Sachs disease?*

- *Which of the following genetic diseases is/are inherited as an autosomal dominant trait: sickle cell anemia, hemophilia A, Huntington's disease?*
- *Which of the following genetic diseases is/are inherited as an X-linked recessive trait: hemophilia A, cystic fibrosis, Tay-Sachs disease?*

16.4 GENE THERAPY

LEARNING OBJECTIVE

8 Briefly discuss the process of gene therapy, including some of its technical challenges.

Because serious genetic diseases are difficult to treat, scientists have dreamed of developing actual cures. One strategy is **gene therapy,** which aims to compensate for a defective, mutant allele by adding a normal, therapeutic allele (and its expressed protein) to certain cells. The rationale is that although a particular allele may be present in all cells, it is expressed only in some. Expression of the normal allele in only the cells that require it may be sufficient to yield a normal phenotype (**FIG. 16-12**).

This approach presents several technical problems. The solutions to these problems must be tailored to the nature of the gene itself as well as to its product and the types of cells in which it is expressed. First, the gene is cloned and the DNA introduced into the appropriate cells.

One of the most successful techniques is packaging the normal allele in a viral vector, a virus that moves the normal allele into target cells that currently have a mutant allele. Ideally, the virus should infect a high percentage of the cells. Most important, the virus should do no harm, especially over the long term. Early gene therapy trials used an adenovirus, which causes the common cold. However, some individuals had a strong immune reaction to the virus, so many current trials use an adeno-associated virus that does not cause side effects.

Gene therapy has had a greater than 90% success rate in restoring the immune systems of children with *severe combined immunodeficiency (SCID)*. (SCID is a group of inherited disorders that seriously compromise the immune system.) This success rate is significantly better than the 50% success rate of the older therapy involving bone marrow transplants. However, serious safety concerns have impeded progress in gene therapy. The death of a young man in a gene therapy trial in 1999 and five cases of cancer (leukemia) in children, one of whom died, led to a temporary shutdown of many trials in 2003, pending the outcome of investigations about health risks. Although the ban was lifted later that year, researchers have continued to proceed cautiously. The main safety concern is the potential toxicity of viral vectors. The vector used in the young man who died was an adenovirus that was required in large doses to transfer enough copies of the normal alleles for effective therapy. Unfortunately, the high viral doses triggered a fatal immune response in the patient's body. The children who developed leukemia were being treated for SCID. The vector in these cases was a retrovirus that inserted itself into and activated an oncogene that can cause childhood leukemia.

WHY IS IT USED? Mice are a model system for the development of gene therapy, the use of normal genes to correct or alleviate the symptoms of a genetic disease caused by defective copies of a particular gene. This procedure is performed on humans for certain types of genetic disease.

HOW IS IT DONE?

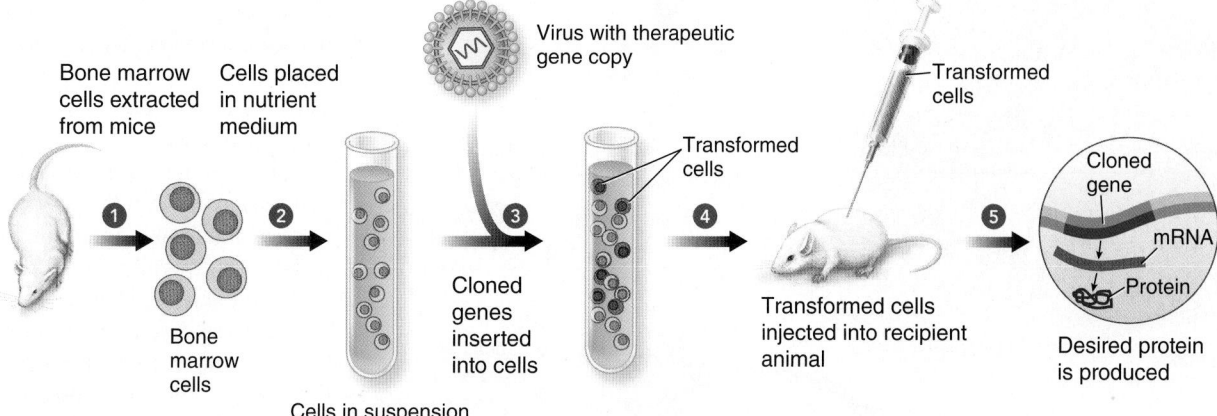

Figure 16-12 **Gene therapy in bone marrow cells of a mouse**
© Cengage Learning

Performing clinical trials on humans always has inherent risks

Researchers carefully select patients and thoroughly explain the potential benefits and risks, as far as they are known, so that the patient—or, in the case of children, the parents—can give informed consent for the procedure. However, the problems in gene therapy trials in recent years have researchers busy developing safer alternatives to viral vectors.

CHECKPOINT 16.4

- *What are the potential concerns regarding the use of viral vectors in gene therapy?*

16.5 GENETIC TESTING AND COUNSELING

LEARNING OBJECTIVES

9   State the relative advantages and disadvantages of amniocentesis, chorionic villus sampling, and preimplantation genetic diagnosis in the prenatal diagnosis of human genetic abnormalities.

10 Distinguish between genetic screening programs for newborns and adults, and discuss the scope and implications of genetic counseling.

Geneticists have made many advances in detecting genetic disorders in individuals in recent years, including in prenatal diagnosis and genetic screening. With these advances comes increased information for couples at risk of having children with genetic diseases. Helping them understand and deal with the genetic information now available is part of the rapidly expanding field of genetic counseling.

Prenatal diagnosis detects chromosome abnormalities and gene defects

Health care professionals are increasingly successful at diagnosing genetic diseases prenatally. In the diagnostic technique called **amniocentesis,** a physician obtains a sample of the *amniotic fluid* surrounding the fetus by inserting a needle through the pregnant woman's abdomen, into the uterus, and then into the amniotic sac surrounding the fetus. Some of the amniotic fluid is withdrawn from the amniotic cavity into a syringe (**FIG. 16-13**). The fetus is normally safe from needle injuries because **ultrasound imaging** helps determine the positions of the fetus, placenta, and the needle. (Figure 51-17 shows an ultrasound of a human fetus.) However, there is a 0.5%, or 1 in 200, chance that amniocentesis will induce a miscarriage.

Amniotic fluid contains living cells sloughed off the body of the fetus and hence genetically identical to the cells of the fetus. After cells grow in culture in the lab, technicians karyotype dividing cells to detect chromosome abnormalities. Other

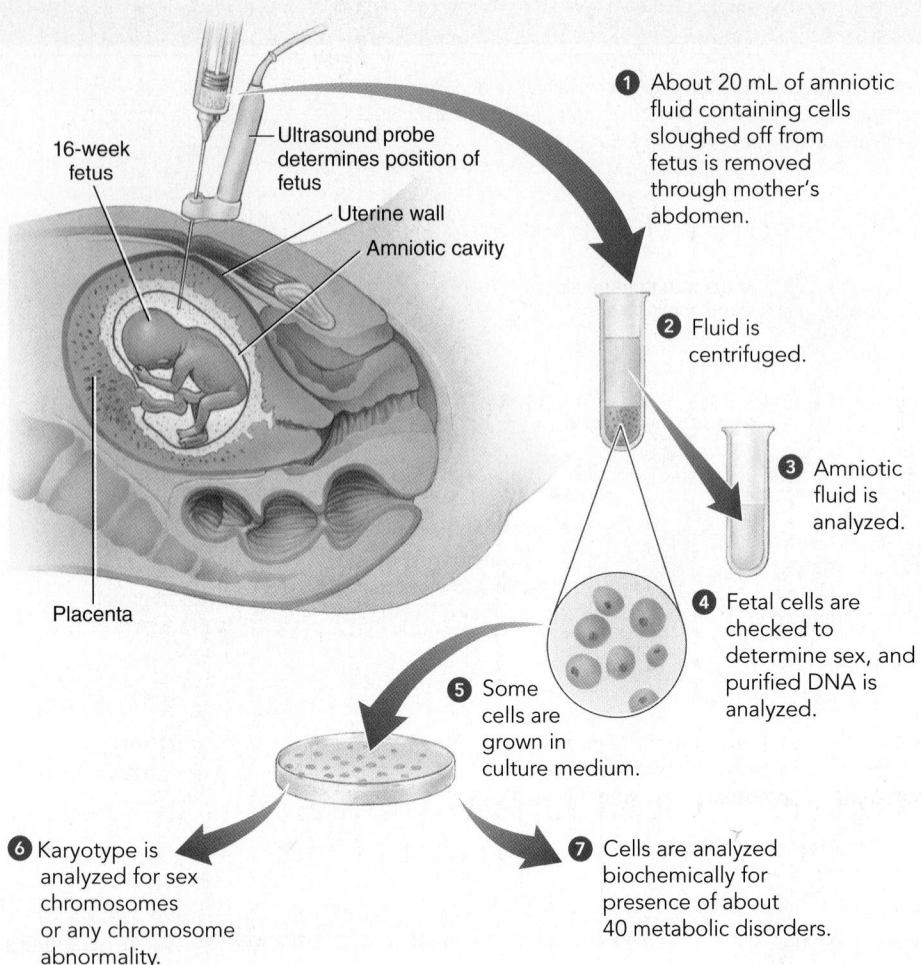

① About 20 mL of amniotic fluid containing cells sloughed off from fetus is removed through mother's abdomen.

16-week fetus

Ultrasound probe determines position of fetus

Uterine wall

Amniotic cavity

Placenta

② Fluid is centrifuged.

③ Amniotic fluid is analyzed.

④ Fetal cells are checked to determine sex, and purified DNA is analyzed.

⑤ Some cells are grown in culture medium.

⑥ Karyotype is analyzed for sex chromosomes or any chromosome abnormality.

⑦ Cells are analyzed biochemically for presence of about 40 metabolic disorders.

Figure 16-13 Amniocentesis

In amniocentesis the fluid surrounding the developing fetus is sampled, usually during the 16th week of pregnancy, to detect genetic and developmental disorders.

© Cengage Learning

DNA tests have also been developed to identify most chromosome abnormalities. Amniocentesis, which has been performed since the 1960s, is routinely offered for pregnant women older than age 35 because their fetuses have a higher-than-normal risk of Down syndrome.

Researchers have developed prenatal tests to detect many genetic disorders with a simple inheritance pattern, but these disorders are rare enough that physicians usually order the tests performed only if they suspect a particular problem. Enzyme deficiencies can often be detected by incubating cells recovered from amniotic fluid with the appropriate substrate and measuring the product; this technique has been useful in prenatal diagnosis of disorders such as Tay-Sachs disease. The tests for several other diseases, including sickle cell anemia, Huntington's disease, and cystic fibrosis, involve directly testing the individual's DNA for the mutant allele.

Amniocentesis is also useful in detecting a condition known as *spina bifida,* in which the spinal cord does not close properly during development. A relatively common malformation (about 1 in 3000 births), this birth defect is associated with abnormally high levels of a normally occurring protein, *a-fetoprotein,* in the amniotic fluid. Some of this protein crosses the placenta into the mother's blood, which is tested for *maternal serum a-fetoprotein (MSAFP)* as a screen for spinal cord defects. If an elevated level of MSAFP is detected, the physician performs diagnostic tests, such as ultrasound imaging and amniocentesis. (Interestingly, abnormally *low* levels of MSAFP are associated with Down syndrome and other trisomies.)

One problem with amniocentesis is that most of the conditions it detects are unpreventable and incurable, and the results are generally not obtained until well into the second trimester, when terminating the pregnancy is both psychologically and medically more difficult than earlier. Therefore, researchers have developed tests that yield results earlier in the pregnancy. **Chorionic villus sampling (CVS)** involves removing and studying cells that will form the fetal contribution to the placenta (**FIG. 16-14**). CVS, which has been performed in the United States since about 1983, is associated with a slightly greater risk of infection or miscarriage than amniocentesis, but its advantage is that results are obtained earlier in the pregnancy than in amniocentesis, usually within the first trimester.

A relatively new embryo screening process, known as **preimplantation genetic diagnosis (PGD),** is available for potential parents who carry alleles for Tay-Sachs disease, hemophilia, sickle cell anemia, and dozens of other inherited genetic conditions. PGD is an adjunct to assisted reproductive technology. Conception is by *in vitro fertilization (IVF),* in which gametes are collected, eggs are fertilized in a dish in the laboratory, and the resulting embryo is then implanted in the uterus for development (see *Inquiring About: Novel Origins,* in Chapter 50). Prior to implantation, the physician screens single cells of early embryos for one or more genetic diseases before placing a healthy embryo into the woman's uterus. PGD differs from amniocentesis and CVS in that the test is performed *before* a woman is pregnant, so it eliminates the decision of whether or not to terminate the pregnancy if an embryo has a genetic abnormality. However, PGD is not as accurate as amniocentesis or CVS, and it is more expensive. Moreover, PGD is sometimes controversial because some potential parents may use it to choose the gender of their offspring, not to screen for genetic diseases.

Although using amniocentesis, CVS, and PGD can help physicians diagnose certain genetic disorders with a high degree of accuracy, the tests are not foolproof, and many disorders cannot be diagnosed at all. Therefore, the lack of an abnormal finding is no guarantee of a normal baby.

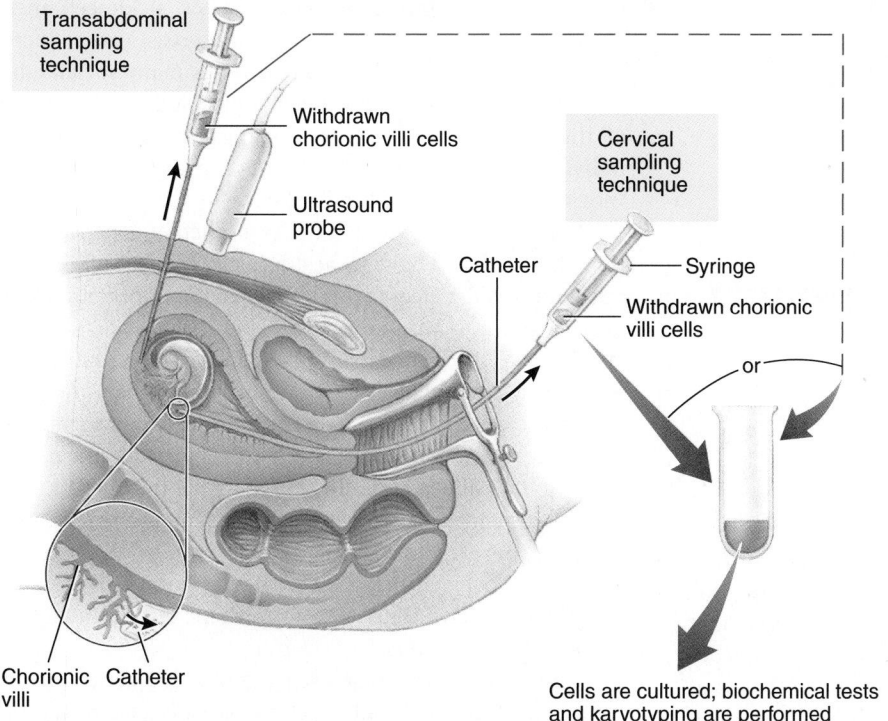

Transabdominal
sampling
technique

Withdrawn
chorionic villi cells

Ultrasound
probe

Cervical
sampling
technique

Catheter

Syringe

Withdrawn chorionic
villi cells

or

Chorionic   Catheter
villi

Cells are cultured; biochemical tests
and karyotyping are performed

Figure 16-14 *Animation* **Chorionic villus sampling (CVS)**

In chorionic villus sampling, chorionic cells genetically identical to the embryo are sampled, usually during the eighth or ninth week of pregnancy, to detect genetic disorders. This test allows the early diagnosis of some genetic abnormalities. Samples may be obtained by inserting a needle through the uterine wall or the cervical opening.

© Cengage Learning

Genetic screening searches for genotypes or karyotypes

Genetic screening is a systematic search through a population for individuals with a genotype or karyotype that might cause a serious genetic disease in themselves or their offspring. There are two main types of genetic screening—one for newborns and one for adults—and each serves a different purpose. Newborns are screened primarily as the first step in preventive medicine, and adults are screened to help them make informed reproductive decisions.

Newborns are screened to detect and treat certain genetic diseases before the onset of serious symptoms. The routine screening of infants for PKU began in 1962 in Massachusetts. Laws in all 50 states of the United States and the District of Columbia, as well as in many other countries, currently require PKU screening. Sickle cell anemia is also more effectively treated with early diagnosis. Screening newborns for sickle cell anemia reduces infant mortality by about 15% because doctors can administer daily doses of antibiotics, thereby preventing bacterial infections common to newborns with the disease. The number of genetic disorders that can be screened in newborns is rapidly increasing, and the March of Dimes organization currently recommends screening newborns for 31 disorders, most of which are genetic.

Genetic screening of adults identifies carriers (heterozygotes) of recessive genetic disorders. If both prospective parents are heterozygous, the carriers are counseled about the risks involved in having children. Since the 1970s, about one million young Jewish adults in the United States, Israel, and other countries have been screened voluntarily for Tay-Sachs disease, and about 1 in 30 has been identified as a carrier. Tay-Sachs screening programs have reduced the incidence of Tay-Sachs disease to almost zero.

Genetic counselors educate people about genetic diseases

Prospective parents who are concerned about the risk of abnormality in their children because they have either an abnormal child or a relative affected by a hereditary disease may seek **genetic counseling** for medical and genetic information as well as support and guidance. Genetic clinics, available in most major metropolitan centers, are usually affiliated with medical schools.

Genetic counselors, who have received training in counseling, medicine, and human genetics, provide people with the information they need to make reproductive decisions. They offer advice, tempered with respect and sensitivity, in terms of risk estimates, that is, the *probability* that any given offspring will inherit a particular condition. The counselor analyzes family histories, and a clinical geneticist (a physician who specializes in genetics) may screen for the detection of heterozygous carriers of certain conditions.

When a disease involves only a single gene locus, probabilities can usually be easily calculated. For example, if one prospective parent is affected with a trait that is inherited as an autosomal dominant disorder, such as Huntington's disease, the probability that any given child will have the disease is 0.5, or 50%. The birth to phenotypically normal parents of a child affected with an autosomal recessive trait, such as albinism or PKU, establishes that both parents are heterozygous carriers, and the probability that any subsequent child will be affected is therefore 0.25, or 25%. For a disease inherited through a recessive allele on the X chromosome, such as hemophilia A, a homozygous normal woman and an affected man will have daughters who are carriers and sons who are phenotypically normal. The probability that the son of a carrier mother and a normal father will be affected is 0.5, or 50%; the probability that their daughter will be a carrier is also 0.5, or 50%.

It is important for identified carriers to receive appropriate genetic counseling. A genetic counselor is trained not only to provide information pertaining to reproductive decisions but also to help individuals understand their situation and avoid feeling stigmatized.

16.6 HUMAN GENETICS, SOCIETY, AND ETHICS

LEARNING OBJECTIVE

11 Discuss the controversies of genetic discrimination.

Many misconceptions exist about genetic diseases and their effects on society. Some people erroneously think of certain individuals or populations as genetically unfit and thus responsible for many of society's ills. They argue, for example, that medical treatment of people affected with genetic diseases, especially those who reproduce, increases the frequency of abnormal alleles in the population. However, such notions are incorrect. Genetic disorders are so rare that modern medical treatments will have only a negligible effect on their incidence.

Recessive mutant alleles are present in *all* individuals and *all* ethnic groups; no one is exempt. According to one estimate, each of us is heterozygous for several (3 to 15) harmful recessive alleles, any of which could cause debilitating illness or death in the homozygous state. Why, then, are genetic diseases relatively uncommon? Each of us has many thousands of essential genes, any of which can be mutated. It is very unlikely that the abnormal alleles that one individual carries are also carried by the other parent of that individual's children. Of course, this possibility is more likely if the harmful allele is a relatively common one, such as the one responsible for cystic fibrosis.

Relatives are more likely than nonrelatives to carry the same harmful alleles, having inherited them from a common ancestor. A greater-than-normal frequency of a particular genetic disease among offspring of *consanguineous matings,* matings between genetically related individuals, is often the first clue that the mode of inheritance is autosomal recessive. The offspring of consanguineous matings have a small but significantly increased risk of genetic disease. In fact, they account for a disproportionately high percentage of those individuals in the population with autosomal recessive disorders. Because of this perceived social cost, marriages of close relatives, including first cousins, are prohibited by about half the states in the United States. However, consanguineous marriages are still relatively common in many other countries.

Genetic discrimination provokes heated debate

One of the fastest-growing areas of medical diagnostics is genetic screening and testing, and new genetic tests that screen for diseases such as cystic fibrosis, sickle cell anemia, Huntington's disease, colon cancer, and breast cancer increase each year. However, genetic testing raises many social, ethical, and legal issues that we as a society must address. Not the least of these is *genetic discrimination,* discrimination against an individual or family member because of differences from the "normal" genome in that individual.

One of the most difficult issues is whether genetic information should be available to health insurance and life insurance companies. Many people think that genetic information should not be given to insurance companies, but others, including employers, insurers, and many organizations representing people affected by genetic disorders, say such a view is unrealistic. If people use genetic tests to help them decide when to buy insurance and how much, insurers insist that they should also have access to this information. Insurers say that they need access to genetic data to help calculate equitable premiums (insurance companies average risks over a large population). Physicians argue that people at risk for a particular genetic disease might delay being tested because they fear genetic discrimination from insurers and employers.

Complicating the issue even more is that genetic tests are sometimes difficult to interpret, in part because of the many complex interactions between genes and the environment. If a woman tests positive for an allele that has been linked to breast cancer, for example, she is at significant risk, but testing positive does not necessarily mean that she will develop breast cancer. These uncertainties also make it hard to decide what form of medical intervention—from frequent mammograms to surgical removal of healthy breasts—is appropriate.

The Ethical, Legal, and Social Implications (ELSI) Research Program of the National Human Genome Research Institute has developed principles designed to protect individuals against genetic discrimination. In 2008, the *Genetic Information Nondiscrimination Act (GINA)* extended significant protection against workplace and health discrimination. This law prohibits employers and insurance companies from discriminating on the basis of information derived from genetic tests. For example, neither employers nor insurance companies are permitted to request a genetic test. Employers cannot base employment decisions on genetic information. Also, insurance companies cannot cancel or deny coverage or increase the price of premiums based on genetic information.

Many ethical issues related to human genetics must be addressed

Genetic discrimination is only one example of ethical issues arising from our expanding knowledge of human genetics. Consider the following questions, all of which deal with the broad ethical issue of individual rights: What is the youngest age at which genetic testing should be permitted for adult-onset diseases, such as Huntington's disease? What are the emotional and psychological effects on individuals who are told that they have tested positive for an incurable genetic disease? Should testing be performed when some family members want testing and others do not?

Should parents be able to test their minor children? Should unexpected, but medically significant, results be reported to patients (or to their parents) when they are uncovered in the course of DNA sequencing for an unrelated medical condition? Should access to genetic test data be permitted in cases of paternity or kinship testing? Should states be able to collect genetic data on their residents? Should school administrators or law enforcement agencies have access to genetic data? These questions are only a sample of the many issues that both ethicists and society must consider now and in the future. As human genetics assumes an increasingly important role in society, issues of genetic privacy and the confidentiality of genetic information must be addressed.

CHECKPOINT 16.6

- *Why is it incorrect to assume that certain individuals or populations carry most of the abnormal alleles found in humans?*
- CONNECT *To be expressed, an autosomal recessive genetic disease must be homozygous. What relationship does this fact have to consanguineous matings?*
- *For what reason do health and life insurance companies want genetic information about their clients?*

SUMMARY: FOCUS ON LEARNING OBJECTIVES

16.1 Studying Human Genetics *(page 337)*

1 Distinguish between karyotyping and pedigree analysis.

- Studies of an individual's *karyotype*, the number and kinds of chromosomes present in the nucleus, enable researchers to identify various chromosome abnormalities.
- A **pedigree** is a "family tree" that shows the transmission of genetic traits within a family over several generations. Pedigree analysis is useful in detecting autosomal dominant mutations, autosomal recessive mutations, X-linked recessive mutations, and defects due to **genomic imprinting,** which is the expression of a gene based on its parental origin.

2 Discuss how gene databases and genomic methods are used to study human genetic diseases.

- The database of the **Human Genome Project** contains the entire sequence of human chromosomal DNA. Large international collaborative efforts such as the **1000 Genomes Project** and the **ENCODE Project** have allowed investigators to construct maps of the human genome showing the location of natural variations in DNA in human populations, such as **single nucleotide polymorphisms (SNPs),** and the locations of DNA sequences involved in biological functions. Researchers use these polymorphisms in **genome-wide association (GWAS) studies** to identify genetic variations associated with complex human diseases. Investigators then use gene database information to identify affected genes and understand the role of each gene, how each gene interacts with other genes, and how the expression of each gene is regulated in different tissues.

3 Discuss the importance of comparative genomics to the study of human genetics.

- Comparative genomics examines the relationships among genes, genomic structures, and functions among different species. The identification of genes and genomic regions in other species that are highly conserved with those in humans allows investigators to study their structure and function in model organisms such as the mouse, *Drosophila, C. elegans,* yeast and *E. coli.* These studies provide important insights into the roles of those genes in humans.

16.2 Abnormalities in Chromosome Number and Structure *(page 340)*

4 Explain how nondisjunction in meiosis is responsible for chromosome abnormalities such as Down syndrome, Klinefelter syndrome, and Turner syndrome.

- In **aneuploidy** there are either missing or extra copies of certain chromosomes. Aneuploidies include **trisomy,** in which an individual's cells contain an extra chromosome, and **monosomy,** in which one member of a pair of chromosomes is missing.
- **Trisomy 21,** the most common form of **Down syndrome,** and **Klinefelter syndrome** (XXY) are examples of trisomy. **Turner syndrome** (X0) is an example of monosomy.
- Trisomy and monosomy are caused by meiotic **nondisjunction,** in which sister chromatids or homologous chromosomes fail to move apart properly during meiosis.

5 Distinguish among the following structural abnormalities in chromosomes: translocations, deletions, and fragile sites.

- In a **translocation** part of one chromosome becomes attached to another. About 4% of individuals with Down syndrome have a translocation in which the long arm of chromosome 21 is attached to the long arm of one of the larger chromosomes, such as chromosome 14.
- A **deletion** can result in chromosome breaks that fail to rejoin. The deletion may range in size from a few base pairs to an entire chromosome arm. One deletion disorder in humans is **cri du chat syndrome,** in which part of the short arm of chromosome 5 is deleted.
- **Fragile sites** may occur at specific locations on both chromatids of a chromosome. In **fragile X syndrome,** a fragile site occurs near the tip on the X chromosome, where the nucleotide triplet CGG is repeated many more times than is normal. Fragile X syndrome is the most common cause of inherited intellectual disability.

6 Explain how genomic imprinting influences inheritance patterns.

- Genomic imprinting can affect the expression of a gene based on its parental origin. An allele can be repressed or expressed, without any changes to the DNA base sequence, depending on the parent from which it was inherited.

16.3 Genetic Diseases Caused by Single-Gene Mutations *(page 347)*

7 State whether each of the following genetic defects is inherited as an autosomal recessive, autosomal dominant, or X-linked recessive trait: phenylketonuria (PKU), sickle cell anemia, cystic fibrosis, Tay-Sachs disease, Huntington's disease, and hemophilia A.

- Most human genetic diseases that show a simple inheritance pattern are transmitted as autosomal recessive traits.

Phenylketonuria (PKU) is an autosomal recessive disorder in which toxic phenylketones damage the developing nervous system. **Sickle cell anemia** is an autosomal recessive disorder in which abnormal hemoglobin (the protein that transports oxygen in the blood) is produced. **Cystic fibrosis** is an autosomal recessive disorder in which abnormal secretions are produced primarily in organs of the respiratory and digestive systems. **Tay-Sachs disease** is an autosomal recessive disorder caused by abnormal lipid metabolism in the brain.

- **Huntington's disease** has an autosomal dominant inheritance pattern and results in mental and physical deterioration, usually beginning in adulthood.
- **Hemophilia** is an X-linked recessive disorder that results in a defect in a blood component required for clotting.

16.4 Gene Therapy (page 350)

8 Briefly discuss the process of gene therapy, including some of its technical challenges.

- In **gene therapy** the normal allele is cloned, and the DNA is introduced into certain human cells where its expression may be sufficient to yield a normal phenotype.
- One technical challenge in gene therapy is finding a safe, effective vector, usually a virus, to deliver the gene of interest into the cells.

16.5 Genetic Testing and Counseling (page 351)

9 State the relative advantages and disadvantages of amniocentesis, chorionic villus sampling, and preimplantation genetic diagnosis in the prenatal diagnosis of human genetic abnormalities.

- In **amniocentesis** a physician samples the amniotic fluid surrounding the fetus and then cultures and screens the fetal cells suspended in the fluid for genetic defects. Amniocentesis provides results in the second trimester of pregnancy.

- In **chorionic villus sampling (CVS),** a physician removes and studies some of the fetal cells. CVS provides results in the first trimester of pregnancy but is associated with a slightly greater risk of infection and miscarriage than amniocentesis.
- Couples who conceive by in vitro fertilization may elect to have **preimplantation genetic diagnosis (PGD),** in which a physician screens the embryos for one or more genetic diseases before placing a healthy embryo into the woman's uterus. PGD is not as accurate as amniocentesis or CVS, and it is more expensive.

10 Distinguish between genetic screening programs for newborns and adults, and discuss the scope and implications of genetic counseling.

- **Genetic screening** identifies individuals who might carry a serious genetic disease. Screening of newborns is the first step in preventive medicine, and screening of adults helps them make informed reproductive decisions.
- Couples who are concerned about the risk of abnormality in their children may seek **genetic counseling.** A genetic counselor provides medical and genetic information pertaining to reproductive decisions and helps individuals understand their situation and avoid feeling stigmatized.

16.6 Human Genetics, Society, and Ethics (page 354)

11 Discuss the controversies of genetic discrimination.

- *Genetic discrimination* is discrimination against an individual or family member because of differences from the "normal" genome in that individual.
- One of the most difficult issues in avoiding genetic discrimination is whether genetic information should be available to employers and to health and life insurance companies. The *Genetic Information Nondiscrimination Act (GINA)* prohibits employers and insurance companies from discriminating on the basis of information derived from genetic tests.

TEST YOUR UNDERSTANDING

Know and Comprehend

1. A diagram of a pedigree shows (a) controlled matings between members of different true-breeding strains (b) the total genetic information in human cells (c) a comparison of DNA sequences among genomes of humans and other species (d) the subtle genetic differences among unrelated people (e) the expression of genetic traits in the members of two or more generations of a family

2. An abnormality in which there is one more or one fewer than the normal number of chromosomes is called (a) a karyotype (b) a fragile site (c) an aneuploidy (d) trisomy (e) a translocation

3. The failure of chromosomes to separate normally during cell division is called (a) a fragile site (b) an inborn error of metabolism (c) a satellite knob (d) a translocation (e) nondisjunction

4. The chromosome composition of an individual or cell is called its (a) karyotype (b) nucleotide triplet repeat (c) pedigree (d) DNA microarray (e) translocation

5. An inherited disorder caused by a defective or absent enzyme is called (a) a karyotype (b) trisomy (c) a reciprocal translocation (d) an inborn error of metabolism (e) an aneuploidy

6. In ___, a genetic mutation codes for an abnormal hemoglobin molecule that is less soluble than usual and more likely than normal to deform the shape of the red blood cell. (a) Down syndrome (b) Tay-Sachs disease (c) sickle cell anemia (d) PKU (e) hemophilia A

7. During this procedure, a sample of the fluid that surrounds the fetus is obtained by inserting a needle through the walls of the abdomen and uterus. (a) DNA marking (b) chorionic villus sampling (c) ultrasound imaging (d) preimplantation genetic diagnosis (e) amniocentesis

Apply and Analyze

8. Which pattern of inheritance is associated with a trait that (1) is not usually expressed in the parents, (2) is expressed in about one-fourth of the children, and (3) is expressed in both male and female children? (a) autosomal recessive (b) autosomal dominant (c) X-linked recessive (d) X-linked dominant (e) Y-linked

Examine the following pedigrees. Which is the most likely mode of inheritance of each disorder? (a) autosomal recessive (b) autosomal dominant (c) X-linked recessive (d) a, b, or c (e) a or c

9. 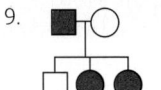   10. 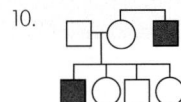   11.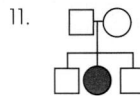

Evaluate and Synthesize

12. **SCIENCE, TECHNOLOGY, AND SOCIETY** Imagine that you are a genetic counselor. What advice or suggestions might you give in the following situations?

    a. A couple has come for advice because the woman had a sister who died of Tay-Sachs disease.

    b. A young man and woman who are not related are engaged to be married. However, they have learned that the man's parents are first cousins, and they are worried about the possibility of increased risk of genetic defects in their own children.

    c. A young woman's paternal uncle (her father's brother) has hemophilia A. Her father is free of the disease, and there has never been a case of hemophilia A in her mother's family. Should she be concerned about the possibility of hemophilia A in her own children?

    d. A 20-year-old man is seeking counseling because his father was recently diagnosed with Huntington's disease.

    e. A 45-year-old woman has just been diagnosed with Huntington's disease. She says she will not tell her college-age sons because of the burden it will place on them. Given that the woman, not her sons, is your client, do you have a duty to inform the sons? Explain your reasoning.

13. A common belief about human genetics is that an individual's genes alone determine his or her destiny. Explain why this idea is a misconception.

14. **CONNECT** Is a chromosome deletion equivalent to a frameshift mutation (discussed in Chapter 13)? Why or why not?

15. **EVOLUTION LINK** Explain some of the evolutionary implications that one can conclude from mice and humans having about five hundred DNA segments that are completely identical.

16. **INTERPRET DATA** Examine Figure 16-11 and estimate the age at which half of individuals carrying a Huntington's disease allele will have developed symptoms. At what age will three-fourths of these individuals have symptoms?

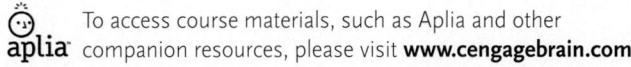 To access course materials, such as Aplia and other companion resources, please visit **www.cengagebrain.com.**

17 | Developmental Genetics

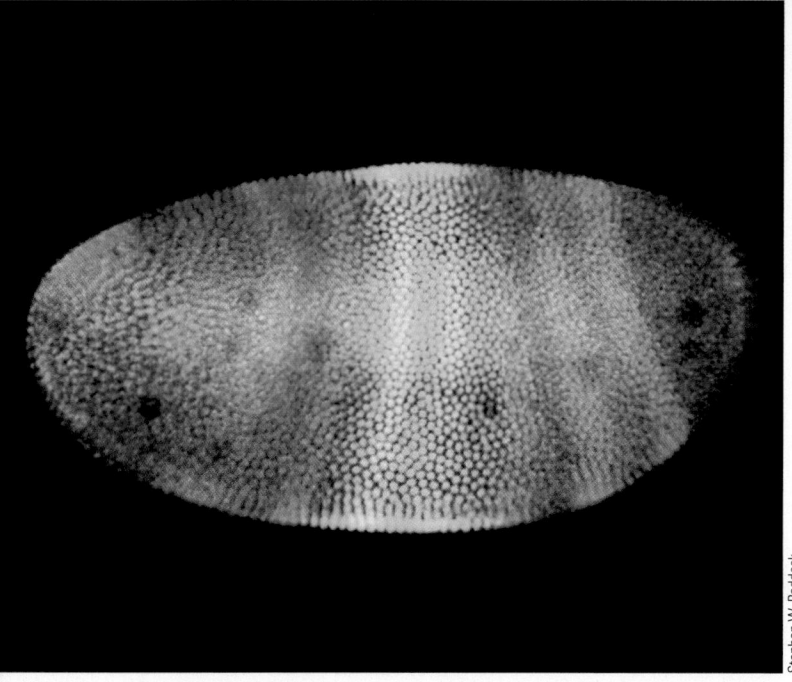

Differential gene expression in an embryo of the model organism *Drosophila melanogaster.*

Stephen W. Paddock

KEY CONCEPTS

17.1 Differential gene expression controls the development of specialized tissues and organs.

17.2 Many genes that regulate development are quite similar in a wide range of organisms, from fruit flies to humans. Mutations in genes that regulate development have provided insights into how those genes function.

17.3 Mutations in oncogenes and tumor suppressor genes may lead to cancer.

Developmental genetics is the study of the genes involved in cell differentiation and development of an organism. Until the late 1970s, biologists knew little about how genes interact to control development. Unraveling the genetic interactions that take place during development was an intractable problem using traditional methods. However, rapid progress in recombinant DNA research led scientists to search for developmental mutants and to apply sophisticated techniques to study them. (A *mutant* is an individual with an abnormal phenotype caused by a gene mutation.)

The organism in the photograph is a developing embryo of the fruit fly *Drosophila melanogaster.* Geneticists used **immunofluorescence,** in which a fluorescent dye is joined to an antibody that binds to a specific protein to localize the protein. In this case, researchers bound three different antibodies—one red, one blue, and one yellow—to three specific proteins. The patterns of colored bands indicate that different cells of the embryo have *differential gene expression;* that is, different genes are active at the same time.

Work with *Drosophila* and other organisms has profound implications for understanding both normal human development (including aging) and malfunctions that lead to birth defects and cancer. Striking similarities among genes that govern development in widely different species suggest that developmentally important genetic mechanisms are deeply rooted in the evolutionary history of multicellular organisms. There are also differences in species' developmental patterns that reflect their separate evolutionary paths. As you will learn in Chapter 18, developmental genes have played a role in reshaping organisms during the course of their evolution.

Biologists are now studying how genes are activated, inactivated, and modified to control development. Eventually, scientists expect to understand how a single cell—a fertilized egg—develops into a multicellular organism as complex as a human.

17.1 CELL DIFFERENTIATION AND NUCLEAR EQUIVALENCE

LEARNING OBJECTIVES

1 Distinguish between cell determination and cell differentiation, and between nuclear equivalence and totipotency.

2 Describe the classic experiments of Steward, Gurdon, and Wilmut.

3 Define *stem cells*, distinguish between embryonic stem cells and pluripotent stem cells, and describe some of the promising areas of research involving stem cells.

The study of **development,** broadly defined as all the changes that occur in the life of an individual, encompasses some of the most fascinating and difficult problems in biology today. Of particular interest is the process by which cells specialize and organize into a complex organism. During the many cell divisions required for a single cell to develop into a multicellular organism, groups of cells become gradually committed to specific patterns of gene activity through the process of **cell determination.** As cell determination proceeds, it restricts an embryonic cell's developmental pathway so that its fate becomes more and more limited. The final step leading to cell specialization is **cell differentiation.** A differentiated cell, which has a characteristic appearance and characteristic activities, appears to be irreversibly committed to its fate.

Another intriguing part of the developmental puzzle is the building of the body. In **morphogenesis,** the development of form, cells in specific locations differentiate and become spatially organized into recognizable structures. Morphogenesis proceeds through the multistep process of **pattern formation,** the organization of cells into three-dimensional structures. Pattern formation includes signaling between cells, changes in cell shapes, and cell migrations. Depending on their location, cells are exposed to different concentrations of signaling molecules that specify positional information. Thus, *where* a given cell is located often determines *what* it will become when it matures.

The human body, like that of other vertebrates, contains about 250 recognizably different types of cells (FIG. 17-1). Combinations of these specialized cells, known as **differentiated cells,** are organized into diverse and complex structures—such as the eye, hand, and brain—each capable of carrying out many sophisticated activities. Most remarkable of all is that all the structures of the body and the different cells within them descend from a unicellular **zygote,** a fertilized egg.

All multicellular organisms undergo complex patterns of development. The root cells of plants, for example, have structures and functions very different from those of the various types of cells located in leaves. Diversity is also found at the molecular level; most strikingly, each type of plant or animal cell makes a highly specific set of proteins. In some cases, such as the protein hemoglobin in red blood cells, one cell-specific protein may make up more than 90% of the cell's total mass of protein. Other cells may have a complement of many cell-specific

proteins, each of which is present in small amounts but still plays an essential role. However, because certain proteins are required in every type of cell (e.g., all cells require the same enzymes for glycolysis), cell-specific proteins usually make up only a fraction of the total number of different kinds of proteins.

When researchers first discovered that each type of differentiated cell makes a unique set of proteins, some scientists hypothesized that each group of cells loses the genes it does not need and retains only those required. However, this hypothesis does not generally seem true. According to the principle of **nuclear equivalence,** the nuclei of essentially all differentiated adult cells of an individual are genetically (although not necessarily metabolically) identical to one another and to the nucleus of the zygote from which they descended. Virtually all *somatic cells* in an adult have the same genes, but different cells express different subsets of these genes.

Somatic cells are all the cells of the body other than **germ line cells,** which ultimately give rise to a new generation. In animals germ line cells—whose descendants ultimately undergo meiosis and differentiate into gametes—are generally set aside early in development. In plants the difference between somatic cells and germ line cells is not as distinct, and the determination that certain cells undergo meiosis is made much later in development.

The evidence for nuclear equivalence comes from cases in which differentiated cells or their nuclei have been found to retain the potential of directing the development of the entire organism. Such cells or nuclei are said to be **totipotent.**

Most cell differences are due to differential gene expression

Because genes do not seem to be lost regularly during development (and thus nuclear equivalence is present in different cell types), differences in the molecular composition of cells must be regulated by the activities of different genes. The process of developmental gene regulation is often referred to as **differential gene expression.**

As discussed in Chapter 14, the expression of eukaryotic genes is regulated in many ways and at many levels. For example, a particular enzyme may be produced in an inactive form and then be activated later. However, much of the regulation that is important in development occurs at the transcriptional level. The transcription of certain sets of genes is repressed, whereas that of other sets is activated. Even the expression of genes that are *constitutive*—that is, constantly transcribed—is regulated during development so that the *quantity* of each product varies from one tissue type to another.

We can think of differentiation as a series of pathways leading from a single cell to cells in each of the different specialized tissues, arranged in an appropriate pattern. At times, a cell makes genetic commitments to the developmental path its descendants will follow. These commitments gradually restrict the development of the descendants to a limited set of final tissue types. Determination, then, is a progressive fixation of the fate of a cell's descendants.

As development proceeds, somatic cells that previously had the potential to develop into a variety of cells become increasingly committed to a specific fate.

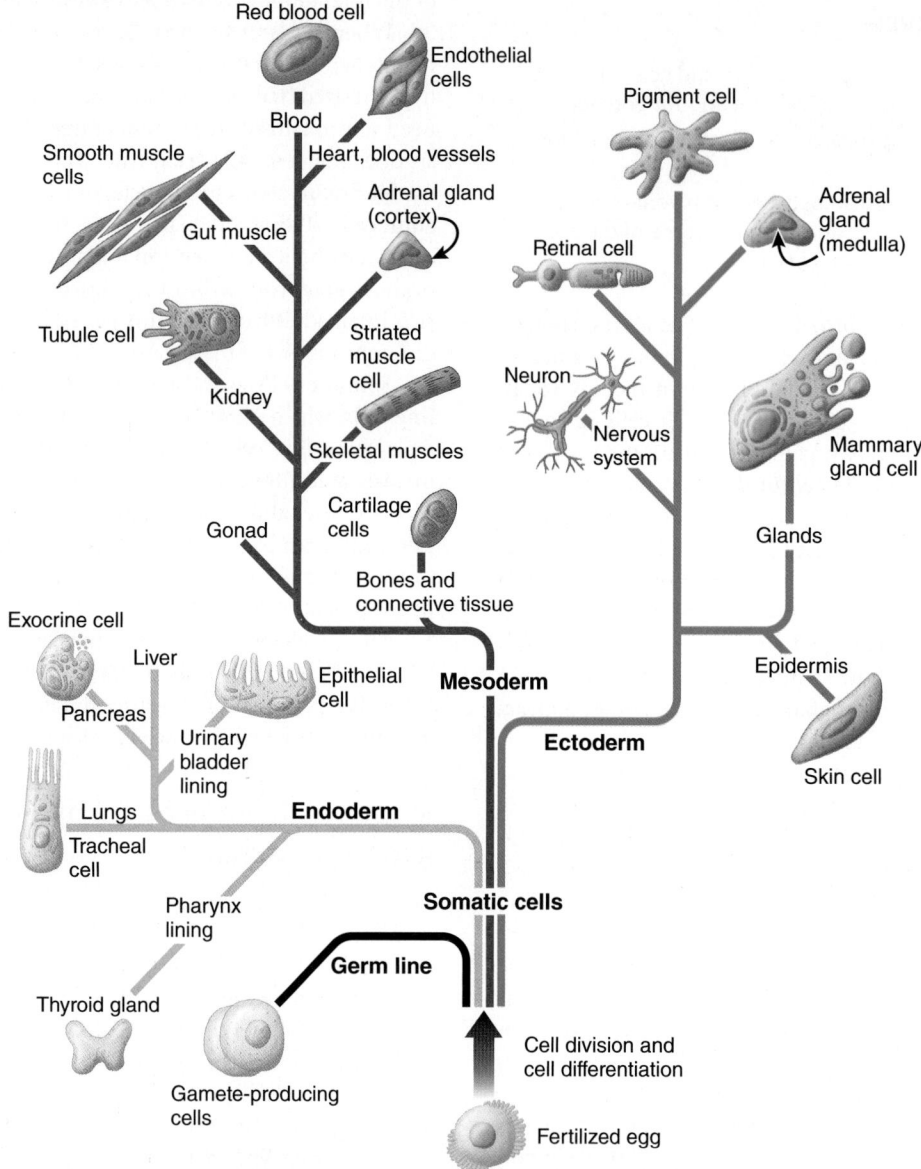

Figure 17-1 Vertebrate cell lineages

Repeated divisions of the fertilized egg (*bottom*) result in the establishment of tissues containing groups of specialized cells. Germ line cells (cells that produce the gametes) are set aside early in development. Somatic cells progress along various developmental pathways undergoing a series of commitments that progressively determine their fates.

CONNECT Can a vertebrate organ contain cells derived from more than one lineage?

© Cengage Learning

As the development of a cell becomes determined along a differentiation pathway, its physical appearance may or may not change significantly. Nevertheless, when a stage of determination is complete, the changes in the cell usually become self-perpetuating and are not easily reversed. Cell differentiation is usually the last stage in the developmental process. At this stage, a precursor cell becomes structurally and functionally recognizable as a bone cell, for example, and its pattern of gene activity differs from that of a neuron (nerve cell) or any other cell type.

A totipotent nucleus contains all the instructions for development

In plants some differentiated cells can be induced to become the equivalent of embryonic cells. Biologists use *tissue culture techniques* to isolate individual cells from certain plants and to allow them to grow in a nutrient medium.

In the 1950s, F.C. Steward and his co-workers at Cornell University conducted some of the first experiments investigating

Are differentiated somatic plant cells totipotent?

HYPOTHESIS: Differentiated somatic carrot cells can be induced to develop into an entire plant.

EXPERIMENT: F.C. Steward and his co-workers cultured carrot root tissues in a liquid nutrient medium. These cells divided to form clumps of undifferentiated cells. The clumps were then transferred to a solid growth medium.

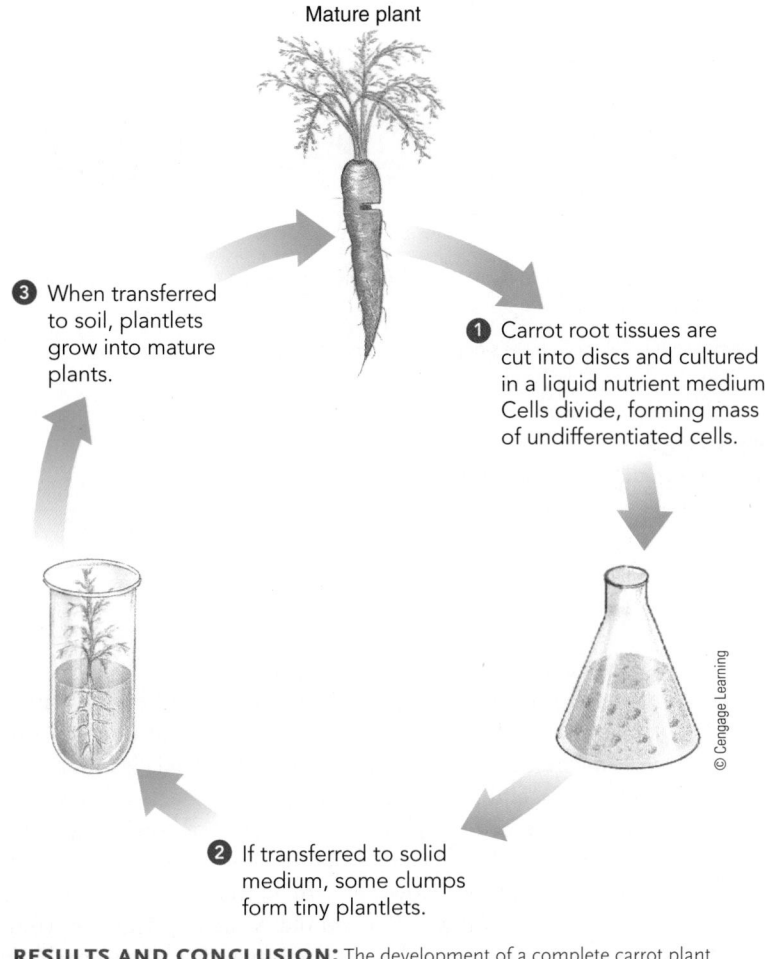

Mature plant

❸ When transferred to soil, plantlets grow into mature plants.

❶ Carrot root tissues are cut into discs and cultured in a liquid nutrient medium. Cells divide, forming mass of undifferentiated cells.

❷ If transferred to solid medium, some clumps form tiny plantlets.

© Cengage Learning

RESULTS AND CONCLUSION: The development of a complete carrot plant from differentiated somatic cells demonstrated the totipotency of these cells.

SOURCE: Shantz, E.M., and F.C. Steward. "Investigations on Growth and Metabolism of Plant Cells VII: Sources of Nitrogen for Tissue Cultures under Optimal Conditions for Their Growth." *Annals of Botany*, Vol. 23, 371–390, 1959. By permission of Oxford University Press.

Figure 17-2 Steward's experiment on cell totipotency in carrots

INTERPRET DATA Does this experiment demonstrate that every differentiated plant cell is totipotent?

cell totipotency in plants (**FIG. 17-2**). Totipotent cells have the potential to give rise to all parts of an organism because they contain a complete set of genetic instructions required to direct the normal development of an entire organism. Steward and his colleagues induced root cells from a carrot to divide in a liquid nutrient medium, forming groups of cells called *embryoid*

(embryo-like) *bodies.* These clumps of dividing cells were then transferred to an agar medium, which provided nutrients and a solid supporting structure for the developing plant cells. Some of the cells of the embryoid bodies gave rise to roots, stems, and leaves. The resulting small plants, called *plantlets* to distinguish them from true seedlings, were then transplanted to soil, where they ultimately developed into adult plants capable of producing flowers and viable seeds.

Because these plants are all derived from the same parent plant, they are genetically alike and therefore constitute a clone. As mentioned in Chapter 15, a **clone** consists of individual organisms, cells, or DNA molecules that are genetically identical to another individual, cell, or DNA molecule from which it was derived. The methods of plant tissue culture are now extensively used to produce genetically engineered plants because they enable researchers to regenerate whole plants from individual cells that have incorporated recombinant DNA molecules (see Chapter 38).

In the 1950s, researchers began testing whether steps in the process of determination are reversible in animal cells by transplanting the *nucleus* of a cell in a relatively late stage of development into an egg cell that had been *enucleated* (i.e., its own nucleus had been destroyed). Robert Briggs and Thomas J. King of the Institute for Cancer Research in Pennsylvania pioneered *nuclear transplantation experiments.* They transplanted nuclei from frog cells at different stages of development into egg cells whose nuclei had been removed. Some of the transplants proceeded normally through several developmental stages, and a few even developed into normal tadpoles. As a rule, the nuclei transplanted from cells at earlier stages were most likely to support development to the tadpole stage. As the fate of the cells became more and more determined, the probability quickly declined that a transplanted nucleus could control normal development.

British biologist John B. Gurdon carried out experiments on nuclear transplantation in frogs during the 1960s. In a few cases, he demonstrated that nuclei isolated from the intestinal epithelial cells of a tadpole directed development up to the tadpole stage (**FIG. 17-3**). This result occurred infrequently (about 1.5% of the time); in these kinds of experiments, however, success counts more than failure. Therefore, he could safely conclude that at least some nuclei of differentiated animal cells are, in fact, totipotent.

For many years, because these successes with frogs could not be repeated with mammalian embryos, many developmental biologists concluded that some fundamental feature of mammalian reproductive biology might be an impenetrable barrier to mammalian cloning. This perception changed markedly in 1996 and 1997 with the first reports of the birth of cloned mammals.

Are nuclei in differentiated animal cells totipotent?

HYPOTHESIS: Nuclei from differentiated cells contain the information required for normal development.

EXPERIMENT: John Gurdon injected the nuclei of differentiated cells (tadpole intestinal cells) into eggs whose own nuclei were destroyed by ultraviolet radiation.

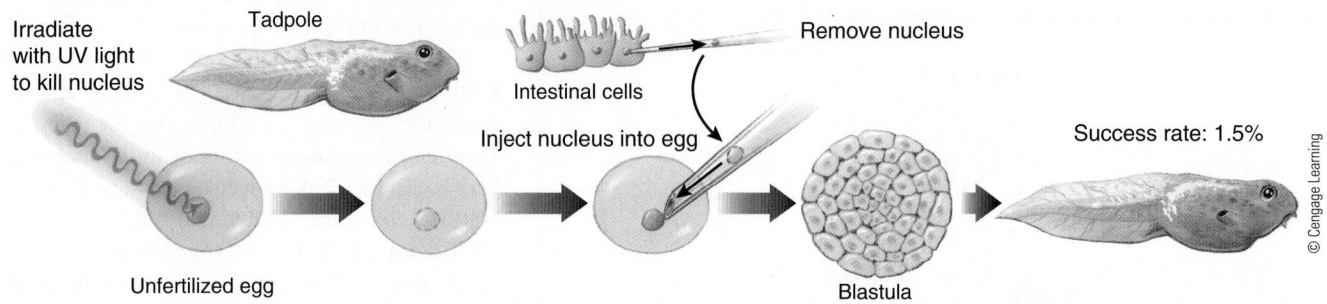

RESULTS AND CONCLUSION: Normal development proceeded to the tadpole stage in about 1.5% of trials, indicating that the genes for programming development up to that point were still present and could be appropriately activated.

SOURCE: Gurdon, J.B. "The Developmental Capacity of Nuclei Taken from Intestinal Epithelium Cells of Feeding Tadpoles." *Journal of Embryology and Experimental Morphology,* Vol. 10, Dec. 1962.

Figure 17-3 Gurdon's experiment on nuclear totipotency in frogs

PREDICT What might have happened in this experiment if the nucleus of the unfertilized egg had not been killed?

The first cloned mammal was a sheep

In 1996, Ian Wilmut, Keith Campbell, and their co-workers at the Roslin Institute in Edinburgh, Scotland, reported that they had succeeded in cloning sheep by using nuclei from an early stage of sheep embryos (the *blastocyst* stage; see Chapter 51). These scientists received worldwide attention in early 1997 when they announced the birth of a lamb named Dolly (after singer Dolly Parton). Dolly's genetic material was derived from a cultured sheep mammary gland cell that was fused with an enucleated sheep's egg. The resulting cell divided and developed into an embryo that was then cultured in vitro until it reached a stage at which it could be transferred to a host mother (**FIG. 17-4**). Not surprisingly, the overall success rate was low: of 277 fused cells, only 29 developed into embryos that could be transferred, and Dolly was the only live lamb produced.

Why did Wilmut's team succeed when so many other researchers had failed? Applying the basic principles of cell biology, they recognized that the *cell cycles* (see Chapter 10) of the egg cytoplasm and the donor nucleus were not synchronous. The egg cell is arrested at metaphase II of meiosis, whereas the actively growing donor somatic cell is usually in the DNA synthesis phase (S), or in G_2. By withholding certain nutrients from the mammary gland cells used as donors, the researchers caused these cells to enter a nondividing state referred to as G_0. This step had the effect of synchronizing the cell cycles of the donor nucleus and the egg. They then used an electric shock to fuse the donor cell with the egg and initiate embryo development.

Although an extremely high level of technical expertise is required, these and other researchers have modified and extended these techniques to produce cloned calves, pigs, horses, rats, mice, dogs, and cats, among others. The list of mammalian species successfully cloned continues to grow. However, the success rate for each set of trials is low, around 1% to 2%, and the incidence of genetic defects is high. Dolly was euthanized at age six because she was suffering from a virus-induced lung cancer that infected several sheep where she was housed. However, she developed arthritis when she was five-and-a-half years old, which is relatively young for a sheep to have this degenerative disease. Some biologists speculate that using adult genetic material to produce a clone might produce an animal with prematurely old cells (see discussion of telomeres and cell aging in Chapter 12). Further research may provide some answers to this potential problem.

The main focus of cloning research is the production of *transgenic* organisms, in which foreign genes have been incorporated (see Chapter 15). Researchers are actively pursuing new techniques to improve the efficiency of the cloning process. Only then will it be possible to produce large numbers of cloned transgenic animals for a variety of uses, such as increasing the populations of endangered species. For example, the first healthy clone of an endangered species, a wild relative of cattle known as a *banteng,* was born in 2003. The nucleus for this clone came from a frozen skin cell of a banteng that died in 1980 at the San Diego Zoo.

Are nuclei in differentiated mammalian cells totipotent?

HYPOTHESIS: The nucleus of a differentiated cell from an adult mammal fused with an enucleated egg can provide the genetic information to direct normal development.

EXPERIMENT: Ian Wilmut and his colleagues produced a sheep embryo by fusing a cultured adult sheep mammary cell with an enucleated sheep's egg. He then implanted the embryo into the uterus of a host mother.

RESULTS AND CONCLUSION: Normal development proceeded, and a female lamb—the world's first cloned mammal—was born. To provide additional evidence that the cloned sheep was fully functioning, when she matured, she was bred and gave birth to a normal offspring.

SOURCE: Campbell, K.H., J. McWhir, W.A. Ritchie, and I. Wilmut. "Sheep Cloned by Nuclear Transfer from a Cultured Cell Line." *Nature*, Vol. 380, 1996.

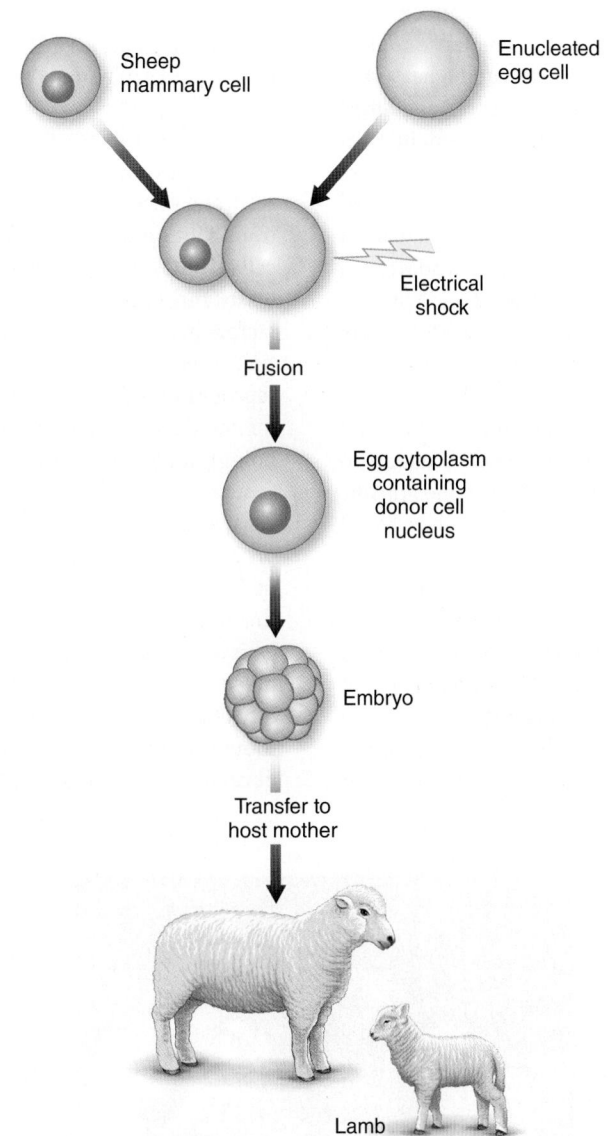

Figure 17-4 *Animation* **Wilmut's experiment on mammalian cloning**

CONNECT Was the egg cell nucleus haploid or diploid prior to enucleation? What was the ploidy of the nucleus derived from the cultured mammary cell?
© Cengage Learning

Stem cells divide and give rise to differentiated cells

Stem cells are undifferentiated cells that can divide to produce a daughter cell that remains as a stem cell and a daughter cell that can differentiate into specialized cells such as muscle cells, neurons, or blood cells. What happens depends on the presence or absence of an array of biochemical signals. One of the most challenging areas of stem cell research today involves determining the identity, order, and amounts of the chemical signals that will result in a specific type of cell differentiation.

The most versatile stem cells—zygotes—are totipotent and can give rise to all tissues of the body and placenta. Stem cells can also be derived from embryos or adult cells. (The term *adult* is somewhat misleading because **adult stem cells** can be harvested from umbilical cord tissue, infants, and children as well as adults.) Embryonic and adult stem cells are known as **pluripotent** stem cells because they can give rise to many, but not all, of the types of cells in an organism. **Embryonic stem cells (ES cells),** formed after a zygote has undergone several rounds of cell division to form a five- or six-day-old blastocyst, are more versatile than adult stem cells (FIG. 17-5). For example, ES cells are pluripotent and have the potential to develop into any type of cell in the body; ES cells are not totipotent because they cannot form cells of the placenta.

Adult stem cells have been found in the human brain, retina, heart, bone marrow, dental pulp, intestines, and other sites. Adult neural stem cells in the brain are pluripotent and differentiate to form neurons and glial cells. Adult stem cells in the bone marrow form both red blood cells and the various types of white blood cells of the immune system. The range of cells that adult stem cells can differentiate into is more limited than that of ES cells. However, recent studies suggest that even specialized stem cells may be more versatile than once thought. For example, adult neural stem cells form blood cells when transplanted into bone marrow.

Stem cells are potential sources for cell transplantation into patients to treat serious degenerative conditions. For example, stem cells may become a source of insulin-producing cells for transplantation into the pancreas of individuals with diabetes mellitus. Stem cells might also provide replacement neurons in people with spinal cord injury or other types of neurological damage.

ES cells are extremely valuable for research because they can give unique insights into the molecular details of development from its earliest stages to full differentiation. Thus far, the only known source of ES cells is early human embryos left over from *in vitro fertilization* (see Chapter 50). Some countries have government restrictions on public funding due to ethical considerations related to the origins of ES cells. Other countries, notably some Asian countries, have no restrictions on this type of research. In the United States, the Obama administration lifted restrictions on stem cell research that were established during the George W. Bush administration, and public funding became available in 2009.

Induced pluripotent stem cells (iPSCs) can be produced from mature cells In 2007, Dr. Shinya Yamanaka and his laboratory reported that the induced expression of only four transcription factors could produce **induced pluripotent stem cells (iPSCs)** from mature mouse and human cells such as fibroblasts and skin cells. (Recall from Chapter 14 that transcription factors are DNA-binding proteins that regulate gene transcription in eukaryotic cells.) This revolutionary ability to produce reprogrammed cells that function as stem cells has the potential for the development of *regenerative medicine*. Using this technology, a patient's own cells could be used to replace diseased tissues, such as heart muscle or insulin-producing pancreatic cells, without being rejected by his or her immune system. This process is commonly referred to as *human therapeutic cloning*. In 2012, Yamanaka, along with John B. Gurdon, was awarded the Nobel Prize in Physiology or Medicine for these discoveries, which showed a way to reprogram mature cells to make them pluripotent.

The process of reprogramming involves the reactivation of thousands of genes that were "locked" in an inactive state during development through DNA methylation, histone modifications,

Blastocyst

① Human embryonic stem (ES) cells are derived from 5- to 6-day-old embryo (blastocyst).

② ES cells are present in liquid drops of this stem cell culture.

Smooth muscle cells

Neuron

Red blood cells

③ It may be possible to induce ES cells to differentiate into any of approximately 250 cell types of the human body.

Volker Steger/Science Photo Library/Science Source

Figure 17-5 Human embryonic stem (ES) cells
© Cengage Learning

and the production of inhibitor microRNAs. One of the primary mechanisms by which the transcription factors reprogram the mature cells to pluripotent cells involves *epigenetic reprogramming* by removing methyl groups from key genes that are normally turned off in adult cells but are active in embryonic cells. (Recall from Chapter 14 that epigenetic inheritance consists of gene inactivation, such as by DNA methylation.) One of the remaining difficulties with the use of either iPSCs or ES cells for regenerative medicine is the interaction of the reprogrammed cells with the older surrounding cells. Cells that are differentiated from the reprogrammed tissue tend to revert to the stem cell state, differentiating into tumors. Because of the promise of these studies, this area of research is highly active.

Ethical questions exist relating to human cloning Cloning research continues to fuel an ongoing debate regarding the potential for human cloning and its ethical implications. In the United States, the National Bioethics Advisory Commission has been established to study this and other questions. In considering these issues, it is important to recognize that *cloning* is a broad term that includes several different processes involved in producing biological cells, tissues, organs, or organisms.

Human reproductive cloning has the goal of producing a newborn human that is genetically identical to another, usually adult, human. It would involve placing a human embryo produced by a process other than fertilization into a woman's body. There is wide opposition to human reproductive cloning.

In contrast, *human therapeutic cloning* discussed earlier would involve duplication of human ES cells or iPSCs for scientific study or medical purposes; no newborn human would develop. Many people support research in therapeutic cloning because of the potential benefits in treating disease. Other people have ethical objections if ES cells are used.

CHECKPOINT 17.1

- *What lines of evidence support the principle of nuclear equivalence?*
- **CONNECT** *Why was an understanding of the cell cycle crucial to the success of Wilmut's team in the first demonstration of mammalian cloning?*
- **CONNECT** *What does the ability to produce iPSCs tell scientists about the differences between ES cells and differentiated cells? Do these findings argue against the principle of nuclear equivalence? Explain your answer.*

17.2 THE GENETIC CONTROL OF DEVELOPMENT

LEARNING OBJECTIVES

4 Indicate the features of *Drosophila melanogaster, Caenorhabditis elegans, Mus musculus,* and *Arabidopsis thaliana* that have made these organisms valuable models in developmental genetics.

5 Distinguish among maternal effect genes, segmentation genes, and homeotic genes in *Drosophila.*

6 Explain the relationship between transcription factors and genes that control development.

7 Define *induction* and *apoptosis,* and give examples of the roles they play in development.

Development has been an important area of research for many years, and biologists have spent considerable time studying the development of invertebrate and vertebrate animals. By investigating patterns of morphogenesis in different species, researchers have identified both similarities and differences in the basic plan of development from a zygote to an adult in organisms ranging from the sea urchin to mammals (see Chapter 51).

In addition to descriptive studies, many classic experiments have demonstrated how groups of cells differentiate and undergo pattern formation. Researchers have developed elaborate screening programs to detect mutations that let them identify many developmental genes in both plants and animals. They then use molecular genetic techniques and other sophisticated methodologies to determine how those genes work and how they interact to coordinate developmental processes.

A variety of model organisms provide insights into basic biological processes

In studies of the genetic control of development, the choice of organism to use as an experimental model is important. A **model organism** is a species chosen for biological studies because it has characteristics that allow for the efficient analysis of biological processes. Because most model organisms are small organisms with short generation times, they are easy to grow and study under controlled conditions. For example, mice are a better model organism than kangaroos.

One of the most powerful approaches in developmental genetics involves isolating mutants of a model organism with abnormal development. Not all organisms have useful characteristics that allow researchers to isolate and maintain developmental mutants for future study. Geneticists so thoroughly understand the genetics of the fruit fly, *Drosophila melanogaster,* that this organism has become one of the most important systems for such studies. Other organisms—the yeast *Saccharomyces cerevisiae;* the nematode worm, *Caenorhabditis elegans;* the zebrafish, *Danio rerio;* the laboratory mouse, *Mus musculus;* and certain plants, including *Arabidopsis thaliana,* a tiny weed— have also become important models in developmental genetics. Each of these organisms has attributes that make it particularly useful for examining certain aspects of development (FIG. 17-6).

In the 1990s, developmental geneticists working on *C. elegans* discovered RNA interference, which has become a powerful research tool. In *RNA interference,* certain small RNA molecules interfere with the expression of genes or their RNA transcripts (see Chapter 13). One way these molecules work is to silence genes by selectively cleaving mRNA molecules that have base sequences complementary to the small RNA molecules. The use of RNAi in a wide variety of organisms makes it possible to silence the expression of a specific gene in an organism during its development, thereby deducing the purpose of the gene.

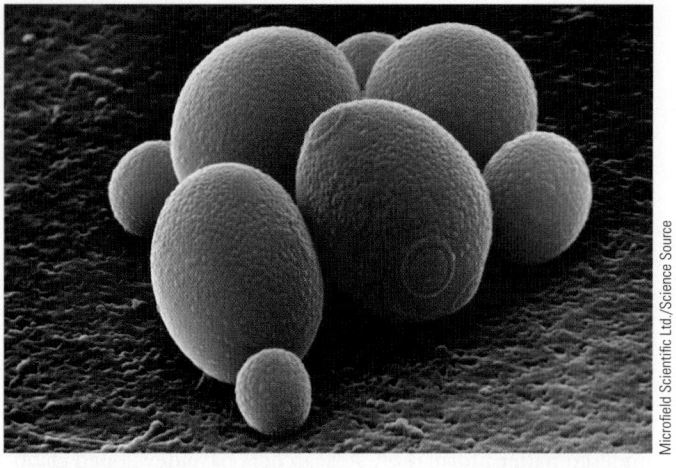

5 μm

(a) **Yeast (*Saccharomyces cerevisiae*)** is a model organism for studying developmental genetics in haploid, unicellular eukaryotes.

(b) **Fruit fly (*Drosophila melanogaster*)** development involves an anterior–posterior body plan that is common to many invertebrates and vertebrates.

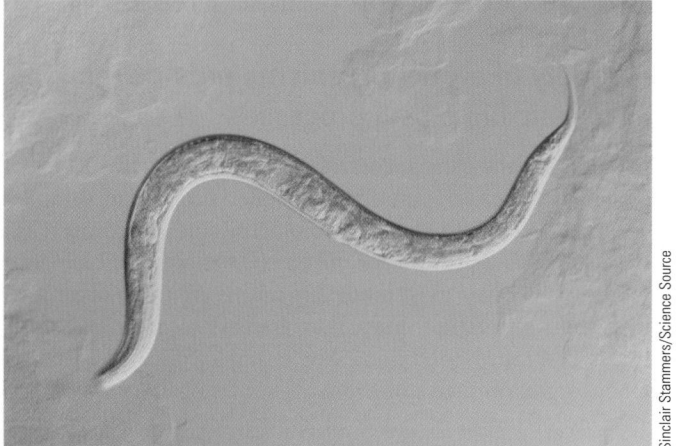

0.25 mm

(c) The **nematode (*Caenorhabditis elegans*)** adult body consists of only 959 cells, and its transparency allows researchers to trace the development of each cell.

(d) **Zebrafish (*Danio rerio*)** are small (2 to 4 cm long) and easy to grow, providing an ideal organism for studying the genetic basis of development in vertebrates.

(e) **Mouse (*Mus musculus*)** is a model organism for studying developmental genetics, including the genetics of cancer, in mammals.

(f) **Mouse-ear cress (*Arabidopsis thaliana*)** is a small plant with a small genome that is used to study development in flowering plants.

Figure 17-6 Model organisms in developmental genetics

Many genes that control development have been identified in the fruit fly

The fruit fly (*Drosophila*) genome sequence, which was completed in late 1999, includes about 13,600 protein-coding genes. At least 1200 of them are essential for embryonic development.

Two of the traditional advantages of *Drosophila* as a research organism are the abundance of mutant alleles, including those of developmental genes, available for study and the relative ease with which a new mutation is mapped on the chromosomes. U.S. biologist Edward B. Lewis (1918–2004), a pioneering developmental geneticist, began working with *Drosophila* mutants in the 1940s. The work of German researcher Christiane Nüsslein-Volhard and American Eric Wieschaus extended our understanding of development in fruit flies. Lewis, Nüsslein-Volhard, and Wieschaus shared the 1995 Nobel Prize in Physiology or Medicine for their decades of painstaking research on the genetics of *Drosophila* development. Many of the genes they discovered in the fruit fly are now known to be important in the growth and development of all animals.

The *Drosophila* life cycle includes egg, larval, pupal, and adult stages Development in *Drosophila* consists of several distinct stages (FIG. 17-7). After the egg is fertilized, a period of embryogenesis occurs during which the zygote develops into a sexually immature form known as a **larva** (pl., *larvae*). After hatching from the egg, each larva undergoes several molts (shedding of the external covering, or cuticle). The periods between molts are called *instars*. Each molt results in a size increase until the larva is ready to become a **pupa.** Pupation involves a molt and the hardening of the new external cuticle so that the pupa is completely encased. The insect then undergoes **metamorphosis,** a complete change in form. During that time, most of the larval tissues degenerate, and other tissues differentiate to form the body parts of the sexually mature adult fly.

The larvae are wormlike in appearance and look nothing like the adult flies. However, very early in embryogenesis of the developing larvae, precursor cells of many of the adult structures are organized as relatively undifferentiated, paired structures called **imaginal discs.** The name comes from *imago,* the adult form of the insect. Each imaginal disc occupies a definite position in the larva and will form a specific structure, such as a wing or a leg, in the adult body (FIG. 17-8). The discs are formed by the time embryogenesis is complete and the larva is ready to begin feeding. In some respects, the larva is a developmental stage that feeds and nurtures the precursor cells that give rise to the adult fly, which is the only form that reproduces.

The organization of the precursors of the adult structures, including the imaginal discs, is under genetic control. Thus far, more than 50 genes have been identified that specify the formation of the imaginal discs, their positions within the larva, and their ultimate functions within the adult fly. Those genes were identified through mutations that either prevent certain discs from forming or alter their structure or ultimate fate.

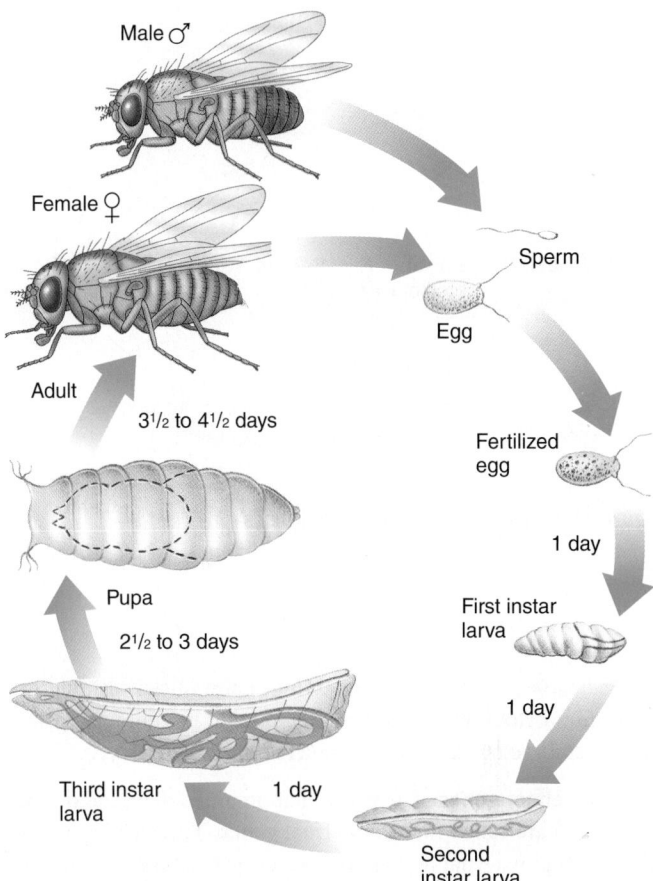

Figure 17-7 The life cycle of *Drosophila*

As it develops from a zygote to a sexually mature adult fly, a fruit fly passes through several stages. It takes about 12 days, at 25°C, to complete the life cycle. The dotted lines within the pupa represent the animal undergoing metamorphosis.

© Cengage Learning

***Drosophila* developmental mutations affect the body plan** Scientists have identified many developmental mutations in *Drosophila*. Researchers have examined their effects on development in various combinations and have studied them extensively at the molecular level. In our discussion we pay particular attention to mutations that affect the segmented body plan of the organism, in both the larva and the adult.

In early development of *Drosophila,* the structure of the egg becomes organized as it develops in the ovary of the female. Stores of mRNA, along with yolk proteins and other cytoplasmic molecules, pass from the surrounding maternal cells into the egg. Immediately after fertilization, the zygote nucleus divides, beginning a series of 13 mitotic divisions.

Each division takes only 5 or 10 minutes, which means that the DNA in the nuclei replicates constantly and at a very rapid rate. During that time, the nuclei do not synthesize RNA. Cytokinesis does not take place, and the nuclei produced by the first seven divisions remain at the center of the embryo until the eighth division occurs. At that time, most of the nuclei migrate out from the center and become localized at the periphery of the embryo. This step is known as the *syncytial blastoderm* stage because the nuclei are not surrounded by individual plasma membranes.

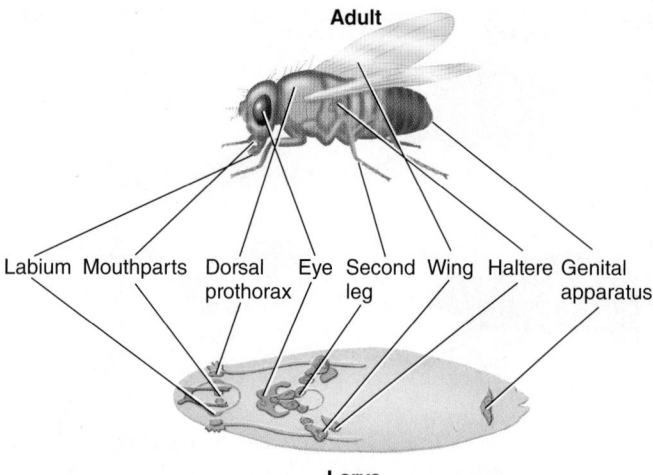

Adult

Labium Mouthparts Dorsal Eye Second Wing Haltere Genital
prothorax leg apparatus

Larva

Figure 17-8 The location of imaginal discs

Each pair of discs in a *Drosophila* larva (*bottom*) develops into a specific pair of structures in the adult fly.
© Cengage Learning

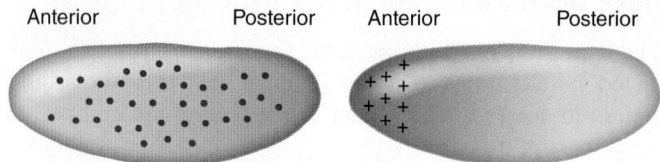

Stage of development   **Gene activity**

Anterior   Posterior   Anterior   Posterior

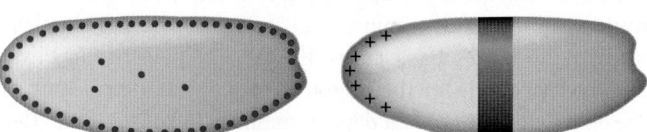

(a) At 1.25 hours after fertilization, the embryo consists of a common cytoplasm with about 128 nuclei (*purple circles*). Maternal effect genes divide the embryo into anterior and posterior sections. The crosses in the anterior region represent mRNA molecules transcribed from one maternal effect gene. The pink shading represents a different maternal mRNA molecule that is more concentrated in the anterior region.

(b) At 2 hours after fertilization, about 1500 nuclei have migrated into the periphery of the embryo and started to make their own mRNA molecules. The gap genes divide the embryo into anterior, middle, and posterior sections. The maternal RNA (*shown in pink*) is now transcribed from segmentation genes (*crosses*) in nuclei in the anterior region. The mRNA from another gap gene is transcribed in the middle of the embryo (*black region*).

Figure 17-9 Early development in *Drosophila*

Longitudinal sections on the *left* show two early stages of development of the *Drosophila* embryo. They are matched with the simplified patterns of gene activity at each stage on the *right*. (Adapted from Akam, M.E., "The Molecular Basis for Metameric Pattern in the Drosophila Embryo," *Development*, Vol. 101, 1987.)

(A *syncytium* is a structure containing many nuclei residing in a common cytoplasm.) Subsequently, plasma membranes form, and the embryo becomes known as a *cellular blastoderm*.

Maternal effect genes The genes that organize the structure of the egg cell are called **maternal effect genes.** These genes in the surrounding maternal tissues are transcribed to produce mRNA molecules that are transported into the developing egg. Analysis of mutant flies with defective maternal effect genes has revealed that many of these genes are involved in establishing the polarity of the embryo, such as which part of the embryo will become the head. Polarity dictates those parts of the egg that are dorsal or ventral and those that are anterior or posterior (see Chapter 30); thus, these genes are known as *egg polarity genes.*

FIGURE 17-9a illustrates concentration gradients for two specific maternal mRNA molecules in the very early embryo. These mRNA transcripts of some of the maternal effect genes are identified by their ability to hybridize with radioactive DNA probes derived from cloned genes (discussed in Chapter 15). Alternatively, researchers use fluorescently tagged antibodies (as in the chapter-opening photograph) to bind to specific protein products of the maternal effect genes. These protein gradients organize the early pattern of development in the embryo by determining anterior and posterior regions.

A combination of protein gradients may provide positional information that specifies the fate—that is, the developmental path—of each nucleus within the embryo. For example, mutations in certain maternal effect genes result in an embryo with two heads or two posterior ends.

In many cases, injecting normal maternal mRNA into the mutant embryo reverses the phenotype associated with a mutation in a given maternal effect gene. The fly subsequently develops normally, indicating that the gene product is needed only for a short time in the earliest stages of development.

Segmentation genes As the nuclei start to migrate to the periphery of the embryo, **segmentation genes** in those nuclei begin to produce embryonic mRNA. Thus far, geneticists have identified at least 24 segmentation genes that are responsible for generating a repeating pattern of body segments within the embryo and adult fly. Based on the study of mutant phenotypes, researchers group the segmentation genes into three classes: gap genes, pair–rule genes, and segment polarity genes.

Gap genes are the first set of segmentation genes to act. These genes interpret the maternal anterior–posterior information in the egg and begin organization of the body into anterior, middle, and posterior regions (FIG. 17-9b). A mutation in one of the gap genes usually causes the absence of one or more body segments in an embryo (FIG. 17-10a).

The other two classes of segmentation genes do not act on small groups of body segments; instead, they affect all segments. Mutations in *pair–rule genes* delete every other segment, producing a larva with half the normal number of segments (FIG. 17-10b and FIG. 17-11). Mutations in *segment polarity genes* produce segments in which one part is missing and the remaining part is duplicated as a mirror image (FIG. 17-10c). The effects of the different classes of mutants are summarized in TABLE 17-1.

Each segmentation gene has distinctive times and places in the embryo in which it is most active. The observed pattern of expression of maternal effect genes and segmentation

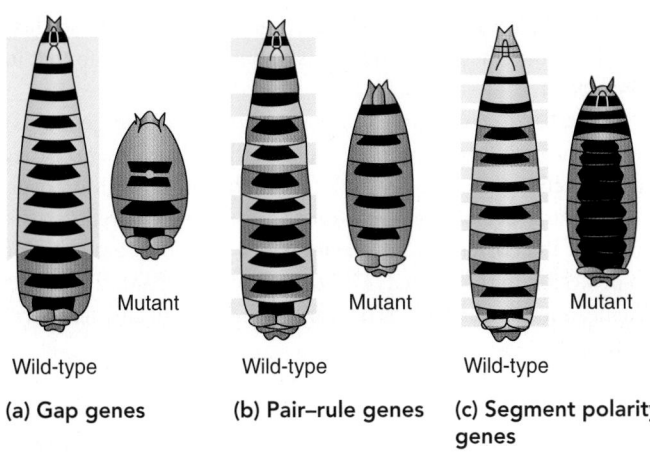

Mutant

Mutant

Mutant

Wild-type

Wild-type

Wild-type

(a) Gap genes

(b) Pair–rule genes

(c) Segment polarity genes

Figure 17-10 A comparison of mutations in *Drosophila* segmentation genes

Gap genes, pair–rule genes, and segment polarity genes control the pattern of body segments in a *Drosophila* embryo. The *blue bands* mark the regions in which the protein products of these genes are normally expressed in wild-type embryos. These same regions are absent in embryos in which the gene is mutated; that is, the mutant lacks the blue-shaded parts. The resulting phenotype is characteristic of the class to which the gene belongs. (Adapted from Nüsslein-Volhard, C., and E. Wieschaus, "Mutations Affecting Segment Number and Polarity in Drosophila," *Nature*, Vol. 287, 1980.)

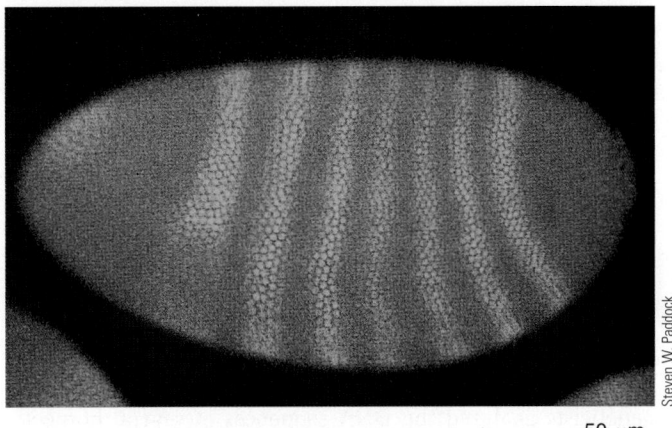

50 μm

Steven W. Paddock

Figure 17-11 Segmentation gene activity

The *bright bands* in this fluorescence LM reveal the presence of mRNA transcribed from one of the pair–rule genes known as *fushi tarazu* (Japanese for "not enough segments"). When this locus is mutated, the segments of the larva that are normally derived from these bands are absent.

genes indicates that a progressive series of developmental events determines cells destined to form adult structures. First, maternal effect genes that form gradients of morphogens in the egg determine the anterior–posterior axis and the dorsal and ventral regions of the embryo. A **morphogen** is a chemical agent that affects the differentiation of cells and the development of form. Morphogen gradients are signals that help cells determine their location within the embryo and promote their eventual differentiation into specialized tissues and organs.

Next, segmentation genes respond to the morphogens at each location to regulate the production of segments from the head to the posterior region. Within each segment other genes are then activated to specify which body part that segment will become. Every cell's position is further specified with a specific "address" designated by combinations of the activities of the regulatory genes.

The segmentation genes act in sequence, with the gap genes acting first, then the pair–rule genes, and finally the segment polarity genes. In addition, the genes of the three groups interact. Each time a new group of genes acts, cells for that group become more determined in their development. As the embryo develops, each region is progressively subdivided into smaller regions.

Most segmentation genes code for transcription factors. For example, some segmentation genes code for a "zinc-finger" type of DNA-binding regulatory protein. (See Figure 14-12b to review zinc fingers.) *Homeotic genes,* discussed next, also code for transcription factors. That many of the genes involved in controlling developmental processes code for transcription factors indicates that those proteins act as genetic switches to regulate the expression of other genes.

Once researchers have identified proteins that function as transcription factors, they can use the purified proteins to determine the DNA target sequences to which the proteins bind. This approach has been increasingly useful in identifying additional parts of the regulatory pathway involved in different stages of

TABLE 17-1	Classes of Genes Involved in Establishing Segmented Body Plan in *Drosophila*	
CLASS OF GENE	**SITE OF GENE ACTIVITY**	**FUNCTIONS OF GENES**
MATERNAL EFFECT GENES	Maternal tissues surrounding egg	Initiate anterior–posterior regions in embryo and regulate expression of segmentation genes
SEGMENTATION GENES		
Gap genes	Embryo	Divide embryo into broad regions and influence activity of pair–rule genes
Pair–rule genes	Embryo	Divide embryo into strips, each about two segments wide, and influence activity of segment polarity genes
Segment polarity genes	Embryo	Define all segments of body and divide each segment into anterior and posterior halves
HOMEOTIC GENES	Embryo	Control identities of segments; expression of homeotic gene is controlled by segmentation genes

© Cengage Learning

development. Transcription factors also play a role in cancer (discussed later in the chapter).

Homeotic genes After segmentation genes have established the basic pattern of segments in the fly body, **homeotic genes** specify the developmental plan for each segment. Mutations in homeotic genes cause one body part to be substituted for another and therefore produce some peculiar changes in the adult. A striking example is the *Antennapedia* mutant fly, which has legs that grow from the head where the antennae would normally be (**FIG. 17-12**).

Homeotic genes in *Drosophila* were originally identified by the altered phenotypes produced by mutant alleles. When geneticists analyzed the DNA sequences of several homeotic genes, they discovered a short DNA sequence of approximately 180 base pairs, which characterizes many homeotic genes as well as some other genes that play a role in development. This DNA sequence is called the **homeobox.** Each homeobox codes for a protein functional region called a **homeodomain,** consisting of 60 amino acids that form four α-helices. One of them serves as a recognition helix that binds to specific DNA sequences and thereby affects transcription. Thus, the products of homeotic genes, like those of the earlier-acting segmentation genes, are transcription factors. In fact, some segmentation genes also contain homeoboxes.

Studies of ***Hox* genes,** clusters of homeobox-containing genes that specify the anterior–posterior axis during development, provide insights about evolutionary relationships. *Hox* genes were initially discovered in *Drosophila,* where they are arranged in two adjacent groups on the chromosome: the *Antennapedia complex* and the *bithorax complex.* As *Hox* genes have been identified in other animals, such as other arthropods, sea anemones, annelids (segmented worms), roundworms, and vertebrates, researchers have found that these genes are also clustered and that their organization is remarkably similar to that in *Drosophila.*

FIGURE 17-13 compares the organization of the *Hox* gene clusters of the fruit fly and mouse. The fruit fly and mouse *Hox* genes are located in the same order along the chromosome. Moreover, the linear order of the genes on the chromosome reflects the order of the corresponding regions they control, from anterior to posterior, in the animal. This organization apparently reflects the need for these genes to be transcribed in a specific temporal sequence.

Drosophila has only one *Antennapedia–bithorax* complex. However, mice, humans, and many other vertebrates have four similar *Hox* gene clusters, each located in a different chromosome. These complexes probably arose through gene duplication. The presence of extra copies of these genes helps explain why mutations causing homeotic-like transformations are seldom seen in vertebrate animals. One does not see an extra eye growing where a mouse leg should be, for example. However, one particular type of *Hox* mutation that has been described in both mice and humans causes abnormalities in the limbs and genitalia. The involvement of the genitalia provides a further

(a) The head of a normal fly and a fly with an *Antennapedia* mutation.

(b) SEM of the head of a fly with an *Antennapedia* mutation.

Figure 17-12 The *Antennapedia* locus

Antennapedia mutations cause homeotic transformations in *Drosophila* in which legs or parts of legs replace the antennae.

© Cengage Learning

Hox genes are arranged in the same order on the chromosome as they are expressed along the anterior–posterior axis of the embryo.

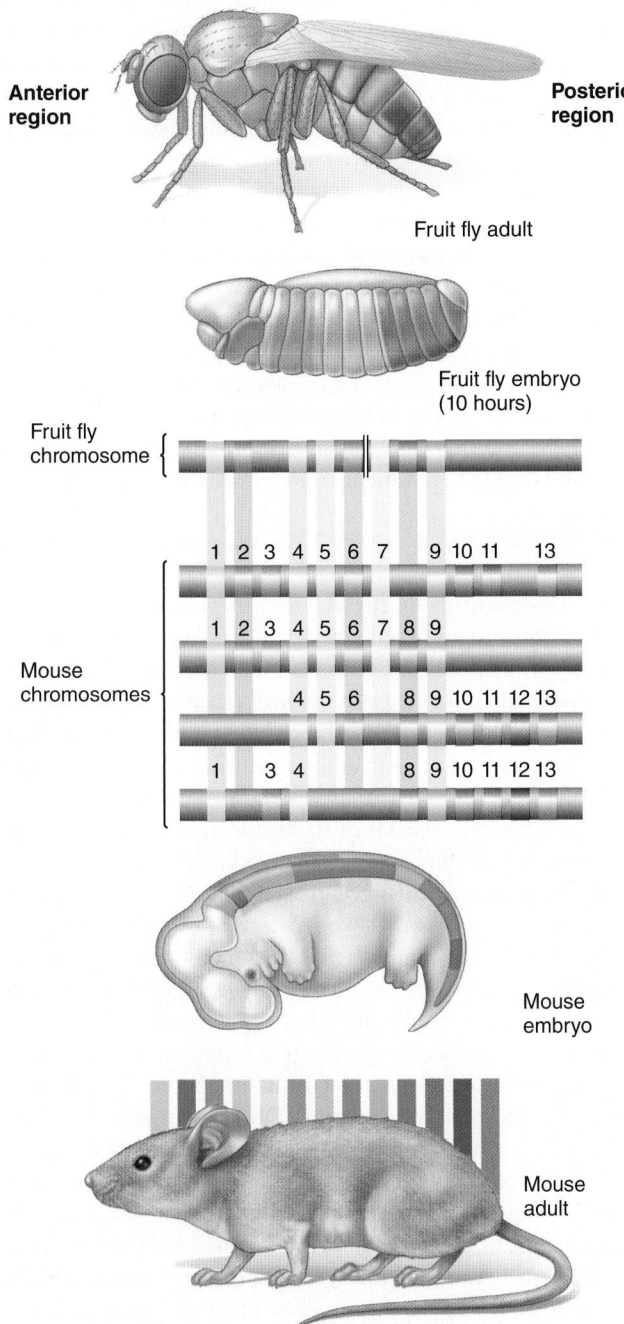

Anterior region

Posterior region

Fruit fly adult

Fruit fly embryo (10 hours)

Fruit fly chromosome

1 2 3 4 5 6 7   9 10 11   13

1 2 3 4 5 6 7 8 9

4 5 6   8 9 10 11 12 13

1   3 4   8 9 10 11 12 13

Mouse chromosomes

Mouse embryo

Mouse adult

Figure 17-13 *Hox* gene clusters

Clusters of *Hox* genes are found in all animal groups except sponges. Note that in each organism, the linear sequence of these developmental genes (*color-coded bands*) on the chromosome(s) reflects their spatial order of expression in the embryo.

CONNECT Which *Hox* genes that are active in the mouse embryo are lacking in the *Drosophila* embryo?

© Cengage Learning

explanation for the rarity of these mutant alleles because affected individuals are unlikely to reproduce.

The very similar developmental controls seen in organisms as diverse as insects and vertebrates (including humans) indicates that the basic mechanism evolved early and has been highly conserved in all animals that have an anterior–posterior axis, even those that are not segmented. Clearly, once a successful way of regulating groups of genes and integrating their activities evolved, it was retained, although it has apparently been modified to provide alterations of the body plan.

The finding of homeobox-like genes in plants suggests that these genes originated early during eukaryotic evolution. With further investigations, researchers hope to develop an overall model of how the rudiments of morphogenesis are controlled in all multicellular eukaryotes. These systems of master genes that regulate development are a rich source of "molecular fossils" that are illuminating evolutionary history in new and exciting ways. Such studies have led to an ongoing synthesis of evolution and developmental biology that has come to be known as **Evo Devo**.

Caenorhabditis elegans has a relatively rigid developmental pattern

The nematode worm *C. elegans* is an ideal model organism because its system for the genetic control of development is relatively easy to study. Sydney Brenner, a British molecular geneticist, began studying molecular development in this animal at Cambridge University in the 1960s. He selected *C. elegans* because it is small, has a short life cycle (its normal lifespan is about three weeks), and is genetically well characterized. Consisting of about 19,700 protein-coding genes, it was the first animal genome sequenced. Today, *C. elegans* is an important tool for answering basic questions about the development of individual cells within a multicellular organism.

When it is an adult, *C. elegans* is only 1.5 mm long and contains only 959 somatic cells (**FIG. 17-14**). Individuals are either males or **hermaphrodites,** organisms with both sexes in the same individual. Hermaphroditic *C. elegans* are capable of self-fertilization, which makes it easy to obtain offspring that are homozygous for newly induced recessive mutations. The availability of males that can reproduce sexually with the hermaphrodites makes it possible to perform genetic crosses as well.

Because the worm's body is transparent, researchers can follow the development of literally every one of its somatic cells using a Nomarski differential interference microscope, which provides contrast in transparent specimens. As a result of efforts by several research teams, the lineage of each somatic cell in the adult has been determined. Those studies have shown that the nematode has a very rigid, or fixed, developmental pattern. After fertilization, the egg undergoes repeated divisions, producing about 550 cells that make up the small, sexually immature larva. After the larva hatches from the egg case, further cell divisions give rise to the adult worm.

The lineage of each somatic cell in the adult can be traced to a single cell in a small group of **founder cells,** which are formed early in development (**FIG. 17-15a**). If a particular founder cell is destroyed or removed, the adult structures that would normally

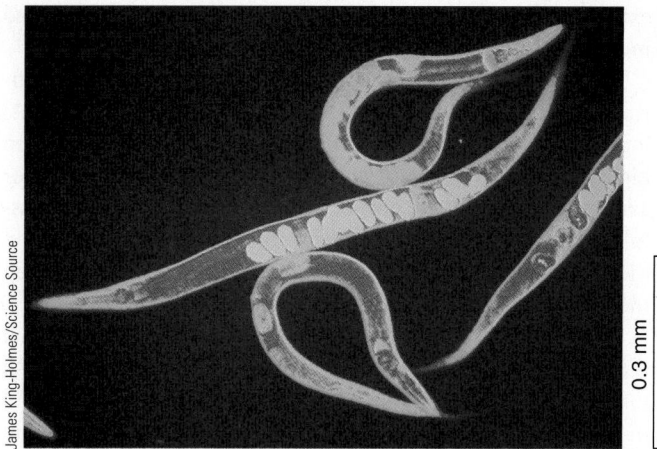

(a) False-color scanning optical micrograph of adult hermaphrodite nematodes. The oval structures are eggs.

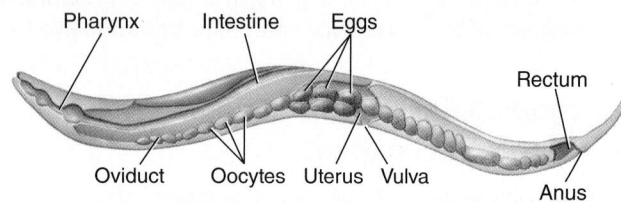

(b) Structures in the adult hermaphrodite. The sperm-producing structures are not shown.

Figure 17-14 *Caenorhabditis elegans*

This transparent organism has a fixed number of somatic cells. (Part **b** adapted from Fig. 22-6a in V. Walbot and N. Holder, *Developmental Biology.* © 1987. Used with permission from McGraw-Hill Companies.)

develop from that cell are missing. Such a rigid developmental pattern, in which the fates of the cells become restricted early in development, is referred to as **mosaic development.** Each cell has a specific fate in the embryo, just as each tile in a mosaic design forms part of the pattern.

Scientists initially hypothesized that every organ system in *C. elegans* might be derived from only one founder cell. Detailed analysis of cell lineages, however, reveals that many of the structures found in the adult, such as the nervous system and the musculature, are, in fact, derived from more than one founder cell (**FIG. 17-15b**). A few lineages have been identified in which a neuron and a muscle cell are derived from the division of a single cell.

A researcher who uses microscopic laser beams small enough to destroy individual cells can determine the influence that one cell has on the development of a neighboring cell. Consistent with the rigid pattern of cell lineages, the destruction of an individual cell in *C. elegans* results, in most cases, in the absence of all the structures derived from that cell but with the normal differentiation of all the neighboring somatic cells. This observation suggests that development in each cell is regulated through its own internal program.

However, the developmental pattern of *C. elegans* is not entirely mosaic. In some cases, cell differentiation is influenced by interactions with particular neighboring cells, a phenomenon known as **induction.** One example is the formation of the vulva (pl., *vulvae*), the reproductive structure through which the eggs are laid. A single nondividing cell, called the *anchor cell,* is a part of the ovary, the structure in which the germ line cells undergo meiosis to produce the eggs. The anchor cell attaches to the ovary and to a point on the outer surface of the animal, triggering the formation of a passage through which the eggs pass to the outside. When the anchor cell is present, it induces cells on the surface to form the vulva and its opening. If the anchor cell is destroyed by a laser beam, however, the vulva does not form, and the cells that would normally form the vulva remain as surface cells (**FIG. 17-16**).

The analysis of worm mutations has contributed to our understanding of inductive interactions. For example, several types of mutations cause more than one vulva to form. In such mutant animals, multiple vulvae form even if the anchor cell is destroyed. Thus, the mutant cells do not require an inductive signal from an anchor cell to form a vulva. Evidently, in these mutants the gene or genes responsible for vulva formation are constitutive. Conversely, mutants lacking a vulva are also known. In some, the cells that would normally form the vulva apparently do not respond to the inducing signal from the anchor cell.

C. elegans **is a model system to study apoptosis** During normal development in *C. elegans,* there are instances in which cells are destined to die shortly after they are produced. **Apoptosis,** or genetically programmed cell death, has been observed in a wide variety of organisms, both plant and animal. For example, the human hand is formed as a webbed structure, but the fingers become individualized when the cells between them undergo apoptosis. In *C. elegans,* as in other organisms, apoptosis is under genetic control. The worm embryos undergo mitosis to produce a total of 1090 cells, but 131 of them undergo apoptosis during development, resulting in adult worms with 959 cells.

In 1986, U.S. molecular geneticist Robert Horvitz discovered several mutant worms that were defective in some way relating to apoptosis. He characterized the mutations and the genes responsible for them at the molecular level. For example, one of the genes codes for a protein that regulates the release from mitochondria of molecules that trigger apoptosis (mitochondria are now known to have a crucial role in apoptosis in animals ranging from *C. elegans* to humans). Other genes were identified that code for a family of proteins known as **caspases,** proteolytic enzymes active in the early stages of apoptosis. Homologous genes were subsequently identified in other organisms, including humans; some of these loci in mammals also code for caspases. The molecular events that trigger apoptosis remain an area of intense research interest, the results of which will shed considerable light on the general processes of cell aging and cancer.

Horvitz and British scientists Sydney Brenner and John Sulston shared the Nobel Prize in Physiology or Medicine in 2002 for their work on the genetic regulation of organ development and apoptosis in *C. elegans.* Sulston was one of Brenner's students. He traced the lineage of how a single zygote gives rise to the 959 cells in the nematode adult and also observed that some cells die during normal development. Continuing Sulston's work on apoptosis, Horvitz, as mentioned earlier, was the first to discover genes involved in apoptosis.

The nematode worm *C. elegans* has a fixed developmental pattern in which the fate of each adult cell can be traced to one of several founder cells in the embryo.

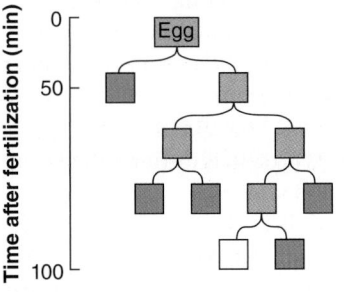

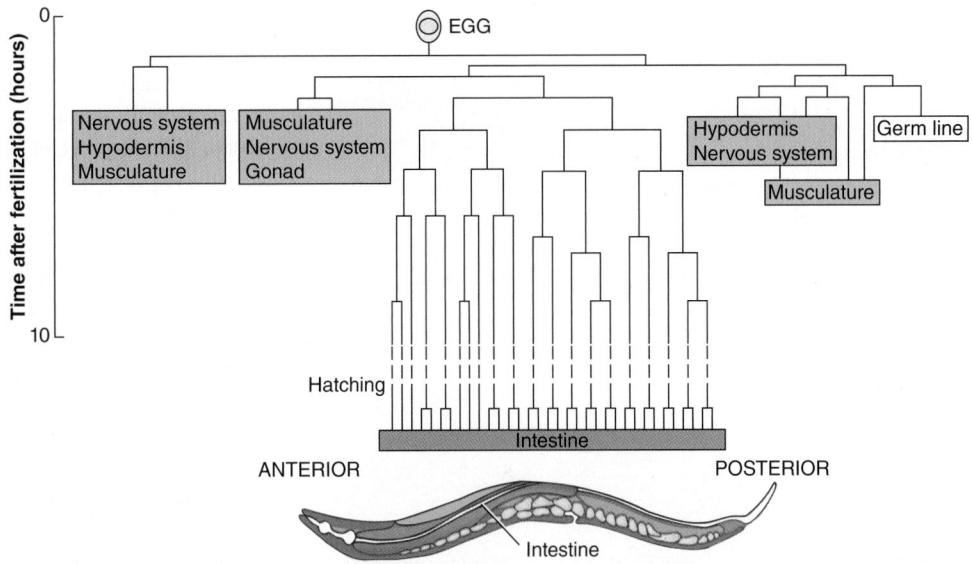

(a) All somatic cells of *C. elegans* are derived from five somatic founder cells (*shown in blue*) produced during the early cell divisions of the embryo. The cell shown in *white* gives rise to germ line cells.

(b) The nervous system, hypodermis, and musculature derive from several founder cells. This lineage map traces the development of the cells that form the intestine. (The dashed lines represent many cell divisions of a particular lineage.)

Figure 17-15 Cell lineages of *Caenorhabditis elegans*

CONNECT Does the cell that gives rise to the *C. elegans* germ line give rise to any other types of cells?
© Cengage Learning

The mouse is a model for mammalian development

Mammalian embryos develop in markedly different ways from the embryos of *Drosophila* and *C. elegans*. The laboratory mouse, *Mus musculus,* is the best-studied example of mammalian development. Researchers have identified numerous genes affecting development in the mouse. The mouse genome sequence, which was published in 2002, contains 27,000 to 30,000 genes, similar to the number of protein-coding genes in the human genome. Indeed, 99% of the genes in the mouse have counterparts in humans.

The early development of the mouse and other mammals is similar in many ways to human development. During the early developmental period, the embryo lives free in the reproductive tract of the female. It then implants in the wall of the uterus, after which the mother meets its nutritional and respiratory needs. Consequently, mammalian eggs are very small and contain little in the way of food reserves. Almost all research on mouse development has concentrated on the stages leading to implantation because in those stages the embryo is free-living and can be experimentally manipulated. During that period, developmental commitments that have a significant effect on the future organization of the embryo take place.

After fertilization, a series of cell divisions gives rise to a loosely packed group of cells. Research shows that all cells are equivalent in the very early mouse embryo. For example, if at the two-cell stage of mouse embryogenesis researchers destroy one cell and implant the remaining cell into the uterus of a surrogate mother, a normal mouse usually develops.

Conversely, if two embryos at the eight-cell stage of development are fused and implanted into a surrogate mother, a normal-sized mouse develops (**FIG. 17-17**). By using two embryos with different genetic characters, such as coat color, researchers demonstrated that the resulting mouse has four genetic parents. These chimeric mice have fur with patches of different colors derived from clusters of genetically different cells. A **chimera** is an organism containing two or more kinds of genetically dissimilar cells arising from different zygotes. (The term is derived from the name of a mythical beast that was said to have the head of a lion, the body of a goat, and a snakelike tail.) Chimeras let researchers use genetically marked cells to trace the fates of certain cells during development.

The responses of mouse embryos to such manipulations contrast with the mosaic or predetermined nature of early *C. elegans* development, in which destroying one of the founder cells results in the loss of a significant portion of the embryo. For this reason, biologists say that the mouse has highly **regulative development;** in other words, the early embryo acts as a self-regulating whole that accommodates missing or extra parts.

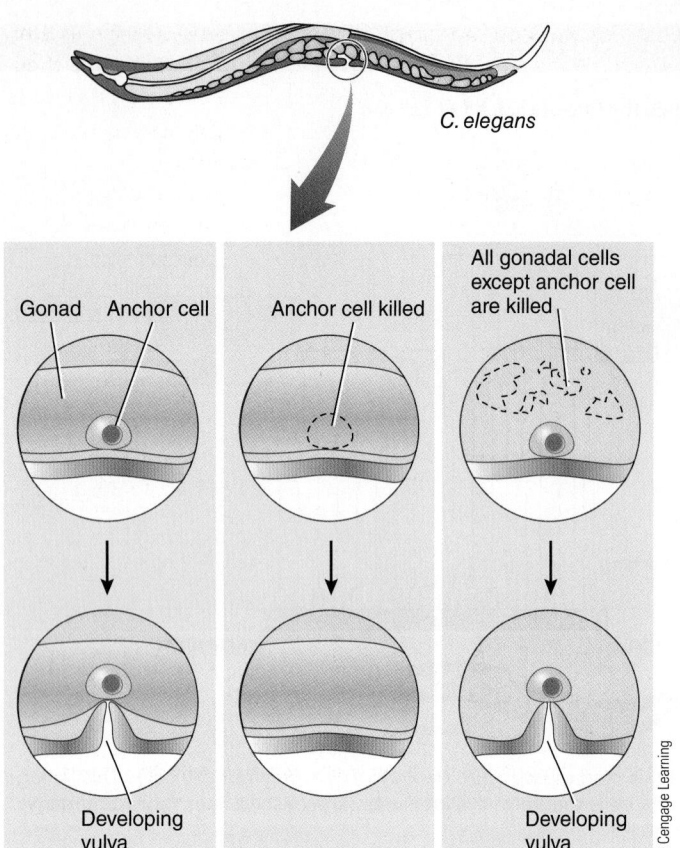

C. elegans

Gonad Anchor cell

Anchor cell killed

All gonadal cells
except anchor cell
are killed

Developing
vulva

Developing
vulva

© Cengage Learning

**(a) Normal
development
of vulva**

**(b) No vulva
develops**

**(c) Normal
development
of vulva**

Figure 17-16 Induction

A single anchor cell induces neighboring cells to form the vulva in *C. elegans*.
This diagram shows how laser destruction of single cells or groups of cells
influences neighboring cells.

Transgenic mice are used in studies of developmental regulation In transformation experiments foreign DNA injected into fertilized mouse eggs is incorporated into the chromosomes and expressed (**FIG. 17-18**). (Also see Figure 15-14 for an example of a transgenic mouse.) The resulting transgenic mice provide insights into how genes are activated during development. Scientists can identify a *transgene* (foreign gene) that has been introduced into a mouse and determine whether it is active by marking the gene in several ways. Sometimes a similar gene from a different species is used; its protein is distinguished from the mouse protein by specific antibodies.

It is also possible to construct a hybrid gene consisting of the regulatory elements of a mouse gene and part of another, non-mouse, gene that codes for a "reporter" protein. For example, the reporter protein could be an enzyme not normally found in mice. Such studies have been important in showing which DNA sequences of a mouse homeobox gene determine where the gene is expressed in the embryo.

Many developmentally controlled genes introduced into mice have yielded important information about gene regulation. When researchers introduce developmentally controlled genes from other species, such as humans or rats, into mice, the genes are regulated the same way they normally are in the donor animal. When introduced into the mouse, for example, human genes encoding insulin, globin, and crystallin—which are normally expressed in cells of the pancreas, blood, and eye lens, respectively—are expressed only in those same tissues in the mouse. That these genes are correctly expressed in their appropriate tissues indicates that the signals for tissue-specific gene expression are highly conserved through evolution. Information on the regulation of genes controlling development in one organism can have valuable applications to other organisms, including humans.

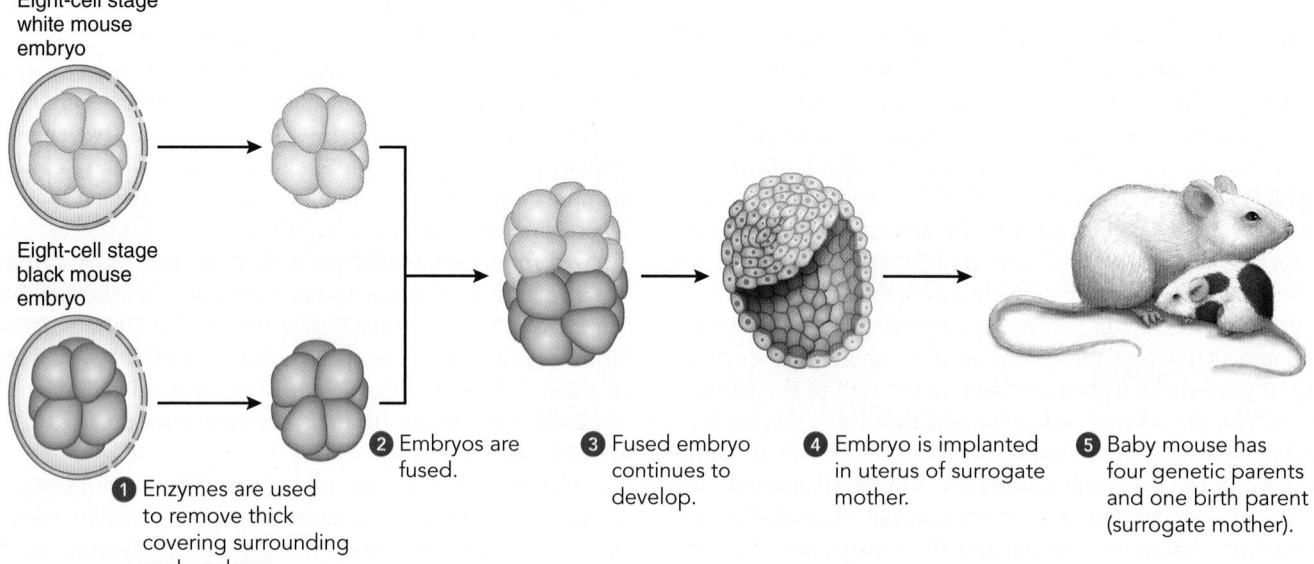

Eight-cell stage
white mouse
embryo

Eight-cell stage
black mouse
embryo

1 Enzymes are used
to remove thick
covering surrounding
each embryo.

2 Embryos are
fused.

3 Fused embryo
continues to
develop.

4 Embryo is implanted
in uterus of surrogate
mother.

5 Baby mouse has
four genetic parents
and one birth parent
(surrogate mother).

Figure 17-17 Chimeric mouse

Researchers remove embryos from females of two different strains and combine the cells in vitro. The resulting aggregate
embryo continues to develop and is implanted in the uterus of a surrogate mother. The offspring has four different genetic
parents. Although the surrogate mother is the birth mother, she is not genetically related.
© Cengage Learning

WHY IS IT USED? This method is used to establish a transgenic line of mice that contain genes with useful traits for scientific and medical research.

HOW IS IT DONE?

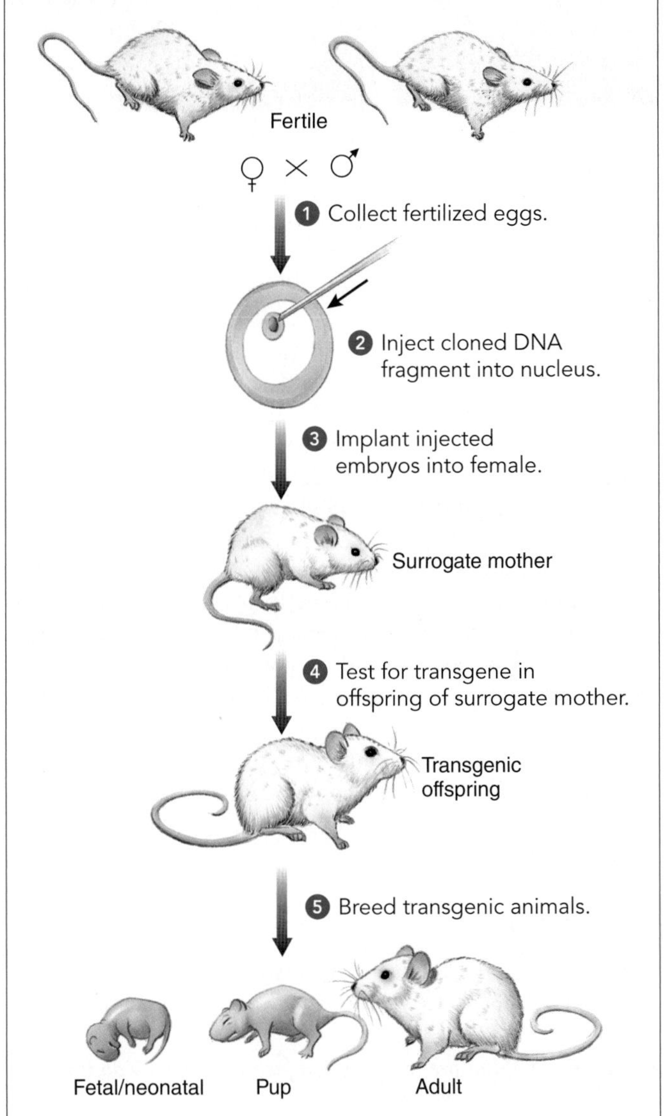

Fertile

♀ × ♂

1 Collect fertilized eggs.

2 Inject cloned DNA fragment into nucleus.

3 Implant injected embryos into female.

Surrogate mother

4 Test for transgene in offspring of surrogate mother.

Transgenic offspring

5 Breed transgenic animals.

Fetal/neonatal   Pup   Adult

Figure 17-18 Producing a transgenic mouse
© Cengage Learning

Mice and other model organisms are providing insights into the aging process Aging, which is defined as a progressive decline in the performance of various parts of the body, is an important field of developmental biology. The study of aging has great practical potential because age-related diseases represent one of the biggest challenges of biomedical research today. Scientists have demonstrated that many environmental factors influence aging. For example, severe calorie restriction in rodents and other mammals delays aging. However, how the human genome interacts with the cell environment during the aging process is not well understood.

Researchers working with model organisms, such as *Drosophila* and *C. elegans,* have found that more than one thousand protein-coding genes affect aging to variable degrees. In the nematode worm *C. elegans,* most aging genes are under the control of three genes (*ELT-3, ELT-5,* and *ELT-6*), all of which code for transcription factors.

Many of the genes that slow aging in organisms as diverse as yeasts, nematodes, and mammals are associated with cells' responses to nutrients. Dietary restrictions are known to extend lifespan for many organisms, from yeasts to dogs and possibly even to primates. Two genes (the *PHA-4* gene and the *SKN-1* gene) that extend survival in *C. elegans* code for transcription factors.

One of the most potent lifespan-extending genes, present in *C. elegans,* codes for a protein similar to the membrane receptor that allows cells to respond to the peptide **insulin-like growth factor (IGF)** in humans and other mammals. The binding of IGF to its receptor triggers signal transduction (see Chapter 6) within the cell, which regulates the expression of many genes.

Researchers wondered if the IGF receptor influences aging in mammals, including humans. To test this hypothesis, they produced *knockout mice* with an inactivated gene for the insulin-like growth factor receptor (IGF-1R). (Knockout mice were first discussed in Chapter 15.) Because mice are diploid organisms, they have two copies of the gene, designated *igf1r.* When both alleles of *igf1r* are inactivated, the mice develop many abnormalities and die at birth. However, when one allele of *igf1r* is inactivated and the other is left in its normal state, the mice thrive and live about 25% longer than control mice. In all other respects, the heterozygous mice are normal. They are virtually indistinguishable from control mice in their rate of development, metabolic rate, ability to reproduce, and body size. Mice heterozygous at the *igf1r* locus produce fewer IGF receptors than the control mice; therefore, the cells are not exposed to as much signaling from IGF.

Research on the IGF receptor and aging in mice suggests that some of the hundreds of other mutant genes associated with aging in *C. elegans* may also be involved in mammalian aging. Thus, the genetic control of mammalian aging has enough research topics to occupy molecular geneticists for many years.

Arabidopsis is a model for studying plant development, including transcription factors

Mouse-ear cress (*Arabidopsis thaliana*), the organism most widely used to study the genetic control of development in plants, is a member of the mustard family. Although *Arabidopsis* itself is a weed of no economic importance, it has several advantages for research. The plant completes its life cycle in just a few weeks and is small enough to be grown in a petri dish, yielding thousands of individuals in limited space. Botanists use chemical mutagens to produce mutant strains and have isolated many developmental mutants. When they insert cloned foreign genes into *Arabidopsis* cells, the genes integrate into the chromosomes and are expressed. The researchers then induce these transformed cells to differentiate into transgenic plants.

In 2000, the *Arabidopsis* genome became the first plant genome sequenced. Although its genome is relatively small

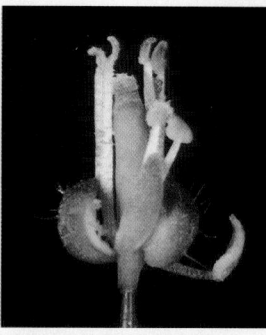

(a) A normal *Arabidopsis* flower has four outer leafy green sepals, four white petals, six stamens (the male reproductive structures), and a central carpel (the female reproductive structure).

(b) This homeotic mutant is missing petals.

(c) This homeotic mutant has only sepals and carpels.

(d) This homeotic mutant has sepals and petals but no other floral structures.

Figure 17-19 Homeotic mutants in *Arabidopsis* flowers

(the rice genome is about four times larger), it includes about 26,000 protein-coding genes. In comparison, *Drosophila* has about 13,600 genes and *C. elegans* about 19,700. Many genes in *Arabidopsis* are functionally equivalent to genes in *Drosophila*, *C. elegans*, and other animal species.

Of particular importance to development are the more than 1500 genes that code for transcription factors in *Arabidopsis* (compared with 635 at last count for *Drosophila*). Not surprisingly, many genes known to specify identities of parts of the *Arabidopsis* flower code for transcription factors. During the development of *Arabidopsis* flowers, four distinct flower parts differentiate: sepals, petals, stamens, and carpels (**FIG. 17-19a**; also see Fig. 33-1). Sepals cover and protect the flower when it is a bud, petals help attract animal pollinators to the flower, stamens produce pollen grains, and the carpel produces ovules, which develop into seeds following fertilization (discussed in Chapter 28).

The *ABC model* explains how three classes of homeotic genes control the development of the four floral organs and may help explain the thousands of different forms of flowers that have evolved. Class *A* genes are needed to specify sepals; both *A* and *B* genes are needed to specify petals; both *B* and *C* genes are needed to specify stamens; and class *C* genes are needed to specify the carpel. Mutations in the *A, B,* or *C* organ-identity genes, all of which code for transcription factors, cause one flower part to be substituted for another. For example, class *C* homeotic mutants (which have an inactive *C* gene) have petals in place of stamens and sepals in place of the carpel. Therefore, the entire flower consists of only sepals and petals. **FIGURES 17-19b, c,** and **d** show three homeotic mutants of *Arabidopsis*.

The ABC model is not the entire explanation for floral development. Another class of genes, designated *SEPALLATA*, interacts with the *B* and *C* genes to specify the development of petals, stamens, and carpels. Remarkably, when *SEPALLATA* genes are turned on permanently in *Arabidopsis*, the resulting plants have white petals growing where the leaves should be. (Because these plants do not photosynthesize, they are grown on a nutrient medium.)

These findings in *Arabidopsis* vastly increase the number of molecular probes available in plants. Researchers use these probes to identify other genes that control development in various plant species and to compare them with genes from a wide range of organisms. The success of the *Arabidopsis* sequencing project led to an international initiative by plant biologists to understand the functions of all genes in *Arabidopsis*. This functional genomic information will lead to a far deeper understanding of plant development and evolutionary history.

CHECKPOINT 17.2

- *What are the relative merits of* Drosophila, C. elegans, M. musculus, *and* Arabidopsis *as model organisms for the study of development?*
- **CONNECT** *What are transcription factors, and how do they influence development?*
- *What role does induction play in development?*
- *What is the importance of apoptosis in normal development?*

17.3 CANCER AND CELL DEVELOPMENT

LEARNING OBJECTIVE

8 Discuss the relationship between cancer and changes in gene expression that affect cell developmental processes.

Cancer cells lack normal biological inhibitions. Normal cells are tightly regulated by control mechanisms that prompt them to divide when necessary and prevent them from growing and dividing inappropriately. Cells of many tissues in the adult are normally prevented from dividing; they reproduce only to replace a neighboring cell that has died or become damaged. Cancer cells have escaped such controls and can divide continuously.

As a consequence of their abnormal growth pattern, some cancer cells eventually form a mass of tissue called a **tumor,** or *neoplasm.* If the tumor remains at the spot where it originated, it can usually be removed by surgery. One of the major problems with certain forms of cancer is that the cells can escape from the controls that maintain them in one location. **Metastasis** is the spreading of cancer cells to different parts of the body. Metastatic cancer cells invade other tissues and form multiple tumors. Lung cancer, for example, is particularly deadly because its cells enter the bloodstream and migrate to form tumors in other parts of the lungs and in other organs, such as the liver and the brain.

Today a 100-fold reduction in the cost of genome-wide sequencing studies has permitted researchers to sequence the genomes of 100 or more tumors of each of any given type of cancer (e.g., breast, colon, brain, pancreas, etc.). This ongoing research is yielding deep insights, particularly through the identification of **cancer driver genes.** When the expression of any of these driver genes is altered, either by a specific mutation or by certain epigenetic changes, the cell will have a selective growth advantage compared to normal cells. Some cancer drivers, called **oncogenes,** function as cancer-activating genes. This activation usually occurs because the normal products of these genes are components of about 12 different signal transduction pathways that are involved in regulating growth, cell division, and development. The abnormal gene products short-circuit these pathways in ways that lead to inappropriate growth. Oncogenes are usually combined with the loss of the expression of other cancer drivers called **tumor suppressor genes** that prevent abnormally growing cells from completing cell division. Oncogenes are often referred to as cancer "accelerators," whereas tumor suppressor genes serve as the "brakes" to cancer cell development. Genomic sequencing studies of over 3000 tumors revealed only 125 driver genes of which 54 were classified as oncogenes and 71 as tumor suppressor genes. These findings also showed that in the cells of common adult tumors (breast, pancreas, brain, colorectal) there are often 3 to 6 mutated driver genes.

Oncogenes are usually altered components of cell signaling pathways that control growth and differentiation

Oncogenes arise from changes in the expression of certain genes called **proto-oncogenes,** which are usually *normal* genes found in all cells. When proto-oncogenes are abnormally expressed, they may inappropriately activate a signaling pathway that triggers cell division.

The current list of known proto-oncogenes includes genes that code for various growth factors or growth factor receptors. Also included are genes that respond to stimulation by growth factors and genes that code for certain transcription factors.

When a proto-oncogene is expressed inappropriately (i.e., when it becomes an oncogene), the cell may misinterpret the signal and respond by growing and dividing.

One of the first cellular oncogenes researchers identified was isolated from a tumor in a human urinary bladder. In the cell that gave rise to the tumor, a proto-oncogene called *RAS* had undergone a single base-pair mutation; the result was that the amino acid valine replaced the amino acid glycine in the Ras protein product of the gene. Ras is a G-protein that serves as an "on/off" switch for a number of different signaling pathways (**FIG. 17-20;** also see Fig. 6-12). This subtle change inappropriately activates Ras so that it is stuck in the "on" state, converting the normal cell into a cancer cell. *RAS* is the most prevalent oncogene in human cancer. Mutations that permanently activate the Ras protein are found in 20% to 25% of all human cancers and in 90% of pancreatic cancers.

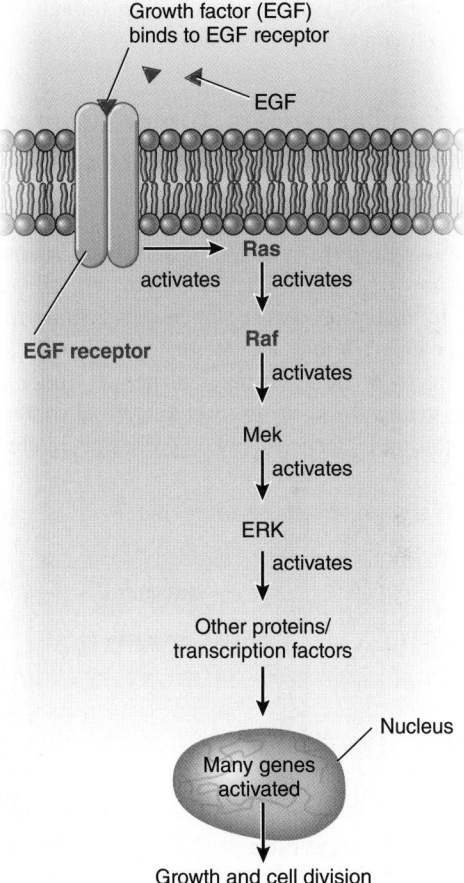

Figure 17-20 Oncogenes inappropriately activate signaling pathways that regulate cell growth and differentiation

Many cancers arise from the activation of proto-oncogenes in signaling pathways that control cell growth and differentiation. Mutations that cause the proto-oncogenes that code for the EGF receptor, Ras, or Raf to function as oncogenes are found in many different types of cancer cells. An activating mutation in any of these three proto-oncogenes that encode components of the MAP kinase pathway can result in the production of an *oncoprotein* (*shown in red*) that would permanently switch the pathway "on," continuously stimulating growth and cell division in a newly formed cancer cell and its progeny. (See Figure 6-12 for a more detailed view of the MAP kinase pathway).

© Cengage Learning

Figure 17-20 illustrates how proto-oncogenes normally code for different components of a signaling pathway. The MAP kinase pathway is normally activated when epidermal growth factor (EGF) binds to its EFG receptor associated with the cell surface, initiating a cascade of events inside the cell leading to growth and cell division. The growth factor–receptor complex acts as a **protein kinase,** an enzyme that activates the membrane-bound Ras protein. Activated Ras then phosphorylates the protein kinase Raf, which triggers the activation of a cascade of protein kinases that catalyze the activation of nuclear proteins, including transcription factors. The activated transcription factors bind to their DNA targets and stimulate the transcription of specific sets of genes that initiate growth and cell division. In this pathway the EGF receptor, Ras, and Raf are encoded by proto-oncogenes. Certain types of mutations affecting the EGF receptor can permanently turn it "on," activating the pathway in the absence of the EFG growth factor. Similarly, mutations affecting either Ras or Raf have been found that turn these proteins "on" in cancer cells, activating the signaling pathway even though the earlier components remain in their "off" state.

In many familial cancers, tumor suppressor genes must be inactivated before cells progress to cancer

A change in a single proto-oncogene is often insufficient to cause a cell to become malignant. The development of cancer is usually a slow, multistep process in which cells acquire mutations or epigenetic changes that activate oncogenes and also mutations that inactivate tumor suppressor genes. Tumor suppressor genes, also known as *anti-oncogenes,* normally interact with growth-inhibiting factors to block cell division in abnormally growing cells. The inactivation of tumor suppressor genes alone usually does not activate the cancer phenotype, but their inactivation is often essential in allowing cells activated by oncogenes to progress toward a malignant state.

For example, *BRCA1* and *BRCA2* are tumor suppressor genes that are found to be in an inactivated state in many cases of breast and ovarian cancer. These proteins function in the cell cycle to detect DNA damage. When mutated, they lose their ability to stop the cell cycle until the breaks can be repaired. Mutant forms of *BRCA1* and *BRCA2* are connected to the increased risks of breast cancer in families who have a high frequency of the disease (see Chapter 50). Persons homozygous for the normal forms of *BRCA1* (on chromosome 17) and *BRCA2* (on chromosome 13) have about a 2% risk of having breast cancer during their lifetime. Individuals who are heterozygous for a mutant allele at either locus have about a 60% risk of developing breast cancer. **FIGURE 17-21** shows how individuals with two normal *BRCA1* alleles must acquire two separate somatic mutations before a breast cell will progress to cancer, whereas a heterozygous individual (*BRCA1/brca1*) requires only single somatic mutation to reach the same state.

CHECKPOINT 17.3

- **CONNECT** *Discuss how oncogenes and tumor suppressor genes are related to genes involved in the control of normal growth and development.*
- **CONNECT** *All four genes that are used to reprogram differentiated cells to function as pluripotent stem cells are transcription factors that are also known to be proto-oncogenes. Many of the early adult stem cell transplant experiments showed the development of cancers in the transplanted tissues. Explain.*

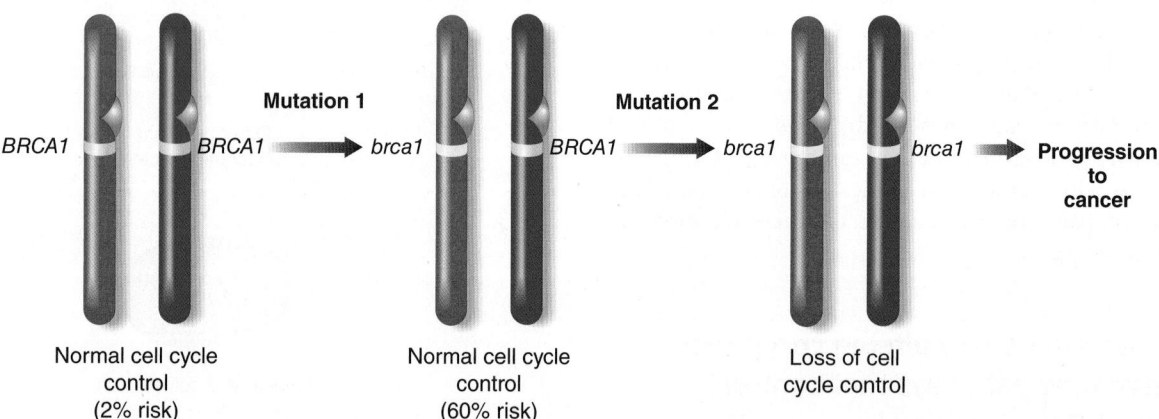

Figure 17-21 Mutations in the *BRCA1* tumor suppressor gene lead to increased risk for breast cancer

BRCA1 is a tumor suppressor gene that encodes a protein that blocks the cell cycle in cells undergoing abnormal growth. If both alleles of this locus become inactivated by mutations in a breast cell (i.e., that cell becomes *brca1/brca1*), it may progress to a cancer phenotype. The actual progression to cancer requires additional activating mutations in certain other genes, such as an oncogene. Many families with a high incidence of breast cancer include individuals carrying an inherited, inactive mutant allele of *BRCA1*. These heterozygous *brca1/BRCA1* individuals have a greatly increased risk of developing breast cancer compared to other individuals because a single somatic mutation (converting *BRCA1* to *brca1*) can cause a breast cell to lose control of the cell cycle.
© Cengage Learning

17.1 Cell Differentiation and Nuclear Equivalence (page 359)

1 Distinguish between cell determination and cell differentiation, and between nuclear equivalence and totipotency.

- An organism contains many types of cells that are specialized to carry out specific functions. These cells are the product of a process of gradual commitment, called **cell determination,** which ultimately leads to the final step in cell specialization, called **cell differentiation.**

- **Nuclear equivalence** is the concept that, with a few exceptions, all the nuclei of the differentiated **somatic cells** of an organism are identical to one another and to the nucleus of the single cell from which they descended. **Totipotency** is the capability of cells to direct the development of an entire organism.

2 Describe the classic experiments of Steward, Gurdon, and Wilmut.

- Steward induced mature carrot root cells to become undifferentiated and express their totipotency, forming an entire plant.

- Gurdon injected nuclei from tadpole intestinal cells into enucleated eggs; a small fraction of these eggs developed into tadpoles, demonstrating the totipotency of the injected nucleus.

- Wilmut fused a differentiated cell from an adult sheep cell with an enucleated egg and implanted it into the uterus of a host mother, where it developed into a normal lamb.

3 Define *stem cells,* distinguish between embryonic stem cells and pluripotent stem cells, and describe some of the promising areas of research involving stem cells.

- **Stem cells** can divide to produce differentiated descendants yet retain the ability to divide to maintain the stem cell population. Totipotent stem cells give rise to all cell types of the body and placenta, whereas **pluripotent stem cells** give rise to many, but not all, types of cells in an organism.

- **Embryonic stem (ES) cells,** formed after a zygote has undergone several rounds of cell division to become a five- or six-day-old blastocyst, are pluripotent and have the potential to develop into any type of cell in the body; ES cells are not totipotent because they cannot form cells of the placenta.

- Stem cells show promise in treating diseases, such as Parkinson's disease and diabetes mellitus.

17.2 The Genetic Control of Development (page 365)

4 Indicate the features of *Drosophila melanogaster, Caenorhabditis elegans, Mus musculus,* and *Arabidopsis thaliana* that have made these organisms valuable models in developmental genetics.

- Developmental mutations have been identified in the fruit fly, *Drosophila,* many of which affect the organism's segmented body plan. Many developmental genes discovered in the fruit fly are now known to be important in the growth and development of all animals.

- *Caenorhabditis elegans* is a roundworm with **mosaic development,** a rigid developmental pattern in which the fates of cells are restricted early in development. The lineage of every somatic cell in the adult is known, and each can be traced to a single **founder cell** in the early embryo.

- The laboratory mouse, *M. musculus,* is extensively used in studies of mammalian development. The mouse shows **regulative development;** the very early embryo is a self-regulating whole and can develop normally even if it has extra or missing cells. **Transgenic** mice, in which foreign genes have been incorporated, have helped researchers determine how genes are activated and regulated during development.

- Genes affecting development have also been identified in certain plants, such as *A. thaliana.* The *ABC model* of interactions among three kinds of genes hypothesizes how floral organs develop in *Arabidopsis.* Mutations in these homeotic genes cause one flower part to be substituted for another.

5 Distinguish among maternal effect genes, segmentation genes, and homeotic genes in *Drosophila.*

- The earliest developmental events to operate in the egg are established by **maternal effect genes** in the surrounding maternal tissues, which are active prior to fertilization. Some produce gradients of **morphogens,** chemical agents that affect the differentiation and development of form. Maternal effect genes establish polarity in the embryo.

- **Segmentation genes** generate a repeating pattern of body segments within the embryo. *Gap genes* are segmentation genes that begin organizing the body into anterior, middle, and posterior regions; *pair–rule genes* and *segment polarity genes* affect all segments.

- The later-acting **homeotic genes** are responsible for specifying the identity of each segment.

6 Explain the relationship between transcription factors and genes that control development.

- **Transcription factors** are DNA-binding proteins that regulate transcription in eukaryotes. Some genes that code for transcription factors contain a DNA sequence called a **homeobox,** which codes for a protein with a DNA-binding region called a **homeodomain.**

- Some homeobox genes are organized into complexes that appear to be systems of master genes specifying an organism's body plan. Parallels exist between the homeobox complexes of *Drosophila* and those of other animals.

7 Define *induction* and *apoptosis,* and give examples of the roles they play in development.

- **Induction** refers to developmental interactions with neighboring cells. During development in *C. elegans,* the anchor cell induces cells on the surface to organize to form the vulva, the structure through which the eggs are laid.

- **Apoptosis** is programmed cell death. During human development, the human hand forms as a webbed structure, but the fingers become individualized when the cells between them die.

17.3 Cancer and Cell Development (page 376)

8 Discuss the relationship between cancer and changes in gene expression that affect cell developmental processes.

- The traits of cancer cells are due to accumulations of mutations and epigenetic changes in **cancer driver genes** that typically include **oncogenes** and **tumor suppressor genes.** Oncogenes are growth-activating genes that arise from changes in the expression of normal genes called **proto-oncogenes,** which exist in all cells and are involved in the control of growth and development. Tumor suppressor genes are normal genes that prevent cells from undergoing inappropriate growth and cell division. These are typically inactivated in cancer cells.

Know and Comprehend

1. Morphogenesis occurs through the multistep process of (a) differentiation (b) determination (c) pattern formation (d) totipotency (e) selection

2. The cloning experiments carried out on frogs demonstrated that (a) all differentiated frog cells are totipotent (b) some differentiated frog cells are totipotent (c) all nuclei from differentiated frog cells are totipotent (d) some nuclei from differentiated frog cells are totipotent (e) cell differentiation always requires the loss of certain genes

3. The anterior–posterior axis of a *Drosophila* embryo is first established by certain (a) homeotic genes (b) maternal effect genes (c) segmentation genes (d) proto-oncogenes (e) pair–rule genes

4. Most segmentation genes code for (a) transfer RNAs (b) enzymes (c) transcription factors (d) histones (e) transport proteins

5. Homeobox genes (a) are found in fruit flies but no other animals (b) tend to be expressed in the order that they appear on a chromosome (c) contain a characteristic DNA sequence (d) b and c (e) a, b, and c

6. *Arabidopsis* is useful as a model organism for the study of plant development because (a) it is of great economic importance (b) it has a very long generation time (c) many developmental mutants have been isolated (d) it contains a large amount of DNA per cell (e) it has a rigid developmental pattern

7. Pluripotent stem cells (a) lose genetic material during development (b) give rise to many, but not all, types of cells in an organism (c) organize into recognizable structures through pattern formation (d) cannot grow in tissue culture (e) have been used to clone a sheep and several other mammals

8. Which of the following statements about cancer is *false*? (a) oncogenes arise from mutations in proto-oncogenes (b) tumor suppressor genes normally interact with growth-inhibiting factors to block cell division (c) more than 120 cancer-driving genes have been discovered (d) oncogenes were first discovered in mouse models for cancer (e) the development of cancer is usually a multistep process involving both oncogenes and mutated tumor suppressor genes

9. Proto-oncogenes code for (a) morphogens (b) antibodies for immune responses (c) growth factor receptors and other components of the growth control cascade (d) enzymes such as reverse transcriptase (e) ES cells

Apply and Analyze

10. You discover a new *Drosophila* mutant in which mouth parts are located where the antennae are normally found. You predict that the mutated gene is most likely a (a) homeotic gene (b) gap gene (c) pair–rule gene (d) maternal effect gene (e) segment polarity gene

11. The developmental pattern of *C. elegans* is said to be mosaic because (a) development is controlled by gradients of morphogens (b) part of the embryo fails to develop if a founder cell is destroyed (c) some individuals are self-fertilizing hermaphrodites (d) all development is controlled by maternal effect genes (e) apoptosis never occurs

12. Which of the following illustrates the regulative nature of early mouse development? (a) the mouse embryo is free-living prior to implantation in the uterus (b) it is possible to produce a transgenic mouse (c) it is possible to produce a mouse in which a specific gene has been knocked out (d) genes related to *Drosophila* homeotic genes have been identified in mice (e) a chimeric mouse can be produced by fusing two mouse embryos

Evaluate and Synthesize

13. **CONNECT** Why is an understanding of gene regulation in eukaryotes crucial to an understanding of developmental processes? Explain your answer.

14. What is the reason that scientists study development in more than one type of organism?

15. Could a gene be involved in the growth of both stem cells and some kinds of cancer? Explain your answer.

16. **EVOLUTION LINK** How are the striking similarities among genes that govern development in widely differing species strong evidence for evolution?

17. **EVOLUTION LINK** What is the common ground between evolutionary biologists and developmental biologists who have adopted the perspective known as Evo Devo?

18. **INTERPRET DATA** Flower parts are arranged in four concentric circles: sepals, petals, stamens, and carpel. According to the ABC model of floral organ development in *Arabidopsis,* class *A* genes are needed to specify sepals, the *A* and *B* genes to specify petals, the *B* and *C* genes to specify stamens, and class *C* genes to specify the carpel. If a mutation occurs in one of the *B* genes, rendering it inactive, what will the resulting flowers consist of?

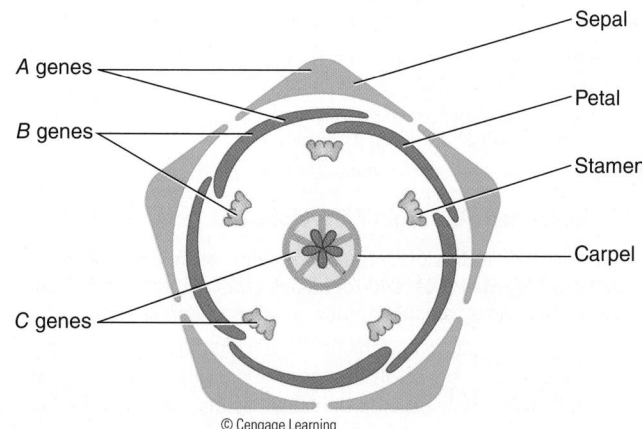

© Cengage Learning

To access course materials, such as Aplia and other companion resources, please visit **www.cengagebrain.com.**

Introduction to Darwinian Evolution

A great deal of evidence suggests that the biological diversity represented by the millions of species currently living on our planet evolved from a single ancestor during Earth's long history. Thus, organisms that are radically different from one another are in fact distantly related, linked through numerous intermediate ancestors to a single, common ancestor. British naturalist Charles Darwin (1809–1882) developed a simple, scientifically testable mechanism to explain the relationship among Earth's diversity of organisms. He argued persuasively that all the species that exist today, as well as the countless extinct species that existed in the past, arose from earlier ones by a process of gradual *divergence* (splitting into separate evolutionary pathways), or *evolution*.

The concept of evolution is the cornerstone of biology because it links all fields of the life sciences into a unified body of knowledge. As stated by U.S. geneticist Theodosius Dobzhansky, "Nothing in biology makes sense except in the light of evolution."[1] Biologists seek to understand both the remarkable variety and the fundamental similarities of organisms within the context of evolution. The science of evolution allows biologists to compare common threads among organisms as seemingly different as bacteria, whales, lilies, slime molds, and tapeworms. Animal behavior, developmental biology, genetics, evolutionary ecology, systematics, and molecular evolution are examples of some of the biological disciplines that are grounded in evolution.

This chapter discusses Charles Darwin and the development of his scientific theory of evolution through the mechanism of natural selection. It also presents evidence that supports evolution, including fossils, biogeography, comparative anatomy, molecular biology, developmental biology, and experimental studies of ongoing evolutionary change in both the laboratory and nature.

Science Source

Charles Darwin. This portrait was made shortly after Darwin returned to England from his voyage around the world.

KEY CONCEPTS

18.1 Evolution is the accumulation of inherited changes within populations over time.

18.2 Ideas about evolution originated long before Darwin's time.

18.3 Darwin's scientific theory of evolution, natural selection, explained how natural forces in the environment could cause evolution. Natural selection occurs because individuals with traits that make them better adapted to local conditions are more likely to survive and produce offspring than are individuals that are not as well adapted. The modern synthesis combines Darwin's scientific theory with the scientific underpinnings of modern genetics.

18.4 The evidence that evolution has taken place and is still occurring is overwhelming. This evidence includes fossils, biogeography, comparative anatomy, molecular biology, developmental biology, and evolutionary experiments with living organisms.

---

[1] Dobzhansky, T. "Nothing in Biology Makes Sense Except in the Light of Evolution," *American Biology Teacher*, Vol. 35, No. 125, pp. 125–129 (1973).

18.1 WHAT IS EVOLUTION?

LEARNING OBJECTIVE

1 Define the scientific theory of *evolution*.

In beginning our study of evolution, we define **evolution** as the accumulation of genetic changes within populations over time. A **population** is a group of individuals of one species that live in the same geographic area at the same time. Just as the definition of a *gene* changed as you studied genetics, you will find that the definition of *evolution* will become more precise in later chapters.

The term *evolution* does not refer to changes that occur in an individual within its lifetime. Instead, it refers to changes in the characteristics of populations over the course of generations. These changes may be so small that they are difficult to detect or so great that the population differs markedly from its ancestral population.

Eventually, two populations may diverge to such a degree that we refer to them as different species. The concept of species is developed extensively in Chapter 20. For now, a simple working definition is that a **species** is a group of organisms, with similar genetic information, structure, function, and behavior, that are capable of interbreeding with one another.

Evolution has two main perspectives. The minor evolutionary changes of populations (*microevolution,* discussed in Chapter 19) are usually viewed over a few generations, and the major evolutionary events (*macroevolution,* discussed in Chapter 20), such as formation of different species from common ancestors, are usually viewed over a long period.

Evolution has important practical applications. Agriculture must deal with the evolution of pesticide resistance in insects and other pests. Likewise, medicine must respond to the rapid evolutionary potential of disease-causing organisms such as bacteria and viruses (**FIG. 18-1**). (Significant evolutionary change occurs in a very short time period in insects, bacteria, and other organisms with short lifespans.) Medical researchers use evolutionary principles to predict which flu strains are evolving more quickly than others, information that scientists need to make the next year's flu vaccine. Also, researchers developing effective treatment strategies for the human immunodeficiency virus (HIV) must understand its evolution, both within and among hosts (see Fig. 23-12).

The conservation management of rare and endangered species makes use of the evolutionary principles of population genetics. The rapid evolution of bacteria and fungi in polluted soils is used in the field of **bioremediation,** in which microorganisms are employed to clean up hazardous-waste sites. Evolution even has applications beyond biology. For example, certain computer applications make use of algorithms that mimic natural selection in biological systems.

CHECKPOINT 18.1

- **CONNECT** *How is microevolution related to macroevolution?*
- *Do individuals evolve? Explain your answer.*

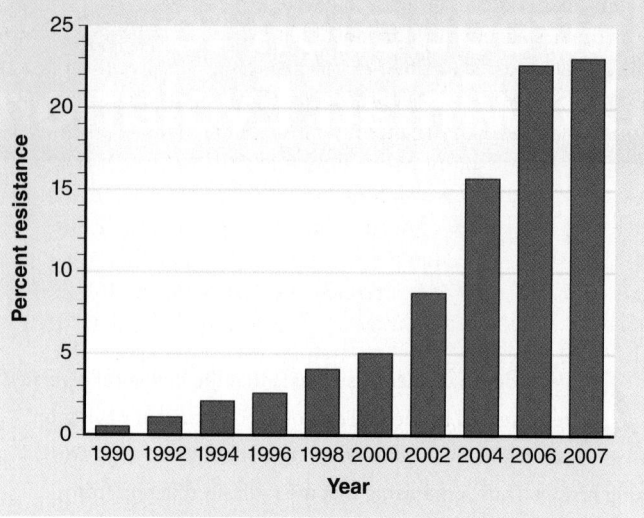

Figure 18-1 *Animation* **Evolution of antibiotic resistance to ciprofloxacin in** *Escherichia coli*

These data show an increased resistance to ciprofloxacin in *E. coli* isolated from blood and cerebrospinal infections in patients in England, Wales, and Northern Ireland, 1990 to 2007. (SOURCE: Livermore, D. "Zietgeist of Resistance." *The Journal of Antimicrobial Chemotherapy*, Vol. 60, i59–i61, 2007. By permission of Oxford University Press.)

18.2 PRE-DARWINIAN IDEAS ABOUT EVOLUTION

LEARNING OBJECTIVE

2 Discuss the historical development of evolutionary thought.

Although Darwin is universally associated with evolution, ideas of evolution predate Darwin by centuries. Aristotle (384–322 BCE) saw much evidence of natural affinities among organisms, which led him to arrange all the organisms he knew in a "scale of nature" that extended from the exceedingly simple to the most complex. Aristotle visualized organisms as being imperfect but "moving toward a more perfect state." Some scientific historians have interpreted this idea as a forerunner of evolutionary thought, but Aristotle was vague on the nature of this "movement toward perfection" and certainly did not propose that natural processes drove the process of evolution. Furthermore, modern scientific evolutionary theory now recognizes that evolution does not move toward more "perfect" states or even necessarily toward greater complexity.

Long before Darwin's time, fossils had been discovered embedded in rocks. Some of these fossils corresponded to parts of familiar species, but others were strangely unlike any known species. Fossils were often found in unexpected contexts; for example, marine invertebrates (sea animals without backbones) were sometimes discovered in rocks high on mountains. Leonardo da Vinci (1452–1519) was among the first to correctly interpret these unusual finds as the remains of animals that had existed in previous ages but had become extinct.

French naturalist Jean Baptiste de Lamarck (1744–1829) was the first scientist to propose that organisms undergo change over time as a result of some natural phenomenon rather than divine intervention. According to Lamarck, a changing environment caused an organism to alter its behavior, thereby using some organs or body parts more and others less. Over several generations, a given organ or body part would increase in size if it was used a lot or would shrink and possibly disappear if it was used less. Lamarck's hypothesis required that organisms pass traits that they acquired during their lifetimes to their offspring. For example, Lamarck suggested that the long neck of the giraffe developed when a short-necked ancestor stretched its neck to browse on the leaves of trees. Its offspring inherited the longer neck, which stretched even more as they ate. This process, repeated over many generations, resulted in the long necks of modern giraffes. Lamarck also thought that all organisms were endowed with a vital force that drove them to change toward greater complexity and "perfection" over time.

Lamarck's proposed mechanism of evolution is quite different from the mechanism later proposed by Darwin. However, Lamarck's hypothesis remained a reasonable explanation for evolution until Mendel's basis of heredity was rediscovered at the beginning of the 20th century. At that time, Lamarck's ideas were largely discredited.

CHECKPOINT 18.2

- CONNECT *How is Aristotle's "scale of nature" idea linked to early evolutionary thought? In what ways does it differ from modern scientific evolutionary theory?*
- CONNECT *How does Jean Baptiste de Lamarck's proposed mechanism for evolution differ from that of Charles Darwin?*

18.3 DARWIN AND EVOLUTION

LEARNING OBJECTIVES

3 Explain the four premises of evolution by natural selection as proposed by Charles Darwin.
4 Compare the modern synthesis with Darwin's original view of evolution.

Darwin, the son of a prominent physician, was sent at the age of 15 to study medicine at the University of Edinburgh. Finding himself unsuited for medicine, he transferred to Cambridge University to study theology. During that time, he became the protégé of the Reverend John Henslow, who was a professor of botany. Henslow encouraged Darwin's interest in the natural world. Shortly after receiving his degree, Darwin embarked on the HMS *Beagle,* which was taking a five-year exploratory cruise around the world to prepare navigation charts for the British navy.

The *Beagle* left Plymouth, England, in 1831 and cruised along the east and west coasts of South America (FIG. 18-2). While other members of the crew mapped the coasts and harbors, Darwin spent many weeks ashore studying the animals, plants, fossils, and geologic formations of both coastal and inland regions, areas that had not been extensively explored. He collected and cataloged thousands of plant and animal specimens and kept notes of his observations, information that became essential in the development of his ideas.

The *Beagle* spent almost two months at the Galápagos Islands, 965 km (600 mi) west of Ecuador, where Darwin continued his observations and collections. He compared the animals and plants of the Galápagos with those of the South

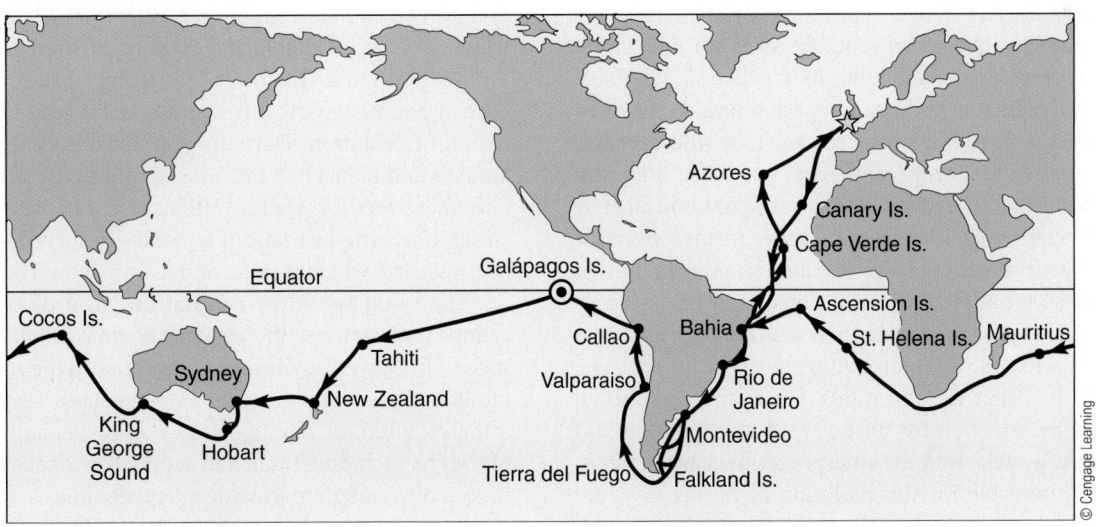

Figure 18-2 The voyage of HMS *Beagle*

The five-year voyage began in Plymouth, England (*star*), in 1831. Observations made in the Galápagos Islands (*bull's-eye*) off the western coast of South America helped Darwin develop a satisfactory mechanism to explain how a population of organisms could change over time.

American mainland. He was particularly impressed by their similarities and wondered why the organisms of the Galápagos should resemble those from South America more than those from other islands in different parts of the world. Moreover, although there were similarities between Galápagos and South American species, there were also distinct differences. There were even recognizable differences in the reptiles and birds from one island to the next. Darwin wondered why these remote islands should have such biological diversity. After he returned home, Darwin pondered these observations and attempted to develop a satisfactory explanation for the distribution of species among the islands.

Darwin drew on several lines of evidence when considering how species might have originated. Despite the work of Lamarck, the general notion in the mid-1800s was that Earth was too young for organisms to have changed significantly since they had first appeared. During the early 19th century, however, geologists advanced the idea that mountains, valleys, and other physical features of Earth's surface did not originate in their present forms. Instead, these features developed slowly over long periods by the geologic processes of volcanic activity, uplift, erosion, and glaciation. On his voyage Darwin took with him *Principles of Geology*, published by English geologist Charles Lyell in 1830, and studied it carefully. Lyell provided an important concept for Darwin: that the slow pace of geologic processes, which still occur today, indicated that Earth was extremely old.

Other important evidence that influenced Darwin was that breeders and farmers could develop many varieties of domesticated animals in just a few generations. They did so by choosing certain traits and breeding only individuals that exhibited the desired traits, a procedure known as **artificial selection.** Breeders, for example, have produced numerous dog varieties—such as bloodhounds, golden retrievers, Chihuahuas, Great Danes, and Pekinese—by artificial selection.

Many plant varieties were also produced by artificial selection. For example, cabbage, broccoli, brussels sprouts, cauliflower, collard greens, kale, and kohlrabi are distinct vegetable crops that are all members of the same species, *Brassica oleracea* (**FIG. 18-3**). Selective breeding of the colewort, or wild cabbage, a leafy plant native to Europe and Asia, produced all seven vegetables. Beginning more than 4000 years ago, some farmers artificially selected wild cabbage plants that formed overlapping leaves. Over time, these leaves became so prominent that the plants, which resembled modern cabbages, became recognized as separate and distinct from their wild cabbage ancestor. Other farmers selected different features of the wild cabbage, giving rise to the other modifications. For example, kohlrabi was produced by selection for an enlarged storage stem and brussels sprouts by selection for enlarged axillary buds. Thus, humans are responsible for the evolution of *B. oleracea* into seven distinct vegetable crops. Darwin was impressed by the changes induced by artificial selection and hypothesized that a similar selective process occurred in nature. He therefore used artificial selection as a model when he developed the concept of natural selection.

John Arnaldi

Figure 18-3 Artificial selection in *Brassica oleracea*

An enlarged terminal bud (the "head") was selected in cabbage (*lower left*), flower clusters in broccoli (*upper left*) and cauliflower (*middle right*), axillary buds in brussels sprouts (*bottom middle*), leaves in collards (*upper right*) and kale (*lower right*), and stems in kohlrabi (*middle*).

The ideas of Thomas Malthus (1766–1834), a British clergyman and economist, were another important influence on Darwin. In *An Essay on the Principle of Population as It Affects the Future Improvement of Society,* published in 1798, Malthus noted that population growth is not always desirable, a view contrary to the beliefs of his day. He observed that populations have the capacity to increase geometrically (1 → 2 → 4 → 8 → 16) and thus outstrip the food supply, which only has the capacity to increase arithmetically (1 → 2 → 3 → 4 → 5). In the case of humans, Malthus suggested that the conflict between population growth and food supply generates famine, disease, and war, which serve as inevitable brakes on population growth.

Malthus's idea that there is a strong and constant check on human population growth strongly influenced Darwin's explanation of evolution. Darwin's years of observing the habits of animals and plants had introduced him to the struggle for existence described by Malthus. It occurred to Darwin that in this struggle inherited variations favorable to survival would tend to be preserved, whereas unfavorable ones would be eliminated.

The result would be **adaptation,** an evolutionary modification that improves the chances of survival and reproductive success in a given environment. Eventually, the accumulation of modifications might result in a new species. Time was the only thing required for new species to originate, and the geologists of the era, including Lyell, had supplied evidence that Earth was indeed old enough to provide adequate time.

Darwin had at last developed a workable scientific explanation of evolution, that of **natural selection,** in which better adapted organisms are more likely to survive and become the parents of the next generation. As a result of natural selection, the population changes over time; the frequency of favorable traits increases in successive generations, whereas less-favorable traits become scarce or disappear. Darwin spent the next 20 years formulating his arguments for natural selection, accumulating an immense body of evidence to support this explanatory mechanism, and corresponding with other scientists.

As Darwin was pondering his ideas, Alfred Russel Wallace (1823–1913), a British naturalist who had studied the plants and animals of the Malay Archipelago for eight years, was similarly struck by the diversity of species and the peculiarities of their distribution. He wrote a brief essay on this subject and sent it to Darwin, by then a world-renowned biologist, asking his opinion. Darwin recognized his own ideas and realized that Wallace had independently arrived at the same conclusion: that evolution occurs by natural selection. Darwin's colleagues persuaded him to present Wallace's manuscript along with an abstract of his own work, which he had prepared and circulated to a few friends several years earlier. Both papers were presented in July 1858 at a London meeting of the Linnaean Society. Darwin's monumental book, *On the Origin of Species by Natural Selection; or, The Preservation of Favored Races in the Struggle for Life,* was published in 1859. In 1870, Wallace's book, *Contributions to the Theory of Natural Selection,* was published, eight years after he had returned from the Malay Archipelago.

Darwin proposed that evolution occurs by natural selection

Darwin's mechanism of evolution by natural selection consists of observations on four aspects of the natural world: variation; overproduction; limits on population growth, or a struggle for existence; and differential reproductive success.

1. *Variation.* The individuals in a population exhibit variation (FIG. 18-4). Each individual has a unique combination of traits, such as size, color, ability to tolerate harsh environmental conditions, and resistance to certain parasites or infections. Some traits improve an individual's chances of survival and reproductive success, whereas others do not. Remember that the variation necessary for evolution by natural selection must be inherited. Although Darwin recognized the importance to evolution of inherited variation, he did not know the mechanism of inheritance.

2. *Overproduction.* The reproductive ability of each species has the potential to cause its population to geometrically increase over time. A female frog lays about 10,000 eggs, and a female cod produces perhaps 40 million eggs! In each case, however, only about two offspring survive to reproduce. Thus, in every generation each species has the capacity to produce more offspring than can survive.

3. *Limits on population growth, or a struggle for existence.* There is only so much food, water, light, growing space, and other resources available to a population, so organisms compete with one another for these limited resources. Because there are more individuals than the environment can support, not all survive to reproduce. Other limits on population growth include predators, disease organisms, and unfavorable weather conditions.

4. *Differential reproductive success.* Those individuals that have the most favorable combination of characteristics (those that make individuals better adapted to their environment) are more likely to survive and reproduce. Offspring tend to resemble their parents because the next generation inherits the parents' genetically based traits. Successful reproduction is the key to natural selection: the best-adapted individuals produce the most offspring, whereas individuals that are less well adapted die prematurely or produce fewer or inferior offspring.

Over time, enough changes may accumulate in geographically separated populations (often with slightly different environments) to produce new species. Darwin noted that the Galápagos finches appeared to have evolved in this way. The 14 species are closely related. All descended from a common ancestor, a single species that found its way from the South American mainland two million to three million years ago. (The closest genetic relatives of the Galápagos finches are small seed-eating birds known as grassquits that live in western South America.)

During this two-million- to three-million-year period, the number of islands increased, the climate changed, and the plant life and food supply changed. The different islands of the Galápagos kept the finches isolated from one another, thereby allowing them to diverge into separate species in response to varying conditions (FIG. 18-5).

Peter Grant, Rosemary Grant, and their colleagues have documented natural selection in the Galápagos finches in their natural environment since the early 1970s. As an example of the evolutionary process in action, consider the sharp-beaked ground finch (*Geospiza difficilis*). This species lives on several different islands, and different beak shapes and sizes have evolved in each population depending on the diet available on the island where it lives.

We revisit the Galápagos finches in later chapters. Some long-term research by Peter Grant, Rosemary Grant, and their colleagues on the microevolution of Galápagos finches when

© Biosphoto

Figure 18-4 Genetic variation in emerald tree boas

These snakes, all the same species (*Corallus caninus*), were caught in a small section of forest in French Guiana. Many snake species exhibit considerable variation in their coloration and patterns.

(a) The cactus finch (*Geospiza scandens*), which feeds on the fleshy parts of cacti such as their flowers, has a long, pointed beak.

(b) The large ground finch (*Geospiza magnirostris*) has an extremely heavy, nutcracker-type beak adapted for eating thick, hard-walled seeds.

(c) The warbler finch (*Certhidia olivacea*) has a slender beak for eating insects.

(d) The woodpecker finch (*Camarhynchus pallidus*) digs insects out of bark and crevices by using spines, twigs, or even dead leaves.

Figure 18-5 *Animation* **Galápagos finches**

Darwin inferred that these birds are derived from a common ancestral population of seed-eating birds from South America. Variation in their beaks is the result of adaptation to the availability of different kinds of food.

droughts affect the food supply is described in Chapter 19; these studies have demonstrated that environmental change can drive natural selection. *Character displacement,* an aspect of evolutionary ecology, is described in Galápagos finches in Chapter 54.

The modern synthesis combines Darwin's scientific theory of evolution with genetics

One premise on which Darwin based his scientific theory of evolution by natural selection is that individuals transmit traits to the next generation. However, Darwin was unable to explain *how* this occurs or *why* individuals vary within a population. As discussed in Chapter 11, Gregor Mendel elucidated the basic patterns of inheritance. Darwin, who was a contemporary of Mendel, was apparently not acquainted with Mendel's work. Indeed, the scientific community did not recognize Mendel's work until the early part of the 20th century.

Beginning in the 1930s and 1940s, biologists experienced a conceptual breakthrough when they combined the principles of Mendelian inheritance with Darwin's principle of natural selection. The result was a unified explanation of evolution known as the **modern synthesis.** In this context, *synthesis* refers to combining parts of several scientific theories to form a unified whole. Some of the founders of the modern synthesis were U.S. geneticist Theodosius Dobzhansky, British geneticist and statistician Ronald Fisher, British geneticist J.B.S. Haldane, British biologist Julian Huxley, U.S. biologist Ernst Mayr, U.S. paleontologist George Gaylord Simpson, U.S. botanist G. Ledyard Stebbins, and U.S. geneticist Sewell Wright.

Today, the modern synthesis incorporates our expanding knowledge in genetics (which on its own makes an irrefutable case for evolution), systematics, paleontology, developmental biology, behavior, and ecology. The modern synthesis explains Darwin's observation of variation among offspring in terms of **mutation,** or changes in DNA, such as nucleotide substitutions. Mutations provide the genetic variability on which natural selection acts during evolution. The modern synthesis, which emphasizes the genetics of populations as the central focus of evolution, has held up well since it was developed. It has dominated the thinking and research of biologists working in many areas and has resulted in an enormous accumulation of new discoveries that validate evolution by natural selection.

Most biologists not only accept the basic principles of the modern synthesis but also try to better understand the causal processes of evolution. For example, what is the role of chance in evolution? How rapidly do new species evolve? These and other questions have arisen in part from a re-evaluation of the fossil record and in part from new discoveries in molecular aspects of inheritance. Such critical analyses are an integral part of the scientific process because they stimulate additional observation and experimentation along with re-examination of previous evidence. Science is an ongoing process, and information obtained in the future may require modifications to certain parts of the modern synthesis.

We now consider one of the many evolutionary questions currently being addressed by biologists: What are the relative effects of chance and natural selection on evolution?

Biologists study the effect of chance on evolution

Biologists have wondered whether we would get the same results if we were able to repeat evolution by starting with similar organisms exposed to similar environmental conditions. That is, would the same kinds of changes evolve as a result of natural selection, or would the organisms be quite different as a result of random events? Studies of evolution in action suggest that chance may not be as important as natural selection, at least at the population level.

A fruit fly species (*Drosophila subobscura*) native to Europe inhabits areas from Denmark to Spain. Biologists noted that the northern flies have larger wings than southern flies (FIG. 18-6). The same fly species was accidentally introduced to North America in the late 1970s. Ten years after its introduction,

Figure 18-6 Wing size in female fruit flies

In Europe female fruit flies (*Drosophila subobscura*) in southern countries have smaller wings than flies in northern countries. Shown are two flies: one from Spain (*left*) and the other from Denmark (*right*). The same evolutionary pattern emerged in North America after the accidental introduction of *D. subobscura* to the Americas.

George Gilchrist, College of William & Mary

biologists determined that no statistically significant changes in wing size had occurred in the different regions of North America. However, 20 years after its introduction, the fruit flies in North America exhibited the same type of north–south wing changes as in Europe. (It is not known why larger wings evolve in northern areas and smaller wings in southern climates.)

A study of the evolution of fishes known as *sticklebacks* in three coastal lakes of western Canada yielded intriguingly similar results to the fruit fly study. Molecular evidence indicates that when the lakes first formed several thousand years ago, they were populated with the same ancestral species. (Analysis of the mitochondrial DNA of sticklebacks in the three lakes supports the hypothesis of a common ancestor.) In each lake the same two species have evolved from the common ancestral fish. One species is large and consumes invertebrates along the bottom of the lake, whereas the other species is smaller and consumes plankton at the lake's surface. Members of the two species within a single lake do not interbreed with each other, but individuals of the larger species from one lake interbreed in captivity with individuals of the larger species from the other lakes. Similarly, smaller individuals from one lake interbreed in captivity with smaller individuals from the other lakes.

In these examples, natural selection appears to be a more important agent of evolutionary change than chance. If chance were the most important factor influencing the direction of evolution, fruit fly evolution would not have proceeded the same way on two different continents and stickleback evolution would not have proceeded the same way in three different lakes. However, just because we have many examples of the importance of natural selection in evolution, it does not necessarily follow that random events should be discounted as a factor in evolutionary change. Proponents of the role of chance think that it is more important in the evolution of major taxonomic groups (macroevolution) than in the evolution of populations (microevolution). It also may be that random events take place but that their effects on evolution are harder to demonstrate than natural selection.

18.4 EVIDENCE FOR EVOLUTION

LEARNING OBJECTIVES

5 Summarize the evidence for evolution obtained from the fossil record.
6 Define *biogeography* and describe how the distribution of organisms supports evolution.
7 Describe the evidence for evolution derived from comparative anatomy.
8 Briefly explain how molecular biology and developmental biology provide insights into the evolutionary process.
9 Give an example of how evolutionary hypotheses are tested experimentally.

A vast body of scientific evidence supports evolution, including observations from the fossil record, biogeography, comparative anatomy, molecular biology, and developmental biology. In addition, evolutionary hypotheses are increasingly being tested experimentally. Taken together, this evidence confirms the scientific theory that life unfolded on Earth by the process of evolution.

The fossil record provides strong evidence for evolution

Perhaps the most direct evidence for evolution comes from the discovery, identification, and interpretation of **fossils,** which are the remains or traces typically left in sedimentary rock by previously existing organisms. (The term *fossil* comes from the Latin word *fossilis,* meaning "something dug up.") Sedimentary rock forms by the accumulation and solidification of particles (pebbles, sand, silt, or clay) produced by the weathering of older rocks, such as volcanic rocks. The sediment particles, which are usually deposited on a riverbed, lake bottom, or the ocean floor, accumulate over time and exhibit distinct layers (FIG. 18-7). In an undisturbed rock sequence, the oldest layer is at the bottom, and upper layers are successively younger. The study of sedimentary rock layers, including their composition, arrangement, and similarity from one location to another, enables geologists to place events recorded in rocks in their correct sequence.

The fossil record shows a progression from the earliest unicellular organisms to the many unicellular and multicellular organisms living today (see Table 21-1). The fossil record therefore demonstrates that life has evolved through time. To date, paleontologists (scientists who study extinct species) have

Figure 18-7 Exposed layers of sedimentary rock at the Burgess shale fossil bed

This site, located in the Canadian Rockies, was formed about 500 million years ago when an avalanche of mud buried and preserved diverse and unusual marine animals.

Alan Sirulnikoff/Science Source

found, for example, because their remains decay extremely rapidly on the forest floor, before fossils can develop. Another reason for bias in the fossil record is that organisms with hard body parts such as bones and shells are more likely to form fossils than are those with soft body parts. Also, rocks of different ages are unequally exposed at Earth's surface; some rocks of certain ages are more accessible to paleontologists for fossil study than are rocks of other ages.

Because of the nature of the scientific process, each fossil discovery represents a separate "test" of scientific evolutionary theory. If any of the tests fail, the theory would have to be modified to fit the existing evidence. The verifiable discovery, for example, of fossil remains of modern humans (*Homo sapiens*) in Precambrian rocks, which are more than 570 million years old, would falsify evolutionary theory as currently proposed. However, Precambrian rocks examined to date contain only fossils of simple organisms, such as algae and small, soft-bodied animals, that evolved early in the history of life. The earliest fossils of *H. sapiens* with anatomically modern features do not appear in the fossil record until approximately 195,000 years ago (see Chapter 22).

Fossils provide a record of ancient organisms and some understanding of where and when they lived. Using fossils of organisms from different geologic ages, scientists can sometimes infer the lines of descent (evolutionary relationships) that gave rise to modern-day organisms. In many instances, fossils provide direct evidence of the origin of new species from pre-existing species, including many transitional forms.

Transitional fossils document whale evolution Over the past century, biologists have found evidence suggesting that whales and other cetaceans (an order of marine mammals) evolved from land-dwelling mammals. During the 1980s and 1990s, paleontologists discovered several fossil intermediates in whale evolution that help document the whales' transition from land to water.

Fossils of *Ambulocetus natans,* a 50-million-year-old transitional form discovered in Pakistan, have many features of modern whales but also possess hind limbs and feet (FIG. 18-9a). (Modern whales do not have hind limbs, although *vestigial* pelvic and hind-limb bones persist. Vestigial structures are discussed later in the chapter.) The vertebrae of *Ambulocetus*'s lower back were very flexible, allowing the back to move dorsoventrally (up and down) during swimming and diving, as with modern whales. This ancient whale could swim but also moved about on land, perhaps as sea lions do today.

Rodhocetus is a fossil whale found in more recent rocks in Pakistan (FIG. 18-9b). The vertebrae of *Rodhocetus* were even more flexible than those found in *Ambulocetus*. The flexible vertebrae allowed *Rodhocetus* a more powerful dorsoventral movement during swimming. *Rodhocetus* may have been totally aquatic.

By 40 million years ago (mya), the whale transition from land to ocean was almost complete. Egyptian fossils of *Basilosaurus* show a whale with a streamlined body and front flippers for steering, like those of modern-day whales (FIG. 18-9c). *Basilosaurus* retained vestiges of its land-dwelling ancestors—a pair of reduced hind limbs that were disjointed from the backbone

described and named about 300,000 fossil species, and others are still being discovered.

Although most fossils are preserved in sedimentary rock, some more recent remains have been exceptionally well preserved in bogs, tar, amber (ancient tree resin), or ice (FIG. 18-8). For example, the remains of a woolly mammoth deep-frozen in Siberian ice for more than 25,000 years were so well preserved that part of its DNA could be analyzed.

Few organisms that die become fossils. The formation and preservation of a fossil require that an organism be buried under conditions that slow or prevent the decay process. These conditions are most likely to occur if an organism's remains are covered quickly by a sediment of fine soil particles suspended in water. In this way, the remains of aquatic organisms may be trapped in bogs, mudflats, sandbars, or deltas. Remains of terrestrial organisms that lived on a floodplain may also be covered by waterborne sediments or, if the organism lived in an arid region, by windblown sand. Over time, the sediments harden to form sedimentary rock, and minerals usually replace the organism's remains so that many details of its structure, even cellular details, are preserved.

The fossil record is not a random sample of past life but instead is biased toward aquatic organisms and those living in the few terrestrial habitats conducive to fossil formation. Relatively few fossils of tropical rainforest organisms have been

(a) Although some fossils contain traces of organic matter, all that remains in this fossil of a 500-million-year-old jellyfish is an impression, or imprint, in the rock.

(b) Petrified wood from the Petrified Forest National Park in Arizona consists of trees that were buried and infiltrated with minerals. The logs were exposed by erosion of the mudstone layers in which they were buried.

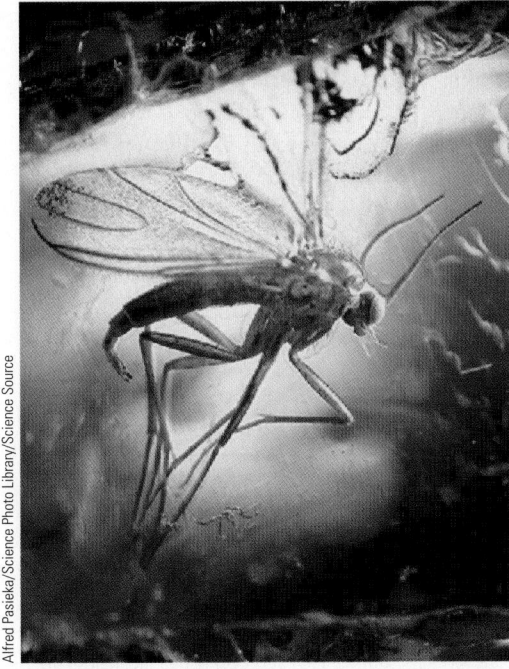

(c) A 2-million-year-old insect fossil (a midge) was embedded in amber.

(d) A cast fossil of ancient echinoderms called *crinoids* formed when the crinoids decomposed, leaving a mold that later filled with dissolved minerals that hardened.

(e) Dinosaur footprints, each 75 to 90 cm (2.5 to 3 ft) in length, provide clues about the posture, gait, and behavior of these extinct animals.

Figure 18-8 Fossils are formed in different ways

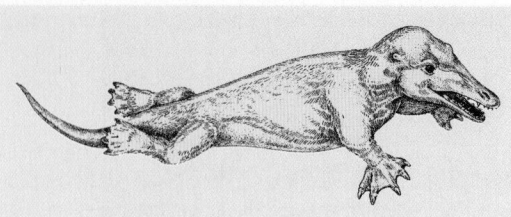

(a) *Ambulocetus natans,* a transitional form between modern whale descendants and their terrestrial ancestors, possessed several recognizable whale features. It retained the hind limbs of its four-legged ancestors, however.

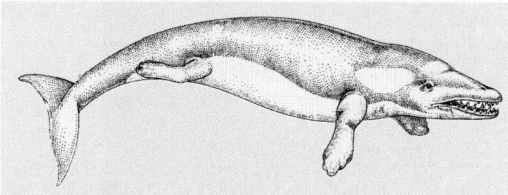

(b) The more recent *Rodhocetus* had flexible vertebrae that permitted a powerful dorsoventral movement during swimming.

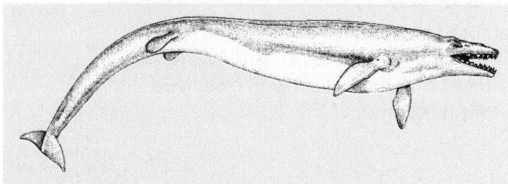

(c) *Basilosaurus* was more streamlined and possessed tiny, nonfunctional hind limbs.

(d) *Balaenoptera,* the modern blue whale, contains vestiges of pelvis and leg bones embedded in its body.

Figure 18-9 Fossil intermediates in whale evolution

Biologists hypothesize that the ancestors of whales had four legs, as shown in these reconstructions of whale intermediates based on fossil evidence. Fossils indicate that ankle bones in whale intermediates match those found in present-day hippos, cows, pigs, and camels. Figures are not drawn to scale. (**a–c:** Adapted with permission from Futuyma, D.J., *Science on Trial: The Case for Evolution,* Fig. 2, pp. 260–261, Sinauer Associates, Sunderland, MA, 1995; **d:** © Cengage Learning)

and probably not used in locomotion—and this reduction in the hind limbs continued to the present. The modern blue whale has vestigial pelvis and femur bones embedded in its body (**FIG. 18-9d**).

Various methods determine the age of fossils Because layers of sedimentary rock occur naturally in the sequence of their deposition, with the more recent layers on top of the older, earlier ones, most fossils are dated by their relative position in

sedimentary rock. However, geologic events occurring after the rocks were initially formed have occasionally changed the relationships of some rock layers. Geologists identify specific sedimentary rocks not only by their positions in layers but also by features such as mineral content and by the fossilized remains of certain organisms, known as **index fossils,** that characterize a specific layer over large geographic areas. Index fossils are fossils of organisms that existed for a relatively short geologic time but were preserved as fossils in large numbers. With this information, geologists can arrange rock layers and the fossils they contain in chronological order and identify comparable layers in widely separated locations.

Radioactive isotopes, also called **radioisotopes,** present in a rock provide a means to accurately measure its age (see Chapter 2). Radioisotopes emit invisible radiations. As radiation is emitted, the nucleus of a radioisotope changes into the nucleus of a different element through a process known as **radioactive decay.** The radioactive nucleus of uranium-235, for example, decays into lead-207.

Each radioisotope has its own characteristic rate of decay. The time required for one-half of the atoms of a radioisotope to change into a different atom is known as its **half-life** (**FIG. 18-10**). Radioisotopes differ significantly in their half-lives. For example, the half-life of iodine-132 is only 2.4 hours, whereas the half-life of uranium-235 is 704 million years. The half-life of a particular radioisotope is constant and does not vary with temperature, pressure, or any other environmental factor.

The age of a fossil in sedimentary rock is usually estimated by measuring the relative proportions of the original radioisotope and its decay product in volcanic rock intrusions that penetrate the sediments. For example, the half-life of potassium-40 is 1.3 billion years, meaning that in 1.3 billion years half of the radioactive potassium will have decayed into its decay product, argon-40. The radioactive clock begins ticking when the magma solidifies into volcanic rock. The rock initially contains

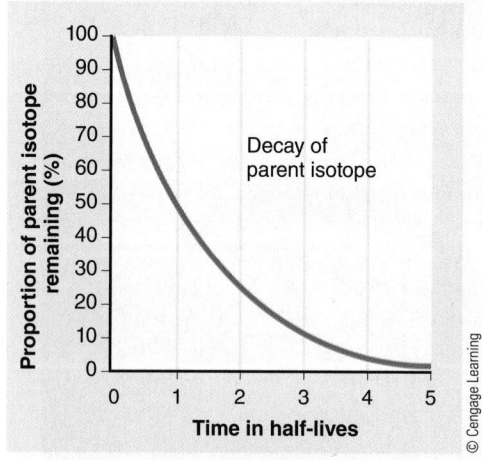

Figure 18-10 *Animation* **Radioisotope decay**

At time zero, the sample is composed entirely of the radioisotope, and the radioactive clock begins ticking. After one half-life, only 50% of the original radioisotope remains. During each succeeding half-life, half of the remaining radioisotope is converted to decay product(s).

some potassium but no argon. Because argon is a gas, it escapes from hot rock as soon as it forms, but when potassium decays in rock that has cooled and solidified, the argon accumulates in the crystalline structure of the rock. If the ratio of potassium-40 to argon-40 in the rock being tested is 1:1, the rock is 1.3 billion years old.

Several radioisotopes are commonly used to date fossils, including potassium-40 (half-life 1.3 billion years), uranium-235 (half-life 704 million years), and carbon-14 (half-life 5730 years). Radioisotopes other than carbon-14 are used to date the *rock* in which fossils are found, whereas carbon-14 is used to date the *carbon remains* of anything that was once living, such as wood, bones, and shells. Whenever possible, the age of a fossil is independently verified using two or more different radioisotopes.

Carbon-14, which is continuously produced in the atmosphere from nitrogen-14 (by cosmic radiation), subsequently decays to nitrogen-14. Because the formation and the decay of carbon-14 occur at constant rates, the ratio of carbon-14 to carbon-12 (a more abundant, stable isotope of carbon) is constant in the atmosphere. Organisms absorb carbon from the atmosphere either directly (by photosynthesis) or indirectly (by consuming photosynthetic organisms). Because each organism absorbs carbon from the atmosphere, its ratio of carbon-14 to carbon-12 is the same as that in the atmosphere. When an organism dies, however, it no longer absorbs carbon, and the proportion of carbon-14 in its remains declines as carbon-14 decays to nitrogen-14. Because of its relatively short half-life, carbon-14 is useful for dating fossils that are 50,000 years old or less. It is particularly useful for dating archaeological sites.

The distribution of plants and animals supports evolution

The study of the past and present geographic distribution of organisms is called **biogeography.** The geographic distribution of organisms affects their evolution. Darwin was interested in biogeography, and he considered why the species found on ocean islands tend to resemble species of the nearest mainland, even if the environment is different. He also observed that species on ocean islands do not tend to resemble species on islands with similar environments in other parts of the world.

Darwin studied the plants and animals of two sets of arid islands: the Cape Verde Islands, nearly 640 km (400 mi) off western Africa; and the Galápagos Islands, about 965 km (600 mi) west of Ecuador, South America. On each group of islands, the plants and terrestrial animals were indigenous (native), but those of the Cape Verde Islands resembled African species and those of the Galápagos resembled South American species. The similarities of Galápagos species to South American species were particularly striking considering that the Galápagos Islands are dry and rocky and that the nearest part of South America is humid and has a lush, tropical rain forest. Darwin concluded that species from the neighboring continent migrated or were carried to the islands, where they

subsequently adapted to the new environment and, in the process, evolved into new species.

If evolution were not a factor in the distribution of species, we would expect to find a given species everywhere it could survive. However, the actual geographic distribution of organisms makes sense in the context of evolution. For example, Australia, which has been a separate landmass for millions of years, has distinctive organisms. Australia has populations of egg-laying mammals (monotremes) and pouched mammals (marsupials) not found anywhere else. Two hundred million years ago, Australia and the other continents were joined in a major landmass. Over the course of millions of years, the Australian continent gradually separated from the others. The monotremes and marsupials in Australia continued to thrive and diversify. The isolation of Australia also prevented placental mammals, which arose elsewhere at a later time, from competing with its monotremes and marsupials. In other areas of the world where placental mammals evolved, most monotremes and marsupials became extinct.

We now consider how Earth's dynamic geology has affected biogeography and evolution.

Biogeography and evolution are related to Earth's geologic history In 1915, German scientist Alfred Wegener, who had noted a correspondence between the geographic shapes of South America and Africa, proposed that all the landmasses had at one time been joined into one huge supercontinent, which he called Pangaea (FIG. 18-11a). He further suggested that Pangaea had subsequently broken apart and that the various landmasses had separated in a process known as **continental drift.** Wegener did not know of any mechanism that could have caused continental drift, so his idea, although debated initially, was largely ignored.

In the 1960s, scientific evidence provided the explanation for continental drift. Earth's crust is composed of seven large plates (plus a few smaller ones) that float on the mantle, which is the mostly solid layer of Earth lying beneath the crust and above the core. (Because of its hotter temperature, the solid rock of the mantle is more plastic than the solid rock of the crust above it.)

The landmasses are situated on some of these plates. As the plates move, the continents change their relative positions (FIGS. 18-11b, c, and d). The movement of the crustal plates is called *plate tectonics.*

Any area where two plates meet is a site of intense geologic activity. Earthquakes and volcanoes are common in such a region. Both San Francisco, noted for its earthquakes, and the Mount St. Helens volcano are situated where two plates meet. If landmasses lie on the edges of two adjacent plates, mountains may form. The Himalayas formed when the plate carrying India rammed into the plate carrying Asia. When two plates grind together, one of them is sometimes buried under the other in a process known as *subduction.* When two plates move apart, a ridge of lava forms between them. The Atlantic Ocean is increasing in size because of the expanding zone of lava along the Mid-Atlantic Ridge, where two plates are separating.

Knowledge that the continents were at one time connected and have since drifted apart is useful in explaining certain

Geologists hypothesize that the breakup of Pangaea is only the latest in a series of continental breakups and collisions that have taken place since early in Earth's history.

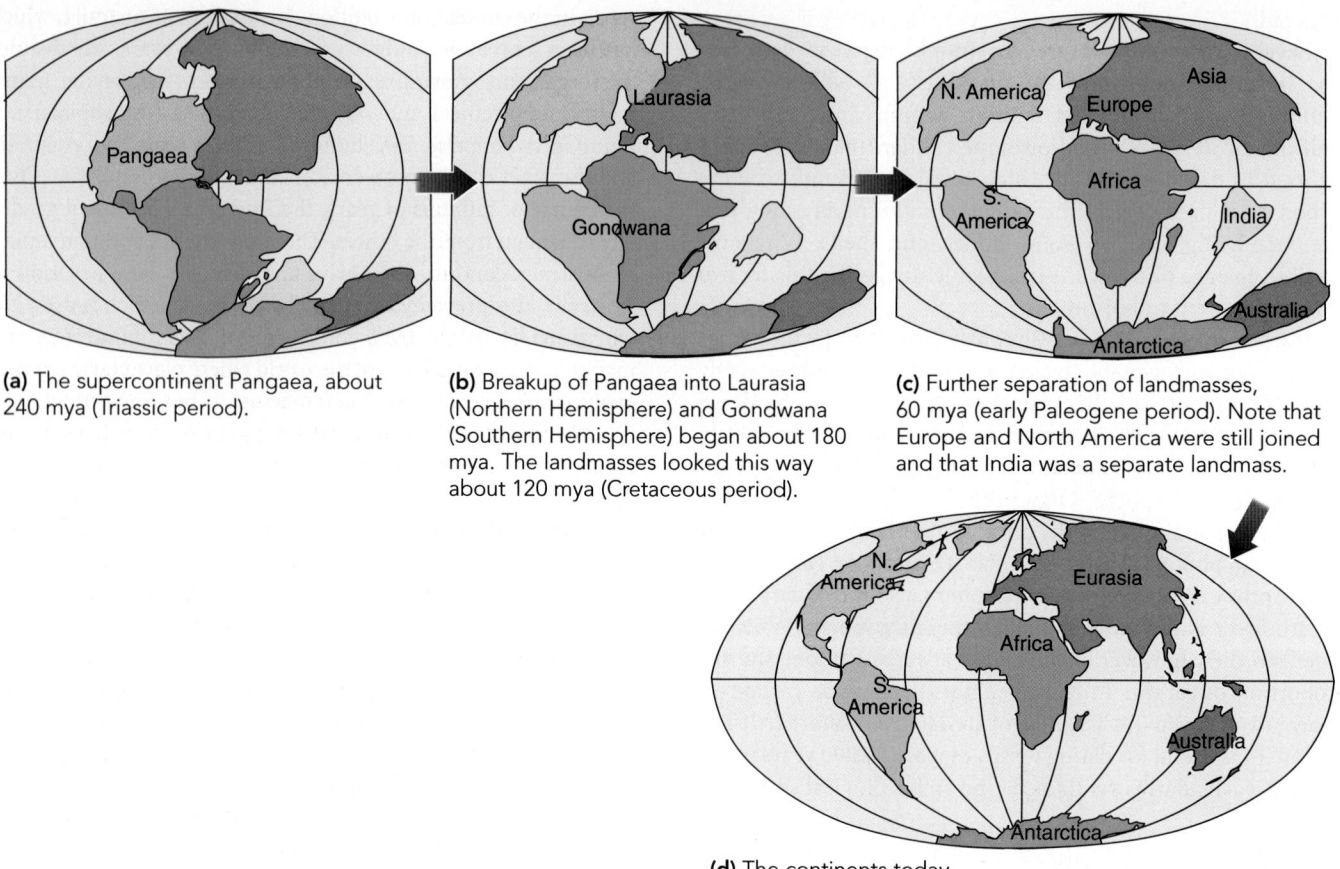

(a) The supercontinent Pangaea, about 240 mya (Triassic period).

(b) Breakup of Pangaea into Laurasia (Northern Hemisphere) and Gondwana (Southern Hemisphere) began about 180 mya. The landmasses looked this way about 120 mya (Cretaceous period).

(c) Further separation of landmasses, 60 mya (early Paleogene period). Note that Europe and North America were still joined and that India was a separate landmass.

(d) The continents today.

Figure 18-11 Continental drift

CONNECT If you were to find 60-million-year-old fossils of a species of land plants on the east coast of the northern part of North America, on what part of which continent would you begin a quest for similar fossils?
© Cengage Learning

aspects of biogeography (FIG. 18-12). Likewise, continental drift has played a major role in the evolution of different organisms. When Pangaea originally formed during the late Permian period, it brought together terrestrial species that had evolved separately from one another, leading to competition and some extinction. Marine life was adversely affected, in part because, with the continents joined as one large mass, less coastal area existed. (Shallow coastal areas contain a greater diversity of marine species than deep-water environments.)

Pangaea separated into several landmasses approximately 180 mya. As the continents began to drift apart, populations became geographically isolated, were exposed to different environmental conditions, and began to diverge along separate evolutionary pathways. As a result, the plants, animals, and other organisms of previously connected continents—South America and Africa, for example—differ. Continental drift also caused gradual changes in ocean and atmospheric currents that have profoundly influenced the biogeography and evolution of organisms. (Biogeography is discussed further in Chapter 56.)

Comparative anatomy of related species demonstrates similarities in their structures

Comparing the structural details of features found in different but related organisms reveals a basic similarity. Such features that are derived from the same structure in a common ancestor are called **homologous features;** the condition is known as **homology.** For example, consider the limb bones of mammals. A human arm, a cat forelimb, a whale front flipper, and a bat wing, although quite different in appearance, have strikingly similar arrangements of bones, muscles, and nerves. FIGURE 18-13 shows a comparison of their skeletal structures. Each has a single bone (the humerus) in the part of the limb nearest the trunk of the body, followed by the two bones (radius and ulna) of the forearm, a group of bones (carpals) in the wrist, and a variable number of digits (metacarpals and phalanges). This similarity is particularly striking because arms, forelimbs, flippers, and wings are used for different types of locomotion, and there is no overriding mechanical

Knowledge that the continents were united at one time explains the unique distribution of certain fossil plants and animals.

(a) *Cynognathus* was a carnivorous reptile found in Triassic rocks in South America and Africa.

(b) *Lystrosaurus* was a herbivorous reptile found in Triassic rocks in Africa, India, and Antarctica.

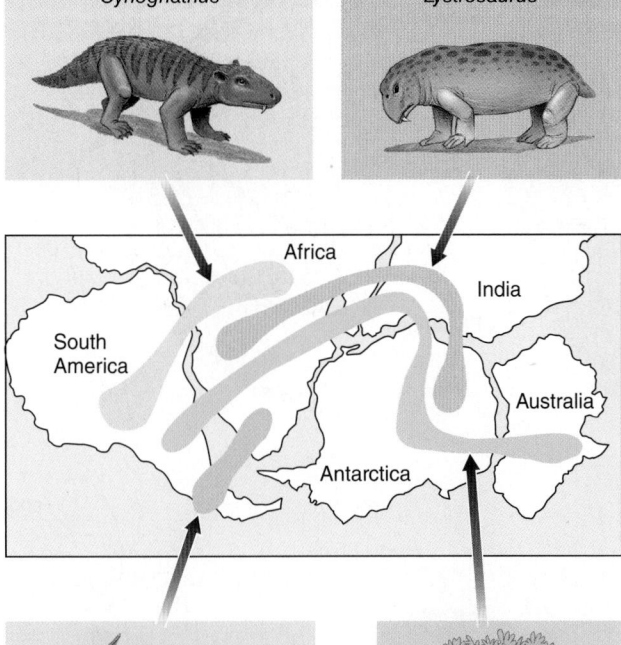

(c) *Mesosaurus* was a freshwater reptile found in Permian rocks in South America and Africa.

Glossopteris

(d) *Glossopteris* was a seed-bearing tree found in Permian rocks in South America, Africa, India, Antarctica, and Australia.

Adapted from E.H. Colbert, *Wandering Lands and Animals*, Hutchinson, London, 1973.

Figure 18-12 Distribution of fossils on continents that were joined during the Permian and Triassic periods (286 mya to 213 mya)

CONNECT Examine Figure 18-11. With which continent is Antarctica most likely to share 60-million-year-old fossils of a particular species of land animals?

reason for them to be so similar structurally. Similar arrangements of parts of the forelimb are evident in ancestral birds, reptiles, and amphibians and even in the first fishes that came out of water onto land hundreds of millions of years ago (see Fig. 32-17).

Leaves provide an example of homology in plants. In many plant species, leaves have been modified for functions other than photosynthesis. A cactus spine and a pea tendril, although quite different in appearance, are homologous because both are modified leaves (**FIG. 18-14**). The spine protects the succulent stem tissue of the cactus, whereas the tendril, which winds around a small object once it makes contact, helps support the climbing stem of the pea plant. Such modifications in organs used in different ways are the expected outcome of a common evolutionary origin. The basic structure present in a common ancestor was modified in different ways for different functions as various descendants subsequently evolved.

Not all species with "similar" features have descended from a recent common ancestor, however. Sometimes similar environmental conditions result in the evolution of similar adaptations. Such independent evolution of similar structures in distantly related organisms is known as **convergent evolution.** Aardvarks, anteaters, and pangolins are an excellent example of convergent evolution (**FIG. 18-15**). They resemble one another in lifestyle and certain structural features. All have strong, sharp claws to dig open ant and termite mounds and elongated snouts with long, sticky tongues to catch these insects. Yet aardvarks, anteaters, and pangolins evolved from three distantly related orders of mammals. (See Figure 32-28, which shows several examples of convergent evolution in placental and marsupial mammals.)

Structurally similar features that are not homologous but have similar functions that evolved independently in distantly related organisms are said to be **homoplastic features.** Such similarities in different species that are independently acquired by convergent evolution and not by common descent are called **homoplasy.**[2]

For example, the wings of various distantly related flying animals, such as insects and birds, resemble one another superficially; they are homoplastic features that evolved over time to meet the common function of flight, although they differ in more fundamental aspects. Bird wings are modified forelimbs supported by bones, whereas insect wings may have evolved from gill-like appendages present in the aquatic ancestors of insects.

Spines, which are modified leaves, and thorns, which are modified stems, are an example of homoplasy in plants. Spines and thorns resemble each other superficially but are homoplastic features that evolved independently to solve the common need for protection from herbivores (**FIG. 18-16**).

Like homology, homoplasy offers crucial evidence of evolution. Homoplastic features are of evolutionary interest because they demonstrate that organisms with separate ancestries may adapt in similar ways to similar environmental demands.

Comparative anatomy also reveals the existence of **vestigial** structures. Many organisms contain organs or parts of organs that are seemingly nonfunctional and degenerate, often undersized or lacking some essential part. Vestigial structures are remnants of more developed structures that were present and functional in ancestral organisms. In the human body, more than one hundred structures are considered vestigial, including

---

[2] An older, less precise term that some biologists still use for nonhomologous features with similar functions is *analogy*.

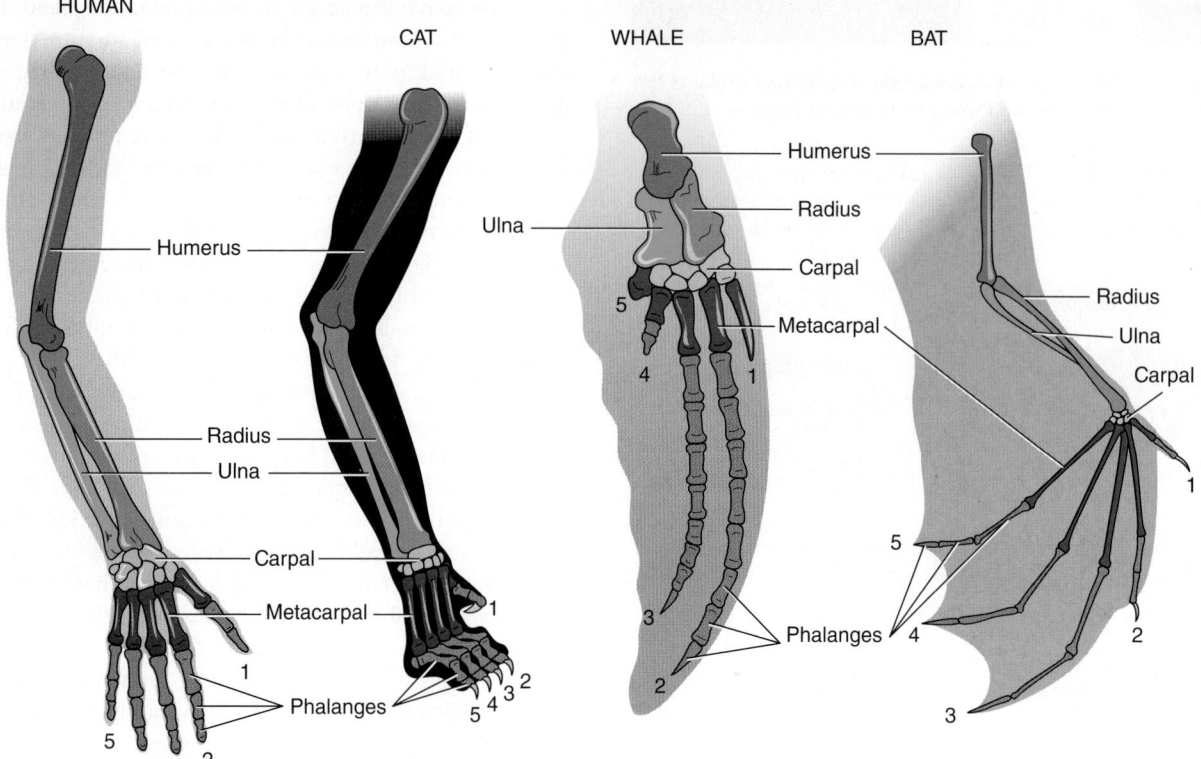

Figure 18-13 *Animation* **Structural homology in vertebrates**

The human arm, cat forelimb, whale flipper, and bat wing have a basic under-lying similarity of structure because they are derived from a common ancestor. The five digits are numbered in each drawing.

© Cengage Learning

the coccyx (fused tailbones), third molars (wisdom teeth), and the muscles that move our ears. Whales and pythons (FIG. 18-17) have vestigial hind-limb bones; pigs have vestigial toes that do not touch the ground; wingless birds such as the kiwi have vestigial wing bones; and many blind, burrowing, or cave-dwelling animals have nonfunctioning, vestigial eyes.

The occasional presence of a vestigial structure is to be expected as a species adapts to a changing mode of life. Some structures become much less important for survival and may end up as vestiges. When a structure no longer confers a selective advantage, it may become smaller and lose all or much of its function with the passage of time. Because the presence of the vestigial structure is usually not harmful to the organism, however, selective pressure for completely eliminating it is weak, and the vestigial structure is found in many subsequent generations.

Molecular comparisons among organisms provide evidence for evolution

Fossils, biogeography, and comparative anatomy provided Darwin with important clues about the evolutionary history of life. Today, similarities and differences in the biochemistry and molecular biology of various organisms provide additional compelling evidence for evolutionary relationships. Molecular evidence for evolution includes the universal genetic code and the conserved sequences of amino acids in proteins and of nucleotides in DNA.

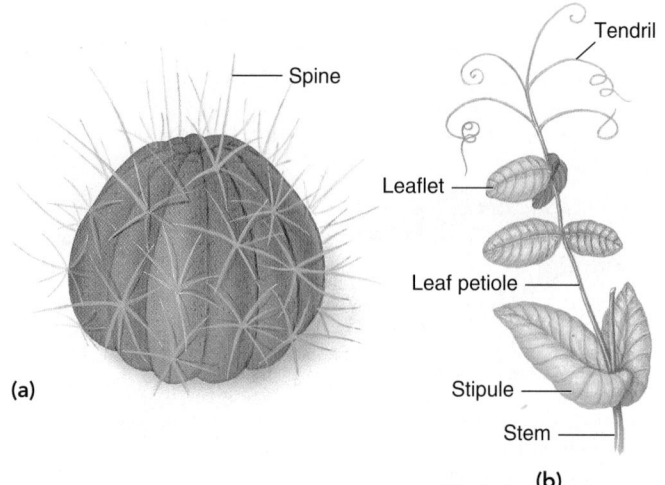

Figure 18-14 **Homology in plants**

(a) The spines of the fishhook cactus (*Ferocactus wislizenii*) are modified leaves, as are **(b)** the tendrils of the garden pea (*Pisum sativum*). Leaves of the garden pea are compound; that is, they are divided into smaller components called leaflets. The terminal leaflets are modified into tendrils that are frequently branched.

© Cengage Learning

The genetic code is virtually universal Organisms owe their characteristics to the types of proteins they possess, which in turn are determined by the sequence of nucleotides in their messenger ribonucleic acid (mRNA) as specified by the order of nucleotides in their DNA. Evidence that all life is related comes from all organisms using a genetic code that is virtually identical. (There is some minor variation in the genetic code. For example, some mitochondria have several deviations from the standard code.)

(a) The aardvark (*Orycteropus afer*) is native to central, southern, and eastern Africa.

(b) A giant anteater (*Myrmecophaga tridactyla*) at a termite mound. The anteater is native to Latin America, from southern Mexico to northern Argentina.

(c) The pangolin (*Manis crassicaudata*) is native to Africa and to southern and southeastern Asia.

Figure 18-15 Convergent evolution

Three distantly related mammals adapted independently to eat ants and termites in similar grassland/forest environments in different parts of the world.

Recall from Chapter 13 that the genetic code specifies a triplet (a sequence of three nucleotides in DNA) that codes for a particular codon (a sequence of three nucleotides in mRNA). The codon then codes for a particular amino acid in a polypeptide chain. For example, AAA in DNA codes for UUU in mRNA, which codes for the amino acid phenylalanine in organisms as diverse as shrimp, humans, bacteria, and tulips. In fact, AAA codes for phenylalanine in all organisms examined to date.

The universality of the genetic code—no other code has been found in any organism—is convincing evidence that all organisms arose from a common ancestor. The genetic code has been maintained and transmitted through all branches of the evolutionary tree since its origin in some extremely early (and successful) organism.

Proteins and DNA contain a record of evolutionary change Researchers have carried out thousands of comparisons of protein and DNA sequences from various species since the late 1970s. Moreover, the recent explosion in DNA sequencing information (more than 1000 species have now had their genomes sequenced) is beginning to provide a wealth of evolutionary data as these genomes are compared. Sequence-based relationships generally agree with earlier studies, which based evolutionary relationships on similarities in structure among living organisms and on fossil data of extinct organisms.

Investigations of the sequence of amino acids in proteins that play the same roles in many species have revealed both great similarities and certain specific differences. Even organisms that are only remotely related share some proteins, such as cytochrome *c*, which is part of the electron transport chain in aerobic respiration. To survive, all aerobic organisms need a respiratory protein with the same basic structure and function as the cytochrome *c* of their common ancestor. Consequently, not all amino acids that confer the structural and functional features of cytochrome *c* are free to change. Any mutations that changed the amino acid sequence at structurally important sites of the cytochrome *c* molecule would have been harmful, and natural selection would have prevented such mutations from being passed to future generations. However, in the course of the long, independent evolution of different organisms, mutations have resulted in the substitution of many amino acids at less important locations in the cytochrome *c* molecule. The greater the differences in the amino acid sequences of their cytochrome *c* molecules, the longer it has been since two species diverged.

Because a protein's amino acid sequences are coded in DNA, the differences in amino acid sequences indirectly reflect the nature and number of underlying DNA base-pair changes that must have occurred during evolution. Of course, not all DNA codes for proteins (witness the regulatory RNAs summarized in Table 13-1, as well as ribosomal RNA and transfer RNA genes). **DNA sequencing,** i.e., determining the order of nucleotide bases in both protein-coding DNA and non-protein-coding DNA, is proving invaluable in determining evolutionary relationships.

Generally, the more closely species are considered related on the basis of other scientific evidence, the greater the percentage of nucleotide sequences that their DNA molecules have in common. By using the DNA sequence data in **FIGURE 18-18,** for example, you can conclude that the closest living relative of humans is the chimpanzee (because its DNA has the lowest percentage of differences in the sequence examined). (Primate evolution is discussed in Chapter 22.)

In some cases, molecular evidence challenges traditional evolutionary ideas that were based on structural comparisons among living species and/or on studies of fossil skeletons.

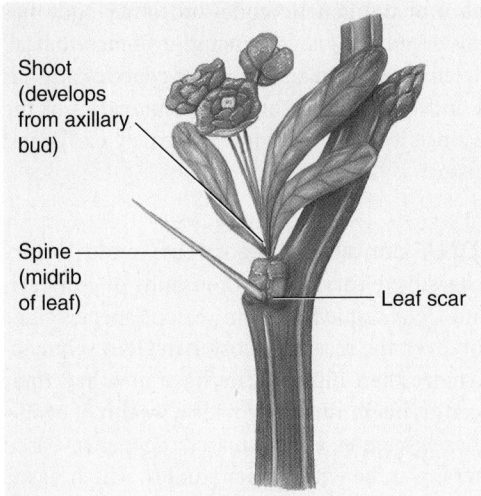

(a) A spine of Japanese barberry (*Berberis thunbergii*) is a modified leaf. (In this example, the spine is actually the midrib of the original leaf, which has been shed.)

(b) Thorns of downy hawthorn (*Crataegus mollis*) are modified stems that develop from axillary buds.

Figure 18-16 **Homoplasy in plants**

Consider even-toed hoofed mammals such as pigs, camels, deer, antelope, cattle, and hippos. Whales lack toes and were traditionally not classified as close relatives of hoofed mammals.

FIGURE 18-19 depicts a **cladogram,** or **phylogenetic tree,** based on molecular data for whales and selected hoofed mammals. Such cladograms—diagrams showing lines of descent—can be derived from differences in a given DNA nucleotide sequence. This diagram indicates that whales are more closely related to hippos than to any other hoofed mammal. The branches representing whales and hippos probably diverged relatively recently because of the close similarity of DNA

(a) An Indian python (*Python molurus*).

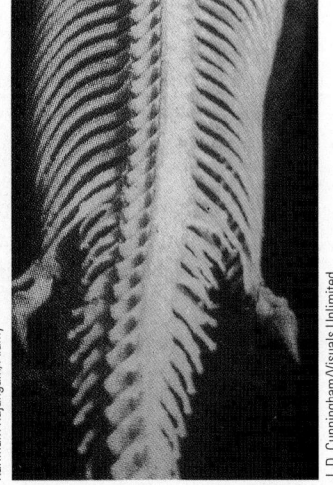

(b) Close-up of part of a python skeleton showing the hind-limb bones.

Figure 18-17 *Animation* **Vestigial structures**

All pythons have remnants of hind-limb bones embedded in their bodies.

sequences in these species. In contrast, camels, which have DNA sequences that are less similar to those of whales, diverged much earlier. The molecular evidence indicates that whales and hippos share a recent common ancestor.

The molecular evidence linking whales and even-toed hoofed mammals has led to a re-examination of early fossil whales. Like today's hoofed mammals, early fossil whales had an even number of toes on their appendages. Fossil whales also had the same type of specialized ankle bone (called a double pulley) as hippos and other hoofed mammals.

Developmental biology helps unravel evolutionary patterns

As early as 1975, biologists were suggesting that regulatory changes in gene expression, particularly of genes involved in development, were responsible for many differences between closely related species. Today much evidence supports the idea that regulatory changes—how genes are switched on and off during development—help explain the diversity of form in species with similar genes. Mutations in genes that regulate development often result in dramatically different structures (see Fig. 17-12).

Increasingly, developmental biology, particularly at the molecular level, is providing answers to such questions as how snakes became elongated and lost their limbs. In many cases, evolutionary changes such as the loss of limbs in snakes occur as a result of changes in genes that regulate the orderly sequence of events that occurs during development. In pythons, for example, the loss of forelimbs and elongation of the body are linked to mutations in several *Hox* genes that affect the expression of body patterns and limb formation in a wide variety of animals. (See the discussion of *Hox* gene clusters in Chapter 17.) The hind limbs may not develop because python embryonic tissue does not respond to internal signals that trigger leg elongation.

Developmental geneticists at Harvard Medical School and Princeton University are studying the developmental basis for the different beak shapes of the Galápagos finches. They have determined that a gene that codes for an important signaling molecule, bone morphogenic protein 4 (BMP4), affects the development of the birds' craniofacial skeletons. The gene for BMP4 is turned on earlier in development and has a higher level of expression in finch species with larger, thicker beaks than in finches with smaller beaks.

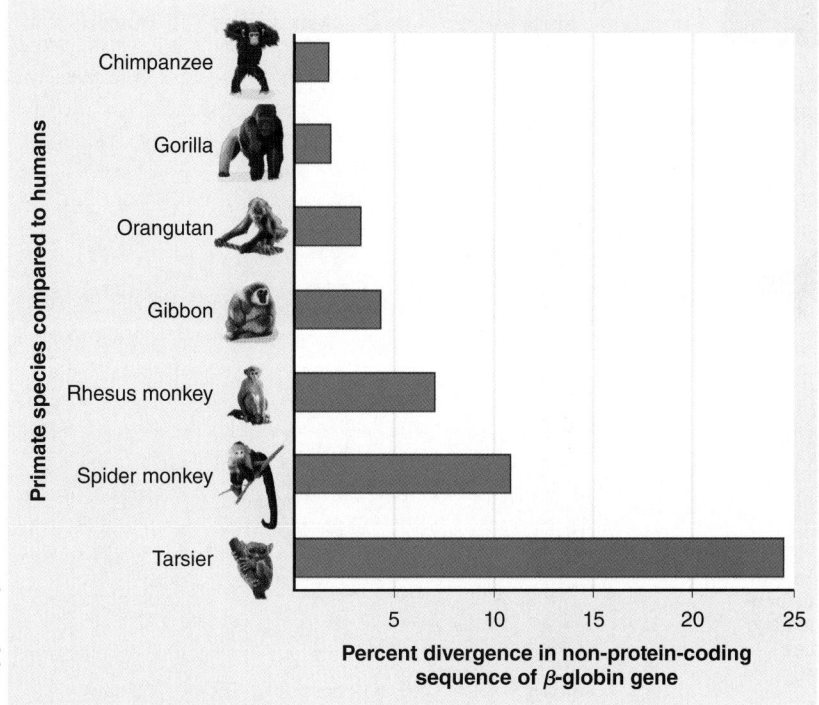

Figure 18-18 Differences in DNA nucleotide sequences as evidence of evolutionary relationships

Comparing the same gene in different organisms provides a window into evolution. Here the differences in the non-protein-coding region of the β-globin gene are compared between humans and other primates. Evolutionary biologists are rapidly expanding such studies from comparing one or several genes to comparing entire genomes.

Scientific evidence overwhelmingly demonstrates that development in different animals is controlled by the same kinds of genes; these genetic similarities in a wide variety of organisms reflect a shared evolutionary history. For example, vertebrates have similar patterns of embryological development that indicate that they share a common ancestor. All vertebrate embryos have segmented muscles, pharyngeal (throat) pouches, a tubular heart without left and right sides, a system of arteries known as *aortic arches* in the pharyngeal region, and many other shared features. All these structures are necessary and functional in the developing fish. The small, segmented muscles of the fish embryo give rise to the segmented muscles used by the adult fish in swimming. The pharyngeal pouches break through to the surface as gill slits. The adult fish heart remains undivided and pumps blood forward to the gills that develop in association with the aortic arches.

Because none of these embryonic features persists in the adults of reptiles, birds, or mammals, why are these fishlike structures present in their embryos? Evolution is a conservative process, and natural selection builds on what has come before rather than starting from scratch. The evolution of new features often does not require the evolution of new developmental genes but instead depends on a modification in developmental genes that already exist (see discussion of pre-adaptations in Chapter 20). Terrestrial vertebrates are thought to have evolved from fishlike ancestors; therefore, they share some of the early stages of development still found in fishes

today. The accumulation of genetic changes over time in these vertebrates has modified the basic body plan laid out in fish development (**FIG. 18-20**).

Evolutionary hypotheses are tested experimentally

Increasingly, biologists are designing imaginative experiments, often in natural settings, to test evolutionary hypotheses. David Reznick from the University of California, Santa Barbara and John Endler from James Cook University in Australia have studied evolution in guppy populations in Venezuela and in Trinidad, a small island in the southern Caribbean.

Reznick and Endler observed that different streams have different kinds and numbers of fishes that prey on guppies. Predatory fishes that prey on larger guppies are present in all streams at lower elevations; these areas of intense predation pressure are known as *high-predation habitats*. Predators are often excluded from tributaries or upstream areas by rapids and waterfalls. The areas above such barriers are known as *low-predation habitats* because they contain only one species of small predatory fish that occasionally eats smaller guppies.

Differences in predation are correlated with many differences in the guppies, such as male coloration, behavior, and attributes known as *life history traits* (discussed in more detail in Chapter 53). These traits include age and size at sexual maturity, the number of offspring per reproductive event, the size of the offspring, and the frequency of reproduction. For example, guppy adults are larger in streams found at higher elevations and smaller in streams found at lower elevations.

Do predators actually cause these differences to evolve? Reznick and his colleagues tested this evolutionary hypothesis by conducting field experiments in Trinidad. Taking advantage of waterfalls that prevent the upstream movement of guppies, guppy predators, or both, they moved either guppies or guppy predators over such barriers. For example, guppies from a high-predation habitat were introduced into a low-predation habitat by moving them over a barrier waterfall into a section of stream that was free of guppies and large predators. The only fish species that lived in this section of stream before the introduction was the small predatory fish.

Eleven years later, the researchers captured adult females from the introduction site (low-predation habitat) and the control site below the barrier waterfall (high-predation habitat). They bred these females in their laboratory and compared the life history traits of succeeding generations. The descendants of guppies introduced into the low-predation habitat matured at a larger size than did the descendants of guppies from the control site below the waterfall (**FIG. 18-21**). They also produced fewer, but larger, offspring. The introduced fish populations had evolved to have life histories similar to those of fishes typically

This branching diagram, called a *cladogram*, shows proposed evolutionary relationships based on available data. The organisms depicted here share a common ancestor.

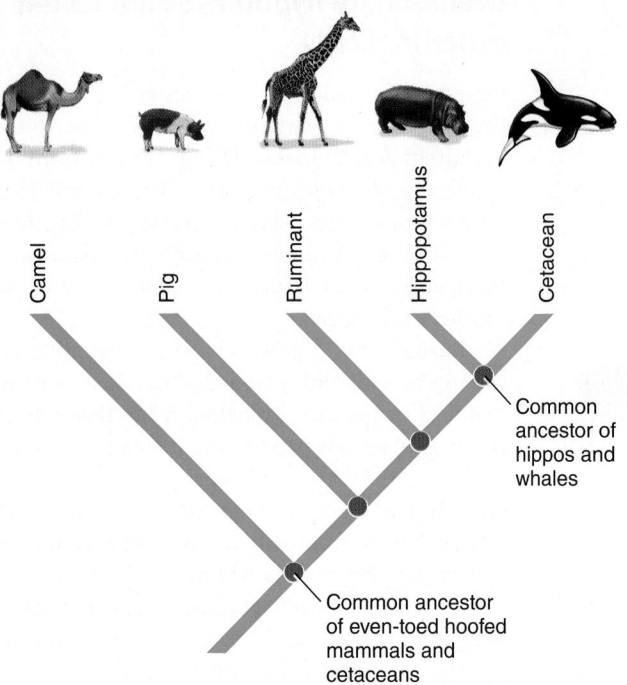

Common ancestor of hippos and whales

Common ancestor of even-toed hoofed mammals and cetaceans

Adapted from Nikaido, M., et al., "Phylogenetic Relationships among Cetartiodactyls Based on Insertions of Short and Long Interspersed Elements: Hippopotamuses Are the Closest Extant Relatives of Whales," *Proceedings of the National Academy of Sciences*, Vol. 96, Aug. 31, 1999.

Figure 18-19 Cladogram of whales and their closest living relatives

DNA sequence differences among selected mammals suggest that hoofed mammals such as hippos and giraffes are the closest living relatives of whales. The hippopotamus is probably the closest living relative of whales and other cetaceans. The nodes (*circles*) represent branch points where a species splits into two or more lineages. (Ruminants are mammals such as cows, sheep, and giraffes that have a multichambered stomach and chew regurgitated plant material to make it more digestible.)

CONNECT According to the cladogram, which animals other than the cetaceans are the closest relatives of ruminants?

found in such low-predation habitats. Similar studies have demonstrated that predators have played an active role in the evolution of other traits, such as the average number of offspring produced during the lifetime of an individual female (fecundity), male coloration, and behavior.

Rapid evolution on a scale of years has also been observed in organisms as diverse as marine snails, mussels, soapberry bugs, mayflies, anole lizards, 'i'iwis (scarlet honeycreepers), and wild rabbits. These and other experiments and observations demonstrate not only that evolution is real but also that it is occurring now, driven by selective environmental forces, such as predation, that can be experimentally manipulated. Darwin incorrectly

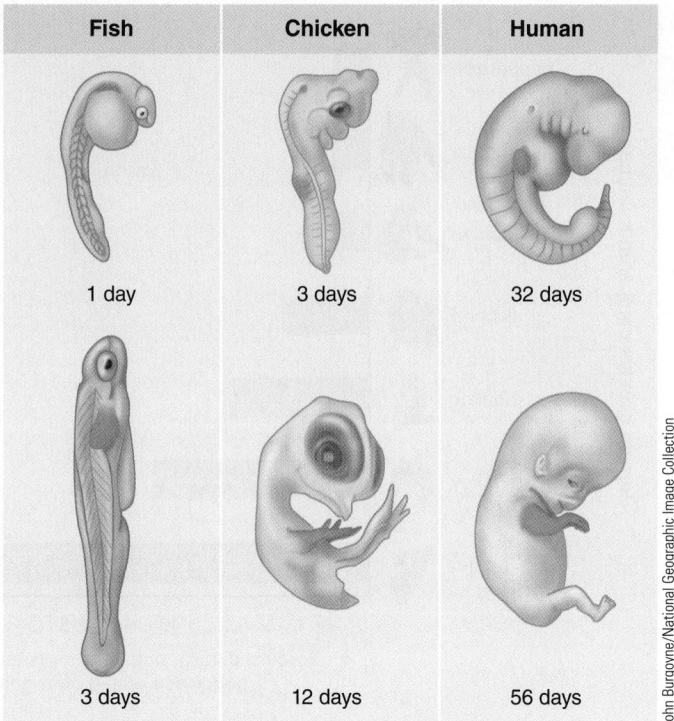

Fish	Chicken	Human
1 day	3 days	32 days
3 days	12 days	56 days

John Burgoyne/National Geographic Image Collection

Figure 18-20 Development of fish fins, chicken wings, and human limbs

Fish, chickens, and humans are vertebrates with strikingly similar genes. Although the early embryos of these organisms are much alike, the areas colored *orange* follow different developmental pathways, resulting in fins, wings, or limbs. (Figures are not to scale.)

assumed evolution to be so gradual that humans cannot observe it. As Jonathan Weiner, author of *The Beak of the Finch: A Story of Evolution in Our Time,* notes, "Darwin did not know the strength of his own theory. He vastly underestimated the power of natural selection. Its action is neither rare nor slow. It leads to evolution daily and hourly, all around us, and we can watch."[3]

CHECKPOINT 18.4

- *How do scientists date fossils?*
- *How can we explain that fossils of* Mesosaurus, *an extinct reptile that could not swim across open water, are found in the southern parts of both Africa and South America?*
- *How do homologous and homoplastic features provide evidence of evolution?*
- **CONNECT** *How does developmental biology provide evidence of a common ancestry for vertebrates as diverse as reptiles, birds, pigs, and humans?*
- *How do predator preferences drive the evolution of size in guppies?*

[3] Weiner, J. *The Beak of the Finch: A Story of Evolution in Our Time,* p. 9. Alfred A. Knopf. New York (1994).

Can natural selection be observed in a natural population?

HYPOTHESIS: A natural population will respond adaptively to environmental change.

EXPERIMENT: Male and female guppies from a stream in which the predators preferred large adult guppies as prey (*brown bars*) were transferred to a stream in which the predators preferred juveniles and small adults.

RESULTS AND CONCLUSION: After 11 years, the descendants of the transferred guppies (*purple bars*) were measurably larger than their ancestors, indicating that larger guppies had a selective advantage in the new environment.

SOURCE: Data used with permission from Reznick, D.N., et al., "Evaluation of the Rate of Evolution in Natural Populations of Guppies [*Poecilia reticulata*]," *Science*, Vol. 275, Mar. 28, 1997.

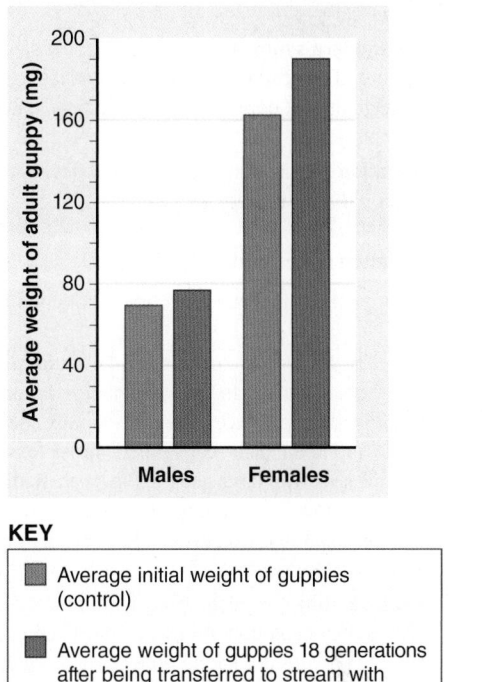

KEY

■ Average initial weight of guppies (control)

■ Average weight of guppies 18 generations after being transferred to stream with different selective predation (experimental)

Figure 18-21 D.N. Reznick's experiment in guppy evolution

CONNECT What do you think would happen next if larger predators were introduced into this low-predation habitat?

18.1 What Is Evolution? *(page 382)*

1 Define the scientific theory of *evolution*.

- **Evolution** is the accumulation of inherited changes within a **population** over time. Evolution is the unifying concept of biology because it links all fields of the life sciences into a coherent body of knowledge.

18.2 Pre-Darwinian Ideas about Evolution *(page 382)*

2 Discuss the historical development of evolutionary thought.

- Jean Baptiste de Lamarck was the first scientist to propose that organisms undergo change over time as a result of some natural phenomenon rather than divine intervention. Lamarck thought that organisms were endowed with a vital force that drove them to change toward greater complexity over time. He thought that organisms could pass traits acquired during their lifetimes to their offspring.

- Charles Darwin's observations while voyaging on the HMS *Beagle* were the basis for his evolutionary theory. Darwin tried to explain the similarities between animals and plants of the arid Galápagos Islands and the humid South American mainland.

- Darwin was influenced by **artificial selection,** in which breeders develop many varieties of domesticated plants and animals

in just a few generations. Darwin applied Thomas Malthus's ideas on the natural increase in human populations to natural populations. Darwin was influenced by the idea that Earth was extremely old, an idea promoted by Charles Lyell and other geologists.

18.3 Darwin and Evolution *(page 383)*

3 Explain the four premises of evolution by natural selection as proposed by Charles Darwin.

- Charles Darwin and Alfred Russel Wallace independently proposed evolution by **natural selection,** which is based on four observations. First, genetic variation exists among the individuals in a population. Second, the reproductive ability of each species causes its populations to have the potential to geometrically increase in number over time. Third, organisms compete with one another for the resources needed for life, such as food, living space, water, and light. Fourth, offspring with the most favorable combination of inherited characteristics are most likely to survive and reproduce, passing those genetic characteristics to the next generation.

- Natural selection results in **adaptations,** evolutionary modifications that improve the chances of survival and reproductive success in a particular environment. Over time, enough changes may accumulate in geographically separated populations to produce new species.

4 Compare the modern synthesis with Darwin's original view of evolution.

- The **modern synthesis** combines Darwin's evolutionary theory by natural selection with modern genetics to explain why individuals in a population vary and how species adapt to their environment.
- **Mutation** provides the genetic variability that natural selection acts on during evolution.

18.4 Evidence for Evolution *(page 387)*

5 Summarize the evidence for evolution obtained from the fossil record.

- Direct evidence of evolution comes from **fossils,** the remains or traces of ancient organisms. Layers of sedimentary rock normally occur in their sequence of deposition, with the more recent layers on top of the older, earlier ones. **Index fossils** characterize a specific layer over large geographic areas. **Radioisotopes** present in a rock provide a way to accurately measure the rock's age.

6 Define *biogeography* and describe how the distribution of organisms supports evolution.

- **Biogeography,** the geographic distribution of organisms, affects their evolution. Areas that have been separated from the rest of the world for a long time contain organisms that have evolved in isolation and are therefore unique to those areas.
- At one time the continents were joined to form a supercontinent. **Continental drift,** which caused the various landmasses to break apart and separate, has played a major role in evolution.

7 Describe the evidence for evolution derived from comparative anatomy.

- **Homologous features** have basic structural similarities even though the structures may be used in different ways because homologous features derive from the same structure in a common ancestor. Evolutionary affinities exist among the organisms that have homologous features.
- **Homoplastic features** evolved independently to have similar functions in distantly related organisms. Homoplastic features demonstrate **convergent evolution,** in which organisms with separate ancestries adapt in similar ways to comparable environmental demands.

- **Vestigial** structures are nonfunctional or degenerate remnants of structures that were present and functional in ancestral organisms. Structures occasionally become vestigial as species adapt to different modes of life.

8 Briefly explain how molecular biology and developmental biology provide insights into the evolutionary process.

- Molecular evidence for evolution includes the universal genetic code and the conserved sequences of amino acids in proteins and of nucleotides in DNA.
- Evolutionary changes are often the result of mutations in genes that affect the orderly sequence of events during development. Development in different animals is controlled by the same kinds of genes, which indicates that these animals have a shared evolutionary history.
- The accumulation of genetic changes since organisms diverged, or took separate evolutionary pathways, has modified the pattern of development in more complex vertebrate embryos.

9 Give an example of how evolutionary hypotheses are tested experimentally.

- David Reznick and John Endler have studied the effects of predation intensity on the evolution of guppy populations in the laboratory and in nature. Such experiments are a powerful way for investigators to test the underlying processes of natural selection.

TEST YOUR UNDERSTANDING

Know and Comprehend

1. Evolution is based on which of the following concepts? (a) organisms share a common origin (b) over time, organisms have diverged from a common ancestor (c) an animal's body parts can change over its lifetime, and these acquired changes are passed to the next generation (d) a and b (e) a, b, and c

2. Evolution is the accumulation of genetic changes within ___ over time. (a) individuals (b) populations (c) communities (d) a and b (e) a and c

3. Charles Darwin proposed that evolution could be explained by the differential reproductive success of organisms that resulted from their naturally occurring variation. Darwin called this process (a) coevolution (b) convergent evolution (c) natural selection (d) artificial selection (e) homoplasy

4. Which of the following is *not* part of Darwin's mechanism of evolution? (a) differential reproductive success (b) variation in a population (c) inheritance of acquired (nongenetic) traits (d) overproduction of offspring (e) struggle for existence

5. The evolution of beak size in the various species of Galápagos finches is associated with their (a) songs (b) diets (c) body sizes (d) predators (e) none of the preceding

6. The fossil record (a) usually occurs in sedimentary rock (b) sometimes appears fragmentary (c) is relatively complete for tropical rainforest organisms but incomplete for aquatic organisms (d) a and b (e) a, b, and c

7. In ___ the selecting agent is the environment, whereas in ___ the selecting agent is humans. (a) natural selection; convergent evolution (b) mutation; artificial selection (c) homoplasy;

homology (d) artificial selection; natural selection (e) natural selection; artificial selection

8. Aardvarks, anteaters, and pangolins are only distantly related but are similar in structure and form as a result of (a) homology (b) convergent evolution (c) biogeography (d) vestigial structures (e) artificial selection

9. The species of the Galápagos Islands (a) are similar to those on other islands at the same latitude (b) are similar to those on the South American mainland (c) are identical to those on other islands at the same latitude (d) are identical to those on the South American mainland (e) are similar to those on both the African and South American mainlands

Apply and Analyze

10. **CONNECT** In what way does the modern synthesis strengthen scientific understanding of evolution? (a) is based on the sequence of fossils in rock layers (b) uses genetics to explain the source of hereditary variation that is essential to natural selection (c) was first proposed by ancient Greek scholars (d) considers the influence of the geographic distribution of organisms on their evolution (e) is reinforced by homologies that are explained by common descent

11. **CONNECT** What types of gene changes are most associated with the evolution of new anatomical features in a population?

Evaluate and Synthesize

12. **EVOLUTION LINK** The use of model organisms such as the mouse for research and biomedical testing of human diseases is based on the assumption that all organisms share a common ancestor. On what evidence is this assumption based?

13. **EVOLUTION LINK** Charles Darwin once said, "It is not the strongest of the species that survive, nor the most intelligent, but the one most responsive to change." Explain what he meant.

14. **EVOLUTION LINK** Write short paragraphs explaining each of the following statements:
    a. Natural selection chooses from among the individuals in a population those most suited to *current* environmental conditions. It does not guarantee survival under future conditions.
    b. Individuals do not evolve, but populations do.
    c. The organisms that exist today do so because their ancestors had traits that allowed them and their offspring to thrive.

d. At the molecular level, evolution can take place by the replacement of one nucleotide by another.

e. Evolution is said to have occurred within a population when measurable genetic changes are detected.

15. **EVOLUTION LINK** Although most salamanders have four legs, a few species that live in shallow water lack hind limbs and have extremely tiny forelimbs (*see photograph*). Develop a hypothesis to explain how limbless salamanders came about according to Darwin's mechanism of evolution by natural selection. How could you test your hypothesis?

Suzanne L. & Joseph T. Collins/Science

The narrow-striped dwarf siren (*Pseudobranchus striatus axanthus*) is an aquatic salamander that resembles an eel. It is native to Florida.

16. **INTERPRET DATA** Which of the primates in Figure 18-18 is the most distantly related to humans? Explain your answer.

17. **SCIENCE, TECHNOLOGY, AND SOCIETY** Farmers often face a predicament, known as the pesticide treadmill, in which the cost of applying pesticides increases because the pesticides have to be applied more frequently or in larger doses, while their effectiveness decreases. Based on what you have learned in this chapter, offer an explanation for why the pesticides lose their effectiveness.

aplia To access course materials, such as Aplia and other companion resources, please visit **www.cengagebrain.com.**

Genetic variation in snail shells. Shown are the shell patterns and colors in a single snail species (*Cepaea nemoralis*), native to Scotland. Variation in shell color may have adaptive value in these snails because some colors predominate in cooler environments, whereas other colors are more common in warmer habitats.

John Prior Images/Alamy

KEY CONCEPTS

19.1 The genotype, phenotype, and allele frequencies of a population's gene pool (all alleles of all loci) can be calculated.

19.2 The Hardy–Weinberg principle predicts allele and genotype frequencies for a population that is not evolving.

19.3 Natural selection, as well as nonrandom mating, mutation, genetic drift, and gene flow, are forces that drive changes in a population's gene pool over successive generations (microevolution). Modes of natural selection include stabilizing selection, directional selection, and disruptive selection.

19.4 Populations typically exhibit genetic variation, which may include genetic polymorphism, balanced polymorphism, or geographic variation.

As you learned in Chapter 18, evolution occurs in populations, not individuals. Although natural selection acts on individuals, causing differential survival and reproduction, individuals themselves do not evolve during their lifetimes. Evolutionary change, which includes modifications in structure, physiology, ecology, and behavior, is inherited from one generation to the next. Although Darwin recognized that evolution occurs in populations, he did not understand how the attributes of organisms are passed to successive generations. One of the most significant advances in biology since Darwin's time has been the demonstration of the genetic basis of evolution.

Recall from Chapter 18 that a **population** consists of all individuals of the same species that live in a particular place at the same time. Individuals within a population vary in many recognizable characters. A population of snails, for example, may vary in shell size, weight, or color (see photograph). Some of this variation is due to heredity, and some is due to environment (nonheritable variation), such as the individual differences observed in the pink color of flamingos, which is partly attributable to differences in diet.

Biologists study variation in a particular character (characteristic) by taking measurements of that character in a population. By comparing the character in parents and offspring, it is possible to estimate the amount of observed variation that is genetic, as represented by the number, frequency, and kinds of alleles in a population. (Recall from Chapter 11 that an **allele** is one of two or more alternate forms of a gene. Alleles occupy corresponding positions, or **loci,** on homologous chromosomes.)

This chapter will help you develop an understanding of the importance of genetic variation as the raw material for evolution and of the basic concepts of **population genetics,** the study of genetic variability within a population and of the evolutionary forces that act on it. Population genetics represents an extension of Gregor Mendel's principles of inheritance (see Chapter 11). You will learn how to distinguish genetic equilibrium from evolutionary change and how to assess the roles of the five factors responsible for evolutionary change: nonrandom mating, mutation, genetic drift, gene flow, and natural selection.

19.1 GENOTYPE, PHENOTYPE, AND ALLELE FREQUENCIES

LEARNING OBJECTIVES

1 Define what is meant by a population's gene pool.
2 Distinguish among genotype, phenotype, and allele frequencies.

Each population possesses a **gene pool,** which includes all the alleles for all the loci present in the population. Because diploid organisms have a maximum of two different alleles at each genetic locus, a single individual typically has only a small fraction of the alleles present in a population's gene pool. The genetic variation that is evident among individuals in a given population indicates that each individual has a different subset of the alleles in the gene pool.

The evolution of populations is best understood in terms of genotype, phenotype, and allele frequencies. Suppose, for example, that all 1000 individuals of a hypothetical population have their genotypes tested with respect to a single locus, with the following results:

GENOTYPE	NUMBER	GENOTYPE FREQUENCY
AA	490	0.49
Aa	420	0.42
aa	90	0.09
Total	1000	1.00

Each **genotype frequency** is the proportion of a particular genotype in the population. Genotype frequency is usually expressed as a decimal fraction, and the sum of all genotype frequencies is 1.0 (somewhat like probabilities, which were discussed in Chapter 11). For example, the genotype frequency for the *Aa* genotype is 420 ÷ 1000 = 0.42.

A **phenotype frequency** is the proportion of a particular phenotype in the population. If each genotype corresponds to a specific phenotype, the phenotype and genotype frequencies are the same. If allele *A* is dominant over allele *a,* however, the phenotype frequencies in our hypothetical population would be the following:

PHENOTYPE	NUMBER	PHENOTYPE FREQUENCY
Dominant	910	0.91
Recessive	90	0.09
Total	1000	1.00

In this example, the dominant phenotype is the sum of two genotypes, *AA* and *Aa,* so the number 910 is obtained by adding 490 and 420.

An **allele frequency** is the proportion of a specific allele (i.e., of *A* or *a*) in a particular population. As mentioned earlier, each individual, being diploid, has two alleles at each genetic locus. Because we started with a population of 1000 individuals, we must account for a total of 2000 alleles. The 490 *AA* individuals have 980 *A* alleles, and the 420 *Aa* individuals have 420 *A* alleles, for a total of 1400 *A* alleles in the population. The total number of *a* alleles in the population is 420 + 90 + 90 = 600. Now it is easy to calculate allele frequencies:

ALLELE	NUMBER	ALLELE FREQUENCY
A	1400	0.7
a	600	0.3
Total	2000	1.0

CHECKPOINT 19.1

- *Does the term* gene pool *apply to individuals, populations, or both?*
- **CONNECT** *Can the frequencies of all genotypes in a population be determined directly with respect to a locus that has only two alleles: one dominant and the other recessive?*
- **INTERPRET DATA** *In a human population of 1000, 840 are tongue rollers (360 TT and 480 Tt), and 160 are not tongue rollers (tt). What is the frequency of the dominant allele (T) in the population?*

19.2 THE HARDY–WEINBERG PRINCIPLE

LEARNING OBJECTIVES

3 Discuss the significance of the Hardy–Weinberg principle as it relates to evolution and list the five conditions required for genetic equilibrium.
4 Use the Hardy–Weinberg principle to solve problems involving populations.

In the example just discussed, we observe that only 90 of the 1000 individuals in the population exhibit the recessive phenotype characteristic of the genotype *aa.* The remaining 910 individuals exhibit the dominant phenotype and are either *AA* or *Aa.* You might assume that after many generations, genetic recombination during sexual reproduction would cause the dominant allele to become more common in the population. You might also assume that the recessive allele would eventually disappear altogether. These assumptions were common among many biologists early in the 20th century. However, they are incorrect because the frequencies of alleles and genotypes do not change from generation to generation unless influenced by outside factors (discussed later in this chapter).

A population whose allele and genotype frequencies do not change from generation to generation is said to be at **genetic equilibrium.** Such a population, with no net change in allele or genotype frequencies over time, is not undergoing evolutionary change. A population that is at genetic equilibrium is not evolving with respect to the locus being studied. However, if allele frequencies change over successive generations, evolution is occurring.

The explanation for the stability of successive generations in populations at genetic equilibrium was provided independently by Godfrey Hardy, an English mathematician, and Wilhelm Weinberg, a German physician, in 1908. They pointed out that the expected frequencies of various genotypes in a population can be described mathematically. The resulting **Hardy–Weinberg principle** shows that if the population is large, the process of inheritance does not by itself cause changes in allele frequencies. It also explains why dominant alleles are not necessarily more common than recessive ones. The Hardy–Weinberg principle represents an ideal situation that seldom occurs in the natural world. However, it is useful because it provides a model to help us understand the real world. Knowledge of the Hardy–Weinberg principle is essential to understanding the mechanisms of evolutionary change in sexually reproducing populations.

We now expand our original example to illustrate the Hardy–Weinberg principle. Keep in mind as we go through these calculations that in most cases we know only the phenotype frequencies. When alleles are dominant and recessive, it is usually impossible to visually distinguish heterozygous individuals from homozygous dominant individuals. The Hardy–Weinberg principle lets us use phenotype frequencies to calculate the expected genotype frequencies and allele frequencies, assuming that we have a clear understanding of the genetic basis for the character under study.

As mentioned earlier, the frequency of either allele, A or a, is represented by a number that ranges from 0 to 1. An allele that is totally absent from the population has a frequency of zero. If all the alleles of a given locus are the same in the population, the frequency of that allele is 1.

Because only two alleles, A and a, exist at the locus in our example, the sum of their frequencies must equal 1. If we let p represent the frequency of the dominant (A) allele in the population and q the frequency of the recessive (a) allele, we can summarize their relationship with a simple binomial equation, $p + q = 1$. When we know the value of either p or q, we can calculate the value of the other: $p = 1 - q$, and $q = 1 - p$.

Squaring both sides of $p + q = 1$ results in $(p + q)^2 = 1$. This equation can be expanded to describe the relationship of the allele frequencies to the genotypes in the population. When it is expanded, we obtain the frequency of the offspring genotypes:

p^2	+	$2pq$	+	q^2	=	1
Frequency of AA		Frequency of Aa		Frequency of aa		All individuals in the population

We always begin Hardy–Weinberg calculations by determining the frequency of the homozygous recessive genotype. We had 90 homozygous recessive individuals in our population of 1000, so we can infer that the frequency of the aa genotype, q^2, is 90/1000, or 0.09. Because q^2 equals 0.09, q (the frequency of the recessive a allele) is equal to the square root of 0.09, or 0.3. From the relationship between p and q, we conclude that the frequency of the dominant A allele, p, is $1 - q = 1 - 0.3 = 0.7$.

Genotypes	AA	Aa	aa
Frequency of genotypes in population	0.49	0.42 (0.21 + 0.21)	0.09
Frequency of alleles in gametes	$A = 0.49 + 0.21$ = 0.7		$a = 0.21 + 0.09$ = 0.3

(a) Genotype and allele frequencies. The figure illustrates how to calculate frequencies of the alleles A and a in the gametes produced by each genotype.

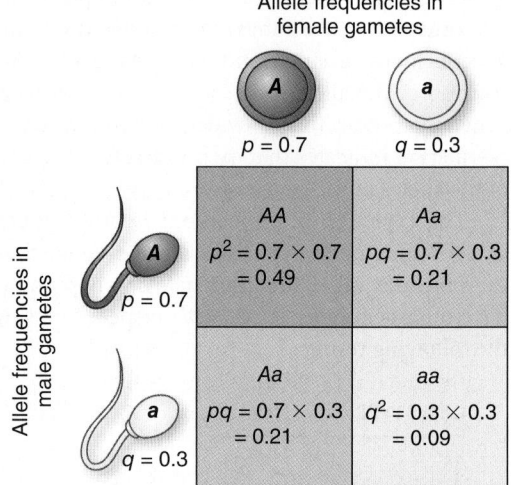

(b) Segregation of alleles and random fertilization. When eggs and sperm containing A or a alleles unite randomly, the frequency of each of the possible genotypes (AA, Aa, aa) among the offspring is calculated by multiplying the frequencies of the alleles A and a in eggs and sperm.

Figure 19-1 *Animation* **The Hardy–Weinberg principle**
© Cengage Learning

Given this information, we can calculate the frequency of homozygous dominant (AA) individuals: $p^2 = 0.7 \times 0.7 = 0.49$ (**FIG. 19-1**). The expected frequency of heterozygous individuals (Aa) would be $2pq = 2 \times 0.7 \times 0.3 = 0.42$. Thus, approximately 490 individuals are expected to be homozygous dominant, and approximately 420 are expected to be heterozygous. Note that the sum of homozygous dominant and heterozygous individuals equals 910, the number of individuals showing the dominant phenotype with which we started.

Any population in which the distribution of genotypes conforms to the relation $p^2 + 2pq + q^2 = 1$, whatever the absolute values for p and q may be, is at genetic equilibrium. The Hardy–Weinberg principle allows biologists to calculate allele frequencies in a given population if we know the genotype frequencies and vice versa. These values provide a basis of comparison with a population's allele or genotype frequencies in succeeding generations. During that time, if the allele or genotype frequencies deviate from the values predicted by the Hardy–Weinberg principle, the population is evolving.

Genetic equilibrium occurs if certain conditions are met

The Hardy–Weinberg principle of genetic equilibrium tells us what to expect when a sexually reproducing population is not evolving. The relative proportions of alleles and genotypes in successive generations will always be the same provided that the following five conditions are met:

1. *Random mating.* In unrestricted random mating, each individual in a population has an equal chance of mating with any individual of the opposite sex. In our example, the individuals represented by the genotypes *AA, Aa,* and *aa* must mate with one another at random and must not select their mates on the basis of genotype or any other factors that result in nonrandom mating.
2. *No net mutations.* There must be no mutations that convert *A* into *a* or vice versa. That is, the frequencies of *A* and *a* in the population must not change because of mutations.
3. *Large population size.* Allele frequencies in a small population are more likely to change by random fluctuations (i.e., by genetic drift, which is discussed later) than are allele frequencies in a large population.
4. *No migration.* There can be no exchange of alleles with other populations that might have different allele frequencies. In other words, there can be no migration of individuals into or out of a population.[1]
5. *No natural selection.* If natural selection is occurring, certain phenotypes (and their corresponding genotypes) are favored over others. These more successful genotypes are said to have greater **fitness,** which is the relative ability to make a genetic contribution to subsequent generations. Consequently, the allele frequencies will change, and the population will evolve.

Human MN blood groups are a valuable illustration of the Hardy–Weinberg principle

Humans have dozens of antigens on the surfaces of their blood cells. (An *antigen* is a molecule, usually a protein or carbohydrate, that is recognized as foreign by cells of another organism's immune system.) One group of antigens, designated the MN blood group, stimulates the production of antibodies when injected into rabbits or guinea pigs. However, humans do not produce antibodies for M and N, so the MN blood group is not medically important, for example, when giving blood transfusions. (Recall the discussion of the medically important ABO alleles in Chapter 11.) The MN blood group is of interest to population geneticists because the alleles for the MN blood group, usually designated *M* and *N*, are codominant (genotype *MM* produces antigen M only, genotype *NN* produces antigen

---

[1] Note that evolutionary biologists use the term *migration,* not in its ordinary sense of periodic or seasonal movement of individuals from one location to another but in reference to a movement of individuals that results in a transfer of alleles from one population to another.

N only, and the heterozygous genotype *MN* produces both antigens). This property allows population geneticists to directly observe all three possible genotype frequencies and compare them with calculated frequencies. The following data are typical of the MN blood group in people in the United States:

GENOTYPE	OBSERVED
MM	320
MN	480
NN	200
Total	1000

Because 1000 diploid individuals are in the sample, there are a total of 2000 alleles. The frequency of *M* alleles in the population is $p = (2 \times 320 + 480) \div 2000 = 0.56$. The frequency of *N* alleles in the population is $q = (2 \times 200 + 480) \div 2000 = 0.44$. As a quick check, the sum of the frequencies should equal 1. Does it?

If this population is in genetic equilibrium, the expected *MM* genotype frequency is $p^2 = (0.56)^2 = 0.31$. The expected *MN* genotype frequency is $2pq = 2 \times 0.56 \times 0.44 = 0.49$. The expected *NN* genotype frequency is $q^2 = (0.44)^2 = 0.19$. As a quick check, the sum of the three genotype frequencies should equal 1. Does it? You can use the calculated genotype frequencies to determine how many individuals in a population of 1000 should have the expected genotype frequencies. By comparing the expected numbers with the actual results observed, you see how closely the population is to genetic equilibrium. Simply multiply each genotype frequency by 1000:

GENOTYPE	OBSERVED	EXPECTED
MM	320	313.6
MN	480	492.8
NN	200	193.6
Total	1000	1000.0

The expected numbers closely match the observed numbers, indicating that the *MN* blood groups in the human population are almost at genetic equilibrium. This finding is not surprising because the lack of medical significance suggests that the *MN* characteristic is not subject to natural selection and that it does not produce a visible phenotype that might affect random mating.

CHECKPOINT 19.2

- **INTERPRET DATA** *In a population at genetic equilibrium, the frequency of the homozygous recessive genotype (tt) is 0.16. What are the allele frequencies of T and t, and what are the expected frequencies of the TT and Tt genotypes?*

- **INTERPRET DATA** *In a population at genetic equilibrium, the frequency of the dominant phenotype is 0.96. What are the frequencies of the dominant (A) and recessive (a) alleles, and what are the expected frequencies of the AA, Aa, and aa genotypes?*

- **INTERPRET DATA** *The genotype frequencies of a population are determined to be 0.6 BB, 0.0 Bb, and 0.4 bb. Is it likely that this population meets all the conditions required for genetic equilibrium?*

19.3 MICROEVOLUTION

LEARNING OBJECTIVES

5 Define *microevolution.*

6 Discuss how each of the following microevolutionary forces alters allele frequencies in populations: nonrandom mating, mutation, genetic drift, gene flow, and natural selection.

7 Distinguish among stabilizing selection, directional selection, and disruptive selection.

Evolution represents a departure from the Hardy–Weinberg principle of genetic equilibrium. The degree of departure between the observed allele or genotype frequencies and those expected by the Hardy–Weinberg principle indicates the amount of evolutionary change. This type of evolution—generation-to-generation changes in allele or genotype frequencies *within* a population—is sometimes referred to as **microevolution** because it often involves relatively small or minor changes, usually over a few generations.

Changes in the allele frequencies of a population result from five microevolutionary processes: nonrandom mating, mutation, genetic drift, gene flow, and natural selection. These microevolutionary processes are the opposite of the conditions that must be met if a population is in genetic equilibrium. When one or more of these processes acts on a population, allele or genotype frequencies change from one generation to the next.

Nonrandom mating changes genotype frequencies

When individuals select mates on the basis of phenotype (thereby selecting the corresponding genotype), they bring about evolutionary change in the population. Two examples of nonrandom mating are inbreeding and assortative mating.

In many populations, individuals mate more often with close neighbors than with more distant members of the population. As a result, neighbors tend to be more closely related—that is, genetically similar—to one another. The mating of genetically similar individuals who are more closely related than if they had been chosen at random from the entire population is known as **inbreeding.** Although inbreeding does not change the overall allele frequency, the frequency of homozygous genotypes increases with each successive generation of inbreeding. The most extreme example of inbreeding is self-fertilization, which is particularly common in certain plants.

Inbreeding does not appear to be detrimental in some populations, but in others it causes **inbreeding depression,** in which inbred individuals have lower fitness than those not inbred. Fitness is usually measured as the average number of surviving offspring of one genotype compared with the average number of surviving offspring of competing genotypes. Inbreeding depression, as evidenced by fertility declines and high juvenile mortality, is thought to be caused by the expression of harmful recessive alleles because homozygosity increases with inbreeding.

Several studies have provided direct evidence of the deleterious consequences of inbreeding in nature. For example, white-footed mice (*Peromyscus leucopus*) were taken from a field and used to develop both inbred and non-inbred populations in the laboratory. When these laboratory-bred populations were returned to nature, their survivorship was estimated from release–recapture data. The non-inbred mice had a statistically significant higher rate of survival than the inbred mice (FIG. 19-2). It is not known why the inbred mice had a lower survival rate. Some possibilities include higher disease susceptibility, poorer ability to evade predators, less ability to find food, and less ability to win fights with other white-footed mice.

Assortative mating, in which individuals select mates by their phenotypes, is another example of nonrandom mating. For example, biologists selected two phenotypes—high bristle number and low bristle number—in a fruit fly (*Drosophila melanogaster*) population. Although the researchers made no effort to control mating, they observed that the flies preferentially mated with those of similar phenotypes. Females with high bristle number tended to mate with males with high bristle number, and females with low bristle number tended to mate with males with low bristle number. Such selection of mates with the same phenotype is known as *positive assortative mating* (as opposed to the less common phenomenon, *negative assortative mating,* in which mates with opposite phenotypes are selected).

Positive assortative mating is practiced in many human societies, in which men and women tend to marry individuals like themselves in such characteristics as height or intelligence. Like inbreeding, assortative mating usually increases homozygosity at the expense of heterozygosity in the population and does not change the overall allele frequencies in the population. However, assortative mating changes genotype frequencies only at the loci involved in mate choice, whereas inbreeding affects genotype frequencies in the entire genome.

Mutation increases variation within a population

Variation is introduced into a population through **mutation,** which is an unpredictable change in deoxyribonucleic acid (DNA). As discussed in Chapter 13, mutations, which are the source of all new alleles, result from (1) a change in the nucleotide base pairs of a gene, (2) a rearrangement of genes within chromosomes so that their interactions produce different effects, or (3) a change in chromosome structure. Mutations occur unpredictably and spontaneously. A particular locus may have a DNA sequence that causes certain types of mutations to occur more frequently than others. The rate of mutation appears relatively constant for a particular locus but may vary by several orders of magnitude among genes within a single species and among different species.

Not all mutations pass from one generation to the next. Those occurring in somatic (body) cells are not inherited. When an individual with such a mutation dies, the mutation is lost. Some mutations, however, occur in reproductive cells. These mutations may or may not overtly affect the offspring because at least some of the DNA in a cell appears to be nonfunctional.

Does inbreeding affect survival?

HYPOTHESIS: Non-inbred white-footed mice (*Peromyscus leucopus*) will have a survival advantage over inbred mice in a natural environment.

EXPERIMENT: Field-captured mice were used to establish inbred and non-inbred laboratory populations. The mice were marked and then released back into the field population. The population was then sampled by capturing mice, noting the number of inbred and non-inbred mice, and re-releasing the mice. The x-axis represents the 10-week period during which the samples were collected (note that no samples were collected in weeks 4 through 7). Values on the y-axis are the estimated proportion of mice that survived since the last time the population was sampled. For example, the value of 0.2 in week 3 for the inbred mice means that 20% of the inbred mice that were alive at the beginning of the week survived through that week.

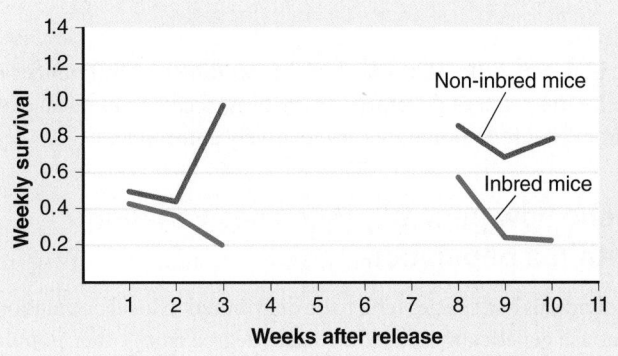

RESULTS AND CONCLUSION: Non-inbred mice (*red*) had a consistently higher survival rate than inbred mice (*blue*). Thus, inbreeding had a negative effect on survival in this species in nature.

SOURCE: Adapted with permission from Jiménez, J.A., et al., "An Experimental Study of Inbreeding Depression in a Natural Habitat," *Science*, Vol. 266, Oct. 14, 1994. Copyright © 1994 American Association for the Advancement of Science.

Figure 19-2 Survival of inbred and non-inbred mice

CONNECT If the researchers estimated there were 200 inbred mice in the population at the beginning of week 3, how many would be estimated to be alive at the beginning of week 4?

Even if a mutation occurs in the DNA that codes for a polypeptide, it may still have little effect on the structure or function of that polypeptide (we discuss such *neutral variation* later in this chapter). However, when a polypeptide is sufficiently altered to change its function, the mutation is usually harmful. By acting against seriously abnormal phenotypes, natural selection eliminates or reduces to low frequencies the most harmful mutations. Mutations with small phenotypic effects, even if slightly harmful, have a better chance of being incorporated into the population, where at some later time, under different environmental conditions, they may produce phenotypes that are useful or adaptive.

The "needs" of a population do not determine what mutations will occur. Consider a population living in an increasingly dry environment. A mutation producing a new allele that helps an individual adapt to dry conditions is no more likely to occur than one for adapting to wet conditions or one with no relationship to the changing environment. The production of new mutations simply increases the genetic variability that is acted on by natural selection and therefore increases the potential for new adaptations.

Mutation by itself causes small deviations in allele frequencies from those predicted by the Hardy–Weinberg principle. Although allele frequencies may be changed by mutation, these changes are typically several orders of magnitude smaller than changes caused by other evolutionary forces, such as genetic drift. Mutation is usually negligible as an evolutionary force, but it is essential to the evolutionary process because it is the ultimate source of genetic variation.

In genetic drift random events change allele frequencies

The size of a population has important effects on allele frequencies because random events, or chance, tend to cause changes of relatively greater magnitude in a small population. If a population consists of only a few individuals, an allele present at a low frequency in the population could be completely lost by chance. Such an event would be unlikely in a large population. For example, consider two populations, one with 10,000 individuals and one with 10 individuals. If an uncommon allele occurs at a frequency of 10%, or 0.1, in both populations, then 1900 individuals in the large population have the allele.[2] That same frequency, 0.1, in the smaller population means that only about 2 individuals have the allele.[3] From this exercise, it is easy to see that there is a greater likelihood of losing the rare allele from the smaller population than from the larger one. Predators, for example, might happen to kill one or two individuals possessing the uncommon allele in the smaller population purely by chance, so these individuals would leave no offspring.

The production of random evolutionary changes in small breeding populations is known as **genetic drift.** Genetic drift results in changes in allele frequencies in a population from one generation to another. One allele may be eliminated from the population purely by chance, regardless of whether that allele is beneficial, harmful, or of no particular advantage or disadvantage. Thus, genetic drift decreases genetic variation *within* a population, although it tends to increase genetic differences *among* different populations.

When bottlenecks occur, genetic drift becomes a major evolutionary force Because of fluctuations in the environment, such as depletion in food supply or an outbreak of disease, a population may rapidly and markedly decrease from time to time. The population is said to go through a bottleneck

---

[2] $2pq + q^2 = 2(0.9)(0.1) + (0.1)^2 = 0.18 + 0.01 = 0.19; 0.19 \times 10,000 = 1900$
[3] $0.19 \times 10 = 1.9$

during which genetic drift can occur in the small population of survivors. As the population again increases in size, many allele frequencies may be quite different from those in the population preceding the decline.

Scientists hypothesize that genetic variation in the cheetah was considerably reduced by a bottleneck at the end of the last Ice Age, some 10,000 years ago. Cheetahs nearly became extinct, perhaps from overhunting by humans. The few surviving cheetahs had greatly reduced genetic variability, and as a result, the cheetah population today is so genetically uniform that unrelated cheetahs can accept skin grafts from one another. (Normally, only identical twins accept skin grafts so readily.)

The founder effect occurs when a few "founders" establish a new colony When one or a few individuals from a large population establish, or found, a colony (as when a few birds separate from the rest of the flock and fly to a new area), they bring with them only a small fraction of the genetic variation present in the original population. As a result, the only alleles among their descendants will be those of the colonizers. Typically, the allele frequencies in the newly founded population are quite different from those of the parent population. The genetic drift that results when a small number of individuals from a large population found a new colony is called the **founder effect.**

The Finnish people may illustrate the founder effect (**FIG. 19-3**). Geneticists who sampled DNA from Finns and from the European population at large found that Finns exhibit considerably less genetic variation than other Europeans. This evidence supports the hypothesis that Finns are descended from a small group of people who settled about 4000 years ago in the area that is now Finland and, because of the geography, remained separate from other European societies for centuries. The subpopulation in eastern Finland is especially homogeneous because it existed in relative isolation after being established by only several hundred founders in the 1500s.

The founder effect and population bottlenecks have also apparently affected the genetic composition of the population of Iceland. Most of Iceland's 322,000 citizens are descended from a small group of Norse and Celtic people who settled that island in the ninth century. Isolation and population bottlenecks due to disease and natural disasters have contributed to the relative genetic homogeneity of the Icelandic people.

The founder effect can be of medical importance. For example, by chance one of the approximately two hundred founders of the Amish population of Pennsylvania carried a recessive allele that, when homozygous, is responsible for a form of dwarfism, Ellis–van Creveld syndrome. Although this allele is rare in the general population (frequency about 0.001), today it is relatively common in the Amish population (frequency about 0.07).

The Finnish and Icelandic populations, as well as certain Amish populations, are being studied by researchers who employ sophisticated technologies such as genome-wide association studies (GWAS; see Chapters 15 and 16) in the quest to identify genetic contributions to a large number of diseases. The task is simplified in these studies because confounding factors

Figure 19-3 Finns and the founder effect
The Finnish people are thought to have descended from a small founding population that remained separate from the rest of Europe for centuries.

such as variability in the rest of the genome and differences in such environmental factors as nutrition, access to medical care, and exposure to pollutants are relatively controlled.

Gene flow generally increases variation within a population

Individuals of a species tend to be distributed in local populations that are genetically isolated to some degree from other populations. For example, the bullfrogs of one pond form a population separated from those in an adjacent pond. Some exchanges occur by migration between ponds, but the frogs in one pond are much more likely to mate with those in the same pond. Because each population is isolated to some extent from other populations, each has distinct genetic traits and gene pools.

The migration of breeding individuals between populations causes a corresponding movement of alleles, or **gene flow,** that has significant evolutionary consequences. As alleles flow from one population to another, they usually increase the amount of genetic variability within the recipient population. If sufficient gene flow occurs between two populations, they become more similar genetically. Because gene flow reduces the amount of variation between two populations, it tends to counteract the effects of natural selection and genetic drift, both of which often cause populations to become increasingly distinct.

If migration by members of a population is considerable and if populations differ in their allele frequencies, significant genetic changes occur in local populations. For example, by ten thousand years ago modern humans occupied almost all Earth's major land areas except a few islands and Antarctica. Because the population density was low in most locations, the small, isolated human populations underwent random genetic drift and natural selection. More recently (during the past three hundred years or so), major migrations have increased gene flow, significantly altering allele frequencies within previously isolated human populations.

Stabilizing, directional, and disruptive selection can change the distribution of phenotypes in a population.

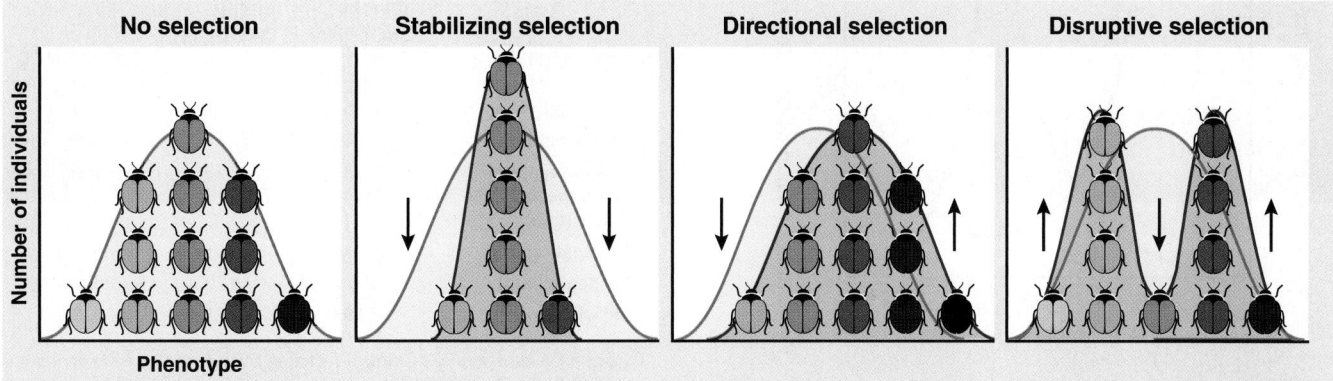

| No selection | Stabilizing selection | Directional selection | Disruptive selection |

(a) A character that is under polygenic control (in this example, wing colors in a hypothetical population of beetles) exhibits a normal distribution of phenotypes in the absence of selection.

(b) As a result of stabilizing selection, which trims off extreme phenotypes, variation about the mean is reduced.

(c) Directional selection shifts the curve in one direction, changing the average value of the character.

(d) Disruptive selection, which trims off intermediate phenotypes, results in two or more peaks.

Figure 19-4 *Animation* **Modes of selection**

The *blue screen* represents the distribution of individuals by phenotype (in this example, color variation) in the original population. The *purple screen* represents the distribution by phenotype in the evolved population. The *arrows* represent the pressure of natural selection on the phenotypes.

PREDICT If a population is experiencing directional selection with respect to wing color, what type of selection would you expect it to experience with respect to antenna length?

© Cengage Learning

Natural selection changes allele frequencies in a way that increases adaptation

Natural selection is the mechanism of evolution first proposed by Darwin in which members of a population that are more successfully adapted to the environment have greater fitness; that is, they are more likely to survive and reproduce (see Chapter 18). Over successive generations, the proportion of favorable alleles increases in the population. In contrast with other microevolutionary processes (nonrandom mating, mutation, genetic drift, and gene flow), natural selection leads to adaptive evolutionary change. Natural selection not only explains why organisms are well adapted to the environments in which they live, but it also helps account for the remarkable diversity of life. Natural selection enables populations to change, thereby adapting to different environments and different ways of life.

Natural selection results in the differential reproduction of individuals with different traits, or phenotypes (and therefore different genotypes), in response to the environment. Natural selection preserves individuals with favorable phenotypes and eliminates those with unfavorable phenotypes. Individuals that survive and produce fertile offspring have a selective advantage.

The mechanism of natural selection does not develop a "perfect" organism. Rather, it weeds out those individuals whose phenotypes are less adapted to environmental challenges, while allowing better-adapted individuals to survive and pass their alleles to their offspring. By reducing the frequency of alleles that result in the expression of less favorable traits, the probability is increased that favorable alleles responsible for an adaptation will come together in the offspring.

Natural selection operates on an organism's phenotype
Natural selection does not act directly on an organism's genotype. Instead, it acts on the phenotype, which is, at least in part, an expression of the genotype. The phenotype represents an interaction between the environment and all the alleles in the organism's genotype. It is rare that alleles of a single locus determine the phenotype, as Mendel originally observed in garden peas. Much more common is the interaction of alleles of several loci for the expression of a single phenotype. Many plant and animal characteristics are under this type of *polygenic* control (see Chapter 11).

When characters are under polygenic control (as is human height), a range of phenotypes occurs, with most of the population located in the median range and fewer at either extreme. This arrangement is a normal distribution or standard bell curve (**FIG. 19-4a**; see also Fig. 11-20). Three kinds of selection—stabilizing, directional, and disruptive—cause changes in the normal distribution of phenotypes in a population. Although we consider each process separately, in nature their influences generally overlap.

Stabilizing Selection The process of natural selection associated with a population well adapted to its environment is known as **stabilizing selection.** Most populations are probably

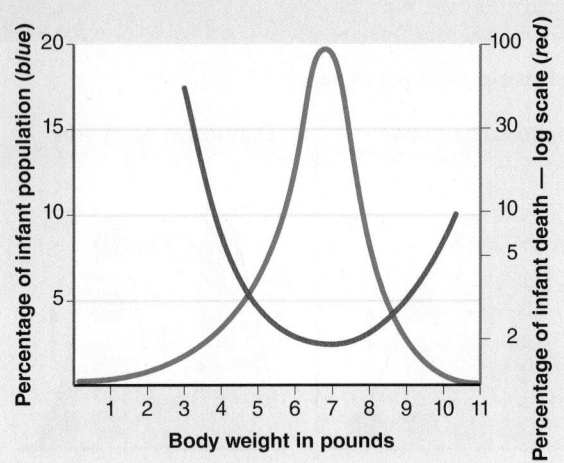

Figure 19-5 Stabilizing selection

The *blue curve* shows the distribution of birth weights in a sample of 13,730 infants. The *red curve* shows mortality (death) at each birth weight. Infants with very low or very high birth weights have higher death rates than infants of average weight. The optimum birth weight—that is, the one with the lowest mortality—is close to the average birth weight (about 7 lb). (Adapted from Cavalli-Sforza, L.L., and W.F. Bodmer, *The Genetics of Human Populations,* W.H. Freeman and Company, San Francisco, 1971.)

influenced by stabilizing forces most of the time. Stabilizing selection selects against phenotypic extremes. In other words, individuals with average, or intermediate, phenotypes are favored.

Because stabilizing selection tends to decrease variation by favoring individuals near the mean of the normal distribution at the expense of those at either extreme, the bell curve narrows (**FIG. 19-4b**). Although stabilizing selection decreases the amount of variation in a population, variation is rarely eliminated by this process because other microevolutionary processes act against a decrease in variation. For example, mutation is slowly but continually adding to the genetic variation within a population.

One of the most widely studied cases of stabilizing selection involves human birth weight, which is under polygenic control and is also influenced by environmental factors. Extensive data from hospitals have shown that infants born with intermediate weights are the ones most likely to survive (**FIG. 19-5**). Infants at either extreme (too small or too large) have higher rates of mortality. When newborn infants are too small, their body systems are immature, and when they are too large, they have difficult deliveries because they cannot pass as easily through the cervix and vagina. Stabilizing selection operates to reduce variability in birth weight so that it is close to the weight with the minimum mortality rate.

Directional Selection If an environment changes over time, **directional selection** may favor phenotypes at one of the extremes of the normal distribution (**FIG. 19-4c**). Over successive generations, one phenotype gradually replaces another. So, for example, if greater size is advantageous in a new environment, larger individuals will become increasingly common in the population. However, directional selection occurs only if alleles favored under the new circumstances are already present in the population.

Darwin's Galápagos finches provide an excellent example of directional selection. Since 1973, Peter Grant and Rosemary

TABLE 19-1 Population Changes in *Geospiza fortis* before and after the 1976–1977 Drought

CHARACTER	AVERAGE BEFORE DROUGHT (634)*	AVERAGE AFTER DROUGHT (135)*	DIFFERENCE
Weight (g)	16.06	17.13	+1.07
Wing length (mm)	67.88	68.87	+0.99
Tarsus (leg, just above the foot) length (mm)	19.08	19.29	+0.21
Bill length (mm)	10.63	10.95	+0.32
Bill depth (mm)	9.21	9.70	+0.49
Bill width (mm)	8.58	8.83	+0.25

From P.R. Grant and B.R. Grant, "Predicting Microevolutionary Responses to Directional Selection on Heritable Variation," *Evolution,* Vol. 49, 1995.

*Number of birds in sample.

Grant of Princeton University have studied the Galápagos finches. The Grants did a meticulous analysis of finch eating habits and beak sizes on Isla Daphne Major during three extended droughts (1976–1977, 1980, and 1982), one of which was followed by an extremely wet El Niño event (1983). During the droughts, the number of insects and small seeds declined, and large, heavy seeds became the finches' primary food source.

Many finches died during this time, and most of the survivors were larger birds whose beaks were larger and deeper. In a few generations, these larger birds became more common in the population (**TABLE 19-1**). After the El Niño event, however, smaller seeds became the primary food source, and smaller finches with average-sized beaks were favored. In this example, natural selection is directional: during the drought, natural selection operated in favor of the larger phenotype; whereas after the wet period, selection occurred in the opposite direction, favoring the smaller phenotype. The guppy populations studied in Venezuela and Trinidad (see Chapter 18) are another example of directional selection.

Directional selection by hunters has been observed in big-horn sheep (*Ovis canadensis*) in Canada. Sport hunters target large rams with rapidly growing horns for their trophy value (FIG. 19-6). Because these rams are typically killed prior to reaching their full reproductive potential, they contribute less to the gene pool; the result was a 25% decline in average ram weight and horn length over a period of 30 years. Paradoxically, the rams that would normally have the greatest fitness because of their size and ability to use their horns in competition with other males for mates become less fit when these attributes cause them to be singled out by hunters.

Disruptive Selection Sometimes extreme changes in the environment may favor two or more different phenotypes at the expense of the mean. That is, more than one phenotype may be favored in the new environment. **Disruptive selection** is a special type of directional selection in which there is a trend in several directions rather than just one (FIG. 19-4d). It results in a divergence, or splitting apart, of distinct groups of individuals within a population. Disruptive selection, which is relatively rare, selects against the average, or intermediate, phenotype.

Limited food supply during a severe drought caused a population of finches on another island in the Galápagos to experience disruptive selection. The finch population initially exhibited a variety of beak sizes and shapes. Because the only foods available on this island during the drought were wood-boring insects and seeds from cactus fruits, natural selection favored birds with beaks suitable for obtaining these types of food. Finches with longer beaks survived because they could open cactus fruits, and those with wider beaks survived because they could strip off tree bark to expose insects. However, finches with intermediate beaks could not use either food source efficiently and had a lower survival rate.

Natural selection induces change in the types and frequencies of alleles in populations only if there is pre-existing inherited variation. Genetic variation is the raw material for evolutionary change because it provides the diversity on which natural selection acts. Without genetic variation, evolution cannot occur. In the next section, we explore the genetic basis for variation that is acted on by natural selection.

CHECKPOINT 19.3

- Which microevolutionary force leads to adaptive changes in allele frequencies?
- CONNECT Why is mutation important to evolution if it is the microevolutionary force that generally has the smallest effect on allele frequencies?
- Which microevolutionary forces are most associated with an increase in variation within a population? among populations?
- Which microevolutionary force typically changes genotype frequencies without changing allele frequencies? Explain.

19.4 GENETIC VARIATION IN POPULATIONS

LEARNING OBJECTIVE

8 Describe the nature and extent of genetic variation, including genetic polymorphism, balanced polymorphism, neutral variation, and geographic variation.

Populations contain abundant genetic variation that was originally introduced by mutation. Sexual reproduction—with its associated crossing-over, independent assortment of chromosomes during meiosis, and random union of gametes—also contributes to genetic variation. The sexual process allows the variability introduced by mutation to be combined in new ways, which may be expressed as new phenotypes.

Genetic polymorphism can be studied in several ways

Genetic polymorphism, which is genetic variation among individuals in a population, is extensive in populations, although many of the genetic variants are present at low frequencies. Much of genetic polymorphism is not evident because it does not produce distinct phenotypes.

One way biologists estimate the total amount of genetic polymorphism in populations is by comparing the different forms of a particular protein. Each form consists of a slightly different amino acid sequence that is coded for by a different allele. For example, tissue extracts containing a particular enzyme may be analyzed for different individuals by gel electrophoresis. In *gel electrophoresis,* the enzymes are placed in slots on an agarose gel and an electric current is applied, causing each enzyme to migrate across the gel (see Fig. 15-8). Slight variations in amino acid sequences in the different forms of a particular enzyme cause each to migrate at a different rate, which can be detected using special stains or radioactive labels. Depending on the species studied, the percentage of enzymes that are polymorphic can range from about 15% to as much as 70%. In general,

Figure 19-6 Bighorn sheep
This ram exhibits the massive horns that characterize this species.

Harry Engels/Science Source

polymorphism of enzyme-coding loci tends to be greater in plants than in animals.

Determining the sequence of nucleotides in DNA from individuals in a population provides a *direct* estimate of genetic polymorphism that is sensitive enough to detect alleles that differ by only one nucleotide (**single nucleotide polymorphisms, or SNPs**). (DNA sequencing is discussed in Chapter 15). DNA sequencing of specific alleles in an increasing number of organisms, including humans, indicates that SNPs are common in most populations.

Recent studies have revealed that genetic changes in which segments of DNA have been gained or lost (compared to a reference genome) are common in model organisms such as mice and *Drosophila*, as well as humans. These **copy number variations, or CNVs,** can involve DNA segments that range from as few as five hundred to one thousand base pairs up to about two million base pairs. It is currently estimated that the average human may have a dozen or more CNVs in their genome. In most cases, there does not appear to be an associated phenotype, or the phenotype is benign, such as an effect on eye color. However, as more sophisticated technologies have been developed to support ongoing research, evidence is accumulating that links specific CNVs to increased risk for certain medically important conditions, such as autism, Alzheimer's disease, and susceptibility to HIV infection.

Balanced polymorphism exists for long periods

Balanced polymorphism is a special type of genetic polymorphism in which two or more alleles persist in a population over many generations as a result of natural selection. Heterozygote advantage and frequency-dependent selection are mechanisms that preserve balanced polymorphism.

Genetic variation may be maintained by heterozygote advantage We have seen that natural selection often eliminates unfavorable alleles from a population, whereas favorable alleles are retained. However, natural selection sometimes helps maintain genetic diversity. In some cases, natural selection may even maintain a population's alleles that are unfavorable in the homozygous state. That happens, for example, when the heterozygote *Aa* has a higher degree of fitness than either homozygote *AA* or *aa*. This phenomenon, known as **heterozygote advantage,** is demonstrated in humans by the selective advantage of heterozygous carriers of the sickle cell allele.

The mutant allele (Hb^S) for sickle cell anemia produces an altered hemoglobin that deforms or sickles the red blood cells, making them more likely to form dangerous blockages in capillaries and to be destroyed in the liver, spleen, or bone marrow (discussed in Chapter 16). People who are homozygous for the sickle cell allele ($Hb^S Hb^S$) usually die at an early age if medical treatment is not available.

Heterozygous individuals carry alleles for both normal (Hb^A) and sickle cell hemoglobin. The heterozygous condition ($Hb^A Hb^S$) makes a person more resistant to a type of severe malaria (caused by the parasite *Plasmodium falciparum*) than people who are homozygous for the normal hemoglobin allele ($Hb^A Hb^A$). In a heterozygous individual, each allele produces its own specific kind of hemoglobin, and the red blood cells contain the two kinds in roughly equivalent amounts. Such cells sickle much less readily than cells containing only the Hb^S allele. They are more resistant to infection by the malaria-causing parasite, which lives in red blood cells, than are the red blood cells containing only normal hemoglobin.

Where malaria is a problem, each of the two types of homozygous individuals is at a disadvantage. Those homozygous for the sickle cell allele are likely to die of sickle cell anemia, whereas those homozygous for the normal allele may die of malaria. The heterozygote is therefore more fit than either homozygote. In parts of Africa, the Middle East, and southern Asia where falciparum malaria is prevalent, heterozygous individuals survive in greater numbers than either homozygote (**FIG. 19-7**). The Hb^S allele is

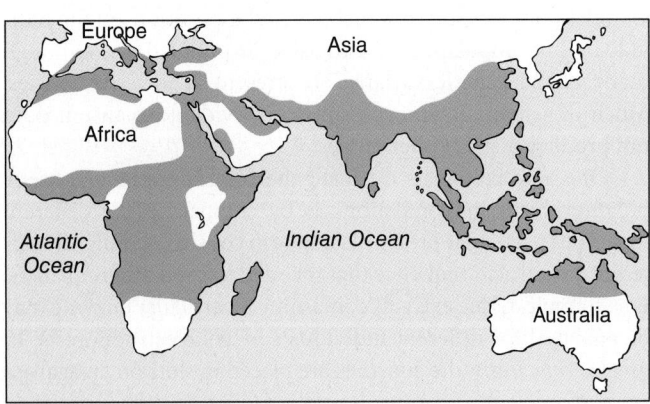

(a) The geographic distribution of falciparum malaria (*green*).

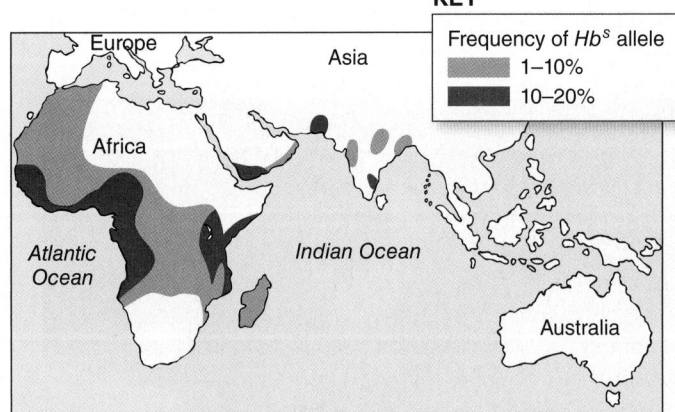

(b) The distribution of the sickle cell anemia allele (*red* and *orange*).

Figure 19-7 Heterozygote advantage

Over large geographic areas, the distribution of individuals heterozygous for the sickle cell (Hb^S) allele is similar to the distribution of falciparum malaria. This evidence for greater fitness of heterozygous individuals in malarial regions supports the hypothesis of heterozygote advantage. (Adapted from Allison, A.C., "Protection Afforded by Sickle-Cell Traits against Subtertian Malarial Infection," *British Medical Journal,* Vol. 1, 1954.)

maintained at a high frequency in the population, even though the homozygous recessive condition is almost always lethal.

What happens to the frequency of Hb^S alleles in Africans and others who possess it when they migrate to the United States and other countries with few cases of malaria? As might be expected, the frequency of the Hb^S allele gradually declines in such populations, possibly because it confers a selective disadvantage by causing sickle cell anemia in homozygous individuals but no longer confers a selective advantage by preventing malaria in heterozygous individuals. The Hb^S allele never disappears from the population, however, because it is "hidden" from strong selection in heterozygous individuals and because it is reintroduced into the population by gene flow from the African population.

Genetic variation may be maintained by frequency-dependent selection Thus far in our discussion of natural selection, we have assumed that the fitness of particular phenotypes (and their corresponding genotypes) is independent of their frequency in the population. However, in cases of **frequency-dependent selection,** the fitness of a particular phenotype depends on how frequently it appears in the population. Often a phenotype has a greater selective value when rare than when common in the population. Such phenotypes lose their selective advantage as they become more common.

Frequency-dependent selection often acts to maintain genetic variation in populations of prey species. In this case, the predator catches and consumes the more common phenotype but may ignore the rarer phenotypes. Consequently, the less common phenotype produces more offspring and therefore makes a greater relative contribution to the next generation.

Frequency-dependent selection is demonstrated in scale-eating fish (cichlids of the species *Perissodus microlepis*) from Lake Tanganyika in Africa (**FIG. 19-8**). The scale-eating fish, which obtain food by biting scales off other fish, have either left-pointing or right-pointing mouths. A single locus with two alleles determines this characteristic; the allele for right-pointing mouth is dominant over the allele for left-pointing mouth. These fish attack their prey from behind and from a single direction, depending on mouth morphology. Those with left-pointing mouths always attack the right flanks of their prey, whereas those with right-pointing mouths always attack the left flanks.

The prey species are more successful at evading attacks from the more common form of scale-eating fish. For example, if the cichlids with right-pointing mouths are more common than those with left-pointing mouths, the prey are attacked more often on their left flanks. They therefore become more wary against such attacks, conferring a selective advantage to the less common cichlids with left-pointing mouths. The cichlids with left-pointing mouths would be more successful at obtaining food and would therefore have more offspring. Over time, the frequency of fish with left-pointing mouths would increase in the population until their abundance causes frequency-dependent selection to work against them and confer an advantage on the now less common fish with right-pointing mouths. Thus, frequency-dependent selection maintains both populations of fish at approximately equal numbers.

KEY EXPERIMENT

Can evidence for frequency-dependent selection be found in natural populations?

HYPOTHESIS: Equal proportions of scale-eating cichlids with right-pointing mouths versus left-pointing mouths in a population are maintained by frequency-dependent selection.

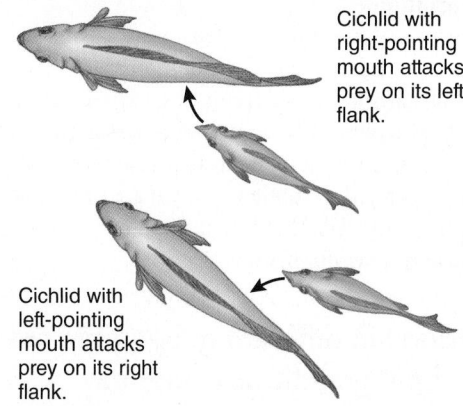

Cichlid with right-pointing mouth attacks prey on its left flank.

Cichlid with left-pointing mouth attacks prey on its right flank.

EXPERIMENT: The relative frequencies of scale-eating cichlids with left-pointing mouths and right-pointing mouths in Lake Tanganyika in Africa were monitored over a 10-year period.

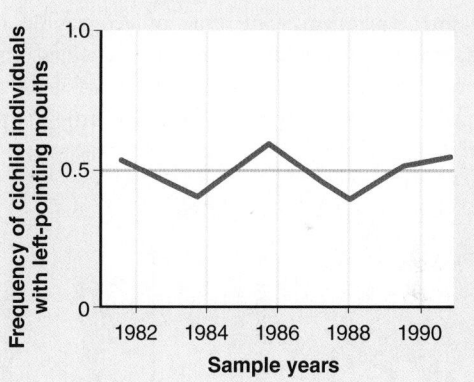

RESULTS AND CONCLUSION: Relative frequencies of left-pointing cichlids and right-pointing cichlids varied slightly from year to year, but remained close to 0.5. This approximate 1:1 ratio is consistent with frequency-dependent selection because prey fish become especially wary to attacks by the cichlid type that is currently most common; therefore, either type is at a selective disadvantage if its frequency rises above 0.5.

SOURCE: Adapted with permission from Hori, M., "Frequency-Dependent Natural Selection in the Handedness of Scale-Eating Cichlid Fish," *Science*, Vol. 260, April 9, 1993. Copyright © 1993 American Association for the Advancement of Science.

Figure 19-8 Frequency-dependent selection in scale-eating cichlids

INTERPRET DATA Recall that the allele for right-pointing mouth (*A*) is dominant over that for left-pointing mouth (*a*). Use the Hardy–Weinberg principle to calculate the allele frequencies that would correspond to approximately equal frequencies of the two phenotypes. Does your answer surprise you?

Neutral variation may give no selective advantage or disadvantage

Some of the genetic variation observed in a population may confer no apparent selective advantage or disadvantage in a particular environment. For example, random changes in DNA that do not alter protein structure usually do not affect the phenotype. Variation that does not alter the ability of an individual to survive and reproduce—and is therefore not adaptive—is called **neutral variation.**

The extent of neutral variation in organisms is difficult to determine. It is relatively easy to demonstrate that an allele is beneficial or harmful if its effect is observable, but the variation in alleles that involves only slight differences in the proteins they code for may or may not be neutral. These alleles may be influencing the organism in subtle ways that are difficult to measure or assess. Also, an allele that is neutral in one environment may be beneficial or harmful in another.

Populations in different geographic areas often exhibit genetic adaptations to local environments

In addition to the genetic variation among individuals within a population, genetic differences often exist among different populations within the same species, a phenomenon known as **geographic variation.** One type of geographic variation is a **cline,** which is a gradual change in a species' phenotype and genotype frequencies through a series of geographically separate populations as a result of an environmental gradient.

A cline exhibits variation in the expression of such an attribute as color, size, shape, physiology, or behavior. Clines are common among species with continuous ranges over large geographic areas. For example, the body sizes of many widely distributed birds and mammals increase gradually as the latitude increases, presumably because larger animals are better able to withstand the colder temperatures of winter.

The common yarrow (*Achillea millefolium*), a wildflower that grows in a variety of North American habitats from lowlands to mountain highlands, exhibits clinal variation in height in response to different climates at different elevations. Although substantial variation exists among individuals within each population, individuals in populations at higher elevations are, on average, shorter than those at lower elevations. The genetic basis of these clinal differences was experimentally demonstrated in a set of classical experiments in which series of populations from different geographic areas were grown in the same environment (FIG. 19-9). Despite being exposed to identical environmental conditions, each experimental population exhibited the traits characteristic of the elevation from which it was collected.

CHECKPOINT 19.4

- *How does the sickle cell allele illustrate heterozygote advantage?*
- *How does frequency-dependent selection affect genetic variation within a population over time?*
- CONNECT *How can researchers test the hypothesis that clinal variation among populations of a particular species has a genetic basis?*

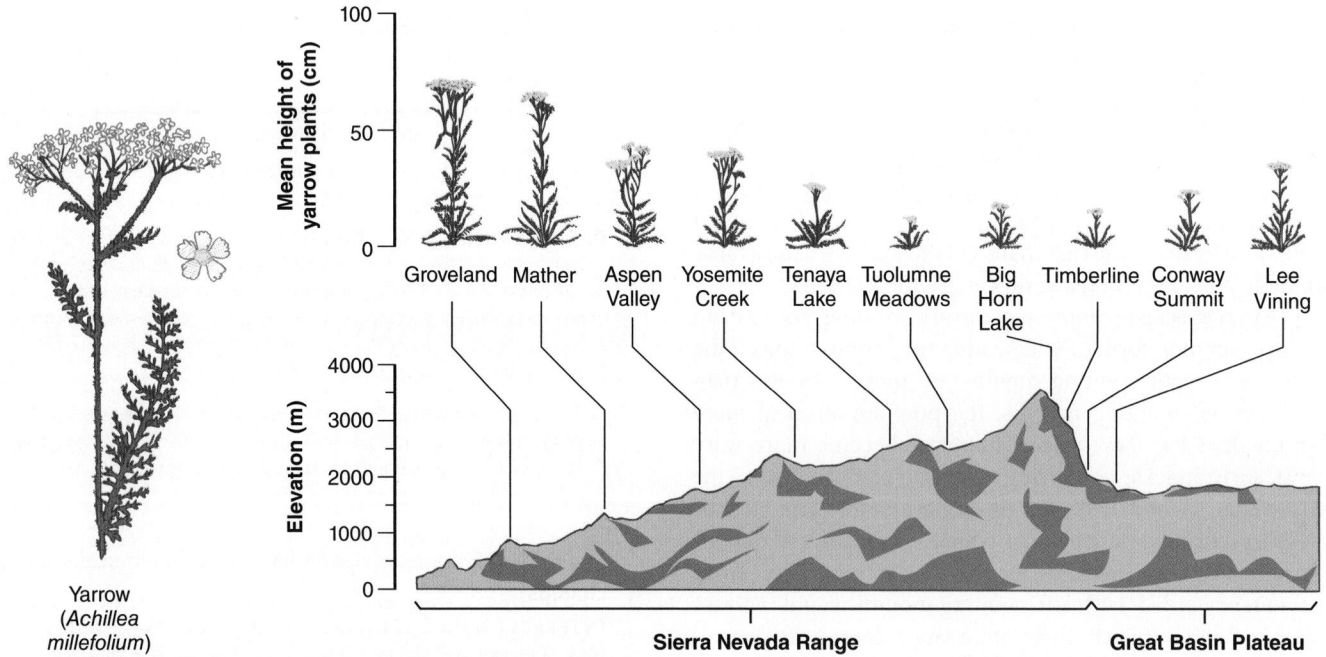

Figure 19-9 Clinal variation in yarrow (*Achillea millefolium*)

(After Clausen, J., D.D. Keck, and W.M. Hiesey, "Experimental Studies on the Nature of Species: III. Environmental Responses of Climatic Races of *Achillea*," Carnegie Institute Washington Publication, Vol. 58, 1948.)

19.1 Genotype, Phenotype, and Allele Frequencies
(page 403)

1 Define what is meant by a population's gene pool.

- All the individuals that live in a particular place at the same time make up a **population.** Each population has a **gene pool,** which includes all the **alleles** for all the **loci** present in the population.

- **Population genetics** is the study of genetic variability within a population and of the forces that act on it.

2 Distinguish among genotype, phenotype, and allele frequencies.

- A **genotype frequency** is the proportion of a particular genotype in the population. A **phenotype frequency** is the proportion of a particular phenotype in the population. An **allele frequency** is the proportion of a specific allele of a given genetic locus in the population.

19.2 The Hardy–Weinberg Principle *(page 403)*

3 Discuss the significance of the Hardy–Weinberg principle as it relates to evolution and list the five conditions required for genetic equilibrium.

- The **Hardy–Weinberg principle** states that allele and genotype frequencies do not change from generation to generation (no evolution is occurring) in a population at **genetic equilibrium.**

- The Hardy–Weinberg principle applies only if mating is random in the population, there are no net mutations that change the allele frequencies, the population is large, individuals do not migrate between populations, and natural selection does not occur.

4 Use the Hardy–Weinberg principle to solve problems involving populations.

- In the Hardy–Weinberg equation, p = the frequency of the dominant allele, q = the frequency of the recessive allele, and $p + q = 1$.

- The genotype frequencies of a population are described by the relationship $p^2 + 2pq + q^2 = 1$, where p^2 is the frequency of the homozygous dominant genotype, $2pq$ is the frequency of the heterozygous genotype, and q^2 is the frequency of the homozygous recessive genotype.

19.3 Microevolution *(page 406)*

5 Define *microevolution.*

- **Microevolution** is a change in allele or genotype frequencies within a population over successive generations.

6 Discuss how each of the following microevolutionary forces alters allele frequencies in populations: nonrandom mating, mutation, genetic drift, gene flow, and natural selection.

- In nonrandom mating individuals select mates on the basis of phenotype, indirectly selecting the corresponding genotype(s). **Inbreeding** is the mating of genetically similar individuals that are more closely related than if they had been chosen at random from the entire population. Inbreeding in some populations causes **inbreeding depression,** in which inbred individuals have lower **fitness** (relative ability to make a genetic contribution to the next generation) than non-inbred

individuals. In **assortative mating** individuals select mates by their phenotypes. Both inbreeding and assortative mating increase the frequency of homozygous genotypes.

- **Mutations,** unpredictable changes in DNA, are the source of new alleles. Mutations increase the genetic variability acted on by natural selection.

- **Genetic drift** is a random change in the allele frequencies of a small population. Genetic drift decreases genetic variation within a population, and the changes caused by genetic drift are usually not adaptive. A sudden decrease in population size caused by adverse environmental factors is known as a bottleneck. The **founder effect** is genetic drift that occurs when a small population colonizes a new area.

- **Gene flow,** a movement of alleles caused by the migration of individuals between populations, causes changes in allele frequencies.

- **Natural selection** causes changes in allele frequencies that lead to adaptation. Natural selection operates on an organism's phenotype, but it changes the genetic composition of a population in a favorable direction for a particular environment.

7 Distinguish among stabilizing selection, directional selection, and disruptive selection.

- **Stabilizing selection** favors the mean at the expense of phenotypic extremes. **Directional selection** favors one phenotypic extreme over another, causing a shift in the phenotypic mean. **Disruptive selection** favors two or more phenotypic extremes.

19.4 Genetic Variation in Populations *(page 411)*

8 Describe the nature and extent of genetic variation, including genetic polymorphism, balanced polymorphism, neutral variation, and geographic variation.

- **Genetic polymorphism** is the presence of genetic variation within a population.

- **Balanced polymorphism** is a special type of genetic polymorphism in which two or more alleles persist in a population over many generations as a result of natural selection.

- **Heterozygote advantage** occurs when the heterozygote exhibits greater fitness than either homozygote. In **frequency-dependent selection,** a genotype's selective value varies with its frequency of occurrence.

- **Neutral variation** is genetic variation that confers no detectable selective advantage.

- **Geographic variation** is genetic variation that exists among different populations within the same species. A **cline** is a gradual change in a species' phenotype and genotype frequencies through a series of geographically separate populations.

Know and Comprehend

1. The genetic description of an individual is its genotype, whereas the genetic description of a population is its (a) phenotype (b) gene pool (c) genetic drift (d) founder effect (e) changes in allele frequencies

2. In a diploid species, each individual possesses (a) one allele for each locus (b) two alleles for each locus (c) three or more alleles for each locus (d) all the alleles found in the gene pool (e) half of the alleles found in the gene pool

3. The MN blood group is of interest to population geneticists because (a) people with genotype *MN* cannot receive blood transfusions from either *MM* or *NN* people (b) the *MM, MN,* and *NN* genotype frequencies can be observed directly and compared with calculated expected frequencies (c) the *M* allele is dominant to the *N* allele (d) people with the *MN* genotype exhibit frequency-dependent selection (e) people with the *MN* genotype exhibit heterozygote advantage

4. If a population's allele and genotype frequencies remain constant from generation to generation, (a) the population is undergoing evolutionary change (b) the population is said to be at genetic equilibrium (c) microevolution has taken place (d) directional selection is occurring, but only for a few generations (e) genetic drift is a significant evolutionary force

5. Analysis of DNA sequences for a particular locus among individuals within a population provides biologists with one way to estimate (a) genetic drift (b) genetic polymorphism (c) gene flow (d) heterozygote advantage (e) frequency-dependent selection

6. The continued presence of the allele that causes sickle cell anemia in areas where falciparum malaria is prevalent demonstrates which of the following phenomena? (a) inbreeding depression (b) frequency-dependent selection (c) heterozygote advantage (d) genetic drift (e) a genetic bottleneck

7. According to the Hardy–Weinberg principle, (a) allele frequencies are not dependent on dominance or recessiveness but remain essentially unchanged from generation to generation (b) the sum of allele frequencies for a given locus is always greater than 1 (c) if a locus has only one allele, the frequency of that allele is zero (d) allele frequencies change from generation to generation (e) the process of inheritance, by itself, causes changes in allele frequencies

8. The Hardy–Weinberg principle may be applicable if (a) the population size is small (b) migration occurs only at the beginning of the breeding season (c) mutations occur at a constant rate (d) matings occur exclusively between individuals of the same genotype (e) natural selection does not occur

9. Mutation (a) leads to adaptive evolutionary change (b) adds to the genetic variation of a population (c) is the result of genetic drift (d) almost always benefits the organism (e) a and b

10. Which of the following is *not* true of natural selection? (a) natural selection acts to preserve favorable traits and eliminate unfavorable traits (b) the offspring of individuals that are better adapted to the environment will make up a larger proportion of the next generation (c) natural selection directs the course of evolution by preserving the traits acquired during an individual's lifetime (d) natural selection acts on a population's genetic variability, which arises through mutation (e) natural selection may result in changes in allele frequencies in a population

Apply and Analyze

11. If all copies of a given locus have the same allele throughout the population, the allele frequency is (a) 0 (b) 0.1 (c) 0.5 (d) 1.0 (e) 10.0

12. **VISUALIZE** Draw four simple graphs, each with a range of phenotypes (e.g., height) on the *x*-axis and numbers of individuals on the *y*-axis. Each graph should illustrate one of the following: (a) no selection (b) stabilizing selection (c) directional selection (d) disruptive selection. Consult Figure 19-4 to check your answers.

Evaluate and Synthesize

13. **EVOLUTION LINK** Given that mutations are almost always neutral or harmful, why are mutations nevertheless essential to evolution? Explain your answer.

14. **EVOLUTION LINK** Explain this apparent paradox: Scientists discuss evolution in terms of *genotype* fitness (the selective advantage that a particular genotype confers on an individual), yet natural selection acts on an organism's *phenotype*.

15. **EVOLUTION LINK** Why is it easier for researchers to study the genetic contribution to disease in the population of Finland or Iceland as opposed to that of the United States?

16. **EVOLUTION LINK** Evolution is sometimes characterized as "survival of the fittest." Is this wording consistent with an evolutionary biologist's definitions of *fitness* and *natural selection*? Is it a good way to think about evolution or a poor one?

17. **INTERPRET DATA** The recessive allele that causes Ellis–van Creveld syndrome when homozygous has a frequency of about 0.07 in the Amish population of Pennsylvania, although its frequency is only about 0.001 in the general population. How many persons out of one thousand in the Amish population would be expected to have the disease? How many out of one million in the general population?

18. **PREDICT** You study males in populations of a certain species of minnows in a series of lakes at different latitudes. You find that they exhibit clinal variation in average weight at maturity and hypothesize that the weight differences are due to genetic factors. What might you predict about the average weights at maturity of representatives of each population reared in aquaria if your hypothesis is correct?

19. **SCIENCE, TECHNOLOGY, AND SOCIETY** Automated techniques are providing new information on the extent of genetic polymorphism among humans, particularly with respect to single nucleotide polymorphisms (SNPs) and copy number variations (CNVs). In what ways might this information prove useful? In what ways might it present challenges?

aplia To access course materials, such as Aplia and other companion resources, please visit **www.cengagebrain.com**.

Speciation and Macroevolution

A Brazilian rain forest contains thousands of species of insects, amphibians, reptiles, birds, and mammals. The Great Barrier Reef off the coast of Australia has thousands of species of sponges, corals, mollusks, crustaceans, sea stars, and fishes. The various environments on Earth abound in rich assemblages of species. Scientists have currently named about 1.8 million species.

We do not know exactly how many species exist today, but many biologists estimate the number to be on the order of 4 million to 100 million! Often this diversity has been portrayed as many twig tips on a "tree of life." As one follows the twigs from the tips toward the main body of the tree, they form connections that represent common ancestors that are now extinct. Indeed, more than 99.9% of all species that ever existed are now extinct. Tracing the branches of the tree of life to larger and larger connections eventually leads to a single trunk that represents the few simple unicellular organisms that evolved early in Earth's history and became the ancestors of all species living today.

Thus, life today is the product of three billion to four billion years of evolution. How did all these species diversify throughout Earth's history? Darwin's studies led him to conclude that the Galápagos Islands were the birthplace of many new species (see photograph). Our understanding of how species evolve has advanced significantly since Darwin published *On the Origin of Species by Means of Natural Selection*.

In this chapter we consider the reproductive barriers that isolate species from one another and the possible evolutionary mechanisms that explain how the millions of species that live today or lived in the past originated from ancestral species. We then examine the rate of evolutionary change and the evolution of higher taxonomic categories (species, genus, family, order, class, and phylum), which is the focus of macroevolution.

© javarman/Shutterstock.com

Blue-footed boobies. Blue-footed boobies are large sea birds that nest on the Galápagos Islands and several other arid islands in the eastern Pacific Ocean. Male and female pairs are monogamous and typically nest on black lava. The largest blue-footed booby breeding population is found in the Galápagos Islands.

KEY CONCEPTS

20.1 According to the biological species concept, a species consists of individuals that can successfully interbreed with one another but not with individuals from other species.

20.2 The evolution of different species begins with reproductive isolation, in which two populations are no longer able to interbreed successfully.

20.3 In allopatric speciation populations diverge into different species due to geographic isolation, or physical separation. In sympatric speciation populations become reproductively isolated from one another despite living in the same geographic area.

20.4 Speciation may require millions of years but sometimes occurs much more quickly.

20.5 The evolution of species and higher taxa is known as macroevolution.

20.1 WHAT IS A SPECIES?

The concept of distinct kinds of organisms, known as **species** (from Latin, meaning "kind"), is not new. However, there are several different definitions of species, and each definition has some sort of limitation. Linnaeus was the 18th-century biologist who founded modern taxonomy: the science of naming, describing, and classifying organisms. He classified plants and other organisms into separate species based on their visible structural differences, such as feathers or number of flower parts. This method, known as the *morphological species concept,* is still used to help characterize species, but structural differences alone are not adequate to explain what constitutes a species.

The biological species concept is based on reproductive isolation

Population genetics did much to clarify the concept of species. According to the **biological species concept,** first expressed by evolutionary biologist Ernst Mayr in 1942, a species consists of one or more populations whose members interbreed in nature to produce fertile offspring and do not interbreed with—that is, are reproductively isolated from—members of different species. Individuals belonging to the same species tend to share common features because their genes are derived from a common *gene pool* (see Chapter 19). Gene flow occurs between individuals of a given species when they mate, but reproductive barriers restrict individuals of one species from interbreeding, or exchanging genes, with members of other species. As a result, reproductive barriers keep a given species genetically distinct from other species. An extension of the biological species concept holds that new species evolve when formerly interbreeding populations become reproductively isolated from one another.

The biological species concept is currently the most widely accepted definition of species, but it has several shortcomings. One problem is that it applies only to sexually reproducing organisms. Bacteria reproduce asexually, although they may have considerable exchange of genetic material by means of *horizontal gene transfer* among bacterial individuals that are classified as different species. Thus, reproductive isolation is not a good criterion for defining species of bacteria.

Small marine organisms known as bdelloid rotifers are an example of animals that reproduce asexually (**FIG. 20-1**). All bdelloid rotifers are females, and they produce offspring without needing males. Interestingly, these animals engage in massive horizontal gene transfer, and their genomes include DNA sequences that appear

to have originated in bacteria, fungi, and even plants! These organisms and also extinct organisms must be classified based on their structural and biochemical characteristics; the biological species concept does not apply to them. (Horizontal gene transfer is discussed in more detail in Chapter 23.)

Another potential problem with the biological species concept is that although individuals assigned to different species do not normally interbreed, they may *sometimes* successfully interbreed. For example, where their ranges overlap, some coyotes have been known to mate with wolves. As a result, some wolves carry coyote genes and vice versa. Most biologists agree that wolves and coyotes are separate species, however.

The phylogenetic species concept defines species based on such evidence as molecular sequencing

Some biologists prefer the **phylogenetic species concept,** also called the *evolutionary species concept,* in which a population is declared a separate species if it has undergone evolution long enough for statistically significant differences in diagnostic traits to emerge. If a population has a unique diagnostic trait—a set of spines running down the back, for example—that can be demonstrated to be inherited, this population is considered a separate species. This approach has the advantage of being testable by comparing gene sequences between two groups. However, many biologists do not want to abandon the biological species concept, in part because the phylogenetic species concept cannot be applied if the evolutionary history (as determined by DNA sequencing) of a taxonomic group has not been carefully studied, and most groups have not been rigorously

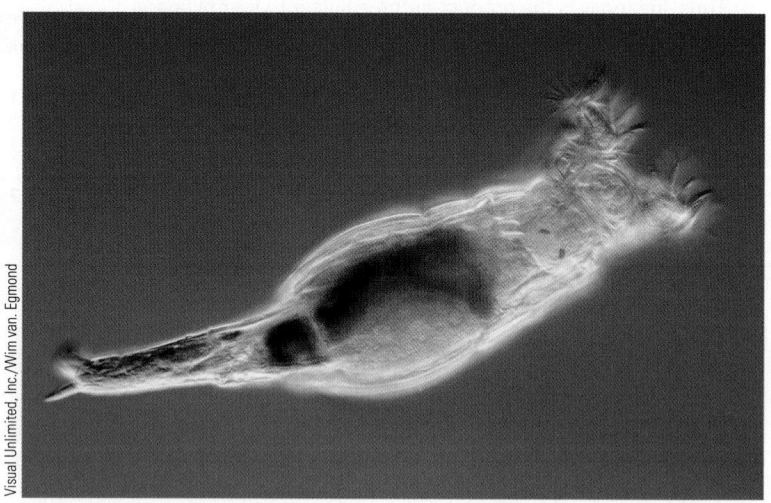

150 μm

Figure 20-1 Bdelloid rotifer (*Philodina roseola*)

Bdelloid rotifers are all females. Because they do not reproduce sexually, it is difficult to classify these small marine animals as species under the biological species concept.

analyzed. Also, if the phylogenetic species concept were universally embraced, the number of named species would probably double. This increase would occur because many closely related populations that are classified as *subspecies* (a taxonomic subdivision of a species) under the biological species concept would fit the requirements of separate species under the phylogenetic species concept.

Thus, the exact definition of a species remains fuzzy. Unless we clearly state otherwise, when we mention *species* in this text, we are referring to the biological species concept.

CHECKPOINT 20.1

- *What is a species, according to the biological species concept?*
- *Is the biological species definition likely to work well for classifying insects in amber? plants that reproduce asexually? fish in an aquarium?*
- CONNECT *If a population meets all the criteria of the biological species concept, is it likely to meet the definition of a morphological species? Explain your answer.*

20.2 REPRODUCTIVE ISOLATION

LEARNING OBJECTIVE

2 Explain the significance of reproductive isolating mechanisms and distinguish among the various prezygotic and postzygotic barriers.

Various **reproductive isolating mechanisms** prevent interbreeding between two different species whose *ranges* (areas where each lives) overlap. These mechanisms preserve the genetic integrity of each species because gene flow between the two species is prevented.

Most species have two or more mechanisms that block a chance occurrence of individuals from two closely related species overcoming a single reproductive isolating mechanism. Although most mechanisms operate before mating or fertilization occurs (*prezygotic*), others work after fertilization has taken place (*postzygotic*).

Prezygotic barriers interfere with fertilization

Prezygotic barriers are reproductive isolating mechanisms that prevent fertilization from taking place. Because male and female gametes never come into contact, an *interspecific zygote*—that is, a fertilized egg formed by the union of an egg from one species and a sperm from another species—is never produced. Prezygotic barriers include temporal isolation, habitat isolation, behavioral isolation, mechanical isolation, and gametic isolation.

Sometimes genetic exchange between two groups is prevented because they reproduce at different times of the day, season, or year. Such examples demonstrate **temporal isolation.** For example, two very similar species of fruit flies, *Drosophila pseudoobscura* and *D. persimilis,* have ranges that overlap to a great extent, but they do not interbreed. *Drosophila pseudoobscura* is sexually active only in the afternoon; and *D. persimilis,* only in the morning. Similarly, two frog species have overlapping ranges in eastern Canada and the United States. The wood frog (*Rana sylvatica*) usually mates in late March or early April, when the water temperature is about 7.2°C (45°F), whereas the northern leopard frog (*R. pipiens*) usually mates in mid-April, when the water temperature is 12.8°C (55°F) (FIG. 20-2).

Although two closely related species may be found in the same geographic area, they usually live and breed in different habitats in that area, providing **habitat isolation** between the two species. For example, the five species of small birds known as flycatchers are nearly identical in appearance and have overlapping ranges in the eastern part of North America. They exhibit habitat isolation because during the breeding season, each species stays in a particular habitat within its range, so potential mates from different species do not meet. The least flycatcher (*Empidonax minimus*) frequents open woods, farms, and orchards; whereas the acadian flycatcher (*E. virescens*) is found in deciduous forests, particularly in beech trees, and swampy woods. The alder flycatcher (*E. alnorum*) prefers wet thickets of alders, the yellow-bellied flycatcher (*E. flaviventris*)

(a) The wood frog (*Rana sylvatica*) mates in early spring, often before the ice has completely melted in the ponds.

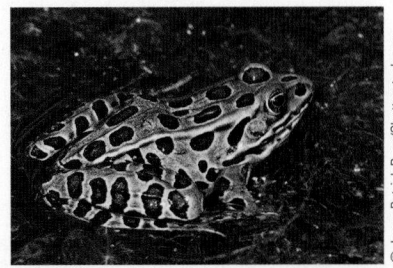

(b) The leopard frog (*Rana pipiens*) typically mates a few weeks later.

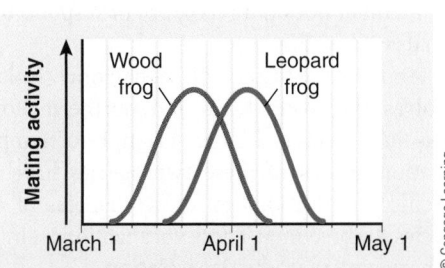

(c) Graph of peak mating activity in wood and leopard frogs. In nature, wood and leopard frogs do not interbreed.

Figure 20-2 *Animation* **Temporal isolation in wood and leopard frogs**

Figure 20-3 Behavioral isolation in bowerbirds

Each bowerbird species has highly specialized courtship patterns that prevent its mating with another species. The male satin bowerbird (*Ptilonorhynchus violacens*) constructs an enclosed place, or bower, of twigs to attract a female. (The bower is the dark "tunnel" on the left.) Note the parrot feather in his beak and blue decorations, including human-made objects such as bottle caps, that he has arranged at the entrance to his bower.

Tim Laman/National Geographic Creative

nests in conifer woods, and the willow flycatcher (*E. traillii*) frequents brushy pastures and willow thickets.

Many animal species exchange a distinctive series of signals before mating. Such courtship behaviors illustrate **behavioral isolation** (also known as **sexual isolation**). Bowerbirds, for example, exhibit species-specific courtship patterns. The male satin bowerbird of Australia constructs an elaborate bower of twigs, adding decorative blue parrot feathers and white flowers at the entrance (**FIG. 20-3**). When a female approaches the bower, the male dances around her, holding a particularly eye-catching decoration in his beak. While dancing, he puffs his feathers, extends his wings, and sings a courtship song that consists of a variety of sounds, including loud buzzes and hoots. These specific courtship behaviors keep closely related bird species reproductively isolated from the satin bowerbird. If a male and a female of two different bowerbird species begin courtship, it stops when one member does not recognize or respond to the signals of the other.

Another example of behavioral isolation involves the wood frogs and northern leopard frogs just discussed as an example of temporal isolation. Males of these two species have very specific vocalizations to attract females of their species for breeding. These vocalizations reinforce each species' reproductive isolation.

Sometimes members of different species court and even attempt copulation, but the incompatible structures of their genital organs prevent successful mating. Structural differences that inhibit mating between species produce **mechanical isolation.** For example, many flowering plant species have physical differences in their flower parts

that help them maintain their reproductive isolation from one another. In such plants the flower parts are adapted for specific insect pollinators. Two species of sage, for example, have overlapping ranges in southern California. Black sage (*Salvia mellifera*), which is pollinated by small bees, has a floral structure different from that of white sage (*S. apiana*), which is pollinated by large carpenter bees (**FIG. 20-4**). Interestingly, black sage and white sage are also prevented from mating by a temporal barrier: black sage flowers in early spring, and white sage flowers in late spring and early summer. Presumably, mechanical isolation prevents insects from cross-pollinating the two species should they happen to flower at the same time.

If mating takes place between two species, their gametes may still not combine. Molecular and chemical differences between species cause **gametic isolation,** in which the egg and sperm of different species are incompatible. In aquatic animals that release their eggs and sperm into the surrounding water simultaneously, interspecific fertilization is extremely rare. The surface of the egg contains specific proteins that bind only to complementary molecules on the surface of sperm cells of the same species (see Chapter 51). A similar type of molecular recognition often occurs between pollen grains and the stigma (receptive surface of the female part of the flower) so that pollen does not germinate on the stigma of a different plant species.

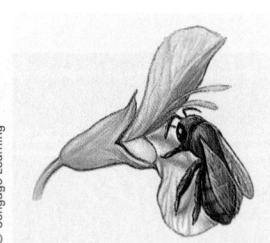

© Cengage Learning

(a) The petal of the black sage functions as a landing platform for small bees. Large bees cannot fit on this platform.

(b) The larger landing platform and longer stamens of white sage allow pollination by larger California carpenter bees (a different species). If smaller bees land on white sage, their bodies do not brush against the stamens. (In the figure, the upper part of the white sage flower has been removed.)

Figure 20-4 Mechanical isolation in black sage and white sage

The differences in their floral structures allow black sage and white sage to be pollinated by different insects. Because the two species exploit different pollinators, they cannot interbreed.

Postzygotic barriers prevent gene flow when fertilization occurs

Fertilization sometimes occurs between gametes of two closely related species despite the existence of prezygotic barriers. When such fertilization happens, **postzygotic barriers** that increase the likelihood of reproductive failure come into play. Generally, the embryo of an interspecific hybrid spontaneously aborts. **Development,** the orderly sequence of events that occurs as an organism grows and matures, is a complex process requiring the precise interaction and coordination of many genes. Apparently, the genes from parents belonging to different species do not interact properly in regulating the mechanisms for normal development. In this case, reproductive isolation occurs by **hybrid inviability.** For example, in crosses between different species of irises, the embryos die before reaching maturity. Similarly, nearly all the hybrids die in the embryonic stage when the eggs of a bullfrog are fertilized artificially with sperm from a leopard frog.

If an interspecific hybrid does live, it may not reproduce for any of several reasons. Hybrid animals may exhibit courtship behaviors incompatible with those of either parental species, and as a result they will not mate. More often, **hybrid sterility** occurs when problems during meiosis cause the gametes of an interspecific hybrid to be abnormal. Hybrid sterility is particularly common if the two parental species have different chromosome numbers. For example, a mule is the offspring of a female horse ($2n = 64$) and a male donkey ($2n = 62$) (FIG. 20-5). This type of union almost always results in sterile offspring ($2n = 63$) because synapsis (the pairing of homologous chromosomes during meiosis) and segregation of chromosomes cannot occur properly.

Occasionally, a mating between two F_1 hybrids produces a second hybrid generation (F_2). The F_2 hybrid may exhibit **hybrid breakdown,** the inability of a hybrid to reproduce because of some defect. For example, hybrid breakdown in the F_2 generation of a cross between two sunflower species was 80%. In other words, 80% of the F_2 generation were defective in some way and could not reproduce successfully. Hybrid breakdown also occurs in the F_3 and later generations.

TABLE 20-1 summarizes the various prezygotic and postzygotic barriers that prevent interbreeding between two species.

Biologists are discovering genes responsible for reproductive isolating mechanisms

Progress has been made in identifying some of the genes involved in reproductive isolation. For example, scientists have determined the genetic basis for prezygotic isolation in species of abalone, large mollusks found along the Pacific coast of North America. In abalone the fertilization of eggs by sperm requires *lysin,* a sperm protein that attaches to a lysin receptor protein located on the egg envelope. After attachment, the lysin produces a hole in the egg envelope that permits the sperm to penetrate the egg. Scientists cloned the lysin receptor gene and demonstrated that this gene varies among abalone species. Differences in the lysin receptor protein in various abalone species determine sperm compatibility with the egg. Sperm of one abalone species do not attach to a lysin receptor protein of an egg of a different abalone species.

Other genes related to reproductive isolating mechanisms—loosely referred to as "speciation genes"—have been identified in a diverse group of organisms, such as Florida beach mice, cactus finches, fruit fly (*Drosophila*) species, and flycatcher birds in the South Pacific.

CHECKPOINT 20.2

- What barriers prevent wood frogs and leopard frogs from interbreeding in nature?
- How is temporal isolation different from behavioral isolation?
- How is mechanical isolation different from gametic isolation?
- Of which postzygotic barrier is the mule an example?

20.3 SPECIATION

LEARNING OBJECTIVES

3 Explain the mechanism of allopatric speciation and give an example.
4 Explain the mechanisms of sympatric speciation and give both plant and animal examples.

We are now ready to consider how entirely new species may arise from previously existing ones. The evolution of a new species is called **speciation.** The formation of two species from a single species occurs when a population becomes reproductively isolated from other populations of the species and the **gene pools** of the two separated populations begin to diverge in genetic composition. When a population is sufficiently different from its ancestral species that there is no genetic exchange

Figure 20-5 Hybrid sterility in mules

Mules are interspecific hybrids formed by mating a female horse (*left*) with a male donkey (*right*). Although the mule (*center*) exhibits valuable characteristics of each of its parents, it is sterile.

TABLE 20-1	Reproductive Isolating Mechanisms	
MECHANISM	**HOW IT WORKS**	**EXAMPLE**

PREZYGOTIC BARRIERS PREVENT FERTILIZATION

Temporal isolation	Similar species reproduce at different times. 	Two species of fruit flies of the genus *Drosophila* live in the same geographic area, but one is sexually active only in the afternoon and the other only in the morning.
Habitat isolation	Similar species reproduce in different habitats. 	Several flycatcher species of the genus *Empidonax* live in the same geographic area but in different habitats; for example, the least flycatcher lives in open woods, farms, and orchards, and the acadian flycatcher lives in deciduous forests and swampy woods.
Behavioral isolation	Similar species have distinctive courtship behaviors. 	The male satin bowerbird of Australia constructs an elaborate bower, dances around the female, and sings a specific courtship song that keeps closely related bowerbird species reproductively isolated from the satin bowerbird.
Mechanical isolation	Similar species have structural differences in their reproductive organs. 	Black sage (genus *Salvia*) is pollinated by small bees, whereas white sage is pollinated by large carpenter bees. Because of differences in the flower structures, neither type of bee can successfully pollinate the other sage species.
Gametic isolation	Gametes of similar species are chemically incompatible. 	In aquatic animals such as sponges that release their eggs and sperm into the water simultaneously, interspecific fertilization rarely occurs because the egg surface is compatible only with the sperm surface of the same species.

POSTZYGOTIC BARRIERS REDUCE VIABILITY OR FERTILITY OF HYBRID

Hybrid inviability	Interspecific hybrid dies at early stage of embryonic development. 	In crosses between different species of irises (genus *Iris*), fertilization occurs, but the embryos in the seeds die before reaching maturity.
Hybrid sterility	Interspecific hybrid survives to adulthood but is unable to reproduce successfully. 	A mule, the offspring of a female horse and a male donkey, has sterile offspring because synapsis and segregation of chromosomes do not occur properly during gamete formation.
Hybrid breakdown	Offspring of interspecific hybrid are unable to reproduce successfully. 	In a cross between two sunflower species, most members of the F_2 generation are defective in some way and cannot reproduce successfully. Hybrid breakdown also occurs in later generations.

between them, we consider speciation to have occurred. Speciation happens in two ways: allopatric speciation and sympatric speciation.

Long physical isolation and different selective pressures result in allopatric speciation

Speciation that occurs when one population becomes geographically separated from the rest of the species and subsequently evolves by natural selection and/or genetic drift is known as **allopatric speciation** (from the Greek *allo,* "different," and *patri,* "native land"). Recall from Chapter 18 that *natural selection* occurs as individuals that possess favorable adaptations to the environment survive and become parents of the next generation. *Genetic drift* is a random change in allele frequency resulting from the effects of chance on the survival and reproductive success of individuals in a small breeding population (see Chapter 19). Both natural selection and genetic drift result in changes in allele frequencies in a population, but only in natural selection is the change in allele frequency adaptive.

Allopatric speciation is the most common method of speciation and accounts for almost all evolution of new animal species. The geographic isolation required for allopatric speciation may occur in several ways. Earth's surface is in a constant state of change. Such change includes rivers shifting their courses; glaciers migrating; mountain ranges forming; land bridges forming that separate previously united aquatic populations; and large lakes diminishing into several smaller, geographically separated pools.

What may be an imposing geographic barrier to one species may be of no consequence to another. Birds and cattails, for example, do not become isolated when a lake subsides into smaller pools; birds easily fly from one pool to another, and cattails disperse their pollen and fruits by air currents. Fishes, in contrast, usually cannot cross the land barriers between the pools and so become reproductively isolated.

In the Death Valley region of California and Nevada, large interconnected lakes formed during wetter climates of the last Ice Age. These lakes were populated by one or several species of pupfishes. Over time, the climate became drier, and the large lakes dried up, leaving isolated pools. Presumably, each pool contained a small population of pupfish that gradually diverged from the common ancestral species by genetic drift and natural selection in response to habitat differences such as the high temperatures, high salt concentrations, and low oxygen levels characteristic of desert springs. Today, there are 20 or so distinct species, subspecies, and populations of pupfish in the area of Death Valley, California, and Ash Meadows, Nevada (**FIG. 20-6**). Many, such as the Devil's Hole pupfish and the Owens pupfish, are restricted to one or two isolated springs.

Allopatric speciation also occurs when a small population migrates or is dispersed (such as by a chance storm) and colonizes a new area away from the range of the original species. This colony is geographically isolated from its parental species, and the small *microevolutionary* changes that accumulate in the isolated gene pool over many generations may eventually be sufficient to form a new species. Because islands provide the geographic isolation required for allopatric speciation, they offer excellent opportunities to study this evolutionary mechanism. A few individuals of a few species probably colonized the Galápagos Islands and the Hawaiian Islands, for example. The hundreds of unique species presently found on each island presumably descended from these original colonizers (**FIG. 20-7**).

Speciation is more likely to occur if the original isolated population is small. Recall from Chapter 19 that genetic drift, including the founder effect, is more consequential in small populations than larger ones. Genetic drift tends to result in rapid changes in allele frequencies in a small, isolated population. The different selective pressures of the new environment to which the population is exposed further accentuate the divergence caused by genetic drift.

(a) Devil's Hole pupfish (*Cyprinodon diabolis*), photographed at Devil's Hole, Ash Meadows National Wildlife Refuge, Nevada.

(b) Ash Meadows pupfish (*Cyprinodon nevadensis ssp. mionectes*), photographed at Point of Rocks Springs, Ash Meadows National Wildlife Refuge, Nevada.

Figure 20-6 Allopatric speciation of pupfishes (*Cyprinodon*)

Shown are two pupfish species that evolved when larger lakes in southern Nevada dried up about 10,000 years ago, leaving behind small, isolated desert pools fed by springs. The pupfish's short, stubby body is characteristic of fish that live in springs; fish that live in larger bodies of water are more streamlined.

Figure 20-7 Allopatric speciation of the Hawaiian goose (the nene)

Nene (pronounced "nay-nay"; *Branta sandvicensis*) are geese originally found only on volcanic mountains on the geographically isolated islands of Hawaii and Maui, which are some 4200 km (2600 mi) from the nearest continent. Compared with those of other geese, the feet of Hawaiian geese are not completely webbed, their toenails are longer and stronger, and their foot pads are thicker; these adaptations enable Hawaiian geese to walk easily on lava flows. Nenes probably evolved from a small population of geese that originated in North America.

The Kaibab squirrel is an example of allopatric speciation in progress About 10,000 years ago, when the American Southwest was less arid than it is today, the ponderosa pine forests in the area supported a tree squirrel with conspicuous tufts of hair sprouting from its ears. A small tree squirrel population living on the Kaibab Plateau of the Grand Canyon became geographically isolated when the climate changed, thus causing areas to the north, west, and east to become desert. Just a few miles to the south lived the rest of the squirrels, which we know as Abert squirrels, but the two groups were separated by the Grand Canyon. With changes over time in both its appearance and its ecology, the Kaibab squirrel is on its way to becoming a new species.

During its many years of geographic isolation, the small population of Kaibab squirrels has diverged from the widely distributed Abert squirrels in several ways. Perhaps most evident are changes in fur color. The Kaibab squirrel now has a white tail and a gray belly in contrast to the gray tail and white belly of the Abert squirrel (**FIG. 20-8**). Biologists think that these striking changes arose in Kaibab squirrels as a result of genetic drift.

Scientists have not reached a consensus about whether the two squirrel populations are separate species, and most biologists today consider the Kaibab squirrel (*Sciurus aberti ssp. kaibabensis*) and the Abert squirrel (*S. aberti*) to be distinct subspecies. Because the Kaibab and Abert squirrels remain reproductively isolated from each other, it is assumed that they will continue to evolve into separate species.

Porto Santo rabbits diverged rapidly from European rabbits Allopatric speciation has the potential to occur quite rapidly. Early in the 15th century, a small population of rabbits was released on Porto Santo, a small island off the coast of Portugal. Because there were no other rabbits or competitors and no predators on the island, the rabbits thrived. By the 19th century, these rabbits were markedly different from their European ancestors. They were only half as large (weighing slightly more than 500 g, or 1.1 lb), with a different color pattern and a more nocturnal lifestyle. More significant, attempts to mate Porto Santo rabbits with mainland European rabbits failed. Many biologists concluded that within four hundred years, an extremely brief period in evolutionary history, a new species of rabbit had evolved.

Not all biologists agree that the Porto Santo rabbit is a new species. The objection stems from a more recent breeding experiment and is based on biologists' lack of a consensus about the definition of a species. In the experiment, foster mothers of the wild Mediterranean rabbit raised newborn Porto Santo rabbits. When they reached adulthood, these Porto Santo rabbits mated successfully with Mediterranean rabbits to produce healthy, fertile offspring. To some biologists, this experiment clearly demonstrated that Porto Santo rabbits are not a separate species but a subspecies. These biologists cite the Porto Santo rabbits as an example of speciation in progress, much like the Kaibab squirrels just discussed.

Many other biologists think that the Porto Santo rabbit is a separate species because it does not interbreed with the other rabbits under natural conditions. They point out that the breeding experiment was successful only after the baby Porto Santo

(a) The Kaibab squirrel, with its white tail and gray belly, is found north of the Grand Canyon.

(b) The Abert squirel, with its gray tail and white belly, is found south of the Grand Canyon.

Figure 20-8 Allopatric speciation in progress

rabbits were raised under artificial conditions that probably modified their natural behavior.

Two populations diverge in the same physical location by sympatric speciation

Although geographic isolation is an important factor in many cases of evolution, it is not an absolute requirement. In **sympatric speciation** (from the Greek *sym,* "together," and *patri,* "native land"), a new species evolves within the same geographic region as the parental species. The divergence of two populations in the same geographic range occurs when reproductive isolating mechanisms evolve at the *start* of the speciation process. Sympatric speciation is especially common in plants. The role of sympatric speciation in animal evolution is probably much less important than allopatric speciation; until recently, sympatric speciation in animals has been difficult to demonstrate in nature.

Sympatric speciation occurs in at least two ways: a change in **ploidy** (the number of chromosome sets making up an organism's genome) and a change in ecology. We now examine each of these mechanisms.

Allopolyploidy is an important mechanism of sympatric speciation in plants As a result of reproductive isolating mechanisms discussed earlier, the union of two gametes from different species rarely forms viable offspring; if offspring are produced, they are usually sterile. Synapsis cannot usually occur in interspecific hybrid offspring because not all the chromosomes are homologous. However, if the chromosome number doubles *before* meiosis, homologous chromosomes can undergo synapsis. Although not common, this spontaneous doubling of chromosomes has been documented in a variety of plants and a few animals. It produces nuclei with multiple sets of chromosomes.

Polyploids, which possess more than two sets of chromosomes, have played a major role in plant evolution. Reproductive isolation occurs in a single generation when a polyploid species with multiple sets of chromosomes arises from diploid parents. There are two kinds of polyploidy: autopolyploidy and allopolyploidy. An **autopolyploid** contains multiple sets of chromosomes from a single species, and an **allopolyploid** contains multiple sets of chromosomes from two or more species. We discuss only allopolyploidy because it is much more common in nature than autopolyploidy.

Allopolyploidy occurs in conjunction with **hybridization,** which is sexual reproduction between individuals from closely related species. Allopolyploidy produces a fertile interspecific hybrid because the polyploid condition provides the homologous chromosome pairs necessary for synapsis during meiosis. As a result, gametes may be viable (**FIG. 20-9**). An allopolyploid— that is, an interspecific hybrid produced by allopolyploidy— reproduces with itself (self-fertilization) or with a similar individual. However, allopolyploids are reproductively isolated from both parents because their gametes have a different number of chromosomes from those of either parent.

Changes in chromosome number have often led to speciation in plants.

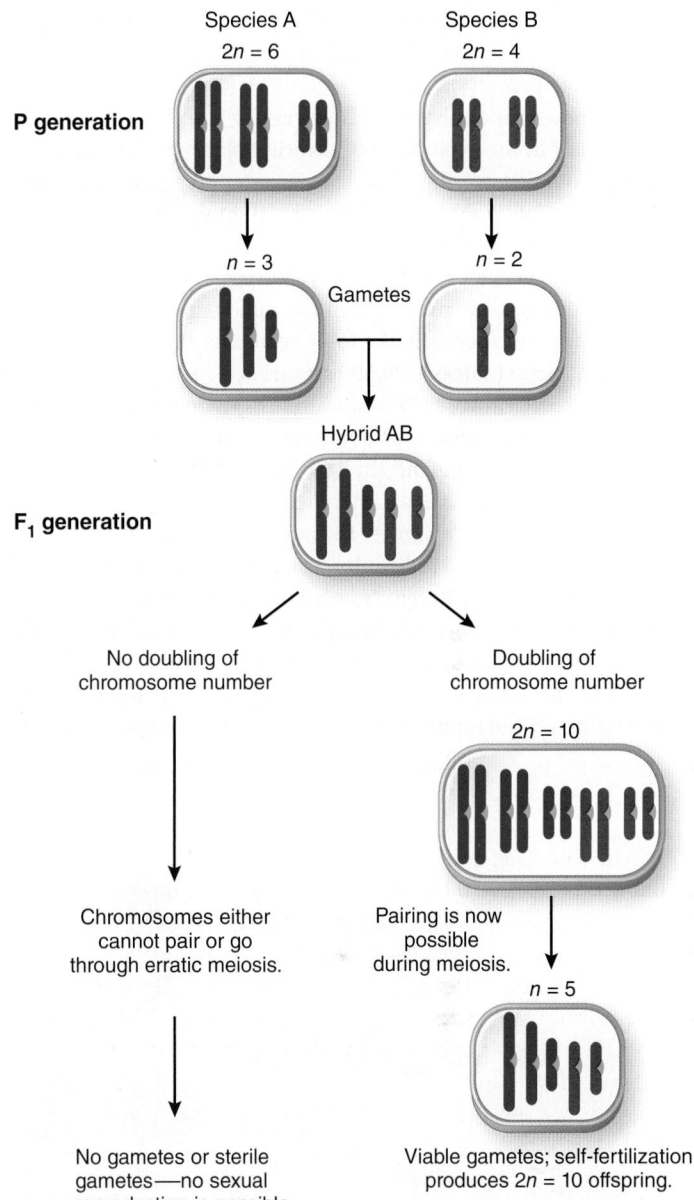

Figure 20-9 *Animation* **Sympatric speciation by allopolyploidy in plants**

When two species (designated the P generation) successfully interbreed, the interspecific hybrid offspring (the F₁ generation) are almost always sterile (*bottom left*). If the chromosomes double, proper synapsis and segregation of the chromosomes can occur, and viable gametes may be produced (*bottom right*). (Unduplicated chromosomes are shown for clarity.)

PREDICT Do you think that the 2n = 10 hybrids could produce fertile offspring when crossed with species A? with species B?

© Cengage Learning

If a population of allopolyploids (i.e., a new species) becomes established, selective pressures cause one of three outcomes. First, the new species may not compete successfully against species that are already established, so it becomes extinct. Second, the allopolyploid individuals may assume a new role in the environment and so coexist with both parental species. Third, the new species may successfully compete with either or both of its parental species. If it has a combination of traits that confers greater fitness than one or both parental species for all or part of the original range of the parent(s), the hybrid species may replace the parent(s).

Although allopolyploidy is extremely rare in animals, it is significant in the evolution of flowering plant species. As many as 80% of all flowering plant species are polyploids, and most of those are allopolyploids. Moreover, allopolyploidy provides a mechanism for extremely rapid speciation. A single generation is all that is needed to form a new, reproductively isolated species. Allopolyploidy may explain the rapid appearance of many flowering plant species in the fossil record and their remarkable diversity today (at least 300,000 species).

The kew primrose (*Primula kewensis*) is an example of sympatric speciation that was documented at the Royal Botanic Gardens at Kew, England, in 1898 (FIG. 20-10). Plant breeders developed *P. kewensis* as an interspecific hybrid of two primrose species, *P. floribunda* ($2n = 18$) and *P. verticillata* ($2n = 18$). *Primula kewensis* had a chromosome number of 18 but was sterile. Then, at three different times, it was reported to have spontaneously formed a fertile branch, which was an allopolyploid ($2n = 36$) that produced viable seeds of *P. kewensis*.

The mechanism of sympatric speciation has been experimentally verified for many plant species. One example is a group of species, collectively called *hemp nettles,* that occur in temperate parts of Europe and Asia. One hemp nettle, *Galeopsis tetrahit* ($2n = 32$), is a naturally occurring allopolyploid hypothesized to have formed by the hybridization of two species, *G. pubescens* ($2n = 16$) and *G. speciosa* ($2n = 16$). This process occurred in nature but was experimentally reproduced. *Galeopsis pubescens* and *G. speciosa* were crossed to produce F_1 hybrids, most of which were sterile. Nevertheless, both F_2 and F_3 generations were produced. The F_3 generation included a polyploid plant with $2n = 32$ that self-fertilized to yield fertile F_4 offspring that could not mate with either of the parental species. These allopolyploid plants had the same appearance and chromosome number as the naturally occurring *G. tetrahit*. When the experimentally produced plants were crossed with the naturally occurring *G. tetrahit,* a fertile F_1 generation was formed. Thus, the experiment duplicated the speciation process that occurred in nature.

Changing ecology causes sympatric speciation in animals Biologists have observed the occurrence of sympatric speciation in animals, but its significance—how often it happens and under what conditions—is still debated. Many examples of sympatric speciation in animals involve parasitic insects and rely on genetic mechanisms other than polyploidy. For example, in the 1860s in the Hudson River valley of New York, a population of fruit maggot flies (*Rhagoletis pomonella*) parasitic on the small red fruits of native hawthorn trees was documented to have switched to a new host, domestic apples, which had been introduced from Europe. Although the sister populations (hawthorn maggot flies and apple maggot flies) continue to occupy the same geographic area, no gene flow occurs between them because they eat, mate, and lay their eggs on different hosts (FIG. 20-11). In other words, because the hawthorn and apple maggot flies have diverged and are reproductively isolated from each other, they have effectively become separate species. Most entomologists, however, still recognize hawthorn and apple maggot flies as a single species because their appearance is virtually identical.

In situations such as that of the fruit maggot flies, a mutation arises in an individual and spreads through a small group of insects by sexual reproduction. The particular mutation leads to *disruptive selection* (see Chapter 19), in which both the old and new phenotypes are favored. Selection is disruptive because the original population is still favored when it parasitizes the original host, and the mutants are favored by a new ecological opportunity, which in this case is to parasitize a different host species. Additional mutations may occur that cause the sister populations to diverge even further.

Cichlids in East Africa intrigue evolutionary biologists Biologists have studied the speciation of colorful fishes known as *cichlids* (pronounced "sik-lids") in several East African lakes. The different species of cichlids in a given lake have remarkably different eating habits, which are reflected in the shapes of their jaws and teeth. Some graze on algae; some consume dead organic material at the bottom of the lake; and others are predatory and eat plankton (microscopic aquatic organisms), insect larvae, the scales off fish, or even other cichlid species.

| Primula floribunda | Primula kewensis | Primula verticillata |

Figure 20-10 Sympatric speciation of a primrose

A new species of primrose, *Primula kewensis,* arose in 1898 as an allopolyploid derived from the interspecific hybridization of *P. floribunda* and *P. verticillata.* Today, *P. kewensis* is a popular houseplant.

© Cengage Learning

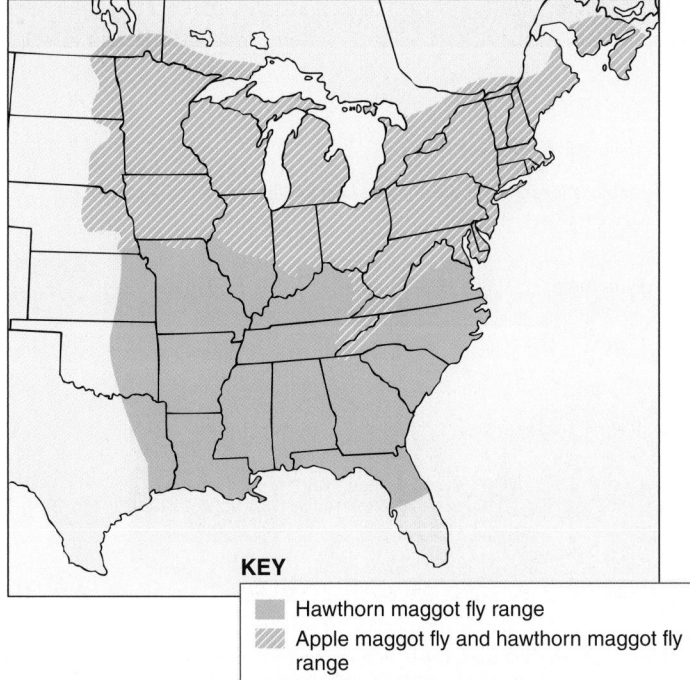

KEY

◼ Hawthorn maggot fly range

▨ Apple maggot fly and hawthorn maggot fly range

Figure 20-11 Ranges of apple and hawthorn maggot flies

Apple and hawthorn maggot flies are sympatric throughout the northern half of the hawthorn maggot fly's range. (Adapted from G.L. Bush, "Sympatric Host Race Formation and Speciation in Frugivorous Flies of the Genus *Rhagoletis* [Diptera, Tephritidae]," *Evolution*, Vol. 23, No. 2, June 1969.)

In some cichlids food preferences are related to size (smaller cichlids consume plankton), which in turn is related to mating preference (small, plankton-eating cichlids mate only with other small, plankton-eating cichlids).

In other cichlids, related species do not differ in size but only in color and mating habits. Each species has a distinct male coloration, which matches the preferences of females of that species in choosing mates (FIG. 20-12).

Charles Darwin called choosing a mate based on its color or some other characteristic **sexual selection.** He recognized that female preference for showy colors or elaborate ornaments on males, if it was based on *inherited* variation, could result in these male sexual traits becoming more common and more pronounced over time. Female preference for a particular male trait confers a selective advantage to the males exhibiting that trait because they have an improved chance to mate and pass on their genes. Darwin suggested that sexual selection, like natural selection, could eventually result in the evolution of new species. (Sexual selection is discussed in greater detail in Chapter 52.)

DNA sequence data indicate that the cichlid species within a single lake are more closely related to one another than to fishes in nearby lakes or rivers. These molecular data suggest that cichlid species evolved sympatrically, or at least within the confines of a lake, rather than by repeated colonizations by fish populations in nearby rivers. How rapidly did sympatric speciation occur in cichlids? Seismic and drill core data provide evidence that Lake Victoria dried up completely during the Late Pleistocene, when much of northern and equatorial Africa was arid. The diverse cichlid species may have arisen after the climate became wetter and Lake Victoria refilled. Further, the more than five hundred **endemic species** (species found nowhere else) of cichlids in Lake Victoria may have evolved from one or a few common ancestors in a remarkably short period, as little as 12,000 years. Other biologists have offered competing hypotheses about the timing and geographic origin of Lake Victoria's cichlids. Regardless of the uncertainties of certain details, Lake Victoria's cichlids are widely recognized as exemplifying the fastest rate of evolution known for such a large group of vertebrate species.

Loss of cichlid diversity in Lake Victoria In recent years, many of the cichlid species living in Lake Victoria have become extinct. The Nile perch, a voracious predator that was deliberately introduced into the lake in 1960 to stimulate the local fishing economy and whose population exploded in the 1980s, has been blamed for the extinction of most of the cichlid species that have disappeared.

Biologists from the University of Leiden in the Netherlands observed, however, that species were also disappearing in habitats in the lake where the Nile perch is known to have little or no

(a) *Pundamilia pundamilia* males have bluish-silver bodies.

(b) *Pundamilia nyererei* males have red backs.

(c) *Pundamilia* "red head" males have a red "chest." (The "red head" species has not yet been scientifically named.)

Figure 20-12 Color variation in Lake Victoria cichlids

Some evidence suggests that changes in male coloration may be the first step in speciation of Lake Victoria cichlids. Later, other traits, including ecological characteristics, diverge. Female cichlids generally have cryptic coloration; their drab colors help them blend into their surroundings.

Why is cichlid diversity declining in Lake Victoria?

HYPOTHESIS: Changes in their habitat are causing closely related cichlid species to interbreed in Lake Victoria.

EXPERIMENT: Seehausen and his colleagues took field measurements in 22 different locations in Lake Victoria and later verified those field observations with laboratory experiments in which two cichlid species were placed in aquaria that were either well-lit or not.

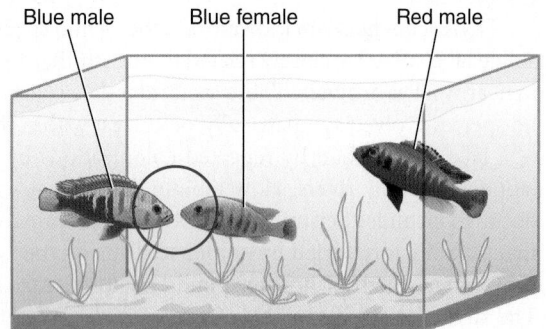

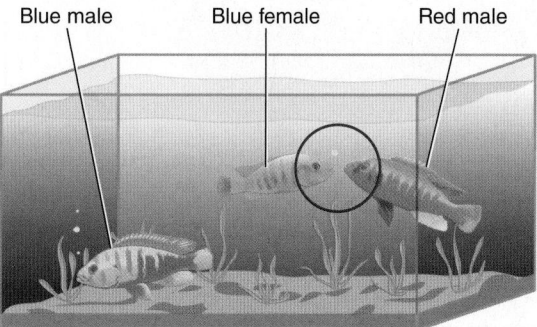

RESULTS AND CONCLUSION: When aquaria were well-lit (*left*), females of two closely related cichlid species (one in which the males are red, the other in which the males are blue) consistently chose mates of their own species (*circled area*) over those of other species. When light was blocked to simulate turbid conditions found in parts of Lake Victoria (*right*), females often chose males of other species.

Thus, increasing turbidity of Lake Victoria's water is preventing females of cichlid species from using color to select mates of their own species. As a result, females are interbreeding with males of other species, and the number of cichlid species is declining as the distinctive gene pools of different species are merging.

SOURCE: Seehausen, O., et al. "Speciation through Sensory Drive in Cichlid Fish," *Nature*, Vol. 455, October 2, 2008.

Figure 20-13 Explaining the rapid loss of cichlid diversity in Lake Victoria

PREDICT You conduct an experiment in which you test mate preferences of F_1 hybrid females for red males versus blue males in well-lit aquaria. Predict the results that would support the hypothesis that hybrid females do not use color cues to select mates.

© Cengage Learning

effect. Ole Seehausen and his colleagues measured environmental features at 22 localities with such habitats and found strong correlations between bright and diverse male colors and clear, well-lit water. Each species has its own distinct male coloration, which is used by females of that species to choose mates. In clear water members of each species do not interbreed with members of other species.

As nearby forests have been cut, soil erosion has made the water of Lake Victoria more turbid (cloudy). Agricultural practices in the area have also contributed fertilizer and sediment pollution. Seehausen demonstrated that as the water became more turbid from pollution, light could not penetrate it effectively. He hypothesized that as a result, females could not distinguish males of their own species from males of closely related species.

As already discussed, cichlids have evolved during the past 12,000 years. Their relative youth as species means that reproductive isolating mechanisms other than mate preference based on color may not have evolved. It also means that closely related species can be expected to interbreed without loss of fertility. With increasing turbidity, males lose their bright colors because they are no longer favored by sexual selection, and females have mated with males of other

species (**FIG. 20-13**). Thus, many species are being replaced with a few hybrids.

The study of hybrid zones has made important contributions to what is known about speciation

When two populations have significantly diverged as a result of geographic separation, there is no easy way to determine if the speciation process is complete (recall the disagreement about whether Porto Santo rabbits are a separate species or a subspecies). If such populations, subspecies, or species come into secondary contact before strong prezygotic isolating mechanisms have evolved, they may hybridize, producing fertile offspring, where they meet. A **hybrid zone** is an area of overlap between two recently diverged populations in which interbreeding takes place and hybrid offspring are common. Hybrid zones are typically narrow, presumably because the hybrids are not well adapted for either parental environment. The hybrid population is typically very small compared with the parental populations.

On the Great Plains of North America, red-shafted and yellow-shafted flickers (types of woodpeckers) meet and interbreed. The red-shafted flicker, named for the bird's red underwings and tail, is found in the western part of North America, from the Great Plains to the Pacific Ocean. Yellow-shafted flickers, which have yellow underwings and tails, range

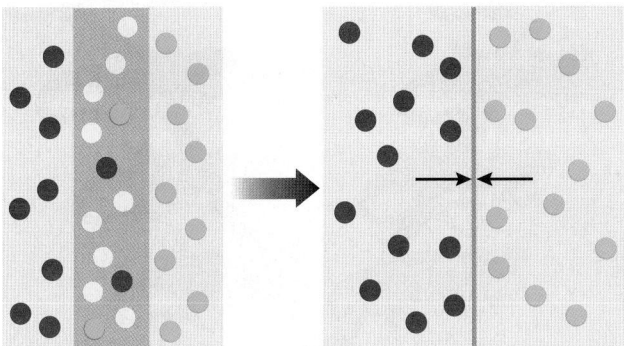

(a) Reinforcement. Hybrids are not as fit as the parents. Over time, the differences between the two parental populations increase, and hybrids are no longer produced. As the hybrid zone shrinks, the resulting outcome is two species.

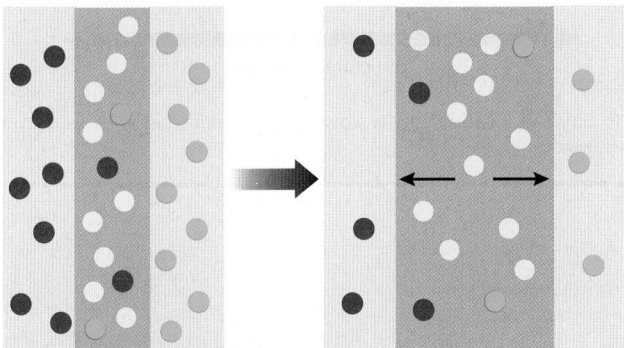

(b) Fusion. Hybrids are as fit as the two parental populations. Over time, differences between the two parental populations weaken, and the hybrid zone enlarges. The essential result is a single species.

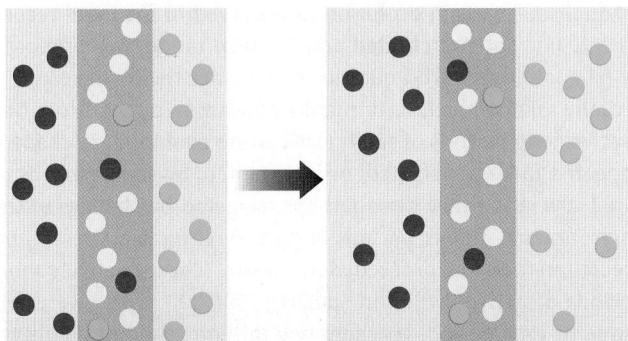

(c) Stability. Hybrids are more fit than either parental population. The fitness of the hybrids results in a stable hybrid zone. Eventually, a third species may form in the hybrid zone.

Figure 20-14 How hybrid zones between two closely related species or diverging populations can change over time
© Cengage Learning

east of the Rockies. Hybrid flickers, which form a stable hybrid zone from Texas to southern Alaska, are varied in appearance, although many have orange underwings and tails.

According to the biological species concept, if red-shafted flickers and yellow-shafted flickers are two species, they should maintain their reproductive isolation. Because a hybrid zone has formed, most biologists think that the red-shafted flickers and yellow-shafted flickers are geographic subspecies instead of separate species; red-shafted, yellow-shafted, and hybrid flickers are collectively called northern flickers.

Three evolutionary processes may occur in hybrid zones When two diverging populations or closely related species come into contact and form a hybrid zone, one of three possibilities can occur: reinforcement, which helps maintain two separate species; fusion of the two species into one; or stability of the hybrid zone (**FIG. 20-14**). **Reinforcement** occurs when the hybrids in the hybrid zone are less fit than the two parental populations. Reinforcement is the increase in reproductive isolation that often occurs over time in a hybrid zone; that is, natural selection strengthens and increases the number of prezygotic barriers between the two species. As a result, fewer and fewer hybrids are formed, and the hybrid zone gradually disappears.

In *fusion* the hybrids in a hybrid zone are as fit as either of the parental populations. Over time, differences in the reproductive barriers between the two parental populations weaken, and the hybrid zone enlarges. The eventual result is a fusing of two species (or subspecies) into one. Hybridization of the Lake Victoria cichlids (discussed earlier in the chapter) is an example of species loss due to fusion.

Stability is an evolutionary process in which the hybrids are more fit than either parental population. Hybrids continue to be formed, resulting in a stable hybrid zone. Stable hybrid zones often occur in **ecotones,** areas of transition between two different environments. In this case, the two incipient species are adapted to two different communities, and the hybrids' combination of characters from both parental populations allows them to flourish in the ecotone. Eventually, a third species may form in such a hybrid zone. (Ecotones are discussed further in Chapter 56.)

In the example of the flicker hybrid zone, reinforcement does not appear to be occurring. At the same time, the two kinds of flickers are not fusing together into one distinct group. The flicker hybrid zone has not expanded; the two types of flickers have maintained their distinctiveness and have not rejoined into a single, freely interbreeding population. Thus, at this time, flicker hybrids have formed a stable hybrid zone.

CHECKPOINT 20.3

- *What are five geographic barriers that might lead to allopatric speciation?*
- *How do hybridization and polyploidy cause a new plant species to form in as little as one generation?*
- *What is the likely mechanism of speciation for pupfishes? for cichlids?*

Punctuated equilibrium and phyletic gradualism represent opposite ends of a spectrum of speciation patterns. Speciation often proceeds in an intermediate or mixed pattern.

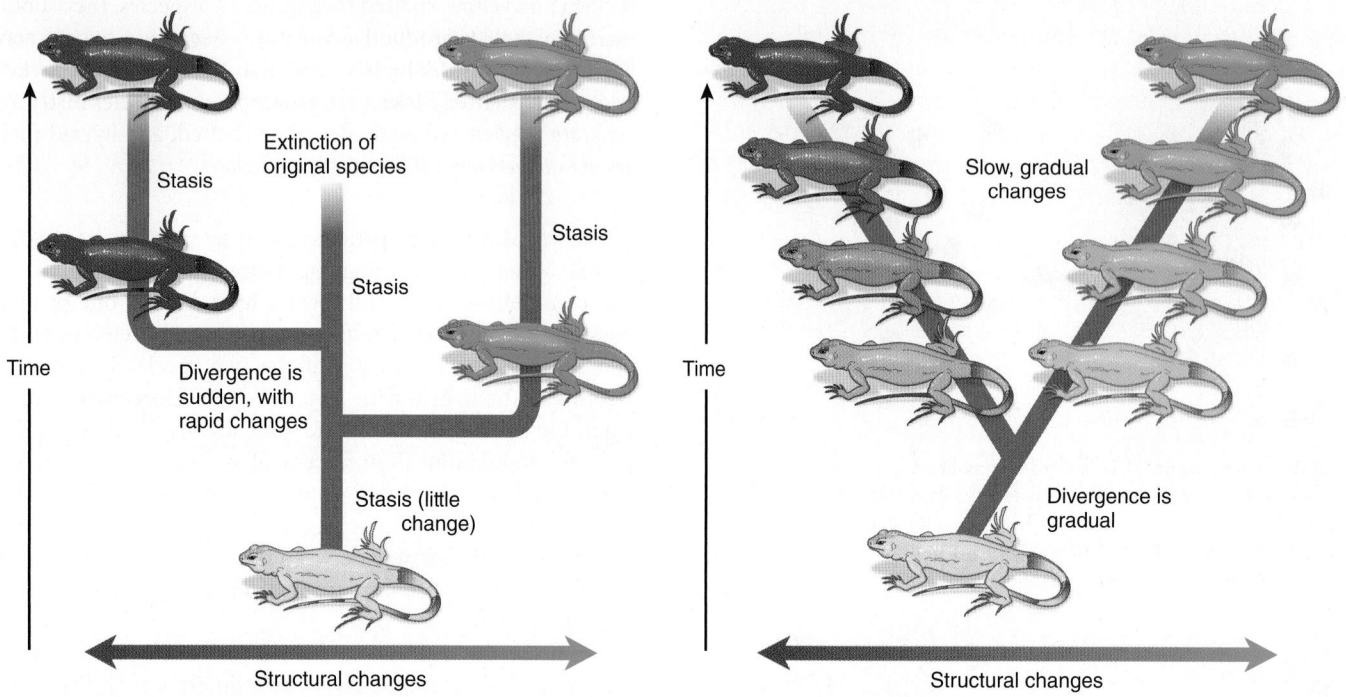

(a) In **punctuated equilibrium**, long periods of stasis are interrupted by short periods of rapid speciation.

(b) In **phyletic gradualism**, a slow, steady change in species occurs over time.

Figure 20-15 Punctuated equilibrium and gradualism

In this figure structural changes in the lizards are represented by changes in skin color.

CONNECT Many scientists think that tropical rain forests have been regions of long-term environmental stability, at least until modern times. If you had the means to follow the evolution of a particular clade of tropical rain forest organisms over the past one million years, would you expect to observe a pattern that would be closer to punctuated equilibrium or to phyletic gradualism?

© Cengage Learning

20.4 THE RATE OF EVOLUTIONARY CHANGE

LEARNING OBJECTIVE

5 Discuss the pace of evolution by describing punctuated equilibrium and phyletic gradualism.

We have seen that speciation is hard for us to directly observe as it occurs. Does the fossil record provide clues about how rapidly new species arise? Biologists have long recognized that the fossil record lacks many transitional forms; the starting points (ancestral species) and the end points (new species) are present, but the intermediate stages in the evolution from one species to another are often lacking. This observation has traditionally been explained by the incompleteness of the fossil record, although scientists have sometimes made new fossil discoveries that fill in the missing parts.

Two different models—punctuated equilibrium and phyletic gradualism—explain evolution as observed in the fossil record (FIG. 20-15). The **punctuated equilibrium** model was proposed by paleontologists who questioned whether the fossil record is actually as incomplete as it initially appeared. In the history of a species, long periods of **stasis** (little or no evolutionary change) appeared to be punctuated, or interrupted, by short periods of rapid speciation that were perhaps triggered by changes in the environment—that is, periods of great evolutionary stress—and speciation therefore proceeded in "spurts." These relatively short periods of active evolution (perhaps 100,000 years) were followed by long periods (perhaps two million years) of stability.

Punctuated equilibrium accounts for the abrupt appearance of a new species in the fossil record, with little or no evidence of intermediate forms. Proponents hypothesize that few transitional forms appear in the fossil record because few transitional forms occurred during speciation.

In contrast, the traditional Darwinian view of evolution espouses the **phyletic gradualism** model, in which evolution proceeded continuously over long periods. Gradualism is rarely observed in the fossil record because the record is incomplete. (Recall from Chapter 18 that precise conditions are required for fossil formation. Most organisms leave no trace of their existence because they decompose when they die.) Occasionally, a complete fossil record of transitional forms has been discovered and cited as a strong case for gradualism. The gradualism model maintains that populations slowly diverge from one another by the gradual accumulation of adaptive characteristics within each population. These adaptive characteristics accumulate as a result of different selective pressures encountered in different environments.

Scientists supporting phyletic gradualism maintain that any periods of stasis evident in the fossil record are the result of *stabilizing selection* (see Chapter 19). They also emphasize that stasis in fossils is deceptive because fossils do not reveal all aspects of evolution. Although fossils display changes in external structure and skeletal structure, other genetic changes— in physiology, internal structure, resistance to infection, and behavior—all of which also represent evolution, are not evident. Gradualists recognize rapid evolution only when strong *directional selection* occurs.

Many biologists today embrace both models to explain the fossil record. They contend that the pace of evolution may be abrupt in certain instances and gradual in others and that neither punctuated equilibrium nor gradualism exclusively characterizes the changes associated with evolution. Many biologists do not view the distinction between punctuated equilibrium and phyletic gradualism as real. They suggest that genetic changes occur gradually and at a roughly constant pace and that the majority of these mutations do not cause speciation. When the mutations that do cause speciation occur, they are dramatic and produce a pattern consistent with the punctuated equilibrium model.

CHECKPOINT 20.4

- *Are the phyletic gradualism and punctuated equilibrium models mutually exclusive? Explain your answer.*

20.5 MACROEVOLUTION

LEARNING OBJECTIVES

6 Define *macroevolution*.

7 Discuss macroevolution in the context of novel features, including preadaptations, allometric growth, and paedomorphosis.

8 Discuss the macroevolutionary significance of adaptive radiation and extinction.

Macroevolution concerns large-scale phenotypic changes in populations that generally warrant their placement in taxonomic groups at the species level and higher, that is, new species, genera, families, orders, classes, and even phyla, kingdoms, and domains. One concern of macroevolution is to explain evolutionary novelties, which are large phenotypic changes such as the appearance of jointed limbs during the evolution of arthropods (crustaceans, insects, and spiders). These phenotypic changes are so great that the new species possessing them are assigned to different genera or higher taxonomic categories. Studies of macroevolution also seek to discover and explain major changes in species diversity through time such as occur during *adaptive radiation,* when many species appear, and *mass extinction,* when many species disappear. Thus, evolutionary novelties, adaptive radiation, and mass extinction are important aspects of macroevolution.

Evolutionary novelties originate through modifications of pre-existing structures

New designs arise from structures already in existence. A change in the basic pattern of an organism produces something unique, such as wings on insects, flowers on plants, or feathered wings on birds. Usually, these evolutionary novelties are variations of some pre-existing structures, called **preadaptations,** that originally fulfilled one role but subsequently changed in a way that was adaptive for a different role. Feathers, which evolved from reptilian scales and may have originally provided thermal insulation in primitive birds and some dinosaurs, represent a preadaptation for flight. With gradual modification feathers evolved to function in flight as well as to fulfill their original thermoregulatory role. (Interestingly, a few feather-footed bird species exist; this phenotype is the result of a change in gene regulation that alters scales, normally found on bird feet, into feathers.)

How do such evolutionary novelties originate? Many are probably due to changes during development. *Regulatory genes* exert control over hundreds of other genes during development, and very slight genetic changes in regulatory genes could ultimately cause major structural changes in the organism (see Chapters 17 and 18).

For example, during development most organisms exhibit varied rates of growth for different parts of the body, known as **allometric growth** (from the Greek *allo,* "different," and *metr,* "measure"). The size of the head in human newborns is large in proportion to the rest of the body. As a human grows and matures, the torso, hands, and legs grow more rapidly than the head. Allometric growth is found in many organisms, including the male fiddler crab with its single, oversized claw and the ocean sunfish with its enlarged tail (FIG. 20-16). If growth rates are altered even slightly, drastic changes in the shape of an organism may result, changes that may or may not be adaptive. For example, allometric growth may help explain the extremely small and relatively useless forelegs of the dinosaur *Tyrannosaurus rex* compared with those of its ancestors.

Sometimes novel evolutionary changes occur when a species undergoes changes in the *timing* of development. Consider, for example, the changes that would occur if juvenile characteristics were retained in the adult stage, a phenomenon known as **paedomorphosis** (from the Greek *paed,* "child," and *morph,* "form"). Adults of some salamander species have external gills

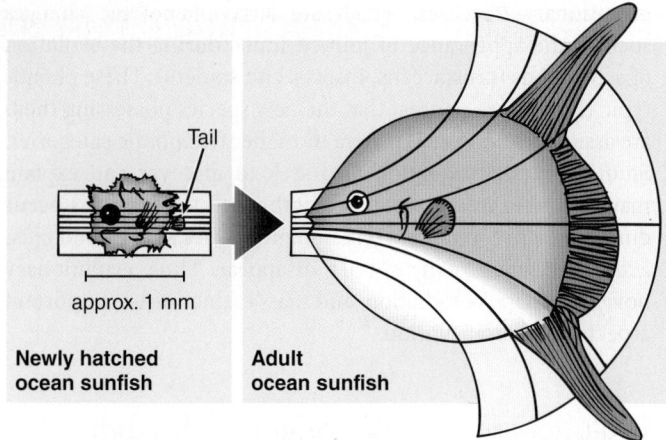

(a) A newly hatched ocean sunfish, only 1 mm long, has an extremely small tail.

(b) This allometric transformation is visualized by drawing rectangular coordinate lines through a picture of the juvenile fish and then changing the coordinate lines mathematically.

(c) An ocean sunfish swims off the coast of southern California. The adult ocean sunfish may reach a length of 4 m (13 ft) and weigh about 1500 kg (3300 lb).

Figure 20-16 Allometric growth in the ocean sunfish

The tail end of an ocean sunfish (*Mola mola*) grows faster than the head end, resulting in the unique shape of the adult.

© Cengage Learning

and tail fins, features found only in the larval (immature) stages of other salamanders.

Retention of external gills and tail fins throughout life obviously alters the axolotl salamander's behavioral and ecological characteristics (**FIG. 20-17**). Perhaps such salamanders succeeded because they had a selective advantage over "normal" adult salamanders; that is, by remaining aquatic, they did not have to compete for food with the terrestrial adult forms of related species. The paedomorphic forms also escaped the typical predators of terrestrial salamanders (although they had other predators in their aquatic environment). Studies suggest that paedomorphosis in salamanders is probably the result of mutations in genes that block the production of hormones that stimulate metamorphic changes. When paedomorphic salamanders receive hormone injections, they develop into adults lacking external gills and tail fins.

Figure 20-17 Paedomorphosis in a salamander

An adult axolotl salamander (*Ambystoma mexicanum*) retains the juvenile characteristics of external gills (feathery structures protruding from the neck) and a tail fin (*not visible*). Paedomorphosis allows the axolotl to remain permanently aquatic and to reproduce without developing typical adult characteristics.

Adaptive radiation is the diversification of an ancestral species into many species

Some ancestral species have given rise to far more species than have other evolutionary lineages. **Adaptive radiation** is the evolutionary diversification of many related species from one or a few ancestral species in a relatively short period.

Evolutionary innovations such as those we considered in the last section may trigger an adaptive radiation. Alternatively, an adaptive radiation may occur because the ancestral species happens to have been in the right place at the right time to exploit numerous ecological opportunities.

Biologists developed the concept of adaptive zones to help explain this type of adaptive radiation. **Adaptive zones** are new ecological opportunities that were not exploited by an ancestral organism. At the species level, an adaptive zone is essentially identical to one or more similar *ecological niches* (the functional roles of species within a community; see Chapter 54). Examples of adaptive zones include nocturnal flying to catch small insects, grazing on grass while migrating across a savanna, and swimming at the ocean's surface to filter out plankton. When many adaptive zones are empty, as they were in Lake Victoria when it refilled some 12,000 years ago (discussed earlier in the chapter), colonizing species such as the cichlids may rapidly diversify and exploit them.

Because islands have fewer species than do mainland areas of similar size, latitude, and topography, vacant adaptive zones are more common on islands than on continents. Consider the Hawaiian honeycreepers, a group of related birds found in the Hawaiian Islands. When the honeycreeper ancestors reached Hawaii, few other birds were present. The succeeding generations of honeycreepers quickly diversified into many new species and, in the process, occupied the many available adaptive zones that on the mainland are occupied by finches, honeyeaters, treecreepers, and woodpeckers. The diversity of honeycreeper bills is a particularly good illustration of adaptive radiation (**FIG. 20-18**). Some honeycreeper bills are curved

Adaptive radiation occurs when a single ancestral species diversifies into a variety of species, each adapted to a different ecological niche.

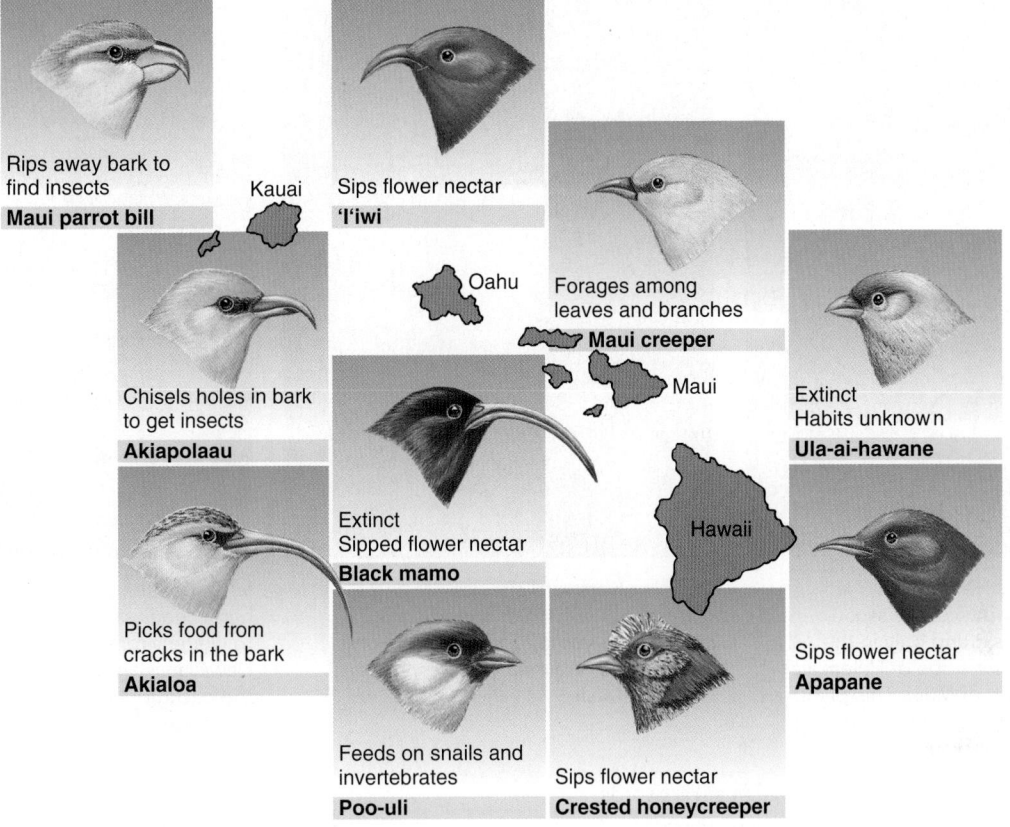

Figure 20-18 Adaptive radiation in Hawaiian honeycreepers

Compare the various beak shapes and methods of obtaining food. Many honeycreeper species are now extinct or nearing extinction as a result of human activities, including the destruction of habitat and the introduction of predators such as rats, dogs, and pigs.

© Cengage Learning

CONNECT Do you think that adaptive radiation in the Hawaiian honeycreepers will continue in the future? Explain you answer.

to extract nectar out of tubular flowers, whereas others are short and thick for ripping away bark in search of insects.

Another example of vacant adaptive zones involves the Hawaiian silverswords, 28 species of closely related plants found only in the Hawaiian Islands. When the silversword ancestor, a California plant related to daisies, reached the Hawaiian Islands, many diverse environments—for example, cool, arid mountains; exposed lava flows; dry woodlands; shady, moist forests; and wet bogs—were present and more or less unoccupied. The succeeding generations of silverswords quickly diversified in structure and physiology to occupy the many adaptive zones available to them. The diversity in their leaves, which changed during the course of natural selection as different populations adapted to various levels of light and moisture, is a particularly good illustration of adaptive radiation (FIG. 20-19). For example, leaves of silverswords that are adapted to shady, moist forests are large, whereas those of silverswords living in arid areas are small. The leaves of silverswords living on exposed volcanic slopes are covered with dense, silvery hairs that may reflect some of the intense ultraviolet radiation off the plant.

Adaptive radiation appears more common during periods of major environmental change, but it is difficult to determine how these changes are related to adaptive radiation. It is possible that major environmental change indirectly affects adaptive radiation by increasing the extinction rate. Extinction produces empty adaptive zones, which provide new opportunities for species that remain. Mammals, for example, diversified and exploited a variety of adaptive zones relatively soon after the dinosaurs' extinction. Flying bats, running gazelles, burrowing moles, and swimming whales all originated from early mammals that had coexisted with reptiles for millions of years.

Extinction is an important aspect of evolution

Extinction, the end of a lineage, occurs when the last individual of a species dies. The loss is permanent, for once a species is extinct, it never reappears. Extinctions have occurred continually since the origin of life; by one estimate, only one species is alive today for every 2000 that have become extinct. Extinction

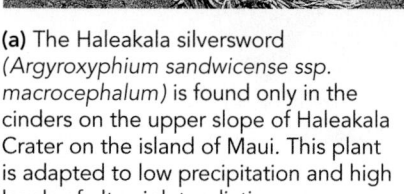

(a) The Haleakala silversword (*Argyroxyphium sandwicense ssp. macrocephalum*) is found only in the cinders on the upper slope of Haleakala Crater on the island of Maui. This plant is adapted to low precipitation and high levels of ultraviolet radiation.

(b) This silversword species (*Wilkesia gymnoxiphium*), which superficially resembles a yucca, is found only along the slopes of Waimea Canyon on the island of Kauai.

(c) *Daubautia scabra* is a small, herbaceous silversword found in moist to wet environments on several Hawaiian islands. (The fern fronds in the background give an idea of the small size of *D. scabra*.)

Figure 20-19 Adaptive radiation in Hawaiian silverswords
The 28 silversword species, which are classified in three closely related genera, live in a variety of habitats.

is the eventual fate of all species in the same way that death is the eventual fate of all individual organisms.

Although extinction has a negative short-term effect on the number of species, it facilitates evolution over a period of thousands to millions of years. As mentioned previously, when species become extinct, their adaptive zones become vacant. Consequently, surviving species are presented with new evolutionary opportunities and may diverge, filling in some of the unoccupied zones. In other words, the extinct species may eventually be replaced by new species.

During the long history of life, extinction appears to have occurred at two different rates. The continuous, low-level extinction of species is sometimes called **background extinction**. In contrast, five or possibly six times during Earth's history **mass extinctions** of numerous species and higher taxonomic groups have taken place in both terrestrial and marine environments. The most recent mass extinction, which occurred about 65 million years ago (mya), killed off many marine organisms, terrestrial plants, and vertebrates, including the last of the dinosaurs. The time span over which a mass extinction occurs may be several million years, but that is relatively short compared with the 3.5 billion years or so of Earth's history of life. Each period of mass extinction has been followed by a period of adaptive radiation of some of the surviving groups.

The causes of past episodes of mass extinction are not well understood. Both environmental and biological factors seem to have been involved. Major changes in climate could have adversely affected those plants and animals that lacked the genetic flexibility to adapt. In particular, marine organisms are adapted to a steady, unchanging climate. If Earth's temperature were to increase or decrease by just a few degrees overall, many marine species would probably perish.

It is also possible that past mass extinctions were due to changes in the environment induced by catastrophes. If a large comet or small asteroid collided with Earth, the dust ejected into the atmosphere on impact could have blocked much of the sunlight. In addition to disrupting the food chain by killing many plants (and therefore terrestrial animals), this event would have lowered Earth's temperature, leading to the death of many marine organisms. Evidence exists that the extinction of dinosaurs was caused by an extraterrestrial object's collision with Earth (see Chapter 21).

Biological factors also trigger extinction. Competition among species may lead to the extinction of species that cannot compete effectively. The human species in particular has had a profound effect on the rate of extinction. The habitats of many animal and plant species have been altered or destroyed by humans, and habitat destruction can result in a species' extinction. Some biologists think that we have entered the greatest mass extinction episode in Earth's history. (Extinction is discussed further in Chapter 57.)

Is microevolution related to speciation and macroevolution?

The concepts presented in Chapters 18 and 19 represent the **modern synthesis,** in which mutation provides the genetic variation on which natural selection acts. The modern synthesis combines Darwin's scientific theory with important aspects of genetics. Many aspects of the modern synthesis have been tested and verified at the population and subspecies levels. Many biologists contend that microevolutionary processes explain macroevolution. They assert that natural selection, mutation, genetic drift, and gene flow account for not only the genetic variation within populations but also the origin of new species and higher taxa.

A considerable body of data from many fields supports the modern synthesis as it relates to speciation and macroevolution. Consider, for example, the *fish–tetrapod transition* (the evolution of four-legged amphibians from fishes), which was a major macroevolutionary event in the history of vertebrates (see Chapter 32). The fish–tetrapod transition is well documented in the fossil record. Studies of fossil intermediates show that the transition from aquatic fishes to terrestrial amphibians occurred as evolutionary novelties, such as changes in the limbs and skull roof, were added. These novelties accumulated as a succession of small changes over a period of 9 million to 14 million years. This time scale is sufficient to have allowed natural selection and other microevolutionary processes to produce the novel characters.

Although few biologists doubt the role of natural selection and microevolution in generating specific adaptations, some question the *extent* of microevolution's role in the overall pattern of life's history. These biologists ask whether speciation and macroevolution have been dominated by microevolutionary processes or by external, chance events, such as an impact by an asteroid. Chance events do not "care" about adaptive superiority but instead lead to the random extinction or survival of species. In the case of an asteroid impact, for example, certain species may survive because they were "lucky" enough to be in a protected environment at the time of impact. If chance events have been the overriding factor during life's history, microevolution cannot be the exclusive explanation for the biological diversity present today (see discussion of the role of chance in evolution in Chapter 18).

CHECKPOINT 20.5

- *Why are evolutionary novelties important to scientists studying macroevolution?*
- *In what ways does preadaptation facilitate macroevolution?*
- *What are some potential selective advantages of paedomorphosis?*
- CONNECT *What is the relationship between mass extinction and adaptive radiation?*

SUMMARY: FOCUS ON LEARNING OBJECTIVES

20.1 What Is a Species? *(page 418)*

1 Describe the biological species concept and list two potential problems with the concept.

- According to the **biological species concept,** a species consists of one or more populations whose members interbreed in nature to produce fertile offspring and do not interbreed with members of different species.
- One problem with the biological species concept is that it applies only to sexually reproducing organisms. Also, although individuals assigned to different species do not normally interbreed, they may occasionally successfully interbreed.

20.2 Reproductive Isolation *(page 419)*

2 Explain the significance of reproductive isolating mechanisms and distinguish among the various prezygotic and postzygotic barriers.

- **Reproductive isolating mechanisms** restrict gene flow between species.
- **Prezygotic barriers** are reproductive isolating mechanisms that prevent fertilization from taking place. **Temporal isolation** occurs when two species reproduce at different times of the day, season, or year. In **habitat isolation** two closely related species live and breed in different habitats in the same geographic area. In **behavioral isolation** distinctive courtship behaviors prevent mating between species. **Mechanical**

isolation is due to incompatible structural differences in the reproductive organs of similar species. In **gametic isolation,** gametes from different species are incompatible because of molecular and chemical differences.

© Cengage Learning

- **Postzygotic barriers** are reproductive isolating mechanisms that prevent gene flow after fertilization has taken place. **Hybrid inviability** is the death of interspecific embryos during development. **Hybrid sterility** prevents interspecific hybrids that survive to adulthood from reproducing successfully. **Hybrid breakdown** prevents the offspring of hybrids that survive to adulthood and successfully reproduce from reproducing beyond one or a few generations.

20.3 Speciation *(page 421)*

3 Explain the mechanism of allopatric speciation and give an example.

- **Speciation** is the evolution of a new species from an ancestral population. **Allopatric speciation** (e.g., in Death Valley pupfishes) occurs when one population becomes geographically isolated from the rest of the species and subsequently diverges. Speciation is more likely to occur if the original isolated population is small because genetic drift is more significant in small populations than in large populations.

4 Explain the mechanisms of sympatric speciation and give both plant and animal examples.

- **Sympatric speciation** does not require geographic isolation. In plants sympatric speciation can result when a **polyploid** individual (one with more than two sets of chromosomes) is an **allopolyploid** hybrid derived from two species. One example of sympatric speciation by allopolyploidy is the hemp nettle.

- Sympatric speciation occurs in animals, such as fruit maggot flies and cichlids, but how often it occurs and under what conditions remain to be determined.

20.4 The Rate of Evolutionary Change *(page 430)*

5 Discuss the pace of evolution by describing punctuated equilibrium and phyletic gradualism.

- According to the **punctuated equilibrium** model, evolution of species proceeds in spurts. Short periods of active speciation are interspersed with long periods of **stasis.** According to the **phyletic gradualism** model, populations slowly diverge from one another by the accumulation of adaptive characteristics within a population.

20.5 Macroevolution *(page 431)*

6 Define *macroevolution*.

- **Macroevolution** concerns large-scale phenotypic changes in populations that typically warrant the placement of the populations in taxonomic groups at the species level and higher, that is, new species, genera, families, orders, classes, and even phyla, kingdoms, and domains.

7 Discuss macroevolution in the context of novel features, including preadaptations, allometric growth, and paedomorphosis.

- Evolutionary novelties may be due to changes during **development,** the orderly sequence of events that occurs as an organism grows and matures. Slight genetic changes in regulatory genes could cause major structural changes in an organism.

- Evolutionary novelties may originate from **preadaptations,** structures that originally fulfilled one role but changed in a way that was adaptive for a different role. Feathers are an example of a preadaptation.

- **Allometric growth,** varied rates of growth for different parts of the body, results in overall changes in the shape of an organism. Examples include the expanded tail region of the ocean sunfish and the enlarged claw of the male fiddler crab.

- **Paedomorphosis,** the retention of juvenile characteristics in the adult, occurs because of changes in the timing of development. Adult axolotl salamanders, with external gills and tail fins, are an example of paedomorphosis.

8 Discuss the macroevolutionary significance of adaptive radiation and extinction.

- **Adaptive radiation** is the process of diversification of an ancestral species into many new species. **Adaptive zones** are new ecological opportunities that were not exploited by an ancestral organism. When many adaptive zones are empty, colonizing species may rapidly diversify and exploit them. Hawaiian honeycreepers and silverswords both underwent adaptive radiation after their ancestors colonized the Hawaiian Islands.

© Cengage Learning

- **Extinction** is the death of a species. When species become extinct, the adaptive zones that they occupied become vacant, allowing other species to evolve and fill those zones. **Background extinction** is the continuous, low-level extinction of species. **Mass extinction** is the extinction of numerous species and higher taxonomic groups in both terrestrial and marine environments.

TEST YOUR UNDERSTANDING

Know and Comprehend

1. A prezygotic barrier prevents (a) the union of egg and sperm (b) reproductive success by an interspecific hybrid (c) the development of the zygote into an embryo (d) allopolyploidy from occurring (e) changes in allometric growth

2. The reproductive isolating mechanism in which two closely related species live in the same geographic area but reproduce at different times is (a) temporal isolation (b) behavioral isolation (c) mechanical isolation (d) gametic isolation (e) hybrid inviability

3. Interspecific hybrids, if they survive, are (a) always sterile (b) always fertile (c) usually sterile (d) usually fertile (e) never sterile

4. The first step leading to allopatric speciation is (a) hybrid inviability (b) hybrid breakdown (c) adaptive radiation (d) geographic isolation (e) paedomorphosis

5. The pupfishes in the Death Valley region are an example of which evolutionary process? (a) background extinction (b) allopatric speciation (c) sympatric speciation (d) allopolyploidy (e) paedomorphosis

6. Which of the following evolutionary processes is associated with allopolyploidy? (a) gradualism (b) allometric growth (c) sympatric speciation (d) mass extinction (e) preadaptation

7. According to the punctuated equilibrium model, (a) populations slowly diverge from one another (b) the evolution of species occurs in spurts interspersed with long periods of stasis (c) evolutionary novelties originate from preadaptations (d) reproductive isolating mechanisms restrict gene flow between species (e) the fossil record, being incomplete, does not accurately reflect evolution as it actually occurred

8. The evolutionary conversion of reptilian scales into feathers is an example of (a) allometric growth (b) paedomorphosis (c) gradualism (d) hybrid breakdown (e) a preadaptation

9. Adaptive radiation is common following a period of mass extinction, probably because (a) the survivors of a mass extinction are remarkably well adapted to their environment (b) the unchanging environment following a mass extinction drives the evolutionary process (c) many adaptive zones are empty (d) many ecological niches are filled (e) the environment induces changes in the timing of development for many species

10. The Hawaiian silverswords are an excellent example of which evolutionary process? (a) allometry (b) preadaptation (c) micro-evolution (d) adaptive radiation (e) extinction

Apply and Analyze

11. **VISUALIZE** Use two different colors to depict the unduplicated chromosomes of species C with larger chromosomes ($2n = 8$) and species D with slightly smaller chromosomes ($2n = 10$), and of their F_1 hybrid. Is the hybrid likely to be fertile?

Evaluate and Synthesize

12. **EVOLUTION LINK** Is allopatric speciation more likely to occur if the original isolated population is small or large? Explain your answer.

13. **EVOLUTION LINK** Could hawthorn and apple maggot flies be considered an example of assortative mating, which was discussed in Chapter 19? Explain your answer.

14. **EVOLUTION LINK** Because mass extinction is a natural process that may facilitate evolution during the period of thousands to millions of years that follow it, should humans be concerned about the current mass extinction we are causing? Why or why not?

15. **INTERPRET DATA** Examine Figure 20-2c and predict which date is likeliest for researchers to have collected wood frogs and leopard frogs and interbred them in the lab: March 10, April 1, or April 15. Explain your answer.

aplia To access course materials, such as Aplia and other companion resources, please visit **www.cengagebrain.com.**

The Origin and Evolutionary History of Life

Fossil of an Early Permian reptile. This well-preserved fossil of *Orobates pabsti* was matched to well-preserved fossil footprints found nearby. Although the limbs are spread out in death, *Orobates* walked with an upright gait with its limbs under its body. Discovered and photographed in Germany.

© Phil Degginger/Carnegie Museum/Alamy

KEY CONCEPTS

21.1 Although there is no direct fossil evidence of the origin of life, biochemical experiments have demonstrated how the complex organic molecules that are found in all living organisms may have formed.

21.2 The first cells were probably heterotrophic, anaerobic, and pro-karyotic. Photosynthesis, aerobic respiration, and eukaryotic cell structure represent several major advances that occurred during the early history of life.

21.3 The fossil record tells us much of what we know about the history of life, such as what kinds of organisms existed and where and when they lived. Scientists identify and demonstrate relationships among fossils in rock layers from different periods of geologic time.

The preceding three chapters were concerned with biological evolution. However, we have not dealt with what many regard as a fundamental question: How did life begin? Although biologists generally accept the hypothesis that life developed from nonliving matter, exactly how this process, called **chemical evolution,** occurred is not certain. Current models suggest that small organic molecules first formed spontaneously and accumulated over time. Large organic macromolecules such as proteins and nucleic acids could have then assembled from the smaller molecules.

The macromolecules interacted, combining into more-complicated structures that could eventually metabolize and replicate, passing their genetic information on to their descendants, which eventually became the first true cells. After the first cells originated, they diverged over several billion years into the rich biological diversity that characterizes our planet today.

Initially, unicellular prokaryotes (archaea and bacteria) predominated, followed by unicellular eukaryotes. The first multicellular eukaryotes were soft-bodied marine animals that did not leave many fossils. Shelled animals and other marine invertebrates (animals without backbones) appeared next, as exemplified by trilobites, primitive aquatic arthropods. Marine invertebrates were followed by the first vertebrates (animals with backbones). The first fishes with jaws appeared and diversified, with some of them giving rise to amphibians, the first vertebrates with limbs that were capable of moving about on land. Amphibians gave rise to reptiles, which diversified and populated the land (see photograph). Reptiles, in turn, gave rise independently to birds and to mammals. Plants underwent a comparable evolutionary history and diversification.

In this chapter we survey life over a vast span of time, starting when our planet was relatively young. We first examine proposed models about how life began and then trace life's long evolutionary history.

21.1 CHEMICAL EVOLUTION ON EARLY EARTH

LEARNING OBJECTIVES

1 Describe the conditions that scientists think existed on early Earth.
2 Compare the prebiotic soup hypothesis with the iron–sulfur world hypothesis.

Scientists generally agree that life's beginnings occurred under environmental conditions quite different from those of today. Therefore, we must examine the conditions of early Earth to understand the origin of life. Evidence from many sources provides us with clues that help us formulate plausible scenarios of the steps in which life originated. Study of the origin of life is an active area of scientific research today, and many important contributions are adding to our understanding of how life began.

Astrophysicists and geologists estimate that Earth is about 4.6 billion years old. The atmosphere of early Earth apparently included carbon dioxide (CO_2), water vapor (H_2O), carbon monoxide (CO), and nitrogen (N_2). It may also have contained some ammonia (NH_3), hydrogen sulfide (H_2S), and methane (CH_4), although ultraviolet radiation from the sun most likely broke these reduced molecules down rapidly. The early atmosphere probably contained little or no free oxygen (O_2).

Four requirements must have existed for the chemical evolution of life: little or no free oxygen, a source of energy, the availability of chemical building blocks, and time. First, life could

have begun only in the absence of free oxygen. Oxygen is quite reactive and would have oxidized the organic molecules that are necessary building blocks in the origin of life. Earth's early atmosphere was probably strongly reducing, which means that any free oxygen would have reacted with other elements to form oxides. Thus, oxygen would have been tied up in compounds.

The origin of life would also have required energy to do the work of building biological molecules from simple inorganic chemicals. Early Earth was a place of high energy with violent thunderstorms; widespread volcanic activity; bombardment from meteorites and other extraterrestrial objects; and intense radiation, including ultraviolet radiation from the sun (FIG. 21-1). The young sun probably produced more ultraviolet radiation than it does today, and ancient Earth had no protective ozone layer to filter it.

A third requirement would have been the presence of the chemical building blocks needed for chemical evolution, including water, dissolved inorganic minerals (present as ions), and the gases present in the early atmosphere. A final requirement for the origin of life was time for molecules to accumulate and react with one another. Earth is approximately 4.6 billion years old, and evidence suggests that life arose early in Earth's history.

Organic molecules formed on primitive Earth

Because organic molecules are the building materials for organisms, it is reasonable to first consider how they may have originated. Two main models seek to explain how the organic precursors of life originated: the **prebiotic soup hypothesis** proposes that these molecules formed near Earth's surface, whereas

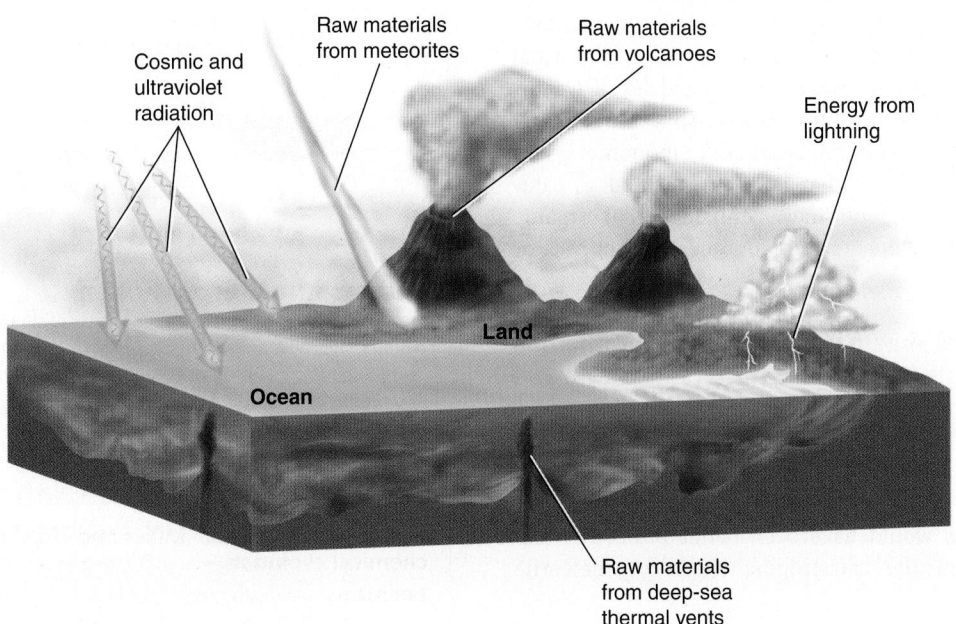

Figure 21-1 Conditions on early Earth

Volcanoes erupted, spewing gases that contributed to the atmosphere, and violent thunderstorms triggered torrential rainfall that washed molecules out of the atmosphere and eroded the land. Meteorites and other extraterrestrial objects bombarded Earth, cataclysmically changing the crust, ocean, and atmosphere. Deep-sea thermal vents, heated by Earth's internal heat, spewed forth chemicals. High-energy radiation from cosmic and ultraviolet rays blasted the planet.

© Cengage Learning

the **iron–sulfur world hypothesis** proposes that organic precursors formed at cracks in the ocean's floor.

Organic molecules may have been produced at Earth's surface

The concept that simple organic molecules such as sugars, nucleotide bases, and amino acids could form spontaneously from simpler raw materials was first advanced in the 1920s by two scientists working independently: A.I. Oparin, a Russian biochemist, and J.B.S. Haldane, a Scottish physiologist and geneticist.

Their hypothesis was tested in the 1950s by U.S. biochemists Stanley Miller and Harold Urey, who designed a closed apparatus that simulated conditions that presumably existed on early Earth (FIG. 21-2). They exposed an atmosphere rich in H_2O, CH_4, NH_3, and H_2 to an electric discharge that simulated lightning. Their analysis of the chemicals produced during one week revealed that amino acids and other organic molecules had formed. Although more recent data suggest that Earth's early atmosphere was not rich in methane, ammonia, or hydrogen, similar experiments using different combinations of gases have produced a wide variety of organic molecules that are important in contemporary organisms. They include all 20 amino acids, several sugars, lipids, the nucleotide bases of RNA and DNA, and ATP (when phosphate is present). Thus, before life began, its chemical building blocks were probably accumulating as a necessary step in chemical evolution.

Oparin envisioned that the organic molecules would, over vast spans of time, accumulate in the shallow seas to form a "sea of organic soup." He envisioned smaller organic molecules (monomers) combining to form larger ones (polymers) under such conditions. Evidence gathered since Oparin's time indicates that organic polymers may have formed and accumulated on rock or clay surfaces rather than in the primordial seas. Clay, which consists of microscopic particles of weathered rock, is particularly intriguing as a possible site for early polymerizations because it binds organic monomers and contains zinc and iron ions that may have served as catalysts. Laboratory experiments have confirmed that organic polymers form spontaneously from monomers on hot rock or clay surfaces.

Organic molecules may have formed at hydrothermal vents

Some biologists hypothesize that early polymerizations leading to the origin of life may have occurred in cracks in the deep-ocean floor where hot water, carbon monoxide, and minerals such as sulfides of iron and nickel spew forth; this view is known as the iron–sulfur world hypothesis. Such **hydrothermal vents** would have been better protected than Earth's surface from the catastrophic effects of meteorite bombardment.

Today, these anaerobic hot springs produce precursors of biological molecules and of energy-rich "food," including the highly reduced compounds hydrogen sulfide and methane. These chemicals support a diverse community of microorganisms, clams, crabs, tube worms, and other animals (see *Inquiring About: Life without the Sun,* in Chapter 55).

Could organic molecules have formed in the conditions of early Earth?

HYPOTHESIS: Organic molecules can form in a reducing atmosphere similar to that thought to be present on early Earth.

EXPERIMENT: The apparatus that Miller and Urey used to simulate the reducing atmosphere of early Earth contained water, methane, ammonia, and hydrogen. An electrical spark was produced in the upper right flask to simulate lightning.

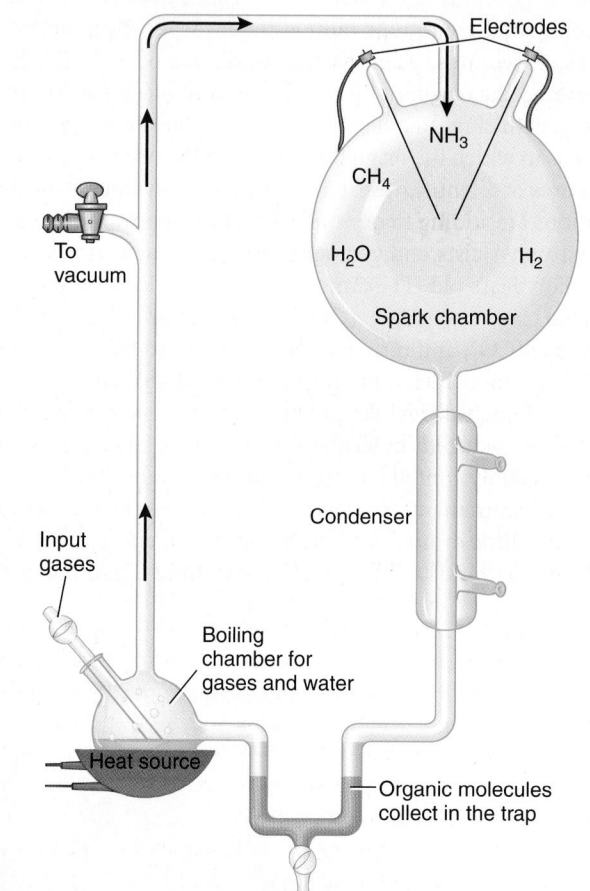

RESULTS AND CONCLUSION: The gases present in the flask reacted together, and in one week a variety of simple organic compounds, such as amino acids, accumulated in the trap at the bottom. Thus, the formation of organic molecules—the first step in the origin of life—can be produced from simple precursors.

SOURCE: Miller, S.L., and H.C. Urey, "Organic Compound Synthesis on the Primitive Earth," *Science*, Vol. 130, July 1959.

Figure 21-2 *Animation* **Miller and Urey's experiment in chemical evolution**

PREDICT Could organic molecules have been produced in this experiment if methane had been omitted?

© Cengage Learning

Testing the iron–sulfur world hypothesis at hydrothermal vents is difficult, but laboratory experiments simulating the high pressures and temperatures at the vents have yielded intriguing results. For example, iron and nickel sulfides catalyze reactions between carbon monoxide and hydrogen sulfide, producing acetic acid and other simple organic compounds. Also, experiments show that ammonia, one of the precursors of proteins and nucleic acids, is produced in abundance, suggesting that vents were ammonia-rich environments in the prebiotic world and that the earliest organisms could have formed in this seemingly inhospitable environment.

CHECKPOINT 21.1

- *What are the four requirements for chemical evolution, and why is each essential?*
- CONNECT *How does the iron–sulfur world hypothesis differ from the prebiotic soup hypothesis? What do these hypotheses have in common?*

21.2 THE FIRST CELLS

LEARNING OBJECTIVES

3 Outline the major steps hypothesized to have occurred in the origin of cells.

4 Explain how the evolution of photosynthetic autotrophs affected both the atmosphere and other organisms.

5 Describe the hypothesis of serial endosymbiosis.

After the first polymers formed, could they have assembled spontaneously into more complex structures with an outer membranous boundary? Scientists have synthesized several different **protobionts,** which are vesicle-like assemblages of abiotically produced (i.e., not produced by organisms) organic polymers. These protobionts resemble living cells in several ways, thus providing clues as to how aggregations of complex nonliving molecules took that "giant leap" and became living cells. These protobionts exhibit many functional and structural attributes of living cells. They often divide in half (binary fission) after they have sufficiently "grown." Protobionts maintain an internal chemical environment that is different from the external environment (homeostasis), and some of them show the beginnings of metabolism (catalytic activity). They are highly organized, considering their relatively simple composition.

Microspheres are a type of protobiont formed by adding water to abiotically formed polypeptides (FIG. 21-3). Some microspheres are excitable: they produce an electrical potential across their surfaces, reminiscent of electrochemical gradients in cells. Microspheres can also absorb materials from their surroundings (selective permeability) and respond to changes in osmotic pressure as though membranes enveloped them, even though they contain no lipid.

Scientists have constructed protobionts in which membranes of fatty acids and monoglycerides surround larger DNA-like molecules. These protobionts are impermeable to the genetic molecules but allow smaller organic molecules—that is, "nutrients"—to enter. This study is significant because it suggests how early cells could have held on to their "genes" and absorbed needed nutrients from their environment without the sophisticated membrane structure associated with contemporary cells.

The study of protobionts shows that relatively simple nonliving structures have some of the properties of life. However, it is a major step (or several steps) to go from simple molecular aggregates such as protobionts to living cells.

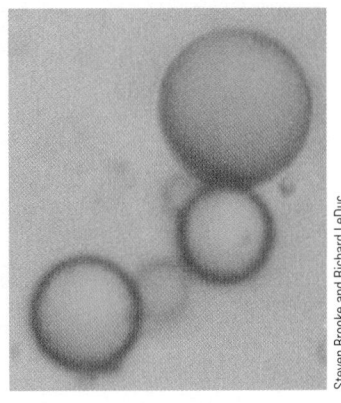

2 μm

Figure 21-3 Microspheres
These tiny protobionts exhibit some of the properties of life.

The origin of a simple metabolism within a membrane boundary may have occurred early in the evolution of cells

Although we have learned many things about how organic molecules may have formed on primitive Earth, the problem of how pre-cells evolved into living cells remains to be solved. Let's examine one possible scenario—the **metabolism first hypothesis**—supported by many chemists, including Robert Shapiro at New York University. Shapiro envisions life beginning as a self-sustaining, organized system consisting of chemical reactions between simple molecules that are enclosed within a boundary, that is, separated from the external environment (FIG. 21-4a).

A source of energy must be available to drive these chemical reactions (thereby maintaining an internal organization different from the external environment). To be self-sustaining, the energy source must be coupled to the chemical reaction sequence (FIG. 21-4b). (Recall how ATP links exergonic and endergonic reactions, shown in Figure 7-6.)

As the pre-cells continue to evolve, their organization increases (FIG. 21-4c). Finally, to survive, the pre-cell system must grow and reproduce. One of the most significant parts of the origin of living cells from pre-cells was the evolution of molecular reproduction (FIG. 21-4d).

Molecular reproduction was a crucial step in the origin of cells

In living cells genetic information is stored in the nucleic acid DNA, which is transcribed into messenger RNA (mRNA), which in turn is translated into the proper amino acid sequence in proteins. All three macromolecules in the DNA → RNA → protein sequence contain precise information. Of these macromolecules, only DNA and RNA are capable of self-replication (although only in the presence of the proper enzymes in cells today). Because both RNA and DNA can form spontaneously on clay in much the same way that other organic polymers

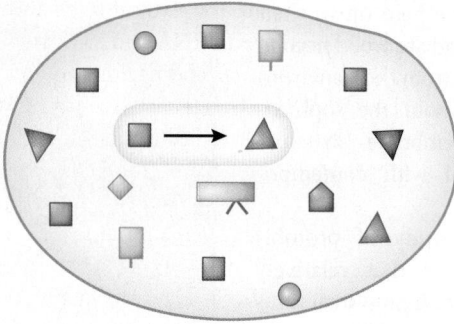

(a) Within a boundary, chemical reactions occur among simple molecules.

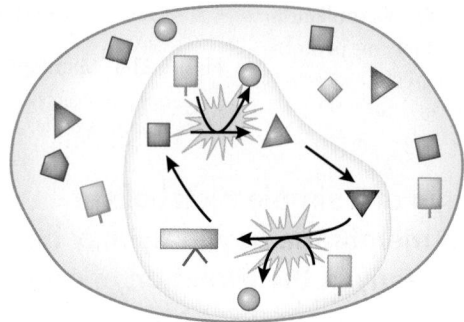

(b) A source of energy is coupled to the chemical reaction sequence.

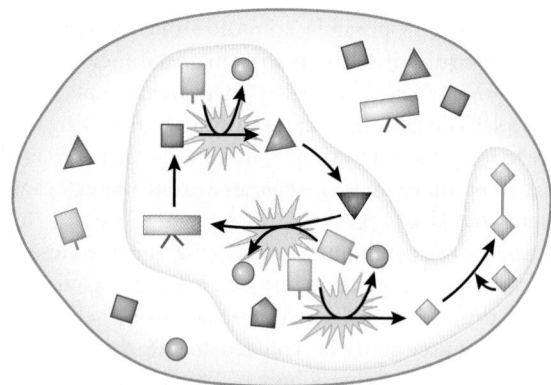

(c) As the pre-cell system continues to evolve, its size and organization increase.

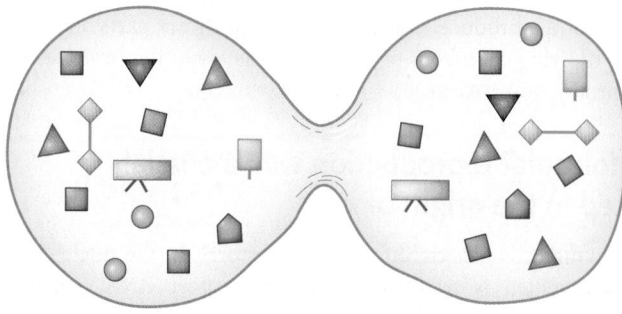

(d) The pre-cell system develops the ability to reproduce.

Figure 21-4 *Animation* **Origin of life: the metabolism first scenario**
© Cengage Learning

do, the question becomes which molecule, DNA or RNA, first appeared in the prebiotic world.

Many scientists have suggested that RNA was the first informational molecule to evolve in the progression toward a self-sustaining, self-reproducing cell and that proteins and DNA came along later. According to a model known as the **RNA world,** the chemistry of prebiotic Earth gave rise to self-replicating RNA molecules that functioned as both enzymes and substrates for their own replication. We represent the replication of RNA in the RNA world scenario as a circular arrow:

As discussed in Chapter 13, RNA often has catalytic properties; such enzymatic RNAs are called **ribozymes.** Before the evolution of true cells, ribozymes may have catalyzed their own replication in the clays, shallow rock pools, or hydrothermal vents where life originated. When RNA strands are added to a test tube containing RNA nucleotides but no enzymes, the nucleotides combine to form short RNA molecules.

The occurrence of an RNA world early in the history of life can never be proven, but experiments with **in vitro evolution,** also called **directed evolution,** have shown that it is feasible. These experiments address an important question about the RNA world, namely, could RNA molecules have catalyzed the many different chemical reactions needed for life? In directed evolution a large pool of RNA molecules with different sequences is mixed, and molecules are selected for their ability to catalyze a single biologically important reaction (**FIG. 21-5**). Those molecules that have at least some catalytic ability are selected by researchers, who then amplify them into many copies. The researchers use certain chemicals to cause mutations and then expose the molecules to another round of selection. After this cycle is repeated several times, the RNA molecules at the end of the selection process function efficiently as catalysts for the reaction. In vitro evolution studies have shown that RNA has a large functional repertoire—that is, RNA can catalyze a variety of biologically important reactions.

Biologists hypothesize that in the RNA world, ribozymes initially catalyzed protein synthesis and other important biological reactions; only later did protein enzymes catalyze these reactions.

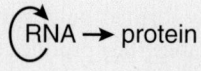

Interestingly, in test-tube experiments RNA molecules that direct protein synthesis by catalyzing peptide bond formation have evolved. Some single-stranded RNA molecules fold back on themselves as a result of interactions among the nucleotides composing the RNA strand. Sometimes the conformation (shape) of the folded RNA molecule is such that it weakly binds to an amino acid (see discussion of peptidyl transferase in Chapter 13). Amino acids held close to one another by RNA molecules may bond, forming a polypeptide.

Can in vitro natural selection result in chemical evolution of catalysts?

HYPOTHESIS: Self-replicating RNA molecules can undergo changes that make them more-efficient catalysts.

EXPERIMENT: RNA molecules are selected from a large pool, based on their ability to catalyze a specific reaction, and then amplified (*see figure*). Mutations occur as this process is repeated 7 to 20 additional times.

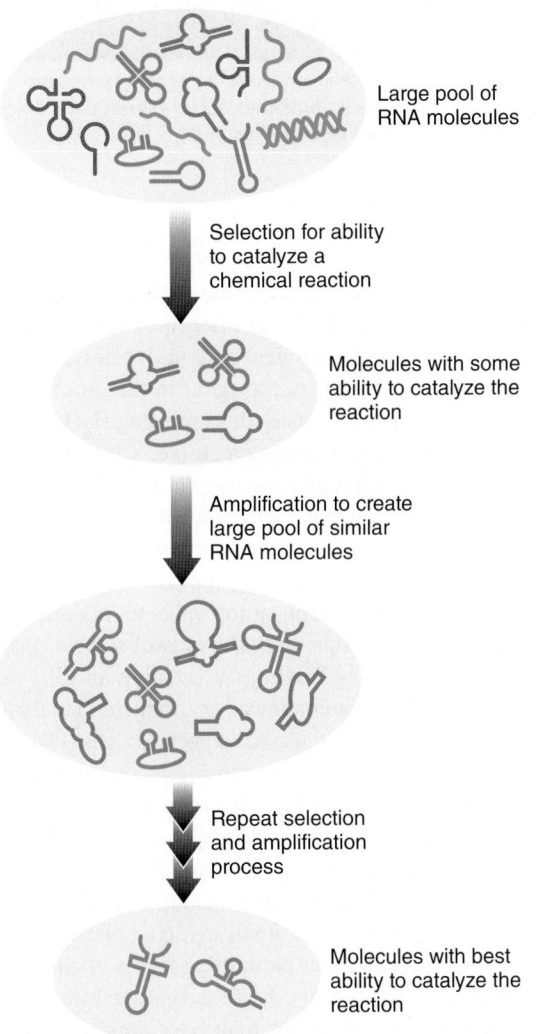

Large pool of
RNA molecules

Selection for ability
to catalyze a
chemical reaction

Molecules with some
ability to catalyze the
reaction

Amplification to create
large pool of similar
RNA molecules

Repeat selection
and amplification
process

Molecules with best
ability to catalyze the
reaction

RESULTS AND CONCLUSION: The final group of RNA molecules is the most efficient at catalyzing the chemical reaction selected for. Scientists have developed more than two dozen synthetic RNA catalysts by in vitro evolution.

SOURCE: Wright, M.C., and G.F. Joyce, "Continuous in Vitro Evolution of Catalytic Function," *Science* 246: 614–617, 1997.

Figure 21-5 In vitro evolution of RNA molecules

PREDICT Would the same result have been obtained if no mutations had occurred during the amplification process?

© Cengage Learning

We have considered how the evolution of informational molecules may have given rise to RNA and later to proteins. If a self-replicating RNA capable of coding for proteins appeared prior to DNA, how did DNA, the universal molecule of heredity in cells, become involved? Perhaps RNA made double-stranded copies of itself that eventually evolved into DNA.

$$DNA \leftarrow RNA \rightarrow protein$$

The incorporation of DNA into the information transfer system was advantageous because the double helix of DNA is more stable (less reactive) than the single strand of RNA. Such stability in a molecule that stores genetic information would have provided a decided advantage in the prebiotic world (as it does today).

Thus, in the DNA/RNA/protein world, DNA became the information storage molecule, RNA remained involved in protein synthesis and various regulatory activities, and protein enzymes catalyzed most cell reactions, including DNA replication and RNA synthesis.

$$\circlearrowleft DNA \rightarrow RNA \rightarrow protein$$

RNA is still a necessary component of the information transfer system because DNA is not catalytic. Thus, natural selection at the molecular level favored the DNA → RNA → protein information sequence. Once DNA was incorporated into this sequence, RNA molecules assumed their present role as an intermediary in the transfer of genetic information.

Several additional steps had to occur before a true living cell could evolve from macromolecular aggregations. For purposes of discussion, scientists have coined the acronym **LUCA** to refer to the hypothesized *last universal common ancestor*. The self-replicating genetic code that all known living organisms possess is certainly a legacy from LUCA, but how did this code originate? Also, how did an encapsulating membrane of lipid and protein evolve? These kinds of questions remain for scientists to address.

Biological evolution began with the first cells

No one knows exactly when or where the first cells appeared on Earth. Did they live at hydrothermal vents or at Earth's surface? Were they bacteria? We cannot answer these questions, in part because no fossils exist that trace the transition from nonlife to life. Nonfossil evidence—isotopic "fingerprints" of organic carbon in ancient rocks in Greenland—suggests to some researchers that life existed as early as 3.8 billion years ago (bya). However, other scientists vigorously debate this conclusion.

Microfossils, ancient remains of microscopic life, suggest that cells may have been thriving as long as 3.5 bya. Rich deposits of microfossils appear to exist in the Pilbara rocks of northwestern Australia and the Barberton rocks in South Africa; both of these deposits are about 3.3 billion to 3.5 billion years old. However, some scientists have challenged the interpretation of the oldest carbon-rich "squiggles" in ancient rocks as microfossils of prokaryotic organisms. They suggest that microfossils are not fossils at all but are formed by natural geologic processes.

(a) These living stromatolites at Hamlin Pool in Western Australia consist of mats of cyanobacteria (unicellular photosynthetic prokaryotes) and minerals such as calcium carbonate. They are several thousand years old.

(b) Cutaway view of a fossil stromatolite showing the layers of microorganisms and sediments that accumulated over time. This stromatolite, also from Western Australia, is about 3.5 billion years old.

Figure 21-6 Stromatolites

Evidence suggests that the earliest cells were prokaryotic. **Stromatolites,** another type of fossil evidence of the earliest cells, are rocklike columns composed of many minute layers of prokaryotic cells, that is, microbial biofilms (**FIG. 21-6**). Over time, sediment collects around the cells and mineralizes. Meanwhile, a new layer of living cells grows over the older, dead cells. Fossil stromatolite reefs are found in several places in the world, including Great Slave Lake in Canada and the Gunflint Iron Formations along Lake Superior in the United States. Some fossil stromatolites are extremely ancient. One group in Western Australia, for example, has been dated at more than 3 billion years. Living stromatolite reefs are rare but are found in certain hot springs and warm, shallow pools of fresh and salt water.

The first cells were probably heterotrophic

The earliest cells probably did not synthesize the organic molecules they needed but instead obtained them from the environment. These primitive **heterotrophs** may have consumed many types of organic molecules that had formed spontaneously, such as sugars, nucleotides, and amino acids. By fermenting these organic compounds, they obtained the energy they needed to support life. Of course, *fermentation,* discussed in Chapter 8, is an anaerobic process (performed in the absence of oxygen), and the first cells were almost certainly **anaerobes.**

When the supply of spontaneously generated organic molecules gradually declined, only certain organisms could survive. Mutations had probably already occurred that permitted some cells to obtain energy directly from sunlight, perhaps by using sunlight to make ATP. These cells, which did not require the energy-rich organic compounds that were now in short supply in the environment, had a distinct selective advantage.

Photosynthesis requires both light energy and a source of electrons, which are used to reduce CO_2 to form organic molecules such as glucose. (Recall the discussion of photosynthesis in Chapter 9.) Most likely, the first photosynthetic

autotrophs—organisms that produce their own food from simple raw materials—used the energy of sunlight to obtain electrons by splitting hydrogen-rich molecules such as H_2S, releasing elemental sulfur (not oxygen) in the process. Splitting H_2S is energetically much easier than splitting H_2O because sulfur is not electronegative like oxygen (see Chapter 9). Indeed, the green sulfur bacteria and the purple sulfur bacteria still use H_2S as a hydrogen (electron) source for photosynthesis. (Members of a third bacterial group, the purple nonsulfur bacteria, use other organic molecules or hydrogen gas as a hydrogen source.)

The first photosynthetic autotrophs to obtain hydrogen electrons by splitting *water* were the **cyanobacteria.** Water was quite abundant on early Earth, as it is today, and the selective advantage of splitting water allowed cyanobacteria to thrive. The process of splitting water released oxygen as a gas (O_2). Initially, the oxygen released during photosynthesis oxidized minerals in the ocean and in Earth's crust, and oxygen did not begin to accumulate in the atmosphere for a long time. Eventually, however, oxygen levels increased in the ocean and the atmosphere.

Scientists estimate the timing of the events just described on the basis of geologic and fossil evidence. Fossils from that period, which include rocks containing traces of chlorophyll as well as the fossil stromatolites discussed earlier, indicate that the first photosynthetic organisms may have appeared as early as 3.5 bya. This evidence suggests that heterotrophic forms may have existed even earlier.

Aerobes appeared after oxygen increased in the atmosphere

Based on sulfur isotope data from ancient rocks in South Africa, it appears that cyanobacteria had produced enough oxygen to begin significantly changing the composition of the atmosphere by 2.4 bya. This date may be changed as more data accumulate. Geoscientists at Pennsylvania State University have found strong isotopic evidence in Australia that oxygen was present in the oceans as early as 3.46 bya, and scientists are now looking for the same kind of evidence in other sedimentary layers known to be older than 2.4 bya.

The increase in atmospheric oxygen affected life profoundly. The oxygen poisoned *obligate anaerobes* (organisms that cannot use oxygen for cellular respiration), and many species undoubtedly perished. Some anaerobes, however, survived in environments where oxygen did not penetrate; adaptations evolved in others that neutralized the oxygen so that it could not harm them. In some organisms, called **aerobes,** a respiratory pathway that *used* oxygen to extract more energy from food evolved. Aerobic respiration (see Chapter 8) was joined with the existing anaerobic process of glycolysis.

The evolution of organisms that could use oxygen in their metabolism had several consequences. Organisms that respire aerobically gain much more energy from a single molecule of glucose than anaerobes gain by fermentation. (Recall the comparison of fermentation and aerobic respiration in Table 8-2.) As a result, the newly evolved aerobic organisms were more efficient and more competitive than anaerobes. Coupled with the poisonous nature of oxygen to many anaerobes, the efficiency of aerobes forced anaerobes into relatively minor roles. Today, the vast majority of organisms—including plants, animals, and most fungi, protists, archaea, and bacteria—use aerobic respiration; only a few archaea and bacteria and even fewer protists and fungi are anaerobic.

The evolution of aerobic respiration stabilized both oxygen and carbon dioxide levels in the biosphere. Photosynthetic organisms used carbon dioxide as a source of carbon for synthesizing organic compounds. This raw material would have been depleted from the atmosphere in a relatively brief period without the advent of aerobic respiration, which released carbon dioxide as a waste product from the complete breakdown of organic molecules. Carbon thus started cycling in the biosphere, moving from the nonliving physical environment to photosynthetic organisms to heterotrophs that ate the photosynthetic organisms (see Chapter 55). Aerobic respiration released carbon back into the physical environment as carbon dioxide, and the carbon cycle continued. In a similar manner, molecular oxygen was produced by photosynthesis and used during aerobic respiration.

Another significant consequence of photosynthesis occurred in the upper atmosphere, where molecular oxygen reacted to form **ozone (O_3)** (FIG. 21-7). A layer of ozone eventually blanketed Earth, preventing much of the sun's ultraviolet radiation from penetrating to the surface. With the ozone layer's protection from the mutagenic effect of ultraviolet radiation, organisms could live closer to the surface in aquatic environments and eventually move onto land. Because the energy in ultraviolet radiation may have been necessary to form organic molecules, however, their abiotic synthesis decreased.

Eukaryotic cells descended from prokaryotic cells

Eukaryotes may have appeared in the fossil record as early as 2.2 bya, and geochemical evidence suggests that eukaryotes were present much earlier. *Steranes,* molecules derived from steroids, have been discovered in Australian rocks dated at 2.7 billion

Ozone (O_3) forms in the upper atmosphere when ultraviolet radiation from the sun breaks the double bonds of oxygen molecules.

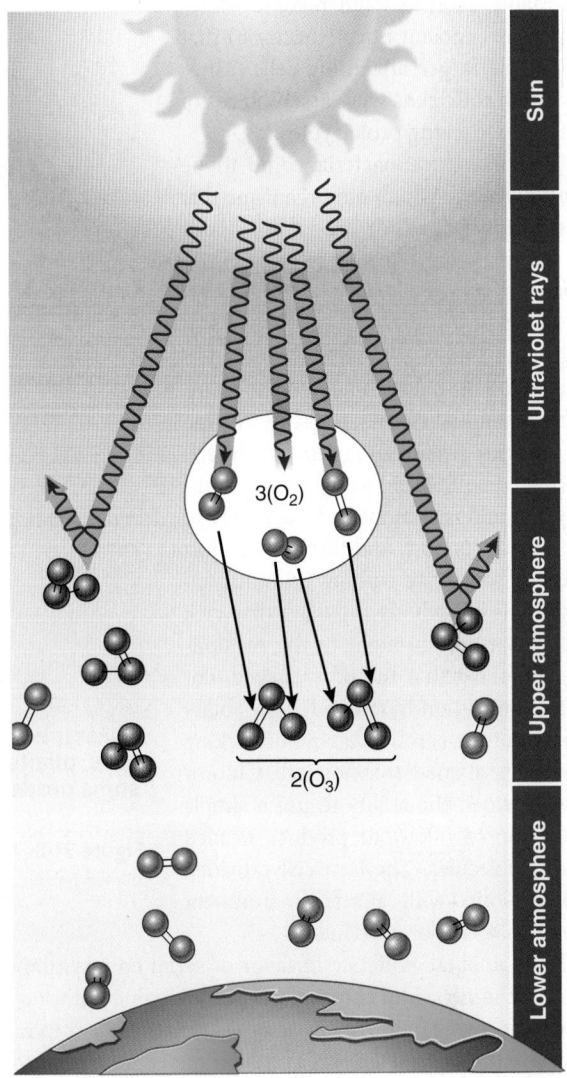

Figure 21-7 The formation of ozone

CONNECT As discussed in Chapter 57, certain human-produced chemicals rise to the stratosphere and react chemically with ozone. What consequences would you expect this reaction to have for organisms?

© Cengage Learning

years old. Because bacteria are not known to produce steroids, the steranes may be biomarkers for eukaryotes. These ancient rocks lack fossil traces of ancient organisms because the rocks have since been exposed to heat and pressure that would have destroyed any fossilized cells. Steranes, however, are very stable in the presence of heat and pressure.

Eukaryotes arose from prokaryotes. Recall from Chapter 4 that archaeal and bacterial cells lack nuclear envelopes as well as other membranous organelles such as mitochondria and chloroplasts. How did these organelles arise? According to the hypothesis of **serial endosymbiosis,** organelles such as mitochondria and chloroplasts may have originated from mutually advantageous symbiotic relationships between two prokaryotic organisms (FIG. 21-8). Chloroplasts apparently evolved from photosynthetic bacteria (cyanobacteria) that lived inside larger heterotrophic cells, whereas mitochondria presumably evolved from aerobic bacteria (perhaps ancient purple bacteria) that lived inside larger anaerobic cells. Thus, early eukaryotic cells were assemblages of formerly free-living prokaryotes.

How did these bacteria come to be **endosymbionts**, which are organisms that live symbiotically inside a host cell? They may originally have been ingested, but not digested, by a host cell. Once incorporated, they could have survived and reproduced along with the host cell so that future generations of the host also contained endosymbionts. A mutualistic relationship evolved between these two organisms in which each contributed something to the other. Eventually, the endosymbiont lost the ability to exist outside its host, and the host cell lost the ability to survive without its endosymbionts. This hypothesis stipulates that each of these partners brought to the relationship something that the other lacked. For example, mitochondria provided the ability to carry out the aerobic respiration lacking in the original anaerobic host cell. Chloroplasts provided the ability to use a simple carbon source (CO_2) to produce needed organic molecules. The host cell provided endosymbionts with a safe environment and raw materials or nutrients.

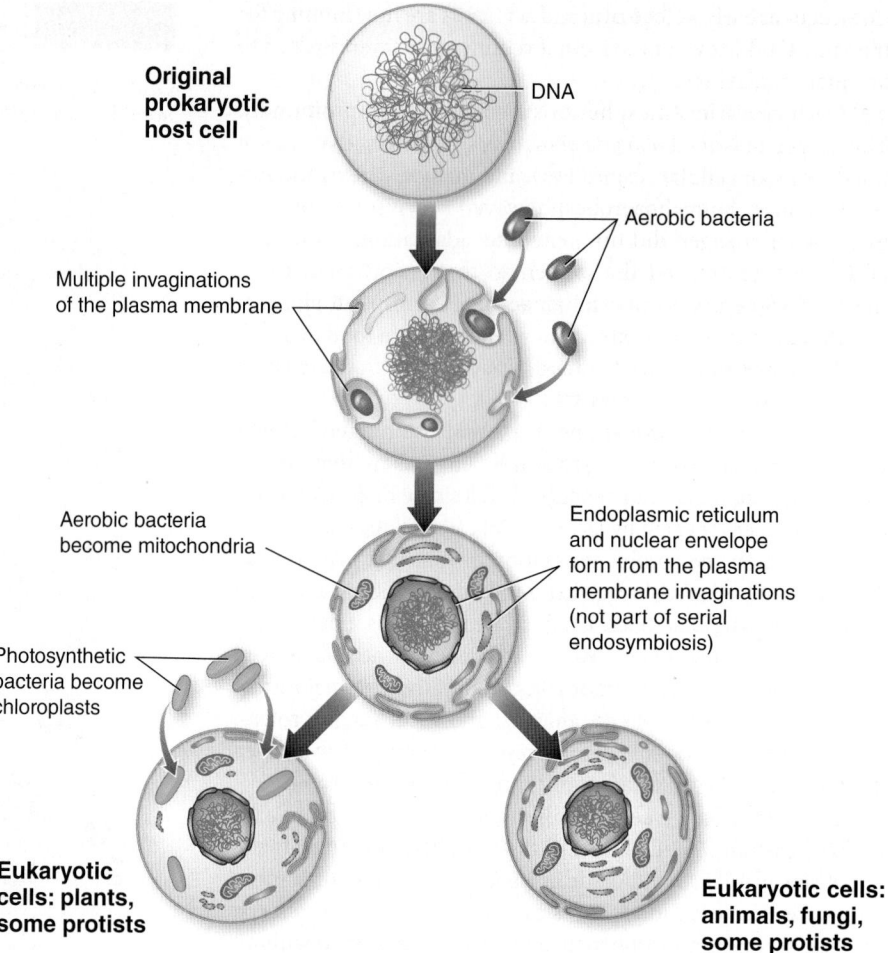

Figure 21-8 *Animation* **Serial endosymbiosis**
© Cengage Learning

The principal evidence in favor of serial endosymbiosis is that mitochondria and chloroplasts possess some (although not all) of their own genetic material and translational components. They have their own DNA (as a circular molecule similar to that of archaea and bacteria; discussed in Chapter 25) and their own ribosomes (which resemble bacterial rather than eukaryotic ribosomes). Mitochondria and chloroplasts also possess some of the machinery for protein synthesis, including tRNA molecules, and conduct protein synthesis on a limited scale independent of the nucleus. Furthermore, it is possible to poison mitochondria and chloroplasts with an antibiotic that affects bacteria but not eukaryotic cells. As discussed in Chapter 4, double membranes envelope mitochondria and chloroplasts. The outer membrane apparently developed from the invagination (infolding) of the host cell's plasma membrane, whereas the inner membrane is derived from the endosymbiont's plasma membrane. (Serial endosymbiosis is discussed in greater detail in Chapter 26.)

Many endosymbiotic relationships exist today. Many corals have algae living as endosymbionts within their cells (see Fig. 54-12). Protozoa (*Myxotricha paradoxa*) live symbiotically in the gut of termites; in turn, several different endosymbionts, including spirochete bacteria, attach to the protozoa and function as whiplike flagella, allowing the protozoa to move.

Although much evidence supports serial endosymbiosis, it does not completely explain the evolution of eukaryotic cells from prokaryotes. For example, serial endosymbiosis does not currently explain how a double membranous envelope came to surround the genetic material in the nucleus.

CHECKPOINT 21.2

- *What major steps probably occurred in the origin of cells?*
- CONNECT *How did the presence of molecular oxygen in the atmosphere affect early life?*
- VISUALIZE *Draw a simple sketch illustrating the way in which aerobic bacteria are hypothesized to have become incorporated into an original prokaryotic host cell.*

21.3 THE HISTORY OF LIFE

LEARNING OBJECTIVE

6 Briefly describe the distinguishing organisms and major biological events of the Ediacaran period and the Paleozoic, Mesozoic, and Cenozoic eras.

The sequence of biological, climate, and geologic events that make up the history of life is recorded in rocks and fossils. The sediments of Earth's crust consist of five major rock strata (layers), each subdivided into minor strata, lying one on top of the other. Very few places on Earth show all layers, but the strata present typically occur in the correct order, with younger rocks on top of older ones. These sheets of rock formed from the accumulation of mud and sand at the bottoms of the ocean, seas, and lakes. Each layer contains certain characteristic **index fossils** that identify deposits made at about the same time in different parts of the world.

Geologists divide Earth's history into units of time based on major geologic, climate, and biological events. **Eons** are the largest divisions of the geologic time scale. Eons are divided into **eras;** where fossil evidence exists, these divisions are based primarily on organisms that characterized each era. Eras are subdivided into **periods,** which in turn are composed of **epochs.**

Relatively little is known about Earth from its beginnings approximately 4.6 bya up to 2.5 bya. Life originated on Earth during the **Archaean eon,** the period between 4.0 bya when Earth's crust formed and 2.5 bya (TABLE 21-1). Signs of life date as early as 3.5 bya. Not much physical evidence is available because the rocks of the Archaean eon, being extremely ancient, are deeply buried in most parts of the world. Ancient rocks are exposed in a few places, including the bottom of the Grand Canyon and along the shores of Lake Superior. Many Archaean rock formations have revealed what appear to be microfossils of cyanobacteria and other bacteria.

Rocks from the Ediacaran period contain fossils of cells and simple animals

That part of time from 2500 mya to 541 mya is known as the **Proterozoic eon.** This enormous span of time is easier to study than the preceding Archaean eon because the rocks are less altered by heat and pressure. Life at the beginning of the Proterozoic eon consisted of prokaryotes such as cyanobacteria. Stromatolites were abundant. About 2.2 bya, the first eukaryotic cells appeared. By the end of the Proterozoic eon, multicellularity was evident in the abundant fossils of small, soft-bodied, invertebrate animals.

The **Ediacaran period** (pronounced "ee-dee-ack′-uh-run"), from 635 mya to 541 mya, is the last (most recent) period of the Proterozoic eon. It is named for the fossil deposits in the Ediacara Hills in South Australia, although Ediacaran fossils are also found in Newfoundland, the Mackenzie Mountains of northwestern Canada, the Doushantuo Formation of China, and other locations. Currently, more than 270 Ediacaran species have been identified and described.

Ediacaran fossils are the oldest known fossils of multicellular animals (FIG. 21-9). Experts have not yet identified all the simple, soft-bodied animals found in the Ediacara Hills and at other sites. Some paleontologists interpret many of these fossils as early sponges, jellyfish, corals, and comb jellies; however, other scientists think that the Ediacaran animals were not ancestral to modern-day species but instead became extinct at the end of the Ediacaran period.

Figure 21-9 Reconstruction of life in an Ediacaran sea
The organisms shown here are based on fossils from the Ediacara Hills of South Australia, although similar associations have been found in Ediacaran rocks from every continent except Antarctica. Shown are organisms that some scientists have interpreted as jellyfish, flatworms, algae, and paddle-shaped organisms similar in appearance to soft corals.

A diversity of organisms evolved during the Paleozoic era

The **Paleozoic era** began approximately 541 mya and lasted about 289 million years. It is divided into six periods: Cambrian, Ordovician, Silurian, Devonian, Carboniferous, and Permian.

Rocks rich in fossils represent the oldest subdivision of the Paleozoic era, the **Cambrian period.** For about 40 million years, evolution was in such high gear, with the sudden appearance of many new animal body plans, that this period is called the **Cambrian radiation,** or more informally, **Cambrian explosion.** Fossils of all contemporary animal phyla are present, along with many bizarre, extinct phyla, in marine sediments. The seafloor was covered with sponges, corals, sea lilies, sea stars, snails, clamlike bivalves, primitive squidlike cephalopods, lampshells (brachiopods), and trilobites (see Fig. 31-26). In addition, small vertebrates—cartilaginous fishes—became established in the marine environment.

Scientists have not determined the factors responsible for the Cambrian radiation, a period unmatched in the evolutionary history of life. There is some evidence that oxygen concentrations, which had continued to gradually increase in the atmosphere, passed some critical threshold late in the Proterozoic eon. Scientists who advocate the *oxygen enrichment hypothesis* note that until late in the Proterozoic eon, Earth did not have enough oxygen to support larger animals. The most important fossil sites that document the Cambrian radiation are the *Chengjiang site* in China (for Early Cambrian fossils) and the *Burgess Shale* in British Columbia (for Middle Cambrian fossils) (FIG. 21-10).

According to geologists, seas gradually flooded the continents during the Cambrian period. In the **Ordovician period,** much land was covered by shallow seas, in which there was another burst of evolutionary diversification, although not as dramatic as the Cambrian radiation. The Ordovician seas were inhabited by giant cephalopods, squidlike animals with straight shells 5 to 7 m (16 to 23 ft) long and 30 cm (12 in.) in diameter.

TABLE 21-1 Some Important Biological Events in Geologic Time

EON	ERA	PERIOD	EPOCH	TIME*	SOME IMPORTANT BIOLOGICAL EVENTS
Phanerozoic eon	Cenozoic era	Quaternary period	Holocene epoch	0.01 (10,000 years ago)	Decline of some woody plants; rise of herbaceous plants; age of *Homo sapiens*
			Pleistocene epoch	2.6 mya	Extinction of some plant species; extinction of many large mammals at end
		Neogene period	Pliocene epoch	5 mya	Expansion of grasslands and deserts; many grazing animals
			Miocene epoch	23 mya	Flowering plants continue to diversify; diversity of songbirds and grazing mammals
		Paleogene period	Oligocene epoch	34 mya	Spread of forests; apes appear; present mammalian families are represented
			Eocene epoch	56 mya	Flowering plants dominant; modern mammalian orders appear and diversify; modern bird orders appear
			Paleocene epoch	66 mya	Semitropical vegetation (flowering plants and conifers) widespread; primitive mammals diversify
	Mesozoic era	Cretaceous period		145 mya	Rise of flowering plants; dinosaurs reach peak, then become extinct at end; toothed birds become extinct
		Jurassic period		201 mya	Gymnosperms common; large dinosaurs; first toothed birds
		Triassic period		252 mya	Gymnosperms dominant; ferns common; first dinosaurs; first mammals
	Paleozoic era	Permian period		299 mya	Conifers diversify; cycads appear; modern insects appear; mammal-like reptiles; extinction of many invertebrates and vertebrates at end of Permian
		Carboniferous period		359 mya	Forests of ferns, club mosses, horsetails, and gymnosperms; many insect forms; spread of ancient amphibians; first reptiles
		Devonian period		419 mya	First forests; gymnosperms appear; many trilobites; wingless insects appear; fishes with jaws appear and diversify; amphibians appear
		Silurian period		444 mya	Vascular plants appear; coral reefs common; jawless fishes diversify; terrestrial arthropods
		Ordovician period		485 mya	Fossil spores of terrestrial plants (bryophytes?); invertebrates dominant; coral reefs appear; first fishes appear
		Cambrian period		541 mya	Bacteria and cyanobacteria; algae; fungi; age of marine invertebrates; first chordates
Proterozoic eon		Ediacaran period		635 mya	Algae and soft-bodied invertebrates diversify
		Early Proterozoic period		2500 mya	Eukaryotes evolve
Archaean eon				4000 mya	Oldest known rocks; prokaryotes evolve; atmospheric oxygen begins to increase

*Time from beginning of eon, period, or epoch to present (millions of years). Dates: Walker, J.D., Geissman, J.W, Bowring, S.A., and Babcock, L.E., compilers, 2012, Geologic Time Scale v.4:0: Geological Society of America.

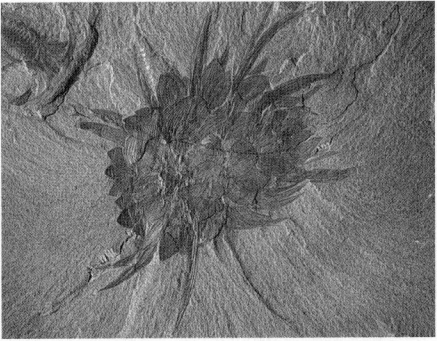

(a) *Marrella splendens* was a small arthropod with 26 body segments. It is the most abundant fossil arthropod found in the Burgess Shale.

(b) *Wiwaxia* was a bristle-covered marine worm that was distantly related to earthworms. It had scaly armor and needlelike spines for protection.

Figure 21-10 Fossils from the Cambrian radiation

These fossils were discovered in the Burgess Shale in the Canadian Rockies of British Columbia.

Coral reefs first appeared during this period, as did small, jawless, bony-armored fishes called *ostracoderms* (**FIG. 21-11**). Lacking jaws, these fishes typically had round or slitlike mouth openings that may have sucked in small food particles from the water or scooped up organic debris from the bottom. Ordovician deposits also contain fossil spores of terrestrial (land-dwelling) plants, suggesting that the colonization of land had begun.

During the **Silurian period,** jawless fishes diversified considerably, and jawed fishes first appeared. Definitive evidence of two life-forms of great biological significance appeared in the Silurian period: terrestrial plants and air-breathing animals. The evolution of plants allowed animals to colonize the land because plants provided the first terrestrial animals with food and shelter. All air-breathing land animals discovered in Silurian rocks were arthropods: millipedes, spiderlike arthropods, and possibly centipedes. From an ecological perspective, the energy flow from plants to animals probably occurred via detritus, which is organic debris from decomposing organisms, rather than directly from living plant material. Millipedes eat plant detritus today, and spiders and centipedes prey on millipedes and other animals.

The **Devonian period** is frequently called the Age of Fishes. This period witnessed the explosive radiation of fishes with jaws, an adaptation that lets a vertebrate chew and bite. Armored *placoderms,* an extinct group of jawed fishes, diversified to exploit varied lifestyles (see Fig. 32-10). Appearing in Devonian deposits are sharks and the two predominant types of bony fishes: lobe-finned fishes and ray-finned fishes, which gave rise to the major orders of modern fishes (see Chapter 31).

Tiktaalik was a transitional form between fishes and *tetrapods,* which are vertebrates with four limbs (see Fig. 32-17). *Tiktaalik* is considered a fish because it had scales and fins. Like tetrapods, however,

Tiktaalik had a movable neck and ribs that enclosed lungs. Upper (more recent) Devonian sediments contain fossil remains of salamander-like amphibians (*labyrinthodonts*) that were often quite large, with short necks and heavy, muscular tails. Wingless insects also originated in the Late Devonian period.

The early vascular plants (plants with specialized tissue to conduct water and nutrients) diversified during the Devonian period in a burst of evolution that rivaled that of animals during the Cambrian radiation. With the exception of flowering plants, all major plant groups appeared during the Devonian period. Forests of ferns, club mosses, horsetails, and seed ferns (an extinct group of ancient plants that had fernlike foliage but reproduced by forming seeds) flourished.

The **Carboniferous period** is named for the great swamp forests whose remains persist today as major coal deposits. Much of the land during this time was covered with low swamps filled with horsetails, club mosses, ferns, seed ferns, and gymnosperms, which are seed-bearing plants such as conifers (**FIG. 21-12**). Amphibians, which underwent an **adaptive radiation** and exploited both aquatic and terrestrial ecosystems, were the dominant terrestrial carnivores of the Carboniferous period. Reptiles first appeared and diverged to form two major lines at this time. One line consisted of mostly small and midsized insectivorous (insect-eating) lizards; this line later led to lizards, snakes, crocodiles, dinosaurs, and birds. The other reptilian line led to a diverse group of Permian and Early Mesozoic mammal-like reptiles. Two groups of winged insects, cockroaches and dragonflies, appeared in the Carboniferous period.

Amphibians continued in importance during the **Permian period,** but they were no longer the dominant carnivores in

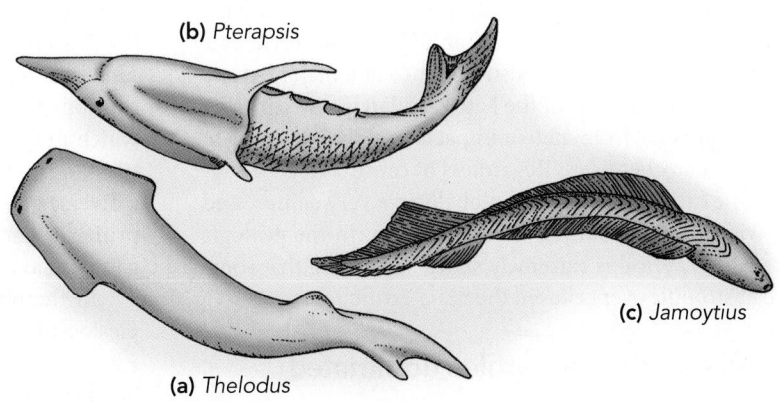

(b) *Pterapsis*

(c) *Jamoytius*

(a) *Thelodus*

Figure 21-11 Ostracoderms

Ostracoderms, primitive jawless fishes that lived in the Ordovician, Silurian, and Devonian periods, ranged from 10 to 50 cm (4 to 20 in.) in length.

© Cengage Learning

Figure 21-12 Reconstruction of a Carboniferous forest
Plants of this period included giant ferns, horsetails, club mosses, seed ferns, and early gymnosperms.

terrestrial ecosystems. During the Permian period, reptiles diversified and dominated both carnivorous and herbivorous terrestrial lifestyles. One important group of mammal-like reptiles, originating in the Permian and extending into the Mesozoic era, were the *therapsids,* a group that included the ancestor of mammals (see Fig. 32-24). During the Permian period, seed plants diversified and dominated most plant communities. Cone-bearing conifers were widespread, and cycads (plants with crowns of fernlike leaves and large, seed-containing cones) and ginkgoes (trees with fan-shaped leaves and exposed, fleshy seeds) appeared.

The greatest **mass extinction** of all time occurred at the end of the Paleozoic era, between the Permian and Triassic periods, 252 mya. More than 90% of all existing marine species became extinct at this time, as did more than 70% of the vertebrate genera living on land. There is also evidence of a major extinction of plants. Many causes for the Late Permian mass extinction have been suggested, from meteor impacts to global warming to changes in ocean chemistry. Regardless of cause, evidence suggests that the extinction occurred globally in a very compressed period, within a few hundred thousand years. In the geologic time scale, this period is extremely short, suggesting that some sort of catastrophic event caused the mass extinction.

Dinosaurs and other reptiles dominated the Mesozoic era

The **Mesozoic era** began about 252 mya and lasted some 186 million years. It is divided into the Triassic, Jurassic, and Cretaceous periods. Fossil deposits from the Mesozoic era occur worldwide.

Notable sites include the *Yixian Formation* in northeastern China, the Solnhofen Limestone in Germany, northwestern Patagonia in Argentina, the Sahara in central Niger, the Badlands in South Dakota, and other sites in western North America.

The outstanding feature of the Mesozoic era was the origin, differentiation, and ultimately the extinction of a large variety of reptiles. For this reason, the Mesozoic era is commonly called the Age of Reptiles. Most of the modern orders of insects appeared during the Mesozoic era. Snails and bivalves (clams and their relatives) increased in number and diversity, and sea urchins reached their peak diversity. From a botanical viewpoint, the Mesozoic era was dominated by gymnosperms until the Mid-Cretaceous period, when the flowering plants first diversified.

During the **Triassic period,** reptiles underwent an adaptive radiation leading to the formation of many groups. On land, the dominant Triassic groups were the mammal-like therapsids, which ranged from small-sized insectivores (insect-eating reptiles) to moderately large herbivores (plant-eating reptiles), and a diverse group of *thecodonts,* early "ruling reptiles," which were primarily carnivores. Thecodonts are the ancestral reptiles that gave rise to crocodilians, flying reptiles, dinosaurs, and birds.

In the ocean several important marine reptile groups, the plesiosaurs and ichthyosaurs, appeared in the Triassic period. *Plesiosaurs* had bodies up to 15 m (about 49 ft) long and paddle-like fins. *Ichthyosaurs* had body forms superficially resembling those of sharks or porpoises, with short necks, large dorsal fins, and shark-type tails (**FIG. 21-13**). Ichthyosaurs had very large eyes, which may have helped them see at diving depths of 500 m (1650 ft) or more.

Figure 21-13 Ichthyosaurs

Brachypterygius (also known by the genus *Grendelius*) was an ichthyosaur that superficially resembled a shark or porpoise. It had large teeth and strong jaws and was about 4 m (13 ft) long.

Figure 21-14 Pterosaurs

Shown are *Peteinosaurus* (*left*), with a wingspan of 60 cm (2 ft), and *Eudimorphodon* (*right foreground*), with a wingspan of 75 cm (2.5 ft). Both species had long, sharp teeth for catching insects or fishes while flying.

Pterosaurs, the first flying reptiles, appeared and underwent considerable diversification during the Mesozoic era (**FIG. 21-14**). This group produced some quite spectacular forms, most notably the giant *Quetzalcoatlus,* known from fragmentary Cretaceous fossils in Texas to have had a wingspan of 11 to 15 m (36 to 49 ft). Pterosaur wings were leathery membranes of skin that were supported by an elongated fourth finger bone. Claws extended from the other finger bones.

The first mammals to appear in the Triassic period were small insectivores that evolved from the mammal-like therapsids. Mammals diversified into a variety of mostly small, nocturnal insectivores during the remainder of the Mesozoic era, with marsupial and placental mammals appearing later in the Mesozoic.

During the **Jurassic** and **Cretaceous periods,** crocodiles, lizards, snakes, and birds appeared, and the dinosaurs diversified dramatically (**FIG. 21-15**). The *mosasaurs,* one group of lizards, were large, voracious marine predators during the Late Cretaceous period. The mosasaurs, which are now extinct, attained lengths of 10 m (33 ft) or more.

The evolutionary radiation of the dinosaurs expanded from one lineage to several dozen that ecologically filled a variety of adaptive zones. Dinosaurs are placed in two main groups based on their pelvic bone structure: the *saurischians* and the *ornithischians* (**FIG. 21-16**). Some saurischians

were fast, bipedal (walking on two feet) forms ranging from those the size of a dog to the ultimate representatives of this group, the gigantic carnivores of the Cretaceous period: *Tyrannosaurus, Giganotosaurus,* and *Carcharodontosaurus.* Other saurischians were huge, quadrupedal (walking on four feet) dinosaurs that ate plants. Some of these dinosaurs were the largest terrestrial animals that have ever lived, including *Argentinosaurus,* with an estimated length of 30 m (98 ft) and an estimated weight of 72 to 90 metric tons (80 to 100 tons).

The other group of dinosaurs, the ornithischians, was entirely herbivorous. Although some ornithischians were bipedal, most were quadrupedal. Some had no front teeth and possessed stout, horny, birdlike beaks. In some species these

Figure 21-15 Dinosaurs

Three *Deinonychus* dinosaurs attack a larger *Tenontosaurus.* The name *Deinonychus* means "terrible claw" and refers to the enlarged, sharp claw on the second digit of its hind feet. *Deinonychus* dinosaurs were small (3 m, or 10 ft in length) but fearsome predators that hunted in packs. *Tenontosaurus* adults were as long as 7.5 m (24 ft).

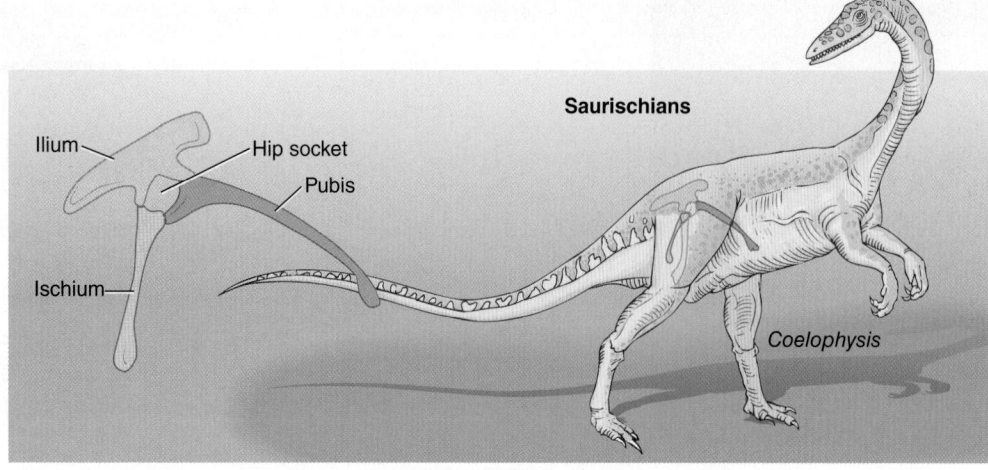

(a) **The saurischian pelvis**. Note the opening (hip socket), a trait possessed by no quadrupedal vertebrates other than dinosaurs.

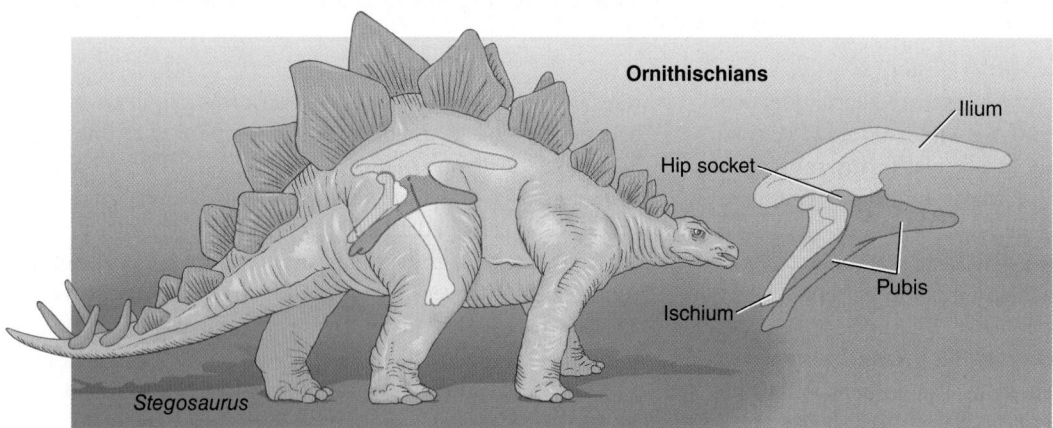

(b) **The ornithischian pelvis**. Note that it has the hole in the hip socket but differs from the saurischian pelvis in that it has a backward-directed extension of the pubis.

Figure 21-16 Saurischian and ornithischian dinosaurs

In each dinosaur figure, the pale yellow femur is shown relative to the pelvic bone.

© Cengage Learning

beaks were broad and ducklike, hence the common name *duck-billed dinosaurs*. Other ornithischians had great armor plates, possibly as protection against carnivorous saurischians. *Ankylosaurus*, for example, had a broad, flat body covered with armor plates (actually bony scales embedded in the skin) and large, laterally projecting spikes.

Over the past few decades, scientists have reconsidered many traditional ideas about dinosaurs and no longer think that they were all cold-blooded, slow-moving monsters living in swamps. Recent evidence suggests that at least some dinosaurs were warm-blooded, agile, and able to move extremely fast. Many dinosaurs appear to have had complex social behaviors, including courtship rituals and parental nurturing of their young. Some species lived in social groups and hunted in packs.

Birds appeared by the Late Jurassic period, and fossil evidence indicates that they evolved directly from saurischian dinosaurs (see *Inquiring About: The Origin of Flight in Birds*).

Archaeopteryx, the oldest known bird in the fossil record, lived about 150 mya (see Fig. 32-22b). It was about the size of a pigeon and had rather feeble wings that it used to glide rather than actively fly. Although *Archaeopteryx* is considered a bird, it had many reptilian features, including a mouthful of teeth and a long, bony tail.

Thousands of well-preserved bird fossils have been found in Early Cretaceous deposits in China. They include *Sinornis*, a 135-million-year-old sparrow-sized bird capable of perching; and the magpie-sized *Confuciusornis*, the earliest known bird with a toothless beak. *Confuciusornis* may date back as far as 142 mya.

At the end of the Cretaceous period, 66 mya, dinosaurs, pterosaurs, and many other animals abruptly became extinct. Many gymnosperms, with the exception of conifers, also perished. Evidence suggests that a catastrophic collision of a large extraterrestrial body with Earth dramatically changed the climate at the end of the Cretaceous period. Part of the evidence is a thin band of dark clay, with a high concentration of iridium, located between Mesozoic and Cenozoic sediments at more than two hundred sites around the world. Iridium is rare on Earth but abundant in meteorites. The force of the impact would have driven the iridium into the atmosphere, to be deposited later on the land by precipitation.

The Chicxulub crater, buried under the Yucatán Peninsula in Mexico, is the apparent site of this collision. The impact produced giant tsunamis (tidal waves) that deposited materials from the extraterrestrial body around the perimeter of the Gulf of Mexico, from Alabama to Guatemala. It may have caused global forest fires and giant smoke and dust clouds that lowered global temperatures for many years.

Although scientists widely accept that a collision with an extraterrestrial body occurred 66 mya, they have reached no consensus about the effects of such an impact on organisms. The extinction of many marine organisms at or immediately

Do we know how birds evolved? The evolution of birds is arguably one of the most interesting chapters in Earth's history of life. Given the substantial fossil evidence, most paleontologists have concluded that the ancestors of birds were dinosaurs, specifically the *dromaeosaurs,* a group of ground-dwelling, bipedal theropods (carnivorous saurischians). Beginning in 1997, paleontologists made several discoveries of fossil dinosaurs with feathers, indicating that feathers appeared before birds. Scientists think that feathers evolved as one or a series of evolutionary novelties, or **preadaptations** (see Chapter 20). The first feathers may have provided thermal insulation but were subsequently modified for flight.

Once dinosaurs and early birds had feathers, how did they fly? Did tree-dwelling animals glide as an intermediate step in bird flight, or did ground-dwelling animals flap their wings and run to provide thrust and lift for a takeoff? The question about how flight originated in birds has intrigued biologists for more than a century. In 1915, U.S. zoologist William Beebe hypothesized that the ancestors of birds were probably tree-dwelling gliders that had feathers on all four limbs.[1] Because no fossil evidence supported his suggestion, scientists did not consider it seriously at the time.

Almost a century later, in 2003, a group of Chinese paleontologists announced the discovery of two nearly complete fossils of the organisms that Beebe had hypothesized.[2] The fossils of a small, feathered dromaeosaur dinosaur

[1] Beebe, W.H. "A Tetrapteryx Stage in the Ancestry of Birds." *Zoologica,* Vol. 2, 1915.
[2] Xu, X., Z. Zhou, X. Wang, X. Kuang, F. Zhang, and X. Du. "Four-Winged Dinosaurs from China." *Nature,* Vol. 421, Jan. 23, 2003.

were found in Liaoning Province in northeastern China. The dinosaur, *Microraptor gui,* had feathers on both forelimbs and hind limbs as well as on its long tail (see figure). It was small—77 cm (about 30 in.) in length, including the tail—and appeared to be adapted to life in the trees. The feathers on *M. gui* were similar to those of modern-day birds. Downy feathers covered the body, and each limb had about 12 "primary" flight feathers and about 18 shorter, "secondary" feathers. *Microraptor gui*'s flight feathers were asymmetrical, a characteristic associated with flight or gliding in modern birds. The primary and secondary feathers followed a similar pattern on both the forelimbs and hind limbs, and this pattern resembles that on modern birds. In part because *M. gui*'s breastbone was not structured to attach large flight muscles, the dinosaur probably glided rather than flapped its wings.

Microraptor gui, which is about 126 million years old, is not a direct ancestor of birds. Birds had already evolved when *M. gui* existed. The earliest known bird, *Archaeopteryx,* lived about 150 mya and therefore predates *M. gui* by about 25 million years. *Microraptor gui* is considered to have evolved from a *basal member*—that is, an earlier evolutionary branch—of the most recent ancestor of *Archaeopteryx* and other birds. Like *M. gui,* earlier feathered dinosaurs may have been four-winged organisms that lived and glided in the trees.

Portia Sloan/Getty Images

During the course of bird evolution, the feathered hind limbs may have become reduced and eventually lost. However, an alternative hypothesis is that feathered hind limbs may have been a failed evolutionary experiment restricted to dromaeosaurs and not important as an intermediate step in bird evolution. Further analysis of *M. gui* and future discoveries of earlier dromaeosaur fossils may shed some light on the importance of feathered hind limbs.

Microraptor gui has given paleontologists and biologists much to consider in the evolutionary transition from dinosaurs to birds. Scientists will continue to study and debate the evolution of flight in birds for many years.

---

after the time of the impact was probably the result of the environmental upheaval that the collision produced. However, many clam species associated with the mass extinction at the end of the Cretaceous period seem to have gone extinct *before* the impact, suggesting that other factors caused some of the massive extinctions occurring then.

The Cenozoic era is the age of mammals

With equal justice the Cenozoic era could be called the Age of Mammals, the Age of Birds, the Age of Insects, or the Age of Flowering Plants. This era is marked by the appearance of all these forms in great variety and numbers of species. The Cenozoic era extends from 66 mya to the present. It is subdivided into three periods: the **Paleogene period,** encompassing some 43 million years; the **Neogene period,** which was 20.4 million years in length; and the **Quaternary period,**

which covers the last 2.6 million years. The Paleogene period is subdivided into three epochs, named from earliest to latest: Paleocene, Eocene, and Oligocene. The Neogene period is subdivided into two epochs: the Miocene and Pliocene. The Quaternary period is subdivided into the Pleistocene and Holocene epochs.

Flowering plants, which arose during the Cretaceous period, continued to diversify during the Cenozoic era. During the Paleocene and Eocene epochs, fossils indicate that tropical and semitropical plant communities extended to relatively high latitudes. Palms, for example, are found in Eocene deposits in Wyoming. Later in the Cenozoic era, there is evidence of more open habitats. Grasslands and savannas spread throughout much of North America during the Miocene epoch, with deserts developing later in the Pliocene and Pleistocene epochs. During the Pleistocene epoch, plant communities changed dynamically in response to the fluctuating climates associated with the multiple advances and retreats of continental glaciers.

During the Paleocene epoch, an explosive radiation of primitive mammals occurred. Most of them were small forest dwellers that are not closely related to modern mammals. During the Eocene epoch, mammals continued to diverge, and all the modern orders first appeared. Again, many of the mammals were small, but there were also some larger herbivores.

During the Eocene epoch, there was an explosive radiation of birds, which acquired adaptations for different habitats. Paleontologists hypothesize that the jaws and beak of the flightless giant bird *Diatryma,* for example, may have been adapted primarily for crushing and slicing vegetation in Eocene forests, marshes, and grasslands. Other paleontologists hypothesize that these giant birds were carnivores that killed or scavenged mammals and other vertebrates (**FIG. 21-17**).

During the Oligocene epoch, many modern families of mammals evolved, including the first apes in Africa. Many lineages showed adaptations that suggest a more open type of habitat, such as grassland or savanna, than previously. Many mammals were larger than earlier mammals and had longer legs for running, specialized teeth for chewing coarse vegetation or for preying on animals, and increases in their relative brain sizes. The *indricotheres,* for example, are extinct relatives of the rhinoceros. These mammals, which lived on the grassless plains of Eurasia, became progressively larger during the Oligocene epoch (**FIG. 21-18**).

Human ancestors appeared in Africa during the Late Miocene and Early Pliocene epochs. *Homo,* the genus to

Figure 21-18 A mammal from the Oligocene epoch

Paraceratherium was an indricothere, a hornless relative of the rhinoceros. This huge land mammal was about 8 m (26 ft) long and weighed about 15 to 20 tons. It probably ate leaves and branches of deciduous trees, much as a modern-day giraffe does.

which humans belong, appeared approximately 2.5 mya. (Primate evolution, including human evolution, is discussed in Chapter 22.)

The Pliocene and Pleistocene epochs witnessed the introduction of spectacular North and South American large-mammal fauna, including mastodons, saber-toothed cats, camels, giant ground sloths, and giant armadillos. However, many of the large mammals became extinct at the end of the Pleistocene epoch. This extinction was possibly due to climate change—the Pleistocene epoch was marked by several ice ages—or to the influence of humans, which had spread from Africa to Europe and Asia, and later to North and South America by crossing a land bridge between Siberia and Alaska. Archaeological evidence indicates that this mass extinction event was concurrent with the appearance of human hunters.

CHECKPOINT 21.3

- **CONNECT** *What is the correct order of appearance in the fossil record, starting with the earliest: eukaryotic cells, multicellular organisms, prokaryotic cells?*

- **CONNECT** *What is the correct order of appearance in the fossil record, starting with the earliest: reptiles, mammals, amphibians, fishes?*

- **CONNECT** *What is the correct order of appearance in the fossil record, starting with the earliest: flowering plants, ferns, gymnosperms?*

Figure 21-17 A bird from the Eocene epoch

The flightless bird *Diatryma,* which stood 2.1 m (7 ft) tall and weighed about 175 kg (385 lb), may have been an herbivore or a formidable predator. In this picture *Diatryma* has captured a small, horselike perissodactyl.

21.1 Chemical Evolution on Early Earth *(page 439)*

1 Describe the conditions that scientists think existed on early Earth.

- Biologists generally agree that life originated from nonliving matter by **chemical evolution.** Hypotheses about chemical evolution are testable.

- Four requirements for chemical evolution are (1) the absence of oxygen, which would have reacted with and oxidized abiotically produced organic molecules; (2) energy to form organic molecules; (3) chemical building blocks, including water, minerals, and gases present in the atmosphere; and (4) sufficient time for molecules to accumulate and react.

2 Compare the prebiotic soup hypothesis with the iron–sulfur world hypothesis.

- During chemical evolution, small organic molecules formed spontaneously and accumulated. The **prebiotic soup hypothesis** proposes that organic molecules formed near Earth's surface in a "sea of organic soup" or on rock or clay surfaces. The **iron–sulfur world hypothesis** suggests that organic molecules were produced at **hydrothermal vents,** cracks in the deep-ocean floor.

21.2 The First Cells *(page 441)*

3 Outline the major steps hypothesized to have occurred in the origin of cells.

- After small organic molecules formed and accumulated, macromolecules assembled from the small organic molecules. Macromolecular assemblages called **protobionts** formed from macromolecules. Cells arose from the protobionts.

- According to a model known as the **RNA world,** RNA was the first informational molecule to evolve in the progression toward a self-sustaining, self-reproducing cell. Natural selection at the molecular level eventually resulted in the information sequence DNA → RNA → protein.

4 Explain how the evolution of photosynthetic autotrophs affected both the atmosphere and other organisms.

- The first cells were prokaryotic **heterotrophs** that obtained organic molecules from the environment. They were almost certainly **anaerobes.** Later, **autotrophs**—organisms that produce their own organic molecules by photosynthesis—evolved.

- The evolution of oxygen-generating photosynthesis ultimately changed early life. The accumulation of molecular oxygen in the atmosphere permitted the evolution of **aerobes,** organisms that could use oxygen for a more efficient type of cellular respiration.

5 Describe the hypothesis of serial endosymbiosis.

- Eukaryotic cells arose from prokaryotic cells. According to the hypothesis of **serial endosymbiosis,** certain eukaryotic organelles (mitochondria and chloroplasts) evolved from prokaryotic **endosymbionts** incorporated within larger prokaryotic hosts.

21.3 The History of Life *(page 446)*

6 Briefly describe the distinguishing organisms and major biological events of the Ediacaran period and the Paleozoic, Mesozoic, and Cenozoic eras.

- Life at the beginning of the **Proterozoic eon** (2500 mya to 541 mya) consisted of prokaryotes. About 2.2 bya, the first eukaryotic cells appeared. The **Ediacaran period,** from 635 mya to 541 mya, is the last period of the Proterozoic eon. Ediacaran fossils are the oldest known fossils of multicellular animals. Ediacaran fauna were small, soft-bodied invertebrates.

- During the **Paleozoic era,** which began about 541 mya and lasted approximately 289 million years, all major groups of plants, except flowering plants, and all animal phyla appeared. Fishes and amphibians flourished, and reptiles appeared. The greatest mass extinction of all time occurred at the end of the Paleozoic era, 252 mya. More than 90% of marine species and 70% of land-dwelling vertebrate genera as well as many plant species became extinct.

- The **Mesozoic era** began about 252 mya and lasted some 186 million years. Flowering plants appeared, and reptiles diversified. Dinosaurs, which descended from early reptiles, dominated. Insects flourished, and birds and early mammals appeared. At the end of the Cretaceous period, 66 mya, many species abruptly became extinct. A collision of a large extraterrestrial body with Earth may have resulted in dramatic climate changes that played a role in this mass extinction.

- In the **Cenozoic era,** which extends from 66 mya to the present, flowering plants, birds, insects, and mammals diversified greatly. Human ancestors appeared in Africa during the Late Miocene and Early Pliocene epochs.

Know and Comprehend

1. Energy, the absence of molecular oxygen, chemical building blocks, and time were the requirements for (a) chemical evolution (b) biological evolution (c) the Cambrian radiation (d) the mass extinction episode at the end of the Cretaceous period (e) directed evolution

2. Many scientists think that ___ was the first information molecule to evolve. (a) DNA (b) RNA (c) a protein (d) an amino acid (e) a lipid

3. The first cells were probably (a) heterotrophs (b) autotrophs (c) anaerobes (d) a and c (e) b and c

4. According to the hypothesis of serial endosymbiosis, (a) life originated from nonliving matter (b) the pace of evolution quickened at the start of the Cambrian period (c) chloroplasts, mitochondria, and possibly other organelles originated from intimate relationships among prokaryotic cells (d) banded iron formations reflect the buildup of sufficient oxygen in the atmosphere to oxidize iron at Earth's surface (e) the first photosynthetic organisms appeared 3.1 bya to 3.5 bya

5. During the Early ___, life consisted of prokaryotic cells, but by the end of this geologic time span, multicellular eukaryotic organisms had evolved. (a) Cenozoic era (b) Paleozoic era (c) Mesozoic era (d) Archaean eon (e) Proterozoic eon

6. Geologists divide the eons into (a) periods (b) epochs (c) eras (d) millennia (e) none of the preceding

7. Ediacaran fossils (a) are the oldest known fossils of multicellular animals (b) come from the Burgess Shale in British Columbia (c) contain remains of large salamander-like organisms (d) are the oldest fossils of early vascular plants (e) contain a high concentration of iridium

8. The time of greatest evolutionary diversification in the history of life occurred during the (a) Cambrian period (b) Ordovician period (c) Silurian period (d) Carboniferous period (e) Permian period

9. The greatest mass extinction episode in the history of life occurred at what boundary? (a) Pliocene–Pleistocene (b) Permian–Triassic (c) Mesozoic–Cenozoic (d) Cambrian–Ordovician (e) Triassic–Jurassic

10. The Age of Reptiles corresponds to the (a) Paleozoic era (b) Mesozoic era (c) Cenozoic era (d) Pleistocene epoch (e) Permian period

11. Evidence exists that a catastrophic collision between Earth and a large extraterrestrial body occurred 66 mya, resulting in the extinction of (a) worms, mollusks, and soft-bodied arthropods (b) jawless ostracoderms and jawed placoderms (c) dinosaurs, pterosaurs, and many gymnosperm species (d) mastodons, saber-toothed cats, and giant ground sloths (e) ferns, horsetails, and club mosses

12. Flowering plants and mammals diversified and became dominant during the (a) Paleozoic era (b) Mesozoic era (c) Cenozoic era (d) Devonian period (e) Cambrian period

Apply and Analyze

13. Which are thought to have evolved first, aerobic bacteria or photosynthetic bacteria? Explain.

Evaluate and Synthesize

14. **EVOLUTION LINK** If you were studying how protobionts evolved into cells and you developed a protobiont that was capable of self-replication, would you consider it a living cell? Why or why not?

15. **EVOLUTION LINK** If living cells were produced in a test tube from nonbiological components by chemical processes, would this accomplishment prove that life evolved in a similar manner billions of years ago? Explain your answer.

16. **EVOLUTION LINK** Why did the evolution of multicellular organisms such as plants and animals have to be preceded by the evolution of oxygen-producing photosynthesis?

17. **INTERPRET DATA** Evidence for the oldest cyanobacterial fossils (stromatolites) dates to about 2 billion years, but molecular evidence inferred from lipids in ancient rocks puts the date for living cyanobacteria and eukaryotes at 2.7 billion years. Develop two opposing hypotheses to explain the discrepancy between the dates.

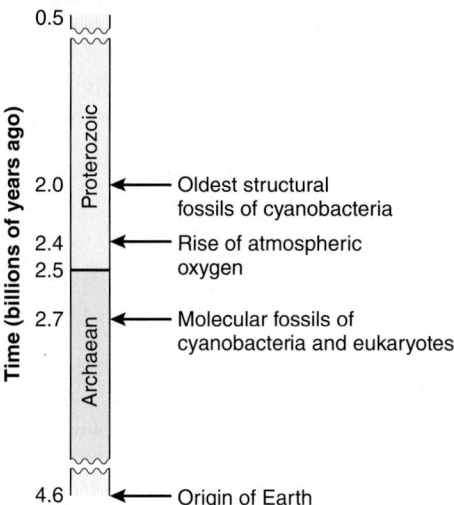

aplia To access course materials, such as Aplia and other companion resources, please visit **www.cengagebrain.com**.

The Evolution of Primates

Twelve years after Darwin wrote *On the Origin of Species by Means of Natural Selection,* he published another controversial book, *The Descent of Man,* which addressed human evolution. In this later book, Darwin hypothesized that humans and apes share a common ancestry. For nearly a century after Darwin's studies, fossil evidence of human ancestry remained fairly incomplete. However, research during the last several decades, especially in Africa, has yielded a rapidly accumulating set of fossils that gives an increasingly clear answer to the question, Where did we come from? (see photograph).

Fossil evidence from **paleoanthropology,** the study of human evolution, has allowed scientists to infer not only the structure but also the habits of early humans and other primates. Teeth and bones are the main fossil evidence that paleoanthropologists study. Much information can be obtained by studying teeth, which have changed dramatically during the course of primate and human evolution. Because tooth enamel is more mineralized (harder) than bone, teeth are more likely to fossilize. The teeth of each primate species, living or extinct, are distinctive enough to identify the species, approximate age, diet, and even sex of the individual.

Based on fossil evidence, paleontologists hypothesize that the first primates descended from small, shrewlike *placental mammals* that lived in trees and ate insects. (Placental mammals are the largest and most successful group of mammals. They have a **placenta,** an organ that exchanges materials between the mother and the embryo/fetus developing in the uterus.) Many traits of the 233 generally recognized living primate species are related to their **arboreal** (tree-dwelling) past. This chapter focuses on humans and their ancestors, who differ from most other primates because they did not remain in the trees but instead adapted to a terrestrial way of life.

Des Bartlett

Studying human evolution. The late Drs. Mary Leakey and Louis Leakey studied fossil teeth from *Australopithecus boisei,* an early hominin (human) that lived in Africa. This photo was taken in Olduvai Gorge in Tanzania. The Leakey family has contributed much of what we know about human evolution. Son Richard, daughter-in-law Meave, and granddaughter Louise have followed in the footsteps of Mary and Louis.

KEY CONCEPTS

22.1 Humans, along with lemurs, tarsiers, monkeys, and apes, are classified in the order Primates. This classification is based on close evolutionary ties.

22.2 The study of living primates provides clues to help scientists reconstruct the adaptations and lifestyles of early primates, some of which were ancestors of humans.

22.3 Fossil evidence indicates that the earliest human ancestors arose in Africa and shared many features with their apelike ancestors. The human brain did not begin to enlarge to its present size and complexity until long after bipedal locomotion had evolved.

22.4 Human culture began when human ancestors started making stone tools.

22.1 PRIMATE ADAPTATIONS

1 Describe the structural adaptations that primates have for life in treetops.

Humans and other **primates**—including lemurs, tarsiers, monkeys, and apes—are mammals that share such traits as flexible hands and feet with five digits, a strong social organization, and front-facing eyes, which permit depth perception. Mammals (class Mammalia) evolved from mammal-like reptiles known as *therapsids* more than 200 million years ago (mya), during the Mesozoic era (see Fig. 32-24). These early mammals remained a minor component of life on Earth for almost 150 million years and then rapidly diversified during the Cenozoic era (from 66 mya to the present).

Fossil evidence indicates that the first primates with traits characteristic of modern primates appeared by the Early Eocene epoch about 56 mya. These early primates had digits with nails, and their eyes were directed somewhat forward on the head.

Several novel adaptations evolved in early primates that allowed them to live in trees. One of the most significant primate features is that each limb has five highly flexible digits: four lateral digits (fingers) plus a partially or fully opposable first digit (thumb and, in many primates, big toe; FIG. 22-1). An **opposable thumb** positions the fingers opposite the thumb, enabling primates to grasp objects such as tree branches with precision. Nails (instead of claws) provide a protective covering for the tips of the digits, and the fleshy pads at the ends of the digits are sensitive to touch. Another arboreal feature is long, slender limbs that rotate freely at the hips and shoulders, giving primates full mobility to climb and search for food in the treetops.

Having eyes located in the front of the head lets primates integrate visual information from both eyes simultaneously; they have *stereoscopic* (three-dimensional) vision, which is important in judging distance and in depth perception. Stereoscopic vision is vital in an arboreal environment, especially for species that leap from branch to branch, because an error in depth perception may cause a fatal fall. In addition to having sharp sight, primates hear acutely.

Primates share several other characteristics, including a relatively large brain. Biologists have suggested that the increased sensory input associated with primates' sharp vision and greater agility favored the evolution of larger brains. Primates are generally very social and intelligent animals that reach sexual maturity relatively late in life. They typically have long lifespans. Females usually bear one offspring at a time; the baby is helpless and requires a long period of nurturing and protection.

CHECKPOINT 22.1

- CONNECT *What features of primate hands and feet are adaptations to an arboreal existence?*
- CONNECT *What anatomical feature confers stereoscopic vision on primates?*

KEY POINT

Primates have five grasping digits at the end of each limb, and the thumb or first toe is often partially or fully opposable.

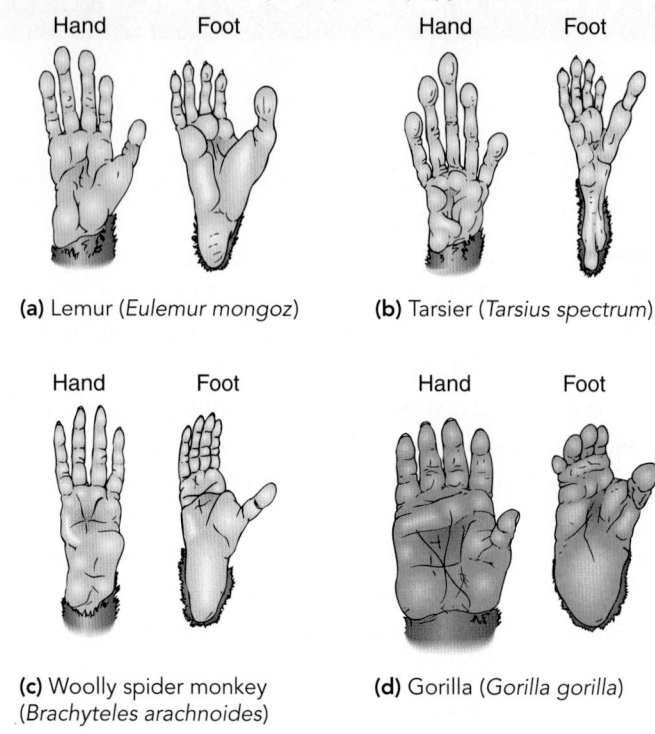

(a) Lemur (*Eulemur mongoz*) **(b)** Tarsier (*Tarsius spectrum*)

(c) Woolly spider monkey (*Brachyteles arachnoides*) **(d)** Gorilla (*Gorilla gorilla*)

Figure 22-1 Right hands and feet of selected primates

Figures not drawn to scale. Adapted from Schultz, A.H. *The Life of Primates,* Weidenfeld and Nicholson, London, 1969.

CONNECT Compare the palm of your right hand to those illustrated here. To which is it most similar?

22.2 PRIMATE CLASSIFICATION

2 List the three suborders of primates and give representative examples of each.

3 Distinguish among anthropoids, hominoids, and hominins.

Now that we have surveyed the general characteristics of primates, let us look at how they are classified. Many biologists currently divide the order Primates into three groups, or suborders (FIG. 22-2). The suborder Prosimii includes lemurs, galagos, and lorises; the suborder Tarsiiformes includes tarsiers; and the suborder Anthropoidea includes **anthropoids** (monkeys, apes, and humans).

All lemurs are restricted to the island of Madagascar off the coast of Africa. Because of extensive habitat destruction and hunting, they are very endangered. Lorises, which live in tropical areas of Southeast Asia and Africa, resemble lemurs in many respects, as do galagos, which live in sub-Saharan

Several kinds of data support the hypothesis that chimpanzees are the closest living relative of humans.

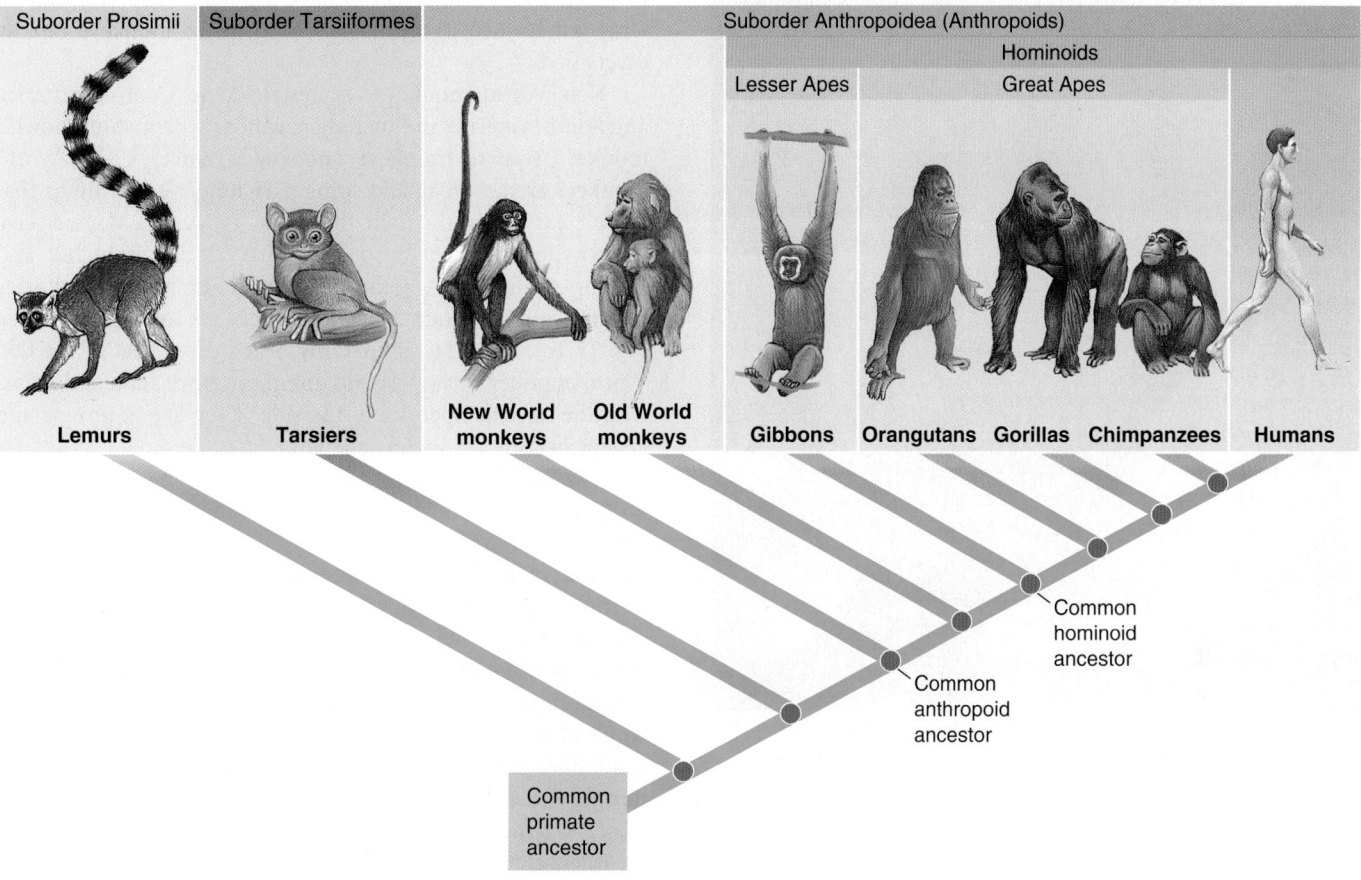

Figure 22-2 *Animation* **Primate evolution**

This branching diagram, called a cladogram, shows evolutionary relationships among living primates, based on current scientific evidence. The nodes (*circles*) represent branch points where a species splits into two or more lineages. The divergence of orangutans from the ape/hominin line occurred some 12 mya to 16 mya. Gorillas separated from the chimpanzee/hominin line an estimated 8 mya, and the hominin (human) lineage diverged from that of chimpanzees about 6 mya. *Figures not drawn to scale.*

CONNECT According to the cladogram, which non-human primates are the closest living relatives of chimpanzees?
© Cengage Learning

Africa. Lemurs, lorises, and galagos have retained several early mammalian features, such as elongated, pointed faces.

Tarsiers are found in rain forests of Indonesia and the Philippines (**FIG. 22-3**). They are small primates (about the size of a small rat) and are very adept leapers. These nocturnal primates resemble anthropoids in several ways, including their shortened snouts and forward-pointing eyes.

Suborder Anthropoidea includes monkeys, apes, and humans

Anthropoid primates arose during the Middle Eocene epoch, at least 45 mya. Several different fossil anthropoids have been identified, from Asia and North Africa, and there is a growing consensus about the relationships of these fossil groups to one another and to living anthropoids. Evidence indicates that anthropoids originated in Africa or Asia. The oldest known

anthropoid fossils, such as 42-million-year-old *Eosimias,* have been found in China and Myanmar. Given details about their dentition and the few bones that have been discovered, scientists infer that *Eosimias* and other ancestral anthropoids were small, insect-eating arboreal primates that were active during the day. Once they evolved, anthropoids quickly spread throughout Europe, Asia, and Africa and arrived in South America much later. (Paleoanthropologists date the oldest known South American primate, *Branisella,* from Bolivia, at 26 million years old.)

One significant difference between anthropoids and other primates is the size of their brains. The cerebrum in particular is more developed in monkeys, apes, and humans, where it functions as a highly complex center for learning, voluntary movement, and interpretation of sensation.

Monkeys are generally diurnal (active during the day) tree dwellers. They tend to eat fruit and leaves, with nuts, seeds, buds,

Figure 22-3 A tarsier
The huge eyes of the tarsier help it find insects, lizards, and other prey when it hunts at night. When a tarsier sees an insect, it pounces on it and grasps the prey with its hands. Tarsiers live in the rain forests of Indonesia and the Philippines.

The ancestors of New World monkeys may have rafted from Africa to South America on floating masses of vegetation. The South Atlantic Ocean would have been about half as wide as it is today, and islands could have provided "stepping stones." Once established in the New World, these monkeys rapidly diversified.

New World monkeys are restricted to Central America and South America and include marmosets, capuchins, howler monkeys, squirrel monkeys, and spider monkeys. New World monkeys are arboreal, and some have long, slender limbs that permit easy movement in the trees (**FIG. 22-4a**). A few have **prehensile tails** capable of wrapping around branches and serving as fifth limbs. Some New World monkeys have shorter thumbs, and in certain cases the thumbs are totally absent (see Fig. 22-1c). Their facial anatomy differs from that of the Old World monkeys; New World monkeys have flattened noses with the nostrils opening to the side. They live in groups and engage in complex social behaviors.

Old World monkeys are distributed in tropical parts of Africa and Asia. In addition to baboons and macaques (pronounced "muh-kacks'"), the Old World group includes guenons, mangabeys, langurs, and colobus monkeys. Most Old World monkeys are arboreal, although some, such as baboons and macaques, spend much of their time on the ground (**FIG. 22-4b**). The ground dwellers, which are **quadrupedal** (four footed; they walk on all fours), arose from arboreal monkeys. None of the Old World monkeys has a prehensile tail, and some have extremely short tails. They have a fully opposable thumb, and unlike the New World monkeys, their nostrils are closer together and directed downward. Old World monkeys are intensely social animals.

insects, spiders, birds' eggs, and even small vertebrates playing a smaller part in their diets. The two main groups of monkeys, New World monkeys and Old World monkeys, are named for the hemispheres where they diversified. Monkeys in South America and Central America are called New World monkeys, whereas monkeys in Africa, Asia, and Europe are called Old World monkeys. New World and Old World monkeys have been evolving separately for tens of millions of years.

One of the most important and interesting questions in anthropoid evolution concerns *how* monkeys arrived in South America. Africa and South America had already drifted apart. (Continental drift is presented in Figure 18-11.)

(a) New World monkey. The white-faced monkey (*Cebus capucinus*) has a prehensile tail and a flattened nose with nostrils directed to the side.

(b) Old World monkey. The Anubis baboon (*Papio anubis*) is native to Africa. Note that its nostrils are directed downward.

Figure 22-4 New World and Old World monkeys

Apes are our closest living relatives

Old World monkeys shared a common ancestor with the **hominoids,** a group composed of apes and **hominins** (humans and their ancestors; also called *hominids*). A fairly primitive anthropoid, discovered in Egypt, was named *Aegyptopithecus* (**FIG. 22-5a**). A cat-sized, forest-dwelling arboreal monkey with a few apelike characteristics, *Aegyptopithecus* lived during the Oligocene epoch, approximately 34 mya.

Fossil evidence reported in 2010 indicates that apes and Old World monkeys had diverged between 29 mya and 28 mya. During the Miocene epoch, as many as one hundred species of apes lived in Africa, Asia, and Europe.

Paleoanthropologists have discovered the oldest fossils with hominoid features in East Africa, mostly in Kenya. *Proconsul,* for example, appeared early in the Miocene epoch, about 20 mya (**FIG. 22-5b**). It had a larger brain than that of monkeys and apelike teeth and diet (fruits), but a monkeylike body. At least 30 other early hominoid species lived during the Miocene epoch, but most of them became extinct and were not the common ancestor of modern apes and humans.

Miocene fossils of forest-dwelling, chimpanzee-sized apes called *dryopithecines,* which lived about 15 mya, are of special interest because this hominoid lineage may have given rise to modern apes as well as to humans (**FIG. 22-5c**). The dryopithecines, such as *Dryopithecus, Kenyapithecus,* and *Morotopithecus,* were distributed widely across Europe, Africa, and Asia. As the climate gradually cooled and became drier, their range became more limited. These apes had highly modified bodies for swinging through the branches of trees, although there is also evidence that some of them may have left the treetops for the ground as dense forest gradually changed into open woodland. Many questions about the relationships among the various early apes have been generated by the discovery of these and other Miocene hominoids. As future fossil finds are evaluated, they may lead to a rearrangement of ancestors in the hominoid family tree.

Five genera of hominoids exist today: gibbons (*Hylobates*), orangutans (*Pongo*), gorillas (*Gorilla*), chimpanzees (*Pan*), and humans (*Homo*). Gibbons are known informally as lesser apes; orangutans, gorillas, and chimpanzees are known as great apes. Molecular evidence indicates a close relationship between humans and the great apes, particularly chimpanzees.

Gibbons are natural acrobats that can *brachiate,* or arm-swing, with their weight supported by one arm at a time (**FIG. 22-6a**). Orangutans are also tree dwellers, but chimpanzees and especially gorillas have adapted to life on the ground (**FIGS. 22-6b–d**). Gorillas and chimpanzees have retained long arms typical of brachiating primates but use them to assist in quadrupedal walking, sometimes known as *knuckle walking* because of the way they fold (flex) their digits when moving. Like humans, apes lack tails, a characteristic that makes them easy to distinguish from monkeys. They are generally much larger than monkeys, although gibbons are a notable exception.

Molecular data provide clues about evolution of anthropoid primates
Evidence of the close relatedness of orangutans, gorillas, chimpanzees, and humans is abundant at the molecular

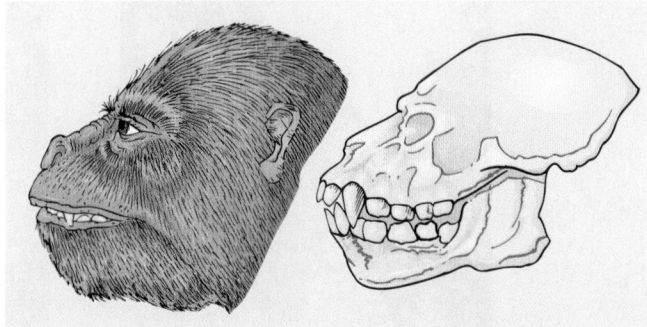

(a) Fossils of *Aegyptopithecus*, a fairly primitive anthropoid, were discovered in Egypt.

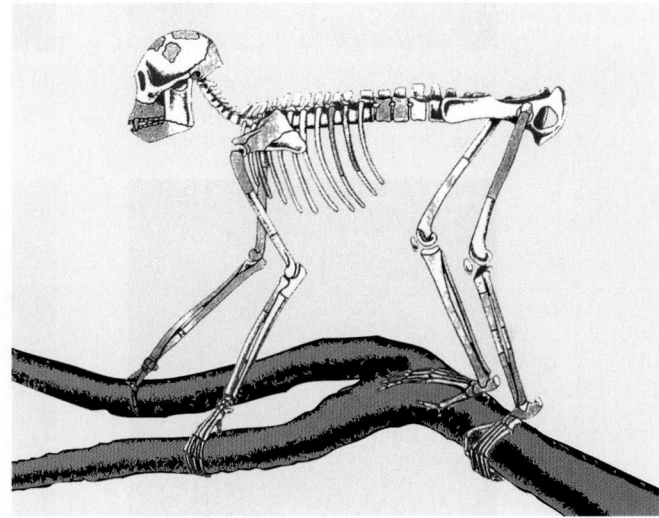

(b) Skeletal reconstruction of *Proconsul*. (The reconstructed parts are *white*.) This anthropoid had the limbs and body proportions of a monkey but lacked a tail, like all apes.

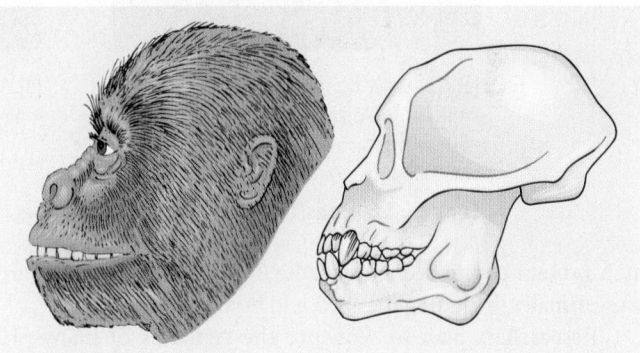

(c) *Dryopithecus*, a more advanced ape, may have been ancestral to modern hominoids.

Figure 22-5 Ape evolution
Figures not drawn to scale.
© Cengage Learning

level. The amino acid sequence of the chimpanzee's hemoglobin is identical to that of the human; hemoglobin molecules of the gorilla and rhesus monkey differ from the human's by 2 and 15 amino acids, respectively. DNA sequence analyses indicate that chimpanzees are likely to be our nearest living relatives among the apes. (Figure 18-18 shows the percent divergence in a

(a) A white-handed gibbon (*Hylobates lar*). Gibbons are extremely acrobatic and often move through the trees by brachiation.

(b) Orangutan (*Pongo pygmaeus*) anatomy is adapted to living in trees.

(c) A mother bonobo chimpanzee (*Pan paniscus*) holds her baby. Bonobos are endemic to a single country, the Democratic Republic of Congo.

(d) A female lowland gorilla (*Gorilla gorilla*), shown with her baby, maintains a knuckle-walking stance.

Figure 22-6 Apes

non-protein-coding sequence on the β-globin gene between various primates, including the apes, and humans.)

Researchers plan to sequence the genomes of many primates, including all the apes. The chimpanzee genome was sequenced in 2005 and the macaque (an Old World monkey) in 2007. Based on this and additional molecular data, the lineage leading to the macaques is thought to have split off about 25–33 mya, so comparing their genome to those of apes helps researchers identify the genetic changes that evolved only in the hominoid lineage.

Molecular evidence suggests that orangutans (genome sequenced in 2010) may have diverged from the gorilla, chimpanzee, and hominin lines about 12–16 mya. Gorillas (genome sequenced in 2012) may have diverged from the chimpanzee and hominin lines some 10 mya, whereas chimpanzee and hominin lines probably separated about 4 mya.

CHECKPOINT 22.2

- *What are the three suborders of primates?*
- **CONNECT** *How can you distinguish between monkeys and hominoids?*

22.3 HOMININ EVOLUTION

LEARNING OBJECTIVES

4 Describe skeletal and skull differences between apes and hominins.

5 Briefly describe the following early hominins: *Orrorin, Ardipithecus, Australopithecus anamensis, Au. afarensis, Au. africanus,* and *Au. sediba.*

6. Distinguish among the following members of genus *Homo*: *H. habilis, H. ergaster, H. erectus, H. antecessor, H. heidelbergensis, H. neanderthalensis,* and *H. sapiens.*

7. Discuss the origin of modern humans.

Scientists have a growing storehouse of hundreds of hominin fossils, which provide useful data about general trends in the body design, appearance, and behavior of ancestral humans. Despite the wealth of fossil evidence and a growing body of molecular data, scientists continue to vigorously debate interpretations of hominin characteristics, classification, and evolution. New discoveries will continue to raise new questions.

Evolutionary changes from the earliest hominins to modern humans are evident in some of the characteristics of the skeleton and skull. For example, before their brains enlarged, early hominins clearly adopted a **bipedal** (two-footed) posture. Compared with the ape skeleton, the human skeleton shows distinct differences that reflect humans' ability to stand erect and walk on two feet (**FIG. 22-7**). These differences also reflect the habitat change for early hominins, from an arboreal existence in the forest to a life spent at least partly on the ground.

The curvature of the human spine provides better balance and weight distribution for bipedal locomotion than the spine of the ape. The human pelvis is shorter and broader than the ape pelvis, allowing better attachment of muscles used for upright walking. (The shape of the human pelvis also allows the birth of offspring with large brains.) In apes the **foramen magnum,** the hole in the base of the skull for the spinal cord, is located in the rear of the skull. In contrast, the human foramen magnum is centered in the skull base, positioning the head for erect walking. An increase in the length of the legs relative to the arms and alignment of the first toe with the rest of the toes further adapted the early hominins for bipedalism.

Another major trend in hominin evolution was an increase in brain size relative to body size (**FIG. 22-8**). The ape skull has prominent **supraorbital ridges** above the eye sockets, whereas modern human skulls lack these ridges. Human faces are flatter than those of apes, and the jaws are different. The arrangement of teeth in the ape jaw is somewhat rectangular, compared with a rounded, or U-shaped, arrangement in humans. Apes have larger front teeth (canines and incisors)

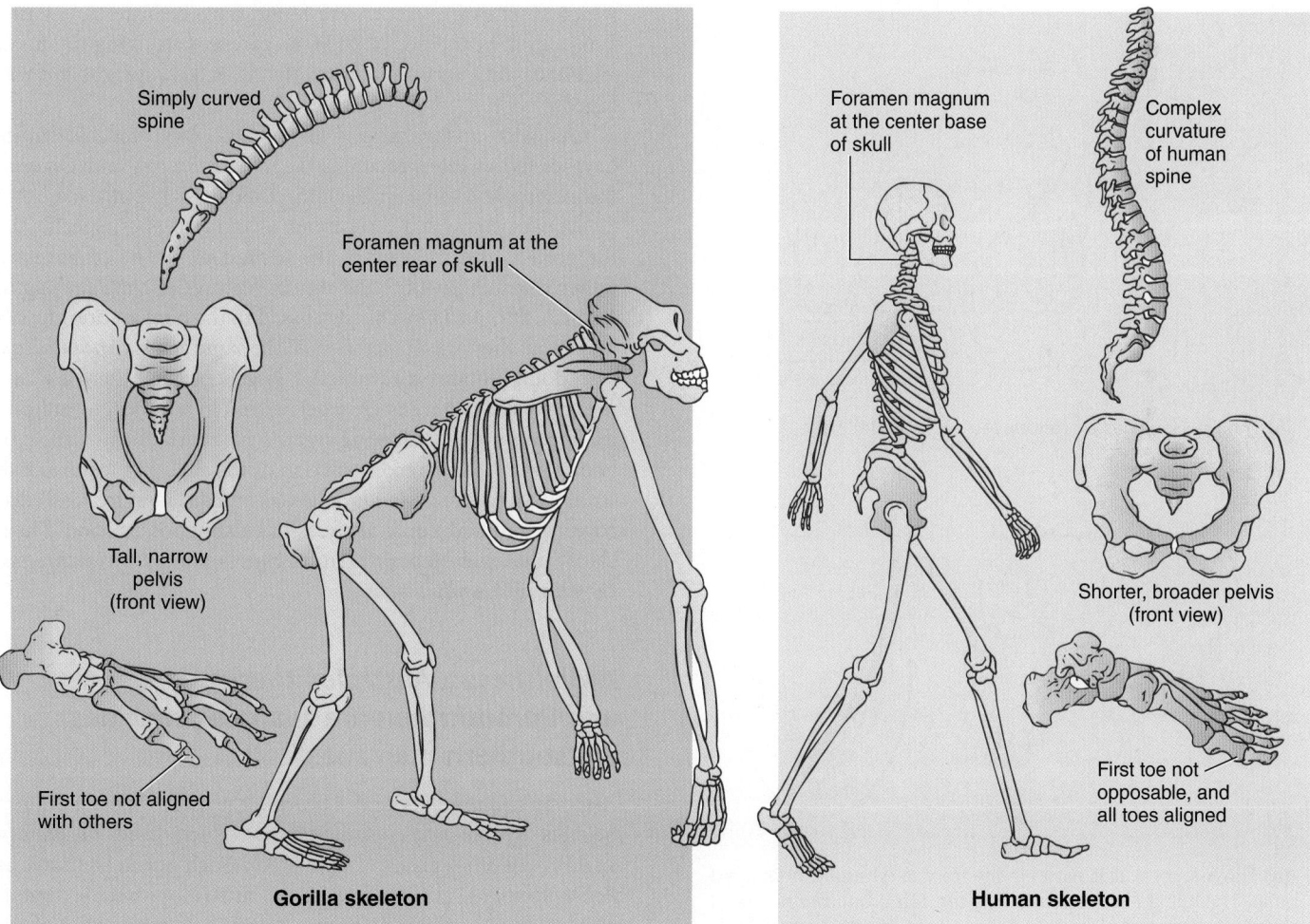

Figure 22-7 *Animation* **Gorilla and human skeletons**

When gorilla and human skeletons are compared, the skeletal adaptations for bipedalism in humans become apparent.

© Cengage Learning

than do humans, and their canines are especially large. Gorillas and orangutans also have larger back teeth (premolars and molars) than do humans.

We now examine some of the increasing number of fossil hominins in the human lineage. As you read the following descriptions of human evolution, keep in mind that much of what is discussed is open to reinterpretation and major revision as additional discoveries are made. Although we present human evolution in a somewhat linear fashion, from ancient hominins to anatomically and behaviorally modern humans, the human family tree is not a single trunk; rather, it has several branches (FIG. 22-9). Perhaps it is most useful to think of the known hominin fossils as a sampling of human evolution instead of a continuous sequence.

Homo sapiens is the only species of hominin in existence today, but more than one hominin species coexisted at any given time for most of the past 4 million years. In addition, do not make the mistake of thinking that your smaller-brained ancestors were inferior to yourself. Ancestral hominins were evolutionarily successful in that they were well adapted to their environment and survived for millions of years.

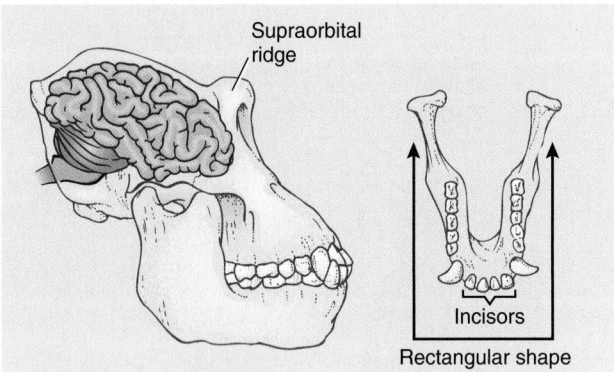

(a) The ape skull has a pronounced supraorbital ridge.

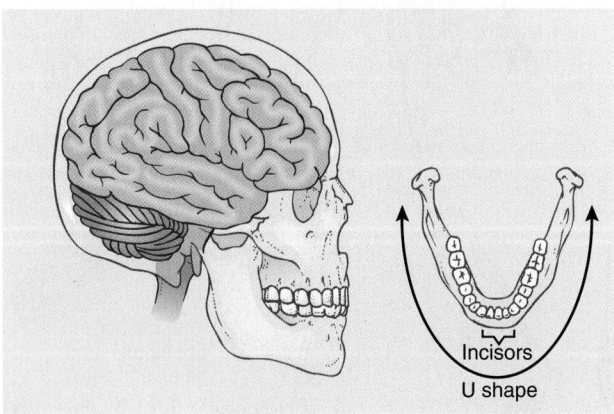

(b) The human skull is flatter in the front and has a pronounced chin. The human brain, particularly the cerebrum (*purple*), is larger than that of an ape, and the human jaw is structured so that the teeth are arranged in a U shape. Human canines and incisors are also smaller than those of apes.

Figure 22-8 Gorilla and human heads
© Cengage Learning

The earliest hominins may have lived 6 mya to 7 mya

Hominin evolution began in Africa. Although most hominin fossils have been discovered in Ethiopia and Kenya, French paleontologist Michel Brunet and an international team made a stunning discovery in 2002 in a dry lake bed in Chad, which is in central Africa. The fossil, which has been reliably dated at 6 million to 7 million years old, may be the earliest known hominin. Viewed from the back, the skull of *Sahelanthropus*, with its small braincase, resembles that of a chimpanzee. However, viewed from the front, the face and teeth have many characteristics of larger-brained human ancestors. Most paleoanthropologists place *Sahelanthropus* close to the base of the human family tree, that is, close to when genetic analyses suggest that the last common ancestor of hominins and chimpanzees existed. The discovery of *Sahelanthropus* is important in its own right, but it is also significant because it shows that early hominins exhibited more variation and lived over a larger area of Africa than had been originally hypothesized.

Orrorin, discovered in 2000 in Kenya, may also represent one of the earliest known hominins. The sediments in which fossils of *Orrorin* were found have been reliably dated at 6 million years old. In 2008, researchers studying the fossil leg bones of *Orrorin* concluded that it walked upright and was bipedal.

As with many aspects of human evolution, scientists have differing interpretations of *Sahelanthropus* and *Orrorin*. For example, some paleoanthropologists hypothesize that *Sahelanthropus* was a forerunner of modern apes, specifically the gorilla, and not one of the earliest ancestors of humans. Other researchers think the similar features of *Sahelanthropus*, *Orrorin*, and *Ardipithecus* (discussed in the next section) fossils mean that they are all members of the same genus, *Ardipithecus*. This point remains controversial, largely because there are currently no skeletal bones (leg, pelvis, and foot bones) to indicate if *Sahelanthropus* walked upright, a hallmark characteristic of hominins. (A 2005 reconstruction of the skull has features that strongly suggest *Sahelanthropus* was bipedal.) Future fossil discoveries and additional analyses of existing fossils should help clarify the evolutionary relationships among *Sahelanthropus*, *Orrorin*, and *Ardipithecus*.

Ardipithecus, Australopithecus, and *Paranthropus* are australopithecines, or "southern man apes"

After *Sahelanthropus* and *Orrorin*, the next-oldest hominin belongs to the genus *Ardipithecus*, which lived in eastern Africa between 4 mya and 6 mya. Although not as primitive as *Sahelanthropus*, *Ardipithecus* is an **australopithecine** that is close to the "root" of the human family tree. The shape of the toe bones suggests that *Ardipithecus* walked upright.

Two species of *Ardipithecus* have been described. The canine teeth of the earliest species, *Ar. kadabba*, have certain primitive features similar to those of *Sahelanthropus* and *Orrorin*.

Paleoanthropologists do not completely agree about certain specific details of the human family tree and hold many possible interpretations of classification lines of descent.

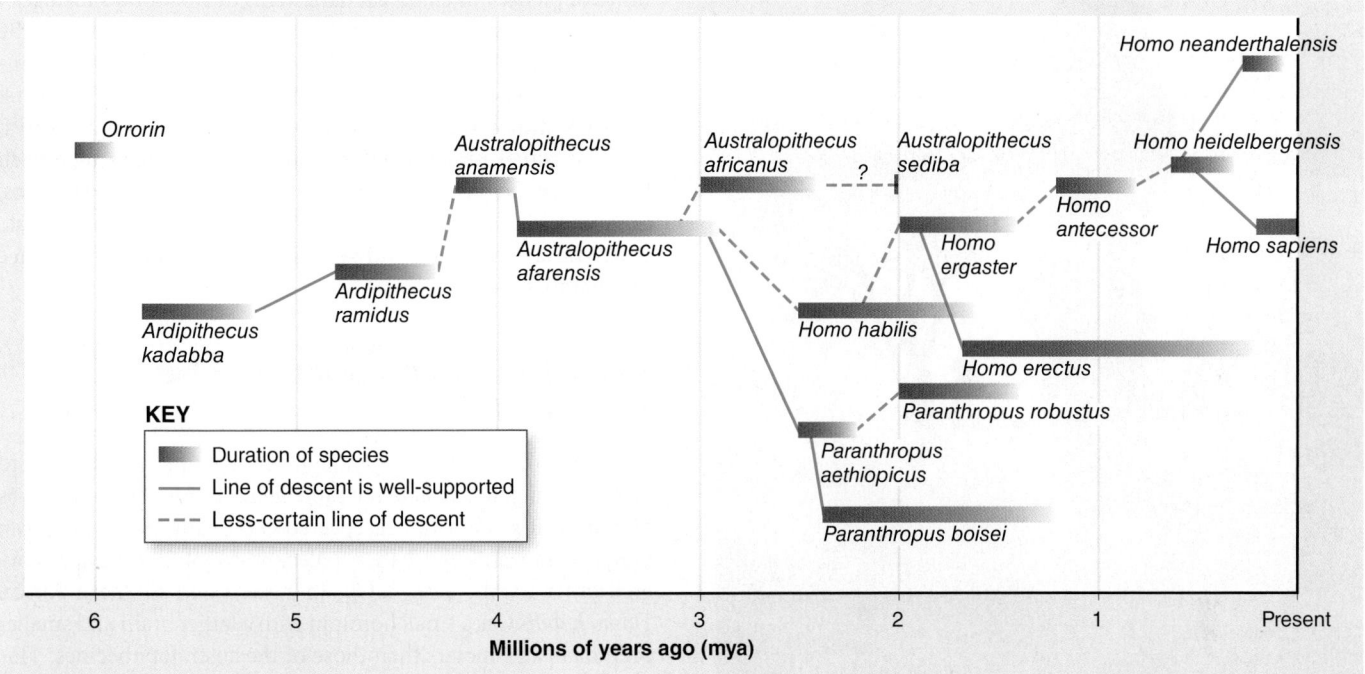

Figure 22-9 One interpretation of human evolution

This interpretation, compiled by the author from multiple sources, will likely change as new evidence comes to light.

CONNECT If you were to add *Homo floresiensis* and the Denisovans to this chart, where would you place them? Justify your answers.

© Cengage Learning

The specific epithet *kadabba* comes from Afar words meaning "basal family ancestor." ***Ardipithecus ramidus*** appeared about 4.8 mya. A remarkably complete fossil of a female *Ar. ramidus,* nicknamed Ardi, was described in a series of scientific papers in 2009. Based on details of her anatomy, Ardi has been described as a "facultative biped," capable of walking upright but also able to climb.

Ardipithecus gave rise to *Australopithecu*s, a genus that includes several species that lived between 4 mya and 1 mya. Both *Ardipithecus* and *Australopithecus* had longer arms, shorter legs, and smaller brains relative to those of modern humans.

Hominins that existed between 4.2 mya and 3.9 mya are assigned to the species ***Australopithecus anamensis,*** first named by paleoanthropologist Meave Leakey and her co-workers in 1995 from fossils discovered in eastern Africa. This hominin species, which has a mixture of apelike and humanlike features, presumably evolved from *Ardipithecus ramidus.* A comparison of male and female *Australopithecus anamensis* body sizes and canine teeth reveals **sexual dimorphism,** marked phenotypic differences between the two sexes of the same species. (The modern-day gorilla is sexually dimorphic.) The back teeth and jaws of *Au. anamensis* are larger than those of modern chimpanzees, whereas the front teeth are smaller and more like those of later hominins. A fossil

leg bone, the tibia, indicates that *Au. anamensis* had an upright posture and was bipedal, although it also may have foraged in the trees. Thus, bipedalism occurred early in human evolution and appears to have been the first human adaptation.

Australopithecus afarensis, another primitive hominin, probably evolved directly from *Au. anamensis.* Many fossils of *Au. afarensis* skeletal remains have been discovered in Africa, including a remarkably complete 3.2-million-year-old skeleton, nicknamed Lucy, found in Ethiopia in 1974 by a team led by U.S. paleoanthropologist Donald Johanson (**FIG. 22-10**). In 1978, British paleoanthropologist Mary Leakey and her co-workers discovered beautifully preserved fossil footprints of three *Au. afarensis* individuals who walked more than 3.6 mya. In 1994, paleoanthropologists found the first adult skull of *Au. afarensis.* The skull, characterized by a relatively small brain, pronounced supraorbital ridges, a jutting jaw, and large canine teeth, is an estimated 3 million years old. It is probable that *Au. afarensis* did not construct tools or make fires because no evidence of tools or fire has been found at fossil sites.

Many paleoanthropologists think that *Au. afarensis* gave rise to several australopithecine species, including ***Australopithecus africanus,*** which may have appeared as early as 3 mya. The first *Au. africanus* fossil was discovered in South Africa in

Figure 22-10 *Australopithecus afarensis*

This artist's rendition of a female australopithecine and child is based on the fossils of Lucy, discovered in 1974.

Lionel Bret/Science Source

1924, and since then hundreds have been found. This hominin walked erect and possessed hands and teeth that were distinctly humanlike. Given the characteristics of the teeth, paleoanthropologists think that *Au. africanus* ate both plants and animals. Like *Au. afarensis,* it had a small brain, more like that of its primate ancestors than of present-day humans.

In 2010 and 2013, a series of papers in the journal *Science* described **Australopithecus sediba** based on studies of the well-preserved partial skeletons of an adult female and a young boy found in South Africa. Dated at 1.95 mya, these fossils exhibit a mixture of apelike and humanlike features. Like apes, they have small brains, long arms, and, in the female, a small birth canal. However, among their humanlike features are long legs, a pelvis consistent with bipedal posture, and long, opposable thumbs. Most perplexing of all, they have apelike feet, but their ankle joints are like those of humans. Many scientists think that *Au. sediba* was probably a descendant of *Au. africanus,* but there is considerable disagreement on other issues. Was *Au. sediba* part of the line leading to humans or an evolutionary dead end?

Three australopithecine species (*Paranthropus robustus* from South Africa and *P. aethiopicus* and *P. boisei,* both from East Africa) were larger than *Au. africanus* and had extremely large molars, very powerful jaws, relatively small brains, and bony skull crests. Most females lacked the skull crests and had substantially smaller jaws than *Au. africanus,* another example of sexual dimorphism in early hominins. The teeth and jaws

suggest a diet, perhaps of tough roots, nuts, and seeds, that would require powerful grinding. (However, a 2008 study of fossil teeth of *P. boisei* found that the wear on the teeth suggested a softer diet, such as fruits. *P. boisei* apparently ate soft food during times of plenty but could eat hard items during food shortages.) These so-called *robust australopithecines* may or may not be closely related but are generally thought to represent evolutionary offshoots, or side branches, of human evolution. The first robust australopithecine, *P. aethiopicus,* appeared about 2.5 mya.

The actual number of australopithecine species for which fossil evidence has been found is under debate. In some cases, differences in the relatively few skeletal fragments could indicate either variation among individuals within a species or evidence of separate species.

Homo habilis is the oldest member of genus *Homo*

The first hominin to have enough uniquely human features to be placed in genus *Homo,* the same genus as modern humans, is **Homo habilis.** It was first discovered in the early 1960s at Olduvai Gorge in Tanzania. Since then, paleoanthropologists have discovered other fossils of *H. habilis* in eastern and southern Africa. *Homo habilis* was a small hominin with a larger brain and smaller premolars and molars than those of the australopithecines. This hominin appeared approximately 2.5 mya and persisted for about 900,000 years. Fossils of *H. habilis* have been found in numerous areas in Africa. These sites contain primitive tools, stones that had been chipped, cracked, or hammered to make sharp edges for cutting or scraping. *Oldowan* pebble choppers and flakes, for example, were probably used to cut through animal hides to obtain meat and to break bones for their nutritious marrow.

The relationship between the australopithecines and *H. habilis* is not clear. Using physical characteristics of their fossilized skeletons as evidence, many paleoanthropologists have inferred that the australopithecines were ancestors of *H. habilis.* Some researchers do not think that *H. habilis* belongs in the genus *Homo,* and they suggest that it should be reclassified as *Australopithecus habilis.* The discovery of *Au. sediba,* described in the previous section, has added fuel to this debate. Discoveries of additional fossils may help clarify these relationships.

Homo ergaster may have arisen from *H. habilis*

Initially, fossils that are now classified as *Homo ergaster* were considered by some scientists as early *H. erectus* (discussed in next section). With the discovery of additional fossils, however, many scientists hypothesized that the fossils classified as *H. erectus* really represent two species, **Homo ergaster,** an earlier African species, and *H. erectus,* a later eastern Asian offshoot. The best-known fossils of *H. ergaster,* which existed from 2.0 to 1.4 mya, come from the Lake Turkana region in Kenya. Researchers who support the split speculate that *H. ergaster* may be the direct ancestor of later humans, whereas *H. erectus* may be an evolutionary dead end.

Homo erectus probably evolved from *H. ergaster*

Investigators found the first fossil evidence of *Homo erectus* in Indonesia in the 1890s. Since then, searchers have found numerous fossils of *H. erectus* throughout Africa and Asia. Paleoanthropologists think that *H. erectus* originated in Africa about 1.7 mya and then spread to Europe and Asia. Peking man and Java man, discovered in Asia, were later examples of *H. erectus,* which existed until at least 200,000 years ago; some populations of *H. erectus* may have persisted more recently.

Homo erectus was taller than *H. habilis.* Its brain, which was larger than that of *H. habilis,* got progressively larger during the course of its evolution. Its skull, although larger, did not possess totally modern features, retaining the heavy supraorbital ridge and projecting face that are more characteristic of its ape ancestors (**FIG. 22-11**). *Homo erectus* is the first hominin to have fewer differences between the sexes.

The increased intelligence associated with a larger brain enabled these early humans to make more advanced stone tools, known as *Acheulean* tools, including hand axes and other implements that scientists have interpreted as choppers, borers, and scrapers (**FIG. 22-12**). Their intelligence also allowed these humans to survive in cold areas. *Homo erectus* obtained food by hunting or scavenging and may have worn clothing, built fires,

Figure 22-12 Acheulean tool

This stone tool is a typical Acheulean tool. It was made about 800,000 years ago by striking both sides of the biface (*shown*) to make a teardrop shape. The tool was found with other fossils and tools in Gran Dolina, Sierra de Atapuerca, Spain.

and lived in caves or shelters. Evidence of weapons (spears) has been unearthed at *H. erectus* sites in Europe. (For the remarkable story of an evolutionary offshoot from *H. erectus,* see *Inquiring About: The Smallest Humans.*)

Archaic humans date from about 1.2 mya to 200,000 years ago

Archaic humans are regionally diverse descendants of *H. ergaster* that lived in Africa, Asia, and Europe from about 1.2 mya to 200,000 years ago. The brains of archaic humans were essentially the same size as our brains, although their skulls retained some ancestral characters. Populations of archaic humans had rich and varied cultures (learned traditions) that included making many kinds of tools and objects with symbolic and ceremonial meaning.

Some researchers classify the oldest archaic human fossils discovered in Europe (in Spain) as *Homo antecessor. Homo antecessor* existed from about 1.2 mya to 800,000 years ago. Sites where *H. antecessor* fossils have been found show numerous cuts on human bones that suggest that these early humans practiced cannibalism.

Homo heidelbergensis, which appeared about 600,000 years ago and existed until about 300,000 years ago, may have descended

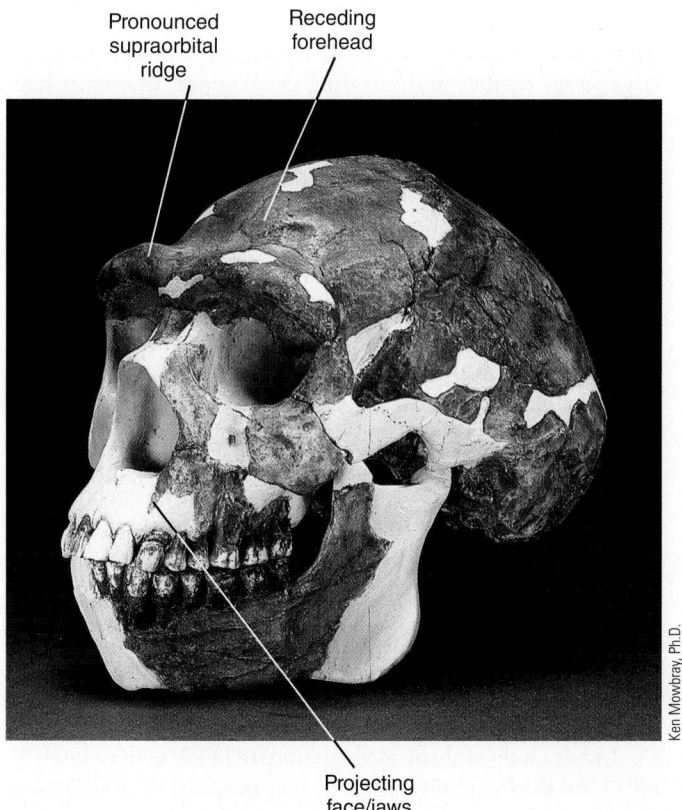

Pronounced supraorbital ridge

Receding forehead

Projecting face/jaws

Figure 22-11 *Homo erectus* skull from China

The reconstructed parts are *white.* Note the receding forehead, pronounced supraorbital ridge, and projecting face and jaws.

Evolutionarily speaking, what are "hobbits"? In 2004, paleoanthropologists reported a startling discovery of fossils of seven humans in a cave on the Indonesian island of Flores. The finding was completely unexpected because the fossil bones and teeth were from adult humans that were about 1 m (about 3 ft) tall.

The tiny humans, which have been dubbed "hobbits" by the media, may represent a new species, *Homo floresiensis*. Although these hominins had small, ape-sized brains, sophisticated stone tools, fireplaces, and charred bones found with the fossils indicate that *H. floresiensis* was capable of complex thinking and activities. The charred bones were primarily from pygmy elephants (*Stegodon*) and Flores giant rats (*Papagomys*), and they provide clues about the food that *H. floresiensis* hunted and killed (see figure). Based on several dating methods, *H. floresiensis* is thought to have existed from about 38,000 years ago to as recently as 12,000 years ago, when it went extinct.

Detailed studies of the *H. floresiensis* braincase, published in 2005, revealed that it was similar in many ways to the larger brain of *H. erectus*, which is known to have lived on nearby islands. Many researchers have concluded from this and other evidence that *H. floresiensis* was an evolutionary offshoot of *H. erectus*.

Why was *H. floresiensis* so small? Biologists have often observed two evolutionary trends of mammals living on remote islands: large mammals tend to evolve into much smaller species, and small mammals tend to evolve into much larger species. It is not unreasonable to assume that the small population of *H. erectus* ancestors that colonized Flores, perhaps by rafting to the island on a log, underwent evolutionary dwarfing over the thousands of years that they were isolated on Flores. (If *H. erectus* had been a frequent visitor to Flores, gene flow between the indigenous population and the visitors would have prevented the indigenous population from becoming measurably smaller.)

Not everyone agrees with the hypothesis that *H. floresiensis* is a separate species of tiny humans. Some researchers think that the fossils have not been identified properly. In a study published in 2006, these researchers suggest that the fossils are from *H. sapiens* individuals suffering from a rare genetic defect (microcephaly) that causes small brains and bodies.

The research team that unearthed the fossils has returned to the cave and will continue excavations. It is anticipated that future discoveries will help scientists answer the many questions we have about Earth's smallest humans. Meanwhile, the debate about the origin of *H. floresiensis* continues.

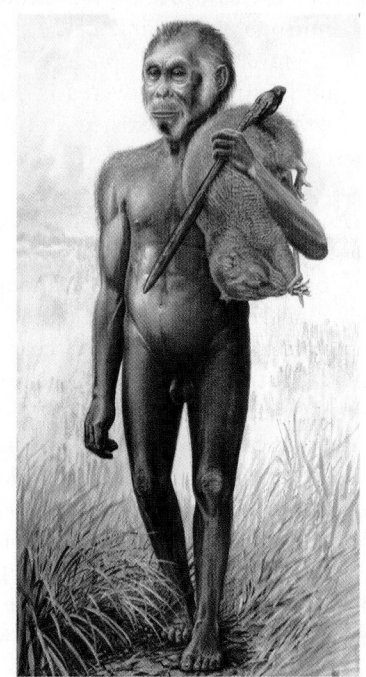

Peter Schouten

---

from *H. antecessor*. Fossils of *H. heidelbergensis* were first discovered in Germany about one hundred years ago. This extinct hominin had a larger brain—similar in size to that of modern humans—than *H. ergaster* or *H. antecessor*.

Some scientists do not recognize *H. antecessor* and suggest that *H. antecessor* fossils should be classified as *H. heidelbergensis*. Classification of *H. heidelbergensis* is also controversial because many (but not all) of the scientists studying its fossil remains think that both Neandertals and modern humans descended from it.

Neandertals appeared approximately 250,000 years ago

Fossils of **Neandertals** were first discovered in the Neander Valley in Germany.[1] They lived throughout Europe and western Asia from about 250,000 to 28,000 years ago. Although at one time many scientists considered Neandertals as a subspecies of *H. sapiens*, increasing evidence suggests that Neandertals were a separate species from modern humans, carrying the name *Homo neanderthalensis.*

These early humans had short, sturdy builds. Their faces projected slightly, their chins and foreheads receded, they had heavy supraorbital ridges and jawbones, and their brains and front teeth were larger than those of modern humans. They had large nasal cavities and receding cheekbones. Scientists have suggested that the large noses provided larger surface areas in Neandertal sinuses, enabling them to better warm the cold air of Ice Age Eurasia as inhaled air traveled through the head to the lungs.

Neandertal tools, known as *Mousterian* tools, were more sophisticated than those of *H. erectus* (**FIG. 22-13**). Studies of Neandertal sites indicate that Neandertals hunted large animals. The existence of skeletons of elderly Neandertals and of some with healed fractures may demonstrate that Neandertals cared for the aged and the sick, an indication of advanced social cooperation. They apparently had rituals, possibly of religious significance, and sometimes buried their dead.

The disappearance of the Neandertals about 28,000 years ago is a mystery that has sparked debate among paleoanthropologists. Groups of *H. sapiens* with more modern features coexisted with the Neandertals for several thousand years. Perhaps the other humans outcompeted or exterminated the Neandertals, leading to their extinction. It is also possible that the Neandertals interbred with these humans, diluting their features beyond recognition.

Increasingly sophisticated methods of DNA analysis contribute useful data for such controversies. Studies of nuclear DNA of Neandertals reported in 2010 indicated that between 1% and 4% of Neandertal DNA is shared with modern human populations outside of Africa, supporting the hypothesis that some human–Neandertal interbreeding did occur. In 2012, scientists reported

---

[1] *Neandertal* was formerly spelled *Neanderthal*. The silent *h* has been dropped in modern German but not in the scientific name.

Figure 22-13 Mousterian tool

Mousterian tools are named after a Neandertal site in Le Moustier, France. Mousterian tools include a variety of skillfully made stone tools; this tool, with its sharp edges, was probably used to scrape hides and/or to form the point of a spear. Although Neandertals had spears, they did not develop bows and arrows and so had to get close to their prey when hunting.

on the genome of a Neandertal relative. DNA was extracted from fossil remains of a young girl excavated from Denisova Cave in Siberia. The fossils, consisting of a finger bone and two teeth, are from sediments that have been dated to 50,000 years (and some genetic evidence has been interpreted to mean that the fossils may be as old as 80,000 years). Scientists consider these archaic humans, provisionally known as the Denisovans, a sister group to the Neandertals. Again there is evidence for inbreeding between archaic humans and *H. sapiens*. Denisovan DNA sequences ranging from traces to as much as 3% have been found in some present-day Asian and Oceanic island populations.

Scientists have reached a near consensus on the origin of modern *H. sapiens*

Homo sapiens with anatomically modern features existed in Africa about 195,000 years ago. As these anatomically modern humans dispersed around the world, they displaced other hominins such as *H. erectus* and *H. neanderthalensis*. By about 30,000 years ago, anatomically modern humans were the only members of genus *Homo* remaining, excluding small, isolated populations.

Both recent fossil finds and extensive molecular analyses have promoted the **(recent) out-of-Africa model** (also called *Recent Africa Origin,* or *RAO*) as the main explanation for the origin of modern humans living around the world. New data include the discovery of the earliest fossilized remains of modern *H. sapiens* in Africa and analyses of DNA

from mitochondria, the Y chromosome, and autosomes. In 2005, fossils of the earliest known modern *H. sapiens* were reliably dated at 195,000 years old; these fossils were found in southern Ethiopia. No *H. sapiens* fossils of comparable age have been found in Europe or Asia. The earliest fossils of anatomically modern *H. sapiens* in Europe and most parts of Asia date to 45,000 to 40,000 years ago. One exception is in Israel, where *H. sapiens* fossils have been dated at 100,000 years before present.

Molecular anthropology, the comparison of biological molecules from present-day individuals of regional human populations, provides clues that help scientists unravel the origin of modern humans and trace human migrations. A series of recent genetic studies of mtDNA, the Y chromosome, and autosomal DNA has strengthened the case for Africa as the birthplace of modern humans (**FIG. 22-14**). These studies provide detailed comparisons of DNA in present-day populations around the world as well as ancient DNA extracted from Neandertal and early *H. sapiens* remains.

CHECKPOINT 22.3

- **CONNECT** *How do the skulls of apes and humans differ?*
- **CONNECT** *How does an ape skeleton differ from a human skeleton?*
- **CONNECT** *How do australopithecines and* Homo *differ?*
- **CONNECT** *How can you distinguish between* H. habilis *and* H. erectus?
- **CONNECT** *How do* H. neanderthalensis *and* H. sapiens *differ?*

22.4 CULTURAL CHANGE

LEARNING OBJECTIVE

8 What are the generally recognized cultural stages that human populations have experienced?

Genetically speaking, humans are not very different from other primates. At the level of our DNA sequences, we are roughly 98% identical to gorillas and 99% identical to chimpanzees. Our relatively few genetic differences, however, give rise to several important distinguishing features, such as greater intelligence and the ability to capitalize on it by transmitting knowledge from one generation to the next. Human culture is not inherited in the biological sense but instead is learned, largely through the medium of language.[2] It is dynamic and is modified as people obtain new knowledge. Human culture is generally divided into three stages: (1) the development of hunter–gatherer societies, (2) the development of agriculture, and (3) the Industrial Revolution.

---

[2] Humans are not the only animals to have culture. Chimpanzees have primitive cultures that include tool-using techniques, hunting methods, and social behaviors that vary from one population to another. These cultural traditions are passed to the next generation by teaching and imitation; see Figure 1-17.

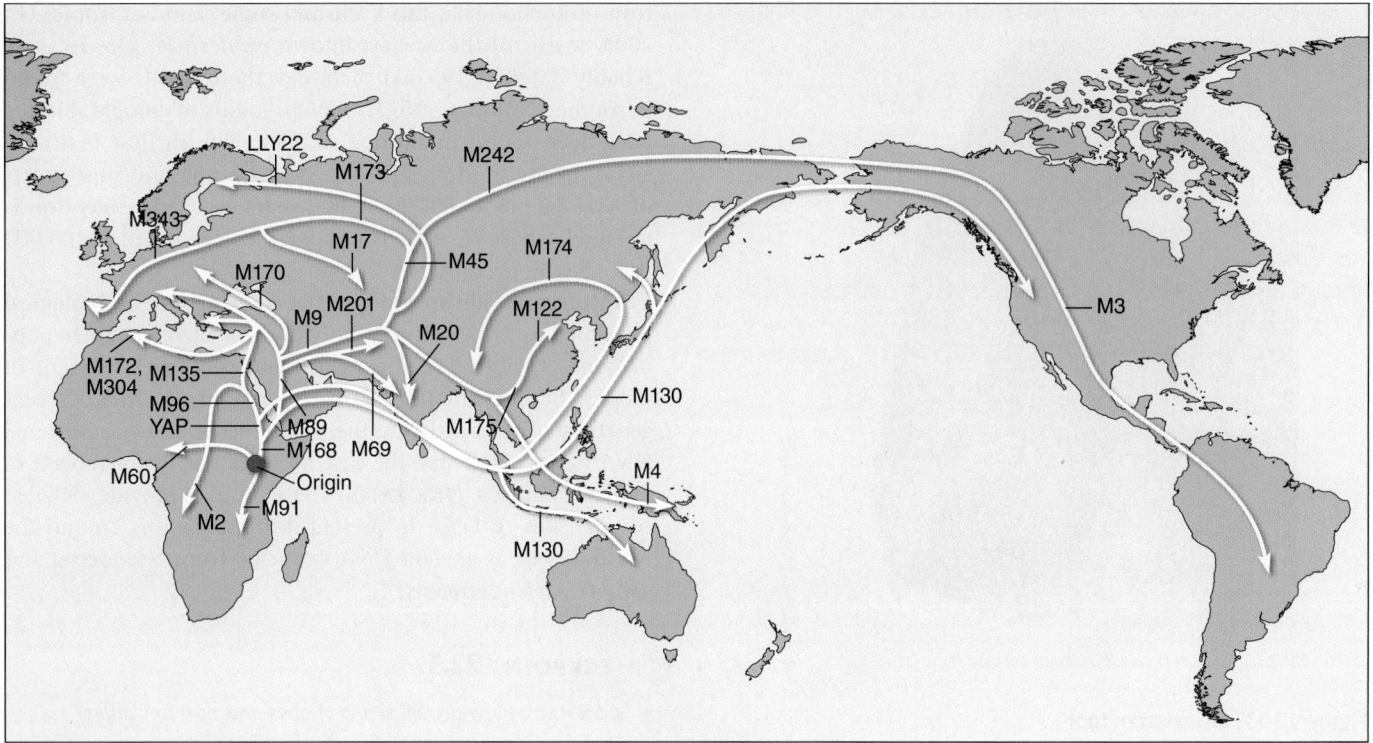

Figure 22-14 The path of human migrations, based on Y chromosome data

Homo sapiens originated in eastern Africa and migrated from there throughout the world. (SOURCE: Stix, G. "Traces of a Distant Past." *Scientific American*, p. 59, July 2008. Author credits information in *National Geographic* maps.)

Early humans were hunters and gatherers who relied on what was available in their immediate environment. They were nomadic, and as the resources in a given area were exhausted or as the population increased, they migrated to a different area. These societies required a division of labor and the ability to make tools and weapons, which were needed to kill game, scrape hides, dig up roots and tubers, and cook food. Although scientists are not certain when hunting was incorporated into human society, they do know that it declined in importance approximately 15,000 years ago. This decline may have been due to a decrease in the abundance of large animals, triggered in part by overhunting. A few isolated groups of hunter–gatherer societies, such as the Inuit of the northern polar region and the Mbuti of Africa, survived into the 21st century.

Development of agriculture resulted in a more dependable food supply

Evidence that humans had begun to cultivate crops approximately 10,000 years ago includes the presence of agricultural tools and plant material at archaeological sites. Agriculture resulted in a more dependable food supply. Archaeological evidence suggests that agriculture arose in several steps. Although there is variation from one site to another, plant cultivation, in combination with hunting, usually occurred first.

Animal domestication generally followed later, although in some areas, such as Australia, early humans did not domesticate animals.

Agriculture, in turn, often led to more permanent dwellings because considerable time was invested in growing crops in one area. Villages and cities often grew up around the farmlands, but the connection between agriculture and the establishment of villages and towns is complicated by certain discoveries. For example, Abu Hureyra in Syria was a village founded *before* agriculture arose. The villagers subsisted on the rich plant life of the area and the migrating herds of gazelle. Once people turned to agriculture, however, they seldom went back to hunting and gathering to obtain food.

Other advances in agriculture include the domestication of animals, which people kept to supply food, milk, and hides. Archaeological evidence indicates that wild goats and sheep were probably the first animals to be domesticated in southwestern Turkey, northern Iraq, and Iran. In the Old World, people also used animals to prepare fields for planting. Another major advance in agriculture was irrigation, which began more than 5000 years ago in Egypt.

Producing food agriculturally was more time-consuming than hunting and gathering, but it was also more productive. In hunter–gatherer societies, everyone shares the responsibility for obtaining food, but in agricultural societies, fewer people are needed to provide food for everyone. Thus, agriculture freed some people to pursue other endeavors, including religion, art, and various crafts.

Human culture has had a profound effect on the biosphere

Human culture has had far-reaching effects on both human society and on other organisms. The Industrial Revolution, which began in the 18th century, drew populations to concentrate in urban areas near centers of manufacturing. Advances in agriculture encouraged urbanization because fewer and fewer people were needed in rural areas to produce food for everyone. The spread of industrialization increased the demand for natural resources to supply the raw materials for industry. Chapter 53 discusses increases in human population size, and Chapter 57 investigates the effect of humans on the environment.

CHECKPOINT 22.4

- CONNECT *Describe some of the transitions from hunter–gatherer societies to agricultural societies to the Industrial Revolution. Have all societies gone through these stages in this order?*

SUMMARY: FOCUS ON LEARNING OBJECTIVES

22.1 Primate Adaptations *(page 458)*

1 Describe the structural adaptations that primates have for life in treetops.

- **Primates** are *placental mammals* that arose from small, arboreal (tree-dwelling), shrewlike mammals. Primates possess five grasping digits, including an **opposable thumb** or toe; long, slender limbs that move freely at the hips and shoulders; and eyes located in front of the head.

22.2 Primate Classification *(page 458)*

2 List the three suborders of primates and give representative examples of each.

- Primates are divided into three suborders. The suborder Prosimii includes lemurs, galagos, and lorises. The suborder Tarsiiformes includes tarsiers. The suborder Anthropoidea includes anthropoids: monkeys, apes, and humans.

3 Distinguish among anthropoids, hominoids, and hominins.

- **Anthropoids** include monkeys, apes, and humans. Early anthropoids branched into two groups: New World monkeys and Old World monkeys.
- **Hominoids** include apes and humans. Hominoids arose from the Old World monkey lineage. There are four modern genera of apes: gibbons, orangutans, gorillas, and chimpanzees.
- The **hominin** line consists of humans and their ancestors.

22.3 Hominin Evolution *(page 462)*

4 Describe skeletal and skull differences between apes and hominins.

- Unlike ape skeletons, hominin skeletons have adaptations that reflect the ability to stand erect and walk on two feet. These adaptations include a complex curvature of the spine; a shorter, broader pelvis; repositioning of the **foramen magnum** to the base of the skull; and a first toe that is aligned with the other toes.
- The human skull lacks a pronounced **supraorbital ridge,** is flatter than ape skulls in the front, and has a pronounced chin. The human brain is larger than ape brains, and the jaw is structured so that the teeth are arranged in a U shape.

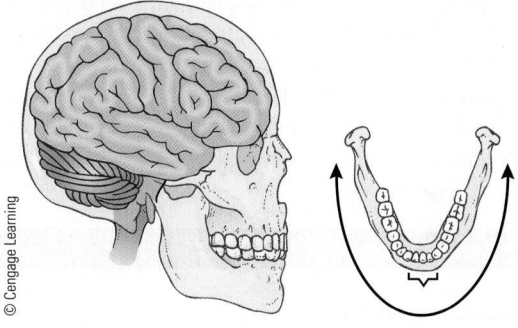
© Cengage Learning

5 Briefly describe the following early hominins: *Orrorin, Ardipithecus, Australopithecus anamensis, Au. afarensis, Au. africanus,* and *Au. sediba.*

- *Orrorin* is an early hominin that arose about 6 mya. Researchers studying the fossil leg bones of *Orrorin* think that it walked upright and was **bipedal** (walked on two feet).
- *Ardipithecus, Australopithecus,* and *Paranthropus* species are often referred to as **australopithecines.** Australopithecines were bipedal, a hominin feature. The first *Ardipithecus* species, *Ar. kadabba,* appeared about 5.8 mya. **Ardipithecus ramidus,** which appeared about 4.8 mya, may have given rise to **Australopithecus anamensis,** which in turn probably gave rise to another primitive hominin, **Australopithecus afarensis.** Many paleoanthropologists think that *Au. afarensis* gave rise to several australopithecine species, including **Australopithecus africanus,** *Paranthropus* spp., and possibly *Homo habilis.* **Australopithecus sediba** may have descended from *Au. africanus.*

6 Distinguish among the following members of genus *Homo: H. habilis, H. ergaster, H. erectus, H. antecessor, H. heidelbergensis, H. neanderthalensis,* and *H. sapiens.*

- **Homo habilis** was the earliest known hominin with some of the human features lacking in the australopithecines, including a slightly larger brain. *Homo habilis* fashioned crude tools from stone.
- The best-known fossils of **H. ergaster,** which existed from 2.0 to 1.4 mya, come from Kenya. *Homo ergaster* may be the direct ancestor of later humans.

- **Homo erectus** had a larger brain than *H. habilis;* made more sophisticated tools; and may have worn clothing, built fires, and lived in caves or shelters. *Homo erectus*, which is probably a later Asian offshoot of *H. ergaster,* appears to be an evolutionary dead end.

- Archaic humans are regionally diverse descendants of *H. ergaster* that lived in Africa, Asia, and Europe from about 1.2 mya to 200,000 years ago. The brains of archaic humans were essentially the same size as our brains, although their skulls retained some ancestral characters, and they had rich and varied cultures.

- Some researchers classify the oldest archaic human fossils discovered in Europe (in Spain) as **Homo antecessor.** *Homo antecessor* existed from about 1.2 mya to 800,000 years ago.

- **Homo heidelbergensis,** which appeared about 600,000 years ago and existed until about 300,000 years ago, may have descended from *H. antecessor*. This extinct hominin had a larger brain—similar in size to that of modern humans—than either *H. ergaster* or *H. antecessor*. Many scientists studying *H. heidelbergensis* think that both Neandertals and modern humans descended from it.

- **Neandertals** lived from about 250,000 to 28,000 years ago. Neandertals had short, sturdy builds; receding chins and foreheads; heavy supraorbital ridges and jawbones; larger front teeth; and nasal cavities with unusual triangular bony projections. Many scientists think that Neandertals were a separate species, **Homo neanderthalensis.**

- **Homo sapiens,** anatomically modern humans, existed in Africa about 195,000 years ago. By about 30,000 years ago, anatomically modern humans were the only members of genus *Homo* remaining, excluding small, isolated populations.

7 Discuss the origin of modern humans.

- Recent fossil finds and extensive molecular analyses have promoted the **(recent) out-of-Africa model** as the main explanation for the origin of modern humans living around the world. New data include the discovery of the earliest fossilized remains of modern *H. sapiens* in Africa and analyses of DNA from mitochondria, the Y chromosome, and autosomes.

22.4 Cultural Change *(page 469)*

8 What are the generally recognized cultural stages that human populations have experienced?

- Large human brain size makes possible the transmission of knowledge from one generation to the next. Significant advances in human culture were the transition from a hunter–gatherer way of life to the development of agriculture and the Industrial Revolution.

TEST YOUR UNDERSTANDING

Know and Comprehend

1. The first primates evolved from (a) shrewlike monotremes (b) therapsids (c) shrewlike placental mammals (d) tarsiers (e) shrewlike marsupials

2. The anthropoids are more closely related to ___ than to ___. (a) tarsiers; lemurs (b) lemurs; monkeys (c) tree shrews; tarsiers (d) lemurs; tarsiers (e) tree shrews; monkeys

3. With what group do hominoids share the most recent common ancestor? (a) Old World monkeys (b) New World monkeys (c) tarsiers (d) lemurs (e) lorises and galagos

4. The ___ in humans is centered at the base of the skull, positioning the head for erect walking. (a) supraorbital ridge (b) foramen magnum (c) pelvis (d) bony skull crest (e) femur

5. Scientists collectively call humans and their *immediate* ancestors (a) mammals (b) primates (c) anthropoids (d) hominoids (e) hominins

6. The earliest hominin that most paleoanthropologists place in genus *Homo* is (a) *H. habilis* (b) *H. ergaster* (c) *H. erectus* (d) *H. heidelbergensis* (e) *H. neanderthalensis*

7. Some scientists now think that fossils identified as *Homo erectus* represent which two different species? (a) *H. habilis* and *H. erectus* (b) *H. ergaster* and *H. erectus* (c) *H. heidelbergensis* and *H. ergaster* (d) *H. neanderthalensis* and *H. erectus* (e) *H. neanderthalensis* and *H. sapiens*

8. Archaic humans appeared as early as ___ years ago. (a) 5 million (b) 1.2 million (c) 250,000 (d) 100,000 (e) 5000

9. ___ were an early group of humans with short, sturdy builds and heavy supraorbital ridges that lived throughout Europe and western Asia from about 250,000 to 28,000 years ago. (a) australopithecines (b) dryopithecines (c) archaic humans (d) Neandertals (e) Denisovans

10. The modern human skull *lacks* (a) small canines (b) a foramen magnum centered in the base of the skull (c) pronounced supraorbital ridges (d) a U-shaped arrangement of teeth on the jaw (e) a large cranium (braincase)

11. The comparison of genetic material from individuals of regional populations of humans, used to help unravel the origin and migration of modern humans, is known as (a) paleoarchaeology (b) cultural anthropology (c) molecular anthropology (d) cytogenetics (e) genetic dimorphism

Apply and Analyze

12. Place the following hominins in chronological order of appearance in the fossil record, beginning with the earliest: 1. *H. ergaster* 2. *H. sapiens* 3. *H. neanderthalensis* 4. *H. erectus* (a) 1, 4, 2, 3 (b) 4, 1, 2, 3 (c) 4, 1, 3, 2 (d) 3, 1, 4, 2 (e) 1, 4, 3, 2

Evaluate and Synthesize

13. If you were evaluating whether other early humans exterminated the Neandertals, what kinds of archaeological evidence might you look for?

14. The remains of *H. sapiens* have been found in southern Europe alongside reindeer bones, but reindeer currently exist only in northern Europe and Asia. Explain the apparent discrepancy.

15. **EVOLUTION LINK** What was the common ancestor of chimpanzees and humans: a chimpanzee, a human, or neither? Explain your answer.

16. **EVOLUTION LINK** Some scientists say that modern medicine and better sanitation are slowing down or altering the course of human evolution in highly developed countries today. As a result, evolution in humans today has gone from survival of the

fittest to survival of almost everyone. Do you think that this idea is valid? Why or why not?

17. **INTERPRET DATA** Using the information in Figure 22-9, draw a simple cladogram that represents a reasonable hypothesis of human evolution at the genus level. (*Hint:* Your diagram will have three branches.)

18. **INTERPRET DATA** The four chromosomes depicted here are Y chromosomes from men in different parts of the world. The *purple bands* represent specific genes that are identical in all four men. The *red bands* represent genetic mutations (M) that appeared at different times; M168 appeared about 50,000 years ago, M9 about 40,000 years ago, and M3 about 10,000 years ago. Based on what you have learned in this chapter, identify the geographic source of men carrying each of these

chromosomes: African, Eurasian, Amerindian, and first migrants out of Africa. (You may wish to refer back to Figure 22-14.)

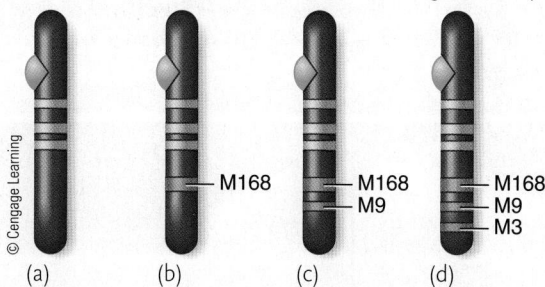

© Cengage Learning

(a)   (b)   (c)   (d)

To access course materials, such as Aplia and other companion resources, please visit **www.cengagebrain.com.**

23 Understanding Diversity: Systematics

Kayan loris (*Nycticebus kayan*) eating a cicada. Identified as a new species in 2012, this primate inhabits the jungles of Borneo.

© Ch'ien Lee/Minden Pictures

KEY CONCEPTS

23.1 Traditionally, biologists have named organisms using a binomial system (in which each species has a genus name followed by a specific epithet) and have classified organisms in taxonomic categories arranged in a hierarchy from most inclusive (domain) to least inclusive (species).

23.2 Based on current knowledge, the three main branches of the tree of life are the three domains: Bacteria, Archaea, and Eukarya.

23.3 Modern systematics seeks to reconstruct phylogeny (evolutionary history of a group of organisms) based on common ancestry inferred from shared characters (characteristics)—including structural, developmental, behavioral, and molecular similarities—as well as from fossil evidence.

23.4 Cladograms (a type of phylogenetic tree) are diagrams of hypothetical evolutionary relationships; they are constructed by analyzing shared derived characters.

23.5 Knowledge of modern systematics informs research in many other areas, including ecology and medicine.

The total number of living eukaryotic species on Earth is estimated to be between about 8.7 million and 100 million, but only about 1.9 million species have been described. Biologists estimate that to date they have identified less than 15% of terrestrial species and less than 10% of marine species. Many regions on our planet have not yet been carefully inventoried, including the forest canopy, soil, and the deep regions of the ocean.

Biologists discover thousands of new species each year. The Kayan loris (*Nycticebus kayan*) was identified as a new species of slow loris in 2012 (see photo). The slow loris is a primate genus related to the lemur. Slow lorises are the only known primates with a toxic bite. The toxin is secreted from glands in their elbow. Slow lorises lick this toxin, mixing it with saliva. When they bite, the toxin delivered into their victim is powerful enough to cause fatal anaphylactic shock in humans.

The variety of living organisms and the ecosystems they are part of are referred to as **biological diversity,** or **biodiversity.** The totality of life on Earth represents our biological heritage, and the quality of life for all organisms depends on the health and balance of this worldwide web of life-forms. For example, we depend on organisms to maintain the life-sustaining composition of gases in the atmosphere, form soil, break down wastes, recycle nutrients, and provide food for one another.

We humans exploit many species for economic benefit. One very anthropocentric (human-centered) example of the practical importance of biodiversity is that more than 40% of the prescriptions that pharmacists dispense in the United States derive from living organisms. Investigators are just learning how to effectively screen organisms for potential drugs. Unfortunately, human activity is seriously reducing biodiversity, and species are becoming extinct faster than researchers can study them. (See the discussion on declining biodiversity in Chapter 57.)

Alarmed by these extinctions, biologists are mobilizing to more rapidly identify new species and to conserve biodiversity. Many biologists agree that describing and classifying all the surviving species of the world should be a major scientific goal of the 21st century. Important steps toward this goal are being taken by international research endeavors. For example, the *Catalogue of Life* is the work of a collaborative international research team that manages a comprehensive global database made available on the Internet. This team categorizes and lists all the world's known species. The *Encyclopedia of Life (EOL)*, a project launched in 2007, is an electronic database that will include a home page for each species. There are currently more than one million pages in this online resource. Data include alternative names for each organism, classification, habitat, distribution, and diet. The Convention on Biological Diversity is an international treaty committed to reducing the "rate of loss of biodiversity." The 190 countries that make up this convention work to develop strategies for the conservation and sustainable use of Earth's biological diversity.

In this chapter we explore some of the approaches and methods used by biologists to classify organisms and infer their evolutionary history and relationships. (The evolution of species is discussed in Chapter 20.)

Information comes from many sources, including fossils, and biogeography as well as homologies in structure, physiology, behavior, and patterns of development. Advances in DNA sequencing have generated huge amounts of data about molecular similarities and differences.

Systematics is the scientific study of the diversity of organisms and their evolutionary relationships. Recall from earlier chapters (e.g., see Chapter 18) that **evolution** is the accumulation of inherited changes within populations over time. **Systematists,** biologists who study systematics, organize information about groups of organisms ranging from a small number of populations to all life on Earth. They generate hypotheses regarding patterns of evolutionary relationships for these groups of organisms. As more sophisticated molecular methods and discovery of new fossils provide new data, old data are reinterpreted. As we gain greater understanding of how organisms are related, we revise the way we classify organisms. We can use the phylogenetic information for naming species as well as for investigating specific hypotheses about directions of evolutionary change and about the particular factors that may have caused certain evolutionary changes.

23.1 CLASSIFYING ORGANISMS

LEARNING OBJECTIVES

1 State two justifications for the use of scientific names and classifications of organisms.
2 Describe the binomial system of naming organisms and arrange the Linnaean categories in hierarchical fashion from most inclusive to least inclusive.

Taxonomy, the science of naming, describing, and classifying organisms is an important aspect of systematics. In biology the term **classification** means arranging organisms into groups based on similarities that reflect evolutionary relationships among lineages.

People have always categorized organisms into groups to convey information. For example, groups like edible versus inedible species, or dangerous versus not dangerous species, are useful in that they can convey some important information about species without knowing the details about every species you encounter. For hundreds of years, scientists have realized that certain groups of organisms share many features. Saying something like "Species X is some kind of cat," tells a person that species X falls somewhere between tigers and house cats. It also suggests that species X has many features shared with other cats, ranging from details of the heart, skin, muscles, and skeleton, to molecular traits like amino acid sequences in particular proteins and nitrogenous base sequences in DNA, and even to complex traits like behaviors and aspects of their physiology.

After the work of Charles Darwin, we came to realize that organisms within such groups share so many traits because they have all inherited those traits from the same ancestor. The living species are "related" via shared ancestors, and the traits they share are **homologous traits** (see Chapter 18). Modern systematists seek to name and classify organisms according to their genealogical (evolutionary) relationships. This work involves some careful investigations as we attempt to determine the patterns of relationship.

Organisms are named using a binomial system

In the mid-18th century, Carolus Linneaus developed a system of naming organisms to allow scientists to communicate more clearly. We will discuss his contributions to classification in the next section, but here we focus on his idea for naming species.

Before the mid-18th century, each species had a lengthy descriptive name, sometimes consisting of ten or more Latin words! Linnaeus simplified scientific classification, developing a **binomial system of nomenclature** in which each species is assigned a unique two-part name. The first part of a binomial scientific name is a noun that designates the **genus** (pl., *genera*), and the second part, an adjective modifying that noun, is called the **specific epithet.**

The genus name is always capitalized, whereas the specific epithet is usually not. Both names are underlined or italicized. The genus, or generic, name can be used alone to designate all species in the genus (e.g., the genus *Quercus* includes all oak species). Note that the specific epithet alone is *not* the name of the species. In fact, the same specific epithet can be used as the second name of species in different genera. For example, *Quercus alba* is the scientific species name for the white oak, and *Salix alba* is the species name for the white willow (*alba* comes from a Latin word meaning "white"). Thus, both parts of the name must be used to identify the species accurately. The specific epithet is never used alone; it must always follow the full or abbreviated genus name, such as *Quercus alba* or *Q. alba.*

Scientific names are generally derived from Greek or Latin roots or from Latinized versions of the names of persons, places, or characteristics. For example, the generic name for the bacterium *Escherichia coli* is based on the name of the scientist, Theodor Escherich, who first described it. The specific epithet *coli* reminds us that *E. coli* lives in the colon (large intestine).

Scientific names permit biology to be a truly international science. Even though the common names of an organism may vary in different locations and languages, an organism can be universally identified by its scientific name. A researcher in Puerto Rico knows exactly which organisms were used in a study published by a Russian scientist and therefore can repeat or extend that scientist's experiments using the same species.

Each taxonomic level is more general than the one below it

Linnaeus devised a system for assigning species to a hierarchy of increasingly broader groups. As you move up the hierarchy, each group is more inclusive; that is, it includes the groups below it. When he set up his system, Linnaeus did not have a theory of evolution in mind, nor did he have any idea of the vast number of extant (living) and extinct organisms that would later be discovered. His system nonetheless ended up providing important evidence for evolution, and although some substantial modifications have been necessary, Linnaeus's basic system still forms a framework for much of our modern classifications.

The range of taxonomic categories from species to domain forms a hierarchy (**TABLE 23-1** and **FIG. 23-1**). Closely related species are assigned to the same genus, and closely related genera are grouped in a single **family.** Families are grouped into **orders,** orders into **classes,** classes into **phyla,** phyla into **kingdoms,** and kingdoms into **domains.** A **taxon** (pl., *taxa*) is a formal grouping of organisms at any given level, such as species, genus, or phylum. Today we realize that classifying organisms requires many more classification levels than

TABLE 23-1	Classification of Corn
DOMAIN	Eukarya
	Organisms that have nuclei and other membrane-enclosed organelles
KINGDOM	Plantae
	Terrestrial, multicellular, photosynthetic organisms
PHYLUM	Anthophyta
	Vascular plants with flowers, fruits, and seeds
CLASS	Monocotyledones
	Monocots: Flowering plants with one seed leaf (cotyledon) and flower parts in threes
ORDER	Commelinales
	Monocots with reduced flower parts, elongated leaves, and dry one-seeded fruits
FAMILY	Poaceae
	Grasses with hollow stems; fruit is a grain; and abundant endosperm in seed
GENUS	*Zea*
	Tall annual grass with separate female and male flowers
SPECIES	*Zea mays*
	Corn

© Cengage Learning

envisioned by Linnaeus, so between any two categories there could be additional levels. For example, the order Coleoptera (beetles) contains more than 400,000 species, and the class is divided into subclasses, with each subclass containing multiple infraorders, each infraorder containing multiple superfamilies, and each superfamily containing numerous families. Sometimes even these additional levels are not sufficient to categorize important groups.

Another difference between the current classifications and those of Linnaeus is even more substantial. Linnaeus grouped species into particular taxa because they shared certain *features.* In light of our modern understanding of shared ancestry and evolution, we seek to place organisms into particular taxa because they share particular *ancestry.* For example, a modern biologist would place a species within the taxon Mammalia if all the available evidence showed that the species evolved from the same recent ancestor as the rest of the mammals. This decision would not be based on whether or not members of the species had any one particular feature, like hair or mammary glands.

CHECKPOINT 23.1

- *Eating a single "death cap" mushroom can be fatal. The specific epithet of this mushroom is* phalloides. *Its genus name is* Amanita. *Write its scientific name.*
- *What are the key features of the system of hierarchical classification?*
- VISUALIZE *Write a list of the following taxonomic categories in order from most inclusive to least inclusive: genus, order, domain, family, phylum, species, kingdom, class.*

23.2 DETERMINING THE MAJOR BRANCHES IN THE TREE OF LIFE

LEARNING OBJECTIVES

3 Describe the three domains and argue for and against classifying organisms in these domains. (Compare this approach with using kingdoms as the main rank of classification.)

4 Interpret a cladogram, describing the meaning of its specific nodes and branches.

The history of taxonomy at the kingdom and domain levels is a good example of the process of science. In this section we will review some highlights of this process.

Systematics is an evolving science

From the time of Aristotle to the mid-19th century, biologists divided organisms into two kingdoms: **Plantae** and **Animalia.** After the development of microscopes, it became increasingly obvious that many organisms could not be easily assigned to either the plant or the animal kingdom. For example, the unicellular organism *Euglena* was classified at various times in the

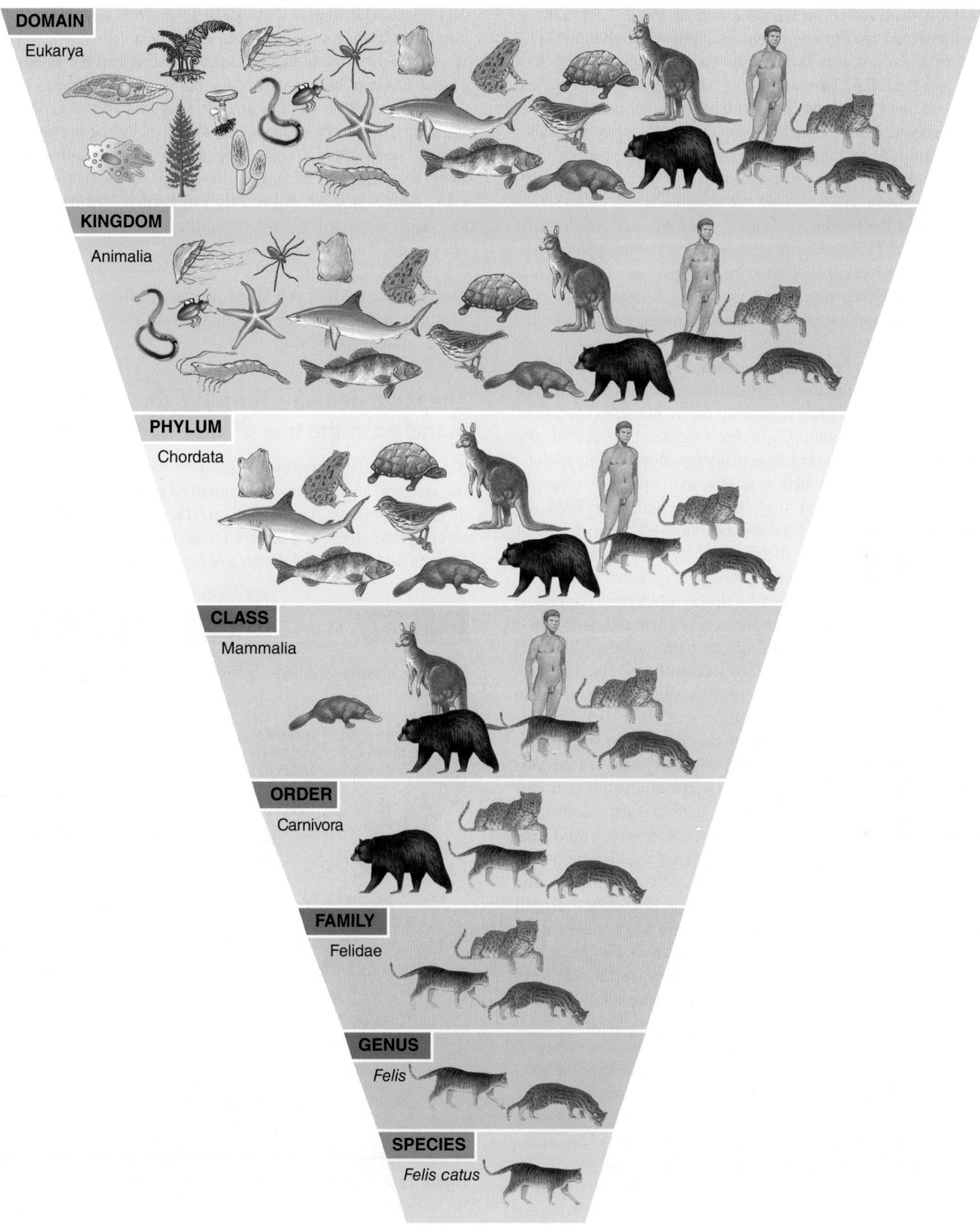

Figure 23-1 The principal categories used in classification

The domestic cat (*Felis catus*) is classified here to illustrate the hierarchical organization of the Linnaean taxonomic system. Each level is more inclusive than the one below it, meaning that it includes more groups of organisms.

© Cengage Learning

plant kingdom and in the animal kingdom, but did not really fit in either kingdom. *Euglena* carries on photosynthesis in the light, but in the dark, it uses its flagellum to move about in search of food (see Fig. 26-5). In 1866, a German biologist, Ernst Haeckel, proposed that a third kingdom, **Protista,** be established to accommodate bacteria and other microorganisms. However, biologists largely ignored this kingdom for about a hundred years.

In 1937, French marine biologist Edouard Chatton suggested the term *procariotique* ("before nucleus") to describe bacteria and the term *eucariotique* ("true nucleus") to describe all other cells. This dichotomy between prokaryotes and eukaryotes is now universally accepted by biologists as a fundamental evolutionary divergence.

In the 1960s, advances in electron microscopy and biochemical techniques revealed further cell differences that inspired many new proposals for classifying organisms. In 1969, R. H. Whittaker proposed a five-kingdom classification based mainly on cell structure and the way that organisms derive nutrition from their environment. (Although Whittaker is generally credited, the concept of five kingdoms was actually first proposed in 1949 by T.L. Jahn and F. Jahn in their book *How to Know the Protozoa*.)

Whittaker suggested that the fungi (which include the mushrooms, molds, and yeasts) be removed from the plant kingdom and classified in their own kingdom, **Fungi.** After all, fungi are not photosynthetic. They obtain their nutrition by absorbing nutrients produced by other organisms. Fungi also differ from plants in the composition of their cell walls, their body structures, and their modes of reproduction. Kingdom **Prokaryotae** was established to accommodate the bacteria, which are fundamentally different from all other organisms in that they do not have distinct nuclei and other membranous organelles and do not undergo mitotic division (see Fig. 4-8).

In the late 1970s, Carl Woese (pronounced "woes") of the University of Illinois and his colleagues began to study the evolutionary relationships among organisms by analyzing genes coding for their ribosomal RNA. Using sequence analysis, Woese used variations in this universal molecule to challenge the long-held view that all prokaryotes were closely related and very similar to one another. The analysis of ribosomal RNA is an important tool used by systematists. Woese showed that there are two fundamentally different groups of prokaryotes, archaea and bacteria. He and his colleagues proposed not only that the archaea are different from the bacteria, but also that the archaea are genealogically more closely related to the Eukaryotes than they are to the bacteria.

Woese's hypothesis gained support in 1996 when Carol J. Bult of the Institute for Genomic Research in Rockville, Maryland, reported in the journal *Science* that she and her colleagues had sequenced the complete genome of a methane-producing archaea, *Methanococcus jannaschii*. When these researchers compared gene sequences with those of two previously sequenced bacteria, they found that fewer than half of the genes matched. Gene sequencing indicates that the archaea have a combination of bacteria-like and eukaryote-like genes. Biologists have identified other important differences between bacteria and archaea. For example, bacteria are characterized by the presence of a compound called peptidoglycan in their cell walls, whereas this compound is not present in archaea. Based on molecular evidence, biologists now divide the prokaryotes into two major groups: *bacteria* and *archaea*.

Kingdom Protista has had an interesting history. Biologists placed *Euglena,* along with other unicellular organisms traditionally referred to as protozoa, in kingdom Protista. At various times, they have also assigned algae (including multicellular forms), water molds, and slime molds to kingdom Protista. Thus, this kingdom became a diverse group of mainly unicellular, mainly aquatic eukaryotic organisms. During the past few years, systematists have established that the protist groups did not descend from one recent common ancestor. As we will discuss in Chapter 26, many biologists have abandoned kingdom Protista (as well as a kingdom classification for plants and animals) and now assign the eukaryotes to five "supergroups" based on molecular data.

The three domains form the three main branches of the tree of life

Based on fundamental molecular differences among the bacteria, archaea, and eukaryotes, biologists now classify organisms in three domains: **Archaea, Bacteria,** and **Eukarya** (eukaryotes). Systematists have inferred that the three domains are the three main branches of the tree of life (**FIG. 23-2**). In Chapter 25

KEY POINT

Biologists classify organisms in three major groups called domains.

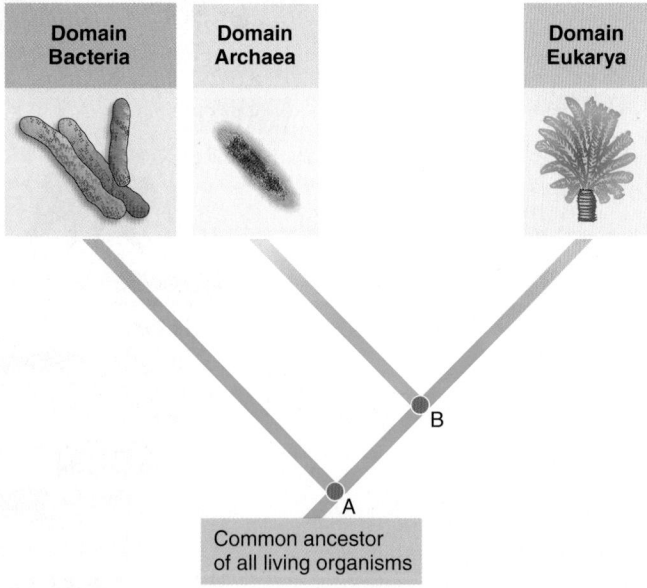

Figure 23-2 *Animation* **The three domains**

This cladogram, a type of evolutionary tree, illustrates the evolutionary relationships among organisms in the three domains. Each branch of the cladogram represents a clade, a group of organisms with a recent common ancestor. Each node (*circle*) represents the point at which two groups diverged from each other.

PREDICT According to this cladogram, are the Archaea more closely related to the Bacteria or to the Eukarya?

© Cengage Learning

we will discuss characters that distinguish these domains. As we will discuss in Chapter 24, the viruses are a special case and are not classified in any of the three domains.

Some biologists are moving away from Linnaean categories

Because modern biologists are interested in having the classifications of living organisms reflect their patterns of relatedness as closely as possible, some of the terminology that has been used for decades is either being put aside or redefined. In addition, our increased ability to investigate the DNA sequences of organisms and improved techniques for analyzing these molecular data are providing many new discoveries regarding the patterns of relatedness among living species. We are in the midst of a time of transition.

For students learning biology during this time when even the experts are disagreeing, it is important to note that biology textbooks vary in the classification and terminology they present. For example, most biologists are now using the classification system with three domains (Bacteria, Archaea, and Eukarya), with the Eukarya divided into a number of large taxa reflecting shared ancestry. Many biologists no longer recognize the kingdom Protista because many of the organisms within that group are actually more closely related to either plants or animals or fungi than they are to one another (see TABLE 23-2 and FIG. 23-3).

In this edition of *Biology*, we classify organisms using the three domains and also use many of the traditional taxonomic names. We discuss throughout the book instances where the traditional taxonomy is undergoing revision to better reflect our knowledge concerning patterns of common ancestry.

Phylogenetic trees show hypothesized evolutionary relationships

Systematists use **phylogenetic trees** to graphically represent hypothesized evolutionary relationships among organisms that have a common ancestor. In Figure 23-2 we use a phylogenetic tree to illustrate the relationship of the three domains, and the phylogenetic tree in Figure 23-3 illustrates the main branches that make up each of the three domains. This tree is one of the current versions of the tree of life.

TABLE 23-2	Domains and Kingdoms		
DOMAIN	**KINGDOM**	**CHARACTERISTICS**	**ECOLOGICAL ROLE AND COMMENTS**
Bacteria	Bacteria	Prokaryotes (lack distinct nuclei and other membranous organelles); unicellular; microscopic; cell walls generally composed of peptidoglycan.	Most are decomposers; some parasitic (and pathogenic); some chemosynthetic autotrophs; some photosynthetic; important in recycling nitrogen and other elements; some used in industrial processes.
Archaea	Archaea	Prokaryotes; unicellular; microscopic; peptidoglycan absent in cell walls; differ biochemically from bacteria.	Decomposers; important in recycling carbon, nitrogen, and other nutrients; methanogens are anaerobes that inhabit sewage, swamps, and animal digestive tracts; extreme halophiles inhabit salty environments; extreme thermophiles inhabit hot, sometimes acidic environments.
	Protists formerly classified in kingdom Protista; now assigned to a number of "supergroups"	Eukaryotes; mainly unicellular or simple multicellular.	Protozoa are an important part of zooplankton. Algae are important producers, especially in marine and freshwater ecosystems; important oxygen source. Some protists cause diseases (e.g., malaria).
	Plantae	Eukaryotes; multicellular; photosynthetic; possess multicellular reproductive organs; alternation of generations; cell walls of cellulose.	Terrestrial biosphere depends on plants in their role as primary producers; important source of oxygen in Earth's atmosphere.
Eukarya*	Fungi	Eukaryotes; heterotrophic; absorb nutrients; do not photosynthesize; body composed of threadlike hyphae that form tangled masses that infiltrate food or habitat; cell walls of chitin	Decomposers; some parasitic (and pathogenic); some form important symbiotic relationships with plant roots (mycorrhizae) or algae (lichens); some used as food; yeast used in making bread and alcoholic beverages; some used to make industrial chemicals or antibiotics; responsible for much spoilage and crop loss.
	Animalia	Eukaryotes; multicellular heterotrophs; many exhibit tissue differentiation and complex organ systems; most able to move about by muscular contraction; nervous tissue coordinates responses to stimuli.	Consumers; some specialized as herbivores, predators, or detritus feeders.

*For another approach to classification of eukarya, see Figure 23-3.

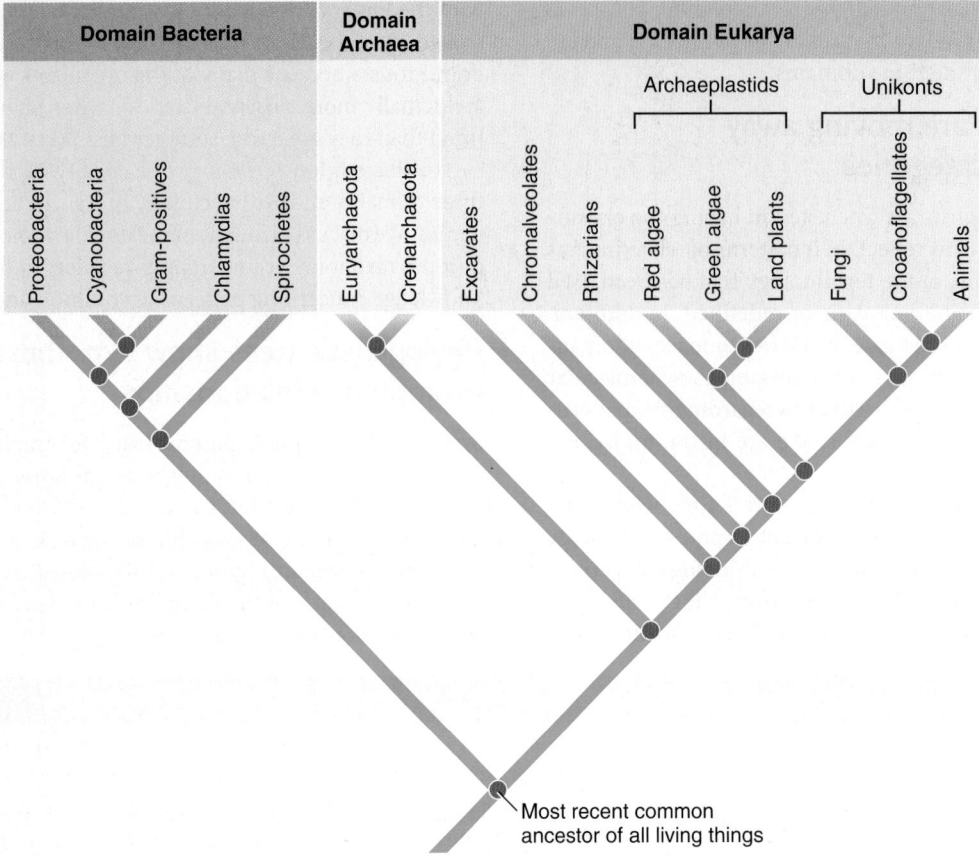

Figure 23-3 The tree of life, a work in progress

Some major branches of each of the three domains are shown. As new data are analyzed, the relationships of some of the branches will change.

© Cengage Learning

The type of phylogenetic tree we use in this book is called a **cladogram.** Each **branch** in a cladogram represents a **clade,** a group of organisms with a common ancestor. Each branching point, referred to as a **node** (depicted by a circle), represents the divergence, or splitting, of two or more new groups from a common ancestor. Thus, the node represents the *most recent common ancestor* of each clade depicted by the branches. In this way, a cladogram uses the positions of branch points to illustrate the hypothesized evolutionary relationships among taxa. Each of the branches is formed based on evidence in terms of observable traits that are shared by the organisms at the end of the branch but are not found in organisms on other branches. (*Shared derived characters* are discussed in the next section). These shared characteristics can be indicated by labels or by bars across the branches.

Cladograms are rooted if the most recent common ancestor is known. The **root,** or node at the base of the cladogram, represents the most recent common ancestor of all the clades depicted in the tree. We will discuss more about constructing cladograms as we proceed through this chapter.

Although a cladogram depicts the evolutionary relationships of each group through time, the *lengths* of the branches do not tell us *when* a particular species evolved. Other types of phylogenetic trees—for example, *phylograms*—can be constructed to indicate time, or rate of evolution, as well as the relationships

among taxa. In a phylogram the length of the branches is proportional to the amount of inferred change in characteristics.

Systematists continue to consider other hypotheses

When we think of how organisms acquire their genes, we think of **vertical gene transfer** in which genes are transmitted from parent to offspring within the same species. We also usually attribute similarities among species in their heritable traits to be the result of shared ancestry, or **homology.** For example, if two species share the same sequence of nitrogenous bases in their DNA for a particular gene, we hypothesize that this sharing is due to the two species having inherited that particular gene from the same ancestor. As biologists have continued to investigate the evolution of genes, they have discovered that in some cases inheritance can occur outside the parent to offspring pathway.

Throughout the history of life, genes were not only passed down "vertically" from one generation to the next, but sometimes were also exchanged "laterally." Such gene swapping between organisms in one taxon and unrelated organisms in another taxon is called **horizontal gene transfer.** In this process, genes move from one species to another species in the same generation.

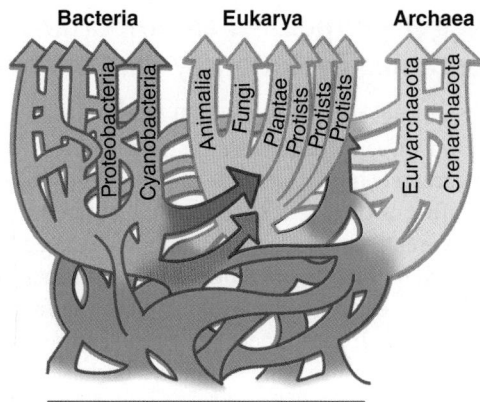

(a) The three-domain approach drawn to show horizontal gene transfer as a continuous process between domains and also among groups within each domain. Only some of the branches are labeled. The *diagonal brown arrow* represents mitochondrial endosymbiosis; the *diagonal green arrow* represents chloroplast endosymbiosis.

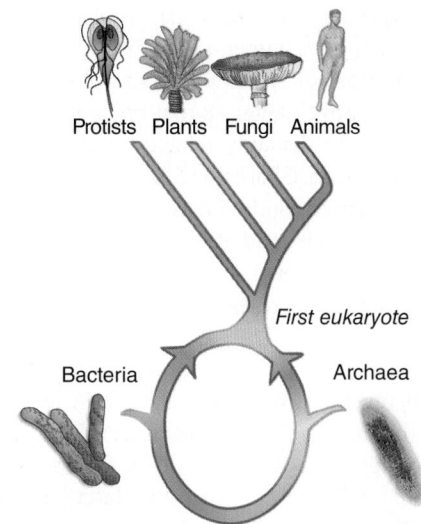

(b) The ring of life is based on the hypothesis that lateral gene transfer between bacteria and archaea gave rise to the eukaryotes.

Figure 23-4 Other ways to represent the tree of life
(Part **a** based on Doolittle, W.F., *Science*, Vol. 284, 2124–2128, 1999; part **b** based on Rivera, M.C., and J.A. Lake. *Nature*, Vol. 431, 152–155, Sept. 9, 2004.)

Horizontal gene transfer can occur in several ways, such as by exchange of DNA among different populations or species of bacteria, or by interbreeding between closely related groups. One way that horizontal gene transfer has occurred in eukaryotes is by **endosymbiosis,** a process in which one organism lives inside the cell of another organism and the two become dependent on each other. Over time, such relationships may result in a complete loss of the separate identity of the symbionts. In fact, eukaryotic cells most likely evolved from prokaryotic cells that lived symbiotically, one within another. Recall that considerable evidence has been found supporting the hypothesis that mitochondria and chloroplasts originated from prokaryotes that lived as endosymbionts inside other cells (see Chapter 21). Horizontal

gene transfer also occurs between free-living organisms classified in different domains. In 2013, a team of researchers reported that horizontal gene transfers from free-living bacteria and archaea enabled the unicellular red alga *Galdieria sulphuraria* to adapt to harsh environments that are too hot, acidic, or toxic for most other eukaryotes. The study suggests that horizontal gene transfer is a very important mechanism in evolution.

Based on what is now known about horizontal gene transfer, some systematists argue that the common ancestor of all living things may have been a *community* of species that traded their genes. Some systematists suggest that there is no simple tree of life and have proposed alternative hypotheses. One group hypothesizes a complex bush with many connecting branches (**FIG. 23-4a**). Researchers at the University of California at Los Angeles have proposed another approach: the tree of life may actually be a ring of life (**FIG. 23-4b**). Based on their analysis of hundreds of genes, these systematists hypothesize that horizontal gene transfer between bacteria and archaea gave rise to the eukaryotes. Horizontal gene transfer appears to be a continuous process among members of the three domains.

The evolution of systematics reflects the creative and dynamic process of science. Systematists are very responsive to new data; consequently, classification of organisms at all levels is a continuously changing process. Although many biologists are moving away from a hierarchical system, some suggest adding new groups so that the classical approach keeps up with new findings. Whatever approach is used, the goal of systematics is to base classification on evolutionary change and relationships.

CHECKPOINT 23.2

- *What are the major groups of organisms that belong to each of the three domains? In which domain and kingdom would you classify each of the following: white willow, the bacterium* Escherichia coli, *tapeworm, and black bread mold?*

- **CONNECT** *In what specific ways does a cladogram give us information about the evolutionary relationships of groups of organisms?*

23.3 RECONSTRUCTING EVOLUTIONARY HISTORY

LEARNING OBJECTIVES

5 Critically review the difficulties encountered in choosing taxonomic criteria.

6 Apply the concept of shared derived characters to the classification of organisms.

7 Describe how analyses of molecular homologies contribute to the science of systematics.

8 Contrast monophyletic, paraphyletic, and polyphyletic taxa.

A major goal of modern systematics is to reconstruct **phylogeny** (literally, "production of phyla"), the evolutionary history of any group of organisms from a common ancestor. In their efforts, systematists must employ a wide range of evidence, from shared traits (homologous traits) to geographic distributions to positions in strata (for fossils), and must incorporate our best understandings of evolutionary processes. As systematists determine evolutionary relationships among species and higher taxa, they build classifications based on the patterns of shared ancestry. Systematics is at the center of how we understand and explain the patterns of similarities and relationships of life-forms on Earth.

Once phylogenies are established, they help us answer other questions in biology. For example, phylogenies help us understand evolutionary patterns that might provide clues to the origin and spread of HIV and other pathogens (discussed later in this chapter). Phylogenies also help biologists identify new species and predict their characteristics. Systematists may classify a new species based on specific characters that it shares with other organisms in a particular taxon. They may then infer that the new species shares many other characters with organisms in that group.

As you read the following sections, remember that phylogenies are testable hypotheses. They are supported or falsified by the available data. Systematics proceeds by constantly re-evaluating data, hypotheses, and theoretical constructs. As new data are discovered and old data are reinterpreted, systematists modify their hypotheses. As a result, our understanding of how organisms are related and the way we classify organisms are continuously being revised. Systematics is a dynamic science that changes as biologists discover new species and use ever more sophisticated techniques to investigate the evolutionary relationships among organisms.

Recall that a **population** is made up of all the individuals of the same species that live in a particular area. A population has a dimension in space—its geographic range—and also a dimension in time. Each population extends backward in time. Somewhat like branches of a tree, a population may diverge from other populations enough to become a new species (which may be depicted by a new tree branch; see Chapters 19 and 20). Species have various degrees of evolutionary relationships with one another, depending on the degree of genetic divergence since their populations branched from a common ancestor.

Homologous structures are important in determining evolutionary relationships

The evidence suggests that some 380 to 400 million years ago a population of early fishlike vertebrates evolved. These animals, the early **tetrapods,** had four limbs instead of the pectoral and pelvic fins of their fish ancestors. Because all the modern amphibians, reptiles, birds, and mammals evolved from that group of early tetrapods, members of the modern taxa share many traits. For example, the wings of bats, the front legs and paws of tigers, the front flippers of whales, and the front limbs of frogs all share similar bone structures because the genes for these bones were inherited from the same earliest tetrapods. We say that the bones are *homologous* among these animals (see Fig. 18-13). Based on their shared ancestry, the tetrapods are now classified in the group, **Tetrapoda.**

Although homologous similarities are very important to scientists for inferring phylogeny, similarities among species can occur for reasons other than homology. Similar structures sometimes evolve when unrelated or distantly related species become adapted to similar environmental conditions. Thus, wings may occur in two or more species not derived from a recent common ancestor. Recall from Chapter 18 that independent evolution of similar structures in distantly related organisms is known as **convergent evolution.** Sharks and dolphins have similar, but independently derived, body forms because they have become adapted to similar environments (aquatic) and lifestyles (predatory). Sometimes it is obvious that convergent evolution has occurred. In other cases, researchers must test their hypotheses using all available data.

Another challenge in deciding on homology is **reversal,** in which a trait reverts to its ancestral state. A reversal removes a similarity that had evolved. A characteristic that superficially appears homologous but is actually independently acquired by convergent evolution or reversal is described as exhibiting **homoplasy.** Distinguishing between homology and homoplasy can be challenging (**FIG. 23-5**).

Figure 23-5 Homology and homoplasy

These two salamander species (*Notophthalmus viridescens*, red eft stage, and *Plethodon cinereus*, red-backed salamander) share many skeletal features because they inherited these features from the same early salamander species that was their shared ancestor. This is an example of homology. However, the bright red color of the two species has independently evolved, which is an example of homoplasy.

Shared derived characters provide clues about phylogeny

Systematists rely on homologous traits for evidence of shared ancestry. However, they also distinguish among those homologous traits, determining which traits will help most for studying particular relationships. In making decisions about taxonomic relationships, the systematist first examines the homologous characteristics in the largest group (such as phylum or class) of organisms being studied and interprets them as indicating the most *remote* common ancestry. These **shared ancestral characters** are features that were present in an ancestral species and remain present in all groups descended from that ancestor. For example, the vertebrae, present in all vertebrates (with the exception of the hagfishes), is an ancestral character for study of classes within the subphylum Vertebrata. Studying the presence or absence of the vertebrae does not help us discriminate among various classes of vertebrates (e.g., between amphibians and mammals) because all individuals in these classes (except the hagfishes) have vertebrae.

When two populations become separated and begin to evolve independently, some of their homologous traits change as a result of mutation, natural selection, and genetic drift. The novel traits that evolve and remain present in the descendant taxa are referred to as **shared derived characters,** or *synapomorphies*. Note that shared derived characters originate in a *recent* common ancestor and are present in its descendants. Species that share derived characters form a clade. Systematists use shared derived characters (recently evolved homologies) to identify points where groups diverged from one another. A trait viewed as a *derived character* in a more inclusive (broader) taxon may also be considered an *ancestral character* in a less inclusive (narrower) taxon. For example, vertebrae are a shared derived character among vertebrates, but an ancestral character among mammals.

More recent common ancestry is indicated by classification into less and less inclusive taxonomic groups with more and more specific shared derived characters. For example, the small bones in the middle ear are useful in identifying a branch point between reptiles and mammals. All tetrapods (amphibians, reptiles, birds, and mammals) have one bone, the stapes, present in the middle ear, but mammals have two additional bones, the incus and malleus, in the middle ear. The evolution of this derived character (three bones rather than one bone) was a unique event, and only mammals have these bones. However, if we compare mammals with one another, the three ear bones are a shared ancestral character because all mammals have them. Consequently, they have no value for distinguishing among mammalian taxa. Other characters must be used to establish branch points among the mammals.

If we compare dogs, goats, and dolphins (which are all mammals), we find that dogs and goats have two nostrils at the tip of their snouts, whereas dolphins have their nostrils at the tops of their heads as a single or paired blowhole. Having nostrils at the tip of the snout is an ancestral trait in mammals and therefore cannot be used as evidence that dogs and goats share a more recent common ancestor. In contrast, the presence of a blowhole in dolphins is a derived character within mammals. When we compare dogs, dolphins, and whales, we find that dolphins and whales share this derived character, providing evidence that these animals evolved from a common ancestor not shared by dogs.

Systematists base taxonomic decisions on recent shared ancestry

Both dolphins and many fishes have streamlined body forms, but this similarity is homoplastic and does not indicate close evolutionary relationships. In contrast, dolphins share important homologous derived characters (synapomorphies) with mammals: mammary glands, which produce milk for the young; three small bones in the middle ear; and a muscular diaphragm that helps move air into and out of the lungs, to name just a few traits. Thus, our evidence suggests that dolphins evolved from earlier mammals, and we classify dolphins as mammals.

Historically, people have thought of higher taxonomic groups as being comprised of species that share certain defining features. For example, previously someone might say that mammals are defined as vertebrates that possess hair. Modern systematics, with its emphasis on recognizing taxonomic groups based on shared genealogy (ancestry), requires a different perspective. Systematists do use traits like mammary glands, muscular diaphragm, and hair as evidence that dolphins should be classified as mammals. Dolphins are included within the Mammalia because all the available evidence suggests that they evolved from the same recent ancestor as the rest of the mammals. Even if dolphins evolved to completely lose all body hair, they would still be classified within the Mammalia because of their ancestry.

Another example was mentioned earlier in this chapter. The tetrapods are all those vertebrates that evolved from the first fishlike vertebrates that had limbs with joints and digits instead of fins. As the name suggests, tetrapods are the "four-legged" vertebrates. However, the taxonomic group Tetrapoda is not defined by the presence of four legs; in fact, many groups of animals that are classified as tetrapods have either no limbs (e.g., snakes and limbless lizards) or greatly modified limbs (e.g., whales and birds).

Molecular homologies help clarify phylogeny

When a new species evolves, it does not always exhibit obvious phenotypic differences when compared to closely related species. For example, two distinct species of fruit flies may appear identical to us. Some of their DNA, proteins, and other molecules, however, are different. Such variations in the structure of specific macromolecules among species, just like differences in anatomical structure, result from mutations. Advances in molecular biology have provided the tools for biologists to compare the macromolecules of various organisms. In fact, organisms can now be identified by their molecular structure.

In 2003, scientists at the University of Guelph in Canada proposed in the British *Proceedings of the Royal Society* that we identify all living things by their unique sequences of nitrogenous bases in particular regions of DNA rather than by their physical structure. The DNA bar-code project is based on the concept that scientists should be able to find a particular region of DNA (1) that can be found in all species, (2) that is relatively short and easy to sequence, and (3) for which each species has a unique sequence of nitrogenous bases. Currently, some scientists have agreed that the gene for the protein cytochrome oxidase subunit I may fit the requirements for eukaryotes, and they are attempting to determine the sequences for as many species as possible. The hope is that the relatively easy-to-determine DNA sequences will work like product barcodes, uniquely identifying each species at any stage of life, including eggs or seeds.

The science of **molecular systematics** focuses on molecular structure to clarify evolutionary relationships. DNA, RNA, and amino acid sequencing are used to compare the macromolecules of organisms being studied. Macromolecules that are functionally similar in two different types of organisms are considered homologous if their subunit sequence is similar. Such comparisons provide systematists with valuable information about the degree of relatedness among organisms. The more closely the subunit sequences of two species correspond, the more closely related the species are considered to be. The number of differences in certain DNA nucleotide sequences in two groups of organisms may approximately reflect how much time has passed since the groups branched from a common ancestor. (This can be true only when the mutations in DNA occur at a steady rate.) Thus, specific DNA sequences may be useful as **molecular clocks.**

Many systematists look to ribosomal RNA (rRNA) structure and the DNA sequences that code for the RNA in ribosomes (ribosomal DNA) to help determine phylogenies. All known organisms have ribosomes that function in protein synthesis, and the DNA nucleotide sequences coding for rRNA have been highly conserved during evolution. Recall that the division of organisms into three domains was based in large part on the comparison of rRNA by Carl Woese and his research team.

The ribosomes of archaea and bacteria contain three subunits of RNA, named in order of increasing size: 5S, 16S, and 23S. (These numbers are sedimentation coefficients, measures of relative size, used to characterize behavior of a particle when centrifuged.) The DNA sequences for the 5S and 16S RNAs have been extensively used to determine evolutionary relationships among bacteria. These gene sequences for rRNA are useful because the number of base pairs is manageable, the DNA has shown little change over long evolutionary periods, and the genes show some species-specific variability among bacteria and archaea.

Researchers have also used comparison of the DNA sequences for rRNA to challenge the once widely accepted idea that fungi are closely related to plants. According to rRNA analysis, fungi are more closely related to animals than to plants. Animals and fungi share a more recent common ancestor, perhaps a flagellate, a one-celled protist that has one or more long, whiplike flagella (used for locomotion).

Because different genes can evolve at different rates, researchers now typically compare nitrogenous base sequences from many different genes of the organisms under study. Molecular sequence data have been sampled from more than 10% of all known species. The very precise tools provided by molecular

How is the dog related to other canids?

HYPOTHESIS: Analysis of molecular data from a variety of canids will clarify the evolutionary relationships of canids.

EXPERIMENT: Researcher Kerstin Lindblad-Toh and her colleagues sequenced the genome of the domestic dog. They compared coding regions of more than 13,000 dog genes with corresponding human and mouse genes. Then they selected 12 exons and 4 introns and sequenced them in 30 of 34 living canids.

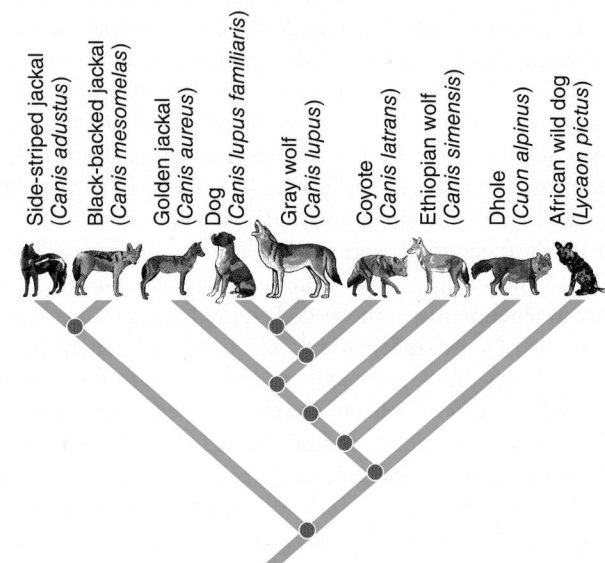

RESULTS AND CONCLUSION: Analysis of the comparisons of exon and intron nucleotide sequences allowed these researchers to build a cladogram of Canidae species. The data indicated that the dog is most closely related to the gray wolf. (Sequence divergences of nuclear exon and intron sequences are 0.04% and 0.21%, respectively.) The two African jackals (*branch at left*) are the sister group of this clade, suggesting an African origin for the canids.

SOURCE: Lindblad-Toh, K., et al. "Genome Sequence, Comparative Analysis and Haplotype Structure of the Domestic Dog." *Nature*, Vol. 438, Dec. 8, 2005.

Figure 23-6 Molecular phylogeny

Molecular similarities indicate that the domestic dog is closely related to the gray wolf. Note that some biologists list the scientific name for the domestic dog as *Canis familiaris*, but others, including the Smithsonian Institution and the American Society of Mammalogists, consider the dog a subspecies of the gray wolf (*Canis lupus*) and list its name as *Canis lupus familiaris*.

CONNECT According to the cladogram, which animal is most closely related to the wolf/domestic dog clade?

biology have put systematics on the cutting edge of biological research.

Scientists at Broad's Genome Sequencing and Analysis program at the Massachusetts Institute of Technology (MIT) and at Harvard provided another example of applied molecular systematics (**FIG. 23-6**). They sequenced the genome of domestic dogs

Groups of organisms can be described as monophyletic, paraphyletic, or polyphyletic. Only monophyletic groups are considered clades.

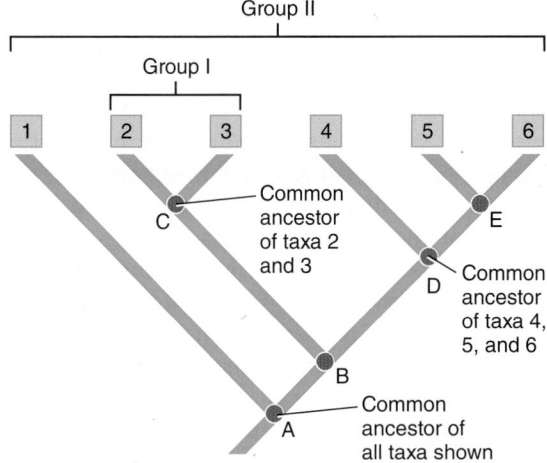

(a) Taxa 2 and 3 are a monophyletic group or clade (Group I). Taxa 1 through 6 also make up a monophyletic group (Group II). Each group includes a common ancestor and all its descendants.

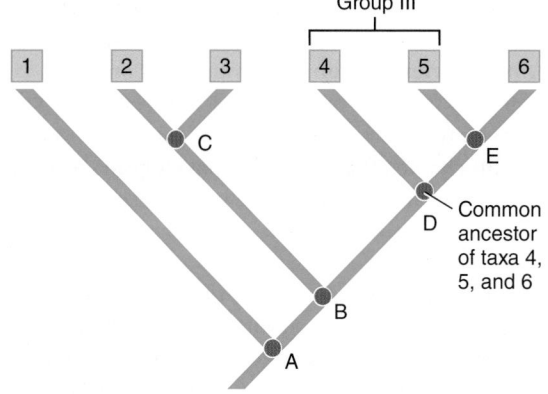

(b) Taxa 4 and 5 make up a paraphyletic group (Group III). This group includes some of, but not all, the descendants of the recent common ancestor indicated at node D. Which taxon is not included?

Figure 23-7 Evolutionary relationships

In all three figures, the branch point at node A represents the common ancestor of all taxa shown. Node B represents the common ancestor of taxa 2 through 6. Node C represents the common ancestor of taxa 2 and 3. Node D represents the common ancestor of taxa 4, 5, and 6. Node E represents the common ancestor of taxa 5 and 6. Common ancestry is the basis for making decisions about classification.

CONNECT Look at part (b). Consider the organisms that evolved from the ancestor at node D. Which of the taxa shown form a monophyletic group?

© Cengage Learning

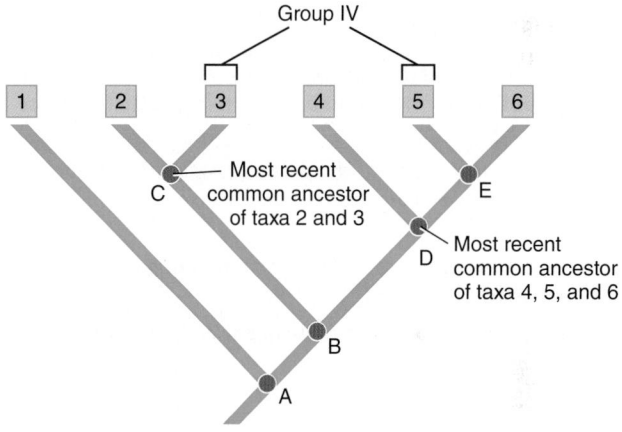

(c) Group IV is polyphyletic. Members of this group (taxa 3 and 5) do not share the same recent common ancestor.

and compared 16 specific coding regions with the sequences found in 30 other living canids (members of the family Canidae to which the domestic dog belongs). The data they gathered enabled these investigators to construct a cladogram of the phylogenetic relationships of canids. They confirmed earlier research based on other DNA sequences as well as morphological traits showing that domestic dogs are most closely related to gray wolves.

Taxa are grouped based on their evolutionary relationships

Most known organisms were classified by biologists before the days of phylogenetics. Modern systematists use phylogenetic analysis to test these hypothesized relationships among groups of organisms. Based increasingly on molecular data, they either confirm or modify the work of earlier biologists. Using their data, systematists construct cladograms to reflect relationships among clades. Cladograms show that previous classifications reflect three different types of taxonomic relationships: monophyletic groups, paraphyletic groups, and polyphyletic groups.

A **monophyletic group** includes an ancestral species and all its descendants (FIG. 23-7a). It is defined by shared derived characters. Mammals, for example, have mammary glands that produce milk for the young, three small bones in the middle ear, and a muscular diaphragm. All mammals are thought to have evolved from a common ancestral mammal that had these shared derived characters, and all descendants of this ancestor are mammals. Monophyletic taxa are clades. They are referred to as "natural" groupings because they represent true evolutionary relationships and include all close relatives. **Sister taxa,** or sister groups, share a more recent common ancestor with one another than either taxon does with any other group shown on a cladogram. In Figure 23-7, taxon 2 and taxon 3 are sister taxa to each other; taxon 5 and taxon 6 are also sister taxa.

A **paraphyletic group** is a group that contains a common ancestor and some of, but not all, its descendants (FIG. 23-7b). Members of the group share ancestral characters. As discussed later in this chapter, the traditional class Reptilia is paraphyletic because it does not include all descendants of the most recent common ancestor of reptiles. Birds share a recent common ancestor with reptiles.

A **polyphyletic group** consists of several evolutionary lines that do not share the same recent common ancestor (FIG. 23-7c). Biologists might have mistakenly classified the members of such a group together because these organisms share similar (homoplastic) features arising from convergent evolution. Systematists try to avoid constructing polyphyletic taxa because they are unnatural and misrepresent evolutionary relationships. Polyphyletic taxa are sometimes accepted temporarily until research provides additional data.

CHECKPOINT 23.3

- *How are shared ancestral characters and shared derived characters different? How is the concept of homology related to these concepts?*
- *Why don't shared ancestral characters provide evidence for relationships between organisms within a taxon that has those traits? Give an example.*
- **CONNECT** *How is molecular biology contributing to the science of systematics?*
- *Systematists prefer to recognize monophyletic taxa rather than polyphyletic taxa. Why?*

23.4 CONSTRUCTING PHYLOGENETIC TREES

LEARNING OBJECTIVES

9 Contrast the traditional classification with the current classification of reptiles and birds.

10 Describe the construction of a cladogram by using outgroup analysis.

11 Apply the principle of parsimony to cladogram construction.

In the second half of the 20th century, several methodological and philosophical schools of systematics were in a sense competing for supremacy among biologists. Three major schools emerged—*phenetics* (also known as numerical taxonomy), *evolutionary taxonomy,* and **phylogenetic systematics,** also known as **cladistics.** Their approaches differed in terms of broad goals, the methodologies used, and the accepted naming systems. The

debate among followers of these three approaches was at times quite forceful and animated. What has emerged is a loosely defined consensus that borrows modified techniques from all three approaches but largely adopts many of the concepts from phylogenetic systematics.

Biologists today generally agree that ideally the goal of systematics should be to reconstruct the genealogical (evolutionary) relationships among species and that the best evidence for such relationships can be found in the patterns of traits shared by organisms (homologous traits). For many traits, the differences between shared derived characters and shared ancestral characters can be distinguished, and these traits can be analyzed as outlined below following many approaches derived from phylogenetic systematics. Aspects of these approaches are described in the next few sections. In some cases, however, systematists cannot clearly separate ancestral and derived states for certain traits; there, the more statistical approaches derived from phenetics are employed to construct phylogenies.

One conflict from the 20th century that has persisted into the present concerns the naming of taxonomic groups. Most modern systematists agree that scientific names should reflect monophyletic groups because such names will generally carry the most meaning. For example, when someone uses a taxonomic name that reflects a monophyletic group, we immediately understand that the name reflects an ancestor and all its descendants. We understand that because of their relatedness, the species included in the monophyletic group are likely to share many homologous similarities. Although this approach seems almost intuitive, it is not without controversy, mostly because full adoption of this approach will require changes to traditional taxonomy.

Some important monophyletic groups were not named in the past and now need to be recognized. At the same time, some traditional group names do not reflect monophyletic groups and will now need to be discarded or redefined. An example of a new taxon required is Tetrapoda, discussed earlier in this chapter. The tetrapods include amphibians, reptiles (including birds), and mammals and their ancestors tracing back to the first vertebrate with four legs. All available evidence suggests that the tetrapods are a monophyletic group and should be recognized as the group Tetrapoda.

More controversial is how traditional names that do not reflect monophyletic groups should be handled. Traditionally, the class Reptilia has been defined to include turtles, lizards, snakes, crocodilians, and dinosaurs. The phylogenetic evidence suggests that this group is paraphyletic because birds share a common ancestor with the saurischian dinosaurs yet are not included in the traditional Reptilia (FIG. 23-8a).

One solution would be to redefine the group Reptilia to include birds. Another solution is to abandon Reptilia as a taxon. Reptiles, birds, and mammals are amniotes, vertebrates that have an **amniotic egg.** The amniotic egg contains an **amnion,** a membrane that forms a fluid-filled sac around the embryo. The amniotic egg permits development in terrestrial environments. Animals with an amniotic egg make up a monophyletic group known as Amniota (FIG. 23-8b).

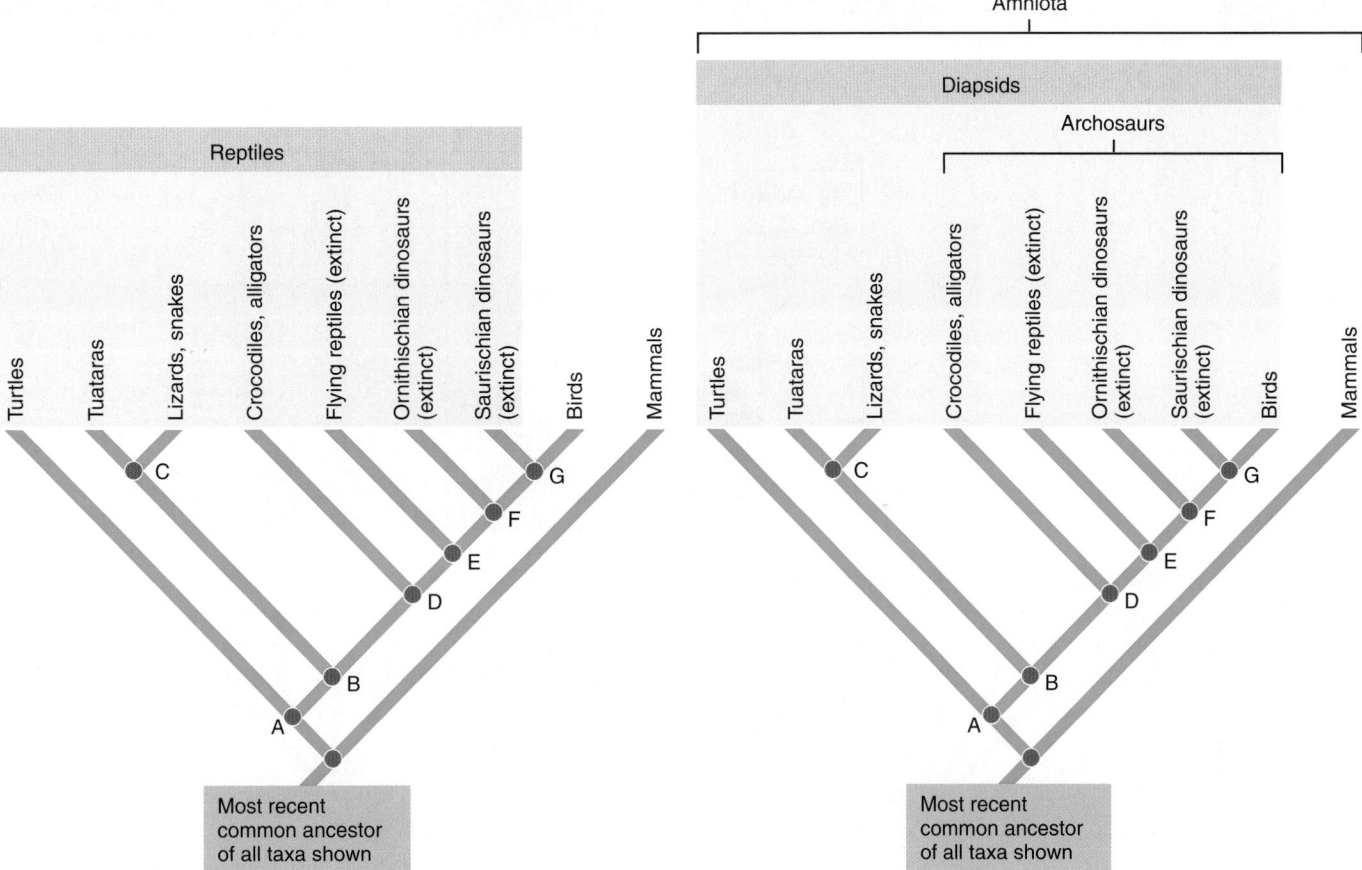

(a) The traditional approach to classification recognized the common ancestry of birds and reptiles, but assigned them to different classes. In this approach reptiles are a paraphyletic group.

b) Phylogenetic systematists (cladists) classify birds and reptiles together because they have a recent common ancestor and are a monophyletic group. The cladogram shows the branching points in the evolution of the major groups of reptiles. Lizards, snakes, and crocodiles are phenotypically similar, but crocodiles, dinosaurs, and birds are most closely related because they evolved most recently from a common ancestor (node D). Node F represents the branching of two clades of dinosaurs from a common ancestor and the subsequent branching of birds from the lineage of the saurischian dinosaurs.

Figure 23-8 Two approaches to the classification of reptiles and birds

Phylogenetic systematics, also known as cladistics, recognizes reptiles and birds as a monophyletic group.
© Cengage Learning

Two large monophyletic groups within the Amniota are the **diapsids** (traditional reptiles and birds) and the mammals. Within the diapsid group, the **archosaurs** include the crocodiles, flying reptiles (the extinct pterosaurs), the extinct dinosaurs, and the birds. These groups are discussed in more detail in Chapter 32.

Outgroup analysis is used in constructing and interpreting cladograms

A crucial step in most cladistic analyses is **outgroup analysis,** a research method for estimating which attributes are shared derived characters in a given group of organisms (**FIG. 23-9**).

An **outgroup** is a taxon that is considered to have branched off earlier than the taxa under investigation, the **ingroup.** An ideal outgroup is the closest relative of the group being studied, its sister taxon, and has not been highly modified since its origin. Recall that sister taxa evolved from the same recent common ancestor. Systematists argue that an outgroup is likely to retain the ancestral state for characters being used in the analysis, allowing the researchers to identify the evolutionary changes leading to derived characters. To help you understand outgroup analysis, we describe the specific example illustrated in Figure 23-9.

The first step in constructing a cladogram is to select the taxa, which may consist of individuals, species, genera, or other taxonomic levels. Here we use a representative group of eight

WHY IS IT USED? Outgroup analysis is used to organize character states into their evolutionary order, that is, to reconstruct phylogeny.

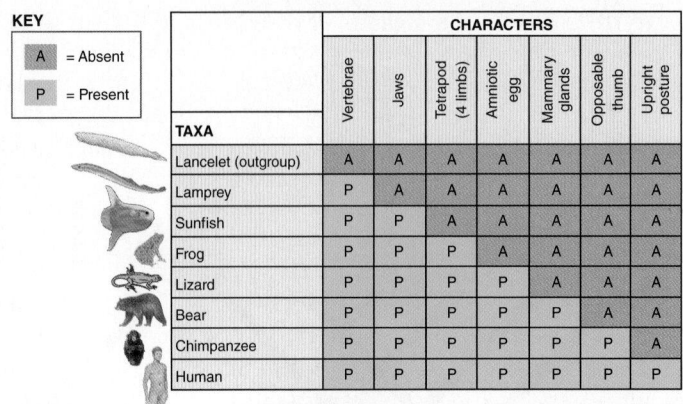

KEY		
A	= Absent	
P	= Present	

	CHARACTERS						
TAXA	Vertebrae	Jaws	Tetrapod (4 limbs)	Amniotic egg	Mammary glands	Opposable thumb	Upright posture
Lancelet (outgroup)	A	A	A	A	A	A	A
Lamprey	P	A	A	A	A	A	A
Sunfish	P	P	A	A	A	A	A
Frog	P	P	P	A	A	A	A
Lizard	P	P	P	P	A	A	A
Bear	P	P	P	P	P	A	A
Chimpanzee	P	P	P	P	P	P	A
Human	P	P	P	P	P	P	P

Jaws

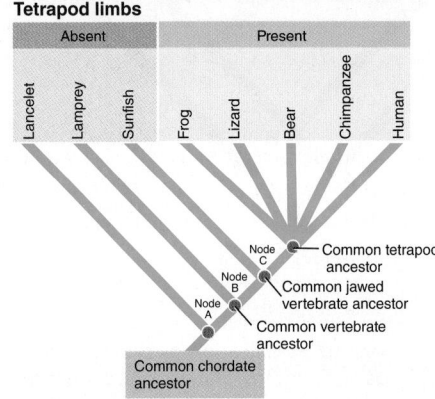

(a) All seven vertebrate taxa shown here have jaws, except the lamprey. Jaws are a shared derived character for these six taxa.

Tetrapod limbs

(b) Tetrapod limbs are a shared derived character for all vertebrate taxa shown here except the lamprey and sunfish.

Amniotic egg

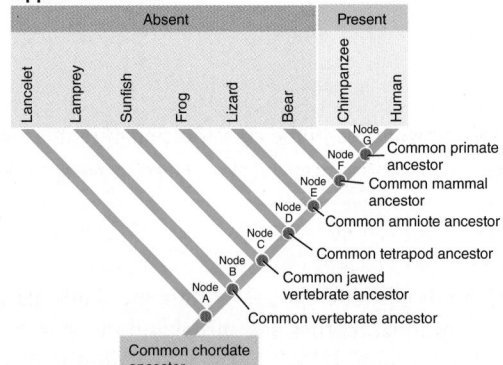

(c) Lizards, bears, chimpanzees, and humans have amniotic eggs, a shared derived character. Therefore, they are members of the clade Amniota.

Opposable thumb

(d) Of the vertebrate taxa shown here, only the chimpanzee and human share the derived character opposable thumb.

HOW IS IT DONE?
1 Select the taxa of interest and the homologous characters to be analyzed.
2 Define the possible conditions, for example, present (P) or absent (A).
3 Here, the lancelet is selected as the outgroup, a taxon that diverged earlier than any of the other taxa being considered. The lancelet represents an approximation of the ancestral condition.
4 Refer to the table as you follow steps (a) through (d) to build a cladogram.

Figure 23-9 *Animation* Constructing a cladogram using outgroup analysis
© Cengage Learning

chordates: the lancelet, lamprey, sunfish, frog, lizard, bear, chimpanzee, and human. (Chordates are animals that, at some time in their lives, have a cartilaginous, dorsal skeletal structure called a notochord; a dorsal, tubular nerve cord; and grooves in the wall of the pharynx.) The next step is to select the homologous characters to be analyzed. In our example we use seven characters. For each character, we must define all the different conditions, or states, as they exist in our taxa. For simplicity, we consider our characters to have only two different states: *present* or *absent*. Keep in mind that many characters used in cladistics have more than two states. For example, black, brown, yellow, and red may be only a few of the many possible states for the character of hair color.

The last, and often the most difficult, step in preparing the data is to organize the character states into their correct evolutionary order. For this step, we use outgroup analysis. In our example the lancelet, a small marine chordate with a fishlike appearance, is the chosen outgroup. It belongs to a taxon that is considered to have diverged earlier than any of the other taxa under investigation but to be closely related to vertebrates; thus, the lancelet represents an approximation of the ancestral condition for vertebrates. Therefore, the character state "absent" for a particular character state such as vertebrae is the ancestral condition, and the character state "present" is the derived condition for the characters in Figure 23-9.

A cladogram is constructed by considering shared derived characters

Our objective is to construct a cladogram that requires the fewest number of evolutionary changes in the characters. Taxa are grouped by the presence of shared derived characters. To form a valid monophyletic group, all members must share at least one derived character. Membership in a clade cannot be established by shared ancestral characters.

In our example notice that all taxa, except the outgroup lancelet, have vertebrae. We can therefore conclude that these seven vertebrate taxa form a valid clade. Next, among the seven vertebrate taxa, note that jaws are present in all groups except lampreys (a group of jawless vertebrates). Using these data, we construct a preliminary cladogram (see Fig. 23-9a).

Recall that the root, or base, of the cladogram represents the common ancestor for all taxa being analyzed. In Figure 23-9a, node A represents the common chordate ancestor from which the outgroup (lancelet) and the seven vertebrate taxa evolved. Similarly, node B represents the common ancestor of the vertebrates, and node C represents the common ancestor of the jawed vertebrates. Continuing with this procedure, note that among the six jawed taxa, all but sunfish have four limbs with digits (see Fig. 23-9b). Among the five groups with limbs, all but frogs have amniotic eggs in which the embryo is surrounded by an amnion (see Fig. 23-9c). The branching process is continued, using the data in the table in the top center of the figure, until all clades are established (see Fig. 23-9d).

Each branch point represents a major evolutionary step

Notice that humans and chimpanzees share a recent common ancestor at node G in Figure 23-9d. This hypothesis is supported by the derived characters they share (synapomorphies). In the same way, bears are more closely related to the human–chimpanzee clade than to any other clade considered in our example, as indicated by the common ancestor at node F.

Note that shared derived characters are nested. As you trace the tree from its root to its tips, each branch reflects the addition of one or more shared derived characters. When we compare the nodes, the order of divergence (branching) is indicated by relative distance from the base of the diagram. The farther a node is located from the base of the cladogram, the more recently the group diverged. In our example node G represents the most recent divergence, and node A represents the most ancient divergence. Thus, in our example humans are closely related to chimpanzees (through node G) but more distantly related to bears (through node F). Therefore, biologists assign humans and chimpanzees to a less inclusive taxon (order Primates), whereas humans, chimpanzees, and bears are assigned to a broader, more inclusive taxon (class Mammalia). In addition, the cladogram reveals that lizards are more closely related to the mammal clade than to frog, sunfish, or any other clade in our example. Can you explain why?

Two important concepts guide interpretation of cladograms. First, the relationships among taxa are determined only by tracing along the branches back to the most recent common ancestor and *not* by the relative placement of the branches along the horizontal axis. It is possible to represent the same relationships with many different types of branching diagrams (FIG. 23-10). Note that all three cladograms in this figure are equivalent. (Verify this fact by comparing the nodes and checking the relationships described earlier.) Thus, a tree can be rotated around its nodes. The branching relationships, not the order of the taxa or the size of the branches, are important.

A second important concept is that the cladogram tells us which taxa shared a common ancestor and how recently they shared a common ancestor. The ancestor itself remains unspecified. The cladogram does not establish direct ancestor–descendant relationships among taxa. In other words, a cladogram does not suggest that a taxon gave rise to any other taxon. Very specifically, the cladograms in Figures 23-9 and 23-10 do not suggest that the lampreys evolved before the other vertebrates or that the other vertebrates evolved from the lampreys, only that the lineage that eventually evolved into the lampreys separated from the rest of the vertebrates before the separation of the tetrapods from the fishes.

Note that any particular cladogram, such as those shown in Figures 23-8, 23-9, and 23-10, depicts only some of the nodes and branches in the tree of life. For example, in Figure 23-10a the single branch from node C that leads toward sunfish could be drawn to show thousands of nodes and tens of thousands of branches representing all the species of modern bony fishes. Cladograms used throughout this book will by necessity

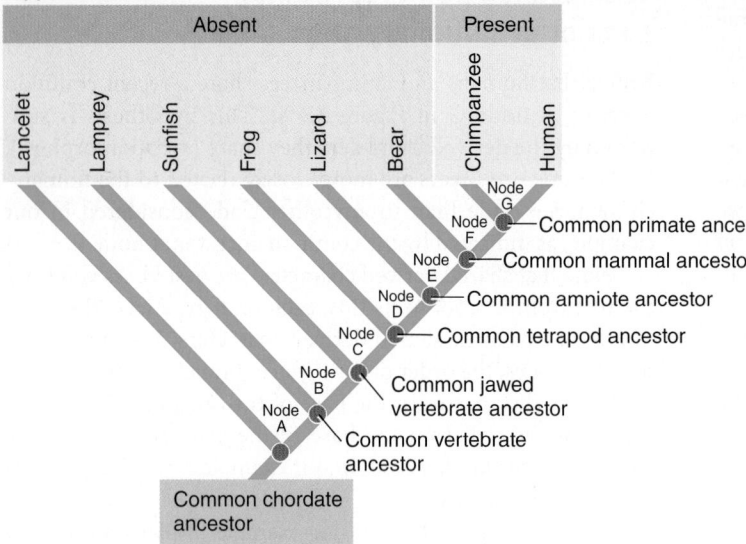

(a) This cladogram, with its diagonal branches, illustrates the style most commonly used in this book.

(b) This cladogram is another way to depict the same relationships shown in (a).

illustrate only particular nodes and branches of interest. The reader is always encouraged to think about which organisms have been left off any cladograms and where their favorite organisms might be placed. Where on Figure 23-10a should salamanders be placed?

Systematists use the principles of parsimony and maximum likelihood to make decisions

When systematists consider the evolutionary relationships of organisms, they must choose between multiple, competing cladograms. How do they choose the most accurate branching pattern? The most common criterion is the principle of **parsimony:** they choose the simplest explanation to interpret the data. Parsimony, a guiding principle in many areas of research, is based on the experience that the simplest explanation is probably the correct one. Applied to choice of cladograms, parsimony requires that the cladogram with the fewest changes in characters (the one with the fewest homoplasies) be accepted as most probable (**FIG. 23-11**).

In actual practice it is often possible to generate several cladograms that are equally parsimonious. The cladograms that are selected are the ones with more shared derived characters. The cladograms with more homoplasies are excluded because homoplasies are less likely to evolve than shared derived characters.

Systematists also use **maximum likelihood** to make decisions, especially when analyzing molecular data. Maximum likelihood is a statistical method that depends on probability (e.g., the probability that nucleotide sequences in DNA change at a constant rate over time). Complex computer programs analyze large data sets and report the probability of a particular tree.

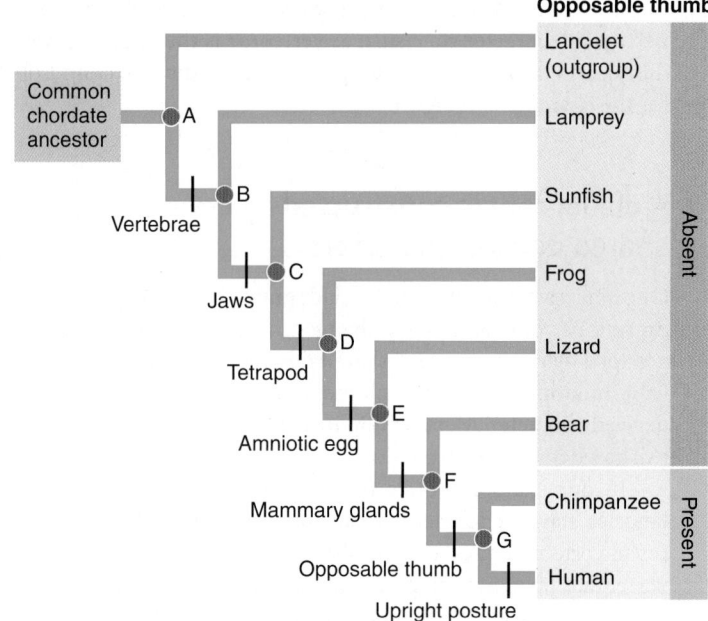

(c) This cladogram has rectangular branches and is rotated 90 degrees. However, it illustrates the same relationships as the cladograms in (a) and (b).

Figure 23-10 Cladogram styles

These three cladograms look very different, but they depict the same relationships. Remember that the relationships are depicted by which groups branch at the nodes. The length of the branches is not important.

© Cengage Learning

CHECKPOINT 23.4

- **CONNECT** *In what way do systematists use shared derived characters in their work?*
- *What is outgroup analysis?*
- *How do systematists apply the principle of parsimony?*

WHY **IS IT USED?** Cladogram construction is used to choose between alternative hypotheses of evolutionary relationships by selecting the most accurate branching pattern.

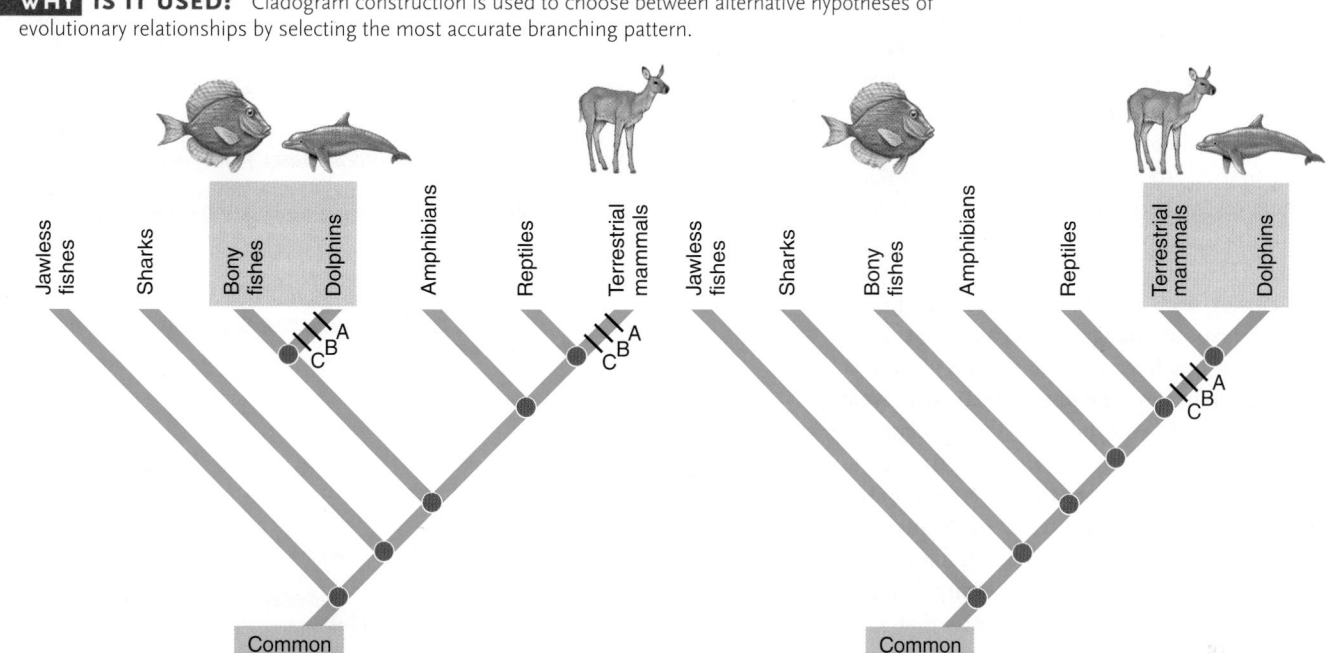

(a) Hypothesis 1: Dolphins and bony fishes are close relatives.

(b) Hypothesis 2: Dolphins and terrestrial mammals are close relatives.

HOW **IS IT DONE?** Systematists choose the simplest explanation to interpret the data.

  1 Draw the possible cladograms.
  2 Examine the relevant characters. In this example character A = hair, B = mammary glands, and C = middle ear bones.
  3 Note that in the first cladogram (hypothesis 1) all three characters examined had to evolve independently twice.
  4 Hypothesis 2 is chosen because it requires that each character evolve only once.

Figure 23-11 **Applying the principle of parsimony to cladogram construction**
© Cengage Learning

23.5 APPLYING PHYLOGENETIC INFORMATION

LEARNING OBJECTIVE

12 Describe how phylogenetic information can be applied to problems in other disciplines.

Modern systematics offers much more than simply a set of consistent species names and organized categories. The phylogenetic trees that are constructed contain useful information about the relationships among the organisms involved. This information can be applied to problems in many disciplines, including in ecology, evolution, and medicine.

When combined with other data, phylogenetic trees can be used to investigate the evolutionary histories of entire groups of organisms as well as the evolutionary changes that have occurred in specific traits. For example, phylogenetic trees based on the DNA sequences of modern humans combined with information on the geographic homelands of those humans are helping us understand the pathways used by humans as they originally dispersed out of Africa to inhabit the rest of the world (see Chapter 22).

Phylogenetic information has proven important in medical research and often has public health implications. **Human immunodeficiency virus (HIV)**, the virus that causes **acquired immunodeficiency syndrome (AIDS)** in humans, has had a devastating effect on humans since it first became prevalent in the 1980s. A major medical question concerned the origin of this virus. HIV is a **lentivirus,** a group of retroviruses (viruses that use RNA as their genetic material). Scientists have been able to determine the sequences of nucleotides in the RNA of many strains of HIV that have infected humans around the world. Researchers have sequenced somewhat similar viruses, called SIV viruses, that infect other primate species. Systematists have used these nucleotide sequences to generate phylogenetic trees for the viruses (FIG. 23-12).

When researchers mapped the primate host species onto the phylogeny for the viruses, they found that HIV-2, the somewhat less virulent virus, is most closely related to an SIV virus in a species of monkey (sooty mangabey). The more virulent HIV-1 strains have evolved from a virus that infected chimpanzees. The phylogenetic data were critically important in showing that precursors to HIV probably jumped from other primates to humans several times before the onset of the HIV epidemic. (HIV is discussed further in Chapter 45.)

This type of phylogenetic research has important public health implications. Many primate species are hosts for human pathogens that have spread into human populations, causing epidemics. Understanding the source and spread of these infections may lead to prevention measures that mitigate or prevent future epidemics.

Many evolutionary questions remain. New hypotheses about how organisms are related suggest new taxonomic schemes. Systematics offers tools to help us test these hypotheses. In Chapters 24 through 32, we explore many evolutionary puzzles and discuss current hypotheses about relationships and classifications of organisms.

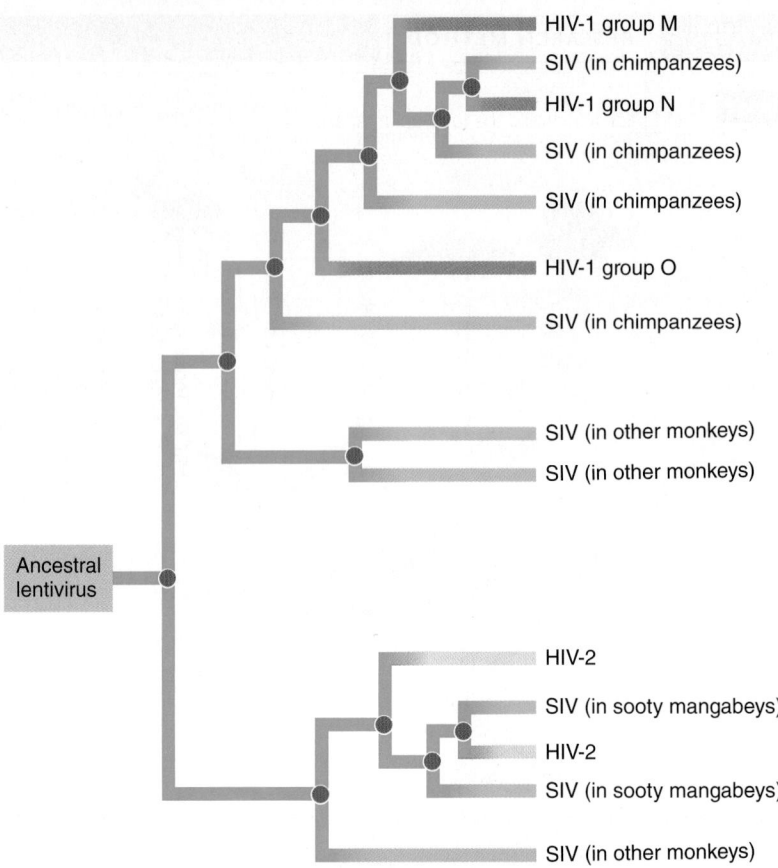

Figure 23-12 Understanding the origins of HIV infection in humans

The human immunodeficiency viruses (HIVs) and the related simian immunodeficiency viruses (SIVs) are lentiviruses. A phylogenetic analysis of nucleotide sequences for this group of viruses gives insight into the origins of HIV infection in humans. There are two major types of HIV, HIV-1 and HIV-2, with multiple strains within each of these types. Groups M, N, and O of HIV-1 are shown in the cladogram. HIV-1, group M, has affected millions of people around the world. Although all the HIV types and strains are from humans, different strains of SIV have been isolated from other species of primates. The cladogram shows four different strains of SIV isolated from chimpanzees, two SIV strains isolated from sooty mangabey monkeys, and three SIV strains isolated from other species of monkeys. The branching pattern derived from this phylogenetic analysis shows that the different strains of HIV-1 and HIV-2 have evolved separately from the various strains of SIV. Chimpanzees were the original source of the HIV-1 viral strains and sooty mangabey monkeys were the original source of the HIV-2 viral strains.

© Cengage Learning

CHECKPOINT 23.5

- **CONNECT** *How does modern systematics help medical researchers?*

SUMMARY: FOCUS ON LEARNING OBJECTIVES

23.1 Classifying Organisms *(page 475)*

1 State two justifications for the use of scientific names and classifications of organisms.

- **Systematics** is the scientific study of the diversity of organisms and their evolutionary relationships. **Taxonomy** is the branch of systematics devoted to naming, describing, and classifying organisms. **Classification** is the process of arranging organisms into groups based on similarities that reflect evolutionary relationships.

- Scientific names allow biologists from different countries with different languages to communicate about organisms. Biologists in distant locations must know with certainty that they are studying the same (or different) organisms.

- Classifications help biologists organize their knowledge.

2 Describe the binomial system of naming organisms and arrange the Linnaean categories in hierarchical fashion from most inclusive to least inclusive.

- In the **binomial system of nomenclature,** the basic unit of classification is the species.

- The name of each species has two parts: the **genus** name followed by the **specific epithet.**

- The hierarchical system of classification includes **domain, kingdom, phylum, class, order, family, genus,** and **species.** Each formal grouping at any given level is a **taxon.**

23.2 Determining the Major Branches in the Tree of Life *(page 476)*

3 Describe the three domains and argue for and against classifying organisms in these domains. (Compare this approach with using kingdoms as the main rank of classification.)

- The three-domain classification system assigns organisms to domains **Archaea, Bacteria,** or **Eukarya.** The domain classification is based on molecular data.
- Domain Eukarya includes the fungi, plants, animals, and protists.
- Some systematists recognize the kingdoms **Archaea, Bacteria, Fungi, Plantae,** and **Animalia.** Several "supergroups" composed of mainly unicellular, mainly aquatic eukaryotic organisms were formerly classified as kingdom Protista. Members of Archaea and Bacteria are prokaryotes. The fungi, which include molds, yeasts, and mushrooms, absorb nutrients produced by other organisms. Kingdoms Plantae and Animalia both consist of multicellular eukaryotes. For additional characteristics of the kingdoms, refer to Table 23-2. As new data are interpreted, organisms assigned to these kingdoms often must be reclassified.

4 Interpret a cladogram, describing the meaning of its specific nodes and branches.

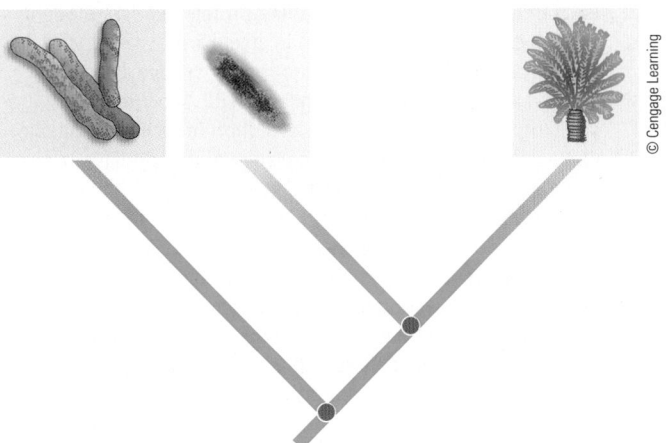

- In a **cladogram** each **branch** represents a **clade,** a group of organisms with a common ancestor. Each **node,** or branching point, represents the splitting of two or more new groups from a common ancestor. The node represents the *most recent common ancestor* of the clade represented by the branches. The **root** represents the most recent common ancestor of all the clades shown in the tree.
- We can determine the relationships among taxa by tracing along the branches back to the nodes. The cladogram indicates which taxa shared a recent common ancestor and how recently they shared that ancestor compared to other groups.

23.3 Reconstructing Evolutionary History *(page 481)*

5 Critically review the difficulties encountered in choosing taxonomic criteria.
- Systematists seek to determine evolutionary relationships, or **phylogeny,** based on shared characteristics. **Homology,** the presence in two or more species of a trait derived from a common ancestor, implies evolution from a common ancestor.
- Some seemingly homologous characters are acquired independently by **convergent evolution,** independent evolution of similar structures in distantly related organisms, or by **reversal,** reversion of a trait to its ancestral state. The term **homoplasy** refers to superficially similar characters that are not homologous.

6 Apply the concept of shared derived characters to the classification of organisms.
- **Shared ancestral characters** suggest a distant common ancestor.
- **Shared derived characters** (synapomorphies) indicate a more recent common ancestor and can be used as evidence for constructing cladograms.

7 Describe how analyses of molecular homologies contribute to the science of systematics.
- **Molecular systematics** depends on molecular structure to clarify phylogeny. Comparisons of nucleotide sequences in DNA and RNA, and of amino acid sequences in proteins, provide important information about how closely organisms are related.

8 Contrast monophyletic, paraphyletic, and polyphyletic taxa.

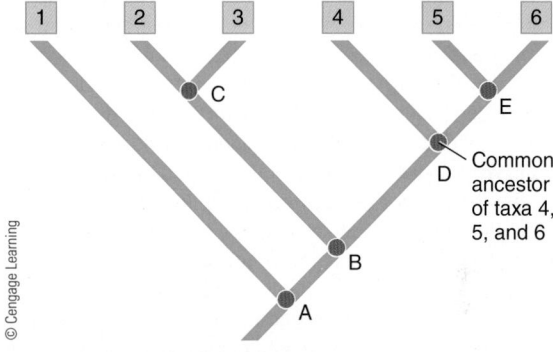

- A **monophyletic group,** or clade, includes all the descendants of the most recent common ancestor.
- A **paraphyletic group** consists of a common ancestor and some of, but not all, its descendants.
- A **polyphyletic group** consists of organisms that evolved from different recent ancestors.

23.4 Constructing Phylogenetic Trees *(page 486)*

9 Contrast the traditional classification with the current classification of reptiles and birds.
- Modern systematists agree that taxa must be monophyletic. Each monophyletic group consists of a common ancestor and all its descendants. Phylogenetic systematists (cladists) use shared derived characters to determine relationships among groups of organisms.
- Contemporary systematists classify reptiles and birds in a single clade.

10 Describe the construction of a cladogram by using outgroup analysis.
- Cladists use shared derived characters to reconstruct evolutionary relationships and diagram these relationships in cladograms. They use **outgroup analysis** to determine which characters in a given group of taxa are ancestral and which are derived.
- An **outgroup** is a taxon that diverged earlier than any of the other taxa being investigated.

11 Apply the principle of parsimony to cladogram construction.
- **Systematists** choose the simplest explanation to interpret the data—the principle of **parsimony.**

23.5 Applying Phylogenetic Information *(page 491)*

12 Describe how phylogenetic information can be applied to problems in other disciplines.

- Understanding how species are related can help scientists answer questions and solve problems in other disciplines. For example, phylogenetic information has helped us better understand the origin and transmission of HIV.

TEST YOUR UNDERSTANDING

Know and Comprehend

1. The mold that produces penicillin is *Penicillium notatum*. *Penicillium* is the name of its (a) genus (b) order (c) family (d) species (e) specific epithet

2. Decomposers such as molds and mushrooms belong to which domain and kingdom? (a) Eukarya, Plantae (b) Bacteria, Archaea (c) Archaea, Fungi (d) Eukarya, Archaea (e) Eukarya, Fungi

3. Each branching point in a cladogram (a) is called a root (b) represents a clade (c) represents the divergence of two or more groups from a common ancestor (d) represents horizontal gene transfer (e) marks the divergence of two kingdoms

4. The presence of homologous structures in two different groups of organisms suggests that (a) the organisms evolved from a common ancestor (b) convergent evolution has occurred (c) they belong to a polyphyletic group (d) homoplasy has occurred (e) independently acquired characters may evolve when organisms inhabit similar environments

5. Dolphins and humans both have the ability to nurse their young, whereas the less closely related sharks do not. The ability to nurse their young is (a) a shared derived character of fish and mammals (b) a shared ancestral character of all vertebrates (c) a shared derived character of mammals (d) an example of homoplastic behavior (e) an example of reversal

6. A group of organisms that includes a recent common ancestor and all its descendants is (a) polyphyletic (b) paraphyletic (c) monophyletic (d) an example of horizontal gene transfer (e) a sister group

7. In cladistic analysis (a) ancestral characters are used to reconstruct phylogenies (b) characters must be homoplastic (c) polyphyletic groups are preferred (d) outgroup analysis is rarely used (e) shared derived characters are the preferred evidence for relatedness

8. When systematists apply the principle of parsimony, they (a) choose the cladogram that requires the fewest changes in characters as the most probable explanation to interpret the data (b) select multiple hypotheses to explain each relationship (c) do not use outgroups (d) typically use a polyphyletic approach (e) hypothesize that the most complex explanation is most probably the correct one

Apply and Analyze

9. In interpreting a cladogram, (a) we can identify the specific ancestor of each taxon by tracing each branch back to the node closest to the root (b) taxa on the right side of a cladogram have evolved from the taxa on the left side (c) the relative placement of smaller branches allows us to determine the number of years since a particular taxon has evolved (d) we can determine relationships by tracing along the branches back to the most recent common ancestor (e) we must first identify horizontal gene transfer

10. What are some of the criteria and tools biologists might use to assign the different types of protists to monophyletic groups?

11. What are the advantages of using domains, rather than kingdoms, as the highest rank of classification?

Evaluate and Synthesize

12. **EVOLUTION LINK** Are members of a clade similar because they share a common ancestor, or do they belong to the same clade because they are similar?

13. **EVOLUTION LINK** The TATA-binding protein (TBP) is thought to be necessary for transcription in all eukaryotic cell nuclei. Studies show that archaea, but not bacteria, have a protein structurally and functionally similar to TBP. What does this similarity suggest regarding the evolution of archaea, bacteria, and eukaryotes? How might knowledge of this similarity affect how systematists classify these organisms?

14. **VISUALIZE** Construct a cladogram based on the following data. Mosses are plants with no vascular tissue. Horsetails, ferns, gymnosperms (pines and other plants with naked seeds), and angiosperms (flowering plants) are all vascular plants. Seeds are absent in all but the gymnosperms and angiosperms. Angiosperms are the only seed plants with flowers. (*Hint*: To help you construct the cladogram, draw a simple table showing which characters are present in each group. See Fig. 23-9.)

15. **INTERPRET DATA** What kind of grouping is represented by the bracketed area in the illustration? What type of group is formed by taxon 2 and taxon 3? What type of group is formed by taxa 2, 4, and 6?

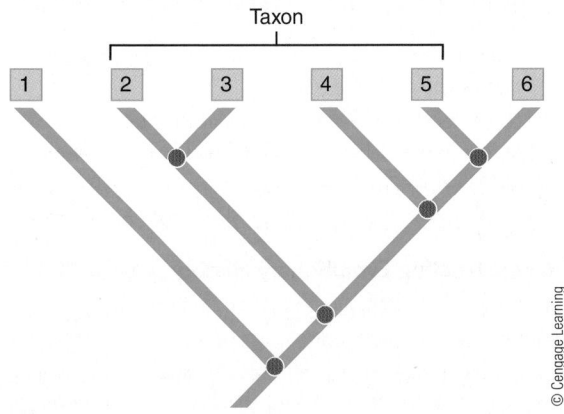

![aplia] To access course materials, such as Aplia and other companion resources, please visit **www.cengagebrain.com.**

Viruses and Subviral Agents

24

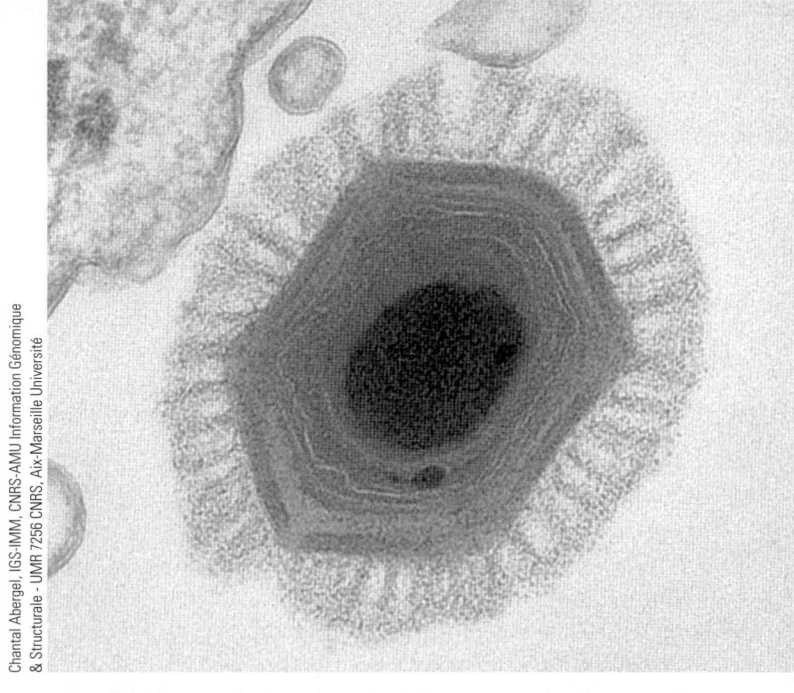

EM of *Megavirus chilensis.* Megavirus is a giant virus, the size of a very small bacterium (mycoplasma). The giant viruses discovered to date infect amoebas.

During the late 1800s, botanists searched for the cause of tobacco mosaic disease, which stunts the growth of tobacco plants and gives the infected tobacco leaves a spotted, mosaic appearance. The investigators found that they could transmit this disease to healthy plants by daubing their leaves with the sap of diseased plants. In 1892, Dmitri Ivanowsky, a Russian botanist, showed that the sap was still infective after it had been passed through porcelain filters designed to filter out all known bacteria. A few years later, in 1898, Martinus Beijerinck, a Dutch microbiologist, provided evidence that the agent that caused tobacco mosaic disease had many characteristics of a living organism. However, it appeared that the infective agent could reproduce only within a living cell. Beijerinck named the infective agent "virus" (from the Latin word *virus,* which means "poison"). Tobacco mosaic virus was the first virus to be recognized.

Early in the 20th century, scientists discovered other infective agents, like those responsible for tobacco mosaic disease, that could cause disease in animals or kill bacteria. These pathogens were so small that they could not be seen with the light microscope. They also passed through filters that removed bacteria. Curiously, they could not be grown in laboratory cultures unless living cells were present. The development of the electron microscope in the 1930s made it possible to see viruses for the first time.

Most of the viruses that infect animals, plants, and bacteria were identified during the second half of the 20th century. Viruses cause many plant diseases and are responsible for billions of dollars in crop losses each year. You may recognize some of the many human viral diseases, such as rabies, influenza, herpes, and acquired human immunodeficiency syndrome (AIDS). You have probably received vaccinations for many common childhood viral diseases, including polio, rubella (German measles), mumps, and chickenpox.

Viruses make up a large component of Earth's biomass. Their huge numbers make up for their small size and weight. Viruses influence many ecological processes. For example, they kill large amounts of marine biomass each day, contributing significantly to the recycling of nutrients. Viruses adapt to their environments through evolution. They also influence the biodiversity of other organisms. Viruses can transfer their own genes into the genetic material of their hosts. They can also transfer eukaryotic genes from

KEY CONCEPTS

24.1 A virus is a small particle consisting of a nucleic acid core surrounded by a protein coat; to reproduce, a virus must infect a living cell.

24.2 Viruses can be classified based on their host range, what type of nucleic acid they have, and whether the nucleic acid is single-stranded or double-stranded.

24.3 In a lytic reproductive cycle, a virus uses the host cell's molecular machinery to replicate itself, destroying the host cell in the process. In a lysogenic cycle, the viral genome becomes integrated into the host DNA and is then called a "provirus."

24.4 Viruses infect the cells of all kinds of organisms; they cause serious diseases in plants and animals.

24.5 Three different hypotheses suggest how viruses may have originated and evolved.

24.6 Subviral agents, which include satellites, viroids, and prions, are smaller and simpler than viruses.

one organism to another, crossing species boundaries. The discovery of giant viruses (see photograph) has raised new questions about the structure, genomes, origin, and phylogenetic relationships of viruses.

In this chapter we examine the characteristics and diversity of viruses as well as the smaller subviral agents, viroids and prions. We include viruses in the diversity unit of *Biology* because they have some characteristics of living things and because they impact all living organisms.

24.1 THE STATUS AND STRUCTURE OF VIRUSES

LEARNING OBJECTIVES

1 Contrast a virus with a cellular organism.
2 Describe the structure of a virus.

A **virus** is a very small infective agent that consists of a core of nucleic acid and is dependent on a living host. The study of viruses is called *virology,* and biologists who study viruses are *virologists.* Most biologists view viruses as nonliving particles because they are not composed of cells and they cannot carry on metabolic activities or reproduce on their own. They do not have the components necessary to carry on cellular respiration or to synthesize proteins and other molecules.

Viruses do contain the nucleic acids necessary to make copies of themselves. They replicate by invading living cells and commandeering their metabolic machinery. To multiply, a virus must *infect* a cell in which it can replicate. Thus, viruses are *obligate intracellular parasites;* they survive only by using the resources of a *host cell,* the cell the virus invades. Viruses infect all types of organisms, including bacteria, archaea, protists, plants, fungi, and animals. Some viruses even infect other viruses. Interestingly, viruses evolve by natural selection. For these reasons, most biologists view viruses as being at the edge of life, although a few argue that they are very simple life-forms.

Viruses are very small

Most known viruses are very small, ranging in size from 20 to 300 nm. (Recall that a nanometer is one-thousandth of a micrometer.) The poliovirus is about 30 nm in diameter (about the size of a ribosome). If we could line these viruses up end to end, it would take almost a million of them to span 1 inch! A poxvirus that causes smallpox can measure up to 300 nm long and 200 nm wide.

The paradigm that all viruses are very small and simple, with very small genomes, has been challenged by the discovery of *giant* viruses. *Mimivirus* was discovered in 2004, followed by *Megavirus* in 2011, and Pandoraviruses in 2013. One of the Pandoraviruses approaches 1 μm in length and is about 0.5 μm wide, larger than many bacteria and some eukaryotic cells. These giant viruses were discovered by French researchers in seawater off the coast of Chile and in a freshwater pond in Australia.

A virus consists of nucleic acid surrounded by a protein coat

The nucleic acid core of the virus is surrounded by a protein coat called a **capsid.** Microbiologists use the term *virion* to refer to the complete virus particle that is outside a host cell in a dormant state. The virion is the form in which the virus moves from the cell in which it was produced to a new host cell in which it can replicate its genome. However, here we will use the term *virus* in a broad sense, even when referring to virions.

A typical virus contains *either* deoxyribonucleic acid (DNA) or ribonucleic acid (RNA), not both. In 2012, however, researchers discovered an RNA–DNA hybrid virus in a high temperature acidic California lake. Further studies have found three other similar RNA–DNA hybrids in marine environments.

The nucleic acid of a virus can be single-stranded or double-stranded. Thus, a virus can have single-stranded (ss) DNA, double-stranded (ds) DNA, ssRNA, or dsRNA. As we will discuss, the type of nucleic acid is important in classifying viruses. The virus genome typically consists of 5000 to more than 100,000 bases or base pairs (depending on whether it is single-stranded or double-stranded). Again, the giant viruses are exceptions. *Megavirus* has seven transfer RNAs and some metabolic genes never before identified in any other virus. Researchers have sequenced the double-stranded DNA genomes of two species of *Pandoravirus* and reported that they range from 1.9 to 2.5 million base pairs. The species with the largest genome has 2556 proposed protein-coding genes (only 6% of which match known proteins).

The capsid is a protective protein coat

The capsid consists of protein subunits called *capsomers.* The capsomers determine the shape of the virus. Capsids generally are helical, polyhedral, or a combination of both shapes. Helical viruses, such as the tobacco mosaic virus (TMV), appear as long rods or threads (**FIG. 24-1a**). Its capsid is a hollow cylinder made up of proteins that form a groove into which the RNA fits.

Polyhedral viruses, such as the adenoviruses (which cause a number of human illnesses, including some respiratory infections), appear somewhat spherical (**FIG. 24-1b**). Its capsomers are organized in equilateral triangles. The capsid of a large virus can consist of several hundred capsomers. The most common polyhedral structure is an *icosahedron,* a structure with 20 identical surface faces (each face is a triangle). The human immunodeficiency virus (HIV) that causes AIDS is an enveloped virus (see next section) (**FIG. 24-1c**).

Some viruses have both helical and polyhedral components. Viruses that infect bacteria are called *bacteriophages,* or simply *phages* (**FIG. 24-1d**). The T4 phage that infects the bacterium *Escherichia coli* consists of a polyhedral "head" attached to a helical "tail." Many phages have this shape. Many phages also have tail fibers that attach to the host cell.

Some viruses are surrounded by an envelope

Some viruses, called *enveloped viruses,* have an outer membranous envelope that surrounds the capsid. Typically, the virus

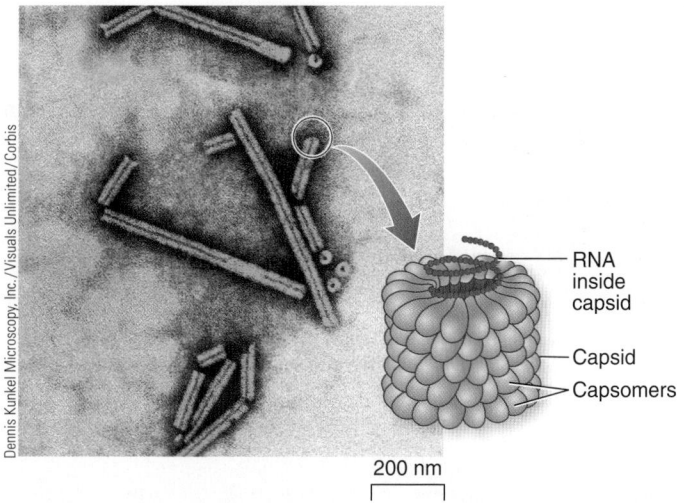

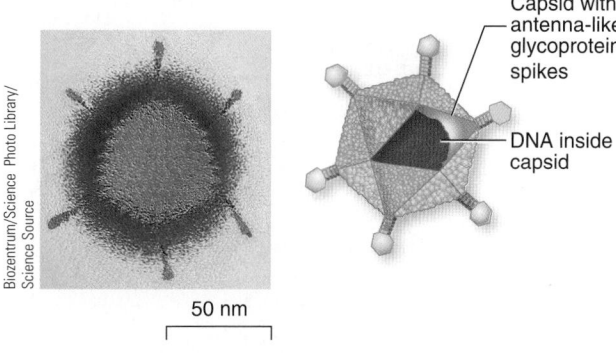

(b) **Color-enhanced TEM of an adenovirus.** The capsid is composed of 252 subunits (visible as tiny ovals) arranged into a 20-sided polyhedron. Twelve of the subunits have projecting glycoprotein spikes that permit the virus to recognize host cells.

(a) **TEM of tobacco mosaic virus.** This rod-shaped virus has a helical arrangement of capsid proteins.

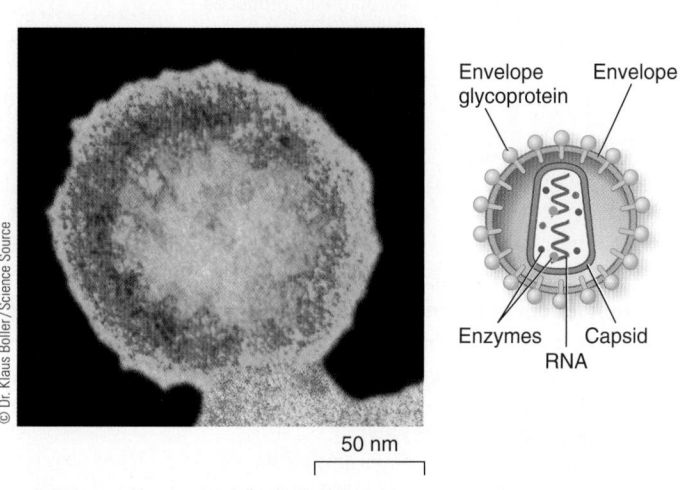

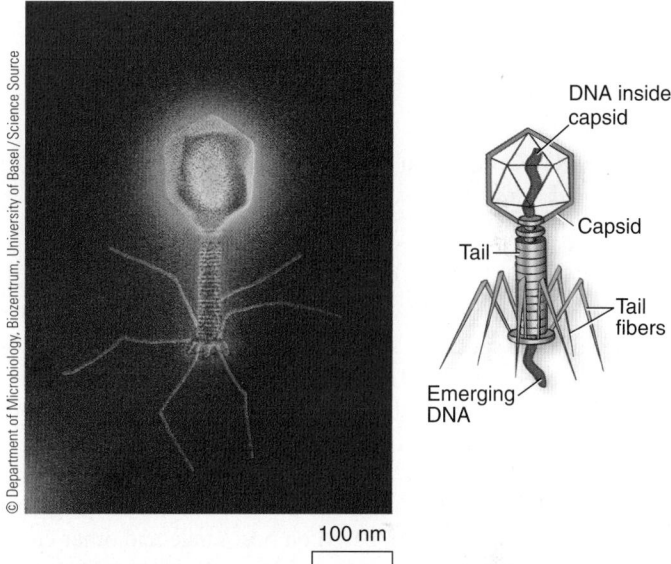

(c) **Color-enhanced TEM of HIV, the virus that causes AIDS.** The virus is leaving a host cell (*pink*). The virus is enclosed in an envelope (*green*) derived from the host's plasma membrane. Viral proteins project from the envelope. The capsid of the virus is shown in *yellow*.

(d) **Color-enhanced TEM of T4 bacteriophage.** This virus has a polyhedral head and a helical tail. The virus attaches to the cell wall of the bacterial host by its tail fibers.

Figure 24-1 *Animation* **The structure of a virus**

A virus consists of a DNA or RNA core surrounded by a protein coat called a capsid. The capsid is made up of protein subunits called capsomers. Some viruses have an outer membranous envelope surrounding the capsid.
© Cengage Learning

acquires the envelope from the host cell's plasma membrane as it leaves the host cell (**FIG. 24-2**). Interestingly, while inside the host cell, the virus synthesizes certain proteins and inserts them into the host's plasma membrane. Thus, the viral envelope consists of phospholipids and proteins of the host's plasma membrane as well as distinctive proteins produced by the virus itself. Some viruses produce envelope glycoproteins that extend out from the envelope as spikes. As we will discuss, these glycoprotein spikes can be very important in the virus's interaction with the host cell.

CHECKPOINT 24.1

- **CONNECT** *What characteristics of a living organism are absent in a virus?*

- *What are the structural components of a virus?*

24.2 CLASSIFICATION OF VIRUSES

LEARNING OBJECTIVE

3 Identify three characteristics used to classify viruses.

Viruses present a taxonomic challenge to biologists because they do not have the characteristics that define living organisms (see Chapter 1). They are not cellular, they do not carry on metabolic activities, and they reproduce only by taking over the reproductive machinery of other cells. Viruses do not manufacture proteins and so do not have distinctive rRNA. For these reasons, viruses are not classified in any of the three domains.

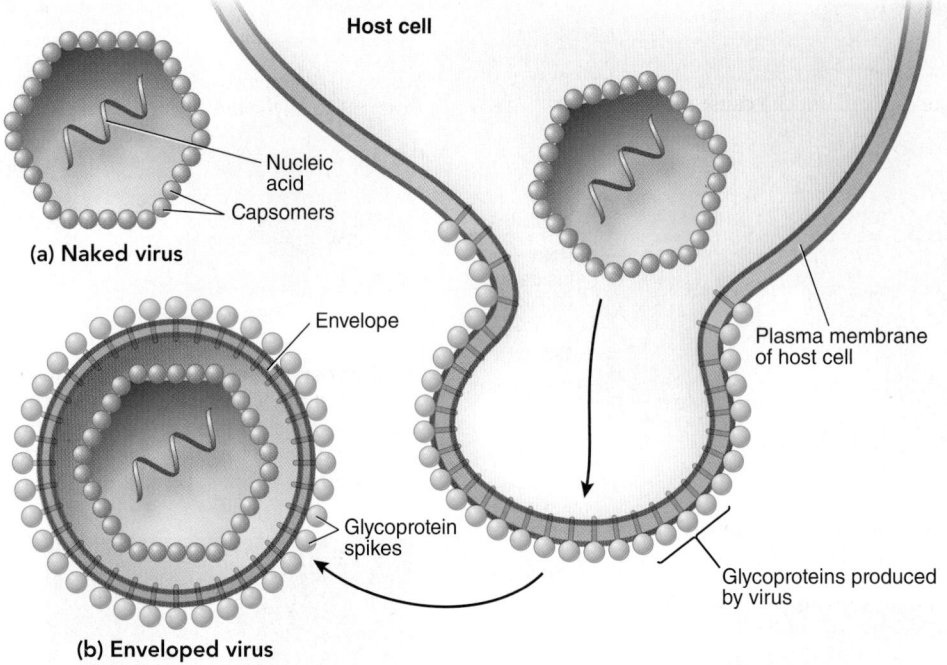

Host cell

Nucleic acid
Capsomers

(a) Naked virus

Envelope

Plasma membrane of host cell

Glycoprotein spikes

Glycoproteins produced by virus

(b) Enveloped virus

Figure 24-2 Comparison of naked and enveloped viruses

As the virus leaves the host cell, the host's plasma membrane wraps around the virus, forming the envelope. The envelope contains proteins produced by the virus.

© Cengage Learning

Viruses can be classified based on their **host range,** the types of host species a specific virus (or other type of infectious agent) can infect. They may be referred to as plant viruses, animal viruses, bacterial viruses, and so on. Viruses are more formally classified into taxa from species to orders. The International Committee on Taxonomy of Viruses (ICTV), a group of virologists, decides on specific criteria for classifying and naming viruses. Recently, based on host range and other characteristics, the ICTV classified viruses into 7 orders, 96 families, 420 genera, and more than 2600 species. Virus family names include the suffix *-viridae.* Note that this classification system is not a traditional Linnaean system; it does not assign viruses to domains, kingdoms, or phyla.

The Baltimore classification system classifies viruses based on the type of nucleic acid the virus contains, whether the nucleic acid is single-stranded or double-stranded, and how mRNA is produced. Other traits considered in viral classification are the size and shape of the virus, the presence of an envelope, and the method by which the virus is transmitted from host to host.

CHECKPOINT 24.2

- *What are three characteristics used to classify viruses?*

24.3 VIRAL REPLICATION

LEARNING OBJECTIVES

4   Characterize bacteriophages.

5   Contrast a lytic cycle with a lysogenic cycle.

As we have discussed, viruses reproduce, but only within the complex environment of the living host cells they infect. Viruses use their genetic information to force their host cells to replicate their viral nucleic acid and make the proteins they need. They take over the transcriptional and translational mechanisms of the host cell.

Bacteriophages infect bacteria

Much of our knowledge of viruses has come from studying **bacteriophages** ("bacteria eaters"), more simply referred to as **phages.** These viruses can be cultured easily within living bacteria in the laboratory. Microbiologists have identified more than 5000 phages. Bacteriophages are among the most complex viruses (see Fig. 24-1d). Their most common structure consists of a long nucleic acid molecule (usually dsDNA) coiled within a polyhedral head. Most have a tail, which may be contractile and may function in penetration of the host cell.

Phages have been clinically used to treat infection for close to a century. Although abandoned in Western countries after the discovery of sulfa drugs and antibiotics in the 1940s, they have continued to be commonly used in member countries of the former Soviet Union to treat patients. Antibiotics, which have long been more dependable and easier to use, are increasingly less effective than in the past due to the widespread and escalating problem of bacterial resistance. Phage therapy, alone and paired with antibiotics, is again a focus of research in the United States as well as in Eastern Europe. Phages are being engineered to specifically destroy particular bacteria due to their ability to target specific host species. Scientists are also genetically engineering phages so that bacteria will be slower to evolve resistance to them.

Phage therapy also has applications in dentistry, veterinary medicine, agriculture, and food science. For example, certain phages can kill deadly strains of *E. coli* in cattle. These bacteria do not appear to make cattle ill but can cause illness and death in people who eat contaminated, undercooked ground beef.

Viruses replicate inside host cells

Viruses infect bacterial, plant, and animal cells in basically similar ways. We will focus on phage infection of bacteria because this process is best understood. The reproductive cycle of viruses begins with a virus coming into contact with a host cell. The virus typically attaches to the surface of the host cell. The viral nucleic acid must enter the host cell and synthesize the components it needs to reproduce itself. Then viral components are assembled, and viruses are released from the cell, ready to invade other cells. Two types of viral reproductive cycles are lytic and lysogenic cycles.

Lytic reproductive cycles destroy host cells. In a **lytic cycle,** the virus lyses (destroys) the host cell. When the virus infects a susceptible host cell, it forces the host to use its metabolic machinery to replicate viral particles. Viruses that have only a lytic cycle are described as *virulent,* which means that they cause disease and often death.

Five steps are typical in lytic viral reproduction (**FIG. 24-3**):

1. **Attachment (or adsorption).** The virus attaches to specific receptors on the host cell. This process ensures that the virus infects only its specific host.
2. **Penetration.** The virus penetrates the host plasma membrane and moves into the cytoplasm. Many viruses that infect animal cells enter the host cell intact. Some phages inject only their nucleic acid into the cytoplasm of the host cell; the capsid remains on the outside.
3. **Replication and synthesis.** The viral genome contains all the information necessary to produce new viruses. Once inside a host cell, the virus degrades the host-cell nucleic acid. The virus then uses the molecular machinery of the host cell to replicate its own nucleic acid and produce viral proteins. Many antiviral drugs interfere with replication of viral nucleic acid.
4. **Assembly.** The newly synthesized viral components are assembled into new viruses.
5. **Release.** Assembled viruses are released from the cell. Generally, lytic enzymes, produced by the phage late in the replication process, destroy the host plasma membrane. Phage release typically occurs all at once and results in rapid cell lysis. In contrast, animal viruses are often released slowly or bud off from the plasma membrane.

Once released, the viruses infect other cells, and the process begins anew. The time required for viral reproduction, from attachment to the release of new viruses, varies from less than 20 minutes to more than 1 hour.

How do bacteria protect themselves from phage infection? You may recall from Chapter 15 that bacteria produce *restriction enzymes,* enzymes that cut up the foreign DNA of the phage. This action prevents the phage DNA from duplicating. The

Viruses reproduce by seizing control of the metabolic machinery of a host cell. In a lytic infection, the viruses destroy the host cell.

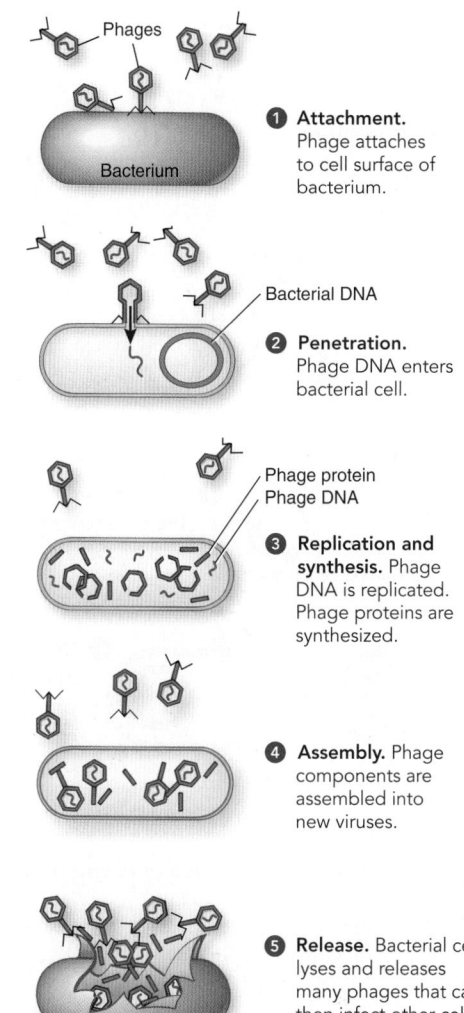

1 Attachment. Phage attaches to cell surface of bacterium.

2 Penetration. Phage DNA enters bacterial cell.

3 Replication and synthesis. Phage DNA is replicated. Phage proteins are synthesized.

4 Assembly. Phage components are assembled into new viruses.

5 Release. Bacterial cell lyses and releases many phages that can then infect other cells.

(a) The sequence of events in a lytic infection.

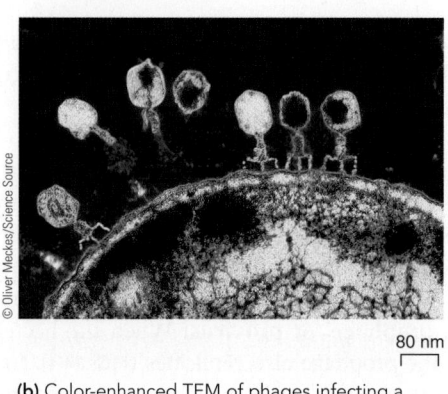

80 nm

(b) Color-enhanced TEM of phages infecting a bacterium, *Escherichia coli.*

Figure 24-3 *Animation* **The sequence of events in a lytic infection**

Virulent bacteriophages kill their host cell during each cycle of infection.

PREDICT Would the phage components be able to infect other bacteria if they were released prior to assembly?

© Cengage Learning

In a lysogenic cycle, temperate phages integrate their nucleic acid into the host genome. The prophage replicates when the bacterial DNA replicates.

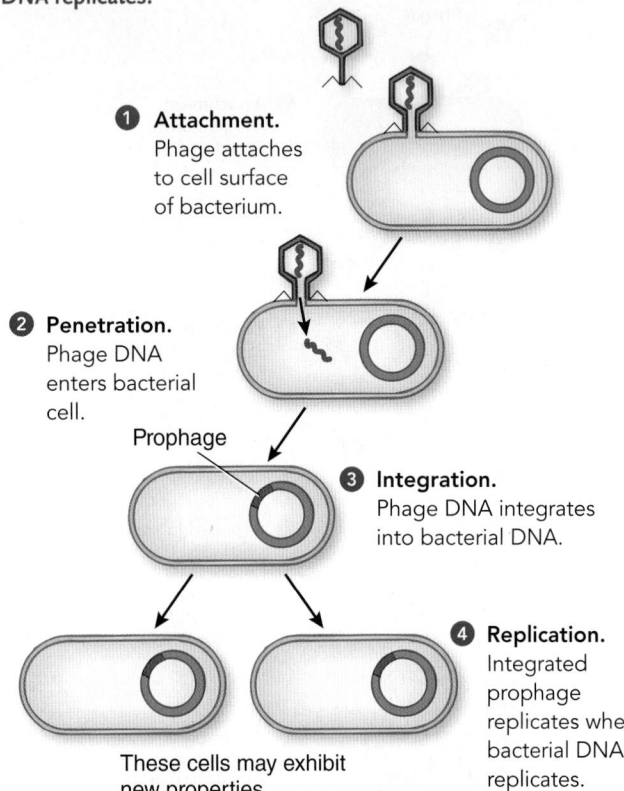

① **Attachment.** Phage attaches to cell surface of bacterium.

② **Penetration.** Phage DNA enters bacterial cell.

Prophage

③ **Integration.** Phage DNA integrates into bacterial DNA.

④ **Replication.** Integrated prophage replicates when bacterial DNA replicates.

These cells may exhibit new properties.

Figure 24-4 *Animation* **The sequence of events in a lysogenic cycle**

Temperate phages integrate their nucleic acid into the DNA of their bacterial host. The integrated viral DNA is called a prophage. The bacterium is called a lysogenic cell.

PREDICT How do temperate phages benefit from integrating into the bacterial DNA?

© Cengage Learning

bacterial cell protects its own DNA by slightly modifying it after replication so that the restriction enzyme does not recognize the sites it would cut.

In lysogenic cycles temperate viruses integrate into the host DNA **Temperate phages** do not immediately destroy their hosts. These phages alternate between a lytic cycle and a lysogenic cycle. In a **lysogenic cycle,** the viral genome becomes integrated into the host bacterial DNA. The integrated virus is called a **prophage,** or **provirus.** When the bacterial DNA replicates, the prophage also replicates (**FIG. 24-4**). In this way, the prophage is passed on to new generations of the original infected cell. The viral genes that code for viral structural proteins may be repressed indefinitely. The temperate phage lambda (designated by the Greek letter λ) is a model organism used to study temperate viruses.

Certain external conditions (such as ultraviolet light and X-rays) cause temperate viruses to revert to a lytic cycle and then destroy their host. Sometimes temperate viruses become lytic spontaneously.

Bacterial cells carrying prophages are called *lysogenic cells.* Such bacterial cells may exhibit new properties. This type of change is called **lysogenic conversion.** An interesting example involves the bacterium *Corynebacterium diphtheriae,* which causes diphtheria. Two strains of this species exist, one that produces a toxin (and causes diphtheria) and one that does not. The only difference between these two strains is that the toxin-producing bacteria are infected by a specific temperate phage. The phage DNA codes for the powerful toxin that causes the symptoms of diphtheria. Similarly, the bacterium *Clostridium botulinum,* which causes botulism, a serious form of food poisoning, is harmless unless it contains certain prophage DNA that induces synthesis of the toxin.

CHECKPOINT 24.3

- **VISUALIZE** *What are the steps in a lytic cycle? Draw diagrams to illustrate your answer.*
- *How is a lysogenic cycle different from a lytic cycle?*

24.4 VIRAL DISEASES

LEARNING OBJECTIVES

6 Contrast viral infection of plants and animals, and identify specific diseases caused by viruses.
7 Describe the reproductive cycle of a retrovirus, such as human immunodeficiency virus (HIV).

Sir Peter Medawar, a winner of the 1960 Nobel Prize in Physiology or Medicine, defined a *virus* as "a piece of bad news wrapped in protein." Viruses that are **pathogens** are responsible for some very serious diseases, including Ebola hemorrhagic fever, rabies, influenza (flu), and AIDS. However, although viruses must infect cells to reproduce, most known viruses do *not* cause disease. Here we focus on viral infection of plants and animals.

Viruses cause serious plant diseases

Viruses cause many important plant diseases and are responsible for billions of dollars in agricultural losses and lower crop quality worldwide each year. Infected crops almost always produce lower yields than noninfected crops.

Viral infections do not usually kill plants, but they do stunt their growth; cause changes in the shape of the foliage; and may cause spots, streaks, or mottled patterns on leaves, flowers, or fruits (**FIG. 24-5**). Most plant viruses have capsids, but not envelopes. The genome of most plant viruses consists of ssRNA. Plant viruses are typically named according to the type of host plant they infect and their effects on the plant. Tobacco mosaic virus, described at the beginning of this chapter, was the first virus ever discovered.

Insects are important vectors of plant disease. As they feed on plant tissues, aphids, leaf-hoppers, and many other insects spread viral diseases among plants. Soil-borne nematodes and protozoa are vectors of viral infections in plant roots.

Because plants have thick cell walls, viruses cannot penetrate plant cells unless the cells are damaged. Plant viruses can be transmitted from one generation to the next through infected seeds or by asexual propagation (Fig. 24-5). Once a plant is infected, the virus spreads through the plant body by passing through *plasmodesmata,* cytoplasmic channels that connect adjacent plant cells.

There are no known cures for most viral diseases of plants, so infected plants are commonly burned. Many agricultural scientists are focusing their efforts on preventing viral disease by developing virus-resistant strains of important crop plants.

(a) Virus-streaked tulip. The virus that causes this relatively harmless disease affects pigment formation in the petals.

(b) Pepper leaves infected with tobacco mosaic virus. The leaf is characteristically mottled with light green areas.

Figure 24-5 Plant diseases caused by viruses

Viruses cause serious diseases in animals

Hundreds of different viruses infect animals, including humans. Animal viruses cause hog cholera, foot-and-mouth disease, canine distemper, and certain types of cancer (such as feline leukemia and cervical cancer). Viruses cause chickenpox, herpes simplex (one type causes genital herpes), mumps, rubella (German measles), rubeola (measles), rabies, warts, infectious mononucleosis, influenza, viral hepatitis, Ebola hemorrhagic fever, and AIDS (**TABLE 24-1** and *Inquiring About: Influenza and Other Emerging Diseases*). Most humans suffer from two to six viral infections each year, including common colds and gastrointestinal disorders.

Norwalk virus is responsible for most worldwide viral gastroenteritis outbreaks and epidemics, affecting more than 200 million and killing more than 200,000 people each year. Norwalk virus is transmitted directly from person to person and indirectly by fecal contamination of water and food. The virus is also spread when infected people sneeze or cough, sending particles into the air; that is, virus particles become aerosolized. Infected individuals can shed viral particles for many weeks after symptoms abate. Norwalk virus infections are a serious concern, especially in closed communities, such as health care institutions, prisons, and cruise ships where people are in prolonged close contact. The genetically diverse strains of the Norwalk virus species make up the noroviruses.

Human activity, including social factors such as urbanization, global travel, and war, contributes to epidemics of infectious disease. Living conditions, including sanitation, nutrition, physical stress, level of health care, and sexual practices, are important factors in the spread of disease. In the United States and other highly developed countries, infectious disease accounts for about 4% to 8% of deaths compared with death rates of 30% to 50% in developing regions. Even at the level of our current knowledge

about viruses and epidemiology, just how prepared are we to contain a particularly virulent virus? For example, how well are we containing HIV (the virus that causes AIDS), which continues to infect and kill millions of people worldwide every year?

Since the September 11, 2001, terrorist attacks in the United States, bioterrorism has become a critical concern worldwide. **Bioterrorism** is the intentional use of microorganisms or toxins derived from living organisms to cause death or disease in humans, animals, or plants on which humans depend. Terrorists could conceivably initiate epidemics of smallpox, anthrax, plague, yellow fever, Ebola hemorrhagic fever, and other potentially fatal diseases. (Note that anthrax and plague are caused by bacteria; see Table 25-5.) The quest for rapid identification and effective treatments and vaccines for these diseases has taken on new urgency.

Animal viruses have varied structure and life cycles

Most viruses cannot survive very long outside a living host cell, so their survival depends on their being transmitted from one animal to another. However, their host range may be quite limited because attachment to a host cell is very specific. The type of attachment proteins on the surface of a virus determines what type of cell it can infect. Receptors typically vary with each species and sometimes with each type of tissue. Thus, many human viruses can infect only humans because the viral attachment proteins combine only with receptor sites found on human cell membranes. The measles virus and poxviruses infect many types of human tissue because their attachment proteins combine with receptor sites on a variety of cells. In contrast, the poliovirus attaches to specific types of human cells, such as those that line the digestive tract and motor neurons of the brain and spinal cord.

Some viruses, such as the adenoviruses, have fibers that project from the capsid and adhere to complementary receptors

TABLE 24-1 | Some Viruses That Infect Vertebrates

GROUP	DISEASES CAUSED	CHARACTERISTICS
DNA VIRUSES WITH ENVELOPE		
Poxviruses	Smallpox, cowpox,* monkeypox, and economically important diseases of domestic fowl	**dsDNA;** large, complex viruses; replicate in the cytoplasm of the host cell
Herpesviruses	Cold sores (herpes simplex virus type 1); genital herpes, a sexually transmitted disease (herpes simplex virus type 2); chickenpox and shingles (herpes varicella–zoster virus); infectious mononucleosis and Burkitt's lymphoma (Epstein–Barr virus)	**dsDNA;** medium to large, enveloped viruses; replicate in the host nucleus[†]
DNA VIRUSES WITH NO ENVELOPE		
Adenoviruses	Respiratory tract disorders (e.g., sore throat, tonsillitis), conjunctivitis, and gastrointestinal disorders are caused by more than 40 types of adenoviruses in humans; other varieties infect other animals	**dsDNA;** replicate in the host nucleus
Papovaviruses	Human warts and some degenerative brain diseases; some cancers, including cervical cancer[‡‡]	**dsDNA**
Parvoviruses	Infections in dogs, swine, arthropods, rodents; gastroenteritis in humans (transmitted by consumption of infected shellfish)	**ssDNA;** some require a helper virus to multiply
RNA VIRUSES WITH ENVELOPE		
Togaviruses	Rubella (German measles)	**ssRNA** that can serve as mRNA; large diverse group of medium-sized enveloped viruses; many transmitted by arthropods
Orthomyxoviruses	Influenza (flu) in humans and other animals	**ssRNA** that serves as template for mRNA synthesis; medium-sized viruses that often exhibit projecting glycoprotein spikes
Paramyxoviruses	Rubeola (measles) and mumps in humans; distemper in dogs	**ssRNA;** resemble orthomyxoviruses but somewhat larger
Rhabdoviruses	Rabies	**ssRNA**
Coronaviruses	Upper respiratory infections; SARS	**ssRNA;** largest known RNA virus
Flaviviruses	Yellow fever; West Nile virus; hepatitis C (the most common reason for liver transplants in the United States)	**ssRNA**
Filoviruses	Hemorrhagic fever, including that caused by the Ebola virus	**ssRNA**
Bunyaviruses	St. Louis encephalitis; hantavirus pulmonary syndrome (caused by Sin Nombre virus, a hantavirus)	**ssRNA**
Retroviruses	AIDS; some types of cancer	**ssRNA** viruses that contain reverse transcriptase for transcribing the RNA genome into DNA; two identical molecules of ssRNA
RNA VIRUSES WITH NO ENVELOPE		
Picornaviruses	Polio (poliovirus); hepatitis A (hepatitis A virus); intestinal disorders (enteroviruses); common cold (rhinoviruses); aseptic meningitis (coxsackievirus, echovirus)	**ssRNA** that can serve as mRNA; diverse group of small viruses
Reoviruses	Vomiting and diarrhea; encephalitis	**dsRNA**
Noroviruses (This group made up of strains of Norwalk virus)	Most common cause of viral gastroenteritis in humans	**ssRNA**

*The vaccinia (cowpox) virus is used to produce genetically engineered vaccines.
[†]These viruses frequently cause latent infections; some cause tumors.
[‡‡]The virus SV40 has been used as a vector to transport genes into cells.

© Cengage Learning

on the host cell. Other viruses, such as those that cause herpes and rabies, are surrounded by a lipoprotein envelope with projecting glycoprotein spikes that attach to a host cell. The influenza virus has rodlike, glycoprotein spikes that project from its envelope. These spikes bind with specific receptors present only on cells lining the respiratory tract of certain vertebrates, including wild and domestic birds, horses, swine, and humans. Influenza vaccine prevents attachment by stimulating the host's antibodies to cover the glycoprotein spikes projecting from the virus.

Viruses have several ways to penetrate animal cells (**FIG. 24-6**). After attachment to a host-cell receptor, some enveloped viruses fuse with the animal cell's plasma membrane. The entire virus, including the capsid and nucleic acid, is released into the animal cell. Other viruses enter the host cell by *endocytosis*. In this process the plasma membrane of the animal cell invaginates, forming a membrane-enclosed vesicle that contains the virus. Endocytosis is advantageous to the virus because the endocytotic vesicle delivers the virus deep into the cytosol.

How do new strains of the influenza virus re-emerge? **Emerging diseases,** those new to the human population—such as AIDS, severe acute respiratory syndrome (SARS), Ebola hemorrhagic fever, eastern equine encephalitis, and West Nile virus—often appear suddenly. Many new, continual, or re-emerging pathogens can strike globally.

Re-emerging diseases are those that have been almost eradicated and then suddenly recur, causing an epidemic, sometimes in a new geographic area. Many familiar diseases, such as influenza, malaria, tuberculosis, and bacterial pneumonias, continue to infect large numbers of people and increasingly reappear in forms that are resistant to drug therapy. Drug-resistant pathogens are a major challenge to public health and other health care professionals.

Historically, new viral strains have claimed many human lives. In 1918, for example, an influenza (flu) pandemic killed more than 20 million people throughout the world. Flu pandemics also occurred in 1957 and 1968. Each year new strains of influenza virus evolve and become infectious. If the new combination of viral genes is unfamiliar to the human immune system, the virus may spread easily, resulting in a pandemic. According to the Centers for Disease Control and Prevention, influenza kills about 36,000 individuals in the United States every year.

New strains of influenza are thought to infect one species, such as a species of birds. At first, they are not able to spread among mammals. However, because viruses can evolve, the avian strain can mutate and become virulent. The avian virus can also exchange RNA with a virus that has the genetic material necessary to spread among swine or other mammals. Swine can be susceptible to both avian and human strains of the virus. Inside a swine's body, the viruses can exchange bits of RNA and develop the ability to infect humans. When an animal disease crosses the species barrier and infects humans, it is called a **zoonotic disease.** (We discuss zoonosis further in Chapter 25.) Reassortment of genetic material can also occur inside a human host, producing a viral strain that can spread from person to person (see figure). H1N1 influenza evolved in this way.

1. Influenza virus may mutate to forms that can infect human cells.

2. Influenza viruses from two different strains infect human cell by releasing their RNA into cell. Neither of these viruses can spread from human to human.

3. RNA from both viruses is duplicated in nucleus (duplicated RNA strands not shown).

4. New viruses assembled with reassorted RNA; new viruses have RNA from both strains.

5. Viruses of new strain leave host cell and now are highly infectious to humans.

An influenza virus such as H1N1 evolves as it mutates and its RNA reassorts in the bodies of birds, swine, and humans. Eventually, it becomes virulent.

© Cengage Learning

In DNA animal viruses, replication of viral DNA and protein synthesis are similar to the processes by which the host cell would normally carry out its own DNA replication and protein synthesis. In most RNA viruses, RNA synthesis takes place with the help of an RNA-dependent RNA polymerase.

Retroviruses are RNA viruses that have a DNA polymerase called **reverse transcriptase,** which transcribes the RNA genome into a DNA intermediate (**FIG. 24-7**). This DNA becomes integrated into the host DNA by an enzyme also carried by the virus. Copies of the viral RNA are synthesized as the incorporated DNA is transcribed by host RNA polymerases. The **human immunodeficiency virus (HIV)** that causes AIDS is a retrovirus. Some retroviruses can cause cancer directly or by interacting with other cancer-causing viruses.

After viral genes are transcribed, the viral structural proteins are synthesized. The capsid is produced, and then assembly of new virus particles takes place. Finally, release occurs.

Viruses bind to specific receptor proteins on the animal cell's plasma membrane. Enveloped viruses then enter the cell by membrane fusion.

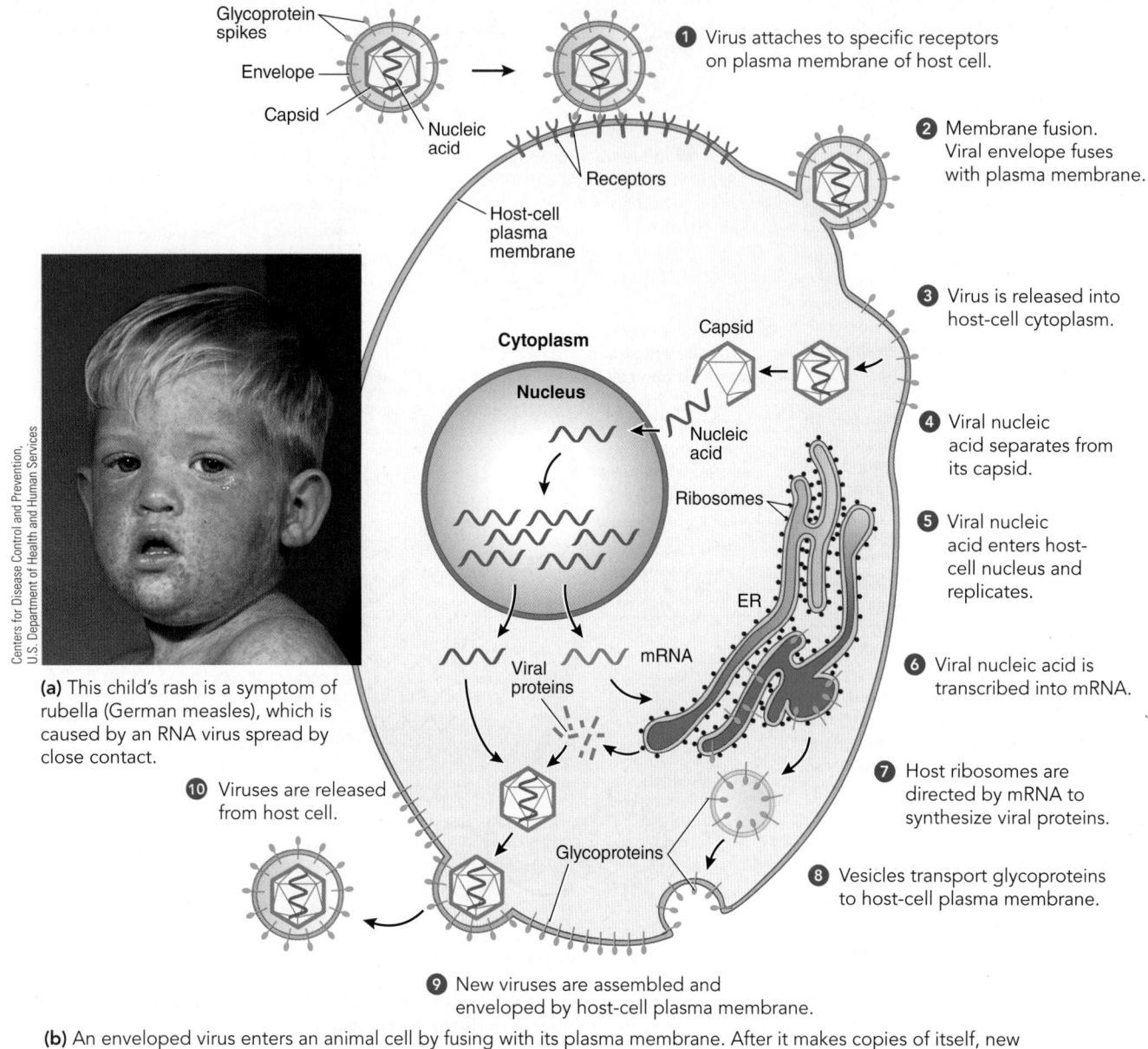

Glycoprotein spikes

Envelope

Capsid

Nucleic acid

Receptors

Host-cell plasma membrane

Cytoplasm

Nucleus

Capsid

Nucleic acid

Ribosomes

ER

mRNA

Viral proteins

Glycoproteins

① Virus attaches to specific receptors on plasma membrane of host cell.

② Membrane fusion. Viral envelope fuses with plasma membrane.

③ Virus is released into host-cell cytoplasm.

④ Viral nucleic acid separates from its capsid.

⑤ Viral nucleic acid enters host-cell nucleus and replicates.

⑥ Viral nucleic acid is transcribed into mRNA.

⑦ Host ribosomes are directed by mRNA to synthesize viral proteins.

⑧ Vesicles transport glycoproteins to host-cell plasma membrane.

⑨ New viruses are assembled and enveloped by host-cell plasma membrane.

⑩ Viruses are released from host cell.

(a) This child's rash is a symptom of rubella (German measles), which is caused by an RNA virus spread by close contact.

Centers for Disease Control and Prevention, U.S. Department of Health and Human Services

(b) An enveloped virus enters an animal cell by fusing with its plasma membrane. After it makes copies of itself, new viruses are released. A new envelope forms around each virus as it leaves the host cell.

Figure 24-6 *Animation* **Viral infection in animals**

CONNECT Why can't new viruses be completely assembled in the nucleus of the host?

© Cengage Learning

Viruses that do not have an outer envelope exit by cell lysis. The plasma membrane ruptures, releasing many new viral particles. Enveloped viruses obtain their lipoprotein envelopes by picking up a fragment of the host plasma membrane as they leave the infected cell (see Figs. 24-6b and 24-7).

Viral proteins damage the host cell in several ways. These proteins may alter the permeability of the plasma membrane or may inhibit synthesis of host nucleic acids or proteins. Viruses sometimes damage or kill their host cells by their sheer numbers. A poliovirus can produce 100,000 new viruses within a single host cell!

Viruses quickly become resistant to antiviral drugs
Antibiotics are specific for fighting bacteria; they do not kill viruses. Fortunately, researchers have developed antiviral drugs. These drugs inhibit the development, or replication, of many types of RNA and DNA viruses. Unfortunately, viruses rapidly develop resistance, making antiviral drug design and therapy a

Retroviruses use reverse transcriptase to transcribe their RNA into DNA; the viral DNA becomes part of the host DNA. When activated, viral DNA uses host enzymes to transcribe viral RNA.

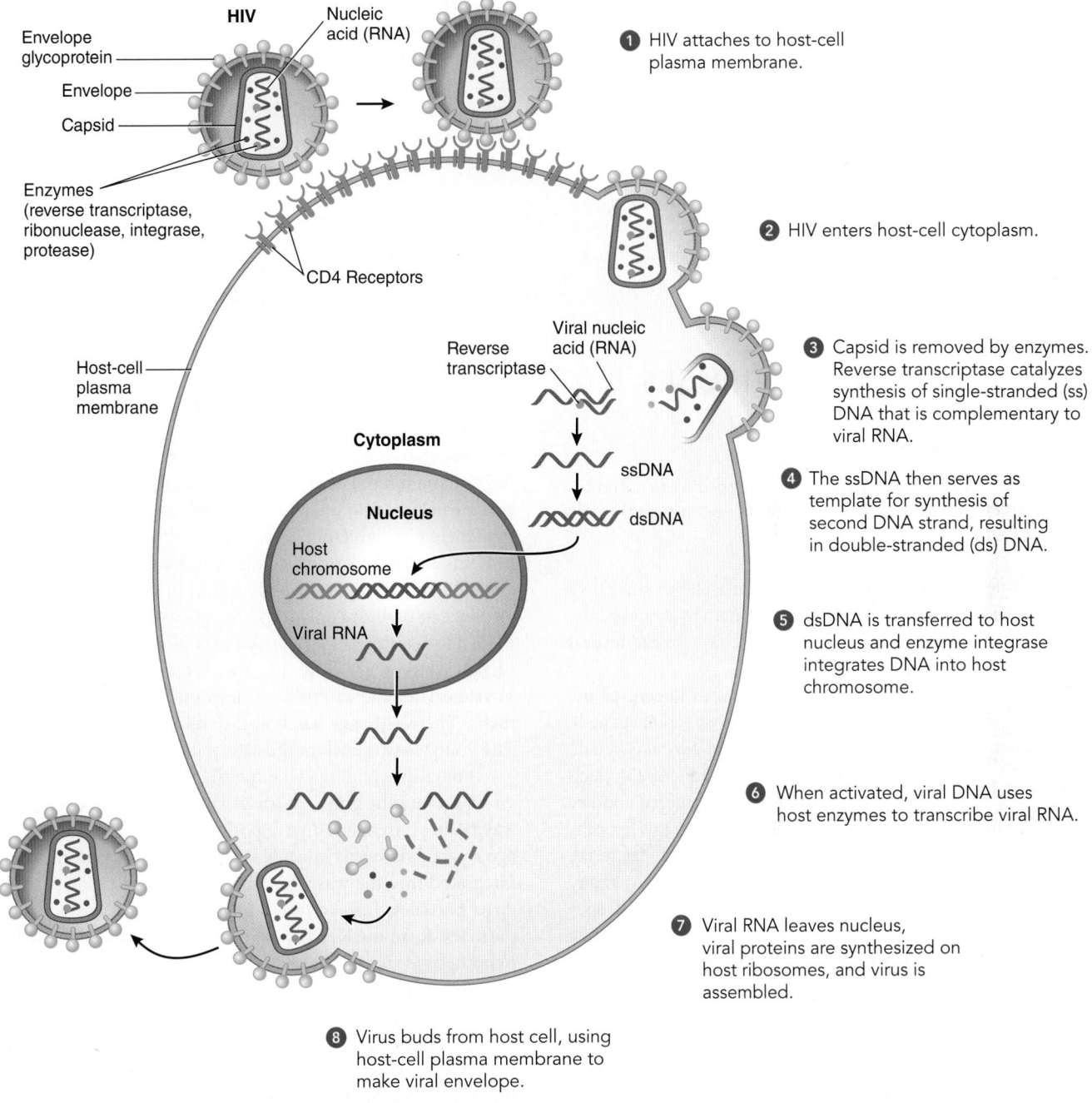

Figure 24-7 *Animation* **Life cycle of HIV, the retrovirus that causes AIDS**

HIV infects T helper cells, specialized cells of the host's immune system. The virus attaches to protein receptors, known as CD4, on the plasma membrane of T helper cells. HIV has two identical single-stranded RNA molecules.

CONNECT What advantage does the retrovirus that causes AIDS have compared to an RNA virus?

© Cengage Learning

constantly changing, competitive endeavor. For example, amantadine, which inhibits penetration or uncoating of viral nucleic acids, has been effective in treating patients with influenza. However, most strains of influenza virus have become resistant to amantadine, perhaps as a result of common use of the drug in poultry feed in certain areas of Asia, such as China.

Another type of antiviral drug (e.g., Tamiflu) inhibits a viral enzyme (neuraminidase) necessary for the virus to leave the host cell. Certain viral strains are becoming resistant to Tamiflu. A single mutation that occurs spontaneously is apparently responsible for this resistance. Many more antiviral drugs are being designed or are currently in clinical trials. Some inhibit viral attachment to host cells, and others interfere with replication of viral nucleic acid.

CHECKPOINT 24.4

- *How do viruses infect plant cells?*
- *What are the mechanisms by which viruses enter animal cells?*
- VISUALIZE *Draw a diagram illustrating how retroviruses, such as HIV, make copies of themselves.*

24.5 EVOLUTION OF VIRUSES

LEARNING OBJECTIVE

8 Trace the evolutionary origin of viruses according to current hypotheses and describe research regarding the evolution of polydnaviruses.

Scientists have been debating the origin of viruses ever since they were discovered. Currently, there are three main hypotheses: the progressive (or escape) hypothesis, the regressive (or reduction) hypothesis, and the virus-first hypothesis.

According to the *progressive hypothesis,* also known as the *escape hypothesis,* viruses may have originated as mobile genetic elements such as *transposons* (see Chapter 13) or *plasmids* (small, circular DNA fragments discussed in Chapters 15 and 25). Such fragments could have escaped from one cell and entered another cell through damaged cell membranes. According to this hypothesis, viruses may trace their origin to animal, plant, bacterial, or archaeal cells. Their multiple origins may explain why many viruses are species specific; perhaps they infect only those species that are closely related to the organisms from which they originated. This hypothesis is supported by the genetic similarity between some viruses and their host cells—a closer similarity than exists between one type of virus and another.

The *regressive hypothesis,* also called the *reduction hypothesis,* asserts that viruses are remnants of cellular organisms and evolved from small cells that were parasites in larger cells. Genes that they did not need, like those for protein synthesis, were gradually lost through evolution. This hypothesis is supported by certain bacteria (chlamydia and rickettsia) that are able to reproduce only inside the cells of their hosts (see Table 25-4).

The regressive hypothesis is also supported by the discovery that some giant viruses have genes that encode components for protein translation. Proponents of the regressive hypothesis suggest that these giant viruses evolved from a free-living, complex cellular ancestor. Later, they became parasitic and gradually lost some of the genes necessary for protein synthesis. This hypothesis explains how viruses could have existed before their present-day hosts evolved.

The *virus-first hypothesis* states that viruses predate or coevolved with their current cellular hosts even before the life-forms assigned to the three domains diverged. One research team recently suggested that viruses initially existed in a pre-cellular world as self-replicating units. Gradually, these small units became more organized and more complex. Eventually they evolved to produce enzymes for the synthesis of membranes and cell walls, leading to the origin of the first cells.

Evidence for this hypothesis comes from similarities found in the protein structures of some viral capsids and in genetic similarities between some viruses that infect archaea and some that infect bacteria. Molecular biologists studying this hypothesis consider it improbable that these similarities evolved independently. This evidence suggests that viruses diverged very early—before archaea, bacteria, or eukaryotes.

Recent studies on the origin of polydnaviruses contribute to our understanding of viral evolution. **Polydnaviruses** are particles that consist of multiple circles of dsDNA encased in capsid proteins and an envelope. Each circle of DNA contains part of the virus genome. These viruses are found in ovary cells of many species of parasitic wasps. The polydnaviruses are unusual in that their circular DNA does not have genes for making the proteins needed to replicate and produce new viruses. Genes needed for viral replication are found in the wasp genome; the viruses can replicate only in the wasp ovary cells.

The wasp injects polydnaviruses along with her eggs into certain caterpillars. The polydnaviruses express toxins that interfere with the caterpillar's immune defenses and development. The wasp eggs hatch and develop inside the caterpillar. The young wasps feed on the caterpillar.

Biologists wondered whether the polydnaviruses were really viruses. Were the polydnaviruses simply particles derived from wasp genes? An alternative hypothesis was that millions of years ago a particular virus infected wasps. The virus genome became integrated into the wasp genome and lost the capacity to enter virus particles. Instead, wasp DNA was incorporated into the particles. Read the Key Experiment, FIGURE 24-8, to learn which hypothesis is supported by the research of Jean-Michel Drezen, Annie Bézier, and their colleagues.

Today's viruses may have originated multiple times, through many mechanisms, or by a mechanism not yet discovered. No matter how viruses came to be and how they continue to evolve, they have important roles in evolution. They mutate constantly, producing many new alleles. They replicate rapidly, and some of their genes are incorporated into the genomes of host cells through horizontal gene transfer. In fact, viruses may provide a major source of new genes for their host cells.

CHECKPOINT 24.5

- CONNECT *How could viruses have evolved before their hosts? Explain your answer.*
- CONNECT *In what way are polydnaviruses mutualistic partners with certain wasps?*

24.6 SUBVIRAL AGENTS

What is the origin of polydnaviruses?

HYPOTHESIS: Polydnaviruses evolved from ancient viruses that infected wasps.

EXPERIMENT: Annie Bézier, Jean-Michel Drezen, and their colleagues searched for virus-related genes in the wasp genome. The team analyzed DNA from three different wasp species. They compared nucleotide sequences with sequences of public databases containing sequences of all organisms and viruses studied.

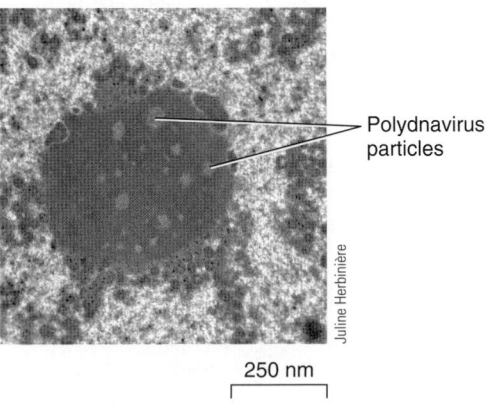

Polydnavirus particles

Juline Herbinière

250 nm

RESULTS AND CONCLUSION: The researchers identified 22 genes expressed in wasp ovaries where viral particles are produced. These genes are present in the wasp genome and resemble key genes of nudiviruses, a family of insect viruses. The investigators found that many of these genes code for proteins that are part of the polydnavirus particles. These genes have been conserved in several different wasp families that have polydnaviruses.

The research team concluded that the polydnavirus evolved from a nudivirus that infected wasps millions of years ago. In time, the virus genome became incorporated into the wasp genome. Proteins needed for viral replication are now part of the wasp's DNA, and the virus can replicate only in the wasp's ovaries.

At the same time, the wasps depend on the virus. The wasps inject their eggs into certain caterpillars along with polydnaviruses. The viruses depress the caterpillar's immune defenses, allowing the wasp larvae to develop in the caterpillar. Thus, a mutualistic relationship has evolved between the wasp and the polydnavirus.

The investigators suggested that the polydnavirus acts as a gene vector, transporting large chunks of DNA to the caterpillar. Might learning more about this mechanism have clinical applications?

SOURCE: Bézier, Annie, et al., "Polydnaviruses of Braconid Wasps Derive from an Ancestral Nudivirus." *Science*, Vol. 323, No. 5916, Feb. 13, 2009, pp. 926–930.

Figure 24-8 Exploring the evolution of polydnaviruses

This TEM shows polydnavirus particles in the nucleus of a cell in a wasp ovary. The polydnavirus particles are located at the periphery of an electron-dense region thought to be a "viral factory."

PREDICT What would happen if the polydnavirus mutated and no longer depressed the caterpillar's immune defenses?

LEARNING OBJECTIVE

9 Compare satellites, viroids, and prions.

Subviral agents are infective agents that are smaller and simpler than viruses. Subviral agents include satellites, viroids, and prions.

Satellites depend on helper viruses

Satellites are subviral agents that depend on co-infection of a host cell with a helper virus. The satellite, which consists of either DNA or RNA, cannot reproduce without the helper virus. The agent that causes hepatitis D is a satellite that can reproduce only when the hepatitis B virus is also present.

Sputnik is a satellite that infects Mimivirus, which in turn infects an amoeba. Sputnik is dependent on the replication and assembly machinery set up by Mimivirus. Some researchers now refer to Sputnik as a *virophage*, a word not yet completely defined by many virologists. The term implies that Sputnik, and perhaps other satellites, impairs the reproduction of Mimivirus, its helper virus.

Viroids are short, single strands of naked RNA

In 1971, Theodor Otto Diener, a plant pathologist, discovered an infective agent in potatoes that he named a **viroid.** Since then, about 30 species of viroids have been identified. Smaller than a virus, a viroid consists of a very short, circular, single strand of naked RNA (only 250 to 400 nucleotides). The viroid has no protective protein coat and no associated proteins to assist in duplication. Its RNA serves as a template that is copied by host RNA polymerases. Viroids are extremely hardy and are able to resist heat and ultraviolet radiation because of the condensed folding of their RNA.

Many viroids that cause plant diseases—including potato spindle tuber viroid, apple scar skin viroid, avocado sunblotch viroid, and tomato chlorotic dwarf viroid—have been identified. Viroids cause stunted or distorted growth and sometimes kill the plant. For example, cadang-cadang disease has killed more than 30 million coconut palm trees in the Philippines. Viroids are transmitted by infected pollen or seeds.

Viroids are generally found within the host-cell nucleus and appear to interfere with gene regulation. Molecular biologists have hypothesized that viroids mirror gene sequences in their hosts and silence critically important host genes. The viroid's ssRNA replicates, forming dsRNA. The plant's defense response cleaves this viroid RNA, producing small interfering RNAs (siRNAs), discussed in Chapters 13 and 14. However, the viroid siRNAs then cause host ribonucleases to selectively cleave host mRNAs with complementary base sequences. This action inactivates host mRNA and silences specific host genes. The viroids themselves are resistant to RNA silencing.

Prions are protein particles

Motivated by the death of a patient from Creutzfeldt–Jakob disease (CJD), a degenerative brain disease, Stanley Prusiner, professor of neurology and biochemistry at the University of California, San Francisco, began his studies of prions in the early 1970s. Prusiner discovered that the infective agent was not affected by radiation (which typically mutates nucleic acids), and he could not find DNA or RNA in the particles. In 1982, he named the infective agent **prion,** for "proteinaceous infectious particle."

Prusiner's hypothesis that an infectious agent could cause disease without nucleic acids was not readily accepted because there was no known mechanism by which this action could occur. However, Prusiner and other researchers continued to study prions and discovered how they can cause cells to malfunction. In 1997, Prusiner was awarded the Nobel Prize in Physiology or Medicine for his discovery of prions.

Prusiner and others have shown that animals have a gene that encodes a normally harmless protein known as PrP. Normal forms of the protein are found on the surfaces of brain cells and many other types of cells. The PrP protein, which consists of 208 amino acids, helps neurons transport copper and may play a role in the sense of smell in animals. Researchers have discovered that normal PrP interacts with the peptides associated with Alzheimer's disease.

Sometimes, the PrP protein folds into a different shape, an insoluble form that can cause disease. This misfolded protein is the prion. Mutations in the gene that encodes the PrP protein increase the risk that the protein will misfold and become a prion. The prion then somehow induces other PrP molecules to misfold into the pathogenic form (FIG. 24-9). Prions apparently can aggregate and accumulate in the brain and in certain other tissues and cause serious damage.

Prions are found in the brains of patients with *transmissible spongiform encephalopathies* (*TSEs*). This group of fatal degenerative brain diseases has been identified in birds and mammals. These diseases are called TSEs because when infected, the brain appears to develop holes and becomes somewhat spongelike. The TSEs are in some ways like viral illnesses. However, there is a long latency period (about 5 years in cows and up to 10 years in humans) between contracting the infection and developing symptoms of the illness. Also, the TSEs have proven very difficult to treat. Genetically engineered mice that lack the prion protein gene are immune to TSE infection.

The oldest known prion disease is *scrapie* in sheep and goats. When infected, animals lose coordination, become irritable, and itch so severely that they scrape off their wool or hair. Bovine spongiform encephalopathy (BSE) is a related prion disease popularly referred to as "mad cow disease" because some diseased cattle become aggressive. In the 1990s, BSE became epidemic in cattle in the United Kingdom.

More than 200 people have died from a human variety of BSE, providing evidence that the disease is transmissible from cow to human. The human disease is called vCJD because it is a variant of CJD, which is caused by the transformation of PrP proteins into prions. The infective agent of vCJD has

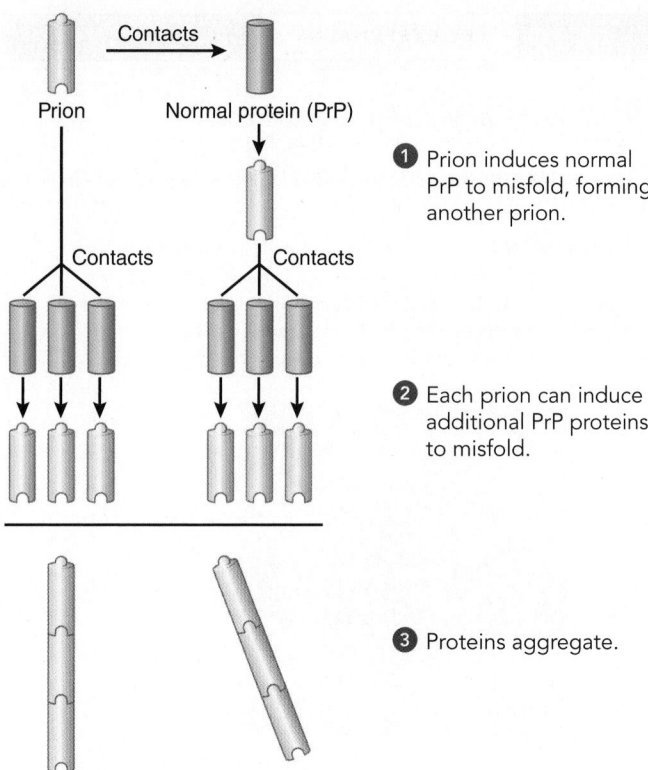

1 Prion induces normal PrP to misfold, forming another prion.

2 Each prion can induce additional PrP proteins to misfold.

3 Proteins aggregate.

Figure 24-9 A model for how prion populations expand
Prions contact normal PrP proteins and induce them to misfold and become prions. Each new prion can then contact additional PrP molecules and induce them to misfold, thus expanding the prion population. Prions can form clumps by aggregating.
© Cengage Learning

been recovered from infected human neural tissue and appears similar to the prion that causes scrapie in sheep. New regulations have been instituted in many countries to safeguard the food supply. Human-to-human transmission of vCJD has been associated with tissue and organ transplants and with transfusion with contaminated blood.

Chronic wasting disease, an illness related to mad cow disease, has spread among deer and elk populations in North America. Studies are under way to determine whether chronic wasting disease can infect livestock or humans. Feline spongiform encephalopathy is a prion disease of domestic and captive felines. With this disease, large deposits of a misfolded protein accumulate in the liver and spleen.

Like viruses, prions exhibit variability. Different strains of prions may differ from one another in the conformation of their proteins, in their properties, and in the diseases they cause. Prions are different from viruses in that they can arise spontaneously, mostly as a result of mutation. Researchers are studying ways to block prions from forming. They are also looking for ways to stimulate cells to destroy prions. Early diagnosis is critical, and research for an assay, or test, to detect prion disease is ongoing.

CHECKPOINT 24.6

- **CONNECT** *What do viroids and prions have in common? How are they different?*

24.1 The Status and Structure of Viruses *(page 496)*

1 Contrast a virus with a cellular organism.

- A **virus** is a very small infective agent that consists of a core of nucleic acid and is dependent on a living host. Viruses contain the nucleic acids necessary to make copies of themselves, but to reproduce they must invade living cells and commandeer their metabolic machinery.

2 Describe the structure of a virus.

- A virus is a subcellular particle consisting of a DNA or RNA genome surrounded by a protein coat, called a **capsid.** In some viruses the capsid is surrounded by an outer envelope.

24.2 Classification of Viruses *(page 497)*

3 Identify three characteristics used to classify viruses.

- Viruses are classified based on **host range,** the types of host species they can infect; the type of nucleic acid they contain; and whether the nucleic acid is single-stranded or double-stranded. Other factors used to classify viruses include size and shape, presence of an envelope, and method of transmission from host to host.

24.3 Viral Replication *(page 498)*

4 Characterize bacteriophages.

- **Bacteriophages,** or **phages,** are complex viruses that infect bacteria. They typically consist of a long molecule of dsDNA coiled within a polyhedral head.

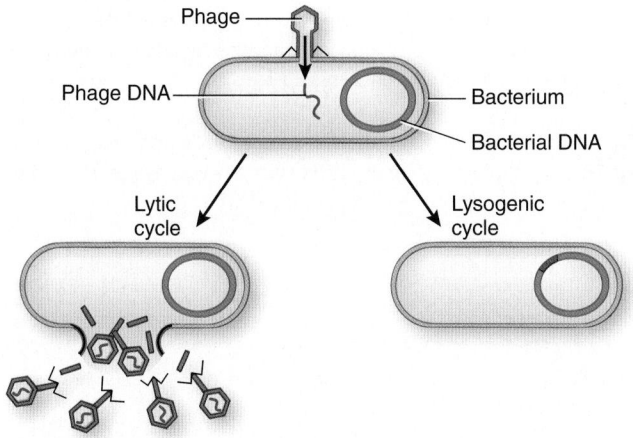

© Cengage Learning

5 Contrast a lytic cycle with a lysogenic cycle.

- In a **lytic cycle,** the virus destroys the host cell. The five steps in a lytic cycle are *attachment* to the host cell, *penetration* of viral nucleic acid into the host cell, *replication* of the viral nucleic acid, *assembly* of newly synthesized components into new viruses, and *release* from the host cell.
- The nucleic acid of some phages becomes integrated into the DNA of its bacterial host. The phage is then called a **prophage.** A **temperate virus** alternates between a lytic and a **lysogenic cycle.** In a lysogenic cycle, the genome of a temperate virus is replicated along with replication of the host's DNA without causing death of the host cell.

- Bacterial cells that carry prophages are lysogenic cells. In **lysogenic conversion** bacterial cells containing certain temperate viruses exhibit new properties.

24.4 Viral Diseases *(page 500)*

6 Contrast viral infection of plants and animals, and identify specific diseases caused by viruses.

- Most plant viruses are ssRNA viruses that do not have envelopes. Plant viruses penetrate plants through damaged cells and are also transmitted by infected seeds. Viruses can be spread among plants by insect vectors. Viruses spread through the plant via plasmodesmata.
- Viruses enter animal cells by membrane fusion or by endocytosis. Diseases caused by DNA viruses include herpes, respiratory infections, and gastrointestinal disorders. RNA viruses cause influenza, upper respiratory infections, AIDS, and some types of cancer.

7 Describe the reproductive cycle of a retrovirus, such as human immunodeficiency virus (HIV).

- **Retroviruses,** such as HIV, use **reverse transcriptase** to transcribe their RNA genome into a DNA intermediate that becomes integrated into the host DNA. Copies of the viral RNA are then synthesized.

24.5 Evolution of Viruses *(page 506)*

8 Trace the evolutionary origin of viruses according to current hypotheses and describe recent research regarding the evolution of polydnaviruses.

- According to the *progressive,* or *escape, hypothesis,* viruses may have originated as mobile genetic elements that could have escaped and moved from one cell and entered another through damaged cell membranes. Viruses may trace their origin to animal cells, plant cells, bacterial cells, or archaeal cells.
- The *regressive,* or *reduction, hypothesis* asserts that viruses are remnants of cellular organisms; they evolved from free-living, complex cellular ancestors. As they became parasitic, the viruses gradually lost some of the genes necessary for protein synthesis. Genes they no longer needed, like those for protein synthesis, were gradually lost through evolution.
- According to the *virus-first hypothesis,* viruses predate or coevolved with their current cellular hosts. Viruses may have initially existed as self-replicating units. As they evolved, viruses produced enzymes for the synthesis of membranes and cell walls, leading to the formation of the first cells.
- Evidence suggests that **polydnaviruses** evolved from viruses that infected wasps. They have become mutualistic partners with their wasp hosts.

24.6 Subviral Agents *(page 507)*

9 Compare satellites, viroids, and prions.

- Satellites, viroids, and prions are smaller than viruses. Satellites reproduce only with the help of a helper virus.
- A **viroid** consists of a short strand of RNA with no protein coat. Many viroids cause plant diseases. A **prion** consists only of protein. Prions cause fatal degenerative brain diseases, such as *transmissible spongiform encephalopathies (TSEs).*

Know and Comprehend

1. The genome of a virus consists of (a) DNA (b) RNA (c) prions (d) DNA and RNA (e) DNA or RNA

2. The capsid of a virus consists of (a) protein subunits (b) nucleic acid (c) helical lipids (d) a carbohydrate envelope (e) RNA and lipid

3. Viruses that kill host cells are (a) lysogenic (b) lytic (c) viroids (d) prophages (e) temperate

4. In lysogenic conversion (a) bacterial cells may exhibit new properties (b) the host cell dies (c) prions sometimes convert to viroids (d) reverse transcriptase transcribes DNA into RNA (e) lytic viruses become temperate

5. The types of host species a particular virus can infect are referred to as its (a) assembly (b) commensal cohort (c) host range (d) prophage factor (e) virulence factor

6. Plant viruses (a) typically cause rapid death of infected plants (b) are mainly retroviruses (c) can be killed with antibiotics (d) are transmitted throughout the plant body by insects (e) can be transmitted from one generation to the next through infected seeds

7. According to the progressive, or escape, hypothesis, viruses (a) appeared before the three domains diverged (b) came from mobile genetic elements and had multiple origins (c) evolved from early plant cells (d) are bits of nucleic acid that escaped from animal cells (e) evolved from cells that were parasites in larger cells

8. Prions (a) consist of RNA with no protein coat (b) are misfolded proteins (c) cause several important plant diseases (d) are the infective agents of several emerging diseases (e) consist of proteins that stimulate host RNA to produce DNA

Apply and Analyze

9. Arrange the following list into the correct sequence for viral reproduction:
   1. penetration 2. assembly 3. replication 4. attachment 5. release
   (a) 1, 2, 3, 4, 5 (b) 5, 2, 3, 4, 1 (c) 4, 1, 3, 2, 5 (d) 4, 1, 2, 3, 5 (e) 3, 1, 2, 4, 5

10. Arrange the following list into the correct sequence for part of the cycle of a retrovirus:
    1. dsDNA integrated into host DNA 2. viral proteins synthesized on host ribosomes 3. viral DNA uses host enzymes to transcribe viral RNA 4. reverse transcriptase catalyzes synthesis of ssDNA 5. synthesis of second DNA strand
    (a) 5, 2, 1, 3, 4 (b) 5, 2, 3, 4, 1 (c) 4, 5, 1, 3, 2 (d) 4, 1, 2, 3, 5 (e) 2, 1, 3, 4, 5

11. **VISUALIZE** What does this diagram illustrate? Complete the labels.

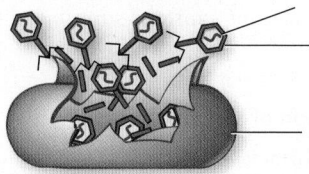

© Cengage Learning

12. **CONNECT** How are viral genomes different from those of plant or animal cells?

13. **CONNECT** How do viroids and prions differ from viruses?

Evaluate and Synthesize

14. **EVOLUTION LINK** Discuss the coevolution of polydnaviruses and their mutualistic partners. (See the discussion and Key Experiment on polydnaviruses.)

15. **EVOLUTION LINK** Based on what you have learned about viruses, present an argument for one of the three hypotheses of virus origin.

16. **SCIENCE, TECHNOLOGY, AND SOCIETY** What is the continuous challenge that the influenza virus presents to scientists and public health officials? How can they use technology and work together to meet this challenge?

aplia To access course materials, such as Aplia and other companion resources, please visit **www.cengagebrain.com**.

Bacteria and Archaea

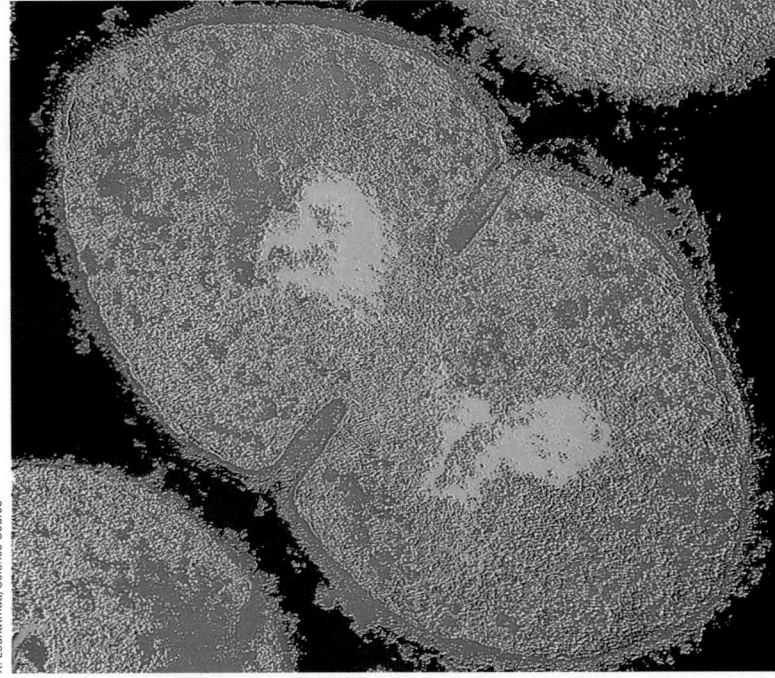

K. Lounatmaa/Science Source

In this chapter we examine the diversity and characteristics of the members of Domain Bacteria and Domain Archaea, two of the three main branches of the tree of life. These organisms, informally called prokaryotes, have inhabited our planet for more than 3.5 billion years, much longer than eukaryotes, which evolved at least 2.2 billion years ago. Although prokaryotes are microscopic, they are so numerous that they probably account for more than half of Earth's biomass, the mass of living material. Thus, the biomass of prokaryotes is greater than that of all the eukaryotes: fungi, plants, and animals. As scientists discover new prokaryotes and explore new niches and habitats, the percentage of bacteria and archaea in the Earth's total biomass continues to increase.

Anton van Leeuwenhoek, a Dutch microscopist, discovered bacteria and other microorganisms in 1674 when he looked at a drop of lake water through a glass lens. During the late 1800s, many microorganisms, including some bacteria, fungi, and protozoa, were identified as **pathogens,** agents that cause disease (see photograph). Bacteria cause many diseases, such as tuberculosis, tetanus, respiratory infections, and food poisoning in humans. However, only a small minority of bacterial species are pathogens. (No pathogenic archaea have been identified.)

Both bacteria and archaea play essential roles in the biosphere. As decomposers they break down organic molecules into their components. Along with fungi, prokaryotes are nature's chief recyclers. Without these microorganisms, elements such as carbon, nitrogen, phosphorus, and sulfur would remain locked up in the wastes and dead bodies of plants, animals, and other organisms and would be unavailable for the synthesis of new cells and organisms.

Some prokaryotes are producers that carry on photosynthesis. Others convert atmospheric nitrogen to ammonia and then to nitrates, forms that are used by plants (see Fig. 55-8). This conversion enables plants and animals (because they eat plants) to manufacture essential nitrogen-containing compounds such as proteins and nucleic acids.

We begin this chapter with a description of the structure of bacteria and archaea, and then we discuss prokaryote reproduction and evolution. We outline some of the nutritional and metabolic

Color-enhanced TEM of a bacterium (*Streptococcus pyogenes*) dividing by binary fission. This bacterium, a pathogen that inhabits the human nose and throat, can cause scarlet fever and inflammation of the heart tissue. The strain shown here is resistant to antibiotics, and infection can be fatal.

KEY CONCEPTS

25.1 Bacteria and archaea are single-celled organisms that, in contrast to eukaryotic cells, do not have membrane-enclosed organelles; most have a cell wall that surrounds the plasma membrane, and most have a single, circular DNA molecule.

25.2 Evolution occurs rapidly in prokaryotes; natural selection acts on the genetic variation provided by mutations and genetic recombination and is facilitated by rapid reproduction.

25.3 The great diversity that has evolved in their metabolism and mode of nutrition allows prokaryotes to thrive in all kinds of habitats.

25.4 Prokaryotes make up two of the three domains: Bacteria and Archaea.

25.5 Prokaryotes play critical roles in ecology, commerce, and technology.

25.6 The many adaptations that have evolved in pathogenic bacteria contribute to their great success.

adaptations of these organisms and describe the phylogeny of the two prokaryote domains. We then present an overview of the effect of prokaryotes on ecology, technology, and commerce, and end the chapter with a discussion of bacteria and disease.

25.1 THE STRUCTURE OF BACTERIA AND ARCHAEA

LEARNING OBJECTIVES

1 Describe the structure and common shapes of prokaryotic cells.
2 Compare the bacterial cell wall in gram-positive and gram-negative bacteria.
3 Describe movement in prokaryotes and describe the structure of the bacterial flagellum.

In contrast to viruses, viroids, and prions, which consist only of nucleic acid and/or protein, **prokaryotes,** assigned to domain Archaea or Bacteria, are cellular organisms. However, recall from Chapters 4 and 23 that the cell structure of prokaryotes is fundamentally different from that of the eukaryotic cells of other living organisms.

Most prokaryotic cells are very small. Typically, their diameter ranges from 0.5 to 1.0 μm, and their length ranges from 1.0 to 5.0 μm. Their cell volume is only about one-thousandth that of small eukaryotic cells, and their length is only about one-tenth. Most prokaryotes are unicellular, but some form colonies or filaments containing specialized cells.

Prokaryotes have several common shapes

Two basic prokaryote shapes are spherical and rod-shaped. Spherical prokaryotes, known as **cocci** (sing., *coccus*), occur singly in some species and in groups of independent cells in others (FIG. 25-1a). Cells may be grouped in twos (*diplococci*), in long chains (*streptococci*), or in irregular clumps that look like bunches of grapes (*staphylococci*). Rod-shaped prokaryotes, called **bacilli** (sing., *bacillus*), may occur as single rods or as long chains of rods (FIG. 25-1b). Some prokaryotes form spirals. If the spiral-shaped cell is flexible, it is a **spirochete;** if rigid, it is a **spirillum** (pl., *spirilla*) (FIG. 25-1c). A spirillum shaped like a comma is called a **vibrio.** Some archaea have unusual shapes, such as triangular or square-shaped cells.

Prokaryotic cells do not have membrane-enclosed organelles

In contrast to eukaryotic cells, prokaryotic cells do not have a nucleus or other membrane-enclosed organelles (FIG. 25-2; see Chapter 4). Although the prokaryotic cell does not have a membrane-enclosed nucleus, it does have a **nuclear area,** also referred to as the **nucleoid,** which contains DNA.

The dense cytoplasm of the prokaryotic cell contains ribosomes (smaller than those found in eukaryotic cells) and storage granules that hold glycogen, lipid, and phosphate compounds. Enzymes needed for metabolic activities may be located in the cytoplasm. Although the membranous organelles of eukaryotic cells are absent, in some prokaryotic cells the plasma membrane is extensively folded inward. Enzymes needed for cellular respiration and photosynthesis may be associated with the plasma membrane or its folds.

A cell wall protects most prokaryotes

Most prokaryotic cells have a **cell wall** surrounding the plasma membrane. The cell wall provides a rigid framework that supports the cell and maintains its shape. Bacterial cells have a high concentration of dissolved solutes. The cell wall keeps the cell from bursting under hypotonic conditions (see Chapter 5). Thus, most bacteria are adapted to hypotonic surroundings. When wall-less forms of bacteria are produced experimentally, they must be

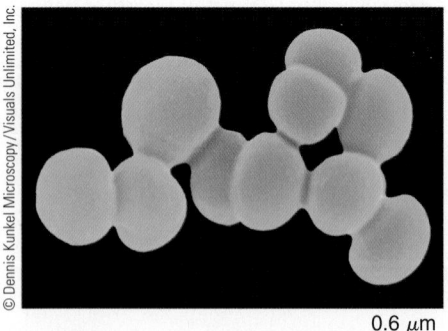

(a) **Cocci bacteria.** Colorized SEM of *Staphylococcus aureus.* These cocci cause skin and wound infections, food poisoning, and toxic shock syndrome.

0.6 μm

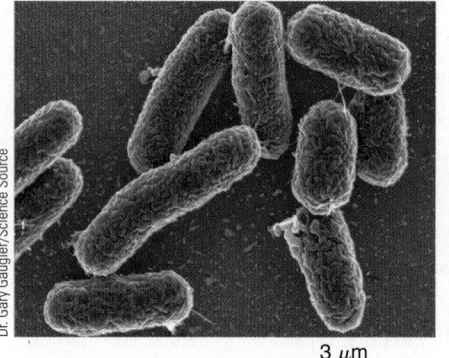

(b) **Bacilli bacteria.** Colorized SEM of *Salmonella.* These bacilli cause food poisoning.

3 μm

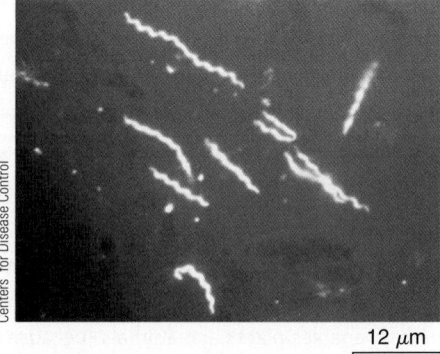

(c) **Spirochete bacteria.** Colorized SEM of *Borrelia burgdorferi.* These spirochetes cause Lyme disease, transmitted by infected deer ticks.

12 μm

Figure 25-1 Common shapes of prokaryotes

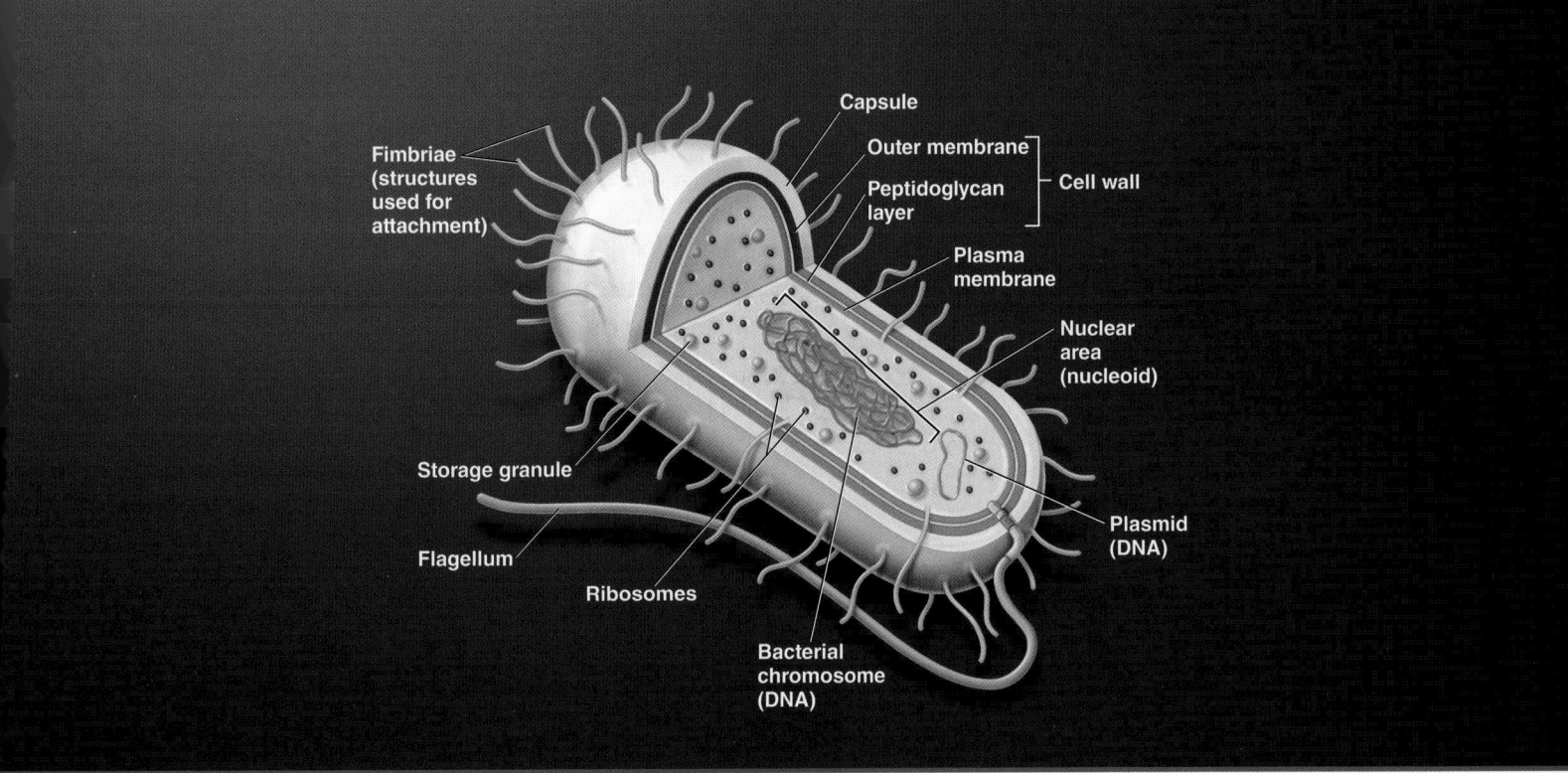

Figure 25-2 *Animation* **Structure of a prokaryotic cell**

This bacillus is a gram-negative bacterium (discussed in text). Prokaryotic cells typically have a nuclear area with a single, circular DNA molecule. They may also have one or more plasmids, small rings of DNA that replicate independently.

PREDICT Why might the absence of a nucleus be an advantage for prokaryotes?

© Cengage Learning

maintained in isotonic solutions to keep them from bursting. However, cell walls are of little help when a bacterium is in a hypertonic environment, as in food preserved by a high sugar or salt content. For this reason, most bacteria grow poorly in jellies, jams, salted fish, and other foods preserved in these ways.

The bacterial cell wall includes **peptidoglycan,** a complex polymer that consists of two unusual types of sugars (amino sugars) linked with short polypeptides. The sugars and polypeptides are cross-linked to form a crystalline lattice that surrounds the entire plasma membrane. Peptidoglycan is absent in the archaeon cell wall.

Differences in bacterial cell wall composition are of great interest to microbiologists and are important clinically. In 1888, Danish physician Christian Gram developed the Gram staining procedure. Bacteria that absorb and retain crystal violet stain in the laboratory are referred to as **gram-positive bacteria.** Bacteria that do not retain the stain when rinsed with alcohol are **gram-negative bacteria.** The cell walls of gram-positive bacteria are very thick and consist primarily of peptidoglycan. The cell walls of gram-negative bacteria have two layers: a thin

peptidoglycan layer and a thick outer membrane. The outer membrane resembles the plasma membrane but contains polysaccharides bonded to lipids (**FIG. 25-3**).

Distinguishing between gram-positive and gram-negative bacteria is important in treating certain diseases. For example, the antibiotic penicillin interferes with peptidoglycan synthesis, ultimately resulting in a fragile cell wall that cannot protect the cell (see Chapter 7 discussion of drugs that are enzyme inhibitors). Predictably, penicillin works most effectively against gram-positive bacteria.

Some bacteria produce capsules or slime layers

Many prokaryote species produce a **capsule** or **slime layer** that surrounds the cell wall. A slime layer is more loosely attached to the cell wall than a capsule. These outer layers are made of polysaccharide or protein.

In free-living species, the outer covering may provide the cell with added protection against phagocytosis (engulfment;

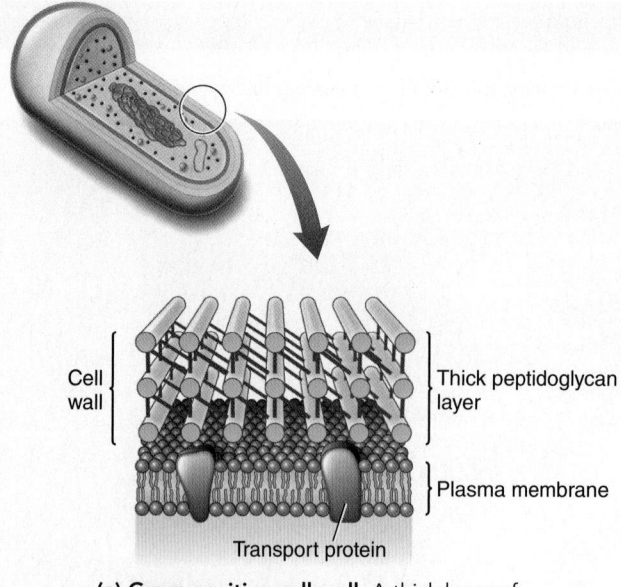

(a) **Gram-positive cell wall.** A thick layer of peptidoglycan molecules is held together by amino acids.

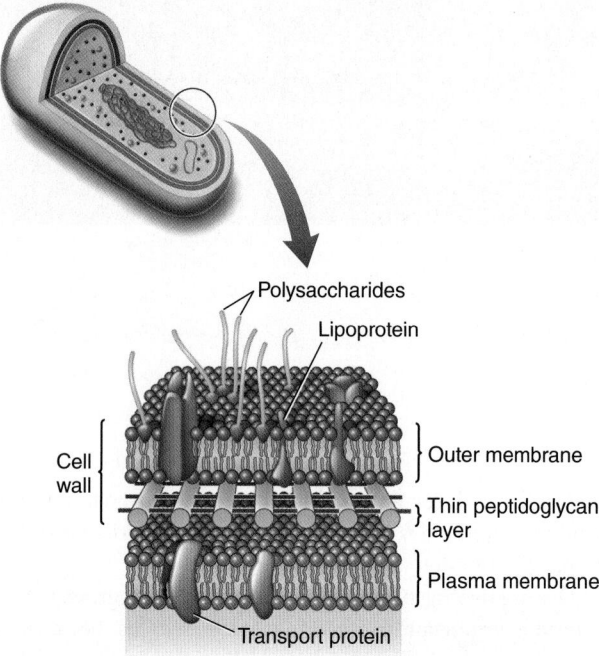

(b) **Gram-negative cell wall.** A thin layer of peptidoglycan is covered by an outer membrane.

Figure 25-3 *Animation* **Bacterial cell walls**

If present, the capsule (shown only in the locater cells) is outside the cell wall.
© Cengage Learning

see Chapter 5) by other microorganisms. In disease-causing bacteria, a capsule or slime layer may protect against phagocytosis by the host's white blood cells. The ability of *Streptococcus pneumoniae* to cause bacterial pneumonia depends on its capsule. A strain of *S. pneumoniae* that lacks a capsule does not cause the disease. Bacteria also use their capsules to attach to surfaces such as rocks, plant roots, and human teeth (where they cause dental plaque).

Some prokaryotes have fimbriae or pili

Some prokaryotes have hundreds of hair-like appendages called **fimbriae** (sing., *fimbria*). Fimbriae, which are made of protein, are shorter than flagella. **Pili** (sing., *pilus*) are usually longer than fimbriae. There are typically fewer pili on the cell surface than fimbriae. Prokaryotes use fimbriae and pili to attach to cell surfaces and, in the case of pathogenic bacteria, to the surfaces of cells they infect. These protein structures also help bacteria adhere to one another. Some elongated pili, called *sex pili*, are important in transmitting DNA between bacteria.

Some bacteria survive unfavorable conditions by forming endospores

When the environment becomes unfavorable—for example, when nutrients are limited or the environment becomes very dry or hot—some types of bacteria become dormant. The cell loses water, shrinks slightly, and remains inactive until water is again available. Certain types of bacteria form dormant, extremely durable cells called **endospores.**

After an endospore forms, the cell wall of the original cell lyses, releasing the endospore. Formation of endospores is not a type of reproduction in bacteria; endospores are not comparable to the reproductive spores of fungi and plants. Only one endospore is formed per original cell, so the total number of individuals does not increase. Archaea do not form endospores; these prokaryotes produce unique enzymes on the cell surface that protect them from cold, heat, and desiccation.

Endospores survive in very dry, hot, or frozen environments or at times when food is scarce. Some endospores are so resistant that they can survive an hour or more of boiling or centuries of freezing. When environmental conditions are again suitable for growth, the endospore germinates, forming an active, growing bacterial cell.

Several types of bacteria that form endospores cause disease (FIG. 25-4). The endospore of *Bacillus anthracis,* the bacterium that causes anthrax, is so hardy that this pathogen has become a concern as an agent of biological warfare. The bacteria (*Clostridium tetani*) that cause tetanus and the bacteria (*C. perfringens*) that cause gas gangrene typically enter the body with soil when a person suffers a deep cut or puncture wound. Patients may also be exposed to these serious diseases when surgical instruments are not effectively sterilized, allowing endospores to survive.

Many types of prokaryotes are motile

Can you imagine trying to swim through molasses? Water has the same relative viscosity to prokaryotes that molasses has to humans. Most motile prokaryotes manage to move by means of rotating **flagella.** The number and location of flagella are important in classifying some bacterial species.

Unlike eukaryotic flagella, prokaryotic flagella do not consist of microtubules (see Chapter 4). A bacterial flagellum is a long, thin appendage consisting of three parts: a basal body, a hook, and a single filament (FIG. 25-5). The basal body is a complex

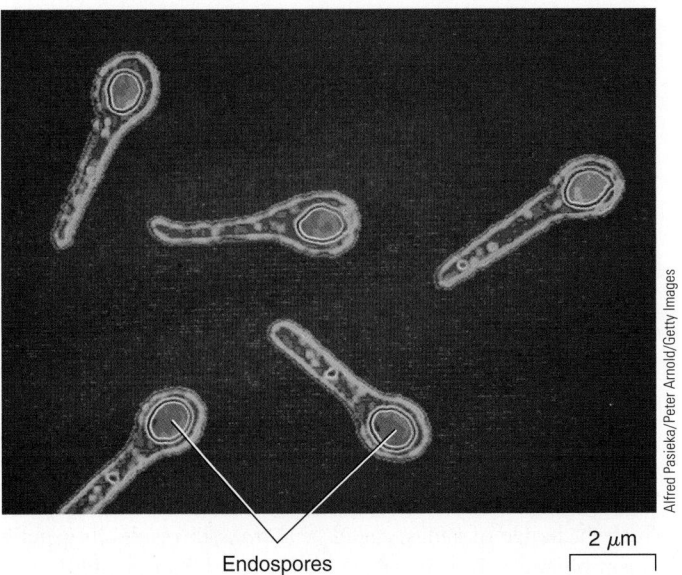

Endospores

2 μm

Figure 25-4 Endospores

Color-enhanced TEM of *Clostridium tetani,* the bacterium that causes tetanus. Each bacterial cell (*blue*) contains one endospore (*orange*), a resistant, dehydrated cell that develops within the original cell.

structure that anchors the flagellum into the cell wall by disc-shaped plates. The curved hook connects the basal body to the long, hollow filament that extends into the outside environment.

The basal body is a motor, somewhat like the propeller of a boat motor. The bacterium uses energy from ATP to pump protons out of the cell. Diffusion of these protons back into the cell powers the motor that spins the flagellum. The rotary motion produced pushes the cell much like a propeller pushes a boat through the water. The flagellum rotates counterclockwise, pushing the cell forward.

Archaea also have flagella. These flagella are thinner than those of bacteria and are more similar to a type of bacterial pili than to bacterial flagella. Biologists have determined that the structure of the motor and the assembly of the flagellum in archaea evolved differently from those of bacteria.

In addition to swimming, prokaryotic flagella perform other functions. Some prokaryotes use their flagella to swarm, to adhere to surfaces, and to participate in biofilm formation. Researchers have documented that bacteria use flagella to sense wetness. A recent study reported that some archaea can interact with one another by flagella, resulting in the formation of a biofilm composed of more than one species. Prokaryotes that lack flagella may move by gliding or twitching.

Scientists have recently discovered two extracellular structures that are unique to archaea: *cannulae* and *hami.* Cannulae are hollow glycoprotein tubes that extend from cells forming a network. Cannulae are extremely resistant to heat. They are produced by *Pyrodictium,* a genus of archaea that live in marine hydrothermal habitats, Hami are complex archaeal cell appendages that have been discovered entwined with bacterial filaments in sulfurous springs. Hami are helical, have barbs along the length of the filament, and have a three-part hook at the distal end, which facilitates attachment to surfaces. Cannulae and hami are adaptations to the extreme environments that many archaea inhabit.

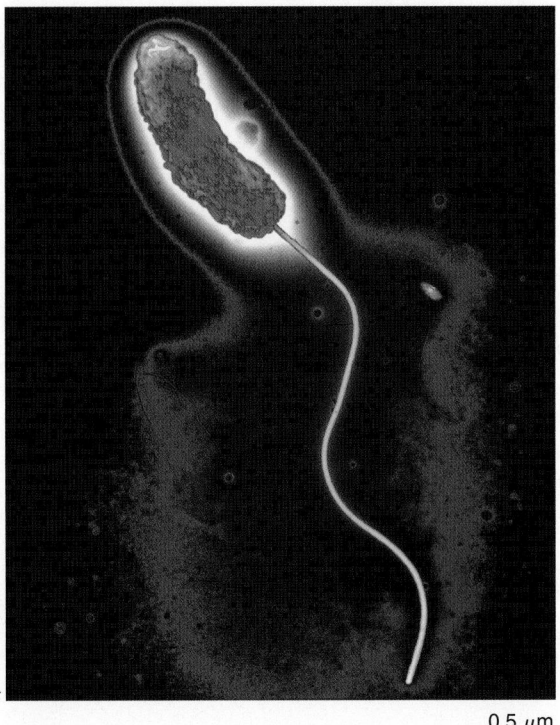

0.5 μm

(a) Color-enhanced TEM of *Vibrio cholerae.* This flagellate bacterium causes cholera in humans.

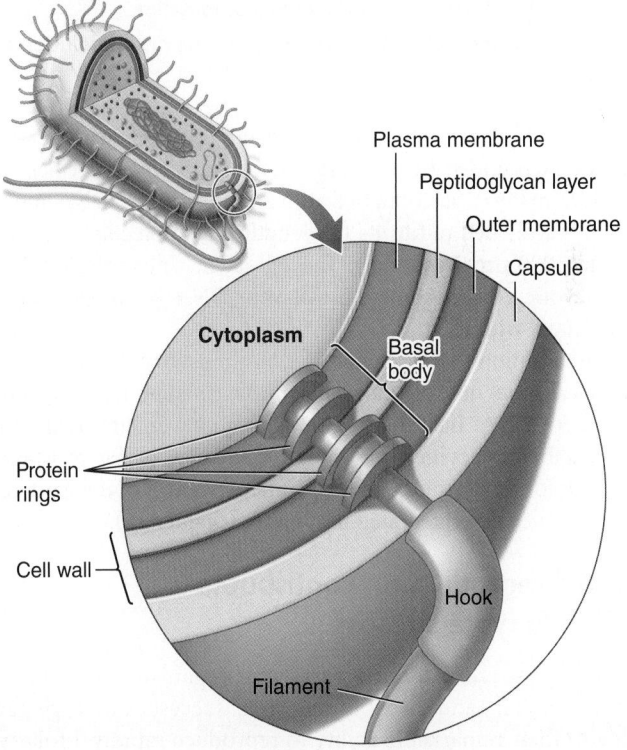

(b) Structure of a bacterial flagellum. The basal body is the motor. It consists of a series of disc-shaped plates that anchor the flagellum to the cell wall and plasma membrane. These plates spin the hook and filament of the flagellum.

Figure 25-5 Bacterial flagella

Flagellated prokaryotes do not move aimlessly. Many prokaryotes exhibit *chemotaxis,* movement in response to chemicals in the environment. For example, prokaryotes move toward food and may also move toward one another using diverse signaling mechanisms. Some bacteria and archaea move away from certain harmful chemicals. In addition, many prokaryotes exhibit *phototaxis,* movement in response to light in the environment.

CHECKPOINT 25.1

- CONNECT *How do prokaryotic cells differ from eukaryotic cells? How do bacteria and archaea differ from each other?*

- CONNECT *How do gram-positive and gram-negative bacteria differ? Why is this difference important to humans?*

- CONNECT *How does a bacterial flagellum differ from a eukaryotic flagellum?*

25.2 PROKARYOTE REPRODUCTION AND EVOLUTION

LEARNING OBJECTIVES

4 Describe asexual reproduction in prokaryotes and summarize three mechanisms (transformation, transduction, and conjugation) that may lead to genetic recombination.

5 State specific factors that contribute to the rapid evolution of bacteria and archaea.

The genetic material of a prokaryote lies in the nuclear area but is not surrounded by a nuclear envelope. In most species the genetic material is contained in a single, circular DNA molecule. If stretched out to its full length, this molecule would be about 1000 times as long as the cell itself. Unlike the DNA in eukaryotic chromosomes, prokaryote DNA has little protein associated with it.

In addition to their genomic DNA, most bacteria and many archaea have one or more **plasmids,** smaller circular fragments of DNA. Bacterial plasmids often have genes that code for catabolic enzymes, for genetic exchange, or for resistance to antibiotics. Plasmids replicate independently of the genomic DNA or become integrated into it (see Chapter 15).

Rapid reproduction contributes to prokaryote success

Prokaryotes are extremely successful organisms in terms of their numbers and distribution. Their success is in large part due to their remarkable ability to reproduce rapidly. Prokaryotes reproduce asexually, generally by **binary fission,** a process in which one cell divides into two similar cells, as in the chapter-opening photograph (also see Fig. 10-12). First the circular DNA replicates, and then an ingrowth of both the plasma membrane and the cell wall forms a transverse wall.

Binary fission occurs with remarkable speed. Under ideal conditions, some bacterial species divide in less than 20 minutes.

At this rate, if nothing interfered, one bacterium could give rise to more than one billion bacteria within 10 hours! Prokaryotes cannot reproduce at this rate for very long, however, before lack of food or the accumulation of waste products slows their population expansion.

Bacteria and archaea also commonly reproduce by **budding.** In budding a cell develops a bulge, or bud, that enlarges, matures, and eventually separates from the mother cell. Some prokaryotes divide by **fragmentation.** In this process, walls develop within the cell, which then separates into several new cells.

Prokaryotes transfer genetic information

Although sexual reproduction involving the fusion of gametes does not occur in prokaryotes, genetic material can be exchanged among bacteria and among archaea and also between domains. Such exchange of genes, called *gene transfer,* results in genetic recombination. Transfer of genetic material from parent to offspring is called *vertical gene transfer.* **Horizontal gene transfer** occurs when an organism (or virus) transfers genetic material to another organism that is not its offspring. Gene transfer among prokaryotes takes place by three different mechanisms: transformation, transduction, and conjugation.

1. When prokaryotes die, they release DNA that can be taken in by certain other prokaryotes. In **transformation** a prokaryotic cell takes up fragments of foreign DNA (or RNA) released by another prokaryotic cell. The foreign DNA must bind to DNA-binding proteins on the surface of the host cell. Once it enters the host cell, segments of the foreign DNA may be exchanged with homologous segments of the host DNA (reciprocal recombination) (FIG. 25-6). Recall

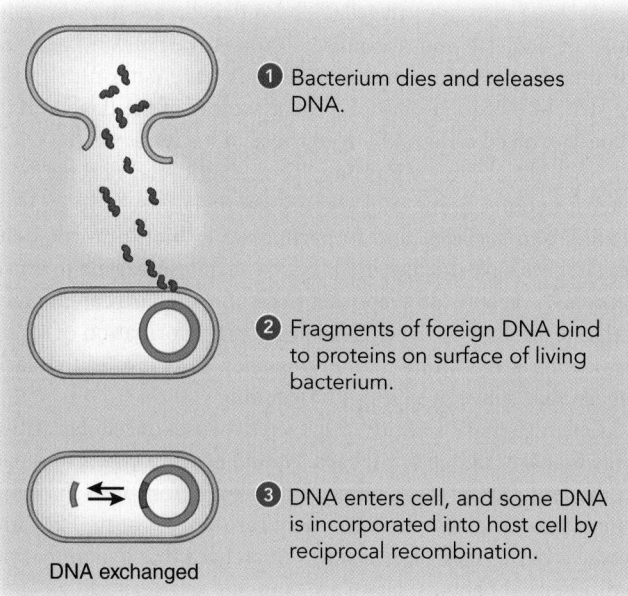

1 Bacterium dies and releases DNA.

2 Fragments of foreign DNA bind to proteins on surface of living bacterium.

3 DNA enters cell, and some DNA is incorporated into host cell by reciprocal recombination.

DNA exchanged

Figure 25-6 Transformation

In transformation a prokaryote takes in foreign DNA from its environment. The host cell exchanges some of its own DNA with homologous segments of the foreign DNA, resulting in a recombinant cell.

© Cengage Learning

from Chapter 12 that Oswald T. Avery and his colleagues identified DNA as the agent that transformed bacterial cells experimentally and showed that DNA is the chemical basis of heredity.

Foreign DNA can also be taken up as plasmids. When that occurs, DNA does not undergo recombination; instead, it remains as a plasmid separate from the prokaryotic chromosome.

2. In a different process of horizontal gene transfer, **transduction,** a phage carries bacterial or archaeal genes from one bacterial or archaeal cell into another (**FIG. 25-7**). Normally, a phage contains only its own DNA. However, sometimes a phage incorporates some of the DNA of its host. Then, when the phage infects another bacterium or archaeon, it transfers that DNA to its new host. The chromosome of this new host then becomes a recombination of its own original DNA and DNA from another bacterium or archaeon.

3. In **conjugation** two cells of different mating types come together, and genetic material is transferred from one to the other (**FIG. 25-8**). In contrast to transformation and transduction, conjugation involves contact between two cells.

Conjugation has been most extensively studied in the bacterium *Escherichia coli*. In the *E. coli* population, there are *donor cells,* or F^+ *cells,* which have DNA that can be transmitted to *recipient cells,* or F^- *cells.* F^+ cells have a DNA sequence known as the *F factor* (F stands for fertility) that is necessary for a bacterium to serve as a donor during conjugation. The F factor can be in the form of a plasmid, or it can be part of the DNA in the bacterial chromosome.

F genes encode enzymes essential for transferring DNA. Certain F genes encode **sex pili,** long, hair-like extensions that project from the cell surface. The sex pilus of an F^+ cell recognizes and binds to the surface of an F^- cell, forming a cytoplasmic conjugation bridge between the two cells. The F plasmid replicates itself, and DNA is transferred from donor to recipient bacterium through the conjugation bridge. F plasmids may also have other types of genes, including those that determine resistance to antibiotics.

Evolution proceeds rapidly in bacterial populations

Because prokaryotes reproduce rapidly by binary fission, mutations are quickly passed on to new generations. Mutations that confer some advantage spread through the population, and the effects of natural selection are quickly evident.

Horizontal gene transfer greatly contributes to the rapid evolution that takes place in prokaryotes. Acquisition of new DNA and genetic recombination are important sources of the genetic variation required for diversification and adaptation. New DNA introduced into a prokaryote's genome represents raw material for evolution. New genes are subject to mutation and are acted on by the forces of natural selection. The changes in the genetic material are passed to succeeding generations by binary fission. Changes that result in adaptation

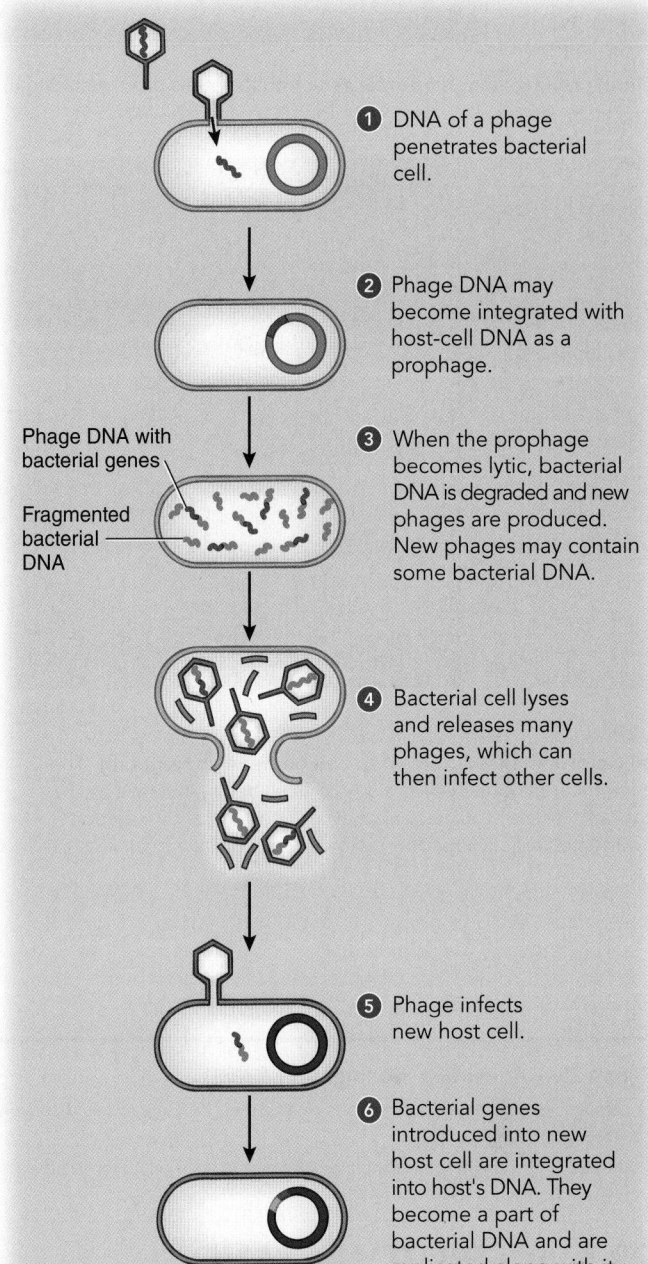

1 DNA of a phage penetrates bacterial cell.

2 Phage DNA may become integrated with host-cell DNA as a prophage.

Phage DNA with bacterial genes

Fragmented bacterial DNA

3 When the prophage becomes lytic, bacterial DNA is degraded and new phages are produced. New phages may contain some bacterial DNA.

4 Bacterial cell lyses and releases many phages, which can then infect other cells.

5 Phage infects new host cell.

6 Bacterial genes introduced into new host cell are integrated into host's DNA. They become a part of bacterial DNA and are replicated along with it.

Figure 25-7 Transduction

In transduction a phage transfers bacterial DNA from one bacterium to another, resulting in genetic recombination. Transduction is an important means of horizontal gene transfer.
© Cengage Learning

can quickly spread through future bacterial and archaeal populations.

CHECKPOINT 25.2

- *How do prokaryotes reproduce? What mechanisms result in gene transfer?*
- **CONNECT** *How does transduction contribute to the rapid evolution of bacterial populations?*
- **VISUALIZE** *Draw the steps that take place during conjugation.*

During conjugation, horizontal gene transfer takes place resulting in genetic recombination.

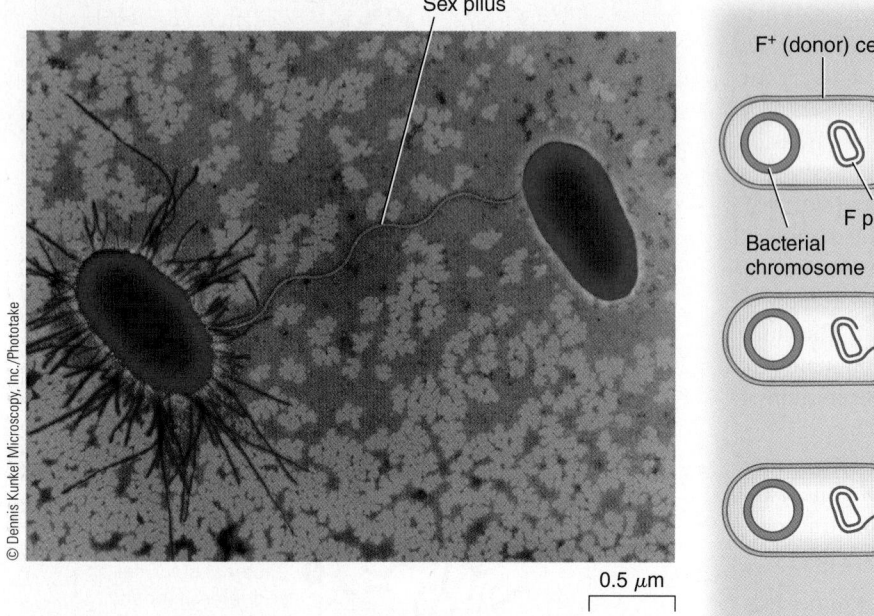

Sex pilus

0.5 μm

(a) **Color-enhanced SEM of *E. coli* bacteria conjugating.** The bacteria are connected by a sex pilus. When stimulated by the contact, the cells pull close together and form a conjugation bridge between donor and recipient cells (*shown in* **b**).

© Dennis Kunkel Microscopy, Inc./Phototake

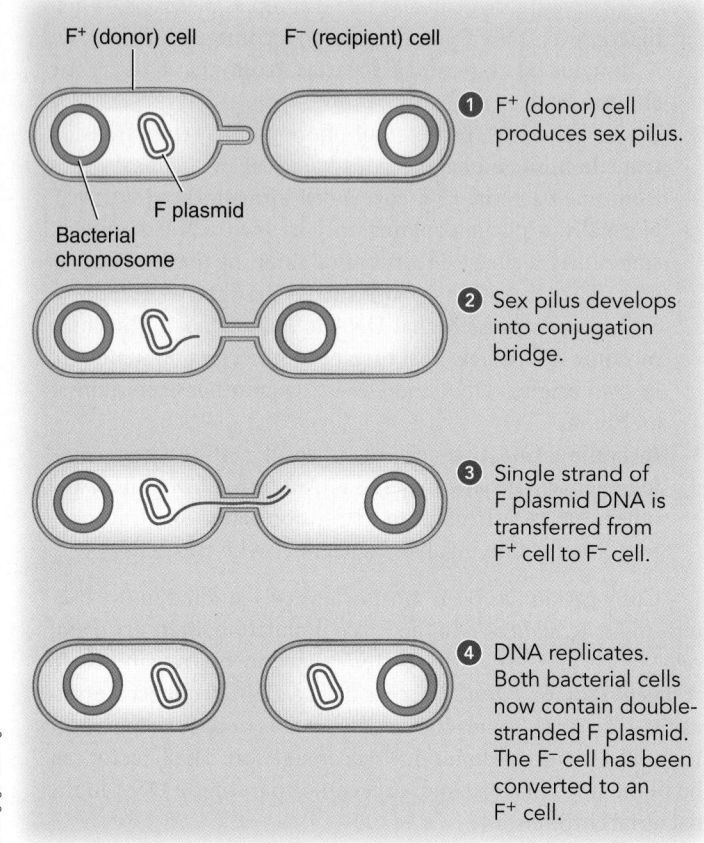

F⁺ (donor) cell   F⁻ (recipient) cell

Bacterial chromosome   F plasmid

❶ F⁺ (donor) cell produces sex pilus.

❷ Sex pilus develops into conjugation bridge.

❸ Single strand of F plasmid DNA is transferred from F⁺ cell to F⁻ cell.

❹ DNA replicates. Both bacterial cells now contain double-stranded F plasmid. The F⁻ cell has been converted to an F⁺ cell.

© Cengage Learning

(b) **The process of conjugation.**

Figure 25-8 *Animation* **Conjugation**

In conjugation a donor bacterium transfers plasmid DNA to a recipient bacterium. Conjugation requires cell-to-cell contact.

CONNECT Why might it be advantageous for a bacterial cell to receive the F plasmid from another cell?

25.3 NUTRITIONAL AND METABOLIC ADAPTATIONS

LEARNING OBJECTIVE

6   Describe the principal modes by which prokaryotes carry on nutrition and energy capture, and compare their requirements for oxygen.

Some prokaryotes are *autotrophs* (-*troph* comes from a Greek word that means "nutrition"), and others are *heterotrophs*. **Autotrophs** are able to use inorganic compounds, such as carbon dioxide, as a source of carbon for manufacturing their organic molecules. Most prokaryotes are **heterotrophs** that obtain carbon atoms from the organic compounds of other organisms.

Based on *two principal ways of capturing energy,* an autotroph is classified as a chemotroph or a phototroph. **Chemotrophs** obtain their energy from chemical compounds. **Phototrophs** capture energy from light. As early as 3.5 billion years ago the ability to use the sun as an energy source evolved in early prokaryotes. These early phototrophs used hydrogen sulfide to reduce carbon dioxide. They released sulfur as a waste product. About 2.7 billion years ago, the ability to use water rather than hydrogen sulfide to reduce carbon dioxide evolved in early cyanobacteria, and oxygen was released as a waste product. When we consider the source of carbon *and* the source of energy, we can classify prokaryotes into four main groups (**TABLE 25-1**):

1. **Photoautotrophs,** such as cyanobacteria, use the energy from sunlight to synthesize organic compounds from carbon dioxide and other inorganic compounds.

2. The majority of archaea and some bacteria are **chemoautotrophs,** which use carbon dioxide as a carbon source, but do not use sunlight as their energy source. Instead, they obtain energy by oxidizing inorganic chemical substances such as ammonia (NH_3) and hydrogen sulfide (H_2S).

3. **Photoheterotrophs,** such as the purple nonsulfur bacteria, obtain their carbon from other organisms but use

TABLE 25-1	Modes of Nutrition and Energy Capture		
MODE OF NUTRITION	ENERGY SOURCE	CARBON SOURCE	EXAMPLES OF ORGANISMS
AUTOTROPH			
Photoautotroph	Sunlight	CO_2	Cyanobacteria; purple sulfur bacteria
Chemoautotroph	Inorganic chemicals (e.g., NH_3, H_2S, Fe^{2+})	CO_2	Certain proteobacteria; most archaea (e.g., methanogens, extreme halophiles)
HETEROTROPH			
Photoheterotroph	Sunlight	Organic compounds	Purple and green nonsulfur bacteria
Chemoheterotroph	Organic compounds	Organic compounds	Free-living decomposers; most bacterial pathogens

© Cengage Learning

chlorophyll and other photosynthetic pigments to trap energy from sunlight.

4. The majority of bacteria are **chemoheterotrophs.** They depend on organic molecules for both their carbon and energy. Many prokaryote chemoheterotrophs are free-living **decomposers** that obtain their carbon and energy from dead organic matter. These organisms are sometimes called *saprotrophs*. Some bacterial chemoheterotrophs are pathogens, which obtain their nourishment from the organisms they infect. They harm their hosts by causing diseases. Other heterotrophic bacteria benefit their hosts. For example, some of the bacteria that inhabit the human large intestine produce vitamin K and certain B vitamins for their hosts.

Most prokaryotes require oxygen

Whether they are heterotrophs or autotrophs, most bacterial cells and many archaeal cells are **aerobic** and require oxygen for cellular respiration. Many are **facultative anaerobes** that use oxygen for cellular respiration if it is available but can carry on metabolism anaerobically when necessary. Other prokaryotes are **obligate anaerobes** that carry on anaerobic respiration; they respire with terminal electron acceptors other than oxygen, such as sulfate (SO_4^{2-}), nitrate (NO_3^{-}), or iron (Fe^{2+}). Some obligate anaerobes, including certain archaea, are actually killed by even low concentrations of oxygen.

Some prokaryotes fix and metabolize nitrogen

All organisms require nitrogen to manufacture amino acids and nucleic acids. Some bacteria (e.g., certain cyanobacteria) and archaea (e.g., methanogens) can reduce nitrogen in the atmosphere to ammonia. This process is called **nitrogen fixation.** Ammonia produced by nitrogen fixation is converted to ammonium ions (NH_4^{+}). Nitrogen-fixing prokaryotes can use these simple forms of nitrogen to produce organic compounds.

Certain prokaryotes convert ammonia or ammonium ions to nitrite (NO_2^{-}), and others convert nitrite to nitrate (NO_3^{-}). This process, called **nitrification,** converts nitrogen to a form that can be used by plants and fungi. Animals obtain nitrogen from organic compounds when they eat other organisms. As we will discuss in a later section, all other organisms ultimately depend on nitrogen fixation and nitrification by prokaryotes for their survival (discussed further in Chapter 55).

CHECKPOINT 25.3

- *How do chemoheterotrophs obtain energy?*
- CONNECT *How do facultative anaerobes differ from obligate anaerobes? How do they differ from aerobes?*
- *How do prokaryotes obtain nitrogen needed to produce amino acids and nucleic acids?*

25.4 THE PHYLOGENY OF THE TWO PROKARYOTE DOMAINS

LEARNING OBJECTIVES

7 Compare characteristics of the three domains: Archaea, Bacteria, and Eukarya.

8 Distinguish between the four main groups of archaea and identify specific types of archaea belonging to each group.

9 Describe the main groups of bacteria discussed in this chapter.

Under a microscope, most prokaryotes appear rather similar in size and form. However, using sequence analysis of small subunit 16S ribosomal RNA (SSU rRNA), Carl Woese and his co-workers demonstrated that there are two fundamentally different groups of prokaryotes (see Chapter 23). Each group has regions of SSU rRNA that have unique nucleotide sequences. The explanation is that after they diverged, prokaryote populations diversified, and mutations occurred that affect RNA sequences. Using such analyses, Woese hypothesized that ancient prokaryotes split into two lineages early in the history of life.

Based on Woese's work and on other recent data, systematists now classify the modern descendants of these two ancient lines in two domains: **Archaea** and **Bacteria** (FIG. 25-9). These groups are thought to have diverged from a common ancestor about four billion years ago. Archaea and bacteria were the only living organisms on our planet for about two billion years. Thus, archaea and bacteria have had a very long time to evolve and adapt to all types of environments. Their diversity is staggering. Horizontal gene transfer has contributed to the diversity of these organisms. As a result of gene transfer, sometimes from distantly related species, the genomes of archaea and bacteria are actually a mix of genes from many prokaryotes.

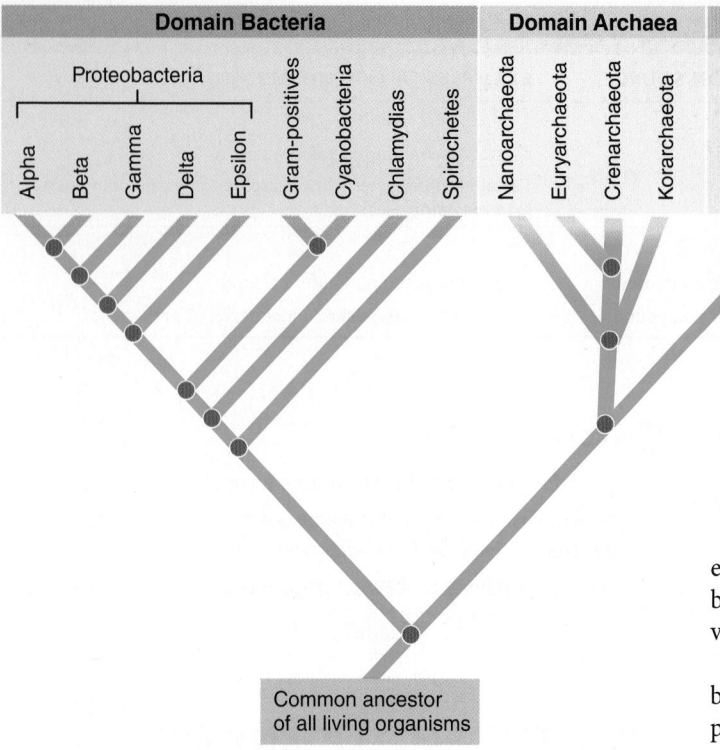

Figure 25-9 Three domains

This highly simplified diagram depicts some representative taxa of domain Bacteria and domain Archaea. The relationships illustrated are based on molecular data. As taxonomists consider additional data, these relationships will be modified.

© Cengage Learning

Key characters distinguish the three domains

Several key characters distinguish archaea from bacteria. In contrast to bacteria, archaea do not have peptidoglycan in their cell walls. Although their plasma membranes are structurally similar, they are chemically unique. In the plasma membranes of bacteria and eukaryotes, straight-chain fatty acids are linked to glycerol molecules by *ester linkages*. In contrast, fatty acid components are not found in archaea. Instead, branched-chain hydrocarbons (synthesized from isoprene units) are bonded to glycerol by *ether linkages* (TABLE 25-2).

Ester linkage   Ether linkage

The absence of a second electronegative oxygen atom makes ether linkages stronger than ester linkages. This unique membrane structure may contribute to the ability of archaea to survive and thrive in harsh environments.

In some ways, archaea are more like eukaryotes than like bacteria. For example, archaea do not have the simple RNA polymerase found in bacteria. Like eukaryotes, their translation process begins with methionine, whereas in bacteria, translation begins with formylmethionine (see Chapter 13). In addition, several antibiotics that affect bacteria do not affect archaea or eukaryotes.

Bacteria also share some characteristics (that are absent in archaea) with eukaryotes. For example, as stated above, bacteria and eukarya both have ester-linked membrane lipids, whereas archaea have ether-linked membrane lipids. Some microbiologists hypothesize that eukaryotes are a product of fusion between an archaeon, which contributed components for transcription and translation, and a bacterium, which contributed enzymes necessary for energy metabolism.

TABLE 25-2	Comparison of the Three Domains		
CHARACTERISTIC	**BACTERIA**	**ARCHAEA**	**EUKARYA**
Nuclear envelope	Absent	Absent	Present
Membrane-enclosed organelles	Absent	Absent	Present
Circular chromosome	Present (linear in some species)	Present	Absent
Number of chromosomes	Typically one (may also have plasmids)	Typically one (may also have plasmids)	Typically many
Histones associated with DNA	Absent	Present	Present
Peptidoglycan in cell wall	Present	Absent	Absent
Structure of lipids in plasma membrane	Straight-chain fatty acids bonded to glycerol by ester linkages	Branched-chain hydrocarbons linked to glycerol by ether linkages	Straight-chain fatty acids bonded to glycerol by ester linkages
Size of ribosomes	70S*	70S	80S except in mitochondria and chloroplasts
RNA polymerase	One relatively simple RNA polymerase	One relatively complex RNA polymerase	Several relatively complex RNA polymerases
Translation	Begins with formylmethionine	Begins with methionine	Begins with methionine
Growth above 70°C	Yes	Yes	No

*The numbers 70S and 80S refer to the sedimentation coefficient (a measure of relative size) when centrifuged.

© Cengage Learning

Taxonomy of archaea and bacteria continuously changes

Prokaryote taxonomy is based largely on molecular data, mainly RNA sequencing and more recently on sequencing entire genomes. Groups that branched off earlier had more time to accumulate mutations in their SSU rRNA. Their nucleotide sequences are less similar than those of groups that diverged more recently. Microbiologists who are developing phylogenetic trees based on sequencing entire genomes argue that there may be 1000 genes that code proteins for every 1 gene that codes an rRNA. These researchers prefer to consider the proportions of genes (or proteins) that genomes of various groups have in common. Taxonomy of archaea and bacteria continuously changes as systematists study new molecular data that provide new clues to the phylogeny of these groups.

Although prokaryotic taxonomy is controversial and continuously changing, about 10,000 species of prokaryotes have been classified. Many thousands of additional species are thought to exist. The editors of *Bergey's Manual of Systematic Bacteriology,* considered the definitive reference text by microbiologists, have divided archaea into 4 phyla and bacteria into more than 30 phyla based on 16S rRNA analyses.

Many archaea inhabit harsh environments

Domain Archaea consists of four major phyla: *Crenarchaeota, Euryarchaeota, Nanoarchaeota,* and *Korarchaeota.* This phylogeny is based on 16S rRNA and on sequencing of entire genomes. Based on continued detection of previously unidentified archaea and sequencing of archaeal genomes, systematists estimate that more than a dozen additional phyla may be added.

The **Crenarchaeota** include **extreme thermophiles,** archaea that require a very high temperature or very low temperature for growth. The optimum temperature for many is greater than 80°C (176°F) and some thrive at temperatures greater than 100°C. Some crenarcheotes have the highest growth temperatures known of any organisms. Some species inhabit acidic environments. One species is found in the hot sulfur springs of Yellowstone National Park at temperatures near 80°C and pH values of 1 to 2, the pH of concentrated sulfuric acid (**FIG. 25-10a**). Other crenarchaeotes inhabit volcanic areas under the sea. One species, found near deep-sea hydrothermal vents on the seafloor of the Pacific Ocean, lives at temperatures ranging from 80°C to 120°C. In contrast, some Crenarchaeota species live in very cold environments (1.8°C).

Studies have identified many diverse species of Crenarchaeota in soil and fresh water, demonstrating that members of this phylum are common to most environments. Crenarchaeotes are a main contributor to carbon fixation. They are also an important part of the plankton in cold, oxygen-rich seas. A few are photoheterotrophs.

The **Euryarchaeota** also include many archaea that inhabit extreme environments. This group includes methanogens, extreme aerobic and anaerobic thermophiles, acidophiles, and

(a) **Extreme thermophiles.** Orange and yellow colonies of extreme thermophiles thrive in the Grand Prismatic Spring in Yellowstone National Park in Wyoming.

(b) **Extreme halophiles.** These seawater salt-evaporation ponds are colored from the large number of extreme halophiles that inhabit them. The colors are from pigments in the cell membranes. These archaea are harmless, and the ponds are used to produce salt commercially.

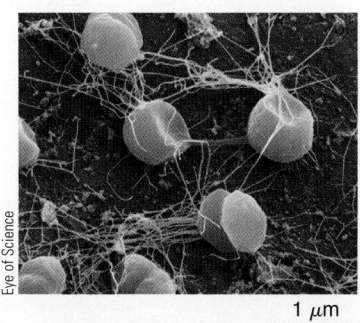

1 μm

(c) **SEM of *Pyrococcus furiosus,* an anaerobe that inhabits marine sediments.** This methanogen is highly resistant to heat; its optimum temperature is 100°C. *Pyrococcus* is classified as a euryarchaeote.

Figure 25-10 Archaea that inhabit extreme environments

halophiles (some extreme) (**FIG. 25-10b**). The **methanogens** (methane producers) are a large, diverse group that inhabit oxygen-free environments in sewage, swamps, and the digestive tracts of humans and other animals. They are obligate anaerobes that produce methane gas from simple carbon compounds. The methanogens are important in recycling components of organic products of organisms that inhabit swamps. Methanogens that inhabit the digestive tracts of cows and other grazing animals produce methane, which is belched out by the animals. Methanogens produce more than 80% of the methane (more than 2 billion tons each year) in Earth's atmosphere. Methanogens are also found in marine sediments (**FIG. 25-10c**). Methane is an important greenhouse gas (discussed in Chapter 57).

Extreme halophiles are heterotrophs that require large amounts of Na^+ for their growth. They live in saturated brine solutions such as salt ponds, the Dead Sea, and Great Salt Lake. The extreme halophiles use aerobic respiration to make ATP. However, they also carry out a form of the Calvin cycle (a part of photosynthesis) in which they capture the energy of sunlight using a purple pigment (*bacteriorhodopsin*). This pigment is very similar to the pigment rhodopsin involved in animal vision.

Korarchaeota is a phylum of Archaea that appears to have branched before the Crenarchaeota and Euryarchaeota branches separated. Korarchaeota have some characteristics of Crenarchaeota and some features of Euryarchaeota. Korarchaeotes have been found in terrestrial hot springs.

To date, only one genus of **Nanoarchaeota** has been identified. *Nanoarchaeum equitans* was discovered in a hydrothermal vent in 2002. This microbe is a very small (400 nm) anaerobic, extreme thermophile with a very small genome (less than 500,000 nucleotides). It lives attached to another archaeon, an autotroph, and depends on its host for many of its metabolic needs.

As more archaea have been discovered and studied, it has become evident that there are likely more archaea in marine and soil environments than in extreme habitats. The archaea are important in biogeochemical cycles and in marine food chains. Although archaea continue to be discovered on and in humans, no pathogenic archaea have been identified.

Bacteria are the most familiar prokaryotes

Bacteria have been known and studied much longer than the archaea. Bacteria are also widely distributed in the environment. Five major groups are considered here: proteobacteria (gram-negative), cyanobacteria (gram-negative), gram-positive bacteria, chlamydias (gram-negative), and spirochetes (gram-negative). These groups are summarized in **TABLE 25-3**.

CHECKPOINT 25.4

- *What types of environments are inhabited by the four phyla of archaea? Why are methanogens important?*
- **CONNECT** *How does each of the following groups of bacteria (described in Table 25-3) affect the biosphere: alpha proteobacteria, cyanobacteria, and chlamydias?*

25.5 IMPACT ON ECOLOGY, TECHNOLOGY, AND COMMERCE

LEARNING OBJECTIVES

10 Identify the critical ecological roles played by prokaryotes.
11 Describe some of the important roles played by prokaryotes in commerce and technology.

Prokaryotes inhabit virtually every environment on Earth and are vital members of the biosphere. They affect other organisms directly and by the ecological roles they play. Prokaryotes produce nitrogen in forms usable by other organisms, and they are a reservoir for nutrients. These microscopic organisms recycle nutrients and are key players in biogeochemical cycles.

Prokaryotes form intimate relationships with other organisms

Prokaryotes interact with other organisms in both beneficial and harmful ways. An intimate relationship between members of two or more species is called **symbiosis.** The partners in a symbiotic relationship are called **symbionts.** Symbiotic relationships arise by coevolution. Three forms of symbiosis are mutualism, commensalism, and parasitism.

Mutualism is a symbiotic relationship in which both partners benefit. Cows and other ruminants (cud-chewing animals) have mutualistic relationships with bacteria and archaea that inhabit their digestive tracts. Ruminants lack enzymes for digesting cellulose. They provide the prokaryotes with a nutrient-rich home, and the prokaryotes digest the cellulose for them.

Trillions of bacteria and some archaea inhabit the nutrient-rich human intestine. Some of them are mutualistic bacteria. For example, in exchange for nutrients and a place to live, *Bacteroides* break down indigestible complex carbohydrates into sugars that their human host can absorb. These bacteria also produce certain vitamins that their host absorbs and uses. In addition, *Bacteroides* promote proliferation of blood vessels that improve intestinal function. Investigators have reported that these bacteria can turn on specific genes in cells of the host's intestine; apparently, they can induce synthesis of a compound that may kill competing bacteria.

In **commensalism** one partner benefits, and the other is neither harmed nor helped. Many prokaryotes that inhabit the human intestine are commensals that live on unused food. In **parasitism** one partner lives on or in the other. The **parasite** benefits, and the **host** is harmed. Disease-causing bacteria are usually not considered obligate parasites because these *pathogens* typically can survive in other ways.

Some types of bacteria may influence evolution of other species The proteobacteria *Wolbachia* infect many invertebrates, including insects, spiders, crustaceans, and nematodes (roundworms). *Wolbachia* are transmitted from generation to generation in the eggs of their hosts, so male hosts are not useful to them. Consequently, these parasites limit or eradicate males from the population. In some insect species, infected males can reproduce only when they mate with infected females carrying the same *Wolbachia* strain. In some other species, these parasites convert male insects into females. In some wasp species, *Wolbachia* induces *parthenogenesis,* the development of unfertilized eggs into adult organisms. Because they affect reproduction in their hosts, *Wolbachia* and other reproductive parasites may influence evolutionary divergence and even extinction in some species. *Wolbachia* may also affect evolution by horizontal gene transfer.

Many prokaryotes form biofilms Many types of bacteria and archaea that inhabit watery environments form **biofilms,** dense communities of microorganisms that attach to solid surfaces. The prokaryotes secrete a slimy, gluelike substance rich in polysaccharides and become embedded in this matrix. A biofilm may consist of layers up to 200 μm thick. Biofilms are communities of many microorganisms and may consist of many species of bacteria, archaea, fungi, and protists. Scientists have found evidence of ancient biofilms in sediments from coastal marine environments 3.5 billion years old.

The dental plaque that forms on teeth is a familiar example of a biofilm (**FIG. 25-11**). Dental plaque consists of several hundred different types of bacteria and archaea. Biofilms also commonly form on the surfaces of contact lenses and catheters. They sometimes develop on surgical implants such as pacemakers and joint replacements. Biofilms, which form on and inside plants, cause considerable crop loss.

Prokaryotes play key ecological roles

Bacteria, especially actinomycetes and myxobacteria, are the most numerous inhabitants of soil. As described earlier in this chapter, many prokaryotes are essential decomposers, chemoheterotrophs that break down dead organic matter and wastes. They use the products of decomposition as an energy source. When prokaryotes break down organic compounds, many of their components, including nitrogen, oxygen, carbon, phosphorus, sulfur, and certain trace elements, are recycled. (The roles of bacteria and archaea in biogeochemical cycles, particularly the nitrogen cycle, are discussed in Chapter 55.)

Nitrogen is constantly removed from the soil by plants and other natural processes as well as by human activities such as agriculture. Plant growth depends on the availability of usable nitrogen, so it must be continually added to the soil. Several types of bacteria, including cyanobacteria, and some archaea transform atmospheric nitrogen to forms that can be used by plants (see Table 25-3). Marine archaea that carry on nitrification are important in the ocean nitrogen cycle.

Figure 25-11 A familiar biofilm: dental plaque
Colorized SEM of dental plaque, which consists of a film of bacteria (*red*) embedded in a matrix of glycoprotein (*blue*). Bacteria in the plaque can produce acids that erode tooth enamel, leading to tooth decay.

Rhizobial prokaryotes, motile inhabitants of the soil, form mutualistic relationships with the roots of legumes, a large family of plants that includes important crops such as beans, peas, and peanuts. The infected plant cells form tumorlike nodules in which the microbes reside and fix nitrogen (see Fig. 55-9). The prokaryotes supply the plant with the nitrogen it requires, and the plant provides the prokaryotes with organic compounds, including sugar needed for cellular respiration. Because many soils are deficient in nitrogen, legumes that have formed mutualistic associations with rhizobial microbes have a decided advantage over other plants. When the legumes die and are decomposed, the fixed nitrogen is released and enriches the soil.

Many prokaryotes, such as cyanobacteria, carry on photosynthesis, using water as the electron source and generating oxygen. During this process they fix huge amounts of carbon dioxide into organic molecules.

Microbiologists are only beginning to unravel the mysteries of prokaryote ecology. For example, alpha proteobacteria in the SAR11 clade are among the most successful organisms on Earth. Although they are one of the most abundant organisms in the Atlantic Ocean, they were not successfully cultured until 2002, and very little is known about their ecological role. There is evidence that they are important in cycling carbon, nitrogen, and sulfur in the ocean.

Prokaryotes are important in many commercial processes and in technology

Some microorganisms produce *antibiotics*. These compounds limit competition for nutrients by inhibiting or destroying other microorganisms. By the 1950s, antibiotics had become

| TABLE 25-3 | Major Groups of Bacteria |

PROTEOBACTERIA (GRAM-NEGATIVE)

A large, very diverse clade of gram-negative bacteria. Based on rRNA sequences, the group is divided into five subgroups designated as alpha, beta, gamma, delta, and epsilon.

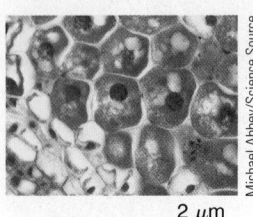

2 μm

Rhizobia (*Rhizobium leguminosarum*) in root nodule
Rhizobia are nitrogen-fixing bacteria.

ALPHA PROTEOBACTERIA

Includes many symbionts of plants and animals, and some pathogens. *Rhizobium* species live symbiotically in root nodules of legumes (e.g., beans) and convert atmospheric nitrogen to a form usable by plants (nitrogen fixation). **Rickettsias** are very small, rod-shaped bacteria. A few species are pathogenic to humans and other animals; transmitted by arthropods through bites or through contact with their excretions. Rickettsias cause typhus (transmitted by fleas and lice) and Rocky Mountain spotted fever (transmitted by ticks). Members of the SAR11 clade are extremely abundant in the ocean.

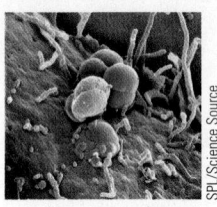

1 μm

SEM of bacteria (*Neisseria gonorrhoeae*) that cause gonorrhea
In this colorized SEM, these beta proteobacteria (*blue*) are infecting a human epithelial cell (*purple*).

BETA PROTEOBACTERIA

Several diverse groups, including *Nitrosomonas*, which oxidizes ammonia. Pathogenic bacteria in this group include *Neisseria gonorrhoeae*, which causes gonorrhea.

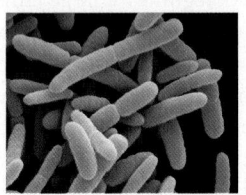

2 μm

SEM of *Escherichia coli* colony
E. coli are gamma proteobacteria.

GAMMA PROTEOBACTERIA

Includes the **enterobacteria,** decomposers that live on decaying plant matter, pathogens, and a variety of bacteria that inhabit humans. Although *Escherichia coli* is a normal inhabitant of the animal intestinal tract, certain strains can cause moderate to severe diarrhea. One species of *Salmonella* infects food and produces a toxin that causes a form of food poisoning; another species causes typhoid fever.

Vibrios are mainly marine; some are bioluminescent. *Vibrio cholerae* causes cholera.

Pseudomonads are heterotrophs that produce nonphotosynthetic pigments; cause disease in plants and animals, including humans.

Purple sulfur bacteria are photoautotrophs that do not produce oxygen.

DELTA PROTEOBACTERIA

Includes the **myxobacteria** (slime bacteria), which secrete slime and glide or creep along. When nutrients are exhausted, these bacteria aggregate into stalked, multicellular reproductive structures called fruiting bodies. Bacterial cells within the fruiting body enter a resting stage. When conditions are favorable, resting cells become active.

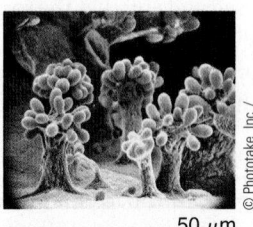

50 μm

SEM of fruiting body of the myxobacterium *Stigmatella aurantiaca*
Protective resting cells within the fruiting bodies are very resistant to heat and drying. Myxobacteria are delta proteobacteria.

EPSILON PROTEOBACTERIA

A small group of bacteria that inhabit the animal digestive tract. *Helicobacter* can cause peptic ulcers.

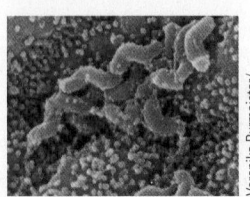

2 μm

SEM of *Helicobacter pylori* attached to the epithelial lining of the stomach

Continued

TABLE 25-3 | Major Groups of Bacteria (*continued*)

GRAM-POSITIVE BACTERIA

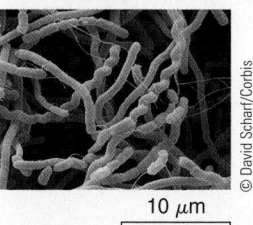

10 μm

SEM of *Actinomycetes naeslundi*, a soil-dwelling bacterium that forms filamentous colonies

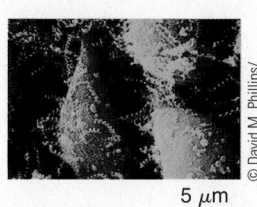

5 μm

SEM of *Mycoplasma* on fibroblast cells

Actinomycetes superficially resemble fungi. However, they have peptidoglycan in their cell walls, lack nuclear envelopes, and have other prokaryotic characteristics. Most actinomycetes are saprotrophs that decompose organic materials in soil. Some are anaerobic. Several species of the genus *Streptomyces* produce antibiotics such as streptomycin, erythromycin, chloramphenicol, and the tetracyclines. Some actinomycetes cause serious lung disease and other infections in humans and other animals.

Lactic acid bacteria ferment sugar, producing lactic acid as the main end product. Inhabit decomposing plant material, milk, and other dairy products; responsible for the characteristic taste of yogurt, pickles, sauerkraut, and green olives. Among the normal inhabitants of the human mouth and vagina.

Mycobacteria are slender, irregular rods; contain a waxy substance in their cell walls. One species causes tuberculosis; another causes leprosy.

Streptococci inhabit the mouth and digestive tract of humans and some other animals. Among the harmful species are those that cause "strep throat," dental caries, a form of pneumonia, scarlet fever, and rheumatic fever (see chapter-opening photograph).

Staphylococci normally live in the nose and on skin. Opportunistic pathogens that cause disease when the immunity of the host is lowered. *Staphylococcus aureus* causes boils and skin infections (some extremely serious); may infect wounds. Certain strains of *S. aureus* cause a form of food poisoning; other strains cause toxic shock syndrome (see Fig. 25-14).

Clostridia are anaerobic. One species causes tetanus; another causes gas gangrene. *Clostridium botulinum* can cause botulism, an often fatal type of food poisoning.

The **mycoplasmas** are a group of very small bacteria that lack cell walls. They may have evolved from bacteria with gram-positive cell walls. They inhabit soil and sewage; some are parasitic on plants or animals. Some inhabit human mucous membranes but do not generally cause disease; one species causes a mild type of bacterial pneumonia in humans.

CYANOBACTERIA (GRAM-NEGATIVE)

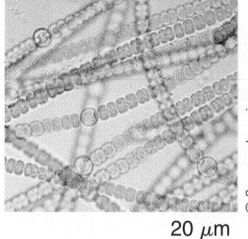

20 μm

LM of *Anabaena*, a filamentous cyanobacterium that fixes nitrogen
Nitrogen fixation takes place in the rounded cells, called *heterocysts*.

Cyanobacteria contain chlorophyll *a* and are the only prokaryotes that, like plants and algae, carry on photosynthesis that generates oxygen. Cyanobacteria were the first organisms that carried on oxygen-generating photosynthesis; very important in the evolution of life-forms because their photosynthesis changed the early reducing atmosphere on Earth to an oxidizing atmosphere. Chloroplasts are thought to have evolved from endosymbiotic cyanobacteria. Inhabit ponds, lakes, swimming pools, moist soil, dead logs, and tree bark. Some form filaments; other species are solitary. As primary producers, they are an important source of food for marine and freshwater organisms. Some species have special structures that fix nitrogen.

CHLAMYDIAS (GRAM-NEGATIVE)

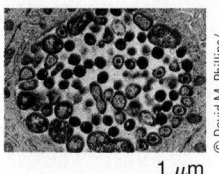

1 μm

TEM of *Chlamydia trachomatis* in human oviduct cell

Chlamydias lack peptidoglycan in their cell walls. They are energy parasites, completely dependent on their hosts for ATP. Infect almost every species of bird and mammal. A strain of *Chlamydia* causes trachoma, the leading cause of blindness in the world. Sexually transmitted chlamydias are the major cause of pelvic inflammatory disease in women.

SPIROCHETES (GRAM-NEGATIVE)

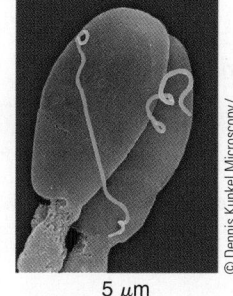

5 μm

LM of *Treponema pallidum*, the spirochete that causes syphilis

Spirochetes are spiral-shaped bacteria with flexible cell walls; move by means of unique internal flagella called *axial filaments*. Some species are free-living, whereas others form symbiotic associations; a few are parasitic. *Treponema pallidum* causes syphilis.

Figure 25-12 Bioremediation

As they feed on the gasoline and certain other waste products in contaminated soil, certain microorganisms convert hydrocarbons in these pollutants to carbon dioxide and water. Photographed at an oil refinery and chemical plant in the United Kingdom.

Paul Rapson/Science Photo Library/Science Source

important clinical tools that transformed the treatment of infectious disease. Today, more than 100 clinically useful antibiotics are available, and literally tons of antibiotics are produced annually. Pharmaceutical companies obtain most antibiotics from three groups of microorganisms: a large group of gram-positive soil bacteria, the *actinomycetes;* gram-positive bacteria of the genus *Bacillus;* and molds (eukaryotes belonging to kingdom Fungi).

Because of their prolific reproduction rates, bacteria are ideal "factories" for the production of biomolecules. Microbiologists have genetically engineered bacteria to produce certain vaccines, human growth hormone, insulin, and many other clinically important compounds (see Chapter 15). Researchers are developing genetically engineered bacteria for production of many other medically and agriculturally useful products.

Microbial fermentation helps produce many foods and beverages. Lactic acid bacteria are used in producing acidophilus milk, yogurt, pickles, olives, and sauerkraut. Several types of bacteria are used in the production of cheese. Bacteria are involved in making fermented meats such as salami and in producing vinegar, soy sauce, chocolate, and certain B vitamins (B_{12} and riboflavin).

Bacteria are used in **bioremediation,** the process of using microorganisms (and sometimes other organisms) to detoxify or remove oil, gasoline, and other pollutants or toxic chemicals from the environment. Microorganisms break down certain toxins, leaving behind harmless metabolic byproducts such as carbon dioxide and chlorides (FIG. 25-12). More than 1000 different species of bacteria and fungi have been used to clean up various forms of pollution, and microbiologists are searching for others. For example, bacteria and other microorganisms are used in oil spills to break down oil, oxidizing it to CO_2. Microorganisms are also used in sewage treatment and to break down solid wastes in landfills.

Archaea are also economically important. Archaea that are adapted to high temperatures or extremely acidic conditions, for example, are a source of enzymes that can be used under these extreme conditions. Archaeal enzymes have been added to laundry and industrial detergents and to organic solvents to increase performance at higher temperatures and pH levels. Methanogens are important in the biogas production industry and in sewage treatment. Another archaeal enzyme has been useful in the food industry to convert cornstarch to dextrins (low molecular weight carbohydrates produced by hydrolysis of starch or glycogen).

Molecular biology and forensic biology have greatly benefited from heat-resistant DNA polymerase, derived from the archaeon *Pyrococcus furiosus* (see Fig. 25-10c). Like heat-resistant Taq polymerase, derived from a bacterium, it is used in the polymerase chain reaction (PCR) discussed in Chapter 15. Some archaea have antimicrobial properties and may someday be used in the production of new antibiotics. Acidophilic archaeons are promising as resources for extracting metal from ores and remediating toxic mining sites.

CHECKPOINT 25.5

- CONNECT *In what ways do bacteria form relationships with other organisms?*
- *Where would you expect to find biofilms? Name organisms you would expect to find in biofilms.*
- CONNECT *Describe how prokaryotes are ecologically important. Give specific examples.*
- *Describe the process of bioremediation.*

25.6 BACTERIA AND DISEASE

LEARNING OBJECTIVES

12 Describe the roles played by Louis Pasteur and Robert Koch in understanding infectious disease; list Koch's postulates.

13 Identify adaptations that have contributed to pathogen success.

Some prokaryote species have coevolved with eukaryotes and are interdependent with them. All plants and animals harbor a community of microorganisms referred to as **microbiota.** (This term is replacing a traditional designation, *microflora,* which is a misnomer because "flora" refers to plants.) An estimated 700 trillion symbiotic bacteria, archaea, and (a few) eukaryotic microorganisms normally inhabit the human body! This number greatly exceeds the number of the body's own cells (about 70 trillion).

Normal microbial populations have been shown to have multiple effects, including preventing harmful microorganisms from flourishing. The term **microbiome** refers to the community of these microorganisms, including their genomes and all of their interactions. Rapid DNA sequencing has greatly facilitated microbiome research because there are thousands of species of these organisms and many are difficult to grow in culture. Through the *Human Microbiome Project* supported by the National Institutes of Health since 2009, investigators are studying the **human microbiome.** Through their research, we are developing an ever-increasing understanding of the critical effects of these symbionts on health and disease.

A small percentage of bacterial species are important pathogens of plants and animals. Some of the normal bacterial inhabitants are opportunistic pathogens that cause disease only under certain conditions. For example, when the immune system is compromised, opportunistic bacteria increase in number and cause disease. Some important bacterial diseases and the pathogens that cause them are briefly described in **TABLE 25-4.**

Many scientists have contributed to our understanding of infectious disease

The idea that some unknown agent caused disease was debated long before Leeuwenhoek discovered microorganisms with his microscope in the late 1600s. However, not until much later did scientists develop the tools and methods needed to accurately understand the relationships between bacteria and disease. During the late 19th century, several physicians, microbiologists, and chemists working independently laid the foundations for the science of microbiology. French chemist Louis Pasteur disproved the prevailing views of spontaneous generation by demonstrating that sterilization of a sugar and protein culture prevented bacterial growth. Pasteur also developed a rabies vaccine, showing that people can be stimulated to develop immunity to disease.

German physician Robert Koch was the first to clearly demonstrate that bacteria cause infectious disease. In 1876, he showed that *Bacillus anthracis* caused anthrax. Using a microscope, Koch observed the bacteria in the blood and spleens of dead sheep. When he inoculated mice with the infected sheep blood, he was able to identify *B. anthracis* in the blood of the mice. He also cultured *B. anthracis* and showed that when he injected the bacteria into healthy mice they developed anthrax.

Koch proposed a set of guidelines, now known as **Koch's postulates,** that are still used to demonstrate that a specific pathogen causes specific disease symptoms: (1) the pathogen must be present in every individual with the disease, (2) a sample of the microorganism taken from the diseased host can be grown in pure culture, (3) a sample of the pure culture causes the same disease when injected into a healthy host, and (4) the microorganism can be recovered from the experimentally infected host. Sometimes these guidelines cannot be met, as, for example, when certain microorganisms cannot be grown in pure culture. In those situations, other criteria must be used.

Many adaptations contribute to pathogen success

Pathogenic microorganisms can enter the body in food, dust, or droplets or through wounds. Many diseases are transmitted by insect or animal bites. To cause disease, a pathogen must adhere to a specific cell type, multiply, and produce toxic substances. Adherence and multiplication occur only when the pathogen competes successfully with the normal microbiota and counteracts the host's defenses against invasion.

Helicobacter pylori, the most common cause of peptic ulcers (ulcers of the stomach and duodenum), is an extremely successful pathogen (**FIG. 25-13**). It is also associated with chronic gastritis (stomach inflammation) and with stomach cancer, the second most common type of cancer in the world. *Helicobacter pylori* inhabits the gastrointestinal tracts of an estimated 40% of adults in developed countries and 80% of adults in developing countries. Among its many adaptations is its ability to produce an alkaline shield around itself that protects it from stomach acid. Also contributing to its success are several powerful flagella used to propel the pathogen through the thick mucus lining the stomach.

Pathogens produce a variety of substances that increase their success. Some bacteria produce **exotoxins,** strong poisons that are either secreted from the cell or leak out when the bacterial cell is destroyed. The toxin, not the presence of the bacteria themselves, is responsible for the disease. Diphtheria is caused by a gram-positive bacillus (*Corynebacterium diphtheriae*) that produces a toxin only when lysogenized by a phage. The diphtheria toxin kills cells and causes inflammation.

Botulism, a type of food poisoning that can lead to paralysis and sometimes death, can result from eating improperly canned food. Botulism is caused by an exotoxin released by the

TABLE 25-4 | Important Bacterial Diseases and Their Causative Agents

DISEASE	PATHOGEN	EPIDEMIOLOGY/COMMENTS
Anthrax	*Bacillus anthracis*	Most commonly occurs in domestic animals such as cattle. Can be transmitted to humans from infected animals or animal products. Endospores can live in soil for many years. Anthrax is not spread from person to person. Infection can occur in three ways: cutaneous, by inhalation, and gastrointestinal.
Antibiotic-associated diarrhea; and inflammation of colon	*Clostridium difficile* (common name: *C. dif*)	Risk greatest for older people taking antibiotics, people with compromised immune systems, and patients in hospitals and health care facilities for extended periods
Botulism	*Clostridium botulinum*	Contracted by eating foods that contain the exotoxin or from infected wound. Infant botulism is caused by ingesting endospores. Causes muscle paralysis and can cause death from respiratory failure.
Chlamydia	*Chlamydia trachomatis*	One of the most frequently reported sexually transmitted infections in the U.S. About 75% of infected women and 50% of infected men have no symptoms. If untreated, infection spreads and can lead to infertility. *Chlamydia* can also infect eyes; cause millions of cases of blindness worldwide each year.
Cholera	*Vibrio cholerae*	Contracted by eating food or drinking water contaminated with the bacteria. Common in areas with inadequate sewage treatment and impure water. Infects intestine and can cause severe diarrhea. Rapid fluid loss can lead to dehydration and death.
Diphtheria	*Corynebacterium diphtheriae*	Transmitted from person to person by intimate respiratory and physical contact. Endemic in developing countries. Not common in U.S. since vaccine became available in 1920s. Infects the heart muscle and respiratory passageways.
Epidemic typhus	*Rickettsia prowazekii*	Transmitted by infected body lice. After an 8- to 12-day incubation period, symptoms include fever, severe headache, muscle aches, and chills. Several days later a rash appears. About 40% of untreated patients die.
Gonorrhea	*Neisseria gonorrhoeae*	Common sexually transmitted disease.
Hansen disease (leprosy)	*Mycobacterium leprae*	Thought to be spread from person to person in nasal secretions. Worldwide, this disease has disabled up to 2 million people.
Lyme disease	*Borrelia burgdorferi*	Transmitted to humans by bite of infected blacklegged ticks. Symptoms include skin rash, headache, fever, and fatigue. If untreated, infection can spread to joints, heart, and nervous system.
Peptic ulcer disease	*Helicobacter pylori*	Causes peptic ulcer, a lesion in the lining of the stomach or duodenum (upper part of small intestine).
Pertussis (whooping cough)	*Bordetella pertussis*	Highly communicable from person to person. Causes spasms of severe coughing. Vaccination available.
Plague	*Yersinia pestis*	Transmitted from wild rodents, squirrels, and cats to people by infected fleas. If untreated, can be fatal. Killed millions of people in Europe during the Middle Ages.
Pneumonia	*Streptococcus pneumoniae*	Transmitted from person to person. Strains of *S. pneumoniae* are resistant to some antibiotics. Incidence has decreased since introduction of a vaccine.
Salmonella (salmonellosis)	*Salmonella* sp.	Transmitted via contaminated chicken, eggs, or other food; also from feces of infected animals. Symptoms include fever, diarrhea, and stomach pain.
Syphilis	*Treponema pallidum*	Sexually transmitted disease passed through direct contact with a syphilis sore. If untreated, eventually damages brain, liver, bone, and spleen; can cause death.
Travelers' diarrhea	*Escherichia coli* is most common cause	Ingested in contaminated food and water. Affects 30% to 50% of travelers in high-risk areas (Central America, South America, Africa, Middle East, and most of Asia). Infecting bacteria produce toxins in gastrointestinal tract.
Tuberculosis	*Mycobacterium tuberculosis*	Transmitted from person to person through inhalation of air containing the pathogen. Symptoms include fatigue, cough, fever, and weight loss. Not common in U.S., but multidrug-resistant form is a growing threat.
Typhoid fever	*Salmonella typhi*	Transmitted from person to person in food or water contaminated with feces. Risk greatest for travelers in developing countries, but vaccine available. Symptoms include high fever, headache, and loss of appetite. Can cause death if not treated.

© Cengage Learning

gram-positive, endospore-forming *Clostridium botulinum*. During the canning process, food must be heated sufficiently to kill any highly heat-resistant endospores that may be present. If not, the endospores can germinate. The resulting bacterial population grows and releases an exotoxin so powerful that 1 g could kill one million humans! Like many exotoxins, the one that causes botulism is inactivated by heating. (Food must be heated to 80°C for 10 minutes or boiled for 3 to 4 minutes to inactivate the exotoxin.)

The botulism exotoxin, marketed under the trade name Botox, is used in extremely tiny amounts to treat several medical conditions involving spasms (involuntary muscle contractions). Because Botox is a neurotoxin that works by paralyzing muscles, it can also relax facial wrinkles caused by contraction

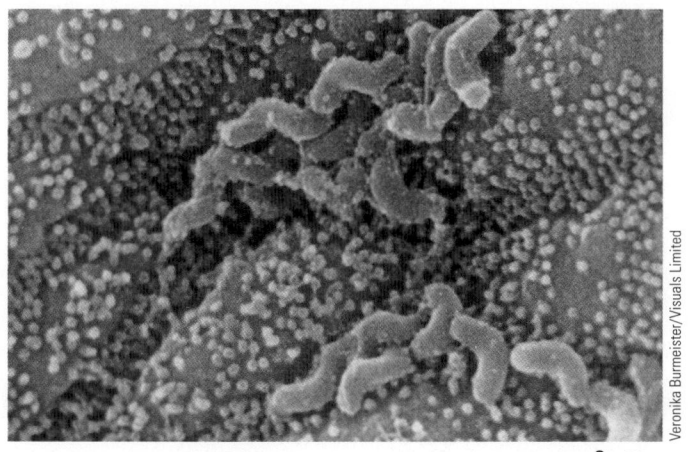

Figure 25-13 SEM of *Helicobacter pylori* attached to the epithelial lining of the stomach

Helicobacter pylori causes peptic ulcers.

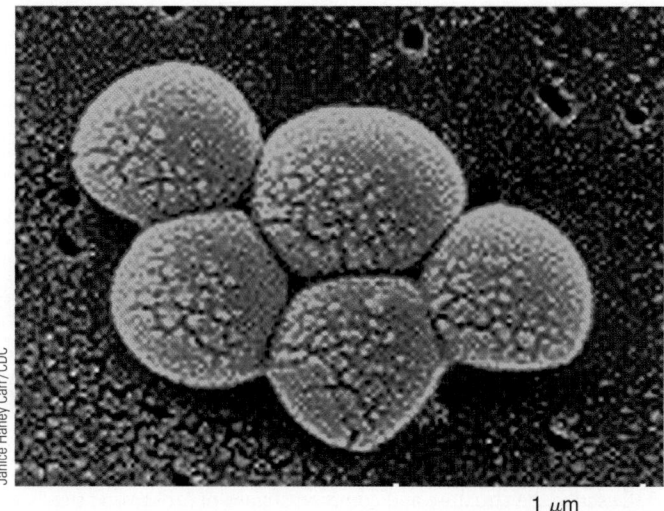

Figure 25-14 Colorized SEM of methicillin-resistant *Staphylococcus aureus* (MRSA)

These bacteria also show increased resistance to vancomycin.

of the underlying muscles. However, its effects last for only three to eight months.

Another gram-positive, endospore-forming bacterium, *Clostridium difficile,* is an anaerobe that produces two exotoxins. It is a common cause of antibiotic-associated diarrhea (AAD), accounting for 15% to 25% of all episodes of AAD. *C. difficile* has recently become more virulent and more resistant to antibiotics. It has been linked to more than 14,000 deaths in the United States each year (see Table 25-4).

Endotoxins are not secreted by pathogens but instead are components of the cell walls of most gram-negative bacteria. These compounds affect the host only when they are released from dead bacteria. Endotoxins bind to the host's macrophages (large phagocytic cells of the immune system) and stimulate them to release substances that cause fever and other symptoms of infection. Unlike exotoxins, which cause specific symptoms, endotoxins appear to affect the entire body. Endotoxins are not destroyed by heating.

Antibiotic resistance is a major public health problem

Bacteria mutate frequently and reproduce rapidly, often developing resistance to antibiotics. Many antibiotics target protein synthesis in bacteria. For example, streptomycin and related antibiotics block the initiation of protein synthesis. The tetracyclines block aminoacyl tRNA from binding to the A site on the ribosome.

Drug resistance may result from an accumulation of mutations in plasmid or chromosomal DNA. Plasmids that have genes for antibiotic resistance are called *R factors.* They have genes for resistance to a specific drug and for transferring the resistance to other bacteria. Some R factors have several genes for drug resistance, each encoding resistance to a different drug.

Methicillin-resistant *Staphylococcus aureus* (SA), referred to as *MRSA,* and vancomycin-resistant *Staphylococcus aureus,* or *VRSA,* have been directly linked to the horizontal transfer of antibiotic resistance genes by plasmids during conjugation. An estimated 1 in every 100 healthy people in the United States now carries MRSA (FIG. 25-14). This "superbug" can cause infection in individuals with compromised immune systems. The Centers for Disease Control estimates that MRSA causes more than 95,000 serious infections each year, resulting in more than 18,000 deaths annually.

Overuse of antibiotics is the principal contributing factor to drug resistance. In any bacterial population, there are likely at least a few bacteria that are genetically resistant to a particular antibiotic. The bacteria that are *not* resistant are killed by the antibiotic, leaving the resistant bacteria to multiply and produce a resistant population. Note that this process is a common, present-day example of natural selection.

The practice of feeding low-dose antibiotics to farm animals to promote growth has also resulted in many types of antibiotic-resistant bacteria that survive and multiply after susceptible bacteria are killed. Many countries have banned some antibiotics in animal feed. Antibiotic-resistance has significantly reduced the effectiveness of most common, inexpensive antibiotics. Scientists and pharmaceutical companies are challenged to rapidly develop new antibiotics to treat bacterial infections that were formerly easily cured.

Another type of drug resistance explains why some infections, such as urinary tract infections, are difficult to cure. When *E. coli* infect the bladder, the immune system launches a powerful defense. The bacteria subvert the attack by forming *biofilms* (discussed earlier in this chapter). Each biofilm is surrounded by a matrix rich in polysaccharides and by a protective shell. The bacteria in the biofilm are resistant to antibiotics as well as to host defenses. By one estimate, bacteria growing in a biofilm are up to 1000 times as resistant to antibiotics as are the same type of bacteria that have not formed a biofilm. Within biofilms differential gene expression and mutation lead to diverse bacterial colonies. The biofilm strategy also explains some other types of chronic or recurrent infections. A high percentage of infections acquired in hospitals involve biofilms.

CHECKPOINT 25.6

- **CONNECT** *Why is each of Koch's postulates important in determining whether a particular pathogen causes specific symptoms?*
- *Contrast endotoxins with exotoxins.*

SUMMARY: FOCUS ON LEARNING OBJECTIVES

25.1 The Structure of Bacteria and Archaea *(page 512)*

1 Describe the structure and common shapes of prokaryotic cells.

- Prokaryotic cells are very small and do not have membrane-enclosed organelles such as nuclei and mitochondria.
- Prokaryotic cells have several common shapes: spherical **(cocci),** rod-shaped **(bacilli),** and spiral. Spiral bacteria include the **spirillum,** which is a rigid helix, and **spirochete,** which is a flexible helix.

2 Compare the bacterial cell wall in gram-positive and gram-negative bacteria.

- Most bacteria have **cell walls** composed of **peptidoglycan.** The walls of **gram-positive** bacteria are very thick and consist mainly of peptidoglycan. The cell walls of **gram-negative** bacteria consist of a thin peptidoglycan layer and an outer membrane resembling the plasma membrane. Some species of bacteria produce a **capsule** or **slime layer** that surrounds the cell wall.
- Some prokaryotes have hairlike appendages called **fimbriae. Pili** also extend from the surface of some prokaryotes. Both fimbriae and pili help cells adhere to one another or to certain other surfaces, including cells they infect. *Cannulae* and *hami* are recently discovered hairlike appendages unique to archaea.

3 Describe movement in prokaryotes and describe the structure of the bacterial flagellum.

- Bacterial **flagella** are structurally different from eukaryotic flagella; each flagellum consists of a basal body, hook, and filament. They produce a rotary motion.

25.2 Prokaryote Reproduction and Evolution *(page 516)*

4 Describe asexual reproduction in prokaryotes and summarize three mechanisms (transformation, transduction, and conjugation) that may lead to genetic recombination.

- The genetic material of a bacterium typically consists of a circular DNA molecule and one or more **plasmids,** smaller circular fragments of DNA.
- Prokaryotes reproduce asexually by **binary fission** (the cell divides, forming two cells), **budding** (a bud forms and separates from the mother cell), or **fragmentation** (walls form inside the cell, which then separates into several cells).
- In prokaryotes genetic material can be exchanged by transformation, transduction, or conjugation. In **transformation** a prokaryotic cell takes in foreign DNA released by another

cell. Homologous segments of foreign and host DNA are exchanged. In **transduction** a phage carries bacterial DNA from one bacterial cell into another. In **conjugation** a donor cell transfers plasmid DNA to a recipient cell.

© Cengage Learning

5 State specific factors that contribute to the rapid evolution of bacteria and archaea.

- Rapid reproduction ensures that mutations are rapidly passed to new generations. Horizontal gene transfer—by transformation, transduction, or conjugation—contributes to rapid evolution in prokaryotes.

25.3 Nutritional and Metabolic Adaptations *(page 518)*

6 Describe the principal modes by which prokaryotes carry on nutrition and energy capture, and compare their requirements for oxygen.

- Most prokaryotes are **heterotrophs** that obtain carbon from other organisms; some are **autotrophs** that make their own organic molecules from simple raw materials.
- **Chemotrophs** obtain energy from chemical compounds; **phototrophs** capture energy from light.
- Autotrophs may be **photoautotrophs,** which obtain energy from sunlight, or **chemoautotrophs,** which obtain energy by oxidizing inorganic chemicals such as ammonia.
- **Photoheterotrophs** obtain carbon from other organisms but use chlorophyll and other photosynthetic pigments to trap energy from sunlight. The majority of bacteria are **chemoheterotrophs.** They are mainly free-living **decomposers** that obtain both carbon and energy from dead organic matter.
- Most bacteria are **aerobic;** that is, they require oxygen for cellular respiration. Some prokaryotes are **facultative anaerobes** that metabolize anaerobically when necessary; others are **obligate anaerobes** that can carry on metabolism *only* anaerobically.
- Some bacteria and archaea carry on **nitrogen fixation;** that is, they reduce nitrogen in the atmosphere to ammonia. Other prokaryotes convert ammonia to nitrite or nitrate in a process called **nitrification.**

25.4 The Phylogeny of the Two Prokaryote Domains
(page 519)

7 Compare characteristics of the three domains: Archaea, Bacteria, and Eukarya.

- Prokaryotes are assigned to domain **Archaea** and domain **Bacteria.**
- Unlike those of bacteria, the cell walls of archaea do not have peptidoglycan. The translational mechanisms of eukaryotes more closely resemble those of archaea than those of bacteria.

8 Distinguish between the four main groups of archaea and identify specific types of archaea belonging to each group.

- The **Crenarchaeota** include many **extreme thermophiles,** archaea that can inhabit very hot, sometimes acidic, environments and archaea that are marine dwellers. **Euryarchaeota** include methanogens, extreme halophiles, and some extreme thermophiles. **Methanogens** are obligate anaerobes that produce methane gas from simple carbon compounds. **Extreme halophiles** inhabit saturated salt solutions. **Korarchaeota** have been found in terrestrial hot springs. **Nanoarchaeota** has only one member to date, *Nanoarchaeum equitans,* a very small, extreme thermophile discovered in a hydrothermal vent.

9 Describe the main groups of bacteria discussed in this chapter.

- Major groups of bacteria include proteobacteria, gram-positive bacteria, cyanobacteria, chlamydias, and spirochetes (see Table 25-3).

25.5 Impact on Ecology, Technology, and Commerce
(page 522)

10 Identify the critical ecological roles played by prokaryotes.

- Many bacteria are symbiotic with other organisms. **Mutualism** is a symbiotic relationship in which both partners benefit. In **commensalism** one partner benefits, and the other is neither harmed nor helped. In **parasitism** the **parasite** benefits, and the **host** is harmed. Bacterial **pathogens** cause disease, but are usually not considered obligate parasites. **Biofilms** are dense communities of microorganisms, in which cells adhere to one another on a surface. Biofilms may include bacteria, archaea, protists, and fungi.
- Prokaryotes play essential ecological roles as decomposers and are important in recycling nitrogen and other nutrients. Some bacteria carry out photosynthesis.

11 Describe some of the important roles played by prokaryotes in commerce and technology.

- Some prokaryotes produce antibiotics. We have developed the technology for using certain bacteria to produce vaccines, insulin, and other important compounds. We use bacteria in the production of many foods, including cheese, yogurt, vinegar, and chocolate. We also use microbes in sewage treatment and in bioremediation.

25.6 Bacteria and Disease *(page 527)*

12 Describe the roles played by Louis Pasteur and Robert Koch in understanding infectious disease; list Koch's postulates.

- Louis Pasteur demonstrated that sterilization prevented bacterial growth. **Koch's postulates** are a set of guidelines developed by Robert Koch to demonstrate that a specific pathogen causes specific disease symptoms: (1) the pathogen must be present in every individual with the disease, (2) a sample of the microorganism taken from the diseased host can be grown in pure culture, (3) a sample of the pure culture causes the same disease when injected into a healthy host, and (4) the microorganism can be recovered from the experimentally infected host.

13 Identify adaptations that have contributed to pathogen success.

- Some pathogenic bacteria release strong poisons called **exotoxins;** others produce **endotoxins,** poisonous components of their cell walls that are released when bacteria die. Many bacteria have become resistant to antibiotics. Plasmids that have genes for antibiotic resistance are called *R factors.* See Table 25-4 for a description of some pathogenic bacteria.

TEST YOUR UNDERSTANDING

Know and Comprehend

1. Peptidoglycan is a chemical compound found in the cell walls of (a) most viroids (b) most archaea (c) all prokaryotes (d) most bacteria (e) most eukarya
2. Bacterial flagella (a) are homologous with eukaryotic flagella (b) exhibit a rotary motion (c) consist of a basal body and nine pairs of microtubules (d) are important in transduction (e) are characteristic of gram-positive bacteria
3. Endospores (a) are formed by some viruses (b) are extremely durable cells (c) are comparable to the reproductive spores of fungi and plants (d) cause fever and other symptoms in the host (e) are vulnerable to infection by archaea
4. In conjugation (a) two bacterial cells of different mating types come together, and genetic material is transferred from one to another (b) a bacterial cell develops a bulge that enlarges and eventually separates from the mother cell (c) fragments of DNA released by a broken cell are taken in by another bacterial cell (d) a phage carries bacterial genes from one bacterial cell into another (e) walls develop in the cell, which then divides into several new cells
5. The majority of heterotrophic bacteria are (a) free-living chemoheterotrophs (b) photoautotrophs (c) chemoautotrophs (d) facultative anaerobes (e) obligate anaerobes
6. Bacteria that are autotrophs (a) do not require atmospheric oxygen for cellular respiration (b) must obtain organic compounds from other organisms (c) manufacture their own organic molecules from simple raw materials (d) get their nourishment from dead organisms (e) produce endospores when oxygen levels are too low for active growth
7. Bacteria that thrive in puncture wounds are likely to be (a) aerobes (b) photoautotrophs (c) chemoautotrophs (d) endospores (e) obligate anaerobes

8. Which of the following do *not* belong to domain Archaea? (a) prokaryotes that produce methane from carbon dioxide and hydrogen (b) thermophiles (c) halophiles (d) bacteriophages (e) prokaryotes with cell walls that lack peptidoglycan

9. Rhizobial bacteria (a) are used in the production of yogurt (b) produce antibiotics (c) cause peptic ulcers (d) are the pathogens that cause syphilis (e) form mutualistic relationships with the roots of legumes and fix nitrogen

10. Robert Koch (a) proposed a set of guidelines to demonstrate that a specific pathogen causes specific disease symptoms (b) discovered *Helicobacter* (c) showed that biofilms consist of microorganisms (d) proposed a hypothesis for antibiotic resistance (e) demonstrated that people can be stimulated to develop immunity to disease

11. Which group of bacteria contains the gram-positive, anaerobic bacterium that causes botulism? (a) clostridia (b) actinomycetes (c) enterobacteria (d) spirochetes (e) streptococci

Apply and Analyze

12. **VISUALIZE** Label the diagram.

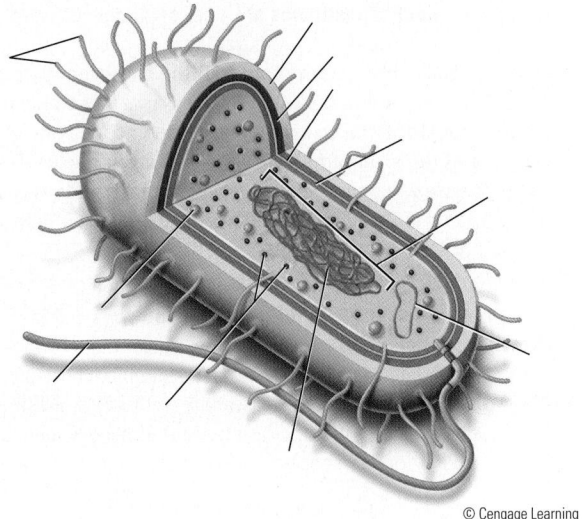

© Cengage Learning

13. Imagine that you discover a new microorganism. After careful study you determine that it should be classified in domain Archaea. What characteristics might lead you to this decision?

Evaluate and Synthesize

14. What would be the consequences for eukaryotes if all prokaryotes suddenly became extinct?

15. **EVOLUTION LINK** In what way does the use of antibiotics impose selective pressure on bacteria?

16. **SCIENCE, TECHNOLOGY, AND SOCIETY** The Centers for Disease Control considers antibiotic resistance to be one of the world's most pressing health problems. How can scientists help solve this problem? (Consider both technology and the general public.)

To access course materials, such as Aplia and other companion resources, please visit **www.cengagebrain.com.**

Protists

Protists are an informal group of primarily aquatic eukaryotic organisms with diverse body forms, types of reproduction, modes of nutrition, and lifestyles. Protists, which include algae, water molds, slime molds, and protozoa, are unicellular, colonial, or simple multicellular organisms that have a eukaryotic cell organization (see photograph). The word *protist,* from the Greek for "the very first," reflects the idea that protists were the first eukaryotes to evolve.

Protists are members of domain Eukarya, the third domain on the tree of life. (Recall the three domains depicted in Figure 23-2.) Eukaryotic cells are characteristic of protists as well as of complex multicellular organisms belonging to the kingdoms Fungi, Animalia, and Plantae. However, having a eukaryotic cell structure clearly differentiates protists from members of the prokaryotic domains Bacteria and Archaea. Recall from Chapter 4 that unlike prokaryotic cells, eukaryotic cells have nuclei and other membrane-enclosed organelles such as mitochondria and plastids, 9 + 2 flagella, and multiple chromosomes in which DNA and proteins form a complex called chromatin. Sexual reproduction, meiosis, and mitosis are also characteristic of eukaryotes.

Until recently, all protists were classified as the kingdom Protista, but this classification was by default: protists included any eukaryotes that were not land plants, fungi, or animals. Based largely on recent molecular data that have clarified many evolutionary relationships among eukaryotes, kingdom Protista is no longer a recognized clade. Although we have made great progress in understanding the evolutionary relationships among protists, many relationships remain uncertain.

Biologists currently recognize dozens of protist taxa, which, along with other eukaryotes, are classified into five "supergroups." Consideration of all protist taxa is beyond the scope of this text, but we discuss several representative examples of each supergroup and provide insights into how such diverse eukaryotes may have evolved.

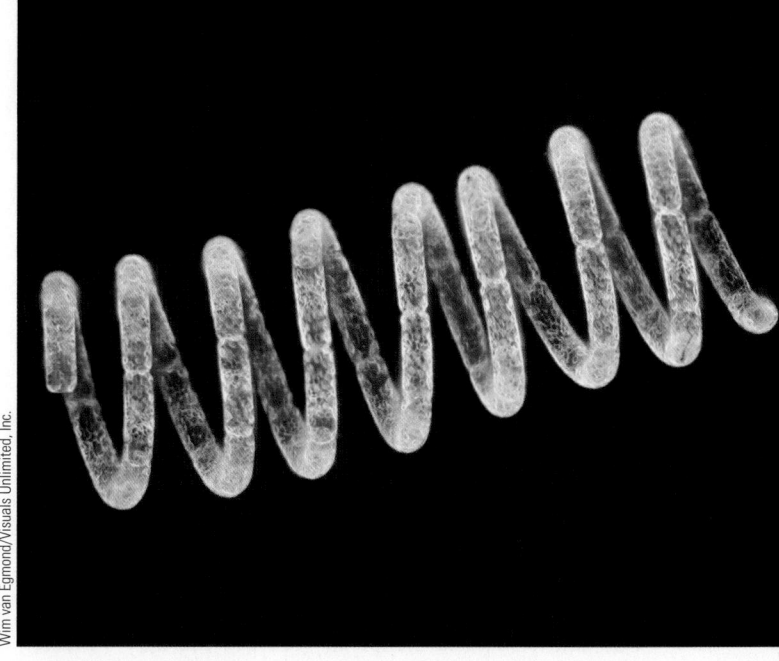

Wim van Egmond/Visuals Unlimited, Inc.

Light micrograph (LM) of a marine diatom. The widely distributed diatom *Guinardia striata* forms colonies composed of spiraling chains of cells.

KEY CONCEPTS

26.1 Protists are a diverse group of eukaryotic organisms that vary in body plan (unicellular, colonial, coenocytic, multicellular), method of motility (pseudopodia, cilia, flagella), nutrition type (autotrophic, heterotrophic), and mode of reproduction (asexual, sexual).

26.2 Much uncertainty surrounds eukaryote evolution, but eukaryote organelles such as chloroplasts probably descended from engulfed cells that survived and became organelles. Current scientific evidence supports splitting the protists and other eukaryotes (land plants, fungi, and animals) into five informal "supergroups."

26.3 Excavates are a supergroup of unicellular protists with atypical, greatly modified mitochondria; they get their name because many have a deep, or excavated, oral groove.

26.4 Chromalveolates are a supergroup of diverse protists that may have originated as a result of secondary endosymbiosis in which an ancestral cell engulfed a red alga.

26.5 Rhizarians are a supergroup of amoeboid cells that often have tests (shells) through which cytoplasmic projections extend; forams, actinopods, and certain shell-less amoebas are rhizarians.

26.6 Archaeplastids are a supergroup that includes red algae, green algae, and land plants, all of which have plastids enclosed by two external membranes.

26.7 Unikonts are a supergroup that includes amoebozoa and choanoflagellates as well as fungi and animals.

26.1 DIVERSITY IN THE PROTISTS

1 Discuss in general terms the diversity inherent in protists, including means of locomotion, modes of nutrition, interactions with other organisms, habitats, and modes of reproduction.

Because of their huge numbers, protists are crucial to the natural balance of the living world. Protists are an important source of food for other organisms, and photosynthetic protists supply oxygen to aquatic and terrestrial ecosystems. Certain protists are economically important, and others cause devastating diseases such as malaria.

Body plan varies considerably among protists. Most protists are *unicellular,* with each cell forming a complete organism capable of performing all the functions characteristic of life. Some protists form **colonies,** loosely connected groups of cells; some are **coenocytes,** consisting of a multinucleate mass of cytoplasm; and some are *multicellular,* composed of many cells. Unlike animals, plants, and many fungi, most multicellular protists have relatively simple body forms without specialized tissues.

Size and structural complexity are not the only variable features of protists. During the course of the long evolutionary history of protists, diversity has evolved in their means of locomotion, ways of obtaining nutrients, interactions with other organisms, habitats, and modes of reproduction.

Protists, most of which are motile at some point in their life cycle, have various means of locomotion. Some move by pushing out cytoplasmic extensions (*pseudopodia*) along the leading edge and retracting the cytoplasm that trails behind, as an amoeba does. Other protists move by flexing individual cells; by gliding over surfaces; by waving *cilia,* short, hairlike organelles; or by lashing *flagella,* long, whiplike organelles. Some protists have two or more means of locomotion, such as both flagella and pseudopodia.

Methods of obtaining nutrients differ widely among protists. Most algae are autotrophic and photosynthesize as plants do (**FIG. 26-1a**). Some heterotrophic protists obtain their nutrients by absorption, as fungi do, whereas others resemble animals in that they ingest food. Some protists switch their modes of nutrition and are autotrophic at certain times and heterotrophic at others.

Although many protists are free-living, others form stable symbiotic associations with unrelated organisms. These intimate associations range from **mutualism,** a more or less equal partnership where both partners benefit; to **commensalism,** where one partner benefits and the other is unaffected; to **parasitism,** where one partner (the parasite) lives on or in another (the host) and metabolically depends on it (see Chapter 54). Some parasitic protists are important *pathogens* (disease-causing agents) of plants or animals. Throughout this chapter, we describe specific examples of symbiotic associations involving protists.

Most protists are aquatic and live in the ocean or in freshwater streams, lakes, and ponds (**FIG. 26-1b**). They make up most of the **plankton,** the floating, often microscopic organisms that inhabit surface waters and are the base of the food web in aquatic ecosystems. Other aquatic protists attach to rocks or other surfaces in the water. Even parasitic protists are aquatic because they live in the watery environments of other organisms' body fluids. Terrestrial protists are restricted to damp places such as soil, cracks in bark, and leaf litter.

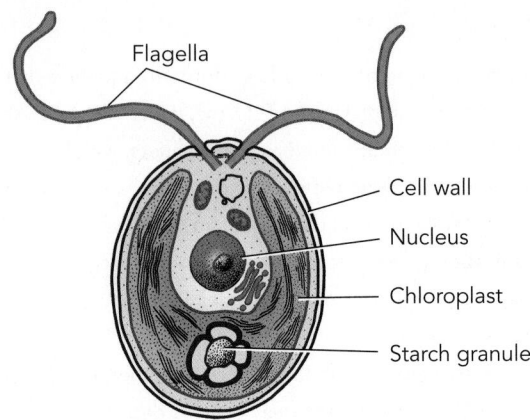

(a) *Chlamydomonas* is a photosynthetic organism with two flagella and a cup-shaped chloroplast.

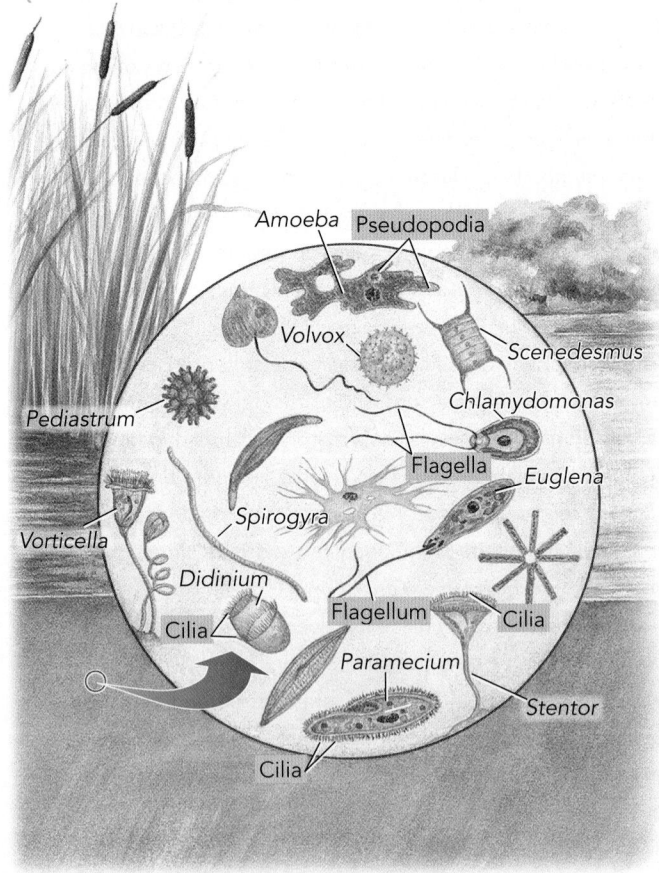

(b) Protists in a drop of pond water. Several modes of locomotion are shown.

Figure 26-1 Protists
© Cengage Learning

Reproduction is varied among protists. Almost all protists reproduce asexually, and many also reproduce sexually. However, most protists do not develop multicellular reproductive organs, nor do they form embryos the way more complex organisms do.

CHECKPOINT 26.1

- *How do protists vary in their means of obtaining nutrients?*
- *What are some of the ways protists interact with other organisms?*

26.2 HOW DID EUKARYOTES EVOLVE?

LEARNING OBJECTIVES

2 Discuss the hypothesis of serial endosymbiosis and briefly explain some of the evidence that supports it.

3 Describe the kinds of data biologists use to classify eukaryotes.

For many years biologists have hypothesized that protists were the first eukaryotic cells and that they evolved from ancestral prokaryotes. However, the more we study eukaryote origins, the more uncertain we are. One thing we can say with absolute certainty is that the evolution of eukaryotes was a complex process.

Eukaryotes may have appeared in the fossil record as early as 2.2 billion years ago. Other than a few protists with hard shells, such as diatoms and forams, most ancient protists did not leave many fossils because their bodies were too soft to leave permanent traces. Evolutionary studies of protists focus primarily on molecular and structural comparisons of present-day organisms, which contain many clues about their evolutionary history.

Mitochondria and chloroplasts probably originated from endosymbionts

Throughout evolutionary history, one organism has engulfed another to the mutual benefit of both. According to the hypothesis of **serial endosymbiosis,** certain eukaryotic organelles, particularly mitochondria and chloroplasts, arose from symbiotic relationships between larger cells and smaller bacteria that were incorporated and lived within them. (You might want to review serial endosymbiosis in Figure 21-8.) Cell biologists hypothesize that mitochondria originated from aerobic bacteria. Studies of mitochondrial DNA suggest that it is a remnant from the mitochondrion's past, when it was an independent organism. Ribosomal RNA (rRNA) sequences from mitochondria closely match rRNAs found in purple bacteria, suggesting that ancient purple bacteria were the ancestors of mitochondria.

Chloroplast evolution is more complex given that there were probably several endosymbiotic events (FIG. 26-2). Molecular evidence supports the view that incorporation of an ancient cyanobacterium within a host cell, known as *primary endosymbiosis,* resulted in the chloroplasts in today's red algae, green algae, and land plants. (The host cell was eukaryotic because mitochondria almost certainly evolved before chloroplasts.) Biologists hypothesize that these chloroplasts, which are enclosed by two external membranes (known as the outer and inner chloroplast membranes; see Chapter 9), later provided other eukaryotes with their chloroplasts during *secondary endosymbiosis.*

Secondary endosymbiosis occurred frequently in eukaryote evolution, as evidenced by the presence of additional chloroplast membranes. For example, *three* membranes envelop the chloroplasts of euglenoids and dinoflagellates, and *four* membranes surround the chloroplasts of diatoms, golden algae, and brown algae. Understanding how these membranes originated is an essential aspect of serial endosymbiosis, and many researchers are studying the origin of chloroplasts in different organisms.

Even non-photosynthetic protists may contain chloroplast relics from secondary endosymbiotic events. Apicomplexans—protists such as *Plasmodium,* which causes malaria—have a non-photosynthetic chloroplast derived from a red alga, surrounded by four external membranes. Because this plastid carries out certain functions essential to the survival of the parasite, it has become a target of ongoing research focused on the development of antimalarial drugs.

A consensus in eukaryote classification is beginning to emerge

Scientists re-evaluate evolutionary relationships among the eukaryotes as additional evidence becomes available. Two types of modern research, molecular analysis and ultrastructural studies, have contributed substantially to scientific understanding of the phylogenetic relationships among protists. Molecular data were initially obtained for the gene that codes for small subunit ribosomal RNA in different eukaryotes (SSU rRNA; see Chapter 25). More recently, biologists have compared other nuclear genes, many of which code for proteins, in different protist taxa.

Ultrastructure is the fine details of cell structure revealed by electron microscopy. In many cases, ultrastructure data complement molecular data. Electron microscopy reveals similar structural patterns among those protist taxa that comparative molecular evidence suggests are **monophyletic;** that is, they evolved from a common ancestor (see Chapter 23). For example, molecular and ultrastructure data suggest that water molds, diatoms, golden algae, and brown algae—protist taxa that at first glance seem to share few characteristics—are a monophyletic group.

Given the diversity in protist ultrastructure and molecular data, biologists regard the protists as a **paraphyletic group;** that is, protists contain some, but not all, of the descendants of a common eukaryote ancestor. Molecular and ultrastructural analyses continue to help biologists clarify relationships among the various protist phyla and among protists and the other eukaryotic kingdoms.

Biologists use these data to develop various classification schemes. A current scheme that is used in this edition splits the protists and other eukaryotes (land plants, fungi, and animals)

Scientists hypothesize that plastids have evolved through both primary and secondary endosymbiosis.

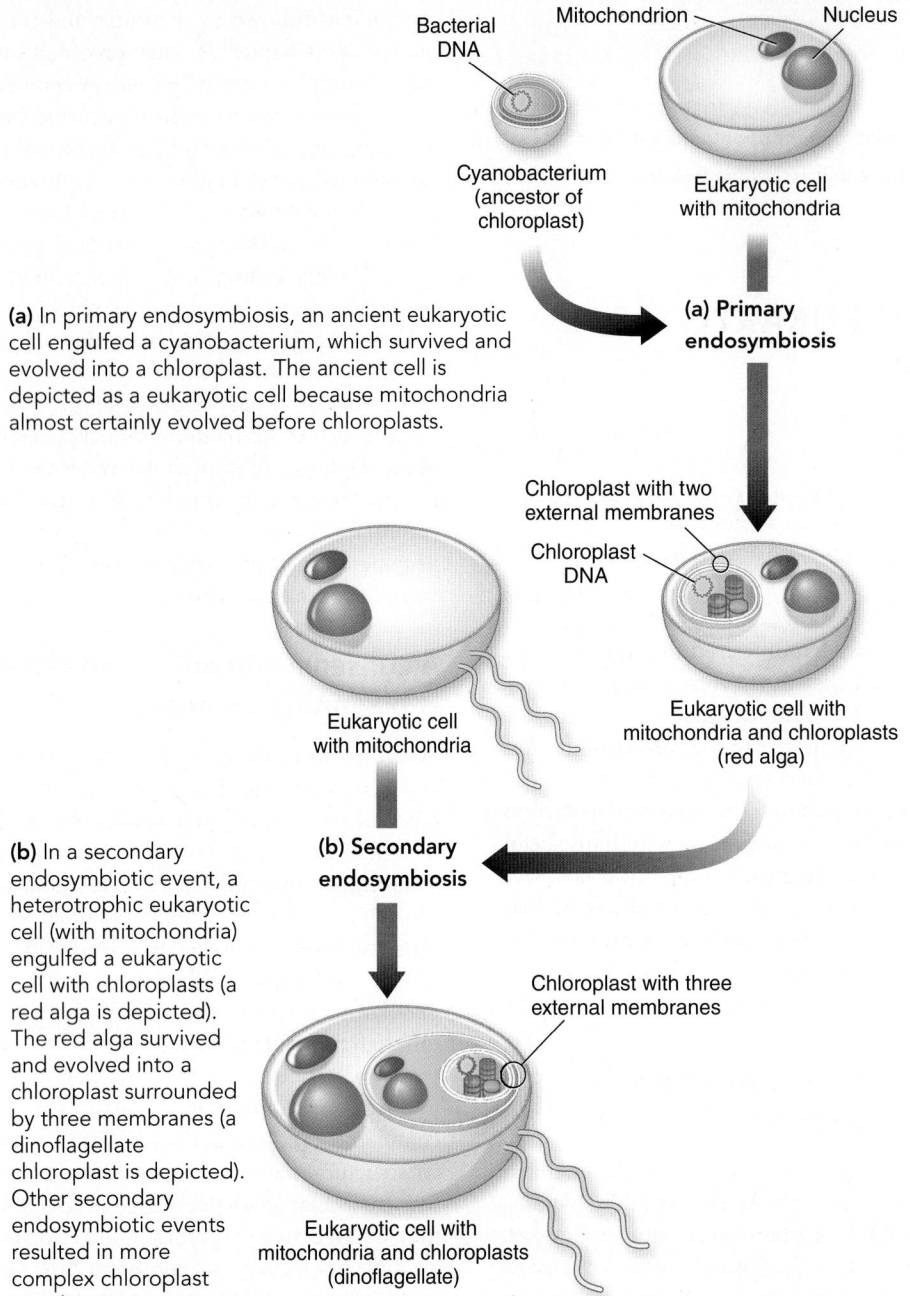

(a) In primary endosymbiosis, an ancient eukaryotic cell engulfed a cyanobacterium, which survived and evolved into a chloroplast. The ancient cell is depicted as a eukaryotic cell because mitochondria almost certainly evolved before chloroplasts.

(b) In a secondary endosymbiotic event, a heterotrophic eukaryotic cell (with mitochondria) engulfed a eukaryotic cell with chloroplasts (a red alga is depicted). The red alga survived and evolved into a chloroplast surrounded by three membranes (a dinoflagellate chloroplast is depicted). Other secondary endosymbiotic events resulted in more complex chloroplast membrane structures.

Figure 26-2 Chloroplast evolution by primary and secondary endosymbiosis

CONNECT How many external membranes are present in a land plant chloroplast? Use Figure 9-4b to check your answer.

© Cengage Learning

into five informal supergroups (**FIG. 26-3** and **TABLE 26-1**). This classification scheme will almost certainly be modified as new information becomes available.

Now that we have a basic understanding of evolution of the protists and other eukaryotes, let's examine representative protists within the five supergroups.

CHECKPOINT 26.2

- *How does serial endosymbiosis explain the origin of chloroplasts?*
- *What kinds of scientific evidence support the hypothesis that protists are a paraphyletic group?*

Unraveling phylogenetic relationships among the protists has been difficult, but progress is occurring. Many biologists currently classify eukaryotes into five major "supergroups."

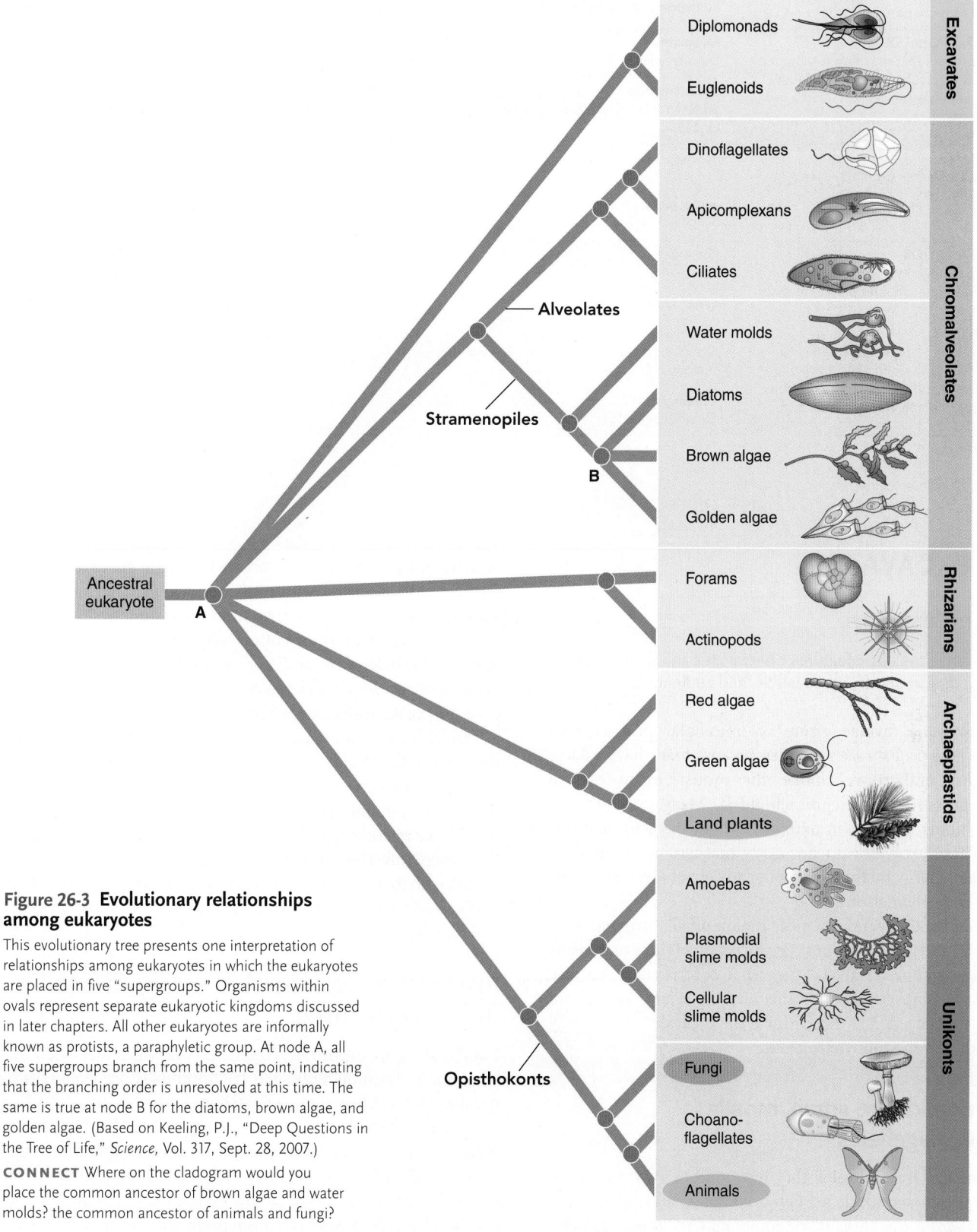

Figure 26-3 Evolutionary relationships among eukaryotes

This evolutionary tree presents one interpretation of relationships among eukaryotes in which the eukaryotes are placed in five "supergroups." Organisms within ovals represent separate eukaryotic kingdoms discussed in later chapters. All other eukaryotes are informally known as protists, a paraphyletic group. At node A, all five supergroups branch from the same point, indicating that the branching order is unresolved at this time. The same is true at node B for the diatoms, brown algae, and golden algae. (Based on Keeling, P.J., "Deep Questions in the Tree of Life," *Science*, Vol. 317, Sept. 28, 2007.)

CONNECT Where on the cladogram would you place the common ancestor of brown algae and water molds? the common ancestor of animals and fungi?

TABLE 26-1 | "Protist" Clades

EUKARYOTE "SUPERGROUP"	REPRESENTATIVE "PROTIST" CLADES	KEY CHARACTERS
EXCAVATES Unicellular protists with atypical, greatly modified mitochondria; bikonts	**Diplomonads and parabasalids**	Two or more flagella; ventral oral (feeding) groove
	Euglenoids and trypanosomes	Some with plastids; crystalline rod in flagella
CHROMALVEOLATES Diverse protists that may have originated as a result of secondary endosymbiosis in which an ancestral cell engulfed a red alga; bikonts	**Alveolates** Dinoflagellates, ciliates, apicomplexans	Alveoli (flattened vesicles) just inside the plasma membrane
	Stramenopiles Water molds, diatoms, brown algae, golden algae	Most have two flagella, one with hairs; no flagella in some
RHIZARIANS Amoeboid cells that often have tests (shells); bikonts	**Forams**	Porous tests (hard shells) through which cytoplasmic projections (pseudopods) extend
	Actinopods	Endoskeletons (internal shells) through which axopods (filamentous pseudopods) extend
ARCHAEPLASTIDS Plastids bounded by outer and inner membranes; include land plants; bikonts	**Red algae**	Chloroplast pigments include phycoerythrin (red pigment) and phycocyanin (blue pigment)
	Green algae	Chloroplast pigments identical to those in land plants
UNIKONTS Cells that have a single flagellum or are amoebas with no flagella; have a triple-gene fusion that is lacking in other eukaryotes; include animals and fungi	**Amoebozoa** Amoebas, plasmodial slime molds, cellular slime molds	Naked amoebas (no tests) with lobelike pseudopods
	Opisthokonts Choanoflagellates	No flagella or single posterior flagellum on motile cells

© Cengage Learning

26.3 EXCAVATES

○LEARNING OBJECTIVE

4 Summarize the basic features of excavates and distinguish among diplomonads, parabasilids, and euglenoids.

Excavates are a diverse group of unicellular protists with flagella. These protists are so named because many have a deep, or *excavated,* oral groove. Unlike other protists, excavates have atypical, greatly modified mitochondria. Many excavates are endosymbionts and live in anoxic (without oxygen) environments. These excavates do not carry out aerobic respiration; they obtain energy by the anaerobic pathway of glycolysis (presumably by fermentation).

Currently, excavates include diplomonads, parabasilids, euglenoids, and trypanosomes. The inclusion of these organisms into a single superfamily is somewhat controversial because their relationships to one another are uncertain. Additional studies will be needed to determine if the excavates as currently presented are a monophyletic group.

Diplomonads are small, mostly parasitic flagellates

Diplomonads are excavates that have one or two nuclei, no functional mitochondria, no Golgi complex, and up to eight flagella. *Giardia* is a parasitic diplomonad (FIG. 26-4a). Interestingly, *Giardia* has two haploid nuclei, each of which contains a complete copy of *Giardia's* genome. *Giardia* lacks functional mitochondria, although it contains certain genes that code for proteins associated with mitochondria in other organisms. *Giardia* also has reduced structures that somewhat resemble mitochondria. This information suggests to some biologists that an early eukaryotic ancestor of *Giardia* may have possessed mitochondria, which were somehow lost or reduced at a later time during its evolutionary history.

Giardia intestinalis is a major cause of water-borne diarrhea throughout the world. *Giardia* is eliminated as a resistant cyst in the feces of many vertebrate animals. These cysts are a common contaminant in untreated drinking water, even in environments considered relatively pristine, such as isolated mountain streams. In a heavy infection, much of the wall of the small intestine is coated with these flagellates, which interfere with the absorption of digested nutrients and cause weight loss, abdominal cramps, and diarrhea.

Parabasilids are anaerobic endosymbionts that live in animals

Parabasilids are anaerobic, flagellated excavates that often live in animals. Trichonymphs and trichomonads are examples of parabasilids. *Trichonymphs,* which have hundreds of flagella, live in the guts of termites and wood-eating cockroaches (FIG. 26-4b). Trichonymphs ingest wood chips from the wood that termites or roaches eat. The trichonymphs rely on endosymbiotic bacteria to digest cellulose in the wood. Thus, in an excellent example of mutualism, the insects, trichonymphs, and bacteria all obtain their nutrients from cellulose.

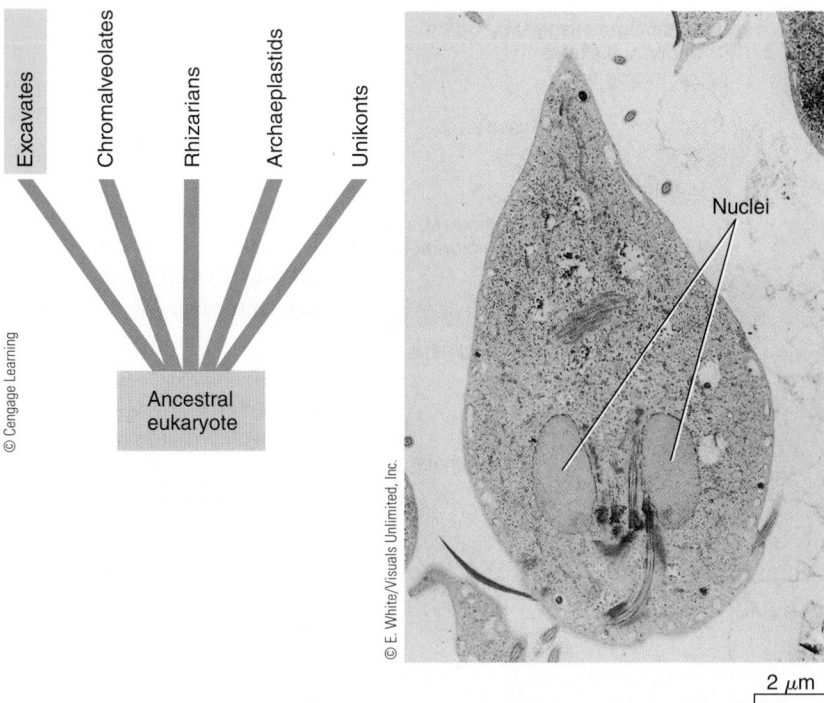

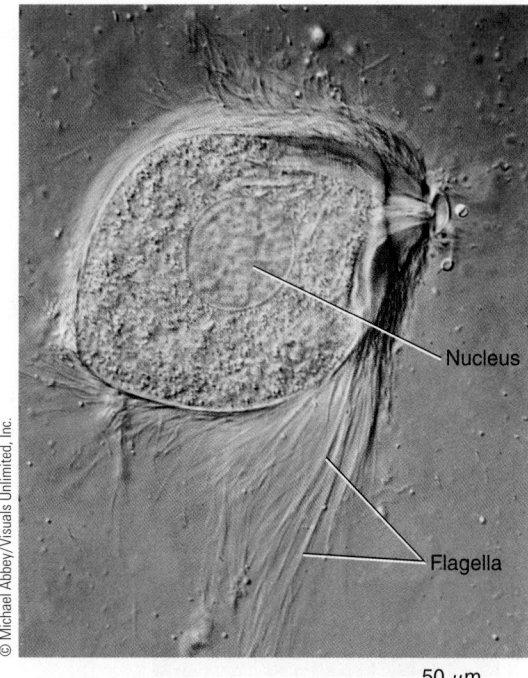

(a) **Diplomonad.** This colorized TEM of *Giardia intestinalis,* a parasitic diplomonad, reveals two nuclei.

(b) **Parabasalid.** LM of *Trichonympha,* a parabasalid that lives in the gut of wood-eating termites and cockroaches. *Trichonympha* has hundreds of flagella.

Figure 26-4 Diplomonads and parabasalids

The most well-known trichomonad is probably *Trichomonas vaginalis,* which causes trichomoniasis, a curable sexually transmitted disease (STD) in humans. Trichomoniasis affects both men and women, although the symptoms are more obvious in women. According to the Centers for Disease Control and Prevention, about 7.4 million new cases occur in the United States each year.

Euglenoids and trypanosomes include both free-living species and parasites

Euglenoids and *trypanosomes* are characterized by an unusual flagellum: in addition to the 9 + 2 arrangement of microtubules characteristic of all eukaryotic flagella, these excavates have a crystalline rod in their flagella; the function of this rod is unknown. Like other excavates, euglenoids and trypanosomes also have atypical mitochondria.

Most **euglenoids** are unicellular flagellates, and about one-third of them are photosynthetic (**FIGS. 26-5a and b**). They generally have two flagella: one long and whiplike and one that is often so short that it does not extend outside the cell. Some euglenoids, such as *Euglena,* change shape continually as they move through the water because their **pellicle,** or outer covering, is flexible.

Autotrophic euglenoids have chloroplasts with the same photosynthetic pigments that green algae and plants have. However, these chloroplasts were acquired by secondary endosymbiosis; euglenoids are not closely related to either

group, as shown in Figure 26-3. Some photosynthetic euglenoids lose their chlorophyll when grown in the dark, and they obtain their nutrients heterotrophically, by ingesting organic matter. Other euglenoids are always colorless and heterotrophic. Some heterotrophic species absorb organic compounds from the surrounding water, whereas others engulf bacteria and protists by **phagocytosis;** they digest the prey within food vacuoles.

Trypanosomes are excavates with a single mitochondrion that has an organized deposit of DNA called a **kinetoplastid.** Trypanosomes are colorless, and many are parasitic and cause disease. In vertebrates, including humans, trypanosomes live in the blood. For example, Trypanosoma *brucei* is a human parasite that causes African sleeping sickness (**FIG. 26-5c**). It is transmitted by the bite of infected tsetse flies. Early symptoms include recurring attacks of fever. Later, when the trypanosomes have invaded the central nervous system, infected people are lethargic, have difficulty speaking or walking, and may die if the disease is untreated. Control efforts are showing some success; the incidence of the disease dropped from at least 50,000 new cases in 2005 to an estimated 30,000 in 2010.

CHECKPOINT 26.3

- *What cell organelle is atypical in excavates?*
- *Give an example of a human disease caused by each of the following: diplomonads, parabasilids, and trypanosomes.*

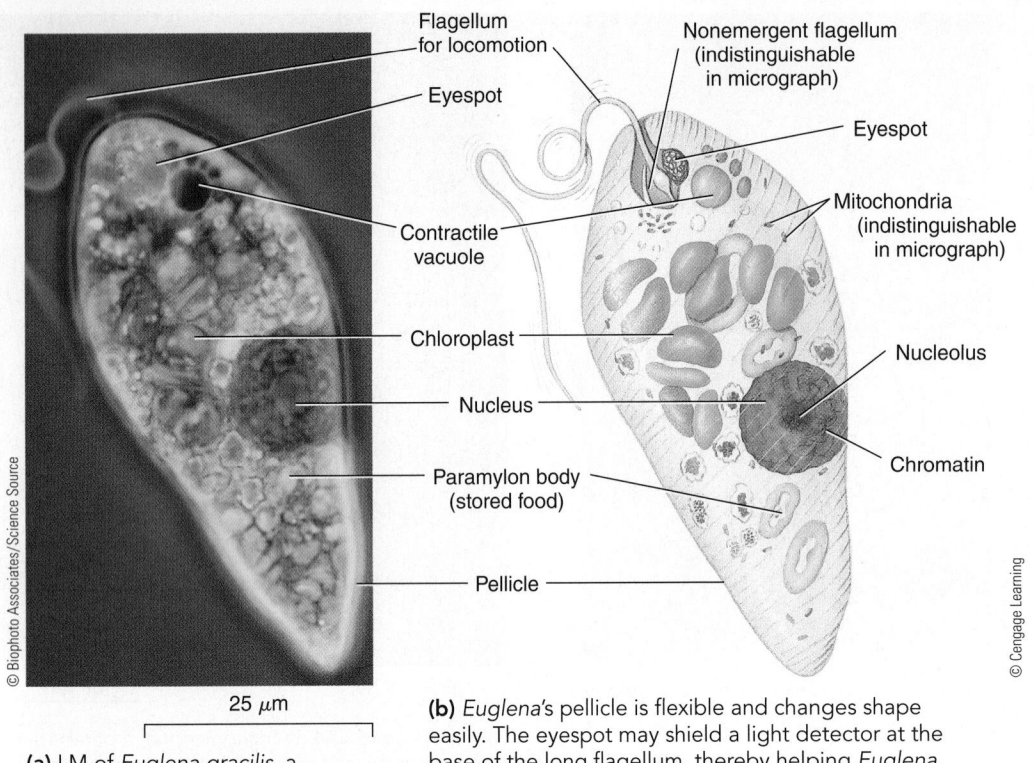

Flagellum for locomotion

Eyespot

Contractile vacuole

Chloroplast

Nucleus

Paramylon body (stored food)

Pellicle

Nonemergent flagellum (indistinguishable in micrograph)

Eyespot

Mitochondria (indistinguishable in micrograph)

Nucleolus

Chromatin

© Biophoto Associates/Science Source

25 μm

(a) LM of *Euglena gracilis*, a unicellular, flagellate euglenoid.

© Cengage Learning

(b) *Euglena*'s pellicle is flexible and changes shape easily. The eyespot may shield a light detector at the base of the long flagellum, thereby helping *Euglena* move to light of an appropriate intensity.

© Eye of Science/Science Source

Red blood cells

Trypanosome with undulating membrane

Flagellum

10 μm

(c) SEM of the flagellate *Trypanosoma brucei* among human red blood cells. *Trypanosoma brucei* causes African sleeping sickness in humans.

Figure 26-5 *Animation* **Euglenoids and trypanosomes**

26.4 CHROMALVEOLATES

LEARNING OBJECTIVES

5 Contrast the two main groups of chromalveolates: alveolates and stramenopiles.

6 Distinguish among the alveolates: dinoflagellates, apicomplexans, and ciliates.

7 Distinguish among the stramenopiles: water molds, diatoms, golden algae, and brown algae.

The **chromalveolates** are a supergroup composed of extremely diverse protists with few shared characters. Chromalveolates probably originated as a result of secondary endosymbiosis in which an ancestral cell engulfed a red alga (which itself was the result of primary endosymbiosis). Most chromalveolates are photosynthetic, and evidence suggests that heterotrophic chromalveolates, such as the water molds and ciliates, descended from autotrophic ancestors. Classification of the chromalveolates as a monophyletic supergroup is controversial because some DNA sequence data have indicated that they are not monophyletic. The chromalveolates are divided into two main groups: alveolates and stramenopiles.

The unifying features of protists classified as **alveolates** include similar ribosomal DNA sequences and **alveoli** (sing., *alveolus*), flattened vesicles located just inside the plasma membrane. In some alveolates the vesicles contain plates of cellulose. Alveolates include the dinoflagellates, apicomplexans, and ciliates.

The **stramenopiles** include water molds, diatoms, golden algae, and brown algae. At first glance stramenopiles appear too diverse to classify together. However, most stramenopiles have motile cells with two flagella, one of which has tiny hairlike projections extending from the shaft. The word *stramenopile* comes from Latin words referring to "straw" (i.e., the shaft of the flagellum) and "hairs."

Most dinoflagellates are a part of marine plankton

Dinoflagellates are generally unicellular, although a few are colonial. Their alveoli contain interlocking cellulose plates impregnated with silicates. The typical dinoflagellate has two flagella. One flagellum wraps around a transverse groove in the center of the cell like a belt, and the other lies in a longitudinal groove (perpendicular to the transverse groove), projecting behind the cell (**FIG. 26-6a**). The undulation of these flagella propels the dinoflagellate through the water like a spinning top. Indeed, the name is derived from the Greek *dinos*, meaning "whirling." Many marine dinoflagellates are bioluminescent.

Many dinoflagellates are photosynthetic, but others are heterotrophic and ingest other microorganisms for food. Some dinoflagellates are endosymbionts that live in the bodies

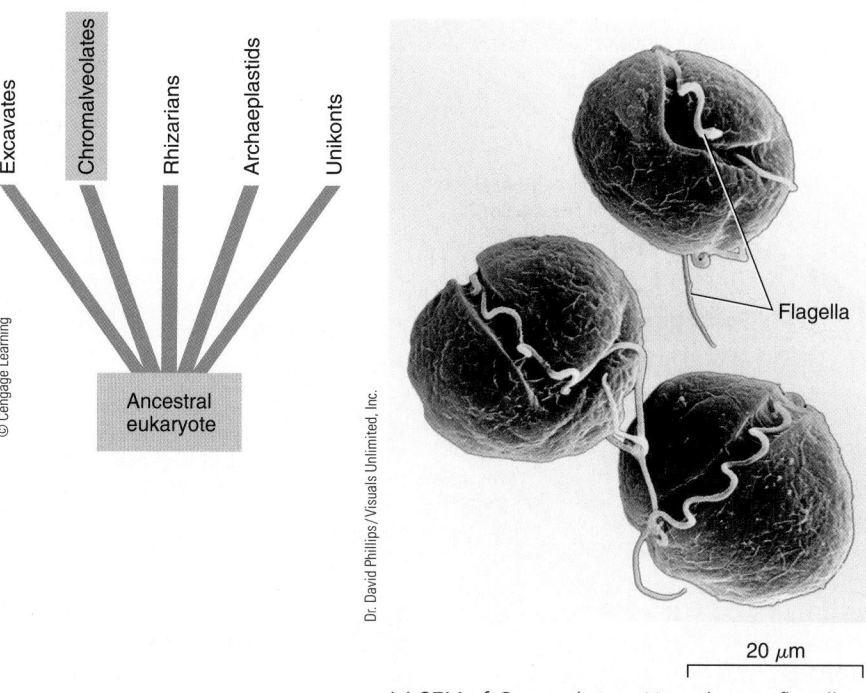

Flagella

20 μm

(a) SEM of *Gymnodinium*. Note the two flagella, which are located in grooves.

(b) Red tide in Mexico. Billions of dinoflagellates produce the orange cloudiness in the water.

Figure 26-6 Dinoflagellates

of marine invertebrates such as mollusks, jellyfish, and corals (see Fig. 54-12). These symbiotic dinoflagellates, called **zooxanthellae,** photosynthesize and provide carbohydrates for their invertebrate partners. Zooxanthellae contribute substantially to the productivity of coral reefs. Other dinoflagellates that are endosymbionts lack pigments and are parasites that live off their hosts.

Ecologically, dinoflagellates are important producers in marine ecosystems. A few dinoflagellates are known to have occasional population explosions, or blooms. These blooms, known as **red tides,** frequently color coastal waters orange, red, or brown (**FIG. 26-6b**). Dinoflagellate blooms are particularly common in warm, nutrient-enriched water. Some dinoflagellate species that form red tides produce a toxin that attacks the nervous systems of fishes, leading to fish kills. Birds sometimes die after eating contaminated fish. Research has also linked red tides to manatee and dolphin deaths in Florida.

Apicomplexans are spore-forming parasites of animals

Apicomplexans are a large group of parasitic, spore-forming alveolates, some of which cause serious diseases in humans. As discussed earlier in the chapter, they contain the unpigmented remnant of a chloroplast derived from a red alga. Apicomplexans lack specific structures for locomotion (cilia, flagella, or pseudopodia) and move by flexing.

Apicomplexans have an *apical complex* of microtubules that attaches the parasite to its host cell; the apical complex is visible only by using electron microscopy. They also have the ability to form a structure known as a *moving junction,* which enables them to form a vacuole that encloses and protects them as they invade a host cell. At some stage in their life cycle, apicomplexans produce **sporozoites,** small infective agents transmitted to the next host. Many apicomplexans spend part of their complex life cycle in one host species and part in a different host species.

Malaria is caused by an apicomplexan (**FIG. 26-7**). According to the World Health Organization, there were approximately 219 million cases of malaria worldwide in 2010; about 660,000 people, mostly children in developing countries, died from the disease. Although for centuries Chinese, Greek, Arabic, and Roman writings described the disease, its causative agent and mode of transmission by means of mosquitoes were not identified until the end of the 19th century. British scientist Ronald Ross received the Nobel Prize in Physiology or Medicine in 1902 for his role in elucidating the life cycle of *Plasmodium,* the apicomplexan that causes malaria.

Control measures that have lost effectiveness due to the evolution of chemotherapy resistance on the part of *Plasmodium* and pesticide resistance on the part of the mosquitoes are currently being replaced by newer methods, but the battle is ongoing. For example, chemotherapies that include the drug artemisinin in combination with other drugs have emerged as the standard of care, but there are signs that resistance of the parasite to artemisinin is beginning to evolve in some countries.

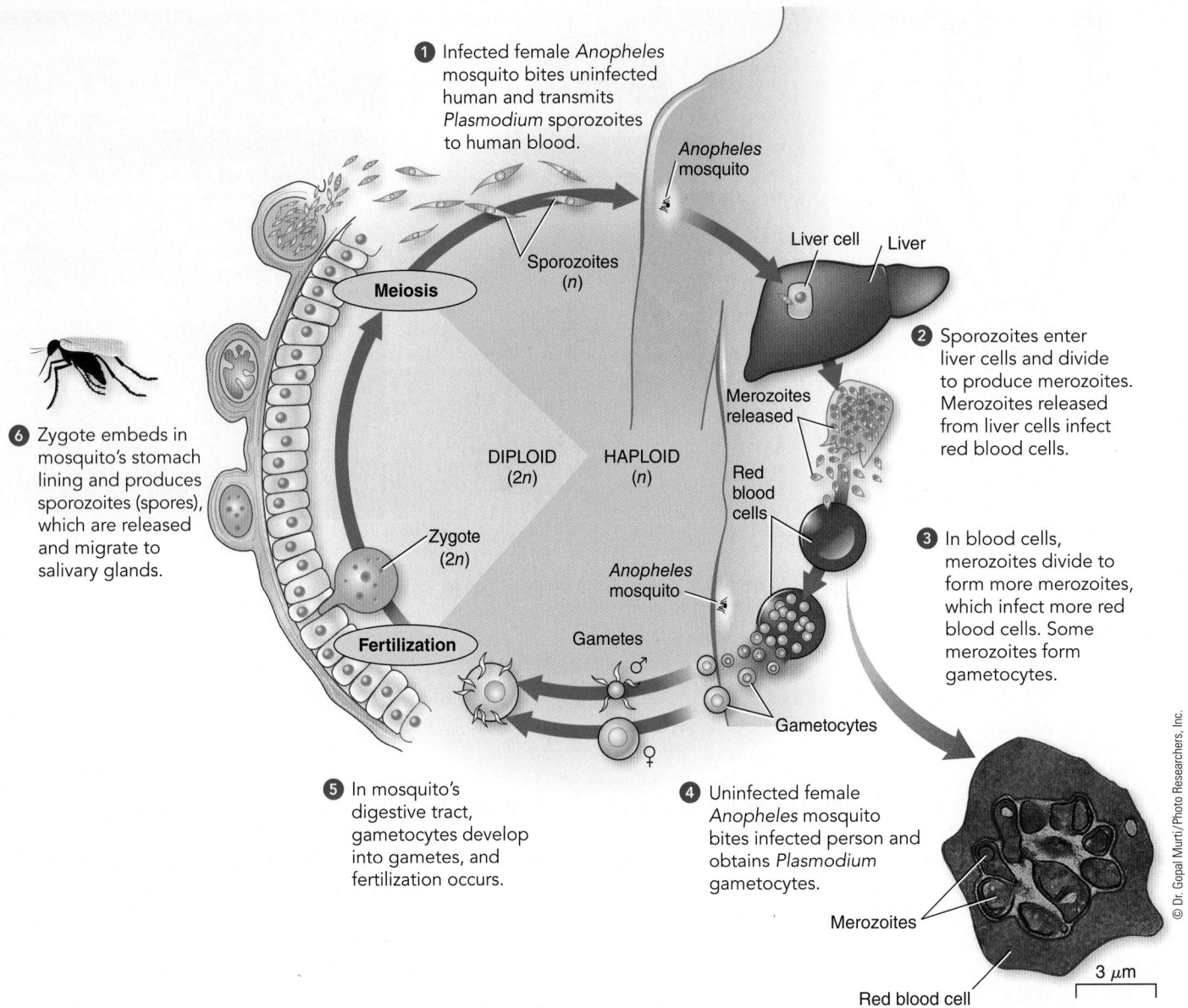

Figure 26-7 Animation Life cycle of *Plasmodium*, the causative agent of malaria

Plasmodium, an apicomplexan, lives in two hosts: mosquitoes and humans. The inset shows a TEM of a colorized human red blood cell filled with merozoites.

© Cengage Learning

The figure contains the following labels:

1. Infected female *Anopheles* mosquito bites uninfected human and transmits *Plasmodium* sporozoites to human blood.

2. Sporozoites enter liver cells and divide to produce merozoites. Merozoites released from liver cells infect red blood cells.

3. In blood cells, merozoites divide to form more merozoites, which infect more red blood cells. Some merozoites form gametocytes.

4. Uninfected female *Anopheles* mosquito bites infected person and obtains *Plasmodium* gametocytes.

5. In mosquito's digestive tract, gametocytes develop into gametes, and fertilization occurs.

6. Zygote embeds in mosquito's stomach lining and produces sporozoites (spores), which are released and migrate to salivary glands.

Meiosis · Sporozoites (*n*) · *Anopheles* mosquito · Liver cell · Liver · Merozoites released · Red blood cells · DIPLOID (2*n*) · HAPLOID (*n*) · Zygote (2*n*) · *Anopheles* mosquito · Fertilization · Gametes · Gametocytes · Merozoites · Red blood cell · 3 μm

© Dr. Gopal Murti/Photo Researchers, Inc.

Researchers are currently testing new antimalarial drugs and several possible vaccines against malaria. The sequencing of the genomes of *P. falciparum* and the *Anopheles* mosquito may lead to new diagnostics, drugs, and vaccines.

Ciliates use cilia for locomotion

Ciliates are among the most complex of eukaryotic cells. These unicellular alveolates have a pellicle that gives them a definite but changeable shape. In *Paramecium* the surface of the cell is covered with several thousand fine, short, hairlike **cilia** that extend through pores in the pellicle to facilitate movement (FIGS. 26-8a and b; also see Fig. 4-5c). The cilia beat with such precise coordination that the organism can back up and turn around as well as move forward.

Not all ciliates are motile. Some sessile forms have stalks, and others, although capable of some swimming, are more likely to remain attached to a rock or other surface at one spot. Their cilia set up water currents that draw food toward them.

Ciliates differ from other protists in having two kinds of nuclei: one or more small, diploid **micronuclei** that function in reproduction; and a larger, polyploid **macronucleus** that controls cell metabolism and growth. Most ciliates are capable of a sexual process called **conjugation,** in which two individuals come together and exchange genetic material (FIG. 26-8c). Conjugation results in two "new" cells that are genetically identical to each other but different from what they were before conjugation. Mitosis and cell division need not follow immediately after conjugation. Ciliates usually divide perpendicularly to their longitudinal axis.

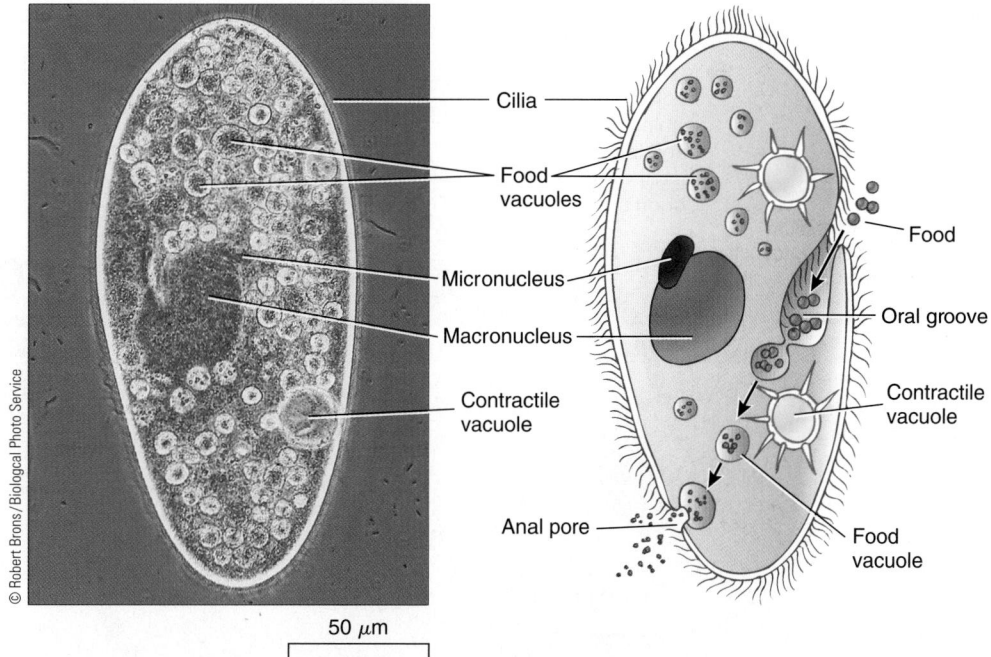

Cilia

Food vacuoles

Micronucleus

Macronucleus

Contractile vacuole

Food

Oral groove

Contractile vacuole

Anal pore

Food vacuole

50 μm

© Robert Brons / Biological Photo Service

(a) Note the complex cell structure in this phase-contrast LM of *Paramecium,* a freshwater ciliate. Like many ciliates, *Paramecium* has multiple nuclei: a macronucleus and one or more smaller micronuclei.

(b) Food particles are swept into *Paramecium's* ciliated oral groove and incorporated into food vacuoles. Lysosomes fuse with the food vacuoles, and the food is digested and absorbed; undigested wastes are eliminated through the anal pore.

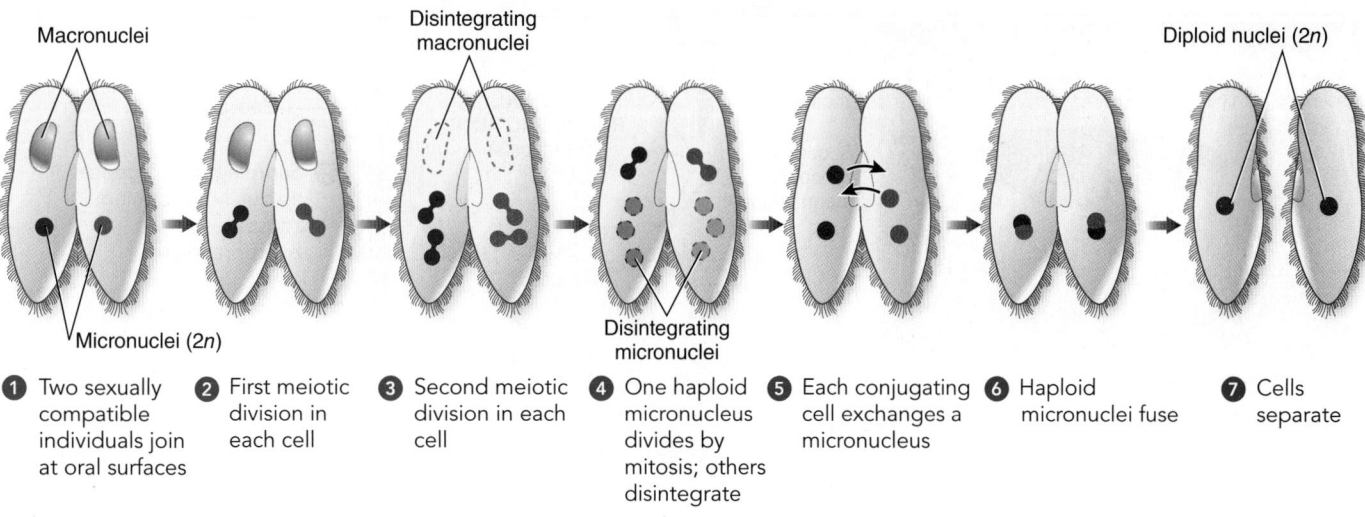

Macronuclei

Disintegrating macronuclei

Diploid nuclei (2*n*)

Micronuclei (2*n*)

Disintegrating micronuclei

① Two sexually compatible individuals join at oral surfaces

② First meiotic division in each cell

③ Second meiotic division in each cell

④ One haploid micronucleus divides by mitosis; others disintegrate

⑤ Each conjugating cell exchanges a micronucleus

⑥ Haploid micronuclei fuse

⑦ Cells separate

(c) Conjugation in *Paramecium caudatum.*

Figure 26-8 *Animation* **Ciliates**

© Cengage Learning

Water molds produce biflagellate reproductive cells

Water molds are stramenopiles that were once classified as fungi because of their superficial resemblance. Both water molds and fungi have a body, called a **mycelium,** that grows over organic material, digesting it and then absorbing the predigested nutrients (FIG. 26-9). The threadlike **hyphae** that make up

the mycelium in water molds are *coenocytic,* meaning that there are no cross walls; the body consists of a single multinucleate cell. The cell walls of water molds are composed of cellulose (as in plants), chitin (as in fungi), or both.

When food is plentiful and environmental conditions are favorable, water molds reproduce asexually. A hyphal tip swells, and a cross wall is formed, separating the hyphal tip from the rest of the mycelium. Within this structure, called

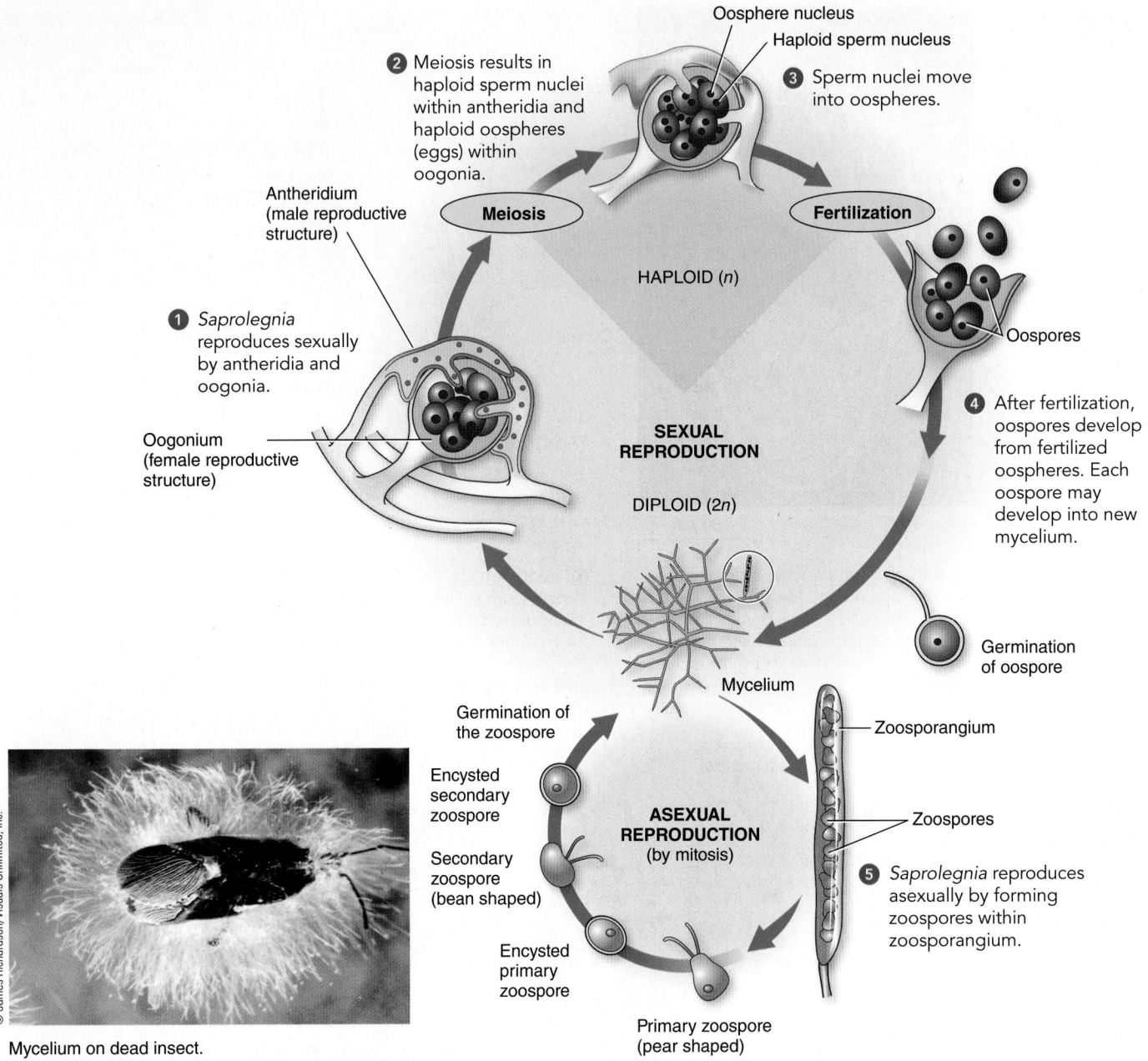

② Meiosis results in haploid sperm nuclei within antheridia and haploid oospheres (eggs) within oogonia.

Oosphere nucleus
Haploid sperm nucleus

③ Sperm nuclei move into oospheres.

Meiosis

Fertilization

HAPLOID (*n*)

Antheridium (male reproductive structure)

① *Saprolegnia* reproduces sexually by antheridia and oogonia.

Oogonium (female reproductive structure)

SEXUAL REPRODUCTION

DIPLOID (*2n*)

Oospores

④ After fertilization, oospores develop from fertilized oospheres. Each oospore may develop into new mycelium.

Germination of oospore

Mycelium

Zoosporangium

Germination of the zoospore

Encysted secondary zoospore

ASEXUAL REPRODUCTION (by mitosis)

Zoospores

Secondary zoospore (bean shaped)

⑤ *Saprolegnia* reproduces asexually by forming zoospores within zoosporangium.

Encysted primary zoospore

Primary zoospore (pear shaped)

© James Richardson/Visuals Unlimited, Inc.

Mycelium on dead insect.

Figure 26-9 Water molds

The life cycle of *Saprolegnia*, a water mold. *Inset:* A mycelium of *Saprolegnia* radiates from a dead insect.
© Cengage Learning

a **zoosporangium,** tiny biflagellate **zoospores** form, each of which swims about, lands and encysts, and eventually develops into a new mycelium. When environmental conditions worsen, water molds initiate sexual reproduction. After fusion of male and female nuclei, thick-walled **oospores** develop from the oospheres (female gametes). Water molds often spend the winter as oospores.

Some water molds have played a profound role in human history. For example, the Irish potato famine of the 19th century was precipitated by the water mold *Phytophthora infestans,* which causes late blight of potatoes. (The genus *Phytophthora* is named from Greek words meaning "plant destruction.") During several

rainy, cool summers in Ireland in the 1840s, the water mold multiplied unchecked, causing potato tubers to rot in the fields. Because potatoes were the staple of the Irish peasants' diet, as many as one million people starved. The famine prompted a mass migration out of Ireland to the United States and other countries.

A close relative of the late blight water mold, *P. ramorum,* causes sudden oak death, which is killing oak forests in several western states. Plant pathologists are concerned that the disease may spread to midwestern and eastern forests. This particular water mold also attacks redwoods, Douglas firs, bay trees, maples, and several other plant species, but most have only twig and leaf infections, not the rapid death observed in oaks.

Diatoms are stramenopiles with shells composed of two parts

Most **diatoms** are unicellular, although a few exist as colonies (see chapter-opener figure). The cell wall of each diatom consists of two shells that overlap where they fit together, much like a petri dish. Silica is deposited in the shell, and this glasslike material is laid down in intricate patterns (FIG. 26-10a). There are two basic groups of diatoms: those with radial symmetry (wheel shaped) and those with bilateral symmetry (boat shaped or needle shaped). Although some diatoms are part of the floating **plankton,** others live on rocks and sediments, where they move by gliding. This gliding movement is facilitated by the secretion of a slimy material from a small groove along the shell.

Diatoms most often reproduce asexually by mitosis. When a diatom divides, the two halves of its shell separate, and each becomes the larger half of a new diatom shell (FIG. 26-10b). Because the glass shell cannot grow, some diatom cells get progressively smaller with each succeeding generation. When a diatom reaches a fraction of its original size, sexual reproduction occurs, with the production of shell-less gametes. Sexual reproduction restores the diatom to its original size because the resulting *zygote,* a 2*n* cell that results from the fusion of *n* gametes, grows substantially before producing a new shell.

Diatoms are common in fresh water but are especially abundant in relatively cool ocean water. They are ecologically significant as major producers in aquatic ecosystems. At least one species is toxic and linked to shellfish poisonings, marine mammal strandings, and the deaths of sea lions along the central California coast.

When diatoms die, their shells trickle to the ocean floor and accumulate in layers that eventually become sedimentary rock. After millions of years, geologic upheaval has exposed some of these deposits on land. Called *diatomaceous earth,* these deposits are mined and used as filtering, insulating, and soundproofing materials. As a filtering agent, diatomaceous earth is used to refine raw sugar and process vegetable oils. Because of its abrasive properties, diatomaceous earth is a common ingredient in scouring powders and metal polishes; it is no longer added to most toothpastes because it is too abrasive for tooth enamel. The intricately detailed diatom shells are often used to test microscope resolution down to 1 μm.

Brown algae are multicellular stramenopiles

Brown algae are the largest and most complex of all algae commonly called seaweeds. All brown algae are multicellular and range in size from a few centimeters (about an inch) to 75 m (about 260 ft). Their body forms are branched filaments; tufts; fleshy "ropes"; or thick, flattened branches. The largest brown algae, called *kelps,* are tough and leathery in appearance.

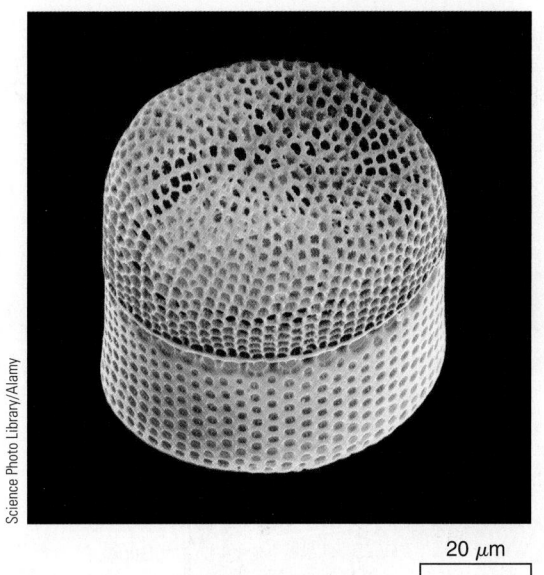

20 μm

(a) False color SEM of a unicellular diatom. Note the striking shell that contains silica.

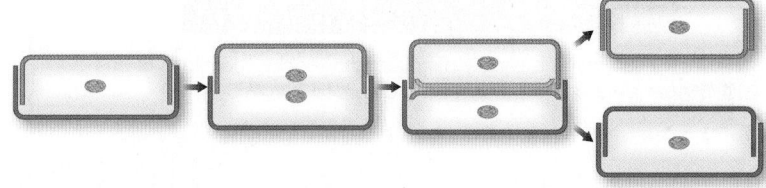

(b) Asexual reproduction in diatoms. After cell division, each new cell retains half of the original shell. The newly synthesized half of the shell always fits *inside* the original half. As a result, one of the new cells is slightly smaller than the other.

Figure 26-10 Diatoms

© Cengage Learning

Many kelps have leaflike **blades** in which most photosynthesis occurs, stemlike **stipes,** and rootlike anchoring **holdfasts** (FIG. 26-11a). They often have gas-filled *bladders* that provide buoyancy. (The blades, stipes, and holdfasts of brown algae are not homologous to the leaves, stems, and roots of plants. Brown algae and plants arose from different unicellular ancestors, as shown in Figure 26-3.)

Reproduction is varied and complex in the brown algae. Their reproductive cells, both asexual zoospores and sexual gametes, are usually biflagellate. Most have a life cycle that exhibits **alternation of generations,** in which they spend part of their life as multicellular haploid organisms and part as multicellular diploid organisms (see Fig. 10-19c).

Brown algae are commercially important for several reasons. Their cell walls contain a polysaccharide called *algin* that is harvested from kelps such as *Macrocystis* and used as a thickening and stabilizing agent in ice cream, toothpaste, shaving cream, hair spray, and hand lotion. Brown algae are an important human food, particularly in eastern Asia, and they are rich sources of certain vitamins and minerals such as iodine.

Blade

Stipe

Holdfast

(a) *Laminaria* is widely distributed on rocky coastlines of temperate and polar seas. It grows to 2 m (6.5 ft).

(b) A kelp (*Macrocystis pyrifera*) bed is ecologically important to aquatic organisms, including the sea lion shown here. Photographed off the coast of California.

Figure 26-11 Brown algae

Brown algae are common in cooler marine waters, especially along rocky coastlines, where they live mainly in the intertidal zone or relatively shallow offshore waters. Kelps form extensive underwater "forests," or kelp beds (FIG. 26-11b). They are essential in that ecosystem as important food producers, and they provide habitat for many marine invertebrates, fishes, and mammals. The diversity of life supported by kelp beds rivals that found in coral reefs.

of golden algae gives them a golden or golden brown color. A few species ingest bacteria and other particles of food. Ecologically, golden algae are important producers in marine environments. They compose a significant portion of the ocean's **nanoplankton,** extremely minute algae (2 to 10 μm) that are major producers because of their great abundance.

Most golden algae are unicellular biflagellates

Golden algae are found in both freshwater and marine environments. Most species are biflagellate, unicellular organisms, although some are colonial (FIG. 26-12a). A few of these stramenopiles lack flagella and are similar to amoebas in appearance except that golden algae contain chloroplasts. Tiny scales of either silica or calcium carbonate may cover the cells. Reproduction in golden algae is primarily asexual and involves the production of biflagellate, motile *zoospores*.

Most golden algae are photosynthetic, and the pigment composition

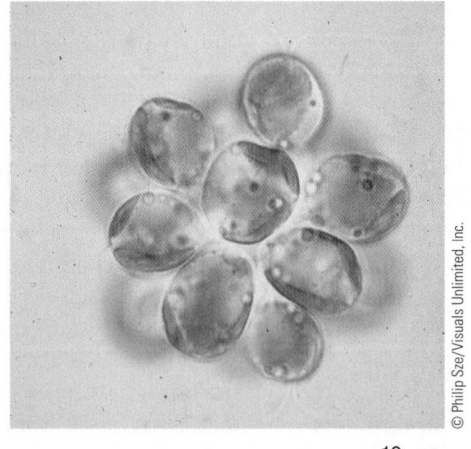

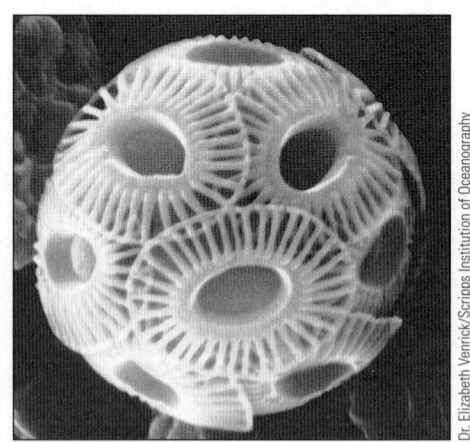

10 μm

1 μm

(a) LM of a colonial golden alga (*Synura*) found in freshwater lakes and ponds.

(b) SEM of a coccolithophorid (*Emiliania huxleyi*). Note the overlapping scales of calcium carbonate.

Figure 26-12 Golden algae

Classification of golden algae is controversial. Some biologists lump diatoms and golden algae in a single phylum, whereas others classify both groups as brown algae. At the other extreme, some biologists divide the golden algae into two phyla by placing many of the marine species, such as *coccolithophorids* (FIG. 26-12b), in a separate phylum.

CHECKPOINT 26.4

- *What are the two main groups of chromalveolates? What are their distinguishing characters?*
- *Why do some biologists think that the apicomplexans descended from dinoflagellates?*
- *Which water mold has influenced human history? Explain your answer.*
- *What is the ecological significance of the diatoms? the brown algae?*

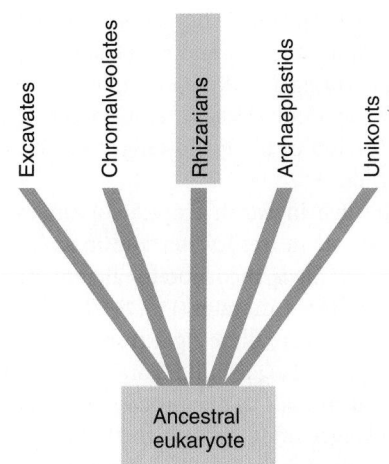

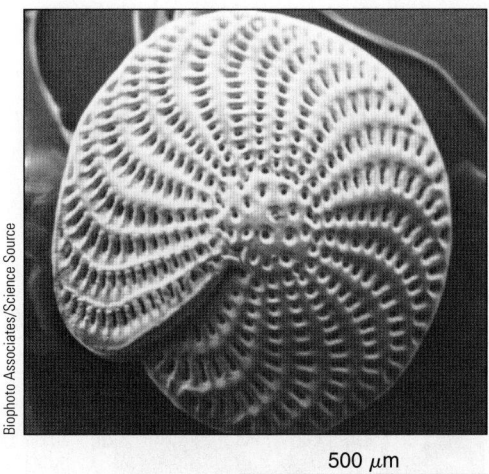

(a) SEM of a foram test. Note the pores through which cytoplasm extrudes.

Figure 26-13 Rhizarians

26.5 RHIZARIANS

LEARNING OBJECTIVE

8 Describe the forams and actinopods, and explain why many biologists classify them in the rhizarian supergroup.

Rhizarians are a diverse supergroup of amoeboid cells that often have hard outer shells, called **tests,** through which cytoplasmic projections extend. The threadlike cytoplasmic projections suggest the name *rhizarian,* from the Greek *rhiza,* meaning "root." Forams and actinopods are rhizarians, as are certain shell-less amoebas. Not all amoebas are rhizarians, however, and many amoeba species are more closely related to other eukaryotic clades. Current molecular evidence indicates that the rhizarian supergroup is monophyletic.

Forams extend cytoplasmic projections that form a threadlike, interconnected net

Almost all **foraminiferans (forams)** are marine rhizarians that produce elaborate tests (FIG. 26-13a). The ocean contains enormous numbers of forams, which secrete chalky, many-chambered tests with pores through which cytoplasmic projections are extended. (*Foraminifera* is derived from the Latin for "bearing openings.") The cytoplasmic projections

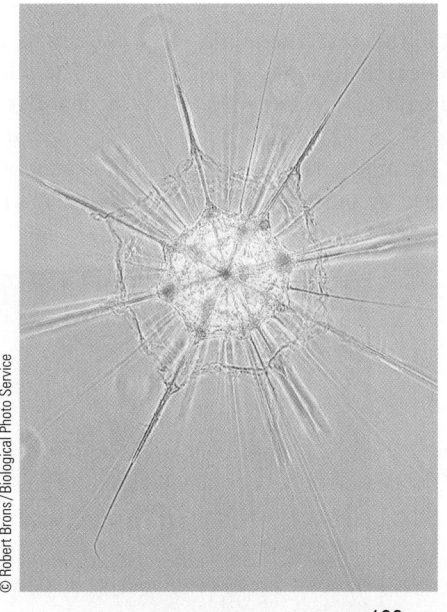

(b) LM of an unidentified living actinopod from the Red Sea. Many slender axopods project from the cell. The shell is not visible because it is an endoskeleton and cytoplasm covers it on all sides.

form a sticky, interconnected net that entangles prey. Many forams contain unicellular algal endosymbionts (green algae, red algae, or diatoms) that provide food by photosynthesis. Many foram species live on the ocean floor, but others are part of the plankton.

Dead forams settle on the bottom of the ocean, where their tests form a gray mud that is gradually transformed into chalk. With geologic uplifting, these chalk formations become part of the land, as in the White Cliffs of Dover in England. (The White Cliffs of Dover are the remains of a variety of carbonate organisms, not only forams.) Because foram tests often appear in rock layers covering oil deposits, geologists exploring for oil look for foram tests in rock strata. Forams are well preserved in the fossil record, and biologists use some as **index fossils,** markers to help identify ancient sedimentary rock layers (see Chapter 18).

Actinopods project slender axopods

Actinopods are mostly marine plankton rhizarians with long, filamentous cytoplasmic projections called **axopods** that protrude through pores in their shells (FIG. 26-13b). A cluster of microtubules strengthens each axopod. Unicellular algae and other prey become entangled in these axopods and are engulfed outside the main body of the actinopod; cytoplasmic streaming carries the prey inside the body. Many actinopods contain algal endosymbionts that provide them with the products of photosynthesis.

Some actinopods, called **radiolarians,** secrete elaborate, beautiful glassy shells made of silica. Radiolarians are an important constituent of marine plankton. When radiolarians and other actinopods die, their shells settle and become an ooze (sediment) that may be several meters thick on the ocean floor.

CHECKPOINT 26.5

- *What character does the term* rhizarian *indicate about the forams and actinopods?*

26.6 ARCHAEPLASTIDS

LEARNING OBJECTIVE

9 Describe evidence supporting the hypothesis that red algae and green algae should be included in a monophyletic group with land plants.

In the classification scheme adopted in this text, the monophyletic group of **archaeplastids** includes red algae and green algae, which are discussed here, and land plants, which are in a separate kingdom (see Chapters 27 and 28). Biologists classify these groups together based on molecular data and on the presence of chloroplasts bounded by outer and inner membranes, suggesting that they developed directly from a cyanobacterial endosymbiont. All photosynthetic protists other than archaeplastids have plastids surrounded by three or four membranes.

Red algae do not produce motile cells

The vast majority of **red algae** are multicellular organisms, although there are a few unicellular species. The multicellular body form of red algae commonly consists of complex, interwoven filaments that are delicate and feathery (FIG. 26-14a); a few red algae are flattened sheets of cells. Most multicellular red algae attach to rocks or other substrates by a basal holdfast. Reproduction in the red algae is remarkably complex, with an alternation of sexual and asexual stages. No flagellate cells develop during the life cycle. Some unicellular red algae may exhibit unusual features. For example, in 2013 a team of scientists working in the United States, Germany, and France reported on the genome of a unicellular extremophilic red alga adapted to toxic hot sulfur springs. The genome includes at least 75 genes acquired from archaea and bacteria needed for this extreme lifestyle, providing a striking example of horizontal gene transfer involving a eukaryote. (See Chapter 23 and Figure 23-4a for a discussion of the evolutionary significance of horizontal gene transfer.)

Red algae primarily live in warm tropical ocean waters, although a few species occur in fresh water and in soil. Some red algae, known as *coralline algae,* incorporate calcium carbonate in their cell walls from the ocean water (FIG. 26-14b). The hard calcium carbonate may protect coralline algae from the rigors of wave action. These coralline red algae build "coral" reefs and are perhaps as crucial as coral animals in this process.

The cell walls of red algae often contain thick, sticky polysaccharides that have commercial value. For example, *agar* is a polysaccharide extracted from certain red algae used as a food thickener and culture medium, a substrate on which to grow microorganisms and propagate some plants, such as orchids.

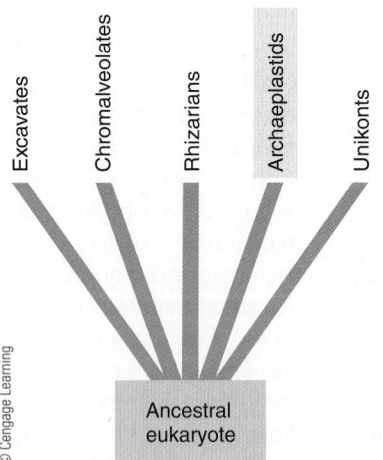

Excavates | Chromalveolates | Rhizarians | Archaeplastids | Unikonts

Ancestral eukaryote

© Cengage Learning

© Philip Sze/Visuals Unlimited, Inc.

(a) *Polysiphonia,* which is widely distributed throughout the world, has a highly branched body of interwoven filaments.

© D. Gotshall/Visuals Unlimited, Inc.

(b) *Bossiella* is a coralline red alga encrusted with calcium carbonate. It lives in the Pacific Ocean.

Figure 26-14 *Animation* **Red algae**

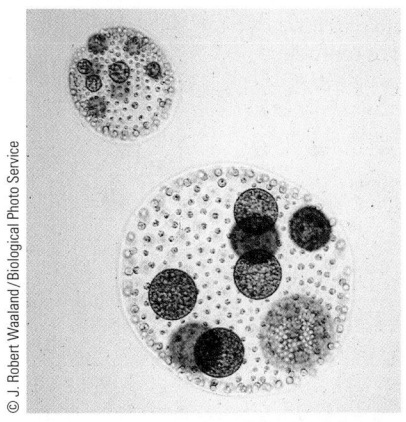

100 μm

(a) LM of two *Volvox* colonies, each composed of up to 50,000 cells. New colonies are inside the parental colonies, which eventually break apart.

(b) *Ulva's* thin, sheetlike form suggests its common name, sea lettuce.

(c) *Chara*, a green alga commonly called a stonewort, is closely related to land plants. *Chara* is widely distributed in fresh water.

Figure 26-15 Green algae

Another polysaccharide extracted from red algae, *carrageenan*, is a food additive used to stabilize chocolate milk; it also provides a thick, creamy texture to ice cream and other soft processed foods. Carrageenan is also used to stabilize paints and cosmetics. Red algae are a source of vitamins (particularly A and C) and minerals, especially in Japan and other eastern Asian countries where people eat red algae fresh, dried, or toasted in such traditional foods as sushi and nori.

Green algae share many similarities with land plants

Green algae have pigments, energy reserve products, and cell walls that are chemically identical to those of land plants. Green algae are photosynthetic, with chloroplasts of a wide variety of shapes. Most green algae have cell walls with cellulose, although some lack walls. Because of these and other similarities, biologists generally accept that land plants arose from ancestral green algae (see Fig. 26-3). Based on recent molecular and ultrastructure data, some biologists classify this diverse group in the plant kingdom.

Green algae exhibit a variety of body types, from single cells to colonial forms, to coenocytic algae (multinucleate), to multicellular filaments and sheets (FIG. 26-15). The multicellular forms do not have cells differentiated into tissues, a characteristic that separates them from land plants. Most green algae have, or produce, flagellate cells during their life cycle, although a few are totally nonmotile.

Reproduction in the green algae is as varied as their body forms, with both sexual and asexual reproduction. Many green algae have life cycles with an alternation of multicellular haploid and multicellular diploid generations. Asexual reproduction is by mitosis and cell division in single cells or by fragmentation in multicellular forms. Many green algae produce spores asexually by mitosis; if these spores have flagella and are motile, they are called *zoospores* (FIG. 26-16). Sexual reproduction in the green algae involves gamete formation in unicellular **gametangia** (sing., *gametangium*), reproductive structures in which gametes are produced. Green algae are found in both aquatic and terrestrial environments. Aquatic green algae primarily inhabit fresh water, although there are many marine species. Terrestrial green algae are restricted to damp soil, cracks in tree bark, and other moist places. Many green algae are symbionts with other organisms; some live as endosymbionts in body cells of invertebrates, and a few grow together with fungi as "dual organisms" called *lichens* (discussed in Chapter 29). Regardless of where they live, green algae are ecologically important as producers.

CHECKPOINT 26.6

- *Why do many biologists classify red algae and green algae with land plants?*

26.7 UNIKONTS

LEARNING OBJECTIVE

10 Briefly describe and compare the following unikonts: amoebas, plasmodial slime molds, cellular slime molds, and choanoflagellates.

Unikonts are a supergroup composed of certain amoebas, plasmodial slime molds, cellular slime molds, choanoflagellates, fungi (discussed in Chapter 29), and animals (discussed in Chapters 30 through 32). Unikonts share a single posterior flagellum in flagellate cells such as sperm and motile spores, although some extant organisms in this supergroup have lost the flagellum. Many unikonts also have a single centriole. Examine

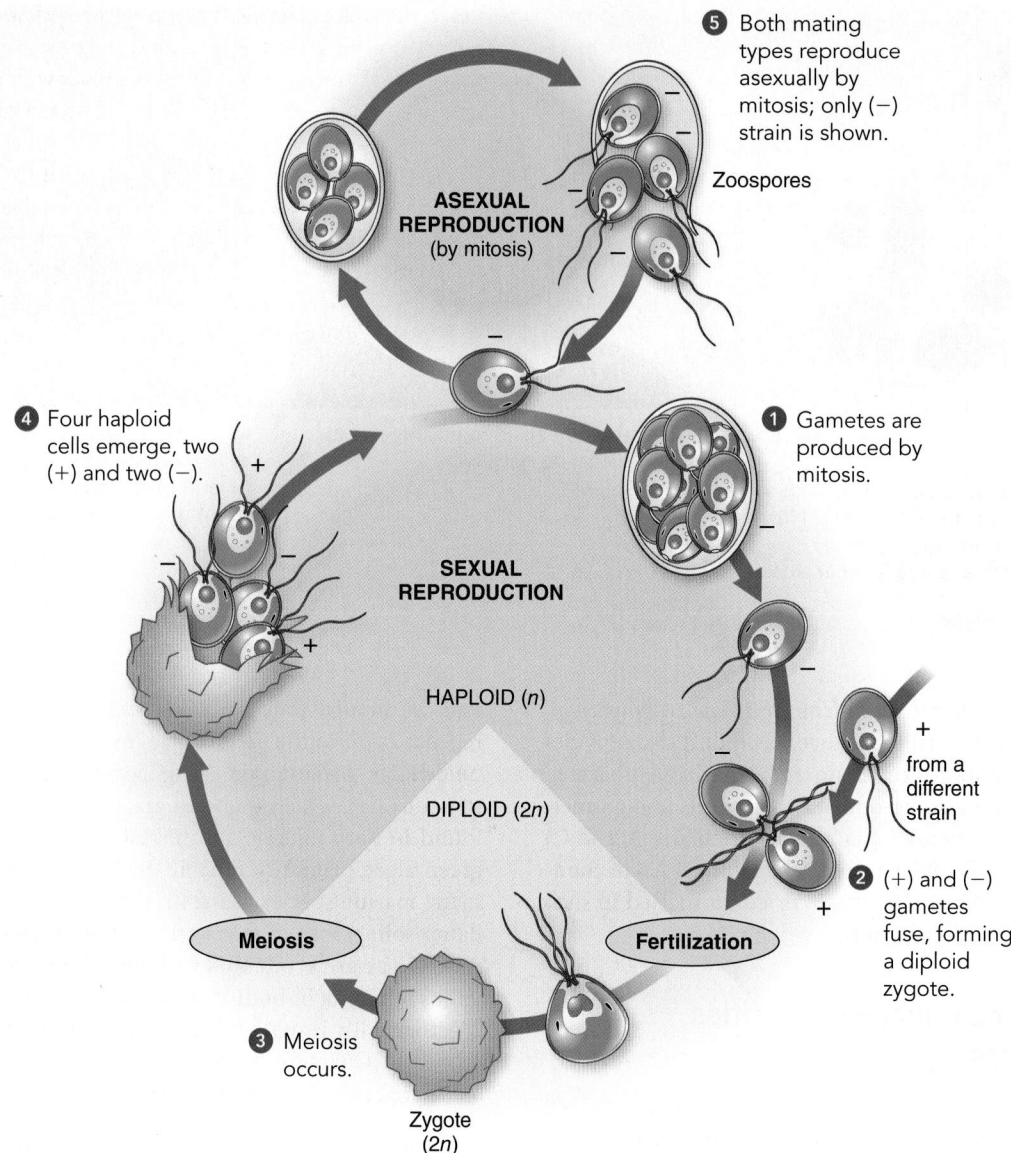

5 Both mating types reproduce asexually by mitosis; only (−) strain is shown.

Zoospores

ASEXUAL REPRODUCTION (by mitosis)

4 Four haploid cells emerge, two (+) and two (−).

SEXUAL REPRODUCTION

1 Gametes are produced by mitosis.

+ from a different strain

2 (+) and (−) gametes fuse, forming a diploid zygote.

HAPLOID (*n*)

DIPLOID (2*n*)

Meiosis

Fertilization

3 Meiosis occurs.

Zygote (2*n*)

Figure 26-16 *Animation* **The life cycle of *Chlamydomonas***

Chlamydomonas is a unicellular, haploid green alga with two mating types, (+) and (−). The only diploid cell in the life cycle is the zygote.

© Cengage Learning

Figure 26-3 again and note that the unikonts are divided into two clades. One clade is the **opisthokonts,** which consist of fungi, choanoflagellates, and animals. You will learn more about the opisthokonts in Chapters 29 and 30.

Unikonts also have a **triple-gene fusion** that has major evolutionary significance. In a triple-gene fusion, three separate genes fused into a single unit early in the course of eukaryote evolution; the fused gene codes for a multi-enzyme protein. You may recall that Figure 26-3 showed all five supergroups branching from the same point, indicating that the branching order is unresolved. However, triple-gene fusion provides evidence of a bifurcation, or branch of the eukaryotes into two main clades, close to the root of the common ancestor of all eukaryotes (**FIG. 26-17**). The two branches consist of (1) unikonts, which had a common ancestor with

a single posterior flagellum; and (2) all other eukaryotes, collectively called **bikonts,** which had a common ancestor with two flagella.

Amoebozoa are unikonts with lobose pseudopodia

Most **amoebozoa** produce temporary cytoplasmic projections called **pseudopodia** (sing., *pseudopodium,* meaning "false foot") at some point in their life cycle. The pseudopodia of amoebozoa are *lobose*—that is, rounded and wide—as opposed to the slender cytoplasmic projections characteristic of rhizarians. Many biologists currently classify amoebas, plasmodial slime molds, and cellular slime molds as amoebozoa.

How do we deduce details of eukaryote evolution near the root of the evolutionary tree?

HYPOTHESIS: Derived gene fusions, which have occurred rarely during the course of evolution, can provide insights into eukaryote evolutionary relationships.

EXPERIMENT: Alexandra Stechmann and Thomas Cavalier-Smith of Oxford University sequenced a double-gene fusion that codes for two enzymes: dihydrofolate reductase (DHFR) and thymidylate synthase (TS). In a second experiment, they sequenced a triple-fused gene involved in the enzymatic synthesis of pyrimidine nucleotides. They tested for the presence of these two gene fusions in various eukaryote groups as well as in bacteria and archaea.

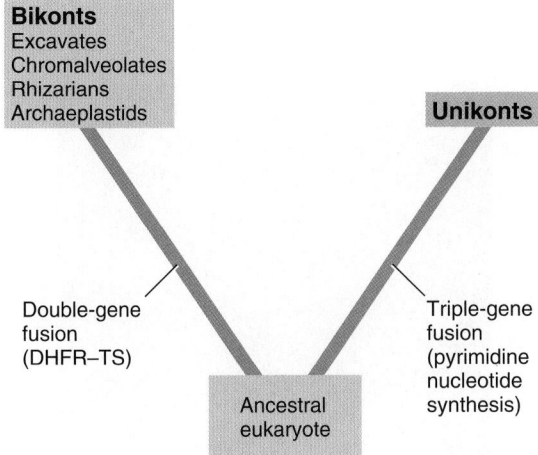

RESULTS AND CONCLUSION: The double-gene fusion is present in all eukaryotes examined except for unikonts; bacteria and archaea also lack the double-gene fusion. The unikonts have the triple-gene fusion that is lacking in all other eukaryotes, bacteria, and archaea. These gene fusions suggest a fundamental division of eukaryotes into two main clades, the unikonts and all other eukaryotes, which are collectively called bikonts.

SOURCE: Stechmann, A. and T. Cavalier-Smith. "The Root of the Eukaryote Tree Pinpointed." *Current Biology*, Vol. 13, Sept. 2, 2003.

Figure 26-17 Stechmann and Cavalier-Smith's research on evolution of the protists

CONNECT Are you a bikont or a unikont?

Amoebas move by forming pseudopodia Amoebas are unicellular amoebozoa found in soil, fresh water, the ocean, and other organisms (as parasites). Because of the extreme flexibility of their outer plasma membrane, many members of this group have an asymmetrical body form and continually change shape as they move. (The word *amoeba* derives from a Greek word meaning "change.") An amoeba moves by pushing out lobose pseudopodia from the surface of the cell. More cytoplasm flows into the pseudopodia, enlarging them until all the cytoplasm has entered and the organism as a whole has

moved. Pseudopodia also capture and engulf food by surrounding and forming a vacuole around it (FIG. 26-18). Food particles are digested when the food vacuole fuses with a lysosome containing digestive enzymes. Digested materials are absorbed from the food vacuole, which gradually shrinks as it empties. Amoebas reproduce asexually, splitting into two equal parts after mitotic division of the nucleus; sexual reproduction has not been observed.

Parasitic amoebas include *Entamoeba histolytica,* which causes *amoebic dysentery,* a serious human intestinal disease characterized by severe diarrhea, bloody stools, and ulcers in the intestinal wall. In especially severe cases the organism spreads from the large intestine and causes abscesses in the liver, lungs, or brain. *Entamoeba histolytica* is transmitted as cysts in contaminated drinking water. A *cyst* is a thick-walled, resistant, resting stage in the life cycle of some protists. Other amoebas, such as *Acanthamoeba,* are usually free-living but produce opportunistic infections such as eye infections in wearers of contact lenses.

Plasmodial slime molds feed as multinucleate plasmodia The feeding stage of a **plasmodial slime mold** is a **plasmodium,** a multinucleate mass of cytoplasm that can grow up to 30 cm (1 ft) in diameter (FIG. 26-19a). The slimy plasmodium streams over damp, decaying logs and leaf litter, often forming a network of channels that covers a large surface area. As it creeps along, it ingests bacteria, yeasts, spores, and decaying organic matter.

When the food supply dwindles or there is insufficient moisture, the plasmodium crawls to an exposed surface and starts reproducing. Stalked structures of intricate complexity and beauty usually form from the drying plasmodium (FIG. 26-19b). Within these structures, called **sporangia,** meiosis produces haploid spores that are extremely resistant to adverse environmental conditions.

When conditions become favorable, the spores germinate, and a haploid reproductive cell emerges from each. This haploid cell is either a biflagellate *swarm cell* or an amoeboid *myxamoeba,* depending on available moisture; flagellate cells form in wet conditions. Swarm cells and myxamoebas act as gametes, which fuse to form a zygote with a diploid nucleus. The resultant diploid nucleus divides many times by mitosis, but the cytoplasm does not divide, so the result is a multinucleate plasmodium.

The plasmodial slime mold *Physarum polycephalum* is a model organism that researchers use to study many fundamental biological processes, such as growth, cytoplasmic streaming, and the function of the cytoskeleton.

Cellular slime molds feed as individual amoeboid cells The **cellular slime molds** are amoebozoa with close affinities to amoebas and plasmodial slime molds. During its feeding stage, each cellular slime mold is an individual amoeboid cell that behaves as a separate, solitary organism (FIG. 26-20). Each cell creeps over rotting logs and soil or swims in fresh water, ingesting bacteria and other particles of food as

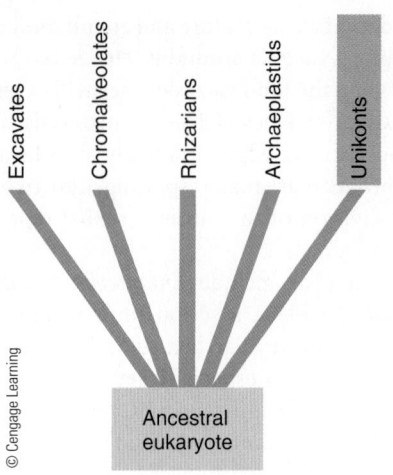

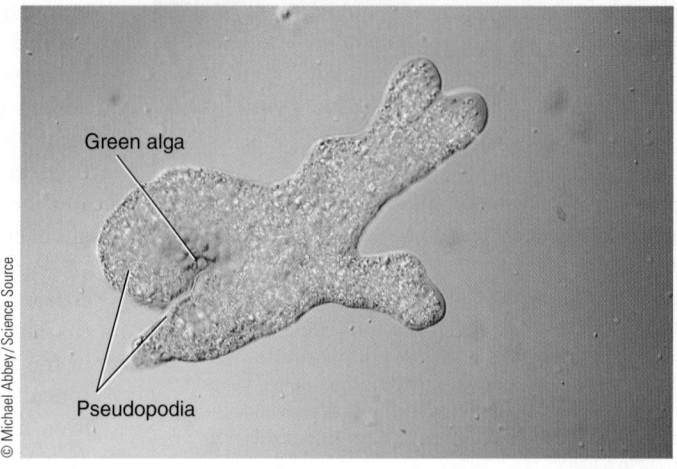

Green alga

Pseudopodia

100 μm

Figure 26-18 *Animation* **Amoeba**

LM of a giant amoeba (*Chaos carolinense*). This unicellular protist, which moves and feeds by pseudopodia, is surrounding and ingesting a colonial green alga. *Chaos* amoebas are generally scavengers that feed on debris in freshwater habitats, but they ingest living organisms when the opportunity arises.

it goes. Each amoeboid cell has a haploid nucleus and reproduces by mitosis, as a true amoeba does.

When moisture or food becomes inadequate, certain cells send out a chemical signal, cyclic adenosine monophosphate (cAMP; see Fig. 3-26), that causes them to aggregate by the hundreds or thousands. During this stage the cells creep about for short distances as a single multicellular aggregate, or slug. Each cell of the slug retains its plasma membrane and individual identity. Eventually, the slug settles and reorganizes, forming a stalked fruiting body containing spores. After being released each spore opens, and a single haploid amoeboid cell—the feeding stage—emerges. The spore-forming reproductive cycle is asexual, although sexual reproduction is observed occasionally. The life cycles of most cellular slime molds lack a flagellate stage.

The cellular slime mold *Dictyostelium discoideum* is a model organism for the study of cell differentiation, cell communication, and cell motility and adhesion. Its biology has been studied intensively, particularly as it relates to **cell signaling** molecules, such as cAMP, which are found in many organisms in addition to the cellular slime molds (see Chapter 6).

Choanoflagellates are opisthokonts closely related to animals

Choanoflagellates are collared flagellates in the opisthokont clade, which also includes fungi and animals. These small, inconspicuous unikonts are found globally in both freshwater and marine environments. Choanoflagellates include both free-swimming and **sessile** species that are permanently attached by a thin stalk to bacteria-rich debris. Their single flagellum is surrounded at the base by a delicate collar of microvilli that trap food (**FIG. 26-21**).

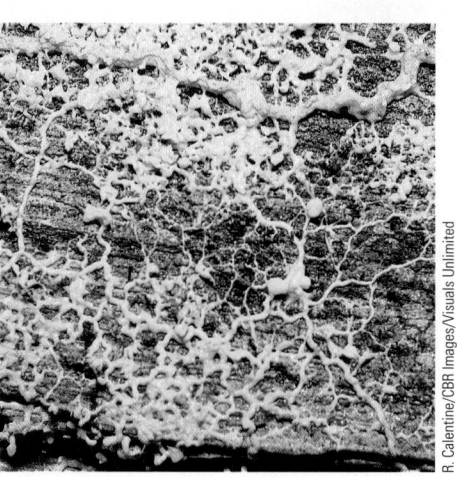

(a) The brightly pigmented plasmodium, shown on a dead log, feeds on bacteria and other microorganisms.

250 μm

(b) The reproductive structures are sporangia on stalks.

Figure 26-19 **The plasmodial slime mold** *Physarum polycephalum*

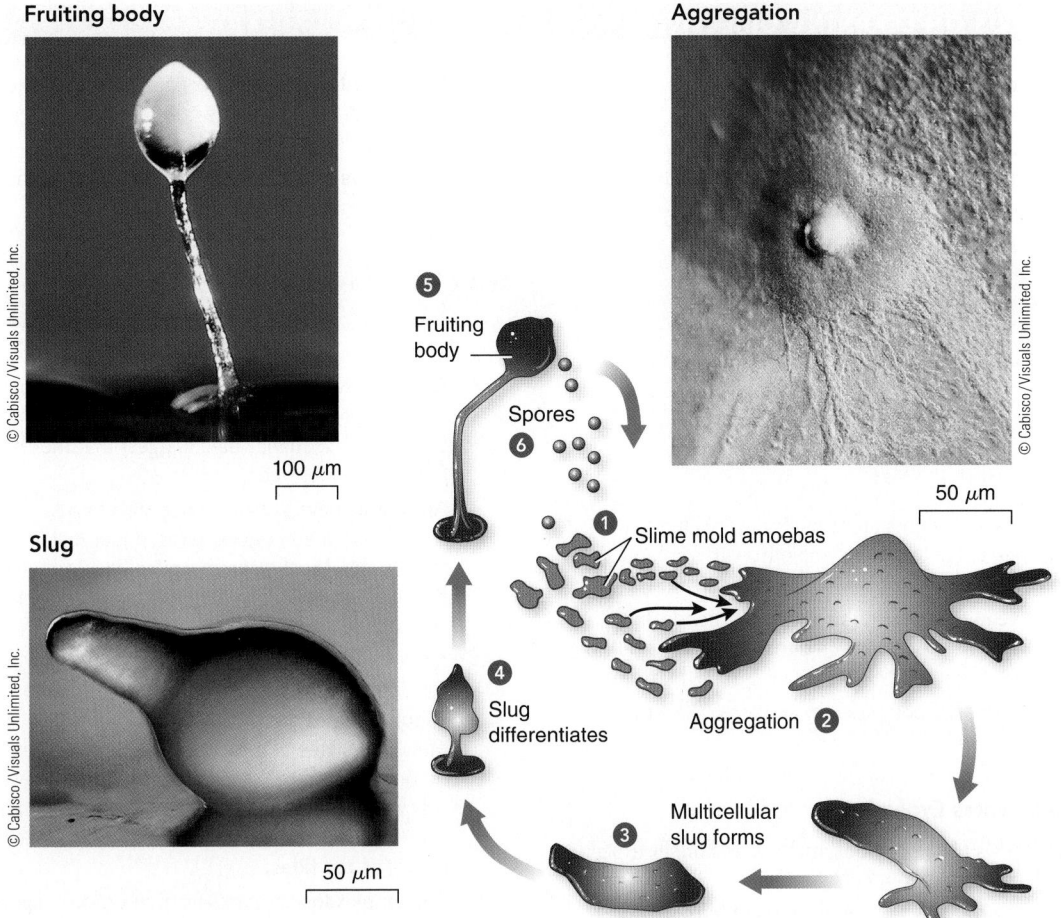

Fruiting body

100 μm

Slug

50 μm

Aggregation

50 μm

⑤ Fruiting body

Spores ⑥

① Slime mold amoebas

④ Slug differentiates

Aggregation ②

③ Multicellular slug forms

Figure 26-20 *Animation* **The cellular slime mold *Dictyostelium discoideum***

Slime mold amoebas ingest food, grow, and reproduce by cell division. After their food is depleted, hundreds of amoeboid cells stream together (*inset in upper right*) and form a migrating sluglike aggregate (*inset in lower left*). When it stops migrating, the aggregate forms a fruiting body on a stalk (*inset in upper left*). The fruiting body releases spores, each of which opens in a favorable environment to liberate an amoeboid cell.

© Cengage Learning

Choanoflagellates are of special interest because of their striking resemblance to collar cells in sponges (see Fig. 31-1b). Other animal phyla, such as cnidarians, flatworms, and echinoderms, also contain choanoflagellate-like cells, but no other group of protists has been observed to possess these cells. Evidence supporting the close relationship between choanoflagellates and animals includes comparative DNA sequence data of mitochondrial and nuclear genes. Given the similarities in structure and molecular genomics that has accumulated in recent years, many biologists hypothesize that choanoflagellates are the closest living nonanimal relative of animals. Thus, living choanoflagellates and animals probably share a common choanoflagellate-like ancestor.

CHECKPOINT 26.7

- *What features distinguish the amoebozoa from the rhizarians?*
- CONNECT *Which group of unikonts is structurally and genetically most similar to animals?*

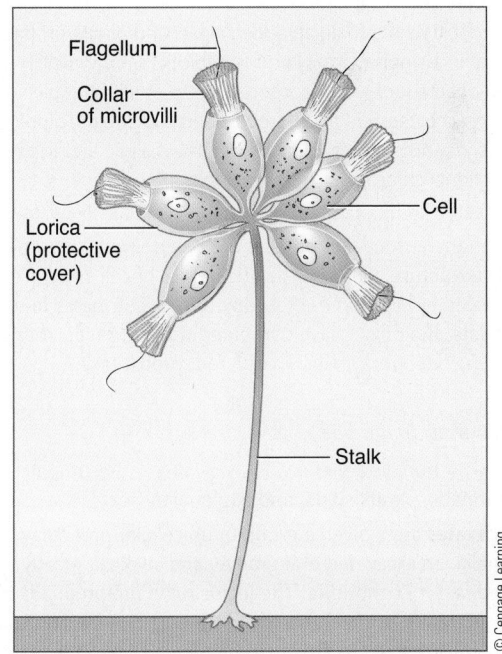

Flagellum

Collar of microvilli

Lorica (protective cover)

Cell

Stalk

Figure 26-21 Choanoflagellates

Choanoflagellates are free-living flagellates that obtain food by waving their flagella, causing water currents to carry bacteria and other small particles of food into the collar of microvilli. A colonial form is shown. Each cell is 5 to 10 μm long, not including the flagellum.

26.1 Diversity in the Protists (page 534)

1 Discuss in general terms the diversity inherent in protists, including means of locomotion, modes of nutrition, interactions with other organisms, habitats, and modes of reproduction.

- Protists have various means of locomotion, including pseudopodia, flagella, and cilia; a few are nonmotile.

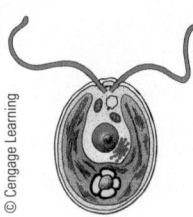

© Cengage Learning

- Protists obtain their nutrients autotrophically or heterotrophically. Protists are free-living or symbiotic, with symbiotic relationships ranging from **mutualism** to **parasitism.**
- Most protists live in the ocean or in freshwater ponds, lakes, and streams. Parasitic protists live in the body fluids or cells of their hosts.
- Many protists reproduce both sexually and asexually; others reproduce only asexually.

26.2 How Did Eukaryotes Evolve? (page 535)

2 Discuss the hypothesis of serial endosymbiosis and briefly explain some of the evidence that supports it.

- According to the hypothesis of **serial endosymbiosis,** mitochondria and chloroplasts arose from symbiotic relationships between larger cells and the smaller bacteria that were incorporated and lived within them.
- Chloroplasts of red algae, green algae, and plants probably arose in a single primary endosymbiotic event in which a cyanobacterium was incorporated into a cell. Multiple secondary endosymbioses led to chloroplasts in euglenoids, dinoflagellates, diatoms, golden algae, and brown algae and to the nonfunctional chloroplasts in apicomplexans.

3 Describe the kinds of data biologists use to classify eukaryotes.

- Relationships among protists are determined largely by **ultrastructure,** which is the fine details of cell structure revealed by electron microscopy, and by comparative molecular data. Biologists have compared nuclear genes, many of which code for proteins, in different protist taxa.

26.3 Excavates (page 538)

4 Summarize the basic features of excavates and distinguish among diplomonads, parabasilids, and euglenoids.

- **Excavates** are a diverse group of unicellular protists with flagella, an excavated oral groove, and atypical, greatly modified mitochondria. The inclusion of diplomonads, parabasilids, and euglenoids in the excavate superfamily is controversial.
- **Diplomonads** are excavates with one or two nuclei, no functional mitochondria, no Golgi complex, and up to eight flagella.

- **Parabasilids** are anaerobic, flagellated excavates that often live in animals. Trichonymphs and trichomonads are examples of parabasilids.
- **Euglenoids** are unicellular and flagellate. Some euglenoids are photosynthetic.

26.4 Chromalveolates (page 540)

5 Contrast the two main groups of chromalveolates: alveolates and stramenopiles.

- **Chromalveolates** probably originated as a result of secondary endosymbiosis in which an ancestral cell engulfed a red alga. Some DNA sequence data suggest that the chromalveolates are not monophyletic.
- **Alveolates** have similar ribosomal DNA sequences and alveoli, flattened vesicles located just inside the plasma membrane. Most **stramenopiles** have motile cells with two flagella, one of which has tiny hairlike projections off the shaft.

6 Distinguish among the alveolates: dinoflagellates, apicomplexans, and ciliates.

- **Dinoflagellates** are mostly unicellular, biflagellate, photosynthetic alveolates of great ecological importance as producers in marine ecosystems. Their **alveoli,** flattened vesicles under the plasma membrane, often contain cellulose plates impregnated with silicates. Some dinoflagellates produce toxic blooms known as **red tides.**
- **Apicomplexans** are parasites that produce **sporozoites** and are nonmotile. An apical complex of microtubules attaches the apicomplexan to its host cell. The apicomplexan *Plasmodium* causes malaria.
- **Ciliates** are alveolates that move by hairlike **cilia,** have **micronuclei** (for sexual reproduction) and **macronuclei** (for controlling cell metabolism and growth), and undergo a sexual process called **conjugation.**

7 Distinguish among the stramenopiles: water molds, diatoms, golden algae, and brown algae.

- **Water molds** have a coenocytic **mycelium.** They reproduce asexually by forming biflagellate **zoospores** and sexually by forming **oospores.**
- **Diatoms** are mostly unicellular, with shells containing silica. Some diatoms are part of floating **plankton,** and others live on rocks and sediments where they move by gliding.
- **Brown algae** are multicellular stramenopiles that are ecologically important in cooler ocean waters. The largest brown algae (kelps) possess leaflike **blades,** stemlike **stipes,** anchoring **holdfasts,** and gas-filled bladders for buoyancy.
- **Golden algae** are mostly unicellular, biflagellate freshwater and marine stramenopiles that are of ecological importance as a component of the ocean's extremely minute **nanoplankton.**

26.5 Rhizarians (page 547)

8 Describe the forams and actinopods, and explain why many biologists classify them in the rhizarian supergroup.

- **Rhizarians** are amoeboid cells that often have hard outer shells, called **tests,** through which cytoplasmic projections extend; molecular evidence indicates that this group is monophyletic.
- **Forams** secrete many-chambered tests with pores through which cytoplasmic projections extend to move and obtain food.
- **Actinopods** are mostly marine plankton that obtain food by means of **axopods,** slender cytoplasmic projections that extend through pores in their shells.

26.6 Archaeplastids (page 548)

9 Describe evidence supporting the hypothesis that red algae and green algae should be included in a monophyletic group with land plants.
- Red algae, green algae, and land plants, collectively called **archaeplastids,** are considered a monophyletic group based on molecular data and on the presence of chloroplasts bounded by outer and inner membranes.
- **Red algae,** which are mostly multicellular seaweeds, are ecologically important in warm tropical ocean waters.

- **Green algae** exhibit a wide diversity in size, structural complexity, and reproduction. Botanists hypothesize that ancestral green algae gave rise to land plants.

26.7 Unikonts (page 549)

10 Briefly describe and compare the following unikonts: amoebas, plasmodial slime molds, cellular slime molds, and choanoflagellates.
- **Unikonts** have a single posterior flagellum in flagellate cells.
- **Amoebas** move and obtain food using cytoplasmic extensions called **pseudopodia.**
- The feeding stage of **plasmodial slime molds** is a multinucleate **plasmodium.** Reproduction is by haploid spores produced within **sporangia.**
- **Cellular slime molds** feed as individual amoeboid cells. They reproduce by aggregating into an aggregate (slug) and then forming asexual spores.
- **Choanoflagellates** are unikonts that are probably the closest living nonanimal relative of animals. A collar of microvilli surrounds their single flagellum at the base. Choanoflagellates are included with animals in the **opisthokont** clade, which also includes fungi.

TEST YOUR UNDERSTANDING

Know and Comprehend

1. Which of the following is *not* true of the protists? (a) they are unicellular, colonial, coenocytic, or simple multicellular organisms (b) their cilia and flagella have a 9 + 2 arrangement of microtubules (c) they are prokaryotic, as bacteria and archaea are (d) some are free-living, and some are endosymbionts (e) most are aquatic and live in the ocean or in freshwater ponds
2. Molecular evidence supports the view that all plastids evolved from an ancient (a) cyanobacterium (b) archaean (c) diplomonad (d) apicomplexan (e) unikont
3. Forams (a) are endosymbionts that live in many marine invertebrates (b) were responsible for the Irish potato famine in the 19th century (c) secrete many-chambered tests with pores through which cytoplasmic extensions project (d) have numerous axopods that may aid in trapping and holding prey (e) have numerous cilia to direct food into the oral groove
4. *Paramecium* and other ciliates often display a sexual phenomenon called (a) oogamy (b) conjugation (c) serial endosymbiosis (d) red tide (e) alternation of generations
5. Parasitic alveolates that form spores at some stage in their life belong to which group? (a) actinopods (b) ciliates (c) coccolithophorids (d) apicomplexans (e) dinoflagellates
6. Malaria (a) is transmitted by the bite of a female tsetse fly (b) is caused by a parasitic flagellate, *Giardia intestinalis* (c) is a serious form of amoebic dysentery caused by *Entamoeba histolytica* (d) is caused by an apicomplexan that spends part of its life cycle in the *Anopheles* mosquito and part in humans (e) is transmitted when people drink water tainted by a red tide

7. Photosynthetic protists with shells composed of two halves that fit together like a petri dish are (a) golden algae (b) diatoms (c) euglenoids (d) brown algae (e) forams
8. The pigments, energy reserve products, and cell walls found in land plants are also characteristic of (a) green algae (b) brown algae (c) golden algae (d) diatoms (e) euglenoids and diatoms
9. Kelps are ___ with multicellular bodies differentiated into blades, stipes, holdfasts, and gas-filled floats. (a) golden algae (b) diatoms (c) euglenoids (d) brown algae (e) red algae
10. Cellular slime molds (a) include *Physarum* and *Phytophthora* (b) are more closely related to bacteria than are any other protists (c) are responsible for late blight of potatoes, which led to starvation in Ireland in the 1840s (d) have double-gene fusion (e) form a slug when cells aggregate in response to cyclic AMP
11. Water molds reproduce asexually by forming ___ and sexually by forming ___. (a) oospores; holdfasts (b) zoospores; zooxanthellae (c) zoospores; oospores (d) holdfasts; oospores (e) oospores; zoospores

Apply and Analyze

12. **VISUALIZE** Draw simple sketches of a *Euglena* and a *Paramecium.* Choose the appropriate structures for each organism from the following list and label them in your drawings: flagellum, cilia, chloroplast, mitochondrion, nucleus, macronucleus, micronucleus, contractile vacuole, pellicle, food vacuole.

Evaluate and Synthesize

13. **EVOLUTION LINK** Why are the protists considered paraphyletic? Use Figure 26-3 to help explain your answer.

14. **INTERPRET DATA** In the corresponding darkfield LM of unicellular choanoflagellates, actin has been stained *red,* tubulin *green,* and DNA *blue.* Identify the cell parts that contain each of these chemical materials.

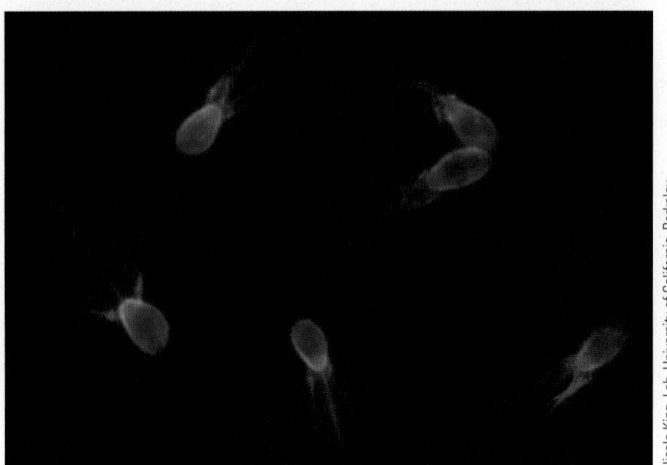

Nicole King Lab, University of California, Berkeley

15. **SCIENCE, TECHNOLOGY, AND SOCIETY** Use examples associated with the malaria parasite, *Plasmodium,* to illustrate why an understanding of evolution is indispensable to modern medical research. (*Hint:* Be sure to include serial endosymbiosis and drug resistance in your answer.)

aplia To access course materials, such as Aplia and other companion resources, please visit **www.cengagebrain.com.**

Seedless Plants

About 445 million years ago (mya), planet Earth would have seemed an inhospitable place because even though life abounded in the ocean, it did not yet exist in abundance on land. Occasionally, perhaps, an animal would crawl out of the water onto land, but it never stayed there permanently because there was little to eat on land: not a single blade of grass, no fruit, and no seeds.

During the next 30 million years, a time corresponding roughly to the Silurian period in geologic time (see Table 21-1), plants appeared in abundance and colonized the land. Where did they come from? Although plants living today exhibit great diversity in size, form, and habitat (see photograph), biologists hypothesize that they all evolved from a common ancestor that was an ancient green alga.

Modern green algae share many biochemical and metabolic traits with modern plants. Both green algae and plants contain the same photosynthetic pigments: chlorophylls *a* and *b* and carotenoids, including xanthophylls (yellow pigments) and carotenes (orange pigments). Both store excess carbohydrates as starch and have cellulose as a major component of their cell walls. In addition, plants and some green algae share certain details of cell division, including formation of a cell plate during cytokinesis (see Chapter 10).

What exactly is a plant? A **plant** is a complex multicellular eukaryote that has cellulose cell walls, chlorophylls *a* and *b* in plastids, and starch as a storage product and that may have cells with two anterior flagella. In addition, all plants develop from multicellular *embryos* that are enclosed in maternal tissues; this last character is one that distinguishes plants from green algae. Many botanists refer to land plants as *embryophytes* because of this character.

Plants range in size from minute, almost microscopic duckweeds and water-meal to massive giant sequoias, some of the largest organisms that have ever lived. Plants include hundreds of thousands of species that live in varied habitats, from frozen arctic tundra to lush tropical rain forests to harsh deserts to moist stream banks.

© Josef Zima/Shutterstock.com

Land plants. Note the ferns and the moss-covered dead tree. Both mosses and ferns are seedless plants.

KEY CONCEPTS

27.1 Adaptations of land plants include a waxy cuticle to prevent water loss; multicellular gametangia; stomata; and for most plants, vascular tissues containing lignin. Plants undergo an alternation of generations between multicellular gametophyte and sporophyte generations.

27.2 Mosses and other bryophytes lack vascular tissues and do not form true roots, stems, or leaves.

27.3 In club mosses and ferns, lignin-hardened vascular tissues that transport water and dissolved substances throughout the plant body have evolved.

27.1 ADAPTATIONS OF PLANTS TO LIFE ON LAND

LEARNING OBJECTIVES

1 Discuss some environmental challenges of living on land and describe how several plant adaptations meet these challenges.

2 Name the green algal group from which plants are hypothesized to have descended and describe supporting evidence.

How were plants modified during the transition from life in the water to life on land? What are some features of plants that have let them colonize so many types of environments? One important difference between plants and algae is that a waxy **cuticle** covers the aerial portion of a plant. Essential for existence on land, the cuticle helps prevent desiccation, or drying out, of plant tissues by evaporation. Plants that are adapted to moister habitats may have a very thin layer of wax, whereas those adapted to drier environments often have a thick, crusty cuticle. (Many desert plants also have a reduced surface area, particularly of leaves, which minimizes water loss.)

Plants obtain the carbon they need for photosynthesis from the atmosphere as carbon dioxide (CO_2). To be fixed into organic molecules such as sugar, CO_2 must first diffuse into the chloroplasts that are inside green plant cells. Because a waxy cuticle covers the external surfaces of leaves and stems, however, gas exchange through the cuticle between the atmosphere and the interior of cells is negligible. Tiny pores called **stomata** (sing., *stoma*), which dot the surfaces of leaves and stems of almost all plants, facilitate gas exchange.

The sex organs, or **gametangia** (sing., *gametangium*), of most plants are multicellular, whereas the gametangia of algae are unicellular (**FIG. 27-1**). Each plant gametangium has a layer of sterile (nonreproductive) cells that surrounds and protects the delicate gametes (eggs and sperm cells). In plants the fertilized egg develops into a multicellular **embryo** (young plant) within the female gametangium. Thus, the embryo is protected during its development. In algae the fertilized egg develops away from its gametangium; in some algae the gametes are released before fertilization, whereas in others the fertilized egg is released.

The plant life cycle alternates between haploid and diploid generations

Plants have a clearly defined **alternation of generations** in which they spend part of their lives in a multicellular haploid stage and part in a multicellular diploid stage (**FIG. 27-2**).[1] The haploid portion of the life cycle is called the **gametophyte generation** because it gives rise to

haploid gametes by mitosis. When two gametes fuse, the diploid portion of the life cycle, called the **sporophyte generation,** begins. The sporophyte generation produces haploid spores by the process of meiosis; these spores represent the first stage in the gametophyte generation.

Let us examine alternation of generations more closely. The haploid gametophytes produce male gametangia, known as **antheridia** (sing., *antheridium*), in which sperm cells form (**FIG. 27-3a**). Gametophytes also produce female gametangia, known as **archegonia** (sing., *archegonium*), each bearing a single egg (**FIG. 27-3b**). Sperm cells reach the female gametangium in a variety of ways, and one sperm cell fertilizes the egg to form a **zygote,** or fertilized egg.

The diploid zygote is the first stage in the sporophyte generation. The zygote divides by mitosis and develops into a multicellular embryo, the young sporophyte plant. Embryo development takes place within the archegonium; thus, the embryo is protected as it develops. Eventually, the embryo grows into a mature sporophyte plant. The mature sporophyte has special cells called *sporogenous cells* (spore-producing cells, also called *spore mother cells*) that divide by meiosis to form haploid **spores.**

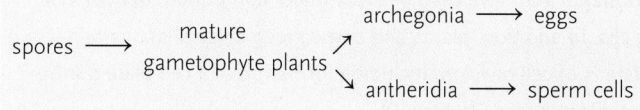

All plants produce spores by meiosis, in contrast with algae and fungi, which may produce spores by meiosis or mitosis. The spores represent the first stage in the gametophyte generation. Each spore divides by mitosis to produce a multicellular gametophyte, and the cycle continues. Plants therefore alternate between a haploid gametophyte generation and a diploid sporophyte generation.

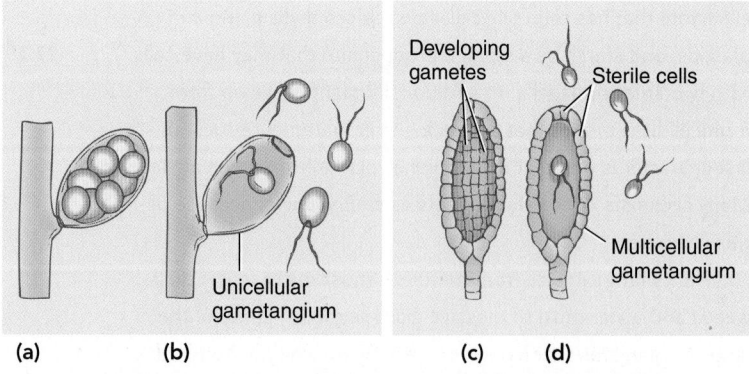

Figure 27-1 **Generalized reproductive structures of algae and plants**

(a, b) In algae gametangia are generally unicellular. When the gametes are released, only the wall of the original cell remains. **(c, d)** In plants the gametangia are multicellular, but only the inner cells become gametes. The gametes are surrounded by a protective layer of sterile (nonreproductive) cells.

© Cengage Learning

---

[1] For convenience, we limit our discussion to plants that are not polyploid, although polyploidy is very common in land plants. We therefore use the terms *diploid* and *2n* (and *haploid* and *n*) interchangeably, but these terms are not actually synonymous.

Plants have an alternation of generations, spending part of the cycle in a haploid gametophyte stage and part in a diploid sporophyte stage.

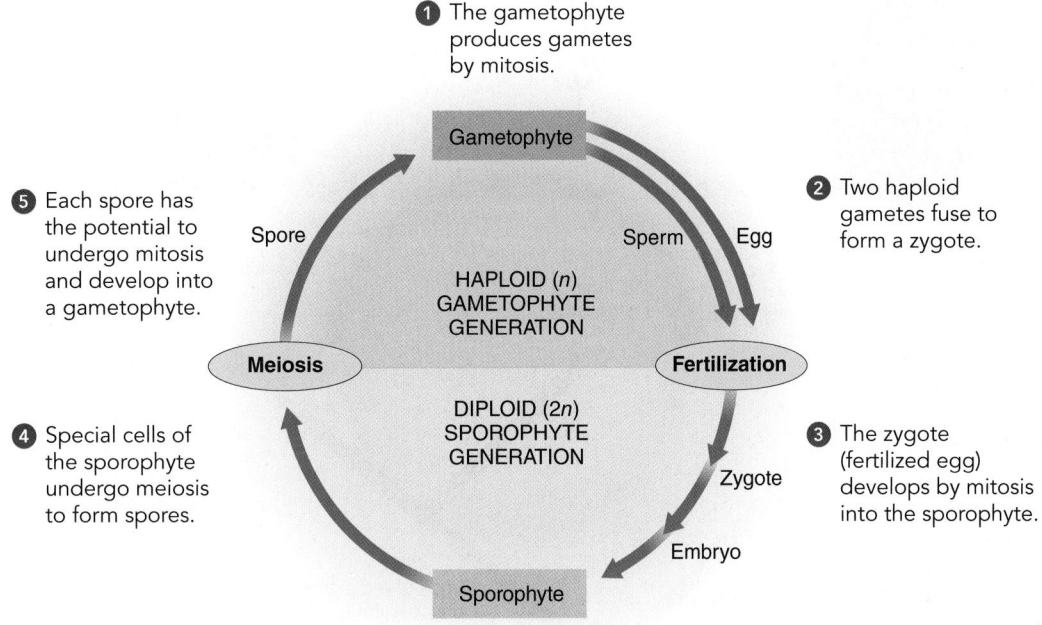

1 The gametophyte produces gametes by mitosis.

2 Two haploid gametes fuse to form a zygote.

3 The zygote (fertilized egg) develops by mitosis into the sporophyte.

4 Special cells of the sporophyte undergo meiosis to form spores.

5 Each spore has the potential to undergo mitosis and develop into a gametophyte.

HAPLOID (*n*) GAMETOPHYTE GENERATION

DIPLOID (*2n*) SPOROPHYTE GENERATION

Meiosis

Fertilization

Gametophyte

Sporophyte

Spore

Sperm

Egg

Zygote

Embryo

Figure 27-2 *Animation* **The basic plant life cycle**

Depending on the plant group, the haploid or the diploid stage may be greatly enlarged or reduced; as you will learn, this fact has evolutionary significance.

CONNECT Do diploid cells undergo mitosis in this life cycle? haploid cells?

© Cengage Learning

Four major groups of plants exist today

Structural and molecular data indicate that land plants probably descended from a group of green algae called **charophytes** or *stoneworts* (see Figure 26-15c, which shows a charophyte). Recall from Chapter 26 that red algae, green algae, and land plants are

collectively classified as **archaeplastids.** Molecular comparisons, particularly of DNA and RNA sequences, provide compelling evidence that green algae are closely allied to plants (**FIG. 27-4**). These comparisons among plants and various green algae include sequences of chloroplast DNA, certain nuclear genes, and ribosomal RNA. In each case, the closest match occurs between

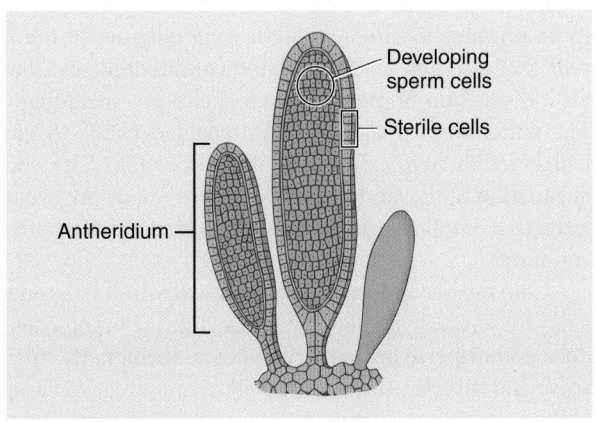

Developing sperm cells

Sterile cells

Antheridium

(a) Each antheridium, the male gametangium, produces numerous sperm cells.

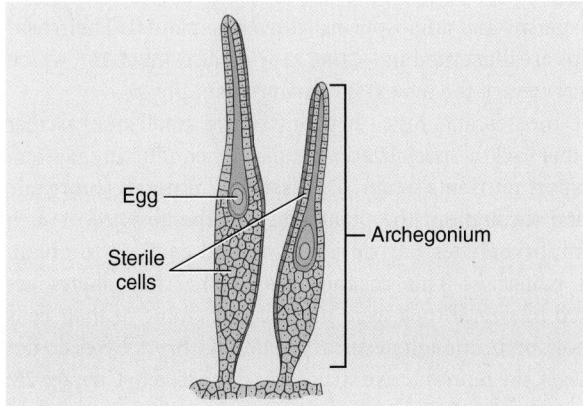

Egg

Sterile cells

Archegonium

(b) Each archegonium, the female gametangium, produces a single egg.

Figure 27-3 **Plant gametangia**

Shown are generalized moss gametangia.

© Cengage Learning

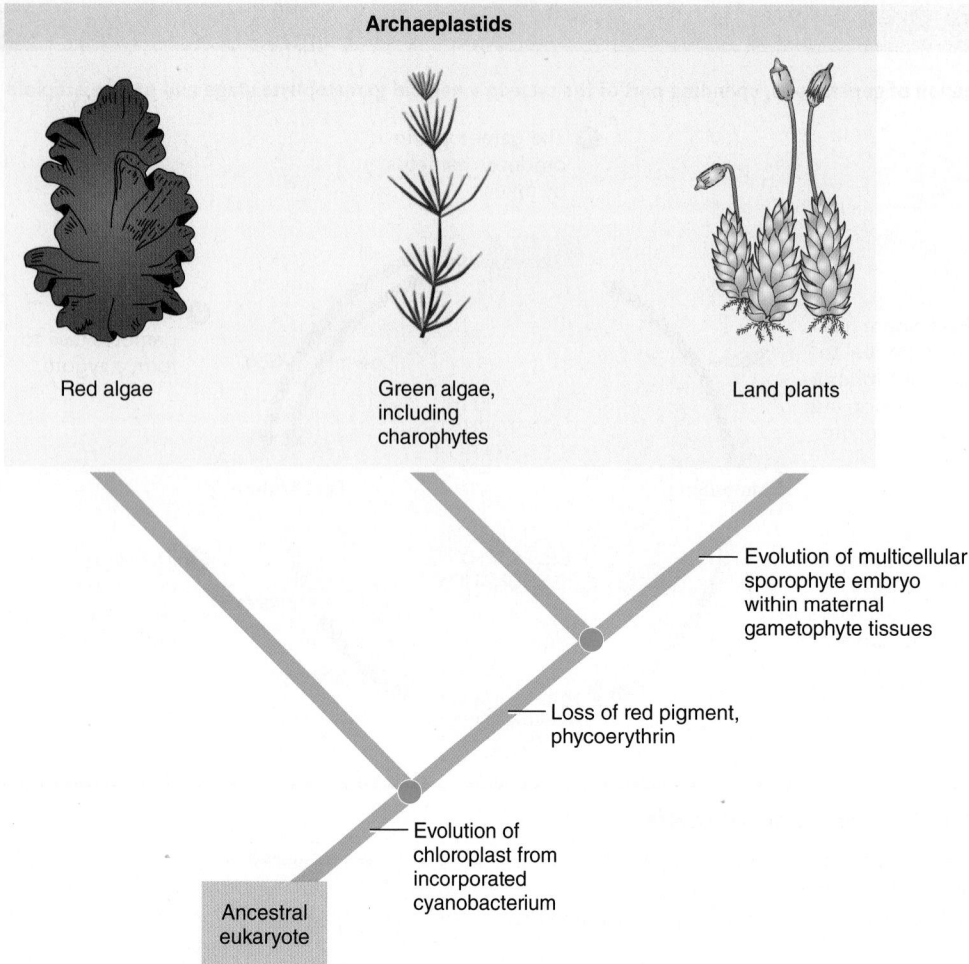

Archaeplastids

Red algae

Green algae, including charophytes

Land plants

Evolution of multicellular sporophyte embryo within maternal gametophyte tissues

Loss of red pigment, phycoerythrin

Evolution of chloroplast from incorporated cyanobacterium

Ancestral eukaryote

Figure 27-4 Evolution of land plants

Cladograms such as this one represent an emerging consensus that is open to change as new discoveries are made.
© Cengage Learning

charophytes and plants, indicating that modern charophytes and plants probably share a recent common ancestor.

Plants consist of four major groups: bryophytes; seedless vascular plants; and two groups of seeded vascular plants, the gymnosperms and angiosperms (flowering plants). Their relationships are illustrated in FIGURE 27-5; see also TABLE 27-1, which is an overview of the ten extant (living) plant phyla.

The mosses and other bryophytes are small nonvascular plants that lack a specialized vascular, or conducting, system to transport nutrients, water, and essential minerals (inorganic nutrients) throughout the plant body. In the absence of such a system, bryophytes rely on diffusion and osmosis to obtain needed materials. This reliance means that bryophytes are restricted in size; if they were much larger, some of their cells could not obtain enough necessary materials. Bryophytes do not form seeds, the reproductive structures discussed in Chapter 28. Bryophytes reproduce and disperse primarily via haploid spores. Recent molecular and fossil evidence, discussed later in this chapter, suggests that bryophytes may have been the earliest plants to colonize land.

The other three groups of plants—seedless vascular plants, gymnosperms, and flowering plants—have vascular tissues and

are thus known as vascular plants. The two vascular tissues are **xylem**, for conducting water and dissolved minerals, and **phloem**, for conducting dissolved organic molecules such as sugar. A key step in the evolution of vascular plants was the ability to produce **lignin**, a strengthening polymer in the walls of cells that function for support and conduction (see Chapter 33 for a discussion of plant cell wall chemistry, including lignin). The stiffening property of lignin enabled plants to grow tall, which let them maximize light interception. The successful occupation of the land by plants, in turn, made the evolution of terrestrial animals possible by providing them with both habitat and food.

Club mosses and ferns (which include whisk ferns and horsetails) are seedless vascular plants that, like the bryophytes, reproduce and disperse primarily via spores. Seedless vascular plants arose and diversified during the Silurian and Devonian periods of the Paleozoic era, between 444 mya and 359 mya. Club mosses and ferns extend back more than 420 million years and were of considerable importance as Earth's dominant plants in past ages. Fossil evidence indicates that many species of these plants were the size of immense trees. Most living representatives of club mosses and ferns are small.

Based on recent evidence, scientists are beginning to reach a consensus about the evolutionary relationships among living plants.

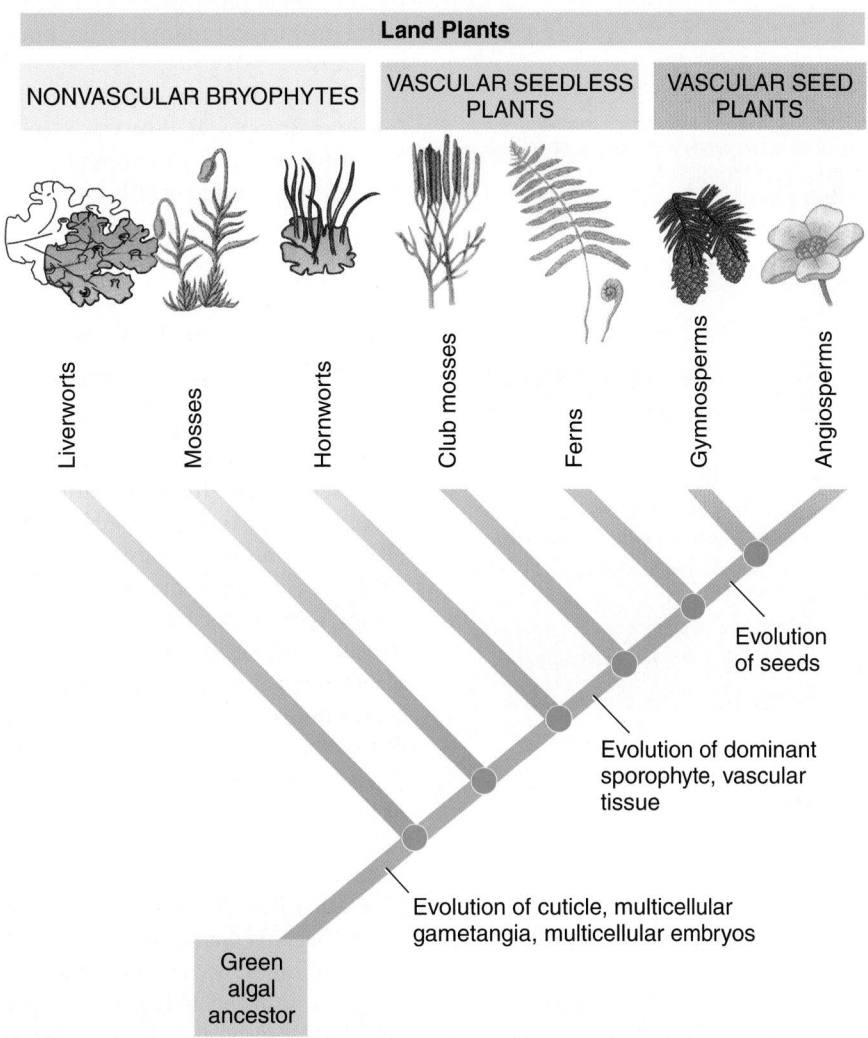

Figure 27-5 *Animation* **Plant evolution**

The four major groups of plants are bryophytes, seedless vascular plants, and two groups of seed plants: gymnosperms and angiosperms. The seed plants are covered in Chapter 28.

CONNECT How does the cladogram relate to many botanists referring to land plants as embryophytes?

© Cengage Learning

The gymnosperms are vascular plants that reproduce by forming seeds. Gymnosperms produce seeds borne exposed (unprotected) on a stem or in a cone. Plants with seeds as their primary means of reproduction and dispersal first appeared about 359 mya, at the end of the Devonian period. These early seed plants diversified into many varied species of gymnosperms.

The most recent plant group to appear is the flowering plants, or angiosperms, which arose during the early Cretaceous period of the Mesozoic era, about 130 mya. Like gymnosperms, flowering plants reproduce by forming seeds. Flowering plants, however, produce seeds enclosed within a fruit.

CHECKPOINT 27.1

- *What are the most important environmental challenges that plants face living on land?*

- *What adaptations do plants have to meet these environmental challenges?*

- *What types of evidence support the hypothesis that land plants descended from the group of green algae known as charophytes?*

- **VISUALIZE** *Draw a simple diagram illustrating alternation of generations in plants, including the sporophyte and gameto-phyte generations, spores, gametes (eggs and sperm), meiosis, and fertilization. Be sure to indicate whether each generation or kind of cell is haploid or diploid.*

TABLE 27-1 | Ten Extant Plant Phyla

NONVASCULAR PLANTS WITH A DOMINANT GAMETOPHYTE GENERATION (BRYOPHYTES)

Phylum Hepatophyta (liverworts)

Phylum Bryophyta (mosses)

Phylum Anthocerophyta (hornworts)

VASCULAR PLANTS WITH A DOMINANT SPOROPHYTE GENERATION

Seedless plants

Phylum Lycopodiophyta (club mosses)

Phylum Pteridophyta (ferns and their allies, the whisk ferns and horsetails)

Seed plants

Plants with naked seeds (gymnosperms)

Phylum Coniferophyta (conifers)

Phylum Cycadophyta (cycads)

Phylum Ginkgophyta (ginkgoes)

Phylum Gnetophyta (gnetophytes)

Seeds enclosed within a fruit (angiosperms)

Phylum Anthophyta (angiosperms or flowering plants)

Class Eudicotyledones (eudicots)

Class Monocotyledones (monocots)

© Cengage Learning

27.2 BRYOPHYTES

LEARNING OBJECTIVES

3 Summarize the features that distinguish bryophytes from other plants.

4 Name and briefly describe the three phyla of bryophytes.

5 Describe the life cycle of mosses and compare their gametophyte and sporophyte generations.

The **bryophytes** (from the Greek words meaning "moss plant") consist of about 16,000 species of mosses, liverworts, and hornworts; bryophytes are the only living nonvascular plants (**TABLE 27-2**). Because they have no means for extensive internal transport of water, sugar, and essential minerals, bryophytes are typically small. They generally require a moist environment for active growth and reproduction, but some bryophytes tolerate dry areas.

The bryophytes are divided into three distinct phyla: mosses (phylum Bryophyta), liverworts (phylum Hepatophyta), and hornworts (phylum Anthocerophyta). These three groups differ in many ways and may or may not be closely related. They are usually studied together because they lack vascular tissues and have similar life cycles.

Moss gametophytes are differentiated into "leaves" and "stems"

Mosses (phylum Bryophyta), which include about 9900 species, usually live in dense colonies or beds (**FIG. 27-6**). Each individual gametophyte plant has tiny, hairlike absorptive structures called *rhizoids* and an upright, stemlike structure that bears leaflike blades, each normally consisting of a single layer of undifferentiated cells except at the midrib. Because mosses lack vascular tissues, they do not have true roots, stems, or leaves; the moss structures are not homologous to roots, stems, or leaves in vascular plants. Some moss species have water-conducting cells and sugar-conducting cells, although these cells are not lignified, nor are they as specialized or effective as the conducting cells of vascular plants.

Alternation of generations is clearly defined in the life cycle of mosses (**FIG. 27-7**). The green moss gametophyte often bears its gametangia at the top of the plant. Many moss species have separate sexes: male plants that bear antheridia and female plants that bear archegonia. Other mosses produce antheridia and archegonia on the same plant.

Fertilization occurs when one of the sperm cells fuses with the egg within the archegonium. Sperm cells, which have flagella, are transported from antheridium to archegonium by flowing water, such as splashing rain droplets. A raindrop lands on the top of a male gametophyte, and sperm cells are released into it from the antheridia. Another raindrop landing on the male plant may splash the sperm-laden droplet into the air and

TABLE 27-2 | A Comparison of Major Groups of Seedless Plants

PLANT GROUP	DOMINANT STAGE OF LIFE CYCLE	REPRESENTATIVE GENERA
NONVASCULAR; REPRODUCE BY SPORES (BRYOPHYTES)		
Liverworts (phylum Hepatophyta)	Gametophyte: thalloid or leafy plant	*Marchantia*
Mosses (phylum Bryophyta)	Gametophyte: leafy plant	*Polytrichum, Sphagnum, Physcomitrella*
Hornworts (phylum Anthocerophyta)	Gametophyte: thalloid plant	*Anthoceros*
VASCULAR; REPRODUCE BY SPORES		
Club mosses (phylum Lycopodiophyta)	Sporophyte: roots, rhizomes, erect stems, and leaves (microphylls)	*Lycopodium, Selaginella*
Ferns (phylum Pteridophyta)	Sporophyte: roots, rhizomes, and leaves (megaphylls)	*Pteridium, Polystichum, Azolla, Platycerium*
Whisk ferns (phylum Pteridophyta)	Sporophyte: rhizomes and erect stems; no true roots or leaves	*Psilotum*
Horsetails (phylum Pteridophyta)	Sporophyte: roots, rhizomes, erect stems, and leaves (reduced megaphylls)	*Equisetum*

© Cengage Learning

(a) **Moss gametophytes.** Close-up of haircap moss (*Polytrichum commune*) gametophytes. The haircap moss is a popular ground cover in rock gardens, particularly in Japan.

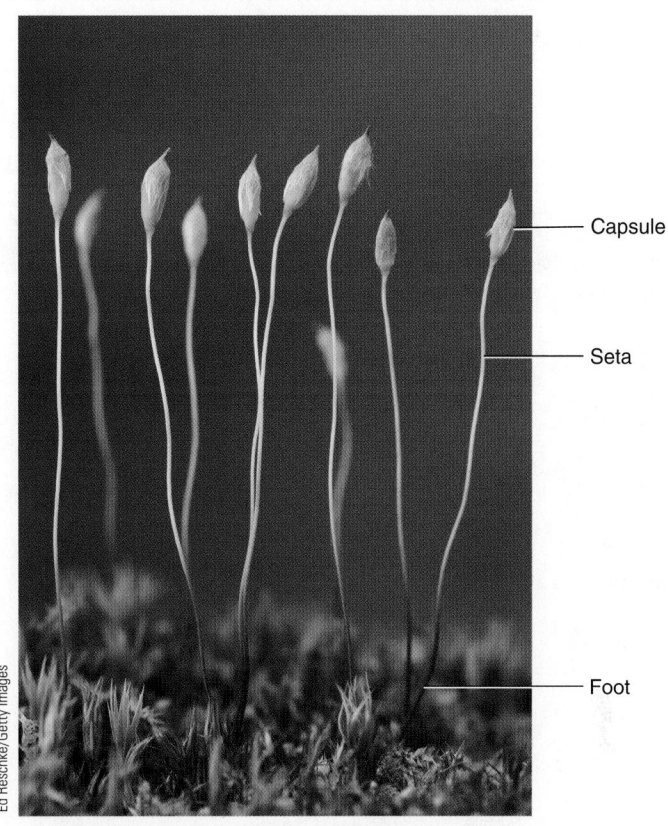

Capsule

Seta

Foot

(b) **Moss sporophytes.** Each consisting of a foot, seta, and capsule, the haircap moss sporophytes grow out of the top of the gametophytes. Spores are produced within the capsule.

Figure 27-6 Bryophytes: mosses

onto the top of a nearby female plant. Alternatively, arthropods such as insects and mites may touch the sperm-laden fluid and inadvertently carry it for considerable distances. Once in a film of water on the female moss, a sperm cell swims into the archegonium, which secretes chemicals to attract and guide the sperm cells, and fuses with the egg.

The diploid zygote, formed by fertilization, grows by mitosis into a multicellular embryo that develops into a mature moss sporophyte. This sporophyte grows out of the top of the female gametophyte and remains attached and nutritionally dependent on the gametophyte throughout its existence (see Fig. 27-6b). Initially green and photosynthetic, the sporophyte becomes golden brown at maturity. It consists of three main parts: a *foot,* which anchors the sporophyte to the gametophyte and absorbs minerals and nutrients from it; a *seta,* or stalk; and a *capsule,* which contains sporogenous cells (spore mother cells). The capsule of some species is covered by a caplike structure, the *calyptra,* which is derived from the archegonium.

The sporogenous cells undergo meiosis to form haploid spores. When the spores are mature, the capsule opens and releases them, and they are then transported by wind or rain. If a moss spore lands in a suitable spot, it germinates and grows into a filament of cells called a **protonema.** The protonema, which superficially resembles a filamentous green alga, forms buds, each of which grows into a green gametophyte, and the life cycle continues.

Biologists consider the haploid gametophyte generation the dominant generation in mosses because it lives independently of the diploid sporophyte. In contrast, the moss sporophyte is attached to and nutritionally dependent on the gametophyte.

Mosses make up an inconspicuous but significant part of their environment. They play an important role in forming soil. Mosses, which form mats that cover the rock, eventually die, forming a thin layer of organic matter in which grasses and other plants can grow. Because they grow tightly packed in dense colonies, mosses hold soil in place and help prevent erosion. At the same time, they retain moisture that they and other organisms need. Waxwings and other birds use moss, along with twigs and grass, as nesting material.

Commercially, the most important mosses are the peat mosses in the genus *Sphagnum.* One of the distinctive features of *Sphagnum* "leaves" is the presence of many large, empty cells that absorb and hold water. This feature makes peat moss a useful packing material for shipping live plants as well as a good soil conditioner. Added to sandy soils, for example, peat moss helps absorb and retain moisture.

The acidic and anaerobic conditions of a peat bog retard the growth of bacterial and fungal decomposers. As a result,

The dominant phase in the life cycle of mosses, like that of other bryophytes, is the gametophyte.

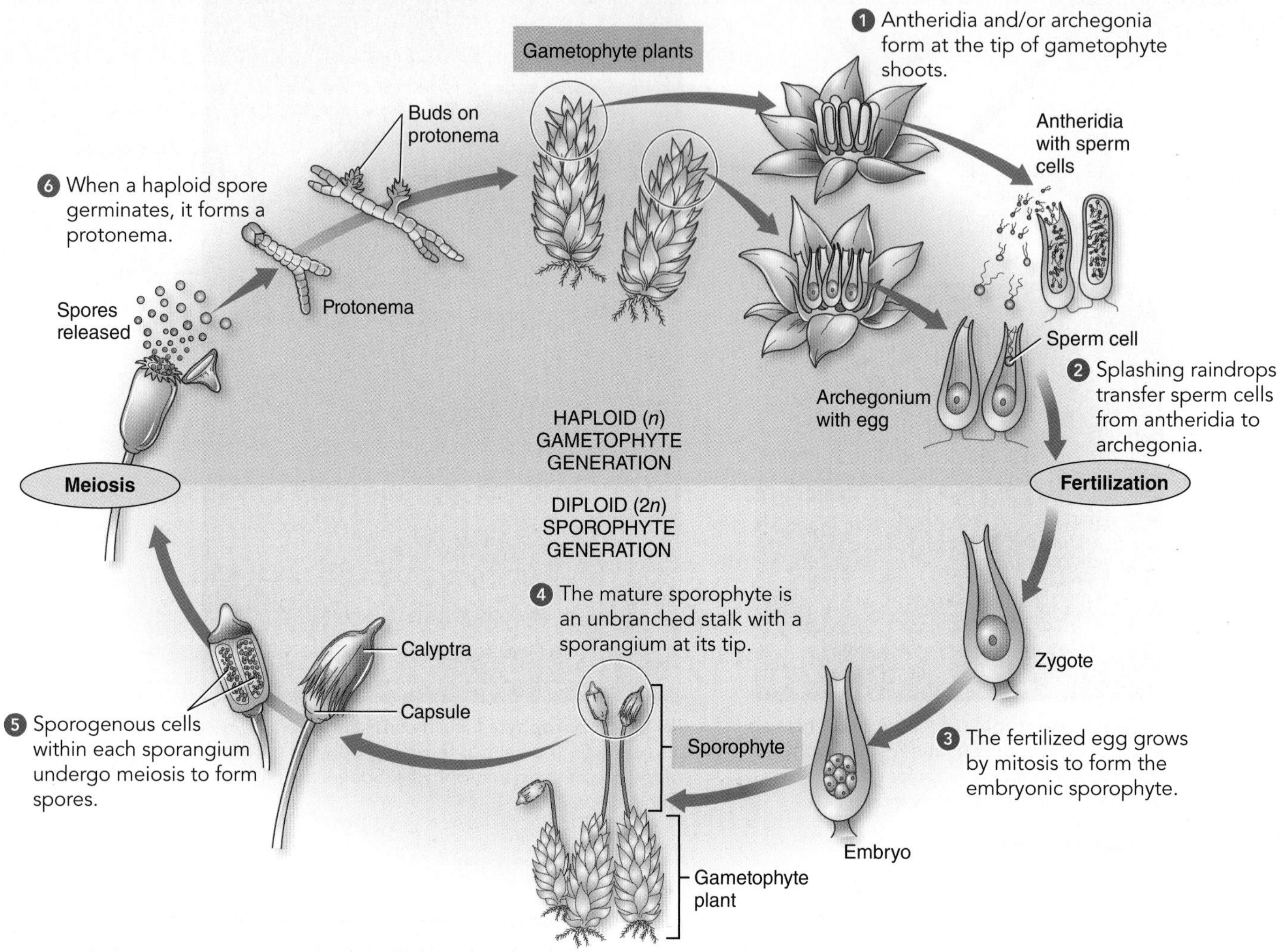

1 Antheridia and/or archegonia form at the tip of gametophyte shoots.

Gametophyte plants

Buds on protonema

6 When a haploid spore germinates, it forms a protonema.

Spores released

Protonema

Antheridia with sperm cells

Sperm cell

2 Splashing raindrops transfer sperm cells from antheridia to archegonia.

Archegonium with egg

Fertilization

HAPLOID (*n*) GAMETOPHYTE GENERATION

Meiosis

DIPLOID (2*n*) SPOROPHYTE GENERATION

Zygote

4 The mature sporophyte is an unbranched stalk with a sporangium at its tip.

Calyptra

Capsule

5 Sporogenous cells within each sporangium undergo meiosis to form spores.

Sporophyte

3 The fertilized egg grows by mitosis to form the embryonic sporophyte.

Embryo

Gametophyte plant

Figure 27-7 *Animation* **The life cycle of mosses**

The gametophyte generation is dominant in the moss life cycle. After sexual reproduction, the sporophyte grows out of the gametophyte.

© Cengage Learning

VISUALIZE Sketch an embryo developing in an archegonium. Indicate to what generation each belongs and whether the tissue is haploid or diploid.

dead peat mosses accumulate as thick deposits—some several meters in depth—under the growing mat of living peat mosses. Over time, the organic material compresses to form *peat*. In some countries, such as Ireland and Scotland, people cut out blocks of peat that has accumulated for hundreds of years in peat bogs, dry them, and burn them for fuel. Occasionally, the remains of well-preserved humans have been uncovered during the excavations of old peat bogs in Ireland and other parts of Europe (**FIG. 27-8**).

Figure 27-8 **Preserved human remains in a peat bog in Denmark**

The clothing and features of the Tollund man, estimated to be about two thousand years old, are remarkably well preserved because the bog's acidic conditions inhibited decay.

© Robin Weaver/Alamy

The name *moss* is often misused to refer to plants that are not truly mosses. For example, reindeer "moss" is a lichen that is a dominant form of vegetation in the arctic tundra, Spanish "moss" is a flowering plant, and club "moss" (discussed later in this chapter) is a relative of ferns.

Liverwort gametophytes are either thalloid or leafy

Liverworts (phylum Hepatophyta) consist of about 6000 species of nonvascular plants with a dominant gametophyte generation, but the gametophytes of some liverworts are quite different from those of mosses. Their body form is often a flattened, lobed structure called a **thallus** (pl., *thalli*) that is not differentiated into leaves, stems, or roots. The common liverwort, *Marchantia polymorpha,* is thalloid (**FIGS. 27-9a** and b). Liverworts are so named because the lobes of their thalli superficially resemble the lobes of the human liver; *wort* is derived from the Old English word *wyrt,* meaning "plant." On the underside of the liverwort thallus are hairlike rhizoids that anchor the plant to the soil. Other liverworts, known as *leafy liverworts,* superficially resemble mosses, with leaflike blades, "stems," and rhizoids rather

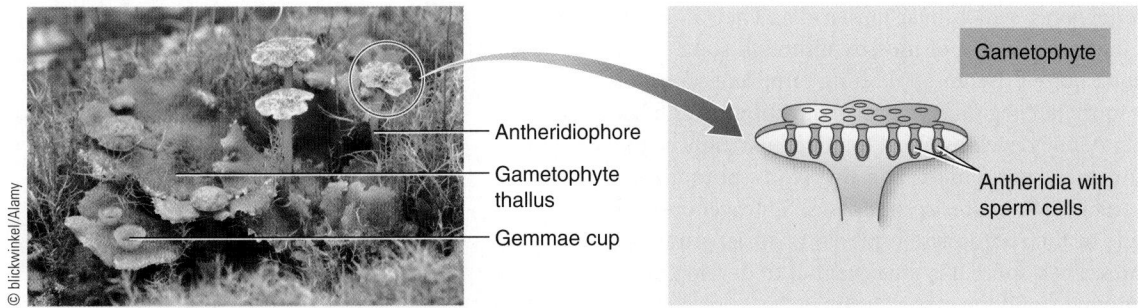

(a) Flattened, ribbon-like lobes characterize the gametophyte of the common liverwort (*Marchantia polymorpha*). This male gametophyte thallus has both asexual gemmae cups and sexual antheridiophores, which produce sperm-bearing antheridia.

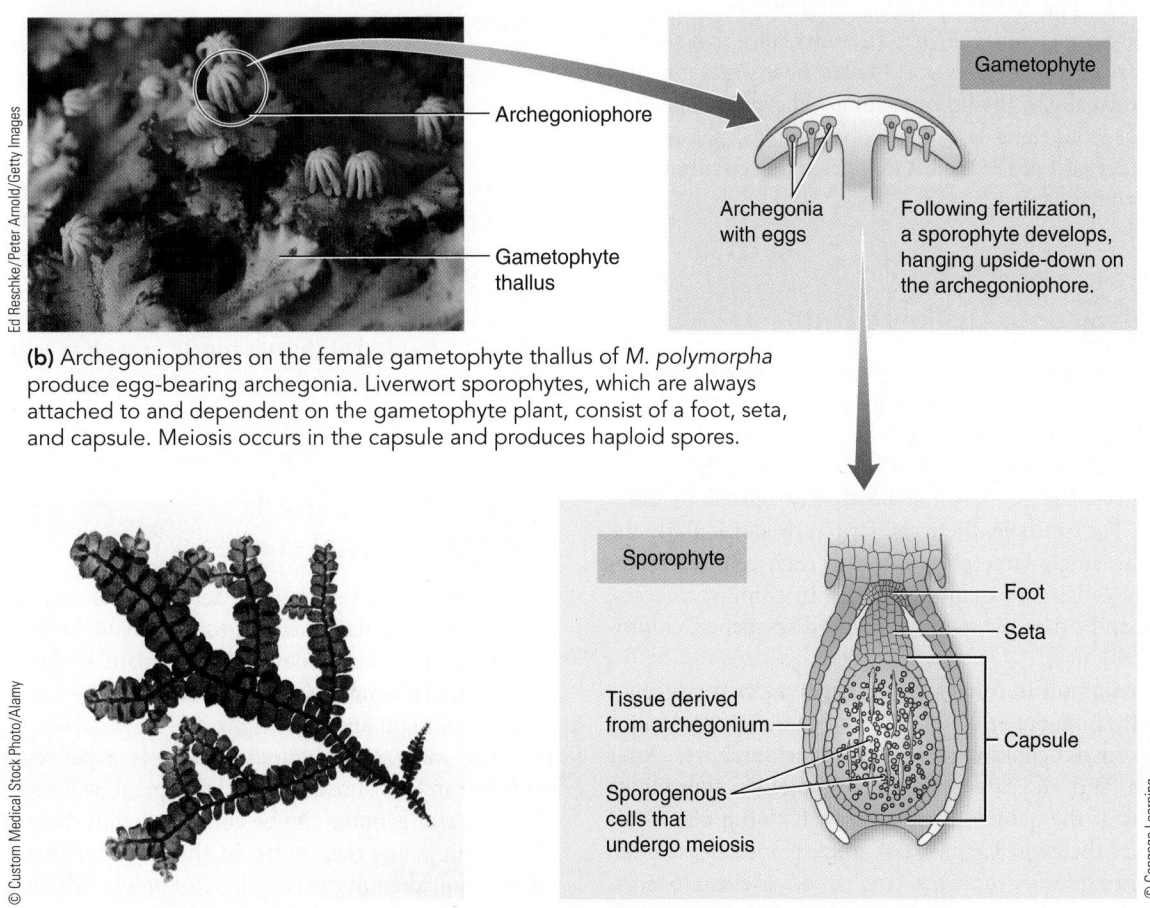

(b) Archegoniophores on the female gametophyte thallus of *M. polymorpha* produce egg-bearing archegonia. Liverwort sporophytes, which are always attached to and dependent on the gametophyte plant, consist of a foot, seta, and capsule. Meiosis occurs in the capsule and produces haploid spores.

(c) *Porella* is a leafy liverwort. The leafy plant is the gametophyte. It bears antheridia and archegonia on special branches that look quite similar to the nonreproductive branches. After fertilization, a small sporophyte develops that produces spores following meiosis.

Figure 27-9 *Animation* **Bryophytes: liverworts**

than a lobed thallus (FIG. 27-9c). As in the mosses, leafy liverwort "leaves" consist of a single layer of undifferentiated cells. Like other bryophytes, liverworts are small, generally inconspicuous plants that are largely restricted to damp environments. Unlike mosses, hornworts, and other plants, liverworts lack stomata, although some liverworts have surface pores thought to be analogous to stomata.

Liverworts reproduce both sexually and asexually (see Figs. 27-9a and b). Their sexual reproduction involves production of archegonia and antheridia on the haploid gametophyte. In some liverworts these gametangia are borne on stalked structures called *archegoniophores,* which bear archegonia, and *antheridiophores,* which bear antheridia. Their life cycle is basically the same as that of mosses, although some of the structures look quite different. Splashing raindrops transport sperm cells to the archegonia, where fertilization takes place. The resulting zygote develops into a multicellular embryo that becomes a mature sporophyte. The liverwort sporophyte is attached to the gametophyte, as in mosses. Sporogenous cells in the capsule of the sporophyte undergo meiosis, producing haploid spores. Each spore has the potential to develop into a green gametophyte, and the cycle continues.

Some liverworts reproduce asexually by forming tiny balls of tissue called **gemmae** (sing., *gemma*), which are borne in a saucer-shaped structure, the gemmae cup, directly on the liverwort thallus (see Fig. 27-9a). Splashing raindrops and small animals help disperse gemmae. When a gemma lands in a suitable place, it grows into a new liverwort thallus. Liverworts may also reproduce asexually by thallus branching and growth. The individual thallus lobes elongate, and each becomes a separate plant when the older part of the thallus that originally connected the individual lobes dies.

Hornwort gametophytes are inconspicuous thalloid plants

Hornworts (phylum Anthocerophyta) are a small group of about 100 species of bryophytes whose gametophytes superficially resemble those of the thalloid liverworts. Hornworts live in disturbed habitats such as fallow fields and roadsides.

Hornworts may or may not be closely related to other bryophytes. For example, their cell structure, particularly the presence of a single large chloroplast in each cell, resembles certain algal cells more than plant cells. In contrast, mosses, liverworts, and other plants have many disc-shaped chloroplasts per cell.

In the common hornwort (*Anthoceros natans*), archegonia and antheridia are embedded in the gametophyte thallus rather than on archegoniophores and antheridiophores. After fertilization and development, the needlelike sporophyte projects out of the gametophyte thallus, forming a spike or "horn," hence the name *hornwort.* A single gametophyte often produces multiple sporophytes (FIG. 27-10). Meiosis occurs, forming spores within each **sporangium** (pl., *sporangia*), or spore case. The sporangium splits open from the top to

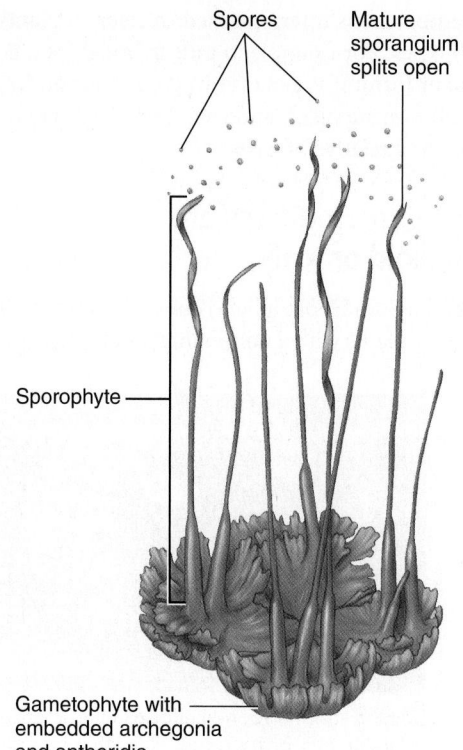

Figure 27-10 Bryophytes: hornworts

The gametophyte of the common hornwort (*Anthoceros natans*) is a small thallus with unicellular rhizoids on the lower (ventral) surface. After fertilization, sporophytes project up out of the gametophyte, forming "horns."
© Cengage Learning

release the spores; each spore can give rise to a new gametophyte thallus. A unique feature of hornworts is that the sporophytes, unlike those of mosses and liverworts, continue to grow from their bases for the remainder of the gametophyte's life, a characteristic known as **indeterminate growth.** Some botanists think that indeterminate growth may indicate that hornworts evolved from plants with larger, more complex sporophytes.

Bryophytes are used for experimental studies

Botanists use certain bryophytes as experimental models to study many fundamental aspects of plant biology, including genetics, growth and development, plant ecology, plant hormones, and *photoperiodism,* which is plant responses to varying periods of night and day length.

The moss *Physcomitrella patens* is a particularly important research organism for studying plant evolution because its features and genome can be compared with those of algae and flowering plants (FIG. 27-11). In this regard *Physcomitrella* is a plant equivalent of the fruit fly *Drosophila,* which is an important model organism for studies of animal inheritance, development, and evolution. As experimental organisms, *Physcomitrella*

Does the Physcomitrella *genome reveal any insights in plant evolution?*

HYPOTHESIS: Comparing the genome of the model organism *Physcomitrella* with genomes of other land plants as well as green algae will provide details of genes that were probably associated with the colonization of land.

EXPERIMENT: Several dozen scientists in Germany, the United States, Japan, the United Kingdom, Australia, Canada, and Belgium collaborated to produce the draft genome sequence for the model organism *Physcomitrella* (see photograph). This bryophyte was selected because researchers have used *Physcomitrella* in a wide variety of experiments and have a good understanding of its biology.

After sequencing the genome of *Physcomitrella*, it was compared with available genomes of green algae (*Ostreococcus* and *Chlamydomonas*) and land plants: the model flowering plant *Arabidopsis*, rice (*Oryza sativa*), and western balsam poplar (*Populus trichocarpa*).

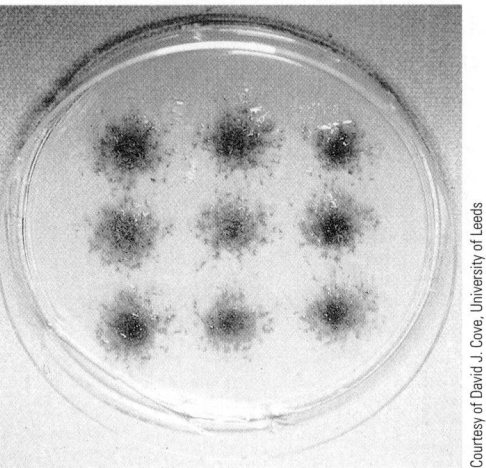

Courtesy of David J. Cove, University of Leeds

RESULTS AND CONCLUSION: Researchers found that *Physcomitrella* has lost some genes needed for life in the water and gained new genes needed for life on land. Every land plant has to meet the same environmental challenges, such as obtaining enough water and preventing excessive water loss, tolerating high and low temperatures, and adapting to increased levels of solar radiation. These challenges have resulted in many modifications in plant body plans and physiological processes.

Bryophytes and vascular plants living today represent remnants of the first plants that colonized land some 450 mya. Compared with green algae, *Physcomitrella* lacks the genes needed to survive in aquatic environments but has gained genes involving plant signaling molecules (auxin, abscisic acid, and cytokinins) needed to adapt to drought, extremes of temperature, and light reception. Compared with *Physcomitrella*, other land plants have gained additional signaling molecules (gibberellins, jasmonic acid, and brassinosteroids) that enable a more complex tolerance of environmental stressors.

Because of its intermediary position in evolution (i.e., between green algae and vascular land plants), *Physcomitrella*'s genome allows biologists to reconstruct which genes may have been acquired and lost during the colonization of land by the last common ancestor of all land plants.

SOURCE: Rensing, S.A., et al. "The *Physcomitrella* Genome Reveals Evolutionary Insights into the Conquest of Land by Plants." *Science*, Vol. 319, pp. 64–69, Jan. 4, 2008.

PREDICT Would you expect *Physcomitrella* to possess genes required to produce xylem or phloem?

Figure 27-11 The moss *Physcomitrella*, the first bryophyte to have its genome sequenced

and other bryophytes are easy to grow on artificial media and do not require much space because they are so small.

Recap: details of bryophyte evolution are based on fossils and on structural and molecular evidence

Plants are a **monophyletic group;** that is, all plants probably evolved from a common ancestral green alga. Fossil evidence indicates that bryophytes are ancient plants, probably the first group of plants to arise from the common plant ancestor. The fossil record of ancient bryophytes is incomplete, consisting mostly of spores and small tissue fragments, and can be interpreted in different ways. As a result, it does not provide a definite answer on bryophyte evolution.

The oldest known recognizable plant fossils are dated at about 425 million years old. These fossils resemble modern liverworts in many respects, but the spores are virtually identical to those in 470-million-year-old rocks. Fossil fragments of tiny liverwort-like plants associated with ancient spores have been discovered in Oman. This evidence suggests that liverwort-like plants may have been the earliest plants to colonize land.

CHECKPOINT 27.2

- *Which of the following are parts of the gametophyte generation in mosses: antheridia, zygote, embryo, capsule, archegonia, sperm cells, egg cell, spores, and protonema?*

- **CONNECT** *How are mosses, liverworts, and hornworts similar? How is each group distinctive?*

27.3 SEEDLESS VASCULAR PLANTS

6 Discuss the features that distinguish seedless vascular plants from algae and bryophytes.
7 Name and briefly describe the two phyla of seedless vascular plants.
8 Describe the life cycle of ferns and compare their sporophyte and gametophyte generations.
9 Compare the generalized life cycles of homosporous and heterosporous plants.

The most important adaptation found in seedless vascular plants, although absent in algae and bryophytes, is specialized vascular tissues—xylem and phloem—for support and conduction. This system of conduction lets vascular plants grow larger than bryophytes because water, minerals, and sugar are transported to all parts of the plant. Although seedless vascular plants in temperate environments are relatively small, tree ferns in the tropics may grow to heights of 18 m (60 ft). All seedless vascular plants have true stems with vascular tissues, and most also have true roots and leaves.

Botanists have extensively studied the evolution of the leaf as the main organ of photosynthesis. The two basic types of true leaves—microphylls and megaphylls—evolved independently of each other (FIG. 27-12). The **microphyll,** which is usually small and has a single vascular strand, probably evolved from small, projecting extensions of stem tissue (*enations*). Only one group of living plants, the club mosses, has microphylls.

In contrast, **megaphylls** probably evolved from stem branches that gradually filled in with additional tissue (*webbing*) to form most leaves as we know them today. Megaphylls have more than one vascular strand, as we would expect if they evolved from branch systems. Ferns (with the exception of whisk ferns, discussed later in this chapter), gymnosperms, and flowering plants have megaphylls.

Recent evidence suggests megaphylls evolved over a 40-million-year period in the Late Paleozoic era in response to a gradual decline in the level of atmospheric CO_2. As CO_2 declined, plants developed a flattened blade with more stomata for gas exchange. (More stomata allowed cells inside the leaf to get enough CO_2.)

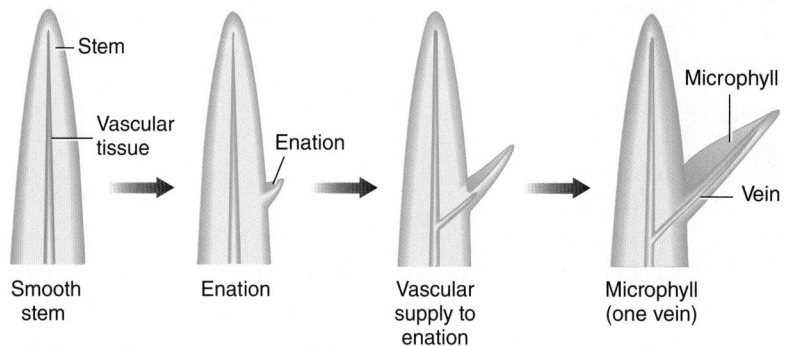

(a) Microphyll evolution. Microphylls probably originated as outgrowths (enations) of stem tissue that developed a single vascular strand later. Club mosses have microphylls.

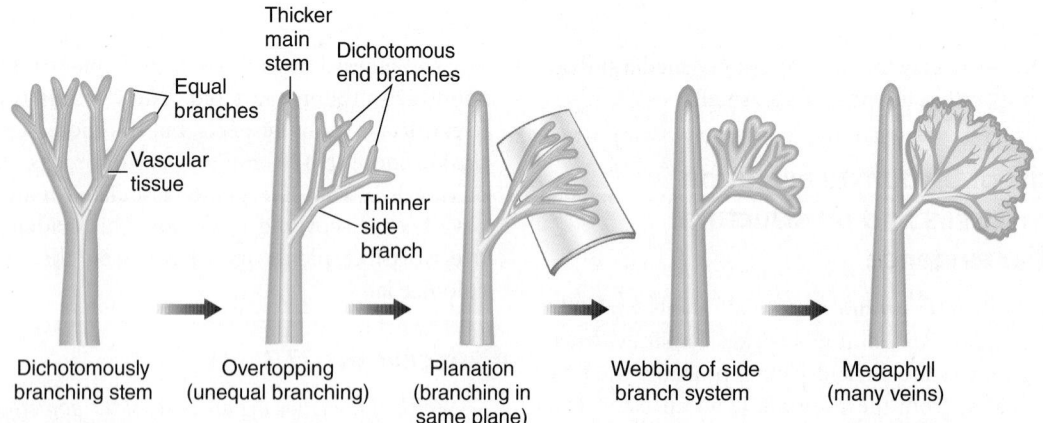

(b) Megaphyll evolution. Megaphylls probably evolved from the evolutionary modification of side branches. Webbing is the evolutionary process in which the spaces between close branches become filled with chlorophyll-containing cells. Ferns, horsetails, gymnosperms, and flowering plants have megaphylls.

Figure 27-12 Evolution of microphylls and megaphylls
© Cengage Learning

There are two main clades of seedless vascular plants: the club mosses and the ferns. Biologists originally considered horsetails and whisk ferns distinct enough to be classified in separate phyla. However, many kinds of evidence, such as DNA comparisons and similarities in sperm structure, have resulted in their being reclassified as ferns. As shown in FIGURE 27-13, ferns, including horsetails and whisk ferns, are a monophyletic group and the closest living relatives of seed plants (also see Table 27-2).

Club mosses are small plants with rhizomes and short, erect branches

Club mosses (phylum Lycopodiophyta) were important plants millions of years ago, when species that are now extinct often reached great size (FIG. 27-14a). These large, treelike plants were major contributors to our present-day coal deposits (see *Inquiring About: Ancient Plants and Coal Formation*).

The 1200 or so species of club mosses living today, such as *Lycopodium* (FIG. 27-14b), are small (less than 25 cm, or 10 in., tall), attractive plants common in temperate woodlands. They possess true roots, both rhizomes and erect aerial stems, and small, scalelike leaves (microphylls). Sporangia are borne on reproductive leaves that are either clustered in conelike strobili at the tips of stems or scattered in reproductive areas along the stem. Club mosses are evergreen and often fashioned into wreaths and other decorations. In some areas they are endangered by overharvesting.

That common names are sometimes misleading in biology is vividly evident in this group of plants. The most common names for the phylum Lycopodiophyta are "club mosses" and "ground pines," yet these plants are neither mosses, which are nonvascular, nor pines, which are seed plants.

Ferns are a diverse group of spore-forming vascular plants

Most of the 11,000 species of **ferns** (phylum Pteridophyta) are terrestrial, although a few have adapted to aquatic habitats. Ferns range from the tropics to the Arctic Circle, with most species living in tropical rain forests, where they perch high in the branches of trees. In temperate regions ferns commonly inhabit swamps, marshes, stream banks, and moist woodlands (FIG. 27-15a). Some species grow in fields, rocky crevices on cliffs or mountains, or even deserts.

The life cycle of ferns involves a clearly defined alternation of generations. The ferns grown as houseplants (e.g., Boston fern, maiden-hair fern, and staghorn fern) represent the larger, more conspicuous sporophyte generation.

The fern sporophyte consists of a horizontal underground stem, or *rhizome,* that bears leaves, called *fronds,* and true roots. As each young frond first emerges from the ground, it is tightly coiled and resembles the top of a violin, hence the name *fiddlehead* (FIG. 27-15b). As fiddleheads grow, they unroll and expand to form fronds. Fern fronds are usually compound (the blade is divided into leaflets), with the leaflets forming

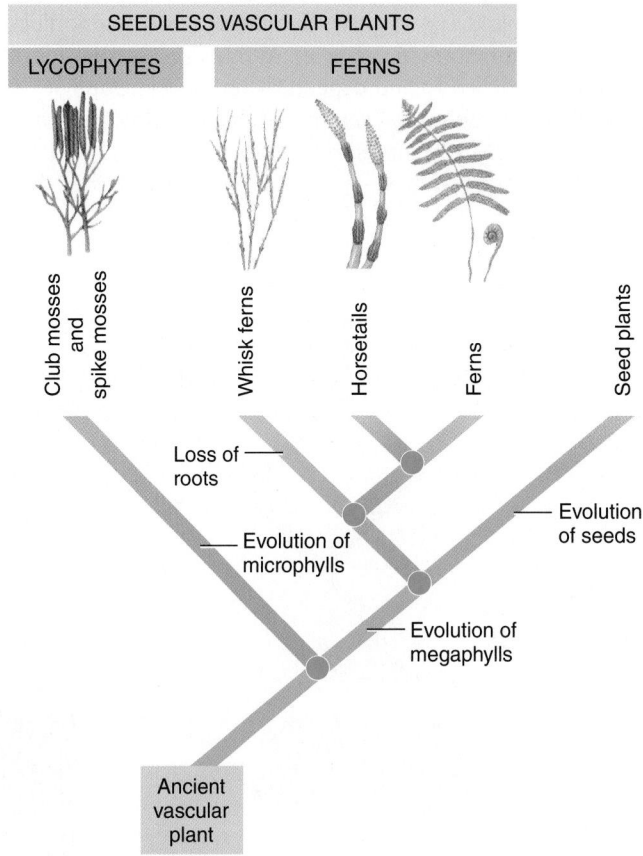

KEY POINT

Seedless vascular plants include lycophytes and ferns.

SEEDLESS VASCULAR PLANTS

LYCOPHYTES — FERNS

Club mosses and spike mosses — Whisk ferns — Horsetails — Ferns — Seed plants

Loss of roots

Evolution of microphylls

Evolution of seeds

Evolution of megaphylls

Ancient vascular plant

Figure 27-13 Evolutionary relationships among extant seedless vascular plants

These relationships are based on structural and molecular comparisons.

CONNECT According to the cladogram, which group, the lycophytes or the ferns, is most closely related to the seed plants?

© Cengage Learning

beautifully complex leaves. Fronds, roots, and rhizomes all contain vascular tissues.

Spore production usually occurs in certain areas on the fronds, which develop sporangia. Many species bear the sporangia in clusters, called **sori** (sing., *sorus*) (FIGS. 27-15c and d). Within sporangia, sporogenous cells (spore mother cells) undergo meiosis to form haploid spores. The sporangia burst open and discharge spores that may germinate and grow by mitosis into gametophytes.

The mature fern gametophyte, which bears no resemblance to the sporophyte, is a tiny (less than half the size of one of your fingernails), green, often heart-shaped structure that grows flat against the ground. Called a **prothallus** (pl., *prothalli*), the fern gametophyte lacks vascular tissues and has tiny, hairlike absorptive rhizoids to anchor it (FIG. 27-15e). The prothallus usually produces both archegonia and antheridia on its underside. Each archegonium contains a single egg, whereas numerous sperm cells are produced in each antheridium.

Ferns use water as a transport medium. The flagellate sperm cells swim, usually from a nearby prothallus, to the neck of an archegonium through a thin film of water on the ground underneath the prothallus. After one of the sperm cells fertilizes the egg, a diploid zygote grows by mitosis into a multicellular embryo (an immature sporophyte). At this stage the sporophyte embryo is attached to and dependent on the gametophyte, As the embryo matures, however, the prothallus withers and dies, and the sporophyte becomes free-living.

The fern life cycle alternates between the dominant, diploid sporophyte with its rhizome, roots, and fronds and the haploid gametophyte (prothallus) (FIG. 27-16). The sporophyte generation is dominant not only because it is larger than the gametophyte but also because it persists for an extended period (most fern sporophytes are perennials), whereas the gametophyte dies soon after reproducing.

Whisk ferns are classified as reduced ferns

Only about 12 species of **whisk ferns** (phylum Pteridophyta) exist today, and the fossil record contains several extinct species. Whisk ferns, which live mainly in the tropics and subtropics, are relatively simple in structure and lack true roots and leaves but have vascularized stems. *Psilotum nudum,* a representative whisk fern, has both a horizontal underground rhizome and vertical aerial stems (FIG. 27-17a). Whenever the stem forks, or branches, it always divides into two equal halves. Botanists call this forking **dichotomous branching.** In contrast, when most plant stems branch, one stem is more vigorous and becomes the main trunk.

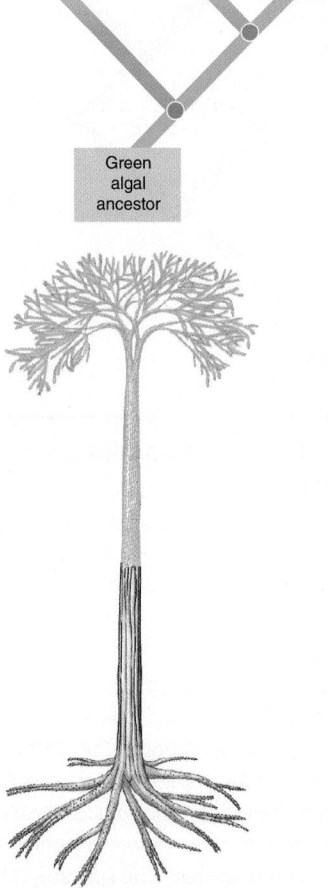

Nonvascular bryophytes

Vascular seedless plants

Vascular seed plants

Green algal ancestor

© Cengage Learning

(a) Reconstruction of *Lepidodendron,* an ancient club moss the size of a large tree. Numerous fossils of *Lepidodendron* were preserved in coal deposits, particularly in Great Britain and the central United States. (Redrawn from Hirmer, M., *Handbuch der Paläobotanik,* R. Olderbourg, Munich, 1927.)

Strobilus

Leaves (microphylls)

Ed Reschke/Getty Images

(b) The sporophyte of *Lycopodium,* a club moss, has small, scalelike, evergreen leaves (microphylls). Spores are produced in sporangia on reproductive leaves clustered in a conelike strobilus (*shown*) or, in other species, scattered along the stem.

Figure 27-14 Seedless vascular plants: club mosses

How did the coal that industrial society depends on for energy form? Coal, which is burned to produce electricity and to manufacture items of steel and iron, is one of the most important fossil fuels. Although mined, coal is not an inorganic mineral like gold or aluminum but rather is an organic material formed from the remains of ancient vascular plants, particularly those of the Carboniferous period, approximately 320 mya. Four main groups of plants contributed to coal formation. Two were seedless vascular plants: the club mosses and ferns (see figure), including horsetails. The other two were seed plants: seed ferns (now extinct) and early gymnosperms.

It is hard to imagine that relatives of the small, relatively inconspicuous club mosses, ferns, and horsetails of today were so significant in forming vast beds of coal. However, many members of these groups that existed during the Carboniferous period were giants compared with their modern counterparts and formed immense forests. (Figure 21-12 shows a reconstruction of a Carboniferous forest.)

The climate during the Carboniferous period was warm, moist, and mild. Plants in most locations could grow year-round because of the favorable conditions. Forests of these plants often grew in low-lying, swampy areas that periodically flooded when the sea level rose. As the sea level receded, these plants would re-establish.

When these large plants died or were blown over in storms, they decomposed incompletely because they were covered by swamp water. (The anaerobic conditions of the water prevented wood-rotting fungi from decomposing the plants, and anaerobic bacteria do not decompose wood rapidly.) Thus, over time the partially decomposed plant material accumulated and consolidated.

Layers of sediment formed over the plant material each time the water level rose and flooded the low-lying swamps. With time, heat and pressure built up in these accumulated layers and converted the plant material to coal and the sediment layers to sedimentary rock. Much later, geologic upheavals raised the layers of coal and sedimentary rock. Coal is usually found in seams, underground layers that vary in thickness from 2.5 cm (1 in.) to more than 30 m (100 ft).

© David Lyons/Alamy

This piece of Carboniferous coal contains a fossilized fern. The coal and fossil are about 300 million years old.

The various grades of coal (lignite, the lowest grade; subbituminous; bituminous; and anthracite, the highest grade) formed as a result of the different temperatures and pressures to which the layers were exposed. Coal exposed to high heat and pressure during its formation is drier, is more compact (and therefore harder), and has a higher heating value (i.e., a higher energy content) than coal exposed to lower heat and temperature conditions.

The upright stems of *Psilotum* are green and are the main organs of photosynthesis. Tiny, round sporangia, borne directly on the erect, aerial stems, contain sporogenous cells that undergo meiosis to form haploid spores. After dispersal, the spores germinate to form haploid prothalli. Because they grow underground, the prothalli of whisk ferns are difficult to study (**FIG. 27-17b**). They are nonphotosynthetic as a result of their subterranean location, and they apparently have a symbiotic relationship with mycorrhizal fungi that provides them with sugar and essential minerals (see Chapter 29).

Botanists have carefully studied whisk ferns in recent years. Molecular data, including comparisons of nucleotide sequences of ribosomal RNA, chloroplast DNA, and mitochondrial DNA in living species, support the hypothesis that the whisk ferns should be classified as reduced ferns rather than as surviving representatives of extinct vascular plants (see discussion of rhyniophytes later in this chapter).

Horsetails are an evolutionary line of ferns

About 300 mya, the **horsetails** (phylum Pteridophyta) were among the dominant plants and grew as large as modern trees (**FIG. 27-18a**). Because they contributed to Earth's vast coal deposits, these ancient horsetails, like ancient club mosses, are still significant today.

The few surviving horsetails, about 15 species in the genus *Equisetum,* grow mostly in wet, marshy habitats and are less than 1.3 m (4 ft) tall, but extremely distinctive (**FIG. 27-18b**). They are widely distributed on every continent except Australia. Traditionally classified in their own phylum, horsetails are now grouped with ferns. This reclassification is based on molecular similarities between horsetails and other ferns.

Horsetails have true roots, stems (both rhizomes and erect aerial stems), and small leaves. The hollow, jointed stems are impregnated with silica, which gives them a gritty texture. Small leaves, interpreted as reduced megaphylls, are fused in whorls at each node (the area on the stem where leaves attach). The green stem is the main organ of photosynthesis. Horsetails are so named because certain vegetative (nonreproductive) stems have whorls of branches that give the appearance of a bushy horse's tail. In the past, horsetails were called "scouring rushes" and were used to scrub pots and pans along stream banks.

Each reproductive branch of a horsetail bears a terminal conelike **strobilus** (pl., *strobili*). The strobilus consists of several stalked, umbrella-like structures, each of which bears five to ten sporangia in a circle around a common axis.

The horsetail life cycle is similar in many respects to the fern life cycle. In horsetails, as in ferns, the sporophyte is the conspicuous plant, whereas the gametophyte is a minute, lobed thallus ranging in width from the size of a pinhead to about 1 cm across. The sporophyte and gametophyte are both photosynthetic and nutritionally independent at maturity. Like ferns, horsetails require water as a medium for flagellate sperm cells to swim to the egg.

(a) Fronds. The Christmas fern (*Polystichum acrostichoides*), photographed in the Great Smoky Mountains in Tennessee, has fronds that grow to 0.6 m (2 ft) in length.

(b) Fiddleheads. Some fiddleheads are edible.

(c) Sori. These round sori of rabbit's foot fern (*Polypodium aureum*) are arranged in two prominent rows on the leaf's underside.

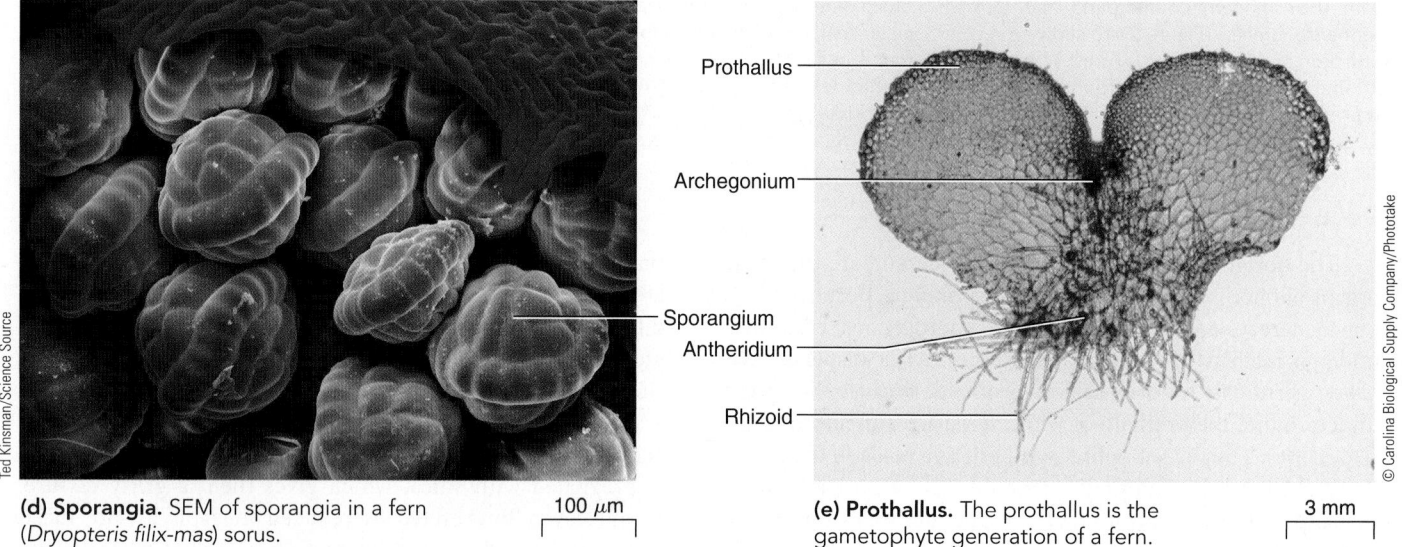

(d) Sporangia. SEM of sporangia in a fern (*Dryopteris filix-mas*) sorus.

100 μm

(e) Prothallus. The prothallus is the gametophyte generation of a fern.

3 mm

Prothallus
Archegonium
Sporangium
Antheridium
Rhizoid

Figure 27-15 Seedless vascular plants: ferns

Some ferns and club mosses are heterosporous

In the life cycles examined thus far, plants produce only one type of spore as a result of meiosis. This condition, known as **homospory,** is characteristic of bryophytes, horsetails, whisk ferns, and most ferns and club mosses. However, certain ferns and club mosses (known as spike mosses) exhibit **heterospory,** in which they produce two types of spores: microspores and megaspores.

FIGURE 27-19 illustrates the generalized life cycle of a heterosporous plant. The sporophyte plant produces sporangia within a conelike strobilus. Each strobilus usually bears two kinds of sporangia: microsporangia and megasporangia. *Microsporangia* are sporangia that produce *microsporocytes* (also called *microspore mother cells*), which undergo meiosis to form microscopic, haploid **microspores.** Each microspore develops into a male gametophyte that produces sperm cells within antheridia.

Megasporangia produce *megasporocytes* (also called *megaspore mother cells*). When megasporocytes undergo meiosis, they form haploid **megaspores,** each of which develops into a female gametophyte that produces eggs in archegonia.

The development of male gametophytes from microspores and of female gametophytes from megaspores occurs within their respective spore walls, using stored food provided by the sporophyte. As a result, and unlike the gametophytes of other seedless vascular plants, the male and female gametophytes are not truly free-living. Fertilization is followed by the development of a new sporophyte.

Heterospory evolved several times during the history of land plants. It was a significant development in plant evolution because it was the forerunner of the evolution of seeds. Heterospory characterizes the two most successful groups of plants existing today, the gymnosperms and the flowering plants, both of which produce seeds.

The dominant phase in the life cycle of ferns, like that of all other vascular plants, is the sporophyte.

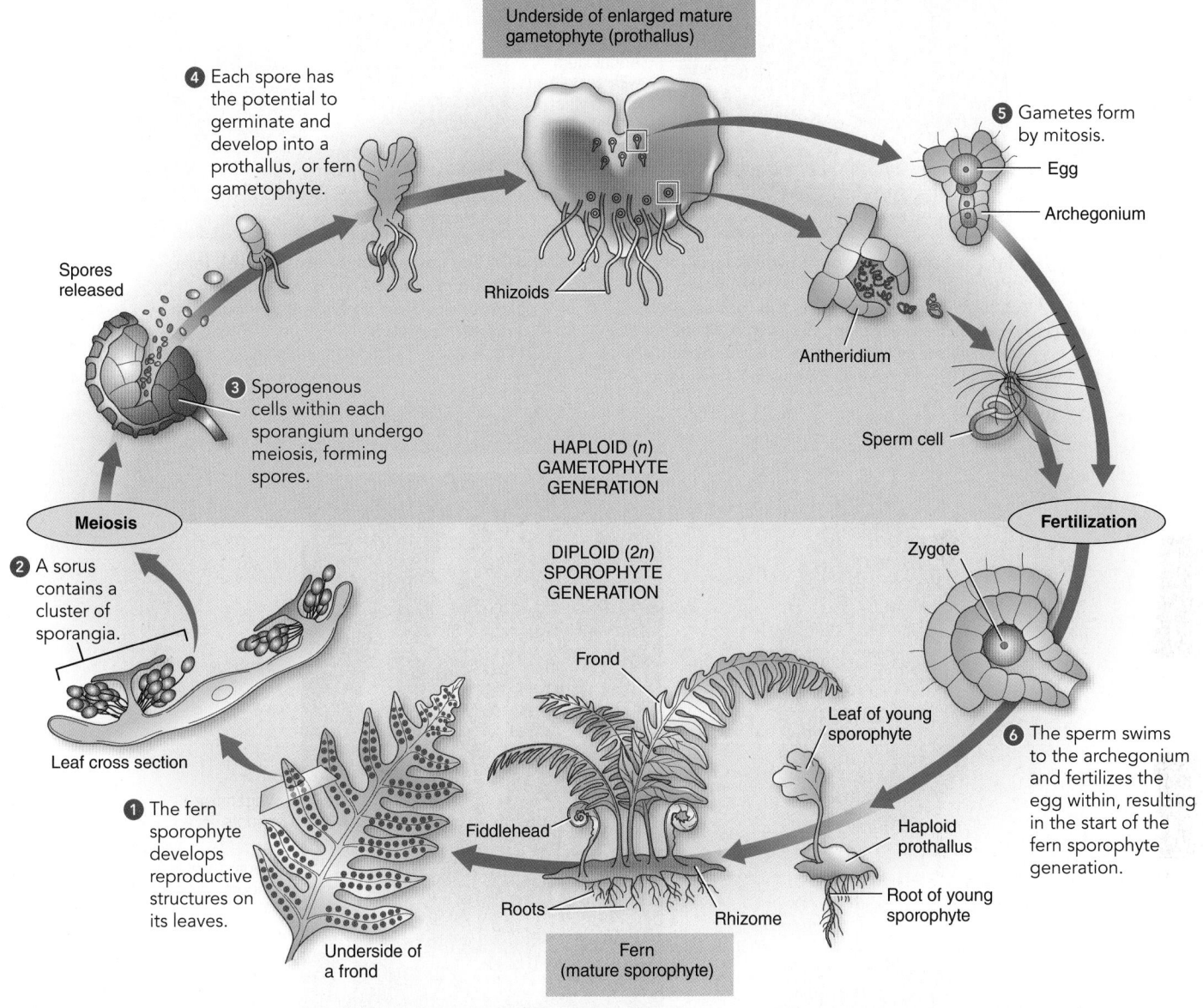

Figure 27-16 *Animation* **The life cycle of ferns**

Note the clearly defined alternation of generations between the gametophyte (prothallus) and sporophyte (leafy plant) generations.

CONNECT What is the most fundamental difference between the fern life cycle and that of mosses?

© Cengage Learning

Seedless vascular plants are used for experimental studies

Botanists use many seedless vascular plants as experimental models to study certain aspects of plant biology, such as physiology, growth, and development. Ferns and other seedless vascular plants are useful in studying how apical meristems give rise to plant tissues. An **apical meristem** is the area at the tip (apex) of a root or shoot where growth—cell division,

elongation, and differentiation—occurs. Ferns and other seedless vascular plants have a single large *apical cell* located at the center tip of the apical meristem. This apical cell is the source, by mitosis, of all the cells that eventually make up the root or shoot. The apical cell divides in an orderly fashion, and the smaller daughter cells produced by the apical cell, in turn, divide and give rise to different parts of the root or shoot. It is possible to trace mature cells in the root or shoot back to their origin from the single apical cell.

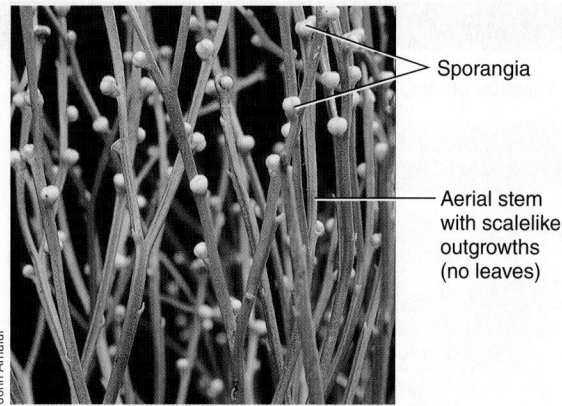

(a) **The sporophyte of the whisk fern** *Psilotum nudum.* The stem is the main organ of photosynthesis in this rootless, leafless, vascular plant. Sporangia, which turn yellow as they mature, are borne on short lateral branches directly on the stems.

(b) **The gametophyte of the whisk fern** *Psilotum nudum.* The gametophyte (prothallus) lives underground, nourished by mycorrhizal fungi.

Figure 27-17 Whisk ferns

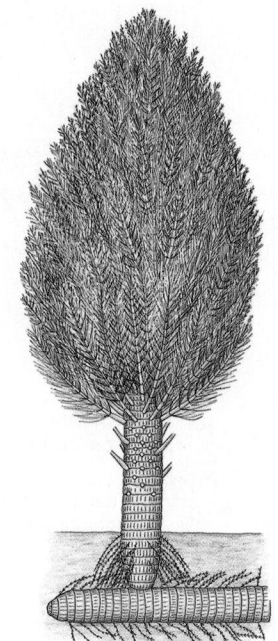

(a) Reconstruction of *Calamites*, an ancient horsetail that grew as tall as many modern trees, to 20 m (about 65 ft). *Calamites* had an underground rhizome where roots and aerial shoots originated. (Redrawn from Emberger, L., *Les Plantes Fossiles*, Masson et Cie, Paris, 1968.)

(b) *Equisetum telematia,* a horsetail with a wide distribution in Eurasia, Africa, and North America, has unbranched reproductive shoots bearing conelike strobili and separate, highly branched vegetative (nonreproductive) shoots. In some horsetail species, both reproductive and vegetative shoots are unbranched.

Figure 27-18 Horsetails

Ferns are interesting research plants for studies in genetics because they are **polyploids** and have multiple sets of chromosomes. (Many ferns have hundreds of chromosomes.) However, gene expression in ferns is exactly what one would expect of a *diploid* plant. Apparently, genes in the extra sets of chromosomes are gradually silenced and therefore not expressed.

Seedless vascular plants arose more than 420 mya

Currently, the oldest known megafossils of early vascular plants are from Silurian (420 mya) deposits in Europe. (Plant *megafossils* are fossilized roots, stems, leaves, and reproductive structures.)

Heterospory—the production of two types of spores—was the forerunner of the evolution of seeds.

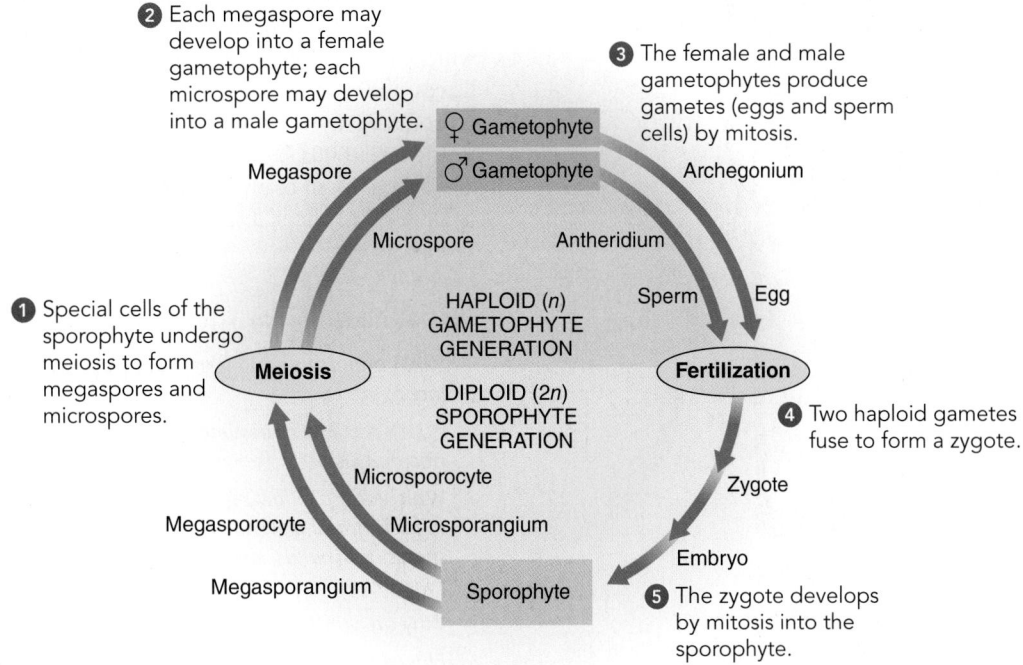

Figure 27-19 The basic life cycle of heterosporous plants

Two types of spores, microspores and megaspores, are produced during the life cycle of heterosporous plants.

CONNECT What fundamental similarities does a heterosporous life cycle share with the basic plant life cycle depicted in Figure 27-2?

© Cengage Learning

Megafossils of several kinds of small, seedless vascular plants also occur in Silurian deposits in Bolivia, Australia, and north-western China. Microscopic spores of early vascular plants appear in the fossil record earlier than megafossils, suggesting that even older megafossils of simple vascular plants may be discovered.

Botanists assign the oldest vascular plants to phylum Rhy-niophyta, which, according to the fossil record, arose some 420 mya and became extinct about 380 mya. The rhyniophytes are so named because many fossils of these extinct plants were found in fossil beds near Rhynie, Scotland. *Rhynia gwynne-vaughanii* is an example of an early vascular plant that superficially resem-bled whisk ferns in that it consisted of leafless upright stems that branched dichotomously from an underground rhizome (**FIG. 27-20**). *Rhynia* lacked roots, although it had absorptive rhi-zoids. Sporangia formed at the ends of short branches. The inter-nal structure of its rhizome contained a central core of xylem cells for conducting water and minerals.

For many years, botanists considered *Rhynia major*, a plant that grew about 50 cm (20 in.) tall and probably lived in marshes, a classic example of a rhyniophyte. Fossils indicate that this plant had rhizoids, dichotomously branching rhi-zomes, and upright stems that terminated in sporangia. How-ever, recent microscopic studies of fossil rhizomes indicate that the central core of tissue lacked the xylem cells characteristic

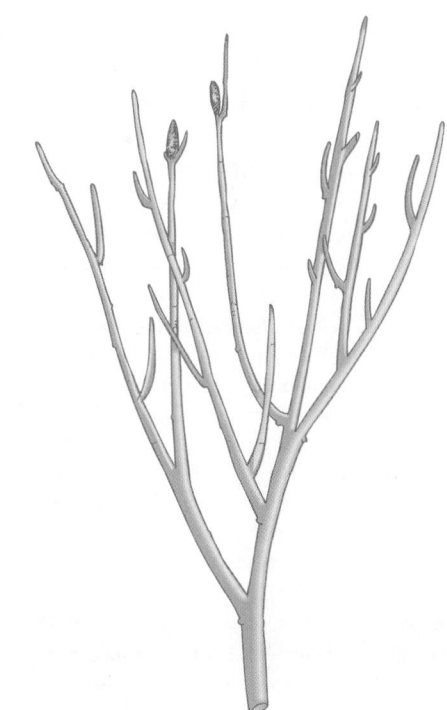

Figure 27-20 Reconstruction of *Rhynia gwynne-vaughanii*

This leafless plant, one of Earth's earliest vascular plants, is now extinct. It grew about 18 cm (7 in.) tall. (Redrawn from Edwards, D., "Evidence for the Sporophytic Status of the Lower Devonian Plant *Rhynia gwynne-vaughanii*," *Review of Palaeobotany and Palynology*, Vol. 29, 1980.)

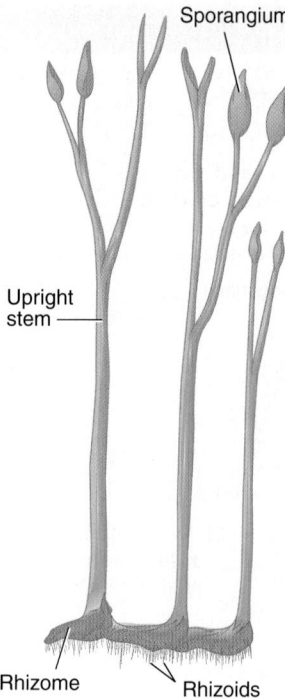

Figure 27-21 Reconstruction of *Aglaophyton major*

Recent evidence indicates that this plant, although superficially similar to other early vascular plants, lacked conducting tissues that are characteristic of vascular plants. For this reason, it has been reclassified into a new genus and is no longer considered a rhyniophyte. (Redrawn from Mauseth, J.D., *Botany: An Introduction to Plant Biology*, 2nd ed., Saunders College Publishing, Philadelphia, 1995.)

of vascular plants. For that reason, *R. major* was reclassified into a new genus, *Aglaophyton,* and is no longer considered a rhyniophyte (FIG. 27-21).

The oldest known megafossils of fernlike trees, discovered in New York, are dated at 380 million years old. These trees, which were about 8 m (26 ft) tall, had vascular tissues but no leaves and a minimal root system. They grew vertically and reproduced by forming spores. The trunks were discovered in the late 1800s (and named *Eospermatopteris*), but not until 2007 was a fossil of an entire tree pieced together. The treetops had previously been found separate from the trunks and named *Wattieza*.

CHECKPOINT 27.3

- *What adaptations do ferns have that both algae and bryophytes lack?*
- CONNECT *How does one distinguish between megaphylls and microphylls?*
- *Which of the following are parts of the sporophyte generation in ferns: frond, sperm cells, egg cell, roots, sorus, sporangium, spores, prothallus, rhizome, antheridium, archegonium, and zygote?*
- *Why are whisk ferns and horsetails now classified as ferns?*
- *How does heterospory modify the plant life cycle?*

SUMMARY: FOCUS ON LEARNING OBJECTIVES

27.1 Adaptations of Plants to Life on Land *(page 558)*

1 Discuss some environmental challenges of living on land and describe how several plant adaptations meet these challenges.

- The colonization of land by plants required the evolution of many anatomical, physiological, and reproductive adaptations. Plants have a waxy **cuticle** to protect against water loss and **stomata** for gas exchange needed for photosynthesis.

- Plant life cycles have an **alternation of generations** in which they spend part of their life cycle in a multicellular haploid **gametophyte generation** and part in a multicellular diploid **sporophyte generation.** The gametophyte plant produces gametes by mitosis. During fertilization these gametes fuse to form a **zygote,** the first stage of the sporophyte generation. The zygote develops into a multicellular **embryo** that the gametophyte protects and nourishes. The mature sporophyte plant develops from the embryo and produces sporogenous cells (spore mother cells). These cells undergo meiosis to form **spores,** the first stage in the gametophyte generation.

- Most plants have multicellular **gametangia** with a protective jacket of sterile cells surrounding the gametes. **Antheridia** are gametangia that produce sperm cells, and **archegonia** are gametangia that produce eggs.

- Ferns and other vascular plants have **xylem** to conduct water and dissolved minerals and **phloem** to conduct dissolved sugar.

2 Name the green algal group from which plants are hypothesized to have descended and describe supporting evidence.

- Plants probably arose from a group of green algae called **charophytes.** This conclusion is based in part on molecular comparisons of DNA and RNA sequences, which show the closest match between charophytes and plants.

27.2 Bryophytes *(page 562)*

3 Summarize the features that distinguish bryophytes from other plants.

- Unlike other land plants, **bryophytes** are nonvascular and lack xylem and phloem. Bryophytes are the only plants with a dominant gametophyte generation. Their sporophytes remain permanently attached and nutritionally dependent on the gametophytes.

4 Name and briefly describe the three phyla of bryophytes.

- **Mosses** (phylum Bryophyta) have gametophytes that are green plants that grow from a filamentous **protonema.**

- Many **liverworts** (phylum Hepatophyta) have gametophytes that are flattened, lobelike **thalli;** others are leafy.

- **Hornworts** (phylum Anthocerophyta) have thalloid gametophytes.

5 Describe the life cycle of mosses and compare their gametophyte and sporophyte generations.

- The green moss gametophyte bears archegonia, antheridia, or both at the top of the plant. During fertilization, a sperm cell fuses with an egg cell in the archegonium. The zygote grows into an embryo that develops into a moss sporophyte, which is attached to the gametophyte. Meiosis occurs within the capsule of the sporophyte to produce spores. When a spore germinates, it grows into a protonema that forms buds that develop into gametophytes.

27.3 Seedless Vascular Plants (page 568)

6 Discuss the features that distinguish seedless vascular plants from algae and bryophytes.

- Seedless vascular plants have several adaptations that algae and bryophytes lack, including vascular tissues and a dominant sporophyte generation. As in bryophytes, reproduction in seedless vascular plants depends on water as a transport medium for motile sperm cells.

7 Name and briefly describe the two phyla of seedless vascular plants.

- Sporophytes of **club mosses** (phylum Lycopodiophyta) consist of roots, rhizomes, erect branches, and leaves that are **microphylls.**
- **Ferns** (phylum Pteridophyta) are the largest and most diverse group of seedless vascular plants. The fern sporophyte consists of a rhizome that bears fronds and true roots. Phylum Pteridophyta also includes whisk ferns and horsetails. Sporophytes of **whisk ferns** have **dichotomously branching** rhizomes and

erect stems; they lack true roots and leaves. **Horsetail** sporophytes have roots, rhizomes, aerial stems that are hollow and jointed, and leaves that are reduced **megaphylls.**

8 Describe the life cycle of ferns and compare their sporophyte and gametophyte generations.

- Fern sporophytes have roots, rhizomes, and leaves that are megaphylls. Their leaves, or fronds, bear sporangia in clusters called **sori.** Meiosis in sporangia produces haploid spores. The fern gametophyte, called a **prothallus,** develops from a haploid spore and bears both archegonia and antheridia.

9 Compare the generalized life cycles of homosporous and heterosporous plants.

- **Homospory,** the production of one kind of spore, is characteristic of bryophytes, most club mosses, and most ferns, including whisk ferns and horsetails. In homospory spores give rise to gametophyte plants that produce both egg cells and sperm cells.
- **Heterospory,** the production of two kinds of spores (microspores and megaspores), occurs in certain club mosses, certain ferns, and all seed plants. **Microspores** give rise to male gametophytes that produce sperm cells. **Megaspores** give rise to female gametophytes that produce eggs. The evolution of heterospory was an essential step in the evolution of seeds.

Know and Comprehend

1. Plants probably descended from a group of green algae called (a) rhyniophytes (b) *Calamites* (c) epiphytes (d) charophytes (e) club mosses
2. Which of the following is *not* a characteristic of plants? (a) cuticle (b) unicellular gametangia (c) stomata (d) multicellular embryo (e) alternation of generations
3. In plant life cycles (a) the first products of meiosis are gametes (b) spores are part of the diploid sporophyte generation (c) the embryo gives rise to a zygote (d) the first stage in the diploid sporophyte generation is the zygote (e) the first stage in the haploid gametophyte generation is the prothallus
4. The bryophytes (a) include mosses, liverworts, and hornworts (b) include whisk ferns, horsetails, and club mosses (c) are small plants that lack a vascular system (d) a and c (e) b and c
5. The waxy layer that covers aerial parts of plants is the (a) cuticle (b) archegonium (c) protonema (d) stoma (e) thallus
6. A strengthening compound found in cell walls of vascular plants is (a) chitin (b) lignin (c) cutin (d) cellulose (e) carotenoid
7. Stomata (a) help prevent desiccation of plant tissues (b) transport water and minerals through plant tissues (c) allow gas exchange for photosynthesis (d) strengthen cell walls (e) produce male gametes
8. The green, gametangia-bearing moss plant (a) is the haploid gametophyte generation (b) is the diploid sporophyte generation (c) is called a protonema (d) contains cells with single large chloroplasts (e) b and c
9. ___ is a leaf that arose from a branch system. (a) An antheridium (b) A microphyll (c) A megaphyll (d) A sorus (e) A microspore
10. These plants have vascularized stems but lack true roots and leaves. (a) mosses (b) club mosses (c) horsetails (d) whisk ferns (e) hornworts

11. These plants have hollow, jointed stems that are impregnated with silica. (a) mosses (b) club mosses (c) horsetails (d) whisk ferns (e) hornworts
12. Which of the following statements about ferns is *not* true? (a) ferns have motile sperm cells that swim through water to the egg-containing archegonium (b) ferns are vascular plants (c) ferns are the most economically important group of bryophytes (d) the fern sporophyte consists of a rhizome, roots, and fronds (e) the diversity of ferns is greatest in the tropics

Apply and Analyze

13. **VISUALIZE** Draw a simple diagram illustrating a heterosporous life cycle. Include the sporophyte generation, megaspore, female gametophyte, egg, microspore, male gametophyte, sperm, meiosis, and fertilization. Be sure to indicate whether each generation or kind of cell is haploid or diploid.

Evaluate and Synthesize

14. **EVOLUTION LINK** How may the following trends in plant evolution be adaptive to living on land?
    a. dependence on water for fertilization → no need for water as a transport medium
    b. homospory → heterospory
15. **INTERPRET DATA** According to the cladogram in Figure 27-5, which plants evolved first: nonvascular bryophytes, seedless vascular plants, or seed plants?
16. **EVOLUTION LINK** Where would you position the rhyniophytes on Figure 27-13? Would the line for rhyniophytes extend to the tips of the rest of the cladogram? Why or why not?

28 | Seed Plants

Hairy vetch seed pods. The seeds are developing within each pod. Hairy vetch (*Vicia villosa*) is native to Europe and Asia but has become naturalized in the United States.

KEY CONCEPTS

28.1 Seed plants, which produce young sporophytes enclosed within seeds, include gymnosperms and angiosperms.

28.2 Gymnosperms produce exposed ovules that, following fertilization, develop into seeds that are usually borne in cones on the sporophytes. Conifers are the most diverse and numerous of the four living gymnosperm phyla.

28.3 Angiosperms produce ovules enclosed within carpels; following fertilization, seeds develop from the ovules, and the ovaries of carpels become fruits. Angiosperms dominate the land and exhibit great diversity in both vegetative and reproductive structures.

28.4 Gymnosperms evolved from ancestral seedless vascular plants; angiosperms evolved from ancestral gymnosperms, possibly an ancient conifer.

Chapter 27 focused on plants that reproduce by means of *spores,* haploid cells that disperse and germinate to produce gametophytes. Although gymnosperms and angiosperms also produce spores, their primary means of reproduction and dispersal is by **seeds,** which represent an important adaptation for life on land (see photograph). Each seed consists of an embryonic sporophyte, nutritive tissue, and a protective coat. Seeds develop from the fertilized egg cell, the female gametophyte, and its associated tissues. The two groups of seed plants, gymnosperms and angiosperms (flowering plants), exhibit the greatest evolutionary complexity of land plants and are the dominant plants in most terrestrial environments.

Seeds are reproductively superior to spores for several reasons. First, a seed is further along in its development before it is released to survive on its own: a seed contains a multicellular young plant with embryonic root, stem, and one or more leaves already formed, whereas a spore is a single cell. Second, a seed contains an abundant food supply. After germination, food stored in the seed nourishes the plant embryo until it becomes self-sufficient. Because a spore is a single cell, few food reserves exist for the plant that develops from a spore. Third, a seed is protected by a multicellular **seed coat** that is very thick and hard in some plants, as, for example, in lima beans. Like spores, seeds live for extended periods at reduced rates of metabolism and germinate when conditions become favorable.

Seeds and seed plants are intimately connected with the development of human civilization. From prehistoric times, early humans collected and used seeds for food. Seeds often contain a concentrated source of proteins, oils, carbohydrates, and vitamins, which are nourishing for humans as well as for germinating plants. Seeds are easy to store (if kept dry), so humans can collect them during times of plenty to save for times of need. Few other foods are stored as conveniently or for as long. Although flowering plants produce most seeds that humans consume, the seeds of certain gymnosperms—the piñon pine, for example—are edible.

In this chapter we present the diversity in seed plants—both gymnosperms and angiosperms—followed by an examination of what we know about how seed plants evolved.

28.1 AN INTRODUCTION TO SEED PLANTS

1 Compare the features of gymnosperms and angiosperms.

Like the bryophytes and seedless vascular plants introduced in Chapter 27, seed plants have life cycles with an **alternation of generations;** they spend part of their lives in the multicellular diploid sporophyte stage and part in the multicellular haploid gametophyte stage. The sporophyte generation is the dominant stage in seed plants, and the gametophyte generation is significantly reduced in size and entirely dependent on the sporophyte generation. Unlike the bryophytes and ferns, seed plants do not have free-living gametophytes. Instead, the female gametophyte is attached to and nutritionally dependent on the sporophyte generation.

In Chapter 27 you learned that some seedless vascular plants are heterosporous. (Figure 27-19 shows a generalized life cycle for heterosporous plants.) However, *all* seed plants are heterosporous and produce two types of spores: microspores and megaspores. In fact, heterospory is a requirement of seed production.

Seed plants produce **ovules,** each of which is a *megasporangium* surrounded by **integuments,** layers of sporophyte tissue that enclose the megasporangium. After fertilization takes place, the ovule develops into a seed, and the integuments develop into the seed coat (**FIG. 28-1a**).

Botanists divide seed plants into two groups based on whether or not an ovary wall surrounds their ovules (an *ovary* is a structure that contains one or more ovules). The two groups of seed plants are the **gymnosperms** and the **angiosperms** (**TABLE 28-1**). The word *gymnosperm* is derived from the Greek for "naked seed." Gymnosperms produce seeds that are totally exposed or borne on the scales of cones (**FIG. 28-1b**). In other words, an ovary wall does not surround the ovules of gymnosperms. Pine, spruce, fir, hemlock, and ginkgo are examples of gymnosperms.

The term *angiosperm* is derived from the Greek expression that means "seed enclosed in a vessel or case." Angiosperms are flowering plants that produce their seeds within a *fruit* (a mature ovary) (**FIG. 28-1c**). Thus, the ovules of angiosperms are protected. Flowering plants, which are extremely diverse, include corn, oaks, water lilies, cacti, apples, grasses, palms, and buttercups.

Both gymnosperms and flowering plants have vascular tissues: **xylem,** for conducting water and dissolved minerals (inorganic nutrients), and **phloem,** for conducting dissolved sugar.

CHECKPOINT 28.1

- *What is an ovule?*
- *What kinds of seeds are surrounded by an ovary wall?*

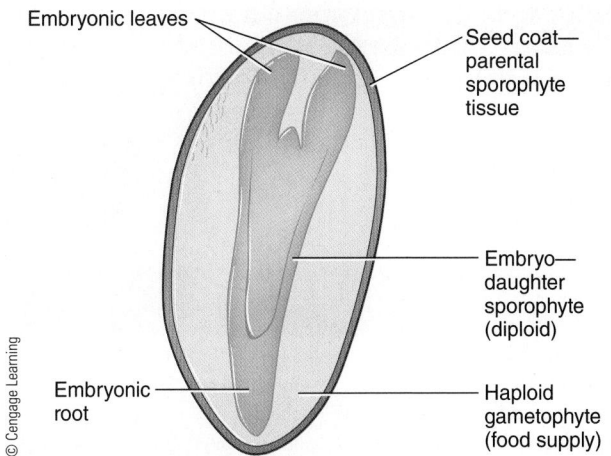

(a) Gymnosperm seed. Longitudinal section through a pine seed.

Embryonic leaves

Seed coat— parental sporophyte tissue

Embryo— daughter sporophyte (diploid)

Embryonic root

Haploid gametophyte (food supply)

© Cengage Learning

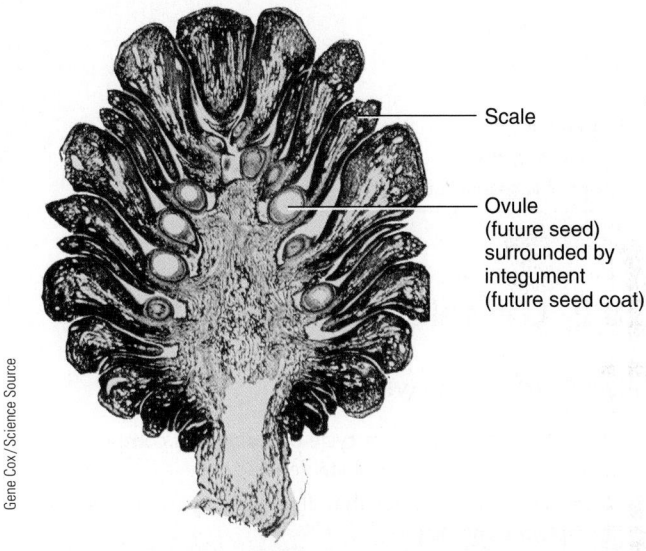

Scale

Ovule (future seed) surrounded by integument (future seed coat)

Gene Cox / Science Source

(b) Gymnosperm cone. Longitudinal section through a female pine cone, showing the ovules (which develop into seeds) borne on scales. Note the absence of an ovary wall.

Fruit (ovary wall)

Seed

C Squared Studios / Photodisc / Getty Images

(c) Angiosperm fruit. Longitudinal section through an avocado fruit, showing the seed surrounded by ovary tissue of the maternal sporophyte.

Figure 28-1 Seeds

TABLE 28-1	A Comparison of Gymnosperms and Angiosperms	
CHARACTERISTIC	**GYMNOSPERMS**	**ANGIOSPERMS**
	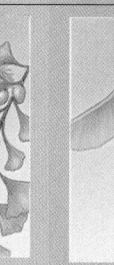	
Growth habit	Woody trees and shrubs	Woody or herbaceous
Conducting cells in xylem	Tracheids	Vessel elements and tracheids
Reproductive structures	Cones (usually)	Flowers
Pollen grain transfer	Wind (usually)	Animals or wind
Fertilization	Egg and sperm → zygote	Double fertilization: egg and sperm → zygote; two polar nuclei and sperm → endosperm
Seeds	Exposed or borne on scales of cones	Enclosed within fruit derived from ovary
Nutritive material	Female gametophyte	Endosperm
Number of species	About 840	More than 300,000
Geographic distribution	Worldwide	Worldwide

© Cengage Learning

28.2 GYMNOSPERMS

LEARNING OBJECTIVES

2 Trace the steps in the life cycle of a pine and compare its sporophyte and gametophyte generations.

3 Summarize the features that distinguish gymnosperms from bryophytes and ferns.

4 Name and briefly describe the four phyla of gymnosperms.

The gymnosperms include some of the most interesting plants. For example, a giant sequoia (*Sequoiadendron giganteum*) known as the General Sherman Tree, in Sequoia National Park in California, is one of the world's most massive organisms. It is 84.2 m (275 ft) tall and has a girth of 24.1 m (79 ft) measured 1.5 m (5 ft) above ground level. Another gymnosperm, a coast redwood (*Sequoia sempervirens*) nicknamed "Hyperion" after a giant in Greek mythology, is possibly the world's tallest tree, measuring 115.7 m (379 ft) in 2007. Botanists using tree-ring analysis determined that one of Earth's oldest living organisms, a bristlecone pine (*Pinus longaeva*) in the White Mountains of California, is about 5,000 years old.

Gymnosperms are usually classified into four phyla, which branch from a single evolutionary line (FIG. 28-2). Numbering 630 species, the largest phylum of gymnosperms is Coniferophyta, commonly called *conifers*. Two phyla of gymnosperms, Ginkgophyta (the ginkgoes) and Cycadophyta (the cycads), are evolutionary remnants of groups that were more significant in the past. The fourth phylum, Gnetophyta (gnetophytes), is a collection of some unusual plants that share certain traits not found in other gymnosperms; until recently, the gnetophytes were thought to be more closely related to flowering plants than were other gymnosperm clades. Current evidence, however, suggests that gnetophytes are probably most closely related to conifers (see Fig. 28-2).

Conifers are woody plants that produce seeds in cones

The **conifers** (phylum Coniferophyta), which include pines, spruces, hemlocks, and firs, are the most familiar group of gymnosperms (FIG. 28-3a). These 630 species of woody trees or shrubs produce annual additions of secondary tissues (wood and bark); there are no herbaceous (nonwoody) conifers. The wood (*secondary xylem*) consists of **tracheids,** which are long, tapering cells with pits through which water and dissolved minerals move from one cell to another.

Many conifers produce **resin,** a viscous, clear or translucent substance consisting of several organic compounds that may protect the plant from attack by fungi or insects. The resin collects in resin ducts, tubelike cavities that extend throughout the roots, stems, and leaves. Cells lining the resin ducts produce and secrete resin.

Conifers generally have leaves called *needles* that are long, narrow, tough, and leathery (FIG. 28-3b). Most pines bear clusters of two to five needles, depending on the species. In a few conifers, such as American arborvitae, the leaves are scalelike and cover the stem (FIG. 28-3c). Most conifers are evergreen and bear their leaves throughout the year. Only a few, such as the dawn redwood, larch, and bald cypress, are deciduous and shed their needles at the end of each growing season.

Most conifers are **monoecious:** they have separate male and female reproductive parts in different locations on the same plant. These reproductive parts are generally borne in *strobili* (commonly called *cones*), hence the name *conifer,* which means "cone-bearing."

Conifers occupy extensive areas, ranging from the Arctic to the tropics, and are the dominant vegetation in the forested regions of Alaska, Canada, northern Europe, and Siberia. In addition, they are important in the Southern Hemisphere, particularly in wet, mountainous areas of temperate and tropical regions in South America, Australia, New Zealand, and Malaysia. Southwestern China, with more than 60 species of conifers, has the greatest regional diversity of conifer species in the world. California, New Caledonia (an island east of Australia), southeastern China, and Japan also have considerable diversity of conifer species. Ecologically, conifers contribute food and shelter to animals and other organisms, and their roots hold the soil in place and help prevent soil erosion.

Humans use conifers for their wood (for building materials as well as paper products), medicinal value (such as the anticancer drug Taxol from the Pacific yew), turpentine, and resins.

Seed plants include four gymnosperm phyla and one phylum of flowering plants (angiosperms).

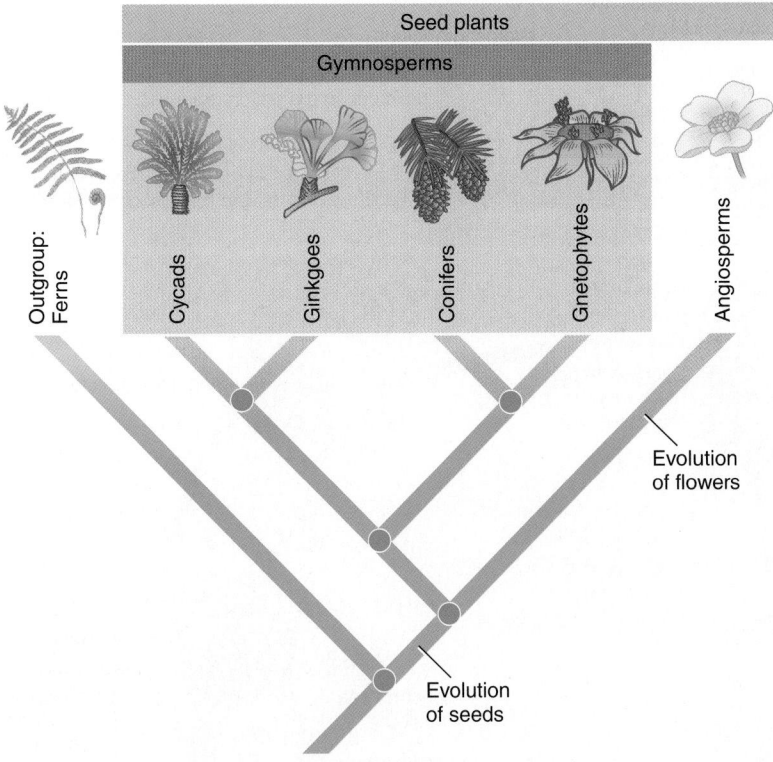

Figure 28-2 Gymnosperm and angiosperm evolution

This cladogram shows a current hypothesis of phylogenetic relationships among living seed plants, based on structural evidence, molecular comparisons, and fossils. Relationships among extant gymnosperm clades and angiosperms remain controversial. The arrangement of the phyla shown here may change as future analyses help clarify relationships.

CONNECT According to the cladogram, which group is most closely related to the conifers?

© Cengage Learning

(a) White fir (*Abies concolor*). Photographed in Milford, Pennsylvania, at the historic home of Gifford Pinchot, the first Chief Forester of the U.S. Forest Service (under T. R. Roosevelt).

(b) In white pine (*Pinus strobus*), leaves are long, slender needles that occur in clusters of five.

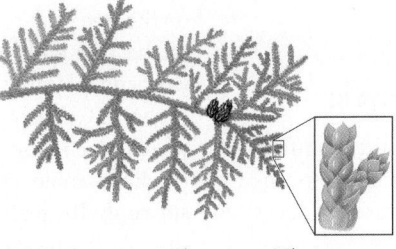

(c) In American arborvitae (*Thuja occidentalis*), leaves are small and scalelike.

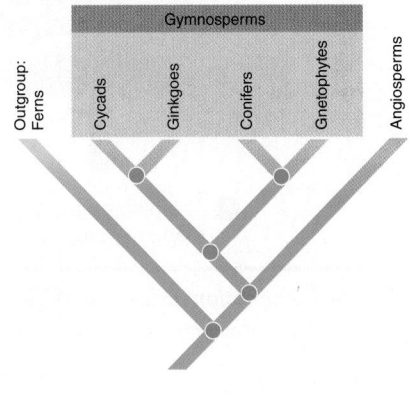

Figure 28-3 Conifers
© Cengage Learning

Gymnosperms, like all seed plants, are heterosporous.

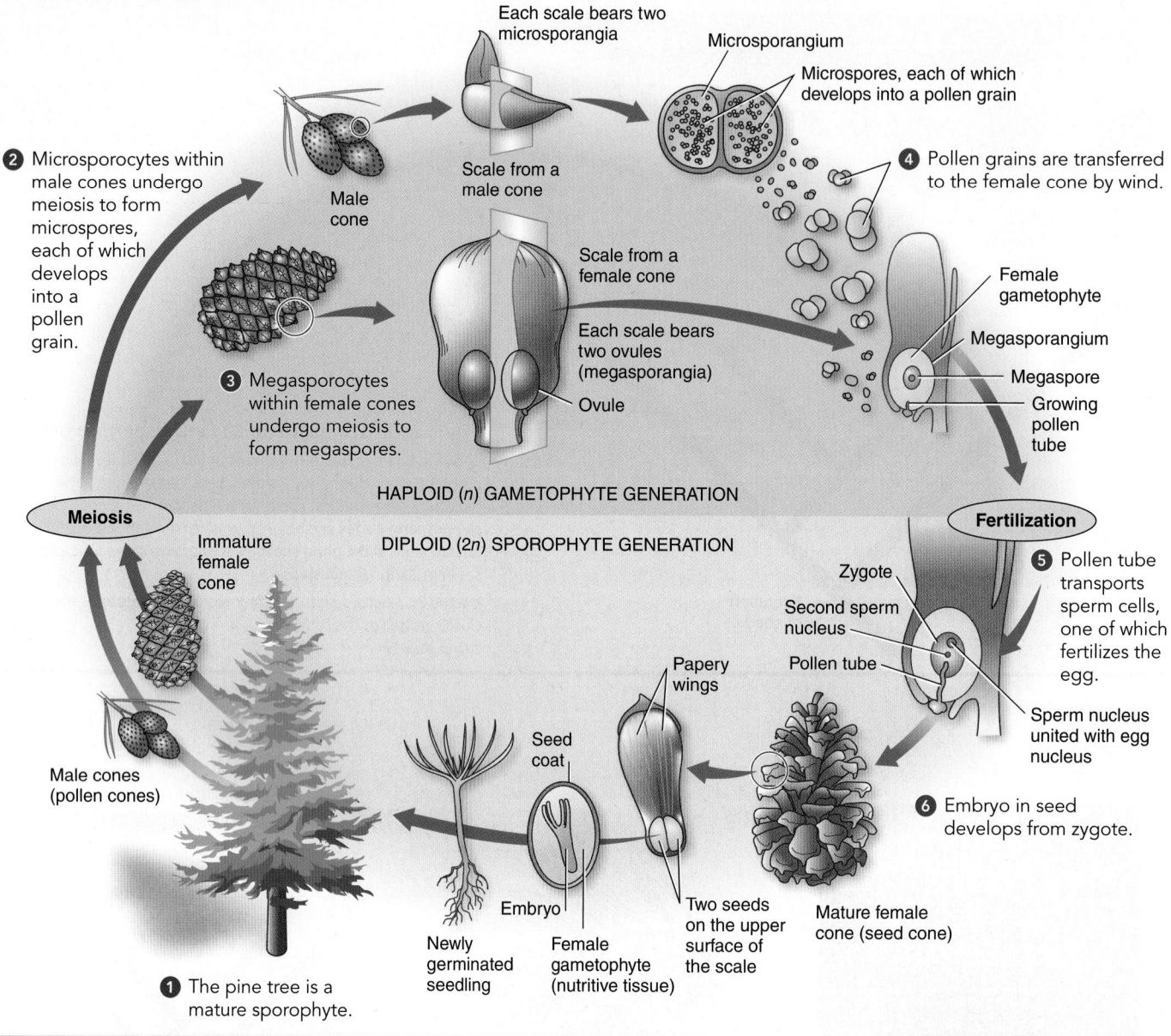

Figure 28-4 *Animation* **Life cycle of pine**

One major advantage of gymnosperms over the seedless vascular plants is the production of wind-borne pollen grains.

© Cengage Learning

CONNECT Is a pine pollen grain haploid or diploid? Is it a gamete? Explain your answer.

Because of their attractive appearance, conifers such as firs, spruces, pines, and cedars are grown for landscape design and decorative holiday trees and wreaths.

Pines represent a typical conifer life cycle

The genus *Pinus,* by far the largest genus in the conifers, consists of about 100 species. A pine tree is a mature sporophyte (FIG. 28-4). Pine is heterosporous and therefore produces microspores and megaspores in separate cones. Male cones, usually 1 cm or less in length, are smaller than female cones

and are generally produced on the lower branches each spring (FIG. 28-5). The more familiar, woody female cones, which are on the tree year-round, are usually found on the upper branches of the tree and bear seeds after reproduction. Female cones vary considerably in size. The sugar pine (*P. lambertiana*) that grows in California produces the world's longest female cones, which reach lengths of 60 cm (2 ft).

Each male cone, also called a *pollen cone,* consists of **sporophylls,** leaflike scales that bear sporangia on the underside. At the base of each sporophyll are two **microsporangia,** which contain numerous **microsporocytes,** also called

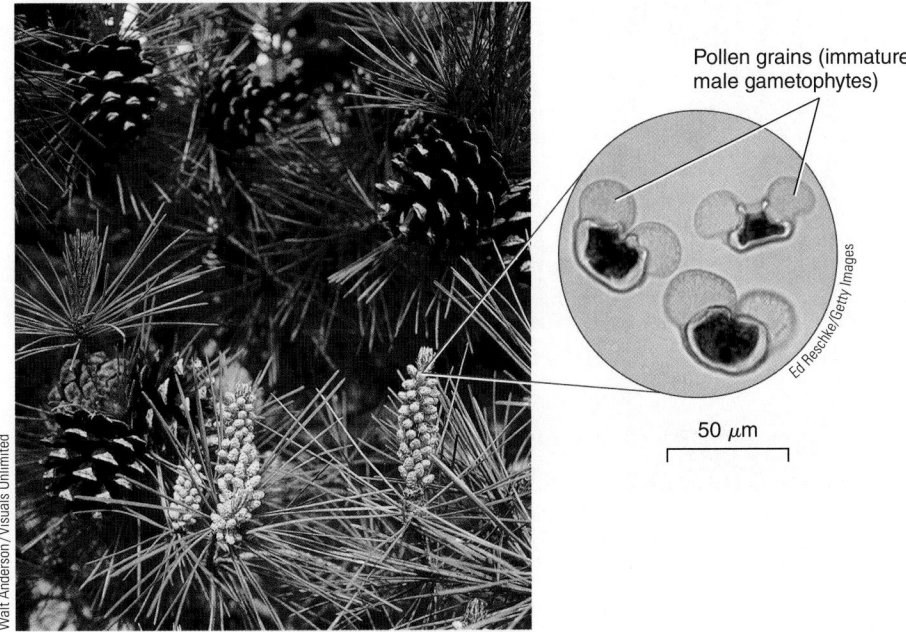

Figure 28-5 Male and female cones in lodgepole pine (*Pinus contorta*)
Mature woody female cones (*top*) have opened to shed their seeds. Clusters of male cones (*bottom*) produce copious amounts of pollen grains in the spring. (*Inset*) Each pollen grain develops from a microspore.

Pollen grains (immature male gametophytes)

50 μm

microspore mother cells. In Figure 28-4, each microsporocyte undergoes meiosis to form four haploid microspores. **Microspores** then develop into extremely reduced male gametophytes. Each immature male gametophyte, also called a **pollen grain,** consists of four cells, two of which—a *generative cell* and a *tube cell*—are involved in reproduction. The other two cells soon degenerate. Two large air sacs on each pollen grain provide buoyancy for wind dissemination. Male cones shed pollen grains in great numbers, and wind currents carry some to the immature female cones.

Many botanists think that the female cones (also called *seed cones*) are modified branch systems. Each cone scale bears two **ovules,** or **megasporangia,** on its upper surface. Within each megasporangium, meiosis of a **megasporocyte,** or *megaspore mother cell,* produces four haploid **megaspores.** One of them divides mitotically, developing into the female gametophyte, which produces an egg within each of several archegonia. The other three megaspores are nonfunctional and soon degenerate.

When the ovule is ready to receive pollen, it produces a sticky droplet at the opening where the pollen grains land. **Pollination,** the transfer of pollen to the female cones, occurs in the spring for a week or ten days, after which the pollen cones wither and drop off the tree. One of the many pollen grains that adhere to the sticky female cone grows a **pollen tube,** an outgrowth that digests its way through the megasporangium to the egg within the archegonium. The germinated pollen grain with its pollen tube is the mature male gametophyte.

Ultimately, two nonmotile (nonflagellate) sperm cells form within the germinated pollen grain. When it reaches the female gametophyte, the pollen tube discharges the two sperm cells near the egg. One of these sperm cells fuses with the egg, in the process of **fertilization,** to form a zygote, or fertilized egg, which subsequently grows into a young pine embryo in the seed. The other sperm cell degenerates.

The developing embryo consists of an embryonic root and an embryonic shoot with several cotyledons (embryonic leaves). Haploid female gametophyte tissue surrounds the embryo and becomes the nutritive tissue in the mature pine seed. A tough, protective seed coat derived from the integuments encloses the embryo and nutritive tissue. The seed coat forms a thin, papery wing at one end that enables dispersal by air currents. Some seeds remain within the female cones for several years before being shed.

In the pine life cycle, the sporophyte generation is dominant, and the gametophyte generation is restricted in size to microscopic structures in the cones. Although the female gametophyte produces archegonia, the male gametophyte is so reduced that it does not produce antheridia. The gametophyte generation in pines, as in all seed plants, depends totally on the sporophyte generation for nourishment.

A major adaptation in the pine life cycle is elimination of the need for external water as a sperm transport medium. Instead, air currents carry pine pollen grains to female cones, and nonflagellate sperm cells move through a pollen tube to the egg. Pine and other conifers are plants whose reproduction is totally adapted for life on land.

Cycads have seed cones and compound leaves

Cycads (phylum Cycadophyta) were very important during the Triassic period, which began about 251 million years ago (mya) and is sometimes referred to as the "Age of Cycads." Most cycad species are now extinct, and the few surviving ones, about 140 species, are tropical and subtropical plants with stout, trunklike stems and compound leaves that resemble those of palms or tree ferns (**FIG. 28-6**). Many cycads are endangered, primarily because they are popular as ornamentals and are gathered from the wild and sold to collectors. In the United States, cycads can be found in Florida, Texas, and California.

Cycad reproduction is similar to that in pines except that cycads are **dioecious** and therefore have seed cones on female plants and pollen cones on male plants. Their seed structure is most like that of the earliest seeds found in the fossil record. Cycads have also retained motile sperm cells, each of which has many hairlike flagella. Motile sperm cells are a vestige retained from the ancestors of cycads, in which sperm cells swam from antheridia to archegonia. In cycads specialized insects (almost exclusively beetles) carry pollen grains to the female plants and

Figure 28-6 Cycads

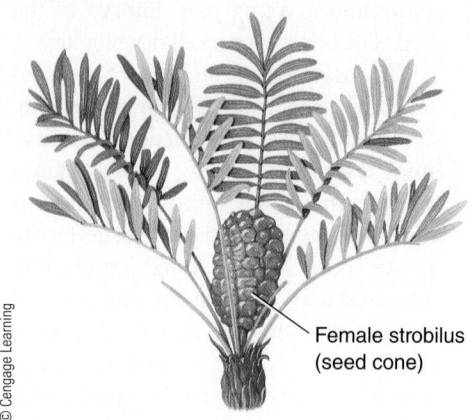

Female strobilus (seed cone)

© Cengage Learning

(a) A female Florida arrowroot (*Zamia integrifolia*) produces seed cones. This plant is the only cycad native to the United States. Like *Zamia*, most cycads are short plants, less than 2 m (6.5 ft) tall.

Martin Harvey/Getty Images

(b) This female cycad (*Encephalartos princeps*) native to South Africa bears three seed cones.

outer coverings give off a foul odor that smells like rancid butter. In China and Japan, where people eat the seeds, the female trees are more common.

Ginkgo has been an important medicinal plant for centuries and is still a common herbal remedy today. Extracts from the leaves may enhance neurological functioning by increasing blood flow to the brain, although several recent studies have concluded that ginkgo does not improve memory or prevent dementia in elderly people. In 2013 scientists from the National Toxicology Program of the National Institutes of Health reported the development of cancers in rats and mice fed high doses of *Gingko biloba* extract.

their cones. Following transfer to the female cone, the pollen grain germinates and grows a pollen tube. The sperm cells are released into this tube and swim to the egg.

Ginkgo biloba is the only living species in its phylum

Ginkgo (phylum Ginkgophyta) is represented by a single *extant* (living) species, the maidenhair tree, *Ginkgo biloba* (FIG. 28-7). This native of eastern China grew in the wild in only two locations, although people had cultivated it for its edible seeds in China and Japan for centuries. *Ginkgo* is the oldest genus (and species) of extant trees. Fossil ginkgoes 200 million years old have been discovered that are strikingly similar to the modern ginkgo.

People often plant ginkgo in North America and Europe today, particularly in parks and along city streets, because it is hardy and somewhat resistant to air pollution. Its leaves are deciduous and turn a beautiful yellow before being shed in the fall.

Like cycads, ginkgo is dioecious, with separate male and female trees. It has flagellate sperm cells, an evolutionary vestige that is not required because ginkgo produces airborne pollen grains. Ginkgo seeds are completely exposed rather than contained within cones. Male trees are usually planted because the female trees bear seeds whose fleshy

Gnetophytes include three unusual genera

The **gnetophytes** (nee´toe phites) (phylum Gnetophyta) consist of about 70 species in three diverse and obscure genera (*Gnetum, Ephedra,* and *Welwitschia*). Gnetophytes share certain features that make them unique among the gymnosperms. For example, gnetophytes have more efficient water-conducting cells, called *vessel elements,* in their xylem (see Chapter 33). Flowering plants also have vessel elements in their xylem, but of the gymnosperms, only the gnetophytes do. In addition, the cone clusters that some gnetophytes produce resemble flower clusters, and certain details in their life cycles resemble those of flowering plants. Despite these similarities, most botanists now think that gnetophytes represent evolutionary diversity in gymnosperms and are not in a direct line to flowering plants (see Fig. 28-2).

Marion Lobstein

(a) Close-up of a branch from a female ginkgo, showing the exposed seeds and the distinctive, fan-shaped leaves.

Joseph Malcolm Smith/Science Source

(b) Ginkgo (*Ginkgo biloba*) tree in a formal garden in northern England.

Figure 28-7 Ginkgo, or maidenhair tree

The genus *Gnetum* contains tropical vines, shrubs, and trees with broad leaves (**FIG. 28-8a**). Species in the genus *Ephedra* include many shrubs and vines that grow in deserts and other dry temperate and tropical regions. Some *Ephedra* species resemble horsetails in that they have jointed green stems with tiny leaves (**FIG. 28-8b**). Commonly called *joint fir*, *Ephedra* has been used medicinally for centuries. An Asiatic *Ephedra* is the source of ephedrine, which stimulates the heart and raises blood pressure. At one time, ephedrine was commonly sold over the counter in weight-control medications and herbal energy boosters; several deaths were reported from chronic use or overdose of products containing ephedrine, so its use has been restricted.

(a) The leaves of *Gnetum gnemon* resemble those of flowering plants. Note the exposed seeds. The species is native to southern Asia and the Maldives.

(b) A male joint fir (*Ephedra*) has pollen cones clustered at the nodes. In the 19th century, European pioneers used species native to the American Southwest to make a beverage, Mormon tea.

The third gnetophyte genus, *Welwitschia,* contains a single species found in deserts of southwestern Africa (**FIG.28-8c**). Most of *Welwitschia*'s body—a long taproot—grows underground. Its short, wide stem forms a shallow disc, up to 0.9 m (3 ft) in diameter, from which two ribbon-like leaves extend. These two leaves continue to grow from the stem throughout the plant's life, but their ends are usually broken and torn by the wind, giving the appearance of numerous leaves. Each leaf grows to about 2 m (6.5 ft) in length. When *Welwitschia* reproduces, cones form around the edge of its disclike stem.

CHECKPOINT 28.2

- *What is the dominant generation in the pine life cycle? How does pollination occur in gymnosperms?*
- **CONNECT** *What features distinguish gymnosperms from seedless plants?*
- *What are the four groups of gymnosperms?*
- *What features distinguish cycads from ginkgo? from gnetophytes?*

(c) *Welwitschia mirabilis* is native to deserts in southwestern Africa. It survives on moisture-laden fogs that drift inland from the ocean. Photographed in the Namib Desert, Namibia.

Figure 28-8 Gnetophytes

28.3 FLOWERING PLANTS

LEARNING OBJECTIVES

5 Summarize the features that distinguish flowering plants from other plants.
6 Briefly explain the life cycle of a flowering plant and describe double fertilization.
7 Contrast monocots and eudicots, the two largest classes of flowering plants.
8 Discuss the evolutionary adaptations of flowering plants.

Flowering plants, or angiosperms (phylum Anthophyta), are the most successful plants today, surpassing even the gymnosperms in importance. They have adapted to almost every habitat and, with at least 300,000 species, are Earth's dominant plants. Flowering plants come in a wide variety of sizes and forms, from herbaceous violets to massive eucalyptus trees. Some flowering plants—tulips and roses, for example—have large, conspicuous flowers; others, such as grasses and oaks, produce small, inconspicuous flowers.

Flowering plants are vascular plants that reproduce sexually by forming flowers and, following a unique double fertilization process, seeds within fruits. The fruit protects the developing seeds and often aids in their dispersal (see Chapter 37). Flowering plants have efficient water-conducting cells called **vessel elements** in their xylem and efficient sugar-conducting cells called **sieve tube elements** in their phloem (see Chapter 33).

Flowering plants are extremely important to humans because our survival as a species literally depends on them. All our major food crops are flowering plants; examples include rice, wheat, corn, potatoes, tomatoes, beans, apples, and citrus fruits. Woody flowering plants such as oak, cherry, and walnut provide valuable lumber. Flowering plants give us fibers, such as cotton and linen, and medicines, such as digitalis and codeine. Products as diverse as rubber, tobacco, coffee, chocolate, wine, and aromatic oils for perfumes come from flowering plants. *Economic botany* is the subdiscipline of botany that deals with plants of economic importance.

TABLE 28-2	Distinguishing Features of Monocots and Eudicots	
FEATURE	MONOCOTS	EUDICOTS
Flower parts	Usually in threes	Usually in fours or fives
Pollen grains	One furrow or pore	Three furrows or pores
Leaf venation	Usually parallel	Usually netted
Vascular bundles in stem cross section	Usually scattered or more complex arrangement	Arranged in a circle (ring)
Roots	Fibrous root system	Taproot system
Seeds	Embryo with one cotyledon	Embryo with two cotyledons
Secondary growth (wood and bark)	Absent	Often present

© Cengage Learning

Monocots and eudicots are the two largest classes of flowering plants

Phylum Anthophyta is divided into several classes with only a few members each and two very large classes: the monocots (class Monocotyledones) and the eudicots (class Eudicotyledones). The smaller classes will be discussed later in this chapter in the context of their evolutionary significance. For now, we restrict our discussion of flowering plants to the monocots and eudicots, which collectively represent about 97% of all flowering plant species. Eudicots are more diverse and include many more species (at least 200,000) than the monocots (at least 90,000). **TABLE 28-2** provides a comparison of some of the general features of the two classes.

Monocots include palms, grasses, orchids, irises, onions, and lilies. Monocots are mostly herbaceous plants with long, narrow leaves that have parallel veins (the main leaf veins run parallel to one another). The parts of monocot flowers usually occur in threes (**FIG. 28-9a**). For example, a flower may have three sepals, three petals, six stamens, and a compound pistil consisting of three fused carpels (these flower parts are

discussed shortly). Monocot seeds have a single **cotyledon,** or embryonic seed leaf; **endosperm,** a nutritive tissue, is usually present in the mature seed.

Eudicots include oaks, roses, mustards, cacti, blueberries, and sunflowers. Eudicots are either herbaceous (such as a tomato plant) or woody (such as a hickory tree). Their leaves vary in shape but usually are broader than monocot leaves, with netted (finely branched) veins. Flower parts usually occur in fours or fives or multiples thereof (**FIG. 28-9b**). Two cotyledons are present in eudicot seeds, and endosperm is usually absent in the mature seed, having been absorbed by the two cotyledons during seed development.

Sexual reproduction takes place in flowers

Flowers are reproductive shoots usually composed of four parts—sepals, petals, stamens, and carpels—arranged in whorls (circles) on the end of a flower stalk, or **peduncle** (**FIG. 28-10**). The peduncle may terminate in a single flower or a cluster of flowers known as an **inflorescence.** The tip of the flower stalk that bears the flower parts is known as the **receptacle.**

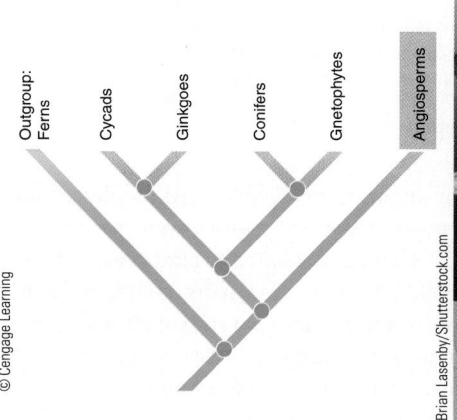

© Cengage Learning

© Brian Lasenby/Shutterstock.com

(a) Monocot. *Trillium erectum*, like most monocots, has floral parts in threes. Note the three green sepals, three red petals, six stamens, and three stigmas (the compound pistil consists of three fused carpels).

Glenn Harper/AgeFotostock

(b) Eudicot. Most eudicots, such as this *Tacitus*, have floral parts in fours or fives. Note the five petals, ten stamens, and five separate pistils. Five sepals are also present but barely visible against the background.

Figure 28-9 Flowering plants

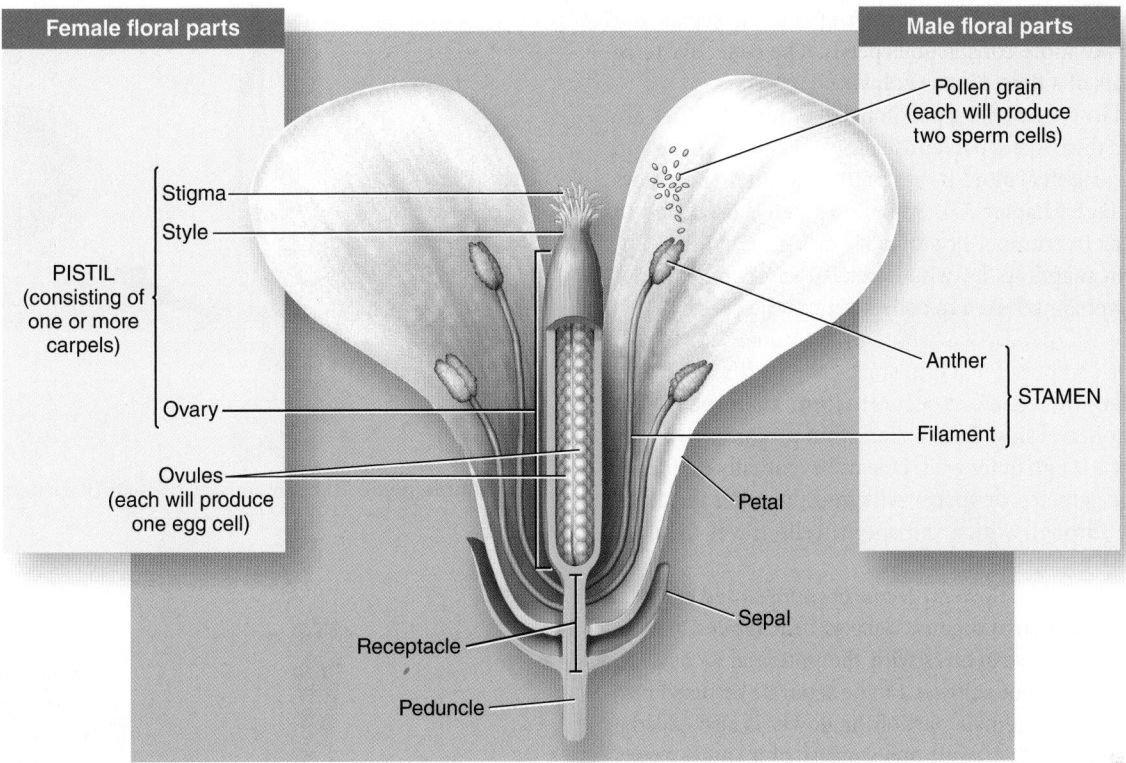

Figure 28-10 *Animation* **Floral structure**

This cutaway view of an *Arabidopsis* flower shows the details of basic floral structure. Each flower has four sepals (two are shown), four petals (two are shown), six stamens, and one long pistil. Four of the stamens are long, and two are short (two long and two short are shown). Pollen grains develop within sacs in the anthers. In *Arabidopsis,* the compound pistil consists of two carpels that each contain numerous ovules.
© Cengage Learning

All four floral parts are important in the reproductive process, but only the stamens (the "male" organs) and carpels (the "female" organs) produce gametes. A flower that has all four parts is a **complete flower,** whereas an **incomplete flower** lacks one or more of these four parts. A flower with both stamens and carpels is a **perfect flower,** whereas an **imperfect flower** has stamens or carpels, but not both.

Sepals, which make up the lowermost and outermost whorl on a floral shoot, are leaflike in appearance and often green (**FIG. 28-11a**). Sepals cover and protect the other flower parts

(a) The leaflike sepals of a rosebud (*Rosa*) enclose and protect the inner flower parts.

Figure 28-11 Parts of a flower

(b) A twinleaf (*Jeffersonia diphylla*) flower has eight yellow stamens. Note the simple pistil with its green ovary in the center of the flower.

when the flower is a bud. As the flower opens, the sepals fold back to reveal the more conspicuous petals. The collective term for all the sepals of a flower is the **calyx.**

The whorl just above the sepals consists of **petals,** which are broad, flat, and thin (like sepals and leaves) but vary in shape and are frequently brightly colored. Petals attract animal pollinators to the flower (see Chapter 37). Sometimes petals are fused to form a tube (as in trumpet honeysuckle flowers) or other floral shape (as in snapdragons, whose petals form two lips). The petals of a flower are referred to collectively as the **corolla.**

Just inside the petals is a whorl of **stamens** (FIG. 28-11b). Each stamen is composed of a thin stalk, called a **filament,** and a saclike **anther,** where meiosis occurs to form microspores that develop into pollen grains. Each pollen grain produces two cells surrounded by a tough outer wall. One cell eventually divides to form two male gametes, or sperm cells, and the other produces a pollen tube through which the sperm cells travel to reach the ovule.

In the center of most flowers are one or more closed **carpels,** the "female" reproductive organs. Carpels bear ovules, which, as you may recall, are structures with the potential to develop into seeds. The carpels of a flower can be separate or fused into a single structure. The female part of the flower is also called a **pistil** (see Fig. 28-11b). A pistil may consist of a single carpel (a simple pistil) or a group of fused carpels (a compound pistil) (FIG. 28-12). Each pistil generally has three sections: a **stigma,** on which the pollen grain lands; a **style,** a necklike structure through which the pollen tube grows; and an **ovary,** an enlarged structure that contains one or more ovules. Each young ovule contains a female gametophyte that forms one female gamete (an egg), two *polar nuclei,* and several other haploid cells. After fertilization, the ovule develops into a seed, and the ovary develops into a fruit.

The life cycle of flowering plants includes double fertilization

Flowering plants undergo alternation of generations in which the sporophyte generation is larger and nutritionally independent (FIG. 28-13). The gametophyte generation in flowering plants is microscopic and nutritionally dependent on the sporophyte. Flowering plants, like gymnosperms and certain other vascular plants, are heterosporous and produce two kinds of spores: microspores and megaspores. Sexual reproduction occurs in the flower.

Each young ovule within an ovary contains a megasporocyte (megaspore mother cell) that undergoes meiosis to produce four haploid megaspores. Three of them usually disintegrate, and one divides mitotically and develops into a mature female gametophyte, also called an **embryo sac.**

Embryo sacs in the vast majority of angiosperms contain seven cells with eight haploid nuclei. Six of these cells, including the egg cell, contain a single nucleus each, and a central cell has two nuclei, called **polar nuclei.** The egg and the central cell with two polar nuclei are directly involved in fertilization; the other five cells in the embryo sac apparently have no direct role in

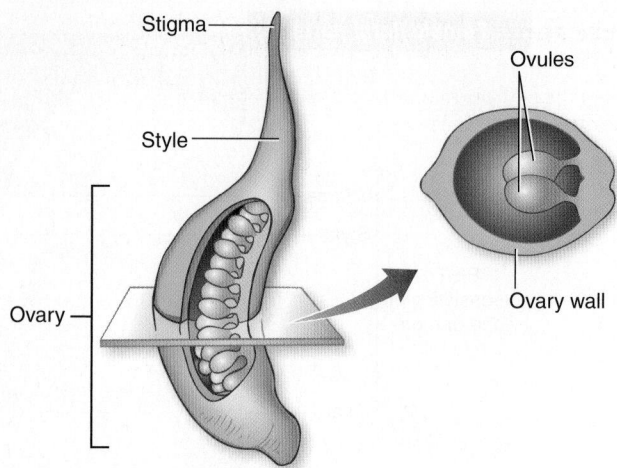

(a) Simple pistil. This simple pistil consists of a single carpel.

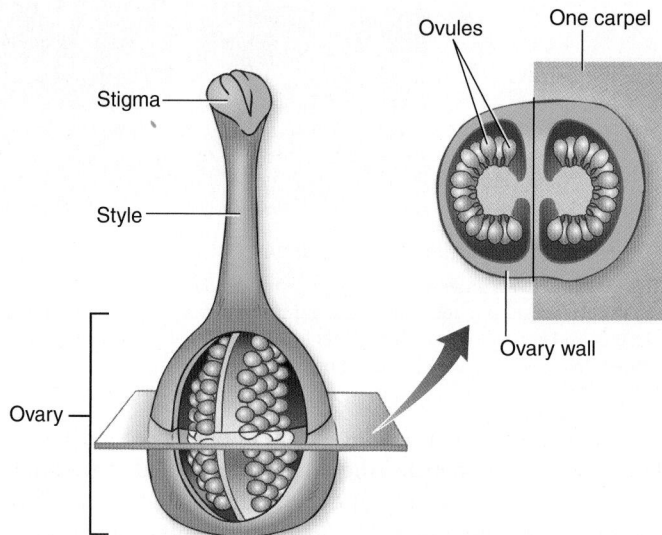

(b) Compound pistil. This compound pistil has two united carpels. In most flowers with single pistils, the pistils are compound, consisting of two or more fused carpels.

Figure 28-12 Simple and compound pistils
© Cengage Learning

the fertilization process and disintegrate. As the *synergids* (the two cells flanking the egg) disintegrate, however, they release chemicals that may affect the direction of pollen tube growth.

Each pollen sac, or microsporangium, of the anther contains numerous microsporocytes (microspore mother cells), each of which undergoes meiosis to form four haploid microspores. Every microspore develops into an immature male gametophyte, also called a *pollen grain.* Pollen grains are small, and each consists of two cells: the *tube cell* and the *generative cell.*

The anthers split open and begin to shed pollen. A variety of agents—including wind, water, insects, and other animal pollinators—transfer pollen grains to the stigma (see

A significant feature of the flowering plant life cycle is double fertilization, in which one sperm cell unites with the egg, forming a zygote; the other sperm cell unites with the two polar nuclei, forming a triploid cell that gives rise to endosperm.

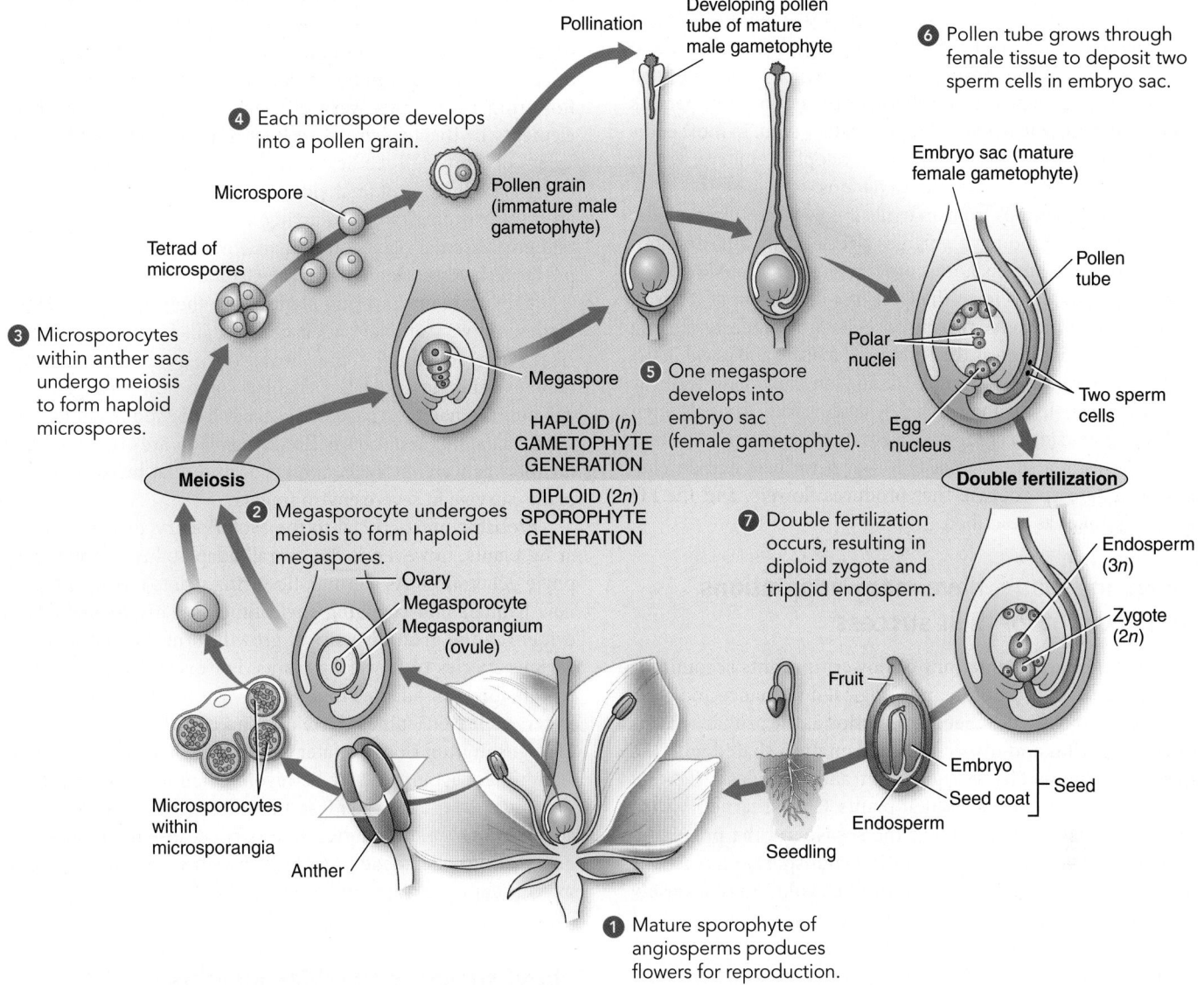

Figure 28-13 *Animation* **Life cycle of flowering plants**

CONNECT Are microspores and megaspores haploid or diploid? Are they gametes? Explain your answer.

© Cengage Learning

Chapter 37). If compatible with the stigma, the pollen grain germinates, and a pollen tube grows down the style and into the ovary. The germinated pollen grain with its pollen tube is the mature male gametophyte. Next, the generative cell divides to form two nonflagellate sperm cells. The sperm cells move down the pollen tube and are discharged into the embryo sac.

When the two sperm cells enter the embryo sac, *both* participate in fertilization. One sperm cell fuses with the egg, forming a zygote that divides by mitosis and develops into a multicellular embryo in the seed. The second sperm cell fuses with the two haploid polar nuclei of the central cell to form a triploid ($3n$) cell that divides by mitosis and develops into **endosperm,** a nutrient tissue rich in lipids, proteins, and carbohydrates that nourishes the growing embryo. This fertilization process, which involves two separate nuclear fusions, is called **double fertilization** and is, with two exceptions, unique to flowering plants. (Double fertilization has been reported in the gymnosperms *Ephedra nevadensis* and *Gnetum gnemon.* This process differs from double fertilization in flowering plants in that an additional zygote, rather than endosperm, is produced. The second zygote later disintegrates.)

Seeds and fruits develop after fertilization

As a result of double fertilization and subsequent growth and development, each seed contains a young plant embryo and nutritive tissue (the endosperm), both of which are surrounded by a protective seed coat. In monocots the endosperm persists and is the main source of food in the mature seed. In most eudicots the endosperm nourishes the developing embryo, which subsequently stores food in its cotyledons.

As a seed develops from an ovule following fertilization, the ovary wall surrounding it enlarges dramatically and develops into a **fruit.** In some instances, other tissues associated with the ovary also enlarge to form the fruit (see discussion of fruits in Chapter 37). Fruits serve two purposes: to protect the developing seeds from desiccation as they grow and mature and to aid in the dispersal of seeds. For example, dandelion fruits have feathery plumes that are lifted and carried by air currents. Animals often assist in dispersing seeds found in edible fruits (**FIG. 28-14**). Once a seed lands in a suitable place, it may germinate and develop into a mature sporophyte that produces flowers, and the life cycle continues as described.

Flowering plants have many adaptations that account for their success

The evolutionary adaptations of flowering plants account for their success in terms of their ecological dominance and their great number of species. Seed production as the primary means of reproduction and dispersal, an adaptation shared with the gymnosperms, is clearly significant and provides a definite advantage over seedless vascular plants. Closed carpels, which give rise to fruits surrounding the seeds, and the process of double fertilization with its resulting endosperm increase the likelihood of reproductive success. The evolution of a variety of interdependencies with many types of insects, birds, and bats, which disperse pollen from one flower to another of the same species, is another reason for angiosperm success. Pollen transfer results in cross-fertilization, which mixes the genetic material and promotes genetic variation among the offspring.

Several distinctive features have contributed to the success of flowering plants in addition to their highly successful reproduction involving flowers, fruits, and seeds. Recall that most flowering plants have very efficient water-conducting vessel elements in their xylem, as well as tracheids. In contrast, the xylem of almost all seedless vascular plants and gymnosperms consists exclusively of tracheids. Most flowering plants also have efficient carbohydrate-conducting sieve tube elements in their phloem. Vascular plants other than flowering plants and gnetophytes lack vessel elements and sieve tube elements.

The leaves of flowering plants, with their broad, expanded blades, are very efficient at absorbing light for photosynthesis. Abscission (shedding) of these leaves during cold or dry periods reduces water loss and has enabled some flowering plants to expand into habitats that would otherwise be too harsh for survival. The stems and roots of flowering plants are often modified for food or water storage, another feature that helps flowering plants survive in severe environments.

Probably most crucial to the evolutionary success of flowering plants, however, is the overall adaptability of the sporophyte generation. As a group, flowering plants readily adapt to new habitats and changing environments. This adaptability is evident in the great diversity of growth forms exhibited by the various species of flowering plants. For example, the cactus is remarkably well-adapted to desert environments. Its stem stores water; its leaves (spines) have a reduced surface area available for transpiration (loss of water vapor; see Chapter 34) and may also protect against thirsty herbivorous animals; and its thick, waxy cuticle reduces water loss. In contrast, the water lily is well-adapted for wet environments, in part because it has air channels that provide adequate oxygen to stems and roots living in oxygen-deficient water and mud.

Floral structure provides insights into the evolutionary process

In evolution new structures or organs often originate by modification of previously existing structures or organs. (See Chapter 20 discussion of preadaptations.) Much evidence supports the classical interpretation that the four organs of a flower—sepals, petals, stamens, and carpels—arose from highly modified leaves. This evidence includes comparisons of the arrangement of vascular tissues in both flowers and leafy stems and of the developmental stages of floral parts and leaves.

Sepals are the most leaflike of the four floral organs, and botanists generally agree that sepals are specialized leaves. Although petals of many flowering plant species are leaflike in appearance, botanists generally view petals as modified stamens that later became sterile and leaflike. Cultivated roses and camellias provide evidence supporting this hypothesis; in some varieties the stamens have been transformed into petals,

Figure 28-14 Guava fruit and seeds
Animals eat fleshy fruits such as guava. The seeds are frequently swallowed whole and pass unharmed through the animals' digestive tracts.

© irabel8/Shutterstock.com

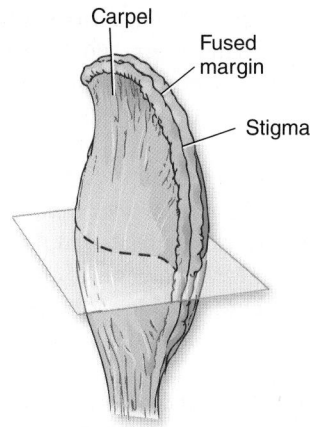

Carpel
Fused margin
Stigma

(a) The carpel resembles a folded leaf in which the ovules borne on its upper surface are enclosed.

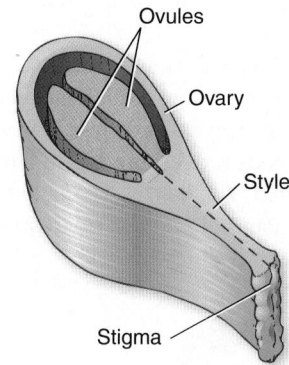

Ovules
Ovary
Style
Stigma

(b) A cross section of the carpel, cut along the dashed line in **(a)**.

Figure 28-15 Carpel of *Drimys piperita*
© Cengage Learning

forming showy flowers with large numbers of petals.

The remarkable leaflike stamens and carpels of certain tropical trees and other species support the origin of stamens and carpels from leaves or leaflike organs. Consider, for example, the carpel of *Drimys,* a genus of flowering trees and shrubs native to Southeast Asia, Australia, and South America. This carpel resembles a leaf that is folded inward along the midrib, thereby enclosing the ovules, and joined along the entire length of the leaf's margin (**FIG. 28-15**).

The fundamental question is whether these leaflike stamens and carpels are shared ancestral characters that were conserved (retained) during the course of evolution or instead are highly specialized organs (i.e., shared derived characters) that do not resemble early stamens and carpels. Many botanists who have studied this question have concluded that stamens and carpels are probably derived from leaves. Not all botanists accept the origin of stamens and carpels from highly modified leaves, however.

During the course of more than 130 million years of angiosperm evolution, floral structure diversified as floral organs fused or became reduced in size or number. These changes led to greater complexity in floral structure in some species and greater simplicity in other species. Interpreting the floral structures of so many different angiosperm species is sometimes difficult, but it is important because correct interpretations are essential to devising a phylogenetic classification scheme.

CHECKPOINT 28.3

- *How do nonreproductive adaptations of flowering plants differ from those of gymnosperms?*
- **CONNECT** *How does the flowering plant life cycle differ from that of the gymnosperms?*
- *What are the two major classes of flowering plants, and how can one distinguish between them?*
- **CONNECT** *How does fertilization differ in gymnosperms and flowering plants?*

28.4 THE EVOLUTION OF SEED PLANTS

LEARNING OBJECTIVE

9   Summarize the evolution of gymnosperms from seedless vascular plants and trace the evolution of flowering plants from gymnosperms.

One group that descended from ancestral seedless vascular plants was the **progymnosperms,** all of which are now extinct. Progymnosperms had two derived features: leaves with branching veins (*megaphylls*) and woody tissue (*secondary xylem*) similar to that of modern gymnosperms. Progymnosperms, however, reproduced by spores, not seeds. *Archaeopteris,* a progymnosperm that lived about 370 mya, is one of the earliest known trees with "modern" woody tissue (**FIG. 28-16a**).

Fossils of several progymnosperms with reproductive structures intermediate between those of spore plants and seed plants have been discovered. For example, the evolution of microspores into pollen grains and of megasporangia into ovules (seed-producing structures) can be traced in fossil progymnosperms. Plants producing seeds appeared during the late Devonian period, more than 359 mya. The fossil record indicates that different groups of seed plants apparently arose independently several times.

As mentioned previously, fossilized remains of ginkgo are found in 200-million-year-old rocks, and other groups of gymnosperms were well established by 160 mya to 100 mya. Although the gymnosperms are an ancient group, some questions persist about the exact pathways of gymnosperm evolution. The fossil record indicates that progymnosperms probably gave rise to conifers and to another group of extinct plants called **seed ferns,** which were seed-bearing woody plants with fernlike leaves (**FIG. 28-16b**). The seed ferns, in turn, probably gave rise to cycads and ginkgo as well as to several gymnosperm groups now extinct. The origin of gnetophytes remains unclear, although molecular data indicate that they are closely related to conifers.

Our understanding of the evolution of flowering plants has made great progress in recent years

Flowering plants are the most recent clade of plants to evolve. The fossil record, although incomplete, suggests that flowering plants descended from gymnosperms. By the middle of the Jurassic period, about 180 mya, several gymnosperm lines existed with some features resembling those of flowering plants. Among other traits, these derived gymnosperms possessed leaves with broad, expanded blades and the first modified seed-bearing leaves, which nearly enclose the ovules. Beetles were evidently visiting these plants, and biologists have suggested that perhaps this relationship was the beginning of **coevolution,** a mutual adaptation between plants and their animal pollinators (see Chapter 37).

One important task facing paleobotanists (biologists who study fossil plants) is determining which of the ancient gymnosperms are in the direct line of evolution leading to the flowering plants. Given the structural data, most botanists hypothesize that flowering plants arose only once; that is, they think that there is only one line of evolution from the gymnosperms to the flowering plants.

Based largely on structural data, many evolutionary botanists previously thought that the gnetophytes were the gymnosperm clade most closely related to flowering plants. However, new hypotheses of angiosperm origin have been proposed in recent years with advances in molecular comparisons, genetic studies (particularly as gene function relates to reproductive development), and additional fossil discoveries. Currently, many botanists think that the closest extant gymnosperms related to flowering plants are probably conifers, but this hypothesis is far from certain, and the origin and early evolution of flowering plants continue to challenge botanists.

The oldest definitive trace of flowering plants in the fossil record consists of ovules enclosed in tiny podlike fruits interpreted as carpels in Jurassic and Lower Cretaceous rocks some 125 million to 145 million years old (FIG. 28-17a). The oldest fossilized flowers are about 118 million to 120 million years old.

(a) Progymnosperm. *Archaeopteris,* which existed about 370 mya, had some features in common with modern seed plants but did not produce seeds.

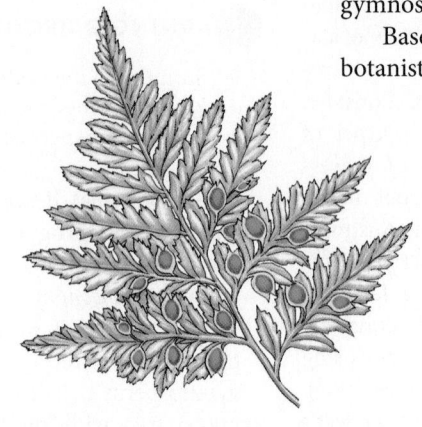

(b) Seed fern. *Emplectopteris* produced seeds on fernlike leaves. Seed ferns existed from about 360 mya to 250 mya.

Figure 28-16 Evolution of seed plants

(**a,** Redrawn from Beck, C.B., "Reconstructions of *Archaeopteris* and Further Consideration of Its Phylogenetic Position," *American Journal of Botany,* Vol. 49, 1962. **b,** Redrawn from Andrews, H.N., *Ancient Plants and the World They Lived In,* Comstock, New York, 1947.)

David Dilcher and Ge Sun

Carpel

Ovule

5 mm

(a) The oldest known fossil angiosperm. This fossil of *Archaefructus* shows a carpel-bearing stem. Discovered in northeastern China, it is about 125 million years old.

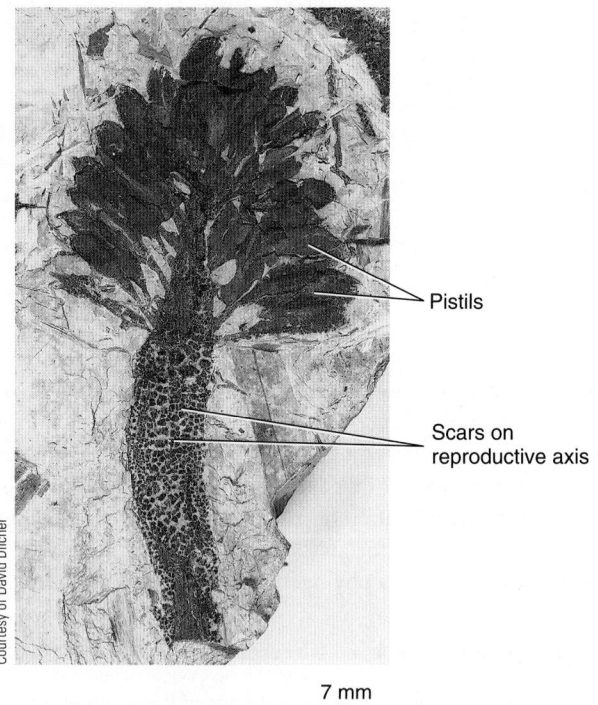

Courtesy of David Dilcher

Pistils

Scars on reproductive axis

7 mm

(b) Fossil flower. The fossilized flower of the extinct plant *Archaeanthus linnenbergeri,* which lived about 100 mya. The scars on the reproductive axis (receptacle) may show where stamens, petals, and sepals were originally attached but abscised (fell off). Many spirally arranged pistils were still attached at the time this flower was fossilized.

Figure 28-17 Fossil angiosperms

Flowering plants evolved from ancient gymnosperms. Basal angiosperms are early diverging clades that still survive today.

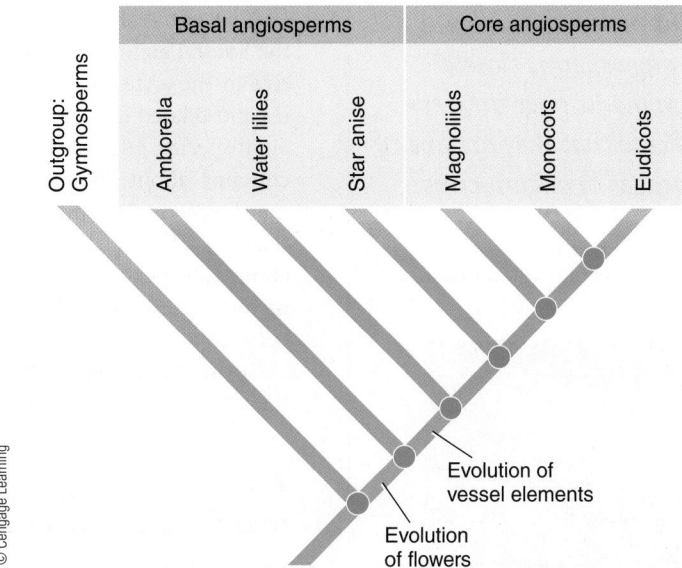

(a) One hypothesis of relationships among the flowering plants, based on fossil and molecular evidence. *Amborella*, water lilies, and star anise are living plants whose ancestors apparently branched off of the angiosperm family tree early. These early clades were followed by the magnoliids, the monocots, and the eudicots.

(b) *Amborella trichopoda*, a basal angiosperm.

(c) Water lily (*Nymphaea*), a basal angiosperm.

(d) Star anise (*Illicium verum*), a basal angiosperm.

(e) *Magnolia*, a core angiosperm.

Figure 28-18 Evolution of flowering plants

CONNECT According to the cladogram, do all basal angiosperms possess vessel elements?

By 90 mya, during the Cretaceous period, flowering plants had diversified and had begun to replace gymnosperms as Earth's dominant plants. Fossils of flowering plant leaves, stems, flowers, fruits, and seeds are numerous and diverse. They outnumber fossils of gymnosperms and ferns in Late Cretaceous deposits, indicating the rapid success of flowering plants once they appeared (FIG. 28-17b). Many angiosperm species apparently arose from changes in chromosome number. (See discussion of allopolyploidy and sympatric evolution in Chapter 20.)

The basal angiosperms comprise three clades

Beginning in 1999, a series of cladistic analyses of structural features and molecular comparisons have helped clarify relationships among the angiosperm classes. In the process many long-held assumptions about early angiosperm lineages have been discarded. The evidence indicates that several clades of basal (early-diverging) angiosperms evolved before the divergence of core angiosperms (FIG. 28-18a). The 170 species of

Plants commonly known as hydatellas were incorrectly classified as grasses until recent molecular evidence indicated that they are basal angiosperms closely related to water lilies. Do these plants have any reproductive features that may indicate that they are one of the most ancient surviving lineages of angiosperms?

EXPERIMENT: *Trithuria inconspicua* (formerly *Hydatella inconspicua*), a small aquatic plant, was collected at a lake in Northland, New Zealand, and its embryo sacs were carefully analyzed.

© Justin Goh, School of Biological Sciences, University of Auckland. Auckland, New Zealand

RESULTS AND CONCLUSION: As in gymnosperms, the mother plant turns over nutrients to the seed *before* fertilization has occurred. This gymnosperm-like reproductive feature may represent a shared ancestral character that was retained during the transition from gymnosperms to flowering plants, thus supporting the view that hydatellas are representative of an ancient angiosperm lineage. Alternatively, this feature may represent a shared derived character not found in earlier hydatella ancestors.

SOURCE: Friedman, W.E. "Hydatellaceae Are Water Lilies with Gymnospermous Tendencies." *Nature*, Vol. 453, May 1, 2008.

Figure 28-19 The basal angiosperm *Trithuria*

Trithuria inconspicua belongs to an ancient clade that consists of about 10 species native to Australia, New Zealand, and India.

CONNECT How is the nutritive tissue in the seeds of gymnosperms formed?

basal angiosperms represent groups thought to be ancestral to all other flowering plants.

The oldest surviving clade of basal angiosperms is represented by a single extant species, *Amborella trichopoda* (**FIG. 28-18b**). A shrub native to New Caledonia, an island in the South Pacific, *Amborella* may be the nearest living relative to the ancestor of all flowering plants. The water lilies and related families compose the second clade of basal angiosperms (**FIG. 28-18c** and **FIG. 28-19**). This clade, which contains about 70 species of aquatic or wetland herbs, may be the second-oldest surviving lineage. Star anise and relatives—the third clade of basal angiosperms—consist of about 100 species of vines, trees, and shrubs found mostly in warmer climates. Star anise is important economically because it is a source of spice used by confectioners and anise oil (**FIG. 28-18d**). Star anise is also used to make Tamiflu, a treatment for influenza.

The core angiosperms comprise magnoliids, monocots, and eudicots

Most angiosperm species belong to a clade of **core angiosperms,** which is divided into three subclades: magnoliids, monocots, and eudicots. **Magnoliids** include species in the magnolia, laurel, and black pepper families as well as several related families (**FIG. 28-18e**). Although magnoliids were traditionally classified with the eudicots as "dicots," molecular evidence such as DNA sequence comparisons indicates that the magnoliids are neither eudicots nor monocots. Native to tropical or warm temperate regions, magnoliids include several economically important plants, such as avocado, black pepper, nutmeg, and bay laurel.

CHECKPOINT 28.4

- *What features distinguish progymnosperms from seed ferns?*
- *Describe the significant features of the oldest known fossil angiosperm.*
- *Are monocots considered basal or core angiosperms? Explain your answer.*

SUMMARY: FOCUS ON LEARNING OBJECTIVES

28.1 An Introduction to Seed Plants *(page 579)*

1 Compare the features of gymnosperms and angiosperms.

- The two groups of seed plants are the **gymnosperms** and the **angiosperms.** Gymnosperms produce seeds that are totally exposed or borne on the scales of cones; an ovary wall does not surround the ovules of gymnosperms. Angiosperms are flowering plants that produce their seeds within a fruit (a mature ovary).

28.2 Gymnosperms *(page 580)*

2 Trace the steps in the life cycle of a pine and compare its sporophyte and gametophyte generations.

- A pine tree is a mature sporophyte; pine gametophytes are extremely small and nutritionally dependent on the sporophyte generation. Pine is heterosporous and produces microspores and megaspores in separate cones.

- Male cones produce **microspores** that develop into **pollen grains** (immature male gametophytes) that are carried by air currents to female cones.
- Female cones produce **megaspores.** One of each four megaspores produced by meiosis develops into a female gametophyte within an **ovule (megasporangium).**
- After **pollination,** the transfer of pollen to the female cones, a **pollen tube** grows through the megasporangium to the egg within the archegonium. After **fertilization,** the zygote develops into an embryo encased inside a seed adapted for wind dispersal.

3 Summarize the features that distinguish gymnosperms from bryophytes and ferns.
- Unlike bryophytes, gymnosperms are vascular plants. Unlike bryophytes and ferns, gymnosperms produce seeds. Gymnosperms also produce wind-borne pollen grains, a feature absent in ferns and other seedless vascular plants.

4 Name and briefly describe the four phyla of gymnosperms.
- **Conifers** (phylum Coniferophyta), the largest phylum of gymnosperms, are woody plants that bear *needles* (slender leaves that are usually evergreen) and produce seeds in cones. Most conifers are **monoecious** and have male and female reproductive parts in separate cones on the same plant.
- **Cycads** (phylum Cycadophyta) are palmlike or fernlike in appearance. They are **dioecious**—they have male and female reproductive structures on separate plants—but reproduce with pollen and seeds in conelike structures.
- *Ginkgo biloba,* the only surviving species in phylum Ginkgophyta, is a deciduous, dioecious tree. The female **ginkgo** produces fleshy seeds directly on branches.
- **Gnetophytes** (phylum Gnetophyta) are an obscure clade of gymnosperms that has a few traits associated with angiosperms.

28.3 Flowering Plants (page 585)

5 Summarize the features that distinguish flowering plants from other plants.
- **Flowering plants,** or angiosperms (phylum Anthophyta), constitute the phylum of vascular plants that produce flowers and seeds enclosed within a **fruit.** They are the most diverse and most successful group of plants.
- The flower, which may contain **sepals, petals, stamens,** and **carpels,** functions in sexual reproduction. Unlike those of gymnosperms, the ovules of flowering plants are enclosed within an **ovary.** After fertilization, the ovules become seeds, and the ovary develops into a fruit.

6 Briefly explain the life cycle of a flowering plant and describe double fertilization.
- The sporophyte generation is dominant in flowering plants; gametophytes are extremely reduced in size and nutritionally dependent on the sporophyte generation. Flowering plants are heterosporous and produce microspores and megaspores within the flower.
- Each microspore develops into a pollen grain (immature male gametophyte). One of each four megaspores produced by meiosis develops into an **embryo sac** (female gametophyte). Within the embryo sac, the egg cell and the central cell with two **polar nuclei** participate in fertilization.
- **Double fertilization,** which results in the formation of a diploid zygote and triploid **endosperm,** is characteristic of flowering plants.

7 Contrast monocots and eudicots, the two largest classes of flowering plants.
- Most **monocots** (class Monocotyledones) have floral parts in threes, and their seeds each contain one **cotyledon.** The nutritive tissue in their mature seeds is endosperm.
- **Eudicots** (class Eudicotyledones) usually have floral parts in fours or fives or multiples thereof, and their seeds each contain two cotyledons. The nutritive organs in their mature seeds are usually the cotyledons, which have absorbed the nutrients in the endosperm.

8 Discuss the evolutionary adaptations of flowering plants.
- Flowering plants reproduce sexually by forming flowers. After double fertilization, seeds form within fruits. Flowering plants have efficient water-conducting **vessel elements** in their xylem and efficient carbohydrate-conducting **sieve tube elements** in their phloem. Wind, water, insects, or other animals transfer pollen grains in various flowering plants.

28.4 The Evolution of Seed Plants (page 591)

9 Summarize the evolution of gymnosperms from seedless vascular plants and trace the evolution of flowering plants from gymnosperms.
- Seed plants arose from seedless vascular plants. **Progymnosperms** were seedless vascular plants that had megaphylls and "modern" woody tissue. Progymnosperms probably gave rise to conifers as well as to **seed ferns,** which in turn likely gave rise to cycads and ginkgo.
- The evolution of the gnetophytes is unclear, although molecular data indicate that they are closely related to conifers.
- Flowering plants probably descended from ancient gymnosperms that had specialized features, such as leaves with broad, expanded blades and closed carpels. Flowering plants likely arose only once; that is, there is only one line of evolution from the gymnosperms to the flowering plants.

TEST YOUR UNDERSTANDING

Know and Comprehend

1. Seed plants *lack* which of the following structure(s)? (a) ovules surrounded by integuments (b) microspores and megaspores (c) vascular tissues (d) a large, nutritionally independent sporophyte (e) a large, nutritionally independent gametophyte

2. Conifers, cycads, ginkgo, and gnetophytes are collectively called (a) club mosses (b) gymnosperms (c) angiosperms (d) eudicots (e) seedless vascular plants

3. The immature male gametophytes of pine are called (a) ovules (b) stamens (c) seed cones (d) pollen grains (e) polar nuclei

4. The transfer of pollen grains from the male to the female reproductive structure is known as (a) pollination (b) fertilization (c) embryo sac development (d) seed development (e) fruit development

5. Motile sperm cells are found as vestiges in these two gymnosperm groups: (a) monocots, eudicots (b) gnetophytes, conifers (c) gnetophytes, flowering plants (d) cycads, conifers (e) cycads, ginkgo

6. There are at least _____ species of flowering plants worldwide. (a) 300 (b) 3000 (c) 30,000 (d) 300,000 (e) 3,000,000

7. A simple pistil consists of a single (a) calyx (b) carpel (c) ovule (d) filament (e) petal

8. A flower that lacks stamens is both _____ and _____. (a) complete; imperfect (b) incomplete; perfect (c) complete; perfect (d) incomplete; imperfect

9. After fertilization, the _____ develop(s) into a fruit and the _____ develop(s) into a seed. (a) ovary; ovule (b) polar nuclei; ovule (c) ovary; endosperm (d) ovule; ovary (e) ovule; polar nuclei

10. The female gametophyte in flowering plants is also called the (a) polar nuclei (b) anther (c) embryo sac (d) endosperm (e) sporophyll

11. This flowering plant may be the nearest living relative to the ancestor of all flowering plants. (a) *Amborella* (b) *Archaeopteris* (c) *Gnetum* (d) water lily (e) magnolia

Apply and Analyze

12.  This cross section through an ovary reveals that the pistil is (a) simple, with one carpel (b) compound, with two fused carpels (c) compound, with three fused carpels (d) compound, with six separate carpels (e) compound, with six fused carpels

13. You are given a plant that you have never seen before (see figure). Is it a gymnosperm or angiosperm? A monocot or eudicot? What are the features that helped you make these determinations?

14. **VISUALIZE** Sketch a seed of a gymnosperm and of a monocot. Label the embryo, seed coat, and nutritive tissue (giving its specific name). Indicate the ploidy of each structure (haploid, diploid, or triploid).

15. **CONNECT** How do the life cycles of seedless plants (see Chapter 27) and seed plants differ? In what fundamental ways are they alike?

Evaluate and Synthesize

16. **EVOLUTION LINK** Most flowers contain both male and female reproductive structures, in contrast to the cones of gymnosperms, which are either male or female. Explain how bisexual flowers might be an advantageous evolutionary adaptation to the flowering plants that possess them.

17. **EVOLUTION LINK** Contrast the algae, mosses, ferns, gymnosperms, and angiosperms with respect to their dependence on water as a transport medium for reproductive cells. Suggest a hypothesis to explain how the differences might be adaptive to living on land.

18. **EVOLUTION LINK** Where would you place the progymnosperms on Figure 28-2? Explain your reasoning.

aplia To access course materials, such as Aplia and other companion resources, please visit **www.cengagebrain.com.**

Carlyn Iverson

The Fungi

Mushrooms, morels, and truffles, delights of the gourmet, share a recent common ancestry with baker's yeast, the black mold that forms on stale bread, and the mildew that collects on damp shower curtains. The science of **mycology,** the study of fungi, is concerned with all these diverse life-forms. Mycologists have described about 100,000 species, most of which are terrestrial, but they estimate that there are more than 1.5 million species. Biologists who use the kingdom as a taxon group these organisms in kingdom Fungi, one of the eukaryotic kingdoms.

Fungi grow best in moist habitats, but they are found universally wherever organic material is available. They require moisture to grow, and they can obtain water from a humid atmosphere as well as from the medium on which they live. When the environment becomes dry, fungi survive by going into a resting stage or by producing spores (see photograph) that resist desiccation (drying out).

Some fungi grow to enormous size. In Washington State a fungal clone (*Armillaria ostoyae*) has been identified that covers more than 1500 acres. This giant fungus, which is mainly underground, developed from a single spore that germinated more than a thousand years ago. The fungus has fragmented and is no longer one continuous body.

Like prokaryotes, most fungi are decomposers that obtain nutrients and energy from dead organic matter. They are vital members of ecosystems because they break down the organic compounds found in dead organisms, leaves, garbage, sewage, and other waste. When they decompose organic compounds, carbon and other elements are released into the environment, where they are recycled.

Many fungi form vital symbiotic associations. For example, most terrestrial plants have fungal partners that live in close association with their roots. The fungi help the plants obtain phosphate ions and other needed minerals from the soil. In exchange, the plants provide the fungi with organic nutrients. Some fungi live symbiotically with algae and cyanobacteria as lichens. Others are parasites and pathogens that cause disease in animals or plants.

© Nature Photographers Ltd/Alamy

Fungal spores. The collared earthstar (*Geastrum triplex*) releases a puff of microscopic spores after the sac, which is about 1.3 cm (0.5 in.) wide, is hit by a raindrop. This fungus is common in leaf litter under trees throughout North America.

KEY CONCEPTS

29.1 Fungi are eukaryotic heterotrophs that absorb nutrients from their surroundings; most grow as multicellular filaments called hyphae that form a tangled mass called a mycelium.

29.2 Most fungi have complex life cycles and reproduce both asexually and sexually by means of spores.

29.3 According to current hypotheses, fungi evolved from a unicellular, flagellate protist and diverged into several groups.

29.4 Fungi are of major ecological importance; they are decomposers, and they form symbiotic relationships with other organisms.

29.5 Fungi are of major economic, biological, and medical importance.

29.1 CHARACTERISTICS OF FUNGI

LEARNING OBJECTIVES

1 Describe the distinguishing characteristics of fungi.
2 Describe the body plan of a fungus.

All **fungi** (sing., *fungus*) are eukaryotes; their cells contain membrane-enclosed nuclei, mitochondria, and other membranous organelles. Although they vary strikingly in size and shape, fungi share certain key characters, including their way of obtaining nutrition.

The optimum pH for most fungal species is about 5.6, but various fungi can tolerate and grow in environments where the pH ranges from 2 to 9. Many fungi are less sensitive to high osmotic pressures than are bacteria. As a result, they can grow in concentrated salt solutions or in sugar solutions such as jelly, which discourage or prevent bacterial growth. Fungi also thrive over a wide temperature range. Even refrigerated food may be invaded by fungi.

Fungi absorb food from the environment

Like animals, fungi are heterotrophs. For their nutritional and energy needs, they depend on preformed carbon molecules produced by other organisms. However, fungi do not ingest food and then digest it in the body as animals do. Instead, they infiltrate a food source and secrete digestive enzymes onto it. Digestion takes place outside the body. When complex molecules are broken down into smaller compounds, the fungus absorbs the predigested food into its body.

The fungus is very efficient at absorbing nutrients and growing. It rapidly converts nutrients into new cell material.

If excessive amounts of nutrients are available, fungi store them, usually as lipid droplets or glycogen.

Fungi have cell walls that contain chitin

Like the cells of bacteria, certain protists, and plants, fungal cells are enclosed by cell walls during at least some stage in their life cycle. Fungal cell walls, however, have a different chemical composition from cell walls of other organisms. In most fungi the cell wall consists of complex carbohydrates, including **chitin,** a polymer that consists of subunits of a nitrogen-containing sugar (see Fig. 3-11). Chitin is also a component of the external skeletons of insects and other arthropods. It is resistant to breakdown by most microorganisms.

Most fungi consist of a network of filaments

The simplest fungi are the **yeasts,** which are unicellular, with a round or oval shape. Yeasts are widely distributed in the soil; on leaves, fruits, and cured meats; and on and in our bodies. The importance of some yeasts in medicine, biological research, and the food industry is discussed later in this chapter.

Most fungi are multicellular. The body consists of long, branched, threadlike filaments called **hyphae** (sing., *hypha*) (**FIGS. 29-1a** and b). Hyphae consist of tubular cell walls surrounding the plasma membranes of the fungal cells. They are an adaptation to the fungal mode of nutrition. Growth occurs at the tips of the hyphae; as the hyphae elongate, the fungus grows into and infiltrates food sources. The fungus absorbs nutrients through its very large surface area.

As hyphae grow they form a tangled mass or tissuelike network, called a **mycelium** (pl., *mycelia*). Fungi that form mycelia are commonly called *molds.* The cobweblike mold sometimes seen on bread is the mycelium of a fungus. What is not seen is

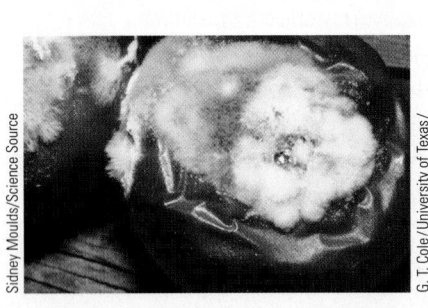

Hyphae

(a) A mycelium. A mass of hyphae has formed a mycelium (*white area*). Fruiting bodies, called sporangia, are visible as *gray areas*.

(b) SEM of a mycelium. The fungus *Blumeria graminis* growing on a leaf (darker area underneath the mycelium).

25 μm

Sidney Moulds/Science Source

G. T. Cole/University of Texas/ Biological Photo Service

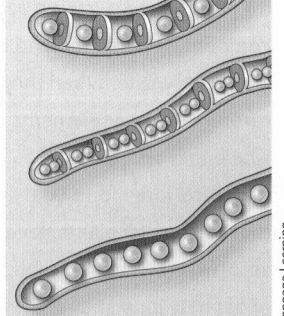

(c) A hypha divided into cells by septa. Each cell is monokaryotic (has one nucleus). In some taxa the septa are perforated, as shown.

(d) A septate hypha. In this hypha, each cell is dikaryotic (has two nuclei).

(e) A coenocytic hypha.

© Cengage Learning

Figure 29-1 *Animation* **Fungus body plan**

Spores carried by the air settle on food. They germinate and produce a mass of threadlike filaments called hyphae. The hyphae penetrate the food to obtain nourishment and produce a mass of hyphae called the mycelium. Eventually, specialized hyphae grow from the mycelium, giving rise to erect stalks with fruiting bodies (sporangia) at their tops. When the walls of the sporangia rupture, new spores are released into the air.

the extensive mycelium that grows into the bread. Depending on environmental conditions, some fungi can alternate between a yeast phase and a phase in which they produce hyphae.

In most fungi hyphae are divided by cross walls, called **septa** (sing., *septum*), into individual cells containing one or more nuclei (FIGS. 29-1c and d). As we will discuss, the presence of septa is an important character in the two largest fungal phyla (which include the most complex fungi). The septa of many fungi are perforated by a pore that may be large enough to permit organelles to flow from cell to cell. Some fungi are **coenocytes,** which lack septa. In these species nuclear division is not followed by cytoplasmic division. As a result, a coenocytic fungus is one elongated, multinucleated, giant cell (FIG. 29-1e).

CHECKPOINT 29.1

- **CONNECT** *What characters distinguish fungi from other organisms?*
- *How does the body of a yeast differ from that of a mold?*

29.2 FUNGAL REPRODUCTION

LEARNING OBJECTIVE

3   Describe the life cycle of a typical fungus, including sexual and asexual reproduction.

Most fungi reproduce by means of microscopic **spores,** reproductive cells that can develop into new organisms. In most groups spores are nonmotile. They are dispersed by wind, water, or animals. The air we breathe is filled with hundreds of thousands of fungal spores. When a spore germinates, it gives rise to a hypha, which then develops into a mycelium (FIG. 29-2).

Fungi produce spores either sexually or asexually. With asexual reproduction new individuals are produced quickly, but there is little genetic variability. Sexual reproduction involves meiosis and generates new genotypes.

Spores are usually produced on specialized aerial hyphae or in fruiting structures. When positioned up above the ground,

spores can be easily dispersed. Structures in which spores are produced are called **sporangia** (sing., *sporangium*). The aerial hyphae of some fungi produce spores in large, complex reproductive structures, referred to as **fruiting bodies.** The familiar part of a mushroom is a large fruiting body. We do not normally see the bulk of the fungus, a nearly invisible mycelium buried out of sight in the rotting material or soil on which it grows.

Many fungi reproduce asexually

Yeasts reproduce asexually, primarily by forming buds that pinch off from the parent cell (FIG. 29-3). Many species of multicellular fungi also reproduce asexually. Spores are produced by mitosis and then released into the air or water. **Conidiophores** (from the Greek, meaning "dust-bearers") are specialized hyphae that produce asexual spores called **conidia** (sing., *conidium*). The arrangement of conidia on conidiophores varies from species to species.

Most fungi reproduce sexually

Fungi are a diverse group with many variations in their life cycles. Most fungi (but not all) reproduce both asexually and sexually. FIGURE 29-4 illustrates a generalized life cycle. Many fungal species reproduce sexually when they come into contact with other mating types. In contrast to the majority of animal and plant cells, most fungal cells contain haploid nuclei. In sexual reproduction the hyphae of two genetically compatible mating types come together, and their cytoplasm fuses,

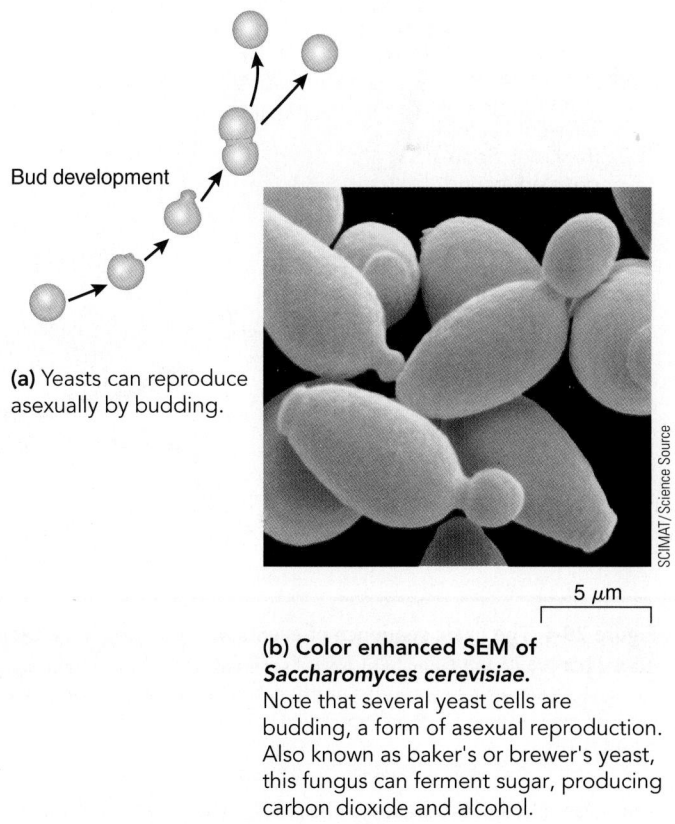

(a) Yeasts can reproduce asexually by budding.

Bud development

5 μm

(b) Color enhanced SEM of *Saccharomyces cerevisiae.*
Note that several yeast cells are budding, a form of asexual reproduction. Also known as baker's or brewer's yeast, this fungus can ferment sugar, producing carbon dioxide and alcohol.

Figure 29-3 Yeasts are unicellular fungi
© Cengage Learning

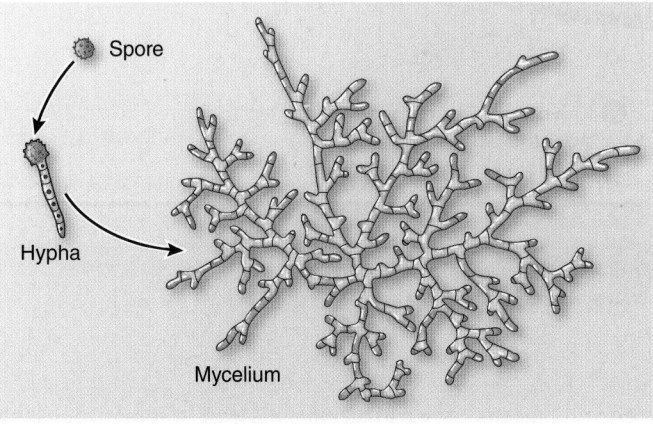

Spore

Hypha

Mycelium

Figure 29-2 Germination of a spore to form a mycelium
© Cengage Learning

Most fungi can reproduce both asexually (which allows rapid proliferation) and sexually (which produces new genotypes).

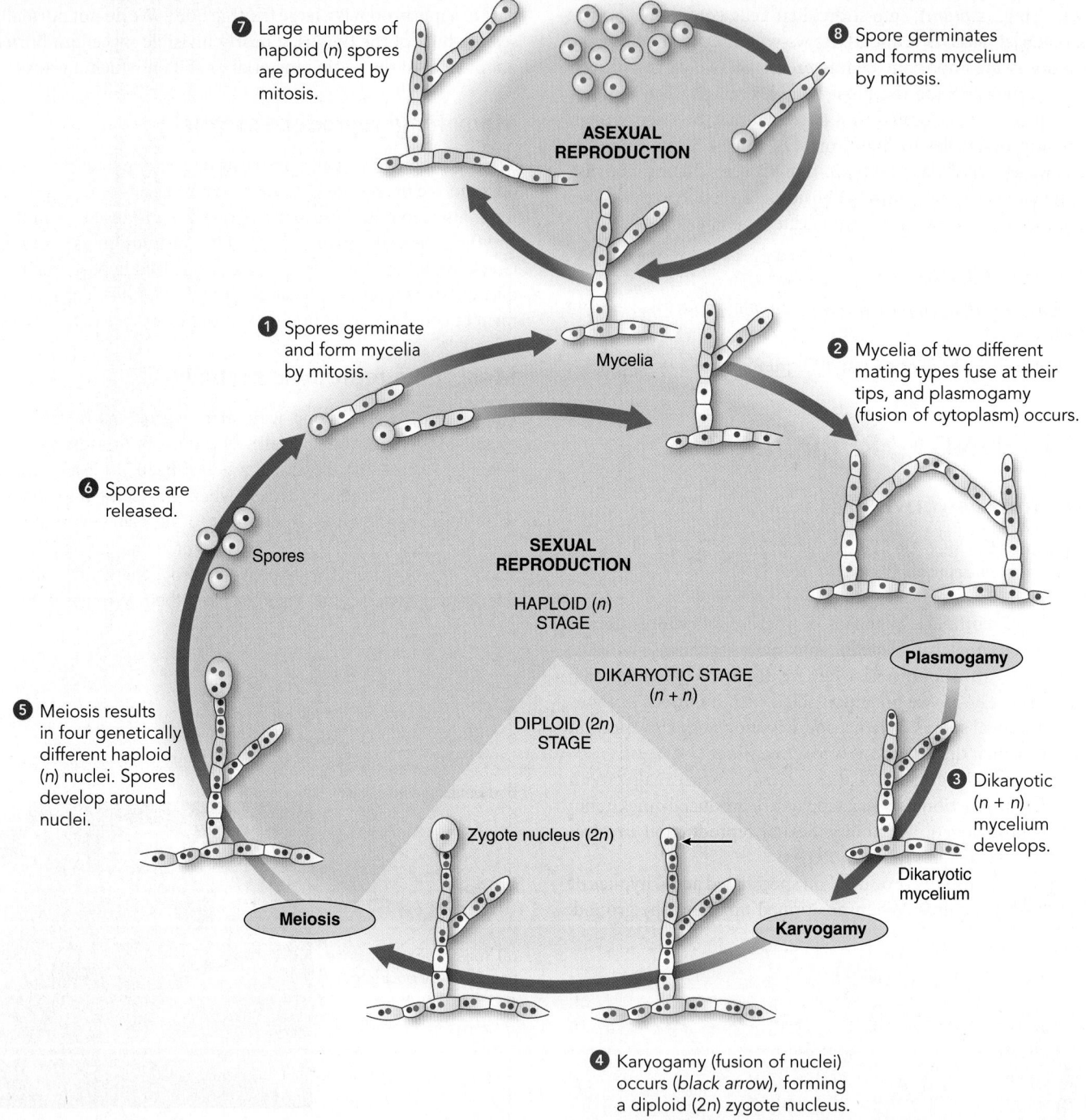

7 Large numbers of haploid (*n*) spores are produced by mitosis.

8 Spore germinates and forms mycelium by mitosis.

ASEXUAL REPRODUCTION

1 Spores germinate and form mycelia by mitosis.

Mycelia

2 Mycelia of two different mating types fuse at their tips, and plasmogamy (fusion of cytoplasm) occurs.

6 Spores are released.

Spores

SEXUAL REPRODUCTION

HAPLOID (*n*) STAGE

DIKARYOTIC STAGE (*n* + *n*)

DIPLOID (2*n*) STAGE

Plasmogamy

5 Meiosis results in four genetically different haploid (*n*) nuclei. Spores develop around nuclei.

3 Dikaryotic (*n* + *n*) mycelium develops.

Dikaryotic mycelium

Zygote nucleus (2*n*)

Meiosis

Karyogamy

4 Karyogamy (fusion of nuclei) occurs (*black arrow*), forming a diploid (2*n*) zygote nucleus.

Figure 29-4 The basic sequence of events in most fungal life cycles

CONNECT Beadle and Tatum used asexual spores of the fungus *Neurospora* in their experiments described in Figure 13-2. How did they obtain strains that differed in their nutritional requirements, given that asexual spores derived from the same parent are expected to be genetically identical?

© Cengage Learning

a process called **plasmogamy.** The resulting cell has two haploid nuclei, one from each fungus. This cell gives rise by mitosis to other cells with two nuclei. At some point the two haploid nuclei fuse. This process, called **karyogamy,** results in a cell containing

a diploid nucleus known as a *zygote nucleus.* In some groups the zygote nucleus is the only diploid nucleus.

In the two largest fungal phyla, the ascomycetes and basidiomycetes (discussed later in this chapter), plasmogamy occurs (hyphae fuse), but karyogamy (fusion of the two different nuclei) does not follow immediately. For a time, the nuclei remain separate within the fungal cytoplasm. Hyphae that contain two genetically distinct, sexually compatible nuclei within each cell are described as **dikaryotic** (see Fig. 29-1d). This condition is referred to as $n + n$ rather than $2n$ because there are two separate haploid nuclei. Hyphae that contain only one nucleus per cell are described as **monokaryotic.** The presence of a dikaryotic stage is an important defining character of the ascomycetes and basidiomycetes.

Fungi communicate chemically by secreting signaling molecules called **pheromones.** At least one pheromone has been identified in each major fungal group. The pheromone binds with a compatible receptor on a different mating type. For example, in the zygomycetes, a pheromone induces the formation of specialized aerial hyphae. Another pheromone causes the tips of aerial hyphae of opposite mating types to grow toward each other and fuse prior to sexual reproduction.

CHECKPOINT 29.2

- CONNECT *How is a diploid cell different from a dikaryotic cell?*
- VISUALIZE *Draw a generalized life cycle of a fungus. Include asexual and sexual reproduction as well as haploid, diploid, and dikaryotic stages.*

29.3 FUNGAL DIVERSITY

LEARNING OBJECTIVES

4 Give arguments to support the hypothesis that fungi are opisthokonts, more closely related to animals than to plants.

5 Give arguments to support the hypothesis that chytrids may have been the earliest fungal group to evolve from the most recent common ancestor of fungi.

6 List distinguishing characteristics, describe a typical life cycle, and give examples of each of the following fungal groups: chytridiomycetes, zygomycetes, glomeromycetes, ascomycetes, and basidiomycetes.

For centuries biologists classified fungi in the plant kingdom. Like plants, fungi have cell walls and vacuoles and are sessile; that is, they cannot move around from place to place. Also, like plants, many types of fungi inhabit the soil. However, systematists began to question this classification, and in 1969, R.H. Whittaker proposed that fungi be assigned to a separate kingdom, Fungi.

Unlike plants, fungal cell walls do not contain cellulose. Rather, they contain chitin, a polysaccharide found in insect skeletons. The fungal mode of nutrition is also very different from that of plants. Unlike plants, fungi cannot produce their own organic materials from a simple carbon source (carbon dioxide). Like animals, fungi are heterotrophs.

As with systematics for other kingdoms, fungal systematics is a challenging and continuously changing process. For example, slime molds and water molds were formerly classified as fungi but are now included among the protists (see Chapter 26).

Fungi are assigned to the opisthokont clade

Systematists hypothesize that the common ancestor of all plants, fungi, and animals was an ancient flagellate protist. As discussed in Chapter 26, amoebozoa (a group of amoebas), fungi, animals, and a few protists, including the choanoflagellates (a group of flagellate protists), form a monophyletic "supergroup," the **unikonts.** Within this supergroup, the clade **opisthokonts** includes the choanoflagellates, the animals, and the fungi. Both genetic and structural similarities support this grouping. Like animals, fungi have platelike cristae in their mitochondria. Another key character shared by members of this clade is that flagellate cells propel themselves with a single posterior flagellum. In other eukaryote groups, flagellate cells move by means of one or more anterior flagella. Based on structural characters and on molecular data, systematists now view fungi as more closely related to animals than to plants.

Diverse groups of fungi have evolved

Fossil evidence has not been very helpful to systematists studying evolutionary relationships among fungal groups. Most fossilized fungi recovered to date have been microscopic. For example, fossilized fungal spores have been found in amber more than 225 million years old, and fossils of hyphae associated with cyanobacteria or algae have been dated as more than 550 million years old. Few large fungal fossils, such as mushrooms, have been found.

Historically, fungi have been classified mainly on the characteristics of their sexual spores and fruiting bodies. More recently, molecular data, such as comparative DNA and RNA sequences, have helped clarify relationships among fungal groups. Currently, many mycologists assign fungi to five main groups: Chytridiomycota, Zygomycota, Glomeromycota, Ascomycota, and Basidiomycota (FIG. 29-5 and TABLE 29-1). Some biologists consider each of these groups to be a phylum. However, some of these groups are not monophyletic (see Chapter 23), and mycologists are in the process of assigning fungi to additional clades.

In this edition of *Biology,* we will discuss fungi in the five phyla shown in Figure 29-5 and listed in Table 29-1. Microsporidia, a group of intracellular parasites, are classified in this text with the zygomycetes, although in the future they may be assigned to their own taxon. About 95% of all named fungi have been assigned to phyla Ascomycota and Basidiomycota. These phyla are considered *sister taxa* because they share a more recent common ancestor with each other than either does with any other group. Fungi of phyla Ascomycota and Basidiomycota have septate hyphae and a dikaryotic stage during the sexual part of their life cycle.

Until recently, mycologists had assigned about 25,000 species of fungi that did not fit into the major groups to a group called deuteromycetes (phylum Deuteromycota). It was a polyphyletic group (members did not share a recent common ancestor)

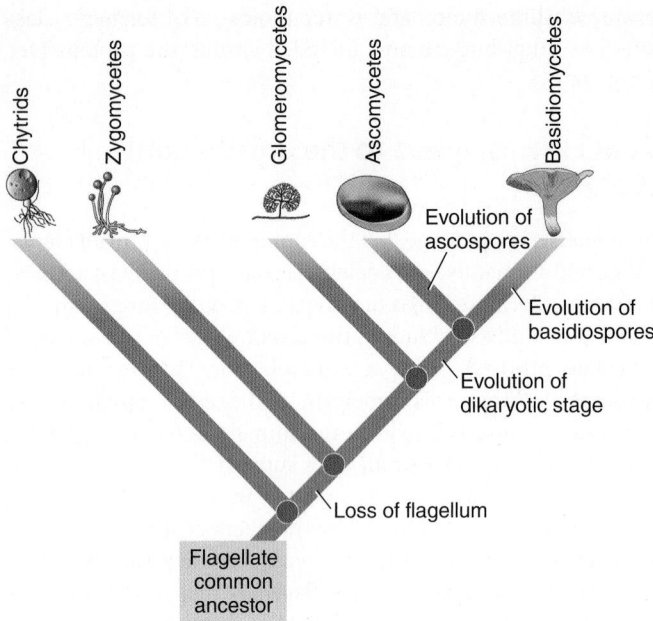

Figure 29-5 Cladogram of currently recognized groups of fungi

This cladogram shows phylogenetic relationships among living fungi, based on comparisons of ribosomal and nuclear gene sequence data for many species. The chytrids were the lineage that branched off first during fungal evolution. Note that ascomycetes and basidiomycetes are sister clades. Remember that the phylogeny of fungi is a work in process.

© Cengage Learning

lumped together simply as a matter of convenience. Mycologists classified fungi as deuteromycetes if no sexual stage had been observed for them at any point during their life cycle. Some of these fungi have lost the ability to reproduce sexually, whereas others reproduce sexually only rarely. Most fungi classified as deuteromycetes reproduce only by means of asexual spores, as ascomycetes do. Mycologists have identified relationships between deuteromycetes and their sexually reproducing relatives, based on DNA comparisons among various species. Most of the deuteromycetes have been reassigned to phylum Ascomycota, and a few have been reassigned to phylum Basidiomycota. (By convention, the asexual stage is still identified as a deuteromycete.)

Chytrids have flagellate spores

At one time, biologists considered the **chytrids,** also known as **chytridiomycetes** (phylum Chytridiomycota), to be funguslike protists, similar in many respects to the water molds. However, both structural and molecular characters indicate that the approximately 1000 species of chytridiomycetes are members of kingdom Fungi. Like fungi, their cell walls contain chitin, and molecular comparisons—particularly of DNA and RNA sequences—have provided compelling evidence that chytrids are indeed fungi. However, recent comparisons of rRNA sequences suggest that chytrids are not a monophyletic group, and they may be divided into four clades.

TABLE 29-1	Characteristics of Currently Recognized Phyla of Fungi			
	PHYLUM AND COMMON TYPES	**ASEXUAL REPRODUCTION**	**SEXUAL REPRODUCTION**	**OTHER KEY CHARACTERS**
	Chytridiomycota (chytrids or chytridiomycetes) *Allomyces*	Flagellate, diploid zoospores produced by mitosis in zoosporangia	Flagellate, haploid gametes in some species	Haploid zoospores produced in resting sporangia; form haploid thallus
	Zygomycota (zygomycetes) Black bread mold. Microsporidia are classified with the zygomycetes	Haploid spores produced in sporangia	Zygospores develop in zygosporangia	Important decomposers; some are insect parasites. Microsporidia are opportunistic pathogens that infect animals.
	Glomeromycota (glomeromycetes)	Large, multinucleate blastospores	Has not been observed	Form arbuscular mycorrhizae with plant roots
	Ascomycota (ascomycetes) Yeasts, powdery mildews, molds, morels, truffles	Conidia pinch off from conidiophores	Ascospores develop in asci	Have a dikaryotic stage; form important symbiotic relationships as lichens and mycorrhizae
	Basidiomycota (basidiomycetes or club fungi) Mushrooms, bracket fungi, puffballs, rusts, smuts	Uncommon	Basidiospores develop on club-shaped basidia	Have a dikaryotic stage; many form mycorrhizae with tree roots

© Cengage Learning

Chytrids are small, relatively simple fungi that inhabit ponds and damp soil. A few species have been found in salt water. Most chytrids are decomposers that degrade organic matter. However, a few species cause disease in plants and animals. A parasitic chytrid has been partly responsible for declining amphibian populations. Infected frogs have been identified in many parts of the world (see *Inquiring About: Declining Amphibian Populations,* in Chapter 57).

Most chytrids are unicellular or composed of a few cells that form a simple body, called a **thallus.** The term *thallus* describes the simple body plan of certain algae, fungi, and plants. The thallus may have slender extensions, called *rhizoids,* that anchor it to a food source and absorb food (**FIG. 29-6**). Chytrids are the only fungi that have flagellate cells. Their spores bear a single, posterior flagellum. Sexual reproduction has not been identified in most chytrids. Species that do reproduce sexually have flagellate gametes.

Some chytrids produce branched, coenocytic mycelia. *Allomyces,* a large, common chytrid, has an unusual life cycle compared with that of most fungi. It undergoes an **alternation of generations** (common in plants, but rare in fungi), spending part of its life as a multicellular haploid (*n*) thallus and part as a multicellular diploid (*2n*) thallus (**FIG. 29-7**). The haploid and diploid thalli are similar in appearance. At the tips of its branches, the haploid thallus bears two types of sporangia, structures in which gametes form by mitosis. Each sporangium produces a different type of flagellate gamete.

Each type of gamete secretes a pheromone that attracts the other type. The two gametes fuse, and plasmogamy and karyogamy occur, resulting in a motile zygote. Each zygote can develop into a diploid thallus. The thallus bears two kinds of spore cases: zoosporangia and resting sporangia. Zoosporangia produce flagellate diploid **zoospores** that develop into new diploid thalli. Meiosis occurs within resting sporangia, producing haploid zoospores. Each zoospore has the potential to develop into a haploid thallus.

Molecular evidence suggests that chytrids were probably the earliest fungal group to evolve. Chytrids produce flagellate cells at some stage in their life history, a character that systematists trace to the protist that was the common ancestor of all opisthokonts. No other fungal group has flagellate cells. At some point in their evolutionary history, other fungal groups apparently lost the ability to produce motile cells, perhaps during the transition from aquatic to terrestrial habitats.

Zygomycetes reproduce sexually by forming zygospores

Mycologists have named more than 1100 species of **zygomycetes** (phylum Zygomycota). Of all the fungi, members of this taxon appear most closely related to the chytrids. However, zygomycetes are not a monophyletic group, and as mycologists learn more about fungal relationships, they may divide this phylum into several taxa or reassign its members to other existing phyla.

Most zygomycetes are decomposers that live in the soil on decaying plant or animal matter (**FIG. 29-8**). Some zygomycetes form a type of symbiotic association (mycorrhizal relationship) with plant roots. (Recall that a symbiotic association is an intimate relationship between organisms of different species.) A few species cause disease in plants and animals, including humans.

During sexual reproduction, zygomycetes produce sexual spores, called **zygospores.** The zygospores are typically produced in spore sacs called **zygosporangia.** The hyphae in zygomycetes are coenocytic; that is, they lack regularly spaced septa. However, septa do form to separate the hyphae from reproductive structures.

Perhaps the most familiar zygomycete is the black bread mold, *Rhizopus stolonifer,* a decomposer that breaks down bread and other foods. If preservatives are not added, bread left at room temperature for a few days often becomes covered with a black, fuzzy growth. Bread becomes moldy when a spore falls on it and then germinates and grows into a mycelium (**FIG. 29-9** on page 606). Hyphae penetrate the bread and absorb nutrients. Eventually, certain hyphae grow upward and develop sporangia at their tips. Clusters of more than 50,000 black asexual spores can develop within each sporangium. The spores are released when the delicate sporangium ruptures. The spores give the black bread mold its characteristic color.

Sexual reproduction in the black bread mold occurs when the hyphae of two different mating types, designated plus (+) and minus (−), grow into contact with one another. The bread mold is **heterothallic,** meaning that an individual fungal hypha

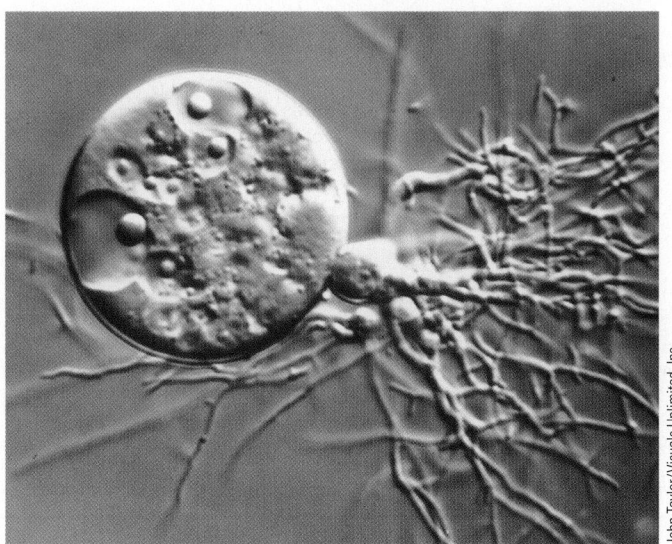

5 μm

Figure 29-6 Chytrid

Nomarski differential interference micrograph of a common chytrid (*Chytridium convervae*). Many chytrids have a microscopic body form consisting of a rounded, coenocytic thallus and branched rhizoids that superficially resemble roots. The rhizoids may anchor the chytrid thallus and absorb predigested food.

John Taylor/Visuals Unlimited, Inc.

Chytrids have flagellate reproductive cells.

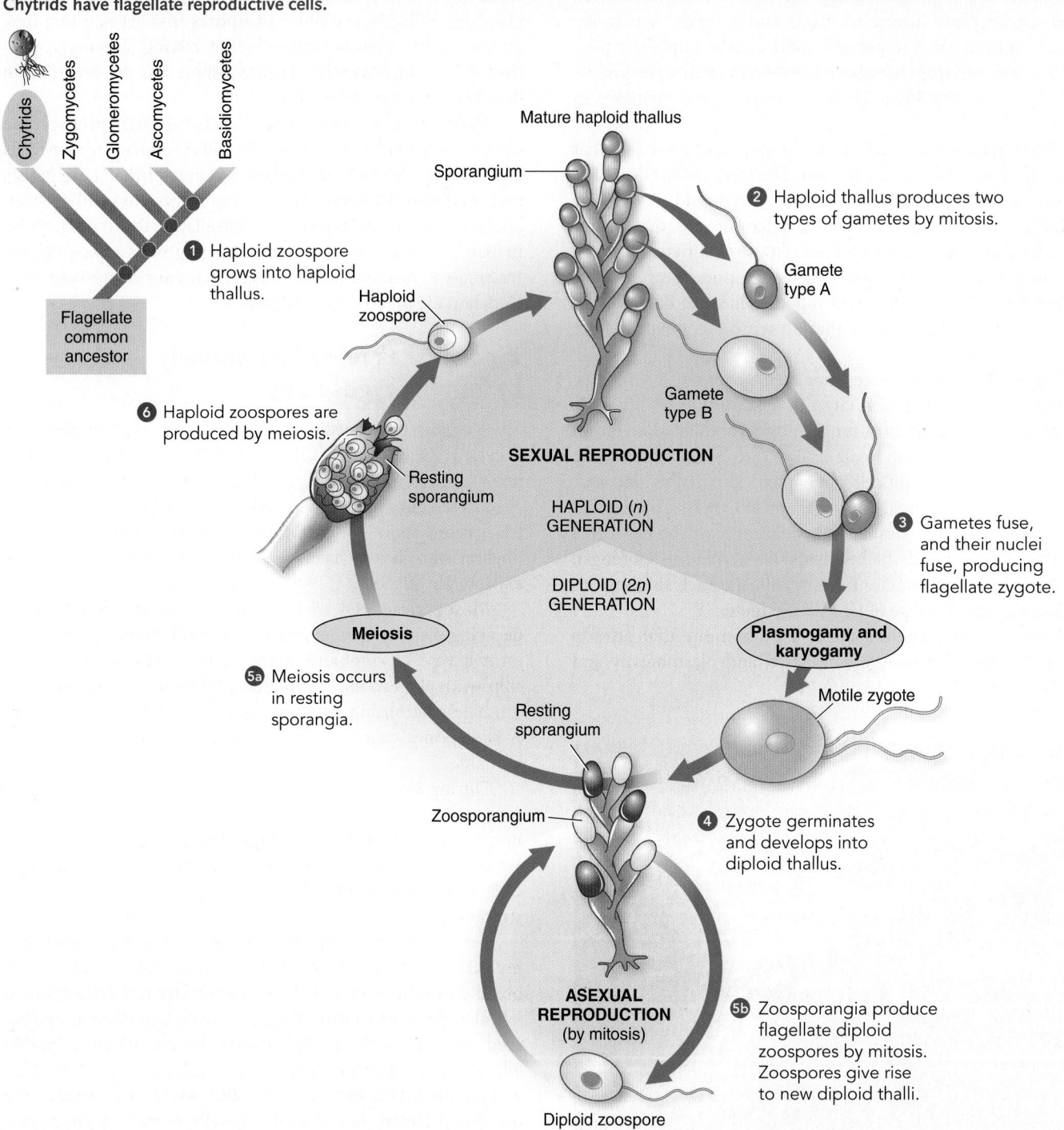

Figure 29-7 Life cycle of the chytrid *Allomyces arbuscula*

Allomyces alternates between multicellular haploid and multicellular diploid stages, which are similar in appearance.

CONNECT *Allomyces* produces gametes by mitosis. By what type of cell division do animals produce gametes?

© Cengage Learning

mates only with a hypha of a different mating type. That is, sexual reproduction occurs only between a member of a (+) strain and one of a (−) strain, not between members of two (+) strains or members of two (−) strains. Because there are

no physical differences between the two mating types, it is not appropriate to refer to them as "male" and "female."

When hyphae of opposite mating types grow in close proximity, they signal one another with pheromones. In response

Figure 29-8 *Pilobolus*, a zygomycete that grows in animal dung
Stalked sporangia of *Pilobolus* protruding from a pile of dung, which contains an extensive mycelium of the fungus. The stalked sporangia, which are 5 to 10 mm tall, act like shotguns and forcefully discharge sporangia (the black tips) away from the dung onto nearby grass. When animals such as cattle or horses eat the grass, the spores pass unharmed through the animal's digestive tract and are deposited in a fresh pile of dung.

to these chemical signals, the tips of the hyphae come together and form **gametangia,** which serve as gametes. Plasmogamy occurs as the gametangia fuse. Then karyogamy occurs as the (+) and (−) nuclei fuse to form the diploid zygote nucleus. The zygote develops into a zygospore. Zygospores are encased in a thick protective zygosporangium. The zygospore may lie dormant for several months. It can survive desiccation and extreme temperatures. Meiosis probably occurs at or just before germination of the zygospore.

When the zygospore germinates, an aerial hypha develops with a sporangium at the tip. Mitosis within the sporangium produces haploid spores. These spores may be all (+) spores, all (−) spores, or a mixture of (+) and (−) spores. When released, the spores germinate to form new hyphae. Only the zygote and zygospore of a black bread mold are diploid; all the hyphae and the asexual spores are haploid.

Microsporidia have been a taxonomic mystery

Microsporidia are small, unicellular parasites that infect eukaryotic cells. They are opportunistic pathogens that infect animals. For example, microsporidia infect people with compromised immune systems, such as HIV-infected individuals. Microsporidia cause a variety of diseases involving many organ systems, and some species cause lethal infections. Some species of microsporidia appear to be host-specific.

Microsporidia have two developmental stages inside their host: a feeding stage and a reproductive stage. Some microsporidian species divide into two cells by binary fission, and others divide into several cells. Some species undergo nuclear fusion and meiosis before producing spores. The spores, which have thick protective walls, can pass from cell to cell inside the host

or can be excreted in urine or through the skin. The spores, the only stage with distinct characters, are used to identify groups.

Each spore is equipped with a unique structure, a long, threadlike *polar tube*. When the spore enters the gut of a new host, it discharges its polar tube and penetrates the lining of the gut. Acting as a hypodermic needle, the polar tube injects the contents of the spore into the host cell (**FIG. 29-10**).

Microbiologists estimate that there may be more than one million species of microsporidia, but only about 1500 species have been named. Microsporidia were originally classified with yeasts and bacteria. In 1976, they were assigned to the protozoa. In the 1980s, biologists viewed microsporidia as the most primitive example of a eukaryote. They were the smallest and simplest known eukaryotes, and the genomes of some species are smaller than most bacterial genomes. Microsporidia lack mitochondria, flagella, and Golgi complexes. Their ribosomes resemble those of prokaryotes.

In 1998, British biologist Thomas Cavalier-Smith reassigned microsporidia to kingdom Fungi. Molecular studies show that microsporidia have gene sequences that indicate that they originally had mitochondria. Microbiologists now generally agree that these organisms have become simpler as they adapted to their parasitic way of life. Other molecular studies have provided additional evidence for their taxonomic relationship with fungi. Establishing that they are closely related to fungi is important in developing medications that will be effective in treating microsporidian infections. Recent genome studies suggest that microsporidia descended from a zygomycete ancestor. For now, we classify the microsporidia with the zygomycetes, but in the future they may be assigned to a separate taxon.

Glomeromycetes have a symbiotic relationship with plant roots

Glomeromycetes (phylum Glomeromycota) have coenocytic (no septa) hyphae. They reproduce asexually with large, multinucleate spores called *blastospores*. Sexual reproduction has not been documented. About 230 species of glomeromycetes have been described. Glomeromycetes were previously considered zygomycetes, but taxonomists using molecular data have determined that they form a separate monophyletic group.

Glomeromycetes are symbionts that form intracellular associations with the roots of most trees and herbaceous plants. These symbiotic associations between the hyphae of certain fungi and the roots of plants are called **mycorrhizae** (from Greek words meaning "fungus roots") (**FIG. 29-11**). Glomeromycetes extend their hyphae through the cell walls of root cells but often may not penetrate the plasma membrane. As each hypha pushes forward, the plasma membrane of the root cell surrounds it. Thus, the hyphae can be thought of as fingers pushing into a glove formed by the plasma membrane. Because they penetrate the cell wall, these fungi are referred to as **endomycorrhizal fungi.**

The most widespread endomycorrhizae are called *arbuscular mycorrhizae* because the hyphae inside the root cells form branched, tree-shaped structures known as **arbuscules** (see Fig. 29-11 and Fig. 36-11b). The arbuscules are the sites of

Like most fungi, most zygomycetes reproduce both asexually and sexually.

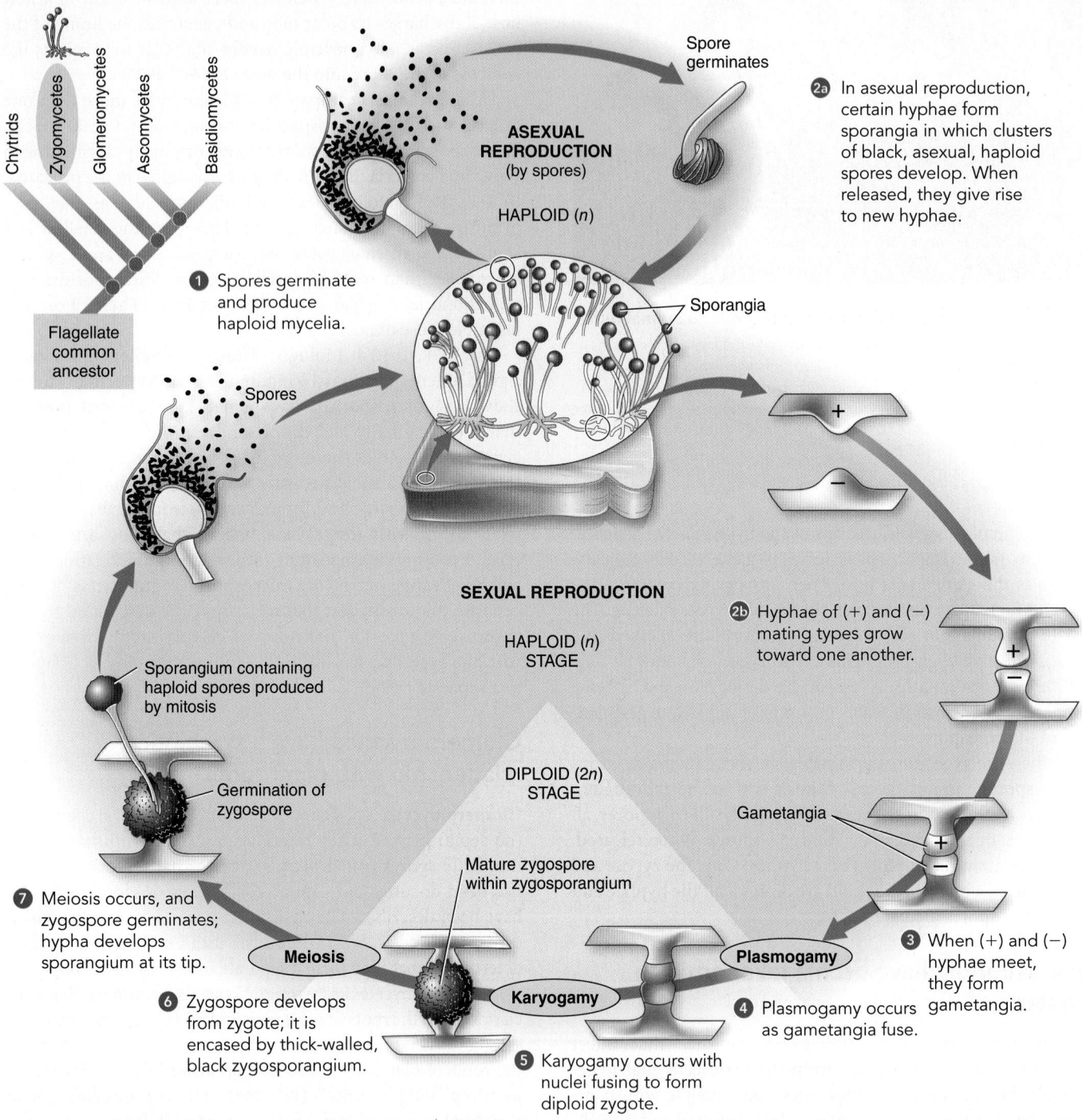

1 Spores germinate and produce haploid mycelia.

2a In asexual reproduction, certain hyphae form sporangia in which clusters of black, asexual, haploid spores develop. When released, they give rise to new hyphae.

Spore germinates

ASEXUAL REPRODUCTION (by spores)

HAPLOID (*n*)

Spores

Sporangia

2b Hyphae of (+) and (−) mating types grow toward one another.

SEXUAL REPRODUCTION

HAPLOID (*n*) STAGE

DIPLOID (2*n*) STAGE

Sporangium containing haploid spores produced by mitosis

Germination of zygospore

Gametangia

3 When (+) and (−) hyphae meet, they form gametangia.

Plasmogamy

4 Plasmogamy occurs as gametangia fuse.

Mature zygospore within zygosporangium

Meiosis

Karyogamy

7 Meiosis occurs, and zygospore germinates; hypha develops sporangium at its tip.

6 Zygospore develops from zygote; it is encased by thick-walled, black zygosporangium.

5 Karyogamy occurs with nuclei fusing to form diploid zygote.

Chytrids · Zygomycetes · Glomeromycetes · Ascomycetes · Basidiomycetes

Flagellate common ancestor

Figure 29-9 *Animation* **Life cycle of a zygomycete, the black bread mold (*Rhizopus stolonifer*)**

CONNECT Identify the structures that function as the equivalent of gametes in the zygomycete life cycle. Are they haploid, diploid, or dikaryotic?

© Cengage Learning

nutrient exchange between the plant and the fungus. Arbuscular mycorrhizae live entirely underground.

In mycorrhizal relationships the roots supply the fungus with sugars, amino acids, and other organic substances. The mycorrhizal fungus decomposes organic material in the soil and also benefits the plant by extending the reach of its roots. The slender mycelia are far thinner than roots and can extend into narrow spaces, absorbing nutrients that the plant could not

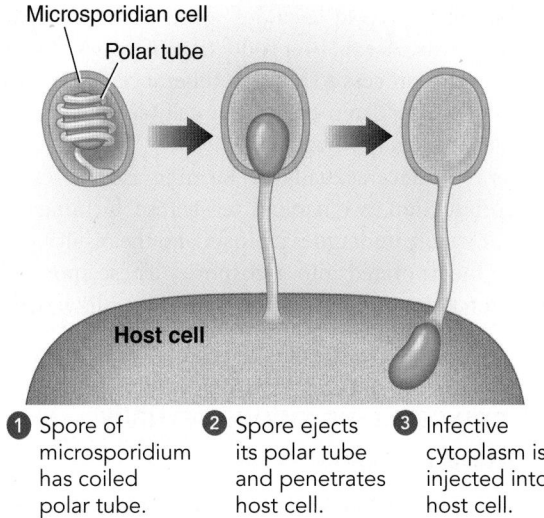

Figure 29-10 Infection by a microsporidium
① Spore of microsporidium has coiled polar tube.
② Spore ejects its polar tube and penetrates host cell.
③ Infective cytoplasm is injected into host cell.
© Cengage Learning

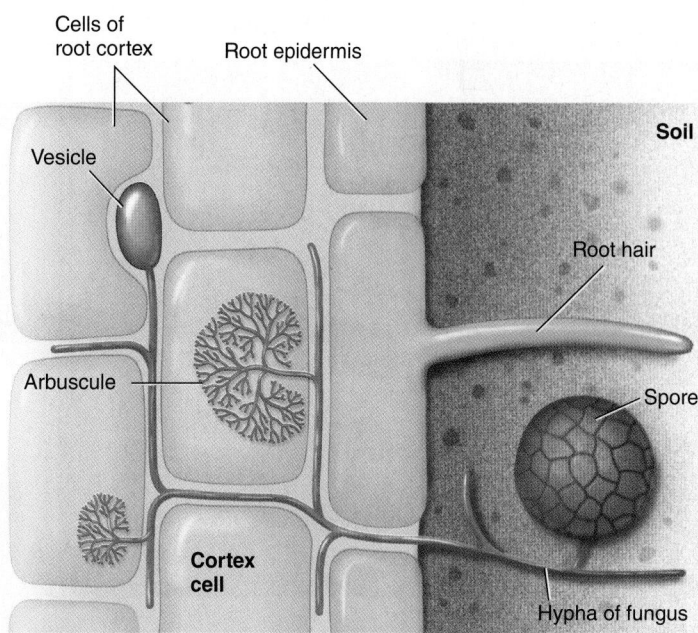

Figure 29-11 Arbuscular mycorrhizae
This mycelium has grown into a plant root. Its hyphae branch between the cells of the root. Hyphae have penetrated through the cell walls of two root cells and have branched extensively to form arbuscules. The tip of one hypha between root cells has enlarged and serves as a vesicle that stores food. The tip of a hypha in the soil has enlarged, forming a spore. The spaces between the root cells have been magnified for clarity.
© Cengage Learning

capture on its own. Thus, with the help of the mycorrhizal fungus, the plant can also take in more nutrient minerals such as phosphorus and nitrogen.

What we have just described is a mutualistic symbiotic association: both partners benefit. Studies show that if a plant grows in phosphate-deficient soil or has a limited root system, its growth is enhanced by having a fungal partner. However, a plant in a phosphate-rich soil with a well-developed root system may not need fungal partners. For these plants, the fungus may be a parasite.

Much remains to be learned about mycorrhizal associations. Other, yet unknown, benefits of the symbiosis may further define the association as mutualistic. For example, some fungi release alkaloids that protect the plants from herbivores and pathogens. Plants also exchange nutrients with one another through fungi that connect them.

Scientists have discovered mycorrhizal fungi within ancient plant fossils in rocks that are about 400 million years old. These findings suggest that when plants moved onto the land, their fungal partners moved with them. In fact, the fungal partners may have been critical for early vascular plants to colonize the land because the fungal hyphae may have provided plants with water and minerals before their own root systems evolved.

Ascomycetes reproduce sexually by forming ascospores

Ascomycetes (phylum Ascomycota) comprise a large group of fungi consisting of more than 32,000 described species. The diverse ascomycetes include most yeasts; the powdery mildews; most of the blue-green, pink, and brown molds that cause food to spoil; decomposer cup fungi; and the edible morels and truffles.

The ascomycetes, more than any other group of fungi, affect humans. As we will discuss in a later section, ascomycetes are used to flavor cheeses, to bake bread (yeast), and to ferment alcohol. Some are enjoyed as foods (morels and truffles). Ascomycetes are used to produce antibiotics. They have also served as valuable model organisms for biologists studying cellular processes, including protein synthesis. Many fungi in this group form mycorrhizae with tree roots, and about 40% join with green algae or cyanobacteria to form lichens. On the negative side, ascomycetes cause most fungal diseases of plants and animals, including humans. For example, ascomycetes cause serious plant diseases such as Dutch elm disease, ergot disease on rye, powdery mildew on fruits and ornamental plants, and chestnut blight.

Ascomycetes are sometimes referred to as *sac fungi* because their sexual spores are produced in microscopic sacs called **asci** (sing., *ascus*). Their hyphae usually have septa, but these cross walls have pores so that cytoplasm is continuous from one cell compartment to another.

In most ascomycetes asexual reproduction involves production of spores called **conidia,** which form at the tips of certain specialized hyphae known as *conidiophores* (FIG. 29-12). Production of these spores is a means of rapidly propagating new mycelia when environmental conditions are favorable. Conidia occur in various shapes, sizes, and colors in different species. The color of the conidia produces the characteristic blue-green, pink, brown, or other tints of many of these molds.

Some species of ascomycetes are heterothallic. (Recall that heterothallic means that an individual fungal hypha mates only with a hypha of a different mating type.) Others are **homothallic,** which means that they are self-fertile and

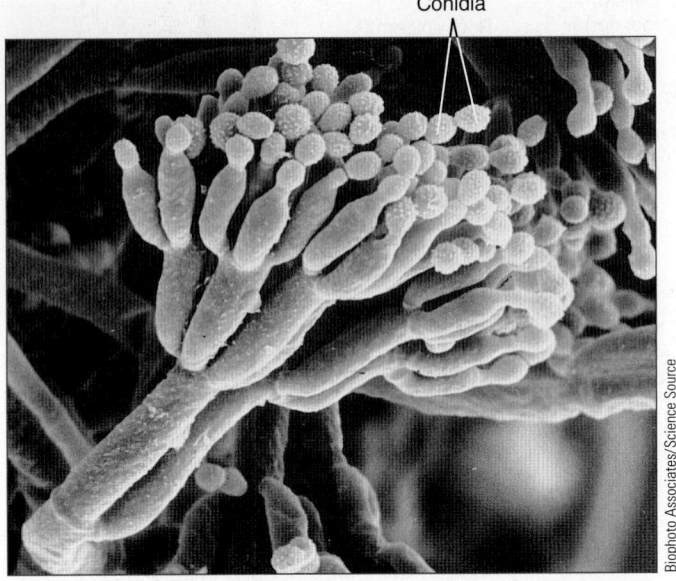

Conidia

10 μm

Biophoto Associates/Science Source

Figure 29-12 Conidia

SEM of *Penicillium* conidiophores, which resemble paintbrushes. Note the conidia pinching off from the tips of the "brushes." Conidia are asexual reproductive cells produced by ascomycetes and some basidiomycetes. Biologists use the arrangement of conidia on conidiophores to identify species of these fungi.

have the ability to mate with themselves. In both heterothallic and homothallic ascomycetes, sexual reproduction takes place after two gametangia come together and their cytoplasm mingles.

Let us examine the life cycle of a typical ascomycete (FIG. 29-13). In our example plasmogamy takes place as hyphae of two different mating types come together and fuse. Within this fused structure, pairs of haploid nuclei, one from each parent hypha, associate but do not fuse. New hyphae, with dikaryotic cells, develop from the fused structure. The hyphae branch repeatedly until the hyphal tips reach the site where asci will be produced. As the many sac-shaped asci develop, each containing two dissimilar nuclei (one from each parent), they are surrounded by intertwining haploid (monokaryotic) hyphae. These hyphae help make a fruiting body known as an **ascocarp** (FIG. 29-14a).

Karyogamy occurs in each ascus. The two nuclei fuse and form a diploid zygote nucleus. The zygote nucleus then undergoes meiosis to form four haploid nuclei with different genotypes. One mitotic division of each of the four nuclei usually follows, resulting in eight haploid nuclei. Each haploid nucleus becomes incorporated into a thick-walled **ascospore**; thus, there are typically eight haploid ascospores within the ascus (FIG. 29-14b). The ascospores are usually released through a pore, slit, or hinged lid at the tip of the ascus. Air currents carry individual ascospores, often for long distances. If one lands in a suitable location, it germinates and forms a new mycelium. The fungus can reproduce asexually by producing conidia that can develop into new mycelia.

Phylum Ascomycota includes more than 300 species of unicellular yeasts. Asexual reproduction of yeasts is mainly by **budding;** in this process a small protuberance (bud) grows and eventually separates from the parent cell (see Fig. 29-3). Each bud can grow into a new yeast cell.

Yeasts reproduce sexually by forming ascospores. During sexual reproduction, two haploid yeasts fuse, forming a diploid zygote. The zygote undergoes meiosis, and the resulting haploid nuclei are incorporated into ascospores. These spores remain enclosed for a time within the original cell wall, which corresponds to an ascus.

Basidiomycetes reproduce sexually by forming basidiospores

The more than 30,000 species of **basidiomycetes** (phylum Basidiomycota) include the largest and most familiar of the fungi: the mushrooms, bracket fungi, and puffballs (FIG. 29-15). Many basidiomycetes are decomposers that obtain nutrients by breaking down organic matter. Some species cause great economic loss because they cause dry rot in buildings. Certain basidiomycetes form mycorrhizae. Others, such as wheat rust and corn smut, infect important crops. A few basidiomycetes cause human disease.

Sometimes called *club fungi,* basidiomycetes derive their name from their microscopic club-shaped **basidia** (sing., *basidium*). Basidia are comparable in function to the asci of ascomycetes. Each basidium is an enlarged hyphal cell that undergoes meiosis to form four **basidiospores** (FIG. 29-16). Note that basidiospores develop on the *outside* of a basidium, whereas ascospores develop *within* an ascus.

Each individual fungus produces millions of basidiospores, and each basidiospore has the potential to give rise to a new **primary mycelium.** Hyphae of a primary mycelium consist of monokaryotic cells. The mycelium of a basidiomycete, such as the commonly cultivated mushroom *Agaricus brunnescens,* consists of a mass of white, branching, threadlike hyphae that live mostly underground. Septa divide the hyphae into cells, but as in ascomycetes, the septa are perforated and allow cytoplasmic streaming between cells.

Let us examine the life cycle of a typical basidiomycete. Asexual reproduction is less common in basidiomycetes than in other groups, so we will focus here on sexual reproduction. We begin with two compatible primary mycelia (FIG. 29-17 on page 612). When in the course of its growth a hypha of a primary mycelium encounters a compatible monokaryotic hypha, typically of a different mating type, the two hyphae fuse (plasmogamy). As in the ascomycetes, the two haploid nuclei remain separate within each cell. In this way a **secondary mycelium** with dikaryotic hyphae, in which each cell contains two haploid nuclei, is produced. The $n + n$ hyphae of the secondary mycelium grow rapidly and extensively.

When environmental conditions are favorable, the hyphae form compact masses, called *buttons,* along the mycelium. Each button grows into a fruiting body that we know as a mushroom. A mushroom is more formally referred to as a **basidiocarp.**

Ascomycetes produce asexual spores called conidia and sexual spores called ascospores.

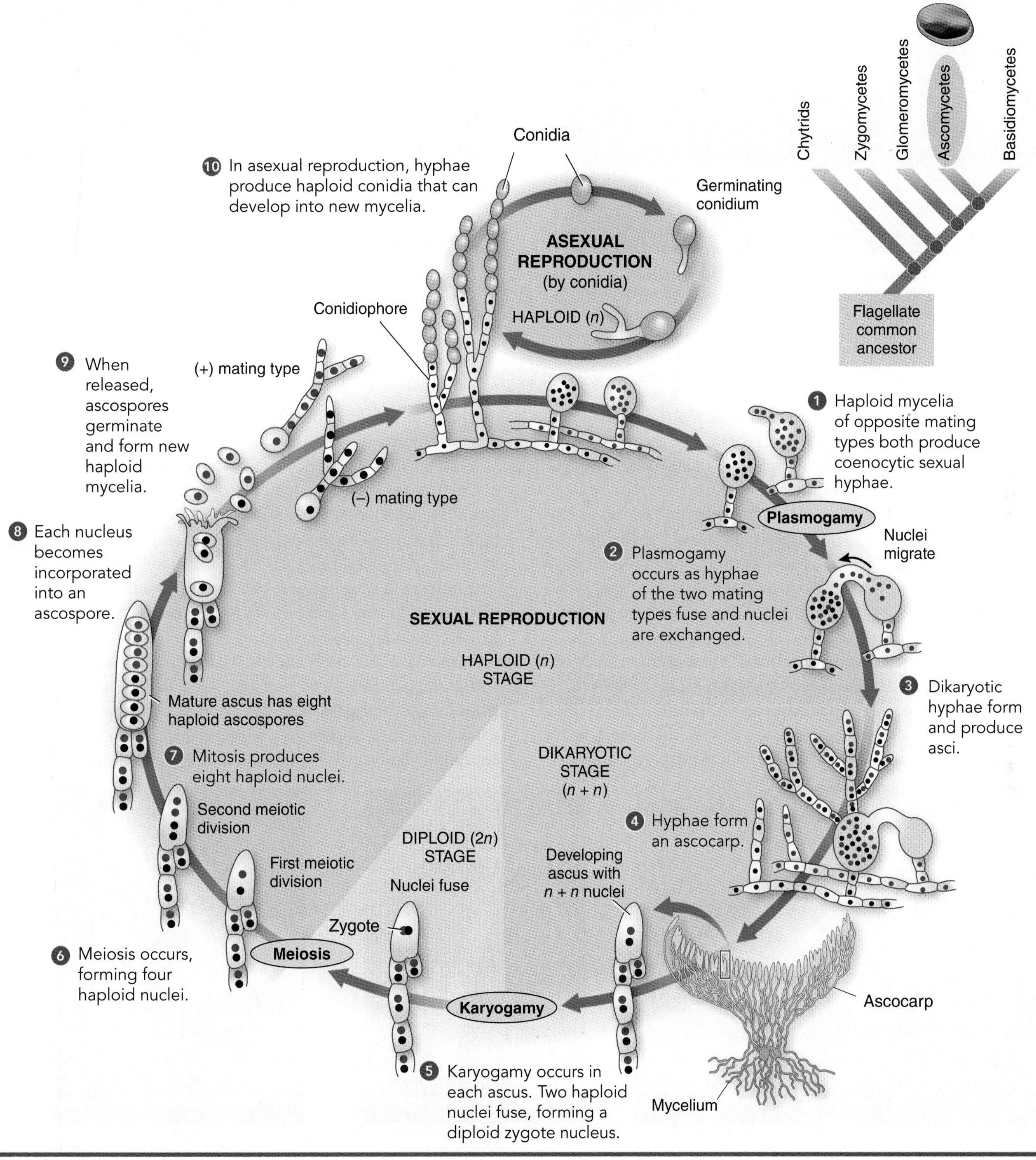

10 In asexual reproduction, hyphae produce haploid conidia that can develop into new mycelia.

Conidia

Germinating conidium

ASEXUAL REPRODUCTION (by conidia)

HAPLOID (*n*)

Conidiophore

Chytrids
Zygomycetes
Glomeromycetes
Ascomycetes
Basidiomycetes

Flagellate common ancestor

9 When released, ascospores germinate and form new haploid mycelia.

(+) mating type

(−) mating type

1 Haploid mycelia of opposite mating types both produce coenocytic sexual hyphae.

Plasmogamy

Nuclei migrate

8 Each nucleus becomes incorporated into an ascospore.

SEXUAL REPRODUCTION

HAPLOID (*n*) STAGE

2 Plasmogamy occurs as hyphae of the two mating types fuse and nuclei are exchanged.

3 Dikaryotic hyphae form and produce asci.

Mature ascus has eight haploid ascospores

7 Mitosis produces eight haploid nuclei.

Second meiotic division

First meiotic division

DIKARYOTIC STAGE (*n* + *n*)

4 Hyphae form an ascocarp.

Developing ascus with *n* + *n* nuclei

DIPLOID (2*n*) STAGE

Nuclei fuse

Zygote

Meiosis

6 Meiosis occurs, forming four haploid nuclei.

Karyogamy

Ascocarp

Mycelium

5 Karyogamy occurs in each ascus. Two haploid nuclei fuse, forming a diploid zygote nucleus.

Figure 29-13 Life cycle of a typical heterothallic ascomycete

Sexual reproduction requires haploid mycelia of different mating types. Note the dikaryotic stage and the separation of plasmogamy and karyogamy. Steps 5 through 8 take place within an ascus in the ascocarp.

CONNECT Is each of the following haploid, diploid, or dikaryotic: conidium, ascocarp, ascospore, zygote?

© Cengage Learning

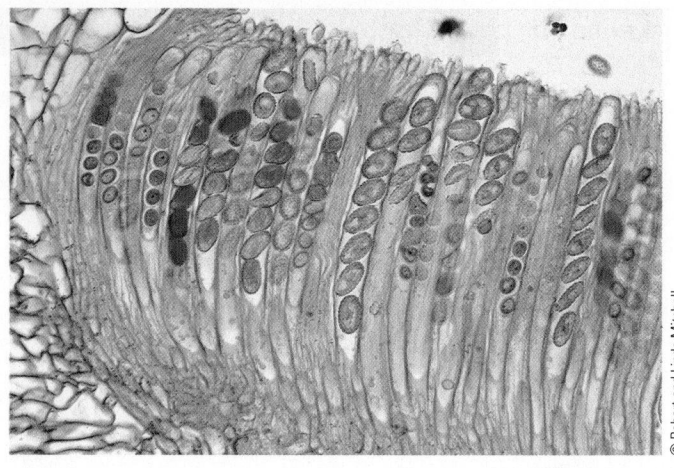

100 μm

(a) The ascocarp (fruiting body) of the common brown cup (*Peziza badio-confusa*). This ascocarp is shaped like a saucer or bowl and is 3 to 10 cm (1 to 4 in.) wide. It is found on damp soil in woods throughout North America. Photographed in Muskegon, Michigan.

(b) Asci. Each ascus contains eight ascospores. Asci line the inner portion of the ascocarp.

Figure 29-14 Sexual reproduction in the ascomycetes

Each basidiocarp consists of intertwined, matted hyphae and has a stalk and a cap. The lower surface of the cap usually consists of many thin, perpendicular plates called **gills** that radiate from the stalk to the edge of the cap.

Karyogamy takes place within the young basidia on the gills of the mushroom. The haploid nuclei fuse in the dikaryotic cells, forming diploid zygote nuclei. They are the only diploid cells that form during a basidiomycete's life cycle. Meiosis then takes place, forming four haploid nuclei with different genotypes. These nuclei move to the outer edge of the basidium. Fingerlike extensions of the basidium develop, into which the nuclei and some cytoplasm move; each of these extensions becomes a basidiospore. A septum forms that separates the basidiospore from the rest of the basidium by a delicate stalk that breaks when the basidiospore is forcibly discharged. Each basidiospore can germinate and give rise to a primary mycelium.

Many basidiomycetes produce "fairy rings" in lawns and forests (FIG. 29-18 on page 613). A fairy ring may first appear as a dark green ring surrounding an inner brown circle. The size of the ring ranges from a few centimeters to more than 15 m (about 51 ft) in diameter. The green ring consists of grass, well nourished by the nutrients released as the fungi decompose organic material. Grass dies, producing the inner brown circle, because the mass of mycelia decreases the movement of water into the area. As the fungi grow outward, the circle widens. The rings grow a few centimeters to more than a meter per year. After rainfall or irrigation, a ring of mushrooms may appear just outside the green circle. The name "fairy ring" comes from a legend that a ring of mushrooms appeared where fairies had danced in a circle the night before.

(a) Basidia line the gills of the Jack-o'-lantern mushroom (*Omphalotus olearius*). The gills of this poisonous species produce a greenish glow in the dark. Each cap is about 15 cm (6 in.) wide. The Jack-o'-lantern occurs throughout eastern North America and California.

(b) The elegant stinkhorn (*Phallus ravenelii*) has a foul smell that attracts flies. The flies help disperse the slimy mass of basidiospores. Fruiting bodies of elegant stinkhorns grow to 18 cm (7 in.) tall. Photographed in Pennsylvania.

(c) Turkey-tail (*Trametes versicolor*) is a common bracket fungus. Bracket fungi grow on both dead and living trees and produce shelf-like fruiting bodies. Basidiospores are produced in pores located underneath each shelf.

Figure 29-15 Basidiomycete fruiting bodies

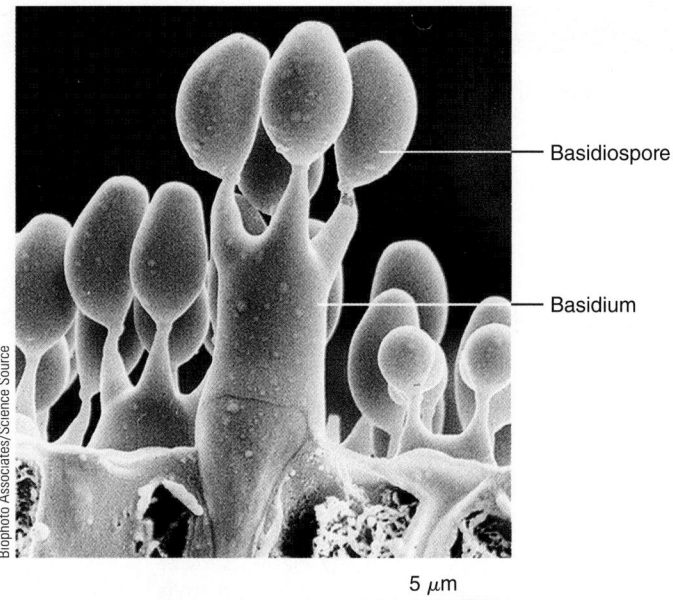

— Basidiospore

— Basidium

5 μm

Figure 29-16 SEM of a basidium
Each basidium produces four basidiospores.

CHECKPOINT 29.3

- *What evidence supports the hypothesis that chytrids were the earliest fungal group to evolve from the common ancestor of fungi?*

- *What are the distinguishing characteristics of each of the following fungal groups: zygomycetes, glomeromycetes, ascomycetes, and basidiomycetes?*

- VISUALIZE *Use simple diagrams to illustrate how the life cycle of a typical basidiomycete differs from that of a typical ascomycete.*

- *Distinguish among (1) ascocarp, ascus, and ascospore and among (2) basidiocarp, basidium, and basidiospore.*

29.4 ECOLOGICAL IMPORTANCE OF FUNGI

LEARNING OBJECTIVES

7 Summarize the ecological significance of fungi as decomposers.

8 Describe the important ecological role of mycorrhizae.

9 Characterize the unique nature of a lichen.

Fungi make vital contributions to the ecological balance of our planet. Like bacteria, most fungi are free-living decomposers, chemoheterotrophs that absorb nutrients from organic wastes and dead organisms. For example, many fungal decomposers degrade cellulose and lignin, the main components of plant cell walls. When fungi degrade wastes and dead organisms, they release water, carbon (as CO_2), and mineral components of organic compounds, and these elements are recycled (see biogeochemical

cycles in Chapter 55). Without this continuous decomposition, essential nutrients would remain locked up in huge mounds of animal carcasses, feces, branches, logs, and leaves. These nutrients would be unavailable for use by new generations of organisms, and life would eventually cease.

Fungi form important symbiotic relationships with animals, plants, bacteria, and protists. A three-way symbiotic relationship involving a fungus, a grass, and a virus has recently been reported. In the geothermal hot spots of Yellowstone National Park, a fungus infects the roots of the host grass. When infected with a specific virus, the fungus is heat tolerant and confers heat tolerance to the grass. When not infected by the virus, the fungus does not confer heat tolerance. The symbiotic relationships of fungi with other organisms have major effects on ecosystems.

Fungi form symbiotic relationships with some animals

Because animals do not have the enzymes necessary to digest cellulose and lignin, cattle and other grazing animals cannot, by themselves, obtain needed nutrients from the plant material they eat. Their survival depends on fungi that inhabit their guts because fungi, like many other microorganisms, do have the enzymes that break down these organic compounds. The fungi benefit by living in a nutrient-rich environment.

Fungi also form symbiotic associations with ants and termites. More than 200 species of ants farm fungi. Leaf-cutting ants bring leaves to their fungi and protect them from competitors and predators. The ants also disperse the fungi to new locations. In exchange, the fungi digest the leaves, providing nutrients for the ants. This symbiosis can involve other organisms. The farmed fungi can be infested by fungal parasites. In response, the ants culture bacteria (actinomycetes) that produce antibiotics to control these parasites. These symbiotic relationships, the most complex known, are the product of 50 million years of coevolution.

Mycorrhizae are symbiotic associations between fungi and plant roots

Mycorrhizae occur in about 80% of plants (and more than 90% of all plant families). As discussed in the section on glomeromycetes, mycorrhizal fungi decompose organic material in the soil and increase the surface area of a plant's roots so that the plant can absorb more water and mineral nutrients. In exchange, the roots supply the fungus with organic nutrients.

To establish and maintain a symbiotic relationship, cells of the fungi and cells of the plant roots must communicate. For example, signaling molecules from the plant root cells stimulate fungal cells to shift to a presymbiotic growth phase in which their energy metabolism increases and their hyphae branch. The fungal cells then signal the root cells, activating a signaling pathway that activates gene expression in the root cells.

The importance of mycorrhizae first became evident when horticulturalists observed that orchids do not grow unless an appropriate fungus lives with them. Similarly, many

Basidiomycetes produce sexual basidiospores on the gills of basidiocarps (fruiting bodies).

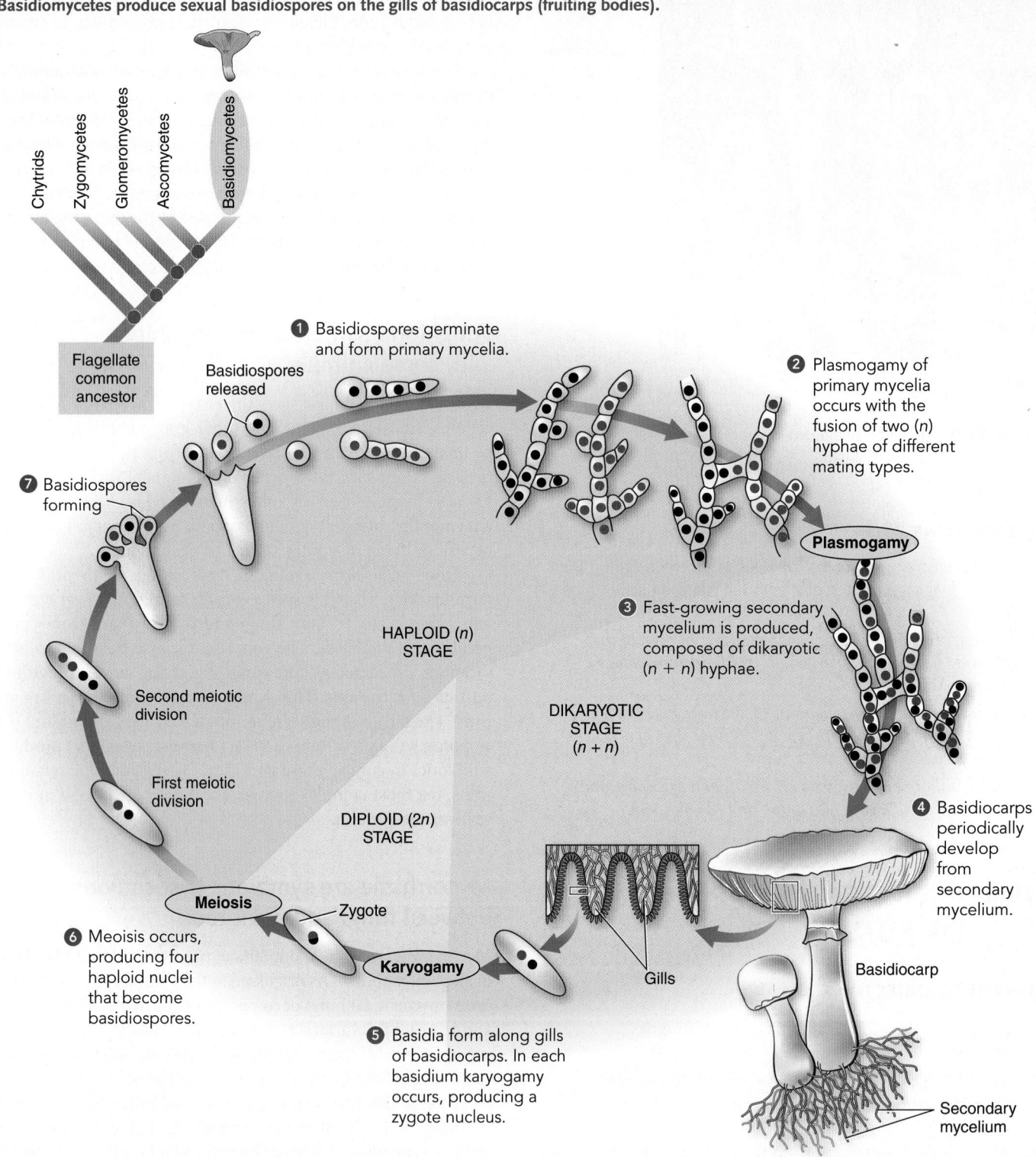

1 Basidiospores germinate and form primary mycelia.

2 Plasmogamy of primary mycelia occurs with the fusion of two (n) hyphae of different mating types.

7 Basidiospores forming

Basidiospores released

Second meiotic division

First meiotic division

HAPLOID (n) STAGE

DIKARYOTIC STAGE (n + n)

DIPLOID (2n) STAGE

3 Fast-growing secondary mycelium is produced, composed of dikaryotic (n + n) hyphae.

Plasmogamy

4 Basidiocarps periodically develop from secondary mycelium.

Basidiocarp

Secondary mycelium

Gills

Meiosis

Zygote

6 Meoisis occurs, producing four haploid nuclei that become basidiospores.

Karyogamy

5 Basidia form along gills of basidiocarps. In each basidium karyogamy occurs, producing a zygote nucleus.

Chytrids

Zygomycetes

Glomeromycetes

Ascomycetes

Basidiomycetes

Flagellate common ancestor

Figure 29-17 *Animation* Life cycle of a typical basidiomycete

Note the dikaryotic stage and the separation of plasmogamy and karyogamy. Steps 5 and 6 take place within the basidia of the basidiocarp. Asexual reproduction is uncommon in this group.

CONNECT Is each of the following haploid, diploid, or dikaryotic: basidiocarp, basidiospore, primary mycelium, zygote, secondary mycelium?

© Cengage Learning

Figure 29-18 A fairy ring

forest trees, such as pines, decline and eventually die from mineral deficiencies when transplanted to mineral-rich grassland soils that lack the appropriate mycorrhizal fungi. When forest soil containing the appropriate fungi or their spores is added to the soil around these trees, they quickly resume normal growth. Studies performed with various types of plants, including cedar, have confirmed the role of mycorrhizae in plant growth (FIG. 29-19).

As we have discussed, glomeromycetes form *endomycorrhizal* connections; they infiltrate the cells of plant roots. At least 5000 species of ascomycetes and basidiomycetes also form mycorrhizal connections, but their hyphae coat the plant root rather than penetrate its cells. These species are referred to as **ectomycorrhizal fungi.** Interestingly, researchers have shown that some mycorrhizal fungi harbor bacteria in their cytoplasm. Although the role of the bacteria is not yet clear, their presence suggests that they may be members of a three-way partnership: fungus, plant, and bacteria.

Mycorrhizal fungi connect plants, allowing nutrient transfer among them. Scientists have measured the movement of organic materials from one tree species to another through shared mycorrhizal connections. Mycorrhizal fungi also release chemicals that protect the plant against herbivores and pathogens.

Mycorrhizae improve the soil by decreasing water loss and erosion. Ecologists are studying the role of mycorrhizal fungi in reclaiming soils damaged by pollution. For example, mycorrhizae can modify toxic heavy metals, such as cadmium, so that plants cannot absorb them.

A lichen consists of two components: a fungus and a photoautotroph

Although a **lichen** looks like a single organism, it is actually a dual organism, a combination of a fungus and a *photoautotroph* (FIG. 29-20a). Almost one-fifth of all known fungal species form these symbiotic relationships. About 14,000 kinds of lichens have been described. Fossils suggest that fungi developed symbiotic partnerships with photoautotrophs before the evolution of vascular plants.

The photoautotrophic component of a lichen is a green alga, a cyanobacterium, or both. The fungus is most often an ascomycete, although a basidiomycete is the fungal partner in some tropical lichens. Most photoautotrophic organisms found in lichens also occur as free-living species in nature, but the fungal components are generally found only as a part of the lichen. Typically, the fungus forms most of the lichen thallus (body). The fungus surrounds hundreds of photosynthetic partners and holds them in place. Lichens are named for the fungal component.

In the laboratory researchers can isolate the fungal and photoautotrophic components of some lichens and grow them separately in appropriate culture media. The photoautotroph grows more rapidly when separated, whereas the fungus grows more slowly and requires many complex carbohydrates. Neither organism resembles a lichen in appearance when grown separately. The photoautotroph and fungus can be reassembled as a lichen thallus, but only if they are placed in a culture medium under conditions that cannot support either of them independently.

What is the nature of this partnership? The lichen was originally considered a definitive example of mutualism. The photoautotroph carries on photosynthesis, producing energy-rich carbon compounds for both members of the lichen. It is unclear how the photoautotroph benefits from the relationship. Some biologists have suggested that the photoautotroph obtains water and nutrient minerals from the fungus as well as protection against desiccation. More recently, researchers have suggested that the lichen partnership is not really a case of mutualism but one of controlled parasitism of the photoautotroph by the fungus.

Lichens typically exhibit one of three different growth forms (FIG. 29-20b). *Crustose lichens* are flat and grow tightly against their substrate (the surface they are growing on). *Foliose lichens* are also flat, but they have leaflike lobes and are not as tightly pressed to the substrate. *Fruticose lichens* grow erect and have many branches.

Able to tolerate extremes of temperature and moisture, lichens grow in almost all terrestrial environments except polluted cities. They exist farther north than any plants of the arctic region and are equally at home in the steaming equatorial rain forest. They grow on tree bark, leaves, and exposed rock surfaces, from solidified lava to tombstones. In fact, lichens are often the first organisms to inhabit rocky areas, and their growth in these areas is important in forming soil from rock. They secrete acid that gradually etches tiny cracks in the rock, releasing minerals. This process sets the stage for further disintegration of the rock by wind and rain. When the lichens themselves die and are decomposed, they become part of the soil.

Reindeer mosses of the arctic region, which serve as the main source of food for migrating herds of caribou, are actually lichens, not mosses. Some lichens produce colored pigments. One of them, orchil, is used to dye wool; another, litmus, is widely used in chemistry laboratories as an acid–base (pH) indicator.

Lichens vary greatly in size. Some are almost invisible, whereas others, such as the reindeer mosses, may cover many

Is plant growth affected by fungi in the soil?

HYPOTHESIS: In low-phosphorus soils, plants that form mycorrhizal associations with fungi exhibit enhanced growth.

EXPERIMENT: Western red cedar seedlings (*Thuja plicata*) were selected for this study. Control seedlings were grown in low-phosphorus soil in the absence of the fungus. Experimental seedlings of the same age as the control plants were grown under the same conditions as the controls except that their roots formed mycorrhizal associations.

(a) No mycorrhizal associations. Control seedlings were grown in low-phosphorus soil in the absence of the fungus.

(b) Mycorrhizal associations. Experimental seedlings were grown under the same conditions as the controls, except that the fungus was present. The seedlings formed mycorrhizal associations with the fungus.

RESULTS AND CONCLUSION: Growth of the plants that formed mycorrhizal associations was significantly enhanced. Mycorrhizal associations enhance the growth of western red cedar plants. Many similar studies using other types of plants and other types of soil have confirmed the importance of mycorrhizal association to plant growth.

SOURCE: Kough, J.L., R. Molina, and R.G. Linderman. "Mycorrhizal Responsiveness of *Thuja, Calocedrus, Sequoia*, and *Sequoiadendron* Species of Western North America," *Canadian Journal of Forest Research*, Vol. 15 (1985): 1049–1054.

Figure 29-19 **The effect of mycorrhizae on western red cedar (*Thuja plicata*) seedlings**

CONNECT Biologists have discovered that many mycorrhizal fungi are sensitive to a low pH. What human-caused environmental problem may prove catastrophic for these fungi? How may this problem affect their plant partners? What measures could we take to decrease the problem?

square kilometers of land with an ankle-deep growth. Growth proceeds slowly; the radius of a lichen may increase by less than 1 mm each year. Some mature lichens are thought to be thousands of years old.

Lichens absorb minerals from the air, rainwater, and the surface on which they grow. They cannot excrete the elements they absorb, and perhaps for this reason they are extremely sensitive to toxic compounds. This sensitivity was first reported in 1866 by a Finnish biologist who observed that lichens growing on tree trunks in Paris were poorly developed or sterile. He deduced that lichens could be used to measure air purity. Today, reduction in lichen growth is used as a sensitive indicator of air pollution, particularly from sulfur dioxide. In one study investigators demonstrated a relationship between lung cancer and air pollution by comparing the locations of low lichen biodiversity (and therefore of air pollution) with the locations of lung cancer deaths in young males. The return of lichens to an area indicates an improvement in air quality.

Lichens reproduce mainly by asexual means, usually by fragmentation, a process in which special dispersal units of the lichen, called **soredia** (sing. *soredium*)**,** break off and, if they land on a suitable surface, establish themselves as new lichens. Soredia contain cells of both partners. In some lichens the fungus produces ascospores, which may be dispersed by wind and find an appropriate algal partner only by chance.

CHECKPOINT 29.4

- **CONNECT** *What is the ecological importance of fungal decomposers?*
- *What are some ways in which the relationship between a plant root and a mycorrhizal fungus is mutualistic?*
- *Many biologists consider a lichen an example of controlled parasitism. In this view, which component is the likely parasite, and which is the likely host?*

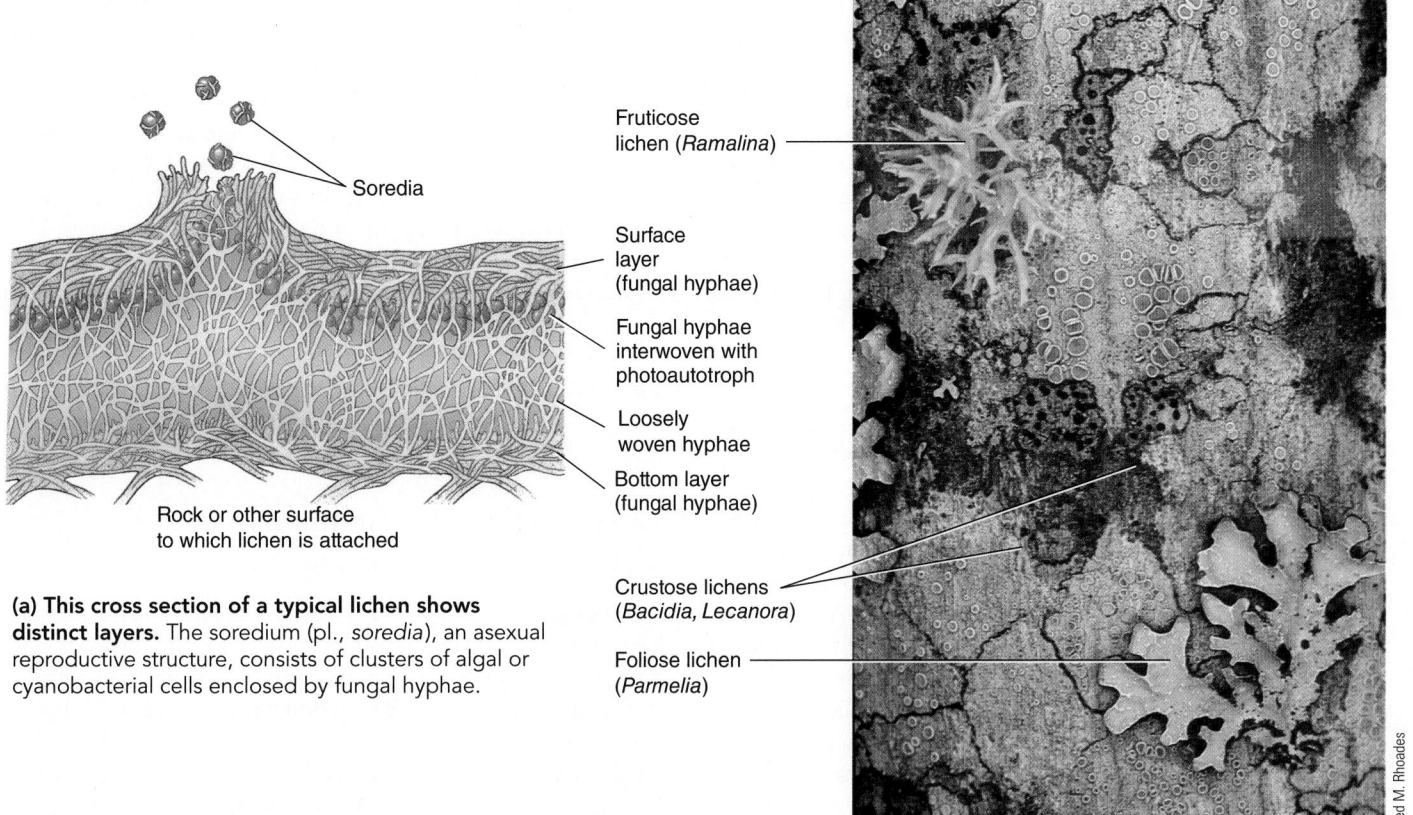

Soredia

Surface
layer
(fungal hyphae)

Fungal hyphae
interwoven with
photoautotroph

Loosely
woven hyphae

Bottom layer
(fungal hyphae)

Rock or other surface
to which lichen is attached

(a) This cross section of a typical lichen shows distinct layers. The soredium (pl., *soredia*), an asexual reproductive structure, consists of clusters of algal or cyanobacterial cells enclosed by fungal hyphae.

Fruticose
lichen (*Ramalina*)

Crustose lichens
(*Bacidia, Lecanora*)

Foliose lichen
(*Parmelia*)

Fred M. Rhoades

(b) Lichens vary in color, shape, and overall appearance. Three growth forms—crustose, foliose, and fruticose—are shown on a maple branch in Washington State.

Figure 29-20 Lichens

These organisms are a combination of a fungus and an alga or cyanobacterium.
© Cengage Learning

29.5 ECONOMIC, BIOLOGICAL, AND MEDICAL IMPACT OF FUNGI

LEARNING OBJECTIVES

10 Summarize some specific ways that fungi affect humans economically.

11 Summarize the importance of fungi to biology and medicine; describe how fungi infect plants and humans, describing at least one fungal animal disease and one fungal plant disease.

The same powerful digestive enzymes that fungi use to decompose wastes and dead organisms can also be used with great efficiency to reduce wood, fiber, and food to their basic components. Many species of basidiomycetes have enzymes that break down the lignin in wood. (Lignin is the second most abundant organic compound on Earth, second only to cellulose.) From the human perspective, various fungi cause incalculable damage to stored goods and building materials each year. Bracket fungi,

for example, cause enormous losses by decaying wood, both in living trees and in stored lumber.

Some fungi cause serious diseases in animals and plants, yet fungi also contribute to our quality of life. They are responsible for economic gains as well as losses. People eat them and grow them to make various medications, such as penicillin. We use them to make certain industrial chemicals and for bioremediation. Renewable fuel companies are actively searching for fungi and other microbes that can produce fuel. For example, a fungus has been discovered that synthesizes 55 hydrocarbons, perhaps to inhibit the growth of other organisms. Researchers may be able to use the genes of this fungus to engineer other microorganisms to produce fuel more efficiently.

Fungi provide beverages and food

Humans exploit the ability of yeasts to produce bread and alcoholic beverages. Yeasts produce ethyl alcohol and carbon dioxide from glucose and other sugars by fermentation (see Chapter 8).

Yeast species of the genus *Saccharomyces* (ascomycetes) are used to produce wine, beer, and other fermented beverages.

Wine is produced when yeasts ferment fruit sugar, and beer results when yeasts ferment sugar derived from starch in grains (usually barley).

Saccharomyces cerevisiae, referred to as baker's yeast, is used to bake bread, pizza, and other wheat products. During the process of making bread, carbon dioxide produced by yeast becomes trapped in dough as bubbles, causing the dough to rise and giving leavened bread its light quality. Both the carbon dioxide and the alcohol produced by the yeast escape during baking.

The unique flavor of cheeses such as Roquefort, Brie, Gorgonzola, and Camembert is produced by species of *Penicillium*. For example, *P. roquefortii*, found in caves near the French village of Roquefort, is used to make Roquefort cheese. By French law, only cheeses produced in this area can be called Roquefort cheese. (The blue spots in Roquefort and certain other cheeses are masses of conidia.)

Aspergillus tamarii and certain other fungi are used to produce traditional soy sauce by fermenting soybeans with the fungi for at least three months. Soy sauce enriches other foods with more than just its special flavor. It also adds vital amino acids from both the soybeans and the fungi themselves, which, in some parts of the world, supplement a low-protein rice diet.

Among the basidiomycetes, there are some 200 kinds of edible mushrooms and about 70 species of poisonous ones. Many edible mushrooms are cultivated commercially. The mushroom *Agaricus brunnescens* is the principal fungal species grown extensively for food. About 30 other mushroom species, such as oyster, shiitake, portobello, and straw mushrooms, are available in supermarkets. Morels, which superficially resemble mushrooms, and truffles, which produce underground fruiting bodies, are ascomycetes (**FIG. 29-21**). Truffles are now cultivated as mycorrhizal fungi on the roots of tree seedlings.

Edible and poisonous mushrooms can look very much alike and may even belong to the same genus. There is no simple way to tell them apart; an expert must identify them. Some of the most poisonous mushrooms belong to the genus *Amanita* (**FIG. 29-22**). Toxic species of this genus have been appropriately called such names as "destroying angel" (*A. virosa*) and "death cap" (*A. phalloides*). Eating a single mushroom of either species can be fatal.

Certain species of mushrooms cause intoxication and hallucinations. The sacred mushrooms of the Aztecs—*Conocybe* and *Psilocybe*—are still used in religious ceremonies by native peoples of Central America for their hallucinogenic properties. The chemical ingredient *psilocybin* is responsible for the trances and visions experienced by those who eat these mushrooms. Ingestion of psychoactive mushrooms is dangerous because negative reactions vary considerably, from mild indigestion, sweating, and heart palpitations, to death. In addition, the possession and use of such mushrooms are illegal in the United States and some other countries.

Fungi are important to modern biology and medicine

As discussed in Chapter 17, yeast *Saccharomyces cerevisiae* has served as a model eukaryotic cell (see Fig. 29-3). It was the first eukaryote whose genome was sequenced, and with its 6000 genes, it has the smallest genome of any eukaryotic model organism. Molecular biologists are in the process of determining the functions of the proteins encoded by its genes. Biologists have used *S. cerevisiae* to study molecular genetics, including how genes regulate cell division. Researchers continue to use this yeast to study such problems as genetic recombination and the correlation between cell age and cancer. *Saccharomyces cerevisiae* is also being used to study the mechanism of action of antifungal drugs and resistance to these drugs.

Biologists have used the ascomycete *Aspergillus nidulans*, an opportunistic pathogen of humans, to study mitosis and other cell processes. This fungus has provided valuable knowledge about the genetics of microtubules. Biologists are using

(a) The yellow morel (*Morchella esculenta*). This morel grows 6 to 10 cm (2.5 to 4 in.) tall. It is found throughout North America. Photographed in Michigan.

(b) The Oregon white truffle (*Tuber gibbosum*). This truffle, which is found underground near Douglas firs and possibly oak trees in British Columbia and northern California, is 1 to 5 cm (0.4 to 2 in.) wide. People find these subterranean ascocarps with the help of trained dogs or pigs. Here, truffles are shown whole and sectioned to show the conspicuous white, marbled tissue.

Figure 29-21 Edible ascomycetes

Figure 29-22 Poisonous mushrooms

The destroying angel (*Amanita virosa*) is an extremely poisonous mushroom that is distinguished, as are other amanitas, by the ring of tissue around its stalk and by the underground cup from which the stalk protrudes. About 50 g (2 oz) of this mushroom can kill an adult man. The destroying angel, which is 7.5 to 20 cm (3 to 8 in.) tall, is found in grass or near trees throughout North America.

recombinant DNA techniques to manipulate yeasts and certain filamentous fungi to produce important biological molecules, such as hormones. Among the many genes that have been cloned in yeast are those for insulin, human growth hormone, and molecules important in immune function. These procedures allow researchers to produce unlimited amounts of these compounds for study and eventual medical use.

Fungi produce useful drugs and chemicals. Discovered in 1928 by British bacteriologist Alexander Fleming, penicillin, produced by the mold *Penicillium notatum,* is still among the most widely used and effective antibiotics (see Chapter 1). Other drugs derived from fungi include the cephalosporin antibiotics (produced by *Cephalosporium*), statins (used to lower blood cholesterol levels), and cyclosporine (used to suppress immune responses in patients who receive organ transplants). Fumagillin, a chemical produced by the ascomycete *Aspergillus fumigatus,* inhibits the formation of new blood vessels. Because solid tumors need a rich blood supply, fumagillin shows promise as an anticancer agent. Fumagillin is also used to treat diseases caused by microsporidia. Researchers have identified several other promising compounds produced by fungi that are antiviral or that destroy cancer cells.

The ascomycete *Claviceps purpurea* infects the flowers of rye plants and other cereals. It produces a structure called an *ergot* where a seed would normally form in the grain head. When livestock eat this grain or when humans eat bread made from ergot-contaminated rye flour, they may be poisoned by the extremely toxic substances in the ergot. However, some ergot compounds are now used clinically in small quantities as drugs to induce labor, to stop uterine bleeding, to treat high blood pressure, and to relieve one type of migraine headache.

Fungi are used in bioremediation and to biologically control pests

Some fungi can biodegrade pesticides, herbicides, coal tars, and petroleum. Fungi convert these products into carbon dioxide and the basic elements of which they are composed. These fungi can be used along with certain bacteria to decontaminate farm land and to clean up oil spills.

Researchers are investigating fungi—for example, certain species of microsporidia—for the biological control of pathogens and insect pests. Some of these species are already being used to parasitize insect pests. In some cases, they interfere with reproduction in their insect host. It should be noted that some microsporidia may pose a threat to beneficial insects. For example, a microsporidian has been implicated as one factor in the die-off of honeybee colonies.

Some fungi cause diseases in humans and other animals

Certain ascomycetes cause superficial infections in which only the skin, hair, or nails are infected. Ringworm, athlete's foot, and jock itch are examples of superficial fungal infections. Because these fungi infect dead layers of skin that are not fed by capillaries, the immune system cannot launch an effective response.

Many pathogenic fungi are opportunists that cause infections only when the body's immune system is compromised, such as in patients infected with HIV. Cancer patients and organ transplant recipients who are given medication to suppress their immune systems are also at risk. *Candida* is an ascomycete that inhabits the human mouth and vagina. The immune system and the normal bacteria of these regions normally prevent this yeast from causing infection. However, when the immune system is compromised, *Candida* multiplies, causing thrush, a painful yeast infection of the mouth, throat, and vagina.

The ascomycete *Aspergillus fumigatus* is usually harmless but causes aspergillosis in people with lowered immune function. During the course of aspergillosis, the fungus can invade the lungs, heart, brain, kidneys, and other vital organs and cause death.

Other fungi also infect internal tissues and organs and may spread through many regions of the body. Histoplasmosis, for example, is a lung infection caused by inhaling spores of a fungus common in soil contaminated with bird feces. Most people in the eastern and midwestern parts of the United States have been exposed to this fungus at some time, and an estimated 40 million Americans have had mild infections. Fortunately, the infection is usually confined to the lungs and is of short duration, but if the infection spreads through the blood to the heart, brain, or other parts of the body, it can be serious and sometimes fatal.

Some fungi produce poisonous compounds collectively called **mycotoxins.** A few species of *Aspergillus,* for example, produce potent mycotoxins called *aflatoxins* that harm the liver and are known carcinogens. Foods on which aflatoxin-producing fungi commonly grow include peanuts, pecans, corn, and other grains. Other foods that may contain traces of aflatoxins include animal products such as milk, eggs, and meat (from animals that consumed feed contaminated by aflatoxin). Avoiding aflatoxin in the diet is impossible, but exposure should be minimized as much as possible. Any human food or animal forage product that has become moldy should be suspected of aflatoxin contamination and should be discarded.

Fungi contribute to sick-building syndrome, a situation in which occupants of a building experience acute adverse health effects linked to the time they spend in a given building. Mold-related insurance claims amount to hundreds of millions of dollars each year. When conditions are moist, molds can grow on carpets, leather, cloth, wood, insulation, and food. Mold spores, fragments, and aerosol mold products make their way into the air, and people are exposed through inhalation as well as by skin contact.

Exposure to molds and their toxins has been linked to depressed immune function, irritation of the throat and respiratory passageways, infection, and toxicity. The most common responses to mold exposure are reactions that range from mild to severe illnesses, including hay fever, sinusitis, asthma, and dermatitis.

Fungi cause many important plant diseases

Fungi are more destructive to plants than any other disease-causing organism. They are responsible for about 70% of all major crop diseases. Fungi cause serious epidemic diseases that spread rapidly and often result in complete crop failure. Fungal plant diseases costs billions of dollars in agricultural damage yearly.

All plants are apparently susceptible to some fungal infection. Damage may be localized in certain tissues or structures of the

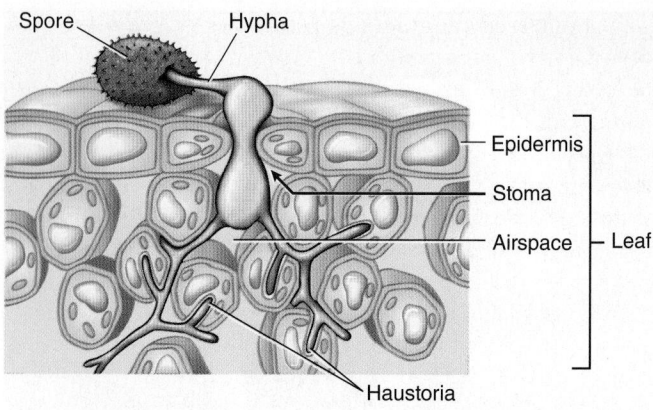

Figure 29-23 How a fungus parasitizes a plant
In this example, the hypha enters the leaf through a stoma. The hypha grows, branching extensively through the internal air spaces, and penetrates plant cells with specialized hyphal extensions called haustoria.
© Cengage Learning

plant, or the disease may be systemic and spread throughout the entire plant. Fungal infections may cause stunting of plant parts or of the plant and may cause wartlike growths or kill the plant.

A plant often becomes infected after hyphae enter through stomata (pores) in the leaf or stem or through wounds in the plant body (FIG. 29-23). Alternatively, the fungus may produce *cutinase,* an enzyme that dissolves the waxy cuticle that covers the surface of leaves and stems. After dissolving the cuticle, the fungus easily invades the plant tissues. As the mycelium grows, it may stay mainly between the plant cells, or it may penetrate the cells. Parasitic fungi often produce special hyphal branches called **haustoria** (sing., *haustorium*) that penetrate the host cells and obtain nourishment from the cytoplasm.

(a) Brown rot of peaches. This disease is caused by *Monilinia fruticola,* an ascomycete. Photographed in Oregon.

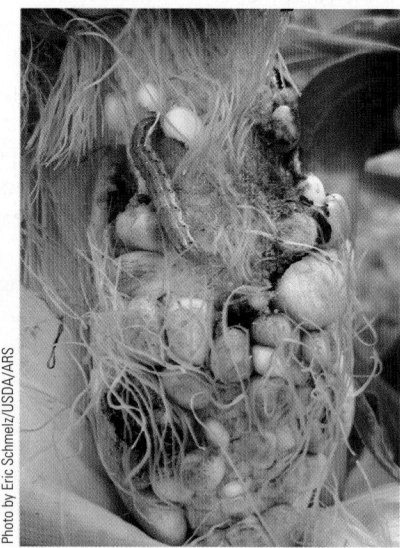

(b) Corn smut on an ear of sweet corn. This fungal disease is caused by *Ustilago maydis,* a basidiomycete.

(c) Black stem rust infection on wheat (*Triticum* sp.) stem. This plant disease is caused by *Puccinia graminis,* a basidiomycete.

Figure 29-24 Fungi that cause plant diseases

Ascomycetes cause serious plant diseases, including powdery mildew, chestnut blight, Dutch elm disease, apple scab, wilt on potatoes, and brown rot, which attacks cherries, peaches, plums, and apricots (FIG. 29-24a).

Basidiomycetes cause smuts and rusts that attack corn, wheat, oats, and other grains (FIGS. 29-24b and c). Some fungal parasites, such as the basidiomycete *Puccinia graminis,* which causes black stem rust of wheat, have complex life cycles that involve two or more different host plants and the production of several kinds of spores. Before the late 1950s, black stem rust outbreaks occurred every few years somewhere in the world, destroying entire wheat crops. By the early 1960s, Norman Borlaug, winner of the Nobel Peace Prize in 1970, and other researchers had developed rust-resistant varieties of wheat, heralding in the green revolution. However, new mutations have put *Puccinia* and wheat rust back in the news. The new strain has spread through parts of Africa and into the Middle East. Each fungus releases billions of spores, which the wind can blow for hundreds of miles, spreading the fungus to new regions of the world. Its spread may lead to widespread food shortages. The genomes of several *Puccinia* species have been sequenced, and researchers are developing methods to quickly diagnose the diseases they cause. They are also working to develop new varieties of disease-resistant wheat.

CHECKPOINT 29.5

- **CONNECT** *In what ways (both positive and negative) are fungi important in modern biology and medicine?*
- **CONNECT** *Can fungi be beneficial to plants? In what ways are they harmful?*

SUMMARY: FOCUS ON LEARNING OBJECTIVES

29.1 Characteristics of Fungi (page 598)

1 Describe the distinguishing characteristics of fungi.
- **Fungi** are eukaryotic heterotrophs that secrete digestive enzymes onto their food source and then absorb the predigested food. Fungi are characterized by cell walls that contain **chitin.**

2 Describe the body plan of a fungus.
- A fungus may be a unicellular **yeast** or a filamentous, multicellular *mold*. The body of most multicellular fungi consists of long, threadlike filaments called **hyphae** that branch and form a tangled mass called a **mycelium.**
- In most fungi perforated **septa**, or cross walls, divide the hyphae into individual cells. In some fungi the hyphae are **coenocytes** that form an elongated, multinuclear cell.

29.2 Fungal Reproduction (page 599)

3 Describe the life cycle of a typical fungus, including sexual and asexual reproduction.
- Most fungi reproduce both sexually and asexually by means of **spores.** Spores are produced on aerial hyphae. When fungal spores land in a suitable spot, they germinate.
- When fungi of two different mating types meet, their hyphae fuse, a process called **plasmogamy.** The cytoplasm fuses, but the nuclei remain separate. The fungi enter a **dikaryotic** (n + n) stage in which each new cell formed has one nucleus of each type.
- **Karyogamy,** fusion of the nuclei, takes place in the hyphal tip and results in a diploid (2n) *zygote nucleus.*
- Meiosis produces four genetically different haploid (n) nuclei. Each nucleus becomes part of a spore. When the spores germinate, they form new mycelia by mitosis.
- Genetically similar asexual spores are produced by mitosis. When these spores germinate, they also develop into mycelia.

29.3 Fungal Diversity (page 601)

4 Give arguments to support the hypothesis that fungi are opisthokonts, more closely related to animals than to plants.
- Like animals, some fungi have flagellate cells—for example, chytrid gametes and spores—and the flagellate cells propel themselves with a single posterior flagellum. Also like animal cells, fungal mitochondria have platelike cristae.
- Based on chemical and structural characters, fungi are classified, along with animals and choanoflagellates, as **opisthokonts.**

5 Give arguments to support the hypothesis that chytrids may have been the earliest fungal group to evolve from the most recent common ancestor of fungi.
- **Chytrids,** or **chytridiomycetes,** produce flagellate spores at some stage in their life cycle. No other fungi have flagella. Thus, chytrids probably were the earliest fungi to evolve; the most recent common ancestor of all fungi was a flagellate protist.

6 List distinguishing characteristics, describe a typical life cycle, and give examples of each of the following fungal groups: chytridiomycetes, zygomycetes, glomeromycetes, ascomycetes, and basidiomycetes.

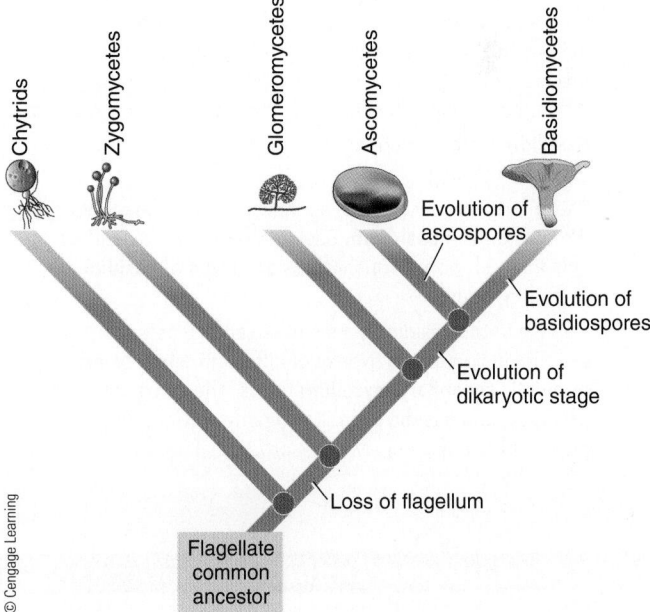

- Chytrids reproduce both asexually and sexually. Their gametes and zoospores are flagellate. *Allomyces,* a common chytrid, spends part of its life as a multicellular haploid **thallus** and part as a multicellular diploid thallus. The haploid thallus produces

The Fungi   **619**

two types of flagellate gametes that fuse. Both plasmogamy and karyogamy occur, producing a flagellate zygote. The diploid thallus bears zoosporangia that produce diploid **zoospores** and resting sporangia in which haploid zoospores form by meiosis. The haploid zoospores form new haploid thalli.

- **Zygomycetes,** such as the black bread mold, *Rhizopus,* form a haploid thallus that produces both asexual spores and sexual spores. Asexual spores germinate and form new thalli. In sexual reproduction hyphae of two different haploid mating types form gametangia. Plasmogamy occurs as the gametangia fuse. Karyogamy occurs, and a diploid zygote is formed; the zygote develops into a **zygospore.** Meiosis produces recombinant haploid zygospores. When zygospores germinate, each hypha develops a sporangium at its tip. Spores are released and develop into new hyphae.

- **Microsporidia,** currently classified as zygomycetes, are opportunistic pathogens that penetrate and infect animal cells with their long, threadlike *polar tubes.*

- **Glomeromycetes** have coenocytic hyphae. They reproduce asexually with large, multinucleate spores called blastospores. Glomeromycetes are symbionts that form intracellular associations called **mycorrhizae** with the roots of plants. Because they extend their hyphae into root cells, glomeromycetes are **endomycorrhizal fungi.** The most common endomycorrhizae are called *arbuscular mycorrhizae* because the hyphae inside the root cells form branched, tree-shaped structures known as **arbuscules.**

- **Ascomycetes** include yeasts, cup fungi, morels, truffles, and blue-green, pink, and brown molds. Some ascomycetes form mycorrhizae; others form lichens. Ascomycetes produce asexual spores called **conidia;** they produce sexual spores called **ascospores** in saclike **asci.** The asci line a fruiting body called an **ascocarp.**

- In ascomycetes haploid mycelia of opposite mating types produce septate hyphae. Plasmogamy occurs, and nuclei are exchanged. A dikaryotic ($n + n$) stage occurs in which hyphae form and produce asci and an ascocarp. Karyogamy occurs, followed by meiosis. The recombinant nuclei divide by mitosis, producing eight haploid nuclei that develop into ascospores. When the ascospores germinate, they can form new mycelia.

- **Basidiomycetes** include mushrooms, puffballs, bracket fungi, rusts, and smuts.

- These fungi produce sexual spores called **basidiospores** on the outside of a **basidium.** Basidia develop on the surface of **gills** in mushrooms; mushrooms are a type of **basidiocarp** (a fruiting body).

- Hyphae in the basidiomycetes have septa. Plasmogamy occurs with the fusion of two hyphae of different mating types. A dikaryotic **secondary mycelium** forms. Then a basidiocarp develops, and basidia form. Karyogamy occurs, producing a diploid zygote nucleus. Meiosis produces four haploid nuclei

that become basidiospores. When basidiospores germinate, they form haploid **primary mycelia.**

29.4 Ecological Importance of Fungi *(page 611)*

7 Summarize the ecological significance of fungi as decomposers.

- Most fungi are decomposers that break down organic compounds in dead organisms, leaves, garbage, and wastes into simpler nutrients that can be recycled.

8 Describe the important ecological role of mycorrhizae.

- Mycorrhizae are mutualistic associations between fungi and the roots of plants. The fungus supplies water and nutrient minerals to the plant, and the fungus obtains organic compounds from the plant. Glomeromycetes form endomycorrhizae with roots. Some ascomycetes and basidiomycetes are **ectomycorrhizal fungi** that form mycorrhizae when their hyphae coat tree roots, but do not penetrate the root cells.

9 Characterize the unique nature of a lichen.

- A **lichen** is a combination of a fungus and a photoautotroph (an alga or cyanobacterium). In this symbiotic relationship, the photoautotroph provides the fungus with organic compounds. The fungus may provide the photoautotroph with shelter, water, and minerals. Lichens have three main forms: crustose, foliose, and fruticose.

29.5 Economic, Biological, and Medical Impact of Fungi *(page 615)*

10 Summarize some specific ways that fungi affect humans economically.

- Fungi cause huge economic losses by damaging food and crops. On the other hand, some fungi, like mushrooms, are foods; others, like yeasts, are used to produce beer, wine, and bread; and still others are used to produce cheeses or industrial chemicals.

11 Summarize the importance of fungi to biology and medicine; describe how fungi infect plants and humans, describing at least one fungal animal disease and one fungal plant disease.

- Biologists use the yeast *Saccharomyces cerevisiae* and other fungi as model organisms for research in molecular biology and genetics. Fungi are also being investigated for the biological control of insects.

- Fungi are used to make many medications, including penicillin and other antibiotics; they are used in bioremediation and to control pests biologically.

- Fungi are opportunistic pathogens in humans. They cause human diseases, such as histoplasmosis; some fungi produce **mycotoxins,** such as *aflatoxins,* which can cause liver damage and cancer.

- Fungal hyphae infect plants through stomata. Hyphal branches called **haustoria** penetrate plant cells and obtain nourishment from the cytoplasm. Fungi cause many important plant diseases, including brown rot, corn smut, and wheat rust.

TEST YOUR UNDERSTANDING

Know and Comprehend

1. Fungi are (a) eukaryotes and opisthokonts (b) prokaryotes and opisthokonts (c) flagellate and dikaryotic (d) autotrophic eukaryotes (e) heterotrophs with cellulose cell walls

2. Which of the following fungi does *not* have a mycelium? (a) black bread mold (b) yeast (c) decomposer cup fungus (d) cultivated mushroom (e) *Penicillium*

3. A cell described as *n* + *n* is (a) monokaryotic (b) diploid (c) haploid (d) coenocytic (e) dikaryotic

4. With the exception of chytridiomycetes, fungi are generally disseminated by (a) water currents (b) fragmentation of hyphae (c) soredia (d) airborne spores (e) flagellate zoospores

5. Which statement is *not* true of the chytrids? (a) they are simple aquatic fungi (b) they produce motile cells with single, posterior flagella (c) they undergo both sexual and asexual reproduction (d) their cells are dikaryotic (e) they were the earliest fungi to evolve

6. Which statement is *not* true of the zygomycetes? (a) many members of this group form endomycorrhizae with tree roots (b) their sexual spores are called zygospores (c) they undergo both sexual and asexual reproduction (d) plasmogamy and karyogamy occur (e) they have coenocytic hyphae

7. Glomeromycetes (a) reproduce mainly by sexual spores called glomerospores (b) are characterized by unique structures called polar tubes (c) associate with cyanobacteria to form lichens (d) include many opportunistic pathogens that cause human disease (e) form arbuscular endomycorrhizae with tree roots

8. The ascomycete life cycle typically includes (a) mainly diploid thalli (b) the formation of a thick zygosporangium (c) the production of eight haploid ascospores within an ascus (d) the production of microsporidia (e) the production of ascospores, zoospores, and conidia at different stages

9. Which statement is *not* true of the basidiomycetes? (a) they produce a secondary mycelium with *n* + *n* hyphae (b) their sexual spores are called basidiospores (c) they have a diploid thallus that produces zoospores (d) reproduction is mainly sexual (e) they have microscopic basidia

10. A combination organism consisting of a photoautotroph and a fungus is called (a) an arbuscular endomycorrhiza (b) an ectomycorrhiza (c) a lichen (d) a pathogenic agent (e) an aflatoxin

11. Mutualistic associations between fungi and the roots of plants are called (a) lichens (b) mycorrhizae (c) pathogenic associations (d) parasitic haustoria (e) mycotoxic symbioses

12. When a fungus infects a plant, it (a) infiltrates leaves with lichens (b) forms relationships by attaching mycorrhizae to stems (c) secretes powerful digestive juices onto the leaves (d) uses haustoria to dissolve roots (e) enters leaves or stems through stomata

Apply and Analyze

13. **VISUALIZE** The secondary mycelium of a basidiomycete is shown here. (a) How would you describe its cells? (b) What stage of its life cycle is it in? (c) Draw the next steps in this stage of its life cycle.

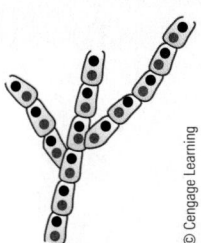
© Cengage Learning

14. Explain the statement "Mushrooms are like the tips of icebergs." If you do not see mushrooms in your lawn, can you conclude that no fungi live there? Why or why not?

Evaluate and Synthesize

15. **EVOLUTION LINK** Justify (a) classifying fungi as opisthokonts, (b) classifying microsporidia as fungi, and (c) grouping ascomycetes and basidiomycetes as sister clades.

16. **CONNECT** The development of safe and effective antifungal agents presents a daunting challenge to researchers. Explain. (*Hint:* Drugs that combat bacterial diseases typically target differences between bacterial cells and those of the host.)

17. **SCIENCE, TECHNOLOGY, AND SOCIETY** Develop an argument for or against the use of scientific resources, including the latest technology, to sequence the genomes of fungi.

aplia To access course materials, such as Aplia and other companion resources, please visit **www.cengagebrain.com.**

30 | An Introduction to Animal Diversity

The tube sponge (*Callyspongia vaginalis*). This animal, sometimes mistaken for a plant, ranges in color from purple to blue to gray. It is common on coral reefs in the Caribbean, from Florida to Mexico.

Daniela Dirscherl/Getty Images

KEY CONCEPTS

30.1 Animals are multicellular, eukaryotic heterotrophs composed of cells specialized to perform specific functions. Most animals are diploid organisms that reproduce sexually, and most have a nervous system and a muscular system.

30.2 Animals evolved in marine environments, and members of most animal phyla still inhabit marine environments. However, many animals are adapted to life in fresh water and others to terrestrial habitats.

30.3 The common ancestors of animals are choanoflagellates; choanoflagellates, fungi, and animals are a monophyletic group known as opisthokonts.

30.4 Biologists classify animals based on many characteristics, including their morphology (structure), features of their early development, and molecular data; they generally agree that bilateral animals split into at least three major clades.

Although members of most animal species are readily recognizable as animals, the identity of some others is less obvious. Early naturalists thought sponges were plants because they did not move from place to place. Some people still mistake certain marine animals, such as sponges and corals, for plants (see photograph; also see the image of the sea anemone on the cover of your textbook). Locomotion is not a requirement for being classified as an animal.

Animal phylogeny is an exciting and rapidly changing field of study. Biologists have described and named more than 1.5 million species of animals, and 15,000 to 20,000 new species are named each year. Millions more probably remain to be discovered and classified. Interestingly, an estimated 99% of all animal species that ever inhabited our planet are extinct. Taxonomists have assigned the extant (living) animals to about 35 phyla. Molecular studies have confirmed that many of these groups are **monophyletic;** that is, they consist of *all* the descendants and *only* the descendants of a common ancestor. (Recall from Chapter 23 that a monophyletic group is called a **clade.**)

Animal groups that are not monophyletic have been split or reorganized, with some members of the groups being reassigned to other taxa. It is important to remember that the classification of animals and the relative positions of animal groups are a work in progress. As they consider new data, systematists redraw the tree of animal life.

In this chapter we discuss the characteristics of animals and their habitats. We then explore animal origins and some of the criteria biologists use to determine evolutionary relationships and to classify animals. Finally, we introduce the major animal groups, including the three major clades of bilateral animals.

30.1 ANIMAL CHARACTERISTICS

LEARNING OBJECTIVE

1 Describe several characteristics common to most animals.

Animals are so diverse that for almost any definition we can find exceptions. We can best describe animals by the characteristics they share:

1. Animals are multicellular eukaryotes. In contrast to plants, algae, and fungi, animal cells lack cell walls. Instead, structural support depends on an **extracellular matrix,** which the cells secrete (see Chapter 4). Collagen, the main structural protein in the extracellular matrix, forms very tough fibers. Collagen is an important *shared derived character* in animals (see Chapter 23).
2. Animals are **heterotrophs.** As consumers, they depend on producers for their raw materials and energy. In contrast to the fungi, most animals ingest their food first and then digest it inside the body, usually within a digestive system.
3. Cells that make up the animal body are specialized to perform specific functions. In all but the simplest animals, cells are organized to form tissues, and tissues are organized to form organs. In small animals with simple body plans, life processes such as gas exchange, circulation of materials, and waste disposal can take place by diffusion of gases and other substances directly to and from the environment. In larger animals, specialized organ systems perform these functions.
4. Animals have diverse body plans. The term *body plan* refers to the basic structure and functional design of the body. An animal's body plan and lifestyle are adapted to its methods of obtaining food and reproducing.
5. Most animals are capable of locomotion at some time during their life cycle. Some animals (such as sponges and corals) move about as larvae (immature forms) but are **sessile** (firmly attached to the ground or some other surface) as adults (see chapter-opening photograph).
6. Most animals have nervous systems and muscle systems that enable them to respond rapidly to stimuli in their environment.
7. Most animals are diploid organisms that reproduce sexually, with large, nonmotile eggs and small, flagellate sperm. A haploid sperm unites with a haploid egg, forming a diploid **zygote** (fertilized egg).
8. Animals go through a period of embryonic development. The zygote undergoes **cleavage,** a series of mitotic cell divisions. During cleavage the zygote develops into a hollow ball of cells called a **blastula.** Although some animals develop directly into adults, the majority first develop into a **larva,** a sexually immature form that may look very different from the adult (FIG. 30-1). The larva differs from the adult in many ways, including where it lives (its habitat), how it moves, and what it eats. Larvae typically go through **metamorphosis,** a developmental process that converts the immature animal into a juvenile form that can then grow into an adult.

CHECKPOINT 30.1

- **CONNECT** *For centuries, scientists classified sponges as plants, but now they are classified as animals. What characteristics do sponges share with other animals?*

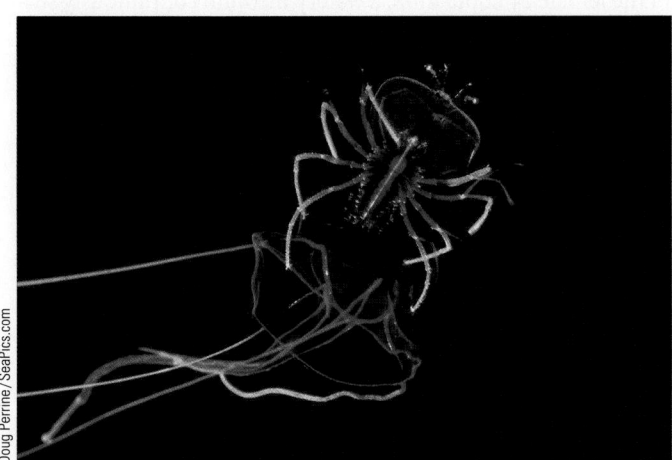

(a) Spiny lobster larva. This larva is hitching a ride on a jellyfish. Over a period of months, the larva passes through several stages before becoming an adult. Photographed in open ocean at night, Hawaii.

Doug Perrine/SeaPics.com

(b) Adult spiny lobster. The adult has two long antennae used to sense movement. Spiny lobsters lack large claws. The spines on their back help protect them. In this image you can see the two large spines above the animal's eyes. Spiny lobsters leave their hiding places at night to hunt.

Dr. Magnus Kjaergaard

Figure 30-1 Larva and adult spiny lobster (*Panulirus* sp.)

Most animals go through a larva stage before developing into an adult. The larva usually differs from the adult in size, appearance, and lifestyle.

30.2 ADAPTATIONS TO OCEAN, FRESHWATER, AND TERRESTRIAL HABITATS

LEARNING OBJECTIVE

2 Compare the advantages and disadvantages of life in the ocean, in fresh water, and on land.

Fossil evidence suggests that animals first evolved in shallow, marine environments during the Proterozoic eon, at least 600 million years ago (mya; see Chapter 21). Although animals are now distributed in virtually every environment, at least some members of most animal phyla still inhabit marine environments.

Marine habitats offer many advantages

The buoyancy of sea water provides support, and its large volume keeps the water temperature relatively stable. The body fluids of most invertebrates have about the same osmotic concentration as sea water, so fluid and salt balance are more easily maintained there than in fresh water. **Plankton,** which consists of the mainly microscopic animals and protists that are suspended in water and float with its movement, provides a ready source of food for many aquatic animals.

Life in the ocean also presents some challenges. Although the continuous motion of water brings nutrients to animals and washes their wastes away, animals must be able to cope with the water's movements and the currents that could sweep them away. Squids, fishes, and marine mammals have evolved as strong swimmers, usually able to direct their movements and maintain their location. However, most invertebrates and young vertebrates cannot swim strongly, and they have adapted in various ways to the tides and currents.

Some sessile animals attach permanently to a stable structure such as a rock. Others burrow in the sand and silt that cover the sea bottom. Many invertebrates have adapted by maintaining a small body size and becoming part of the plankton. As they are tossed about, their food supply continues to surround them.

Some animals are adapted to freshwater habitats

Far fewer kinds of animals make their homes in fresh water than in the ocean because living in this habitat is more difficult. Fresh water is hypotonic to the tissue fluids of animals, so water tends to move into the animal by osmosis. To survive in this habitat, freshwater species must have mechanisms for removing excess water while retaining salts. This *osmoregulation* requires an expenditure of energy.

Fresh water offers a much less constant environment than sea water. Animals that inhabit fresh water must have adaptations for surviving variations in oxygen content, temperature, turbidity (because of sediments suspended in the water), and even water volume. In addition, fresh water generally contains less food than the sea.

Terrestrial living requires major adaptations

Living on land is even more difficult than living in fresh water, and the evolution of terrestrial animals involved major adaptations. Analyzing the fossil record, many biologists hypothesize that the first air-breathing terrestrial animals were scorpionlike arthropods that came ashore in the Silurian period about 444 mya. The first vertebrates to inhabit terrestrial environments, the amphibians, did not appear until the Devonian period, about 30 million years later.

The chief problem facing all terrestrial organisms is desiccation (drying out). Water is constantly lost by evaporation and is often difficult to replace. A body covering adapted to minimize fluid loss helps solve this problem in many terrestrial animals (FIG. 30-2). Location of the respiratory surface deep within the animal also helps prevent fluid loss. Thus, the gills of aquatic animals are typically located externally, but lungs and tracheal tubes of terrestrial animals are typically found deep within the body.

Reproduction on land also poses challenges to protecting gametes and the developing offspring from desiccation. Aquatic animals typically shed their gametes in the water, where fertilization occurs. Some land animals, including most amphibians, return to the water for reproduction, and their larval forms develop in the water.

The evolution of internal fertilization has permitted many terrestrial animals, including land planarians, earthworms, land snails, insects, reptiles (including birds), and mammals, to meet the desiccation challenge. Because these terrestrial animals transfer sperm from the body of the male directly into the body of the female by copulation, a watery medium continuously surrounds the sperm. Another important adaptation to reproduction on land is the tough, protective shell that surrounds the eggs of many species (see Fig. 30-2). Secreted by the female, this

Figure 30-2 Adaptations to terrestrial life
The tough, horny skin of the green iguana (*Iguana iguana*) has scales and is water resistant. Leathery eggs protect the embryos from drying out.

shell protects the developing embryo from drying out. An alternative adaptation for terrestrial reproduction is development of the embryo within the moist body of the mother.

Water has buoyancy that helps support animals that inhabit aquatic environments. Air is less dense than water, and to inhabit the land, animals must have structures, such as a skeletal system and muscles, that support the body. The temperature extremes of terrestrial habitats also present challenges. In later chapters we discuss behavioral and physiological adaptations for maintaining body temperature.

CHECKPOINT 30.2

- *What are some advantages of marine environments over freshwater and terrestrial habitats?*
- CONNECT *What are some animal adaptations to the terrestrial environment?*

30.3 ANIMAL EVOLUTION

LEARNING OBJECTIVE

3   Use current hypotheses to trace the early evolution of animals.

Biologists generally agree that animals share a common ancestor with a group of protists known as *choanoflagellates* (see Fig. 26-21). The cells of these colonial flagellates became specialized to perform specific functions, such as movement, feeding, or reproduction. As this division of labor evolved, a colony of flagellates reached the level of cooperation and coordination that qualified it to be considered a single organism, the first animal. The choanoflagellates, fungi, and animals are a monophyletic group known as **opisthokonts.** Recall from Chapter 26 that opisthokonts are characterized by a posterior flagellum on motile cells.

Historically, biologists depended on fossils, on similarities in body plan (i.e., structure), and on patterns of development to determine evolutionary relationships among various groups of animals. **Molecular systematics,** the science that focuses on molecular structure to clarify evolutionary relationships, has provided additional data that are critical in answering questions about phylogeny. In many cases, molecular data have confirmed hypotheses that were based on morphology (structure).

Complex genomes were apparently present early during animal evolution. Molecular studies suggest that the ancestor of animals had more than 1500 genes not found in other eukaryotes. Some of these genes may be traced to horizontal gene transfer from other domains, followed by modification of the genes. Complex genomes have been described in the sea anemone and in other animals that are relatively simple morphologically.

Molecular analyses indicate that the structure of genes that control development, RNA molecules, and many other molecules are very similar among all animal groups that have been studied. According to the *principle of parsimony,* such complex molecules are unlikely to have evolved multiple times

(see Chapter 23). Thus, these data support the hypothesis that animals evolved only once. Animals are a monophyletic group.

Molecular systematics helps biologists interpret the fossil record

Because early animals were soft-bodied forms that left few fossils, the evolutionary history of animals has been vigorously debated. The scarcity of fossils has made it difficult to determine the age, rate of divergence, and number of branches of animal groups. In 2009, a research team found fossil traces in an oil field on the Arabian Peninsula that are thought to date back more than 635 million years. The fossil traces are steroids found only in the skeletal structures of certain sponges (demosponges). Before this discovery, the earliest known animal fossils were the *Ediacaran biota* from the **Ediacaran period** (635 mya to 541 mya). These fossils of small, simple animals suggest that sponges, jellyfish, and comb jellies were present during this period (see Fig. 21-9).

Paleontologists have discovered many large, complex animal fossils in Chengjiang, an Early Cambrian (542 mya to 520 mya) fossil site in China, and in the Burgess Shale in British Columbia, a Middle Cambrian fossil site (520 mya to 515 mya). Fossils of most extant phyla (and also many extinct animals) have been found at these sites. The rapid appearance of an amazing variety of body plans during this time is known as the **Cambrian radiation,** or less formally as the **Cambrian explosion** (see Fig. 21-10). According to the Cambrian radiation hypothesis, which is based on the fossil record, major modifications in body plan that occurred during this time account for many branches of the animal tree.

Studies of large molecular data sets suggest that most animal clades actually diverged over a very long period during the Proterozoic eon (2.5 bya to 541 mya). Thus, the animal phyla that first left fossils during the Cambrian radiation may have evolved several hundred million years *before* they appear in the fossil record. Biologists estimate that certain groups are about twice as old as the oldest fossils found to date. According to this view, the Cambrian radiation was a rapid evolution of new animal body plans among clades that already existed. Perhaps fossils of these early animals remain to be discovered in Proterozoic rocks. Another hypothesis holds that a change in environmental conditions that occurred prior to the Cambrian radiation allowed fossils to form.

Biologists develop hypotheses about the evolution of development

Changes in animal body plans are linked to changes in patterns of embryonic development. Biologists have long used similarities and differences in embryonic development to hypothesize how animal groups are related. Traditionally, biologists depended mainly on structural changes to compare the process of development in various groups.

Today, researchers are focusing on the molecular basis of developmental processes. They have identified the genes that direct the early development of the body plan and have discovered that many of these genes have been conserved during animal evolution. The same basic set of genes controls early development in all animal groups. Furthermore, the same genes are used in the same ways to regulate development.

Evolutionary developmental biology, sometimes referred to as **Evo Devo,** has become an important approach to studying animal relationships. Biologists compare molecular events, such as gene regulation during development, in various animal groups.

Recall from Chapter 17 that *Hox* **genes** are a group of regulatory genes that specify the anterior–posterior axis during development (see Fig. 17-13). The presence and number of *Hox* genes provide insights about evolutionary relationships. These genes have been identified in all the bilateral animal groups that have been studied, suggesting that the last common ancestor of all bilateral animals had similar *Hox* genes. These genes have been identified in a sea anemone (*Nematostella vectensis*), which is a cnidarian. Cnidarians are marine animals, for example, jellyfish, with radial symmetry. This finding suggests that the cnidarians share a common ancestor with the bilateral animals.

Investigators think that all the *Hox* gene groups had evolved by the beginning of the Cambrian period. Mutations in *Hox* genes could have resulted in rapid changes in animal body plans. For example, regulation by *Hox* gene groups has been linked with the development of wings or legs. Similarities in molecular development among different animal groups suggest that they had a common ancestor.

CHECKPOINT 30.3

- *What was the Cambrian radiation?*
- *According to the current hypothesis, when did most major groups of animals evolve?*
- CONNECT *How has the discovery of* Hox *genes helped biologists understand animal evolution?*

30.4 RECONSTRUCTING ANIMAL PHYLOGENY

LEARNING OBJECTIVES

4 Describe how biologists use morphology (including variations in body symmetry, number of tissue layers, and type of body cavity) and patterns of early development to infer relationships among animal phyla.

5 Cite specific examples of how data from molecular systematics have confirmed or modified traditional animal phylogeny and identify the three major clades of bilateral animals.

Because some animal body plans have been highly conserved throughout evolutionary history, variations in body plans can provide clues to animal relationships. For example, biologists compare variations in body symmetry, number of tissue layers,

types of body cavity, and pattern of development. Biologists also use similarities and differences in embryonic development to infer evolutionary relationships among animal groups. In addition to these traditional methods, researchers now have molecular tools to enhance our understanding of animal phylogeny. Because of technological improvements over the last few decades, comparisons of nucleic acid (DNA and RNA) and protein structure provide critical data for biologists seeking to interpret and reconstruct animal phylogeny.

Animals exhibit two main types of body symmetry

Symmetry refers to the arrangement of body structures in relation to the body axis. Most sponges are not symmetrical, so when a sponge is cut in half, the two halves are not similar to each other. Most other animals exhibit either radial or bilateral body symmetry.

Cnidarians (jellyfish, sea anemones, and their relatives) and adult echinoderms (sea stars and their relatives) have **radial symmetry.** The body has the general form of a wheel or cylinder, and similar structures are regularly arranged as spokes from a central axis (FIG. 30-3a). Multiple planes can be drawn through the central axis, each dividing the organism into two mirror images. An animal with radial symmetry receives stimuli equally from all directions in the environment. Some animals have modified radial symmetry. For example, sea anemones and ctenophores (comb jellies) have *biradial symmetry,* in which parts of the body have become specialized so that only two planes can divide the body into similar halves.

Most animals exhibit **bilateral symmetry,** at least in their larval stages. A bilaterally symmetrical animal can be divided through only one plane (which passes through the midline of the body) to produce roughly equivalent right and left halves that are mirror images (FIG. 30-3b).

As bilateral symmetry evolved, natural selection led to **cephalization,** the development of a head where sensory structures are concentrated. In these groups, concentrations of nerve cells in the head form a brain, and one or two nerve cords extend from the brain toward the rear end of the animal. Bilateral symmetry and cephalization are adaptations for locomotion. The head end of the animal meets its environment first and is best equipped to capture food or respond to danger.

Some definitions of basic terms and directions will help in locating body structures in bilaterally symmetrical animals. The back surface of an animal is its **dorsal** surface; the underside (belly) is its **ventral** surface. **Anterior** (or *cephalic*) means toward the head end of the animal; **posterior,** or *caudal,* means toward the tail end. A structure is said to be *medial* if it is located toward the midline of the body, and it is *lateral* if it is toward one side of the body; for example, the human ear is lateral to the nose. In human anatomy the term *superior* refers to a structure located above some point of reference, or toward the head end of the body. The term *inferior* is used in human anatomy to mean located below some point of reference, or toward the feet.

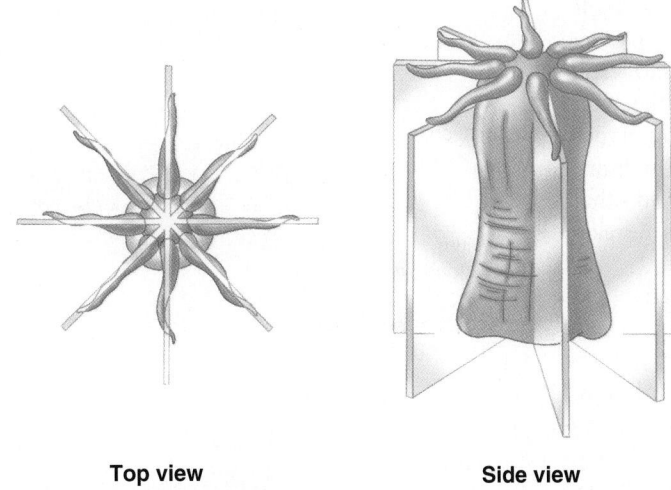

Top view   **Side view**

(a) Radial symmetry. Multiple planes can be drawn through the central axis; each divides the animal into two mirror images.

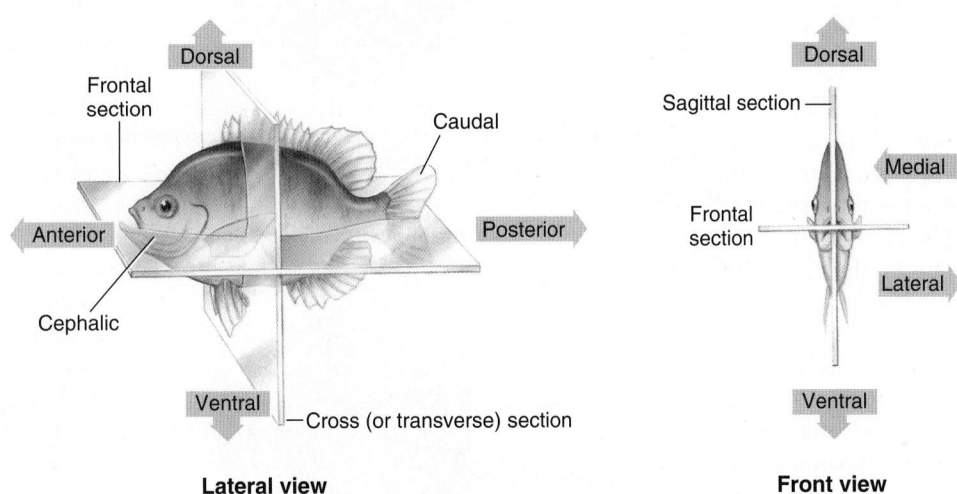

Lateral view   **Front view**

(b) Bilateral symmetry. The head of the animal is its anterior end, and the opposite end is its posterior end. The back of the animal is its dorsal surface, and the belly is its ventral surface. The diagrams illustrate various ways the body can be sectioned (cut) to study its internal structure. A sagittal section (lengthwise vertical cut) divides the animal into right and left parts. A frontal, or longitudinal, cut (lengthwise horizontal) divides the body into dorsal and ventral parts.

Figure 30-3 *Animation* **Radial and bilateral symmetry**

Sections of animals are used in illustrations throughout this book to show the structure and arrangement of tissues and organs.

© Cengage Learning

A bilaterally symmetrical animal has three axes, each at right angles to the other two: an anterior–posterior axis extending from head to tail, a dorsal–ventral axis extending from back to belly, and a left–right axis extending from side to side. We can distinguish three planes or sections that divide the body into specific parts. A *sagittal plane* divides the body into right and left parts. A sagittal plane passes from anterior to posterior and from dorsal to ventral. A *frontal plane* divides a bilateral body into dorsal and ventral parts. A *transverse section,* or *cross section,* cuts at right angles to the body axis and separates anterior and posterior parts.

Animal body plans are linked to the level of tissue development

Sponges have several types of cells, but their cells are not organized into **tissues,** which are groups of closely associated, similar cells that work together to carry out specific functions.

In the early development of all animals except sponges, cells form layers, called **germ layers.** The outer germ layer, or **ectoderm,** gives rise to the tissues that form the outer covering of the body as well as to nervous tissue. The inner layer, or **endoderm,** forms the lining of the digestive tube and other digestive structures. These layers develop into specific types of tissues.

Biologists describe cnidarians and ctenophores as **diploblastic** because they have only two germ layers. Other animals are **triploblastic.** They have a third germ layer, the **mesoderm,** which gives rise to most other body structures, including muscles, skeletal structures, and circulatory system (when present).

Most bilateral animals have a body cavity lined with mesoderm

The vast majority of bilateral animals have a fluid-filled body cavity, or **coelom** (pronounced "see´-lum"), between the outer wall of the body and the digestive tube (**FIG. 30-4**). The flatworms and ribbon worms are exceptions. They are bilateral and triploblastic but have a solid body; that is, they have no body cavity. They are referred to as **acoelomates** (*a-*, "without"; and *coelom,* "cavity").

Most animals have a body cavity that is completely lined with mesoderm. Such a body cavity is a *true* coelom. An animal with a true coelom is referred to as a **coelomate.** The coelom was one of the most important early animal adaptations. Evolution of the coelom was a critical step in the evolution of larger, more complex animals.

With the evolution of the coelom came a new body design, the *tube-within-a-tube* body plan. The coelom is a space that separates the body wall, the outer tube, from the digestive tube (gut), which is the inner tube. The digestive tube is attached to the body wall at its ends. Typically, the digestive tube has a mouth at one end for taking in food and an anus at the other end for eliminating wastes. Because the coelom separates the muscles of the body wall from those in the wall of the digestive tract, the digestive tube can move food along independently of body movements.

Because it is an enclosed compartment (or series of compartments) of fluid under pressure, the coelom can serve as a **hydrostatic skeleton** in which contracting muscles push against a tube of fluid. The hydrostatic skeleton also shapes the body of soft animals. The evolution of various shapes and divisions of the coelom provided the opportunity for animals to become

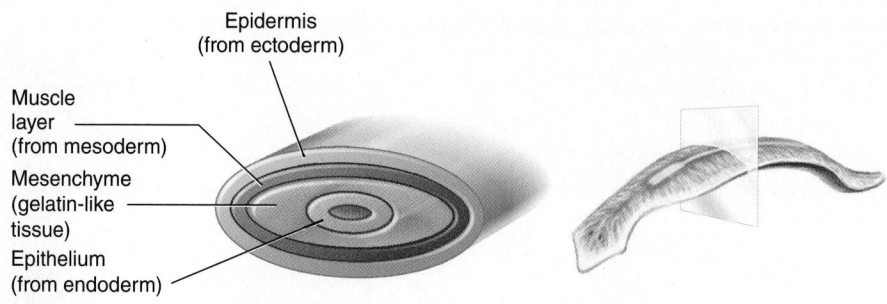

(a) Acoelomate: flatworm (planarian worm). Acoelomate animals have no body cavity.

Epidermis (from ectoderm)
Muscle layer (from mesoderm)
Mesenchyme (gelatin-like tissue)
Epithelium (from endoderm)

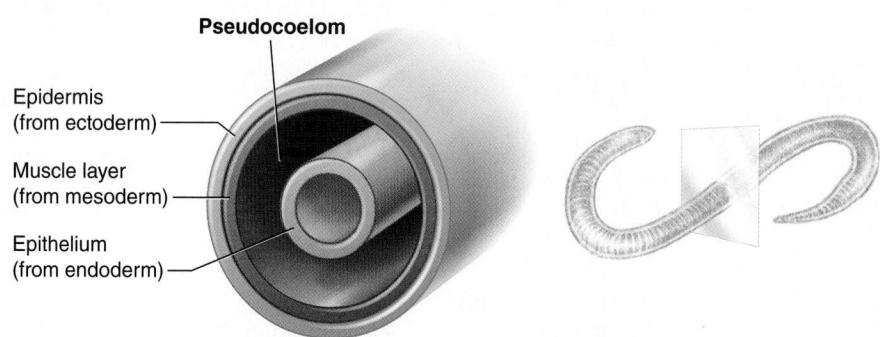

(b) Pseudocoelomate: nematode. Pseudocoelomate animals have a body cavity that is not completely lined with mesoderm.

Pseudocoelom
Epidermis (from ectoderm)
Muscle layer (from mesoderm)
Epithelium (from endoderm)

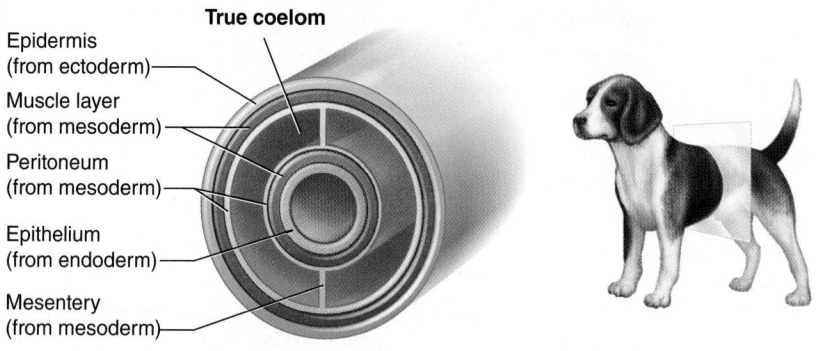

(c) True coelomate: vertebrate. Coelomate animals have a coelom, a body cavity completely lined with tissue that develops from mesoderm. Many internal organs are located in the coelom.

True coelom
Epidermis (from ectoderm)
Muscle layer (from mesoderm)
Peritoneum (from mesoderm)
Epithelium (from endoderm)
Mesentery (from mesoderm)

Figure 30-4 *Animation* **Three basic body plans in triploblastic animals**

The germ layer from which each tissue was derived is indicated in parentheses. Ectoderm is shown in *blue,* mesoderm in *red,* and endoderm in *yellow.*

© Cengage Learning

specialized in swimming, crawling, or walking. Animals that can move quickly to capture food or avoid predators are more likely than slower-moving animals to survive.

Evolution of the coelom provided a space where internal organs could develop and function. For example, the pumping action of the heart is possible because of the surrounding space the coelom provides. The fluid-filled coelom also protects internal organs by cushioning them. (Other functions of the coelom will be discussed in Chapter 31.)

Some (typically, small) animals have a body cavity that is not completely lined with mesoderm. This type of body cavity is called a **pseudocoelom** ("false coelom"). Animals with a pseudo-coelom, such as nematodes (roundworms) and rotifers, are called **pseudocoelomates.** Because the pseudocoelom appears to have evolved independently in different animal taxa, the presence of a pseudocoelom is not considered a useful characteristic for determining phylogeny. Instead, it is an example of homoplasy.

Bilateral animals form two main clades based on differences in development

Embryonic development begins as the zygote undergoes cleavage, the first several cell divisions of the embryo. During cleavage, the embryo develops into a hollow ball of cells, the blastula. Cells of the blastula undergo **gastrulation,** a process that forms and segregates the three germ layers.

Basic differences in the pattern of early development distinguish two main evolutionary lines of bilateral animals: **protostomes** (pro´-tuh-stomes), assigned to the clade *Protostomia,* and **deuterostomes** (doo´-ter-uh-stomes), assigned to the clade *Deuterostomia.* The protostomes include mollusks (e.g., snails, clams, squids), annelids (e.g., earthworms), arthropods (e.g., crabs, insects), and several other groups. Deuterostomes include the echinoderms (such as sea stars and sea urchins) and chordates (which include the vertebrates).

One important difference in the development of protostomes and deuterostomes is the pattern of cleavage. In many protostomes the early cell divisions are diagonal to the polar axis (the long axis of the embryo), resulting in a somewhat spiral arrangement of cells; any one cell lies between the two cells above or below it (FIG. 30-5a). This pattern of division is known as **spiral cleavage.** In **radial cleavage,** characteristic of the deuterostomes, the early divisions are either parallel or at right angles to the polar axis. The resulting cells lie directly above or below one another (FIG. 30-5b).

In the protostomes the developmental fate of each embryonic cell is typically fixed very early. For example, if the first four cells of an annelid embryo are separated, each cell develops into only a fixed quarter of the larva; this pattern of cleavage is called **determinate cleavage.** In contrast, deuterostomes typically undergo **indeterminate cleavage.** During early cleavage, each cell has the potential of developing into a complete embryo.

For example, if the first four cells of a sea star embryo are separated, each cell can form a complete, although small, larva. If a few cells are removed from a blastula undergoing indeterminate cleavage, other cells compensate, and the embryo develops normally. In contrast, if a few cells are removed from the blastula of an embryo undergoing determinate cleavage, some structure, such as a limb, does not develop.

During gastrulation, a group of cells moves inward, forming a sac that becomes the embryonic gut. The opening to the outside is called the **blastopore.** In most protostomes the blastopore develops into the mouth. The word *protostome* comes from Greek words meaning "first" and "the mouth." In deuterostomes the blastopore does not give rise to the mouth but generally develops into the anus. A second opening that forms later in development gives rise to the mouth. The word *deuterostome* derives from words meaning "second" and "the mouth."

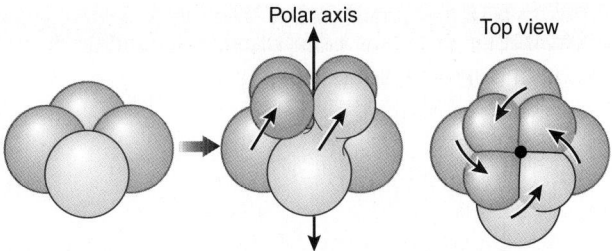

(a) Spiral cleavage is characteristic of protostomes. Note the spiral arrangement, with the upper cells centered between the lower cells.

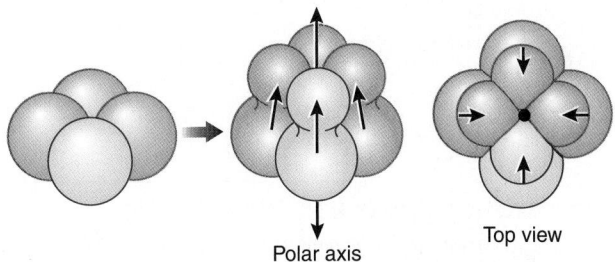

(b) Radial cleavage is characteristic of deuterostomes. The early divisions are either parallel to the polar axis or at right angles to it. The cells are stacked, with the upper cells centered directly above the lower cells.

Figure 30-5 Spiral and radial cleavage

The pattern of cleavage can be appreciated by comparing the positions of the *purple cells* in **(a)** and **(b)**.

© Cengage Learning

Biologists have identified major animal clades based on structure, development, and molecular data

Biologists have long inferred evolutionary relationships among animals based on the structural variations and patterns of development we have just discussed. The development of sophisticated new technology, such as new types of microscopy and cell labeling, has helped systematists in their work. As discussed earlier, techniques for determining nucleotide sequences in DNA and RNA and other molecular tools have helped systematists clarify evolutionary relationships among animal groups. Molecular data have also challenged some traditional conclusions, such as *when* various animal groups diverged in relation to other groups. Such questions continue to be the topic of much debate.

Animals are referred to as *metazoa.* Animals with two or three germ layers (diploblastic or triploblastic) make up the *eumetazoa.* As we have described, some animals exhibit asymmetry (sponges), radial symmetry (cnidarians), or biradial symmetry (ctenophores). The *Bilateria,* the lineage of animals with bilateral symmetry, are triploblastic. (They have three germ layers: ectoderm, endoderm, and mesoderm.) As explained in the last section, biologists classify bilateral animals into protostome and deuterostome clades.

Molecular data show that the protostomes split into two major clades: **Lophotrochozoa** (pronounced "lo-fah-tro-kah-zo´-ah") and **Ecdysozoa** (pronounced "ek-dah-so-zo´-ah") (TABLE 30-1 and FIG. 30-6). Animals assigned to the *Lophotrochozoa*

TABLE 30-1 Overview of the Animal Kingdom

MAJOR GROUPS AND SUBGROUPS

SOME KEY CHARACTERS/COMMENTS

 **Poriferans** (sponges)

Collar cells, flagellated cells that trap food; cells loosely associated and do not form true tissues; sponge larvae have flagella and can swim about; poriferans are probably not a monophyletic group

 Cnidarians

Hydrozoans: hydras

Scyphozoans: jellyfish

Cubozoans: box jellyfish

Anthozoans: corals, sea anemones

Radial symmetry; cnidocytes (stinging cells); polyp and medusa body forms; tentacles surround mouth; mainly marine

 **Ctenophores** (comb jellies)

Biradial symmetry; eight rows of cilia that resemble combs; tentacles with adhesive glue cells; marine predators

---

PROTOSTOMES: LOPHOTROCHOZOAN BRANCH

Bilateral symmetry; triploblastic

Flatworms (platyhelminths)

Turbellarians: planarians

Trematodes and monogeneans: flukes

Cestodes: tapeworms

Gastrovascular cavity with one opening; no coelom; cephalization

 **Nemerteans** (ribbon worms)

Proboscis (long, muscular tube that can be everted to capture prey); complete digestive tube; mainly marine carnivores

 **Mollusks**

Chitons

Gastropods: snails, slugs, nudibranchs

Bivalves: clams, oysters

Cephalopods: squids, octopods

Soft body usually covered by dorsal shell; muscular foot; mantle covers visceral mass; most have radula (belt of teeth)

 **Annelids**

Polychaetes: sandworms, tubeworms

Oligochaetes: earthworms, freshwater worms

Hirudinids: leeches

Segmented body; most have bristles, called setae, that provide traction during crawling

 **Lophophorates**

Brachiopods

Phoronids

Bryozoans (ectoprocts)

Lophophore (ring of ciliated tentacles around mouth); mainly sessile, marine

 **Rotifera** (wheel animals)

Crown of cilia at anterior end; microscopic; freshwater and marine

Continued

---

are characterized by (1) a *lophophore,* a ciliated ring of tentacles surrounding the mouth that serves as a feeding organ; or (2) a type of larva called a *trochophore larva.* The Lophotrochozoa include the flatworms, ribbon worms, mollusks, annelids, and three groups sometimes referred to as the lophophorate phyla.

The name *Ecdysozoa* is derived from the animals in this group molting, a process called *ecdysis.* The Ecdysozoa include the nematodes and arthropods. Note that the phylogeny we have described assigns bilateral animals to three major clades: Lophotrochozoa, Ecdysozoa, and Deuterostomia.

TABLE 30-1 | Overview of the Animal Kingdom *(continued)*

MAJOR GROUPS AND SUBGROUPS	SOME KEY CHARACTERS/COMMENTS
PROTOSTOMES: ECDYSOZOAN BRANCH	Protostomes with cuticle that is molted and replaced as animal grows
 **Nematodes** (roundworms)	Fluid-filled pseudocoelom serves as hydrostatic skeleton; important decomposers; many are predators
Onycophorans (velvet worms)	Unjointed, paired appendages; sister group of arthropods?
Tardigrades ("water bears")	Unjointed, clawed legs; sister group of arthropods?
Arthropods	Segmented; exoskeleton of chitin; paired, jointed appendages; insects and many crustaceans have compound eyes
 Myriapods (centipedes, millipedes)	
Chelicerates (horseshoe crabs, arachnids)	
Crustaceans (lobsters, crabs, barnacles, copepods)	
Hexapods (insects)	
DEUTEROSTOMES	Radial, indeterminate cleavage; pharyngeal slits
Echinoderms	Water vascular system; tube feet; endoskeleton with spines; larvae bilateral, ciliated; adult, pentaradial symmetry; marine
 Crinoids (sea lilies, feather stars)	
Asteroids (sea stars)	
Ophiuroids (basket stars, brittle stars)	
Echinoids (sea urchins, sand dollars)	
Holothuroids (sea cucumbers)	
Hemichordates (acorn worms)	Proboscis, collar, and trunk
Chordates	Notochord; dorsal, tubular nerve cord; postanal tail; endostyle; segmented body
Cephalochordates (lancelets)	Notochord extends from anterior to posterior tip
Urochordates (tunicates)	Larvae have chordate characters
 Vertebrates (hagfishes, lampreys, cartilaginous fishes, ray-finned fishes, coelacanths, lungfishes, amphibians, reptiles [including birds], mammals)	Vertebral column, cranium, neural crest cells; endoskeleton

© Cengage Learning

Segmentation apparently evolved three times

Over millions of years, evolutionary forces acting on the basic animal body plan have produced changes resulting in a remarkable diversity of body forms. One very important innovation has been *segmentation,* a body plan in which certain structures are repeated, producing a series of body structures and compartments. In some cases, each compartment can be regulated somewhat independently of the others, which allows various parts of the body to become specialized to perform specific functions.

Perhaps the most obvious example of segmentation can be found in earthworms. However, as will be discussed in the following chapters, arthropods and vertebrates also have segmented body plans. Thus, segmented animals are found within each of the three major clades of bilateral animals. Molecular data suggest that segmentation evolved independently three times and is therefore another example of homoplasy. This view is reflected in the cladogram in Figure 30-6. In each independent origin of segmentation, natural selection apparently acted on many of the same genes (e.g., *Hox* genes).

In this chapter we have briefly discussed characteristics common to animals and the early evolution of animals. We described animal body plans and examined some of the criteria that biologists use to reconstruct phylogenetic relationships. The cladogram shown in Figure 30-6 depicts some of the current hypotheses regarding the relationships among major animals groups, and Table 30-1 summarizes these relationships. In the next two chapters, we survey these animal groups.

In Chapter 31 we describe three groups—sponges, cnidarians, and ctenophores—traditionally viewed as diverging early in the evolutionary history of animals. We then discuss one of the major clades of animals: the protostomes. Then, in Chapter 32, we discuss the deuterostomes, which include the echinoderms and chordates, the clade to which we humans belong. Many hypotheses are presented in these chapters, and we will discuss many examples of how systematists revise the relationships of the branches of the animal phylogenetic tree in response to new data.

CHECKPOINT 30.4

- *How are animals classified based on type of symmetry?*
- *What are some differences between protostomes and deuterostomes?*
- **VISUALIZE** *Draw a simple cladogram illustrating the evolutionary relationships among the three main clades of bilateral animals. How do they differ from one another?*

Biologists recognize three major clades of bilateral animals: Lophotrochozoa, Ecdysozoa, and Deuterostomia.

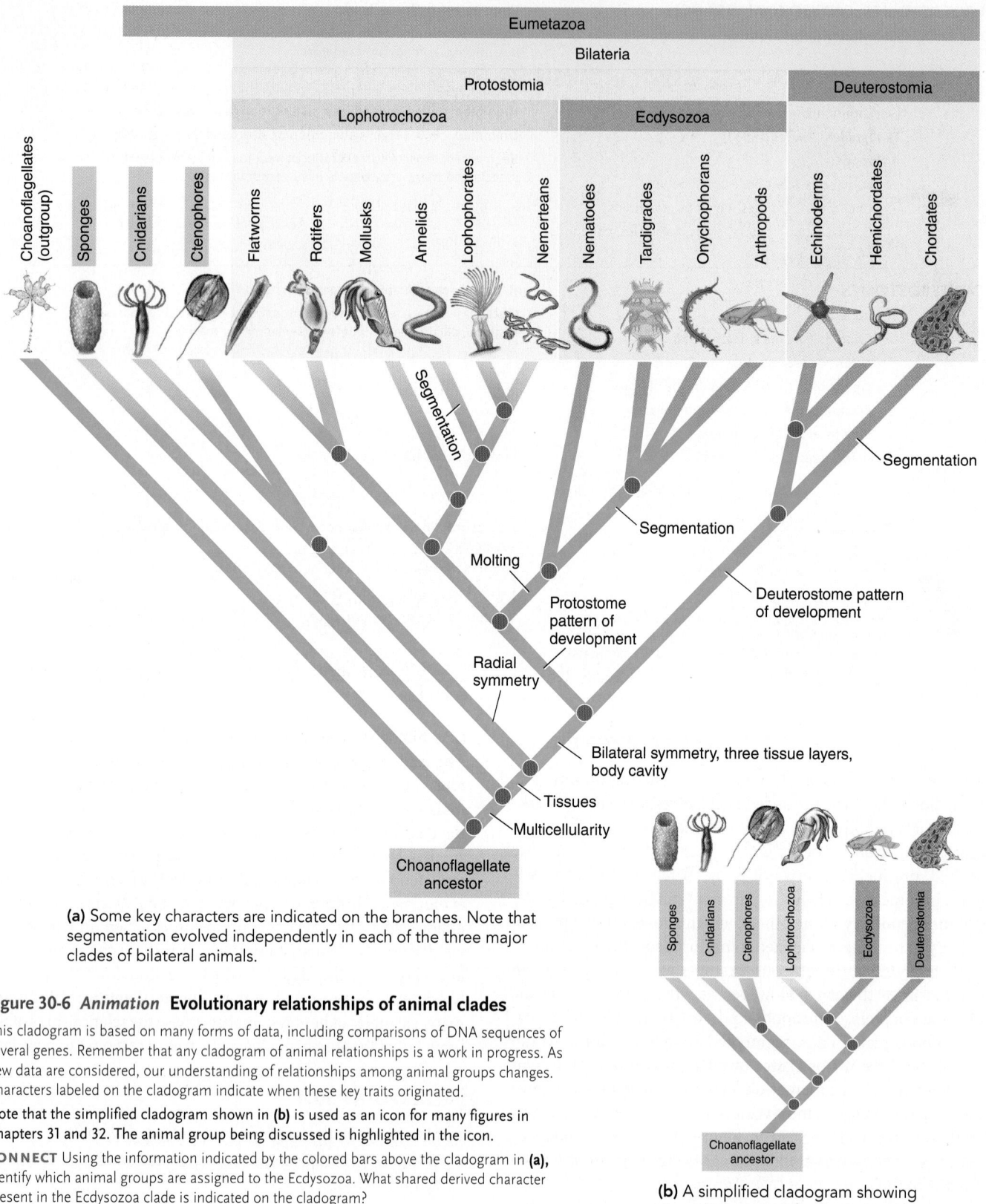

(a) Some key characters are indicated on the branches. Note that segmentation evolved independently in each of the three major clades of bilateral animals.

Figure 30-6 *Animation* **Evolutionary relationships of animal clades**

This cladogram is based on many forms of data, including comparisons of DNA sequences of several genes. Remember that any cladogram of animal relationships is a work in progress. As new data are considered, our understanding of relationships among animal groups changes. Characters labeled on the cladogram indicate when these key traits originated.

Note that the simplified cladogram shown in **(b)** is used as an icon for many figures in Chapters 31 and 32. The animal group being discussed is highlighted in the icon.

CONNECT Using the information indicated by the colored bars above the cladogram in **(a)**, identify which animal groups are assigned to the Ecdysozoa. What shared derived character present in the Ecdysozoa clade is indicated on the cladogram?

© Cengage Learning

(b) A simplified cladogram showing six major groups of animals.

30.1 Animal Characteristics (page 623)

1 Describe several characteristics common to most animals.

- Animals are eukaryotic, multicellular, heterotrophic organisms with cells specialized to perform specific functions. Animals have diverse body plans. The *body plan* is the basic structure and functional design of the body.

- Most animals are capable of locomotion at some time during their life cycle, can respond adaptively to external stimuli, and can reproduce sexually.

- In sexual reproduction sperm and egg unite to form a **zygote.** The zygote undergoes **cleavage,** a series of cell divisions that produce a hollow ball of cells called a **blastula.** Most animals develop into a **larva,** a sexually immature form that may appear and behave differently from the adult. Larvae typically go through **metamorphosis,** a developmental process that converts the immature animal into a juvenile form that grows into an adult.

30.2 Adaptations to Ocean, Freshwater, and Terrestrial Habitats (page 624)

2 Compare the advantages and disadvantages of life in the ocean, in fresh water, and on land.

- Marine environments have relatively stable temperatures, provide buoyancy, and provide readily available food. Fluid and salt balance are more easily maintained in sea water than in fresh water. Currents and other water movements can be a disadvantage.

- Fresh water offers a less constant environment and less food than sea water. Because fresh water is hypotonic to tissue fluid, animals must osmoregulate.

- Terrestrial animals must have adaptations that protect them from drying out and from temperature changes, and that protect their gametes and embryos.

30.3 Animal Evolution (page 625)

3 Use current hypotheses to trace the early evolution of animals.

- Based on molecular data, biologists hypothesize that most animal clades actually diverged over a long period during the Proterozoic eon. During the **Cambrian radiation,** new animal body plans rapidly evolved among clades that already existed.

- *Hox* **genes** control early development in animal groups. These genes had evolved by the beginning of the Cambrian period, and mutations in these genes could have resulted in rapid changes in animal body plans.

30.4 Reconstructing Animal Phylogeny (page 626)

4 Describe how biologists use morphology (including variations in body symmetry, number of tissue layers, and type of body cavity) and patterns of early development to infer relationships among animal phyla.

- Biologists hypothesize that cnidarians (which have radial symmetry) and ctenophores (which have biradial symmetry) are more closely related to each other than to animals that exhibit

bilateral symmetry. **Cephalization,** the development of a head, evolved along with bilateral symmetry.

- Biologists have also inferred relationships based on level of tissue development and type of body cavity. Embryonic tissues, called **germ layers,** include the outer layer, **ectoderm,** which gives rise to the body covering and the nervous system; the inner layer, **endoderm,** which lines the gut and other digestive organs; and a middle layer, **mesoderm,** which gives rise to muscle, skeletal structures, and most other body structures.

- In bilateral animals the type of body cavity has been used to classify animals. **Acoelomate** animals have no body cavity, and **coelomate** animals have a *true* coelom, a body cavity completely lined with mesoderm. Some animals have a **pseudocoelom** (literally, a "false cavity"), a body cavity that is not completely lined with mesoderm.

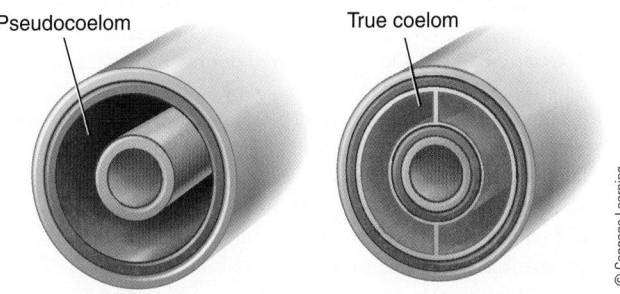

Pseudocoelom   True coelom

© Cengage Learning

- Two major evolutionary branches of bilateral animals are **protostomes** (mollusks, annelids, and arthropods) and **deuterostomes** (echinoderms and chordates).

- Protostomes undergo **spiral cleavage,** in which early cell divisions are diagonal to the polar axis. Deuterostomes undergo **radial cleavage,** in which the early cell divisions are either parallel or at right angles to the polar axis, so the cells lie directly above or below one another.

- Protostomes undergo **determinate cleavage,** in which the fate of each embryonic cell is fixed very early. Deuterostomes undergo **indeterminate cleavage,** in which early in development each cell has the potential to develop into a complete organism.

- In protostomes the **blastopore,** the opening from the embryonic gut to the outside, develops into the mouth; in deuterostomes the blastopore typically becomes the anus.

5 Cite specific examples of how data from molecular systematics have confirmed or modified traditional animal phylogeny and identify the three major clades of bilateral animals.

- Molecular systematics has confirmed much of animal phylogeny that was originally based on structural characters, including the axiom that animal body plans usually evolved from simple to complex. However, molecular systematics has also provided evidence for exceptions.

- Based on molecular data, biologists now subdivide the protostomes into two clades: **Lophotrochozoa** and **Ecdysozoa.** The Lophotrochozoa include the flatworms, ribbon worms, mollusks, annelids, rotifers, and animals that have a lophophore, a ciliated ring of tentacles surrounding the mouth. The Ecdysozoa, animals that molt, include the nematodes and arthropods. The third clade of animals, Deuterostomia, includes the echinoderms, hemichordates, and chordates. These animals have radial, indeterminate cleavage and pharyngeal slits.

Eumetazoa		
Bilateria		
Protostomia		Deuterostomia
Lophotrochozoa	Ecdysozoa	

© Cengage Learning

TEST YOUR UNDERSTANDING

Know and Comprehend

1. Which of the following is *not* a characteristic of all animals? (a) heterotrophic (b) multicellular (c) eukaryotic (d) presence of a coelom (e) formation of a zygote that undergoes cleavage

2. Which of the following is *not* an adaptation to terrestrial living? (a) internal fertilization (b) shell surrounding egg (c) adaptations for maintaining body temperature (d) surface for gas exchange deep in body (e) ability to maintain location

3. The Cambrian radiation (a) occurred during the late Cambrian period (b) was a rapid evolution of new animal body plans during the middle Cambrian (c) was a result of prokaryotic and eukaryotic migration to many new regions (d) is supported by the great variety of Ediacaran fossils (e) b and c

4. The germ layer that gives rise to the outer covering of the body and the nervous system is the (a) blastula (b) ectoderm (c) gastrula (d) endoderm (e) mesoderm

5. Radial symmetry is characteristic of (a) protostomes (b) chordates (c) mollusks (d) cnidarians (e) sponges

6. A coelom is a body cavity completely lined with (a) epithelium (b) ectoderm (c) mesoderm (d) endoderm (e) epidermis

7. Protostomes are characterized by (a) spiral cleavage (b) indeterminate cleavage (c) an acoelomate body plan (d) radial symmetry (e) a and c

8. The evolution of animals (a) followed an orderly progression from simple to complex (b) will be better understood as biologists continue to collect molecular and other types of data (c) has been determined by studying classification (d) began with the cnidarians (e) began with their common ancestor, a protostome

9. Which of the following is an example of a deuterostome? (a) lophotrochozoan (b) coral (c) chordate (d) planarian flatworm (e) insect

Apply and Analyze

10. **VISUALIZE** Label the branches of the diagram.

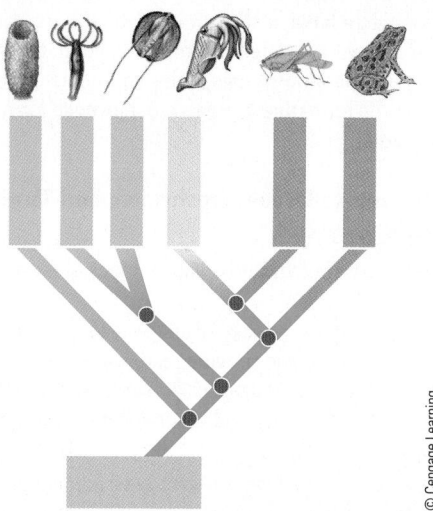

© Cengage Learning

11. Imagine that you discover a new organism in a lake. How would you decide whether it is an animal? What are some characteristics that might contribute to your decision?

12. **EVOLUTION LINK** Examine the cladogram in Figure 30-6a. Based on the discussion in this chapter, what were some of the types of data that biologists used to determine these phylogenetic relationships?

Evaluate and Synthesize

13. **EVOLUTION LINK** Suggest and support a hypothesis for the view that segmentation evolved in three different animal clades.

14. **INTERPRET DATA** Imagine that a biologist discovers evidence that a newly discovered animal evolved from a very early protostome. Where would you place the branch (clade) for that animal on Figure 30-6a?

Sponges, Cnidarians, Ctenophores, and Protostomes | 31

I n Chapter 30 we introduced the animal kingdom and explored animal phylogeny. We examined how systematists use fossils, morphology, developmental patterns, and molecular data to determine animal relationships. In this chapter we begin with an introduction to the sponges, cnidarians, and ctenophores, groups characterized by asymmetry, radial symmetry, and biradial symmetry, respectively. The phylogeny of these animals is the focus of much current research. Although sponges, cnidarians, and ctenophores are mainly small animals with simple body structures, they are of great ecological importance. They are important members of marine food chains, many provide shelter for other organisms, and some form symbiotic relationships with other animals. Corals that produce reefs are among the most ecologically important animals in the world.

Following our discussion of the sponges, cnidarians, and ctenophores, we begin our survey of the bilateral animals, the most familiar and by far the most numerous animals on Earth. More than 99% of animal species belong to the Bilateria. These diverse forms have adapted to life in almost every imaginable habitat in the ocean and in freshwater and terrestrial environments. The bearded fireworm shown in the photograph is an example of the diverse animals found among the Bilateria.

The remarkable success of the bilateral animals can be attributed to the evolution of adaptations that facilitated food capture, escape from predators, and reproduction. Among their key adaptations are cephalization (formation of a head), a central nervous system, muscles, a coelom, and compartmentalization of the body. Large animals developed body systems for gas exchange, waste disposal, and internal transport of nutrients, respiratory gases, wastes, and other materials.

Recall that based on molecular data, pattern of early development, and other data, biologists divide the bilateral animals into two main groups: protostomes and deuterostomes. The protostomes form two clades: Lophotrochozoa and Ecdysozoa. In this chapter we focus on the innovations in body plans, phylogenetic relationships, and life histories of several phyla of these two clades.

Marty Snyderman/Visuals Unlimited, Inc.

A bearded fireworm (*Hermodice carunculata*). The fireworm is a segmented worm (an annelid) that can grow up to 31 cm (11.8 in.) long. It lives in shallow marine waters: in coral reefs, in beds of turtle grass, or under rocks. Fireworm bristles, which are filled with venom, can break off in the skin and cause irritation.

KEY CONCEPTS

31.1 Animals with asymmetry, radial symmetry, or biradial symmetry include sponges, cnidarians (hydras, jellyfish, sea anemones), and ctenophores (comb jellies). Sponges are characterized by collar cells and by loosely associated cells that do not form true tissues; cnidarians are characterized by radial symmetry, two tissue layers, and cells that contain stinging organelles; and ctenophores have biradial symmetry, two tissue layers, eight rows of cilia, and tentacles with adhesive glue cells.

31.2 Protostomes are a large monophyletic group that includes two major clades: Lophotrochozoa and Ecdysozoa. The lophotrochozoans include the annelids (segmented worms), mollusks, and several smaller groups.

31.3 The ecdysozoans include the nematodes (roundworms) and arthropods (horseshoe crabs, spiders, crustaceans, insects). The remarkable biological success of the arthropods can be attributed to the evolution of complex body plans and life cycles, including their exoskeleton, segmentation, specialized jointed appendages, ability to fly (among insects), and metamorphosis.

31.1 SPONGES, CNIDARIANS, AND CTENOPHORES

LEARNING OBJECTIVES

1 Identify important characteristics of poriferans.
2 Identify distinguishing characteristics of cnidarians, describe four groups, and give examples of animals that belong to each group.
3 Identify characteristics of ctenophores.

Recall from Chapter 30 that based on available evidence, animals and choanoflagellates (a group of unicellular and colonial protists) share a common choanoflagellate ancestor. Many biologists assign sponges to a basal group, *Parazoa* (*para*, "alongside"; and *zoa*, "animals"). This classification is based on the asymmetry and simple body plan of sponges. The cells of sponges are loosely associated and do not form true tissues. Other animals have true tissues and are classified as **Eumetazoa** (*eu*, "true"; *meta*, "later"; and *zoa*, "animals"). Thus, these biologists divide the animal kingdom into two major animal groups: Parazoa and Eumetazoa.

According to this view, sponges are a monophyletic group. Based on molecular data, some biologists hypothesize that sponges are a polyphyletic group. There is some evidence that at least one group of sponges (calcareous sponges) became simplified as they evolved from animals with more complex body plans.

Systematists also debate about which animal group is the oldest. Some systematists have proposed that comb jellies (marine animals that swim by means of cilia), rather than sponges, are the oldest group of animals. Systematists continue to search for new clues that will settle the controversies over which branch of animals is the oldest and over the exact pattern of relationships among the major animal groups.

Sponges have collar cells and other specialized cells

Sponges, or **poriferans,** are aquatic, mainly marine animals that are most abundant in warm waters. Their phylum name, *Porifera* (paw-rif′-er-ah), means "to have pores." This name aptly describes these animals, whose bodies are perforated by tiny holes. Biologists have identified about 10,000 species of sponges. They range in size from a few millimeters to more than a meter in height and diameter (loggerhead sponges). Many sponges are asymmetrical, but they vary in shape from flat, encrusting growths to balls, cups, fans, or vases. Living sponges may be brightly colored—green, orange, red, yellow, blue, or purple—or they may be white or drab (FIG. 31-1). Some species are inhabited by symbiotic bacteria or algae that give them color.

Although they are multicellular and can be large, sponges function much like choanoflagellates. These protists are characterized by a single flagellum surrounded by a collar of microvilli (see Fig. 26-21). Sponges have flagellate cells called **collar cells,** or **choanocytes** (ko-an′-uh-sites), which are strikingly similar to choanoflagellates.

Sponge larvae have flagella and can swim about. Adult sponges attach to some solid object and have long been described as sessile. However, biologists have observed adults of several species moving slowly (about 4 mm per day), perhaps by the cumulative movement of cells along the sponge's lower surface.

Although sponges are multicellular, their cells are loosely associated and do not form true tissues. However, a division of labor exists among the several types of cells that make up the sponge, with certain cells specializing in nutrition, support, contraction, or reproduction. Many sponge cells are extremely versatile and can change form and function.

Collar cells make up the inner layer of certain sponges. Each cell is equipped with a tiny collar surrounding the base of the flagellum. The collar is an extension of the plasma membrane and consists of microvilli. Collar cells create the water current that brings food and oxygen to the cells and carries away carbon dioxide and other wastes. Collar cells also trap and phagocytize food particles. Together, the collar cells of some sponges can pump a volume of water equal to the volume of the sponge each minute!

Sponges have three types of canal systems through which water circulates. In the simple *asconoid* canal system, the beating flagella of collar cells create a current that pulls water through hundreds of tiny pores, called *ostia*. Specialized tubelike cells, called *porocytes*, form the pores. These cells regulate the diameter of the pores by contracting. Water passes into the central cavity, or **spongocoel** (spon′-jo-seel) (not a digestive cavity), and then flows out through the sponge's open end, the **osculum** (os′-kyuh-lum). The asconoid system is illustrated in Figure 31-1.

In the *syconoid* system, water enters through canals lined by collar cells. Most types of sponges have a *leuconoid* system. The body wall is extensively folded, and complex systems of canals provide increased surface area for food capture. *Epidermal cells* form the outer layer of the sponge and line the canals. The canals lead into small chambers that are lined with collar cells.

The skeletal framework of a sponge can be both rigid and fibrous. Between the outer and inner cell layers, the sponge body has a gelatin-like layer, the *mesohyl*, which is supported by slender skeletal spikes, or **spicules** (spik′-yuls). *Amoeboid cells*, which wander about in the mesohyl, secrete the spicules. The spicules are made of calcium carbonate or of silica. The fibrous part of the sponge skeleton consists of *spongin*, a form of collagen.

Sponges are *suspension feeders*, adapted for trapping and eating whatever food is suspended in the water. As water circulates through the body, bacteria, algae, and other organic particles are trapped along the sticky collars of the collar cells. Thus, the collars *filter* the suspended food particles out of the water. Suspension feeders that filter the suspended food are known as *filter feeders*. Food particles are either digested within the collar cell or transferred to an amoeboid cell for digestion. The amoeboid cell transports nutrients to epidermal cells. Undigested food passes out through the osculum and is simply eliminated into the water.

Gas exchange and excretion of wastes depend on diffusion into and out of individual cells. Although cells of the sponge can react to stimuli, sponges do not have specialized nerve cells and

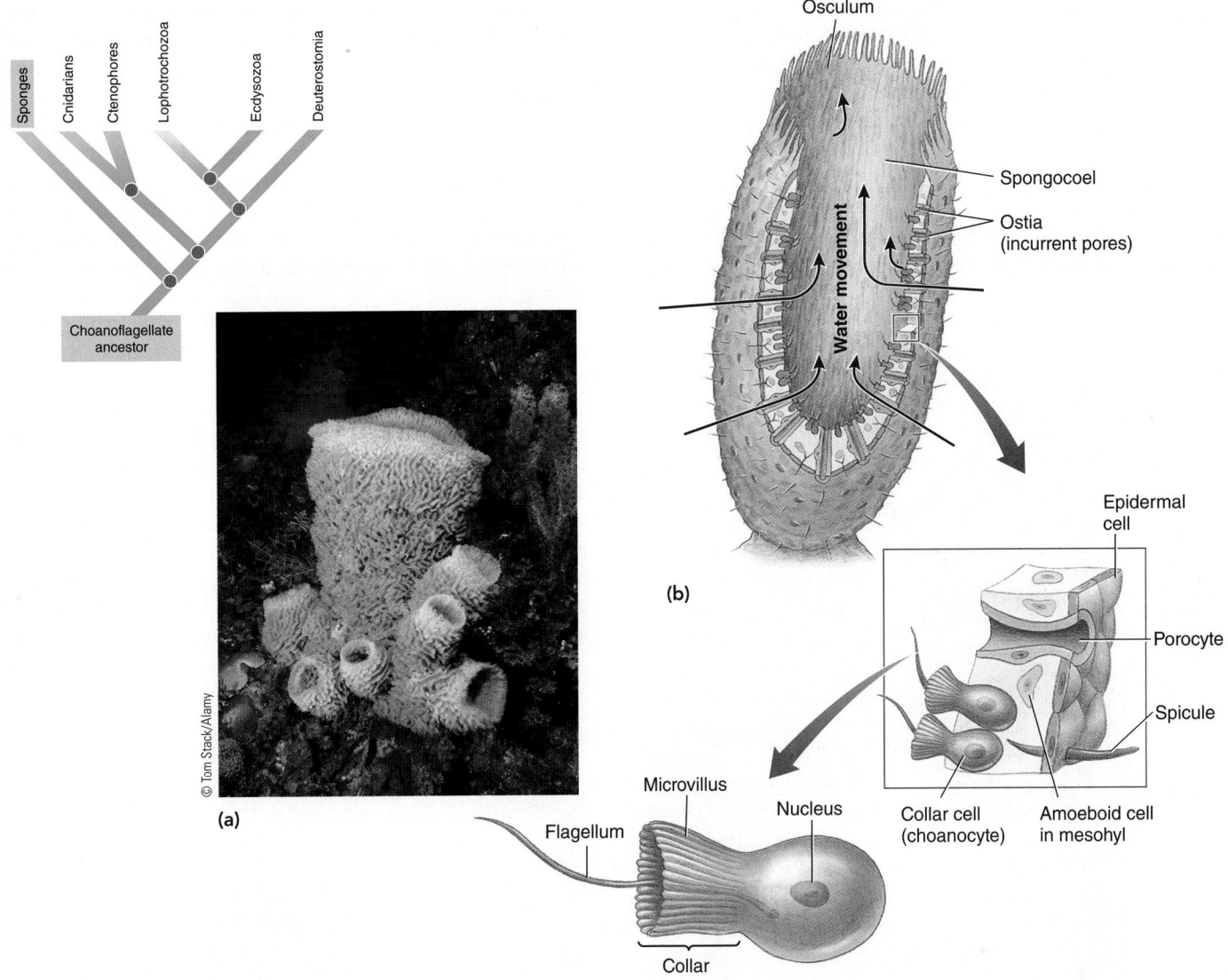

Figure 31-1 *Animation* **Sponge structure**

(a) Tube sponges (*Spinosella plicifera*) from the Caribbean attached to the coral reef substrate. **(b)** Sponge cut to expose its organization. This sponge has an asconoid type of canal system.

© Cengage Learning

so cannot react as a whole. However, electrical signals have been identified in certain glass sponges. Behavior appears limited to basic metabolic necessities such as capturing food and regulating the flow of water through the body.

Sponges reproduce both asexually and sexually. In asexual reproduction, a small fragment or bud may break free from the parent sponge and give rise to a new sponge. Such fragments may attach to the parent sponge, forming, or becoming part of, a colony. Most sponges are **hermaphrodites,** meaning that the same individual can produce both eggs and sperm. Some of the amoeboid cells develop into sperm cells and others into egg cells. However, hermaphroditic sponges usually produce eggs and sperm at different times, and they cross-fertilize with other sponges. They release mature sperm into the water, which are taken in by other sponges of the same species.

Fertilization and early development take place within the jellylike mesohyl. Zygotes develop into flagellate larvae that leave the parent along with the stream of outflowing water. After swimming for a while, a larva finds a solid object, attaches to it, and settles down to a sessile life.

Sponges have a remarkable ability to repair themselves when injured and to regenerate lost parts. When the cells of a sponge are separated experimentally, they recognize one another and their place in the whole and aggregate to re-form a complete sponge.

There are three main groups of sponges: the *calcareous sponges* have spicules composed of calcium carbonate, whereas the *glass sponges* and the *demosponges* have spicules of silicon. About 95% of living sponges are demosponges. In some of the demosponges, spicules are bound together by spongin. Some demosponges do not have spicules. These soft sponges have been used for thousands of years as bath sponges and for cleaning. Unfortunately, as a result of overfishing, bath sponges are now in short supply.

Sponges play significant roles in marine habitats. Larger species are important structural components in both shallow and deep water, providing homes to many species of annelids, crustaceans, and echinoderms. Sponges also serve as food for some invertebrates, fish, and sea turtles.

Cnidarians have unique stinging cells

More than 10,000 species of **cnidarians** (ni-dah'-ree-ans; phylum Cnidaria) have been described. Most are marine.

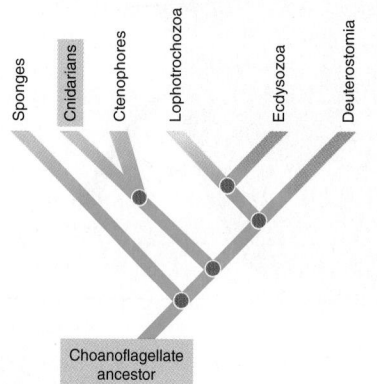

Choanoflagellate ancestor

The radially symmetrical cnidarian body is organized as a hollow sac with the mouth and surrounding tentacles located at one end. Some cnidarians live a solitary existence, whereas many others, such as corals, form colonies.

Cnidarians have two body shapes: polyp and medusa (FIG. 31-2). The **polyp** form, represented by *Hydra*, typically has a dorsal mouth surrounded by tentacles. In the **medusa** (pl., *medusae*), or jellyfish form, the mouth is located in the lower concave, or *oral*, surface; the convex upper surface is the *aboral* surface. Some cnidarians have the polyp shape during one stage of their life cycle and the medusa form during another stage. The Portuguese man-of-war and some other cnidarians consist of colonies of many individuals, some of which are polyps and others medusae.

Cnidarians get their name from specialized cells, called **cnidocytes** (from a Greek word meaning "nettle cells"), that contain stinging organelles. Cnidocytes are located mainly in the epidermis, especially on the tentacles. The cnidocytes contain stinging "thread capsules," called **nematocysts** (FIG. 31-3). Each cnidocyte has a small, projecting trigger (*cnidocil*) on its outer surface and a coiled, hollow thread inside. We can compare the cnidocyte to a loaded gun, ready to fire. When stimulated by touch or certain chemicals dissolved in the water, the

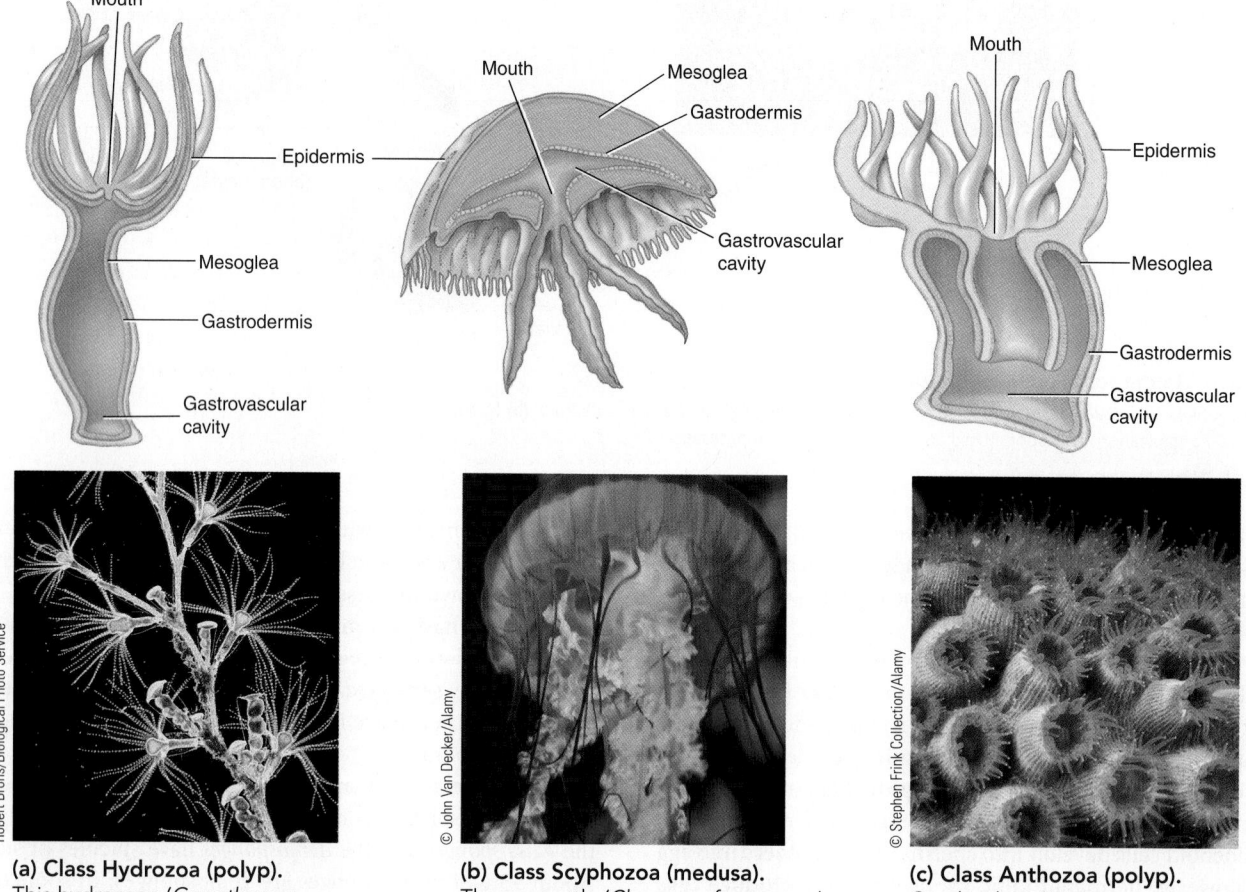

(a) Class Hydrozoa (polyp).
This hydrozoan (*Gonothyraea loveni*) forms a colony of polyps. The drawing illustrates a single polyp.

(b) Class Scyphozoa (medusa).
The sea nettle (*Chrysaora fuscescens*) uses its tentacles equipped with cnidocytes to capture small animals (zooplankton) suspended in the water.

(c) Class Anthozoa (polyp).
Coral polyps (*Montastrea cavernosa*) extended for feeding.

Figure 31-2 *Animation* **Polyp and medusa body forms of cnidarians**
© Cengage Learning

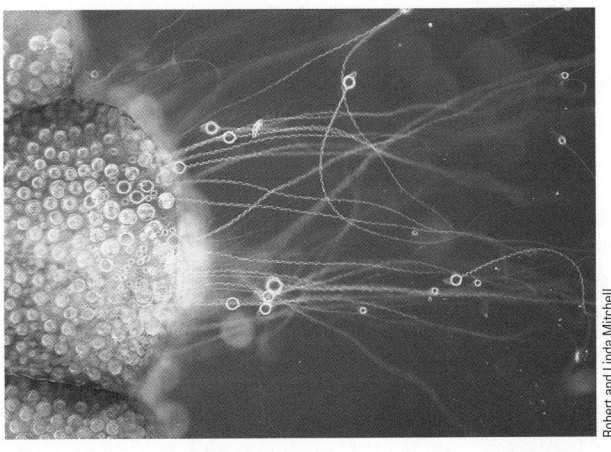

250 μm

(a) LM of discharged nematocysts of a Portuguese man-of-war (*Physalia physalis*). Photographed in the Gulf of Mexico.

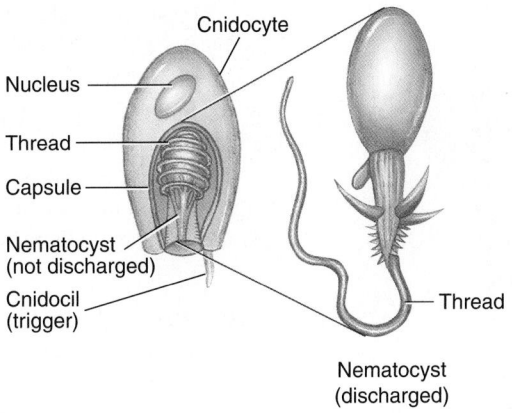

(b) Undischarged and discharged nematocyst. The cnidocil, or trigger, is a mechanoreceptor that discharges the nematocyst when it senses contact with an object.

Figure 31-3 *Animation* **Nematocysts**

When cnidarian stinging cells (cnidocytes) are stimulated, the nematocyst discharges, ejecting a thread that may entangle or penetrate the prey. Some nematocysts secrete a toxic substance that immobilizes the prey.
© Cengage Learning

nematocyst fires its thread. Some types of nematocyst threads are sticky. Others are long and coil around prey. A third type bears barbs or spines that can inject a protein toxin. This toxin paralyzes prey animals, such as small crustaceans.

Cnidarians use their tentacles to capture prey and push it into the mouth. The mouth leads into the **gastrovascular cavity,** where digestion takes place. The mouth is the only opening into the gastrovascular cavity and so must serve for both ingestion of food and expulsion of undigested material. Gas exchange and excretion occur by diffusion. The body wall is thin enough that no cell is far from the surface.

More highly organized than sponges, cnidarians are diploblastic; that is, they have two definite tissue layers. The ectoderm gives rise to the outer epidermis, a protective layer covering the body. The endoderm gives rise to the inner *gastrodermis,* which

lines the gastrovascular cavity and functions in digestion. These thin layers are separated by a thick, jellylike **mesoglea,** which is mainly acellular.

The nerve cells of a cnidarian form a **nerve net** that connects sensory cells in the body wall to contractile cells and gland cells. An impulse set up by one sensory cell passes in all directions more or less equally. Nerve cells are not organized to form a brain or nerve cord. In some types of medusae, sense organs are positioned around the edge of the body. The sense organs include simple "eyes," called eyespots; organs of balance (statocysts); and chemoreceptors (cells that sense certain chemicals).

Both the epidermis and gastrodermis have cells specialized to contract. (However, they are not true muscle cells). Contractile fibers in the epidermal cells are arranged lengthwise, and those in the gastrodermis are arranged in a circular pattern. These two sets of contractile cells act on the water-filled gastrovascular cavity, forming a **hydrostatic skeleton.** This skeleton supports the body and allows movement. By contracting one set of contractile cells or the other, the hydra can shorten, lengthen, or bend its body (see Fig. 40-2). We consider four groups of cnidarians: hydrozoans, scyphozoans, cubozoans, and anthozoans (TABLE 31-1).

Most hydrozoans form colonies Hydrozoans include hydras and *hydroids,* such as *Obelia* and the Portuguese man-of-war. Although not really typical, the solitary *Hydra* is the cnidarian that beginning biology students most often study (FIG. 31-4). To the naked eye, *Hydra* looks like a bit of frayed string. This tiny hydrozoan is found in freshwater ponds. Because it has a remarkable ability to regenerate, biologists named *Hydra* after the multiheaded monster of Greek mythology that could grow two new heads for each head cut off. When *Hydra* is cut into several pieces, each piece can regenerate all the missing parts and become a whole animal.

Hydra lives in fresh water and typically attaches to a rock, aquatic plant, or detritus by a disc of cells at its base. Hydras

TABLE 31-1	Major Classes of Phylum Cnidaria
CLASS AND REPRESENTATIVE ANIMALS	**CHARACTERISTICS**
HYDROZOA *Hydra, Obelia,* Portuguese man-of-war	Mainly marine, but some freshwater species; alternation of polyp and medusa stages in most species (polyp form only in *Hydra*); some form colonies
SCYPHOZOA Jellyfish	Mainly marine; typically inhabit coastal water, free-swimming medusa most prominent form; polyp stage often reduced
CUBOZOA "Box jellyfish"	Inhabit tropical and subtropical waters; have polyp stage, but medusa form most prominent; square shape when viewed from above; actively hunt prey; complex eyes form blurred images
ANTHOZOA Sea anemones, corals, sea fans	Marine; solitary or colonial polyps; no medusa stage in most; gastrovascular cavity divided by partitions into chambers, increasing area for digestion

© Cengage Learning

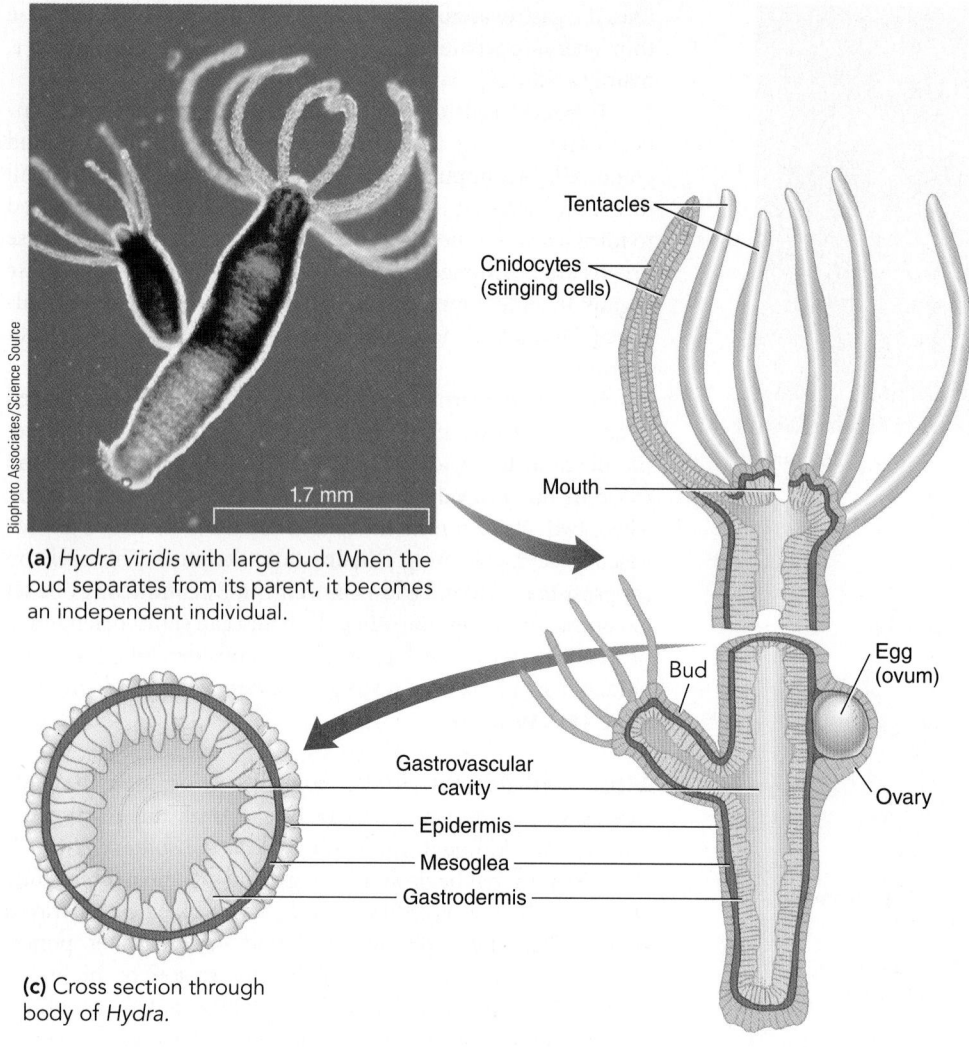

(a) *Hydra viridis* with large bud. When the bud separates from its parent, it becomes an independent individual.

1.7 mm

Biophoto Associates/Science Source

Tentacles

Cnidocytes (stinging cells)

Mouth

Bud

Egg (ovum)

Ovary

Gastrovascular cavity

Epidermis

Mesoglea

Gastrodermis

(c) Cross section through body of *Hydra*.

(b) *Hydra* cut longitudinally to show internal structure. Asexual reproduction by budding is represented on *left*; sexual reproduction is represented by ovary on *right*. Male hydras develop testes that produce sperm.

Figure 31-4 *Hydra*, a freshwater hydrozoan
© Cengage Learning

reproduce asexually by budding during periods when environmental conditions are optimal. However, they differentiate as males and females and reproduce sexually in the fall or when pond water becomes stagnant. The zygote may become covered with a shell that protects it through the winter or until conditions become more favorable.

Many hydrozoans form colonies consisting of hundreds or thousands of individuals. A colony begins with a single polyp that reproduces asexually by budding. However, instead of separating from the parent, the bud remains attached and eventually forms additional buds. Several types of individuals may develop in the same colony, some specialized for feeding, some for reproduction, and others for defense.

Some marine hydrozoans are remarkable for their alternation of sessile polyp and motile medusa stages. The life cycle of the colonial marine hydrozoan *Obelia* illustrates alternation of polyp

and medusa stages (FIG. 31-5). In *Obelia* polyps reproduce asexually by budding, whereas the medusae reproduce sexually, using meiosis to produce gametes. Both polyp and medusa are diploid; only sperm and eggs are haploid.

Scyphozoans are the "true" jellyfish In *scyphozoans* the medusa is the dominant body form. Jellyfish medusae are generally larger than hydrozoan medusae, and they have a thick, viscous mesoglea that gives firmness to the body. In scyphozoans the polyp stage is small and inconspicuous or may even be absent. The largest jellyfish, *Cyanea*, may be more than 2 m (6.5 ft) in diameter and have tentacles 30 m (98 ft) long. These orange and blue "monsters," among the largest invertebrates, can cause painful stings to swimmers in the North Atlantic and Pacific Oceans.

Cubozoans include the "box jellyfish" When viewed from above, *cubozoans* have a square shape. They have four tentacles, or groups of tentacles. These fast-swimming jellyfish have complex eyes that form blurred images, and they actively hunt for prey. Found in waters off the northern Australian coast, the sea wasp (*Chironex fleckeri*) has long tentacles that can extend more than 3 m (10 ft). The sea wasp produces one of the deadliest venoms in the animal kingdom. Its venom contains toxins that can stun or kill fish and other prey. These toxins can also cause respiratory failure and cardiac arrest in humans. Fatalities are mainly associated with severe stings caused by contact with tentacles over a large area of the body. Interestingly, sea turtles feed on box jellyfish, apparently unaffected by their venom.

Anthozoans are polyps *Anthozoans*, which include the sea anemones and corals, are either individual or colonial polyps. There is no free-swimming medusa stage. The polyp produces eggs and sperm, and the fertilized egg develops into a small, ciliated **planula larva.** This larval form may swim to a new location before attaching to develop into a polyp.

Anthozoans differ from hydrozoans in that a series of vertical partitions partially divides the gastrovascular cavity into

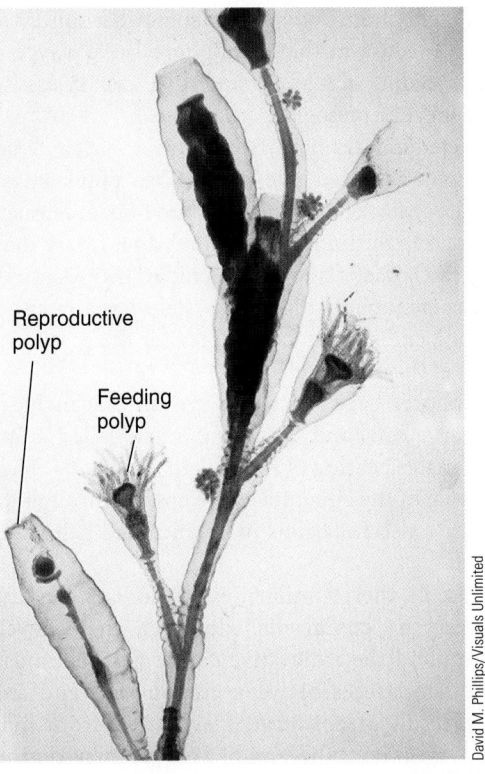

Reproductive polyp

Feeding polyp

David M. Phillips/Visuals Unlimited

250 μm

(a) LM of *Obelia*. Some polyps have tentacles and are specialized for feeding, whereas others are specialized for reproduction.

connected chambers. The partitions increase the surface area for digestion, enabling an anemone to digest an animal as large as a crab. Although corals can capture prey, many tropical species depend for nutrition on photosynthetic algae (*zooxanthellae*) that live within cells lining the coral's digestive cavity (see Chapter 26 and Fig. 54-12). The relationship between coral and zooxanthellae is symbiotic and mutually beneficial. The coral supplies the algae with waste products such as ammonia from which the algae make nitrogen compounds for both partners. In exchange, the algae provide the coral with oxygen and with carbon and nitrogen compounds. In warm, shallow seas, much of the bottom is covered with coral or anemones, most of them brightly colored. Coral reefs consist of colonies of millions of corals and of certain algae (mainly coralline red algae). Living colonies occur only in the uppermost regions of such reefs, adding their own skeletons to the forming rock. Coral reefs are among the most productive of all ecosystems, rivaling tropical rain forests in species diversity (see Fig. 56-21). A single reef can serve as home for more than 3000 species of fishes and other marine organisms, and an estimated one-fourth of all marine species depend on coral reefs.

Many species of reef-building corals are considered endangered or threatened and are at increased risk for extinction. Human activities are responsible for much of this damage. In 2012, the National Oceanic and Atmospheric Administration reported that the three major threats to coral reefs are rising ocean temperatures (due to global climate change), acidification of the ocean (caused by increase in carbon dioxide in the

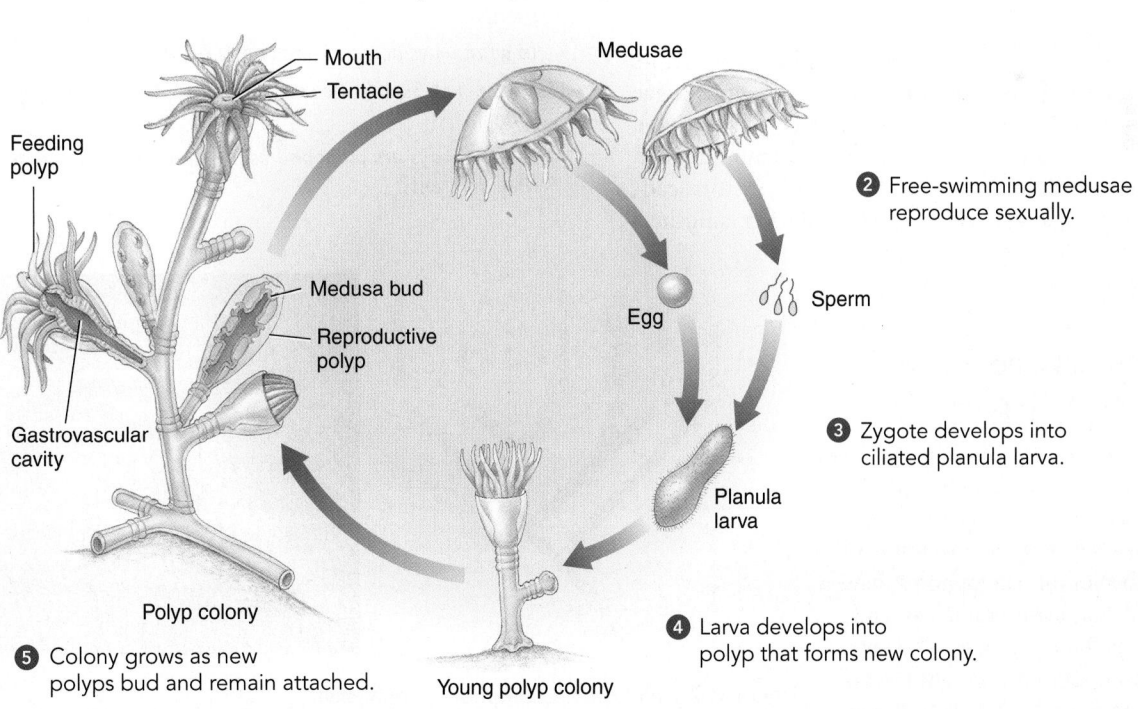

① Reproductive polyps produce medusae by budding asexually.

Mouth

Tentacle

Medusae

Feeding polyp

② Free-swimming medusae reproduce sexually.

Medusa bud

Reproductive polyp

Sperm

Egg

Gastrovascular cavity

③ Zygote develops into ciliated planula larva.

Planula larva

Polyp colony

④ Larva develops into polyp that forms new colony.

⑤ Colony grows as new polyps bud and remain attached.

Young polyp colony

(b) Life cycle of *Obelia*.

Figure 31-5 *Animation* **Obelia, a marine colonial hydrozoan**

© Cengage Learning

ways. The ctenophore body plan is somewhat similar to that of a cnidarian medusa in that ctenophores have a type of radial symmetry, feeding tentacles, and two cell layers separated by a thick, jellylike mesoglea.

Some ctenophores have two tentacles. Their tentacles are equipped with adhesive glue cells. When prey comes in contact with the tentacle, the glue cells burst open, releasing sticky threads that trap the prey. Ctenophores do not have the stinging nematocysts characteristic of the cnidarians. Also unlike cnidarians, the ctenophore digestive system has a mouth for food intake at one end and two anal pores for the egestion of water and wastes at the other end.

Ctenophores have a nervous system that includes a network of nerve cells and a sensory organ called a "statocyst" that coordinates beating of the cilia in the combs. The coordinated beating of the cilia moves the animal through the water. The statocyst also functions in balance and helps the animal orient itself.

Because of their similarities, biologists classify ctenophores near the cnidarians. However, their development is different, and their digestive cavity has openings at both ends. The similarities between ctenophores and cnidarians may be a result of convergent evolution from living in a similar environment, the ocean. Where, then, do the ctenophores belong on our cladogram? As systematists gather new data, the position of the ctenophores on the animal tree will become clearer.

CHECKPOINT 31.1

- **CONNECT** *What is the significance of choanocytes in terms of the evolution of sponges? Explain their function in sponges.*
- *In what ways do cnidarians differ from sponges?*
- **VISUALIZE** *Sketch the major events of the life cycle of Obelia.*
- *In what ways are ctenophores like cnidarians? In what ways are they different?*

Figure 31-6 Bleached staghorn coral

This staghorn coral (*Acropora*) was photographed off Heron Island on the Great Barrier Reef off the coast of Australia.

atmosphere), and disease. The stress caused by these factors can result in the coral expelling the colorful symbiotic algae that inhabit their cells. This process is called **coral bleaching** (FIG. 31-6). Without their algae, coral become malnourished and may die if conditions do not improve. Other factors leading to the decline of coral reefs include pollution, smothering coral with the silt washed downstream from clear-cut forests, and overfishing.

Comb jellies have adhesive glue cells that trap prey

The fragile, luminescent *comb jellies* are marine animals known as **ctenophores** (ten′-oh-forz). They are assigned to phylum *Ctenophora* (teh-nof′-er-uh). Approximately 150 species of ctenophores have been described. Some are as small as a pea; others are larger than a tomato. The outer surface of a ctenophore bears eight rows of cilia that resemble combs (FIG. 31-7). Ctenophores are biradially symmetrical, meaning you could obtain equal halves by cutting through the body axis in two different

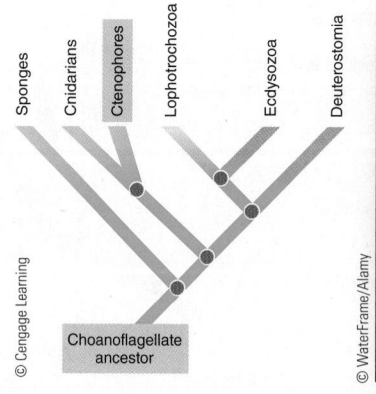

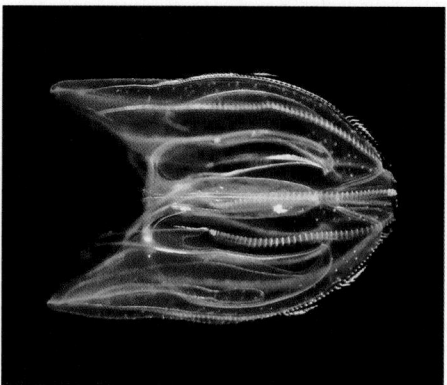

Figure 31-7 *Animation* Ctenophore (comb jelly)

Ctenophores are free-swimming, bioluminescent hermaphrodites capable of self-fertilization. The warty comb jelly (*Mnemiopsis leidyi*; also known as the sea walnut) refracts light so that it appears to have colors running down the track of its cilia. *Mnemiopsis* is also bioluminescent, and flashes when disturbed. This animal is a predator that consumes crustaceans, eggs and larvae of fishes, and zooplankton.

31.2 THE LOPHOTROCHOZOA

LEARNING OBJECTIVES

4 Summarize what is known about the phylogeny of the lophotrochozoans and cite the adaptive advantages of having a coelom and of cephalization.

5 Identify distinguishing characteristics of flatworms and nemerteans, and describe the main groups of flatworms, giving examples of animals that belong to each class.

6 Describe the main characteristics of mollusks and of the four main groups of mollusks discussed, giving examples of each group.

7 Describe the principal characteristics of annelids and of the three main groups of annelids discussed, giving examples of each group.

8 Describe the distinguishing characteristics of the lophophorates and of the rotifers.

We begin our discussion of the bilateral animals with the **Lophotrochozoa,** one of the two major clades of protostomes. The Lophotrochozoa includes the flatworms (platyhelminths), rotifers (wheel animals), nemerteans (ribbon worms), mollusks, annelids (segmented worms), and the lophophorate phyla (see Fig. 30-6). Recall from Chapter 30 that the name *Lophotrochozoa* comes from two important characters of some animals in this clade: (1) the *lophophore,* a ciliated ring of tentacles surrounding the mouth in three small groups of animals; and (2) the *trochophore larva,* a type of larva that characterizes two major groups: the mollusks and annelids. Some systematists argue that flatworms and rotifers should be assigned to a separate group, the *platyzoa.* Here, we include them with the lophotrochozoa.

Most lophotrochozoans have **bilateral symmetry,** at least in their larval stages. They are **triploblastic;** that is, they have three definite tissue layers. In addition to the outer epidermis derived from ectoderm and an inner endodermis derived from endoderm, lophotrochozoans have a middle tissue layer that develops from mesoderm.

Most lophotrochozoans have a true **coelom** and a *tube-within-a-tube body plan* (see Chapter 30). Most coelomate animals have well-developed circulatory, excretory, and nervous systems. Many internal organs are suspended within folds of the tissue lining the coelom and can move independently of the outer body wall. In addition, the coelom provides space for the gonads to develop. During the breeding season of many animals, such as birds, the gonads enlarge within the coelom as they fill with ripe gametes.

In some animals fluid within the coelom helps transport materials such as food, oxygen, and wastes. Cells bathed by the coelomic fluid can exchange materials with it. The cells receive nutrients and oxygen from the coelomic fluid and excrete wastes into it. Some coelomates have excretory structures that remove wastes directly from the coelomic fluid.

Animals that move about typically have an elongated body with definite anterior and posterior ends. They exhibit **cephalization,** the evolution of a head with a concentration of sense organs at the anterior end. A simple brain and paired sense organs are concentrated in the "head" region. An animal with a front end generally moves forward. With a concentration of sense organs in the part of the body that first meets its environment, the animal can actively search for food, shelter, and mates or can detect enemies quickly.

Flatworms are bilateral acoelomates

Platyhelminths (plat-ee-hel′-minths), or *flatworms,* (phylum *Platyhelminthes* are soft-bodied animals that left few fossils to provide clues to their evolutionary history. Their flat, elongated bodies are solid; that is, they are **acoelomate** (have no body cavity). Historically, biologists considered flatworms to be the simplest bilaterally symmetrical, triploblastic animals. However, it is unclear whether the Platyhelminthes evolved from ancestors that possessed a body cavity. A group of flatworm-like animals, the Acoelomorpha, were once considered a part of the phylum Platyhelminthes, but modern evidence, specifically DNA and RNA sequences, suggests that they form a separate lineage branching from the beginnings of the bilateral animals.

Based on our current understanding of the evidence, the 20,000 or so species within the Platyhelminthes fall within four classes. Class *Turbellaria* consists of the free-living flatworms, including planarians and their relatives. Classes *Trematoda* and *Monogenea* include the flukes, which are either internal or external parasites. Class *Cestoda* includes the tapeworms, which as adults are intestinal parasites of vertebrates (TABLE 31-2).

Flatworms typically have a simple **nervous system.** The brain consists of two masses of nervous tissue, called **ganglia,** in the head region. In many species the ganglia connect to two nerve cords that extend the length of the body. This nervous system is sometimes referred to as a "ladder-type nervous system" because a series of nerves connects the cords like the rungs of a ladder (see Fig. 42-2). Sense organs include simple "eyes," called eyespots, organs of balance (statocysts), and numerous chemoreceptors over their bodies for locating food and mates.

TABLE 31-2	Classes of Phylum Platyhelminthes
CLASS AND REPRESENTATIVE ANIMALS	**CHARACTERISTICS**
TURBELLARIA* Planarians	Mainly free-living; marine, freshwater, and terrestrial; body covered by ciliated epidermis; typically predatory on other invertebrates
TREMATODA AND MONOGENEA Flukes	Parasites with a wide range of vertebrate and invertebrate hosts; may require intermediate hosts; adults have suckers for attachment to host
CESTODA Tapeworms	Parasites of vertebrates; complex life cycle usually with one or two intermediate hosts; larval host may be invertebrate; typically have suckers and sometimes hooks for attachment to host; eggs produced within proglottids, which are shed; no digestive or nervous systems

*Systematists view Turbellaria as a paraphyletic group that will probably be divided into at least three groups.

© Cengage Learning

Flatworms have no organs for circulation or gas exchange. Gas exchange depends largely on diffusion through the body wall. The activities of some organs are coordinated and form simple organ systems, such as the digestive and nervous systems. As in the cnidarians, the digestive system is a gastrovascular cavity with only one opening, a mouth. The gastrovascular cavity is often extensively branched.

Parasitic flatworms—flukes and tapeworms—are highly adapted to and modified for their parasitic lifestyle. They have suckers or hooks for holding on to their hosts. The bodies of those that live in digestive tracts resist the digestive enzymes secreted by their hosts. Many have complicated life cycles and produce large numbers of eggs. Other adaptations include the loss of certain structures such as sense organs. Tapeworms have also lost the digestive system.

Turbellarians are free-living flatworms Most members of class Turbellaria are free-living flatworms. They inhabit marine, freshwater, and terrestrial habitats. Turbellarian flatworms typically have a muscular pharynx that takes in food and is connected with a branching gastrovascular cavity. They have a simple brain, multiple eyespots, and other sensory organs in the head; **protonephridia,** structures that function in osmoregulation (fluid balance) and metabolic waste disposal (excretion); and reproductive organs.

Planarians are turbellarian flatworms found in ponds, quiet streams, and moist terrestrial habitats throughout the world. One common genus of American planarians, *Dugesia*, contains aquatic worms that reach about 15 mm (0.6 in.) in length. They have what appear to be crossed "eyes" and flapping lateral projections called *auricles* (FIG. 31-8). The auricles actually serve as organs of chemoreception, which is important in locating food.

Most planarians are predators, capable of capturing and killing small animals. Their digestive system consists of a single opening (the mouth); a tubelike, muscular **pharynx** (the first portion of the digestive tube) and a branched gastrovascular cavity. Many planarians evert (turn inside out) their pharynx out of their mouth and release digestive enzymes onto their captured prey. The liquefied tissues of the prey are then swept by ciliary action into the gastrovascular cavity. The branches of the gastrovascular cavity distribute the nutrients throughout the body of the flatworm. Because planarians lack an anus, digestive wastes leave the body through the mouth.

Some metabolic wastes leave the body by diffusion. Other metabolic wastes are excreted by the protonephridia, blind tubules that end in **flame cells.** The flame cells are collecting cells equipped with cilia. The beating of the cilia channels waste through the system of tubules and eventually out of the body through excretory pores. Osmoregulation and protonephridia are discussed further in Chapter 48 (see Fig. 48-2).

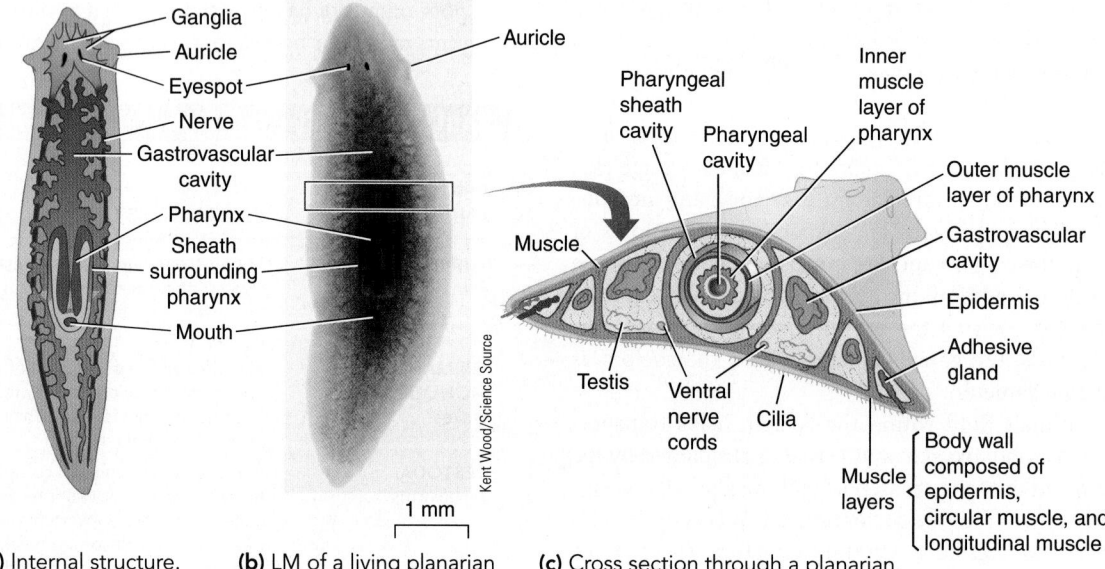

(a) Internal structure.

(b) LM of a living planarian (*Dugesia dorotocephala*). Note the auricles, used to locate food.

(c) Cross section through a planarian.

Figure 31-8 *Animation* **The common planarian, *Dugesia***
© Cengage Learning

Planarians reproduce either asexually or sexually. In asexual reproduction an individual constricts its body and divides into two planarians. Each regenerates its missing parts. Sexually, these animals are hermaphrodites. During the warm months of the year, each is equipped with a complete set of male organs and female organs. Two planarians come together in copulation and exchange sperm cells so that their eggs are cross-fertilized.

Flukes parasitize other animals Although their body plan generally resembles that of the free-living flatworms, the *flukes*, members of classes Trematoda and Monogenea, have structures, such as hooks and suckers, for attachment to their host. Flukes also have extremely prolific reproductive organs.

Flukes that are parasitic in humans include blood flukes, which are widespread in tropical areas of the world, and liver flukes, which are common in Asia, particularly in areas where humans use their own feces for fertilizing crops. Blood flukes of the genus *Schistosoma* cause the human disease schistosomiasis,

the second most debilitating parasitic disease in the world. (Malaria is the first.) An estimated 200 million people are infected, most of them children. The worms live in blood vessels for many years and can damage the liver and kidneys.

Both blood flukes and liver flukes go through complicated life cycles involving several different forms. There is an alternation of sexual and asexual stages as well as parasitism on one or more intermediate hosts, such as snails and fishes (**FIG. 31-9**). The aquatic snails that serve as intermediate hosts thrive in ponds, rice paddies, and the marshy areas that develop when dams are built.

Tapeworms inhabit the intestine of vertebrates Adult members of the more than 5000 species of class Cestoda live as parasites in the intestine of probably every kind of vertebrate, including humans. Tapeworms are long, flat, ribbonlike animals strikingly specialized for their parasitic mode of life. The body consists of a *scolex* for attachment to a host and a long chain of segments called *proglottids*. The scolex is equipped

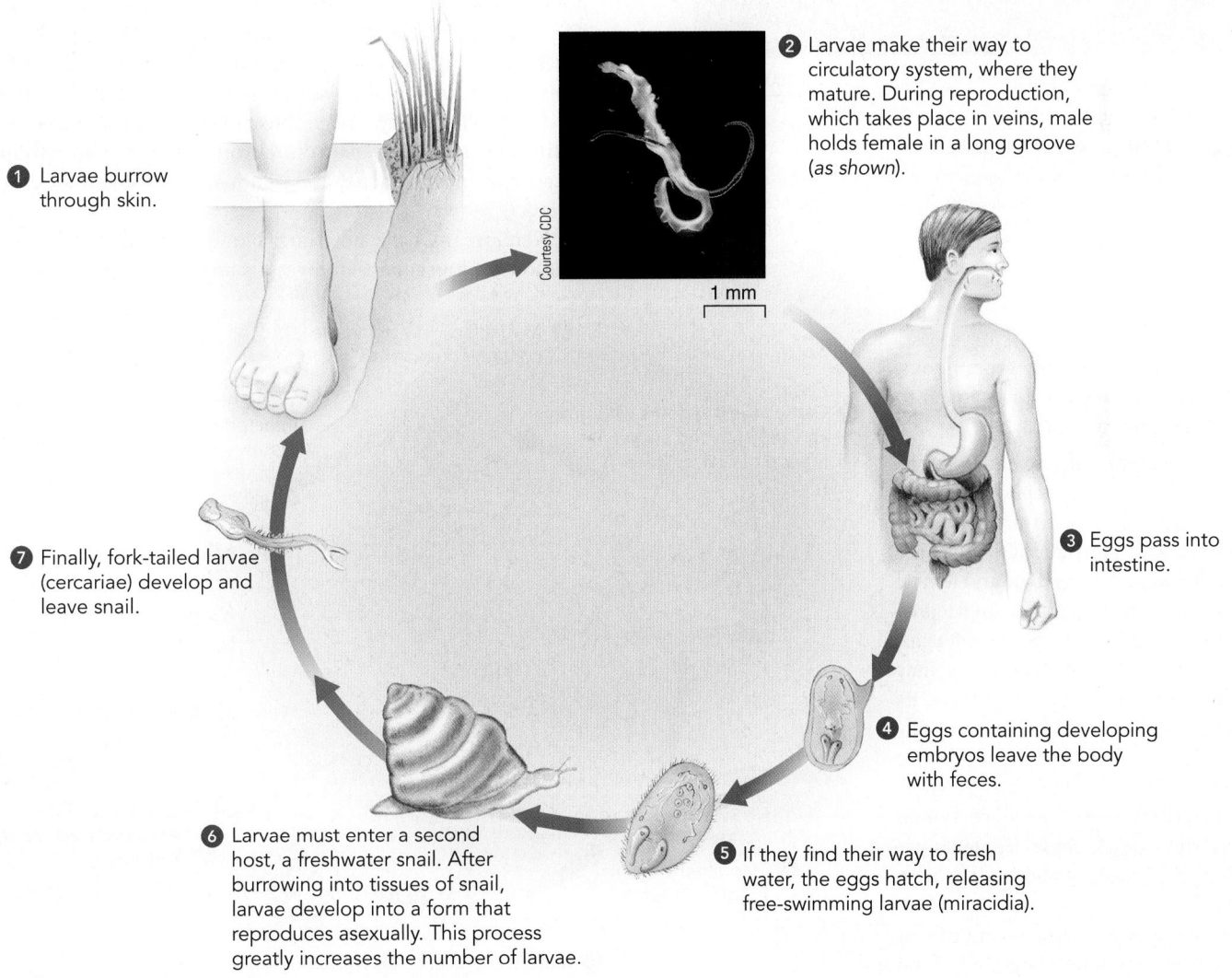

1 Larvae burrow through skin.

2 Larvae make their way to circulatory system, where they mature. During reproduction, which takes place in veins, male holds female in a long groove (*as shown*).

Courtesy CDC

⊢ 1 mm ⊣

3 Eggs pass into intestine.

4 Eggs containing developing embryos leave the body with feces.

5 If they find their way to fresh water, the eggs hatch, releasing free-swimming larvae (miracidia).

6 Larvae must enter a second host, a freshwater snail. After burrowing into tissues of snail, larvae develop into a form that reproduces asexually. This process greatly increases the number of larvae.

7 Finally, fork-tailed larvae (cercariae) develop and leave snail.

Figure 31-9 *Animation* Life cycle of the blood fluke (*Schistosoma*)
The LM shows an adult male enfolding a smaller female.
© Cengage Learning

250 μm

Figure 31-10 Scolex of a tapeworm

The small tapeworm (*Acanthrocirrus retrisrostris*) reaches maturity in the intestine of wading birds that eat barnacles. The color-enhanced SEM shows the pistonlike cluster of hooks that can be withdrawn into the head or thrust out and buried in the host's tissue. You can see two of the four powerful suckers beneath the hooks.

organs. Some tapeworms have complex life cycles, spending their larval stage within the body of an intermediate host and their adult life within the body of a different, final host. **FIGURE 31-11** illustrates the life cycle of the beef tapeworm, which can infect humans when they eat undercooked beef containing the larvae.

Nemerteans are characterized by their proboscis

The **nemerteans** (neh-mur′-tee-uns; phylum *Nemertea*) include the ribbon worms, a relatively small group of about 1200 species of free-living animals (**FIG. 31-12**). Most burrow in marine sediments, but a few species inhabit deep sea water, fresh water, or damp soil. Many nemerteans are predators, feeding on crustaceans and annelids. Nemerteans have long, narrow bodies, either cylindrical or flattened, ranging in length from 1 mm to more than 30 m (100 ft). Some are a vivid orange, red, or green, with black or colored stripes.

Their most remarkable organ, the *proboscis,* is a long, hollow, muscular tube that can be rapidly everted from the anterior end of the body. Used to capture prey, the sticky proboscis can be wrapped around a small animal. In some species the proboscis is sharp, and in various species it secretes toxic fluid that immobilizes prey. The proboscis is a *derived character* that distinguishes nemerteans from all other invertebrate groups. Because of this character, these animals are sometimes called *proboscis worms.*

Nemerteans have no heart; blood is circulated by contractions of muscular blood vessels and by movements of the

with suckers, and sometimes hooks, that enable the parasite to attach to the host's intestine (**FIG. 31-10**).

The reproductive adaptations and abilities of tapeworms are extraordinary. Each proglottid is an entire reproductive machine equipped with both male and female reproductive organs. A single proglottid contains up to 100,000 eggs. Because an adult tapeworm may have as many as 2000 segments, its reproductive potential is staggering. A single tapeworm can produce 600 million eggs per year! Proglottids farthest from the scolex contain the ripest eggs; these segments are shed from the host's body along with the feces.

A tapeworm has no mouth or digestive system. Digested food from the host is absorbed across the worm's body wall. Tapeworms also lack well-developed sense

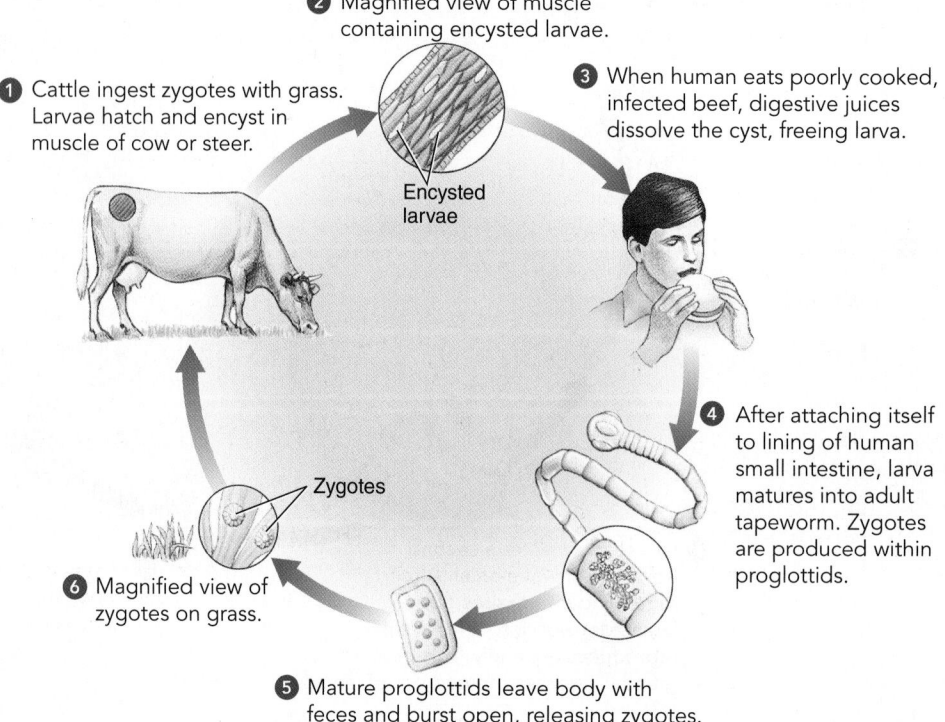

1 Cattle ingest zygotes with grass. Larvae hatch and encyst in muscle of cow or steer.

2 Magnified view of muscle containing encysted larvae.

Encysted larvae

3 When human eats poorly cooked, infected beef, digestive juices dissolve the cyst, freeing larva.

4 After attaching itself to lining of human small intestine, larva matures into adult tapeworm. Zygotes are produced within proglottids.

Zygotes

5 Mature proglottids leave body with feces and burst open, releasing zygotes.

6 Magnified view of zygotes on grass.

Figure 31-11 *Animation* **Life cycle of the beef tapeworm,** *Taenia saginata*
© Cengage Learning

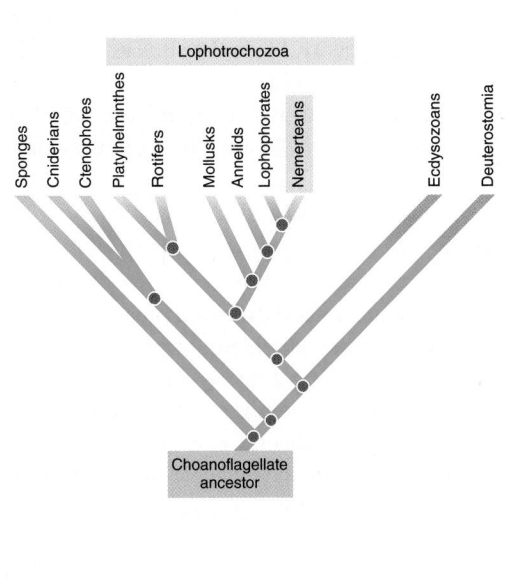

(a) Ribbon worm (*Lineus*) from the Pacific coast of Panama.

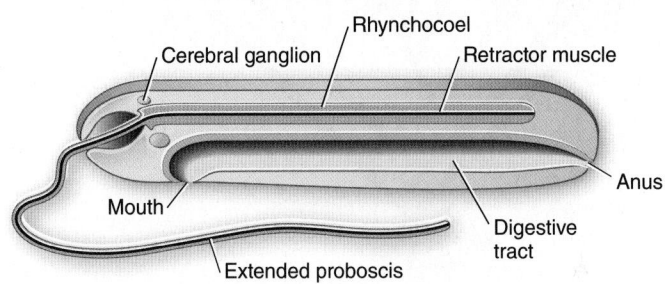

(b) Lateral view of a typical nemertean. Note the complete digestive tract that extends from mouth to anus.

Figure 31-12 Nemerteans (ribbon worms)
© Cengage Learning

body. The chamber surrounding the proboscis is a coelomic space known as a *rhynchocoel*. It develops in the embryo as the coelom does in coelomate protostomes (see Chapter 30). However, it differs in position and function from the coelom of other animals. Biologists view the rhynchocoel as a remnant of the coelom of a coelomate ancestor. Because of its presence and recent molecular evidence, biologists classify nemerteans with the lophotrochozoans. Like the flatworms, nemerteans evolved from coelomate animals and became simpler over time.

Mollusks have a muscular foot, visceral mass, and mantle

The **mollusks** (phylum *Mollusca*) include clams, oysters, snails, slugs, octopuses, and the largest of all the invertebrates, the giant squid, which averages 9 to 16 m (about 30 to 53 ft) in length, including its tentacles. More than 80,000 living species and 35,000 fossil species (second only to the arthropods in number) have been described. Here we discuss four of eight recognized classes. We illustrate representatives of these classes in FIGURE 31-13 and describe them in TABLE 31-3.

Although most mollusks are marine, many snails and clams live in fresh water, and some species of snails and slugs inhabit the land. Mollusks probably evolved early in the history of the protostome clade, soon after the evolution of the coelom but

before the origin of the segmented body that is characteristic of annelids. Although mollusks vary widely in outward appearance, most share six basic characteristics:

1. A soft body, usually covered by a dorsal shell composed mainly of calcium carbonate.
2. A broad, flat, muscular foot, located ventrally, which is used for locomotion.
3. The body organs (viscera) are concentrated as a **visceral mass** located above the foot.
4. The dorsal body wall forms a pair of folds called the **mantle.** The mantle is a thin sheet of tissue that generally overhangs the visceral mass, forming a mantle cavity. The mantle may contain glands that secrete a shell. The mantle cavity contains gills or a lung.
5. A rasplike structure called a **radula,** which is a belt of teeth in the mouth region. (The radula is not present in clams or their relatives, which are filter feeders.)
6. A coelom, generally reduced to small compartments around certain organs, including the heart and excretory organs (*metanephridia*). The main body cavity is typically a **hemocoel,** a space containing blood (see the following discussion of open circulatory systems). The hemocoel is not a coelom.

Mollusks have all the organ systems typical of complex animals. The digestive system is a tube, often coiled, consisting of a mouth, buccal cavity (mouth cavity), esophagus, stomach, intestine, and anus. The mollusk radula, located within the buccal cavity, projects out of the mouth and shows variation in structure among species related to feeding behaviors. In herbivorous species the radula is used to scrape algae from the surface of rocks or to remove tissue from live plants. The radulas of predatory mollusks may be used to inject toxins into prey, drill holes in the shells of prey animals, or rip apart flesh held in the beaks of cephalopods.

Most mollusks have an **open circulatory system** in which the blood, called **hemolymph,** bathes the tissues directly. The heart pumps blood into a single blood vessel, the aorta, which may branch into other vessels. Eventually, blood flows into a network of large spaces called *sinuses,* bringing the blood into direct contact with the tissues. This network makes up the hemocoel, or blood cavity. From the sinuses, blood drains into vessels that conduct it to the gills, where it is recharged with oxygen. After passing through the gills, the blood returns to the heart. Thus, blood flow in a mollusk follows this pattern:

heart $\longrightarrow$ aorta $\longrightarrow$ smaller blood vessels $\longrightarrow$ blood sinuses (hemocoel) $\longrightarrow$ blood vessels to gills $\longrightarrow$ heart

In open circulatory systems, blood pressure tends to be low, and tissues are not very efficiently oxygenated. Because most mollusks are slow-moving animals with low metabolic rates, this type of circulatory system is adequate. The active cephalopods (the class that includes the squids and octopuses) have a **closed circulatory system** in which blood flows through a complete circuit of blood vessels.

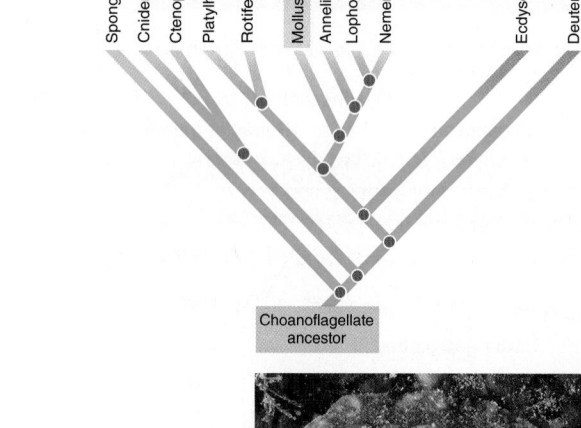

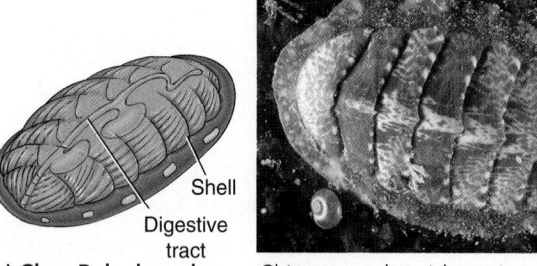

(a) Class Polyplacophora Chitons are sluggish marine animals with shells composed of eight overlapping plates. This sea cradle chiton (*Tonicella lineata*), which reaches about 5 cm (2 in.) in length, inhabits rocks in coastal waters off the U.S. Pacific Northwest.

Scott Leslie/Getty Images

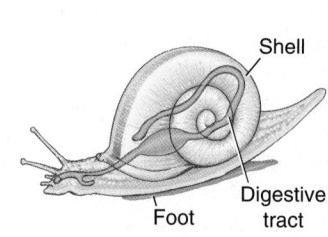

(b) Class Gastropoda The broad, flat foot of the gastropod is an adaptation to its mobile lifestyle. The mystery snail (*Pomacea bridgesi*) inhabits fresh water and can be found burrowing in the mud.

E. R. Degginger/Dembinsky Photo Associates

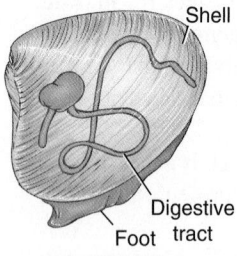

(c) Class Bivalvia The compressed body of the horseneck clam (*Tresus capax*) is adapted for burrowing in the North Pacific mud. Its shell grows to about 20 cm (8 in.) in length.

Tom McHugh/Science Source

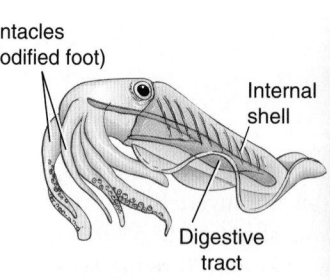

(d) Class Cephalopoda The squid body is streamlined for swimming. To avoid being seen by potential predators, a squid can change color to blend with its background. This northern short-fin squid (*Illex illecebrosus*), which grows to a mantle length of up to about 31 cm (about 1 ft), inhabits the North Atlantic Ocean.

Jeffrey L. Rotman/Getty Images

Figure 31-13 *Animation* The mollusk body plan
© Cengage Learning

TABLE 31-3 | Major Classes of Phylum Mollusca

CLASS AND REPRESENTATIVE ANIMALS	CHARACTERISTICS
POLYPLACOPHORA Chitons	Marine; dorsal shell consisting of eight separate transverse plates; head reduced; broad foot used for locomotion; uses radula to scrape algae and other small organisms off rocks
GASTROPODA Snails, slugs, nudibranchs	Marine, freshwater, or terrestrial; coiled shell in many species; torsion of visceral mass; well-developed head with tentacles and eyes; predators, herbivores, and detritus feeders
BIVALVIA Clams, oysters, mussels	Marine or freshwater; body laterally compressed; two-part shell hinged dorsally; hatchet-shaped foot; filter feeders
CEPHALOPODA Squids, octopuses	Marine; foot modified into tentacles, usually bearing suckers; well-developed eyes; closed circulatory system; predatory

© Cengage Learning

Most marine mollusks pass through one or more larval stages. The first stage is typically a **trochophore larva,** a free-swimming, top-shaped larva with two bands of cilia around its middle (**FIG. 31-14**). In many mollusks (gastropods such as snails and bivalves such as clams) the trochophore larva develops into a **veliger larva,** which has a shell, foot, and mantle. The veliger larva is unique to the mollusks.

Chitons may be similar to ancestral mollusks

Polypla-cophorans (pol-ee-plah-kof'-o-rans; meaning "many plates") are *chitons,* marine animals with flattened bodies (see Fig. 31-13a). Their most distinctive feature is a shell composed of eight separate but overlapping dorsal plates. The head is reduced, and there are no eyes or tentacles.

Chitons inhabit rocky intertidal zones, using a broad, flat foot to move and to hold on firmly to rocks. By pressing its mantle against the substratum and lifting the inner edge of the mantle, a chiton can produce a partial vacuum. The resulting suction lets the animal adhere powerfully to its perch. Using the radula for grazing, chitons scrape algae and other small organisms off rocks and shells.

Gastropods are the largest group of mollusks

The *gastropods*—the snails, slugs, conchs, sea slugs, and their relatives—are the largest and most diverse group of mollusks (see Fig. 31-13b). In fact, with more than 70,000 extant species, gastropods constitute the second-largest class in the animal kingdom, second only to insects. Most gastropods inhabit marine waters, but others make their homes in brackish or fresh water or on land. We think of snails as having a single, spirally coiled shell into which they can withdraw the body, and many do. However, other gastropods, such as limpets, have shells like flattened dunce caps. Still others, such as garden slugs and the beautiful marine slugs known as *nudibranchs,* have no shell at all (**FIG. 31-15**).

Many gastropods have a well-developed head with tentacles. Two simple eyes may be located on stalks that extend from the head. The gastropod uses its broad, flat foot for creeping. Most land snails do not have gills. Instead, the mantle is highly vascularized and functions as a lung. Biologists describe these garden snails and slugs as *pulmonate* ("having a lung").

Torsion, a twisting of the visceral mass, is a unique feature of gastropods. This twisting is unrelated to the coiling of the shell. As the bilateral larva develops, one side of the visceral mass grows more rapidly than the other side. This uneven growth rotates the visceral mass. The visceral mass and mantle twist permanently

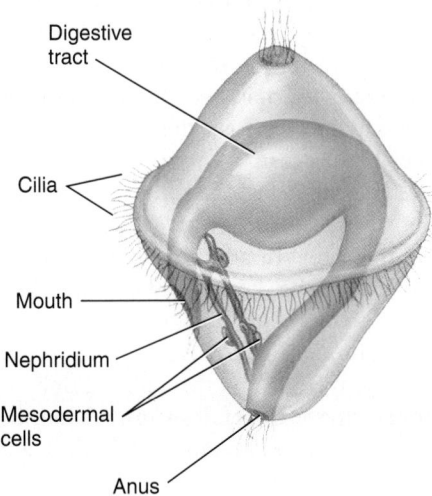

Figure 31-14 A trochophore larva

The first larval stage of a marine mollusk, the trochophore larva, is also characteristic of annelids. Just above the mouth a band of ciliated cells functions as a swimming organ and, in some species, collects suspended food particles.
© Cengage Learning

Figure 31-15 Nudibranch (*Felimare tricolor*)

Nudibranchs (known as sea slugs) are a group of gastropods that have no shell. They feed on sponges and other small marine animals. The species shown here is small, about 25 mm, but other nudibranchs grow to about 600 mm (24 in.).

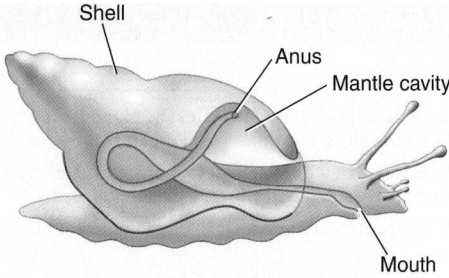

Figure 31-16 *Animation* **Torsion in a gastropod**

As the bilateral larva develops, the visceral mass twists 180 degrees relative to the head. The digestive tube coils, and the anus becomes relocated near the mouth.

© Cengage Learning

up to 180 degrees relative to the head. As a result, the digestive tract becomes somewhat U-shaped, and the anus comes to lie above the head and gill (**FIG. 31-16**). Subsequent growth is dorsal and usually in a spiral coil. Torsion limits space in the body, and typically the gill, excretory organ (metanephridium), and gonads are absent on one side. Some biologists hypothesize that torsion is an adaptation that protects the head by allowing it to enter the shell first during withdrawal from potential predators. Without torsion, the foot would be withdrawn first.

Most bivalves are filter feeders The *bivalves* include clams, oysters, mussels, scallops, and their relatives (see Fig. 31-13c). Typically, the soft body is laterally compressed and completely enclosed by a two-part shell that hinges dorsally and opens ventrally (**FIG. 31-17**). This arrangement allows the hatchet-shaped foot to protrude ventrally for locomotion and for burrowing in mud. The two parts, or *valves,* of the shell are connected by an elastic ligament. Stretching of the ligament opens the shell. Large, strong adductor muscles attached to the shell permit the animal to close its shell.

The inner, pearly layer of the bivalve shell is made of calcium carbonate secreted in thin sheets by the epithelial cells of the mantle. Known as *mother-of-pearl,* this material is valued for making jewelry and buttons. Should a bit of foreign matter lodge between the shell and the mantle, the epithelial cells covering the mantle secrete concentric layers of calcium carbonate around the intruding particle. Many species of oysters and clams form pearls in this way.

Some bivalves, such as oysters, attach permanently to the substrate. Others burrow slowly through rock or wood, seeking protected dwellings. The shipworm, *Teredo,* a clam that damages dock pilings and other marine installations, is just looking for a home. A few bivalves, such as scallops, swim rapidly by clapping their two shells together with the contraction of a large adductor muscle (the part of the scallop that humans eat).

Clams and oysters are filter feeders that trap food particles suspended in sea water. They take water in through an extension of the mantle called the *incurrent siphon.* As the water passes over the gills, mucus secreted by the gills traps food particles in the stream of water. Cilia move the food to the mouth. Water and digestive wastes leave by way of an *excurrent siphon.* An oyster can filter more than 30 L (about 32 qt) of water per hour! As filter feeders, bivalves have no need for a radula, and indeed they are the only group of mollusks that lack this structure.

Cephalopods are active predators In contrast with most other mollusks, members of class *Cephalopoda* (meaning "head-foot") are fast-swimming predators (see Fig. 31-13d). The mouth of cephalopods is surrounded by tentacles, or arms: 8 in octopuses, 10 in squids, and as many as 90 in the chambered nautilus. The large cephalopod head has well-developed eyes that form images. Although they develop differently, their complex eyes are structurally similar to vertebrate eyes and function in much the same way. Octopuses have no shell, and the shells of squid, located inside the body, are greatly reduced.

Nautilus has a coiled shell consisting of many chambers built up over time. Each year, the animal lives in the newest and largest chamber of the series. *Nautilus* secretes a mixture of gases similar to air into the other chambers. By regulating the amount of gas in the chambers, the animal controls its buoyancy in the ocean.

The tentacles of squids, octopuses, and cuttlefish are covered with suckers for seizing and holding prey. In addition to a radula, the mouth has two strong, horny beaks used to kill prey and tear it to bits. The thick, muscular mantle is fitted with a funnel-like *siphon.* By filling the cavity with water and ejecting it through the siphon, cephalopods achieve forceful jet propulsion. Squids and cuttlefish have streamlined bodies adapted for efficient swimming.

In addition to speed, a cephalopod has two other adaptations that facilitate escape from its predators, which include certain whales, seals, large fish, and other cephalopods. It can confuse the enemy by rapidly changing colors. By expanding and

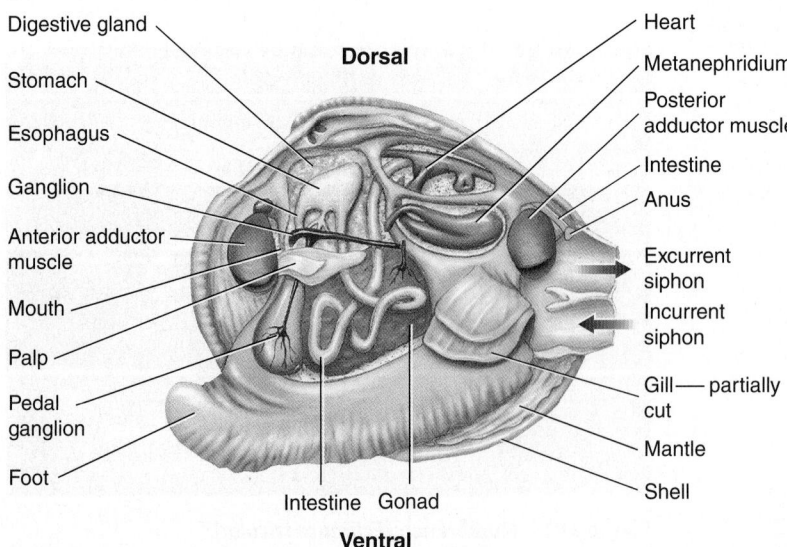

Figure 31-17 *Animation* **Internal anatomy of a clam**

The two shells of a bivalve hinge dorsally and open ventrally.

© Cengage Learning

contracting *chromatophores,* cells in the skin that contain pigment granules, a cephalopod can display an impressive variety of mottled colors. Another defense mechanism is its *ink sac,* which produces a thick, black liquid the animal releases in a dark cloud when alarmed. While its enemy pauses, temporarily blinded and confused, the cephalopod escapes. The ink inactivates the chemical receptors of some predators, making them incapable of detecting their prey.

Octopuses feed on crabs and other arthropods, catching and killing them with a poisonous secretion of their salivary glands. During the day, an octopus usually hides among the rocks; in the evening, it emerges to hunt for food. Its motion is incredibly fluid, giving little hint of the considerable strength in its eight arms.

Small octopuses survive well in aquariums and have been studied extensively. Because octopuses are relatively intelligent and can make associations among stimuli, researchers have used them as models for studying learning and memory. Their highly adaptable behavior more closely resembles that of the vertebrates than the more stereotypic patterns of behavior observed in other invertebrates (see Chapter 52).

Annelids are segmented worms

Annelids (an'-eh-lids; phylum *Annelida*) are segmented worms with bilateral symmetry and a tubular body that may be partitioned into more than 100 ringlike segments. This phylum, composed of about 15,000 species, includes three main groups: the polychaetes, a group of mainly marine worms; the earthworms and their relatives; and the leeches (**TABLE 31-4** and **FIG. 31-18**).

The term *Annelida* (from a Latin word meaning "little rings") refers to the series of rings, or segments, that make up the annelid body. Both the body wall and many of the internal organs are segmented. Some structures, such as the digestive tract and certain nerves, extend the length of the body, passing through successive segments. Other structures, such as excretory organs, are repeated in each segment. In polychaetes and earthworms, segments are separated from one another internally by transverse partitions called **septa** (sing., *septum*).

An important advantage of **segmentation** is that it facilitates locomotion. The coelom is divided into segments, and

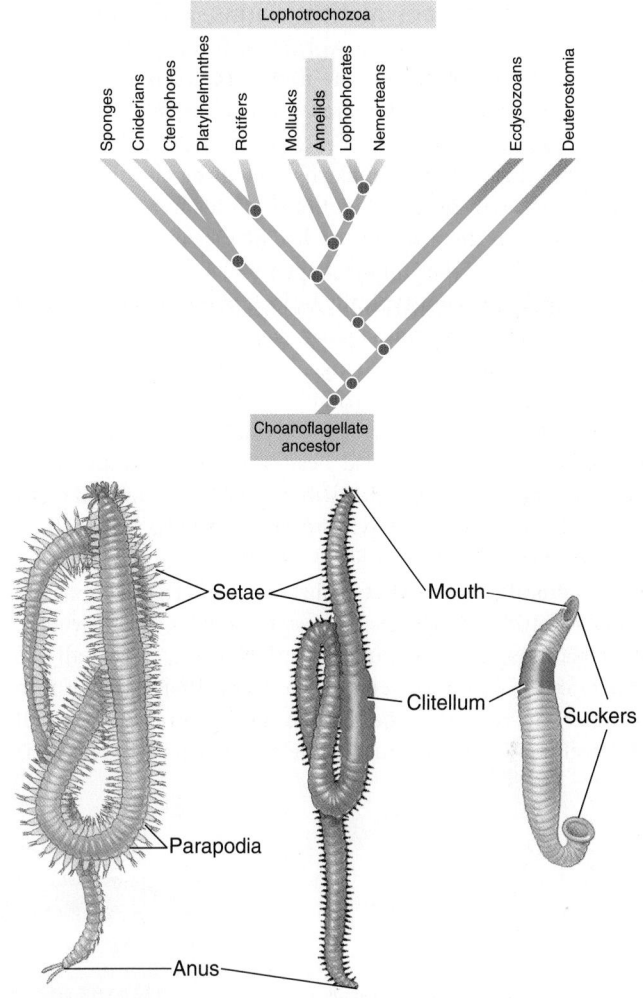

(a) Class Polychaeta. Polychaetes are marine worms with paddle-shaped parapodia.

(b) Class Oligochaeta. Oligochaetes, which include the earthworms, inhabit fresh water and moist terrestrial areas.

(c) Class Hirudinida. Many leeches are parasites that feed on blood.

Figure 31-18 Three major groups of annelids
© Cengage Learning

TABLE 31-4	Major Classes of Phylum Annelida
CLASS AND REPRESENTATIVE ANIMALS	**CHARACTERISTICS**
POLYCHAETA Sandworms, tubeworms	Mainly marine; each segment bears a pair of parapodia with many setae; well-developed head; separate sexes; trochophore larva
OLIGOCHAETA Earthworms	Terrestrial and freshwater worms; few setae per segment; lack well-developed head; hermaphroditic
HIRUDINIDA Leeches	Most are blood-sucking parasites that inhabit fresh water; appendages and setae absent; prominent muscular suckers

© Cengage Learning

each segment has its own muscles. This arrangement allows the animal to elongate one part of its body while shortening another part. The annelid's hydrostatic skeleton is important in movement. Polychaetes and earthworms have bristlelike structures called **setae** (sing., *seta*) located on each segment. Setae provide traction as the worm moves along by alternating contraction of its longitudinal and circular muscles. In earthworms setae consist mainly of chitin.

The annelid nervous system typically consists of a ventral nerve cord and a simple brain consisting of a pair of ganglia. Each segment has a pair of ganglia and lateral nerves. Annelids have a large, well-developed coelom; a closed circulatory system; and a complete digestive tract extending from mouth to anus. Respiration is *cutaneous*—that is, through the skin—or by gills. Typically, a pair of excretory tubules called **metanephridia** is located in each segment (described in Chapter 48).

Most polychaetes are marine worms with parapodia
Most *polychaetes* are marine worms, and many swim freely in the sea. Some polychaetes are part of the plankton and are important in marine food chains. Others burrow in the mud near the shore or in crevices of rocks or coral reefs. Still others live in tubes they secrete or make by cementing bits of shell and sand together with mucus. Giant tube worms have been found in deep vents in the floor of the Pacific Ocean (see *Inquiring About: Life without the Sun* in Chapter 55).

Each body segment typically bears a pair of paddle-shaped appendages called **parapodia** (sing., *parapodium*) that function in locomotion and in gas exchange. These fleshy structures bear many stiff setae (the term *polychaete* means "many bristles"; see the fireworm in the chapter-opening photograph). Most polychaetes have a well-developed head with eyes and antennae. The head may also be equipped with sensory tentacles and palps (feelers). Polychaetes develop from free-swimming trochophore larvae similar to those of mollusks.

Behavioral patterns that ensure fertilization have evolved in many polychaete species. By responding to certain rhythmic variations, or cycles, in the environment, nearly all the females and males of a given species release their gametes into the water at the same time. For example, more than 90% of reef-dwelling *Palolo* worms of the South Pacific shed their eggs and sperm within a 2-hour period on the same night of the year. The posterior portion of the *Palolo* worm, loaded with eggs or sperm, breaks off from the rest of the body. It comes to the surface and bursts, releasing its gametes. This mass reproductive event occurs in October or November at the beginning of the last lunar quarter. Local islanders gather great numbers of the swarming polychaetes, which they bake or eat raw.

Earthworms help maintain fertile soil The approximately 3100 species of *oligochaetes* live almost exclusively in fresh water and in moist terrestrial habitats. These worms lack parapodia, have only a few bristles per segment (the term *oligochaete* means "few bristles"), and lack a well-developed head. All oligochaetes are hermaphroditic.

Lumbricus terrestris, a common earthworm, can reach 20 cm (8 in.) or more in length. Its body has more than 100 segments, separated externally by grooves that indicate the internal position of the septa (**FIG. 31-19**). The earthworm's body is somewhat protected from drying by a thin, transparent **cuticle** secreted by the cells of the epidermis. Mucus secreted by gland cells of the epidermis forms an additional protective layer over the body surface. The body wall has an outer layer of circular muscles and an inner layer of longitudinal muscles.

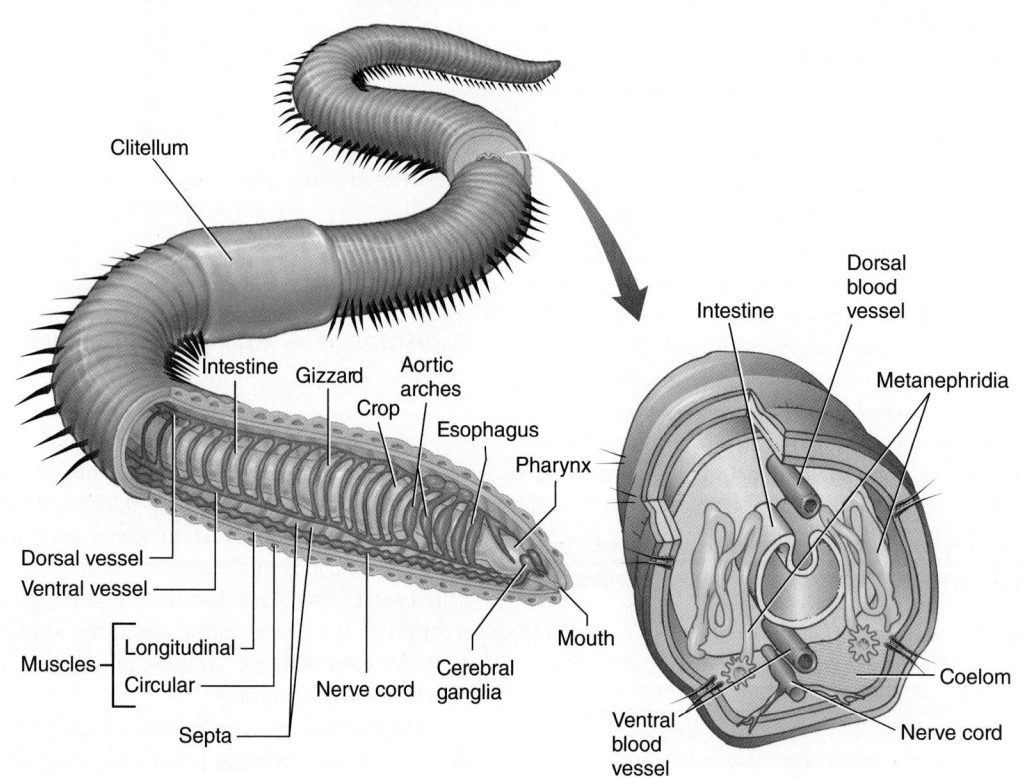

(a) The internal structure has been exposed at the anterior end of an earthworm.

(b) Cross section of an earthworm.

Figure 31-19 *Animation* **Body plan of an earthworm**

(Reproductive structures are not shown.)

© Cengage Learning

Earthworms eat living and dead plant material; some species feed mainly above the soil surface and other species feed within the ground, consuming mineral soil particles and organic matter. Burrowing and feeding activity by earthworms moves nutrients through the soil and creates pathways for oxygen and water movement, greatly influencing suitability of the soil for plant growth.

Earthworms process their meals in complex digestive systems. Food is swallowed through the muscular **pharynx** and passes through the **esophagus,** which is modified to form a thin-walled *crop* where food is stored, and a thick-walled, muscular *gizzard*. In the gizzard food is ground to bits by sand grains consumed along with the meal. The rest of the digestive system is a long, straight *intestine* that chemically digests food and absorbs nutrients. Wastes pass out of the intestine to the exterior through the **anus.**

The efficient closed circulatory system consists of two main blood vessels that extend longitudinally. The dorsal blood vessel, just above the digestive tract, collects blood from vessels in the segments. The dorsal vessel functions as a heart; it contracts to pump blood anteriorly. In the region of the esophagus, five pairs of circumesophageal blood vessels, referred to as *aortic arches,* conduct blood from the dorsal to the ventral blood vessel. The ventral blood vessel receives blood from the aortic arches. Small vessels branch from it and deliver blood to the various structures in each segment and to the body wall. Within these structures, blood flows through tiny capillaries before returning to the dorsal blood vessel.

Gas exchange takes place through the moist skin. Oxygen is transported by *hemoglobin,* a respiratory pigment in the blood. The excretory system consists of paired metanephridia, repeated in almost every segment of the body, that remove nitrogenous wastes from the coelomic fluid.

The nervous system consists of a pair of *cerebral ganglia* (a simple brain) just above the pharynx and a subpharyngeal ganglion just below the pharynx. A ring of nerve fibers connects these ganglia. From the lower ganglion, a ventral nerve cord extends beneath the digestive tract to the posterior end of the body. A pair of fused ganglia is present in each segment along the nerve cord. Nerves extend laterally from the ganglia to the muscles and other structures of that segment. The ganglia coordinate muscle contractions of the body wall, allowing the worm to creep along.

Like most annelids, earthworms are hermaphroditic. During copulation, two worms, headed in opposite directions, press their ventral surfaces together. These surfaces become glued together by the thick mucous secretions of each worm's *clitellum,* a thickened ring of epidermis. Sperm are then exchanged and stored in the *seminal receptacles* (small sacs) of the other worm. A few days later, each clitellum secretes a membranous cocoon containing a sticky fluid. As the cocoon is slipped forward, eggs are laid in it. Sperm are added as the cocoon passes over the openings of the seminal receptacles. As the cocoon slips free over the worm's head, its openings constrict so that a spindle-shaped capsule is formed. The fertilized eggs develop into tiny worms within this capsule. This reproductive pattern is an adaptation to terrestrial life; the cocoon protects the delicate gametes and young worms from drying out.

Many leeches are blood-sucking parasites Leeches are members of a group called *Hirudinida*. Most leeches inhabit fresh water, but some live in the sea or in moist areas on land. Some leeches are nonparasitic predators that capture small invertebrates such as earthworms and snails. However, about 75% of the known species of leeches are blood-sucking parasites. Leeches differ from other annelids in having neither setae nor parapodia.

Leeches have muscular suckers at both the anterior and posterior ends of their body. Most parasitic leeches attach themselves to a vertebrate host, bite through the skin, and suck out a quantity of blood, which is stored in pouches in the digestive tract. *Hirudin,* an anticoagulant secreted by glands in the crop, ensures leeches a full meal of blood. In 31 minutes, a leech can suck out as much as 10 times its own weight in blood! Some leeches feed only about twice each year because they digest their food slowly over several months.

Leeches have been used since ancient times for drawing blood from areas swollen by poisonous stings and bites. During the 19th century, they were widely used to remove "bad blood," thought to be the cause of many diseases. Leeches are sometimes used in modern medicine to remove excess fluid and blood that accumulate within body tissues as a result of injury, disease, or surgery (**FIG. 31-20**). The leech attaches its sucker near the site of injury, makes an incision, and secretes hirudin. The hirudin prevents the blood from clotting and dissolves clots that already exist.

The lophophorates are distinguished by a ciliated ring of tentacles

With a few exceptions, **lophophorates** are marine animals adapted for life on the ocean floor. These coelomates are distinguished by their *lophophore,* a ring of ciliated tentacles that surrounds the mouth. The lophophore is an adaptation for capturing suspended particles in the water. Lophophorates do not have a distinct head.

The three lophophorate groups are Brachiopoda, Phoronida, and Bryozoa (also known as Ectoprocta). *Brachiopods,* or lampshells, are solitary marine animals that inhabit cold water. Until the mid-19th century, they were thought to be mollusks because they are suspension feeders that superficially resemble clams and other bivalve mollusks. Their body is enclosed between two shells and has a mantle and mantle cavity (**FIG. 31-21a**). However, brachiopods differ from bivalve mollusks in that the shells are dorsal and ventral rather than lateral; each shell is symmetrical about the midline, and the two shell valves are typically of unequal size.

Brachiopods attach to the substrate by a long stalk. The action of cilia on the lophophore brings water with suspended food into the slightly opened shell. Although there are now only about 325 living species, brachiopods were abundant and diversified during the Paleozoic era. Paleobiologists have described about 12,000 fossil species.

Only about 20 extant species of *phoronids* are known. They are wormlike, sessile animals found in coastal marine sediments.

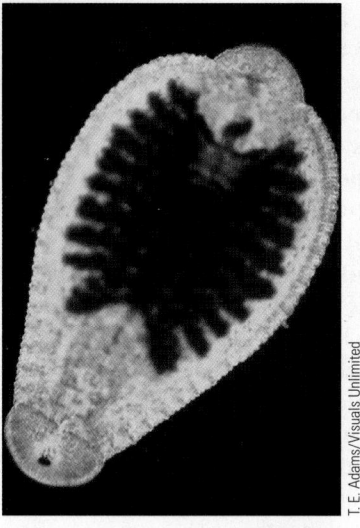

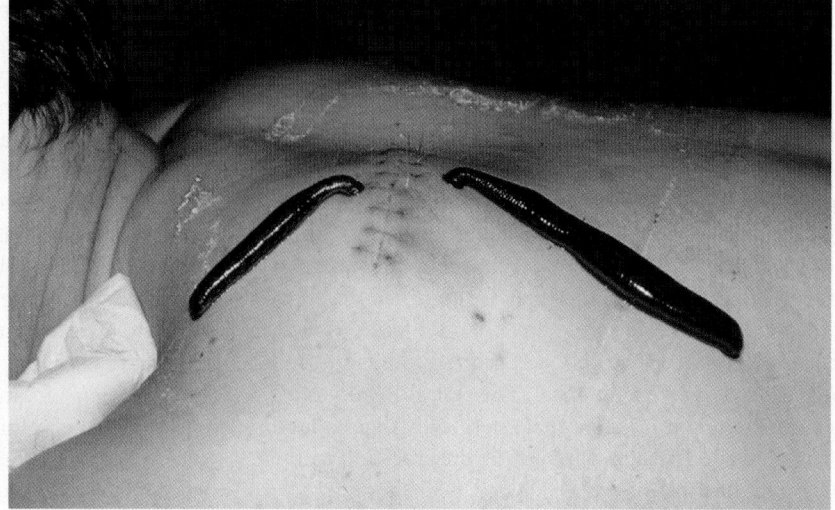

(a) LM of a blood-sucking leech (*Helobdella stagnalis*) that feeds on mammals. The digestive tract (*dark area*) of its swollen body is filled with ingested blood.

(b) Medicinal leeches (*Hirudo medicinalis*) are used here to treat hematoma, an accumulation of blood within tissues that results from injury or disease.

Figure 31-20 Leeches

Adult phoronids secrete chitinous tubes in which they live (**FIG. 31-21b**). They extend their lophophores from their tubes for feeding. Phoronids vary in length from 2 cm to more than 20 cm (less than 1 in. to about 8 in.), but they are only 1 to 3 mm in diameter.

Bryozoans, also known as *ectoprocts* or "moss animals," are microscopic aquatic animals that form sessile colonies by asexual budding (**FIG. 31-21c**). Each colony can consist of millions of individuals and can extend in length up to about 31 cm (about 1 ft). The colonies typically appear plantlike, but some have the appearance of coral. Approximately 4500 living species are known.

Biologists are still debating the evolutionary position of the lophophorates. They were traditionally considered a monophyletic group and, based on certain structural similarities, a sister group to the deuterostomes. However, the lophophorate groups also have several morphological characters that suggest a relationship with the lophotrochozoans.

(a) **Brachiopods.** Like other brachiopods, these northern lampshells (*Terebratulina septentrionalis*) superficially resemble clams.

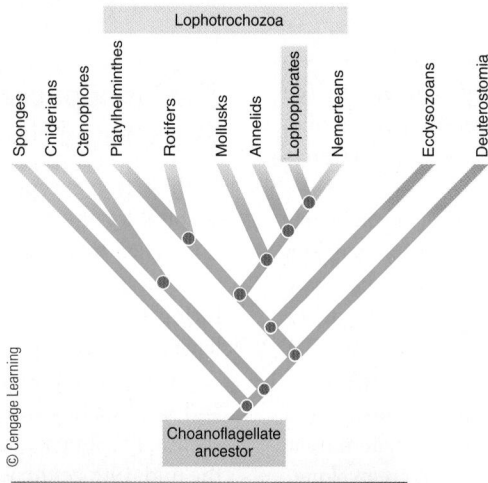

(b) **Phoronids.** A single lophophore of a phoronid (*Phoronopsis viridis*). This animal is about 20 cm (8 in.) long, but only 1–3 mm in diameter.

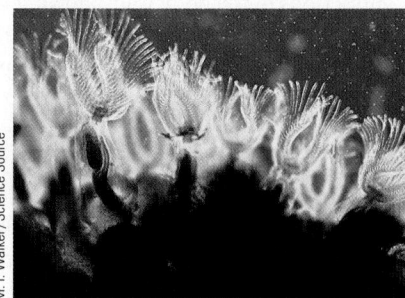

(c) **Bryozoans.** Like many bryozoans, this freshwater species (*Cristatella mucedo*) forms colonies by asexual budding.

Figure 31-21 Representative lophophorates

Based on molecular data (18S rDNA nucleotide sequences), the lophophorate phyla are currently classified as Lophotrochozoa, although their phylogeny is still being debated.

Rotifers have a crown of cilia

Rotifers (phylum *Rotifera*), or "wheel animals," are among the less familiar invertebrates. Although no larger than many protozoa, these aquatic, microscopic animals are multicellular with complex organ systems. Most of the 2000 described species inhabit fresh water, but some live in marine environments or damp soil. Rotifers are important in cycling nutrients in aquatic ecosystems. These animals do not have a true coelom and were formerly classified as pseudocoelomates. Current evidence suggests that rotifers evolved from animals with a true coelom, and we classify them as lophotrochozoans.

Rotifers have a characteristic crown of cilia on their anterior end. The cilia beat rapidly during swimming and feeding, giving the appearance of a spinning wheel (FIG. 31-22; also see Fig. 20-1). Rotifers feed on tiny organisms suspended in the stream of water drawn into the mouth by the cilia. A muscular organ posterior to the mouth grinds the food. The complete digestive tract ends in an anus through which wastes are eliminated.

Rotifers have a nervous system with a "brain" and sense organs, including eyespots. Protonephridia with flame cells remove excess water from the body and may also excrete metabolic wastes.

Biologists have discovered that rotifers, like some other animals with a pseudocoelom, are "cell constant," meaning that each member of a given species has exactly the same number of cells. Indeed, each part of the body has a precisely fixed number of cells arranged in a characteristic pattern. Cell division does not take place after embryonic development, and mitosis cannot be induced. Growth results from an increase in cell size. Do you think that rotifers could develop cancer? One of the challenging research problems of our time is to discover the difference between such nondividing cells and the dividing cells of other animals.

Another interesting characteristic of rotifers is their ability to survive very dry conditions and temperature extremes. These animals can remain in a dormant state for months or years. When conditions become favorable, they recover and become active once again.

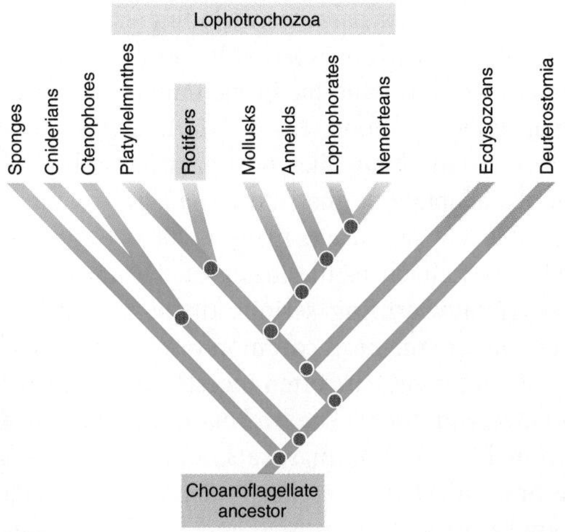

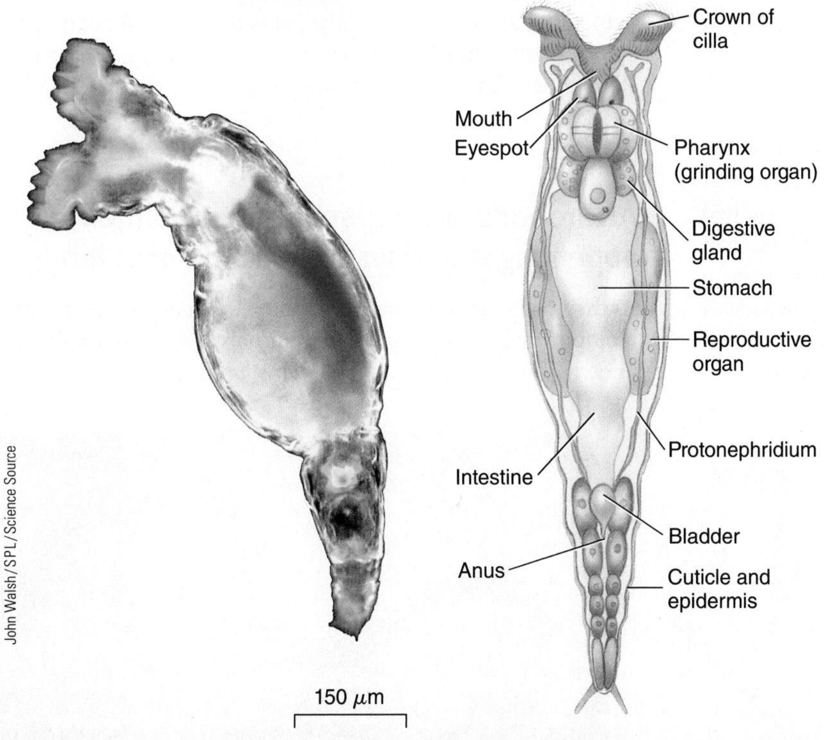

CHECKPOINT 31.2

- *What are some advantages of cephalization and a coelom?*
- CONNECT *On what basis have biologists classified nemerteans as lophotrochozoans?*
- CONNECT *How does the lifestyle of a gastropod differ from that of a cephalopod? Identify adaptations that have evolved in each for its particular lifestyle.*
- *What are two distinguishing characteristics for each of the following: flatworms, mollusks, and annelids?*

(a) LM of an Antarctic rotifer (*Philodina gregaria*) that survives the winter by forming a cyst. It reproduces in great numbers, sometimes coloring the water red.

150 μm

John Walsh/SPL/Science Source

(b) Longitudinal section showing rotifer anatomy. The motion of its cilia draws particles of food such as algae into the mouth.

Figure 31-22 Rotifers (wheel animals)
© Cengage Learning

31.3 THE ECDYSOZOA

LEARNING OBJECTIVES

9   Summarize what is known about the phylogeny of the ecdysozoans.

10 Describe characteristics of nematodes and identify four parasitic nematodes.

11 Describe six key characteristics of arthropods, summarize proposed phylogeny of arthropods, and distinguish among arthropod subphyla and classes; give examples of animals that belong to each group.

12 Identify adaptations that have contributed to the biological success of insects.

The **Ecdysozoa** include the nematodes (roundworms) and arthropods, animals characterized by a **cuticle,** a noncellular body covering secreted by the epidermis. The name *Ecdysozoa* refers to the process of **ecdysis,** or **molting,** characteristic of animals in this group. During ecdysis, an animal sheds its outer covering (e.g., cuticle, exoskeleton, or skin), which is then replaced by the growth of a new one. The taxonomic validity of the Ecdysozoa has been supported by many types of data, including molecular data.

Roundworms are of great ecological importance

Nematodes (nem′-uh-todes; phylum *Nematoda*), or *round-worms,* play key ecological roles as decomposers, parasites, and predators of smaller organisms. Many soil nematodes eat bacteria. Nematodes are numerous and are widely distributed in soil and in marine and freshwater sediments. More than 25,000 species have been described, and many more remain to be discovered. The nematodes rival the insects in number of individuals. A spadeful of soil may contain thousands of these mainly microscopic worms, which thrash around, coiling and uncoiling. Greater understanding of soil nematode diversity and ecology will contribute to solving some important agricultural problems. *Caenorhabditis elegans,* a free-living soil nematode, is an important model research organism for biologists studying the genetic control of development (see Chapter 17).

The elongated, cylindrical, threadlike nematode body is pointed at both ends and covered with a tough, flexible cuticle (**FIG. 31-23**). Secreted by the underlying epidermis, the thick cuticle gives the nematode body shape and offers some protection. The epidermis is unusual in that it does not consist of distinct cells.

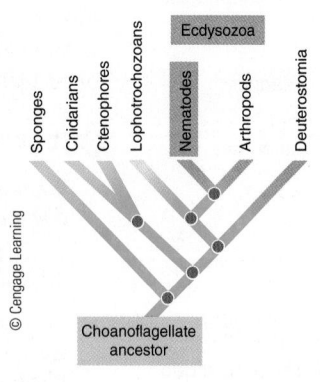

Nematodes have a fluid-filled pseudocoelom that serves as a hydrostatic skeleton. It transmits the force of muscle contraction to the enclosed fluid. Movement of fluid in the pseudocoelom is also important in transporting nutrients and wastes. Like ribbon worms, nematodes exhibit bilateral symmetry, a complete digestive tract, three definite tissue layers, and definite organ systems; however, they lack specific circulatory structures. The sexes are usually separate, and the male is generally smaller than the female.

Although most nematodes are free-living, others are important parasites in plants and animals. More than 50 species of roundworms are human parasites, including *Ascaris,* hookworms, pinworms, and the trichina worm. The common intestinal parasite *Ascaris* spends its adult life in the human intestine, where it ingests partly digested food (**FIG. 31-24**). *Ascaris* is a white worm up to 25 cm (10 in.) long. Like most parasites, it has remarkable reproductive adaptations that ensure survival within hosts. A mature female may produce as many as 200,000 eggs per day!

Hookworms attach to the lining of the intestine and suck blood, potentially causing serious tissue damage and blood loss. *Pinworms* are the most common worms found in children. The tiny pinworm eggs are often ingested by eating with hands contaminated with them. The *trichina worm* lives inside a variety of animals, including pigs, rats, and bears. Humans typically become infected by eating undercooked, infected meat. Larvae encyst in skeletal muscle and nervous tissue. Filiarial nematodes are small, threadlike parasitic worms that are spread among hosts by the bites of insects. Members of this group cause elephantiasis in humans. The dog heartworm infects cats, dogs, and their relatives.

Arthropods are characterized by jointed appendages and an exoskeleton of chitin

Arthropods (ar′-thro-pods; phylum *Arthropoda*) are the most biologically successful group of animals. More than 80% of all known animals are arthropods! They are more diverse and live

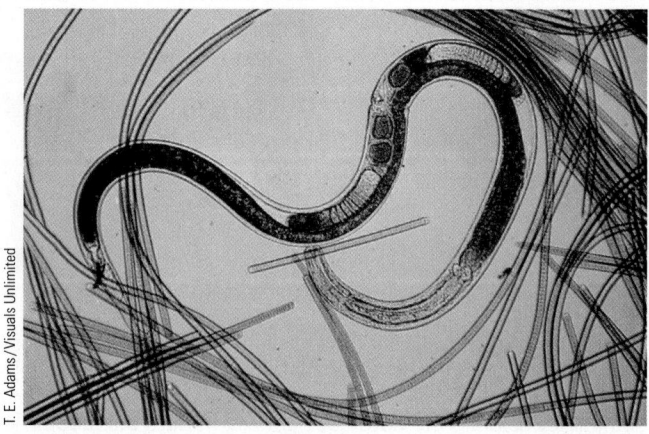

Figure 31-23 LM of a free-living nematode
This unidentified aquatic nematode is shown among the cyanobacterium *Oscillatoria,* which it eats.

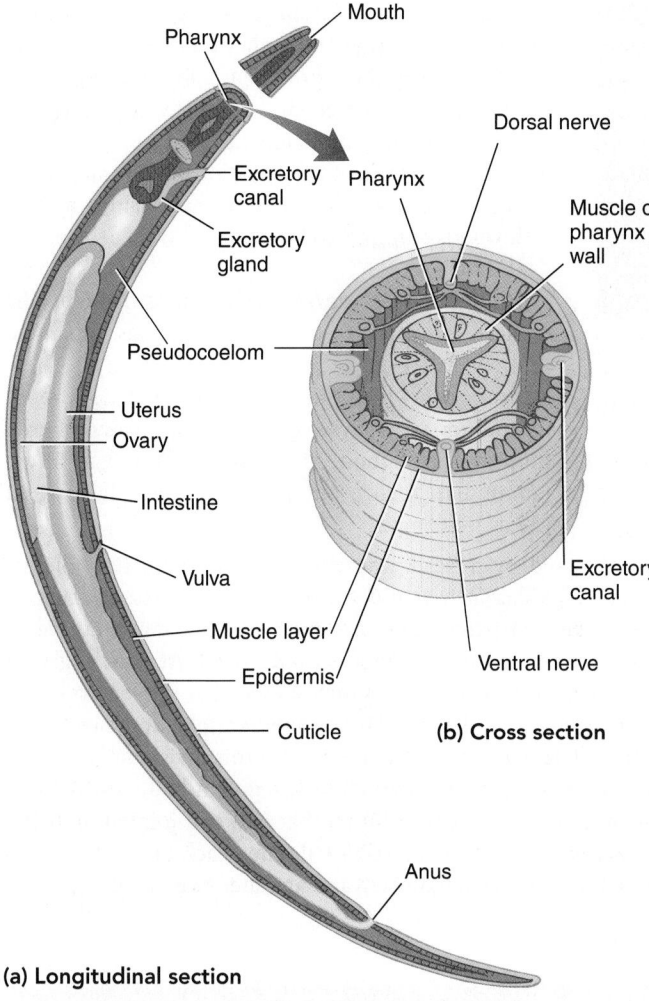

Mouth
Pharynx
Excretory canal
Excretory gland
Pseudocoelom
Uterus
Ovary
Intestine
Vulva
Muscle layer
Epidermis
Cuticle

(a) Longitudinal section

Pharynx
Dorsal nerve
Muscle of pharynx wall
Excretory canal
Ventral nerve

(b) Cross section

Anus

Figure 31-24 *Animation* **The roundworm *Ascaris***
(a) Note the complete digestive tract that extends from mouth to anus.
(b) This cross section through *Ascaris* shows the tube-within-a-tube body plan. The protective cuticle that covers the body helps this parasite resist host defenses.
© Cengage Learning

in a greater range of habitats than the members of any other animal phylum. More than one million species have been described, and biologists predict millions more will be identified. The following adaptations have greatly contributed to their success:

1. The arthropod body, like that of the annelid, is segmented. *Segmentation* is important from an evolutionary perspective because it provides the opportunity for specialization of body regions. In arthropods groups of segments are specialized to perform particular functions. Segments differ in shape, muscles, or the appendages they bear.

2. A hard **exoskeleton,** composed of chitin and protein, covers the entire body and appendages. The exoskeleton serves as a coat of armor that protects against predators and helps prevent excessive loss of moisture. It also supports the underlying soft tissues. Arthropods move effectively because distinct muscles attach to the inner surface of the exoskeleton and operate the joints of the body and appendages.

A disadvantage of the exoskeleton is that it is nonliving, and the arthropod periodically outgrows it. Recall that *molting* is the process of shedding an old exoskeleton, which is then replaced by a larger one. The shed exoskeleton represents a net metabolic loss, and molting also leaves the arthropod temporarily vulnerable to predators.

3. *Paired, jointed appendages,* from which this group gets its name (*arthropod* means "jointed foot"), are modified for many functions. They serve as swimming paddles, walking legs, mouthparts for capturing and manipulating food, sensory structures, or organs for transferring sperm.

4. The nervous system, which resembles that of the annelids, consists of a "brain" (cerebral ganglia) and a ventral nerve cord with ganglia. In some arthropods successive ganglia may fuse. Arthropods have a variety of very effective sense organs. Many have organs of hearing and **antennae** that sense taste and touch. Most insects and many crustaceans have **compound eyes** composed of many light-sensitive units called **ommatidia** (see Fig. 43-17). The compound eye can form an image and is especially adapted for detecting movement.

Arthropods have an *open circulatory system*. A dorsal, tubular heart pumps **hemolymph** into a dorsal artery, which may branch into smaller arteries. From the arteries, hemolymph flows into large spaces that collectively make up the hemocoel. Eventually, hemolymph re-enters the heart through openings, called *ostia,* in its walls.

The exoskeleton presents a barrier to diffusion of oxygen and carbon dioxide through the body wall, necessitating the evolution of specialized respiratory systems for gas exchange. Most aquatic arthropods have gills that function in gas exchange, whereas many terrestrial forms have a system of internal branching air tubes called **tracheae,** or **tracheal tubes** (see Fig. 46-1b). Other terrestrial arthropods have platelike *book lungs* (see Fig. 46-1d).

The arthropod digestive system is a tube similar to that of earthworms. Excretory structures vary somewhat among groups. The coelom is small and filled chiefly by the organs of the reproductive system.

Arthropod evolution and classification are controversial

Two groups of animals thought to be closely related to arthropods are the onychophorans and the tardigrades (**FIG. 31-25**). These three groups make up clade *Panarthropoda*. Like arthropods, onychophorans and tardigrades have a thick cuticle and must molt to grow. The body is segmented in all three groups, and all have legs, claws, and a ventral nervous system. Panarthropods have an open circulatory system, and the coelom is reduced to a hemocoel.

The caterpillar-like *onychophorans,* known as velvet worms, probably branched early from the arthropod line. They have paired appendages, but these are not jointed. As in arthropods, their jaws are derived from appendages. Some biologists place the *tardigrades,* known as "water bears," as the sister taxon of the arthropods. These tiny animals (typically less than 1 mm in length) have short, unjointed, clawed legs (see Fig. 2-12).

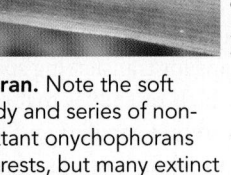

(a) Onychophoran. Note the soft segmented body and series of non-jointed legs. Extant onychophorans live in humid forests, but many extinct species inhabited the ocean.

(b) Tardigrades. Commonly known as water bears, tardigrades normally live in moist habitats, such as thin films of water on mosses. Most are less than 1 mm long.

Figure 31-25 Onychophorans and tardigrades

These animals are thought to be close relatives of arthropods.

Arthropod fossils have been identified dating back to the Proterozoic eon. Early in their evolutionary history, more than 540 million years ago (mya), arthropods diverged into several groups. Arthropods evolved rapidly during the Cambrian radiation, as evidenced by many diverse fossils in the Burgess Shale and other ancient sites.

We discuss five main arthropod groups, based on molecular and other data: the extinct trilobites and the extant Myriapoda, Chelicerata, Crustacea, and Hexapoda (TABLE 31-5). Some systematists consider each of these groups a phylum; others view them as subphyla or classes. In this edition we classify these groups as subphyla of phylum Arthropoda. However, as you study the following sections, bear in mind that arthropod systematics is a work in progress. Both the relationship of arthropods to other protostomes and the relationships among arthropods continue to be topics of lively debate.

Trilobites were early arthropods Among the earliest arthropods to evolve, *trilobites* inhabited shallow Paleozoic seas more than 500 mya. These arthropods, which have been extinct for about 250 million years, lived on the sea bottom and filtered mud to obtain food. Most ranged from 3 to 10 cm (about 1 to 4 in.), but a few reached almost 1 m (39 in.) in length.

Covered by a hard, segmented exoskeleton, the trilobite body was a flattened oval divided into three parts: an anterior head bearing a pair of antennae and a pair of compound eyes, a thorax, and a posterior abdomen (FIG. 31-26). At right angles to these divisions, two dorsal grooves extended the length of the animal, dividing the body into a median lobe and two lateral lobes. (The name *trilobite* derives from this division of the body into three longitudinal parts.) Each segment had a pair of segmented **biramous appendages,** that had two jointed branches extending from their base. In trilobites each appendage consisted of an inner walking leg and an outer branch with gills.

TABLE 31-5	Extant Arthropod Subphyla					
SUBPHYLUM AND SELECTED GROUPS	**BODY DIVISIONS**	**APPENDAGES: ANTENNAE, MOUTHPARTS**	**APPENDAGES: LEGS**	**GAS EXCHANGE**	**DEVELOPMENT**	**MAIN HABITAT**
MYRIAPODA Chilopoda (centipedes) Diplopoda (millipedes)	Head with segmented body	Uniramous. Antennae: one pair Mouthparts: mandibles, maxillae	Chilopods: one pair/segment Diplopods: usually two pairs/ segment	Tracheae	Direct	Terrestrial
CHELICERATA Merostoma (horseshoe crabs) Arachnida (spiders, scorpions, ticks, mites)	Cephalothorax and abdomen	Uniramous. Antennae: none Mouthparts: chelicerae, pedipalps	Merostomes: four pairs of walking legs Arachnids: four pairs on cephalothorax	Merostomes: gills Arachnids: book lungs or tracheae	Direct, except mites and ticks	Merostomes: marine Arachnids: mainly terrestrial
CRUSTACEA Malacostraca (lobsters, crabs, shrimp, isopods) Cirripedia (barnacles) Copepoda (copepods)	Head, thorax, and segmented abdomen	Biramous. Antennae: two pairs Mouthparts: mandibles, two pairs of maxillae (for handling food)	Typically one pair/segment	Gills	Usually larval stages (nauplius larva)	Marine or freshwater; a few are terrestrial
HEXAPODA Insecta (bees, ants, grasshoppers, roaches, flies, beetles)	Head, thorax, and abdomen (some multi-segmented)	Uniramous. Antennae: one pair Mouthparts: mandibles, maxillae	Three pairs on thorax	Tracheae; gills in aquatic species	Typically, larval stages; most with complete metamorphosis	Mainly terrestrial

© Cengage Learning

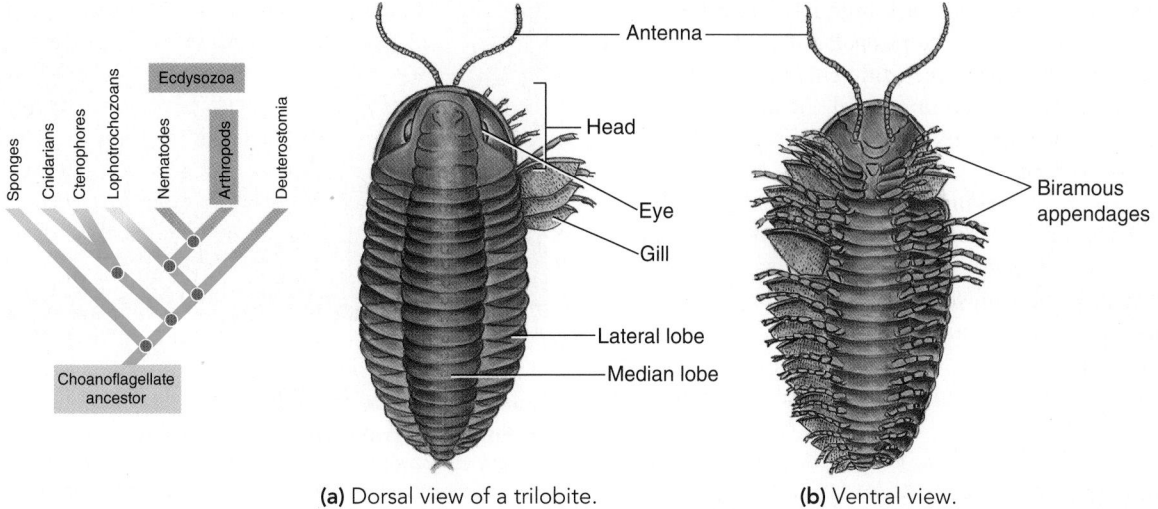

(a) Dorsal view of a trilobite. **(b)** Ventral view.

Figure 31-26 Fossil trilobites

These extinct arthropods (*Phacops rana*) were about 3 cm (1.2 in.) long. Note the large, well-developed eyes visible on either side of the head region. Biologists consider these extinct marine arthropods the most primitive members of phylum Arthropoda. Trilobites flourished in the ocean during the Paleozoic era.

© Cengage Learning

Subphylum Myriapoda includes the centipedes and millipedes Members of subphylum *Myriapoda* are characterized by unbranched appendages called **uniramous appendages,** jawlike mandibles, and a single pair of antennae. Centipedes (*Chilopoda*) and millipedes (*Diplopoda*) are terrestrial and are typically found beneath stones or wood in the soil in both temperate and tropical regions. Both centipedes and millipedes have a head and an elongated trunk with many segments, each bearing uniramous legs (FIG. 31-27).

Centipedes ("hundred-legged") have one pair of legs on each segment behind the head. Most centipedes do not have enough legs to merit their name; typically, they have approximately 30, although a few species have 100 or more. Because centipedes have long legs, they are able to run rapidly. They are predators that feed on other animals, mostly insects. Larger centipedes eat snakes, mice, and frogs. Using their poison claws located just behind the head on the first trunk segment, centipedes capture and kill their prey.

Millipedes ("thousand-legged") have two pairs of legs on most body segments. They are not as agile as chilopods, and most species can crawl only slowly over the ground. However, they can powerfully force their way through earth and rotting wood. Millipedes are generally herbivorous and feed on both living and dead vegetation.

Chelicerates do not have antennae Subphylum *Chelicerata* includes the merostomes (horseshoe crabs) and the arachnids. The chelicerate body consists of a *cephalothorax* (fused head and thorax) and an abdomen. Chelicerates are the only arthropods without antennae. They have no chewing mandibles. Instead, the first appendages, located immediately anterior to the mouth, are a pair of **chelicerae** (sing., *chelicera*), fanglike feeding

(a) Centipede (*Lithobius*), a member of class Chilopoda. Centipedes have one pair of uniramous appendages per segment.

(b) Millipede (*Diplopoda pachydesmus*), a member of class Diplopoda. Note the two pairs of uniramous appendages per segment.

Figure 31-27 Animation Myriapods

appendages. Most chelicerates suck fluid food from their prey. The second appendages are the *pedipalps*. The chelicerae and pedipalps are modified to perform different functions in various groups, including manipulation of food, locomotion, defense, and copulation. The four pairs of legs on the cephalothorax are specialized for walking.

Almost all the *merostomes* are extinct. Only the horseshoe crabs have survived, essentially unchanged for more than 350 million years. *Limulus polyphemus,* the species common along the Atlantic shore of North America, is horseshoe-shaped (**FIG. 31-28a**). Its long, spikelike tail is used in locomotion, not for defense or offense. Horseshoe crabs feed at night, capturing mollusks, worms, and other invertebrates that they find on the sandy ocean floor.

Arachnids include spiders, scorpions, ticks, harvestmen (daddy longlegs), and mites (**FIGS. 31-28b** and **c**). Most of the approximately 80,000 named species are predators that feed on insects and other small arthropods. Most arachnids have six pairs of jointed appendages. In spiders the first appendages, the chelicerae, are used to penetrate prey; some species use the chelicerae to inject venom into their prey. Spiders use the second pair of appendages, the pedipalps, to manipulate food. Many arachnid species have pedipalps modified as sense organs for tasting food or for reproduction (for sperm transfer and courtship displays). Scorpions use their very large pedipalps as pincers for capturing prey. Chelicerates use the remaining four pairs of appendages for walking.

Typically, spiders have eight eyes arranged in two rows of four each along the anterior dorsal edge of the cephalothorax. The eyes detect movement, locate objects, and in some species form a relatively sharp image.

Gas exchange in arachnids takes place by tracheal tubes, book lungs, or both. A *book lung* consists of 15 to 20 parallel plates (like pages of a book) that contain tiny blood vessels. Air enters the body through abdominal slits and circulates between the plates. As air passes over the blood vessels in the plates, oxygen diffuses into the blood and carbon dioxide diffuses out of the blood into the air. As many as four pairs of book lungs provide an extensive surface area for gas exchange.

Spiders have unique glands in the abdomen that secrete silk, an elastic protein that is spun into fibers by organs called *spinnerets.* The silk is liquid as it emerges from the spinnerets but hardens after it leaves the body. Spiders use silk to build nests, to encase their eggs in a cocoon, and in some species to trap prey in a web. Many spiders lay down a silken dragline that serves both as a safety line and a means of communication between members of a species. From the type of dragline, another spider can determine the sex and maturity level of the spinner.

Although all spiders (and scorpions) have poison glands useful in capturing prey, only a few produce poison toxic to humans. The most widely distributed venomous spider in the United States is the black widow (see Fig. 31-28b). Its venom is a neurotoxin that interferes with transmission of messages from nerves to muscles. The shiny black female is about 3.8 cm (1.5 in.) long with a red or orange hourglass-shaped marking on the underside of its abdomen. Adult male black widow spiders

(a) Horseshoe crabs (*Limulus polyphemus*). The only living merostomes are a few closely related species of horseshoe crabs. Seasonally, horseshoe crabs return to beaches for mating.

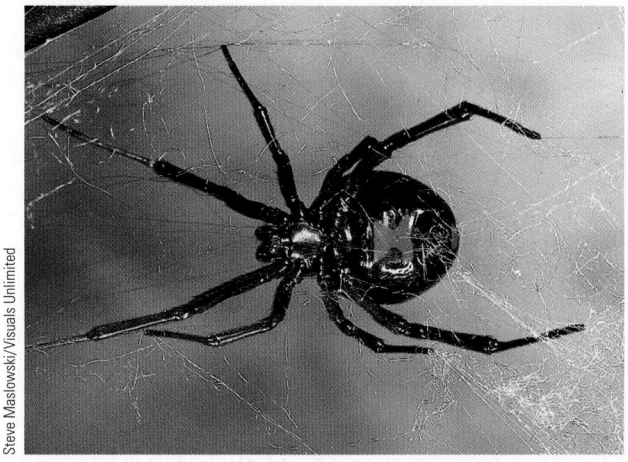

(b) A female black widow spider (*Latrodectus mactans*) rests on her web. The venom of the black widow spider is a neurotoxin.

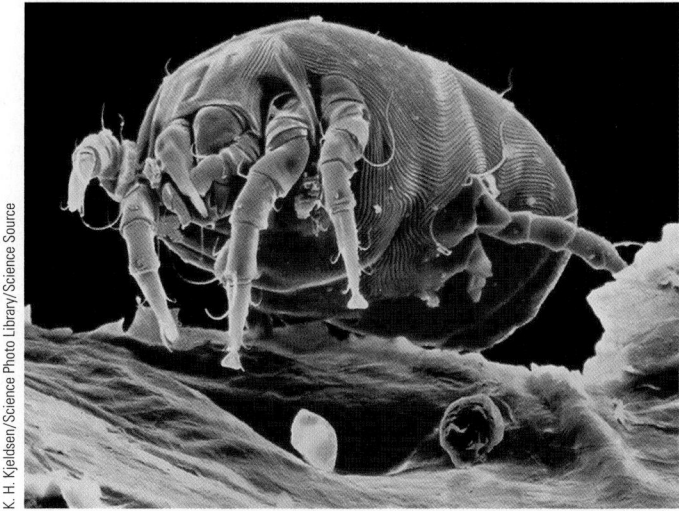

100 μm

(c) The house dust mite *Dermatophagoides.* This mite, a common inhabitant of homes, is associated with house dust allergies (color-enhanced SEM).

Figure 31-28 Chelicerates

are smaller and are harmless. After mating, the female sometimes kills and eats the male.

The brown recluse spider (*Loxosceles reclusa*) is more slender than the black widow and has a violin-shaped stripe on its dorsal cephalothorax. Its venom destroys the tissues surrounding the bite. Although painful, spider bites cause fewer than three fatalities per year in the United States.

Mites are incredibly abundant, but are rarely noticed because of their small size. Although some mites contribute to soil fertility by breaking down organic matter, many are serious nuisances. They eat crops, infest livestock and pets, and are external parasites. Some species have modified body shapes and reduced appendages, adaptations for their parasitic lifestyles. Chiggers (red bugs), the larval form of some red mite species, attach themselves to the skin and secrete an irritating digestive fluid that may cause itchy red welts. Certain mites cause mange in dogs and other domestic animals.

Larger than mites, ticks are parasites on dogs, deer, and many other animals. They transmit the bacteria that cause diseases such as Rocky Mountain spotted fever and Lyme disease (see Fig. 54-2).

Crustaceans are vital members of marine food webs

Crustaceans, members of Subphylum **Crustacea,** include lobsters, crabs, shrimp, and barnacles plus tens of thousands of smaller species with great diversity (**FIG. 31-29**). Many crustaceans are primary consumers of algae and detritus. Countless billions of microscopic crustaceans are part of marine zooplankton, the free-floating, mainly microscopic organisms in the upper layers of the oceans. These crustaceans are food for many fishes and other marine animals, such as certain baleen whales. Paleobiologists have dated crustacean fossils to the Cambrian period (more than 505 mya).

A distinctive feature of many crustaceans is the *nauplius larva,* which is often the first stage after hatching. This larva has only the most anterior three pairs of appendages. Crustaceans are also characterized by mandibles, biramous appendages, and two pairs of **antennae.** Their antennae, which are the first and second pairs of appendages, serve as sensory organs for touch and taste. Most adult crustaceans have compound eyes, and many crustaceans have **statocysts,** sense organs that detect gravity (see Fig. 43-6).

The hard **mandibles,** used for biting and grinding food, are the third pair of appendages; they lie on each side of the ventral mouth. Crustaceans show great diversity in the number of appendages posterior to the mandibles. Some species have several pairs of appendages near the mouth, called **maxillae** and maxillipeds, that manipulate and hold food. More posteriorly along the thorax and along the segmented abdomen, other appendages are modified for swimming, walking, defense, sensation, mating, or carrying eggs or young.

Crustaceans are primarily aquatic; not surprisingly, they typically have gills for gas exchange. Two large glands located in the head excrete metabolic wastes and regulate salt balance. Most crustaceans have separate sexes. During copulation, the male uses specialized appendages to transfer sperm into the female. Many crustaceans hatch into a nauplius larva

Colin Milkins/Getty Images

(a) Goose barnacles (*Lepas* sp.) filter feeding. These stalked barnacles are found in large numbers on intertidal rocks.

© Wolfgang Polzer/Alamy

(b) Sponge crab (*Criptodromia octodenta*) wearing a sponge on its back. The sponge provides camouflage and shelter. The sponge crab uses its claws to cut out a fragment from a sponge and then trims the sponge to conform to its own shape. The last two pairs of the crab's legs bend upward over the crab's body and hold the sponge in place. The sponge grows along with the crab.

Figure 31-29 Crustaceans

and then molt through a series of larval stages before developing the adult body form. However, some species resemble small adults immediately upon hatching. Newly hatched animals may resemble adults, or they may undergo successive molts as they pass through a series of larval stages before developing an adult body.

Barnacles, the only sessile crustaceans, differ markedly in external anatomy from other members of the subphylum. They are marine suspension feeders that secrete limestone cups within which they live (see Fig. 31-29a). The larvae of barnacles are free-swimming forms that go through several molts. They eventually become sessile and develop into the adult form. Louis Agassiz, a 19th-century naturalist, described the barnacle as "nothing more than a little shrimplike animal standing on its head in a limestone house and kicking food into its mouth."

The bane of marine boaters, barnacles can proliferate on ship bottoms in great numbers. Barnacles can reduce the speed of a ship by more than 10% and increase fuel consumption by as much as 40%. The U.S. Office of Naval Research has been testing an underwater robot that cleans barnacles and other marine life from the hulls of ships when they are in port.

Isopods are mainly tiny (5 to 15 mm in length) marine crustaceans that inhabit the ocean floor. However, this group also includes some terrestrial animals: pill bugs and sow bugs. The most abundant crustaceans are the *copepods*, which are typically microscopic. Marine copepods are the most numerous component of zooplankton.

The largest and most familiar order of crustaceans, Decapoda, contains more than 10,000 species of lobsters, crayfish, crabs, and shrimp. Most **decapods** are marine, but a few, such as crayfish, certain shrimp, and a few crabs, live in fresh water. The crustaceans in general and the decapods in particular show striking specialization and differentiation of parts in the various regions of the animal. For example, the appendages in the different parts of the body differ markedly in form and function (**FIG. 31-30**).

In the lobster the five segments of the head and the eight segments of its thorax are fused into a cephalothorax, which is covered on the top and sides by a shield, the *carapace*, composed of chitin impregnated with calcium salts. The two pairs of antennae serve as chemoreceptors and tactile sense organs. The mandibles are short and heavy, with opposing surfaces used in grinding and biting food. Behind the mandibles are two pairs of accessory feeding appendages: the first and second maxillae.

The appendages of the first three segments of the thorax are the maxillipeds, which aid in chopping up food and passing it to the mouth. The fourth segment of the thorax has a pair of large *chelipeds*, or pinching claws. The last four thoracic segments bear walking legs. Gills for respiration branch dorsally from each of the thoracic appendages.

The appendages of the first abdominal segment are part of the reproductive system and are used by the male to transfer sperm. The following four abdominal segments bear paired *swimmerets,* small paddlelike structures used by some decapods for swimming and by the females of all species for holding eggs. Each branch of the sixth abdominal appendages consists of a large flattened structure. Together with the flattened posterior end of the abdomen, they form a tail fan used for swimming backward.

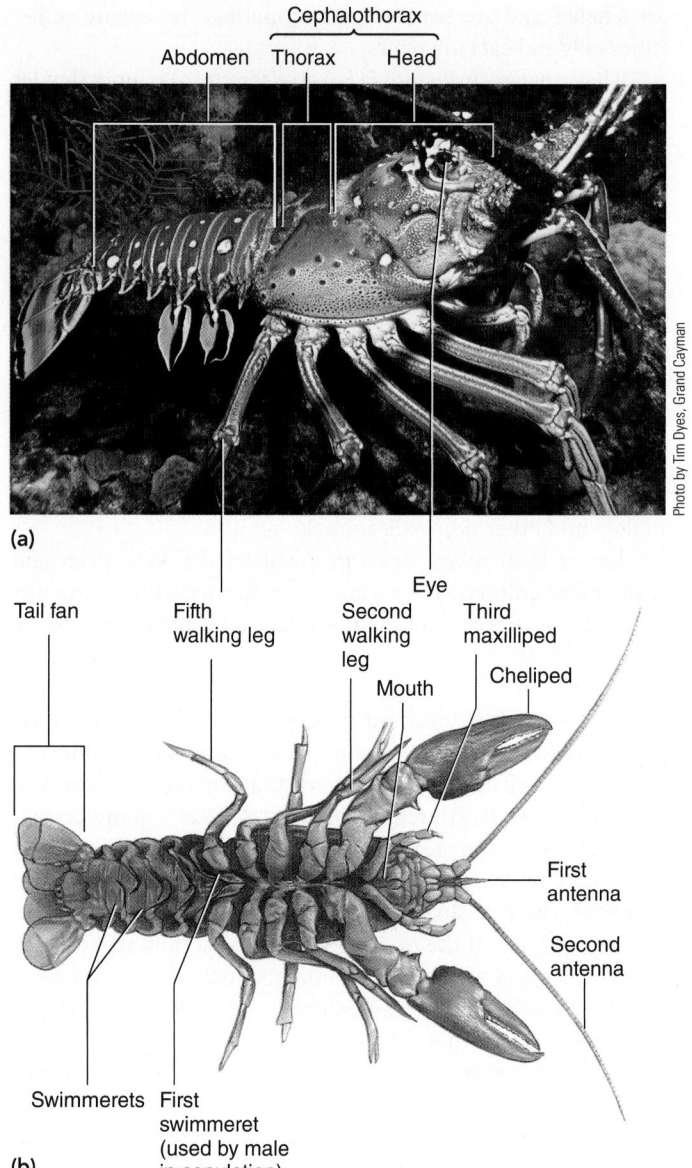

(a)

(b)

Figure 31-30 Anatomy of the lobster

(a) Like other decapods, the spiny lobster (*Panulirus argus*) has five pairs of walking legs. The first pair of walking legs is modified as chelipeds (*large claws*). Photographed in Grand Cayman. **(b)** Ventral view of a lobster. Note the variety of specialized appendages.
© Cengage Learning

Insects are articulated, tracheated hexapods With more than one million described species (and perhaps millions more not yet identified), *Insecta* of subphylum *Hexapoda* is the most successful group of animals on our planet in terms of number of individuals, number of species, diversity, and geographic distribution (**TABLE 31-6**). What they lack in size, insects make up in sheer numbers. If we could weigh all the insects in the world, their weight would exceed that of all the remaining terrestrial animals. Although primarily terrestrial, some insect species live in fresh water, a few are truly marine, and others inhabit the shore between the tides.

| TABLE 31-6 | Some Orders of Insects |

ORDER, NUMBER OF SPECIES, SOME CHARACTERISTICS	REPRESENTATIVE ADULT	ORDER, NUMBER OF SPECIES, SOME CHARACTERISTICS	REPRESENTATIVE ADULT

HOMOPTERA (43,000)*

Leafhoppers, scale insects
Typically, two pairs of membranous wings; piercing–sucking mouthparts form beak; parasites of plants; vectors of plant diseases; incomplete metamorphosis

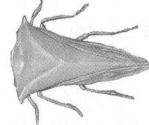

Buffalo treehopper
Stictocephala bubalus

ODONATA (6000)

Dragonflies, damselflies
Two pairs of long membranous wings; chewing mouthparts; large, compound eyes; active predators; larvae very different from adult; incomplete metamorphosis

Damselfly
Ischnura

ORTHOPTERA (20,000)

Grasshoppers, crickets
Forewings leathery, hindwings membranous; chewing mouthparts; most herbivorous, some cause crop damage; some predatory; incomplete metamorphosis

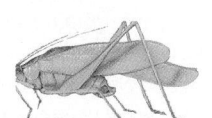

Fork-tailed bush katydid
Scudderia furcata

BLATTODEA (4000)

Cockroaches
When wings present, forewings leathery, hindwings membranous; chewing mouthparts; legs adapted for running; incomplete metamorphosis

American cockroach
Periplaneta americana

ISOPTERA (2500)

Termites
Two pairs of wings, or none; wings shed by sexual forms after mating; chewing mouthparts; social insects, form large colonies; eat wood; incomplete metamorphosis

Eastern subterranean termite
Reticulitermes flavipes

PHTHIRAPTERA (3000)

Lice, sucking lice
No wings; piercing–sucking mouthparts; ectoparasites of mammals; head and body louse and crab louse are human parasites; vectors of typhus fever; incomplete metamorphosis

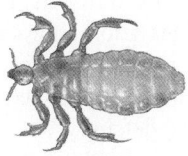

Human head and body louse
Pediculus humanus

HEMIPTERA (75,000)

Chinch bugs, bedbugs, water striders; cicadas and aphids are now assigned to this order
Hindwings membranous; forewings smaller; piercing–sucking mouthparts form beak; most herbivorous; some parasitic; incomplete metamorphosis

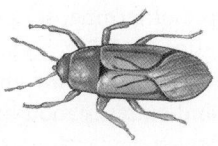

Chinch bug
Blissus leucopterus

LEPIDOPTERA (160,000)

Moths, butterflies
Usually two pairs of membranous, colorful, scaled wings; larvae are wormlike caterpillars with chewing mouthparts for eating plants; adult mouthparts adapted for sucking nectar; complete metamorphosis

Luna moth
Aetias luna

DIPTERA (150,000)

Houseflies, mosquitoes, fruit flies
Only forewings functional in flying; hindwings reduced to small, knob-like balancing organs (halteres); mouthparts usually adapted for sucking (and piercing in some); larvae are maggots that damage domestic animals or food; adults transmit diseases such as sleeping sickness, yellow fever, and malaria; complete metamorphosis

Deerfly
Chrysops vittatus

SIPHONAPTERA (2000)

Fleas
No wings; piercing–sucking mouthparts; legs adapted for clinging and jumping; parasites on birds and mammals; vectors of bubonic plague and typhus; complete metamorphosis

Dog flea
Ctenocephalides canis

COLEOPTERA (360,000)

Beetles, weevil
Forewings modified as protective coverings for membranous hindwings (which are sometimes absent); chewing mouthparts; largest order of insects; most herbivorous; some aquatic; many larvae are grubs; complete metamorphosis

Colorado potato beetle
Leptinotarsa decemlineata

HYMENOPTERA (more than 100,000)

Ants, bees, wasps
Usually two pairs of membranous wings; mouthparts may be modified for sucking or lapping nectar; many are social insects; some sting; important pollinators; complete metamorphosis

Bald-faced hornet
Vespula maculata

THYSANURA (600)

Silverfish
No wings; biting–chewing mouthparts; two to three "tails" extend from posterior tip of abdomen; inhabit dead leaves; eat starch in books; direct development

Silverfish
Lepisma saccharina

*The number of species listed for each order is approximate. Estimates vary widely.

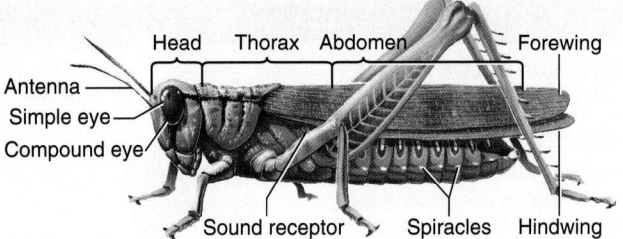

Head   Thorax Abdomen   Forewing

Antenna
Simple eye
Compound eye

Sound receptor   Spiracles   Hindwing

(a) External structure. Note the three pairs of segmented legs.

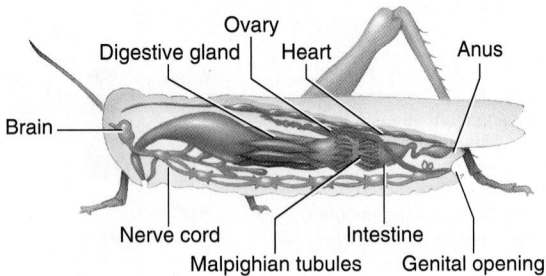

Ovary
Digestive gland   Heart   Anus

Brain

Nerve cord   Intestine
        Malpighian tubules   Genital opening

(b) Internal anatomy.

Figure 31-31 Anatomy of the grasshopper

Insects, the most successful of all animals, are articulated, tracheated hexapods.
© Cengage Learning

The earliest fossil insects—primitive, wingless species—date to the Devonian period more than 360 mya. Insect fossils from the Carboniferous period more than 310 mya include both wingless and primitive winged species. Cockroaches, mayflies, and cicadas are among the insects that have survived relatively unchanged from the Carboniferous period to the present day.

Molecular and developmental studies suggest that hexapods are a monophyletic group that evolved from within the crustaceans. Many biologists today classify crustaceans along with hexapods in clade *Pancrustacea*. We describe an insect as an *articulated* (jointed), *tracheated* (having tracheal tubes for gas exchange) *hexapod* (having six feet). The insect body consists of three distinct parts: head, thorax, and abdomen (FIG. 31-31). An insect's uniramous (unbranched) appendages include three pairs of legs that extend from the adult thorax and, in many orders, one or two pairs of wings. One pair of antennae protrudes from the head, and the sense organs include both simple and compound eyes. The complex mouthparts, which include mandibles and maxillae, are adapted for piercing, chewing, sucking, or lapping.

The tracheal system of insects has made an important contribution to their diversity. Air enters the tracheal tubes, or tracheae, through **spiracles,** tiny openings in the body wall. Oxygen passes directly to the internal organs via the tubes branching throughout the body. This effective oxygen delivery system permits insects to have the high metabolic rate necessary for activities such as flight, even though they have an open circulatory system.

Excretion is accomplished by two or more **Malpighian tubules,** which receive metabolic wastes from the blood, concentrate the wastes, and discharge them into the intestine. Unique to terrestrial arthropods, Malpighian tubules also perform the extremely important function of conserving water (see Fig. 48-4).

The sexes are separate, and fertilization takes place internally. Several molts occur during development. The immature stages between molts are called *instars.* Primitive insects with no wings, such as silverfish, have direct, or simple, development: the young hatch as juveniles that resemble the adult form. Other insects, such as grasshoppers and cockroaches, undergo **incomplete metamorphosis** in which the egg gives rise to a larva that resembles the adult in many ways but lacks functional wings and reproductive structures. The larva goes through a series of molts during which it becomes more and more like the adult.

Most insects, including bees, butterflies, and fleas, undergo **complete metamorphosis** with four distinct stages in the life cycle: *egg, larva, pupa,* and *adult* (FIG. 31-32). The wormlike larva does not look at all like the adult. For example, caterpillars have a body form entirely different from that of butterflies. Typically, an insect spends most of its life as a larva, which has chewing mouthparts. Eventually, the larva stops feeding, molts, and enters a pupal stage, usually within a protective cocoon or underground burrow. The **pupa** does not feed and typically cannot defend itself. Energy reserves stored during the larval stage are spent remodeling its body. When it emerges as an adult, it is equipped with functional wings and reproductive organs.

Certain species of bees, ants, and termites live as colonies or societies made up of several different types of individuals, each adapted for some particular function. The members of some insect societies communicate with one another by "dances" and by means of **pheromones,** substances secreted to the external environment. Social insects and their communication are discussed in Chapter 52.

Adaptations that contribute to the biological success of insects What are the secrets of insect success? The tough exoskeleton protects insects from predators and helps reduce water loss. Segmentation allows mobility and flexibility and permits regional specialization of the body. For example, cephalization and highly developed sense organs are important in both offense and defense.

The insect body plan has been modified and specialized in so many ways that insects are adapted to a remarkable number of lifestyles. The jointed appendages are specialized for various types of feeding, sensory functions, and different types of locomotion (walking, jumping, swimming). For example, grasshopper mouthparts are adapted for biting and chewing leaves. Moth and butterfly mouthparts are adapted for sucking nectar from flowers, and mosquito mouthparts are adapted for sucking blood. Aphids and leafhoppers have mouthparts specialized to pierce plants and feed on plant juices.

Another adaptation that contributes to insect success is the ability to fly. Unlike other terrestrial invertebrates, which creep slowly along on or under the ground, many insects fly rapidly through the air. Their wings and small size facilitate their wide

Complete metamorphosis reduces competition among insects within the same species; larval forms do not compete with adults for food or habitats.

Matt Meadows/Getty Images

James H. Robinson/Science Source

(b) Caterpillar feeding on sweet gum leaf.

Adam Jones/Science Source

Millard H. Sharp/Science Source

(a) Polyphemus moth caterpillars hatching from eggs.

(c) Cocoon cut open to show developing pupa.

(d) Adult on tree. Note eyespots.

Figure 31-32 Complete metamorphosis in a Polyphemus moth

Insects, the most successful of all animals, are articulated, tracheated hexapods. This moth (*Antheraea polyphemus*) belongs to the family of giant silk moths.

CONNECT Which of the following stages is haploid? Diploid? Embryo developing in egg, larva, pupa, adult?

distribution and, in many instances, their immediate survival. For example, when a pond dries up, adult aquatic insects can fly to another habitat.

The reproductive capacity of insects is remarkable. Under ideal conditions, the fruit fly *Drosophila* can produce 25 generations in a single year! Insect eggs are protected by a thick membrane. In addition, several eggs may be enclosed in a protective egg case. By dividing the insect life cycle into different stages, complete metamorphosis reduces competition among members of the same species (intraspecific competition). For example, larval forms have different lifestyles, so they do not compete with adults for food or habitats.

Insects have many interesting adaptations for offense and defense (see Fig. 7-8). We all are familiar with the stingers of bees and wasps, which are specialized egg-laying structures (ovipositors). Many insects have **cryptic coloration,** which allows them to blend into the background of their habitat; some look like dead twigs or leaves. Others mimic poisonous insects, deriving protection from this resemblance. Still others "play dead" by remaining motionless.

Insects communicate by tactile, auditory, visual, or chemical signals. For example, certain ant species use pheromones to mark trails and to warn of danger. Many insects use pheromones to attract mates.

Impact of insects on humans Not all insects compete with humans for food or cause us to scratch, swell up, or recoil from their presence. Bees, wasps, beetles, and many other insects pollinate the flowers of crops and fruit trees. Some insects destroy other insects that are harmful to humans. For example, dragonflies eat mosquitoes. And many organic farmers and home gardeners buy ladybird beetles, which are adept at ridding plants of aphids and other insect pests. Insects are important members of many food webs. Many birds, mammals, amphibians, reptiles, and some fishes depend on insects for food. Some beetles and the larvae (maggots) of flies are detritus feeders: they break down dead plants and animals and their wastes, permitting nutrients to be recycled.

Various insect products are useful to humans. Bees produce honey as well as beeswax, which we use to make candles, lubricants, and other products. Shellac is made from lac, a substance given off by certain scale insects that feed on the sap of trees. And the labor of silkworms provides us with beautiful fabric.

On the negative side, each year insects destroy crops worth billions of dollars. Fire ants not only inflict painful stings but cause farmers serious economic loss because of their large mounds, which damage mowers and other farm equipment. Moths damage clothing, and termites destroy buildings. Bloodsucking flies, screwworms, lice, fleas, and other insects annoy and transmit disease to humans and domestic animals. Mosquitoes are vectors of yellow fever and malaria (see Fig. 26-7). Fleas and lice transmit the rickettsia that cause typhus, and houseflies sometimes transmit typhoid fever and dysentery. Tsetse flies transmit African sleeping sickness, and fleas may be vectors of bubonic plague.

In this chapter we have surveyed the major groups of protostomes. TABLE 31-7 summarizes these groups. Chapter 32 focuses on the deuterostomes, which include the echinoderms and chordates.

CHECKPOINT 31.3

- **CONNECT** *What characteristics distinguish nematodes from flatworms?*

- **CONNECT** *Describe four key arthropod characteristics, and explain how each contributes to arthropod success.*

- *How do crustaceans differ from chelicerates?*

- **CONNECT** *Describe four adaptations that have contributed to insect success.*

TABLE 31-7	Overview of Sponges, Cnidarians, Ctenophores, and Protostomes	
MAJOR GROUPS	**BODY PLAN**	**SOME KEY CHARACTERISTICS**
Poriferans (sponges)	Asymmetrical; cells loosely arranged and do not form true tissues; body is sac with pores, central cavity, and osculum	Collar cells (choanocytes); aquatic, mainly marine
Cnidarians (hydras, jellyfish, corals)	Radial symmetry; diploblastic; gastrovascular cavity with one opening	Tentacles with cnidocytes (stinging cells) that discharge nematocysts; two body forms: polyp and medusa; some form colonies; mainly marine
Ctenophores (comb jellies)	Biradial symmetry; diploblastic; digestive system with mouth and anal pores	Eight rows of cilia that resemble combs; tentacles with adhesive glue cells; marine predators
PROTOSTOMES: LOPHOTROCHOZOAN CLADE		
Flatworms (planarians, tapeworms, flukes)	Bilateral symmetry; triploblastic; simple organ systems; gastrovascular cavity with one opening	No body cavity; some cephalization; some are free-living predators; many are parasites
Nemerteans (ribbon worms)	Bilateral symmetry; triploblastic; organ systems; complete digestive tube*	Proboscis (long, muscular tube that can be everted to capture prey); true coelom reduced; mainly marine predators
Mollusks (clams, snails, squids)	Bilateral symmetry; triploblastic; organ systems; complete digestive tube; true coelom	Soft body usually covered by dorsal shell; flat, muscular foot; mantle covers visceral mass; most have a radula (belt of teeth)
Annelids (some marine worms, earthworms, leeches)	Bilateral symmetry; triploblastic; organ systems; complete digestive tube; true coelom	Segmented body; most have setae, bristles that provide traction during crawling
Lophophorates (brachiopods, phoronids, bryozoans)	Bilateral symmetry; triploblastic; organ systems; complete digestive tube; true coelom	Lophophore (ciliated ring of tentacles) surrounds mouth, captures suspended particles in water; mainly sessile, marine animals
Rotifers (wheel animals)	Bilateral symmetry; triploblastic; organ systems; complete digestive tube; pseudocoelom	Crown of cilia at anterior end; cell number constant; microscopic, aquatic animals
PROTOSTOMES: ECDYSOZOAN CLADE		
Nematodes (roundworms, e.g., *Ascaris*, hookworms, trichina worms)	Bilateral symmetry; triploblastic; organ systems; complete digestive tube; pseudocoelom	Cylindrical, threadlike body; widely distributed in soil and aquatic sediments; important as decomposers; many are predators; some are parasites
Arthropods (centipedes, spiders, crabs, lobsters, insects)	Bilateral symmetry; triploblastic; organ systems; complete digestive tube; true coelom	Segmented body; exoskeleton; paired, jointed appendages; insects and many crustaceans have compound eyes; insects have tracheal tubes for gas exchange; wings

*A complete digestive tube has a mouth for food intake and an opening (anus) for elimination of wastes.

© Cengage Learning

31.1 Sponges, Cnidarians, and Ctenophores (page 636)

1 Identify important characteristics of poriferans.

- The **poriferans,** or sponges, are characterized by flagellate **collar cells (choanocytes),** which generate a water current that brings food and oxygen to the cells. Collar cells also trap and phagocytize food particles. The sponge body is a sac with tiny openings through which water enters; a central cavity, or **spongocoel;** and an open end, or **osculum,** through which water exits. The cells of sponges are loosely associated; they do not form true tissues.

2 Identify distinguishing characteristics of cnidarians, describe four groups, and give examples of animals that belong to each group.

- **Cnidarians** are characterized by radial symmetry, two tissue layers, and **cnidocytes,** cells containing stinging organelles called **nematocysts.** The **gastrovascular cavity** has a single opening that serves as both mouth and anus. Nerve cells form irregular, nondirectional **nerve nets** that connect sensory cells with contractile and gland cells.

- The life cycle of many cnidarians includes a sessile **polyp** stage (a form with a dorsal mouth surrounded by tentacles) and a free-swimming **medusa** (jellyfish) stage.

- Phylum Cnidaria includes four groups. *Hydrozoa* (hydras, hydroids, and the Portuguese man-of-war) are typically polyps and may be solitary or colonial. *Scyphozoa* (jellyfish) are generally medusae. *Cubozoa,* the "box jellyfish," have complex eyes that form blurred images. *Anthozoa* (sea anemones and corals) are polyps and may be solitary or colonial; anthozoans differ from hydrozoans in the organization of the gastrovascular cavity.

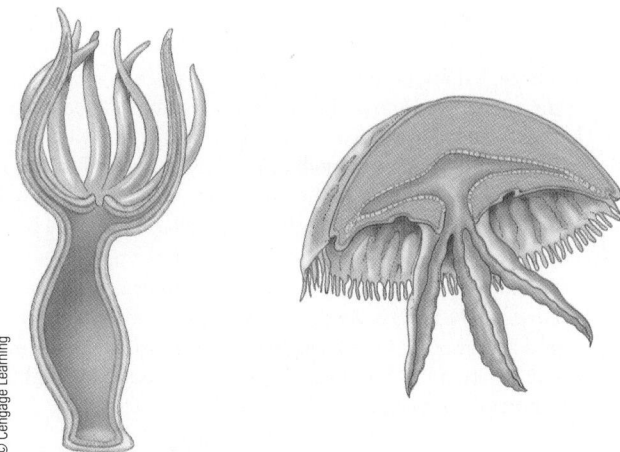

© Cengage Learning

3 Identify characteristics of ctenophores.

- **Ctenophores,** or *comb jellies,* are fragile, luminescent marine predators with biradial symmetry. Ctenophores have eight rows of cilia that resemble combs. They are diploblastic and have tentacles with adhesive glue cells.

31.2 The Lophotrochozoa (page 643)

4 Summarize what is known about the phylogeny of the lophotrochozoans and cite the adaptive advantages of having a coelom and of cephalization.

- The **Lophotrochozoa** make up a clade that includes some of the flatworms, nemerteans, mollusks, annelids, the lophophorate phyla, and rotifers.

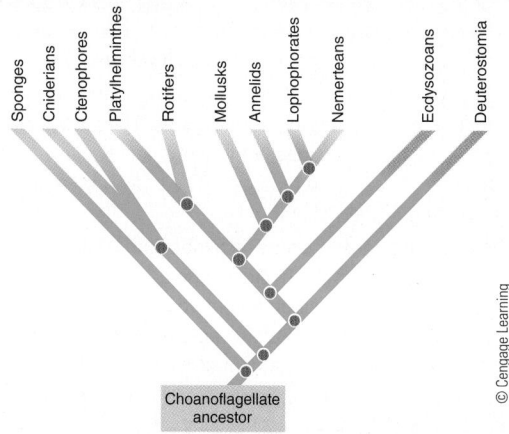

© Cengage Learning

- The true **coelom** is a fluid-filled body cavity completely lined by mesoderm that lies between the digestive tube and the outer body wall. The coelom brings about the *tube-within-a-tube body plan.* The body wall is the outer tube. The inner tube is the digestive tube. The coelom can serve as a **hydrostatic skeleton** in which contracting muscles push against a tube of fluid. The coelom is a space in which internal organs, including gonads, can develop; it helps transport materials and protects internal organs.

- **Cephalization,** the evolution of a head with the concentration of sense organs and nerve cells at the anterior end, increases the effectiveness of a bilateral animal to actively find food, shelter, and mates and to detect enemies.

5 Identify distinguishing characteristics of flatworms and nemerteans, and describe the main groups of flatworms, giving examples of animals that belong to each class.

- The *flatworms* are **acoelomate** (have no coelom) animals with bilateral symmetry, cephalization, three definite tissue layers, and well-developed organs. Many flatworms are **hermaphrodites:** a single animal produces both sperm and eggs. Flatworms have a ladder-type nervous system, typically consisting of sense organs and a simple brain composed of two **ganglia.** The ganglia are connected to two nerve cords that extend the length of the body. **Protonephridia** function in osmoregulation and disposal of metabolic wastes.

- Four groups of flatworms are recognized: class *Turbellaria* comprises free-living flatworms, including planarians; classes *Trematoda* and *Monogenea* include the parasitic *flukes;* and class *Cestoda* includes the parasitic tapeworms. The parasitic flukes and tapeworms typically have suckers or hooks for holding on to their hosts; they have complicated life cycles with intermediate hosts and produce large numbers of eggs.

- **Nemerteans** (ribbon worms) are characterized by the *proboscis,* a muscular tube used in capturing food and in defense. Nemerteans have a complete digestive tract with mouth and anus, and a circulatory system. The coelom is reduced.

6 Describe the main characteristics of mollusks and of the four main groups of mollusks discussed, giving examples of each group.

- **Mollusks** are soft-bodied animals typically covered by a shell. They have a ventral *foot* for locomotion and a pair of folds called the **mantle** that covers the **visceral mass,** a concentration of body organs.

- Mollusks have an **open circulatory system** except for cephalopods, which have a **closed circulatory system.** A rasplike **radula** functions as a scraper in feeding in all groups except the bivalves, which are filter feeders. Typically, marine mollusks have a free-swimming, ciliated **trochophore larva.**
- *Polyplacophorans* are chitons, mollusks with shells consisting of eight overlapping dorsal plates. The *gastropods,* which include the snails, slugs, and their relatives, have a well-developed head with tentacles. The body undergoes **torsion,** a twisting of the visceral mass.
- *Bivalves* are aquatic clams, scallops, and oysters. A two-part shell, hinged dorsally, encloses the bodies of these filter feeders. *Cephalopods* include the squids, octopuses, and *Nautilus.* These active, predatory swimmers have tentacles surrounding the mouth, which is located in the large head.

7 Describe the principal characteristics of annelids and of the three main groups of annelids discussed, giving examples of each group.

- The **annelids,** the segmented worms, include many aquatic worms, earthworms, and leeches. Annelids have long bodies with **segmentation** both internally and externally; their large, compartmentalized coelom serves as a hydrostatic skeleton.
- *Polychaetes* are marine annelids characterized by **parapodia,** appendages used for locomotion and gas exchange. The parapodia bear many bristlelike structures called **setae.** Polychaetes also differ from other annelids in having a well-defined head with sense organs.
- *Oligochaetes,* the group that includes the earthworms, are characterized by a few short setae per segment. The body is divided into more than one hundred segments separated internally by **septa.**
- Leeches belong to the group *Hirudinida.* Setae and appendages are absent. Parasitic leeches are equipped with suckers for holding on to their host.

8 Describe the distinguishing characteristics of the lophophorates and of the rotifers.

- The **lophophorates,** marine animals that have a lophophore, include the *brachiopods, phoronids,* and *bryozoans.* The *lophophore,* a ciliated ring of tentacles surrounding the mouth, is specialized for capturing suspended particles in the water.
- *Rotifers* are pseudocoelomates that are thought to have evolved from animals with a true coelom. They have a crown of cilia at their anterior end.

31.3 The Ecdysozoa *(page 656)*

9 Summarize what is known about the phylogeny of the ecdysozoans.

- **Ecdysozoa** is one of the three major animal clades; its validity is based on many types of evidence, including molecular data. Members of this group go through the process of **ecdysis,** or **molting,** during which an animal sheds its outer covering; the covering is then replaced by the growth of a new one.

10 Describe characteristics of nematodes and identify four parasitic nematodes.

- **Nematodes,** or *roundworms,* have a pseudocoelom. The body is covered by a tough **cuticle** that helps prevent desiccation. Parasitic nematodes that infect humans include *Ascaris, hookworms, trichina worms,* and *pinworms.*

11 Describe six key characteristics of arthropods, summarize proposed phylogeny of arthropods, and distinguish among arthropod subphyla and classes; give examples of animals that belong to each group.

- **Arthropods** are segmented animals with *paired, jointed appendages* and an armorlike **exoskeleton** of chitin. Molting is necessary for the arthropod to grow. Arthropods have an open circulatory system with a dorsal heart that pumps **hemolymph.** Aquatic forms have gills for gas exchange; terrestrial forms have either **tracheae** or *book lungs.*
- The arthropods along with the *onychophorans* (velvet worms) and *tardigrades* (water bears) make up the clade *Panarthropoda.* Based on molecular and other data, arthropods are currently assigned to five main groups: extinct trilobites and extant Myriapoda, Chelicerata, Crustacea, and Hexapoda.
- The *trilobites* are extinct marine arthropods covered by a hard, segmented shell. Each segment had a pair of **biramous appendages,** appendages with two jointed branches: an inner walking leg and an outer gill branch.
- Subphylum *Myriapoda* includes *Chilopoda,* the centipedes, and *Diplopoda,* the millipedes. Members of this subphylum have **uniramous appendages,** that is unbranched appendages, and a single pair of antennae.
- Subphylum *Chelicerata* includes the *merostomes* (horseshoe crabs) and the **arachnids** (spiders, mites, and their relatives). The chelicerate body consists of a cephalothorax and abdomen; there are six pairs of uniramous, jointed appendages, of which four pairs serve as legs. The first appendages are **chelicerae,** and the second are *pedipalps.* These appendages are adapted for manipulation of food, locomotion, defense, or copulation. Chelicerates have no antennae and no mandibles.
- **Crustaceans** include lobsters, crabs, shrimp, pill bugs, and barnacles, and their many relatives. The body typically consists of a cephalothorax and abdomen. Crustaceans vary greatly in the appearance and in the number of biramous appendages. Crustaceans have two pairs of **antennae** that sense taste and touch, and a pair of **mandibles** used for chewing. Two pairs of **maxillae,** posterior to the mandibles, manipulate and hold food. The decapod crustaceans typically have five pairs of walking legs.
- Subphylum *Hexapoda* includes *Insecta.* An insect is an *articulated, tracheated hexapod;* its body consists of head, thorax, and abdomen. Insects have uniramous appendages, a single pair of antennae, tracheae for gas exchange, and **Malpighian tubules** for excretion.

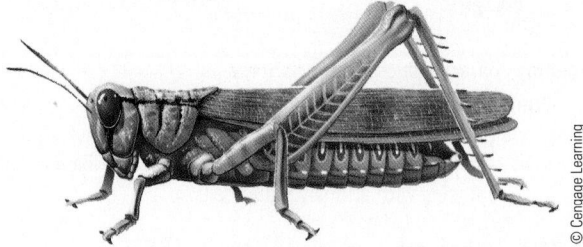

12 Identify adaptations that have contributed to the biological success of insects.

- The biological success of the insects can be attributed to their many adaptations, including a versatile exoskeleton, segmentation, specialized jointed appendages, highly developed sense organs, and ability to fly. **Complete metamorphosis,** transition during the life cycle from one developmental stage to another, includes egg, larva, pupa, and adult stages. Complete metamorphosis reduces competition within the same species.

Effective reproductive strategies, effective mechanisms for defense and offense, and the ability to communicate have evolved in insects.

Know and Comprehend

1. Collar cells (choanocytes) are characteristic of (a) poriferans (b) cnidarians (c) most coelomates (d) lophotrochozoans (e) ecdysozoans

2. Cnidocytes (a) are characteristic of sponges (b) are found among cells lining the gastrovascular cavity (c) contain stinging organelles (d) are lined with mesoderm (e) have two main shapes

3. Which of the following is associated with the evolution of the coelom? (a) radial symmetry (b) tube-within-a-tube body plan (c) incomplete metamorphosis (d) bilateral symmetry (e) development of ganglia

4. Trochophore larvae are characteristic of (a) arthropods (b) cnidarians (c) flatworms (d) mollusks (e) crustaceans

5. Which of the following is *not* an adaptation to parasitic life? (a) production of a few well-protected eggs (b) hooks (c) suckers (d) reduced digestive system (e) intermediate host

6. An open circulatory system (a) is characteristic of squids and other active mollusks (b) permits contracting muscles to recover quickly (c) has no blood vessels (d) is a unique characteristic of annelids (e) has hemolymph, which bathes tissues directly

7. Which of the following characteristics is associated with mollusks? (a) mandibles (b) mantle (c) pedipalps (d) chelipeds (e) setae

8. Trilobites (a) were early mollusks (b) are onychophorans (c) are characterized by parapodia and setae (d) were early arthropods (e) are an evolutionary link between annelids and arthropods

9. Which of the following is *not* characteristic of arthropods? (a) exoskeleton (b) pseudocoelom (c) paired, jointed appendages (d) chitin (e) segmentation

10. Which of the following is characteristic of insects? (a) biramous appendages (b) two pairs of antennae (c) chelicerae (d) eight legs (e) mandibles

Apply and Analyze

11. The correct sequence of complete metamorphosis of an insect is
    (a) egg $\longrightarrow$ immature form $\longrightarrow$ adult
    (b) egg $\longrightarrow$ trochophore larva $\longrightarrow$ adult
    (c) egg $\longrightarrow$ pupa $\longrightarrow$ larva $\longrightarrow$ adult
    (d) egg $\longrightarrow$ larva $\longrightarrow$ pupa $\longrightarrow$ adult
    (e) adult $\longrightarrow$ larva $\longrightarrow$ egg $\longrightarrow$ pupa

12. **VISUALIZE** Draw a cross section through a hydra. Label the epidermis, gastrodermis, gastrovascular cavity, and mesoglea.

13. **EVOLUTION LINK** Discuss important adaptations made possible by the development of the coelom.

14. **EVOLUTION LINK** Discuss the idea that every evolutionary adaptation has both advantages and disadvantages, using each of the following as an example: (a) cephalization, (b) the arthropod exoskeleton, and (c) segmentation with specialization.

Evaluate and Synthesize

15. **INTERPRET DATA** Imagine that you discover a new animal in a rain forest. It has an elongated, segmented body with bristles, and it has no obvious head. Where would you place this animal in the accompanying cladogram? Support your decision. Look at the more detailed cladogram in Figure 30-6a (in Chapter 30). What additional characteristics would help you place your animal in this cladogram? Explain your decision-making process.

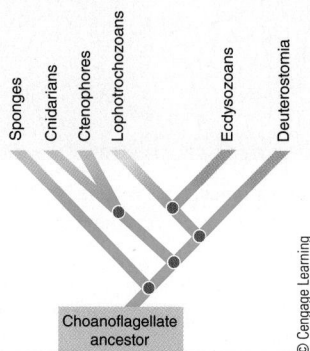

16. **EVOLUTION LINK** Why have biologists stopped using type of body cavity as a major criterion for inferring relationships among animal groups?

17. **EVOLUTION LINK** Animal body plans have generally evolved from simple to more complex. How can we explain the exceptions?

18. **EVOLUTION LINK** Insects that undergo complete metamorphosis outnumber those that do not by more than 10 to 1. Hypothesize an explanation.

19. **SCIENCE, TECHNOLOGY, AND SOCIETY** Several international monitoring projects are gathering data to help us understand coral reef destruction. In your opinion, why is it important to take action to protect coral reefs? What role might technology play?

aplia To access course materials, such as Aplia and other companion resources, please visit **www.cengagebrain.com.**

Representative deuterostomes. This marine habitat was photographed in Hawaii above a coral reef. Echinoderms are represented by the pencil sea urchins (*Heterocentrotus mammilatus*). Chordates are represented by the fish, a Moorish idol (*Zanchus cornutus*).

© David Fleetham/Alamy

KEY CONCEPTS

32.1 Echinoderms and chordates are the two most successful deuterostome lineages in terms of diversity, number of species, and number of individuals.

32.2 Echinoderms are characterized by radial symmetry in adults, a water vascular system, tube feet, and spiny skin.

32.3 At some time in its life, a chordate has a notochord; a dorsal, tubular nerve cord; a muscular postanal tail; and an endostyle (or thyroid gland).

32.4 Invertebrate chordates include tunicates and lancelets.

32.5 Shared derived characters of vertebrates include a vertebral column, a cranium, neural crest cells, and an endoskeleton of cartilage or bone.

32.6 Among the earliest vertebrates, ostracoderms were armored, jawless fishes. The extant hagfishes and lampreys have neither jaws nor paired fins.

32.7 Jaws and fins were key adaptations that contributed to the success of jawed fishes; and the evolution of limbs was important in the transition to terrestrial life.

32.8 The evolution of the amniotic egg allowed animals to become fully terrestrial; the amniotic egg was the key adaptation leading to the evolution of reptiles and, later, to birds and mammals.

What does a sea star have in common with a fish, frog, hawk, or human? You may think it strange to group the echinoderms—the sea stars, sea urchins, and sand dollars—with the chordates, the phylum to which we humans and other animals with a backbone belong. However, even though these animals look and behave very differently from one another, fossil evidence and morphological, developmental, and molecular data suggest that chordates and echinoderms share a common ancestor and are closely related.

In Chapter 31 we discussed the two major protostome branches of the animal kingdom: the lophotrochozoans and the ecdysozoans. In this chapter we focus on the third major branch of the animal kingdom, the deuterostomes. Some biologists speculate that the last common ancestor of the deuterostomes was an animal that obtained food by filtering ocean water. The two major groups (phyla) of living animals assigned to the deuterostomes are the echinoderms and chordates. The largest chordate subphylum is Vertebrata, which includes the animals with which we are most familiar: fishes, amphibians, reptiles (including birds), and mammals. Echinoderms are represented in the photograph by the pencil urchins, and chordates are represented by the fish.

We begin this chapter with an introduction to the deuterostomes and a survey of the echinoderms. We then turn our attention to the chordates. We describe the key chordate characteristics and discuss the invertebrate chordates. After an introduction to the vertebrates, we discuss jawless fishes and the evolution of jaws, and later the origin of limbed vertebrates. We discuss evolution of the adaptations that made the transition to life on land possible. We then describe the evolution of amniotes, which encompasses reptiles (including birds) and mammals.

32.1 WHAT ARE DEUTEROSTOMES?

LEARNING OBJECTIVE

1 Identify shared derived characters of deuterostomes and briefly describe the hemichordates.

The **deuterostomes** (clade Deuterostomia) make up the third major branch of the animal kingdom. Deuterostomes include the *echinoderms*—the sea stars, sea urchins, and sand dollars—and the *chordates,* the phylum to which we humans and other *vertebrates* (animals with a backbone) belong. Biologists also classify the **hemichordates,** a small group of wormlike marine animals, as deuterostomes. Hemichordates have a three-part body made up of a proboscis, collar, and trunk. These animals also have a characteristic ring of cilia surrounding the mouth. The most familiar of the hemichordates are the *acorn worms,* animals that live buried in mud or sand.

Deuterostomes evolved from a common ancestor during the Proterozoic eon more than 550 million years ago (mya). The major groups were present during the Early Cambrian period. Deuterostomes are characterized by several **shared derived characters** (synapomorphic characters), evolutionary novelties present in their most recent common ancestor (see Chapter 23). Evidence for the relatedness of the chordates, hemichordates, and echinoderms comes from molecular data and patterns of embryonic development. For example, deuterostomes are characterized by radial, rather than spiral, cleavage (see Chapter 30). Their cleavage is indeterminate, which means that the fate of their cells is fixed later in development than is the case in protostomes. In deuterostomes the mouth does not develop from the blastopore as in protostomes. The blastopore of deuterostomes becomes the anus (or is located near the future site of the anus), and the mouth develops from a second opening at the anterior end of the embryo, thus the name *deuterostome,* which is derived from the Greek words for "second mouth."

At some time in their life cycle, most deuterostomes develop *pharyngeal slits,* openings that connect the pharynx with the outside environment. Although there is evidence that some early echinoderms had pharyngeal slits, the extant (living) echinoderms have lost this character. *Basal deuterostomes* (members of the earliest deuterostome group to evolve) have a type of larva with a loop-shaped band of cilia used for locomotion. Deuterostomes have a true coelom.

CHECKPOINT 32.1

- *What are three shared derived characters of deuterostomes?*

32.2 ECHINODERMS

LEARNING OBJECTIVE

2 Identify three shared derived characters of echinoderms and describe the main classes of echinoderms.

The **echinoderms** (phylum *Echinodermata*) have one of the most highly derived body plans in the animal kingdom. With their unusual symmetry, spiny skins, water vascular system, and tube feet, echinoderms are one of the most unique animal groups. Echinoderm larvae are bilaterally symmetrical, ciliated, and free-swimming. However, during development the body reorganizes, and the adult exhibits *pentaradial symmetry,* in which the body is arranged in five parts around a central axis. Biologists have hypothesized that early echinoderms were sessile and that radial symmetry evolved as an adaptation to that lifestyle. Their radial symmetry allows these animals to respond effectively in every direction of their surrounding environment.

The most unique derived character of echinoderms is the **water vascular system,** a network of fluid-filled canals and chambers. In the sea star, for example, sea water enters through small pores in the *madreporite,* a sievelike structure on the body surface. Cilia lining the canals of the system move the water along. Branches of the water vascular system lead to numerous tiny **tube feet** that extend when filled with fluid. Each tube foot receives fluid from the main system of canals. A rounded muscular sac, or **ampulla,** at the base of the foot stores fluid and is used to operate the tube foot. A valve separates each tube foot from other parts of the system. When the valve shuts, the ampulla contracts, forcing fluid into the tube foot. The fluid causes the tube foot to extend. At the bottom of the foot, a suction-type structure presses against and adheres to whatever surface the tube foot is on. The water vascular system functions in feeding and gas exchange, and it serves as a hydrostatic skeleton important in locomotion.

Another unique echinoderm character is the **endoskeleton,** an internal skeleton, covered by a thin, ciliated epidermis. The endoskeleton consists of calcium carbonate ($CaCO_3$) plates and spines. The name *Echinodermata,* derived from words meaning "spiny skinned," was inspired by the spines that project outward from the endoskeleton. Some groups have pincerlike, modified spines called *pedicellariae* on the body surface. These structures, found only among the echinoderms, keep the surface of the animal free of debris.

Echinoderms have a well-developed coelom, containing coelomic fluid that transports materials. Although its structure varies in different groups, the complete digestive system is the most prominent body system. A variety of respiratory structures are found in the various classes. No excretory organs are present. The nervous system is simple, generally consisting of a nerve ring with nerves that extend out from it. Echinoderms have no brain. The sexes are usually separate, and eggs and sperm are generally released into the water, where fertilization takes place.

The fossil evidence suggests that echinoderms evolved from bilaterally symmetrical ancestors, probably during the Early Cambrian period. They achieved maximum diversity by the middle of the Paleozoic era, about 400 mya. By the beginning of the Mesozoic era 252 mya, they had declined, leaving five main groups that have survived to the present day. Biologists have identified about 7000 living and more than 13,000 extinct species. All echinoderms inhabit marine environments. They are found in the ocean at all depths. We consider the five extant groups: class Crinoidea, sea lilies and feather stars; class Asteroidea, sea stars; class Ophiuroidea, basket stars and brittle stars;

TABLE 32-1 Classes of Echinoderms

CLASS AND REPRESENTATIVE ANIMALS		CHARACTERISTICS
CRINOIDEA Feather stars, sea lilies		Feather stars motile; sea lilies, sessile, attach to ocean floor by stalk; oral surface on upper side of disc; tube feet along feathery arms trap microscopic organisms
ASTEROIDEA Sea stars		Arms (rays) extend from central disc; tube feet on undersurface of each arm; mouth in center of underside of disc; carnivorous predators and scavengers
OPHIUROIDEA Basket stars, brittle stars		Resemble sea stars, but arms long, slender, and more set off from central disc; tube feet used to collect and handle food
ECHINOIDEA Sea urchins, sand dollars		Skeletal plates flattened and fused to form a test; sand dollars burrow in sand and feed on organic particles; sea urchins graze on algae
HOLOTHUROIDEA Sea cucumbers		Body is elongated, muscular sac; mouth surrounded by tentacles that are modified tube feet; sluggish animals that live on sea bottom; many species eject organs when conditions are poor; for defense, can eject tubules out of anus

© Cengage Learning

class Echinoidea, sea urchins and sand dollars; and class Holothuroidea, sea cucumbers (TABLE 32-1 and FIG. 32-1).

Feather stars and sea lilies are suspension feeders

Feather stars and sea lilies (class *Crinoidea*) are the oldest living echinoderms (see Fig. 32-1a). Although many extinct crinoids are known, relatively few living species have been identified. The feather stars are motile, although they often remain in the same location for long periods. Sea lilies are sessile and remain attached to the ocean floor by a stalk.

Crinoids remove suspended food from the water. Several branched, feathery arms extend upward. Numerous tube feet shaped like small tentacles are located along the feathery arms. These tube feet are coated with mucus that traps microscopic organisms. The *oral surface* (location of the mouth) is on the upper side of the disc. In all other echinoderms, the mouth is located on the underside of the disc toward the substratum.

Many sea stars capture prey

Sea stars (commonly called starfish) are asteroids (class *Asteroidea;* see Fig. 32-1b). Their bodies consist of a central disc from which extend 5 to more than 20 arms, or rays (FIG. 32-2). The undersurface of each arm has hundreds of pairs of tube feet. The endoskeleton consists of a series of calcareous plates that permit some movement of the arms. Delicate dermal gills, small extensions of the body wall, carry on gas exchange. The mouth lies in the center of the underside of the disc.

Most sea stars are predators and scavengers that feed on cnidarians, crustaceans, mollusks, annelids, and even other echinoderms. A few species may catch small fish. The sea star's water vascular system does not permit rapid movement, so its prey usually consists of stationary or slow-moving animals, such as clams.

To attack a clam or other bivalve mollusk, the sea star mounts it and assumes a humped position as it straddles the edge opposite the hinge. Then, holding itself in position with its tube feet, the sea star slides its thin, flexible stomach out through its mouth and between the closed, or slightly gaping, valves (shell parts) of the clam. While the clam is still in its own shell, the sea star secretes enzymes that digest the soft parts of the clam to the consistency of a thick soup. When the partly digested meal is taken into the sea star body, it is further digested by enzymes secreted by digestive glands located in each arm.

The circulatory system in sea stars is poorly developed and probably of little help in transporting materials. Instead, the coelomic fluid, which fills the large coelom and bathes the internal tissues, assumes this function. Metabolic wastes pass to the outside by diffusion across the tube feet and dermal gills. The nervous system consists of a ring of nervous tissue encircling the mouth and a nerve extending from this ring into each arm.

Sea daisies are a curious group of asteroids characterized by small, disc-shaped, flat bodies (less than 1 cm in diameter) with no arms or mouth. They inhabit bacteria-rich wood sunk in deep water and apparently absorb bacteria through their body surface.

Basket stars and brittle stars make up the largest group of echinoderms

Basket stars and brittle stars (serpent stars) make up the largest group of echinoderms (class *Ophiuroidea*) both in number of species and individuals (see Fig. 32-1c). These animals resemble

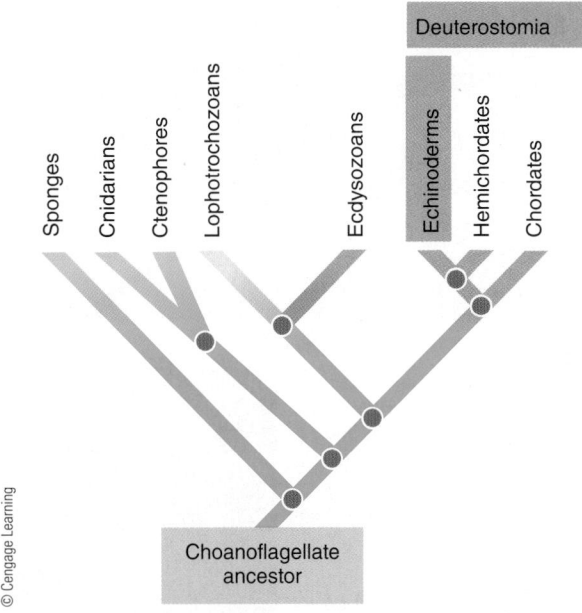

(a) Class Crinoidea. Feather stars use their slender, jointed appendages to cling to the surface of a rock or coral reef. They can creep away to escape predators.

(b) Class Asteroidea. Orange and red sea star (*Fromia monilis*) on a colonial tunicate.

(c) Class Ophiuroidea. Daisy brittle star (*Ophiopholis aculeata*), photographed in Muscongus Bay, Maine.

(d) Class Echinoidea. With its flattened, circular body, the sand dollar (*Dendraster excentricus*) is adapted for burrowing on the ocean floor.

(e) Class Holothuroidea. A sea cucumber (*Thelonota*) raises its body to spawn.

Figure 32-1 Representative echinoderms

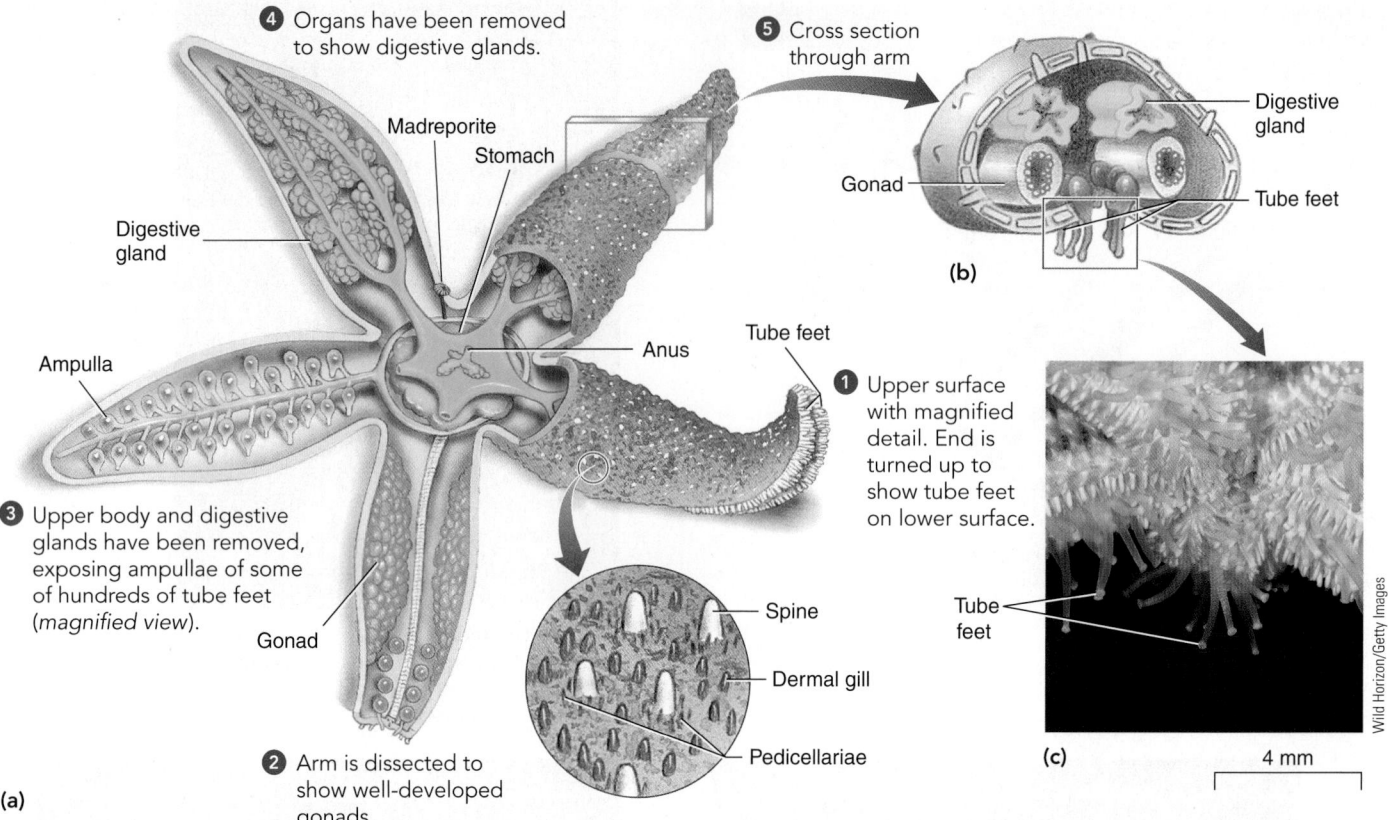

④ Organs have been removed to show digestive glands.

Madreporite

Stomach

Digestive gland

Ampulla

③ Upper body and digestive glands have been removed, exposing ampullae of some of hundreds of tube feet (*magnified view*).

Gonad

Anus

Tube feet

② Arm is dissected to show well-developed gonads.

(a)

⑤ Cross section through arm

Digestive gland

Gonad

Tube feet

(b)

① Upper surface with magnified detail. End is turned up to show tube feet on lower surface.

Spine

Dermal gill

Pedicellariae

Tube feet

(c)

4 mm

Wild Horizon/Getty Images

Figure 32-2 Body plan of a sea star

(a) A sea star viewed from above, with its arms in various stages of dissection. Similar structures are present in each arm, but some organs are not shown so as to highlight certain other structures. The two-part stomach is in the central disc with the anus on the aboral (*upper*) surface and the mouth beneath on the oral surface. **(b)** Cross section through arm and tube feet. **(c)** LM of underside of a sea star, showing its tube feet.

© Cengage Learning

sea stars in that their bodies consist of a central disc with arms. However, the arms are long and slender and more sharply set off from the central disc than those of sea stars. Some brittle stars use their arms for locomotion, making rowing movements. Their tube feet lack suckers; they may be used to collect and handle food and may also serve a sensory function, perhaps that of taste.

Sea urchins and sand dollars have movable spines

Sea urchins and sand dollars (class *Echinoidea*) have no arms (see Fig. 32-1d and chapter-opening photograph). Their skeletal plates are flattened and fused to form a solid shell called a test. The flattened body of the sand dollar is adapted for burrowing in the sand, where it feeds on tiny organic particles. Sand dollars have smaller spines than sea urchins.

Many sea urchins graze on algae, scraping the sea floor with their calcareous teeth. Sea urchins use their tube feet for locomotion. They also push themselves along with their movable spines. In some species the spines covering the body can penetrate flesh and are difficult to remove. So threatening are these spines that swimmers on tropical beaches are often cautioned to wear shoes when venturing offshore, where these living pincushions may dwell in abundance.

Sea cucumbers are elongated, sluggish animals

Sea cucumbers (class *Holothuroidea*) are appropriately named, for some species are about the size and shape of a cucumber. The elongated body is a flexible, muscular sac (see Fig. 32-1e). The mouth is usually surrounded by a circle of tentacles that are modified tube feet. The endoskeleton consists of microscopic plates embedded in the body wall. The circulatory system, which is more highly developed than in other echinoderms, functions to transport oxygen and perhaps nutrients.

Sea cucumbers are sluggish animals that usually live on the bottom of the sea, sometimes burrowing in the mud. Some graze with their tentacles, whereas others stretch their branched tentacles out in the water and wait for dinner to float by. Algae and other morsels are trapped in mucus along the tentacles.

An interesting habit of some sea cucumbers is evisceration, in which the digestive tract, respiratory structures, and gonads are ejected from the body, usually when environmental conditions are unfavorable. When conditions improve, the lost parts are regenerated. Even more curious, when certain sea cucumbers are irritated or attacked, they direct their rear end toward the enemy and shoot red tubules out of their anus! These unusual weapons become sticky in sea water, and the attacking animal may become hopelessly entangled in the spaghetti-like tangle of tubules. Some of the tubules release a toxic substance.

CHECKPOINT 32.2

- *What are three derived characters of echinoderms? Describe each.*
- **CONNECT** *How do sea stars differ from crinoids? from sea urchins?*

32.3 THE CHORDATES: MAJOR CHARACTERISTICS

LEARNING OBJECTIVE

3 Describe characteristics of chordates, including four shared derived characters.

Biologists currently assign the chordates (phylum Chordata) to three groups, or subphyla: urochordates, marine animals called *tunicates;* cephalochordates, marine animals called *lancelets;* and vertebrates, animals with backbones. Although the structures of living species may suggest otherwise, recent molecular data support that urochordates, and not cephalochordates, are the sister group to the vertebrates (**FIG. 32-3**).

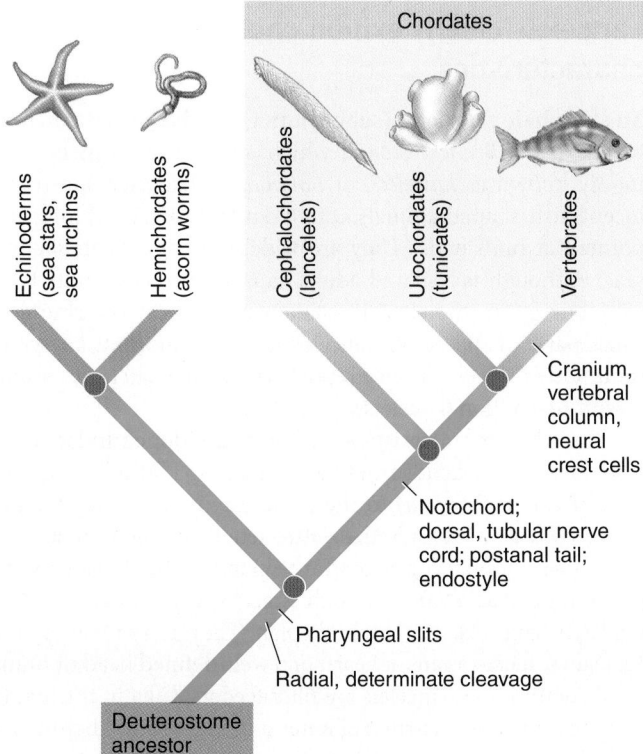

Figure 32-3 Evolutionary relationships of the chordates

This cladogram shows hypothesized phylogenetic relationships among deuterostomes based on structural and DNA data.

© Cengage Learning

Chordates are deuterostomes. They are coelomates with bilateral symmetry, a tube-within-a-tube body plan, and three well-developed germ layers. Four shared derived characters distinguish the chordates from all other groups of animals. These characters, which evolved in connection with their evolving methods of locomotion and obtaining food, are the notochord; the dorsal, tubular nerve cord; a postanal tail; and an endostyle or thyroid gland (**FIG. 32-4**).

1. All chordates have a **notochord** during some time in their life cycle. The notochord is a dorsal, longitudinal rod composed of spongy connective tissue cells surrounded by a tough fibrous sheath. The notochord is firm but flexible, and it supports the body. It also plays an important role in embryonic development of the vertebrates.
2. At some time in their life cycle, chordates have a **dorsal, tubular nerve cord.** The chordate nerve cord differs from the nerve cord of most other animals in that it is located dorsally rather than ventrally, is hollow rather than solid, and is single rather than double.
3. Chordates have a larva or embryo with a muscular **postanal tail,** an appendage that extends posterior to the anus.
4. The **endostyle** is a groove in the floor of the pharynx that secretes mucus and traps food particles in the sea water passing through the pharynx. The endostyle is present in the urochordates, cephalochordates, and lamprey larvae. The thyroid gland evolved from the endostyle and is present in all other chordates.

Pharyngeal slits are another important characteristic of chordates. During chordate embryonic development, a series of grooves develop in the body wall in the region of the pharynx (the part of the digestive tract just posterior to the mouth). A series of outpocketings from the lateral sides of the pharynx (pharyngeal pouches) extend to the grooves. In aquatic chordates the tissue between the grooves and pharyngeal pouches breaks through and forms **pharyngeal slits,** passageways that connect the inside of the pharynx with the surrounding environment.

The perforated pharynx first evolved as an adaptation for filtering food. As water flowed through the mouth and passed out of the pharynx through the pharyngeal slits, suspended food particles were trapped in mucus. Later, modifications evolved, including highly vascularized gills in the passageways. These additions adapted the pharyngeal slits for gas exchange in fishes and some other aquatic vertebrates. Early in vertebrate evolution, anterior pharyngeal arches (supporting tissue between the pharyngeal pouches) evolved into jaws. Although pharyngeal slits are found in all chordates, they are not considered a derived character of chordates because they are also present in hemichordates.

Typically, chordates have an endoskeleton and a closed circulatory system with a ventral heart. Chordates have segmented bodies, but specialization is so pronounced that the basic segmentation of the body plan may not be apparent. The segmentation is most obvious in the serially repeated body muscles and skeletal structures (like vertebrae), and in the nerves, which supply these tissues.

The shared derived characters of chordates include a notochord; dorsal, tubular nerve cord; postanal tail; and endostyle, a groove in the floor of the pharynx that traps food particles in sea water.

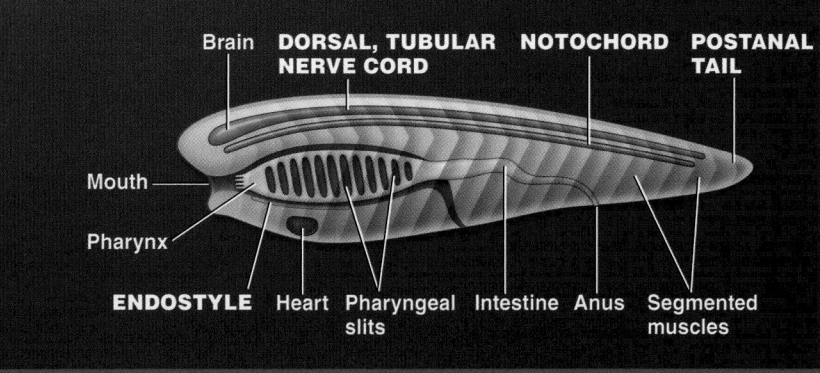

Brain   **DORSAL, TUBULAR NERVE CORD**   **NOTOCHORD**   **POSTANAL TAIL**

Mouth

Pharynx

ENDOSTYLE   Heart   Pharyngeal slits   Intestine   Anus   Segmented muscles

Figure 32-4 *Animation* **Generalized chordate body plan**

Identify the shared derived characters: notochord; dorsal, tubular nerve cord; postanal tail; and endostyle. Other key characteristics include pharyngeal slits and segmented muscles.

PREDICT What structures do you think the dorsal, tubular nerve cord gives rise to during the development of vertebrate chordates?

© Cengage Learning

CHECKPOINT 32.3

- *What are four shared derived characters of chordates?*
- **CONNECT** *How is the chordate nerve cord different from the nerve cord of most other animals?*
- **CONNECT** *What is the relationship between your jaws and the pharyngeal slits of early chordates?*

32.4 INVERTEBRATE CHORDATES

LEARNING OBJECTIVE

4   Compare tunicates and lancelets, and summarize the phylogeny of chordates.

Some chordates are not vertebrates. The invertebrate chordates include tunicates (urochordates) and lancelets (cephalochordates).

Tunicates are common marine animals

The **tunicates,** or **urochordates** (subphylum Urochordata), include the sea squirts and their relatives (**FIG. 32-5**). Larval tunicates have typical chordate characteristics and superficially resemble tadpoles. The expanded body has a pharynx with slits, and the long muscular tail contains a notochord and a dorsal, tubular nerve cord. Some tunicates (*appendicularians*) retain their chordate features and ability to swim. These animals are common members of the zooplankton.

Most tunicates are *ascidians,* commonly known as sea squirts (class Ascidiacea). A sea squirt larva swims for a time and then attaches itself to a rock, a piling, or the sea bottom.

It loses its tail, notochord, and much of its nervous system. Adult sea squirts are barrel-shaped, sessile marine animals unlike other chordates. Indeed, they are often mistaken for sponges or cnidarians. Only the pharyngeal slits, endostyle, and the structure of its larva indicate that the sea squirt is a chordate.

Adult tunicates develop a protective covering, or **tunic,** that may be soft and transparent or quite leathery. Curiously, the tunic consists of a carbohydrate much like cellulose. The tunic has two openings: the incurrent siphon, through which water and food enter; and the excurrent siphon, through which water, waste products, and gametes pass to the outside. Sea squirts get their name from their practice of forcefully expelling a stream of water from the excurrent siphon when irritated.

Tunicates are filter feeders that remove plankton suspended in the stream of water passing through the pharynx. Food particles are trapped in mucus secreted by cells of the endostyle, a groove that extends the length of the pharynx. Ciliated cells of the pharynx move the stream of food-laden mucus into the esophagus. Much of the water entering the pharynx passes out through the pharyngeal slits into an *atrium* (a chamber) and is discharged through the excurrent siphon.

Some species of tunicates form large colonies in which members share a common tunic and excurrent siphon. Colonial forms often reproduce asexually by budding. Sexual forms are usually hermaphroditic.

Lancelets clearly exhibit chordate characteristics

Most **cephalochordates** (subphylum **Cephalochordata**) belong to the genus *Branchiostoma,* which consists of animals commonly known as *lancelets,* or *amphioxus.* Lancelets are translucent, fish-shaped animals, 3 to 8 cm (1.3 to 3.2 in.) long and pointed at both ends. They are widely distributed in shallow seas. Although larvae and adults can swim freely, adults typically burrow in the sand in shallow water near the shore. In some parts of the world, lancelets are an important source of food. One Chinese fishery reports an annual catch of 35 tons (about one billion lancelets).

Chordate characteristics are highly developed in lancelets. The notochord extends from the anterior tip ("head"; hence the name *Cephalochordata*) to the posterior tip. A dorsal, tubular nerve cord also extends the entire length of the animal, and many pairs of pharyngeal slits are evident in the large pharyngeal region (**FIG. 32-6**). Although superficially similar to fishes, lancelets have a far simpler body plan. They do not have paired fins, jaws, sense organs, a heart, or a well-defined head or brain.

Like tunicates, lancelets are filter feeders. Cilia in the mouth and pharynx draw a current of water into the mouth. Microscopic organisms in the water are trapped in mucus secreted by the endostyle and are then moved back to the intestine by beating cilia.

Water passes through the pharyngeal slits into the atrium, a chamber with a ventral opening (the *atriopore*) located anterior to the anus. Metabolic wastes are excreted by segmentally

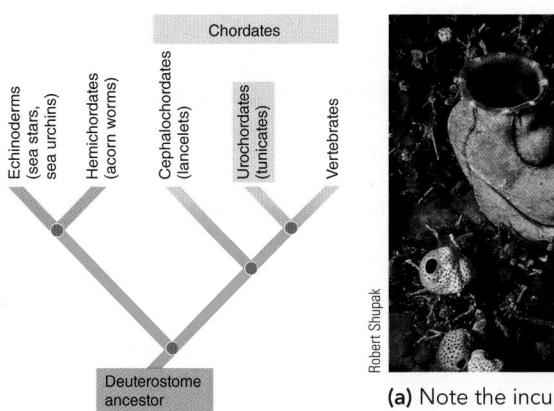

(a) Note the incurrent (*top*) and excurrent (*side*) siphons of this sea peach (*Halocynthia aurantium*), a solitary tunicate.

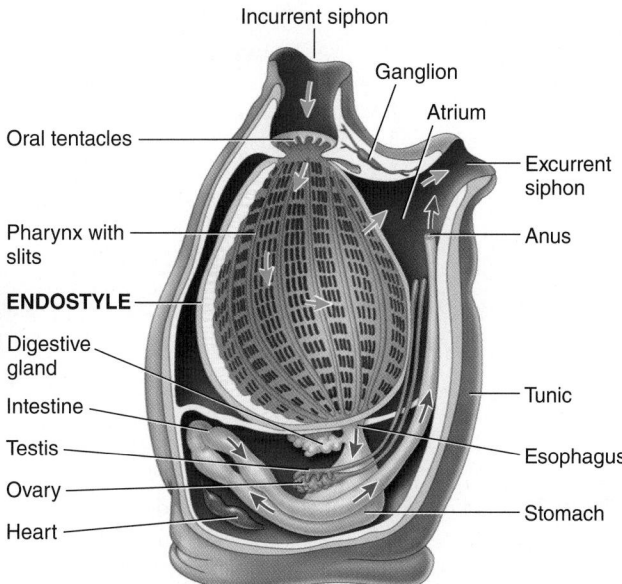

(b) Lateral view of an adult tunicate. The *blue arrows* represent the flow of water, and the *red arrows* represent the path of food.

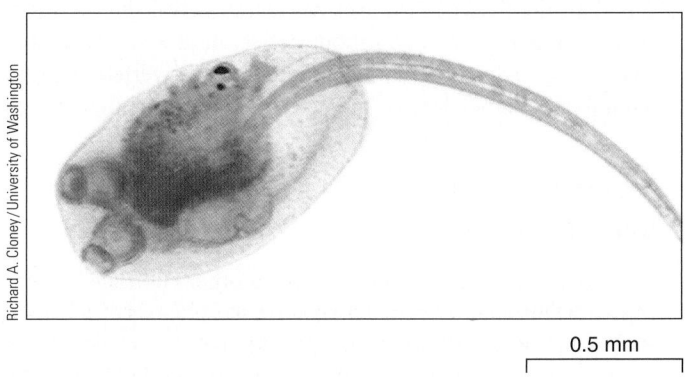

(c) Swimming larval stage of a colonial species, *Distaplia occidentalis*.

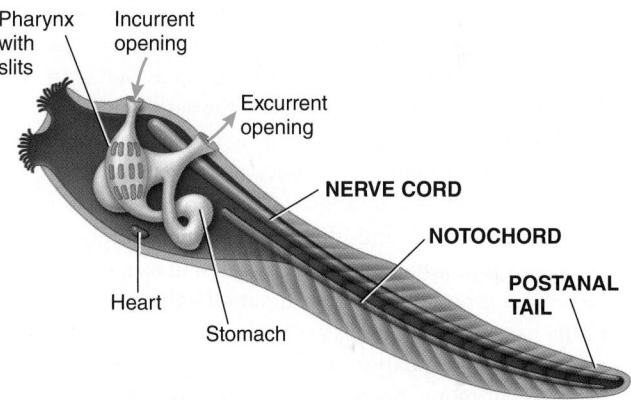

(d) Internal structure of a larval tunicate (*lateral view*).

Figure 32-5 *Animation* **Tunicate body plan**
© Cengage Learning

---

arranged, ciliated *protonephridia* that open into the atrium. Unlike circulation in other invertebrates, the blood flows anteriorly in the ventral vessel and posteriorly in the dorsal vessel in lancelets. This circulatory pattern is similar to that of fishes.

Systematists debate chordate phylogeny

Fossils of early chordates contribute to our understanding of chordate phylogeny. Invertebrate chordates are soft-bodied and did not leave many fossils. However, some well-preserved, key fossils that date back to the Middle Cambrian have been found in the Burgess Shale of British Columbia, Canada. Early Cambrian fossil sites have been discovered in Chengjiang and Haikou, China. At Chengjiang, fine-grained rocks about 530 million years old have preserved soft-bodied animals, including *Haikouella*, in some detail. *Haikouella* was about 2 to 3 cm (approximately 1 in.) long and had a nerve cord, a notochord, pharyngeal slits,

muscle segments, and a brain. Many biologists view *Haikouella* as an early chordate. Some investigators hypothesize that this fossil represents a transitional form between the earliest chordates and the first vertebrates.

Pikaia, a lancelet-like fossil discovered in the Burgess Shale, had a primitive notochord with muscles attached to it that allowed for locomotion. About 4 cm (1.5 in.) in length, *Pikaia* had a tail fin and probably filtered food from the water. *Pikaia* is also considered an early chordate, possibly sharing many similarities with modern cephalochordates.

Are cephalochordates or urochordates the sister group of the vertebrates? Although structural similarities implicated cephalochordates, rather than urochordates, as the sister group to the vertebrates, the predominance of recent molecular data supports that urochordates are the closest relatives of the vertebrates. The molecular data combined with information gathered from the available fossils suggest that cephalochordates, urochordates,

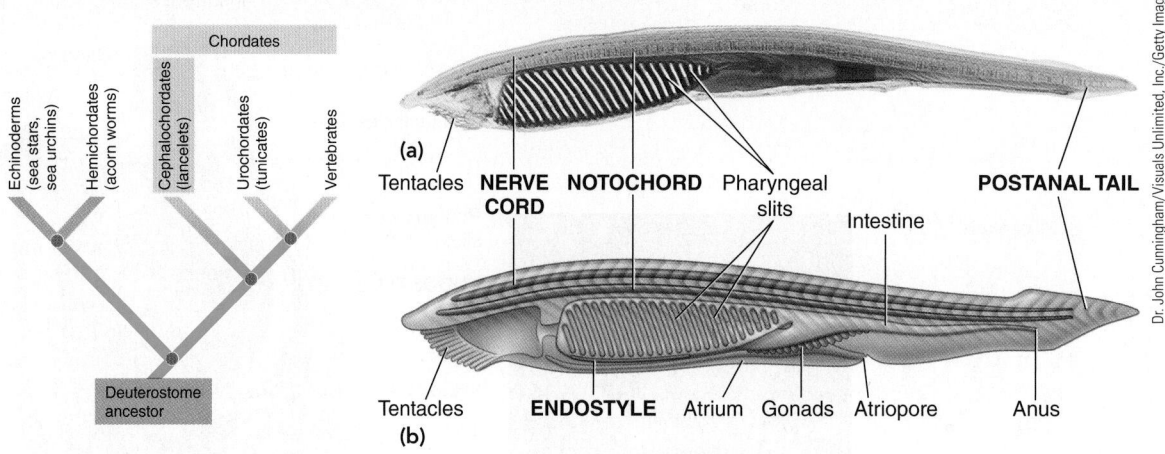

Figure 32-6 *Animation* **Cephalochordate body plan**

(a) Photograph of a lancelet, *Branchiostoma* (amphioxus). Note the prominent pharyngeal slits. **(b)** Longitudinal section showing derived characters and other internal structures.

© Cengage Learning

and vertebrates all evolved from early chordates that had elongated bodies. Some of these very early chordates retained the same general body form and gave rise to the cephalochordates. In others, an adult that was sessile, rather than swimming, gave rise to the urochordates. A third lineage added a number of specialized features while giving rise to the vertebrates.

The tunicate *Ciona intestinalis* is an excellent model organism for studying chordate genetics and development because its genome consists of only about 16,000 protein-coding genes, the basic set of genes found in chordates. Interestingly, about 80% of *Ciona*'s genes are found in vertebrates. In fact, *Ciona* has most vertebrate gene families, but in simplified form. For example, *Ciona* has only a single copy of each gene family involved in cell signaling and regulation of development, whereas vertebrates have two or more copies of these gene families. (A gene family is a group of genes that evolved over time from a single ancestral gene through duplication and divergence.) Some of the "extra" copies probably evolved to code for new structures or functions. *Ciona* also has genes for some vertebrate structures and processes, even though it does not express these genes.

CHECKPOINT 32.4

- **CONNECT** *How are the main derived chordate characters evident in a tunicate larva and in an adult tunicate?*
- **CONNECT** *If you found a small fishlike animal along the shoreline, how would you determine whether or not it was a lancelet?*

32.5 INTRODUCING THE VERTEBRATES

LEARNING OBJECTIVES

5 Describe four shared derived characters of vertebrates.
6 Describe the major taxa of extant vertebrates.

Numbering about 50,000 described species, the **vertebrates** (subphylum Vertebrata) are less diverse and much less numerous than the insects. However, vertebrates rival the insects in their adaptations to an enormous variety of lifestyles (**FIG. 32-7**). In addition to the basic chordate characteristics, vertebrates have a number of shared derived characters not found in other groups.

The vertebral column is a derived vertebrate character

The vertebrates are distinguished from other chordates in having a **vertebral column** that forms the skeletal axis of the body. Although sometimes called the "backbone," the vertebral column actually consists of a series of separate skeletal elements, the *vertebrae,* which are made of cartilage or bone. This flexible support develops around the notochord, and in most species it largely replaces the notochord during embryonic development. Dorsal projections of the vertebrae enclose the nerve cord along its length. Anterior to the vertebral column, a cartilaginous or bony **cranium,** or braincase, encloses and protects the brain, the enlarged anterior end of the nerve cord.

The cranium and vertebral column are part of the *endoskeleton.* In contrast with the nonliving exoskeleton of many invertebrates, the vertebrate endoskeleton is a living tissue that grows with the animal. Some vertebrates (jawless and cartilaginous fishes) have skeletons made of cartilage. However, in most vertebrates the skeleton is mainly bone, a tissue that contains fibers made of the protein collagen. The hard matrix of bone consists of the compound hydroxyapatite, composed mainly of calcium phosphate. Muscles are attached to the endoskeleton.

Many characters common to vertebrates have been derived from a group of embryonic cells called **neural crest cells.** These cells, found only in vertebrates, appear early in development and migrate to various parts of the embryo. Neural crest cells give rise to or influence the development of many structures, including nerves, head muscles, cranium, and jaws.

This cladogram provides a framework for understanding the evolution of vertebrate diversity and adaptations to aquatic and terrestrial environments.

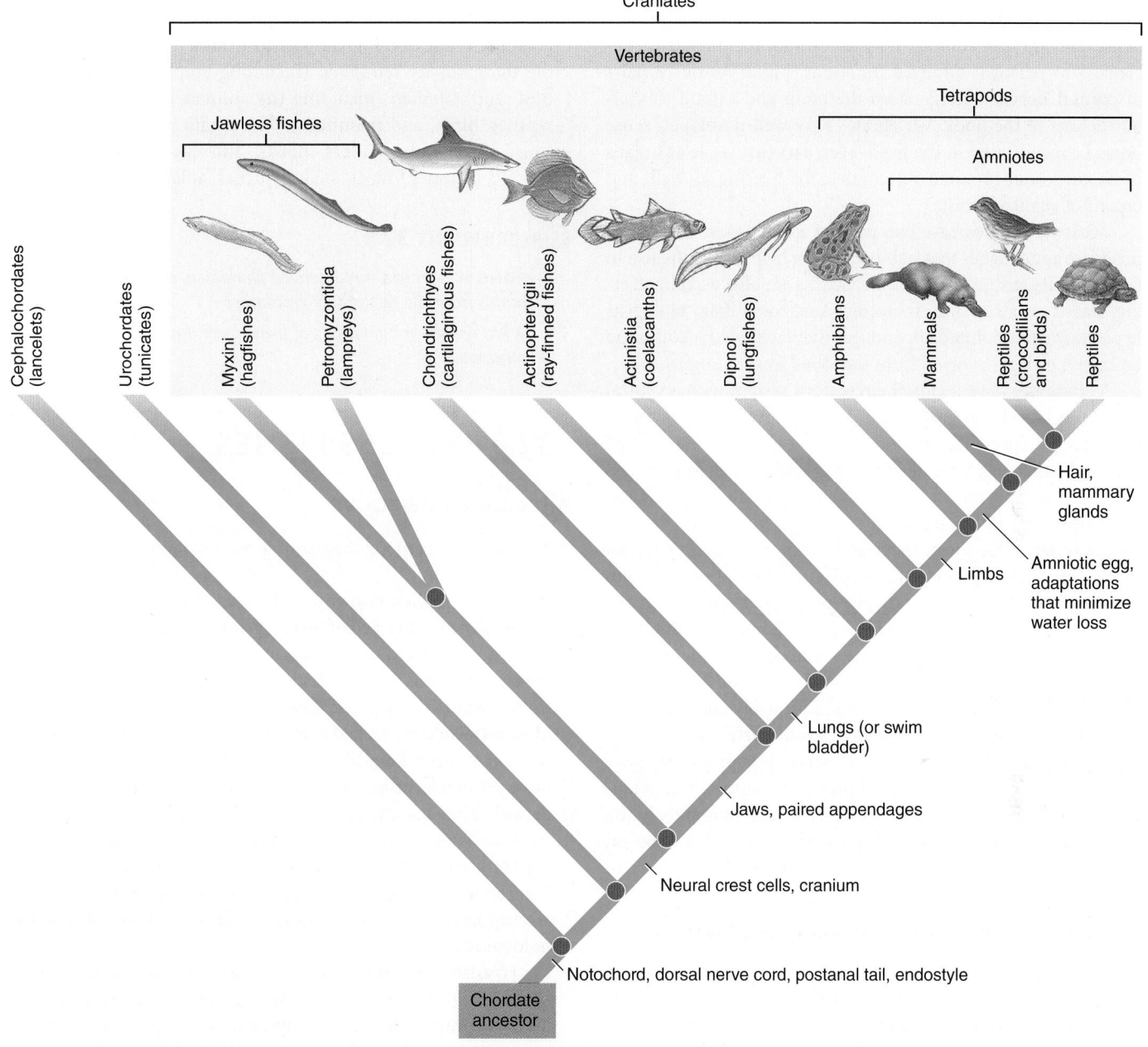

Figure 32-7 *Animation* **Evolutionary relationships of extant vertebrates**

This cladogram represents one phylogenetic interpretation of vertebrate phylogeny. The evolution of certain key characters is indicated. As new data are collected and considered, details of the cladogram will likely change. Amniote relationships are shown in more detail in Figure 32-20.
© Cengage Learning

CONNECT What are some advantages of the pronounced cephalization that occurred as vertebrates evolved from jawless fish to tetrapods?

Recall that *Hox* genes are important in determining pattern development in embryos (see Chapter 17). Specifically, *Hox* genes determine the fate of cells in the anterior–posterior axis. These genes occur in clusters on specific chromosomes. All invertebrates and non-vertebrate chordates that have been studied have one *Hox* gene cluster; vertebrates have four duplicated clusters. Other gene clusters, including some encoding transcription factors and others coding globin proteins for transporting oxygen, were also duplicated during the evolution of the early vertebrates. The significance of these genetic changes is the focus of current investigation, but it is reasonable to hypothesize that the increase in genetic complexity contributed to the development of more complex and varied bodies.

Recall that some invertebrate groups show an evolutionary trend toward **cephalization,** the concentration of nerve cells and sense organs in a definite head. The vertebrates are characterized by *pronounced* cephalization. The brain has become larger and more elaborate, and its various regions have specialized to perform different functions. Either 10 or 12 pairs of **cranial nerves** emerge from the brain and extend to various organs of the body. Vertebrates have well-developed sense organs concentrated in the head: eyes; ears that serve as organs of balance and, in some vertebrates, for hearing as well; and organs of smell and taste.

Most vertebrates have two pairs of appendages. The fins of fishes are appendages that stabilize and/or help propel the fish in the water. Paired pectoral and pelvic fins are also used in steering. Based on structural, molecular, and fossil data, biologists hypothesize that jointed appendages that facilitated locomotion on land (i.e., legs) evolved from the lobed fins of lungfishes.

Vertebrates have a closed circulatory system with a ventral heart and blood containing hemoglobin. The complete digestive tract has specialized regions and large digestive glands (the liver and pancreas). Several *endocrine glands,* which are ductless glands, secrete hormones. Paired kidneys regulate fluid balance. The sexes are typically separate.

Early vertebrates were probably marine animals. In contrast to the tunicates and lancelets, which use cilia to beat a stream of water into the mouth and filter particles of food from the water, vertebrates use muscles for feeding. Some vertebrates use muscles to draw in a current of water from which both food and oxygen can be extracted. Muscles in the wall of the digestive tube are more powerful than cilia and probably contributed to an increase in both size and activity of early vertebrates. A muscular pharynx with firm skeletal elements for support may have permitted animals to crush small prey after capture. The evolution of more effective sensory systems, a more complex brain, and organs that could support increased activity was important in the shift from filter-feeding to a more active lifestyle.

Vertebrate taxonomy is a work in progress

The study of evolutionary relationships of vertebrates is an important focus of research. The earliest vertebrate fossils include *Haikouichthys,* conodonts, and a group known as *ostracoderms* (discussed in the next section). *Haikouichthys* was a fishlike animal about 2.5 cm (1 in.) long. It had several characters found in vertebrates, and some researchers think it was an early jawless fish.

Conodonts were simple fishlike chordates with gill arches (tissues that support the gills), muscular segments, and fins. They had large eyes and complex tooth-like hooks that were probably used in capturing prey. Conodonts were abundant from the Precambrian to the Late Triassic period; they have been identified in several fossil beds around the world. Cladistic analyses suggest that conodonts were early vertebrates.

By the Late Cambrian, the chordates included a number of different taxa with skeletal elements (a rudimentary braincase,

or cranium) protecting the brain. Some of these animals also showed evidence of vertebrae. These earliest vertebrates lacked jaws and are sometimes referred to as **agnathans** (*a,* "without"; and *gnathos,* "jaw").

The extant vertebrates can be assigned to nine classes: six classes of fishes and three of **tetrapods** (four-limbed vertebrates) (TABLE 32-2; see also Fig. 32-7). Most modern biologists classify the tetrapods as *Tetrapoda* (including amphibians and amniotes) and *Amniota* (including the animals traditionally called reptiles, birds, and mammals). The extant Amphibia includes frogs, toads, salamanders, newts, and caecilians; the extant Reptilia includes lizards, snakes, turtles, alligators, and birds.

CHECKPOINT 32.5

- CONNECT *What shared derived characters distinguish the vertebrates from the rest of the chordates?*
- *What are the main classes of fishes? the main groups of tetrapods?*

32.6 JAWLESS FISHES

LEARNING OBJECTIVE

7 Distinguish among the major groups of jawless fishes.

Some of the earliest known vertebrates, collectively referred to as *ostracoderms,* consisted of several groups of small, armored, jawless fishes that lived on the bottom and strained their food from the water (see Fig. 21-11). Thick bony plates protected their heads from predators, and thick scales covered their trunks and tails. Most ostracoderms lacked paired fins. Fragments of ostracoderm scales have been found in rocks from the Cambrian period, but most ostracoderm fossils are from the Ordovician and Silurian periods. They became extinct by the end of the Devonian period.

Like ostracoderms, present-day hagfishes and lampreys have neither jaws nor paired fins. They are eel-shaped animals, up to 1 m (about 3 ft) long. Their smooth skin lacks scales, and they are supported by a cartilaginous skeleton and well-developed notochord.

Hagfishes, assigned to class *Myxini* (mik-sin'-y), are marine scavengers (FIG. 32-8). They burrow for worms and other invertebrates or prey on dead and disabled fishes. Hagfishes use tooth-like projections from their tongue to pull off flesh from prey. For leverage, a hagfish can tie itself into a knot. As a defense mechanism, hagfishes secrete large amounts of fluid that forms sticky slime in sea water. The hagfish becomes so slippery that predators cannot grasp it.

Some systematists have argued that the hagfishes do not quite qualify as vertebrates. Although they have many vertebrate characteristics, including a cranium, hagfishes have no trace of vertebrae. The notochord is their only axial support. Some biologists use the term *Craniata,* based on the presence of a cranium, to designate a clade that includes the vertebrates plus the hagfishes. They view the Myxini as the sister group of all other living craniates. Molecular analyses, however, now support

TABLE 32-2   Extant Vertebrate Classes   681

CLASS		EXAMPLES	CHARACTERISTICS
MYXINI		Hagfishes	Jawless, marine fishes that lack paired appendages; the notochord is their only axial support; lack vertebrae
PETROMYZONTIDA		Lampreys	Jawless, freshwater and marine fishes with skeleton of cartilage; complete cranium and rudimentary vertebrae; gills; specialized sense organs
CHONDRICHTHYES		Sharks, rays, skates, chimaeras	Jawed marine and freshwater fishes with skeleton of cartilage; vertebrae present; gills; placoid scales; two pairs of fins; oviparous, ovoviparous, or viviparous (a few species); well-developed sense organs
ACTINOPTERYGII (ray-finned fishes)		Perch, salmon, tuna, trout	Bony, marine and freshwater fishes; gills; swim bladder; generally oviparous
ACTINISTIA (fleshy-finned fishes)		Coelacanths	Bony fishes; marine, nocturnal predators on fish; lobed fins
DIPNOI (fleshy-finned fishes)		Lungfishes	Bony freshwater fishes; have both functional gills and lungs
AMPHIBIA		Salamanders, frogs and toads, caecilians	Aquatic larva typically undergoes metamorphosis into terrestrial adult; gas exchange through lungs and/or moist skin; two atria and single ventricle; systemic and pulmonary circulation
REPTILIA		Turtles, lizards, snakes	Amniotes with horny scales; adapted for reproduction on land (internal fertilization, leathery shell, amnion); lungs; ventricles of heart partly divided
		Crocodilians and birds	Amniotes with two complete ventricles; care for their young. Birds have feathers; anterior limbs modified as wings; compact, endothermic; vocal calls and complex songs
MAMMALIA		Monotremes (Protheria), marsupials (Metatheria), Eutheria (mammals with well-developed placentas)	Amniotes with hair; females nourish young with mammary glands; differentiation of teeth; three middle-ear bones; diaphragm; heart with two separate atria and two separate ventricles; endothermic; highly developed nervous system

© Cengage Learning

classifying hagfishes as vertebrates. Hagfishes may have evolved from vertebrates that had vertebrae but lost them during the course of their evolution.

Lampreys are jawless vertebrates assigned to class Petromyzontida (pet-tro-my-zon'-tih-da) (from the Greek *petros,* "stone," and *myzon,* "sucking"; lampreys hold onto stones by their mouths to prevent being washed away by the currents). Some lampreys spend their adult lives in the ocean and return to fresh water to reproduce. Many species of adult lampreys are parasites on other fishes (**FIG. 32-9**). Adult parasitic lampreys have a circular sucking disc around the mouth, which lies on the ventral side of the anterior end of the body. Using this disc to attach to a fish, the lamprey bores through the skin of its host

with horny teeth (made of keratin) on the disc and tongue. Then the lamprey injects an anticoagulant into its host and sucks out blood and soft tissues.

As in hagfish, the notochord persists throughout life and is not replaced by a vertebral column. However, lampreys have rudiments of vertebrae, cartilaginous segments that extend dorsally around the spinal cord.

CHECKPOINT 32.6

- **CONNECT** *How do lampreys and hagfishes resemble ostracoderms?*
- **CONNECT** *In what ways do hagfishes differ from other fishes?*

Figure 32-8 Pacific hagfish (*Eptatretus stoutii*)

The hagfish has a flexible body, which it can tie into knots. It knots its tail and then slides the knot toward its head. Being knotted gives it leverage when tearing flesh from its prey. Knotting itself also helps the hagfish remove excess slime from its body.

32.7 EVOLUTION OF JAWS AND LIMBS: JAWED FISHES AND TETRAPODS

LEARNING OBJECTIVE

8   Trace the evolution of jawed fishes and early tetrapods, and describe modern amphibians.

Imagine trying to hunt for food and then eat without jaws or limbs. Such was the challenge for early vertebrates, which helps explain why the jawless fishes are mainly scavengers and

parasites. The evolution of jaws from a portion of the gill arch skeleton and the development of fins allowed fishes to become active predators. With jaws an animal can grasp and hold on to live prey while eating it. The evolution of fins allowed fishes to swim faster and with more control. Fishes with jaws and fins had many new opportunities for capturing food.

Fossil evidence suggests that jaws and paired fins evolved during the Late Silurian and Devonian periods. Two early groups of jawed fishes, now extinct, were the **acanthodians,** armored fishes with paired spines and pectoral and pelvic fins, and **placoderms,** armored fishes with paired fins (**FIG. 32-10**). Vertebrates with jaws are referred to as **gnathostomes** (nath'-o-stomes). The success of the jawed vertebrates probably contributed to the extinction of the ostracoderms.

When the fins of certain fishes evolved into limbs about 370 mya, a different type of locomotion was possible. These limbed vertebrates (early tetrapods) were able to move about in shallow waters and wetlands in search of food. Some of these early tetrapods were able to move on land and gained access to new food and habitats.

Most cartilaginous fishes inhabit marine environments

The cartilaginous fishes, members of class **Chondrichthyes** (kon-drik'-thee-eez), appeared as successful marine forms in the Devonian period. Chondrichthyes are a monophyletic group that includes the sharks, rays, skates, and ratfishes (**FIG. 32-11**). Most species are ocean dwellers, but a few have invaded fresh water. With the exception of whales, the sharks are the largest living vertebrates. Some whale sharks (*Rhincodon*) exceed 15 m (49 ft) in length.

Most rays and skates are flattened creatures that live partly buried in the sand. Their enormous pectoral fins propel them along the bottom, where they feed on mussels and clams. The

(a) Three lampreys attached to a carp by their suction-cup mouths. Note the absence of jaws and paired fins.

(b) Suction-cup mouth of adult lamprey (*Entosphenus japonicus*). Note the rasplike teeth.

Figure 32-9 Lampreys

Parasitic lampreys attach to fish and suck out body fluids. They leave wounds in the fish that may be fatal.

(a) An acanthodian. *Climatius* was a spiny-skinned acanthodian with large fin spines and five pairs of accessory fins between the pectoral and pelvic pairs. *Climatius* was a small fish that reached a length of 8 cm (3 in.).

(b) A placoderm. *Dunkleosteus* was a giant placoderm that grew to a length of 8 m (26 ft). [Most placoderms were only about 20 cm (8 in.) long.] Its head and thorax were covered by bony armor, but the rest of the body and tail were naked.

Figure 32-10 *Animation* **Early jawed fishes**
Acanthodians and placoderms flourished in the Devonian period.
© Cengage Learning

(a) Blue-spotted stingray (*Taeniura lymma*). Stingrays typically feed on shellfish and bottom-dwelling fishes.

(b) The great white shark (*Carcharodon carcharias*). Photographed in Australia, this shark is considered the most dangerous shark to humans. It is actually white only on its ventral aspect; the rest of the body is brownish gray or bluish gray.

Figure 32-11 Cartilaginous fishes

stingray has a whiplike tail with a barbed spine at its base that can inflict a painful wound. The electric ray has electric organs on either side of the head. These modified muscles can discharge enough electric current (up to 200 volts) to stun fairly large fishes as well as human swimmers.

Chondrichthyes retain their cartilaginous embryonic skeleton. Although this skeleton is not replaced by bone, in many species calcium salts are added to the cartilage for strength. All chondrichthyes have jaws and two pairs of fins. The skin contains **placoid scales,** which are toothlike structures (**FIG. 32-12**). The lining of the mouth contains larger, but essentially similar, scales that serve as teeth. The teeth of other vertebrates are homologous with these scales. Shark teeth are embedded in the flesh and not attached to the jawbones; new teeth develop continuously in rows behind the functional teeth and migrate forward to replace any that are lost.

The streamlined shark body is adapted for rapid swimming. Lift is provided by body shape and fins. The shark stores a great deal of oil in its large liver (which may account for up to 30% of its body weight). Fats and oils decrease the overall density of fishes and contribute to buoyancy. Even so, the shark body is denser than water, so sharks tend to sink unless they are actively swimming.

Most sharks are predators that swim actively and prey on other fishes as well as on crustaceans and mollusks. The largest sharks and rays, like the largest whales, are filter feeders that strain plankton from the water. They gulp water through the mouth. As the water passes through the pharynx and out the gill slits, food particles are trapped in a sievelike structure.

Predatory sharks are attracted to blood, so a wounded swimmer or a skin diver towing speared fish is a target. However, despite the common portrayal of sharks in books and films as monstrous enemies, most do not go out of their way to attack humans. In fact, of the approximately 350 known shark species, fewer than 30 are known to attack humans.

Sharks have a complex brain and a spinal cord that is protected by vertebrae. Their well-developed sense organs effectively locate prey in the water. Sharks may detect other animals electrically in addition to sensing them by sight or smell. **Electroreceptors** on the shark's head sense weak electric currents generated by the muscle activity of animals. The **lateral line organ,** found in all fishes and many amphibians, is a groove along each side of the body with many tiny openings to the outside. Sensory cells in the lateral line organ detect vibrations caused by waves and other movement in the water, including movements by predators or prey (see Fig. 43-7).

Cartilaginous fishes have no lungs. Gas exchange takes place through their five to seven pairs of gills. A current of water enters the mouth and passes over the gills and out the pharyngeal slits, constantly providing the fish with a fresh supply of dissolved oxygen. Sharks that actively swim depend on their motion to enhance gas exchange. Rays, skates, and sharks that spend time on the ocean floor use muscles of the jaw and pharynx to pump water over their gills.

The digestive tract of sharks consists of the mouth cavity; a long pharynx leading to the stomach; a short, straight intestine; and a **cloaca,** which opens on the underside of the body and is characteristic of many vertebrates (see Fig. 32-12). The liver and pancreas discharge digestive juices into the intestine. The cloaca

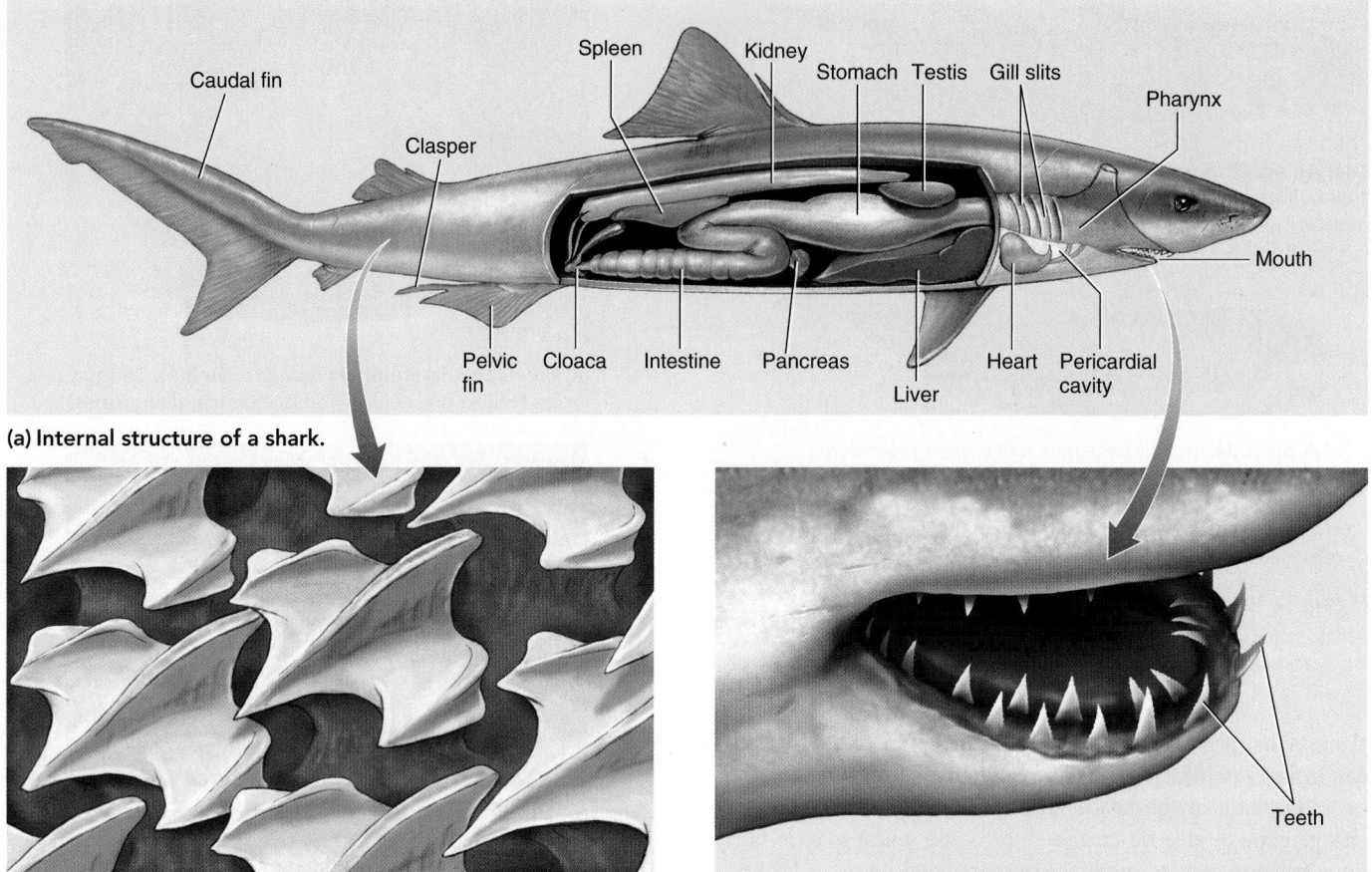

(a) Internal structure of a shark.

(b) Placoid scales. These tough scales protect the shark against predators. The base of each placoid scale is embedded in the skin. Placoid scales are toothlike structures. They are covered with hard enamel.

(c) Teeth. Placoid scales are embedded in the flesh along the inner surface of the jaw cartilage where they function as teeth. Sharks have several rows of teeth. Old teeth are shed and quickly replaced.

Figure 32-12 *Animation* **Anatomy of a shark**
© Cengage Learning

receives digestive wastes as well as metabolic wastes from the urinary system. In females the cloaca also serves as a reproductive organ.

The sexes are separate, and fertilization is internal. In the mature male, each pelvic fin has a slender, grooved section, known as a *clasper*, used to transfer sperm into the female's cloaca. The eggs are fertilized in the upper part of the female's oviducts. Part of the oviduct is modified as a shell gland, which secretes a protective coat around each egg.

Skates and some species of sharks are **oviparous;** that is, they lay eggs. Many species of sharks, however, are **ovoviviparous,** meaning that their young are enclosed in eggs and incubated within the mother's body. During development, the young depend on stored yolk for their nourishment rather than on transfer of materials from the mother. The young are born after hatching from the eggs.

A few species of sharks are **viviparous.** Not only do the embryos develop within the uterus, but much of their nourishment is delivered to them by the mother's blood. Nutrients are transferred between the blood vessels in the lining of the uterus and the yolk sac surrounding each embryo.

The ray-finned fishes gave rise to modern bony fishes

Although bony fishes appear earlier in the fossil record than cartilaginous fishes, both groups may have evolved about the same time, during the Late Silurian or Early Devonian period. The two groups share many characteristics (such as continuous tooth replacement), but they also differ in important ways.

Most bony fishes are characterized by a bony skeleton with many vertebrae. Bone has advantages over cartilage because it provides excellent support and effectively stores calcium. Most species have flexible median and paired fins supported by long rays made of cartilage or bone. Overlapping, bony dermal scales cover the body. A lateral bony flap, the **operculum,** extends posteriorly from the head and protects the gills.

Most bony fishes are oviparous. Most species lay an impressive number of eggs and fertilize them externally. The ocean sunfish, for example, lays more than 300 million eggs! Of course, most of the eggs and young become food for other animals. The probability of survival is increased by certain behavioral adaptations. For example, many species of fishes build nests for their eggs and protect them. Other species have internal fertilization and give birth to live young.

During the Devonian period, the bony fishes diverged into two major groups: the fleshy-finned fishes, *Sarcopterygii;* and the **ray-finned fishes,** class *Actinopterygii.* Fossils of the earliest sarcopterygians date back to the Devonian period, about 400 mya. Lungs and fleshy, *lobed fins* characterized these fishes. The flexible fins were supported by thin, bony rays radiating out from a thick base of multiple bones and muscles. Lobed fins may have evolved as an adaptation for pushing along the bottom of any body of water.

Early sarcopterygians evolved along three separate lines: the *lungfishes* (class *Dipnoi*), the *coelacanths* (class *Actinistia*), and the *Tetrapodomorpha,* a clade consisting of tetrapods and their most recent ancestors. All three lineages had fins with fleshy bases of muscle and bone. Surviving from these groups today are three genera of lungfishes, two species of *coelacanths,* and the many thousands of species of tetrapods (amphibians and amniotes) in habitats worldwide.

Although the earliest ray-finned fishes also had both lungs and gills, they differed from the sarcopterygians in having more flexible fins with less bone and muscle at the base. The ray-finned fishes (actinopterygians) underwent two important adaptive radiations. The first gave rise during the Late Paleozoic era to a group of fishes that are now mostly extinct. The second radiation began during the Early Mesozoic era and gave rise to the very successful modern bony fishes.

In addition to gills, the common ancestor of the bony fishes had primitive lungs that could exchange gases in air. Lungs for gas exchange were retained by the tetrapods and lungfishes. In the modern ray-finned fishes, the lungs became modified as a **swim bladder,** an air sac that helps regulate buoyancy (FIG. 32-13). Bones and muscles are heavier than water, and without the swim bladder the fish would sink. By regulating gas exchange between its blood and swim bladder, a fish can control the amount of gas in the swim bladder and thus change the overall density of its body. This ability allows a bony fish, in contrast to a shark, to hover at a given depth of water without much muscular effort.

There are more species of bony fishes than of any other vertebrate group. Biologists have identified tens of thousands of living species of freshwater and saltwater bony fishes, of many shapes and colors (FIG. 32-14). Bony fishes range in size from that of *Paedocypris progenetica,* which has been measured at less than 8 mm (about 0.3 in) long, to that of the ocean sunfish (or *Mola;* see Fig. 20-16c), which may reach 4 m (13 ft) and weigh about 1500 kg (about 3300 lb). *Paedocypris progenetica,* native to Indonesia, is the smallest known vertebrate.

The diversity of bony fishes may have resulted from the action of *Hox* genes. During the radiation of ray-finned fishes, entire chromosomes duplicated, resulting in one or more additional clusters of *Hox* genes. The zebrafish (see Fig. 17-6d) has seven *Hox* clusters. These additional genes could have provided the genetic material for the evolution of the diverse species of ray-finned fishes.

Tetrapods evolved from sarcopterygian ancestors

Biologists thought that the **coelacanths** were extinct by the end of the Paleozoic era, so in 1938 the scientific community was very excited when a commercial fisherman caught one off the coast of South Africa. Since that time, more than 200 specimens of these giant "living fossils" have been found in the deep waters off the southeastern coast of Africa, the Comores Islands, Madagascar, and Indonesia (FIG. 32-15). These coelacanths, which measure nearly 2 m (about 6 ft) in length, are nocturnal predators on other fishes.

The relationships among the three living groups of sarcopterygians (coelocanths, lungfishes, and tetrapods) have been investigated for many years. Although it was known that the tetrapods evolved from the fleshy-finned fishes, it was not known whether the lungfishes or the coelacanths were the

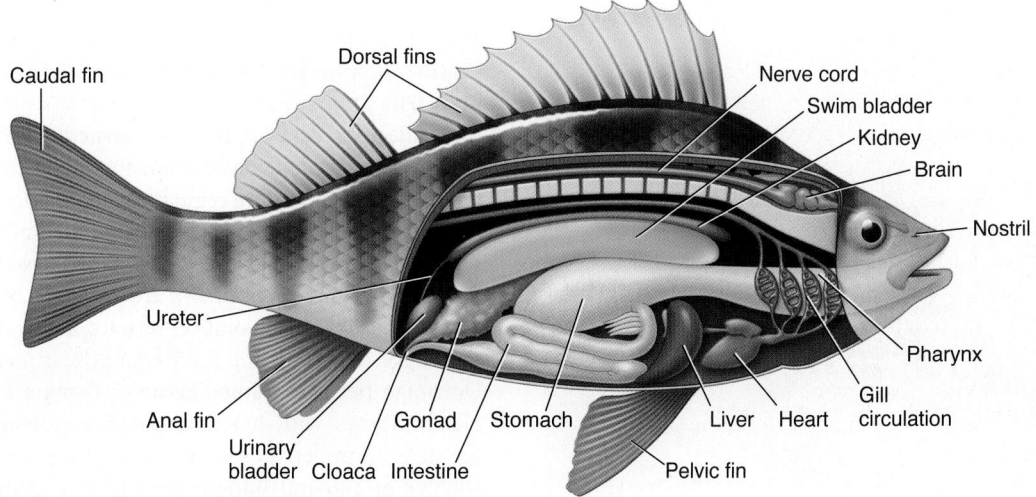

Figure 32-13 *Animation* **Perch, a representative bony fish**

The swim bladder is a hydrostatic organ that enables the fish to change the density of its body and remain stationary at a given depth. Pectoral fins (*not shown*) and pelvic fins are paired.

© Cengage Learning

(a) The porcupine fish (*Diodon hystrix*). This fish swallows air or water to inflate its body, a strategy that discourages potential predators. Photographed in the Virgin Islands.

(b) The parrotfish (*Scarus gibbus*). The parrotfish feeds on coral, grinds it in its digestive tract, and extracts the coralline algae. The fish eliminates a fine white sand. These fishes contribute to white sand beaches in many parts of the world. Parrotfish begin life as females and later become males.

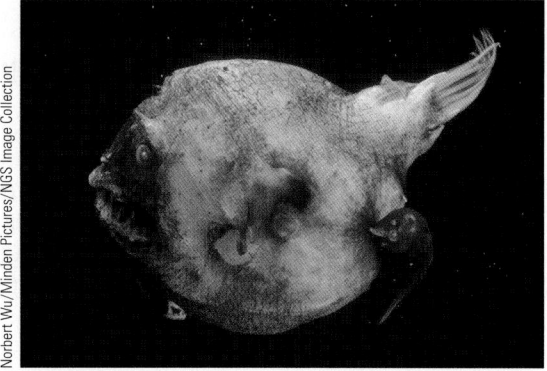

(c) Anglerfishes. In some clades of anglerfishes, the much smaller male attaches himself to a female with his sharp teeth and becomes a lifelong parasitic mate. Most species have at least one long spine projecting from the middle of the head. This "lure" is used as bait to attract prey.

(d) The leafy sea dragon (*Phycodurus eques*). The leafy sea dragon inhabits kelp-covered rocky reefs and seaweed beds in the waters off the southern and western coast of Australia. This fish is a striking example of camouflage.

Figure 32-14 Modern bony fishes

Figure 32-15 Coelacanth
Although they were more diverse and widespread during their early history, the coelacanths, like this *Latimeria chalumnae*, are restricted today to certain deep sea environments.

nearest relatives of this group. Fossil evidence and DNA analyses now support the lungfishes as the sister group to the tetrapods (FIG. 32-16).

Molecular data support the structural evidence that the limbs of tetrapods evolved from the lobed fins of fishes. Proteins encoded by certain regulatory genes, including *Hox* genes, have been found at the same developmental times and in the same locations in both limbs and fins. These findings indicate that limb and fin development are governed by the same genes.

What drove the evolution of tetrapod limbs? One hypothesis suggests that during the frequent seasonal droughts of the Devonian period, swamps became stagnant or dried up completely. Fishes with lobed fins had a tremendous advantage for survival under those conditions. They were strikingly preadapted for moving onto the land. They had lungs for breathing air, and their sturdy, fleshy fins allowed them to "walk" along in shallow water. These fins could support the fish's weight, so it could emerge onto dry land and make its way to another pond or stream.

Figure 32-16 South American lungfish (*Lepidosiren paradoxus*)

Lungfishes were common in Devonian and Carboniferous times (400 mya to 300 mya). South American lungfish have paired lungs on either side of the throat and can survive for long periods if the river they inhabit dries up.

A more recent hypothesis holds that limbs evolved in a fully aquatic environment. Animals with more-developed limbs could move along in shallow water or creep through dense aquatic vegetation more efficiently than their lobe-finned ancestors. *Tiktaalik* was a transitional form between fishes and tetrapods (**FIG. 32-17**). Biologists consider *Tiktaalik,* which lived about 375 mya (during the Devonian period), a fish because it had scales and fins. However, *Tiktaalik* also had tetrapod features such as a movable neck and ribs that supported lungs.

Acanthostega was one of the earliest known tetrapods. Fossils from the Devonian period show that this animal had four legs with well-formed digits. However, because its limbs were not properly positioned for walking effectively on land and because it possessed gills and a tail fin, *Acanthostega* was a fully aquatic tetrapod. The ichthyostegids, tetrapods known from somewhat later in the Devonian, had more robust limbs and girdles, suggesting that they could move more easily on land.

Figure 32-17 *Tiktaalik*

This extinct fish, which grew to 2.75 m (9 ft), had limblike fins and other tetrapod features.

However, the evidence suggests that these early tetrapods were also very aquatic.

Those early tetrapods that could explore shallow wetlands and make their way onto dry land had access to new food sources. Terrestrial plants were already established, and terrestrial insects and arachnids were rapidly evolving. A vertebrate that could survive on land had less competition for food than one that was solely aquatic. However, success on land required the evolution of several major adaptations in addition to legs.

As already discussed, the early sarcopterygians had lungs for gas exchange in addition to gills. Lungs were important because gills cannot function in air. Life on land also required changes in the muscles and skeleton to support the body's weight in air. Body coverings and other mechanisms were needed to protect animals against the drying effect of air.

Evolution of ears that could hear sound transmitted through air and olfactory mechanisms for detecting airborne odors contributed to the ability to find food and mates, and to avoid predators. It is likely that early tetrapods needed to return to the water for reproduction, but eventually adaptations evolved that allowed some animals to reproduce in terrestrial environments. The early tetrapods were very successful and eventually gave rise to the amphibians and the amniotes (including reptiles, birds, and mammals).

Amphibians were the first successful land vertebrates

Biologists classify modern **amphibians** (class *Amphibia*) in three orders. Order Caudata ("with tail") includes salamanders, mud puppies, and newts, all animals with long tails; order *Anura* ("no tail") is made up of frogs and toads, most with legs adapted for hopping; and order *Gymnophiona* contains the limbless caecilians (**FIG. 32-18**). Although some adult amphibians are quite successful as land animals and live in dry environments, most must return to the water to reproduce. Eggs and sperm are typically released in the water.

Many amphibians undergo **metamorphosis,** a transition from larva to adult. The embryos of most frogs and toads develop into larvae called *tadpoles*. These larvae have tails and gills, and most feed on aquatic plants. After a time, the tadpole undergoes metamorphosis, which is regulated by hormones secreted by the *thyroid gland*. During metamorphosis, gills and gill slits disappear, the tail is resorbed, and limbs emerge. The digestive tract shortens, and food preference shifts from plant material to a carnivorous diet; the mouth widens; a tongue develops; the tympanic membrane (eardrum) and eyelids appear; and the eye lens changes shape. Many biochemical changes accompany the transformation from a completely aquatic life to a semiterrestrial one.

Several salamanders, such as the mudpuppy *Necturus,* do not undergo complete metamorphosis; they retain many larval characteristics even when sexually mature adults. Recall from Chapter 20 that this retention of juvenile features is called **paedomorphosis** (see Fig. 20-17). This type of development

(a) Red dart frog (*Dendrobates histrionicus*). Poison dart frogs, native to Central America and South America, are active during the day. Most are brightly colored. Many secrete toxins through their skin as a chemical defense against predators.

(b) Fire salamander (*Salamandra salamandra*). Fire salamanders are well known in Europe. Their poison glands release neurotoxins.

Figure 32-18 Modern amphibians

permits these salamanders to remain aquatic rather than having to compete on land.

The coloration of amphibians may conceal them in their habitat or may be very bright and striking. Many of the brightly colored species are poisonous (see Fig. 32-18). Their distinctive colors warn predators that they are not encountering an ordinary amphibian. Some frogs camouflage themselves by changing color.

Adult amphibians do not depend solely on their primitive lungs for the exchange of respiratory gases. Their moist, glandular skin, which lacks scales and is plentifully supplied with blood vessels, also serves as a respiratory surface. The numerous mucous glands within the skin help keep the body surface moist, which is important in gas exchange. The mucus also makes the animal slippery, facilitating its escape from predators. Most amphibians have glands in their skin that secrete toxic and/or foul-tasting substances that repel predators.

The amphibian heart contains two **atria** (sing. *atrium*), receiving blood, and a single **ventricle,** pumping blood into the arteries. A double circuit of blood vessels keeps oxygen-rich and oxygen-poor blood partially separate. Blood passes through the **systemic circulation** to the various tissues and organs of the

body. Then, after returning to the heart, it is directed through the **pulmonary circulation** to the lungs and skin, where it is recharged with oxygen. The oxygen-rich blood returns to the heart to be pumped out into the systemic circulation again. We discuss the comparative anatomy of the heart and circulation of various vertebrate classes in Chapter 44.

CHECKPOINT 32.7

- **CONNECT** *From an evolutionary perspective, what is the significance of each of the following: (1) placoderms, (2) lungfishes, (3) ray-finned fishes, and (4) Tiktaalik?*
- *What changes allow amphibians to move from aquatic life as a tadpole to terrestrial life as an adult?*

32.8 AMNIOTES: TERRESTRIAL VERTEBRATES

LEARNING OBJECTIVES

9 Describe three vertebrate adaptations to terrestrial life.
10 Describe the reptiles (including the birds) and argue for including the birds in the reptile clade.
11 Describe five key characters of mammals and contrast protherian (monotremes), metatherian (marsupials), and eutherian mammals, giving examples of animals that belong to each group.

The evolution of **reptiles** from ancestral amphibians required many adaptations that allowed them to become completely terrestrial. Evolution of the **amniotic egg** was an extremely important event because it allowed terrestrial vertebrates to complete their life cycles on land. The amniotic egg contains an **amnion,** a membrane that forms a fluid-filled sac around the embryo. The amnion provides the embryo with its own private "pond," permitting independence from a watery external environment. In addition to keeping the embryo moist, the amniotic fluid serves as a shock absorber that cushions the developing embryo.

The evolution of the amniotic egg is so important to the success of terrestrial vertebrates—reptiles (including birds) and mammals—that biologists refer to these animals as **amniotes**. Amniotes are a monophyletic group because they have a recent common ancestor that was itself an amniote.

In addition to the amnion, the amniotic egg has three other extraembryonic (not part of the developing body itself) membranes: *yolk sac, chorion,* and *allantois* (FIG. 32-19). These membranes protect the developing embryo, store nutrients (*yolk sac*), carry on gas exchange (chorion, allantois, and yolk sac), and store wastes (allantois). We discuss development of the extraembryonic membranes in Chapter 51. Although most extant mammals do not lay eggs, their embryos have an amnion and other extraembryonic membranes.

Another important adaptation to terrestrial life is a body covering that minimizes water loss. The amniote body covering is typically thick and contains **keratin,** a water-insoluble

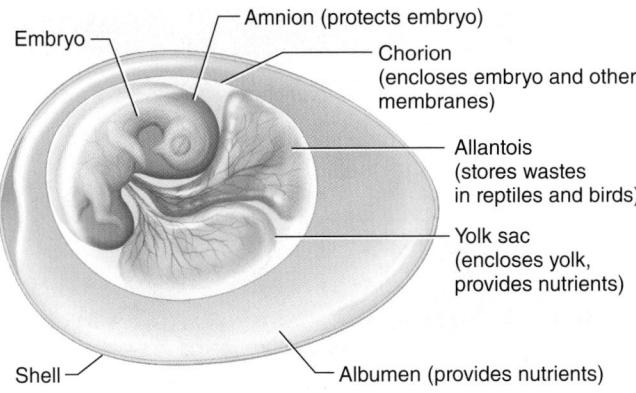

Figure 32-19 *Animation* **An amniotic egg**

The amnion, a fluid-filled sac surrounding the embryo, keeps the embryo moist and protects it from mechanical shock. Other extraembryonic membranes include the yolk sac, which functions in nutrition; the allantois, which stores metabolic wastes; and the chorion, which along with the allantois, functions in gas exchange.

© Cengage Learning

protein that helps protect the animal from injury. Keratin is found in the epidermis and in derivatives of the epidermis, such as scales, nails, feathers, hair, and horns. A body covering containing keratin presents another challenge: it severely decreases gas exchange across the body surface. This challenge has been met by the evolution of efficient lungs and circulatory systems for exchange of oxygen and carbon dioxide. Amniotes also have physiological mechanisms for conserving water. For example, much of the water filtered from the blood by the kidneys is reabsorbed in the kidney tubules to decrease fluid loss during excretion of metabolic wastes.

Our understanding of amniote phylogeny is changing

Biology instructors once taught that the three classes of amniotes were Reptilia, Aves (birds), and Mammalia. However, cladistic analysis determined that class Reptilia was not a monophyletic group without the birds. Because it included some, but not all, of its descendants, it was paraphyletic (see Fig. 23-7). For this reason, most biologists now classify the birds as a group of reptiles.

Biologists hypothesize that the earliest amniotes resembled lizards. By the Late Carboniferous period, about 290 mya, amniotes had undergone an impressive adaptive radiation and had diverged into two main branches: **diapsids** and **synapsids** (FIG. 32-20 and TABLE 32-3). The term *diapsid* refers to the two pairs of openings in the temporal bones that characterize the skulls of these animals. Synapsid skulls have one pair of temporal openings. The diapsids comprise all extant reptilian groups, including the birds and most extinct reptiles. The synapsids include the extinct **therapsids** and the mammals.

A second great adaptive radiation of amniotes occurred during the Mesozoic era, which ended about 66 mya. During that time reptiles were the dominant terrestrial vertebrates (see Chapter 21). In fact, the Mesozoic era is known as the "Age of Reptiles." These animals had radiated into an impressive variety of ecological lifestyles (see Figs. 21-13, 21-14, and 21-15). Some could fly, others became marine, and many filled terrestrial habitats. Among the major groups were the *pterosaurs* (ter'-uh-sawr), the flying reptiles; the *saurischian dinosaurs* (saw-ris'-kee-un), a group that included *Tyrannosaurus* and *Diplodocus*; and the *ornithischian dinosaurs* (awr-nuh-this'-kee-un), including *Triceratops*.

Some of the dinosaurs were among the largest land animals to ever walk on Earth. Some dinosaurs apparently traveled in social groups and took care of their young. Fossil evidence supports the hypothesis that at least some dinosaurs may have been **endotherms,** meaning that they used metabolic energy to maintain a constant body temperature despite changes in the temperature of the environment. An advantage of endothermy is that it allows animals to be more active. The natural history and evolution of dinosaurs and other early reptiles are discussed in more detail in Chapter 21.

The reptiles were the dominant land vertebrates for almost 200 million years. Then, toward the end of the Mesozoic era, many reptiles, including all the dinosaurs and pterosaurs, disappeared from the fossil record. In fact, more than half of all animal species became extinct at that time (see Chapter 21).

Reptiles have many terrestrial adaptations

Many reptilian characters are adaptations to terrestrial life. The female reptile secretes a protective leathery shell around the egg, which helps prevent the developing embryo from drying out. The shell presents a challenge for reproduction because sperm cannot penetrate it. Fertilization must take place within the body of the female before the shell is added. In this process, the male uses a copulatory organ (penis) to transfer sperm into the female reproductive tract. An amnion surrounds the embryo as it develops within the protective shell.

The hard, dry, keratin scales that are part of the reptile's skin retard drying in yet another adaptation to life on land, This scaly protective armor, which also protects the reptile from predators, is shed periodically. The dry reptilian skin does not allow effective gas exchange. Reptilian lungs are better developed than the saclike lungs of amphibians. Divided into many chambers, the reptilian lung provides an increased surface area for gas exchange.

The hearts of amniotes contain two atria. The ventricle of reptiles is either partially or completely divided. The division into right and left ventricles enhances the separation of oxygen-rich and oxygen-poor blood. The more efficient circulatory and respiratory systems of reptiles are critical for animals with a keratinized epidermis.

Like fishes and amphibians, many extant reptiles lack metabolic mechanisms for regulating body temperature. They are **ectotherms,** meaning that their body temperature fluctuates with the temperature of the surrounding environment. Some reptiles have behavioral adaptations that help them maintain a body temperature higher than that of their environment. For

The diapsid reptiles gave rise to the birds, and the synapsids gave rise to the mammals. Note the five main branches of extant diapsids.

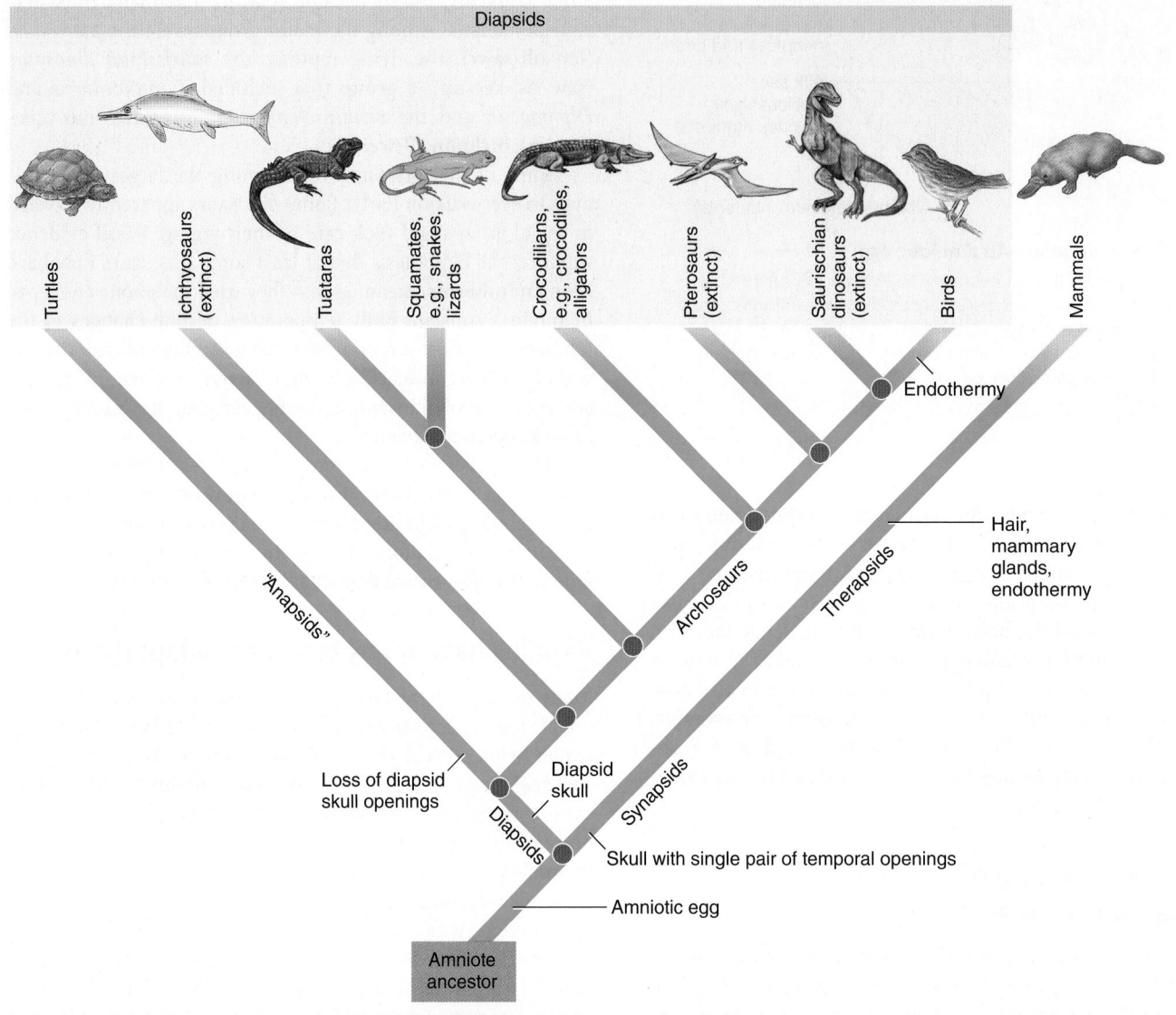

Figure 32-20 Evolutionary relationships of some amniotes

This cladogram shows proposed evolutionary relationships among reptiles, birds, and mammals. Systematists continue to study these relationships and may reinterpret some of them. Identify five main branches of extant diapsids on the cladogram. (*Hint:* Recall that turtles are diapsids even though they have "lost" their temporal openings during their evolution.)

CONNECT Are birds more closely related to extinct branches of reptiles or to mammals?

© Cengage Learning

example, you may have observed a lizard basking in the sun, which raises its body temperature and so increases its metabolic rate. This increased rate permits the lizard to hunt actively for food. When the body of a reptile is cold, its metabolic rate is low and the animal tends to be sluggish. Ectothermy helps explain why lizards, snakes, and turtles are more successful in warm than in cold climates.

Many reptiles are predators. Their paired limbs, usually with five toes, are well adapted for running and climbing in search of

prey. In addition, their well-developed sense organs enable them to locate prey.

Biologists assign reptiles to several major clades

Order Testudines includes the turtles, terrapins, and tortoises (see Table 32-3). The **lepidosaurs** make up a superorder of diapsid vertebrates. This clade includes order Squamata—snakes,

TABLE 32-3 Some Major Groups of Amniotes

CLADES	IMPORTANT CHARACTERS
SUBCLASS DIAPSIDA	**Skull with two pairs of temporal openings**
Order Testudines (Chelonia) Turtles, terrapins, tortoises	Enclosed in bony shell; beak of keratin instead of teeth; some can withdraw head and legs into shell; temporal openings lost ("anapsids")
SUPERORDER LEPIDOSAURIA	Overlapping scales; a very successful group of modern reptiles
Order Squamata Snakes, lizards, amphisbaenians (worm lizards)	Flexible armor of overlapping, horny scales, which are shed Snakes: elongated body with no limbs; flexible, loosely jointed jaw; forked tongue protrusible Lizards: slender body, typically with four legs; some with no legs
Order Sphenodonta Tuataras	Vertebrae concave at both ends; only two extant species
SUPERORDER ICHTHYOSAURIA (extinct) Marine reptiles of Mesozoic	Body shaped like that of modern dolphins; large eyes; vertical tails
SUPERORDER ARCHOSAURIA	Mainly terrestrial; some specialized for flight; ventricle completely divided in living forms
Order Crocodilia Crocodiles, alligators, caimans, gavials	Heart with two atria and two ventricles; closest extant relative to birds
Pterosauria (extinct) Flying reptiles of Mesozoic	Membranous wings
Saurischia (extinct) Mesozoic dinosaurs (*Tyrannosaurus, Diplodocus*) **Birds descended from this lineage**	Some were two-legged carnivores; others were four-legged herbivores; some had feathers
Ornithischia (extinct) Mesozoic dinosaurs (*Triceratops*)	Bipedal and quadrupedal herbivores; social behavior and parental care in many
SUBCLASS SYNAPSIDA	**Skull with one pair of temporal openings**
Therapsida (extinct) **Mammals descended from this lineage**	Many mammal-like characters; became dominant land animals during Middle Permian

© Cengage Learning

lizards, and amphisbaenians (worm lizards)—and order Sphenodonta, which includes the tuataras, lizardlike animals that live in burrows. **Archosaurs** make up another superorder of diapsid vertebrates. Archosaurs include the extinct **pterosaurs**—the flying reptiles and dinosaurs—and the extant crocodiles, alligators, caimans, and gavials (order Crocodilia) and birds (order Aves). (**FIG. 32-21**).

Turtles have protective shells

Members of order Testudines (te-stood′-n-eez) are enclosed in a protective shell made of bony plates overlaid by keratin scales. Some terrestrial species can withdraw their heads and legs completely into their shells. Their keratin beak covers the jaws. Turtles do not have teeth. Turtles are diapsid reptiles. However, they are referred to as "anapsids" because during the course of evolution, they have "lost" the temporal openings.

The size of adult turtles ranges in length from about 8 cm (3 in.) to more than 2 m (6.5 ft) in leatherback turtles, which are the largest marine species. The weight of a leatherback turtle can exceed 500 kg (more than 1000 lb). The forelimbs of marine turtles are modified into flippers. Sea turtles migrate hundreds of miles from the beaches where they hatch to feeding grounds. Females return to the same beaches where they hatched to mate and nest. Most species of sea turtles are endangered as a result of human activities.

Lizards and snakes are common modern reptiles

Lizards and snakes are assigned to order Squamata (squa-ma′-tah). These animals have rows of scales that overlap like shingles on a roof, forming a continuous, flexible armor that is shed periodically. Lizards range in size from certain geckos, which weigh as little as 1 g (less than 0.1 oz), to the Komodo dragon of Indonesia, which may weigh 100 kg (220 lb). Their body sizes and shapes vary greatly. Some lizards—for example, the "worm lizards" and the glass lizards—are legless.

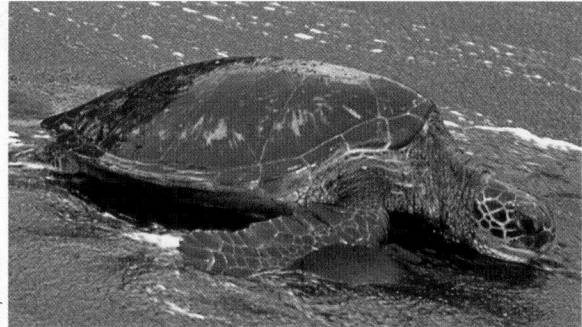

Carlyn Iverson

(a) Green turtle (*Chelonia mydas*). The paddlelike appendages are adapted for swimming.

Pete Oxford/Minden Pictures

(b) The emerald tree boa (*Corallus canina*). This boa inhabits tropical South American forests. It rarely leaves the trees to go to the ground. Emerald boas use their heat-sensitive pit organs to locate prey.

Betty and Nathan Cohen/Visuals Unlimited

(c) Tuatara (*Sphenodon punctatus*). These nocturnal predators reach a length of 60 cm (more than 2 ft) or longer. There are two species of tuataras. Both are classified as endangered.

© Trevor Kelly/Shutterstock.com

(d) Nile crocodiles (*Crocodilus niloticus*) hatching from their eggs. Both parents guard the leathery eggs. When the hatchlings emerge, the mother leads them to the water. She may carry them in her mouth.

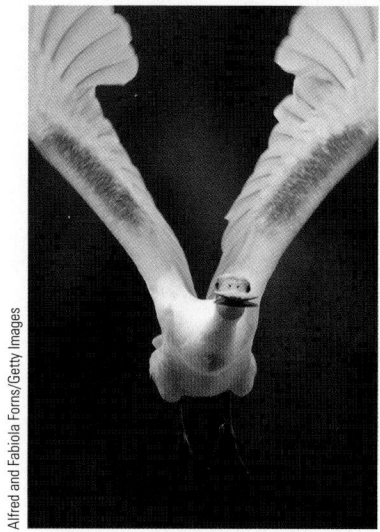

Alfred and Fabiola Forns/Getty Images

(e) Roseate spoonbill (*Ajaia ajaja*) in its full breeding colors. The roseate spoonbill is a long-legged wading bird nearly three feet tall with a wingspan of more than four feet. This bird is taking off for flight in the Alafia river banks in central Florida. The future of spoonbills in Florida depends on protection of their habitats—coastal marshes, estuaries, and mangrove areas.

Figure 32-21 Representative extant reptiles

Snakes are characterized by a flexible, loosely jointed jaw structure that lets them swallow animals larger than the diameter of their own jaws. These reptiles have elongated bodies with no legs, although pythons have vestigial hindlimb bones (see Fig. 18-17). Remember that although not all the tetrapods have four limbs, all evolved from four-limbed ancestors. Snake eyes are covered by a transparent scale. They do not have movable eyelids. Also absent are an external ear opening, a tympanic membrane (eardrum), and a middle-ear cavity.

Snakes use their forked tongues as accessory sensory organs for touch and smell. Chemicals from the ground or air adhere to the tongue. Snakes rub the tips of their tongues across the opening of a sense organ in the roof of the mouth that detects odors. Pit vipers and some boas also have *pit organs* that detect heat from endothermic prey (see Fig. 43-3). These snakes use their pit organs to locate and capture birds and small nocturnal mammals.

Some snakes, such as king snakes, pythons, and boa constrictors, kill their prey by rapidly wrapping themselves around the animal and squeezing so it cannot breathe. Others have fangs, which are hollow teeth connected to venom glands. When the snake bites, it pumps venom through the fangs into the prey. Some snake venoms cause the breakdown of red blood cells; others, such as that of the coral snake, are neurotoxins, which interfere with nerve function. Venomous snakes of the United States include rattlesnakes, copperheads, cottonmouths, and coral snakes. Except for the coral snakes, all are pit vipers.

Amphisbaenians (am-fus-bee′-nee-uns), or worm lizards, are considered an additional group of modified lizards. They are well adapted for their burrowing lifestyle. They have elongated bodies with either no legs or a single pair of legs. In many species the eyes are hidden under the skin. Most are less than 15 cm (6 in.) long. Most species of amphisbaenians inhabit South America and southern Africa. Only one species (known as the Florida worm lizard) is found in the United States.

Tuataras superficially resemble lizards

Tuataras look somewhat like iguanas but have certain distinct characters. For example, they are the only amniotes with vertebrae that are concave at both ends; such vertebrae are characteristic of fish and some amphibians. The two extant species of tuataras inhabit New Zealand and several small islands off its coast. Tuataras have long been considered endangered species. When humans brought rats, cats, and dogs to New Zealand, these non-native animals decimated the population of tuataras by eating their eggs. Tuataras became extinct on the New Zealand mainland. In 2005, these animals were reintroduced into a sanctuary in New Zealand, and they are now protected.

Crocodilians have an elongated skull

The extant members of order Crocodilia (crok-uh-dil′-ee-uh), along with the birds, are the surviving reptiles of the archosaur lineage. This lineage gave rise to the **pterosaurs** (flying reptiles) and the dinosaurs that dominated the Mesozoic era. Modern crocodilians include three groups: (1) the crocodiles of Africa, Asia, and the Americas; (2) the alligators of the southern United States and China, plus the caimans of Central America; and (3) the gavials of South Asia. Most species live in swamps, in rivers, or along seacoasts, feeding on various kinds of animals.

Crocodiles are the largest living reptiles; some exceed 7 m (23 ft) in length. The cranial skeleton is adapted for aquatic life. In the Americas, crocodiles can be distinguished from alligators or caimans by their more tapered snouts and by the large fourth tooth on the lower jaw that is visible when the mouth is closed.

How do we know that birds are really dinosaurs?

Cladistic analyses indicate that birds evolved from the lineage of saurischian dinosaurs, specifically from the **theropods** (theer′-uh-pods), a group of bipedal, saurischian dinosaurs that included *Tyrannosaurus* and *Deinonychus* (see Fig. 21-15). This view is supported by the remarkable array of Cretaceous and late Jurassic fossils discovered in the Yixian Formation and in other parts of China as well as in Europe.

Some dinosaurs had feathers Paleontologists have identified many theropod dinosaur fossils that demonstrate the evolution of feathers from very simple, tubular structures to complex modern forms. *Caudipteryx,* a feathered dinosaur discovered in the Yixian Formation, was about the size of a turkey. Fossils of *Caudipteryx* exhibit both dinosaur and bird characteristics. The bones in the foot and the shape and orientation of the pelvis were similar to those of dinosaurs. However, the fossils have well-preserved impressions of complex feathers on the tail and forelimbs. *Caudipteryx* had only a few teeth in the front of its upper jaw, and its bones show many birdlike characteristics. Although some investigators have argued that *Caudipteryx* was a flightless bird, the most recent consensus view is that this animal was a feathered dinosaur. Analyzing fossil evidence and molecular data, many paleontologists have hypothesized that feather evolution took place in terrestrial, bipedal dinosaurs *before* the evolution of birds or flight.

Modern birds use asymmetrical feathers for flight. In 2003, paleontologists reported finding a new species of feathered dinosaur, *Microraptor gui. Microraptor* had asymmetrical feathers on both its forelimbs and hind limbs, and also on its tail. (See the figure in *Inquiring About: The Origin of Flight in Birds,* in Chapter 21.) However, their small size and body structure support the hypothesis that flying evolved from gliding; that is, *Microraptor* climbed trees and used its feathered limbs for gliding. These small dinosaurs, which lived about 126 mya, coexisted with early birds.

Biologists have debated why feathers may have evolved in dinosaurs. Were feathers important in courtship rituals or in camouflage? Were feathers an adaptation that conserved body heat? Insulating feathers could have contributed to the evolution of endothermy (the ability to maintain a constant body temperature), permitting animals to be more active.

In 2010, paleontologists described a nearly complete skeleton of a basal theropod discovered in northwestern China. This new fossil, named *Haplocheirus sollers,* had organelles containing pigment that colored its feathers, suggesting that feathers may have first evolved for color displays in courtship rituals. The presence of feathers and the bone structure of *Haplocheirus* indicate that striking evolutionary convergence occurred between feathered dinosaurs and birds. In 2013 the fossil of a new bird-like theropod from the Jurassic period was discovered. This theropod was named *Eosinopteryx*.

Early birds were transitional forms Many extinct theropods were long-tailed animals that moved about on two feet and had forelimbs with three clawed fingers. Although the bones of birds are fragile and disintegrate quickly, paleobiologists have found a few fossils of early birds. The first birds looked very much like dinosaurs. They had teeth (which modern birds lack), a long tail, and bones with thick walls. Like certain extinct theropods, modern birds have feet with three digits and thin-walled, hollow bones. Both have a *furcula,* or wishbone, which is formed by the two clavicles (collarbones) fusing in the midline. Many biologists classify modern birds as theropods.

Archaeopteryx (meaning "ancient wing") was an early bird that was about the size of a pigeon. More than ten specimens of this genus have been found in Bavaria in Jurassic limestone, which was laid down about 150 mya. Unlike those of extant birds, the jawbones of *Archaeopteryx* were armed with teeth, and its long, reptilian tail was covered with feathers (FIG. 32-22a). Each of its short, broad wings had three claw-bearing, functional digits. Like modern birds, *Archaeopteryx* had wings, feathers, and a furcula. Its feathers were very similar to those of modern birds.

Using computer tomography and computerized 3-D reconstruction, researchers examined the inside of the *Archaeopteryx* skull. They found that *Archaeopteryx* had a birdlike brain with comparatively larger cerebral lobes than those of other reptiles. It had enlarged visual centers and highly developed inner ear canals similar to those of modern birds. These structures would support the coordination and agility needed for flight.

A complete fossil of an early bird from China was described in 2013. This fossil, named *Aurornis xui*, predates *Archaeopteryx* by about 10 million years. Like *Archaeopteryx*, *Aurornis xui* has a furcula, feathers, and teeth (FIG. 32-22b). The *Aurornis* fossil preserved traces of downy feathers. However, it shows no evidence of larger feathers, suggesting that this early bird was not able to fly.

Cretaceous rocks have yielded fossils of other early birds. *Hesperornis,* which lived in North America, was a toothed, aquatic diving bird with small wings and broad, lobed feet for swimming. *Ichthyornis* was a toothed, flying bird about the size of a small tern. Based on fossils, structural similarities, and molecular data, most biologists now view birds as living dinosaurs.

Modern birds are adapted for flight

By the end of the Cretaceous period, the major clades of present-day birds—palaeognaths and neognaths—had evolved. (Some biologists assign birds to class Aves and consider palaeognathae and neognathae as superorders.) *Palaeognaths* (from Greek *palaios,* "ancient," and *gnathos,* "jaw") include flightless birds such as the ostriches, kiwis, cassowaries, and emus. These birds are called **ratites.** Their sternum (breastbone) is flat and does not have a ridge for attachment of wing muscles. They have

(a) **Reconstruction of *Archaeopteryx*, an early bird.** This reconstruction represents the view that *Archaeopteryx* was a climbing animal that had at least some ability to use its wings and feathers for gliding.

(b) **Fossil of *Aurornis xui*, a very early bird.** *Aurornis xui*, about the size of a pheasant, lived during the Middle–Late Jurassic period about 160 million years ago.

Figure 32-22 Fossil evidence supports the evolution of birds from the dinosaur lineage

only vestigial wings, but they have well-developed legs used for running.

Most modern birds are *neognaths* (from Greek *neos,* "new," and *gnathos,* "jaw"). These birds have powerful flight muscles attached to their keeled (ridged) sternum, and most neognaths fly. The penguins are an exception. They use their strong pectoral muscles and small, flipperlike wings to swim.

Biologists have described more than 9000 species of extant birds and classified them into about 23 monophyletic taxa. Birds are a diverse group (FIG. 32-23). They have become adapted to a variety of environments, and various species have different types of beaks, feet, wings, tails, and behavioral patterns. They can be

(a) Southern cassowary (*Casuarius casuarius*), a ratite. The flightless southern cassowary, native to Australia, is the third largest bird in the world, after the ostrich and emu. This is the only bird known to have protective armor, including a hard crest that serves as a helmet atop its head.

(b) Peacock (*Pavo cristatus*). The Indian peacock displays his colorful, fully fanned tail feathers to get a female's attention. Peafowl (peacocks and peahens) belong to the pheasant family.

(c) Atlantic puffin (*Fratercula arctica*). This puffin has caught several fish. The puffin's beak becomes brightly colored during the breeding season.

Figure 32-23 Modern birds
Although they are diverse, most birds are highly adapted for flight, and their basic structure is similar.

found on all continents, most islands, and even the open sea. The largest living birds are the ostriches of Africa, which may be up to 2 m (6.5 ft) tall and weigh 136 kg (300 lb). The great condors of the Americas have wingspans of up to 3 m (10 ft). The smallest known bird is the bee hummingbird of Cuba, with a length of less than 6 cm (about 2 in.) and a weight of less than 2 g (about 0.06 oz).

Like their reptilian ancestors, modern birds lay eggs and have reptilian-type scales on their legs. Birds have evolved remarkable specializations for flight. They are the only extant animals with feathers. Their feathers are an amazing example of biological engineering. They are very light, yet flexible and strong, and present a flat surface to the air. Feathers also protect the body and decrease water and heat loss. The anterior limbs of birds are wings, usually adapted for flight. The posterior limbs are modified for walking, swimming, or perching.

Other adaptations for flight include the compact, streamlined body and the fusion of many bones, which provides the rigidity needed for flying. The bones are strong but very light. Many are hollow, containing large air spaces. The avian jaw is light, and instead of teeth, birds have a lightweight beak of bone and keratin. The breastbone is broad and keeled for the attachment of the large flight muscles.

Birds have lungs with *air sacs,* thin-walled extensions that occupy spaces between the internal organs and within certain bones. The unique "one-way" flow of air through their respiratory system is metabolically efficient because their lungs can extract great amounts of oxygen from the air. (This respiratory system is described in more detail in Chapter 46.) Birds have a four-chambered heart and a double circuit of blood flow. One circuit delivers oxygen-rich blood to the body tissues and returns low-oxygen blood to the heart; the other circuit takes blood to and from the lungs where it is oxygenated. The very effective respiratory and circulatory systems provide the cells with enough oxygen to permit a high metabolic rate, which is necessary for the strenuous muscular activity that flying requires. Some of the heat a bird generates by metabolic activities helps it maintain a constant body temperature. Because they are endotherms, birds can remain active in cold climates.

Birds excrete nitrogenous wastes mainly as semisolid uric acid. Because birds typically do not have a urinary bladder, these solid wastes are delivered into the cloaca. They leave the body with the feces, which are dropped frequently. This adaptive mechanism helps maintain a light body weight.

Bills are specifically adapted for the type of food the bird eats (see Fig. 1-13). Birds must eat frequently because they have a high metabolic rate and typically do not store much fat. Although the choice of food varies widely among species, most birds eat energy-rich foods such as seeds, fruits, worms, mollusks, or arthropods. Warblers and some other species eat mainly insects. Owls and hawks eat small mammals and other birds. Some hawks catch snakes and lizards. Vultures feed on dead animals. Pelicans, gulls, terns, and kingfishers catch fish.

An interesting feature of the bird digestive system is the *crop,* an expanded, saclike portion of the digestive tract below the esophagus in which food is temporarily stored. The stomach is divided into a *proventriculus*, which secretes gastric juices; and a thick, muscular *gizzard*, which grinds food. The bird swallows small bits of gravel that act as "teeth" in the gizzard to mechanically break down food.

Birds have a well-developed nervous system, with a brain that is proportionately larger than that of other reptiles. Birds rely heavily on vision, and their eyes are relatively larger than those of other vertebrates. Hearing is also well developed.

In striking contrast with the relatively silent non-avian reptiles, birds are very vocal. Most have short, simple *calls* that signal danger or influence feeding, flocking, or interaction between parent and young. *Songs* are usually more complex than calls and are performed mainly by males. Birds sing songs to attract and keep a mate and to claim and defend a territory.

One of the most fascinating aspects of bird behavior is the annual migration made by many species. Some birds, such as the golden plover and Arctic tern, fly from Alaska to Patagonia, South America, and back each year, covering perhaps 40,250 km (25,000 mi) en route. Migration and navigation are discussed in Chapter 52.

Many birds have beautiful, striking colors. The colors are due partly to pigments deposited during the development of the feathers and partly to reflection and refraction of light of certain wavelengths. Many birds, especially females, are protectively colored by their plumage. During the breeding season, the male often assumes brighter colors, which help attract a mate.

Mammals have hair and mammary glands

Hair is a key derived character of **mammals** (class Mammalia). All mammals have at least a few hairs at some time in their life, and no other organism has true hair. Hair insulates the body, helping maintain the high constant body temperature required for endothermy. The benefit of endothermy is a high metabolic rate, which allows a high level of activity even in low winter temperatures.

Other derived characters of mammals are **mammary glands,** which produce milk for the young; one pair of temporal openings in the skull; *differentiation of teeth* into incisors, canines, premolars, and molars; and three *middle-ear bones* (malleus, incus, and stapes) that transmit vibrations from the tympanic membrane (eardrum) to the inner ear. The other tetrapods have only a single bone, the stapes, in the middle ear. The evolution of the **cochlea,** the organ of hearing in the inner ear, gives mammals an excellent sense of hearing.

Contributing significantly to the success of the mammals, the nervous system is more highly developed than in any other group of animals. The cerebrum is especially large and complex, with an outer gray region, the **cerebral cortex.** In mammals the highly specialized cerebral cortex, called the *neocortex,* has six layers of neurons. Specific regions in the neocortex are specialized for functions such as vision, hearing, touch, movement, emotional response, and higher cognitive functions.

Fertilization in mammals is internal, and with the exception of the monotremes that lay eggs, mammals are viviparous.

Most mammals develop a **placenta,** an organ of exchange between developing embryo and mother. As the mother's blood passes through blood vessels in the placenta, it delivers nourishment and oxygen to the embryo and carries off wastes. By carrying their developing young internally, mammals avoid the hazards of having their eggs consumed by predators. By nourishing the young and caring for them, the parents offer both protection and an "education" on how to obtain food and avoid being eaten.

A muscular **diaphragm** helps move air into and out of the lungs. Like birds, mammals are endotherms, but mammals appear to have evolved endothermy independently of birds. Some of the adaptations that allow mammals to maintain a constant body temperature include the covering of insulating hair, very efficient lungs with alveoli (air sacs through which gas exchange with the blood takes place), a fully divided ventricle, and complete separation of pulmonary and systemic circulations. Red blood cells without nuclei serve as efficient oxygen transporters.

The limbs of mammals are variously adapted for walking, running, climbing, swimming, burrowing, or flying. In most terrestrial mammals, the limbs are more directly under the body than they are in extant reptiles, which contributes to speed and agility.

New fossil discoveries are changing our understanding of the early evolution of mammals

Mammals descended from **therapsids,** a group of synapsid amniotes, during the Triassic period more than 200 mya. The therapsids were somewhat doglike carnivores with differentiated teeth and legs adapted for running (**FIG. 32-24**). Until

Figure 32-24 Therapsid (*Lycaenops*)
The therapsids were synapsid amniotes. *Lycaenops,* a predator about the size of a coyote, lived in South Africa during the Late Permian period.

recently, early mammals were thought to be small, about the size of a mouse or shrew. However, fossil evidence indicates that mammals diversified earlier than previously thought. Some early mammals were fairly large, with limbs adapted for swimming or burrowing. Some had fur, and some were probably endothermic.

How did mammals manage to coexist with the abundant archosaurs during the approximately 160 million years that reptiles ruled Earth? Many adaptations permitted early mammals to compete for a place on this planet. Perhaps one of the most important was their skill at being inconspicuous. Many were **arboreal** (tree dwelling) and nocturnal (active at night), searching for food (mainly insects and plant material, and perhaps reptile eggs) at night while the reptiles were inactive. Although their eyes may have been small, early mammals had well-developed sense organs for hearing and smell.

As many reptiles became extinct, mammals adapted to the niches (lifestyles) that reptiles abandoned. During this time, the flowering plants, including many trees, underwent an adaptive radiation, providing new habitats, sources of food, and protection from predators. Numerous new varieties of mammals evolved. During the Early Cenozoic era (more than 55 mya), the mammals underwent an adaptive radiation; they became widely distributed and adapted to an impressive variety of ecological lifestyles.

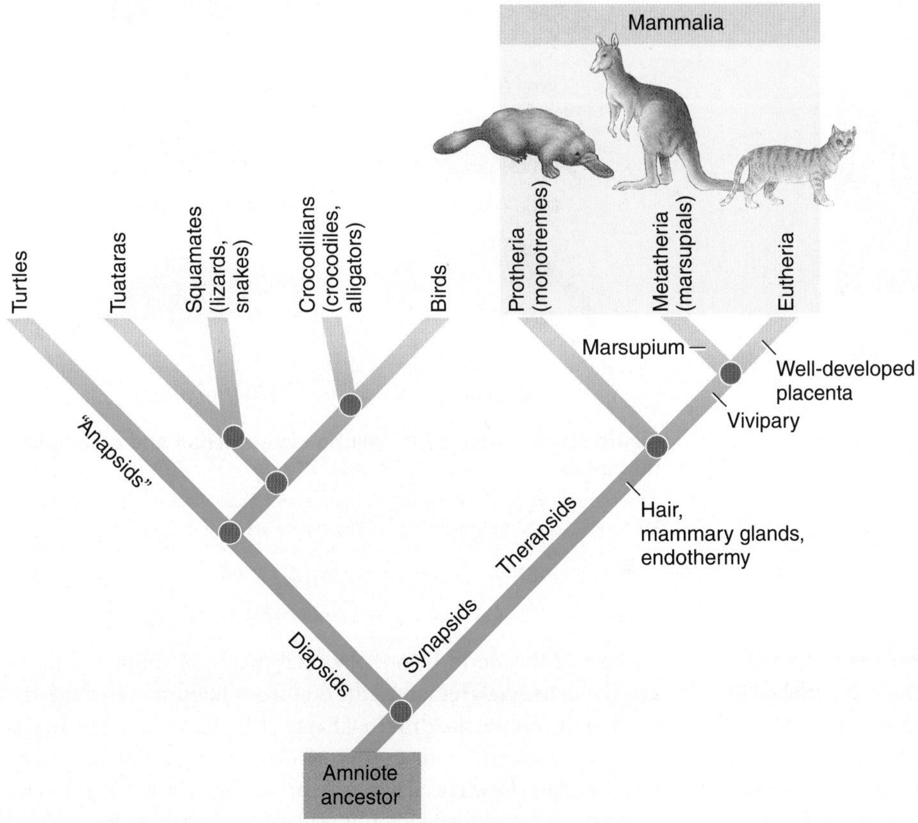

Modern mammals are assigned to three subclasses

By the end of the Cretaceous period, three main groups of mammals had evolved (**FIG. 32-25**). Today, mammals inhabit virtually every corner of Earth; they live on land, in fresh water and salt water, and in the air. They range in size from the tiny pygmy shrew, weighing about 2.5 g (less than 0.1 oz), to the blue whale, which may weigh up to 136,000 kg (150 tons) and which is probably one of the largest animals that has ever lived.

Modern mammals are classified in two main clades: Protheria, which includes the egg-laying mammals, or **monotremes;** and Theria, which includes the mammals that bear their young alive. The therian clade is further divided into two groups: Metatheria, the **marsupials** (pouched mammals); and Eutheria, mammals that are more developed at birth than the marsupials. **Eutherians** are often referred to as placental mammals because they have well-developed placentas. However, some marsupials do have simple placentas.

One group of protherians includes the duck-billed platypus and a second, the spiny echidnas (**FIG. 32-26**). These animals live in Australia and Tasmania; two species of spiny echidnas inhabit New Guinea. Female protherians lay eggs that they carry in a pouch on the abdomen or keep warm in a nest. When the young hatch, they lap up milk secreted by their mother's mammary glands. Unlike other mammals, monotremes do not have nipples. With their long beaks and long, sticky tongues, spiny echidnas capture ants and termites. The duck-billed platypus, which lives in burrows along river banks, preys on freshwater invertebrates. It has webbed feet and a flat, beaverlike tail, which help it swim.

Marsupials include pouched mammals such as kangaroos and opossums (**FIG. 32-27a**). Embryos begin their development in the mother's uterus, where they are nourished by fluid and yolk. After a brief gestation, the young are

Figure 32-25 Evolution of mammals

This cladogram shows the current interpretation of the evolution of mammals.
© Cengage Learning

Figure 32-26 Short-beaked echidna (*Tachyglossus aculeatus*)

Female protherians lay eggs. They have mammary glands, but do not have nipples.

(a) Eastern gray kangaroo (*Macropus giganteus*) with young, known as a joey. The kangaroo is native to Australia.

(b) A kangaroo soon after birth. Marsupials are born in an embryonic state and continue to develop in the safety of the mother's marsupium (pouch).

Figure 32-27 Marsupials

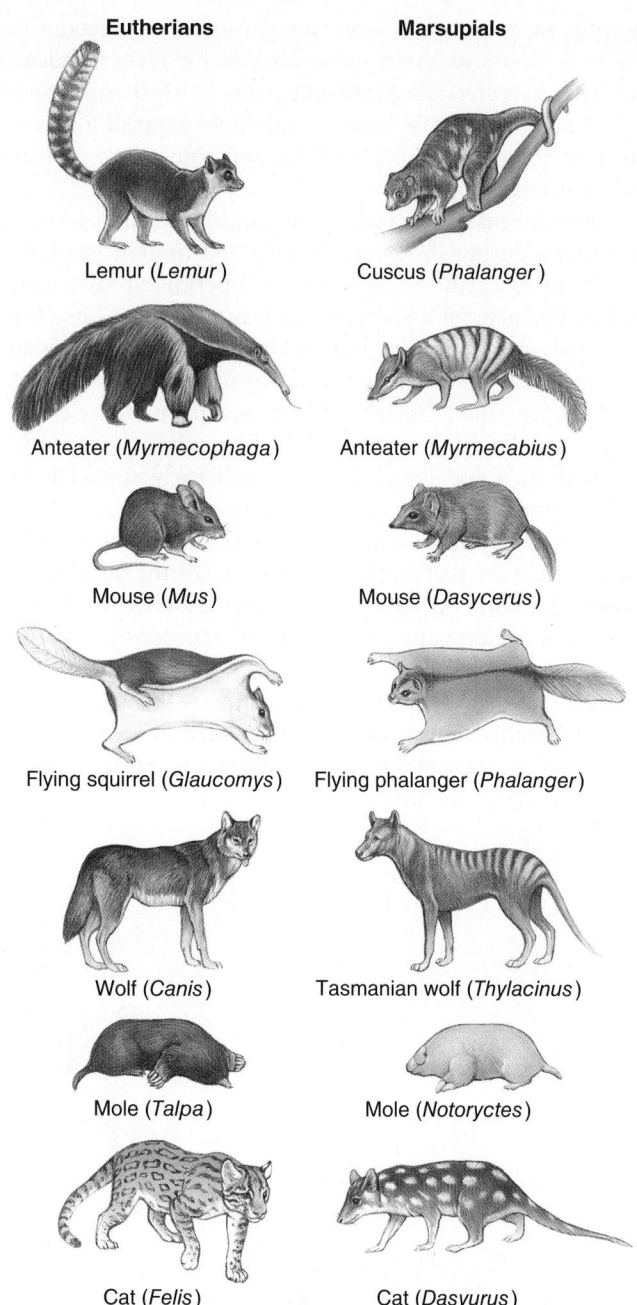

Eutherians	Marsupials
Lemur (*Lemur*)	Cuscus (*Phalanger*)
Anteater (*Myrmecophaga*)	Anteater (*Myrmecabius*)
Mouse (*Mus*)	Mouse (*Dasycerus*)
Flying squirrel (*Glaucomys*)	Flying phalanger (*Phalanger*)
Wolf (*Canis*)	Tasmanian wolf (*Thylacinus*)
Mole (*Talpa*)	Mole (*Notoryctes*)
Cat (*Felis*)	Cat (*Dasyurus*)

Figure 32-28 Convergent evolution in eutherian and marsupial mammals

For each mammal in one group, a counterpart has evolved in the other group. Their similarities include both lifestyles and structural features.

© Cengage Learning

born in an undeveloped stage. The young make their way to the *marsupium* (pouch), where they attach to a nipple. Nourished by the mother's milk, they complete their development (FIG. 32-27b).

At one time, marsupials probably inhabited much of the world, but placental mammals largely outcompeted them. Now marsupials live mainly in Australia, Central America, and South America. The only marsupial species that ranges into the United States is the Virginia opossum. Australia became geographically isolated from

the rest of the world before placental mammals migrated there, and the marsupials remained the dominant mammals (see the discussion of continental drift in Chapter 18). Australian marsupials underwent adaptive radiation, paralleling the evolution of placental mammals elsewhere. Thus, marsupials that live in Australia and adjacent islands independently evolved body forms and natural histories similar to North American placental wolves, bears, rats, moles, flying squirrels, and even cats (FIG. 32-28).

(a) Wildebeest (*Connochaetes taurinus*), photographed in Tanzania. The dominant plains antelope in many areas of eastern and southern Africa, wildebeest migrate long distances during the dry season in search of food and water.

(b) Humpback whales (*Megaptera novaeangliae*) exhibiting a rare double breach.

(c) Wolverine (*Gulo gulo*), a tough, powerful animal. Wolverines are fierce predators that sometimes take down large prey, such as caribou. The wolverine population has declined due to trapping, fragmentation of their habitat, and reduction of their range. Their habitat is also being reduced by climate change. They are protected by the Endangered Species Act.

(d) Egyptian fruit bat (*Rousettus aegyptiacus*). Often called "flying foxes," these small bats cluster in large colonies. They often hang by one foot from the cave ceiling.

Figure 32-29 Eutherian mammals

Most familiar to us are the eutherian mammals, which are born at a more mature stage than marsupials (FIG. 32-29). Indeed, among some species the young can walk and begin to interact with other members of the group within a few minutes after birth. Eutherians have a well-developed **placenta.** The placenta develops from both embryonic membranes and the maternal uterine wall. In the placenta blood vessels of the embryo come very close to the blood vessels of the mother, so materials can be exchanged by diffusion. (The two circulations do not normally mix.) The placenta allows the young to remain within the mother's body until embryonic development is complete.

Biologists assign extant placental mammals to about 19 orders. TABLE 32-4 gives a brief summary of some of these orders.

Remember that aquatic mammals, such as dolphins, whales, and seals, evolved from terrestrial ancestors (see Fig. 18-9). Also refer to TABLE 32-5, which reviews the three main deuterostome groups.

CHECKPOINT 32.8

- CONNECT *Describe three adaptations that have allowed amniotes to become completely terrestrial.*
- CONNECT *Which amniotes are endothermic? What are some advantages of endothermy?*
- *Argue for classifying birds and reptiles in a single clade.*
- *Contrast the three main clades of mammals.*

TABLE 32-4 Some Orders of Extant Eutherian Mammals

ORDER, REPRESENTATIVE MEMBERS, AND SOME CHARACTERISTICS

INSECTIVORA (moles, hedgehogs, shrews)

Nocturnal; eat insects. Shrew is smallest living mammal; some weigh less than 5 g (about 0.2 ounces).

African hedgehog
Atelerix albiventris

ARTIODACTYLA (cattle, sheep, pigs, deer, giraffes)

Hoofed with even numbers of digits per foot; most have two toes, some have four. Most have antlers or horns. Herbivores; most are ruminants, chew a cud, and have a series of stomachs inhabited by bacteria that digest cellulose.

American elk
Cervus elaphus

CHIROPTERA (bats)

Adapted for flying; fold of skin extends from elongated fingers to body and legs, forming wing. Guided in flight by type of biological sonar; emit high-frequency squeaks and are guided by echoes from obstructions. Eat insects and fruit or suck blood of other animals.

Bat
Eptesicus fuscus

XENARTHRA (sloths, anteaters, armadillos)

Teeth reduced or no teeth. Sloths are sluggish animals that hang upside down from branches; often protectively colored by green algae that grow on their hair. Armadillos are protected by bony plates; eat insects and small invertebrates.

Nine-banded armadillo
Dasypus bellus

CARNIVORA (cats, dogs, wolves, foxes, bears, otters, mink, weasels, skunks, seals, sea lions, walruses)

Predators with sharp, pointed canine teeth and molars for shearing; keen sense of smell; complex social interactions; among fastest, strongest, and smartest animals. Seals, sea lions, and walruses are large marine predators with limbs adapted as flippers for swimming.

Wolf
Canis lupus

RODENTIA (squirrels, beavers, rats, mice, hamsters, porcupines, guinea pigs)

Gnawing animals with chisel-like incisors. As they gnaw, teeth are worn down and so must grow continually. Most numerous mammals both in numbers and species.

Flying squirrel
Glaucomys volans

PERISSODACTYLA (horses, zebras, tapirs, rhinoceroses)

Herbivores; hoofed with odd number of digits per foot; one or three toes; teeth adapted for chewing; usually large animals with long legs.

Tapir
Tapirus sp.

PROBOSCIDEA (elephants)

Largest land animals; weigh up to 7 tons; large head; broad ears; long, muscular, flexible trunk (proboscis); thick loose skin is characteristic; two upper incisors are elongated as tusks. Includes extinct mastodons and woolly mammoths.

African elephant
Loxodonta africana

LAGOMORPHA (rabbits, hares, pikas)

Like rodents, have chisel-like incisors; typically have long hind legs adapted for jumping; many have long ears.

Pika
Ochontona sp.

SIRENIA (sea cows, manatees)

Herbivorous, aquatic mammals with fin-like forelimbs and no hind limbs. They are probably the basis for most tales about mermaids. Evolved from Proboscidean ancestors.

Manatee
Trichechus manatus

PRIMATES (lemurs, monkeys, apes, humans)

Highly developed brain and eyes; nails instead of claws; opposable thumb; eyes directed forward; omnivores; most species arboreal. (Primate evolution discussed in Chapter 22.)

Ring-tailed lemur
Lemur catta

CETACEA (whales, dolphins, porpoises)

Adapted for aquatic life with fish-shaped body and broad, paddlelike forelimbs (flippers); posterior limbs absent; many have thick layer of blubber under skin; some are filter feeders; mate and bear young in the water; very intelligent. Evolved from Artiodactylian ancestors.

Humpback whale
Megaptera noveangliae

TABLE 32-5	Overview of the Deuterostomes		
MAJOR GROUPS		**BODY PLAN**	**KEY CHARACTERISTICS**
Echinoderms (sea stars, sea urchins, sand dollars)		Larva bilateral, ciliated; adult pentaradial; triploblastic; organ systems; complete digestive tube	Endoskeleton with spines; water vascular system functions in locomotion, feeding, and gas exchange; tube feet; marine
Hemichordates (acorn worms)		Bilateral symmetry; triploblastic; organ systems; complete digestive tube	Ring of cilia surrounds mouth; three-part body: proboscis, collar, and trunk; pharyngeal slits; wormlike marine animals
Chordates (tunicates, lancelets, vertebrates)		Bilateral symmetry; triploblastic; organ systems; complete digestive tube	Notochord; dorsal, tubular nerve cord; endostyle; postanal tail; pharyngeal slits during some time in life cycle; segmented muscles

© Cengage Learning

32.1 What Are Deuterostomes? *(page 671)*

1 Identify shared derived characters of deuterostomes and briefly describe the hemichordates.

- The **deuterostomes** include echinoderms, hemichordates, and chordates. Shared derived characters include: radial, indeterminate cleavage; the blastopore becomes (or is near the future site of) the anus; and pharyngeal slits at some time in the life cycle. Basal deuterostomes have a larva with a loop-shaped ciliated band used for locomotion.
- **Hemichordates** (acorn worms) are marine deuterostomes with a three-part body, including proboscis, collar, and trunk.

32.2 Echinoderms *(page 671)*

2 Identify three shared derived characters of echinoderms and describe the main classes of echinoderms.

- **Echinoderms** (phylum *Echinodermata*) are marine animals with a spiny "skin," **water vascular system, tube feet,** and **endoskeleton.** The larvae exhibit bilateral symmetry; most of the adults exhibit *pentaradial symmetry*.
- Class Crinoidea includes sea lilies and feather stars. The *oral surface* of crinoids is turned upward; some crinoids are sessile.
- Class Asteroidea consists of the sea stars. They have a central disc with five or more arms, and they use tube feet for locomotion.
- Class Ophiuroidea includes the brittle stars, which resemble sea stars but have longer, more slender arms that are set off more distinctly from the central disc. They use their arms for locomotion. Their tube feet lack suckers and are not used in locomotion.
- Class Echinoidea includes the sea urchins and sand dollars. Echinoids lack arms; they have a solid shell and are covered with spines.
- Class Holothuroidea consists of sea cucumbers, animals with elongated flexible bodies. The mouth is surrounded by a circle of modified tube feet that serve as tentacles.

32.3 The Chordates: Major Characteristics *(page 675)*

3 Describe characteristics of chordates, including four shared derived characters.

- The **chordates** (Phylum *Chordata*) include three subphyla: Urochordata, Cephalochordata, and Vertebrata. At some time in its life cycle, a chordate has a flexible, supporting **notochord;** a **dorsal, tubular nerve cord;** a muscular **postanal tail;** and an **endostyle,** or thyroid gland; they are also characterized by **pharyngeal slits,** but that is a derived character of deuterostomes.

32.4 Invertebrate Chordates *(page 676)*

4 Compare tunicates and lancelets, and summarize the phylogeny of chordates.

- The **tunicates,** which are **urochordates,** are suspension-feeding marine animals with tunics. Larvae have typical chordate characteristics and are free-swimming. Adults of most groups are sessile suspension feeders.
- The *lancelets* are **cephalochordates,** small, segmented, fishlike animals; their chordate characteristics are highly developed.
- The available evidence suggests that urochordates are the sister group of the vertebrates.

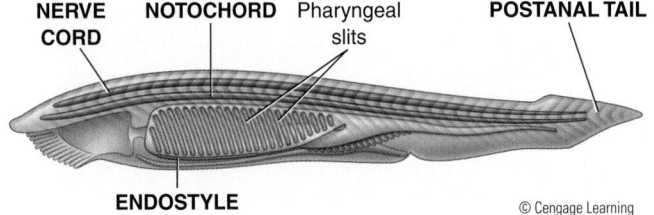

NERVE CORD **NOTOCHORD** Pharyngeal slits **POSTANAL TAIL**

ENDOSTYLE

© Cengage Learning

32.5 Introducing the Vertebrates *(page 678)*

5 Describe four shared derived characters of vertebrates.

- The vertebrates have a *vertebral column* composed of *vertebrae* that forms the chief skeletal axis of the body and a braincase, or **cranium. Neural crest cells** are embryonic cells important in the development of many structures, including the cranium and jaws. Vertebrates have pronounced cephalization, a complex brain, a muscular pharynx, and muscles attached to the endoskeleton.

6 Describe the major taxa of extant vertebrates.

- Vertebrates can be assigned to nine classes. The hagfishes, which make up the *Myxini*, and the lampreys, which make up the *Petromyzontida*, have neither jaws nor paired fins. The *Chondrichthyes* comprise the sharks, rays, and skates; they are jawed fishes with skeletons of cartilage. The extant (living) bony fishes can be assigned to three classes: *Actinopterygii*, ray-finned fishes; *Actinistia*, coelacanths; and *Dipnoi*, lungfishes.

- The **tetrapods** (Tetrapoda) include the amphibians (class Amphibia; salamanders, frogs, and caecilians), many of which have aquatic larvae that undergo metamorphosis, and the amniotes (Amniota), which include reptiles and mammals. **Reptiles** (class Reptilia) include turtles, lizards, snakes, alligators, and birds. Reptiles are amniotes with keratin scales or feathers and reproduction adapted for terrestrial life; mammals (class Mammalia) include monotremes, marsupials, and placental mammals. Mammals are amniotes with hair and mammary glands. (See Table 32-2 for a more detailed review.)

32.6 Jawless Fishes *(page 680)*

7 Distinguish among the major groups of jawless fishes.

- The extant jawless fishes are the hagfishes (Myxini) and the lampreys (Petromyzontida). Jaws and paired fins are absent in both hagfishes and lampreys. Hagfishes are marine scavengers that secrete slime as a defense mechanism. Many lampreys are parasites on other fishes.

32.7 Evolution of Jaws and Limbs: Jawed Fishes and Tetrapods *(page 682)*

8 Trace the evolution of jawed fishes and early tetrapods, and describe modern amphibians.

- Chondrichthyes, the cartilaginous fishes (sharks, rays, and skates), evolved during the Devonian period. They have jaws, two pairs of fins, and **placoid scales.**

- During the Devonian period, bony fishes gave rise to two evolutionary lines: the Actinopterygii, or **ray-finned fishes,** and the *Sarcopterygii,* or *lobe-finned fishes*. The ray-finned fishes gave rise to the modern bony fishes. Their lungs have been modified as a **swim bladder,** an air sac for regulating buoyancy.

- The Sarcopterygii includes the Tetrapodomorpha, the *lung-fishes* (Dipnoi), and the **coelacanths** (Actinistia). Evidence suggests that the Tetropodomorpha gave rise to the **tetrapods,** the land vertebrates. *Tiktaalik* was a transitional form between fishes and tetrapods.

- Early tetrapods were mainly aquatic animals that ventured onto the land to find food or escape predators. These early tetrapods had limbs strong enough to support the weight of their bodies on land.

- Modern **amphibians** (Amphibia) include salamanders and newts, frogs and toads, and caecilians. Most amphibians return to the water to reproduce. Frog embryos develop into tadpoles, which undergo **metamorphosis** to become adults. Amphibians use their moist skin as well as lungs for gas exchange. They have systemic and pulmonary circulations as well as hearts with two atria and one ventricle.

32.8 Amniotes: Terrestrial Vertebrates *(page 688)*

9 Describe three vertebrate adaptations to terrestrial life.

- Terrestrial vertebrates, or **amniotes,** include reptiles (including birds) and mammals. Adaptations for life on land include (1) the evolution of the **amniotic egg** with its shell and **amnion,** a membrane that forms a fluid-filled sac around the embryo; (2) internal fertilization; and (3) a body covering that retards water loss and physiological mechanisms that conserve water.

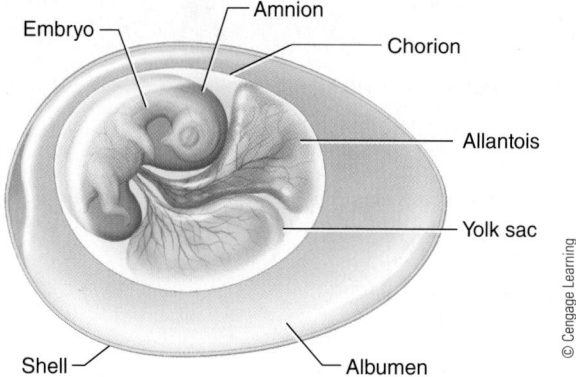

Labels: Embryo, Amnion, Chorion, Allantois, Yolk sac, Shell, Albumen
© Cengage Learning

10 Describe the reptiles (including the birds) and argue for including the birds in the reptile clade.

- Biologists classify amniotes in two main groups: diapsids and synapsids. **Diapsids** include the turtles, squamates (snakes and lizards), tuataras, extinct ichthyosaurs, crocodiles, the extinct pterosaurs, the extinct ornisthischian dinosaurs, the extinct saurischian dinosaurs, and birds. The **synapsids** include the **therapsids,** which gave rise to the mammals. (See Table 32-3.)

- Extant **reptiles** (including birds) can be classified in five clades: (1) turtles, terrapins, and tortoises; (2) lizards, snakes, and amphisbaenians; (3) tuataras; (4) crocodiles, alligators, caimans, and gavials; and (5) birds (avian reptiles).

- Most non-avian reptiles have dry skin with horny scales, lungs with many chambers, and a heart with two completely separated atria and two ventricles that are incompletely separated. (In the crocodiles the two ventricles are completely partitioned.) Birds have many adaptations for powered flight, including feathers; wings; and light, hollow bones containing air spaces. They have completely divided ventricles and very efficient lungs, and they excrete solid metabolic wastes (uric acid). They are **endotherms;** that is, they maintain a constant body temperature. Birds have a well-developed nervous system with excellent vision and hearing.

- Based on fossil evidence and molecular data, feather evolution took place in terrestrial, bipedal dinosaurs before the evolution of birds or flight. Birds are considered feathered reptiles that evolved from the lineage of saurischian dinosaurs, specifically from the **theropods,** a group of bipedal, saurischian dinosaurs. Like theropods, modern birds have feet with three digits; thin-walled, hollow bones; and a *furcula*, or wishbone.

11 Describe five key characters of mammals and contrast protherian (monotremes), metatherian (marsupials), and eutherian mammals, giving examples of animals that belong to each group.

- **Mammals** have *hair*, **mammary glands**, *differentiated teeth*, lungs with alveoli, completely divided ventricles, and three *middle-ear bones*. They are endotherms and have a highly developed nervous system and a muscular **diaphragm.**

- *Protherians*, the **monotremes,** include the duck-billed platypus and spiny anteaters. Monotremes lay eggs.

- *Metatherians*, the **marsupials,** include pouched mammals, such as kangaroos and opossums. The young, born at an immature stage, complete their development in their mother's marsupium, where they are nourished with milk from mammary glands.

- **Eutherians** are more developed at birth than marsupials; they are characterized by a well-developed **placenta,** an organ of exchange that develops between the embryo and the mother.

TEST YOUR UNDERSTANDING

Know and Comprehend

1. Which of the following is *not* a shared derived character of echinoderms? (a) water vascular system (b) notochord (c) tube feet (d) pentaradial symmetry in adult (e) endoskeleton of calcium carbonate plates and spines
2. Which of the following is/are found in tunicates? (a) dorsal, tubular nerve cord (b) tube feet (c) anal gill slits (d) two pairs of appendages (e) vertebral column
3. A shark is characterized by (a) amnion (b) bony skeleton (c) water vascular system (d) placoid scales (e) endothermy *and* amnion
4. Which of the following characteristics is/are associated with amphibians? (a) amnion (b) placoid scales (c) heart with two complete ventricles (d) swim bladder (e) metamorphosis
5. Reptiles (a) are all endotherms (b) are amniotes (c) have a great deal of keratin in their epidermis (d) have external fertilization (e) b and c are both correct
6. Which of the following is *not* characteristic of birds? (a) hollow bones (b) amnion (c) ectothermy (d) high metabolic rate (e) reptilian-like scales on legs
7. Which of the following is true of mammals? (a) they evolved from saurischian dinosaurs (b) they are exotherms (c) they have hair and three middle ear bones (d) mammalian embryos do not have an amnion (e) they all bear their young alive (do not lay eggs).

Apply and Analyze

8. **VISUALIZE** Draw a simple cladogram illustrating the evolutionary relationships among extant mammals (marsupials, eutherians, and monotremes). Include the following characters in your cladogram: well-developed placenta, vivipary, endothermy, marsupium, hair.
9. **EVOLUTION LINK** Sea urchins have radial symmetry. Explain why they are not classified as cnidarians.
10. **EVOLUTION LINK** Most biologists consider birds as living dinosaurs. Justify this position.
11. **EVOLUTION LINK** Discuss the relationships among the echinoderms and chordates, describing shared derived characters that support grouping these animals as deuterostomes.

Evaluate and Synthesize

12. **INTERPRET DATA** Imagine that you discover an interesting new animal. You examine it and gather the following data. It has a dorsal, tubular nerve cord; a cranium; skin with scales; and a heart with two atria and two ventricles. How would you classify the animal? Explain each step in your decision. Where would you place this animal on the cladogram shown in Figure 32-20?
13. **SCIENCE, TECHNOLOGY, AND SOCIETY** Every year hundreds of sea turtles are injured and killed as a result of human activities. All seven species of marine turtles are listed under the Endangered Species Act. What measures would you propose to protect sea turtles?

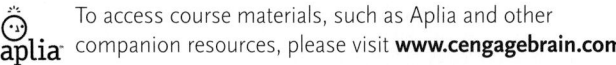 To access course materials, such as Aplia and other companion resources, please visit **www.cengagebrain.com.**

Foxgloves, herbaceous biennials, and oxeye daisies, herbaceous perennials. Foxgloves (*Digitalis purpurea*) and oxeye daisies (*Leucanthemum vulgare*) grow wild in this mountain meadow in Idaho. These non-native species are also popular garden plants. Foxglove is the source of digitalis, which is used to treat heart disease.

KEY CONCEPTS

33.1 The structure of plant cells, tissues, and organs matches their functions. The vascular plant body is differentiated into three tissue systems: the ground tissue system, the vascular tissue system, and the dermal tissue system.

33.2 Primary meristems lengthen roots and shoots throughout the lifespan of most plants. Woody plants have both primary meristems and secondary meristems, which increase roots and shoots in thickness.

33.3 Plant development requires not only cell division and expansion but also cell determination and cell differentiation, and pattern formation and morphogenesis.

The more than 300,000 species of flowering plants that live in and are adapted to the many environments on Earth represent remarkable variety. From desert cacti with enormously swollen stems, to cattails partly submerged in marshes, to orchids growing in the uppermost tree branches of lush tropical rain forests, they are all recognizable as plants. Almost all plants have the same basic body plan, which consists of roots, stems, and leaves.

Plants are either herbaceous or woody. *Herbaceous* plants are nonwoody. In temperate climates the aerial parts (stems and leaves) of herbaceous plants die back to the ground at the end of the growing season. In contrast, the aerial parts of *woody* plants (trees and shrubs) persist. Botanically speaking, woody plants produce hard, lignified secondary tissues (the cell walls of secondary tissues contain *lignin*), and herbaceous plants do not. Lignin and the production of secondary tissues are discussed later in this chapter and in Chapters 35 and 36.

Annual plants are herbaceous plants (such as corn, geraniums, and marigolds) that grow, reproduce, and die in one year or less. Other herbaceous plants (such as carrots, Queen Anne's lace, cabbages, and foxgloves) are **biennial plants;** they take two years to complete their life cycles before dying (see photograph). During their first season, biennials produce extra carbohydrates, which they store and use during their second year, when they typically form flowers and reproduce.

Perennial plants are herbaceous and woody plants that have the potential to live for more than two years. In temperate climates the aerial (aboveground) stems of herbaceous perennials such as irises, rhubarb, onions, asparagus, and oxeye daisies (see photograph) die back each winter. Their underground parts (roots and underground stems) become dormant during the winter and send out new growth each spring. (In **dormancy** an organism reduces its metabolic state to a minimum level to survive unfavorable conditions.) Similarly, in certain tropical climates with pronounced wet and dry seasons, the aerial parts of herbaceous perennials die back, and the underground parts become dormant during the unfavorable dry season. Other tropical plants, such as orchids, are herbaceous perennials that grow year-round.

All woody plants are perennials, and some of them live for hundreds or even thousands of years. In temperate climates the aerial stems of woody plants become dormant during the winter. Many temperate woody perennials are **deciduous;** that is, they shed their leaves before winter and produce new stems with new leaves the following spring. Other woody perennials are **evergreen** and shed their leaves over a long period, so some leaves are always present. Because they have permanent woody stems that are the starting points for new growth the following season, many trees attain massive sizes.

In this chapter we focus on the structure, growth, and development of flowering plants, which are vascular plants characterized by flowers, double fertilization, endosperm, and seeds enclosed within fruits (see Chapter 28). We then examine other aspects of flowering plants in Chapters 34 to 38.

33.1 THE PLANT BODY

LEARNING OBJECTIVES

1 Discuss the functions of various parts of the vascular plant body, including the nutrient- and water-absorbing root system and the photosynthesizing shoot system.

2 Describe the structure and functions of the ground tissue system (parenchyma tissue, collenchyma tissue, and sclerenchyma tissue).

3 Describe the structure and functions of the vascular tissue system (xylem and phloem).

4 Describe the structure and functions of the dermal tissue system (epidermis and periderm).

The plant body of flowering plants (and other vascular plants) is usually organized into a root system and a shoot system (FIG. 33-1). The **root system** is generally underground. The aerial portion, the **shoot system,** usually consists of a vertical stem that bears leaves and, in flowering plants, flowers and fruits that contain seeds.

Each plant typically grows in two different environments: the dark, moist soil and the illuminated, relatively dry air. Plants usually have both roots and shoots because they require resources from both environments. Thus, roots branch extensively through the soil, forming a network that anchors the plant firmly in place and absorbs water and dissolved minerals (inorganic nutrients). Leaves, the flattened organs of photosynthesis, are attached more or less regularly on the stem, where they absorb the sun's light and atmospheric carbon dioxide (CO_2) used in photosynthesis.

The plant body consists of cells and tissues

As in other organisms, the basic structural and functional unit of plants is the cell. Plants have evolved a variety of cell types, each specialized for particular functions. Plant cells are organized into tissues. A **tissue** is a group of cells that forms a structural and functional unit. *Simple tissues* are composed of only one kind of cell, whereas *complex tissues* have two or more kinds of cells.

In vascular plants tissues are organized into three tissue systems, each of which extends throughout the plant body (FIG. 33-2). Each tissue system contains two or more kinds of tissues (TABLE 33-1). Most of the plant body is composed of the **ground tissue system,** which has a variety of functions, including photosynthesis, storage, and support. The **vascular tissue system,** an intricate conducting system that extends throughout the plant body, is responsible for conduction of various substances, including water, dissolved minerals, and food (dissolved sugar). The vascular tissue system also functions in strengthening and supporting the plant. The **dermal tissue system** provides a covering for the plant body.

Roots, stems, leaves, flower parts, and fruits are **organs** because each is composed of all three tissue systems. The tissue systems of different plant organs form an interconnected network throughout the plant. For example, the vascular tissue system of a leaf is continuous with the vascular tissue system of the stem to which it is attached, and the vascular tissue system of the stem is continuous with the vascular tissue system of the root.

The ground tissue system is composed of three simple tissues

The bulk of a herbaceous plant is its ground tissue system, which is composed of three tissues: parenchyma, collenchyma, and sclerenchyma (TABLE 33-2 on pages 708 and 709). These tissues can be distinguished by their cell wall structures. Recall that plant cells are surrounded by a cell wall that provides structural support (see Chapter 4). A growing plant cell secretes a thin *primary cell wall,* which stretches and expands as the cell increases in size. After the cell stops growing, it sometimes secretes a thick, strong *secondary cell wall,* which is deposited *inside* the primary cell wall—that is, between the primary cell wall and the plasma membrane (see Fig. 4-30).

Plant cell walls have several important roles. As discussed later in this chapter, cell walls are involved in growth; the ability of primary cell walls to expand allows cells to increase in size. Scientists are increasingly aware that **signal transduction** takes place in plant cell walls because many carbohydrate and protein molecules in plant cell walls communicate with other molecules both inside and outside the cell. The cell wall is also the first line of defense against disease-causing organisms.

Parenchyma cells have thin primary walls Parenchyma tissue, a simple tissue composed of parenchyma cells, is found throughout the plant body and is the most common type of cell and tissue (FIG. 33-3a on page 710). The soft parts of a plant, such as the edible part of an apple or a potato, consist largely of parenchyma.

Parenchyma cells perform several important functions, such as photosynthesis, storage, and secretion. Parenchyma cells that function in photosynthesis contain green chloroplasts, whereas nonphotosynthetic parenchyma cells lack chloroplasts and are often colorless. Materials stored in parenchyma cells include starch grains, oil droplets, water, and salts, which are sometimes visible as crystals. Resins, tannins, hormones, enzymes, and

During the evolution of plants, the root and shoot systems specialized to obtain resources from the soil and air, respectively.

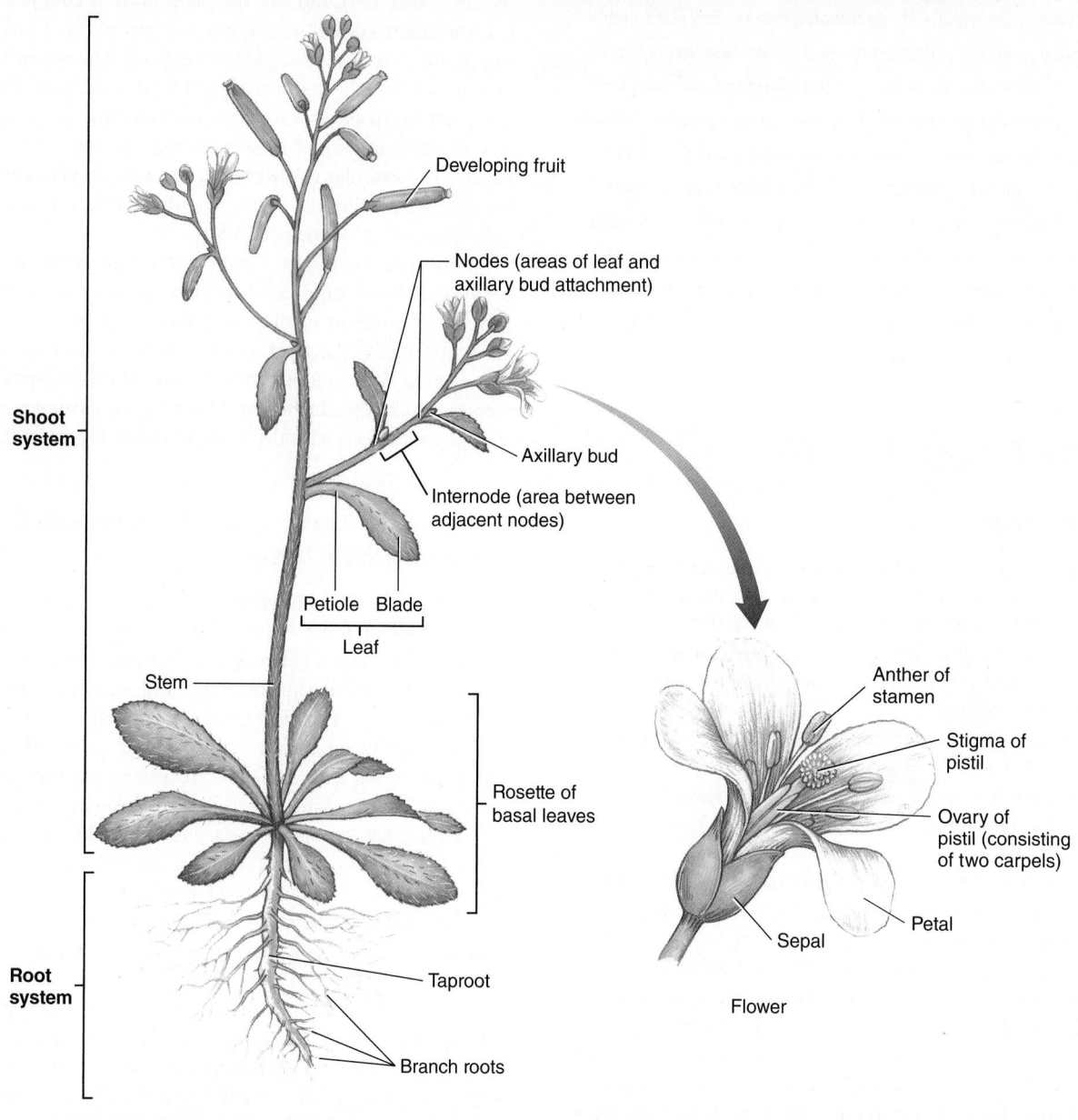

Figure 33-1 *Animation* **The plant body**

A plant body consists of a root system, usually underground, and a shoot system, usually aerial. Shown is mouse-ear cress (*Arabidopsis thaliana*), a small plant in the mustard family that is a model plant for biological research. *Arabidopsis* is native to North Africa and Eurasia and has been naturalized (was introduced and is now growing in the wild) in California and the eastern half of the United States.
© Cengage Learning

CONNECT What resources required for photosynthesis do plants obtain from the air? from the soil?

sugary nectar are examples of substances that may be secreted by parenchyma cells. The various functions of parenchyma require that parenchyma cells be alive and metabolically active.

Parenchyma cells have the ability to *differentiate* into other kinds of cells, particularly when a plant has been injured. If xylem (water-conducting) cells are severed, for example, adjacent parenchyma cells may divide and differentiate into new xylem cells within a few days. (Recall from Chapter 17 that it is possible to induce certain plant cells to become the equivalent of embryonic stem cells that can then differentiate into specialized cells.)

Collenchyma cells have unevenly thickened primary walls Collenchyma tissue, a simple plant tissue composed of collenchyma cells, is an extremely flexible structural tissue that provides much of the support in soft, nonwoody plant organs

The tissue systems are continuous throughout the plant. For example, the vascular tissue system in a leaf is continuous with the vascular tissue system in the stem to which it is attached.

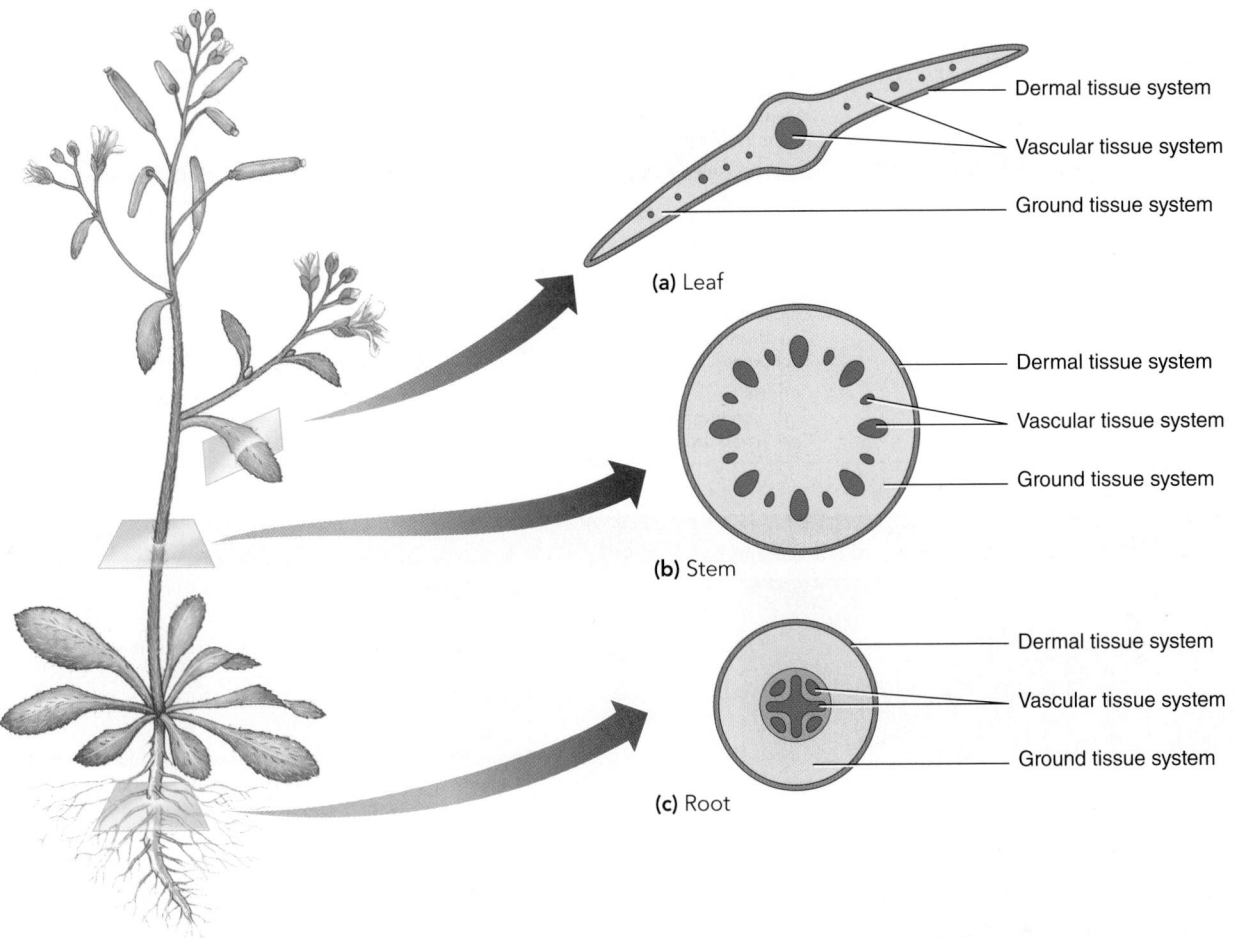

(a) Leaf
- Dermal tissue system
- Vascular tissue system
- Ground tissue system

(b) Stem
- Dermal tissue system
- Vascular tissue system
- Ground tissue system

(c) Root
- Dermal tissue system
- Vascular tissue system
- Ground tissue system

Figure 33-2 *Animation* **The three tissue systems in the plant body**

This figure shows the distribution of the ground tissue system, vascular tissue system, and dermal tissue system in a herbaceous eudicot such as *Arabidopsis*.

© Cengage Learning

PREDICT What do you think would happen if the vascular tissue system were to become discontinuous in the stem but the dermal and ground tissue systems were to remain intact?

(FIG. 33-3b). Support is crucial for plants, in part because it lets them grow upward to compete with other plants for available sunlight in a plant-crowded area. Plants lack the bony skeletal system typical of many animals; instead, individual cells, including collenchyma cells, support the plant body.

Collenchyma cells, which are usually elongated, are alive at maturity. Their primary cell walls are unevenly thickened and are especially thick in the corners. Collenchyma is not found uniformly throughout the plant and often occurs as long strands near stem surfaces and along leaf veins. The "strings" in a celery stalk, for example, consist of collenchyma tissue.

Sclerenchyma cells have both primary walls and thick secondary walls Another simple plant tissue specialized for structural support is **sclerenchyma** tissue, whose cells have

both primary and secondary cell walls. The root of the word *sclerenchyma* is derived from the Greek word *sclero,* meaning "hard." The secondary cell walls of sclerenchyma cells become strong and hard because of extreme thickening. *Pits,* which are thin areas where secondary walls did not develop, allow an exchange of substances between adjacent living sclerenchyma cells. Mature sclerenchyma cells are unable to stretch or elongate. At functional maturity, when sclerenchyma tissue is providing support for the plant body, its cells are often dead.

Sclerenchyma tissue may be located in several areas of the plant body. Two types of sclerenchyma cells are sclereids and fibers. **Sclereids** are cells of variable shape common in the shells of nuts and in the stones of fruits such as cherries and peaches. The slightly gritty texture of pears results from the presence of clusters of sclereids. **Fibers,** which are long, tapered cells that often occur

TISSUE SYSTEM	TISSUE	CELL TYPES	MAIN FUNCTIONS OF TISSUE
Ground tissue system	Parenchyma tissue	Parenchyma cells	Storage, secretion, photosynthesis
	Collenchyma tissue	Collenchyma cells	Support
	Sclerenchyma tissue	Sclerenchyma cells (sclereids or fibers)	Support, strength
Vascular tissue system	Xylem	Tracheids	Conduction of water and nutrient minerals, support
		Vessel elements	Conduction of water and nutrient minerals, support
		Xylem parenchyma cells	Storage
		Fibers (sclerenchyma cells)	Support, strength
	Phloem	Sieve tube elements	Conduction of sugar in solution, support
		Companion cells	May control functioning of sieve tube elements, loading sugar into sieve tube elements
		Phloem parenchyma cells	Storage
		Fibers (sclerenchyma cells)	Support, strength
Dermal tissue system	Epidermis	Epidermal cells	Protective covering over surface of plant body
		Guard cells	Regulate stomata
		Trichomes	Variable functions
	Periderm	Cork cells	Protective covering over surface of plant body
		Cork cambium cells	Meristematic (form new cells)
		Cork parenchyma cells	Storage

© Cengage Learning

TABLE 33-2 | Cell Types in the Ground Tissue System

Parenchyma cell

Description
Living, actively metabolizing; thin primary cell walls

Functions
Storage; secretion, and/or photosynthesis

Location and comments
Throughout the plant body; shown is an LM of a stamen hair of a spiderwort (*Tradescantia virginiana*) flower; note the large pigmented vacuole; the nucleus is not inside the vacuole but is lying on *top* of the vacuole

Parenchyma cell

Description
Living, actively metabolizing; thin primary cell walls

Functions
Storage; secretion; photosynthesis

Location and comments
Throughout the plant body; shown is an LM of leaf cells from an aquatic plant (*Elodea*); note the many chloroplasts in the thin layer of cytoplasm surrounding the large, transparent vacuole

Parenchyma cell

Description
Living, actively metabolizing; thin primary cell walls

Functions
Storage; secretion; photosynthesis

Location and comments
Throughout the plant body; shown is an LM of a cross section of part of a buttercup (*Ranunculus*) root; note the starch grains filling the cells

Continued

TABLE 33-2 | Cell Types in the Ground Tissue System (*continued*)

Primary cell walls are thickened in corners

Ed Reschke/Getty Images

50 μm

Collenchyma cell

Description
Living; unevenly thickened primary cell walls

Function
Elastic support

Location and comments
Just under stem epidermis; shown is an LM of a cross section of an elderberry (*Sambucus*) stem; note the unevenly thickened cell walls that are especially thick in the corners, making the cell contents assume a rounded shape in cross section

Nucleus Secondary cell walls Cytoplasm Pits

James Mauseth, University of Texas

50 μm

Sclereid (sclerenchyma cell)

Description
May be living or dead at maturity; thick secondary cell walls; lacks secondary wall at pits

Functions
Strength; sclereid-rich tissue is hard and inflexible

Location and comments
Shells of walnuts and coconuts; pits of cherries and peaches; shown is an LM of living sclereids in a section through a cherry (*Prunus avium*) pit

Thick secondary cell walls Lumen (location of cell contents when these cells were alive)

James Mauseth, University of Texas

50 μm

Fiber (sclerenchyma cell)

Description
Often dead at maturity; thick secondary cell walls; fewer pits than in sclereids

Functions
Support; provides strength

Location and comments
Throughout the plant body; common in stems and certain leaves; shown is an LM of a cross section through a clump of fibers from a century plant (*Agave*) leaf

© Cengage Learning

in patches or clumps, are particularly abundant in the wood, inner bark, and leaf ribs (veins) of flowering plants (**FIG. 33-3c**).

Cells of the three simple tissues vary in their cell wall chemistry Parenchyma, collenchyma, and sclerenchyma cells can be distinguished by the chemistry of their cell walls. Cell walls may contain cellulose, hemicelluloses, pectin, and lignin. **Cellulose,** the most abundant polymer in the world, accounts for about 40% to 60% of the dry weight of plant cell walls. As discussed in Chapter 3, cellulose is a **polysaccharide** composed

of glucose units joined by β-1,4 bonds. Each cellulose molecule consists of thousands of glucose subunits joined to form a flat, ribbonlike chain. From 40 to 70 of these chains lie parallel to one another and connect by hydrogen bonding to form a *cellulose microfibril,* a strong, tiny strand visible under the electron microscope (see Fig. 3-10a).

Cellulose microfibrils are cemented together by a matrix of hemicelluloses and pectins. *Hemicelluloses* are a group of polysaccharides that are more soluble than cellulose. Hemicelluloses vary in their chemical composition from one species

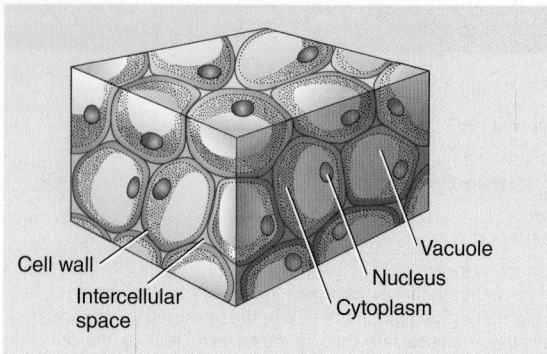

(a) **Parenchyma cells** have thin primary cell walls. They vary in size and structure, depending on their various functions within the plant body.

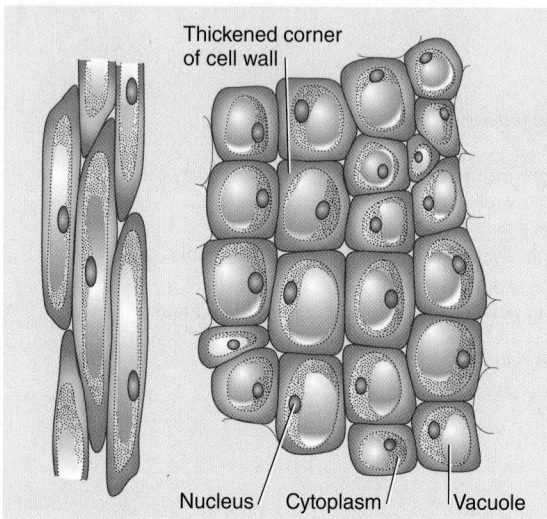

Thickened corner of cell wall

Nucleus   Cytoplasm   Vacuole

(b) **Collenchyma cells** in longitudinal section (*left*) and cross section. Note the elongated cells, evident in longitudinal section, and the unevenly thickened primary cell walls, evident in cross section.

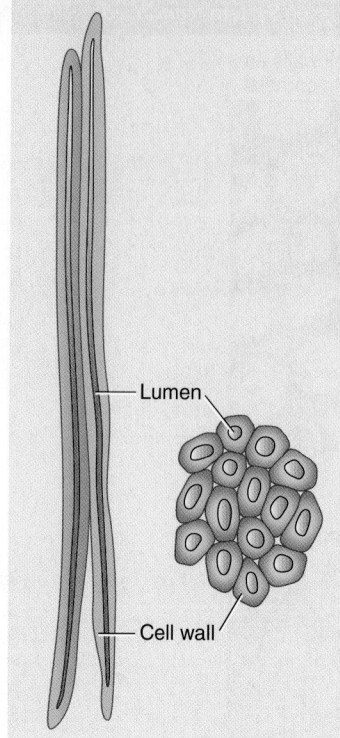

Lumen

Cell wall

(c) **Sclerenchyma cells** (fibers) in longitudinal section (*left*) and cross section. Mature fibers have thick secondary cell walls, are often dead at functional maturity, and therefore lack nuclei and cytoplasm; the lumen is the space formerly occupied by the living cell.

Figure 33-3 Cell types: parenchyma, collenchyma, and sclerenchyma
© Cengage Learning

to another. Some hemicelluloses are composed of *xyloglucan,* which consists of a backbone of β-1,4-glucose molecules to which side chains of *xylose,* a five-carbon sugar, are attached. Despite their name, the chemical structure of the hemicelluloses is significantly different from that of cellulose. *Pectin,* another cementing polysaccharide, is less variable in its monomer composition than are the hemicelluloses. Pectin monomer units are *α-galacturonic acid,* a six-carbon molecule that is a derivative of glucose.

An important component of secondary plant cell walls, particularly those of wood, is **lignin.** Composing as much as 35% of the dry weight of the secondary cell wall, lignin is a strengthening polymer made up of monomers derived from certain amino acids. Scientists have not yet determined the exact chemical structure of lignin because it is hard to remove from cellulose and other cell wall materials to which it is covalently bound.

Having discussed the four main components in plant cell walls, we can now generalize about the cell chemistry of parenchyma, collenchyma, and sclerenchyma cells. The thin primary

cell walls of parenchyma cells contain predominantly cellulose, although they also contain hemicelluloses and pectin. Both parenchyma and collenchyma cells have primary cell walls, but their walls are chemically distinct because the thickened areas of collenchyma walls contain large quantities of pectin in addition to cellulose and hemicelluloses. The thick secondary walls of sclerenchyma cells are chemically different because they are rich in lignin in addition to cellulose, hemicelluloses, and pectin.

The vascular tissue system consists of two complex tissues

The **vascular tissue system,** which is embedded in the ground tissue, transports needed materials throughout the plant via two complex tissues: xylem and phloem (**TABLE 33-3**). Both xylem and phloem are continuous throughout the plant body. (Chapter 35 discusses the mechanisms of transport in xylem and phloem.)

The conducting cells in xylem are tracheids and vessel elements
Xylem conducts water and dissolved minerals from the roots to the stems and leaves and provides structural support. In flowering plants xylem is a complex tissue composed of four cell types: tracheids, vessel elements, parenchyma cells, and fibers. Two of the four cell types found in xylem, the **tracheids** and **vessel elements,** conduct water and dissolved minerals. Xylem also contains fibers that provide support, and parenchyma cells, known as *xylem parenchyma,* that perform storage functions.

Tracheids and vessel elements are highly specialized for conduction. As they develop, both types of cells undergo **apoptosis,** or programmed cell death. As a result, mature tracheids and vessel elements are dead and therefore hollow; only their cell walls remain. Tracheids, the chief water-conducting cells in gymnosperms (such as pines) and seedless vascular plants (such as ferns), are long, tapering cells located in patches or clumps (**FIG. 33-4a**). Water is conducted upward, from roots to shoots, passing from one tracheid into another through wide pits.

In addition to tracheids, flowering plants possess extremely efficient vessel elements (**FIG. 33-4b**). The cell diameters of vessel elements are usually greater than those of tracheids. Vessel elements are hollow, but unlike tracheids, the end walls have holes, known as *perforations,* or are absent. Vessel elements are stacked

TABLE 33-3 | Selected Cell Types in the Vascular Tissue System

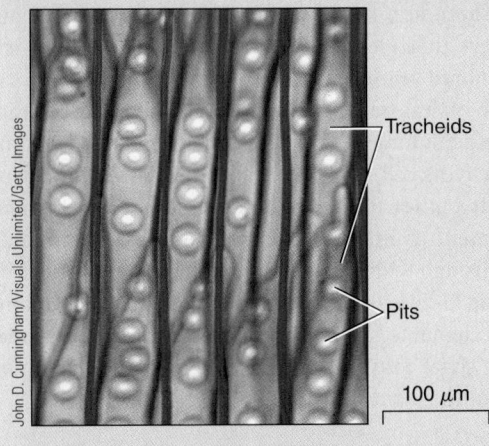

John D. Cunningham/Visuals Unlimited/Getty Images

Tracheids

Pits

100 μm

Tracheid

Description
Dead at maturity; lacks secondary wall at pits

Functions
Conduction of water and nutrient minerals; support

Location and comments
Occurs in clumps in xylem throughout plant body; shown is an LM of a longitudinal section of tracheids from white pine (*Pinus strobus*) wood

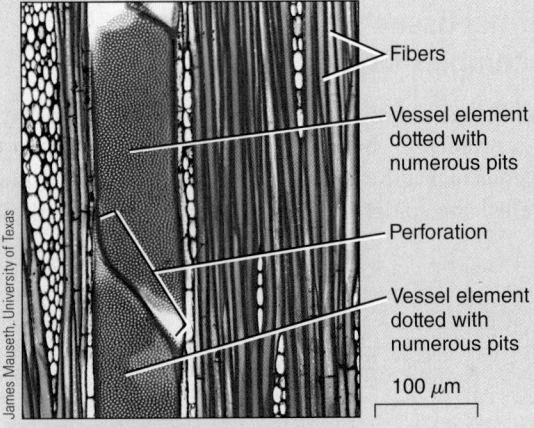

James Mauseth, University of Texas

Fibers

Vessel element dotted with numerous pits

Perforation

Vessel element dotted with numerous pits

100 μm

Vessel element

Description
Dead at maturity; end walls have perforations; lacks secondary wall at pits

Functions
Conduction of water and nutrient minerals; support

Location and comments
Xylem throughout plant body; vessel elements are more efficient than tracheids in conduction; shown is an LM of a longitudinal section of two vessel elements from an unidentified woody eudicot

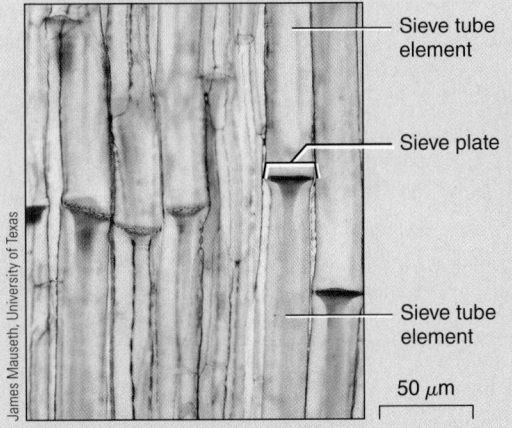

James Mauseth, University of Texas

Sieve tube element

Sieve plate

Sieve tube element

50 μm

Sieve tube element

Description
Living but lacks nucleus and other organelles at maturity; end walls are sieve plates

Function
Conduction of sugar in solution

Location and comments
Phloem throughout plant body; shown is an LM of a longitudinal section through a clump of sieve tube elements in a squash (*Cucurbita*) petiole (leaf stalk)

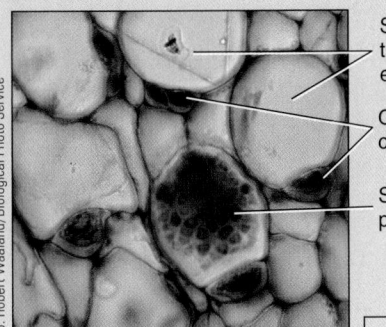

J. Robert Waaland/Biological Photo Service

Sieve tube elements

Companion cells

Sieve plate

50 μm

Companion cell

Description
Living; has cytoplasmic connections with sieve tube element

Function
Assists in moving sugars into and out of sieve tube element

Location and comments
Phloem throughout plant body; shown is an LM of phloem from a squash (*Cucurbita*) petiole (leaf stalk) in cross section

one on top of the other, and water is conducted readily from one vessel element into the next. A stack of vessel elements, called a **vessel,** resembles a miniature water pipe. Vessel elements also have pits in their side walls that permit the lateral transport (sideways movement) of water from one vessel to another.

Sieve tube elements are the conducting cells of phloem

Phloem conducts food materials—that is, carbohydrates formed in photosynthesis—throughout the plant and provides structural support. In flowering plants phloem is a complex tissue composed of four cell types: sieve tube elements, companion cells, fibers, and phloem parenchyma cells (**FIGS. 33-4c and 33-4d**). Fibers, which are usually extensive in the phloem of herbaceous plants, provide additional structural support for the plant body.

Food materials are conducted in solution—that is, dissolved in water—through the **sieve tube elements,** which are among the most specialized plant cells. Sieve tube elements are joined end to end to form long **sieve tubes.** The cells' end walls, called *sieve plates,* have a series of holes through which cytoplasm extends from one sieve tube element into the next. Sieve tube elements are living at maturity, but many of their organelles, including the nucleus, vacuole, mitochondria, and ribosomes, disintegrate or shrink as they mature.

Sieve tube elements are among the few eukaryotic cells that can function without nuclei. These cells typically live for less than a year. There are, however, notable exceptions. Certain palms have sieve tube elements that remain alive for approximately one hundred years.

Adjacent to each sieve tube element is a **companion cell** that assists in the functioning of the sieve tube element. The companion cell is a living cell, complete with a nucleus. This nucleus is thought to direct the activities of both the companion cell and sieve tube element. Numerous **plasmodesmata**—cytoplasmic connections through which cytoplasm extends from one cell to another (see Fig. 5-24)—link a companion cell with its adjoining sieve tube element. The companion cells play an essential role in moving sugar into the sieve tube elements for transport to other parts of the plant.

The dermal tissue system consists of two complex tissues

The **dermal tissue system**—the epidermis and periderm—provides a protective covering over plant parts (**TABLE 33-4**). In herbaceous plants the dermal tissue system is a single layer of cells called the epidermis. Woody plants initially produce

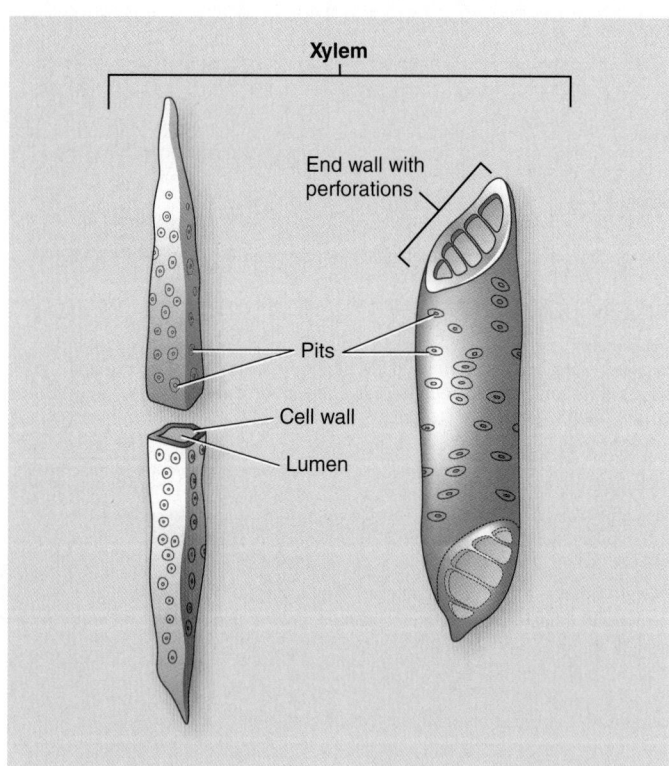

Xylem

End wall with perforations

Pits

Cell wall

Lumen

(a) Tracheid (xylem cell). The tracheid is opened to reveal the cell's appearance in cross section. At maturity, tracheids are usually dead.

(b) Vessel element (xylem cell). The end walls of vessel elements are perforated. Vessel elements are joined end to end, from roots to leaves and other shoot parts.

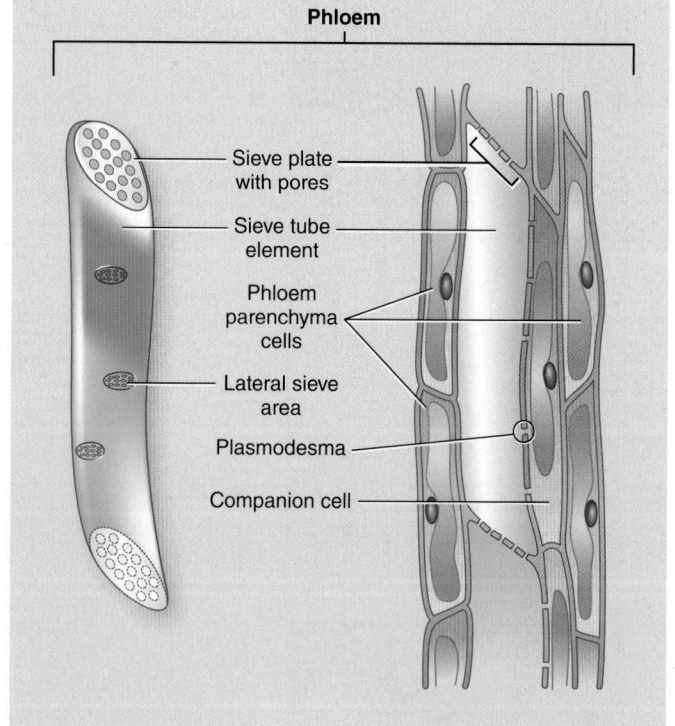

Phloem

Sieve plate with pores

Sieve tube element

Phloem parenchyma cells

Lateral sieve area

Plasmodesma

Companion cell

(c) Sieve tube element, showing the sieve plate. Sieve tube elements are joined end to end to form sieve tubes.

(d) Longitudinal section of phloem tissue. Sieve tube elements, companion cells, and phloem parenchyma cells. Fiber cells are not shown.

Figure 33-4 *Animation* **Longitudinal views of cell types in xylem and phloem tissues**
© Cengage Learning

TABLE 33-4 | Selected Cell Types in the Dermal Tissue System

Cuticle
Epidermis
Collenchyma
100 μm

James Mauseth, University of Texas

Epidermal cell

Description
Relatively unspecialized cell with thin primary wall; outer wall often thicker and covered by a noncellular waxy layer (cuticle)

Functions
Protective covering over surface of plant body; helps reduce water loss

Location and comments
Epidermis is usually one cell thick; shown is an LM of a cross section through epidermis of an ivy (*Hedera helix*) stem

Guard cell
Stomatal pore
Epidermal cell
Guard cell
10 μm

Ed Reschke/Getty Images

Guard cell

Description
Chloroplast-containing cell that occurs in pairs; pair changes shape to open and close stomatal pore

Function
Opens and closes stomatal pore

Location and comments
Epidermis of stems and leaves; shown is an LM of the epidermis of a spiderwort (*Tradescantia virginiana*) leaf

Trichomes
Epidermis
100 μm

Biophoto Associates/Science Source

Trichome

Description
Hair or other epidermal outgrowth; may be unicellular or multicellular; occurs in variety of sizes and shapes

Functions
Varied; absorption; secretion; excretion; protection; reduction of water loss

Location and comments
Epidermis; shown is an SEM of a nettle (*Solanum carolinense*) leaf, which has trichomes that break off inside the skin of animals and inject irritating substances that cause a stinging sensation

Remnants of epidermis
Cork cells
Cork cambium and cork parenchyma
Periderm
200 μm

James Mauseth/University of Texas

Cork cell

Description
Dead at maturity; cell walls impregnated with waterproof material (suberin)

Functions
Reduces water loss and prevents disease-causing organisms from penetrating

Location and comments
Produced in large numbers; cork often forms just under the epidermis; replaces epidermis in older stems and roots; shown is an LM of a cross section through the periderm of a calico flower (*Aristolochia elegans*) stem

© Cengage Learning

an epidermis, but it splits apart as a result of the production of additional woody tissues inside the epidermis as the plant increases in girth. Periderm, which replaces the epidermis in the stems and roots of older woody plants, composes the outer bark.

Epidermis is the outermost layer of a herbaceous plant

The **epidermis** is a complex tissue composed primarily of relatively unspecialized living cells. Dispersed among these cells are more specialized guard cells and outgrowths called trichomes (discussed later). In most plants the epidermis consists of a single layer of flattened cells (FIG. 33-5). Epidermal cells generally contain no chloroplasts and are therefore transparent, so light can penetrate the interior tissues of stems and leaves. In both stems and leaves, photosynthetic tissues lie *beneath* the epidermis.

An important requirement of the aerial parts (stems, leaves, flowers, and fruits) of a plant is the ability to control water loss. Epidermal cells of aerial parts secrete a waxy **cuticle** over the surface of their exterior walls; this waxy layer greatly restricts water loss from plant surfaces.

Although the cuticle is extremely efficient at preventing most water loss through epidermal cells, it also slows the diffusion of CO_2, needed for photosynthesis, from the atmosphere into the leaf or stem. The diffusion of CO_2 is facilitated by **stomata** (sing., *stoma*). Stomata are minute pores in the epidermis that are surrounded by two cells called **guard cells** (see Fig. 34-5). Many gases, including CO_2, oxygen, and water vapor, pass through the stomata by diffusion. Stomata are generally open during the day when photosynthesis is occurring, and the water loss that also takes place when stomata are open provides some evaporative cooling. At night, stomata usually close. In drought conditions the need to conserve water overrides the need to cool the leaves and exchange gases. Thus, during a drought, the stomata close in the daytime. Stomata are discussed in greater detail in Chapter 34.

The epidermis also contains special outgrowths, or hairs, called **trichomes,** which occur in many sizes and shapes and have a variety of functions. Plants that tolerate salty environments such as the seashore often have specialized trichomes on their leaves to remove excess salt that accumulates in the plant. The presence of trichomes on the aerial parts of desert plants may increase the reflection of light off the plants, thereby cooling the internal tissues and decreasing water loss. Other trichomes have a protective function. For example, the trichomes on stinging nettle leaves and stems contain irritating substances that may discourage herbivorous animals from eating the plant. **Root hairs** are simple, unbranched trichomes that increase the surface area of the root epidermis (which comes into contact with the soil) for more effective water and mineral absorption.

Periderm replaces epidermis in woody plants **Periderm,** a tissue that can be anywhere from several to many cell layers thick, forms under the epidermis to provide a new protective covering as the epidermis is destroyed. As a woody plant continues to increase in girth, its epidermis is sloughed off, exposing the periderm. Periderm forms the outer bark of older stems and roots. It is a complex tissue composed mainly of cork cells and cork parenchyma cells. **Cork cells** are dead at maturity, and their walls are heavily coated with a waterproof substance called *suberin,* which helps reduce water loss. **Cork parenchyma** cells function primarily in storage. Periderm is discussed further in Chapters 35 and 36.

CHECKPOINT 33.1

- *What are some of the functions of roots? of shoots?*
- *What are the three tissue systems in plants? Describe the functions of each.*
- *How do parenchyma, collenchyma, and sclerenchyma tissues differ in cell structure and function?*
- *What are the functions of xylem and phloem? Describe the conducting cells that occur in each of these complex tissues.*
- CONNECT *How are epidermis and periderm alike? How are they different?*

33.2 PLANT MERISTEMS

LEARNING OBJECTIVES

5   State how growth in plants differs from growth in animals.
6   Distinguish between primary and secondary growth.
7   Distinguish between apical meristems and lateral meristems.

One difference between plants and animals is the *location* of growth. When plants grow, their cells divide only in specific areas, called **meristems,** which are composed of cells whose

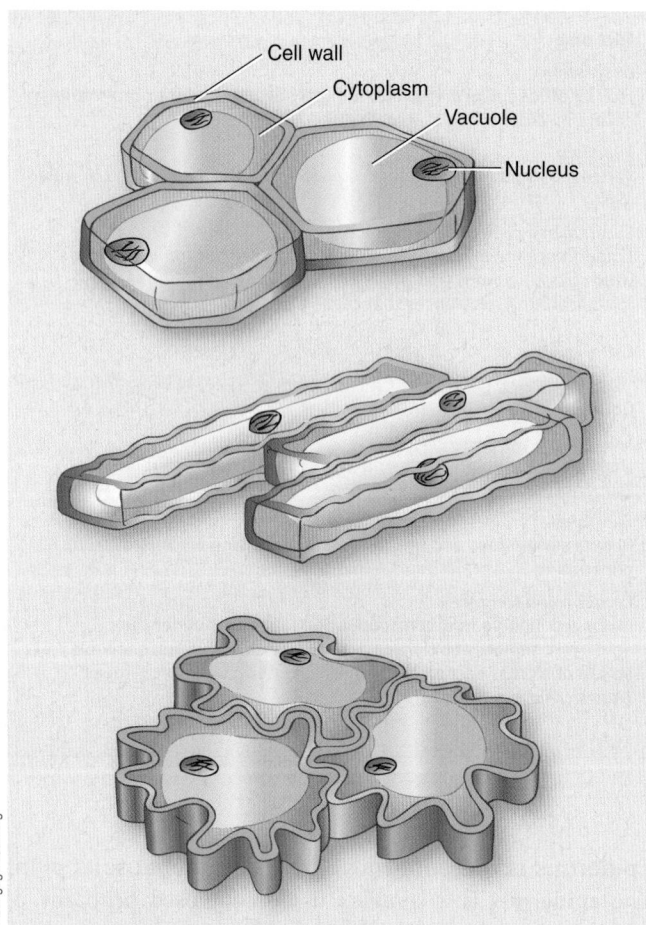

© Cengage Learning

Figure 33-5 Cell types: epidermal cells
Epidermal cells from three different plants. Note that regardless of the cell shape, epidermal cells fit tightly together.

Cell wall
Cytoplasm
Vacuole
Nucleus

primary function is to form new cells by mitotic division. (In contrast, when a young animal is growing, all parts of its body grow, although not necessarily at the same rate.) Meristematic cells, also known as *stem cells,* do not differentiate. Instead, they retain the ability to divide by mitosis, a trait that many differentiated cells lose. The persistence of mitotically active meristems means that plants, unlike most animals, can grow throughout their entire lifespan.

This ability of roots and stems to grow throughout a plant's life is known as **indeterminate growth.** In contrast, many leaves and flowers have **determinate growth;** that is, they stop growing after reaching a certain size. The size of leaves and flowers with determinate growth varies from species to species and from individual to individual, depending on the plant's genetic programming and on environmental conditions such as availability of sunlight, water, and essential minerals.

Two kinds of meristematic growth may occur in plants. **Primary growth** is an increase in stem and root length. All plants have primary growth, which produces the entire plant body in herbaceous plants and the young, soft shoots and roots in woody trees and shrubs. **Secondary growth** is an increase in the girth of a plant. For the most part, only gymnosperms and woody eudicots have secondary growth. (Woody eudicots include such trees and shrubs as oak, sycamore, ash, cherry, apple, beech, and maple.)

Tissues produced by secondary growth compose the wood and bark, which make up most of the bulk of trees and shrubs. A few annuals—geraniums and sunflowers, for example—undergo limited secondary growth even though they lack obvious wood and bark tissues.

Primary growth takes place at apical meristems

Primary growth occurs as a result of the activity of **apical meristems,** areas located at the tips of roots and shoots, including within the buds of stems. **Buds** are dormant embryonic shoots that eventually develop into branches.

Primary growth is evident when a root tip is examined (**FIG. 33-6**). A protective layer of cells called a **root cap** covers the root tip. Directly behind the root cap is the root apical meristem, which consists of meristematic cells. Meristematic cells, which are cube shaped, remain small because they are continually dividing. (In meristems, as daughter cells begin to enlarge, one or both will divide again.)

Farther from the root tip, just behind the area of cell division, is an area of cell elongation where the cells have been displaced from the meristem. Here the cells are no longer dividing but instead are growing longer, pushing the root tip ahead of them, deeper into the soil. Some differentiation also begins to occur in the area of cell elongation, and immature tissues such as differentiating xylem and phloem become evident. The tissue systems continue to develop and differentiate into primary tissues (epidermis, xylem, phloem, and ground tissues) of the adult plant. Farther from the tip, behind the area of cell elongation, most cells have completely differentiated and are fully mature. Root hairs are evident in this area.

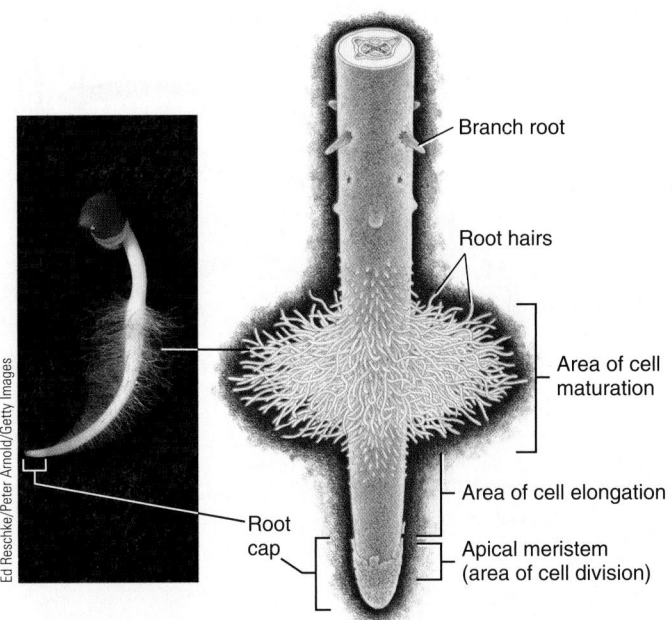

Figure 33-6 A root tip

The root apical meristem (where cells divide and thus increase in number) is protected by a root cap. Farther from the tip is an area of cell elongation where cells enlarge and begin to differentiate. Note the young root hairs in the area of cell maturation, which has fully mature, differentiated cells.

© Cengage Learning

A shoot apex (such as the terminal bud) is quite different in appearance from a root tip (**FIG. 33-7**). A dome of minute, regularly arranged meristematic cells—the shoot apical meristem—is located in each shoot apex. *Leaf primordia* (developing leaves) and *bud primordia* (developing buds) arise from the shoot apical meristem. The tiny leaf primordia cover and protect the shoot apical meristem. As the cells formed by the shoot apical meristem elongate, the shoot apical meristem is pushed upward. Subsequent cell divisions produce additional stem tissue and cause new leaf and bud primordia to appear. Farther from the stem tip, the immature cells differentiate into the three tissue systems of the mature plant body.

Secondary growth takes place at lateral meristems

Trees and shrubs undergo both primary and secondary growth. These plants increase in length by primary growth and increase in girth by secondary growth. The increase in girth, which occurs in areas that are no longer elongating, is due to cell divisions that take place in **lateral meristems,** areas extending along the entire length of the stems and roots except at the tips. Two lateral meristems, the vascular cambium and the cork cambium, are responsible for secondary growth, which is the formation of secondary tissues: secondary xylem, secondary phloem, and periderm (**FIG. 33-8**).

The **vascular cambium** is a layer of meristematic cells that forms a long, thin, continuous cylinder within the stem and

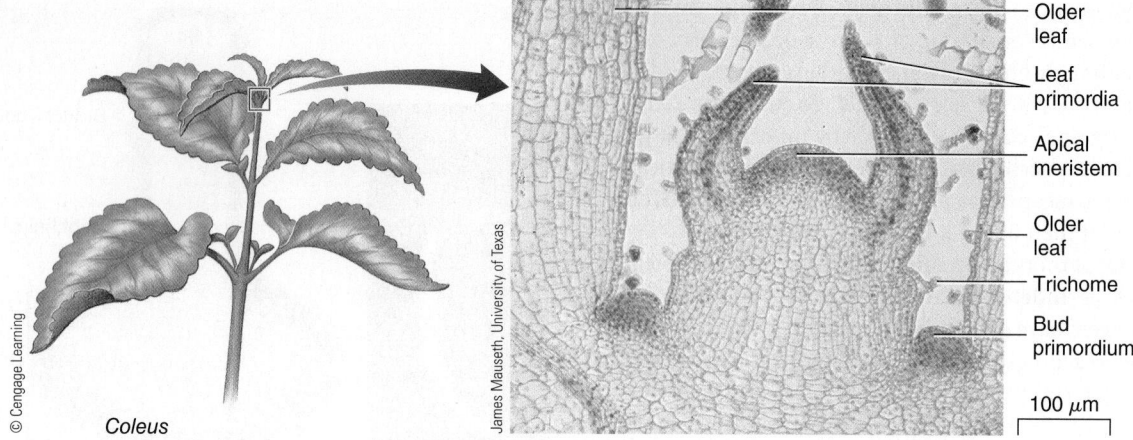

Coleus

Figure 33-7 *Animation* **A shoot apex**

LM of a longitudinal section through a shoot apex of *Coleus*, showing the shoot apical meristem, leaf primordia, and bud primordia.

Older leaf
Leaf primordia
Apical meristem
Older leaf
Trichome
Bud primordium
100 μm

James Mauseth, University of Texas

root. It is located between the wood and bark of a woody plant. Division of cells of the vascular cambium adds more cells to the wood (secondary xylem) and inner bark (secondary phloem).

The **cork cambium,** located in the outer bark, is composed of a thin cylinder or irregular arrangement of meristematic cells. Cells of the cork cambium divide and form **cork cells** toward the outside and one or more underlying layers of **cork parenchyma** cells that function in storage. Collectively, cork cells, cork cambium, and cork parenchyma make up the periderm.

We are now ready to give a more precise definition of *bark*. **Bark,** the outermost covering over woody stems and roots, consists of all plant tissues located outside the vascular cambium. Bark has two regions, a living inner bark composed of

secondary phloem and a mostly dead outer bark composed of periderm. Chapters 35 and 36 present a more comprehensive discussion of secondary growth.

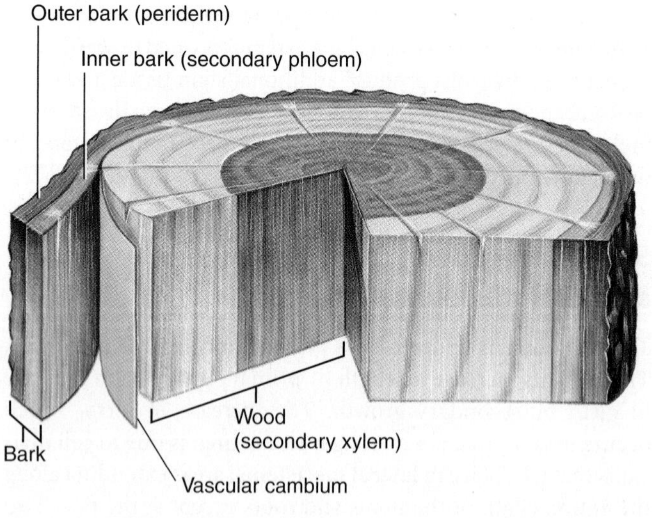

CHECKPOINT 33.2

- **CONNECT** *What is the role of plant meristems? Does animal growth involve meristems?*
- *How does secondary growth differ from primary growth?*
- **CONNECT** *How do the functions of apical and lateral meristems relate to primary and secondary growth?*

33.3 DEVELOPMENT OF FORM

LEARNING OBJECTIVES

8 Distinguish between cell division and cell expansion, and describe the roles of the preprophase band and cellulose microfibrils.

9 Describe the relationship between cell determination and cell differentiation and between pattern formation and morphogenesis.

10 Explain why the model organism *Arabidopsis* is so useful in the study of plant development.

Plants are growing, dynamic organisms. **Development** in plants encompasses all the changes that take place during the entire life of an individual. Of particular interest in developmental biology is the process by which cells specialize and organize into a complex organism. For example, how does a microscopic, unicellular zygote give rise to a multicellular embryo in a seed? (See the stages of embryonic development in a typical eudicot in Figure 37-8.) How does the embryo in the seed give rise to the leaves, stems, and roots of a *juvenile* plant, and what mechanisms underlie the transition from a nonreproductive juvenile plant to an *adult plant* capable of flowering (**FIG. 33-9**)?

Outer bark (periderm)
Inner bark (secondary phloem)
Bark
Wood (secondary xylem)
Vascular cambium

Figure 33-8 Secondary growth

The vascular cambium, a thin layer of cells sandwiched between the wood and bark, produces the secondary vascular tissues: the wood, which is secondary xylem, and the inner bark, which is secondary phloem. The cork cambium produces the periderm, the outer bark tissue that replaces the epidermis in a plant with secondary growth.

© Cengage Learning

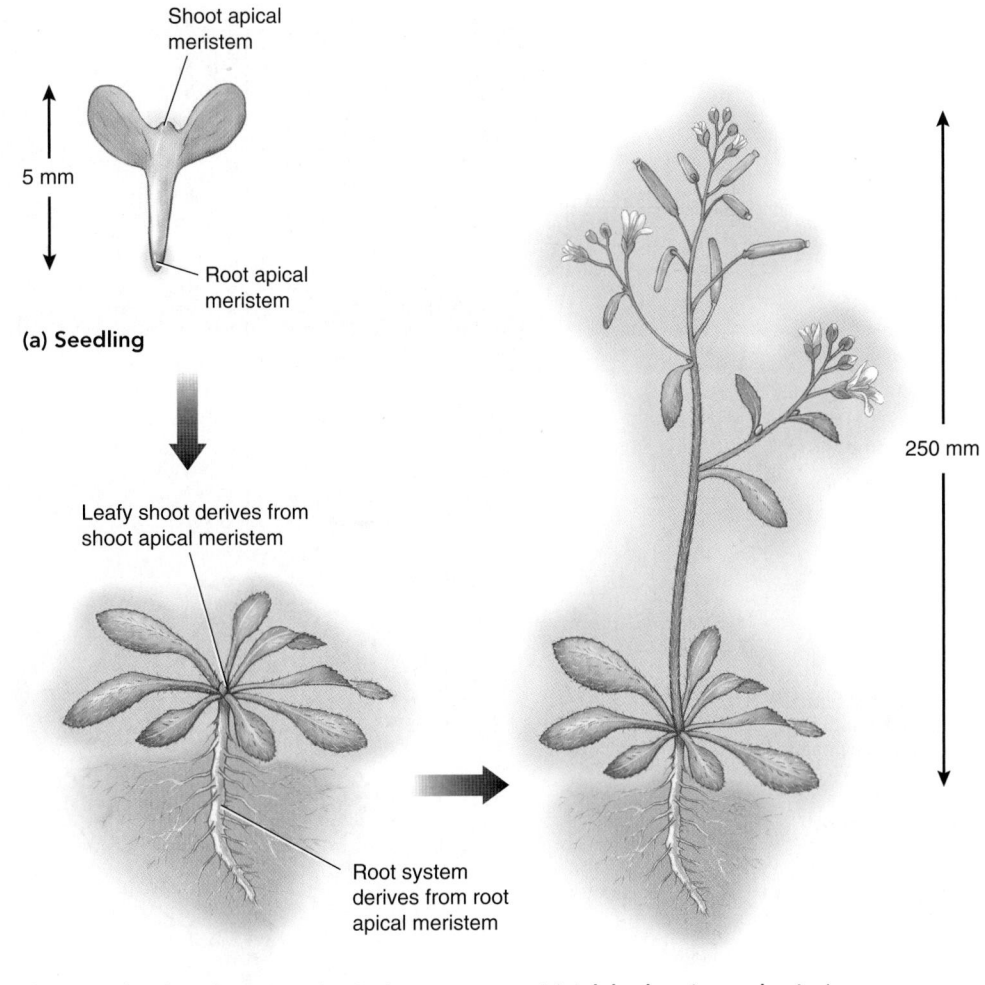

Shoot apical meristem

5 mm

Root apical meristem

(a) Seedling

Leafy shoot derives from shoot apical meristem

Root system derives from root apical meristem

250 mm

(b) Juvenile plant (nonreproductive)

(c) Adult plant (reproductive)

Figure 33-9 Growth of an *Arabidopsis* seedling into an adult plant

Almost all the differentiated tissues formed after a seed is germinated (often referred to as postembryonic growth) come from a small number of stem cells in the root and shoot apical meristems.

© Cengage Learning

We now examine several aspects of plant development that affect form in plants. Flowering is discussed elsewhere in this book (see the ABC model in Chapter 17).

The plane and symmetry of cell division affect plant form

You have seen that plant growth is localized into meristems for cell division. At these sites, the cells undergo cell division in an orderly manner. In the onion root apical meristem, for example, files of cells can be traced back to the meristem region (FIG. 33-10). These cells divided in the same plane to produce each file of cells. Cell divisions in one plane, two planes, or more help establish the tissue patterns that will be produced during development.

How is the plane of cell division for a particular meristematic cell determined? In plant cells a dense array of microtubules

forms a **preprophase band** just inside of the plasma membrane, encircling the nucleus much like a belt. (Chapter 4 contains a discussion of the dynamic nature of microtubules.) The preprophase band appears just prior to mitosis and determines the plane in which the cell will divide (FIG. 33-11). The preprophase band, which is located where the cell plate will form following mitosis and cytokinesis, also helps orient the mitotic spindle. The preprophase band is a plant characteristic; in animal cells, the orientation of the spindle determines the plane in which cells divide, and there is no preprophase band.

The orientation of cellulose microfibrils affects the direction of cell expansion

In plants growth occurs by both an increase in the number of cells and an increase in cell size. Cell division and cell expansion (also called *elongation*) are governed by a genetic program that

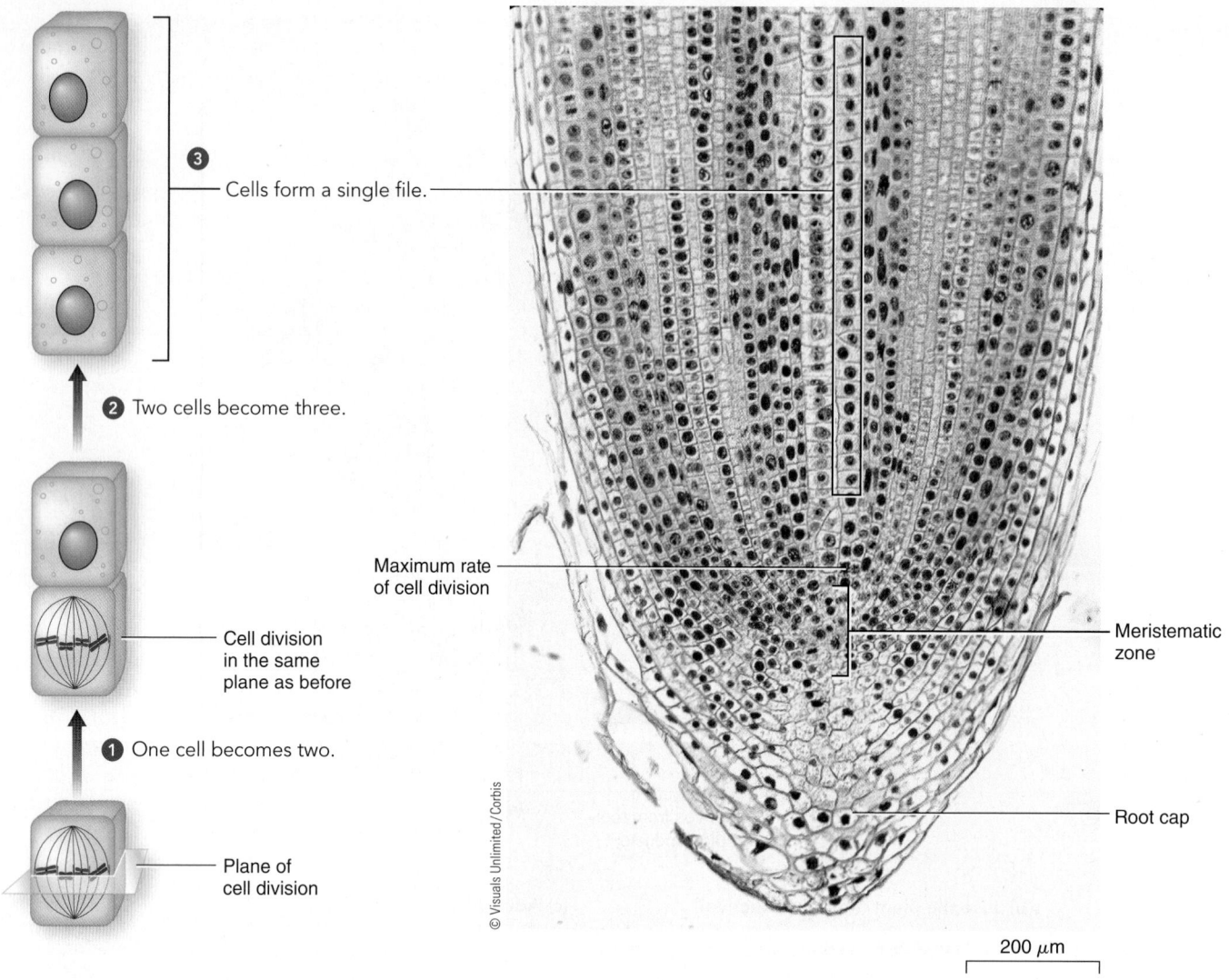

③ Cells form a single file.

② Two cells become three.

Cell division in the same plane as before

① One cell becomes two.

Plane of cell division

Cells form a single file.

Maximum rate of cell division

Meristematic zone

Root cap

© Visuals Unlimited/Corbis

200 μm

(a) Cells that divide in the same plane form a row, or file, of cells.

(b) A single file of cells in the onion root tip is labeled. If you look closely, you can see many other files.

Figure 33-10 The plane of cell division and the onion root tip
The root apical meristem maintains postembryonic root growth.
© Cengage Learning

interacts with environmental signals from both the external and internal environments.

In growth a newly formed plant cell absorbs water into its vacuole by osmosis. The water increases the cell's **turgor pressure,** hydrostatic pressure that develops within a walled cell and presses outward against the plasma membrane. Turgor pressure provides a force that exerts pressure on the cell wall and causes it to expand. In an onion root cell, the vacuole increases in size by some 30 to 150 times during elongation.

In growing tissue, the arrangement of cellulose microfibrils in the wall prevents the cell from expanding equally in all directions due to turgor pressure. When the microfibrils in the cell wall are oriented in the same direction, the wall expands perpendicularly to them (**FIG. 33-12**).

Cell differentiation depends in part on a cell's location

By themselves, cell division and cell expansion would produce only a formless heap of similar cells. The development of plant form is the result of a balanced combination of several fundamental processes that go beyond cell division and cell expansion: cell differentiation and cell determination, and pattern formation and morphogenesis.

As growth proceeds, cell division gives rise to increasing numbers of cells, which serve as the building blocks of development. At various times certain cells become biochemically and structurally specialized to carry out specific functions through a process known as **cell differentiation.** Generally,

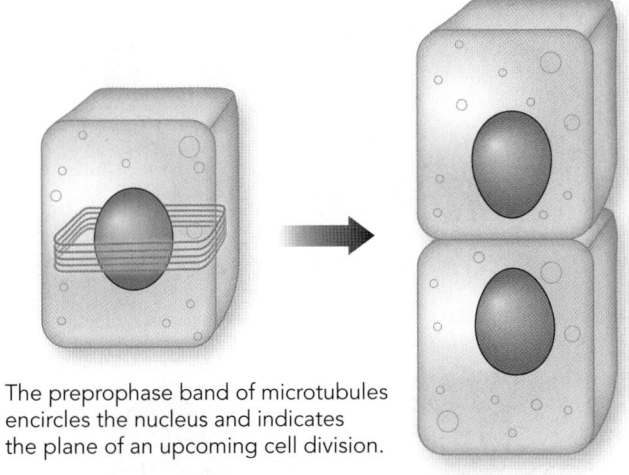

The preprophase band of microtubules encircles the nucleus and indicates the plane of an upcoming cell division.

Figure 33-11 Preprophase band
© Cengage Learning

a differentiated animal cell is irreversibly committed to its fate, but that is not the case with plants. As discussed earlier in this chapter, parenchyma cells can differentiate into other kinds of cells, for example, when the plant body is injured.

Cell differentiation occurs through **cell determination,** a series of molecular events in which the activities of certain genes are altered in ways that cause a cell to progressively commit to a particular differentiation pathway. As cell determination proceeds, a young cell's fate becomes more and more limited, restricting its development to a limited set of tissue types.

During cell differentiation, a cell's physical appearance may or may not change significantly. Nevertheless, when cell determination is complete, a cell becomes structurally and functionally recognizable—as an epidermal cell, for example—and its pattern of gene activity differs from that of other cell types. Thus, differentiated cells of the plant body

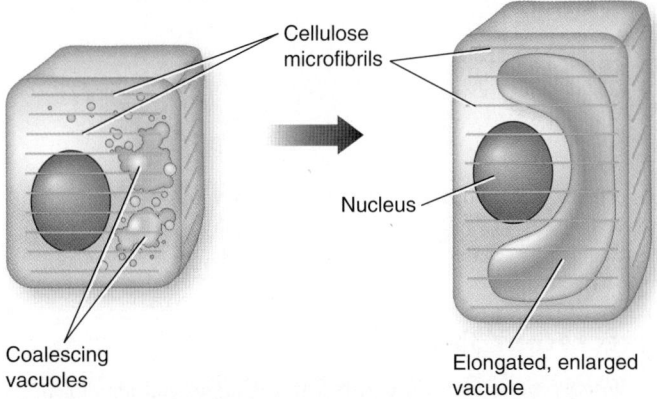

Figure 33-12 Direction of cell expansion

Following cell division, cellulose microfibrils are deposited into new cell walls in an orderly fashion that permits cell expansion. The orientation of the cellulose microfibrils determines the direction of cell expansion. Note that most of the cellulose microfilbrils are oriented transversely in this figure, which results in longitudinal cell expansion.

© Cengage Learning

have **differential gene expression;** that is, different genes are active in different cells at the same time. Differential gene expression is responsible for variations in chemistry, behavior, and structure among cells.

Tissue cultures involve differential gene expression Cells can be isolated from certain plants and grown in a chemically defined, sterile nutrient medium. In initial experiments with such cultures, plant cells could be kept alive, but they did not divide. Researchers later discovered that adding certain natural materials, such as the liquid endosperm of coconut, also known as coconut milk, induced cells to divide in culture. By the late 1950s, plant cells from a variety of sources could be cultured successfully, dividing to produce a mass of disorganized, relatively undifferentiated cells, or **callus.**

In 1958, F.C. Steward, a plant physiologist at Cornell University, succeeded in generating an entire carrot plant from a single callus cell derived from a carrot root (see Fig. 17-2), thus demonstrating that each plant cell contains a genetic blueprint for all features of an entire organism. His work also showed that an entire plant can be grown from a single cell, provided differential gene expression occurs so that the proper genes are expressed at the appropriate times.

Since Steward's pioneering work, biologists have successfully cultured many plants by using a variety of cell sources. Plants have been regenerated from different tissues, organ explants (excised organs or parts such as root apical meristems), and single cells. Cell and tissue culture techniques help answer many fundamental questions involving growth and development in plants. These techniques also have great practical potential. Using tissue culture, researchers can regenerate many genetically identical plants from the cells of a single, genetically superior plant. Many kinds of plants—from orchids and African violets to coastal redwoods—have been cultured in this way.

Morphogenesis occurs through pattern formation

Cell differentiation by itself does not explain overall plant form. The differentiated cells must become progressively organized, shaping the intricate pattern of tissues and organs that characterizes a multicellular plant. This development of form, in which differentiated cells in specific locations become spatially organized into recognizable structures, is known as **morphogenesis.**

Morphogenesis proceeds through the process of **pattern formation,** a series of steps requiring signaling between cells, changes in the shapes and metabolism of certain cells, and precise cell interactions. Pattern formation organizes cells into three-dimensional structures (**FIG. 33-13**). Depending on their location, cells are exposed to different concentrations of signaling molecules that specify **positional information,** that is, where the cell is located relative to the body's axes. This information affects cell differentiation and tissue formation. Thus, *where* a given cell is located often determines *what* it will become when it matures.

(a) Examples of leaf shapes.

(b) Examples of leaf margins.

Figure 33-13 Leaf variation and pattern formation

Leaves are quite variable in form; there are dozens of different leaf shapes and margins. Leaves on juvenile plants are often shaped differently than those on adult plants, and some adult plants have several leaf shapes on the same individual. Pattern formation, which is under genetic control, is largely responsible for leaf variation. (Also see Figure 34-2 for more examples of variation in leaves.)

© Cengage Learning

***Arabidopsis* mutants provide crucial insights into plant development** Until the late 1970s, biologists knew little about how genes interact to control development. However, rapid progress in molecular biology led scientists to search for developmental mutants and to apply sophisticated techniques to study them. (A *mutant* is an individual with an abnormal phenotype caused by a gene mutation.) Biologists are now studying how genes are activated, inactivated, and modified to control development. Eventually, scientists expect to reach a major goal of systems biology: understanding how a single cell (a fertilized egg) develops into a multicellular plant.

You may recall from Chapter 17 that mouse-ear cress (*Arabidopsis thaliana*) is the organism most widely used to study the genetic control of development in plants. When biologists insert cloned foreign genes into *Arabidopsis* cells, the genes integrate into the chromosomes and are expressed. The researchers then induce these transformed cells to differentiate into transgenic plants.

Botanists use chemical mutagens to produce mutant strains and have isolated many developmental mutants of *Arabidopsis*. Consider the *monopteros* (*mp*) gene, which codes for a transcription factor affecting apical-basal pattern formation and the development of vascular tissues. Mutant *mp* seedlings lack basal structures, including a primary root (**FIG. 33-14**). Such mutant phenotypes have helped scientists identify the genes essential for normal development.

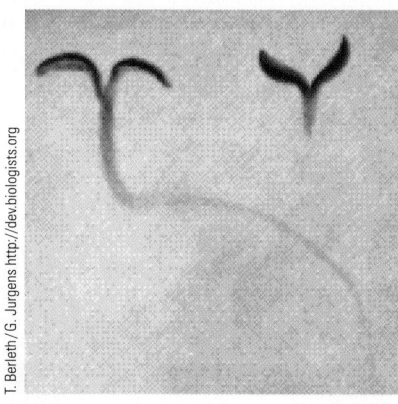

T. Berleth / G. Jürgens http://dev.biologists.org

Figure 33-14 *Monopteros* mutation in *Arabidopsis* seedling

The *monopteros* gene affects development of the root system. The wild type seedling (*left*) has a normal *monopteros* gene, whereas the mutant (*right*) lacks basal structures.

CHECKPOINT 33.3

- *Which process typically occurs first, cell differentiation or cell expansion?*
- *What are some kinds of events that contribute to pattern formation?*
- **CONNECT** *What is the relationship between pattern formation and morphogenesis?*
- *Why is* Arabidopsis *such a useful model organism for plant development?*

33.1 The Plant Body *(page 705)*

1 Discuss the functions of various parts of the vascular plant body, including the nutrient- and water-absorbing root system and the photosynthesizing shoot system.

- The vascular plant body typically consists of a root system and a shoot system. The **root system** is generally underground and obtains water and dissolved minerals for the plant. Roots also anchor the plant firmly in place.

- The **shoot system** is generally aerial and obtains sunlight and exchanges gases such as CO_2, oxygen, and water vapor. The shoot system consists of a vertical stem that bears leaves and reproductive structures. **Buds,** undeveloped embryonic shoots, develop on stems.

2 Describe the structure and functions of the ground tissue system (parenchyma tissue, collenchyma tissue, and sclerenchyma tissue).

- The **ground tissue system** consists of three tissues with a variety of functions. **Parenchyma** tissue is composed of living parenchyma cells that have thin *primary cell walls.* Functions of parenchyma tissue include photosynthesis, storage, and secretion.

- **Collenchyma** tissue consists of collenchyma cells with unevenly thickened primary cell walls. This tissue provides flexible structural support.

- **Sclerenchyma** tissue is composed of sclerenchyma cells—**sclereids** or **fibers**—that have both *primary cell walls* and *secondary cell walls.* Sclerenchyma cells are often dead at maturity but provide structural support.

3 Describe the structure and functions of the vascular tissue system (xylem and phloem).

- The **vascular tissue system** conducts materials throughout the plant body and provides strength and support.

- **Xylem** is a complex tissue that conducts water and dissolved minerals. The actual conducting cells of xylem are **tracheids** and **vessel elements.**

- **Phloem** is a complex tissue that conducts sugar in solution. **Sieve tube elements** are the conducting cells of phloem; they are assisted by **companion cells.**

4 Describe the structure and functions of the dermal tissue system (epidermis and periderm).

- The **dermal tissue system** is the outer protective covering of the plant body.

- The **epidermis** is a complex tissue that covers the herbaceous plant body. The epidermis that covers aerial parts secretes a waxy **cuticle** that reduces water loss. **Stomata** permit gas exchange between the interior of the shoot system and the surrounding atmosphere. **Trichomes** are outgrowths, or hairs, that occur in many sizes and shapes and have a variety of functions.

- The **periderm** is a complex tissue that covers the plant body in woody plants.

33.2 Plant Meristems *(page 714)*

5 State how growth in plants differs from growth in animals.

- Growth in plants, unlike that in animals, is localized in specific regions called **meristems.** Meristems are composed of cells whose primary function is to form new cells by mitotic division.

6 Distinguish between primary and secondary growth.

- **Primary growth** is an increase in stem or root length. Primary growth occurs in all plants.

- **Secondary growth** is an increase in stem or root girth (thickness). Secondary growth typically occurs in long cylinders of meristematic cells throughout the length of older stems and roots.

7 Distinguish between apical meristems and lateral meristems.

- Primary growth results from the activity of **apical meristems** that are localized at the tips of roots and shoots and within the buds of stems.

- The two **lateral meristems** responsible for secondary growth are the **vascular cambium** and the **cork cambium.**

33.3 Development of Form *(page 716)*

8 Distinguish between cell division and cell expansion, and describe the roles of the preprophase band and cellulose microfibrils.

- Growth occurs by both cell division (an increase in the number of cells) and cell expansion (an increase in cell size).

- In plant cells a dense array of microtubules forms a **preprophase band** just inside of the plasma membrane. The preprophase band appears just prior to mitosis and determines the plane in which the cell will divide.

- Absorption of water into a cell's vacuole increases the **turgor pressure,** which exerts pressure on the cell wall. The arrangement of cellulose microfibrils in the wall prevents the wall from expanding equally in all directions. When microfibrils are oriented the same way, the wall expands perpendicular to the microfibrillar orientation.

9 Describe the relationship between cell determination and cell differentiation and between pattern formation and morphogenesis.

- In **cell differentiation** cells become biochemically and structurally specialized to carry out specific functions. Cell differentiation occurs through **cell determination,** a series of molecular events in which the activities of certain genes are altered in ways that cause a cell to progressively commit to a particular differentiation pathway.

- The development of form, in which differentiated cells in specific locations become spatially organized into recognizable structures, is known as **morphogenesis.** Morphogenesis proceeds through the process of **pattern formation,** a series of steps requiring signaling between cells, changes in the shapes and metabolism of certain cells, and precise cell interactions.

10 Explain why the model organism *Arabidopsis* is so useful in the study of plant development.

- Botanists have isolated many developmental mutants of *Arabidopsis.* Such mutant phenotypes have helped scientists identify genes essential for normal development.

Know and Comprehend

1. Most of the plant body consists of the _____ tissue system. (a) ground (b) vascular (c) periderm (d) dermal (e) cortex

2. The cell walls of parenchyma cells (a) contain large quantities of pectin in the thickened corners (b) are rich in lignin but do not contain hemicelluloses and pectin (c) are predominantly cellulose, although they also contain hemicelluloses and pectin (d) contain cellulose, hemicelluloses, and lignin in approximately equal amounts (e) contain hemicelluloses, pectin, and lignin but no cellulose

3. Which tissue system provides a covering for the plant body? (a) ground (b) vascular (c) periderm (d) dermal (e) cortex

4. The two simple tissues that are specialized for support are (a) parenchyma and collenchyma (b) collenchyma and sclerenchyma (c) sclerenchyma and parenchyma (d) parenchyma and xylem (e) xylem and phloem

5. Sclereids and fibers are examples of which plant tissue? (a) parenchyma (b) collenchyma (c) sclerenchyma (d) xylem (e) epidermis

6. Which of the following statements about the vascular tissue system is *not* true? (a) xylem and phloem are continuous throughout the plant body (b) xylem not only conducts water and dissolved minerals but also provides support (c) four cell types occur in phloem: sieve tube elements, companion cells, tracheids, and vessel elements (d) sieve tube elements lack nuclei (e) vessel elements are hollow, and their end walls have perforations or are entirely dissolved away

7. Minute pores known as _____ dot the surface of the epidermis of leaves and stems; each pore is bordered by two _____. (a) stomata; guard cells (b) stomata; fibers (c) sieve tube elements; companion cells (d) sclereids; guard cells (e) cuticle; guard cells

8. Primary growth, an increase in the length of a plant, occurs at the (a) cork cambium (b) apical meristem (c) vascular cambium (d) lateral meristem (e) periderm

9. The preprophase band (a) appears just prior to mitosis in both plant and animal cells (b) determines the plane in which the cell will divide (c) attaches to cellulose microfibrils in the cell wall (d) is an essential part of pattern formation in plants (e) increases a cell's turgor pressure

10. Cell differentiation occurs through _____, whereas morphogenesis involves _____. (a) pattern formation; cell determination (b) cell division; cell expansion (c) cell expansion; pattern formation (d) cell determination; cell division (e) cell determination; pattern formation

11. The *monopteros* mutant (a) lacks a primary root (b) develops normally until the plant becomes reproductive (c) has abnormal leaf morphogenesis (d) was discovered in rice after its genome was sequenced (e) is one of only two developmental mutants known in *Arabidopsis*.

Apply and Analyze

12. Label the three tissue systems in this leaf cross section.

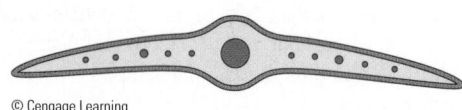

© Cengage Learning

13. **VISUALIZE** Sketch a roughly cuboidal cell preparing to divide. Indicate the orientation of the preprophase band and the site where the new cell walls of the daughter cells will form.

14. **VISUALIZE** Sketch a cell preparing to expand. Indicate the orientation of the cellulose microfibrils and use a double-headed arrow to indicate the direction of subsequent cell elongation.

Evaluate and Synthesize

15. A couple carved a heart with their initials into a tree trunk 4 ft above ground level; the tree was 25 ft tall at the time. Twenty years later the tree was 50 ft tall. How far above the ground were the initials then? Explain your answer.

16. Sclerenchyma in plants is the functional equivalent of bone in humans (both sclerenchyma and bone provide support). However, sclerenchyma is dead, whereas bone is living tissue. What are some of the advantages of a plant having dead support cells? Can you think of any disadvantages? Explain your answer.

17. **EVOLUTION LINK** Flowering plants have both tracheids and vessel elements in their xylem. Conifers, which evolved earlier than flowering plants, have tracheids only. Do these differences in xylem structure help explain the difference in the success of these two plant groups? Explain your answer.

18. **SCIENCE, TECHNOLOGY, AND SOCIETY** Why is knowledge gained from sequencing the *Arabidopsis* genome important even though *Arabidopsis* has no commercial value?

To access course materials, such as Aplia and other aplia companion resources, please visit **www.cengagebrain.com**.

Leaf Structure and Function | 34

Plants allocate many resources into the production of leaves that are often short-lived. Each year, a large maple tree, for example, may produce 47 m² (500 ft²) of leaves, which may weigh more than 113 kg (250 lb). The metabolic cost of producing so many leaves is high, but necessary. Simply put, leaves are essential to the tree's survival, so plants must invest in their production.

In most vascular plants, leaves are the primary organs to effectively collect solar energy and convert it into chemical energy. That is, leaves gather the sunlight necessary for **photosynthesis,** the biological process that converts radiant energy into the chemical energy of carbohydrate molecules. Plants use these molecules as starting materials to synthesize all other organic compounds and as fuel to provide energy for metabolism. During a single summer, the leaves of a maple tree will fix about 454 kg (1000 lb) of carbon dioxide (CO_2) into organic compounds.

The structure of a leaf is superbly adapted for its primary function of photosynthesis. Most leaves are thin and flat, a shape that allows optimal absorption of light energy and the efficient internal diffusion of gases such as CO_2 and O_2. As a result of their ordered arrangement on the stem, leaves efficiently catch the sun's rays with a minimum of "interference" from neighboring leaves. The leaves form an intricate green mosaic, bathed in sunlight and atmospheric gases.

To control water loss, a thin, transparent layer of wax covers the leaf surface. Such structural adaptations are compromises between competing needs, and some features that optimize photosynthesis promote water loss. For example, plants have minute pores that allow gas exchange for photosynthesis, but these openings also let water vapor escape into the atmosphere. Thus, leaf structure represents a trade-off between photosynthesis and water conservation.

In this chapter we examine leaf form and function. Leaf form is highly variable, and thousands of different leaf shapes have evolved over millions of years. Despite this diversity, most leaves "work" in the same way, and their internal structures (see photograph) support this similarity in function.

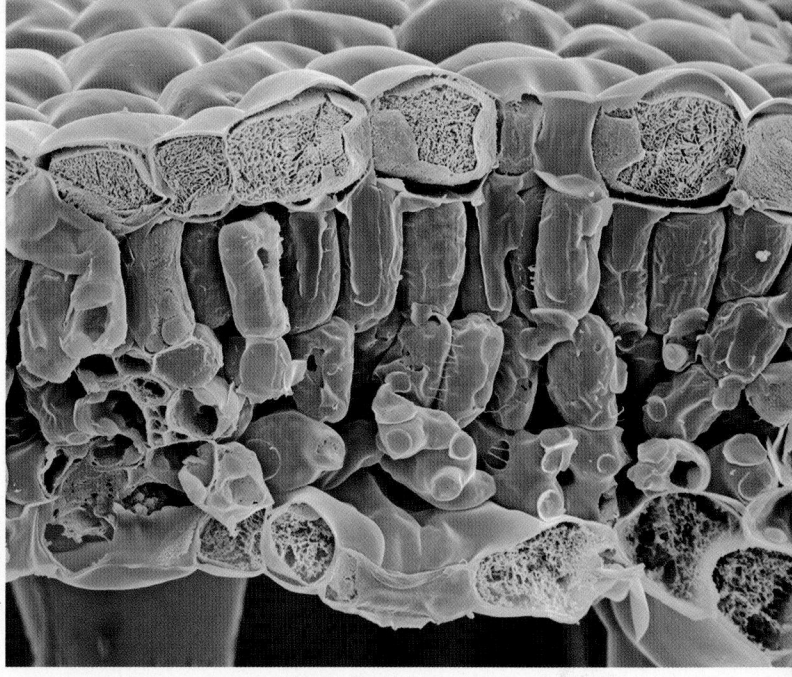

Peter Bond/Science Source

Colorized SEM of the interior of a stinging nettle leaf. The loose arrangement of internal cells in stinging nettle (*Urtica dioica*) optimizes the diffusion of necessary raw materials such as carbon dioxide and water into photosynthetic cells.

KEY CONCEPTS

34.1 Leaf structure reflects its primary function of photosynthesis.

34.2 Opening and closing of stomata affect carbon dioxide availability in the daily cycle of photosynthesis.

34.3 Transpiration, the loss of water vapor from leaf surfaces, keeps leaves from becoming overheated and promotes water transport in the plant. Guttation is the exudation of liquid water from leaf margins.

34.4 Leaf abscission is the seasonal removal of deciduous leaves.

34.5 Some leaves are modified for special functions in addition to photosynthesis and transpiration.

34.1 LEAF FORM AND STRUCTURE

LEARNING OBJECTIVES

1 Discuss variation in leaf form, including simple versus compound leaves, leaf arrangement on the stem, and venation patterns.
2 Describe the major tissues of the leaf (epidermis, photosynthetic ground tissue, xylem, and phloem) and sketch how they are arranged in a leaf cross section.
3 Compare leaf anatomy in eudicots and monocots.
4 Relate leaf structure to its function of photosynthesis.

Foliage leaves are the most variable of plant organs, so much so that plant biologists have developed specific terminology to describe their shapes, margins (edges), vein patterns, and the way they attach to stems. Because each leaf is characteristic of the species on which it grows, many plants can be identified by their leaves alone. Leaves may be round, needlelike, scalelike, cylindrical, heart-shaped, fan-shaped, or thin and narrow. They vary in size from those of the raffia palm (*Raphia ruffia*), whose leaves often grow more than 20 m (65 ft) long, to those of water-meal (*Wolffia*), whose leaves are so small that 16 of them laid end to end measure only 2.5 cm (1 in.).

The broad, flat portion of a leaf is the **blade;** the stalk that attaches the blade to the stem is the **petiole.** Some leaves also have **stipules,** which are leaflike outgrowths usually present in pairs at the base of the petiole (FIG. 34-1). Some leaves do not have petioles (i.e., the blades are attached directly to the stem) or stipules; leaves without a petiole are said to be *sessile.*

Leaves may be *simple* (having a single blade) or *compound* (having a blade divided into two or more leaflets) (FIG. 34-2a). Sometimes it is difficult to tell whether a plant has formed one compound leaf or a small stem bearing several simple leaves. One easy way to determine if a plant has simple or compound leaves is to look for axillary buds, so-called because each develops in a leaf *axil* (the angle between the stem and petiole). Axillary buds form at the base of a leaf, whether it is simple or compound. However, axillary buds never develop at the base of leaflets. Also, the leaflets of a compound leaf lie in a single plane (you can lay a compound leaf flat on a table), whereas simple leaves usually are not arranged in one plane on a stem.

Leaves are arranged on a stem in one of three possible ways (FIG. 34-2b). Plants such as beeches and walnuts have an *alternate leaf arrangement,* with one leaf at each **node,** the area of the stem where one or more leaves are attached. In an *opposite leaf arrangement,* as occurs in maples and ashes, two leaves grow at each node. In a *whorled leaf arrangement,* as in catalpa trees, three or more leaves grow at each node.

Leaf blades may possess *parallel venation,* in which the primary **veins**—strands of vascular tissue—run approximately parallel to one another (generally characteristic of monocots), or *netted venation,* in which veins are branched in such a way that they resemble a net (generally characteristic of eudicots;

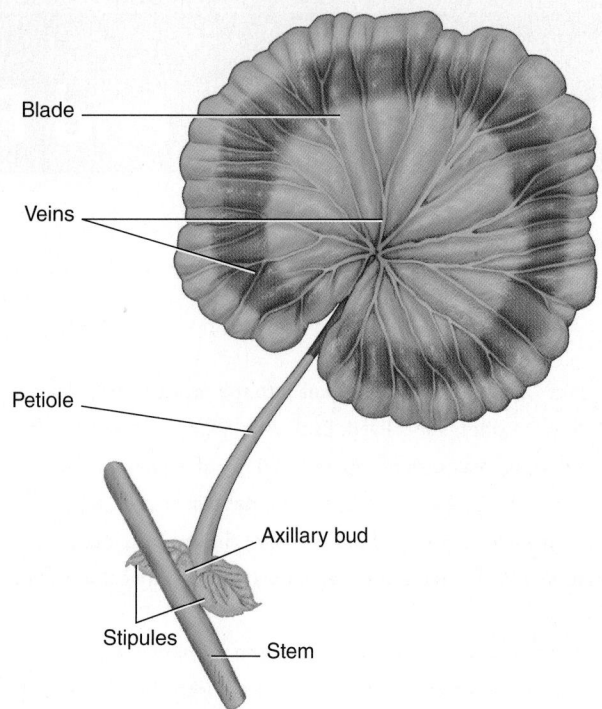

Figure 34-1 Parts of a leaf

A geranium leaf consists of a blade, a petiole, and two stipules at the base of the leaf. Note the axillary bud in the leaf axil.

© Cengage Learning

FIG. 34-2c).[1] Netted veins can be *pinnately netted,* with major veins branching off in succession along the entire length of the **midvein** (main or central vein of a leaf), or *palmately netted,* with several major veins radiating out from one point.

Plant scientists use sophisticated molecular techniques to study the genetic basis of various aspects of plant growth and development, including leaf form. Many of the genes under study code for transcription factors. As discussed in Chapter 14, *transcription factors* are DNA-binding proteins that regulate the synthesis of RNA from a DNA template. An exciting example of this type of research is described in FIGURE 34-3. Eventually, such research will help plant biologists to understand the roles of all genes involved in plant development, an important goal of systems biology.

Leaf structure is adapted for maximum light absorption

The leaf is a complex organ composed of several tissues organized to optimize photosynthesis (FIG. 34-4). The leaf blade has upper and lower surfaces, each consisting of a layer of **epidermis.** The *upper epidermis* covers the upper surface, and the *lower epidermis* covers the lower surface. Most cells in these layers lack chloroplasts and are relatively transparent. One interesting feature of leaf epidermal cells is that the cell wall facing toward the outside environment is somewhat thicker than the cell wall facing inward. This extra thickness may provide the plant with additional protection against injury or water loss.

[1] Recall that flowering plants, the focus of this chapter, are divided into two main groups, informally called *eudicots* and *monocots* (see Chapter 28). Examples of eudicots include beans, petunias, oaks, cherry trees, roses, and snapdragons; monocots include corn, lilies, grasses, palms, tulips, orchids, and bananas.

Simple

California white oak
(*Quercus lobata*)

Pinnately compound

White ash
(*Fraxinus americana*)

Palmately compound

Ohio buckeye
(*Aesculus glabra*)

(a) Leaf form: simple and compound.

Alternate

American beech
(*Fagus grandifolia*)

Opposite

Sugar maple
(*Acer saccharum*)

Whorled

Southern catalpa
(*Catalpa bignonioides*)

(b) Leaf arrangement on a stem.

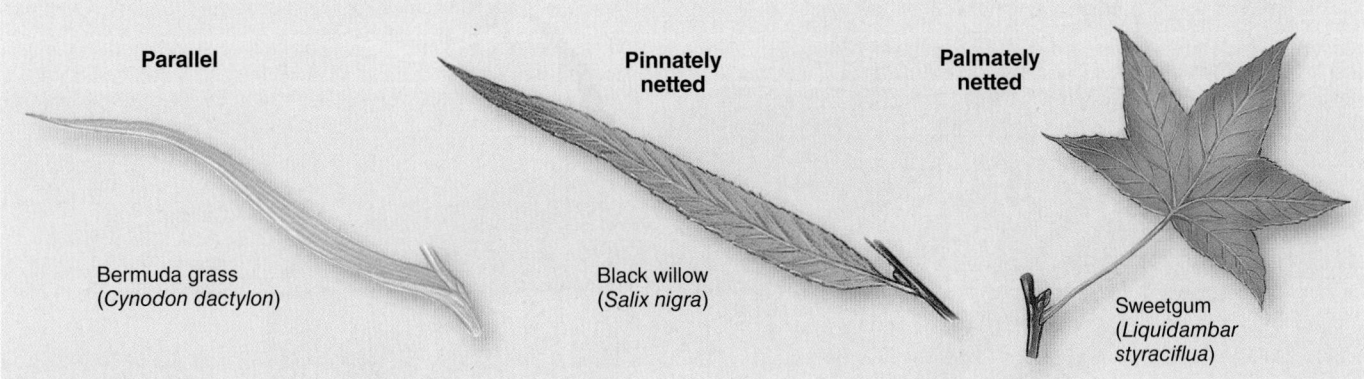

Parallel

Bermuda grass
(*Cynodon dactylon*)

Pinnately netted

Black willow
(*Salix nigra*)

Palmately netted

Sweetgum
(*Liquidambar
styraciflua*)

(c) Venation patterns.

Figure 34-2 Leaf morphology

All leaves shown are woody eudicot trees from North America, except Bermuda grass, which is a herbaceous monocot native to Europe and Asia.
© Cengage Learning

Because leaves have such a large surface area exposed to the atmosphere, water loss by evaporation from the leaf's surface is unavoidable. However, epidermal cells secrete a waxy layer, the **cuticle,** that reduces water loss from their exterior walls (see Table 33-4). The cuticle, which consists primarily of the waxy substance *cutin,* varies in thickness in different plants, in part as a result of environmental conditions. As one might expect, the leaves of plants adapted to hot, dry climates have extremely thick cuticles. Furthermore, a leaf's exposed (and warmer) upper epidermis generally has a thicker cuticle than its shaded (and cooler) lower epidermis.

The epidermis of many leaves is covered with various hairlike structures called **trichomes,** which have several functions (see Table 33-4). Trichomes of some plants help reduce water loss from the leaf surface by retaining a layer of moist air next to the leaf and by reflecting sunlight, thereby protecting the plant from overheating. Some trichomes secrete stinging irritants for deterring *herbivores,* animals that feed on plants. In addition, a leaf covered with trichomes is difficult for an insect to walk over or eat. Other trichomes excrete excess salts absorbed from a salty soil.

The leaf epidermis contains minute openings, or **stomata** (sing., *stoma*), for gas exchange between leaf cells and the environment; the stomata are evenly spaced to optimize this gas exchange. Each stoma is flanked by two specialized epidermal **guard cells,** which are responsible for opening and closing the stoma (FIG. 34-5). Guard cells are usually the only epidermal cells with chloroplasts.

What molecular controls determine if a developing leaf will become simple or compound?

EXPERIMENT: *NAM/CUC* (NO APICAL MERISTEM/CUP-SHAPED COTYLEDON) genes were cloned and genetically manipulated in four distantly related flowering plants that normally produce compound leaves: *Aquilegia caerulea* (columbine), *Solanum lycopersicum* (tomato), *Cardamine hirsuta* (hairy bittercress), and *Pisum sativum* (garden pea). *NAM/CUC* genes were already known to code for plant transcription factors involved in the establishment and function of the shoot apical meristem.

RESULTS AND CONCLUSION: European scientists Thomas Blein, Patrick Laufs, and colleagues reported that the *timing* of gene silencing of *NAM/CUC* genes had the potential to produce less elaborate, simpler leaf structures. If a young leaf at the shoot apical meristem of a columbine plant had already developed deeply lobed leaflets, silencing the *NAM/CUC* genes would not affect its further development (leaf 1). The leaf margins of leaves 2 and 3 developed with fewer lobes because the *NAM/CUC* genes had been silenced before much leaf development had taken place. In leaf 4, very early silencing resulted in a simple leaf with an entire margin. Similar leaf simplifications occurred due to the timing of silencing of the *NAM/CUC* genes in the developing leaves of tomato, hairy bittercress, and garden pea.

SOURCE: Blein, T., et al. "A conserved molecular framework for compound leaf development." *Science*, Vol. 322, Dec. 19, 2008. (Photo is taken from Figure 2 on page 1837.)

Figure 34-3 Molecular controls and the development of leaf form

CONNECT Studies by other researchers have provided evidence that genes other than the ones described here also influence leaf form. Based on what you have learned in Chapters 14 and 17, would you expect them all to code for transcription factors? Explain your answer.

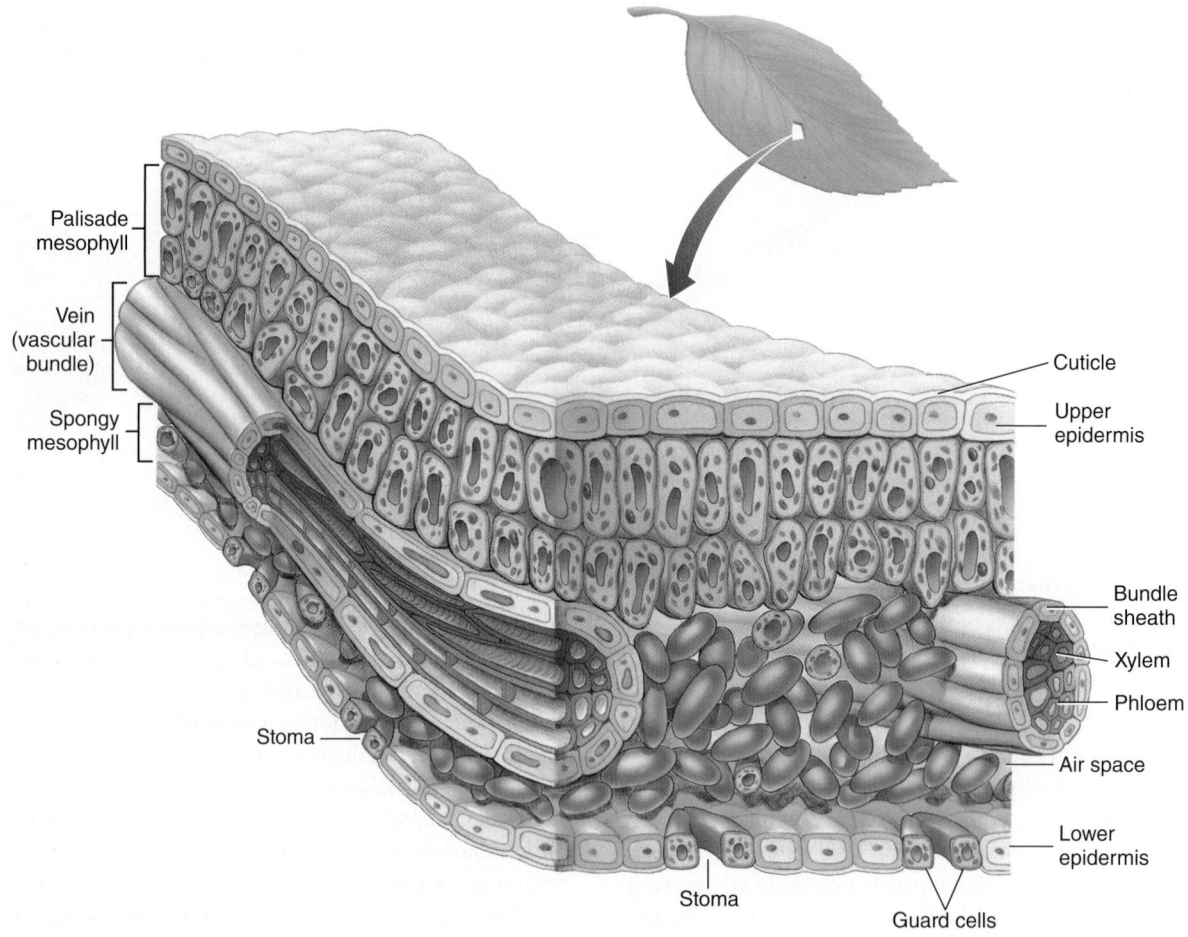

Figure 34-4 *Animation* The three plant tissue systems in a typical leaf blade

The dermal tissue system is represented by the upper epidermis and lower epidermis, which cover the blade. The photosynthetic ground tissue, called *mesophyll,* is often arranged into palisade and spongy layers. The vascular tissue system is represented by the xylem and phloem in the veins, which branch throughout the mesophyll.

© Cengage Learning

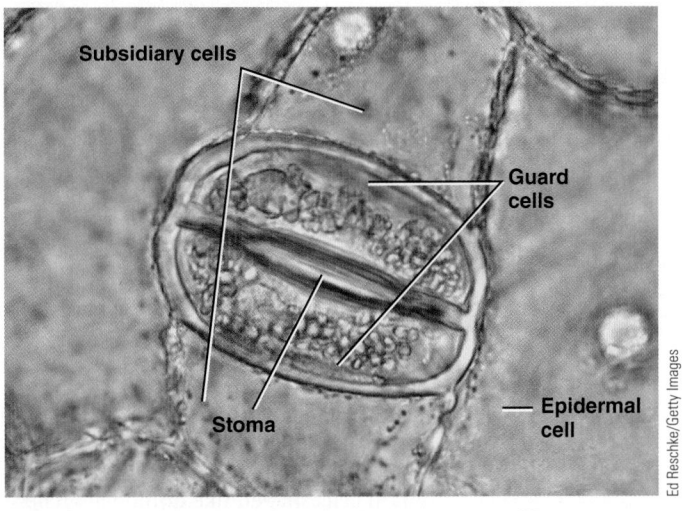

Figure 34-5 Stoma

LM of an open stoma from the leaf epidermis of a spiderwort plant (*Tradescantia virginiana*). Note the chloroplasts present in the guard cells and the thicker inner wall of each guard cell.

Stomata are especially numerous on the lower epidermis of horizontally oriented leaves (an average of about 100 stomata per square millimeter) and in many species are located *only* on the lower surface. The lower epidermis of apple (*Malus sylvestris*) leaves, for example, has almost 400 stomata per square millimeter, whereas the upper epidermis has none. This adaptation reduces water loss because stomata on the lower epidermis are shielded from direct sunlight and are therefore cooler than those on the upper epidermis. In contrast, floating leaves of aquatic plants, such as water lilies, have stomata only on the upper epidermis.

Guard cells are associated with special epidermal cells called **subsidiary cells** that are often structurally different from other epidermal cells. Subsidiary cells provide a reservoir of water and ions that move into and out of the guard cells as they change shape during stomatal opening and closing (discussed later in this chapter).

The photosynthetic ground tissue of the leaf, called the **mesophyll** (from the Greek *meso,* "the middle of," and *phyll,* "leaf"), is sandwiched between the upper epidermis and the lower epidermis. Mesophyll cells, which are parenchyma cells (see Chapter 33) packed with chloroplasts, are loosely arranged, with many air spaces between them that facilitate gas exchange. These intercellular air spaces account for as much as 70% of the leaf's volume.

In many plants the mesophyll is divided into two sublayers. Toward the upper epidermis, the columnar cells are stacked closely together in a layer called **palisade mesophyll.** In the lower portion, the cells are more loosely and irregularly arranged in a layer called **spongy mesophyll.** The two layers have different functions. Palisade mesophyll is specialized for light capture and is the main site of photosynthesis in the leaf. Photosynthesis also occurs in the spongy mesophyll, but the primary function of the spongy mesophyll is to allow diffusion of gases, particularly CO_2, within the leaf.

Palisade mesophyll may be organized into one, two, three, or even more layers of cells. The presence of additional layers is at least partly an adaptation to environmental conditions. Leaves exposed to direct sunlight contain more layers of palisade mesophyll than do shaded leaves on the same plant. In direct sunlight the light is strong enough to effectively penetrate multiple layers of palisade mesophyll, allowing all layers to photosynthesize efficiently.

The veins, or *vascular bundles,* of a leaf extend through the mesophyll. Branching is extensive, and no mesophyll cell is more than two or three cells away from a vein. Therefore, the slow process of diffusion does not limit the movement of needed resources between mesophyll cells and veins. Each vein contains two types of vascular tissue: xylem and phloem. **Xylem,** which conducts water and dissolved minerals (inorganic nutrients), is usually located in the upper part of a vein, toward the upper epidermis. **Phloem,** which conducts dissolved sugars, is usually confined to the lower part of a vein.

One or more layers of nonvascular cells surround the larger veins and make up the **bundle sheath.** Bundle sheaths are composed of parenchyma or sclerenchyma cells (see Chapter 33). Frequently, the bundle sheath has support columns, called **bundle sheath extensions,** that extend through the mesophyll from the upper epidermis to the lower epidermis (**FIG. 34-6**). Bundle sheath

Figure 34-6 Bundle sheath extension

LM of a wheat (*Triticum aestivum*) midvein in cross section. Note the bundle sheath extensions to both the upper epidermis and lower epidermis. Photographed using fluorescence microscopy.
© Cengage Learning

Upper epidermis
Bundle sheath extension
Midvein
Bundle sheath
Bundle sheath extension
Lower epidermis

25 μm

250 μm

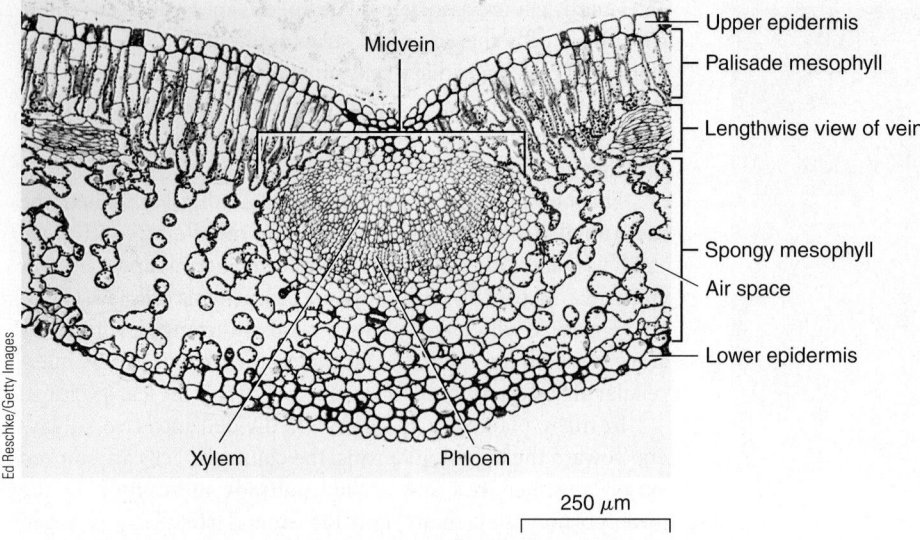

(a) Eudicot leaf. LM of part of a leaf cross section of privet (*Ligustrum vulgare*). The mesophyll has distinct palisade and spongy sections.

Labels (top figure): Midvein, Upper epidermis, Palisade mesophyll, Lengthwise view of vein, Spongy mesophyll, Air space, Lower epidermis, Xylem, Phloem

250 μm

Ed Reschke/Getty Images

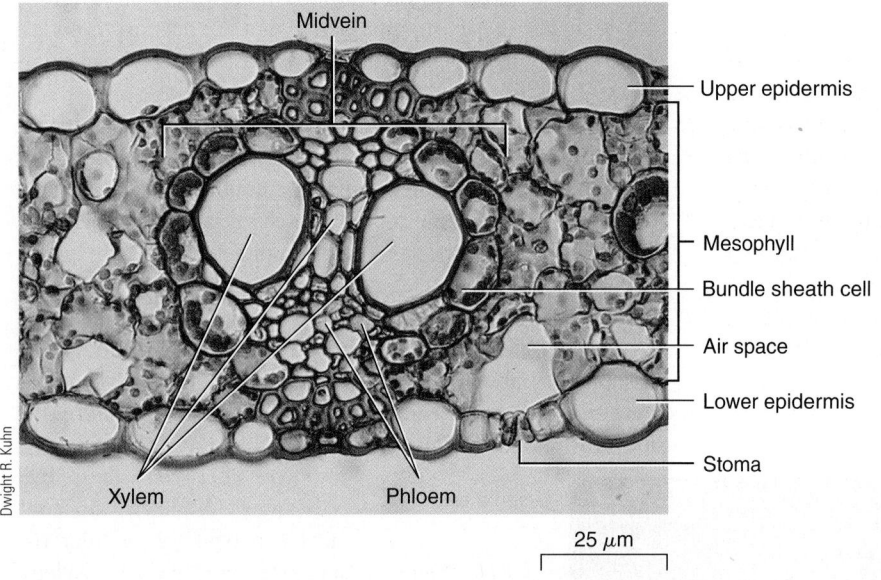

(b) Monocot leaf. LM of part of a leaf cross section of corn (*Zea mays*). Corn leaves lack distinct regions of palisade and spongy mesophyll.

Labels (bottom figure): Midvein, Upper epidermis, Mesophyll, Bundle sheath cell, Air space, Lower epidermis, Stoma, Xylem, Phloem

25 μm

Dwight R. Kuhn

Figure 34-7 Eudicot and monocot leaf cross sections

extensions may be composed of parenchyma, collenchyma, or sclerenchyma cells.

Leaf structure differs in eudicots and monocots Each eudicot leaf is usually composed of a broad, flattened blade and a petiole. As mentioned previously, eudicot leaves typically have netted venation. In contrast, many monocot leaves lack a petiole; they are narrow, and the base of the leaf often wraps around the stem to form a sheath. Parallel venation is characteristic of monocot leaves.

Eudicots and certain monocots also differ in internal leaf anatomy (FIG. 34-7). Although most eudicots and monocots have both palisade and spongy layers, some monocots (corn and other grasses) do not have mesophyll differentiated into distinct palisade and spongy layers. Because eudicots have netted veins, a cross section of a eudicot blade often shows veins in both cross-sectional and lengthwise views. In a cross section of a monocot leaf, in contrast, the parallel venation pattern produces evenly spaced veins, which all appear in cross section.

Differences between the guard cells in eudicot and certain monocot leaves also occur (FIG. 34-8). The guard cells of eudicots and many monocots are shaped like kidney beans. Other monocot leaves (those of grasses, reeds, and sedges) have guard cells shaped like dumbbells. These structural differences affect how the cells swell or shrink to open or close the stoma.

Leaf structure is related to function
How is leaf structure related to its primary function of photosynthesis? The epidermis of a leaf is relatively transparent and allows light to penetrate to the interior of the leaf where the photosynthetic ground tissue, the mesophyll, is located. Stomata, which dot the leaf surfaces, permit the exchange of gases between the atmosphere and the leaf's internal tissues. CO_2, a raw material of photosynthesis, diffuses into the leaf through stomata, and the oxygen produced during photosynthesis diffuses rapidly out of the leaf through stomata. Other gases, including air pollutants, also enter the leaf's interior via stomata.

Water required for photosynthesis is obtained from the soil and transported in the xylem to the leaf, where it diffuses into the mesophyll and moistens the surfaces of mesophyll cells. The loose arrangement of the mesophyll cells, with air spaces between cells, allows for rapid diffusion of CO_2 to the mesophyll cell surfaces; there it dissolves in a film of water before diffusing into the cells.

The veins not only supply the photosynthetic ground tissue with water and minerals (from the roots, by way of the xylem) but also carry (in the phloem) dissolved sugar produced during photosynthesis to all parts of the plant. Bundle sheaths and bundle sheath extensions associated with the veins provide additional support to prevent the leaf, which is structurally weak because of the large amount of air space in the mesophyll, from collapsing under its own weight.

Leaves are adapted to help a plant survive in its environment Leaf structure reflects the environment to which a

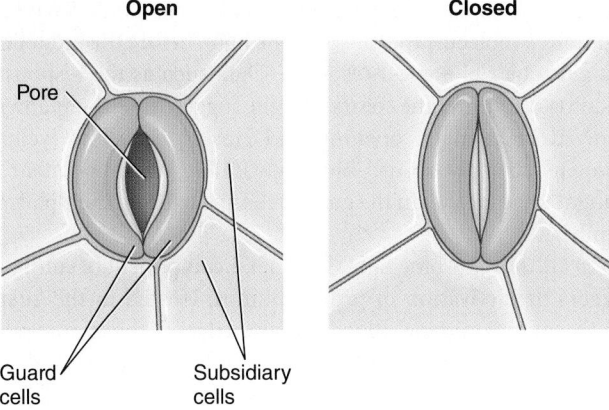

Open   **Closed**

Pore

Guard cells   Subsidiary cells

(a) Guard cells of eudicots and many monocots are bean-shaped.

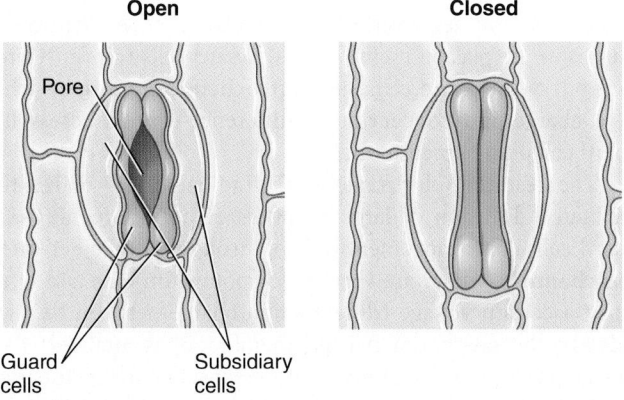

Open   **Closed**

Pore

Guard cells   Subsidiary cells

(b) Some monocot guard cells are narrow in the center and thicker at each end.

Figure 34-8 Variation in guard cells

Guard cells are associated with epidermal cells called subsidiary cells.

© Cengage Learning

particular plant is adapted. Although both aquatic plants and those adapted to dry conditions perform photosynthesis and have the same basic leaf anatomy, their leaves are modified to enable them to survive different environmental conditions. The leaves of water lilies have petioles long enough to allow the blade to float on the water's surface. Large air spaces in the mesophyll make the floating blade buoyant. The petioles and other submerged parts have an internal system of air ducts; oxygen moves through these ducts from the floating leaves to the underwater roots and stems, which live in a poorly aerated environment.

Conifers are an important group of woody trees and shrubs that includes pine, spruce, fir, redwood, and cedar.[2] Most conifers are evergreen, which means that they lose leaves throughout the year rather than during certain seasons. Conifers dominate a large portion of Earth's land area, particularly in northern forests and mountains. The leaves of most conifers are waxy needles. Their needles have structural adaptations that help them survive winter, the driest part of the year. (Winter is arid even in areas of heavy snows because roots cannot absorb water from soil when the soil temperature is very low.) Indeed, many of the structural features of needles are also found in many desert plants.

FIGURE 34-9 shows a cross section of a pine needle. Note that the needle is somewhat thickened rather than thin and bladelike. The needle's relative thickness, which results in less surface area exposed to the air, reduces water loss. Other features that help conserve water include the thick, waxy cuticle and sunken stomata; these permit gas exchange while

---

[2] As discussed in Chapter 28, conifers are gymnosperms, one of the two groups of seed plants (the other group is the flowering plants, or angiosperms). Unlike flowering plants, whose seeds are enclosed in fruits, conifers bear "naked" seeds on the scales of female cones.

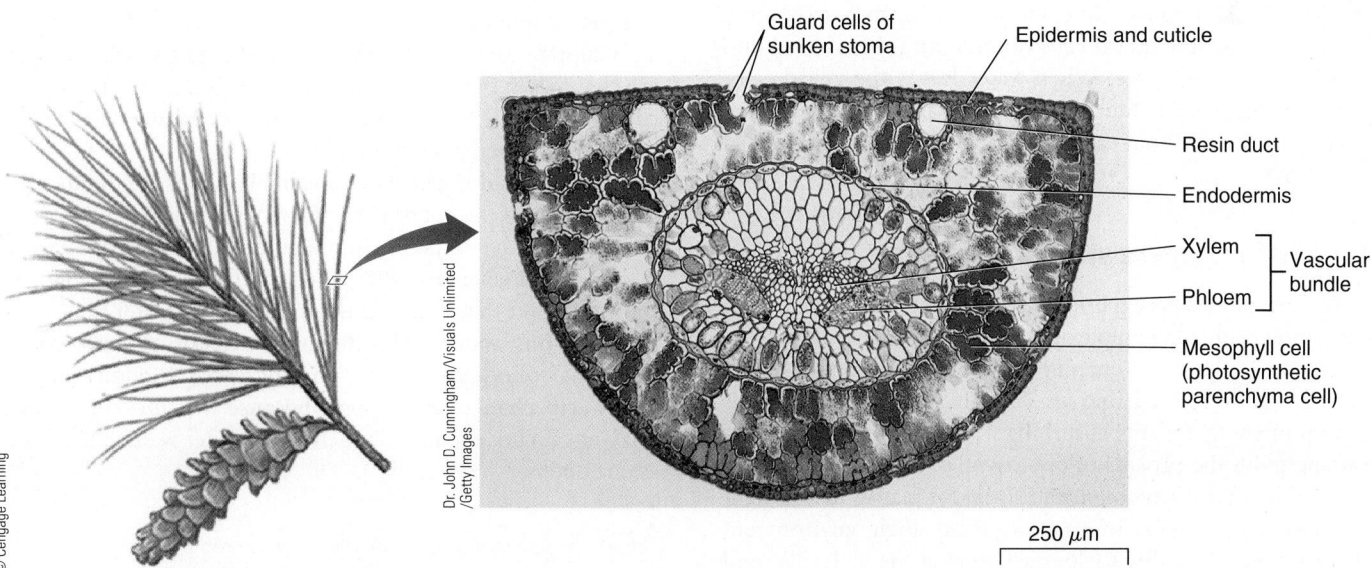

Guard cells of sunken stoma

Epidermis and cuticle

Resin duct

Endodermis

Xylem — Vascular bundle

Phloem

Mesophyll cell (photosynthetic parenchyma cell)

250 μm

Dr. John D. Cunningham/Visuals Unlimited/Getty Images

© Cengage Learning

Figure 34-9 LM of a pine needle in cross section

The thick, waxy cuticle and sunken stomata are two structural adaptations that enable pine (*Pinus*) to retain its needles throughout the winter.

minimizing water loss. Thus, needles help conifers tolerate the dry winds that occur during winter (*dry* refers here to low relative humidity). With the warming of spring, soil water again becomes available, and the needles quickly resume photosynthesis.

CHECKPOINT 34.1

- *How are leaves adapted to conserve water?*
- *What is the photosynthetic ground tissue of a leaf called? What are its two sublayers?*
- *What are the two types of vascular tissue in a vascular bundle? Which vascular tissue is usually located in the upper part of the vascular bundle?*
- *How is the leaf organized to deliver the raw materials and remove the products of photosynthesis?*

34.2 STOMATAL OPENING AND CLOSING

LEARNING OBJECTIVES

5 Explain the role of blue light in the opening of stomata.
6 Outline the physiological changes that accompany stomatal opening and closing.

Stomata are adjustable pores that are usually open during the day when CO_2 is required for photosynthesis and closed at night when photosynthesis is shut down (see the section on CAM photosynthesis in Chapter 9 for an interesting exception). The opening and closing of stomata are controlled by changes in the shape of the two guard cells that surround each pore. The guard cells' shape is determined by their rigidity. When water moves into guard cells from surrounding nonguard cells, the guard cells become turgid (swollen) and bend, producing a pore. When water leaves the guard cells, they become flaccid (limp) and collapse against one another, closing the pore. What causes water to move into and out of guard cells?

Blue light triggers stomatal opening

Data from numerous experiments and observations are beginning to explain the complexities of stomatal movements. Let us begin with stomatal opening, which occurs when the plant detects light from the rising sun. You already know that light is a form of energy; plants absorb light and convert it to chemical energy in the process of photosynthesis. However, light is also an important *environmental signal* for plants; that is, light provides plants with information about their environment that they use to modify various activities at the molecular and cellular levels.

In stomatal opening and several other plant responses, *blue light,* which has wavelengths from 400 to 500 nm, is an environmental signal. Any plant response to light must involve a **pigment,** a molecule that absorbs the light, before the induction of a particular biological response. Data such as the responses of stomata to different colors of light suggest that the pigment involved in stomatal opening and closing is yellow (yellow pigments strongly absorb blue light). The yellow pigment is thought to be located in the guard cells, probably in their plasma membranes.

In **FIGURE 34-10** blue light, which is a component of sunlight, triggers the activation of proton pumps, located in the guard cell plasma membrane. Blue light also triggers the synthesis of malic acid and the hydrolysis (splitting) of starch (discussed later).

The proton pumps use ATP energy to actively transport protons (H^+) out of the guard cells. The H^+ that are pumped are formed when malic acid produced in the guard cells ionizes to form H^+ and negatively charged malate ions. As the proton pumps in the plasma membranes of guard cells transport protons out of the guard cells, an **electrochemical gradient**—that is, a charge and concentration difference—forms across the guard cell plasma membrane.

The resulting electrochemical gradient of H^+ drives the facilitated diffusion of large numbers of potassium ions *into* guard cells. This movement occurs through **voltage-activated ion channels,** which are specific for potassium ions and open when a certain voltage (difference in charge between the two sides of the guard cell plasma membrane) is attained. This movement of potassium ions has been experimentally measured by the **patch clamp technique,** in which researchers seal the tip of a micropipette to a tiny patch of membrane that contains a single ion channel. As illustrated in Figure 41-5, these scientists then measure the flow of ions through that channel between the cytoplasm and the solution in the micropipette. (Interestingly, in animals, voltage-activated ion channels are found in the plasma membranes of nerve cells and are involved in transmitting neural impulses.)

Chloride ions are also taken into the guard cells through ion channels in the guard cell plasma membrane. The negatively charged chloride and malate ions help electrically balance the positively charged potassium ions.

The potassium, chloride, and malate ions accumulate in the vacuoles of the guard cells, thus increasing the solute concentration in the vacuoles. You may recall from the discussion of osmosis in Chapter 5 that when a cell has a solute concentration greater than that of surrounding cells, water flows *into* the cell. Thus, water enters the guard cells from surrounding epidermal cells by osmosis. The increased turgidity of the guard cells changes their shape because the thickened inner cell walls do not expand as much as the outer walls, and so the stoma opens.

---

blue light activates proton pumps ⟶ proton pumps move H^+ out of guard cells ⟶ K^+ and Cl^- diffuse into guard cells through voltage-activated ion channels ⟶ water diffuses by osmosis into guard cells ⟶ guard cells change shape and stoma opens

The accumulation of osmotically active ions (K⁺ and Cl⁻) in the guard cells is driven by a proton (H⁺) gradient.

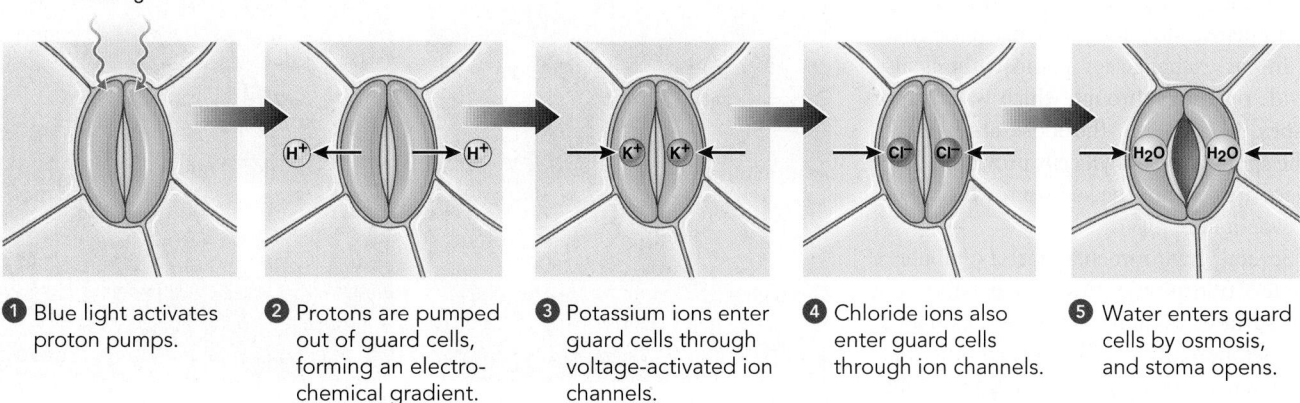

Blue light

① Blue light activates proton pumps.

② Protons are pumped out of guard cells, forming an electro-chemical gradient.

③ Potassium ions enter guard cells through voltage-activated ion channels.

④ Chloride ions also enter guard cells through ion channels.

⑤ Water enters guard cells by osmosis, and stoma opens.

Figure 34-10 Mechanism of stomatal opening

CONNECT Guard cells are usually the only epidermal cells that are photosynthetic. How does that feature relate to the mechanisms of stomatal opening and closing?

© Cengage Learning

Stomata close in the late afternoon or early evening, but not by an exact reversal of the opening process. Recent studies have demonstrated that the concentration of potassium ions in guard cells slowly decreases during the day. However, the concentration of sucrose, another osmotically active substance, increases during the day—maintaining the open pore—and then slowly decreases as evening approaches. This sucrose comes from the hydrolysis of the polysaccharide starch, which is stored in the guard cell chloroplasts. As evening approaches, the sucrose concentration in the guard cells declines as sucrose is converted back to starch (which is osmotically inactive), water leaves by osmosis, the guard cells lose their turgidity, and the pore closes.

To summarize, different mechanisms appear to regulate the opening and closing of stomata. The uptake of potassium and chloride ions is mainly associated with stomatal opening, and the declining concentration of sucrose is mainly associated with stomatal closing. In Chapter 38 we discuss other blue-light responses in plants.

Additional factors affect stomatal opening and closing

Although light and darkness trigger the opening and closing of stomata, other environmental factors are also involved, including CO_2 concentration. A low concentration of CO_2 in the leaf induces stomata to open, even in the dark. The effects of light and CO_2 concentration on stomatal opening are interrelated. Photosynthesis, which occurs in the presence of light, reduces the internal concentration of CO_2 in the leaf, triggering sto-matal opening. Another environmental factor that affects sto-matal opening and closing is water stress. During a prolonged drought, stomata remain closed, even during the day. Stomatal opening and closing are under hormonal control, particularly by the plant hormone *abscisic acid* (see Table 38-1).

The opening and closing of stomata are also regulated by an internal biological clock that in some way measures time. For example, after plants are placed in continual dark-ness, their stomata continue to open and close at more or less the same time each day. Such biological rhythms that fol-low an approximate 24-hour cycle are known as **circadian rhythms.** Other examples of circadian rhythms are provided in Chapter 38.

CHECKPOINT 34.2

- How does blue light trigger stomatal opening?
- What physiological changes occur in guard cells during stoma-tal opening? during stomatal closing?

34.3 TRANSPIRATION AND GUTTATION

LEARNING OBJECTIVES

7 Discuss transpiration and its effects on plants.

8 Distinguish between transpiration and guttation.

Despite leaf adaptations such as the cuticle, approximately 99% of the water that a plant absorbs from the soil is lost by evaporation from the leaves and, to a lesser extent, the stems. Loss of water vapor by evaporation from aerial plant parts is called **transpiration.**

The cuticle is extremely effective in reducing water loss by transpiration. It is estimated that only 1% to 3% of the water lost from a plant passes directly through the cuticle. Most transpiration occurs through open stomata. The numerous stomatal pores that are so effective in gas exchange for photosynthesis also provide openings through which water vapor escapes. In addition, the loose arrangement of the spongy mesophyll cells provides a large surface area within the leaf from which water can evaporate.

Several environmental factors influence the rate of transpiration. More water is lost from plant surfaces at higher temperatures than at lower temperatures. Light increases the transpiration rate, in part because it triggers stomatal opening and in part because it increases the leaf's temperature. Wind and dry air increase transpiration, but humid air *decreases* transpiration because the air is already saturated, or nearly so, with water vapor.

Although transpiration may seem wasteful, particularly to farmers in arid lands, it is an essential process that has adaptive value. Transpiration is responsible for water movement in plants, and without it water would not reach the leaves from the soil (see discussion of the tension–cohesion model in Chapter 35). The large amount of water that plants lose by transpiration may provide some additional benefits. Transpiration, like sweating in humans, cools the leaves and stems. When water passes from a liquid state to a vapor, it absorbs a great deal of heat. When the water molecules leave the plant as water vapor, they carry this heat with them. Thus, the cooling effect of transpiration may prevent leaves from overheating, particularly in direct sunlight.

Another benefit of transpiration is that it distributes essential minerals throughout the plant. The water that a plant transpires is initially absorbed from the soil, where it is present not as pure water but as a dilute solution of dissolved mineral salts. The water and dissolved minerals are then transported in the xylem throughout the plant, including its leaves. Water moves from the plant to the atmosphere during transpiration, but minerals remain in plant tissues. Many of these minerals are required for the plant's growth. It has been suggested that transpiration enables a plant to take in sufficient water to provide enough essential minerals; it has also been suggested that plants cannot satisfy their mineral requirements if the transpiration rate is not high enough.

There is no doubt, however, that under certain circumstances excessive transpiration can be harmful to a plant. On hot summer days, plants frequently lose more water by transpiration than they can take in from the soil. Their cells experience a loss of turgor, and the plants wilt (**FIG. 34-11**). If a plant is able to recover overnight because of the combination of negligible transpiration (recall that stomata are closed) and absorption of water from the soil, the plant is said to have experienced *temporary wilting*. Most plants recover from temporary wilting with no ill effects. In cases of prolonged drought, however, the soil

(a) In the late afternoon of a hot day, the leaves have wilted because of water loss. Note that wilting helps reduce the surface area from which transpiration occurs.

(b) The next morning, water in the leaves has been replenished. Transpiration is negligible during the night, and the plants recover by absorbing water from the soil.

Figure 34-11 Temporary wilting in squash (*Cucurbita pepo*) leaves

may not contain sufficient moisture to permit recovery from wilting. A plant that cannot recover is said to be *permanently wilted* and will die.

Transpiration is an important part of the **hydrologic cycle,** in which water cycles from the ocean and land to the atmosphere and then back to the ocean and land (see Fig. 55-11). As a result of transpiration, water evaporates from leaves and stems to form clouds in the atmosphere. Thus, transpiration eventually results in precipitation. As you may expect, forest trees release substantial amounts of moisture into the air by transpiration. Researchers have determined that at least half the rain that falls in the Amazon rainforest basin is recycled again and again by transpiration and precipitation.

Some plants exude liquid water

Many leaves have special structures at vein endings through which liquid water is literally forced out. This loss of liquid water, known as **guttation,** occurs when transpiration is negligible and available soil moisture is high. Guttation typically occurs in low-growing plants at night because the stomata are closed, but water continues to move into the roots by osmosis. People sometimes mistake the early morning water droplets formed on leaf margins by guttation for dew (**FIG. 34-12**). Unlike dew, which condenses from cool night air, guttation droplets come from within the plant. (The mechanism for guttation is discussed in Chapter 35.)

CHECKPOINT 34.3

- **CONNECT** *How is leaf structure related to transpiration?*
- *How do environmental factors (sunlight, temperature, humidity, and wind) influence the rate of transpiration?*

Figure 34-12 Guttation

Shown is a compound leaf of strawberry (*Fragaria*) with water droplets formed by guttation.

- **CONNECT** *In what ways is transpiration detrimental to plants? What are its benefits?*
- *How does guttation differ from transpiration?*

34.4 LEAF ABSCISSION

LEARNING OBJECTIVE

9 Define leaf *abscission,* explain why it occurs, and describe the physiological and anatomical changes that precede it.

All trees shed leaves, a process known as **abscission.** Many conifers shed their needles in small numbers year-round. The leaves of deciduous plants turn color and abscise, or fall off, once a year as winter approaches in temperate climates or at the beginning of the dry period in tropical climates with pronounced wet and dry seasons. In temperate forests most woody plants with broad leaves shed their leaves to survive the low temperatures of winter. During winter, the plant's metabolism, including its photosynthetic machinery, slows down or halts temporarily.

Another reason for abscission is related to a plant's water requirements, which become critical during the physiological drought of winter. As mentioned previously, as the ground chills, absorption of water by the roots is inhibited. When the ground freezes, *no* absorption occurs. If the broad leaves were to stay on the plant during the winter, the plant would continue to lose water by transpiration but would be unable to replace it with water absorbed from the soil.

Leaf abscission is a complex process that involves many physiological changes, all initiated and orchestrated by changing levels of plant hormones, particularly *ethylene.* Briefly, the process is as follows. As autumn approaches, sugars, amino acids, and many essential minerals (such as nitrogen, phosphorus, and possibly potassium) are mobilized and transported from the leaves to other plant parts; these minerals will be reused during the following spring. Chlorophyll breaks down, allowing the **carotenoids** (orange carotenes and yellow xanthophylls), some of the accessory pigments in the chloroplasts of leaf cells, to become evident (see Fig. 3-14b). Accessory pigments are always present in the chloroplasts but are masked by the green of the chlorophyll. In addition, red water-soluble pigments called **anthocyanins** may accumulate in the vacuoles of epidermal leaf cells in some species; anthocyanins may protect leaves against damage by ultraviolet radiation. The various combinations of carotenoids and anthocyanins are responsible for the brilliant colors found in autumn landscapes in temperate climates.

In many leaves abscission occurs at an abscission zone near the base of the petiole

The area where a petiole detaches from the stem is a structurally distinct area called the **abscission zone.** Composed primarily of thin-walled parenchyma cells, it is anatomically weak because it contains few fibers (**FIG. 34-13**). A protective layer of cork cells develops on the stem side of the abscission zone. These cells have a waxy, waterproof material impregnated in their walls. Enzymes then dissolve the **middle lamella** (the "cement" that holds the primary cell walls of adjacent cells together) in the abscission zone (see Fig. 4-30). Once this process is completed, nothing holds the leaf to the stem but a few xylem cells. A sudden breeze is enough to make the final break, and the leaf detaches. The protective layer of cork remains, sealing off the area and forming a leaf scar.

Although the structure of abscission zones and the physiological changes associated with abscission have been well known for many years, only recently have botanists investigated the genes responsible for the development of an abscission zone and the physiological changes of leaf abscission. Using **DNA microarrays,** for example, botanists have identified about 35 genes that are expressed in leaves only as winter approaches. As shown in Figure 15-12, a DNA microarray contains hundreds of different DNA molecules placed on a glass slide or chip. Some of these genes code for degradative enzymes involved in breaking down proteins and other molecules; others have functions not yet identified.

CHECKPOINT 34.4

- *Why do many woody plants living in temperate zones lose their leaves in autumn?*
- *What physiological changes occur during leaf abscission?*
- *What is the abscission zone?*

Figure 34-13 Abscission zone

This LM of a longitudinal section through a silver maple (*Acer saccharinum*) branch shows the abscission zone at the base of a petiole. An axillary bud with its protective bud scales is also evident above the petiole.

Labels on figure: Axillary bud; Bud scales; Petiole; Abscission zone; Stem; 0.5 mm

James Mauseth, University of Texas

34.5 MODIFIED LEAVES

LEARNING OBJECTIVE

10 List at least four examples of modified leaves and give the function of each.

Although photosynthesis is the main function of leaves, certain leaves have special modifications for other functions. Some plants have leaves specialized for deterring herbivores. **Spines,** modified leaves that are hard and pointed, are found on many desert plants, such as cacti (**FIG. 34-14a**). In the cactus the main organ of photosynthesis is the stem rather than the leaf. Spines discourage animals from eating the succulent stem tissue.

Vines are climbing plants whose stems cannot support their own weight, so they often possess **tendrils** that help keep the vine attached to the structure on which it is growing. The tendrils of many vines, such as peas, cucumbers, squash, and sweet peas, are specialized leaves (**FIG. 34-14b**). However, some tendrils, such as those of ivy, Virginia creeper, and grape, are specialized stems.

The winter buds of a dormant woody plant are covered by **bud scales.** These modified leaves protect the delicate meristematic tissue of the bud from injury, freezing, or drying out (**FIG. 34-14c**).

Some leaves associated with flower clusters (inflorescences) are modified as **bracts.** In flowering dogwood the inconspicuous flowers are clustered in the center of each inflorescence, and what appears to be four white or pink petals are actually bracts.

Similarly, the red "petals" of poinsettia are not petals at all but bracts (**FIG. 34-14d**).

Leaves may also be modified for storage of water or food. For example, a **bulb** is a short, underground stem to which large, fleshy leaves are attached (**FIG. 34-14e**). Onions and tulips form bulbs. Many plants adapted to arid conditions, such as jade plants, medicinal aloes, and string-of-beads, have succulent leaves for water storage (**FIG. 34-14f**). These leaves are usually green and function in photosynthesis.

Some leaves are modified for either sexual reproduction (see discussion of flower evolution in Chapter 28, including Figure 28-15) or asexual reproduction (see Fig. 37-17).

Modified leaves of carnivorous plants capture insects

Carnivorous plants are plants that capture insects. Most carnivorous plants grow in poor soil that is deficient in certain essential minerals, particularly nitrogen. These plants meet some of their mineral requirements by digesting insects and other small animals. The leaves of carnivorous plants are adapted to attract, capture, and digest their animal prey.

Some carnivorous plants have passive traps. The leaves of a pitcher plant, for example, are shaped so that rainwater collects and forms a reservoir that also contains acid secreted by the plant (**FIG. 34-15**). Some pitchers are quite large; in the tropics pitcher plants may be large enough to hold 1 L (approximately 1 qt) or more of liquid. An insect attracted by the odor or nectar of the pitcher may lean over the edge and fall in. Although it may

(a) The leaves of this pincushion cactus (*Mammilaria*) are spines for deterring herbivores.

(b) Tendrils of a sweet pea (*Lathyrus odoratus*) are modified leaves that wind around objects and aid in climbing.

(c) A terminal bud and two axillary buds of a maple (*Acer*) twig have overlapping bud scales to protect the buds.

(d) Showy red bracts surround each poinsettia (*Euphorbia pulcherrima*) inflorescence; the inconspicuous flowers are in the center.

(e) The leaves of bulbs such as the onion (*Allium cepa*) are fleshy for storage of food materials and water.

(f) The succulent leaves of the string-of-beads plant (*Senecio rowleyanus*) are spherical to minimize surface area, thereby conserving water.

Figure 34-14 Leaf modifications

Figure 34-15 A common pitcher plant
This species (*Sarracenia purpurea*), whose pitchers grow to 30.5 cm (12 in.), is widely distributed in acidic bogs and marshes in eastern North America. Young pitchers are green but turn red as they age. Note the dead beetle in the pitcher.

make repeated attempts to escape, the insect is prevented from crawling out by the slippery sides and the rows of stiff hairs that point downward around the lip of the pitcher. The insect eventually drowns, and part of its body disintegrates and is absorbed.

Most insects are killed in pitcher plants. However, the larvae of several insects (certain flies, midges, and mosquitoes) and a large community of microorganisms live inside the pitchers. These insect species obtain their food from the insect carcasses, and the pitcher plant digests what remains. It is not known how these insects survive the acidic environment inside the pitcher.

The Venus flytrap is a carnivorous plant with active traps. Its leaf blades resemble tiny bear traps (see Fig. 1-3). Each side of the leaf blade contains three small, stiff hairs. If an insect alights and brushes against two of the hairs or against the same hair twice in quick succession, the trap springs shut with amazing rapidity, about 100 milliseconds. The spines along the margins of the blades fit closely together to prevent the insect from escaping. After the leaf initially traps the insect, the leaf continues to slowly close for the next several hours. Digestive glands on the surface of the trap secrete enzymes in response to the insect pressing against them. Days later, after the insect has died and been digested, the trap reopens and the indigestible remains fall out.

CHECKPOINT 34.5

- *What are the primary functions of each of the following modified leaves: spines, tendrils, and bud scales?*
- *What are the functions of bulbs? of succulent leaves?*
- *What are some of the specialized features of the leaves of carnivorous plants?*

SUMMARY: FOCUS ON LEARNING OBJECTIVES

34.1 Leaf Form and Structure (*page 724*)

1 Discuss variation in leaf form, including simple versus compound leaves, leaf arrangement on the stem, and venation patterns.

- Leaves may be simple (having a single **blade**) or compound (having a blade divided into two or more leaflets).
- Leaf arrangement on a stem may be alternate (one leaf at each **node**), opposite (two leaves at each node), or whorled (three or more leaves at each node).
- Leaves may have parallel or netted venation. Netted venation may be palmately netted, with several major **veins** radiating from one point, or pinnately netted, with veins branching along the entire length of the midvein.

2 Describe the major tissues of the leaf (epidermis, photosynthetic ground tissue, xylem, and phloem) and sketch how they are arranged in a leaf cross section.

- Upper and lower surfaces of the leaf blade are covered by an **epidermis.** A waxy **cuticle** coats the epidermis, enabling the plant to survive the dry conditions of a terrestrial existence.
- **Stomata** are small pores in the epidermis that permit gas exchange needed for photosynthesis. Each pore is surrounded by two **guard cells** that are often associated with special epidermal cells called **subsidiary cells.** Subsidiary cells provide a reservoir of water and ions that move into and out of the guard cells as they change shape during stomatal opening and closing.

- **Mesophyll** consists of photosynthetic parenchyma cells. Mesophyll is divided into **palisade mesophyll,** which functions primarily for photosynthesis, and **spongy mesophyll,** which functions primarily for gas exchange.
- Leaf veins have **xylem** to conduct water and essential minerals to the leaf and **phloem** to conduct sugar produced by photosynthesis to the rest of the plant.

3 Compare leaf anatomy in eudicots and monocots.

- Monocot leaves have parallel venation, whereas eudicot leaves have netted venation. Some monocots (corn and other grasses) do not have mesophyll differentiated into distinct palisade and spongy layers. Some monocots (grasses, reeds, and sedges) have guard cells shaped like dumbbells, unlike the more common bean-shaped guard cells.

4 Relate leaf structure to its function of photosynthesis.

- Leaf structure is adapted for its primary function of **photosynthesis.** Most leaves have a broad, flattened blade that efficiently collects the sun's radiant energy. The transparent epidermis allows light to penetrate into the middle of the leaf, where photosynthesis occurs.
- Stomata generally open during the day for gas exchange needed during photosynthesis and close at night to conserve water when photosynthesis is not occurring.
- Air spaces in mesophyll tissue permit the rapid diffusion of CO_2 and water into, and oxygen out of, mesophyll cells.

34.2 Stomatal Opening and Closing *(page 730)*

5 Explain the role of blue light in the opening of stomata.

- *Blue light,* which is a component of sunlight, triggers the activation of proton pumps located in the guard cell plasma membrane. Blue light also triggers the synthesis of malic acid and the hydrolysis of starch.

6 Outline the physiological changes that accompany stomatal opening and closing.

- Protons (H^+), produced when malic acid ionizes, are pumped out of the guard cells. As protons leave the guard cells, an **electrochemical gradient** (a charge and concentration difference) forms on the two sides of the guard cell plasma membrane.
- The electrochemical gradient drives the uptake of potassium ions (K^+) through specific **voltage-activated ion channels** into the guard cells. Chloride ions (Cl^-) are also taken into the guard cells through ion channels. These osmotically active ions increase the solute concentration in the guard cell vacuoles. The resulting osmotic movement of water into the guard cells causes them to become turgid, forming a pore.
- As the day progresses, potassium ions slowly leave the guard cells, and starch is hydrolyzed to sucrose, an osmotically active solute, which increases in concentration in the guard cells. Stomata close when water leaves the guard cells as a result of a decline in the concentration of sucrose, which occurs as evening approaches. The sucrose is converted to starch, which is osmotically inactive.

34.3 Transpiration and Guttation *(page 731)*

7 Discuss transpiration and its effects on plants.

- **Transpiration** is the loss of water vapor from aerial parts of plants. Transpiration, which occurs primarily through the stomata, is affected by environmental factors such as temperature, wind, and relative humidity.
- Transpiration appears to be both beneficial and harmful to the plant; that is, transpiration represents a trade-off between the CO_2 requirement for photosynthesis and the need for water conservation.

8 Distinguish between transpiration and guttation.

- **Guttation,** the release of liquid water from leaf margins of some plants, occurs through special structures when transpiration is negligible and available soil moisture is high. In contrast, transpiration is the loss of water vapor and occurs primarily through the stomata.

34.4 Leaf Abscission *(page 733)*

9 Define leaf *abscission,* explain why it occurs, and describe the physiological and anatomical changes that precede it.

- Leaf **abscission** is the loss of leaves that often occurs as winter approaches in temperate climates or at the beginning of the dry period in tropical climates with wet and dry seasons.
- Abscission involves physiological and anatomical changes that occur prior to leaf fall. An **abscission zone** develops where the petiole detaches from the stem. Sugars, amino acids, and many essential minerals are transported from the leaves to other plant parts. Chlorophyll breaks down, and **carotenoids** and **anthocyanins** become evident.

34.5 Modified Leaves *(page 734)*

10 List at least four examples of modified leaves and give the function of each.

- **Spines** are leaves adapted to deter herbivores. Some **tendrils** are leaves modified for grasping and holding on to other structures (to support weak stems). **Bud scales** are leaves modified to protect delicate meristematic tissue or dormant buds. **Bracts** are modified leaves associated with some inflorescences. **Bulbs** are short, underground stems with fleshy leaves specialized for storage. Many plants adapted to arid conditions have succulent leaves for water storage. Carnivorous plants have leaves modified to trap insects.

TEST YOUR UNDERSTANDING

Know and Comprehend

1. Plants with an alternate leaf arrangement have (a) blades divided into two or more leaflets (b) major veins that radiate out from one point (c) one leaf at each node (d) major veins branching off along the entire length of the midvein (e) two leaves at each node

2. The primary function of the spongy mesophyll is (a) reducing water loss from the leaf surface (b) changing the shape of the guard cells (c) supporting the leaf to prevent it from collapsing under its own weight (d) diffusing gases within the leaf (e) deterring herbivores

3. Gas exchange occurs through microscopic pores formed by two (a) subsidiary cells (b) abscission cells (c) mesophyll cells (d) guard cells (e) stipules

4. The thin, noncellular layer of wax secreted by the epidermis of leaves is the (a) stoma (b) abscission zone (c) trichome (d) bundle sheath (e) cuticle

5. Which of the following is *not* an adaptation of pine needles to conserve water? (a) less surface area exposed to the air than thin-bladed leaves (b) a relatively thick cuticle (c) sunken stomata (d) netted veins instead of parallel veins (e) both c and d are not adaptations of pine needles

6. When transpiration is negligible, plants such as grasses exude excess water by (a) guttation (b) circadian rhythm (c) abscission (d) pumping H^+ out of and K^+ into guard cells (e) photosynthesis

7. At sunrise, the accumulation in the guard cells of the osmotically active substance _____ causes an inflow of water and the opening of the pore. (a) protons (b) starch (c) ATP synthase (d) sucrose (e) potassium ions

8. Stomatal opening is most pronounced in response to _____ light. (a) green (b) yellow (c) blue (d) ultraviolet (e) infrared

9. Anatomically, the abscission zone where a petiole detaches from a stem consists of (a) thin-walled parenchyma cells with few fibers (b) thick-walled cork parenchyma cells (c) clusters of fibers and collenchyma strands (d) hard, pointed stipules (e) epidermal cells with sunken stomata

10. Modified leaves that enable a stem to climb are called _____, whereas modified leaves that cover the winter buds of a dormant woody plant are called _____ (a) spines; bud scales

(b) bud scales; tendrils (c) tendrils; bud scales (d) tendrils; spines (e) carnivorous leaves; spines

11. There is a trade-off between photosynthesis and transpiration in leaves because (a) numerous stomatal pores provide both gas exchange for photosynthesis and openings through which water vapor escapes (b) a waxy layer, the cuticle, reduces water loss (c) blue light triggers an influx of potassium ions (K^+) into the guard cells (d) leaves of deciduous plants abscise as winter approaches in temperate climates (e) stomata are closed at night, although water continues to move into the roots by osmosis

Apply and Analyze

12. Suppose that you are asked to observe a micrograph of a leaf cross section and distinguish between the upper and lower epidermis. How would you make this decision?

13. **CONNECT** Given that (a) xylem is located toward the upper epidermis in leaf veins and phloem is toward the lower epidermis and (b) the vascular tissue of a leaf is continuous with that of the stem, suggest one possible arrangement of vascular tissues in the stem that might account for the arrangement of vascular tissue in the leaf.

14. **VISUALIZE** Draw a simple cross section of a leaf. Label the upper epidermis, lower epidermis, guard cells, spongy mesophyll, palisade mesophyll, and vascular tissue. Which cells contain chloroplasts?

Evaluate and Synthesize

15. What might be some of the advantages of a plant having a few large leaves? What might be some disadvantages? What might be some advantages of having many small leaves? What disadvantages might that entail? How would your answers differ for plants growing in a humid environment compared with those in a desert?

16. **EVOLUTION LINK** What is the selective advantage of seasonal leaf abscission in woody flowering plants living in colder climates? What adaptations enable conifers to survive these climates without leaf abscission?

17. **SCIENCE, TECHNOLOGY, AND SOCIETY** Briefly explain why research on the molecular mechanism of stomatal closure might be of future use in agriculture.

aplia To access course materials, such as Aplia and other companion resources, please visit **www.cengagebrain.com.**

Stem Structure and Transport | 35

A vegetative (not sexually reproductive) vascular plant has three parts: roots, leaves, and stems. As discussed in Chapter 33, roots serve to anchor the plant and absorb materials from the soil, whereas leaves are primarily for photosynthesis, converting radiant energy into the chemical energy of carbohydrate molecules. Stems, the focus of this chapter, link a plant's roots to its leaves and are usually located aboveground, although many plants have underground stems. Stems exhibit varied forms, ranging from ropelike vines to massive tree trunks (see photograph). They can be herbaceous, with soft, nonwoody tissues, or they can be woody, with extensive hard tissues of wood and bark.

Stems perform three main functions in plants. First, stems of most species support leaves and reproductive structures. The upright position of most stems and the arrangement of the leaves on them allow each leaf to absorb light for use in photosynthesis. Reproductive structures (such as flowers and fruits) are located on stems in areas accessible to insects, birds, and air currents, which transfer pollen from flower to flower and help disperse seeds and fruits.

Second, stems provide internal transport. They conduct water and dissolved minerals (inorganic nutrients) from the roots, where these materials are absorbed from the soil, to leaves and other plant parts. Stems also conduct the sugar produced in leaves by photosynthesis to roots and other parts of the plant. Remember, however, that stems are not the only plant organs that conduct materials. The vascular system is continuous throughout all parts of a plant, and conduction occurs in roots, stems, leaves, and reproductive structures.

Third, stems produce new living tissue. They continue to grow throughout a plant's life, producing *buds* that develop into stems with new leaves and/or reproductive structures. In addition to the main functions of support, conduction, and production of new stem tissues, stems of some species are modified for asexual reproduction (see Chapter 37) or, if green, to manufacture sugar by photosynthesis.

Giant sequoias. Giant sequoias (*Sequoiadendron giganteum*) are the most massive trees on Earth. Their huge stems grow to 75 m (250 ft) or more. Photographed in Sequoia National Park, California.

KEY CONCEPTS

35.1 Primary tissues (epidermis, cortex, pith, xylem, and phloem) of stems develop from shoot apical meristems. Secondary tissues (wood and bark) of stems develop from two lateral meristems: vascular cambium and cork cambium.

35.2 The concept of water potential explains the direction of water transport into, through, and out of a plant. According to the tension–cohesion model, transpiration pulls water up through the stem as water evaporates from leaves by transpiration.

35.3 According to the pressure–flow model, sucrose is translocated in phloem sap from the source, where the sugar is loaded into phloem, to the sink, where the sugar is removed from phloem.

35.1 STEM GROWTH AND STRUCTURE

LEARNING OBJECTIVES

1 Sketch cross sections of herbaceous eudicot and monocot stems, and describe the functions of each tissue.
2 Name the two lateral meristems and describe the tissues that arise from each.
3 Outline the transition from primary growth to secondary growth in a woody stem.

You may recall from Chapter 33 that plants have two different types of growth. **Primary growth** is an increase in the length of a plant and occurs at **apical meristems** located at the tips of roots and shoots and also within the buds of stems. **Secondary growth** is an increase in the girth (thickness) of a plant as a result of the activity of **lateral meristems** located within stems and roots. The new tissues formed by the lateral meristems are called *secondary tissues* to distinguish them from *primary tissues* produced by apical meristems.

All plants have primary growth; some plants have both primary and secondary growth. Stems with only primary growth are herbaceous, whereas those with both primary and secondary growth are generally woody. (Certain herbaceous stems, such as geranium and sunflower, also have a limited amount of secondary growth.) A woody plant increases in length by primary growth at the tips of its stems and roots, whereas its older stems and roots farther back from the tips increase in girth by secondary growth. In other words, at the same time that primary growth is increasing the length of the stem, secondary growth is adding wood and bark, thereby causing the stem to thicken.

Herbaceous eudicot and monocot stems differ in internal structure

Although considerable structural variation exists in stems, they all possess an outer protective covering (epidermis or periderm), one or more types of ground tissue, and vascular tissues (xylem and phloem). Let us first consider the structure of herbaceous eudicot stems and then the structure of monocot stems.

Vascular bundles of herbaceous eudicot stems are arranged in a circle in cross section A young sunflower stem is a representative herbaceous eudicot stem that exhibits primary growth (FIG. 35-1). Its outer covering, the **epidermis,** provides protection in herbaceous stems, as it does in leaves and herbaceous roots (see Table 33-4 and Fig. 33-5). The **cuticle,** a waxy layer of *cutin,* covers the stem epidermis and reduces water loss from the stem surface. **Stomata** permit gas exchange. (Recall from Chapter 34 that a cuticle and stomata are also associated with the leaf epidermis.)

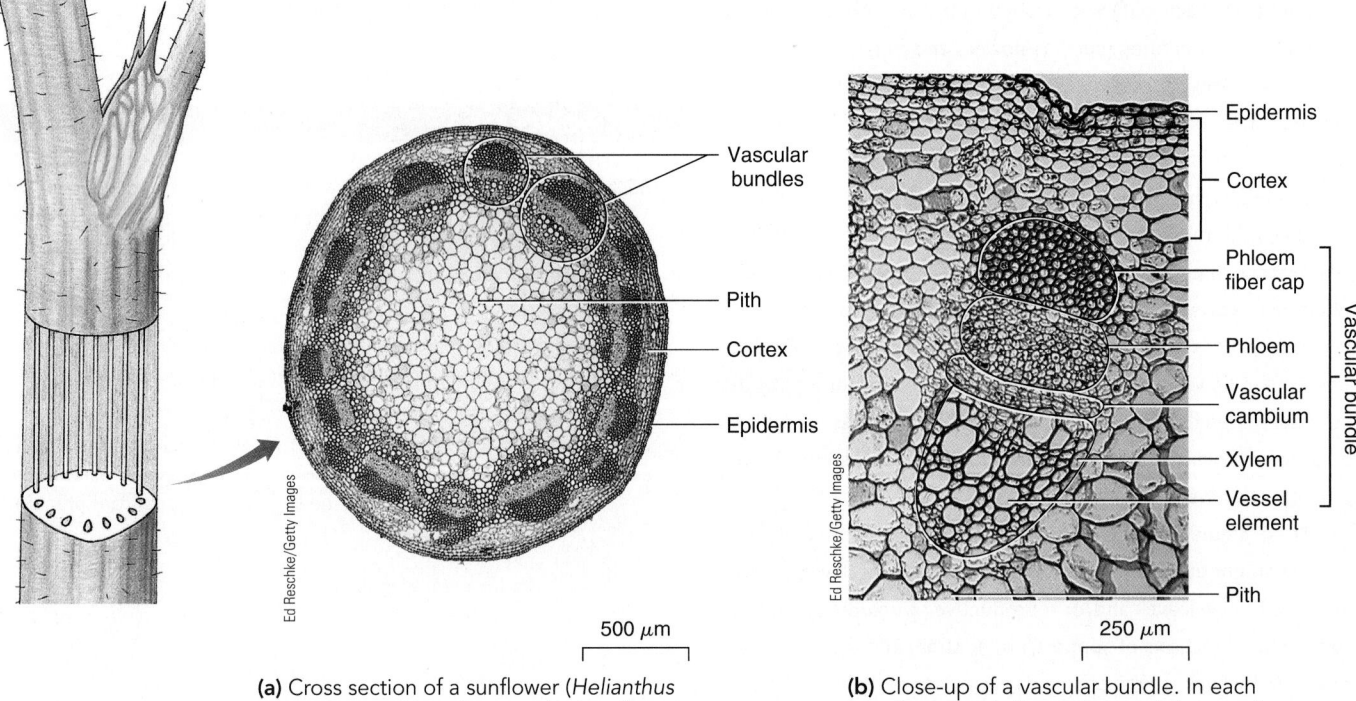

Ed Reschke/Getty Images

500 μm

(a) Cross section of a sunflower (*Helianthus annuus*) stem. Note the vascular bundles arranged in a circle around a central core of pith.

Ed Reschke/Getty Images

Epidermis
Cortex
Phloem fiber cap
Phloem
Vascular cambium
Xylem
Vessel element
Pith

Vascular bundle

250 μm

(b) Close-up of a vascular bundle. In each bundle, xylem is located toward the stem's interior, and phloem toward the exterior. Each vascular bundle is "capped" by a batch of fibers for additional support.

Vascular bundles
Pith
Cortex
Epidermis

Figure 35-1 LMs of a herbaceous eudicot stem

© Cengage Learning

Inside the epidermis is the **cortex,** a cylinder of ground tissue that may contain parenchyma, collenchyma, and sclerenchyma cells (see Table 33-2 and Fig. 33-3). As might be expected from the various types of cells that it contains, the cortex in herbaceous eudicot stems can have several functions, such as photosynthesis, storage, and support. If a stem is green, photosynthesis occurs in chloroplasts of cortical parenchyma cells. Parenchyma in the cortex also stores starch (in amyloplasts) and crystals (in vacuoles). Collenchyma and sclerenchyma in the cortex confer strength and structural support for the stem.

The vascular tissues provide conduction and support. In herbaceous eudicot stems, the vascular tissues are located in bundles that, when viewed in cross section, are arranged in a circle. However, viewed lengthwise, these bundles extend as long strands throughout the length of a stem and are continuous with vascular tissues of both roots and leaves.

Each vascular bundle contains both **xylem,** which transports water and dissolved minerals from roots to leaves, and **phloem,** which transports dissolved sugar (see Table 33-3 and Fig. 33-4). Xylem is located on the inner side of the vascular bundle, and phloem is found toward the outside. Sandwiched between xylem and phloem in some herbaceous stems is a single layer of cells called the *vascular cambium,* a lateral meristem responsible for secondary growth (discussed later).

Because most stems support the aerial plant body, they are much stronger than roots. The thick walls of tracheids and vessel elements in xylem help support the plant. Fibers also occur in both xylem and phloem, although they are usually more extensive in phloem. These fibers add considerable strength to the herbaceous stem. In sunflowers and certain other herbaceous eudicot stems, phloem contains a cluster of fibers toward the outside of the vascular bundle, called a *phloem fiber cap,* that helps strengthen the stem. The phloem fiber cap is not present in all herbaceous eudicot stems.

The **pith** is a ground tissue at the center of the herbaceous eudicot stem that consists of large, thin-walled parenchyma cells that function primarily in storage. Because of the arrangement of the vascular tissues in bundles, there is no distinct separation of cortex and pith between the vascular bundles. The areas of parenchyma between the vascular bundles are often referred to as *pith rays.*

Vascular bundles are scattered throughout monocot stems An epidermis with its waxy cuticle covers monocot stems, such as the herbaceous stem of corn. As in herbaceous eudicot stems, the vascular tissues run in strands throughout the length of a stem. In cross section the vascular bundles contain xylem toward the inside and phloem toward the outside. In contrast with herbaceous eudicots, however, vascular bundles of monocots are not arranged in a circle but are scattered throughout the stem (**FIG. 35-2**). Each vascular bundle is enclosed in a bundle sheath of supporting sclerenchyma

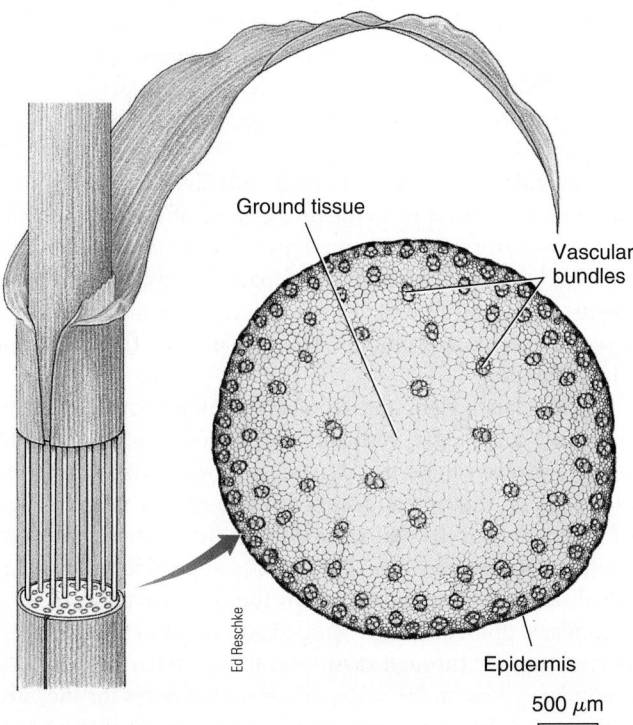

Ground tissue

Vascular bundles

Epidermis

500 μm

(a) Cross section of a corn (*Zea mays*) stem shows vascular bundles scattered throughout ground tissue.

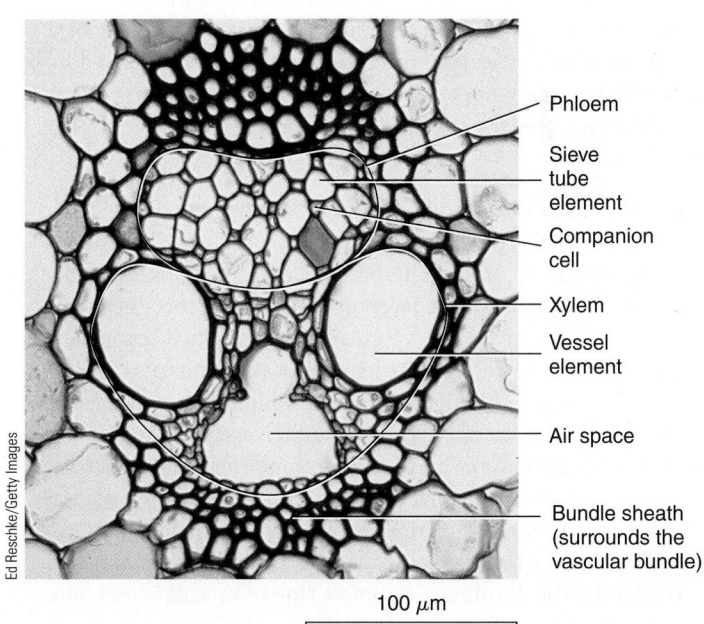

Phloem

Sieve tube element

Companion cell

Xylem

Vessel element

Air space

Bundle sheath (surrounds the vascular bundle)

100 μm

(b) Close-up of a vascular bundle. The air space is the site where the first xylem elements were formed and later disintegrated. The entire bundle is enclosed in a bundle sheath of sclerenchyma for additional support.

Figure 35-2 LMs of a monocot stem
© Cengage Learning

cells. The monocot stem does not have distinct areas of cortex and pith. The ground tissue in which the vascular tissues are embedded performs the same functions as cortex and pith in herbaceous eudicot stems.

Monocot stems do not possess lateral meristems (vascular cambium and cork cambium) that give rise to secondary growth. Monocots have primary growth only and do not produce wood and bark. Although some treelike monocots (such as palms) attain considerable size, they do so by a modified form of primary growth in which parenchyma cells divide and enlarge. Stems of some monocots (such as bamboo and palm) contain a great deal of sclerenchyma tissue, which makes them hard and woodlike in appearance.

Woody plants have stems with secondary growth

Woody plants undergo secondary growth, an increase in the girth of stems and roots. Secondary growth occurs as a result of the activity of two lateral meristems: vascular cambium and cork cambium. Among flowering plants, only woody eudicots (such as apple, hickory, and maple) have secondary growth. Cone-bearing gymnosperms (such as pine, juniper, and spruce) also have secondary growth.

Cells in the **vascular cambium** divide and produce two conducting and supporting tissues: secondary xylem (wood) to replace primary xylem and secondary phloem (inner bark) to replace primary phloem. Primary xylem and primary phloem are not able to transport materials indefinitely and so are replaced in plants that have extended lifespans.

Cells of the outer lateral meristem, called cork cambium, divide and produce cork cells and cork parenchyma. **Cork cambium** and the tissues it produces are collectively referred to as **periderm** (outer bark), which functions as a replacement for the epidermis (see the last LM in Table 33-4).

Vascular cambium gives rise to secondary xylem and secondary phloem Primary tissues in woody eudicot stems are organized like those in herbaceous eudicot stems, with the vascular cambium a thin layer of cells sandwiched between xylem and phloem in the vascular bundles. Once secondary growth begins, the internal structure of a stem changes considerably (**FIG. 35-3**). Although vascular cambium is not initially a continuous cylinder of cells (because the vascular bundles are separated by pith rays), it becomes continuous when production of secondary tissues begins. This continuity develops because certain parenchyma cells in each pith ray retain the ability to divide. These cells connect to vascular cambium cells in each vascular bundle, forming a complete ring of vascular cambium in cross section.

Cells in the vascular cambium divide and produce daughter cells in two directions. The cells formed from the dividing vascular cambium are located either *inside* the ring of vascular cambium (to become secondary xylem, or wood) or *outside* it (to become secondary phloem, or inner bark) (**FIG. 35-4**). When a cell in the vascular cambium divides tangentially (inward or outward), one daughter cell remains meristematic; that is, it remains part of the vascular cambium. The other cell may divide again several times, but eventually it stops dividing and develops into mature secondary tissue. Thus, vascular cambium is a thin layer of cells sandwiched between the wood and inner bark, the two tissues it produces (**FIG. 35-5**).

As the stem increases in circumference, the number of cells in the vascular cambium also increases. This increase in cells occurs by an occasional radial division of a vascular cambium cell, at right angles to its normal direction of division. In this case, both daughter cells remain meristematic.

What happens to the original primary tissues of a stem once secondary growth develops? As a stem increases in thickness, the orientation of the original primary tissues changes. For example, secondary xylem and secondary phloem are laid down between the primary xylem and primary phloem within each vascular bundle. Therefore, as vascular cambium forms secondary tissues, the primary xylem and primary phloem in each vascular bundle become separated from each other. The primary tissues located outside the cylinder of secondary growth (i.e., primary phloem, cortex, and epidermis) are subjected to the mechanical pressures produced by secondary growth and are gradually crushed or torn apart and sloughed off.

Secondary tissues replace the primary tissues in function. Secondary xylem conducts water and dissolved minerals from roots to leaves in the woody plant. It contains the same types of cells found in primary xylem: water-conducting tracheids and vessel elements in addition to xylem parenchyma cells and fibers. The arrangement of the different cell types in secondary xylem produces the distinctive wood characteristics of each species.

Secondary phloem conducts dissolved sugar from its place of manufacture or of breakdown of starch to a place of use and storage. The same types of cells found in primary phloem (sieve tube elements, companion cells, phloem parenchyma cells, and fibers) are also found in secondary phloem, although secondary phloem usually has more fibers than primary phloem.

Secondary xylem and secondary phloem transport water, minerals, and sugar vertically throughout the woody plant body. However, materials must also move horizontally (laterally). Lateral movement occurs through *rays,* which are chains of parenchyma cells that radiate from the center of the woody stem or root (see Fig. 35-5). The vascular cambium forms rays, which are often continuous from the secondary xylem to the secondary phloem. Water and dissolved minerals are transported laterally through rays, from the secondary xylem to the secondary phloem. Likewise, rays form pathways for the lateral transport of dissolved sugar, from the secondary phloem to the secondary xylem, and of waste products to the center, or heart, of the tree (discussed later).

Cork cambium produces periderm Cork cambium, which usually arises from parenchyma cells in the outer

The stems of many plants undergo secondary growth to increase in girth.

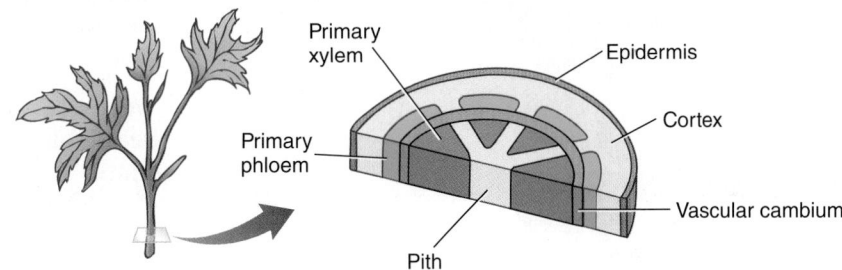

1 At the onset of secondary growth, vascular cambium arises in the parenchyma between the vascular bundles (i.e., in the pith rays), forming a cylinder of meristematic tissue (*blue circle in cross section*).

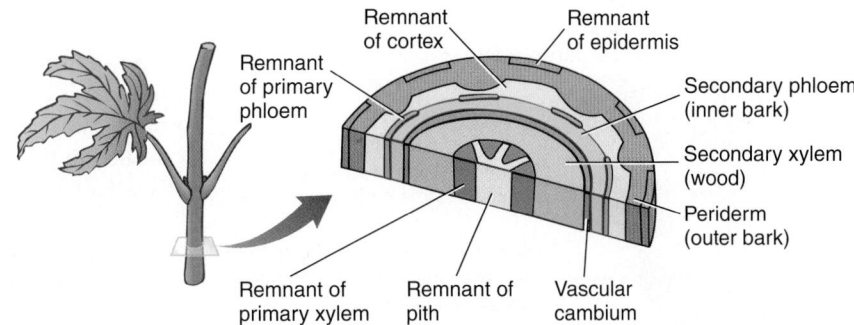

2 Vascular cambium begins to divide, forming secondary xylem on the inside and secondary phloem on the outside.

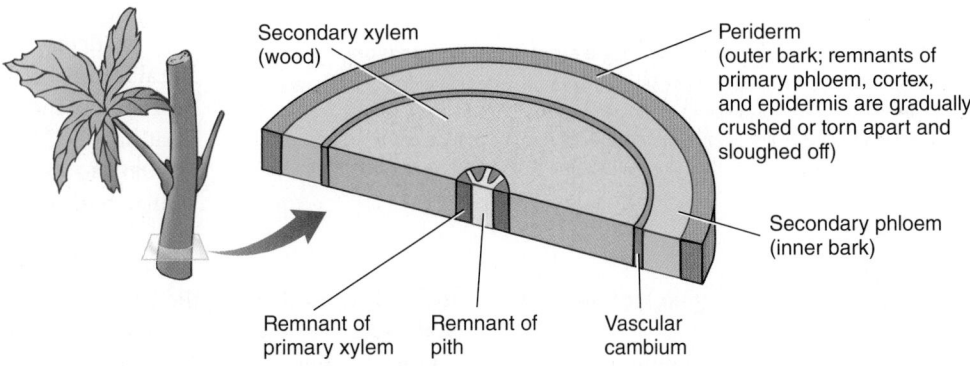

3 A young woody stem. Vascular cambium produces significantly more secondary xylem than secondary phloem.

Figure 35-3 *Animation* **Development of secondary growth**

Vascular cambium and the tissues it produces are shown in cross section; cork cambium is not depicted. (The figures change in scale because of space limitations; pith and primary xylem are actually the same size in all three diagrams, but the change in scale makes those tissues appear to shrink in each succeeding diagram.)

© Cengage Learning

CONNECT How is the arrangement of the vascular bundles in a monocot stem unsuited to the functioning of a vascular cambium?

cortex, produces *periderm,* the functional replacement for the epidermis. Cork cambium is either a continuous cylinder of dividing cells (similar to vascular cambium) or a series of overlapping arcs of meristematic cells that form from parenchyma cells in successively deeper layers of the cortex and, eventually, secondary phloem. Variation in cork cambia and their rates of division explain why the outer bark of some tree species is fissured (as in bur oak), rough and

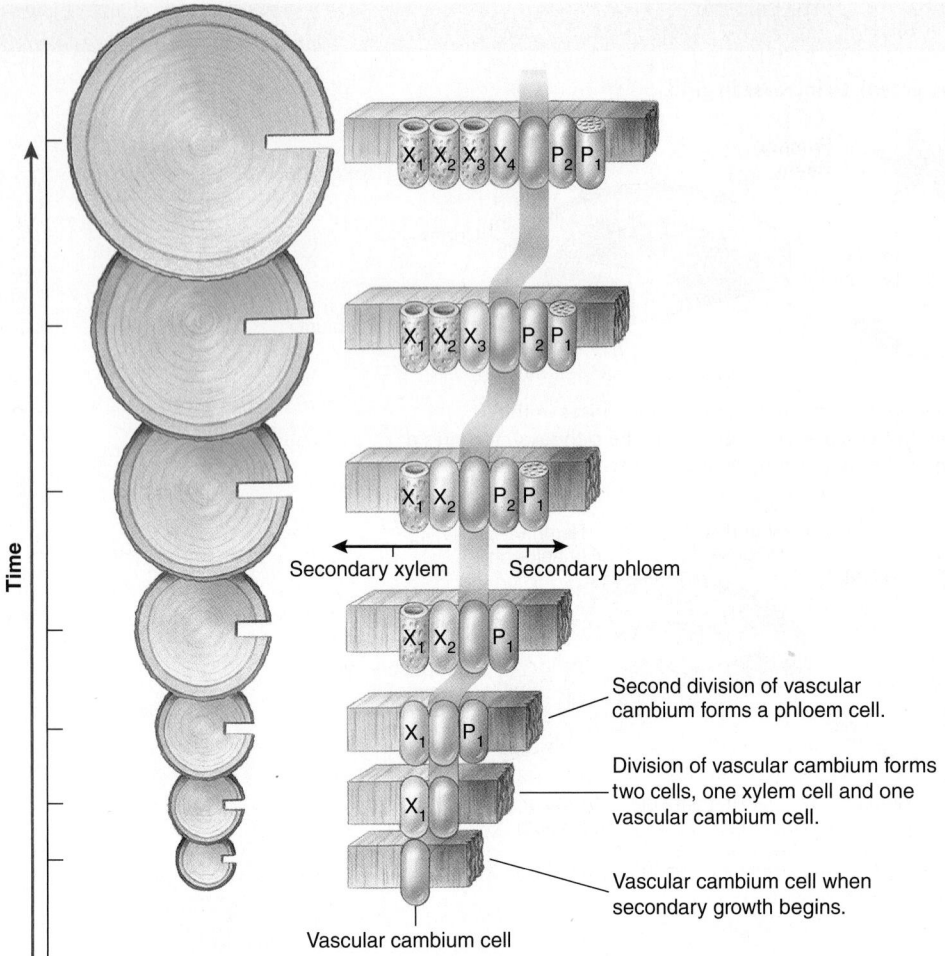

Secondary xylem ← → Secondary phloem

Second division of vascular cambium forms a phloem cell.

Division of vascular cambium forms two cells, one xylem cell and one vascular cambium cell.

Vascular cambium cell when secondary growth begins.

Vascular cambium cell

Figure 35-4 Divisions of a vascular cambium cell during a single growing season

To study the figure, start at the bottom and move up. Note that vascular cambium (*blue cell*) divides in two directions, forming secondary xylem (X) to the inside and secondary phloem (P) to the outside. These cells, which are numbered in the order in which they are produced, differentiate into the mature cell types associated with xylem and phloem. As secondary xylem accumulates, vascular cambium "moves" outward, and the woody stem increases in diameter.

© Cengage Learning

---

stomata that exchange gases for the herbaceous stem, dies. Stomata are replaced by **lenticels,** which permit gas exchange through the periderm (**FIG. 35-6**).

Woody twigs of deciduous trees bear characteristic features in winter A woody *twig* is that part of a branch produced during the current growing season. Twigs have **buds,** which are embryonic shoots. A **terminal bud** is the embryonic shoot located at the tip of a stem. The dormant (not actively growing) apical meristem of a terminal bud is covered and protected by an outer layer of **bud scales,** which are modified leaves (see Fig. 34-14c). **Axillary buds,** also called *lateral buds,* are located in the axils of a plant's leaves (see Fig. 34-1). An *axil* is the upper angle between a leaf and the stem to which it is attached. When terminal and axillary buds grow, they form branches that bear leaves and/or flowers. The area on a stem where each leaf is attached is called a **node,** and the region between two successive nodes is an **internode.**

To demonstrate certain structural features of the stem, we can use a woody branch of a deciduous tree that has shed its leaves, as shown in **FIGURE 35-7.** Bud scales cover the terminal bud and protect its delicate apical meristem during dormancy. When the bud resumes growth, the bud scales covering the terminal bud fall off, leaving **bud scale scars** on the stem where they were attached. Because temperate-zone woody plants form terminal buds at the end of each year's growing season, the number of sets of bud scale scars on a twig indicates its age. A **leaf scar** shows where each leaf was attached on the stem; the pattern of leaf scars can be used to determine leaf arrangement on a stem: alternate, opposite, or whorled (see Fig. 34-2b). The vascular (conducting) tissue that extends from the stem out into the leaf forms **bundle scars** within a leaf scar. Axillary buds may be found above the leaf scars. Also, the bark of a woody twig has lenticels, which look like tiny specks on the bark of a twig.

Common terms associated with wood are based on plant structure If you have ever examined different types of lumber, you may have noticed that some trees have wood with two different colors (**FIG. 35-8**). The functional secondary

---

shaggy (shagbark hickory), scaly (Norway pine), or smooth and peeling (paper birch).

As is true of vascular cambium, cork cambium divides to form new tissues in two directions: to its inside and its outside. Cork cells, formed to the outside of cork cambium, are dead at maturity and have walls that contain layers of *suberin* and waxes, making them waterproof. These cork cells protect the woody stem against mechanical injury, mild fires, attacks by insects and fungi, temperature extremes, and water loss. To its inside, cork cambium sometimes forms cork parenchyma cells that store water and starch granules. Cork parenchyma is only one to several cells thick, much thinner than the cork cell layer.

Cork cells are impermeable to water and gases, yet the living internal cells of the woody stem require oxygen and must be able to exchange gases with the surrounding atmosphere. As a stem thickens from secondary growth, the epidermis, including

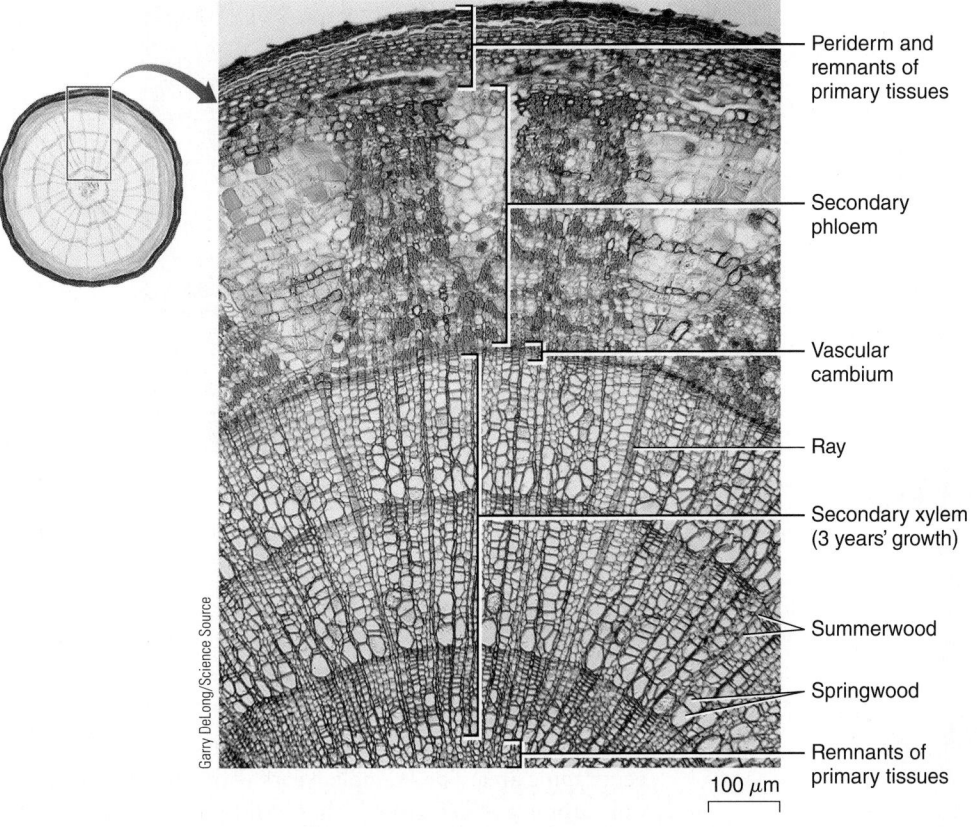

Figure 35-5 **LM of portion of a three-year-old basswood (*Tilia americana*) stem in cross section**
© Cengage Learning

Labels on figure:
- Periderm and remnants of primary tissues
- Secondary phloem
- Vascular cambium
- Ray
- Secondary xylem (3 years' growth)
- Summerwood
- Springwood
- Remnants of primary tissues
- 100 μm

Garry DeLong/Science Source

are tracheids. These cell differences generally make conifer wood softer than the wood of flowering plants, although there is a substantial variation from one species to another. The balsa tree, for example, is a flowering plant whose extremely light, soft wood is used to fashion airplane models.

Woody plants that grow in temperate climates where there is a growing period (during spring and summer) and a dormant period (during winter) exhibit *annual rings,* concentric circles found in cross sections of wood. To determine the age of a woody stem in the temperate zone, simply count the annual rings. In the tropics environmental conditions, particularly seasonal or year-round precipitation patterns, determine the presence or absence of rings, so rings are not a reliable method of determining the ages of most tropical trees.

Examination of annual rings with a magnifying lens reveals no actual "ring," or line, separating one year's growth from the next. The appearance of a ring in cross section is due to differences in cell size and cell wall

xylem—the part that conducts water and dissolved minerals—is the *sapwood,* a thin layer of younger, lighter-colored wood that is closest to the bark. *Heartwood,* the older wood in the center of the tree, is typically a brownish red. A microscopic examination of heartwood reveals that its vessels and tracheids are plugged with pigments, tannins, gums, resins, and other materials. Therefore, heartwood no longer functions in conduction but instead functions as a storage site for waste products. Because it is denser than sapwood, heartwood provides structural support for trees. Some evidence suggests that heartwood is also more resistant to decay than sapwood.

Almost everyone has heard of hardwood and softwood. Botanically speaking, *hardwood* is the wood of flowering plants, and *softwood* is the wood of conifers (cone-bearing gymnosperms). The wood of pine and other conifers typically lacks fibers (with their thick secondary cell walls) and vessel elements; the conducting cells in gymnosperms

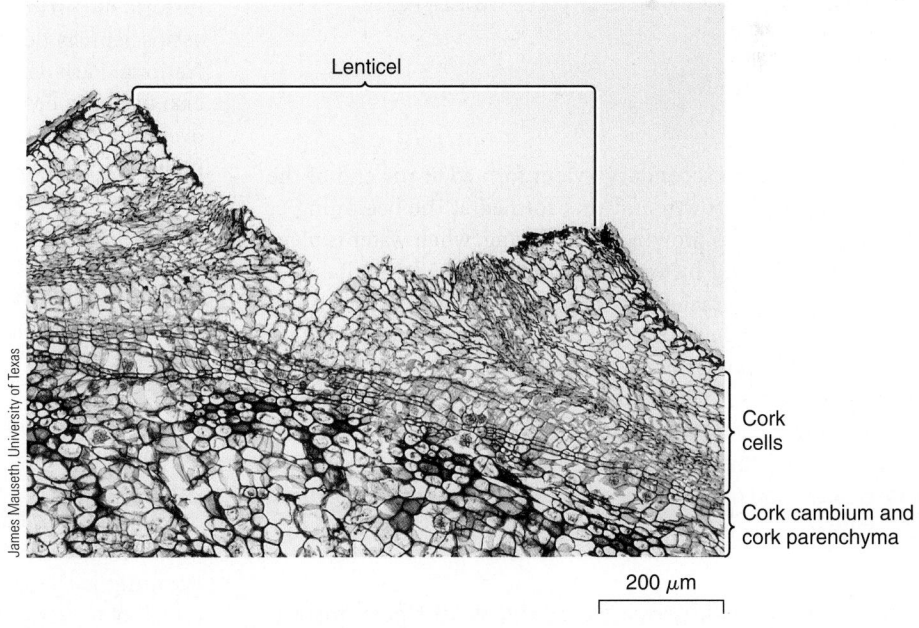

Labels on figure:
- Lenticel
- Cork cells
- Cork cambium and cork parenchyma
- 200 μm

James Mauseth, University of Texas

Figure 35-6 **LM of stem periderm, showing a lenticel**
The epidermis has ruptured because of the proliferation of loosely arranged cork cells in the lenticel. From the bark of a calico flower (*Aristolochia elegans*) stem.

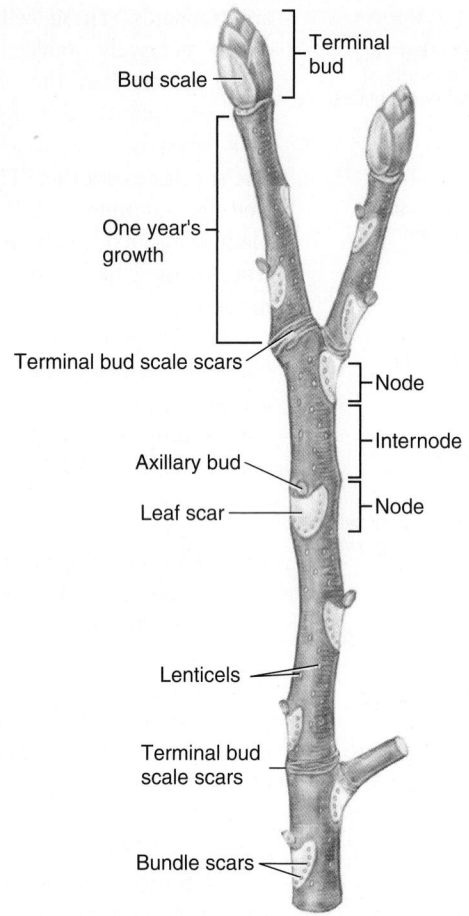

Figure 35-7 External structure of a woody branch in its winter condition

The age of a woody branch can be determined by the number of sets of bud scale scars (do not count side branches).

CONNECT How old is this branch? Can you point out any twigs?

© Cengage Learning

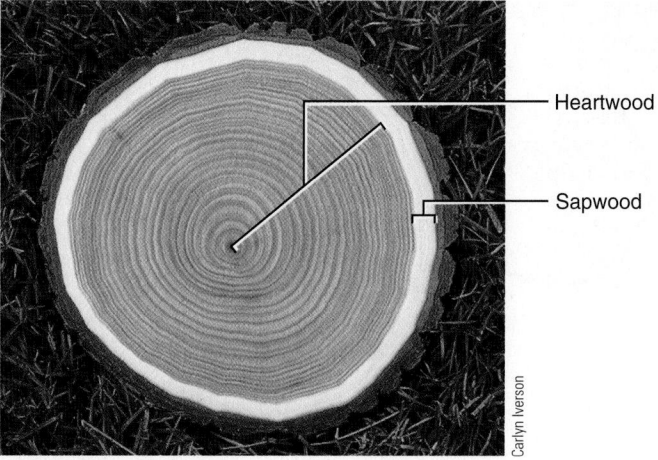

Figure 35-8 Heartwood and sapwood

The wood of older trees consists of a dense central heartwood and an outer layer of sapwood. The sapwood is the functioning xylem that conducts water and dissolved minerals. The annual rings in the heartwood are very conspicuous.

thickness between secondary xylem formed at the end of the preceding year's growth and that formed at the beginning of the following year's growth. In the spring, when water is plentiful, wood formed by vascular cambium has large-diameter conducting cells (tracheids and vessel elements) and few fibers and is appropriately called *springwood* or *early wood* (see Fig. 35-5). As summer progresses and water becomes less plentiful, the wood formed, known as *summerwood* or *late wood*, has narrower conducting cells and many fibers. It is this difference in cell size between the summerwood of one year and the springwood of the following year that gives the appearance of rings.

Tree-ring analysis provides useful scientific information In temperate climates the number of annual rings indicates the age of a tree. The size of each ring varies depending on local weather conditions. Sometimes the variation can be attributed to a single environmental factor, and similar patterns appear in the rings of different species over a large geographic area. For example, years with adequate precipitation produce wider growth rings, and years of drought produce narrower rings. It is possible to study ring sequences that are several thousand years old by constructing a *master chronology,* a complete sample of rings dating back as far as possible (FIG. 35-9).

Dendrochronology, the study of both visible and microscopic details of tree rings, has been used extensively in several fields. Tree-ring analysis has helped date Native American prehistoric sites in the Southwest. For example, using tree-ring analysis, scientists determined that the Cliff Palace in Mesa Verde National Park dates back to the year 1073. Tree-ring analysis is also useful in ecology (to study changes in a forest community over time), environmental science (to study the effects of air pollution on tree growth), and geology (to date earthquakes and volcanic eruptions).

Climatologists are increasingly using tree-ring data to study past climate patterns. Annual rings of certain tree species that grow at high elevations are sensitive to yearly temperature variations; the rings of these trees are wider in warm years and narrower in cool years. Studying tree rings over long periods helps researchers determine the natural pattern of global temperature fluctuations. This information is particularly important because of concerns about the human influence on global climate.

Although scientists generally agree that Earth has warmed in recent decades, they are not sure how much of the recent warming is the result of human influence rather than natural climate variability. To help answer this vital question, European scientists have been developing a 10,000-year master chronology that will help reconstruct annual temperatures across northern Europe and Asia since the end of the last Ice Age.

WHY IS IT USED? By matching the rings of a wood sample of unknown age to the master chronology, investigators can accurately determine the age of the sample. Tree-ring analysis provides useful information in a variety of fields, such as to date prehistoric sites, study past climate patterns, and date major prehistoric earthquakes.

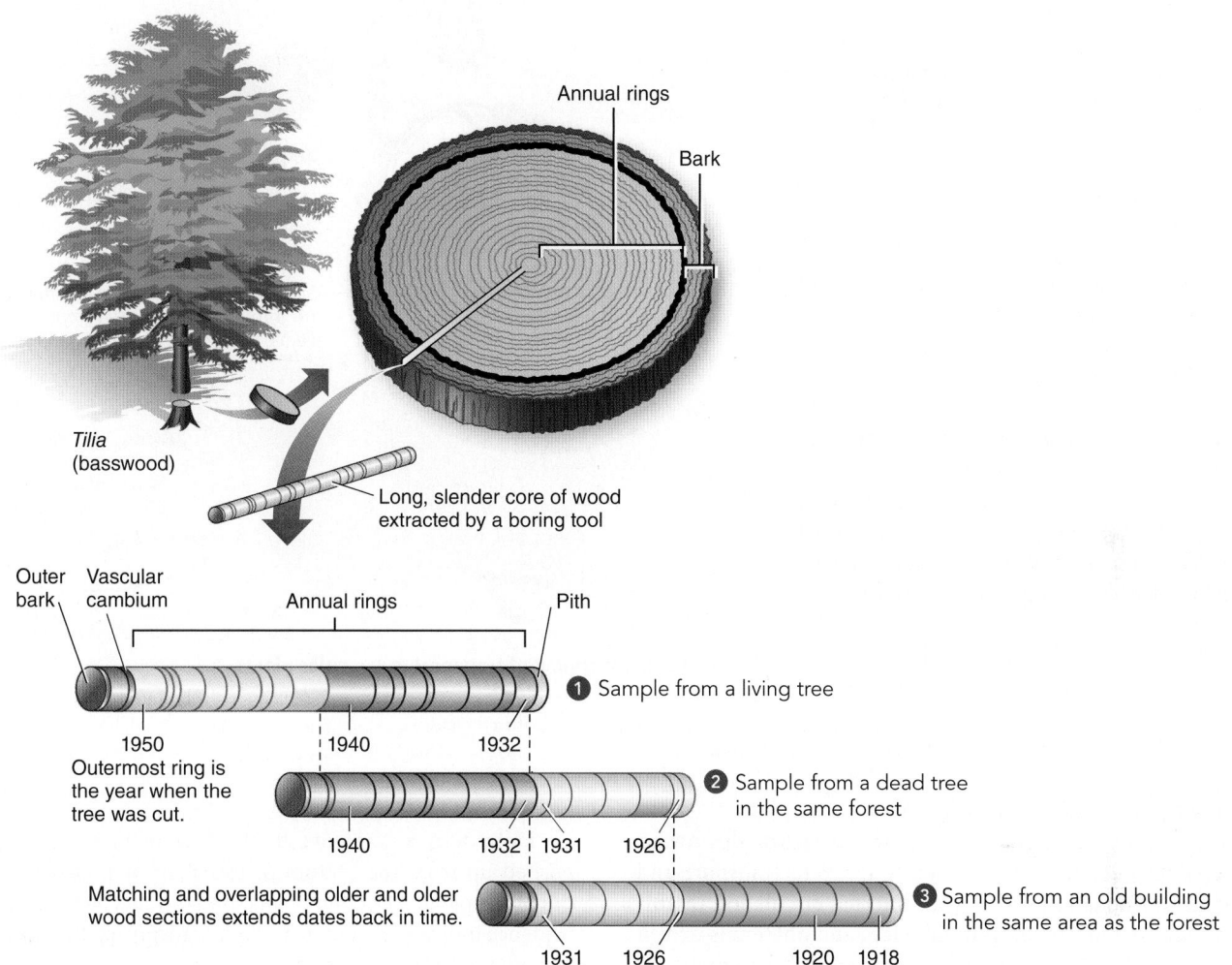

Annual rings

Bark

Tilia (basswood)

Long, slender core of wood extracted by a boring tool

Outer bark   Vascular cambium

Annual rings   Pith

❶ Sample from a living tree

1950
Outermost ring is the year when the tree was cut.

1940   1932

❷ Sample from a dead tree in the same forest

1940   1932 1931   1926

Matching and overlapping older and older wood sections extends dates back in time.

❸ Sample from an old building in the same area as the forest

1931   1926   1920 1918

HOW IS IT DONE? A master chronology is developed using progressively older pieces of wood from the same geographic area. A small core of wood is bored out of the trunk of a living tree. The oldest rings (toward the center of the tree) are matched with the youngest rings (toward the outside) of an older tree, perhaps a dead one in the forest. The master chronology is extended back in time by overlapping matching ring sequences of successively older and older sections of wood, even those found in prehistoric dwellings. Currently, the longest master chronology—of bristlecone pines in the western United States—goes back almost 9000 years. Ring matching, once a tedious, painstaking job, is performed by computers today.

Figure 35-9 How tree rings are analyzed
© Cengage Learning

CHECKPOINT 35.1

- *How does the arrangement of vascular tissue in cross section differ in primary eudicot stems and monocot stems?*
- *What is the difference between vascular cambium and cork cambium? What tissues arise from vascular cambium? from cork cambium?*
- *What happens to the primary tissues of a stem when secondary growth occurs?*
- *How do growth rings form in woody stems?*
- *What is the difference between terminal and axillary buds?*

35.2 WATER TRANSPORT

LEARNING OBJECTIVES

4 Describe the pathway of water movement in plants.

5 Define *water potential*.

6 Compare tension–cohesion and root pressure as mechanisms responsible for the rise of water and dissolved minerals in xylem.

Now that we have discussed stem structure and primary and secondary growth, we examine internal transport in the vascular system of the plant (**FIG. 35-10**). Roots obtain water and dissolved minerals from the soil. Once inside roots, these materials are transported upward to stems, leaves, flowers, fruits, and seeds. Furthermore, sugar molecules are transported from their place of manufacture or of breakdown of starch throughout the plant body, including into the subterranean roots. Water and dissolved minerals are transported from roots to other parts of the plant in xylem, whereas dissolved sugar is *translocated* in phloem.

Xylem transport and phloem translocation do not resemble the movement of materials in animals because in plants nothing *circulates* in a system of vessels. Water and minerals, transported in xylem, travel in one direction only (upward), whereas translocation of dissolved sugar may occur upward or downward in separate phloem cells. In addition, xylem transport and phloem translocation differ from internal circulation in animals because movement in both xylem and phloem is driven largely by natural physical processes rather than by a pumping organ, or heart.

How do materials travel in the continuous system of the plant's vascular tissues? We first examine water and its movement through the plant, and later we discuss the translocation of dissolved sugar.

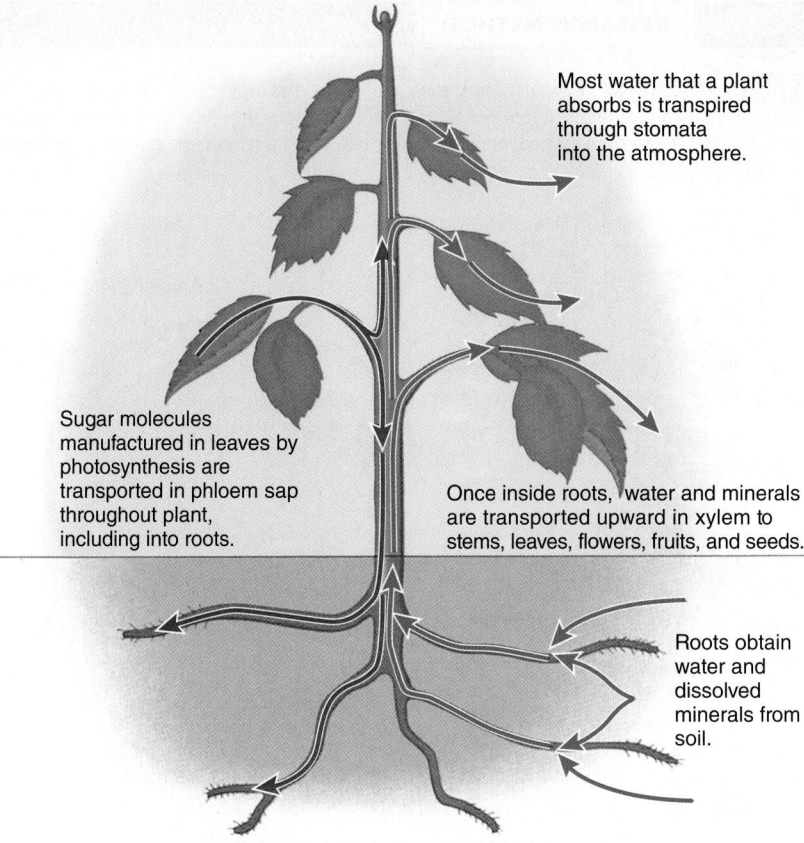

Figure 35-10 Overview of transport in vascular plants

Xylem transport is explained on the *right side* of the figure (start at the bottom and work upward), and phloem transport is on the *left side*.

© Cengage Learning

Most water that a plant absorbs is transpired through stomata into the atmosphere.

Sugar molecules manufactured in leaves by photosynthesis are transported in phloem sap throughout plant, including into roots.

Once inside roots, water and minerals are transported upward in xylem to stems, leaves, flowers, fruits, and seeds.

Roots obtain water and dissolved minerals from soil.

Water and minerals are transported in xylem

Water initially moves horizontally into roots from the soil, passing through several tissues until it reaches xylem. Once the water moves into the tracheids and vessel elements of root xylem, it travels upward through a continuous network of these hollow, dead cells from root to stem to leaf. Dissolved minerals are carried along passively in the water. The plant does not expend any energy of its own to transport water, which moves as a result of natural physical processes. The transport of xylem sap is the most rapid of any movement of materials in plants (**TABLE 35-1**).

How does water move to the tops of plants? It is either pushed up from the bottom of the plant or pulled up by the top of the plant. Although plants use both mechanisms, current evidence indicates that most water is transported through xylem by being *pulled* to the top of the plant.

TABLE 35-1	Xylem and Phloem Transport Rates in Selected Plants	
PLANT	**MAXIMUM RATE IN XYLEM (cm/min)**	**MAXIMUM RATE IN PHLOEM (cm/min)**
Conifer	2	0.8
Woody eudicot	73	2
Herbaceous eudicot or monocot	100	2.8–11
Herbaceous vine	250	1.2

Note: Xylem and phloem rates are from different plants within each general group and should be used for comparative purposes only.

Source: Adapted from Mauseth, J.D., *Botany: An Introduction to Plant Biology*, 2nd ed. Saunders College Publishing, Philadelphia, 1995.

Water movement can be explained by a difference in water potential

To understand how water moves, it is helpful to introduce **water potential,** which is defined as the **free energy** of water (see Chapter 7). Water potential (Ψ_w), represented by the Greek letter psi, is important in plant physiology because it is a measure of a cell's ability to absorb water by osmosis. (Recall from Chapter 5 and Figure 5-11 that **osmosis** is a special kind of diffusion that involves the net movement of water through a selectively permeable membrane from a region of higher effective concentration of water to a region of lower concentration.) Water potential also provides a measure of water's tendency to evaporate from cells.

The water potential of pure water is conventionally set at 0 megapascal (MPa) because it cannot be measured directly. (A megapascal is a unit of pressure equal to about 10.2 kg/cm^2, or 145.1 lb/in.^{2}.) However, botanists can measure differences in the free energy of water molecules in different situations. When solutes dissolve in water, the free energy of water decreases. Solutes induce **hydration,** in which water molecules surround ions and polar molecules, keeping them in solution by preventing them from coming together (see Fig. 2-10). The association of water molecules with hydrated molecules and ions reduces the motion of water molecules, decreasing their free energy. Thus, dissolved solutes lower the water potential to a negative number. Water moves from a region of higher (less negative) water potential to a region of lower (more negative) water potential.

The water potential of the soil varies, depending on how much water it contains. When a soil is extremely dry, its water potential is very low (very negative). When a soil is moister, its water potential is higher, although it still has a negative value because dissolved minerals are present in dilute concentrations (TABLE 35-2).

The water potential in root cells is also negative because of the presence of dissolved solutes. Roots contain more dissolved materials than soil water does, unless the soil is extremely dry. This means that under normal conditions the water potential of the root is more negative than the water potential of the soil. Thus, water moves by osmosis from the soil into the root.

TABLE 35-2	Representative Water Potential (Ψ_w) Values
PLANT PARTS AND THE NEARBY ENVIRONMENT	**WATER POTENTIALS IN MEGAPASCALS (MPa)**
Atmosphere near leaves	−80.0
Leaf interior	−1.5
Stem xylem	−0.7
Root interior	−0.4
Soil water	−0.1

Note: Actual values vary widely, depending on environmental factors such as solute concentration, hydrostatic pressure, and gravity.

© Cengage Learning

According to the tension–cohesion model, water is pulled up a stem

In 1896, Irish botanist Henry Dixon proposed the tension–cohesion model to explain the ascent of water against the force of gravity. Although initially met with widespread skepticism in the scientific community, this model has withstood the test of time. According to the **tension–cohesion model,** also known as the **transpiration–cohesion model,** water is pulled up the plant as a result of a *tension* produced at the top of the plant (FIG. 35-11). This tension is caused by the evaporative pull of transpiration. Recall from Chapter 34 that **transpiration** is the evaporation of water vapor from plants.

Most water loss from transpiration takes place through stomata, the numerous microscopic pores present on leaf and stem surfaces. Water vapor diffuses through the stomata into a *boundary layer* of quiet air adjacent to the leaf surface. The boundary layer is thicker when the wind is calm than when it is gusty; a thicker boundary layer results in a slower rate of transpiration.

The tension extends from leaves, where most transpiration occurs, down the stems and into the roots. It draws water up stem xylem to leaf cells that have lost water as a result of transpiration and pulls water from root xylem into stem xylem. As water is pulled upward, additional water from the soil is drawn into the roots. Thus, the pathway of water movement is as follows:

soil ⟶ root tissues (epidermis, cortex, etc.) ⟶ root xylem ⟶ stem xylem ⟶ leaf xylem ⟶ leaf mesophyll ⟶ atmosphere

This upward pulling of water is possible only as long as there is an unbroken column of water in xylem throughout the plant. Water forms an unbroken column in xylem because water molecules exhibit **cohesion**—that is, they are strongly attracted to one another—due to **hydrogen bonding.** In addition, the **adhesion** of water to the walls of xylem cells, also the result of hydrogen bonding, is an important factor in maintaining an unbroken column of water. Thus, the cohesive and adhesive properties of water enable it to form a continuous column that can be pulled up through the xylem.

The movement of water in xylem by the tension–cohesion mechanism can be explained in terms of water potential. The atmosphere has an extremely negative water potential. For example, air with a relative humidity of 50% has a water potential of −100 MPa; even moist air at a relative humidity of 90% has a negative water potential of −13 MPa. Thus, there is a water potential gradient from the least negative (the soil) up through the plant to the most negative (the atmosphere). This gradient literally pulls the water from the soil up through the plant.

Although the tension–cohesion model was first proposed toward the end of the 19th century, conclusive experimental evidence to support this mechanism was not obtained until the late 1990s and early 2000s. At that time, several research groups made direct measurements of the large negative pressure that exists in xylem, indicating that the water potential gradients in root, stem, and leaf xylem are adequate to explain the observed movement of water.

Transpiration is the driving force of the tension–cohesion model.

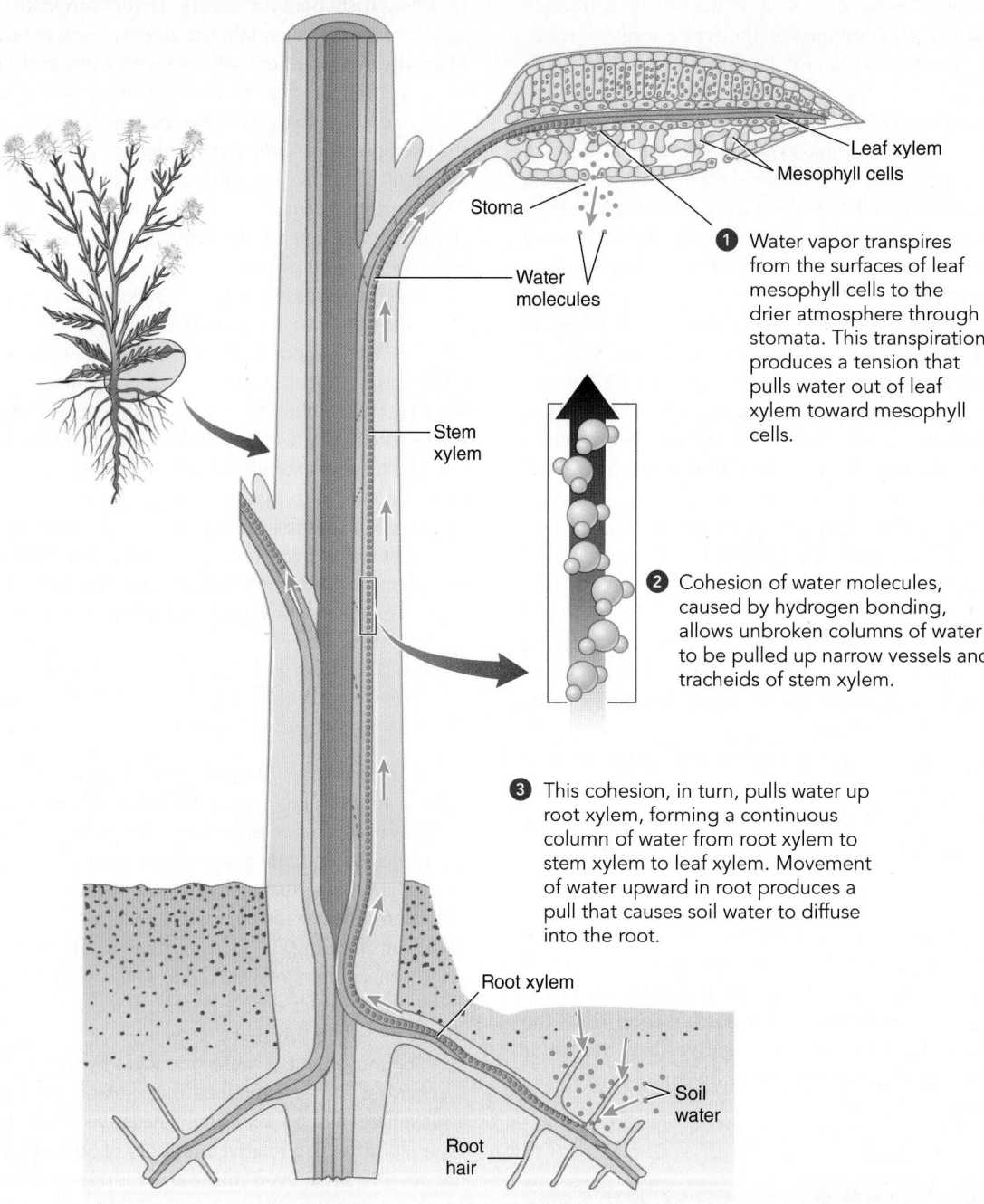

Leaf xylem
Mesophyll cells

Stoma

Water molecules

1 Water vapor transpires from the surfaces of leaf mesophyll cells to the drier atmosphere through stomata. This transpiration produces a tension that pulls water out of leaf xylem toward mesophyll cells.

Stem xylem

2 Cohesion of water molecules, caused by hydrogen bonding, allows unbroken columns of water to be pulled up narrow vessels and tracheids of stem xylem.

3 This cohesion, in turn, pulls water up root xylem, forming a continuous column of water from root xylem to stem xylem to leaf xylem. Movement of water upward in root produces a pull that causes soil water to diffuse into the root.

Root xylem

Soil water

Root hair

Figure 35-11 *Animation* **The tension–cohesion model**

PREDICT What do you think would happen if solutes added to the soil water were to make its water potential more negative than that of the root tissue?

© Cengage Learning

Is the tension–cohesion model powerful enough to explain the rise of water in the tallest plants? Plant biologists have studied this question for years. For example, a 2004 study of five of the eight tallest trees in the world (all redwoods, or *Sequoia sempervirens*) concluded that the tension produced by transpiration is strong enough to pull water upward to a maximum height of 130 m (422 ft). Because the height of the tallest known living tree, a coast redwood in California, measured 115.7 m (379 ft) in 2007, the tension–cohesion model easily accounts for the transport of water. Botanists generally consider the tension–cohesion model to be the dominant mechanism of xylem transport in most plants.

Root pressure pushes water from the root up a stem

In the less important mechanism for water transport, known as **root pressure,** water that moves into roots from the soil is *pushed* up through xylem toward the top of the plant. Root pressure occurs because mineral ions that are actively absorbed from the soil are pumped into the xylem, decreasing its water potential. This accumulation of ions has an osmotic effect, causing water to move into xylem cells from surrounding root cells. In turn, water moves into roots by osmosis because of the difference in water potential between the soil and root cells. The accumulation of water in root tissues produces a positive pressure (as high as +0.2 MPa) that forces the water up through the root xylem into the shoot.

Guttation, a phenomenon in which liquid water is forced out through special openings in the leaves (see Fig. 34-12), results from root pressure. However, root pressure is not strong enough to explain the rise of water to the tops of tall trees. Root pressure exerts an influence in much smaller plants, particularly in the spring when the soil is moist, but it clearly does not cause water to rise to the tops of the tallest plants. Furthermore, root pressure does not occur to any appreciable extent in summer (when water is often not plentiful in soil), yet water transport is greatest during hot summer days.

CHECKPOINT 35.2

- **CONNECT** *How is the movement of water in a plant related to water potential?*
- *How does the tension–cohesion model explain the rise of water in the tallest trees?*

35.3 TRANSLOCATION OF SUGAR IN SOLUTION

LEARNING OBJECTIVES

7 Describe the pathway of sugar translocation in plants.
8 Discuss the pressure–flow model of sugar translocation in phloem.

The sugar produced during photosynthesis is converted into sucrose (common table sugar), a disaccharide composed of one molecule of glucose and one of fructose (see Fig. 3-8b), before being loaded into phloem and translocated to the rest of the plant. Sucrose is the predominant photosynthetic product carried in phloem. Phloem sap also contains much smaller amounts of other materials, such as amino acids, organic acids, proteins, hormones, and certain minerals (e.g., potassium, chloride, phosphate, and magnesium). Sometimes disease-causing plant viruses are transported throughout the plant in phloem. Translocation of phloem sap is not as rapid as xylem transport (see Table 35-1).

Fluid within phloem tissue moves both upward and downward. Sucrose is translocated in individual sieve tubes from a *source,* an area of excess sugar supply (usually a leaf), to a *sink,* an area of storage (as insoluble starch) or of sugar use, such as roots, apical meristems, fruits, and seeds.

The pressure–flow model explains translocation in phloem

Current experimental evidence supports the **translocation** of dissolved sugar in phloem by the **pressure–flow model,** which was first proposed in 1926 by German scientist Ernst Münch. The pressure–flow model states that solutes (such as dissolved sugars) move in phloem by means of a pressure gradient, that is, a difference in pressure. The pressure gradient exists between the source, where the sugar is loaded into phloem, and the sink, where the sugar is removed from phloem.

At the source, the dissolved sucrose is moved from a leaf's mesophyll cells, where it was manufactured, into the companion cells, which load sucrose into the sieve tube elements of phloem. Sucrose loading occurs by active transport, a process that requires adenosine triphosphate (ATP) (**FIG. 35-12**). The ATP supplies energy to pump protons (H^+) out of the sieve tube elements, producing a proton gradient that drives the uptake of sugar through specific channels by the **cotransport** of protons back into the sieve tube elements. (Recall the discussion of a cotransport system in Chapter 5.) In the cotransport system involved in phloem loading, sugar is moved from a region of low concentration to a region of high concentration by coupling its transport to the transport of protons down their concentration gradient.

The sugar therefore accumulates in the sieve tube element. The increase in dissolved sugars in the sieve tube element at the source—a concentration that is two to three times as great as in surrounding cells—decreases (makes more negative) the water potential of that cell. As a result, water moves by osmosis from nearby xylem cells into the sieve tubes, increasing the **turgor pressure** (hydrostatic pressure) inside them. Thus, phloem loading at the source occurs as follows:

proton pump moves H^+ out of sieve tube element $\longrightarrow$ sugar is actively transported into sieve tube element $\longrightarrow$ water diffuses from xylem into sieve tube element $\longrightarrow$ turgor pressure increases within sieve tube

At its destination (the sink), sugar is unloaded by various mechanisms, both active and passive, from the sieve tube elements. With the loss of sugar, the water potential in the sieve tube elements at the sink increases (becomes less negative). Therefore, water moves out of the sieve tubes by osmosis and into surrounding cells where the water potential is more negative. Most of this water diffuses back to the xylem to be transported upward. This water movement decreases the turgor

In phloem solutes move from sources to sinks. The pressure gradient within the sieve tube causes translocation from the area of higher turgor pressure (the source) to the area of lower turgor pressure (the sink).

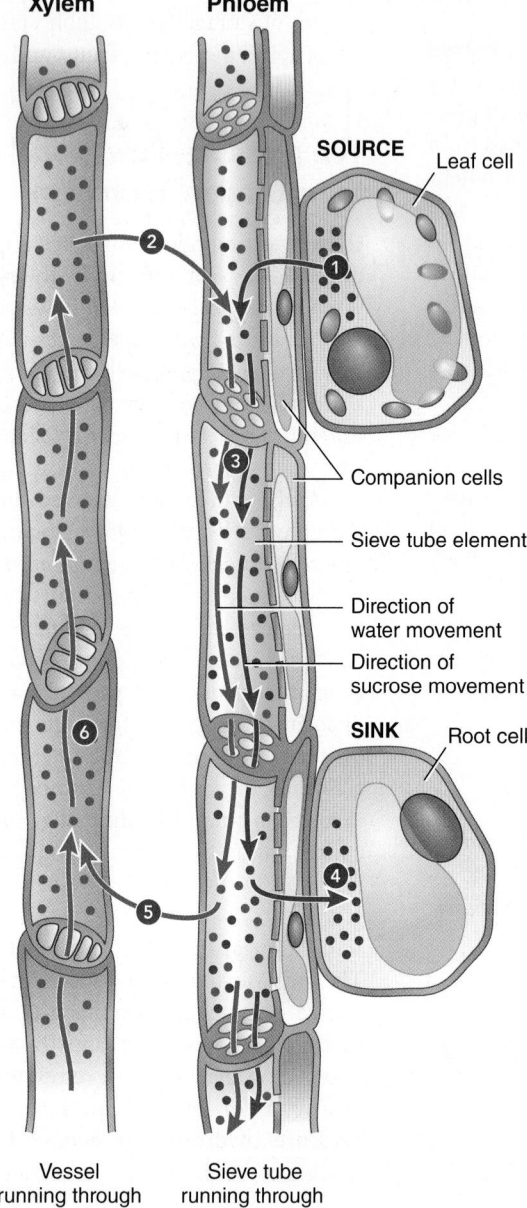

Xylem   **Phloem**

❶ Sucrose is actively loaded into sieve tube elements (requires ATP).

❷ Water diffuses from xylem as a result of decreased (more negative) water potential in sieve tube.

❸ The increase in pressure forces the fluid through the sieve tube to the root.

❹ Sucrose is actively and passively unloaded into sink cell, such as parenchyma cell in the root cortex. (Active unloading requires ATP.)

❺ Water diffuses from phloem to xylem as a result of increased (less negative) water potential in sieve tube.

❻ Once water is in the xylem, transpiration pulls water to leaves again.

SOURCE Leaf cell

Companion cells

Sieve tube element

Direction of water movement
Direction of sucrose movement

SINK Root cell

Vessel running through length of plant

Sieve tube running through length of plant

Figure 35-12 *Animation* **The pressure–flow model**

PREDICT In what direction would transport occur in the phloem if the source were starch being broken down in a storage root in the spring and the sink were a developing leaf?
© Cengage Learning

pressure inside the sieve tubes at the sink. Thus, phloem unloading at the sink proceeds as follows:

sugar is transported out of sieve tube element ⟶ water diffuses out of sieve tube element and into xylem ⟶ turgor pressure decreases within sieve tube

The pressure–flow model explains the movement of dissolved sugar in phloem by means of a pressure gradient. The difference in sugar concentrations between the source and the sink causes translocation in phloem as water and dissolved sugar flow along the pressure gradient. This pressure gradient pushes the sugar solution through phloem much as water is forced through a hose.

The actual translocation of dissolved sugar in phloem does not require metabolic energy. However, the loading of sugar at the source and the active unloading of sugar at the sink require energy derived from ATP to move the sugar across cell membranes by active transport.

Although the pressure–flow model adequately explains current data on phloem translocation, much remains to be learned about this complex process. Phloem translocation is difficult to study in plants. Because phloem cells are under pressure, cutting into phloem to observe it releases the pressure and causes the contents of the sieve tube elements (the phloem sap) to exude and mix with the contents of other severed cells that are also unavoidably cut. In the 1950s, scientists developed a unique research tool to avoid contaminating the phloem sap: aphids, which are small insects that insert their mouthparts into phloem sieve tubes for feeding (**FIG. 35-13**). The pressure in the punctured phloem drives the sugar solution through the aphid's mouthpart into its digestive system. When the aphid's mouthpart is severed from its body by a laser beam, the sugar solution continues to flow through the mouthpart at a rate proportional to the pressure in phloem. This rate can be measured, and the effects on phloem transport of different environmental conditions—varying light intensities, darkness, and mineral deficiencies, for example—can be ascertained.

The identity and proportions of translocated substances can also be determined using severed aphid mouthparts. This technique has verified that in most plant species the sugar sucrose is the primary carbohydrate transported in phloem; however, some species transport other sugars, such as raffinose, or sugar alcohols, such as sorbitol.

CHECKPOINT 35.3

- *How does the direction of transport differ in xylem and phloem?*
- *How does the pressure–flow model explain sugar movement in phloem? Include in your answer the activities at source and sink.*

How can phloem sap be studied without cutting non-phloem cells that would contaminate the sap?

HYPOTHESIS: An aphid mouthpart can be used to penetrate a single sieve tube.

EXPERIMENT: After allowing aphids to insert their mouthparts into phloem of a stem, researchers anesthetized the feeding aphids with CO_2 and used a laser to cut their bodies away from their mouthparts. The mouthparts remained in the phloem and functioned like miniature pipes.

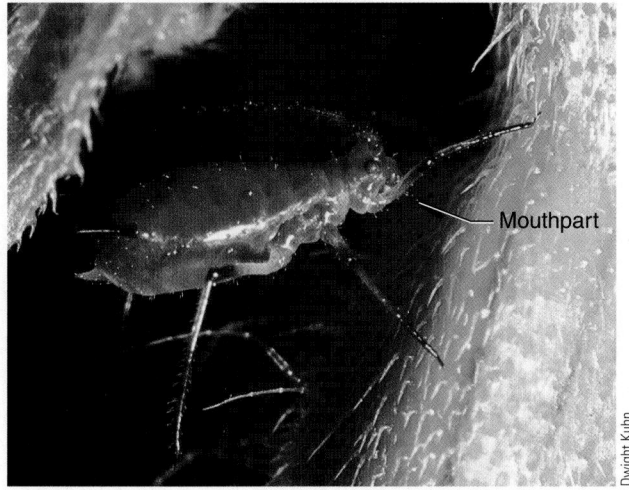

Mouthpart

Dwight Kuhn

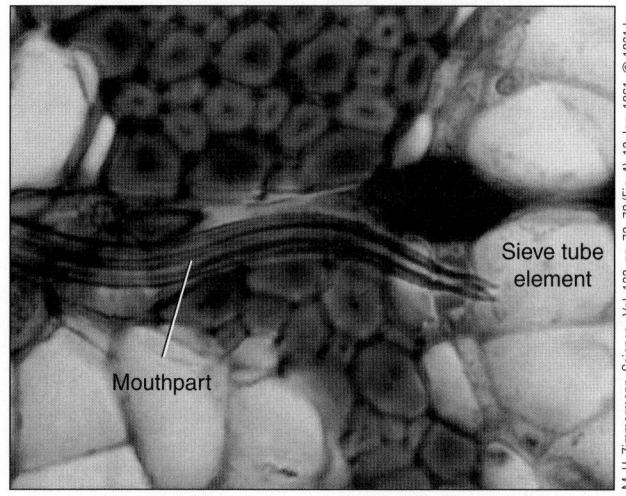

Sieve tube element

Mouthpart

M. H. Zimmermann, *Science*, Vol. 133, pp. 73–79 (Fig. 4), 13 Jan. 1961. © 1961 by the American Association for the Advancement of Science

25 μm

(a) Aphid feeding on a stem.

(b) LM of phloem cells, showing an aphid mouthpart penetrating a sieve tube element.

RESULTS AND CONCLUSION: About 1 mm³ of phloem sap exuded from each severed mouthpart per hour for several days, so researchers were able to collect and analyze the composition of the sap.

SOURCE: M.H. Zimmermann, *Science*, Vol. 133, pp. 73–79 (Fig. 4), 13 Jan. 1961. © 1961 by the American Association for the Advancement of Science

Figure 35-13 Collecting and analyzing phloem sap

This method, first used in the 1950s by insect physiologists studying aphids, was quickly adopted by plant physiologists studying phloem translocation.

PREDICT Would you expect to observe outward movement of fluid through the aphid's mouthpart if the stem xylem of a large plant were penetrated?

SUMMARY: FOCUS ON LEARNING OBJECTIVES

35.1 Stem Growth and Structure *(page 740)*

1 Label cross sections of herbaceous eudicot and monocot stems, and describe the functions of each tissue.

- Herbaceous stems possess an epidermis, vascular tissue, and either ground tissue or cortex and pith. The **epidermis** is a protective layer covered by a water-conserving **cuticle. Stomata** permit gas exchange. **Xylem** conducts water and dissolved minerals, and **phloem** conducts dissolved sugar. The **cortex, pith,** and ground tissue function primarily for storage.

- All herbaceous stems have the same basic tissues, but their arrangement varies. Herbaceous eudicot stems have the vascular bundles arranged in a circle (in cross section) and have a distinct cortex and pith. Monocot stems have vascular bundles scattered in ground tissue.

2 Name the two lateral meristems and describe the tissues that arise from each.

- **Vascular cambium** is the lateral meristem that produces secondary xylem (wood) and secondary phloem (inner bark).

- **Cork cambium** produces **periderm,** which consists of cork parenchyma and cork cells. Cork cells are the functional replacement for epidermis in a woody stem. Cork parenchyma functions primarily for storage in a woody stem.

3 Outline the transition from primary growth to secondary growth in a woody stem.

- **Secondary growth** (the production of the secondary tissues, wood and bark) occurs in some flowering plants (woody eudicots) and in all cone-bearing gymnosperms. During secondary growth, the vascular cambium divides in two directions to form secondary xylem (to the inside) and secondary phloem (to the outside). The primary xylem and primary phloem in the original vascular bundles become separated as secondary growth proceeds.

35.2 Water Transport (page 748)

4 Describe the pathway of water movement in plants.

- Water and dissolved minerals move from the soil into root tissues (epidermis, cortex, etc.). Once in root xylem, water and minerals move upward, from root xylem to stem xylem to leaf xylem. Much of the water entering the leaf exits leaf veins and passes into the atmosphere.

5 Define *water potential*.

- **Water potential** (Ψ_w) is a measure of the **free energy** of water. Pure water has a water potential of 0 megapascal (MPa), whereas water with dissolved solutes has a negative water potential. Water moves from an area of higher (less negative) water potential to an area of lower (more negative) water potential.

6 Compare tension–cohesion and root pressure as mechanisms responsible for the rise of water and dissolved minerals in xylem.

- The **tension–cohesion model** explains the rise of water in even the tallest plants. The evaporative pull of **transpiration** causes tension at the top of the plant. This tension is the result of a water potential gradient that ranges from the slightly negative water potentials in the soil and roots to the very negative water potentials in the atmosphere. As a result of the **cohesion** of water molecules to each other and their **adhesion** to the xylem cell walls, the column of water pulled up through the plant remains unbroken.

- **Root pressure,** caused by the movement of water into roots from the soil as a result of the active absorption of mineral ions from the soil, helps explain the rise of water in smaller plants, particularly when the soil is wet. Root pressure pushes water up through xylem.

35.3 Translocation of Sugar in Solution (page 751)

7 Describe the pathway of sugar translocation in plants.

- **Translocation** of dissolved sugar occurs upward or downward in phloem, from a source (an area of excess sugar, usually a leaf) to a sink (an area of storage or of sugar use, such as roots, apical meristems, fruits, and seeds). Sucrose is the predominant sugar translocated in phloem.

8 Discuss the pressure–flow model of sugar translocation in phloem.

- Movement of materials in phloem is explained by the **pressure–flow model.** Companion cells actively load sugar into the sieve tubes at the source; ATP is required for this process. The ATP supplies energy to pump protons out of the sieve tube elements. The proton gradient drives the uptake of sugar by the cotransport of protons back into the sieve tube elements. Sugar therefore accumulates in the sieve tube element, causing the movement of water into the sieve tubes by **osmosis.**

- Companion cells actively (requiring ATP) and passively (not requiring ATP) unload sugar from the sieve tubes at the sink. As a result, water leaves the sieve tubes by osmosis, decreasing the **turgor pressure** (hydrostatic pressure) inside the sieve tubes.

- The flow of materials between source and sink is driven by the turgor pressure gradient produced by water entering phloem at the source and water leaving phloem at the sink.

TEST YOUR UNDERSTANDING

Know and Comprehend

1. Axillary buds are located (a) at the tips of stems (b) in unusual places, such as on roots (c) in the region between two successive nodes (d) in the upper angle between a leaf and the stem to which it is attached (e) within the loosely arranged cells of lenticels

2. Ground tissue in monocot stems performs the same functions as _____ and _____ in herbaceous eudicot stems. (a) phloem; xylem (b) cork cambium; vascular cambium (c) epidermis; periderm (d) primary xylem; secondary xylem (e) cortex; pith

3. Which of the following statements is *false?* (a) primary growth is an increase in the length of a plant (b) primary growth occurs at both apical and lateral meristems (c) all plants have primary growth (d) herbaceous stems have primary growth, whereas woody stems have both primary and secondary growth (e) buds are embryonic shoots that contain apical meristems

4. The two lateral meristems responsible for secondary growth are (a) phloem and xylem (b) cork cambium and vascular cambium (c) epidermis and periderm (d) primary xylem and secondary xylem (e) cortex and pith

5. Cork cambium and the tissues it produces are collectively called (a) periderm (b) lenticels (c) cortex (d) epidermis (e) wood

6. Each annual ring in a section of wood represents one year's growth of (a) primary xylem (b) secondary xylem (c) primary xylem or secondary xylem in alternate years (d) primary phloem (e) secondary phloem

7. Water potential is (a) the formation of a proton gradient across a cell membrane (b) the transport of a watery solution of sugar in phloem (c) the transport of water in both xylem and phloem (d) the removal of sucrose at the sink, causing water to move out of the sieve tubes (e) the free energy of water in a particular situation

8. Which of the following is a mechanism of water movement in xylem that is responsible for guttation? (a) pressure–flow (b) tension–cohesion (c) root pressure (d) active transport of potassium ions into guard cells (e) transpiration

9. Which of the following is a mechanism of water movement in xylem that combines the evaporative pull of transpiration with the cohesive and adhesive properties of water? (a) pressure–flow (b) tension–cohesion (c) root pressure (d) active transport of potassium ions into guard cells (e) guttation

10. Which of the following is a mechanism of phloem transport in which dissolved sugar is moved by means of a pressure gradient that exists between the source and the sink? (a) pressure–flow (b) tension–cohesion (c) root pressure (d) active transport of potassium ions into guard cells (e) guttation

11. How does increasing solute concentration affect water potential? (a) water potential becomes more positive (b) water potential becomes more negative (c) water potential becomes

more positive under certain conditions and more negative under other conditions (d) water potential is not affected by solute concentration (e) water potential is always zero when solutes are dissolved in water

Apply and Analyze

12. **VISUALIZE** Sketch side by side two daughter cells formed after mitosis of a meristematic cell in the vascular cambium of a eudicot stem. Label the cell formed toward the interior of the stem a meristematic cell. Should the daughter cell formed toward the outside be labeled primary xylem, primary phloem, secondary xylem, or secondary phloem?

Evaluate and Synthesize

13. When a strip of bark is peeled off a tree branch, what tissues are usually removed?
14. Could a tree grow to a height of 150 m (488 ft)? Why or why not?

15. **EVOLUTION LINK** Like stems in general, some vines are herbaceous and others are woody. Tropical rain forests have a greater diversity of vines than in any other environment on Earth, and most of these vines are woody. Develop a hypothesis to explain why natural selection has favored the evolution of more species of woody vines (as opposed to herbaceous vines) in tropical rain forests.
16. **INTERPRET DATA** A plant cell with a water potential $\Psi_w = -1.5$ MPa is adjacent to a second cell with $\Psi_w = -1.8$ MPa. Will water flow by osmosis from one cell to the other? If so, in which direction?

aplia To access course materials, such as Aplia and other companion resources, please visit **www.cengagebrain.com.**

36 Roots and Mineral Nutrition

Storage roots. Cassava (*Manihot esculenta*) plants have been dug up to show the growth habit of the storage roots. Cassava roots are an important staple crop in the tropics. Some cassava roots contain a poison (prussic acid), which is destroyed during cooking.

© Inga Spence/Alamy

KEY CONCEPTS

36.1 Root structure is related to function. Primary tissues (epidermis, cortex, endodermis, pericycle, xylem, and phloem) of roots develop from root apical meristems. Secondary tissues (wood and bark) of woody roots develop from lateral meristems.

36.2 Roots associate and interact with many organisms, including soil bacteria and fungi.

36.3 Soil, the layer of Earth's crust that has been modified by contact with weather, wind, water, and organisms, contains most of the essential elements required by plants.

Branching underground root systems are often more extensive than a plant's aerial parts. The roots of a corn plant, for example, may grow to a depth of 2.5 m (about 8 ft) and spread outward 1.2 m (4 ft) from the stem. Desert-dwelling tamarisk (*Tamarix*) trees reportedly have roots that grow to a depth of 50 m (163 ft) to tap underground water. The extent of a plant's root depth and spread varies considerably among different species and even among different individuals in the same species.

Because roots are usually underground and out of sight, people do not always appreciate the important functions they perform. First, as anyone who has ever pulled weeds can attest, roots anchor a plant securely in the soil. A plant needs a solid foundation from which to grow. Firm anchorage is essential to a plant's survival so that the stem remains upright, enabling leaves to absorb sunlight effectively. Second, roots absorb water and dissolved minerals (inorganic nutrients) such as nitrates, phosphates, and sulfates, which are necessary for synthesizing important organic molecules. These dissolved minerals are then transported throughout the plant in the xylem.

Many roots perform the function of storage. Carrots, sweet potatoes, cassava, and other root crops are important sources of human food (see photograph). Surplus sugars produced in the leaves by photosynthesis are transported in the phloem to the roots for food storage (usually as starch or sucrose) until needed. Other plants, particularly those living in arid regions, possess storage roots adapted to store water.

In certain species roots are modified for functions other than anchorage, absorption, conduction, and storage. Roots specialized to perform uncommon functions are discussed later in the chapter.

36.1 ROOT STRUCTURE AND FUNCTION

LEARNING OBJECTIVES

1 Distinguish between taproot and fibrous root systems.
2 Compare cross sections of a primary eudicot root and a monocot root, and describe the functions of each tissue.
3 Trace the pathway of water and mineral ions from the soil through the various root tissues and distinguish between the symplast and apoplast.
4 Discuss the structure of roots with secondary growth.
5 Describe at least three roots that are modified to perform uncommon functions.

Two types of root systems, a taproot system and a fibrous root system, occur in plants (**FIG. 36-1**). A **taproot system** consists of one main root that formed from the seedling's enlarging **radicle,** or embryonic root. Many lateral roots of various sizes branch out of a taproot. Taproots are characteristic of many eudicots and gymnosperms. A dandelion is a good example of a common herbaceous plant with a taproot system. A few trees, such as hickory, retain their taproots, which become quite massive as the plants age. Most trees, however, have taproots when young and later develop large, shallow, lateral roots from which other roots branch off and grow downward.

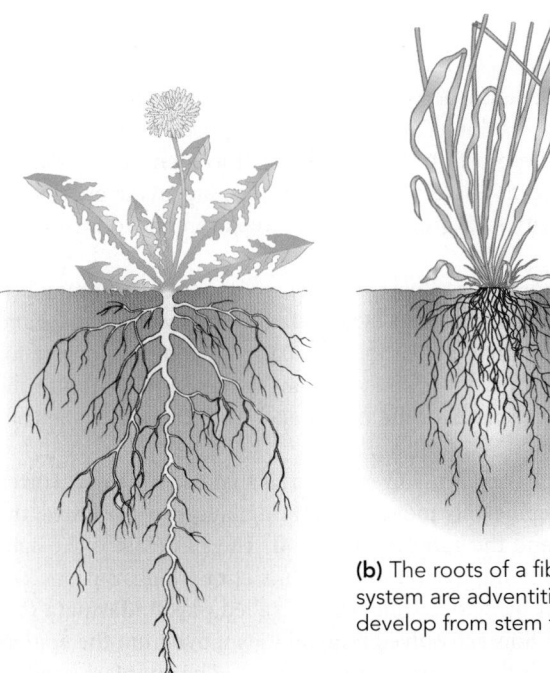

(b) The roots of a fibrous root system are adventitious and develop from stem tissue.

(a) A taproot system develops from the embryonic root in the seed.

Figure 36-1 *Animation* **Root systems**
© Cengage Learning

A **fibrous root system** has several to many roots of similar size developing from the end of the stem, with lateral roots branching off these roots. Fibrous root systems form in plants that have a short-lived embryonic root. The roots originate first from the base of the embryonic root and later from stem tissue. The main roots of a fibrous root system do not arise from pre-existing roots but from the stem; such roots are called **adventitious** roots. Adventitious organs occur in an unusual location, such as roots that develop on a stem or buds that develop on roots. Onions, crabgrass, and other monocots have fibrous root systems.

Taproot and fibrous root systems are adapted to obtain water in different sections of the soil. Taproot systems often extend down into the soil to obtain water located deep underground, whereas fibrous root systems, which are located relatively close to the soil surface, are adapted to obtain rainwater from a larger area as it drains into the soil.

Roots have root caps and root hairs

Because of the need to adapt to the soil environment instead of the atmospheric environment, roots have several structures, such as root caps and root hairs, that shoots lack. Although stems and leaves have various types of hairs, they are distinct from root hairs in structure and function.

Each root tip is covered by a **root cap,** a protective, thimblelike layer many cells thick that covers the delicate root **apical meristem** (**FIG. 36-2a**; also see Fig. 33-6). As the root grows, pushing its way through the soil, cells of the root cap are sloughed off by the frictional resistance of the soil particles and replaced by new cells formed by the root apical meristem. The root cap cells secrete lubricating polysaccharides that reduce friction as the root passes through the soil. The root cap also appears to be involved in orienting the root so that it grows downward (see discussion of gravitropism in Chapter 38). When a root cap is removed, the root apical meristem grows a new cap. However, until the root cap has regenerated, the root grows randomly rather than in the direction of gravity.

Root hairs are short-lived tubular extensions of epidermal cells located just behind the growing root tip. Root hairs continually form in the area of cell maturation closest to the root tip to replace those that are dying off at the more mature end of the root-hair zone (**FIG. 36-2b**; also see Fig. 33-6). Each root hair is short (typically less than 1 cm, or 0.4 in., in length), but they are quite numerous. Root hairs greatly increase the absorptive capacity of roots by increasing their surface area in contact with moist soil. Soil particles are coated with a microscopically thin layer of water in which minerals are dissolved. The root hairs establish an intimate contact with soil particles, which allows efficient absorption of water and minerals. They also modify the composition of the soil by exuding a diverse array of organic molecules.

Scientists are actively studying the control of root hair development. Through investigations using the model angiosperm *Arabidopsis thaliana,* it has been learned that the spatial arrangement of root hairs in the root epidermis is a consequence

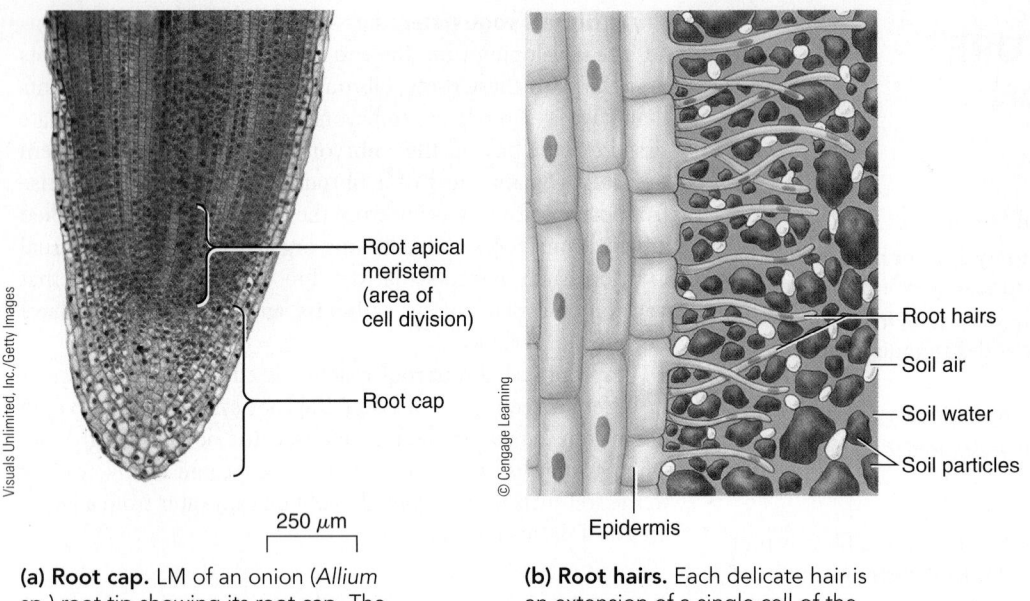

(a) **Root cap.** LM of an onion (*Allium* sp.) root tip showing its root cap. The root apical meristem is protected by the root cap.

Root apical meristem (area of cell division)

Root cap

250 μm

Root hairs
Soil air
Soil water
Soil particles

Epidermis

(b) **Root hairs.** Each delicate hair is an extension of a single cell of the root epidermis. Root hairs increase the surface area of the root in contact with the soil.

Figure 36-2 *Animation* **Structures unique to roots**

of complex interactions between cells that form root hairs and adjacent cells that do not. Many of the genes involved code for **transcription factors.** (You may recall from Chapter 14 that transcription factors are DNA-binding proteins that regulate the synthesis of RNA from a DNA template.) Some of these genes have been shown to have an ancient origin with evolutionary significance. Researchers have demonstrated that two genes that code for transcription factors involved in the control of root hair formation in the sporophyte stage of *Arabidopsis* are essentially the same as two genes that control rhizoid formation in the gametophyte stage of *Physcomitrella patens,* the first moss to have its genome sequenced (see Fig. 27-11). Mosses such as *Physcomitrella* represent an ancient lineage of land plants thought to have diverged from the line leading to vascular plants about 420 million years ago. It therefore appears that these genes were conserved during the evolution from a dominant gametophyte in mosses to a dominant sporophyte in flowering plants.

The arrangement of vascular tissues distinguishes the roots of herbaceous eudicots and monocots

Although considerable variation exists in herbaceous eudicot and monocot roots, they all have an outer protective covering (epidermis), a cortex for storage of starch and other organic molecules, and vascular tissues for conduction. Let us first consider the structure of herbaceous eudicot roots.

In most herbaceous eudicot roots, the central core of vascular tissue lacks pith The buttercup root is a representative eudicot root with primary growth (**FIG. 36-3**). Like other parts of this herbaceous eudicot, a single layer of protective

tissue, the **epidermis,** covers its roots. The root hairs are a modification of the root epidermis that enables it to absorb more water from the soil. The root epidermis does not secrete a thick, waxy cuticle in the region of root hairs because this layer would impede the absorption of water from the soil. Both the lack of a cuticle and the presence of root hairs increase absorption.

Beginning with the root hairs, most of the water that enters the root moves along the cell walls rather than enters the cells. A major component of cell walls is cellulose, which absorbs water as a sponge does. An example of the absorptive properties of cellulose is found in cotton balls, which are almost pure cellulose.

The root **cortex,** which is made up primarily of loosely packed **parenchyma** cells, composes the bulk of a herbaceous eudicot root. Roots usually lack supporting **collenchyma** cells, probably because the soil supports the root, although roots as they age may develop some **sclerenchyma** (another supporting tissue; see Chapter 33). The primary function of the root cortex is storage. A microscopic examination of the parenchyma cells that form the cortex often reveals numerous amyloplasts (see Figs. 36-3b and 3-9a), which store starch. Starch, an insoluble carbohydrate composed of glucose subunits, is the most common form of stored energy in plants. When used at a later time, these reserves provide energy for such activities as growth and cell replacement following an injury.

The large intercellular (between-cell) spaces, a common feature of the root cortex, provide a pathway for water uptake and allow for aeration of the root. The oxygen that root cells need for aerobic respiration diffuses from air spaces in the soil into the intercellular spaces of the cortex and from there into the cells of the root.

The water and dissolved minerals that enter the root cortex from the epidermis move in solution along two pathways: the symplast and the apoplast (**FIG. 36-4**). The **symplast** is the continuum of living cytoplasm, which is connected from one cell to the next by cytoplasmic bridges called **plasmodesmata** (see Fig. 5-24). Some dissolved mineral ions move from the epidermis through the cortex via the symplast. The **apoplast** consists of the interconnected porous cell walls of a plant along which water and mineral ions move freely. The water and mineral ions can diffuse across the cortex without ever crossing a plasma membrane to enter a living cell. Together the symplast and apoplast make up the entire plant body.

The inner layer of the cortex, the **endodermis,** regulates the movement of water and minerals that enter the xylem in the

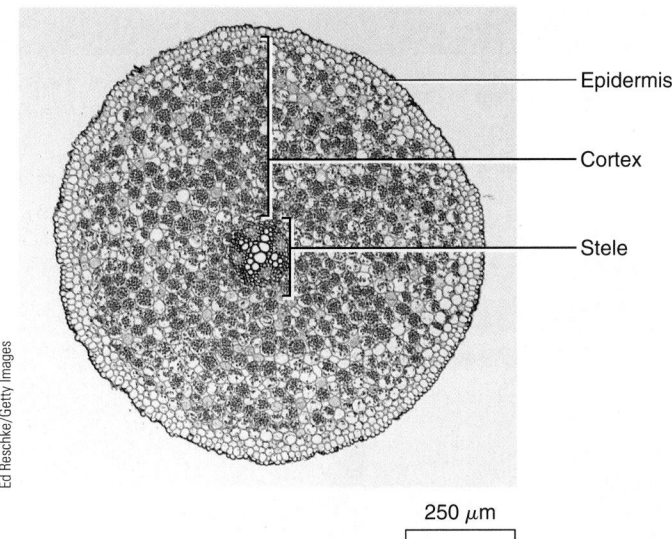

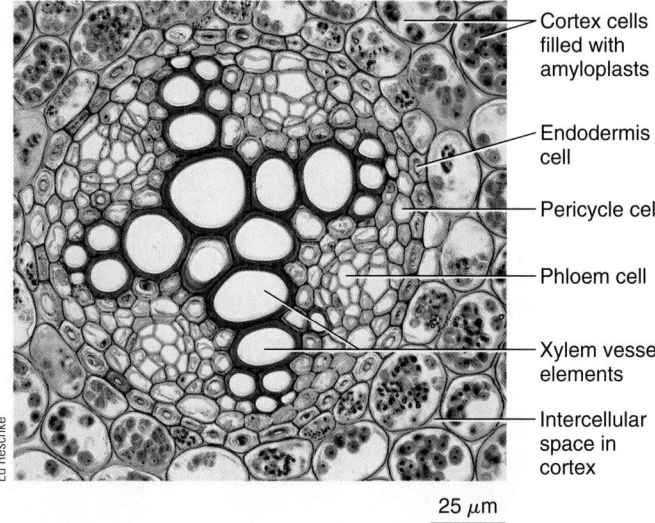

Epidermis

Cortex

Stele

Cortex cells
filled with
amyloplasts

Endodermis
cell

Pericycle cell

Phloem cell

Xylem vessel
elements

Intercellular
space in
cortex

Ed Reschke/Getty Images

Ed Reschke

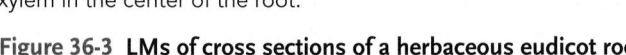

250 μm

25 μm

(a) Eudicot root. Cortex composes the bulk of herbaceous eudicot roots. Note the X-shaped xylem in the center of the root.

(b) Stele of eudicot root. Surrounding the solid core of vascular tissues is a single layer of pericycle, which is meristematic in growing roots.

Figure 36-3 LMs of cross sections of a herbaceous eudicot root

Shown is a buttercup (*Ranunculus*) root.

KEY POINT

The symplast pathway (*blue*) consists of living cells interconnected by plasmodesmata. The apoplast pathway (*red*) uses cell walls and intercellular spaces.

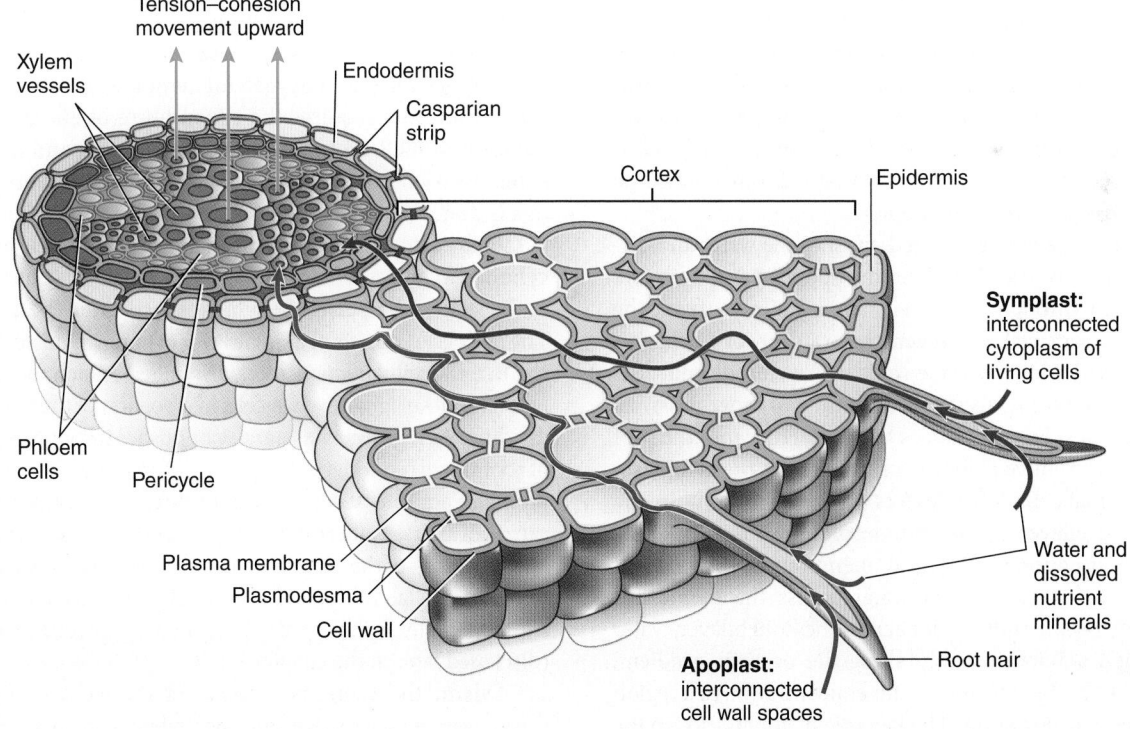

Xylem
vessels

Tension–cohesion
movement upward

Endodermis

Casparian
strip

Cortex

Epidermis

Symplast:
interconnected
cytoplasm of
living cells

Phloem
cells

Pericycle

Plasma membrane

Plasmodesma

Cell wall

Apoplast:
interconnected
cell wall spaces

Water and
dissolved
nutrient
minerals

Root hair

Figure 36-4 *Animation* **Pathways of water and dissolved minerals in the root**

Water and dissolved minerals travel from cell to cell along the interconnected porous cell walls (the apoplast) or from one cell's cytoplasm to another through plasmodesmata (the symplast). On reaching the endodermis, water and minerals can move into the root's center only if they pass through a plasma membrane and enter the cytoplasm of an endodermal cell.

© Cengage Learning

PREDICT Would you expect to find minerals not needed by the plant in the apoplast? in the symplast? Explain your answer.

The endodermis regulates the kinds of minerals that are absorbed from the soil and conducted to the rest of the plant body.

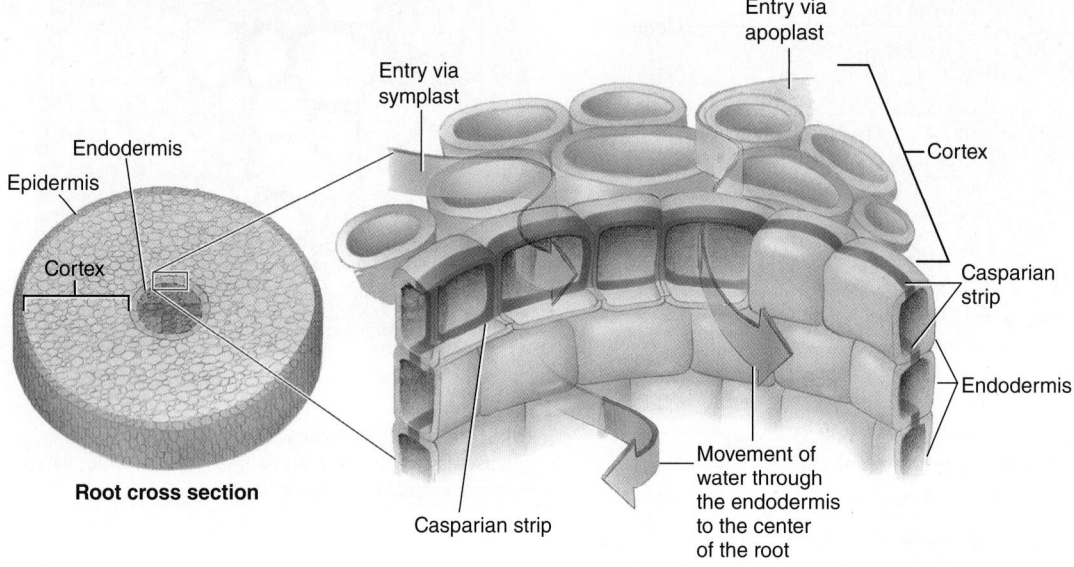

Figure 36-5 *Animation* **Endodermis and mineral uptake**

Note the Casparian strip around the radial and transverse walls that prevents water and dissolved minerals from passing into the stele along endodermal cell walls. To reach the vascular tissues, water and dissolved minerals must pass through the plasma membranes of endodermal cells.

© Cengage Learning

CONNECT Does the entry of water from the apoplast into the endodermal cells require a direct or indirect input of metabolic energy? Explain your answer.

root's center. Structurally, the endodermis differs from the rest of the cortex. Endodermal cells fit snugly against one another, and each has a special bandlike region, called a **Casparian strip** (**FIG. 36-5**), on its radial (side) and transverse (upper and lower) walls. Casparian strips contain *suberin,* a fatty material that is waterproof. (Recall from Chapter 33 that suberin is also the waterproof material in cork cell walls.)

Until the endodermis is reached, most of the water and dissolved minerals have traveled along the apoplast and therefore have not passed through a plasma membrane or entered the cytoplasm of a root cell. However, the waterproof Casparian strip on the radial and transverse walls of the endodermal cells prevents water and minerals from continuing to move passively along the cell walls. For substances to pass farther into the root interior, they must move from the cell walls into the cytoplasm of the endodermal cells. Water enters by osmosis, both directly across the lipid bilayer of the endodermal plasma membrane and through **aquaporins,** integral membrane proteins that facilitate the rapid transport of water across the membrane (**FIG. 36-6a**). As in water movement across the lipid bilayer, water moves through aquaporins only along the osmotic gradient, from a region of higher effective water concentration to a region of lower water concentration. The key role of aquaporins in the uptake of water by root cells is currently under active study.

Minerals enter endodermal cells by passing through carrier proteins in their plasma membranes. In **carrier-mediated active transport,** the mineral ions are pumped *against* their concentration gradient—that is, from an area of *low* concentration of that mineral in the soil solution to an area of *high* concentration in the plant's cells (**FIG. 36-6b**). One of many reasons root cells require sugar and oxygen for aerobic respiration is that this active transport requires the expenditure of energy, usually in the form of ATP. From the endodermis, water and mineral ions enter the root xylem (botanists do not yet know precisely how this is done) and are conducted to the rest of the plant.

At the center of a eudicot primary root is the **stele,** or **vascular cylinder,** a central cylinder of vascular tissues. The outermost layer of the stele is the **pericycle,** which is just inside the endodermis. The pericycle consists of a single layer of parenchyma cells that give rise to multicellular *lateral roots,* also called branch roots (**FIG. 36-7**). Lateral roots originate when cells in a portion of the pericycle start dividing. As it grows, the lateral root pushes through several layers of root tissue (endodermis, cortex, and epidermis) before entering the soil. Each lateral root has all the structures and features—root cap, root hairs, epidermis, cortex, endodermis, pericycle, xylem, and phloem—of the larger root from which it emerges. In addition to producing lateral roots, the pericycle is involved in forming the lateral meristems that produce secondary growth in woody roots (discussed later in the chapter).

Xylem, the centermost tissue of the stele, often has two, three, four, or more extensions, or "xylem arms" (see Fig. 36-3b). **Phloem** is located in patches between the xylem arms. The xylem and phloem of the root have the same functions and kinds of cells as in the rest of the plant: water and dissolved minerals are conducted in **tracheids** and **vessel elements** of xylem, and dissolved sugar (sucrose) is conducted in **sieve tube elements** of phloem.

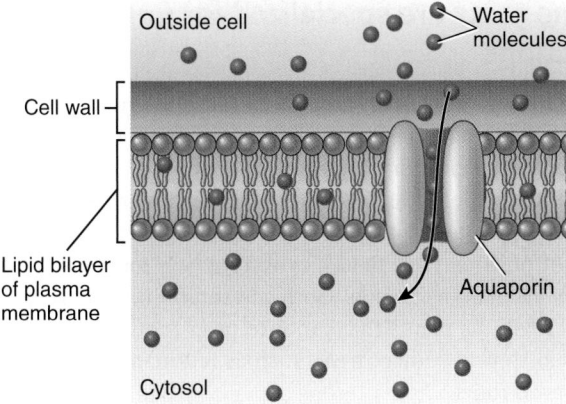

(a) Movement of water. Water crosses membranes by diffusion through the lipid layer (*left*) and, more rapidly, through aquaporins (*right*).

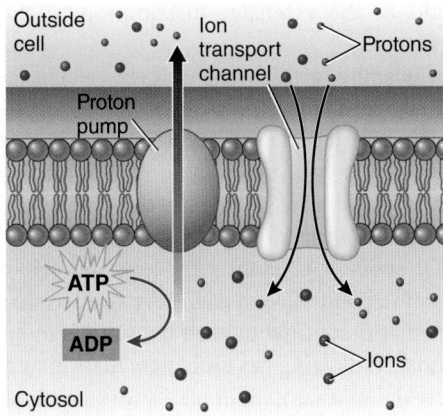

(b) Movement of mineral ions by carrier-mediated active transport. The lipid bilayer of plant membranes is impermeable to mineral ions, which can only pass through ion transport channels. The hydrolysis of ATP pumps protons outside the cell, forming an energy-charged proton gradient. Ion transport channels allow mineral ions to enter the cell with protons.

Figure 36-6 **Transport of water and minerals across plant membranes**
© Cengage Learning

After passing through the endodermal cells, water enters the root xylem, often at one of the xylem arms. Up to this point, the pathway of water has been horizontal from the soil into the center of the root:

root hair/epidermis ⟶ cortex ⟶ endodermis ⟶ pericycle ⟶ root xylem

Once water enters the xylem, it is transported upward through root xylem into stem xylem and from there to the rest of the plant.

One direction of phloem conduction is from the leaves, where sugar is made by photosynthesis, to the root, where sugar

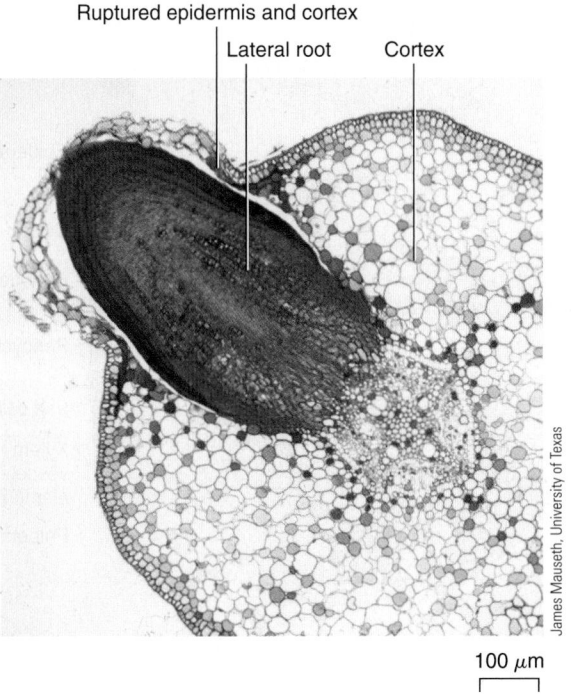

100 μm

Figure 36-7 **LM of a lateral root**
Lateral roots originate at the pericycle.

is used for the growth and maintenance of root tissues or stored, usually as starch. Another direction of phloem conduction is from the root, where sugar is stored as starch, to other parts of the plant, where sugar is used for growth and maintenance of tissues.

The *vascular cambium*, which gives rise to secondary tissues in woody plants, is sandwiched between the xylem and phloem. Because it has an inner core of vascular tissue, the primary eudicot root lacks **pith,** a ground tissue found in the centers of many stems and roots.

Xylem does not form the central tissue in some monocot roots Monocot roots vary considerably in internal structure compared with eudicot roots. The layers found in some monocot roots, starting from the outside, are epidermis, cortex, endodermis, and pericycle (**FIG. 36-8**). Unlike the xylem in herbaceous eudicot roots, the xylem in many monocot roots does not form a solid cylinder in the center. Instead, the phloem and xylem are in separate, alternating bundles arranged around the central pith, which consists of parenchyma cells.

Because virtually no monocots have secondary growth, no vascular cambium exists in monocot roots. Despite their lack of secondary growth, long-lived monocots, such as palms, may have thickened roots produced by a modified form of primary growth in which parenchyma cells in the cortex divide and enlarge.

Woody plants have roots with secondary growth

Plants that produce stems with secondary growth also produce roots with secondary growth. These plants—gymnosperms

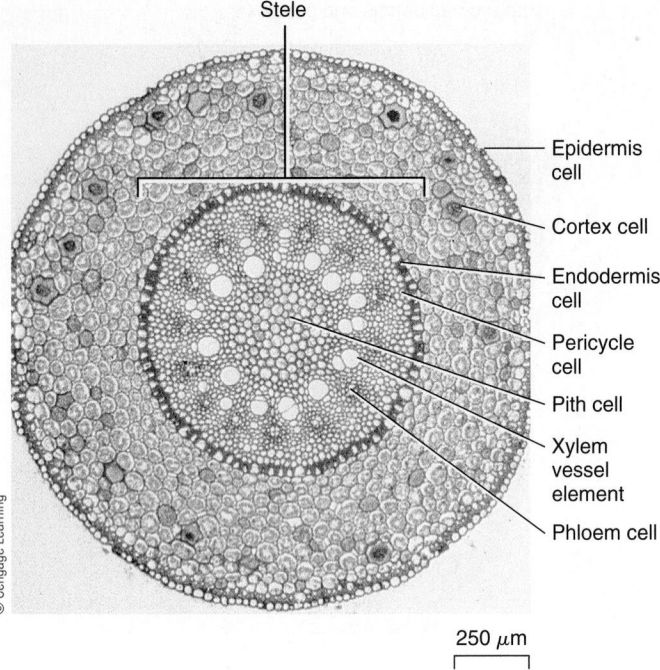

Stele

Epidermis cell

Cortex cell

Endodermis cell

Pericycle cell

Pith cell

Xylem vessel element

Phloem cell

© Cengage Learning

250 μm

Figure 36-8 LM of a cross section of a monocot root

Shown is a greenbrier (*Smilax*) root. As in herbaceous eudicot roots, the cortex of a monocot root is extensive.

and woody eudicots—have primary growth at apical meristems and secondary growth at lateral meristems. The production of secondary tissues occurs some distance back from the root tips and results from the activity of the same two lateral meristems found in woody stems: the vascular cambium and the cork cambium. Major roots of trees are often massive and have both wood and bark. In temperate climates the wood of both roots and stems exhibits annual rings in cross section.

Before secondary growth starts in a root, the **vascular cambium** is sandwiched between the primary xylem and the primary phloem (**FIG. 36-9**). At the onset of secondary growth, the vascular cambium extends out to the pericycle, which develops into vascular cambium opposite the xylem arms. As a result, the pericycle links the separate sections of vascular cambium so that the vascular cambium becomes a continuous, noncircular loop of cells in cross section. As the vascular cambium divides to produce secondary tissues, it eventually forms a cylinder of vascular cambium that continues to divide, producing secondary xylem (wood) to the inside and secondary phloem (inner bark) to the outside. The root increases in girth (thickness), and the vascular cambium continues to move outward.

The epidermis, cortex, endodermis, and primary phloem are gradually torn apart as the root increases in girth. The root epidermis is replaced by **periderm,** composed of cork cells and cork parenchyma, both produced by the **cork cambium** (the last figure in Table 33-4 shows an LM of periderm). The cork cambium in the root initially arises from regions in the pericycle.

Some roots are specialized for unusual functions

Adventitious roots often arise from the nodes of stems. Many aerial adventitious roots are adapted for functions other than anchorage, absorption, conduction, or storage. **Prop roots** are adventitious roots that develop from branches or a vertical stem and grow downward into the soil to help support the plant in an upright position (**FIG. 36-10a**). Prop roots are more common in monocots than in eudicots. Corn and sorghum, both monocots, are herbaceous plants that produce prop roots. Many tropical and subtropical eudicot trees, such as red mangrove, also produce prop roots.

The roots of many tropical rainforest trees are shallow and concentrated near the surface in a mat only a few centimeters (an inch or so) thick. The root mat catches and absorbs almost all minerals released from leaves by decomposition. Swollen bases or braces called **buttress roots** hold the trees upright and aid in the extensive distribution of the shallow roots (**FIG. 36-10b**).

In swampy or tidal environments where the soil is flooded or waterlogged, some roots grow upward until they are above the high-tide level. Even though roots live in the soil, they still require oxygen for aerobic respiration. Flooded soils are depleted of oxygen, so these aerial "breathing" roots, known as **pneumatophores,** may assist in getting oxygen to the submerged roots (**FIG. 36-10c**). Pneumatophores, which also help anchor the plant, have a well-developed system of internal air spaces that is continuous with the submerged parts of the root, presumably allowing gas exchange. Black mangroves and white mangroves are examples of plants with pneumatophores.

Climbing plants and **epiphytes,** which are plants that grow attached to other plants, have aerial roots that anchor the plant to the bark, branch, or other surface on which they grow. Some epiphytes have aerial roots specialized for functions other than anchorage. Certain epiphytic orchids, for example, have photosynthetic roots (**FIG. 36-10d**). Epiphytic roots may absorb moisture as well.

Some parasitic epiphytes, such as mistletoe, have modified roots that penetrate the host-plant tissues and absorb water. Another plant that starts its life as an epiphyte is the strangler fig, which produces long roots that eventually reach the ground and anchor the plant (now a tree rather than an epiphyte) in the soil. The tree on which the strangler fig originally grew is often killed as the strangler fig grows around it, competing with it for light and other resources and crushing its secondary phloem.

CHECKPOINT 36.1

- *What are the advantages of a taproot system? of a fibrous root system?*
- *If you were examining a cross section of a primary root of a flowering plant, how would you determine whether it was a eudicot or a monocot?*
- *By what mechanism(s) do water and dissolved minerals move through the symplast? the apoplast?*

The production of secondary tissues in roots, like that in stems, is the result of the activity of two lateral meristems: the vascular cambium and the cork cambium.

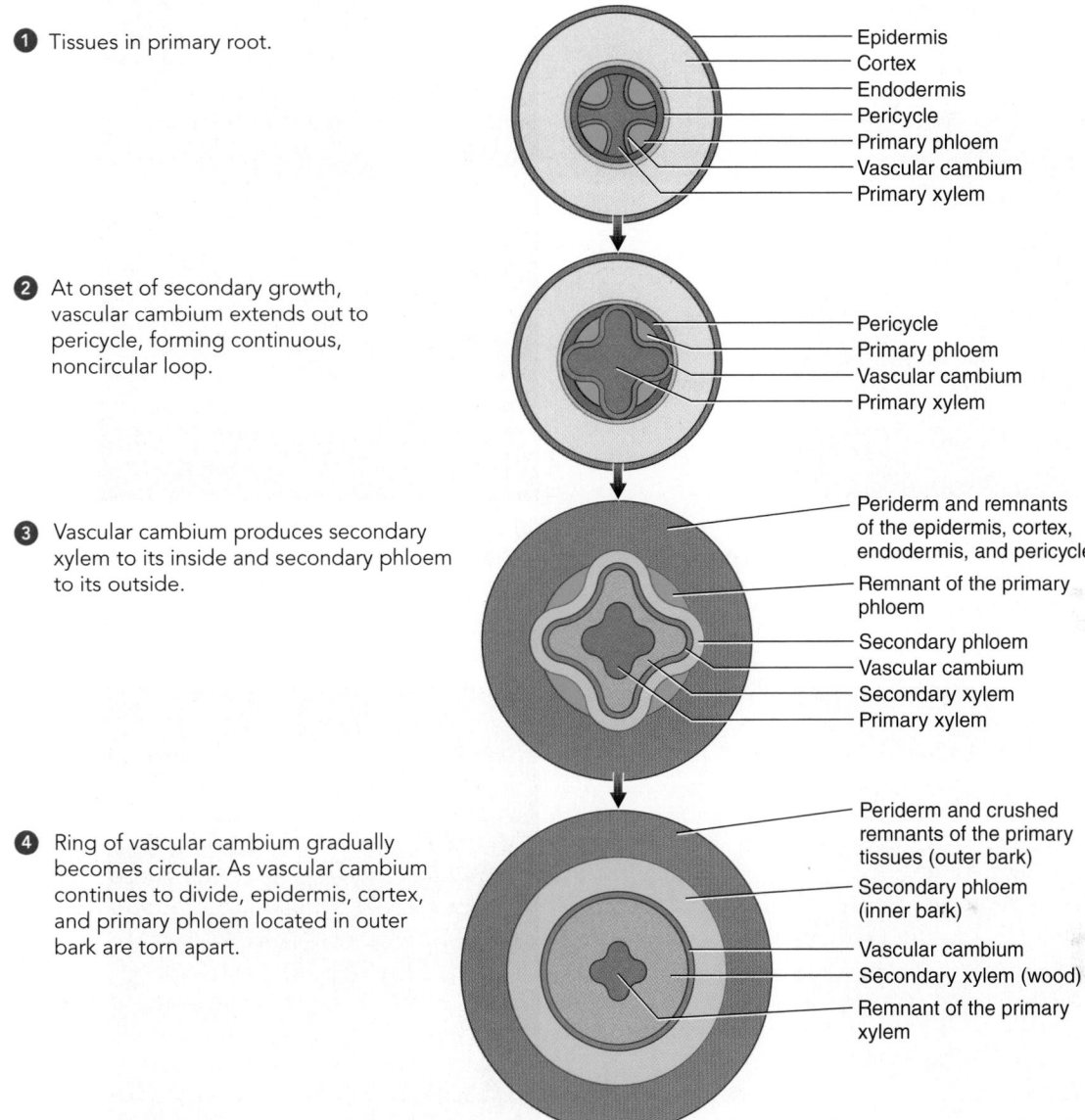

1 Tissues in primary root.

- Epidermis
- Cortex
- Endodermis
- Pericycle
- Primary phloem
- Vascular cambium
- Primary xylem

2 At onset of secondary growth, vascular cambium extends out to pericycle, forming continuous, noncircular loop.

- Pericycle
- Primary phloem
- Vascular cambium
- Primary xylem

3 Vascular cambium produces secondary xylem to its inside and secondary phloem to its outside.

- Periderm and remnants of the epidermis, cortex, endodermis, and pericycle
- Remnant of the primary phloem
- Secondary phloem
- Vascular cambium
- Secondary xylem
- Primary xylem

4 Ring of vascular cambium gradually becomes circular. As vascular cambium continues to divide, epidermis, cortex, and primary phloem located in outer bark are torn apart.

- Periderm and crushed remnants of the primary tissues (outer bark)
- Secondary phloem (inner bark)
- Vascular cambium
- Secondary xylem (wood)
- Remnant of the primary xylem

Figure 36-9 *Animation* **Development of secondary vascular tissues in a primary root**
The figures are not drawn to scale because of space limitations; the primary xylem is actually the same size in all four diagrams, but differences in scale make the xylem appear different sizes.
© Cengage Learning

CONNECT In what directions does the stem vascular cambium produce secondary xylem and secondary phloem?

- *In what way does the possession of a Casparian strip enable endodermal cells to carry out their essential function?*
- *How do the vascular cambium and the cork cambium of a herbaceous eudicot root function in the production of secondary tissues?*
- *What are the functions of each of the following: prop roots, buttress roots, and pneumatophores?*

36.2 ROOT ASSOCIATIONS AND INTERACTIONS

LEARNING OBJECTIVE

6 List and describe two mutualistic relationships between roots and other organisms.

Linda R. Berg

(a) Prop roots are adventitious roots that arise near the base of the stem and provide additional support. Screw pine (*Pandanus*) has an elaborate set of aerial prop roots. Photographed in Kauai, Hawaii.

Prop roots

John Arnaldi

Buttress roots

(b) Tropical rainforest trees typically possess buttress roots that support them in the shallow, often wet soil. Shown are buttress roots on Australian banyan (*Ficus macrophylla*). Photographed at Selby Gardens in Sarasota, Florida.

Paul Vinten/Alamy

Pneumatophores

(c) White mangrove (*Laguncularia racemosa*) produces pneumatophores, shown protruding from the water in the foreground. Pneumatophores may provide oxygen for roots buried in anaerobic (oxygen-deficient) soil.

Carlyn Iverson

Aerial roots

(d) This Ascodenda orchid has photosynthetic aerial roots.

Figure 36-10 Specialized roots

As roots of an individual plant extend through the soil, they often encounter the roots of related or unrelated plants. The nature of these interactions is largely unknown, but several studies suggest that roots are involved in complex ways,

sometimes recognizing the identity of neighboring roots and modifying their growth accordingly. For example, wild strawberry roots have been demonstrated to grow more vigorously when they encounter ground ivy roots; in contrast,

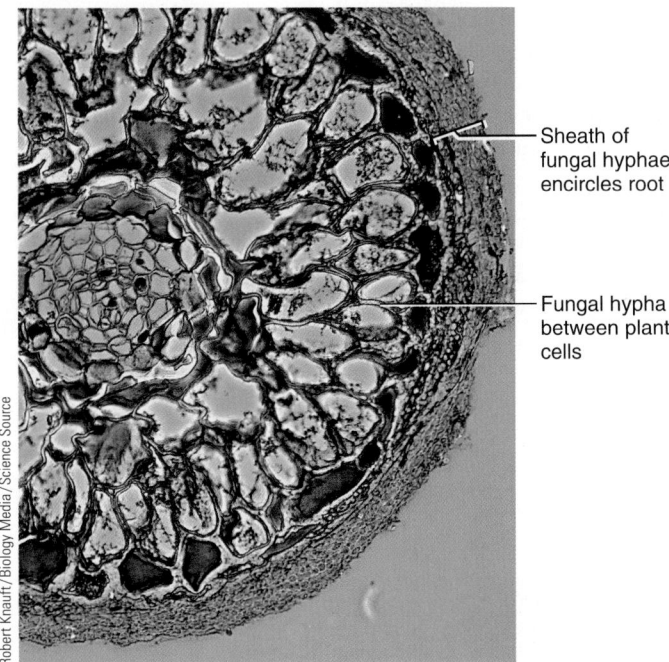

Sheath of
fungal hyphae
encircles root

Fungal hypha
between plant
cells

Robert Knauft/Biology Media/Science Source

250 μm

(a) LM of ectomycorrhizae, fungal associations that form a sheath around the root. The fungal hyphae penetrate the root between cortical cells but do not enter the cells.

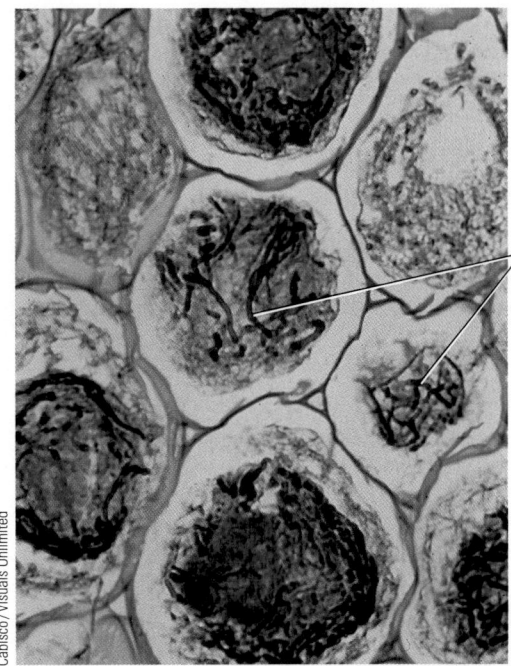

Fungal hyphae
within plant
cortical cells

Cabisco/Visuals Unlimited

100 μm

(b) LM of endomycorrhizae, fungal associations in which the fungal hyphae penetrate root cells. The hyphae aid in delivering and receiving nutrients. Roots of most vascular plant species are colonized by endomycorrhizae.

Mycorrhizae facilitate the uptake of essential minerals by roots

Subterranean associations between roots and certain fungi, known as **mycorrhizae,** are found in virtually all terrestrial ecosystems. Mycorrhizae permit the transfer of materials (such as carbon compounds produced by photosynthesis) from roots to the fungus. At the same time, essential minerals, such as phosphorus, move from the fungus to the roots of the host plant. The threadlike body of the fungal partner extends into the soil, extracting minerals well beyond the reach of the plant's roots. In some mycorrhizae the fungal mycelium encircles the root like a sheath; in others the fungus penetrates cell walls of the root cortex and forms branching **arbuscules** (FIG. 36-11; also see Fig. 29-11).

The mycorrhizal relationship is an ancient one: some of the earliest known plant fossils (about 400 million years old) contain mycorrhizal fungi in the roots. The mycorrhizal relationship is mutually beneficial because when mycorrhizae are not present, neither the fungus nor the plant grows as well (see Fig. 29-19). The hyphal network of mycorrhizae appears to interconnect the roots of different plant species in the community so that carbon compounds may flow from one plant to another through their mutual fungal partner.

The genome of an ectomycorrhizal fungus, *Laccaria bicolor,* has been sequenced, and scientists have identified many genes that code for *small secreted proteins* that may be involved in establishing the mutualistic relationship between the fungus and its plant partner (see Fig. 36-11c). For example, it has been learned that one such protein is secreted by the fungus in response to signals that diffuse from plant roots. The fungal protein is then

U.S. Department of Energy

(c) Fruiting body of the mycorrhizal fungus *Laccaria bicolor*, which is colonizing the roots of nearby Douglas fir seedlings.

Figure 36-11 Mycorrhizae

Mycorrhizae enhance plant growth by providing essential minerals to the roots.

root growth of ground ivy is inhibited by strawberry roots. Such root associations affect community dynamics in natural ecosystems and may affect the productivity of agricultural ecosystems.

The roots of terrestrial plant species may form *mutualistic—* that is, mutually beneficial—relationships with certain soil microorganisms: *mycorrhizal fungi* and *rhizobial bacteria.* These mutualistic relationships provide plants with access to essential minerals that are in limited supply in the soil.

taken up by plant root cells by endocytosis and enters the nuclei where it affects the transcription of plant genes that promote the mutualistic mycorrhizal relationship.

Rhizobial bacteria fix nitrogen in the roots of leguminous plants

Certain nitrogen-fixing bacteria, collectively called **rhizobia,** form associations with the roots of leguminous plants such as clover, peas, and soybeans. **Nodules** (swellings) that house millions of the rhizobia develop on the roots (see Fig. 55-9). Evidence indicates that the rhizobial relationship evolved much later than the mycorrhizal relationship; unlike mycorrhizae, which infect most plant species, rhizobia almost exclusively infect plants of the legume

family, which evolved about 70 million years ago. (A few other plant–bacteria associations that fix nitrogen have also evolved.)

As with mycorrhizae, the association between nitrogen-fixing bacteria and the roots of legumes is mutually beneficial. Bacteria receive photosynthetic products from plants while helping plants meet their nitrogen requirements by producing ammonia (NH_3) from atmospheric nitrogen.

For several decades, biologists have been studying the molecular basis of associations between plants and nitrogen-fixing bacteria. The two partners in this association use **cell signaling**—that is, a molecular dialogue back and forth between them—to establish contact and develop nodules. When soil contains a low level of nitrogen, legume roots secrete chemical attractants (flavonoids) toward which the rhizobial bacteria swim (**FIG. 36-12**).

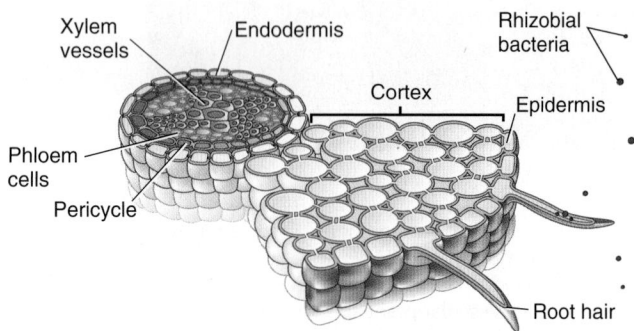

❶ Legume root secretes chemical signal that attracts rhizobia.

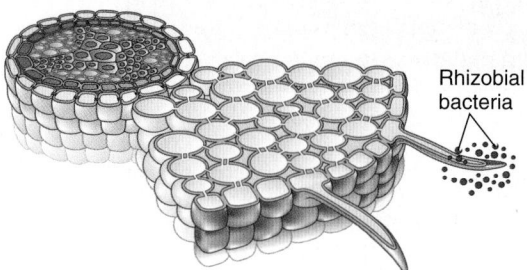

❷ Rhizobia congregate around root hair.

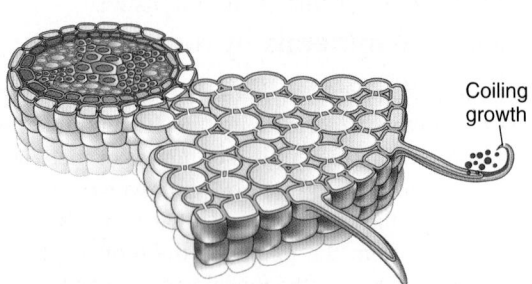

❸ Rhizobia secrete chemicals (nodulation, or nod, factors) that cause root hair to coil around them.

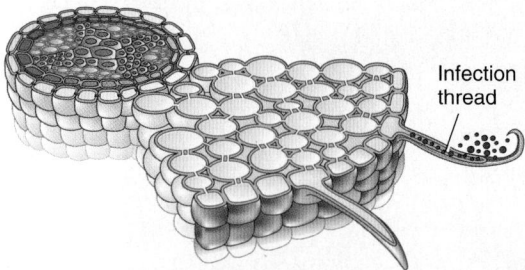

❹ Rhizobia enter root hair and move through infection thread (tube surrounded by plasma membrane) that extends down root-hair cell.

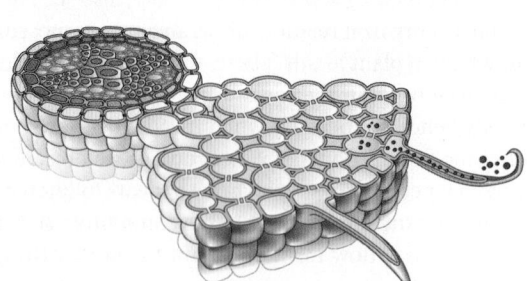

❺ Rhizobia move from infection thread into cells in cortex; each bacterium is surrounded by a membrane.

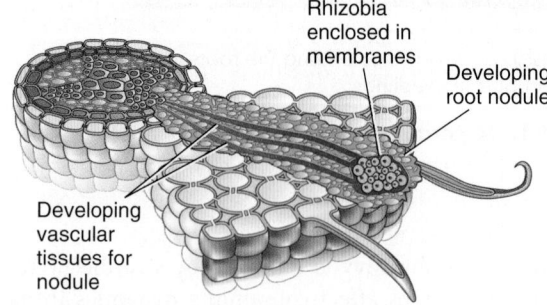

❻ Chemicals produced by rhizobia induce cell division and formation of root nodule. Some nodule cells are "infected."

Figure 36-12 Establishment of a relationship between rhizobial bacteria and leguminous plants

Much of the nitrogen produced by nitrogen-fixing bacteria in legume root nodules is transported in the vascular tissues throughout the plant body, including the seeds.

© Cengage Learning

The initial signal molecules produced by rhizobial bacteria are called *nodulation factors,* or *nod factors.* These nod factors help root-hair cells in the roots of leguminous plants recognize the presence of the bacteria; recognition is the first step in the infection by rhizobia and the development of root nodules. The nod factor apparently attaches to a plant receptor on the root-hair plasma membrane. This attachment triggers a calcium-dependent **signal transduction** cascade within root cells that alters gene expression and metabolism, resulting in nodule formation. The plant hormone **cytokinin,** which triggers cell division (see Chapter 38), is involved in root nodule formation.

Some genes are involved in the initiation of both mycorrhizal and rhizobial associations Because both mycorrhizae and rhizobia infect plant roots, some biologists have hypothesized that the genes and signal transduction components involved in mycorrhizal relationships may have been recruited for the rhizobial relationships that evolved much later. There is experimental evidence to support this hypothesis. Some of the same genes that trigger infection and changes in root development to form rhizobial nodules are also involved in establishing mycorrhizal infections. Evidence includes several legume mutants that biologists have identified; these mutants lack the ability to develop nodules for rhizobial bacteria and cannot form mycorrhizal associations.

CHECKPOINT 36.2

- *How are mycorrhizae and root nodules similar? How are they different?*

36.3 THE SOIL ENVIRONMENT

LEARNING OBJECTIVES

7 Describe the roles of weathering, organisms, climate, and topography in soil formation.

8 List the four components of soil and give the ecological significance of each.

9 Describe how roots absorb positively charged mineral ions by the process of cation exchange.

10 Distinguish between macronutrients and micronutrients.

11 Explain the effects of mineral depletion and soil erosion on plant growth.

We now examine the soil environment in which most roots live. *Soil* is a relatively thin layer of Earth's crust that has been modified by the natural actions of weather, wind, water, and organisms. It is easy to take soil for granted. We walk on and over it throughout our lives but rarely stop to think about how important it is to our survival. Vast numbers and kinds of organisms colonize soil and depend on it for shelter and food. Most plants anchor themselves in soil, and from it they receive water and minerals. Almost all the elements essential for plant growth are obtained directly from the soil. With few exceptions, plants cannot survive on their own without soil, and because we depend on plants for our food, neither could humans exist without soil.

Soils are generally formed from rock (called "parent material") that is gradually broken down, or fragmented, into smaller and smaller particles by biological, chemical, and physical **weathering processes** (FIG. 36-13). Two important factors that work together in the weathering of rock are climate and organisms. When plant roots and other organisms living in the soil respire, they produce carbon dioxide (CO_2), which diffuses into the soil and reacts with soil water to form carbonic acid (H_2CO_3). Soil organisms such as lichens produce other kinds of acids. Recall from Chapter 29 that a lichen is a dual organism composed of a fungus and a phototroph (photosynthetic organism). These acids etch tiny cracks, or fissures, in the rock surface, and water then seeps into these cracks. If the parent material is located in a temperate climate, the alternate freezing and thawing of the water during winter causes the cracks to enlarge, breaking off small pieces of rock. Small plants can then become established and send their roots into the larger cracks, fracturing the rock further.

Topography, a region's surface features such as the presence or absence of mountains and valleys, is also involved in soil formation. Steep slopes often have little or no soil on them because the soil and rock are continually transported down the slopes by gravity. Runoff from precipitation tends to amplify erosion on steep slopes. Moderate slopes, in contrast, may encourage the formation of deep soils.

The disintegration of solid rock into finer and finer mineral particles in the soil takes an extremely long time, sometimes thousands of years. Soil forms constantly as the weathering of parent material beneath soil that is already formed continues to add new soil.

Soil comprises inorganic minerals, organic matter, air, and water

Soil comprises four distinct components: inorganic mineral particles (which make up about 45% of a typical soil), organic matter (about 5%), water (about 25%), and air (about 25%). The plants, animals, fungi, and microorganisms that inhabit soil interact with it, and minerals are continually cycled from the soil to organisms, which use them in their biological processes. When the organisms die, bacteria and other soil organisms decompose the remains, returning the minerals to the soil.

The inorganic mineral particles, which come from weathered rock, provide anchorage and essential minerals for plants as well as pore space for water and air. Because different rocks consist of different minerals, soils vary in mineral composition and chemical properties. Rocks rich in aluminum form acidic soils, for example, whereas rocks that contain silicates of magnesium and iron form soils that may be deficient in calcium, nitrogen, and phosphorus. Also, soils formed from the same kind of parent material may not develop in the same way because other factors, such as weather, topography, and organisms, differ.

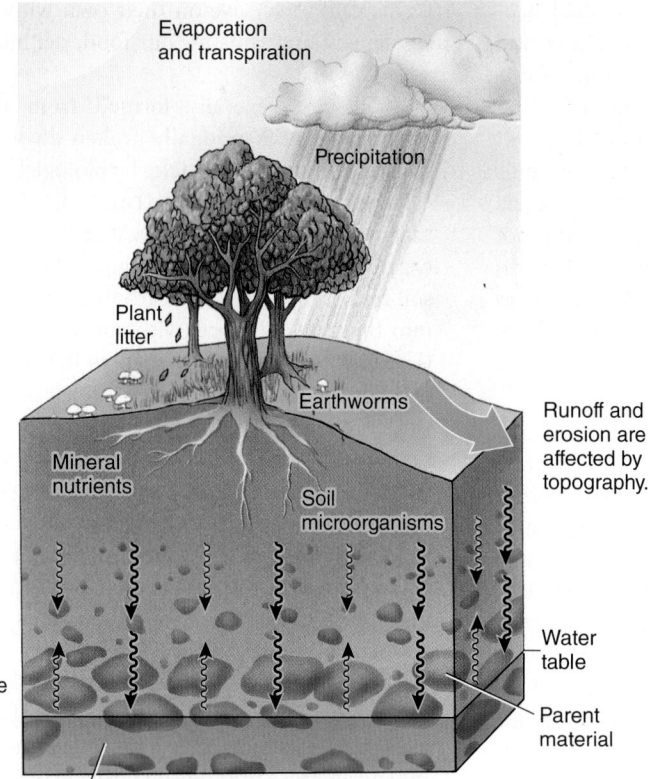

Figure 36-13 The dynamics of soil formation

Weather, climate, topography, and organisms interact with Earth's crust to form soil, the material that supports plants and many other organisms on land.

© Cengage Learning

Labels in figure:

Evaporation and transpiration

Precipitation

Plant litter

Earthworms

Mineral nutrients

Soil microorganisms

Runoff and erosion are affected by topography.

Water table

Parent material

Soil organisms add organic additions underground.

Fine particles and solubles are washed downward.

Capillary action and evaporation cause some materials to rise.

Loss of soluble materials in groundwater

The *texture,* or structural characteristic, of a soil is determined by the percentages (by weight) of the different-sized inorganic mineral particles—sand, silt, and clay—in it. The size assignments for sand, silt, and clay give soil scientists a way to classify soil texture. Particles larger than 2 mm in diameter, called gravel or stones, are not considered soil particles because they do not have any direct value to plants. The largest soil particles are called "sand" (0.02 to 2 mm in diameter); the medium-sized particles, "silt" (0.002 to 0.02 mm in diameter); and the smallest particles, "clay" (less than 0.002 mm in diameter). Sand particles are large enough to be seen easily with the eye, and silt particles (about the size of flour particles) can be seen with an ordinary light microscope. Most individual clay particles can be seen only under an electron microscope.

A soil's texture affects many of that soil's properties, in turn influencing plant growth. The clay component of a soil is particularly important in determining many of its characteristics because clay particles have the greatest surface area of all soil particles. If the surface areas of about 450 g (1 lb) of clay particles were laid out side by side, they would occupy 1 hectare (2.5 acres).

Each clay particle has predominantly negative electric charges on its outer surface that attract and reversibly bind **cations,** which are positively charged mineral ions such as potassium (K^+) and magnesium (Mg^{2+}). Because many cations are essential for plant growth, cation adhesion to soil particles is an important aspect of soil fertility. Roots secrete protons (H^+), which are exchanged for other positively charged mineral ions adhering to the surface of soil particles, in a process known as **cation exchange.** These "freed" ions and the water that forms a film around the soil particles are absorbed by the plant's roots (**FIG. 36-14**).

In contrast, **anions,** which are negatively charged mineral ions, are repelled by the negative surface charges of clay particles and tend to remain in solution. Anions such as nitrate (NO_3^-) are often washed out of the root zone by water moving through the soil.

Soil always contains a mixture of different-sized particles, but the proportions vary from one soil to another. A *loam,* which is an ideal agricultural soil, has an optimum combination of different soil particle sizes: it contains approximately 40% each of sand and silt and about 20% of clay. Generally, the larger particles provide structural support, aeration, and permeability to the soil, whereas the smaller particles bind into aggregates, or clumps, and hold minerals and water. Soils with larger proportions of sand are not as desirable for most plants because they do not hold water and mineral ions well. Plants grown in such soils are more susceptible to drought and mineral deficiencies. Soils with larger proportions of clay are also not desirable for most plants because

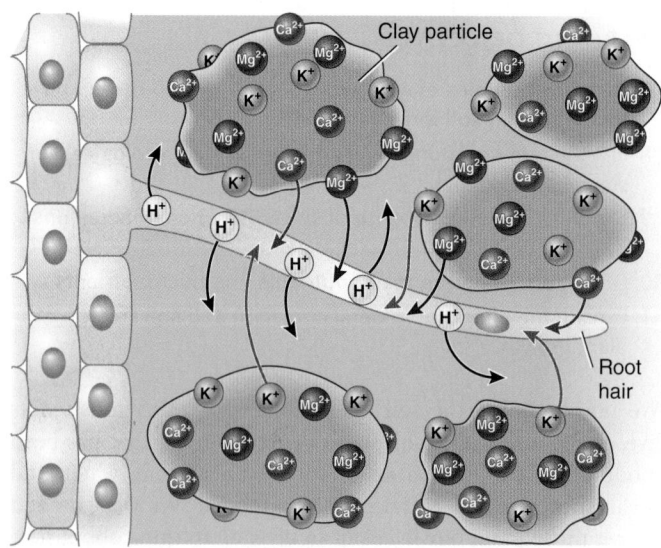

Figure 36-14 *Animation* **Cation exchange**

Negatively charged clay particles bind to positively charged mineral cations, holding them in the soil. Roots pump out protons (H^+), which are exchanged for the cations, facilitating their absorption.

© Cengage Learning

they provide poor drainage and often do not provide enough oxygen. Clay soils used in agriculture tend to compact, which reduces the number of soil spaces that can be filled by water and air.

A soil's organic matter consists of the wastes and remains of soil organisms The organic matter in soil is composed of litter (dead leaves and branches on the soil's surface); droppings (animal dung); and the dead remains of plants, animals, and microorganisms in various stages of decomposition. Organic matter is decomposed by microorganisms, particularly bacteria and fungi, that inhabit the soil. During decomposition, essential mineral ions are released into the soil, where they may be bound by soil particles or absorbed by plant roots. Organic matter increases the soil's water-holding capacity by acting much like a sponge. For this reason, gardeners often add organic matter to soils, especially sandy soils, which are naturally low in organic matter.

The partly decayed organic portion of the soil is referred to as **humus.** Humus, which is not a single chemical compound but a mix of many organic compounds, binds mineral ions and holds water. On average, humus persists in agricultural soil for about 20 years. Certain components of humus may persist in the soil for hundreds of years. Although humus is somewhat resistant to decay, a succession of microorganisms gradually reduces it to carbon dioxide, water, and minerals.

About 50% of soil volume is composed of pore spaces Soil has numerous pore spaces of different sizes around and among the soil particles. Pore spaces occupy roughly 50% of a soil's volume and are filled with varying proportions of air and water. Both soil air and soil water are necessary to produce a moist but aerated soil that sustains plants and other soil-dwelling organisms. Water is usually held in the smaller pores, whereas air is found in the larger pores. After a prolonged rain, almost all the pore spaces may be filled with water, but water drains rapidly from the larger pore spaces, drawing air from the atmosphere into those spaces.

Soil air contains the same gases as atmospheric air, although they are usually present in different proportions. As a result of respiration by soil organisms, there is less oxygen and more carbon dioxide in soil air than in atmospheric air. (Aerobic respiration uses oxygen and produces carbon dioxide.) Among the important gases in soil are oxygen (O_2), required by soil organisms for aerobic respiration; nitrogen (N_2), used by nitrogen-fixing bacteria; and carbon dioxide (CO_2), involved in soil weathering.

Soil water originates as precipitation, which drains downward, or as groundwater (water stored in porous underground rock), which rises upward from the water table (the uppermost level of groundwater). Soil water contains low concentrations of dissolved minerals that enter the roots of plants when they absorb water. Water not bound to soil particles or absorbed by roots percolates (moves down) through soil, carrying dissolved minerals with it. The removal of dissolved materials from soil by percolating water is called **leaching.** The deposition of leached material in the lower layers of soil is known as **illuviation.** Iron

and aluminum compounds, humus, and clay are some illuvial materials that can gather in the subsurface portion of the soil. Some substances completely leach out of the soil because they are so soluble that they migrate down into the groundwater. It is also possible for water that is moving *upward* through the soil to carry dissolved materials with it.

Soil organisms form a complex ecosystem

A single teaspoon of fertile agricultural soil may contain millions of microorganisms, such as bacteria, fungi, algae, protozoa, as well as microscopic nematodes and other worms. Other organisms also colonize the soil ecosystem, including plant roots, earthworms, insects, moles, snakes, and groundhogs (**FIG. 36-15**). Most numerous in soil are bacteria, which number in the hundreds of millions per gram of soil. Scientists have identified about 170,000 species of soil organisms, but thousands remain to be identified. Little is known about the roles and interactions of soil organisms, in part because it is hard to study their activities under natural conditions.

Worms are some of the most important organisms living in soil. Earthworms, probably one of the most familiar soil inhabitants, ingest soil and obtain energy and raw materials by digesting humus. *Castings,* bits of soil that have passed through the gut of an earthworm, are deposited on or near the soil surface. In this way, minerals from deeper layers are brought to upper layers. Earthworm tunnels serve to aerate the soil, and the worms' waste products and corpses add organic material to the soil.

Ants live in the soil in enormous numbers, constructing tunnels and chambers that help aerate it. Members of soil-dwelling ant colonies forage on the surface for bits of food, which they carry back to their nests. Not all this food is eaten, however, and its eventual decomposition helps increase the organic matter in the soil. Many ants are also indispensable in plant reproduction because they bury seeds in the soil (discussed in Chapter 37).

Soil pH affects soil characteristics and plant growth

Acidity is measured using the pH scale, which extends from 0 (extremely acidic) through 7.0 (neutral) to 14.0 (extremely alkaline). The pH of most soils ranges from 4.0 to 8.0, but some soils are outside this range. The soil of the Pygmy Forest in Mendocino County, California, is extremely acidic (pH 2.8 to 3.9). At the other extreme, certain soils in Death Valley, California, have a pH of 10.5.

Plants are affected by soil pH, partly because the solubility of certain minerals varies with differences in pH. Soluble minerals can be absorbed by the plant, whereas insoluble forms cannot. At a low pH, for example, aluminum and manganese in soil water are more soluble than at higher pH and are sometimes absorbed by the roots in toxic concentrations. At a higher pH, certain mineral salts essential for plant growth, such as calcium

Soil provides most of the minerals found in plants

More than 90 naturally occurring elements exist on Earth, and more than 60 of them, including elements as common as carbon and as rare as gold, have been found in plant tissues. Not all these elements are essential for plant growth, however.

Nineteen elements have been found essential for most, if not all, plants (**TABLE 36-1**). Ten of them are required in fairly large quantities (greater than 0.05% dry weight) and are therefore known as **macronutrients.** Macronutrients are carbon, hydrogen, oxygen, nitrogen, potassium, calcium, magnesium, phosphorus, sulfur, and silicon. The remaining nine **micronutrients** are needed in trace amounts (less than 0.05% dry weight) for normal plant growth and development. Micronutrients are chlorine, iron, boron, manganese, sodium, zinc, copper, nickel, and molybdenum.

Four of the 19 elements—carbon, oxygen, hydrogen, and nitrogen—come directly or indirectly from soil water or from gases in the atmosphere. Carbon is obtained from carbon dioxide in the atmosphere during photosynthesis. Oxygen is obtained from atmospheric oxygen (O_2) and water (H_2O). Water also supplies hydrogen to the plant. Plants absorb their nitrogen from the soil as ions of nitrogen salts, but the nitrogen in nitrogen salts ultimately comes from atmospheric nitrogen (N_2). The remaining 15 essential elements are obtained from the soil as dissolved mineral ions. Their ultimate source is the parent material from which the soil was formed.

Let us examine the main functions of essential elements. Carbon, hydrogen, and oxygen are found as part of the structure of all biologically important molecules, including lipids, carbohydrates, nucleic acids, and proteins. Nitrogen is part of proteins, nucleic acids, and chlorophyll.

Potassium, which plants use in fairly substantial amounts, is not found in a specific organic compound in plant cells. Instead, it remains as free K^+ and plays a key physiological role in maintaining the turgidity of cells because it is osmotically active. The presence of K^+ in cytoplasm causes water to pass through the plasma membrane into the cell by osmosis. Potassium is also involved in the opening of stomata.

Calcium plays a key structural role as a component of the middle lamella (the cementing layer between cell walls of adjacent plant cells). Calcium ions (Ca^{2+}) are also important in

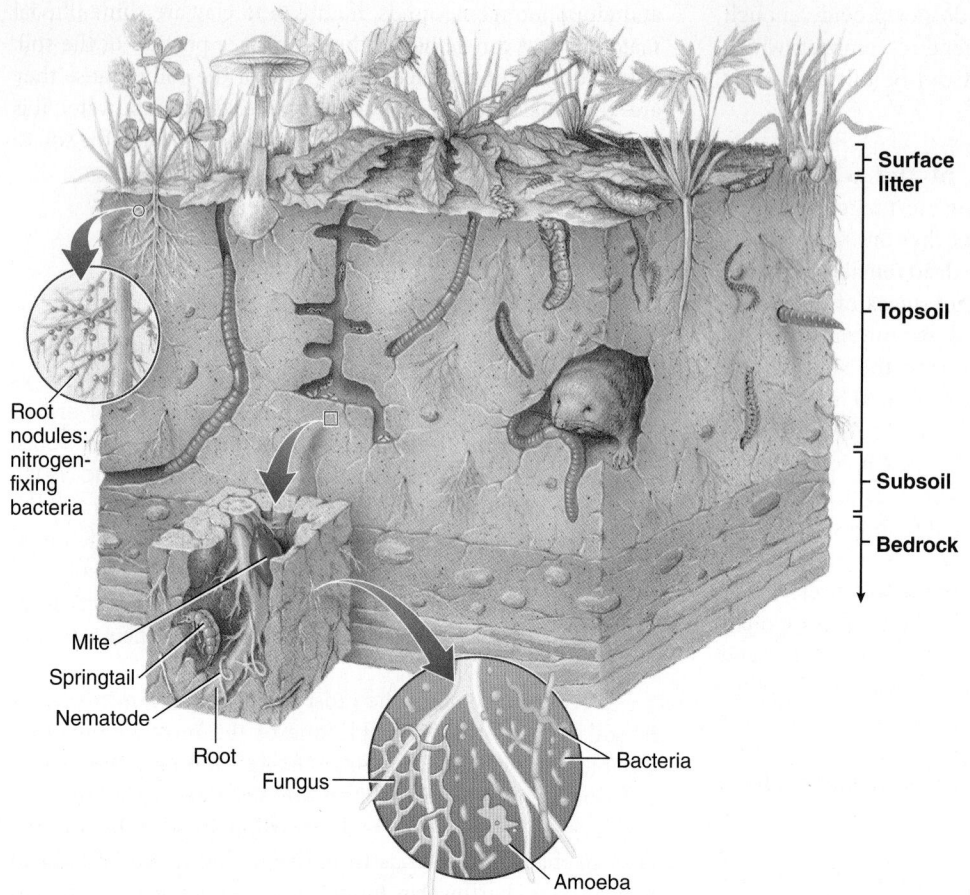

Figure 36-15 Diversity of life in fertile soil

Plants, algae, protozoa, fungi, bacteria, earthworms, flatworms, roundworms, insects, spiders and mites, and burrowing animals such as moles and groundhogs live in soil.

© Cengage Learning

phosphate, become less soluble and thus less available to plants than at lower pH.

Soil pH also affects the leaching of minerals. An acidic soil has less ability than other soil to bind positively charged ions to it because the soil particles also bind the abundant protons (**FIG. 36-16**). As a result, certain mineral ions essential for plant growth, such as potassium (K^+), are leached more readily from acidic soil than other soil. The optimum soil pH for most plant growth is 6.0 to 7.0 because most essential minerals are available to plants in that pH range.

Acid precipitation, a type of air pollution in which sulfuric and nitric acids produced by human activities fall to the ground as acid rain, sleet, snow, or fog, can seriously decrease soil pH. Acid precipitation is one of several factors implicated in *forest decline,* the gradual deterioration, and often death, of trees that has been observed in many European and North American forests in recent decades. Forest decline may be partly the result of soil changes, such as leaching of essential cations, caused by acid precipitation. In central European forests that have experienced forest decline, for example, a strong correlation has been demonstrated between forest damage and soil chemistry altered by acid precipitation.

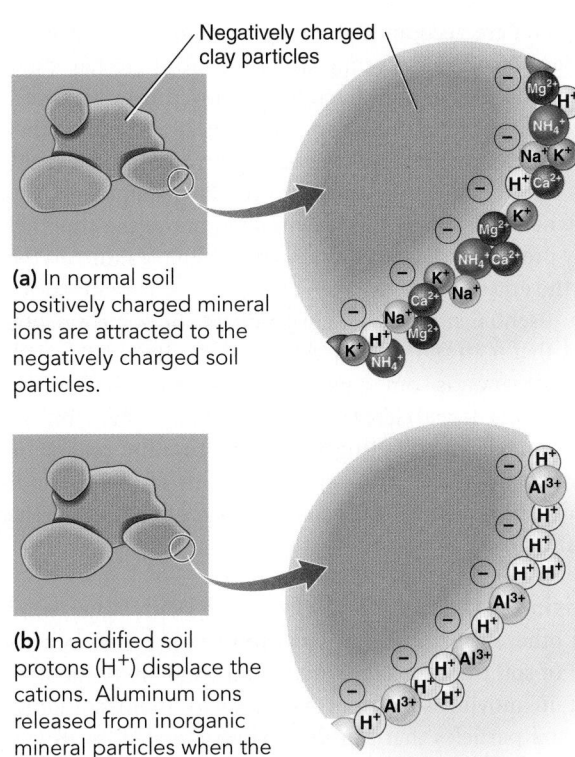

(a) In normal soil positively charged mineral ions are attracted to the negatively charged soil particles.

(b) In acidified soil protons (H^+) displace the cations. Aluminum ions released from inorganic mineral particles when the soil becomes acidified also adhere to soil particles.

Figure 36-16 How acid alters soil chemistry

© Cengage Learning

physiological roles in plants, such as altering membrane permeability and as **second messengers** in various cell signaling responses (see Chapter 6).

Magnesium is critical for plants because it is part of the chlorophyll molecule. Phosphorus is a component of nucleic acids, phospholipids (part of cell membranes), and energy transfer molecules such as ATP. Sulfur is essential because it is found in certain amino acids and vitamins. Many plants accumulate silicon in their cell walls and intercellular spaces. Silicon enhances the growth and fertility of some species and may help reinforce cell walls.

Chlorine and sodium are micronutrients that help maintain cell turgor. In addition to this osmotic role, chloride (Cl^-) and sodium (Na^+) ions are essential for photosynthesis. Six of the micronutrients (iron, manganese, zinc, copper, nickel, and molybdenum) are associated with various plant enzymes, often as **cofactors,** and are involved in certain enzymatic reactions. Boron, present in cell walls, is also involved in nucleic acid metabolism and in cell growth.

How do biologists determine whether an element is essential? It is impossible to conduct mineral nutrition experiments by growing plants in soil because soil is too complex and contains too many elements. Therefore, nutritional studies require special methods. One of the most useful techniques to test whether an element is essential is **hydroponics,** which is the growing of plants in aerated water to which mineral salts have

TABLE 36-1	Functions of Elements Required by Most Plants	
ELEMENT	**TAKEN UP AS**	**MAJOR FUNCTIONS**
MACRONUTRIENTS		
Carbon	CO_2	Component of carbohydrate, lipid, protein, and nucleic acid molecules
Hydrogen	H_2O	Component of carbohydrate, lipid, protein, and nucleic acid molecules
Oxygen	O_2, H_2O	Component of carbohydrate, lipid, protein, and nucleic acid molecules
Nitrogen	NO_3^-, NH_4^+	Component of proteins, nucleic acids, chlorophyll, and certain coenzymes
Potassium	K^+	Osmotic and ionic balance; opening of stomata; enzyme activator (for 40 enzymes)
Calcium	Ca^{2+}	In cell walls; involved in membrane permeability; enzyme activator; second messenger in metabolism
Magnesium	Mg^{2+}	In chlorophyll; enzyme activator in carbohydrate metabolism (phosphate transfer)
Phosphorus	$HPO_4^{2-}, H_2PO_4^-$	In nucleic acids, phospholipids, ATP (energy transfer compound), and coenzymes
Sulfur	SO_4^{2-}	In certain amino acids and vitamins
Silicon	SiO_3^{2-}	In cell walls; increased resistance to pests and diseases; important primarily in grasses and sedges
MICRONUTRIENTS		
Chlorine	Cl^-	Ionic balance; involved in photosynthesis (production of oxygen)
Iron	Fe^{2+}, Fe^{3+}	In enzymes and electron transport molecules involved in photosynthesis, respiration, and nitrogen fixation
Boron	$H_2BO_3^-$	In cell walls; involved in nucleic acid metabolism and in cell elongation
Manganese	Mn^{2+}	In enzymes involved in respiration and nitrogen metabolism; required for photosynthesis (production of oxygen)
Sodium	Na^+	Involved in photosynthesis (in C_4 and CAM plants); substitutes for potassium in osmotic and ionic balance
Zinc	Zn^{2+}	In enzymes involved in respiration and nitrogen metabolism
Copper	Cu^+, Cu^{2+}	In enzymes involved in photosynthesis
Nickel	Ni^{2+}	In enzymes (urease) involved in nitrogen metabolism; required by nitrogen-fixing bacteria
Molybdenum	MoO_4^{2-}	In enzymes involved in nitrogen metabolism

© Cengage Learning

WHY IS IT USED? Using hydroponics, it is possible to grow plants in which only the mineral under investigation is absent from a nutrient solution. (It would be impossible to design such an experiment using soil.) Growing plants hydroponically in the absence of an essential mineral produces recognizable deficiency symptoms.

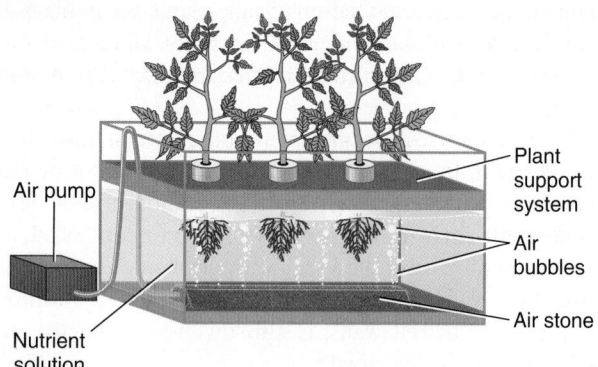

HOW IS IT DONE? The experimental and control plants are grown in two liquid solutions of mineral nutrients through which air is bubbled to allow the roots to respire. The liquid solution for the experimental plants contains all nutrient minerals known to be essential for plant growth except for one. The liquid solution for the control plants contains all known essential elements, including the one under study.

Figure 36-17 Plant nutritional studies in hydroponic growth systems
© Cengage Learning

been added (**FIG. 36-17**). Hydroponics has commercial applications in addition to its scientific use. Lettuce is one of several crops that is grown hydroponically.

If biologists suspect that a particular element is essential for plant growth, they grow plants in a nutrient solution that contains all known essential elements *except* the one in question. If plants grown in the absence of that element cannot develop normally or complete their life cycle, the element may be essential. Additional criteria are used to confirm whether an element is essential. For example, it must be demonstrated that the element has a direct effect on the plant's metabolism and that the element is essential for a wide variety of plants.

Soil can be damaged by human mismanagement

Soil is a valuable natural resource on which humans depend for food. Many human activities generate or aggravate soil problems, including mineral depletion, soil erosion, and accumulation of salt.

Mineral depletion may occur in agricultural soils In a natural ecosystem, the essential minerals removed from the soil by plants are returned when the plants or the animals that eat

them die and are decomposed. An agricultural system disrupts this pattern of nutrient cycling when the crops are harvested. Plant material containing minerals is removed from the nutrient cycle, and the harvested crops fail to decay and release their nutrients back to the soil. Over time, farmed soil inevitably loses its fertility, that is, its ability to produce abundant crops. Homeowners often mow their lawns and remove the clippings, similarly preventing decomposition and cycling of minerals that were in the grass blades.

The essential material (water, sunlight, or some essential element) that is in shortest supply usually limits plant growth. This phenomenon is sometimes called the concept of **limiting resources.** The three elements that are most often limiting resources for plants are nitrogen, phosphorus, and potassium. To sustain the productivity of agricultural soils, farmers periodically add fertilizers to depleted soils to replace the minerals that limit plant growth.

Soil erosion is the loss of soil from the land Water, wind, ice, and other agents cause **soil erosion,** the wearing away or removal of soil from the land. Water and wind are particularly effective in moving soil from one place to another. Rainfall loosens soil particles that can then be transported by moving water (**FIG. 36-18**). Wind loosens soil and blows it away, particularly if the soil is barren and dry. Soil erosion is a natural process that can be greatly accelerated by human activities.

Soil erosion is a national and international problem that does not make the headlines very often. The U.S. Department of Agriculture estimates that approximately 4.2 billion metric tons (4.6 billion tons) of topsoil are eroded each year from U.S. croplands and rangelands. Erosion causes a loss in soil fertility because essential minerals and organic matter that are a part of the soil are removed. As a result of these losses, the productivity

Figure 36-18 Soil erosion in an open field
Removal of plant cover exposes bare soil to erosion in heavy rains. Precipitation can cause gullies to enlarge rapidly because they provide channels for the runoff of water.

of eroded agricultural soils declines, and more fertilizer must be used to replace the nutrients lost to erosion.

Humans often accelerate soil erosion through poor soil management practices. Agriculture is not the only culprit; the removal of natural plant communities during surface mining, unsound logging practices (such as clear-cutting large areas of forest), and the construction of roads and buildings also accelerate erosion.

Soil erosion has an impact on other natural resources. Sediment that enters streams, rivers, and lakes degrades water quality and fish habitats. Pesticides and fertilizer residues in sediment may add pollutants to the water. Also, when forests are removed within the watershed of a hydroelectric power facility, accelerated soil erosion fills the reservoir behind the dam with sediment much faster than usual. This process reduces electricity production at that facility.

Sufficient plant cover reduces the amount of soil erosion: leaves and stems cushion the effect of rainfall, and roots help hold the soil in place. Although soil erosion is a natural process, abundant plant cover makes it negligible in many natural ecosystems.

Salt accumulates in soil that is improperly irrigated

Although irrigation improves the agricultural productivity of arid and semiarid lands, it sometimes causes salt to accumulate in the soil, a process called *salinization*. In a natural scenario, as a result of precipitation runoff, rivers carry dissolved salts away. Irrigation water, however, normally soaks into the soil and does not run off the land into rivers, so when water evaporates, the salt remains behind and accumulates in the soil. Salty soil results in a decline in productivity and, in extreme cases, renders the soil completely unfit for crop production.

Most plants cannot obtain all the water they need from salty soil because a water balance problem exists: water moves by osmosis *out* of plant roots and into the salty soil. Obviously, most plants cannot survive under these conditions (see Fig. 5-13). Plant species that thrive in saline soils have special adaptations that enable them to tolerate the high amount of salt. Black mangroves, for example, excrete excess salt through their leaves.

Unless they have been genetically selected to tolerate high salt, most crops are not productive in saline soil. Research on the molecular aspects of salt tolerance suggests that mutations in membrane transport proteins confer salt tolerance. Biologists genetically engineered a salt-tolerant variety of *Arabidopsis* to overexpress a single gene that codes for a sodium transport protein in the vacuolar membrane. These plants can store large quantities of sodium in their vacuoles, thereby tolerating saline soils.

CHECKPOINT 36.3

- What are the four components of soil, and how is each important to plants?
- How does cation exchange make mineral ions available for absorption by plant roots?
- In what ways do weathering processes convert rock into soil?
- What criteria are used to distinguish macronutrients from micronutrients?
- What are three ways in which human mismanagement can cause soil damage?

SUMMARY: FOCUS ON LEARNING OBJECTIVES

36.1 Root Structure and Function (page 757)

1 Distinguish between taproot and fibrous root systems.

- A **taproot system** has one main root (formed from the **radicle**) from which many lateral roots extend. A **fibrous root system** has several to many **adventitious** roots of the same size developing from the end of the stem. Lateral roots branch from these adventitious roots.

2 Compare cross sections of a primary eudicot root and a monocot root, and describe the functions of each tissue.

- Primary roots have an epidermis, ground tissues (cortex and, in certain roots, pith), and vascular tissues (xylem and phloem). Each root tip is covered by a **root cap**, a protective layer that covers the delicate root **apical meristem** and may orient the root so that it grows downward.
- **Epidermis** protects the root. **Root hairs,** short-lived extensions of epidermal cells, aid in absorption of water and dissolved minerals.
- **Cortex** consists of parenchyma cells that often store starch. **Endodermis,** the innermost layer of the cortex, regulates the movement of water and minerals into the root xylem. Cells of the endodermis have a **Casparian strip** around their radial and transverse walls that is impermeable to water and dissolved minerals.

- Pericycle, xylem, and phloem collectively make up the root's **stele**, or **vascular cylinder. Pericycle** gives rise to lateral roots and lateral meristems. **Xylem** conducts water and dissolved minerals; **phloem** conducts dissolved sugar.
- Xylem of a herbaceous eudicot root forms a solid core in the center of the root. The center of a monocot root often consists of pith surrounded by a ring of alternating bundles of xylem and phloem.

3 Trace the pathway of water and mineral ions from the soil through the various root tissues and distinguish between the symplast and apoplast.

- As water and dissolved mineral ions move from the soil into the root, they pass through the following tissues: root hair/epidermis ⟶ cortex ⟶ endodermis ⟶ pericycle ⟶ root xylem.
- Water and dissolved minerals move through the epidermis and cortex along one of two pathways: the **apoplast** (along the interconnected porous cell walls) or the **symplast** (from one cell's cytoplasm to the next through **plasmodesmata**).

4 Discuss the structure of roots with secondary growth.

- Roots of gymnosperms and woody eudicots develop secondary tissues (wood and bark). The production of secondary tissues is the result of the activity of two lateral meristems: the vascular cambium and cork cambium. The **vascular cambium** produces secondary xylem (wood) and secondary phloem (inner bark). The **cork cambium** produces **periderm** (outer bark).

5 Describe at least three roots that are modified to perform uncommon functions.

- **Prop roots** develop from branches or from a vertical stem and grow downward into the soil to help support certain plants in an upright position. **Buttress roots** are swollen bases or braces that support certain tropical rainforest trees that have shallow root systems. **Pneumatophores** are aerial "breathing" roots that may assist in getting oxygen to submerged roots.

36.2 Root Associations and Interactions *(page 763)*

6 List and describe two mutualistic relationships between roots and other organisms.

- **Mycorrhizae,** which are mutually beneficial associations between roots and soil fungi, enhance plant growth by providing essential minerals such as phosphorus to the roots.
- Root **nodules** are swellings that develop on roots of leguminous plants and house millions of **rhizobia** (nitrogen-fixing bacteria).

36.3 The Soil Environment *(page 767)*

7 Describe the roles of weathering, organisms, climate, and topography in soil formation.

- Factors influencing soil formation include parent material (type of rock), climate, organisms, the passage of time, and topography. Most soils are formed from parent material that is broken into smaller and smaller particles by **weathering processes.** Climate and organisms work together in weathering rock.
- Soil organisms such as plants, algae, fungi, worms, insects, spiders, and bacteria are important not only in forming soil but also in cycling minerals.
- Topography, a region's surface features, affects soil formation. Steep slopes have little or no soil on them, whereas moderate slopes often have deep soils.

8 List the four components of soil and give the ecological significance of each.

- Soil comprises inorganic minerals, organic matter, air, and water. Inorganic minerals provide anchorage and essential minerals for plants. Organic matter increases the soil's water-holding capacity and, as it decomposes, releases essential minerals into the soil. Soil air provides oxygen for soil organisms to use during aerobic respiration. Soil water provides water and dissolved minerals to plants and other organisms.

9 Describe how roots absorb positively charged mineral ions by the process of cation exchange.

- **Cations,** positively charged mineral ions, are attracted and reversibly bound to clay particles, which have predominantly negative charges on their outer surfaces. In **cation exchange** roots secrete protons (H^+), which are exchanged for other positively charged mineral ions, freeing them into the soil water to be absorbed by roots.

10 Distinguish between macronutrients and micronutrients.

- Plants require 19 essential elements for normal growth. Ten elements are **macronutrients:** carbon, hydrogen, oxygen, nitrogen, potassium, calcium, magnesium, phosphorus, sulfur, and silicon. Macronutrients are required in fairly large quantities.
- Nine elements are **micronutrients:** chlorine, iron, boron, manganese, sodium, zinc, copper, nickel, and molybdenum. Micronutrients are needed in trace amounts.

11 Explain the effects of mineral depletion and soil erosion on plant growth.

- Mineral depletion may occur in soils that are farmed because the natural pattern of nutrient cycling is disrupted when crops are harvested (and not allowed to decompose into the soil).
- **Soil erosion** is the removal of soil from the land by agents such as water and wind. Erosion causes a loss in soil fertility because essential minerals and organic matter that are a part of the soil are removed.

TEST YOUR UNDERSTANDING

Know and Comprehend

1. One main root, formed from the enlarging embryonic root, with many smaller lateral roots branching off it is (a) a fibrous root system (b) an adventitious root system (c) a taproot system (d) a storage root system (e) a prop root system

2. Root hairs (a) cover and protect the delicate root apical meristem (b) increase the absorptive capacity of roots (c) secrete a waxy cuticle (d) orient the root so that it grows downward (e) store excess sugars produced in the leaves

3. Certain plants adapted to flooded soil produce aerial "breathing" roots known as (a) fibrous roots (b) pneumatophores (c) mycorrhizae (d) nodules (e) prop roots

4. The waterproof region around the radial and transverse walls of endodermal cells is the (a) Casparian strip (b) pericycle (c) apoplast (d) symplast (e) pneumatophore

5. The apoplast is (a) a layer of cells that surround the vascular region in roots (b) the layer of cells just inside the endodermis (c) a system of interconnected plant cell walls through which water moves (d) the central cylinder of the root that comprises the vascular tissues (e) a continuum of cytoplasm of many cells, all connected by plasmodesmata

6. Plants obtain positively charged mineral ions from clay particles in the soil by cation exchange, in which (a) roots passively absorb the positively charged mineral ions (b) mineral ions flow freely along porous cell walls (c) roots secrete protons, which free other positively charged mineral ions to be absorbed by roots (d) the Casparian strip effectively blocks the passage of water and mineral ions along the endodermal cell wall (e) a well-developed system of internal air spaces in the root allows both gas exchange and cation exchange

7. The cell layer from which lateral roots originate is the (a) epidermis (b) cortex (c) endodermis (d) pericycle (e) vascular cambium

8. Mutually beneficial associations between certain soil fungi and the roots of most plant species are called (a) mycorrhizae (b) pneumatophores (c) nodules (d) rhizobia (e) humus

9. Which of the following statements about soil is *true*? (a) pore spaces are always filled with about 50% air and 50% water

(b) a single teaspoon of fertile agricultural soil may contain up to several hundred living microorganisms (c) the texture of a soil is determined by the soil's pH (d) a soil's organic matter includes litter, droppings, and the dead remains of plants, animals, and microorganisms (e) soil formation is unaffected by a region's climate or topography

10. Carbon, hydrogen, oxygen, nitrogen, potassium, calcium, magnesium, phosphorus, sulfur, and silicon are collectively known as (a) micronutrients (b) microvilli (c) micronuclei (d) macronuclei (e) macronutrients

Apply and Analyze

11. **VISUALIZE** Draw a simple sketch illustrating the relationships of the following in a cross section of a primary root: remnant of primary xylem, vascular cambium, secondary phloem, secondary xylem, and periderm.

Evaluate and Synthesize

12. A mesquite root is found penetrating a mine shaft about 46 m (150 ft) below the surface of the soil. How could you determine *when* the root first grew into the shaft? (*Hint:* Mesquite is a woody plant.)

13. You are given a plant part that was found growing in the soil and are asked to determine whether it is a root or an underground stem. How would you identify the structure without a microscope? with a microscope?

14. Explain why, once secondary growth has occurred, that portion of the root is no longer involved in absorption. Where does absorption of water and dissolved minerals occur in plants that have roots with secondary growth?

15. **EVOLUTION LINK** A barrel cactus that is 60 cm tall and 30 cm in diameter has roots more than 3 m long. However, all the plant's roots are found in the soil at a depth of 5 to 15 cm. What possible adaptive value does such a shallow root system confer on a desert plant?

16. **SCIENCE, TECHNOLOGY, AND SOCIETY** What is the agricultural value of research to understand the control of root hair development? of the interactions leading to the development of mycorrhizal associations?

17. **INTERPRET DATA** When soil has become polluted with hazardous materials such as heavy metals, it is sometimes cost-effective to construct an artificial marsh around the site. Water is pumped through the contaminated soil, picking up the hazardous material, and then the water is passed through the artificial wetland. In one experiment scientists compared the effectiveness of the roots of two different marsh plants (cattails and common reeds) in absorbing the heavy metal selenium. Which plant's roots removed the most selenium? What should be done to the plants after they have absorbed selenium?

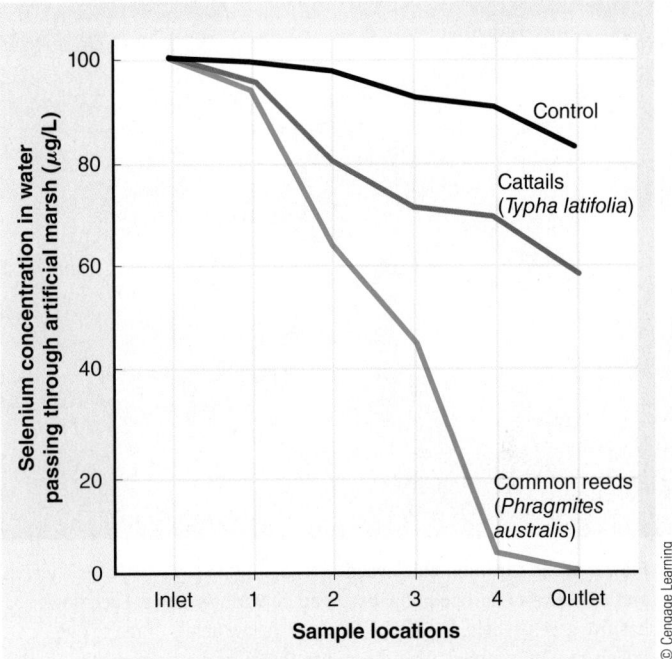

© Cengage Learning

To access course materials, such as Aplia and other companion resources, please visit **www.cengagebrain.com**.

37 | Reproduction in Flowering Plants

Dr. Jeremy Burgess/Science Source

Flower of the common chickweed. This plant (*Stellaria media*) is a weedy annual native to Europe but widespread in North America. Each flower has five green sepals, five deeply notched yellow petals, three to five pollen-bearing stamens (this flower has three), and a single pistil.

KEY CONCEPTS

37.1 The flower is the site of sexual reproduction in angiosperms. A typical flower consists of four whorls: sepals, petals, stamens, and carpels.

37.2 Pollen grains are transported to stigmas by a variety of agents, such as animals and wind.

37.3 Double fertilization results in a plant embryo and endosperm. The seed is a mature ovule, and the fruit is a mature ovary.

37.4 Following a period of dormancy, a seed germinates, or sprouts. The young seedling becomes anchored in the ground and begins to photosynthesize.

37.5 Many vegetative organs (roots, stems, and leaves) are modified for asexual reproduction.

37.6 Sexual reproduction in flowers results in genetic variability in the offspring, whereas asexual reproduction in vegetative structures generally results in individuals that are genetically identical.

Flowering plants, or **angiosperms,** include about 300,000 species and are the largest, most successful group of plants. You may have admired flowers for their fragrances as well as their appealing colors and varied shapes (see photograph). The biological function of flowers is sexual reproduction. Their colors, shapes, and fragrances increase the likelihood that pollen grains, which produce sperm cells, will be carried from one plant to another. Sexual reproduction in plants includes *meiosis* and the fusion of reproductive cells: egg and sperm cells, collectively called **gametes.** The union of gametes, called fertilization, occurs within the flower's ovary.

Sexual reproduction offers the advantage of new gene combinations, not found in either parent, that may make an individual plant better suited to its environment. These new combinations result from the crossing-over and independent assortment of chromosomes that occur during meiosis, before the production of egg and sperm cells (see Chapter 11).

Many flowering plants also reproduce asexually. Asexual reproduction often does not involve the formation of flowers, seeds, and fruits. Instead, offspring generally form when a vegetative organ (such as a stem, root, or leaf) expands, grows, and then becomes separated from the rest of the plant, often by the death of tissues. Because asexual reproduction requires only one parent and no meiosis or fusion of gametes occurs, the offspring of asexual reproduction are virtually genetically identical to one another and to the parent plant.[1]

This chapter examines both sexual and asexual reproduction in flowering plants, including floral adaptations that are important in pollination, seed and fruit structure and dispersal, germination and early growth, and several kinds of asexual reproduction. We conclude with a discussion of the evolutionary advantages and disadvantages of sexual and asexual reproduction.

---

[1] Somatic mutations can result in some variability among asexually derived offspring.

37.1 THE FLOWERING PLANT LIFE CYCLE

LEARNING OBJECTIVES

1 Describe the functions of each part of a flower.
2 Identify where eggs and pollen grains are formed within the flower.

In Chapters 27 and 28 you learned that angiosperms and other plants undergo a cyclic **alternation of generations** in which they spend a portion of their life cycle in a multicellular haploid stage and a portion in a multicellular diploid stage. The haploid portion, called the **gametophyte generation,** gives rise to gametes by mitosis. When two gametes fuse during **fertilization,** the diploid portion of the life cycle, called the **sporophyte generation,** begins. The sporophyte generation produces haploid spores by meiosis. Each spore has the potential to give rise to a gametophyte plant, and the cycle continues.

In flowering plants the diploid sporophyte generation is larger and nutritionally independent. The haploid gametophyte generation, which is located in the flower, is microscopic and nutritionally dependent on the sporophyte. We say more about alternation of generations in flowering plants after our discussion of the role of flowers as reproductive structures. (It may be helpful for you to review Figure 28-13, which shows the main stages in the flowering plant life cycle.)

Flowers develop at apical meristems

How does a plant "know" that it is time to start forming flowers? Correct timing in the switch from vegetative to reproductive development is crucial to ensure reproductive success. A variety of environmental cues, such as temperature and day length, ensure proper timing, and different species are adapted to respond to distinct environmental cues. These environmental signals interact with a variety of plant hormones and developmental pathways.

In recent years some biologists have focused on the molecular aspects of developmental pathways that initiate flowering in plants such as the model organism *Arabidopsis.* When environmental conditions induce flowering, many genes are activated or inactivated. For example, one gene, the *Flowering Locus C (FLC)* gene, codes for a *transcription factor* that represses flowering. (You may recall from Chapter 14 that transcription factors are DNA-binding proteins that regulate the synthesis of RNA from a DNA template.) Another gene, called *Flowering Locus D (FLD),* codes for a protein that removes acetyl groups from histones in the chromatin where the *FLC* gene is located. When deacetylation occurs, the *FLC* gene is not transcribed (i.e., the repressive transcription factor is not produced), and the shoot apical meristem undergoes a transition from vegetative growth to reproductive growth. It is intriguing that the plant FLD protein is homologous to a mammalian protein that also removes acetyl groups from chromatin.

Other genes are also involved in the initiation of flowering, and this area remains a focus of active research interest. In Chapter 38 we discuss the initiation of flowering further.

Each part of a flower has a specific function

Flowers are reproductive shoots, usually consisting of four kinds of organs—sepals, petals, stamens, and carpels—arranged in whorls (circles) on the end of a flower stalk (**FIG. 37-1**; also see Fig. 28-11). In flowers with all four organs, the normal order of whorls from the flower's periphery to the center (or from the flower's base upward) is as follows:

sepals $\longrightarrow$ petals $\longrightarrow$ stamens $\longrightarrow$ carpels

The tip of the stalk enlarges to form a **receptacle** on which some or all of the flower parts are borne. All four floral parts are important in the reproductive process, but only the stamens (the "male" organs) and carpels (the "female" organs) participate directly in sexual reproduction; sepals and petals are sterile.

Sepals, which constitute the outermost and lowest whorl on a floral shoot, cover and protect the flower parts when the flower is a bud. Sepals are leaflike in shape and form and are often green. Some sepals, such as those in lily flowers, are colored and resemble petals (**FIG. 37-2**). The collective term for all the sepals of a flower is **calyx.**

The whorl just inside and above the sepals consists of **petals,** which are broad, flat, and thin (like sepals and leaves) but widely varied in shape and frequently brightly colored, which attracts pollinators. Petals play an important role in ensuring that sexual reproduction will occur. Sometimes petals fuse to form a tube or other floral shape. The collective term for all the petals of a flower is **corolla.**

Just inside and above the petals are the **stamens,** the male reproductive organs. Each stamen has a thin stalk, called a **filament,** at the top of which is an **anther,** a saclike structure in which **pollen grains** form. For sexual reproduction to occur, pollen grains must be transferred from the anther to the female reproductive structure (the carpel), usually of another flower of the same species. At first, each pollen grain consists of two cells surrounded by a tough outer wall. One cell, the *generative cell,* divides mitotically to form two nonflagellate male gametes, known as *sperm cells.* The other cell, the *tube cell,* produces a **pollen tube** through which the sperm cells travel to reach the ovule.

One or more **carpels,** the female reproductive organs, are located in the center or top of most flowers. Carpels bear **ovules,** which are structures with the potential to develop into seeds. The carpels of a flower may be separate or fused into a single structure. The female part of the flower, often called a **pistil,** may be a single carpel (a *simple pistil*) or a group of fused carpels (a *compound pistil*) (see Fig. 28-12). Each pistil has three sections: a **stigma,** on which the pollen grains land; a **style,** a necklike structure through which the pollen tube grows; and an **ovary,** a juglike structure that contains one or more ovules and can develop into a fruit.

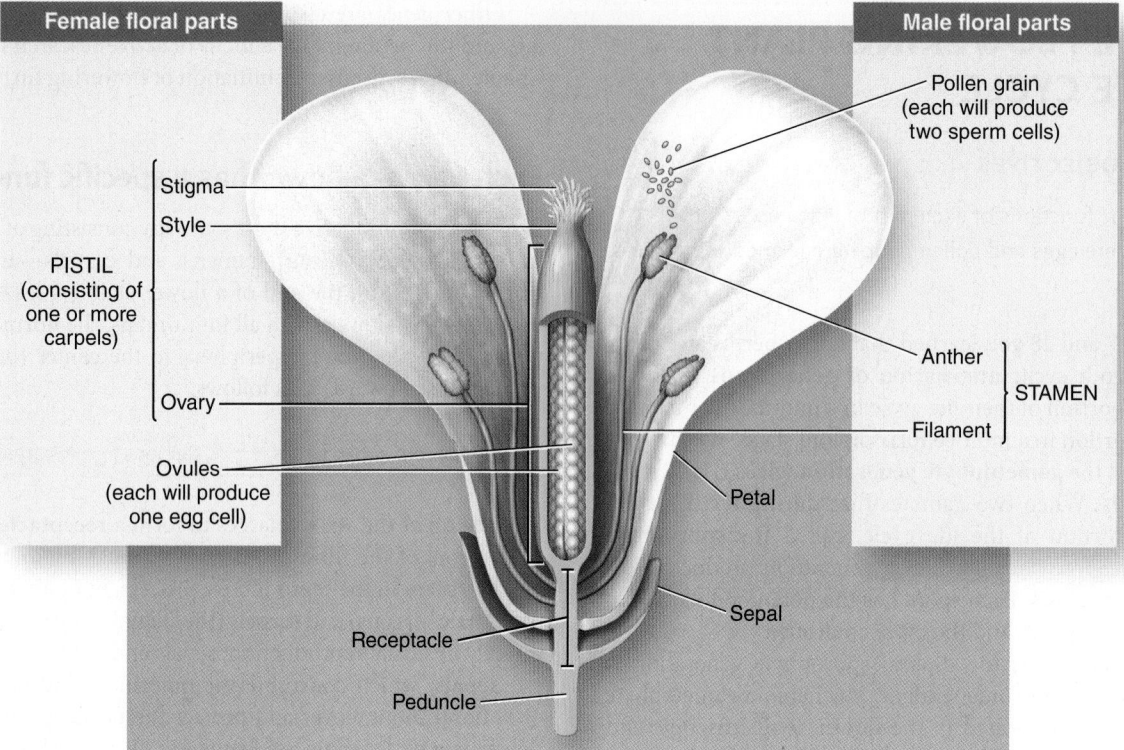

Pollen grain
(each will produce
two sperm cells)

Stigma

Style

PISTIL
(consisting of
one or more
carpels)

Ovary

Anther

STAMEN

Filament

Ovules
(each will produce
one egg cell)

Petal

Sepal

Receptacle

Peduncle

Figure 37-1 *Animation* **Floral structure**

An *Arabidopsis* flower has four sepals (two are shown in this cutaway view), four petals (two are shown), six stamens, and one long pistil. Four of the stamens are long, and two are short (two long and two short are shown). Pollen grains develop within sacs in the anthers. In *Arabidopsis* the compound pistil consists of two carpels, each containing numerous ovules.

© Cengage Learning

Considerable variation in flower structure occurs from one species to another, but flower structure is virtually constant within a given species. For that reason, flower structure is important in plant identification, and if flowers are present, flowering plant species are easy to evaluate in the field.

© Holmes Garden Photos / Alamy

Figure 37-2 Lily flowers

In lily (*Lilium*) the three sepals and three petals are similar in size and color.

Female gametophytes (embryo sacs) are produced in the ovary, male gametophytes in the anther Before we proceed, it may be helpful to relate the stages in alternation of generations to floral structure. As discussed in Chapters 27 and 28, angiosperms and certain other plants are heterosporous and produce two kinds of spores: megaspores and microspores (FIG. 37-3).

Each young ovule within an ovary contains a diploid cell, the *megasporocyte,* which undergoes meiosis to produce four haploid **megaspores.** Three of them usually disintegrate, and the fourth, the functional megaspore, divides mitotically to produce a multicellular **embryo sac,** which is the *female gametophyte.* The embryo sac, which is embedded in the ovule, typically contains seven cells with eight haploid nuclei. Six of these cells, including the egg cell (the female gamete), contain a single nucleus each; a large central cell has two nuclei, called **polar nuclei.** The egg and both polar nuclei participate directly in fertilization.

Pollen sacs within the anther contain numerous diploid cells called *microsporocytes,* each of which undergoes meiosis to produce four haploid cells called **microspores.** Each microspore divides mitotically to produce an immature male gametophyte, also called a pollen grain, that consists of two cells: the tube cell and the generative cell. The pollen grain becomes mature when its generative cell divides to form two nonmotile sperm cells.

In alternation of generations in flowering plants, the female and male gametophytes are microscopic and nutritionally dependent on the sporophyte plant.

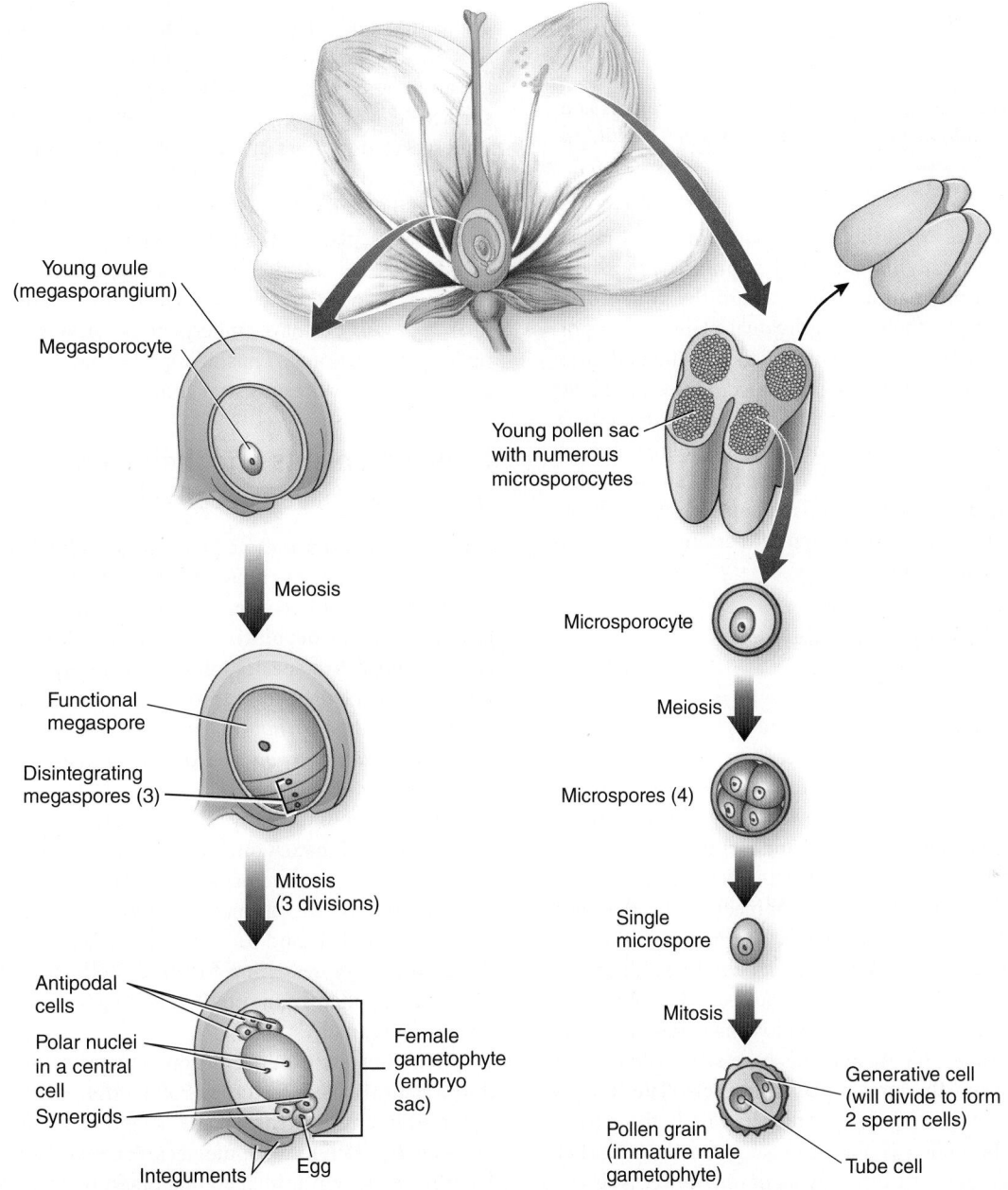

Figure 37-3 *Animation* **Development of female and male gametophytes**

(*Left side*) The embryo sac (female gametophyte) develops within the ovule. (*Right side*) The immature male gametophytes, or pollen grains, develop within pollen sacs in the anthers.

CONNECT State if each of the following is haploid, diploid, or triploid: megasporocyte, megaspore, polar nucleus, egg, microsporocyte, microspore, generative cell, and sperm cell.

© Cengage Learning

CHECKPOINT 37.1

- *How do petals differ from sepals? How are they similar?*
- *How do stamens differ from carpels? How are they similar?*

- **CONNECT** *What are the female gametophytes of flowering plants, and where are they formed?*
- **CONNECT** *What are the male gametophytes of flowering plants, and where are they formed?*

37.2 POLLINATION

3 Compare the evolutionary adaptations that characterize flowers pollinated in different ways (by insects, birds, and bats).

4 Define *coevolution* and give examples of ways that plants and their animal pollinators have affected each other's evolution.

Before fertilization can occur, pollen grains must travel from the anther (where they form) to the stigma. The transfer of pollen grains from anther to stigma is known as **pollination.** Plants are *self-pollinated* if pollination occurs within the same flower or a different flower on the same individual plant. When pollen grains are transferred to a flower on another individual of the same species, the plant is *cross-pollinated.* Flowering plants accomplish pollination in a variety of ways. Beetles, bees, flies, butterflies, moths, wasps, and other insects pollinate many flowers. Animals such as birds, bats, snails, and small nonflying mammals (rodents, primates, and marsupials) also pollinate plants. Wind is an agent of pollination for certain flowers; and water, for a few aquatic flowers.

Many plants have mechanisms that prevent self-pollination

In plant sexual reproduction, the two gametes that unite to form a zygote may be from the same parent or from two different parents. The combination of gametes from two different parents increases the variation in offspring, and this variation may confer a selective advantage. Some offspring, for example, may be able to survive environmental changes better than either parent can.

Evolution has resulted in a variety of mechanisms that prevent self-pollination and thus prevent **inbreeding,** which is the mating of genetically similar individuals. Inbreeding can increase the concentration of harmful genes in the offspring. Some plants, such as asparagus and willow, have separate male and female individuals; the male plants have staminate flowers that lack carpels, and the female plants have pistillate flowers that lack stamens. Other species have flowers with both stamens and pistils, but the pollen is shed from a given flower either before or after the time when the stigma of that flower is receptive to pollen. These characteristics promote **outbreeding** (also called *outcrossing*), which is the mating of dissimilar individuals.

Many species have genes for **self-incompatibility,** a genetic condition in which the pollen is ineffective in fertilizing the same flower or other flowers on the same plant. In other words, an individual plant can identify and reject its own pollen. Genes for self-incompatibility usually inhibit the growth of the pollen tube in the stigma and style, thereby preventing delivery of sperm cells to the ovules. Self-incompatibility, which is more common in wild species than in cultivated plants, ensures that reproduction occurs only if the pollen comes from a genetically different individual.

In plants such as oilseed rape, self-incompatibility is based on a high degree of variation at a particular locus called the *S* locus; the many alleles at this locus are designated S_1, S_2, S_3, S_4, and so on. Here is an example of how the *S* locus blocks self-fertilization. A plant with the genotype S_1S_2 produces pollen grains that land on a stigma of another plant with the genotype S_1S_3. In this case, the presence of the S_1 allele in both the pollen and stigma triggers a self-recognition signaling cascade in surface cells of the stigma. As a result, the stigma cells do not undergo changes that allow the pollen grain to grow a pollen tube. Therefore, fertilization does not occur.

The molecular basis of self-incompatibility in *Arabidopsis* and related plants is an area of active scientific interest. *Arabidopsis* can self-pollinate. Biologists have determined that the genes involved in self-incompatibility and outcrossing exist in *Arabidopsis* but have undergone mutations so that they are no longer functional. Thus, it appears that self-incompatibility in these plants is the ancestral condition.

Flowering plants and their animal pollinators have coevolved

Animal pollinators and the plants they pollinate have had such close, interdependent relationships over time that they have affected the evolution of certain physical and behavioral features in one another. **Coevolution** describes such reciprocal adaptation, in which two species interact so closely that they become increasingly adapted to each other as they each undergo evolutionary change by *natural selection.* We now examine some of the features of flowers and their pollinators that may be the products of coevolution.

Flowers pollinated by animals have various features to attract their pollinators, including showy petals (a visual attractant) and scent (an olfactory attractant) (**FIG. 37-4** and **TABLE 37-1**). One reward for the animal pollinator is food. Pollen grains are a protein-rich food for many animals. As they move from flower to flower searching for food, pollinators inadvertently carry along pollen grains on their body parts, helping the plants reproduce sexually.

Some flowers produce nectar, a sugary solution in special floral glands called *nectaries.* Pollinators use nectar, which is also rich in amino acids, alkaloids, and other floral chemicals, as an energy-rich food. Some nectars contain alkaloids such as nicotine to prevent animal pollinators from drinking all the nectar in a single visit. (Nicotine is poisonous.) Thus, pollinators are encouraged to visit multiple flowers, and each flower has multiple pollinator visitors.

Biologists estimate that insects pollinate about 70% of all flowering plant species. Bees are particularly important as pollinators of crop plants. Crops pollinated by bees provide about 30% of human food. Plants pollinated by insects often have blue or yellow petals. The insect eye does not see color the same way the human eye does. Most insects see well in the violet, blue, and yellow range of visible light but do not perceive red as a distinct color. So, flowers pollinated by insects are not usually red. Insects also see in the ultraviolet range, wavelengths that

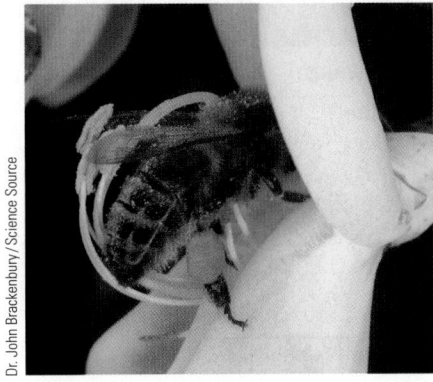

(a) A honeybee (*Apis mellifera*) pollinates a Scotch broom flower (*Cytisus scoparius*). When the bee presses in to obtain nectar, the stamens spring out, dropping pollen grains on the bee's body.

(b) An eastern tiger swallowtail butterfly (*Pterourus glaucas*) obtains nectar from butterfly weed (*Asclepias tuberosa*) flowers.

(c) An Australian long-nosed lycid beetle (*Metriorrhynchus rhipidus*) obtains nectar from an Australian tea-tree (*Leptospermum laevigatum*).

(d) The carrion flower (*Stapelia* sp.), which has petals that resemble dried blood, emits a putrid scent. This desert plant is pollinated by flies. Photographed in the Fairchild Tropical Garden in Miami, Florida.

(e) A broad-billed hummingbird obtains nectar from a desert flower, ocotillo (*Fouquieria splendens*), in Arizona. The pollen grains on the bird's feathers are carried to the next plant.

(f) An orchid flower (*Ophrys scolopax*) that mimics female bees attracts male bees, thereby accomplishing pollination.

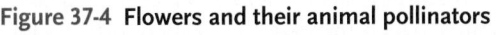

(g) Ultraviolet markings on the evening primrose (*Oenothera* sp.) draw attention to the center of the flower, where insects find nectar and pollen grains. The human eye perceives the flower as solid yellow (*left*). The same flower viewed under ultraviolet radiation (*right*) provides clues about how the insect eye perceives it. The light blue portions of the petals may appear purple to a bee, whereas the dark blue inner parts may appear yellow.

(h) A lesser long-nosed bat (*Leptonycteris curasoae*) obtains nectar from a cardon cactus flower (*Pachycereus*) and transfers pollen as it moves from flower to flower. Bats pollinate several hundred species of plants.

Figure 37-4 Flowers and their animal pollinators

TABLE 37-1   Floral Characteristics Associated with Various Animal Pollinators

ANIMAL POLLINATOR	COMMON FLOWER COLOR	SCENT	POLLINATOR REWARD	TIME OF FLOWERING
Bee	Yellow, blue, purple	Strong, floral	Nectar, pollen	Day
Butterfly	Red, pink	Weak, floral	Nectar	Day
Moth	White	Strong, sweet	Nectar	Dusk, night
Carrion fly	Reddish-brown, purple	Strong, rotten	No reward	Day, night
Beetle	Green, white	Strong, various scents	Nectar, pollen	Day, night
Bird	Red, pink	No scent	Nectar	Day
Bat	White	Strong, musklike or sweet	Nectar, pollen	Night

© Cengage Learning

are invisible to the human eye. Insects see ultraviolet radiation as a color called *bee's purple*. Many flowers have dramatic UV markings that may or may not be visible to humans but that direct insects to the center of the flower where the pollen grains and nectar are.

Insects have a well-developed sense of smell, and many insect-pollinated flowers have a strong scent that may be pleasant or foul to humans. The carrion plant, for example, is pollinated by flies and smells like the rotting flesh in which flies lay their eggs. As flies move from one reeking flower to another looking for a place to lay their eggs, they transfer pollen grains.

Birds such as hummingbirds are important pollinators. Flowers pollinated by birds are usually red, orange, or yellow because birds see well in this range of visible light. Because most birds do not have a strong sense of smell, bird-pollinated flowers usually lack a scent.

Bats, which feed at night and do not see well, are important pollinators, particularly in the tropics. Bat-pollinated flowers bloom at night and have dull white petals and a strong nighttime output of scent that usually smells like fermented fruit. Nectar-feeding bats are attracted to the flowers by the scent; they lap up the nectar with their long, extendible tongues. As they move from flower to flower, they transfer pollen grains. An unusual adaptation that encourages pollination by bats has evolved in the tropical vine *Mucana holtonii*. When the pollen in a given flower is mature, a concave petal lifts up. The petal bounces the echo from the bat's echo-locating calls back to the bat, helping it find the flower.

Specialized features such as petals, scent, and nectar to attract pollinators have coevolved with adaptations of animal pollinators. Specialized body parts and behaviors adapted animals to aid pollination as they obtain nectar and pollen grains as a reward. For example, coevolution has selected for bumblebees' hairy bodies, which catch and hold the sticky pollen grains for transport from one flower to another.

Coevolution may have led to the long, curved beaks of the 'i'iwi, one of the Hawaiian honeycreepers, that inserts its beak into tubular flowers to obtain nectar (see Figs. 1-13b and 20-18). The long, tubular corolla of the flowers that 'i'iwis visit probably also came about through coevolution. During the 20th century, some of the tubular flower species (such as lobelias) became rare, largely as a result of grazing by nonnative cows and feral goats, and about 25% of lobelia species have become extinct. The 'i'iwi now feeds largely on the flowers of the 'ohi'a

tree, which lacks petals, and the 'i'iwi bill appears to be adapting to this change in feeding preference. Comparison of the bills of 'i'iwi museum specimens collected in 1902 with the bills of live birds captured in the 1990s showed that 'i'iwi bills were about 3% shorter in the 1990s than in 1902.

Plants and the behavior of their animal pollinators have coevolved, sometimes in bizarre and complex ways. The flowers of certain orchids (*Ophrys*) resemble female bees in coloring and shape. The unpollinated flowers secrete a scent similar to that produced by female bees, and the males are irresistibly attracted to it. The resemblance between *Ophrys* flowers and female bees is so strong that male bees mount the flowers and try to copulate with them. During this *pseudocopulation,* a pollen sac usually attaches to the back of the bee. When the frustrated bee departs and tries to copulate with another orchid flower, pollen grains are transferred to that flower. Interestingly, once a flower has been pollinated, it emits a scent like that released by female bees that have already mated. Male bees have no interest in visiting flowers that have been pollinated (just as they lose interest in female bees that have already been inseminated).

Biologists think that hummingbird-pollinated flowers arose from insect-pollinated flowers many times during the course of flowering plant evolution. In one experiment researchers demonstrated that a single gene mutation affecting petal color could have been partly responsible for the pollinator shift that resulted in the evolution of two closely related species, *Mimulus lewisii* and *Mimulus cardinalis,* from their insect-pollinated common ancestor (FIG. 37-5).

Some flowering plants depend on wind to disperse pollen

Some flowering plants, such as grasses, ragweed, maples, and oaks, are pollinated by wind. Wind-pollinated plants produce many small, inconspicuous flowers (FIG. 37-6). They do not produce large, colorful petals, scent, or nectar. Some have large, feathery stigmas, presumably to trap wind-borne pollen grains. Because wind pollination is a hit-or-miss affair, the likelihood of a particular pollen grain landing on a stigma of the same species of flower is slim. Wind-pollinated plants produce large quantities of pollen grains, which increase the likelihood that some pollen grains will land on the appropriate stigma.

Can a single-gene mutation in flower color affect animal pollinators?

HYPOTHESIS: If flower color is changed by a mutation, pollinator preferences for the resulting flower may change.

EXPERIMENT: *Mimulus lewisii* has violet-pink flowers and is pollinated by bumblebees; the closely related *M. cardinalis* has orange-red flowers and is pollinated by hummingbirds. Alleles at the same locus in both species affect flower color.

Researchers transferred the allele for orange-red flowers from *M. cardinalis* to *M. lewisii*, which then produced yellow-orange petals. They also transferred the allele for pink flowers from *M. lewisii* to *M. cardinalis*, which then produced dark pink petals.

Wild-type *M. lewisii*

Altered *M. lewisii*

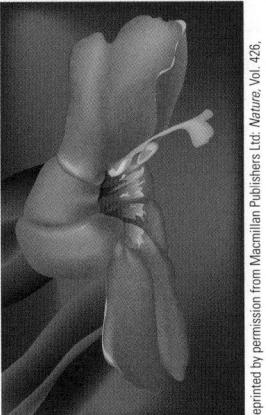

Wild-type *M. cardinalis*

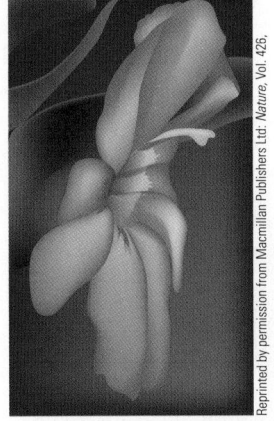

Altered *M. cardinalis*

Reprinted by permission from Macmillan Publishers Ltd: *Nature*, Vol. 426, November 13, 2003

RESULTS AND CONCLUSION: The modified *M. lewisii* flowers were yellow-orange in color and had 68 times as many hummingbird visits as the unaltered *M. lewisii* flowers. The modified *M. cardinalis* flowers had dark pink petals and were visited by 74 times as many bumblebees as the unaltered *M. cardinalis* flowers. Thus, a single mutation in a gene for flower color resulted in a shift in animal pollinators.

SOURCE: Bradshaw, H.D. Jr. and D.W. Schemske, "Allele Substitution at a Flower Colour Locus Produces a Pollinator Shift in Monkeyflowers," *Nature*, Vol. 426, pp. 176–178, Nov. 13, 2003.

Figure 37-5 Studying how pollinator preferences may have evolved

Biologists from the University of Washington and Michigan State University used monkey flowers to determine if a single gene mutation would have any effect on animal pollinators.

PREDICT What other floral traits in monkey flowers are likely to influence whether they are pollinated by hummingbirds or bumblebees?

Dr. Jeremy Burgess/Science Source

Figure 37-6 Wind pollination

Each cluster of male oak flowers (*Quercus*) dangles from a tree branch and sheds a shower of pollen when the wind blows. These flowers lack petals.

CHECKPOINT 37.2

- **PREDICT** *A flower has yellow petals, a pleasant scent, and sugary nectar. How is it probably pollinated?*
- **PREDICT** *A flower has small, inconspicuous flowers without petals, a scent, or nectar. How is it probably pollinated?*
- *Give three examples in which the relationship of a plant and its pollinators demonstrates coevolution.*

37.3 FERTILIZATION AND SEED AND FRUIT DEVELOPMENT

LEARNING OBJECTIVES

5 Distinguish between pollination and fertilization.
6 Trace the stages of embryo development in flowering plants and list and define the main parts of seeds.
7 Explain the relationships among ovules, ovaries, seeds, and fruits.
8 Distinguish among simple, aggregate, multiple, and accessory fruits.

After pollen grains have been transferred from anther to stigma, the tube cell—one of the two cells in the pollen grain—grows a thin pollen tube down through the style and into an ovule in the ovary. How does the pollen tube "know" where to grow? Scientists have identified molecular signals from the synergid cells located adjacent to the egg that guide the growing pollen tube toward the ovule. Two attracting signals (called LUREs), both small polypeptides, have been studied in wishbone flowers (*Torenia fournieri*), and there are ongoing investigations to detect others and to understand their mechanisms of action.

Once a pollen tube penetrates the ovule, the attracting signals are no longer produced. As a result, only one pollen tube enters each embryo sac. The second cell within the pollen grain divides to form two male gametes (the sperm cells), which move down the pollen tube and enter the ovule (**FIG. 37-7**).

A unique double fertilization process occurs in flowering plants

The egg within the ovule unites with one of the sperm cells, forming a zygote (fertilized egg) that develops into an embryonic plant contained in the future seed. The two polar nuclei in the central cell of the ovule fuse with the second sperm cell to form the first cell of the triploid (*3n*) **endosperm,** the tissue with

nutritive and hormonal functions that surrounds the developing embryonic plant in the seed. This process, in which two separate cell fusions occur, is called **double fertilization.** It is, with few exceptions, unique to flowering plants. (A type of double fertilization has been reported in two gymnosperm species, *Ephedra nevadensis* and *Gnetum gnemon.*)

After double fertilization, the ovule develops into a *seed,* and the surrounding ovary develops into a *fruit* (see discussion in Chapter 28).

Embryonic development in seeds is orderly and predictable

Flowering plants package a young plant embryo, complete with stored nutrients, in a compact **seed,** which develops from the ovule after fertilization. The nutrients in seeds are not only used by germinating plant embryos but also eaten by animals, including humans. Development of the embryo and endosperm following fertilization is possible because of the constant flow of nutrients into the developing seed from the parent plant.

Cell divisions of the fertilized egg to form a multicellular embryo proceed in several ways in flowering plants. We describe eudicot embryonic development; monocot embryonic development is similar in the early stages.

The two cells (basal cell and apical cell) that are formed as a result of the first division of the fertilized egg establish *polarity,* or direction, in the embryo. The large *basal cell* (located toward the outside of the ovule) typically develops into a **suspensor,** an embryonic tissue

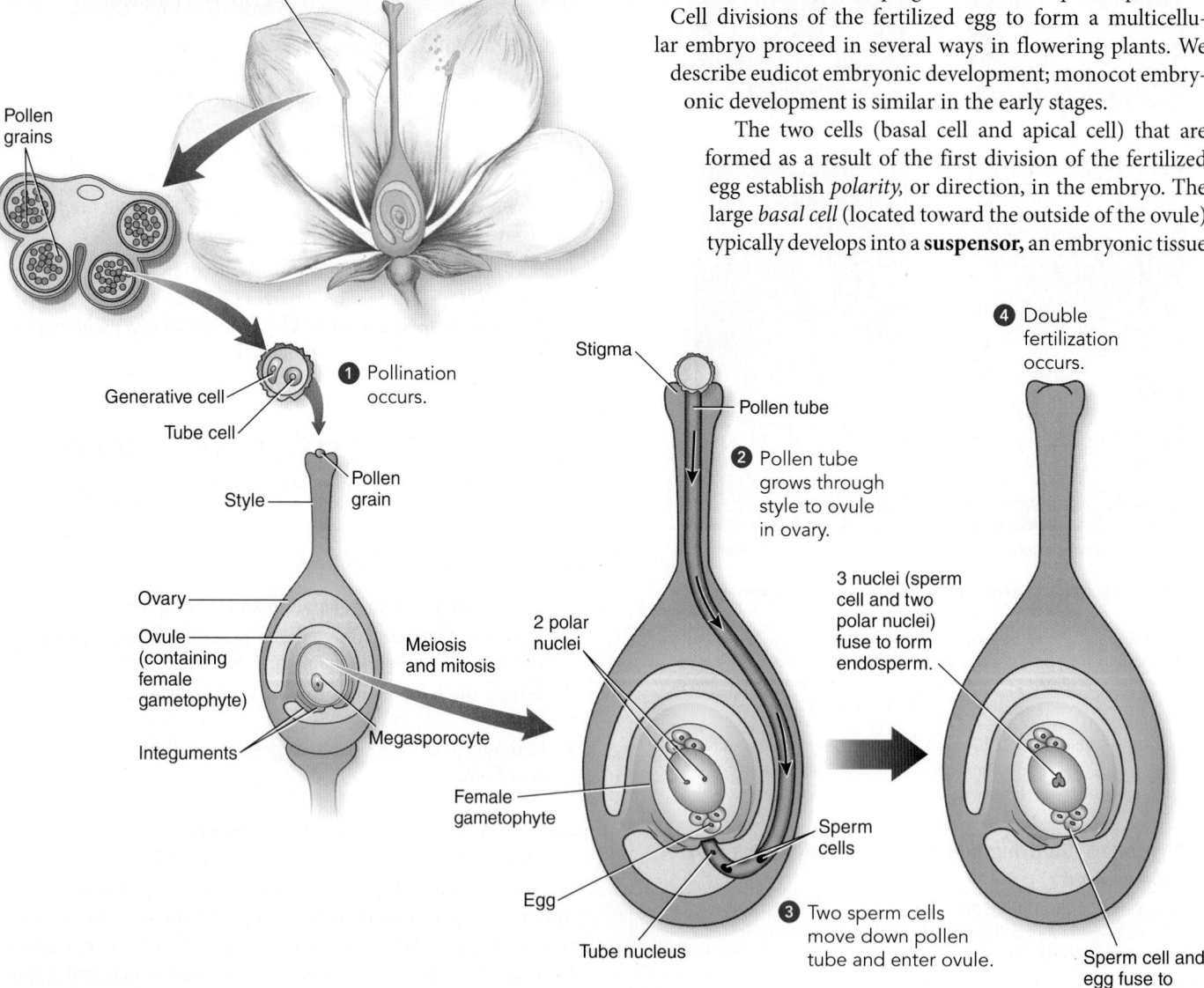

Figure 37-7 Pollination, pollen tube growth, and double fertilization

© Cengage Learning

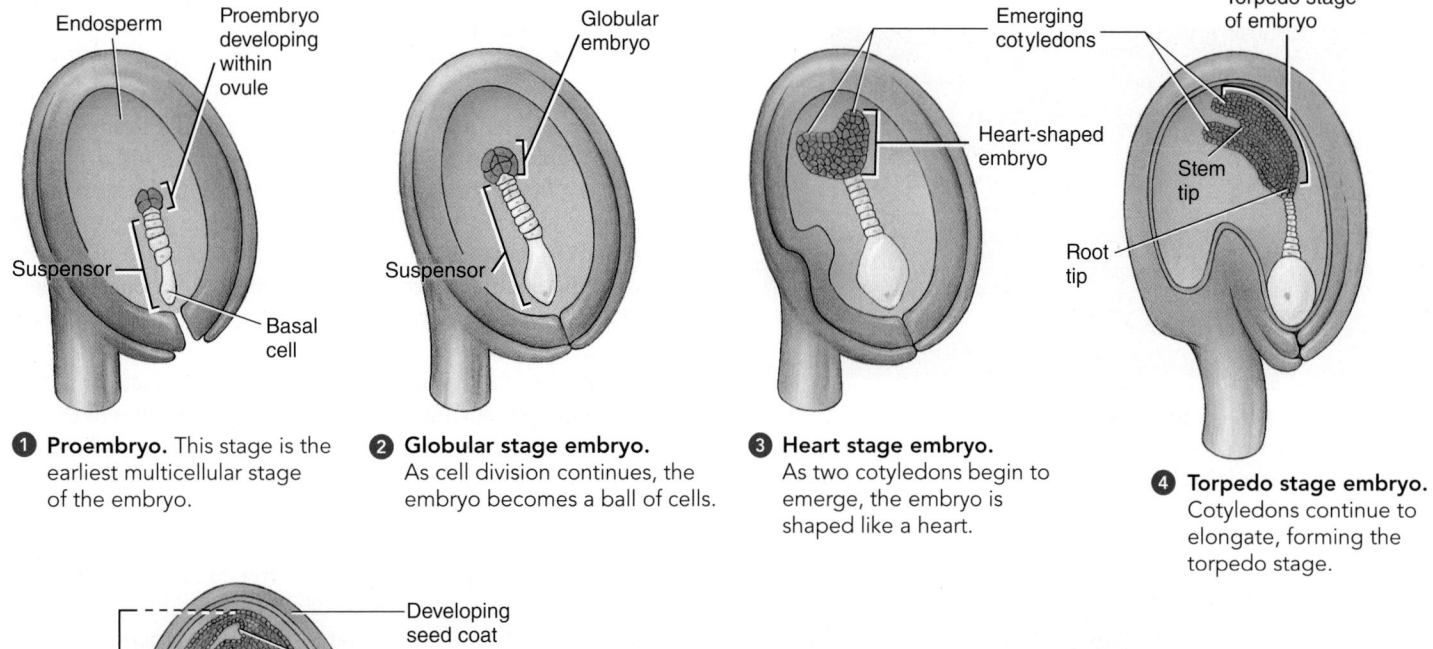

1 **Proembryo.** This stage is the earliest multicellular stage of the embryo.

2 **Globular stage embryo.** As cell division continues, the embryo becomes a ball of cells.

3 **Heart stage embryo.** As two cotyledons begin to emerge, the embryo is shaped like a heart.

4 **Torpedo stage embryo.** Cotyledons continue to elongate, forming the torpedo stage.

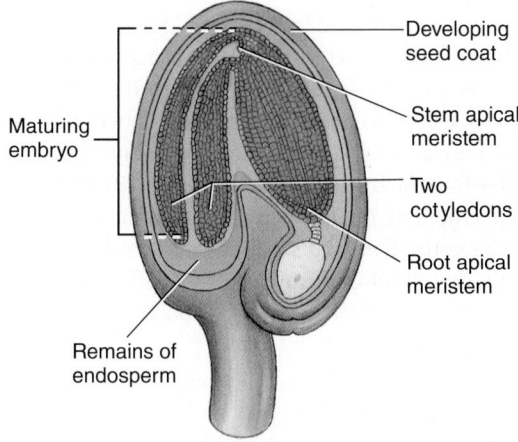

5 **Maturing embryo within seed.** Food originally stored in endosperm has been almost completely depleted during embryonic growth and development, and most of the food for the embryonic plant is now stored in the cotyledons.

6 **Fruit.** Longitudinal section through a fruit of shepherd's purse reveals numerous tiny seeds, each containing a mature embryo. Each seed developed from an ovule.

Figure 37-8 *Animation* **Embryonic development in shepherd's purse (*Capsella bursa-pastoris*)**

The ovule is shown apart from the ovary in parts 1 through 5.
© Cengage Learning

that anchors the developing embryo and aids in nutrient uptake from the endosperm. The *apical cell* (toward the inside of the ovule) develops into the plant embryo. Scientists are currently studying molecular differences between the apical cell and basal cell to determine the initial molecular signals that are involved in establishing polarity.

Initially, the apical cell divides to form a small cluster of cells, called a *proembryo* (**FIG. 37-8**). As cell division continues, a sphere of cells, often called a *globular embryo*, develops. Cells begin to develop into specialized tissues during this stage. When the eudicot embryo starts to develop its two cotyledons (seed leaves), it has two lobes and resembles a heart (the *heart stage*). During the *torpedo stage*, the embryo continues to grow as the cotyledons elongate. As the embryo enlarges, it often curves back on itself and crushes the suspensor.

The mature seed contains an embryonic plant and storage materials

A mature seed contains an embryonic plant and food (stored in the cotyledons or endosperm), surrounded by a tough, protective **seed coat** derived from the **integuments,** which are the outermost layers of an ovule. The seed, in turn, is enclosed within a fruit.

The mature embryo within the seed consists of a short embryonic root, or **radicle;** an embryonic shoot; and one or two seed leaves, or **cotyledons** (**FIG. 37-9**). Monocots have a single cotyledon, whereas eudicots have two. The short portion of the embryonic shoot connecting the radicle to one or two cotyledons is the **hypocotyl.** The shoot apex, or terminal bud, located above the point of attachment of the cotyledon(s) is the **plumule.** After

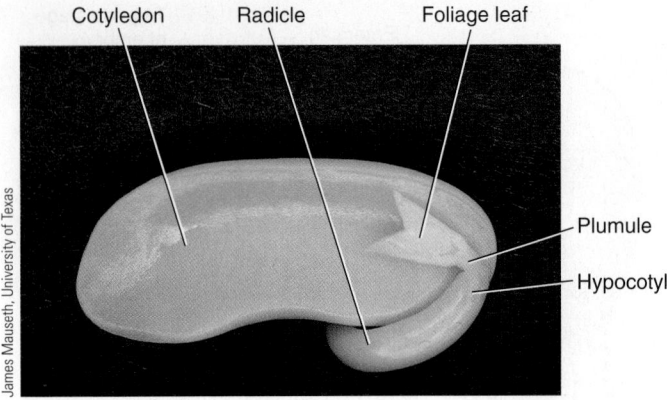

Figure 37-9 Seed structure

A bean seed has been dissected—its seed coat and one of its two cotyledons were removed—to show the radicle, hypocotyl, remaining cotyledon, and plumule with its foliage leaf at the shoot apex. This particular seed had begun germinating, so its radicle is larger than in an ungerminated seed.

the radicle, hypocotyl, cotyledon(s), and plumule have formed, the young plant's development is arrested, usually by desiccation (drying out) or **dormancy** (a temporary state of arrested physiological activity). When conditions are right for continuing the developmental program, the seed *germinates,* and the embryo resumes growth.

Many seeds are known to remain dormant for decade or so. Only a few instances of germination after hundreds of years of dormancy have been scientifically documented. In 2008, a team of biologists from Israel and Switzerland successfully germinated a date seed (*Phoenix dactylifera*) that was radiocarbon dated at about 2000 years old. Before this report, the oldest documented instance of germination was a 1200-year-old lotus seed (*Nelumbo nucifera*).

Because the embryonic plant is nonphotosynthetic, it must be nourished during germination until it becomes photosynthetic and therefore self-sufficient. The cotyledons of many plants function as storage organs and become large, thick, and fleshy as they absorb the food reserves (starches, oils, and proteins) initially produced as endosperm. Seeds that store nutrients in cotyledons have little or no endosperm at maturity. Examples of such seeds are peas, beans, squashes, sunflowers, and peanuts. Other plants—wheat and corn, for example—have thin cotyledons that function primarily to help the young plant digest and absorb food stored in the endosperm.

Fruits are mature, ripened ovaries

After double fertilization takes place within the ovule, the ovule develops into a seed, and the ovary surrounding it develops into a **fruit.** For example, a pea pod is a fruit, and the peas within it are seeds. A fruit may contain one or more seeds; some orchid fruits contain several thousand to a few million seeds! Fruits provide protection for the enclosed seeds and sometimes aid in their dispersal.

There are several types of fruits; their differences result from variations in the structure or arrangement of the flowers from which they were formed. The four basic types of

fruits—simple fruits, aggregate fruits, multiple fruits, and accessory fruits—are summarized in **FIGURE 37-10**.

Most fruits are simple fruits. A **simple fruit** develops from a single ovary, which may consist of a single carpel or several fused carpels. At maturity, simple fruits may be fleshy or dry. Two examples of simple, fleshy fruits are berries and drupes. A **berry** is a fleshy fruit that has soft tissues throughout and contains few to many seeds; a blueberry is a berry, as are grapes, cranberries, bananas, and tomatoes. Many so-called berries do not fit the botanical definition. Strawberries, raspberries, and mulberries, for example, are not berries; these three fruits are discussed shortly.

A **drupe** is a simple, fleshy or fibrous fruit that contains a hard stone (pit) surrounding a single seed. Examples of drupes are peaches, plums, olives, and almonds. The almond shell is actually the stone, which remains after the rest of the fruit has been removed.

Many simple fruits are dry at maturity; some of them split open, usually along seams, called *sutures,* to release their seeds. A milkweed pod is an example of a **follicle,** a simple, dry fruit that develops from a single carpel and splits open along one suture to release its seeds. A **legume** is a simple, dry fruit that develops from a single carpel and splits open along two sutures. Pea pods are legumes, as are green beans, although both are generally harvested before the fruit has dried out and split open. Pea seeds are usually removed from the fruit and consumed, whereas in green beans the entire fruit and seeds are eaten. A **capsule** is a simple, dry fruit that develops from two or more fused carpels and splits open along two or more sutures or pores. Iris, poppy, and cotton fruits are capsules.

Other simple, dry fruits, such as **caryopses** (sing., *caryopsis*), or **grains,** do not split open at maturity. Each caryopsis contains a single seed. Because the seed coat is fused to the fruit wall, a caryopsis looks like a seed rather than a fruit. Kernels of corn and wheat are fruits of this type.

An **achene** is similar to a caryopsis in that it is simple and dry, does not split open at maturity, and contains a single seed. However, the seed coat of an achene is not fused to the fruit wall. Instead, the single seed is attached to the fruit wall at one point only, permitting an achene to be separated from its seed. The sunflower fruit is an example of an achene. One can peel off the fruit wall (the shell) to reveal the sunflower seed within.

Nuts are simple, dry fruits that have a stony wall and do not split open at maturity. Unlike achenes, nuts are usually large, single seeded, and often derived from a compound ovary. Examples of nuts are chestnuts, acorns, and hazelnuts. Many so-called nuts do not fit the botanical definition. Peanuts and Brazil nuts, for example, are seeds, not nuts.

Aggregate fruits are a second main type of fruit. An **aggregate fruit** is formed from a single flower that contains several to many separate (free) carpels (FIG. 37-11). After fertilization, each ovary from each individual carpel enlarges. As they enlarge, the ovaries may fuse to form a single fruit. Raspberries, blackberries, and magnolia fruits are examples of aggregate fruits.

A third type is the **multiple fruit,** formed from the ovaries of many flowers that grow in proximity on a common floral

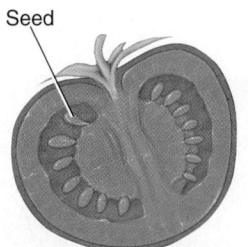

Berry (simple fruit)
A simple, fleshy fruit in which the fruit wall is soft throughout.

Tomato (*Lycopersicon lycopersicum*)

Seed

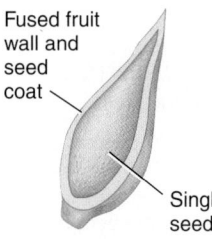

Fused fruit wall and seed coat

Single seed

Caryopsis (simple fruit)
A simple, dry fruit in which the fruit wall is fused to the seed coat.

Wheat (*Triticum*)

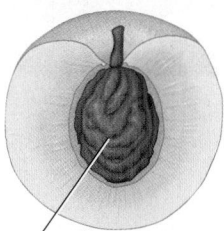

Drupe (simple fruit)
A simple, fleshy fruit in which the inner wall of the fruit is a hard stone.

Peach (*Prunus persica*)

Single seed inside stone

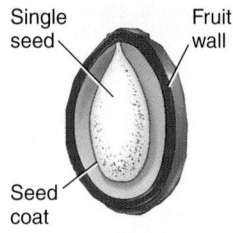

Single seed

Fruit wall

Seed coat

Achene (simple fruit)
A simple, dry fruit in which the fruit wall is separate from the seed coat.

Sunflower (*Helianthus annuus*)

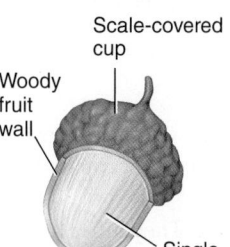

Scale-covered cup

Woody fruit wall

Single seed

Nut (simple fruit)
A simple, dry fruit that has a stony wall, is usually large, and does not split open at maturity.

Oak (*Quercus*)

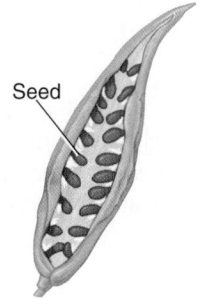

Follicle (simple fruit)
A simple, dry fruit that splits open along one suture to release its seeds; fruit is formed from ovary that consists of a single carpel.

Milkweed (*Asclepias syriaca*)

Seed

Seed

Aggregate fruit
A fruit that develops from a single flower with several to many pistils (i.e., carpels are not fused into a single pistil).

Blackberry (*Rubus*)

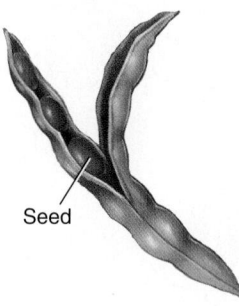

Legume (simple fruit)
A simple, dry fruit that splits open along two sutures to release its seeds; fruit is formed from ovary that consists of a single carpel.

Green bean (*Phaseolus vulgaris*)

Seed

Seed

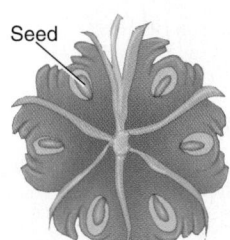

Seed

Multiple fruit
A fruit that develops from the ovaries of a group of flowers.

Mulberry (*Morus*)

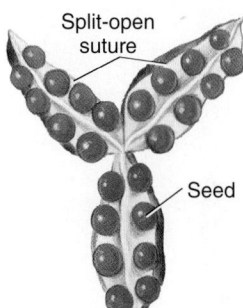

Split-open suture

Seed

Capsule (simple fruit)
A simple, dry fruit that splits open along two or more sutures or pores to release its seeds; fruit is formed from ovary that consists of two or more carpels.

Iris (*Iris*)

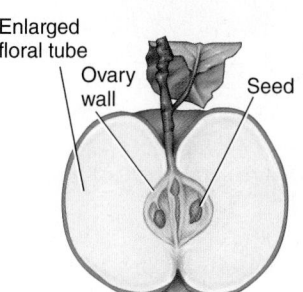

Enlarged floral tube

Ovary wall

Seed

Accessory fruit
A fruit composed primarily of nonovarian tissue (such as the receptacle or floral tube).

Apple (*Malus sylvestris*)

Figure 37-10 Fruit types

Fruits are botanically classified into four groups—simple, aggregate, multiple, and accessory—based on their structure and mechanism of seed dispersal.

© Cengage Learning

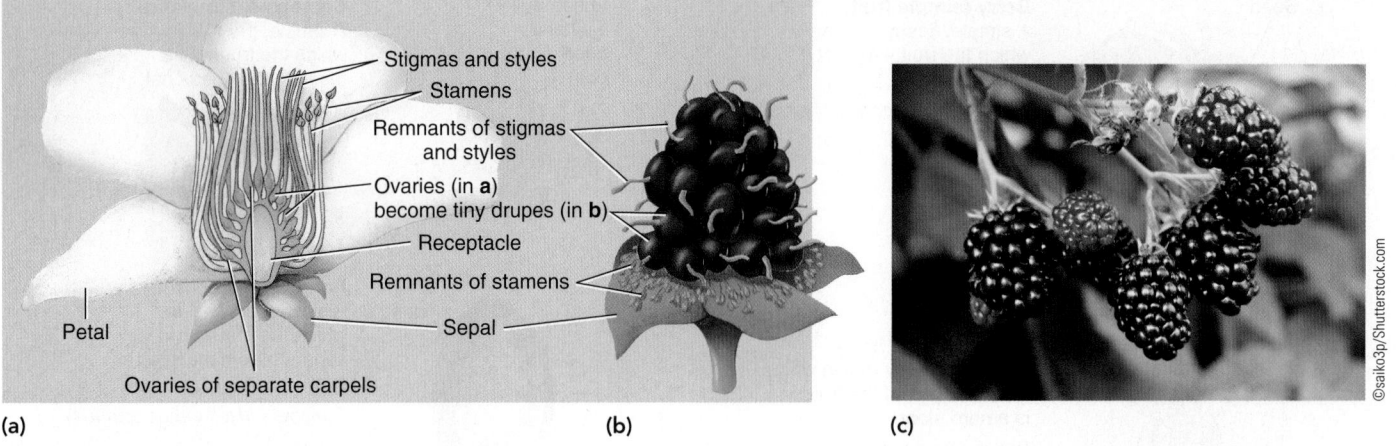

Figure 37-11 An aggregate fruit

(a) Cutaway view of a blackberry flower (*Rubus*), showing the many separate carpels in the center of the flower.
(b) A developing blackberry fruit is an aggregate of tiny drupes. The little "hairs" on the blackberry are remnants of stigmas and styles. **(c)** Developing blackberry fruits at various stages of maturity.
© Cengage Learning

stalk. The ovary from each flower fuses with nearby ovaries as it develops and enlarges after fertilization. Pineapples, figs, and mulberries are multiple fruits.

Accessory fruits are the fourth type. They differ from other fruits in that plant tissues in addition to ovary tissue make up the fruit. For example, the edible portion of a strawberry is the red, fleshy receptacle. Apples and pears are also accessory fruits; the outer part of each of these fruits is an enlarged *floral tube,*

consisting of receptacle tissue along with portions of the calyx, that surrounds the ovary (**FIG. 37-12**).

Seed dispersal is highly varied

Wind, animals, water, and explosive dehiscence disperse the various seeds and fruits of flowering plants (**FIG. 37-13**). Effective methods of seed dispersal have made it possible for certain plants

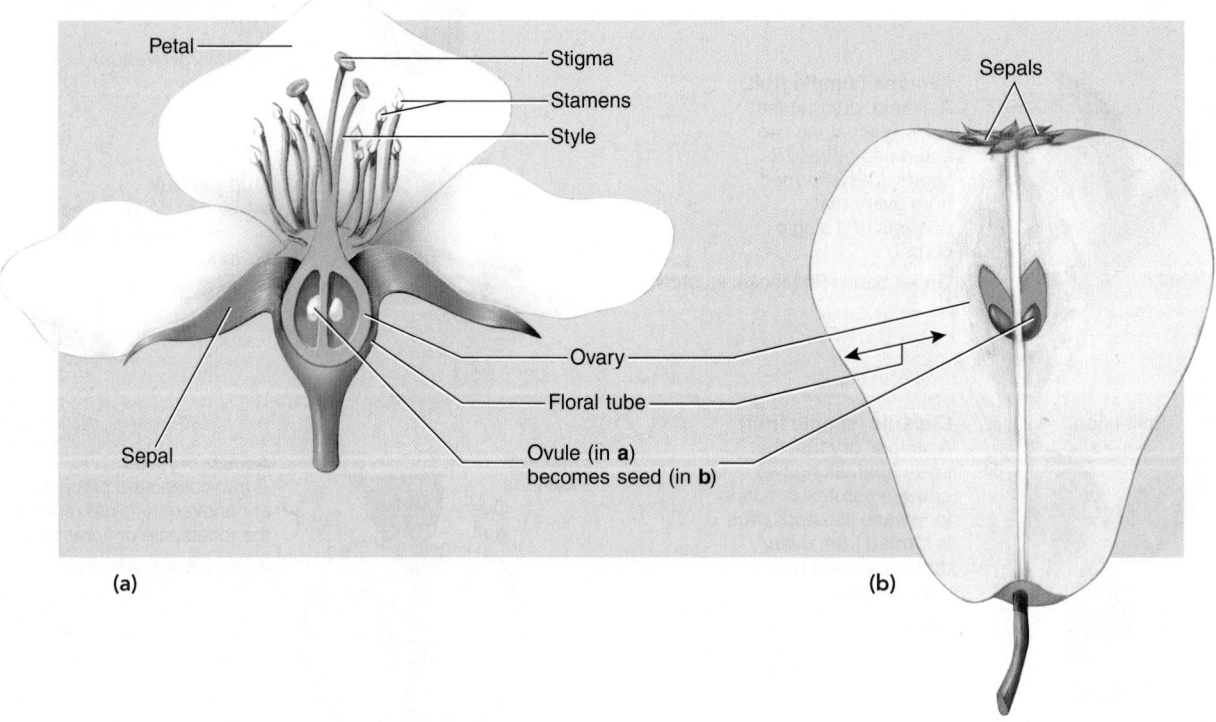

Figure 37-12 An accessory fruit

(a) Note the floral tube surrounding the ovary in the pear flower (*Pyrus communis*). This tube becomes the major edible portion of the pear. **(b)** Longitudinal section through a pear showing the fruit tissue, which derives from both the floral tube and the ovary.
© Cengage Learning

(a) The feathery plumes of milkweed (*Asclepias syriaca*) seeds make them buoyant for dispersal by wind.

(b) Burdock (*Arctium minus*) burs (the hooked fruits) are carried away from the parent plant after sticking to bird feathers, mammal fur, or human clothing.

(c) Dispersal of seeds by ants. The brown part of each bloodroot (*Sanguinaria canadensis*) seed is the seed proper, and the white part is the elaiosome, or oil body. The seeds have been placed along the midvein of an oak leaf to indicate scale.

(d) Fleshy fruits such as blackberries (*Rubus*) are eaten by animals such as this white-footed mouse. The seeds are frequently swallowed whole and pass unharmed through the animal's digestive tract.

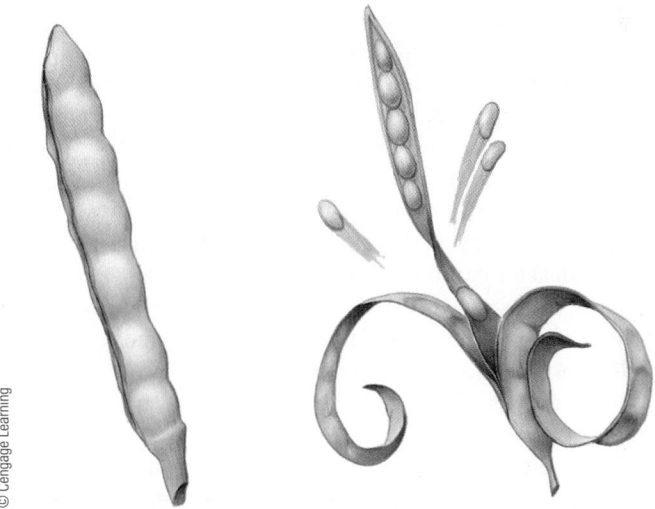

(e) (*Left*) An intact fruit of bitter cress (*Cardamine pratensis*), before it has opened. (*Right*) The fruit splits open with explosive force, flinging the seeds some distance from the plant.

Figure 37-13 Seed (and fruit) dispersal

to expand their geographic range. In some cases, the seed is the actual agent of dispersal; in others, the fruit performs this role. In tumbleweeds, such as Russian thistle, the entire plant is the agent of dispersal because it detaches and blows across the ground, scattering seeds as it bumps along. Tumbleweeds are lightweight and are sometimes blown many kilometers by the wind.

Wind is responsible for seed dispersal in many plants. Plants such as maple trees have winged fruits adapted for wind dispersal. Light, feathery plumes enable other seeds or fruits to be transported by wind, often for considerable distances. Both dandelion fruits and milkweed seeds have this type of adaptation.

Some plants have special structures that aid in dispersal of their seeds and fruits by animals. The spines and barbs of burdock burs and similar fruits catch in animal fur and fall off as the animal moves about.

Fleshy, edible fruits are also adapted for animal dispersal. Birds, bats, primates, grazing ruminants, and ants are common dispersal agents. These animals are attracted to the fruit by its color, location, odor, and taste. As the animal eats these fruits, it either discards or swallows the seeds. Many seeds that are swallowed have thick seed coats and are not digested; instead, they pass through the digestive tract and are deposited in the animal's feces some distance from the parent plant. In fact, some seeds do not germinate unless they have passed through an animal's digestive tract; the animal's digestive juices probably aid germination by helping break down the seed coat. Some edible fruits apparently contain chemicals that function as laxatives to speed seeds through an animal's digestive tract; the less time these seeds spend in the gut, the more likely they will germinate.

Animals such as squirrels and many bird species also help disperse acorns and other fruits and seeds by burying them for winter use. Many buried seeds are never used by the animal and germinate the following spring. Ants collect the seeds of many plants and take them underground to their nests. Ants disperse and bury seeds for hundreds of plant species in almost every terrestrial environment, from northern coniferous forests to tropical rain forests to deserts.

Both ants and flowering plants benefit from their association. The ants ensure the reproductive success of the plants whose seeds they bury, and the plants supply food to the ants. A seed that is collected and taken underground by ants often contains a special structure called an *elaiosome,* or *oil body,* that protrudes from the seed. Elaiosomes are a nutritious food for ants, which carry seeds underground before removing the elaiosome. Once an elaiosome is removed from a seed, the ants discard the undamaged seed in an underground refuse pile, which happens to be rich in organic material (such as ant droppings and dead ants) and minerals (inorganic nutrients) required by young seedlings. Thus, ants not only bury the seeds away from animals that might eat them but also place the seeds in rich soil that is ideal for germination and seedling growth.

The coconut is an example of a fruit adapted for dispersal by water. The coconut has air spaces that make it buoyant and capable of being carried by ocean currents for thousands of kilometers. When it washes ashore, the seed may germinate and grow into a coconut palm tree.

Some seeds are not dispersed by wind, animals, or water. Such seeds are found in fruits that use *explosive dehiscence,* in which the fruit bursts open suddenly and quite often violently, forcibly discharging its seeds. These fruits burst open due to pressure caused by differences in *turgor* (hydrostatic pressure) as the cells in the fruit dry out. The fruits of plants such as touch-me-not and bitter cress split open so explosively that seeds are scattered a meter or more.

CHECKPOINT 37.3

- *What is the difference between pollination and fertilization? Which process occurs first?*
- *What are the main stages of eudicot embryonic development?*
- *What is a fruit?*
- *How are the following fruits distinguished: simple, aggregate, multiple, and accessory?*
- *What are some characteristics of animal-dispersed seeds and fruits?*

37.4 GERMINATION AND EARLY GROWTH

LEARNING OBJECTIVE

9 Summarize the influence of internal and environmental factors on the germination of seeds.

In flowering plants, pollination and fertilization are followed by seed and fruit development. Each seed develops from an ovule and contains an embryonic plant and food to provide nourishment for the embryo during germination. **Germination**—the process of a seed sprouting—and the growth of young seedlings into mature plants are aspects of growth and development. Within a given species, the precise requirements for seed germination represent evolutionary adaptations that protect the young seedlings from adverse environmental conditions. Environmental cues, such as the presence of water and oxygen, proper temperature, and sometimes the presence of light penetrating the soil surface, influence whether or not a seed germinates.

No seed germinates unless it has absorbed water. The embryo in a mature seed is dehydrated, and a watery environment is necessary for active metabolism. When a seed germinates, its metabolic machinery is turned on, and numerous materials are synthesized and others degraded. Therefore, water is an absolute requirement for germination. The absorption of water by a dry seed is **imbibition.** As a seed imbibes water, it often swells to several times its original, dry size (**FIG. 37-14**). Cells imbibe water by adhesion of water onto and into materials such as cellulose, pectin, and starches within the seed.

Germination and subsequent growth require a great deal of energy. Because young plants obtain this energy by converting the energy of food molecules stored in the seed's endosperm or cotyledons to ATP during aerobic respiration, much oxygen is usually needed during germination. (Some plants, such as rice, grow in flooded soil where oxygen is absent and carry out

Marion Lobstein

Figure 37-14 Imbibition

Pinto bean seeds (*Phaseolus vulgaris*) before imbibition (*left*) and after (*right*). Dry seeds imbibe water before they germinate.

alcohol fermentation during the early stages of germination and seedling growth.)

Temperature is another environmental factor that affects germination. Each species has an optimal, or ideal, temperature at which the germination percentage is highest. For most plants, the optimal germination temperature is between 25°C and 30°C (77°F and 86°F). Some seeds, such as those of apples, require prolonged exposure to low temperatures before their seeds break dormancy and germinate. Some of the environmental factors needed for germination help ensure the survival of the young plant. The requirement of a prolonged low-temperature period ensures that seeds adapted to temperate climates germinate in the spring rather than the fall.

Some plants, especially those with tiny seeds, such as lettuce, require light for germination. A light requirement ensures that a tiny seed germinates only if it is close to the soil surface. If such a seed were to germinate several centimeters below the soil surface, it might not have enough food reserves to grow to the surface. If this light-dependent seed remains dormant until the soil is disturbed and it is brought to the surface, however, it has a much greater likelihood of survival.

Some seeds do not germinate immediately

A mature seed is often dormant and may not germinate immediately even if growing conditions are ideal. In certain seeds internal factors, which are under genetic control, prevent germination even when all external conditions are favorable. Many seeds are dormant because certain chemicals are present or absent or because the seed coat restricts germination. For example, the seeds of many desert plants contain high concentrations of *abscisic acid* (discussed in Chapter 38), which inhibits germination under unfavorable conditions. Abscisic acid is washed out only when rainfall is sufficient to support the plant's growth after the seed germinates.

Some seeds, such as certain legumes, have extremely hard, thick seed coats that prevent water and oxygen from entering, thereby inducing dormancy. *Scarification,* the process of scratching or scarring the seed coat (physically with a knife or chemically with an acid) before sowing it, induces germination in these plants. Scarification in nature occurs, for example, when these seeds pass through the digestive tracts of animals or when the seed coats are partially digested by soil bacteria.

Eudicots and monocots exhibit characteristic patterns of early growth

Once conditions are right for germination, the first part of the plant to emerge from the seed is the radicle, or embryonic root. As the root grows and forces its way through the soil, it encounters considerable friction from soil particles. A root cap protects the delicate apical meristem of the root tip.

The shoot (the stem with its leaves and reproductive structures) is next to emerge from the seed. Stem tips are not protected by a structure comparable to a root cap, but plants have ways to protect the delicate stem tip as it grows through the

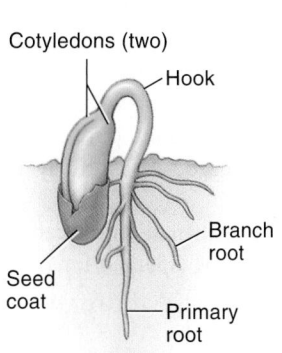
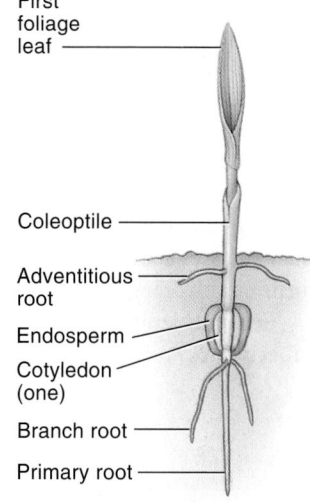

(a) Common bean, a eudicot. The hook in the stem protects the delicate stem tip as it grows through the soil. Once the shoot emerges from the soil, the hook straightens.

(b) Corn, a monocot. The coleoptile, a sheath of cells, emerges first from the soil. The shoot and leaves grow through the middle of the coleoptile.

Figure 37-15 *Animation* **Germination and seedling growth**
© Cengage Learning

soil to the surface. The stem of a bean seedling (a eudicot), for instance, curves over to form a hook so that the stem tip and cotyledons are actually *pulled up* through the soil (**FIG. 37-15a**). Corn and other grasses (monocots) have a special sheath of cells called a **coleoptile** that surrounds and protects the young shoot (**FIG. 37-15b**). First, the coleoptile pushes up through the soil, and then the leaves and stem grow through the tip of the coleoptile.

CHECKPOINT 37.4

- *Give three examples of environmental factors that influence seed germination in certain plants. What is an advantage of each of these germination requirements?*

- **CONNECT** *How does early seedling development differ in eudicots and monocots?*

37.5 ASEXUAL REPRODUCTION IN FLOWERING PLANTS

LEARNING OBJECTIVES

10 Explain how the following structures may be used to propagate plants asexually: rhizomes, tubers, bulbs, corms, stolons, plantlets, and suckers.

11 Define *apomixis*.

Flowering plants have many kinds of asexual reproduction, several of which involve modified stems: rhizomes, tubers, bulbs, corms, and stolons. A **rhizome** is a horizontal underground stem that may or may not be fleshy. Fleshiness indicates that the rhizome is used for storing food materials, such as starch (FIG. 37-16a). Although rhizomes resemble roots, they are actually stems, as indicated by the presence of scalelike leaves, buds, nodes, and internodes. (Roots have none of these features.) Rhizomes frequently branch in different directions. Over time, the old portion of the rhizome dies, and the two branches eventually separate to become distinct plants. Irises, bamboos, ginger, and many grasses are examples of plants that reproduce asexually by forming rhizomes.

Some rhizomes produce greatly thickened ends called **tubers,** which are fleshy underground stems enlarged for food storage. When the attachment between a tuber and its parent plant breaks, often as a result of the death of the parent plant, the tuber grows into a separate plant. Potatoes and elephant's ear (*Caladium*) are examples of plants that produce tubers (FIG. 37-16b). The "eyes" of a potato are axillary buds, evidence that the tuber is an underground stem rather than a storage root such as a sweet potato or carrot.

A **bulb** is a modified underground bud in which fleshy storage leaves are attached to a short stem (FIG. 37-16c). A bulb is globose (round) and covered by paperlike bulb scales, which are modified leaves. It frequently forms axillary buds that develop into small daughter bulbs (bulblets). These new bulbs are

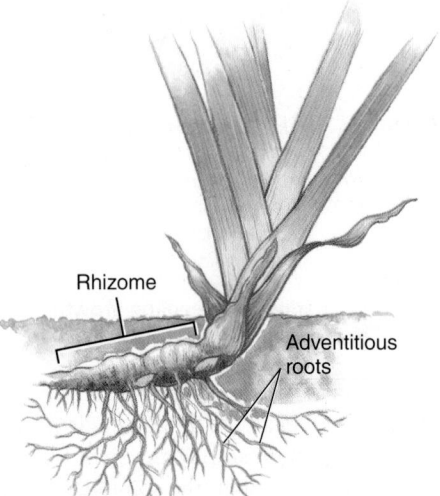

(a) Rhizome. Irises have horizontal underground stems called rhizomes. New aerial shoots arise from buds that develop on the rhizome.

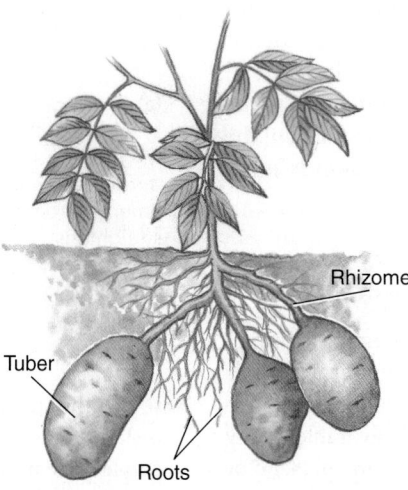

(b) Tuber. Potato plants form rhizomes, which enlarge into tubers (the potatoes) at the ends.

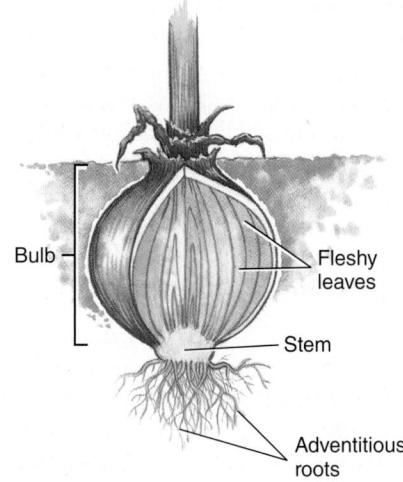

(c) Bulb. A bulb is a short underground stem to which overlapping, fleshy leaves are attached; most of the bulb consists of leaves.

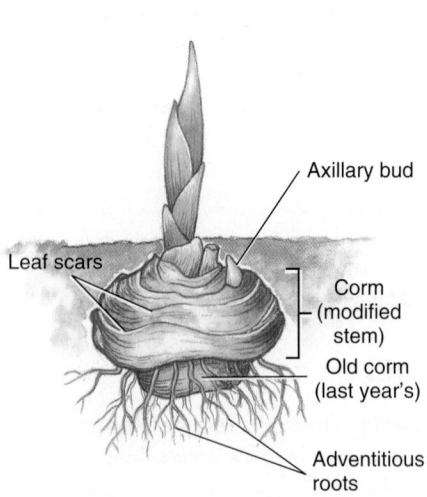

(d) Corm. A corm is an underground stem that is almost entirely stem tissue surrounded by a few papery scales.

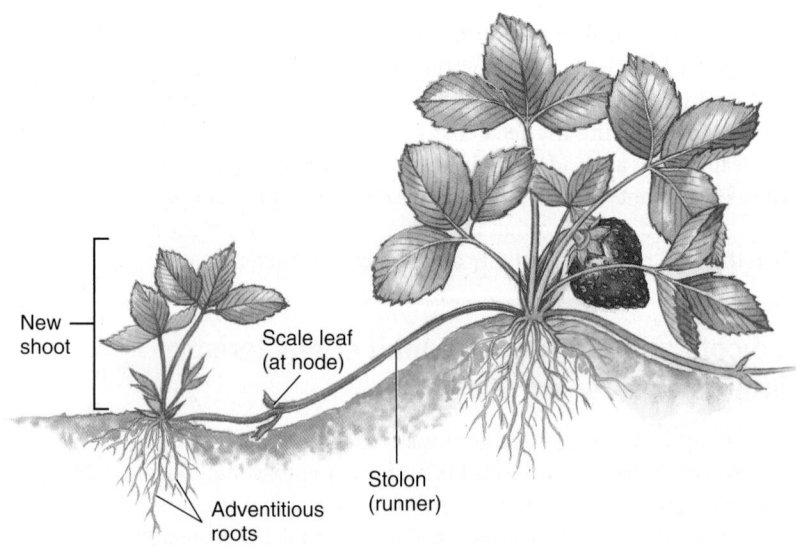

(e) Stolon. Strawberries reproduce asexually by forming stolons, or runners. New plants (shoots and roots) are produced at every other node.

Figure 37-16 Modified stems

© Cengage Learning

initially attached to the parent bulb, but when the parent bulb dies and rots away, each daughter bulb can become established as a separate plant. Lilies, tulips, onions, and daffodils are some plants that form bulbs.

A **corm** is a short, erect underground stem that superficially resembles a bulb (FIG. 37-16d). Unlike the bulb, whose food is stored in underground leaves, the corm's storage organ is a thickened underground stem covered by papery scales (modified leaves). Axillary buds frequently give rise to new corms; the death of the parent corm separates these daughter corms, which then become established as separate plants. Familiar garden plants that produce corms include crocus, gladiolus, and cyclamen.

Stolons, or **runners,** are horizontal aboveground stems that grow along the surface and have long internodes (FIG. 37-16e). Buds develop along the stolon, and each bud gives rise to a new shoot that roots in the ground. When the stolon dies, the daughter plants live separately. The strawberry plant produces stolons.

Some plants form detachable *plantlets* (small plants) in notches along their leaf margins. *Kalanchoe,* whose common name is "mother of thousands," has meristematic tissue that gives rise to an individual plantlet at each notch in the leaf (FIG. 37-17). When these plantlets reach a certain size, they drop to the ground, root, and grow.

Some plants reproduce asexually by producing **suckers,** aboveground shoots that develop from adventitious buds on roots (FIG. 37-18). Each sucker grows additional roots and becomes an independent plant when the parent plant dies. Examples of plants that form suckers are black locust, pear, apple, cherry, blackberry, and aspen. A quaking aspen (*Populus tremuloides*) colony in the Wasatch Mountains of Utah contains at least 47,000 tree trunks formed from suckers that can be traced back to a single individual; this massive "organism" occupies almost 43 hectares (106 acres). Some weeds, such as field bindweed, produce many suckers. These plants are difficult to control because pulling the plant out of the soil seldom

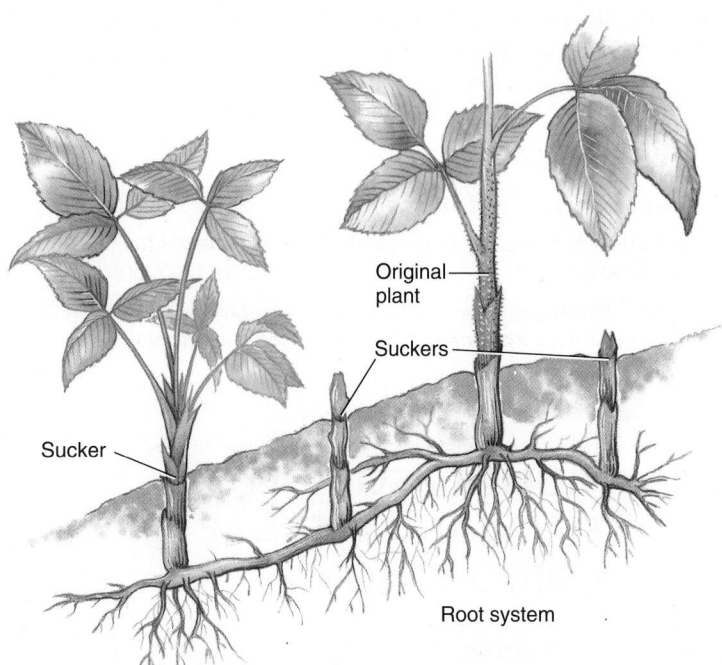

Figure 37-18 Suckers in various stages of development
All the suckers are connected to one another through the root system.
© Cengage Learning

removes all the roots, which can grow as deep as 3 m (10 ft). In fact, in response to wounding, the roots produce additional suckers, which can be a considerable nuisance to humans.

Apomixis is the production of seeds without the sexual process

Some flowering plants produce embryos in seeds without meiosis and the fusion of gametes. Seed production by **apomixis** is a form of asexual reproduction in which the offspring are maternal clones. For example, an embryo may develop from a diploid cell in the ovule rather than from a diploid zygote that forms from the union of two haploid gametes. Because there is no fusion of gametes, the embryo is virtually genetically identical to the maternal genotype. However, the advantage of apomixis over other methods of asexual reproduction is that the seeds and fruits produced by apomixis can be dispersed by methods associated with sexual reproduction.

Apomixis, which occurs in various species of more than 40 angiosperm families, is often found in plants that are polyploid with varying degrees of sterility. Apomixis allows these sterile plants to reproduce and therefore survive. Examples of plants that can reproduce by apomixis are dandelions, citrus trees, mango, blackberries, garlic, and certain grasses.

Plant biologists are actively engaged in identifying the genes involved in asexual reproduction by apomixis. For example, it has been demonstrated that mutation of a single gene in *Arabidopsis* can result in production of diploid female gametes, a first step toward apomixis. If researchers can transfer this or similar genes to crop plants that possess superior traits such as

Figure 37-17 Plantlets

The "mother of thousands" (*Kalanchoe*) produces detachable plantlets along the margins of its leaves. The young plantlets drop off and root in the ground.

high yield, they may be able to develop apomictic crop plants whose traits do not get lost or diluted during the genetic shuffling of sexual reproduction.

CHECKPOINT 37.5

- *How are rhizomes and tubers involved in asexual reproduction?*
- CONNECT *How are corms and bulbs similar? How are they different?*
- *What is apomixis? How is the ability to reproduce by apomixis advantageous compared with other forms of asexual reproduction?*

37.6 A COMPARISON OF SEXUAL AND ASEXUAL REPRODUCTION

LEARNING OBJECTIVE

12 State the differences between sexual reproduction and asexual reproduction, and discuss the evolutionary advantages and disadvantages of each.

Sexual reproduction and asexual reproduction are suited for different environmental circumstances. As you know, sexual reproduction results in offspring that are genetically different from the parents; that is, the parental genotypes are not preserved in the offspring. This genetic diversity of offspring may be selectively advantageous, particularly in an unstable, or changing, environment. (We define *environment* broadly to include all external conditions, both living and nonliving, that affect an organism.) If a plant species that reproduces sexually (and is therefore genetically diverse) is exposed to increasing annual temperatures as a result of climate change, for example, some of the individuals may be more fit than the parents or other offspring. The genetic diversity from sexual reproduction may also let the individuals of a species exploit new environments, thereby expanding their range.

You have seen that asexual reproduction results in offspring that are virtually genetically identical to the parent; that is, the parental genotype is preserved. Assuming that the parent is well adapted to its environment (i.e., the parent has a favorable combination of alleles), this genetic similarity may be selectively advantageous if the environment remains stable (unchanging) for several generations. None of the offspring of asexual reproduction is more fit than the parent, but neither is any less fit.

Despite the apparent advantages of asexual reproduction, most plant species whose reproduction is primarily asexual occasionally reproduce sexually. Even in what appears to be a stable environment, plants are exposed to changing selective pressures, such as changes in the number and kinds of herbivores and parasites, the availability of food, competition from other species, and climate. Sexual reproduction permits species whose reproduction is primarily asexual to increase their genetic variability so that at least some individuals are adapted to the changing selective pressures in a stable environment.

Sexual reproduction has some disadvantages

Although the genetic diversity produced by sexual reproduction is advantageous to a species' long-term survival, sexual reproduction is a costly form of reproduction. Sexual reproduction requires both male and female gametes, which must meet for reproduction to occur. The many adaptations of flowers for different modes of pollination represent one cost of sexual reproduction.

Sexual reproduction produces some individuals with genotypes that are well-adapted to the environment, but it also produces some individuals that are less well-adapted. Therefore, sexual reproduction is usually accompanied by high death rates among offspring, particularly when selective pressures are strong. (As discussed in Chapter 18, however, this aspect of sexual reproduction is an important part of evolution by natural selection.)

Every biological process involves trade-offs, and sexual reproduction is no exception. Although sexual reproduction has its costs, it also has adaptive advantages over asexual reproduction.

CHECKPOINT 37.6

- *How does sexual reproduction in plants differ from asexual reproduction?*
- CONNECT *What is an adaptive advantage of sexual reproduction? Why is sexual reproduction considered costly?*

37.1 The Flowering Plant Life Cycle *(page 777)*

1 Describe the functions of each part of a flower.

- **Sepals** cover and protect the flower parts when the flower is a bud. **Petals** play an important role in attracting animal pollinators to the flower.

- **Stamens** produce pollen grains. Each stamen consists of a thin stalk (the **filament**) attached to a saclike structure (the **anther**).

- The **carpel** is the female reproductive unit. A **pistil** may consist of a single carpel or a group of fused carpels. Each pistil has three sections: a **stigma,** on which the pollen grains land; a **style,** through which the pollen tube grows; and an **ovary** that contains one or more **ovules.**

2 Identify where eggs and pollen grains are formed within the flower.

- **Pollen grains** form within pollen sacs in the anther.
- An egg and two **polar nuclei,** along with several other nuclei, are formed in the ovule. Both egg and polar nuclei participate directly in fertilization.

37.2 Pollination *(page 780)*

3 Compare the evolutionary adaptations that characterize flowers pollinated in different ways (by insects, birds, and bats).

- Flowers pollinated by insects are often yellow or blue and have a scent. Bird-pollinated flowers are often yellow, orange, or red and do not have a strong scent. Bat-pollinated flowers often have white petals and are scented.

4 Define *coevolution* and give examples of ways that plants and their animal pollinators have affected each other's evolution.

- **Coevolution** is reciprocal adaptation caused by two different species (such as flowering plants and their animal pollinators) forming an interdependent relationship and affecting the course of each other's evolution. For example, flowers with large, showy petals and scent have evolved in some plants, whereas hairy bodies that catch and hold sticky pollen grains have evolved in bees.

37.3 Fertilization and Seed and Fruit Development *(page 783)*

5 Distinguish between pollination and fertilization.

- **Pollination** is the transfer of pollen grains from anther to stigma. After pollination, **fertilization,** the fusion of gametes, occurs.
- Flowering plants undergo **double fertilization.** The egg fuses with one sperm cell, forming a zygote (fertilized egg) that eventually develops into a multicellular embryo in the **seed.** The two polar nuclei fuse with the second sperm cell, forming a triploid nutritive tissue called **endosperm.**

6 Trace the stages of embryo development in flowering plants and list and define the main parts of seeds.

- A eudicot embryo develops in the seed in an orderly fashion, from proembryo to globular embryo to the heart stage to the torpedo stage.
- A mature seed contains both a young plant embryo and nutritive tissue—stored in the endosperm or in the **cotyledons** (seed leaves)—for use during germination. A tough, protective **seed coat** surrounds the seed.

7 Explain the relationships among ovules, ovaries, seeds, and fruits.

- Ovules are structures with the potential to develop into seeds, whereas ovaries are structures with the potential to develop into fruits. Seeds are enclosed within **fruits,** which are mature, ripened ovaries.

8 Distinguish among simple, aggregate, multiple, and accessory fruits.

- **Simple fruits** develop from a single ovary that consists of one carpel or several fused carpels. **Aggregate fruits** develop from a single flower with many separate ovaries.

Multiple fruits develop from the ovaries of many flowers growing in close proximity on a common axis. In **accessory fruits** the major part of the fruit consists of tissue other than ovary tissue.

37.4 Germination and Early Growth *(page 790)*

9 Summarize the influence of internal and environmental factors on the germination of seeds.

- **Germination** is the process of seed sprouting. Internal factors affecting whether a seed germinates include the maturity of the embryo; the presence or absence of chemical inhibitors; and the presence or absence of hard, thick seed coats.
- External environmental factors that may affect germination include requirements for oxygen, water, temperature, and light.

37.5 Asexual Reproduction in Flowering Plants *(page 791)*

10 Explain how the following structures may be used to propagate plants asexually: rhizomes, tubers, bulbs, corms, stolons, plantlets, and suckers.

- A **rhizome** is a horizontal underground stem. A **tuber** is a fleshy underground stem enlarged for food storage. A **bulb** is a modified underground bud with fleshy storage leaves attached to a short stem. A **corm** is a short, erect underground stem covered by papery scales. A **stolon** is a horizontal aboveground stem with long internodes.
- Some leaves have meristematic tissue along their margins and give rise to detachable *plantlets*.
- Roots may develop adventitious buds that develop into **suckers.** Suckers produce additional roots and may give rise to new plants.

11 Define *apomixis*.

- **Apomixis** is the production of seeds and fruits without sexual reproduction.

37.6 A Comparison of Sexual and Asexual Reproduction *(page 794)*

12 State the differences between sexual reproduction and asexual reproduction, and discuss the evolutionary advantages and disadvantages of each.

- Sexual reproduction involves the union of two gametes; the offspring produced by sexual reproduction are genetically variable. Asexual reproduction involves the formation of offspring without the fusion of gametes; the offspring are virtually genetically identical to the single parent.
- Genetic diversity among offspring produced by sexual reproduction may let certain individuals survive in a changing environment or exploit new environments. Sexual reproduction is costly because both male and female gametes must be produced and must meet.
- Genetic similarity among offspring produced by asexual reproduction may be advantageous if the environment is stable. Most plant species whose reproduction is primarily asexual occasionally reproduce sexually, thereby increasing their genetic variability.

Know and Comprehend

1. The normal order of whorls from the flower's periphery to the center is (a) sepals, petals, carpels, stamens (b) stamens, carpels, sepals, petals (c) sepals, petals, stamens, carpels (d) petals, carpels, stamens, sepals (e) carpels, stamens, petals, sepals

2. A pistil consists of (a) stigma, style, and stamen (b) anther and filament (c) sepal and petal (d) stigma, style, and ovary (e) radicle, hypocotyl, and plumule

3. The transfer of pollen grains from anther to stigma is (a) fertilization (b) double fertilization (c) pollination (d) germination (e) apomixis

4. The observation that insects with long mouthparts pollinate long, tubular flowers and insects with short mouthparts pollinate flowers with short corollas is explained by (a) coevolution (b) germination (c) double fertilization (d) apomixis (e) explosive dehiscence

5. The process of _____ in flowering plants involves one sperm cell fusing with an egg cell and one sperm cell fusing with two polar nuclei. (a) coevolution (b) germination (c) double fertilization (d) apomixis (e) pollination

6. The nutritive tissue in the seeds of flowering plants that is formed from the union of a sperm cell and two polar nuclei is called the (a) plumule (b) endosperm (c) cotyledon (d) hypocotyl (e) radicle

7. After fertilization, the ovule develops into a _____, and the ovary develops into a _____. (a) fruit; seed (b) seed; fruit (c) calyx; corolla (d) corolla; calyx (e) follicle; legume

8. In plants that lack endosperm in their mature seeds, the cotyledons function to (a) enclose and protect the seed (b) aid in seed dispersal (c) serve as an absorptive embryonic root (d) store food reserves (e) attach the embryo within the ovule

9. _____ fruits develop from many ovaries of a single flower, whereas _____ fruits develop from the ovaries of many separate flowers. (a) multiple; accessory (b) simple; accessory (c) aggregate; multiple (d) accessory; aggregate (e) simple; multiple

10. A horizontal underground stem that may or may not be fleshy and that is often specialized for asexual reproduction is called a (a) stolon (b) bulb (c) corm (d) rhizome (e) tuber

Apply and Analyze

11. Place the following events in the correct order.

    1. pollen tube grows into ovule 2. insect lands on flower to drink nectar 3. embryo develops within the seed 4. fertilization occurs 5. pollen carried by insect contacts stigma

    (a) 2, 5, 1, 4, 3 (b) 1, 4, 2, 5, 3 (c) 3, 2, 5, 1, 4 (d) 5, 1, 3, 4, 2 (e) 2, 5, 4, 3, 1

12. **VISUALIZE** Draw a simple sketch of a eudicot flower with a single pistil and label a petal, anther, ovule, stigma, ovary, filament, stamen, style, and sepal.

13. **VISUALIZE** Sketch the kinds of flowers that form simple, aggregate, multiple, and accessory fruits.

Evaluate and Synthesize

14. Using what you have learned in this chapter, speculate whether it is more likely that offspring of asexual reproduction develop in close proximity to or widely dispersed from the parent plant. Explain your reasoning. How could you test your hypothesis?

15. Which type of reproduction, sexual or asexual, might be more beneficial in the following circumstances and why? (a) a perennial (plant that lives more than two years) in a stable environment (b) an annual (plant that lives one year) in a rapidly changing environment (c) a plant adapted to an extremely narrow climate range

16. **EVOLUTION LINK** Is seed dispersal by ants an example of coevolution? Why or why not?

17. **SCIENCE, TECHNOLOGY, AND SOCIETY** What is the value to society of research on genes that control flowering? of those that control seed germination?

To access course materials, such as Aplia and other companion resources, please visit **www.cengagebrain.com.**

aplia

Plant Developmental Responses to External and Internal Signals

38

The ultimate control of plant growth and development, which includes all the changes that take place during the entire life of an individual, is genetic. If the genes required for development of a particular trait, such as the shape of a leaf, the color of a flower, or the type of root system, are not present, that characteristic does not develop. When a particular gene is present, its expression—that is, how it exhibits itself as an observable feature of an organism—is determined by several factors, including regulatory signals from other genes and from the environment. The location of a cell in the young plant body also has a profound effect on gene expression during development. Chemical signals from adjacent cells help a cell "perceive" its location within the plant body. Each cell's spatial environment helps determine what that cell ultimately becomes.

Growth and development, including a plant's responses to various changes in its environment, are controlled by plant *hormones,* organic molecules that are present in very low concentrations in plant tissues and that act as chemical signals between cells. Environmental cues, such as changing day length and variations in precipitation and temperature, exert an important influence on gene expression and hormone production, as they do on all aspects of plant growth and development.

For example, the initiation of sexual reproduction is often under environmental control, particularly in temperate latitudes, and plants switch from vegetative to reproductive growth after receiving appropriate signals from the environment. Many flowering plants are sensitive to changes in the relative amounts of daylight and darkness that accompany the changing seasons, and these plants flower in response to those changes (see photograph). In this chapter we will see that plants use light as an external signal to help regulate many aspects of growth and development, such as tropisms (directional growth), time of flowering, shade avoidance, and circadian rhythms.

Black-eyed Susan. This plant (*Rudbeckia hirta*) produces flowers in response to the shortening nights of spring and early summer.

KEY CONCEPTS

38.1 A plant may respond to an external stimulus, such as light, gravity, or touch, by a tropism, a directional growth response. Study of tropisms has helped biologists elucidate links between the external environment and internal signals such as hormones.

38.2 Hormones are chemical signals responsible for coordinating and regulating many aspects of plant development. Although there are many plant hormones, six have been well characterized: auxins, gibberellins, cytokinins, ethylene, abscisic acid, and brassinosteroids.

38.3 Plants have different receptors that detect various colors of light. Phytochrome detects red light and far-red light, which affect several aspects of development, including the timing of flowering.

38.4 The exposure of a plant to a herbivore or pathogen causes a localized hypersensitive response that seals off the affected area. The plant also sends signals from the site of localized infection, leading to systemic acquired resistance and long-lasting, enhanced defenses throughout the plant.

Other plants have temperature requirements that induce sexual reproduction. Thus, plants continually perceive information from the environment and use this information to help regulate normal growth and development. We consider these various aspects of growth and development in this chapter.

38.1 TROPISMS

LEARNING OBJECTIVE

1 Describe phototropism, gravitropism, and thigmotropism.

Plants exhibit movements in response to environmental stimuli such as light, gravity, and touch. A plant may respond to such an external stimulus by directional growth; that is, the direction of growth depends on the direction of the stimulus. Such a directional growth response, called a **tropism,** results in a change in the position of a plant part. Tropisms are irreversible and may be positive or negative, depending on whether the plant grows toward the stimulus (a positive tropism) or away from it (a negative tropism). Tropisms are under hormonal control, which is discussed later in the chapter.

Phototropism is the directional growth of a plant caused by light (**FIG. 38-1**). Most growing shoot tips exhibit positive phototropism by bending (growing) toward light, something you may have observed if you place houseplants near a sunny window. This growth response increases the likelihood that stems and leaves receive adequate light for photosynthesis. The bending response of phototropism is triggered by blue light with wavelengths less than 500 nm. (You may recall from Chapter 34 that blue light also induces stomata to open.)

For a plant or any organism to have a biological response to light, it must contain a light-sensitive substance, called a *photoreceptor,* to absorb the light. The photoreceptor that absorbs blue light and triggers the phototropic response and other blue-light responses (such as stomatal opening) is a

Figure 38-1 Phototropism

A bean seedling (*Phaseolus vulgaris*) grows in the direction of light and therefore exhibits positive phototropism. The bending is caused by greater elongation on the shaded side of the stem than on the lighted side.

family of yellow pigments called **phototropins.** Phototropins are light-activated **kinases,** enzymes that transfer phosphate groups. There is evidence that phototropins become phosphorylated—that is, a phosphate group is added—in response to blue light. Thus, phosphorylation is an early step in the blue-light-signaling pathway.

Growth in response to the direction of gravity is called **gravitropism.** Most stem tips exhibit negative gravitropism by growing away from Earth's center, whereas most root tips exhibit positive gravitropism (**FIG. 38-2**). The root cap is the site of gravity perception in roots; when the root cap is removed, the root continues to grow, but it loses any ability to perceive gravity. Special cells in the root cap possess starch-containing **amyloplasts** (see Fig. 3-9a) that collect toward the bottom of the cells in response to gravity, and these amyloplasts may initiate at least some of the gravitropic response. If the root is put in a different position, as when a potted plant is laid on its side,

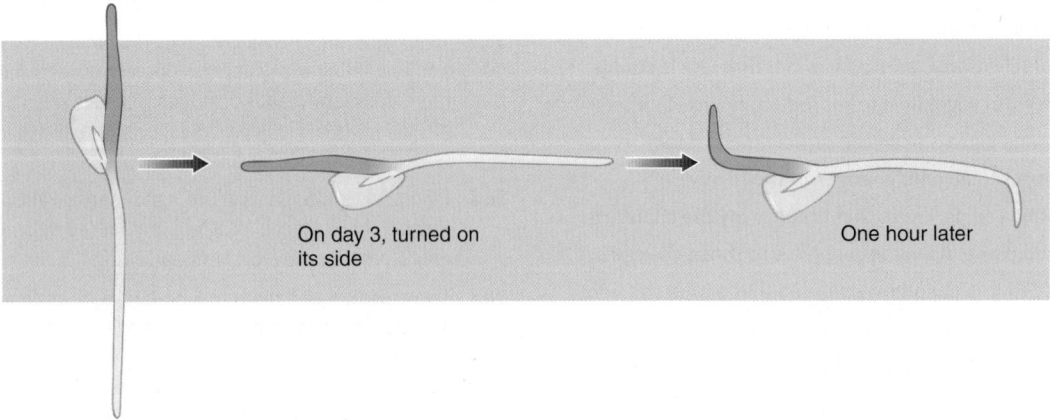

On day 3, turned on its side

One hour later

Figure 38-2 Gravitropism

A corn (*Zea mays*) seedling was turned on its side three days after germinating. In 1 hour the root and shoot tips had curved.

© Cengage Learning

the amyloplasts tumble to a new position, always settling in the direction of gravity. The gravitropic response (bending) occurs shortly thereafter and involves the hormone *auxin* (discussed later in the chapter).

Despite the movement of amyloplasts in response to gravity, researchers question their role in gravitropism. A mutant *Arabidopsis thaliana* plant that lacks amyloplasts in its root cap cells still responds gravitropically when placed on its side, indicating that roots do not necessarily need amyloplasts to respond to gravity. Ongoing research may clarify how roots perceive gravity.

Thigmotropism is growth in response to a mechanical stimulus, such as contact with a solid object. The twining or curling growth of tendrils or stems, which helps attach a climbing plant such as a vine to some type of support, is an example of thigmotropism (see Fig. 34-14b).

CHECKPOINT 38.1

- **CONNECT** *How does the yellow color of phototropins relate to the wavelengths of light to which they respond?*
- *How do phototropism, gravitropism, and thigmotropism differ?*

38.2 PLANT HORMONES AND DEVELOPMENT

LEARNING OBJECTIVES

2 Describe a general mechanism of action for plant hormones, using auxin as your example.

3 Describe early auxin experiments involving phototropism.

4 List several ways that each of these hormones affects plant growth and development: auxins, gibberellins, cytokinins, ethylene, abscisic acid, and brassinosteroids.

A plant **hormone** is an organic compound that acts as a chemical signal eliciting a variety of responses that regulate growth and development. The study of plant hormones is challenging

because hormones are effective in extremely small concentrations (less than 10^{-6} mol/L). In addition, the effects of different plant hormones overlap, and it is difficult to determine which hormone, if any, is the primary cause of a particular response. Plant hormones may also stimulate a response at one concentration and inhibit that same response at a different concentration.

For many years, biologists studied five major classes of plant hormones: auxins, gibberellins, cytokinins, ethylene, and abscisic acid. More recently, researchers have uncovered compelling evidence for brassinosteroids as a sixth plant hormone. Additional plant hormones will probably be characterized in the future. (**TABLE 38-1** summarizes the six plant hormones.)

Plant hormones act by signal transduction

Researchers have used molecular genetic techniques to better understand the biology of plant hormones. *Arabidopsis* mutants are particularly useful. For example, different mutants have defects in hormone synthesis, hormone transport, signal reception, or signal transduction (**FIG. 38-3**). In **signal transduction** a receptor converts an extracellular signal into an intracellular signal that causes one or more cellular responses.

Such mutants enable plant biologists to identify and clone genes involved in these aspects of hormone biology. Studying the mutant phenotypes helps plant biologists establish connections between the mutant genes and specific physiological activities involved in growth and development. Using this research, biologists are elucidating the general mechanisms of plant hormone action, an important goal of systems biology.

As discussed in Chapter 6, it appears that many plant hormones bind to *enzyme-linked receptors;* the hormone binds to the receptor, where it triggers an enzymatic reaction of some sort. Let us consider a specific example that involves the plant hormone *auxin,* which is known to cause rapid changes in gene expression. One receptor for auxin is located in either the cytosol or the nucleus and has a three-dimensional shape that binds to the auxin molecule. The binding of auxin to its receptor (designated the *TIR1 receptor*) catalyzes the attachment of the molecule

TABLE 38-1	Plant Hormones	
HORMONE	**SITE OF PRODUCTION**	**PRINCIPAL ACTIONS**
Auxins (e.g., IAA)	Shoot apical meristem, young leaves, seeds	Stem elongation, apical dominance, root initiation, fruit development
Gibberellins	Young leaves and shoot apical meristems, embryo in seed	Seed germination, stem elongation, flowering, fruit development
Cytokinins (e.g., zeatin)	Various sites in plants	Cell division, delay of leaf senescence, inhibition of apical dominance, flower development, embryo development, seed germination
Ethylene	Stem nodes, ripening fruit, damaged or senescing tissue	Fruit ripening, responses to environmental stressors, seed germination, maintenance of apical hook on seedlings, root initiation, senescence and abscission in leaves and flowers
Abscisic acid	Almost all cells that contain plastids (leaves, stems, roots)	Seed dormancy, responses to water stress
Brassinosteroids (e.g., brassinolide)	Shoots (leaves and flower buds), seeds, fruits	Light-mediated gene expression, cell division, cell elongation, seed germination, vascular development

© Cengage Learning

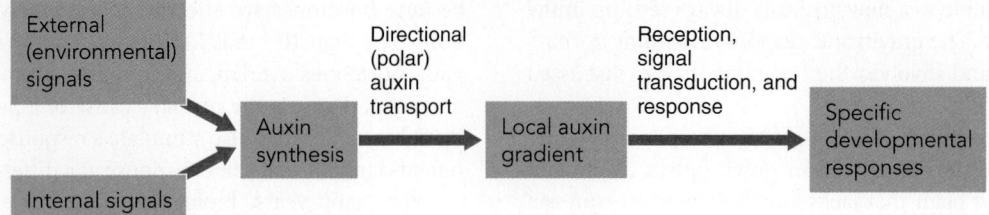

Figure 38-3 General mechanism of action for the hormone auxin

Various external and/or internal signals trigger the synthesis of auxin. Once synthesized, auxin is transported directionally, causing local auxin gradients to form in the plant. In turn, these gradients trigger specific developmental responses in certain cells by reception, signal transduction, and response (discussed shortly). *Arabidopsis* mutants have been identified with defects involving auxin synthesis, auxin transport, or other steps in the process.

© Cengage Learning

ubiquitin to repressor proteins that inhibit the transcription of certain genes, designated *auxin response genes* (FIG. 38-4).

The ubiquitinylated protein is therefore targeted for degradation into peptide fragments in a proteasome. (See Figure 14-17, which shows the ubiquitin–proteasome degradation of proteins.)

As a result, transcription of the auxin response gene occurs, resulting in changes in cell growth and development.

The *TIR1* receptor for auxin is an *F-box protein*. The F-box is a short sequence of amino acids found in molecules that catalyze the addition of ubiquitin tags to proteins targeted for destruction.

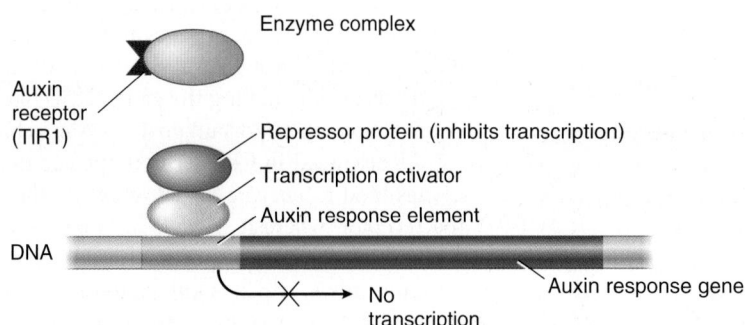

(a) Auxin absent. A repressor protein binds to a transcription activator, inhibiting transcription of the auxin response gene.

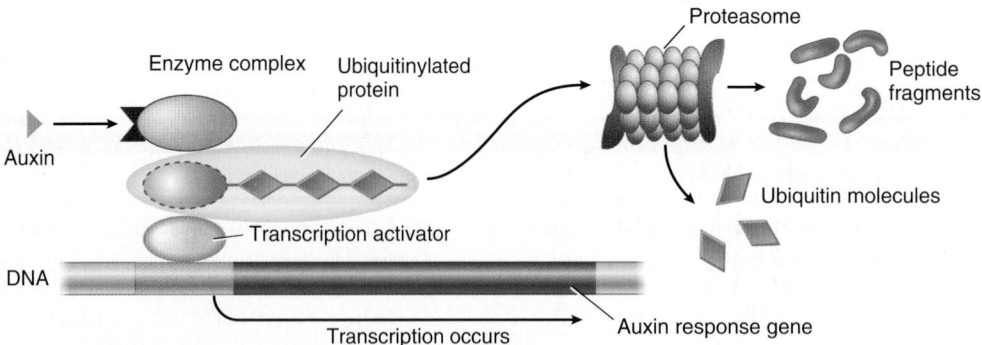

(b) Auxin present. Auxin binds to the TIR1 receptor, which is part of an enzyme complex that attaches ubiquitin to the repressor protein. The ubiquitinylated repressor protein is then degraded into peptide fragments in a proteasome. With the repressor protein no longer bound to the transcription activator, transcription of the auxin response gene occurs.

Figure 38-4 Model of a signal transduction pathway for auxin

© Cengage Learning

What part of the grass coleoptile depends on light for phototropic growth?

HYPOTHESIS: The coleoptile tip is required for the growth (bending) of the coleoptile shaft that accompanies exposure to light coming from one direction.

EXPERIMENT: Some canary grass coleoptiles were uncovered, some were covered only at the tip, some had the tip removed, and some were covered everywhere but at the tip (*left*). The covers were impervious to light.

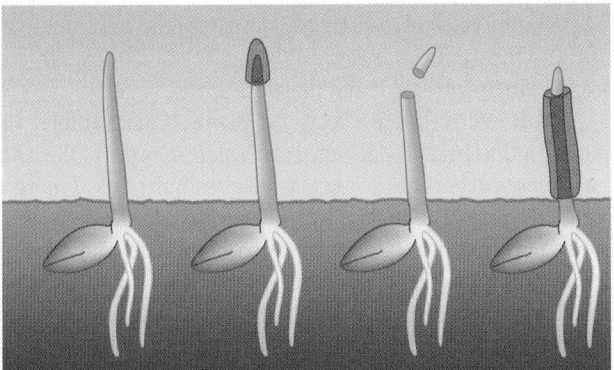 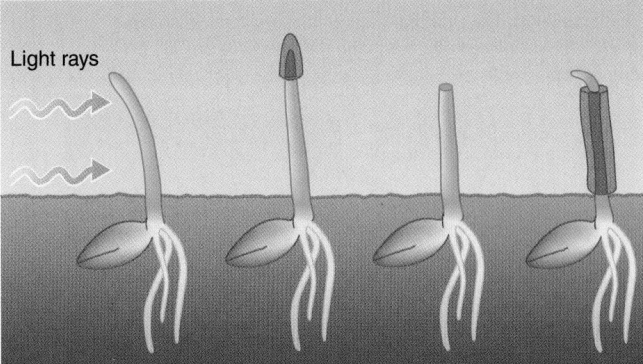

Light rays

RESULTS AND CONCLUSION: After exposure to light coming from one direction, the uncovered plants and the plants with uncovered tips grew toward the light (*right*). The plants with tips covered or removed did not bend toward light. The Darwins concluded that some substance was produced in the tip and transmitted to the lower part, causing it to bend.

SOURCES: Darwin, C.R. *The Power of Movement in Plants.* John Murray, London, 1880; Boysen-Jensen, P. "Concerning the Performance of Phototropic Stimuli on the *Avena* Coleoptile." *Berichte der Deutschen Botanischen Gesellschaft* (Reports of the German Botanical Society), Vol. 31, pp. 559–566, 1913.

Figure 38-5 The Darwins' phototropism experiments

PREDICT Devise an experiment to test the hypothesis that a minimum period of light exposure is required to produce bending.

© Cengage Learning

Interestingly, all animals, plants, and fungi examined to date use ubiquitin to target proteins for destruction, but bacteria do not, which suggests that the use of ubiquitin tags evolved early in eukaryote evolution and was conserved as the various groups of eukaryotes diversified. Plants have more than 700 different F-box proteins. Although not much is known about most of them, it is significant that researchers have identified other plant hormones and signaling molecules that use F-box proteins as receptors, including gibberellin, ethylene, and abscisic acid.

Auxins promote cell elongation

Charles Darwin, the British naturalist best known for developing the scientific theory of natural selection to explain evolution, also provided the first evidence for the existence of auxins. The experiments that Darwin and his son Francis performed in the 1870s involved positive phototropism, the directional growth of plants toward light. The plants they used were newly germinated canary grass seedlings. As in all grasses, the first part of a canary grass seedling to emerge from the soil is the coleoptile, a protective sheath that encircles the stem. When coleoptiles are exposed to light from only one direction, they bend toward the light. The bending occurs below the tip of the coleoptile.

The Darwins tried to influence this bending in several ways (**FIG. 38-5**). For example, they covered the tip of the coleoptile as soon as it emerged from the soil. When they covered that part of the coleoptile above where the bend would be expected to occur, the plants did not bend. On other plants, they removed the coleoptile tip and found that bending did not occur. When the bottom of the coleoptile where the curvature would occur was shielded from the light, the coleoptile bent toward light. From these experiments, the Darwins concluded that "some influence is transmitted from the upper to the lower part, causing it to bend."

In the 1920s, Frits Went, a young Dutch scientist, isolated the phototropic hormone from oat coleoptiles. He removed the coleoptile tips and placed them on tiny blocks of agar for a period of time. When he put one of these agar blocks squarely on a decapitated coleoptile, normal growth resumed. When he placed one of these agar blocks to one side of the tip of a decapitated coleoptile in the dark, bending occurred (**FIG. 38-6**). The results indicated that the substance had diffused from the coleoptile tip into the agar and, later, from the agar into the decapitated coleoptile. Went named this substance **auxin** (from the Greek *aux,* "enlarge" or "increase").

Auxin is a group of several natural (and artificial) plant hormones; the most common and physiologically important auxin

Is there a chemical substance responsible for elongation of coleoptiles, and can it be isolated?

HYPOTHESIS: The factor responsible for coleoptile growth is a diffusible chemical that can be isolated from coleoptile tips.

EXPERIMENT: Coleoptile tips were placed on agar blocks for a period of time **(a).** The agar block was transferred to a decapitated coleoptile. It was placed off-center, and the coleoptile was left in darkness **(b).**

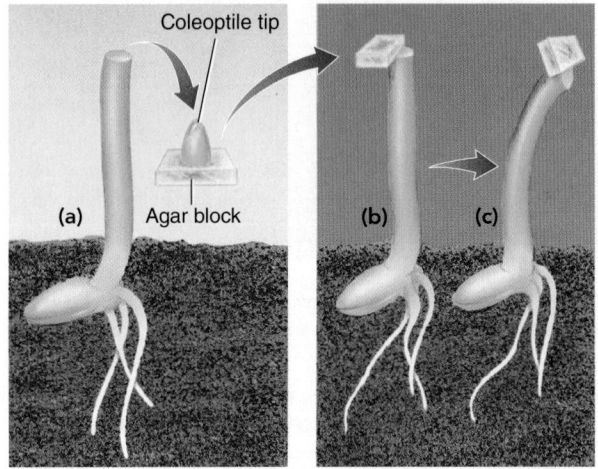

RESULTS AND CONCLUSION: The coleoptile bent **(c),** indicating that a chemical moved from the original coleoptile tip to the agar block and from there to the same side of the decapitated coleoptile, causing that side to elongate.

SOURCE: Went, F. "A Growth Substance and Growth." *Recueils Travaux Botaniques Néerlandais* (Collections of Dutch Botanical Works), Vol. 25, pp. 1–116, 1928.

Figure 38-6 *Animation* **Isolating auxin from coleoptiles**

PREDICT Devise an experiment to test the hypothesis that the auxin that has diffused into the agar blocks is sensitive to heat.

© Cengage Learning

is **indoleacetic acid (IAA).** The purification and elucidation of the primary auxin's chemical structure were accomplished in the mid-1930s by a research team led by Kenneth Thimann at the California Institute of Technology.

The movement of auxin in the plant is said to be *polar,* or unidirectional. Auxin moves downward along the shoot–root axis from its site of production, usually the shoot apical meristem. (Young leaves and seeds are also sites of auxin production.) This directional cell-to-cell transport results in an *auxin gradient* (a spatial difference in auxin concentration) that triggers specific responses. Special auxin transport proteins in the plasma membranes of plant cells control the movement of auxin out of cells; this process is responsible for the directional movement of auxin. Mutations in auxin transport proteins inhibit the transmission of auxin, which normally moves about 1 cm per hour.

Auxin's most characteristic action is promotion of cell elongation in stems and coleoptiles. This effect, apparently exerted by acidification of cell walls, increases their plasticity and thus enables them to expand under the force of the cell's internal turgor pressure. Auxin's effect on cell elongation also provides an explanation for phototropism. When a plant is exposed to a light from only one direction, some of the auxin migrates laterally to the shaded side of the stem before moving down the stem by polar transport. Because of the greater auxin concentration on the shaded side of the stem, the cells there elongate more than the cells on the light side, and the stem bends toward the light (**FIG. 38-7**). Auxin is also involved in gravitropism and thigmotropism.

Auxin exerts other effects on plants. For example, some plants tend to branch out very little when they grow. Growth in these plants occurs almost exclusively from the apical meristem rather than from axillary buds, which do not develop as long as the terminal bud is present. Such plants are said to exhibit **apical dominance,** the inhibition of axillary bud growth by the apical meristem. In plants with strong apical dominance, auxin produced in the apical meristem inhibits axillary buds near the apical meristem from developing into actively growing shoots. When the apical meristem is pinched off, the auxin source is removed, and axillary buds grow to form branches. Apical dominance is often quickly re-established, however, as one branch begins to inhibit the growth of others. Other hormones (ethylene and cytokinin, both discussed later) are also involved in apical dominance. As with many physiological activities, the

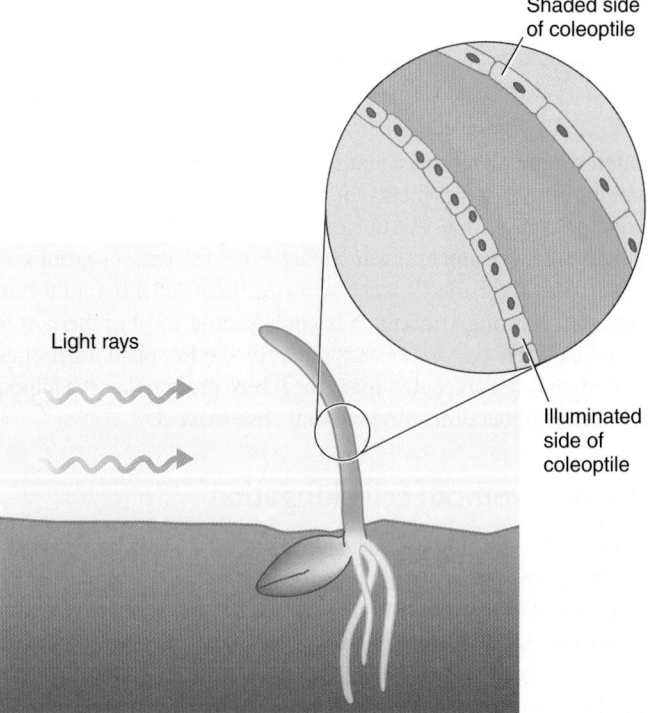

Figure 38-7 Phototropism and the unequal distribution of auxin
Auxin travels down the side of the stem or coleoptile *away* from the light, causing cells on the shaded side to elongate. Therefore, the stem or coleoptile bends toward light.

© Cengage Learning

Figure 38-8 Auxin and root development on stem cuttings

Many adventitious roots developed on a honeysuckle (*Lonicera fragrantissima*) cutting placed in a solution with a high concentration of synthetic auxin (*left*). Fewer roots developed in a lower auxin concentration (*middle*). The cutting placed in water (no auxin) served as a control and did not form roots in the same period (*right*).

changing ratios of these hormones may be the factor responsible for apical dominance.

Auxin produced by developing seeds stimulates the development of the fruit. When auxin is applied to certain flowers in which fertilization has not occurred (and therefore in which seeds are not developing), the ovary enlarges and develops into a seedless fruit. Seedless tomatoes are produced in this manner.[1] Auxin is not the only hormone involved in fruit development, however.

Some manufactured, or synthetic, auxins have structures similar to that of IAA. The synthetic auxin naphthalene acetic acid stimulates root development on stem cuttings and is used for asexual propagation, particularly of woody plants with horticultural importance (FIG. 38-8). The synthetic auxins 2,4-D and 2,4,5-T have been used as selective herbicides (weed killers). These compounds kill plants with broad leaves but, for reasons not completely understood, do not kill grasses. Both herbicides are similar in structure to IAA and disrupt the plants' normal growth processes. Because many of the world's most important crops are grasses (such as wheat, corn, and rice), 2,4-D and 2,4,5-T can kill broadleaf weeds that compete with these crops. The use of 2,4,5-T is no longer allowed in the United States, however, because of its association with dioxins, a group of mildly to very toxic compounds formed as byproducts when 2,4,5-T is manufactured.

Gibberellins promote stem elongation

In the 1920s and 1930s, a Japanese biologist was studying a disease of rice in which the young rice seedlings grew extremely tall and spindly, fell over, and died. The cause of the disease

was a fungus (*Gibberella fujikuroi*) that produces a chemical substance named **gibberellin (GA).** Not until after World War II did scientists in Europe and North America learn of the work done by this Japanese scientist and others. During the 1950s and 1960s, studies in the United States and Great Britain showed that gibberellins are produced by healthy plants as well as by the fungus.

Gibberellins are hormones involved in many normal plant functions. The symptoms of the seedling disease are caused by an abnormally high GA concentration in the plant tissue (because both plant and fungus produce gibberellin). Currently, dozens of naturally occurring GAs are known, although only about four function as plant hormones; the rest are inactive precursors. There are no synthetic gibberellins.

Gibberellins promote stem elongation in many plants. When GA is applied to a plant, particularly certain dwarf varieties, this elongation may be spectacular. Some corn and pea plants that are dwarfs as a result of one or more mutations grow to a normal height when treated with GA. Short-stemmed, high-yielding varieties of wheat have short stems because they have a reduced response to GA. These varieties put less of their resources into stem height and more resources into grain production. Gibberellins are also involved in **bolting,** the rapid elongation of a floral stalk that occurs naturally in many plants when they initiate flowering (FIG. 38-9).

Gibberellins cause stem elongation by stimulating cells to divide as well as to elongate. The actual mechanism of cell elongation differs from that caused by auxin, however. Recall that IAA-induced cell elongation involves the acidification of the cell wall. In GA-induced cell elongation, cell wall acidification does

Figure 38-9 Gibberellin and stem elongation

Like many biennials, Indian blanket (*Gaillardia pulchella*) grows as a rosette, which is a circular cluster of leaves close to the ground, during its first year (*left*). It then bolts when it initiates flowering in the second year (*right*). The plant in bloom grows to 0.6 m (2 ft).

[1] Not all seedless fruits are produced by treatment with auxin. In Thompson seedless grapes, fertilization occurs but the embryos abort; therefore, the seeds fail to develop. Thompson seedless grapes are sprayed with the hormone gibberellin to increase berry size.

not occur; biologists do not yet know how gibberellins modify cell wall properties to cause cell elongation.

Gibberellins affect several reproductive processes in plants. They stimulate flowering, particularly in long-day plants (discussed later in this chapter). In addition, GAs substitute for the low temperature that biennials require before they begin flowering. If GAs are applied to biennials during their first year of growth, flowering occurs without exposure to a period of low temperature. Gibberellins, like auxin, affect fruit development. Agriculturalists apply GAs to several varieties of grapes to produce larger fruits.

Gibberellins are also involved in seed germination in certain plants. In a classic experiment involving barley seed germination, researchers showed that the release of GA from the embryo triggers the synthesis of α-amylase, an enzyme that digests starch in the endosperm. As a result, glucose becomes available for absorption by the embryo. Although enzymes mobilize starch reserves in many types of seeds, GA control of seed enzymes appears restricted to cereals and other grasses. In addition to mobilizing food reserves in newly germinated grass seeds, application of gibberellins substitutes for low-temperature or light requirements for germination in seeds of plants such as lettuce, oats, and tobacco.

Scientists are actively conducting research in gibberellin signaling and have identified several GA receptors in the nucleus, repressor proteins designated *DELLA proteins,* as well as an F-box protein. When GA is not bound to the receptor, a DELLA protein inhibits transcription of *GA response genes.* When GA is present, it binds to the receptor, which triggers an enzyme that attaches ubiquitin to the DELLA protein. This protein is targeted for degradation into peptide fragments in a proteasome. With the DELLA protein no longer present, transcription of the GA response genes occurs. This signaling process may sound familiar: the GA signal transduction process shares similarities with the auxin signal transduction pathway described earlier in this chapter.

Cytokinins promote cell division

During the 1940s and 1950s, researchers were trying to find substances that might induce plant cells to divide in **tissue culture,** a technique in which cells are isolated from plants and grown in a nutrient medium. (Plant tissue culture is discussed in Chapter 33.) They discovered that cells would not divide without a substance found in coconut milk. Because coconut milk has a complex chemical composition, investigators did not chemically identify the division-inducing substance for some time. Finally, researchers isolated an active substance from a different source, aged DNA from herring sperm. They called it **cytokinin** because it induces cytokinesis, or cytoplasmic division. In 1963, researchers identified the first plant cytokinin, zeatin, in corn. Since then, similar molecules have been identified in other plants. Biologists have also synthesized several cytokinins.

Cytokinins are structurally similar to adenine, a purine base that is part of DNA and RNA molecules. Biologists are beginning to elucidate the cytokinin signal transduction pathway, which is more complex than those of auxin and gibberellin. The cytokinin pathway has multiple steps, each under the control of a number of genes. The presence of feedback loops and links to other signaling pathways adds another level of complexity to the regulation of cytokinin activity.

Cytokinins promote cell division and differentiation of young, relatively unspecialized cells into mature, more specialized cells in intact plants. They are a required ingredient in any plant tissue culture medium and must be present for cells to divide. In tissue culture cytokinins interact with auxin during the formation of plant organs such as roots and stems (**FIG. 38-10**). For example, in tobacco tissue culture a high ratio of auxin to cytokinin induces root formation, whereas a low ratio of auxin to cytokinin induces shoot formation.

Cytokinins and auxin also interact in the control of apical dominance. Here their relationship is antagonistic: auxin inhibits the growth of axillary buds, and cytokinin promotes their growth.

One effect of cytokinins on plant cells is to delay the aging process. Plant cells, like all living cells, go through a natural aging process known as **senescence.** Senescence is accelerated in cells of plant parts that are cut, such as flower stems. Botanists think that plants must have a continual supply of cytokinins from the roots. Cut stems, of course, lose their source of cytokinins and therefore age rapidly. When cytokinins are sprayed on leaves of a cut stem of many species, they remain green, whereas unsprayed leaves turn yellow and die.

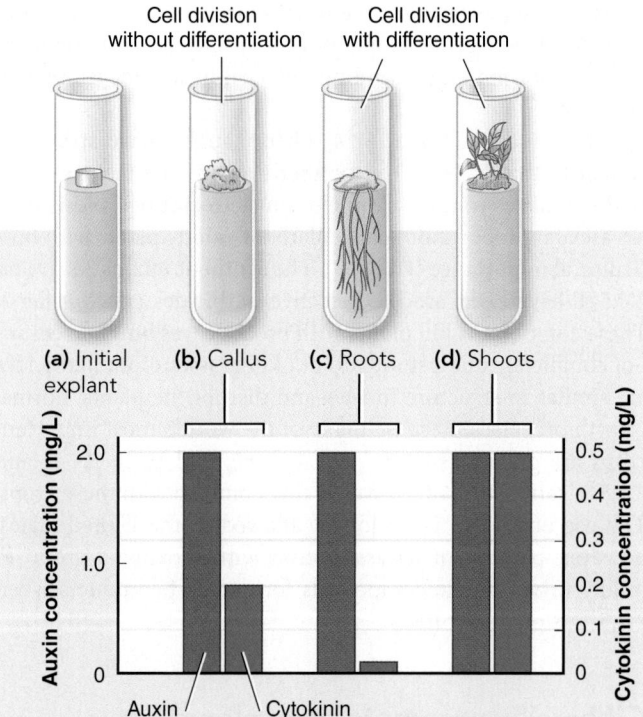

Figure 38-10 Auxin–cytokinin interactions in tissue culture

Varying amounts of auxin and cytokinin in the culture media produce different growth responses. **(a)** The initial explant is a small piece of sterile tissue from the pith of a tobacco stem, which is placed on nutrient agar. **(b)** Nutrient agar containing 2.0 mg/L of auxin and 0.2 mg/L of cytokinin causes cells to divide and form a clump of undifferentiated tissue called a callus. **(c)** Agar with 2.0 mg/L of auxin and 0.02 mg/L of cytokinin (high ratio of auxin to cytokinin) stimulates root growth. **(d)** Agar with 2.0 mg/L of auxin and 0.5 mg/L of cytokinin (low ratio of auxin to cytokinin) stimulates shoot growth.
© Cengage Learning

Despite their involvement in regulating many aspects of plant growth and development, cytokinins currently have few commercial applications other than plant tissue culture. However, molecular biologists at the University of Wisconsin have combined a promoter from a gene activated during normal senescence with a gene that encodes an enzyme involved in cytokinin synthesis. (Recall from Chapter 13 that the **promoter** is the nucleotide sequence in DNA to which RNA polymerase attaches to begin transcription.) The leaves of transgenic tobacco plants that contained this recombinant DNA produced more cytokinin and therefore lived longer and continued to photosynthesize (FIG. 38-11). This type of application of molecular technology may have the potential to increase the longevity and productivity of certain crops.

Ethylene promotes abscission and fruit ripening

During the early 20th century, scientists observed that the gas **ethylene** (C_2H_4) has several effects on plant growth, but not until 1934 did they demonstrate that plants produce ethylene. This natural hormone influences many plant processes. Ethylene inhibits cell elongation, promotes seed germination, promotes apical dominance, and is involved in plant responses to wounding or invasion by disease-causing microorganisms.

Ethylene also has a major role in many aspects of senescence, including fruit ripening. As a fruit ripens, it produces ethylene, which triggers an acceleration of the ripening process and induces the fruit to produce more ethylene, which further accelerates ripening. The expression "one rotten apple spoils the barrel" is true. A rotten apple is one that is overripe and produces large amounts of ethylene, which diffuses and triggers the ripening process in nearby apples. Humans use ethylene commercially to uniformly ripen bananas and tomatoes. These fruits are picked while green and shipped to their destination, where they are exposed to ethylene before they are delivered to grocery stores (FIG. 38-12).

Plants growing in a natural environment encounter rain, hail, wind, and contact with passing animals. All these mechanical stressors alter their growth and development, making them shorter and stockier than plants grown in a greenhouse. Ethylene regulates such developmental responses to mechanical stimuli, known as **thigmomorphogenesis.** Plants that are mechanically disturbed produce additional ethylene, which in turn inhibits stem elongation and enhances cell wall thickening in supporting cells (collenchyma and sclerenchyma). These changes are adaptive because shorter, thicker stems are less likely than longer, thinner stems to be damaged by mechanical stressors.

Ethylene is involved in leaf abscission, which is actually influenced by two antagonistic hormones, ethylene and auxin. As a leaf ages (as autumn approaches, for deciduous trees in temperate climates), the level of auxin in the leaf decreases. Concurrently, cells in the abscission layer at the base of the petiole (where the leaf will break away from the stem) begin producing ethylene.

Researchers studying ethylene signaling have identified a similar mechanism to that already described for auxin and gibberellin. When ethylene is present, it binds to a receptor. This

Figure 38-11 Cytokinin synthesis and delay of senescence

The tobacco (*Nicotiana tabacum*) plant on the *left* was genetically engineered to produce additional cytokinin as it aged, whereas the tobacco plant of the same age on the *right* served as a control. Note the extensive senescence and death of older leaves on the control plant. Depending on the variety, tobacco grows 0.9 to 3 m (3 to 10 ft) tall.

Courtesy of Dr. Richard M. Amasino, University of Wisconsin

Illustration Services

Figure 38-12 Ethylene and fruit ripening

Both boxes of tomatoes were picked at the same time, while green. The tomatoes in the box on the *right* were exposed to an atmosphere containing 100 ppm of ethylene for three days.

binding triggers an enzyme that attaches ubiquitin to a protein targeted for degradation into peptide fragments in a proteasome. At least two F-box proteins are involved in the protein degradation that occurs in ethylene signal transduction.

Abscisic acid promotes seed dormancy

In 1963, two different research teams discovered **abscisic acid (ABA).** Despite its name, abscisic acid does not induce abscission in most plants. Instead, ABA is involved in a plant's response to stress and in seed **dormancy,** a temporary state of arrested physiological activity. As an environmental stress hormone, ABA particularly promotes changes in plant tissues that are water-stressed. (Recall that ethylene also affects plant responses to certain stressors, such as mechanical stressors and wounding.)

Botanists best understand the effect of abscisic acid on plants suffering from water stress. The level of ABA increases dramatically in the leaves of plants exposed to severe drought conditions. The high level of ABA in the leaves activates a signal transduction process that leads to the closing of stomata. The closing of stomata saves the water that the plant would normally lose by transpiration, thereby increasing the plant's likelihood of survival. As knowledge of ABA signaling in guard cells increases, botanists hope to use this information to engineer crops and horticultural plants that are resistant to drought.

The onset of winter is also a type of stress on plants. A winter adaptation that involves abscisic acid is dormancy in seeds. Many seeds have high levels of ABA in their tissues and do not germinate until the ABA washes out. In a corn mutant unable to synthesize ABA, the seeds germinate as soon as the embryos are mature, even while attached to the ear (FIG. 38-13).

Abscisic acid is not the only hormone involved in seed dormancy. For example, addition of gibberellin reverses the effects of dormancy. In seeds the level of ABA decreases during the winter, and the level of GA increases. Cytokinins are also implicated in breaking dormancy. Once again, we see that a single physiological activity such as seed dormancy may be controlled in plants by the interaction of several hormones. The plant's actual response may result from changing ratios of hormones rather than the effect of each individual hormone.

Brassinosteroids are plant steroid hormones

Although steroid hormones have long been known to have crucial roles in animals, the roles of steroid hormones in plants were not well characterized until the mid-1990s. The **brassinosteroids (BRs)** are a group of steroids that function as plant hormones. BRs are involved in several aspects of growth and development. *Arabidopsis* mutants that cannot synthesize BRs are dwarf plants with small curly leaves and reduced fertility; researchers can reverse this defect by applying BR to the plant. Studies of these mutants suggest that brassinosteroids are involved in multiple developmental processes, such as cell division, cell elongation, light-induced differentiation, vascular development, and seed germination (FIG. 38-14).

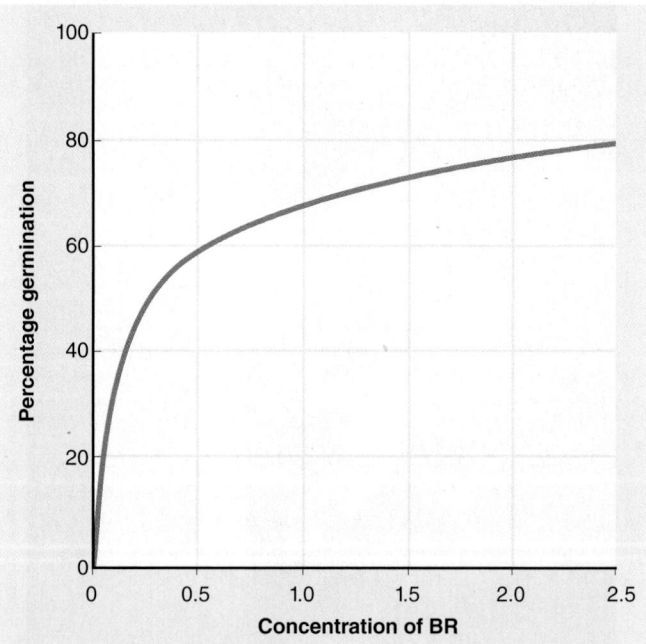

Figure 38-14 Brassinosteroids and seed germination in *Arabidopsis*

Researchers used a GA-insensitive *Arabidopsis* mutant that produced seeds that did not germinate. (Recall that GA promotes seed germination, so a plant that does not respond to endogenous GA is required.) The scientists overcame the inability of the mutant seeds to germinate by treating them with a brassinosteroid (BR). Note how the percentage of germination increases as the concentration of BR to which the seeds were exposed increases.

Original source: Steber, C.M., and P. McCourt. "A Role for Brassinosteroids in Germination in *Arabidopsis.*" *Plant Physiology,* Vol. 125, pp. 287–297, 2001.

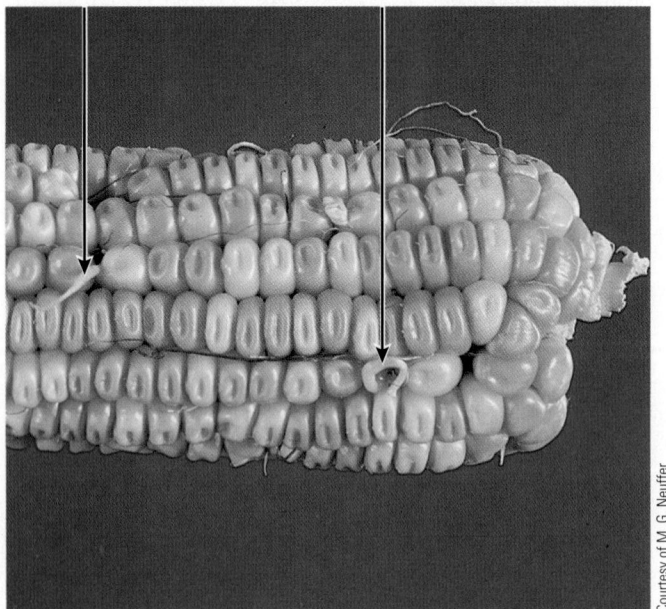

Courtesy of M. G. Neuffer

Figure 38-13 The absence of abscisic acid causes precocious germination

In a corn (*Zea mays*) mutant that does not produce abscisic acid, some of the kernels have germinated while still on the ear, producing roots (*arrows*).

Many steps of the BR signaling pathway have been characterized. BRs bind to a receptor kinase in the plant plasma membrane (unlike steroid hormones in animals, which enter the cell and bind to receptors in the cytoplasm or nucleus). When BR binds to its receptor, the receptor phosphorylates molecules in the cytosol, which in turn triggers a signaling cascade that ultimately alters gene expression in the cell. Part of the signal transduction process involves attaching ubiquitin to nuclear proteins targeted for degradation in a proteasome. As a result, the expression of BR target genes is adjusted.

Identification of a universal flower-promoting signal remains elusive

Experiments in which different tobacco species are grafted together indicate that unidentified flower-promoting and flower-inhibiting substances may exist. *Nicotiana silvestris* is a *long-day* tobacco plant (it requires a short night to flower); a variety of *N. tabacum* is a *day-neutral* tobacco plant (seasonal changes do not affect when it flowers). When a long-day tobacco is grafted to a day-neutral tobacco and exposed to short nights, both plants flower (FIG. 38-15). The day-neutral tobacco plant flowers sooner than it normally would.

Biologists hypothesize that a flower-promoting substance, *florigen,* may be induced in the long-day plant and transported to the day-neutral plant through the graft union, causing the day-neutral tobacco plant to flower sooner than expected. An intact plant may produce florigen in the leaves and transport it in the phloem to the shoot apical meristem. There, it induces a transition from vegetative to reproductive development, that is, to a meristem that produces flowers.

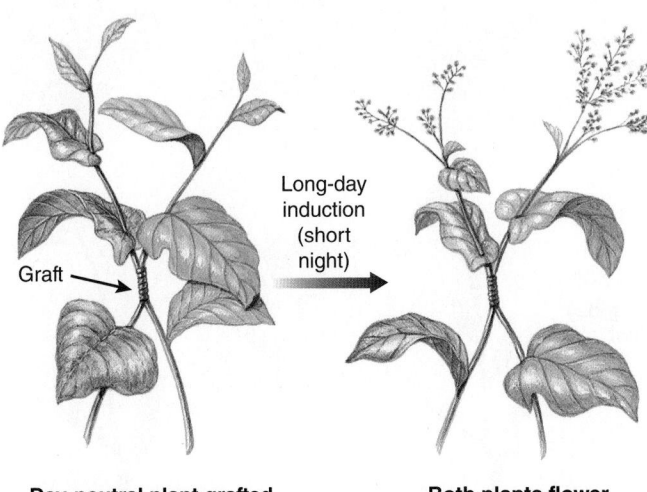

Day-neutral plant grafted to long-day plant

Both plants flower

Figure 38-15 Evidence for the existence of a flower-promoting substance

When a long-day tobacco plant (*Nicotiana silvestris*) is grafted to a day-neutral tobacco plant (*N. tabacum*) and both plants are exposed to a long-day/short-night regimen, they both flower. The day-neutral plant flowers sooner than it normally would, presumably because a flower-promoting substance passes from the long-day plant to the day-neutral one through the graft.

© Cengage Learning

When a botanist grafts a long-day tobacco to a day-neutral tobacco and exposes them to long nights, neither plant flowers. As long as these conditions continue, the day-neutral plants do not flower even when they would normally do so. In this case, the long-day tobacco may produce a flower-inhibiting substance that is transported to the day-neutral tobacco through the graft union. This substance prevents the day-neutral tobacco from flowering.

CHECKPOINT 38.2

- *How are signal reception, signal transduction, and altered cell activity part of the general mechanism of action for auxin?*
- *How is auxin involved in phototropism?*
- *Which hormones are involved in each of the following physiological processes: (1) seed germination, (2) stem elongation, (3) fruit ripening, (4) leaf abscission, and (5) seed dormancy?*

38.3 LIGHT SIGNALS AND PLANT DEVELOPMENT

LEARNING OBJECTIVES

5 Explain how varying amounts of light and darkness induce flowering.
6 Describe the role of phytochrome in flowering, including a brief discussion of phytochrome signal transduction.
7 Define *circadian rhythm* and give an example.
8 Distinguish between phytochrome and cryptochrome.

Photoperiodism is any response of a plant to the relative lengths of daylight and darkness. Initiation of flowering at the shoot apical meristem is one of several physiological activities that are photoperiodic in many plants. Plants are classified into four main groups—short day, long day, intermediate day, and day neutral—on the basis of how photoperiodism affects their transition from vegetative growth to flowering.

Short-day plants (also called **long-night plants**) flower when the night length is equal to or greater than some critical period (FIG. 38-16a). Thus, short-day plants detect the lengthening nights of late summer or fall. There are two kinds of short-day plants: qualitative and quantitative. In *qualitative short-day plants,* flowering occurs only in short days; in *quantitative short-day plants,* flowering is accelerated by short days. The initiation of flowering in short-day plants is not due to the shorter period of daylight but to the long, uninterrupted period of darkness. The minimum critical night length varies considerably from one plant species to another but falls between 12 and 14 hours for many. Examples of short-day plants are florist's chrysanthemum, cocklebur, and poinsettia, which typically flower in late summer or fall. Poinsettias, for example, typically initiate flower buds in early October in the Northern Hemisphere and flower about 8 to 10 weeks later; hence, they are traditionally associated with Christmas.

Long-day plants (also called **short-night plants**) flower when the night length is equal to or less than some critical period (FIG. 38-16b). In *qualitative long-day plants,* flowering occurs only in long days; in *quantitative long-day plants,* long days accelerate flowering. Long-day plants, such as spinach, black-eyed Susan (see chapter-opening photograph), and the model research plant *Arabidopsis,* flower when they detect the shortening nights of spring and early summer.

Intermediate-day plants do not flower when night length is either too long or too short (FIG. 38-16c). Sugarcane and coleus are intermediate-day plants. These plants flower when exposed to days and nights of intermediate length.

Some plants, called **day-neutral plants,** do not initiate flowering in response to seasonal changes in the period of daylight and darkness but instead respond to some other type of stimulus, external or internal. Cucumber, sunflower, corn, and onion are examples of day-neutral plants. Many of these plants originated in the tropics, where day length does not vary appreciably during the year. In contrast, short-day, long-day, and intermediate-day plants are temperate or subtropical (mid-latitude species).

Plant biologists have experimented with the effects of various light regimens on flowering. FIGURE 38-17 shows how flowering in long-day and short-day plants is affected by different light treatments, including a short-day/long-night regimen with a *night break,* a short burst of light in the middle of the night.

Phytochrome detects day length

The main photoreceptor for photoperiodism and many other light-initiated plant responses (such as germination and seedling establishment) is **phytochrome,** a family of about five blue-green pigment proteins, each of which is coded for by a different gene. A mixture of phytochrome proteins is present in cells of all vascular plants examined so far. For example, five members of the phytochrome family—designated phyA, phyB, phyC, phyD, and phyE—occur in *Arabidopsis.* Phytochrome is also a photoreceptor in certain bacteria, fungi, and slime molds.

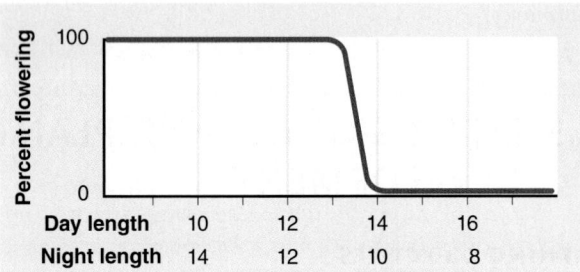

(a) Short-day plants flower when the night length is equal to or exceeds a certain critical length in a 24-hour period.

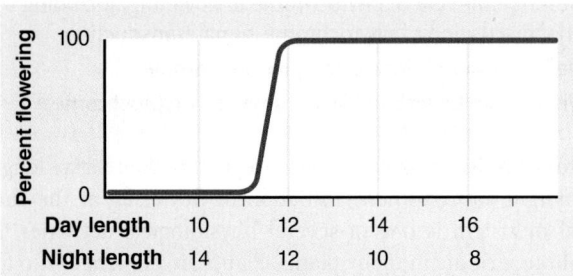

(b) Long-day plants flower when the night length is equal to or less than a certain critical length in a 24-hour period. The critical length varies among different species; short-day plants may require a shorter night length than long-day plants, as shown.

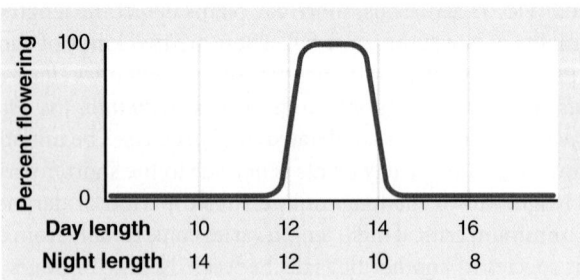

(c) Intermediate-day plants have a narrow night-length requirement.

Figure 38-16 Generalized photoperiodic responses in short-day, long-day, and intermediate-day plants

© Cengage Learning

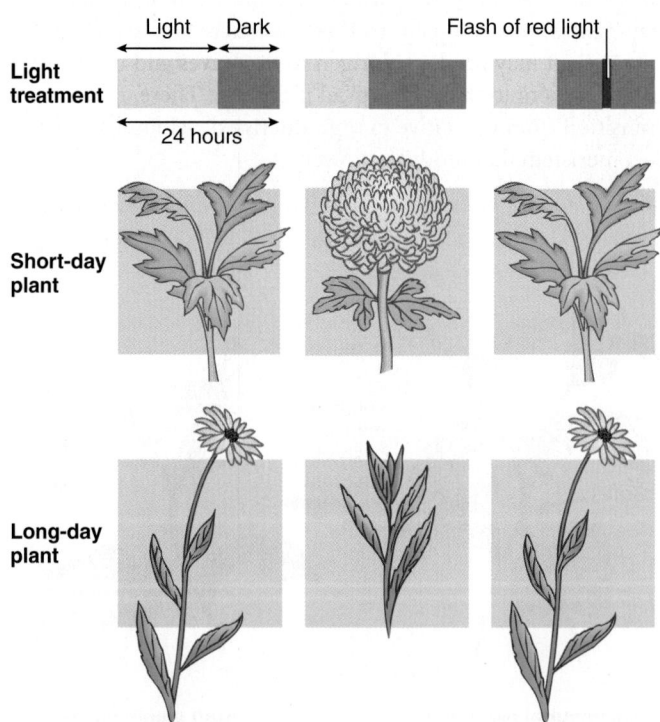

Figure 38-17 *Animation* Photoperiodic responses of short-day and long-day plants

A short-day plant flowers when it is grown under long-night conditions (*top middle*), but it does not flower when exposed to a long night interrupted with a brief flash of red light (*top right*). A long-day plant does not flower when grown under long-night conditions (*bottom middle*) unless the long night is interrupted with a brief flash of red light (*bottom right*).

© Cengage Learning

Much of the current knowledge of phytochrome in *Arabidopsis* is based on various mutant plants that do not express a specific phytochrome gene, such as the gene that codes for phyA. By studying the physiological response of plants that do not produce an individual phytochrome, biologists have concluded that the individual forms of phytochrome have both unique and overlapping functions. PhyB appears to exert its influence at all stages of the plant life cycle, whereas the other forms of phytochrome have narrower functions at specific stages in the life cycle.

PhyA and phyB may have antagonistic (opposite) effects on flowering. In long-day plants phyB may inhibit flowering, and phyA may induce flowering. Flowering occurs more rapidly in long-day plants with mutations in the gene that codes for phyB; such mutations reduce or eliminate the production of phyB. In contrast, flowering is delayed or prevented in long-day plants with mutations in the gene that codes for phyA.

Each member of the phytochrome family exists in two forms and readily converts from one form to the other after absorption of light of specific wavelengths. One form, designated **Pr** (for *r*ed-absorbing *p*hytochrome), strongly absorbs light with a relatively short red wavelength (660 nm). In the process, the shape of the molecule changes to the second form of phytochrome, **Pfr,** so designated because it absorbs *far-r*ed light, which is light with a relatively long red wavelength (730 nm; FIG. 38-18). When it absorbs far-red light, Pfr reverts to the original form, Pr. Pfr is the active form of phytochrome, triggering or inhibiting physiological responses such as flowering.

What does a pigment that absorbs red light and far-red light have to do with daylight and darkness? Sunlight consists of various amounts of the entire spectrum of visible light in addition to ultraviolet and infrared radiation. Sunlight contains more red than far-red light, however, so when a plant is exposed to sunlight, the level of Pfr increases. During the night,

the level of Pfr slowly decreases as Pfr, which is less stable than Pr, is degraded.

The importance of phytochrome to plants cannot be overemphasized. Timing of day length and darkness is the most reliable way for plants to measure the change from one season to the next. This measurement, which synchronizes the stages of plant development, is crucial for survival, particularly in environments where the climate has an annual pattern of favorable and unfavorable seasons.

Competition for sunlight among shade-avoiding plants involves phytochrome

Plants sense the proximity of nearby plants, which are potential competitors, and react by changing the way they grow and develop. Many plants, from small herbs to large trees, compete for light, a response known as **shade avoidance,** in which plants tend to grow taller when closely surrounded by other plants. If successful, the shade-avoiding plant projects its new growth into direct sunlight and thereby increases its chances of survival.

Since the 1970s, botanists have recognized the environmental factor that triggers shade avoidance: plants perceive changes in the ratio of red to far-red light that result from the presence of nearby plants. The leaves of neighboring plants absorb much more red light than far-red light. (The green pigment chlorophyll strongly absorbs red light during photosynthesis.) In a densely plant-populated area, the ratio of red light to far-red light (r/fr) decreases, affecting the equilibrium between the Pr and Pfr forms, particularly of phyB. This signal triggers a series of responses that cause the shade-avoiding plant, which is adapted to full-light environments, to grow taller or flower earlier than

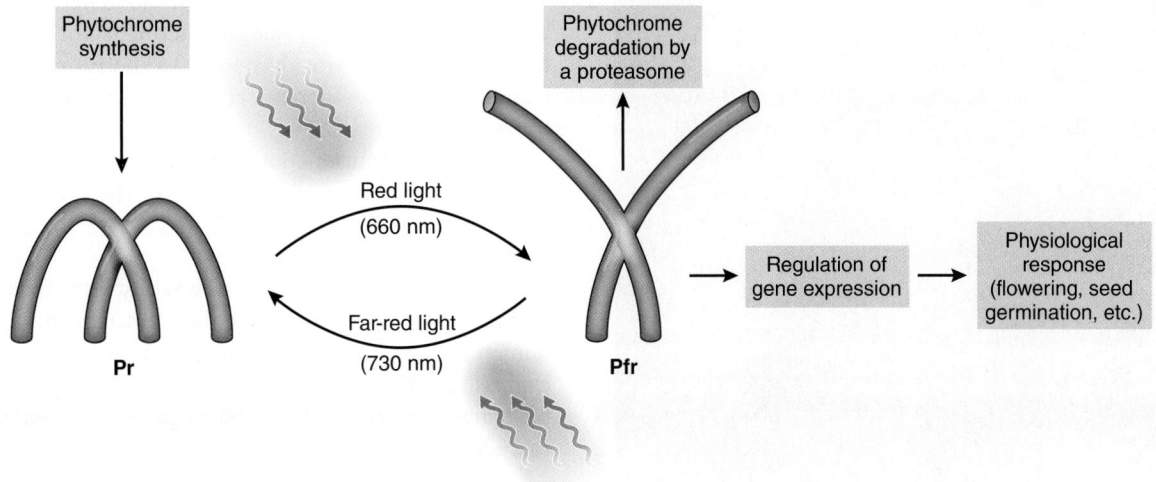

Figure 38-18 *Animation* **Phytochrome**

Each member of the phytochrome family occurs in two forms, designated Pr and Pfr, and readily converts from one form to the other. Red light (660 nm) converts Pr to Pfr, and far-red light (730 nm) converts Pfr to Pr.

© Cengage Learning

it would in a less densely populated area. **FIGURE 38-19** shows an interesting application of additional (reflected) far-red light without the reduction in incoming sunlight that occurs in shade avoidance.

When a plant is using many of its resources for stem elongation, it has fewer resources for new leaves and branches, storage tissues, and reproductive tissues. However, for a shade-avoiding plant that is shaded by its neighbors, a rapid increase in stem length is advantageous because once this plant is taller than its neighbors, it obtains a larger share of unfiltered sunlight.

Phytochrome is involved in other responses to light, including germination

Phytochrome is involved in the light requirement that some seeds have for germination. Seeds with a light requirement must be exposed to light containing red wavelengths. Exposure to red light converts Pr to Pfr, and germination occurs. Many temperate forest species with small seeds require light for germination. (Larger seeds generally do not have a light requirement.) This adaptation enables the seeds to germinate at the optimal time. During early spring, sunlight, including red light, penetrates the bare branches of overlying deciduous trees and reaches the soil between the trees. As spring temperatures warm, the seeds on the soil absorb red light and germinate. During their early growth, the newly germinated seedlings do not have to compete with the taller trees for sunlight.

Cary Wolinsky

Figure 38-19 Colored mulches and plant growth processes
The red mulch reflects far-red light from the ground to the tomato plants. The plants, absorbing the extra far-red light, react as if nearby plants are present. Their aboveground growth is greater, and their fruits ripen earlier. Manipulating mulch colors also modifies yield, flavor, and nutrient content.

Other physiological functions under the influence of phytochrome include sleep movements in leaves (discussed later); shoot dormancy; leaf abscission; and pigment formation in flowers, fruits, and leaves.

Phytochrome acts by signal transduction

Some phytochrome-induced responses are very rapid and short term, whereas others are slower and long term. The rapid responses probably involve reversible changes in the properties of membranes, altering cell ionic balances. Red light, for example, causes potassium ion (K^+) channels to open in cell membranes involved in plant movements caused by changes in turgor. Exposure to far-red light causes the K^+ channels to close. Slower phytochrome-induced responses involve the selective regulation of transcription of numerous genes. For example, phytochrome activates transcription of the gene for the small subunit of *rubisco*, an enzyme involved in photosynthesis (see Chapter 9).

Each phytochrome molecule consists of a protein attached to a light-absorbing photoreceptor. Biologists hypothesize that the absorption of light by the photoreceptor portion of phytochrome elicits a change in the shape of the protein portion of the phytochrome molecule. This change, in turn, triggers one or more **signal transduction** pathways.

One important research tool in studying phytochrome signal transduction is mutant plants that respond as if they were exposed to a particular light stimulus even if they were not. Research indicates that the active form of phytochrome moves from the cytoplasm into the nucleus, where it affects gene expression by activating a **transcription factor.** When activated, the transcription factor, which binds to the promoter of a gene, either turns on or represses transcription that leads to protein synthesis.

In one such pathway, inactive phytochrome (Pr) in the cytoplasm absorbs red light and is converted into the active form, Pfr, which moves into the nucleus. There, phytochrome binds to PIF3 (phytochrome-interacting factor 3), a transcription factor that is already bound to the promoters of light-responsive genes, thereby activating (or repressing) their transcription.

The signal transduction pathway is shut down by far-red light, which is absorbed by the Pfr in the Pfr–PIF3 complex in the nucleus. When Pfr is converted to Pr, it dissociates from PIF3.

The signal transduction pathway just described is not the end of the story. Phytochrome does not regulate just one but numerous light-responsive signaling pathways that affect gene expression. Also, we do not understand the specifics of how the transition from Pr to Pfr participates in so many signaling cascades.

Light influences circadian rhythms

Almost all organisms, including plants, animals, fungi, eukaryotic microorganisms, and many bacteria, appear to have an internal timer, or biological clock, that approximates

(a) Leaf position at noon in a bean (*Phaseolus vulgaris*) seedling.

(b) Leaf position at midnight.

Figure 38-20 *Animation* **Sleep movements**

a 24-hour cycle, the time it takes Earth to rotate around its own axis. These internal molecular cycles are known as **circadian rhythms** (from the Latin *circum,* "around," and *diurn,* "daily"). Circadian rhythms help an organism detect the time of day, whereas photoperiodism enables a plant to detect the time of year.

Why do plants and other organisms exhibit daily circadian rhythms? Predictable environmental changes, such as sunrise and sunset, occur during the course of each 24-hour period. These predictable changes may be important to an individual organism, causing it to change its physiological activities or its behavior (in the case of animals). Researchers think that circadian rhythms help an organism synchronize repeated daily activities so that they occur at the appropriate time each day. If, for example, an insect-pollinated flower does not open at the time of day that pollinating insects are foraging for food, reproduction will be unsuccessful.

Circadian rhythms in plants affect such biological events as gene expression, the timing of photoperiodism (i.e., of seasonal reproduction), the rate of photosynthesis, and the opening and closing of stomata. *Sleep movements* observed in the common bean and other plants are another example of a circadian rhythm (FIG. 38-20). During the day, bean leaves are horizontal, possibly for optimal light absorption, but at night the leaves fold down or up, a movement that orients them perpendicular to their daytime position. The biological significance of sleep movements is unknown at this time.

When constant environmental conditions are maintained, circadian rhythms such as sleep movements repeat every 20 to 30 hours, at least for several days. In nature the rising and setting of the sun reset the biological clock so that the cycle repeats every 24 hours. The functioning of the circadian clock is under genetic control.

Mutants of the model plant *Arabidopsis* in which the circadian clock is not synchronized to match the external day–night cycle have been found to contain less chlorophyll, fix less carbon by photosynthesis, grow more slowly, and have less of a competitive advantage than *Arabidopsis* plants with a normal circadian clock.

For *Arabidopsis* and many other plants, two photoreceptors—the red light–absorbing phytochrome and the blue/ultraviolet-A light–absorbing **cryptochrome**—are implicated in resetting the biological clock. Certain amino acid sequences of the protein portion of phytochrome are homologous with amino acid

sequences of clock proteins in fruit flies, fungi, mammals, and bacteria; this molecular evidence strongly supports the circadian-clock role of phytochrome. The evidence for cryptochrome as a clock protein is also convincing. First discovered in plants, cryptochrome counterparts are found in the fruit fly and mouse biological-clock proteins. Possibly both photoreceptors are involved in resetting the biological clock in plants; researchers have evidence that phytochrome and cryptochrome sometimes interact to regulate similar responses.

CHECKPOINT 38.3

- *What is phytochrome? What are two roles of phytochrome?*
- *What are the steps in phytochrome signal transduction?*
- *In what process is cryptochrome involved?*

38.4 RESPONSES TO HERBIVORES AND PATHOGENS

LEARNING OBJECTIVE

9 Distinguish between the hypersensitive response and systemic acquired resistance.

Plants lack an immune system such as is found in humans and other mammals, but plants can defend themselves from attack by pathogenic and herbivorous enemies. Each plant cell has an innate immune system to fight against local infections. When a molecule produced by a pathogen or herbivore binds to a receptor in a plant cell, it triggers a signal transduction pathway. As a result, a *hypersensitive response* occurs that seals off the infected area, causing it to die. These *necrotic lesions,* or dead areas, prevent or slow the spread of the initial infection (FIG. 38-21).

Figure 38-21 Hypersensitive response
The necrotic lesions on these oak leaves were produced in response to invasion by a fungal pathogen.

The plant also sends signals from the site of localized infection, leading to **systemic acquired resistance (SAR)**, a buildup of long-lasting, enhanced resistance throughout the plant. The signal travels in the phloem. Systemic acquired resistance is broad spectrum; that is, SAR is a heightened defense against many types of pathogens, not just the one that caused the initial infection.

Plant biologists continue to discover new hormonelike signaling molecules. Many of these signaling molecules are involved in the defensive responses of plants to disease organisms and insects. Here we briefly consider two groups: jasmonic acid and methyl salicylate.

Jasmonic acid activates several plant defenses

Jasmonic acid affects several plant processes, such as pollen development, root growth, fruit ripening, and senescence. This lipid-derived plant hormone, which is structurally similar to *prostaglandins* in animals (see Chapter 49), is also produced in response to the presence of insect pests and disease-causing organisms.

Jasmonic acid triggers the production of enzymes that confer an increased resistance against herbivorous (plant-eating) insects. For example, caterpillar-infested tomato plants release volatile jasmonic acid into the air that attracts natural caterpillar enemies, such as parasitic wasps; the wasps lay their eggs on the caterpillars' bodies. When the eggs hatch, the wasp larvae consume, and ultimately kill, the host insects. In one study, treating plants with jasmonic acid increased parasitism on caterpillar pests twofold over control fields in which the plants did not have jasmonic acid applications. Jasmonic acid may be of practical value in controlling certain insect pests without the use of chemical pesticides.

Biologists have made progress in identifying key aspects of the jasmonic acid signaling pathway. Interestingly, the jasmonic acid signaling pathway includes ubiquitin, which was discussed earlier in the chapter in relation to auxin (see Fig. 38-4). As in the auxin signaling pathway, the ubiquitin tags in the jasmonic acid signaling pathway cause degradation of transcription repressors, thereby altering gene expression.

Methyl salicylate may induce systemic acquired resistance

For centuries, people chewed willow (*Salix*) bark to treat headaches and other types of pain. **Salicylic acid** was first extracted from willow bark and is chemically related to aspirin (acetylsalicylic acid). More recently, biologists have shown that salicylic acid helps plants defend against insect pests and pathogens such as viruses. When a plant is under attack, the concentration of salicylic acid increases, and it spreads systemically throughout the plant.

Based on current research, biologists think that *methyl salicylate*, a derivative of salicylic acid also known as *oil of wintergreen*, binds to a cell receptor, triggering a signal transduction pathway that switches on genes. These genes code for proteins that establish systemic acquired resistance to fight infection and promote wound healing.

Plants also use methyl salicylate, which is volatile, to signal nearby plants. Tobacco plants infected with tobacco mosaic virus release methyl salicylate into the air. When nearby healthy plants receive the airborne chemical signal, they begin synthesizing antiviral proteins that enhance their resistance to the virus.

CHECKPOINT 38.4

- *You observe necrotic lesions on the leaves of a plant in your garden. Are they more likely due to a hypersensitive response or to systemic acquired resistance (SAR)? Explain your answer.*

SUMMARY: FOCUS ON LEARNING OBJECTIVES

38.1 Tropisms *(page 798)*

1 Describe phototropism, gravitropism, and thigmotropism.

- **Tropisms** are directional growth responses. **Phototropism** is growth in response to the direction of light. **Gravitropism** is growth in response to the influence of gravity. **Thigmotropism** is growth in response to contact with a solid object.

38.2 Plant Hormones and Development *(page 799)*

2 Describe a general mechanism of action for plant hormones, using auxin as your example.

- Plants produce and respond to **hormones,** organic molecules that act as highly specific chemical signals to elicit a variety of responses that regulate growth and development.
- Many plant hormones bind to *enzyme-linked receptors* located in the plasma membrane; the hormone binds to the receptor, triggering an enzymatic reaction of some sort. The receptor for auxin, which is located in the cytosol or nucleus, binds to auxin. This catalyzes the attachment of ubiquitin to a repressor protein, targeting it for destruction. When genes repressed by

those proteins are then expressed, changes in cell growth and development result.

3 Describe early auxin experiments involving phototropism.

- Charles Darwin and his son Francis performed phototropism experiments in the 1870s on grass seedlings. When they covered the coleoptile tip, the plant did not bend. When they removed the coleoptile tip, bending did not occur. When the bottom of the coleoptile was shielded from the light, the coleoptile bent. The Darwins concluded that some substance is transmitted from the upper to the lower part that causes the plant to bend.
- In the 1920s, Frits Went isolated the phototropic hormone from oat coleoptiles by removing the coleoptile tips and placing them on agar blocks. Normal growth resumed when he put one of these agar blocks squarely on a decapitated coleoptile; the substance had diffused from the coleoptile tip into the agar and, later, from the agar into the decapitated coleoptile. Went named this substance auxin.

4 List several ways that each of these hormones affects plant growth and development: auxins, gibberellins, cytokinins, ethylene, abscisic acid, and brassinosteroids.

- **Auxin** is involved in cell elongation; tropisms; **apical dominance,** the inhibition of axillary buds by the apical meristem; and fruit development. Auxin also stimulates root development on stem cuttings.
- **Gibberellins (GA)** are involved in stem elongation, flowering, and germination.
- **Cytokinins** promote cell division and differentiation; delay **senescence,** the natural aging process; and interact with auxin and ethylene in apical dominance. Cytokinins induce cell division in **tissue culture,** a technique in which cells are isolated from plants and grown in a nutrient medium.
- **Ethylene** plays a role in ripening fruits; apical dominance; leaf abscission; wound response; **thigmomorphogenesis,** a developmental response to mechanical stressors such as wind; and senescence.
- **Abscisic acid (ABA)** is an environmental stress hormone involved in stomatal closure caused by water stress and in seed **dormancy,** a temporary state of reduced physiological activity.
- Plant steroids known as **brassinosteroids (BRs)** are involved in several aspects of plant growth and development, such as cell division, cell elongation, light-induced differentiation, seed germination, and vascular development.

38.3 Light Signals and Plant Development *(page 807)*

5 Explain how varying amounts of light and darkness induce flowering.
- **Photoperiodism** is any response of plants to the duration and timing of light and dark. Flowering is a photoperiodic response in many plants. **Short-day plants** detect the lengthening nights of late summer or fall and flower at that time. **Long-day plants** detect the shortening nights of spring and early summer and flower at that time. **Intermediate-day plants** flower when exposed to days and nights of intermediate length.

6 Describe the role of phytochrome in flowering, including a brief discussion of phytochrome signal transduction.

- The photoreceptor in photoperiodism is **phytochrome,** a family of about five blue-green pigments. Each type of phytochrome has two forms, **Pr** and **Pfr,** named by the wavelength of light they absorb. Pfr is the active form, triggering or inhibiting physiological responses such as flowering, **shade avoidance,** and a light requirement for germination.
- The pathway in phytochrome **signal transduction** begins when inactive phytochrome in the cytoplasm absorbs red light and is converted into the active form, Pfr, which moves into the nucleus. There, phytochrome binds to the transcription factor PIF3 (phytochrome-interacting factor) and activates or represses the transcription of light-responsive genes.

7 Define *circadian rhythm* and give an example.
- A **circadian rhythm** is a regular period in an organism's growth or activities that approximates the 24-hour day and is reset by the rising and setting of the sun. Two examples of circadian rhythms are the opening and closing of stomata and *sleep movements.*

8 Distinguish between phytochrome and cryptochrome.
- Both phytochrome and **cryptochrome** are photoreceptors that sometimes interact to regulate similar responses, such as resetting the biological clock. Phytochrome strongly absorbs red light, whereas cryptochrome absorbs blue and ultraviolet-A light.

38.4 Responses to Herbivores and Pathogens *(page 811)*

9 Distinguish between the hypersensitive response and systemic acquired resistance.
- Each plant cell has an innate immune system to fight against local infections. A *hypersensitive response* seals off the infected area, causing it to die. These dead areas prevent or slow the spread of the initial infection.
- **Systemic acquired resistance (SAR)** is a buildup of long-lasting, enhanced resistance throughout the plant. Systemic acquired resistance is broad spectrum (against many types of pathogens).

TEST YOUR UNDERSTANDING

Know and Comprehend

1. In the signal transduction process for the hormone auxin, the molecule ubiquitin (a) absorbs blue light (b) becomes phosphorylated (c) tags certain proteins for destruction (d) interacts antagonistically with gibberellins (e) binds to a receptor in the plant cell's plasma membrane
2. When you prune shrubs to make them "bushier"—that is, to prevent apical dominance—you are affecting the distribution and action of which plant hormone? (a) auxin (b) jasmonic acid (c) salicylic acid (d) ethylene (e) abscisic acid
3. Research on a fungal disease of rice provided the first clues about the plant hormone (a) auxin (b) gibberellin (c) cytokinin (d) ethylene (e) abscisic acid
4. This plant hormone interacts with auxin during the formation of plant organs in tissue culture. (a) florigen (b) gibberellin (c) cytokinin (d) ethylene (e) abscisic acid
5. The stress hormone that helps plants respond to drought is (a) auxin (b) gibberellin (c) cytokinin (d) ethylene (e) abscisic acid
6. This hormone promotes seed dormancy. (a) auxin (b) gibberellin (c) cytokinin (d) ethylene (e) abscisic acid

7. The three classes of photoreceptors that enable plants to sense the presence and duration of light are (a) phytochrome, cryptochrome, and circadian rhythms (b) chlorophyll, photosynthesis, and photoperiodism (c) phytochrome, photoperiodism, and chlorophyll (d) phytochrome, cryptochrome, and phototropin (e) phototropin, photosynthesis, and photoperiodism
8. Pfr, the active state of _____, forms when red light is absorbed. (a) photosystem I (b) phytochrome (c) far-red light (d) phototropin (e) cryptochrome
9. Which photoreceptor(s) is/are implicated in resetting the biological clock? (a) phytochrome only (b) cryptochrome only (c) phototropin only (d) cryptochrome and gibberellin (e) phytochrome and cryptochrome
10. Which signaling molecule triggers the release of volatile substances that attract parasitic wasps to plant-eating caterpillars? (a) phytochrome (b) jasmonic acid (c) auxin (d) methyl salicylate (e) gibberellin

Apply and Analyze

11. Arrange the following steps in the correct order in the phytochrome signal transduction pathway: 1. red light; 2. light-responsive gene is switched on (or off); 3. movement of Pfr to nucleus; 4. conversion of Pr to Pfr; 5. formation of Pfr–PIF3 complex that is bound to promoter region

12. **PREDICT** State whether flowering in a short-day plant with a minimum critical night length of 14 hours would be expected to occur in the following situations. Explain each answer.
    (a) The plant is exposed to 15 hours of daylight and 9 hours of uninterrupted darkness.
    (b) The plant is exposed to 9 hours of daylight and 15 hours of darkness.
    (c) The plant is exposed to 9 hours of daylight and 15 hours of darkness, with a 10-minute exposure to red light in the middle of the night.

13. **PREDICT** If you transplanted the short-day plant discussed in Question 12 to the Equator, would it flower? Explain your answer.

Evaluate and Synthesize

14. **EVOLUTION LINK** What adaptive advantages are conferred on a plant whose stems are positively phototropic and whose roots are positively gravitropic?

15. **EVOLUTION LINK** Explain why a plant with a mutation in one of its phytochrome genes that affects flowering time would be at an evolutionary disadvantage.

16. **INTERPRET DATA** Examine the figure. When the day length is 10 hours and the night length is 14 hours, what percentage of the plants flower? When the day length is 16 hours and the night length is 8 hours, what percentage of the plants flower? What is the critical night length for this plant?

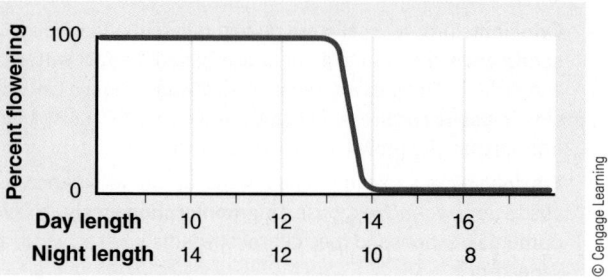

17. **SCIENCE, TECHNOLOGY, AND SOCIETY** Do you think that our expanding knowledge of the molecular basis of plant signaling molecules could improve agricultural productivity in the future? Explain your answer.

To access course materials, such as Aplia and other **aplia** companion resources, please visit **www.cengagebrain.com.**

Animal Structure and Function: An Introduction

39

© Francois Gagnon/Shutterstock.com

Larger body size does not mean bigger cells. The cells of the adult elephant, the young elephant, and the bird on the adult's head are all about the same size. The adult elephant is larger than the bird on its head because its genes specify that its body consists of a larger number of cells. The cells of the young elephant will continue to multiply until it reaches adult size.

Animal groups are dramatically diverse, with radically different body structures. For example, consider how different the elephant and the bird are, not only in size but also in body form and lifestyle (see photograph). Despite their differences, animal groups share many characteristics, including their relatively large size.

Why are most animals larger than bacteria, archaea, protists, and fungi? The answer may be related to *ecological niches,* which are the functional roles of a species within a community. By the time animals evolved, other organisms already occupied most available ecological niches. For new species to succeed, they had to displace others from a niche or adapt to a new one. Success in a new niche required a new body plan, and new body plans often required and accommodated larger size. Increased size also provided more opportunity for capturing food. Predators are typically larger than their prey.

To grow larger than their bacterial and protist competitors, animals had to be multicellular. Recall that the size of a single cell is limited by the ratio of its surface area (plasma membrane) to its volume (see Chapter 4). The plasma membrane needs to be large enough relative to the cell's volume to permit passage of materials into and out of the cell so that the conditions necessary for life can be maintained. In a multicellular animal, each cell has a large enough ratio of surface area to volume to effectively regulate its internal environment. Individual cells live and die, and they are replaced while the organism continues to maintain itself and thrive. The number of cells, not their individual sizes, is mainly responsible for the size of an animal.

In unicellular organisms such as bacteria and many protists, the single cell carries on all the activities necessary for life. Recall that unicellular and small, flat organisms depend on diffusion for many life processes, including gas exchange and disposal of metabolic wastes. One reason they can be small is that they do not require complex organ systems.

In this chapter we focus on the basic form and function of the animal body. **Anatomy** is the study of an organism's structure. **Physiology** is the study of how the body functions. Remember that in biological systems, structures are adapted to function at every level of organization. As we describe the types and

KEY CONCEPTS

39.1 Cells make up tissues, various types of tissues make up organs, and tissues and organs working together make up organ systems. The main types of tissues found in animals are epithelial, connective, muscle, and nervous tissues.

39.2 Homeostatic mechanisms are mainly negative feedback systems that maintain a relatively stable internal environment.

39.3 Thermoregulation is the process of maintaining homeostasis of body temperature despite changes in surrounding (or internal) temperature.

functions of tissues and the principal organ systems of animals, note the many examples of this basic principle. In this chapter we also discuss the important concept of *homeostasis*, using regulation of body temperature as an example. In the following chapters, we discuss how organ systems work together to maintain homeostasis as the animal carries out its many life processes.

39.1 TISSUES, ORGANS, AND ORGAN SYSTEMS

LEARNING OBJECTIVES

1 Compare the structure and function of the four main kinds of animal tissues: epithelial, connective, muscle, and nervous tissues.
2 Compare the main types of epithelial tissue and describe their functions.
3 Compare the main types of connective tissue and describe their functions.
4 Contrast the three types of muscle tissue and describe their functions.
5 Relate the structure of the neuron to its function.
6 Briefly describe the organ systems of a mammal and summarize the functions of each organ system.

In a multicellular organism, cells specialize to perform specific tasks. Recall from Chapter 1 that cells organize to form *tissues,* and tissues associate to form *organs* such as the heart or stomach. Groups of tissues and organs make up the *organ systems* of a complex *organism*. Billions of cells organize to form the tissues, organs, and organ systems of a butterfly, a crocodile, or an elephant.

A **tissue** consists of a group of closely associated, similar cells that carry out specific functions. Biologists classify animal tissues as epithelial, connective, muscle, or nervous tissue. Classification of tissues depends on their structure and origin. Each kind of tissue is composed of cells with characteristic sizes, shapes, and arrangements, and each type of tissue is specialized to perform a specific function or group of functions. For example, some tissues are specialized to transport materials, whereas others contract, enabling the animal to move. Still others secrete hormones that regulate metabolic processes. Structure and function are closely linked at every level of organization. As we discuss each tissue type, notice the relationship between its form and its function. Notice also how biological systems continuously interact.

Epithelial tissues cover the body and line its cavities

Epithelial tissue (also called *epithelium*) consists of cells fitted tightly together to form a continuous layer, or sheet, of cells. One surface of the sheet is typically exposed because it covers the body (outer layer of the skin) or lines a cavity, such as the *lumen* (the cavity in a hollow organ) of the intestine. The other surface of an epithelial layer attaches to the underlying tissue by a noncellular **basement membrane** consisting of tiny *fibers* and nonliving polysaccharide material that the epithelial cells produce.

Epithelial tissue forms the outer layer of the skin and the linings of the digestive, respiratory, excretory, and reproductive tracts. As a result, everything that enters or leaves the body must cross at least one layer of epithelium. Food taken into the mouth and swallowed is not really "inside" the body until it is absorbed through the epithelium of the digestive tract and enters the blood. To a large extent, the permeabilities of the various epithelial tissues regulate the exchange of substances between the different parts of the body as well as between the animal and the external environment.

Epithelial tissues perform many functions, including protection, absorption, secretion, and sensation. The epithelial layer of the skin, the **epidermis,** covers the entire body and protects it from mechanical injury, chemicals, bacteria, and fluid loss. The epithelial tissue lining the digestive tract absorbs nutrients and water into the body. Some epithelial cells form **glands** that secrete cell products such as hormones, enzymes, or sweat. Other epithelial cells are sensory receptors that receive information from the environment. For example, epithelial cells in taste buds and in the nose specialize as chemical receptors.

TABLE 39-1 illustrates the main types of epithelial tissue, indicates their locations in the body, and describes their functions (pages 818–819). We can distinguish three types of epithelial cells on the basis of shape. *Squamous* epithelial cells are thin, flat cells shaped like flagstones. *Simple squamous epithelium* lines the blood vessels and the air sacs in the lungs. *Cuboidal* epithelial cells are short cylinders that from the side appear cube-shaped, like dice. Actually, each cuboidal cell is typically hexagonal in cross section, making it an eight-sided polyhedron. *Simple cuboidal epithelium* lines the kidney tubules.

When viewed from the side, *columnar* epithelial cells look like columns or cylinders. The nucleus is usually located near the base of the cell. Viewed from above or in cross section, these cells often appear hexagonal. On its free surface, a columnar epithelial cell may have cilia that beat in a coordinated way, moving materials over the tissue surface. Most of the upper respiratory tract is lined with ciliated *columnar epithelium* that moves particles of dust and other foreign material away from the lungs.

Epithelial tissue is also classified by number of layers. *Simple epithelium* is composed of one layer of cells. It is usually located where substances are secreted, excreted, or absorbed, or where materials diffuse between compartments. For example, simple squamous epithelium lines the air sacs in the lungs. The structure of this thin tissue allows diffusion of gases in and out of air sacs.

Stratified epithelium, which has two or more layers, protects underlying tissues. For example, stratified squamous epithelium, which makes up the outer layer of your skin, is continuously sloughed off during normal wear and tear. It must also continuously regenerate. The cells of *pseudostratified epithelium* appear layered, but they are not. Although all its cells rest on a basement membrane, not every cell extends to the exposed surface of the tissue. This arrangement gives the impression of two or more cell layers. Some of the respiratory passageways are lined with pseudostratified epithelium equipped with cilia.

The lining of blood and lymph vessels is called endothelium. Endothelial cells have a different embryonic origin from "true" epithelium. However, these cells are structurally similar to squamous epithelial cells and can be included in that category.

Glands are made of epithelial cells

A **gland** consists of one or more epithelial cells specialized to produce and secrete a product such as sweat, milk, mucus, wax, saliva, hormones, or enzymes (**FIG. 39-1**). Epithelial tissue lining the cavities and passageways of the body typically has some specialized mucus-secreting cells called **goblet cells.** The mucus lubricates these surfaces, offers protection, and facilitates the movement of materials.

Glands are classified as exocrine or endocrine. **Exocrine glands,** like goblet cells and sweat glands, secrete their products onto a free epithelial surface, typically through a duct (tube). **Endocrine glands** lack ducts. These glands release their products, called **hormones,** into the **interstitial fluid** (tissue fluid) or blood. Hormones are typically transported by the cardiovascular system. (Endocrine glands are discussed in Chapter 49.)

Epithelial cells form membranes

An *epithelial membrane* consists of a sheet of epithelial tissue and a layer of underlying connective tissue. Types of epithelial membranes include mucous membranes and serous membranes. A **mucous membrane,** or *mucosa,* lines a body cavity that opens to the outside of the body, such as the digestive tract or respiratory tract. Goblet cells in the epithelial layer secrete mucus that lubricates the tissue and protects it from drying.

A **serous membrane** lines a body cavity that does not open to the outside of the body. It consists of simple squamous epithelium over a thin layer of loose connective tissue. This type of membrane secretes fluid into the cavity it lines. Examples of serous membranes are the pleural membranes lining the pleural cavities around the lungs and the pericardial membranes lining the pericardial cavity around the heart.

Connective tissues support other body structures

Almost every organ in the body has a framework of **connective tissue** that supports and cushions it. Compared with epithelial tissues, connective tissues contain relatively few cells. Its cells are embedded in an extensive **intercellular substance** consisting of threadlike, microscopic **fibers** scattered throughout a **matrix,** a thin gel of polysaccharides that the cells secrete. The structure and properties of the intercellular substance help determine the nature and function of each kind of connective tissue.

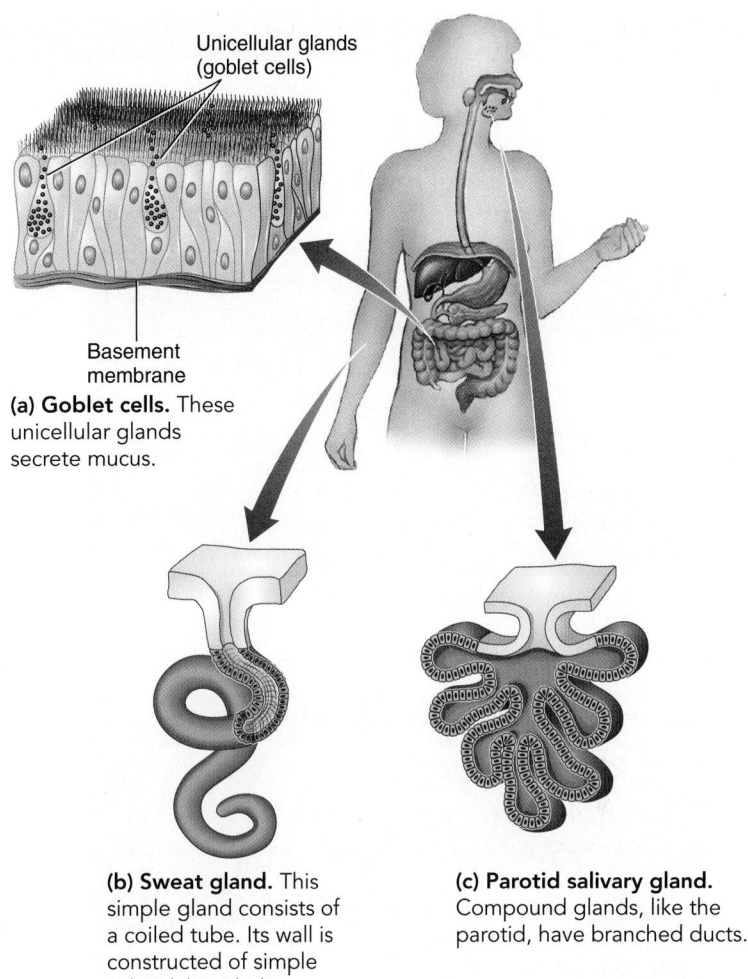

(a) **Goblet cells.** These unicellular glands secrete mucus.

(b) **Sweat gland.** This simple gland consists of a coiled tube. Its wall is constructed of simple cuboidal epithelium.

(c) **Parotid salivary gland.** Compound glands, like the parotid, have branched ducts.

Figure 39-1 Glands

A gland consists of one or more epithelial cells.

© Cengage Learning

Connective tissue typically contains three types of fibers: collagen, elastic, and reticular. *Collagen fibers,* the most numerous type, are made of **collagens,** a group of fibrous proteins found in all animals (see Fig. 3-23b). Collagens are the most abundant proteins in mammals, accounting for about 25% of their total protein mass. Collagen is very tough (meat is tough because of its collagen content). The tensile strength (ability to stretch without tearing) of collagen fibers is comparable to that of steel. Collagen fibers are wavy and flexible, allowing them to remain intact when tissue is stretched.

Elastic fibers branch and fuse to form networks. They can be stretched by a force and then (like a stretched rubber band) return to their original size and shape when the force is removed. Elastic fibers, composed of the protein elastin, are an important component of structures that must stretch.

Reticular fibers are very thin, branched fibers that form delicate networks joining connective tissues to neighboring tissues. Reticular fibers consist of collagen and some glycoprotein.

The cells of various kinds of connective tissues differ in their shapes and structures and in the kinds of fibers and matrices they

TABLE 39-1 | Epithelial Tissues

Nuclei of squamous epithelial cells

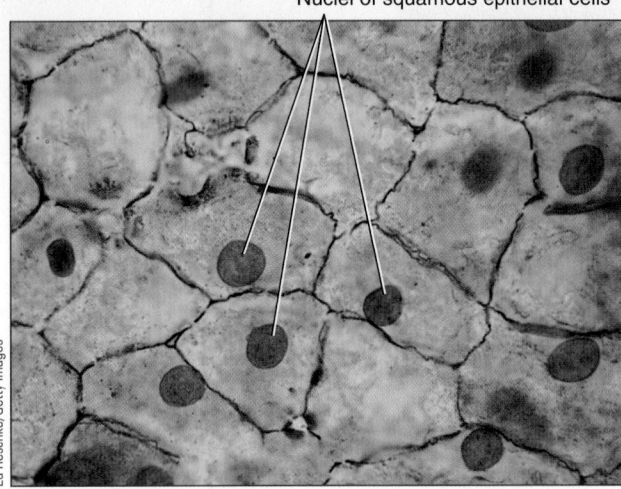

LM of simple squamous epithelium.

25 μm

Simple Squamous Epithelium

Main Locations
Air sacs of lungs; lining of blood vessels

Functions
Passage of materials where little or no protection is needed and where diffusion is major form of transport

Description and Comments
Cells are flat and arranged as single layer

Nuclei of cuboidal epithelial cells   Lumen of tubule

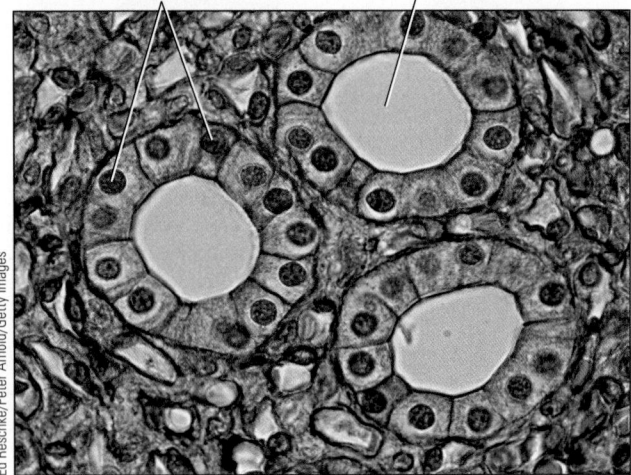

LM of simple cuboidal epithelium.

25 μm

Simple Cuboidal Epithelium

Main Locations
Linings of kidney tubules; gland ducts

Functions
Secretion and absorption

Description and Comments
Single layer of cells; LM shows cross section through tubules; from the side each cell looks like a short cylinder; some have microvilli for absorption

Goblet cell   Nuclei of columnar cells

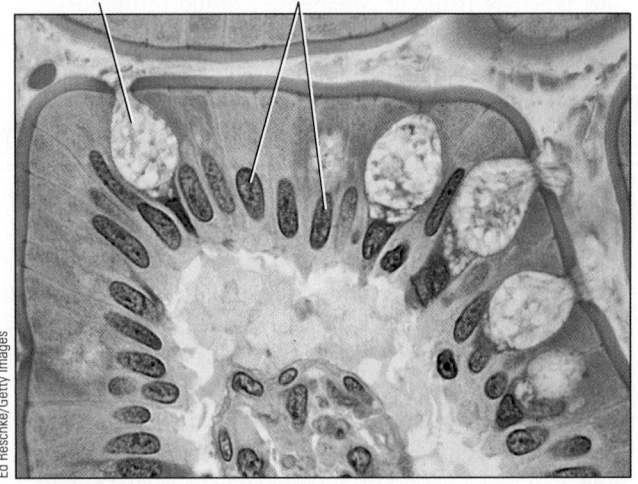

LM of simple columnar epithelium.

25 μm

Simple Columnar Epithelium

Main Locations
Linings of much of digestive tract and upper part of respiratory tract

Functions
Secretion, especially of mucus; absorption; protection; moves layer of mucus

Description and Comments
Single layer of columnar cells; highly developed Golgi complex; often ciliated; goblet cells secrete mucus

Continued

TABLE 39-1 | Epithelial Tissues (*continued*)

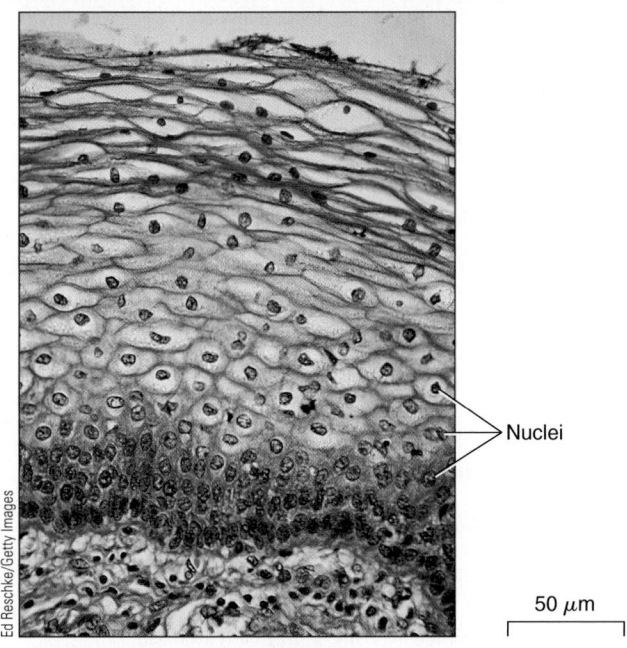

LM of stratified squamous epithelium.

Nuclei

50 μm

Ed Reschke/Getty Images

Stratified Squamous Epithelium

Main Locations
Skin; mouth lining; vaginal lining

Functions
Protection only; little or no absorption or transit of materials; outer layer continuously sloughed off and replaced from below

Description and Comments
Several layers of cells, with only the lower ones columnar and metabolically active; division of lower cells causes older ones to be pushed upward toward surface, becoming flatter as they move

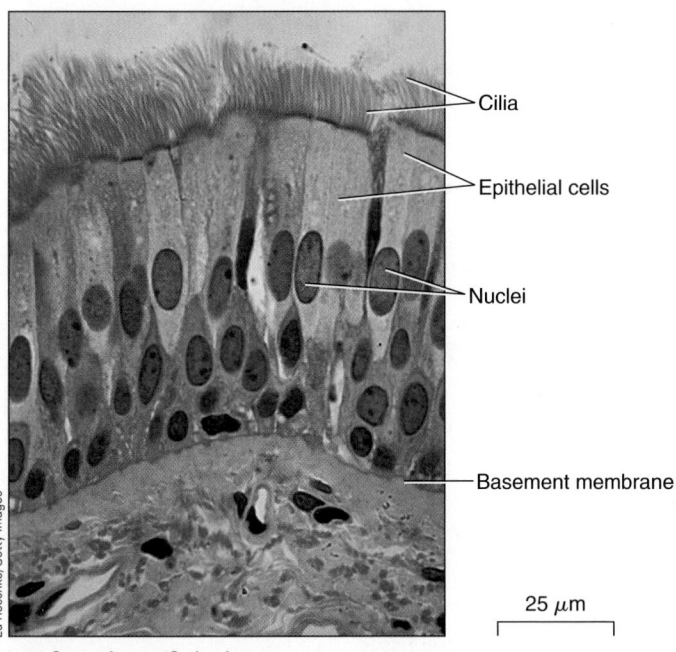

LM of pseudostratified columnar epithelium, ciliated.

Cilia

Epithelial cells

Nuclei

Basement membrane

25 μm

Ed Reschke/Getty Images

Pseudostratified Epithelium

Main Locations
Some respiratory passages; ducts of many glands

Functions
Secretion; protection; moves layer of mucus

Description and Comments
Ciliated, mucus-secreting, or with microvilli; comparable in many ways to columnar epithelium except that not all cells are the same height; so, although all cells contact the same basement membrane, the tissue appears stratified

© Cengage Learning

secrete. **Fibroblasts** are connective tissue cells that produce the fibers, as well as the protein and carbohydrate complexes, of the matrix. Fibroblasts release protein components that become arranged to form the characteristic fibers. These cells are especially active in developing tissues and are important in healing wounds. As tissues mature, the number of fibroblasts decreases, and they become less active. **Macrophages,** the body's scavenger cells, commonly wander through connective tissues, cleaning up cell debris and phagocytosing foreign matter, including bacteria.

Some of the main types of connective tissue are (1) loose and dense connective tissues; (2) elastic connective tissue; (3) reticular connective tissue; (4) adipose tissue; (5) cartilage; (6) bone; and (7) blood, lymph, and tissues that produce blood cells. These tissues vary widely in their structural details and in the functions they perform (**TABLE 39-2**).

TABLE 39-2 | Connective Tissues

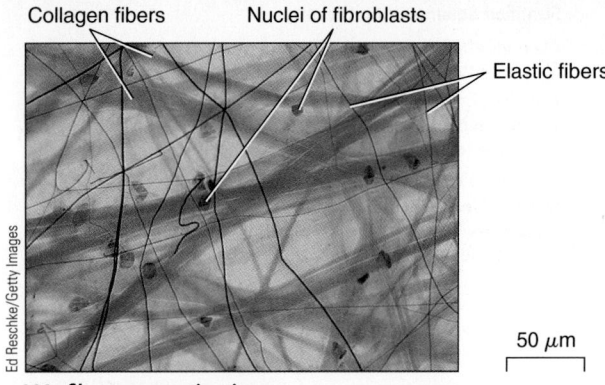

Collagen fibers   Nuclei of fibroblasts

Elastic fibers

50 μm

LM of loose connective tissue.

Loose Connective Tissue

Main Locations
Everywhere that support must be combined with elasticity, such as subcutaneous tissue (the layer of tissue beneath the dermis of the skin)

Functions
Support; reservoir for fluid and salts

Description and Comments
Fibers produced by fibroblast cells embedded in semifluid matrix; other types of cells, e.g., macrophages, present

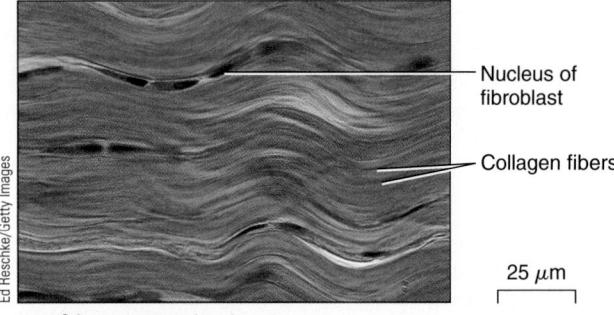

Nucleus of fibroblast

Collagen fibers

25 μm

LM of dense connective tissue.

Dense Connective Tissue

Main Locations
Tendons; many ligaments; dermis of skin

Functions
Support; transmits mechanical forces

Description and Comments
Collagen fibers may be regularly or irregularly arranged

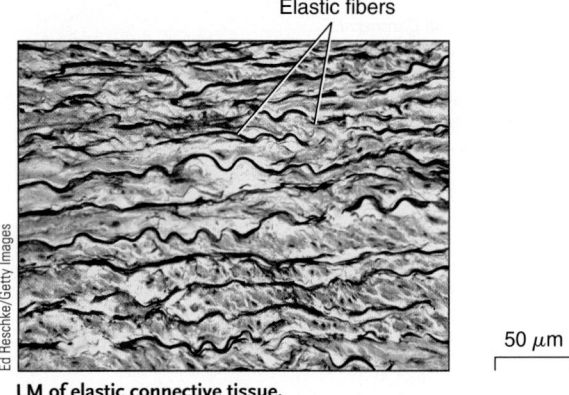

Elastic fibers

50 μm

LM of elastic connective tissue.

Elastic Connective Tissue

Main Locations
Structures that must both expand and return to their original size, such as lung tissue and large arteries

Function
Confers elasticity

Description and Comments
Branching elastic fibers interspersed with fibroblasts

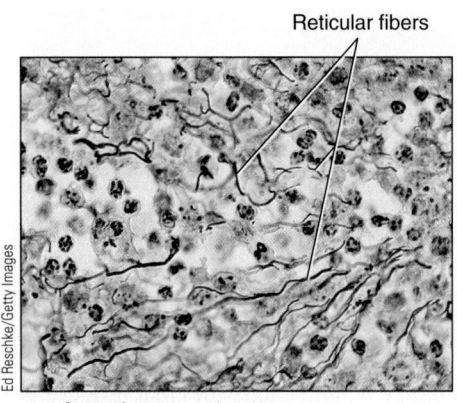

Reticular fibers

50 μm

LM of reticular connective tissue.

Reticular Connective Tissue

Main Locations
Framework of liver; lymph nodes; spleen

Function
Support

Description and Comments
Consists of interlacing reticular fibers

Continued

Ed Reschke/Getty Images

TABLE 39-2 | Connective Tissues (*continued*)

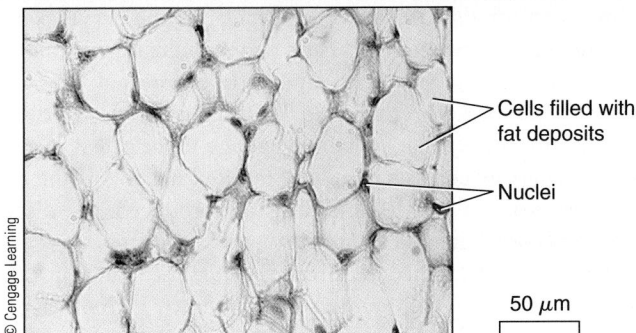

Cells filled with fat deposits

Nuclei

50 μm

LM of adipose tissue.

Adipose Tissue

Main Locations
Subcutaneous layer; forms protective pads around certain internal organs

Functions
Stores fat; insulation; supports organs such as mammary glands, kidneys

Description and Comments
Fat cells are star shaped at first; fat droplets accumulate until typical ring-shaped cells are produced

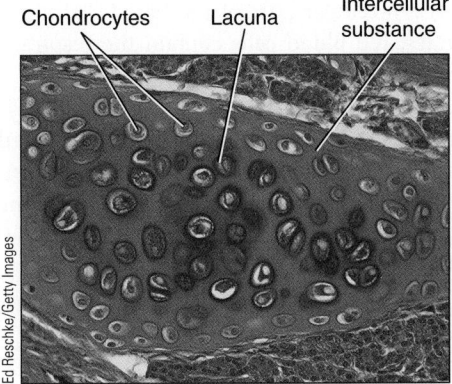

Chondrocytes   Lacuna   Intercellular substance

50 μm

LM of cartilage.

Cartilage

Main Locations
Supporting skeletons in sharks and rays; ends of bones in mammals and some other vertebrates; supporting rings in wall of trachea; tip of nose; external ear

Function
Flexible support

Description and Comments
Cells (chondrocytes) separated from one another by intercellular substance; cells occupy lacunae

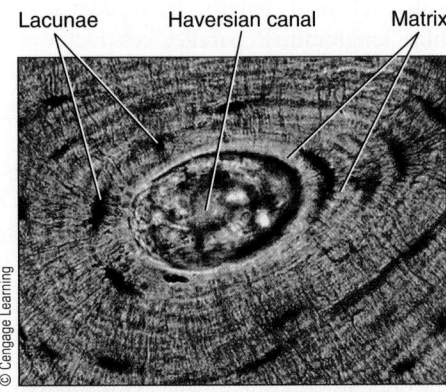

Lacunae   Haversian canal   Matrix

50 μm

LM of bone.

Bone

Main Locations
Forms skeletal structure in most vertebrates

Functions
Supports and protects internal organs; calcium reservoir; skeletal muscles attach to bones

Description and Comments
Cells (osteocytes) in lacunae; in compact bone lacunae embedded in lamellae, concentric circles of matrix surrounding Haversian canals

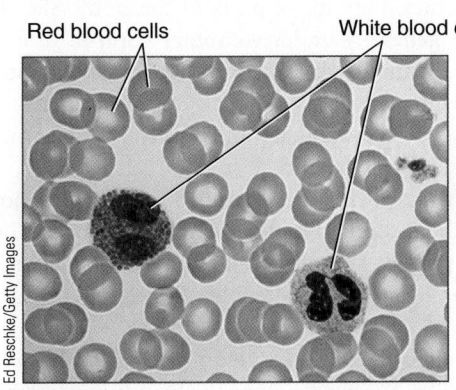

Red blood cells   White blood cells

25 μm

LM of blood.

Blood

Main Locations
Within heart and blood vessels of circulatory system

Functions
Transports oxygen, nutrients, wastes, and other materials; white blood cells defend against disease organisms

Description and Comments
Consists of red blood cells, white blood cells, and platelets dispersed in fluid matrix (plasma)

Loose connective tissue is the most widely distributed connective tissue in the vertebrate body. Together with adipose tissue, loose connective tissue forms the subcutaneous (below the skin) layer that attaches skin to the muscles and other structures beneath. Nerves, blood vessels, and muscles are wrapped in loose connective tissue. This tissue also forms a thin filling between body parts and serves as a reservoir for fluid and salts. Loose connective tissue consists of fibers running in all directions through a semifluid matrix. Its flexibility permits the parts it connects to move.

Dense connective tissue, found in the dermis (lower layer) of the skin, is very strong, but is somewhat less flexible than loose connective tissue. Collagen fibers predominate. **Tendons,** the cords that connect muscles to bones, and **ligaments,** the cables that connect bones to one another, consist of dense connective tissue in which collagen bundles are arranged in a definite pattern.

Elastic connective tissue consists mainly of bundles of parallel elastic fibers. This tissue is found in structures that must expand and then return to their original size, such as lung tissue and the walls of large arteries.

Reticular connective tissue is composed mainly of interlacing reticular fibers. It forms a supporting internal framework in many organs, including the liver, spleen, and lymph nodes.

The cells of **adipose tissue** store fat and release it when fuel is needed for cellular respiration. Adipose tissue is found in the subcutaneous layer and in tissue that cushions internal organs.

The supporting skeleton of a vertebrate is made of cartilage or of both cartilage and bone. **Cartilage** is the supporting skeleton in the embryonic stages of all vertebrates. In most vertebrates bone replaces cartilage during development. However, cartilage remains in some supporting structures. In humans, for example, cartilage is found in the external ear, the supporting rings in the walls of the respiratory passageways, the tip of the nose, the ends of some bones, and the discs that serve as cushions between the vertebrae.

Cartilage is firm yet elastic. Its cells, called **chondrocytes,** secrete a hard, rubbery matrix that surrounds them. They also secrete collagen fibers, which become embedded in the matrix and strengthen it. Chondrocytes eventually come to lie, singly or in groups of two or four, in small cavities in the matrix called *lacunae.* These cells remain alive and are nourished by nutrients and oxygen that diffuse through the matrix. Cartilage tissue lacks nerves, lymph vessels, and blood vessels.

Bone, the main vertebrate skeletal tissue, is like cartilage in that it consists mostly of matrix material. The bone cells, called **osteocytes,** are contained within lacunae. Osteocytes secrete and maintain the matrix (**FIG. 39-2**). Unlike cartilage, however, bone is a highly vascular tissue, with a substantial blood supply. Osteocytes communicate with one another and with capillaries by tiny channels (*canaliculi*) that contain long cytoplasmic extensions of the osteocytes.

A typical bone has an outer layer of **compact bone** surrounding a filling of *spongy bone.* Compact bone consists of spindle-shaped units called **osteons.** Within each osteon, osteocytes are arranged in concentric layers of matrix called *lamellae.* In turn, the lamellae surround central microscopic channels known as **Haversian canals,** through which capillaries and nerves pass.

Bones are amazingly light and strong. Calcium salts of bone render the matrix very hard, and collagen prevents the bony matrix from being overly brittle. Most bones have a large, central *marrow cavity* that contains a spongy tissue called *marrow.* Yellow marrow consists mainly of fat. Red marrow is the connective tissue in which blood cells are produced. We discuss bone in more detail in Chapter 40.

Blood and **lymph** are circulating tissues that help other parts of the body communicate and interact. Like other connective tissues, they consist of specialized cells dispersed in an intercellular substance. In mammals blood consists of **red blood cells, white blood cells,** and **platelets,** all suspended within **plasma,** the liquid, noncellular part of the blood. In humans and other vertebrates, red blood cells contain the respiratory pigment that transports oxygen. White blood cells defend the body against disease-causing microorganisms (see Chapter 45). Platelets, small fragments broken off from large cells in the bone marrow, play a key role in blood clotting. Plasma consists of water, proteins, salts, and a variety of soluble chemical messengers such as hormones that it transports from one part of the body to another. We discuss blood in Chapter 44.

Muscle tissue is specialized to contract

Most animals move by contracting the long, cylindrical or spindle-shaped cells of **muscle** tissue. Muscle cells are called muscle fibers because of their length. Each **muscle fiber** contains many thin, longitudinal, parallel contractile units called **myofibrils.** Two proteins, **myosin** and **actin,** are the chief components of myofibrils. Myosin and actin play a key role in contraction of muscle fibers.

Many invertebrates have skeletal and smooth muscle. Vertebrates have three types of muscle tissue: skeletal, cardiac, and smooth (**TABLE 39-3** on page 824). **Skeletal muscle** makes up the large muscle masses attached to the bones of the body. Skeletal muscle fibers are very long, and each fiber has many nuclei. The nuclei of skeletal muscle fibers lie just under the plasma membrane, which frees the entire central part of the skeletal muscle fiber for the myofibrils. This adaptation appears to increase the efficiency of contraction. When skeletal muscles contract, they move parts of the body. Skeletal muscle fibers are generally under voluntary control. In contrast, you do not normally contract your cardiac and smooth muscle fibers at will.

Light microscopy shows that both skeletal and cardiac fibers have alternating light and dark transverse stripes, or *striations*, that change their relative sizes during contraction. Striated muscle fibers contract rapidly but cannot remain contracted for a long period. They must relax and rest momentarily before contracting again. (Muscle contraction is discussed in Chapter 40.)

Cardiac muscle is the main tissue of the heart. When cardiac muscle contracts, the heart pumps the blood. The fibers of

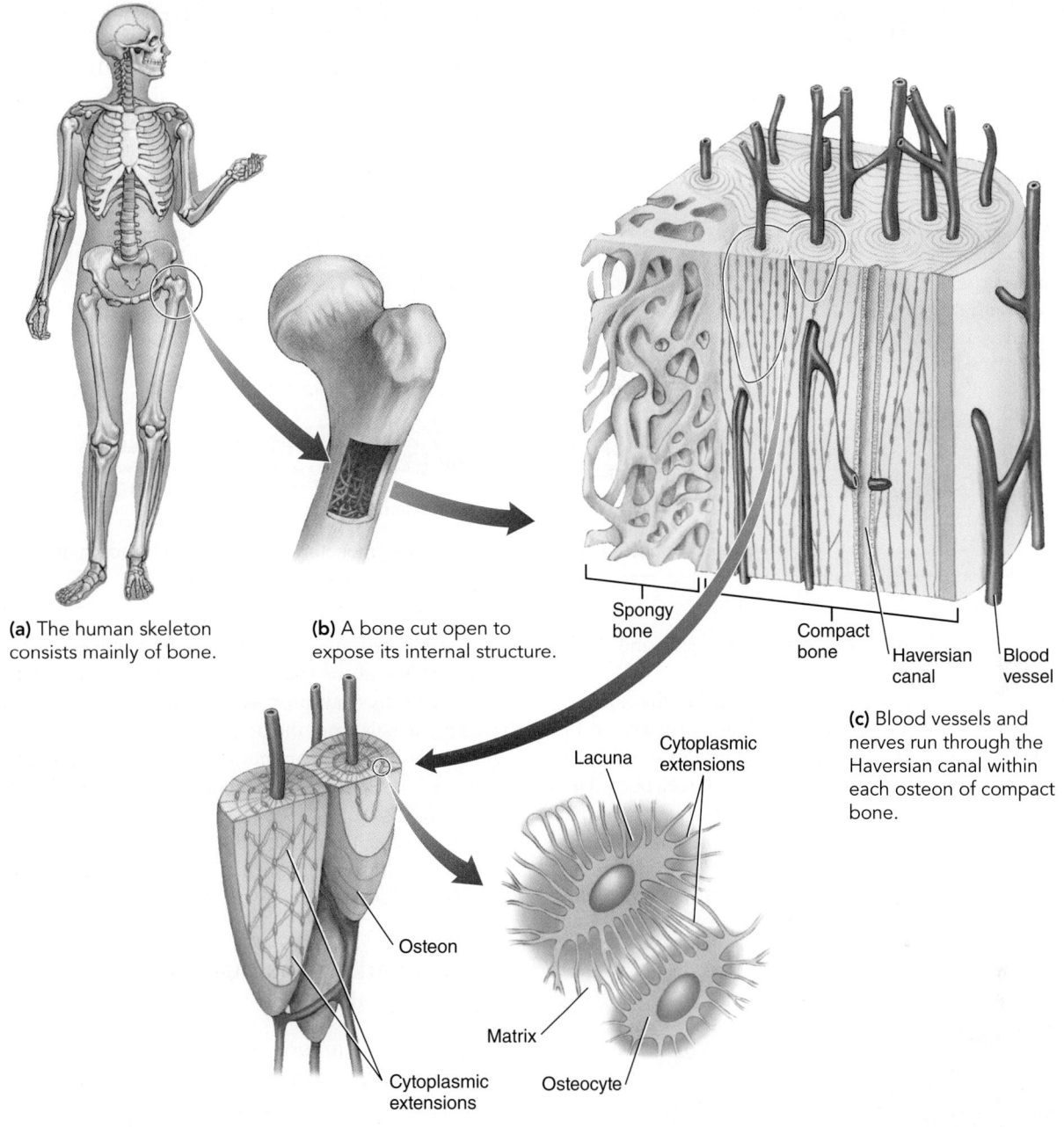

(a) The human skeleton consists mainly of bone.

(b) A bone cut open to expose its internal structure.

Spongy bone

Compact bone

Haversian canal

Blood vessel

(c) Blood vessels and nerves run through the Haversian canal within each osteon of compact bone.

Lacuna

Cytoplasmic extensions

Osteon

Matrix

Cytoplasmic extensions

Osteocyte

(d) The bone matrix is rigid and hard. Osteocytes become trapped within lacunae but communicate with one another by way of cytoplasmic extensions that extend through tiny canals.

Figure 39-2 *Animation* **Bone**
© Cengage Learning

cardiac muscle join end to end, and they branch and rejoin to form complex networks. One or two nuclei lie within each fiber. A characteristic feature of cardiac muscle tissue is the presence of *intercalated discs,* specialized junctions where the fibers join.

Smooth muscle occurs in the walls of the digestive tract, uterus, blood vessels, and many other internal organs. Contraction of smooth muscle is necessary for these organs to perform certain functions. For example, smooth muscle contraction in the wall of the digestive tract moves food through the digestive

tract. When smooth muscle in the walls of arterioles (small arteries) contracts, the blood vessels constrict, raising blood pressure. Each spindle-shaped smooth muscle fiber contains a single, central nucleus.

Nervous tissue controls muscles and glands

Nervous tissue consists of neurons and glial cells. **Neurons** are specialized for receiving and transmitting signals. **Glial cells**

TABLE 39-3 | Muscle Tissues

	SKELETAL	CARDIAC	SMOOTH
Location	Attached to skeleton	Walls of heart	Walls of stomach, intestines, blood vessels, uterus
Type of control	Voluntary	Involuntary	Involuntary
Shape of fibers	Elongated, cylindrical, blunt ends	Elongated, cylindrical, fibers that branch and fuse	Elongated, spindle shaped, pointed ends
Striations	Present	Present	Absent
Number of nuclei per fiber	Many	One or two	One
Position of nuclei	Peripheral	Central	Central
Speed of contraction	Most rapid	Intermediate (varies)	Slowest
Resistance to fatigue (with repetitive contraction)	Least	Intermediate	Greatest

Nuclei Striations Nuclei Nuclei

Intercalated discs

Skeletal muscle fibers **Cardiac muscle fibers** **Smooth muscle fibers**

© Cengage Learning

support and nourish the neurons, destroy pathogens, and modulate transmission of impulses (**FIG. 39-3**).

A typical neuron has a **cell body** containing the nucleus as well as two types of cytoplasmic extensions (discussed in Chapter 41). **Dendrites** are cytoplasmic extensions specialized for receiving signals and transmitting them to the cell body. The single **axon** transmits signals, called *nerve impulses,* away from the cell body. Axons range in length from 1 or 2 mm to more than a meter. Those extending from the spinal cord down the arm or leg in a human, for example, may be a meter or more in length.

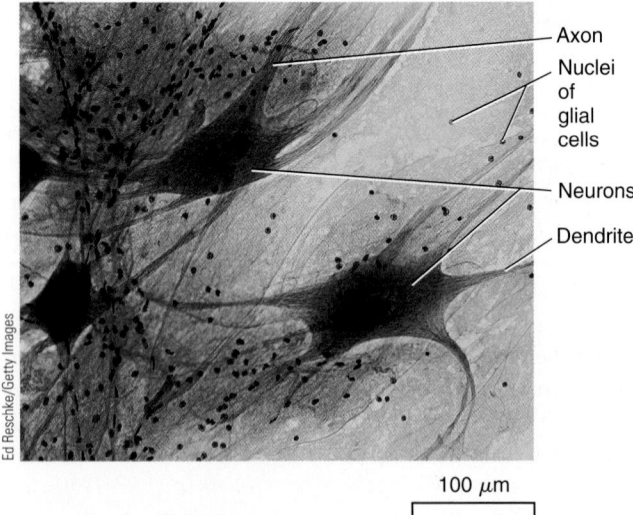

Axon
Nuclei of glial cells
Neurons
Dendrite

100 μm

Figure 39-3 LM of nervous tissue
Neurons transmit information in the form of electrochemical signals. Glial cells support, protect, and nourish neurons. They also communicate and help regulate neural function.

Certain neurons receive signals from the external or internal environment and transmit them to the spinal cord and brain. Other neurons relay, process, or store information. Still others transmit signals from the brain and spinal cord to the muscles and glands. Neurons communicate at junctions called **synapses.** A **nerve** consists of a great many neurons bound together by connective tissue.

In this chapter we have focused on normal tissues. For a discussion of some abnormal tissues, see *Inquiring About: Unwelcome Tissues: Cancers.*

Tissues and organs make up the organ systems of the body

Tissues associate to form **organs.** Although an animal organ may be composed mainly of one type of tissue, other types are needed to support, protect, provide a blood supply, and transmit information. For example, the heart is mainly cardiac muscle tissue, but its chambers are lined with endothelium, and its walls contain blood vessels made of endothelium, smooth muscle, and connective tissue. The heart also has nerves that transmit information and help regulate the rate and strength of its contractions.

An organized group of tissues and organs that together perform a specialized set of functions make up an **organ system.** Working together in a very coordinated way, organ systems perform the functions required by the **organism.** We can identify 11 major organ systems that work together to carry out the physiological processes of a mammal: **integumentary system, skeletal system, muscular system, nervous system, endocrine system, cardiovascular system, lymphatic system,** which functions as the **immune system, respiratory system, digestive system, urinary system,** and **reproductive system. FIGURE 39-4** summarizes the principal organs and functions of each organ system.

A **neoplasm** ("new growth"), or **tumor,** is an abnormal growth of cells. A neoplasm may be benign ("kind") or malignant (cancerous). A benign neoplasm tends to grow slowly, and its cells stay together. Because benign tumors form masses with distinct borders, they can usually be removed surgically. A **malignant** ("wicked") **neoplasm,** or **cancer,** typically grows much more rapidly and invasively than a benign tumor.

In Chapter 17 you learned that cancer results from abnormal expression of specific genes critical for cell division (see Fig. 17-20). Most cancer cells divide in an uncontrolled way. Unlike normal cells, which respect one another's boundaries and form tissues in an orderly, organized manner, cancer cells grow helter-skelter and infiltrate normal tissues. They apparently no longer receive or respond appropriately to signals from surrounding cells; communication is lacking (see figure).

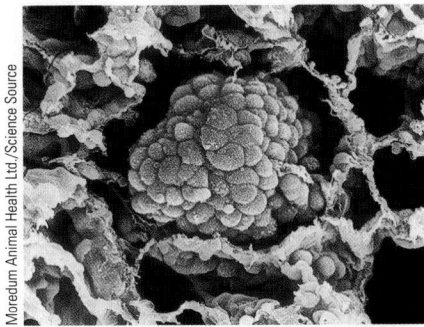

50 μm

When cancer cells multiply, they invade normal tissues and interfere with their functions. This SEM shows a mass of malignant cells (*pink*) in the air sac in the center. Cancer cells that have separated from the main tumor can be seen in other air sacs. Microvilli on the surface of the cancer cells give them a fuzzy appearance.

When a cancer cell multiplies, all the cells derived from it are also abnormal. Unlike the cells of benign neoplasms, cancer cells do not retain normal structural features. Most human cancers originate in epithelial tissue and are called *carcinomas*. This group includes breast, prostate, colon, lung cancer, and most ovarian cancers. Cancers that develop from connective tissues or muscle are referred to as *sarcomas*.

Death from cancer typically results from **metastasis,** migration of cancer cells through blood or lymph channels to other parts of the body. Once there, cancer cells multiply, forming new malignant neoplasms that interfere with the normal functions of the tissues being invaded. Cancer often spreads so rapidly and extensively that surgeons cannot locate or remove all the malignant masses.

Solid tumors, which account for more than 85% of cancer deaths, require blood vessels to ensure delivery of nourishment and oxygen. Some tumors grow to several millimeters in diameter and then enter a dormant stage, which may last for months or even years. Eventually, cancer cells release a chemical substance that stimulates nearby blood vessels to develop new capillaries. These blood vessels grow into the abnormal mass of cells. Nourished by its new blood supply, the neoplasm may grow rapidly. Newly formed blood vessels have leaky walls that provide a route for metastasis. Malignant cells enter the blood through these walls and are transported to new sites.

Worldwide, cancer causes more than seven million deaths each year. In the United States, cancer is the second leading cause of death. One in three people in the United States develops cancer at some time in his or her life. Currently, the key to survival is early diagnosis and treatment with some combination of surgery, hormonal treatment, radiation therapy, chemotherapy, inhibitors that suppress development of new blood vessels, immunotherapy, and targeted therapies. Many treatments are under investigation, including new agents that inhibit the development of new blood vessels. Immunotherapy for cancer will be discussed in Chapter 45. Cancer is a large family of closely related diseases (there are hundreds of distinct varieties), and treatment must be tailored to the particular type of cancer.

Most cancers are thought to be triggered by **carcinogens,** cancer-producing agents, in the environment, and by diet and lifestyle. Alleles of some genes appear to affect an individual's level of tolerance to carcinogens. You can decrease your risk of developing cancer by following these recommendations:

1. Do not smoke or use tobacco. Smoking is responsible for more than 80% of lung cancer cases, and it increases the risk for many other cancers.
2. Avoid prolonged exposure to the sun. When in the sun, use sunscreen or sunblock. Exposure to the sun is responsible for almost all of the more than two million cases of skin cancer reported each year in the United States alone.
3. Eat a healthy diet, including fresh, unprocessed fruits, vegetables, and grains. Limit intake of red meat. Avoid smoked, salt-cured, and nitrite-cured foods. Limit intake of alcoholic beverages. Reduce intake of foods and drinks that contribute to weight gain. Obesity increases the risk of cancer.
4. Exercise. Physical inactivity has been linked with increased risk of colon, breast, and other cancers.
5. Avoid unnecessary exposure to X-rays.
6. Self-examination and screening can lead to early diagnosis. Beginning at age 50, both men and women should be screened for colorectal cancer. Detection and removal of polyps (benign growths that can become malignant) can prevent cancer.
7. Self-examination and screening for women includes examining their breasts each month and having regular mammograms after age 40. Cervical cancer can be prevented with regular screening tests, such as the Papanicolaou (Pap) and human papilloma virus (HPV) tests. Certain HPV strains cause cervical cancer; vaccines are available.
8. Men over age 55 should discuss with their doctor the benefits of screening for prostate cancer with the prostate-specific antigen (PSA) blood test.
9. Consider genetic testing. If you have a family history of certain cancers, you may be at higher risk genetically for cancer. Genetic tests are now available to determine if you have mutations in the *BRCA1* or *BRCA2* genes, which increase risk for breast, ovarian, and pancreatic cancers. Mutations in several other genes indicate increased risk for colon, uterine, stomach, and urinary tract cancer. If you are at increased risk for cancer, a genetic counselor can advise you of measures you can take to reduce that risk.

ⒸHECKPOINT 39.1

- **CONNECT** *What are the main differences in structure and function between epithelial tissue and connective tissue?*
- *What type of tissue lines the air sacs of the lungs? How is its structure adapted to its function?*
- *What are some differences between the three types of muscle tissue? How is the structure of each type adapted for its function?*
- *What are the main functions of each of the following organ systems: (1) respiratory, (2) urinary, and (3) endocrine? (Consult Figure 39-4 for help.)*

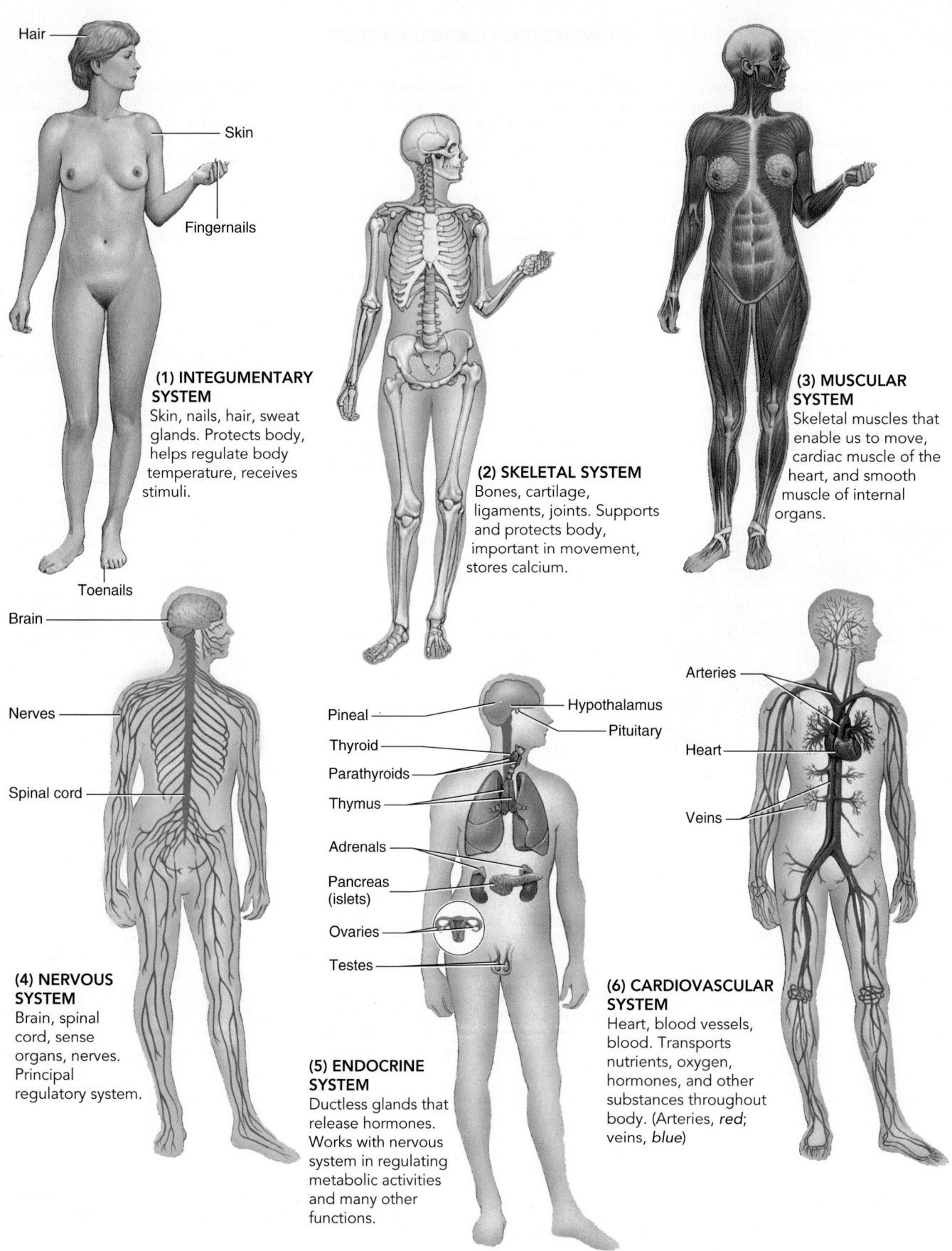

(1) INTEGUMENTARY SYSTEM
Skin, nails, hair, sweat glands. Protects body, helps regulate body temperature, receives stimuli.

Hair
Skin
Fingernails
Toenails

(2) SKELETAL SYSTEM
Bones, cartilage, ligaments, joints. Supports and protects body, important in movement, stores calcium.

(3) MUSCULAR SYSTEM
Skeletal muscles that enable us to move, cardiac muscle of the heart, and smooth muscle of internal organs.

(4) NERVOUS SYSTEM
Brain, spinal cord, sense organs, nerves. Principal regulatory system.

Brain
Nerves
Spinal cord

(5) ENDOCRINE SYSTEM
Ductless glands that release hormones. Works with nervous system in regulating metabolic activities and many other functions.

Pineal
Thyroid
Parathyroids
Thymus
Adrenals
Pancreas (islets)
Ovaries
Testes
Hypothalamus
Pituitary

(6) CARDIOVASCULAR SYSTEM
Heart, blood vessels, blood. Transports nutrients, oxygen, hormones, and other substances throughout body. (Arteries, *red*; veins, *blue*)

Arteries
Heart
Veins

Figure 39-4 *Animation* **The principal organ systems of the human body**
© Cengage Learning

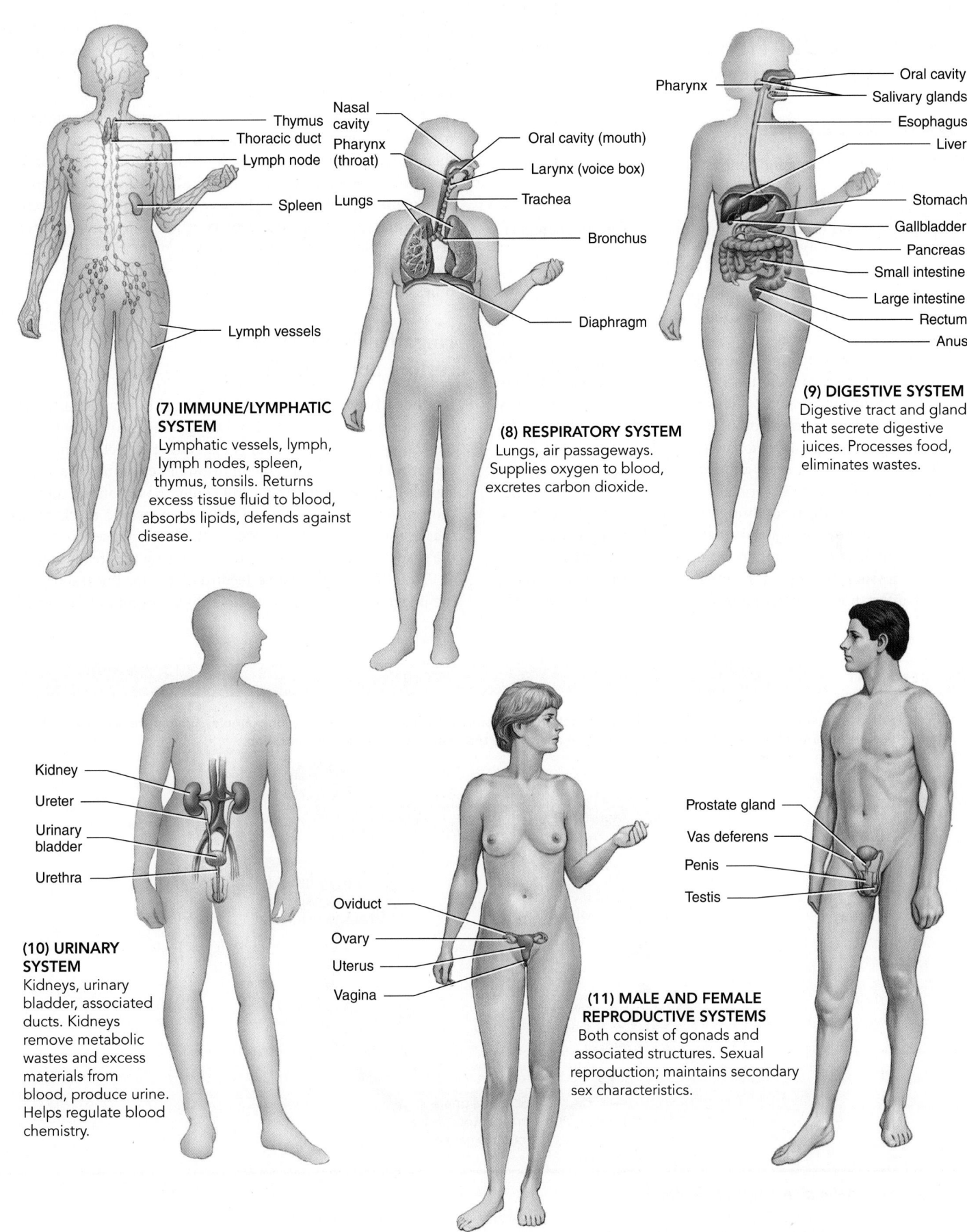

(7) IMMUNE/LYMPHATIC SYSTEM
Lymphatic vessels, lymph, lymph nodes, spleen, thymus, tonsils. Returns excess tissue fluid to blood, absorbs lipids, defends against disease.

Thymus
Thoracic duct
Lymph node
Spleen
Lymph vessels

(8) RESPIRATORY SYSTEM
Lungs, air passageways. Supplies oxygen to blood, excretes carbon dioxide.

Nasal cavity
Pharynx (throat)
Lungs
Oral cavity (mouth)
Larynx (voice box)
Trachea
Bronchus
Diaphragm

(9) DIGESTIVE SYSTEM
Digestive tract and glands that secrete digestive juices. Processes food, eliminates wastes.

Pharynx
Oral cavity
Salivary glands
Esophagus
Liver
Stomach
Gallbladder
Pancreas
Small intestine
Large intestine
Rectum
Anus

(10) URINARY SYSTEM
Kidneys, urinary bladder, associated ducts. Kidneys remove metabolic wastes and excess materials from blood, produce urine. Helps regulate blood chemistry.

Kidney
Ureter
Urinary bladder
Urethra

(11) MALE AND FEMALE REPRODUCTIVE SYSTEMS
Both consist of gonads and associated structures. Sexual reproduction; maintains secondary sex characteristics.

Oviduct
Ovary
Uterus
Vagina

Prostate gland
Vas deferens
Penis
Testis

39.2 REGULATING THE INTERNAL ENVIRONMENT

7   Define *homeostasis* and contrast negative and positive feedback systems.

To survive and function, animals must regulate the composition of the fluids that bathe their cells. They must maintain pH and internal temperature within relatively narrow limits. The body must also maintain the appropriate concentration of nutrients, oxygen, and other gases, ions, and compounds needed for metabolism at all times.

Cells, tissues, organs, and organ systems work together to maintain appropriate conditions in the body. The balanced internal environment is referred to as **homeostasis.** Homeostasis is a basic concept in physiology. First coined by U.S. physiologist Walter Cannon, the word *homeostasis* is derived from the Greek *homoios,* meaning "same," and *stasis,* "standing." Although the internal environment never really stays the same, it is a dynamic equilibrium in which conditions are maintained within narrow limits, which we call a **steady state.** The control processes that maintain these conditions are **homeostatic mechanisms.**

Stressors, changes in the internal or external environment that affect normal conditions within the body, continuously challenge homeostasis. An internal condition that moves out of its homeostatic range (either too high or too low) causes **stress.** An organism functions effectively because homeostatic mechanisms continuously operate to manage stress.

Many animals are **conformers** for certain environmental conditions. Some of their internal states vary with changes in their surroundings. For example, most marine invertebrates conform to the salinity of the surrounding sea water. Mammals are superb **regulators.** They have complex homeostatic mechanisms that maintain relatively constant internal conditions despite changes in the outside environment. How do homeostatic mechanisms work? Many are **feedback systems,** sometimes called "biofeedback systems."

Negative feedback systems restore homeostasis

In a **negative feedback system,** a change in some steady state (e.g., normal body temperature) triggers a response that counteracts, or reverses, the change. A **sensor** detects a change, a deviation from the normal condition or **set point.** The sensor signals an **integrator,** or control center. Based on the input of the sensor, the integrator activates homeostatic mechanisms that restore the steady state (**FIG. 39-5**). The response counteracts the inappropriate change, thereby restoring the steady state.

Note that in a negative feedback system, the response of the integrator is *opposite* (negative) to the output of the sensor.

KEY POINT

In a negative feedback system, the response of the integrator is opposite to the input of the sensor; for example, if the glucose concentration in the blood is too low, alpha cells in the pancreas secrete a hormone that increases glucose concentration.

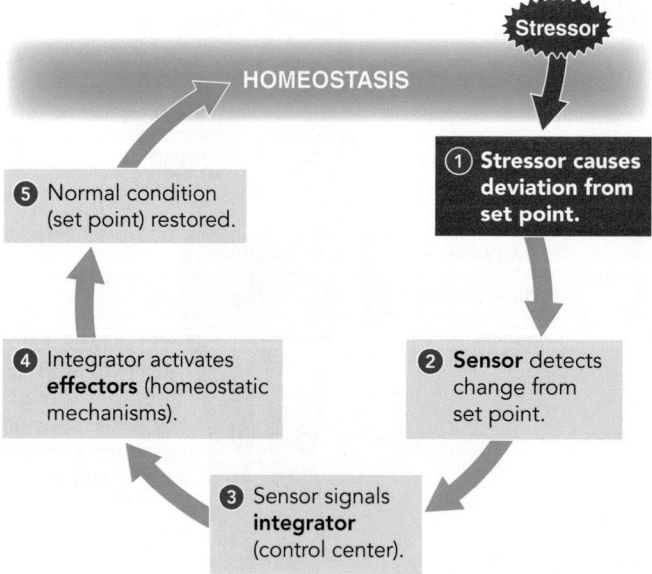

Figure 39-5 *Animation* **Negative feedback**

PREDICT If you drank a soda (high in carbohydrates), what would happen to the glucose concentration in your blood? How do you think your body would respond?

© Cengage Learning

When some condition varies too far from the steady state (either too high or too low), a control system using negative feedback brings the condition back to the steady state. For example, when the glucose concentration in the blood *decreases* below its homeostatic level, negative feedback systems *increase* its concentration. Most homeostatic mechanisms in the body are negative feedback systems.

Let us discuss heat regulation, a specific example of a negative feedback system. The temperature-regulating system in the human body is somewhat similar to how you regulate temperature in your home. You might set a particular room temperature on your thermostat. If the temperature in the room falls, the thermometer in the thermostat acts as a sensor that detects a change, or deviation, from the set point (FIG. 39-6a). The thermometer sends a signal to the thermostat, which then acts as an integrator, or control center. The thermostat compares the sensor's input with the set point. The thermostat then signals the furnace, which is the **effector** in this system. An effector is the device, organ, or process that helps restore the steady state. The furnace increases its heat output, a corrective response that brings the room temperature back to the set point. The thermometer no longer detects a change from the set point, so the thermostat and furnace are switched off.

As shown in FIGURE 39-6b, the negative feedback system that regulates body temperature works like a home thermostat. When body temperature decreases below normal limits, specialized nerve cells (sensors) signal the temperature-regulating center in the hypothalamus of the brain (the integrator). The integrator activates effectors that bring the temperature back to the set point. The return to normal temperature signals the sensors, and the temperature-regulating center switches off the effectors.

A few positive feedback systems operate in the body

In a **positive feedback system,** a change in some steady state sets off a response that intensifies (rather than reverses) the changing condition. Although some positive feedback mechanisms are beneficial, they do not maintain homeostasis. For example, a positive feedback cycle operates during the birth of a baby. As the baby's head pushes against the cervix (lower part of uterus), a reflex action causes the uterus to contract. The contraction forces the head against the cervix again, stimulating another contraction, and the positive feedback cycle repeats again and again until the baby is born. Some positive feedback sequences, such as those that deepen circulatory shock following severe hemorrhage, can disrupt steady states and lead to death. A simplified example is shown in FIGURE 39-7.

Homeostatic mechanisms maintain the internal environment within the physiological limits that support life. As you continue your study of animal processes, you will learn many ways in which organ systems interact to maintain the steady state of the organism. Although the nervous and endocrine systems play major roles, all organ systems participate in these regulatory processes. In the next section, we discuss some specific homeostatic mechanisms that help regulate body temperature.

CHECKPOINT 39.2

- *Contrast negative and positive feedback systems.*

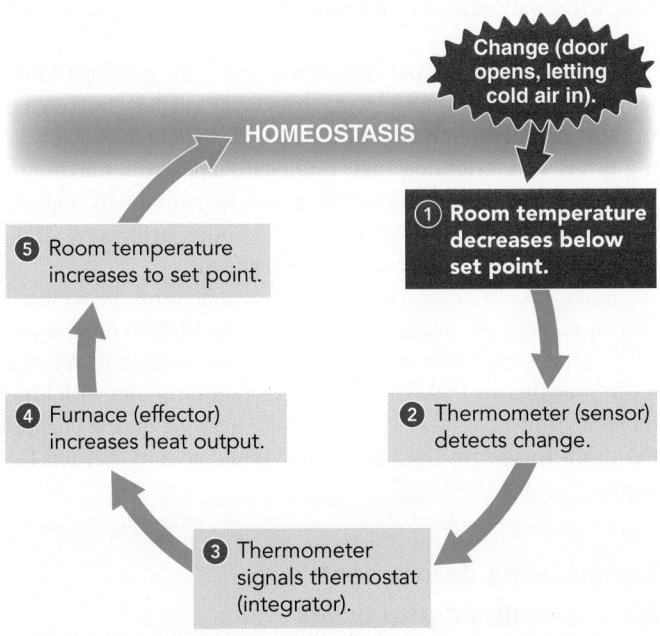

(a) Regulation of room temperature.

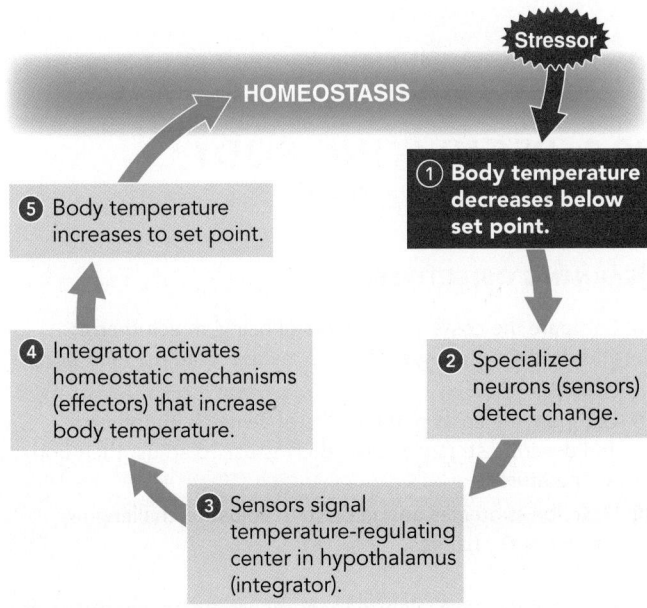

(b) Regulation of body temperature.

Figure 39-6 Negative feedback in temperature regulation

Note that the diagram in **(b)** is highly simplified. Figure 39-9 shows greater detail.

© Cengage Learning

Positive feedback intensifies the change taking place, moving conditions farther away from homeostasis; in some situations such as hemorrhage, the results can be fatal.

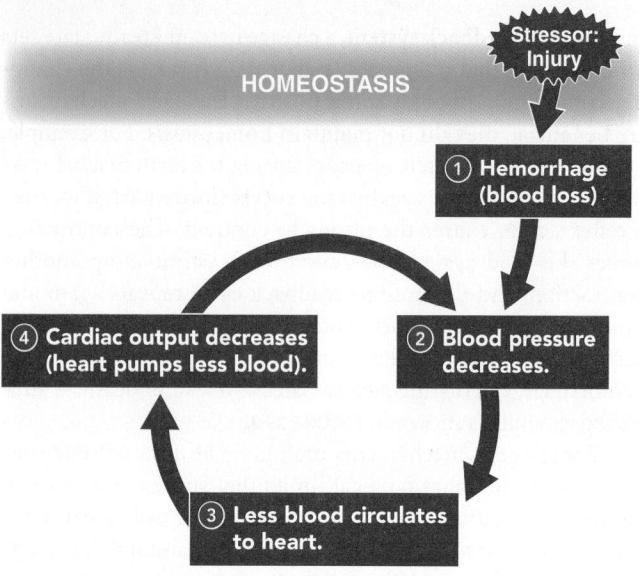

Figure 39-7 Positive feedback

In positive feedback the changes that occur increase the deviation from the set point. Conditions move away from the normal range. Loss of blood decreases blood pressure. Less blood reaches the heart, so heart function decreases. The resulting decrease in cardiac output further decreases blood pressure, bringing conditions farther from homeostasis.

PREDICT What might happen if you could convert this positive feedback system to a negative feedback system?

© Cengage Learning

39.3 REGULATING BODY TEMPERATURE

LEARNING OBJECTIVES

8 Compare the costs and benefits of being an ectotherm; and describe strategies ectotherms use to adjust their body temperature.

9 Compare the costs and benefits of being an endotherm; and describe strategies endotherms use to adjust their body temperature.

10 Describe strategies animals use to adjust to challenging temperature changes.

Many animals have elaborate homeostatic mechanisms for regulating body temperature. Some are physiological, and others are structural or behavioral. **Thermoregulation** is the process of maintaining body temperature within certain limits despite changes in the surrounding temperature. Animals produce heat as a byproduct of metabolic activities. Body temperature is determined by the rate at which heat is produced and the rate at which heat is lost to, or gained from, the outside environment.

The strategies for maintaining body temperature that are available to an animal may restrict the type of environment it can inhabit. Each species has an optimal environmental temperature range. Some animals, such as snowshoe hares, snowy owls, and weasels, can survive in cold arctic regions. Others, such as the Cape ground squirrel, which inhabits South Africa, are adapted to hot tropical climates. Although some animals can survive at temperature extremes, most survive only within moderate temperature ranges.

Ectotherms absorb heat from their surroundings

Ectotherms are animals that depend on the environment for their body heat. Their body temperature is determined mainly by the changing temperature of their surroundings. Most of the heat for their thermoregulation comes from the sun. In contrast, **endotherms** have homeostatic mechanisms that maintain body temperature despite changes in the external temperature. You may be surprised to learn that most animals are ectotherms.

An ectotherm's metabolic rate tends to change with the weather. These animals have a far lower daily energy expenditure than endotherms because they do not maintain a high metabolic rate. They therefore survive on less food and convert more of the energy in their food to growth and reproduction than do endotherms. Ectothermy also has costs. One disadvantage is that daily and seasonal temperature conditions may limit their activity.

Many ectotherms use behavioral strategies to adjust body temperature. For example, lizards may keep warm by burrowing in the soil at night. During the day, lizards take in heat by basking in the sun, orienting their bodies to expose the maximum surface area to the sun's rays (FIG. 39-8a). Many animals migrate to warmer climates during the winter. Another behavioral strategy for regulating temperature is *hibernation*.

Some insects use a combination of structural, behavioral, and physiological mechanisms to regulate body temperature. The "furry" body of the moth helps conserve body heat. When a moth prepares for flight, it contracts its flight muscles with little movement of its wings. The metabolic heat generated enables the moth to sustain the intense metabolic activity needed for flight.

Endotherms derive heat from metabolic processes

Birds and mammals, as well as some species of fish (e.g., tuna and some sharks) and some insects, are endotherms. The most important benefit of endothermy is the high metabolic rate, which can be as much as six times higher than that of ectotherms. The constant body temperature of endotherms allows a higher rate of enzyme activity than is possible for ectotherms living in the same habitat. Endotherms also can respond more rapidly to internal and external stimuli than ectotherms. They can be active even in low winter temperatures, but these animals must pay the high energy cost of thermoregulation even during times when they are inactive. You must maintain your body temperature even when you are asleep.

Endotherms have structural adaptations for maintaining body temperature. For example, the insulating feathers of birds, hair of mammals, and insulating layers of fat in birds and mammals reduce heat loss from the body. Birds and mammals also have behavioral adaptations for maintaining body temperature. The Cape ground squirrel positions its tail to shade its body from the direct rays of the sun. Elephants spray themselves with cool water.

Endotherms have a variety of physiological mechanisms for maintaining temperature homeostasis. They regulate heat production and regulate heat exchange with the environment. Most of their body heat comes from their own metabolic processes (see *Inquiring About: Electron Transport and Heat*, in Chapter 8).

In mammals receptors located in the hypothalamus of the brain and in the spinal cord regulate temperature. Heat from metabolic activities can be increased directly or can be increased indirectly by the action of hormones (such as thyroid hormones) that increase metabolic rate. Heat production is increased by contracting muscles, and in cold weather many animals shiver.

When body temperature rises, birds and many mammals pant, and some mammals sweat (FIG. 39-8b). These processes provide fluid for evaporation: the conversion of a liquid, such as sweat, to water vapor. (Recall from Chapter 2 that when molecules enter the vapor phase, they take their heat energy with them.) Heat transfers from the body to the surroundings, resulting in evaporative cooling.

In the human body, about 2.5 million sweat glands secrete sweat. As sweat evaporates from the skin surface, body temperature decreases. Constriction and dilation of capillaries in the skin are also homeostatic mechanisms for regulating body temperature.

In humans, when body temperature increases above normal, specialized nerve cells signal the temperature-regulating center in the hypothalamus (FIG. 39-9). This center sends messages by way of nerves to the sweat glands, which increase sweat secretion. At the same time, the hypothalamus sends messages to smooth muscle in the walls of blood vessels in the skin that cause them to dilate. More blood circulates through the skin, bringing heat to the body surface. The skin acts as a heat radiator that allows heat to radiate from the body surface into the environment. These homeostatic mechanisms help return body temperature to normal.

When body temperature decreases below normal, the hypothalamus signals the anterior pituitary gland to release a hormone that signals the thyroid gland. Secretion of thyroid hormones increases, which raises metabolic rate. Body tissues increase heat production. The hypothalamus also sends neural signals that cause blood vessels in the skin to constrict. As a result, less heat is brought to the body surface. In addition, nerves signal muscles to shiver or let you move muscles voluntarily to increase body temperature.

(a) **Ectotherm.** The rainbow agama (*Agama agama*) increases its body temperature by sunning itself. This insect-eating lizard was photographed in Serengeti National Park, Tanzania.

(b) **Endotherm.** This young emperor penguin (*Aptenodytes forsteri*) is panting to stay cool. Body temperature decreases as heat leaves the body through its open mouth.

Figure 39-8 Behavioral adaptations for thermoregulation

Figure 39-9 *Animation* Regulation of temperature in the human body

The hypothalamus is the integrator that maintains homeostasis of body temperature by activating mechanisms that **(a)** cool or **(b)** warm the body in response to stressors.

© Cengage Learning

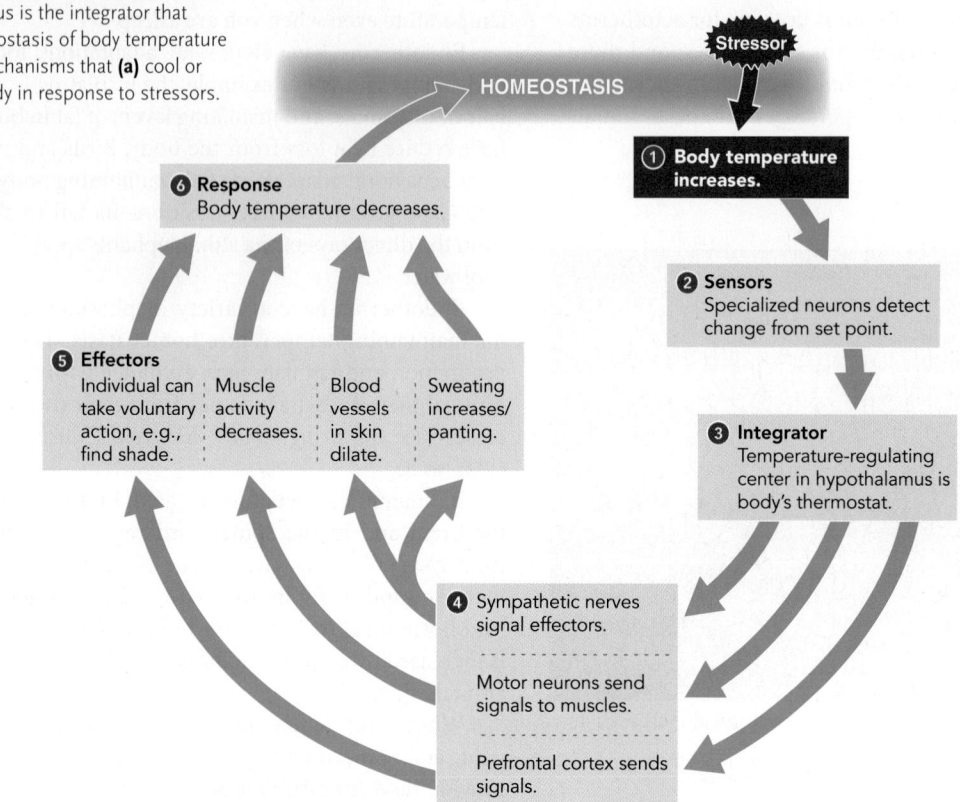

(a) When body temperature increases. Sensors detect any increase in body temperature above its homeostatic range. They send signals to the hypothalamus, which then signals effectors to take corrective action. Temperature decreases, restoring homeostasis.

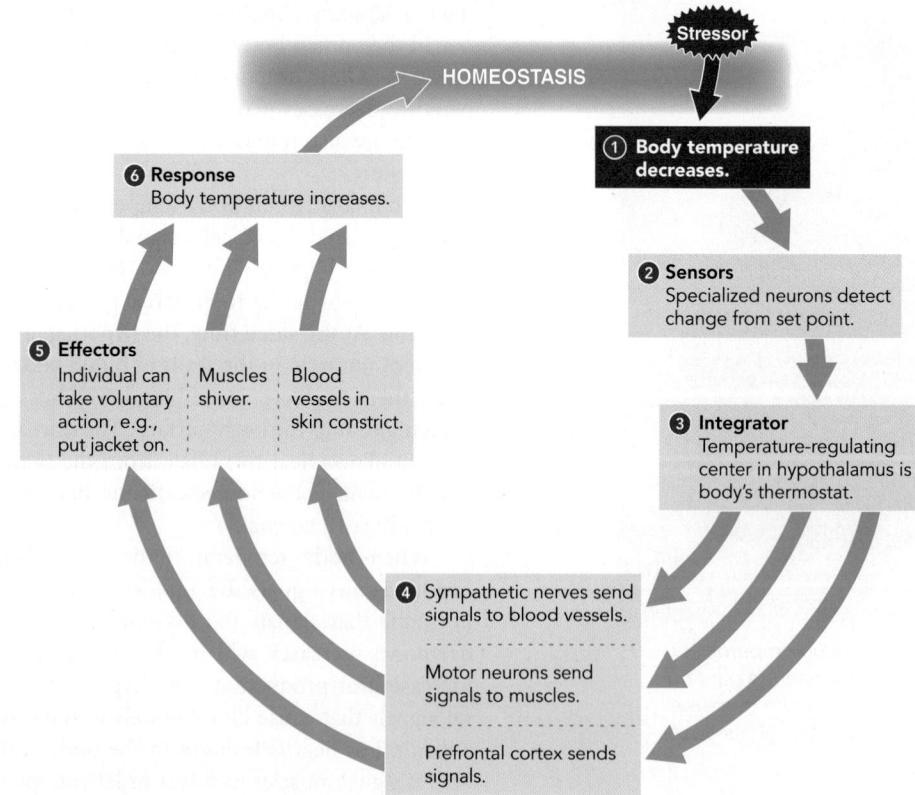

(b) When body temperature decreases. Sensors detect any decrease in body temperature below its homeostatic range. They send signals to the hypothalamus, which then signals effectors to take corrective action. Temperature increases, restoring homeostasis.

Many animals adjust to challenging temperature changes

Animals adjust to seasonal changes, a process called **acclimatization.** A familiar example is the thickening of a dog's coat in the winter. As water temperature decreases during fall and winter, a trout's enzyme systems decrease their level of activity, allowing the trout to remain active at the lowest metabolic cost.

When stressed by cold, many small endotherms sink into **torpor,** a short-term state in which metabolic rate decreases, sometimes dramatically. Torpor saves the energy that the animal would use to maintain a high body temperature. Instead, body temperature decreases below normal levels. The decrease in body temperature is brought about by a decrease in the temperature set point in the hypothalamus. Heart and respiratory rates decrease, and animals are less responsive to external stimulation. Daily torpor occurs in some small endotherms, such as hummingbirds and shrews. We can think of torpor as an adaptive hypothermia that helps animals survive during times of cold temperatures.

Hibernation is long-term torpor in response to winter cold and scarcity of food. Animals that hibernate store unsaturated fats to use as energy sources. **Estivation** is a state of torpor caused by lack of food or water during periods of high temperature. During estivation, some mammals retreat to their burrows. The cactus mouse enters a state of hibernation during the winter in response to cold and scarcity of food. In the summer, it estivates in response to lack of either food or water.

CHECKPOINT 39.3

- *What are some costs and benefits of ectothermy? How do ectotherms adjust body temperature?*
- *What are some costs and benefits of endothermy?*
- **VISUALIZE** *How is body temperature regulated in humans? Draw a diagram to illustrate your answer.*

SUMMARY: FOCUS ON LEARNING OBJECTIVES

39.1 Tissues, Organs, and Organ Systems *(page 816)*

1 Compare the structure and function of the four main kinds of animal tissues: epithelial, connective, muscle, and nervous tissues.
- A **tissue** consists of a group of similarly specialized cells that associate to perform specific functions. **Epithelial tissue** *(epithelium)* forms a continuous layer, or sheet, of cells covering a body surface or lining a body cavity. Epithelial tissue functions in protection, absorption, secretion, and sensation.
- **Connective tissue** consists of relatively few cells separated by an **intercellular substance,** composed of **fibers** scattered throughout a **matrix.** The intercellular substance contains three types of fibers.
- Connective tissue contains specialized cells such as **fibroblasts** and **macrophages.** Connective tissue joins other tissues of the body, supports the body and its organs, and protects underlying organs.
- **Muscle** tissue consists of cells specialized to contract. Each cell is an elongated muscle fiber containing many contractile units called **myofibrils. Nervous tissue** consists of elongated cells called neurons, specialized for transmitting impulses, and **glial cells,** which support and nourish the neurons.

2 Compare the main types of epithelial tissue and describe their functions.
- Epithelial cells may be *squamous, cuboidal,* or *columnar* in shape. Epithelial tissue may be *simple, stratified,* or *pseudostratified* (summarized in Table 39-1).
- *Simple squamous epithelium* lines blood vessels and the air sacs in the lungs; it permits exchange of materials by diffusion. *Simple cuboidal epithelium* and *columnar epithelium* line passageways and are specialized for secretion and absorption. *Stratified squamous epithelium* forms the outer layer of the skin and lines passageways into the body; it provides protection. *Pseudostratified epithelium* also lines passageways and protects underlying tissues.
- Some epithelial tissue is specialized to form **glands. Goblet cells** are unicellular glands that secrete mucus. Goblet cells are **exocrine glands** that secrete their product through a duct onto an exposed epithelial surface. In contrast, **endocrine glands** release hormones into the **interstitial fluid** or blood.
- An *epithelial membrane* consists of a sheet of epithelial tissue and a layer of underlying connective tissue. A **mucous membrane** lines a cavity that opens to the outside of the body. A **serous membrane** lines a body cavity that does not open to the outside.

3 Compare the main types of connective tissue and describe their functions.
- The cells of connective tissue are embedded in an intercellular substance that consists of microscopic *collagen fibers, elastic fibers,* and *reticular fibers* (thin branched fibers) scattered through a **matrix,** a thin gel of polysaccharides. **Loose connective tissue** consists of fibers running in various directions through a semifluid matrix; it forms a covering for nerves, blood vessels, and muscles.
- **Dense connective tissue** is strong but less flexible than loose connective tissue. It forms **tendons,** cords that connect muscles to bones, and **ligaments,** cables that connect bones to one another.
- Elastic connective tissue consists of bundles of parallel elastic fibers; it is found in lung tissue and in walls of large arteries. *Reticular connective tissue,* which consists of interlacing reticular fibers, forms a supporting framework for many organs.
- **Adipose tissue** is made up of fat cells; it is found along with loose connective tissue in subcutaneous tissue.
- **Cartilage** consists of **chondrocytes** that lie in lacunae, small cavities in a hard matrix. **Osteocytes** secrete and maintain the matrix of **bone.** Unlike cartilage, bone is quite vascular. Cartilage and bone form the skeleton of vertebrates.
- **Blood** and **lymph** are circulating tissues that help various parts of an animal communicate with one another. The intercellular substances of blood and lymph are more fluid than those of other tissues.

4 Contrast the three types of muscle tissue and describe their functions.

- **Skeletal muscle** is striated and under voluntary control. Each elongated, cylindrical **muscle fiber** has several nuclei. When skeletal muscles contract, they move parts of the body.

- **Cardiac muscle** is also striated, but its contractions are involuntary. Its elongated, cylindrical fibers branch and fuse; each fiber has one or two central nuclei. When this muscle contracts, the heart pumps the blood.

- **Smooth muscle** contracts involuntarily; its elongated, spindle-shaped fibers lack striations. Each fiber has a single central nucleus. Smooth muscle is responsible for movement of body organs; for example, it pushes food through the digestive tract.

5 Relate the structure of the neuron to its function.

- The elongated **neuron** is adapted for receiving and transmitting information. **Dendrites** receive signals and transmit them to the **cell body.** The **axon** transmits signals away from the cell body to other neurons or to a muscle or gland. A **synapse** is a junction between neurons.

6 Briefly describe the organ systems of a mammal and summarize the functions of each organ system.

- Tissues and **organs** work together, forming **organ systems.** In mammals 11 organ systems work together to carry out the functions required by the **organism:** the **integumentary system, skeletal system, muscular system, nervous system, endocrine system, cardiovascular system, lymphatic system,** which functions as part of the **immune system, respiratory, digestive, urinary,** and **reproductive systems** (summarized in Figure 39-4). Each organ system functions to maintain homeostasis.

39.2 Regulating the Internal Environment *(page 828)*

7 Define *homeostasis* and contrast negative and positive feedback systems.

- **Homeostasis** is the balanced internal environment, or steady state. The control processes that maintain these conditions are **homeostatic mechanisms,** mainly **negative feedback systems.** When a stressor causes a change in some steady state, a response is triggered that counteracts the change.

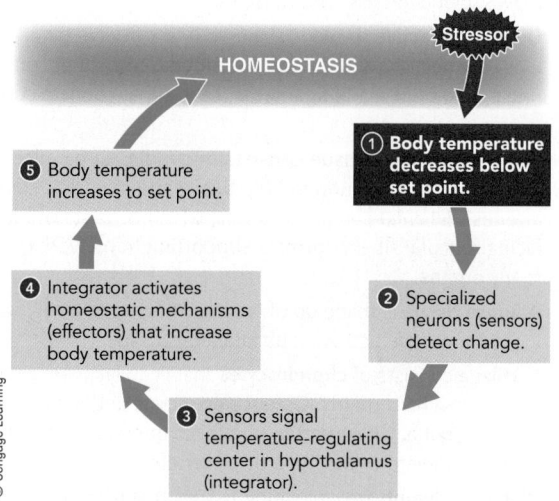

© Cengage Learning

A **sensor** detects the change, a deviation from the normal condition or **set point.** The sensor signals an **integrator,** or control center. Based on the output of the sensor, the integrator activates one or more **effectors,** organs or processes that restore the steady state.

- In a **positive feedback system,** a deviation from the steady state sets off a series of changes that intensify (rather than reverse) the changes.

39.3 Regulating Body Temperature *(page 830)*

8 Compare the costs and benefits of being an ectotherm; and describe strategies ectotherms use to adjust their body temperature.

- Animals have structural, behavioral, and physiological strategies for **thermoregulation,** the process of maintaining body temperature within certain limits despite changes in the surrounding temperature.

- In **ectotherms** body temperature depends to a large extent on the temperature of the environment. A benefit of ectothermy is that very little energy is used to maintain metabolic rate. A disadvantage is that activity may be limited by daily and seasonal temperature conditions. Many ectotherms use behavioral strategies to adjust body temperatures.

9 Compare the costs and benefits of being an endotherm; and describe strategies endotherms use to adjust their body temperature.

- **Endotherms** have homeostatic mechanisms for regulating body temperature within a narrow range. The most important benefit of endothermy is a high metabolic rate. Another advantage is that constant body temperature allows a higher rate of enzyme activity. Many endotherms are active even in low winter temperatures. A disadvantage of endothermy is its high energy cost.

10 Describe strategies animals use to adjust to challenging temperature changes.

- **Acclimatization** is the process of adjustment to seasonal changes.

- When challenged by a drop in surrounding temperature, many small endotherms enter a state of torpor. **Torpor** is an adaptive hypothermia; body temperature is maintained below normal levels. **Hibernation** is long-term torpor in response to winter cold. **Estivation** is torpor caused by lack of food or water during summer heat.

Know and Comprehend

1. Tissue that contains fibroblasts and a great deal of intercellular substance is (a) connective tissue (b) muscle tissue (c) nervous tissue (d) pseudostratified epithelium (e) simple squamous epithelium

2. Tissue that contracts and is striated and involuntary would most likely be found in (a) the leg (b) the wall of the stomach (c) the exocrine glands (d) body structures requiring very rapid contraction (e) the heart

3. Which organ system has the homeostatic function of helping regulate volume and composition of blood and body fluids? (a) integumentary (b) muscular (c) reproductive (d) urinary (e) exocrine

4. Most homeostatic functions in the body are maintained by (a) exocrine glands (b) negative feedback systems (c) set points (d) stressors (e) positive feedback systems

5. Which of the following is *not* true of torpor? (a) it is an adaptive state of low body temperature (b) the temperature set point in the hypothalamus decreases (c) metabolic rate increases (d) it is a strategy used by some endotherms (e) respiratory rate decreases

6. An ectotherm (a) has a higher rate of enzyme activity than a typical endotherm (b) has a variety of homeostatic mechanisms for regulating body temperature (c) depends on sensors in the hypothalamus to regulate temperature (d) may use behavioral strategies to help adjust body temperature (e) must expend more energy on thermoregulation than an endotherm

Apply and Analyze

7. **VISUALIZE** Draw (a) simple cuboidal epithelium lining a kidney tubule and (b) loose connective tissue.

8. **VISUALIZE** Draw a flowchart showing the relationship among organ systems, tissues, cells, and organs.

9. **PREDICT** Jim has lost a great deal of epithelium due to a chemical burn. What effects might this change have on his body and its ability to function?

10. **PREDICT** The intercellular substance of the connective tissue in a dog's skin is decreasing due to a rare disease. What effect might this change have on the dog's body?

11. **PREDICT** A high concentration of carbon dioxide in the blood and interstitial fluid results in more rapid breathing. What effect would rapid breathing have on homeostasis?

Evaluate and Synthesize

12. **EVOLUTION LINK** From an evolutionary perspective, why do you think most animals are ectotherms? (*Hint:* What are some benefits of ectothermy?)

13. **SCIENCE, TECHNOLOGY, AND SOCIETY** Imagine that you are a health care professional. If a patient told you that she had a family history of breast cancer, would you advise her to be genetically tested? Why or why not? What if you had a family history of cancer? Would you want to be tested?

To access course materials, such as Aplia and other companion resources, please visit **www.cengagebrain.com.**

40 Protection, Support, and Movement

Stephen Dalton/Science Source

Grasshopper in midjump. The grasshopper (*Tropidacris dux*) uses powerful muscles in its hind legs to catapult itself into the air. Insect muscles attach to the exoskeleton.

KEY CONCEPTS

40.1 Many structures and processes have evolved in animals for protection, support, and movement. Epithelial coverings are adapted for protecting underlying tissues and may be specialized for gas exchange, sensory reception, secretion, or other functions.

40.2 Skeletal systems, whether they are hydrostatic skeletons, exoskeletons, or endoskeletons, are adapted for supporting and protecting the body and for transmitting mechanical forces important in movement.

40.3 Muscles are adapted for contracting; during contraction, energy from ATP is used to move muscle filaments so that the muscle shortens. Body parts move when muscles pull on them.

I n Chapter 39 we described animal tissues, organs, and organ systems, laying the foundation for the more detailed discussions in this and later chapters. Here we focus on epithelial coverings, skeletons, and muscles, structures that are closely interrelated in the functions of protection, support, and movement. In our survey of the animal kingdom (Chapters 30–32), we described many adaptations involving these structures. For example, we discussed how the arthropod *exoskeleton* contributes to the great biological success of insects. In this chapter we compare these systems in several animal groups and then focus on their structure and function in mammals.

The *epithelial coverings* of animals protect underlying tissues and have several other functions as well. One of the most interesting is secretion of peptides and other substances that kill harmful bacteria and certain other pathogens. Frogs and toads, in particular, secrete a variety of potent peptides that protect them against harmful bacteria, fungi, and protozoa. Secretions of the frog *Rana ridibanda* have been shown to kill "superbugs," such as MRSA (methicillin-resistant *Staphylococcus aureus;* see Chapter 25). The skin of amphibians is a promising potential source of new antibiotics and other drugs.

Most animals are also protected by a *skeletal system*. Whether it is a fluid-filled compartment, a shell, or a system of bones, the skeletal system protects internal organs and provides support for the body. The skeletal system and *muscular system* work together to produce movement. Muscle tissue is specialized to contract. Although a few animals remain rooted to one spot, sweeping their surroundings with tentacles, **locomotion**—movement from place to place in the environment—is characteristic of most animals. Muscles responsible for locomotion anchor to the skeleton, which gives them something firm on which to act. In most vertebrates bones serve as levers that transmit the force necessary to move various parts of the body.

During the course of evolution, most animals have developed adaptations for dependable, rapid, and responsive movement. Animals creep, walk, run, jump, swim, or fly (see photograph). These adaptations enable them to find food and escape from predators. In complex animals effective movement depends on the cooperation and interactions of several body systems, including the muscular, skeletal, nervous, circulatory, respiratory, and endocrine systems. The precisely regulated interactions among tissues, organs, and organ systems that make complex function and maintenance of homeostasis possible are a dynamic example of systems biology in action.

40.1 EPITHELIAL COVERINGS

LEARNING OBJECTIVES

1 Compare the functions of the epithelial coverings of invertebrates and vertebrates.
2 Relate the structure of vertebrate skin to its functions.

Epithelial tissue covers all external and internal surfaces of the animal body. The structure and functions of the external epithelial covering are adapted to the animal's environment and lifestyle. In both invertebrates and vertebrates, epithelial coverings protect the body. Epithelial coverings may also be specialized for secretion, gas exchange, or temperature regulation and typically contain receptors that receive sensory signals from the environment.

Invertebrate epithelium may secrete a cuticle

In many invertebrates the outer epithelium is specialized to secrete protective or supportive layers of nonliving material. In insects and many other animals, the secreted material forms an outer covering called a **cuticle.** Some animals, including corals and mollusks, secrete a shell made mainly of calcium carbonate.

In many species the epithelium contains cells that secrete lubricants or adhesives. Such cells may release odorous secretions used for communication or for marking trails. In other species secretory cells produce toxins used for offense or defense. The earthworm epithelium secretes a lubricating mucus that promotes efficient diffusion of gases across the body wall and reduces friction during movement through the soil. Epithelial cells may be modified as sensory cells that are selectively sensitive to light; chemical stimuli; or mechanical stimuli, such as contact with an object, or pressure.

Vertebrate skin functions in protection and temperature regulation

The **integumentary system** of vertebrates includes the skin and structures that develop from it. In many fishes, in some reptiles, and in the ant-eating pangolin (a mammal; see Fig. 18-15c), skin has developed into a set of scales formidable enough to be considered armor. Even human skin has considerable strength. The thin, moist skin of many amphibians is adapted for gas exchange as well as for defense.

Derivatives of skin differ considerably among vertebrates. Fishes have bony or toothlike scales. Amphibians have naked skin covered with mucus, and some species are equipped with poison glands. Reptiles have epidermal scales, mammals have hair, and birds have feathers that provide even more effective insulation than fur. The feathers of birds and the fur of mammals are insulating structures that help maintain body temperature by trapping heat in the body. Skin and its derivatives are often brilliantly colored in connection with courtship rituals, territorial displays, and other kinds of communication.

In mammals structures derived from skin include claws (modified as fingernails and toenails in primates), hair, sweat glands, oil (sebaceous) glands, and several types of sensory receptors that give mammals the ability to feel pressure, temperature, and pain. The skin of mammals also contains mammary glands, specialized in females for secretion of milk.

In humans and other mammals, oil glands empty via short ducts into hair follicles. A *hair follicle* is the part of a hair below the skin surface, together with the surrounding epithelial tissue. Oil glands secrete *sebum,* a mixture of fats and waxes that inhibits the growth of harmful bacteria.

Although the skin of fishes, frogs, snakes, pelicans, and elephants appears very different, the basic structure of skin is the same in all vertebrates. The outer layer of skin, the **epidermis,** is a waterproof protective barrier. The epidermis consists of several strata, or sublayers. The deepest is **stratum basale,** and the most superficial is **stratum corneum** (FIG. 40-1). Pigment cells in stratum basale and in the dermis below produce **melanin,** a pigment that contributes to the color of the skin. In stratum basale epithelial cells continuously divide and are pushed outward as other cells are produced below them. Epidermal cells mature as they move toward the skin surface.

As they move toward the body surface, epidermal cells manufacture **keratin,** an elaborately coiled protein that gives the skin considerable mechanical strength and flexibility. Keratin is insoluble and serves as a diffusion barrier for the body surface. As epidermal cells move through stratum corneum, they die. When they reach the outer surface of the skin, they wear off and must be continuously replaced.

Beneath the epidermis lies the **dermis,** a dense, fibrous connective tissue made up mainly of collagen fibers (see Fig. 40-1). Collagen imparts strength and flexibility to the skin. The dermis also contains blood vessels that nourish the skin and sensory receptors for touch, pain, and temperature. Hair follicles and, in most mammals, sweat glands are embedded in the dermis. In birds and mammals, the dermis rests on a layer of *subcutaneous tissue.* This tissue consists mainly of adipose tissue that insulates the body from outside temperature extremes.

In humans exposure to ultraviolet (UV) radiation—the short, invisible rays from the sun—causes the epidermis to thicken and stimulates pigment cells in the skin to produce melanin at an increased rate. Melanin is an important protective screen against the sun because it absorbs some of the harmful UV rays. An increase in melanin causes the skin to darken. The suntan so prized by sun worshippers is actually a sign that the skin has been exposed to too much UV radiation. When the melanin cannot absorb all the UV rays, the skin becomes inflamed, or sunburned. Dark-skinned people have more melanin and so suffer less sunburn, wrinkling, and skin cancer than lighter-skinned people. However, they still require protection from UV rays.

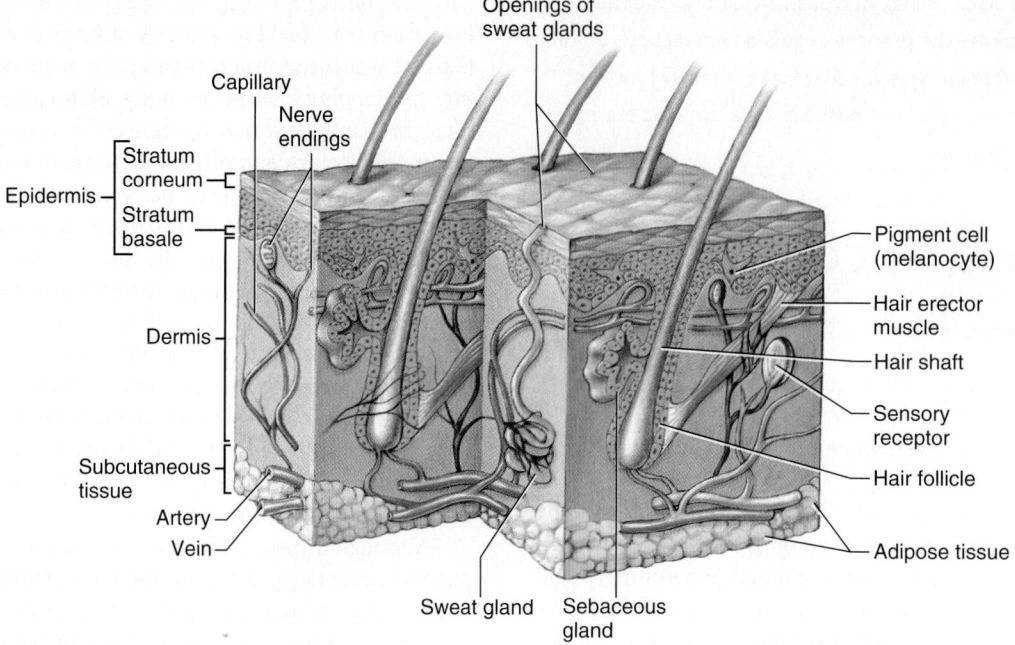

Figure 40-1 *Animation* **Human integumentary system**
The integumentary system of humans and other mammals is a complex organ system consisting of skin and structures, such as hair and nails, that develop from it. The outer layer of the skin, the epidermis, consists of several sublayers, or strata.
© Cengage Learning

UV radiation damages DNA, causing mutations that lead to malignant transformation of cells (see Chapter 17). Skin cancer is the most common type of cancer in humans, and most cases of skin cancer are caused by excessive, chronic exposure to UV radiation. The incidence of *malignant melanoma,* the most lethal type of skin cancer, is increasing faster than any other type of cancer. Greater understanding of the Ras/ERK cell signaling pathway has led to research aimed at developing targeted drugs for treating melanoma. (Recall from Chapter 6 that this pathway is the main signaling cascade for cell division and differentiation. Also see Figure 17-20 in Chapter 17.)

CHECKPOINT 40.1

- **CONNECT** *How are the body coverings of invertebrates and vertebrates alike?*
- **CONNECT** *How is the structure of vertebrate skin related to its specific functions?*
- *What properties does keratin give human skin?*
- *What is the function of melanin?*

40.2 SKELETAL SYSTEMS

LEARNING OBJECTIVES

3 Compare the structure and functions of different types of skeletal systems, including the hydrostatic skeleton, exoskeleton, and endoskeleton.

4 Describe the main divisions of the vertebrate skeleton and identify the bones that make up each division.

5 Describe (or draw and label) the structure of a typical long bone and contrast the roles of osteoblasts and osteoclasts in bone remodeling.

6 Compare the main types of vertebrate joints.

In addition to having an epithelial covering, many animals are protected by a **skeletal system** that forms the framework of the body. The skeleton also functions in locomotion. Muscles typically act on hard structures such as chitin or bone. These skeletal structures *transmit* mechanical forces generated by muscle contraction and *transform* them into the variety of motions that animals make. Most invertebrates have *hydrostatic skeletons* or *exoskeletons.* (You may recall from Chapter 32 that echinoderms are an exception.) Vertebrates have *endoskeletons* of cartilage or bone.

In hydrostatic skeletons body fluids transmit force

Many soft-bodied invertebrates have **hydrostatic skeletons** made of fluid-filled body compartments. Imagine an elongated balloon full of water. If you pull on the balloon, it lengthens and becomes thinner. It does the same if you squeeze it. Conversely, if you push the ends toward the center, the balloon shortens and thickens. Many invertebrates, including most cnidarians, flatworms, annelids, and roundworms, have hydrostatic skeletons that work something like a balloon filled with water. Fluid in a closed compartment of the body is held under pressure. When muscles in the compartment wall contract, they push against the tube of fluid. Because fluids cannot be compressed, the force is transmitted through the fluid, changing the shape and movement of the body.

In *Hydra* and other cnidarians, cells of the two body layers can contract. The contractile cells in the outer epidermal layer lie longitudinally, whereas the contractile cells of the inner layer (the gastrodermis) are arranged circularly around the central body axis. The two groups of cells work in *antagonistic* fashion: what one can do, the other can undo. When the inner (circular) layer contracts, the hydra thins, and its fluid contents force it to lengthen (FIG. 40-2a). In contrast, when the epidermal (longitudinal) layer contracts, the hydra shortens. Due to the fluid in the gastrovascular cavity, force is transmitted so that the hydra thickens as well (FIG. 40-2b).

Mechanically, the hydra is a bag of fluid. The fluid acts as a hydrostatic skeleton because it transmits force when the contractile cells contract against it. (Although technically not a closed compartment, the gastrovascular cavity functions as a hydrostatic skeleton because its opening is small.) Hydrostatic skeletons permit only crude mass movements of the body or its appendages. Delicate movements are difficult because force tends to be transmitted equally in all directions throughout the entire fluid-filled body of the animal. For example, it is not easy for the hydra to thicken one part of its body while thinning another.

The more sophisticated hydrostatic skeleton of the annelid worm lets it move with increased flexibility. An earthworm's body consists of a series of segments divided by transverse partitions, or septa (see Fig. 31-19). The septa isolate portions of the body cavity and the coelomic fluid they contain. This arrangement allows the hydrostatic skeleton of each segment to be largely

independent. As a result, contraction of the circular muscle in the elongating anterior end does not interfere with the action of the longitudinal muscle in the segments at the posterior end.

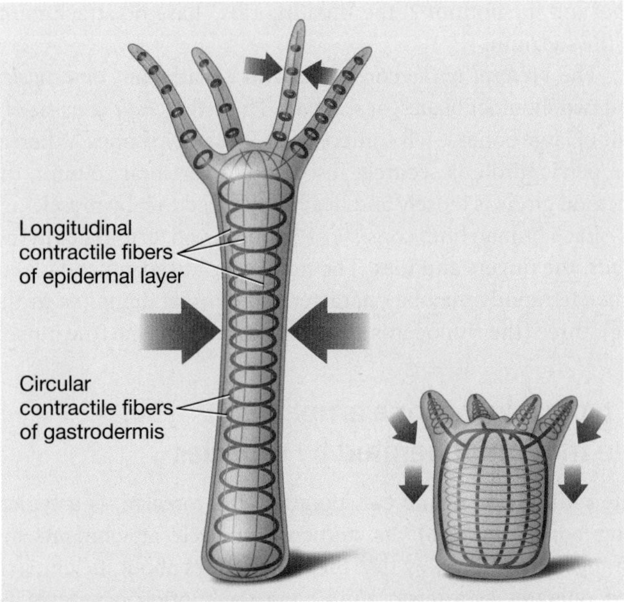

You can find some examples of hydrostatic skeletons in invertebrates equipped with shells or endoskeletons and even in vertebrates with endoskeletons of cartilage or bone. For example, sea stars and sea urchins have an endoskeleton, but they move their tube feet by an ingenious adaptation of a hydrostatic skeleton (see Chapter 32). And the human penis becomes erect and stiff because of the turgidity of pressurized blood in its internal spaces.

Mollusks and arthropods have nonliving exoskeletons

In most animals the skeleton is a lifeless shell, or **exoskeleton,** deposited atop the outer epithelial covering. In mollusks the exoskeleton is a calcium carbonate shell secreted by the mantle, a thin sheet of epithelial tissue that extends from the body wall. The exoskeleton provides protection, a retreat used in emergencies, with the bulk of the tasty, naked body exposed at other times.

Exoskeletons of arthropods serve not only to protect but also to transmit forces. In this respect they are comparable to the skeletons of vertebrates. The arthropod exoskeleton is a nonliving *cuticle* that contains the polysaccharide **chitin.** This exoskeleton consists of large, thick, inflexible plates that completely cover the body. These plates are separated from one another by thin, flexible joints arranged segmentally. Enough joints are present to make the arthropod's body as flexible as those of many vertebrates. The arthropod exoskeleton is adapted to a vast variety of lifestyles, and parts of it are modified to function as specialized tools or weapons.

A disadvantage of the rigid arthropod exoskeleton is that to accommodate growth an arthropod must shed its exoskeleton and replace it with a new, larger one (FIG. 40-3). Recall from Chapter 31 that this process, called **molting,** or **ecdysis,** is characteristic of the **Ecdysozoa,** a major branch of invertebrates. During ecdysis, the animal is weak and vulnerable to predators.

Internal skeletons are capable of growth

Echinoderms and chordates have internal skeletons, or **endoskeletons.** Composed of living tissue, the endoskeleton grows along with the animal as a whole. This skeleton consists of plates or shafts of calcium-impregnated tissue (such as cartilage or bone). The echinoderm endoskeleton consists of spines and plates of calcium salts embedded in the body wall, beneath an epidermis that covers the body. This endoskeleton forms what amounts to an internal shell that provides support and protection (FIG. 40-4). Many echinoderm endoskeletons bear spines that project to the outer surface.

You are probably most familiar with the endoskeleton of vertebrates. This endoskeleton provides support and protection and transmits muscle forces. Members of class Chondrichthyes (sharks and rays) have skeletons of cartilage, but in most vertebrates the skeleton consists mainly of bone. Many bones form

(a) Contraction of circular contractile fibers (*red*) elongates the body.

(b) Contraction of longitudinal fibers (*red*) shortens the body.

Figure 40-2 *Animation* **Hydrostatic skeleton**

In *Hydra* fluid in the gastrovascular cavity transmits force when contractile cells in the body wall contract against it. The longitudinally arranged contractile cells of *Hydra* are antagonistic to the cells arranged circularly around the body axis.
© Cengage Learning

Longitudinal contractile fibers of epidermal layer

Circular contractile fibers of gastrodermis

Figure 40-3 **Ecdysis**

A greengrocer cicada (*Magicicada*) requires 13 years to mature. It then emerges from the soil, climbs a tree, and sheds its exoskeleton prior to reproducing.

systems of levers that transmit muscle forces. Bones store calcium and are important in maintaining homeostatic levels of calcium in the blood.

The vertebrate skeleton has two main divisions

The two main divisions of the vertebrate skeleton are the axial and appendicular skeletons (FIG. 40-5). The *axial skeleton,* located along the central axis of the body, consists of the skull, vertebral column, ribs, and sternum (breastbone). The *appendicular skeleton* consists of the bones of the limbs (arms and legs) plus the bones making up the girdles that connect the limbs to the axial skeleton, the pectoral (shoulder) girdle, and most of the pelvic (hip) girdle.

The *skull,* the bony framework of the head, consists of the cranial and facial bones. In the human skull, 8 cranial bones enclose the brain, and 14 bones make up the facial portion of the skull. Several cranial bones that are single in the adult human result from the fusion of 2 or more bones that are separate in the fetus or newborn.

The vertebrate spine, or **vertebral column,** supports the body and bears its weight. In humans it consists of 24 **vertebrae** and 2 bones composed of fused vertebrae: the *sacrum* and *coccyx.* The vertebral column consists of the *cervical* (neck) region, with 7 vertebrae; the *thoracic* (chest) region, with 12 vertebrae; the *lumbar* (back) region, with 5 vertebrae; the *sacral* (pelvic) region,

Figure 40-4 **The echinoderm endoskeleton**

As in other echinoderms, the endoskeleton of the sunflower sea star (*Pycnopodia helianthoides*) is composed of spines and plates of nonliving calcium salts embedded in the body wall. The endoskeleton provides support and protection.

with 5 fused vertebrae; and the *coccygeal* region, also composed of fused vertebrae.

The *rib cage* is a bony "basket" formed by the *sternum* (breastbone), thoracic vertebrae, and, in mammals, 12 pairs of ribs. The rib cage protects the internal organs of the chest, including the heart and lungs. It also supports the chest wall, preventing it from collapsing as the diaphragm contracts with each breath. Each pair of ribs is attached dorsally to a separate vertebra. Of your 12 pairs of ribs, the first 7 are attached ventrally to the sternum; the next 3 are attached indirectly by cartilage; and the bottom 2, the "floating ribs," have no attachments to the sternum.

The *pectoral girdle* consists of two collarbones, or *clavicles,* and two shoulder blades, or *scapulas.* The *pelvic girdle* consists of a pair of large bones, each composed of 3 fused hip bones. Whereas the pelvic girdle is securely fused to the vertebral column, the pectoral girdle is loosely and flexibly attached to it by muscles.

Each human limb consists of 30 bones and terminates in five *digits,* the fingers and toes. The more specialized appendages of other tetrapods may be characterized by four digits (as in the pig), three (the rhinoceros), two (the camel), or one (the horse).

A typical long bone amplifies the motion generated by muscles

The radius, one of the two bones of the forearm, is a typical long bone (FIG. 40-6). Its numerous muscle attachments are arranged in such a way that the bone rotates about its long axis and operates as a lever, amplifying the motion generated by the muscles. By themselves, muscles cannot shorten enough to produce large movements of the body parts to which they are attached.

Like other bones, the radius is covered by a connective tissue membrane, the *periosteum,* to which muscle tendons and ligaments attach. The periosteum can produce new layers of bone, thus increasing the bone's diameter. The main shaft of a long bone is its *diaphysis,* and each expanded end is an *epiphysis.*

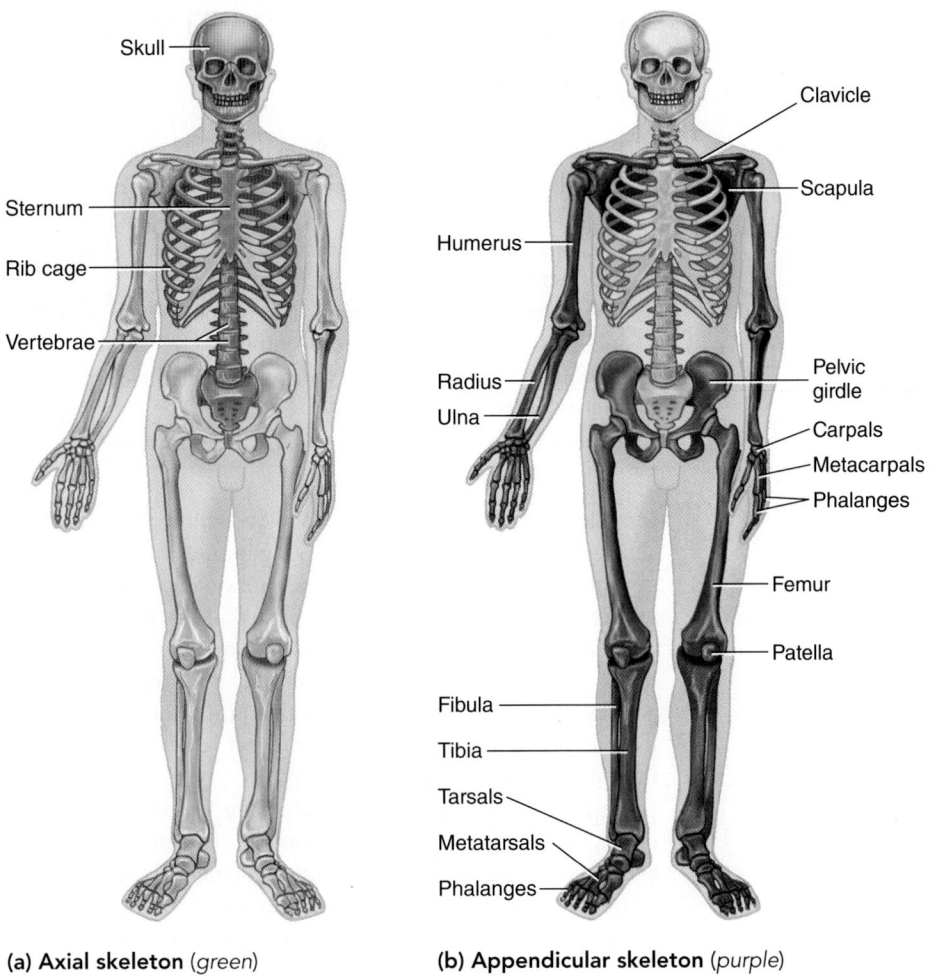

Skull

Sternum

Rib cage

Vertebrae

Clavicle

Scapula

Humerus

Radius

Ulna

Pelvic girdle

Carpals

Metacarpals

Phalanges

Femur

Patella

Fibula

Tibia

Tarsals

Metatarsals

Phalanges

(a) Axial skeleton (*green*)

(b) Appendicular skeleton (*purple*)

Figure 40-5 *Animation* **The human skeletal system**

Only a few of the 206 named bones are labeled here.

© Cengage Learning

Bones are remodeled throughout life

Bones are modeled during growth and remodeled continuously throughout life in response to physical stresses and other changing demands. As muscles develop in response to physical activity, the bones to which they are attached thicken and become stronger. **Osteoblasts** are bone-building cells. They secrete the protein collagen, which forms the strong fibers of bone. The compound hydroxyapatite, composed mainly of calcium phosphate, is present in the interstitial fluid (tissue fluid). It automatically crystallizes around the collagen fibers to form the hard matrix of bone. As the matrix forms around the osteoblasts, they become isolated within the lacunae. The trapped osteoblasts are then called **osteocytes.**

Osteoclasts are large, multinucleated cells that resorb (break down) bone. The osteoclasts move about, secreting hydrogen ions that dissolve the crystals and enzymes that digest the collagen. Osteoclasts and osteoblasts are synergistic; together they shape bones. The remodeling process is an extensive process of breaking down and rebuilding bone. The adult human skeleton is completely replaced every 10 years!

In *osteoporosis,* the most common progressive and degenerative bone disease, bone is broken down more rapidly than it is rebuilt. In women the decrease in estrogen after menopause accelerates bone resorption. Osteoporosis patients lose so much bone mass that their bones become fragile, which greatly increases their risk for fracture. Most drugs used to treat osteoporosis inhibit osteoclast activity, thus slowing the bone breakdown process. Medications being developed stimulate osteoblasts to promote bone formation.

Joints are junctions between bones

Joints, or articulations, are junctions between two or more bones. Joints facilitate flexibility and movement. At the joint, the outer surface of each bone consists of *articular cartilage.* One way to classify joints is according to the degree of movement they allow. The *sutures* between bones of the human skull are *immovable joints.* In a suture bones are held together by a thin layer of dense fibrous connective tissue, which may be replaced by bone in the adult. *Slightly movable* joints, found between vertebrae, are made of cartilage. These joints help absorb shock.

Most joints are *freely movable joints.* Each is enclosed by a joint capsule of connective tissue and lined with a membrane

In children a disc of cartilage, the *metaphysis,* lies between the epiphyses and the diaphysis. Metaphyses are growth centers that disappear at maturity, becoming vague *epiphyseal lines.* Long bones have a central cavity that contains **bone marrow.** Yellow marrow consists mainly of a fatty connective tissue; the red marrow in certain bones produces blood cells.

The radius has a thin outer shell of **compact bone,** which is very dense and hard. Compact bone lies primarily near the surfaces of a bone, where it provides great strength. Recall from Chapter 39 that compact bone consists of interlocking spindle-shaped units called **osteons** (see Fig. 39-2). Within an osteon, **osteocytes** (bone cells) lie in small cavities called *lacunae* (sing., *lacuna*). The lacunae are arranged in concentric circles around central **Haversian canals.** Blood vessels that nourish the bone tissue pass through the Haversian canals. Osteocytes are connected by threadlike extensions of their cytoplasm that extend through narrow channels (called *canaliculi*).

Interior to the thin shell of compact bone is a filling of **spongy bone** that provides mechanical strength. Spongy bone consists of a network of thin strands of bone. Its spaces are filled with bone marrow.

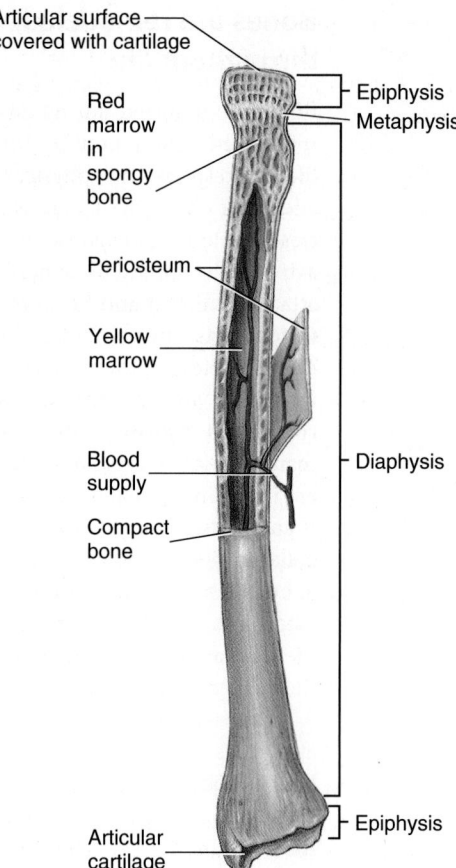

Articular surface covered with cartilage

Red marrow in spongy bone

Periosteum

Yellow marrow

Blood supply

Compact bone

Articular cartilage

Epiphysis
Metaphysis

Diaphysis

Epiphysis

Figure 40-6 A typical long bone

A long bone has a thin shell of compact bone and a filling of spongy bone that contains marrow.

© Cengage Learning

that secretes a lubricant called *synovial fluid.* This viscous fluid reduces friction during movement and absorbs shock. The joint capsule is typically reinforced by **ligaments,** bands of fibrous connective tissue that connect bones and limit movement at the joint.

With time and use, joints wear down. In *osteoarthritis,* a group of common joint disorders, cartilage repair does not keep up with degeneration, and the articular cartilage wears out. Inflammation of the joint capsule may occur, and bone outgrowths, or spurs, may develop. In *rheumatoid arthritis,* an autoimmune disease, the synovial membrane thickens and becomes inflamed. Synovial fluid accumulates, causing pressure, pain, stiffness, and progressive deformity, leading to loss of function.

CHECKPOINT 40.2

- **CONNECT** *How do the septa in the annelid worm contribute to the effectiveness of its hydrostatic skeleton?*
- *What are some advantages of an exoskeleton? some disadvantages?*
- *What are the main bones in each division of the human skeleton?*
- *How do osteoblasts and osteoclasts remodel bone?*

40.3 MUSCLE CONTRACTION

LEARNING OBJECTIVES

7 Relate the structure of insect flight muscles to their function.
8 Describe the structure of skeletal muscles and their antagonistic actions.
9 List, in sequence, the events that take place during muscle contraction.
10 Compare the roles of glycogen, creatine phosphate, and ATP in providing energy for muscle contraction.
11 Compare the structures and functions of the three types of skeletal muscle fibers.
12 Describe factors that influence muscle contraction.

Most animals have the ability to move from place to place. Throughout most of the animal kingdom, a *muscular system* generates the mechanical forces and motion necessary for locomotion and for manipulation of objects. Muscle also powers many physiological actions necessary to maintain homeostasis. Many animals have digestive systems that push food along with peristaltic contractions and hearts that pump internal circulating fluids. Some animals also have blood vessels that maintain their pressure with gentle squeezing.

Animals with very simple body plans do not have muscle tissue, but all eukaryotic cells do contain the contractile protein **actin.** The major component of microfilaments (see Fig. 4-27), actin is important in many cell processes, including amoeboid movement and attachment of cells to surfaces. In most cells actin is functionally associated with the contractile protein **myosin.** Actin and myosin are most highly organized in muscle fibers.

Invertebrate muscle varies among groups

Earlier in this chapter, we discussed the contractile cells of the hydrostatic skeleton of *Hydra* and other cnidarians. In most other animal groups, muscle is a specialized tissue organized into definite layers, or straplike bands. Some invertebrate phyla have skeletal and smooth muscle. Bivalve mollusks, such as clams, have two sets of muscles for opening and closing the shell. Their specialized smooth muscle, which is capable of slow, sustained contraction at low energy cost, keeps the two shells tightly closed for long periods, even weeks at a time. Striated muscle, which contracts rapidly, is used to swim and to shut the shell quickly when the mollusk is threatened. Arthropod muscles are striated, even in the walls of the digestive tract.

Insect flight muscles are adapted for rapid contraction

Insects were the first animals in which flight evolved, an adaptation that has contributed to their impressive biological success (see Chapter 31). Just how insects fly has been an aerodynamic mystery, and scientists are working to understand their remarkable ability to maneuver. Insect flight muscles contract more rapidly than any other known muscle, up to 1000 contractions per second! Not surprisingly, insect flight muscles in action have the highest known metabolic rate of any muscle tissue.

As you might imagine, insect flight muscles are structurally well adapted to their function. They contain more mitochondria than any other known variety of muscle. Their mitochondria also have more cristae (folds of their inner membranes) and can consume oxygen twice as fast as mammalian mitochondria. Insect muscles are also elaborately infiltrated with *tracheae,* tiny air-filled tubes that carry oxygen directly to each muscle fiber.

Many insects must warm up before they fly because an increase in body temperature increases the rate of ATP synthesis. Butterflies and some other insects bask in the sun before they fly. Honeybees "shiver" their muscles to increase body temperature.

In the common blowfly, the wings may beat at 120 cycles *per second,* yet in the same blowfly, the *motor neurons* that innervate those furiously contracting flight muscles are delivering impulses to them at the astonishingly low frequency of three per second. In fact, about 75% of flying insect species beat their wings too rapidly for each contraction to be controlled by a signal from a motor neuron. Instead, they have **asynchronous,** or indirect, muscle contractions in which muscle contraction is not synchronized with signals from motor neurons.

In insects with asynchronous muscle contractions, the striated flight muscles do not attach directly to the wings; rather, they attach to the flexible portions of the exoskeleton that articulate with the wings. The mechanical properties of this musculoskeletal arrangement provide the stimuli for contraction by stretching the muscle fibers at a high frequency. Each contraction of the muscles "dimples" the exoskeleton in association with a downstroke and sometimes, depending on the exact arrangement of the muscles, on the upstroke as well. When the dimple springs back into resting position, the muscles attached to it stretch. The stretching immediately initiates another contraction, and the cycle repeats. Thus, the deformation of the exoskeleton is transmitted as a force to the wings, which beat so quickly that we may perceive the sounds as a musical tone. However, nerve impulses are needed to maintain contraction.

Somehow, insects create lift that is 20 times or more their body weight. A central question has been how their flapping wings generate enough force to keep them airborne. Insect flight involves much more than just flapping the wings up and down. The flapping motion of the insect wing changes direction and speed, and upstrokes alternate with downstrokes at very high rates. At each shift of stroke, the wing rotates about its long axis and tilts to just the correct angle for the new direction of motion.

Vertebrate skeletal muscles act antagonistically to one another

Skeletal muscles produce movements by pulling on **tendons,** tough cords of connective tissue that anchor muscles to bone. Tendons then pull on bones. Skeletal muscles, or their tendons, pass across a joint and are attached to the bones on each side of the joint. When the muscle contracts, it pulls one bone toward or away from the bone with which it articulates.

Because muscles can only contract, they can only pull; they cannot push. Muscles act **antagonistically** to one another, which means that the movement produced by one can be reversed by another. For example, contraction of the biceps muscle flexes (bends) your arm, whereas contraction of the triceps muscle extends it (**FIG. 40-7**).

The muscle that contracts to produce a particular action is known as the *agonist.* The muscle that produces the opposite movement is the *antagonist.* When the agonist is contracting, the antagonist is relaxed. Generally, movements are accomplished by groups of muscles working together, so several agonists and several antagonists may take part in any action. Note that muscles that are agonists in one movement may serve as antagonists in another. Some of the superficial skeletal muscles of the human body are shown in **FIGURE 40-8.**

A vertebrate muscle may consist of thousands of muscle fibers

Skeletal muscle is the most abundant tissue in the vertebrate body. Its elongated cells, called **muscle fibers,** are organized in bundles wrapped by connective tissue. The biceps in your arm, for example, consists of thousands of individual muscle fibers and their connective tissue coverings.

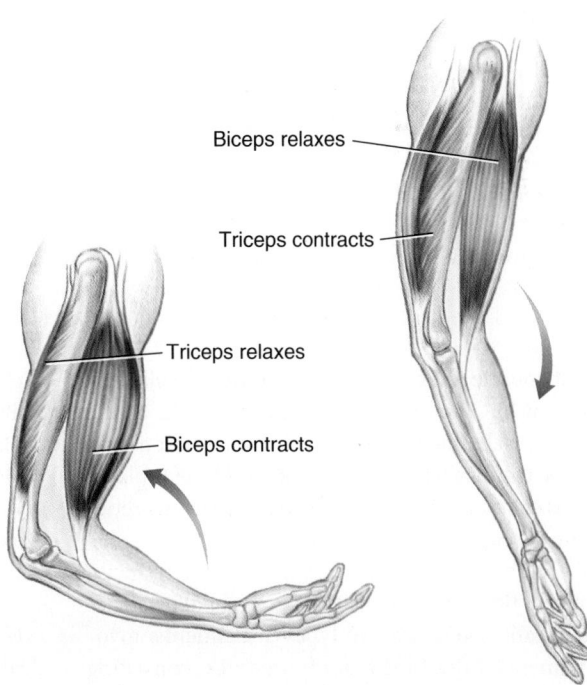

Biceps relaxes

Triceps contracts

Triceps relaxes

Biceps contracts

(a) Flexion. When the biceps muscle contracts, the arm flexes (bends).

(b) Extension. When the triceps muscle contracts, the arm extends.

Figure 40-7 *Animation* **Muscle action**

The biceps and triceps muscles function antagonistically.

© Cengage Learning

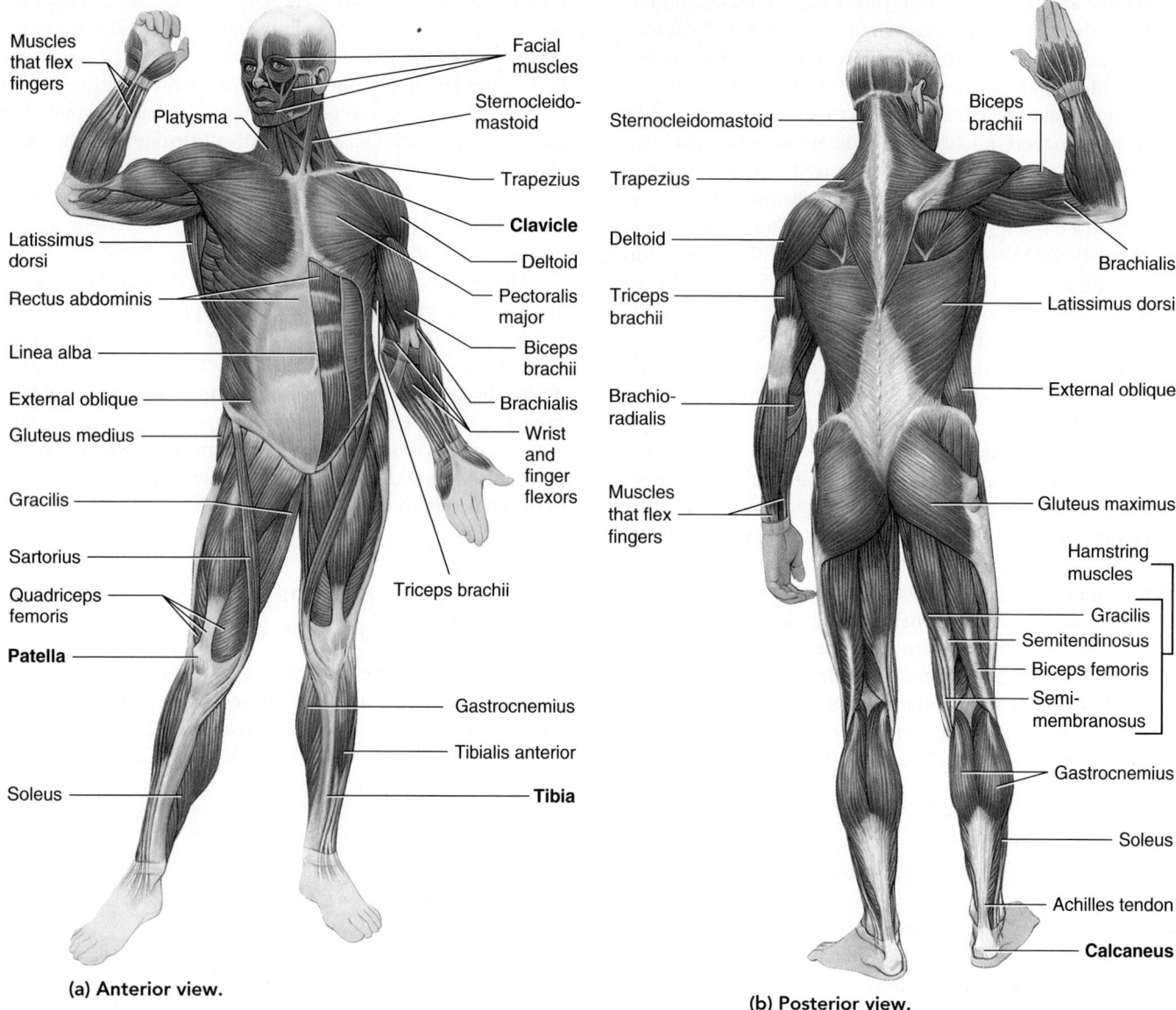

Figure 40-8 Some superficial muscles of the human body
The labels in boldface are bones.
© Cengage Learning

Each striated muscle fiber is a long, cylindrical cell with many nuclei (**FIG. 40-9**). The plasma membrane, known as the **sarcolemma** in a muscle fiber, has multiple inward extensions that form a set of **T tubules** (transverse tubules). The cytoplasm of a muscle fiber is called *sarcoplasm,* and the endoplasmic reticulum is called the **sarcoplasmic reticulum.**

Myofibrils are threadlike structures that run lengthwise through the muscle fiber. They consist of even smaller structures called **filaments.** There are two types of filaments: myosin and actin filaments. **Myosin filaments** are thick, consisting mainly of the protein myosin. The thin **actin filaments** consist mostly of the protein actin; they also contain the proteins **tropomyosin** and **troponin,** which regulate the actin filament's interaction with myosin filaments.

Myosin and actin filaments are organized into repeating units called **sarcomeres,** the basic units of muscle contraction.

Hundreds of sarcomeres connected end to end make up a myofibril. Sarcomeres are joined at their ends by an interweaving of filaments called the *Z line.* Each sarcomere consists of overlapping myosin and actin filaments. The filaments overlap lengthwise in the muscle fibers, producing the pattern of transverse bands or striations characteristic of striated muscle (**FIG. 40-10**; also see Fig. 40-9).

The bands in striated muscle are designated by the letters *A, H,* and *I.* The *I band* consists of parts of actin filaments of two adjacent sarcomeres. The *A band* is the wide, dark region that includes overlapping myosin and actin filaments. Within the A band is a narrow, light area, the *H zone,* made up exclusively of myosin filaments; the actin filaments do not extend into this region. A system of proteins holds each stack of myosin filaments together. These proteins are visible as a thin line, the *M line,* that extends down the center of the H zone within the A band.

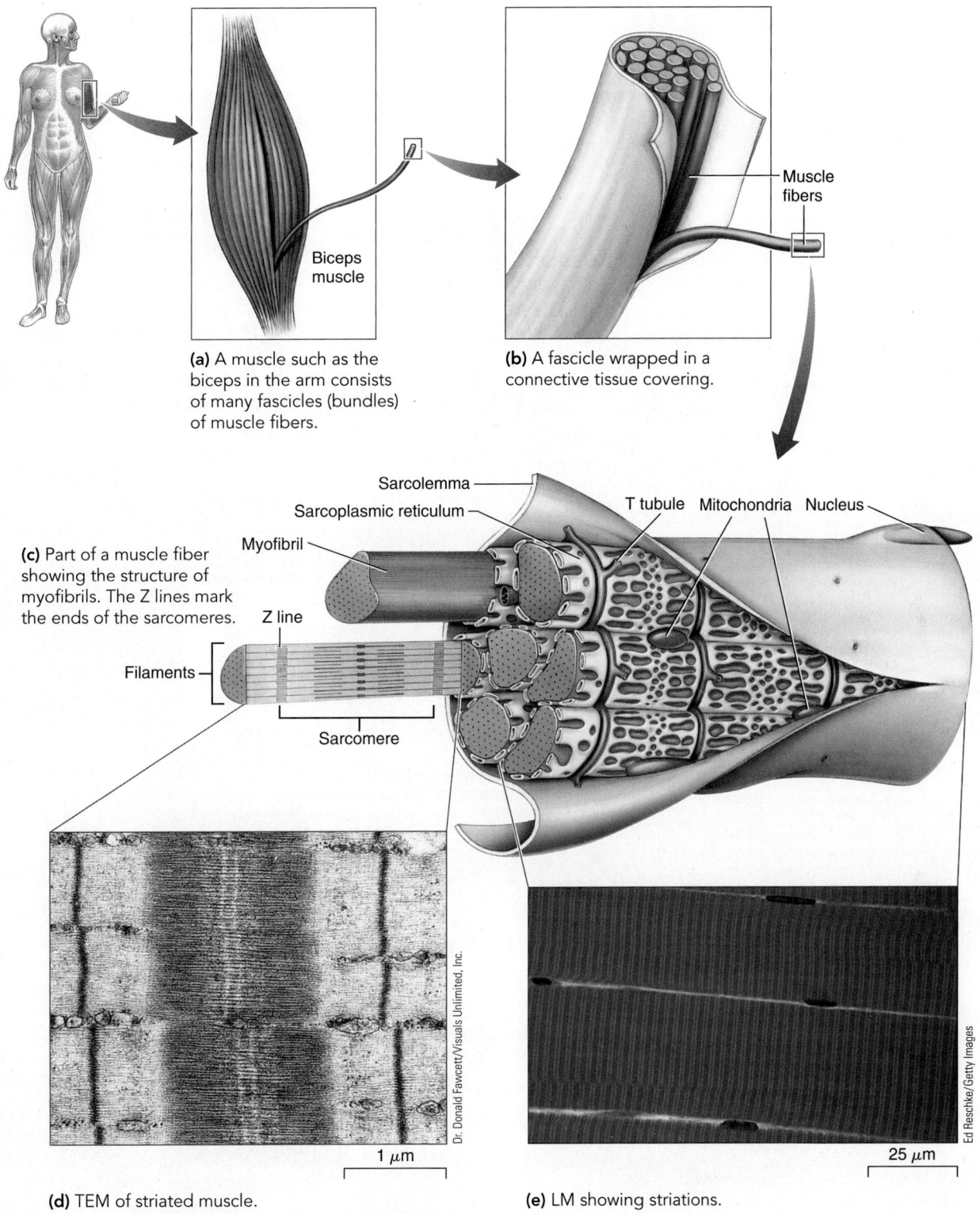

(a) A muscle such as the biceps in the arm consists of many fascicles (bundles) of muscle fibers.

(b) A fascicle wrapped in a connective tissue covering.

Muscle fibers

Sarcolemma
Sarcoplasmic reticulum
Myofibril
T tubule   Mitochondria   Nucleus

(c) Part of a muscle fiber showing the structure of myofibrils. The Z lines mark the ends of the sarcomeres.

Z line
Filaments
Sarcomere

Dr. Donald Fawcett/Visuals Unlimited, Inc.

1 μm

(d) TEM of striated muscle.

Ed Reschke/Getty Images

25 μm

(e) LM showing striations.

Figure 40-9 *Animation* **Muscle structure**

A skeletal muscle fiber contains threadlike myofibrils that consist of actin and myosin filaments. These filaments are organized into repeating units called sarcomeres. The regular pattern of overlapping filaments gives skeletal and cardiac muscle their striated appearance.

© Cengage Learning

A muscle contracts when actin and myosin filaments move past one another. The length of each filament remains the same.

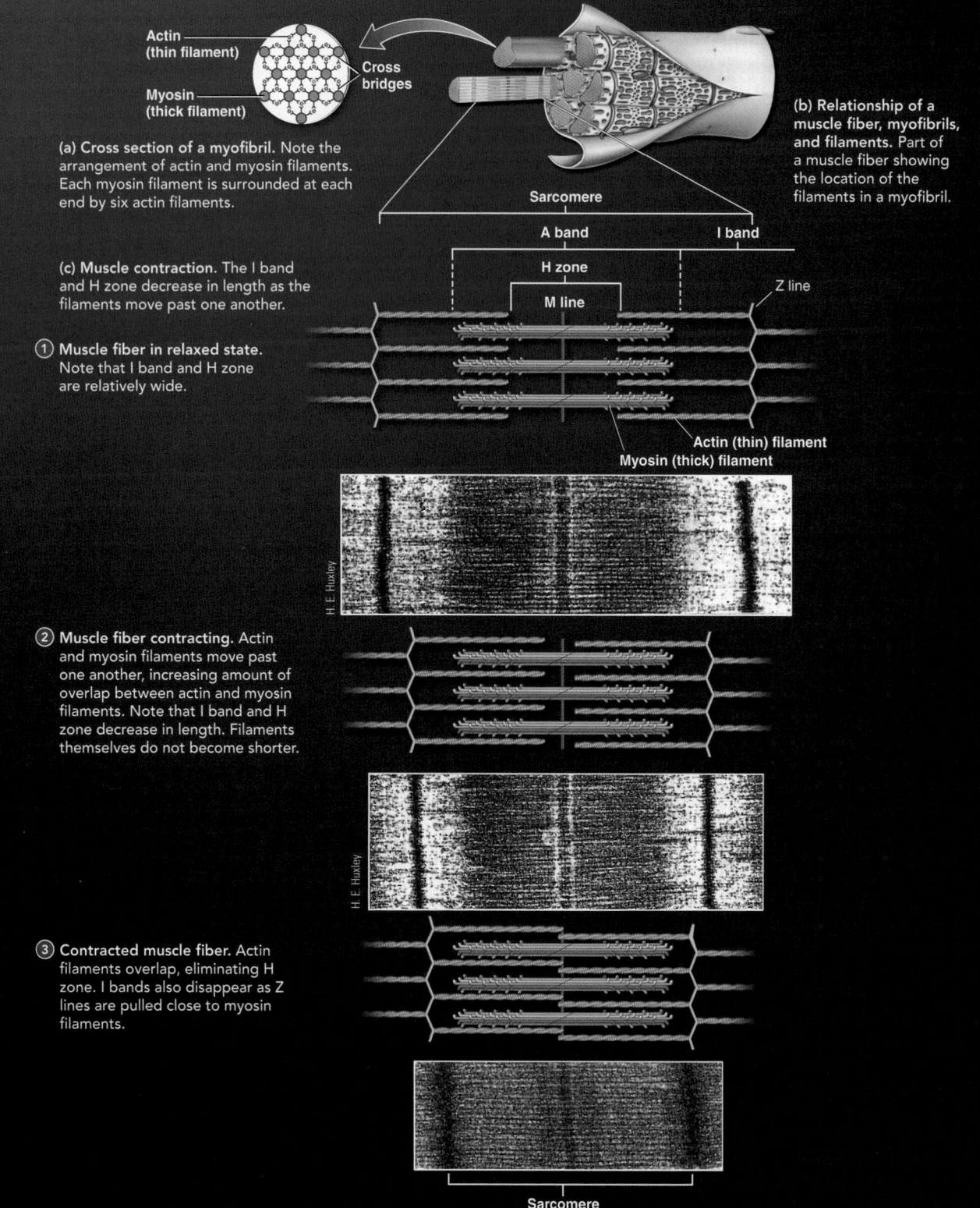

Actin (thin filament)

Cross bridges

Myosin (thick filament)

(a) Cross section of a myofibril. Note the arrangement of actin and myosin filaments. Each myosin filament is surrounded at each end by six actin filaments.

(b) Relationship of a muscle fiber, myofibrils, and filaments. Part of a muscle fiber showing the location of the filaments in a myofibril.

Sarcomere

A band

I band

(c) Muscle contraction. The I band and H zone decrease in length as the filaments move past one another.

H zone

M line

Z line

① **Muscle fiber in relaxed state.** Note that I band and H zone are relatively wide.

Actin (thin) filament

Myosin (thick) filament

H. E. Huxley

② **Muscle fiber contracting.** Actin and myosin filaments move past one another, increasing amount of overlap between actin and myosin filaments. Note that I band and H zone decrease in length. Filaments themselves do not become shorter.

H. E. Huxley

③ **Contracted muscle fiber.** Actin filaments overlap, eliminating H zone. I bands also disappear as Z lines are pulled close to myosin filaments.

Sarcomere

Figure 40-10 *Animation* **The sliding filament model of muscle contraction**
Myosin and actin filaments are organized into repeating units called sarcomeres. As the sarcomeres become shorter, the muscle fiber shortens, but the size of the filaments does not change. (The I band consists of actin filaments of two adjacent sarcomeres.)

PREDICT How would the volume of myofibrils in a muscle fiber affect the force that the muscle can generate?

© Cengage Learning

Contraction occurs when actin and myosin filaments move past one another

Muscle contraction occurs when the muscle fibers, and thus the sarcomeres, shorten. This explanation of muscle contraction, known as the **sliding filament model,** was developed in the 1950s by two British biologists, Hugh Huxley and Andrew Huxley. (The Huxleys were not related. They worked separately, and they independently came to the same general conclusions at about the same time.) Thanks to the Huxleys, we now understand that the muscle shortens as the actin and myosin filaments use a ratcheting mechanism to move past one another, increasing their overlap. You might think of an extension ladder. The overall ladder length changes as the ends get closer or farther apart, but the length of each ladder section stays the same. The I band and H zone decrease in length, but neither the actin nor myosin filaments themselves shorten.

Motor neurons transmit messages from the brain or spinal cord to muscle fibers. When a motor neuron transmits a message, it releases the neurotransmitter **acetylcholine** into the **synaptic cleft,** a small space between the motor neuron and each muscle fiber. Acetylcholine binds with receptors on each muscle fiber, causing **depolarization,** a change in the distribution of electric charge across its sarcolemma. Depolarization can generate an **action potential,** an electrical signal that travels along the membrane of the muscle fiber (or neuron).

In a muscle fiber, an action potential is a wave of depolarization that travels along the sarcolemma and into the system of T-tubule membranes. Depolarization of the T tubules opens calcium channels in the sarcoplasmic reticulum, and stored calcium ions are released into the myofibrils. Calcium ions bind to the protein troponin on the actin filaments, which changes the shape of the troponin. This change results in the troponin pushing tropomyosin away from the *active sites* on the actin filament (FIG. 40-11). These active sites, also called *myosin-binding sites,* are now exposed.

One end of each myosin molecule is folded into two globular structures called *heads.* The rounded heads of the myosin molecules extend away from the body of the myosin filament. Each myosin molecule also has a long tail that joins other myosin tails to form the body of the thick filament. ATP is bound to the myosin when the muscle fiber is at rest (not contracting). Myosin is an adenosine triphosphatase (ATPase), an enzyme that splits ATP to form ADP and inorganic phosphate (P_i). Myosin converts the chemical energy of ATP into the mechanical energy of sliding filaments.

According to the current model of muscle contraction, when ATP is split, the ADP and P_i initially remain attached to the myosin head. The myosin head (with ADP and P_i still bound to it) is in an energized state; it is "cocked." The myosin head binds to an exposed active site on the actin filament, forming a **cross bridge** linking the myosin and actin filaments. The inorganic phosphate is then released, which triggers a conformational change in the myosin head. The myosin head bends about 45 degrees, in a flexing motion. This movement, the *power stroke,* pulls the actin filament closer to the center of the sarcomere. During the power stroke, the ADP is released.

A new ATP must bind to the myosin head before the myosin can detach from the actin. If sufficient calcium ions are present,

the cycle begins anew. Energized once again, myosin heads contact a second set of active sites on the actin filament. This next set is farther down the filament, closer to the end of the sarcomere. The process repeats with a third set of active sites, and so on. This series of stepping motions pulls the actin filaments toward the center of the sarcomere.

Each time the myosin heads attach, move 45 degrees, detach, and then re-attach farther along the actin filament, the muscle shortens. One way to visualize this process is to imagine the myosin heads engaging "hand over hand" with the actin filaments. When many sarcomeres contract simultaneously, they contract the muscle as a whole. The sequence of events in muscle contraction can be summarized as follows:

1. A motor neuron releases acetylcholine, which binds with receptors on the muscle fiber; binding causes depolarization and generation of an action potential.
2. The action potential spreads through T tubules, triggering Ca^{2+} release from the sarcoplasmic reticulum.
3. Ca^{2+} bind to troponin, causing troponin to change shape, which exposes the active sites on the actin filaments.
4. ATP (attached to myosin) is split, and the energized myosin head is cocked; it binds to the active site on the actin filament, forming a cross bridge.
5. The release of P_i from the myosin head triggers the power stroke.
6. During the power stroke, myosin heads pull actin filament toward the center of the sarcomere, shortening the muscle; ADP is released.
7. The myosin head binds a new ATP and detaches from actin. If the concentration of Ca^{2+} is high enough, the cycle repeats from step 3.

ATP powers contraction through repeated cycles of cross bridge formation followed by myosin heads pulling actin filaments toward the center of the sarcomere.

When impulses from the motor neuron cease, the enzyme acetylcholinesterase in the synaptic cleft inactivates acetylcholine. Muscle fibers return to their resting state. Calcium ions are pumped back into the sarcoplasmic reticulum by active transport, a process that requires ATP. Calcium removal is the key step in muscle relaxation. Without Ca^{2+}, tropomyosin once again covers active sites on the actin filaments. The actin filaments move back to their original position, and the muscle relaxes. This entire series of events happens in milliseconds.

ATP powers muscle contraction

Muscle cells often perform strenuously and need large amounts of energy. As noted, the immediate source of energy for muscle contraction is ATP. Energy stored in ATP molecules powers attachment of the myosin heads to the actin filaments, bending of the cross bridges, and release of the myosin heads. Note that energy is needed not only for the pull exerted by the cross bridges, but also for their release from each active site.

Rigor mortis, the temporary but very marked muscle stiffening that occurs after death, results from ATP depletion. When death occurs, cellular respiration ceases, and ATP is used up. However,

The energy of ATP powers muscle contraction. During the power stroke, myosin heads move actin filaments toward the center of the sarcomere, which shortens the muscle.

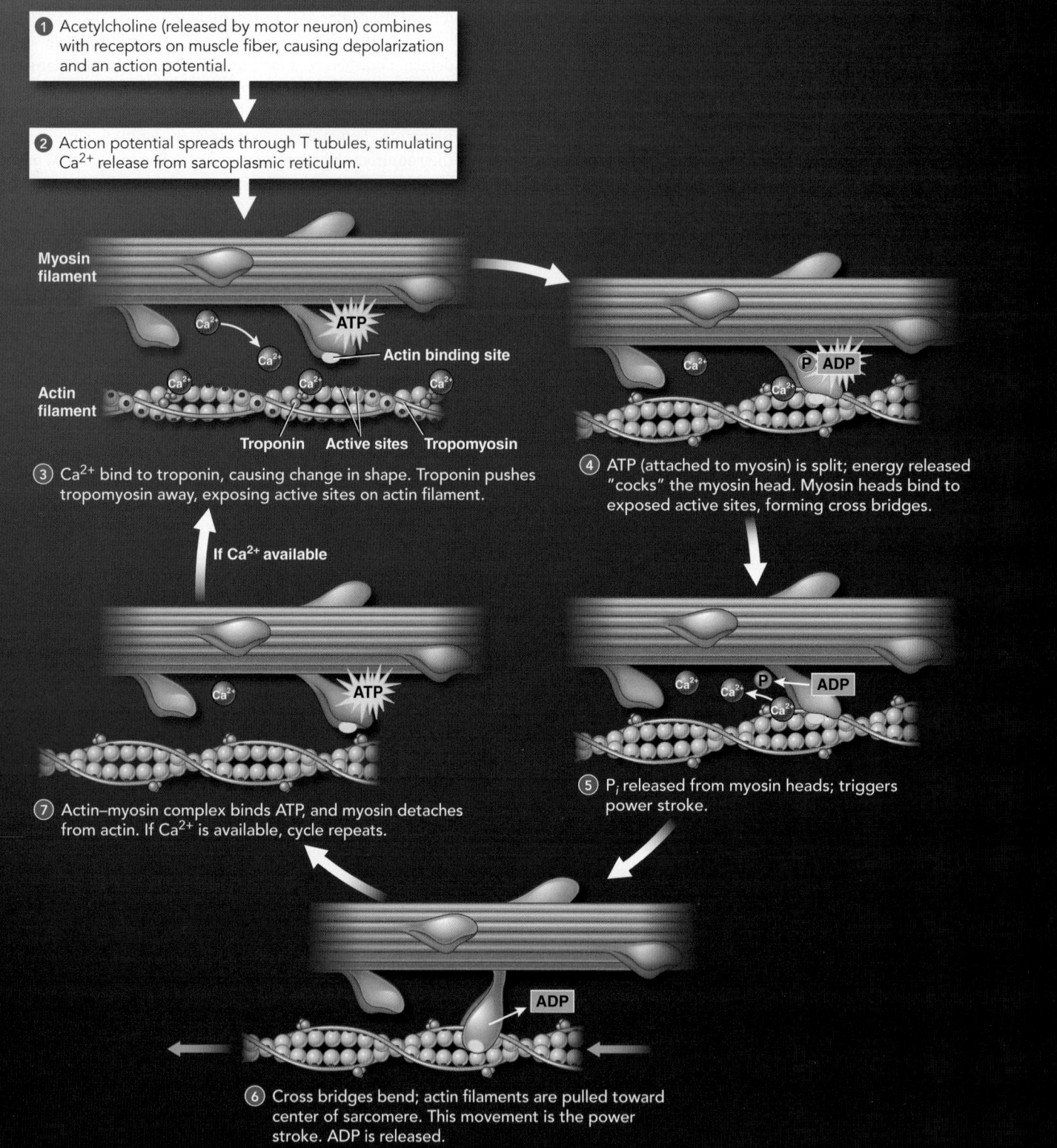

① Acetylcholine (released by motor neuron) combines with receptors on muscle fiber, causing depolarization and an action potential.

② Action potential spreads through T tubules, stimulating Ca^{2+} release from sarcoplasmic reticulum.

Myosin filament

ATP

Actin binding site

Actin filament

Troponin   Active sites   Tropomyosin

③ Ca^{2+} bind to troponin, causing change in shape. Troponin pushes tropomyosin away, exposing active sites on actin filament.

If Ca^{2+} available

ATP

⑦ Actin–myosin complex binds ATP, and myosin detaches from actin. If Ca^{2+} is available, cycle repeats.

P ADP

④ ATP (attached to myosin) is split; energy released "cocks" the myosin head. Myosin heads bind to exposed active sites, forming cross bridges.

P ADP

⑤ P$_i$ released from myosin heads; triggers power stroke.

ADP

⑥ Cross bridges bend; actin filaments are pulled toward center of sarcomere. This movement is the power stroke. ADP is released.

Figure 40-11 *Animation* **Muscle contraction: model of actin and myosin interactions**

Contraction results when actin filaments move toward the center of individual sarcomeres of a myofibril. After step 7 the cycle repeats from step 3 if calcium ions are available. In the absence of calcium ions, the muscle relaxes.

PREDICT How would muscle contraction be affected if the body stopped producing acetylcholine?

© Cengage Learning

at the time of death, many muscle fibers are in the process of contraction, and cross bridges have formed. Because ATP is essential for release of the myosin heads, cross bridges remain intact. Rigor mortis does not persist indefinitely because the entire contractile apparatus of the muscles eventually decomposes, restoring pliability. The phenomenon is temperature dependent, so given the prevailing temperature, a medical examiner can estimate the time of death of a cadaver from its degree of rigor mortis.

ATP molecules can provide energy for only a few seconds of strenuous activity. Muscle fibers have a backup energy storage compound, **creatine phosphate,** that can be stockpiled. The energy stored in creatine phosphate is transferred to ATP as needed, but during vigorous exercise the supply of creatine phosphate is quickly depleted. Muscle cells must replenish their supplies of these energy-rich compounds.

Fortunately, there is yet another backup energy storage compound. Muscle fibers store chemical energy in **glycogen,** a large polysaccharide formed from hundreds of glucose molecules. Glycogen can be degraded, yielding glucose, which is then degraded in cellular respiration (see Chapters 3 and 8). When sufficient oxygen is available, enough energy is captured from glucose to produce needed quantities of ATP and creatine phosphate.

During a burst of strenuous exercise, the circulatory system cannot deliver enough oxygen to keep up with the demand of the rapidly metabolizing muscle fibers. The result is an **oxygen debt.** Under these conditions, muscle fibers break down fuel molecules anaerobically (without oxygen) for short periods. Lactic acid fermentation is a method of generating ATP anaerobically, but not in great quantity (see Fig. 8-14c). ATP depletion results in weaker contractions and muscle fatigue. Accumulation of the waste product **lactic acid** also contributes to muscle fatigue. Well-conditioned athletes develop the ability to tolerate the high levels of lactic acid generated during high-performance activity. The period of rapid breathing that generally follows strenuous exercise pays back the oxygen debt by consuming lactic acid.

The type of muscle fibers determines strength and endurance

Three major types of skeletal muscle fibers have been identified in vertebrates: slow-oxidative, fast-oxidative, and fast-glycolytic fibers (**TABLE 40-1**). **Slow-oxidative fibers** are well adapted for endurance activities such as swimming and running long distances and for maintaining posture. They contract slowly and fatigue slowly. These fibers require a steady supply of oxygen. They derive most of their energy from aerobic respiration and are rich in mitochondria and capillaries. Slow-oxidative fibers are rich in **myoglobin,** a red pigment similar to hemoglobin, which stores oxygen in red blood cells. Myoglobin enhances rapid diffusion of oxygen from blood into muscles during strenuous muscle exertion. Because of their myoglobin, slow-oxidative fibers are red and are sometimes called red fibers.

TABLE 40-1	Characteristics of Skeletal Muscle Fibers		
CHARACTERISTICS	SLOW-OXIDATIVE FIBERS	FAST-OXIDATIVE FIBERS	FAST-GLYCOLYTIC FIBERS
Contraction speed	Slow	Fast	Fast
Rate of fatigue	Slow	Intermediate	Fast
Major pathway for ATP synthesis	Aerobic respiration	Aerobic respiration	Glycolysis
Mitochondria	Many	Many	Few
Intensity of contraction	Low	Intermediate	High
Myoglobin content	High	High	Low
Color of fiber	Red	Red	White

© Cengage Learning

Fast-oxidative fibers contract rapidly and have an *intermediate* rate of fatigue. Like slow-oxidative fibers, they are rich in mitochondria and obtain most of their ATP from aerobic respiration. They have a high myoglobin content and a red color.

Fast-glycolytic fibers generate a great deal of power and carry out rapid movements but can sustain that activity for only a short time. They are important in activities such as sprinting and weight lifting. Fast-glycolytic fibers contract rapidly, have few mitochondria, and obtain most of their energy from glycolysis. When their glycogen supply is depleted, they fatigue rapidly. These fibers have a low myoglobin content and are white. (They are sometimes referred to as white fibers.) People who are sedentary have more fast-glycolytic fibers than do physically fit individuals. With physical training, these fibers change to fast-oxidative fibers.

Entire muscles may be specialized for quick or slow responses. In chickens, for instance, the white breast muscles are efficient for quick responses, perhaps because a short flight is an escape mechanism for chickens. However, they walk about on the ground all day; the dark (red) meat of the leg and thigh is muscle specialized for more sustained activity. Birds that fly have red breast muscles specialized to support sustained activity.

In humans most muscles have a mixture of different types of fibers. The proportions of the types of fibers vary among individuals and from muscle to muscle in the same person. Aerobic exercise, such as jogging or bicycling, increases endurance. *Aerobic training* increases capillary density, the number of mitochondria, and the myoglobin content of muscle fibers. In *strength training*—for example, weight lifting—specific muscles are repeatedly contracted under heavy loads. This action increases the number of filaments in the muscle fibers, which increases the size of the muscle.

Several factors influence the strength of muscle contraction

In addition to the types of muscle fibers making up the muscle, contraction of a whole muscle depends on the number of muscle fibers contracting, the tension developed by each fiber, frequency of stimulation, condition of the muscle (e.g., whether or not it is fatigued), and the load on the muscle. The more muscle fibers that contract, the greater the tension in the muscle.

In a **motor unit,** a motor neuron is functionally connected with an average of about 150 muscle fibers (**FIG. 40-12**). However, some motor units have fewer than a dozen motor neurons, and others have several hundred. Each junction of a motor neuron with a muscle fiber is called a *neuromuscular junction.* Messages from the brain or spinal cord activate motor units. The more motor units *recruited,* the stronger the contractions.

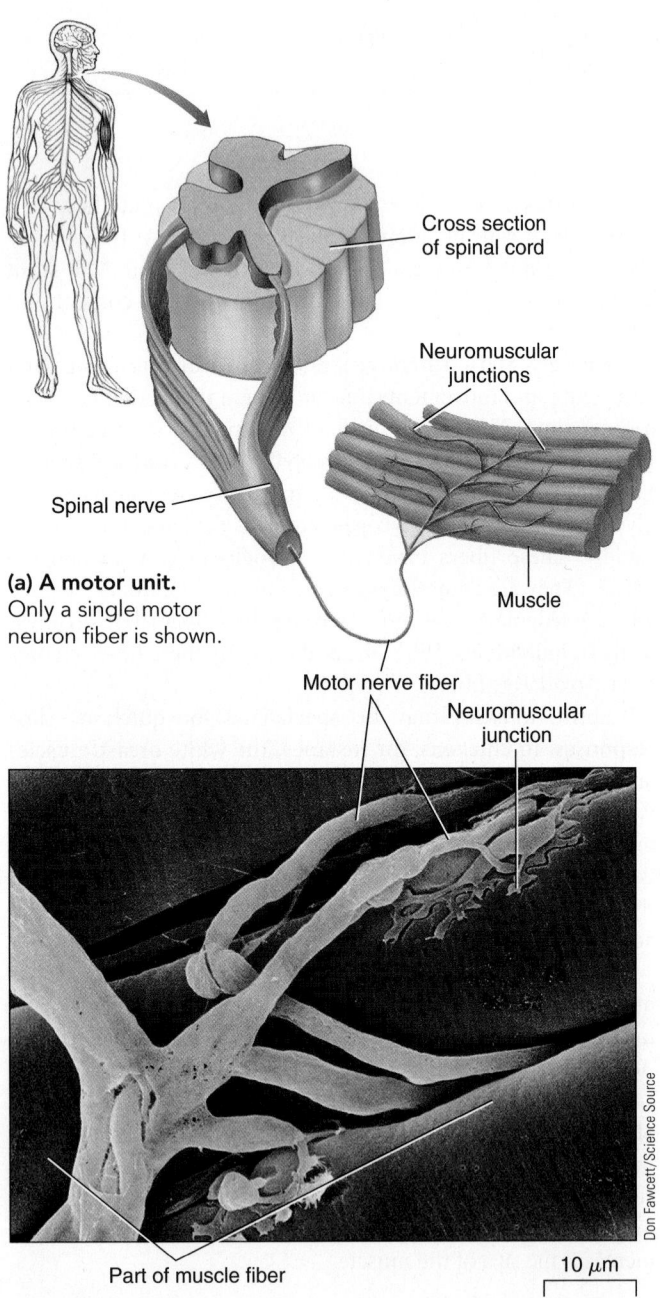

(a) A motor unit.
Only a single motor
neuron fiber is shown.

Cross section
of spinal cord

Neuromuscular
junctions

Spinal nerve

Muscle

Motor nerve fiber

Neuromuscular
junction

Part of muscle fiber

10 μm

Don Fawcett/Science Source

(b) SEM of some of the motor neuron fibers in a motor unit.
Note how neurons branch to innervate all muscle fibers in the
motor unit.

Figure 40-12 *Animation* **A motor unit**

A motor unit consists of a motor neuron and the muscle fibers that it innervates.
© Cengage Learning

An important factor affecting the tension of contracting muscle fibers is the frequency of stimulation. In response to a single, brief electrical stimulus, skeletal muscle contracts with a single, quick contraction called a *simple twitch.* When a second stimulus is received before the first contraction is complete, the two twitches may add together, a process called **summation** (**FIG. 40-13**). Typically, skeletal muscle receives a series of separate stimuli timed very closely together. Summation results in a smooth, sustained contraction called *tetanus.* The identity and number of our muscle fibers that are tetanically contracting help determine whether we can play the piano, draw a picture, rock a baby, drive a car, or run a mile.

When you lift a heavy object, you place a load on the muscles affected. As the weight of the load increases, the muscles contract more strongly (within limits).

Even when you are not moving, your muscles are in a state of partial contraction known as **muscle tone.** At any given moment, some muscle fibers are contracted, stimulated by messages from motor neurons. Muscle tone is an unconscious process that keeps muscles prepared for action. When the motor nerve to a muscle is cut, the muscle becomes limp (completely relaxed, or flaccid) and eventually atrophies (decreases in size).

Smooth muscle and cardiac muscle are involuntary

The three types of vertebrate muscle—skeletal, smooth, and cardiac—are compared in Table 39-3. All three types of muscle have thin actin filaments that move along thick myosin filaments in response to an increase in calcium ion concentration. All three muscle types use ATP to power contraction. However, each type of muscle is specialized for particular types of responses. Neither smooth muscle nor cardiac muscle is under voluntary control.

Smooth muscle is not attached to bones but instead forms tubes that squeeze like the muscle tissue in the body wall of the earthworm. Smooth muscle often contracts in response to simple stretching, and its contraction tends to be sustained. It is well adapted to performing such tasks as regulating blood pressure by sustained contraction of the walls of the arterioles. Although smooth muscle contracts slowly, it shortens much more than striated muscle does; it squeezes impressively.

Smooth muscle is not striated because its actin and myosin filaments are not organized into myofibrils or into sarcomeres. The fibers of smooth muscle tissue function as a unit because they are connected by *gap junctions* (see Fig. 5-23c). Gap junctions permit electrical signals to pass rapidly from fiber to fiber. Although smooth muscle contraction is similar to skeletal muscle contraction (occurring by a sliding filament mechanism), the cross bridges in smooth muscle remain in the attached state longer. For this reason, less ATP is required to maintain a high level of force in smooth muscle than in skeletal muscle.

Cardiac muscle contracts and relaxes in alternating rhythm, propelling blood with each contraction. Sustained contraction of cardiac muscle would be disastrous! Cardiac muscle fibers are electrically coupled by junctions called **intercalated discs.** Gap

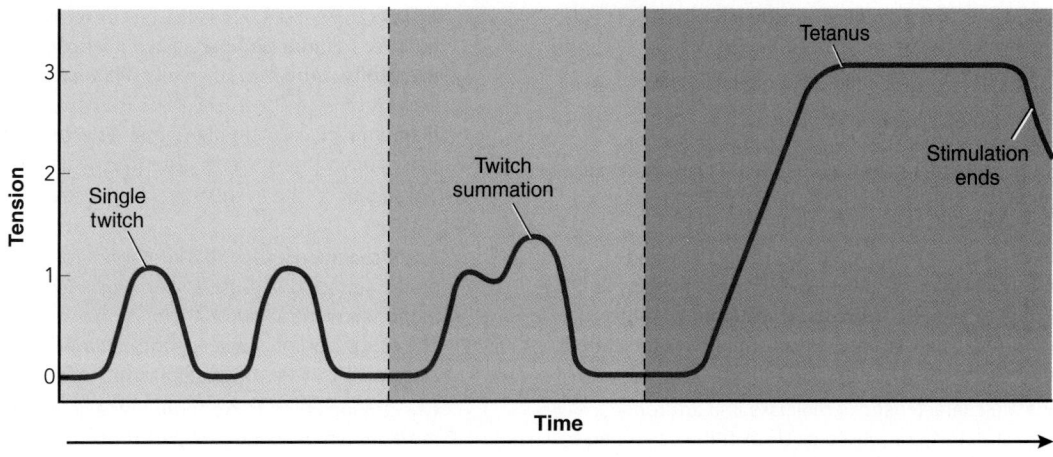

(a) When a muscle fiber is allowed to completely relax and is then stimulated a second time, the second twitch is the same magnitude as the first twitch.

(b) When a muscle fiber is stimulated a second time before it has completely relaxed, the second twitch is added on to the first twitch. The result is summation.

(c) A volley of rapid stimuli does not allow the muscle fiber to relax between stimuli. The result is a sustained contraction called tetanus.

Figure 40-13 Summation and tetanus
© Cengage Learning

junctions within the intercalated discs allow action potentials to spread rapidly from one cardiac fiber to another.

Each heartbeat is initiated by a *pacemaker,* a mass of specialized cardiac muscle. A conduction system transmits the signal throughout the cardiac muscle. The cardiac fibers within the atria contract as a functional unit, and the cardiac fibers within the ventricles also function as a functional unit. The heart beats in a regular rhythm independent of a nerve supply. However, the heart rate is regulated, according to the needs of the body, by neurons in the cardiac centers in the medulla of the brain (discussed in Chapter 44).

CHECKPOINT 40.3

- VISUALIZE *Sketch a longitudinal section of a sarcomere in a relaxed state. Label the Z lines, thin filaments, and thick filaments. Which filaments are actin? myosin?*
- *What is the sequence of events that occur when a muscle fiber contracts (begin with release of acetylcholine and include cross-bridge action)?*
- *Compare and contrast the functions of ATP, creatine phosphate, and glycogen in muscle contraction.*
- *Compare skeletal, smooth, and cardiac muscle.*

SUMMARY: FOCUS ON LEARNING OBJECTIVES

40.1 Epithelial Coverings *(page 837)*

1 Compare the functions of the epithelial coverings of invertebrates and vertebrates.

- **Epithelial tissue** protects underlying tissues and may be specialized for sensory or respiratory functions. The outer epithelium may be specialized to secrete antibacterial or poisonous substances.
- In many invertebrates the outer epithelium is specialized to secrete a protective **cuticle,** or shell.
- The **integumentary system** of vertebrates includes the skin and structures that develop from it. Mammalian skin includes hair, claws or nails, sweat glands, oil glands, and sensory receptors.

2 Relate the structure of vertebrate skin to its functions.

- The feathers of birds and the hair of mammals form an insulating layer that helps maintain a constant body temperature. Mammalian skin protects the body from the wear and tear that occurs as it interacts with the outer environment. **Stratum corneum,** the most superficial layer of the **epidermis,** consists of dead cells filled with **keratin,** a protein that gives mechanical

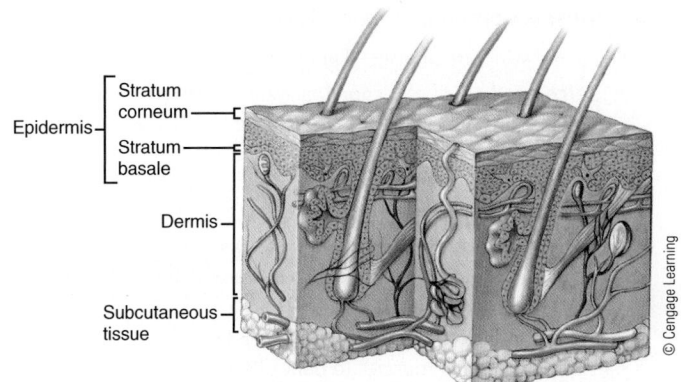

strength to the skin and reduces water loss. Cells in **stratum basale,** the deepest layer of the epidermis, divide and are pushed upward toward the skin surface. These cells mature, produce keratin, and eventually die and slough off.

- The **dermis** consists of dense, fibrous connective tissue. In birds and mammals, the dermis rests on a layer of *subcutaneous tissue* composed largely of insulating fat.

40.2 Skeletal Systems *(page 838)*

3 Compare the structure and functions of different types of skeletal systems, including the hydrostatic skeleton, exoskeleton, and endoskeleton.

- The **skeletal system** supports and protects the body and transmits mechanical forces generated by muscles. Many soft-bodied invertebrates have a **hydrostatic skeleton;** fluid in a closed body compartment is used to transmit forces generated by contractile cells or muscle.
- **Exoskeletons** are characteristic of mollusks and arthropods. The arthropod skeleton, composed partly of **chitin,** is jointed for flexibility and adapted for many lifestyles. This nonliving skeleton does not grow, making it necessary for arthropods to undergo **ecdysis (molting)** periodically.
- The **endoskeletons** of echinoderms and chordates consist of living tissue and therefore can grow.

4 Describe the main divisions of the vertebrate skeleton and identify the bones that make up each division.

- The vertebrate skeleton consists of an *axial skeleton* and an *appendicular skeleton*. The axial skeleton consists of the *skull,* the **vertebral column,** the *rib cage,* and the *sternum*. The appendicular skeleton consists of the bones of the limbs, the *pectoral girdle,* and the *pelvic girdle*.

5 Describe (or draw and label) the structure of a typical long bone and contrast the roles of osteoblasts and osteoclasts in bone remodeling.

- A typical long bone consists of a thin outer shell of **compact bone** surrounding the inner **spongy bone** and a central cavity that contains **bone marrow.**
- **Osteoblasts,** cells that produce bone, and **osteoclasts,** cells that break down bone, work together to shape and remodel bone.

6 Compare the main types of vertebrate joints.

- **Joints** are junctions of two or more bones. **Ligaments** are connective tissue bands that connect bones and limit movement at the joint. The sutures of the skull are *immovable joints*. Joints between vertebrae are *slightly movable joints*. A *freely movable joint* is enclosed by a joint capsule lined with a membrane that secretes *synovial fluid*.

40.3 Muscle Contraction *(page 842)*

7 Relate the structure of insect flight muscles to their function.

- Large numbers of mitochondria and tracheae (air tubes) present in insect flight muscles support the high metabolic rate required for flight.

8 Describe the structure of skeletal muscles and their antagonistic actions.

- A *muscular system* is found in most invertebrate phyla and in vertebrates. As muscle tissue contracts (shortens), it moves body parts by pulling on them.
- Vertebrate skeletal muscles pull on **tendons,** connective tissue cords that attach muscles to bones. When a muscle contracts, it pulls a bone toward or away from the bone with which it articulates.
- Skeletal muscles act *antagonistically* to one another. The muscle that produces a particular action is the *agonist;* the *antagonist* produces the opposite movement.

- A skeletal muscle such as the biceps is an organ made up of hundreds of **muscle fibers.** Each fiber consists of threadlike **myofibrils** composed of smaller **filaments.** The striations of skeletal muscle fibers reflect the overlapping of their **actin filaments** and **myosin filaments.** A **sarcomere** is a contractile unit of actin (thin) and myosin (thick) filaments.

9 List, in sequence, the events that take place during muscle contraction.

- **Acetylcholine** released by a motor neuron binds with receptors on the surface of a muscle fiber. This may cause **depolarization** of the sarcolemma and transmission of an **action potential.** The action potential spreads through the **T tubules,** releasing calcium ions from the **sarcoplasmic reticulum.**
- Calcium ions bind to **troponin** in the actin filaments, causing the troponin to change shape. Troponin pushes **tropomyosin** away from the binding sites on the actin filaments.
- ATP binds to myosin; ATP is split, putting the myosin head in a high-energy state (it is "cocked"). Energized myosin heads attach to the exposed binding sites on the actin filaments, forming **cross bridges** that link the myosin and actin filaments.
- After myosin attaches to the actin filament, the cross bridge flexes as phosphate is released. This power stroke pulls the actin filament toward the center of the sarcomere. ADP is released during the power stroke.
- The myosin head binds a new ATP, which lets the myosin head detach from the actin. As long as the calcium ion concentration remains elevated, the new ATP is split, and the sequence repeats. The myosin re-attaches to new active sites so that the filaments are pulled past one another, and the muscle continues to shorten.

10 Compare the roles of glycogen, creatine phosphate, and ATP in providing energy for muscle contraction.

- ATP is the immediate source of energy for muscle contraction. ATP hydrolysis provides the energy to "cock" the myosin. Muscle tissue has an intermediate energy storage compound, **creatine phosphate. Glycogen** is the fuel stored in muscle fibers.

11 Compare the structures and functions of the three types of skeletal muscle fibers.

- **Slow-oxidative fibers** are specialized for endurance activities. They contract slowly, fatigue slowly, are rich in mitochondria, and obtain most of their ATP from aerobic respiration. These fibers have a red color because they have a high content of **myoglobin,** a red pigment that stores oxygen.
- **Fast-oxidative fibers** are specialized for rapid response. They contract quickly, have an intermediate rate of fatigue, are rich in mitochondria, and obtain most of their ATP from aerobic respiration. They have a high myoglobin content and a red color.
- **Fast-glycolytic fibers** generate a great deal of power for a brief period. These white fibers contract rapidly, fatigue quickly, have few mitochondria, and use glycolysis as a major pathway for ATP synthesis.

12 Describe factors that influence muscle contraction.

- Contraction of a whole muscle depends on several factors, including the number of muscle fibers contracting, tension developed by each fiber, frequency of stimulation, condition of the muscle, and load on the muscle.
- A **motor unit** consists of all the skeletal muscle fibers stimulated by a single motor neuron. Messages from the brain

or spinal cord activate motor units. The more motor units *recruited*, the stronger the contractions.

- When activated by a brief electrical stimulus, skeletal muscle responds with a *simple twitch*. When a second stimulus is received before the first contraction is complete, the two twitches can add together, a process called **summation.**

Typically, skeletal muscle is stimulated by a series of separate stimuli timed close together and responds with a smooth, sustained contraction called *tetanus*.

- **Muscle tone** is the state of partial contraction characteristic of muscles.

TEST YOUR UNDERSTANDING

Know and Comprehend

1. An endoskeleton (a) is typically composed of dead tissue (b) is characterized by fluid in a closed compartment (c) is found mainly in echinoderms and vertebrates (d) is typical of echinoderms and arthropods (e) requires the animal to molt

2. Which of the following is part of the vertebrate appendicular skeleton? (a) skull (b) vertebral column (c) pelvic girdle (d) rib cage (e) sternum

3. An energy storage compound that can be stockpiled in muscle cells for short-term use is (a) myoglobin (b) ADP (c) troponin (d) myosin (e) creatine phosphate

4. Calcium ions are released from the sarcoplasmic reticulum. What happens next? (a) acetylcholine is released (b) active sites on the actin filaments are exposed (c) filaments move past one another, and the muscle fiber shortens (d) myosin is activated (e) P_i is released, and the cross bridge flexes

5. Myosin binds to actin, forming a cross bridge. What happens next? (a) acetylcholine is released (b) calcium ions stimulate exposure of active sites (c) myosin filaments move past one another, and the muscle fiber shortens (d) myosin is activated (e) P_i is released, and the cross bridge flexes

6. When skeletal muscle is stimulated by a series of closely timed separate stimuli, (a) white fibers respond (b) a simple twitch occurs (c) the muscle responds with a smooth, sustained contraction called tetanus (d) red fibers respond (e) summation is inhibited

Apply and Analyze

7. Insect flight muscles have a very high metabolic rate. In what way is this characteristic adaptive?

8. **VISUALIZE** Label the diagram.

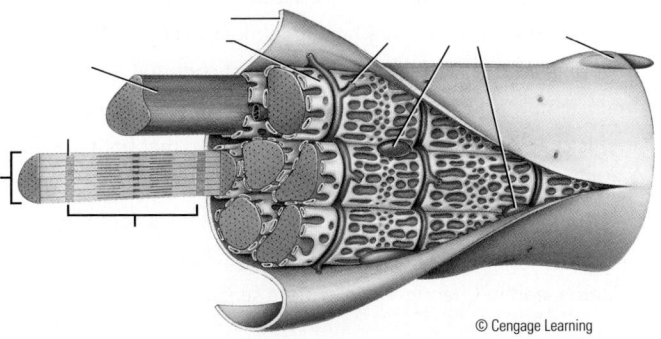
© Cengage Learning

Evaluate and Synthesize

9. **INTERPRET DATA** Examine Figure 40-13. Based on the summation of two stimuli shown in that graph, draw a curve showing summation when the muscle fiber is stimulated three times.

10. Skeletal muscles are able to shift functions, sometimes acting as an agonist and other times acting as an antagonist. How is this ability to shift functions advantageous?

11. **CONNECT** What is the functional relationship between skeletal and muscle tissues? How can muscle damage affect skeletal tissue?

12. **EVOLUTION LINK** Very different skeletal systems have evolved in arthropods and vertebrates. What are some benefits and some disadvantages of each type of skeleton? In what way is the insect skeleton an adaptation to the insect lifestyle?

13. **EVOLUTION LINK** What does human muscle contraction have in common with cellular movement in single-cell organisms?

14. **SCIENCE, TECHNOLOGY, AND SOCIETY** Explore the relationships between society's need for joint replacements and the interconnections of science and technology with this need.

aplia To access course materials, such as Aplia and other companion resources, please visit **www.cengagebrain.com.**

41 | Neural Signaling

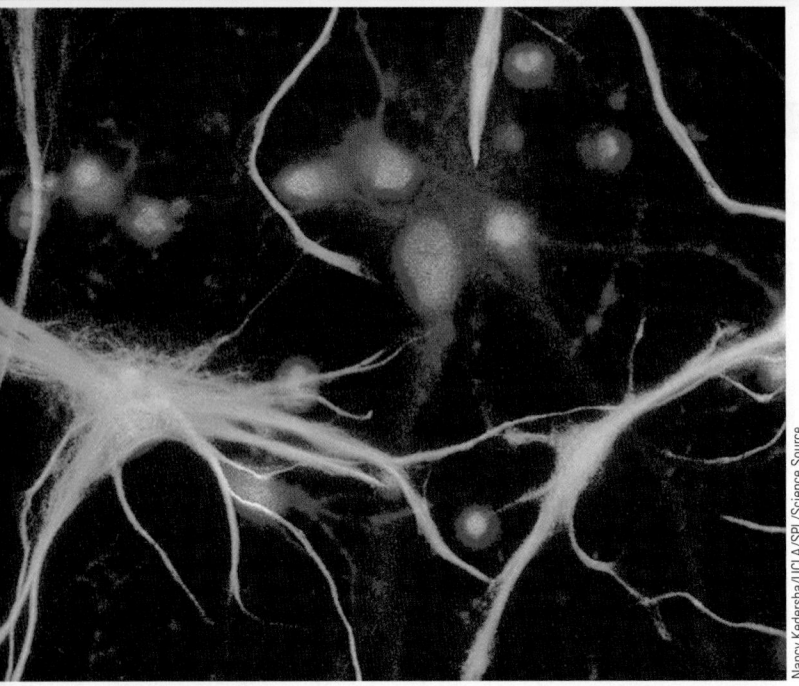

Nancy Kedersha/UCLA/SPL/Science Source

Specialized cells in the nervous system. This immunofluorescent LM of mammalian spinal cord tissue includes several astrocytes (*green*). The *blue dots* are the nuclei of other glial cells. The smaller cells (*dark pink*) are neurons; their cell bodies appear *light pink*.

KEY CONCEPTS

41.1 Neural signaling involves reception, transmission, integration, and action by effectors.

41.2 Neurons are specialized to receive stimuli and transmit signals; glial cells support, protect, and nourish neurons, and they can modify neural signals.

41.3 The resting potential of a neuron is maintained by ion channels and ion pumps. Depolarization of the neuron plasma membrane to threshold level generates action potentials, the electrical signals transmitted by axons.

41.4 Most neurons signal other cells by releasing neurotransmitters at chemical synapses.

41.5 During integration, incoming neural signals are summed; temporal and spatial summation can bring a neuron to threshold level.

41.6 Neurons are organized in specific pathways called neural circuits, which, in turn, are organized to form neural networks.

An animal's ability to survive and to maintain homeostasis depends largely on how effectively it detects and responds to *stimuli,* the changes that take place in its environment. Most animals have a *nervous system* that takes in information, transmits it to the spinal cord and brain where it is integrated, and then responds. The nervous system is composed mainly of two specialized types of cells: *neurons* and *glial cells* (see photograph). Just how animals respond to stimuli depends on how their neurons are organized and connected to one another. A single neuron in the vertebrate brain may be functionally connected to thousands of other neurons.

The nervous system is the principal regulatory system in animals. Regulation requires communication, and the nervous system transmits information to and from all parts of the body. Neurobiology is one of the most exciting areas of biological research. Active areas of research include *neurotransmitters,* the chemical messengers used by neurons to signal other neurons, and the *receptors* that bind with the neurotransmitters.

Neurogenesis, the production of new neurons, has been a very interesting focus of research. One of the most time-honored views about the nervous system was that the brain of mammals cannot generate new neurons after birth. Nobel laureate Santiago Ramon y Cajal promoted this concept at the beginning of the 20th century, and the idea that we are born with all the brain cells we will ever have was widely held for almost a century. Another enduring idea was that glial cells, the most common cells in the brain, simply provide support services for neurons. As we discuss in this chapter, researchers have demonstrated that both of these hypotheses were false.

In this chapter we focus on how information is communicated by neurons and glial cells. In Chapter 42 we discuss nervous systems, and in Chapter 43 we explore sensory reception.

41.1 NEURAL SIGNALING: AN OVERVIEW

LEARNING OBJECTIVE

1 Describe the processes involved in neural signaling: reception, transmission, integration, and action by effectors that brings about the actual response.

Thousands of stimuli constantly bombard an animal. Survival depends on identifying and evaluating these stimuli, and then responding effectively. Information about each stimulus must be transmitted *to* the nervous system, and then information *from* the nervous system must be transmitted to muscles and glands. Animals respond to changes in temperature, sound, light, odors, and movement that may indicate the presence of a predator or of prey. Internal signals, such as hunger, pain, fluctuations in blood pressure, or changes in blood sugar concentration, also demand responses. In most animals responses to stimuli depend on **neural signaling,** the transmission of information by **neurons,** conducting cells of the nervous system. Neural signaling typically involves four processes: reception, transmission, integration, and the action by effectors (muscles or glands) that brings about the actual response.

Reception, the process of detecting a stimulus, is the job of neurons and of specialized *sensory receptors* such as those in the skin, eyes, and ears (**FIG. 41-1**). **Neural transmission** is the process of sending messages along a neuron, from one neuron to another or from a neuron to a muscle or gland. In vertebrates a neural message is transmitted from a receptor to the **central nervous system (CNS),** which consists of the brain and spinal cord. Neurons that transmit information to the CNS are called **afferent neurons** (meaning "to carry toward"), or **sensory neurons.**

Afferent neurons generally transmit information to **interneurons,** or association neurons, in the CNS. Most neurons, perhaps 99% of them, are interneurons. Their function is to integrate input and output. **Integration** involves sorting and interpreting incoming sensory information and determining the appropriate response.

Neural messages are transmitted from the CNS by **efferent neurons** (meaning "to carry away") to **effectors** (muscles and glands). Efferent neurons that signal skeletal muscle are called **motor neurons.** The *action by effectors* brings about the actual **response** to the stimulus. Sensory receptors and afferent and efferent neurons are part of the **peripheral nervous system (PNS).** In summary, information flows through the nervous system in the following sequence:

reception by sensory receptor ⟶ transmission by afferent neuron ⟶ integration by interneurons in CNS ⟶ transmission by efferent neuron ⟶ action by effectors ⟶ actual response

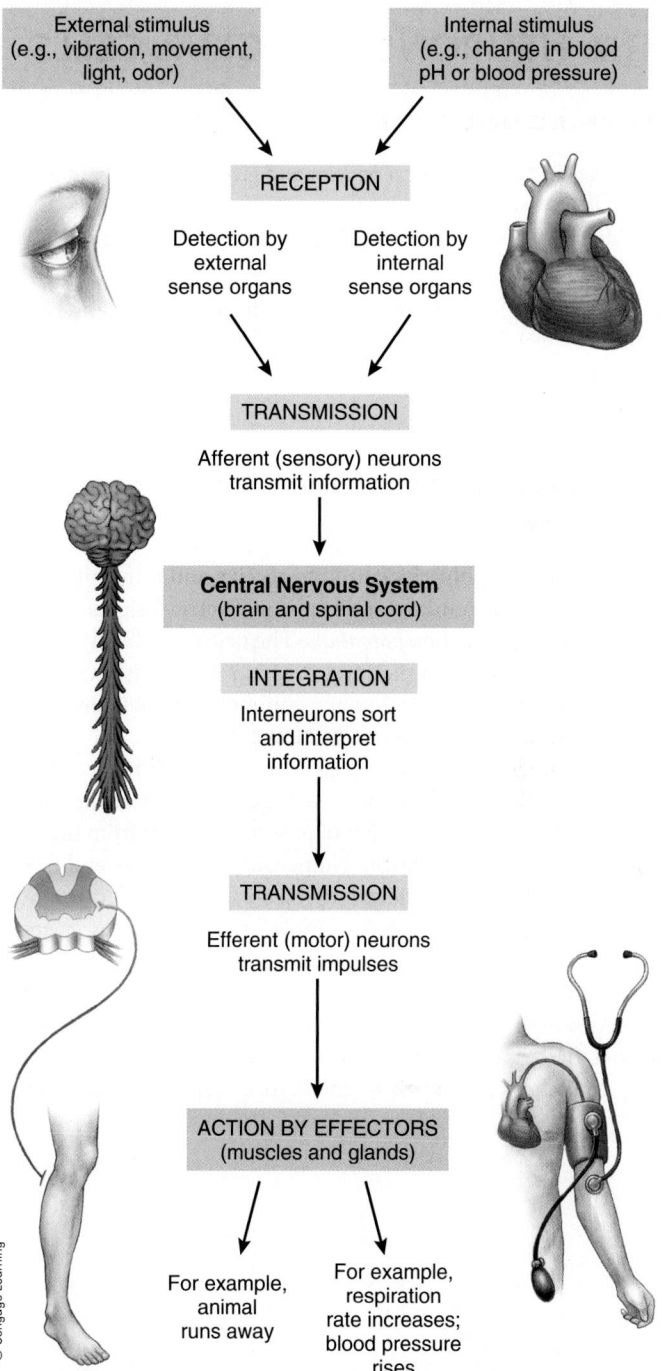

Figure 41-1 Response to a stimulus

Whether a stimulus originates in the outside world or inside the body, information must be received, transmitted to the central nervous system, integrated, and then transmitted to effectors (muscles and glands) that carry out some action, the actual response.

CHECKPOINT 41.1

- **CONNECT** *Imagine that you are swimming and suddenly spot a shark fin moving in your direction. What sequence of processes must take place within your nervous system before you can make your escape?*

- *What happens after neural signals are integrated in the CNS?*

41.2 NEURONS AND GLIAL CELLS

Neurons and glial cells are unique to the nervous system. Neurons are specialized to receive and send information. Glial cells support and protect neurons and carry out many regulatory functions.

Neurons receive stimuli and transmit neural signals

The neuron is highly specialized to receive and transmit information. Neurons produce and transmit electrical signals called *nerve impulses,* or *action potentials.* The neuron is distinguished from all other cells by its long *processes* (cytoplasmic extensions). Examine the structure of a common type of neuron, the multipolar neuron shown in **FIGURE 41-2.**

The largest portion of the neuron, the **cell body,** contains the bulk of the cytoplasm, the nucleus, and most of the other organelles. Typically, two types of processes project from the cell body of a multipolar neuron. Numerous *dendrites* extend from one end, and a long, single *axon* projects from the opposite end. **Dendrites** are typically short, highly branched processes specialized to receive stimuli and send signals to the cell body. The cell body integrates incoming signals.

Although microscopic in diameter, an **axon** may be 1 m (more than 1 yd) or more in length and may divide, forming branches called *axon collaterals.* The axon conducts nerve impulses away from the cell body to another neuron or to a muscle or gland. At its end the axon divides, forming many *terminal branches* that end in *synaptic terminals,* also called *axon terminals.* The synaptic terminals release **neurotransmitters,** chemicals that transmit signals from one neuron to another or from a neuron to an effector. The junction between a synaptic terminal and another neuron (or effector) is called a **synapse.** Typically, a small space exists between these two cells.

In vertebrates the axons of many neurons outside the CNS are surrounded by a series of **Schwann cells.** The plasma membranes of these glial cells contain *myelin,* a white, fatty material. Schwann cells wrap their plasma membranes around the axon, forming an insulating covering called the **myelin sheath.** Gaps in the myelin sheath, called **nodes of Ranvier,** occur between successive Schwann cells. At these points the axon is not insulated with myelin. Axons more than 2 μm in diameter have myelin sheaths and are described as *myelinated.* Those of smaller diameter are generally unmyelinated.

Certain regions of the CNS produce new neurons

At the beginning of this chapter, we described Cajal's idea that **neurogenesis,** the production of new neurons, does not occur in the CNS after birth. Cajal's idea was widely accepted for many years. However, in science prevailing hypotheses are continuously challenged by new findings. Researchers investigating seemingly unrelated problems made observations that shed new light on the neurogenesis issue.

Since the 1970s Fernando Nottebohm at Rockefeller University has studied birdsong, including its neural basis. Nottebohm has found that birds produce new neurons daily in the brain center that regulates their songs. His work was at first

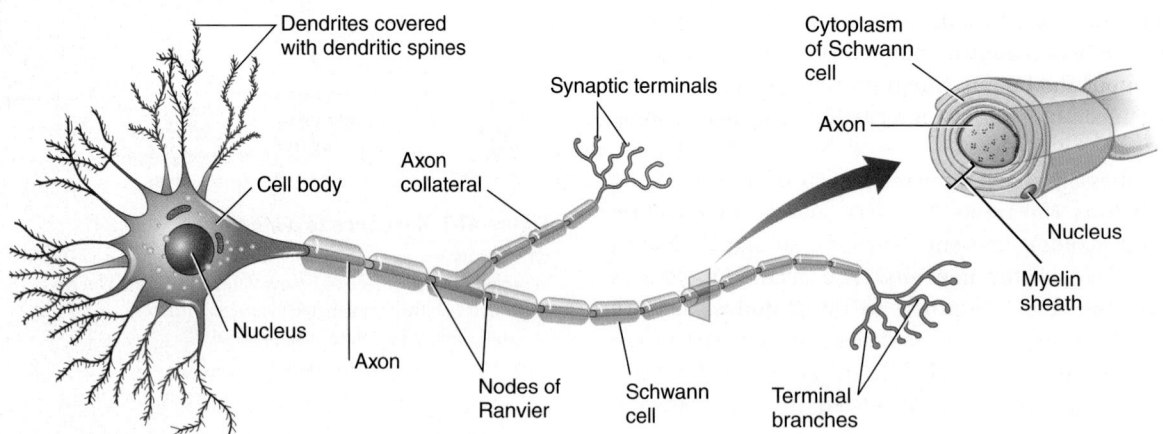

Figure 41-2 *Animation* **Structure of a multipolar neuron**

The cell body contains most of the organelles. Many dendrites and a single axon extend from the cell body. Schwann cells form a myelin sheath around the axon by wrapping their own plasma membranes around the neuron in jelly-roll fashion.

© Cengage Learning

dismissed as a phenomenon peculiar to birds and not relevant to mammals.

However, other investigators, notably psychologist Elizabeth Gould at Princeton University, have provided additional data that together have falsified Cajal's hypothesis. Gould demonstrated neuron proliferation in primates. She and her research team also discovered that stress damages the brain and inhibits mitosis in neurons. They stressed newborn rats by separating them from their mothers for up to 3 hours a day. The stressed rats did not learn how to cope with stress, and as adults they had fewer new neurons in certain areas of their brains than control rats. Neurobiologists now agree that neurogenesis does occur regularly in the adult mammalian brain, at least in certain areas of the hippocampus and in the lining of the lateral ventricles. The paradigm shift that has occurred with regard to Cajal's long accepted hypothesis demonstrates the power and importance of the process of science.

Axons aggregate to form nerves and tracts

A **nerve** consists of hundreds or even thousands of axons wrapped together in connective tissue (FIG. 41-3). We can compare a nerve to a telephone cable. The individual axons correspond to the wires that run through the cable, and the sheaths and connective tissue coverings correspond to the insulation. Within the CNS, bundles of axons are called **tracts** or *pathways* rather than nerves.

Outside the CNS, a mass of neuron cell bodies is called a **ganglion** (plural, *ganglia*). Inside the CNS, a collection of cell bodies is called a **nucleus,** rather than a *ganglion.*

Glial cells play critical roles in neural function

Glial cells collectively make up the *neuroglia* (literally, "nerve glue"). Glial cells were once thought to be passive cells that simply support and protect the neurons. Armed with more sophisticated technology, researchers are discovering that glial cells communicate and carry out major regulatory functions.

More than 75% of the cells in the human CNS are glial cells. They account for only about 50% of the brain volume because they do not branch as extensively as neurons. Four types of glial cells are found in the vertebrate CNS: astrocytes, oligodendrocytes, ependymal cells, and microglia (TABLE 41-1).

Astrocytes are star-shaped glial cells that provide both physical support and nutrients for neurons. Astrocytes help regulate the composition of the extracellular fluid in the CNS by removing excess potassium ions. This action helps maintain normal neuron excitability. Some astrocytes position the ends of their long processes on blood vessels in the brain. In response, endothelial cells lining the blood vessels form *tight junctions* (see Fig. 5-23a) that prevent many potentially harmful substances in the blood from entering the brain tissue. This protective wall is the *blood–brain barrier.* Astrocytes also help

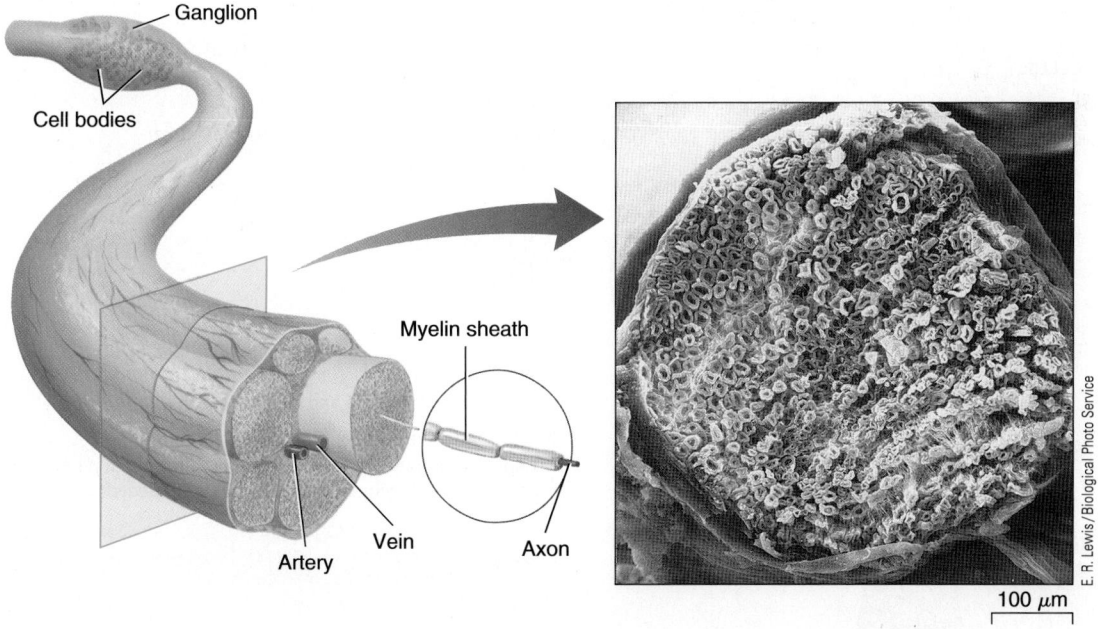

(a) A nerve consists of bundles of axons held together by connective tissue. The cell bodies belonging to the axons of a nerve are grouped in a ganglion.

(b) SEM showing a cross section through a myelinated afferent nerve of a bullfrog.

Figure 41-3 *Animation* **Structure of a myelinated nerve**
© Cengage Learning

| TABLE 41-1 | Functions of Glial Cells |

TYPE OF GLIAL CELL	FUNCTIONS
ASTROCYTES 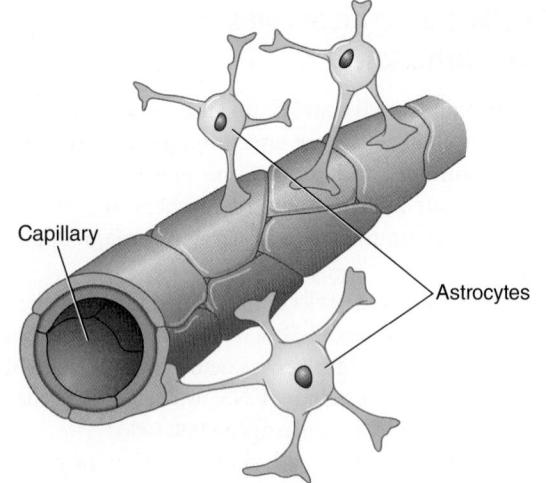 Capillary — Astrocytes	Physically support neurons Help transfer nutrients to neurons Remove excess K^+, which helps regulate composition of extracellular fluid in CNS Induce blood vessels to form blood–brain barrier and help maintain the barrier Communicate with one another and with neurons Induce synapse formation and strengthen synaptic transmission by chemically signaling neurons Respond to neurotransmitters and help regulate neurotransmitter reuptake May be important in memory and learning Guide neurons during embryonic development
OLIGODENDROCYTES 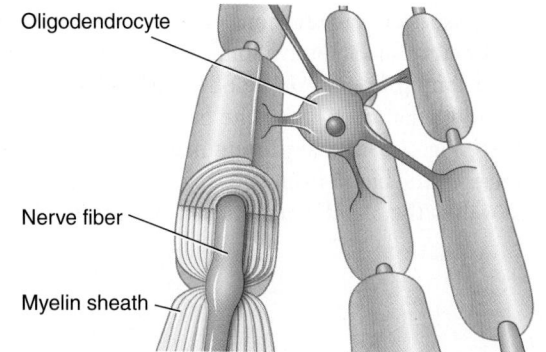 Oligodendrocyte Nerve fiber Myelin sheath	Form myelin sheaths around neurons in CNS
EPENDYMAL CELLS 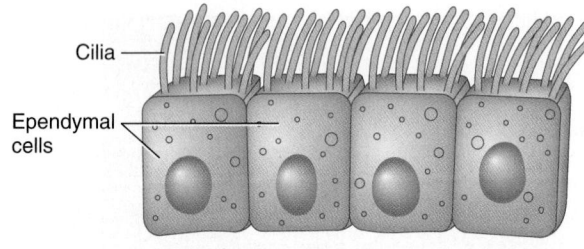 Cilia Ependymal cells	Line cavities of CNS Help produce and circulate cerebrospinal fluid May function as neural stem cells
MICROGLIA 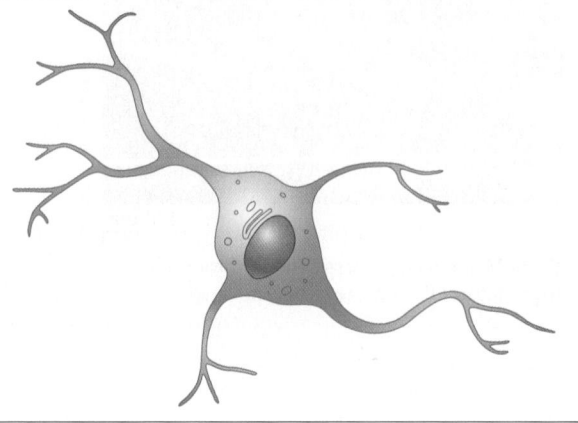	Phagocytosis of bacteria and debris Release signaling molecules that mediate inflammation May play important role in regulating the number of neurons in the developing brain by destroying neurons that are not needed

regulate self-cleaning mechanisms in the brain that involve flushing out wastes with cerebrospinal fluid.

Neurobiologists have shown that astrocytes are functionally connected throughout the brain. These interesting glial cells participate in information signaling by coordinating activity among neurons. They communicate through *gap junctions* and with signaling molecules. Although astrocytes can generate only weak electrical signals, they communicate with one another and with neurons by means of chemical signals. Astrocytes appear to respond to neurotransmitters released by neurons. They also help regulate the reuptake of excess neurotransmitters from synapses. Astrocytes induce synapse formation and can strengthen activity of synapses in the brain. Their actions appear to be important in learning and memory.

Astrocytes play a role in development by guiding neurons to appropriate locations in the body. Certain astrocytes function as stem cells in the brain and spinal cord. These cells can give rise to new neurons, additional astrocytes, and certain other glial cells. Taken out of their normal environment in the adult mouse brain, astrocytes can give rise to cells of all germ layers (the embryonic tissue layers: ectoderm, mesoderm, and endoderm). Human astrocytes may someday be used to produce specific types of cells needed for treating various medical conditions.

Oligodendrocytes are glial cells that envelop neurons in the CNS, forming insulating myelin sheaths around them. Because myelin is an excellent electrical insulator, its presence speeds transmission of neural impulses. In the neurological disease **multiple sclerosis,** patches of myelin deteriorate at irregular intervals along axons in the CNS and are replaced by scar tissue. This damage interferes with conduction of neural impulses, and the victim suffers loss of coordination, tremor, and partial or complete paralysis of parts of the body. Multiple sclerosis affects more than 2 million people worldwide.

Do not confuse oligodendrocytes with Schwann cells. Although sometimes considered glial cells, Schwann cells are located outside the CNS. Schwann cells form myelin sheaths around axons of many neurons that are part of the PNS.

Ependymal cells are ciliated glial cells that line the internal cavities of the CNS. Ependymal cells help produce and circulate the *cerebrospinal fluid* that bathes the brain and spinal cord of vertebrates (discussed in Chapter 42). There is some evidence that ependymal cells function as neural stem cells, and that they have the potential to give rise to new neurons, as well as to glial cells.

Microglia are actually specialized macrophages (phagocytic cells that ingest and digest cell debris and bacteria). Microglia respond to signals from neurons and are important in mediating responses to injury or disease. These cells are found near blood vessels. When the brain is injured or infected, microglia multiply and move to the affected area. There they remove bacteria and cell debris by phagocytosis. In addition, they release signaling molecules (also produced by macrophages and certain other cells in the immune system) that mediate inflammation. Microglia appear to help regulate the number of neurons in the developing brain by destroying neurons that are not needed.

CHECKPOINT 41.2

- **VISUALIZE** Sketch a multipolar neuron and label the cell body, dendrites, axon, and synaptic terminals. Give the function of each structure.
- How is a neuron different from a nerve?
- What are the functions of astrocytes? of microglia?

41.3 TRANSMITTING INFORMATION ALONG THE NEURON

LEARNING OBJECTIVES

4 Explain how the neuron develops and maintains a resting potential.

5 Compare a graded potential with an action potential, describing the production and transmission of each.

6 Contrast continuous conduction with saltatory conduction.

Most animal cells have a difference in electric charge across the plasma membrane: the electric charge inside the cell is more negative than the electric charge of the extracellular fluid. The plasma membrane is said to be electrically *polarized,* meaning that one side, or pole, has a different charge than the other side. When electric charges are separated in this way, a potential energy difference exists across the membrane.

The difference in electric charge across the plasma membrane gives rise to an *electrical gradient.* Voltage is the force that causes charged particles to flow between two points. The voltage measured across the plasma membrane is a form of potential energy called the **membrane potential.** If the charges are permitted to come together, they have the ability to do work. Thus, the cell can be thought of as a biological battery. In excitable cells, such as neurons and muscle cells, the membrane potential can change rapidly. Such changes can transmit signals to other cells.

Ion channels and pumps maintain the resting potential of the neuron

The membrane potential in a resting (not excited) neuron or muscle cell is its **resting potential.** The resting potential is generally expressed in units called *millivolts* (mV). (A millivolt equals one thousandth of a volt.) Like other cells that can produce electrical signals, the neuron has a resting potential of about 70 mV. By convention, this value is expressed as −70 mV because the cytosol close to the plasma membrane is negatively charged relative to the extracellular fluid (FIG. 41-4a).

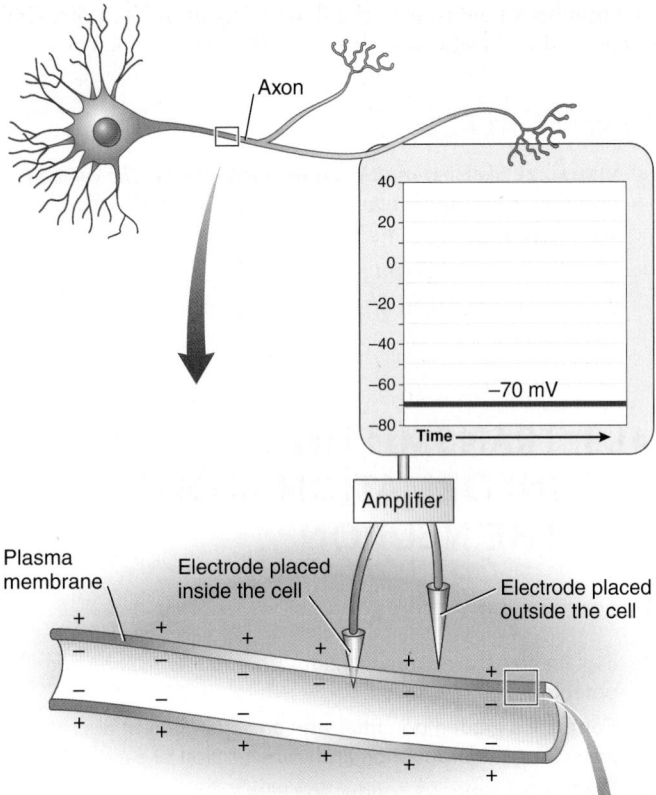

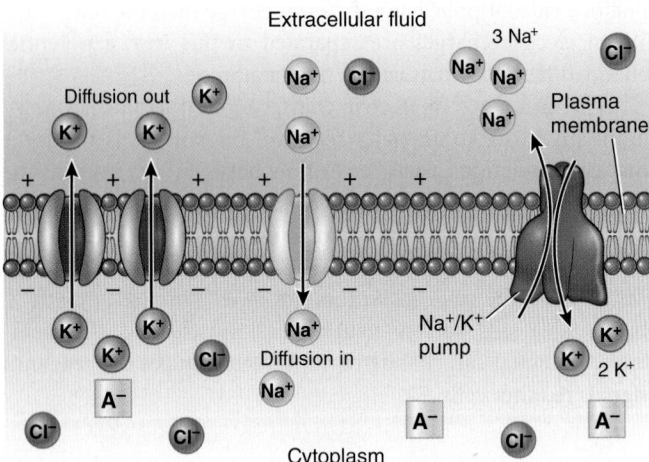

(a) Measuring the resting potential of a neuron. The axon is negatively charged compared with the surrounding extracellular fluid. The difference in electric charge across the plasma membrane is measured by placing an electrode just inside the neuron and a second electrode in the extracellular fluid just outside the plasma membrane.

(b) Permeability of the neuron membrane. The membrane is less permeable to Na^+ than to K^+. Passive ion channels allow K^+ to diffuse out along their concentration gradient. Potassium ions are mainly responsible for the voltage across the membrane. Sodium–potassium pumps in the plasma membrane actively pump Na^+ out of the cell and pump K^+ in. The plasma membrane is permeable to negatively charged Cl^-, and they contribute slightly to the negative charge. Large anions (A^-) such as proteins also contribute to the negative charge.

Figure 41-4 *Animation* **Resting potential**

© Cengage Learning

Biologists can measure the potential across the membrane by placing one electrode inside the cell and a second electrode outside the cell and connecting them through a very sensitive voltmeter or oscilloscope. If you place both electrodes on the outside surface of the neuron, no potential difference between them is registered. (All points on the same side of the membrane have the same charge.) However, once one of the electrodes penetrates the cell, the voltage changes from zero to approximately −70 mV.

Two main factors determine the magnitude of the membrane potential: (1) differences in the concentrations between specific ions inside the cell and those in the extracellular fluid and (2) selective permeability of the plasma membrane to these ions. The distribution of ions inside neurons and in the extracellular fluid surrounding them is like that of most other cells in the body. The potassium ion (K^+) concentration is about 10 times greater inside the cell than outside the cell. In contrast, the sodium ion (Na^+) concentration is about 10 times greater outside than inside. This asymmetric distribution of ions across the plasma membrane at rest is brought about by the action of selective ion channels and ion pumps. In vertebrate neurons (and skeletal muscle fibers), the resting membrane potential depends mainly on the diffusion of ions down their *concentration gradients* (see discussion of diffusion in Chapter 5).

Ions cross the plasma membrane by diffusion through ion channels

Recall that net movement of ions occurs from an area of higher concentration of that type of ion to one of lower concentration. Membrane proteins form ion channels that typically allow only specific types of ions to pass. Neurons have three types of ion channels: *passive ion channels, voltage-activated channels,* and *chemically activated ion channels.* **Passive ion channels** permit the passage of specific ions such as Na^+, K^+, Cl^-, and Ca^{2+} (**FIG. 41-4b**). Unlike voltage-activated and chemically activated ion channels, passive ion channels are not controlled by gates.

Potassium channels are the most common type of passive ion channel in the plasma membrane, and cells are more permeable to potassium than to other ions. In fact, in the resting neuron, the plasma membrane is up to 100 times more permeable to K^+ than to Na^+. When pumped out of the neuron, Na^+ cannot easily pass back into the cell. However, K^+ pumped into the neuron easily diffuses out.

Potassium ions leak out through passive ion channels following their concentration gradient. As K^+ diffuse out of the neuron, they increase the positive charge in the extracellular fluid relative to the charge inside the cell. The resulting change in the electrical gradient across the membrane influences the flow of ions. This electrical gradient forces some of the positively charged potassium ions back into the cell.

The membrane potential at which the flux (flow) of K^+ inward (due to the electrical gradient) equals the flux of K^+ outward (because of the concentration gradient) is the

equilibrium potential for K⁺. The equilibrium potential for any particular ion is a steady state in which opposing chemical and electrical fluxes are equal and there is no net movement of the ion. The equilibrium potential for K⁺ in the typical neuron is −90 mV. The equilibrium potential for Na⁺ is +60 mV; it differs from the equilibrium potential of K⁺ because of the difference in concentration of Na⁺ across the membrane (concentration is high outside and low inside the cell). Because the membrane is much more permeable to K⁺ than to Na⁺, the resting potential of the neuron is closer to the potassium equilibrium potential than to the sodium equilibrium potential. (Remember that the resting potential of the neuron is about −70 mV.)

The resting potential is primarily established by the net flow of K⁺ into the neuron. However, the net flow of Na⁺ outward contributes to the resting potential. Because the plasma membrane is permeable to negatively charged chloride ions, Cl⁻ also contribute slightly. These ions accumulate in the cytosol close to the plasma membrane. Negative charges on large molecules such as proteins add to the negative charge in the cytosol. These large anions cannot cross the plasma membrane.

Ion pumping maintains the gradients that determine the resting potential

The neuron plasma membrane has very efficient **sodium–potassium pumps** that actively transport Na⁺ out of the cell and K⁺ into the cell (see Fig. 41-4b; also see Fig. 5-16). These pumps require ATP to pump Na⁺ and K⁺ against their concentration and electrical gradients. For every three Na⁺ pumped out of the cell, two K⁺ are pumped in. Thus, more positive ions are pumped out than in. The sodium–potassium pumps maintain a higher concentration of K⁺ inside the cell than outside and a higher concentration of Na⁺ outside than inside.

In summary, the development of the resting potential depends mainly on the diffusion of K⁺ out of the cell. However, once the neuron reaches a steady state, net Na⁺ diffusion into the cell is greater than net K⁺ diffusion out of the cell. This situation is offset by the sodium–potassium pumps that pump three Na⁺ out of the cell for every two K⁺ pumped in.

Graded local signals vary in magnitude

Neurons are excitable cells. They respond to stimuli and can convert stimuli into nerve impulses. An electrical, chemical, or mechanical stimulus may alter the resting potential by increasing the membrane's permeability to sodium ions.

When a stimulus causes the membrane potential to become less negative (closer to zero) than the resting potential, that region of the membrane is **depolarized.** Because depolarization brings a neuron closer to transmitting a neural impulse, it is described as *excitatory.* In contrast, when the membrane potential becomes more negative than the resting potential, the membrane is **hyperpolarized.** Hyperpolarization is *inhibitory;* it decreases the neuron's ability to generate a neural impulse.

A stimulus may change the potential in a relatively small region of the plasma membrane. Such a **graded potential** is a local response. It functions as a signal only over a very short distance because it fades out within a few millimeters of its point of origin. A graded potential varies in magnitude; that is, the potential charge varies depending on the strength of the stimulus applied. A larger stimulus causes a larger change in permeability and a greater change in membrane potential.

Axons transmit signals called action potentials

When a stimulus is strong enough, a rapid, large change in membrane potential occurs, depolarizing the membrane to a critical point known as the **threshold level.** At that point, the neuron fires a *nerve impulse,* or **action potential,** an electrical signal that travels rapidly down the axon into the synaptic terminals. All cells can generate graded potentials, but only neurons, muscle cells, and a few cells of the endocrine and immune systems can generate action potentials.

In a series of experiments in the 1940s, pioneering English researchers Alan Hodgkin and Andrew Huxley inserted electrodes into large axons found in squids. By measuring voltage changes as they varied ion concentrations, these investigators showed that passage of Na⁺ ions *into* the neuron and K⁺ ions *out* of the neuron resulted in an action potential. The process of Na⁺ rushing into the cell may be thought of as fueling the action, whereas the process of K⁺ leaving the cell brakes the action potential. In 1963, Hodgkin and Huxley received the Nobel Prize in Physiology or Medicine for this research.

We now know that changes in voltage regulate specific **voltage-activated ion channels** (also called *voltage-gated ion channels*) in the plasma membrane of the axon and cell body. Biologists can study these channels by using the **patch clamp technique** to measure currents across a very small segment of the plasma membrane rather than across the entire membrane (**FIG. 41-5**). The patch of membrane measured is so small that it may contain a single channel.

Biologists have used certain toxins that affect the nervous system to investigate how membrane channels function. For example, the poison tetrodotoxin (TTX) binds to voltage-activated Na⁺ channels, blocking the passage of Na⁺. Using TTX and other toxins, researchers were able to identify the channel protein. TTX was first isolated from the Japanese puffer fish, which is eaten as a delicacy (**FIG. 41-6**). In tiny amounts TTX tingles the taste buds, but a slightly larger amount can prevent breathing. (The Japanese government operates a certification program that teaches puffer fish chefs to carefully remove the organs containing the toxin.)

Investigators have established that voltage-activated ion channels have charged regions that act as gates. Voltage-activated Na⁺ channels have two gates: an *activation gate* and an *inactivation gate* (**FIG. 41-7a**). Voltage-activated K⁺ channels have a single gate (**FIG. 41-7b**).

Local anesthetics, such as procaine (novocaine) and lidocaine (Xylocaine) as well as cocaine, bind to voltage-activated

WHY IS IT USED? The patch clamp technique allows researchers to study the action of a single ion channel over time, even in very small cells. This research method was developed in the mid-1970s by Erwin Neher and Bert Sakmann, both of the Max Planck Institute in Germany. They were awarded a Nobel Prize in Physiology or Medicine in 1991 for this work.

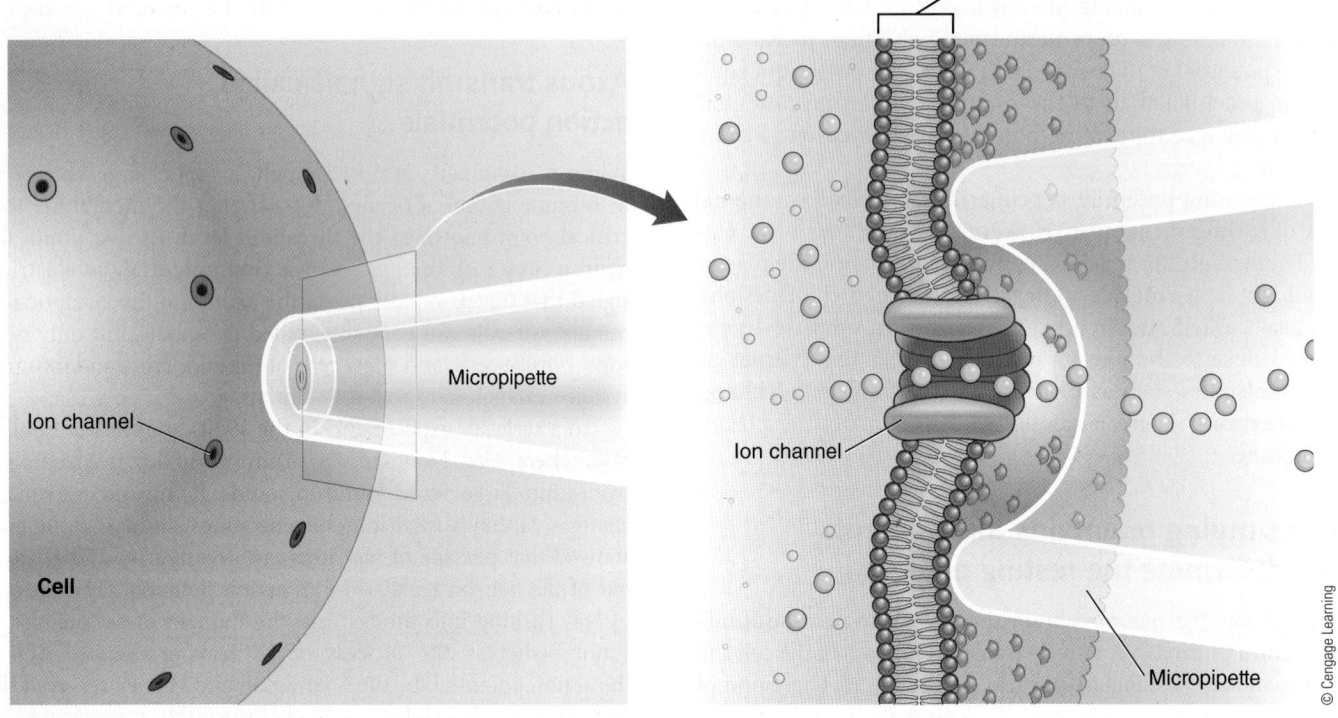

HOW IS IT DONE? The tip of a micropipette is tightly sealed to a patch of membrane so small that it generally contains only a single ion channel. The flow of ions through the channel can be measured using an extremely sensitive recording device. Cell biologists have found that the current flow is intermittent and corresponds to the opening and closing of the ion channel. The permeability of the channel affects the magnitude of the current.

Figure 41-5 The patch clamp technique

The patch clamp technique has been modified in many ways and has been applied to studies of the roles of ion channels in a wide range of cellular processes in both plants and animals. For example, neuroscientists once thought that transmitting action potentials was a very inefficient process. They thought that there was a high degree of overlap between the time that Na$^+$ rush into the cell and K$^+$ move out of the cell. This process would be a waste of energy because it would be like stepping on the gas and brake pedals at the same time. Henrik Alle, a neuroscientist at the Max Planck Institute for Brain Research in Frankfurt, Germany, and his colleagues used the patch clamp technique to record currents in neurons of the learning and memory areas of rat brains. They found that the flows of Na$^+$ and K$^+$ were almost completely separated in time. Na$^+$ have almost finished entering the neuron before K$^+$ ions begin to exit. In contrast to earlier estimates of 30% efficiency for transmission of signals, neurons may transmit action potentials with more than 70% efficiency.

Na$^+$ channels and block them. These blocked channels cannot open in response to depolarization, and the neuron cannot transmit an impulse through the anesthetized region. Signals do not reach the brain, so pain is not experienced.

An action potential is generated when the voltage reaches threshold level

The membrane of most neurons can depolarize by about 15 mV—that is, to a potential of about −55 mV—without initiating an action potential. However, when depolarization is greater than −55 mV, the threshold level is reached, and an action potential is generated. The neuron membrane quickly reaches zero potential and even *overshoots* to +35 mV or more as a momentary reversal in polarity takes place. The sharp rise and fall of the action potential is called a *spike*. **FIGURE 41-8a** illustrates an action potential recorded by placing one electrode inside an axon and one just outside.

At resting potentials, activation gates of the voltage-activated Na$^+$ channels are closed, and Na$^+$ cannot pass into the neuron (**FIG. 41-8b**). Follow the steps illustrated in Figure 41-8b as you read about transmission of action potentials, repolarization, and return to the resting state. When the voltage reaches threshold level, the channel proteins change shape, opening

Figure 41-6 Fish with tetrodotoxin
Like the Japanese puffer fish (*Fuju rubripes*), this closely related porcupine fish has tetrodotoxin in its internal organs. This fish can inflate its body by swallowing air or water. When it is larger and rounder, it is safer from many predators.

the activation gates. Sodium ions can then flow through the voltage-activated Na⁺ channels. As a result, the inside of the neuron becomes positively charged relative to the fluid outside the plasma membrane, generating an action potential. Inactivation gates that are open in the resting neuron close

slowly in response to depolarization. Sodium ions can pass through the channel only during the brief period when both activation and inactivation gates are open.

The generation of an action potential depends on a **positive feedback system.** (Recall from Chapter 39 that in a positive feedback system, a change in some condition triggers a response that *intensifies* the change.) As just discussed, when a neuron becomes depolarized, voltage-activated Na⁺ channels open, increasing the membrane permeability to Na⁺. Sodium ions diffuse into the cell, moving from an area of higher concentration to an area of lower concentration. As Na⁺ flow into the cell, the cytosol becomes positively charged relative to the extracellular fluid. This charge further depolarizes the neuron so that more voltage-activated Na⁺ channels open, which increases membrane permeability to Na⁺.

The neuron repolarizes and returns to a resting state

After a certain period, inactivation gates close in the Na⁺ channels, and the membrane again becomes impermeable to Na⁺. This inactivation begins the process of **repolarization,** during which the membrane potential returns to its resting level. Voltage-activated K⁺ channels slowly open in response to depolarization. When Na⁺ have almost stopped rushing into the neuron, K⁺ channels open, and K⁺ leak out of the neuron. This decrease in intracellular K⁺ returns the interior of the membrane to its relatively negative state, repolarizing the membrane.

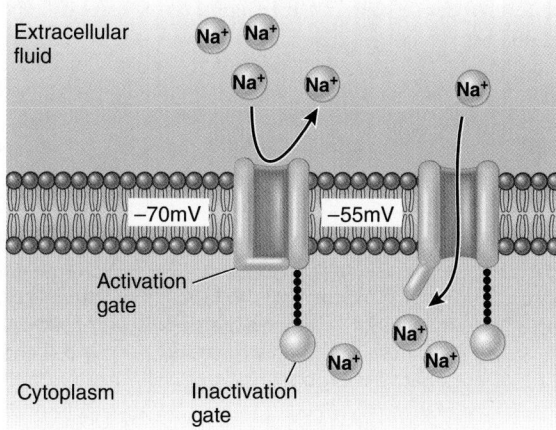

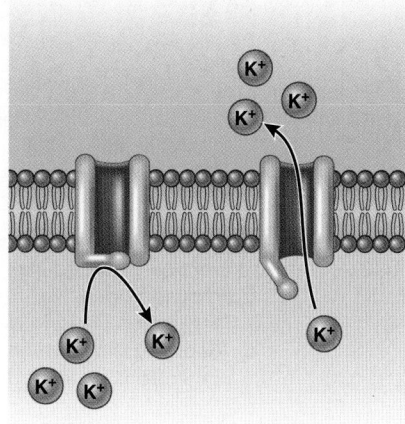

(a) Sodium channels. In the resting state, voltage-activated Na⁺ channels are closed (*left*). When the voltage reaches threshold level, activation gates open quickly, allowing Na⁺ to pass into the cell (*right*). After a certain time, inactivation gates close, blocking the channels (*not shown*). Inactivation gates are open when a neuron is in the resting state; they close slowly in response to depolarization. The ion channel is open only during the brief period when both activation and inactivation gates are open.

(b) Potassium channels. Voltage-activated K⁺ channels have activation gates that open slowly in response to depolarization. The activation gates close after the resting potential has been restored. Potassium channels have no inactivation gates.

Figure 41-7 *Animation* **Voltage-activated ion channels**
© Cengage Learning

When the neuron plasma membrane is depolarized to threshold level, voltage-activated Na⁺ channels open and Na⁺ rush into the neuron, causing greater depolarization; an action potential is generated.

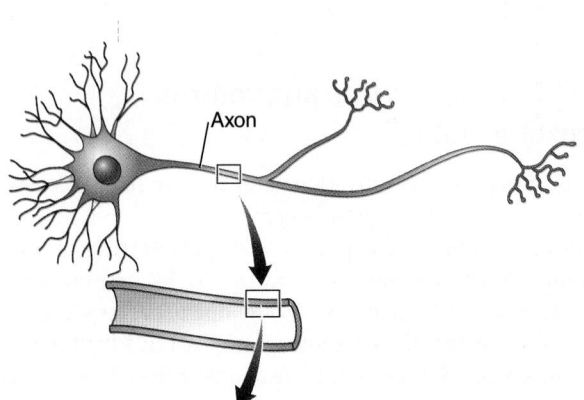

(a) Action potential. When the axon depolarizes to about –55 mV, an action potential is generated. (The numerical values vary for different nerve cells.)

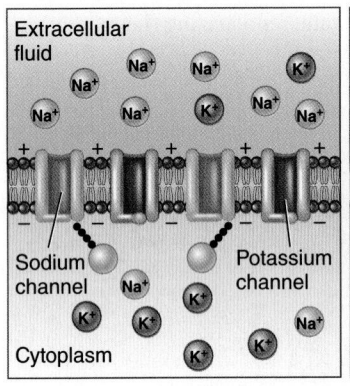

❶ **Resting state.** Voltage-activated Na⁺ and K⁺ channels are closed.

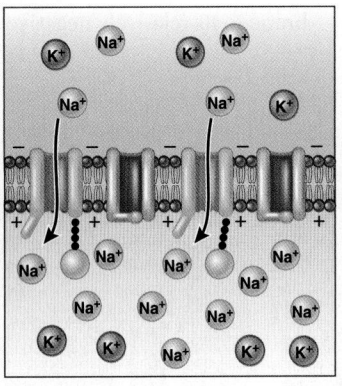

❷ **Depolarization.** At threshold, voltage-activated Na⁺ channels open. Na⁺ rush into the neuron, causing further depolarization. Action potential is generated.

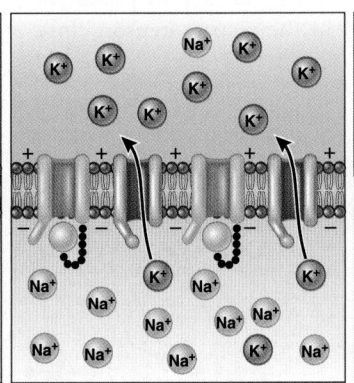

❸ **Repolarization.** Voltage-activated Na⁺ channels become inactivated. Voltage-activated K⁺ channels are open; K⁺ diffuse out of cell, restoring negative charge to inside of cell.

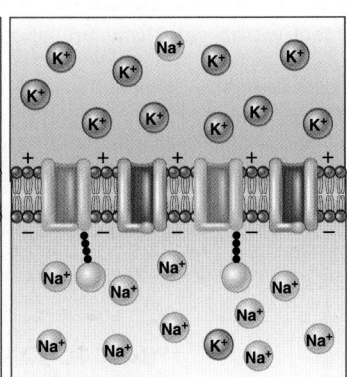

❹ **Return to resting state.** Voltage-activated Na⁺ and K⁺ channels close. Na⁺/K⁺ pumps (*not shown*) move Na⁺ out of the cell and K⁺ into the cell, restoring the original ion distribution.

(b) The action of ion channels in the plasma membrane determines the state of the neuron.

Figure 41-8 *Animation* **Action potentials depend on voltage-gated ion channels**

Generation of an action potential depends on the action of voltage-gated sodium and potassium ion channels in the plasma membrane.

© Cengage Learning

PREDICT One of the toxins (charybdotoxin) produced by scorpions blocks K⁺ channels. What effect would this toxin have on the action of a neuron?

Voltage-activated K⁺ channels remain open until the resting potential has been restored.

As a *wave of depolarization* moves down the membrane of the neuron, the normal polarized state is quickly re-established behind it. The entire process—depolarization and then repolarization—can take place in less than 1 millisecond.

During the millisecond or so in which it is depolarized, the axon membrane is in an *absolute* **refractory period:** it cannot transmit another action potential no matter how great a stimulus is applied because the voltage-activated Na⁺ channels are inactivated. Until their gates are reset, they cannot open again. When enough Na⁺ channel gates have been reset, the neuron enters a *relative* refractory period that lasts for a few more milliseconds. During this period, the axon can transmit impulses, but the threshold is higher (less negative). Even with the limits imposed by their refractory periods, most neurons can transmit several hundred impulses per second.

In summary, neural action proceeds as follows:

neuron in resting state ⟶ stimulus causes depolarization ⟶ threshold reached ⟶ action potential transmits signal ⟶ repolarization and return to resting state

The action potential is an all-or-none response

Any stimulus too weak to depolarize the plasma membrane to threshold level cannot fire the neuron. It merely sets up a local signal that fades and dies within a few millimeters from the point of stimulus. Only a stimulus strong enough to depolarize the membrane to its critical threshold level results in transmission of an impulse along the axon. An action potential is an **all-or-none response** because either it occurs or it does not. No variation exists in the strength of a single impulse.

If the all-or-none law is valid, how can we explain differences in the levels of intensity of sensations? After all, we have no difficulty distinguishing between the pain of a severe toothache and that of a minor cut on the arm. We can explain this apparent inconsistency by understanding that the intensity of sensation depends on several factors, including the *number of neurons stimulated* and their *frequency of discharge*. For example, suppose that you burn your hand. If a large area is burned, more pain receptors are stimulated and more neurons become depolarized. Also, each neuron transmits more action potentials per unit of time.

An action potential is self-propagating

Once initiated, an action potential continues along the neuron. During depolarization, the affected area of the membrane is more positive relative to adjacent regions where the membrane is still at resting potential. The difference in potentials between active and resting membrane regions causes ions to flow between them (an electric current). The flow of Na⁺ into the adjacent region induces the opening of voltage-activated Na⁺ channels in that area of the membrane (**FIG. 41-9**). The process is repeated like a chain reaction until the end of the axon is reached. Thus, an

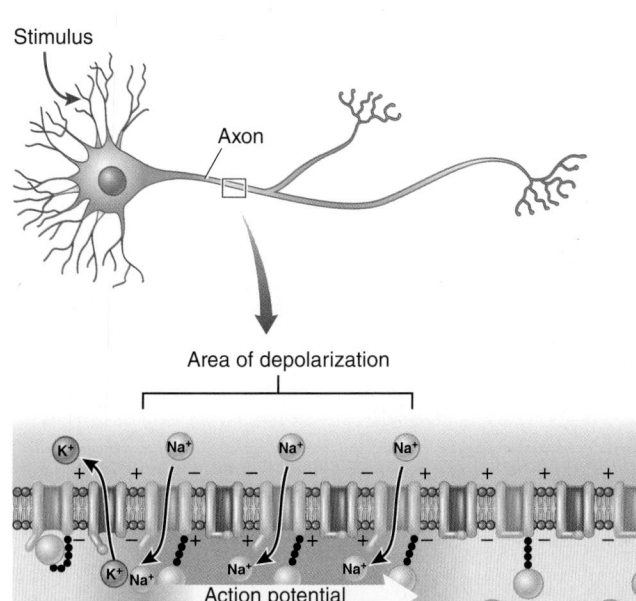

① An action potential is transmitted as a wave of depolarization that travels down the axon. At the region of depolarization, Na⁺ diffuse into the cell.

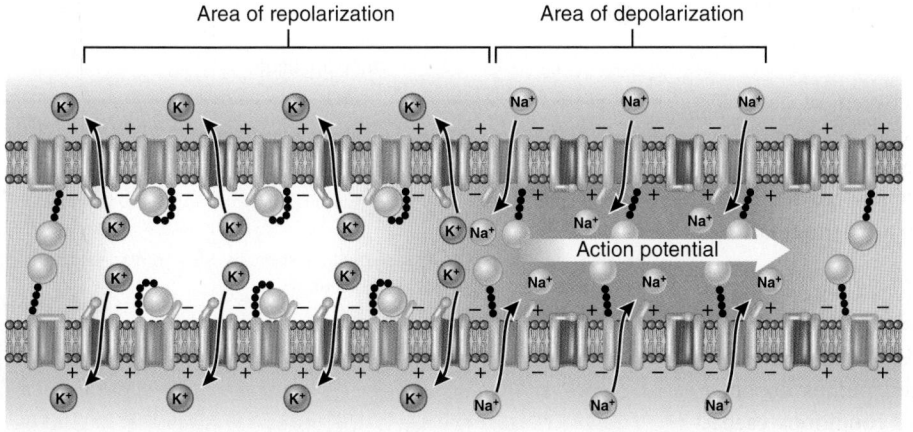

② As an action potential progresses along the axon, repolarization occurs quickly behind it.

Figure 41-9 *Animation* Transmission of an action potential along the axon

When the dendrites (or cell body) of a neuron are stimulated sufficiently to depolarize the membrane to its threshold level, a strong pulse of voltage change travels down the axon. This pulse is the action potential.
© Cengage Learning

action potential is a **wave of depolarization** that travels down the length of the axon.

Transmission of an action potential is sometimes compared with a flame moving along a trail of gunpowder. Once the gunpowder is ignited at one end of the trail, the flame moves steadily to the other end by igniting the powder particles ahead of it. An amazing difference is that after a brief refractory period, the neuron recovers and is able to transmit another action potential.

Several factors determine the velocity of an action potential

Compared with the flow of electrons in electrical wiring or the speed of light, a nerve impulse travels rather slowly. Most axons transmit impulses at about 1 to 10 m per second. The smooth, progressive impulse transmission just described, called *continuous conduction,* occurs in unmyelinated neurons. In these axons the speed of transmission is proportional to the diameter of the axon. Axons with larger diameters transmit faster than those with smaller diameters because an axon with a large diameter presents less internal resistance to the ions flowing along its length. Squids and certain other invertebrates have giant axons, up to 1 mm in diameter, which allow them to respond rapidly when escaping from predators.

In vertebrates another strategy that speeds transmission has evolved—myelinated neurons. Myelin acts as an effective electrical insulator around the axon. However, the axon is not myelinated at the nodes of Ranvier. At these nodes, the plasma membrane of the axon makes direct contact with the surrounding extracellular fluid. Voltage-activated Na^+ and K^+ channels are concentrated at the nodes. In fact, in myelinated neurons ion movement across the membrane occurs only at the nodes.

The ion activity at an active node results in diffusion of ions along the axon that depolarize the next node. The action potential appears to jump from one node of Ranvier to the next (FIG. 41-10). This type of impulse transmission is known as **saltatory conduction** (from the Latin word *saltus,* meaning "to leap").

In myelinated neurons the distance between successive nodes of Ranvier affects speed of transmission. When nodes are farther apart, less of the axon must depolarize, and the axon conducts the impulse faster. Using saltatory conduction, a myelinated axon can conduct an impulse up to 50 times faster than the fastest unmyelinated axon.

Saltatory conduction has another advantage over continuous conduction: it requires less energy. Ions move across the plasma membrane only at the nodes, so fewer Na^+ and K^+ are displaced. As a result, the sodium–potassium pumps do not expend as much ATP to re-establish resting conditions each time an impulse is conducted.

CHECKPOINT 41.3

- *How do ion channels and sodium–potassium pumps contribute to the resting potential?*

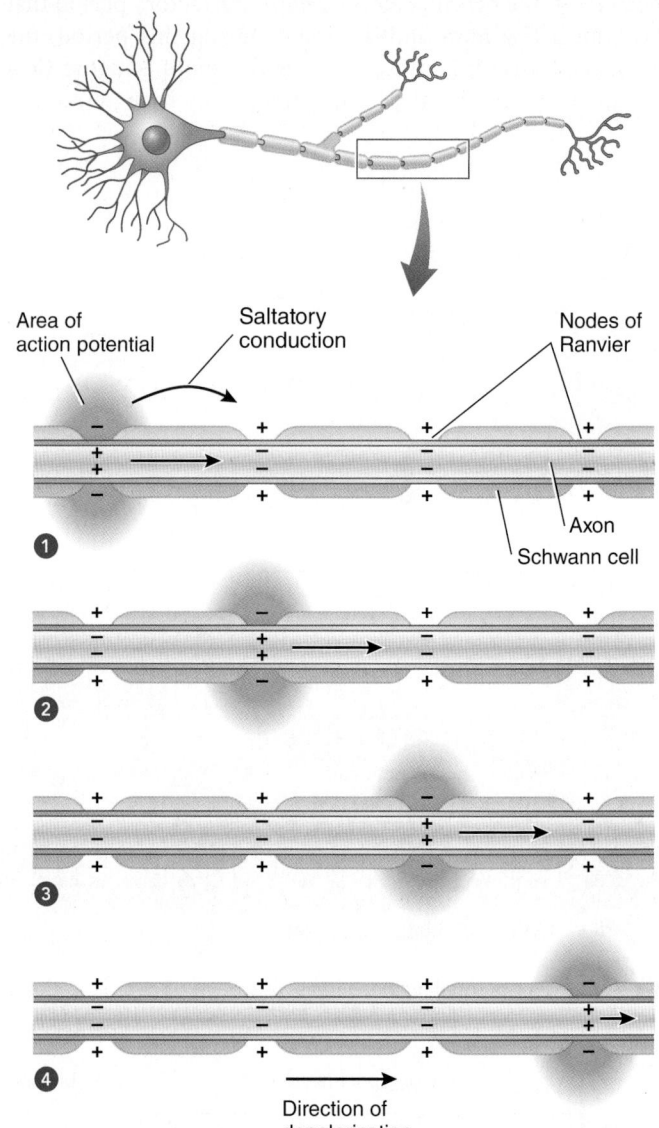

KEY POINT

In saltatory conduction Na^+ rapidly diffuse from one node of Ranvier to the next. Action potentials appear to leap from node to node.

Area of action potential — Saltatory conduction — Nodes of Ranvier

Axon

Schwann cell

Direction of depolarization

Figure 41-10 *Animation* **Saltatory conduction**

This sequence of diagrams (steps 1 through 4) illustrates transmission of an action potential in a myelinated neuron. The *arrows* in the axons indicate diffusion of Na^+.

PREDICT How might the disease multiple sclerosis affect transmission of action potentials along a myelinated neuron?

© Cengage Learning

- *Describe the sequence of events that occurs when threshold level is reached in a neuron.*
- *Contrast an action potential with a graded potential.*
- *What are the benefits of saltatory conduction compared with continuous conduction?*

41.4 TRANSMITTING INFORMATION ACROSS SYNAPSES

A **synapse** is a junction between two neurons or between a neuron and an effector, such as between a neuron and a muscle cell. A neuron that *terminates* at a specific synapse is called a **presynaptic neuron,** whereas a neuron that *begins* at that synapse is a **postsynaptic neuron.** Note that these terms are relative to a specific synapse. A neuron that is postsynaptic with respect to one synapse may be presynaptic to the next synapse in the sequence.

Signals across synapses can be electrical or chemical

Based on how presynaptic and postsynaptic neurons communicate, two types of synapses have been identified: electrical synapses and chemical synapses. In an **electrical synapse** the presynaptic and postsynaptic neurons occur very close together (within 2 nm of one another) and form *gap junctions* (see Fig. 5-23c). The interiors of the two neurons are physically connected by a protein channel. Electrical synapses let ions pass from one cell to another, permitting an impulse to be directly and rapidly transmitted from a presynaptic to a postsynaptic neuron. The escape responses of many animals involve electrical synapses. For example, the "tail-flick" escape response of the crayfish involves giant neurons in the nerve cord that form electrical synapses with large motor neurons. The motor neurons signal muscles in the abdomen to contract. Electrical synapses are important in the mammalian nervous system as well. For example, many interneurons communicate by way of electrical synapses.

The majority of synapses are **chemical synapses.** Presynaptic and postsynaptic neurons are separated by a space, the **synaptic cleft,** which is about 20 nm wide (less than one-millionth of an inch). Depolarization is a property of the plasma membrane, so when an action potential reaches the end of the axon, it cannot jump the gap. The electrical signal must be converted into a chemical one.

When an action potential reaches a synaptic terminal, it triggers the release of a *neurotransmitter,* a chemical messenger that conducts the neural signal across the synapse. Neurotransmitter molecules then bind to chemically activated ion channels in the membrane of the postsynaptic neuron. This binding triggers specific gated ion channels to open (or close), resulting in changes in permeability of the postsynaptic membrane. When a postsynaptic neuron reaches its threshold level of depolarization, it transmits an action potential.

Neurons use neurotransmitters to signal other cells

Neurotransmitters signal other neurons, muscle fibers, or gland cells. We can categorize neurotransmitters into different chemical groups, including acetylcholine, biogenic amines, amino acids, neuropeptides, and gaseous neurotransmitters (TABLE 41-2).

Acetylcholine is a low-molecular-weight neurotransmitter that is released from motor neurons and triggers muscle contraction. Acetylcholine is also released by some neurons in the brain and in the autonomic nervous system (see *Inquiring About: Alzheimer's Disease;* also see Chapter 42). Cells that release acetylcholine are called **cholinergic neurons.**

Neurons that release norepinephrine are called **adrenergic neurons. Norepinephrine, epinephrine,** and the neurotransmitter **dopamine** are **catecholamines.** The catecholamines and the neurotransmitters **serotonin** and **histamine** belong to a class of compounds called **biogenic amines.** Biogenic amines affect mood, and their imbalance has been linked to several disorders, including major depression, anxiety disorders, attention deficit hyperactivity disorder (ADHD), and schizophrenia.

Several amino acids function as neurotransmitters. **Glutamate** is the major excitatory neurotransmitter in the brain. One type of glutamate receptor is the target of several mind-altering drugs, including phencyclidine ("angel dust").

The amino acid **glycine** and a modified amino acid, **gamma-aminobutyric acid (GABA),** are neurotransmitters that inhibit neurons in the spinal cord and brain. Glycine is the main inhibitory neurotransmitter in the spinal cord. GABA is the neurotransmitter used at most rapid inhibitory synapses throughout the brain. Many drugs that reduce anxiety enhance the actions of GABA by permitting lower concentrations of GABA to open Cl^- channels. These drugs include benzodiazepines, such as diazepam (Valium) and alprazolam (Xanax), and barbiturates, such as phenobarbital.

Neuropeptides, such as endorphins and enkephalins, appear to function mainly as *neuromodulators,* messenger molecules that stimulate long-term changes that enhance or inhibit synaptic function. Some neurobiologists consider them a group of neurotransmitters, but others classify them as a separate group. **Endorphins,** such as *beta-endorphin,* and **enkephalins** are the body's own endogenous opioids. Opiate drugs—for example, morphine and codeine—are powerful analgesics, drugs that relieve pain without causing loss of consciousness. Endorphins and enkephalins bind to opioid receptors (the same receptors to which opiate drugs bind) and block pain signals. Opioids modulate the effect of other neurotransmitters, including **substance P,** which activates pathways that transmit pain signals from certain sensory neurons to the CNS. (We discuss pain perception in Chapter 43.)

The signaling molecule **nitric oxide (NO)** is a gas that diffuses across cell membranes. It can transmit information from the postsynaptic neuron to the presynaptic neuron, the opposite direction of transmission from other neurotransmitters. Thus, it acts as a *retrograde messenger* at some synapses.

TABLE 41-2 | Selected Neurotransmitters

NEUROTRANSMITTER	STRUCTURE	SITE OF ACTION	ACTION
ACETYLCHOLINE	$H_2C-\overset{\displaystyle O}{\overset{\|}{C}}-O-CH_2-CH_2-N^+-[CH_3]_3$	CNS, neuromuscular junctions	Excitatory effect on skeletal muscle; inhibitory effect on cardiac muscle; excitatory or inhibitory at other sites; acetylcholine concentration in the brain decreases during the progression of Alzheimer's disease
BIOGENIC AMINES Norepinephrine	(structure)	CNS, PNS	Excitatory or inhibitory; concentration in brain affects mood, sleep, wakefulness, attention
Dopamine	(structure)	CNS	Helps maintain balance between excitation and inhibition of neurons; important in motor functions; concentration in brain affects mood; also involved in motivation and reward
Serotonin (5-hydroxytryptamine, 5-HT)	(structure)	CNS	Excitatory effect on pathways that control muscle action; inhibitory effect on sensory pathways; helps regulate food intake, sleep, and wakefulness; affects mood
AMINO ACIDS Glutamate	$H_2N-CH-CH_2-CH_2-COOH$ 　　　$\|$ 　　$COOH$	CNS, invertebrate neuromuscular junctions	Principal excitatory neurotransmitter in vertebrate brain; functions in learning and memory; excitatory effect at invertebrate neuromuscular junctions
Aspartate	$H_2N-CH-CH_2-COOH$ 　　　$\|$ 　　$COOH$	CNS	Excitatory; functions in learning and memory
Glycine	H_2N-CH_2-COOH	CNS	Inhibitory (mainly in spinal cord)
Gamma-aminobutyric acid (GABA)	$H_2N-CH_2-CH_2-CH_2-COOH$	CNS, invertebrate neuromuscular junctions	Major inhibitory neurotransmitter in mammalian brain
NEUROPEPTIDES Endorphins and enkephalins	Y—G—G—F—M* (Methionine enkephalin)	CNS, PNS	Endogenous opiates; bind with opiate receptors on afferent fibers that transmit pain signals; inhibit release of substance P
Substance P	R—P—K—P—Q—Q—F—F—G—L—M	CNS, PNS	Activates ascending pathways that transmit pain signals
GASEOUS NEUROTRANSMITTER Nitric oxide	NO	Most neurons?	Diffuses across cell membranes; retrograde messenger; can transmit signals from postsynaptic to presynaptic neuron

*Letters represent amino acids; refer to Figure 3-17.

© Cengage Learning

Neurotransmitters bind with receptors on postsynaptic cells

Neurotransmitters are stored in synaptic terminals within hundreds of small membrane-enclosed sacs called *synaptic vesicles* (FIG. 41-11a). Each time an action potential reaches a synaptic terminal, voltage-gated Ca^{2+} channels open. Calcium ions from the extracellular fluid then flow into the synaptic terminal. The Ca^{2+} induce synaptic vesicles to fuse with the presynaptic membrane

What is Alzheimer's disease, and what causes it? **Alzheimer's disease (AD)** is a progressive, degenerative brain disorder in which neural signaling is disrupted and neurons become damaged and die prematurely. Alzheimer's disease is the leading cause of senile dementia, the loss of memory, judgment, and ability to reason that is sometimes associated with aging. The first symptom of AD is mild cognitive impairment, typically a problem with short-term memory. As the disease progresses, the ability to read, write, and think logically becomes compromised, and long-term memory is affected. Eventually, the AD patient loses the ability to carry on routine daily activities, such as dressing, bathing, and eating, and may no longer recognize family members and close friends. Patients typically die within 10 to 12 years of diagnosis.

Worldwide, more than 35 million people are affected by AD, and more than 5 million people in the United States suffer from this disease. One in eight older individuals in the United States has AD, and this disease is one of the top ten leading causes of death. Only 4% of those with AD are under age 65. These early-onset cases are caused by mutations in one of three genes. There is some evidence that you can reduce your risk for AD by engaging in regular physical exercise, pursuing higher education and working in a profession that is cognitively challenging, staying cognitively active, and interacting socially with other people.

AD typically develops over several decades. One startling study demonstrated that a person's writing early in life can predict the disease.

Young adults whose writings had a lower density of ideas were more likely to develop the disease than those whose density of ideas was higher. Early changes in the brain can now be detected by positron emission tomography scans, magnetic resonance imaging, and spinal taps years before AD symptoms are apparent (see Figure 1).

In AD neurons are damaged in the cerebral cortex and hippocampus, areas of the brain important in thinking and remembering. Neurons that secrete the neurotransmitter acetylcholine are especially affected. Three abnormalities that develop in brain tissue as humans age are especially characteristic of AD: *amyloid plaques, neurofibrillary tangles,* and death of neurons. Understanding the biochemistry and genetic basis of both plaques and tangles may provide clues to the causes and cures of AD.

Amyloid plaques were first identified in autopsied brains in 1906 by Alois Alzheimer, a German physician. These plaques are extracellular deposits of a peptide called amyloid-beta (amyloid-β) that are surrounded by degenerating dendrites and axons. Researchers found that amyloid-β is cut from a precursor protein called β-APP, a large protein in neuron plasma membranes. Normal amyloid-β molecules have 40 amino acids. However, about 10% of the peptides produced have two additional amino acids. This longer peptide appears to form the plaques. Neurons near these plaques appear swollen and misshapen. Microglia are typically present, suggesting an attempt to remove the damaged cells or the plaques themselves. The microglia cause inflammation and release toxic compounds that

damage nearby neurons. The abnormal peptide also appears to disrupt calcium regulation, which destroys neurons in the brain. Another hypothesis is that amyloid-β damages mitochondria, releasing harmful free radicals.

Amyloid-β and its toxic effects may lead to formation of **neurofibrillary tangles**. These tangles, which form in the neuron cytoplasm, are another piece of the AD puzzle. The cytoskeletal protein tau normally associates with the protein tubulin, which forms microtubules. According to one current hypothesis, amyloid-β interferes with the way the tau protein binds to microtubules, resulting in abnormal microtubules that cannot effectively transport materials through axons. Abnormal tau proteins accumulate in the cytoplasm, forming fibrous deposits that make up the neurofibrillary tangles (see Figure 2). These tangles interfere with neural signaling.

AD is currently treated with two types of drugs. The first type—for example, Aricept—inhibits the enzyme acetylcholinesterase, slowing breakdown of acetylcholine, a neurotransmitter that plays a role in memory and thought. The second type of drug—for example, Namenda—affects glutamate, a neurotransmitter important in learning and memory. Namenda-type drugs slow cognitive decline. However, none of the available drugs prevents the eventual damage that occurs in the brain. At least part of the problem may be that individuals with AD are not treated until symptoms are evident and the disease has already severely damaged the brain. Several new drugs designed to slow the progress of Alzheimer's disease are in clinical trials.

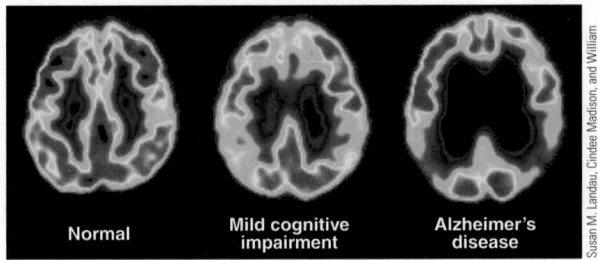

Figure 1 These fluorodeoxyglucose (FDG)-PET scans measure glucose metabolism. FDG-PET scanning is one of the major methods used by the Alzheimer's Disease Neuroimaging Initiative (ADNI) to track progression of Alzheimer's disease. *Orange* and *yellow* colors indicate metabolic activity. Note reduced metabolic activity in the PET scan image of the brain of a patient with Alzheimer's disease.

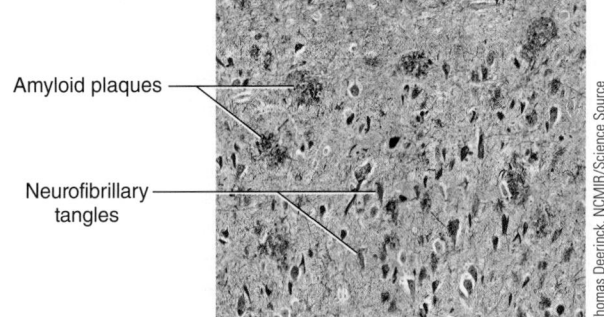

Figure 2 Brain tissue from the cerebral cortex of a patient with Alzheimer's disease, showing characteristic amyloid plaques (*rounded brown areas*) and neurofibrillary tangles (*dark teardrop shapes*).

and release neurotransmitter molecules into the synaptic cleft by exocytosis (**FIG. 41-11b**).

Neurotransmitter molecules diffuse across the synaptic cleft and bind with specific receptors on the dendrites or cell bodies of postsynaptic neurons (or on the plasma membranes of effector cells). Many neurotransmitter receptors are chemically

activated ion channels known as **ligand-gated channels** (see Chapter 6). When the neurotransmitter, the *ligand*, binds with the receptor, the ion channel opens (**FIG. 41-11c**). (Recall that a ligand is a molecule that binds to a specific site in a receptor.) The acetylcholine receptor, for example, is a ligand-gated ion channel that permits the passage of Na^+ and K^+.

At most synapses the action potential cannot jump across the synaptic cleft between the two neurons. This problem is solved by chemical signaling across synapses.

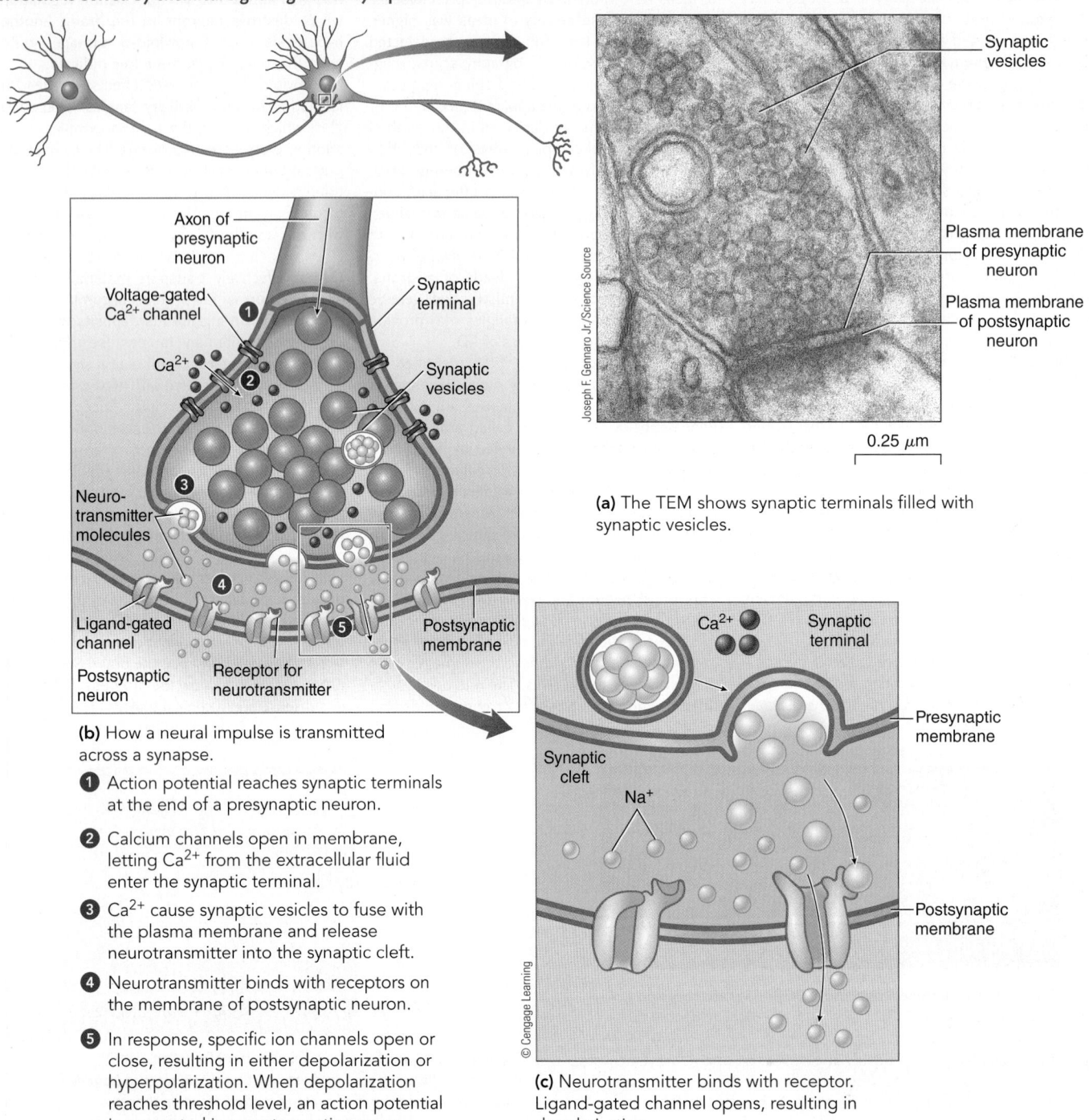

(b) How a neural impulse is transmitted across a synapse.

1 Action potential reaches synaptic terminals at the end of a presynaptic neuron.

2 Calcium channels open in membrane, letting Ca^{2+} from the extracellular fluid enter the synaptic terminal.

3 Ca^{2+} cause synaptic vesicles to fuse with the plasma membrane and release neurotransmitter into the synaptic cleft.

4 Neurotransmitter binds with receptors on the membrane of postsynaptic neuron.

5 In response, specific ion channels open or close, resulting in either depolarization or hyperpolarization. When depolarization reaches threshold level, an action potential is generated in a postsynaptic neuron.

(a) The TEM shows synaptic terminals filled with synaptic vesicles.

(c) Neurotransmitter binds with receptor. Ligand-gated channel opens, resulting in depolarization.

Figure 41-11 *Animation* **Neural signaling across a synapse**

PREDICT What would be the effect of blocking the receptors on the postsynaptic neuron?

Some neurotransmitters, such as serotonin, operate by a different mechanism. They work indirectly through a **second messenger.** Binding of the neurotransmitter with a receptor activates a **G protein.** The G protein then activates an enzyme, such as adenylyl cyclase, in the postsynaptic membrane. Adenylyl cyclase converts ATP to **cyclic AMP (cAMP),** which acts as a second messenger (see Chapters 6 and 49). Cyclic AMP activates a kinase that phosphorylates a protein, which then closes K^+ channels.

For a postsynaptic neuron to repolarize quickly, any excess neurotransmitter in the synaptic cleft must be removed. Some neurotransmitters are inactivated by enzymes. For example, excess acetylcholine is degraded into its chemical components, choline and acetate, by the enzyme acetylcholinesterase. Other neurotransmitters, such as the biogenic amines, are actively transported back into the synaptic terminals, a process known as *reuptake.* These neurotransmitters are repackaged in vesicles and recycled.

Many drugs inhibit the reuptake of neurotransmitters. For example, some antidepressant medications are selective serotonin reuptake inhibitors (SSRIs). SSRIs such as fluoxetine (Prozac) and escitalopram (Lexapro) selectively inhibit the reuptake of serotonin in the brain. This action concentrates serotonin in the synaptic cleft, elevating mood. A few antidepressants inhibit the reuptake of both serotonin and norepinephrine. New antidepressants may target dopamine reuptake.

Activated receptors can send excitatory or inhibitory signals

A postsynaptic neuron may have receptors for several types of neurotransmitters. Some of its receptors may be excitatory, and others may be inhibitory. Depending on the type of postsynaptic receptor with which it combines, the same neurotransmitter can have different effects. For example, acetylcholine has an excitatory effect on skeletal muscle and an inhibitory effect on cardiac muscle.

When neurotransmitter molecules bind to receptors, the activated receptor can directly or indirectly open or close ion channels; this action changes the permeability of the postsynaptic membrane to certain ions. The resulting redistribution of ions changes the electrical potential of the membrane, and the membrane may depolarize. For example, when neurotransmitter molecules combine with receptors that open Na^+ channels, the resulting influx of Na^+ partially depolarizes the membrane. If sufficiently intense, a local depolarization can set off an action potential.

A change in membrane potential that brings the neuron closer to firing is called an **excitatory postsynaptic potential (EPSP).** Imagine that sufficient sodium ions enter to change the membrane potential from −70 to −60 mV. The membrane would be only −5 mV away from threshold. Under such conditions, an additional relatively weak stimulus, causing a small influx of Na^+, can cause the neuron to fire. *Note that unlike action potentials, postsynaptic potentials are graded responses.*

Some neurotransmitter–receptor combinations hyperpolarize the postsynaptic membrane, causing the membrane potential to become more negative. Because such an action takes the neuron farther away from the firing level, a potential change in this direction is called an **inhibitory postsynaptic potential (IPSP).** For example, if the membrane potential changes from −70 to −80 mV, the membrane is farther away from threshold, and a stronger stimulus is required to fire the neuron.

Like an EPSP, an IPSP can be produced in several ways. Two types of GABA receptors are known to produce IPSPs. When GABA binds to a $GABA_B$ receptor, K^+ channels open. As K^+ diffuse out, the neuron becomes more negative, hyperpolarizing the membrane. Activated $GABA_A$ receptors produce IPSPs by opening Cl^- channels. In this case, the influx of negative ions

hyperpolarizes the membrane. Although the mechanisms are different, the inhibiting effect of GABA is the same in both cases.

In summary, neural signaling across synapses involves the following sequence of events:

action potential reaches synaptic terminal ⟶ calcium ions enter synaptic terminal ⟶ synaptic vesicles release neurotransmitter into the synaptic cleft ⟶ neurotransmitter diffuses across synaptic cleft ⟶ neurotransmitter binds with receptors in membrane of postsynaptic neuron ⟶ ion channels open or close, resulting in depolarization or hyperpolarization of postsynaptic membrane

CHECKPOINT 41.4

- *What are some functions of biogenic amines? of GABA?*
- VISUALIZE *Sketch the sequence of events that takes place at a chemical synapse.*
- *How are EPSPs produced? IPSPs?*

41.5 NEURAL INTEGRATION

LEARNING OBJECTIVE

10 Define neural integration and describe how a postsynaptic neuron sums incoming stimuli and "decides" whether to fire.

Each neuron may synapse with hundreds of other neurons. Indeed, thousands of synaptic terminals of presynaptic neurons may cover as much as 40% of a postsynaptic neuron's dendritic surface and cell body. It is the job of the dendrites and the cell body of every neuron to integrate the messages that continually bombard them.

Neural **integration** is the process of summing, or integrating, incoming signals. EPSPs and IPSPs are produced continually in postsynaptic neurons, and IPSPs cancel the effects of some EPSPs. It is important to remember that each EPSP and IPSP is not an all-or-none response. Rather, each is a graded response that does *not* travel like an action potential but may be added to or subtracted from other EPSPs and IPSPs.

As the membrane of the postsynaptic neuron continuously updates its molecular tabulations, the neuron may be inhibited or brought to threshold level. This mechanism integrates hundreds of signals (EPSPs and IPSPs) before an action potential is actually transmitted along the axon of a postsynaptic neuron. Local responses permit the neuron and the entire nervous system a far greater range of response than would be possible if every EPSP generated an action potential.

Postsynaptic potentials are summed over time and space

Each EPSP or IPSP is a graded potential that varies in magnitude depending on the strength of the stimulus applied. One EPSP is usually too weak to trigger an action potential by itself. Its effect

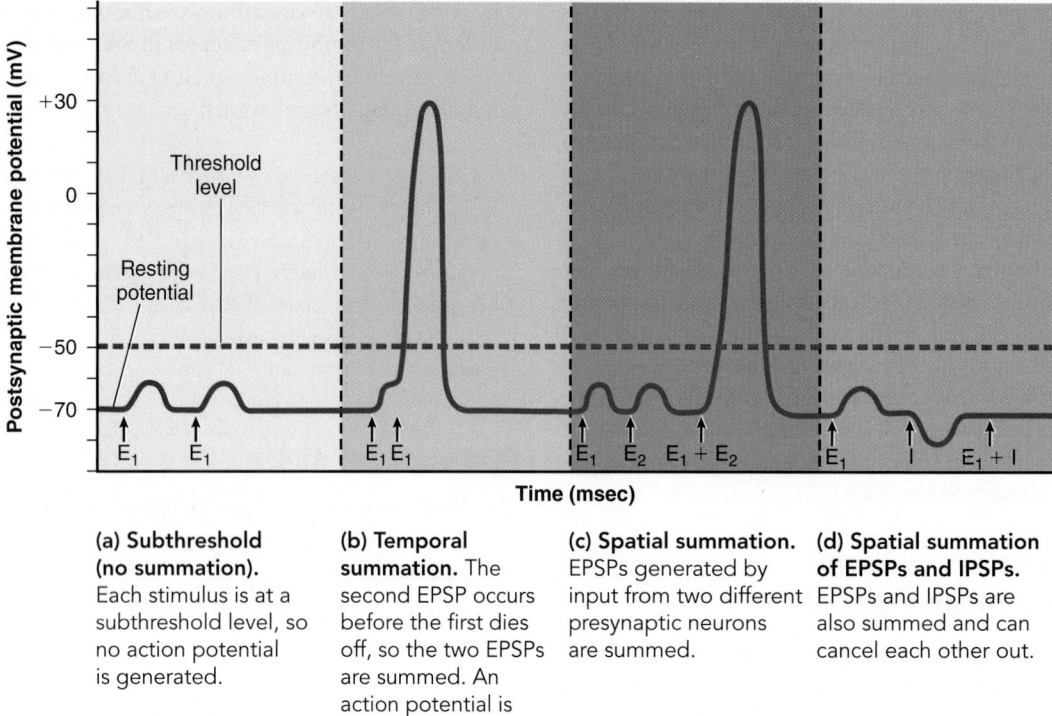

Figure 41-12 Neural integration

E_1 and E_2 are excitatory stimuli. I represents an inhibitory stimulus.

© Cengage Learning

Within the figure:

Postsynaptic membrane potential (mV)

+30

0

Threshold level

Resting potential

−50

−70

E_1 E_1 | E_1 E_1 | E_1 E_2 $E_1 + E_2$ | E_1 I $E_1 + I$

Time (msec)

(a) Subthreshold (no summation). Each stimulus is at a subthreshold level, so no action potential is generated.

(b) Temporal summation. The second EPSP occurs before the first dies off, so the two EPSPs are summed. An action potential is generated.

(c) Spatial summation. EPSPs generated by input from two different presynaptic neurons are summed.

(d) Spatial summation of EPSPs and IPSPs. EPSPs and IPSPs are also summed and can cancel each other out.

---

is below threshold level. However, even though subthreshold EPSPs do not produce an action potential, they do affect the membrane potential. EPSPs can be added together in a process called **summation** (FIG. 41-12).

Temporal summation occurs when a presynaptic neuron fires multiple times in rapid succession, resulting in repeated stimulation of a postsynaptic neuron. New EPSPs develop before previous EPSPs have decayed. The summation of several EPSPs may bring the neuron to the critical firing level.

Spatial summation can also bring the postsynaptic neuron to threshold level. **Spatial summation** is the summing of several postsynaptic potentials resulting when closely spaced presynaptic neurons release neurotransmitter simultaneously. The postsynaptic neuron may be stimulated at multiple regions at the same time.

Neurons sum IPSPs as well as EPSPs. As IPSPs occur, they move the membrane potential farther away from threshold.

Where does neural integration take place?

On a molecular level, every neuron tabulates the thousands of bits of information that continuously bombard it. In vertebrates more than 90% of the neurons in the body are located in the CNS, so most neural integration takes place there, within the brain and spinal cord. These neurons are responsible for making most of the "decisions." The brain and spinal cord are examined in some detail in the next chapter.

CHECKPOINT 41.5

- *What is neural integration?*
- *How do temporal and spatial summation differ?*
- *Where does most neural integration take place in vertebrates?*

41.6 NEURAL CIRCUITS: COMPLEX INFORMATION SIGNALING

LEARNING OBJECTIVE

11 Distinguish between convergence and divergence, and explain why each is important to the functioning of neural circuits.

The CNS contains millions of neurons, but it is not just a tangled mass of nerve cells. Its neurons are functionally arranged in specific pathways, or **neural circuits.** A neural circuit typically consists of one or more afferent neurons, one or more interneurons, and one or more efferent neurons.

Neural circuits are organized into *neural networks.* Although each network has some specific characteristics, the neural circuits in all the networks share many organizational features. For example, convergence and divergence are characteristic of them all.

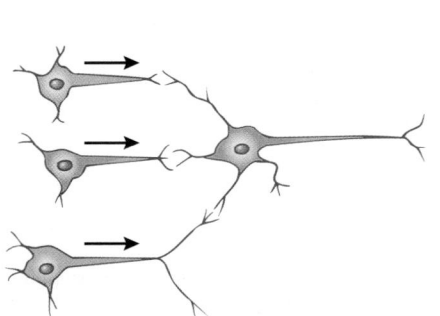

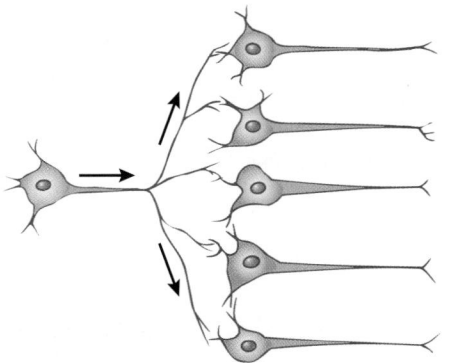

(a) Convergence of neural input.
Several presynaptic neurons synapse with one postsynaptic neuron.

(b) Divergence of neural output.
A single presynaptic neuron synapses with many postsynaptic neurons.

Figure 41-13 Neural circuits
© Cengage Learning

In **convergence** of neural input, a single neuron receives converging signals (inputs) from two or more presynaptic neurons (**FIG. 41-13a**). An interneuron in the spinal cord, for instance, may receive converging information from sensory neurons entering the cord, from neurons bringing information from various parts of the brain, and from neurons coming from different levels of the spinal cord. Information from all these sources must be integrated before an action potential is transmitted and an appropriate motor neuron is stimulated.

In **divergence** of neural input a single presynaptic neuron makes connections with many postsynaptic neurons (**FIG. 41-13b**). Each presynaptic neuron may branch and synapse with thousands of different postsynaptic neurons. For example, a single neuron transmitting an impulse from the motor area of the brain may make connections with hundreds of interneurons in the spinal cord, and each of them may diverge, in turn, so that hundreds of muscle fibers are stimulated.

CHECKPOINT 41.6

- **VISUALIZE** *Draw simple sketches that contrast convergence of neural input with divergence of neural input.*

SUMMARY: FOCUS ON LEARNING OBJECTIVES

41.1 Neural Signaling: An Overview *(page 855)*

1 Describe the processes involved in neural signaling: reception, transmission, integration, and action by effectors that brings about the actual response.

- Neural signaling involves (1) **reception** of information by a sensory receptor; (2) **neural transmission** by an **afferent neuron** to the **central nervous system (CNS),** which consists of the brain and spinal cord; (3) **integration** by **interneurons** in the central nervous system; (4) transmission by an **efferent neuron** to other neurons or to an effector; and (5) action by **effectors,** the muscles and glands that bring about the actual **response** to the stimulus. Sensory receptors and neurons outside the CNS make up the **peripheral nervous system (PNS).**

41.2 Neurons and Glial Cells *(page 856)*

2 Draw and label a typical neuron and give the function of each of its parts.

- **Neurons** are specialized to receive stimuli and transmit electrical and chemical signals. In a typical neuron, a **cell body** contains the nucleus and most of the organelles. Many branched **dendrites** extend from the cell body; they are specialized to receive stimuli and send signals to the cell body.
- The single long **axon** extends from the cell body and forms branches called *axon collaterals*. The axon transmits signals into its *terminal branches*, which end in *synaptic terminals*.
- Many axons are surrounded by an insulating **myelin sheath.** In the PNS the myelin sheath is formed by **Schwann cells. Nodes of Ranvier** are gaps in the sheath between successive Schwann cells.

- A **nerve** consists of several hundred axons wrapped in connective tissue; a **ganglion** is a mass of neuron cell bodies in the PNS.

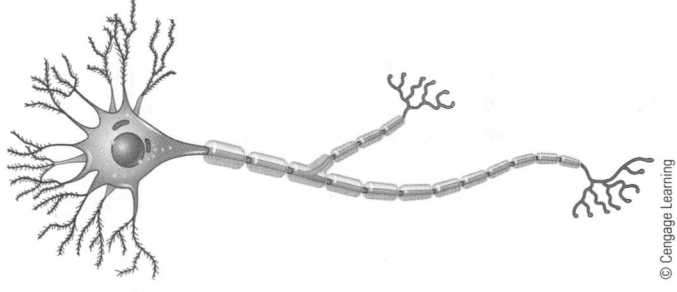

3 Identify the main types of glial cells and describe the functions of each.

- **Glial cells** support and nourish neurons and are important in neural communication. **Astrocytes** are glial cells that physically support neurons. They help regulate the composition of the extracellular fluid in the CNS by taking up excess potassium ions. Astrocytes communicate with one another and with neurons; they also induce and stabilize synapses.
- **Oligodendrocytes** are glial cells that form myelin sheaths around axons in the CNS. **Ependymal cells** line cavities in the CNS and help produce and circulate cerebrospinal fluid. **Microglia** are phagocytic cells.

41.3 Transmitting Information Along the Neuron *(page 859)*

4 Explain how the neuron develops and maintains a resting potential.

- Neurons use electrical signals to transmit information along axons. The plasma membrane of a resting neuron, one that is not transmitting an impulse, is *polarized*. The inner surface of the plasma membrane is negatively charged relative to the extracellular fluid just outside the membrane. The voltage measured across the plasma membrane is the **membrane potential.**
- The **resting potential** is the potential difference of about −70 mV that exists across the membrane. The magnitude of the resting potential is determined by (1) differences in concentrations of specific ions (mainly Na^+ and K^+) inside the cell relative to the extracellular fluid and (2) selective permeability of the plasma membrane to these ions.
- Ions pass through specific **passive ion channels.** K^+ leak out more readily than Na^+ can leak in. The resting potential is mainly due to the net flow of K^+ out of the neuron. Cl^- and large anions inside the neuron contribute negative charges.
- The opposing Na^+ and K^+ gradients are maintained by ATP-requiring **sodium–potassium pumps** that continuously transport three Na^+ out of the neuron for every two K^+ transported in.

5 Compare a graded potential with an action potential, describing the production and transmission of each.
- If a stimulus causes the membrane potential to become less negative, the membrane becomes **depolarized.** If the membrane potential becomes more negative than the resting potential, the membrane is **hyperpolarized.**
- A **graded potential** is a local response that varies in magnitude depending on the strength of the applied stimulus. A graded potential fades out within a few millimeters of its point of origin.
- If voltage across the membrane declines to a critical **threshold level, voltage-activated ion channels** open, allowing Na^+ to flow into the neuron, and an **action potential** is generated. The action potential is a *wave of depolarization* that moves down the axon.
- As the action potential moves down the axon, **repolarization** occurs behind it. During depolarization, the axon enters an *absolute* **refractory period,** a time when it cannot transmit another action potential. When enough gates controlling Na^+ channels have been reset, the neuron enters a *relative* refractory period, a time when the threshold to fire the neuron is higher.
- The action potential is an **all-or-none response;** no variation exists in the strength of a single impulse. Either the membrane potential exceeds threshold level, leading to transmission of an action potential, or it does not. Once begun, an action potential is self-propagating.

6 Contrast continuous conduction with saltatory conduction.
- *Continuous conduction*, which involves the entire axon plasma membrane, takes place in unmyelinated neurons.
- **Saltatory conduction,** which is more rapid and requires less energy than continuous conduction, takes place in myelinated neurons. Depolarization appears to leap along the axon from one node of Ranvier to the next.

41.4 Transmitting Information Across Synapses *(page 867)*

7 Describe the actions of the neurotransmitters identified in this chapter.
- The junction between two neurons or between a neuron and effector is a **synapse.** Most synapses are chemical. In synaptic transmission a **presynaptic neuron** releases a neurotransmitter, a chemical messenger, from its *synaptic vesicles.*
- **Acetylcholine** triggers contraction of skeletal muscle.
- The **biogenic amines,** which include **norepinephrine, serotonin,** and **dopamine,** have multiple functions. They play important roles in regulating mood. Dopamine is also important in motor function.
- Several amino acids function as neurotransmitters, including **glutamate,** the main excitatory neurotransmitter in the brain, and **GABA,** a widespread inhibitory neurotransmitter.
- The neuropeptides, which are opioids, include the **endorphins** (e.g., *beta-endorphin*) and the **enkephalins.**
- **Nitric oxide (NO),** a gaseous neurotransmitter, transmits signals from postsynaptic neurons to presynaptic neurons, the opposite direction from most neurotransmitters.

8 List the sequence of events that take place in synaptic transmission and draw diagrams to support your description.
- Calcium ions induce synaptic vesicles to fuse with the presynaptic membrane and release neurotransmitter into the **synaptic cleft.** The neurotransmitter diffuses across the synaptic cleft and combines with specific receptors on a postsynaptic neuron.
- Many neurotransmitter receptors are proteins that form **ligand-gated channels.** Others activate a G protein and work through a **second messenger** such as **cAMP.**

9 Compare excitatory and inhibitory signals and their effects.
- Binding of a neurotransmitter to a receptor can cause either an **excitatory postsynaptic potential (EPSP)** or an **inhibitory postsynaptic potential (IPSP),** depending on the type of receptor. EPSPs bring the neuron closer to firing. IPSPs move the neuron farther away from its firing level.

41.5 Neural Integration *(page 871)*

10 Define *neural integration* and describe how a postsynaptic neuron sums incoming stimuli and "decides" whether to fire.
- Neural **integration** is the process of summing, or integrating, incoming signals.
- Each EPSP or IPSP is a graded potential that varies in magnitude depending on the strength of the stimulus applied. **Summation** is the process of adding and subtracting incoming signals. By summation of several EPSPs, the neuron may be brought to the critical firing level.
- **Temporal summation** occurs when repeated stimuli cause new EPSPs to develop before previous EPSPs have decayed. **Spatial summation** occurs when several closely spaced synaptic terminals release neurotransmitter simultaneously, stimulating the postsynaptic neuron at several different places.

41.6 Neural Circuits: Complex Information Signaling *(page 872)*

11 Distinguish between convergence and divergence, and explain why each is important to the functioning of neural circuits.
- Complex **neural circuits** are possible because of associations such as convergence and divergence. In **convergence** a single neuron is affected by signals from two or more presynaptic neurons. Convergence allows the CNS to integrate incoming information from various sources. In **divergence** a single presynaptic neuron stimulates many postsynaptic neurons, allowing widespread effect.

Know and Comprehend

1. Summing incoming neural signals is part of (a) reception (b) transmission (c) integration (d) action by effectors (e) afferent neuron transmission

2. The myelin sheath is produced around axons in the PNS by (a) ganglia (b) neuron cell bodies (c) dendrites (d) Schwann cells (e) oligodendrocytes

3. Which of the following occurs first when voltage reaches the threshold level? (a) voltage-activated Na^+ channels open (b) K^+ channels open (c) the membrane hyperpolarizes (d) neurotransmitter is released (e) K^+ channels close

4. Saltatory conduction (a) requires more energy than continuous conduction (b) occurs in unmyelinated neurons (c) occurs when the action potential jumps from one node of Ranvier to the next (d) slows transmission of an impulse (e) depends on the action of GABA

5. Receptors for serotonin and many other neurotransmitters (a) are voltage-activated ion channels (b) permit influx of chloride ions, leading to depolarization (c) inhibit EPSPs (d) are ligand-gated channels (e) are passive ion channels

6. A presynaptic neuron in the cerebrum transmits information to hundreds of other neurons. This process is an example of (a) convergence (b) divergence (c) temporal summation (d) spatial summation (e) a graded potential

Apply and Analyze

7. **VISUALIZE** Describe the action taking place at each of the numbered steps in the figure.

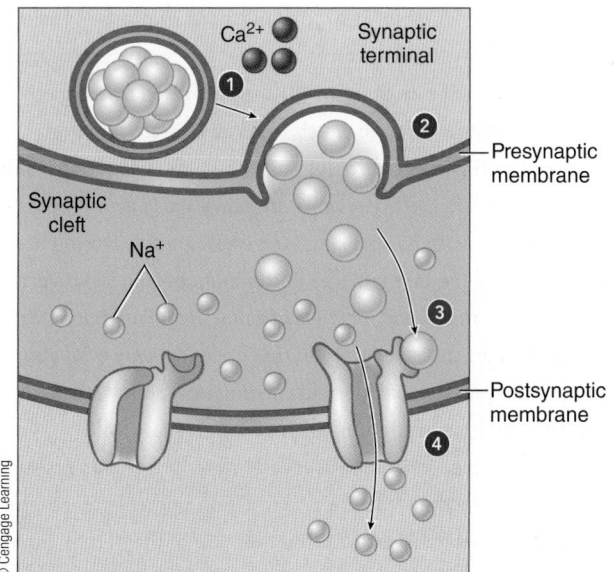

8. **VISUALIZE** Using the diagram in question 7, add several red spheres in the synaptic cleft to represent molecules of an antidepressant medication. Illustrate a potential mechanism of action for the antidepressant molecules. (Extend the diagram if you like.)

9. **INTERPRET DATA** State whether each of the following would result in an EPSP or IPSP (explain your answer). (a) Na^+ enter postsynaptic neuron (b) K^+ diffuse out of postsynaptic neuron (c) Cl^- enter postsynaptic neuron.

Evaluate and Synthesize

10. Develop a hypothesis to explain why acetylcholine has an excitatory effect on skeletal muscle but an inhibitory effect on cardiac muscle.

11. **SCIENCE, TECHNOLOGY, AND SOCIETY** A test for risk of Alzheimer's disease is available. Based on gene markers that can be identified in a saliva sample, a healthy person can learn his or her approximate risk for AD. How might this information be useful to people? Would you be concerned about the psychological consequences of a person knowing that he or she has a high risk of AD? Would you want to know?

12. **SCIENCE, TECHNOLOGY, AND SOCIETY** Investigators have genetically engineered cells to produce acetylcholine, dopamine, GABA, and other neurotransmitters. In what ways might these cells be useful in neurobiological research? How might they be useful in clinical medicine?

13. **EVOLUTION LINK** From the perspective of adaptation, hypothesize the advantages of neural circuits characterized by a great deal of convergence and divergence.

aplia To access course materials, such as Aplia and other companion resources, please visit **www.cengagebrain.com**.

42 | Neural Regulation

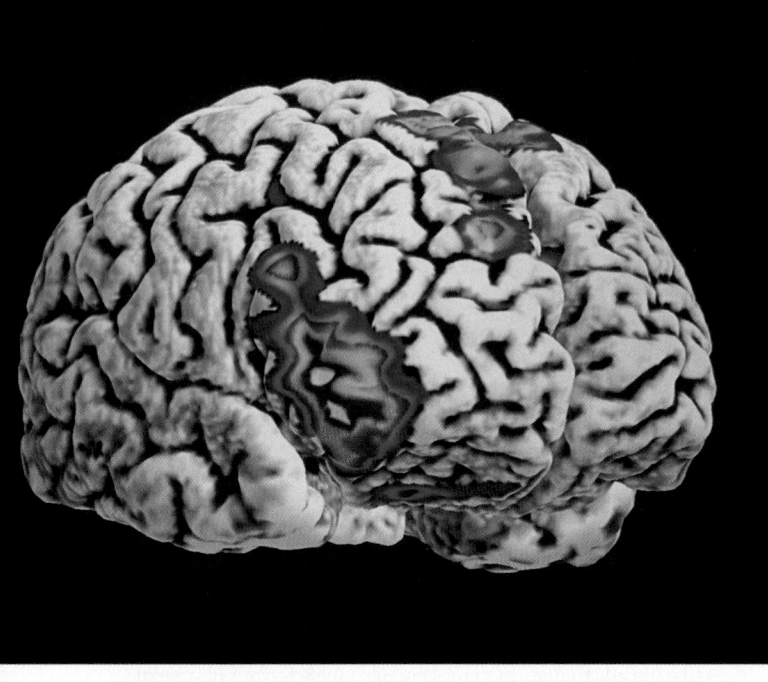

The brain during speech. In this functional magnetic resonance image (fMRI) of the brain, *red* indicates areas of greatest activation during speech, and *yellow* indicates medium activation.

© Collection CNRI/Phototake

KEY CONCEPTS

42.1 The trend in the evolution of invertebrate nervous systems is toward greater complexity, with increased number and concentration of neurons and greater specialization of function.

42.2 The vertebrate nervous system consists of a central nervous system (CNS), which includes the brain and spinal cord, and a peripheral nervous system (PNS), which includes sensory receptors and nerves.

42.3 During the evolution of the vertebrate brain, the cerebrum and cerebellum have become larger and more complex.

42.4 The CNS receives sensory information from both the internal and external environments, integrates it, and determines appropriate responses. The association areas of the brain are responsible for learning, memory, language, thought, and judgment.

42.5 The PNS has two divisions. The somatic division transmits information between the external environment and the CNS. The autonomic division transmits information between the internal environment and the CNS.

42.6 Many prescribed and illicit drugs affect mood by affecting neurotransmitter concentrations in the brain.

A frog flicks out its tongue with lightning speed to capture a fly, a rabbit escapes from a predator, and you learn biology, perhaps not that quickly. In a complex nervous system, millions of neurons work together to transmit information from the external environment to the central nervous system (CNS). After the information is integrated, neural signals are transmitted to the effectors that can produce effective responses.

The nervous system also regulates heart rate, breathing, and hundreds of other internal activities. Nerves must transmit information from all the organs in the body to the CNS. Then, other nerves transmit back the information that will allow organs to make the adjustments necessary to maintain homeostasis. Researchers have focused extensively on just how neurons interact, and we are only beginning to understand the mechanisms that permit the many vital functions of the nervous system.

Neurobiologists use a variety of methods to study the mechanisms of neural function. For example, improved imaging methods have revolutionized the study of the brain. Functional magnetic resonance imaging (fMRI) has provided investigators with a window through which to observe brain function. Functional MRI allows neurobiologists to study the responses of neural networks in the brain while an individual is actually performing a task, such as speaking (see photograph). During the performance of the task, an area of the brain becomes active, and flow of oxygen-rich blood increases to that area. Functional MRI detects changes that take place in response to a very brief stimulus. For example, a visual stimulus lasting only 30 milliseconds stimulates brain activation that can be detected by fMRI.

In this chapter we compare several types of nervous systems. We then examine the structure and function of the vertebrate nervous system, with emphasis on the function of the human brain. We explore some of the frontiers of neurobiology, such as mechanisms involved in information processing and drug addiction.

42.1 INVERTEBRATE NERVOUS SYSTEMS: TRENDS IN EVOLUTION

LEARNING OBJECTIVES

1 Contrast nerve nets and radial nervous systems with bilateral nervous systems.
2 Identify trends in the evolution of invertebrate nervous systems.

An animal's lifestyle is closely related to the organization and complexity of its nervous system. *Hydra* and other cnidarians have a **nerve net** consisting of interconnected neurons with no central control organ. Sensory cells are located in the epidermis, and the cnidarian can respond to stimuli from any direction (see Chapter 31). When a neuron is stimulated, electrical signals are sent from neuron to neuron in all directions (**FIG. 42-1**). Responses may involve large parts of the body.

An advantage of the nerve net is that the cnidarian responds effectively to predator or prey approaching from any direction. *Hydra* responds by discharging nematocysts (stinging structures) and coordinating the movements of its tentacles to capture food. Some cnidarians have two or more nerve nets. In some jellyfish a slow-transmission nerve net coordinates movement of the tentacles, and a second nerve net, which is faster, coordinates swimming.

The *radial nervous system* of the sea star and other echinoderms is a modified nerve net. This system shows some degree of selective organization of neurons into more than a diffuse network. It consists of a nerve ring around the mouth, from which a large radial nerve extends into each arm. Branches of these nerves, which form a network somewhat similar to the nerve net of *Hydra*, coordinate the animal's movement.

Bilaterally symmetrical animals have a *bilateral nervous system*. In planarian flatworms the head region contains concentrations of nerve cells called *cerebral ganglia* (**FIG. 42-2**). These ganglia serve as a primitive brain and have some control over the rest of the nervous system. Typically, two solid, ventral, longitudinal *nerve cords* extend from the ganglia to the posterior end of the body. A **nerve cord** is a bundle of nerves that extends from a central ganglion (or brain) and gives rise to smaller nerves. In planaria, transverse nerves connect the two nerve cords and connect the brain with the eyespots. This arrangement is called a "ladder-type" nervous system.

Annelids and arthropods typically have a solid ventral nerve cord (**FIG. 42-3**). The cell bodies of many of the neurons are massed into **ganglia.** Afferent and efferent neurons lie in lateral nerves that link the ganglia with muscles and other body structures. If an earthworm's brain is removed, the animal can move almost as well as before. However, when it bumps into an obstacle, it persists in a futile effort to move forward rather than turn aside. To respond adaptively to environmental change, the earthworm needs its brain.

The cerebral ganglia of some arthropods differ from those of annelids in that they have specific functional regions. These areas are specialized for integrating information transmitted to the ganglia from sense organs.

Mollusks with inactive lifestyles have relatively simple nervous systems with little cephalization and very simple sense organs. Typically, there are two nerve cords; one extends down each side of the body. Several pairs of ganglia are located along the cords.

The cephalopod mollusks (squids and octopods) have complex nervous systems that include well-developed sense organs. The cephalopod nervous system is adapted to the active, predatory lifestyle of these animals. Neurons concentrated in ganglia are massed in a ring surrounding the esophagus; they make up a brain that contains about 168 million nerve cells. Like the vertebrate brain, the cephalopod brain has lobes and intricate folds, and particular areas are specialized as centers for specific functions. The octopus is capable of considerable learning and can be taught difficult tasks. In fact, cephalopods are considered the most intelligent invertebrates.

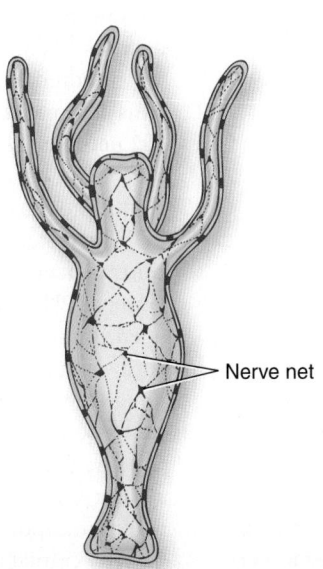

Figure 42-1 *Hydra*'s nerve net

Hydra and other cnidarians have a network of neurons with no central control organ.

© Cengage Learning

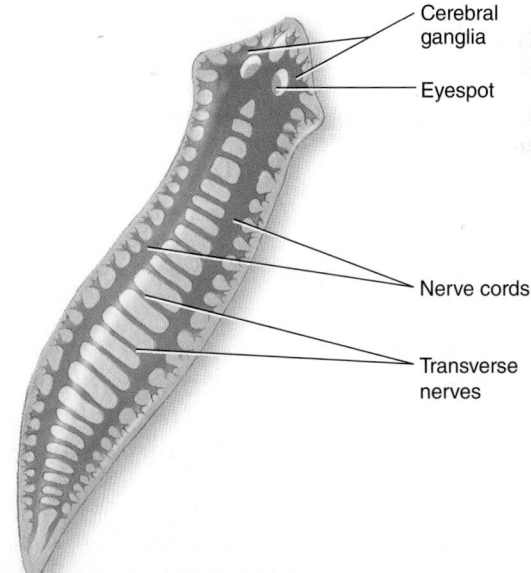

Figure 42-2 Ladder-type nervous system of flatworms

Cerebral ganglia in the head region serve as a simple brain. The longitudinal nerve cords are connected by transverse nerves.

© Cengage Learning

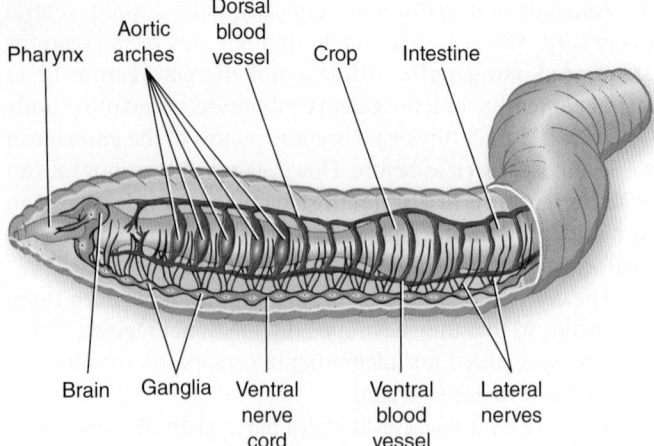

(a) Earthworm nervous system. Annelids have a dorsal anterior brain and one or more ventral nerve cords. Cell bodies of the neurons are located in ganglia that are connected by the ventral nerve cord.

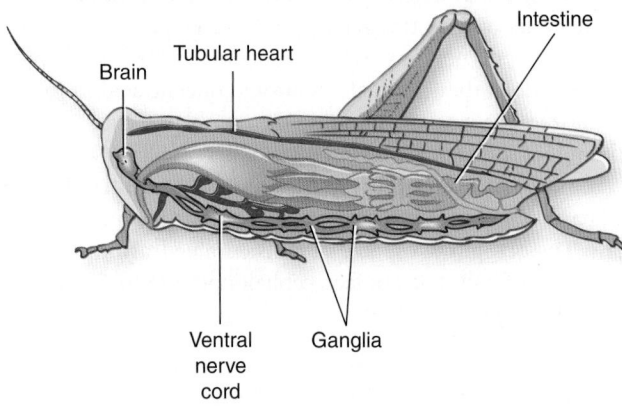

(b) Insect nervous system. The arthropod brain is continuous with the ventral nerve cord. The brain is more specialized than in annelids.

Figure 42-3 *Animation* **The annelid and arthropod nervous systems**
© Cengage Learning

We can identify the following trends in the evolution of nervous systems:

1. *Increased number of nerve cells.*
2. *Concentration of nerve cells.* The nerve cells form masses of tissue that become *ganglia* (concentrations of nerve cell bodies) and *brain,* and they form thick cords of tissue that become *nerve cords* and *nerves.*
3. *Specialization of function.* For example, transmission of nerve impulses in one direction requires both **afferent nerves,** which conduct impulses toward a central nervous system (CNS), and **efferent nerves,** which transmit impulses away from the CNS to the **effectors** (muscles and glands). Certain parts of the CNS become specialized to perform specific functions, and distinct structural and functional regions can be identified.
4. *Increased number of interneurons and more complex synaptic contacts.* These contacts permit greater integration of

incoming messages, provide a greater range of responses, and allow more precise responses.

5. *Cephalization, or formation of a head.* A bilaterally symmetrical animal generally moves in a forward direction. Concentration of sense organs at the front end of the body lets the animal detect an enemy quickly enough to escape or to see, hear, or smell food in time to capture it. Response can be rapid if short pathways link sense organs to decision-making nerve cells nearby. Therefore, nerve cells are typically concentrated in the head region and organized to form ganglia or a **brain.**

CHECKPOINT 42.1

- *Contrast the nervous system of the insect with that of the flatworm.*
- *What are some trends in the evolution of nervous systems?*
- *What are some advantages of cephalization?*

42.2 OVERVIEW OF THE VERTEBRATE NERVOUS SYSTEM

LEARNING OBJECTIVE

3 Describe the two main divisions of the vertebrate nervous system and summarize their functions.

An animal's range of possible responses depends in large part on how many neurons it has and how they are organized in the nervous system. Some invertebrate nervous systems have only a few hundred neurons. As animal groups evolved, nervous systems became increasingly complex. A common roundworm has 302 neurons. Elephant, whale, and human brains have billions. How neurons are organized is critical in determining learning ability and flexibility of behavior.

The vertebrate nervous system has two main divisions: the **central nervous system (CNS)** and the **peripheral nervous system (PNS)** (FIG. 42-4). The CNS consists of a highly developed brain that is continuous with the dorsal, tubular nerve cord, or **spinal cord.** Serving as central control, these organs integrate incoming information and determine appropriate responses.

The PNS is made up of the sensory receptors (e.g., tactile, auditory, and visual receptors) and the nerves, which are the communication lines. Various parts of the body are linked to the CNS by cranial nerves and by spinal nerves. Afferent (sensory) neurons in these nerves continuously inform the CNS of changing conditions. Then efferent neurons transmit the "decisions" of the CNS to appropriate muscles and glands, which make the adjustments needed to maintain homeostasis.

For convenience, the PNS is subdivided into somatic and autonomic divisions. Most of the receptors and nerves concerned with changes in the external environment are somatic.

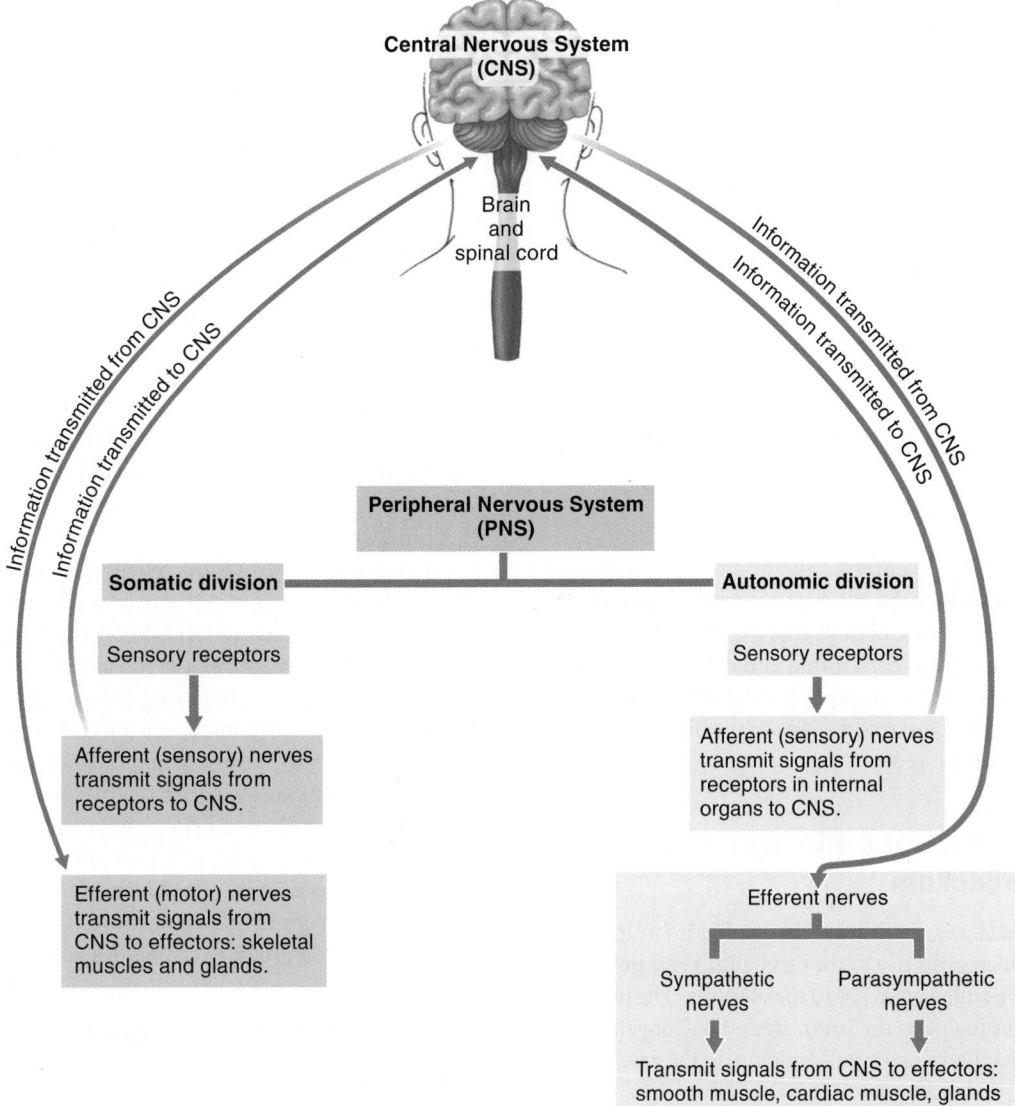

Central Nervous System (CNS)

Brain and spinal cord

Information transmitted from CNS

Information transmitted to CNS

Information transmitted from CNS

Information transmitted to CNS

Peripheral Nervous System (PNS)

Somatic division

Sensory receptors

Afferent (sensory) nerves transmit signals from receptors to CNS.

Efferent (motor) nerves transmit signals from CNS to effectors: skeletal muscles and glands.

Autonomic division

Sensory receptors

Afferent (sensory) nerves transmit signals from receptors in internal organs to CNS.

Efferent nerves

Sympathetic nerves

Parasympathetic nerves

Transmit signals from CNS to effectors: smooth muscle, cardiac muscle, glands

Figure 42-4 *Animation* **Organization of the vertebrate nervous system**

The two main divisions of the vertebrate nervous system are the central nervous system (CNS) and the peripheral nervous system (PNS).

© Cengage Learning

Those that regulate the internal environment are autonomic. Both divisions have afferent nerves, which transmit messages from sensory receptors to the CNS, and efferent nerves, which transmit information back from the CNS to the structures that respond. The autonomic division has two kinds of efferent pathways: sympathetic and parasympathetic nerves (discussed later in the chapter).

CHECKPOINT 42.2

- **VISUALIZE** *Draw a diagram showing the main components of the CNS and the main components of the PNS.*

- *Describe the functions of each component of the CNS and PNS.*

42.3 EVOLUTION OF THE VERTEBRATE BRAIN

LEARNING OBJECTIVE

4 Trace the development of the principal vertebrate brain regions from the hindbrain, midbrain, and forebrain, and compare the brains of fishes, amphibians, reptiles, birds, and mammals.

All vertebrates, from fishes to mammals, have the same basic brain structure; different parts of the brain are specialized in the various vertebrate classes. The evolutionary trend is toward increasing complexity, especially of the cerebrum and cerebellum.

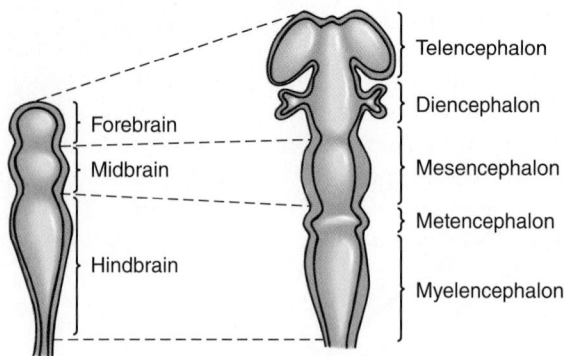

Figure 42-5 Early development of the vertebrate nervous system
Early in the development of the vertebrate embryo, the anterior end of the neural tube differentiates into the hindbrain, midbrain, and forebrain. These primary divisions subdivide and eventually give rise to specific structures of the adult brain.
© Cengage Learning

In the early vertebrate embryo, the brain and spinal cord differentiate from a single tube of tissue, the **neural tube.** Anteriorly, the tube expands and develops into the brain. Posteriorly, the tube becomes the spinal cord. Brain and spinal cord remain continuous, and their cavities communicate. As the brain begins to differentiate, three bulges become visible: the hindbrain, midbrain, and forebrain (**FIG. 42-5**).

The hindbrain develops into the medulla, pons, and cerebellum

As indicated in **TABLE 42-1**, the **hindbrain** subdivides to form the *metencephalon,* which gives rise to the **cerebellum** and **pons,** and the *myelencephalon,* which gives rise to the **medulla.** The medulla, pons, and midbrain make up the **brain stem,** the elongated portion of the brain that looks like a stalk holding up the cerebrum.

The medulla, the most posterior part of the brain, is continuous with the spinal cord. Its cavity, the *fourth ventricle,* is continuous with the central canal of the spinal cord and with a channel that runs through the midbrain. The fourth **ventricle** is one of four interconnected chambers in the brain through which cerebrospinal fluid flows. The walls of the medulla are thick and made up largely of *nerve tracts* (bundles of axons) that connect the spinal cord with various parts of the brain. The

medulla contains centers that regulate life-sustaining functions such as respiration, heartbeat, and blood pressure. Other reflex centers in the medulla regulate activities such as swallowing, coughing, and vomiting.

The cerebellum is responsible for muscle tone, posture, and equilibrium. It also refines and coordinates muscle activity. In humans the cerebellum provides input to motor areas in the cerebral cortex, and in this way it helps plan and initiate voluntary activity. Certain regions of the cerebellum store implicit memories (unconscious memories for perceptual and motor skills, such as swimming and skating).

The size and shape of the cerebellum vary among the vertebrate classes (**FIG. 42-6**). Development of the cerebellum is roughly correlated with the extent and complexity of muscular activity, reflecting the principle that the relative size of a brain part correlates with its importance to the behavior of the species. The cerebellum is not well developed in jawless fishes, amphibians, and reptiles. In jawed fishes, birds, and mammals, the cerebellum is large and well developed.

Injury or removal of the cerebellum results in impaired muscle coordination. A bird without a cerebellum cannot fly, and its wings thrash about jerkily. When the human cerebellum is injured by a blow or by disease, muscular movements are uncoordinated. Any activity requiring delicate coordination, such as threading a needle, becomes very difficult if not impossible.

In mammals the pons contains a thick bundle of fibers that transmits information between the two sides of the cerebellum. The pons also serves as a bridge that connects the medulla and cerebellum with other regions of the brain. The pons contains centers that help regulate respiration and nuclei that relay impulses from the cerebrum to the cerebellum. Recall from Chapter 41 that a **nucleus** is a group of cell bodies within the CNS.

The midbrain is prominent in fishes and amphibians

In fishes and amphibians, the **midbrain,** or *mesencephalon,* is the most prominent part of the brain and serves as the main association area. It receives incoming sensory information, integrates it, and sends decisions to appropriate motor nerves. The dorsal portion of the midbrain is differentiated to some extent. For example, the optic lobes are specialized for visual interpretations.

TABLE 42-1	Differentiation of CNS Structures		
EARLY EMBRYONIC DIVISIONS	**SUBDIVISIONS**	**DERIVATIVES IN ADULT**	**CAVITY**
BRAIN			
Forebrain	Telencephalon	Cerebrum	Lateral ventricles (first and second ventricles)
	Diencephalon	Thalamus, hypothalamus, epiphysis (pineal body)	Third ventricle
Midbrain	Mesencephalon	Optic lobes in fish and amphibians; superior and inferior colliculi in mammals	Cerebral aqueduct
Hindbrain	Metencephalon	Cerebellum, pons	
	Myelencephalon	Medulla	Fourth ventricle
SPINAL CORD		Spinal cord	Central canal

© Cengage Learning

Although different parts of the brain are specialized in various vertebrate groups, there is an evolutionary trend toward larger size and greater complexity of the cerebrum and cerebellum.

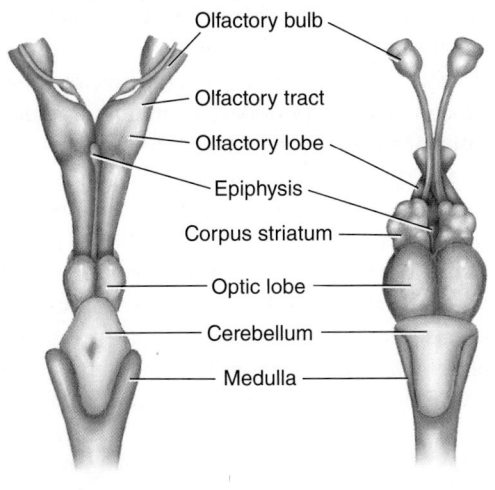

(a) Shark   **(b)** Codfish

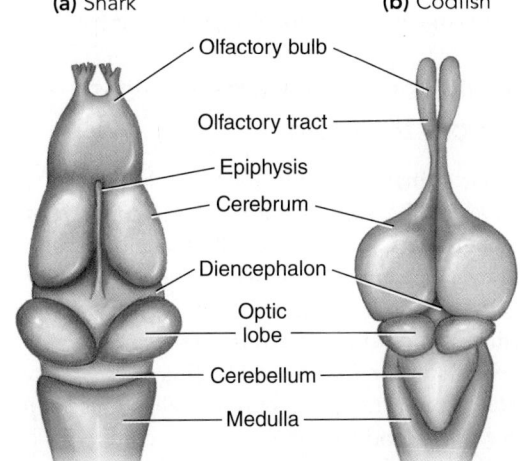

(c) Amphibian (frog)   **(d)** Reptile (alligator)

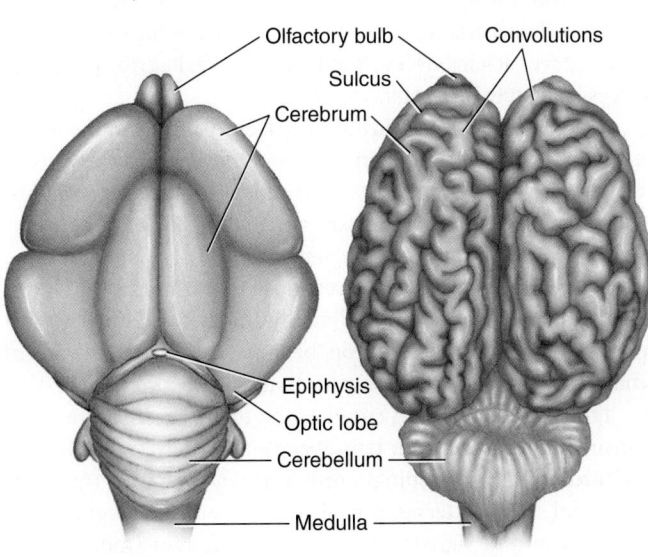

(e) Bird (goose)   **(f)** Mammal (horse)

In reptiles, birds, and mammals, many functions of the optic lobes are assumed by the cerebrum, which develops from the forebrain. In mammals the midbrain consists of the *superior colliculi,* centers for visual reflexes such as pupil constriction, and the *inferior colliculi,* centers for certain auditory reflexes. The inferior colliculi are major integration centers for incoming auditory information. The mammalian midbrain also contains a center (the *red nucleus*) that helps maintain muscle tone and posture.

The forebrain gives rise to the thalamus, hypothalamus, and cerebrum

The **forebrain** subdivides to form the *telencephalon* and *diencephalon.* The *diencephalon* gives rise to the thalamus and hypothalamus.

In all vertebrate classes, the **thalamus** is a relay center for motor and sensory messages. In mammals all sensory messages except those from the olfactory receptors are delivered to the thalamus, where they are integrated before they are relayed to the sensory areas of the cerebrum.

The **hypothalamus,** which lies below the thalamus, forms the floor of the third ventricle. The hypothalamus is a major coordinating center for regulating autonomic and somatic responses. It integrates incoming information and provides input to centers in the medulla and spinal cord that regulate activities such as heart rate, respiration, and digestive system function. In birds and mammals, the hypothalamus controls body temperature. It also contains olfactory centers, regulates appetite and water balance, and is important in emotional and sexual responses. As we discuss in Chapter 49, the hypothalamus links the nervous and endocrine systems and produces certain hormones.

The *telencephalon* gives rise to the **cerebrum.** The *lateral ventricles* (also called the first and second ventricles) lie within the cerebrum. Each lateral ventricle connects with the third ventricle (within the diencephalon) by way of a channel. In most vertebrate groups, the telencephalon also gives rise to the **olfactory bulbs.** These structures are important in the chemical sense of smell, which is the dominant sense in most aquatic and terrestrial vertebrates. In fact, much of brain development in vertebrates appears to focus on integrating olfactory information. In fishes and amphibians, a large part of the cerebrum is

Figure 42-6 *Animation* **Evolution of the vertebrate brain**

Comparison of the brains of members of six vertebrate classes reveals basic similarities and evolutionary trends. (Brains are not drawn to scale.) Note that different parts of the brain are specialized in the various groups. The large olfactory lobes in the shark brain **(a)** are essential to this predator's highly developed sense of smell. **(b–f)** During the course of evolution, the cerebrum and cerebellum have become larger and more complex. In mammals **(f)** the cerebrum is the most prominent part of the brain; the cerebral cortex, the thin outer layer of the cerebrum, is highly convoluted (folded), which greatly increases its surface area. The diencephalon and part of the brain stem are covered by the large cerebrum.

PREDICT Imagine that humans continue to thrive during the next several thousand years. Predict evolutionary changes that might occur in the human brain.

© Cengage Learning

devoted to this function. In birds, centers in the cerebrum control eating, flying, singing, and other action patterns.

In most vertebrates the cerebrum is divided into right and left *cerebral hemispheres.* Most of the cerebrum is made of **white matter,** which consists mainly of myelinated axons that connect various parts of the brain. In mammals and in the majority of reptiles, the **cerebral cortex** makes up the outer portion of the cerebrum. The cerebral cortex is composed of **gray matter,** which consists of cell bodies, dendrites, and unmyelinated axons.

Certain reptiles and all mammals have a type of cerebral cortex called the *neocortex.* Most of the neocortex consists of *association areas,* regions that link sensory and motor functions and that are responsible for higher functions, such as learning and reasoning. The neocortex is very extensive in mammals, making up the bulk of the cerebrum. In humans most of the cerebral cortex is neocortex that consists of six distinct cell layers. Part of the limbic system (discussed later in this chapter) has only three layers. This region is referred to as the *paleocortex* (ancient cortex).

In mammals the cerebrum is the most prominent part of the brain. During embryonic development, it expands and grows backward, covering many other brain structures. The surface area of the human cerebral cortex is greatly expanded by numerous folds called **convolutions.** The furrows between them are called **sulci** (sing., *sulcus*) if shallow and **fissures** if deep. In mammals the cerebrum is responsible for numerous functions performed by other parts of the brain in other vertebrates. In addition, the mammalian cerebrum has highly developed association functions.

CHECKPOINT 42.3

- *What structures are derived from the vertebrate embryonic forebrain? the hindbrain? What are the functions of these structures?*
- *Compare the brain of a mammal with the brain of an amphibian.*

42.4 THE HUMAN CENTRAL NERVOUS SYSTEM

LEARNING OBJECTIVES

5 Describe the structure and functions of the human spinal cord.

6 Describe the structure and functions of the human cerebrum.

7 Summarize the sleep–wake cycle and contrast rapid-eye-movement (REM) sleep with non-REM sleep.

8 Describe the actions of the limbic system; include the role of the amygdala in emotional expression and in response to danger.

9 Summarize how the brain processes information; include a description of synaptic plasticity and the neurobiological changes that take place during learning.

The soft, fragile human brain consists of billions of neurons that are connected by trillions of synapses. This 3-pound mass of tissue perceives and processes information and determines effective responses more capably and efficiently than the most sophisticated computer.

The brain and spinal cord are well protected. Encased within bone, they are covered by three layers of connective tissue, collectively called the **meninges.** The three meningeal layers are the tough, outer **dura mater;** the middle **arachnoid;** and the thin, vascular **pia mater,** which adheres closely to the tissue of the brain and spinal cord (FIG. 42-7). *Meningitis* is a disease in which these coverings become infected and inflamed.

The space between the arachnoid and the pia mater is the subarachnoid space, which contains **cerebrospinal fluid (CSF).** This shock-absorbing fluid cushions the brain and spinal cord against mechanical injury. It also serves as a medium for exchange of nutrients and waste products between the blood and brain. CSF is produced by special networks of capillaries, collectively called the *choroid plexus,* which extend from the pia mater into the ventricles. After circulating through the ventricles, CSF passes into the subarachnoid space. It is then reabsorbed into large blood sinuses within the dura mater.

The spinal cord transmits impulses to and from the brain

The tubular **spinal cord** extends from the base of the brain to the level of the second lumbar vertebra. A cross-section view through the spinal cord reveals a small *central canal* surrounded by an area of gray matter shaped a bit like the letter H (FIG. 42-8). The gray matter consists of large masses of cell bodies, dendrites, unmyelinated axons, and glial cells.

The white matter, found outside the gray matter, consists of myelinated axons arranged in bundles called *tracts* or *pathways.* *Ascending tracts* conduct impulses up the cord to the brain. For example, the spinothalamic tracts in the anterior and lateral columns of the white matter conduct pain and temperature information from sensory neurons in the skin. The pyramidal tracts are *descending tracts* that convey impulses from the cerebrum to motor nerves at various levels in the cord. We describe the spinal nerves later in this chapter.

In addition to transmitting impulses to and from the brain, the spinal cord controls many reflex activities. A **reflex action** is a relatively simple, involuntary motor response to a stimulus. Although most reflex actions are more complex, let us consider a *withdrawal reflex,* in which a neural circuit consisting of three types of neurons carries out a response to a stimulus (FIG. 42-9). Suppose that your hand accidentally touches a flame. Almost instantly, and even before you become consciously aware of what has happened, you jerk your hand away.

The instant your skin contacts the flame, a sensory neuron transmits a message from pain receptors in the skin to the spinal cord. Within the spinal cord, a sensory neuron transmits the signal to an interneuron. The interneuron integrates the information and signals an appropriate efferent (motor) neuron, which conducts the information to groups of muscles. The muscles respond by contracting, jerking the hand away from

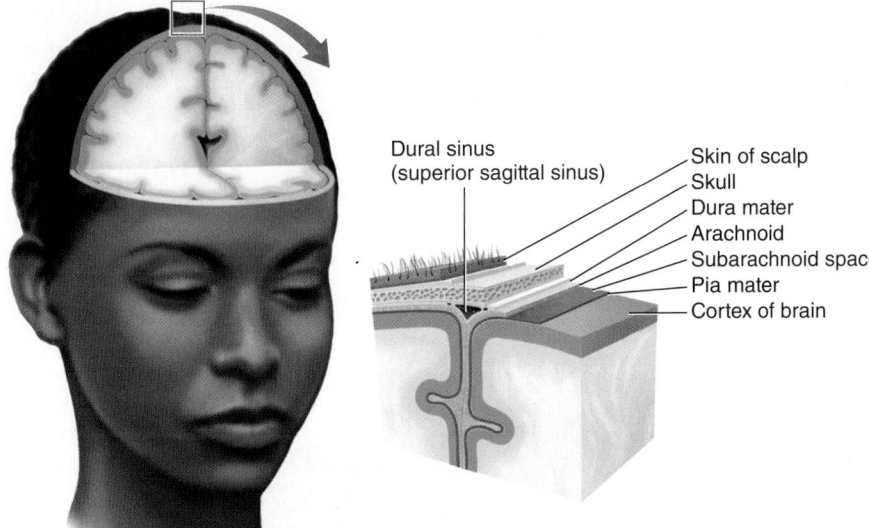

Dural sinus
(superior sagittal sinus)

Skin of scalp
Skull
Dura mater
Arachnoid
Subarachnoid space
Pia mater
Cortex of brain

(a) Frontal section through the superior part of the brain.
Note the dural sinus, a blood sinus, between two layers of
the dura mater. Blood leaving the brain flows into such sinuses
and then circulates to the large jugular veins in the neck.

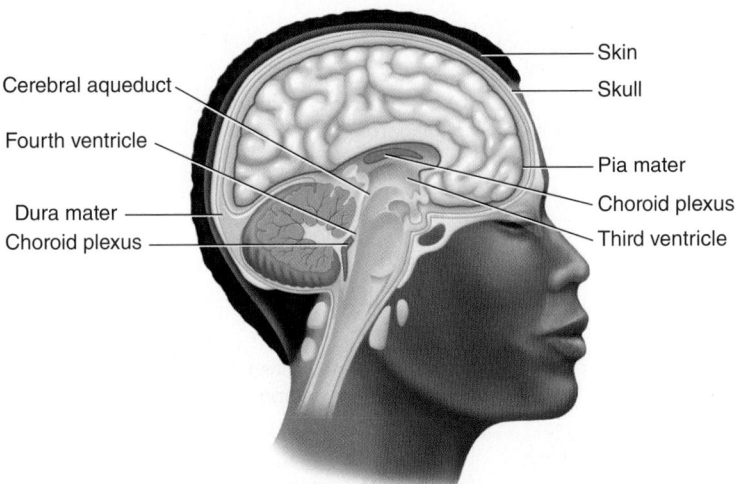

Cerebral aqueduct

Fourth ventricle

Dura mater
Choroid plexus

Skin

Skull

Pia mater

Choroid plexus

Third ventricle

(b) Sagittal section through the brain. The CSF, which cushions the brain and
spinal cord, is produced by the choroid plexi in the walls of the ventricles. This
fluid circulates through the ventricles and subarachnoid space. It is continuously
produced and then reabsorbed into the blood of the dural sinuses.

Figure 42-7 Protection of the brain and spinal cord

The CNS is well protected by the skull and meninges and by cerebrospinal
fluid (CSF).

© Cengage Learning

the flame. Actually, many neurons in sensory, association, and
motor nerves participate in such a reflex action. Generally, we
are not even consciously aware that all these responding muscles
exist. In summary, signals are transmitted through a withdrawal
reflex in the following sequence:

reception by sensory receptor in skin ⟶ sensory neuron
transmits signal to CNS ⟶ interneuron in CNS integrates
information ⟶ efferent (motor) neuron transmits signal to
muscle ⟶ muscle contracts

At the same time that the reflex pathway
is activated, a message is sent up the spinal
cord to the conscious areas of the brain. As
you pull your hand away from the flame, you
become aware of what has happened and
feel the pain. This awareness, however, is not
part of the reflex response.

The spinal cord has some plasticity and
is capable of training. For example, sensory feedback from exercise increases the
strength of neural connections in the spinal cord. More than 200,000 people in the
United States are living with spinal cord
injury. When descending connections with
the brain are intact, some victims of spinal
cord injury can develop at least some limited
ability to walk again.

The most prominent part of the human brain is the cerebrum

The structure and functions of the main parts of the
human brain are summarized in **TABLE 42-2**, and major
regions of the brain are illustrated in **FIGURE 42-10** on
page 886 and **FIGURE 42-11** on page 887. As in other
mammals, the human cerebral cortex consists of right
and left *cerebral hemispheres.*

**Specific areas of the cerebrum are specialized
for performing specific functions** The cerebral cortex is functionally divided into three areas:
(1) *sensory areas* receive incoming signals from the
sense organs; (2) *motor areas* control voluntary movement; and (3) **association areas** link the sensory and
motor areas and are responsible for thought, learning, language, memory, judgment, decision making, and personality. You can see positron-emission
tomography (PET) scans showing some functional
areas of the cerebrum in **FIGURE 42-12** on page 888.

Investigators have mapped the cerebral cortex,
locating the areas responsible for different functions.
The **occipital lobes** contain the visual centers. Stimulation of
these areas, even by a blow on the back of the head, causes the
sensation of light; their removal causes blindness. The centers for
hearing are located in the **temporal lobes** of the brain above the
ear; stimulation by a blow causes a sensation of noise. Removal
of both auditory areas causes deafness. Removal of one does not
cause deafness in one ear; instead, it produces a decrease in the
auditory acuity of both ears. Association areas in the temporal
lobes help us identify incoming sound stimuli.

A groove called the *central sulcus* crosses the top of each
hemisphere from medial to lateral edge. This groove partially
separates the **frontal lobes** from the **parietal lobes.** The frontal lobes have important motor and association areas. The
prefrontal cortex is an association area in each frontal lobe that
is crucial in evaluating information, making decisions, planning,

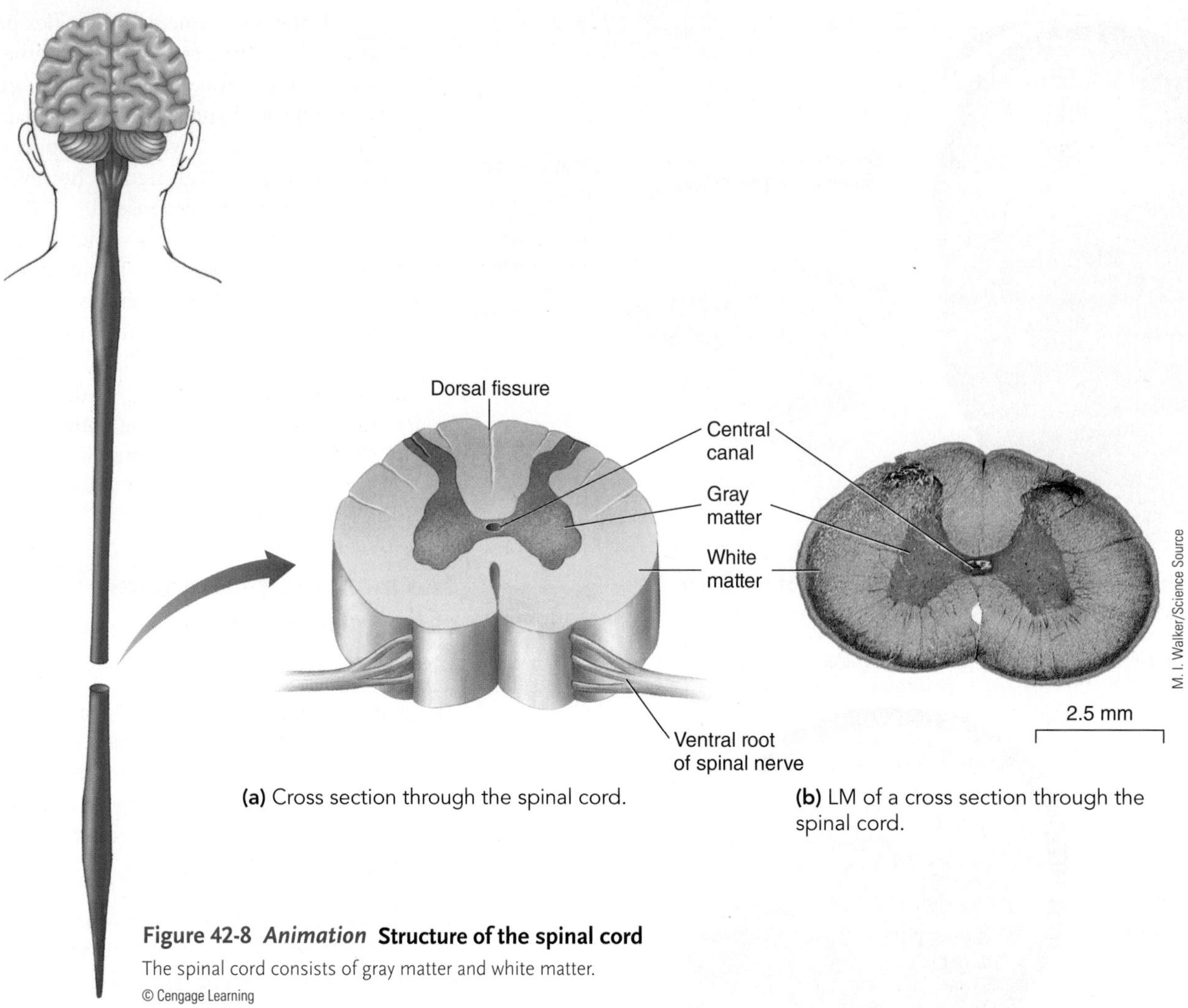

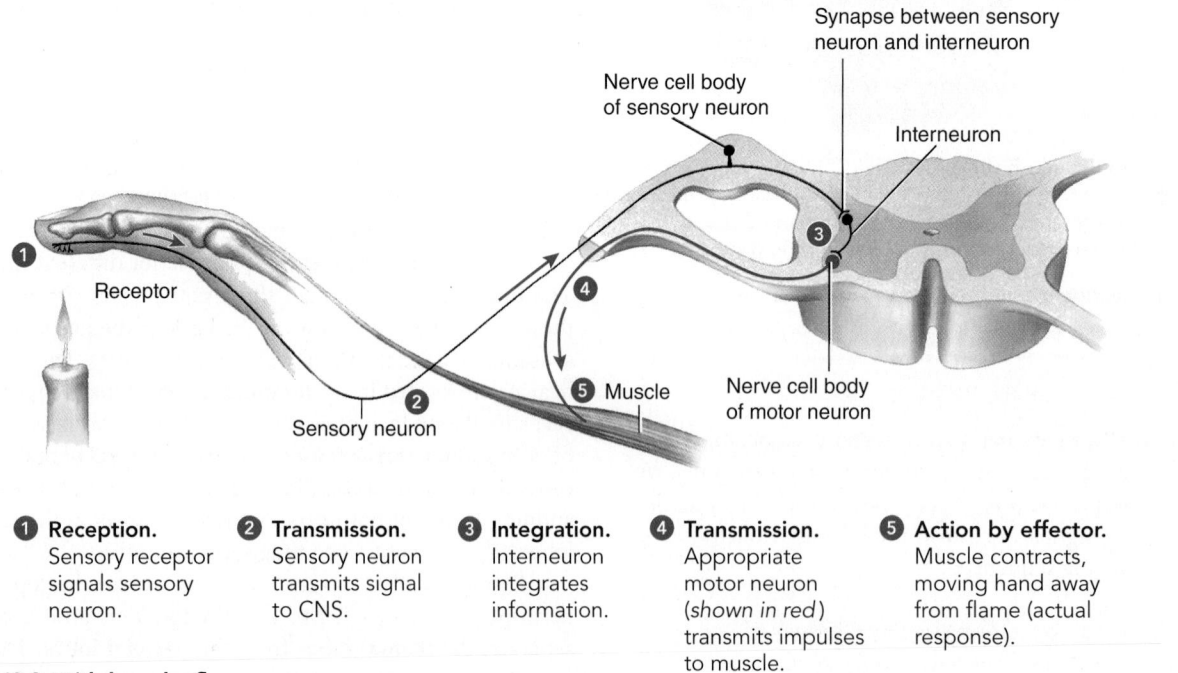

Dorsal fissure

Central canal

Gray matter

White matter

Ventral root of spinal nerve

M. I. Walker/Science Source

2.5 mm

(a) Cross section through the spinal cord.

(b) LM of a cross section through the spinal cord.

Figure 42-8 *Animation* **Structure of the spinal cord**

The spinal cord consists of gray matter and white matter.

© Cengage Learning

Synapse between sensory neuron and interneuron

Nerve cell body of sensory neuron

Interneuron

Receptor

Sensory neuron

Muscle

Nerve cell body of motor neuron

❶ **Reception.**
Sensory receptor signals sensory neuron.

❷ **Transmission.**
Sensory neuron transmits signal to CNS.

❸ **Integration.**
Interneuron integrates information.

❹ **Transmission.**
Appropriate motor neuron (*shown in red*) transmits impulses to muscle.

❺ **Action by effector.**
Muscle contracts, moving hand away from flame (actual response).

Figure 42-9 Withdrawal reflex

A reflex action is a coordinated, involuntary motor response to a stimulus.

© Cengage Learning

TABLE 42-2 The Human Brain

STRUCTURE	DESCRIPTION	FUNCTION
BRAIN STEM		
Medulla	Continuous with spinal cord; primarily made up of nerves passing from spinal cord to rest of brain	Contains vital centers (clusters of neuron cell bodies) that control heartbeat, respiration, and blood pressure; contains centers that control swallowing, coughing, vomiting
Pons	Forms bulge on anterior surface of brain stem	Connects various parts of brain with one another; contains respiratory and sleep centers
Midbrain	Just above pons	Center for visual and auditory reflexes (e.g., pupil reflex, blinking, adjusting ear to volume of sound)
THALAMUS	At top of brain stem	Main sensory relay center for conducting information between spinal cord and cerebrum; neurons in thalamus sort and interpret all incoming sensory information (except olfaction) before relaying messages to appropriate neurons in cerebrum
HYPOTHALAMUS	Just below thalamus; pituitary gland is connected to hypothalamus by stalk of neural tissue	Contains centers for control of body temperature, appetite, and fat metabolism; regulates pituitary gland; important in emotional and sexual responses, and in sleep–wake cycle
CEREBELLUM	Second-largest division of brain	Muscle coordination and refinement of movements; muscle tone, posture, equilibrium; helps plan and initiate voluntary activity; stores implicit memories
CEREBRUM	Largest, most prominent part of human brain; longitudinal fissure divides cerebrum into right and left hemispheres, each divided into frontal, parietal, temporal, and occipital lobes	Center of intellect, memory, consciousness, and language; also controls sensation and motor functions
Cerebral cortex (outer gray matter)	Arranged into convolutions (folds) that increase surface area; functionally, cerebral cortex is divided into:	
	1. Motor areas	1. Control movement of voluntary muscles
	2. Sensory areas	2. Receive incoming information from eyes, ears, touch receptors, and other sensory receptors
	3. Association areas	3. Centers for intellect, memory, language, and emotion; interpret incoming sensory information
White matter	Consists of myelinated axons of neurons that connect various regions of brain; axons are arranged in bundles (tracts)	Connects the following: 1. Neurons within same hemisphere 2. Right and left hemispheres 3. Cerebrum with other parts of brain and spinal cord

© Cengage Learning

organizing responses, and interpreting incoming sensory information. The *motor areas* in the frontal lobes control the skeletal muscles. The *somatosensory area* in the anterior region of the parietal lobes receives information regarding touch, pressure, heat, cold, and pain from sense organs in the skin. This region also receives information regarding position of the body. Association areas in the parietal lobes are important in helping us pay attention to such incoming information.

The size of the motor area in the brain for any given part of the body is proportional to the complexity of movement involved, not to the amount of muscle. Predictably, areas that control the hands and face are relatively large (**FIG. 42-13** on page 889). A similar relationship exists between the sensory area and the sensitivity of the region of the skin from which it receives impulses. These relationships are visually represented as the sensory homunculus ("little person") and the motor homunculus in Figure 42-13.

Neural fibers in the brain cross so that one side of the brain controls the opposite side of the body. As a result of another "reversal," the uppermost part of the cortex controls the lower limbs of the body.

When all the areas of known function are plotted, they cover almost all the rat's cortex, a large part of the dog's, a moderate amount of the monkey's, and only a small part of the total surface of the human cortex. The remaining cortical areas are association areas. Somehow the association regions integrate the diverse impulses reaching the brain so that an appropriate response is made. When disease or accident destroys the functioning of one or more association areas, the ability to recognize certain kinds of symbols may be lost. For example, the names of objects may be forgotten, although their functions are remembered and understood.

The two cerebral hemispheres are specialized for certain functions. As will be discussed later in this chapter, the left hemisphere is specialized for language functions. This part of the brain is also specialized for logical decision making and retrieval of facts. The right hemisphere is specialized for emotional processing and for visual–spatial tasks such as recognizing faces.

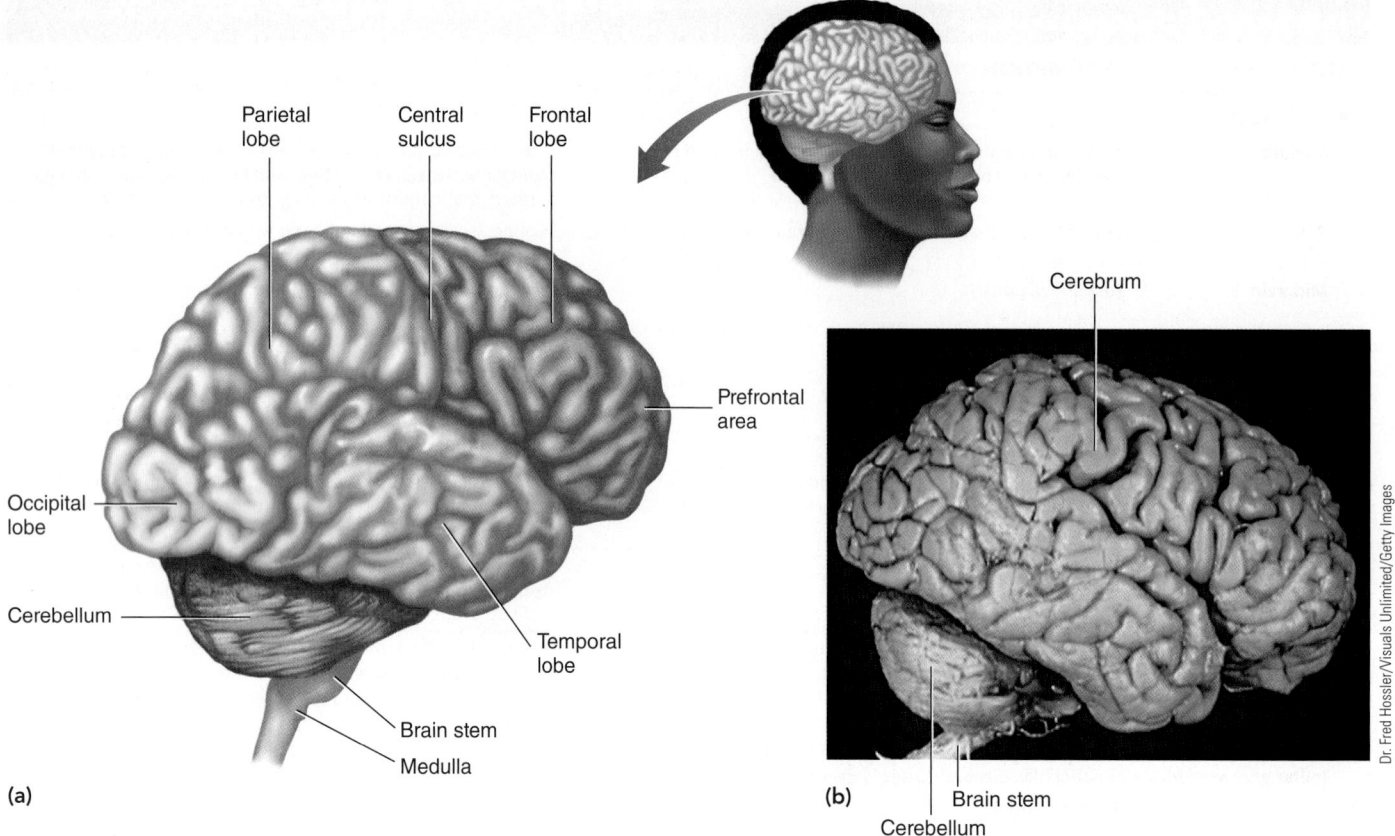

Dr. Fred Hossler/Visuals Unlimited/Getty Images

Parietal lobe — **Central sulcus** — **Frontal lobe**

Prefrontal area

Occipital lobe

Cerebellum

Temporal lobe

Brain stem

Medulla

(a)

Cerebrum

(b) **Brain stem**
Cerebellum

Figure 42-10 The human brain

Human brain, lateral view. Each cerebral hemisphere is divided into four major lobes. The diencephalon and part of the brain stem are covered by the cerebrum. Compare the diagram **(a)** with the photograph of the human brain **(b)**.

© Cengage Learning

Axons in the white matter of the cerebrum connect parts of the brain

The white matter of the cerebrum lies beneath the cerebral cortex. Nerve fibers of the white matter connect the cortical areas with one another and with other parts of the nervous system (see Table 42-2). A large band of white matter, the **corpus callosum,** connects the right and left hemispheres (see Fig. 42-11).

Deep within the white matter of the cerebrum lie the *basal ganglia,* paired groups of nuclei (gray matter). These nuclei play an important role in coordinating movement. The basal ganglia send signals to and receive signals from the *substantia nigra* in the midbrain. Neurons from the substantia nigra that project (extend) to the basal ganglia produce **dopamine,** a neurotransmitter that helps balance inhibition and excitation of neurons involved in motor function. (See the discussion of neurotransmitters in Chapter 41.) Other neurons from the substantia nigra signal nuclei in the thalamus, which in turn relay information to the motor cortex. Neurons from the substantia nigra that signal the thalamus release the neurotransmitter **gamma-aminobutyric acid (GABA).** Just how these areas work together to coordinate motor function is not yet fully understood.

Nearly one million people in the United States alone are affected by Parkinson's disease, a progressive neurological disorder that affects movement. In this disease dopamine-producing neurons die over time, decreasing the available dopamine. This decrease causes a progressive decrease in effective neural messages regarding motor responses. The resulting symptoms include tremors in the limbs and face, a slowing of movement, poor muscle coordination, and impaired walking and balance. Several medications, including a dopamine precursor (l-DOPA), are used to treat symptoms of the disorder.

The body follows a circadian rhythm of sleep and wakefulness

Researchers have determined that many animals, including nematodes and fruit flies, sleep. Predators typically sleep for longer periods than animals that are preyed upon. For example, rabbits sleep only for a few minutes at a time. In some birds and aquatic mammals, one cerebral hemisphere sleeps at a time, allowing the animal to remain aware of its environment. Sleeping with one hemisphere permits eared seals and whales to move to the surface to breathe without awakening.

Humans and many other animals follow a 24-hour **circadian** (daily) rhythm of sleeping and then being awake. When we are awake, we are typically aware of the external world and of ourselves, including our thoughts, perceptions, and emotions.

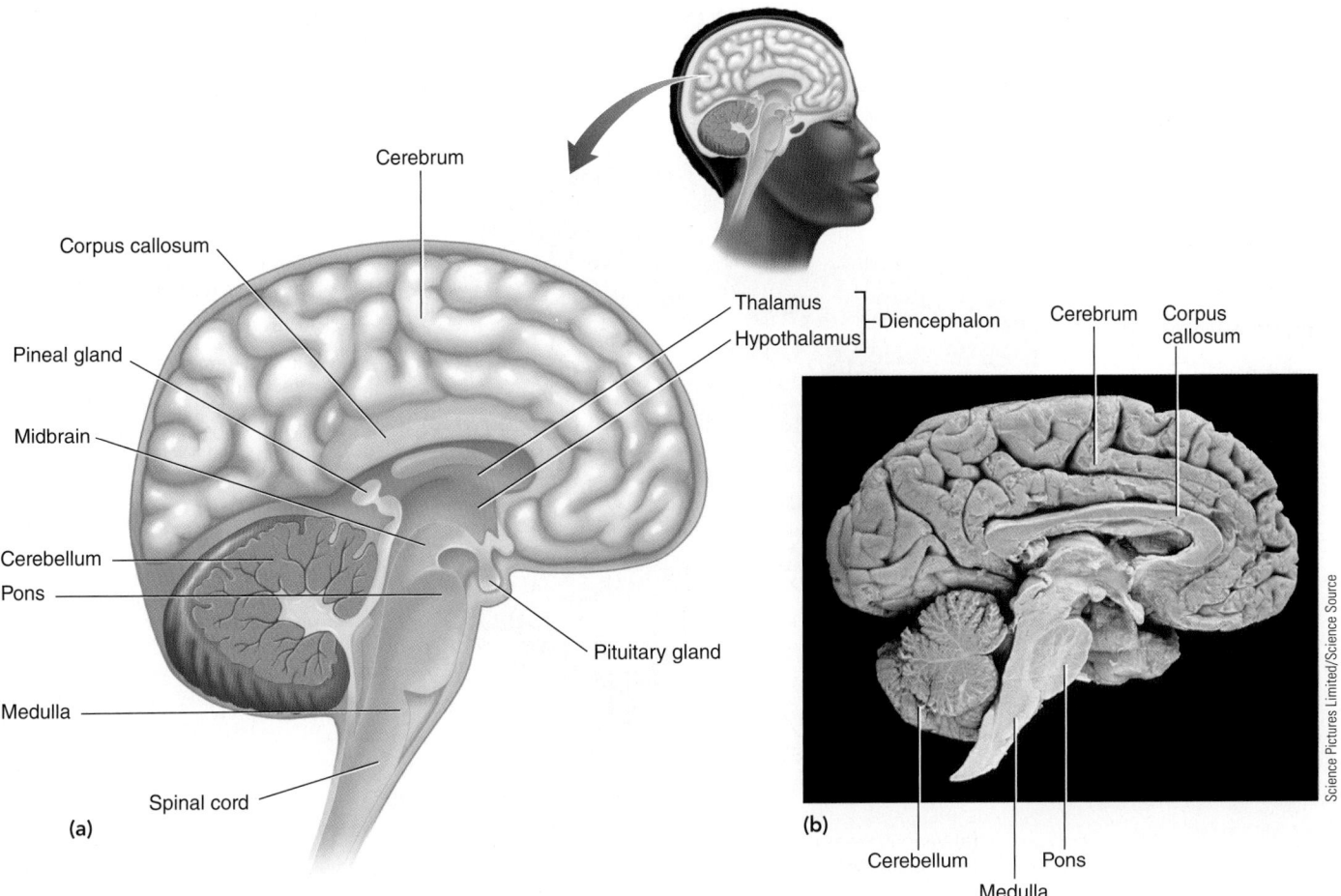

Cerebrum

Corpus callosum

Pineal gland

Midbrain

Cerebellum

Pons

Medulla

Spinal cord

(a)

Thalamus
Hypothalamus ⎫ Diencephalon

Cerebrum   Corpus callosum

Pituitary gland

(b)

Cerebellum   Pons

Medulla

Science Pictures Limited/Science Source

Figure 42-11 *Animation* **Midsagittal section through the human brain**

Half of the brain has been cut away, exposing structures normally covered by the cerebrum. Compare the diagram **(a)** with the photograph of the human brain **(b)**.
© Cengage Learning

In contrast, when we are asleep, the body receives messages from the environment, but we are not aware of them.

The **suprachiasmatic nucleus,** the most important of the body's biological clocks, is located in the hypothalamus. The suprachiasmatic nucleus receives information about the duration of light and dark from the retina of the eye and transmits information to other nuclei and neurons. The suprachiasmatic nucleus signals the **pineal gland** about light and dark. This endocrine gland is located in the midline of the diencephalon.

The pineal gland produces **melatonin,** a hormone that plays a role in regulating the sleep–wake cycle. The pineal gland secretes up to ten times as much melatonin during darkness as during daylight. *Jet lag,* the state of fatigue and decreased physical and mental performance we experience when we fly to a different time zone, occurs because the body is no longer synchronized with its light–dark cycle. Some individuals take melatonin supplements to promote sleep, but their effectiveness has not been proven.

The hypothalamus, thalamus, and brain stem are all involved in regulating the sleep–wake cycle. Neurons in these regions secrete several neurotransmitters important in regulating sleep, including acetylcholine, norepinephrine, and serotonin. The **reticular activating system (RAS)** is a neural pathway within the brain stem and thalamus. The RAS receives messages from neurons in the spinal cord and from many other parts of the nervous system, and it communicates with the cerebral cortex by complex neural circuits. When certain neurons of the RAS bombard the cerebral cortex with stimuli, an individual feels alert and can focus on specific thoughts. If the RAS is severely damaged, the person may pass into a deep, permanent coma.

After many hours of activity, the sleep–wake cycle may be affected by fatigue of the RAS. Sleep centers are then activated, and their neurons release **serotonin.** During *REM sleep,* a stage of sleep characterized by rapid eye movement, neurons in the RAS stimulate heightened activity in certain regions of the brain. After sufficient rest, the inhibitory neurons of the sleep centers become less excitable, and the excitatory neurons of the RAS become more excitable.

Specific types of activity are associated with certain brain wave patterns Brain activity can be studied by measuring and recording the electrical potentials, or "brain waves," generated by thousands of active neurons in various parts of the brain. This electrical activity can be recorded by a device known as an *electroencephalograph.* A recording of this electrical activity,

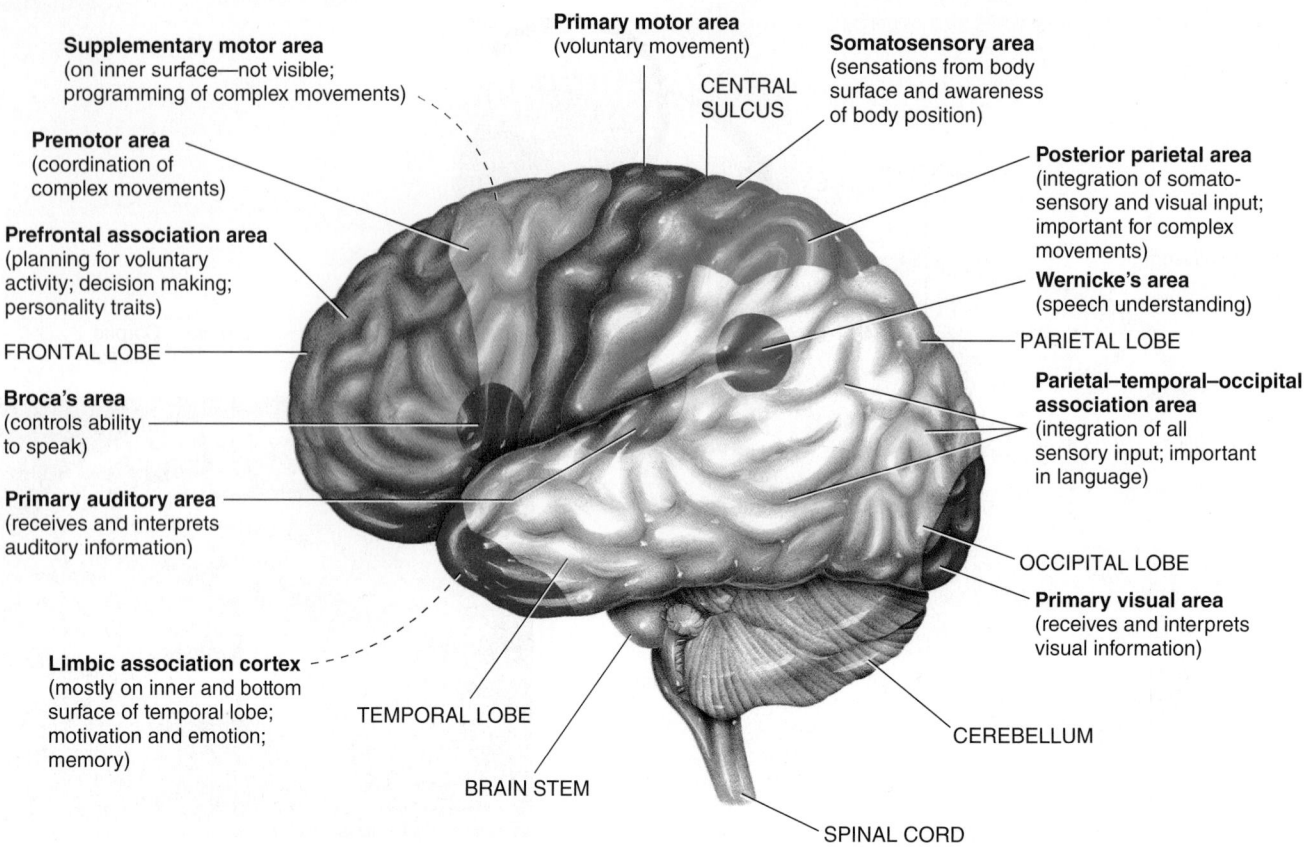

Supplementary motor area
(on inner surface—not visible;
programming of complex movements)

Premotor area
(coordination of
complex movements)

Prefrontal association area
(planning for voluntary
activity; decision making;
personality traits)

FRONTAL LOBE

Broca's area
(controls ability
to speak)

Primary auditory area
(receives and interprets
auditory information)

Limbic association cortex
(mostly on inner and bottom
surface of temporal lobe;
motivation and emotion;
memory)

Primary motor area
(voluntary movement)

CENTRAL
SULCUS

Somatosensory area
(sensations from body
surface and awareness
of body position)

Posterior parietal area
(integration of somato-
sensory and visual input;
important for complex
movements)

Wernicke's area
(speech understanding)

PARIETAL LOBE

**Parietal–temporal–occipital
association area**
(integration of all
sensory input; important
in language)

OCCIPITAL LOBE

Primary visual area
(receives and interprets
visual information)

CEREBELLUM

TEMPORAL LOBE

BRAIN STEM

SPINAL CORD

(a) Various areas of the cerebral cortex are responsible
for specific types of neural processing. (Lateral view of the brain).

called an *electroencephalogram (EEG)*, can be obtained by tap-ing a set of electrodes to standard regions of the scalp and mea-suring the activity of the cerebral cortex (**FIG. 42-14**). The EEG shows that the brain is continuously active.

Certain patterns of brain waves are elicited by specific types of activity. For example, when you are resting quietly with eyes closed, your brain emits *alpha rhythm patterns*. In contrast, as you are reading this biology text, your brain is emitting *beta rhythm patterns,* which have a fast-frequency rhythm. Beta rhythm pat-terns are characteristic of heightened mental activity, such as information processing. During certain stages of sleep, the brain emits slow, high-amplitude *theta waves* and *delta waves.*

Certain brain conditions and diseases change the pattern of brain waves. Individuals with epilepsy, for example, exhibit a recognizable, abnormal wave pattern. The location of a brain tumor or the site of brain damage caused by a blow to the head can sometimes be determined by noting the part of the brain that emits abnormal waves.

Sleep progresses through several stages *Sleep* is an altera-tion of consciousness during which there is decreased electrical activity in the cerebral cortex and from which a person can be aroused. Based on changes in electrical activity, we can identify four stages of *non-REM* sleep (nonrapid eye movement; *REM* is an acronym for *rapid eye movement*) during the first hour of sleep. The deepest sleep is thought to occur during the fourth stage, when it is difficult to awaken the sleeping person. Slow-frequency,

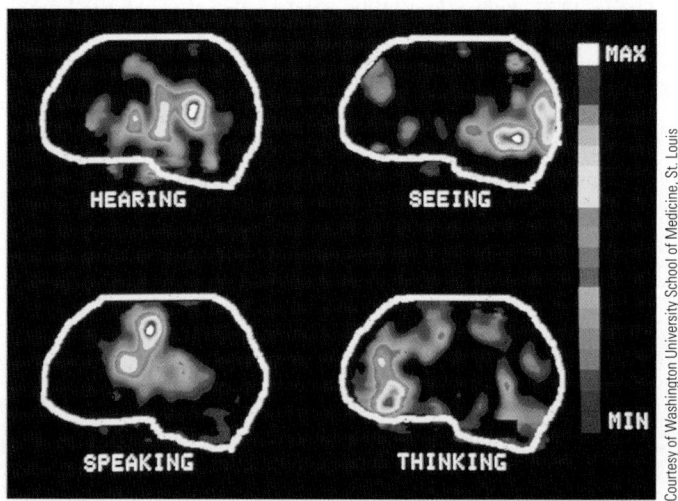

(b) Specific areas of the brain "light up" on positron-emission tomography (PET) scans as an individual performs various tasks. When a specific region of the brain becomes more active, more blood flows to that area. PET scans detect the magnitude of blood flow. Thus, PET scans are pictures of the brain at work, performing specific tasks. The color scale at the right indicates progressively greater activity, from the *purple* color at the bottom to the *white* at the top, which indicates maximum activity.

Figure 42-12 Functional areas of the cerebrum
© Cengage Learning

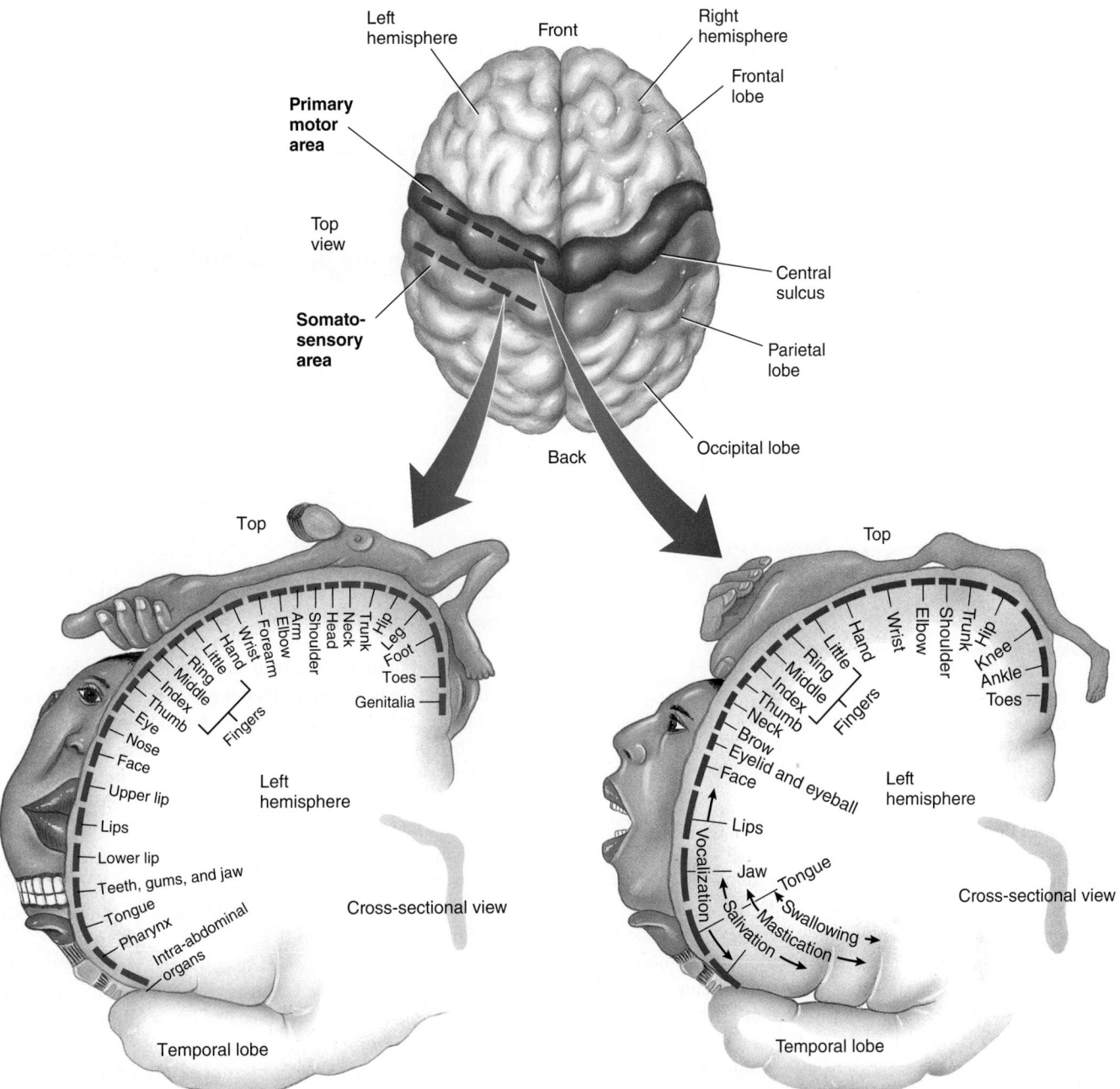

(a) Sensory homunculus. This cross section through the anterior region of the parietal lobe shows the distribution of sensory input to the somatosensory area from various parts of the body. The homunculus is proportioned to reflect the relative amount of cerebral cortex that receives sensory information from each body part.

(b) Motor homunculus. This cross section through the posterior part of the frontal lobe shows which area of the cerebral cortex controls each body part. Note that more of the cerebral cortex is devoted to controlling those body structures capable of skilled, complex movement. The primary motor area is involved in coordinated movement of many muscle groups.

Figure 42-13 Maps of the primary sensory and motor areas
© Cengage Learning

higher-amplitude waves (theta and delta waves) are characteristic of non-REM sleep. This electrical activity is thought to be generated spontaneously by the cerebral cortex when it is not driven by impulses from other parts of the brain. During non-REM sleep, body movement, heart rate, breathing, blood pressure, metabolic rate, and body temperature all decrease.

Every 90 minutes or so, a sleeping person enters the *REM sleep stage* for a time. During this stage (about 20% of total sleep time), the eyes move about rapidly beneath the closed but fluttering lids. Brain waves change to a desynchronized pattern. Everyone dreams, especially during REM sleep. PET scans of sleeping subjects indicate that during REM sleep, blood flow in the frontal

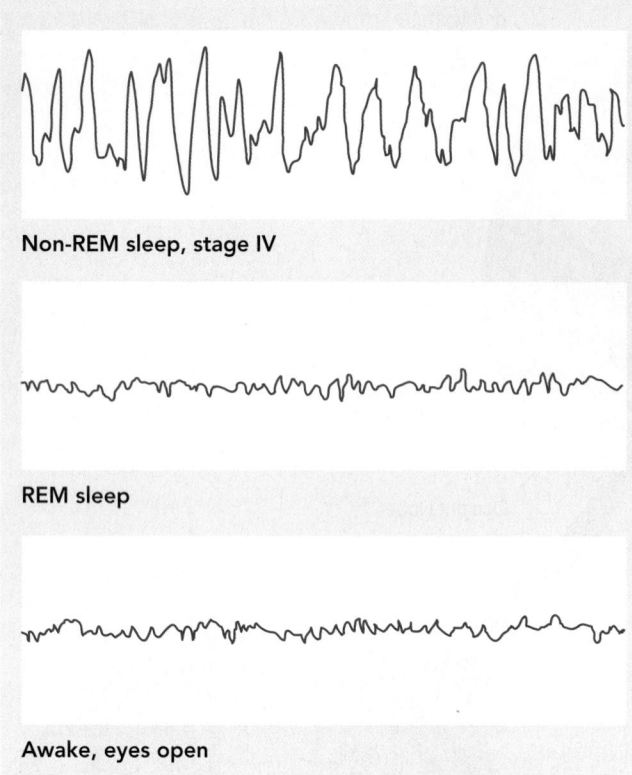

Non-REM sleep, stage IV

REM sleep

Awake, eyes open

(a) EEG recordings show brain wave rhythms during different stages of sleep. During non-REM, stage IV sleep, the brain emits slow-frequency, high-amplitude waves. During REM sleep, the brain emits high-frequency activity similar to that of a person who is awake.

Figure 42-14 Brain wave activity and cyclical pattern of sleep
© Cengage Learning

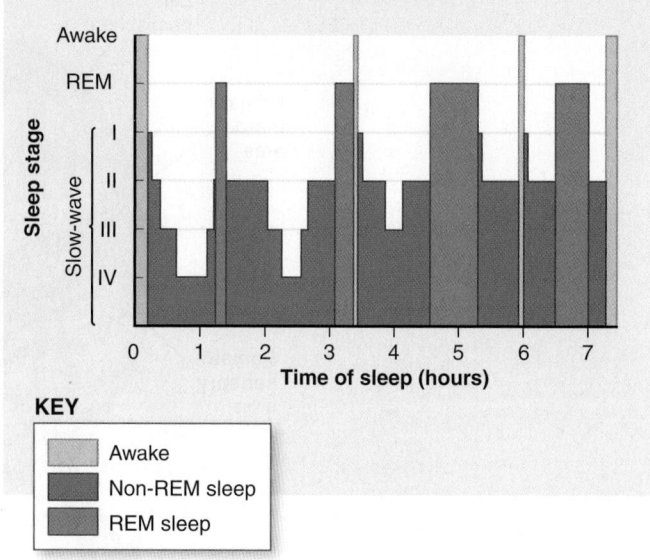

KEY

▢	Awake
▢	Non-REM sleep
▢	REM sleep

(b) A typical sleep pattern in a young adult.

The limbic system affects emotional aspects of behavior

The **limbic system,** present in all mammals, influences the emotional aspects of behavior, evaluates rewards, and is important in motivation. It plays a role in sexual behavior, biological rhythms, and autonomic responses. The limbic system includes regions of the cerebrum (parts of the prefrontal cortex, the cingulate cortex, the temporal lobe, the basal ganglia, the hippocampus, the amygdala, and the olfactory bulb), regions of the thalamus, the hypothalamus, several nuclei in the midbrain, and the neural pathways that connect these structures (**FIG. 42-15**). The limbic system influences the endocrine system and the autonomic division of the nervous system.

Emotions are feeling states that we experience both physiologically and cognitively. Humans are born with the capacity to experience and express emotions, including fear, anger, sadness, and happiness. The **amygdala** filters incoming sensory information and interprets it in the context of emotional needs and survival. When it perceives danger, the amygdala sends information to other parts of the brain so that appropriate responses can be made. This part of the limbic system is important in the experience of fear and aggression. The amygdala often becomes hypersensitive to possible danger following a traumatic experience (see *Inquiring About: The Neurobiology of Traumatic Experience*).

Structures of the limbic system are essential in an infant's bonding with a consistent, nurturing caregiver during the first year of life. The development of a secure attachment bond between infant and caregiver is the foundation for all future emotional and social connections between the child and other human beings. The quality of early attachment between child and caregiver affects brain development and future behavior. The caregiver encourages the baby to express emotions and helps her learn how to regulate them. Attachment also determines how vulnerable (or resilient) the individual will be to traumatic experiences later in life.

lobes is reduced. In contrast, blood flow increases in limbic system areas that produce visual scenes and emotion. Body movement, heart rate, breathing, blood pressure, metabolic rate, and body temperature all increase. Increased secretion of the neurotransmitter GABA by certain neurons in the brain and spinal cord inhibits muscle contraction in some large groups of muscles. This action results in a type of temporary paralysis during REM sleep.

Sleep may have several functions Although most animals have a sleep–wake cycle, neurobiologists are not certain why sleep is necessary. One hypothesis holds that sleep allows certain synaptic connections set during the day to reset. Neurons may strengthen their synaptic contacts with certain neurons and weaken their connections with other neurons. In this way, the brain prunes unneeded synaptic contacts. Memories may be consolidated during REM sleep.

Another hypothesis suggests that non-REM sleep may provide time for the brain to restore itself. When a person stays awake for unusually long periods, fatigue and irritability result, and even routine tasks are not performed well. For example, fatigue is a leading cause of automobile accidents. Sleep also provides the body with the opportunity to conserve energy and replace glycogen stores.

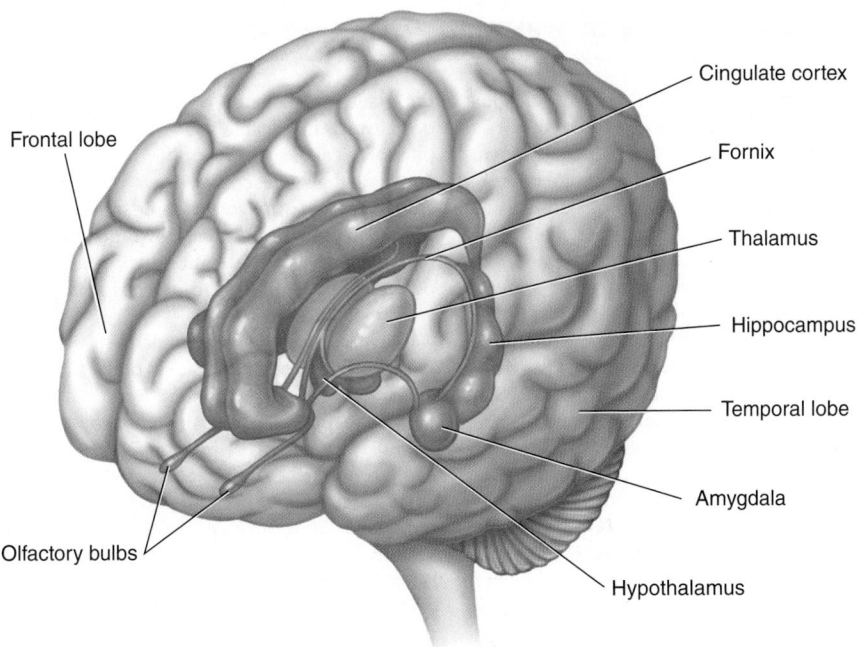

Figure 42-15 The limbic system

The limbic system consists of a ring of forebrain structures that surround the brain stem and are interconnected by complex neural pathways. This system includes the amygdala, hippocampus, parts of the prefrontal cortex, the cingulate cortex, and areas in the thalamus and hypothalamus. The limbic system is important in assessing danger, in emotional expression, sexual behavior, motivation, and learning.

© Cengage Learning

Labels: Frontal lobe, Cingulate cortex, Fornix, Thalamus, Hippocampus, Temporal lobe, Amygdala, Hypothalamus, Olfactory bulbs

Researchers first discovered in the 1950s that the limbic system is part of an important motivational system when they implanted electrodes in certain areas of the brains of laboratory animals. They found that a rat quickly learns to press a lever that stimulates *reward centers* in the brain. In fact, the rat will press the lever as many as 5000 times per hour, choosing this self-stimulation over food and water, until it drops from exhaustion.

The brain has reward centers that give us pleasure when we carry out vital activities such as eating, drinking, and sexual activity. These centers are important in experiencing emotion and in motivation. When stimulated, they activate reward circuits that allow us to feel pleasure in response to certain experiences.

The *mesolimbic dopamine pathway,* a major reward circuit, includes two important midbrain areas: the substantia nigra and the adjacent ventral tegmental area. These areas, which extend into behavioral control centers in the limbic system, contain the largest group of neurons in the brain that release dopamine. Novel stimuli associated with reward or pleasure activate these dopamine neurons. The neurons signal information about surprising events or stimuli that predict rewards. Such signals may motivate us to act when something important is happening. Dopamine pathways are also important in the neurobiology of drug addiction (discussed later in this chapter).

As researchers continue to unravel the mechanisms of dopamine pathways, we may better understand such disorders as attention deficit hyperactivity disorder (ADHD) and schizophrenia, conditions associated with excessive amounts of dopamine and other neurotransmitters in the brain. People with these disorders have difficulty filtering out sensory stimuli. The presence of excess dopamine may drive them to divert their attention to so many sensory stimuli that they have difficulty focusing on the most important stimuli.

Learning and memory involve long-term changes at synapses

Neurobiologists are beginning to understand how the nervous system functions, including how we learn and remember. **Learning** is the process by which we acquire knowledge or skills as a result of experience. For learning to occur, we must be able to remember what we experience. **Memory** is the process of encoding, storing, and retrieving information or learned skills.

An interesting question has been posed: To what extent are the brain's functions **hard-wired,** that is, preset? Many areas of the brain once thought to be hard-wired are now known to be flexible and capable of change. Even animals with very simple nervous systems can learn to repeat behaviors associated with reward and to avoid behaviors that cause pain. Such changes in behavior are possible because of **synaptic plasticity,** the ability of synaptic connections to change in response to experience. Such change involves structure and function.

Familiar examples of plasticity in human motor skills include learning to walk, ride a bicycle, or catch a baseball. At first, you were probably clumsy, but with practice your performance became smoother and more precise. For that to happen, changes must occur at synapses in neural circuits. Similarly, the ability to learn languages, solve problems, and perform scientific experiments depends on synaptic plasticity.

What are the mechanisms by which we learn diverse activities, such as skilled movement and the study of biology? Canadian psychologist Donald O. Hebb proposed in 1949 that when two neurons connected by a synapse (i.e., presynaptic and postsynaptic neurons) are active simultaneously, the synapse is strengthened. We can credit Hebb for the axiom "Neurons that fire together, wire together." Hebb's hypotheses have been supported by experimental studies. Experience and practice do affect how neural circuits in our brains organize themselves.

Learning involves the storage of information and its retrieval **Implicit memory,** also called *procedural memory,* is unconscious memory for perceptual and motor skills, such as running, playing a guitar, or texting a message. Implicit memory is about "how" to do something and involves repeating the behavior until it becomes routine. **Declarative memory,** also called *explicit memory,* involves factual knowledge of people, places, or objects and requires conscious recall of the

Have you been involved in a serious automobile accident, or have you been the victim of a violent crime? Are you a combat veteran of Iraq or Afghanistan? Perhaps you are a survivor of a natural disaster, such as a hurricane or earthquake. Or you may have endured childhood neglect or abuse. If you are a survivor of any of these experiences or of some other terrifying event, you may have experienced the *effects* of trauma. A *traumatic experience* is an event that causes intense fear, helplessness, or horror and that overwhelms normal coping and defense mechanisms.

Most of us have strategies for effectively processing *moderately* disturbing experiences. For example, if you were involved in a minor car accident or were one of the millions of people who watched on television as people were wounded in war or pulled from the rubble of buildings after earthquakes, you may have been disturbed by the events you experienced or witnessed. You probably processed what you saw and heard by thinking about it and talking to your friends and family about what happened. You may have dreamed about your experience. As the brain actively reviews and sorts out a disturbing experience, we try to make sense of what happened and we process the memory. The emotional intensity decreases, and we store the memory along with other memories of more ordinary past events. The memory of the uncomfortable experience fades in importance and intensity.

Traumatic events are more difficult to process than moderately disturbing events. If you survived an experience in which you thought you were going to die or be seriously hurt or if you witnessed someone else being wounded or killed, your body's response to real or perceived danger may be very intense. Years later, your body may still be secreting abnormally high amounts of stress hormones and you may remain hyper-alert, your body poised to deal with any new threat.

How does the brain respond in a traumatic experience, and how does the experience affect the brain? When danger threatens, the amygdala sends messages to both the hypothalamus (which signals the autonomic division of the nervous system and the endocrine system) and the cerebral cortex (which allows us to be aware of our experience). The amygdala is programmed to remember the smells, sounds, and sensations that are part of the experience. Until the memories of the experience are fully processed, similar smells, sounds, and sensations remind us of the traumatic event and trigger the body to

prepare for danger. The traumatized individual may experience fear and anxiety.

Because they are so overwhelming, traumatic memories are very difficult to process. They cause us so much discomfort and anxiety that we tend to avoid them rather than intentionally focus on them and sort them out. As a result, traumatic memories seem to stay "stuck" in the limbic system. When triggered, the experience may be replayed with its original emotional intensity (a flashback). Some trauma survivors develop post-traumatic stress disorder (PTSD), a condition in which they experience (1) intrusive thoughts, images, sensory experiences, memories, and dreams; (2) an urge to avoid reminders of the traumatic event; and (3) physiological hyperarousal, a condition in which the body remains on high-alert status, scanning the environment for potential danger.

Several studies have shown an association between prolonged trauma (e.g., severe childhood neglect or abuse) and long-term changes in the brain. These changes include EEG abnormalities; smaller size of certain regions of the brain; compromised right brain development (the right brain is specialized for processing information related to emotion, social interaction, and physiological states); differences in neural circuits connecting cortical and subcortical areas (which can decrease the individual's sense of self and lead to poor connections with other people); abnormal concentrations of certain neurotransmitters, which affect mood and the ability to inhibit inappropriate behavior; inappropriate modulation of the limbic system, in which its responses are exaggerated; and long-term changes in the sympathetic nervous system and endocrine response to stress, including a faster, more intense response to stress.

Neuroimaging studies have demonstrated changes in brain function, including over-response by the amygdala and decreased response by the medial prefrontal cortex (which normally inhibits the amygdala). When the amygdala over-responds, the individual experiences anxiety, distress, and physiological hyperarousal. These physiological states can lead to intense emotional responses that are rooted in the traumatic experience.

Broca's area, a region in the posterior part of the left frontal cortex, is also affected by trauma. Broca's area is critically important in generating words and thus in expressing language (see Fig. 42-12). Researchers have shown that when subjects are exposed to accounts of their

traumatic experiences, activity in Broca's area decreases. This deactivation appears to be the physiological basis for the difficulty that trauma survivors have describing their experience in words. Trauma survivors have described a state of "speechless terror" that they experience during flashbacks of traumatic experiences. Without words, it is difficult to process and resolve a traumatic experience.

Based on many studies, it is clear that traumatic experiences can cause both structural and functional changes in the brain. The exaggerated responses of the amygdala to harmless stimuli perceived as threatening are disconnected from the influence of the prefrontal cortex that would normally modulate the limbic system. Responses generated by the limbic system are rooted in emotion rather than in rationality and judgment (see figure). PTSD can severely affect and limit one's life. Fortunately, this disorder can be resolved with appropriate treatment.

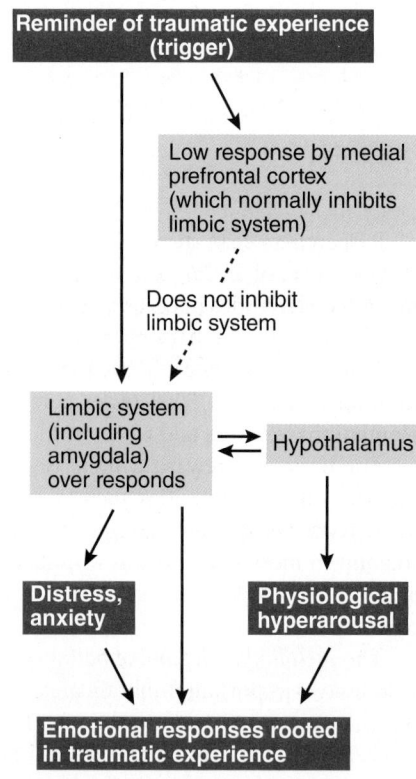

Some limbic system responses to reminders of traumatic experiences.

© Cengage Learning

---

information. The **hippocampus,** part of the limbic system, is critical in forming and retrieving declarative memories.

How long do you remember? You are constantly bombarded with thousands of bits of sensory information. At this very moment, your eyes are receiving information about the

words on this page, the objects around you, and the intensity of the light in the room. At the same time, you may be hearing a variety of sounds, such as music, your friends talking in the next room, or the hum of an air conditioner. Your olfactory epithelium may sense cologne or the smell of coffee. Maybe you are eating as you read. Sensory receptors in your hands may be receiving information regarding the weight and position of your book or notebook computer. We may hold such sensory information in *immediate memory* for fractions of a second. Most sensory stimuli are not important to remember any longer and are filtered out.

Short-term memory lasts only for seconds or for a few minutes. Typically, we can hold only about seven chunks of information (e.g., seven words or numbers) at a time in short-term memory. Short-term memory lets us recall information for a few minutes. When we look up a phone number, for example, we usually remember it only long enough to call it. If we need the same number an hour later, most of us have to look it up again.

When we select information for **long-term memory,** we must process it. The brain rehearses the material and *encodes* it. We recognize patterns and meaningfully *associate* the stimuli to past experience or knowledge. We practice recalling the information. The hippocampus temporarily holds new information and may integrate various aspects of an experience, including odors, sounds, and other information. The hippocampus helps place our experiences in categories so that they can be stored along with similar memories. We integrate new information with other knowledge already stored in the brain.

Memory consolidation allows memories to be transferred to the cerebral cortex and *stored* for long periods. Consolidation involves expression of genes and protein synthesis, and it depends on new neural connections at synapses. Time (several minutes, hours, or even longer) is required for the brain to *consolidate* a new memory.

If a person suffers a brain concussion, memory of what happened immediately before the incident may be completely lost. This condition is known as *retrograde amnesia.* When the hippocampus is damaged, long-term memory may not be impaired, and the patient may be able to recall information stored in the past. However, new experiences (and the short-term memories of them) can no longer be converted to long-term memories.

Where are memories stored? When large areas of the mammalian cerebral cortex are destroyed, information is lost somewhat in proportion to the extent of lost tissue. No specific area can be labeled the "memory bank." Rather, memories seem to be stored within many areas of the brain. Visual memories, for example, may be stored in the visual centers of the occipital lobes; similarly, auditory memories may be stored in the temporal lobes. Memories are integrated in many areas of the brain, including association areas of the cerebral cortex, amygdala, hippocampus, thalamus, and hypothalamus. *Wernicke's area* in the temporal lobe has been identified as an important association area for complex thought processes. Neurons within the association areas form interconnected pathways that permit complicated information transfer.

Retrieval of information stored in long-term memory is of considerable interest, especially to students! The challenge is to find information when you need it. When you seem to forget something, the problem may be that you have not searched effectively for the memory. Information retrieval can be improved by careful encoding, such as by forming strong associations between items you want to remember. On the other hand, unimportant, unused memories are typically forgotten over time. Forgetting can be useful when it clears the brain of unnecessary information.

Neurobiological changes occur during learning Short-term memory involves changes in the neurotransmitter receptors of postsynaptic neurons. These changes strengthen synaptic connections. The receptors are linked by second messengers (e.g., cyclic AMP) to ion channels in the plasma membrane.

In some types of learning, changes take place in presynaptic terminals or postsynaptic neurons that permanently enhance or inhibit the transmission of impulses. In some cases, specific neurons may become more sensitive than usual to a particular neurotransmitter. Neurons typically transmit action potentials in bursts, and the amount of a neurotransmitter released by each action potential may increase or decrease.

Repeated electrical stimulation of neurons causes a functional change at synapses. When a presynaptic neuron continues to transmit action potentials at a high rate for a minute or longer, there is a long-lasting strengthening of the connection between the presynaptic and postsynaptic neurons. This increased strength of the synaptic connection is known as **long-term potentiation (LTP).** (*Potentiation* is the process of "strengthening or making more potent.")

In contrast, low-frequency stimulation of neurons results in a long-lasting *decrease* in the strength of their synaptic connections. This decrease is called **long-term depression (LTD),** which is not related to the mood disorder. Information storage and forgetting depend on the strengthening and weakening of synaptic connections brought about by LTP and LTD.

Induction of LTP and LTD requires activation of two types of glutamate receptors: *NMDA receptors* and *AMPA receptors.* (Each of these receptors is named for the compound that artificially activates it. NMDA is the acronym for N-methyl-D-aspartate, and AMPA stands for α-amino-3-hydroxyl-5-methyl-4-isoxazole-propionate.) NMDA and AMPA receptors are present on the plasma membranes of postsynaptic neurons. They control the passage of calcium ions into neurons. NMDA receptors respond to the neurotransmitter *glutamate* by opening Ca^{2+} channels. However, when the postsynaptic neuron is at its resting potential, NMDA ion channels are blocked by Mg^{2+}.

A model for the mechanism of LTP is illustrated in FIGURE 42-16. A presynaptic neuron releases glutamate, which binds with AMPA receptors. The postsynaptic neuron becomes depolarized. If depolarization of the postsynaptic neuron is sufficiently strong, Mg^{2+} move away from the NMDA receptors, unblocking them. Glutamate can then bind with these receptors, opening Ca^{2+} channels and letting Ca^{2+} move into the cell. Calcium ions appear to be an important trigger for LTP.

Calcium ions act as second messengers that initiate long-term changes ultimately responsible for LTP. For example, calcium ions activate a Ca^{2+}-dependent second-messenger pathway that results in the insertion of more AMPA receptors in the postsynaptic membrane. This step is important because

When glutamate binds to NMDA receptors, Ca^{2+} enter the postsynaptic neuron and activate intracellular signaling systems, leading to changes responsible for LTP.

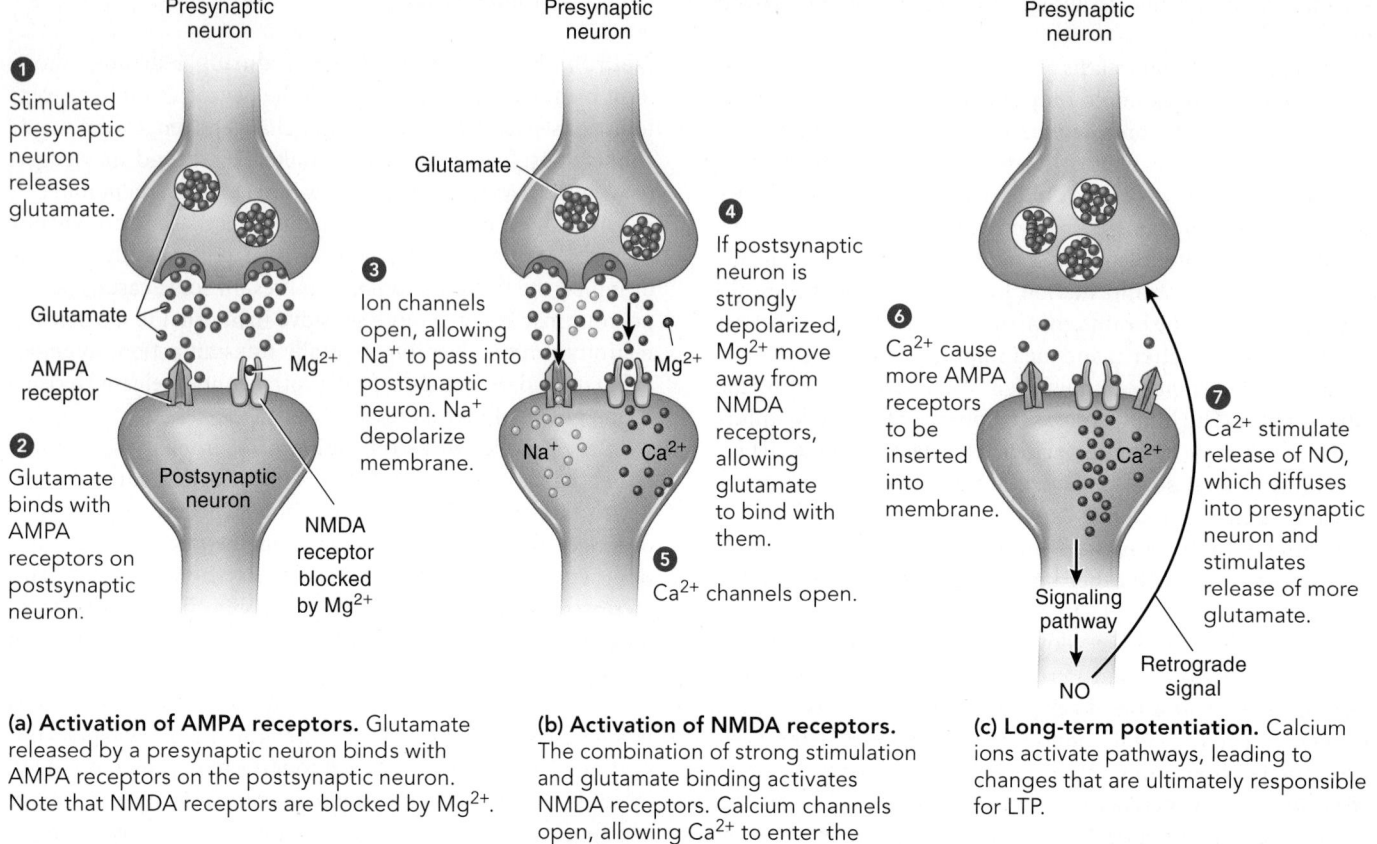

(a) Activation of AMPA receptors. Glutamate released by a presynaptic neuron binds with AMPA receptors on the postsynaptic neuron. Note that NMDA receptors are blocked by Mg^{2+}.

(b) Activation of NMDA receptors. The combination of strong stimulation and glutamate binding activates NMDA receptors. Calcium channels open, allowing Ca^{2+} to enter the postsynaptic neuron.

(c) Long-term potentiation. Calcium ions activate pathways, leading to changes that are ultimately responsible for LTP.

Figure 42-16 A proposed model for the mechanism of long-term potentiation (LTP)
© Cengage Learning

PREDICT Imagine that Ca^{2+} channels did not open (in step 5). How would this affect synaptic connections and LTP?

additional AMPA receptors increase sensitivity to glutamate. More EPSPs (excitatory postsynaptic potentials) are produced, which strengthens the synapse (i.e., helps maintain the LTP). Recall from Chapter 41 that an EPSP is a change in membrane potential that brings the neuron closer to its threshold level.

Calcium ions also activate a pathway that leads to release of a retrograde signal, a signal that moves backward, from the postsynaptic neuron to the presynaptic neuron. The soluble gas *nitric oxide* (NO) has been identified as the retrograde signal. This signal enhances neurotransmitter release by the presynaptic neuron. Note that this positive feedback loop strengthens the connection between the two neurons.

Long-term memory involves gene expression Gene expression and protein synthesis take place during the process of establishing long-term memory. This process involves slower, but longer-lasting, changes in synaptic connections. Long-term memory depends on activated receptors linked to G proteins (see Chapter 6). Cyclic AMP acts as a second messenger.

A high level of cyclic AMP activates a protein kinase that enters the nucleus, leading to gene activation and protein

synthesis. In this process, protein kinase phosphorylates a transcription factor known as *cyclic AMP response element binding protein (CREB)*, which then turns on the transcription process of certain genes. CREB has been shown to be a signaling molecule in the memory pathway in many animals, including fruit flies and mice. The molecules and processes involved in learning and memory have been highly conserved during evolution.

Experience affects development and learning Many studies have demonstrated synaptic plasticity in rats, mice, and other laboratory animals exposed to enriched environments. In contrast with rats housed in standard cages and provided with the basic necessities, those exposed to enriched environments are given toys and other stimulating objects as well as the opportunity to socially interact with other rats. Animals reared in an enriched environment exhibit increased synaptic contacts, and they process and remember information more quickly than animals lacking such advantages. Mice exposed to enriched environments develop significantly greater numbers of neurons in the hippocampus and learn mazes faster than control animals.

Early environmental stimulation can also enhance the development of motor areas in the brain. For example, the brains of rats encouraged to exercise become slightly heavier than those of control animals. Characteristic changes occur within the cerebellum, including the development of larger dendrites.

Apparently, during early life certain critical or sensitive periods of nervous system development occur that are influenced by environmental stimuli. For example, when the eyes of young mice first open, neurons in the visual cortex develop large numbers of dendritic spines (structures on which synaptic contact takes place). If the animals are kept in the dark and deprived of visual stimuli, fewer dendritic spines form. If the mice are exposed to light later in life, some new dendritic spines form, but never as many as develop in a mouse reared in a normal environment.

Studies linking the development of the human brain with environmental experience have demonstrated that early stimulation is important for the sensory, motor, intellectual, and social development of children. Such studies argue for the importance of informed parenting as well as for early childhood education programs, such as Head Start. Investigators have also shown that environmental stimulation is needed to maintain the status of the cerebral cortex in later life.

Language involves comprehension and expression

Language is a form of communication in which we use words to symbolize objects and convey ideas. In about 97% of humans, including most individuals who are left-handed, the areas of the brain responsible for language are located in the left hemisphere (see Fig. 42-12). **Wernicke's area,** located in the temporal lobe, is an important center for language comprehension. This area helps us recognize and interpret both spoken and written words. Wernicke's area also helps us formulate the choice and sequence of words and transfers information to **Broca's area.** Located near motor areas in the left frontal lobe, Broca's area controls our ability to speak. This area is also important in language processing and speech comprehension.

CHECKPOINT 42.4

- *How is the human CNS protected? What is the function of the cerebrospinal fluid?*
- *What are two main functions of the vertebrate spinal cord?*
- VISUALIZE *Draw a lateral view of the human brain, label the major structures, including the lobes; write the functions of each lobe.*
- CONNECT *What role does the limbic system play in emotions? in traumatic experience?*
- CONNECT *In what ways does your success on a biology exam depend on synaptic plasticity? (Include the roles of LTP, gene expression, and NMDA receptors in your answer.)*

42.5 THE PERIPHERAL NERVOUS SYSTEM

LEARNING OBJECTIVES

10 Describe the organization of the peripheral nervous system and compare its somatic and autonomic divisions.

11 Contrast the sympathetic and parasympathetic divisions of the autonomic system and give examples of the effects of these systems on specific organs.

The peripheral nervous system consists of the sensory receptors, the nerves that link these receptors with the central nervous system, and the nerves that link the CNS with effectors (muscles and glands). The somatic division of the PNS helps the body respond to changes in the external environment. The nerves and receptors that maintain homeostasis despite internal changes make up the autonomic division.

The somatic division helps the body adjust to the external environment

The **somatic division** of the PNS includes the receptors that react to changes in the external environment, the sensory neurons that inform the CNS of those changes, and the motor neurons that adjust the positions of the skeletal muscles that help maintain the body's posture and balance. In mammals 12 pairs of **cranial nerves** emerge from the brain. They transmit information to the brain from the sensory receptors for smell, sight, hearing, and taste. For example, cranial nerve II, the optic nerve, transmits signals from the retina of the eye to the brain. The cranial nerves transmit information to the brain from general sensory receptors, especially in the head region.

Cranial nerves bring orders from the CNS to the voluntary muscles that control movements of the eyes, face, mouth, tongue, pharynx, and larynx. Cranial nerve VII, the facial nerve, transmits signals to the muscles used in facial expression and to the salivary glands.

In humans 31 pairs of **spinal nerves** emerge from the spinal cord. Named for the general region of the vertebral column from which they originate, they comprise 8 pairs of cervical, 12 pairs of thoracic, 5 pairs of lumbar, 5 pairs of sacral, and 1 pair of coccygeal spinal nerves. The ventral branches of several spinal nerves form tangled networks called *plexi* (sing., *plexus*). Within a plexus, the fibers of a spinal nerve may separate and then regroup with fibers that originated in other spinal nerves. Thus, nerves emerging from a plexus consist of neurons from several different spinal nerves.

The autonomic division regulates the internal environment

The **autonomic division** of the PNS helps maintain homeostasis in the internal environment. For example, it regulates the heart rate and helps maintain a constant body temperature.

The autonomic system works automatically and without voluntary input. Its effectors are smooth muscle, cardiac muscle, and glands. Like the somatic system, it is functionally organized into reflex pathways. Receptors within the viscera relay information via afferent nerves to the CNS. The information is integrated at various levels. Then the "decisions" are transmitted along efferent nerves to the appropriate muscles or glands.

Afferent and efferent neurons of the autonomic division lie within cranial and spinal nerves. For example, afferent fibers of cranial nerve X, the *vagus nerve,* transmit signals from many organs of the chest and upper abdomen to the CNS. Its efferent fibers transmit signals from the brain to the heart, stomach, small intestine, and several other organs.

The efferent portion of the autonomic division is subdivided into **sympathetic** and **parasympathetic systems** (TABLE 42-3). Many organs are innervated by both types of nerves. In general, sympathetic and parasympathetic systems have *opposite* effects (FIGS. 42-17 and 42-18). For example, the heart rate is speeded up by messages from sympathetic nerve fibers and slowed by impulses from its parasympathetic nerve fibers. In many cases, sympathetic nerves operate to stimulate organs and to mobilize energy, especially in response to stress, whereas the parasympathetic nerves influence organs to conserve and restore energy, particularly during quiet, calm activities.

Instead of using a single efferent neuron, as in the somatic division, the autonomic division uses a relay of two neurons between the CNS and the effector. The first neuron, called the *preganglionic neuron,* has a cell body and dendrites within the CNS. Its axon, part of a peripheral nerve, ends by synapsing with a *postganglionic neuron.* The dendrites and cell body of the postganglionic neuron are in a ganglion outside the CNS. Its axon terminates near or on the effector.

The sympathetic ganglia are paired, and a chain of them— the *paravertebral sympathetic ganglion chain*—runs on each side of the spinal cord from the neck to the abdomen. Some sympathetic preganglionic neurons do not end in these ganglia but pass on to *collateral ganglia* in the abdomen, close to the aorta and its major branches. Parasympathetic preganglionic neurons synapse with postganglionic neurons in *terminal ganglia* near or within the walls of the organs they innervate.

The sympathetic and parasympathetic systems also differ in the neurotransmitters they release at the synapse with the effector. Both preganglionic and postganglionic parasympathetic

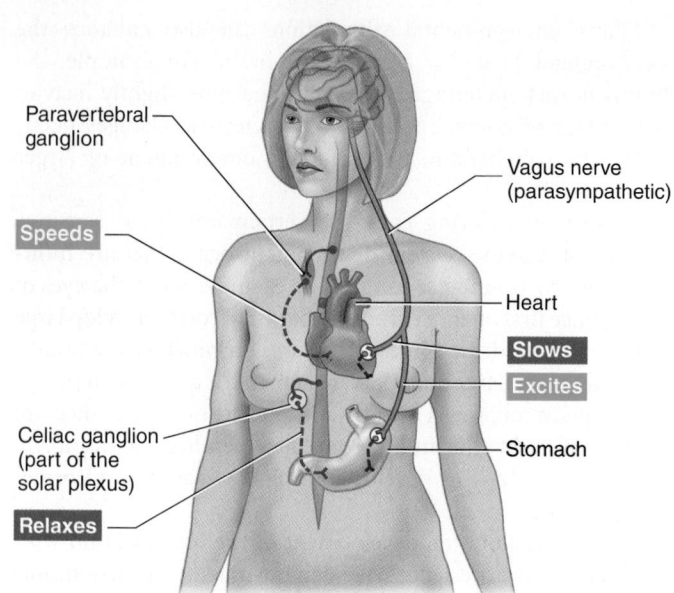

Figure 42-17 Dual innervation of the heart and stomach

Most organs are innervated by both sympathetic and parasympathetic nerves. A sympathetic nerve increases the heart rate, whereas a parasympathetic nerve, the vagus nerve, slows the heartbeat. In contrast, sympathetic nerves slow contractions of the stomach and intestine, whereas the vagus nerve stimulates contractions of these digestive organs. Sympathetic nerves are shown in *red;* postganglionic fibers are shown as *dotted lines.*
© Cengage Learning

neurons secrete acetylcholine. Sympathetic postganglionic neurons release norepinephrine (although preganglionic sympathetic neurons secrete acetylcholine).

The autonomic division got its name from the original idea that it was independent of the CNS, that is, autonomous. Neurobiologists have disproved this concept and shown that the hypothalamus and many other parts of the CNS help regulate the autonomic division. Although the autonomic system usually functions automatically, its activities can be consciously influenced.

Biofeedback provides a person with visual or auditory evidence concerning the status of an autonomic body function.

TABLE 42-3	Comparison of Sympathetic and Parasympathetic Systems	
	SYMPATHETIC SYSTEM	**PARASYMPATHETIC SYSTEM**
General effect	Prepares body to cope with stressful situations	Restores body to resting state after stressful situation; actively maintains normal body functions
Extent of effect	Widespread throughout body	Localized
Duration of effect	Lasting	Brief
Outflow from CNS	Thoracic and lumbar nerves from spinal cord	Cranial nerves and sacral nerves from spinal cord
Location of ganglia	Chain and collateral ganglia	Terminal ganglia
Neurotransmitters released	Preganglionic: acetylcholine	Acetylcholine
	Postganglionic: norepinephrine (usually)	Acetylcholine

© Cengage Learning

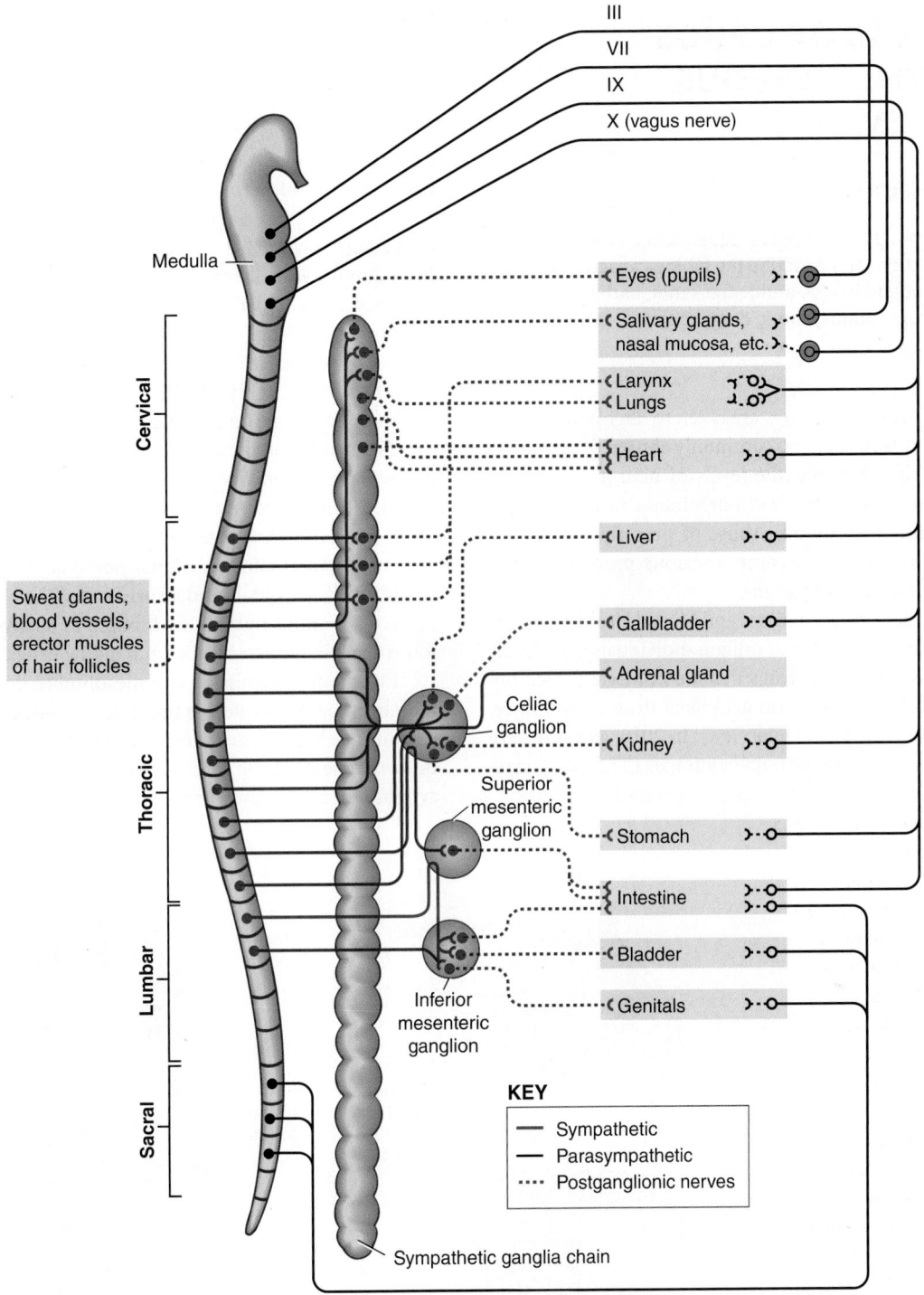

III
VII
IX
X (vagus nerve)

Medulla

Cervical

Eyes (pupils)
Salivary glands, nasal mucosa, etc.
Larynx
Lungs
Heart
Liver

Sweat glands, blood vessels, erector muscles of hair follicles

Gallbladder
Adrenal gland
Celiac ganglion
Kidney
Superior mesenteric ganglion
Stomach
Intestine
Bladder
Genitals

Thoracic

Lumbar

Inferior mesenteric ganglion

KEY

— Sympathetic
— Parasympathetic
···· Postganglionic nerves

Sacral

Sympathetic ganglia chain

Figure 42-18 *Animation* **The sympathetic and parasympathetic systems**

Complex as it appears, this diagram has been greatly simplified. *Red lines* represent sympathetic nerves, *black lines* represent parasympathetic nerves, and *dotted lines* represent postganglionic nerves.
© Cengage Learning

For example, a tone may be sounded when blood pressure reaches a desirable level. Using such techniques, individuals have learned to control autonomic activities such as blood pressure, brain wave pattern, heart rate, and blood sugar level. Even certain abnormal heart rhythms can be consciously modified.

CHECKPOINT 42.5

- *Compare the function of the somatic division of the PNS with that of the autonomic division of the PNS.*
- *Contrast the sympathetic and parasympathetic systems with respect to their general effect and the types of neurotransmitters they release.*

42.6 EFFECTS OF DRUGS ON THE NERVOUS SYSTEM

LEARNING OBJECTIVE

12 Distinguish among psychological dependence, tolerance, and drug addiction, and describe the biological actions and effects on mood of alcohol, antidepressants, barbiturates, anti-anxiety drugs, antipsychotic drugs, opiates, stimulants, hallucinogens, and marijuana.

About 25% of all prescribed drugs are taken to alter psychological conditions, and almost all the commonly abused drugs affect mood. Many act by changing the levels of neurotransmitters within the brain. Changing levels of norepinephrine, serotonin, and dopamine influence mood. Abuse of prescription drugs, particularly painkillers, has become a serious problem, especially among teens and young adults.

According to the Substance Abuse and Mental Health Services Administration, an estimated 22.6 million individuals ages 12 or older in the United States use drugs that are not legal (based on individuals who acknowledged using an illicit drug at least once during the 30 days prior to the interview). In 2010, an estimated 22.1 million persons (8.7% of the population ages 12 or older) were diagnosed as dependent on substances or with substance abuse.

Although legal, alcohol is the most commonly used and abused drug. (See *Inquiring About: Alcohol: The Most Abused Drug.*) More than 9 million children live with a parent who is dependent on alcohol and/or illicit drugs. TABLE 42-4 lists several commonly used and abused drugs and gives their effects.

Habitual use of almost any mood-altering drug can result in *psychological dependence,* in which the user becomes emotionally dependent on the drug. When deprived of it, the user craves the feeling of euphoria (well-being) the drug induces. Some drugs induce *tolerance* after several days or weeks; response to the drug decreases, and greater amounts are required to obtain the desired effect. Tolerance often occurs because the liver cells are stimulated to produce more of the enzymes that metabolize and inactivate the drug. It can also

occur when the number of postsynaptic receptors that bind with the drug decrease (*down-regulation*).

Drug addiction is a serious societal problem that involves compulsive use of a drug despite negative health consequences and negative effects on the ability to function socially, occupationally, and in school. Use of some drugs, such as tobacco, alcohol, barbiturates, and heroin and other opiates (morphine, codeine, oxycodone), can also result in physical dependence; physiological changes occur that make the user dependent on the drug. For example, when heroin or alcohol is withheld, the addict suffers characteristic withdrawal symptoms. Physical addiction can occur when a drug—for example, morphine—has components similar to substances that body cells normally manufacture on their own. Some highly addictive drugs, such as crack cocaine and methamphetamine, do not cause serious withdrawal symptoms.

The neurobiological mechanisms for drug addiction involve the *mesolimbic dopamine pathway,* a major reward circuit (described earlier in this chapter in connection with the limbic system). Many of the neurons in this pathway project from the *ventral tegmental area* in the midbrain and signal neurons in the limbic system, including the amygdala and an area called the *nucleus accumbens.* Projections of this nucleus are thought to be involved in the reinforcement people experience from taking drugs.

Effects on neurons in the mesolimbic dopamine pathway have been demonstrated in nicotine, heroin, amphetamine, and cocaine addiction. For example, amphetamines stimulate dopamine release. Cocaine blocks the reuptake of dopamine into presynaptic neurons, thus prolonging the stimulation (FIG. 42-19).

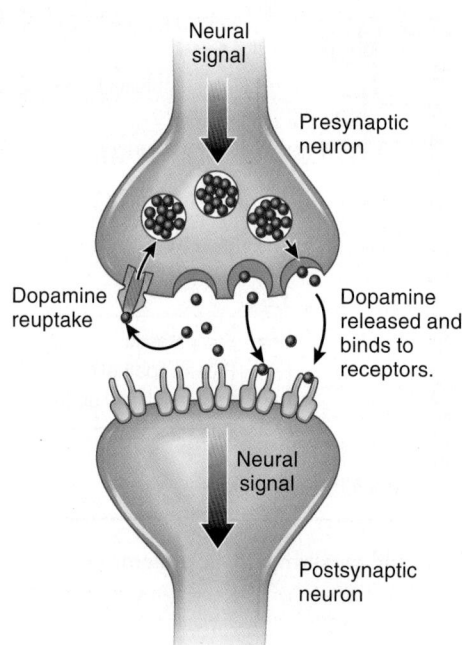

(a) Reuptake of the neurotransmitter dopamine. Dopamine binds with a transporter (*green*) that shuttles it back into the presynaptic terminal.

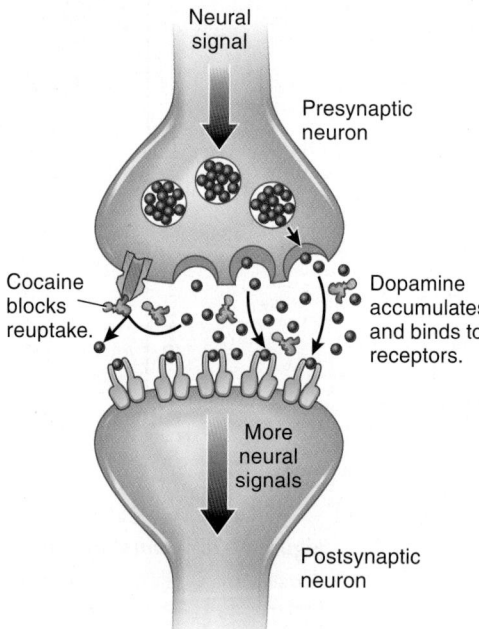

(b) Effect of cocaine. Cocaine blocks the reuptake of dopamine by binding competitively with the dopamine reuptake transporter. As a result, more dopamine is available to bind with postsynaptic receptors.

Figure 42-19 Mechanism of cocaine action
© Cengage Learning

How does alcohol affect health, families, and society? Alcohol is one of the leading preventable risk factors for premature death. According to the World Health Organization, 320,000 young people (15–29 years old) die each year from alcohol-related causes globally. Most alcohol-related deaths result from injuries, cancer, liver cirrhosis, and cardiovascular diseases.

According to the National Survey on Drug Use and Health in 2010, about 18 million individuals in the United States ages 12 or older have been diagnosed with alcohol dependence (also referred to as addiction) or abuse. About 80% of college students drink alcohol. Approximately half of these students also binge drink. According to a 2011 report by the Centers for Disease Control and Prevention (CDC), binge drinking has become a serious problem among women. About 1 in 8 women and 1 in 5 high school girls acknowledge that they binge drink. In fact, nearly 14 million women in the United States binge drink about three times each month. They consume an average of six drinks each time they binge drink. Such excessive alcohol intake increases the risk for breast cancer and other cancers, heart disease, sexually transmitted diseases, and unintended pregnancy.

Almost 3 million students between the ages of 18 and 24 report that they drove a motor vehicle under the influence of alcohol during the past year. According to Mothers against Drunk Driving, teens who drink alcohol kill about 6000 people every year. (The number of deaths is greater than those caused by teen use of all illegal drugs combined.)

Every year, nearly 100,000 college students between the ages of 18 and 24 are victims of alcohol-related sexual assault or date rape. Almost 2000 college students between the ages of 18 and 24 die each year from alcohol-related nonintentional injuries.

Based on many studies, CDC reports that youth who drink alcohol are likely to experience serious problems, including disruption of normal growth and sexual development; abnormal brain development that may have lifelong effects; memory problems; school, social, and legal problems; and higher risk for suicide, homicide, and death from alcohol poisoning. Adults who first used alcohol when they were 14 years old or younger were more than five times as likely to have a problem with alcohol dependence or abuse than adults who had their first drink at age 21 or older. Alcohol abuse and dependence result in physiological, psychological, and social impairment and have serious negative consequences for family, friends, and society.

Each of the following has been linked to alcohol abuse:

- More than 30% of all traffic fatalities (Drugs other than alcohol—e.g., marijuana and cocaine—are involved in about 18% of motor vehicle deaths.)
- Almost 50% of violent crime
- More than 60% of child-abuse and spouse-abuse cases
- More than 50% of college sexual assaults
- About 25% of suicides
- The birth of thousands of babies each year with serious birth defects (Fetal alcohol spectrum disorders are the leading cause of preventable intellectual disability in the United States.)
- Increased risk of mouth, larynx, and throat cancers. Alcohol is also a risk factor for primary liver cancer, breast cancer, and colon cancer. These cancers have been linked to regular alcohol consumption.

A single drink—that is, 12 oz of beer, 5 oz of wine, or 1.5 oz of 72-proof liquor—results in a blood alcohol concentration of approximately 20 mg/dL (milligrams per deciliter). This level represents about 0.5 oz of pure alcohol in the blood. Alcohol accumulates in the blood because absorption occurs more rapidly than oxidation and excretion. Every cell in the body can take in alcohol from the blood.

At first, the drinker may feel stimulated, but alcohol actually depresses the CNS. As blood alcohol concentration rises, information processing, judgment, memory, sensory perception, and motor coordination all become progressively impaired. Depression and drowsiness generally occur. Contrary to popular belief, alcohol decreases sexual performance. Some individuals become loud, angry, or violent.

A blood alcohol concentration of 80 mg/dL (or 0.08 gm/100 mL) legally defines driving under the influence, or DUI. A 170-lb man typically reaches this level by drinking four drinks in an hour on an empty stomach. (Note that the National Transportation Safety Board has recommended that states lower the legal allowable blood alcohol concentration to 50 mg/dl, or 0.05 gm/100 mL.) Alcohol metabolism occurs at a fixed rate in the liver. About one drink per hour is metabolized. Only time, not coffee or other remedies, decreases the effects of alcohol intake. Because alcohol inhibits water reabsorption in the kidneys, more fluid is excreted as urine than is consumed in the alcohol. The result is dehydration that, together with a low blood sugar level, may cause the stupor produced by excessive drinking.

In chronic drinkers cells of the CNS adapt to the presence of the drug, which causes *tolerance* (more and more alcohol is needed to experience the same effect) and physical dependence. Abrupt withdrawal can result in sleep disturbances, severe anxiety, tremors, seizures, hallucinations, and psychosis.

The *A1* allele of the D2 dopamine receptor gene increases the risk for alcoholism. Individuals with this allele have fewer D2 dopamine receptors in their brain and therefore have less dopamine than those without this allele. As a result, they are less able to experience pleasure. Alcohol, like cocaine, heroin, amphetamines, and nicotine, enhances dopamine activity in the *mesolimbic dopamine pathway* of the brain. Individuals with the *A1* allele may compensate for their low dopamine levels by using alcohol and other drugs. However, the mechanisms underlying alcohol addiction are far more complex. Evidence suggests that many neurotransmitters are involved, including glutamate, serotonin, GABA, and opioid peptides.

Treatment for alcohol addiction includes various forms of psychotherapy and use of drugs such as naltrexone and acamprosate, which are anticraving drugs, and disulfiram, which is an aversion drug. The group support offered by Alcoholics Anonymous (AA) has proved effective for many struggling with alcohol dependence. Participation in AA includes attending frequent meetings, having a sponsor within the program, and working on its 12 steps.

Amphetamines or cocaine can increase dopamine concentration in the limbic system by up to 1000 times normal amounts! Prolonged drug use alters gene expression and changes neuron structure.

CHECKPOINT 42.6

- *How do the following drugs affect the CNS: alcohol, antipsychotic drugs, antidepressants, amphetamines, opiates such as oxycodone, and MDMA ("Ecstasy")?*

TABLE 42-4 Effects of Some Commonly Used Drugs

NAME OF DRUG	EFFECT ON MOOD	ACTIONS ON BODY	SIDE EFFECTS/DANGERS ASSOCIATED WITH ABUSE
ANTIDEPRESSANTS			
Selective serotonin reuptake inhibitors (SSRIs) (e.g., Prozac, Zoloft, Lexapro, Pristiq)	Relieve depression; used to treat obsessive–compulsive behavior	Block serotonin reuptake; some also block norepinephrine reuptake	Sometimes cause nausea, headache, insomnia, anxiety
Tricyclic antidepressants (e.g., Elavil, Anafranil)	Elevate mood; relieve depression; used to treat obsessive–compulsive behavior	Most block reuptake of biogenic amines (especially norepinephrine), increasing their concentration in synapses	Sedation, weight gain, sexual dysfunction
MAO (monoamine oxidase) inhibitors (e.g., Parnate, Nardil)	Relieve depression	Block enzymatic breakdown of biogenic amines, increasing their concentration in synapses	Liver toxicity, excessive CNS stimulation; overdose may affect blood pressure, may cause hallucinations
ANTI-ANXIETY MEDICATIONS			
Benzodiazepines (e.g., Xanax, Valium, Librium)	Sedation; induce sleep	Depress CNS; bind to GABA receptors on postsynaptic neuron; increase effectiveness of GABA in opening chloride channels, causing hyperpolarization; cause relaxation of skeletal muscles	Drowsiness, confusion; psychological dependence, addiction; effects additive with alcohol
Barbiturates ("downers") (e.g., phenobarbital)	Sedation; induce sleep; used as anesthesia; used to treat epilepsy	Depress CNS; bind to GABA receptors on chloride channels; the channels open so that more chloride ions move in, causing hyperpolarization	Drowsiness, confusion; psychological dependence, tolerance, addiction; severe CNS depression, resulting in coma and death (especially lethal in combination with alcohol)
ANTIPSYCHOTIC MEDICATIONS			
Phenothiazines (e.g., Thorazine, Mellaril, Stelazine)	Relieve symptoms of schizophrenia; reduce impulsive and aggressive behavior	Block dopamine receptors	Tardive dyskinesia, a neurological syndrome characterized by repetitive involuntary movements, muscle spasms, and shuffling gait
Atypical antipsychotic medications (e.g., Abilify, Risperdal)	Similar to phenothiazines, but also increase motivation	Block dopamine and certain other neurotransmitter receptors	Fewer side effects than with the phenothiazines
NARCOTIC ANALGESICS			
Opioids (e.g., morphine, codeine, meperidine [Demerol], oxycodone [Oxycontin], heroin)	Euphoria; sedation; pain relief	Depress CNS; block pain; mimic actions of endorphins; bind to opiate receptors; like cocaine, inhibit reuptake of dopamine	Depress respiration; constrict pupils; impair coordination; cause structural and functional changes in the brain; tolerance, psychological dependence, addiction; convulsions, death from overdose
COCAINE	Brief, intense high with euphoria, followed by fatigue and depression	Stimulates release and inhibits reuptake of dopamine, leading to CNS stimulation followed by depression; autonomic stimulation; dilation of pupils; local anesthesia	Mental impairment, convulsions, hallucinations, unconsciousness; death from overdose
MARIJUANA	Euphoria	Impairs coordination; impairs depth perception and alters sense of timing; impairs short-term memory (probably by decreasing acetylcholine levels in the hippocampus); inflames eyes; causes peripheral vasodilation	In large doses sensory distortions, hallucinations, evidence of lowered sperm counts and testosterone levels; psychological dependence
AMPHETAMINES			
"Uppers," "pep pills," "crystal meth" (e.g., Dexedrine); methamphetamine is a popular club drug	Euphoria, stimulation, hyperactivity	CNS stimulants; stimulate release and block reuptake of dopamine and norepinephrine; enhance flow of impulses in RAS, leading to increased heart rate and blood pressure; pupil dilation	Tolerance, addiction; hallucinations; increased blood pressure; psychotic episodes; death from overdose

Continued

TABLE 42-4 Effects of Some Commonly Used Drugs (*continued*)

NAME OF DRUG	EFFECT ON MOOD	ACTIONS ON BODY	SIDE EFFECTS/DANGERS ASSOCIATED WITH ABUSE
AMPHETAMINES (*continued*)			
Amphetamine-type prescription stimulants commonly used to treat attention deficit hyperactivity disorder (ADHD) (e.g., methylphenidate: Ritalin, Concerta, Adderall)	CNS stimulation, but has calming effect on many children and adults who have ADHD; increases focus	Increase heart rate, blood pressure, and body temperature; like cocaine, amphetamines block dopamine transporter, which increases dopamine concentration in synapses	Not addictive at prescribed doses; used illegally and abused by some high school and college students who use them in higher doses than would be prescribed; can cause irregular heartbeat and respiration, hallucinations, delusions, and convulsions; abuse can lead to tolerance and physical addiction
MDMA (3,4-methylenedioxymethamphetamine) "Ecstasy," "Molly" (a popular club drug)	Stimulation, sense of pleasure, self-confidence, feelings of closeness to others; mild hallucinogenic	CNS stimulant; damages serotonin pathways in rat and monkey brains	Hyperthermia (excessive body heat); overdose causes rapid heartbeat, high blood pressure, panic attacks, and seizures; long-term neurological changes; cognitive impairment
ROHYPNOL (flunitrazepam)			
A benzodiazepine not legally available in the U.S. "Date rape" drug, "roofies," "roaches"	Sedation, anxiety reduction	Depresses CNS; muscle relaxation; partial amnesia	Psychological dependence; addiction; withdrawal may cause seizures. May be lethal when mixed with alcohol and/or other depressants
PCP (phencyclidine)			
"Angel dust," "ozone;" originally used as general anesthetic for surgery, but discontinued because it caused delirium with psychotic features. Now illegal.	Feelings of strength, power, invulnerability, and a numbing effect on the mind; chronic use can cause depression	Affects several neurotransmitters; can cause psychotic symptoms; increase in blood pressure and heart rate; shallow respiration; flushing, profuse sweating; numbness of extremities; difficulty coordinating muscles; changes in body awareness, similar to those associated with alcohol intoxication	Psychological dependence, craving; in adolescents interferes with hormones related to growth and development; may interfere with learning; causes hallucinations, delusions, disordered thinking; high doses cause drop in blood pressure and heart rate, seizures, coma, and death; chronic use causes memory loss, difficulties with speech and thinking
LSD (lysergic acid diethylamide)			
"acid;" hallucinogen (manufactured from lysergic acid, which is found in ergot, a fungus that grows on rye and other grains)	Effects unpredictable; typically extreme changes in mood; overexcitation	Alters levels of neurotransmitters in brain; potent CNS stimulator; dilates pupils, sometimes unequally; increases heart rate; raises blood pressure	Impaired judgment leading to irrational behavior; sensory distortions, hallucinations, panic attacks; psychosis; flashbacks
METHAQUALONE (e.g., Quaalude, Sopor)			
A synthetic barbiturate-type drug. No longer legally available in U.S.	Sedative–hypnotic; reduces anxiety; causes euphoria	Depresses CNS; depresses certain spinal reflexes	Tolerance, addiction; overdose can lead to seizures, coma, or death
CAFFEINE	Increases mental alertness; decreases fatigue and drowsiness	CNS stimulant; relaxes smooth muscle; stimulates cardiac and skeletal muscle; increases urine volume	Tolerance, psychological dependence, addiction. Large amounts cause anxiety, irritability. Very large doses may slow the heart
NICOTINE Typically inhaled in tobacco smoke which contains more than 19 chemical compounds that cause cancer.	Decreases anxiety; increases alertness; may induce a sense of well-being	Stimulates sympathetic nervous system; increases dopamine in mesolimbic pathway	Tolerance, psychological dependence, addiction; withdrawal can cause anxiety, depression, headaches, difficulty concentrating; nicotine stimulates synthesis of lipid in arterial wall, leading to atherosclerosis
ALCOHOL (see *Inquiring About: Alcohol: The Most Abused Drug*)	Euphoria, relaxation, release of inhibitions	Depresses CNS by affecting GABA receptors; impairs vision, coordination, judgment; lengthens reaction time	Tolerance, psychological dependence, addiction. Damage to pancreas and liver (e.g., cirrhosis); brain damage; increased risk of depression

42.1 Invertebrate Nervous Systems: Trends in Evolution
(page 877)

1 Contrast nerve nets and radial nervous systems with bilateral nervous systems.

- **Nerve nets** and *radial nervous systems* are typical of radially symmetrical invertebrates, such as cnidarians. An animal with a radial nervous system can respond effectively to predator or prey approaching from any direction.

- In planarian flatworms the *bilateral nervous system* includes *cerebral ganglia* and, usually, two solid ventral nerve cords connected by transverse nerves. Annelids and arthropods typically have a ventral nerve cord and numerous ganglia. The cerebral ganglia of arthropods have specialized regions. Octopods and other cephalopod mollusks have highly developed nervous systems with neurons concentrated in a central region.

- The bilateral nervous system is adaptive for bilaterally symmetrical animals that typically move in a forward direction. Nerve cells concentrate to form **nerves, nerve cords, ganglia,** and the **brain;** typically, sense organs are concentrated in the head region.

2 Identify trends in the evolution of invertebrate nervous systems.

- Trends in nervous system evolution include increased number and concentration of nerve cells; specialization of function; increased number of association neurons; more complex synaptic contacts; and **cephalization,** the formation of a head.

42.2 Overview of the Vertebrate Nervous System
(page 878)

3 Describe the two main divisions of the vertebrate nervous system and summarize their functions.

- The **central nervous system (CNS),** which consists of the brain and spinal cord, is specialized to integrate incoming information. The **peripheral nervous system (PNS),** which consists of sensory receptors and nerves, transmits signals between sensory receptors and the CNS, and between the CNS and effectors.

42.3 Evolution of the Vertebrate Brain *(page 879)*

4 Trace the development of the principal vertebrate brain regions from the hindbrain, midbrain, and forebrain, and compare the brains of fishes, amphibians, reptiles, birds, and mammals.

- In the vertebrate embryo, the brain and spinal cord arise from the **neural tube.** The anterior end of the tube differentiates into the **hindbrain, midbrain,** and **forebrain.**

- The hindbrain subdivides into the metencephalon and myelencephalon. The myelencephalon develops into the **medulla,** which contains vital centers and other reflex centers. The fourth ventricle, the cavity of the medulla, communicates with the central canal of the spinal cord. The metencephalon gives rise to the **cerebellum,** which is responsible for muscle tone, posture, and equilibrium, and to the **pons,** which connects various parts of the brain.

- The midbrain is the largest part of the brain in fishes and amphibians. It is their main association area, linking sensory input and motor output. In reptiles, birds, and mammals, the midbrain serves as a center for visual and auditory reflexes. The medulla, pons, and midbrain make up the **brain stem.**

- The forebrain differentiates to form the diencephalon and telencephalon. The diencephalon develops into the thalamus and hypothalamus. The **thalamus** is a relay center for motor and sensory information. The **hypothalamus** controls autonomic functions; links nervous and endocrine systems; controls temperature, appetite, and fluid balance; and is involved in some emotional and sexual responses.

- The telencephalon develops into the cerebrum and **olfactory bulbs.** In most vertebrates the **cerebrum** is divided into right and left cerebral hemispheres. In fishes and amphibians, the cerebrum mainly integrates incoming sensory information. In birds the corpus striatum controls complex behavior patterns, such as flying and singing.

- In mammals the *neocortex* accounts for a large part of the **cerebral cortex,** the **gray matter** of the brain. The cerebrum has complex association functions.

42.4 The Human Central Nervous System *(page 882)*

5 Describe the structure and functions of the human spinal cord.

- The human brain and spinal cord are protected by bone and three layers of connective tissue, the **meninges: the dura mater, arachnoid,** and **pia mater.** The brain and spinal cord are cushioned by **cerebrospinal fluid (CSF).**

- The **spinal cord** transmits impulses to and from the brain and controls many **reflex actions.** The spinal cord consists of *ascending tracts,* which transmit information to the brain, and *descending tracts,* which transmit information from the brain. Its gray matter contains nuclei that serve as reflex centers.

- A *withdrawal reflex* involves sensory receptors; sensory neurons, interneurons, and motor neurons; and effectors, such as muscles.

6 Describe the structure and functions of the human cerebrum.

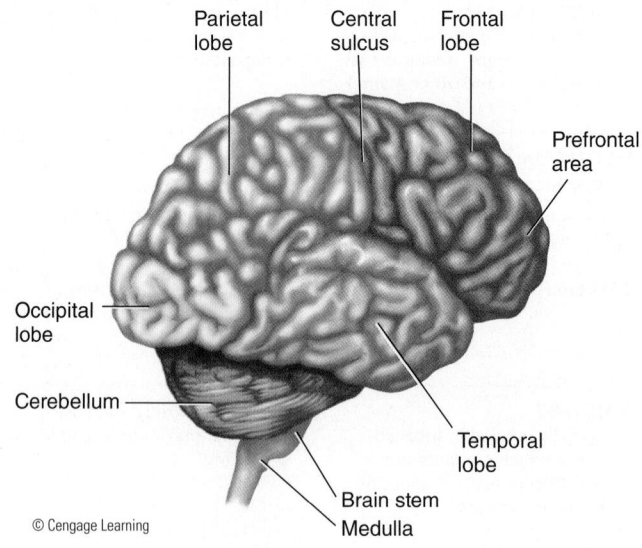

Parietal lobe | Central sulcus | Frontal lobe | Prefrontal area | Occipital lobe | Cerebellum | Brain stem | Medulla | Temporal lobe

© Cengage Learning

- The human cerebral cortex consists of gray matter, which forms folds or **convolutions.** Deep furrows between the folds are called **fissures.**

- The cerebrum is functionally divided into *sensory areas* that receive incoming sensory information; *motor areas* that control voluntary movement; and **association areas** that link sensory and motor areas and are responsible for learning, language, thought, and judgment. The cerebrum consists of lobes, including the **frontal lobes, parietal lobes, temporal lobes,** and **occipital lobes.**

- The **white matter** of the cerebrum lies beneath the cerebral cortex. The **corpus callosum,** a large band of white matter, connects right and left hemispheres. The *basal ganglia,* a cluster of nuclei within the white matter, are important centers for motor function.

7 Summarize the sleep–wake cycle and contrast rapid-eye-movement (REM) sleep with non-REM sleep.
- The body's main biological clock—the **suprachiasmatic nucleus**—receives information about light and dark and transmits it to other nuclei that regulate sleep.
- The **reticular activating system (RAS)** is an arousal system; its neurons filter sensory input, selecting which information is transmitted to the cerebrum.
- The hypothalamus, thalamus, and brain stem regulate the sleep–wake cycle. *Sleep* is an alteration of consciousness with decreased electrical activity in the cerebral cortex.
- Electrical activity of the cerebral cortex and metabolic rate slow during *non-REM sleep;* dreaming characterizes *REM sleep.*

8 Describe the actions of the limbic system; include the role of the amygdala in emotional expression and in response to danger.
- The **limbic system** affects the emotional aspects of behavior, motivation, sexual activity, autonomic responses, and biological rhythms. The **amygdala** evaluates incoming information and signals danger; it is especially important in the experience of fear and aggression.

9 Summarize how the brain processes information; include a description of synaptic plasticity and the neurobiological changes that take place during learning.
- **Synaptic plasticity** is the ability of the nervous system to modify synapses, which allows learning and remembering.
- **Learning** is the process by which we acquire knowledge or skills as a result of experience. **Memory** is the process by which information is encoded, stored, and retrieved.
- *Immediate memory* holds sensory information for fractions of a second. **Short-term memory** holds a small amount of information for a few minutes. Information can be transferred from short-term memory to **long-term memory.** The **hippocampus** is important in categorizing information and consolidating memories.

- Learning and memory involve long-term functional changes at synapses. Known as **long-term potentiation (LTP),** such changes involve increased sensitivity to an action potential by a postsynaptic neuron. **Long-term depression (LTD)** is a long-lasting decrease in the strength of synaptic connections. Long-term memory storage involves gene expression. The transcription factor *CREB* is an important signaling molecule in the memory pathway.

42.5 The Peripheral Nervous System *(page 895)*

10 Describe the organization of the peripheral nervous system and compare its somatic and autonomic divisions.
- The PNS consists of sensory receptors and nerves, including the **cranial nerves** and **spinal nerves** and their branches. The **somatic division** of the PNS responds to changes in the external environment. The **autonomic division** of the PNS regulates the internal activities of the body.

11 Contrast the sympathetic and parasympathetic divisons of the autonomic system and give examples of the effects of these systems on specific organs.
- The **sympathetic system** permits the body to respond to stressful situations. The **parasympathetic system** influences organs to conserve and restore energy. Many organs are innervated by sympathetic and parasympathetic nerves, which function in opposite ways. For example, the sympathetic system increases heart rate, whereas the parasympathetic system decreases heart rate.

42.6 Effects of Drugs on the Nervous System *(page 898)*

12 Distinguish among psychological dependence, tolerance, and drug addiction, and describe the biological actions and effects on mood of alcohol, antidepressants, barbiturates, anti-anxiety drugs, antipsychotic drugs, opiates, stimulants, hallucinogens, and marijuana.
- Many drugs alter mood by increasing or decreasing the concentrations of specific neurotransmitters within the brain. See Table 42-4 for a summary of these drugs.
- Habitual use of mood-altering drugs can result in *psychological dependence* or *drug addiction.* Many drugs induce *tolerance,* in which the body's response to the drug decreases so that greater amounts are needed to obtain the desired effect.

TEST YOUR UNDERSTANDING

Know and Comprehend

1. A radially symmetrical animal such as *Hydra* is likely to have (a) a forebrain (b) a nerve net (c) cerebral ganglia (d) a ventral nerve cord (e) cerebral ganglia and a nerve net
2. Which part of the vertebrate brain maintains posture, muscle tone, and equilibrium? (a) cerebrum (b) medulla (c) cerebellum (d) neocortex (e) thalamus
3. The main association area in the amphibian brain is the (a) cerebrum (b) medulla (c) neocortex (d) midbrain (e) arachnoid
4. The suprachiasmatic nucleus (a) secretes melatonin (b) stimulates long-term potentiation (LTP) in the hippocampus (c) filters incoming sensory information and determines whether it is threatening (d) is critical in memory consolidation (e) is important in regulating the sleep–wake cycle

5. The heart rate is slowed by (a) sympathetic nerves (b) parasympathetic nerves (c) rapid firing in the corpus callosum (d) both sympathetic and parasympathetic nerves (e) the hippocampus working together with sympathetic nerves
6. Amphetamines (a) affect neurons in the mesolimbic dopamine pathway (b) decrease the amount of dopamine secreted (c) affect the CNS in the same way as alcohol (d) are often prescribed to reduce anxiety (e) are CNS depressants

Apply and Analyze

7. Which sequence best describes transmission of information in a withdrawal reflex?
   1. afferent neuron transmits signal to CNS 2. muscle contracts
   3. reception by sensory neuron 4. interneuron in CNS integrates

information 5. efferent neuron transmits signal
(a) 3, 1, 4, 5, 2 (b) 2, 1, 4, 5, 3 (c) 3, 5, 1, 4, 2 (d) 3, 1, 5, 4, 2
(e) 5, 3, 1, 4, 2

8. **VISUALIZE** Label the diagram of the human brain.

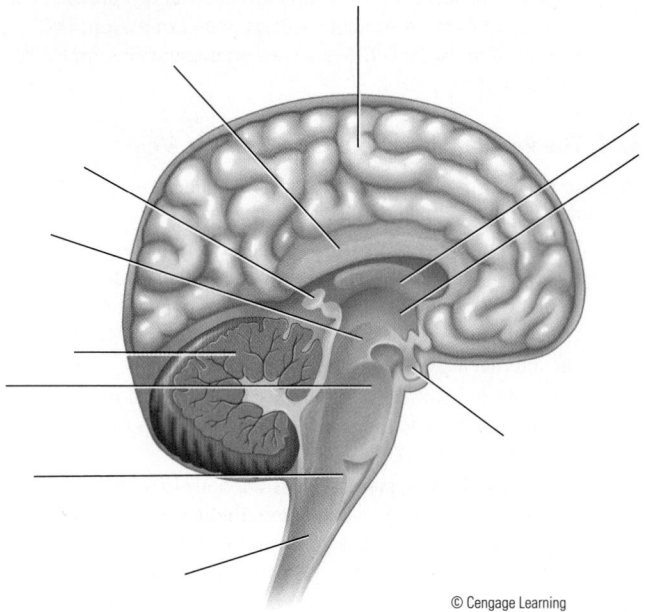

© Cengage Learning

9. **EVOLUTION LINK** What general trends can you identify in the evolution of the vertebrate brain?

Evaluate and Synthesize

10. **SCIENCE, TECHNOLOGY, AND SOCIETY** Imagine that you have just become a parent. What kinds of things can you do to enhance the development of your child's intellectual abilities?

11. What is the adaptive function of the limbic system when an alligator is chasing you?

12. **EVOLUTION LINK** CREB has been shown to be a signaling molecule in the memory pathway in many animals, including fruit flies and mice. What does that suggest about the evolution of learning and memory?

13. Hypothesize a possible relationship between inheritance and intelligence based on gene activation during information processing.

aplia To access course materials, such as Aplia and other companion resources, please visit **www.cengagebrain.com.**

Sensory Systems

Dr. Merlin D. Tuttle/Science Source

Bats are guided by echoes. This California leaf-nosed bat (*Macrotus californicus*) has located an insect.

Have you heard dolphins make clicking sounds? Humans can hear only a limited number of the wide range of sound frequencies that dolphins emit. Dolphins, bats, and a few other vertebrates detect distant objects by *echolocation*, sometimes called biosonar. Echolocation works something like radar but uses sound rather than radio waves. For example, a bat emits high-pitched sounds that bounce off objects in its path and echo back. By rapidly responding to the echo, the bat skillfully avoids obstacles and captures prey (see photograph). Using echolocation, bats and dolphins determine the size, shape, texture, and density of an object as well as its location.

The ability to detect sounds emitted by some foraging bats has evolved in certain moths and other insect prey. Through the course of evolution, some species of bats have adapted to this ability of their prey and now emit even higher-frequency sounds that their prey cannot detect.

Sensory systems transfer information from the outside world to the animal. They inform the animal about predators, food, water, possible mates, temperature, and other changes in the environment. The kinds of sensory receptors an animal has and the way in which its brain *perceives* incoming sensory information determine what the animal's world is like. We humans live in a world of rich colors, numerous shapes, and varied sounds. But we cannot hear the high-pitched whistles that are audible to dogs and cats or the ultrasonic echoes by which bats navigate. Unlike some migratory animals, we cannot use magnetic fields to navigate. Nor do we ordinarily recognize our friends by their distinctive odors. And although vision is our dominant and most refined sense, we are blind to the ultraviolet (UV) hues that light up the world for insects.

We humans are most familiar with the senses of sight, hearing, smell, taste, and touch, but we have other senses. In fact, touch is a compound sense that allows us to detect pressure, pain, and temperature. Some sensory receptors allow us to sense balance, muscle tension, and joint position, providing us with an awareness of our bodies.

In this chapter we first examine how sensory systems function. Next we introduce the major types of sensory receptors. We then discuss several types of sensory systems with emphasis on auditory and visual systems.

KEY CONCEPTS

43.1 Sensory processing includes sensory reception, energy transduction, transmission of signals, and interpretation in the central nervous system (CNS).

43.2 Thermoreceptors respond to heat and cold.

43.3 Some predatory species of sharks, rays, and bony fishes have electroreceptors that detect electric fields generated in the water by the activity of other animals.

43.4 Dendrites of certain sensory neurons are nociceptors (pain receptors).

43.5 Mechanoreceptors respond to touch, pressure, gravity, stretch, or movement.

43.6 Most animals have chemoreceptors for taste and smell that detect chemical substances in food, water, and air.

43.7 Almost all animals have photoreceptors that respond to light, and most animals have eyes that form images; the retina of the vertebrate eye contains photoreceptor cells called rods and cones.

43.1 HOW SENSORY SYSTEMS WORK

LEARNING OBJECTIVES

1   Describe how sensory systems function; include descriptions of sensory receptors, energy transduction, receptor potential, sensory adaptation, and perception.

2   Classify sensory receptors according to the location of the stimuli to which they respond and according to the types of energy they transduce.

A **sensory receptor** detects information about changes in the external or internal environment. Sensory receptors consist of the endings of afferent neurons or cells in close contact with neurons that are specialized to detect specific types of energy in the environment. Along with other types of cells, sensory receptors make up complex **sense organs,** such as eyes, ears, nose, and taste buds. A human taste bud, for example, consists of taste receptor cells that detect chemicals dissolved in saliva, and epithelial cells that support the receptor cells.

Several steps take place in sensory processing, including sensory reception, energy transduction, transmission of the signal, and interpretation in the brain. With minor variations, all sensory systems operate in this way.

Sensory receptors receive information

Sensory receptors receive stimuli from the external or internal environment. In the process of reception, they absorb a small amount of energy from some stimulus. Each kind of sensory receptor is especially sensitive to one particular form of energy. For example, the photoreceptors of the human eye are stimulated by an extremely faint beam of light; similarly, taste receptors are stimulated by a minute amount of a chemical compound.

Sensory receptors transduce energy

Sensory receptors *transduce,* or convert, the energy of the stimulus into electrical signals, the information currency of the nervous system. This process is known as *energy transduction.* When unstimulated, a sensory receptor maintains a *resting potential,* that is, a difference in charge between the inside and outside of the cell (see Chapter 41). Transduction couples a stimulus with the opening or closing of ion channels in the plasma membranes of sensory receptors. The permeability of the plasma membrane to various ions is altered.

A change in ion distribution causes a change in voltage across the membrane. If the difference in charge increases, the receptor becomes hyperpolarized. If the potential decreases, the receptor becomes depolarized. A change in membrane potential is a **receptor potential.** A receptor potential does not directly trigger an action potential. Like an excitatory postsynaptic potential, or EPSP (see Chapter 41), the receptor potential is a **graded potential** in which the magnitude of change depends on the strength of the stimulus.

If the receptor is a separate cell, receptor potentials stimulate the release of a neurotransmitter, which flows across the synapse and binds to receptors on an *afferent neuron,* or **sensory neuron** (FIG. 43-1). When a sensory neuron becomes sufficiently depolarized to reach its threshold level, an action potential is generated. Thus, receptor potentials can generate action potentials that transmit information to the central nervous system (CNS).

Many receptors are specialized neurons rather than separate cells. In these neurons the specialized region of the plasma membrane that transduces energy does not generate action potentials. Current generated by receptor potentials flows to a region along the axon where an action potential can be generated.

We can summarize the process of sensory reception as follows:

stimulus (such as light energy) ⟶ sensory receptor absorbs energy from stimulus ⟶ energy transduction of stimulus into electrical energy ⟶ receptor potential produced ⟶ action potential in sensory neuron ⟶ signal transmitted to CNS

Sensory input is integrated at many levels

Sensory integration begins in the receptor itself. For example, receptor potentials produced by various stimuli are integrated by summation (see Chapter 41). However, integration occurs mainly in the spinal cord and at various locations in the brain.

Sensory receptors adapt to stimuli Many sensory receptors do not continue to respond at the initial rate, even if the stimulus continues at the same intensity. This change in response rate, called **sensory adaptation,** occurs for two reasons. First, during a sustained stimulus, the receptor sensitivity decreases and produces a smaller receptor potential (resulting in a lower frequency of action potentials in the sensory neurons). Second, changes take place at synapses in the neural pathway activated by the receptor. For example, the release of neurotransmitter from a presynaptic terminal may decrease in response to a series of action potentials.

Some receptors, such as those for pain or cold, adapt so slowly that they continue to trigger action potentials as long as the stimulus persists. Other receptors adapt rapidly, permitting us to ignore persistent unpleasant or unimportant stimuli. For example, when you first pull on a pair of tight jeans, your pressure receptors let you know that you are being squished, and you may feel uncomfortable. Soon, though, these receptors adapt, and you hardly notice the sensation of the tight fit. In the same way, people quickly adapt to odors that at first seem to assault their senses. Sensory adaptation enables an animal to discriminate between unimportant background stimuli that can be ignored and new or important stimuli that require attention.

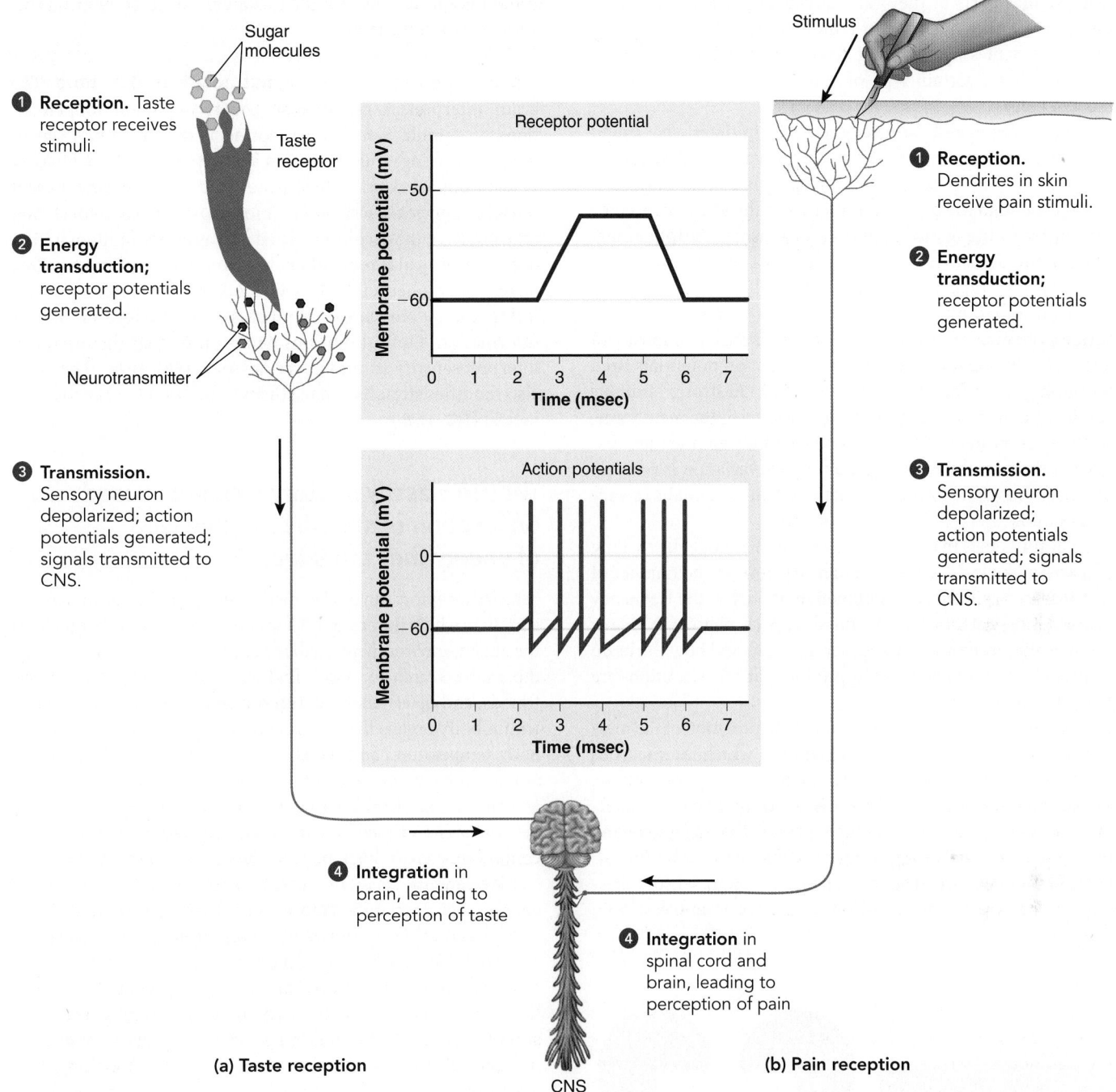

(a) Taste reception

CNS

(b) Pain reception

Figure 43-1 How sensory systems work

Sensory receptors receive stimuli from the environment. They transduce the energy of the stimulus into electrical signals and thus generate receptor potentials. Action potentials are generated in sensory neurons that transmit signals to the CNS, where they are integrated.

© Cengage Learning

Sensation depends on transmission of a coded message How do you know whether you are seeing a blue sky, tasting a doughnut, or hearing a note played on a piano? All action potentials are qualitatively the same. Light of the wavelength 450 nm (blue), sugar molecules (sweet), and sound waves of 440 hertz (Hz) (A above middle C) all cause transmission of similar action potentials. Our ability to differentiate stimuli depends on both the sensory receptor itself and on the brain. We can distinguish the color of a blue sky from the scent of cologne, a sweet taste from a light breeze, or the sound of a piano from the heat of the sun because cells of each sensory receptor are connected to specific neurons in particular parts of the brain. Because a receptor normally responds to only one type of stimulus (e.g., light), the brain interprets a message arriving from a particular receptor as meaning that a certain type of stimulus occurred (such as a flash of color).

When stimulated, a sensory receptor initiates what might be considered a coded message, composed of action potentials transmitted by sensory neurons. This coded message is later decoded in the brain. Impulses from the sensory receptor may differ in the (1) total number of sensory neurons transmitting the signal, (2) specific neurons transmitting action potentials and their targets, (3) total number of action potentials transmitted by a given neuron, and (4) frequency of the action potentials transmitted by a given fiber. The intensity of a stimulus is coded by the frequency of action potentials fired by sensory neurons during the stimulus. A strong stimulus results in a greater depolarization of the receptor membrane, so the sensory neuron fires action potentials with greater frequency than would a weak stimulus. This variation is possible because, as a graded response, the receptor potential can vary in magnitude.

The difference in sound intensity between the gentle rustling of leaves and a loud clap of thunder depends on the number of neurons transmitting action potentials as well as the frequency of the action potentials transmitted by each neuron. Just how the sensory receptor initiates different codes and how the brain analyzes and interprets them to produce various sensations are not yet completely understood.

Sensation takes place in the brain. Interpretation of the message and the type of sensation depends on which interneurons receive the message. The rods and cones of the eye do not see. When stimulated, they send a message to the brain that interprets the signals and translates them into a rainbow, an elephant, or a child. Artificial (e.g., electrical) stimulation of brain centers can also result in sensation. In contrast, many sensory messages never give rise to sensations at all. For example, certain chemoreceptors sense internal changes in the body but never stir our consciousness.

Sensory perceptions are constructed by the mind The brain interprets sensations by converting them to perceptions of stimuli. Sensory **perception** is the process of selecting, interpreting, and organizing sensory information. Sensory perceptions are constructed by the mind by comparing present sensory experience with our memories of past experiences. You may perceive the incoming visual and auditory input of a hissing snake very differently than does your friend who is terrified of snakes or a herpetologist who studies them. Perceptions are influenced by our state of mind at the time we receive sensory information. Optical illusions demonstrate that the brain can interpret sensory stimuli in various ways (**FIG. 43-2a**). The brain also modifies stimuli to make them more complete, familiar, or logical (**FIG. 43-2b**).

We can classify sensory receptors based on location of stimuli or on the type of energy they transduce

Sensory receptors can be classified according to the location of the stimuli to which they respond. **Exteroceptors** receive stimuli from the outside environment, enabling an animal to know and explore the world, search for food, find and attract a mate, recognize friends, and detect enemies. **Interoceptors** are sensory receptors within body organs that detect changes in pH, osmotic pressure, body temperature, and the chemical composition of the blood. You are not usually aware of messages sent to the CNS by these receptors as they work continuously to maintain homeostasis. You become aware of their activity when they signal certain internal conditions such as thirst, hunger, nausea, pain, and orgasm.

We can also classify sensory receptors based on the type of energy they transduce (**TABLE 43-1**). *Thermoreceptors* respond to heat and cold. *Electroreceptors* sense differences in electrical potential. *Electromagnetic receptors* are electroreceptors that are sensitive enough to detect Earth's magnetic field. *Nociceptors* (pain receptors) respond to mechanical, temperature, and other stimuli that could be damaging. *Mechanoreceptors* transduce mechanical energy—touch, pressure, gravity, stretching, and movement—and these receptors convert mechanical forces directly into electrical signals. *Chemoreceptors* transduce the energy of certain chemical compounds, and *photoreceptors* transduce light energy.

CHECKPOINT 43.1

- **CONNECT** *In what way does a receptor potential differ from an action potential?*
- **PREDICT** *Imagine that you are hiking along a bay and notice a really strong, unpleasant odor. How might sensory adaptation be helpful?*
- *How is sensory information integrated?*
- *Identify five kinds of sensory receptors based on the type of energy they transduce. What is the specific function of each?*

(a) Your perception of the same visual input can vary. Do you see a wineglass or the profiles of two faces?

(b) Do you "see" a white square? If so, your brain is completing a familiar pattern suggested by the right-angle wedges taken out of each of the four red circles.

Figure 43-2 Visual perception
© Cengage Learning

TABLE 43-1	Classification of Receptors by Type of Energy They Transduce	
TYPE OF RECEPTOR	**TYPE OF ENERGY TRANSDUCED**	**EXAMPLES**
Thermoreceptors	Heat	Temperature receptors in blood-sucking insects and ticks; pit organs in pit vipers; nerve endings and receptors in skin and tongues of many animals
Electroreceptors and electromagnetic receptors	Electrical; receptors detect differences in electrical potential	Electrical currents in water used to navigate by many fish and some amphibian species; some animals use magnetic fields for orientation and migration
Nociceptors (pain receptors)	Mechanical; physical force such as strong touch, pressure; heat, temperature extremes; damaging chemicals	Neuron endings in skin and other tissues
Mechanoreceptors	Mechanical; change shape as a result of being pushed or pulled	Tactile receptors (free nerve endings, Merkel cells, Meissner corpuscles, Ruffini endings, Pacinian corpuscles); respond to touch and pressure
		Proprioceptors respond to movement and body position Muscle spindles respond to muscle contraction Golgi tendon organs respond to stretch of a tendon Joint receptors respond to movement in ligaments
		Statocysts in invertebrates have hair cells that respond to gravity
		Lateral line organs in fish detect vibrations in the water; respond to waves and currents
		Vestibular apparatus Hair cells in saccule and utricle respond to gravity, linear acceleration Hair cells in semicircular canals respond to angular acceleration
		Hair cells in organ of Corti in cochlea; respond to pressure waves (sound)
Chemoreceptors	Specific chemical compounds	Taste buds; olfactory epithelium
Photoreceptors	Light	Eyespots; ommatidia in compound eye of arthropods; rods and cones in retina of vertebrates

© Cengage Learning

43.2 THERMORECEPTORS

LEARNING OBJECTIVE

3 Identify three functions of thermoreceptors.

Thermoreceptors respond to heat and cold. Mosquitoes, ticks, and other blood-sucking arthropods use thermoreception to find endothermic hosts. Some arthropods have thermoreceptors on their antennae that are sensitive to temperature changes of less than 0.5°C.

Pit vipers (which include rattlesnakes) and some pythons and boas use thermoreceptors to locate their prey, and perhaps, to detect predators (**FIG. 43-3**). These thermoreceptors have heat-sensitive ion channels that detect infrared radiation.

In mammals, which are endothermic, free nerve endings (and, perhaps, specialized receptors) in the skin and tongue detect temperature changes in the outside environment. In addition, pain receptors sense extreme temperatures that pose a threat to the body. Thermoreceptors in the hypothalamus detect internal changes in temperature. They also receive and integrate information from receptors on the body surface. The hypothalamus then initiates homeostatic mechanisms that ensure a constant body temperature.

CHECKPOINT 43.2

- *What are the functions of thermoreceptors in mosquitoes, snakes, and mammals?*

© ephotocorp/Alamy

Figure 43-3 Thermoreception

The pit organ of the bamboo viper (*Trimeresurus gramineus*) is a sense organ located between each eye and nostril. Thermoreceptors in the pit organ can detect the heat emitted by an endothermic animal up to a distance of 1 to 2 m.

43.3 ELECTRORECEPTORS AND MAGNETIC RECEPTION

4 Describe the functions of electroreceptors and identify the functions of magnetic reception.

Electroreceptors sense differences in electrical potential. Some predatory species of sharks, rays, and bony fishes detect electric fields generated in the water by the muscle activity of their prey. Electroreception also appears important in communication, such as in recognizing a potential mate. Males and females have different frequencies of electric discharge.

Several groups of fishes have electric organs, specialized muscle or nerve cells that emit electrical signals and receive a feedback signal. In species that produce a weak electric current, electroreceptors may help in navigation. This mechanism is particularly useful in murky water, where visibility and olfaction are poor.

A few fishes, such as electric eels or electric rays, have electric organs in their heads capable of delivering powerful shocks (up to several hundred volts) that stun prey or predators. Weaker electrical signals are emitted continuously to help in assessing the environment, orienting themselves, and communicating.

Some bacteria, mollusks, crustaceans, insects, bony fishes, amphibians, reptiles, birds, and a few mammals detect Earth's magnetic fields and use this information to orient themselves. Migratory birds and sea turtles are well-known examples of animals that use magnetic fields to navigate (discussed further in Chapter 52).

Earth's magnetic fields penetrate organisms, so specialized receptors may not be required. However, certain receptors of sharks, rays, and skates appear to sense magnetic fields. For example, when a shark swims across one of Earth's magnetic field lines, *electromagnetic receptors* may detect changes that occur in electric currents. This information, which is integrated in the brain, helps the animal orient itself.

CHECKPOINT 43.3

- *What are two functions of electroreceptors?*
- *How do animals use electromagnetic reception?*

43.4 NOCICEPTORS

5 Describe the functions of nociceptors and identify the roles of substance P and endorphins.

Pain receptors, called **nociceptors** (from the Latin *nocere*, meaning "to injure"), are free nerve endings (dendrites) of certain sensory neurons found in almost every tissue. Three main types of nociceptors have been identified: (1) mechanical nociceptors respond to strong tactile stimuli such as cutting, crushing, or pinching; (2) thermal nociceptors respond to temperature extremes (temperatures above 45°C or below 5°C); and (3) polymodal nociceptors respond to a variety of damaging stimuli, including certain chemicals.

When stimulated, nociceptors transmit signals through sensory neurons to interneurons in the spinal cord. These sensory neurons release the neurotransmitters *substance P* and *glutamate*. **Substance P** activates interneurons that transmit the pain message to the opposite side of the spinal cord. Ascending pathways then transmit the message upward to the brain stem and thalamus, where pain perception begins. From the thalamus, impulses are sent into the somatosensory areas that localize the pain. Signals are transmitted to the prefrontal cortex, allowing us to become fully aware of our pain and to evaluate the situation. Information is also sent from the brain stem and thalamus to the limbic system, which generates an emotional response to the injury. **Glutamate** stimulates interneurons in the spinal cord that transmit pain signals to cortical areas of the brain.

Mammals have an analgesic system that inhibits pain signals in the spinal cord. Certain interneurons release **endorphins** and **enkephalins,** neurotransmitters that bind with opiate receptors on the terminals of sensory neurons in the spinal cord. This binding inhibits release of substance P, which stops transmission of the pain signal.

CHECKPOINT 43.4

- *What is the function of each main type of nociceptor?*
- *What is the function of substance P? of endorphins?*

43.5 MECHANORECEPTORS

6 Compare the functions and mechanisms of action of the following mechanoreceptors: tactile receptors, proprioceptors, statocysts, hair cells, and lateral line organs.

7 Compare the structure and function of the saccule and utricle with that of the semicircular canals in maintaining equilibrium.

8 Trace the path taken by sound waves through the structures of the ear and explain how the organ of Corti functions as an auditory receptor.

Mechanoreceptors are activated when they change shape as a result of being mechanically pushed or pulled. They transduce mechanical energy, permitting animals to feel, hear, and maintain balance. These receptors provide information about the shape, texture, weight, and topographic relations of objects in the external environment. Some mechanoreceptors are extremely sensitive. For example, alligators have pressure receptors on their faces that can detect the movement in the surrounding water caused by a single drop of falling water.

Certain mechanoreceptors enable an animal to maintain its body position with respect to gravity (for us, head up and feet down; for a dog, dorsal side up and ventral side down; for a tree sloth, ventral side up and dorsal side down). When displaced from its normal position, the animal quickly adjusts its body to return to its normal orientation. Mechanoreceptors continuously send information to the CNS regarding the position and movements of the body. These receptors also provide information about the operation of internal organs. For example, they inform us about the presence of food in the stomach, feces in the rectum, urine in the bladder, and a fetus in the uterus.

Tactile receptors are located in the skin

The simplest mechanoreceptors are free nerve endings in the skin. They detect touch, pressure, vibration, and pain when stimulated by objects that contact the body surface. In many invertebrates and vertebrates, *tactile* (touch) *receptors* lie at the base of a hair or bristle. These receptors may sense the body's orientation in space with respect to gravity. They may also detect air and water vibrations and contact with other objects. A tactile receptor is stimulated indirectly when the hair is bent or displaced. The stimulus mechanically distorts ion channel proteins in the plasma membrane of the receptor, leading to net entry of Na^+. A receptor potential develops in the afferent neuron, and action potentials may be generated. This type of receptor responds only when the hair is moving. Even though the hair may be maintained in a displaced position, the receptor is not stimulated unless there is motion.

Thousands of specialized tactile receptors are located in mammalian skin (FIG. 43-4). *Merkel cells* form discs in the deep regions of the epidermis that extend down into the dermis. Merkel cells are important in sensing light touch and enable us to distinguish among the softness of a baby's skin, the fuzziness of a peach, or the hardness of a book cover. They adapt slowly, permitting us to know that an object continues to touch the skin.

Three types of mechanoreceptors in the skin—Meissner corpuscles, Ruffini endings, and Pacinian corpuscles—have encapsulated endings. *Meissner corpuscles,* located in the upper dermis, are sensitive to light touch and vibration. They adapt quickly to a sustained stimulus. *Ruffini endings,* located in the dermis, inform us of heavy, continuous pressure and stretch

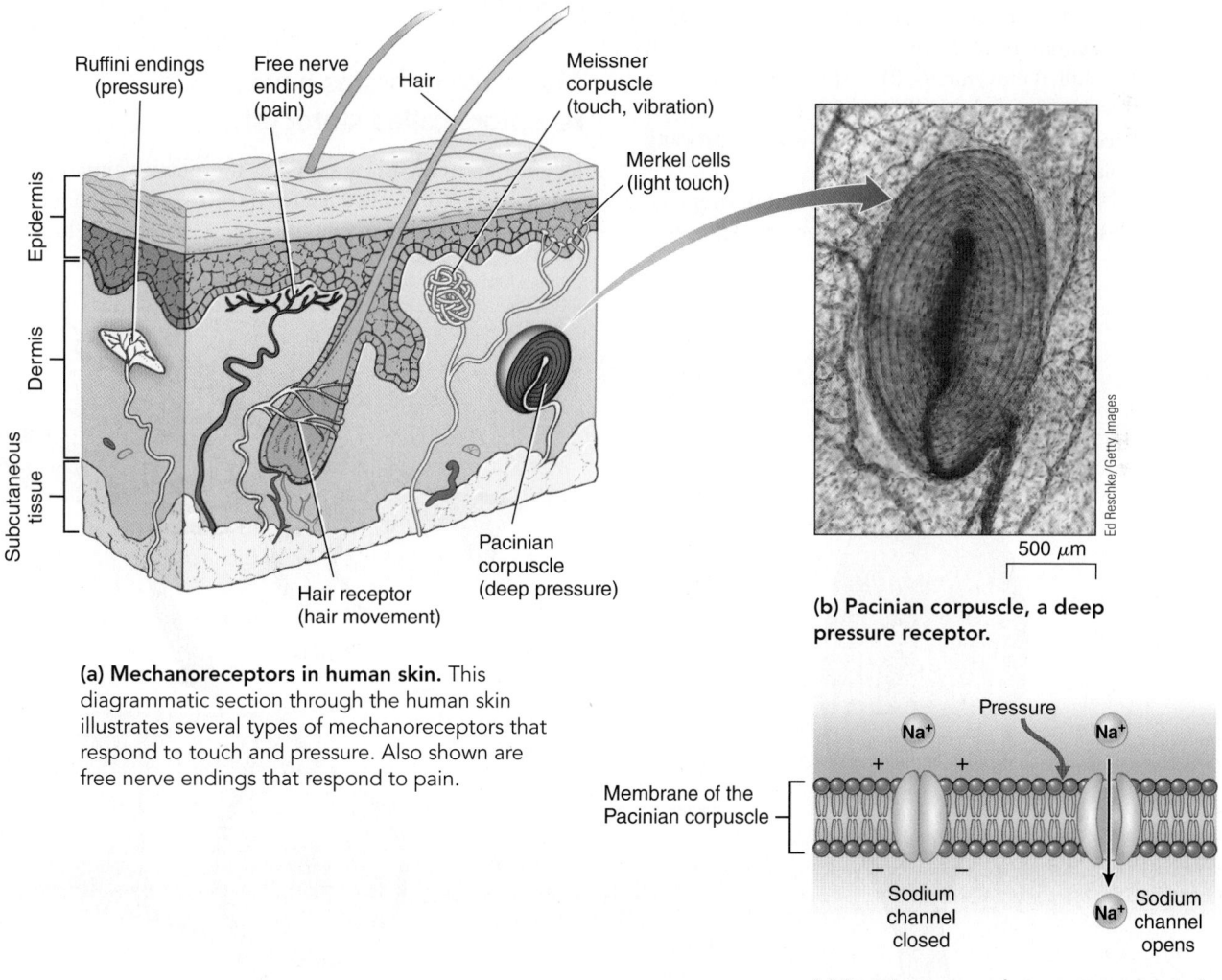

(a) **Mechanoreceptors in human skin.** This diagrammatic section through the human skin illustrates several types of mechanoreceptors that respond to touch and pressure. Also shown are free nerve endings that respond to pain.

(b) **Pacinian corpuscle, a deep pressure receptor.**

(c) **Pacinian corpuscles sense mechanical forces.** Pressure causes Na^+ channels to open, depolarizing the axon.

Figure 43-4 *Animation* **Sensory receptors in human skin**
© Cengage Learning

of the skin. They adapt very slowly and are thought to provide information about the position of our fingers and hand.

Pacinian corpuscles, located deep in the dermis, are sensitive to deep pressure that causes rapid movement of the tissues. They are especially sensitive to stimuli that vibrate. The Pacinian corpuscle consists of a neuron ending surrounded by concentric connective tissue layers interspersed with fluid. Compression causes displacement of the layers, which stimulates the axon. Even though the displacement is maintained under steady compression, the receptor potential rapidly falls to zero, and action potentials cease—an excellent example of sensory adaptation.

Proprioceptors help coordinate muscle movement

Proprioceptors help animals maintain the position of one part of the body with respect to another. Located within muscles, tendons, and joints, proprioceptors continuously respond to tension and movement. With their continuous input, an animal can perceive the positions of its arms, legs, head, and other body parts along with the orientation of its body as a whole. This information is essential for all forms of locomotion and for all coordinated and skilled movements, from spinning a cocoon to completing a reverse one-and-a-half dive with a twist. With the help of proprioceptors, we carry out activities such as dressing or playing the piano even in the dark.

Vertebrates have three main types of proprioceptors: **muscle spindles,** which detect muscle movement (**FIG. 43-5**);

Golgi tendon organs, which respond to tension in contracting muscles and in the tendons that attach muscle to bone; and **joint receptors,** which detect movement in ligaments. Impulses from proprioceptors help coordinate the contractions of the several distinct muscles that may be involved in a single movement. Without such receptors, complicated, skilled acts would be impossible. Proprioceptors are also important in maintaining balance.

Proprioceptors are probably more numerous and more continually active than any of the other sensory receptors, although we are less aware of them than of most of the others. As long as the stimulus is present, receptor potentials are maintained (although not at constant magnitude) and action potentials continue to be generated. Proprioceptors continuously send signals to inform the brain about position.

The mammalian muscle spindle, one of the more versatile stretch receptors, helps maintain muscle tone. It is a bundle of specialized muscle fibers, with a central region encircled by sensory neuron endings. These neurons continue to transmit signals for a prolonged period, in proportion to the degree of stretch.

Many invertebrates have gravity receptors called statocysts

Jellyfish and crayfish are among the many animals that have gravity receptors called **statocysts** (**FIG. 43-6**). Statocysts are the simplest organs of equilibrium. A statocyst is basically an

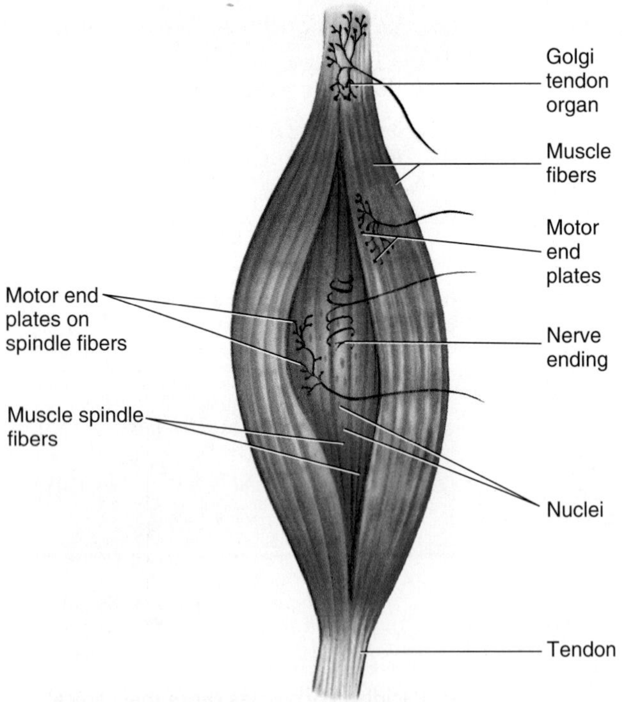

Figure 43-5 Proprioceptors

Muscle spindles detect muscle movement. A muscle spindle consists of an elongated bundle of specialized muscle fibers. Golgi tendon organs respond to tension in muscles and their associated tendons.

© Cengage Learning

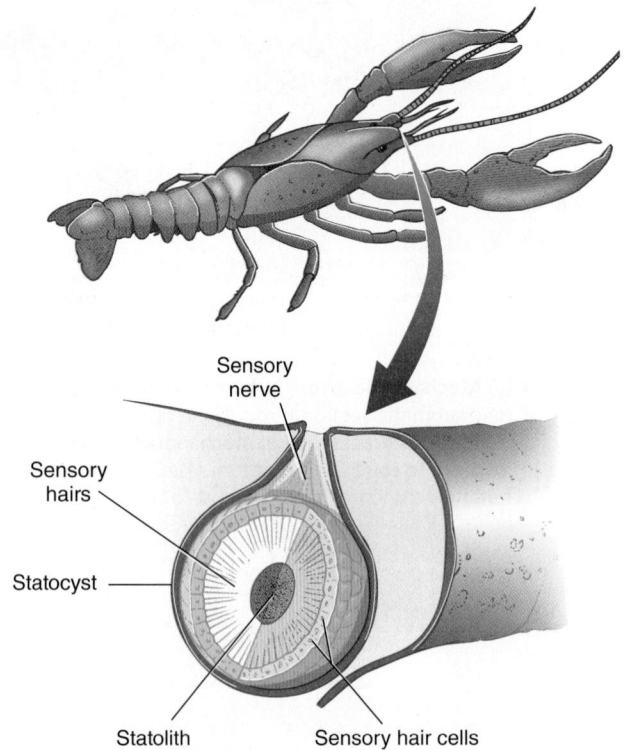

Figure 43-6 A statocyst

Many invertebrates have statocysts, receptors that sense gravitational force and provide information about orientation of the body with respect to gravity.

© Cengage Learning

infolding of the epidermis lined with receptor cells called *sensory hair cells.* These cells are equipped with sensory hairs (not true hairs). The cavity contains one or more **statoliths,** tiny granules of loose sand grains or calcium carbonate held together by an adhesive material secreted by cells of the statocyst. Normally, the particles are pulled downward by gravity and stimulate the sensory hair cells.

When the animal changes position, the statocyst tilts, and the statolith moves in the same direction. This movement bends the sensory hairs. Each sensory hair cell responds maximally when the animal is at a particular position with respect to gravity. The mechanical displacement results in receptor potentials and action potentials that inform the nervous system of the change in position. By "knowing" which sensory cells are firing, the animal knows where "down" is and can correct any abnormal orientation.

In a classic experiment, the function of the statocyst was demonstrated by replacing the sand grains with iron filings in the statocysts of crayfish. The force of gravity was overcome by holding magnets above the animals. The iron filings were attracted upward toward the magnets, and the crayfish began to swim upside down in response to the new information provided by their gravity receptors.

Hair cells are characterized by stereocilia

Vertebrate **hair cells** are mechanoreceptors that detect movement. They are the sensory receptors in the lateral line of fishes that detect water movement (**FIG. 43-7**). Hair cells also help maintain position and equilibrium and are important in hearing.

The surface of a vertebrate hair cell typically has a single long **kinocilium** and many shorter **stereocilia,** hairlike projections that increase in length from one side of the hair cell to the other (see Fig. 43-7c). A kinocilium is a true cilium with a 9 + 2 arrangement of microtubules (see Chapter 4). In contrast, stereocilia are not really cilia. They are microvilli that contain actin filaments.

Mechanical stimulation of the stereocilia causes voltage changes. Mechanical stimulation in one direction causes depolarization and release of neurotransmitter, whereas mechanical stimulation in the opposite direction results in hyperpolarization. However, hair cells have no axons and do not produce their own action potentials. Rather, they release neurotransmitter that depolarizes associated neurons.

Lateral line organs supplement vision in fishes

Lateral line organs of fishes and aquatic amphibians detect vibrations in the water. They inform the animal of obstacles in its way and of moving objects such as prey, enemies, and others in its school. Typically, the lateral line organ is a long canal running the length of the body and continuing into the head. The canal is lined with sensory hair cells. The tips of the stereocilia are enclosed by a **cupula,** a mass of gelatinous material secreted by the hair cells.

Pressure waves, currents, and even slight movement in the water cause vibrations in the lateral line organ. As the water moves the cupula, the stereocilia bend. This motion changes the membrane potential of the hair cell, which may then release neurotransmitter. If an associated sensory neuron is sufficiently stimulated, an action potential is dispatched to the CNS.

The vestibular apparatus maintains equilibrium

When you think of the ear, you probably think of hearing. However, in vertebrates the main function of the ear is to help maintain equilibrium. Many vertebrates do not have outer or middle ears, but all have inner ears (**FIG. 43-8**). In mammals the *inner ear* includes organs of equilibrium equipped with hair cells that sense the position of the body with respect to gravity.

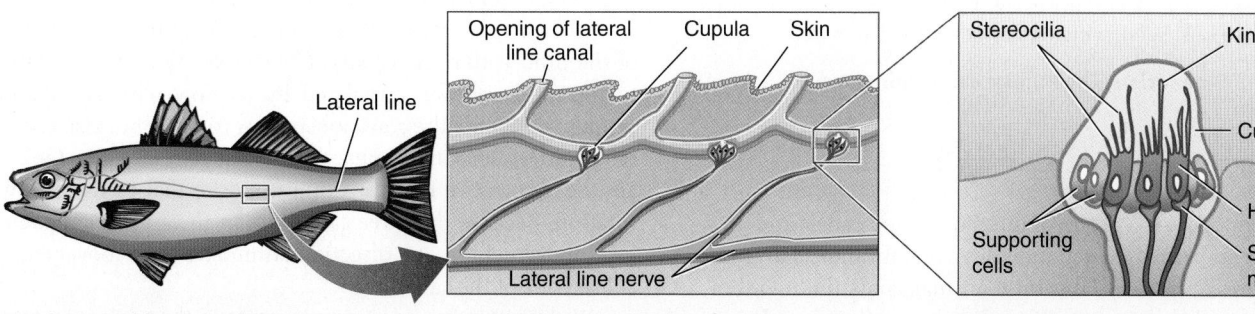

(a) The lateral line organ is a canal that extends the length of the body.

(b) The canal has numerous groupings of hair cells and openings to the outside environment.

(c) The receptor cells are hair cells. Each hair cell has stereocilia and a kinocilium that are covered with a gelatinous cupula. The hair cells respond to waves, currents, and other disturbances in the water.

Figure 43-7 Lateral line organs of fishes
These mechanoreceptors detect vibrations in the water, informing the fish of obstacles or moving objects.
© Cengage Learning

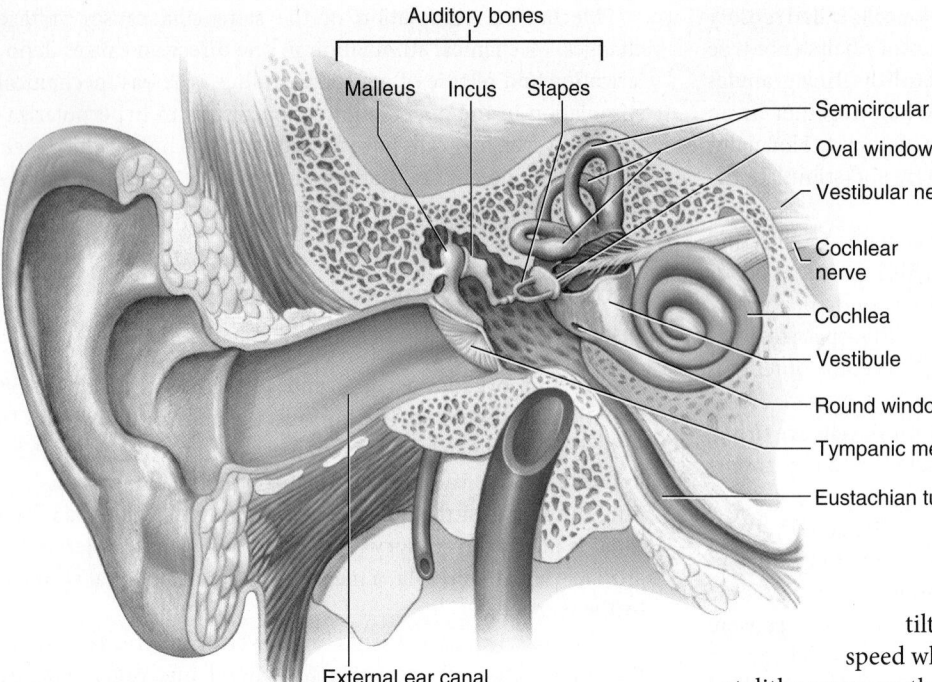

Auditory bones

Malleus   Incus   Stapes

Semicircular canals

Oval window

Vestibular nerve

Cochlear nerve

Cochlea

Vestibule

Round window

Tympanic membrane

Eustachian tube

External ear canal

(a) Structure of the human ear, anterior view.

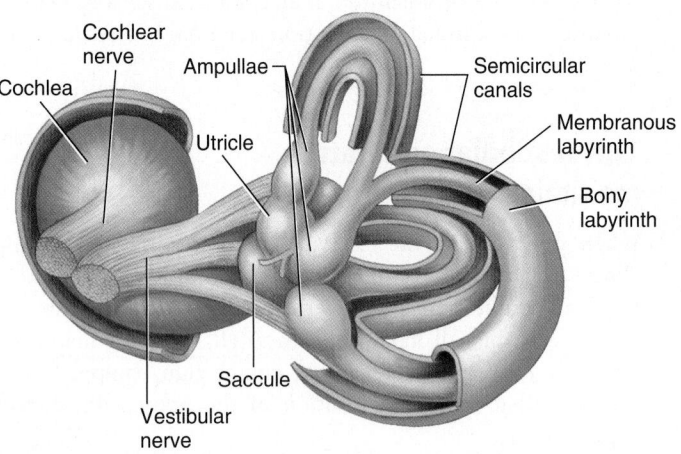

Cochlear nerve

Cochlea

Ampullae

Utricle

Semicircular canals

Membranous labyrinth

Bony labyrinth

Saccule

Vestibular nerve

(b) The vestibular apparatus. The saccule, utricle, and semicircular canals together make up the vestibular apparatus. The utricle and saccule are best seen in the posterior view shown here.

Figure 43-8 *Animation* **The human ear: hearing and equilibrium**
© Cengage Learning

The inner ear is a group of interconnected canals and sacs, called the **labyrinth,** which includes a membranous labyrinth that fits inside a bony labyrinth. In mammals the membranous labyrinth has two saclike chambers—the **saccule** and the **utricle**—and three **semicircular canals.** Collectively, the saccule, utricle, and semicircular canals are called the **vestibular apparatus** (see Fig. 43-8b). Destruction of these organs leads to a considerable loss of the sense of equilibrium. A pigeon whose vestibular apparatus has been destroyed cannot fly but in time can relearn how to maintain equilibrium by using visual stimuli.

The sensory cells of the saccule and utricle are hair cells similar to those of the lateral line organ (FIG. 43-9). The stereocilia projecting from the hair cells are covered by a gelatinous cupula in which calcium carbonate ear stones, called **otoliths,** are embedded. The hair cells in the saccule and utricle lie in different planes.

Normally, the pull of gravity causes the otoliths to press against the stereocilia, stimulating them to initiate impulses. Sensory neurons at the bases of the hair cells transmit these signals to the brain. When the head is tilted or in linear acceleration (increasing speed when the body is moving in a straight line), otoliths press on the stereocilia of different cells, deflecting them. Deflection of the stereocilia toward the kinocilium depolarizes the hair cell. Deflection in the opposite direction hyperpolarizes the hair cell. Thus, depending on the direction of movement, the hair cells release more or less neurotransmitter. When the brain interprets the neural messages, the animal becomes aware of its full body position relative to the ground regardless of how its head is positioned.

Information about turning movements, referred to as angular acceleration, is provided by the three semicircular canals. Each canal, a hollow ring connected with the utricle, lies at right angles to the other two. Each canal is filled with fluid called **endolymph.** At one of the openings of each canal into the utricle there is a small, bulblike enlargement, the **ampulla.** Within each ampulla there is a clump of hair cells called a *crista* (pl., *cristae*), similar to the groups of hair cells in the utricle and saccule. No otoliths are present. The stereocilia of the hair cells of the cristae are stimulated by movements of the endolymph in the canals (FIG. 43-10).

When the head is turned, there is a lag in the movement of the fluid within the canals. The stereocilia move in relation to the fluid and are stimulated by its flow. This stimulation produces not only the consciousness of rotation but also certain reflex movements in response to it. These reflexes cause the eyes and head to move in a direction opposite that of the original rotation. Because the three canals are in three different planes, head movement in any direction stimulates fluid movement in at least one of the canals.

We humans are used to movements in the horizontal plane but not to vertical movements (parallel to the long axis of the upright body). The motion of an elevator or of a ship pitching in a rough sea stimulates the semicircular canals in an unusual way and may cause seasickness or motion sickness, with resultant nausea or vomiting. When a seasick person lies down, the movement stimulates the semicircular canals in a more familiar way, and nausea is less likely to occur.

The utricle and saccule sense changes in the position of the head relative to gravity as well as changes in the rate of motion when moving in a straight line.

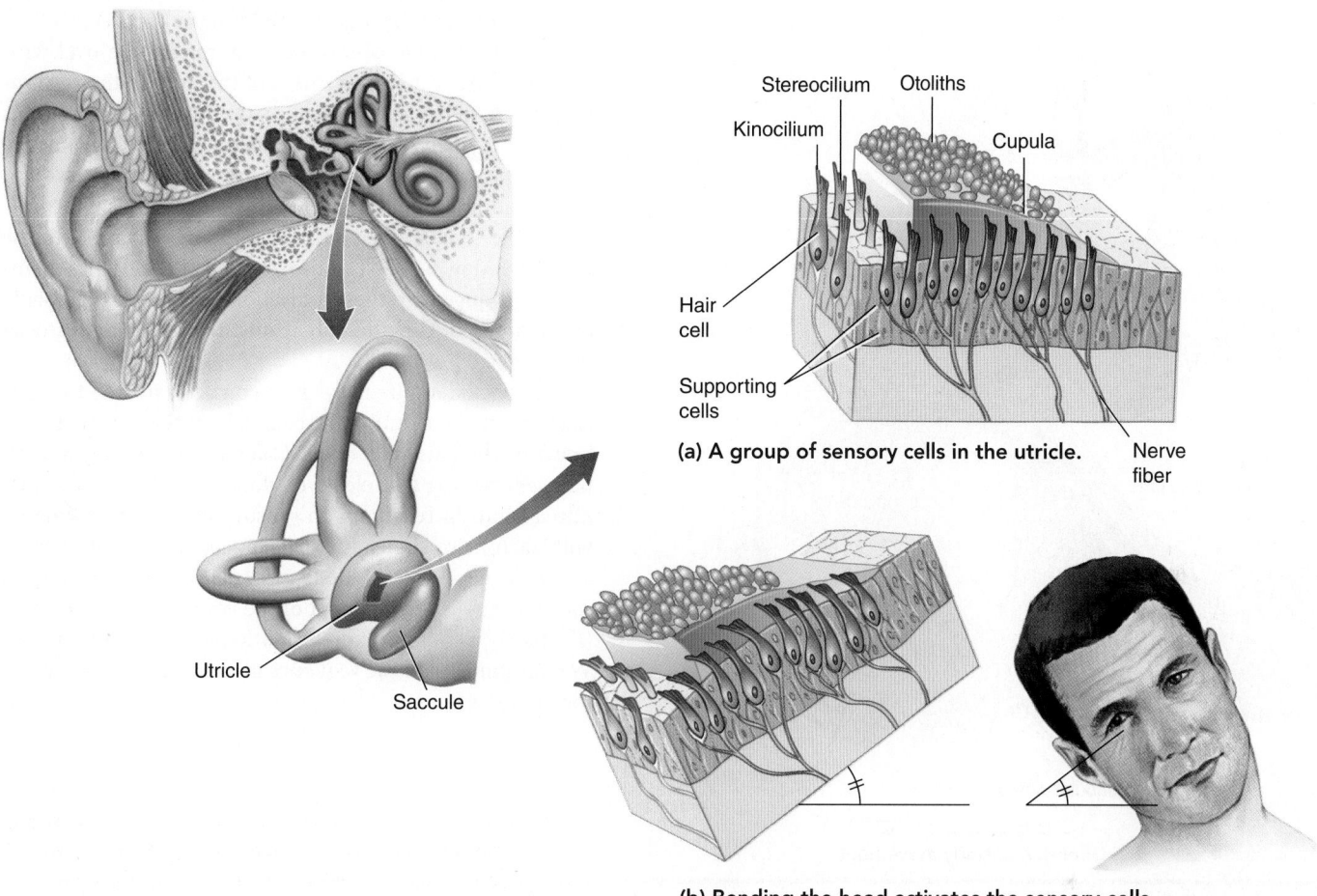

Stereocilium Otoliths

Kinocilium Cupula

Hair cell

Supporting cells

(a) A group of sensory cells in the utricle. Nerve fiber

Utricle

Saccule

(b) Bending the head activates the sensory cells.

Figure 43-9 *Animation* **The saccule and utricle: maintaining equilibrium**
The saccule and utricle sense linear acceleration, allowing us to know our position relative to the ground. Compare the positions of the otoliths and hair cells in **(a)** with those in **(b).** Changes in head position cause the force of gravity to distort the cupula, which in turn distorts the stereocilia of the hair cells. The hair cells respond by releasing neurotransmitters. Impulses are transmitted along the vestibulocochlear nerve to the brain.
© Cengage Learning

PREDICT How would the utricle and saccule function in sensing changes in head position under gravity-free (weightless) conditions?

Auditory receptors are located in the cochlea

Many arthropods and most vertebrates have sound receptors, but for many of them hearing does not seem to be a sensory priority. It is important in tetrapods, however, and both birds and mammals have a highly developed sense of hearing. Their auditory receptors are located in the **cochlea** of the inner ear. These mechanoreceptors contain hair cells that detect pressure waves.

The cochlea is a spiral tube that resembles a snail's shell (**FIG. 43-11**). If the cochlea were uncoiled, you would see that it consists of three canals separated from one another by thin membranes. The canals come almost to a point at the apex. Two of these canals, the *vestibular canal* and the *tympanic canal,* connect at the apex of the cochlea and are filled with a fluid called

perilymph. The middle canal, the *cochlear duct,* is filled with endolymph. The cochlear duct contains the auditory organ, the **organ of Corti.**

The organ of Corti contains about 18,000 hair cells in rows that extend the length of the coiled cochlea. The stereocilia of each hair cell extend into the cochlear duct. The hair cells rest on the **basilar membrane,** which separates the cochlear duct from the tympanic canal. Another membrane, the **tectorial membrane,** overhangs and is in contact with the hair cells.

Sound waves are amplified in the middle ear In terrestrial vertebrates sound waves in the air are transformed into pressure waves in the cochlear fluid. In the human ear, for

The semicircular canals detect angular acceleration or deceleration of the head (changes in the position of the head when you are turning it).

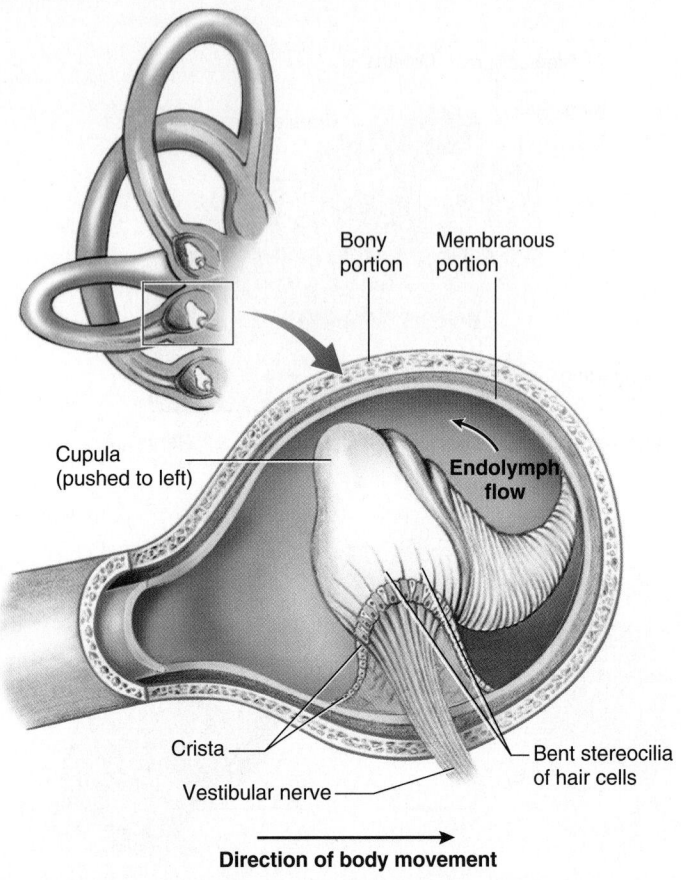

Figure 43-10 Semicircular canals and equilibrium

When the head changes its rate of rotation, endolymph within the ampulla of the semicircular canal distorts the cupula. The stereocilia of the hair cells bend, increasing the frequency of action potentials in sensory neurons. Information is transmitted to the brain via the vestibular nerve.

PREDICT A child spinning in a circle becomes dizzy. The child is advised to spin a few times in the opposite direction. What effect would this have? Explain.

© Cengage Learning

example, sound waves pass through the *external auditory canal* and vibrate the **tympanic membrane,** or eardrum, which separates the outer ear and the middle ear. The vibrations are transmitted across the middle ear by three tiny bones: the **malleus, incus,** and **stapes** (or hammer, anvil, and stirrup, so called because of their shapes). The malleus is in contact with the eardrum, and the stapes is in contact with a thin, membranous region of the cochlea called the *oval window.* The **eustachian tube,** which connects the middle ear with the pharynx (throat region), helps equalize pressure between the middle ear and the atmosphere.

The malleus, incus, and stapes act as three interconnected levers that amplify vibrations. A small movement in the malleus causes a larger movement in the incus and a very large

movement in the stapes. The vibrations pass through the oval window to the perilymph in the vestibular canal. If sound waves were conducted directly from air to the oval window, much energy would be lost. The middle ear functions to couple sound waves in the air with the pressure waves conducted through the fluid in the cochlea.

Because liquids cannot be compressed, the oval window could not cause movement of the fluid in the vestibular canal if there were no escape valve for the pressure. This valve is provided by the membranous *round window* at the end of the tympanic canal. The pressure wave presses on the membranes separating the three ducts and is transmitted to the tympanic canal. The pressure wave makes the round window bulge. The movements of the basilar membrane produced by these pulsations cause the stereocilia of the organ of Corti to rub against the overlying tectorial membrane.

Stereocilia are bent by their contact with the tectorial membrane. As a result, ion channels in the plasma membrane of the hair cells open. Calcium ions move into the hair cell, causing the release of glutamate. This neurotransmitter binds to receptors on sensory neurons that synapse with each hair cell, leading to depolarization of these sensory neurons. Axons of the sensory neurons join to form the *cochlear nerve,* a component of the vestibulocochlear nerve (cranial nerve VIII; also referred to as the *auditory nerve*). We can summarize the sequence of events involved in hearing as follows:

sound waves enter external auditory canal $\longrightarrow$ tympanic membrane vibrates $\longrightarrow$ malleus, incus, and stapes amplify vibrations $\longrightarrow$ oval window vibrates $\longrightarrow$ vibrations are conducted through fluid $\longrightarrow$ basilar membrane vibrates $\longrightarrow$ hair cells in organ of Corti are stimulated $\longrightarrow$ cochlear nerve transmits impulses to brain

Sounds differ in pitch, loudness, and tone quality *Pitch* depends on frequency of sound waves, or number of vibrations per second, and is expressed in hertz (Hz). Low-frequency vibrations result in the sensation of low pitch, whereas high-frequency vibrations result in the sensation of high pitch. Sounds of a given frequency set up resonance waves in the cochlear fluid that cause a particular section of the basilar membrane to vibrate. The brain infers the pitch of a sound from the particular hair cells that are stimulated.

The human ear typically registers sound frequencies between about 20 and 20,000 Hz (also expressed as 20 kilohertz) and is most sensitive to sounds between 1000 and 4000 Hz (**FIG. 43-12** on page 918). Comparing the energy of audible sound waves with the energy of visible light waves, your ear is ten times more sensitive than your eye. Dogs can hear sounds up to 40,000 Hz, and bats can detect frequencies as high as 100,000 Hz.

Loud sounds cause resonance waves of greater amplitude (height). The hair cells are more intensely stimulated, and the

The hair cells in the organ of Corti in the cochlea are the receptors for hearing.

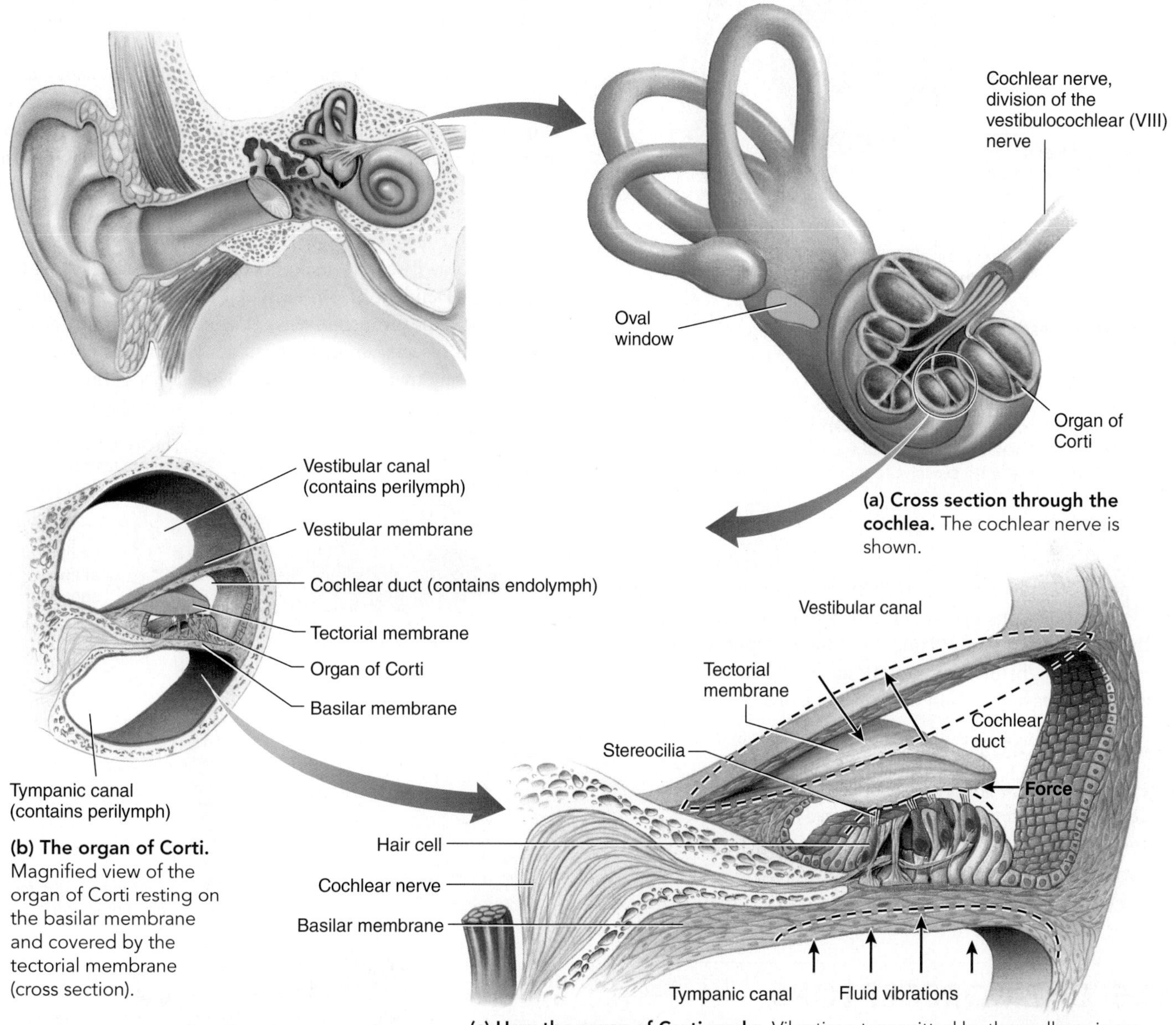

Cochlear nerve, division of the vestibulocochlear (VIII) nerve

Oval window

Organ of Corti

(a) Cross section through the cochlea. The cochlear nerve is shown.

Vestibular canal (contains perilymph)

Vestibular membrane

Cochlear duct (contains endolymph)

Tectorial membrane

Organ of Corti

Basilar membrane

Tympanic canal (contains perilymph)

(b) The organ of Corti. Magnified view of the organ of Corti resting on the basilar membrane and covered by the tectorial membrane (cross section).

Vestibular canal

Tectorial membrane

Stereocilia

Cochlear duct

Force

Hair cell

Cochlear nerve

Basilar membrane

Tympanic canal   Fluid vibrations

(c) How the organ of Corti works. Vibrations transmitted by the malleus, incus, and stapes set the fluid in the tympanic canal in motion. These vibrations (*dotted lines*) are transmitted to the basilar membrane. As the membrane vibrates, hair cells of the organ of Corti rub against the overlying tectorial membrane. This stimulation depolarizes the hair cells, generating action potentials in the sensory neurons of the cochlear nerve.

Figure 43-11 The cochlea, organ of hearing

PREDICT Imagine that the hair cells at one end of the organ of Corti were damaged and could no longer be stimulated by the tectorial membrane. How would that affect hearing?
© Cengage Learning

cochlear nerve transmits a greater number of impulses per second. Variations in tone quality (timbre)—such as when an oboe, a cornet, and a violin play the same note—depend on the number and kinds of overtones, or harmonics, produced. Many hair cells along the basilar membrane are stimulated and vibrate simultaneously in addition to the main vibration common to all three instruments. Thus, differences in tone quality are recognized in the pattern of the hair cells stimulated.

Deafness may be caused by injury to, or malformation of, either the sound-transmitting mechanism of the outer, middle,

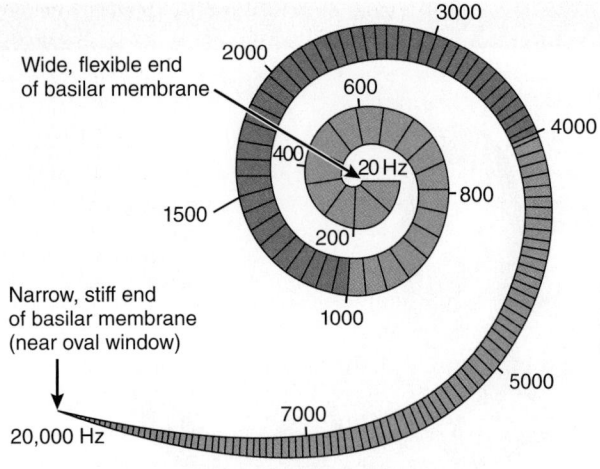

Figure 43-12 Distinguishing pitch

The basilar membrane is shown here coiled as it would be within the cochlea. Various parts of the basilar membrane vibrate in response to specific frequencies. The narrow, stiff end of the basilar membrane responds to higher frequencies, whereas the wide, flexible end responds to lower frequencies. Pitch depends on frequency of sound waves expressed as hertz (Hz). The human ear is most sensitive to sounds between 1000 and 4000 Hz (*darker pink*).

© Cengage Learning

or inner ear or the sound-perceiving mechanism of the inner ear. Efferent neurons from the brain stem protect the hair cells of the organ of Corti by dampening their response. Even so, exposure to high-intensity sound, such as heavily amplified music, damages hair cells in the organ of Corti.

CHECKPOINT 43.5

- **CONNECT** *Which sensory receptors permit you to perform actions such as finding your way into bed in the dark?*
- *What is the function of the vestibular apparatus? Of the semi-circular canals?*
- *List the sequence of events involved in hearing.*

43.6 CHEMORECEPTORS

LEARNING OBJECTIVE

9   Compare the structure and function of the receptors of taste and smell, and describe the function of the vomeronasal organ.

The mechanisms for chemoreception have been highly conserved throughout the animal kingdom. Two highly sensitive types of **chemoreceptors** evolved: one for taste, more formally called *gustation,* and one for smell, or *olfaction.* These chemical senses, found throughout the animal kingdom, allow animals to detect chemical substances in food, water, and air. Evaluating such chemical cues allows animals to find food and mates and to avoid predators. Members of the same species use chemoreception to communicate about these critical issues.

For terrestrial vertebrates, the sense of smell involves gaseous substances, called *odorants,* that reach olfactory receptors through the air. The sense of taste depends on sensing chemicals called *tastants,* dissolved in water (or saliva) in the mouth. For aquatic animals, these distinctions blur. Does a catfish smell or taste the water?

The organs of taste and smell in insects are sensory hairs called *sensilla* (sing., *sensillum*) that project from the cuticle. Olfactory sensilla are typically found on the antennae. In some insects, including ants, bees, and wasps, taste sensilla are also located on the antennae. Taste sensilla typically project from the mouthparts. In butterflies and flies, they may also project from the end of the leg.

In the blowfly each sensillum contains four chemoreceptors (**FIG. 43-13**). Dendrites from each cell extend to an opening at the tip, and axons transmit sensory signals to the brain. Studies suggest that insects can distinguish among various bitter tastes more precisely than mammals can. This ability is important in

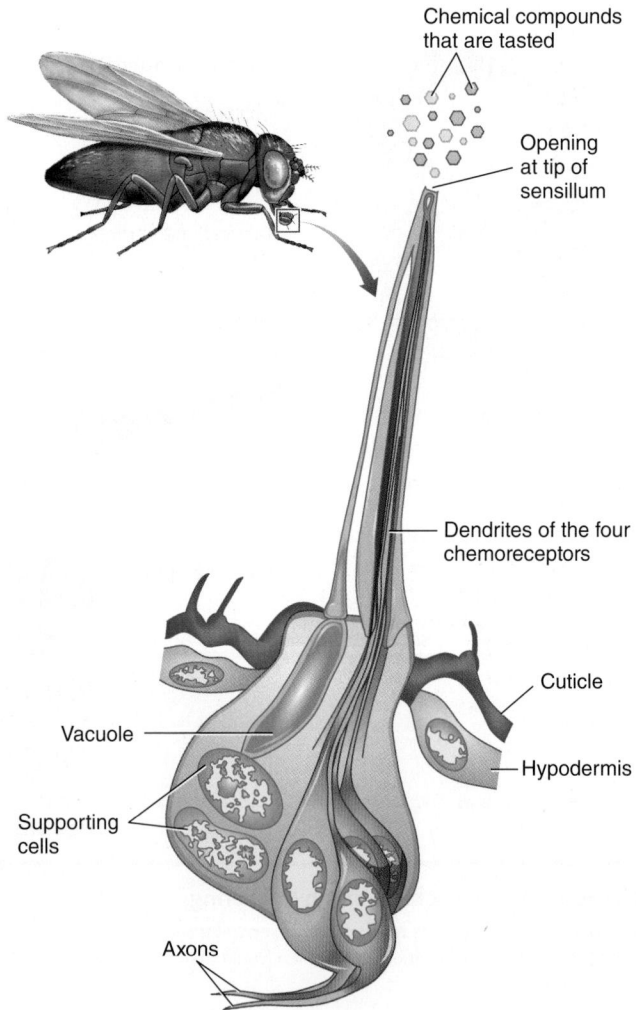

Figure 43-13 An insect taste receptor

Insects respond to a wide variety of chemical substances found in foods. Each sensillum in the blowfly contains four chemoreceptors. Dendrites that extend to an opening at the tip sense taste. The other end of each chemoreceptor is an axon that transmits sensory signals to the brain.

© Cengage Learning

discriminating tastes associated with spoiled, bacteria-laden foods or toxins.

Taste receptors detect dissolved food molecules

As just described for insects, the ability to discriminate among tastes has survival value. Poisons are generally bitter. In contrast, foods with a high caloric value often taste sweet. Mammals sense taste with *taste buds* located in the mouth.

In humans taste buds lie mainly in tiny elevations, or papillae, on the tongue. Each of the 10,000 or so taste buds is an oval epithelial capsule containing about 50 taste receptor cells interspersed with supporting cells (FIG. 43-14). Taste receptor cells are replaced about every ten days. One sensory neuron can innervate several taste buds. The plasma membrane at the tip of each taste receptor cell has microvilli that extend into a taste pore on the surface of the tongue, where they are bathed in saliva. The taste receptors detect tastants dissolved in saliva.

Although mammals can sense thousands of different tastes, most tastes are thought to be combinations of five basic tastes: sweet, sour, salty, bitter, and umami (the Japanese word for delicious). Umami is responsible for the savory flavor of certain aged cheeses, soy sauce, anchovies, and some types of seafood. The amino acid glutamate is the compound that triggers the taste. In the form of its sodium salt, monosodium glutamate, or MSG, this compound is used to flavor foods. Additional basic tastes have been proposed, including *fatty*, which is stimulated by fatty acids.

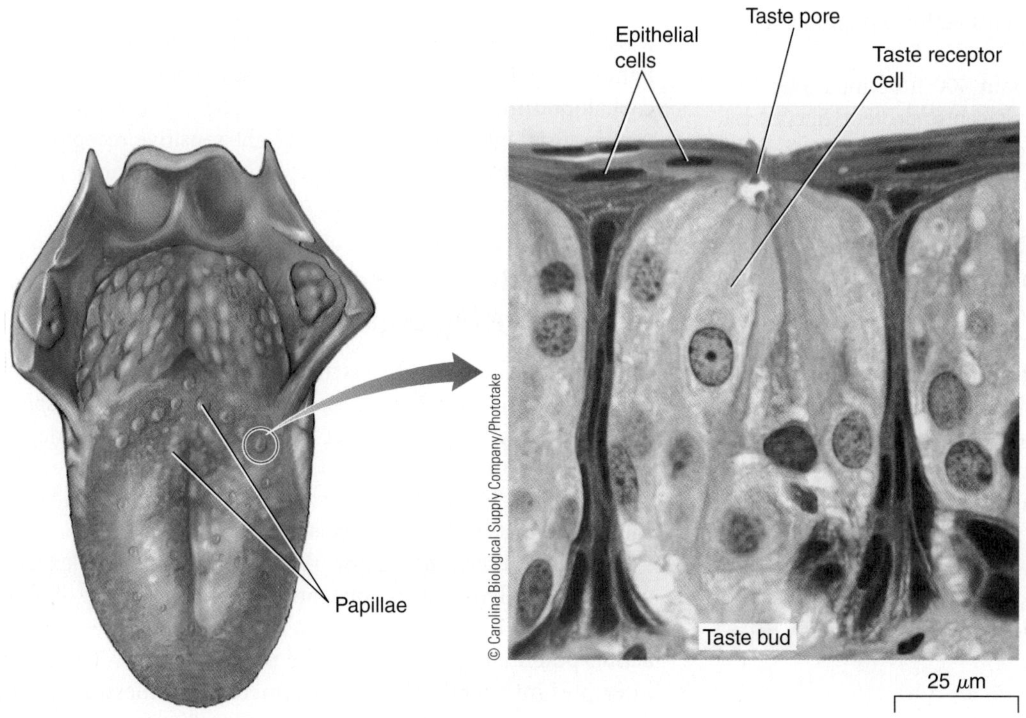

The process of tasting sweet, bitter, or umami involves a signal transduction process in which glucose, or another tastant, binds to a G protein–linked receptor that activates a second messenger system (see Chapter 6). Salty and sour tastants act directly on ion channels. In each case, action potentials are generated. Signals are transmitted (by way of the brain stem and thalamus) to the gustatory area in the parietal lobe as well as to the limbic system.

Flavor depends on the basic tastes in combination with smell, texture, and temperature. Smell affects flavor because odors pass from the nasal chamber to the mouth. You have probably observed that when your nose is congested, food seems to have little "taste." The taste buds are not affected, but the blockage of nasal passages severely reduces the participation of olfactory reception in the composite sensation of flavor.

The olfactory epithelium is responsible for the sense of smell

Most invertebrates and some vertebrates (e.g., rodents) depend on olfaction, the detection of odors, as their main sensory modality. In terrestrial vertebrates olfaction occurs in the nasal epithelium.

In humans the **olfactory epithelium** is found in the roof of the nasal cavity (FIG. 43-15). It contains about 100 million olfactory receptor cells with ciliated tips. Each receptor cell is actually an axon. The long, nonmotile cilia extend into a layer of mucus on the epithelial surface of the nasal passageway. Receptor molecules on the cilia bind with compounds dissolved in the mucus.

Remarkably, the olfactory receptor cells are replaced about every two months in humans, yet another example of the generation of new neurons throughout life.

The olfactory receptor cell axons project directly to the brain. These axons make up the **olfactory nerve** (cranial nerve I), which extends to the **olfactory bulb** in the brain. From there, information is transmitted to the *olfactory cortex* in the limbic system. Recall that the limbic system is also associated with emotions. Odors are often associated with feelings and memories.

When an *odorant*, a molecule that can be smelled, binds with a receptor on a cilium of an olfactory receptor cell, a signal transduction process is initiated.

Figure 43-14 *Animation* **Chemoreception: taste**
© Cengage Learning

Labels: Epithelial cells · Taste pore · Taste receptor cell · Papillae · Taste bud · © Carolina Biological Supply Company/Phototake · 25 μm

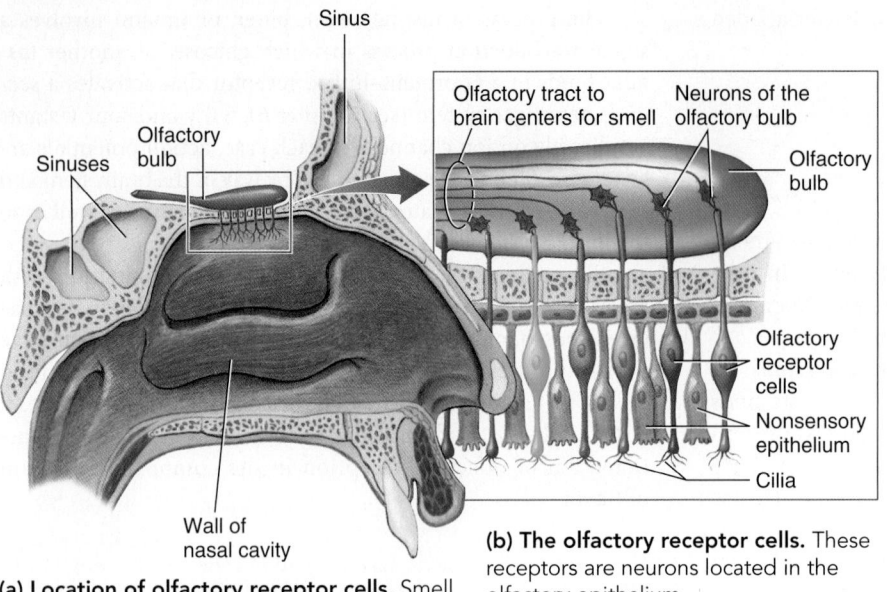

Sinus

Sinuses

Olfactory bulb

Olfactory tract to brain centers for smell

Neurons of the olfactory bulb

Olfactory bulb

Olfactory receptor cells

Nonsensory epithelium

Cilia

Wall of nasal cavity

(a) Location of olfactory receptor cells. Smell depends on thousands of chemoreceptor cells in the roof of the nasal cavity.

(b) The olfactory receptor cells. These receptors are neurons located in the olfactory epithelium.

Figure 43-15 *Animation* **Chemoreception: olfaction**
© Cengage Learning

A G protein is activated, leading to the synthesis of cyclic AMP, which opens ion channels in the plasma membrane. These channels permit Na^+ and Ca^{2+} to enter the cell, causing depolarization. Calcium-gated chloride channels then open, and Cl^- move out of the cell, causing further depolarization. The number of odorous molecules determines the intensity, which in turn determines the magnitude of the receptor potential. The receptor potential generates action potentials in the afferent neuron.

Humans and chimps have about 400 functional olfactory genes, each coding for an olfactory receptor protein. In contrast, dogs have about 1000 functional olfactory genes.

The combination of receptors activated determines the odors perceived. Humans are capable of perceiving about 10,000 scents. Interestingly, studies have shown that dogs can be trained to detect patients with certain types of cancer based on scent.

The olfactory receptors respond to remarkably small amounts of a substance. For example, most people can detect ionone, the synthetic substitute for the odor of violets, when it is present in a concentration of only 1 part to more than 30 billion parts of air. Smell is perhaps the sense that adapts most quickly. The olfactory receptors adapt about 50% in the first second or so after stimulation, so even air that smells awful may seem odorless after only a few minutes.

Many animals communicate with pheromones

Animals within many species communicate with one another by releasing **pheromones,** small, volatile molecules that are secreted into the environment. For example, female moths release pheromones that attract males. Dogs and wolves use pheromones to mark territory.

Mammals have specialized chemoreceptor cells that detect pheromones. These cells make up the **vomeronasal organ** located in the epithelium of the nose (separate from the main olfactory epithelium). The vomeronasal sensory neurons send signals to areas of the hypothalamus, which serves as a link with the endocrine system. (We discuss pheromones further in Chapter 52.)

CHECKPOINT 43.6

- **CONNECT** *In what ways are olfaction and gustation similar?*

43.7 PHOTORECEPTORS

LEARNING OBJECTIVES

10 Contrast simple eyes, compound eyes, and vertebrate eyes.
11 Identify each structure of the vertebrate eye and describe its function.
12 Compare the two types of photoreceptors, describe the signal transduction pathway that is activated when rhodopsin responds to light, and summarize the visual pathway.

Most animals have **photoreceptors** that use pigments to absorb light energy. **Rhodopsins** are the photopigments in the eyes of cephalopod mollusks (squids and octopods), arthropods, and vertebrates. Light energy striking a light-sensitive receptor cell containing these pigments triggers chemical changes in the pigment molecules that result in receptor potentials.

Invertebrates have several types of light-sensing organs

The simplest light-sensitive structures in animals are found in certain cnidarians and flatworms (**FIG. 43-16**). They are *eyespots,* also called **ocelli,** which detect light but do not form images. Eyespots are often bowl-shaped clusters of light-sensitive cells within the epidermis. They may detect the direction of the light source and distinguish light intensity.

Effective image formation requires a more complex *eye,* usually with a lens. A **lens** concentrates light on a group of photoreceptors. Vision also requires a brain that can interpret the action potentials generated by the photoreceptors. The brain integrates information about movement, brightness, location, position, and shape of the visual stimulus. Two fundamentally different types of eyes evolved: the compound eye of arthropods and the camera eye of vertebrates and cephalopod mollusks.

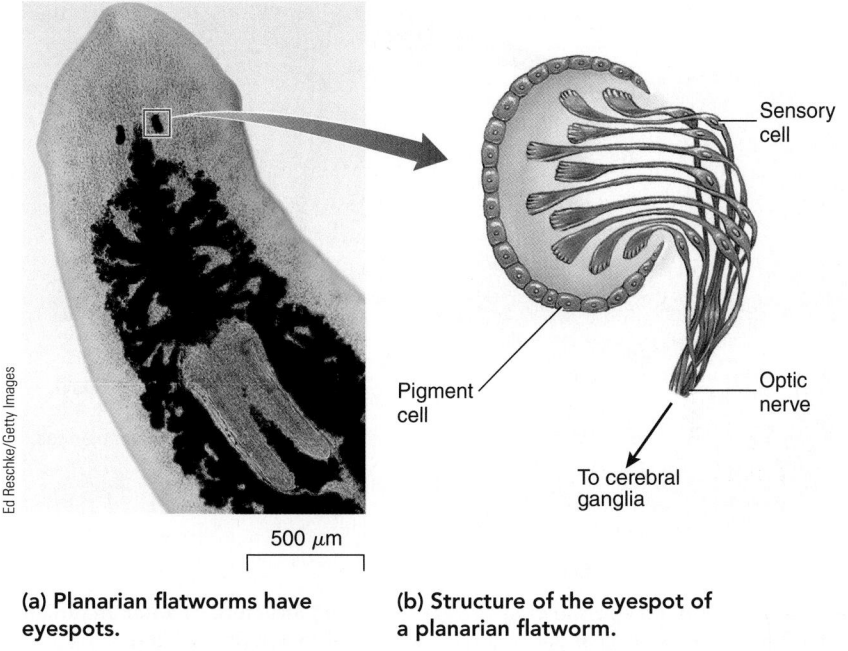

(a) Planarian flatworms have eyespots.

(b) Structure of the eyespot of a planarian flatworm.

Figure 43-16 Photoreception: eyespots

The simplest light-sensitive structures are the ocelli, or eyespots, found in certain invertebrates.
© Cengage Learning

500 μm

Sensory cell

Pigment cell

Optic nerve

To cerebral ganglia

The **compound eyes** of crustaceans and insects differ structurally and functionally from vertebrate eyes. The surface of a compound eye appears faceted, which means "having many faces," like a diamond (**FIG. 43-17**). Each *facet* is the convex *cornea* of one of the eye's visual units, called **ommatidia** (sing., *ommatidium*). The number of ommatidia varies with the species. For example, each eye of certain crustaceans has only 20 ommatidia, whereas the dragonfly eye has as many as 28,000!

The optical part of each ommatidium includes a biconvex lens and a *crystalline cone*. The lens and crystalline cone focus light onto photoreceptor cells called **retinular cells.** These cells have a light-sensitive membrane made up of microvilli that contain rhodopsin. The membranes of several retinular cells may fuse to form a central rod-shaped *rhabdome* that is sensitive to light.

Compound eyes do not perceive form well. Although the lens system of each ommatidium is adequate to focus a small inverted image, there is little evidence that the animal actually perceives them as images. However, all the ommatidia together produce a composite image, or *mosaic* picture. In gathering a point of light from a narrow sector of the visual field, each ommatidium is, in fact, sampling a mean intensity from that sector. All these points of light taken together form a mosaic picture (see Fig. 43-17d).

Arthropod eyes usually adapt to different intensities of light. A sheath of pigmented cells envelops each ommatidium, and screening pigments are present in *iris cells* and retinular cells. In nocturnal and crepuscular (active at dusk) insects and many crustaceans, pigment migrates proximally and distally within each pigmented cell. When the pigment is in the proximal position, each ommatidium is shielded from its neighbor, and only light entering directly along its axis can stimulate the receptors. When the pigment is in the distal position, light striking at any angle can pass through several ommatidia and stimulate many retinal units. As a result, sensitivity is increased in dim light, and the eye is protected from excessive stimulation in bright light. Pigment migration is under neural control in insects and under hormonal control in crustaceans. In some species it follows a daily rhythm.

Although the compound eye can form only coarse images, it compensates by superbly detecting movement. Flies can detect up to about 265 flickers per second. In contrast, the human eye can detect only 45 to 53 flickers per second; for us, flickering lights fuse above these values, so we see light provided by an ordinary bulb as steady and the movement in motion pictures as smooth. To an insect, both room lighting and motion pictures must flicker horribly. The insect's high critical flicker fusion threshold permits immediate detection of even slight movement by prey or enemy. The compound eye is an important adaptation to the arthropod's way of life.

Compound eyes differ from our eyes in another respect. They are sensitive to wavelengths of light in the range from red to ultraviolet (UV). Accordingly, an insect can see UV light well, and its world of color is very different from ours. Because they reflect UV light to various degrees, flowers that appear identically colored to humans may appear strikingly dissimilar to insects (see Fig. 37-4g).

Cephalod mollusks (squids and octopods) and vertebrates have eyes that work somewhat like an old-fashioned camera (described in the next section). Interestingly, the camera eye also evolved in Cubozoa (box jellyfish), a group of cnidarians.

Vertebrate eyes form sharp images

The position of the eyes in the front of the head of primates and certain birds allows both eyes to focus on the same object. The overlap in information they receive results in the same visual information striking the two retinas (light-sensitive areas) at the same time. This *binocular vision* is important in judging distance and in depth perception.

The vertebrate eye can be compared to an old-fashioned camera. An adjustable lens can be focused for different distances, and a diaphragm, called the **iris,** regulates the size of the **pupil,** an opening through which light passes (**FIG. 43-18**). The **retina** corresponds to the light-sensitive film used in a camera. Outside the retina is the **choroid layer,** a sheet of cells filled with black pigment that absorbs extra light. This mechanism prevents light from being reflected back into the photoreceptors,

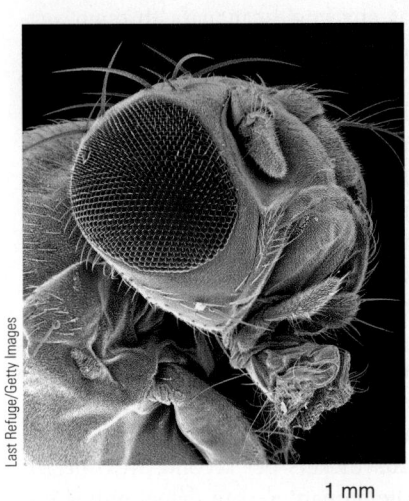

1 mm

(a) SEM of fruit fly (*Drosophila* sp.) showing its prominent compound eyes.

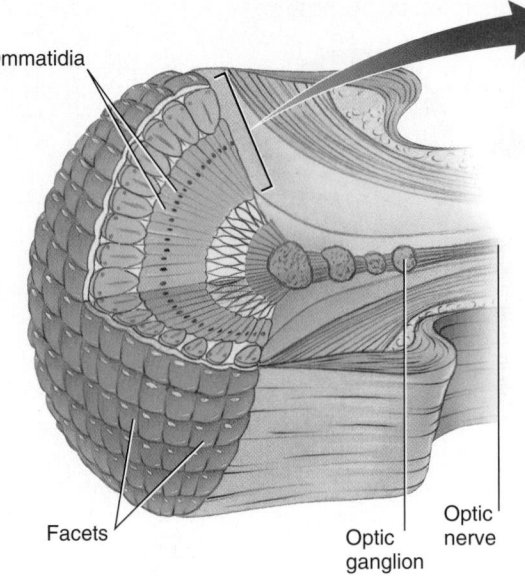

Ommatidia

Facets

Optic ganglion

Optic nerve

(b) Structure of compound eye showing several ommatidia. The eye registers changes in light and shade, permitting the animal to detect movement.

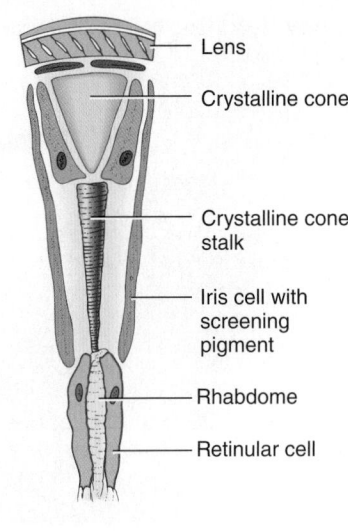

Lens

Crystalline cone

Crystalline cone stalk

Iris cell with screening pigment

Rhabdome

Retinular cell

(c) Structure of an ommatidium. The rhabdome is the light-sensitive core of the ommatidium.

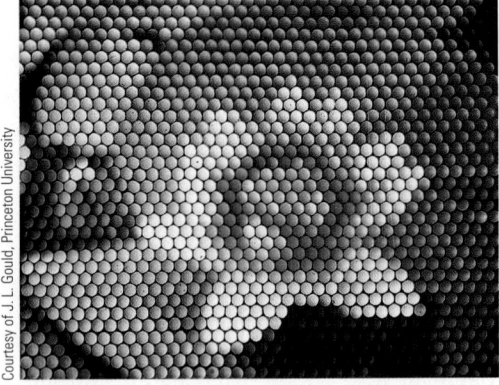

(d) A bee's-eye view of a flower. This photograph was taken using an optical device to achieve an approximate simulation of how the bee might see this flower. However, the bee would see UV light rather than red, and circles would be vertically elongated ellipses.

Figure 43-17 Photoreception: compound eyes
© Cengage Learning

which would cause blurring of images. (Cameras are also black on the inside.) The choroid is rich in blood vessels that supply the retina.

The outer coat of the eyeball, called the **sclera,** is a tough, opaque, curved sheet of connective tissue. The sclera protects the inner structures and helps maintain the rigidity of the eyeball. On the front surface of the eye, this sheet becomes the thinner, transparent **cornea** through which light enters. The cornea serves as a fixed *lens* that focuses light.

The **lens** of the eye is a transparent, elastic ball just behind the iris. It bends the light rays coming in and brings them to

a focus on the retina. The lens is aided by the curved surface of the cornea and by the refractive properties (ability to bend light rays) of the liquids inside the eyeball. The *anterior cavity* between the cornea and the lens is filled with a watery substance, the aqueous fluid. The larger *posterior cavity* between the lens and the retina is filled with a more viscous fluid, the **vitreous body.** By providing an internal hydrostatic pressure, both fluids help maintain the shape of the eyeball.

At its anterior margin, the choroid is thick and projects medially into the eyeball, forming the *ciliary body,* which consists of ciliary processes and the ciliary muscle. The ciliary processes are glandlike folds that project toward the lens and secrete the aqueous fluid.

The eye has the ability to change focus for near or far vision by changing the shape of the lens (**FIG. 43-19**). This process, called **accommodation,** depends on the *ciliary muscle,* a circular muscle that is part of the ciliary body. To focus on close objects, the ciliary muscle contracts, making the elastic lens rounder. To focus on more distant objects, this muscle relaxes and the lens assumes a flattened (ovoid) shape.

The amount of light entering the eye is regulated by the iris, a ring of smooth muscle that appears as blue, green, gray, or brown depending on the amount and nature of pigment present. The iris consists of two mutually antagonistic sets of muscle fibers. One set is arranged circularly and contracts to *decrease* the size of the pupil. The other is arranged radially and contracts to *increase* the size of the pupil.

Each eye also has six muscles that extend from the surface of the eyeball to various points in the bony socket. These muscles enable the eye as a whole to move and be oriented in a given direction. Cranial nerves innervate the muscles in such a way that the eyes normally move together and focus on the same area.

The retina contains light-sensitive rods and cones

The light-sensitive structure in the vertebrate eye is the retina, which lines the posterior two-thirds of the eyeball and covers the choroid. The retina, which is composed of ten layers, contains the photoreceptor cells called, according to shape, **rods** and **cones.** In both rod and cone cells, infoldings of the plasma membrane form stacks of membranous discs that contain the photopigments. The discs greatly increase the light-absorbing surface.

The human eye has about 125 million rods and 6.5 million cones. Rods function in dim light, allowing us to detect shape and movement. They do not discriminate among colors. Because the rods are more numerous in the periphery of the retina, you can see an object better in dim light if you look slightly to one side of it (allowing the image to fall on the rods).

Cones respond to light at higher levels of intensity, such as daylight, and they allow us to perceive fine detail. Cones are responsible for color vision; they are differentially sensitive to different wavelengths (colors) of light. In primates and certain hunting birds (such as eagles), the cones are most concentrated in the fovea, a small, depressed area in the center of the retina. The **fovea** is the region of sharpest vision because it has the greatest density of receptor cells and because the retina is thinner there.

Light must pass through several layers of connecting neurons in the retina to reach the rods and cones (**FIG. 43-20**). This arrangement lets the rods and cones contact a layer of pigmented epithelium that provides **retinal,** a component of rhodopsin. The retina has five main types of neurons.

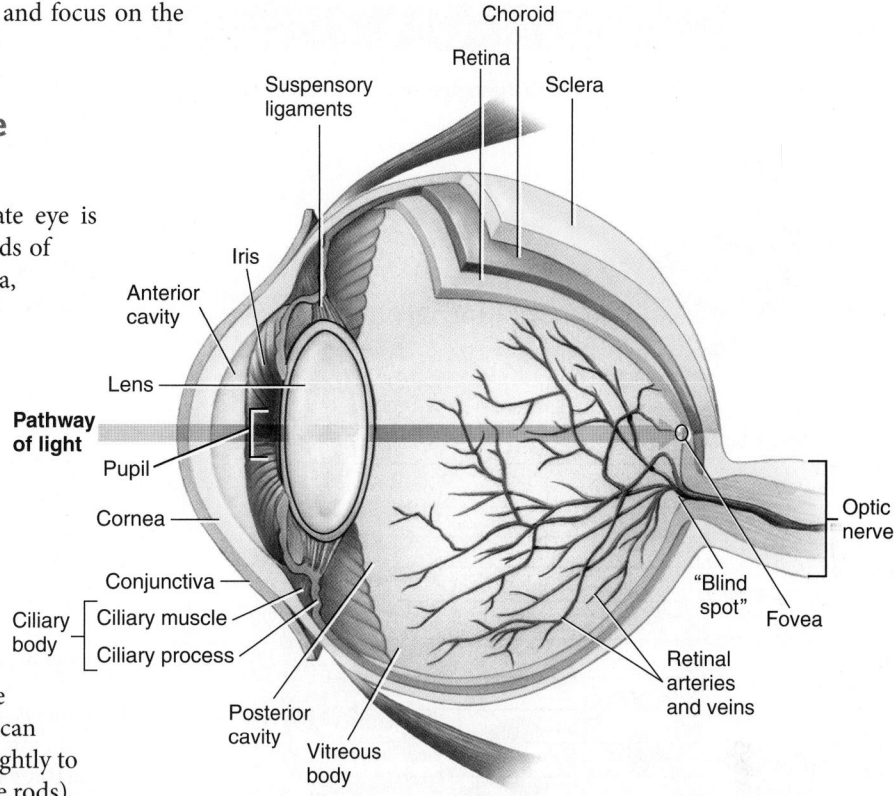

Figure 43-18 Photoreception: structure of the human eye

Light passes through the human eye to photoreceptor cells in the retina. In this lateral view, the eye is shown partly sectioned along the sagittal plane to expose its internal structures.

© Cengage Learning

(1) Photoreceptors (rods and cones) synapse on (2) **bipolar cells,** which make synaptic contact with (3) **ganglion cells.** Two types of lateral interneurons are the (4) **horizontal cells,** which receive information from the photoreceptor cells and send it to bipolar cells, and the (5) **amacrine cells,** which receive messages from the bipolar cells and send signals back to them or to ganglion cells (**FIG. 43-21**). Horizontal and amacrine cells help process visual input within the retina. Interestingly,

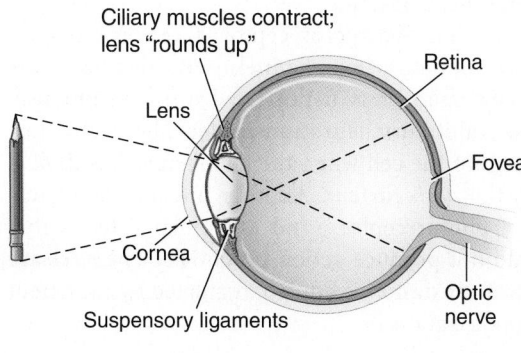

(a) Near vision.

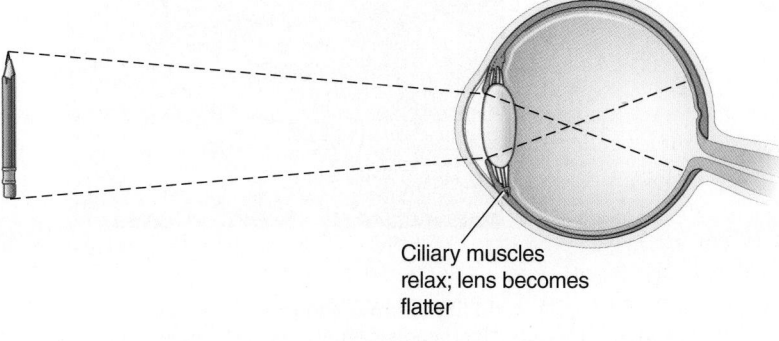

(b) Distant vision.

Figure 43-19 Accommodation

How the eye changes focus for **(a)** near and **(b)** distant vision.

© Cengage Learning

The rods and cones in the retina are the photoreceptors in vertebrates.

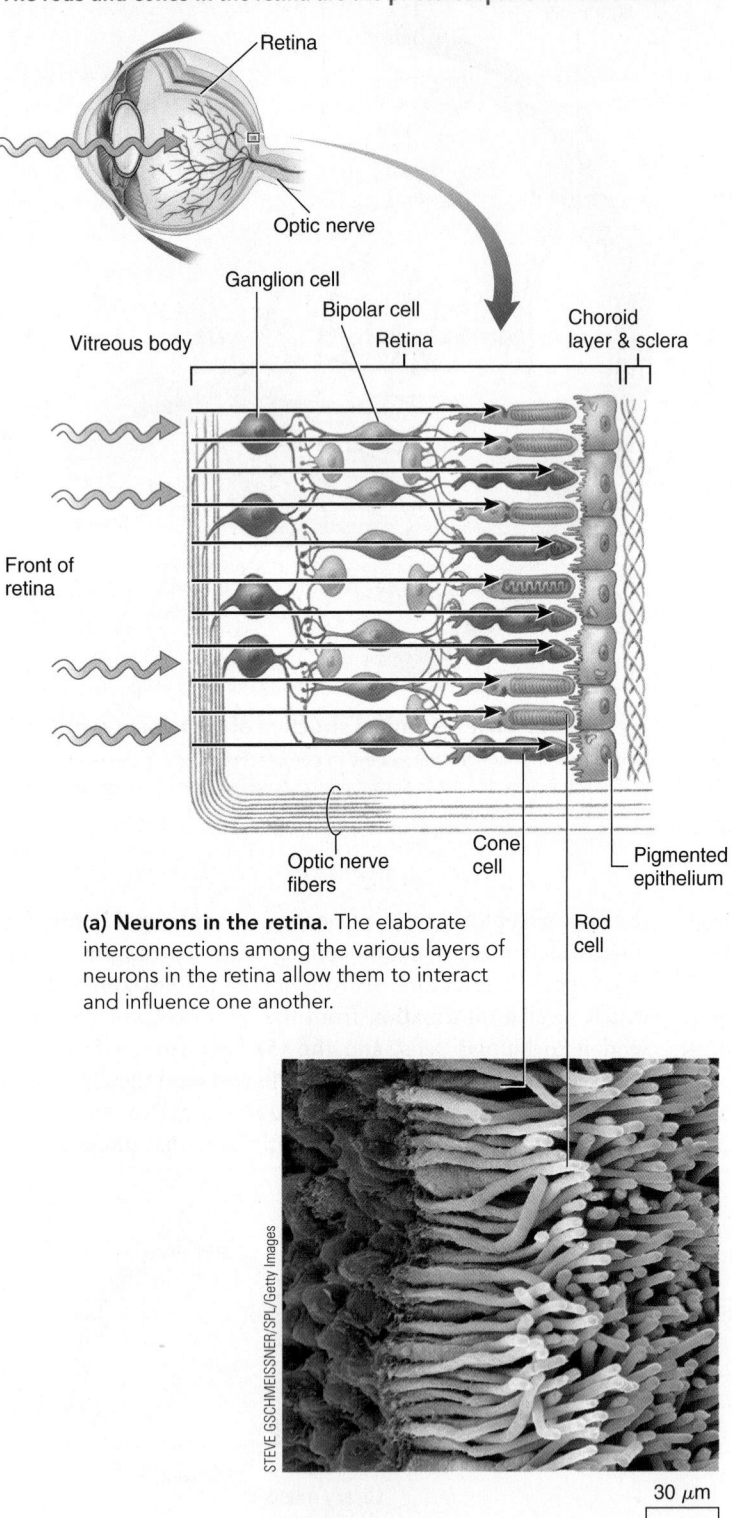

(a) Neurons in the retina. The elaborate interconnections among the various layers of neurons in the retina allow them to interact and influence one another.

STEVE GSCHMEISSNER/SPL/Getty Images

30 μm

(b) SEM of rods and cones. The *tan*, elongated structures are rods. The shorter, thicker, *green* structures are cones. The rods function in dim light and allow you to see shapes and movement. The cones allow you to view the world in color.

one group of ganglion cells *project* to (i.e., have axons that extend to) the *suprachiasmatic nucleus,* the body's main biological clock. These ganglion cells contain a light-sensitive pigment (melanopsin) that is important in nonvisual responses to light.

The axons of the ganglion cells extend across the surface of the retina and unite to form the **optic nerve.** The area where the optic nerve passes out of the eyeball, the *optic disk,* is known as the *blind spot;* because it lacks rods and cones, images falling on it cannot be perceived.

Here is a simplified summary of the visual pathway:

light passes through cornea ⟶ through aqueous fluid ⟶ through lens ⟶ through vitreous body ⟶ image forms on photoreceptor cells in retina ⟶ signal bipolar cells ⟶ signal ganglion cells ⟶ optic nerve transmits signals to thalamus ⟶ integration by visual areas of cerebral cortex

Light activates rhodopsin

How is light converted into the neural signals that transmit information about environmental stimuli into pictures in the brain? Rod cells are so sensitive to light that they can respond to a single photon. Rhodopsin in the rod cells and some very closely related photopigments in the cone cells are responsible for the ability to see. **Rhodopsin** consists of *opsin,* a large protein that is chemically joined with *retinal,* an aldehyde of vitamin A (see Fig. 3-14). Two isomers of retinal exist: the *cis* form, which is folded, and the *trans* form, which is straight.

In the dark, opsin binds to retinal in the *cis* form. Cyclic GMP (cyclic guanosine monophosphate, a molecule similar to cyclic AMP) opens nonspecific channels that permit passage of Na^+ and other cations into the rod cell (**FIG. 43-22**). This process depolarizes the rod cell, which then releases the neurotransmitter glutamate. The glutamate is inhibitory; it hyperpolarizes the membrane of the bipolar cell, preventing it from transmitting messages.

Note that the photoreceptor differs from other neurons in that the ion channels in its membrane are normally open; it is depolarized and continuously releases inhibitory neurotransmitter. The steady flow of ions into the cell when the environment is dark is called the *dark current.* Another unusual characteristic of photoreceptors, and of bipolar cells, is that they do not produce action potentials. Their release of neurotransmitter is graded, regulated by the extent of depolarization.

Figure 43-20 *Animation* **Organization of the retina**

CONNECT Why is it difficult to see colors in dim light?
© Cengage Learning

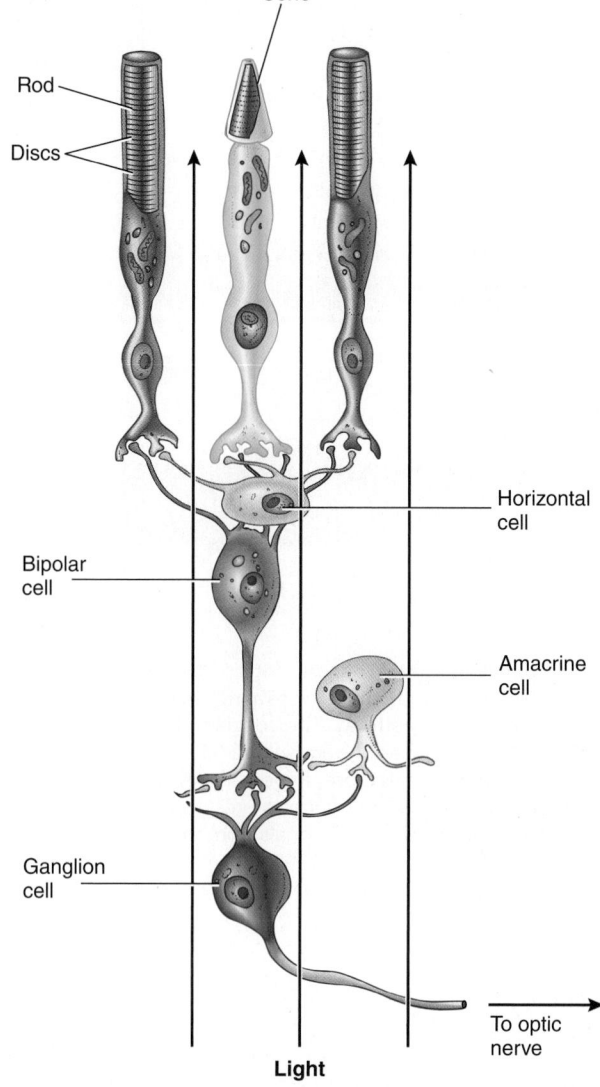

Cone

Rod

Discs

Horizontal cell

Bipolar cell

Amacrine cell

Ganglion cell

Light

To optic nerve

Figure 43-21 Neural pathway in the retina

The photoreceptor cells (rods and cones) are located in the back of the retina. They transmit information to bipolar cells. The bipolar cells synapse with ganglion cells. The horizontal and amacrine cells are important in local processing of visual input within the retina. Axons of ganglion cells make up the optic nerve.

© Cengage Learning

When light strikes rhodopsin, light energy is transduced. This process can be considered in three stages:

1. *Light activates rhodopsin.* Light transforms *cis*-retinal to *trans*-retinal. Rhodopsin changes shape and breaks down into its components: opsin and retinal. This sequence is the light-dependent process in vision.
2. *Rhodopsin activates a signal transduction pathway.* When it changes shape, rhodopsin binds with *transducin,* a G protein. Transducin activates an esterase that hydrolyzes cyclic GMP (cGMP) to GMP, reducing the concentration of cGMP.
3. *A decrease in concentration of cGMP in the rod causes the* Na^+ *channels to close.* Fewer cations pass into the rod cell, and it becomes more negative (hyperpolarized). The rod cell then releases less glutamate. Thus, light decreases the number of neural signals from the rod cells.

The release of glutamate normally hyperpolarizes the membrane of the bipolar cell, so a decrease in glutamate release results in depolarization of the bipolar cell. The depolarized bipolar cell increases its release of a neurotransmitter, which typically stimulates the ganglion cell.

When you move from bright daylight to a dark room, you experience "blindness" for a few moments while your eyes adapt. During exposure to bright light, rhodopsin breaks down, decreasing the sensitivity of your eyes. *Dark adaptation* occurs as enzymes restore rhodopsin to a form in which it can respond to light. In dim light vision depends on the rods. Conversely, when you move from the dark to bright light, your eyes are very sensitive, and you may feel "blinded" by the light. *Light adaptation* occurs as some of the photopigments break down and the sensitivity of your eyes decreases.

Color vision depends on three types of cones

Many invertebrates and at least some animals in each vertebrate class have color vision. Color perception depends on cones. Most mammals have two types of cones, but humans and other primates have three types: blue, green, and red. Each contains a slightly different photopigment. Although the retinal portion of the pigment molecule is the same as in rhodopsin, the opsin protein differs slightly in each type of photoreceptor.

Each type of cone responds to light within a considerable range of wavelengths but is named for the ability of its pigment to absorb a particular wavelength more strongly than other cones do. For example, red light can be absorbed by all three types of cones, but those cones most sensitive to red act as red receptors. By comparing the relative responses of the three types of cones, the brain can detect colors of intermediate wavelengths. Color blindness occurs when there is a deficiency of one or more of the three types of cones. It is usually an inherited X-linked condition (see Fig. 11-15).

Integration of visual information begins in the retina

The size, intensity, and location of light stimuli determine initial processing in the retina. Rods and cones send signals directly to bipolar cells, and bipolar cells send signals directly to ganglion cells. However, as illustrated in Figure 43-21, photoreceptor cells can also transmit signals to horizontal cells, and bipolar cells can transmit signals to amacrine cells. Ganglion cells are inhibited by amacrine cells and can be inhibited indirectly by horizontal cells as a result of their action on bipolar cells. Thus, the horizontal and amacrine cells integrate signals laterally.

Bipolar, horizontal, and amacrine cells combine signals from several photoreceptors. Each ganglion cell has a *receptive field,* a specific group of photoreceptors that light must strike for the ganglion cell to be stimulated. Ganglion cells are activated or inhibited depending on which photoreceptor cells have been stimulated.

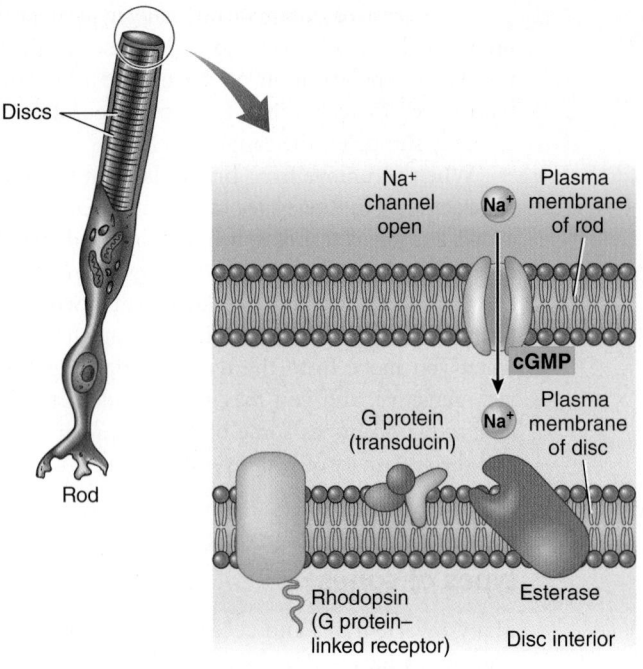

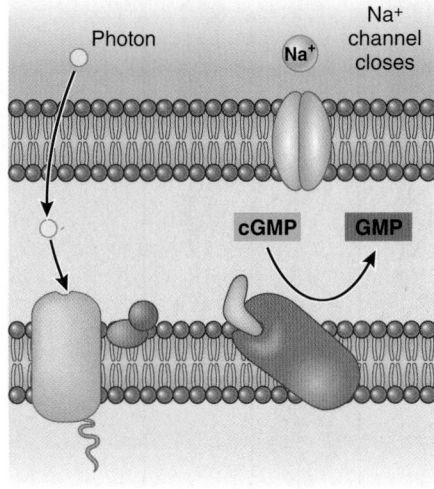

(a) In the dark. Sodium channels are open. The rod cell membrane is depolarized and releases the inhibitory neurotransmitter glutamate (*not shown*).

(b) In the light. When rhodopsin absorbs light, it changes shape, activating a signal transduction pathway. The concentration of cGMP is reduced. As a result cGMP-gated sodium channels close, hyperpolarizing the membrane. Hyperpolarization results in a decrease in inhibitory signals from the rod cell.

Figure 43-22 The biochemistry of vision

Inward foldings of the plasma membrane produce a series of discs in each rod cell. In the dark the rod cell is depolarized; in the light it becomes hyperpolarized.

© Cengage Learning

The pattern of neuron firing in the retina is very important. The signals sent to ganglion cells depend on spatial patterns and timing of the light striking the retina. Ganglion cells receive signals about specific types of visual stimuli, such as color, brightness, and motion, and transmit this information about the characteristics of a visual image. In contrast with rods and cones and most other neurons of the retina, which produce only graded potentials, ganglion cells produce action potentials.

The retina begins the construction of visual images, but those images are interpreted in the brain. Axons of more than one million ganglion cells form the optic nerves. These cranial nerves transmit information to the brain by complex, encoded signals. Axons from the medial half of each retina cross to the opposite side in the floor of the hypothalamus, forming an X-shaped structure, the **optic chiasm** (FIG. 43-23). The axons of the optic nerves that cross over extend to the opposite side of the brain.

Axons of the *optic nerves* end in the *lateral geniculate nuclei* of the thalamus. From there, neurons send signals to the *primary visual cortex* in the occipital lobe of the cerebrum. The lateral geniculate nucleus controls what information is sent to

the visual cortex. Neurons in the reticular activating system are involved in this integration. Information is transmitted from the primary visual cortex to other cortical areas for further processing.

Different types of ganglion cells receive information about different aspects of a visual stimulus. As a result, information from each point in the retina is transmitted in parallel by several types of ganglion cells. These cells project to different kinds of neurons in the lateral geniculate nucleus and primary visual cortex.

Neurobiologists have not yet discovered all the mechanisms by which the brain makes sense of the visual information it receives. We know that a large part of the association areas of the cerebrum is involved in integrating visual input. The neurons of the visual cortex are organized as a map of the external visual field. They are also organized into columns. Each column of neurons receives signals originating from light entering either the right or the left eye. Columns are also organized by orientation of light. Neurons in a column receive signals originally triggered by light with the same orientation. However, these signals originate from different locations in the retina.

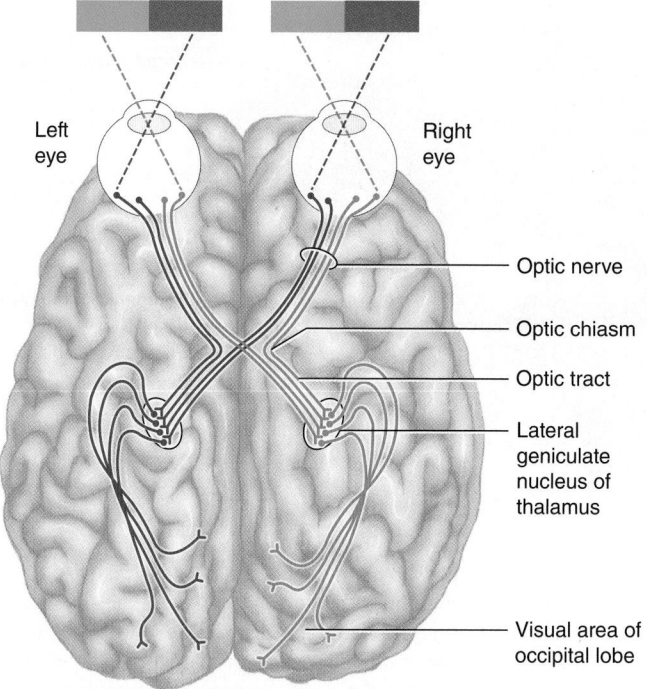

Left eye

Right eye

Optic nerve

Optic chiasm

Optic tract

Lateral geniculate nucleus of thalamus

Visual area of occipital lobe

Figure 43-23 Neural pathway for transmission of visual information

Axons of ganglion cells form the optic nerves. In the brain, fibers from the medial half of each retina cross to the opposite side, forming the optic chiasm. Many optic nerve fibers end in the lateral geniculate nuclei of the thalamus. From there, signals are sent to the visual area in the occipital lobe. (Brain is viewed from above with overlying structures removed.)
© Cengage Learning

CHECKPOINT 43.7

- **CONNECT** *How does the insect's compound eye compare with the vertebrate eye?*
- *What happens when light strikes rhodopsin? (Include a description of the signal transduction pathway in your answer.)*
- *What is the sequence of neural signaling in the retina?*
- *What is meant by the statement "Vision happens mainly in the brain"?*

43.1 How Sensory Systems Work (page 906)

1 Describe how sensory systems function; include descriptions of sensory receptors, energy transduction, receptor potential, sensory adaptation, and perception.

- **Sensory receptors** are neuron endings or specialized receptor cells in close contact with neurons. **Sense organs** consist of sensory receptors and accessory cells.

- A sensory receptor absorbs energy from a stimulus and transduces that energy—*energy transduction*—into electrical energy. In response to stimuli, a sensory receptor produces **receptor potentials,** which are depolarizations or hyperpolarizations of the membrane. Receptor potentials are **graded potentials,** local changes in electrical potential that vary in magnitude depending on the strength of the stimulus. Receptor potentials can generate action potentials in **sensory neurons,** which transmit coded signals to the CNS.

- **Sensory adaptation,** the decrease in frequency of action potentials in a sensory neuron even when the stimulus is maintained, decreases response to that stimulus.

- The brain interprets sensations by converting them to perceptions of stimuli. Sensory **perception** is the process of selecting, interpreting, and organizing sensory information.

2 Classify sensory receptors according to the location of stimuli to which they respond and according to the types of energy they transduce.

- **Exteroceptors** receive stimuli from the outside environment. **Interoceptors** detect changes within the body, such as changes in pH, osmotic pressure, body temperature, and the chemical composition of the blood.

- Sensory receptors can be classified according to the types of energy they transduce. *Thermoreceptors* respond to heat and cold. *Electroreceptors* sense differences in electrical potential. *Nociceptors* (pain receptors) respond to mechanical stimuli, temperature, and other stimuli that could be damaging. *Mechanoreceptors* transduce mechanical energy: touch, pressure, gravity, stretching, and movement. *Chemoreceptors* transduce certain chemical compounds, and photoreceptors transduce light energy.

43.2 Thermoreceptors (page 909)

3 Identify three functions of thermoreceptors.

- **Thermoreceptors** transduce thermal energy; they respond to heat and cold. Some parasites use thermoreceptors to find an endothermic host. Some predators use thermoreceptors to locate endothermic prey. In endothermic animals thermoreceptors provide cues about body temperature.

43.3 Electroreceptors and Magnetic Reception (page 910)

4 Describe the functions of electroreceptors and identify the functions of magnetic reception.

- **Electroreceptors** respond to electrical stimuli. Predatory fishes use electroreceptors to detect prey.

- Some animals detect and use information about Earth's magnetic fields to orient themselves. Many migratory animals use magnetic fields to navigate.

43.4 Nociceptors (page 910)

5 Describe the functions of nociceptors and identify the roles of substance P and endorphins.

- **Nociceptors** are pain receptors; they are free nerve endings of certain sensory neurons that respond to strong tactile stimuli, temperature extremes, and certain chemicals. Sensory neurons that transmit pain signals release **substance P** or **glutamate. Endorphins** and **enkephalins** inhibit release of substance P, ending transmission of the pain signal.

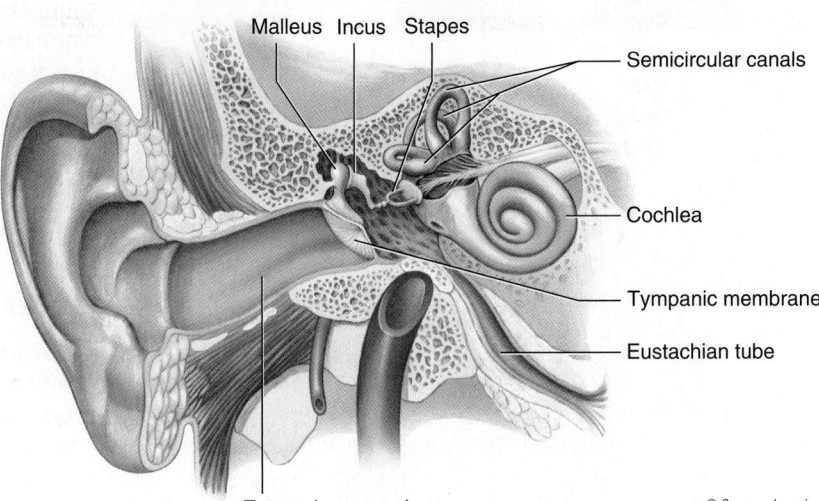

Malleus Incus Stapes
Semicircular canals
Cochlea
Tympanic membrane
Eustachian tube
External ear canal

© Cengage Learning

43.5 Mechanoreceptors *(page 910)*

6 Compare the functions and mechanisms of action of the following mechanoreceptors: tactile receptors, proprioceptors, statocysts, hair cells, and lateral line organs.

- **Mechanoreceptors** transduce mechanical energy, permitting animals to feel, hear, and maintain balance. Mechanoreceptors respond to touch, pressure, gravity, stretch, or movement. They are activated when they change shape as a result of being mechanically pushed or pulled.
- *Tactile receptors* in the skin respond to mechanical displacement of hairs or to displacement of the receptor cells themselves. The **Pacinian corpuscle** is a tactile receptor that responds to touch and pressure.
- **Proprioceptors** allow an animal to perceive orientation of its body and the positions of its parts. **Muscle spindles, Golgi tendon organs,** and **joint receptors** are proprioceptors that continuously respond to tension and movement.
- **Statocysts,** gravity receptors found in many invertebrates, are simple organs of equilibrium; they inform the CNS of change in position. A statocyst consists of a chamber lined with ciliated mechanoreceptors; the chamber contains **statoliths,** granules of sand grains or calcium carbonate that change position as the animal moves. As statoliths move, sensory hairs bend, and different sensory cells are stimulated.
- Vertebrate **hair cells** detect movement; they are found in the lateral line of fishes, the vestibular apparatus, semicircular canals, and cochlea. Each hair cell has a single **kinocilium,** which is a true cilium, and **stereocilia,** which are microvilli.
- **Lateral line organs** supplement vision in fishes and some amphibians by informing the animal of moving objects or objects in its path.

7 Compare the structure and function of the saccule and utricle with that of the semicircular canals in maintaining equilibrium.

- The vertebrate *inner ear* consists of a **labyrinth** of fluid-filled chambers and canals that help maintain equilibrium. The **vestibular apparatus** in the upper part of the labyrinth consists of the **saccule, utricle,** and **semicircular canals.**
- The saccule and utricle contain **otoliths,** calcium carbonate ear stones, that change position when the head is tilted or when the body is moving in a straight line. The otoliths stimulate hair cells that send signals to the brain, enabling the animal to perceive the direction of gravity.
- The semicircular canals inform the brain about turning movements. Clumps of hair cells, called *cristae,* are located within each bulblike enlargement, called an **ampulla.** Cristae are stimulated by movements of the **endolymph,** a fluid that fills each canal.

8 Trace the path taken by sound waves through the structures of the ear and explain how the organ of Corti functions as an auditory receptor.

- In terrestrial vertebrates, the **organ of Corti** within the **cochlea** contains auditory receptors.

- Sound waves pass through the *external auditory canal* and cause the **tympanic membrane** (eardrum) to vibrate. The ear bones—**malleus, incus,** and **stapes**—transmit and amplify the vibrations through the middle ear. Vibrations pass through the *oval window* to fluid within the vestibular canal. Pressure waves press on the membranes that separate the three ducts of the cochlea.
- The bulging of the *round window* serves as an escape valve for the pressure. The pressure waves cause movements of the **basilar membrane.** These movements stimulate the hair cells of the organ of Corti by rubbing them against the overlying **tectorial membrane.**
- Neural impulses are initiated in the dendrites of neurons that lie at the base of each hair cell and are transmitted by the *cochlear nerve* to the brain.

43.6 Chemoreceptors *(page 918)*

9 Compare the structure and function of the receptors of taste and smell, and describe the function of the vomeronasal organ.

- **Chemoreceptors** transduce certain kinds of chemical compounds, allowing *gustation,* or taste, and *olfaction,* or smell.
- In vertebrates taste receptor cells are specialized epithelial cells in *taste buds.* The **olfactory epithelium** contains specialized olfactory receptors with axons that extend to the brain. These axons make up the **olfactory nerve.**
- Receptors that make up the **vomeronasal organ,** located in the nasal epithelium of mammals, detect **pheromones,** small, volatile signaling molecules released by animals.

43.7 Photoreceptors *(page 920)*

10 Contrast simple eyes, compound eyes, and vertebrate eyes.

- **Photoreceptors** transduce light energy and serve as the sensory receptors in eyespots and eyes. *Eyespots,* or **ocelli,** found in cnidarians and flatworms, detect light but do not form images.
- The **compound eye** of insects and crustaceans consists of visual units called **ommatidia,** which collectively produce a *mosaic* image. Each ommatidium has a transparent lens and a *crystalline cone* that focus light onto receptor cells called **retinular cells.**
- Vertebrates have camera eyes that form sharp images.

11 Identify each structure of the vertebrate eye and describe its function.

- In the human eye, light enters through the **cornea,** passes through the **pupil,** is focused by the **lens,** and produces an image on the **retina.** The **iris** regulates the amount of light that can enter. The internal surface of the eye is covered by the **choroid layer,** a black coat that prevents light rays from scattering.

- **Accommodation** is the process by which the lens changes shape, allowing the eye to focus at various distances. The lens is attached to the circular *ciliary muscle.* When the ciliary muscle contracts, the lens becomes rounder. When it relaxes, the lens flattens.

12 Compare the two types of photoreceptors, describe the signal transduction pathway that is activated when rhodopsin responds to light, and summarize the visual pathway.

- The retina contains the photoreceptor cells: **rods,** which function in dim light and form images in black and white, and **cones,** which function in bright light and permit color vision.

- The retina also contains **bipolar cells** that send signals to **ganglion cells.** Two types of lateral interneurons integrate

information: **horizontal cells** receive signals from the rods and cones and send signals to bipolar cells, and **amacrine cells** receive signals from bipolar cells and send signals back to bipolar cells and to ganglion cells.

- When the environment is dark, ion channels in the plasma membranes of rod cells are open, and the cells are depolarized; they release glutamate, which hyperpolarizes the membranes of bipolar cells so that they do not send signals.

- When light strikes the photopigment **rhodopsin** in the rod cells, its retinal portion changes shape and initiates a signal transduction process that involves *transducin.* This G protein activates an esterase that hydrolyzes cGMP, reducing its concentration. As a result, ion channels close, and the membrane becomes hyperpolarized. The rod cells release less glutamate, and fewer signals are transmitted. As a result, bipolar cells become depolarized and release the neurotransmitter that stimulates ganglion cells.

- Axons of the ganglion cells make up the **optic nerves.** The optic nerves transmit information to the *lateral geniculate nuclei* in the thalamus. From there, neurons project to the *primary visual cortex* and then to other cortical areas.

TEST YOUR UNDERSTANDING

Know and Comprehend

1. A sensory receptor absorbs energy from some stimulus. The next step is (a) release of neurotransmitter (b) transmission of an action potential (c) energy transduction (d) transmission of a receptor potential (e) sensory adaptation

2. Which of the following is *not* a correct match? (a) mechanoreceptors—touch, pressure (b) electroreceptors—voltage (c) photoreceptors—light (d) chemoreceptors—extreme temperature (e) nociceptors—pain

3. Which of the following is/are associated with informing the brain about turning movements? (a) semicircular canals (b) saccule (c) lymph (d) otoliths (e) utricle

4. In the process of hearing, the basilar membrane vibrates. Which event occurs next? (a) tympanic membrane vibrates (b) bones in middle ear amplify and conduct vibrations (c) cochlear nerve transmits impulses to organ of Corti (d) hair cells in organ of Corti are stimulated (e) vibrations are conducted to chemoreceptors

5. In the human visual pathway, after light passes through the cornea it next (a) stimulates ganglion cells (b) passes through the lens (c) sends signals through the optic nerve (d) depolarizes horizontal cells (e) hyperpolarizes rod cells

6. When rhodopsin is activated, it (a) concentrates in bipolar cells (b) transmits more action potentials (c) binds with a G protein (d) sends signals to the lateral geniculate nuclei (e) activates substance P

Apply and Analyze

7. **VISUALIZE** Draw a simple diagram illustrating the relationship of the following layers to the path of light in the vertebrate eye: vitreous body, bipolar cells, rods and cones, ganglion cells.

8. **CONNECT** Connoisseurs can recognize many varieties of cheese or wine by "tasting." How can they do so when there are only a few types of taste receptors?

Evaluate and Synthesize

9. **CONNECT** If all neurons transmit the same type of message, how do you know the difference between sound and light? How are you able to distinguish between an intense pain and a mild one? How are these discriminations adaptive?

10. **EVOLUTION LINK** Proteins known as opsins are present in the photoreceptors of all light-sensitive receptors that have been studied. What does this finding suggest about the relationships of organisms as diverse as flatworms, crayfish, and elephants?

11. **EVOLUTION LINK** Once the ability to produce images evolved, eyes evolved relatively rapidly. Hypothesize why image-forming eyes evolved quickly and are now present in most animals.

12. **SCIENCE, TECHNOLOGY, AND SOCIETY** Cochlear implants bring hearing to many children who are born deaf. The prognosis is best when the device is implanted before the child is three years old (during the early years when language is developed). Many individuals in the deaf community (which consists of individuals born deaf or who are affected by deafness) who communicate with sign language oppose cochlear implants. They do not view the inability to hear as a disability. This perspective raises ethical questions for some families with children who are deaf. Argue for and against cochlear implants for very young children.

aplia To access course materials, such as Aplia and other companion resources, please visit **www.cengagebrain.com.**

44 Internal Transport

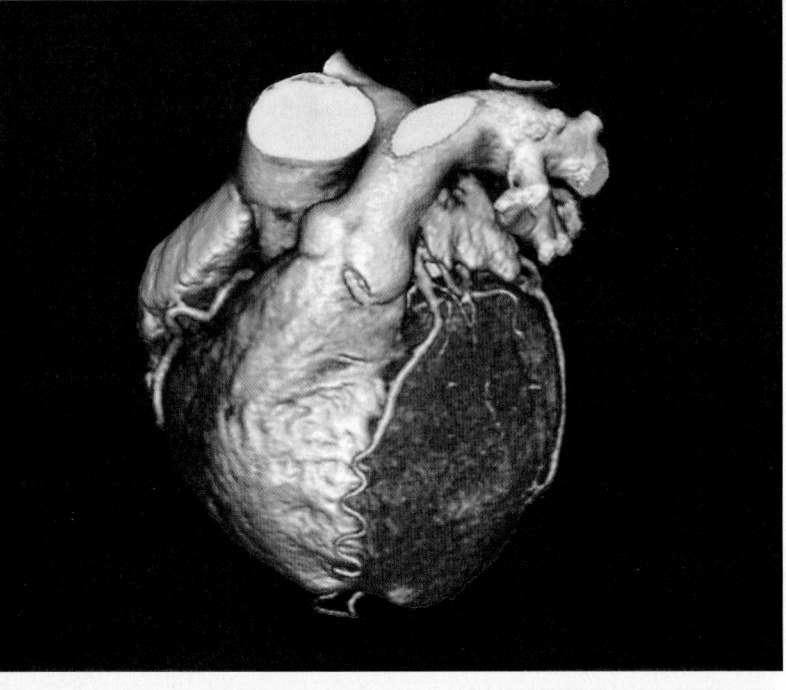

Zephyr/Science Source

Colored 3-D computed tomography (CT) scan of the human heart and some of its blood vessels. The large cream-colored structures at upper center above the heart are blood vessels that carry blood to and from the heart. The fine blood vessels on the surface of the chambers are coronary arteries, which supply the heart muscle itself with oxygen-rich blood and nutrients.

KEY CONCEPTS

44.1 Arthropods and most mollusks have an open circulatory system in which blood bathes the tissues directly; some invertebrates and all vertebrates have a closed circulatory system in which a heart pumps blood that flows through a continuous circuit of blood vessels.

44.2 Vertebrate blood consists of plasma, which transports nutrients, wastes, gases, and hormones; red blood cells, which are specialized to transport oxygen; white blood cells, which defend the body against disease; and platelets, which function in blood clotting.

44.3 Three main types of vertebrate blood vessels are arteries, which carry blood away from the heart; capillaries, which are exchange vessels; and veins, which carry blood back toward the heart.

44.4 During the evolution of terrestrial vertebrates, adaptations in circulatory system structures separated oxygen-rich blood from oxygen-poor blood.

44.5 In amniotes, the pacemaker in the sinoatrial node initiates each heartbeat; one complete heartbeat makes up a cardiac cycle.

44.6 Blood pressure depends on cardiac output, blood volume, and peripheral resistance.

44.7 Along with the evolution of lungs, double circulation evolved in vertebrates. The pulmonary circulation connects the heart and lungs, while the systemic circulation delivers oxygen-rich blood to all the tissues and organs of the body.

44.8 The vertebrate lymphatic system helps maintain fluid homeostasis by returning interstitial fluid to the blood and defends the body against disease organisms.

44.9 Several modifiable risk factors have been identified for cardiovascular disease, which is the leading cause of death in the world.

Most cells require a continuous supply of nutrients and oxygen and removal of waste products. In very small animals, these metabolic needs are met by simple **diffusion,** the net movement of particles from a region of higher concentration of that specific type of particle to a region of lower concentration as a result of random motion. A molecule can diffuse 1 micrometer (μm) in less than 1 millisecond (msec), so diffusion is adequate over microscopic distances. In invertebrates that are only a few cells thick, diffusion is an effective mechanism for distributing materials to and from the cells. Consequently, many small, aquatic invertebrates have no circulatory system. Fluid between the cells, called **interstitial fluid** or tissue fluid, bathes the cells and provides a medium for diffusion of oxygen, nutrients, and wastes.

The time required for diffusion increases with the square of the distance over which diffusion occurs. A cell that is 10 μm away from its oxygen (or nutrient) supply can receive oxygen by diffusion in about 50 msec, but a cell that is 1000 μm (1 mm) away from its oxygen supply would have to wait several minutes! At this distance, a cell could not survive if it had to depend on diffusion alone. The evolution of specialized **circulatory systems** allowed animals to increase in size and become many cells thick. A circulatory system reduces the distance that needed materials must diffuse. It transports oxygen, nutrients, hormones, and other materials to the interstitial fluid surrounding the cells and removes metabolic wastes. In most animals a circulatory system interacts with every organ system in the body.

The computed tomography (CT) scan shown in the opening photograph visualizes the human heart and several blood vessels. The human circulatory system, also known as the **cardiovascular system,** is the focus of extensive research because cardiovascular disease is the leading cause of death in the United States and throughout the world.

44.1 TYPES OF CIRCULATORY SYSTEMS

LEARNING OBJECTIVE

1 Contrast internal transport in animals with no circulatory system, animals with an open circulatory system, and animals with a closed circulatory system.

No specialized circulatory structures are present in sponges, cnidarians (hydras, jellyfish), ctenophores (comb jellies), flatworms, or nematodes (roundworms). In cnidarians the central gastrovascular cavity serves as a circulatory organ as well as a digestive organ (FIG. 44-1a). As the animal stretches and contracts, body movements stir up the contents of the gastrovascular cavity and help distribute nutrients.

The flattened body of the flatworm permits effective gas exchange by diffusion (FIG. 44-1b). Its branched gastrovascular cavity brings nutrients close to all the cells. As in cnidarians, circulation is aided by contractions of the body wall muscles, which move the fluid throughout the gastrovascular cavity. The branching excretory system of planarians provides for transport of metabolic wastes that are then expelled from the body.

Fluid in the body cavity of nematodes and other pseudocoelomate animals helps circulate materials. Nutrients, oxygen, and wastes dissolve in this fluid and diffuse through it to and from the cells. Body movements of the animal move the fluid and thus distribute these materials.

Larger animals require a circulatory system to distribute materials efficiently. A circulatory system typically has the following components: (1) **blood,** a connective tissue consisting of cells and cell fragments dispersed in fluid, usually called *plasma;* (2) a pumping organ, generally a **heart;** and (3) a system of **blood vessels** or spaces through which blood circulates. Two main types of circulatory systems are open and closed systems.

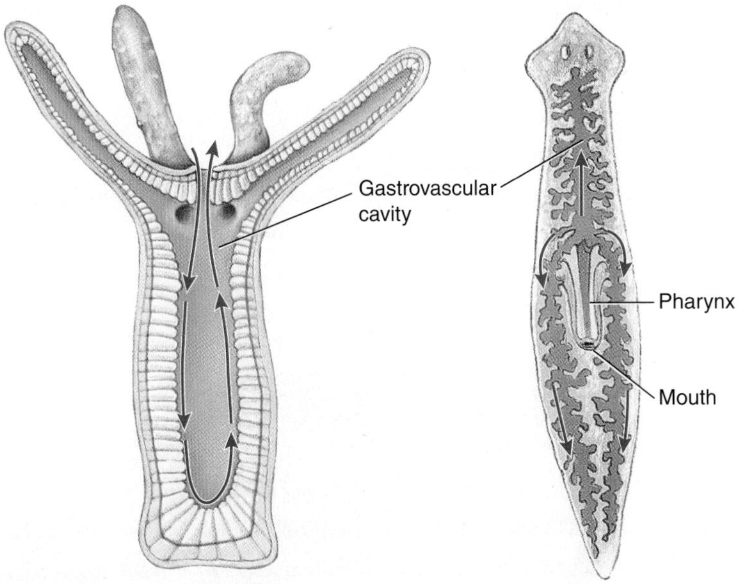

Many invertebrates have an open circulatory system

Arthropods and most mollusks have an **open circulatory system** in which the heart pumps blood into vessels that have open ends. Their blood and interstitial fluid are collectively referred to as **hemolymph.** This fluid spills out of the open ends of the blood vessels, filling large spaces called *sinuses.* The sinuses make up the **hemocoel** (blood cavity), which is not part of the coelom. (In arthropods and mollusks, the coelom is reduced.) The hemolymph bathes the cells of the body directly. Hemolymph re-enters the circulatory system through openings in the heart (in arthropods) or through open-ended vessels that lead to the gills (in mollusks).

In the open circulatory system of most mollusks, the heart has three chambers (FIG. 44-2a). The two *atria* receive hemolymph from the gills. The single *ventricle* then pumps the oxygen-rich hemolymph into blood vessels that conduct it into the large sinuses of the hemocoel. After bathing the body cells, the hemolymph passes into vessels that lead to the gills, where it is recharged with oxygen. The hemolymph then returns to the heart.

Some mollusks as well as arthropods have a hemolymph pigment, **hemocyanin,** containing copper that transports oxygen. When oxygenated, hemocyanin is blue and imparts a bluish color to the hemolymph of these animals (the original blue bloods!).

In arthropods a tubular heart pumps hemolymph into arteries, blood vessels that deliver it to the sinuses of the hemocoel (FIG. 44-2b). Hemolymph then circulates through the hemocoel, eventually returning to the pericardial cavity surrounding the heart. Hemolymph enters the heart through ostia, tiny openings equipped with valves that prevent backflow. The rate of hemolymph circulation increases when the animal moves. Thus, when an animal is active and most needs nutrients for fuel, its own movement ensures effective circulation.

In crayfish and other crustaceans, gas exchange takes place as hemolymph circulates through the gills. However, an open circulatory system cannot provide enough oxygen to maintain the active lifestyle of insects. Hemolymph

(a) In *Hydra* and other cnidarians, nutrients circulate through the gastrovascular cavity and come in contact with the inner layer of body cells. Nutrients diffuse the short distance to the outer layer of cells.

(b) In planarian flatworms, the branched gastrovascular cavity allows nutrients to come within close proximity of most body cells.

Figure labels: Gastrovascular cavity, Pharynx, Mouth

Figure 44-1 Invertebrates with no circulatory system

Red arrows indicate the path of nutrient circulation through the gastrovascular cavity.
© Cengage Learning

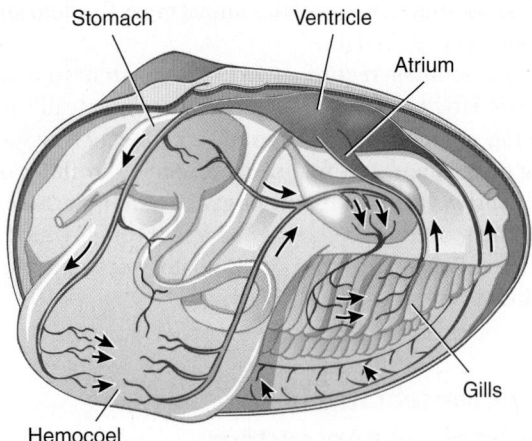

(a) In most mollusks, the heart pumps hemolymph into blood vessels that conduct it to the hemocoel. After bathing the cells, the hemolymph enters vessels that conduct it to the gills. The hemolymph is recharged with oxygen and then returned to the heart.

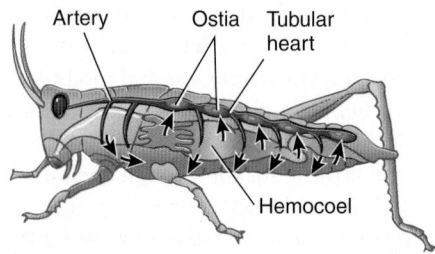

(b) In arthropods, a tubular heart pumps hemolymph into arteries that deliver it to the sinuses of the hemocoel. After circulating, hemolymph re-enters the heart through ostia in the heart wall.

Figure 44-2 *Animation* **Open circulatory systems**

In mollusks and arthropods, a heart pumps the blood into arteries that end in sinuses of the hemocoel. Hemolymph circulates through the hemocoel.

© Cengage Learning

mainly distributes nutrients and hormones in insects. Oxygen diffuses directly to the cells through a system of air tubes (tracheae) that make up the respiratory system (see Fig. 46-2).

Some invertebrates have a closed circulatory system

Annelids, some mollusks (cephalopods), and echinoderms have a **closed circulatory system** in which blood flows through a continuous circuit of blood vessels. The walls of the smallest blood vessels, the *capillaries,* are thin enough to permit diffusion of gases, nutrients, and wastes between blood in the vessels and the interstitial fluid that bathes the cells.

Proboscis worms (phylum Nemertea) have a rudimentary closed circulatory system. Their system consists of a complete network of blood vessels but no heart. Blood flow depends on movements of the animal and on contractions in the walls of large blood vessels.

Earthworms and other annelids have a closed circulatory system in which two main blood vessels extend lengthwise in the body (**FIG. 44-3**). The ventral vessel conducts blood posteriorly, and the dorsal vessel conducts blood anteriorly. Dorsal and ventral vessels are connected by lateral vessels in every segment. Branches of the lateral vessels deliver blood to the surface, where it is oxygenated. In the anterior part of the worm, five pairs of contractile blood vessels (sometimes called "hearts") connect dorsal and ventral vessels. Contractions of these paired vessels and of the dorsal vessel as well as contraction of the body wall muscles circulate the blood. Earthworms have **hemoglobin,** the same red pigment that transports oxygen in vertebrate blood. However, their hemoglobin is not contained within red blood cells; rather, it is dissolved in the blood plasma.

Although other mollusks have an open circulatory system, the fast-moving cephalopods, such as the squid and octopus, require a more efficient means of internal transport. They have a closed system made even more effective by accessory "hearts" at the base of the gills, which speed passage of blood through the gills.

Vertebrates have a closed circulatory system

The vertebrate circulatory system consists of heart, blood vessels, blood, lymph, lymph vessels, and associated organs such as the thymus, spleen, and liver. All vertebrates have a ventral, muscular heart that pumps blood into a closed system of blood vessels. **Capillaries,** the tiniest blood vessels, have very thin walls that permit exchange of materials between blood and interstitial fluid.

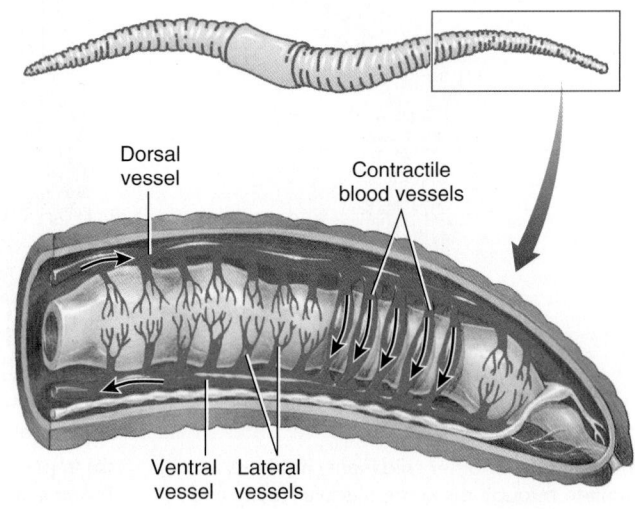

Figure 44-3 Closed circulatory system of the earthworm

Blood circulates through a continuous system of blood vessels. Five pairs of contractile blood vessels deliver blood from the dorsal vessel to the ventral vessel.

© Cengage Learning

The vertebrate circulatory system performs several functions:

1. Transports nutrients from the digestive system and from storage depots to each cell
2. Transports oxygen from respiratory structures (gills or lungs) to the cells
3. Transports metabolic wastes from each cell to organs that excrete them
4. Transports hormones from endocrine glands to target tissues
5. Helps maintain fluid balance
6. Helps distribute metabolic heat within the body, which helps maintain a constant body temperature in endothermic animals
7. Helps maintain appropriate pH
8. Defends the body against invading microorganisms

CHECKPOINT 44.1

- *Compare how oxygen is transported to the body cells in a hydra, flatworm, earthworm, insect, and frog.*
- *Contrast how an open circulatory system differs from a closed circulatory system.*
- *What are five functions of the vertebrate circulatory system?*

44.2 VERTEBRATE BLOOD

LEARNING OBJECTIVES

2 Compare the structure and function of plasma, red blood cells, white blood cells, and platelets.

3 Summarize the sequence of events involved in blood clotting.

In vertebrates blood consists of a pale yellowish fluid called **plasma** in which red blood cells, white blood cells, and platelets are suspended (**FIG. 44-4** and **TABLE 44-1**). In humans the total circulating blood volume is approximately 5 L (5.3 qt) in an adult female and about 5.5 L (5.8 qt) in an adult male. About 55% of the blood volume is plasma. The remaining 45% is made up of blood cells and platelets. Because cells and platelets are heavier than plasma, they can be separated from it by centrifugation. Plasma does not separate from blood cells in the body because the blood is constantly mixed as it circulates through the blood vessels.

Plasma is the fluid component of blood

Plasma consists of water (about 92%), proteins (about 7%), salts, and a variety of materials being transported, such as dissolved gases, nutrients, wastes, and hormones. Plasma is in dynamic equilibrium with the interstitial fluid bathing the cells and with the intracellular fluid. As blood passes through the capillaries, substances continuously move into and out of the plasma. Any deviation from its normal composition signals one or more organs to restore homeostasis.

Plasma contains several kinds of **plasma proteins,** each with specific properties and functions: **fibrinogen;** alpha, beta, and gamma **globulins;** and **albumin.** Fibrinogen is one of the proteins involved in the clotting process. When the proteins involved in blood clotting have been removed from the plasma, the remaining liquid is called **serum.**

Alpha globulins include certain hormones and proteins that transport hormones; prothrombin, a protein involved in blood clotting; and high-density lipoproteins (HDL), which transport fats and cholesterol. Beta globulins include other lipoproteins that transport fats and cholesterol as well as proteins that transport certain vitamins and minerals. The *gamma globulin* fraction of plasma contains many types of antibodies that provide immunity to diseases such as measles and infectious hepatitis. Purified human gamma globulin is sometimes used to treat certain diseases or to reduce the possibility of contracting a disease.

Plasma proteins, especially albumins and globulins, help regulate distribution of fluid between plasma and interstitial fluid. These proteins are too large to pass readily through the walls of blood vessels. They contribute to the blood's osmotic pressure, which helps maintain an appropriate blood volume. Plasma proteins (along with the hemoglobin in the red blood cells) are important acid–base buffers. They help maintain the pH of the blood within a narrow range, near its normal, slightly alkaline pH of 7.4.

Red blood cells transport oxygen

Erythrocytes, informally called **red blood cells (RBCs),** are highly specialized for transporting oxygen. In most vertebrates except mammals, circulating RBCs have nuclei. For example, birds have large, oval, nucleated RBCs. In mammals the nucleus is ejected from the RBC as the cell develops. Each mammalian RBC is a flexible, biconcave disc 7 to 8 μm in diameter and 1 to 2 μm thick. An internal elastic framework maintains the disc shape and permits the cell to bend and twist as it passes through blood vessels even smaller than its own diameter. Its biconcave shape provides a high ratio of surface area to volume, allowing efficient diffusion of oxygen and carbon dioxide into and out of the cell. In an adult human, about 30 trillion RBCs circulate in the blood, approximately 5 million per μL.

Erythrocytes are produced within the red bone marrow of certain bones: vertebrae, ribs, breastbone, skull bones, and long bones. As an RBC develops, it produces large quantities of **hemoglobin,** the oxygen-transporting pigment that gives vertebrate blood its red color. (Oxygen transport is discussed in Chapter 46.) The lifespan of a human RBC is about 120 days. As blood circulates through the liver and spleen, phagocytic cells remove worn-out RBCs from the circulation. These RBCs are then disassembled, and some of their components are recycled. In the human body, more than 2.4 million RBCs are destroyed every second! As you might imagine, an equal number must be produced in the bone marrow to replace them. Red blood cell production is regulated by the hormone **erythropoietin** (eh-rith"-row-poy'-ih-tin), which the kidneys release in response to a decrease in oxygen.

Anemia is a deficiency in hemoglobin (often accompanied by a decrease in the number of RBCs). When hemoglobin is

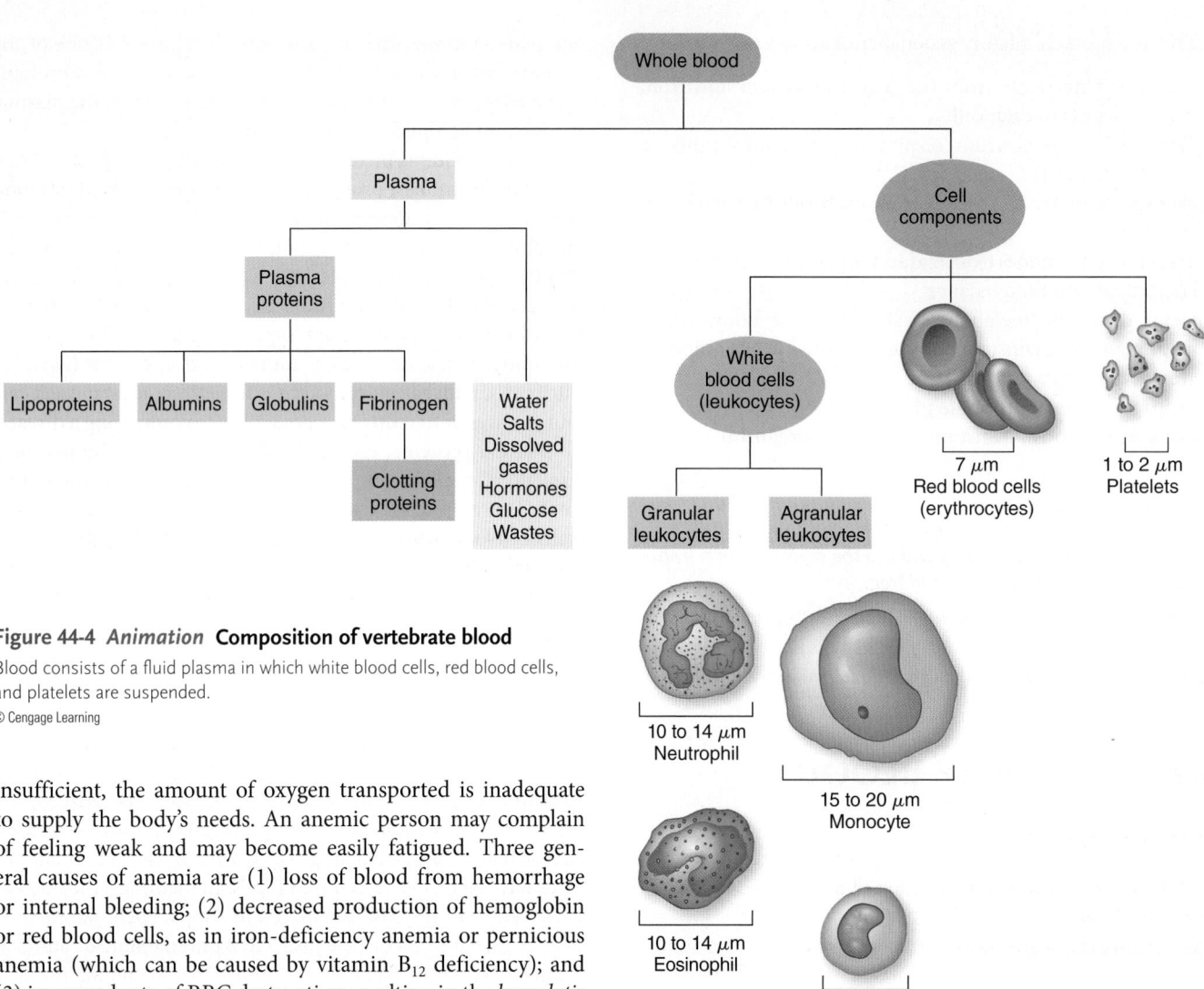

Figure 44-4 *Animation* **Composition of vertebrate blood**
Blood consists of a fluid plasma in which white blood cells, red blood cells, and platelets are suspended.
© Cengage Learning

insufficient, the amount of oxygen transported is inadequate to supply the body's needs. An anemic person may complain of feeling weak and may become easily fatigued. Three general causes of anemia are (1) loss of blood from hemorrhage or internal bleeding; (2) decreased production of hemoglobin or red blood cells, as in iron-deficiency anemia or pernicious anemia (which can be caused by vitamin B_{12} deficiency); and (3) increased rate of RBC destruction resulting in the *hemolytic anemias,* such as sickle cell anemia (see Chapter 16).

White blood cells defend the body against disease organisms

The **leukocytes,** or **white blood cells (WBCs),** are specialized to defend the body against harmful bacteria and other microorganisms. Leukocytes are amoeba-like cells capable of independent movement. Some types routinely slip through the walls of blood vessels and enter the tissues. Human blood contains three kinds of granular leukocytes and two types of agranular leukocytes (see Fig. 44-4). Both types are manufactured in the red bone marrow.

The *granular leukocytes* are characterized by large, lobed nuclei and distinctive granules in their cytoplasm. The three varieties of granular leukocytes are neutrophils, eosinophils, and basophils. **Neutrophils,** the principal phagocytic cells in the blood, are especially adept at seeking out and ingesting bacteria. They also phagocytose dead cells, a cleanup task that is especially demanding after injury or infection. Most granules in neutrophils contain enzymes that digest ingested material.

Eosinophils have large granules that stain bright red with eosin, an acidic dye. Eosinophils increase in number during allergic reactions and during parasitic (e.g., tapeworm) infestations. The lysosomes of these WBCs contain enzymes such as peroxidases that degrade the cell membranes of parasitic worms and protozoa. These substances can also injure normal tissue. Eosiniphils also produce substances that promote an *inflammatory response* (*inflammation*), one of the body's key responses to injury or infection (discussed in Chapter 45).

Basophils exhibit deep blue granules when stained with basic dyes. Like eosinophils, these cells play a role in allergic reactions. Basophils do not contain lysosomes. Granules in their cytoplasm contain **histamine,** a substance that dilates blood vessels and makes capillaries more permeable. Basophils release histamine in injured tissues and in allergic responses. Other

TABLE 44-1 | Cell Components of Blood

	NORMAL RANGE	FUNCTION	PATHOLOGY
RED BLOOD CELLS (RBCs)	Male: 4.2–5.4 million/μL Female: 3.6–5.0 million/μL	Oxygen transport; carbon dioxide transport	Too few: anemia Too many: polycythemia
PLATELETS	150,000–400,000/μL	Essential for clotting	Clotting malfunctions; bleeding; easy bruising
WHITE BLOOD CELLS (WBCs)	5000–10,000/μL		
Neutrophils	About 60% of WBCs	Phagocytosis	Too many: may be due to bacterial infection, inflammation, myelogenous leukemia
Eosinophils	1%–3% of WBCs	Play role in allergic responses; release proteins that are toxic to parasitic worms; release compounds important in immune responses	Too many: may result from allergic reaction, parasitic infestation
Basophils	1% of WBCs	May play role in prevention of inappropriate clotting; release histamine, which is important in inflammatory responses	
Lymphocytes	25%–35% of WBCs	Produce antibodies; destroy foreign cells	Atypical lymphocytes present in infectious mononucleosis; too many may be due to lymphocytic leukemia, certain viral infections
Monocytes	6% of WBCs	Can differentiate to form macrophages and dendritic cells	May increase in monocytic leukemia and fungal infections

© Cengage Learning

basophil granules contain *heparin,* which speeds up fat removal from the blood after you eat a meal high in fat. Heparin is an anticoagulant that may help prevent blood from clotting inappropriately within the blood vessels.

Agranular leukocytes lack large, distinctive granules, and their nuclei are rounded or kidney-shaped. Two types of agranular leukocytes are lymphocytes and monocytes. Some **lymphocytes** are specialized to produce antibodies, whereas others directly attack foreign invaders, such as bacteria or viruses (discussed in Chapter 45).

Monocytes are the largest WBCs, reaching 20 μm in diameter. During infection, monocytes migrate from the blood into the tissues. They phagocytose cells and remove toxic molecules. Monocytes can also differentiate into macrophages or dendritic cells. **Macrophages** are giant scavenger cells that voraciously engulf bacteria, dead cells, and debris. **Dendritic cells** are also important cells of the immune system. For example, some dendritic cells produce compounds that help the body fight viral infection. We discuss macrophages and dendritic cells further in Chapter 45.

Human blood normally has about 7000 WBCs per μL of blood (only 1 for every 700 RBCs). During bacterial infections, the number may rise sharply, so a WBC count is useful in diagnosis. The proportion of each kind of WBC is determined by a differential WBC count. Table 44-1 shows a normal distribution of WBCs.

Leukemia is a form of blood cancer in which any one of the various kinds of WBCs multiplies rapidly within the bone marrow. Many of these cells do not mature, and their large numbers crowd out developing RBCs and platelets, leading to anemia and impaired clotting. A common cause of death from leukemia is internal hemorrhaging, especially in the brain. Another frequent cause of death is infection; although the WBC count may rise dramatically, the cells are immature and abnormal, and they cannot defend the body against disease organisms.

Thanks to new treatments, survival rates for individuals with leukemia and other blood cancers have recently increased sharply, particularly for children. For example, the survival rate for children younger than five years old with acute lymphocytic leukemia is now higher than 90%.

Platelets function in blood clotting

In most vertebrates other than mammals, the blood contains small, oval, nucleated cells called **thrombocytes** that function in blood clotting. Mammals have **platelets,** tiny spherical or disc-shaped bits of cytoplasm that lack nuclei. About 300,000 platelets per μL are present in human blood. Platelets are pinched off from very large cells in the bone marrow. Thus, a platelet is not a whole cell but rather is a fragment of cytoplasm enclosed by a membrane.

Platelets function in blood clotting and may also stimulate the immune system. When a blood vessel is cut, it constricts to reduce blood loss. Platelets physically patch the break by sticking to the rough, cut edges of the vessel. As platelets begin to gather, they release substances that attract other platelets. The platelets become sticky and adhere to collagen fibers in the blood vessel wall. Within about 5 minutes after injury, they form a platelet plug, or temporary clot.

At the same time that the temporary clot forms, a stronger, more permanent clot begins to develop. More than 30 chemical substances interact in this complex process. The series of reactions that leads to clotting is triggered when one of the clotting factors in the blood is activated by contact with the injured tissue. In **hemophilia** one of the clotting factors is absent as a result of an inherited genetic mutation (see Chapter 16).

Prothrombin, a plasma protein manufactured in the liver, requires vitamin K for its production. In the presence of clotting

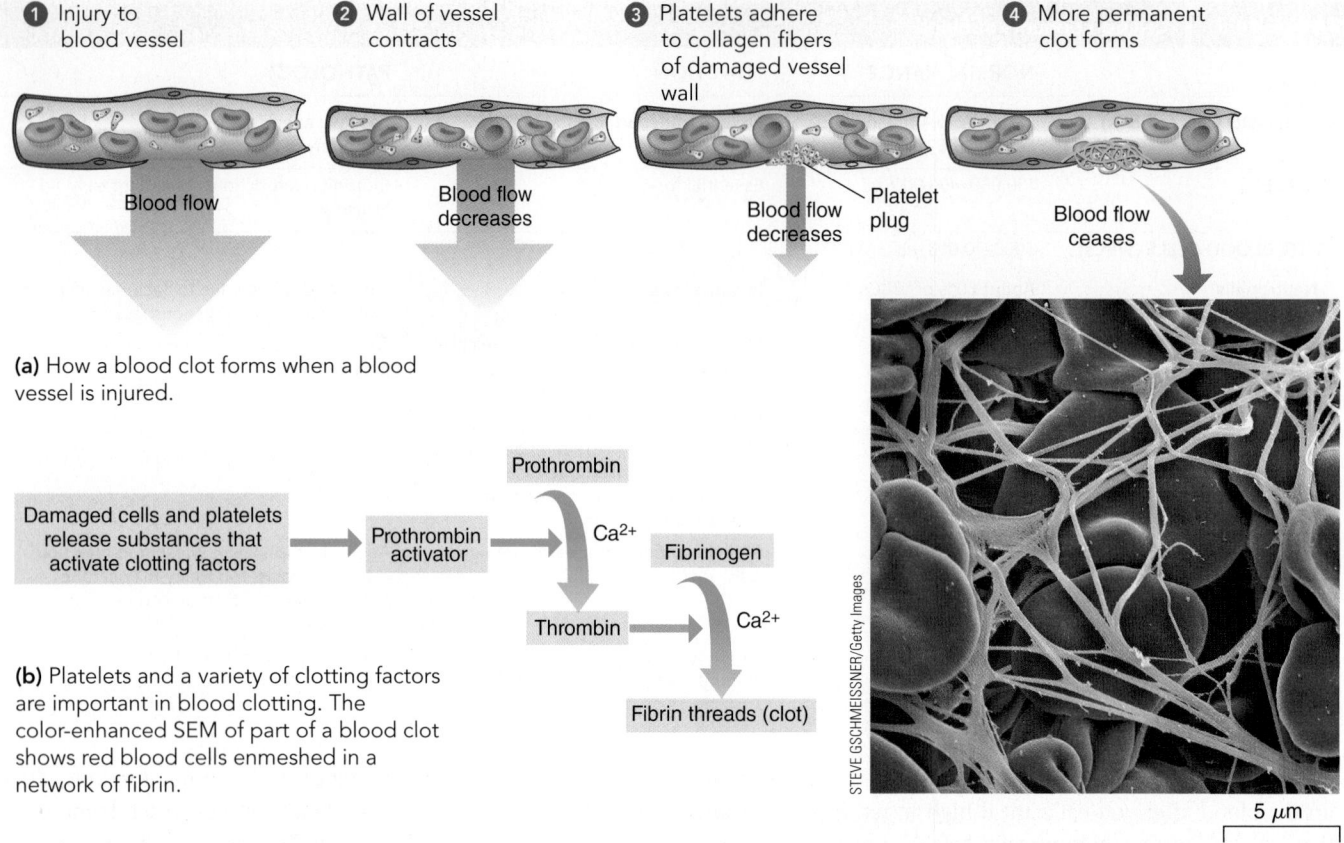

① Injury to blood vessel

② Wall of vessel contracts

③ Platelets adhere to collagen fibers of damaged vessel wall

④ More permanent clot forms

Blood flow

Blood flow decreases

Blood flow decreases — Platelet plug

Blood flow ceases

(a) How a blood clot forms when a blood vessel is injured.

Damaged cells and platelets release substances that activate clotting factors → Prothrombin activator → Prothrombin + Ca^{2+} → Thrombin + Ca^{2+} + Fibrinogen → Fibrin threads (clot)

(b) Platelets and a variety of clotting factors are important in blood clotting. The color-enhanced SEM of part of a blood clot shows red blood cells enmeshed in a network of fibrin.

STEVE GSCHMEISSNER/Getty Images

5 μm

Figure 44-5 *Animation* **Blood clotting**
© Cengage Learning

factors, calcium ions, and compounds released from platelets, prothrombin is converted to *thrombin*. Then thrombin catalyzes the conversion of the soluble plasma protein *fibrinogen* to an insoluble protein, **fibrin**. Once formed, fibrin polymerizes, producing long threads that stick to the damaged surface of the blood vessel and form the webbing of the clot. These threads trap blood cells and platelets, which help strengthen the clot. The clotting process is summarized in FIGURE 44-5.

CHECKPOINT 44.2

- *What are the functions of albumins and globulins?*
- *Contrast the functions of red blood cells and neutrophils.*
- PREDICT *What would happen if there were a shortage of fibrinogen in the blood?*

44.3 VERTEBRATE BLOOD VESSELS

LEARNING OBJECTIVE

4 Compare the structure and function of different types of blood vessels, including arteries, arterioles, capillaries, and veins.

The vertebrate circulatory system includes three main types of blood vessels: arteries, capillaries, and veins (FIG. 44-6). An **artery** carries blood away from a heart chamber toward other tissues. When an artery enters an organ, it divides into many smaller branches called **arterioles.** The arterioles deliver blood into the microscopic **capillaries.** After blood circulates through capillary networks within tissues, capillaries merge to form **veins** that carry the blood back toward the heart.

The wall of an artery or vein has three layers (see Fig. 44-6b). The innermost layer, which lines the blood vessel, consists mainly of **endothelium,** a tissue that resembles squamous epithelium (see Chapter 39). The middle layer is made up of connective tissue and smooth muscle cells, and the outer coat consists of connective tissue rich in elastic and collagen fibers.

Gases and nutrients cannot pass through the thick walls of arteries and veins. Materials are exchanged between the blood and interstitial fluid bathing the cells through the capillary walls, which are only one cell thick. Capillary networks in the body are so extensive that at least one of these tiny vessels is located close to every cell in the body. The total length of all capillaries in the body has been estimated to be more than 96,000 km!

Smooth muscle in the arteriole wall can constrict, a process called **vasoconstriction** or relax, a process called **vasodilation.** Vasoconstriction and vasodilation change the internal diameter of the arteriole. Such changes help maintain

The structure of the wall of each type of blood vessel is adapted for the particular function of the vessel; for example, materials are exchanged through the thin walls of capillaries.

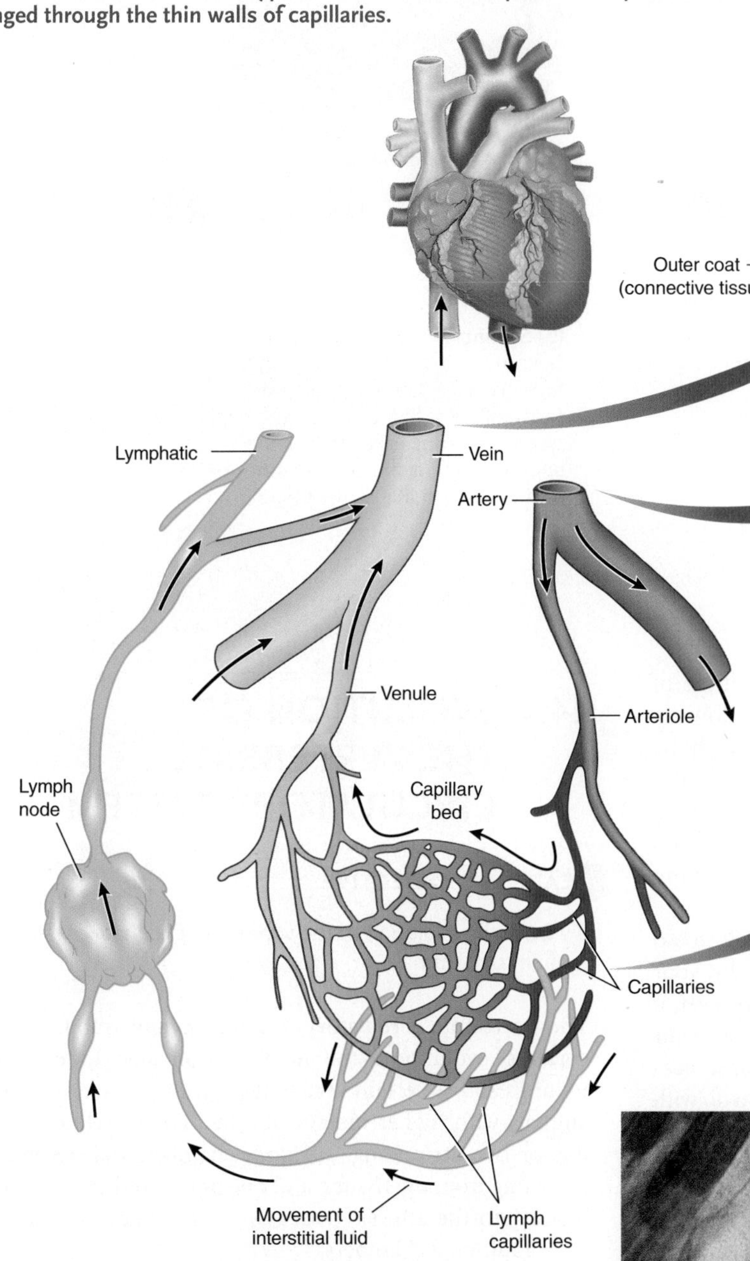

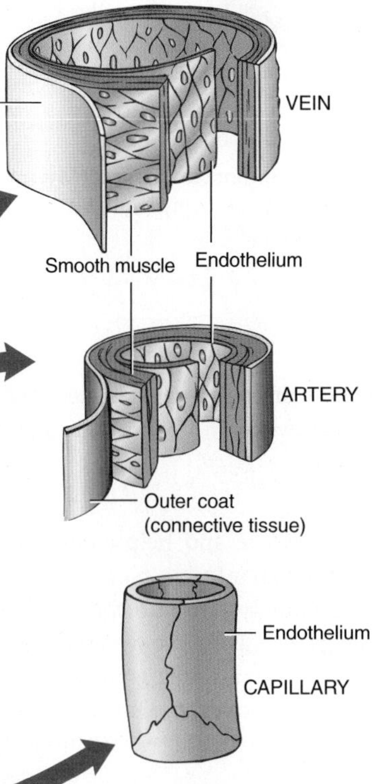

(b) Comparison of the walls of an artery, vein, and capillary. All three vessels are lined with endothelium. The capillary wall is only one cell thick, allowing exchange of materials.

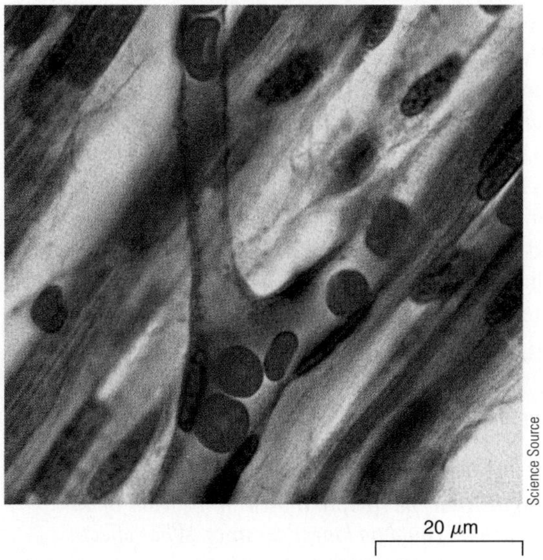

(a) The heart pumps blood into arteries. Blood then flows through arterioles, capillaries, and veins, which return it to the heart. Some plasma leaves the capillaries and becomes interstitial fluid. Lymphatic vessels return excess interstitial fluid to the blood through ducts that lead into large veins in the shoulder region. Blood vessels with oxygen-rich blood are shown in *red*. Those with oxygen-poor blood are shown in *blue*.

(c) LM of red blood cells passing through capillaries almost in single file.

Figure 44-6 *Animation* **Blood and lymphatic vessels**

CONNECT Note that lymph capillaries have dead ends. In what way is the structure of these capillaries connected with their function?

© Cengage Learning

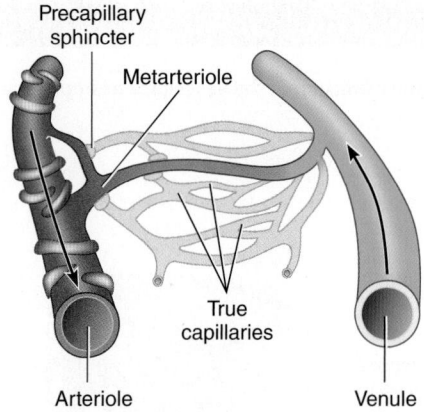

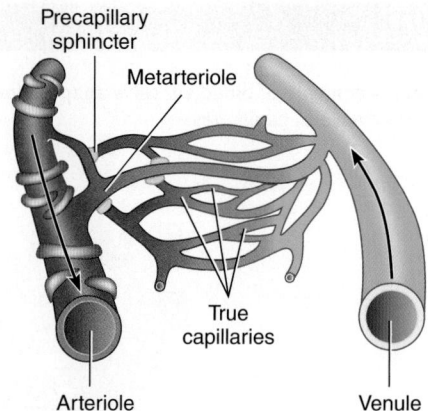

(a) Sphincters closed. When a tissue is inactive, only its metarterioles are open.

(b) Sphincters open. When the tissue becomes active, decreased oxygen tension in the tissue relaxes the precapillary sphincters, and the capillaries open. This process increases the blood supply and thus the delivery of nutrients and oxygen to the active tissue.

Figure 44-7 Blood flow through a capillary network

As a tissue becomes active, the pattern of blood flow through its capillary networks changes.

© Cengage Learning

appropriate blood pressure and help control the volume of blood passing to a particular tissue. Changes in blood flow are regulated by the nervous system in response to the metabolic needs of the tissue as well as by the demands of the body as a whole. For example, when a tissue is metabolizing rapidly, it needs more nutrients and oxygen. During exercise, arterioles within skeletal muscles dilate, increasing blood flow by more than tenfold.

If all your blood vessels dilated at the same time, you would not have enough blood to fill them completely. Normally, your liver, kidneys, and brain receive most of the blood. However, if an emergency suddenly occurred requiring rapid action, your blood would be rerouted quickly in favor of the heart and muscles. At such a time, the digestive system and kidneys can do with less blood because they are not crucial in responding to the crisis.

The small vessels that directly link arterioles with venules (small veins) are **metarterioles.** The true capillaries branch off from the metarterioles and then rejoin them (FIG. 44-7). True capillaries also interconnect with one another. Wherever a capillary branches from a metarteriole, a smooth muscle cell serves as a *precapillary sphincter* that opens and closes continuously, directing blood first to one and then to another section of tissue. These precapillary sphincters (along with the smooth muscle in the walls of arteries and arterioles) regulate the blood supply to each organ and its subdivisions.

CHECKPOINT 44.3

- *Compare the functions of arteries, capillaries, and veins.*

- PREDICT *Imagine that the smooth muscle in the walls of an individual's arterioles could no longer constrict. What effect would that have on homeostasis?*

44.4 EVOLUTION OF THE VERTEBRATE CIRCULATORY SYSTEM

LEARNING OBJECTIVE

5 Trace the evolution of the vertebrate circulatory system from fish to mammal.

The vertebrate circulatory system became modified in the course of evolution as the site of gas exchange shifted from gills to lungs and as certain vertebrates became active, endothermic animals with higher metabolic rates. The vertebrate heart has one or two **atria** (sing., *atrium*), chambers that receive blood returning from the tissues, and one or two **ventricles** that pump blood into the arteries (FIG. 44-8). Vertebrates in some groups have additional chambers.

The fish heart has one atrium and one ventricle. The atrium pumps blood into the ventricle, which pumps blood into a single circuit of blood vessels. Blood is oxygenated as it passes through capillaries in the gills. After blood circulates through the gill capillaries, its pressure is low, so blood passes very slowly to the other organs. The fish's swimming movements facilitate circulation. Blood returning to the heart has a low oxygen content. A thin-walled *sinus venosus* receives blood returning from the tissues and pumps it into the atrium.

In amphibians blood flows through a double circuit: the **pulmonary circulation** and the **systemic circulation.** Oxygen-rich blood and oxygen-poor blood are kept somewhat separate. The amphibian heart has two atria and one

The four-chambered heart and double circuit that separate oxygen-rich from oxygen-poor blood are important adaptations that evolved as vertebrates diversified and some became active, terrestrial, endothermic animals.

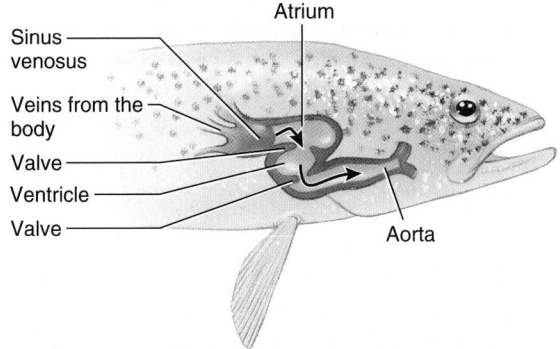

(a) Fishes. The single atrium and ventricle of the fish heart are part of a single circuit of blood flow.

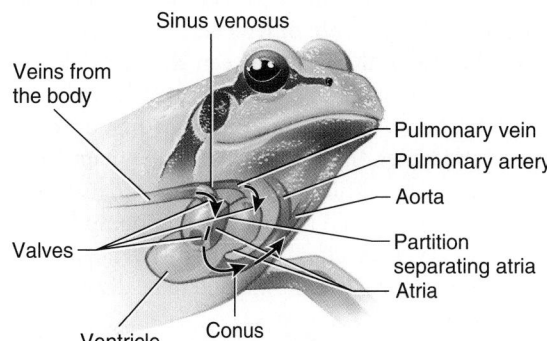

(b) Amphibians. In amphibians, the heart consists of two atria and one ventricle; blood flows through a double circuit, but oxygen-rich blood mixes with oxygen-poor blood.

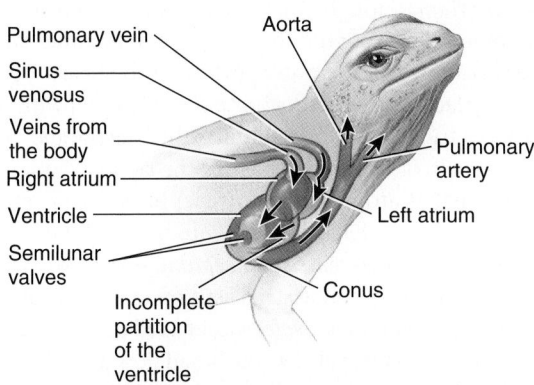

(c) Reptiles (except birds). The reptilian heart has two atria and two ventricles. In most nonavian reptiles, the wall separating the ventricles is incomplete, so blood from the right and left chambers mixes to some extent. In crocodiles and alligators, the septum is complete, and the heart consists of four separate chambers.

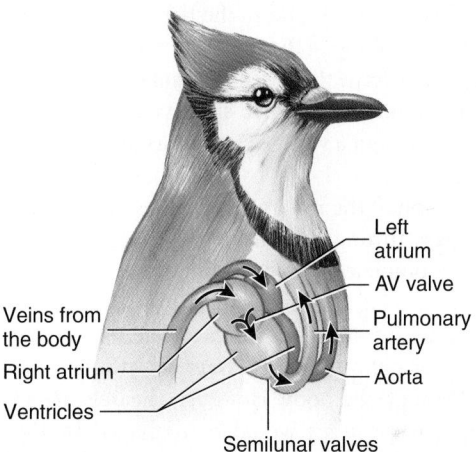

(d) Birds and mammals. These animals have two atria and two ventricles. The interventricular septum is complete; oxygen-rich blood is separated from oxygen-poor blood.

Figure 44-8 *Animation* **Evolution of the vertebrate circulatory system**

CONNECT How does the separation of oxygen-rich blood and oxygen-poor blood in birds and mammals affect endothermy?

© Cengage Learning

ventricle. A sinus venosus collects oxygen-poor blood returning from the veins and pumps it into the right atrium. Blood returning from the lungs passes directly into the left atrium.

Both atria pump into the single ventricle, but oxygen-poor blood is pumped out of the ventricle before oxygen-rich blood enters it. Blood passes into an artery, the *conus arteriosus,*

equipped with a fold that helps keep the blood separate. Much of the oxygen-poor blood is directed into the pulmonary circulation, which delivers it to the lungs and skin, where it is recharged with oxygen. The systemic circulation delivers oxygen-rich blood into arteries that conduct it to the various tissues of the body.

Most nonavian reptiles also have a double circuit of blood flow, made more efficient by a wall that partly divides the ventricles. (In this discussion, birds are considered separately from other reptiles.) Mixing of oxygen-rich and oxygen-poor blood is minimized by the timing of contractions of the left and right sides of the heart and by pressure differences. In crocodilians (crocodiles and alligators), the wall between the ventricles is complete, so the heart consists of two separate atria and two separate ventricles. Thus, a four-chambered heart first evolved among the nonavian reptiles. In contrast to birds and mammals, amphibians and nonavian reptiles do not ventilate their lungs continuously. Therefore, it would be inefficient to pump blood through the pulmonary circulation continuously. The shunts between the two sides of the heart allow blood to be distributed to the lungs as needed.

In all birds and mammals (and in crocodilians), the *septum* (wall) between the ventricles is complete. Biologists hypothesize that the completely divided heart evolved twice during the course of vertebrate evolution, first in the crocodilian–bird clade and then independently in mammals. The septum between the ventricles prevents oxygen-rich blood in the left chamber from mixing with oxygen-poor blood in the right chamber. The conus arteriosus has split and become the base of the **aorta** (the largest artery) and of the pulmonary artery. No sinus venosus is present as a separate chamber, but a vestige remains as the *sinoatrial node* (the pacemaker).

Complete separation of the right and left sides of the heart requires blood to pass through the heart twice each time it tours the body. The complete double circuit allows birds and mammals to maintain relatively high blood pressures in the systemic circulation and modest pressures in the pulmonary circulation. The higher pressure is necessary for efficient circulation of blood through the body. However, the delicate air sacs and capillaries of the lungs would be damaged by this pressure.

The double circulatory circuit delivers materials to the tissues rapidly and efficiently. Because their blood contains more oxygen per unit volume and circulates more rapidly than in other vertebrates, the tissues of birds and mammals receive more oxygen. As a result, these animals can maintain a higher metabolic rate than other animals and a constant high body temperature even in cold surroundings.

The pattern of blood circulation in birds and mammals can be summarized as follows:

veins (conduct blood from organs) ⟶ right atrium ⟶ right ventricle ⟶ pulmonary arteries ⟶ capillaries in the lungs ⟶ pulmonary veins ⟶ left atrium ⟶ left ventricle ⟶ aorta ⟶ arteries (conduct blood to organs) ⟶ arterioles ⟶ capillaries ⟶ veins

CHECKPOINT 44.4

- *What are some of the major adaptations that occurred during the evolution of the vertebrate circulatory system?*

44.5 THE HUMAN HEART

LEARNING OBJECTIVES

6 Describe how the structure and function of the human heart are inter-related; include the heart's conduction system in your answer.

7 Trace the events of the cardiac cycle and relate normal heart sounds to these events.

8 Define *cardiac output,* describe how it is regulated, and identify factors that affect it.

Not much bigger than a fist and weighing less than a pound, the human heart is a remarkable organ. It beats about 2.5 billion times during an average lifetime, pumping about 300 million L (80 million gal) of blood. To meet the body's changing needs, the heart can vary its output from 5 L to more than 20 L of blood per minute.

Your heart is a hollow, muscular organ located in the chest cavity directly under the breastbone. Its wall consists mainly of cardiac muscle attached to a framework of collagen fibers. The *pericardium,* a tough connective tissue sac, encloses the heart. A smooth layer of endothelium covers the inner surface of the pericardium and the outer surface of the heart. Between these two surfaces is a small *pericardial cavity* filled with fluid, which reduces friction to a minimum as the heart beats.

A wall, or **septum,** separates the right atrium and ventricle from the left atrium and ventricle. Between the atria, the wall is known as the *interatrial septum;* between the ventricles, it is known as the *interventricular septum.* A shallow depression, the *fossa ovalis,* on the interatrial septum marks the place where an opening, the *foramen ovale,* was located in the fetal heart. In the fetus the foramen ovale lets the blood flow directly from the right to the left atrium, so very little passes to the nonfunctional lungs. A small muscular pouch, called an *auricle,* lies at the upper surface of each atrium.

To keep blood from flowing backward, the heart has valves that close automatically (FIG. 44-9). The valve between the right atrium and right ventricle is called the *right* **atrioventricular (AV) valve,** or **tricuspid valve.** The *left atrioventricular valve* (between the left atrium and left ventricle) is the **mitral valve,** or *bicuspid valve.* The AV valves are held in place by stout cords, or "heartstrings," the *chordae tendineae.* These cords attach the valves to the papillary muscles that project from the walls of the ventricles.

When blood returning from the tissues fills the atria, blood pressure on the AV valves forces them to open into the ventricles, filling the ventricles with blood. As the ventricles contract, blood is forced back against the AV valves, pushing them closed. Contraction of the papillary muscles and tensing of the chordae tendineae prevent the valves from opening backward into the atria. These valves are like swinging doors that open in only one direction.

Semilunar valves (named for their flaps, which are shaped like half-moons) guard the exits from the heart. The semilunar valve between the left ventricle and the aorta is the *aortic*

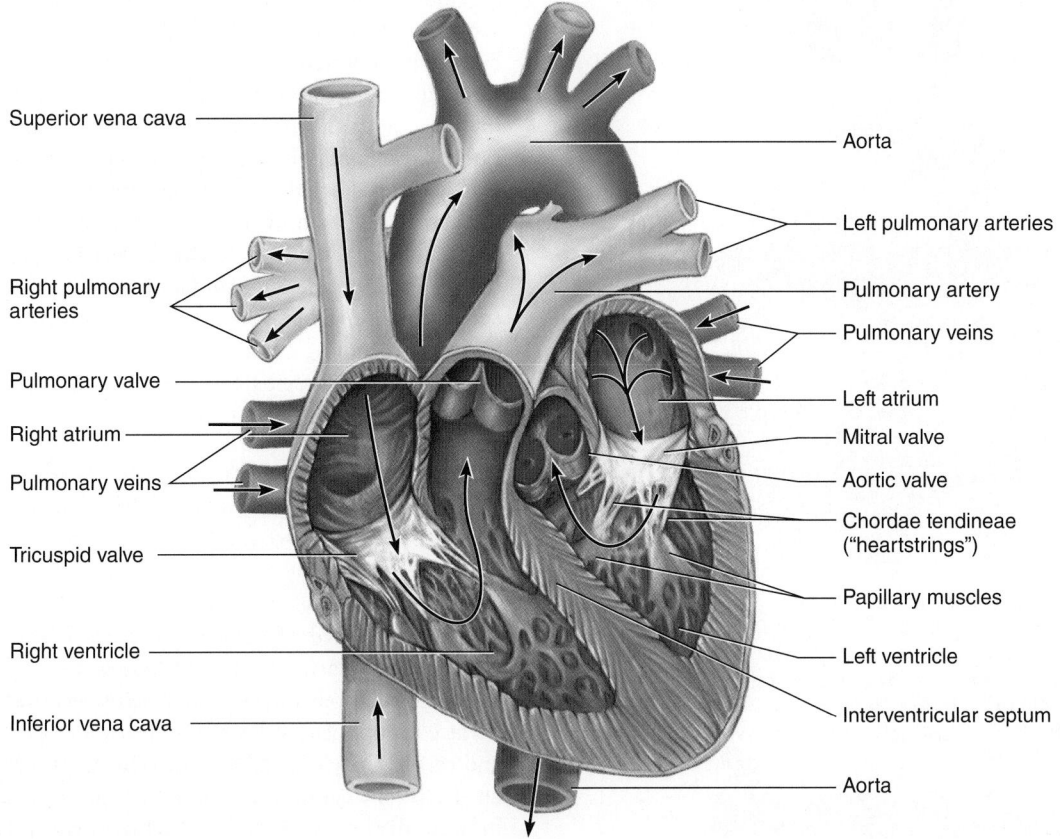

Superior vena cava

Right pulmonary arteries

Pulmonary valve

Right atrium

Pulmonary veins

Tricuspid valve

Right ventricle

Inferior vena cava

Aorta

Left pulmonary arteries

Pulmonary artery

Pulmonary veins

Left atrium

Mitral valve

Aortic valve

Chordae tendineae ("heartstrings")

Papillary muscles

Left ventricle

Interventricular septum

Aorta

Figure 44-9 *Animation* **Section through the human heart showing the valves**

Note the right and left atria, which receive blood, and the right and left ventricles, which pump blood into the arteries. *Arrows* indicate the direction of blood flow.

© Cengage Learning

valve, and the one between the right ventricle and the pulmonary artery is the *pulmonary valve.* When blood passes out of the ventricles, the flaps of the semilunar valves are pushed aside and offer no resistance to blood flow. When the ventricles are relaxing and filling with blood from the atria, however, blood pressure in the arteries is higher than in the ventricles. Blood then fills the pouches of the valves, stretching them across the artery so that blood cannot flow back into the ventricle.

Each heartbeat is initiated by a pacemaker

Horror films sometimes feature a scene in which a heart cut out of a human body continues to beat. Scriptwriters of these tales actually have some factual basis for their gruesome fantasies because a heart carefully removed from the body can continue to beat for many hours if kept in a nutritive, oxygenated fluid. This is possible because the contractions of cardiac muscle begin within the muscle itself and can occur independently of any nerve supply (**FIG. 44-10a**).

At their ends cardiac muscle cells are joined by dense bands called **intercalated discs** (in-ter′-kuh-lay′-ted) (**FIGS. 44-10b** and c). These complex junctions contain *gap junctions.* Recall from Chapter 5 that in a gap junction, two cells connect through pores. Gap junctions are of great physiological importance

because they offer very little resistance to the passage of an action potential. Ions move easily through the gap junctions, allowing the entire atrial (or ventricular) muscle mass to contract as one giant cell.

Compared with skeletal muscle that has action potentials typically lasting 1 to 2 milliseconds (msec), the action potentials of cardiac muscle are much longer, several hundred msec. Voltage-activated Ca^{2+} channels open during depolarization of cardiac muscle fibers. Entry of Ca^{2+} contributes to the longer depolarization time. Another factor is a type of K^+ channel that stays open when the cell is at its resting potential but closes during depolarization, lengthening depolarization time by decreasing the permeability of the membrane to K^+. From experiments on isolated cardiac muscle fibers, investigators determined that spontaneous contraction results from the combination of a slow decrease in K^+ permeability and a slow increase in Na^+ and Ca^{2+} permeability.

A specialized conduction system ensures that the heart beats in a regular and effective rhythm. Each beat is initiated by the **pacemaker,** a small mass of specialized cardiac muscle fibers in the **sinoatrial (SA) node.** The SA node is located in the posterior wall of the right atrium near the opening of a large vein, the superior vena cava. The action potential in the SA node is triggered mainly by the opening of Ca^{2+} channels. Ends of the

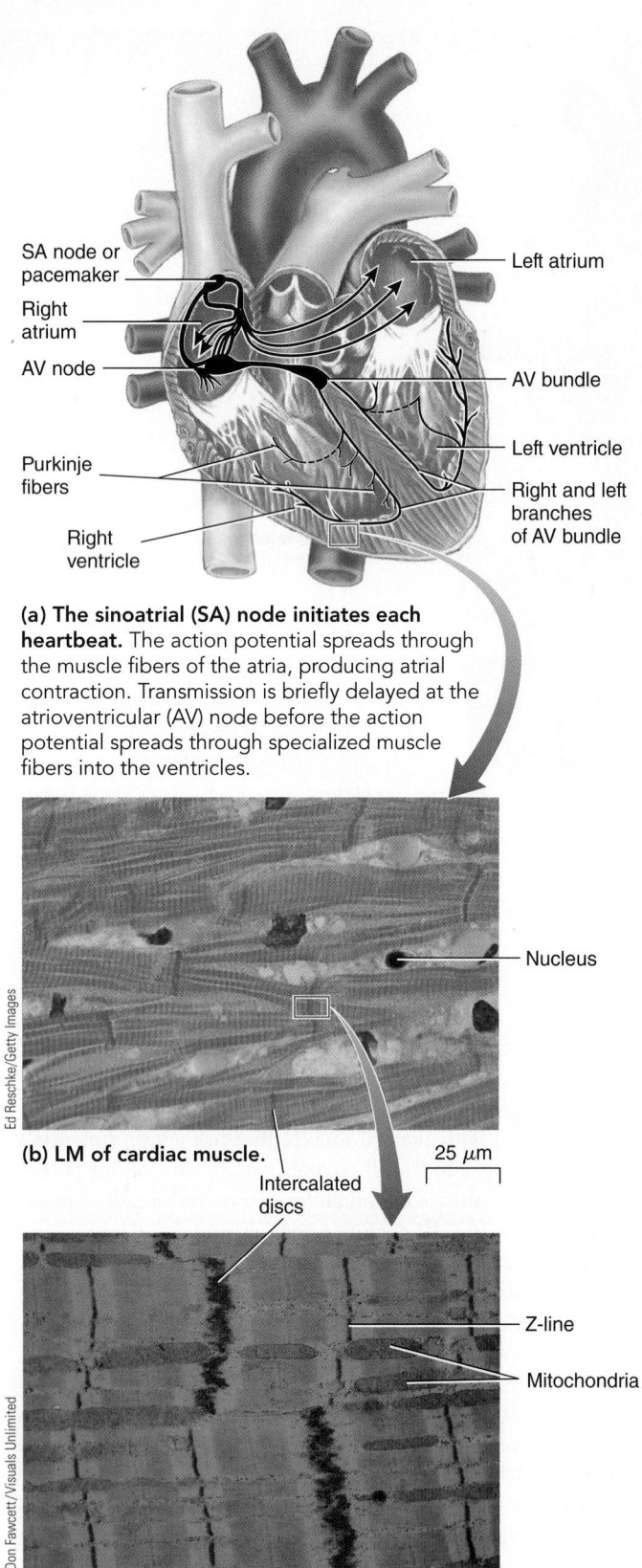

(a) The sinoatrial (SA) node initiates each heartbeat. The action potential spreads through the muscle fibers of the atria, producing atrial contraction. Transmission is briefly delayed at the atrioventricular (AV) node before the action potential spreads through specialized muscle fibers into the ventricles.

(b) LM of cardiac muscle.

25 μm

(c) TEM of cardiac muscle.

1 μm

Figure 44-10 *Animation* **Conduction system of the heart**

When action potentials are initiated by the pacemaker cells in the SA node, gap junctions within the intercalated discs allow the impulses to spread easily from one cardiac muscle fiber to the next within the atria (and within the ventricles). The atria (and then the ventricles) contract as a functional unit.

© Cengage Learning

SA node fibers fuse with ordinary atrial muscle fibers, so each action potential spreads through the cardiac muscle fibers of both atria, producing atrial contraction.

One group of atrial muscle fibers conducts the action potential directly to the **atrioventricular (AV) node,** located in the right atrium along the lower part of the septum. Here transmission is delayed briefly so that the atria finish contracting before the ventricles begin to contract. From the AV node, the action potential spreads into specialized muscle fibers that make up the *AV bundle.* The AV bundle divides, sending branches into each ventricle. Fibers of the bundle branches divide further, eventually forming small *Purkinje fibers.* These fibers conduct impulses to the muscle fibers of both ventricles.

In Summary,

SA node ⟶ atrial muscle fibers (atria contract) ⟶ AV node ⟶ AV bundle ⟶ right and left bundle branches ⟶ Purkinje fibers ⟶ conduct impulses to muscle fibers of both ventricles ⟶ ventricles contract

As each wave of contraction spreads through the heart, electric currents flow into the tissues surrounding the heart and onto the body surface. By placing electrodes on the body surface on opposite sides of the heart, a physician can amplify and record the electrical activity. The graph produced is called an *electrocardiogram* (*EKG* or *ECG*). Abnormalities in the EKG indicate disorders in the heart or its rhythm. For example, in *heart block* impulse transmission is delayed or blocked at some point in the conduction system. Artificial pacemakers can help patients with severe heart block. The pacemaker is implanted beneath the skin, and its electrodes are connected to the heart. This device provides continuous rhythmic impulses that avoid the block and drive the heartbeat.

The cardiac cycle consists of alternating periods of contraction and relaxation

Each minute the heart beats about 70 times. One complete heartbeat takes about 0.8 second and is referred to as a **cardiac cycle.** That portion of the cycle in which contraction occurs is known as **systole;** the period of relaxation is **diastole.** FIGURE 44-11 shows the sequence of events that occur during one cardiac cycle.

When you listen to the heartbeat with a stethoscope, you can hear two main heart sounds, "lub-dup," which repeat rhythmically. These sounds result from the heart valves closing. When the valves close, they cause turbulence in blood flow that sets up vibrations in the walls of the heart chambers. The first heart sound, "lub," is low pitched, not very loud, and fairly long lasting. It is caused mainly by the closing of the AV (mitral and tricuspid) valves and marks the beginning of ventricular systole. The "lub" sound is quickly followed by the higher-pitched, louder, sharper, and shorter "dup" sound. Heard almost as a quick snap, the "dup" marks the closing of the semilunar valves and the beginning of ventricular diastole.

The quality of these sounds tells a discerning physician much about the state of the valves. For example, when a valve does not close tightly, blood may flow backward. When the

The rhythmic changes in electrical activity of the heart are responsible for the cardiac cycle, the pattern of contraction and relaxation that takes place during each heartbeat.

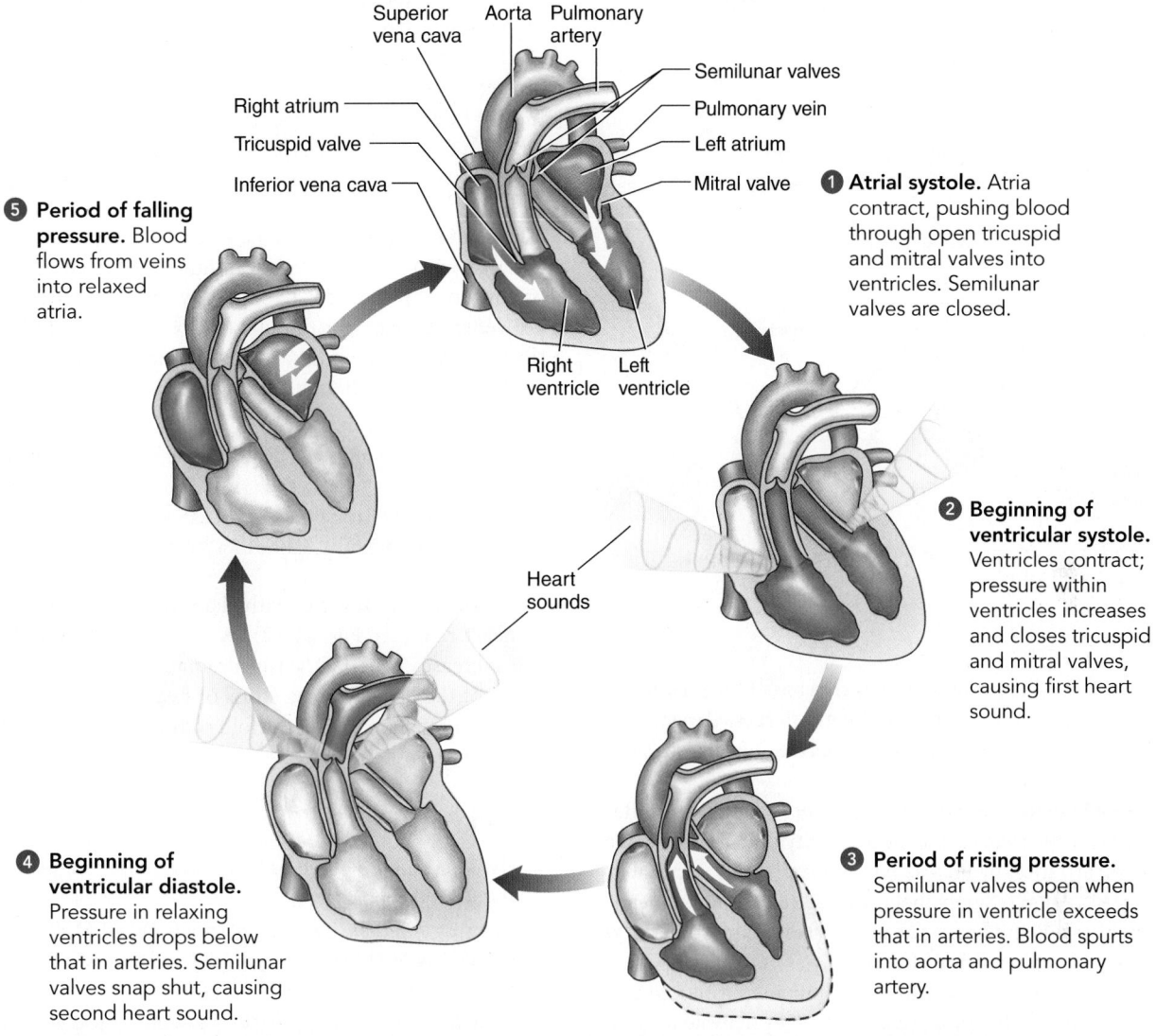

① Atrial systole. Atria contract, pushing blood through open tricuspid and mitral valves into ventricles. Semilunar valves are closed.

② Beginning of ventricular systole. Ventricles contract; pressure within ventricles increases and closes tricuspid and mitral valves, causing first heart sound.

③ Period of rising pressure. Semilunar valves open when pressure in ventricle exceeds that in arteries. Blood spurts into aorta and pulmonary artery.

④ Beginning of ventricular diastole. Pressure in relaxing ventricles drops below that in arteries. Semilunar valves snap shut, causing second heart sound.

⑤ Period of falling pressure. Blood flows from veins into relaxed atria.

Superior vena cava, Aorta, Pulmonary artery, Right atrium, Tricuspid valve, Inferior vena cava, Semilunar valves, Pulmonary vein, Left atrium, Mitral valve, Right ventricle, Left ventricle, Heart sounds

Figure 44-11 *Animation* The cardiac cycle
The cycle comprises contraction of both atria followed by both ventricles. *White arrows* indicate the direction of blood flow; *dotted lines* indicate the change in size as contraction occurs.
© Cengage Learning

PREDICT How is the cardiac cycle affected when the SA node increases its rate of firing?

semilunar valves are injured, a soft, hissing noise ("lub-shhh") known as a *heart murmur* is heard in place of the normal sound. Valve deformities are sometimes present at birth, or they may result from certain diseases, such as rheumatic fever and syphilis. Diseased or deformed valves can be surgically repaired or replaced with artificial valves.

You can measure your heart rate by placing a finger over the radial artery in your wrist or the carotid artery in your neck and counting the pulsations for 1 minute. **Arterial pulse** is the alternate expansion and recoil of an artery. Each time the left ventricle pumps blood into the aorta, the elastic wall of the aorta expands to accommodate the blood. This expansion moves in a wave down the aorta and the arteries that branch from it. When the pressure wave passes, the elastic arterial wall snaps back to its normal size.

The nervous system regulates heart rate

Although the heart can beat independently, its rate is, in fact, carefully regulated by the nervous and endocrine systems (**FIG. 44-12**). Sensory receptors in the walls of certain blood vessels and heart chambers are sensitive to changes in blood pressure. When stimulated, the receptors send messages to *cardiac centers* in the medulla of the brain. These cardiac centers

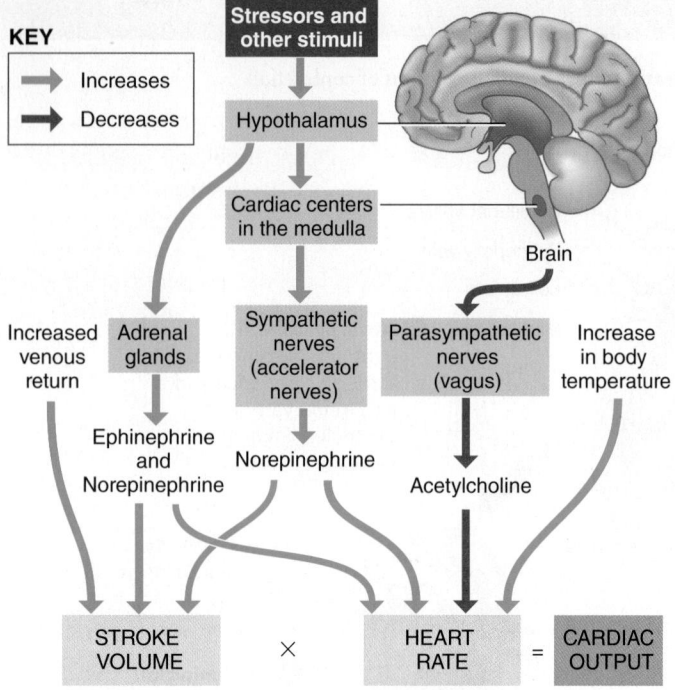

KEY

→ Increases

→ Decreases

Stressors and other stimuli

Hypothalamus

Cardiac centers in the medulla

Brain

Increased venous return

Adrenal glands

Sympathetic nerves (accelerator nerves)

Parasympathetic nerves (vagus)

Increase in body temperature

Ephinephrine and Norepinephrine

Norepinephrine

Acetylcholine

STROKE VOLUME × HEART RATE = CARDIAC OUTPUT

Figure 44-12 Some factors that influence cardiac output
© Cengage Learning

govern two sets of autonomic nerves that pass to the SA node: parasympathetic nerves and sympathetic nerves. Parasympathetic and sympathetic nerves have opposite effects on heart rate (see Fig. 42-17).

Parasympathetic nerves release the neurotransmitter *acetylcholine,* which slows the heart. Acetylcholine slows the rate of depolarization by increasing the membrane's permeability to K$^+$ (**FIG. 44-13a**). Sympathetic nerves release norepinephrine, which speeds the heart rate and increases the strength of contraction. Norepinephrine stimulates Ca^{2+} channel opening during depolarization (**FIG. 44-13b**). Both norepinephrine and acetylcholine act indirectly on ion channels. They activate a *signal transduction* process involving a G protein (see Chapter 6).

Norepinephrine binds to *beta-adrenergic receptors,* one of the two main types of adrenergic receptors. These receptors are targeted by *beta blockers,* drugs that block the actions of norepinephrine on the heart. Beta blockers are used clinically in treating hypertension (high blood pressure) and other types of heart disease.

In response to physical and emotional stressors, the adrenal glands release epinephrine and norepinephrine, which speed the heart. An elevated body temperature also increases heart rate. During fever, the heart may beat more than 100 times per minute. When body temperature is lowered, heart rate decreases.

Stroke volume depends on venous return

The volume of blood one ventricle pumps during one beat is the **stroke volume.** Stroke volume depends mainly on **venous return,** the amount of blood the veins deliver to the heart. According to *Starling's law of the heart,* if the veins deliver more blood to the heart, the heart pumps more blood into the arteries (within physiological limits). Increased venous return stretches the cardiac muscle fibers more, and they contract with greater force, increasing stroke volume. Norepinephrine released by sympathetic nerves and epinephrine released by the adrenal glands during stress also increase the force of contraction of cardiac muscle fibers.

Cardiac output varies with the body's need

The **cardiac output (CO)** is the volume of blood pumped by the left ventricle into the aorta in 1 minute. We can compute the CO by multiplying the stroke volume by the number of times the left ventricle beats per minute. For example, in a resting adult, the heart may beat about 72 times per minute and pump about 70 mL of blood with each contraction.

CO = stroke volume × heart rate (number of ventricular contractions per minute)
= 70 mL/stroke × 72 strokes/min
= 5040 mL/min (about 5 L/min)

Cardiac output varies with changes in either stroke volume or heart rate (see Fig. 44-12). When stroke volume increases, CO increases. The CO varies dramatically with the changing needs of the body. During stress or heavy exercise, the normal heart can increase its CO fivefold so that it pumps 20 to 30 L of blood per minute.

CHECKPOINT 44.5

- List the sequence of events that occurs during cardiac conduction.

- PREDICT A patient's mitral valve (left AV valve) does not close completely at the beginning of ventricular systole. What effect would this have on the cardiac cycle?

- Imagine that you are feeling very anxious as you walk to the podium to deliver a speech. If your heart rate increases to 100 beats per minute and your stroke volume is 100 mL per stroke, what is your cardiac output?

44.6 BLOOD PRESSURE

LEARNING OBJECTIVE

9 Identify factors that determine and regulate blood pressure and compare blood pressure in different types of blood vessels.

Blood pressure is the force exerted by the blood against the inner walls of the blood vessels. It is determined by CO, blood volume, and resistance to blood flow (**FIG. 44-14a**). When CO increases, the increased blood flow causes a rise in blood pressure. When CO decreases, the decreased blood flow causes

① Parasympathetic neuron releases acetylcholine.

② Acetylcholine binds with receptors on plasma membrane of cardiac muscle.

③ Receptor activates G protein.

④ G protein binds with K⁺ channel, opening it.

⑤ K⁺ leaves the cell, hyperpolarizing the membrane. Action potentials occur more slowly. Heart rate decreases.

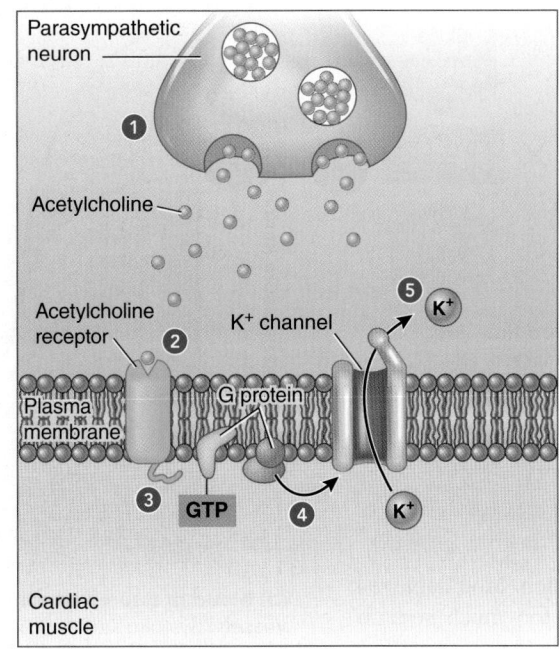

(a) Parasympathetic action on cardiac muscle

① Sympathetic neuron releases norepinephrine.

② Norepinephrine binds with receptors on the plasma membrane of cardiac muscle.

③ Receptor activates G protein.

④ G protein activates adenylyl cyclase, which converts ATP to cyclic AMP (cAMP).

⑤ Cyclic AMP activates protein kinase.

⑥ Protein kinase phosphorylates Ca²⁺ channels so that they open more easily when neuron is depolarized. Action potentials occur more rapidly. Heart rate increases.

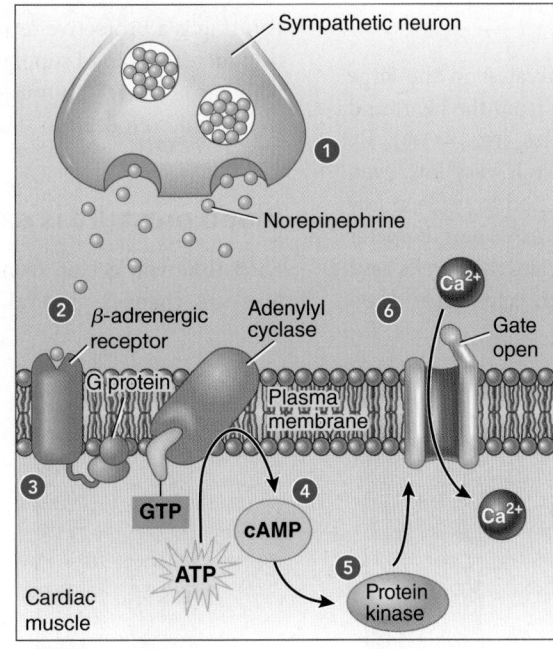

(b) Sympathetic action on cardiac muscle

Figure 44-13 Actions of parasympathetic and sympathetic neurons on cardiac muscle

Parasympathetic neurons release acetylcholine, which slows the heart rate; sympathetic neurons release norepinephrine, which increases both the heart rate and the strength of contraction. Each of these neurotransmitters works through a G protein and initiates a signal transduction process.
© Cengage Learning

a fall in blood pressure. If the volume of blood is reduced by hemorrhage or chronic bleeding, the blood pressure drops. In contrast, an increase in blood volume results in an increase in blood pressure. For example, a high dietary intake of salt causes water retention, which increases blood volume and raises blood pressure.

Blood flow is impeded by resistance; when the resistance to flow increases, blood pressure rises. *Peripheral resistance* is the resistance to blood flow caused by blood viscosity and by friction between blood and the blood vessel wall. In the blood of a healthy person, viscosity remains fairly constant and is only a minor influence on changes in blood pressure. More important is the friction between blood and the blood vessel wall. The length and diameter of a blood vessel determine the amount of surface area in contact with the blood. Blood vessel length does not change, but the diameter, especially of an arteriole, does. A small change in blood vessel diameter causes a big change in blood pressure. Constriction of blood vessels raises blood pressure; dilation lowers blood pressure.

Blood pressure in arteries rises during systole and falls during diastole. The National Institutes of Health defines *normal blood pressure* as a systolic pressure of less than 120 and a diastolic pressure of less than 80. An example of a normal blood pressure, measured in the upper arm with a sphygmomanometer, is 110/73 mm of mercury, abbreviated mm Hg. Systolic pressure is indicated by the first number, diastolic by the second.

If you have a blood pressure of 120 to 139 systolic over 80 to 89 diastolic, you are considered *prehypertensive,* and you need to modify your lifestyle to prevent cardiovascular disease and stroke. Lifestyle changes that will reduce your risk include exercising, losing excess weight, following a heart-healthy diet, reducing salt intake, limiting alcohol, and not smoking.

If your systolic pressure consistently measures 140 mm Hg or higher or your diastolic pressure measures 90 mm Hg or higher, you have **hypertension,** or high blood pressure. Hypertension is a risk factor for atherosclerosis and other cardiovascular disease (discussed later in this chapter).

In hypertension there is usually increased vascular resistance, especially in the arterioles and small arteries. The heart's workload increases because it must pump against this greater resistance. If this condition persists, the left ventricle enlarges and may deteriorate in function. Heredity, aging, race, and

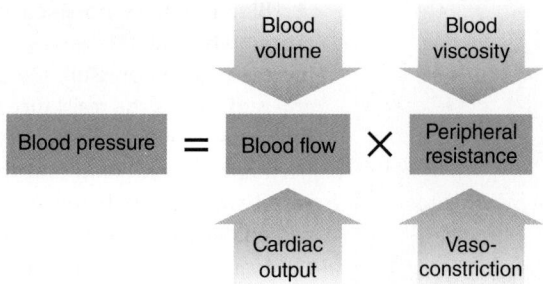

(a) Blood pressure depends on blood flow and resistance to that flow. A variety of factors affect blood flow and resistance.

Figure 44-14 Blood pressure

Area in graph refers to the surface area of blood vessels in contact with blood.
© Cengage Learning

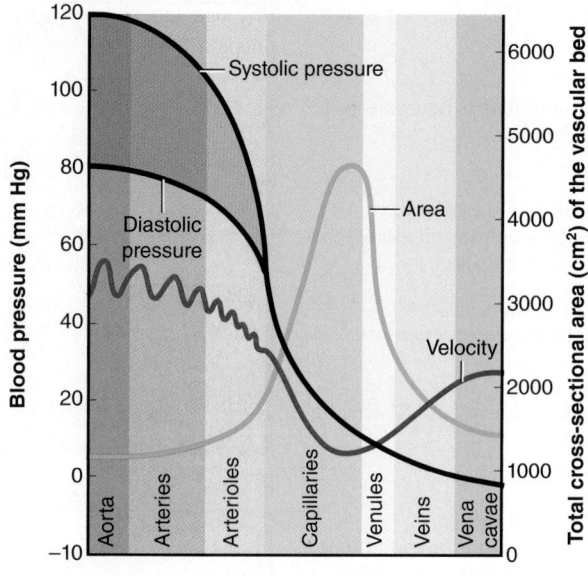

(b) Blood pressure varies in different types of blood vessels. Systolic and diastolic variations in arterial blood pressures are shown. Note that the venous pressure drops below zero (below atmospheric pressure) near the heart.

ethnicity can increase the risk of hypertension. Modifiable risk factors include being overweight, eating too much sodium, smoking, and drinking too much alcohol.

Blood pressure varies in different blood vessels

As you might predict, blood pressure is greatest in the large arteries and decreases as blood flows away from the heart and through the smaller arteries and capillaries (FIG. 44-14b). By the time blood enters the veins, its pressure is very low, even approaching zero. Flow rate can be maintained in veins at low pressure because they are low-resistance vessels. Their diameter is larger than that of corresponding arteries, and their walls have little smooth muscle. Flow of blood through veins depends on several factors, including skeletal muscle movement, which compresses veins. Most veins larger than 2 mm (0.08 in) in diameter that conduct blood against the force of gravity are equipped with *valves* to prevent backflow. Such valves usually consist of two cusps formed by inward extensions of the vein wall (FIG. 44-15).

When a person stands perfectly still for a long time, as when a soldier stands at attention or a store clerk stands at a cash register, blood tends to pool in the veins. When fully distended, veins can accept no more blood from the capillaries. Pressure increases in the capillaries, and large amounts of plasma are forced out of the circulation through the thin capillary walls. Within just a few minutes, as much as 20% of the blood volume can be lost from the circulation, with drastic effect. Arterial blood pressure falls dramatically, reducing blood flow to the brain. Sometimes the resulting lack of oxygen in the brain causes the person to faint.

Fainting is a protective response because lying in a prone position increases blood supply to the brain. In fact, lifting a person who has fainted to an upright position can result in circulatory shock and even death.

Blood pressure is carefully regulated

Each time you get up from a horizontal position, your blood pressure changes. Several mechanisms interact to maintain

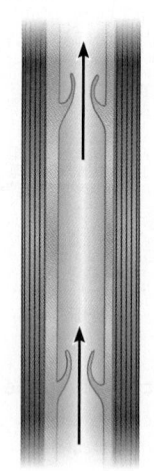

(a) Resting condition.

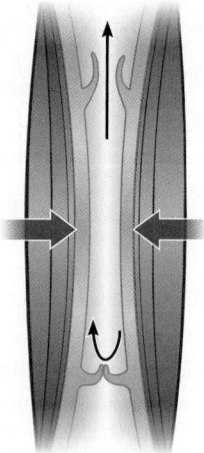

(b) Muscles contract. Muscles bulge, compressing veins and forcing blood toward the heart. The lower valve prevents backflow.

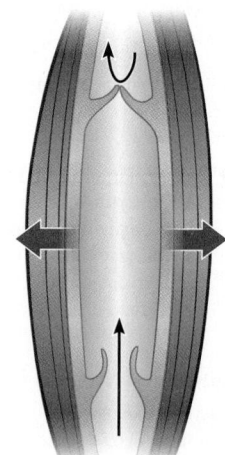

(c) Muscles relax. The vein expands and fills with blood from below. The upper valve prevents backflow.

Figure 44-15 Venous blood flow

Contraction of skeletal muscles helps move blood through the veins.
© Cengage Learning

normal blood pressure so that you do not faint when you get out of bed each morning or change position during the day. When blood pressure decreases, sympathetic nerves to the blood vessels stimulate vasoconstriction, causing the pressure to rise again.

Baroreceptors, specialized receptors in the walls of certain arteries and in the heart wall, are sensitive to changes in blood pressure. When an increase in blood pressure stretches the baroreceptors, messages are sent to the cardiac and vasomotor centers in the medulla of the brain. The cardiac center stimulates parasympathetic nerves that slow the heart, lowering blood pressure. The vasomotor center inhibits sympathetic nerves that constrict arterioles; this action causes vasodilation, which also lowers blood pressure. These neural reflexes continuously work in a complementary way to maintain blood pressure within normal limits.

Several hormones are also involved in regulating blood pressure (discussed in more detail in Chapters 48 and 49). In response to low blood pressure, the kidneys release **renin,** which activates the *renin–angiotensin–aldosterone pathway*. Renin acts on a plasma protein (angiotensinogen), triggering a cascade of reactions that produces the hormone **angiotensin II,** a powerful vasoconstrictor. Vasoconstriction *increases* blood pressure, restoring homeostasis. Angiotensin II also acts indirectly to maintain blood pressure by increasing the synthesis and release of the hormone **aldosterone** by the adrenal glands. Aldosterone increases retention of Na+ by the kidneys, resulting in greater fluid retention and increased blood volume.

When the body becomes dehydrated, the osmotic concentration of the blood increases. In response, the posterior lobe of the pituitary gland releases **antidiuretic hormone (ADH).** This hormone increases reabsorption of water in the kidneys (and only a small volume of concentrated urine is produced). Blood volume increases, raising blood pressure and restoring homeostasis.

When blood volume increases, the atria of the heart release a hormone called **atrial natriuretic peptide (ANP).** This hormone increases sodium excretion. As a result, a large volume of dilute urine is produced, and blood pressure *decreases*. Nitric oxide also helps regulate blood pressure by causing vasodilation, thus decreasing blood pressure.

CHECKPOINT 44.6

- *How does constriction of arterioles affect blood pressure?*
- *How does blood pressure in arteries compare with blood pressure in veins?*

44.7 THE PATTERN OF CIRCULATION

LEARNING OBJECTIVE

10 Trace a drop of blood through the pulmonary and systemic circulations, naming in sequence each structure through which it passes.

The right side of the heart receives oxygen-poor blood returning from the tissues and pumps it into the pulmonary circulation. The left side of the heart receives oxygen-rich blood from the lungs and pumps it into the systemic circulation.

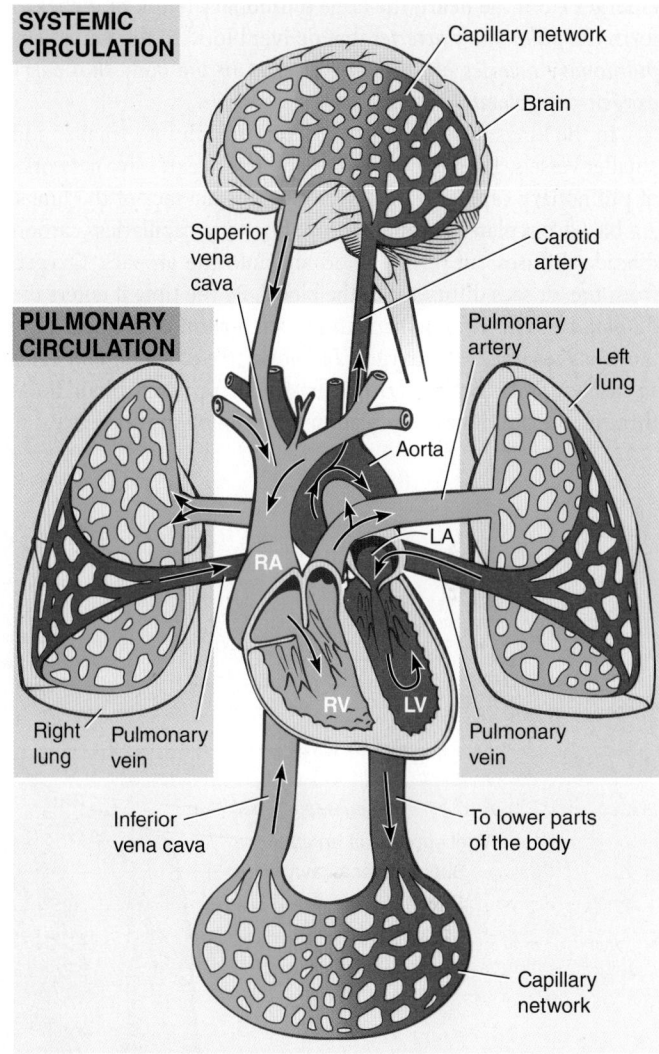

Figure 44-16 *Animation* **Systemic and pulmonary circulation**

In this highly simplified diagram, *red* represents oxygen-rich blood, and *blue* represents oxygen-poor blood. The *blue screen* highlights systemic circulation, and the *red screen* highlights pulmonary circulation.

CONNECT Although not evident in the view depicted in the illustration, the chambers of the left and right ventricles are the same size and pump equal volumes of blood. However, the muscle of the left ventricle is much thicker than that of the right ventricle. How does this difference relate to their functions?

© Cengage Learning

Most vertebrates other than fishes have a double circuit of blood vessels: (1) the **pulmonary circulation** connects the heart and lungs; and (2) the **systemic circulation** connects the heart with all the body tissues. You can trace this general pattern of circulation in **FIGURE 44-16.**

The pulmonary circulation oxygenates the blood

Blood from the tissues returns to the right atrium of the heart. This oxygen-poor blood, loaded with carbon dioxide, is pumped by the right ventricle into the pulmonary circulation. As it emerges from the heart, the large pulmonary trunk branches to form the *pulmonary arteries* that deliver blood to the lungs. *The pulmonary arteries are the only arteries in the body that carry oxygen-poor blood.*

In the lungs the pulmonary arteries branch into smaller and smaller vessels. These blood vessels give rise to extensive networks of pulmonary capillaries that surround the air sacs of the lungs. As blood circulates through the pulmonary capillaries, carbon dioxide diffuses out of the blood and into the air sacs. Oxygen from the air sacs diffuses into the blood. By the time it enters the *pulmonary veins* leading back to the left atrium of the heart, the blood is recharged with oxygen. *Pulmonary veins are the only veins in the body that carry blood rich in oxygen.* In summary, blood flows through the pulmonary circulation in the following sequence:

right atrium ⟶ right ventricle ⟶ pulmonary arteries ⟶ pulmonary capillaries (in lungs) ⟶ pulmonary veins ⟶ left atrium

The systemic circulation delivers blood to the tissues

Blood entering the systemic circulation is pumped by the left ventricle into the **aorta**, the largest artery. Arteries that branch off from the aorta conduct blood to all regions of the body. Some of the principal branches include the *coronary arteries* to the heart wall itself, the *carotid arteries* to the brain, the *subclavian arteries* to the shoulder region, the *mesenteric artery* to the intestine, the *renal arteries* to the kidneys, and the *iliac arteries* to the legs (**FIG. 44-17**). Each of these arteries gives rise to smaller and smaller vessels, somewhat like branches of a tree that divide until they form tiny twigs. Eventually, blood flows into capillary networks within each tissue or organ.

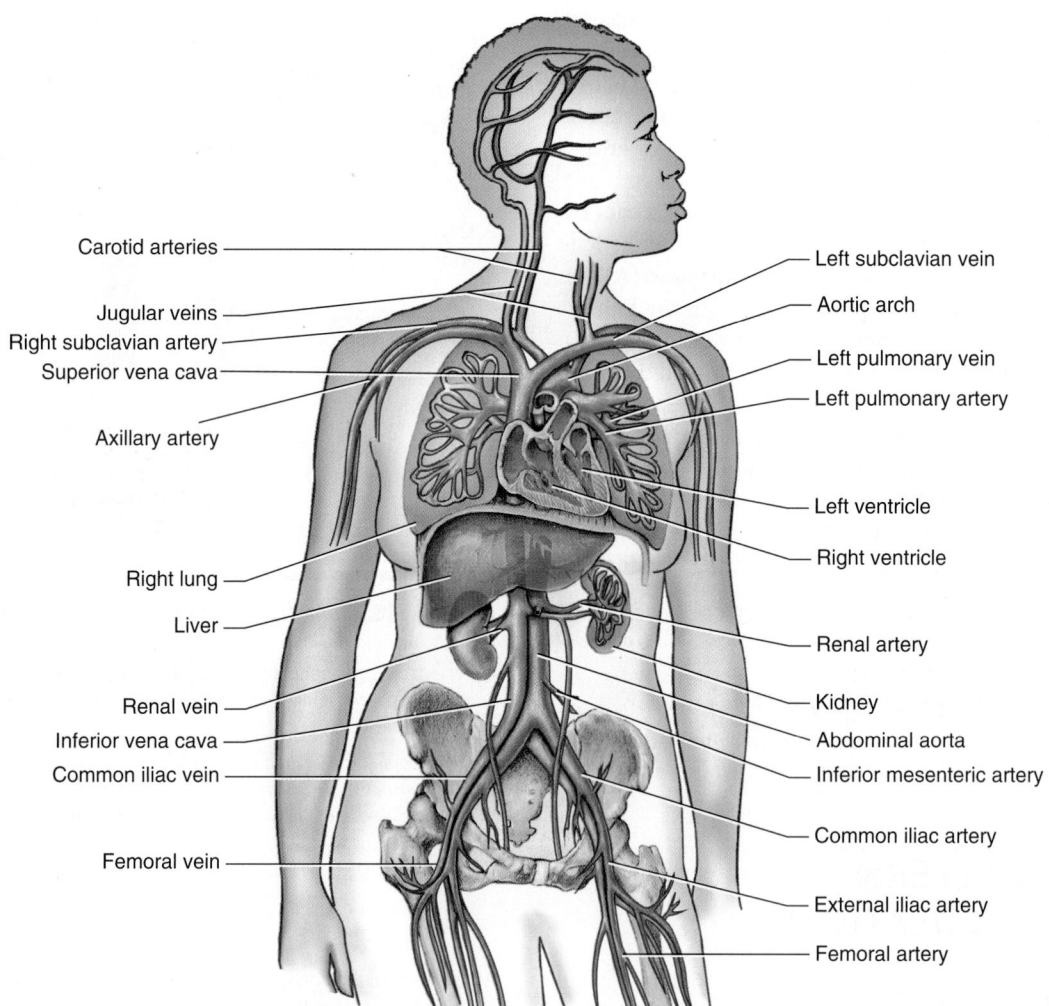

Carotid arteries
Jugular veins
Right subclavian artery
Superior vena cava
Axillary artery
Right lung
Liver
Renal vein
Inferior vena cava
Common iliac vein
Femoral vein

Left subclavian vein
Aortic arch
Left pulmonary vein
Left pulmonary artery
Left ventricle
Right ventricle
Renal artery
Kidney
Abdominal aorta
Inferior mesenteric artery
Common iliac artery
External iliac artery
Femoral artery

Figure 44-17 *Animation* **Blood circulation through some of the principal arteries and veins of the human body**

Blood vessels carrying oxygen-rich blood are *red;* blood vessels carrying oxygen-poor blood are *blue.*

Blood returning from the capillary networks within the brain passes through the *jugular veins*. Blood from the shoulders and arms drains into the *subclavian veins*. These veins and others returning blood from the upper portion of the body merge to form a very large vein that empties blood into the right atrium. In humans this vein is called the **superior vena cava.** The *renal veins* from the kidneys, *common iliac veins* from the lower limbs, *hepatic veins* from the liver, and other veins from the lower portion of the body return blood to the large **inferior vena cava,** which delivers blood to the right atrium.

As an example of blood circulation through the systemic circuit, let us trace a drop of blood from the heart to the right leg and back to the heart:

left atrium ⟶ left ventricle ⟶ aorta ⟶ right common iliac artery ⟶ smaller arteries in leg ⟶ capillaries in leg ⟶ small veins in leg ⟶ common iliac vein ⟶ inferior vena cava ⟶ right atrium

In most fishes, small amphibians, and in some nonavian reptiles, the heart muscle is spongy and receives oxygen directly from the blood as it passes through the heart chambers. However, in other vertebrates, including birds and mammals, the walls of the heart are too thick for nutrients and oxygen to reach all the muscle fibers by diffusion. Instead, the cardiac muscle has its own system of blood vessels, the **coronary circulation.** In humans, *coronary arteries* give rise to a network of capillaries within the heart wall. The *coronary veins* join to form a large vein, the *coronary sinus,* which empties directly into the right atrium.

Blood almost always travels from artery to capillary to vein to the heart. An exception occurs in the **hepatic portal system.** Instead of leading directly back to the heart (as most veins do), the hepatic portal vein delivers nutrients from the intestine to the liver. Within the liver, the hepatic portal vein gives rise to an extensive network of tiny blood sinuses. As blood courses through the hepatic sinuses, liver cells remove nutrients and store them. Eventually, liver sinuses merge to form hepatic veins, which deliver blood to the inferior vena cava.

CHECKPOINT 44.7

- *What sequence of blood vessels and heart chambers would a red blood cell pass through on its way (1) from the carotid artery to the superior vena cava and (2) from the renal vein to the renal artery?*
- *What is the function of the hepatic portal system? How does its sequence of blood vessels differ from that in most other circulatory routes?*

44.8 THE LYMPHATIC SYSTEM

LEARNING OBJECTIVE

11 Describe the inter-relationship between the structures and functions of the lymphatic system.

In addition to the blood circulatory system, vertebrates have an accessory circulatory system, the **lymphatic system** (FIG. 44-18).

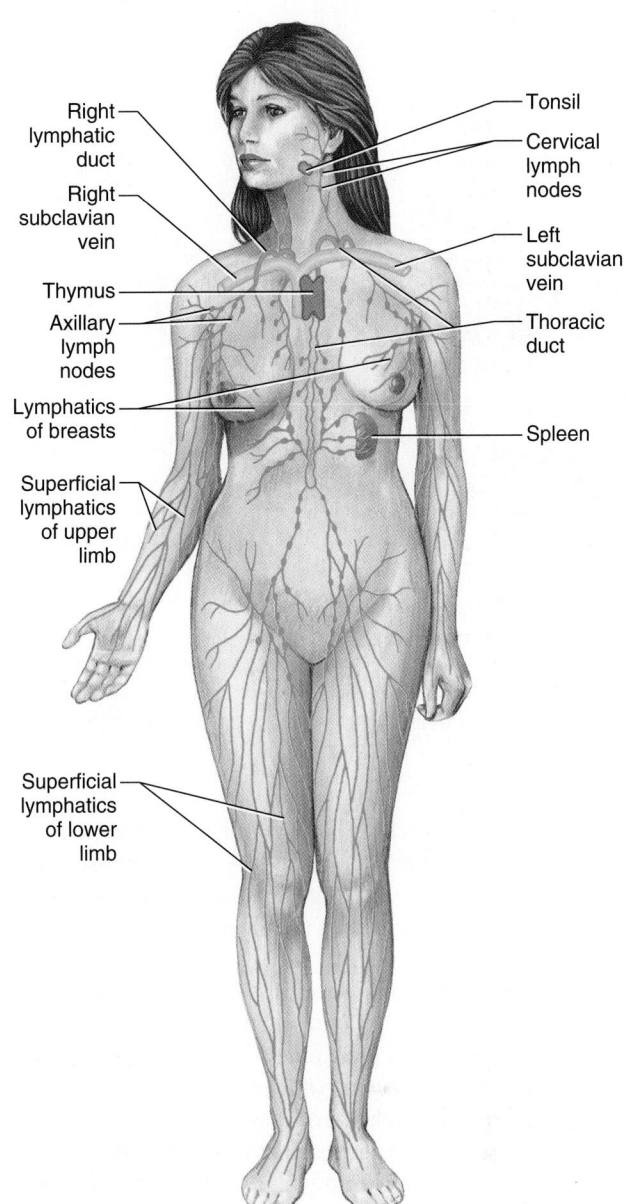

Figure 44-18 *Animation* **Human lymphatic system**

Note that the lymphatic vessels extend into most tissues of the body, but the lymph nodes are clustered in certain regions. The right lymphatic duct drains lymph from the upper right quadrant of the body and returns fluid to the right subclavian vein. The thoracic duct drains lymph from other regions of the body and returns fluid to the left subclavian vein.

© Cengage Learning

The lymphatic system (1) collects and returns interstitial fluid to the blood, (2) launches immune responses that defend the body against disease organisms, and (3) absorbs lipids from the digestive tract. In this section we focus on the first function. We discuss immunity in Chapter 45 and lipid absorption in Chapter 47.

The lymphatic system consists of lymphatic vessels and lymph tissue

The lymphatic system consists of (1) an extensive network of *lymphatic vessels,* or simply *lymphatics,* that conduct **lymph,** the clear, watery fluid formed from interstitial fluid; and (2) *lymph*

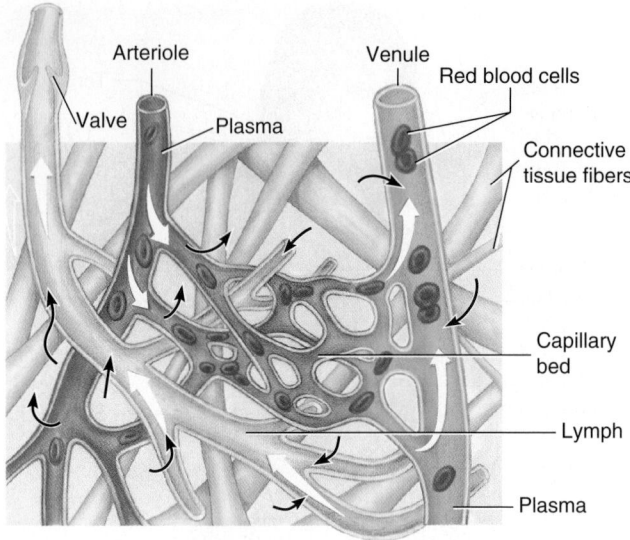

Figure 44-19 Lymph capillaries

Lymph capillaries drain excess interstitial fluid from the tissues. Note that blood capillaries are connected to vessels at both ends, whereas lymph capillaries (*shown in green*) are "dead-end streets." The *white arrows* indicate direction of blood and lymph flow. The *black arrows* indicate direction of interstitial fluid flow.

© Cengage Learning

tissue, a type of connective tissue with large numbers of lymphocytes. Lymph tissue is organized into small masses of tissue called **lymph nodes** and *lymph nodules.* The tonsils, thymus gland, and spleen, which consist mainly of lymph tissue, are also part of the lymphatic system.

Tiny "dead-end" capillaries of the lymphatic system extend into almost all body tissues (**FIG. 44-19**). Interstitial fluid enters lymph capillaries, where it becomes lymph. Lymph capillaries join to form larger lymphatics (which you can think of as lymph veins). There are no lymph arteries.

Lymph passes from lymph capillaries into lymphatics, which at certain locations empty into lymph nodes. As lymph moves through the lymph nodes, phagocytes filter out bacteria and other harmful materials. The lymph then flows into lymphatics that conduct it away from the lymph node. Lymphatics from all over the body conduct lymph toward the shoulder region. These vessels join the circulatory system at the base of the subclavian veins by way of ducts: the *thoracic duct* on the left side and the *right lymphatic duct* on the right.

Tonsils are masses of lymph tissue under the lining of the oral cavity and throat. (When enlarged, the pharyngeal tonsils in back of the nose are called *adenoids.*) Tonsils help protect the respiratory system from infection by destroying bacteria and other foreign matter that enter the body through the mouth or nose. Unfortunately, tonsils are sometimes overcome by invading bacteria and become the site of frequent infection themselves.

Some vertebrates, such as frogs, have lymph "hearts," which pulsate and squeeze lymph along. However, in mammals the walls of lymphatic vessels themselves pulsate. Valves within the lymphatics prevent lymph from flowing backward. When muscles contract or when arteries pulsate, pressure on the lymphatic vessels increases lymph flow. The rate at which lymph flows is slow and variable. The total lymph flow is about 100 mL per hour, which is far slower than the 5 L per minute of blood flowing in the cardiovascular system.

The lymphatic system plays an important role in fluid homeostasis

When blood enters a capillary network, it is under rather high pressure, so some plasma is forced out of the capillaries into the tissues. Once it leaves the blood vessels, this fluid is called *interstitial fluid,* or tissue fluid. It contains no red blood cells or platelets and only a few white blood cells. Its protein content is about one-fourth that found in plasma because proteins are too large to pass easily through capillary walls. However, smaller molecules dissolved in the plasma do pass out with the fluid leaving the blood vessels. Thus, interstitial fluid contains oxygen, glucose, amino acids and other small nutrients, and a variety of salts. This nourishing fluid bathes all the cells.

Three factors are responsible for *net filtration pressure,* which is the tendency for plasma to leave the blood at the arterial end of a capillary and enter the interstitial fluid (**FIG. 44-20**). The main force is *hydrostatic pressure,* the blood pressure against the capillary wall. This effect is augmented by the osmotic pressure of the interstitial fluid. The principal opposing force is the osmotic pressure of the blood, which is greater than the osmotic pressure of the interstitial fluid and restrains fluid loss from the capillary.

At the venous ends of the capillaries, the blood hydrostatic pressure is much lower, and the osmotic pressure of the blood draws fluid back into the capillary. However, not as much fluid is absorbed back into the circulation at the venous ends of capillaries as is filtered out at the arterial ends. Furthermore, protein does not return effectively into the venous capillaries and instead tends to accumulate in the interstitial fluid. These potential problems are so serious that without the lymphatic system, fluid balance in the body would be significantly disturbed within a few hours, and death would occur within about 24 hours. The lymphatic system preserves fluid balance by collecting interstitial fluid, including the protein that accumulates in it, and returning it to the circulation.

The walls of the lymph capillaries consist of endothelial cells that overlap slightly. When interstitial fluid accumulates, it presses against these cells, pushing them inward like tiny swinging doors that swing in only one direction. As fluid accumulates within the lymph capillary, these cell doors are pushed closed.

Obstruction of lymphatic vessels causes *lymphedema,* swelling from excessive accumulation of interstitial fluid. Lymphatic vessels can be blocked as a result of injury, inflammation, surgery, or parasitic infection. For example, when a breast is removed (mastectomy) because of cancer, lymph nodes in the armpit are often removed to prevent the spread of cancer cells. The disrupted lymph circulation may cause the patient's arm to swell. Treatment focuses on minimizing the swelling and controlling discomfort and pain.

CHECKPOINT 44.8

- **VISUALIZE** *Draw a diagram illustrating the relationships among plasma, interstitial fluid, and lymph.*
- *How does the lymphatic system help maintain fluid balance?*

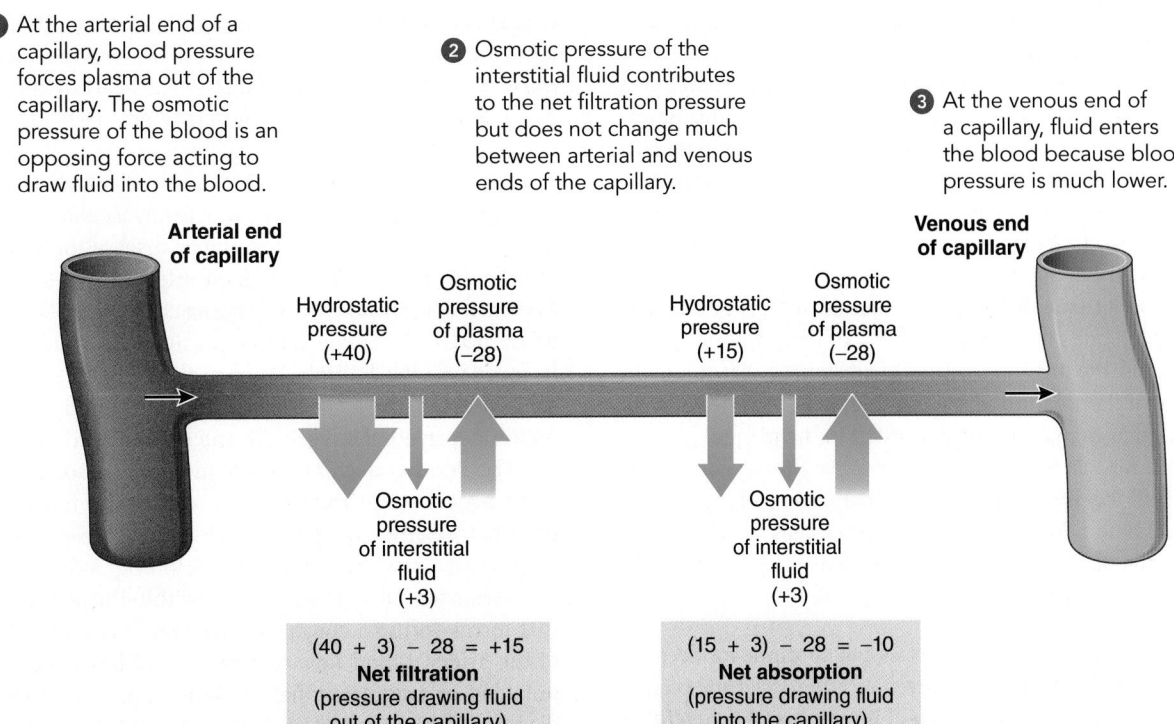

① At the arterial end of a capillary, blood pressure forces plasma out of the capillary. The osmotic pressure of the blood is an opposing force acting to draw fluid into the blood.

② Osmotic pressure of the interstitial fluid contributes to the net filtration pressure but does not change much between arterial and venous ends of the capillary.

③ At the venous end of a capillary, fluid enters the blood because blood pressure is much lower.

Arterial end of capillary

Hydrostatic pressure (+40)

Osmotic pressure of plasma (−28)

Hydrostatic pressure (+15)

Osmotic pressure of plasma (−28)

Venous end of capillary

Osmotic pressure of interstitial fluid (+3)

Osmotic pressure of interstitial fluid (+3)

(40 + 3) − 28 = +15
Net filtration
(pressure drawing fluid out of the capillary)

(15 + 3) − 28 = −10
Net absorption
(pressure drawing fluid into the capillary)

Figure 44-20 Fluid movement between blood and interstitial fluid
The exchange of dissolved materials between blood and interstitial fluid depends on fluid movement between the blood and the interstitial fluid surrounding the capillary. This fluid movement depends on the hydrostatic pressure of the blood and on osmotic pressures. More fluid leaves at the arterial end of each capillary than is returned at the venous end. The lymphatic system collects the excess interstitial fluid and returns it to the blood. The numbers given are hypothetical and represent millimeters of mercury.
© Cengage Learning

44.9 CARDIOVASCULAR DISEASE

LEARNING OBJECTIVE

12 Trace the progression of atherosclerosis and describe at least five risk factors for cardiovascular disease.

Cardiovascular disease includes heart and blood vessel diseases, including cerebrovascular accidents (strokes). Cardiovascular disease is the leading cause of death for both men and women in the United States and in the world. In the United States, cardiovascular disease causes more than 700,000 deaths each year. Death typically results from a combination of *atherosclerosis, thrombosis* (clotting), and *hypertension* (high blood pressure).

Atherosclerosis develops progressively

Atherosclerosis is a progressive disease in which the walls of certain arteries are damaged, inflamed, and narrowed as a result of lipid deposits in their walls. Although it can affect almost any artery, the disease most often develops in the aorta and in the coronary and cerebral arteries. The following sequence of events occurs in the development of atherosclerosis.

1. The earliest events in atherosclerosis begin when excess low-density lipoprotein (LDL) (known as "bad" cholesterol) in the blood enters the inner layer of blood vessels. (LDLs are discussed further in Chapter 47; uptake of LDL by cells is described in Chapter 5 and illustrated in Figure 5-22.) Recall that the inner layer of a blood vessel is made up of endothelial cells; the inner lining is the endothelium. Excess LDLs in the blood vessel wall become oxidized by **free radicals,** highly reactive, toxic compounds with unpaired electrons. (Free radicals bond with other compounds in cells and interfere with normal function.) Oxidized LDLs accumulate between and beneath the endothelial cells.

2. Endothelial cells respond to the oxidized LDLs by releasing compounds that attract monocytes. These white blood cells enter the tissues and trigger an inflammatory response. Monocytes develop into macrophages, which phagocytize the oxidized LDLs. As they become filled with LDLs, the macrophages appear foamy and are called *foam cells.*

3. Large numbers of foam cells lead to the development of fatty streaks, thought to be the initial lesion in the progression of atherosclerosis. (Interestingly, fatty streaks have been identified in the arteries of some children and adolescents.) Cells at the site of inflammation release compounds that trigger underlying smooth muscle fibers to migrate from

the muscle layer of the artery and cover the fatty streaks. The smooth muscle fibers settle just below the endothelium of the artery and multiply there. The deposited lipids together with the smooth muscle fibers form the developing *atherosclerotic plaque.* Inflammation, discussed in Chapter 45, is one of the body's most important responses to infection or injury. However, inflammation can also have harmful effects and appears to be an important factor in the development of atherosclerosis.

4. Connective tissue fibers also proliferate and produce collagen, which adds to the plaque. Cells of the immune system migrate to the area and produce substances that maintain chronic inflammation in the inner layer of the arterial wall. Later, calcium may also be deposited in the plaque.

5. Smooth muscle fibers form a protective *fibrous cap,* which walls off the plaque from the inner lining of the artery and the blood. Macrophages and other cells eventually die within the plaque, recruiting more macrophages to the area to clean up the debris.

The progression of atherosclerosis can continue for decades without causing symptoms. However, as atherosclerotic plaque builds, arteries eventually lose their ability to stretch when filled with blood. As the walls thicken, the diameter of the artery decreases, and it gradually becomes occluded (blocked), as shown in FIGURE 44-21.

Atherosclerosis has many effects

Compared with healthy blood vessels, occluded vessels deliver less blood to the tissues. As a result, tissues become *ischemic,* or lacking in blood, which reduces their oxygen and nutrient supply. In *ischemic heart disease,* the increased need for oxygen during exercise or emotional stress results in the pain characteristic of *angina pectoris.* Nitroglycerin, in the form of pills or a patch, dilates veins, decreasing venous return. Cardiac output is lowered, so the heart is not working as hard and requires less

oxygen. Nitroglycerin also dilates the coronary arteries slightly, allowing more blood to reach the heart muscle. *Calcium ion channel blockers* are drugs that slow the heart by inhibiting the passage of Ca^{2+} into cardiac muscle fibers. They also dilate the coronary arteries.

Myocardial infarction (MI), or heart attack, is a very serious, often fatal, consequence of cardiovascular disease. MI is the first symptom of cardiovascular disease for about 65% of men and almost 50% of women affected by atherosclerosis. Myocardial infarction is the leading cause of death and disability in the United States. It often results from a sudden decrease in coronary blood supply. If only a small region of cardiac muscle is affected, the heart may continue to function. Cells in the region deprived of oxygen die and are replaced by scar tissue.

The sudden decrease in blood supply associated with MI often occurs when inflammation causes the fibrous cap over the atherosclerotic plaque to break open. Platelets adhere to the roughened arterial wall and initiate clotting. A *thrombus* is a clot that forms within a blood vessel or within the heart; a thrombus that forms within a coronary artery can block a sizable branch of the artery. Blood flow to a portion of heart muscle may be impeded or completely halted. When a critical region of cardiac muscle is deprived of its blood supply, the heart may stop beating. This condition, called *cardiac arrest,* can result in death within a few minutes.

An episode of ischemia may trigger a fatal arrhythmia such as *ventricular fibrillation,* a condition in which the ventricles contract very rapidly without actually pumping blood. The pulse may stop, and blood pressure may fall precipitously. Ventricular fibrillation has been linked with about 65% of cardiac arrests. The only effective treatment for fibrillation is *defibrillation* with electric shock. The shock appears to depolarize all the muscle fibers in the heart so that its timing mechanism can reset.

When atherosclerosis develops in the cerebral arteries, less blood is delivered to the brain. This condition can lead to a *cerebrovascular accident (CVA),* commonly referred to as a *stroke.* Most CVAs are caused by ischemia, typically as a result of thrombosis. Others are caused by uncontrolled hypertension, which can cause blood vessels in the brain to leak or rupture.

Cardiovascular disease can be treated

Current treatment for patients with progressive cardiovascular disease includes certain drugs; surgical procedures aimed at restoring and improving blood flow; and most importantly, education regarding lifestyle changes that reduce risk factors. Among the types of medications used are statins, medications that lower cholesterol level; aspirin, which reduces the tendency for blood clotting; and ACE inhibitors (angiotensin-converting enzyme), drugs that decrease blood pressure. By inhibiting ACE, a key enzyme responsible for the synthesis of angiotensin II, these drugs decrease blood pressure. (Angiotensin II is a powerful

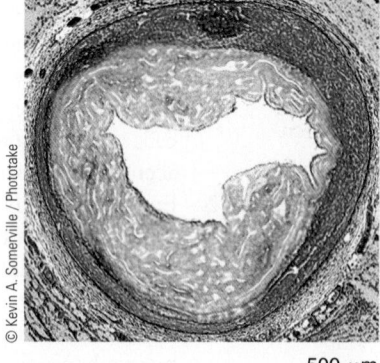

SPL/Science Source

© Kevin A. Somerville / Phototake

500 µm

500 µm

(a) Normal coronary artery.

(b) This artery is severely occluded by the atherosclerotic plaque that has built up within its wall.

Figure 44-21 Progression of atherosclerosis

LMs of cross sections through two arteries showing changes that take place in atherosclerosis.

vasoconstrictor and also promotes production of the hormone aldosterone, which increases sodium reabsorption in the kidneys and thus, retention of water. Angiotensin II and aldosterone are discussed further in Chapter 48.) Drugs that inhibit inflammation are being investigated.

Surgical procedures to treat cardiovascular disease include angioplasty and stent placement, and coronary bypass surgery. *Angioplasty* involves inserting a small balloon into an occluded coronary artery. Inflating the balloon breaks up the plaque in the arterial wall and thus widens the vessel. A stent (a mesh tube) left in the artery helps keep the artery open.

In *coronary bypass surgery,* veins from another location in the patient's body are grafted around occluded coronary arteries. The newly positioned blood vessels allow blood to flow around the blocked arteries, restoring adequate blood flow to the affected area.

Future treatment for damaged hearts may include stem cell therapies. Researchers have demonstrated that patients treated with stem cells harvested from their own bone marrow (in addition to standard treatment) had better recoveries than patients who did not receive stem cell treatment.

The risk of cardiovascular disease can be lowered

Many studies have shown that a healthy lifestyle lowers the risk for cardiovascular disease. For example exercise increases the concentration of HDLs, which protect against heart disease. A balanced diet low in sodium, cholesterol, saturated and *trans* fat, and rich in fruits and vegetables is important. Fruits and vegetables provide antioxidants (such as lycopene). Individuals with higher levels of antioxidant molecules in their body fat appear to have a reduced risk of developing cardiovascular disease.

Several major *modifiable* risk factors for cardiovascular disease have been identified:

1. *Elevated LDL-cholesterol levels in the blood.* These levels are associated with diets rich in total calories, total fats, saturated fats, and cholesterol. Genetic predisposition can also be a factor.

2. *Hypertension.* The higher the blood pressure is, the greater the risk of cardiovascular disease. The hormone angiotensin II, which at abnormally high levels is thought to contribute to hypertension, also stimulates inflammation.

3. *Cigarette smoking.* The risk of developing atherosclerosis is three to six times greater in smokers than in nonsmokers and is directly proportional to the number of cigarettes smoked daily. Nicotine constricts blood vessels, and carbon monoxide in the smoke damages the inner lining of blood vessels. Components of cigarette smoke cause oxidants to form, which may contribute to the inflammation by increasing oxidation of the altered LDL in the arterial wall.

4. *Diabetes mellitus.* In this endocrine disorder, glucose is not metabolized normally. The body shifts to fat metabolism, and there is a marked increase in circulating lipids, leading to atherosclerosis. Impaired glucose tolerance is also a risk factor.

5. *Physical inactivity.* Physical inactivity contributes to more than one-third of the more than 700,000 deaths related to heart disease in the United States each year. Exercise reduces the concentrations of triacylglycerols (triglycerides) and cholesterol in the blood. (These lipids have been associated with heart disease.)

6. *Overweight and obesity.* Being overweight can affect cholesterol levels and increase the risk of hypertension and diabetes.

The risk of developing cardiovascular disease also increases with age. Other probable risk factors currently being studied are hereditary predisposition, hormone levels, stress and behavior patterns, and dietary factors.

CHECKPOINT 44.9

- PREDICT *Over a period of years, Joe pays little attention to his diet. He eats mainly fried foods. He also smokes cigarettes. What outcome would you predict for Joe?*

- CONNECT *How might inflammation contribute to blocked arteries?*

SUMMARY: FOCUS ON LEARNING OBJECTIVES

44.1 Types of Circulatory Systems *(page 931)*

1 Contrast internal transport in animals with no circulatory system, animals with an open circulatory system, and animals with a closed circulatory system.

- Small, simple invertebrates depend on **diffusion** for internal transport. Larger animals require a specialized **circulatory system,** which typically consists of **blood,** a **heart,** and a system of **blood vessels** or spaces through which blood circulates. In all animals **interstitial fluid,** the tissue fluid between cells, brings oxygen and nutrients into contact with cells.

- Arthropods and most mollusks have an **open circulatory system** in which blood flows into a **hemocoel,** bathing the tissues directly.

- Some invertebrates and all vertebrates have a **closed circulatory system** in which blood flows through a continuous circuit of blood vessels.

- The vertebrate circulatory system consists of a muscular heart that pumps blood into a system of blood vessels. The circulatory system transports nutrients, oxygen, wastes, and hormones; helps maintain fluid balance, appropriate pH, and body temperature; and defends the body against disease.

44.2 Vertebrate Blood *(page 933)*

2 Compare the structure and function of plasma, red blood cells, white blood cells, and platelets.

- **Plasma** consists of water, salts, substances in transport, and **plasma proteins,** including albumins, globulins, and fibrinogen.
- **Red blood cells (RBCs),** also called **erythrocytes,** transport oxygen and carbon dioxide. Red blood cells produce large quantities of **hemoglobin,** a red pigment that transports oxygen.
- **White blood cells (WBCs),** also called **leukocytes,** defend the body against disease organisms. **Lymphocytes** and **monocytes** are agranular white blood cells; **neutrophils, eosinophils,** and **basophils** are granular white blood cells. Monocytes can develop into **macrophages** and **dendritic cells.**
- **Platelets** patch damaged blood vessels and release substances essential for blood clotting.

3 Summarize the sequence of events involved in blood clotting.

- Damaged cells and platelets release substances that activate clotting factors. Prothrombin is converted to *thrombin,* which catalyzes the conversion of *fibrinogen* to an insoluble protein, **fibrin.** Fibrin forms long threads that make up the webbing of the clot.

44.3 Vertebrate Blood Vessels *(page 936)*

4 Compare the structure and function of different types of blood vessels, including arteries, arterioles, capillaries, and veins.

- **Arteries** carry blood away from a heart chamber. **Arterioles** are small arteries that regulate blood pressure by constricting (**vasoconstriction**) and dilating (**vasodilation**). **Capillaries** are the thin-walled exchange vessels through which blood and tissues transfer materials. **Veins** return blood to a heart chamber.

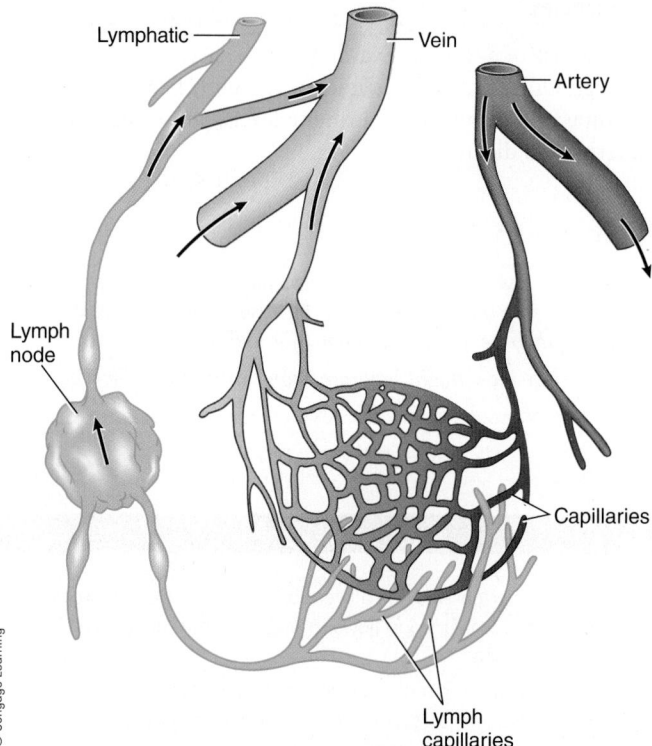

© Cengage Learning

Labels: Lymphatic — Vein — Artery — Lymph node — Capillaries — Lymph capillaries

44.4 Evolution of the Vertebrate Circulatory System *(page 938)*

5 Trace the evolution of the vertebrate circulatory system from fish to mammal.

- The vertebrate heart has one or two **atria,** which receive blood, and one or two **ventricles,** which pump blood into the arteries.

The fish heart consists of a single atrium and ventricle that are part of a single circuit of blood flow.

- In terrestrial vertebrates circulatory systems separate oxygen-rich blood from oxygen-poor blood, which allows the higher metabolic rate needed to support an active terrestrial lifestyle. Amphibians have two atria and a single ventricle. Blood flows through a double circuit so that oxygen-rich blood is partly separated from oxygen-poor blood. Most nonavian reptiles have a wall that partly divides the ventricles, minimizing the mixing of oxygen-rich and oxygen-poor blood.
- The four-chambered hearts of birds and mammals separate oxygen-rich blood from oxygen-poor blood.

44.5 The Human Heart *(page 940)*

6 Describe how the structure and function of the human heart are inter-related; include the heart's conduction system in your answer.

- The heart is enclosed by a *pericardium* and has valves that prevent backflow of blood. The valve between the right atrium and ventricle is the *right* **atrioventricular (AV) valve,** or **tricuspid valve.** The valve between the left atrium and ventricle is the **mitral valve. Semilunar valves** guard the exits from the heart. Cardiac muscle fibers are joined by **intercalated discs.**
- The **pacemaker** in the **sinoatrial (SA) node** initiates each heartbeat. A specialized electrical conduction system coordinates heartbeats.

7 Trace the events of the cardiac cycle and relate normal heart sounds to these events.

- One complete heartbeat makes up a **cardiac cycle.** The period of contraction is called **systole.** The period of relaxation is **diastole.** At the beginning of ventricular systole, the closing of the AV valves makes a low-pitched "lub" sound. At the beginning of ventricular diastole, the closing of the semilunar valves makes a short, loud "dup" sound.

8 Define *cardiac output,* describe how it is regulated, and identify factors that affect it.

- **Cardiac output (CO)** equals **stroke volume** multiplied by heart rate. Stroke volume depends on *venous return* and on neural messages and hormones, especially epinephrine and norepinephrine. According to *Starling's law of the heart,* the more blood delivered to the heart by the veins, the more blood the heart pumps.
- Heart rate is regulated mainly by the nervous system and is influenced by hormones and body temperature.

44.6 Blood Pressure *(page 944)*

9 Identify factors that determine and regulate blood pressure and compare blood pressure in different types of blood vessels.

- **Blood pressure** is the force blood exerts against the inner walls of the blood vessel. Blood pressure is greatest in the arteries and decreases as blood flows through the capillaries.
- Blood pressure depends on cardiac output, blood volume, and resistance to blood flow. *Peripheral resistance* is the resistance to blood flow caused by blood viscosity and by friction between blood and the blood vessel wall.
- **Baroreceptors** are sensitive to blood pressure changes. They send messages to the cardiac and vasomotor centers in the medulla of the brain. When informed of an increase in blood pressure, the cardiac center stimulates parasympathetic nerves that slow the heart rate. The vasomotor center inhibits

sympathetic nerves that constrict blood vessels. These actions reduce blood pressure.

- **Angiotensin II** is a hormone that raises blood pressure. **Aldosterone** helps regulate salt excretion, which affects blood volume and blood pressure.

44.7 The Pattern of Circulation (page 947)

10 Trace a drop of blood through the pulmonary and systemic circulations, naming in sequence each structure through which it passes.

- The **pulmonary circulation** connects the heart and the lungs; the **systemic circulation** connects the heart and the tissues. In the pulmonary circulation, the right ventricle pumps blood into the *pulmonary arteries,* which go to the lungs. Blood circulates through pulmonary capillaries in the lung and then is conducted to the left atrium by a *pulmonary vein.*
- In the systemic circulation, the left ventricle pumps blood into the **aorta,** which branches into arteries leading to the body organs. After flowing through capillary networks within various organs, blood flows into veins that conduct it to the **superior vena cava** or **inferior vena cava.** These large veins return blood to the right atrium.
- The **coronary circulation** supplies the heart muscle with blood. The **hepatic portal system** circulates nutrient-rich blood through the liver.

44.8 The Lymphatic System (page 949)

11 Describe the inter-relationship between the structures and functions of the lymphatic system.

- The **lymphatic system** helps maintain homeostasis of fluids by collecting interstitial fluid and returning it to the blood. The structure of the walls of *lymphatic vessels* allows them to collect excess interstitial fluid. **Lymph** is the clear fluid formed from interstitial fluid.
- Lymphatic vessels conduct lymph to the blood. **Lymph nodes** are small masses of tissue that filter bacteria and harmful materials out of lymph.

44.9 Cardiovascular Disease (page 951)

12 Trace the progression of atherosclerosis and describe at least five risk factors for cardiovascular disease.

- **Atherosclerosis** progresses as the inner walls of certain arteries become inflamed and blocked by the accumulation of fatty deposits and cells that mediate inflammation. The blood supply of tissues served by these vessels becomes increasingly compromised.
- Risk factors include elevated LDL-cholesterol levels in the blood, hypertension, cigarette smoking, diabetes mellitus, physical inactivity, and being overweight.

TEST YOUR UNDERSTANDING

Know and Comprehend

1. Which of the following are most closely associated with oxygen transport? (a) red blood cells (b) platelets (c) neutrophils (d) basophils (e) lymphocytes

2. The sinoatrial (SA) node (a) initiates diastole (b) conducts impulses directly to the muscle fibers of both ventricles (c) briefly delays transmission before the action potential continues into the ventricles (d) transmits action potentials directly into the atrial muscle fibers (e) consists of smooth muscle fibers

3. A cardiac cycle (a) consists of one ventricular heartbeat (b) includes a systole (c) equals stroke volume multiplied by heart rate (d) includes a diastole (e) includes a systole and a diastole

4. Blood pressure can be significantly affected by (a) cardiac output (b) peripheral resistance (c) blood volume (d) a, b, and c (e) a and b only

5. Lymph forms from (a) fluid in the lymph nodes (b) blood serum (c) plasma combined with protein (d) fluid released by lymph nodes (e) interstitial fluid

6. Norepinephrine (a) slows heart rate (b) is released in cardiac muscle by parasympathetic nerves (c) causes K^+ channels in cardiac muscle to open (d) decreases stroke volume (e) causes Ca^{2+} channels in cardiac muscle to open

7. Baroreceptors (a) stimulate renin release (b) activate the renin–angiotensin–aldosterone pathway (c) stimulate sympathetic nerves (d) are stimulated by increased blood pressure (e) send messages to cardiac centers that increase blood pressure

Apply and Analyze

8. Which sequence below describes an accurate sequence of blood flow?
   1. pulmonary artery 2. pulmonary vein 3. left atrium 4. right atrium 5. pulmonary capillaries
   (a) 2, 1, 4, 3 (b) 3, 4, 2, 5 (c) 1, 5, 2, 3 (d) 4, 1, 5, 2 (e) 1, 5, 2, 4

9. **VISUALIZE** Draw a simple cross section of a vein, a capillary, and an artery. Label connective tissue, endothelium, and smooth muscle, if present. How does the structure of each of these vessels relate to its function?

10. When the nerves to the heart are cut, the heart rate increases to about 100 contractions per minute. What does this increase indicate about regulation of the heart rate?

Evaluate and Synthesize

11. **EVOLUTION LINK** A heart divided into four chambers independently evolved in birds and mammals. What are the benefits of this adaptation?

12. **EVOLUTION LINK** How is the heart of a fish specifically adapted to its lifestyle?

13. **SCIENCE, TECHNOLOGY, AND SOCIETY** Explain the association between atherosclerosis and ischemic heart disease. Imagine that you were the Surgeon General of the United States. What types of programs would you advocate to decrease heart disease? Do you think that as a society we should invest resources into the development of stem cell therapies for more effective treatment of myocardial infarction?

aplia To access course materials, such as Aplia and other companion resources, please visit **www.cengagebrain.com.**

45 | The Immune System: Internal Defense

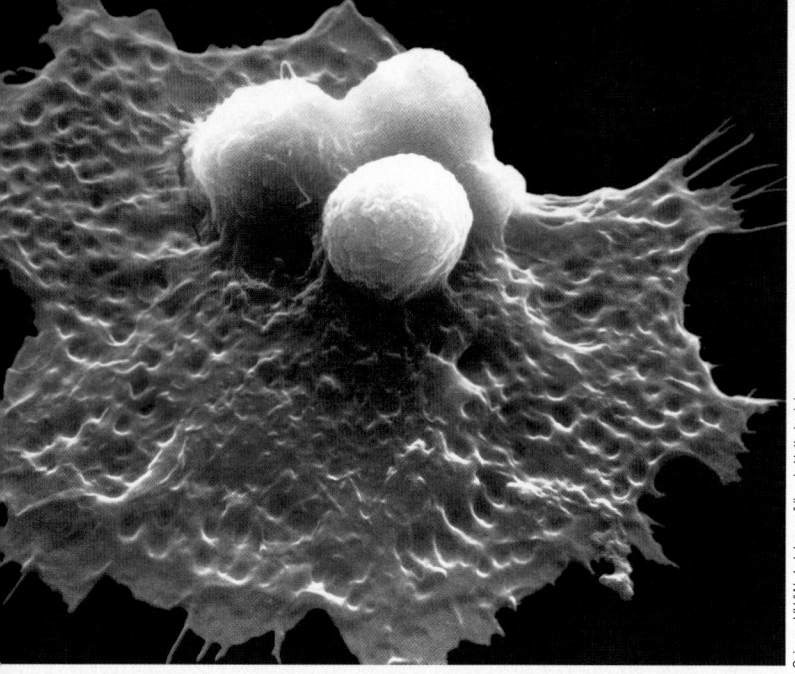

Immune response to cancer cells. This color-enhanced SEM shows a macrophage engulfing three cancer cells.

Science VU/W. J. Johnson/Visuals Unlimited, Inc.

KEY CONCEPTS

45.1 The immune system recognizes and destroys foreign macromolecules, including those associated with pathogens and abnormal cells (such as cancer cells).

45.2 Innate immunity provides general and immediate protection against pathogens, some toxins, and cancer cells.

45.3 Adaptive immunity includes cell-mediated immunity and antibody-mediated immunity.

45.4 In cell-mediated immunity, specific lymphocytes called T cells are activated; they release proteins that destroy cells infected with viruses or other intracellular pathogens.

45.5 In antibody-mediated immunity, specific lymphocytes called B cells are activated; they multiply and differentiate into plasma cells, which produce antibodies.

45.6 Cancer and some other diseases and disease-causing organisms—for example, HIV (the virus that causes AIDS)— compromise or overcome the immune system. Some immune responses have negative effects on health such as allergic responses.

The **immune system,** our internal defense system, is made up of the molecules and cells that protect the body against foreign macromolecules, including disease-causing organisms, certain toxins, and cancer cells. Derived from the Latin for "safe," the word *immune* refers to the early observation that when people recovered from smallpox and other serious infections, they were safe from contracting the same illness again.

Disease-causing organisms, or **pathogens,** include certain viruses, bacteria, fungi, protozoa, and worms. Pathogens enter the body with air, food, and water; during copulation; and through wounds in the skin. The immune system recognizes pathogens and many types of toxins and rapidly responds to eliminate them. For example, *macrophages* act quickly to destroy invading bacteria. Macrophages and other cells of the immune system also destroy cancer cells (see photograph).

The immune system normally functions very effectively. However, some pathogens, and also cancer cells, sometimes evade or overcome the body's internal defenses, resulting in disease. Certain genetic mutations and some pathogens can prevent or compromise immune function. For example, HIV, the retrovirus that causes AIDS, destroys T cells, an important component of the immune system. At times, the immune system over-reacts, as in allergic reactions. Immune responses are sometimes directed against our own tissues, resulting in *autoimmune disease* such as rheumatoid arthritis or multiple sclerosis. At times, the immune system attacks when we would prefer it did not. For example, immune responses can destroy the cells of heart or kidney transplants.

Immunology, the study of internal defense systems of humans and other animals, is one of the most rapidly changing and exciting fields of biomedical research. The immune system is a collection of many types of cells and tissues scattered throughout the body. Effective cell signaling among these cells is critical. Cells of the immune system communicate directly by means of their surface molecules and indirectly by releasing messenger molecules. Understanding the complex signaling systems of the immune system is a major focus of research. Investigators are also working to better understand the evolution of the immune system.

One of the greatest accomplishments of immunologists has been the development of vaccines that prevent disease. Another has been developing techniques for successful tissue and organ transplantation. Sophisticated research strategies, such as gene transfer, have enabled immunologists to expand their knowledge of the cells and molecules that interact to generate immune responses. These techniques have led to the development of new approaches to the prevention and treatment of disease. Much has been learned, and many challenges lie ahead.

45.1 EVOLUTION OF IMMUNE RESPONSES

LEARNING OBJECTIVES

1 Distinguish between innate and adaptive immunity.
2 Compare, in general terms, the immune responses that have evolved in invertebrates and vertebrates.

Immunity is the ability to recognize and destroy foreign or dangerous macromolecules. The immune system distinguishes between *self* and *nonself*. Such recognition is possible because each individual is biochemically unique. Cells have surface proteins different from those on the cells of other species or even other individuals of the same species. An animal's immune system recognizes its own cells and can identify those of other organisms as foreign. Thus, the distinctive macromolecules of an invading pathogen stimulate the animal's defensive responses. A single bacterium may have from 10 to more than 1000 distinctive macromolecules on its surface. The immune system also responds to danger signals from injured tissues, such as proteins released when cell membranes are damaged.

Two main types of immunity protect the body: innate and adaptive immunity (FIG. 45-1). **Innate immunity** consists of defenses that provide immediate, general protection against pathogens, parasites, some toxins and drugs, and cancer cells. Innate immune responses prevent most pathogens from entering the body and rapidly destroy those that do penetrate the outer defenses. For example, the cuticle or skin provides a physical barrier to pathogens that come in contact with an animal's body. *Phagocytosis* and *inflammation* are innate defenses that destroy bacteria when they manage to penetrate the outer defenses.

Adaptive immunity, also called *acquired immunity,* consists of specific defenses that target distinct macromolecules. Any molecule that cells of the immune system specifically recognize as foreign is called an **antigen.** Proteins are the most powerful antigens, but some polysaccharides and lipids are also antigenic. Adaptive immune responses are directed toward particular antigens and typically include the production of **antibodies,** highly specific proteins that recognize and bind to specific antigens. An important characteristic of adaptive immune responses is **immunological memory,** which means that the adaptive immune system "remembers" foreign or dangerous molecules.

Immunological memory enables the body to respond more effectively to repeated encounters with the same molecules.

Immunologists do not yet understand how the immune system is signaled to *not* destroy foreign molecules that are harmless. Animals *manage* bacteria and other microbes that are beneficial, such as certain bacteria in the digestive tract.

Invertebrates launch innate immune responses

An animal's first line of defense is its outer covering. Tough external skeletons, such as cuticles, shells, or chitinous exoskeletons, shield the bodies of nematodes, most mollusks, arthropods, and many other invertebrates (see Chapter 31). Cells in the outer layer of many invertebrates (cnidarians, annelids, and mollusks) produce mucus that traps and kills pathogens.

Recognition of nonself depends on **pattern recognition receptors (PRRs)** in the plasma membrane. PRRs, such as *Toll receptors* in *Drosophila*, recognize certain chemical compounds (or specific parts of compounds) on the surface of bacteria and other microbes as foreign. The chemical compounds that stimulate innate immune responses are called **pathogen-associated molecular patterns,** or **PAMPs.** Each group of microbes—viruses, gram-positive bacteria, and fungi—share certain PAMPs. For example, certain viruses have double-stranded RNA, and gram-positive bacteria have peptidoglycan.

All animal species studied have *phagocytes,* cells that move through the body engulfing and destroying bacteria and other foreign matter. When PRRs on phagocytes bind with PAMPs, innate immune responses are activated. For example, phagocytes then become more effective at taking in and destroying pathogens. These receptors may also transduce signals that stimulate inflammation.

Inflammation, also referred to as the *inflammatory response,* is one of the body's key responses to injury or infection in both invertebrates and vertebrates. During inflammation the body recruits white blood cells and plasma proteins from the blood to the location of an infection. *Phagocytosis* increases at the site of inflammation. Recall that **phagocytosis** is the process by which cells engulf microorganisms, foreign matter, or other cells (see Fig. 5-20). In mollusks substances in the hemolymph (blood) enhance phagocytosis. Annelids are the only invertebrates known to have *natural killer cells,* a type of lymphocyte that destroys tumor cells and cells infected by viruses.

Invertebrates distinguish between their own cells and those of other species. For example, sponge cells have specific glycoproteins on their surfaces that enable them to recognize their own species. When cells of two different species are mixed, they sort themselves out according to species. When two different species of sponges are forced to grow in contact with each other, tissue is destroyed along the region of contact. Cnidarians (e.g., corals), annelids (e.g., earthworms), arthropods (e.g., insects), and echinoderms (e.g., sea stars) reject tissue grafted from other animals, even from the same species.

Pattern recognition receptors activate signal transduction pathways that lead to the production of **antimicrobial peptides.** These compounds inactivate or kill pathogens. Investigators

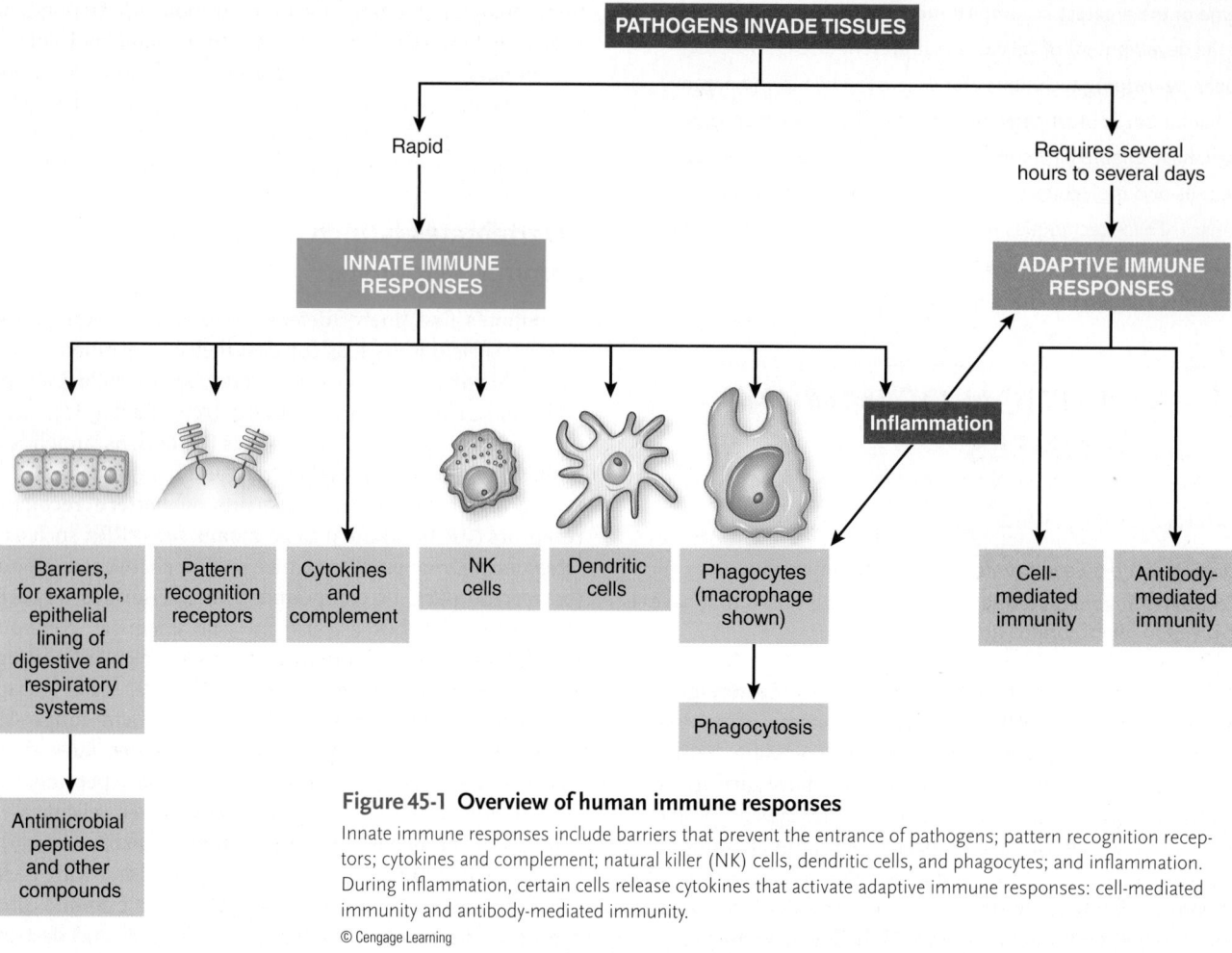

Figure 45-1 Overview of human immune responses

Innate immune responses include barriers that prevent the entrance of pathogens; pattern recognition receptors; cytokines and complement; natural killer (NK) cells, dendritic cells, and phagocytes; and inflammation. During inflammation, certain cells release cytokines that activate adaptive immune responses: cell-mediated immunity and antibody-mediated immunity.

© Cengage Learning

have identified antimicrobial peptides in all eukaryotes (animals, plants, fungi) and prokaryotes (bacteria and archaea) that have been studied, suggesting an early common origin of these molecules. More than 2200 antimicrobial peptides have been described. When researchers inject an antigen into an insect, as many as 15 different antimicrobial peptides are produced within a few hours. These chemical compounds are very effective in part because of their small size (a dozen or fewer amino acids), which facilitates their rapid production and diffusion. Some antimicrobial peptides, called *defensins,* kill bacteria and fungi by punching holes in the plasma membrane. Defensins have been found in many types of animals and in some plants.

Certain cnidarians, arthropods, some echinoderms, and simple chordates (e.g., tunicates) appear to remember antigens for a short period. As mentioned earlier, immunological memory enables the body to respond more effectively when it encounters the same pathogens again. Researchers have discovered that shrimp are able to launch an adaptive immune response, called "specific immune priming." During this response, phagocytic activity increased after the animals were vaccinated with a specific pathogen. Although certain invertebrates demonstrate some specificity and memory, their immune responses are primarily innate responses.

Echinoderms and tunicates are the only invertebrates known to have differentiated white blood cells that perform limited immune functions. Investigators have identified two genes in the purple sea urchin that are very similar to vertebrate genes important in producing antibodies. Some invertebrates appear to have at least a limited capacity to launch adaptive immune responses.

Vertebrates launch both innate and adaptive immune responses

Like invertebrates, vertebrates depend on innate immunity for their initial defense against invading pathogens. However, their more recently evolved adaptive immune responses fine-tune the more general innate responses.

The lymphatic system that evolved in jawed vertebrates includes cells such as **lymphocytes,** white blood cells specialized to carry out immune responses, and organs such as lymph nodes (see Chapter 44). Very specific lymphocytes and antibodies are produced when the body is exposed to specific pathogens. Once exposed to foreign molecules, the adaptive immune system remembers them. In the discussion that follows, we focus on the vertebrate immune system.

45.2 INNATE IMMUNE RESPONSES IN VERTEBRATES

LEARNING OBJECTIVE

3   Describe innate immune responses, including physical barriers; recognition by pattern recognition receptors; actions of phagocytes and other cells; actions of cytokines and complement; and the inflammatory response.

Innate immune responses include physical barriers, such as the skin; action by pattern recognition receptors; action of proteins such as *cytokines* and those that make up the *complement system;* actions by specialized cells, including phagocytes, dendritic cells, and natural killer (NK) cells; and the inflammatory response.

Physical barriers prevent most pathogens from entering the body

A vertebrate's first line of defense is its skin, which blocks the entry of pathogens. In some species the skin is covered by horny scales or plates. The digestive (gastrointestinal) tract, respiratory passageways, urinary tract, and vagina all open to the outside of the body. The epithelial linings of these passageways are barriers between the outside environment and the inside of the body. Various mechanisms within epithelial barriers expel, trap, or destroy pathogens. For example, pathogens that enter the body with inhaled air may be filtered out by hairs in the nose or trapped in the sticky mucous lining of the respiratory tract. Once trapped, they are destroyed by phagocytes.

Physical barriers are also equipped with chemical weapons. For example, defensins are produced by the epithelium of the skin. They are also found in the saliva of many animals, including mammals. Mucus contains a substance called *mucin,* which chemically destroys invaders. **Lysozyme,** an enzyme found in tears and other body fluids, attacks the cell walls of many gram-positive bacteria.

Mucus, tears, urine, and other body fluids flush pathogens out of the body. Invading bacteria must also compete with the populations of microbes that naturally inhabit our skin and other surface tissues. These resident microbes are normally not harmful and are essential to good health.

Pattern recognition receptors called **Toll-like receptors** are expressed on several types of cells. (Toll-like receptors are homologous to the Toll receptors in *Drosophila.*) When activated, these receptors activate transcription factors that stimulate the expression of genes encoding important signaling molecules (e.g., cytokines).

Cells of the innate immune system destroy pathogens

Innate immune responses depend on the action of several types of cells, including phagocytes, natural killer cells, and dendritic cells. **Neutrophils,** the most common type of white blood cell (see Chapter 44), and *macrophages* are the main phagocytes in the body. A neutrophil can phagocytose about 20 bacteria before it becomes inactivated (perhaps by leaking lysosomal enzymes) and dies.

Macrophages are large phagocytes that develop from *monocytes,* a type of nongranular white blood cell. A macrophage can phagocytose about 100 bacteria during its life span (FIG. 45-2). Macrophages also release antiviral agents. Some macrophages patrol the body's tissues, engulfing damaged cells or foreign matter (including bacteria). Other macrophages stay in one place and destroy bacteria that pass by. For example, air sacs in the lungs contain large numbers of macrophages that destroy foreign matter entering with inhaled air.

Vertebrate macrophages have Toll-like receptors that recognize certain PAMPs and respond by producing cytokines. For example, a lipopolysaccharide found on gram-negative bacteria stimulates macrophages. In response, macrophages release cytokines that enhance the inflammatory response.

Can bacteria counteract the phagocyte's attack? Many strategies have evolved in bacteria that protect them from their hosts. For example, the capsules of *Streptococcus pneumoniae* resist the actions of the phagocyte's lysosomal enzymes. Other bacteria release enzymes that destroy lysosomal membranes. The powerful lysosomal enzymes then spill out into the cytoplasm and may destroy the phagocyte.

Natural killer (NK) cells are large, granular lymphocytes that originate in the bone marrow. They account for about 10% of circulating lymphocytes. NK cells are active against tumor cells and cells infected with some types of viruses. They destroy target cells by both innate and adaptive (antibody-mediated) processes.

NK cells release cytokines as well as enzymes that destroy target cells. Some of these enzymes cause pores to form in the plasma membrane of the target cell. Other enzymes enter the cell and activate reactions that cause the cell to self-destruct by **apoptosis,** programmed cell death. Several cytokines stimulate NK cell activity. When NK cell levels are high, resistance to certain cancers may be strengthened. Psychological stressors may decrease NK cell activity and thus enhance tumor growth.

Biologists first identified **dendritic cells** in 1868 and then rediscovered them in 1973, but the small numbers of these cells made them hard to study until sophisticated techniques became available in the 1990s. Their name derives from their long cytoplasmic extensions called *dendrites* (not to be confused with the dendrites of neurons). Dendritic cells develop from precursor cells in the bone marrow and from monocytes.

Dendritic cells are strategically stationed in all tissues of the body that come into contact with the environment: the skin and the epithelial linings of the digestive, respiratory, urinary, and

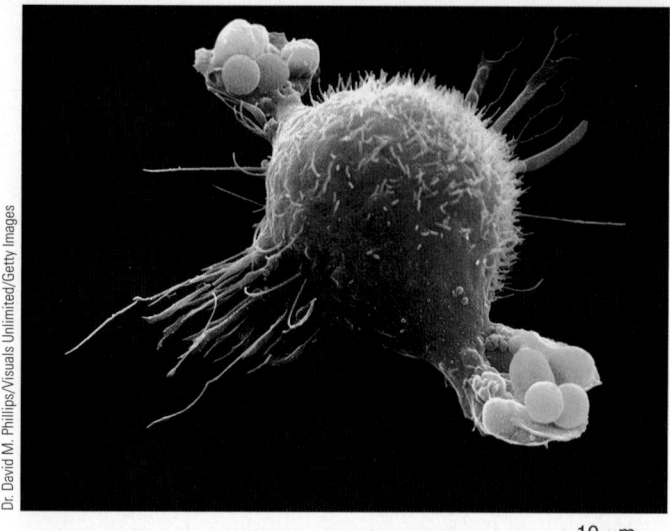

Figure 45-2 Phagocytosis: An innate immune response to infection

Macrophages are efficient warriors. This color-enhanced SEM shows a macrophage (*large, deep blue cell*) engulfing bacteria (*small green cells*).

vaginal passageways. When pathogens infect a tissue, PAMPs activate dendritic cells. One type of dendritic cell responds to viral infection. Dendritic cells produce *interferons,* cytokines that are antiviral. Certain types of activated dendritic cells capture microbial antigens by phagocytosis or by receptor-mediated endocytosis (see Chapter 5).

Cytokines are important signaling molecules

Cells of the immune system secrete a remarkable number of regulatory and antimicrobial peptides and proteins. Two major groups of molecules that are critical in innate defense responses are cytokines and complement. **Cytokines** are a large, diverse group of peptides and proteins (most are glycoproteins) that serve as important signaling molecules and perform regulatory functions. The same kinds of cytokines can be produced by many different types of cells. Also, a cytokine may affect many different kinds of cells. Just as the functions of innate and adaptive immune responses overlap, the actions of cytokines of these subsystems also overlap. For example, cytokines produced by cells such as macrophages can activate lymphocytes involved in adaptive immune responses.

Cytokines help regulate the intensity and duration of immune responses. They are also important in regulating many other biological processes, such as cell growth, repair, and cell activation. Cytokines bind to specific membrane receptors on target cells. These signaling molecules can act as *autocrine agents,* affecting the very cells that produce them, or as *paracrine agents,* regulating the activity of nearby cells (see Chapter 49). When produced in large amounts, some cytokines act as hormones that circulate in the blood and affect distant tissues. Biologists name the types of cytokines for their function or origin. We discuss several examples, including *interferons, tumor necrosis factors, interleukins,* and *chemokines.*

When infected by viruses or other intracellular parasites (some types of bacteria, fungi, and protozoa), certain cells respond by secreting cytokines called **interferons.** *Type I interferons* are produced by either macrophages or *fibroblasts,* cells that produce the fibers of connective tissues. Type I interferons inhibit viral replication and activate NK cells that have antiviral actions. Viruses produced in cells exposed to Type I interferons are not very effective at infecting other cells. Interestingly, even though Type I interferons are produced as part of the innate immune response, once released they can regulate the action of certain cells of the adaptive immune response. This interaction of innate and adaptive immune mechanisms is an example of the power and complexity of the immune system.

Type II interferon (also called IFN-gamma) is critical in both innate and adaptive immunity. In innate immunity Type II interferon stimulates macrophages to destroy tumor cells and host cells that have been infected by viruses. (The interferon activates transcription of genes that encode enzymes needed to produce antimicrobial compounds in lysosomes.)

Since their discovery in 1957, interferons have been the focus of a great deal of research. Pharmaceutical companies now use recombinant DNA techniques to produce large quantities of some interferons. The U.S. Food and Drug Administration has approved interferons for treating several diseases, including hepatitis B and hepatitis C, genital warts, several types of cancer (e.g., leukemia), a type of multiple sclerosis, and AIDS-related Kaposi's sarcoma.

In response to gram-negative bacteria, certain other pathogens, or some tumors, **tumor necrosis factor (TNF)** stimulates immune cells to initiate an inflammatory response. In fact, TNF was so named because it was first identified as a compound that destroyed tumor cells. TNF is secreted mainly by macrophages and by *T cells,* a type of lymphocyte. In severe infections large amounts of TNF are produced. TNF is then circulated in the blood and can have widespread effects. For example, it acts on the hypothalamus, inducing fever.

Sometimes infection by gram-negative bacteria, such as *Salmonella typhi,* results in the release of large amounts of TNF and other cytokines. The result is a cascade of reactions leading to *septic shock,* a potentially fatal condition that may involve high fever and malfunction of the circulatory system. Thus, cytokines can sometimes have harmful effects.

Interleukins are a diverse group of cytokines secreted mainly by macrophages and lymphocytes. Interleukins are numbered according to their order of discovery. They regulate interactions between white blood cells, such as lymphocytes and macrophages, and other cells. *Interleukin-1 (IL-1)* functions with TNF in mediating inflammation. *Interleukin-12 (IL-12)* stimulates NK cells as well as T cells to produce IFN-gamma. Some interleukins have widespread effects. During infection, IL-1 can reset the body's thermostat in the hypothalamus, resulting in fever and its symptoms. The overproduction of certain interleukins has been associated with clinical disease, including atherosclerosis.

Chemokines, a large group of cytokines, are signaling molecules that attract, activate, and direct the movement of various cells of the immune system. For example, they regulate the migration of white blood cells from the blood to the tissues. Some chemokines are produced in response to infection and are mediators of the inflammatory response.

Complement promotes destruction of pathogens and enhances inflammation

Phagocytes secrete cytokines that activate the *complement system.* **Complement,** so named because it "complements" the action of other defensive responses, consists of more than 20 proteins present in plasma and other body fluids. Similarities in complement proteins in many species, including horseshoe crabs, sea urchins, tunicates, and mammals, suggest that these molecules evolved millions of years ago and have been highly conserved.

Normally, complement proteins are inactive until the body is exposed to an antigen. Certain pathogens activate the complement system directly. In other cases, the binding of an antigen and antibody stimulates activation. Complement activation involves a cascade of reactions; each component acts on the next

in the series. Proteins of the complement system then work to destroy pathogens.

Complement proteins are active against many antigens, and their actions are not specific: they (1) lyse viruses, bacteria, and other cells; (2) coat pathogens, making them less "slippery" so that macrophages and neutrophils phagocytose them more easily; (3) attract white blood cells to the site of infection, a process called *chemotaxis;* and (4) bind to specific receptors on cells of the immune system, stimulating specific actions, such as secreting regulatory molecules and enhancing the inflammatory response.

Inflammation is a protective response

The **inflammatory response,** or *inflammation,* is one of the body's key responses to infection or injury. Tissue injury caused by pathogen invasion or physical injury activates a clotting factor in the blood plasma that turns on three different interconnecting molecular cascades in the plasma, including the clotting cascade. Many of these reactions generate molecules—for example, the peptide *bradykinin*—that mediate the inflammatory process. These plasma mediators dilate blood vessels and increase capillary permeability (**FIGS. 45-3** and **45-4**). Many cytokines signal white blood cells to launch the inflammatory response, and some cytokines help regulate the process. Inflammation also activates the complement system. The clinical characteristics of inflammation are heat, redness, edema (swelling), and pain.

The inflammatory response includes three main processes.

1. *Vasodilation.* Macrophages and **mast cells** stationed in the tissues respond rapidly to damaged tissue or infection. Mast cells, which are found in most tissues close to blood vessels, release **histamine,** cytokines, and other compounds that dilate blood vessels in the affected area. Expanded blood vessel diameter brings more blood to the area. The increased blood flow warms the skin and reddens skin that contains little pigment.

2. *Increased capillary permeability.* Histamine and other compounds released by mast cells increase capillary permeability. Fluid and antibodies leave the circulation and enter the tissues. As the volume of interstitial fluid increases, *edema* occurs. The edema, along with the action of certain enzymes in the plasma, causes the pain that characterizes inflammation.

KEY POINT

During the inflammatory response, capillaries dilate and become more permeable, bringing more phagocytes to the infected area; phagocytosis increases.

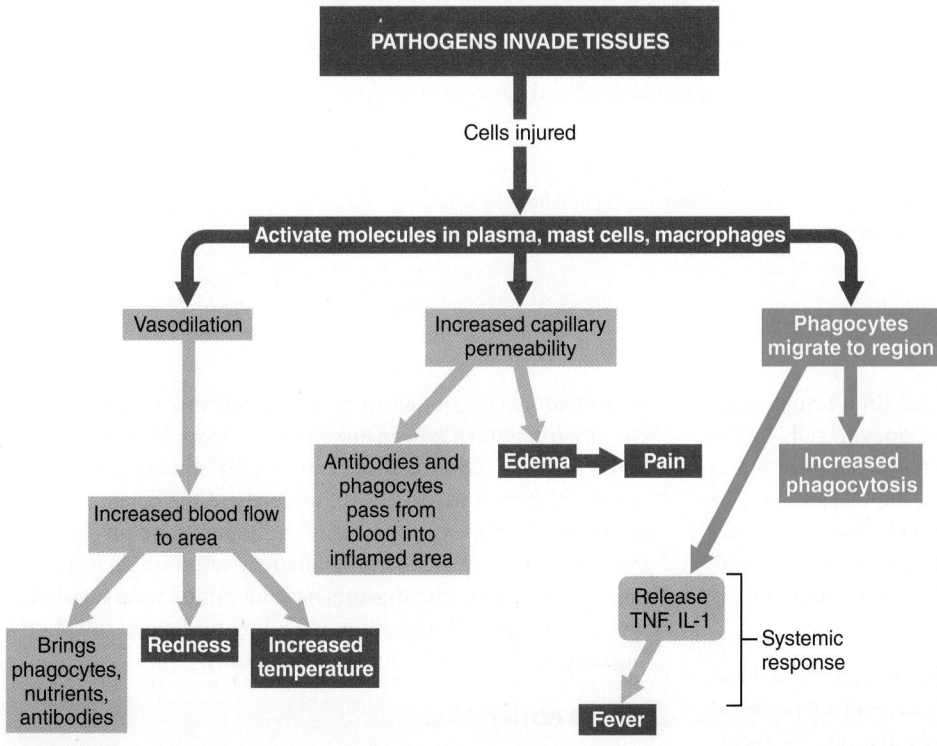

Figure 45-3 Overview of the inflammatory response

When skin is injured, bacteria (and other pathogens) enter the tissues. Tissue injury activates a clotting factor in the blood that turns on three molecular pathways in the plasma. Some of the molecules produced dilate blood vessels and increase capillary permeability.

PREDICT If untreated, how would your foot look the day after stepping on a nail?

© Cengage Learning

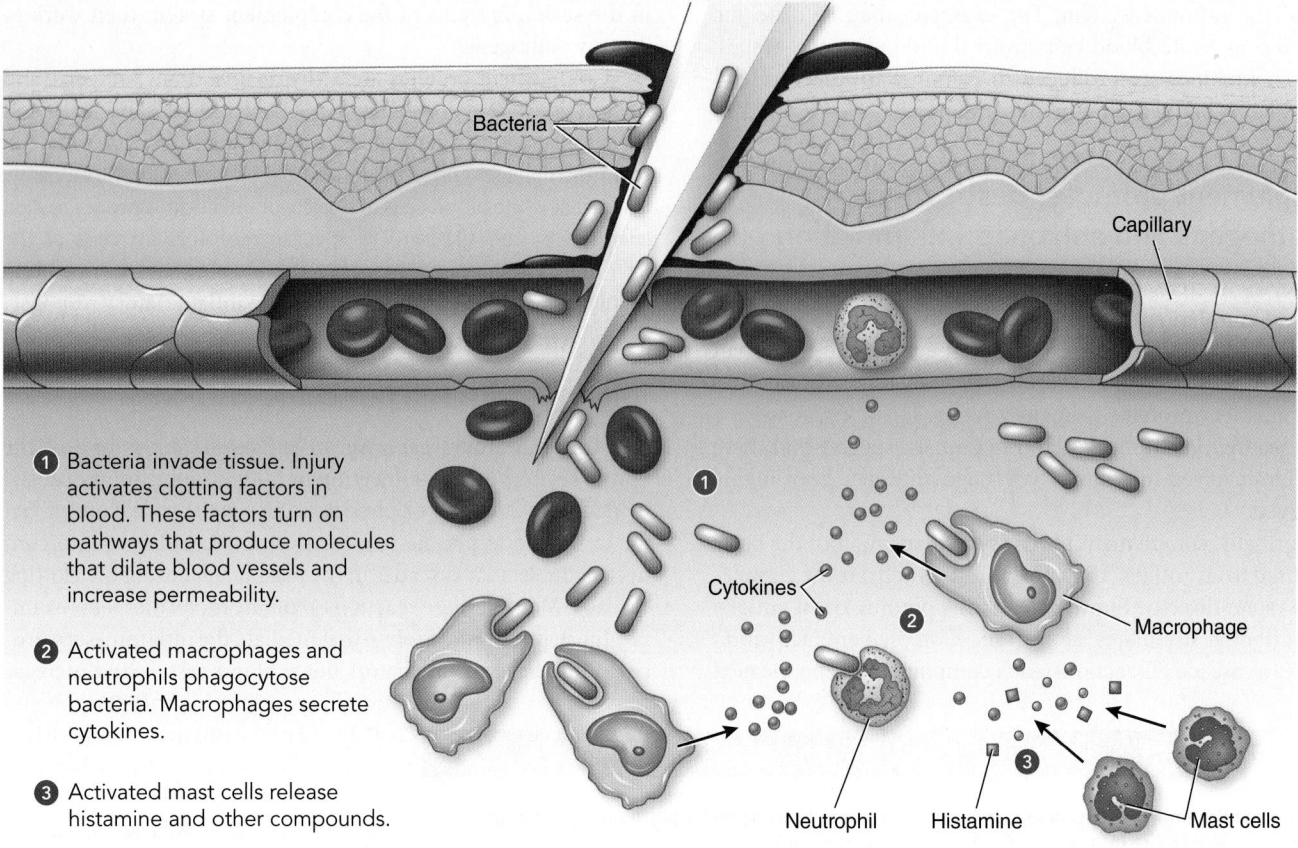

① Bacteria invade tissue. Injury activates clotting factors in blood. These factors turn on pathways that produce molecules that dilate blood vessels and increase permeability.

② Activated macrophages and neutrophils phagocytose bacteria. Macrophages secrete cytokines.

③ Activated mast cells release histamine and other compounds.

(a) Mast cells in the tissue secrete cytokines, histamine, and other compounds that increase capillary permeability and dilate blood vessels. Certain cytokines attract phagocytic cells. Macrophages and neutrophils in the infected area engulf and destroy bacteria. Macrophages also secrete cytokines that recruit additional neutrophils.

Figure 45-4 *Animation* **The inflammatory response**

The body reacts to tissue injury, infection, or exposure to certain toxins with an inflammatory response. Bacteria are represented by *yellow rods*, cytokines by *green dots*, histamines by *orange squares*, and other mast cell secretions by *orange dots*.

© Cengage Learning

3. *Increased phagocytosis.* Increased blood flow brings large numbers of neutrophils and other phagocytic cells to the inflamed region within a few hours of tissue injury or infection. The phagocytes migrate out of the capillaries and into the infected tissues. Macrophages secrete chemokines and other cytokines that recruit and activate additional neutrophils. One of the main functions of inflammation is increased phagocytosis.

Although the inflammatory response begins as a local response, sometimes the entire body may become involved. *Fever,* a common clinical symptom of widespread inflammation, helps the body fight infection. Elevated body temperature increases phagocytosis and interferes with the growth and replication of microorganisms. Fever breaks down lysosomes, destroying cells infected by viruses. Fever also promotes the activity of certain lymphocytes. A short-term, low fever speeds recovery.

The inflammatory response is vital in overcoming infection, but it can also cause tissue damage and disease. *Chronic*

inflammation contributes to many chronic diseases. For example, inflammation has been linked with atherosclerosis, rheumatoid arthritis, Crohn's disease, diabetes, Alzheimer's disease, and cancer. Although we do not yet understand the details, chronic inflammation appears to occur when homeostatic mechanisms that normally turn off the inflammatory response do not function appropriately. Investigators have identified several risk factors for chronic inflammation, including prolonged infections, cigarette smoking, gum disease, and obesity.

ᴄHECKPOINT 45.2

- *What is the function of pattern recognition receptors?*
- PREDICT *What would happen if you had no NK cells? What would happen if you had no macrophages?*
- *Describe two functions of cytokines. What are two functions of complement proteins?*
- VISUALIZE *Draw a diagram illustrating the main steps in the inflammatory response.*

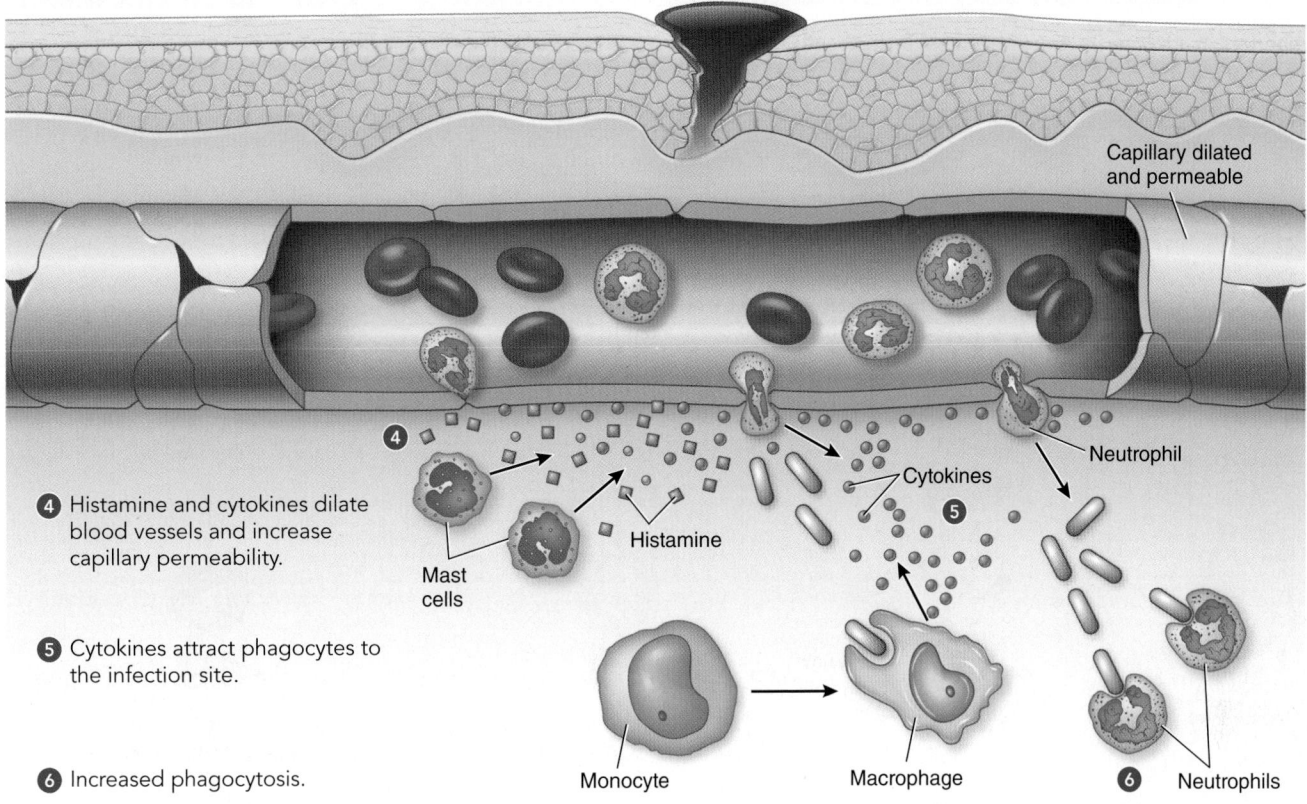

④ Histamine and cytokines dilate blood vessels and increase capillary permeability.

⑤ Cytokines attract phagocytes to the infection site.

⑥ Increased phagocytosis.

Capillary dilated and permeable

Neutrophil

Cytokines

Histamine

Mast cells

Monocyte

Macrophage

Neutrophils

(b) Large numbers of monocytes and phagocytic cells pass through the now dilated, permeable capillaries into the infected area. Monocytes develop into macrophages. Phagocytosis increases.

45.3 ADAPTIVE IMMUNE RESPONSES IN VERTEBRATES

LEARNING OBJECTIVES

4 Contrast cell-mediated and antibody-mediated immunity and give an overview of each process.

5 Describe the functions of the specialized cells of the adaptive immune system and of the major histocompatibility complex.

Adaptive immune responses are also referred to as *acquired immunity.* In contrast to innate immune mechanisms that are always present and ready to respond to infection, adaptive immune responses develop in response to an infection and *adapt* to the specific pathogens invading the body. These immune responses are highly specific for distinct macromolecules.

While innate immune responses destroy pathogens and prevent the spread of infection, the body mobilizes its adaptive immune responses. It takes several days to activate adaptive immune responses, but once in gear, these mechanisms are extremely effective. As discussed earlier, adaptive immune responses are precisely targeted to destroy specific antigens, and they have immunological memory. Two types of adaptive immune responses are *cell-mediated immunity* and *antibody-mediated immunity.*

Many types of cells are involved in adaptive immune responses

The main types of cells of the adaptive immune system are shown in FIGURE 45-5. Two key cell types that participate in adaptive immune responses are lymphocytes and *antigen-presenting cells.* Phagocytes and many other cells important in innate immune responses also participate in specific responses. **Eosinophils,** for example, are white blood cells that release proteins, including cytokines, from their granules. Some of these cytokines are important in adaptive immune responses. Eosinophils also release proteins that are toxic to parasitic worms (the life cycles of some parasitic worms are described in Chapter 31).

Lymphocytes are the principal warriors in adaptive immune responses Lymphocytes are important in both adaptive and innate immunity. Natural killer cells (NK cells) were discussed in the previous discussion of innate immunity. Recall that NK cells are lymphocytes that kill virally infected cells and tumor cells. **T lymphocytes,** also called **T cells,** and **B lymphocytes,** also called **B cells,** recognize antigens and are key cells in adaptive immunity.

T cells are responsible for **cell-mediated immunity.** They are the body's cellular soldiers. T cells travel to the site of infection and attack body cells infected by invading pathogens as well as foreign cells (such as those introduced in tissue grafts or organ transplants). T cells also destroy cells altered by mutation (cancer cells).

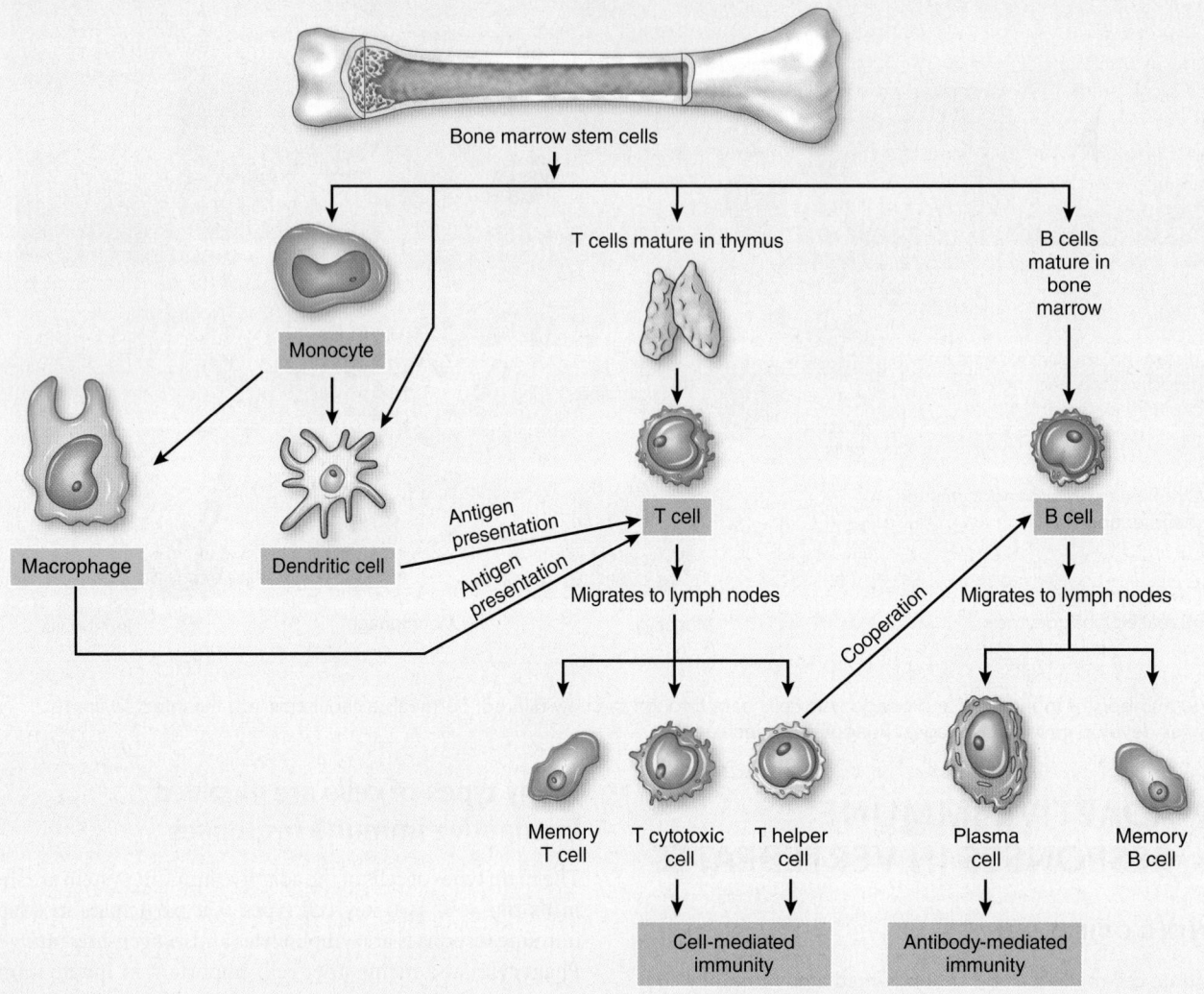

Figure 45-5 *Animation* **Some key cells in adaptive immune responses**
Recall that macrophages and dendritic cells also have vital roles in innate immunity.
© Cengage Learning

PREDICT What effect would multiple myeloma (a bone marrow cancer in which malignant plasma cells accumulate in the bone marrow) have on the production of antigen-presenting cells and lymphocytes?

B cells are responsible for **antibody-mediated immunity.** B cells mature into **plasma cells,** which produce specific antibodies. Antibodies bind to specific antigens, neutralizing them or marking them for destruction. Antibody-mediated immunity is one of the body's chief chemical defense strategies.

All lymphocytes develop from stem cells in the bone marrow. B cells complete their development in the adult bone marrow. T cells mature in the **thymus gland** (from which the *T* in their name derives). Large numbers of mature lymphocytes reside in lymph organs, including the spleen, lymph nodes, tonsils, and other lymph tissues strategically positioned throughout the body.

Although T cells and B cells are similar in appearance when viewed with a light microscope, sophisticated techniques such as fluorescence microscopy demonstrate that these cells can be differentiated by their unique cell-surface macromolecules.

T cells and B cells have different functions and life histories and tend to locate in (or "home" to) separate regions of the spleen, lymph nodes, and other lymph tissues.

Millions of B cells are produced in the bone marrow daily. Each B cell is genetically programmed to encode a glycoprotein receptor that can bind with a specific type of antigen. When a B cell comes into contact with an antigen that binds to its receptors, the B cell becomes activated. B-cell activation is a complex process that typically requires the participation of a particular type of T cell.

Once activated, a B cell divides rapidly to form a clone of identical cells. The cloned B cells differentiate into *plasma cells,* which produce antibodies. A plasma cell can produce more than ten million molecules of antibody per hour! The antibody binds to the antigen that originally activated the B cells. Some activated B cells do not become plasma cells; instead, they become long-living **memory B cells** that continue to produce small amounts of antibody after the body overcomes an infection.

On their way from the bone marrow to the lymph tissues, T cells are processed in the thymus gland. As T cells move through the thymus, they divide many times and develop specific surface proteins with distinctive receptor sites. The thymus makes T cells *immunocompetent,* capable of producing an effective immune response after exposure to an antigen.

The T-cell population undergoes positive and negative selection. In *negative selection* T cells in the thymus that react to self-antigens undergo apoptosis. Immunologists estimate that more than 90% of developing T cells are negatively selected in this way.

In *positive selection* T cells are allowed to mature if they recognize self-antigens and bind with foreign antigens. These T cells differentiate and leave the thymus to take up residence in other lymph tissues or to launch immune responses in infected tissues. By selecting only appropriate T cells, the thymus gland ensures that T cells can distinguish between the body's own molecules and foreign antigens.

Most T cells in the thymus differentiate just before birth and during the first few months of postnatal life. If the thymus is removed before this processing takes place, an animal does not develop cell-mediated immunity. If the thymus is removed after that time, cell-mediated immunity is less seriously impaired.

T cells are distinguished by the **T-cell receptor (TCR),** which recognizes specific antigens. Two main populations of T cells are T cytotoxic cells and T helper cells. **T cytotoxic (T_C) cells** are also known as *CD8 cells* because they have a glycoprotein designated CD8 on their plasma membrane surface. T_C cells recognize and destroy cells with foreign antigens on their surfaces. Their target cells include virus-infected cells, cancer cells, and foreign tissue grafts.

T helper cells (T_H cells), also known as *CD4 cells,* have a surface glycoprotein designated CD4. T_H cells secrete cytokines that activate B cells, T cells, and macrophages. **Regulatory T cells,** known as T_{regs}, are a subpopulation of T_H cells. T_{regs} help regulate immune responses by suppressing the function of certain T cells. After an infection, **memory T cells** (both memory T_C cells and memory T_H cells) remain in the body.

Antigen-presenting cells activate T helper cells Macrophages, dendritic cells, and B cells function as "professional" **antigen-presenting cells (APCs)** that display foreign antigens as well as their own surface proteins. APCs are inactive until their pattern recognition receptors recognize PAMPs on pathogens. Once activated, an APC ingests the pathogen. Lysosomal enzymes inside the APC degrade most bacterial antigens, but not all. The APC displays fragments of the foreign antigens on its cell surface in association with a type of self-molecule (class II MHC, discussed later). The activated APC expresses additional signaling molecules, called *co-stimulatory molecules,* along with the displayed antigen. These fully functional APCs present displayed antigens to T cells.

In addition to their function as APCs, macrophages secrete more than 100 different compounds, including cytokines and enzymes that destroy bacteria. When stimulated by bacteria, macrophages secrete interleukins that activate B cells and certain T cells. We have seen that interleukins also promote innate immune responses such as fever.

Receptors on dendritic cells bind antigens of microbes. Dendritic cells take these antigens, process them, and display antigen fragments on their cell surface. This is how they become APCs. The dendritic cells leave the epithelial lining and, guided by chemokines, migrate to the lymph nodes. Co-stimulatory molecules, along with the displayed antigen, attract and activate specific T cells capable of responding to the antigen. Thus, dendritic cells are specialized to bind, process, transport, and present antigens to T cells.

The major histocompatibility complex is responsible for recognition of self

The ability of the vertebrate immune system to distinguish self from nonself depends largely on a cluster of closely linked genes known as the **major histocompatibility complex (MHC).** The human MHC is also referred to as **human leukocyte antigen complex (HLA).** (We use the designations MHC and HLA interchangeably in this chapter.) The HLA consists of a group of loci on chromosome 6. The HLA genes are polymorphic (variable). Within the population there are multiple alleles for each locus. About 10,000 alleles have been identified. (Note that MHC and HLA refer to both the loci and to the membrane proteins they encode.)

The HLA genes encode HLA membrane proteins referred to as self-antigens. Because there are so many alleles, the HLA membrane proteins differ in chemical structure, function, and tissue distribution. With so many possible combinations, no two people, except identical twins, are likely to have all the same HLA proteins on their cell membranes. The more closely related two individuals are, the more HLA alleles they have in common. You can think of your HLA proteins as a biochemical "fingerprint."

There are three main Class I HLA loci, which encode glycoproteins expressed on the surface of most nucleated cells. These antigens bind with foreign antigens from viruses or other pathogens processed within the cell to form a foreign antigen–class I HLA glycoprotein complex. The cell displays this complex on its surface and presents it to T_C cells. Thus, any infected cell can function as an antigen-presenting cell, which can activate certain T_C cells or can be instructed to self-destruct.

Six main class II HLA loci encode glycoproteins expressed primarily on "professional" APCs: dendritic cells, macrophages, and B cells. These class II HLA antigens combine with foreign antigen from bacteria, and the APC then presents the complex to T_H cells. Class III HLA loci encode many secreted proteins involved in the immune response, including components of the complement system and TNFs.

CHECKPOINT 45.3

- **CONNECT** *How are cell-mediated immunity and antibody-mediated immunity alike? How are they different?*
- *What is the function of antigen-presenting cells? Identify three types of antigen-presenting cells.*
- **PREDICT** *Which kidney transplant would more likely be successful—a transplant from an unknown donor or a transplant between first cousins? Support your answer. (Hint: Consider the HLA.)*

45.4 CELL-MEDIATED IMMUNITY

6 Describe the sequence of events in cell-mediated immunity.

T cells and APCs (mainly dendritic cells and macrophages) are responsible for cell-mediated immunity. T cells destroy cells infected with viruses, as well as macrophages that have ingested "smart" bacteria. These bacteria escape from the vesicles in which they are enclosed during phagocytosis. They escape destruction by lysosomal enzymes by "hiding" in the cytoplasm of macrophages. T cells destroy cells that have been altered in some way, such as cancer cells. They also destroy the cells of foreign grafts such as a transplanted kidney or liver.

How do T cells know which cells to target? T cells do not recognize antigens unless they are presented properly. Recall that when a virus (or other pathogen) infects a cell, some of the viral protein is broken down to peptides and displayed with class I MHC molecules on the cell surface. Only T cells with receptors that bind to the specific antigen–MHC complex become activated.

Recall that there are two general populations of T cells: T_H cells and T_C cells. Each T_H cell expresses molecules of the glycoprotein called CD4, and each T_C cell expresses CD8 molecules. Within each general T cell population there are thousands of different antigen specificities that the T-cell receptors (TCRs) can recognize. Activation of T_H cells, for example, occurs when phagocytic cells digest antigen and parts of the antigen are placed within their surface MHC complexes. They then present these antigens to the TCR on T_H cells as part of a foreign antigen–class II MHC glycoprotein complex.

Once activated by the foreign antigen–MHC class II antigen complex displayed on the surface of APCs, the activated T_H cells secrete interleukin 2. This cytokine is a growth factor that stimulates the activated cells to proliferate. The T_H cells divide several times, giving rise to a clone of T_H cells (FIG. 45-6). This process is called **clonal expansion.**

Some of the T_H cells produced migrate to the site of infection. These cells can produce many different types of cytokines. Some attract macrophages to the site of infection and promote the destruction of intracellular pathogens. Others function mainly in promoting antibody-mediated immunity.

Each T_C cell expresses the CD8 molecule, as well as more than 50,000 identical TCRs, on its surface. The TCRs bind to

Figure 45-6 *Animation* Cell-mediated immunity

When activated by a foreign antigen–MHC complex presented by an APC and by cytokines, a T_H cell divides and gives rise to a clone of T_H cells. T_C cells are also activated by antigens presented on the surfaces of infected cells, and they also form a clone. Both types of T cells migrate to the site of infection. T_H cells release cytokines that stimulate T_C cells and activate macrophages, which then phagocytose pathogens. T_C cells release proteins that destroy infected cells. There are many receptors on the surface of each T cell and each B cell. For simplicity, only one receptor per cell is shown.

CONNECT What types of cells can function as APCs?

© Cengage Learning

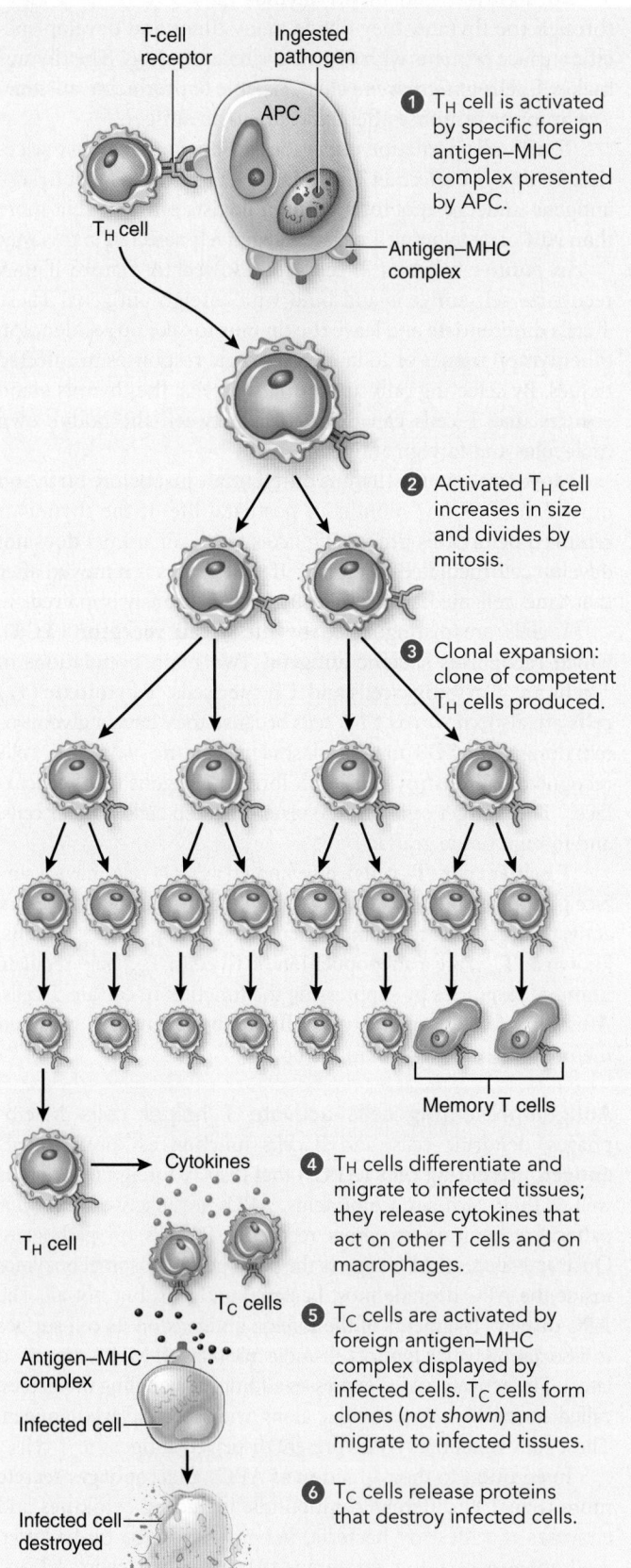

① T_H cell is activated by specific foreign antigen–MHC complex presented by APC.

② Activated T_H cell increases in size and divides by mitosis.

③ Clonal expansion: clone of competent T_H cells produced.

Memory T cells

④ T_H cells differentiate and migrate to infected tissues; they release cytokines that act on other T cells and on macrophages.

⑤ T_C cells are activated by foreign antigen–MHC complex displayed by infected cells. T_C cells form clones (*not shown*) and migrate to infected tissues.

⑥ T_C cells release proteins that destroy infected cells.

one specific type of antigen. T_C cells recognize antigens presented to them as part of a foreign antigen–class I MHC glycoprotein complex. Generally, fewer than 1 in 100,000 T cells have the same antigen specificity and can respond. However, after clonal expansion as many as 1 in 10 T_C cells may be specific for the targeted antigen. T_C-cell activation requires at least two signals in addition to the presented antigen: a co-stimulatory signal and an interleukin signal. Once activated, a T_C cell increases in size and gives rise to a clone of many effector T_C cells.

Effector T_C cells make up the cell infantry. They leave the lymph nodes and make their way to the infected area, where they destroy target cells within seconds after contact. When a T_C cell combines with antigen on the surface of the target cell, it destroys the cell in much the same way that NK cells destroy their target cells. The T_C cell secretes perforins that perforate the plasma membrane of the target cell. Granzymes then induce the cell to self-destruct by apoptosis. After releasing cytotoxic substances, the T cell disengages itself from its victim cell and seeks out a new target.

The sequence of events for production and function of T_C cells in cell-mediated immunity can be summarized as follows:

virus invades body cell $\longrightarrow$ infected cell displays foreign antigen–MHC class I antigen complex $\longrightarrow$ specific T_C cell is activated by this complex $\longrightarrow$ clone of T_C cells produced $\longrightarrow$ effector T_C cells migrate to area of infection $\longrightarrow$ T_C cells are stimulated by cytokines released by T_H cells $\longrightarrow$ T_C cells release enzymes that destroy target cells

CHECKPOINT 45.4

- **CONNECT** *How do APCs, T_C cells, and T_H cells work together in cell-mediated immunity?*

45.5 ANTIBODY-MEDIATED IMMUNITY

LEARNING OBJECTIVES

7 Summarize the sequence of events in antibody-mediated immunity, including the effects of antigen–antibody complexes on pathogens.

8 Describe the basic structure and function of an antibody and explain the basis of antibody diversity.

9 Describe the basis of immunological memory and contrast secondary and primary immune responses.

10 Compare active and passive immunity, and give examples of each.

We have discussed two types of molecules that recognize specific antigens: T-cell receptors and MHC molecules (HLA in humans). Here we turn our attention to the third type, antibodies. B cells are responsible for antibody-mediated immunity.

A given B cell can produce many copies of one specific antibody. Antibody molecules serve as cell-surface receptors that combine with antigens. Only a B cell with a receptor complementary in shape to a particular antigen can bind that antigen. The antigen enters the B cell by receptor-mediated endocytosis (see Chapter 5). The antigen is then degraded into peptide fragments, and the B cell displays peptide fragments together with MHC protein class II on its surface.

In most cases, the activation of B cells involves APCs and T_H cells (FIG. 45-7). T_H cells stimulate B cells to divide and produce antibodies. However, the T_H cell itself must first be activated. T_H cells do not recognize an antigen that is presented alone. An APC must present the antigen as part of a foreign antigen–class II MHC complex on its surface. When an APC displaying a foreign antigen–MHC complex contacts a T_H cell with complementary T-cell receptors, a complicated interaction occurs. Multiple chemical signals are sent back and forth between cells. For example, the APC secretes interleukins, such as IL-1, that activate T_H cells.

B cells serve as APCs to T cells. An activated T_H cell binds with the foreign antigen–MHC complex on the B cell. The T_H cell then releases interleukins, which together with antigen activate the B cell.

Once activated, a B cell divides by mitosis, giving rise to a clone of identical cells (FIG. 45-8). This clonal expansion in response to a specific antigen is known as *clonal selection* (discussed later). The cloned B cells mature into plasma cells that secrete antibodies. These antibodies are specific to the antigen that activated the original B cell. It is important to remember that the specificity of the clone is determined *before* the B cell encounters the antigen.

Unlike T cells, most plasma cells do not leave the lymph nodes. The antibodies they secrete pass out of the lymph tissues and make their way via the lymph and blood to the infected area. The secreted antibody is a soluble form of the glycoprotein receptor of the activated B cell.

The sequence of events in antibody-mediated immunity can be summarized as follows:

a pathogen invades the body $\longrightarrow$ an APC phagocytoses the pathogen $\longrightarrow$ foreign antigen–MHC complex is displayed on APC surface $\longrightarrow$ T_H cell binds with foreign antigen–MHC complex $\longrightarrow$ activated T_H cell interacts with a B cell that displays the same antigen $\longrightarrow$ B cell is activated $\longrightarrow$ clone of B cells produced $\longrightarrow$ B cells differentiate, becoming plasma cells $\longrightarrow$ plasma cells secrete antibodies $\longrightarrow$ antibodies form complexes with pathogen $\longrightarrow$ pathogen is destroyed

Some activated B cells do not differentiate into plasma cells. Instead, they become memory B cells (discussed in a later section).

A typical antibody consists of four polypeptide chains

An antibody molecule, also called **immunoglobulin (Ig)**, has several functions. It combines with antigen and activates

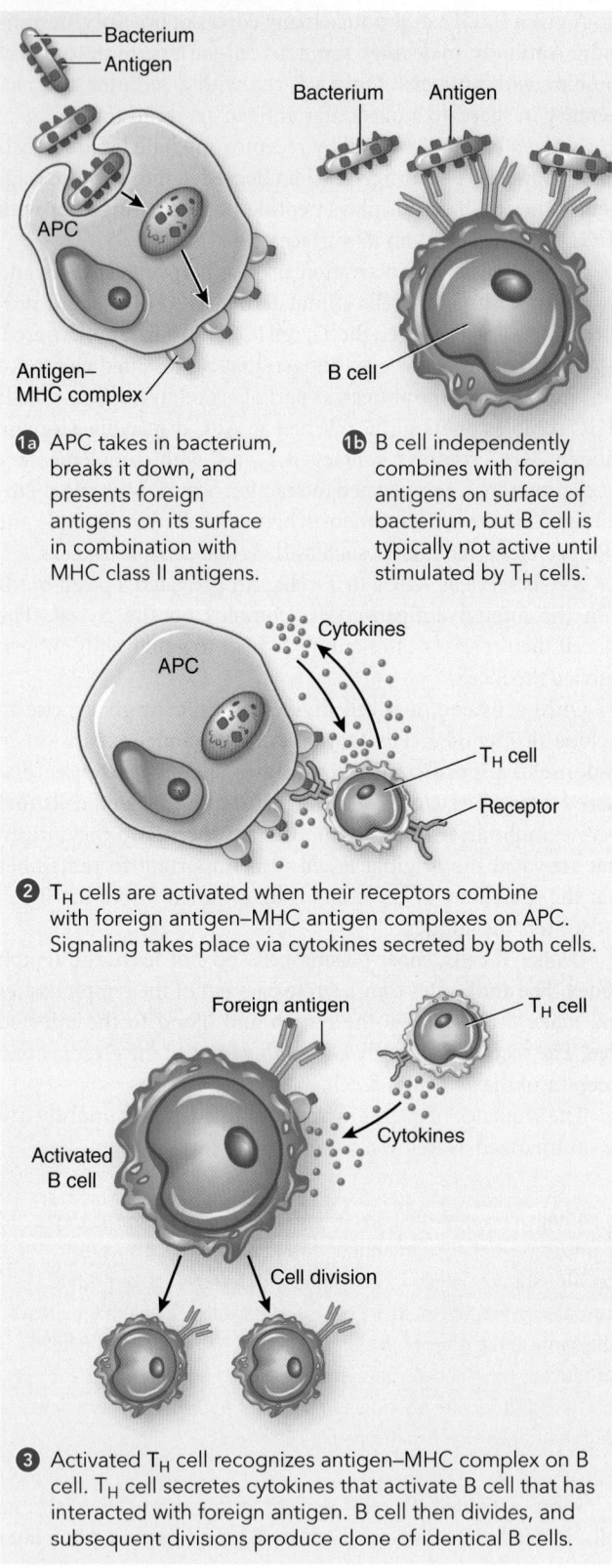

Bacterium
Antigen

Bacterium   Antigen

APC

Antigen–
MHC complex

B cell

1a APC takes in bacterium, breaks it down, and presents foreign antigens on its surface in combination with MHC class II antigens.

1b B cell independently combines with foreign antigens on surface of bacterium, but B cell is typically not active until stimulated by T_H cells.

Cytokines

APC

T_H cell

Receptor

2 T_H cells are activated when their receptors combine with foreign antigen–MHC antigen complexes on APC. Signaling takes place via cytokines secreted by both cells.

Foreign antigen

T_H cell

Activated
B cell

Cytokines

Cell division

3 Activated T_H cell recognizes antigen–MHC complex on B cell. T_H cell secretes cytokines that activate B cell that has interacted with foreign antigen. B cell then divides, and subsequent divisions produce clone of identical B cells.

Figure 45-7 B-cell activation

This figure is a detailed illustration of the first four steps in Figure 45-8. There are many receptors on the surface of each T cell and each B cell. For simplicity, only one receptor per cell is shown.

© Cengage Learning

processes that destroy the bound antigen. For example, an antibody may stimulate phagocytosis. Note that an antibody does not destroy an antigen directly; rather, it *labels* the antigen for destruction. When it is attached to the B-cell surface, an antibody molecule serves as the B-cell receptor for antigen.

In a Y-shaped antibody molecule, the tail of the Y is called the *Fc fragment* (*c* indicates that this fragment crystallizes during cold storage). The Fc fragment binds with receptors on phagocytes and other cells of the immune system. This region of the antibody molecule also binds with molecules of the complement system. The two arms of the Y are referred to as *Fab fragments* (*Fab* stands for *fragment, antigen-binding*). The tips of the Fab fragments bind with antigen (**FIG. 45-9**). The Y shape permits the antibody molecule to combine with two antigen molecules, resulting in the formation of *antigen–antibody complexes.*

Each antibody molecule consists of four polypeptide chains: two long chains, called *heavy chains,* and two short chains, called *light chains.* Each chain has a constant region and a variable region. In the *constant (C) region* of the heavy chains, the amino acid sequence is constant within a particular immunoglobulin class. The C region of the heavy chains forms the tail of the Y part of the antibody molecule. You can think of this C region as the handle portion of a door key.

The *variable (V) region* has a unique amino acid sequence, like the pattern of bumps and notches on the part of a key that fits into a lock. At its variable regions, the antibody folds three-dimensionally, assuming a shape that enables it to combine with a specific antigen. When they meet, antigen and antibody fit together somewhat like a lock and key. They must fit in just the right way for the antibody to be effective (**FIG. 45-10**).

The specific part of the antigen molecule that is recognized by an antibody or T-cell receptor is called an *antigenic determinant,* or *epitope.* An antigen may have many antigenic determinants on its surface; some have hundreds. Antigenic determinants often differ from one another, so several different kinds of antibodies can combine with a single antigen. A given antibody can bind with different strengths, or *affinities,* to different antigens. In the course of an immune response, higher affinity antibodies are generated. This increase in affinity occurs because as activated B cells are replicated, mutations occur in genes that encode the antibodies. Cells with antibodies that bind more tightly are selected during the ongoing immune response—a striking example of microevolution in action.

Antibodies are grouped in five classes

Antibodies are grouped in five classes, defined by unique amino acid sequences in the C regions of the heavy chains. The classes are named using the abbreviation *Ig* for *immunoglobulin* and are designated IgG, IgM, IgA, IgD, and IgE. In humans about 75% of the circulating antibodies belong to the *IgG* class; they are part of the gamma globulin fraction of the plasma. *IgM* is the first antibody that humans make in an immune response. IgG and IgM share some functional properties. Both interact with macrophages and activate the complement system. They defend against many pathogens carried in the blood, including bacteria, viruses, and some fungi. IgG is the antibody that crosses the

B cells, APCs, and T_H cells are responsible for antibody-mediated immunity.

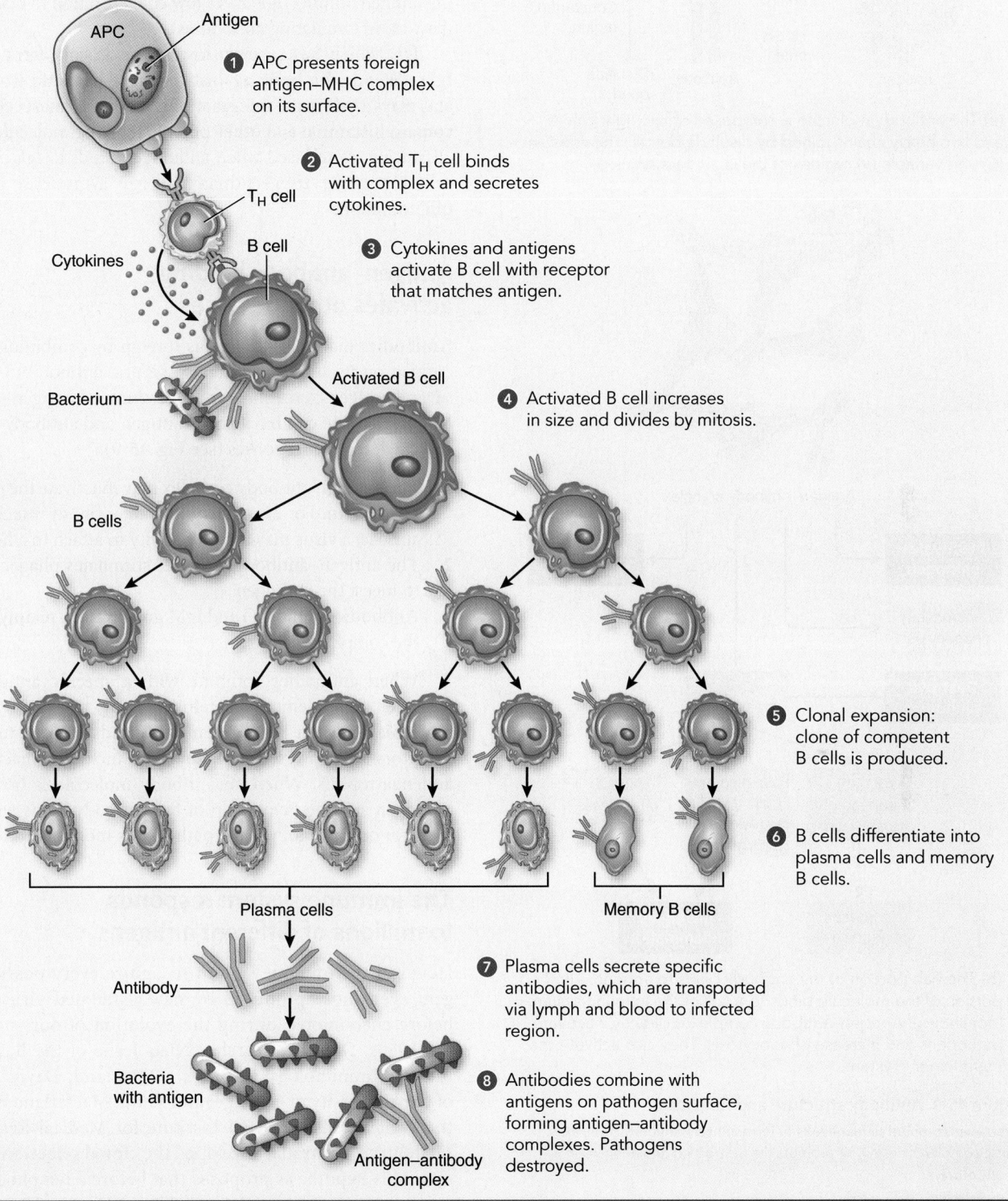

1 APC presents foreign antigen–MHC complex on its surface.

2 Activated T_H cell binds with complex and secretes cytokines.

3 Cytokines and antigens activate B cell with receptor that matches antigen.

4 Activated B cell increases in size and divides by mitosis.

5 Clonal expansion: clone of competent B cells is produced.

6 B cells differentiate into plasma cells and memory B cells.

7 Plasma cells secrete specific antibodies, which are transported via lymph and blood to infected region.

8 Antibodies combine with antigens on pathogen surface, forming antigen–antibody complexes. Pathogens destroyed.

Figure 45-8 *Animation* **Antibody-mediated immunity**

A specific B cell may be activated when a specific antigen binds to receptors on its surface. Typically, the B cell also requires cytokine stimulation from an activated T_H cell (see Fig. 45-7). Once activated, the B cell divides and produces a clone of B cells. These B cells differentiate and become plasma cells that secrete antibodies. Some of the plasma cells remain in the lymph tissues, but antibodies are transported to the site of infection by the blood or lymph. After the infection has been overcome, memory B cells remain in the tissues. There are many receptors on the surface of each T cell and each B cell. For simplicity, only one receptor per cell is shown.

PREDICT What would be the effect on immune responses if T_c cells collected in the lymph tissue as plasma cells do?
© Cengage Learning

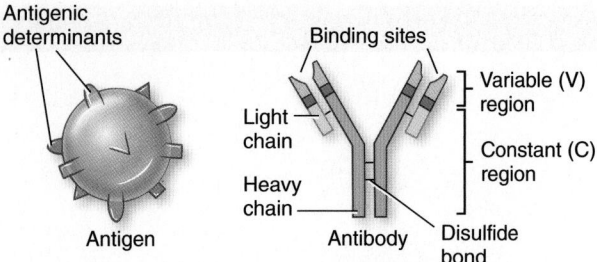

(a) The antibody molecule is composed of two light chains and two heavy chains joined by disulfide bonds. The constant (C) and variable (V) regions of the chains are labeled.

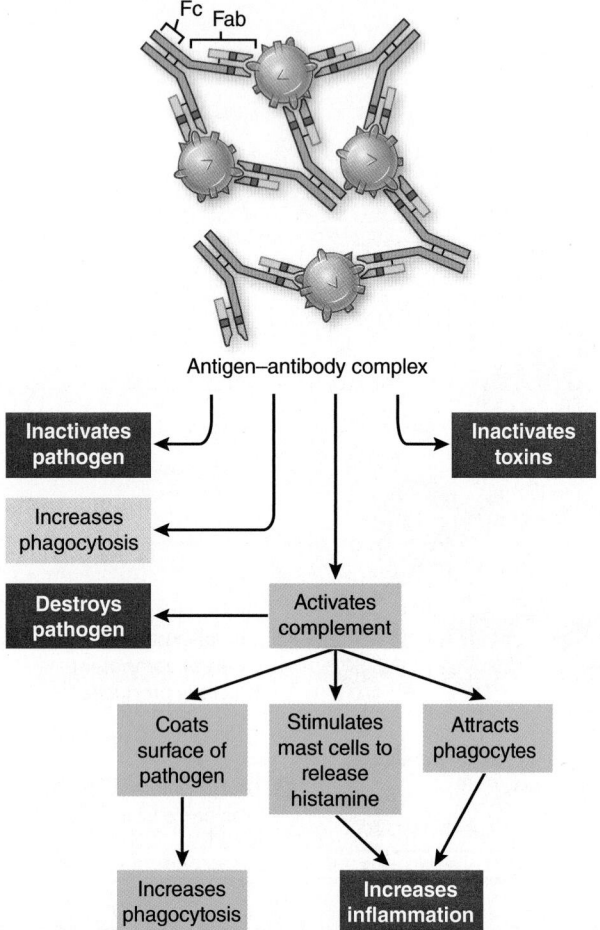

(b) The Fab portion of the antibody binds to antigen. The Fc portion of the molecule binds to a cell of the immune system (*not shown*). Antigen–antibody complexes directly inactivate pathogens and increase phagocytosis. They also activate the complement system.

Figure 45-9 Antibody structure and function

Antibodies combine with antigens to form antigen–antibody complexes. These complexes destroy or promote the destruction of pathogens in a variety of ways.

© Cengage Learning

placenta (the organ of exchange between mother and developing fetus) and protects the fetus and, later, the newborn baby.

IgA, present in mucus, tears, saliva, and breast milk, prevents viruses and bacteria from attaching to epithelial surfaces. This immunoglobulin, which defends against inhaled or ingested pathogens, is strategically secreted into the respiratory, digestive, urinary, and reproductive tracts. Like IgM, *IgD* is an important immunoglobulin on the B-cell surface. It has a critical role in maturation of B cells and helps activate B cells following antigen binding. IgD has a low concentration in plasma (less than 1% of circulating antibodies).

IgE, which has an even lower plasma concentration than IgD, defends the body against invading parasitic worms. IgE also plays a role in allergic reactions. It binds to mast cells, which contain histamine and other potent signaling molecules. These chemicals are released when an antigen binds to IgE on a mast cell. Histamine triggers many allergy symptoms, including inflammation.

Antigen–antibody binding activates other defenses

Antibodies mark a pathogen as foreign by combining with an antigen on its surface. Generally, several antibodies bind with several antigens, creating a mass of clumped antigen–antibody complexes. The combination of antigen and antibody activates several defensive responses (see Fig. 45-9).

1. The antigen–antibody complex may inactivate the pathogen or its toxin. For example, when an antibody attaches to its surface, a virus may lose the ability to attach to a host cell.
2. The antigen–antibody complex stimulates phagocytic cells to ingest the pathogen.
3. Antibodies of the IgG and IgM groups work mainly through the complement system.

When antibodies combine with a specific antigen on a pathogen, complement proteins destroy the pathogen. IgG molecules have an Fc fragment that binds Fc receptors; these receptors are expressed on many cells, including macrophages and neutrophils. When an antibody molecule is bound to a pathogen and the Fc portion of the antibody binds with an Fc receptor on a phagocyte, the pathogen is more easily destroyed.

The immune system responds to millions of different antigens

How can the immune system recognize every possible antigen, even those produced by newly mutated viruses never before encountered during the evolution of our species? In the 1950s, Danish researcher Niels Jerne of the Basel Institute for Immunology in Basel, Switzerland; David Talmage of the University of Chicago; and Frank Macfarlane Burnet of the Walter and Eliza Hall Institute for Medical Research in Melbourne, Australia, developed the **clonal selection** hypothesis. This hypothesis proposes that before a lymphocyte ever encounters an antigen, the lymphocyte has specific receptors for that antigen on its surface. (Jerne and Burnet later were awarded Nobel Prizes in Physiology or Medicine for their work in immunology.)

When an antigen binds to a matching receptor on a lymphocyte (T cell), it activates the lymphocyte, which then gives

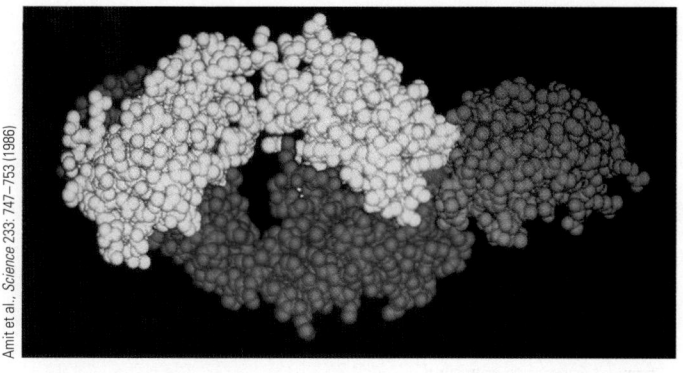

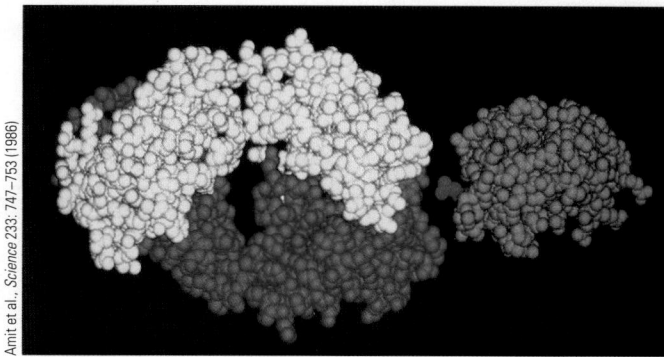

(a) A portion of the antigenic determinant (*red*) fits into a groove in the antibody molecule.

(b) The antigen–antibody complex has been pulled apart to show its structure.

Figure 45-10 An antigen–antibody complex

The components of an antigen–antibody complex fit together as shown in this computer simulation of their molecular structure. Antigen lysozyme, *green;* heavy chain of the antibody, *blue;* light chain, *yellow.*

rise to a clone of cells with identical receptors. A problem with this hypothesis, when it was initially proposed, was its suggestion that our cells must contain millions of separate antibody genes, each coding for an antibody with a different specificity. Although each human cell has a large amount of DNA, it is not enough to provide a different gene to encode each of the millions of possible specific antibody molecules.

An explanation for how the immune system could have so many different specificities came later. In 1965, W. J. Dreyer of the Institute of Biochemistry at the University of Zurich, Switzerland, and Joe Claude Bennett of the University of Alabama at Birmingham suggested that the C region and the V region of an immunoglobulin are encoded by two separate genes. Their hypothesis was at first rejected by many biologists because it contradicted the prevailing hypothesis that one gene codes for one polypeptide.

The technology needed to test Dreyer and Bennett's hypothesis was not immediately available, and it was not until 1976 that Susumu Tonegawa of the Massachusetts Institute of Technology and his colleagues demonstrated that separate genes encode the V and C regions of immunoglobulins. Tonegawa further showed that three separate families of genes code for immunoglobulins and that each gene family contains a large number of DNA segments that code for V regions. Recombination of these DNA segments during the differentiation of B cells is responsible for antibody diversity. Tonegawa was awarded the 1987 Nobel Prize in Physiology or Medicine for his work, which transformed the emerging science of immunology.

We now understand that in undifferentiated B cells, gene segments are present for a number of different V regions, for one or more junction (J) regions, and for one or more different C regions (**FIG. 45-11**). Rearrangement of these DNA segments produces an enormous number of potential combinations. Millions of different types of B (and T) cells are produced. By chance, one of those B cells may produce just the right antibody to destroy a pathogen that invades the body. To appreciate the capacity for antibody diversity, consider the diverse combinations of things you create in your everyday life. A familiar example is making an ice-cream sundae. Imagine the varied combinations that are possible using 10 flavors of ice cream, 6 types of sauce, and 15 kinds of toppings.

Researchers have identified additional sources of antibody diversity. For example, the DNA of mature B cells that codes for the regions of immunoglobulins mutates very readily. These somatic mutations produce genes that code for slightly different antibodies.

We have used antibody diversity here as an example, but similar genetic mechanisms account for the diversity of T-cell receptors. Although we may actually use only a relatively few types of antibodies or T cells in a lifetime, the remarkable diversity of the immune system prepares it to attack most potentially harmful antigens that may invade the body.

Monoclonal antibodies are highly specific

Monoclonal antibodies are identical antibodies produced by cells cloned from a single cell. Each antibody reacts only with the same antigenic determinant (epitope) of the same antigen. Monoclonal antibodies are used in research and in medicine.

One way to produce monoclonal antibodies in the laboratory is to inject mice with the antigen of interest, such as an antigen from a particular bacterium. After the mice have produced antibodies to the antigen, their plasma cells are collected. However, these normal plasma cells survive in culture for only a few generations, limiting the amount of antibody that can be produced. In contrast, cancer cells live and divide in tissue culture indefinitely.

B cells can be suspended in a culture medium together with lymphoma cells from other mice. (Lymphoma is a cancer of lymphocytes.) Then the B cells and lymphoma cells are induced to fuse. They form hybrid cells, known as **hybridomas,** which have properties of the two "parent" cells. Hybridomas can be cultured indefinitely (a cancer cell property) and continue to secrete antibodies (a B-cell property). Researchers select hybrid

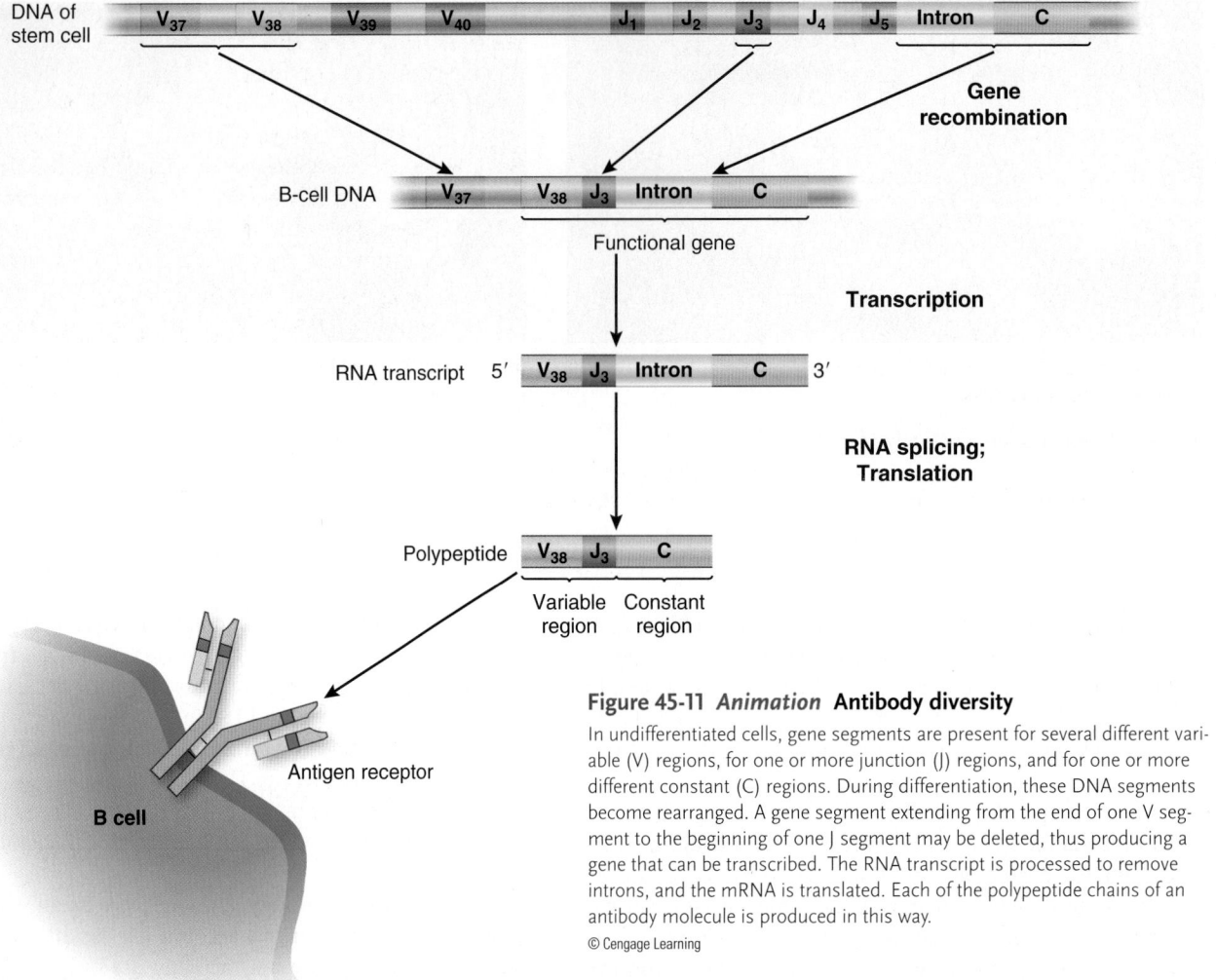

Figure 45-11 *Animation* **Antibody diversity**

In undifferentiated cells, gene segments are present for several different variable (V) regions, for one or more junction (J) regions, and for one or more different constant (C) regions. During differentiation, these DNA segments become rearranged. A gene segment extending from the end of one V segment to the beginning of one J segment may be deleted, thus producing a gene that can be transcribed. The RNA transcript is processed to remove introns, and the mRNA is translated. Each of the polypeptide chains of an antibody molecule is produced in this way.

© Cengage Learning

cells that are manufacturing the specific antibody needed and then clone them in a separate cell culture. Cells of this clone secrete large amounts of the particular antibody. Antibodies produced in this way are called *monoclonal antibodies.*

Immunologists also use recombinant DNA technology to produce monoclonal antibodies. Because of their purity and specificity, monoclonal antibodies have proved to be invaluable tools in modern biology. For example, a researcher may want to detect a specific molecule present in very small amounts in a mixture. The reaction of that molecule (the antigen) with a specific monoclonal antibody makes its presence known. Monoclonal antibodies are used in similar ways in various diagnostic tests. Home pregnancy tests make use of a monoclonal antibody that is specific for human chorionic gonadotropin (hCG), a hormone produced by a developing human embryo (see Chapters 50 and 51). A dozen monoclonal antibodies are currently being used clinically to treat cancer. The drug Herceptin is a monoclonal antibody used in treating a type of breast cancer.

Immunological memory is responsible for long-term immunity

Immunological memory depends on memory B cells and memory T cells. Following an immune response, memory T cells group strategically in many nonlymphatic tissues, including the lung, liver, kidney, and intestine. Many infections occur at such sites, and T cells stationed in those locations can respond quickly. In response to antigen, memory T cells rapidly become T_C cells and produce substances that kill invading cells.

Memory B cells have a "survival gene" that prevents apoptosis, the programmed cell death that is the eventual fate of plasma cells. Memory B cells continue to live and produce small amounts of antibody long after the body has overcome an infection. This circulating antibody is part of the body's arsenal of chemical weapons. If the same pathogen enters the body again, the antibody immediately targets it for destruction. At the same time, specific memory cells are stimulated to divide, producing new clones of plasma cells that produce the

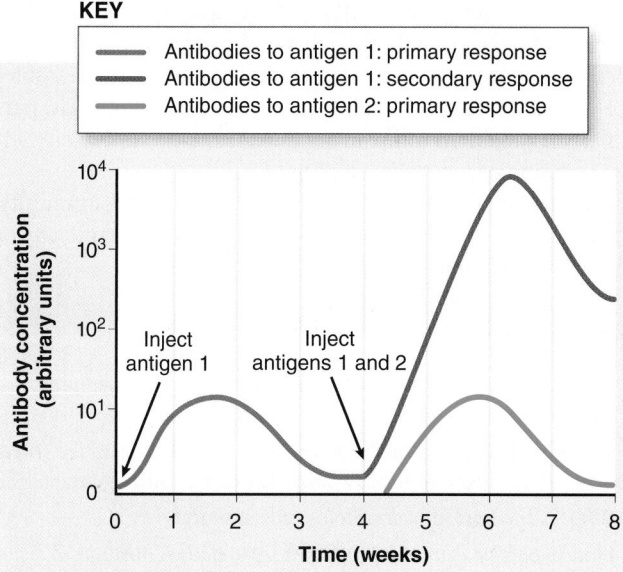

KEY
— Antibodies to antigen 1: primary response
— Antibodies to antigen 1: secondary response
— Antibodies to antigen 2: primary response

Inject antigen 1

Inject antigens 1 and 2

Figure 45-12 Immunological memory
Antigen 1 was injected at day 0, and the immune response was assessed by measuring levels of antibody specific for antigen 1. At week 4, the primary immune response had subsided. Antigen 1 was injected again, together with a new protein, antigen 2. The secondary immune response was greater and more rapid than the primary response. It was also specific to antigen 1. A primary immune response was made to the newly encountered antigen 2.
© Cengage Learning

same antibody. The presence of circulating antibodies is used clinically to detect previous exposure to specific pathogens. For example, some HIV screening tests measure antibody to the virus that causes AIDS.

A secondary immune response is more effective than a primary response The first exposure to an antigen stimulates a **primary immune response.** An infection, or an antigen injected into an animal, causes specific antibodies to appear in the blood plasma in 3 to 14 days. After antigen injection, there is a brief *latent period* during which the antigen is recognized and appropriate lymphocytes begin to form clones. A *logarithmic phase* follows, during which the antibody concentration rises rapidly for several days until it reaches a peak (**FIG. 45-12**). IgM is

the principal antibody synthesized during the primary immune response. Finally, there is a *decline phase,* during which the antibody concentration decreases to a very low level.

A second exposure to the same antigen, even years later, results in a **secondary immune response.** Because memory B cells bearing antibodies to that antigen (and also memory T cells with their specific T-cell receptors) persist for many years, the secondary immune response is much more rapid than the primary immune response and has a shorter latent period. Much less antigen is necessary to stimulate a secondary immune response than a primary response, and more antibodies are produced. In addition, the affinity of antibodies is generally much higher. The predominant antibody in a secondary immune response is IgG.

The body's ability to launch a rapid, effective response during a second encounter with an antigen explains why we do not usually suffer from the same infectious disease several times. For example, a person who contracts measles or chickenpox is unlikely to become reinfected in the future. When someone is exposed a second time, the immune system responds quickly, destroying the pathogens before they have time to multiply and cause symptoms of the disease. Booster shots of vaccine are given to elicit a secondary immune response, thus reinforcing immunological memory.

You may wonder how a person can get influenza (the flu) or a cold more than once. The reason is that there are many varieties of these viral diseases, each caused by a virus with slightly different antigens. For example, more than 200 different viruses cause the common cold, and new varieties of cold and flu viruses evolve continuously by mutation (a survival mechanism for them). As viruses evolve, changes may occur in their surface antigens. Even a slight change may prevent recognition by memory cells. Because the immune system is so specific, each different antigen is treated by the body as a new immunological challenge.

Immunization induces active immunity We have been considering **active immunity,** which develops following exposure to antigens. Active immunity can be naturally or artificially induced (**TABLE 45-1**). If someone with influenza sneezes near you and you contract the disease, you develop active

TABLE 45-1	Active and Passive Immunity		
TYPE OF IMMUNITY	**WHEN DEVELOPED**	**DEVELOPMENT OF MEMORY CELLS**	**DURATION OF IMMUNITY**
ACTIVE			
Naturally induced	After pathogens enter the body through natural encounters (e.g., a person with influenza sneezes on you)	Yes	Many years
Artificially induced	After immunization with a vaccine	Yes	Many years
PASSIVE			
Naturally induced	After transfer of antibodies from mother to developing baby	No	Few months
Artificially induced	After injection with gamma globulin	No	Few months

© Cengage Learning

immunity naturally. Active immunity can also be artificially induced by *immunization*, that is, by exposure to antigens in a *vaccine.* When an effective vaccine is introduced into the body, the immune system actively develops clones of cells, produces antibodies, and develops memory cells.

The first vaccine was prepared in 1796 by British physician Edward Jenner against vaccinia, the cowpox virus. (The term *vaccination* was derived from the name of the cowpox virus.) Jenner's vaccine provided humans with immunity against the deadly disease smallpox, which closely resembles cowpox. Jenner had no knowledge of microorganisms or of immunology, and 100 years passed before French chemist Louis Pasteur began to develop scientific methods for preparing vaccines.

Pasteur showed that inoculations with preparations of attenuated (weakened) pathogens could be used to develop immunity against the virulent (infectious) form of the pathogen. However, not until 20th-century advances in immunology—for example, Burnet's clonal selection theory in 1957 and the discovery of T cells and B cells in 1965—did scientists gain a modern understanding of vaccines. Effective vaccination stimulates the body to launch an immune response against the antigens contained in the vaccine. Memory cells develop, and future encounters with the same pathogen are dealt with rapidly.

Microbiologists prepare effective vaccines in various ways. They can attenuate a pathogen so that it loses its ability to cause disease. When pathogens are cultured for long periods in non-human cells, mutations adapt the pathogens to the nonhuman host so that they no longer cause disease in humans. This method is used to produce the Sabin polio, measles, mumps, and rubella vaccines. Tetanus and diphtheria vaccines are made from toxins secreted by the respective pathogens. The toxin is altered so that it can no longer destroy tissues, but its antigenic determinants are still intact.

Researchers are developing **DNA vaccines** (or RNA vaccines) consisting of plasmids that have been genetically engineered to produce specific antigens from a pathogen. When injected into a patient, the altered DNA is taken up by cells and makes its way to the nucleus. The encoded antigens are manufactured and stimulate both cell-mediated and antibody-mediated immunity. Several DNA vaccines, including vaccines to prevent and treat HIV infection, are in clinical trials.

Passive immunity is borrowed immunity

When individuals have been exposed to certain diseases such as measles or hepatitis A, physicians may inject them with gamma globulin containing antibodies to the pathogen. The individuals injected develop **passive immunity,** boosting the body's defenses against the disease. However, gamma globulin injections offer protection for only a few weeks. Because the body has not actively launched an immune response, it has no memory cells and cannot produce antibodies to the pathogen. Once the injected antibodies are broken down, the immunity disappears.

Pregnant women confer naturally induced passive immunity on their developing babies by manufacturing antibodies for them. These maternal antibodies, of the IgG class, pass through the placenta and provide the fetus and newborn infant with a defense system until its own immune system matures. Babies who are breast-fed continue to receive immunoglobulins, particularly IgA, in their milk during the critical months when their own immune system is developing.

Remember that passive immunity is borrowed immunity. Its effects are temporary, whether the immunity is artificially or naturally induced (see Table 45-1).

CHECKPOINT 45.5

- List the sequence of events that occur in antibody-mediated immunity.
- VISUALIZE Draw an antigen–antibody complex. Describe three ways that antigen–antibody complexes affect pathogens.
- What is the basis of immunological memory?
- How is passive immunity different from active immunity?

45.6 RESPONSE TO DISEASE, IMMUNE FAILURES, AND HARMFUL REACTIONS

LEARNING OBJECTIVES

11 Describe the body's response to cancer cells and HIV.
12 Describe examples of hypersensitivity, including autoimmune diseases, Rh incompatibility, and allergic reactions, and summarize the immunological basis of graft rejection.

The immune system is generally very effective in defending the body against invading pathogens. However, many factors can impair immune function, including genetic mutations, malnutrition, sleep deprivation, pre-existing disease, stress, and virulent pathogens. Ineffective or inappropriately directed immune responses can also result in disease states.

Exaggerated, damaging immune responses are referred to as *hypersensitivity.* For example, sometimes the immune system malfunctions and attacks its own tissues, causing *autoimmune disease.* In allergic reactions individuals are overresponsive to mild environmental antigens. The immune system's ability to distinguish self from nonself sometimes interferes with clinical interventions. Physicians must carefully consider immune system responses when employing such lifesaving procedures as blood transfusions or organ transplants.

Cancer cells evade the immune system

Cancer is a leading cause of death worldwide and the second-leading cause of death in Western countries. An estimated one person in three in the United States will develop cancer, and one in five will die of cancer. Worldwide, lung cancer in men and breast cancer in women are the main types of cancer resulting in death. In our earlier discussions of cancer in Chapters 17 and 39, we explained that cancer is actually a complex group of diseases characterized by dysregulation of normal control mechanisms for cellular growth. Here we focus on the interaction between cancer cells and cells of the immune system and on new treatment strategies based on immune mechanisms.

Cancer cells are body cells that have been transformed in a way that alters their normal growth-regulating mechanisms. Every day a few normal cells may be transformed into precancer cells in each of us in response to the sun's UV rays, X-rays, certain viruses, tobacco smoke, asbestos, and other chemical carcinogens in the environment as well as other unknown factors. How does the immune system respond? And why are immune responses sometimes ineffective?

Cancer cell antigens are classified according to their molecular structure and their source. Some cancer cell antigens are produced by mutant genes that are associated with cancer cells. T cells may respond to them because these antigens are not found in normal cells. Other cancer cell antigens are normal proteins that may stimulate immune responses because they are expressed in abnormally large quantities.

The human papillomavirus (HPV), which has been linked to several types of cancer (including cervical cancer and head and neck cancer), produces antigens that stimulate T cells to respond. Adaptive immune responses sometimes prevent the development of such DNA virus-induced cancers.

Cancer cell antigens induce both cell-mediated and antibody-mediated immunity NK cells and macrophages destroy cancer cells. Macrophages produce cytokines, including TNFs that inhibit tumor growth. Dendritic cells present antigens to T cells, stimulating them to produce interferons, which have an antitumor effect. T_C cells attack cancer cells directly and also produce interleukins, which attract and activate macrophages and NK cells (**FIG. 45-13**). Although the immune system recognizes many cancer cell antigens, these antigens often stimulate only weak immune responses that are not effective.

Sometimes cancer cells evade the immune system and multiply unchecked. Some types of cancer cells can block T_C cells. Others dramatically decrease their expression of class I MHC molecules. Recall that T_C cells recognize only antigens associated with class I MHC, so a low level of these molecules prevents tumor destruction. Some cancer cells do not produce co-stimulatory molecules needed to activate T_C cells.

Future cancer treatment will be more targeted Cancer treatments, such as chemotherapy and radiation, are designed to kill cells that are rapidly dividing. However, these treatments destroy normal cells as well as cancer cells. To treat cancer effectively, treatment strategies must be more specific, or even customized to the particular cancer. For example, cancer researchers are genetically engineering cancer cells to secrete cytokines that would stimulate the patient's immune response.

A few cancer drugs, such as Herceptin, are monoclonal antibodies. Herceptin binds with a growth-factor receptor that is present in excessive numbers on the cells of about 30% of metastatic breast cancers. Herceptin blocks the growth factors that would stimulate proliferation of the cells.

The spread of cancer typically depends on the development of blood vessels that bring nutrients and oxygen to the tumor. *Angiogenesis inhibitors* slow the development of blood vessels and thereby slow the spread of cancer. Many studies under way are attempting to optimize the clinical benefits of these drugs.

Researchers are developing DNA vaccines made from bits of DNA from a patient's cancer cells. When injected into the patient, the vaccine stimulates increased production of the target antigens. In response, the patient's immune system launches cell-mediated responses to tumor-associated antigens. These vaccines differ from traditional vaccines that are used to destroy pathogens *before* they cause disease. Effective new treatment protocols for cancer will use a systems approach and combinations of medications. Cellular mechanisms are complex, and when a drug blocks one signaling pathway, cancer cells often work around the blocked reaction or take advantage of a redundant pathway. When combinations of drugs are used, more than one signaling pathway, or different parts of the same pathway, can be targeted.

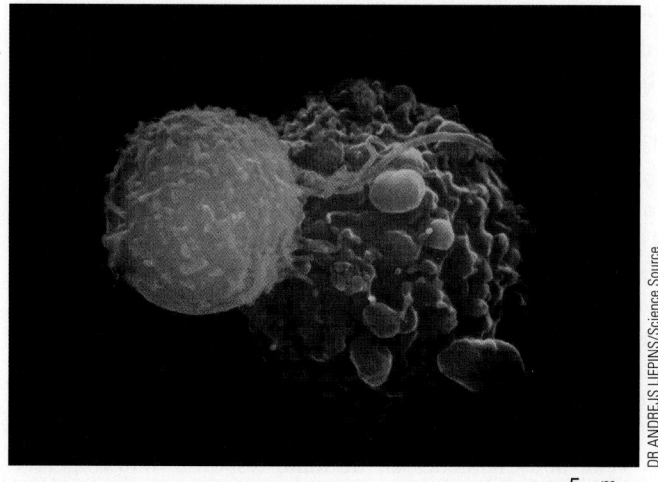

5 μm

DR ANDREJS LIEPINS/Science Source

Figure 45-13 A cytotoxic T cell destroying a cancer cell

This color-enhanced SEM shows a cytotoxic T_C cell (*orange*) destroying a cancer cell (*pink*). The T_C cell has attached itself to a cancer cell and released a compound that induces the cancer cell to undergo apoptosis (programmed cell death).

Immunodeficiency disease can be acquired or inherited

The absence or failure of some component of the immune system can result in **immunodeficiency disease,** a condition that increases susceptibility to infection. Worldwide, the leading cause of *acquired* immunodeficiency in children is protein malnutrition. Lack of protein decreases T-cell numbers and decreases the ability to manufacture antibodies, which impairs the immune system. This condition increases the risk of contracting *opportunistic infections,* infections by microbes that do not typically cause disease. Another important cause of acquired immunodeficiency is chemotherapy administered to cancer patients because chemotherapy is most damaging to cells that divide rapidly, including blood stem cells and activated lymphocytes.

Inherited immunodeficiencies have an estimated incidence of 1 per 10,000 births. *Severe combined immunodeficiency syndromes (SCIDs)* are X-linked and autosomal recessive disorders that profoundly affect both cell-mediated immunity and antibody-mediated immunity, resulting in multiple infections. Babies born with SCID typically die by two years of age unless they are maintained in a protective bubble until they can be effectively treated, typically with bone marrow transplants.

In *DiGeorge syndrome* the thymus is reduced or absent, and the patient is deficient in T cells. Children born with this disorder are prone to serious viral infections. Treatment involves transplanting bone marrow or fetal thymus tissue. Other tissues may take over T-cell production, and function typically improves by age five.

Researchers use several animal models to study immunodeficiency. For example, the nude mouse (a hairless mutant mouse) does not develop a functional thymus. Because nude mice are deficient in mature T cells, they are used to study the effects of compromised cell-mediated immunity as well as possible treatments. *Stem cell research* and advances in genetic engineering suggest new approaches to treating immunodeficiency (see Chapters 16 and 17).

HIV is the major cause of acquired immunodeficiency in adults

The **human immunodeficiency virus (HIV)** was first isolated in 1983 and was shown to be the cause of **acquired immunodeficiency syndrome (AIDS)** in 1984. HIV, which is a retrovirus, has been studied more than any other virus. (Recall from Chapter 24 that a retrovirus is an RNA virus that uses its RNA as a template to make DNA with the help of reverse transcriptase.) Several different strains of the virus are known; HIV-1 is the most virulent form in humans (see Fig. 23-12).

The AIDS pandemic has claimed more than 30 million lives since 1981. HIV kills more than 2 million people every year, making it the sixth-highest cause of death globally. Epidemiologists estimate that more than 34 million people worldwide, including more than 2 million children, are now infected with HIV. More than 2.5 million people are newly infected with HIV each year. The General Assembly of the United Nations has called AIDS a "global emergency" and has outlined measures for mobilizing world resources to combat this disease. HIV/AIDS is more than a medical issue; it is a political, economic, and human rights threat.

HIV is transmitted through body fluids from infected individuals, including semen, vaginal secretions, blood, and breast milk. Transmission occurs mainly during sexual intercourse with an infected person or by direct exposure to infected blood or blood products. Sharing needles and/or syringes (usually for drug injection) with an infected person is a major risk factor. The virus is not spread by casual contact. People do not contract HIV by hugging, casual kissing, sharing food or water, or using the same bathroom facilities. Friends and family members who have contact with AIDS patients are not more likely to become infected than those who do not have such contact.

In the United States, more than 50,000 new HIV infections occur each year. Of these new infections, about 75% occur in men. Sexual contact with persons who have HIV infection is a risk factor that can be modified. Use of a latex condom during sexual intercourse provides some protection against HIV transmission, and use of a spermicide containing nonoxynol-9 may provide additional protection.

As many as one-third of untreated pregnant women infected with HIV will transmit the virus to their babies during pregnancy, childbirth, or breast-feeding. Recent drug treatment protocols have reduced the rate of HIV transmission between mother and fetus. Clinical studies are ongoing to confirm reports that beginning treatment (using triple antiretroviral drugs) of HIV-infected infants within the first 30 hours of life may lead to a cure after about 18 months.

In high-income countries, effective blood-screening procedures now safeguard blood bank supplies, markedly reducing the risk of infection by blood transfusion. In low- and middle-income countries, HIV continues to be a health threat, particularly to women and newborns. As treatment, prevention, and education programs are implemented, the incidence of new infections in many parts of the world is decreasing.

HIV destroys the host immune system HIV typically enters the body through the mucosa that lines passageways into the body. A protein on the outer envelope of the virus attaches to CD4 on the surface of T_H cells, the main target (**FIG. 45-14**). The virus also binds to a co-receptor that normally binds a chemokine. HIV enters the T_H cell and is reverse-transcribed to form DNA. Its DNA is then incorporated into the host genome. The virus reproduces itself, causing death of the host cell. The first phase of HIV infection is the *acute* phase of the infection. Within two weeks of infection, large numbers of T_H and memory T cells are destroyed. HIV also infects dendritic cells and macrophages. Dendritic cells in the mucous membranes transport HIV from their point of entry to the lymph nodes. The virus further spreads through the blood.

The body launches an immune response following HIV infection. Innate responses include production of antimicrobial peptides and action by NK cells, dendritic cells, and the complement system. Adaptive immune responses include activation and mobilization of T_C cells.

An estimated 20% of HIV-infected individuals in the United States do not know that they are infected. During the first few

Figure 45-14 Colorized SEM of HIV infecting T helper cell
HIV (*red particles*) attack and destroy T_H cell (*green*). Depletion of T_H is the main reason for the failure of the immune system in HIV infection.

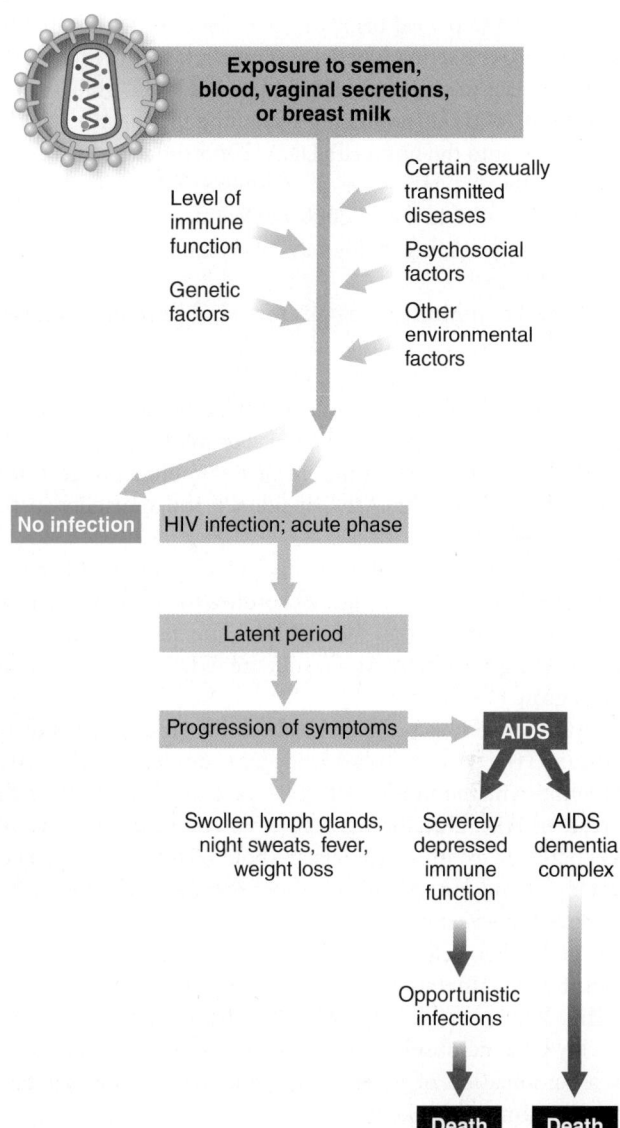

Figure 45-15 The course of HIV infection
Exposure to semen, blood, vaginal secretions, or breast milk containing HIV can lead to HIV infection. The acute (early) infection begins in lymph nodes and other lymph tissues associated with epithelial linings. Most T_H cells may be destroyed within two weeks of infection. During the latent period, chronic infection develops, and the virus replicates at a low level. Eventually, viral replication may increase and lead to a progression of symptoms and to AIDS, the final, often lethal, stage. Although some exposed individuals apparently do not become infected, the risk increases with multiple exposures. Many factors determine whether a person exposed to HIV will develop AIDS.
© Cengage Learning

weeks after infection, some people experience mild flulike symptoms (fever, headache, sore throat, aching muscles). However, some individuals have no symptoms.

The acute phase transitions into the latent period, a *chronic* phase, as the immune system decreases HIV in the blood. Some cells infected by HIV are not destroyed, and the virus may continue to replicate slowly for many years. T_C cells appear to be the main cells that attack HIV, limiting viral replication and delaying the progress of the disease. However, over time HIV continues to destroy T_H cells, resulting in a dramatic decrease in the T_H cell population. This decrease severely impairs the body's ability to resist infection because T_H cells are needed to release the cytokines that activate T_C cells and B cells. After a time, a progression of symptoms occurs, including swollen lymph glands, night sweats, fever, and weight loss (**FIG. 45-15**). Too often, the virus eventually wins the battle.

In most HIV-infected individuals, the disease eventually progresses to AIDS, the final, and often lethal, stage. This process can take more than ten years, and antiretroviral drugs slow the process. By the time the disease progresses to AIDS, lymph tissue has been destroyed and the blood T_H cell count has dropped below a critical level of 200 cells/μL. As a consequence of this immunosuppression, many AIDS patients develop serious opportunistic infections, such as tuberculosis, or cancers, such as lymphomas and Kaposi's sarcoma (a tumor of the endothelial cells lining blood vessels that causes purplish spots on the skin). In about one-third of AIDS patients, the virus infects the nervous system and causes *AIDS dementia complex*. These patients exhibit progressive cognitive, motor, and behavioral dysfunction that typically ends in coma and death.

Researchers are searching for effective treatments
Researchers throughout the world are searching for anti-retroviral drugs that will successfully combat the AIDS virus. Because HIV often infects the central nervous system, an effective drug must cross the blood–brain barrier. **Azidothymidine (AZT),** the first anti-retroviral drug developed to treat HIV infection,

can prolong the period prior to the onset of AIDS symptoms. AZT inhibits the action of reverse transcriptase, the enzyme the retrovirus uses to synthesize DNA. As a result, AZT blocks HIV replication. Without producing DNA, the virus cannot incorporate itself into the host cell's DNA. Unfortunately, the reverse transcriptase that HIV uses to synthesize DNA makes many mistakes: about 1 in every 2000 nucleotides it incorporates is incorrect. By mutating in this way, viral strains resistant to AZT have evolved.

Protease inhibitors block the viral enzyme protease, resulting in viral copies that cannot infect new cells. Protease inhibitors are used in combination with AZT and other reverse transcriptase inhibitors. Triple-combination treatment has been very effective for many AIDS patients, preventing opportunistic infections and prolonging life. In fact, combination treatment has led to a decline in AIDS incidence and mortality in the United States.

Unfortunately, people in many parts of the world cannot afford these expensive drugs. According to the World Health Organization, more than 50% of HIV-infected individuals are not receiving treatment. As a result, the AIDS epidemic continues to grow.

Another serious problem has been the emergence of drug-resistant HIV viruses. Worldwide containment of the AIDS epidemic will require an effective vaccine that prevents the spread of HIV. Developing such a vaccine remains a daunting challenge for immunologists. About 60 vaccines have been developed and tested clinically, but none has proved effective. An effective vaccine would have to counteract HIV's destruction of the key cells needed to launch an immune response. Also, because mutations arise at a very high rate, new viral strains with new antigens evolve quickly. A vaccine might quickly become obsolete. Some optimism has been generated by a combination of vaccine and gene therapy that has been tested in simian monkeys.

While immunologists work to develop a successful vaccine and more effective drugs to treat infected patients, massive educational programs have been launched to slow the spread of HIV. Educating the public that having multiple sexual partners increases the risk of HIV and teaching sexually active individuals the importance of "safe sex" help contain the pandemic. In some countries public health facilities provide free condoms to those who are sexually active and free needles and syringes to those addicted to intravenous drugs. Some HIV prevention programs now provide male circumcision, which reduces the risk of HIV infection through sexual contact. The cost of these measures is far less than the cost of medical care for increased numbers of HIV patients.

In an autoimmune disease, the body attacks its own tissues

Self-tolerance, the ability to recognize self, is established during lymphocyte development. Lymphocytes do not normally attack cells and tissues of their own body. However, some lymphocytes do have the potential to be *autoreactive,* that is, to launch an immune response against self-tissues. In about 5%

of adults in high-income countries, such autoreactivity can lead to a form of hypersensitivity known as *autoimmunity,* or **autoimmune disease,** in which immune cells react against the body's own cells. Some of the diseases that result from such failure in self-tolerance are rheumatoid arthritis, systemic lupus erythematosus, insulin-dependent diabetes, scleroderma, and multiple sclerosis.

In autoimmune diseases antibodies and T cells attack the body's own tissues. In *rheumatoid arthritis* T cells in the area of inflammation produce a cytokine that promotes inflammation. In addition, antibodies to a joint protein may generate antigen–antibody complexes that are deposited in the small joints. In *multiple sclerosis,* an autoimmune disease of the central nervous system, T_H cells attack an individual's own myelin antigens. MRI scans show the loss of myelin sheaths in multiple areas around axons in the brain and spinal cord that are normally myelinated. Scar tissue forms and compromises the ability of these damaged neurons to transmit signals.

Multiple genes, including MHC alleles, contribute to the development of autoimmune disease. Mutations that cause abnormal apoptosis or inadequate clearing of dead cells may result in danger signals that activate the immune system. Another important factor is that viral or bacterial infection often precedes the onset of an autoimmune disease. A tactic known as *molecular mimicry* has evolved in some pathogens. They trick the body by producing molecules that look like self-molecules. For example, an adenovirus that causes respiratory and intestinal illness produces a peptide that mimics myelin protein. When the body launches responses to the adenovirus peptide, it may also begin to attack the similar self-molecule, myelin.

As we have discussed, the numerous types of cells, regulatory and signaling molecules, and the complex actions of the immune system must be carefully regulated to maintain health. For example, specific T cells and B cells proliferate and respond to challenges by pathogens, but after an effective immune response the expanded populations of T cells and B cells must be decreased to normal numbers. Cells that are no longer needed die by apoptosis. Overactivation of critical components can result in inflammation and autoimmune disease. Underactivation allows pathogens to cause serious disease. Either situation can be fatal.

Rh incompatibility can result in hypersensitivity

Named for the rhesus monkeys in whose blood it was first found, the Rh system consists of more than 40 kinds of Rh antigens, each referred to as an **Rh factor.** By far the most important of these factors is *antigen D.* About 85% of U.S. residents who are of Western European descent are Rh positive, meaning that they have antigen D on the surfaces of their red blood cells (in addition to the antigens of the ABO system and other blood group antigens). The 15% or so of this population who are Rh negative have no antigen D.

Unlike the situation discussed in Chapter 11 for the ABO blood group, Rh-negative individuals do not routinely produce antibodies against antigen D (anti-D). However, they do produce anti-D antibodies when they are exposed to Rh-positive blood. The allele coding for antigen D is dominant to the allele for the absence of antigen D. For this reason, Rh-negative people are homozygous recessive, and Rh-positive people are heterozygous or homozygous dominant.

Although several kinds of maternal-fetal blood type incompatibilities are known, **Rh incompatibility** is the most serious (FIG. 45-16). If a woman is Rh negative and the father of the fetus she is carrying is Rh positive, the fetus may also be Rh positive, having inherited the D allele from the father. Ordinarily, the maternal and fetal blood do not mix; nutrients, oxygen, and other substances are exchanged between the two circulatory systems across the placenta. However, late in pregnancy or during the birth process, a few drops of blood from the fetus may pass through a defect in the placenta.

When maternal blood mixes with fetal blood, fetal red blood cells, which bear antigen D, activate the mother's immune system. The mother's B cells then produce antibodies to antigen D. If the woman becomes pregnant again, anti-D antibodies cross the placenta and enter the fetal blood. They combine with antigen D molecules on the surface of the fetal red blood cells, causing hemolysis (cells rupture and release hemoglobin into the circulation). The ability to transport oxygen is reduced, and breakdown products of the released hemoglobin damage organs, including the brain. In extreme cases of this disease, known as *erythroblastosis fetalis,* so many fetal red blood cells are destroyed that the fetus may die.

When Rh-incompatibility problems are suspected, fetal blood can be exchanged by transfusion before birth, but it is a risky procedure. More commonly, Rh-negative women are treated just after childbirth (or at termination of pregnancy by miscarriage or abortion) with a preparation of anti-D antibodies known as RhoGAM. These antibodies clear the Rh-positive fetal red blood cells from the mother's blood very quickly, minimizing the chance for her own white blood cells to become sensitized and develop memory cells. The antibodies are also soon eliminated from her body. As a result, if she becomes pregnant again, her blood does not contain the anti-D that could harm the developing fetus.

Allergic reactions are directed against ordinary environmental antigens

About 20% of the U.S. population is plagued by an allergic disorder, such as allergic asthma or hay fever. A predisposition toward these disorders appears to be inherited. In *allergic reactions* hypersensitivity results in the manufacture of antibodies against mild antigens, called **allergens,** which normally do not stimulate an immune response. Common environmental agents, such as pollen, house-dust mites, roach feces, and pet dander, can trigger allergic reactions in some individuals. In many kinds of allergic reactions, distinctive IgE immunoglobulins are produced.

Let us examine a common allergic reaction, *allergic rhinitis,* commonly known as hay fever. This allergic response is triggered in some individuals by breathing pollen (FIG. 45-17). Exposure to pollen causes *sensitization.* Macrophages degrade the allergen and present fragments of it to T cells. The activated T cells then stimulate B cells to become plasma cells and produce pollen-specific IgE. These antibodies attach to receptors on mast cells. The Fc end of each IgE molecule attaches to a mast cell receptor; its V region is free to combine with the pollen allergen.

When a sensitized, allergic person inhales the microscopic pollen, allergen molecules rapidly attach to the IgE on sensitized mast cells. The IgE antibody then binds with the allergen. In response, mast cells release granules filled with histamine and other molecules that stimulate inflammation. Blood vessels dilate, and capillaries become more permeable, leading to edema and redness. Such responses cause the nasal passages to become swollen and irritated. A runny nose, sneezing, watery eyes, and a general feeling of discomfort are common.

A prolonged allergic response occurs when neutrophils are attracted by chemical compounds released by the mast cells. Neutrophils migrate to the inflamed area and release compounds that lengthen the allergic reaction.

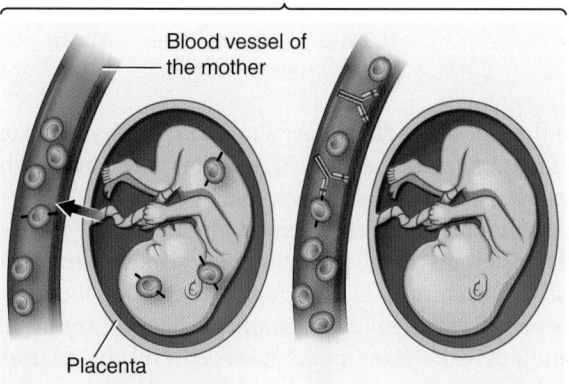

First pregnancy — Subsequent pregnancy

Blood vessel of the mother

Placenta

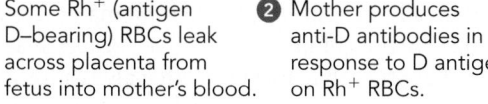

1 Some Rh⁺ (antigen D–bearing) RBCs leak across placenta from fetus into mother's blood.

2 Mother produces anti-D antibodies in response to D antigens on Rh⁺ RBCs.

3 Anti-D antibodies cross placenta and enter blood of fetus. Hemolysis of Rh⁺ blood occurs.

KEY

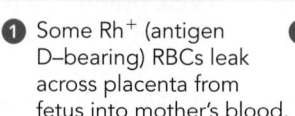

Rh⁻ RBC of mother

Rh⁺ RBC of fetus with antigen D on surface

Anti-D antibody made against Rh⁺ RBC

Hemolysis of Rh⁺ RBC

Figure 45-16 *Animation* **Rh incompatibility**

When an Rh-negative woman produces Rh-positive offspring, her immune system may become sensitized. In subsequent pregnancies she may produce anti-D antibodies that cross the placenta and destroy fetal red blood cells.

© Cengage Learning

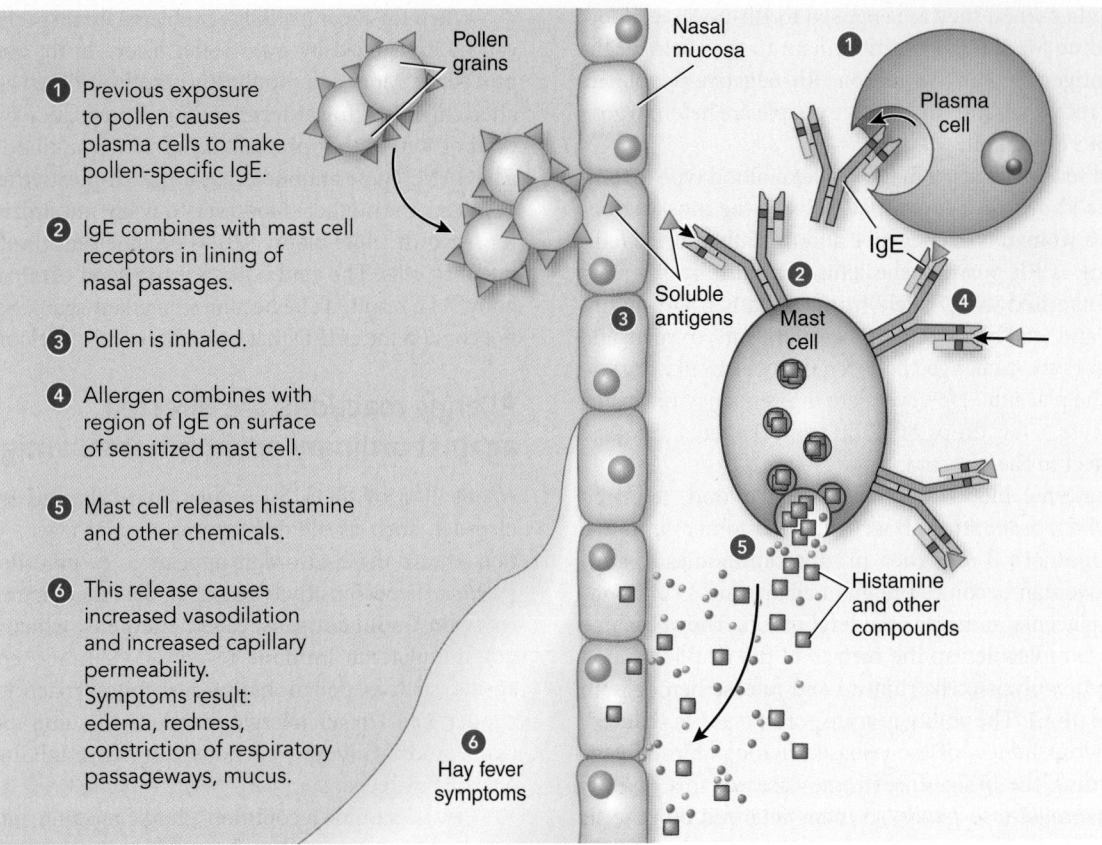

① Previous exposure to pollen causes plasma cells to make pollen-specific IgE.

② IgE combines with mast cell receptors in lining of nasal passages.

③ Pollen is inhaled.

④ Allergen combines with region of IgE on surface of sensitized mast cell.

⑤ Mast cell releases histamine and other chemicals.

⑥ This release causes increased vasodilation and increased capillary permeability. Symptoms result: edema, redness, constriction of respiratory passageways, mucus.

Pollen grains

Nasal mucosa

Plasma cell

IgE

Soluble antigens

Mast cell

Histamine and other compounds

⑥ Hay fever symptoms

Figure 45-17 An allergic reaction

Many people respond to breathing pollen with allergic rhinitis (hay fever), a common type of allergic reaction.

© Cengage Learning

In **allergic asthma** an allergen–IgE response occurs in the bronchioles of the lungs. Mast cells release substances that cause smooth muscle to contract, and the airways in the lungs sometimes constrict for several hours, making breathing difficult. Mast cells also secrete compounds that cause inflammation. Allergic asthma is treated with steroids, which block cytokine production and reduce inflammation. Monoclonal antibodies to IgE are also useful in treating allergic asthma because they reduce IgE levels in the blood.

Certain foods or drugs act as allergens in some people, causing a reaction in the walls of the digestive tract that leads to discomfort and diarrhea. When allergens are absorbed, they can cause mast cells to release granules elsewhere in the body. When the allergen–IgE reaction takes place in the skin, the histamine released by mast cells causes the swollen red welts known as urticaria, or hives.

Systemic anaphylaxis is a dangerous allergic reaction that can occur when a person develops an allergy to a specific drug, such as penicillin; to compounds in the venom injected by a stinging insect; or even to certain foods. Within minutes after the substance enters the body, a widespread allergic reaction takes place. Mast cells release large amounts of histamine and other compounds into the circulation, causing extreme vasodilation and capillary permeability. So much plasma may be

lost from the blood that circulatory shock and death can occur within a few minutes. Epinephrine (adrenaline) must be rapidly injected to reverse anaphylactic symptoms.

The symptoms of allergic reactions are often treated with corticosteroids and antihistamines, drugs that block the effects of histamine. These drugs compete for the same receptors on cells targeted by histamine. When the antihistamine combines with the receptor, it blocks the histamine from binding, thus preventing its harmful effects. Antihistamines are useful clinically in relieving the symptoms of hives and hay fever. They are not completely effective because mast cells release molecules other than histamine that also cause allergic symptoms.

Graft rejection is an immune response against transplanted tissue

Skin can be successfully transplanted from one part of the same body to another or from one identical twin to another. However, when skin is taken from one person and grafted onto the body of a non-twin, it is rejected and sloughs off. Why? Recall that tissues from the same individual or from identical twins have identical HLA alleles and thus the same HLA antigens. Such tissues are compatible. Because there are many alleles for each of the HLA genes, it is difficult to find identical matches among individuals who are not identical twins.

When a tissue or organ is taken from a donor and transplanted to the body of another person, several of the HLA antigens are likely to be different. The recipient's immune system regards the graft as foreign and launches an immune response known as **graft rejection.** In the first stage of rejection, the *sensitization stage,* T_H cells and T_C cells recognize the foreign antigens. In the second stage, the *effector stage,* T_C cells, the complement system, and cytokines secreted by T_H cells attack the transplanted tissue and can destroy it within a week. The sequence of events in graft rejection can be summarized as follows:

tissue from donor transplanted into body of recipient $\longrightarrow$ T_H cells and T_C cells recognize HLA antigens on transplant cells as foreign $\longrightarrow$ T_H cells secrete cytokines, complement system is activated, and T_C cells destroy transplant cells

To avoid graft rejection, tissues from the patient and from potential donors must be typed and matched as closely as possible. Cell typing is somewhat similar to blood typing but is more complex. If all the HLA antigens are matched, the graft has about a 95% chance of surviving the first year. Unfortunately, not many people are lucky enough to have an identical twin to supply spare parts, so perfect matches are difficult to find. Furthermore, some organs, such as the heart, cannot be spared. Most transplanted organs therefore come from unrelated donors, often from patients who have just died.

To prevent graft rejection in less compatible matches, physicians use antirejection drugs such as cyclosporine to suppress the immune system. Cyclosporine prevents T cells from producing cytokines. Unfortunately, immunosuppression makes the transplant patient more vulnerable to pneumonia or other infections and increases the risk of certain types of cancer. After the first few months, immunosuppressant drug dosages may be reduced. Researchers are working to develop more effective immunosuppression techniques.

In the United States, thousands of patients are in need of organ transplants. Because the number of human donors does not meet this need, investigators are developing techniques for culturing stem cells as replacement tissues. Current research is promising for successful human tissue and organ replacement in the future using a patient's own stem cells. Medical engineers also continue to develop effective artificial organs.

Some researchers are working on more effective methods for transplanting animal tissues and organs to humans. Other investigators plan to introduce human stem cells into animal embryos in order to produce human organs in animals. These organs would eventually be harvested and implanted in humans.

The body has a few immunologically privileged locations where foreign tissue is accepted. For example, corneal transplants are highly successful because the cornea has almost no associated blood or lymphatic vessels and is thus out of reach of most lymphocytes. Furthermore, antigens in the corneal graft probably would not find their way into the circulatory system and therefore could not stimulate an immune response.

CHECKPOINT 45.6

- *How do cancer cells evade the immune system?*
- *How does HIV infect the body? How does the immune system respond?*
- *How does an autoimmune disease develop?*
- VISUALIZE *Draw the sequence of immunological events that take place in a common type of allergic reaction, such as hay fever.*
- CONNECT *What is the immunological basis for graft rejection?*

SUMMARY: FOCUS ON LEARNING OBJECTIVES

45.1 Evolution of Immune Responses *(page 957)*

1 Distinguish between innate and adaptive immunity.

- The **immune system** is the internal defense system; it consists of the cells and molecules that recognize foreign or dangerous macromolecules and respond to eliminate them. **Immune responses** are the body's defensive responses against foreign macromolecules, including **pathogens** (disease-causing organisms), cancer cells, toxins, and other harmful agents.
- **Innate immunity** provides general and immediate protection against foreign macromolecules, including pathogens, some toxins and drugs, and cancer cells.
- **Adaptive immunity** targets distinct **antigens,** molecules recognized as foreign or dangerous by cells of the immune system. The adaptive immune system produces **antibodies,** highly specific proteins that recognize and bind to specific antigens.
- Another defining feature of adaptive immunity is **immunological memory;** that is, the system "remembers" foreign or dangerous molecules and responds more strongly to repeated encounters with the same molecules.

2 Compare, in general terms, the immune responses that have evolved in invertebrates and vertebrates.

- Invertebrates depend mainly on innate immune responses, such as physical barriers; **phagocytosis** (engulfing and destroying pathogens); receptors that recognize molecular patterns associated with pathogens; and **antimicrobial peptides** (soluble molecules that destroy pathogens).
- Vertebrates use both innate and adaptive immune responses.

45.2 Innate Immune Responses in Vertebrates *(page 959)*

3 Describe innate immune responses, including physical barriers; recognition by pattern recognition receptors; actions of phagocytes and other cells; actions of cytokines and complement; and the inflammatory response.

- Innate immune responses include physical barriers, such as the skin and the epithelial linings of the respiratory and digestive tracts. Antimicrobial peptides are produced by epithelial membranes.

- When pathogens break through the first-line defenses, **pathogen-associated molecular patterns,** or **PAMPs,** bind with **pattern recognition receptors (PRRs);** this binding activates other innate defenses. PAMPs are sets of common molecular features that are characteristic of each class of pathogen. The best known pattern recognition receptors are **Toll-like receptors,** receptors on phagocytes and certain other types of cells.

- *Phagocytes,* including **neutrophils** and **macrophages,** destroy bacteria. **Natural killer (NK) cells** destroy cells infected with viruses and foreign or altered cells such as tumor cells. **Dendritic cells** release antiviral proteins called **interferons.**

- **Cytokines** are signaling proteins that regulate interactions between cells. Important groups include interferons, tumor necrosis factors, interleukins, and chemokines. Interferons inhibit viral replication and activate natural killer cells. **Tumor necrosis factor (TNF)** stimulates immune cells to initiate an inflammatory response. **Interleukins** help regulate interactions between lymphocytes and other cells of the body, help mediate inflammation, and can mediate fever. *Chemokines* attract, activate, and direct the movement of certain cells of the immune system.

- **Complement** proteins lyse the cell walls of pathogens; coat pathogens, thereby enhancing phagocytosis; and attract white blood cells to the site of infection. These actions enhance the inflammatory response.

- When pathogens invade tissues, they trigger an **inflammatory response,** which includes three main processes: vasodilation, which brings more blood to the area of infection; increased capillary permeability, which allows fluid and antibodies to leave the circulation and enter the tissues; and increased phagocytosis. In response to infection, **mast cells** release **histamine** and other compounds that cause vasodilation and increased capillary permeability.

45.3 Adaptive Immune Responses in Vertebrates *(page 963)*

4 Contrast cell-mediated and antibody-mediated immunity and give an overview of each process.

- In **cell-mediated immunity,** specific T cells are activated; T cells release proteins that destroy cells infected with viruses or other intracellular pathogens.

- In **antibody-mediated immunity,** specific B cells are activated; B cells multiply and differentiate into plasma cells, which produce antibodies.

5 Describe the functions of the specialized cells of the adaptive immune system and of the major histocompatibility complex.

- Two main types of cells important in adaptive immune responses are lymphocytes and antigen-presenting cells. Lymphocytes develop from stem cells in the bone marrow. **T cells** are lymphocytes responsible for cell-mediated immunity. The **thymus gland** confers immunocompetence on T cells by making them capable of distinguishing between self and nonself. The types of T cells include **T cytotoxic cells (T_C cells), T helper cells (T_H), memory T_C cells, memory T_H cells,** and **regulatory T cells (T_{regs}).** T cells are distinguished by their **T-cell receptors (TCRs).**

- **B cells** are lymphocytes responsible for antibody-mediated immunity. B cells differentiate into **plasma cells,** which produce antibodies. Some activated B cells become **memory B cells,** which continue to produce antibodies after the body overcomes the infection.

- **Antigen-presenting cells (APCs),** including dendritic cells, macrophages, and B cells, display foreign antigens as well as their own surface proteins. Dendritic cells are specialized to process, transport, and present antigens.

- Immune responses depend on the **major histocompatibility complex (MHC),** a group of genes that encode MHC proteins. In humans the MHC is called the **human leukocyte antigen complex (HLA).**

45.4 Cell-Mediated Immunity *(page 966)*

6 Describe the sequence of events in cell-mediated immunity.

- In cell-mediated immunity, specific T cells are activated by a foreign antigen–MHC complex on the surface of an infected cell. A *co-stimulatory molecule* and interleukins are also required.

- Activated T_C cells multiply and give rise to a clone, a process called **clonal expansion.** These differentiated cells migrate to the site of infection and destroy pathogen-infected cells. Activated T_H cells give rise to a clone of T_H cells, which secrete cytokines that activate B cells and macrophages.

45.5 Antibody-Mediated Immunity *(page 967)*

7 Summarize the sequence of events in antibody-mediated immunity, including the effects of antigen–antibody complexes on pathogens.

- In antibody-mediated immunity, B cells are activated when they combine with antigen. Activation requires both an APC (such as a dendritic cell or macrophage) that has a foreign antigen–MHC complex displayed on its surface and a T_H cell that secretes interleukins.

- When activated B cells multiply, they give rise to clones of cells. The cloned cells differentiate and form plasma cells. Plasma cells produce specific antibodies, called **immunoglobulins (Ig),** in response to the specific antigens that activated them.

- An antibody combines with a specific antigen to form an *antigen–antibody complex,* which may inactivate the pathogen, stimulate phagocytosis, or activate the complement system.

8 Describe the basic structure and function of an antibody and explain the basis of antibody diversity.

- An antibody molecule consists of four polypeptide chains: two identical heavy chains and two shorter light chains. Each chain has a *constant (C) region* and a *variable (V) region.* In a Y-shaped antibody, the two arms combine with antigen.

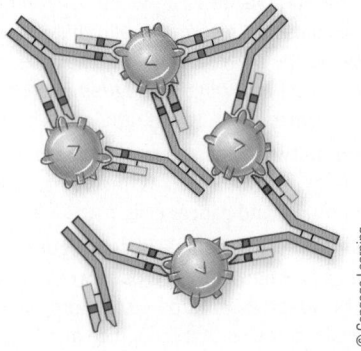

© Cengage Learning

- Rearrangement of DNA segments during the differentiation of B cells is the main factor responsible for antibody diversity; millions of different types of B (and T) cells are produced.

9 Describe the basis of immunological memory and contrast secondary and primary immune responses.

- After an infection, memory B cells and memory T cells remain in the body. These cells are responsible for long-term immunity.
- The first exposure to an antigen stimulates a **primary immune response.** A second exposure to the same antigen evokes a **secondary immune response,** which is more rapid and more intense than the primary response.

10 Compare active and passive immunity, and give examples of each.

- **Active immunity** develops as a result of exposure to antigens; it may occur naturally after recovery from a disease or can be artificially induced by immunization with a vaccine.
- **Passive immunity** is a temporary condition that develops when an individual receives antibodies produced by another person or animal.

45.6 Response to Disease, Immune Failures, and Harmful Reactions *(page 974)*

11 Describe the body's response to cancer cells and HIV.

- NK cells, macrophages, T cells, and other cells of the immune system recognize antigens on cancer cells and launch an immune response against them. Cancer cells evade the immune system by blocking T_C directly or by decreasing the class I MHC molecules on T_C cells.
- **Acquired immunodeficiency syndrome (AIDS)** is caused by the **human immunodeficiency virus (HIV),** a retrovirus. HIV destroys T helper cells, severely impairing immunity and placing the patient at risk for opportunistic infections.

12 Describe examples of hypersensitivity, including autoimmune diseases, Rh incompatibility, and allergic reactions, and summarize the immunological basis of graft rejection.

- In **autoimmune diseases** the body reacts immunologically against its own tissues.
- When an Rh-negative woman gives birth to an Rh-positive baby, she may develop anti-D antibodies. Rh incompatibility can then occur in future pregnancies.
- In an *allergic reaction,* an **allergen** stimulates the production of IgE, which combines with receptors on mast cells. The mast cells release histamine and other molecules that cause inflammation and other symptoms of allergy. **Systemic anaphylaxis** is a rapid, widespread allergic reaction that can lead to death.
- Transplanted tissues have HLA antigens that stimulate **graft rejection,** an immune response in which T cells destroy the transplant.

TEST YOUR UNDERSTANDING

Know and Comprehend

1. Innate immune responses include all but (a) inflammation (b) antigen–antibody complexes (c) binding of PAMPs with pattern recognition receptors (d) complement (e) phagocytosis
2. Cytokines (a) are regulatory Toll-like receptors (b) prevent the inflammatory response (c) include interferons and interleukins (d) are immunoglobulins (e) b and c are correct
3. Which of the following is *not* an action of complement? (a) enhances phagocytosis (b) enhances inflammatory response (c) coats pathogens (d) lyses viruses (e) stimulates production of memory T cells
4. Which of the following cells are antigen-presenting cells? (a) NK cells and monocytes (b) macrophages and plasma cells (c) dendritic cells and macrophages (d) mast cells and B cells (e) memory T cells and plasma cells
5. Which of the following cells are especially adept at destroying tumor cells? (a) NK cells (b) plasma cells (c) neutrophils (d) B cytotoxic cells (e) mast cells
6. The major histocompatibility complex (MHC) (a) consists of Y-shaped molecules (b) encodes certain antibodies (c) encodes Toll-like receptors (d) inhibits complement release from macrophages (e) encodes a group of cell-surface proteins.
7. Immunoglobulin A (a) recognizes pathogen-associated molecular patterns (b) binds with NK cells (c) prevents pathogens from attaching to epithelial surfaces (d) is found mainly on dendritic cell surfaces (e) is found mainly on T-cell surfaces
8. When a person is exposed to the same antigen a second time, the response is (a) less specific than the first time (b) more rapid than the first time (c) mediated by dendritic cells (d) likely to lead to development of an autoimmune disease (e) a hypersensitivity reaction
9. In an allergic reaction, (a) an allergen binds with IgE (b) the body is immunodeficient (c) T helper cells release histamine (d) allergens are produced by plasma cells (e) mast cells are deactivated

Apply and Analyze

10. Which sequence most accurately describes antibody-mediated immunity?
    1. B cell divides and gives rise to clone 2. antibodies produced 3. cells differentiate and form plasma cells 4. activated T helper cell interacts with B cell displaying same antigen complex 5. B cell activated
    (a) 1, 2, 3, 4, 5 (b) 3, 2, 1, 4, 5 (c) 4, 5, 3, 2, 1 (d) 4, 5, 1, 3, 2 (e) 4, 3, 1, 2, 5
11. **VISUALIZE** Label the diagram and explain what this diagram illustrates.

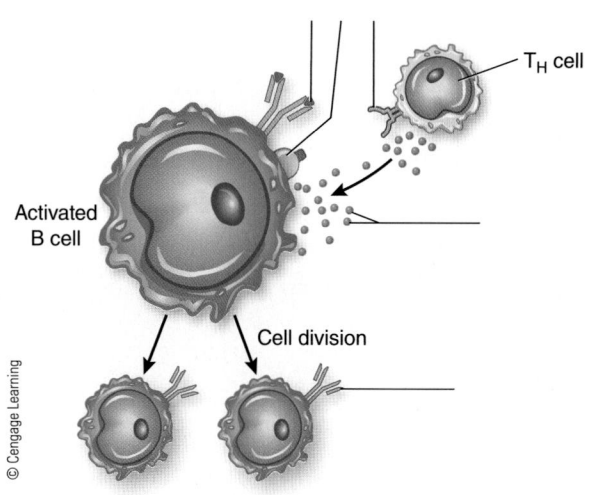

T_H cell

Activated B cell

Cell division

© Cengage Learning

12. Specificity, diversity, and memory are key features of the immune system. Giving specific examples, explain how each of these features is important.

13. **PREDICT** What would be the effects of a significant reduction of macrophages in the body? Which do you think would have a greater effect on the immune system: significant decrease of macrophages or of memory B cells?

14. **PREDICT** A playmate in kindergarten exposes John and Jack to measles. John has been immunized against measles, but Jack has not received the measles vaccine. How are their immune responses different? Five years later, John and Jack are playing together when Judy, who is coming down with measles, sneezes on both of them. Is it probable that either John or Jack will become ill with measles?

Evaluate and Synthesize

15. **EVOLUTION LINK** The enzymes Rag1 and Rag2 cut and rearrange DNA, producing millions of specific antibodies and T-cell receptors. Immunologists thought that the genes (*Rag1* and *Rag2*) that encode these enzymes were found only in jawed vertebrates. However, investigators have found similar genes in the purple sea urchin. What does this finding suggest about the evolution of the adaptive immune system?

16. **SCIENCE, TECHNOLOGY, AND SOCIETY** Imagine that you are a researcher developing new HIV treatments. What approaches might you take? What public policy decisions would you recommend that might help slow the spread of HIV while new treatments or vaccines are being developed? What role might technology play in slowing the spread of HIV?

aplia To access course materials, such as Aplia and other companion resources, please visit **www.cengagebrain.com.**

Gas Exchange

To stay alive, animals must take in oxygen and release excess carbon dioxide. Most animal cells require a continuous supply of oxygen for cellular respiration. Brain cells may be damaged beyond repair if their oxygen supply is cut off for only a few minutes. The exchange of gases between an organism and its environment is called **respiration.**

There are two phases of respiration. During **organismic respiration,** oxygen from the environment is taken up by the animal and delivered to its individual cells. At the same time, carbon dioxide generated during *cellular respiration* is excreted into the environment. **Aerobic cellular respiration** takes place in mitochondria. Oxygen is essential because it serves as the final electron acceptor in the mitochondrial electron transport chain (see Chapter 8). Recall that carbon dioxide is a metabolic waste product of cellular respiration.

Sponges, hydras, flatworms, and many other small, aquatic organisms exchange gases entirely by simple *diffusion.* Recall that **diffusion** is the passive movement of particles (atoms, ions, or molecules) from a region of higher concentration of a particular kind of particle to a region of lower concentration of that particle, that is, down its concentration gradient (see Chapter 5). In these animals most cells are in direct contact with the environment. Dissolved oxygen from the surrounding water diffuses into the cells, while carbon dioxide diffuses out of the cells and into the water. No specialized respiratory structures are needed.

In an animal more than about 1 mm thick, oxygen cannot diffuse quickly enough through the layers of cells to support life. Specialized respiratory structures, such as gills or lungs, deliver oxygen to the cells or to a transport system. These structures also facilitate excretion of carbon dioxide. Lungs are an adaptation for gas exchange in terrestrial environments (see photograph).

In many animal groups, including vertebrates, respiratory systems and circulatory systems are functionally connected. They work together to provide efficient intake and transport of oxygen. Such systems provide sufficient oxygen to support high metabolic rates. As we will discuss, different strategies for gas exchange have evolved in the arthropods.

In this chapter we will examine adaptations for gas exchange in air and water and describe various types of respiratory surfaces.

© FLPA/Alamy

Gas exchange in terrestrial vertebrates. The breath of this juvenile buck (a fallow deer; *Dama dama*) is condensing in cold air.

KEY CONCEPTS

46.1 Air has a higher concentration of molecular oxygen than water does, and animals require less energy to move air than to move water over a gas exchange surface; however, animals that respire in air must have adaptations that reduce water loss so that their respiratory surfaces do not dry out.

46.2 Adaptations for gas exchange include a thin, moist body surface; gills in aquatic animals; and tracheal tubes and lungs in terrestrial animals.

46.3 In mammals oxygen and carbon dioxide are exchanged between alveoli and blood by diffusion; the pressure of a particular gas determines its direction and rate of diffusion. Respiratory pigments combine with oxygen and transport it; carbon dioxide is transported mainly as bicarbonate ions.

46.4 Several defense mechanisms protect the lungs from harmful substances that are inhaled; tobacco smoke and other air pollutants are major causes of disease, disability, and death.

We will then focus on the structure and function of the mammalian respiratory system. We end the chapter with a discussion of the effects of breathing polluted air, including cigarette smoke.

46.1 ADAPTATIONS FOR GAS EXCHANGE IN AIR OR WATER

LEARNING OBJECTIVE

1   Compare the advantages and disadvantages of air and water as mediums for gas exchange and describe adaptations for gas exchange in air.

Gas exchange in air has certain advantages over gas exchange in water. Air contains a much higher concentration of molecular oxygen (at least 20 times more) than water does. In addition, oxygen diffuses about 10,000 times faster through air than through water. Another advantage is that less energy is needed to move air than to move water over a gas exchange surface because air is less dense and less viscous. A mammal uses less than 2% of its energy budget to breathe compared to a fish, which must spend as much as 20% of its energy to obtain its oxygen from water.

Gas exchange in air presents a major challenge: the threat of desiccation. Animals that respire in air struggle continuously with water loss. Respiratory surfaces must be kept moist because oxygen and carbon dioxide are dissolved in the fluid that bathes the cells of these surfaces. Whether an animal makes its home on land or water, gas exchange takes place across a moist surface.

Adaptations have evolved that keep respiratory surfaces moist and minimize desiccation. For example, the lungs of air-breathing vertebrates are located deep within the body, not exposed as gills are. Air is humidified and brought to body temperature as it passes through the moist upper respiratory passageways. Expired air must again pass through these airways (providing opportunity for retaining water) before leaving the body. These adaptations protect the lungs from the drying and cooling effects of air.

CHECKPOINT 46.1

*   *What are some advantages of gas exchange in air over gas exchange in water?*

46.2 TYPES OF RESPIRATORY SURFACES

LEARNING OBJECTIVE

2   Describe the following adaptations for gas exchange: body surface, tracheal tubes, gills, and lungs.

Respiratory structures must be moist and must also have thin walls through which diffusion can easily occur. Four main types of respiratory surfaces have evolved in animals: the animal's own body surface, tracheal tubes, gills, and lungs (FIG. 46-1). Some animals use a combination of these adaptations. Gills are adapted for gas exchange in water, and the tracheal tubes of insects and lungs of vertebrates are respiratory structures adapted for gas exchange in air. With the important exception of insect tracheal tubes, respiratory structures are richly supplied with blood vessels that facilitate exchange and transport of respiratory gases.

Most animals carry on **ventilation;** that is, they actively move air or water over their respiratory surfaces. If the air or water supplying oxygen to the cells can be continuously renewed, more oxygen will be available. Sponges use their flagella to set up a current of water through the channels of their bodies. Most fishes gulp water, which then passes over their gills. Terrestrial vertebrates have lungs and breathe air. When mammals breathe, the diaphragm and other muscles move air in and out of the lungs.

The body surface may be adapted for gas exchange

Gas exchange occurs through the entire body surface in many animals, including nudibranch mollusks, most annelids, and some amphibians. These animals are small, with a high ratio of surface area to volume. They also have a low metabolic rate that requires smaller quantities of oxygen per cell. In aquatic animals the body surface is kept moist by the surrounding water. In some terrestrial animals the body secretes fluids that keep its surface moist. Many animals that exchange gases across the body surface also have gills or lungs.

Tracheal tube systems deliver air directly to the cells

In insects and some other arthropods (such as centipedes and millipedes, some mites, and some spiders), the respiratory system is a network of **tracheal tubes,** also called **tracheae** (FIG. 46-2). This very efficient respiratory system delivers air directly to the cells. Air enters the tracheal tubes through a series of as many as 20 tiny openings called **spiracles** along the body surface. In some insects, especially large, active ones, muscles help ventilate the tracheae by pumping air in and out of the spiracles. For example, the grasshopper draws air in through the first four pairs of spiracles when the abdomen expands. Then the abdomen contracts, forcing air out through the last six pairs of spiracles.

Once inside the body, the air passes through a system of branching tracheal tubes, which extend to all parts of the animal. The tracheal tubes terminate in microscopic, fluid-filled tracheoles. Gases are exchanged between this fluid and the body cells. The tracheal system supplies enough oxygen to support the high metabolic rates required by many insects.

Gills are the respiratory surfaces in many aquatic animals

Found mainly in aquatic animals, **gills** are moist, thin structures that extend outward from the body surface. They are supported by the buoyancy of water but tend to collapse in air. In

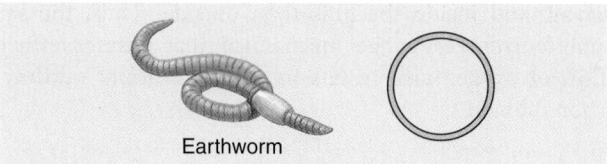

(a) Body surface. Some small, multicellular animals exchange gases through the body surface.

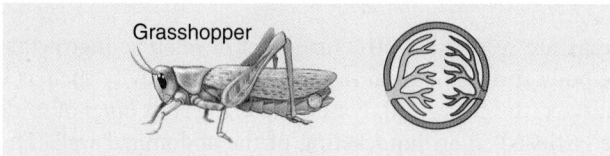

(b) Tracheal tubes. Insects and some other arthropods exchange gases through a system of tracheal tubes, or tracheae.

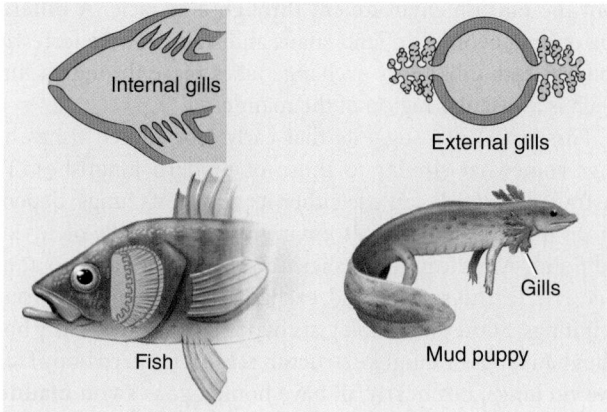

(c) Gills. Most aquatic animals exchange gases through gills, thin structures that extend from the body surface. Gills can be internal or external.

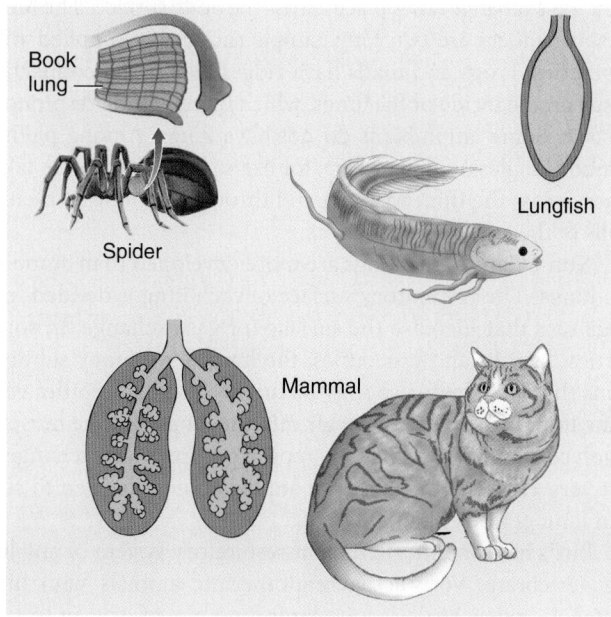

(d) Lungs. Lungs are adaptations for gas exchange in terrestrial habitats.

Figure 46-1 Adaptations for gas exchange

© Cengage Learning

The efficient tracheal tube system of insects delivers air directly to the cells.

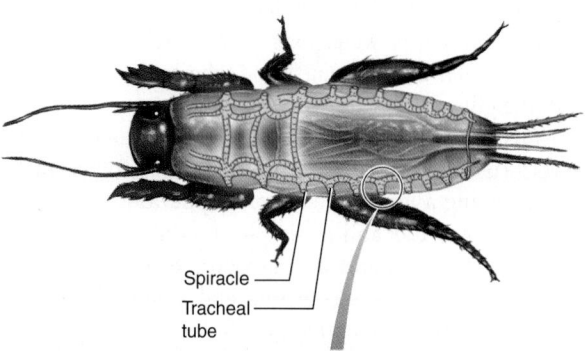

(a) Location of spiracles and tracheal tubes. Air enters the system of tracheal tubes through openings called spiracles.

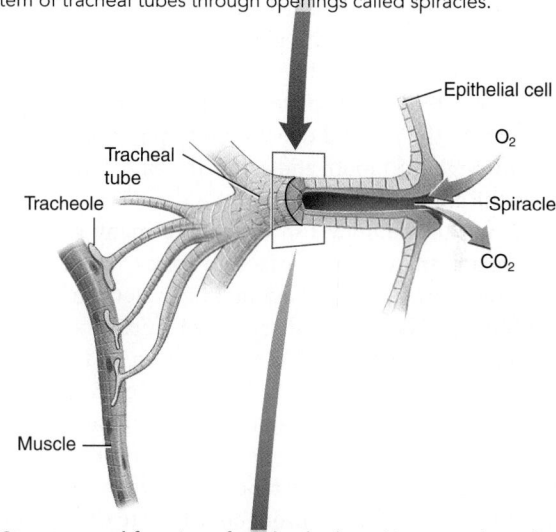

(b) Structure and function of tracheal tubes. Air passes through a system of branching tracheal tubes that conduct oxygen to all cells of the insect.

Courtesy of Dr. James L. Nation and *Stain Technology*, Vol. 58, 1983

100 μm

(c) SEM of a mole cricket trachea. Rings of chitin reinforce the tubes of the tracheal system, preventing their collapse.

Figure 46-2 Gas exchange in insects: tracheal tubes

CONNECT In contrast to vertebrate gas exchange, blood does not play an essential role in insect gas exchange. Why?

© Cengage Learning

many animals the outer surface of the gills is exposed to water, whereas the inner side is in close contact with networks of blood vessels.

Sea stars and sea urchins have *dermal gills* that project from the body wall. Their ciliated epidermal cells ventilate the gills by beating a stream of water over them. Gases are exchanged between the water and the fluid inside the coelom by diffusion through the gills.

Various types of gills are found in some annelids, aquatic mollusks, crustaceans, fishes, and amphibians. Mollusk gills are folded, providing a large surface for respiration. In clams and other bivalve mollusks and in simple chordates, gills may also be adapted for trapping and sorting food. The rhythmic beating of cilia draws water over the gill area, and food is filtered out of the water while gases are exchanged. In mollusks gas exchange also takes place through the mantle.

In chordates gills are usually internal. A series of slits perforates the pharynx, and the gills lie along the edges of these gill slits (see Fig. 32-4). In bony fishes the fragile gills are protected by an external bony plate, the **operculum.** In some fishes movements of the jaw and operculum help pump water rich in oxygen through the mouth and across the gills. The water leaves through the gill slits.

Each gill in the bony fish consists of many *gill filaments,* which provide an extensive surface for gas exchange (FIGS. 46-3a and b). The filaments extend out into the water, which continuously flows over them. A capillary network delivers blood to the gill filaments, facilitating diffusion of oxygen and carbon dioxide between blood and water. This system is extremely efficient because blood flows in a direction opposite to the movement of the water. This arrangement, called a **countercurrent exchange system,** maximizes the difference in oxygen concentration between blood and water throughout the area where the two remain in contact (FIG. 46-3c).

If blood and water flowed in the *same* direction—that is, *concurrent exchange*—the difference between the oxygen concentrations in blood (low) and water (high) would be very large initially and then very small at the end (FIG. 46-3d). The oxygen concentration in the water would decrease as the concentration in the blood increased. When the concentrations in the two fluids became equal, equilibrium would be reached, and the net diffusion of oxygen would stop. Only about 50% of the oxygen dissolved in the water could diffuse into the blood.

In the countercurrent exchange system, however, blood low in oxygen comes in contact with water that is partly depleted of oxygen. Then, as the blood flowing through the capillaries becomes progressively richer in oxygen, the blood comes in contact with water with an increasingly higher concentration of oxygen. Thus, all along the capillaries, the diffusion gradient favors passage of oxygen from the water into the gill. A high rate of diffusion is maintained, ensuring that a very high percentage (more than 80%) of the available oxygen in the water diffuses into the blood.

Oxygen and carbon dioxide do not interfere with each other's diffusion, and they simultaneously diffuse in opposite directions. The reason is that oxygen is more concentrated outside the gills than inside but carbon dioxide is more concentrated inside the gills than outside. Thus, the same countercurrent exchange mechanism that ensures efficient inflow of oxygen also results in equally efficient outflow of carbon dioxide.

Terrestrial vertebrates exchange gases through lungs

Lungs are respiratory structures that develop as ingrowths of the body surface or from the wall of a body cavity such as the pharynx (throat region). For example, the *book lungs* of spiders are enclosed in an inpocketing of the abdominal wall. These lungs consist of a series of thin, parallel plates of tissue (like the pages of a book) filled with hemolymph (see Fig. 46-1d). The plates of tissue are separated by air spaces that receive oxygen from the outside environment through a spiracle. A different type of lung evolved in land snails and slugs. (These terrestrial mollusks lack gills.) Gas exchange takes place through a lung, which is a vascular region of the mantle.

Fossil evidence suggests that early lobe-finned fishes had lungs somewhat similar to those of modern lungfishes. The Australian lungfish can use either its gills or its lungs, depending on the conditions in its environment. The gills of African and South American lungfishes degenerate with age, so adults must rise to the surface and exchange gases entirely through their lungs. Some paleontologists hypothesize that all early bony fishes had lungs or lunglike structures. Most modern bony fishes have no lungs, but nearly all have homologous **swim bladders** (see Chapter 32). By adjusting the amount of gas in its swim bladder, the fish can control its buoyancy.

Most amphibians have lungs (FIG. 46-4). However, most of their gas exchange takes place across the body surface. The lungs of salamanders are two long, simple sacs richly supplied with capillaries. Frogs and toads have ridges containing connective tissue on the inside of the lungs, which increases the respiratory surface. Some amphibians do not have lungs. Among plethodontid (lungless) salamanders, for example, gas exchange takes place across the thin, wet skin and through the moist, vascular walls of the mouth and pharynx.

Non-avian reptile lungs are more developed than amphibian lungs. The respiratory surface of each lung is divided into large sacs that increase the surface for gas exchange. In some turtles, lizards, and crocodiles, the lungs have many subdivisions that give them a spongy texture. Non-avian reptiles ventilate their lungs by drawing air into the lungs, but the muscles involved differ in the various reptile groups. Gas exchange is not very efficient and does not supply enough oxygen to sustain long periods of activity.

Birds have the most efficient respiratory system of any living vertebrate. Very active, endothermic animals with high metabolic rates, birds require large amounts of oxygen to sustain flight and other activities. Their small, bright red lungs have extensions (usually nine) called *air sacs,* which reach into all parts of the body and even connect with air spaces in some of the bones. The air sacs act as bellows, drawing air into the system. Collapse of the air sacs during exhalation forces air out.

Countercurrent flow efficiently oxygenates blood circulating through the gill filaments.

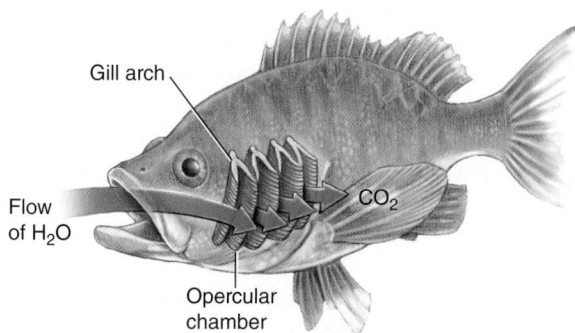

Gill arch

Flow
of H₂O

CO_2

Opercular
chamber

(a) Location of gills. The gills form the lateral wall of the pharyngeal cavity. They lie under a bony plate, the operculum, which has been removed in this side view.

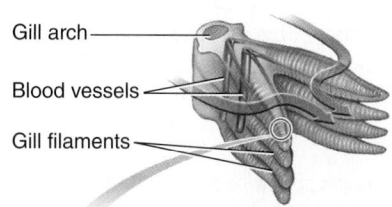

Gill arch

Blood vessels

Gill filaments

(b) Structure of a gill. Each gill consists of a cartilaginous gill arch to which two rows of leaflike gill filaments attach. Each gill filament has many smaller extensions rich in capillaries. As water flows past the gill filament, oxygen passes from the water into blood circulating through the capillaries.

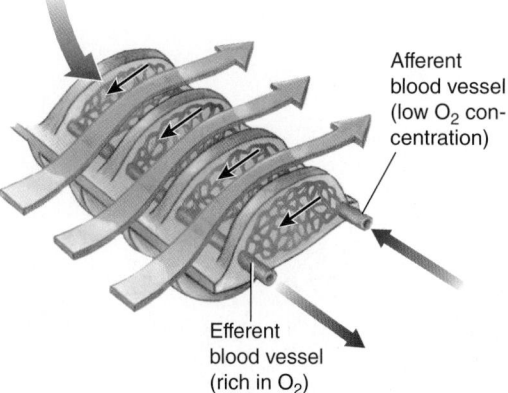

Afferent
blood vessel
(low O₂ con-
centration)

Efferent
blood vessel
(rich in O₂)

(c) Countercurrent flow. Blood entering the capillaries of the gill filament is deficient in oxygen. The blood flows through the capillaries in a direction opposite to that taken by the water. This countercurrent exchange system efficiently charges the blood with oxygen.

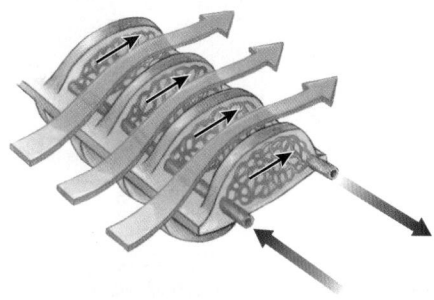

(d) Concurrent flow (hypothetical). If the system were concurrent—that is, if blood flowed through the capillaries in the same direction as the flow of the water—much less of the oxygen dissolved in the water could diffuse into the blood.

Bernard Photo Production / Animals Animals

(e) Gills of the salmon.

Figure 46-3 *Animation* **Gas exchange in bony fishes: gills**

PREDICT In what way would you expect the countercurrent mechanism to affect the removal of CO₂ from the blood?

© Cengage Learning

In birds, gas exchange does not take place across the walls of the air sacs.

The high-performance bird respiratory system is arranged so that air flows in one direction through the lungs and is renewed during a two-cycle process (**FIG. 46-5**). Air entering the body passes into the posterior air sacs and the part of the lungs closest to these air sacs. The posterior air sacs are reservoirs for fresh air. When the bird exhales, that air flows into

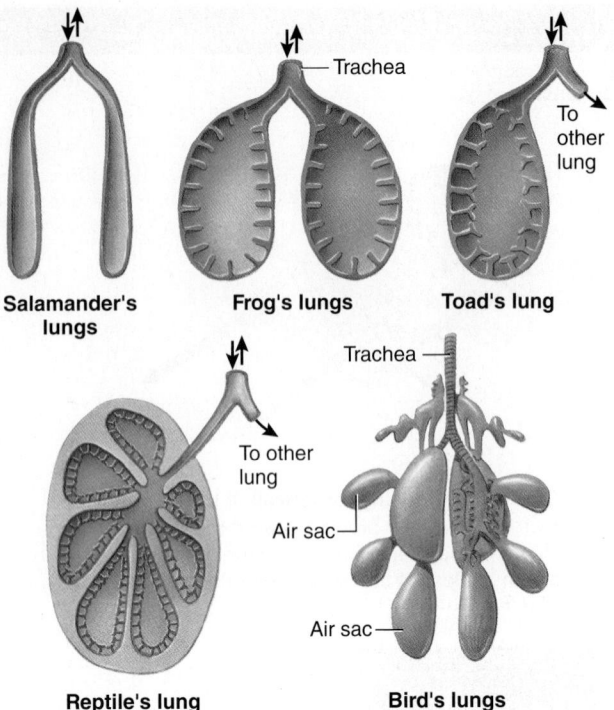

Figure 46-4 *Animation* **Evolution of vertebrate lungs**

During vertebrate evolution, the surface area of the lung increased. Salamander lungs are simple sacs. Other amphibians and reptiles have lungs with small ridges or folds that increase surface area. Birds have an elaborate system of lungs and air sacs. Mammalian lungs have millions of air sacs (alveoli) that increase the surface available for gas exchange (see Fig. 46-7).

© Cengage Learning

the lungs. At the second breath, the air flows from the lungs to the anterior air sacs. Finally, at the second exhalation, the air leaves the body as another breath of air enters the lungs. Thus, a bird gets fresh air across its lungs through both inhalation and exhalation.

In birds the narrowest branches of the airway end in tiny, thin-walled tubes called **parabronchi.** In contrast to the saclike alveoli of mammals, parabronchi are open at both ends and air continuously flows through them. Gas exchange takes place across the walls of these tubes, which are rich in capillaries. Chickens and other weak-flying birds have about 400 parabronchi per lung; pigeons and other strong-flying birds have about 1800 parabronchi per lung.

The direction of blood flow in the lungs is in a different direction from that of airflow through the parabronchi. This arrangement, similar in principle to the countercurrent exchange in the gills of fishes, increases the amount of oxygen that enters the blood. However, in birds the capillaries are oriented at right angles to the parabronchi rather than along their length. For this reason, this arrangement is referred to as *crosscurrent* rather than countercurrent.

CHECKPOINT 46.2

- **CONNECT** *Why are specialized respiratory structures necessary in a tadpole but not in a flatworm?*
- *How does gas exchange differ among the following animals: earthworm, grasshopper, fish, and bird?*
- *How does the countercurrent exchange system increase the efficiency of gas exchange between a fish's gills and blood?*

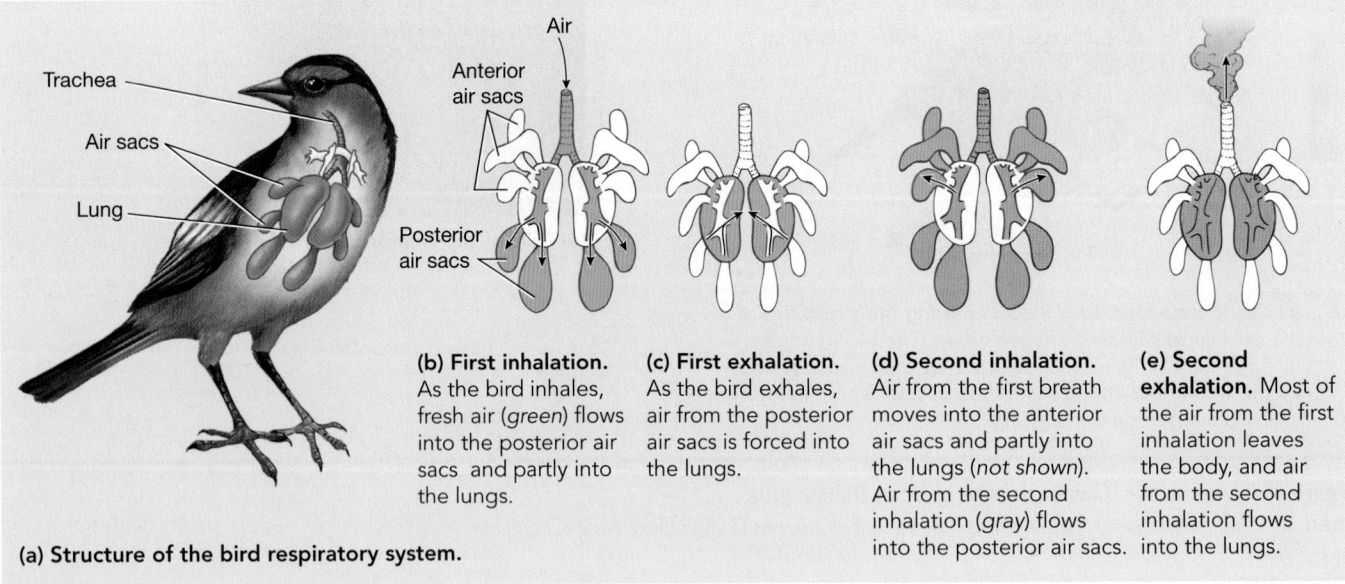

(b) First inhalation. As the bird inhales, fresh air (*green*) flows into the posterior air sacs and partly into the lungs.

(c) First exhalation. As the bird exhales, air from the posterior air sacs is forced into the lungs.

(d) Second inhalation. Air from the first breath moves into the anterior air sacs and partly into the lungs (*not shown*). Air from the second inhalation (*gray*) flows into the posterior air sacs.

(e) Second exhalation. Most of the air from the first inhalation leaves the body, and air from the second inhalation flows into the lungs.

(a) Structure of the bird respiratory system.

Figure 46-5 *Animation* **Gas exchange in birds**

The bird respiratory system includes lungs and air sacs. The bird's breathing process requires two cycles of inhalation and exhalation to support a one-way flow of air through the lungs.

© Cengage Learning

46.3 THE MAMMALIAN RESPIRATORY SYSTEM

LEARNING OBJECTIVES

3 Trace the passage of oxygen through the human respiratory system from nostrils to alveoli.

4 Summarize the mechanics of breathing in humans and describe gas exchange in the lungs and tissues.

5 Summarize the mechanisms by which oxygen and carbon dioxide are transported in the blood and identify factors that determine the oxygen–hemoglobin dissociation curve.

6 Describe the regulation of breathing in humans and summarize the physiological effects of hyperventilation and of sudden decompression when a diver surfaces too quickly from deep water.

The respiratory system of mammals consists of the lungs and a series of tubes through which air passes on its journey from the nostrils to the lungs and back (**FIG. 46-6**). The lungs have an enormous surface area. As in all vertebrates, the respiratory and circulatory systems are functionally connected. The circulatory system delivers oxygen-rich blood to all the cells in the body. In the following sections, we focus mainly on the human respiratory system.

The airway conducts air into the lungs

A breath of air enters the body through the *nostrils* and flows through the *nasal cavities*. Air passing through the nose is filtered, moistened, and brought to body temperature. The nasal cavities are lined with a moist, ciliated epithelium rich in blood vessels. Inhaled dirt, bacteria, and other foreign particles are trapped in the stream of mucus produced by cells within the epithelium and pushed along toward the throat by the cilia. In this way, foreign particles are delivered to the digestive system, which can more effectively dispose of such materials than can the delicate lungs. We normally swallow more than a pint of nasal mucus each day, and even more during an infection or allergic reaction.

The back of the nasal cavities is continuous with the throat region, or **pharynx.** Air finds its way into the pharynx whether we breathe through the nose or mouth. An opening in the floor of the pharynx leads into the **larynx.** Because the larynx contains the vocal cords, it is also referred to as the "voice box." Cartilage embedded in its wall prevents the larynx from collapsing and makes it hard to the touch when felt through the neck.

During swallowing, a flap of tissue called the **epiglottis** automatically closes off the larynx so that food and liquid enter the esophagus rather than the lower airway. If this mechanism fails and foreign matter enters the sensitive larynx, a cough reflex expels the material. Despite these mechanisms, choking sometimes occurs.

From the larynx, air passes into the **trachea,** or windpipe, which is kept from collapsing by rings of cartilage in its wall. The trachea divides into two branches, the **bronchi** (sing., *bronchus*); one bronchus connects to each lung. Both trachea and bronchi are lined by a mucous membrane containing ciliated cells. These cells trap medium-sized particles that escape the cleansing mechanisms of the nose and larynx. Mucus containing these particles is constantly beaten upward by the cilia to the pharynx, where it is periodically swallowed. This mechanism, functioning as a cilia-propelled elevator of mucus, helps keep foreign material out of the lungs.

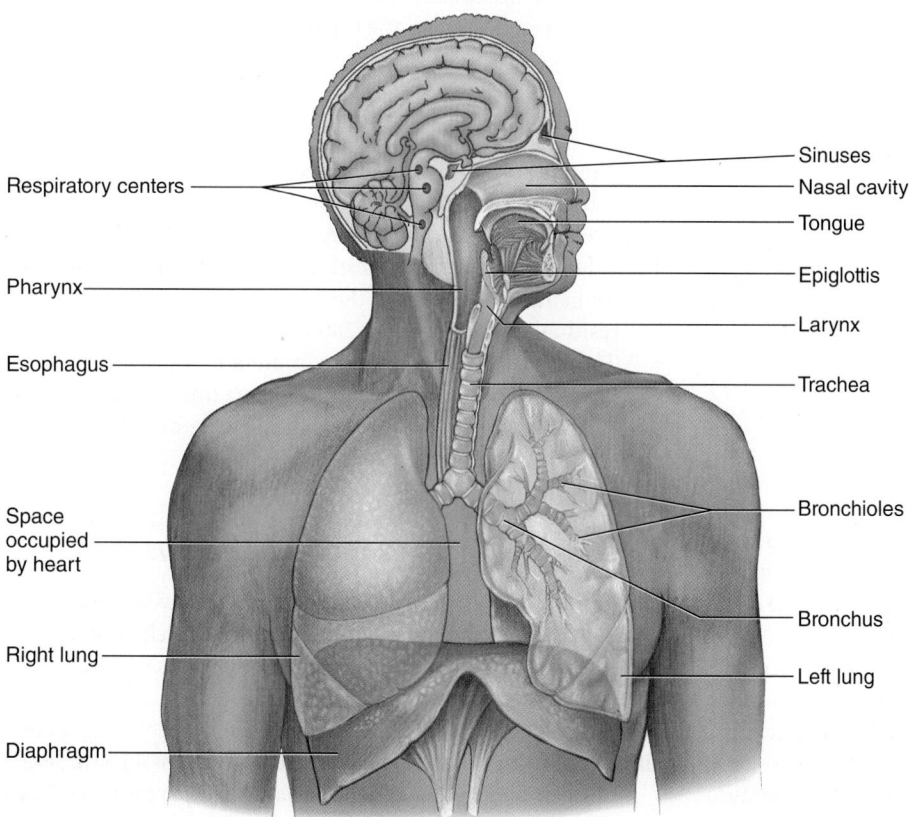

Respiratory centers

Pharynx

Esophagus

Space occupied by heart

Right lung

Diaphragm

Sinuses
Nasal cavity
Tongue
Epiglottis
Larynx
Trachea
Bronchioles
Bronchus
Left lung

Figure 46-6 *Animation* **The human respiratory system**

The internal view of one lung illustrates a portion of its extensive system of air passageways. The muscular diaphragm forms the floor of the thoracic cavity. The respiratory centers in the brain regulate the rate of respiration.
© Cengage Learning

Gas exchange occurs in the alveoli of the lungs

The lungs are large, paired, spongy organs occupying the *thoracic* (chest) *cavity*. The right lung is divided into three lobes, the left lung into two lobes. Each lung is covered with a **pleural membrane,** which forms a continuous sac that encloses the lung and extends outward to become the lining of the thoracic cavity. The *pleural cavity* is the space between the pleural membranes. A film of fluid in the pleural cavity provides lubrication between the lungs and the chest wall.

Because the lung consists largely of air tubes and elastic connective tissue, it is a spongy, elastic organ with a very large internal surface area for gas exchange. Within each lung the bronchi branch and become progressively shorter, narrower, and more numerous, similar to the branching of a tree. Small branches give rise to the even smaller **bronchioles** in each lung. Each bronchiole ends in a cluster of tiny air sacs, the **alveoli** (sing., *alveolus*) (FIG. 46-7). Each human lung contains more than 200 million alveoli, which provide an internal surface area the approximate size of a tennis court!

Each alveolus is lined by an extremely thin, single layer of epithelial cells. Gases diffuse freely through the wall of the alveolus and into the capillaries that surround it. Only two thin cell layers—the endothelium of the alveolar wall and the epithelium of the capillary wall—separate the air in the alveolus from the blood.

In summary, air passes through the following sequence of structures after it enters the body:

nostrils ⟶ nasal cavities ⟶ pharynx ⟶ larynx ⟶ trachea ⟶ bronchi ⟶ bronchioles ⟶ alveoli

Ventilation is accomplished by breathing

Breathing is the mechanical process of moving air from the environment into the lungs and of expelling air from the lungs. Inhaling air is called **inhalation** or *inspiration;* exhaling air is **exhalation** or *expiration.* The thoracic cavity is closed so that air can enter only through the trachea. (When the chest wall is punctured, for example, by a fractured rib or gunshot wound, air enters the pleural space and the lung collapses.)

During inhalation, the volume of the thoracic cavity is increased by the contraction of the **diaphragm,** the dome-shaped muscle that forms its floor. When the diaphragm contracts, it moves downward, increasing the volume of the thoracic cavity (FIG. 46-8). The lungs adhere to the walls of the thoracic cavity, so when the volume of the thoracic cavity increases, the space within each lung also increases. The air in the lungs now has more space in which to move about. As a result, air pressure in the lungs falls by 2 or 3 millimeters of mercury (mm Hg) below the air pressure outside the body. Because of this pressure difference, air from the outside rushes in through the respiratory passageways and fills the lungs until the two pressures are equal once again.

During deep or forced inhalation, the *external intercostal muscles* also contract, pulling the rib cage upward and outward. This action further increases the volume of the thoracic cavity, and more air moves into the lungs.

Exhalation occurs when the diaphragm relaxes. The volume of the thoracic cavity decreases, raising the pressure in the lungs to 2 to 3 mm Hg above atmospheric pressure. The millions of distended air sacs partially deflate and expel the inhaled air. The pressure then returns to normal, and the lung is ready for another inhalation. To summarize: during inhalation, the millions of alveoli fill with air like so many tiny balloons. Then,

during exhalation, the air rushes out of the alveoli, partially deflating them.

During forced exhalation, muscles of the abdominal wall and the *internal intercostal muscles* contract, pushing the diaphragm up and the ribs down. This action decreases the volume of the thoracic cavity, pushing air out of the lungs.

Some of the work required to stretch the thorax and the lungs is necessary to stretch their elastic connective tissue. Work is also required to overcome the cohesive force of water molecules associated with the pleural membranes. The forces between the water molecules produce a surface tension that resists stretching.

The work of breathing is reduced by *pulmonary surfactant,* a detergent-like phospholipid mixture secreted by specialized epithelial cells in the lining of the alveoli. Pulmonary surfactant intersperses between the water molecules, reducing their cohesive force. This action markedly reduces the surface tension of the water, prevents the alveoli from collapsing, and reduces the energy required to stretch the lungs. Premature infants often cannot produce enough surfactant, which causes respiratory distress syndrome. In these infants high surface tension makes it difficult to inflate the lungs, and the alveoli collapse during expiration. Breathing is labored. Respiratory failure as a result of surfactant deficiency is a leading cause of morbidity and mortality in premature infants. Surfactant deficiency can be treated with surfactant replacement therapy in which artificial surfactant is delivered through the trachea.

The quantity of respired air can be measured

The amount of air moved into and out of the lungs with each normal resting breath is called the **tidal volume.** The normal tidal volume is about 500 mL. A great deal of stale air remains in the lungs during normal resting breathing. The volume of air that remains in the lungs at the end of a maximal expiration is the **residual volume** (about 1200 mL).

The **vital capacity** is the maximum amount of air a person can exhale after filling the lungs to the maximum extent (about 4500 mL). Vital capacity is a useful measure of the functional capacity of the lungs.

Gas exchange takes place in the alveoli

The respiratory system delivers oxygen to the alveoli, but if oxygen remained in the lungs, all the other body cells would soon die. The vital link between alveoli and body cells is the circulatory system. Clusters of pulmonary capillaries surround the alveoli, bringing blood close to the alveolar air. Each alveolus serves as a tiny depot from which oxygen diffuses into the blood (FIG. 46-9).

Oxygen molecules efficiently pass by simple diffusion from the alveoli, where they are more concentrated, into the blood in the pulmonary capillaries, where they are less concentrated. At the same time, carbon dioxide moves from the blood, where it is more concentrated, to the alveoli, where it is less concentrated. Each gas diffuses through the single layer of cells lining the alveoli and the single layer of cells lining the capillaries.

Oxygen diffuses through the thin wall of each alveolus and then through the thin walls of the surrounding capillaries to reach the blood. Carbon dioxide diffuses from the blood into the alveoli.

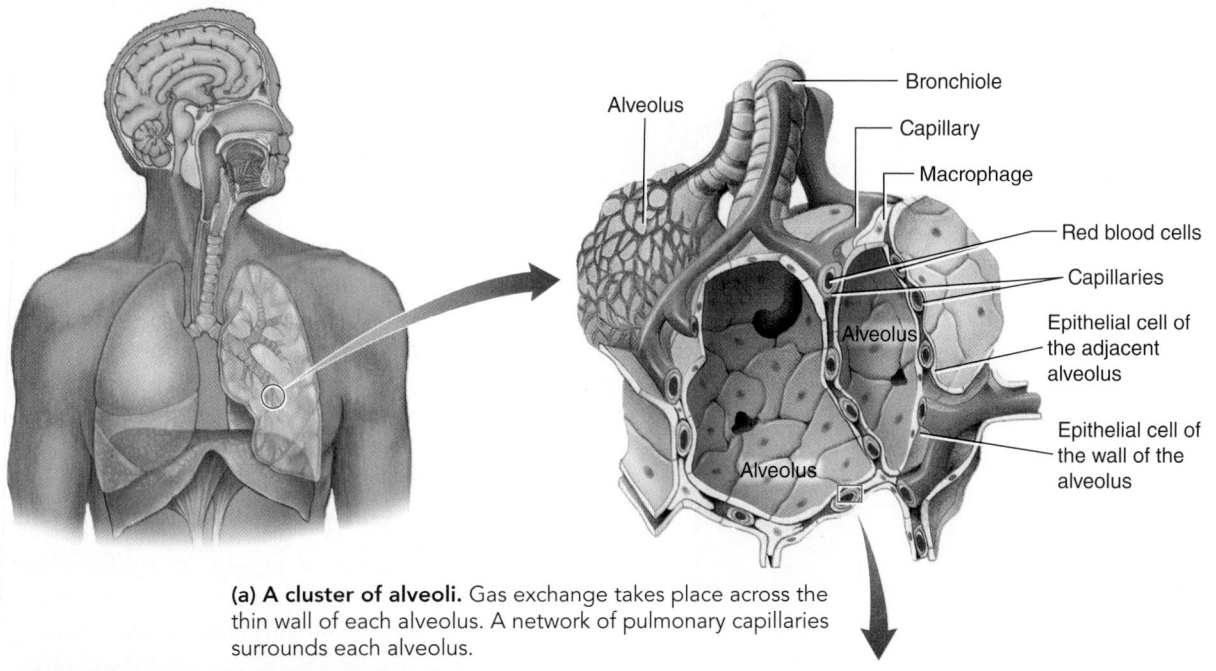

(a) **A cluster of alveoli.** Gas exchange takes place across the thin wall of each alveolus. A network of pulmonary capillaries surrounds each alveolus.

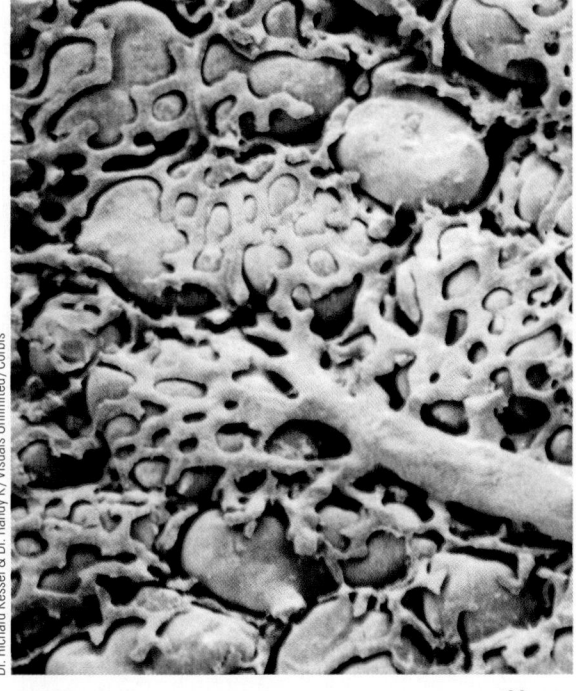

30 μm

(b) **SEM of alveoli and surrounding pulmonary capillaries.** The capillary networks are so dense that each alveolus is surrounded by continuously flowing blood.

1 μm

(c) **Color-enhanced TEM of a portion of a capillary and the wall of an alveolus.** The dark structure extending through the capillary is part of a red blood cell. Note the very short distance that oxygen must diffuse to get from the air within the alveolus to the red blood cells that transport it to the body tissues.

Figure 46-7 *Animation* **Structure and function of alveoli**

Each bronchiole ends in a cluster of alveoli (air sacs). Gases are exchanged between the air in the alveoli and the blood in the capillaries. The lungs contain approximately 500 million alveoli.
© Cengage Learning

CONNECT In what way is the structure of capillaries (discussed in Chapter 44) essential to efficient gas exchange in the alveoli?

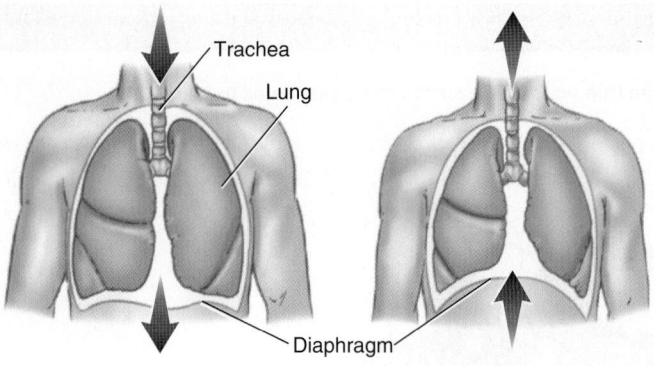

Trachea

Lung

Diaphragm

(a) Inhalation. The diaphragm contracts, increasing the volume of the thoracic cavity. Air moves into the lungs.

(b) Exhalation. The diaphragm relaxes, decreasing the volume of the thoracic cavity. Air moves out of the lungs.

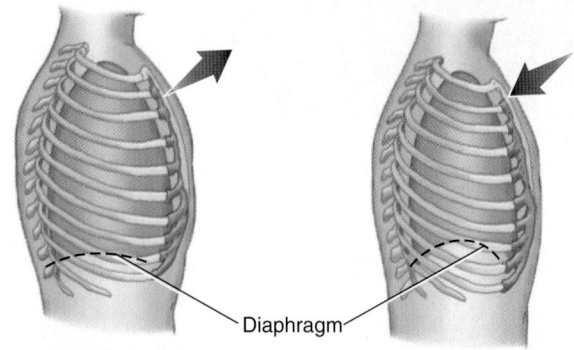

Diaphragm

(c) Forced inhalation. The external intercostal muscles contract, pulling the rib cage upward and outward. This action increases the front-to-back dimension of the chest and correspondingly increases the volume of the thoracic cavity.

(d) Forced exhalation. The internal intercostal muscles contract, pulling the rib cage downward and inward. The volume of the thoracic cavity decreases.

Figure 46-8 *Animation* **Mechanics of breathing**

During inhalation and exhalation, the position of the diaphragm and of the external and internal intercostal muscles changes. These changes increase or decrease the volume of the thoracic cavity.

© Cengage Learning

Recall that in aerobic respiration, oxygen is the final electron acceptor; it accepts two electrons, which are added to two protons to produce water (see Chapter 8). Inhaled (atmospheric) air contains about 20.9% oxygen, but because oxygen is used during cellular respiration, exhaled (alveolar) air contains only 14% oxygen. Carbon dioxide is produced during aerobic respiration. For this reason, exhaled air contains more than one hundred times as much (5.6%) carbon dioxide as inhaled air (about 0.04% carbon dioxide).

The concentration of oxygen in the cells is lower than in the capillaries entering the tissues, and the concentration of carbon dioxide is higher in the cells than in the capillaries. As blood circulates through capillaries of a tissue such as brain or muscle, oxygen moves by simple diffusion from the blood to the cells, and carbon dioxide moves from the cells into the blood.

The factor that determines the direction and rate of diffusion is the pressure or tension of the particular gas. According to *Dalton's law of partial pressures,* in a mixture of gases, the total pressure of the mixture is the sum of the pressures of the individual gases. Each gas exerts, independently of the others, a **partial pressure,** the same pressure it would exert if it were present alone. At sea level, the barometric pressure (the pressure of Earth's atmosphere) typically supports a column of mercury 760 mm high. Because oxygen makes up about 21% of the atmosphere, oxygen's share of that pressure is $0.21 \times 760 = 160$ mm Hg. Thus, 160 mm Hg is the partial pressure of atmospheric O_2, abbreviated P_{O_2}. In contrast, the partial pressure of atmospheric CO_2 is 0.23 mm Hg, abbreviated P_{CO_2}.

Fick's law of diffusion explains that the amount of oxygen or carbon dioxide that diffuses across the membrane of an alveolus depends both on the differences in partial pressure on the two sides of the membrane and on the surface area of the membrane. The gas diffuses faster if the difference in pressure or the surface area increases.

Gas exchange takes place in the tissues

The partial pressure of oxygen in arterial blood is about 100 mm Hg. The P_{O_2} in the tissues is still lower, averaging about 40 mm Hg. Consequently, oxygen diffuses out of the

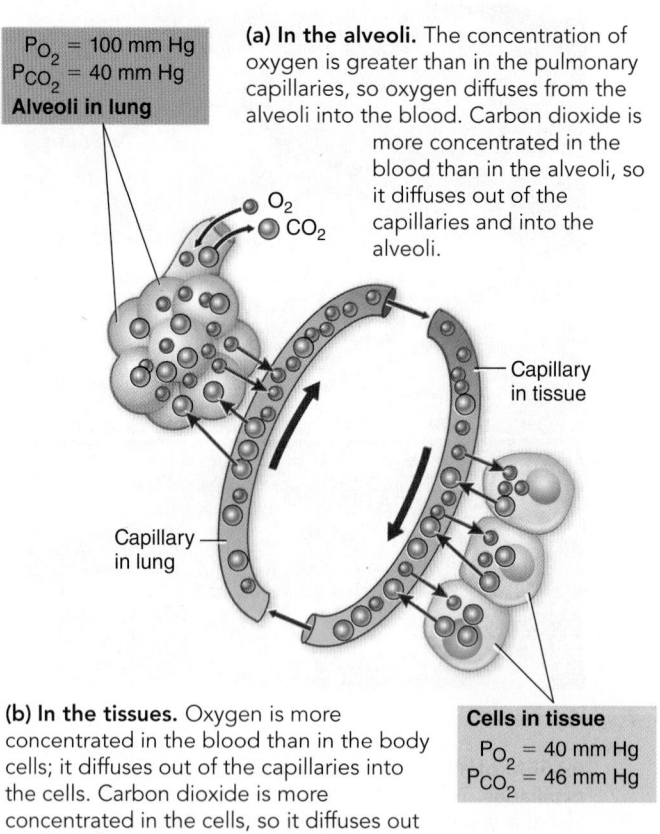

$P_{O_2} = 100$ mm Hg
$P_{CO_2} = 40$ mm Hg
Alveoli in lung

(a) In the alveoli. The concentration of oxygen is greater than in the pulmonary capillaries, so oxygen diffuses from the alveoli into the blood. Carbon dioxide is more concentrated in the blood than in the alveoli, so it diffuses out of the capillaries and into the alveoli.

O_2
CO_2

Capillary in tissue

Capillary in lung

(b) In the tissues. Oxygen is more concentrated in the blood than in the body cells; it diffuses out of the capillaries into the cells. Carbon dioxide is more concentrated in the cells, so it diffuses out of the cells and moves into the blood.

Cells in tissue
$P_{O_2} = 40$ mm Hg
$P_{CO_2} = 46$ mm Hg

Figure 46-9 Gas exchange in the lungs and tissues

Note the differences in partial pressures of oxygen and carbon dioxide before and after gases are exchanged in the tissues.

© Cengage Learning

capillaries and into the tissues. Not all the oxygen leaves the blood, however. The blood passes through the tissue capillaries too rapidly for equilibrium to be reached. As a result, the partial pressure of oxygen in venous blood returning to the lungs is about 40 mm Hg. Thus, exhaled air has had only part of its oxygen removed, which is a good thing for those in need of mouth-to-mouth resuscitation!

Respiratory pigments increase capacity for oxygen transport

Respiratory pigments greatly increase the quantity of oxygen that blood can transport. These molecules combine reversibly with oxygen. **Hemocyanins** are copper-containing proteins dispersed in the hemolymph of many species of mollusks and arthropods. Without oxygen, these pigments are colorless. When oxygen combines with the copper, hemocyanins are blue.

Hemoglobin and **myoglobin** are the most common respiratory pigments in animals. Hemoglobin is the pigment found in the blood of vertebrates. It is also present in many invertebrate species, including annelids, nematodes, mollusks, and arthropods. In some of these animals, the hemoglobin is dispersed in the plasma rather than confined to blood cells. Recall that myoglobin is a form of hemoglobin found in muscle fibers (see Chapter 40).

In humans and other mammals, inhaled oxygen diffuses out of the alveoli and enters the pulmonary capillaries. Plasma in equilibrium with alveolar air can take up only 0.25 mL of oxygen per 100 mL. However, oxygen diffuses into red blood cells (RBCs) and combines with hemoglobin. The properties of hemoglobin permit whole blood to carry some 20 mL of oxygen per 100 mL. Hemoglobin transports almost 99% of the oxygen. The rest is dissolved in the plasma.

The term *hemoglobin* is actually a general name for a group of related compounds, all of which consist of an iron–porphyrin, or heme, group bound to a protein known as a *globin*. The protein portion of the molecule varies in size, amino acid composition, and physical properties among various species.

The protein portion of hemoglobin is composed of four peptide chains, typically two alpha and two beta chains, each attached to a heme (iron–porphyrin) ring (see Fig. 3-23a). An iron atom is bound in the center of each heme ring. Hemoglobin has the remarkable property of forming a weak chemical bond with oxygen. An oxygen molecule can attach to the iron atom in each heme. In the lung (or gill), oxygen diffuses into the RBCs and combines with hemoglobin (Hb) to form **oxyhemoglobin (HbO₂)**. When combined with oxygen, hemoglobin is

bright red; without oxygen, it appears dark red, imparting a purplish color to venous blood.

Because the chemical bond formed between the oxygen and the hemoglobin is weak, the reaction is readily reversible. As blood circulates through tissues where the oxygen concentration is low, the reaction proceeds to the left. Hemoglobin releases oxygen, which diffuses out of the blood and into the cells of the tissues.

$$Hb + O_2 \rightleftharpoons HbO_2$$

The maximum amount of oxygen that hemoglobin can transport is its **oxygen-carrying capacity.** The *actual* amount of oxygen bound to hemoglobin is the *oxygen content.* The ratio of O_2 content to oxygen-carrying capacity is the *percent O_2 saturation* of the hemoglobin. The percent O_2 saturation is highest in the pulmonary capillaries, where the concentration of oxygen is greatest. In the capillaries of the tissues, where there is less oxygen, oxyhemoglobin dissociates, releasing oxygen. There, the percent saturation of hemoglobin is correspondingly lower.

The *oxygen–hemoglobin dissociation curve* shown in **FIGURE 46-10a** illustrates the relationship between partial pressure of oxygen and percent saturation of hemoglobin. As oxygen concentration increases, there is a progressive increase in the percentage of hemoglobin that is combined with oxygen. The ability of oxygen to combine with hemoglobin and be released from oxyhemoglobin is influenced by several factors in addition

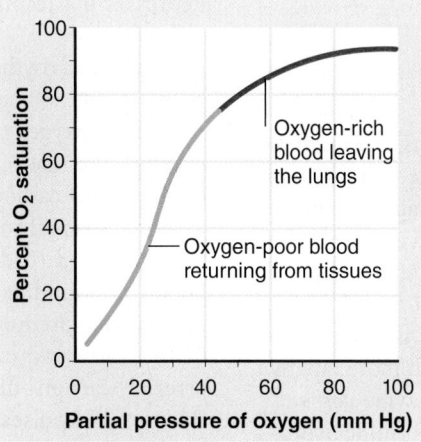

(a) **Normal oxygen–hemoglobin dissociation curve.** Look at the relationship between the partial pressure of oxygen and the percent O_2 saturation of hemoglobin. Oxygen binds to hemoglobin in the lungs and unloads from hemoglobin in the tissues.

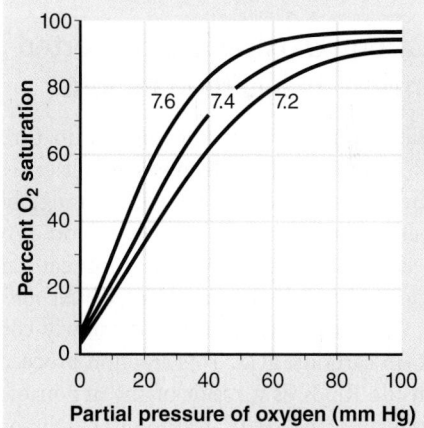

(b) **Effect of pH on the oxygen–hemoglobin curve (Bohr effect).** The normal pH of human blood is 7.4. Find the location on the horizontal axis where the partial pressure of oxygen is 40, and follow the line up through the curves. Notice that the saturation of hemoglobin with oxygen differs among the three curves, even though the partial pressure of oxygen is the same. Oxygen loading increases at higher pH (7.6). At lower pH (7.2), hemoglobin unloads more oxygen.

Figure 46-10 Oxygen–hemoglobin dissociation curves
© Cengage Learning

to percent O_2 saturation. These factors include pH, carbon dioxide concentration, and temperature.

Carbon dioxide produced in respiring tissue reacts with water in the plasma to form carbonic acid, H_2CO_3. For this reason, any increase in carbon dioxide concentration also increases the acidity (lowers the pH) of the blood. Oxyhemoglobin unloads oxygen more readily in an acidic environment than in an environment with normal pH. Displacement of the oxygen–hemoglobin dissociation curve by a change in pH is known as the **Bohr effect** (FIG. 46-10b). Lactic acid released from active muscles also lowers blood pH and has a similar effect on the oxygen–hemoglobin dissociation curve: more oxygen is unloaded and available for muscle contraction.

Some carbon dioxide is transported by the hemoglobin molecule. Although carbon dioxide attaches to the hemoglobin molecule in a different way and at a different site than oxygen does, the attachment of a carbon dioxide molecule releases an oxygen molecule from the hemoglobin. This effect of carbon dioxide concentration on the oxygen–hemoglobin dissociation curve is important. In the capillaries of the lungs (or gills in fishes), carbon dioxide concentration is relatively low and oxygen concentration is high, so oxygen combines with a very high percentage of hemoglobin. In the capillaries of the tissues, carbon dioxide concentration is high and oxygen concentration is low, so hemoglobin readily unloads oxygen.

The oxygen and carbon dioxide bound to hemoglobin in RBCs do not contribute to the partial pressures that govern diffusion. Loading and unloading of gases onto hemoglobin depend on partial pressures of the plasma and interstitial fluid (which is determined by the gases dissolved in these fluids).

Carbon dioxide is transported mainly as bicarbonate ions

Blood transports carbon dioxide in three forms. About 10% of carbon dioxide dissolves in plasma. Another 30% enters the RBCs and combines with hemoglobin. Because the bond between the hemoglobin and carbon dioxide is very weak, the reaction is readily reversible. Most carbon dioxide (about 60%) moves through plasma as *bicarbonate ions* (HCO_3^-).

In plasma carbon dioxide slowly combines with water to form carbonic acid. This reaction proceeds much more rapidly inside RBCs as a result of the action of the enzyme *carbonic anhydrase* (FIG. 46-11). Carbonic acid dissociates, forming hydrogen ions and bicarbonate ions.

$$\underset{\substack{\text{Carbon}\\\text{dioxide}}}{CO_2} + \underset{\text{Water}}{H_2O} \xrightarrow{\substack{\text{Carbonic}\\\text{anhydrase}}} \underset{\substack{\text{Carbonic}\\\text{acid}}}{H_2CO_3} \longrightarrow \underset{\substack{\text{Hydrogen}\\\text{ion}}}{H^+} + \underset{\substack{\text{Bicarbonate}\\\text{ion}}}{HCO_3^-}$$

Most hydrogen ions released from carbonic acid combine with hemoglobin, which is a very effective buffer. Many of the bicarbonate ions diffuse into the plasma. The action of carbonic anhydrase in the RBCs maintains a diffusion gradient for carbon dioxide to move into and then out of RBCs. As the

negatively charged bicarbonate ions move out of RBCs, chloride ions (Cl^-) in the plasma diffuse into the RBCs to replace them, a process known as the *chloride shift*. In the alveolar capillaries, CO_2 diffuses out of the plasma and into the alveoli. As the CO_2 concentration decreases, the reaction sequence just described reverses.

Any condition (such as emphysema) that interferes with the removal of carbon dioxide by the lungs can lead to *respiratory acidosis*. In this situation, carbon dioxide is produced more rapidly than it is excreted by the lungs. As a result, the concentration of carbonic acid in the blood increases. When blood pH falls below 7, the central nervous system becomes depressed, and the individual becomes disoriented. Untreated respiratory acidosis can cause coma and death.

Breathing is regulated by respiratory centers in the brain

Breathing is a rhythmic, involuntary process regulated by **respiratory centers** in the brain stem (see Fig. 46-6). Groups of neurons in the medulla regulate the basic rhythm of breathing. These neurons send a burst of impulses to the diaphragm and external intercostal muscles, which causes them to contract. After several seconds, these neurons become inactive, the muscles relax, and exhalation occurs. Respiratory centers in the pons help control the transition from inspiration to exhalation. These centers can stimulate or inhibit the respiratory centers in the medulla. The cycle of activity and inactivity repeats itself so that at rest you breathe about 14 times per minute. Overdose of certain medications, such as barbiturates, depresses the respiratory centers and may lead to respiratory failure.

The basic rhythm of respiration changes in response to the body's needs. When you are playing a fast game of tennis, you need more oxygen than when you are studying biology. During exercise, the rate of aerobic cellular respiration increases, producing more carbon dioxide. Your body must dispose of this carbon dioxide through increased ventilation.

Carbon dioxide concentration is the most important chemical stimulus for regulating the rate of respiration. Specialized **chemoreceptors** in the medulla and in the walls of the aorta and carotid arteries are sensitive to changes in arterial carbon dioxide concentration. When stimulated, they send impulses to the respiratory centers, which increase breathing rate.

The chemoreceptors in the walls of the aorta, called *aortic bodies,* and those in the walls of the carotid arteries, the *carotid bodies,* are sensitive to changes in hydrogen ion concentration and oxygen concentration as well as to carbon dioxide levels. Recall that an increase in carbon dioxide concentration increases hydrogen ions (from carbonic acid) and lowers blood pH. Even a slight decrease in pH stimulates these chemoreceptors, leading to faster breathing. As the lungs remove carbon dioxide, the hydrogen ion concentration in the blood and other body fluids decreases, and homeostasis is re-established.

Interestingly, oxygen concentration generally does not play an important role in regulating respiration. Only if the partial

The enzyme carbonic anhydrase (in red blood cells) converts carbon dioxide to carbonic acid; carbonic acid dissociates, producing hydrogen ions and bicarbonate ions. About 60% of carbon dioxide is transported in the plasma as bicarbonate ions.

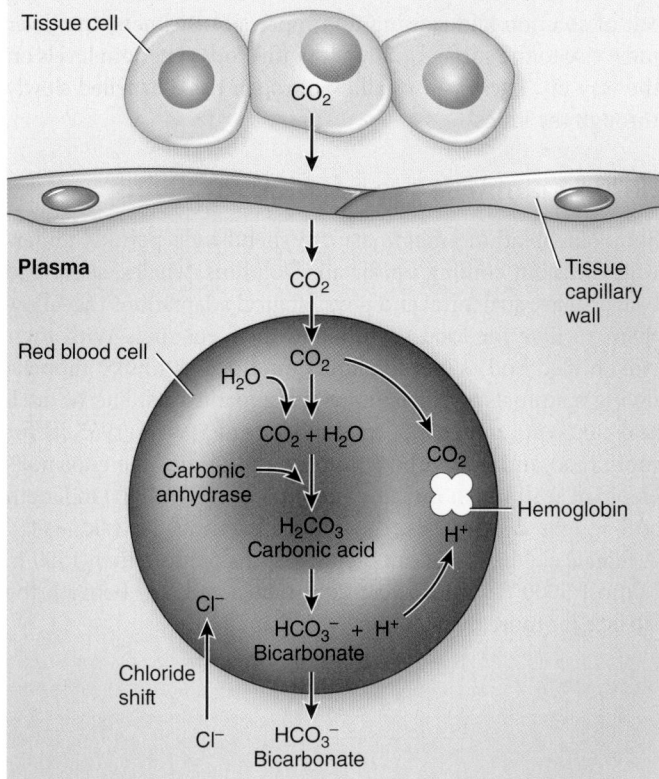

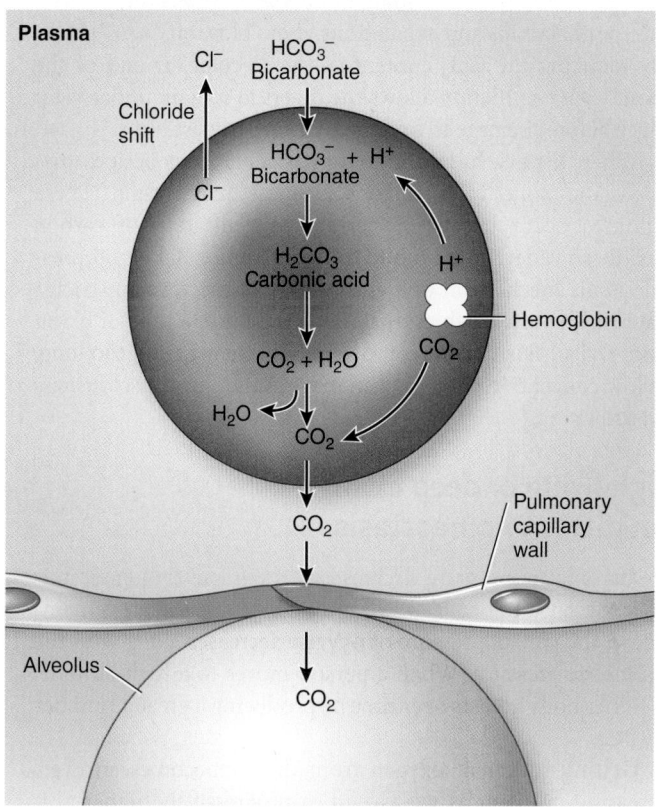

(a) In the tissues, carbon dioxide diffuses from cells into the plasma. Most of the carbon dioxide enters red blood cells, where the enzyme carbonic anhydrase rapidly converts it to carbonic acid. Carbonic acid dissociates, producing bicarbonate ions (HCO_3^-) and H^+. As bicarbonate ions move out into the plasma, chloride ions replace them. Hemoglobin combines with H^+, preventing a decrease in pH.

(b) In the lungs, carbon dioxide diffuses out of the plasma and into the alveoli. The reactions described in (a) are reversed. Bicarbonate ions diffuse from the plasma into the red blood cells. H^+ released from hemoglobin combines with bicarbonate ions, producing carbonic acid. Carbon dioxide produced from the carbonic acid diffuses out of the blood and into the alveoli.

Figure 46-11 Carbon dioxide transport

PREDICT In what way would diffusion of CO_2 into the red blood cells be affected if the carbonic anhydrase enzyme were inhibited?

© Cengage Learning

pressure of oxygen falls markedly do the chemoreceptors in the aorta and carotid arteries become stimulated to send messages to the respiratory centers.

Shallow breathing, which occurs in anxiety and in many respiratory diseases, causes *hypoxia,* a deficiency of oxygen. Even *rapid,* shallow breathing results in hypoxia because we do not clear out the stale air in the airway and ventilate the lungs. Hypoxia causes drowsiness, mental fatigue, headache, and sometimes euphoria. The ability to think and make judgments is impaired, as is the ability to perform tasks requiring coordination.

Although breathing is involuntary, the action of the respiratory centers can be consciously influenced for a short time by stimulating or inhibiting them. For example, you can inhibit respiration by holding your breath. You cannot hold your breath

indefinitely, however, because eventually you feel a strong urge to breathe. Even if you could ignore this urge, you would eventually pass out and resume breathing.

People who have stopped breathing because of drowning, smoke inhalation, electric shock, or cardiac arrest can sometimes be sustained by mouth-to-mouth resuscitation until their own breathing reflexes return. **Cardiopulmonary resuscitation (CPR)** is a method for aiding victims who have suffered respiratory and cardiac arrest. CPR must be started immediately because irreversible brain damage occurs within about 4 minutes of oxygen deprivation. A number of organizations offer training in CPR to the general public. You can also learn to save a life with Hands-Only CPR. Free online instructional materials, including videos, are provided by the American Heart Association and other organizations.

Hyperventilation reduces carbon dioxide concentration

Underwater swimmers and some Asian pearl divers intentionally *hyperventilate* before going underwater. By making a series of deep inhalations and exhalations, they "blow off" CO_2, markedly reducing the CO_2 content of the alveolar air and of the blood. *Hyperventilation* allows the divers to remain underwater longer before the urge to breathe becomes irresistible.

When hyperventilation continues for a long period, dizziness and sometimes unconsciousness may result. These responses occur because a certain concentration of carbon dioxide is needed in the blood to maintain normal blood pressure. (This mechanism operates by way of the vasoconstrictor center in the brain, which maintains the muscle tone of blood vessel walls.) Furthermore, if divers hold their breath too long, the low concentration of oxygen may result in unconsciousness and drowning.

High flying or deep diving can disrupt homeostasis

The barometric pressure decreases at progressively higher altitudes. Because the concentration of oxygen in the air remains at 21%, the partial pressure of oxygen decreases along with the barometric pressure. When a person moves to a high altitude, his or her body adjusts over time by producing a greater number of RBCs.

Getting sufficient oxygen from the air becomes an ever-increasing problem as we ascend to progressively higher altitudes. All high-flying jets have airtight cabins pressurized to the equivalent of the barometric pressure at an altitude of about 2000 m (6600 ft). If a jet flying at 11,700 m (more than 38,000 ft) suddenly decompressed, the pilot would lose consciousness in about 30 seconds and become comatose in about 1 minute.

In addition to the problems of hypoxia, a rapid decrease in barometric pressure causes *decompression sickness* (commonly known as the "bends" because those suffering from it bend over in pain). When the barometric pressure drops below the total pressure of all gases dissolved in the blood and other body fluids, the dissolved gases tend to come out of solution and form gas bubbles. A familiar example occurs each time you uncap a bottle of soda, reducing pressure in the bottle. Carbon dioxide is released from solution and bubbles out into the air. In the body nitrogen has a low solubility in blood and tissues. When it comes out of solution, the bubbles produced may damage tissues and block capillaries, interfering with blood flow. The clinical effects of decompression sickness are pain, dizziness, paralysis, unconsciousness, and even death.

Decompression sickness is more common in scuba diving than in high-altitude flying. As a diver descends, the surrounding pressure increases tremendously at a rate of 1 atmosphere (the atmospheric pressure at sea level, which equals 760 mm Hg) for each 10 m (about 32 ft). To prevent lung collapse, a diver must be supplied with air under pressure, exposing the lungs to very high alveolar gas pressures.

At sea level an adult human has about 1 L of nitrogen dissolved in the body, with about half in the fat and half in the body fluids. After a diver's body has become saturated with nitrogen at a depth of 100 m (325 ft), the body fluids contain about 10 L of nitrogen. To prevent this nitrogen from rapidly bubbling out of solution and causing decompression sickness, the diver must rise to the surface gradually, with stops at certain levels on the way up. These pauses allow nitrogen to be expelled slowly through the lungs.

Some mammals are adapted for diving

Some air-breathing mammals can spend long periods underwater without coming up for air. Dolphins, whales, seals, and beavers have structural and physiological adaptations that allow them to dive for food or to elude their enemies. With their streamlined bodies and forelimbs modified as fins or flippers, diving mammals perform impressive aquatic feats. The Weddell seal can swim under the ice at a depth of 596 m (1955 ft) for more than an hour without coming up for air. The enormous elephant seal, which measures about 5 m (16 to 18 ft) in length and weighs 2 to 4 tons, can plunge even deeper (**FIG. 46-12**). A female elephant seal can dive to depths of more than 1500 m (almost 5000 ft) in less than 20 minutes and stay beneath the surface for more than an hour.

Bruce Watkins/Animals Animals

Figure 46-12 Deep diver

Elephant seals, like this northern elephant seal (*Mirounga angustirostris*), may be the deepest divers on Earth. Researchers have recorded their dives at depths of more than 1500 m (almost 5000 ft).

Physiological adaptations, including ways to distribute and store oxygen, permit some mammals to dive deeply and remain underwater for long periods. Turtles and birds that dive depend on oxygen stored in their lungs. However, diving mammals do not take in and store extra air before a dive. In fact, seals exhale before they dive. With less air in their lungs, they are less buoyant. Their lungs collapse at about 50 to 70 m into their dive and then reinflate as they ascend, so their lungs do not function for most of the dive. These adaptations are thought to reduce the chance of decompression sickness because with less air in the lungs there is less nitrogen in the blood to dissolve during the dive.

Seals have about twice the volume of blood, relative to their body weight, as nondiving mammals. Diving mammals also have high concentrations of myoglobin, which stores oxygen in muscles. These animals have up to ten times as much myoglobin as do terrestrial mammals. The very large spleen typical of many diving mammals stores oxygen-rich red blood cells. During a dive, the increased pressure squeezes the spleen, releasing the stored red blood cells into the circulation.

Diving mammals markedly reduce the energy expended in deep diving (more than 200 m) by gliding. Filmed video sequences of diving seals and whales show that they glide most of the way down. Gliding is possible because the animal's buoyancy decreases as the lungs gradually collapse and the amount of air in the lungs is reduced. Gravity then pulls the animal downward.

When a mammal dives to its limit, physiological mechanisms known collectively as the **diving reflex** are activated. Metabolic rate decreases by about 20%, which conserves oxygen. Breathing stops, and bradycardia (slowing of the heart rate) occurs. The heart rate may decrease to one-tenth the normal rate, reducing the body's consumption of oxygen and energy. Blood is redistributed so that the heart and brain receive the greatest share; skin, muscles, digestive organs, and other internal organs can survive with less oxygen, and less blood is directed to these organs while an animal is submerged.

The diving reflex is present to some extent in humans, where it may act as a protective mechanism during birth, when an infant may be deprived of oxygen for several minutes. Cases of near drownings, especially of young children, have been documented in which the victim was submerged for as long as 45 minutes in very cold water before being rescued and resuscitated. Some of these survivors showed no brain damage. The shock of icy water slows heart rate, increases blood pressure, and shunts blood to internal organs of the body that most need oxygen (blood flow in arms and legs decreases). Metabolic rate decreases, so less oxygen is required.

CHECKPOINT 46.3

- *What is the sequence of inhaled airflow through the respiratory structures in a mammal?*
- *What happens in the alveoli? Why does alveolar air differ in composition from atmospheric air? Explain.*

- CONNECT *How does carbon dioxide concentration affect the oxygen–hemoglobin dissociation curve?*
- CONNECT *Imagine that you are playing basketball. How would your breathing be affected? What mechanisms are activated to restore homeostasis?*

46.4 BREATHING POLLUTED AIR

LEARNING OBJECTIVE

7 Describe the defense mechanisms that protect the lungs and describe the effects of polluted air and cigarette smoking on the respiratory system.

Several defense mechanisms protect the delicate lungs from the harmful substances we breathe (FIG. 46-13). Hairs in the nostrils, the ciliated mucous lining in the nose and pharynx, and the cilia–mucus elevator of the trachea and bronchi trap foreign particles that enter the body in inspired air. Cilia on epithelial cells lining the human airway have chemoreceptors that sense taste. Researchers have shown that when potentially toxic bitter substances are inhaled, the cilia beat more rapidly. This mechanism appears to remove harmful compounds from the airway.

One of the body's most rapid defense responses to breathing dirty air is **bronchial constriction.** In this process, the bronchial tubes narrow, increasing the chance that inhaled particles will land on the sticky mucous lining. Unfortunately, bronchial constriction narrows the airway. Less air can reach the lungs, decreasing the amount of oxygen available to body cells. Fifteen puffs on a cigarette during a 5-minute period increases airway resistance as much as threefold, and this added resistance to breathing lasts more than 30 minutes. Chain-smokers and those who breathe heavily polluted air are in a state of chronic bronchial constriction.

Neither the smallest bronchioles nor the alveoli are equipped with mucus or ciliated cells. Foreign particles that get through other respiratory defenses and find their way into the alveoli

Figure 46-13 Urban air pollution
Industry continues to spew tons of pollutants into the atmosphere. Despite modern technology, air pollution still poses a significant health risk.

Tobacco use is the leading cause of preventable and premature death in the United States. Cigarette smoking kills more people than alcohol, car accidents, suicide, AIDS, homicide, and illegal drugs combined. An estimated 443,000 people die prematurely each year from smoking or exposure to secondhand smoke. More than 8.6 million additional smokers have a serious illness caused by smoking. On average, tobacco smokers die ten years earlier than nonsmokers.[1]

Tobacco smoke is by far the most important risk factor for lung cancer, the most common lethal cancer worldwide. It is also associated with many other types of cancer and accounts for at least 30% of all cancer deaths. Tobacco smoke is also an important risk factor for cardiovascular disease, chronic obstructive pulmonary disease (COPD), and many other diseases. Despite these grim facts, approximately 20% of both high school students and adults in the United States continue to smoke cigarettes—more than 43 million adults and 71 million teens.

Secondhand tobacco smoke is smoke from a burning cigarette, cigar, or pipe, or smoke that has been exhaled by a smoker. Secondhand smoke contains hundreds of toxic chemical compounds, including about 70 known human carcinogens. It causes the deaths of an estimated 49,000 individuals each year in the United States alone. Millions of children and adults are exposed to secondhand smoke in their homes and workplaces. This exposure causes disease and premature death in children and adults who do not smoke. Babies and children exposed to secondhand smoke are at increased risk for sudden infant death syndrome (SIDS), acute respiratory infections, pneumonia, bronchitis, ear infections, and asthma. In adults secondhand smoke causes cardiovascular disease and lung cancer. About 60% of *nonsmokers* in the United States have biological symptoms of exposure to secondhand smoke. Laws passed during the past 20 or so years that ban smoking in workplaces and public places and increased awareness of the dangers of secondhand smoke have led to a decrease in exposure.

[1] Statistics quoted are from the Centers for Disease Control and the American Cancer Society.

Nicotine dependence is the most common type of chemical dependence in the United States. Studies indicate that nicotine may be as addictive as alcohol, cocaine, or heroin. Like cocaine, morphine, and amphetamines, nicotine increases dopamine concentration in the limbic system. This region of the brain helps integrate emotion, and dopamine may facilitate learning an association between the pleasurable effects of the drug and other stimuli, such as the smell of smoke. Nicotine, like cocaine and other abused drugs, causes long-lasting changes in the brain. Cell signaling and other mechanisms underlying drug addiction are very similar to those of learning and memory.

An estimated 70% of smokers (more than 33 million) want to quit, but only 2.5% per year succeed in quitting permanently. Nicotine replacement with gum, patches, nasal spray, inhalers, or lozenges has been shown to be effective as an aid to smoking cessation, especially when used with behavioral therapy. Certain antidepressants are used to reduce the craving for nicotine. One prescription medication (Chantix) works by blocking nicotine receptors in the brain. It reduces symptoms of nicotine withdrawal and also decreases the pleasurable effects of smoking.

Use of electronic cigarettes (also known as e-cigarettes or electronic nicotine delivery systems) is growing rapidly. The effects of e-cigarettes have not yet been thoroughly studied. Brands vary in the amounts of nicotine they deliver and in their toxin content. There is evidence that many individuals who use e-cigarettes become smokers of traditional cigarettes.

Here are some facts about smoking:

- Almost everyone who takes up smoking in the U.S. is a teenager, and nearly 4000 teens smoke their first cigarette every day. About one-third of these teens will die prematurely from a smoking-related disease. The younger you are when you begin to smoke, the more likely you are to be an adult smoker.

- Smoking shortens the life of an adult male smoker by an average of 13.2 years and shortens the life of an adult female smoker by an average of 14.5 years.

- If you smoke more than one pack of cigarettes per day, you are about 20 times as likely as a nonsmoker to develop lung cancer. Cigarette smoking causes almost nine out of ten lung cancer deaths.

- If you smoke, you double your chances of dying from cardiovascular disease.

- If you smoke, you are 20 times more likely as a nonsmoker to develop chronic bronchitis and emphysema. About 90% of all deaths from chronic obstructive pulmonary disease are caused by cigarette smoking.

- If you smoke, you have about 5% less oxygen circulating in your blood (because carbon monoxide binds to hemoglobin) than does a nonsmoker.

- If you smoke when you are pregnant, your baby will weigh about 6 ounces less at birth, and there is a higher risk of miscarriage, premature birth (early delivery), stillbirth, and sudden infant death syndrome.

- Infants whose parents smoke have double the risk of contracting pneumonia or bronchitis in their first year of life.

- There is no risk-free level of exposure to secondhand smoke. Nonsmokers who are exposed to secondhand smoke increase their risk of developing heart disease by 25% to 30% and lung cancer by 20% to 30%.

- When smokers quit smoking, their risk of dying from chronic obstructive pulmonary disease, cardiovascular disease, or cancer gradually decreases. (Precise changes in risk depend on the number of years the person smoked, the number of cigarettes smoked per day, the age of starting to smoke, and the number of years since quitting.)

- The direct medical costs of tobacco use in the U.S. amount to more than $96 billion annually.

- Smokeless tobacco products (snuff, chewing tobacco) are not a safe alternative to tobacco smoking. All forms of these products have chemicals that cause cancer.

---

may be engulfed by macrophages. The macrophages may then accumulate in the lymph tissue of the lungs. Lung tissues of cigarette smokers and those who work in dirty industries burning fossil fuel contain large blackened areas where carbon particles have been deposited (**FIG. 46-14**).

Continued insult to the respiratory system results in disease. Chronic bronchitis, pulmonary emphysema, and lung cancer have been linked to cigarette smoking and breathing polluted air. More than 75% of patients with *chronic bronchitis* have a history of heavy cigarette smoking (see *Inquiring About: The Effects of Smoking*). People with chronic bronchitis often develop **pulmonary emphysema,** a disease also most common in cigarette smokers. In this disorder alveoli lose their elasticity, and walls between adjacent alveoli are destroyed. The surface area of the lung is so reduced that gas exchange is seriously impaired. Air is not expelled effectively, and stale air accumulates in the lungs.

The emphysema victim struggles for every breath, and still the body does not get enough oxygen. To compensate, the right

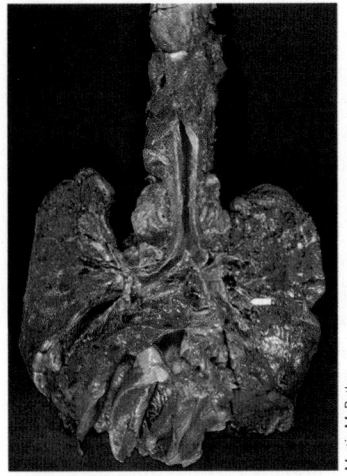

(a) Lungs and major bronchi of a nonsmoker.

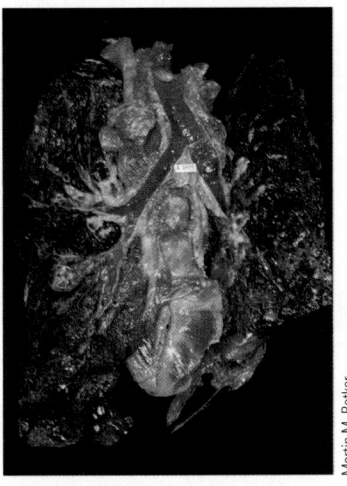

(b) Lungs and heart of a cigarette smoker. The dark spots visible in the lung tissue are particles of carbon, tar, and other substances that passed through the respiratory defenses and lodged in the lungs.

Figure 46-14 Effects of cigarette smoking

ventricle of the heart pumps harder and becomes enlarged. Emphysema patients frequently die of heart failure.

Most patients with *chronic obstructive pulmonary disease (COPD)*, a condition characterized by obstructed airflow, have both chronic bronchitis and emphysema. Asthma also contributes to COPD. People with asthma respond to inhaled stimuli with exaggerated bronchial constriction, and the airway is typically inflamed.

Cigarette smoking is also the main cause of *lung cancer.* At least 70 of the 4800 chemical compounds in tobacco smoke cause cancer in animals, including humans. These carcinogenic substances irritate the cells lining the respiratory passages and alter their metabolic balance. Normal cells are transformed into cancer cells, which may multiply rapidly and invade surrounding tissues.

CHECKPOINT 46.4

- *What mechanisms does the human respiratory system have for getting rid of inhaled particles?*
- **CONNECT** *What happens when so much dirty air is inhaled that the respiratory system's defense mechanisms cannot function effectively?*

SUMMARY: FOCUS ON LEARNING OBJECTIVES

46.1 Adaptations for Gas Exchange in Air or Water *(page 986)*

1 Compare the advantages and disadvantages of air and water as mediums for gas exchange and describe adaptations for gas exchange in air.

- Air contains a higher concentration of molecular oxygen than water, and oxygen diffuses more rapidly through air than through water. Air is less dense and less viscous than water, so less energy is needed to move air over a gas exchange surface.

- Terrestrial animals have adaptations that protect their respiratory surfaces from drying, including location deep in the body. Also, air is humidified as it passes through the moist upper airway.

46.2 Types of Respiratory Surfaces *(page 986)*

2 Describe the following adaptations for gas exchange: body surface, tracheal tubes, gills, and lungs.

- Small aquatic animals exchange gases by diffusion; they require no specialized respiratory structures. Some invertebrates, including most annelids, and a few vertebrates (many amphibians) exchange gases across the body surface.

- In insects and some other arthropods, air enters a network of **tracheal tubes,** or **tracheae,** through openings, called **spiracles,** along the body surface. Tracheal tubes branch and extend to all regions of the body.

- **Gills** are moist, thin projections of the body surface found mainly in aquatic animals. In chordates gills are usually internal, located along the edges of the gill slits. In bony fishes an **operculum** protects the gills. A **countercurrent exchange system** maximizes diffusion of oxygen into the blood and diffusion of carbon dioxide out of blood.

- Animals carry on **ventilation,** the process of actively moving air or water over respiratory surfaces. Terrestrial vertebrates have **lungs** and some means of ventilating them.

- In birds the lungs have extensions, called *air sacs,* that act as bellows, drawing air into the system. Two cycles of inhalation and exhalation support a one-way flow of air through the lungs. Air from the outside flows into the posterior air sacs, to the lung, through the anterior air sacs, and then out of the body. Gas exchange takes place through the walls of the **parabronchi,** small, open tubes in the lungs. A crosscurrent arrangement, in which blood flow is at right angles to the parabronchi, increases the amount of oxygen that enters the blood.

46.3 The Mammalian Respiratory System *(page 991)*

3 Trace the passage of oxygen through the human respiratory system from nostrils to alveoli.

- The human respiratory system includes the lungs and a system of airways. Each breath of air passes in sequence through

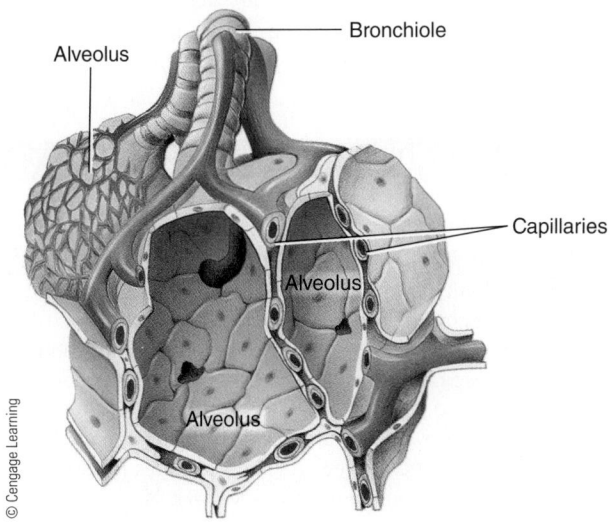

the *nostrils, nasal cavities,* **pharynx, larynx, trachea, bronchi, bronchioles,** and **alveoli.** Each lung occupies a pleural cavity and is covered with a **pleural membrane.**

4 Summarize the mechanics of breathing in humans and describe gas exchange in the lungs and tissues.

- During breathing, the **diaphragm** contracts, expanding the chest cavity. The membranous walls of the lungs move outward along with the chest walls, lowering pressure within the lungs. Air from outside the body rushes in through the air passageways and fills the lungs until the pressure equals atmospheric pressure.

- **Tidal volume** is the amount of air moved into and out of the lungs with each normal breath. The volume of air that remains in the lungs at the end of a maximal expiration is the **residual volume. Vital capacity** is the maximum volume of air that can be exhaled after the lungs fill to the maximum extent.

- Oxygen and carbon dioxide are exchanged between alveoli and blood by diffusion. The pressure of a particular gas determines its direction and rate of diffusion.

- *Dalton's law of partial pressures* explains that in a mixture of gases, the total pressure of the mixture is the sum of the pressures of the individual gases. Thus, each gas in a mixture of gases exerts a **partial pressure,** the same pressure it would exert if it were present alone. The partial pressure of atmospheric oxygen, P_{O_2}, is 160 mm Hg at sea level.

- According to **Fick's law of diffusion,** the greater the difference in pressure on the two sides of a membrane and the larger the surface area, the faster the gas diffuses across the membrane.

5 Summarize the mechanisms by which oxygen and carbon dioxide are transported in the blood and identify factors that determine the oxygen–hemoglobin dissociation curve.

- **Hemoglobin** is the respiratory pigment in the blood of vertebrates. Almost 99% of the oxygen in human blood is transported as **oxyhemoglobin (HbO$_2$).**

- The maximum amount of oxygen that hemoglobin can transport is the **oxygen-carrying capacity.** The actual amount of oxygen bound to hemoglobin is the *oxygen content.* The *percent O$_2$ saturation,* the ratio of oxygen content to oxygen-carrying capacity, is highest in pulmonary capillaries, where oxygen concentration is greatest.

- The *oxygen–hemoglobin dissociation curve* shows that as oxygen concentration increases, there is a progressive increase in the amount of hemoglobin that combines with oxygen. The curve is affected by pH, temperature, and CO$_2$ concentration.

- Oxyhemoglobin dissociates more readily as carbon dioxide concentration increases because carbon dioxide combines

with water and produces carbonic acid, which lowers the pH. Displacement of the oxygen–hemoglobin dissociation curve by a change in pH is called the **Bohr effect.**

- About 60% of the carbon dioxide in the blood is transported as bicarbonate ions. About 30% combines with hemoglobin, and another 10% is dissolved in plasma.

- *Carbonic anhydrase* catalyzes the production of carbonic acid from carbon dioxide and water. Carbonic acid dissociates, producing bicarbonate ions (HCO$_3^-$) and hydrogen ions (H$^+$).

- Hemoglobin combines with H$^+$, which buffers the blood. Many bicarbonate ions diffuse from red blood cells into the plasma; they are replaced by Cl$^-$. This exchange is known as the *chloride shift.*

6 Describe the regulation of breathing in humans and summarize the physiological effects of hyperventilation and of sudden decompression when a diver surfaces too quickly from deep water.

- **Respiratory centers** in the medulla and pons regulate respiration. These centers are stimulated by **chemoreceptors** sensitive to an increase in carbon dioxide concentration. They also respond to an increase in hydrogen ions and to very low oxygen concentration.

- *Hyperventilation* reduces the concentration of carbon dioxide in the alveolar air and in the blood.

- As altitude increases, barometric pressure falls, and less oxygen enters the blood. This situation can lead to *hypoxia,* or oxygen deficiency, which in turn can lead to loss of consciousness and death. Rapid decrease in barometric pressure can also cause *decompression sickness,* especially common among divers who ascend too rapidly from a dive.

- Diving mammals have unusually high concentrations of **myoglobin,** a pigment that stores oxygen, in their muscles. The **diving reflex,** a group of physiological mechanisms that includes a decrease in metabolic rate, is activated when a mammal dives to its limit.

46.4 Breathing Polluted Air *(page 999)*

7 Describe the defense mechanisms that protect the lungs and describe the effects of polluted air and cigarette smoking on the respiratory system.

- The ciliated mucous lining of the airway traps inhaled particles. Inhaling polluted air results in *bronchial constriction,* increased mucus secretion, damage to ciliated cells, and coughing.

- Breathing polluted air or inhaling cigarette smoke can cause many diseases, including *chronic bronchitis, pulmonary emphysema,* and *lung cancer.*

TEST YOUR UNDERSTANDING

Know and Comprehend

1. Which of the following is a benefit of gas exchange in air compared with gas exchange in water? (a) air has a higher concentration of molecular oxygen (b) oxygen diffuses more slowly in air (c) no energy is required for ventilation (d) moist respiratory surface is not needed (e) air is denser than water

2. Which of the following are accurately matched? (a) bony fish—operculum (b) insect—alveoli (c) bird—spiracles (d) aquatic mammal—gill filaments (e) shark—dermal gills

3. The most efficient vertebrate respiratory system is that of (a) amphibians (b) birds (c) reptiles (d) mammals (e) humans

4. In a bird the correct sequence for a breath of air is: 1. anterior air sacs 2. posterior air sacs 3. parabronchi 4. alveoli (a) 1, 2, 4 (b) 2, 1, 3 (c) 2, 3, 1 (d) 2, 4, 1 (e) 1, 3, 2

5. Which sequence most accurately describes the sequence of airflow in the human respiratory system? 1. pharynx 2. bronchus 3. trachea 4. larynx 5. alveolus 6. bronchiole (a) 4, 1, 3, 2, 5, 6 (b) 1, 4, 3, 2, 5, 6 (c) 4, 1, 3, 2, 6, 5 (d) 1, 4, 3, 2, 6, 5 (e) 1, 3, 4, 2, 6, 5

6. The amount of air moved in and out of the lungs with each normal resting breath is the (a) vital capacity (b) residual volume (c) vital volume (d) partial pressure (e) tidal volume

7. According to Fick's law of diffusion, a gas will diffuse faster if the (a) surface is reduced (b) difference in pressure is decreased (c) difference in pressure is increased (d) pH is lowered (e) a and c

8. The concentration of which substance is most important in regulating the rate of respiration? (a) chloride ions (b) oxygen (c) bicarbonate ions (d) nitrogen (e) carbon dioxide

9. When a diver ascends too rapidly, (a) bronchial constriction occurs (b) a diving reflex is activated (c) nitrogen rapidly bubbles out of solution in the body fluids (d) nitrogen hypoxia occurs (e) carbon dioxide bubbles damage the alveoli

10. Which of the following is *not* true of the diving reflex? (a) breathing stops (b) the heart slows (c) less blood is distributed to the muscles (d) metabolic rate increases by about 20% (e) energy consumption decreases

Apply and Analyze

11. **VISUALIZE** Draw and label a diagram illustrating the sequence of structures through which inhaled air travels to reach the alveoli.

12. **EVOLUTION LINK** What are the advantages of having millions of alveoli rather than a pair of simple, balloonlike lungs?

13. **INTERPRET DATA** Look at Figure 46-10b. How would strenuous muscle activity affect the oxygen–hemoglobin dissociation curve? (*Hint:* How does muscular activity affect pH? See Chapter 40 if you need to review this concept.)

14. **PREDICT** What problems would be faced by a terrestrial animal having gills instead of lungs?

Evaluate and Synthesize

15. **EVOLUTION LINK** Under what conditions might it be advantageous for a fish to have lungs as well as gills? What function do the "lungs" of modern fishes serve?

16. **EVOLUTION LINK** Aquatic mammals such as whales and dolphins use lungs rather than gills for gas exchange. Propose a hypothesis to explain why.

17. **SCIENCE, TECHNOLOGY, AND SOCIETY** Tobacco smoking is extremely expensive to our society in economic and health costs as well as in human suffering. How would you solve the tobacco smoking problem? How would you specifically target young people?

aplia To access course materials, such as Aplia and other companion resources, please visit **www.cengagebrain.com.**

47 Processing Food and Nutrition

Jonathan Bird/Getty Images

Atlantic wolffish (*Anarhichas lupus*) eating a sea urchin. Also known as the seawolf, this predator lives on the ocean floor where it hunts hard-shelled prey, such as mollusks, crustaceans, and echinoderms. The Atlantic wolffish, which can weigh up to 40 lb, helps control the density and distribution of populations of sea urchins and certain other invertebrates. The wolffish population has decreased drastically as a result of overfishing, unintentional harm by fishermen (bycatch), and habitat damage. Photographed in Maine.

KEY CONCEPTS

47.1 Food processing includes ingestion, digestion, absorption, and elimination; many animal adaptations are associated with mode of nutrition.

47.2 Various parts of the vertebrate digestive system are specialized to perform specific functions; accessory glands (liver, pancreas, and salivary glands) secrete fluids and enzymes that are important in digestion.

47.3 Most animals require the same basic nutrients: carbohydrates, lipids, proteins, vitamins, and minerals.

47.4 Basal metabolic rate is the body's cost of metabolic living; energy metabolism is regulated by complex signaling pathways. Both overnutrition and undernutrition can result in malnutrition.

Animals as diverse as sponges and baleen whales feed on food particles suspended in oceans. Giraffes eat leaves, and antelope graze on grasses. Frogs, lions, and some insects capture other animals. Although their choices of food and feeding mechanisms are diverse, all animals are **heterotrophs,** organisms that must obtain their energy and nourishment from the organic molecules manufactured by other organisms. What they eat, how they obtain food, and how they process and use food are the focus of this chapter.

Nutrients are substances in food that are used as energy sources to power the systems of the body, as ingredients to make compounds needed for metabolic processes, and as building blocks in the growth and repair of tissues. Obtaining nutrients is of such vital importance that both individual organisms and ecosystems are structured around the central theme of **nutrition,** the process of taking in and using food.

An animal's body plan and its lifestyle are adapted to its particular mode of obtaining food. For example, the Atlantic wolffish lives on the ocean floor at depths of up to 120 m (400 ft). It hunts clams, crabs, sea stars, and sea urchins (see photograph). Both its upper and lower jaw have four to six fanglike teeth that stab and tear its prey. Behind these fangs are three rows of teeth that crush the hard shells of its prey. It also has serrated teeth in its throat.

Nutrition has been an important force in human evolution. Through natural selection, the human diet became more varied than the diets of other primates. Humans also became very efficient at obtaining and modifying (e.g., cooking) food. Their higher-quality diet supported the evolution of a larger, more complex brain.

With only slight variations, all animals require the same basic nutrients: carbohydrates, lipids, proteins, vitamins, and minerals. Carbohydrates, lipids, and proteins are used as energy sources. Eating too much of any of these nutrients can result in weight gain, whereas eating too few nutrients or an unbalanced diet can result in malnutrition and death. **Malnutrition,** or poor nutritional status, results from dietary intake that is either below or above required needs. In human populations both undernutrition (particularly protein deficiency) and obesity (which results from overnutrition) are serious health problems. Food processing and nutrition are active areas of research.

47.1 NUTRITIONAL STYLES AND ADAPTATIONS

LEARNING OBJECTIVE

1 Describe food processing and modes of nutrition, and compare the digestive systems of a cnidarian (such as *Hydra*), a flatworm, an earthworm, and a vertebrate.

Feeding is the selection, acquisition, and ingestion of food. Most animals have a specialized digestive system that processes the food they eat. **Ingestion** is the process of taking food into the digestive cavity. In many animals, including vertebrates, ingestion includes taking food into the mouth and swallowing it.

The process of breaking down food is called **digestion.** Because animals eat the macromolecules made by and for other organisms, they must break down these molecules and synthesize new macromolecules for their own needs. For example, we humans cannot incorporate the proteins and other organic compounds from a hamburger directly into our own cells. We must *mechanically digest* the meat and then *chemically digest* it by enzymatic hydrolysis (see Chapter 3). During digestion, complex organic compounds are degraded into smaller molecular components. For example, proteins are broken down into their component amino acids. These amino acids are then used to synthesize proteins tailored for the needs of that individual at that particular time.

Amino acids and other nutrients pass through the lining of the digestive tract and into the blood by **absorption.** Then the circulatory system transports the nutrients to all the cells of the body. Cells use nutrients to synthesize proteins and other organic compounds or as fuel for cellular respiration. Food that is not digested and absorbed is discharged from the body, a process called *egestion* or **elimination.**

Animals are adapted to their mode of nutrition

We can classify animals based on the type of food they typically eat and on how they obtain food (FIG. 47-1). Many adaptations have evolved for each type of feeding. For example, the beaks of birds and the teeth of many vertebrates are specialized for cutting, tearing, crushing, or chewing food.

Animals that feed directly on producers are **herbivores,** or **primary consumers.** Most of what an herbivore eats is not efficiently digested and is eliminated from the body, almost unchanged, as waste. Herbivores must therefore eat large quantities of food to obtain the nourishment they need. Many herbivores—grasshoppers, manatees, elephants, and cattle, for example—spend a major part of their lives eating.

Animals cannot digest the cellulose of plant cell walls, and many adaptations have evolved for extracting nutrients from the plant material they eat. Many herbivores, including termites, cows, and horses, have a symbiotic relationship with bacteria, fungi, and other microorganisms that inhabit their digestive tracts. *Ruminants* (cattle, sheep, deer, giraffes) are hoofed animals with a stomach divided into four chambers. Symbiotic bacteria and protists living in the first two chambers digest cellulose, splitting some of the molecule into sugars, which are then used by the host and by the bacteria themselves. The bacteria produce fatty acids during their metabolism, some of which are absorbed by the animal and serve as an important energy source.

Solid bits of food clump together to form a cud. The cud is regurgitated into the animal's mouth, where it is mixed with saliva and chewed again. When the cud is reswallowed, the partly digested food is further broken down. Partially digested food passes into the third chamber, where water and minerals are absorbed into the blood. Digestion is completed in the fourth chamber, which corresponds to the stomach of humans and other animals with a single stomach.

Herbivores are sometimes eaten by secondary and higher-level consumers. These **predators** are adapted for capturing and killing prey. Some predators seize their victims and swallow them alive and whole (see Fig. 47-1d). Others paralyze, crush, or shred their prey before ingesting it. Adaptations for this type of feeding include tentacles, claws, fangs, poison glands, and teeth. (Predators, which may also eat one another, are sometimes referred to as *carnivores.*)

Carnivorous mammals have well-developed canine teeth for stabbing their prey during combat. For example, spotted hyenas, which are formidable predators as well as scavengers, have massive jaws that allow them to consume an entire elephant carcass, including bones and hide. Few nutrients are wasted. Meat is more easily digested than plant food, and carnivore digestive tracts are shorter than those of herbivores.

Omnivores, such as pigs, humans, most bears, and some fishes, consume both plants and animals. Earthworms ingest large amounts of soil containing both animal and plant material. The blue whale, the largest animal, filters out tiny algae and animals from the water as it swims. Omnivores often have adaptations that help them distinguish among a wide range of smells and tastes and thereby select a variety of foods.

Animals can also be classified according to the mechanisms they use to feed. Many omnivores are **filter feeders** that remove suspended food particles from the water they inhabit. For example, clams and oysters trap suspended food particles in mucus secreted by their gills. Some echinoderms trap food particles along their tentacles, which are coated with mucus. Baleen whales use rows of hard plates (baleen) suspended from the roof of the mouth to filter small crustaceans.

Some animals feed on fluids by piercing and sucking. Mosquitoes have highly adapted structures for piercing skin and sucking blood. Bats that feed on nectar have a long tongue and a reduced number of teeth. Birds that feed on pollen and nectar have long bills and tongues. The shape, size, and curve of the beak may be specialized for feeding on or capturing a particular type of food (see Fig. 1-13).

Deposit feeders, such as earthworms, fiddler crabs, and sea cucumbers, consume nutrients in soil or sediments. For example, earthworms eat their way through the soil, digesting and absorbing organic material.

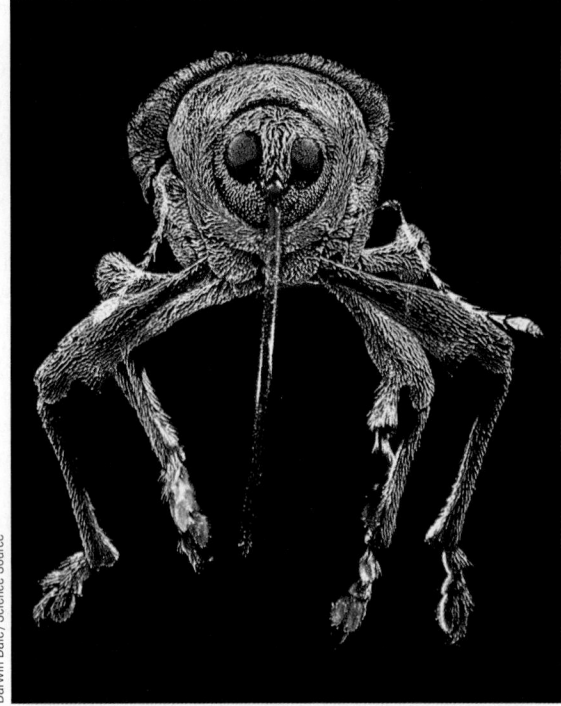

(a) **Unusual adaptation of an acorn-eating insect.** The impressively long "snout" of the herbivorous acorn weevil (*Curculio*) is adapted both for feeding and for making a hole in the acorn through which it deposits an egg. When it has hatched, the larva feeds on the contents of the acorn seed.

(b) **An herbivorous bear, the giant panda (*Ailuropoda melanoleuca*).** The panda's large, flat teeth and well-developed jaws and jaw muscles are adaptations for grinding high-fiber plant food. Most bears are omnivores.

(c) **Long-nose butterfly fish (*Forcipiger longirostris*), a predator.** The mouth (*on right*) of the long-nose butterfly fish is adapted for extracting small worms and crustaceans from crevices in coral reefs. The false eyespot on the fish's tail confuses its predators.

(d) **Snake (*Alsophis* sp.) strangling a lava lizard (*Tropidurus*).** Snakes are predators that can swallow very large prey whole because of the structure of their jaws. The very flexible lower jaw expands during swallowing.

Figure 47-1 Adaptations for obtaining and processing food

Some invertebrates have a digestive cavity with a single opening

Sponges, which are very simple invertebrates, obtain food by filtering microscopic organisms from the surrounding water. Individual cells phagocytose the food particles, and digestion is *intracellular* within food vacuoles. Wastes are egested into the water that continuously circulates through the sponge body.

Most animals have a digestive cavity. Digestion within a cavity is more efficient than intracellular digestion because digestive enzymes can be released into one confined space. Less surface area is required. Cnidarians (such as hydras and jellyfish) and flatworms have a **gastrovascular cavity,** a central digestive cavity with a single opening. Cnidarians capture small aquatic animals with the help of their stinging cells and tentacles (**FIG. 47-2a**). The mouth opens into the gastrovascular cavity. Cells lining this

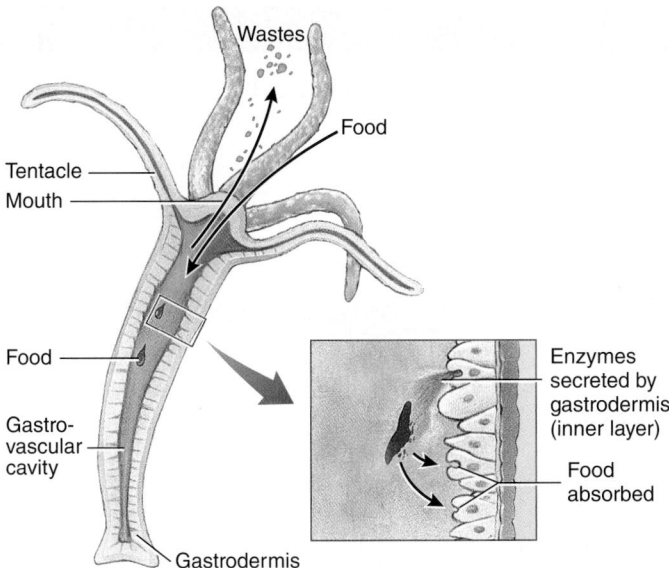

(a) Digestive system of a hydra.

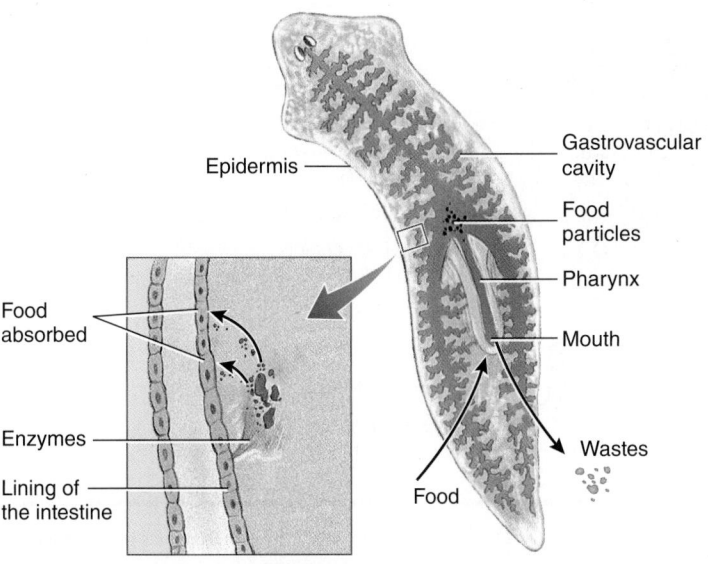

(b) Digestive system of a flatworm.

Figure 47-2 Simple invertebrate digestive systems

(a) Hydras and (b) flatworms (planarians) have a gastrovascular cavity, a digestive tract with a single opening that serves as both mouth and anus.
© Cengage Learning

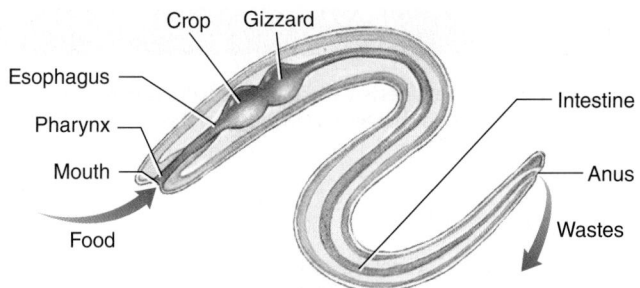

Figure 47-3 *Animation* Digestive tract with two openings

The earthworm, like most animals, has a complete digestive tract extending from mouth to anus. Various regions of the digestive tract are specialized to perform different food-processing functions.
© Cengage Learning

Most animal digestive systems have two openings

Most invertebrates, and all vertebrates, have a tube-within-a-tube body plan. The body wall forms the outer tube. The inner tube is a digestive tract with two openings, sometimes referred to as a complete digestive system (FIG. 47-3). Food enters through the mouth, and undigested food is eliminated through the anus. The mixing and propulsive movements of the digestive tract are referred to as **motility.** The propulsive activity characteristic of most regions of the digestive tract is **peristalsis,** waves of muscular contraction that push the food in one direction. More food can be taken in while previously eaten food is being digested and absorbed farther down the digestive tract. In a digestive tract with two openings, various regions of the tube are adapted to perform specific functions.

CHECKPOINT 47.1

- **CONNECT** *How have predators adapted to their mode of nutrition?*
- *How does food processing differ in earthworms and flatworms?*
- *What are the benefits of a digestive tract with two openings?*

47.2 THE VERTEBRATE DIGESTIVE SYSTEM

LEARNING OBJECTIVES

2 Trace the pathway traveled by an ingested meal through the human digestive system and describe the structure and function of each structure involved.

3 Trace the step-by-step digestion of carbohydrate, protein, and lipid.

4 Describe the structural adaptations that increase the surface area of the digestive tract.

5 Compare lipid absorption with absorption of other nutrients.

digestive cavity secrete enzymes that break down proteins. Digestion continues intracellularly within food vacuoles, and digested nutrients diffuse into other cells. Body contractions promote egestion of undigested food particles through the mouth.

Free-living flatworms begin to digest their prey even before ingesting it. They extend the pharynx out through their mouth and secrete digestive enzymes onto their prey (FIG. 47-2b). After it is ingested, the food enters the branched gastrovascular cavity, where enzymes continue to digest it. Cells lining the gastrovascular cavity phagocytose partly digested food fragments, and digestion is completed within food vacuoles. As in cnidarians, the flatworm digestive cavity has only one opening, so undigested wastes are egested through the mouth.

Various regions of the digestive tract are specialized to perform specific functions. More food can be ingested while the stomach and duodenum are digesting food that was previously swallowed, and food ingested hours earlier is being absorbed.

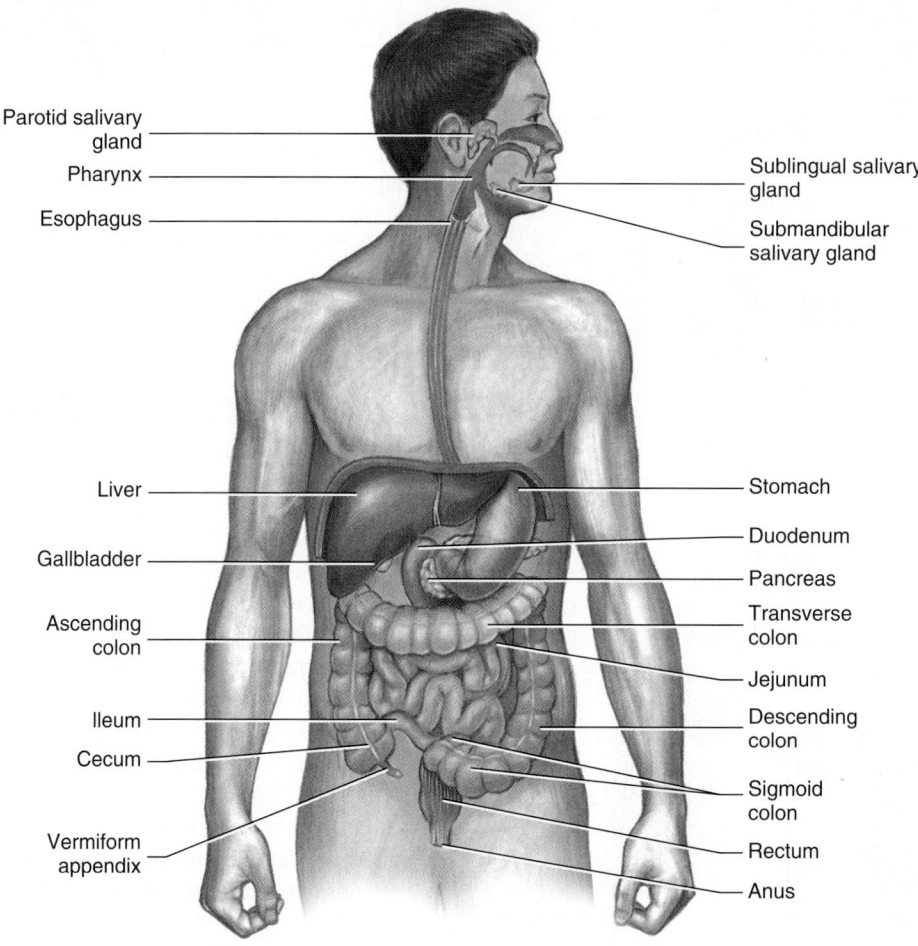

Parotid salivary gland
Pharynx
Esophagus
Sublingual salivary gland
Submandibular salivary gland
Liver
Gallbladder
Ascending colon
Ileum
Cecum
Vermiform appendix
Stomach
Duodenum
Pancreas
Transverse colon
Jejunum
Descending colon
Sigmoid colon
Rectum
Anus

Figure 47-4 *Animation* Human digestive system

The human digestive tract is a long, coiled tube extending from mouth to anus. Food passes from the mouth, through the pharynx and esophagus, and into the stomach. After being partially digested in the stomach, food passes into the small intestine. Digestion continues in the first part of the small intestine. Absorption takes place mainly in the small intestine, which consists of the duodenum, jejunum, and ileum. The large intestine absorbs water and minerals. The large intestine includes the cecum, colon, rectum, and anus. Locate the three types of accessory glands: the liver, pancreas, and salivary glands.

© Cengage Learning

PREDICT A surgeon removes the stomach of a cancer patient. What would the surgeon do to ensure that the patient could continue to digest and absorb food?

Various regions of the vertebrate digestive tract are specialized to perform specific functions (**FIG. 47-4**). Food passes in sequence through the following specialized regions:

mouth ⟶ pharynx (throat) ⟶ esophagus ⟶ stomach ⟶ small intestine ⟶ large intestine ⟶ anus

The liver, pancreas, and salivary glands (in terrestrial vertebrates) are accessory glands that secrete digestive juices into the digestive tract.

The wall of the digestive tract has four layers. Although various regions differ somewhat in structure, the layers are basically similar throughout the digestive tract (**FIG. 47-5**). The **mucosa,** which consists of a layer of epithelial tissue and underlying connective tissue, lines the **lumen** (inner space) of the digestive tract. In the stomach and intestine, the mucosa is greatly folded to increase the secreting and absorbing surface. Surrounding the mucosa is the *submucosa,* a connective tissue layer rich in blood vessels, lymphatic vessels, and nerves.

A *muscle layer,* consisting of two sublayers of smooth muscle, surrounds the submucosa. In the inner sublayer, the muscle fibers are arranged circularly around the digestive tube. In the outer sublayer, the muscle fibers are arranged longitudinally.

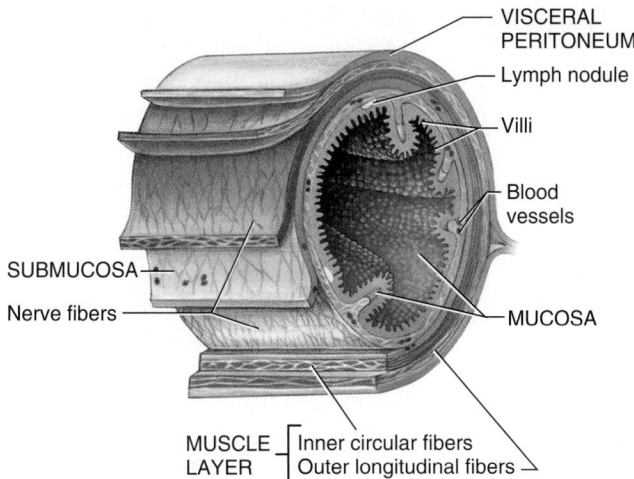

Figure 47-5 Wall of the digestive tract

From inside out, the layers of the wall are the mucosa, submucosa, muscle layer, and visceral peritoneum.

© Cengage Learning

Below the level of the diaphragm, the outer connective tissue coat of the digestive tract is called the *visceral peritoneum.* By various folds it is connected to the *parietal peritoneum,* a sheet of connective tissue that lines the walls of the abdominal and pelvic cavities. The visceral and parietal peritonea enclose part of the coelom, the *peritoneal cavity.* Inflammation of the peritoneum, called *peritonitis,* can be very serious because infection can spread along the peritoneum to most of the abdominal organs.

Food processing begins in the mouth

Imagine that you have just taken a big bite of a hamburger in a bun. The mouth is specialized for ingestion and for beginning the digestive process. Mechanical digestion begins as you bite, grind, and chew the meat and bun with your teeth. Unlike the simple, pointed teeth of fish, amphibians, and reptiles, the teeth of mammals vary in size and shape and are specialized to perform specific functions. The chisel-shaped **incisors** are used for biting and cutting food, whereas the long, pointed **canines** are adapted for piercing prey and tearing food (**FIG. 47-6**). The flattened surfaces of the **premolars** and **molars** are specialized for crushing and grinding.

Each tooth is covered by *enamel,* the hardest substance in the body (**FIG. 47-7**). Most of the tooth consists of *dentin,* which resembles bone in composition and hardness. The *pulp cavity,* which lies beneath the dentin, is a soft connective tissue containing blood vessels, lymph vessels, and nerves.

While food is being mechanically disassembled by the teeth, it is moistened by saliva. As some of the food molecules dissolve, you can taste the food. (Recall from Chapter 43 that taste buds are located on the tongue and other surfaces of the mouth.) Three pairs of **salivary glands** secrete about a liter of saliva into the mouth cavity each day. Saliva contains an enzyme, **salivary amylase,** which begins the chemical digestion of starch into sugar.

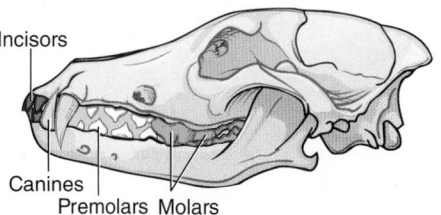

(a) Predator. The skull of a coyote shows the pointed incisors and canines, adaptations for ripping flesh.

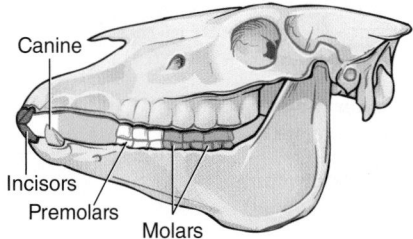

(b) Herbivore. Horses and other herbivores have incisors (and sometimes canines) adapted for cutting off bits of vegetation. Canines are absent in some herbivores. The broad, ridged surfaces of the molars are adapted for grinding plant material.

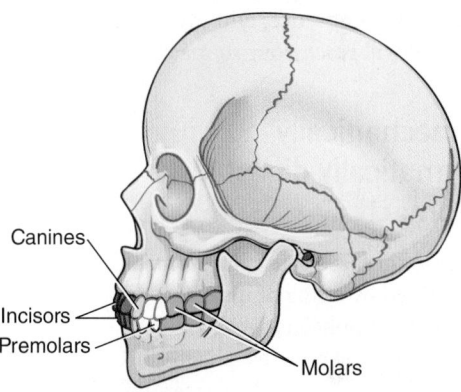

(c) Omnivore. The teeth of humans and other omnivores are adapted for chewing a variety of foods.

Figure 47-6 Teeth and diet

An animal's teeth are adapted for its mode of nutrition.

© Cengage Learning

The pharynx and esophagus conduct food to the stomach

After being chewed and fashioned into a lump called a *bolus,* the bite of food is swallowed, moving through the **pharynx** into the **esophagus.** The pharynx, or throat, is a muscular tube that serves as the hallway of both the respiratory system and the digestive system. During swallowing, a small flap of tissue, the **epiglottis,** closes the opening to the airway.

Waves of peristalsis sweep the bolus through the pharynx and esophagus toward the stomach (**FIG. 47-8**). Circular muscle fibers in the wall of the esophagus contract around the top of the bolus and push it downward. Almost at the same time,

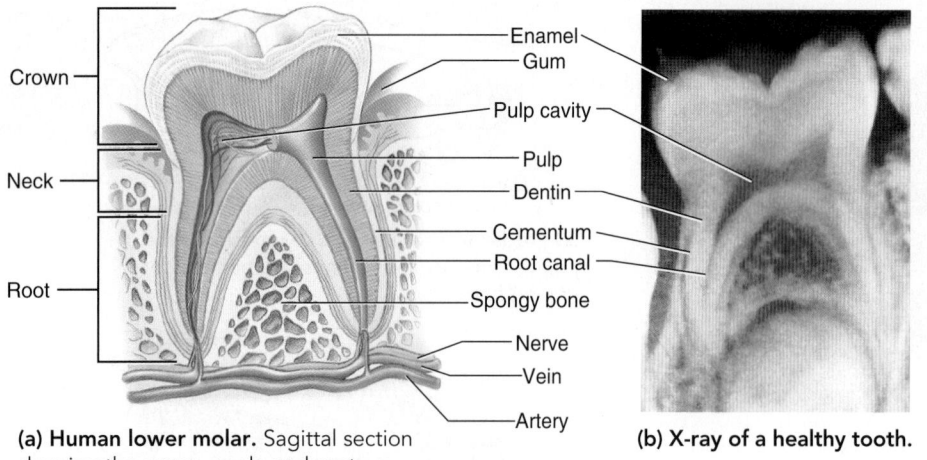

(a) Human lower molar. Sagittal section showing the crown, neck, and root.

(b) X-ray of a healthy tooth.

Figure 47-7 Tooth structure
© Cengage Learning

longitudinal muscles around the bottom of the bolus and below it contract, which shortens the tube.

When the body is in an upright position, gravity helps move the food through the esophagus, but gravity is not essential. Astronauts eat in its absence, and even if you are standing on your head, food will reach your stomach.

Food is mechanically and enzymatically digested in the stomach

The entrance to the large, muscular **stomach** is normally closed by a ring of muscle at the lower end of the esophagus. When a peristaltic wave passes down the esophagus, the muscle relaxes, and the bolus then enters the stomach (**FIG. 47-9**). When empty, the stomach is collapsed and shaped almost like a hot dog. Folds of the stomach wall, called **rugae,** give the inner lining a wrinkled appearance. As food enters, the rugae gradually smooth out, which expands the capacity of the stomach to more than a liter.

The stomach is lined with a simple, columnar epithelium that secretes large amounts of mucus. Tiny pits mark the entrances to the millions of **gastric glands,** which extend deep into the stomach wall. *Parietal cells* in the gastric glands secrete hydrochloric acid and *intrinsic factor,* a substance needed for adequate absorption of vitamin B_{12}. *Chief cells* in the gastric glands secrete **pepsinogen,** an inactive enzyme precursor. When it comes in contact with

the acidic gastric juice in the stomach, pepsinogen is converted to pepsin, the main digestive enzyme of the stomach. **Pepsin** hydrolyzes proteins, converting them to short polypeptides.

Several protective mechanisms prevent the gastric juice from digesting the wall of the stomach. Cells of the gastric mucosa secrete alkaline mucus that coats the stomach wall and neutralizes the acidity of the gastric juice along the lining. In addition, the epithelial cells of the lining fit tightly together, preventing gastric juice from leaking between them and into the tissue beneath. If some of the epithelial cells become damaged, they are quickly replaced. In fact, about a half million of these cells are shed and replaced every minute!

Sometimes, the part of the stomach lining is digested, leaving an open sore, or *peptic ulcer.* Such ulcers also occur in the duodenum and sometimes in the lower part of the esophagus. The bacterium *Helicobacter pylori* is the cause of about 80% of peptic ulcers (see Fig. 25-13). *Helicobacter pylori* infects the mucus-secreting cells of the stomach lining, decreasing the protective mucus, which can lead to peptic ulcers or cancer. This type of infection responds to antibiotic therapy.

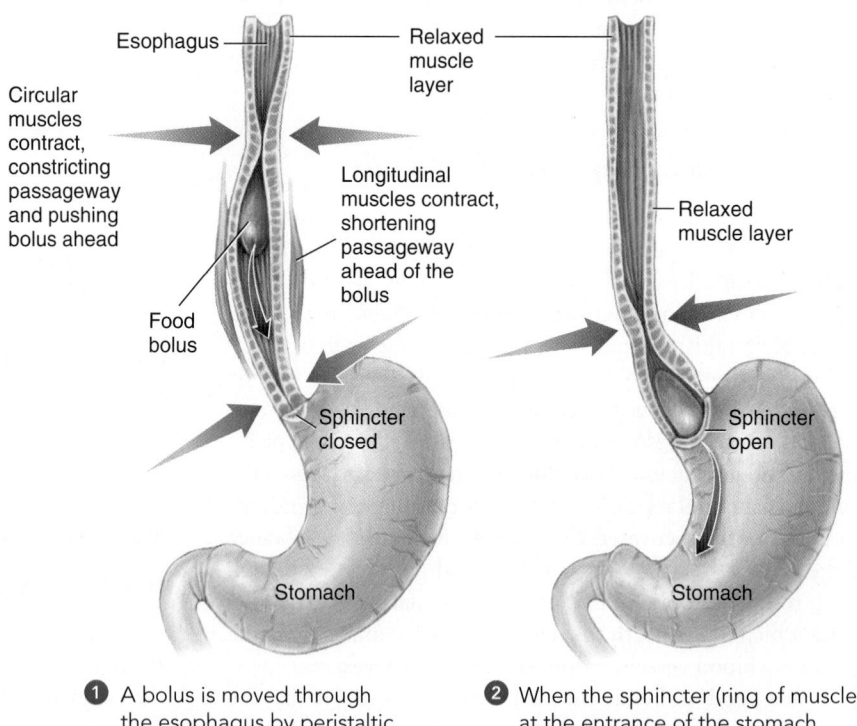

❶ A bolus is moved through the esophagus by peristaltic contractions.

❷ When the sphincter (ring of muscle) at the entrance of the stomach opens, food enters the stomach.

Figure 47-8 *Animation* Peristalsis
© Cengage Learning

The structure of the stomach is adapted for its function of digesting food; gastric glands in its mucosa secrete digestive juices, and the three layers of muscle in the stomach wall are important in mechanically digesting food.

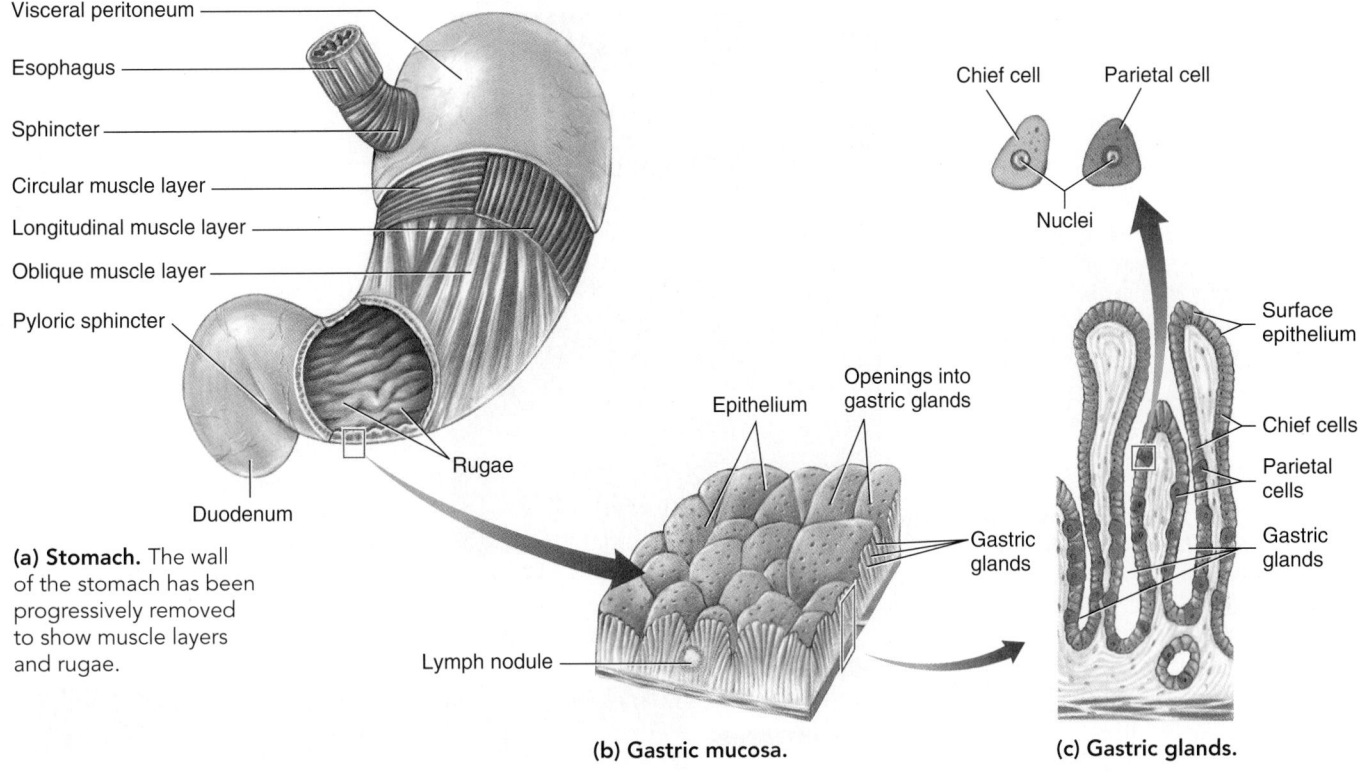

(a) Stomach. The wall of the stomach has been progressively removed to show muscle layers and rugae.

(b) Gastric mucosa.

(c) Gastric glands.

Figure 47-9 *Animation* **Structure of the stomach**
The stomach is adapted for digesting food mechanically and enzymatically.
© Cengage Learning

CONNECT What is the benefit of having lymph nodules within the gastric mucosa?

What changes occur in a bite of hamburger during its 3- to 4-hour stay in the stomach? The stomach churns and chemically degrades the food so that it assumes the consistency of a thick soup; this partially digested food is called *chyme.* Digestion of the starch in the bun to small polysaccharides and maltose continues until salivary amylase is inactivated by the acidic pH of the stomach. Protein digestion degrades much of the hamburger protein to polypeptides. Over a period of several hours, peristaltic waves release the chyme in spurts through the stomach exit, the *pylorus,* and into the small intestine.

Most enzymatic digestion takes place in the small intestine

Digestion of food is completed in the **small intestine,** and nutrients are absorbed through its wall. The small intestine, which is 5 to 6 m (about 17 ft) in length, has three regions: the **duodenum,** *jejunum,* and *ileum.* Most chemical digestion takes place in the duodenum, the first portion of the small intestine, not in the stomach. Bile from the liver and enzymes from the pancreas are released into the duodenum and act on the chyme. Then enzymes produced by the epithelial cells lining the duodenum catalyze the final steps in the digestion of the major types of nutrients.

The lining of the small intestine appears velvety because of its millions of tiny fingerlike projections, the intestinal **villi** (sing., *villus*) (**FIG. 47-10**). The villi increase the surface area of the small intestine for digestion and absorption of nutrients. The intestinal surface is further expanded by **microvilli,** projections of the plasma membrane of the simple columnar epithelial cells of the villi. About 600 microvilli protrude from the exposed surface of each cell, giving the epithelial lining a fuzzy appearance when viewed with the light microscope.

If the intestinal lining were smooth, like the inside of a water pipe, food would zip right through the intestine, and many valuable nutrients would not be digested or absorbed. Folds in the

The digestive and absorptive surface area of the small intestine is vastly increased by millions of fingerlike villi; the surface area is further increased by microvilli of the epithelial cells of the villi.

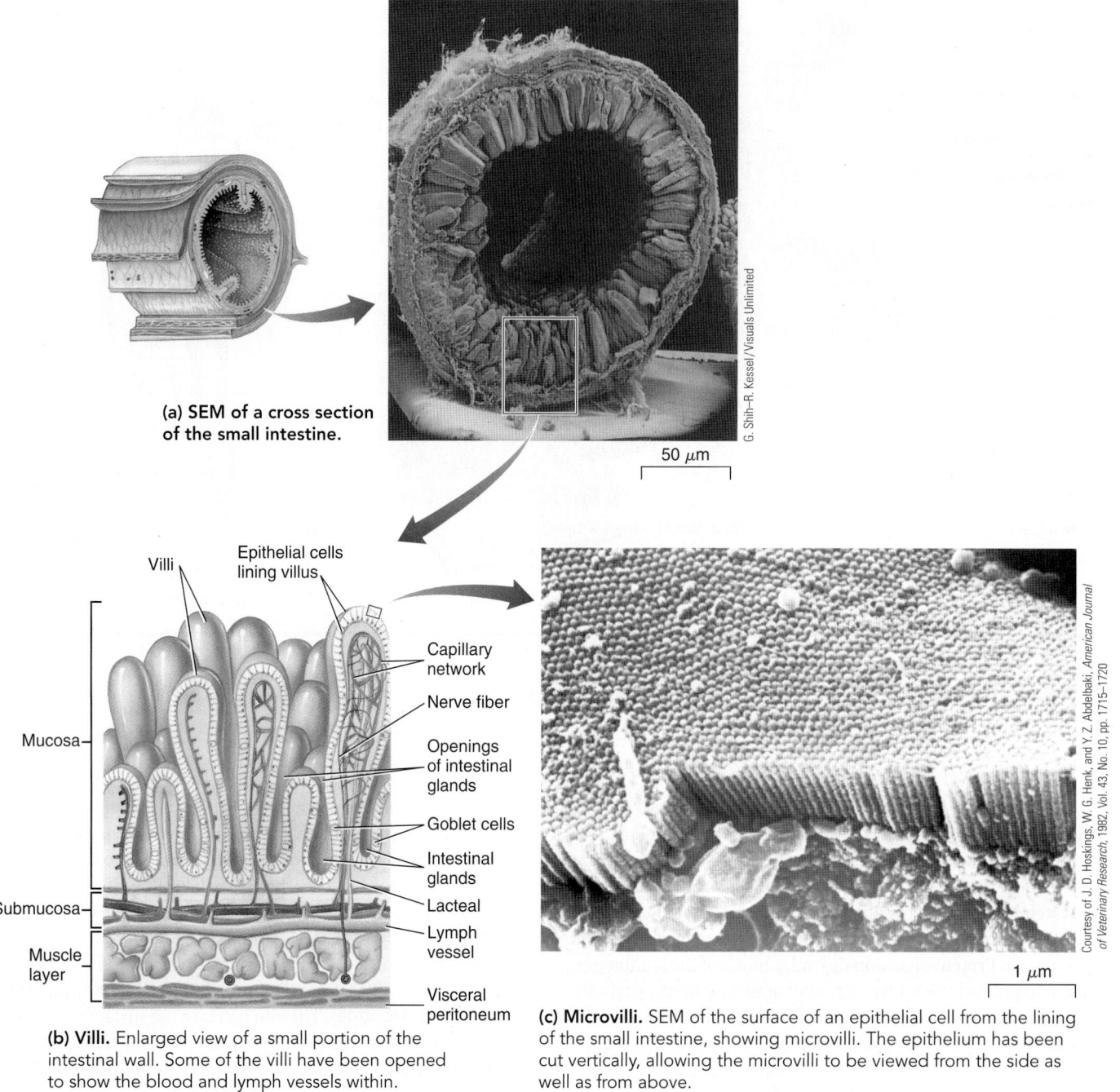

(a) **SEM of a cross section of the small intestine.**

G. Shih–R. Kessel/Visuals Unlimited

50 μm

Villi

Epithelial cells lining villus

Mucosa

Submucosa

Muscle layer

Capillary network

Nerve fiber

Openings of intestinal glands

Goblet cells

Intestinal glands

Lacteal

Lymph vessel

Visceral peritoneum

(b) **Villi.** Enlarged view of a small portion of the intestinal wall. Some of the villi have been opened to show the blood and lymph vessels within.

Courtesy of J. D. Hoskings, W. G. Henk, and Y. Z. Abdelbaki, *American Journal of Veterinary Research*, 1982, Vol. 43, No. 10, pp. 1715–1720

1 μm

(c) **Microvilli.** SEM of the surface of an epithelial cell from the lining of the small intestine, showing microvilli. The epithelium has been cut vertically, allowing the microvilli to be viewed from the side as well as from above.

Figure 47-10 Villi and microvilli

CONNECT In celiac disease, the body launches an immune response to gluten, a protein found in wheat, rye, and barley. How could this disease lead to malnutrition?

© Cengage Learning

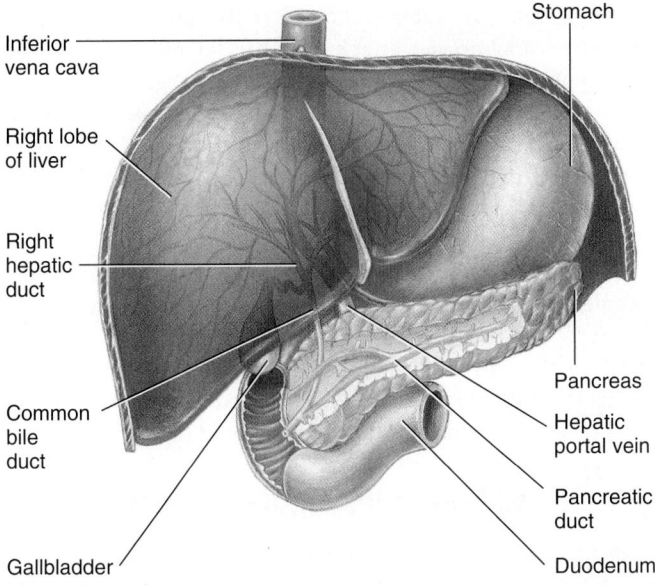

Inferior vena cava

Right lobe of liver

Right hepatic duct

Common bile duct

Gallbladder

Stomach

Pancreas

Hepatic portal vein

Pancreatic duct

Duodenum

Figure 47-11 The liver and pancreas

The gallbladder stores bile from the liver. Note the ducts that conduct bile to the gallbladder and the duodenum. The stomach has been displaced to expose the pancreas, which secretes several digestive enzymes into the duodenum.
© Cengage Learning

wall of the intestine, the villi, and microvilli together increase the surface area of the small intestine by about 600 times. If we could unfold and spread out the lining of the small intestine of an adult human, its surface would approximate the size of a tennis court.

The liver secretes bile

The **liver,** the largest internal organ and also one of the most complex organs in the body, lies in the upper-right part of the abdomen just under the diaphragm (**FIG. 47-11**). The liver secretes **bile,** which mechanically digests fats by a detergent-like action (discussed in a later section). Bile consists of water, bile salts, bile pigments, cholesterol, salts, and lecithin

(a phospholipid). Because it contains no digestive enzymes, bile does not enzymatically digest food. The pear-shaped **gallbladder** stores and concentrates bile and releases it into the duodenum as needed.

A single liver cell can carry on more than 500 separate, specialized metabolic activities! The liver performs these vital functions:

1. Secretes bile that mechanically digests fats
2. Helps maintain homeostasis by removing or adding nutrients to the blood
3. Converts excess glucose to glycogen and stores it
4. Converts excess amino acids to fatty acids and urea
5. Stores iron and certain vitamins
6. Detoxifies alcohol and other drugs and poisons

The pancreas secretes digestive enzymes

The **pancreas** is an elongated gland that secretes both digestive enzymes and hormones that help regulate the level of glucose in the blood. The cells that line the pancreatic ducts secrete an alkaline solution rich in bicarbonate ions. This pancreatic juice neutralizes the stomach acid in the duodenum and provides the optimal pH for action of the pancreatic enzymes. Pancreatic enzymes include **trypsin** and *chymotrypsin,* which digest polypeptides to short peptides; *pancreatic* **lipase,** which degrades fats; *pancreatic amylase,* which breaks down almost all types of complex carbohydrates, except cellulose, to disaccharides; and *ribonuclease* and *deoxyribonuclease,* which split the nucleic acids ribonucleic acid (RNA) and deoxyribonucleic acid (DNA) to free nucleotides.

Nutrients are digested as they move through the digestive tract

Chyme moves through the digestive tract by peristalsis, mixing contractions, and motions of the villi. As nutrients in the chyme move through the small intestine, they come into contact with enzymes that digest them (**TABLE 47-1**).

TABLE 47-1	Summary of Digestion		
	CARBOHYDRATES	**PROTEINS**	**LIPIDS**
Mouth	Polysaccharides ↓ *Salivary amylase* Maltose and small polysaccharides		
Stomach	Action continues until acidic pH inactivates salivary amylase	Protein ↓ *Pepsin* Short polypeptides	
Small intestine	Undigested polysaccharides ↓ *Pancreatic amylase* Maltose and other disaccharides ↓ *Maltase, sucrase, lactase* Monosaccharides	Polypeptides ↓ *Trypsin, chymotrypsin* Short peptides ↓ *Carboxypeptidase, peptidases, dipeptidases* Amino acids	Glob of fat ↓ *Bile salts* Emulsified fat droplets ↓ *Pancreatic lipase* Fatty acids and glycerol

© Cengage Learning

Carbohydrates are digested to monosaccharides Polysaccharides, such as starch and glycogen, are important components of the food ingested by most animals. The glucose units of these large molecules are connected by glycosidic bonds linking carbon 4 (or 6) of one glucose molecule with carbon 1 of the adjacent glucose molecule. These bonds are hydrolyzed by *amylases* that digest polysaccharides to the disaccharide maltose. Although amylase can split the α-glycosidic linkages present in starch and glycogen, it cannot split the β-glycosidic linkages present in cellulose (see Figs. 3-9 and 3-10).

Amylase cannot split the bond between the two glucose units of maltose. Enzymes produced by the cells lining the small intestine break down disaccharides such as maltose to monosaccharides. *Maltase,* for example, splits maltose into two glucose molecules (see Fig. 3-8a). Hydrolysis occurs while the disaccharides are being absorbed through the epithelium of the small intestine.

Proteins are digested to amino acids Several proteolytic enzymes are secreted into the digestive tract. Each breaks peptide bonds at one or more specific locations in a polypeptide chain. Trypsin, secreted in an inactive form by the pancreas, is activated by an enzyme called *enterokinase.* The trypsin then activates chymotrypsin and carboxypeptidase as well as additional trypsin.

Carboxypeptidase removes amino acids with free terminal carboxyl groups from the end of polypeptide chains. Pepsin, trypsin, and chymotrypsin break certain internal peptide bonds of proteins and polypeptides. *Dipeptidases* released by the duodenum then split the small peptides into amino acids.

Fats are digested to fatty acids and monoacylglycerols Lipids are usually ingested as large masses of triacylglycerols (also called triglycerides). They are digested mainly within the duodenum by pancreatic lipase. Like many other proteins, lipase is water soluble, but its substrates are not. Thus, the enzyme can attack only the fat molecules at the surface of a mass of fat. Bile salts act like detergents, reducing the surface tension of fats. Their action, called *emulsification,* breaks large masses of fat into smaller droplets. Emulsification greatly increases the surface area of fat exposed to the action of pancreatic lipase and so increases the rate of lipid digestion.

Conditions in the intestine are usually not optimal for the complete hydrolysis of lipids to glycerol and fatty acids. Consequently, the products of lipid digestion include monoacylglycerols (monoglycerides) and diacylglycerols (diglycerides) as well as glycerol and fatty acids. Undigested triacylglycerols also remain, some of which are absorbed without digestion.

Nerves and hormones regulate digestion

Most digestive enzymes are produced only when food is present in the digestive tract. Salivary gland secretion is controlled entirely by the nervous system, but secretion of other digestive juices is regulated by both nerves and hormones. The wall of the digestive tract contains dense networks of neurons. This *enteric nervous system* continues to regulate many motor and secretory activities of the digestive system even if sympathetic and parasympathetic nerves to these organs are cut. Many neuropeptides present in the brain are also released by neurons in the digestive tract and help regulate digestion. For example, *substance P* stimulates smooth muscle contraction of the digestive tract, and *enkephalin* inhibits it (see Table 41-2).

Several hormones, including **gastrin,** *secretin, cholecystokinin (CCK),* and *glucose-dependent insulinotropic peptide (GIP),* help regulate the digestive system (**TABLE 47-2**). These hormones are small polypeptides secreted by endocrine cells in the mucosa of certain regions of the digestive tract. Investigators are studying

TABLE 47-2	Some Hormones That Regulate Digestion			
HORMONE	**SOURCE**	**TARGET TISSUE**	**ACTIONS**	**FACTORS THAT STIMULATE RELEASE**
Gastrin	Stomach (mucosa)	Stomach (gastric glands)	Stimulates gastric glands to secrete pepsinogen and HCl; stimulates gastric motility	Distention of the stomach by food; certain substances such as partially digested proteins and caffeine
Secretin	Duodenum (mucosa)	Pancreas	Stimulates pancreas to secrete sodium bicarbonate, which neutralizes acid in duodenum	Acidic chyme acting on mucosa of duodenum
		Liver	Stimulates bile secretion	
		Stomach	Inhibits gastric secretion and gastric emptying	
Cholecystokinin (CCK)	Duodenum (mucosa)	Pancreas	Stimulates release of digestive enzymes	Presence of fatty acids and partially digested proteins in duodenum
		Gallbladder	Stimulates emptying of bile	
		Brain	Helps regulate food intake by signaling satiety	
Glucose-dependent insulinotropic peptide (GIP)	Duodenum (mucosa)	Pancreas	Stimulates insulin secretion	Presence of glucose in duodenum

© Cengage Learning

several other messenger peptides that are important in regulating digestive activity.

As an example of the regulation of the digestive system, consider the secretion of gastric juice. Seeing, smelling, tasting, or even thinking about food causes the brain to send neural signals to the gastric glands in the stomach. These signals stimulate these glands to secrete gastric juice. In addition, when food distends the stomach, stretch receptors send neural messages to the medulla. The medulla then sends messages to endocrine cells in the stomach wall that secrete gastrin. After being transported by the circulatory system, gastrin returns to the stomach, where it stimulates the stomach to release gastric juice. Gastrin also stimulates gastric and intestinal motility, and gastric emptying.

Absorption takes place mainly through the villi of the small intestine

Only a few substances—water, simple sugars, salts, alcohol, and certain drugs—are small enough to be absorbed through the stomach wall. Absorption of nutrients is primarily the job of the intestinal villi. As illustrated in Figure 47-10, the wall of a villus is a single layer of epithelial cells. Inside each villus is a network of capillaries and a central lymph vessel, called a **lacteal.**

To reach the blood (or lymph), a nutrient molecule must pass through an epithelial cell lining the intestine and through a cell lining a blood or lymph vessel. Absorption occurs by a combination of simple diffusion, facilitated diffusion, and active transport. Because glucose and amino acids cannot diffuse through the intestinal lining, they must be absorbed by active transport. Absorption of these nutrients is coupled with the active transport of sodium (see Chapter 5). Fructose is absorbed by facilitated diffusion.

Amino acids and glucose are transported directly to the liver by the *hepatic portal vein*. In the liver this vein divides into a vast network of tiny blood sinuses, vessels similar to capillaries. As the nutrient-rich blood moves slowly through the liver, nutrients and certain toxic substances are removed from the circulation.

The products of lipid digestion are absorbed by a different process and different route. After free fatty acids and monoacylglycerols enter an epithelial cell in the intestinal lining, they are reassembled as triacylglycerols in the smooth endoplasmic reticulum. The triacylglycerols, along with absorbed cholesterol and phospholipids, are packaged into protein-covered fat droplets, called **chylomicrons.**

After they are released into the interstitial fluid, chylomicrons enter the lacteal (lymph vessel) of the villus. Chylomicrons are transported in the lymph to the subclavian veins, where the lymph and its contents enter the blood. About 90% of absorbed fat enters the blood circulation in this indirect way. The rest, mainly short-chain fatty acids such as those in butter, are absorbed directly into the blood. After a meal rich in fats, the great number of chylomicrons in the blood may give the plasma a turbid, milky appearance for a few hours. Chylomicrons transport fats to the liver and other tissues.

Most of the nutrients in the chyme are absorbed by the time it reaches the end of the small intestine. What is left (mainly waste) passes through a sphincter, the *ileocecal valve,* into the large intestine.

The large intestine eliminates waste

Undigested material, such as the cellulose of plant foods, along with unabsorbed chyme passes into the **large intestine** (see Fig. 47-4). Although only about 1.3 m (about 4 ft) long, this part of the digestive tract is referred to as "large" because its diameter is greater than that of the small intestine. The small intestine joins the large intestine about 7 cm (2.8 in.) from the end of the large intestine, forming a blind pouch, the **cecum.** The *vermiform appendix* projects from the end of the cecum. (Appendicitis is an inflammation of the appendix.) Herbivores such as rabbits have a large, functional cecum that holds food while bacteria digest its cellulose. In humans the functions of the cecum and appendix are not known, and these structures are generally considered vestigial organs.

From the cecum to the **rectum** (the last portion of the large intestine), the large intestine is known as the **colon.** The regions of the large intestine are the cecum; ascending colon; transverse colon; descending colon; sigmoid colon; rectum; and **anus,** the opening for the elimination of wastes.

As chyme passes slowly through the large intestine, water and sodium are absorbed from it, and it gradually assumes the consistency of normal feces. Bacteria inhabiting the large intestine are nourished by the last remnants of the meal; in exchange, they benefit their host by producing vitamin K and certain B vitamins that can be absorbed and used.

A distinction should be made between elimination and excretion. *Elimination* is the process of getting rid of digestive wastes, materials that have not been absorbed from the digestive tract and did not participate in metabolic activities. In contrast, *excretion* is the process of getting rid of metabolic wastes, which in mammals is mainly the function of the kidneys and lungs. The large intestine, however, does *excrete* bile pigments.

When chyme passes through the intestine too rapidly, *defecation* (expulsion of feces) becomes more frequent, and the feces are watery. This condition, called *diarrhea,* may be caused by anxiety, certain foods, or pathogens that irritate the intestinal lining. Prolonged diarrhea results in loss of water and salts and leads to dehydration, which can be a serious condition, especially in infants. *Constipation* results when chyme passes through the intestine too slowly. Because more water than usual is removed from the chyme, the feces may be hard and dry. Constipation is often caused by a diet deficient in fiber.

Colorectal cancer occurs in the colon or rectum (**FIG. 47-12**). In Western countries, colorectal cancer is the third most common cancer, and in the United States, it is the second leading cancer killer. Several factors contribute to the risk of developing colon cancer, including certain genetic factors and family history, smoking, heavy use of alcohol, physical inactivity, and obesity. A diet high in red meat and processed meat and low in fresh fruit, vegetables, poultry, and fish appears to increase the risk.

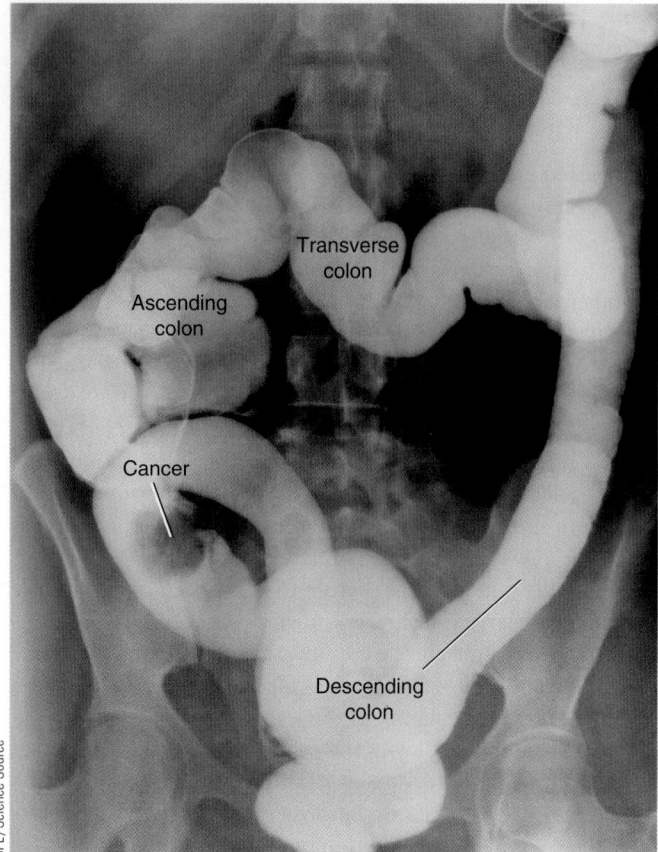

Figure 47-12 Colon cancer

In this radiographic view of the large intestine, cancer is evident as a mass that projects into the lumen of the colon. (The cancer has been color-enhanced.) The large intestine has been filled with a suspension of barium sulfate, which makes irregularities in the wall visible.

CHECKPOINT 47.2

- *Imagine that you have just taken a bite of a chicken sandwich. List in sequence the structures through which that food passes in its journey through the digestive system. What happens in each structure?*

- **CONNECT** *How are villi in the small intestine functionally related to rugae in the stomach?*

- *What are the functions of each of the following: trypsin, bile salts, gastrin, lacteals?*

- **VISUALIZE** *Draw a sketch of the large intestine and label the cecum, transverse colon, descending colon, vermiform appendix, ascending colon, sigmoid colon, and rectum.*

47.3 REQUIRED NUTRIENTS

LEARNING OBJECTIVES

6 Summarize the nutritional requirements for dietary carbohydrates, lipids, and proteins, and trace the fate of glucose, lipids, and amino acids after their absorption.

7 Describe the nutritional functions of vitamins, minerals, and phytochemicals.

Adequate amounts of nutrients are necessary for metabolic processes. Animals require carbohydrates, lipids, proteins, vitamins, minerals, and water. Nutritionists are also investigating plant compounds called *phytochemicals* that are important in nutrition. Water is a necessary dietary component because water is the medium in which all the body's biochemical processes take place. Sufficient water must be ingested to replace water lost in urine, sweat, feces, and breath.

Nutritionists continually consider new data and re-examine the *relative* amounts of various types of nutrients recommended for a healthy diet. According to the U.S. Department of Agriculture's *Dietary Guidelines for Americans,* 45% to 65% of our calories should come from carbohydrates, 20% to 35% from fat, and 10% to 35% from protein.

Recall from Chapter 1 that **metabolism** includes all the chemical processes that take place in the body. Metabolism includes anabolic and catabolic processes. **Anabolism** includes the synthetic aspects of metabolism, such as the production of proteins and nucleic acids. **Catabolism** includes breakdown processes, such as hydrolysis. Nutritionists measure the energy value of food in kilocalories, or simply Calories. A Calorie, spelled with a capital *C,* denotes a **kilocalorie (kcal),** defined as the amount of heat required to raise the temperature of 1 kilogram of water 1°C.

Carbohydrates provide energy

Sugars and starches are important sources of energy in the human diet. Most carbohydrates are ingested in the form of starch and cellulose, both polysaccharides composed of long chains of glucose subunits. (You may want to review the discussion of carbohydrates in Chapter 3.) Nutritionists refer to *polysaccharides* as *complex carbohydrates.* Foods rich in complex carbohydrates include rice, potatoes, corn, and other cereal grains.

When we eat an excess of carbohydrate-rich food, the liver cells become fully packed with glycogen and convert excess glucose to fatty acids and glycerol. Liver cells convert these compounds to triacylglycerols and send them to the fat deposits of the body for storage.

Refined carbohydrates, such as white bread and white rice, are unhealthy because the refining process removes fiber and many vitamins and minerals. The refining process also produces a form of starch that the digestive system rapidly breaks down to glucose. The resulting rapid increase in glucose concentration in the blood stimulates the pancreas to release a large amount of insulin. This hormone lowers blood glucose level by signaling the liver and muscles to remove glucose from the blood. When glucose and insulin levels are high, two things happen: triacylglycerol levels increase, and high-density lipoprotein (HDL; the good cholesterol) concentration decreases. These metabolic events can lead to cardiovascular disease and increase the risk of type 2 diabetes (discussed in Chapter 49).

Dietary fiber decreases cholesterol concentration in the blood and is associated with a lower risk for cardiovascular disease and diabetes. *Fiber* is mainly a mixture of cellulose and other indigestible carbohydrates. We obtain dietary fiber by eating fruits, vegetables, and whole grains. The U.S. diet is low in fiber because of low intake of fruits and vegetables and the use of refined flour. Increasing fiber in the diet has a variety of health benefits. For instance, fiber stimulates the feeling of being satisfied (satiety) after food intake and thus is useful in treating obesity.

Lipids provide energy and are used to make biological molecules

Cells use ingested lipids as a source of energy and to make a variety of lipid compounds, including components of cell membranes, steroid hormones, and bile salts. We ingest about 98% of our dietary lipids in the form of triacylglycerols (triglycerides). (Recall from Chapter 3 that a triacylglycerol is a glycerol molecule chemically combined with three fatty acids; see Fig. 3-12b. Triacylglycerols may be saturated—that is, fully loaded with hydrogen atoms—or their fatty acids may be monounsaturated (containing one double bond) or polyunsaturated (containing two or more double bonds).

Three polyunsaturated fatty acids (linoleic, linolenic, and arachidonic acids) are essential fatty acids that humans must obtain from food. Given these and sufficient nonlipid nutrients, the body can make all the lipid compounds (including fats, cholesterol, phospholipids, and prostaglandins) it needs. The average U.S. diet provides far more cholesterol than the recommended daily maximum of 300 mg. High-cholesterol sources include egg yolks, butter, ice cream, and meat.

How are lipids transported? Recall that chylomicrons transport lipids from the intestine to the liver and to other tissues. As chylomicrons circulate in the blood, the enzyme *lipoprotein lipase* breaks down triacylglycerols. The fatty acids and glycerol can then be taken up by the cells. What is left of the chylomicron, a remnant made up mainly of cholesterol and protein, is taken up by the liver.

Liver cells repackage cholesterol and triacylglycerols. These lipids are bound to proteins and transported as large molecular complexes, called **lipoproteins.** Some plasma cholesterol is transported by **high-density lipoproteins,** or **HDLs** ("good" cholesterol), but most is transported by **low-density lipoproteins,** or **LDLs** ("bad" cholesterol). LDLs deliver cholesterol to the cells. For cells to take in LDLs, a protein (apolipoprotein B) on the LDL surface must bind with a protein *LDL receptor* on the plasma membrane (see Fig. 5-22). After binding, the LDL enters the cell, and its cholesterol and other components are used. When cholesterol levels are high, HDLs collect the excess cholesterol and transport it to the liver. HDLs decrease risk for cardiovascular disease.

When needed, stored fats are hydrolyzed to fatty acids and released into the blood. Before cells can use these fatty acids as fuel, they are broken down into smaller compounds and combined with coenzyme A to form molecules of acetyl coenzyme A (acetyl CoA; **FIG. 47-13**). Acetyl CoA enters the citric acid cycle (see Chapter 8). The conversion of fatty acids to acetyl CoA is accomplished in the liver by a process known as *β-oxidation*.

For transport to the cells, acetyl coenzyme A is converted into one of three types of *ketone bodies* (four-carbon ketones). Normally, the level of ketone bodies in the blood is low, but in

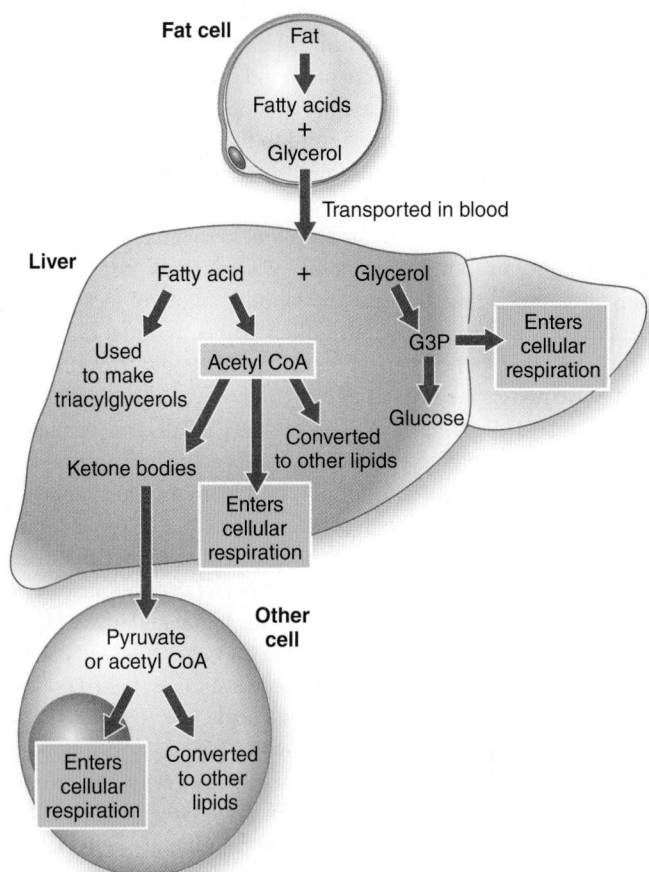

Figure 47-13 How the body uses fat

The liver converts glycerol and fatty acids to compounds that are used as fuel in cellular respiration. Recall from Chapter 8 that G3P is glyceraldehyde-3-phosphate.

© Cengage Learning

certain abnormal conditions, such as starvation and diabetes mellitus, fat metabolism is tremendously increased. Ketone bodies are then produced so rapidly that their level in the blood becomes excessive, which may make the blood too acidic to function normally. Such disrupted pH balance can lead to death.

What is the relationship between fat and cholesterol intake and cardiovascular disease? Lipids play a key role in the development of atherosclerosis, a progressive disease in which the arteries become clogged with fatty material (see Chapter 44). Low-density lipoproteins are the main source of the cholesterol that builds up in the walls of arteries. The type of fat consumed— as well as other dietary and lifestyle factors—is important. You can promote a healthy proportion of HDL to LDL by exercising regularly, following a healthy diet, maintaining appropriate body weight (obesity raises LDL and triacylglycerol levels), and not smoking cigarettes.

In general, animal foods are rich in both saturated fats and cholesterol, whereas most plant foods contain unsaturated fats and no cholesterol. Butter contains mainly saturated fats. Commonly used polyunsaturated vegetable oils are corn, soybean, and safflower oils. Monounsaturated fats raise HDL level, as does stearic acid, the saturated fat in chocolate. Olive, canola, and peanut oils contain large amounts of monounsaturated fats. A few plant oils, including palm oil and coconut oil, are high in saturated fats. Their effects on blood lipids and their role in causing cardiovascular disease are the focus of ongoing investigation. **Omega-3 fatty acids** (found in fish and some plant oils) decrease LDL levels and play other protective roles in decreasing the risk for coronary heart disease.

A diet high in saturated fats and cholesterol increases the cholesterol concentration in the blood by as much as 25%. Because polyunsaturated fats decrease the blood cholesterol level, many people cook with vegetable oils rather than with butter and lard, drink skim milk rather than whole milk, and eat low-fat ice cream. Some use margarine instead of butter. However, during production of many margarines, vegetable oils are partially hydrogenated (some of the carbons accept hydrogen to become fully saturated). During hydrogenation, some double bonds are transformed from a *cis* arrangement to a *trans* arrangement (see Fig. 3-3b), forming *trans fatty acids;* the harder the margarine, the higher the *trans* fatty acid content. Health concerns have been raised because *trans* fatty acids increase LDL cholesterol in the blood and lower HDL. In response there has been a decrease in the use of *trans* fatty acids in commercial products. However, some fast foods, baked goods, and snack foods still contain *trans* fatty acids.

Although some fats, such as *trans* fatty acids and saturated fats, are unhealthy, others, such as omega-3 fatty acids and monounsaturated fat, promote health. The Inuits in Greenland eat diets extremely high in fat. However, because their dietary fat is rich in omega-3 fatty acids, they have a low incidence of heart disease.

Proteins serve as enzymes and as structural components of cells

Proteins are essential building blocks of cells, serve as enzymes, and are used to make essential compounds such as receptors, hemoglobin and myosin. Protein consumption is an index of a country's (or an individual's) economic status because high-quality protein tends to be the most expensive and least available of the nutrients. In many parts of the world, protein poverty is one of the most pressing health problems.

Ingested proteins are degraded in the digestive tract to small peptides and amino acids. Of the 20 or so amino acids important in nutrition, approximately 9 (10 in children) cannot be synthesized by humans at all, or at least not in sufficient quantities to meet the body's needs. The diet must provide these **essential amino acids** (see Chapter 3).

Complete proteins, those that contain the most appropriate distribution of amino acids for human nutrition, are found in fish, meat, nuts, eggs, and milk. Some foods, such as gelatin and legumes (soybeans, beans, peas, peanuts), contain a high proportion of protein. However, they either do not contain all the essential amino acids or do not contain them in proper nutritional proportions. Most plant proteins are deficient in one or more essential amino acids. The healthiest sources of protein are fish, chicken, nuts, and legumes.

Amino acids circulating in the blood are taken up by cells and used for the synthesis of proteins. The liver removes excess amino acids from the circulation. Liver cells deaminate amino acids; that is, they remove the amino group (**FIG. 47-14**). During deamination, ammonia is produced from the amino group. Ammonia, which is toxic at high concentrations, is converted to urea and excreted from the body. The remaining carbon chain of the amino acid (called a *keto acid*) may be converted into carbohydrate or lipid and used as fuel or stored. For this reason, people who eat high-protein diets can gain weight if they eat too much.

Vitamins are organic compounds essential for normal metabolism

Vitamins are organic compounds required in the diet in relatively small amounts for normal biochemical functioning. Many are components of coenzymes (see Chapter 7). Nutritionists divide vitamins into two main groups. *Fat-soluble vitamins* include vitamins A, D, E, and K. *Water-soluble vitamins* are the B and C vitamins. Fruit and vegetables are rich sources of vitamins. **TABLE 47-3** provides the sources, functions, and consequences of deficiency for the vitamins.

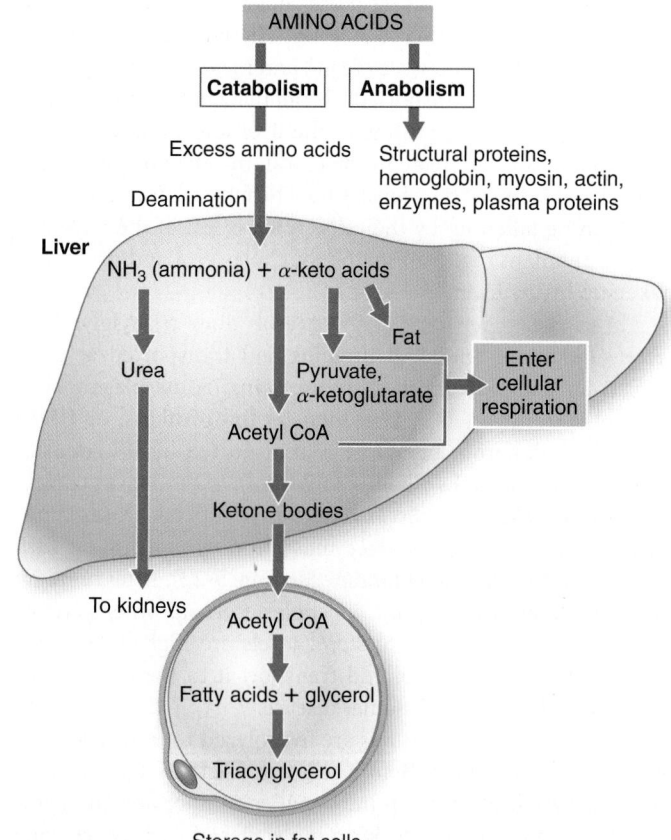

Figure 47-14 How the body uses protein

The liver plays a central role in protein metabolism. Deamination of amino acids and conversion of the amino groups to urea take place there. In addition, many proteins are synthesized in the liver. Note that you can gain weight on a high-protein diet because excess amino acids can be converted to fat.
© Cengage Learning

| TABLE 47-3 | The Vitamins |

VITAMINS AND U.S. RDA	ACTIONS	EFFECTS OF DEFICIENCY	SOURCES
FAT-SOLUBLE VITAMINS			
Vitamin A (retinal)	Converted to retinal; essential for normal vision; essential for normal growth and differentiation of cells; reproduction; immunity	Retarded growth; night blindness; worldwide millions of children are at risk of blindness from vitamin A deficiency; increased risk of infections	Liver; fortified milk and dairy products; orange, yellow, and green vegetables and fruit, such as carrots, broccoli, and cantaloupe
Vitamin D (calciferol) Two major forms: vitamin D_2 is derived mainly from plant foods in diet; vitamin D_3 is derived from animal foods in diet and synthesis in skin	Promotes calcium and phosphorus absorption from digestive tract; essential to normal growth and maintenance of bone	Weak bones; bone deformities; rickets in children, softening of the bones (osteomalacia) in adults	Fish oils, egg yolk, fortified milk, butter, margarine, salmon; synthesized in the skin through action of sunlight
Vitamin E (tocopherols)	Antioxidant; protects unsaturated fatty acids and cell membranes	Increased catabolism of unsaturated fatty acids so that not enough are available for maintenance of cell membranes; prevention of normal growth; nerve damage	Polyunsaturated plant oils, nuts, leafy greens, whole-grain products
Vitamin K	Synthesis of proteins necessary for blood-clotting; important in bone function	Prolonged blood-clotting time; hemorrhage	Synthesized by intestinal bacteria; leafy greens, legumes
WATER-SOLUBLE VITAMINS			
Vitamin C (ascorbic acid)	Synthesis of collagen, some hormones, and neurotransmitters; antioxidant; important in immune function	Scurvy (wounds heal very slowly, and scars become weak and split open; capillaries become fragile; bone does not grow or heal properly); impaired immunity	Citrus fruit, strawberries, tomatoes, leafy vegetables, cabbage-type vegetables
Thiamine (Vitamin B_1)	Active form is a coenzyme in many enzyme systems; important in carbohydrate and amino acid metabolism	Beriberi (weakened heart muscle, enlarged right side of heart, nervous system and digestive tract disorders); deficiency common in alcoholics	Liver, meat, fish, yeast, whole-grain or enriched breads, green leafy vegetables
Riboflavin (Vitamin B_2)	Used to make coenzymes (e.g., FAD) essential in energy metabolism	Dermatitis, inflammation of mouth	Liver, dairy products, meat, green leafy vegetables, whole-grain or enriched breads
Niacin	Component of important coenzymes (NAD^+ and $NADP^+$); essential to energy metabolism	Pellagra (dermatitis, diarrhea, dementia, weakness, fatigue)	Meat, fish, dairy products, nuts, whole-grain or enriched breads
Vitamin B_6 (Pyridoxine)	Part of coenzyme needed in amino acid and fatty acid metabolism	Dermatitis, anemia, depression, convulsions	Meat, fish, whole grains, legumes, soy, green leafy vegetables
Pantothenic acid	Component of coenzyme A (important in energy metabolism)	Deficiency rare; sleep disturbances, fatigue	Widespread in foods; meat, eggs, whole grains, legumes
Folate (folic acid)	Coenzyme needed for nucleic acid synthesis and maturation of red blood cells	Anemia; neural tube birth defects; increased risk of cardiovascular disease and certain cancers; deficiency common in alcoholics	Liver, legumes, dark green leafy vegetables, citrus fruit and juices, whole grains
Biotin	Coenzyme important in metabolism	Deficiency uncommon; anemia, dermatitis	Produced by intestinal bacteria; liver, chocolate, egg yolks
Vitamin B_{12} (cobalamin)	Coenzyme important in DNA synthesis and metabolism; works with folate in production of red blood cells	A type of anemia, fatigue	Liver, meat, fish, dairy products

© Cengage Learning

Health professionals debate the advisability of taking large amounts of certain specific vitamins, such as vitamin C to prevent colds or vitamin E to protect against vascular disease. Some studies suggest that vitamin A (found in yellow and green vegetables) and vitamin C (found in citrus fruits and tomatoes) help protect against certain forms of cancer. We do not yet understand all the biochemical roles played by vitamins or the interactions among various vitamins and other nutrients. We do know, however, that, like vitamin deficiency, large overdoses of some vitamins can be harmful. Moderate overdoses of the B and C vitamins are excreted in the urine, but surpluses of the fat-soluble vitamins are not easily excreted and can accumulate to harmful levels.

Minerals are inorganic nutrients

Minerals are inorganic nutrients ingested in the form of salts dissolved in food and water (**TABLE 47-4**). We need certain minerals, including sodium, chloride, potassium, calcium, phosphorus, magnesium, manganese, and sulfur, in amounts of 100 mg or more daily. These nutrients are referred to as *major minerals.* Several others, such as iron, copper, iodide, fluoride, and selenium, are **trace elements,** minerals that we require in minute amounts (less than 100 mg per day).

Minerals are necessary components of body tissues and fluids. Salt content (about 0.9% in plasma) is vital in maintaining the fluid balance of the body, and because salts are lost from the body daily in sweat, urine, and feces, they must be replaced by dietary intake. Sodium chloride (common table salt) is the salt needed in largest quantity in blood and other body fluids. Deficiency results in dehydration. Iron is the mineral most often deficient in the diet. In fact, iron deficiency is one of the most widespread nutritional problems in the world.

Antioxidants inactivate reactive molecules

Normal cell processes that require oxygen produce **oxidants,** highly reactive molecules such as free radicals, peroxides, and superoxides. **Free radicals,** molecules or ions with one or more unpaired electrons, are also generated by ionizing radiation, tobacco smoke, and other forms of air pollution. Oxidants damage DNA, proteins, and unsaturated fatty acids by snatching

TABLE 47-4	Some Important Minerals and Their Functions		
MINERAL	**FUNCTIONS**	**SOURCES**	**EFFECTS OF DEFICIENCY; COMMENTS**
Calcium	Main mineral of bones and teeth; essential for normal blood clotting, muscle contraction, nerve function, and regulation of cell activities	Milk and other dairy products, fish, green leafy vegetables; bones serve as calcium reservoir	Stunted growth; osteoporosis in adults
Phosphorus	Performs more functions than any other mineral; structural component of bone; component of ATP, DNA, RNA, and phospholipids	Meat, fish, dairy products, cereals	Muscle weakness
Sulfur	Component of many proteins and certain vitamins	High-protein foods such as meat, fish, legumes, nuts	None known
Potassium	Principal positive ion within cells; important in muscle contraction, nerve function, and fluid balance	Meat, milk, fruit, vegetables, grains	Deficiency caused by dehydration; muscle weakness; paralysis; confusion; can cause cardiac arrest
Sodium	Principal positive ion in interstitial fluid; important in fluid balance and acid–base balance; neural transmission	Many foods, table salt; too much ingested in average American diet	Muscle cramps, apathy; excessive amounts may contribute to high blood pressure
Chloride	Principal negative ion of interstitial fluid; important in fluid balance and in acid–base balance; component of hydrochloric acid in stomach	Many foods, table salt	Stunted growth in children; muscle cramps
Magnesium	Needed for normal muscle and nerve function	Nuts, whole grains, legumes, green leafy vegetables, seafood, chocolate	Stunted growth in children; weakness; muscle spasms, confusion
Copper	Component of several enzymes; essential for hemoglobin synthesis	Meats, seafood, whole grains, legumes, seeds	Anemia
Iodine	Component of thyroid hormones (help regulate metabolic rate, growth, and development)	Seafood, iodized salt, vegetables grown in iodine-rich soils	Underactive thyroid gland, goiter (abnormal enlargement of thyroid gland); in children, deficiency affects mental and physical development
Manganese	Activates many enzymes; bone formation	Whole grains, nuts, leafy green vegetables; tea	Poorly absorbed from intestine, but deficiency symptoms rare
Iron	Component of hemoglobin, myoglobin, important respiratory enzymes (cytochromes), and other enzymes essential to oxygen transport and cellular respiration	Good sources: meat (especially liver), fish, nuts, egg yolk, legumes, dried fruit	Iron-deficiency anemia; weakness, fatigue, may impair cognitive function; impaired immunity; mineral most likely to be deficient in diet
Fluoride	Component of bones and teeth; makes teeth resistant to decay	Fluoridated water; seafood; tea	Risk of tooth decay; excess causes tooth mottling
Zinc	Cofactor for at least 70 enzymes; helps regulate synthesis of certain proteins; needed for growth and repair of tissues	Meat, fish, dairy products, whole grains, vegetables	Slowed growth in children; loss of appetite; impaired immune function
Selenium	Antioxidant (cofactor for a peroxidase that protects against oxidation by breaking down peroxides)	Seafood, meat, whole grains	Deficiency is rare

© Cengage Learning

electrons. Damage to DNA causes mutations that may lead to cancer, and injury to unsaturated fatty acids can damage cell membranes. Free radicals are thought to contribute to atherosclerosis by causing oxidation of LDL cholesterol. Oxidative damage to the body over many years contributes to the aging process.

Cells have **antioxidants** that inactivate free radicals and other reactive molecules by donating an electron. Antioxidants in tissues include certain enzymes, such as catalase and peroxidase. Their action requires certain minerals, such as selenium, zinc, manganese, and copper. Vitamin C and vitamin E have strong antioxidant activity. A variety of *phytochemicals* are potent antioxidants (discussed in next section).

Nutritionists recommend that we increase the antioxidants in our diet by eating fruits, vegetables, and other foods high in antioxidants. We do not yet know whether antioxidant supplements are useful. In fact, recent studies suggest that some may be harmful.

Phytochemicals play important roles in maintaining health

A diet that includes all the essential nutrients does not provide the same health benefits as one also rich in fruit and vegetables. The missing ingredients are **phytochemicals,** plant compounds that are biologically active in the body and may promote health. Many phytochemicals are antioxidants. For example, *lycopenes,* which are responsible for the red color of tomatoes, are powerful antioxidants. *Flavonoids,* found in cocoa and especially in dark chocolate, are antioxidants that also appear to lower blood pressure, lower cholesterol, and improve cardiovascular health in other ways.

Nutritionists have begun to intensively investigate phytochemicals. Some important classes of phytochemicals are carotenoids (which include vitamin A and lycopene) in deeply pigmented fruits and vegetables such as carrots and tomatoes; flavonoids in berries, herbs, vegetables, barley, soy, black and green teas, and cocoa; isothiocyanates and indoles in broccoli and other cruciferous vegetables; and organosulfur compounds in garlic and onions. In Asian countries where diets are low in fat and high in soy and green tea, the incidence of breast, prostate, and colorectal cancer is low.

Diets rich in fruits and vegetables lower the incidence of heart disease and of certain types of cancer. Studies comparing diets in various countries suggest that intake of fruits and vegetables may be more important than differences in dietary fat. Yet nutritionists estimate that in the United States only about 1 person in 11 eats the daily recommended 2 to 6.5 cups of fruits and vegetables a day (the equivalent of 4 to 13 servings). In the United States, the National Fruit and Vegetable Program (a public–private partnership that includes government agencies, industry, and not-for-profit groups) promotes increased consumption of fruit and vegetables.

CHECKPOINT 47.3

- *What are the major categories of required nutrients?*
- *What is the function of glucose? of essential amino acids?*
- CONNECT *How are low-density lipoproteins (LDLs) related to cardiovascular disease?*

47.4 ENERGY METABOLISM

LEARNING OBJECTIVES

8 Contrast basal metabolic rate with total metabolic rate; write the basic energy equation for maintaining body weight and describe the consequences of altering it in either direction.
9 Identify components of the regulatory pathways that regulate food intake and energy homeostasis.
10 In general terms, describe the effects of malnutrition, including both overnutrition and undernutrition.

The amount of energy liberated by the body per unit time is a measure of its **metabolic rate.** Much of the energy expended by the body is ultimately converted to heat. Metabolic rate may be expressed either in kilocalories of heat energy expended per day or as a percentage above or below a standard normal level.

The **basal metabolic rate (BMR)** is the rate at which the body releases heat as a result of breaking down fuel molecules. BMR is the body's basic cost of living; that is, it is the rate of energy use during resting conditions. This energy is required to maintain body functions such as heart contraction, breathing, and kidney function. *Total metabolic rate* is the sum of the BMR and the energy used to carry on daily activities. A professional athlete has a greater total metabolic rate than an account executive who is inactive most of the day and who does not exercise regularly.

An average-sized man who does not exercise and who sits at a desk all day expends about 2100 kcal daily (women expend less energy). If the food the individual eats each day also contains about the same number of kilocalories, the body will be in a state of energy balance; that is, energy input will equal energy output. This concept is important because body weight remains constant when

$$\text{energy input} = \text{energy output}$$

When people take in less energy in food than they expend in daily activity, the body burns stored fat, and they lose weight. In this case,

$$\text{energy input} < \text{energy output}$$

When people take in more energy in food than they expend in daily activity, they gain weight. In this case,

$$\text{energy input} > \text{energy output}$$

Body mass index (BMI), used worldwide as a measure of body size, is an index of weight in relation to height. It is calculated by dividing the square of a person's weight (in kilograms) by height (in meters). The equivalent in the United States is 4.89 times the weight (in pounds) divided by the square of the height (in feet). An adult with a BMI between 18.5 and 24.9 typically has a healthy weight. However, BMI is no longer considered a reliable measure because it does not consider important factors such as differences in body composition between males and females, between the average adult and a muscular athlete, or among various races.

Energy metabolism is regulated by complex signaling

The hypothalamus regulates energy metabolism and food intake. Biologists have identified signaling molecules that form complex pathways involving the hypothalamus, digestive system, and adipose tissue. Several appetite regulators signal us to eat or to stop eating. One set of neurons in the hypothalamus produce the appetite-stimulating neurotransmitter **neuropeptide Y (NPY).** Another set of neurons in the hypothalamus release **melanocortins,** which suppress appetite, resulting in decreased food intake and weight loss. NPY and melanocortins also act on other parts of the brain that regulate food intake.

The gastrointestinal tract releases several peptides that signal satiety and suppress food intake. When the stomach is empty, it secretes the peptide hormone **ghrelin** (grell'-in), which stimulates appetite. Ghrelin activates neurons in the hypothalamus to produce NPY. **Peptide YY,** secreted by the small and large intestine, signals the hypothalamus to stop secreting NPY. The decrease in NPY slows gastric emptying, suppressing appetite. *Cholecystokinin* (CCK), released by the duodenum, signals satiety and suppresses appetite.

Fat cells in adipose tissue secrete hormones called *adipokines* that are critical in regulating energy metabolism. **Leptin** (from the Greek word *leptos,* "thin") is an adipokine that helps regulate normal body weight. Leptin signals the hypothalamus about the status of energy stores in adipose tissue. As more fat is stored, more leptin is released. Leptin both slows the release of NPY, which stimulates appetite, and stimulates the release of melanocortins, which decrease appetite.

Another important molecular player in the regulation of energy metabolism is **insulin.** This pancreatic hormone is secreted when the concentration of glucose and other nutrients increases after we eat. Insulin stimulates cells, especially liver, muscle, and fat cells, to take in glucose from the blood and use or store it. The actions of insulin inhibit the hypothalamus from secreting NPY. Thus, insulin inhibits further food intake.

The hypothalamus integrates input from these (and other) signaling molecules with other information, such as the concentration of nutrients circulating in the blood. The brain then sends signals that adjust feeding behavior and metabolism so that homeostasis of stored fat and fuel metabolism is maintained. When food is available and energy stores are adequate, energy use increases, fat stores are mobilized, and energy intake is inhibited. Glucose production in the liver is also inhibited. In contrast, when food intake is low, as during dieting or starvation, the hypothalamus increases secretion of neuropeptide Y. Appetite increases and metabolism slows, two actions that help restore energy homeostasis. The regulation of food intake and energy metabolism continues to be an important focus of research.

Obesity is a serious nutritional problem

Overnutrition causes **obesity,** the excess accumulation of body fat. Approximately 33% of U.S. adults are overweight, and another 33% of U.S. adults (more than 72 million) are obese. An estimated 17% of U.S. children and adolescents are overweight.

Obesity is a serious form of malnutrition that has become a problem of epidemic proportions in affluent societies. The World Health Organization considers obesity to be one of the most important global health problems, and the American Medical Association now considers obesity a disease. Obesity contributes to about 300,000 deaths annually in the United States and is the second leading preventable cause of death (second only to smoking). It is a major risk factor for heart disease, diabetes mellitus, certain types of cancer (including breast and colon cancers), hypertension, osteoarthritis, and liver and gallbladder diseases. In the well-known Framingham study of more than 2000 men, those who were 20% overweight had a significantly higher mortality rate from all causes than those who were not overweight.

Each of us appears to have a **set point,** or steady state, around which body weight is regulated. Dieting can be difficult because when weight decreases below an individual's set point, energy-conserving mechanisms are activated, causing energy expenditure to decrease.

Many factors affect weight In addition to the array of signaling molecules that affect energy metabolism and body weight, researchers have identified several gene mutations that affect food intake and energy homeostasis. For example, the normal *ob* gene product is leptin. Mice with a mutated allele at the *ob* locus become grossly obese. When researchers inject leptin into grossly obese mice, the appetites of the mice decrease and their energy use increases. They lose weight, mainly from loss of body fat (**FIG. 47-15**). Unfortunately, most obese humans are resistant to the actions of leptin.

The number of fat cells in the adult is apparently determined mainly by the amount of fat stored during infancy and childhood. When we are overfed early in life, abnormally large numbers of fat cells are produced. Later in life, these fat cells may be fully stocked with excess lipids or may shrink in size, but they are always there. People with such increased numbers of fat cells are thought to be at greater risk for obesity than are those with normal numbers.

Figure 47-15 *Animation* **Effect of leptin**

Mutant *ob* mice before (*left*) and after (*right*) treatment with leptin.

Eating activates the dopamine reward circuit in the brain. (Recall that this circuit is also activated in drug addiction.) The amount of dopamine released corresponds to the intensity of pleasure experienced by eating the food. High calorie foods rich in fats and sugars strongly stimulate the reward system. These foods also stimulate regions of the brain to produce endorphins. In some people, the pleasurable feelings apparently overpower leptin, insulin, and other hormones that normally interact to suppress appetite and inhibit overeating. As an individual gains weight, more insulin and leptin are released. However, the body apparently develops tolerance to these hormones. As with alcohol, cocaine, and other addictive drugs, the more food the addicted individual consumes, the harder it is to satisfy the craving. One hypothesis holds that some obese individuals may have an underactive dopamine reward system. They overeat to compensate and develop a food addiction in the process.

Although some factors involved in obesity are inherited, behavior is critically important. High-calorie diets, overeating, and underexercising lead to obesity, and psychosocial factors influence these behaviors. A sedentary lifestyle combined with the ready availability of high-energy foods contributes to the problem. One researcher cleverly described the roots of the obesity problem as the combination of "computer chips and potato chips." For every 9.3 kcal of excess food taken into the body, about 1 g of fat is stored. (An excess of about 140 kcal, less than a typical candy bar, per day for one month results in a 1-lb weight gain.)

Researchers have identified targets for obesity treatment

As researchers discover the genetic and biochemical mechanisms that regulate energy metabolism, they provide potential targets for the development of pharmacological treatments for obesity. The following are among the many targets for drug action that are the focus of research: (1) reduce appetite so that food intake is decreased, (2) block absorption of fat, (3) uncouple metabolism from energy production so that food is dissipated as heat (see *Inquiring About: Electron Transport and Heat* in Chapter 8), and (4) block receptors for molecules that send signals leading to increased energy storage.

Undernutrition can cause serious health problems

Millions of people do not have enough to eat or do not eat a balanced diet. Individuals suffering from *undernutrition* are malnourished. Essential amino acids, iron, calcium, iodine, zinc, and vitamin A are nutrients commonly deficient in the diet. Malnutrition results in poor health and lowered resistance to disease.

Worldwide, malnutrition is the leading cause of death among children. According to the World Health Organization, more than one-third of the world's children suffer from malnutrition. Inadequate amounts of protein and other essential nutrients severely affect physical growth and mental development. Because children suffering from malnutrition cannot manufacture antibodies (which are proteins) or cells needed to fight

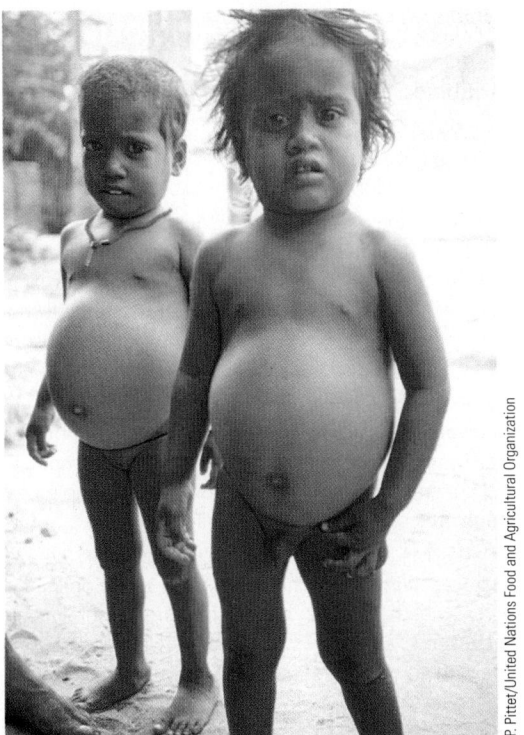

Figure 47-16 Protein deficiency
Millions of children suffer from kwashiorkor, a disease caused by severe protein deficiency. Note the characteristic swollen belly, which results from fluid imbalance.

infection, they often die from common childhood diseases, such as measles, whooping cough, or chickenpox.

In young children severe malnutrition results in the condition known as *kwashiorkor*. This Ashanti (West African) word means "first-second" and refers to the situation in which a first child is displaced from its mother's breast when a younger sibling is born. The older child is then given a diet of starchy cereal or cassava that is deficient in protein. Growth becomes stunted, muscles are wasted, and edema develops (as displayed by a swollen belly); the child is anemic and suffers from apathy and impaired metabolism (**FIG. 47-16**). Without essential amino acids, digestive enzymes cannot be manufactured, so what little protein is ingested cannot be digested. Recent studies suggest that certain gut bacteria may increase the risk for kwashiorkor among malnourished children. In the future, antibiotics may play a role in treatment and recovery.

Another form of severe malnutrition, *marasmus*, results from severe calorie deficiency (as well as protein deficiency). Marasmus causes stunted growth, a wasted body, and apathy. A child with marasmus appears extremely emaciated.

CHECKPOINT 47.4

- **VISUALIZE** *Write an equation to describe energy balance and explain what happens when the equation is altered in either direction.*
- *What is body mass index (BMI)?*
- **CONNECT** *How do melanocortins, neuropeptides Y (NPY), ghrelin, and leptin function to maintain energy balance?*

47.1 Nutritional Styles and Adaptations *(page 1005)*

1 Describe food processing and modes of nutrition, and compare the digestive systems of a cnidarian (such as *Hydra*), a flatworm, an earthworm, and a vertebrate.

- **Nutrition** is the process of taking in and using food. *Feeding* is the selection, acquisition, and **ingestion** of food, that is, taking food into the body. **Digestion** is the process of breaking down food mechanically and chemically. Nutrients pass through the lining of the digestive tract and into the blood by **absorption.** Food that is not digested and absorbed is discharged from the body by *egestion* or **elimination.**

- **Herbivores,** or **primary consumers,** are adapted to feed directly on producers. **Predators** are secondary and higher level consumers adapted for capturing and killing prey. **Omnivores** consume both plants and animals.

- In cnidarians and flatworms, food is digested in the **gastrovascular cavity**. This cavity has a single opening that serves as both mouth and anus. In more complex invertebrates and in all vertebrates, the digestive tract is a complete tube with an opening at each end. Digestion takes place as food passes through the tube. Various parts of the digestive tract are specialized to perform specific functions.

47.2 The Vertebrate Digestive System *(page 1007)*

2 Trace the pathway traveled by an ingested meal through the human digestive system and describe the structure and function of each structure involved.

- Mechanical and enzymatic digestion of carbohydrates begins in the mouth. Mammalian teeth include **incisors,** which are adapted for biting and cutting food; **canines** adapted for tearing food; and **premolars** and **molars** adapted for crushing and grinding. Three pairs of **salivary glands** secrete saliva, which contains the enzyme **salivary amylase** that digests starch.

- As food is swallowed, it is propelled through the **pharynx** and **esophagus.** A *bolus* of food is moved along through the digestive tract by **peristalsis,** waves of muscular contraction that push food along.

- In the **stomach** food is mechanically digested by vigorous churning, and proteins are enzymatically digested by the action of **pepsin** in the gastric juice. **Rugae** are folds in the stomach wall that expand as the stomach fills with food. **Gastric glands** secrete hydrochloric acid and **pepsinogen,** the precursor of pepsin.

- After several hours, a soup of partly digested food, called *chyme,* leaves the stomach through the *pylorus* and enters the **small intestine** in spurts. Most enzymatic digestion takes place in the **duodenum,** which produces several digestive enzymes and receives secretions from the liver and pancreas.

- The **liver** produces **bile,** which emulsifies fats. The **pancreas** releases enzymes that digest carbohydrate, lipid, and protein as well as RNA and DNA. *Pancreatic amylase* digests complex carbohydrates. *Pancreatic* **lipase** degrades fats, and **trypsin** and *chymotrypsin* digest polypeptides to small peptides.

- The **large intestine,** which consists of the **cecum, colon, rectum,** and **anus,** is responsible for eliminating undigested wastes. It also incubates bacteria that produce vitamin K and certain B vitamins.

3 Trace the step-by-step digestion of carbohydrate, protein, and lipid.

- Nutrients in chyme are enzymatically digested as they move through the digestive tract. Polysaccharides are digested to the disaccharide maltose by salivary and pancreatic amylases. *Maltase* in the small intestine splits maltose into glucose, the main product of carbohydrate digestion.

- Proteins are split by pepsin in the stomach and by proteolytic enzymes in the pancreatic juice. *Dipeptidases* split small peptides to amino acids.

- Lipids are emulsified by bile salts and then hydrolyzed by pancreatic lipase.

4 Describe the structural adaptations that increase the surface area of the digestive tract.

- The surface area of the small intestine is greatly expanded by folds in its wall; by the intestinal **villi,** elongated projections of the mucosa; and by **microvilli,** projections of the plasma membrane of the epithelial cells of the villi.

5 Compare lipid absorption with absorption of other nutrients.

- Nutrients are absorbed through the thin walls of intestinal villi. The *hepatic portal vein* transports amino acids and glucose to the liver.

- Fatty acids and monoacylglycerols enter epithelial cells in the intestinal lining, where they are reassembled into triacylglycerols. They are packaged into **chylomicrons,** droplets that also contain cholesterol and phospholipids and are covered with a protein coat. The lymphatic system transports chylomicrons to the blood circulation.

47.3 Required Nutrients *(page 1016)*

6 Summarize the nutritional requirements for dietary carbohydrates, lipids, and proteins, and trace the fate of glucose, lipids, and amino acids after their absorption.

- Most carbohydrates are ingested in the form of polysaccharides: starch and cellulose. Polysaccharides are referred to as *complex carbohydrates.* *Fiber* is mainly a mixture of cellulose and other indigestible carbohydrates. Carbohydrates are used mainly as an energy source. Glucose concentration in the blood is carefully regulated. Excess glucose is stored as glycogen and can also be converted to fat.

- Lipids are used to provide energy, to form components of cell membranes, and to synthesize steroid hormones and other lipid substances. Most lipids are ingested in the form of triacylglycerols. Fatty acids are converted to molecules of acetyl coenzyme A, which enter the citric acid cycle. Excess fatty acids are converted to triacylglycerol and stored as fat.

- Lipids are transported as large molecular complexes called **lipoproteins. Low-density lipoproteins (LDLs)** deliver cholesterol to the cells. **High-density lipoproteins (HDLs)** collect excess cholesterol and transport it to the liver.

- Proteins serve as enzymes and are essential structural components of cells. The best distribution of **essential amino acids** is found in the complete proteins of animal foods. Excess amino acids are deaminated by liver cells. Amino groups are converted to urea and excreted in urine;

the remaining keto acids are converted to carbohydrate and used as fuel or are converted to lipid and stored in fat cells.

7 Describe the nutritional functions of vitamins, minerals, and phytochemicals.

- **Vitamins** are organic compounds required in small amounts for many biochemical processes. Many serve as components of coenzymes. *Fat-soluble vitamins* include vitamins A, D, E, and K. *Water-soluble vitamins* are the B and C vitamins.
- **Minerals** are inorganic nutrients ingested as salts dissolved in food and water. **Trace elements** are minerals required in amounts less than 100 mg per day.
- *Phytochemicals* are plant compounds that promote health. Many phytochemicals are **antioxidants** that destroy **oxidants,** reactive molecules such as **free radicals.** These molecules damage DNA, proteins, and unsaturated fatty acids by snatching electrons.

47.4 Energy Metabolism *(page 1021)*

8 Contrast basal metabolic rate with total metabolic rate; write the basic energy equation for maintaining body weight and describe the consequences of altering it in either direction.

- **Basal metabolic rate (BMR)** is the body's cost of metabolic living. *Total metabolic rate* is the BMR plus the energy used to carry on daily activities.
- When energy (kilocalories) input equals energy output, body weight remains constant. When energy output is greater than energy input, body weight decreases. When energy input exceeds energy output, body weight increases.

9 Identify components of the regulatory pathways that regulate food intake and energy homeostasis.

- The hypothalamus, digestive system, and adipose tissue release signaling molecules that help regulate food intake and energy metabolism.
- The hypothalamus releases **neuropeptide Y (NPY),** which stimulates appetite, and **melanocortins,** which suppress appetite.
- Adipose tissue secretes the hormone **leptin** that signals the brain about the status of energy stores in adipose tissue. Leptin and the pancreatic hormone **insulin** inhibit food intake.
- When empty, the stomach secretes the hormone **ghrelin,** which stimulates appetite. The intestine secretes **peptide YY** and cholecystokinin (CCK), which decrease appetite.

10 In general terms, describe the effects of malnutrition, including both overnutrition and undernutrition.

- In **obesity** an excess amount of fat accumulates in adipose tissues. Obesity is a major risk factor for several major disorders, including heart disease and diabetes mellitus. A person gains weight by taking in more energy, in the form of kilocalories, than is expended in activity.
- Undernutrition can be caused by not eating enough food or by not eating a balanced diet. Nutrients commonly deficient in the diet include essential amino acids, iron, calcium, iodine, zinc, and vitamin A. Malnourishment prevents normal physical and mental development. In severe malnutrition death often results from common diseases as well as the conditions *kwashiorkor* and *marasmus.*

TEST YOUR UNDERSTANDING

Know and Comprehend

1. Teeth adapted for crushing and grinding are (a) incisors (b) canines (c) premolars and molars (d) incisors and premolars (e) canines and molars
2. The layer of tissue that lines the lumen of the digestive tract is the (a) muscle layer (b) visceral peritoneum (c) parietal peritoneum (d) mucosa (e) submucosa
3. Which of the following are accessory digestive glands? (a) salivary glands (b) pancreas (c) liver (d) a, b, and c (e) a and b
4. Arrange the following into the correct sequence.
   1. stomach 2. esophagus 3. pharynx 4. small intestine 5. colon
   (a) 2, 3, 1, 4, 5 (b) 3, 2, 1, 5, 4 (c) 3, 2, 5, 1, 4 (d) 2, 3, 1, 5, 4 (e) 3, 2, 1, 4, 5
5. Which sequence most accurately describes the digestion of protein?
   1. small peptides 2. amino acid 3. protein 4. polypeptide 5. glycerol
   (a) 4, 3, 1, 2 (b) 3, 4, 1, 2 (c) 3, 4, 1, 5 (d) 4, 3, 5, 1, 2 (e) 2, 1, 4, 5
6. The surface area of the small intestine is increased by (a) folds in its wall (b) villi (c) microvilli (d) a, b, and c (e) a and b
7. Lipids are transported from the intestine to the liver by (a) chylomicrons (b) HDLs (c) LDLs (d) glycerol transporters (e) leptin
8. Most vitamins are (a) inorganic compounds (b) components of coenzymes (c) used as fuel (d) electrolytes (e) required by herbivores and carnivores but not omnivores

9. When energy input is greater than energy output, (a) weight loss occurs (b) weight remains stable (c) weight gain occurs (d) leptin secretion is inhibited (e) the *ob* gene is deactivated
10. A hormone that stimulates gastric glands to secrete pepsinogen is (a) secretin (b) GIP (c) gastrin (d) cholecystokinin (CCK) (e) Peptide YY

Apply and Analyze

11. **VISUALIZE** Sketch a small area of the lining of the small intestine. Label intestinal glands, lacteal, capillary network of villus, and epithelial cells of villus.
12. Design an experiment to test the hypothesis that the B vitamin pyridoxine is an essential nutrient in mice.
13. Proteolytic enzymes are produced in an inactive form. What, if any, benefit might that have?

Evaluate and Synthesize

14. **EVOLUTION LINK** If you were presented with an unfamiliar animal and asked to determine its nutritional lifestyle, how would you do so? (*Hint:* Consider adaptations.)
15. **SCIENCE, TECHNOLOGY, AND SOCIETY** Investigators are unraveling the biochemical pathways that help regulate body weight. How might their work lead to a cure for obesity? How might a cure for obesity affect our society?

To access course materials, such as Aplia and other
aplia companion resources, please visit **www.cengagebrain.com.**

48 Osmoregulation and Disposal of Metabolic Wastes

Most terrestrial animals inhabit areas near water sources. Zebra, wildebeest, and a variety of birds drink at a lake in Ngorongoro Crater in Tanzania.

McMurray Photography

KEY CONCEPTS

48.1 Osmoregulation is the process by which animals regulate the concentration of water and salt in the body, allowing them to maintain their body fluids within homeostatic limits.

48.2 During metabolism, most animals produce water, carbon dioxide, and nitrogenous wastes, including ammonia, uric acid, and urea.

48.3 Many invertebrates have nephridial organs that carry on osmoregulation and excretion; insects and spiders have Malpighian tubules that excrete metabolic wastes and conserve water.

48.4 Freshwater, marine, and terrestrial animals have different adaptations for osmoregulation that meet the challenges of these diverse environments.

48.5 The vertebrate kidney maintains water and electrolyte balance and excretes metabolic wastes; the nephron is the functional unit of the vertebrate kidney. Several hormones interact to regulate kidney function.

W ater, the most abundant molecule both in the cell and on Earth's surface, shapes life and is a critical factor in the distribution of organisms on our planet. Like the animals in the photograph, most animals need a dependable source of water with which to replenish their body fluids. For this reason, animals often inhabit areas near water sources.

Ions are dissolved in water, the medium in which most metabolic reactions take place. The volume and ionic composition of an animal's body fluids must be maintained within homeostatic limits. Through natural selection a variety of homeostatic mechanisms have evolved that regulate the volume and composition of fluids in the internal environment.

Many small animals live in the ocean and get their food and oxygen directly from the sea water that surrounds them; they release waste products into the sea water. In larger animals and in most terrestrial animals, an *internal* sea of fluid bathes the cells. Terrestrial animals have a continuous need to conserve water. Water loss from the body must be carefully regulated, and water lost must be replaced. Water is taken in with food and drink, and it is also produced in metabolic reactions.

Two processes that help maintain fluid and electrolyte (salt) homeostasis in animals are *osmoregulation* and *excretion*, the disposal of metabolic wastes. In this chapter we discuss these processes in both invertebrates and vertebrates inhabiting various environments. Then we focus on the structure and functions of the mammalian kidney. We end the chapter with a description of the hormones that regulate kidney function.

1026

48.1 MAINTAINING FLUID AND ELECTROLYTE BALANCE

LEARNING OBJECTIVE

1   Describe how the processes of *osmoregulation* and *excretion* contribute to fluid and electrolyte homeostasis.

Intracellular fluid, the fluid within cells, accounts for most of the body fluid. *Extracellular fluid,* the fluid outside the cells, includes interstitial fluid (the fluid between cells), lymph, and blood plasma (or hemolymph). In vertebrates blood plasma, which is mainly water, transports nutrients, gases, waste products, and other materials throughout the body. **Interstitial fluid,** also called *tissue fluid,* forms from the blood plasma and bathes all the cells. Excess water evaporates from the body surface or is excreted by specialized structures.

Electrolytes are compounds such as inorganic salts, acids, and bases that form ions in solution. Electrolytes are very important solutes in body fluids. Many homeostatic mechanisms regulate fluid and electrolyte balance.

Recall from Chapter 5 that *osmosis* is the diffusion of water through a selectively permeable membrane. The net flow of water is from a more dilute solution (one with a lower solute concentration) to a less dilute solution (one with a higher solute concentration). If the solute concentrations of two solutions are equal, they are *isotonic* to each other. If solution A has a greater solute concentration than solution B, solution A is *hypertonic* to solution B. Solution B, with a lower solute concentration relative to solution A, is *hypotonic* to solution A. Remember that the total solute concentration of a solution is responsible for the direction of water movement. The *osmotic pressure* of a solution is the pressure that must be exerted on the hypertonic side of a selectively permeable membrane to prevent net movement of water from the hypotonic side.

An **osmole** is related to a mole (see Chapter 2), but takes into account the number of particles produced when a solute dissolves. For example, glucose dissolves to give only one kind of particle, so a mole of glucose in solution is one osmole. A mole of NaCl in solution produces two kinds of particles (Na^+ and Cl^-), so a mole of NaCl is two osmoles. Both types of ions affect the osmotic pressure of the solution. One *milliosmole* is 1/1000 of an osmole.

Osmolarity is a measure of the number of osmoles of solute per liter of solution. Solutions with the same osmolarity are described as *iso-osmotic*. If solution X has a higher osmolarity than solution Y, it is described as *hyperosmotic*. Solution Y would be described as *hypo-osmotic* relative to solution X.

Osmoregulation is the process by which animals regulate the concentration of water and salt in the body, allowing them to maintain their body fluids within homeostatic limits. **Excretion** is the process of discharging metabolic wastes from the body. **Excretory systems** have evolved that function in both osmoregulation and in disposal of metabolic wastes. Excretory systems rid the body of excess water and ions, metabolic wastes, and harmful substances. As we will discuss, hormones are important signaling molecules in these regulatory processes.

CHECKPOINT 48.1

- **CONNECT** *How does osmoregulation contribute to homeostasis?*
- *What are the functions of excretory systems?*

48.2 METABOLIC WASTE PRODUCTS

LEARNING OBJECTIVE

2   Contrast the benefits and costs of excreting each of the following: ammonia, uric acid, or urea.

Metabolic wastes must be excreted so that they do not accumulate and reach concentrations that would disrupt homeostasis. The principal metabolic waste products produced by most animals are water, carbon dioxide, and *nitrogenous wastes,* those that contain nitrogen. Carbon dioxide is excreted mainly by respiratory structures (see Chapter 46). In terrestrial animals some water is also lost from respiratory surfaces. However, specialized excretory organs, such as kidneys, remove and excrete most of the water and nitrogenous wastes.

Nitrogenous wastes include ammonia, uric acid, and urea. Recall that amino acids and nucleic acids contain nitrogen. During the metabolism of amino acids, the nitrogen-containing amino group is removed (in a process known as deamination) and converted to *ammonia* (FIG. 48-1). However, ammonia is highly toxic. Some aquatic animals excrete ammonia into the surrounding water before it can build up to toxic concentrations in their tissues. A few terrestrial animals, including some snails and wood lice, vent it directly into the air. But many animals, humans included, convert ammonia to some less toxic nitrogenous waste such as uric acid or urea.

Uric acid is produced both from ammonia and by the breakdown of nucleotides from nucleic acids. Uric acid is insoluble in water and forms crystals that are excreted as a crystalline paste, so little fluid loss results. This excretion is an important water-conserving adaptation in many terrestrial animals, including insects, certain reptiles, and birds. Also, because uric acid is not toxic and can be safely stored, its excretion is an adaptive advantage for species whose young begin their development enclosed in eggs.

Urea, the principal nitrogenous waste product of amphibians and mammals, is synthesized in the liver from ammonia and carbon dioxide by a sequence of reactions known as the urea cycle. As in the production of uric acid, these reactions require specific enzymes and the input of energy by the cells. Compared with the energy cost of producing ammonia, the cost of producing urea and uric acid is high.

Urea has the advantage of being far less toxic than ammonia and can accumulate in higher concentrations without causing tissue damage; thus, it can be excreted in more concentrated form. Because urea is highly soluble, however, it is dissolved in water, and more water is needed to excrete urea than to excrete uric acid.

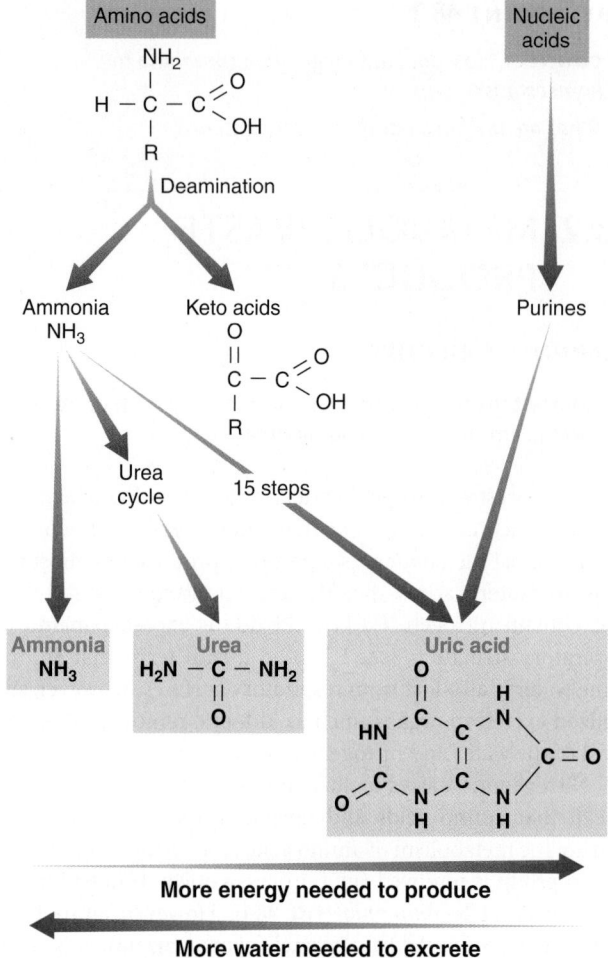

Figure 48-1 Formation of nitrogenous wastes

Deamination of amino acids and metabolism of nucleic acids produce nitrogenous wastes. Ammonia is the first metabolic product of deamination. Many aquatic animals excrete ammonia, but terrestrial animals convert it to urea and uric acid, which are less toxic than ammonia. Most amphibians and mammals convert ammonia to urea via the urea cycle. Insects, many reptiles, and birds convert ammonia to uric acid. Energy is required to convert ammonia to urea and uric acid, but less water is required to excrete these wastes.

© Cengage Learning

CHECKPOINT 48.2

- **CONNECT** *Give examples of animals that excrete each of the principal types of nitrogenous wastes.*

- *What are the benefits of excreting nitrogenous wastes in the form of uric acid? in the form of urea?*

48.3 OSMOREGULATION AND EXCRETION IN INVERTEBRATES

LEARNING OBJECTIVES

3   Contrast osmoconformers and osmoregulators.

4   Compare the structure and function of protonephridia, metanephridia, and Malpighian tubules.

The ocean is a stable environment, and its salt concentration does not vary much. The electrolyte concentration in the cells and body fluids of marine sponges and cnidarians is very similar to sea water, so these animals do not require specialized excretory structures. They expend little or no energy in excreting wastes because wastes simply diffuse from their cells to the external environment and are washed away by water currents. When water stagnates and currents do not wash away wastes, aquatic environments, such as coral reefs, are damaged by the waste accumulation.

The body fluids of most marine invertebrates are in osmotic equilibrium with the surrounding sea water. These animals are known as **osmoconformers** because the concentration of their body fluids varies along with changes in the sea water. However, many osmoconforming marine animals still regulate some ions in their body fluids.

Coastal habitats, such as estuaries that contain brackish water, are much less stable environments than is the open ocean. Salt concentrations change frequently with shifting tides. Many invertebrates and vertebrates that inhabit coastal environments are **osmoregulators.** These animals have homeostatic mechanisms that maintain an optimal salt concentration in their tissues regardless of changes in the salt concentration of their surroundings.

In a coastal environment where fresh water enters the ocean, the water may have a lower salt concentration than the body fluids of an animal. Water osmotically moves into the animal's body, and salt diffuses out. An animal adapted to this environment has excretory structures that remove the excess water. Certain polychaete worms and the blue crab are among the animals that can be either osmoconformers or osmoregulators, depending on environmental conditions.

Dehydration is a constant threat to terrestrial animals. Because their fluid concentration is higher than that of the air around them, they tend to lose water by evaporation from both the body surface and respiratory surfaces. Terrestrial animals may also lose water as wastes are excreted. As animals moved onto the land, natural selection favored the evolution of structures and processes that conserve water.

Excretory systems help maintain fluid and electrolyte homeostasis by selectively adjusting the concentrations of salts and other substances in blood and other body fluids. Typically, an excretory system collects fluid, generally from the blood or interstitial fluid. It then adjusts the composition of this fluid by selectively returning needed substances to the body fluid. Finally, the body releases the adjusted excretory product containing excess or potentially toxic substances.

Nephridial organs are specialized for osmoregulation and/or excretion

Nephridial organs, or nephridia, are structures that function in osmoregulation and excretion in many invertebrates, including flatworms, nemerteans, rotifers, annelids, mollusks, and lancelets. Each nephridial organ consists of simple or branching tubes that typically open to the outside of the body through pores, called *nephridiopores.* Two types of nephridial organs are protonephridia and metanephridia.

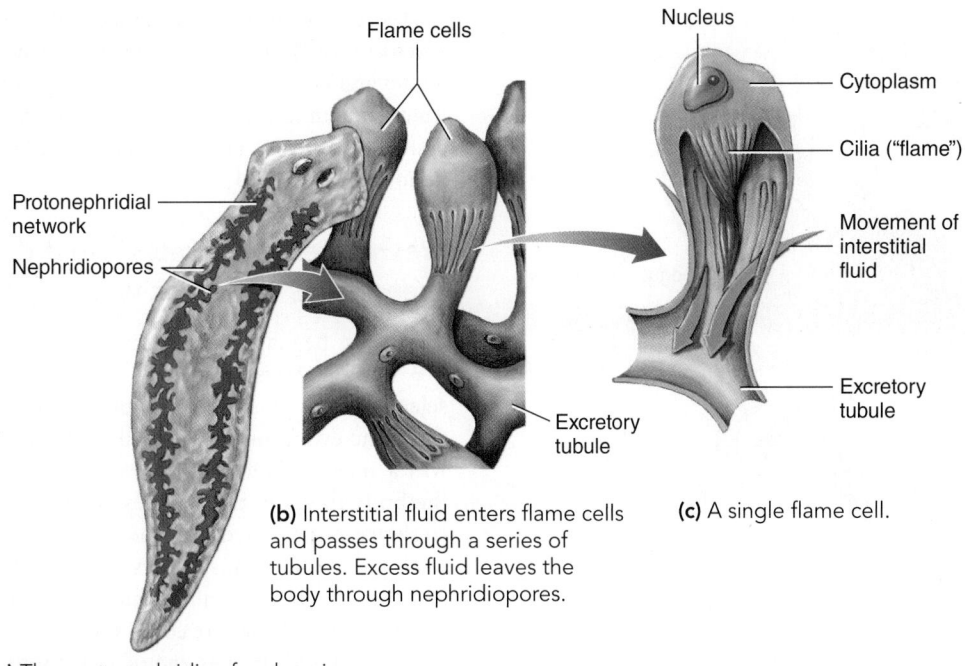

(b) Interstitial fluid enters flame cells and passes through a series of tubules. Excess fluid leaves the body through nephridiopores.

(c) A single flame cell.

(a) The protonephridia of a planarian function mainly in osmoregulation. These organs form a system of branching tubules.

Figure 48-2 Protonephridia of a flatworm

© Cengage Learning

In flatworms and nemerteans, metabolic wastes pass through the body surface by diffusion, but these animals also have **protonephridia** (FIG. 48-2). The protonephridia are composed of tubules with no internal openings. Their enlarged blind ends consist of **flame cells** with brushes of cilia, so named because their constant motion reminded early biologists of flickering flames. The flame cells lie in the interstitial fluid that bathes the body cells. Fluid enters the flame cells, and the beating of the cilia propels the fluid through the tubules. Excess fluid leaves the body through nephridiopores.

Most annelids and mollusks have more complex nephridial organs called **metanephridia** (FIG. 48-3). Each segment of an earthworm has a pair of metanephridia. A metanephridium is a tubule open at both ends. The inner end opens into the coelom as a ciliated funnel, and the outer end opens to the outside through a nephridiopore. A network of capillaries surrounds each tubule. Fluid from the coelom passes into the tubule, bringing with it whatever it contains: glucose, salts, or wastes.

As fluid moves through the tubule, needed materials (e.g., water and glucose) are removed from the fluid by the tubule and are reabsorbed by the capillaries, leaving the wastes behind. In this way, urine that contains concentrated wastes is produced.

Malpighian tubules conserve water

The excretory system of insects and spiders consists of several hundred **Malpighian tubules** (FIG. 48-4). Malpighian tubules are slender extensions of the gut wall. Their blind ends lie in the hemocoel (blood cavity) and are bathed in hemolymph. Cells of the tubule wall actively transport uric acid, potassium ions,

and some other substances from the hemolymph into the tubule lumen. Other solutes and water follow by diffusion. The Malpighian tubules empty into the gut. Water, some salts, and other

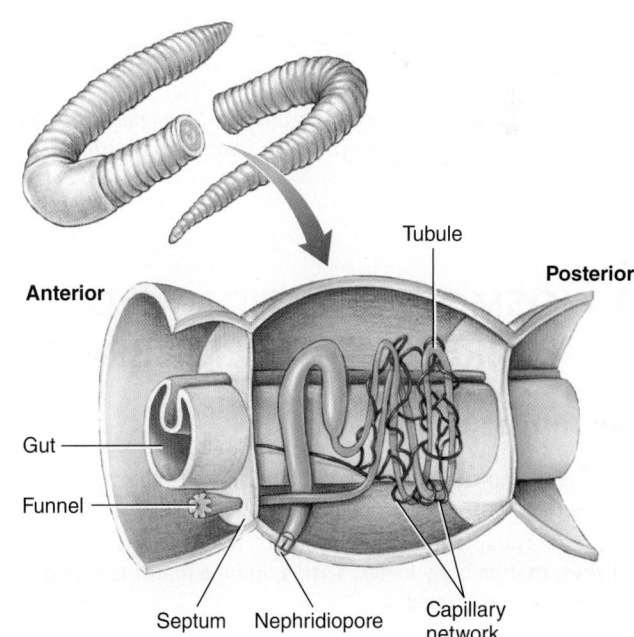

Figure 48-3 Metanephridium of an earthworm

Each metanephridium consists of a ciliated funnel opening into the coelom, a coiled tubule, and a nephridiopore opening to the outside. This 3-D internal view shows parts of three segments of the earthworm body.

© Cengage Learning

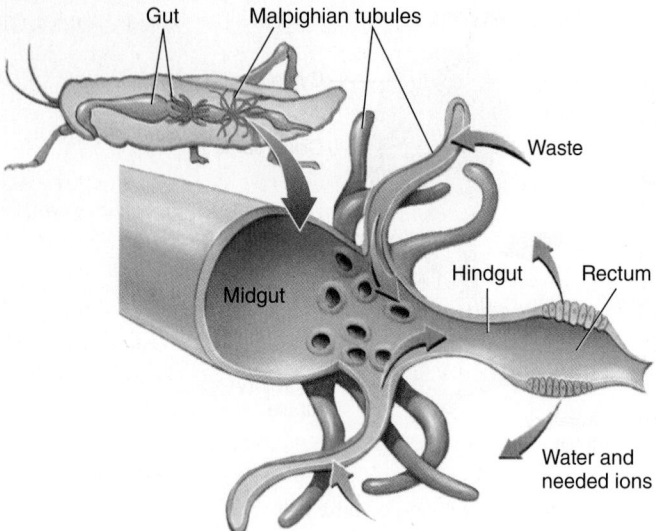

Gut Malpighian tubules

Waste

Midgut

Hindgut Rectum

Water and
needed ions

Figure 48-4 Malpighian tubules of an insect

The slender Malpighian tubules have blind ends that extend into the hemocoel. Their cells transfer uric acid and some ions from the hemolymph into the tubule. Water follows by diffusion. The wastes are discharged into the gut. The epithelium that lines the rectum (part of the hindgut) actively reabsorbs most of the water and needed ions.

© Cengage Learning

solutes are reabsorbed into the hemolymph by a specialized epithelium in the rectum.

Uric acid, the major waste product, is excreted as a semidry paste with a minimum of water loss. Because Malpighian tubules effectively conserve body fluids, they have contributed to the success of insects in terrestrial environments.

CHECKPOINT 48.3

- **CONNECT** *How are nephridial organs and Malpighian tubules alike? How are they different? (Hint: Consider both structure and function.)*

48.4 OSMOREGULATION AND EXCRETION IN VERTEBRATES

LEARNING OBJECTIVES

5 Relate the function of the vertebrate kidney to the success of vertebrates in a wide variety of habitats.

6 Compare adaptations for osmoregulation in freshwater fishes, marine bony fishes, sharks, marine mammals, and terrestrial vertebrates.

Vertebrates live successfully in a wide range of habitats: in fresh water; in the ocean and tidal regions; and on land, even in extreme environments such as deserts. In response to the requirements of these diverse environments, adaptations have evolved for regulating salt and water content and for excreting

wastes. The main osmoregulatory and excretory organ in most vertebrates is the kidney. The **kidney** excretes most nitrogenous wastes and helps maintain fluid balance by adjusting the salt and water content of the urine. The skin, lungs or gills, and digestive system also help maintain fluid balance and dispose of metabolic wastes.

Freshwater vertebrates must rid themselves of excess water

As fishes began to move into freshwater habitats about 470 million years ago (mya), there must have been strong selection for adaptations that promoted effective osmoregulation. The evolution of body fluids more dilute than sea water was a major adaptation. However, the salt concentration of the body fluids of these fishes is still higher than that of the fresh water that surrounds them. They are hypertonic to their watery environment. As a result, water moves into the body, and they are in constant danger of becoming waterlogged.

Freshwater fishes are covered by scales and a mucous secretion that retard the passage of water into the body. However, water constantly enters through the gills and through the mouth (with food). How do these animals meet this challenge? The kidneys of freshwater fishes are adapted to excrete large amounts of very dilute urine (**FIG. 48-5a**). The kidneys have large *glomeruli,* capillary clusters that filter the blood in the first step in producing urine.

Water entry is only part of the challenge of osmoregulation in freshwater fishes. These animals also tend to lose salts by diffusion through the gills into the surrounding water. To compensate, special gill cells have evolved that actively transport salts (mainly sodium chloride) from the water into the body.

The gills of freshwater fishes excrete ammonia, their principal nitrogenous wastes. About 10% of nitrogenous wastes are excreted as urea.

Most amphibians are at least semi-aquatic, and their mechanisms of osmoregulation are similar to those of freshwater fishes. They, too, produce large amounts of dilute urine. In its urine and through its skin, a frog can lose an amount of water equivalent to one-third of its body weight in one day. Active transport of salt into the body by special cells in the skin compensates for loss of salt through skin and in urine. Adult amphibians generally inhabit moist environments. They excrete urea and reabsorb some water from the urinary bladder. When dehydrated, terrestrial frogs can dramatically decrease the process of urine production.

Marine vertebrates must replace lost fluid

Freshwater fishes adapted very successfully to their aquatic habitats. Thus, when some freshwater fishes returned to the sea about 200 mya, their blood and body fluids were less salty than their surroundings; that is, they were hypo-osmotic to their environment. These fishes lose water osmotically and take in salt through their gills and with their food. To compensate for fluid loss, many marine bony fishes drink sea water (**FIG. 48-5b**). They retain the water and excrete salt by the action of specialized

Adaptations for regulating water and salt concentration allow fishes to inhabit diverse environments, from freshwater lakes to salty oceans.

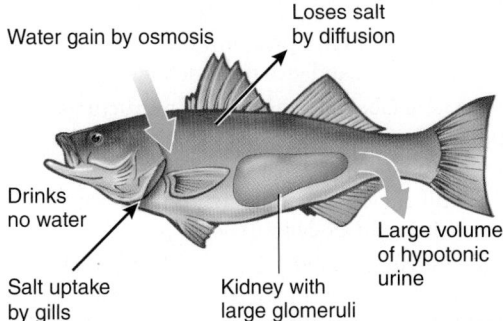

(a) Freshwater fishes live in a hypotonic medium. Water continuously enters the body, and salts diffuse out. These fishes excrete large quantities of dilute urine and actively transport salts in through the gills.

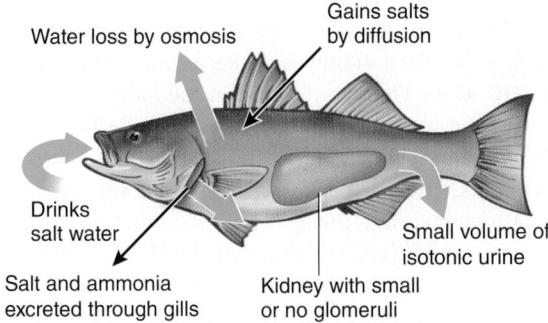

(b) Marine fishes live in a hypertonic medium. They lose water by osmosis. They gain salts from the sea water they drink and by diffusion. To compensate, the fish drinks water, excretes the salt, and produces a small volume of urine.

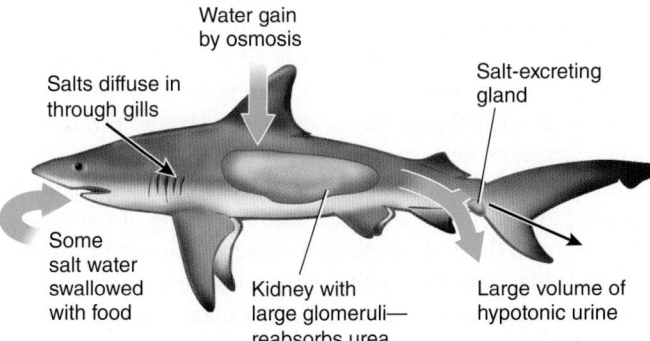

(c) Cartilaginous fishes (sharks and rays). The shark kidney reabsorbs urea in high enough concentration that its tissues become hypertonic to the surrounding medium. As a result, water enters the shark by osmosis. The shark excretes a large quantity of dilute urine.

Figure 48-5 Osmoregulation in fishes

PREDICT A shark is spotted swimming in fresh water. How might this shark's kidney function be affected?

© Cengage Learning

cells in their gills. The gills are also responsible for ammonia excretion. Very little urine is excreted by the kidneys, which have only small (or no) glomeruli. The kidneys excrete certain excess ions (e.g., Mg^{2+}).

Marine cartilaginous fishes (sharks and rays) have different osmoregulatory adaptations that allow them to tolerate the salt concentrations of their environment. These animals accumulate and tolerate urea (**FIG. 48-5c**). Their tissues are adapted to function at concentrations of urea that would be toxic to most other animals. The high urea concentration makes the body fluids slightly hypertonic to sea water, resulting in a net inflow of water. Marine cartilaginous fishes have well-developed kidneys that excrete a large volume of urine. Excess salt is excreted by the kidneys and, in many species, by a rectal gland.

Certain nonavian reptiles and marine birds ingest sea water and take in a lot of salt in their food. Glands in their heads excrete salt. These salt glands are usually inactive; they function in response to osmotic stress. Nonavian reptiles and birds excrete uric acid, which conserves water.

Whales, dolphins, and other marine mammals ingest sea water along with their food. Their kidneys produce concentrated urine, much saltier than sea water. This physiological adaptation is important, especially for marine carnivores. The high-protein diet of these animals results in the production of large amounts of urea, which must be excreted in urine without losing much water.

Terrestrial vertebrates must conserve water

Amniotes (nonavian reptiles, birds, and mammals) have effective adaptations for life on land. Their skin minimizes water loss by evaporation, and many amniotes excrete uric acid, which requires very little water.

Because birds and mammals are endothermic (maintain a constant body temperature) and have a high rate of metabolism, they produce a relatively large volume of nitrogenous wastes. They have numerous adaptations, including very efficient kidneys, for conserving water. An extreme example is the desert-dwelling kangaroo rat, which obtains most of its water from its own metabolism. (Recall from Chapter 8 that aerobic respiration produces water. The kangaroo rat's kidneys are so efficient that it loses little fluid as urine.)

Birds conserve water by excreting nitrogen as uric acid and by efficiently reabsorbing water from the cloaca and intestine. Mammals excrete urea. Their kidneys produce very concentrated (hypertonic) urine.

In terrestrial vertebrates the lungs, skin, and digestive system are important in osmoregulation and waste disposal (**FIG. 48-6**). Most carbon dioxide is excreted by the lungs. In birds and mammals, some water is lost from the body as water vapor in exhaled air. Although primarily concerned with the regulation of body temperature, the sweat glands of humans and some other mammals excrete 5% to 10% of all metabolic wastes.

The liver produces both urea and uric acid, which are transported by the blood to the kidneys. Most of the bile pigments produced by the breakdown of red blood cells are excreted by the liver into the intestine. From the intestine they pass out of the body with the feces.

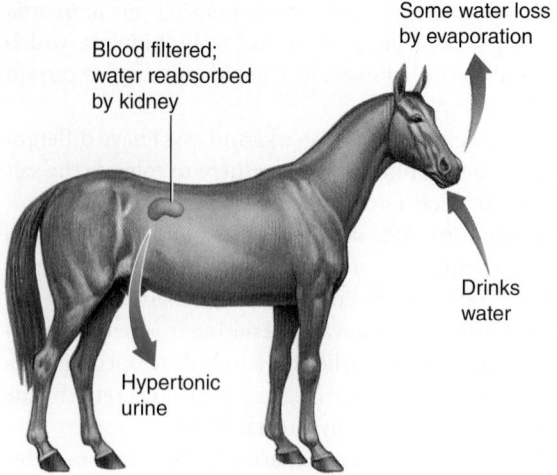

(a) **The kidney of terrestrial vertebrates conserves water by reabsorbing it.** Birds and mammals can produce concentrated urine.

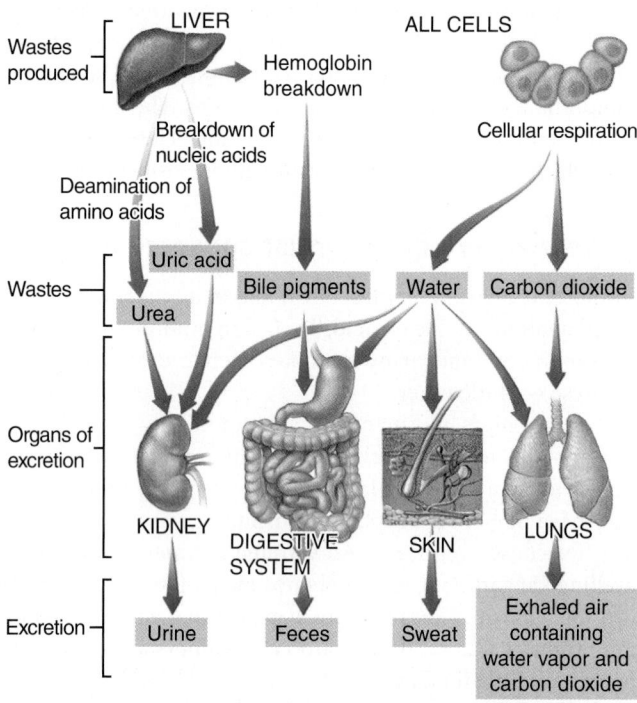

(b) **Disposal of metabolic wastes in humans and other terrestrial mammals.** Nitrogenous wastes are produced by the liver and transported to the kidneys. To conserve water, mammals produce a small amount of hypertonic urine. All cells produce carbon dioxide and some water during cellular respiration. These waste products are excreted by the lungs.

Figure 48-6 Excretory organs in terrestrial vertebrates
© Cengage Learning

CHECKPOINT 48.4

• **CONNECT** *What type of osmoregulatory challenge is faced by marine fishes? by freshwater fishes? What mechanisms have evolved to meet these challenges?*

• **CONNECT** *What adaptations in aquatic birds and mammals help meet the challenges of osmoregulation and metabolic waste disposal? in terrestrial birds and mammals?*

48.5 THE URINARY SYSTEM OF MAMMALS

LEARNING OBJECTIVES

7 Describe (and label on a diagram) the organs of the mammalian urinary system and give the functions of each.

8 Describe (and label on a diagram) the structures of a nephron (including associated blood vessels) and give the functions of each structure.

9 Trace a drop of glomerular filtrate from Bowman's capsule to its release from the body as urine.

10 Describe the hormonal regulation of fluid and electrolyte balance by antidiuretic hormone (ADH), the renin–angiotensin–aldosterone system, and atrial natriuretic peptide (ANP).

The mammalian **urinary system** consists of the kidneys, the urinary bladder, and associated ducts. The overall structure of the human urinary system is shown in **FIGURE 48-7.** Located just below the diaphragm in the "small of the back," the kidneys look like a pair of giant, dark red lima beans, each about the size of a fist. Each kidney is covered by a connective tissue capsule (**FIG. 48-8**). The outer region of the kidney is the **renal cortex,** and the inner region is the **renal medulla.** The renal medulla contains eight to ten cone-shaped structures called *renal pyramids.* The tip of each pyramid is a *renal papilla.* Each papilla has several pores, the openings of **collecting ducts.**

As urine is produced, it flows from collecting ducts through a renal papilla and into the **renal pelvis,** a funnel-shaped chamber. Urine then flows into one of the paired **ureters,** ducts that connect each kidney with the **urinary bladder.** The urinary bladder is a remarkable organ capable of holding (with practice) up to 800 mL (about a pint and a half) of urine. Emptying the bladder changes it from the size of a small melon to that of a pecan. This feat is made possible by the smooth muscle and specialized epithelium of the bladder wall, which is capable of great shrinkage and stretching.

During *urination,* urine is released from the bladder and flows through the **urethra,** a duct leading to the outside of the body. In the male the urethra is lengthy and passes through the penis. Semen as well as urine passes through the male urethra. In the female the urethra is short and transports only urine. Its opening to the outside is just above the opening of the vagina. The length of the male urethra discourages bacterial invasions of the bladder. This length difference helps explain why bladder infections are more common in females than in males. In summary, urine flows through the following structures:

kidney (through renal pelvis) ⟶ ureter ⟶ urinary bladder ⟶ urethra

The nephron is the functional unit of the kidney

The principal function of the kidneys is to help maintain homeostasis by regulating fluid balance and excreting metabolic wastes. However, the kidneys also have other functions. They

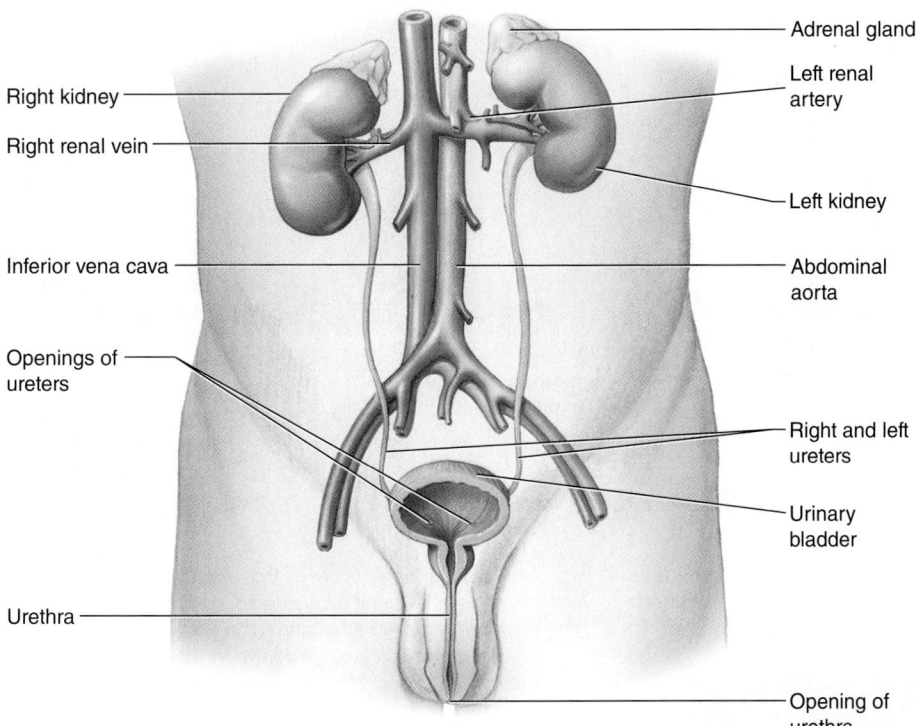

Right kidney

Right renal vein

Inferior vena cava

Openings of
ureters

Urethra

Adrenal gland

Left renal
artery

Left kidney

Abdominal
aorta

Right and left
ureters

Urinary
bladder

Opening of
urethra

Figure 48-7 The human urinary system

The kidneys produce urine, which passes through the ureters to the urinary bladder for temporary storage. During urination, the urethra conducts urine from the bladder to the outside of the body through an opening, the external urethral orifice.

© Cengage Learning

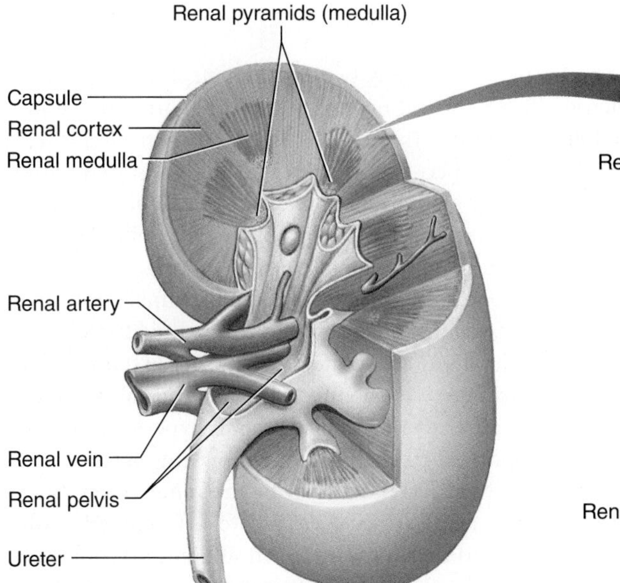

Renal pyramids (medulla)

Capsule
Renal cortex
Renal medulla

Renal artery

Renal vein

Renal pelvis

Ureter

(a) Internal structure of the kidney. The outer region of the kidney is the cortex, and the inner region is the medulla. As urine is produced, it flows into the renal pelvis and leaves the kidney through the ureter.

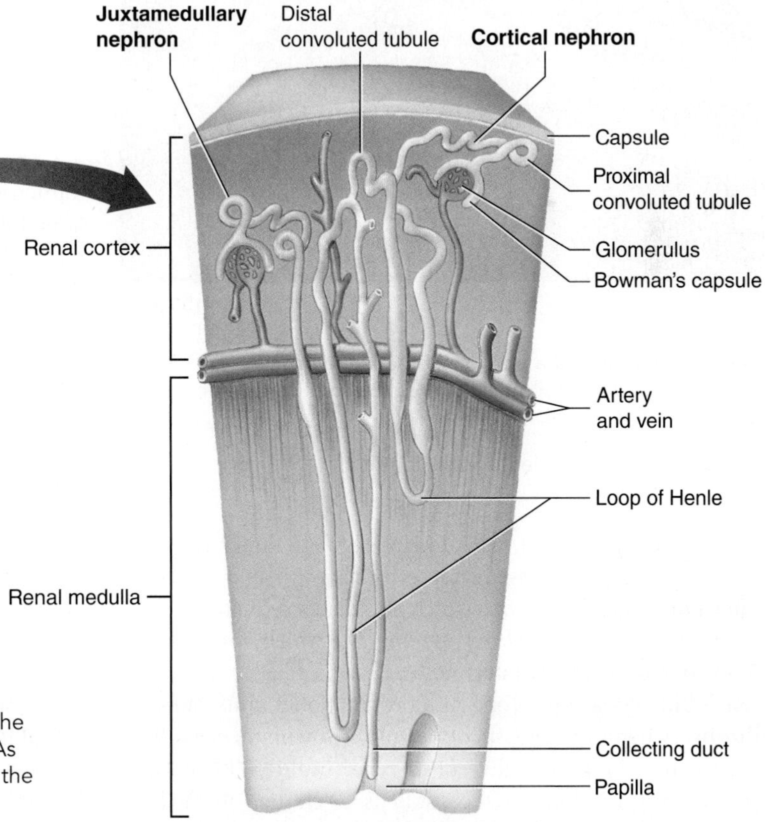

Juxtamedullary nephron   Distal convoluted tubule   **Cortical nephron**

Renal cortex

Renal medulla

Capsule

Proximal convoluted tubule

Glomerulus

Bowman's capsule

Artery and vein

Loop of Henle

Collecting duct

Papilla

(b) Juxtamedullary and cortical nephrons. Longitudinal section showing the location and structure of the two main types of nephrons.

Figure 48-8 Animation Structure of the kidney

© Cengage Learning

Osmoregulation and Disposal of Metabolic Wastes   **1033**

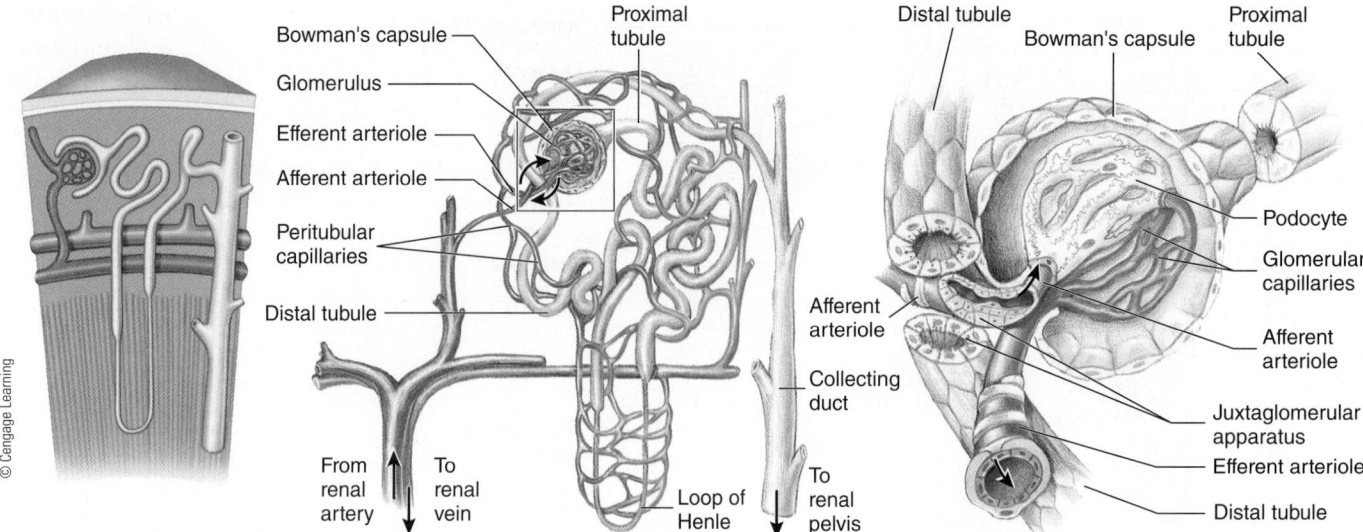

(a) Location and the basic structure of a nephron. Urine forms by filtration of the blood in the glomerulus and by adjustment of the glomerular filtrate as it passes through the series of tubules that drain Bowman's capsule.

(b) Cutaway view of Bowman's capsule. Note that the distal tubule is adjacent to the afferent and efferent arterioles. The juxtaglomerular apparatus (discussed later in the chapter) is a small group of cells located in the walls of the tubule and arterioles.

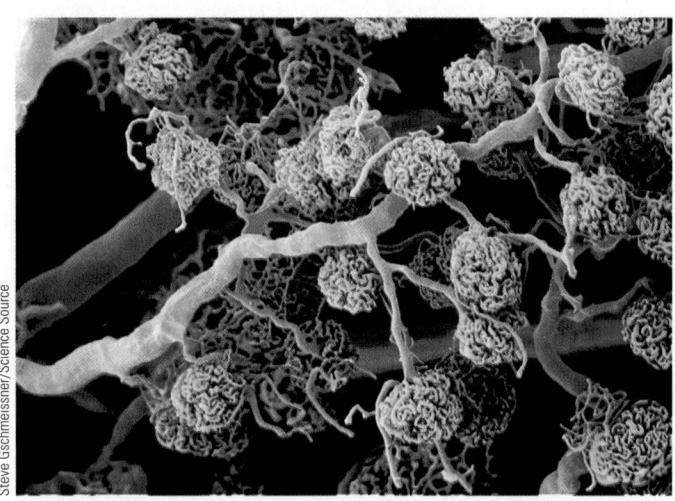

100 μm

(c) Low-power SEM of a portion of the kidney cortex. The tissue was treated to remove some structures and to show glomeruli and associated blood vessels.

Figure 48-9 *Animation* **Nephron structure**

produce the enzyme *renin,* which helps regulate fluid balance and blood pressure (discussed later in the chapter), and at least two hormones: *erythropoietin,* which stimulates red blood cell production, and *1,25-dihydroxy-vitamin D₃,* which stimulates calcium absorption by the intestine.

Each kidney has up to one million functional units called **nephrons.** A nephron consists of a cuplike **Bowman's capsule** connected to a long, partially coiled **renal tubule** (FIG. 48-9). Positioned within Bowman's capsule is a cluster of capillaries called a **glomerulus.** Three main regions of the renal tubule are (1) the **proximal convoluted tubule** (or simply, *proximal tubule*), which conducts the *glomerular filtrate* (the fluid filtered from the blood) from Bowman's capsule; (2) the **loop of Henle,** an elongated, hairpin-shaped portion; and

(3) the **distal convoluted tubule** (or *distal tubule*), which conducts the glomerular filtrate to a collecting duct. Thus, the glomerular filtrate passes through the following structures:

Bowman's capsule ⟶ proximal convoluted tubule ⟶ loop of Henle ⟶ distal convoluted tubule ⟶ collecting duct

The human kidney has two types of nephrons: the more numerous (85%) cortical nephrons and the more internal juxtamedullary nephrons (see Fig. 48-8b). *Cortical nephrons* have relatively small glomeruli and are located almost entirely within the cortex or outer medulla. *Juxtamedullary nephrons* have large glomeruli, and their very long loops of Henle extend deep into the medulla. The loop of Henle consists of a *descending limb* that receives glomerular filtrate from the proximal convoluted tubule and an *ascending limb,* through which the filtrate passes on its way to the distal convoluted tubule. The juxtamedullary nephrons contribute to the ability of the mammalian kidney to concentrate urine. Excretion of urine that is hypertonic to body fluids is an important mechanism for conserving water.

Blood is delivered to the kidney by the **renal artery.** Small branches of the renal artery give rise to **afferent arterioles** (*afferent* means "to carry toward"). An afferent arteriole conducts blood into the capillaries that make up each glomerulus. As blood flows through the glomerulus, some of the plasma is forced into Bowman's capsule.

You may recall that in a typical circulatory route, capillaries deliver blood into veins. Circulation in the kidneys is an exception because blood flowing from the glomerular capillaries next passes into an **efferent arteriole.** Each efferent arteriole conducts blood *away* from a glomerulus. The efferent arteriole delivers blood to a second capillary network, the **peritubular capillaries** surrounding the renal tubule.

Blood is filtered as it flows through the first set of capillaries, those of the glomerulus. The peritubular capillaries receive blood containing materials returned to it by the renal tubule. Blood from the peritubular capillaries enters small veins that eventually lead to the **renal vein.** In summary, blood circulates through the kidney in the following sequence:

renal artery ⟶ an afferent arteriole ⟶ capillaries of a
glomerulus ⟶ an efferent arteriole ⟶ peritubular capillaries
⟶ small veins ⟶ renal vein

Urine is produced by glomerular filtration, tubular reabsorption, and tubular secretion

Urine, the watery discharge of the urinary system, is produced by a combination of three processes: glomerular filtration, tubular reabsorption, and tubular secretion (**FIG. 48-10**).

Glomerular filtration is not selective with regard to ions and small molecules Blood flows through the glomerular capillaries under high pressure, forcing more than 10% of the plasma out of the capillaries and into Bowman's capsule.

Glomerular filtration is similar to the mechanism whereby interstitial fluid is formed as blood flows through other capillary networks in the body. However, blood flow through glomerular capillaries is at much higher pressure, so more plasma is filtered in the kidney.

Several factors contribute to the production of a large amount of *glomerular filtrate.* First, the hydrostatic blood pressure in the glomerular capillaries is higher than in other capillaries. This high pressure is mainly due to the high resistance to outflow presented by the efferent arteriole, which is smaller in diameter than the afferent arteriole (see Fig. 48-9b). A second factor contributing to the large amount of glomerular filtrate is the large surface area for filtration provided by the highly coiled glomerular capillaries. A third factor is the great permeability of the glomerular capillaries. Numerous small pores between the endothelial cells that form their walls make the glomerular capillaries more porous than typical capillaries.

The wall of Bowman's capsule in contact with the capillaries consists of specialized epithelial cells called *podocytes.* These cells have numerous cytoplasmic extensions called *foot processes* that cover most of the surfaces of the glomerular capillaries (**FIG. 48-11**). Foot processes of adjacent podocytes are separated

---

KEY POINT

The glomerulus filters blood. As the glomerular filtrate moves through the renal tubule, its composition is adjusted by selective reabsorption and tubular secretion. The adjusted filtrate is urine.

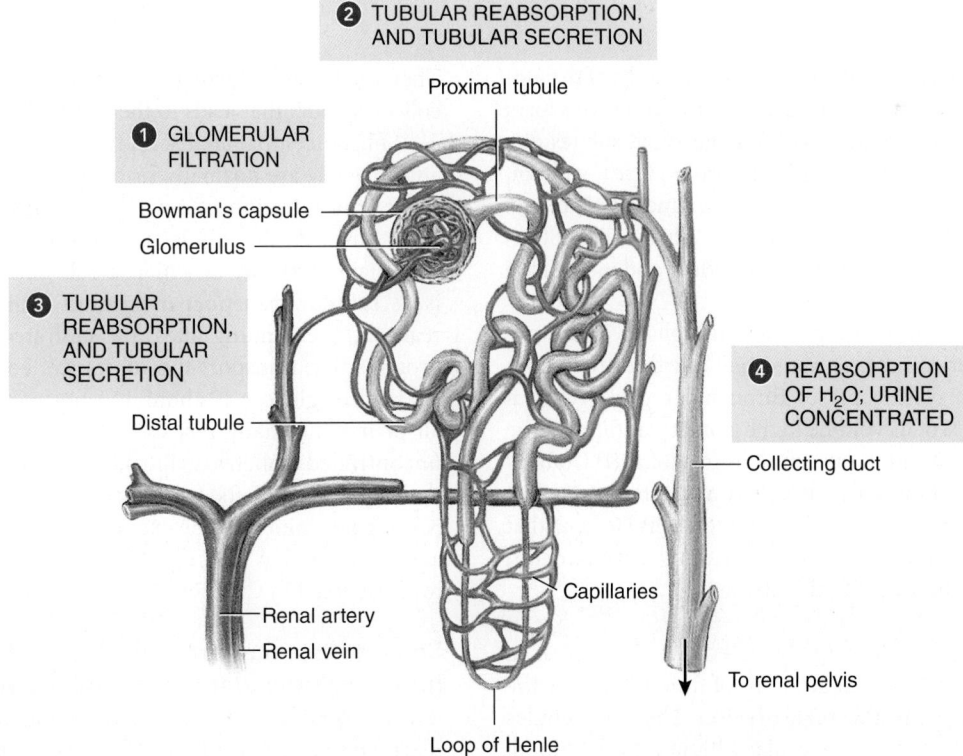

❷ TUBULAR REABSORPTION, AND TUBULAR SECRETION

Proximal tubule

❶ GLOMERULAR FILTRATION

Bowman's capsule

Glomerulus

❸ TUBULAR REABSORPTION, AND TUBULAR SECRETION

Distal tubule

❹ REABSORPTION OF H₂O; URINE CONCENTRATED

Collecting duct

Renal artery

Renal vein

Capillaries

To renal pelvis

Loop of Henle

Figure 48-10 General regions of glomerular filtration, tubular reabsorption, and tubular secretion

© Cengage Learning

PREDICT How would mild dehydration affect kidney function?

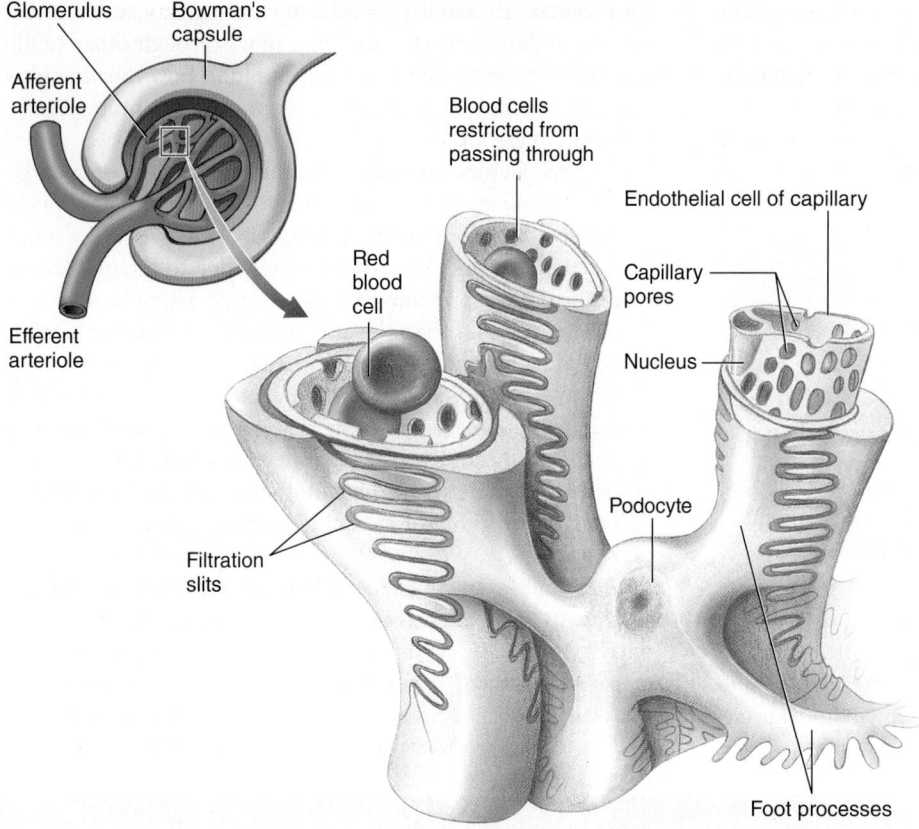

Glomerulus   Bowman's capsule

Afferent arteriole

Efferent arteriole

Blood cells restricted from passing through

Red blood cell

Filtration slits

Endothelial cell of capillary

Capillary pores

Nucleus

Podocyte

Foot processes

Figure 48-11 Filtration membrane of the kidney

The porous walls of the glomerular capillaries and the filtration slits between podocytes form a filtration membrane that is highly permeable to water, small molecules, and ions but restricts the passage of blood cells and large molecules.

© Cengage Learning

by narrow gaps called *filtration slits.* The porous walls of the glomerular capillaries and the filtration slits of the podocytes form a *filtration membrane* that permits fluid and small solutes dissolved in the plasma (including glucose, amino acids, sodium, potassium, chloride, bicarbonate, other salts, and urea) to pass through and become part of the glomerular filtrate. This filtration membrane holds back blood cells, platelets, and most of the plasma proteins.

The total volume of blood passing through the kidneys is about 1200 mL per minute, or about one-fourth of the entire cardiac output. As plasma passes through the glomerulus, it loses more than 10% of its volume to the glomerular filtrate. The normal glomerular filtration rate amounts to about 180 L (about 45 gallons) each 24 hours, which is four and a half times the amount of fluid in the entire body! Common sense tells us that urine could not be excreted at that rate. Within a few moments, dehydration would become life-threatening.

Tubular reabsorption is highly selective The threat to homeostasis posed by the vast amounts of fluid filtered by the kidneys is avoided by **tubular reabsorption.** The renal tubules reabsorb about 99% of the glomerular filtrate into the blood, leaving only about 1.5 L to be excreted as urine during a 24-hour period. Tubular reabsorption permits precise regulation of blood chemistry by the kidneys. Wastes, excess salts, and other

materials remain in the filtrate and are excreted in the urine, whereas needed substances such as glucose and amino acids are returned to the blood. Each day the tubules reabsorb more than 178 L of water, 1200 g (2.6 lb) of salt, and about 250 g (0.5 lb) of glucose. Most of it, of course, is reabsorbed many times over.

The simple epithelial cells lining the renal tubule are well adapted for reabsorbing materials. Their abundant microvilli increase the surface area for tubular reabsorption. These epithelial cells contain numerous mitochondria that provide the energy for running the cell pumps that actively transport materials.

Most (about 65%) of the glomerular filtrate is reabsorbed as it passes through the proximal convoluted tubule. Glucose, amino acids, vitamins, and other substances of nutritional value are entirely reabsorbed there. Many ions, including sodium, chloride, bicarbonate, and potassium, are partially reabsorbed. Some of these ions are actively transported; others are reabsorbed by diffusion. Tubular reabsorption continues as the glomerular filtrate passes through the loop of Henle and the distal convoluted tubule. Then the filtrate is further concentrated as it passes through the collecting duct that leads to the renal pelvis.

Substances that are useful to the body, such as glucose and amino acids, are normally reabsorbed from the renal tubules. If the concentration of a particular substance in the blood is high, however, the tubules may not be able to reabsorb it all. The maximum rate at which a substance can be reabsorbed is its **tubular transport maximum (Tm).** When that rate is reached, the binding sites are saturated on the membrane proteins that transport the substance. For example, the tubular load of glucose is about 125 mg per minute, and almost all of it is normally reabsorbed. However, in a person with uncontrolled diabetes mellitus, the concentration of glucose in the blood exceeds its Tm (glucose concentration in excess of 320 mg per minute). The excess glucose cannot be reabsorbed and is excreted in the urine (glucosuria), a symptom of diabetes (discussed in Chapter 49).

Some substances are actively secreted from the blood into the glomerular filtrate **Tubular secretion** is the selective transfer of substances from the blood in the peritubular capillaries into the renal tubule. Note that in tubular secretion ions and other substances are moved across the tubule epithelium in the opposite direction of tubular reabsorption. Potassium, hydrogen ions, and ammonium ions as well as some organic ions, such as creatinine (a metabolic waste), are secreted into the glomerular filtrate. Certain drugs, such as penicillin, are also removed from the blood by secretion. Secretion occurs mainly in the region of the distal convoluted tubule.

Secretion of hydrogen ions by the collecting ducts is an important homeostatic mechanism for regulating the pH of the blood. Carbon dioxide, which diffuses from the blood into the cells of the distal tubules and collecting ducts, combines with water, which produces carbonic acid. This acid then dissociates to form hydrogen ions and bicarbonate ions. When the blood becomes too acidic, the collecting ducts secrete more hydrogen ions into the urine:

$$CO_2 + H_2O \rightleftharpoons H_2CO_3 \rightleftharpoons H^+ + HCO_3^-$$

Potassium ion secretion is another important homeostatic mechanism. When K^+ concentration is too high, nerve impulses are not effectively transmitted, and the strength of muscle contraction decreases. The heart rhythm becomes irregular, and cardiac arrest can occur. When K^+ concentration exceeds its homeostatic level, K^+ are secreted from the blood into the renal tubules and then excreted in the urine. Secretion results partly from a direct effect of the K^+ on the tubules. In addition, the adrenal cortex (a region of the adrenal gland) increases its output of the hormone *aldosterone*, which further stimulates secretion of K^+.

Urine becomes concentrated as it passes through the renal tubule

We can survive with limited fluid intake because the kidneys can produce highly concentrated urine, more than four times as concentrated as blood. The osmolarity of human blood is about 300 milliosmoles per liter (mOsm/L). The kidneys can produce urine with an osmolarity of about 1200 mOsm/L.

As the glomerular filtrate passes through various regions of the renal tubule, salt (NaCl) is reabsorbed, establishing a salt concentration gradient in the interstitial fluid (**FIGS. 48-12** and **48-13**). The gradient is used to produce concentrated urine.

When glomerular filtrate flows from Bowman's capsule to the proximal tubule, its osmolarity is the same as that of blood (about 300 mOsm/L). Water and salt are reabsorbed from the proximal tubule. Sodium ions are actively transported out of the proximal tubule, and chloride follows passively. As salt passes into the interstitial fluid, water follows osmotically.

The walls of the descending limb of the loop of Henle are relatively permeable to water but relatively impermeable to sodium and urea. The interstitial fluid has a high concentration of Na^+, so as the glomerular filtrate passes down the loop of Henle, water moves out by osmosis. This process concentrates the filtrate inside the loop of Henle.

The loop of Henle is specialized to highly concentrate sodium chloride in the interstitial fluid of the medulla. It maintains a highly hypertonic interstitial fluid in the medulla near the bottom of the loop, which in turn lets the kidneys produce concentrated urine. At the turn of the loop of Henle, the walls become more permeable to salt and less permeable to water. As the concentrated glomerular filtrate begins to move up the ascending limb (the thin region), salt diffuses out into the interstitial fluid. This diffusion contributes to the high salt concentration of the interstitial fluid in the medulla surrounding the loop of Henle. Farther along the ascending limb (the thick region), sodium is actively transported out of the tubule.

Because water passes out of the descending limb of the loop of Henle, the glomerular filtrate at the bottom of the loop has a high salt concentration. However, because salt (but not water) is removed in the ascending limb, by the time the filtrate moves into the distal tubule, its osmolarity may be the same as or even lower than that of blood.

As it passes through the distal tubule, the filtrate may become even more dilute. The distal tubule is relatively impermeable to water but actively transports salt out into the interstitial fluid. The filtrate passes from the renal tubule into a larger collecting duct that eventually empties into the renal pelvis.

Note the *counterflow* of fluid through the two limbs of the loop of Henle. Glomerular filtrate passing down through the descending limb is flowing in a direction opposite that of the filtrate moving upward through the ascending limb. The filtrate becomes concentrated as it moves down the descending limb and diluted as it moves up the ascending limb. This countercurrent mechanism helps maintain a high salt concentration in the interstitial fluid of the medulla. The hypertonic interstitial fluid draws water osmotically from the filtrate in the collecting ducts.

The inner medullary collecting ducts are permeable to urea, so some of the concentrated urea in the filtrate can diffuse out into the interstitial fluid. The urea contributes to the high solute concentration of the inner medulla. This process helps concentrate urine.

The collecting ducts pass through the zone of very salty interstitial fluid. As the glomerular filtrate moves through the collecting duct, water passes osmotically into the interstitial fluid. Excess interstitial fluid continuously enters capillaries and becomes part of the blood. Sufficient water can leave the collecting ducts to produce a highly concentrated urine. Remember that a hypertonic urine conserves water.

Some of the water that diffuses from the filtrate into the interstitial fluid is removed by the *vasa recta*, which are long, straight capillaries that extend from the efferent arterioles of the juxtamedullary nephrons. The vasa recta extend deep into the medulla, only to negotiate a hairpin curve and return to the cortical venous drainage of the kidney. These capillaries parallel the renal tubules.

Blood flows in opposite directions in the ascending and descending regions of the vasa recta, just as glomerular filtrate flows in opposite directions in the ascending and descending limbs of the loop of Henle. As a consequence of this countercurrent flow, much of the salt and urea that enter the blood through the descending region of the vasa recta leave again from the ascending region. As a result, the solute concentration of the blood leaving the vasa recta is only slightly higher than that of the blood entering. This mechanism helps maintain the high solute concentration of the interstitial fluid in the renal medulla.

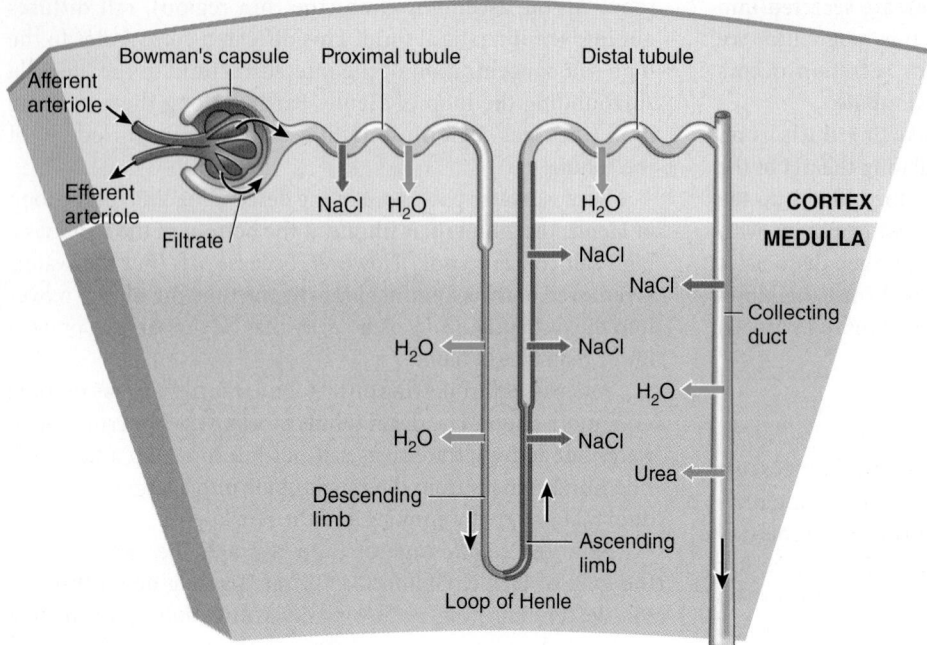

Figure 48-12 *Animation* **Movement of water, ions, and urea through the renal tubule and collecting duct**

Water passes out of the descending limb of the loop of Henle, leaving a more concentrated filtrate inside. The heavy outline along the ascending limb indicates that this region is relatively impermeable to water. NaCl diffuses out from the lower (*thin*) part of the ascending limb. In the upper (*thick*) part of the ascending limb, NaCl is actively transported into the interstitial fluid. The saltier the interstitial fluid becomes, the more water moves out of the descending limb. This process leaves a concentrated filtrate inside, so more salt passes out. Note that this is a positive feedback system. Some urea also moves out into the interstitial fluid through the collecting ducts, contributing to its osmolarity. Water from the collecting ducts moves osmotically out into the hypertonic interstitial fluid and is carried away by capillaries. The concentrated solutes (NaCl and urea) are excreted in the urine.

© Cengage Learning

Urine consists of water, nitrogenous wastes, and salts

By the time the glomerular filtrate reaches the renal pelvis, its composition has been precisely adjusted. The adjusted filtrate, called *urine*, consists of approximately 96% water, 2.5% nitrogenous wastes (mainly urea), 1.5% salts, and traces of other substances, such as bile pigments, that may contribute to the characteristic color and odor. Healthy urine is sterile and has been used to wash battlefield wounds when clean water was not available. However, when exposed to bacterial action, urine swiftly decomposes and forms ammonia and other products. Ammonia produces the diaper rash of infants.

The composition of urine yields many clues to body function and malfunction. *Urinalysis,* the physical, chemical, and microscopic examination of urine, is a very important diagnostic tool that has been used to monitor diabetes mellitus and many other disorders. Urinalysis is also extensively used in drug testing because breakdown products of some drugs can be identified in the urine for several days or weeks after the drugs are ingested.

Hormones regulate kidney function

As we have discussed, the kidneys are vital in maintaining the body's fluid volume and electrolyte concentration. They help maintain the concentration of Na$^+$ and K$^+$ in the blood within narrow limits and also regulate blood pH. The endocrine system regulates these homeostatic functions of the kidneys. Hormones involved include antidiuretic hormone (ADH), aldosterone, angiotensin II, and atrial natriuretic peptide (ANP) (**TABLE 48-1**).

ADH increases water reabsorption The amount of urine produced depends on the body's need to retain or rid itself of water. When fluid intake is low, the body begins to dehydrate, and the blood volume decreases. As blood volume decreases, the concentration of salts dissolved in the blood becomes greater, causing an increase in osmotic pressure. Certain receptors in the hypothalamus are sensitive to this osmotic change. They signal the posterior lobe of the pituitary gland to release **antidiuretic hormone (ADH).** This hormone is actually produced in the hypothalamus, but it is stored in the posterior pituitary gland and is released as needed. A thirst center in the hypothalamus also responds to dehydration, stimulating an increase in fluid intake.

ADH makes the collecting ducts more permeable to water so that more water is reabsorbed. ADH temporarily increases the number of gated water channels in the plasma membranes of the collecting ducts. These channels allow water to pass rapidly through the plasma membrane. As a result, a small volume of concentrated urine is produced (**FIG. 48-14**).

When you drink a large volume of water, your blood becomes diluted, and its osmotic pressure falls. Release of ADH by the pituitary gland decreases, lessening the amount of water reabsorbed from the collecting ducts. The kidneys produce a large volume of dilute urine.

In the disorder *diabetes insipidus* (not to be confused with the more common disorder, diabetes mellitus), water is not efficiently reabsorbed from the ducts, so the body produces a large volume of urine. A person with severe diabetes insipidus may excrete up to 25 quarts of urine each day, a serious water loss. The affected individual becomes dehydrated and must drink almost continually to offset fluid loss. Diabetes insipidus results when the pituitary gland does not release enough ADH.

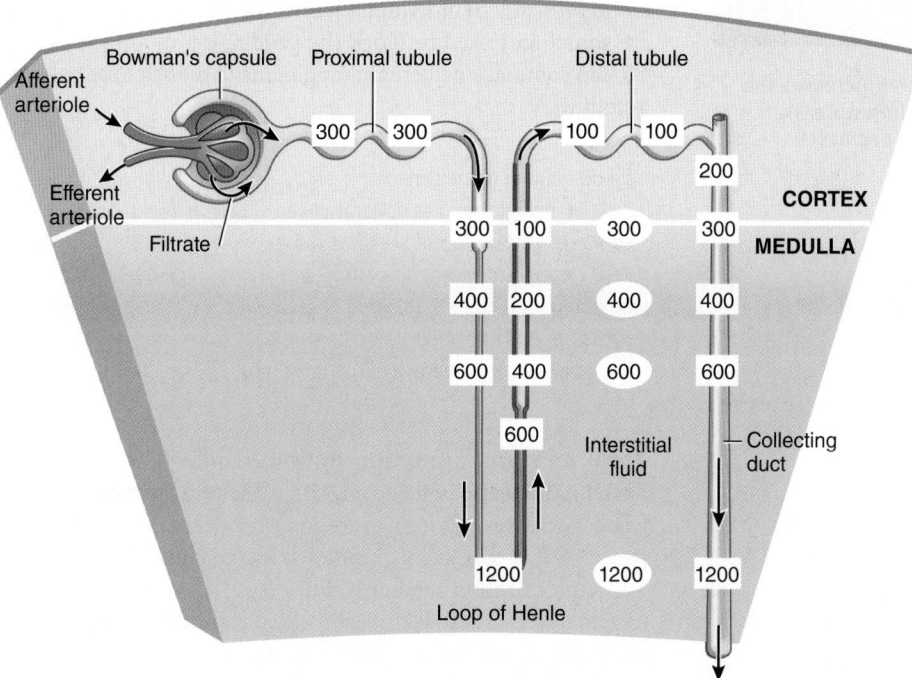

Figure 48-13 Concentration of the glomerular filtrate as it moves through the nephron

This figure shows the relative concentration of ions, mainly Na^+ and Cl^-, during production of a very concentrated urine. Numbers indicate the concentration of salt in the glomerular filtrate and interstitial fluid expressed in milliosmoles per liter. The more hypertonic a solution is, the higher its osmotic pressure. The very hypertonic interstitial fluid near the renal pelvis draws water osmotically from the filtrate in the collecting ducts. The heavy outline along the ascending loop indicates that this region is relatively impermeable to water.

© Cengage Learning

This type of diabetes insipidus can be treated with a synthetic hormone (desmopressin). Diabetes insipidus can also develop from an acquired insensitivity of the kidney to ADH.

The renin–angiotensin–aldosterone pathway increases sodium reabsorption

Sodium is the most abundant extracellular ion, accounting for about 90% of all positive ions in the extracellular fluid. Several hormones work together to regulate sodium concentration. **Aldosterone,** which is secreted by the cortex of the adrenal glands, stimulates the distal tubules and collecting ducts to increase sodium reabsorption. When researchers remove the adrenal glands of experimental animals,

too much sodium is excreted, leading to serious depletion of the extracellular fluid.

Aldosterone secretion is stimulated both by hormones and by a decrease in blood pressure (caused by a decrease in volume of blood and interstitial fluid). When blood pressure falls, cells of the **juxtaglomerular apparatus** secrete the enzyme **renin,** which activates the *renin–angiotensin–aldosterone pathway.* The juxtaglomerular apparatus is a small group of cells in the region where the renal tubule contacts the afferent and efferent arterioles (see Fig. 48-9b). Renin converts the plasma protein angiotensinogen to **angiotensin I.** *Angiotensin-converting enzyme (ACE)* converts angiotensin I

TABLE 48-1	Endocrine Regulation of Kidney Function			
HORMONE	**SOURCE**	**TARGET TISSUE**	**ACTIONS**	**FACTORS THAT STIMULATE RELEASE**
Antidiuretic hormone (ADH)	Produced in hypothalamus; released by posterior pituitary gland	Collecting ducts	Increases permeability of collecting ducts to water, which increases tubular reabsorption and decreases water excretion	Low fluid intake decreases blood volume and increases osmotic pressure of blood; receptors in hypothalamus stimulate posterior pituitary
Angiotensin II	Produced from angiotensin I	Blood vessels and adrenal glands	Constricts blood vessels, which raises blood pressure; stimulates aldosterone secretion; stimulates thirst	Decrease in blood pressure causes renin secretion; renin catalyzes conversion of angiotensinogen to angiotensin I, which is then converted to angiotensin II by ACE
Aldosterone	Adrenal glands (cortex)	Distal tubules and collecting ducts	Increases sodium reabsorption, which raises blood pressure	Angiotensin II; decrease in blood pressure
Atrial natriuretic peptide (ANP)	Atrium of heart	Afferent arterioles; collecting ducts	Dilates afferent arterioles; inhibits sodium reabsorption by collecting ducts; inhibits aldosterone secretion; decreases blood pressure	Stretching of atria due to increased blood volume

© Cengage Learning

When the body is dehydrated, the hormone ADH increases the permeability of the collecting ducts to water. More water is reabsorbed, which increases blood volume and decreases osmotic pressure.

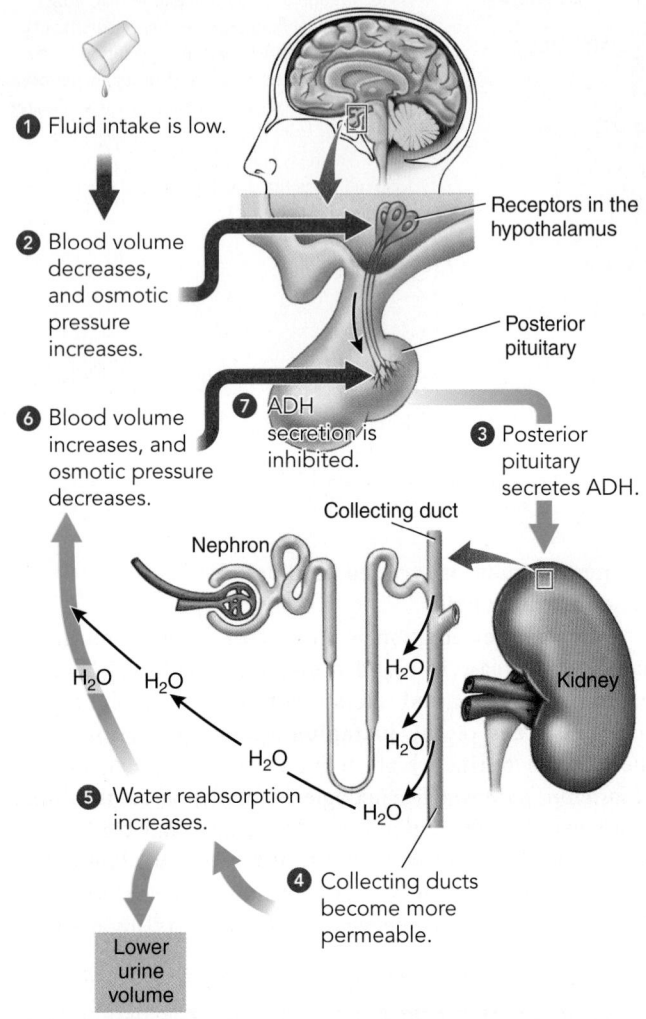

1 Fluid intake is low.

2 Blood volume decreases, and osmotic pressure increases.

Receptors in the hypothalamus

Posterior pituitary

7 ADH secretion is inhibited.

6 Blood volume increases, and osmotic pressure decreases.

3 Posterior pituitary secretes ADH.

Collecting duct

Nephron

H_2O H_2O

H_2O

H_2O

H_2O

Kidney

H_2O

5 Water reabsorption increases.

4 Collecting ducts become more permeable.

Lower urine volume

Figure 48-14 Regulation of urine volume by antidiuretic hormone (ADH)

The hormone ADH helps maintain fluid homeostasis.

CONNECT What effect does ADH have on arterial blood pressure?

© Cengage Learning

into its active form, **angiotensin II,** which is a peptide hormone. Angiotensin II stimulates aldosterone secretion. ACE is produced by the endothelial cells in the walls of pulmonary capillaries.

Angiotensin II increases the synthesis and release of aldosterone. Angiotensin II also raises blood pressure directly by constricting blood vessels. In addition, this hormone stimulates the posterior pituitary to release ADH and stimulates thirst. These actions help increase extracellular fluid volume, which raises

blood pressure. In individuals with hypertension, *ACE inhibitors* are sometimes used to block the production of angiotensin II. We can summarize the renin–angiotensin–aldosterone pathway as follows:

blood volume decreases ⟶ blood pressure decreases ⟶ cells of juxtaglomerular apparatus secrete renin ⟶ renin catalyzes conversion of angiotensinogen to angiotensin I ⟶ ACE catalyzes conversion of angiotensin I to angiotensin II ⟶ angiotensin II constricts blood vessels, which increases blood pressure, and stimulates aldosterone secretion ⟶ aldosterone increases sodium reabsorption ⟶ blood pressure increases

Atrial natriuretic peptide inhibits sodium reabsorption

Atrial natriuretic peptide (ANP), a hormone produced by the heart, increases sodium excretion and decreases blood pressure. ANP is stored in granules in atrial muscle cells. When Na^+ concentration increases, fluid is retained, and blood volume increases. As atrial muscle cells are stretched, they respond by releasing ANP into the circulation.

ANP dilates afferent arterioles, which increases the glomerular filtration rate. This hormone directly inhibits sodium reabsorption by the collecting ducts. ANP also acts indirectly by inhibiting aldosterone secretion. (Recall that aldosterone increases sodium reabsorption.) ANP reduces plasma aldosterone concentration by inhibiting renin release. These actions of ANP increase sodium excretion and urine output, which decreases blood volume and blood pressure. Note that the renin–angiotensin–aldosterone pathway works antagonistically with ANP in regulating fluid balance, electrolyte balance, and blood pressure.

blood volume increases ⟶ blood pressure increases ⟶ atria of heart stretched ⟶ atria release ANP ⟶ ANP directly inhibits sodium reabsorption and inhibits aldosterone secretion (which also inhibits sodium reabsorption) ⟶ urine volume increases ⟶ blood volume decreases ⟶ blood pressure decreases

CHECKPOINT 48.5

- **CONNECT** *Which structure(s) is/are associated with each of the following functions: urea formation, urine formation, temporary storage of urine, and conduction of urine out of the body?*

- **CONNECT** *Which part(s) of the nephron is/are associated with each of the following processes: filtration, reabsorption, and secretion?*

- **VISUALIZE** *Draw a diagram showing the sequence of structures that a drop of glomerular filtrate passes through as it moves from Bowman's capsule to the urinary bladder.*

- *What are the main actions of the renin–angiotensin–aldosterone pathway? Compare the actions of this pathway with the actions of ANP.*

48.1 Maintaining Fluid and Electrolyte Balance (page 1027)

1 Describe how the processes of *osmoregulation* and *excretion* contribute to fluid and electrolyte homeostasis.

- **Osmoregulation** is the active regulation of osmotic pressure of body fluids so that fluid and electrolyte homeostasis is maintained. **Excretion** is the process of disposing of metabolic wastes.

48.2 Metabolic Waste Products (page 1027)

2 Contrast the benefits and costs of excreting each of the following: ammonia, uric acid, and urea.

- The principal waste products of animal metabolism are water; carbon dioxide; and *nitrogenous wastes,* including ammonia, uric acid, and urea. *Ammonia,* which is toxic, is excreted mainly by aquatic animals.

- **Uric acid** and **urea** are far less toxic than ammonia, but their synthesis requires energy. Uric acid can be excreted as a semisolid paste, a water-conserving adaptation. Urea excretion requires water.

48.3 Osmoregulation and Excretion in Invertebrates (page 1028)

3 Contrast osmoconformers and osmoregulators.

- Most marine invertebrates are **osmoconformers,** meaning that the salt concentration of their body fluids varies with changes in the sea water. Some marine invertebrates, especially those inhabiting coastal habitats, are **osmoregulators** that maintain an optimal salt concentration despite changes in salinity of their surroundings.

4 Compare the structure and function of protonephridia, metanephridia, and Malpighian tubules.

- **Nephridial organs** help maintain homeostasis by regulating the concentration of body fluids through osmoregulation and excretion of metabolic wastes.

- **Protonephridia,** nephridial organs found in flatworms and nemerteans, are tubules with no internal openings. Interstitial fluid enters their blind ends, which consist of **flame cells,** cells with brushes of cilia. The cilia propel fluid through the tubules, and excess fluid exits through nephridiopores.

- Most annelids and mollusks have nephridial organs called **metanephridia,** which are tubules open at both ends. As fluid from the coelom moves through the tubule, needed materials are reabsorbed by capillaries. Urine, containing wastes, exits the body through nephridiopores.

- **Malpighian tubules,** extensions of the insect gut wall, have blind ends that lie in the hemocoel. Cells of the tubule actively transport uric acid and some other substances from the hemolymph into the tubule. Water follows by diffusion. The contents of the tubule pass into the gut. Water and some solutes are reabsorbed in the rectum. Malpighian tubules have contributed to the success of insects as terrestrial animals.

48.4 Osmoregulation and Excretion in Vertebrates (page 1030)

5 Relate the function of the vertebrate kidney to the success of vertebrates in a wide variety of habitats.

- The vertebrate **kidney** excretes nitrogenous wastes and helps maintain fluid balance by adjusting the salt and water content of the urine.

- Freshwater, marine, and terrestrial habitats present different problems for maintaining internal fluid balance and for the excretion of nitrogenous wastes. The structure and function of the vertebrate kidney have adapted to meet the various osmotic challenges presented by these different habitats.

6 Compare adaptations for osmoregulation in freshwater fishes, marine bony fishes, sharks, marine mammals, and terrestrial vertebrates.

- Freshwater fishes take in water osmotically, and they excrete a large volume of hypotonic urine.

- Marine bony fishes lose water osmotically. They compensate by drinking sea water and excreting salt through their gills; they produce only a small volume of isotonic urine. Sharks and other marine cartilaginous fishes retain large amounts of urea, allowing them to take in water osmotically through the gills. They excrete a large volume of hypotonic urine. Marine mammals ingest sea water with their food. They produce concentrated urine.

- Terrestrial vertebrates must conserve water. Endotherms have a high metabolic rate and produce a large volume of nitrogenous wastes. They have numerous adaptations for conserving water, including efficient kidneys.

48.5 The Urinary System of Mammals (page 1032)

7 Describe (and label on a diagram) the organs of the mammalian urinary system and give the functions of each.

- The **urinary system** is the principal excretory system in mammals. The kidneys produce urine, which passes through the **ureters** to the **urinary bladder** for storage. During urination, the urine is released from the body through the **urethra.**

- The outer portion of each kidney is the **renal cortex,** and the inner portion is the **renal medulla.** The renal medulla contains eight to ten *renal pyramids.* The tip of each pyramid is a *renal papilla.* As urine is produced, it flows into **collecting ducts,** which empty through a renal papilla into a funnel-shaped chamber, the **renal pelvis.** Each kidney has more than one million functional units called **nephrons.**

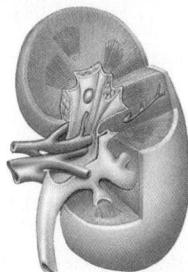

© Cengage Learning

8 Describe (and label on a diagram) the structures of a nephron (including associated blood vessels) and give the functions of each structure.

- Each nephron consists of a cluster of capillaries called a **glomerulus,** which is surrounded by a **Bowman's capsule** that opens into a long, coiled **renal tubule.** The renal tubule consists of a **proximal convoluted tubule, loop of Henle,** and **distal convoluted tubule.**

- *Cortical nephrons*, located almost entirely within the cortex, have small glomeruli. *Juxtamedullary nephrons* have large glomeruli and long loops of Henle that extend deep into the medulla. These nephrons are important in concentrating urine.
- Blood flows from small branches of the **renal artery** to **afferent arterioles** and then to glomerular capillaries. Blood then flows into an **efferent arteriole** that delivers blood into a second set of capillaries, the **peritubular capillaries** that surround the renal tubule. Blood leaves the kidney through the *renal vein*.

9 Trace a drop of glomerular filtrate from Bowman's capsule to its release from the body as urine.
- Urine is produced by **glomerular filtration** of plasma, **tubular reabsorption** of needed materials, and **tubular secretion** of a few substances, such as potassium and hydrogen ions, into the renal tubule.
- Plasma filters through the glomerular capillaries and into Bowman's capsule. The permeable walls of the capillaries and *filtration slits* between *podocytes*, specialized epithelial cells that make up the inner wall of Bowman's capsule, serve as a *filtration membrane*.
- Glomerular filtration is nonselective with regard to small molecules. In addition to metabolic wastes, glucose and other needed substances become part of the filtrate.
- Tubular reabsorption is a highly selective process that returns usable materials to the blood, but leaves wastes and excesses of other substances to be excreted in the urine. The maximum rate at which a substance can be reabsorbed is its **tubular transport maximum (Tm).**
- In tubular secretion hydrogen ions, certain other ions, and some drugs are actively transported into the renal tubule to become part of the urine.
- Production of concentrated urine depends on a high salt and urea concentration in the interstitial fluid of the kidney medulla. The interstitial fluid in the medulla has a concentration gradient in which the salt is most concentrated around the bottom of the loop of Henle. This gradient is maintained, in part, by salt reabsorption from various parts of the renal tubule.
- A counterflow of fluid through the two limbs of the loop of Henle concentrates filtrate as it moves down the descending loop and dilutes it as it moves up the ascending loop.
- Water is drawn by osmosis from the filtrate as it passes through the collecting ducts. This process concentrates urine in the collecting ducts.
- Some of the water that diffuses from the filtrate into the interstitial fluid is removed by the *vasa recta*, a system of capillaries that extend from the efferent arterioles.
- *Urine* is a watery solution of nitrogenous wastes, excess salts, and other unneeded substances.

10 Describe the hormonal regulation of fluid and electrolyte balance by antidiuretic hormone (ADH), the renin–angiotensin–aldosterone system, and atrial natriuretic peptide (ANP).
- When the body needs to conserve water, the posterior pituitary gland increases its release of **antidiuretic hormone (ADH).** The pituitary gland responds to an increase in osmotic concentration of the blood (caused by dehydration). ADH increases the permeability of the collecting ducts to water. As a result, more water is reabsorbed, and only a small volume of urine is produced.
- The renin–angiotensin–aldosterone pathway and atrial natriuretic peptide work antagonistically. When blood pressure decreases, cells of the **juxtaglomerular apparatus** secrete the enzyme **renin,** which activates a pathway leading to production of **angiotensin II.** This hormone stimulates aldosterone release. **Aldosterone** increases sodium reabsorption, and angiotensin II constricts arterioles; both actions raise blood pressure.
- When blood pressure increases, **atrial natriuretic peptide (ANP)** increases sodium excretion and inhibits aldosterone secretion. These actions increase urine output and lower blood pressure.

TEST YOUR UNDERSTANDING

Know and Comprehend

1. The process that maintains homeostasis of body fluids by keeping them from becoming too dilute or too concentrated is (a) excretion (b) elimination (c) osmoregulation (d) glomerular filtration (e) tubular secretion
2. Which of the following is *not* a correct pair? (a) protonephridia; flatworm (b) metanephridia; annelid (c) flame cell; flatworm (d) Malpighian tubule; mollusk (e) kidney; vertebrate
3. To compensate for fluid loss, many marine bony fishes (a) accumulate urea (b) have glands that excrete glucose (c) eat a low-protein diet (d) excrete a large volume of hypertonic urine (e) drink sea water
4. The afferent arteriole delivers blood to the (a) renal artery (b) efferent arteriole (c) renal vein (d) capillaries of the glomerulus (e) peritubular capillaries
5. Which of the following does *not* contribute to the process of filtration? (a) active transport by epithelial cells lining renal tubules (b) large surface area for filtration (c) low permeability of glomerular capillaries (d) high hydrostatic blood pressure in glomerular capillaries (e) podocytes
6. Tubular transport maximum (Tm) is (a) the maximum concentration of a substance in the plasma that can be reabsorbed by the kidney (b) the most rapid rate at which urine can be transported through the ureter (c) the maximum rate at which a substance can be reabsorbed from the glomerular filtrate in the renal tubules (d) the maximum rate at which a substance can pass through the loop of Henle (e) the maximum amount of a substance that can be secreted into the filtrate
7. Which of the following does *not* contribute to the high salt concentration in the interstitial fluid of the kidney medulla? (a) active transport of sodium from the upper part of the ascending limb (b) diffusion of salt from the ascending limb of the loop of Henle (c) reabsorption of salt from various regions of Bowman's capsule (d) counterflow of fluid through the two limbs of the loop of Henle (e) diffusion of urea out of the collecting duct
8. Which is *not* true of ADH? (a) released by posterior lobe of the pituitary gland (b) increases water reabsorption (c) secretion increases when osmotic pressure in body increases (d) increases urine volume (e) secretion decreases when you drink a lot of water

9. Aldosterone (a) is released by the posterior pituitary gland (b) decreases sodium reabsorption (c) secretion is stimulated by an increase in blood pressure (d) is an enzyme that converts angiotensin into angiotensin II (e) secretion increases in response to angiotensin II

Apply and Analyze

10. Arrange the following structures into an accurate sequence through which urine passes.
    1. urethra 2. urinary bladder 3. kidney 4. ureter
    (a) 4, 3, 2, 1 (b) 3, 4, 2, 1 (c) 1, 2, 3, 4 (d) 4, 2, 1, 3 (e) 3, 1, 2, 4
11. Arrange the following structures into an accurate sequence through which glomerular filtrate passes.
    1. proximal convoluted tubule 2. loop of Henle 3. collecting duct 4. distal convoluted tubule 5. Bowman's capsule
    (a) 5, 4, 3, 2, 1 (b) 3, 4, 2, 5, 1 (c) 1, 5, 2, 3, 4 (d) 5, 4, 2, 3, 1 (e) 5, 1, 2, 4, 3
12. **VISUALIZE** Label the diagram.

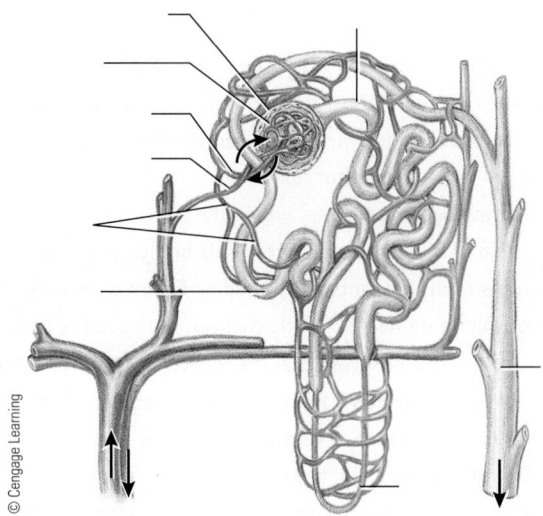

13. **CONNECT** Glucose is normally present in the glomerular filtrate, but not in the urine. Persons with uncontrolled diabetes mellitus are an exception. They have glucose in their urine, as well as increased urine output. Develop a hypothesis to explain these exceptions.
14. What types of osmoregulatory challenges do humans experience? Explain. What mechanisms do we have to meet these challenges?

Evaluate and Synthesize

15. **EVOLUTION LINK** The number of protonephridia in a planarian is related to the salinity of its environment. Planaria inhabiting slightly salty water develop fewer protonephridia than those inhabiting less salty water, but the number of protonephridia quickly increases when the concentration of salt in the environment is lower. How is this increase adaptive? Explain.
16. **EVOLUTION LINK** The kangaroo rat's diet consists of dry seeds, and it drinks no water. What types of osmoregulatory adaptations might help this animal survive?
17. **EVOLUTION LINK** Although the kidney is basically similar in all vertebrates, its structure and function vary somewhat among vertebrate taxa. What are some of the adaptations that have occurred during the evolution of the vertebrate kidney?
18. **SCIENCE, TECHNOLOGY, AND SOCIETY** A person with kidney disease lives 10 to 15 years longer with a kidney transplant than on continued dialysis. However, in the United States many individuals have difficulty being accepted by a transplant program because they do not have adequate insurance. The immunosuppressive drugs required after the surgery are very expensive, costing about $15,000 per year.. Should economic status be a deciding factor about whether a person lives or dies? or about quality of life?

To access course materials, such as Aplia and other companion resources, please visit **www.cengagebrain.com.**

49 Endocrine Regulation

Shark Song/M. Kazmers/Dembinsky Photo Associates

Endocrine regulation of color changes in a crustacean. Hormones regulate the distribution of pigment in the cells, resulting in color changes. This juvenile Puget Sound king crab (*Lopholithodes mandtii*) blends with its background by changing color.

KEY CONCEPTS

49.1 Endocrine glands and tissues secrete hormones, chemical messengers that bind to specific receptors on (or in) target cells and regulate physiological processes. Endocrine regulation depends mainly on negative feedback systems.

49.2 Several types of endocrine signaling are known: in classical endocrine signaling, endocrine glands secrete hormones that are transported to target cells by the blood; in neuroendocrine signaling, neuroendocrine cells produce hormones that are transported down axons and released into interstitial fluid; in autocrine signaling, a hormone acts on the very cells that produce it; and in paracrine signaling, a hormone acts on nearby cells.

49.3 Receptors are responsible for the specificity of the endocrine system. Small, lipid-soluble hormones enter target cells, bind to receptors inside the cell, and then activate or repress specific genes; hydrophilic hormones bind to cell-surface receptors and initiate signal transduction, leading to changes in cell processes.

49.4 Many invertebrate hormones are neurohormones secreted by neuroendocrine cells. Invertebrate hormones regulate metabolism, growth, and development, including molting and metamorphosis.

49.5 In vertebrates many organs and tissues secrete hormones that interact to regulate growth and development, metabolism, fluid balance, response to stress, reproduction, and many other processes.

A caterpillar becomes a butterfly. A crustacean changes color to blend with its background (see photograph). A young girl develops into a woman. An adult copes with chronic stress. The **endocrine system** regulates these physiological processes and many more, including growth and development; metabolism; fluid balance and concentrations of specific ions and chemical compounds in the blood, urine, and other body fluids; reproduction; and response to stress. The tissues and organs of the endocrine system secrete *hormones,* chemical messengers that signal other cells.

Endocrinology, the study of endocrine activity, is an active, exciting field of biomedical research. This branch of biology has its roots in experiments performed by German physiologist A. A. Berthold in the 1840s. Berthold removed the testes from young roosters and observed that the roosters' combs (a male secondary sex characteristic) did not grow as large as those in normal roosters. He then transplanted testes into some of the birds and observed that the combs grew to normal size. Berthold's methods became a model for subsequent studies in endocrinology and are still used by researchers today.

Investigators are actively studying mechanisms of hormone action, including characterizing receptors and identifying the molecules involved in *signal transduction*. Some endocrinologists use a reverse strategy for discovering new hormones and signaling pathways within the cell. They focus on "orphan" nuclear receptors, those for which ligands (the molecules that bind with them) are not yet known. Some of these "orphan" receptors are receptors for hormones that have not yet been identified. Using this strategy, researchers have identified intracellular signaling pathways for steroids, fatty acids, and several other compounds. Interestingly, many signaling pathways share the same molecules even though they respond to different stimuli and activate different responses. Much remains to be learned about how similar pathways lead to many different, and very specific, cell actions.

In this chapter we discuss how the endocrine system regulates life processes. We focus on how hormones maintain homeostasis, and we examine how overproduction or deficiency of various hormones interferes with normal functioning.

49.1 AN OVERVIEW OF ENDOCRINE REGULATION

LEARNING OBJECTIVES

1 Compare endocrine system function with nervous system function and describe how these systems work together to regulate body processes.

2 Summarize the regulation of endocrine action by negative feedback systems.

3 Identify four main chemical groups to which hormones are assigned and give two examples for each group.

The endocrine system is a diverse collection of cells, tissues, and organs including specialized **endocrine glands** that produce and secrete **hormones,** chemical messengers that regulate many physiological processes. The term *hormone* is derived from a Greek word meaning "to excite." Hormones excite, or stimulate, changes in specific tissues. Endocrine glands differ from **exocrine glands** (such as sweat glands and gastric glands) that release their secretions into ducts. Endocrine glands have no ducts and secrete their hormones into the surrounding interstitial fluid or blood.

Endocrinologists have identified about ten discrete endocrine glands. They have also discovered specialized cells in the digestive tract, heart, kidneys, and many other parts of the body that release hormones or hormone-like substances. As a result of these discoveries, the scope of endocrinology now includes the production and actions of chemical messengers produced by a wide variety of organs, tissues, and cells.[1]

Hormones are typically transported by the blood. They produce a characteristic response only after they bind with specific receptors on **target cells,** the cells influenced by a particular hormone. Target cells may be in another endocrine gland or in an entirely different type of organ, such as a bone or kidney. Target cells may be located far from the endocrine gland. For example, the vertebrate thyroid gland secretes hormones that stimulate metabolism in tissues throughout the body.

Several types of hormones may be involved in regulating the metabolic activities of a particular type of cell. In fact, many hormones produce a synergistic effect in which the presence of one hormone enhances the effects of another.

The endocrine system and nervous system interact to regulate the body

The endocrine system works closely with the nervous system to maintain **homeostasis,** the balanced internal environment

---

[1] Pheromones are chemical messengers that animals release to communicate with other animals of the same species. Because pheromones are generally produced by exocrine glands and do not regulate metabolic activities within the animal that produces them, most biologists do not classify them as hormones. Their role in regulating behavior is discussed in Chapter 52.

of the body. Recall from Chapters 41 and 42 that the nervous system responds rapidly to stimuli by transmitting electrical and chemical signals. Neurons signal muscle and gland cells, including endocrine cells. The endocrine system signals a much wider range of target cell types than the nervous system does.

In general, the endocrine system responds more slowly than the nervous system, but its responses may be longer-lasting. As we will learn, the nervous system helps regulate many endocrine responses. For example, when the body is threatened, the hypothalamus signals the adrenal glands to secrete the hormone epinephrine. The hypothalamus, which is the link between the nervous and endocrine systems, also produces several hormones (neurohormones), including hormones that regulate the pituitary gland.

Nervous system and endocrine function further blur when we consider that the same signal molecule can function as either a neurotransmitter or a hormone, depending on its source. (You may recall from Chapters 41 and 42 that norepinephrine is secreted at synapses by neurons in the brain and by the sympathetic system.) Norepinephrine is also secreted as a hormone by the adrenal medulla, an endocrine organ.

Negative feedback systems regulate endocrine activity

Although present in minute amounts, more than 50 different hormones may be circulating in the blood of a vertebrate at any given time. Hormone molecules continuously move out of the circulation and bind with target cells. They are also removed from the blood by the liver, which inactivates some hormones, and by the kidneys, which excrete them.

Most endocrine action is regulated by **negative feedback systems,** discussed in Chapter 39. The parathyroid glands, located in the neck of tetrapod vertebrates, provide a good example of how negative feedback works in endocrine regulation (FIG. 49-1). The parathyroid glands regulate the calcium concentration of the blood. When the calcium concentration is not within homeostatic limits, nerves and muscles cannot function properly. For instance, when too few calcium ions are present, neurons fire spontaneously, causing muscle spasms. When calcium concentration varies too far from the steady state (either too high or too low), negative feedback systems restore homeostasis.

A decrease in calcium concentration in the plasma signals the parathyroid glands to release more parathyroid hormone. This hormone increases the concentration of calcium in the blood. When the calcium concentration rises above normal limits, the parathyroid glands slow their output of hormone. Both responses are negative feedback systems. An increase in calcium concentration results in decreased release of parathyroid hormone, whereas a decrease in calcium leads to increased hormone secretion. In each case, the response counteracts the inappropriate change, restoring the steady state. (Parathyroid action is discussed in greater detail later in the chapter.)

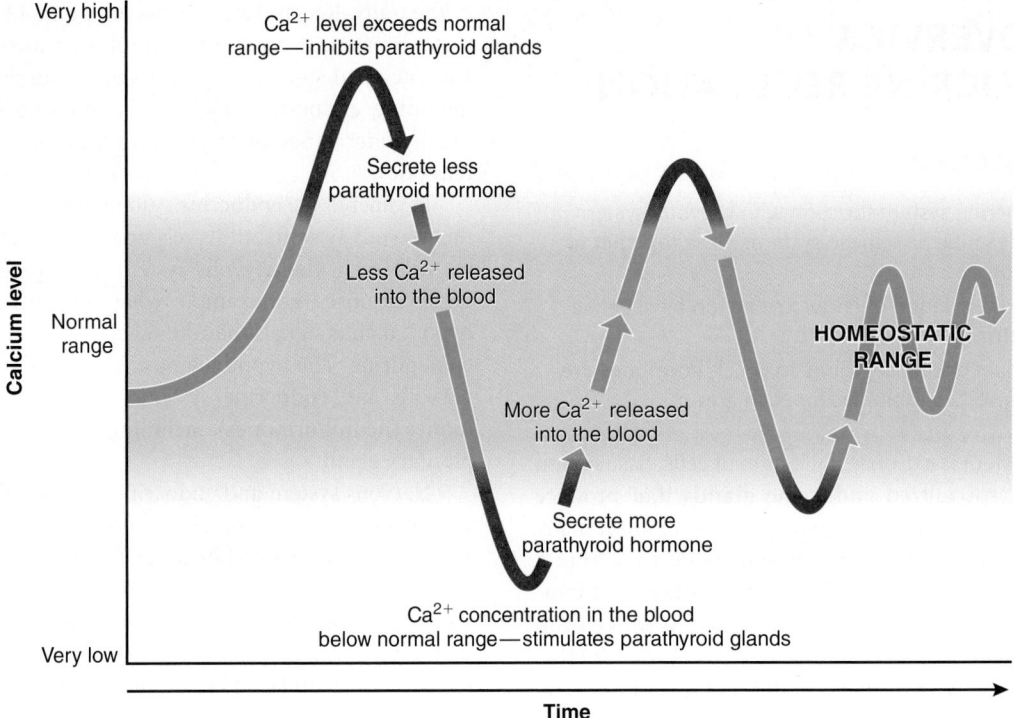

Figure 49-1 *Animation* **Regulation by negative feedback**
Whether the calcium concentration in the blood is too high or too low, negative feedback mechanisms restore homeostasis. Parathyroid hormone increases calcium concentration by stimulating calcium release from bones and calcium reabsorption from the kidney tubules. The *green area* represents the homeostatic range. (See Figure 49-11 for more detailed diagrams of parathyroid hormone regulation.)
© Cengage Learning

Hormones are assigned to four chemical groups

Although hormones are chemically diverse, they generally belong to one of four chemical groups: (1) fatty acid derivatives, (2) steroids, (3) amino acid derivatives, or (4) peptides or proteins. *Prostaglandins* and the juvenile hormones of insects are **fatty acid derivatives** (**FIG. 49-2a**). Prostaglandins are synthesized from arachidonic acid, a 20-carbon fatty acid. A prostaglandin has a five-carbon ring in its structure.

The molting hormone of insects is a **steroid hormone** (**FIG. 49-2b**). In vertebrates the adrenal cortex, testis, ovary, and placenta secrete steroid hormones synthesized from cholesterol. Examples of steroid hormones are cortisol, secreted by the adrenal cortex; the male reproductive hormone testosterone secreted by the testis; and the female reproductive hormones progesterone and estrogen produced by the ovary. Athletes (and others) sometimes abuse synthetic hormones known as anabolic steroids (see *Inquiring About: Anabolic Steroids and Other Performance-Enhancing Drugs*).

Chemically, the simplest hormones are **amino acid derivatives** (**FIG. 49-2c**). The thyroid hormones (T_3 and T_4) are synthesized from the amino acid tyrosine and iodide. Epinephrine (also known as adrenaline) and norepinephrine (also known as noradrenaline), produced by the medulla of the adrenal gland, are also derived from tyrosine. Melatonin is synthesized from the amino acid tryptophan.

The water-soluble **peptide hormones** are the largest hormone group. (Many endocrinologists include protein hormones in this group.) **Neuropeptides** are a large group of signaling molecules produced by neurons. Oxytocin and antidiuretic hormone (ADH), produced in the hypothalamus, are short neuropeptides, each composed of nine amino acids (**FIG. 49-2d**). Seven of the amino acids are identical in the two hormones, but the actions of these hormones are quite different. The hormones glucagon (see Fig. 3-19), secretin, adrenocorticotropic hormone (ACTH), and calcitonin are somewhat longer peptides that each consist of about 30 amino acids.

Insulin is a small protein that consists of two peptide chains joined by disulfide bonds. Growth hormone, thyroid-stimulating hormone, and the gonadotropic hormones, all secreted by the anterior pituitary gland, are large proteins with molecular weights of 25,000 or more.

CHECKPOINT 49.1

- *What are target cells?*
- **CONNECT** *How do the nervous and endocrine systems work together?*
- **PREDICT** *The concentration of calcium in the blood decreases below the homeostatic level. How does the body respond?*
- *What are four major chemical groups of hormones? Give an example of each.*

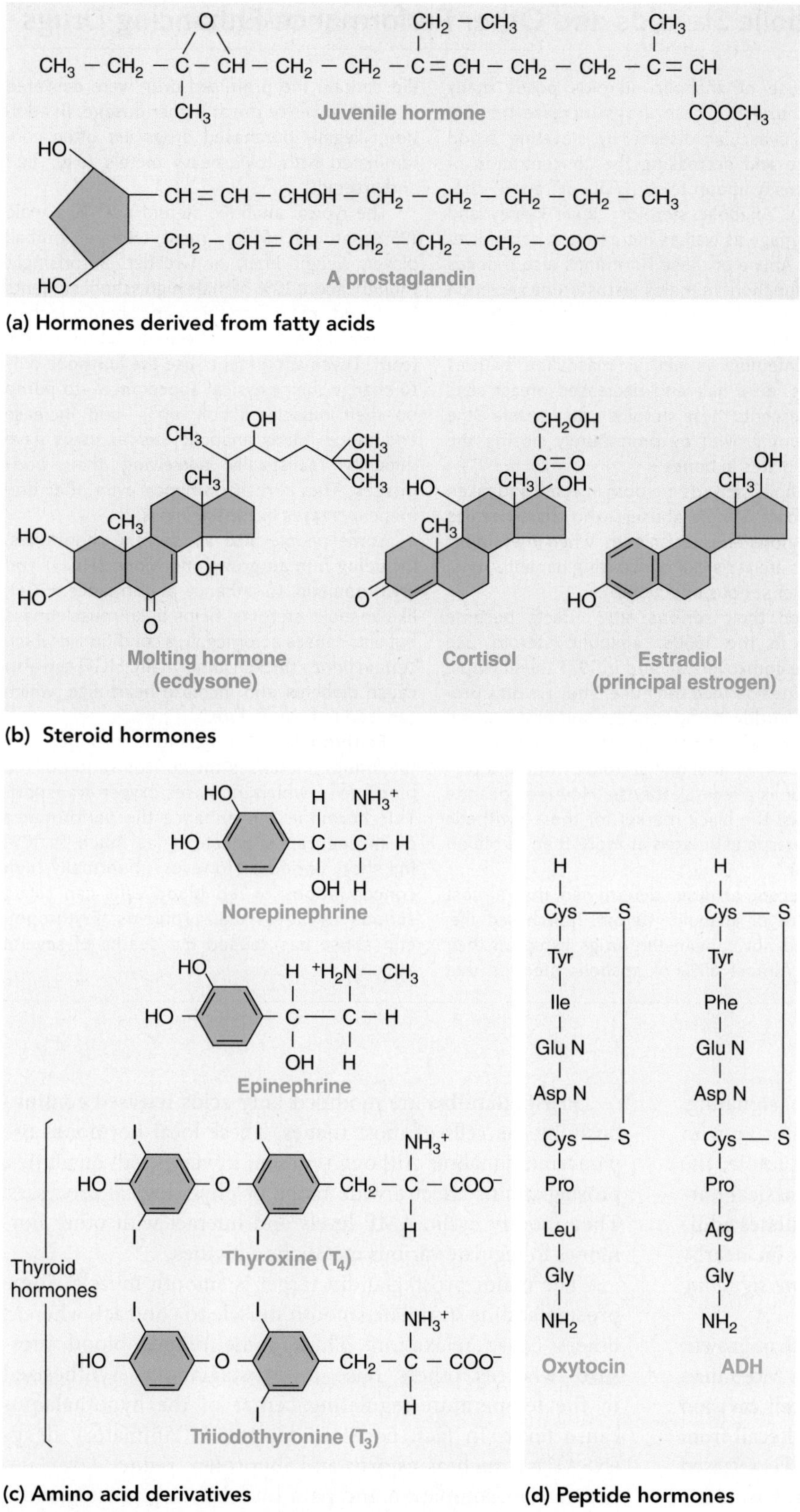

(a) Hormones derived from fatty acids

(b) Steroid hormones

(c) Amino acid derivatives

(d) Peptide hormones

Figure 49-2 Major chemical groups of hormones
© Cengage Learning

49.2 TYPES OF ENDOCRINE SIGNALING

LEARNING OBJECTIVE

4   Compare four types of endocrine signaling.

Hormones signal their target cells by classical endocrine signaling, neuroendocrine signaling, autocrine signaling, and paracrine signaling. In *classical endocrine signaling,* hormones are secreted by endocrine glands and are transported by the blood to target cells (**FIG. 49-3a**). Steroid hormones and thyroid hormones are transported bound to plasma proteins. The water-soluble peptide hormones are dissolved in the plasma. Hormones move out of the blood with plasma and are present in interstitial fluid.

Neurohormones are transported in the blood

Certain neurons, known as **neuroendocrine cells,** are an important link between the nervous and endocrine systems. In *neuroendocrine signaling* neuroendocrine cells produce **neurohormones** that are transported down axons and released into the interstitial fluid. They typically diffuse into capillaries and are transported by the blood (**FIG. 49-3b**). Invertebrate endocrine systems are largely neuroendocrine. In vertebrates the hypothalamus produces neurohormones that link the nervous system with the pituitary gland, an endocrine gland that secretes several hormones.

Some local regulators are considered hormones

A **local regulator** is a signaling molecule that diffuses through the interstitial fluid and acts on nearby cells. Most endocrinologists include at least certain local regulators as hormones. Also, some chemical compounds that act as classical hormones (transported by the blood) function as local regulators under some conditions.

Why are anabolic steroids and some other performance-enhancing drugs abused? An estimated one million people in the United States, half of them adolescents, abuse certain hormones so as to enhance their physical performance. Anabolic steroids, a group of synthetic hormones, are officially called anabolic-androgenic steroids. Anabolic steroids are synthetic androgens (male reproductive hormones) that were developed in the 1930s to prevent muscle atrophy in patients with diseases that prevented them from moving.

In the 1950s, anabolic steroids became popular with professional athletes, who used them to increase muscle mass, physical strength, endurance, and aggressiveness. In truth, their athletic performance was probably enhanced, at least in part, by drug-induced euphoria and increased enthusiasm for training. The term *doping* refers to the use of banned performance-enhancing drugs.

As with other hormones, the concentration of steroids circulating in the body is precisely regulated, so use of anabolic steroids interferes with normal physiological processes. These drugs remain in the body for a long time. Their metabolites (breakdown products) can be detected in the urine for up to six months.

Even during short-term use and at relatively low doses, anabolic steroids have a significant effect on mood and behavior. At higher doses users experience disturbed thought processes, forgetfulness, and confusion and often find themselves easily distracted. The term *roid rage* refers to the mood swings, unpredictable anger, increased aggressiveness, and irrational behavior exhibited by many users.

Misuse of anabolic steroids poses many health hazards. These drugs increase the risk of cardiovascular disease by elevating blood pressure and decreasing the concentration of high-density lipoproteins (HDL, or "good" cholesterol). Anabolic steroids cause kidney and liver damage as well as increase the risk of liver cancer. Abuse of these hormones also reduces sexual function. In males testosterone secretion decreases and the testes shrink, leading to sterility. In females abuse of anabolic steroids can cause infertility as well as masculine features such as facial hair and decreased breast size. In adolescents these steroids cause severe acne and stunt growth by prematurely closing the growth plates in bones.

Anabolic steroids are both injected and taken in pill form. Steroid abusers who share needles or use nonsterile techniques when they inject steroids are at risk for contracting hepatitis, HIV, and other serious infections.

When their serious side effects became known in the 1960s, anabolic steroid use became controversial, and in 1973 the Olympic Committee banned their use. They are now prohibited worldwide by amateur and professional sports organizations. In the United States, possession of anabolic steroids without a prescription is a federal offense. However, despite the risks, the black market for these synthetic hormones is estimated at more than $1 billion per year.

A recent analysis determined that almost one-third of anabolic steroids purchased illegally did not contain the drugs listed on their labels. Almost 50% of anabolic steroids that did contain the promised drug were delivered in much lower or much higher dosage. In addition, illegally purchased drugs are often contaminated with toxic heavy metals (e.g., lead and arsenic).

The typical anabolic steroid user is a male (95%) athlete (65%), most often a football player, weight lifter, or wrestler. Surprisingly, though, about 10% of male high school students have used anabolic steroids, and about one-third of these students are not even on a high school team. These adolescents use the hormone only to change their physical appearance—to pump up their muscles ("bulk up")—and increase endurance. Many anabolic steroid users have difficulty realistically perceiving their body images. They remain unhappy even after dramatic increases in muscle mass.

Some people also abuse other hormones, including human growth hormone (HGH) and erythropoietin, to enhance performance. HGH, like anabolic steroids, helps build muscle mass but also causes acromegaly, a condition in which certain bones thicken abnormally. HGH can also cause diabetes and increase heart size, which can lead to heart failure.

Erythropoietin, a hormone produced by the kidney, increases the concentration of red blood cells, which increases oxygen transport. This hormone can enhance the performance of an endurance athlete by as much as 10% for short periods. However, abnormally high concentrations of red blood cells can cause serious cardiovascular problems. Erythropoietin abuse has caused the deaths of several athletes.

Local regulators use *autocrine* or *paracrine* signaling. In *autocrine signaling* a hormone or other regulator acts on the very cells that produce it (FIG. 49-3c). For example, the female hormone estrogen, which functions as a classical hormone, can also exert an autocrine effect that stimulates additional estrogen secretion. At times, estrogen acts on nearby cells, a type of local regulation known as *paracrine signaling* (FIG. 49-3d).

Local regulators include chemical mediators such as growth factors, histamine, nitric oxide, and prostaglandins. More than 50 **growth factors,** typically peptides, stimulate cell division and normal development in specific types of cells. Recall from Chapter 45 that **histamine** is stored in mast cells and is released in response to allergic reactions, injury, or infection. Histamine causes blood vessels to dilate and capillaries to become more permeable. **Nitric oxide (NO),** another local regulator, is a gas produced by many types of cells, including those lining blood vessels. NO relaxes smooth muscle fibers in the blood vessel wall, causing the blood vessel to dilate. (See discussion of nitric oxide in Chapter 6.)

Prostaglandins are modified fatty acids released continuously by the cells of most tissues. These local hormones use paracrine signaling. Although present in very small quantities, prostaglandins affect a wide range of physiological processes. They modify cyclic AMP levels and interact with other hormones to regulate various metabolic activities.

The major prostaglandin target is smooth muscle. Some prostaglandins stimulate smooth muscle to contract, whereas others cause relaxation. Thus, some reduce blood pressure, whereas others raise it. Prostaglandins synthesized in the temperature-regulating center of the hypothalamus cause fever. In fact, nonsteroidal anti-inflammatory drugs (NSAIDs), such as aspirin and ibuprofen, reduce fever and decrease inflammation and pain by inhibiting prostaglandin synthesis.

Because prostaglandins are involved in the regulation of so many metabolic processes, they have great potential for a variety of clinical uses. Physicians use them to induce labor in pregnant women, to induce abortion, and to promote the healing of ulcers in the stomach and duodenum. Prostaglandins may someday

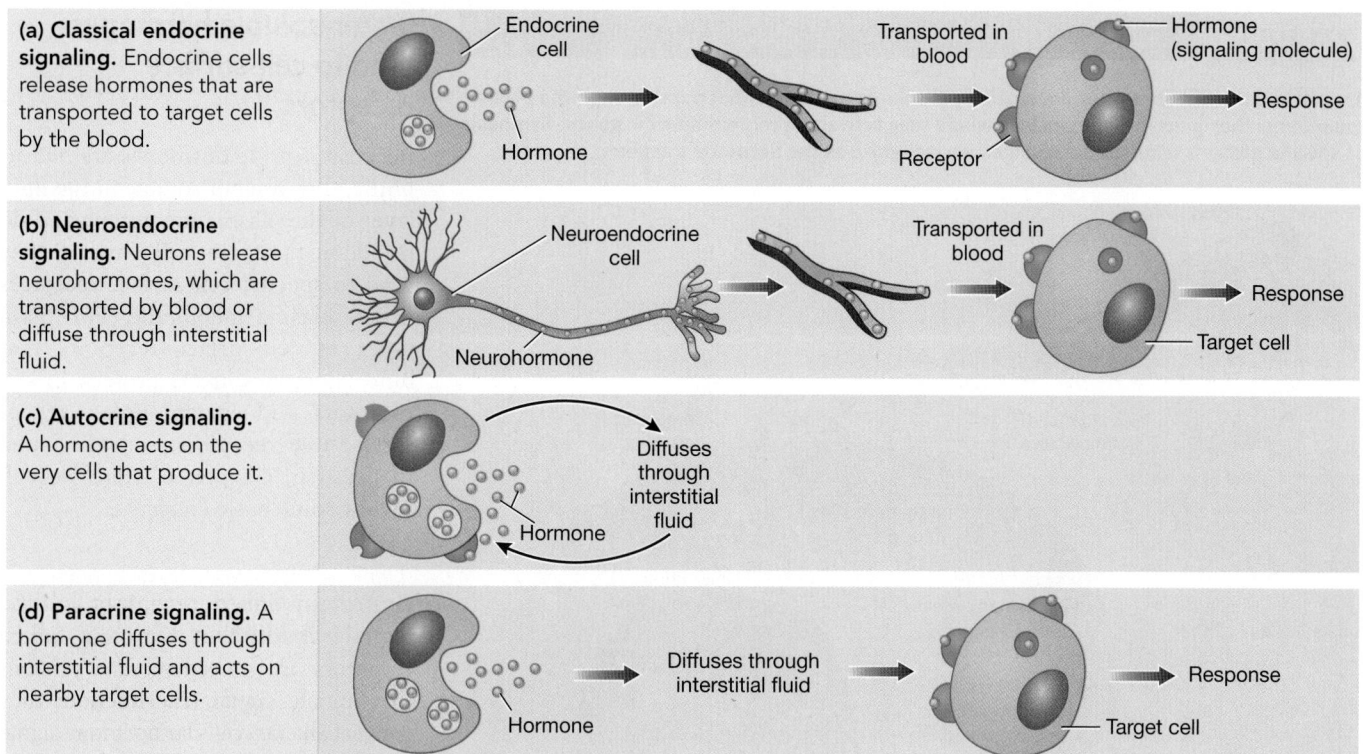

(a) Classical endocrine signaling. Endocrine cells release hormones that are transported to target cells by the blood.

Endocrine cell

Hormone

Transported in blood

Hormone (signaling molecule)

Response

Receptor

(b) Neuroendocrine signaling. Neurons release neurohormones, which are transported by blood or diffuse through interstitial fluid.

Neuroendocrine cell

Neurohormone

Transported in blood

Response

Target cell

(c) Autocrine signaling. A hormone acts on the very cells that produce it.

Diffuses through interstitial fluid

Hormone

(d) Paracrine signaling. A hormone diffuses through interstitial fluid and acts on nearby target cells.

Hormone

Diffuses through interstitial fluid

Response

Target cell

Figure 49-3 Some types of endocrine signaling

Endocrine cells secrete hormones that signal target cells, leading to a change in one or more cell processes.

© Cengage Learning

be used to treat a variety of illnesses, including asthma, arthritis, kidney disease, certain cardiovascular disorders, and some forms of cancer.

CHECKPOINT 49.2

- **VISUALIZE** *Draw diagrams to contrast neuroendocrine signaling with classical endocrine signaling.*
- *How are autocrine and paracrine signaling different?*
- *What are two functions of prostaglandins?*

49.3 MECHANISMS OF HORMONE ACTION

LEARNING OBJECTIVE

5 Compare the mechanism of action of small, lipid-soluble hormones with that of hydrophilic hormones; include the role of second messengers, such as cyclic AMP.

A hormone present in the interstitial fluid remains "unnoticed" until it comes in contact with its target cells and binds to specific receptors on the cell surface or inside the cell. Recall that the receptor site is somewhat similar to a lock, and the hormones are like different keys. Only the hormone that fits the specific receptor can influence the metabolic machinery of the cell (see

Chapter 6). Receptors are responsible for the specificity of the endocrine system.

Receptors are continuously synthesized and degraded. Their numbers are decreased or increased by **receptor down-regulation** and **receptor up-regulation.** For example, when the concentration of a hormone is too high for an extended period, cells decrease the number of receptors for that hormone. Down-regulation decreases the sensitivity of target cells to the hormone, so it dampens the hormone's ability to carry out its function. Up-regulation occurs when the concentration of a particular hormone is too low. Increasing the number of receptors on the plasma membrane increases the hormone's effect on the cell.

Lipid-soluble hormones enter target cells and activate genes

Steroid hormones and thyroid hormones are relatively small, lipid-soluble (hydrophobic) molecules that easily pass through the plasma membrane of the target cell. Specific protein receptors in the cytoplasm or in the nucleus bind with the hormone to form a hormone–receptor complex (**FIG. 49-4**). For example, thyroid hormones bind with receptors in the nucleus that are transcription factors. When activated by the hormone, they activate or repress transcription of messenger RNA coding for specific proteins. Proteins that are synthesized produce changes in structure or metabolic activity.

Steroid and thyroid hormones are small, lipid-soluble molecules that pass through the plasma membrane; they pass into the nucleus where they activate or repress specific genes. Synthesis of specific proteins leads to the changes we recognize as the hormone's actions.

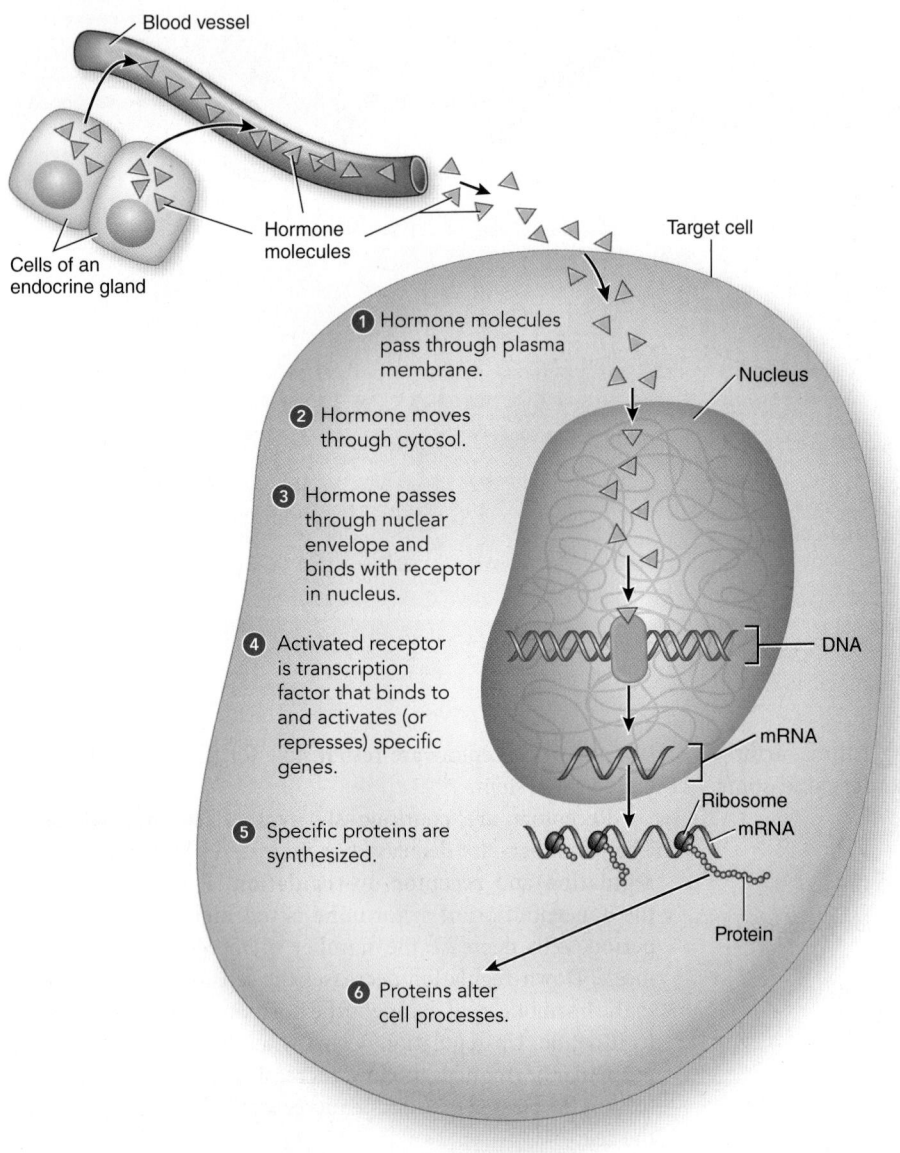

Blood vessel

Hormone molecules

Cells of an endocrine gland

Target cell

1 Hormone molecules pass through plasma membrane.

2 Hormone moves through cytosol.

3 Hormone passes through nuclear envelope and binds with receptor in nucleus.

4 Activated receptor is transcription factor that binds to and activates (or represses) specific genes.

5 Specific proteins are synthesized.

6 Proteins alter cell processes.

Nucleus

DNA

mRNA

Ribosome
mRNA

Protein

Figure 49-4 *Animation* **Mechanism of action of steroid and thyroid hormones**

PREDICT What would happen if the number of receptors in the nucleus were severely decreased?
© Cengage Learning

Gene activation and protein synthesis may take several hours. However, steroid hormones may also have more rapid mechanisms of action that do not require protein synthesis. For example, when an estrogen receptor in the membrane of the endoplasmic reticulum is activated by estrogen binding, it initiates rapid signaling that leads to changes in the cell within minutes or even seconds. Plasma membrane receptors have been described for two steroid hormones, and others will likely be discovered.

Water-soluble hormones bind to cell-surface receptors

Because peptide hormones are hydrophilic, they are not soluble in the lipid layer of the plasma membrane and do not enter the target cell. Instead, they bind to specific cell-surface receptors in the plasma membrane. Two main types of cell-surface receptors that bind hormones are G protein–linked receptors and enzyme-linked receptors. These receptors were discussed in detail in Chapter 6. Here we will review some basic concepts.

G protein–linked receptors initiate signal transduction **G protein–linked receptors** are transmembrane proteins that initiate **signal transduction;** they convert an extracellular hormone signal into an intracellular signal that affects some cell process. The hormone does not enter the cell. It serves as the **first messenger** and relays information to a **second messenger,** or intracellular signal. The second messenger then signals effector molecules that carry out the action.

G protein–linked receptors activate **G proteins,** a group of integral regulatory proteins (**FIG. 49-5**). The G indicates that they bind *guanosine triphosphate (GTP),* which, like ATP, is an important molecule in energy reactions. When the system is inactive, G protein binds to *guanosine diphosphate (GDP),* which is similar to ADP, the hydrolyzed form of ATP. G proteins shuttle a signal between the receptor and a second messenger.

When a hormone-bound receptor binds to it, a stimulatory G protein releases GDP and replaces it with GTP. The G protein undergoes a change in shape that allows it to bind with and activate the enzyme **adenylyl cyclase** on the cytoplasmic side of the plasma membrane. Once activated, adenylyl cyclase catalyzes the conversion of ATP to **cyclic AMP (cAMP).** By coupling the hormone–receptor complex to an enzyme that generates a signal, G proteins amplify hormone effects, and many second-messenger molecules are rapidly produced. One type of G protein, G_s, stimulates adenylyl cyclase, and another, G_i, inhibits it.

Peptide hormones bind with receptors on the plasma membrane. The receptor is a signal transducer that converts the hormone signal to an intracellular signal.

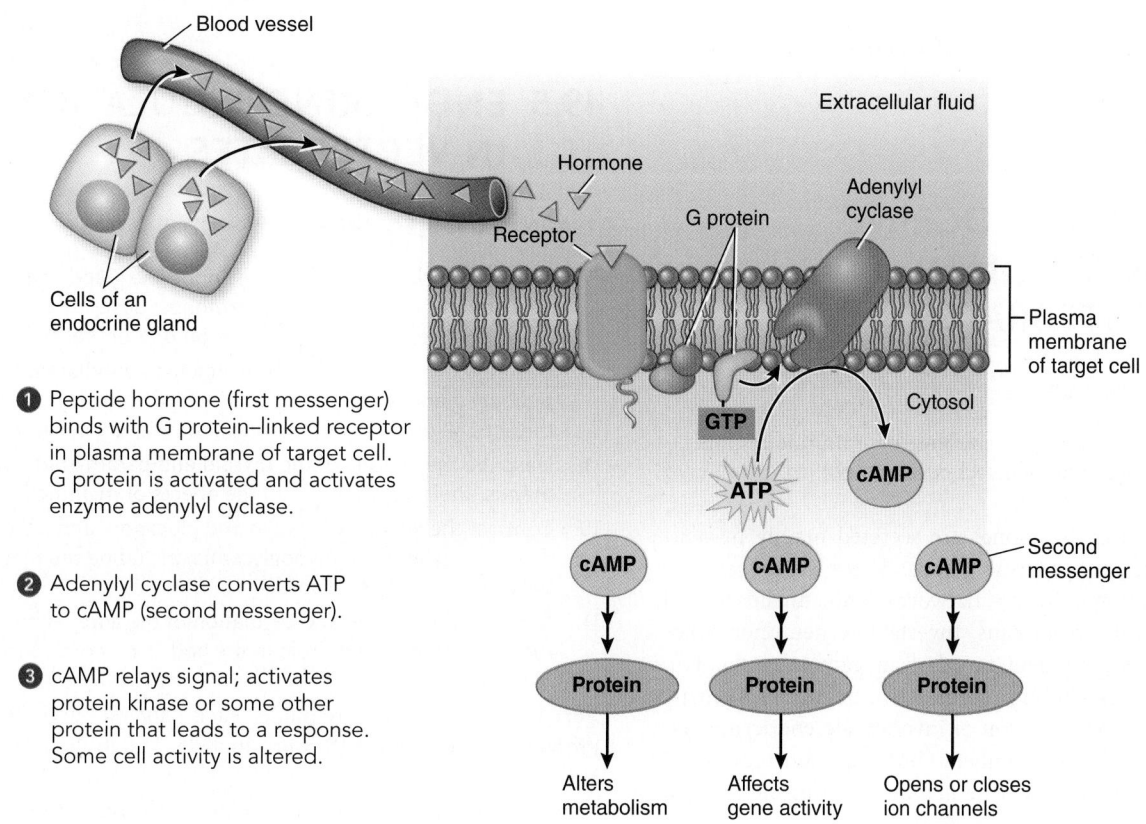

① Peptide hormone (first messenger) binds with G protein–linked receptor in plasma membrane of target cell. G protein is activated and activates enzyme adenylyl cyclase.

② Adenylyl cyclase converts ATP to cAMP (second messenger).

③ cAMP relays signal; activates protein kinase or some other protein that leads to a response. Some cell activity is altered.

Figure 49-5 Mechanism of action of peptide hormones

Many peptide hormones signal target cells by way of a G protein–linked receptor. Cyclic AMP, or some other second messenger, relays the signal by activating a cascade of phosphorylation reactions. The final protein in the pathway changes some cell process.

PREDICT What would happen if adenylyl cyclase were inhibited?

© Cengage Learning

Cyclic AMP activates **protein kinases,** enzymes that phosphorylate (add a phosphate group to) specific proteins. When phosphorylated, protein kinase is activated and may phosphorylate the next protein in the pathway. A chain of reactions leads to some metabolic change. The substrates for protein kinases are different in various cell types, so the effect of cAMP varies. In skeletal muscle cells, a protein kinase triggers the breakdown of glycogen to glucose, whereas in cells of the hypothalamus, a protein kinase activates the gene that encodes a growth-inhibiting hormone.

Recall from Chapter 6 that some G proteins use phospholipids as second messengers. Certain hormone–receptor complexes activate a G protein that then activates phospholipase C, a membrane-bound enzyme (see Fig. 6-9). This enzyme splits a membrane phospholipid into two products: *inositol trisphosphate* (IP_3) and *diacylglycerol* (DAG). Both act as second messengers. DAG (in combination with calcium ions) activates a protein kinase that phosphorylates a variety of proteins, and IP_3 opens calcium channels in the endoplasmic reticulum, releasing calcium ions into the cytosol. Calcium ions bind to certain proteins, such as *calmodulin*. The activated calmodulin then activates other proteins.

Enzyme-linked receptors function directly **Enzyme-linked receptors** are transmembrane proteins with a hormone-binding site outside the cell and an enzyme site inside the cell (see Fig. 6-5c). These receptors are not linked to G proteins. They function directly as enzymes or are directly linked to enzymes. Most enzyme-linked receptors are **tyrosine kinase receptors** that bind growth factors and other signal molecules, including insulin. When the receptor is activated, it phosphorylates the amino acid tyrosine in specific signaling proteins inside the cell.

- **CONNECT** *What are the roles of receptors and second messengers in hormone action?*
- *How do steroid and peptide hormones differ in their mechanisms of action?*
- *What is the mechanism of action of a hormone that uses cyclic AMP as a second messenger?*

49.4 NEUROENDOCRINE REGULATION IN INVERTEBRATES

LEARNING OBJECTIVE

6 Identify four functions of invertebrate neurohormones and describe the regulation of insect development.

Among invertebrates, hormones are secreted mainly by neurons rather than by endocrine glands. These *neurohormones* regulate regeneration in hydras, flatworms, and annelids as well as color changes in crustaceans. Invertebrate neurohormones regulate growth, development, metabolism, gamete production, and reproduction (including reproductive behavior) in many groups. Trends in the evolution of invertebrate endocrine systems include an increasing number of both neurohormones and hormones secreted by endocrine glands and a greater role of hormones in physiological processes.

Insects have endocrine glands and neuroendocrine cells. Their various hormones and neurohormones interact with one another to regulate metabolism, growth, and development, including molting and metamorphosis. Hormonal control of development in insects varies among species. Generally, some environmental factor (e.g., temperature change) activates neuroendocrine cells in the brain. These cells then produce a neurohormone referred to as *brain hormone* (BH), which is transported down axons and stored in the paired *corpora cardiaca* (sing., *corpus cardiacum;* **FIG. 49-6**). When released from the corpora cardiaca, BH stimulates the *prothoracic glands,* endocrine glands in the prothorax, to produce *molting hormone* (MH), also called *ecdysone* (see Fig. 49-2b). MH is a steroid hormone that stimulates growth and molting.

In the immature insect, paired endocrine glands called *corpora allata* (sing., *corpus allatum*) secrete *juvenile hormone* (JH). This hormone suppresses metamorphosis at each larval molt so that the insect increases in size while remaining in its immature state. After the molt, the insect is still in a larval stage. When the concentration of JH decreases below some critical level, metamorphosis occurs, and the insect is transformed into a pupa (see Chapter 31). In the absence of JH, the pupa molts and becomes an adult. The nervous system regulates the secretory activity of the corpora allata, and the amount of JH decreases with successive molts.

49.5 ENDOCRINE REGULATION IN VERTEBRATES

LEARNING OBJECTIVES

7 Identify the classical vertebrate endocrine glands and describe the actions of their hormones; describe the effects of hypersecretion and hyposecretion on homeostasis.

8 Describe the mechanisms by which the hypothalamus and pituitary gland integrate regulatory functions; summarize the actions of the hypothalamic and pituitary hormones.

9 Describe the actions of the thyroid and parathyroid hormones, their regulation, and the effects of malfunction.

10 Contrast the actions of insulin and glucagon, and describe diabetes mellitus and hypoglycemia, including the metabolic effects of these disorders.

11 Describe the actions and regulation of the adrenal hormones, including their role in the body's response to stress.

Vertebrate hormones regulate such diverse activities as growth and development, reproduction, metabolic rate, fluid balance, blood homeostasis, and response to stress. Most vertebrates have similar endocrine glands, although the actions of some hormones may be different in various vertebrate groups. **TABLE 49-1** gives the sources, target tissues, and principal physiological actions of some of the major vertebrate hormones. The hypothalamus and pituitary gland regulate many of these hormones. The principal human endocrine glands are illustrated in **FIGURE 49-7**.

Homeostasis depends on normal concentrations of hormones

When a disorder or disease process affects an endocrine gland, the rate of secretion may become abnormal. If **hyposecretion,** abnormally reduced output, occurs, target cells are deprived of needed stimulation. If **hypersecretion,** abnormally increased output, occurs, the target cells may be overstimulated. In some endocrine disorders, an appropriate amount of hormone is secreted, but the target cell receptors do not function properly. As a result, the target cells cannot respond to the hormone. These failures in regulation threaten homeostasis and can lead to predictable metabolic malfunctions and clinical symptoms (**TABLE 49-2**).

The hypothalamus regulates the pituitary gland

Most endocrine activity is controlled directly or indirectly by the **hypothalamus,** which links the nervous and endocrine

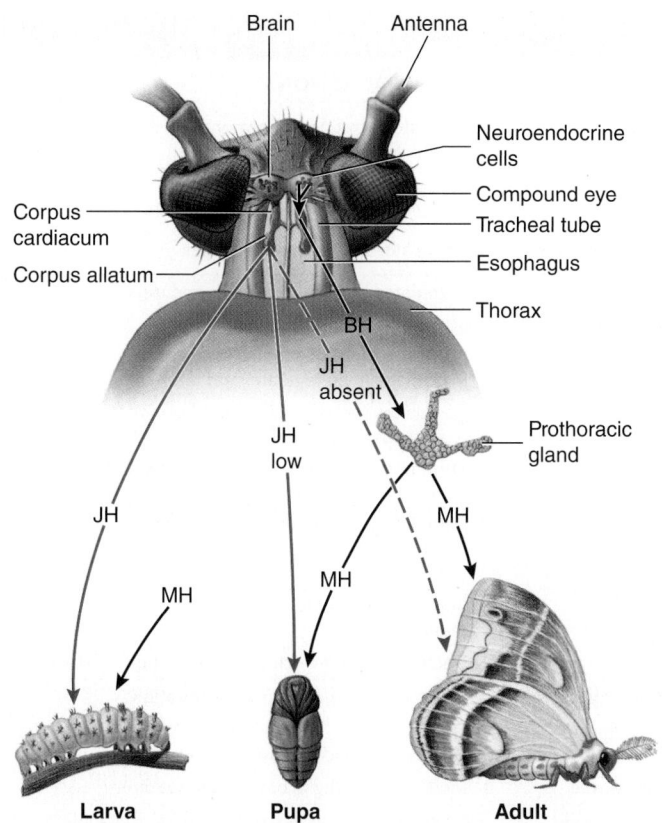

Figure 49-6 Regulation of growth and molting in insects

Hormones regulate insect development. Most insects go through a series of larval stages and then undergo metamorphosis to the adult form. This dissection of an insect head shows the brain, paired corpora allata, and paired corpora cardiaca.

© Cengage Learning

❶ Neuroendocrine cells secrete brain hormone (BH), which is stored in corpora cardiaca.

❷ When released, BH stimulates prothoracic glands to secrete molting hormone (MH), which stimulates growth and molting.

❸ In immature insect, corpora allata secrete juvenile hormone (JH), which suppresses metamorphosis at each larval molt.

❹ Metamorphosis to adult form occurs when MH acts in absence of JH.

systems both anatomically and physiologically. The **pituitary gland** is connected to the hypothalamus by the pituitary stalk. The hypothalamus receives input from other areas of the brain and from hormones in the blood. In response, neurons of the hypothalamus secrete neurohormones that regulate specific physiological processes.

Because secretions of the pituitary gland control the activities of several other endocrine glands, biologists sometimes refer to it as the master gland of the body. Although it is the size of a large pea and weighs only about 0.5 g (0.02 oz), the pituitary gland produces six peptide hormones that exert far-reaching influence over growth, metabolism, reproduction, and many other body activities. The human pituitary gland consists of two main parts: the *anterior pituitary* and the *posterior pituitary.* In some vertebrates, an intermediate lobe is present.

The posterior pituitary gland releases hormones produced by the hypothalamus

The **posterior pituitary,** also called the posterior lobe of the pituitary gland, develops from brain tissue. This neuroendocrine organ *secretes* two peptide hormones, **oxytocin** and **antidiuretic hormone (ADH),** also known as *vasopressin.* Oxytocin and ADH are actually *produced* by neuroendocrine

cells in two distinct areas of the hypothalamus. These hormones are enclosed within vesicles and transported down the axons of the neuroendocrine cells into the posterior pituitary gland (**FIG. 49-8** on page 1056). The vesicles are stored in the axon terminals until the neuron is stimulated. Then the hormones are released and diffuse into surrounding capillaries.

We discussed the function of ADH in Chapter 48. Recall that ADH stimulates reabsorption of water in the kidneys. Oxytocin concentration in the blood rises toward the end of pregnancy, stimulating the strong contractions of the uterus needed to expel a baby. Oxytocin is sometimes administered clinically to initiate or speed labor.

After birth, when an infant sucks at its mother's breast, sensory neurons signal the pituitary to release oxytocin. The hormone stimulates contraction of smooth muscle cells surrounding the milk glands so that milk is let down into the ducts, making it available to the nursing infant. Because oxytocin also stimulates the uterus to contract, breast-feeding promotes recovery of the uterus to nonpregnant size.

Oxytocin also plays a role in social behavior. In voles, for example, males with higher concentrations of oxytocin are more likely to be monogamous (have a single sexual partner) than those with lower concentrations. In humans oxytocin affects maternal behavior. Of critical importance, this hormone facilitates attachment between mother and infant. It appears that oxytocin continues to play a role in social behavior throughout the lifespan. This hormone appears to help us recognize familiar faces and promote trust and cooperation. Oxytocin may also help us notice social cues.

The anterior pituitary gland regulates growth and other endocrine glands

The **anterior pituitary,** also called the anterior lobe of the pituitary, develops from epithelial cells rather than neural cells. The anterior pituitary functions like a classical endocrine gland: it receives signals by way of the blood and releases its hormones into the blood. The hypothalamus produces several **releasing hormones** and **inhibiting hormones** that regulate the production and secretion of specific hormones by the anterior pituitary.

TABLE 49-1 Some Vertebrate Endocrine Glands and Their Hormones

GLAND	HORMONE	TARGET TISSUE	PRINCIPAL ACTIONS
Hypothalamus	Releasing and inhibiting hormones	Anterior pituitary	Regulate secretion of hormones by the anterior pituitary
Posterior pituitary (storage and release of hormones produced by hypothalamus)	Oxytocin	Uterus Mammary glands	Stimulates contraction Stimulates ejection of milk into ducts
	Antidiuretic hormone (ADH)	Kidneys (collecting ducts)	Stimulates reabsorption of water
Anterior pituitary	Growth hormone (GH)	General	Stimulates growth of skeleton and muscle
	Prolactin	Mammary glands	Stimulates milk production
	Melanocyte-stimulating hormones (MSH)	Pigment cells in skin	Stimulate melanin production in some animals; in humans, help regulate food intake
	Thyroid-stimulating hormone (TSH)	Thyroid gland	Stimulates secretion of thyroid hormones; helps regulate bone remodeling
	Adrenocorticotropic hormone (ACTH)	Adrenal cortex	Stimulates secretion of adrenal cortical hormones
	Gonadotropic hormones* (follicle-stimulating hormone [FSH]; luteinizing hormone [LH])	Gonads	Stimulate gonad function and growth
Thyroid gland	Thyroxine (T_4) and triiodothyronine (T_3)	General	Stimulate metabolic rate; regulate energy metabolism
	Calcitonin	Bone	Decreases blood calcium concentration
Parathyroid glands	Parathyroid hormone	Bone, kidneys, digestive tract	Increases blood calcium concentration
Pancreas	Insulin	General	Decreases blood glucose concentration
	Glucagon	Liver, adipose tissue	Increases blood glucose concentration
Adrenal medulla	Epinephrine and norepinephrine	Muscle; blood vessels; liver; adipose tissue	Help body cope with stress; increase metabolic rate; increase blood glucose concentration; increase heart rate and blood pressure
Adrenal cortex	Mineralocorticoids	Kidney tubules	Maintain sodium and potassium balance
	Glucocorticoids	General	Help body cope with long-term stress; increase blood glucose concentration
Pineal gland	Melatonin	Hypothalamus	Regulates biological rhythms
Ovary*	Estrogens	General; uterus	Develop and maintain sex characteristics in females; stimulate growth of uterine lining
	Progesterone	Uterus; breast	Stimulates development of uterine lining
Testis*	Testosterone	General; reproductive structures	Develops and maintains sex characteristics in males; promotes spermatogenesis

*Reproductive hormones are discussed in Chapter 50; see Tables 50-1 and 50-2.

© Cengage Learning

Releasing and inhibiting hormones are produced by specific endocrine cells in the hypothalamus and are stored in vesicles in the axon terminals. When signaled to do so, the cells secrete a particular releasing (or inhibiting) hormone, which then enters a capillary. The hormone is transported in the blood through special portal veins that connect the hypothalamus with the anterior lobe of the pituitary. Within the anterior pituitary, the portal veins divide into a second set of capillaries. Thus, these portal veins, like the hepatic portal vein, do not deliver blood to a larger vein directly; they connect two sets of capillaries.

Releasing and inhibiting hormones pass through the capillary walls into the tissue of the anterior lobe of the pituitary. These hormones control the secretion of anterior pituitary hormones. For example, when the body is stressed, the hypothalamus secretes *corticotropin-releasing factor*. This releasing factor stimulates the anterior pituitary to secrete *adrenocorticotropic hormone (ACTH)*.

The anterior lobe secretes prolactin, melanocyte-stimulating hormones, growth hormone, and several **tropic hormones** that stimulate other endocrine glands (**FIG. 49-9**). **Prolactin** stimulates the cells of the mammary glands to produce milk in a nursing mother. This hormone has diverse roles in various vertebrate groups. In fishes and amphibians, for example, it helps maintain water and electrolyte balance by decreasing Na^+ and K^+ loss in the urine. Prolactin also inhibits metamorphosis in amphibians and stimulates molting in reptiles. This versatile hormone appears to have a role in immune function and in formation of new blood vessels.

Melanocyte-stimulating hormones (MSH) (also referred to as *melanocortins*) are a group of peptide hormones secreted by the anterior pituitary gland and also by certain neurons in the hypothalamus. In some fishes, amphibians, and reptiles, MSH causes the skin to darken, facilitating camouflage. In humans it

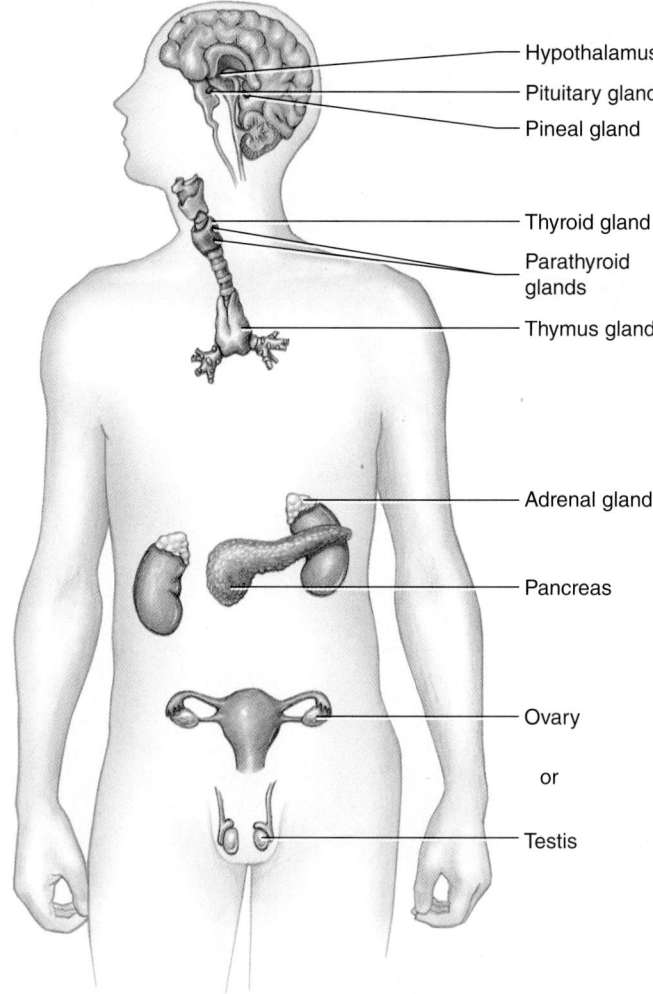

Figure 49-7 *Animation* **The principal human endocrine glands**
Many of the hormones produced by these glands and their actions are discussed in this chapter.
© Cengage Learning

Labels (top to bottom):
- Hypothalamus
- Pituitary gland
- Pineal gland
- Thyroid gland
- Parathyroid glands
- Thymus gland
- Adrenal gland
- Pancreas
- Ovary
- or
- Testis

suppresses appetite, leading to weight loss, and thus is important in regulating body weight and energy.

Growth hormone stimulates protein synthesis Whether someone is tall or short depends on many factors, including genes, diet, and hormone balance. **Growth hormone** (also called

somatotropin) is referred to as an **anabolic hormone** because it promotes tissue growth. (The general acronym for vertebrate growth hormone is GH; human growth hormone is referred to as HGH or hGH.) Many of the effects of GH on skeletal growth are indirect. GH stimulates liver cells (and cells of many other tissues) to produce peptides called **insulin-like growth factors (IGFs).** These growth factors (1) promote the linear growth of the skeleton by stimulating cartilage formation in growth areas of bones and (2) promote protein synthesis and other anabolic processes, which stimulates general tissue growth and increases organ size.

In adults as well as in growing children, GH is secreted in pulses throughout the day. The hypothalamus regulates GH secretion by secreting a **growth hormone–releasing hormone (GHRH)** and a **growth hormone–inhibiting hormone (GHIH),** which is also called *somatostatin.* A high level of GH in the blood signals the hypothalamus to secrete GHIH; in response, the pituitary slows its release of GH. In contrast, a low level of GH in the blood stimulates the hypothalamus to secrete GHRH, which stimulates the pituitary gland to release more GH. Many other factors, including blood sugar level, amino acid concentration, and stress, influence GH secretion.

For children to grow, research supports the age-old notions that they need plenty of sleep, a proper diet, and regular exercise. GH is secreted about 1 hour after the onset of deep sleep and in a series of pulses 2 to 4 hours after a meal. Secretion of GH increases during exercise, probably because rapid metabolism by muscle cells lowers the blood glucose level.

Emotional support is also necessary for proper growth. Growth is retarded in children who are deprived of cuddling, playing, and other forms of nurture, even when their needs for food and shelter are met. In extreme cases, childhood stress can produce a form of retarded development known as *psychosocial dwarfism.* Some emotionally deprived children exhibit abnormal sleep patterns, which may be the basis for decreased secretion of GH.

Other hormones also influence growth. Thyroid hormones appear necessary for normal GH secretion and function and for normal tissue response to IGFs. Sex hormones must be present for the adolescent growth spurt to occur. However, the presence of sex hormones eventually causes the growth centers within the long bones to ossify so that no further increase in height can occur even when GH is present.

TABLE 49-2	Consequences of Endocrine Malfunction	
HORMONE	**HYPOSECRETION**	**HYPERSECRETION**
Growth hormone	Pituitary dwarfism	Gigantism if malfunction occurs in childhood; acromegaly in adult
Thyroid hormones	Cretinism (in children); myxedema, a condition of pronounced adult hypothyroidism; dietary iodine deficiency leads to hyposecretion and goiter	Hyperthyroidism; increased metabolic rate, nervousness, irritability; goiter; Graves' disease
Parathyroid hormone	Spontaneous discharge of nerves; spasms; tetany; death	Weak, brittle bones; kidney stones
Insulin	Diabetes mellitus	Hypoglycemia
Hormones of adrenal cortex	Addison's disease	Cushing's syndrome

© Cengage Learning

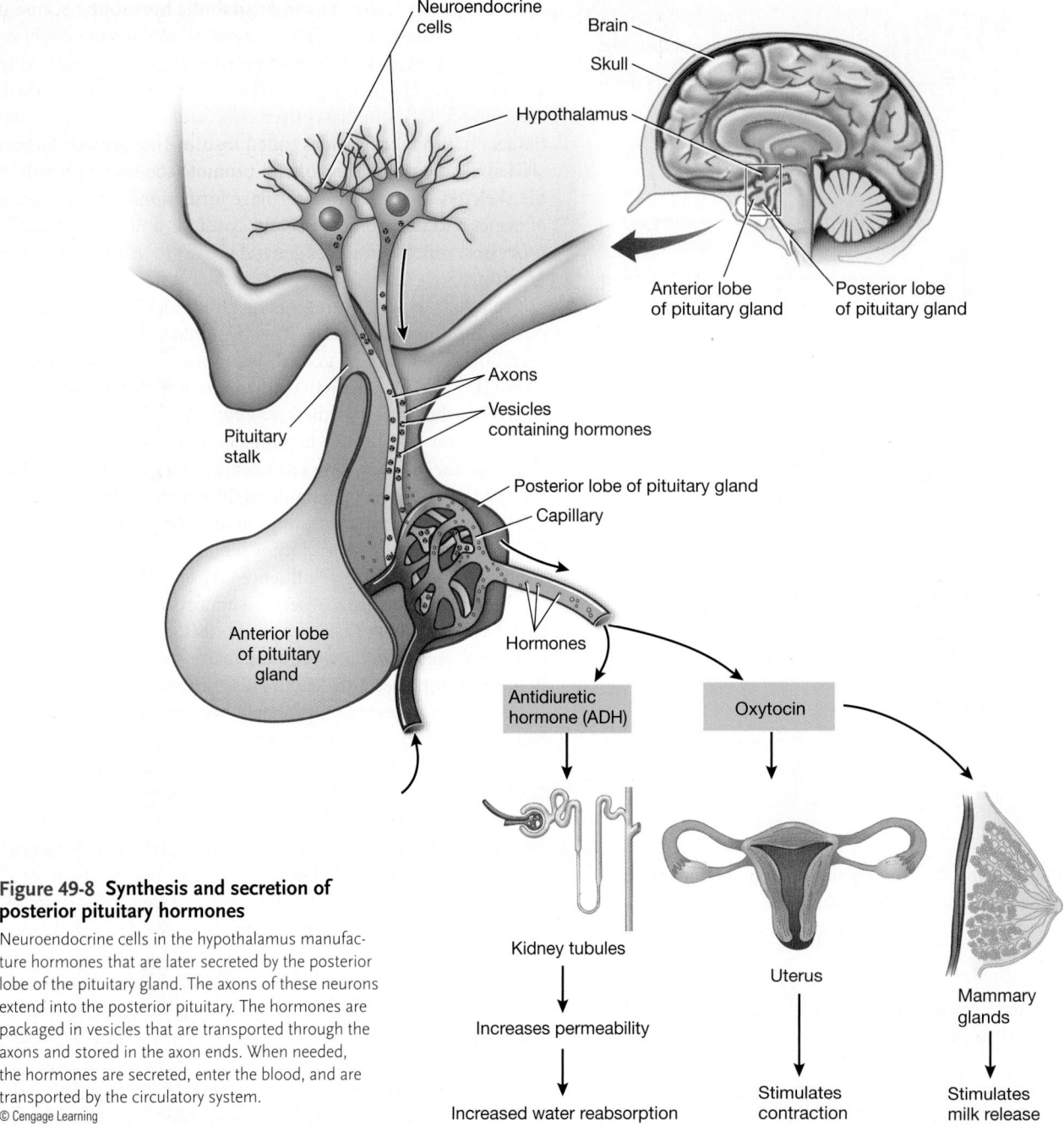

Figure 49-8 Synthesis and secretion of posterior pituitary hormones

Neuroendocrine cells in the hypothalamus manufacture hormones that are later secreted by the posterior lobe of the pituitary gland. The axons of these neurons extend into the posterior pituitary. The hormones are packaged in vesicles that are transported through the axons and stored in the axon ends. When needed, the hormones are secreted, enter the blood, and are transported by the circulatory system.
© Cengage Learning

Inappropriate amounts of growth hormone secretion result in abnormal growth Have you ever wondered why some people fail to grow normally? Extreme deficiency in GH during childhood causes pituitary dwarfism. This condition is usually caused by insufficient production of GH by the pituitary gland. However, dwarfism also occurs when the liver produces insufficient IGF or when target tissues do not respond to IGF. Although miniature, a pituitary dwarf has normal intelligence and is usually well-proportioned. If the growth centers in the long bones are still active when this condition is diagnosed, it can often be effectively treated by HGH injection. Human GH can be synthesized by recombinant DNA technology.

A person grows abnormally tall when the anterior pituitary secretes excessive amounts of HGH during childhood. This condition is called **gigantism.** If hypersecretion of HGH occurs during adulthood, the individual cannot grow taller. Instead, connective tissue thickens, and bones in the hands, feet, and face may increase in diameter. This condition is known as **acromegaly** (*acromegaly* means "large extremities").

Thyroid hormones increase metabolic rate

The **thyroid gland** is located in the neck region, in front of the trachea and below the larynx. The **thyroid hormones—thyroxine,**

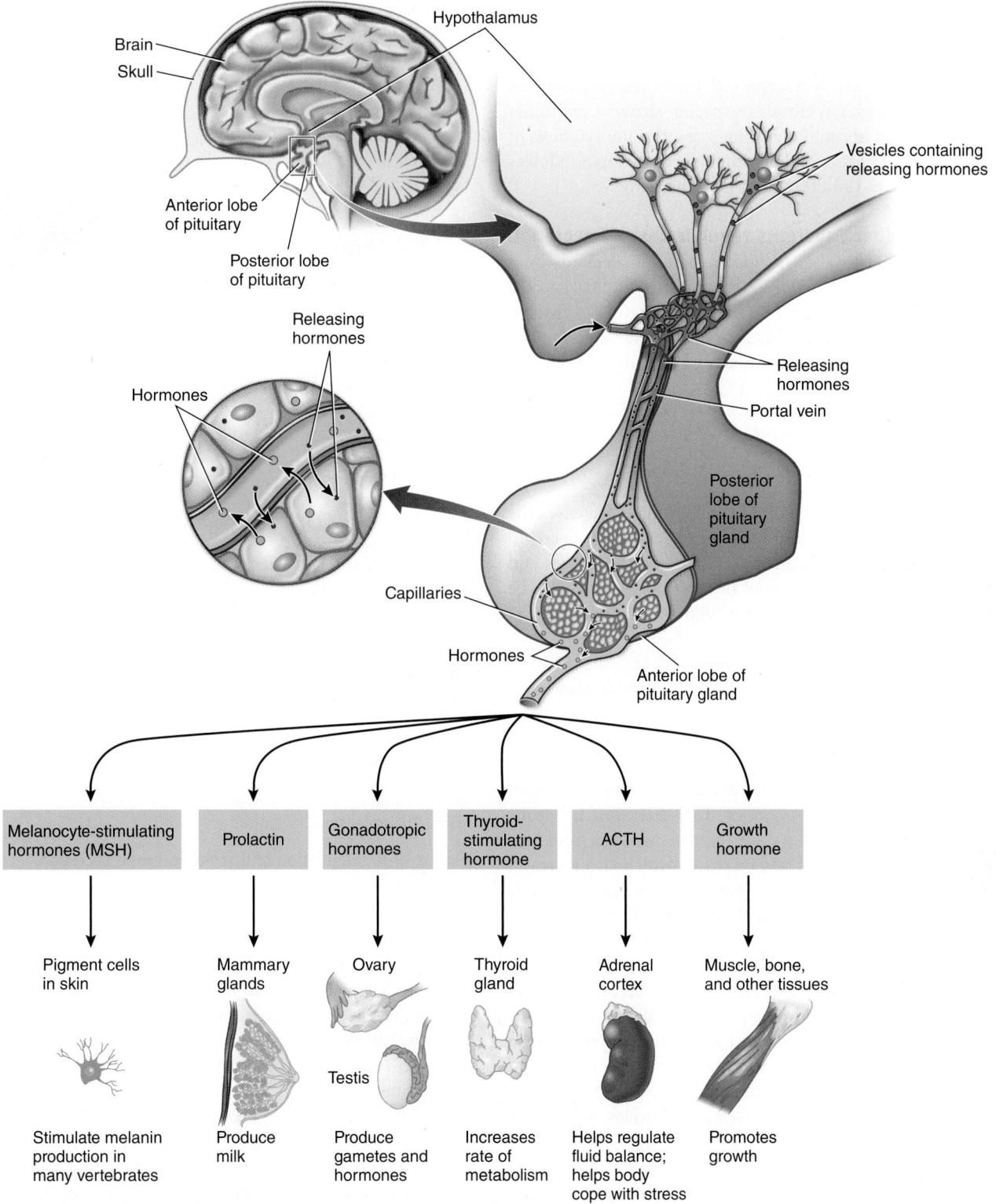

Figure 49-9 *Animation* **Synthesis and secretion of anterior pituitary hormones**

The hypothalamus secretes several specific releasing and inhibiting hormones that are transported to the anterior lobe of the pituitary gland by way of portal veins. These regulatory hormones stimulate or inhibit the release of specific hormones by cells of the anterior pituitary. The anterior pituitary hormones act on a wide variety of target tissues. (ACTH, adrenocorticotropic hormone)

© Cengage Learning

also known as **T₄**, and **triiodothyronine,** or **T₃**—are synthesized from the amino acid tyrosine and iodine. Thyroxine has four iodine atoms attached to each molecule, and T_3 has three. In most target tissues, T_4 is converted to T_3, the more active form (see Fig. 49-2c). We describe the actions of calcitonin, another hormone secreted by the thyroid gland, when we discuss the parathyroid glands.

In vertebrates thyroid hormones are essential for normal growth and development. These hormones increase the rate of metabolism in most body tissues. After T_3 binds with its receptor in the nucleus of a target cell, the T_3 receptor complex induces or suppresses synthesis of specific enzymes and other proteins. Thyroid hormones also help regulate the synthesis of proteins necessary for cell differentiation. For example, tadpoles cannot develop into adult frogs without thyroxine.

Negative feedback systems regulate thyroid secretion

The regulation of thyroid hormone secretion depends on a negative feedback loop between the anterior pituitary and the thyroid gland (**FIG. 49-10**). When the concentration of thyroid hormones in the blood rises above normal, the anterior pituitary secretes less **thyroid-stimulating hormone (TSH):**

> thyroid hormone concentration high $\longrightarrow$ inhibits anterior pituitary $\longrightarrow$ secretes less TSH $\longrightarrow$ thyroid gland secretes less hormone $\longrightarrow$ homeostasis

When the concentration of thyroid hormones decreases, the anterior pituitary secretes more TSH. TSH acts by way of cAMP to promote synthesis and secretion of thyroid hormones and also to promote increased size of the thyroid gland itself. The effect of TSH can be summarized as follows:

> thyroid hormone concentration low $\longrightarrow$ anterior pituitary secretes more TSH $\longrightarrow$ thyroid gland secretes more hormone $\longrightarrow$ homeostasis

In some mammals changes in weather affect the hypothalamus. Exposure to very cold temperature stimulates the hypothalamus to increase secretion of **TSH-releasing hormone (TRH).** This action causes a dramatic increase in thyroid hormone secretion. Metabolic rate increases, which also increases heat production, so the body temperature rises. Increased body temperature has a negative feedback effect, limiting further secretion of thyroid hormones.

Malfunction of the thyroid gland leads to specific disorders

Extreme hypothyroidism during infancy and childhood results in low metabolic rate and can lead to **cretinism,** a condition characterized by intellectual disability and impaired physical development. When diagnosed early enough, hypothyroidism can be treated with thyroid hormones, and cretinism can be prevented.

An adult who feels like sleeping all the time, has little energy, and is mentally slow or confused may be suffering from hypothyroidism. When there is almost no thyroid function, the basal metabolic rate is reduced by about 40% and the patient develops **myxedema,** characterized by a slowing down of physical and mental activity. Hypothyroidism is treated by oral administration of the missing hormone.

Hyperthyroidism does not cause abnormal growth but does increase metabolic rate by 60% or even more. This increase in metabolism results in the rapid use of nutrients, causing the individual to be hungry and to eat more. Increased food intake

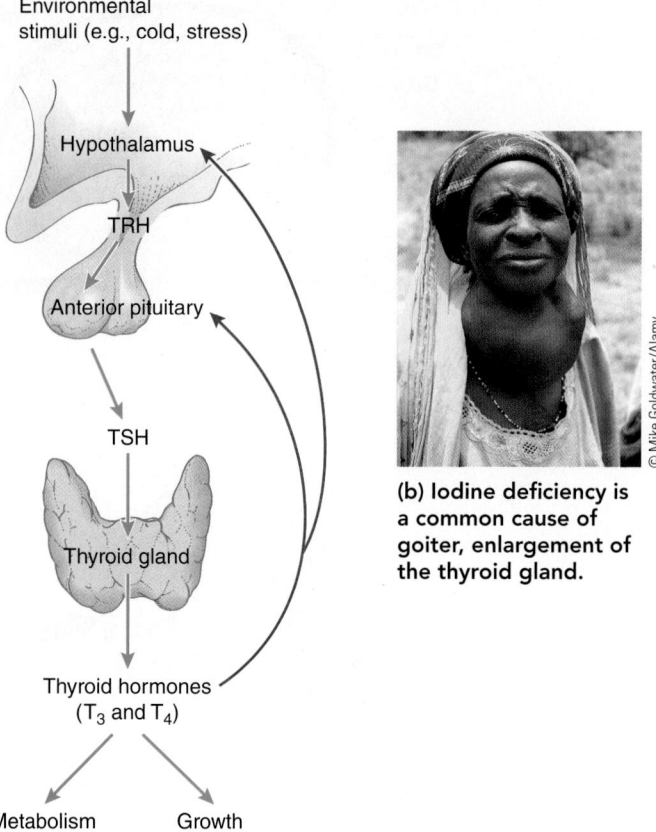

(b) Iodine deficiency is a common cause of goiter, enlargement of the thyroid gland.

© Mike Goldwater/Alamy

(a) Regulation of thyroid hormone secretion by negative feedback. An increase in concentration of thyroid hormones above normal levels signals the anterior pituitary to slow thyroid-stimulating hormone (TSH) production. Thus, thyroid hormones limit their own production by negative feedback. *Green arrows* indicate stimulation; *red arrows* indicate inhibition. (TRH, TSH-releasing hormone)

Figure 49-10 Regulation of thyroid hormone secretion
© Cengage Learning

is not sufficient to meet the demands of the rapidly metabolizing cells, however, and people with this condition often lose weight. They also tend to be nervous, irritable, and emotionally unstable. The most common form of hyperthyroidism is **Graves' disease,** an autoimmune disease. Abnormal antibodies bind to TSH receptors and activate them. Production of thyroid hormones increases.

An abnormal enlargement of the thyroid gland, or **goiter,** may be associated with either hypersecretion or hyposecretion (see Fig. 49-10b). In Graves' disease the abnormal antibodies that activate TSH receptors can cause the development of a goiter. Worldwide, more than 90% of goiters are caused by iodine deficiency. Without iodine the gland cannot make thyroid hormones, resulting in hyposecretion. When the concentration of thyroid hormones in the blood decreases, the anterior pituitary secretes large amounts of TSH. The TSH stimulates growth of the thyroid gland, sometimes to gigantic proportions. However, enlargement of the gland does not increase production of the hormones because the needed ingredient—iodine—is still missing.

Seafood is a rich source of iodine, and iodine is also added to table salt as a nutritional supplement. In fact, thanks to iodized salt, goiter is no longer common in the United States and other highly developed countries. In other parts of the world, however, hundreds of thousands of people still suffer from this easily preventable disorder. Because it results in hypothyroidism, iodine deficiency has a negative effect on intellectual and physical development. *Lancet,* a well-respected British medical journal, has referred to iodine deficiency as "the single greatest preventable cause of mental retardation."

The parathyroid glands regulate calcium concentration

The **parathyroid glands** are typically embedded in the connective tissue surrounding the thyroid gland. These glands secrete **parathyroid hormone (PTH),** a polypeptide that helps regulate the calcium level of the blood and interstitial fluid (**FIG. 49-11a**). Parathyroid hormone acts by way of a G protein–linked receptor and cAMP to stimulate calcium release from bones and to increase calcium reabsorption from the kidney tubules. It also activates vitamin D, which then increases the amount of calcium absorbed from the intestine.

Calcitonin, a peptide hormone secreted by the thyroid gland, works antagonistically to parathyroid hormone (**FIG. 49-11b**). When the concentration of calcium rises above homeostatic levels, calcitonin is released and rapidly inhibits removal of calcium from bone.

The islets of the pancreas regulate blood glucose concentration

In addition to secreting digestive enzymes (see Chapter 47), the pancreas is an important endocrine gland. Its hormones, insulin and glucagon, are secreted by cells that occur in little clusters, the **islets of Langerhans,** throughout the pancreas (**FIG. 49-12**). About one million islets are present in the human pancreas. They consist mainly of **beta cells,** which secrete insulin, and **alpha cells,** which secrete glucagon.

Insulin lowers the concentration of glucose in the blood

Insulin is an anabolic hormone that regulates the use and storage of nutrients. Insulin stimulates cells of many tissues, including liver, muscle, and fat cells, to take up glucose from the blood by facilitated diffusion (see Chapter 5). Once glucose enters muscle cells, it is either used immediately as fuel or stored as glycogen. Insulin also inhibits liver cells from releasing glucose. Thus, insulin activity *lowers* the glucose level in the blood.

Insulin also helps regulate fat and protein metabolism. It reduces the use of fatty acids as fuel and instead stimulates their

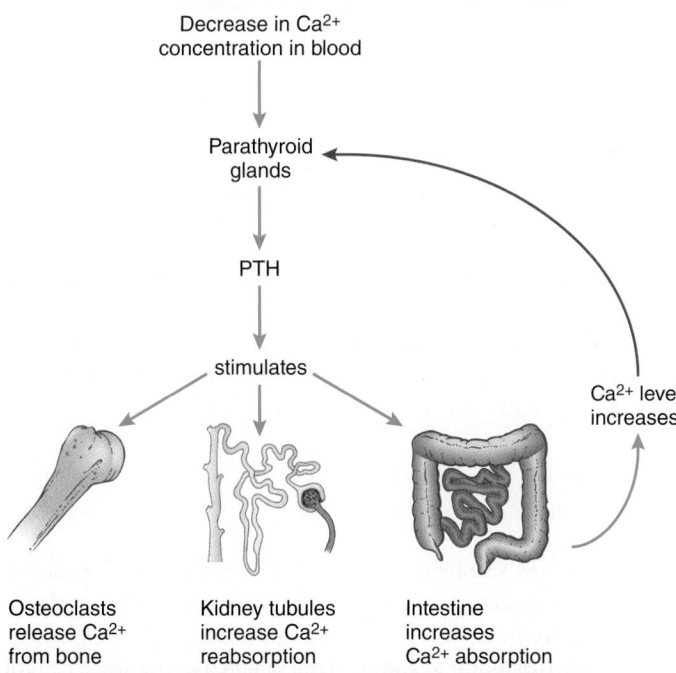

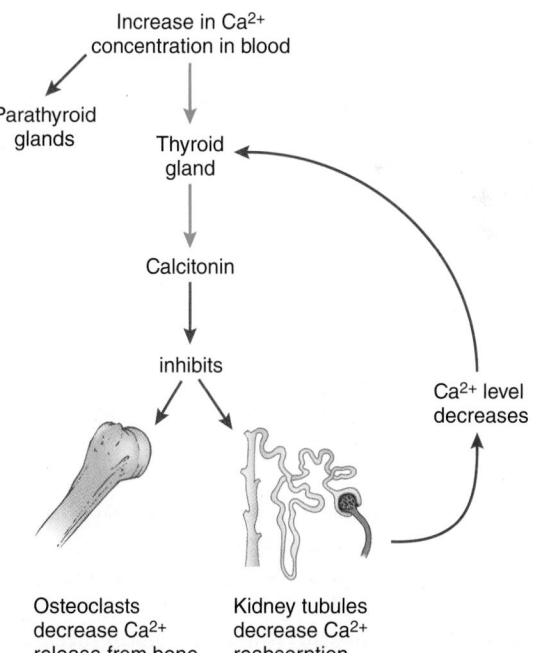

(a) Calcium concentration too low. When calcium concentration decreases below the normal range, the parathyroid glands secrete parathyroid hormone (PTH). PTH stimulates homeostatic mechanisms that restore appropriate calcium concentration.

(b) Calcium concentration too high. When calcium concentration increases above its set point, the parathyroid glands are inhibited. The thyroid gland secretes calcitonin, which inhibits Ca^{2+} release from bone and decreases Ca^{2+} reabsorption in the kidneys.

Figure 49-11 *Animation* **Regulation of calcium homeostasis**

Green arrows indicate stimulation; *red arrows* indicate inhibition.

© Cengage Learning

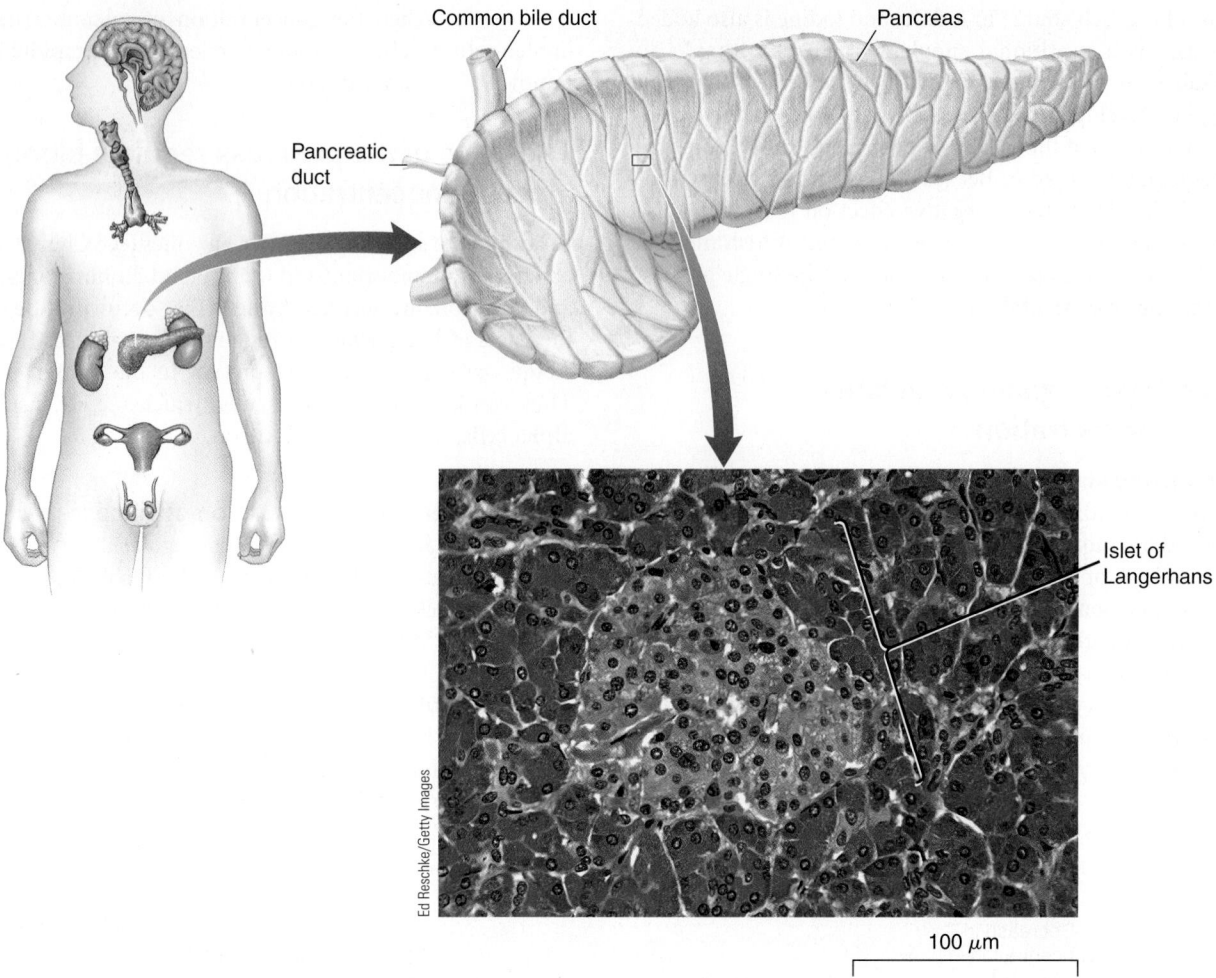

Common bile duct

Pancreatic duct

Pancreas

Islet of Langerhans

Ed Reschke/Getty Images

100 μm

Figure 49-12 *Animation* **An islet of Langerhans in the pancreas**
Scattered throughout the pancreas, clusters of endocrine cells (such as the one shown in this LM) secrete insulin and glucagon.
© Cengage Learning

storage in adipose tissue. Insulin has an anabolic effect on protein metabolism, resulting in a net increase in protein in cells. It promotes protein synthesis by increasing the number of amino acid transporters in the plasma membrane, thus stimulating the transport of certain amino acids into cells. Insulin also promotes transcription and translation.

Glucagon raises the concentration of glucose in the blood The effects of **glucagon** are opposite to those of insulin. Glucagon's main action is to raise blood glucose level. It does so by stimulating liver cells to convert glycogen to glucose. Glucagon also stimulates *gluconeogenesis* (glu-ko-nee-oh-gen′-eh-sis), the production of glucose from noncarbohydrates. Glucagon mobilizes fatty acids by promoting breakdown of fat and inhibiting triacylglycerol synthesis. This hormone also stimulates the liver to degrade proteins and inhibits protein synthesis.

Glucose concentration regulates insulin and glucagon secretion The concentration of glucose in the blood directly controls insulin and glucagon secretion (**FIG. 49-13**). After a meal, when the blood glucose level rises as a result of intestinal absorption, beta cells are stimulated to increase insulin secretion. Then, as cells remove glucose from the blood, decreasing its concentration, insulin secretion decreases.

high blood glucose concentration ⟶ beta cells secrete insulin ⟶ blood glucose concentration decreases ⟶ homeostasis

When you have not eaten for several hours, the concentration of glucose in your blood begins to fall. When it decreases from its normal fasting level of about 90 mg of glucose per 100 mL of blood to about 70 mg of glucose, the alpha cells of the islets increase their secretion of glucagon. Glucose is mobilized from storage in the liver cells, returning blood glucose concentration to normal:

low blood glucose concentration ⟶ alpha cells secrete glucagon ⟶ blood glucose concentration increases ⟶ homeostasis

Insulin and glucagon work antagonistically to regulate the concentration of glucose in the blood.

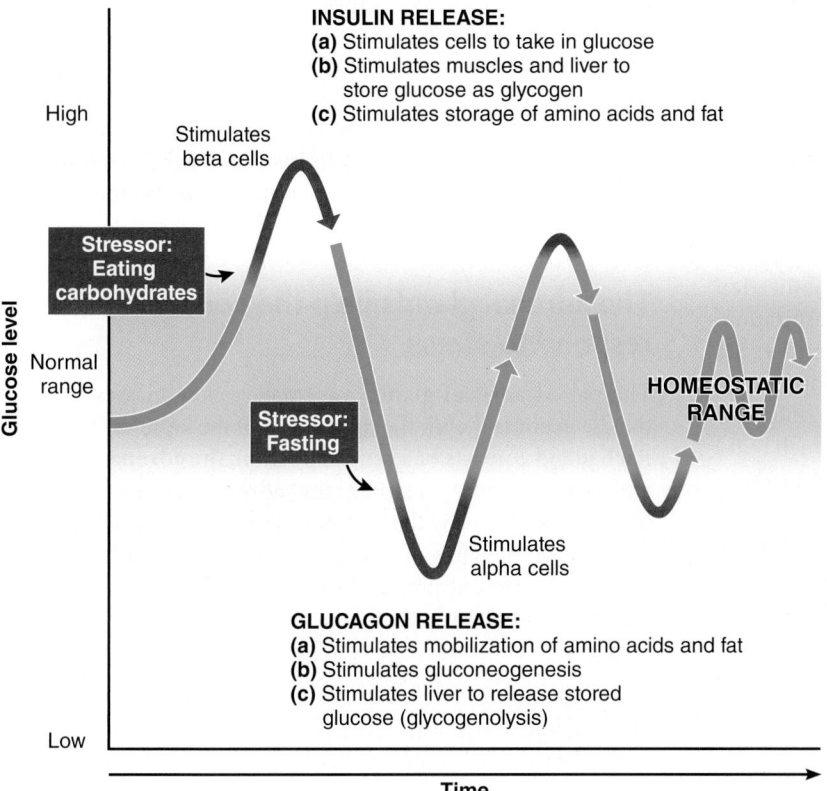

INSULIN RELEASE:
(a) Stimulates cells to take in glucose
(b) Stimulates muscles and liver to store glucose as glycogen
(c) Stimulates storage of amino acids and fat

Stimulates beta cells

Stressor: Eating carbohydrates

HOMEOSTATIC RANGE

Stressor: Fasting

Stimulates alpha cells

GLUCAGON RELEASE:
(a) Stimulates mobilization of amino acids and fat
(b) Stimulates gluconeogenesis
(c) Stimulates liver to release stored glucose (glycogenolysis)

High

Normal range

Low

Glucose level

Time

Figure 49-13 Regulation of blood glucose concentration

When the concentration of glucose in the blood increases above normal, beta cells secrete insulin. This hormone lowers blood glucose concentration, thus restoring homeostasis. When the concentration of glucose in the blood decreases below normal, alpha cells secrete glucagon. This hormone increases blood glucose concentration, restoring homeostasis.

PREDICT What would happen if an autoimmune disease sharply reduced the number of functional alpha cells?

© Cengage Learning

Alpha cells respond to the glucose concentration within their own cytoplasm, which reflects the blood glucose level. When blood glucose level is high, there is generally a high level of glucose within the alpha cells, and glucagon secretion is inhibited.

Note that in these negative feedback systems, insulin and glucagon work antagonistically to keep blood glucose concentration within homeostatic limits. When the glucose level rises, insulin release brings it back to normal; when the glucose concentration falls, glucagon acts to raise it again. The insulin–glucagon system is a powerful, fast-acting mechanism for keeping blood glucose level within normal limits. Why is it important to maintain a constant blood glucose level? Recall that brain cells ordinarily are unable to use other nutrients as fuel, so they depend on a continuous supply of glucose. As we will discuss, several other hormones also affect blood glucose concentration.

Diabetes mellitus is a serious disorder of carbohydrate metabolism **Diabetes mellitus,** the most common endocrine disorder, is a serious health problem, and according to the World Health Organization, its global prevalence is increasing rapidly. About 350 million people worldwide have this disorder. (Note that diabetes mellitus is an entirely different disorder from diabetes insipidus, which was discussed in Chapter 48.) According to the Centers of Disease Control and Prevention (CDC), diabetes is the leading cause of kidney failure, nontraumatic lower limb amputations, and new cases of blindness. Diabetes is also a major risk factor for cardiovascular disease, neurological disease, and premature death. Approximately one-half of the people suffering from diabetes are unaware that they have this disorder; consequently, they do not receive treatment and are at high risk for its serious effects.

Diabetes mellitus is actually a group of related disorders characterized by high blood glucose concentrations. Two main types of diabetes are type 1 and type 2. *Type 1 diabetes* is an autoimmune disease in which antibodies mark the beta cells for destruction. T cells destroy the beta cells, resulting in insulin deficiency. Insulin injections are needed to correct the carbohydrate imbalance that results. This disorder usually develops before age 30, often during childhood. Type 1 diabetes is caused by a combination of genetic predisposition and environmental factors, possibly including infection by a virus.

More than 90% of diabetics have *type 2 diabetes.* This disorder develops gradually, typically in overweight people. **Metabolic syndrome** often precedes type 2 diabetes. The symptoms of metabolic syndrome include obesity, large waist circumference ("apple shape"), elevated blood glucose concentration, high triglyceride and low HDL ("good" cholesterol) concentrations in the blood, and high blood pressure.

More than 20% of obese children have impaired glucose tolerance, an early warning sign of diabetes. Impaired glucose tolerance, which affects more than 200 million people worldwide, is defined as *hyperglycemia* (a blood glucose concentration that is abnormally high) following ingestion of a large amount of glucose (a glucose load).

Type 2 diabetes typically begins as **insulin resistance,** a condition in which cells cannot use insulin effectively. In type 2 diabetes, normal (or greater than normal) concentrations of insulin are present in the blood, but target cells cannot use it. You can decrease your risk of Type 2 diabetes with a healthy diet, regular physical activity (at least 30 minutes of moderate-intensity exercise each day), and by avoiding tobacco use.

Similar metabolic disturbances occur in both types of diabetes mellitus: disruption of carbohydrate, fat, and protein metabolism and electrolyte imbalance.

1. *Glucose use decreases.* In a diabetic, the cells have difficulty taking up glucose from the blood, and its accumulation causes hyperglycemia. Instead of the normal fasting concentration of about 90 mg of glucose per 100 mL of blood, the level may exceed 140 mg per 100 mL and may reach concentrations of more than 500 mg per 100 mL. Recall from Chapter 48 that when glucose concentration exceeds its **tubular transport maximum (Tm),** the maximum rate at which it can be reabsorbed, glucose is excreted in the urine. Despite the large quantities of glucose in the blood, insulin-dependent cells in a diabetic can take in only about 25% of the glucose they require for fuel.

2. *Fat mobilization increases.* Cells turn to fat and protein for energy. The absence of insulin promotes the mobilization of fat stores to provide nutrients for cellular respiration. Unfortunately, however, the blood lipid level may reach five times the normal level, leading to the development of atherosclerosis (see Chapter 44). Increased fat metabolism also increases the formation of ketone bodies, which build up in the blood. This accumulation can cause *ketoacidosis,* a condition in which the body fluids and blood become too acidic. When the ketone level in the blood rises, ketones appear in the urine, another clinical indication of diabetes mellitus. If severe, ketoacidosis can lead to coma and death.

3. *Protein use increases.* Lack of insulin also results in increased protein breakdown relative to protein synthesis, so the untreated diabetic becomes thin and emaciated.

4. *Electrolyte imbalance.* When ketone bodies and glucose are excreted in the urine, water follows by osmosis; as a result, urine volume increases. The resulting dehydration causes thirst and challenges electrolyte homeostasis. When ketones are excreted, they also take sodium, potassium, and some other cations with them, contributing to electrolyte imbalance.

Many type 2 diabetics can keep their blood glucose levels within normal range by diet management, weight loss, and regular exercise. When this treatment approach is not effective, the disorder is treated with oral drugs that stimulate insulin secretion and promote its actions. Approximately one-third of type 2 diabetics eventually need insulin injections.

The dramatic increase in type 2 diabetes is a public health problem resulting from increased prevalence of obesity, rapid increases in people becoming overweight, and lack of physical activity. Because type 2 diabetes is associated with insulin resistance, researchers are studying this metabolic problem in their search for causes of and cures for diabetes. Researchers have found a strong association between increased fat tissue and insulin resistance. When insulin-resistant individuals lose weight, they become more sensitive to insulin.

In hypoglycemia the blood glucose concentration is too low Low blood glucose concentration, or *hypoglycemia,* sometimes occurs in people who later develop diabetes. These individuals may have impaired glucose tolerance in which the islets overreact to glucose challenge. When carbohydrates are ingested, there is a delayed insulin response, followed after about 3 hours by too much insulin secretion. This hypersecretion causes glucose levels to fall, and the individual feels drowsy. If this reaction is severe, the person may become uncoordinated or even unconscious.

Serious hypoglycemia can develop if diabetics receive injections of too much insulin or if the islets, because of a tumor, secrete too much insulin. The blood glucose concentration may then fall drastically, depriving brain cells of their needed fuel supply. These events can lead to **insulin shock,** a condition in which the patient may appear drunk or may become unconscious, suffer convulsions, and even die.

The adrenal glands help the body respond to stress

The paired **adrenal glands** are small, yellow masses of tissue that lie in contact with the upper ends of the kidneys (FIG. 49-14). Each gland consists of a central portion, the **adrenal medulla,** and a larger outer section, the **adrenal cortex.** Although joined anatomically, the adrenal medulla and cortex develop from different types of tissue in the embryo and function as distinctly different glands. However, both secrete hormones that help regulate metabolism and help the body respond to stress.

The adrenal medulla initiates an alarm reaction The adrenal medulla is a neuroendocrine gland that is coupled to

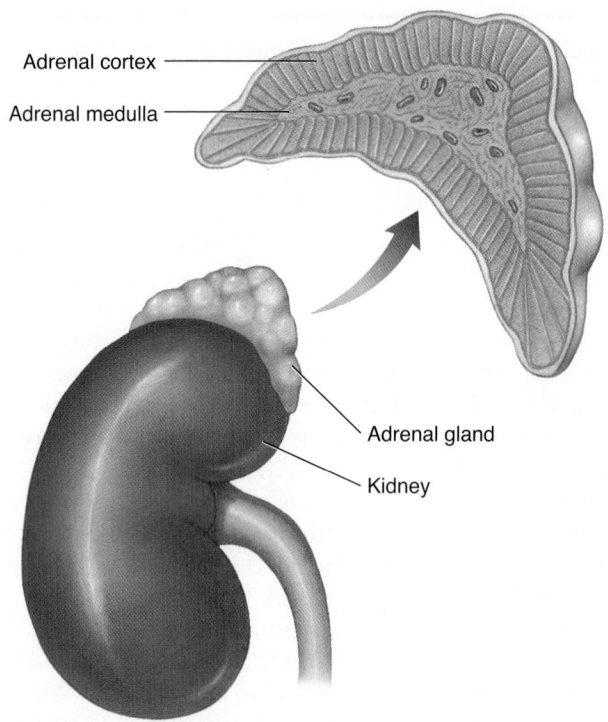

Figure 49-14 Adrenal gland
The adrenal glands are located above the kidneys. Each gland consists of a central adrenal medulla surrounded by an adrenal cortex.
© Cengage Learning

the sympathetic nervous system. It develops from neural tissue, and its secretion is controlled by sympathetic nerves. The adrenal medulla secretes **epinephrine** and **norepinephrine.** Chemically, these hormones are very similar; they are amino acid derivatives that belong to the chemical group known as **catecholamines** (see Fig. 49-2c). Most of the hormone output of the adrenal medulla is epinephrine. Recall that norepinephrine also serves as a neurotransmitter released by sympathetic neurons and by some neurons in the central nervous system (see Chapter 41).

Under normal conditions, the adrenal medulla continuously secretes small amounts of both epinephrine and norepinephrine. Their secretion is under neural control. When anxiety is aroused, the hypothalamus sends signals via sympathetic nerves to the adrenal medulla. Sympathetic neurons release acetylcholine, which triggers the adrenal medulla to release larger amounts of epinephrine and norepinephrine.

During a stressful situation, adrenal medullary hormones initiate an *alarm reaction* that enables you to think more quickly, fight harder, or run faster than usual. Metabolic rate increases by as much as 100%. Blood is rerouted to those organs essential for emergency action (**FIG. 49-15**). Blood vessels going to the brain, muscles, and heart are dilated, whereas those to the skin and kidneys are constricted. Constriction of blood vessels serving the skin has the added advantage of decreasing blood loss in case of hemorrhage (and explains the sudden paling that comes with fear). At the same time, the heart beats faster. Strength of muscle contraction increases. Thresholds in the reticular activating system of the brain are lowered, so you become more alert (see Chapter 42). The adrenal medullary hormones also raise fatty acid and glucose levels in the blood, ensuring needed fuel for extra energy.

The adrenal cortex helps the body deal with chronic stress The adrenal cortex synthesizes steroid hormones from cholesterol (which in turn is made in the liver from acetyl coenzyme A). Although the adrenal cortex produces more than 30 types of steroids, it secretes only three types of hormones in significant amounts: sex hormone precursors, mineralocorticoids, and glucocorticoids.

In both sexes the adrenal cortex secretes sex hormone precursors, such as androgens (masculinizing hormones). Certain tissues convert sex hormone precursors to **testosterone,** the principal male sex hormone, and **estradiol,** the principal female sex hormone. In males androgen production by the adrenal cortex is not significant because the testes produce testosterone. However, in females the adrenal cortex secretes the androgen used to produce most of the testosterone circulating in the blood.

The principal **mineralocorticoid** is **aldosterone.** (Recall from Chapter 48 that aldosterone helps regulate fluid balance by regulating salt balance.) In response to aldosterone, the kidneys reabsorb more sodium and excrete more potassium. As a result of increased sodium, extracellular fluid volume increases, which results in greater blood volume and elevated blood pressure.

When the adrenal glands do not produce enough aldosterone, large amounts of sodium are excreted in the urine. Water leaves the body with the sodium (because of osmotic pressure), and the blood volume may be so markedly reduced that the patient dies of low blood pressure.

Cortisol, also called *hydrocortisone,* accounts for about 95% of the **glucocorticoid** activity of the human adrenal cortex (see Fig. 49-2b). Cortisol helps ensure adequate fuel supplies for the cells when the body is under stress. Its principal action is to stimulate production of glucose from other nutrients in liver cells.

Cortisol helps provide nutrients for glucose production by stimulating amino acid transport into liver cells (see Fig. 49-15). It also promotes mobilization of fats so that glycerol from triacylglycerol molecules is available for conversion to glucose. These actions ensure that glucose and glycogen are produced in the liver, and the concentration of glucose in the blood rises. Thus, the adrenal cortex provides an important backup system for the adrenal medulla, ensuring adequate glucose supplies when the body is under stress and in need of extra energy.

During stress, the brain and the adrenal glands work together to help the body respond effectively (see Fig. 49-15; also see *Inquiring About: The Neurobiology of Traumatic Experience* in Chapter 42). Almost any type of stress stimulates the hypothalamus to secrete **corticotropin-releasing factor (CRF),** which stimulates the anterior pituitary to secrete **adrenocorticotropic hormone (ACTH).** ACTH regulates both glucocorticoid and aldosterone secretion. It is so potent that it can result in a 20-fold increase in cortisol secretion within minutes. When the body is not under stress, high levels of cortisol in the blood inhibit both CRF secretion by the hypothalamus and ACTH secretion by the pituitary.

Addison's disease results when the adrenal cortex produces insufficient amounts of aldosterone and cortisol. This disorder is most commonly caused by autoimmune destruction of the adrenal cortex. Reduction in cortisol prevents the body from regulating the blood glucose concentration because the liver cannot synthesize enough glucose. The cortisol-deficient patient also loses the ability to respond to stress. If cortisol levels are significantly depressed, even the stress of mild infections can cause death.

Glucocorticoids are used clinically to reduce inflammation in allergic reactions, infections, arthritis, and certain types of cancer. These hormones inhibit production of prostaglandins (which are mediators of inflammation). Glucocorticoids also reduce inflammation by decreasing capillary permeability, which reduces swelling. In addition, they reduce the effects of histamine and so are used to treat allergic symptoms.

When used in large amounts over long periods, glucocorticoids can cause serious side effects. Although they help stabilize lysosome membranes so that they do not destroy tissues with their potent enzymes, the ability of lysosomes to degrade foreign molecules is also decreased. Glucocorticoids decrease interleukin-1 production, blocking cell-mediated immunity and reducing the patient's ability to fight infections. Other side effects include ulcers, hypertension, diabetes mellitus, and atherosclerosis.

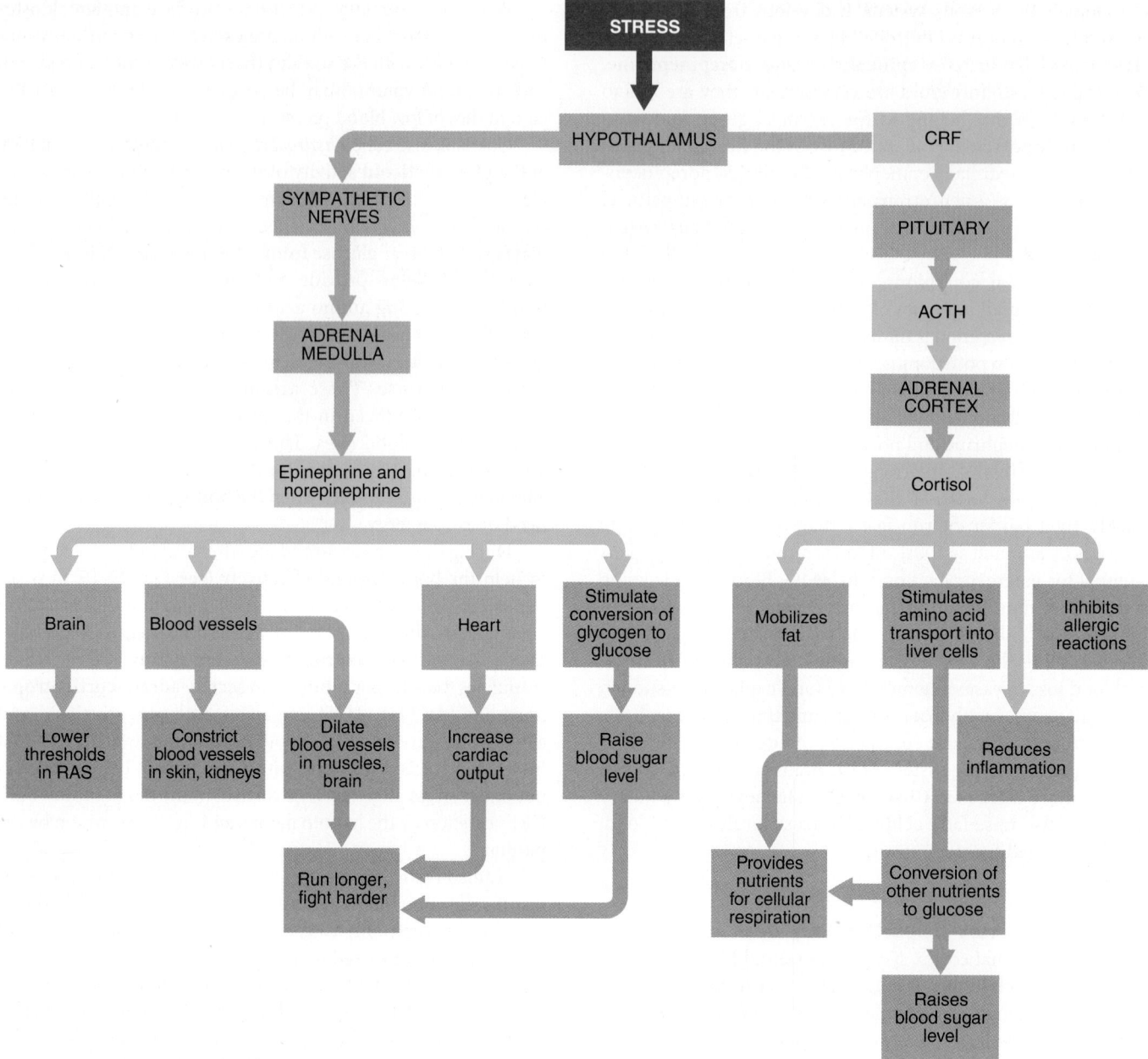

Figure 49-15 Response to stress

The adrenal glands help the body respond to stressful experiences. (RAS, reticular activating system; CRF, corticotropin-releasing factor; ACTH, adrenocorticotropic hormone)

© Cengage Learning

Abnormally large amounts of glucocorticoids, whether due to disease or steroid medication, can result in **Cushing's syndrome.** In this condition, excess glucose is converted to fat and deposited about the trunk. Edema gives the patient's face a full-moon appearance. Blood glucose level may increase by as much as 50% above normal, causing *adrenal diabetes.* If this condition persists for several months, the beta cells in the pancreas may "burn out" from secreting excessive amounts of insulin, leading to permanent diabetes mellitus. Reduction in protein synthesis causes weakness and decreased immune responses.

Many other hormones help regulate life processes

Many other tissues of the body secrete hormones. The **pineal gland,** located in the brain, produces **melatonin,** which is derived from the amino acid tryptophan. Melatonin influences biological rhythms and the onset of sexual maturity. In humans melatonin facilitates the onset of sleep. Interestingly, exposure to light suppresses melatonin secretion.

As we have discussed, several hormones secreted by the digestive tract and by adipose tissue regulate digestive processes (see Chapter 47). The **thymus gland** produces **thymosin,** a hormone that plays a role in immune responses. **Atrial natriuretic peptide,** secreted by the atrium of the heart, promotes sodium excretion and lowers blood pressure (see Chapter 48). We discuss reproductive hormones in Chapter 50.

CHECKPOINT 49.5

- **CONNECT** *Why is the hypothalamus considered the link between the nervous and endocrine systems? What is the role of the anterior pituitary gland? of the posterior pituitary?*

- **VISUALIZE** *Draw a diagram illustrating how thyroid hormone secretion is regulated.*

- **PREDICT** *Imagine that you eat a hot fudge sundae. What would be the response of your pancreas? What are the antagonistic actions of insulin and glucagon in regulating blood glucose level?*

- **PREDICT** *Imagine that you are being chased by an alligator. How would your adrenal glands respond? How would that help? (Be specific.)*

- *How is regulation of the adrenal medulla different from regulation of the adrenal cortex?*

SUMMARY: FOCUS ON LEARNING OBJECTIVES

49.1 An Overview of Endocrine Regulation *(page 1045)*

1 Compare endocrine system function with nervous system function and describe how these systems work together to regulate body processes.

- The **endocrine system** consists of endocrine glands, cells, and tissues that secrete **hormones,** chemical signals that regulate physiological processes. Most endocrine responses are slow but long-lasting. The endocrine system signals a wide range of **target cells.**

- The nervous system responds rapidly to stimuli by transmitting electrical and chemical signals. Neurons signal other neurons, muscle cells, and gland cells, including endocrine cells. The nervous system helps regulate many endocrine responses.

2 Summarize the regulation of endocrine action by negative feedback systems.

- Hormone secretion is typically regulated by **negative feedback systems.** A hormone is released in response to some change in a steady state. The hormone triggers a response that counteracts the changed condition. This process restores homeostasis.

3 Identify four main chemical groups to which hormones are assigned and give two examples for each group.

- Prostaglandins and the juvenile hormone of insects are **fatty acid derivatives.** Hormones secreted by the adrenal cortex, ovary, and testis, as well as the molting hormone of insects, are **steroid hormones.**

- Thyroid hormones and epinephrine are **amino acid derivatives.** Antidiuretic hormone (ADH) and glucagon are examples of **peptide hormones.** Insulin is a small protein.

49.2 Types of Endocrine Signaling *(page 1047)*

4 Compare four types of endocrine signaling.

- In *classical endocrine signaling,* discrete **endocrine glands,** glands without ducts, secrete hormones into the interstitial fluid. Hormones are transported by the blood. They bind with receptors on or in specific **target cells.**

- In *neuroendocrine signaling,* neurons secrete **neurohormones.** These hormones are transported down axons and then secreted; they are typically transported by the blood.

- In *autocrine signaling,* a hormone (or other signal molecule) is secreted into the interstitial fluid and then acts on the very cell that produced it.

- In *paracrine signaling,* a hormone (or other signal molecule) diffuses through interstitial fluid and acts on nearby target cells.

- Some **local regulators** are considered hormones; they use autocrine or paracrine signaling. Local regulators include **growth factors,** peptides that stimulate cell division and development, and **prostaglandins,** a group of local hormones that help regulate many metabolic processes.

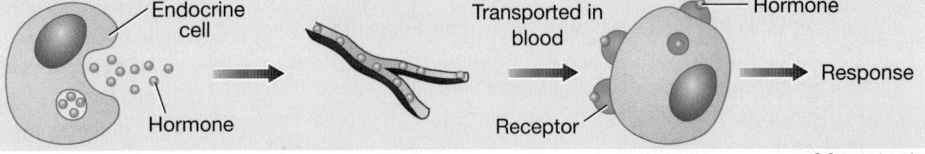

© Cengage Learning

49.3 Mechanisms of Hormone Action *(page 1049)*

5 Compare the mechanism of action of small, lipid-soluble hormones with that of hydrophilic hormones; include the role of second messengers, such as cyclic AMP.

- Steroid hormones and thyroid hormones are small, lipid-soluble molecules that pass through the plasma membrane and combine with receptors within the target cell; the hormone–receptor complex may activate or repress transcription of messenger RNA coding for specific proteins.

- Peptide hormones are hydrophilic and do not enter target cells. They combine with receptors on the plasma membrane of target cells. Many hormones bind to **G protein–linked receptors** that act via **signal transduction.** An extracellular hormone signal is transduced into an intracellular signal by the receptor.

- Most peptide hormones are first messengers that carry out their actions by way of **second messengers,** such as **cyclic AMP (cAMP).** The G protein–linked receptor activates a **G protein.** The G protein either stimulates or inhibits an enzyme that affects the second messenger. For example, G proteins stimulate or inhibit **adenylyl cyclase,** the enzyme that catalyzes the conversion of ATP to cAMP.

- Many second messengers stimulate the activity of *protein kinases,* enzymes that phosphorylate specific proteins that affect some cell process.

- Some G proteins use phospholipid derivatives as second messengers. *Inositol trisphosphate* (IP$_3$) and *diacylglycerol* (*DAG*) are second messengers that increase calcium concentration and activate enzymes. Calcium ions bind with *calmodulin,* which activates certain enzymes.
- **Tyrosine kinase receptors** are **enzyme-linked receptors** that bind growth factors, including insulin and nerve growth factor.

49.4 Neuroendocrine Regulation in Invertebrates *(page 1052)*

6 Identify four functions of invertebrate neurohormones and describe the regulation of insect development.

- Invertebrate hormones and neurohormones help regulate metabolism, growth and development, regeneration, color changes, molting, metamorphosis, reproduction, and behavior.
- Hormones control development in insects. When stimulated by some environmental factor, neuroendocrine cells in the insect brain secrete *brain hormone* (BH). BH stimulates the prothoracic glands to produce *molting hormone* (MH), which stimulates growth and molting.
- In the immature insect, the *corpora allata* secrete *juvenile hormone* (JH), which suppresses metamorphosis at each larval molt. The amount of JH decreases with successive molts.

49.5 Endocrine Regulation in Vertebrates *(page 1052)*

7 Identify the classical vertebrate endocrine glands and describe the actions of their hormones; describe the effects of hypersecretion and hyposecretion on homeostasis.

- Consult Table 49-1 for a description of the main endocrine glands and their actions. Vertebrate hormones regulate growth and development, reproduction, salt and fluid balance, and many aspects of metabolism and behavior.
- **Hyposecretion** (abnormally reduced output) or **hypersecretion** (abnormally increased output) of hormones leads to disruption of homeostasis and specific endocrine disorders.

8 Describe the mechanisms by which the hypothalamus and pituitary gland integrate regulatory functions; summarize the actions of the hypothalamic and pituitary hormones.

- Nervous and endocrine regulation is integrated in the **hypothalamus,** which regulates the activity of the **pituitary gland.**
- The neurohormones **oxytocin** and **antidiuretic hormone (ADH)** are produced by the hypothalamus and released by the **posterior pituitary** of the pituitary gland. Oxytocin stimulates contraction of the uterus and stimulates ejection of milk by the mammary glands. ADH stimulates reabsorption of water by the kidney tubules.
- The hypothalamus secretes **releasing hormones** and **inhibiting hormones** that regulate the hormone output of the **anterior pituitary gland.**
- The anterior pituitary secretes prolactin, melanocyte-stimulating hormones, growth hormone, and several **tropic hormones** that stimulate other endocrine glands. **Melanocyte-stimulating hormones (MSH)** suppress appetite and are important in regulating energy and body weight. **Prolactin** stimulates the mammary glands to produce milk.
- **Growth hormone (GH)** is an **anabolic hormone** that stimulates body growth by promoting protein synthesis. GH stimulates the liver to produce **insulin-like growth factors (IGFs),** which promote skeletal growth and general tissue growth.

9 Describe the actions of the thyroid and parathyroid hormones, their regulation, and the effects of malfunction.

- The **thyroid gland** secretes thyroid hormones: **thyroxine,** or **T$_4$,** and **triiodothyronine,** or **T$_3$.** Thyroid hormones stimulate the rate of metabolism.
- Regulation of thyroid secretion depends mainly on a negative feedback system between the anterior pituitary gland and the thyroid gland.
- Hyposecretion of thyroxine during childhood may lead to **cretinism;** during adulthood it may result in **myxedema. Goiter,** an abnormal enlargement of the thyroid gland, is associated with both hyposecretion and hypersecretion. The most common cause of hyperthyroidism is **Graves' disease,** an autoimmune disease.
- The **parathyroid glands** secrete **parathyroid hormone (PTH),** which regulates the calcium level in the blood. PTH increases calcium concentration by stimulating calcium release from bones, increasing calcium reabsorption by kidney tubules, and increasing calcium reabsorption from the intestine.
- **Calcitonin,** secreted by the thyroid gland, acts antagonistically to parathyroid hormone.

10 Contrast the actions of insulin and glucagon, and describe diabetes mellitus and hypoglycemia, including the metabolic effects of these disorders.

- The **islets of Langerhans** in the pancreas secrete **insulin** and **glucagon.** Insulin and glucagon secretion is regulated directly by glucose concentration. Insulin stimulates cells to take up glucose from the blood, so it lowers blood glucose concentration. Glucagon raises blood glucose concentration by stimulating conversion of glycogen to glucose and by stimulating production of glucose from other nutrients.
- In **diabetes mellitus** either insulin deficiency or **insulin resistance** results in decreased use of glucose, increased fat mobilization, increased protein use, and electrolyte imbalance. In hypoglycemia impaired glucose tolerance results in a delayed insulin response followed by hypersecretion of insulin. The glucose concentration falls, causing drowsiness.

11 Describe the actions and regulation of the adrenal hormones, including their role in the body's response to stress.

- Each **adrenal gland** consists of an inner adrenal medulla and an outer adrenal cortex. The **adrenal medulla** secretes **epinephrine** and **norepinephrine,** hormones that help the body cope with stress. These hormones increase heart rate, metabolic rate, and strength of muscle contraction. They reroute blood to organs needed for fight or flight. The adrenal medulla is regulated by the nervous system.
- The **adrenal cortex** secretes sex hormones; **mineralocorticoids,** such as **aldosterone;** and **glucocorticoids,** such as **cortisol.** Aldosterone increases the rate of sodium reabsorption and potassium excretion by the kidneys.
- During stress, the adrenal cortex ensures adequate fuel supplies for the rapidly metabolizing cells. Cortisol promotes glucose synthesis from other nutrients. During stress, the hypothalamus secretes **corticotropin-releasing factor (CRF),** which stimulates the anterior pituitary gland to secrete **adrenocorticotropic hormone (ACTH).** ACTH regulates both aldosterone and cortisol secretion.

Know and Comprehend

1. A cell secretes a product that diffuses through the interstitial fluid and acts on nearby cells. This process is an example of (a) neuroendocrine secretion (b) autocrine signaling (c) paracrine signaling (d) classical endocrine regulation (e) peptide hormone function

2. Which of the following is/are *true* of steroid hormones? (a) hydrophilic (b) secreted by the posterior pituitary (c) typically work through G proteins and cyclic AMP (d) typically bind with receptor in nucleus and affect transcription (e) a and c are correct

3. Which of the following is *not* a correct pair? (a) neurohormone; insect brain hormone (b) calcium; calmodulin (c) posterior pituitary; releasing hormone (d) anterior pituitary; growth hormone (e) thyroid hyposecretion; cretinism

4. Growth hormone (a) signals the hypothalamus to produce a releasing hormone (b) is regulated mainly by calcium level (c) is a catabolic hormone (d) stimulates metabolic rate (e) stimulates the liver to produce insulin-like growth factors

5. Which of the following occurs in diabetes mellitus? (a) decreased use of glucose (b) decreased fat metabolism (c) decreased protein use (d) increased concentration of TSH-releasing hormone (e) b and c are correct

6. Which of the following is an action of epinephrine? (a) decreases glucose use (b) increases cardiac output (c) constricts blood vessels in brain (d) reduces inflammation (e) mobilizes fat

7. Aldosterone (a) is released by posterior pituitary (b) is an androgen (c) secretion is stimulated by an increase in thyroid-stimulating hormone (d) is an enzyme that converts epinephrine to norepinephrine (e) increases sodium reabsorption

Apply and Analyze

8. Arrange the following events into an appropriate sequence. 1. high thyroid hormone concentration 2. anterior pituitary inhibited 3. homeostasis 4. lower level of thyroid-stimulating hormone 5. thyroid gland secretes less thyroid hormone (a) 1, 2, 4, 5, 3 (b) 5, 4, 3, 2, 1 (c) 1, 2, 5, 4, 3 (d) 4, 5, 2, 3, 1 (e) 1, 4, 2, 5, 3

9. Arrange the following events into an appropriate sequence. 1. blood glucose concentration increases 2. alpha cells in islets stimulated 3. homeostasis 4. low blood glucose concentration 5. glucagon secretion increases (a) 1, 2, 3, 5, 4 (b) 5, 4, 2, 1, 3 (c) 1, 2, 5, 4, 3 (d) 4, 2, 5, 1, 3 (e) 4, 5, 1, 2, 3

10. **VISUALIZE** Draw a diagram illustrating the regulation of blood glucose concentration.

11. Why is it important to maintain a constant blood glucose level? Several hormones discussed in this chapter affect carbohydrate metabolism. Why is it important to have more than one? How do they interact?

12. An injection of too much insulin may cause a diabetic to go into insulin shock, in which the person may appear drunk or may become unconscious, suffer convulsions, and even die. From what you know about the actions of insulin, explain the physiological causes of insulin shock.

Evaluate and Synthesize

13. How do receptors impart specificity within the endocrine system? What might be some advantages of having complex mechanisms for hormone action (such as second messengers)?

14. **EVOLUTION LINK** The receptor for aldosterone appears to have evolved much earlier than aldosterone itself. Propose a hypothesis to explain how that might have occurred.

15. **SCIENCE, TECHNOLOGY, AND SOCIETY** The "thrifty gene" hypothesis holds that the genes that increase risk for type 2 diabetes allow efficient utilization of food (and fat accumulation) when food is available, and this action increases chances of survival in times of food scarcity. Use this hypothesis to explain the rising rate of diabetes in the United States and suggest how science and technology may partner with society to reduce incidence of the disease.

aplia To access course materials, such as Aplia and other companion resources, please visit **www.cengagebrain.com.**

50 | Reproduction

Mating nudibranchs (*Hypselodoris* sp.). Photographed in Bali, Indonesia.

DAVID DOUBILET/National Geographic Creative

The survival of each species requires that its members produce new individuals to replace those that die. The ability to reproduce and perpetuate its species is a basic characteristic of living things. To paraphrase American sociobiologist E. O. Wilson, animal reproduction is an animal's way of making more copies of its genes. How this process is accomplished depends on the relative benefits (and costs) of the reproductive strategies available in any given situation.

The mating nudibranchs (marine slugs) in the photograph are *hermaphrodites*. Each animal has both male and female organs and produces both sperm and eggs. However, they do not self-fertilize. Two nudibranchs come together to mate and exchange sperm through a tube near the head. Each then goes its own way and lays masses containing millions of eggs.

In this chapter we summarize some major features of animal reproduction and discuss the evolutionary advantages of sexual reproduction. We then focus on the process of human reproduction and on the hormones that regulate reproduction. We conclude the chapter with discussions of contraception and sexually transmitted infections.

KEY CONCEPTS

50.1 Most animals reproduce sexually by fusion of sperm and egg; some animals reproduce asexually; and others can reproduce either way, depending on conditions.

50.2 In male vertebrates testes produce sperm and reproductive hormones. In humans and other mammals, reproduction is regulated by hormones produced by the hypothalamus, pituitary gland, and gonads.

50.3 In female vertebrates ovaries produce ova and reproductive hormones. In human females hormones maintain a monthly menstrual cycle that prepares the body for possible pregnancy.

50.4 When a secondary oocyte is fertilized, development begins, and the embryo implants in the wall of the uterus; hormones from the developing embryo, from the corpus luteum, and later from the placenta maintain the pregnancy.

50.5 In both human males and females, sexual response includes four physiological stages: excitement phase, plateau phase, orgasmic phase, and resolution phase.

50.6 A variety of highly reliable contraceptives are available; most hormonal contraceptives prevent ovulation.

50.7 Sexually transmitted infections (STIs) are communicable diseases; the most common STIs in the United States include chlamydia (which can lead to pelvic inflammatory disease), human papillomavirus (HPV), and genital herpes.

50.1 ASEXUAL AND SEXUAL REPRODUCTION

LEARNING OBJECTIVE

1 Compare the benefits of asexual and sexual reproduction, and describe each mode of reproduction, giving specific examples.

Most animals carry on *sexual reproduction,* and some carry on *asexual reproduction.* Some animals reproduce asexually under some conditions and sexually at other times. As we will discuss, many variations of both asexual and sexual reproduction have evolved.

Asexual reproduction is an efficient strategy

In **asexual reproduction** a single parent gives rise to offspring that are genetically identical to the parent (with the exception of mutations). Many invertebrates, including sponges, cnidarians, and some rotifers, flatworms, and annelids, can reproduce asexually. Some vertebrates also reproduce asexually under certain conditions. Asexual reproduction is an adaptation of some sessile animals that cannot move about to search for mates. For animals that do move about, asexual reproduction can be advantageous when the population density is low and mates are not readily available.

In asexual reproduction a single parent may split, bud, or fragment to give rise to two or more offspring. Sponges and cnidarians are among the animals that can reproduce by **budding.** A small part of the parent's body separates from the rest and develops into a new individual (**FIG. 50-1**). Sometimes the buds remain attached and become more or less independent members of a colony.

Oyster farmers learned long ago that when they tried to kill sea stars by chopping them in half and throwing the pieces back into the sea, the number of sea stars preying on the oyster bed doubled! In some flatworms, nemerteans, and annelids, this ability to regenerate is part of a method of reproduction known as **fragmentation.** The body of the parent breaks into several pieces; each piece then regenerates the missing parts and develops into a whole animal.

Parthenogenesis ("virgin development") is a form of asexual reproduction in which an unfertilized egg develops into an adult animal. The adult is typically haploid. Parthenogenesis is common among insects (especially honeybees and wasps) and crustaceans; it also occurs among some other invertebrate and vertebrate groups, including some species of nematodes, gastropods, fishes, amphibians, and reptiles.

Although a few species appear to reproduce solely by parthenogenesis, in most species episodes of parthenogenesis alternate with periods of sexual reproduction. Parthenogenesis may occur for several generations, followed at some point by sexual reproduction in which males develop, produce sperm, and mate with the females to fertilize their eggs. In some species parthenogenesis is a means of rapidly producing individuals when conditions are favorable.

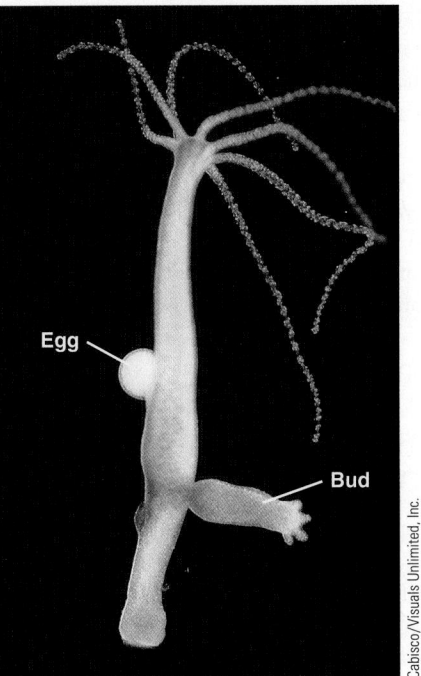

Cabisco/Visuals Unlimited, Inc.

Figure 50-1 Asexual reproduction by budding

A part of the *Hydra*'s body grows outward, then it separates and develops into a new individual. The region of the parent body that buds is not specialized exclusively for reproduction. The *Hydra* shown here is also reproducing sexually, as evidenced by the egg (*left*).

Most animals reproduce sexually

Sexual reproduction in animals involves the production and fusion of two types of **gametes:** sperm and eggs. Typically, two different individuals are required. A male parent contributes a **sperm,** and a female parent contributes an egg, or **ovum** (pl., *ova*). The sperm provides genes coding for some of the male parent's traits, and the egg contributes genes coding for some of the female parent's traits. The egg is typically large and nonmotile, with a store of nutrients that supports the development of the embryo. The sperm is usually small and motile, and is adapted to propel itself by beating its long, whiplike flagellum.

When sperm and egg unite, a **zygote,** or fertilized egg, is produced. The zygote develops into a new animal, similar to both parents but not identical to either. Sexual reproduction typically involves remarkably complex structural, functional, and behavioral processes. In vertebrates hormones secreted by the hypothalamus, pituitary gland, and gonads regulate these processes.

Many aquatic animals practice **external fertilization** in which the gametes meet outside the body (**FIG. 50-2a**). Mating partners usually release eggs and sperm into the water simultaneously. Gametes live for only a short time, and many are lost in the water; some are eaten by predators. However, so many gametes are released that sufficient numbers of sperm and egg cells meet to perpetuate the species.

In **internal fertilization** matters are left less to chance. The male generally delivers sperm cells directly into the body of the female. Her moist tissues provide the watery medium required for the movement of sperm, and the gametes fuse inside the body. Most terrestrial animals, sharks, and aquatic reptiles, birds, and mammals practice internal fertilization (**FIG. 50-2b**).

Hermaphroditism is a form of sexual reproduction in which a single individual produces both eggs and sperm. A few hermaphrodites, such as the tapeworm, are capable of self-fertilization. More typically, two animals come together and fertilize each other's eggs (see the chapter-opening photograph of mating nudibranchs). The common earthworm is also a hermaphrodite. Two animals copulate, and mutual cross-fertilization occurs. Each earthworm inseminates the other. In some hermaphroditic species, self-fertilization is prevented by the development of testes and ovaries at different times.

Sexual reproduction increases genetic variability

Asexual reproduction is the fastest and most efficient way to reproduce. Each cell can produce two new cells. Sexual reproduction is less efficient than asexual reproduction because two cells, rather than just one, are required to make a new organism. (Two parents are typically required.) Sexual reproduction is also more expensive in terms of energy than asexual reproduction because the animal must produce gametes and find mates. Why, then, do most animals reproduce sexually?

A major benefit of sexual reproduction is that it has the potential to increase the *fitness* (reproductive success) of some of the offspring. In contrast to asexual reproduction, in which an animal passes all its genes to its offspring, sexual reproduction has the biological advantage of promoting genetic variety among the members of a species. Each offspring is the product of a particular combination of genes contributed by both parents rather than a genetic copy of a single individual. By combining inherited traits of two parents, sexual reproduction gives rise to at least some offspring that may be better able to survive than either parent. Also,

because the offspring are diploid, they have a backup copy of their genes in case one copy gets damaged by mutation.

Although they agree that sexual reproduction has some selective advantage, biologists do not agree on the details. They are exploring several hypotheses. One hypothesis holds that sexual reproduction provides a mechanism for beneficial mutations from each parent to come together in offspring that can reproduce and spread these mutations through the population. For example, certain beneficial mutations may permit animals to protect themselves from predators or resist parasites.

Another hypothesis holds that sexual reproduction is more efficient than asexual reproduction at removing harmful mutations from a population. As you have learned in earlier chapters, mutations occur constantly, and most mutations are harmful. When animals reproduce asexually, all the offspring inherit all the harmful mutations. As mutations accumulate in a population, individuals carry a bigger and bigger load of harmful genes. In contrast, when animals with different mutations mate, offspring inherit varying numbers and combinations of mutations. Offspring that inherit too many harmful mutations are selected against. They may not live to reproduce, so their harmful mutations are removed from the population.

Wayne Getz, an applied mathematician at the University of California, Berkeley, developed a mathematical model that

(a) External fertilization. Like many aquatic animals, these spawning frogs (*Rana temporaria*) release their gametes into the water. The female lays a mass of eggs, while the male mounts her and simultaneously deposits his sperm in the water.

(b) Internal fertilization. In most terrestrial animals, such as these lions (*Panthera leo*), the male deposits sperm inside the female body. Internal fertilization is also practiced by some fishes and some aquatic reptiles and mammals.

Figure 50-2 Sexual reproduction: external and internal fertilization

predicts whether asexual reproduction or sexual reproduction will exist within a population. According to his model, clones of animals produced by asexual reproduction would be favored in an unchanging environment. The model further predicts that sexual reproduction will be adaptive in an unstable, changing environment and that both types of reproduction can coexist under conditions of moderate change.

Competing hypotheses are an expected part of the scientific process because scientific discovery is rarely a straightforward, linear sequence of question–answer, question–answer. In fact, as scientific knowledge expands, many creative hypotheses are discarded as dead ends. Sometimes researchers show that competing hypotheses may each explain part of the problem under scrutiny.

CHECKPOINT 50.1

- *Distinguish among budding, fragmentation, and parthenogenesis.*
- *What are the advantages and disadvantages of asexual reproduction compared with sexual reproduction?*

50.2 HUMAN REPRODUCTION: THE MALE

LEARNING OBJECTIVES

2   Relate the structure of each organ of the human male reproductive system to its function.
3   Trace the passage of sperm cells through the human male reproductive system from their origin in the seminiferous

tubules to their expulsion from the body in the semen; include a description of spermatogenesis.

4   Describe endocrine regulation of reproduction in the human male.

The human male, like other male mammals, has the reproductive role of producing sperm cells and delivering them into the female reproductive tract. When a sperm combines with an egg, the sperm contributes its genes and determines the sex of the offspring.

The testes produce gametes and hormones

In humans and other vertebrates, **spermatogenesis,** the process of sperm cell production, occurs in the paired male gonads, or **testes** (sing., *testis*) (**FIG. 50-3a**). More specifically, spermatogenesis takes place within a vast tangle of hollow tubules, the **seminiferous tubules,** within each testis (**FIG. 50-3b**). Spermatogenesis begins with undifferentiated cells, the **spermatogonia** (singular, spermatogonium) in the walls of the tubules (**FIG. 50-4**).

The spermatogonia, which are diploid cells, divide by mitosis and produce more spermatogonia. Some enlarge and become **primary spermatocytes,** which undergo **meiosis.** (You may want to review the discussion of meiosis in Chapter 10.) In many animals gamete production occurs only in the spring or fall, but humans have no special breeding season. In the human adult

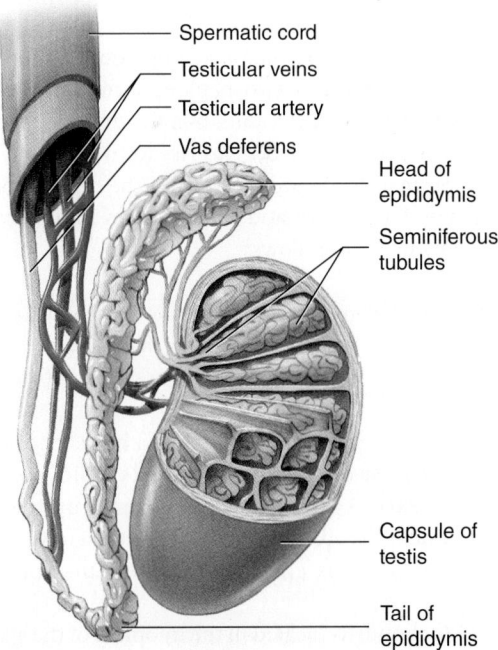

Spermatic cord
Testicular veins
Testicular artery
Vas deferens
Head of epididymis
Seminiferous tubules
Capsule of testis
Tail of epididymis

Seminal vesicle
Ejaculatory duct
Prostate
Rectum
Bulbourethral gland
Epididymis
Scrotum

Bladder
Pubic bone
Vas deferens
Cavernous body
Spongy body
Urethra
Glans penis
Testis

(a) Male reproductive anatomy. The scrotum, penis, and pelvic region of the human male are shown in sagittal section to illustrate their internal structure.

Figure 50-3 *Animation* **Male reproductive system**
© Cengage Learning

(b) Structures that produce and transport sperm. The testis, epididymis, and spermatic cord are shown partly dissected and exposed. The testis is shown in sagittal section to illustrate the arrangement of the seminiferous tubules.

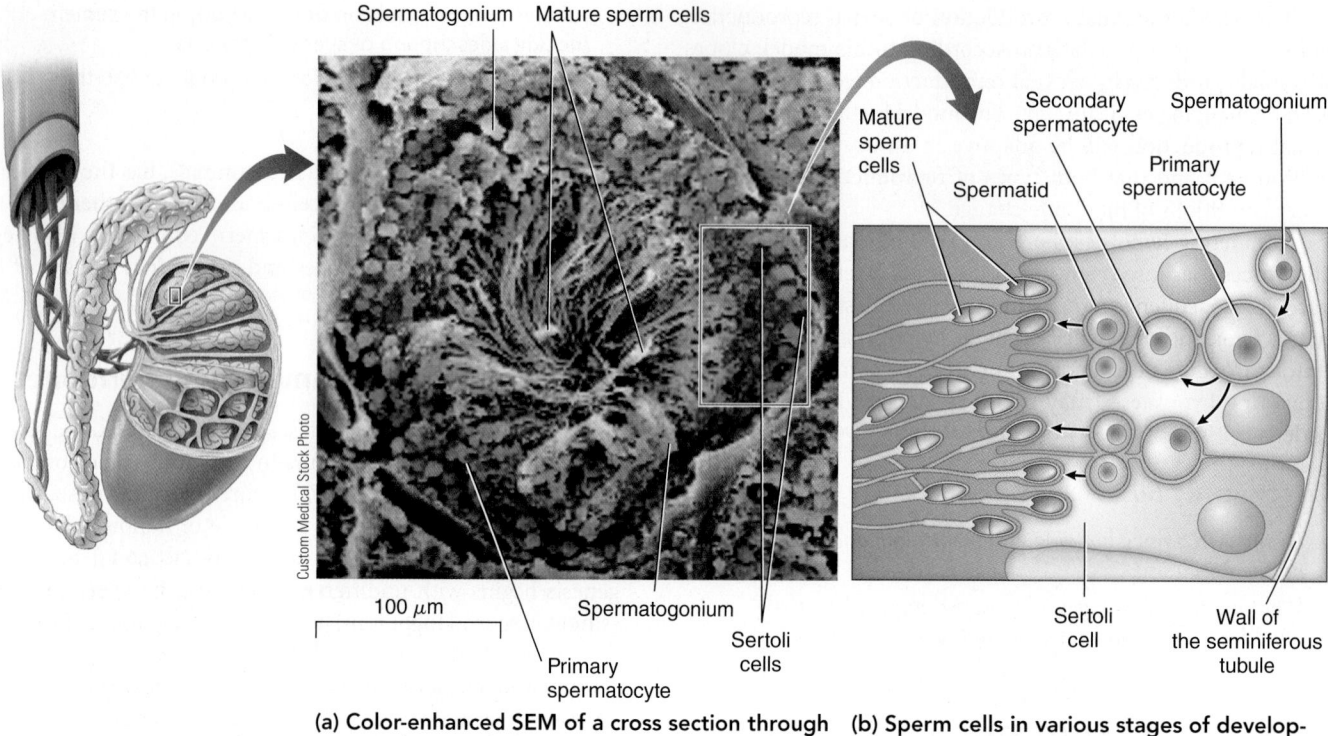

Spermatogonium   Mature sperm cells

Mature sperm cells   Secondary spermatocyte   Spermatogonium

Spermatid   Primary spermatocyte

100 μm

Spermatogonium

Primary spermatocyte

Sertoli cells

Sertoli cell   Wall of the seminiferous tubule

Custom Medical Stock Photo

(a) Color-enhanced SEM of a cross section through a seminiferous tubule.

(b) Sperm cells in various stages of development. Note that the developing sperm cells lie between the large, nutritive Sertoli cells.

Figure 50-4 *Animation* **Spermatogenesis in the seminiferous tubules**
© Cengage Learning

male, spermatogenesis proceeds continuously, and millions of sperm are produced each day.

Each primary spermatocyte undergoes a first meiotic division, which produces two haploid **secondary spermatocytes** (FIG. 50-5). During the second meiotic division, each of the two secondary spermatocytes gives rise to two haploid **spermatids.** Four spermatids are produced from the original primary spermatocyte. Each spermatid differentiates into a mature sperm. The sequence is as follows:

spermatogonium (diploid) ⟶ primary spermatocyte (diploid) ⟶ two secondary spermatocytes (haploid) ⟶ four spermatids (haploid) ⟶ four mature sperm (haploid)

Each mature sperm consists of a head, midpiece, and flagellum (FIG. 50-6). The head consists almost entirely of the nucleus. Part of the nucleus is covered by the *acrosome,* a large vesicle that differentiates from the Golgi complex. The acrosome contains enzymes that help the sperm penetrate the egg.

Mitochondria, located in the midpiece of the sperm, provide the energy for movement of the flagellum. The sperm flagellum has the typical eukaryotic 9 + 2 arrangement of microtubules. During its development, most of the sperm's cytoplasm is discarded and is phagocytosed by the large **Sertoli cells** that ring the fluid-filled lumen of the seminiferous tubule. These cells provide nutrients for the developing sperm cells. Sertoli cells also secrete hormones and other signaling molecules.

Each Sertoli cell extends from the outer membrane of the seminiferous tubule to its lumen. Sertoli cells are joined to one another by *tight junctions* at a place just within the outer membrane of the tubule (see Chapter 5 for discussion of tight junctions). Together, Sertoli cells form a blood–testis barrier that prevents harmful substances from entering the tubule and interfering with spermatogenesis. This barrier also prevents sperm from passing out of the tubule into the blood, where they could stimulate an immune response. The tight junctions between Sertoli cells also form compartments that separate sperm cells in various stages of development.

Human sperm cells cannot develop at body temperature. Although the testes develop within the abdominal cavity of the male embryo, about two months before birth they descend into the **scrotum,** a skin-covered sac suspended from the groin. The scrotum serves as a cooling unit, maintaining sperm below body temperature. In rare cases, the testes do not descend. If this condition is not corrected by surgery or hormone treatment, the seminiferous tubules eventually degenerate and the male becomes *sterile,* unable to produce offspring.

The scrotum is an outpocketing of the pelvic cavity and is connected to it by the *inguinal canals.* As they descend, the testes pull their blood vessels, nerves, and conducting tubes after them. The inguinal region is a weak place in the abdominal wall. Straining the abdominal muscles by lifting heavy objects sometimes tears the inguinal tissue. A loop of intestine can then bulge into the scrotum through the tear, a condition known as an *inguinal hernia.*

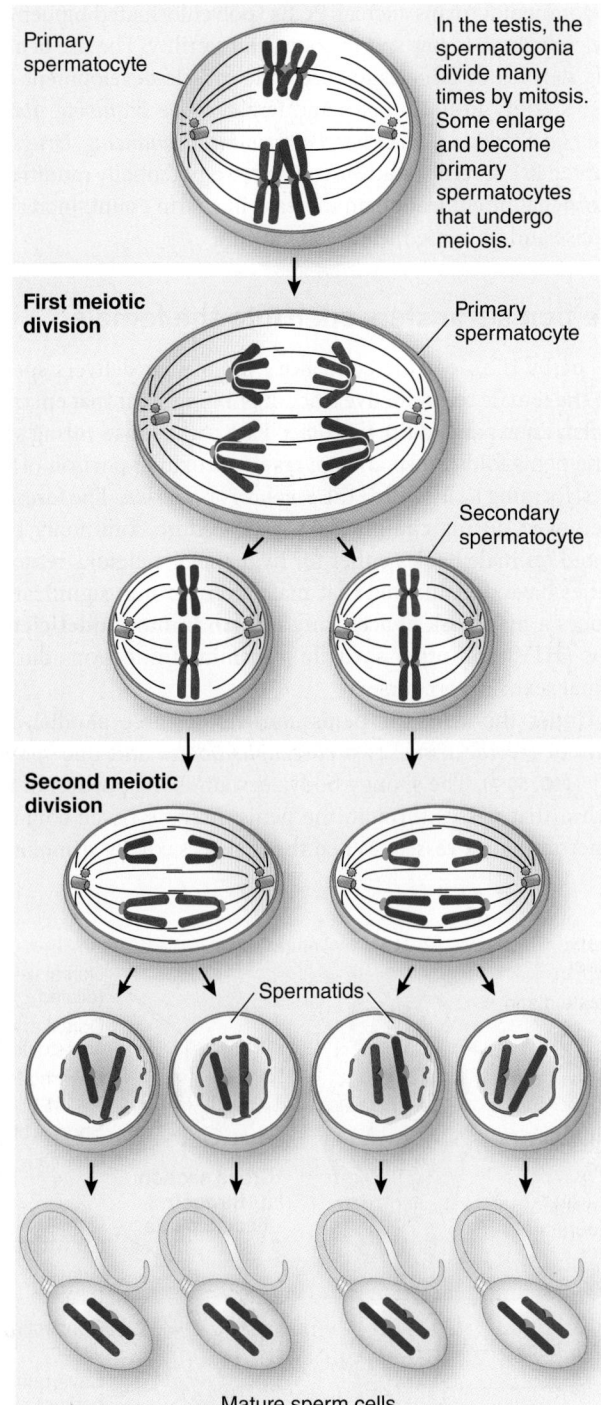

Figure 50-5 Spermatogenesis

Note in this example that four chromosomes are present in the primary spermatocyte (2*n*) and that meiosis produces the haploid (*n*) number (2) in the secondary spermatocytes. The spermatids and mature sperm cells are also haploid.

© Cengage Learning

A series of ducts store and transport sperm

Sperm cells leave the seminiferous tubules of each testis and pass into a larger coiled tube, the **epididymis.** There, sperm finish maturing and are stored. During **ejaculation,** sperm pass from each epididymis into a sperm duct, the **vas deferens** (pl., *vasa*

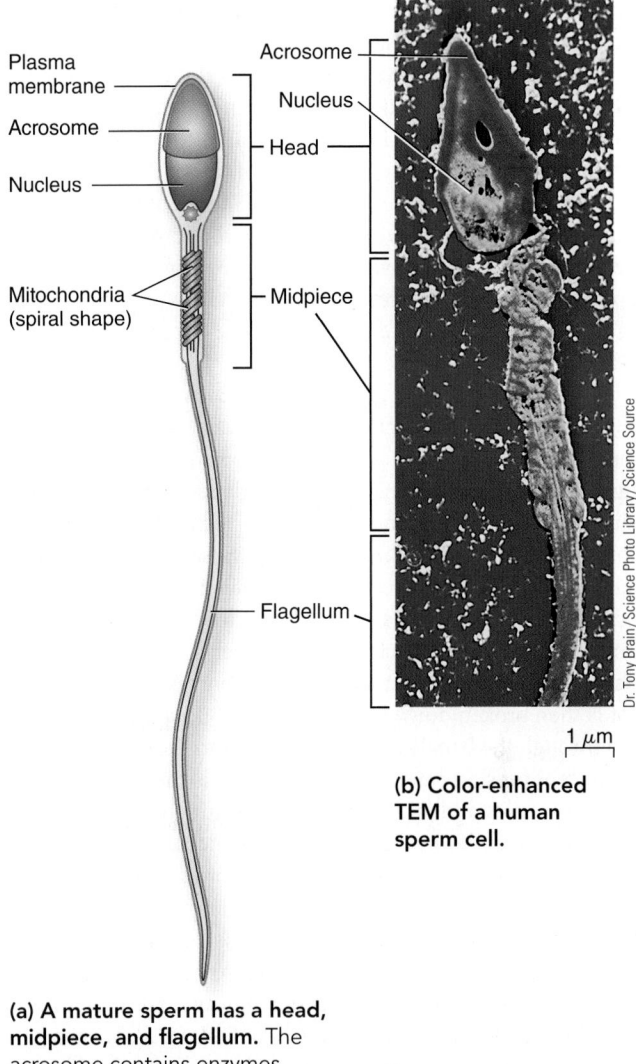

(a) A mature sperm has a head, midpiece, and flagellum. The acrosome contains enzymes important in penetrating the egg.

Figure 50-6 Structure of a mature sperm
© Cengage Learning

deferentia). The vas deferens extends from the scrotum through the inguinal canal and into the pelvic cavity.

Each vas deferens empties into a short **ejaculatory duct,** which passes through the prostate gland and then opens into the single **urethra.** The urethra, which at different times conducts urine and semen, passes through the penis to the outside of the body. Thus, the sperm pass through the following structures:

seminiferous tubules ⟶ epididymis ⟶ vas deferens ⟶ ejaculatory duct ⟶ urethra ⟶ release from body

The accessory glands produce the fluid portion of semen

As sperm travel through the conducting tubes, they mix with secretions from three types of accessory glands. Approximately 3 mL of **semen** is ejaculated during sexual climax. Semen consists of about 200 million sperm cells suspended in the secretions of these glands.

The paired **seminal vesicles** secrete a fluid rich in fructose and prostaglandins into the vasa deferentia (see Fig. 50-3). The fructose provides energy for the sperm after they are ejaculated. Prostaglandins stimulate contractions of smooth muscle in both male and female reproductive tracts. These contractions help transport the sperm down the male reproductive tract and up the female tract to the oviduct where fertilization occurs. The seminal vesicles also secrete fibrinogen, which clots the semen. The fluid secreted by the seminal vesicles accounts for about 60% of the semen volume.

The single **prostate gland** secretes an alkaline fluid that neutralizes the acidic secretions of the vagina. This process is important because sperm are more active and survive longer in a slightly alkaline environment. Prostate gland secretion also contains clotting enzymes and *prostate-specific antigen (PSA)*. The clotting enzymes act on fibrinogen from the seminal vesicles, producing fibrin. You may recall from Chapter 44 that fibrin threads form blood clots. In this case, fibrin clots the semen. When semen is ejaculated, the clotted semen helps keep the sperm in the female reproductive tract while the penis is withdrawn. The clot is then broken down by PSA, and mobile sperm make their way through the female reproductive tract to the ovum.

Except for skin cancer, prostate cancer is the most common cancer in males and is also one of the leading causes of cancer death among males in the United States. According to national guidelines, doctors should talk to their patients about the benefits and risks of PSA-based screening for prostate cancer for men without symptoms. Screening tests for prostate cancer include the digital rectal exam and the prostate specific antigen blood test (PSA). The PSA measures the concentration of PSA in the blood. Elevated concentrations of PSA are associated with prostate cancer, prostate infection, and certain benign changes in the prostate. An elevated PSA indicates a need for further evaluation, typically a prostate biopsy. Prostate cancer can be treated with surgery, hormone therapy, or radiation treatment.

During sexual arousal, the paired *bulbourethral glands,* located on each side of the urethra, release a mucous secretion. This fluid lubricates the penis, facilitating its penetration into the vagina.

A major cause of male infertility is insufficient sperm production. When sperm counts drop below 35 million per milliliter of semen, fertility is impaired, and males with a sperm count lower than 20 million per milliliter are usually considered sterile. When a couple's attempts to produce a child are unsuccessful, a sperm count and analysis may be performed in a clinical laboratory. Sometimes semen is found to contain large numbers of abnormal sperm or, occasionally, no sperm at all.

In the United States in the 1970s, an average healthy young man produced about 100 million sperm per milliliter of semen. Today that average has dropped to about 60 million. Although the cause of this decrease is not known, low sperm counts have been linked to a variety of environmental factors, including chronic marijuana use, alcohol abuse, and cigarette smoking. Studies also show that men who smoke tobacco are more likely than nonsmokers to produce abnormal sperm. Exposure to industrial and environmental toxins such as PCBs (polychlorinated biphenyls) may contribute to low sperm count and sterility. The use of anabolic steroids by athletes to accelerate muscle development can cause sterility in both males and females (see *Inquiring About: Anabolic Steroids and Other Performance-Enhancing Drugs* in Chapter 49). A recent study identified two potentially modifiable factors that might lead to an increase in sperm count: increased exercise and decreased television viewing.

The penis transfers sperm to the female

The **penis** is an erectile copulatory organ that delivers sperm into the female reproductive tract. It is a long shaft that enlarges to form an expanded tip, the *glans.* Part of the loose-fitting skin of the penis folds down and covers the proximal portion of the glans, forming a cuff called the *prepuce,* or *foreskin.* The foreskin is removed during circumcision, a procedure commonly performed on male babies either for hygienic or religious reasons. Studies have documented that male circumcision significantly reduces a man's risk of acquiring **human immunodeficiency virus (HIV)** and other sexually transmitted infections during vaginal sexual intercourse.

Under the skin, the penis consists of three parallel columns of *erectile tissue:* two *cavernous bodies* and one *spongy body* (**FIG. 50-7**). The spongy body surrounds the portion of the urethra that passes through the penis. Erectile tissue contains numerous blood vessels. When the male is sexually stimulated,

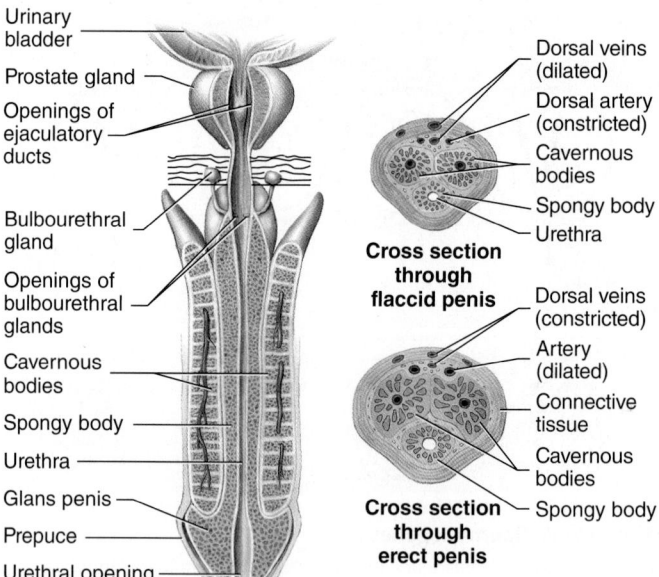

(a) Longitudinal section through the prostate gland and penis. Note the three parallel columns of erectile tissue in the penis: the two cavernous bodies and the spongy body.

(b) Cross sections through a flaccid and an erect penis. The erectile tissues become engorged with blood in the erect penis.

Figure 50-7 Internal structure of the penis
© Cengage Learning

parasympathetic neurons dilate the arteries in the penis. As blood fills the blood vessels of the erectile tissue, the tissue swells. This swelling compresses veins that conduct blood away from the penis, slowing the outflow of blood. Thus, more blood enters the penis than can leave, further engorging the erectile tissue with blood. Erection of the penis occurs as the penis increases in length, diameter, and firmness. Although the human penis contains no bone, penis bones do occur in some other mammals, such as bats, rodents, and some primates.

Erectile dysfunction, the chronic inability to sustain an erection, prevents effective sexual intercourse. Associated with a variety of physical causes and psychological issues, this common disorder is treated with medication, such as sildenafil (Viagra) and related drugs (Levitra, Cialis). These drugs block the action of an enzyme that breaks down the signaling molecule that sustains erection.

Testosterone has multiple effects

Testosterone is the principal **androgen,** or male sex hormone. It is a steroid produced by the **interstitial cells** between the seminiferous tubules in the testes. Testosterone has many functions, beginning during early embryonic development when it stimulates development of the primary male reproductive organs (**TABLE 50-1**). Later during development, testosterone stimulates descent of the testes into the scrotum.

Puberty is the period of sexual maturation during which the secondary sex characteristics begin to develop and the individual becomes capable of reproducing. In males puberty typically begins between ages 10 and 12 and continues until ages 16 to 18. Testosterone directly affects muscle and bone and stimulates the adolescent growth spurt in males at puberty. It produces the male's *primary sex characteristics:* growth of the reproductive organs and spermatogenesis. Testosterone also stimulates the development of the **secondary sex characteristics** at puberty, including growth of facial and

body hair, muscle development, and the increase in vocal cord length and thickness that causes the voice to deepen. Testosterone is necessary for normal sex drive.

In some of its target tissues, testosterone is converted to other steroids. Interestingly, in brain cells testosterone is converted to *estradiol,* the principal female sex hormone. The implications of this transformation are not yet understood.

The hypothalamus, pituitary gland, and testes regulate male reproduction

As you read the following description of male hormone action and regulation, follow the steps in **FIGURE 50-8**. When a boy is about 10 years old, the hypothalamus begins to secrete **gonadotropin-releasing hormone (GnRH).** GnRH stimulates the anterior pituitary to secrete the *gonadotropic hormones* **follicle-stimulating hormone (FSH)** and **luteinizing hormone (LH).** Both FSH and LH are glycoproteins that use cyclic AMP as a second messenger (see Chapter 49). FSH stimulates Sertoli cells to secrete **androgen-binding protein (ABP)** and other signaling molecules that are necessary for spermatogenesis. LH stimulates the interstitial cells to secrete testosterone.

FSH, LH, and testosterone all directly or indirectly stimulate testosterone secretion and spermatogenesis. A high concentration of testosterone in the testes is required for spermatogenesis. Testosterone and FSH stimulate Sertoli cells to produce ABP, which binds to testosterone and concentrates it in the tubules. Testosterone also maintains the male secondary sex characteristics.

Reproductive hormone concentrations are regulated by negative feedback mechanisms (see Fig. 50-8b). Testosterone acts on the hypothalamus, decreasing its secretion of GnRH, which decreases FSH and LH secretion by the pituitary. Testosterone also directly inhibits the anterior lobe of the pituitary by blocking the normal actions of GnRH on LH synthesis and release.

FSH secretion is inhibited mainly by **inhibin,** a peptide hormone secreted by Sertoli cells. FSH itself stimulates inhibin

TABLE 50-1	Principal Male Reproductive Hormones	
ENDOCRINE GLAND AND HORMONES	**PRINCIPAL TARGET TISSUE**	**PRINCIPAL ACTION**
HYPOTHALAMUS		
Gonadotropin-releasing hormone (GnRH)	Anterior pituitary	Stimulates release of FSH and LH
ANTERIOR PITUITARY		
Follicle-stimulating hormone (FSH)	Testes	Stimulates development of seminiferous tubules; stimulates spermatogenesis
Luteinizing hormone (LH)	Testes	Stimulates interstitial cells to secrete testosterone
TESTES		
Testosterone	General	**Before birth:** stimulates development of primary sex organs and descent of testes into scrotum
		At puberty: responsible for growth spurt; stimulates development of reproductive structures and secondary sex characteristics
		In adult: maintains secondary sex characteristics; stimulates spermatogenesis
Inhibin	Anterior pituitary	Inhibits FSH secretion

© Cengage Learning

Hormones from the hypothalamus, anterior pituitary gland, and testes regulate male reproduction by negative feedback systems.

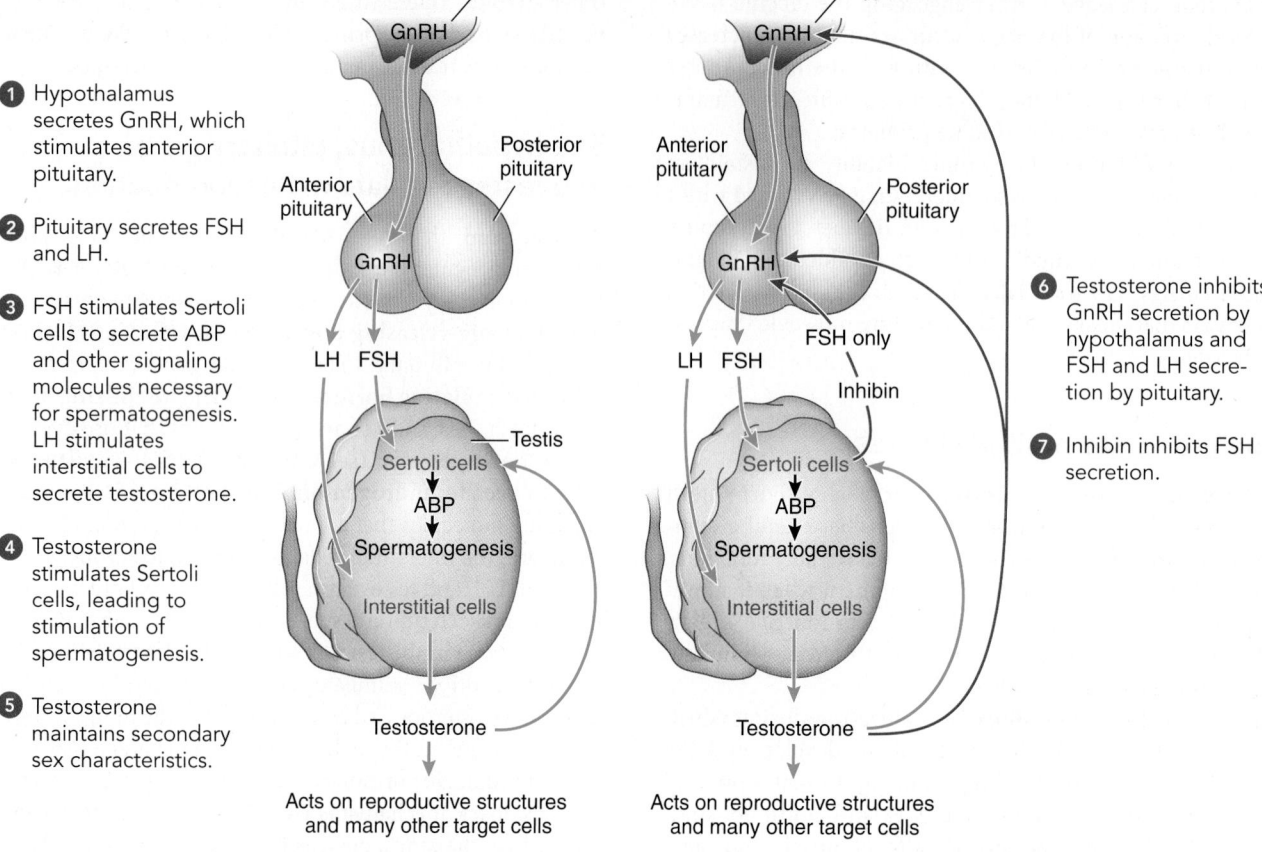

1 Hypothalamus secretes GnRH, which stimulates anterior pituitary.

2 Pituitary secretes FSH and LH.

3 FSH stimulates Sertoli cells to secrete ABP and other signaling molecules necessary for spermatogenesis. LH stimulates interstitial cells to secrete testosterone.

4 Testosterone stimulates Sertoli cells, leading to stimulation of spermatogenesis.

5 Testosterone maintains secondary sex characteristics.

6 Testosterone inhibits GnRH secretion by hypothalamus and FSH and LH secretion by pituitary.

7 Inhibin inhibits FSH secretion.

(a) Overview of hormone action.

(b) Negative feedback systems regulate hormone concentrations.

Figure 50-8 *Animation* **Regulation of male reproduction**

Several hormones interact to maintain male characteristics and reproductive function. *Green arrows* indicate stimulation; *red arrows* indicate inhibition. (GnRH, gonadotropin-releasing hormone; FSH, follicle-stimulating hormone; LH, luteinizing hormone; ABP, androgen-binding protein)

© Cengage Learning

PREDICT Imagine that a man's anterior pituitary stopped producing LH. How would that affect him?

secretion. The endocrine regulation of reproductive function is complex, and it is likely that other hormones and signaling molecules will be identified.

Insufficient testosterone results in sterility. If a male is *castrated*—that is, the testes are removed—before puberty, he is deprived of testosterone and becomes a eunuch. He retains child-like sexual organs and does not develop secondary sexual characteristics. If castration occurs after puberty, increased secretion of male hormones by the adrenal glands helps maintain masculinity.

CHECKPOINT 50.2

- *What are the functions of the testes?*
- **VISUALIZE** *Draw and label a diagram showing the structures through which a sperm passes in its journey from a seminiferous tubule until it leaves the male body during ejaculation.*

- *What are the actions of testosterone?*
- **VISUALIZE** *Draw and label a diagram illustrating the endocrine regulation of male reproduction. (Include LH, FSH, GnRH, ABP, testosterone, and inhibin.)*

50.3 HUMAN REPRODUCTION: THE FEMALE

LEARNING OBJECTIVES

5 Relate the structure of each organ of the human female reproductive system to its function.

6 Trace the development of a human egg and trace its passage through the female reproductive system until it is fertilized.

7 Describe the endocrine regulation of reproduction in the human female and identify key events of the menstrual cycle, such as ovulation and menstruation.

The female reproductive system produces ova (gametes), receives the penis and sperm released from it during sexual intercourse, houses and nourishes the embryo during prenatal development, gives birth, and produces milk for the young (lactation). Hormones secreted by the hypothalamus, pituitary gland, and ovaries interact to regulate and coordinate these processes.

The ovaries produce gametes and sex hormones

Like the male gonads, the female gonads, or **ovaries,** produce both gametes and sex hormones. About the size and shape of large almonds, the ovaries lie close to the lateral walls of the pelvic cavity and are held in position by several connective tissue ligaments (**FIG. 50-9**). Internally, the ovary consists mainly of connective tissue containing scattered ova in various stages of maturation.

The process of ovum production, called **oogenesis,** begins in the ovaries. Before birth, hundreds of thousands of **oogonia** are present in the ovaries. For more than 50 years, biologists agreed that all a female's oogonia were produced during embryonic development and that no new oogonia were produced after birth. Recently, a research team challenged this view, claiming that stem cells in the ovaries of women (and

mice) continue to differentiate into ova throughout adulthood, until menopause.

During prenatal development, the oogonia increase in size and become **primary oocytes.** By the time of birth, they are in the prophase of the first meiotic division. At this stage, they enter a resting phase that lasts throughout childhood and into adult life.

A primary oocyte and the **granulosa cells** surrounding it together make up a **follicle** (**FIG. 50-10**). The granulosa cells are connected by tight junctions that form a protective barrier around the oocyte. With the onset of puberty, a few follicles begin to mature each month in response to FSH secreted by the anterior pituitary gland. As a follicle grows, the granulosa cells proliferate, forming several layers. Connective tissue cells surrounding the granulosa cells differentiate to form a layer of **theca cells.**

As the follicle matures, the primary oocyte completes its first meiotic division. The two haploid cells produced differ in size (**FIG. 50-11**). The smaller one, the first **polar body,** may later divide, forming two polar bodies, but these cells eventually disintegrate. The larger cell, the **secondary oocyte,** proceeds to the second meiotic division but remains in metaphase II until it is fertilized. When meiosis continues, the second meiotic division gives rise to a single ovum and a second polar body. The polar bodies are small and apparently dispose of unneeded chromosomes with a minimal amount of cytoplasm. The sequence is as follows:

oogonium (diploid) ⟶ primary oocyte (diploid) ⟶ secondary oocyte + first polar body (both haploid) ⟶ (after fertilization) ovum + second polar body (both haploid)

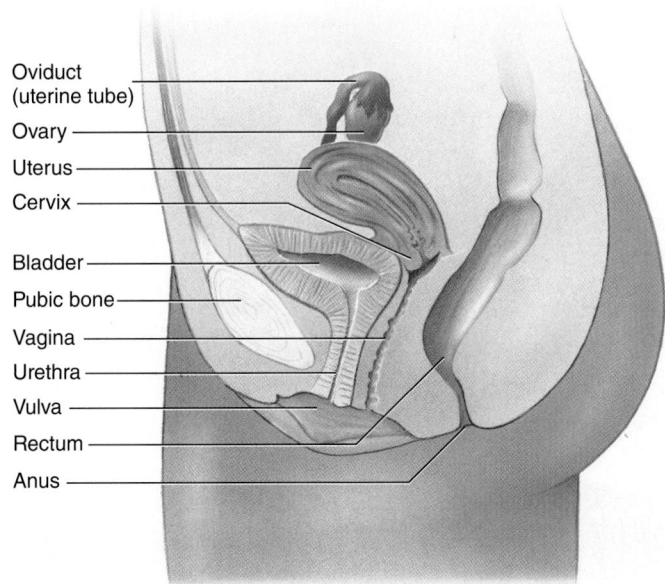

(a) **Midsagittal section through the female pelvis.** Note the position of the uterus relative to the vagina.

Figure 50-9 *Animation* **Female reproductive system**
© Cengage Learning

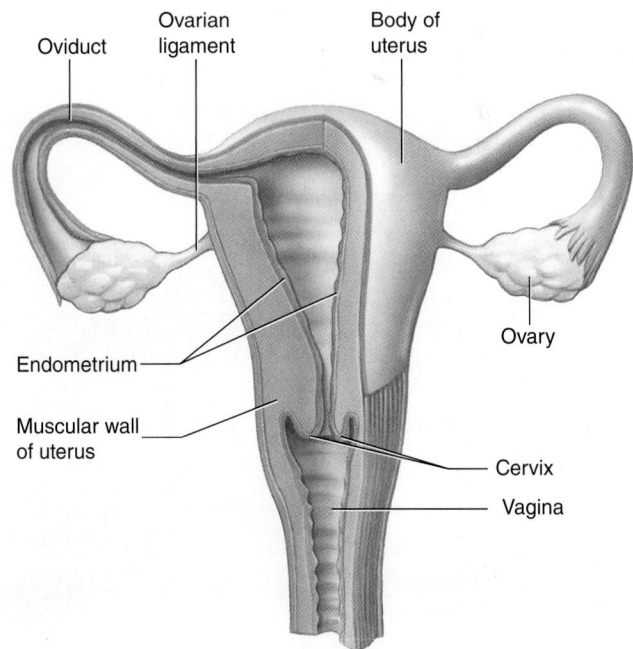

(b) **Anterior view of female reproductive system.** Some organs have been cut open to expose their internal structure. Connective tissue ligaments anchor the reproductive organs in place.

Recall that in the male, each primary spermatocyte gives rise to four functional sperm cells. In contrast, each primary oocyte generates only one ovum.

As an oocyte develops, it becomes separated from its surrounding follicle cells by a layer of glycoproteins called the **zona pellucida.** As the follicle develops, follicle cells secrete fluid, which collects in the antrum (space) between them (see Fig. 50-10). The follicle cells also secrete **estrogens,** female sex hormones. The principal estrogen is *estradiol* (see Fig. 49-2b). Typically, only one follicle fully matures each month. Several others may develop for a while and then deteriorate by apoptosis.

As a follicle matures, it moves closer to the surface of the ovary, eventually resembling a fluid-filled bulge on the ovarian surface. Follicle cells secrete proteolytic enzymes that break down a small area of the ovary wall. During **ovulation,** the secondary oocyte ejects through the ovary wall and into the pelvic cavity. The portion of the follicle that remains in the ovary develops into the **corpus luteum,** a temporary endocrine gland that secretes estrogen and **progesterone.**

The oviducts transport the secondary oocyte

Almost immediately after ovulation, the secondary oocyte is swept into the funnel-shaped opening of the **oviduct,** also called the **uterine tube** or *fallopian tube.* Ciliated epithelial cells lining the oviduct sweep the secondary oocyte into the oviduct and move it along toward the uterus. Fertilization takes place within the oviduct. If fertilization does not occur, the secondary oocyte degenerates there.

Scarring of the oviducts (e.g., by *pelvic inflammatory disease,* caused by a sexually transmitted infection) can block the tubes so that the fertilized ovum cannot pass to the uterus. Women with blocked oviducts may be infertile. Sometimes partial constriction of the oviduct results in *tubal pregnancy.* The embryo cannot progress to the uterus, and it begins to develop in the wall of the oviduct. Because oviducts are not adapted to bear the burden of a developing embryo, the oviduct and the embryo it contains must be surgically removed before it ruptures and endangers the life of the mother.

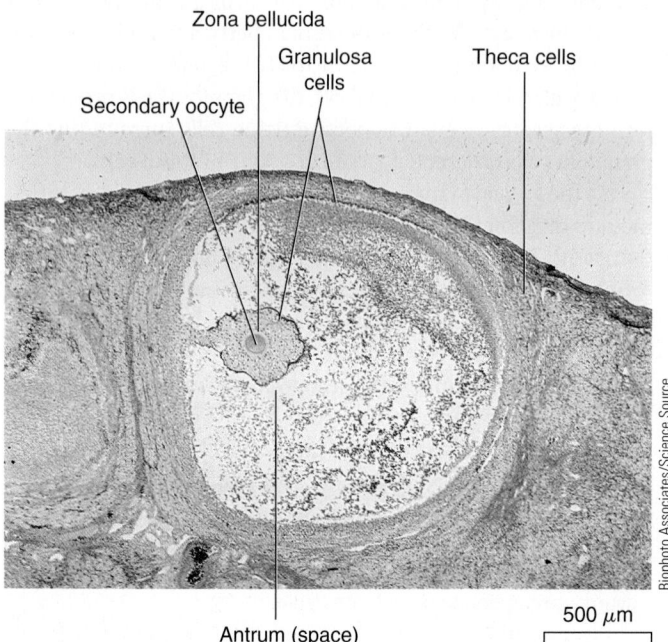

Zona pellucida
Granulosa cells
Theca cells
Secondary oocyte
Antrum (space)
500 μm

Biophoto Associates/Science Source

(a) LM of a developing follicle. The secondary oocyte is surrounded by the zona pellucida (a layer of glycoproteins) and by granulosa cells. Connective tissue cells surrounding the granulosa cells form a layer of theca cells.

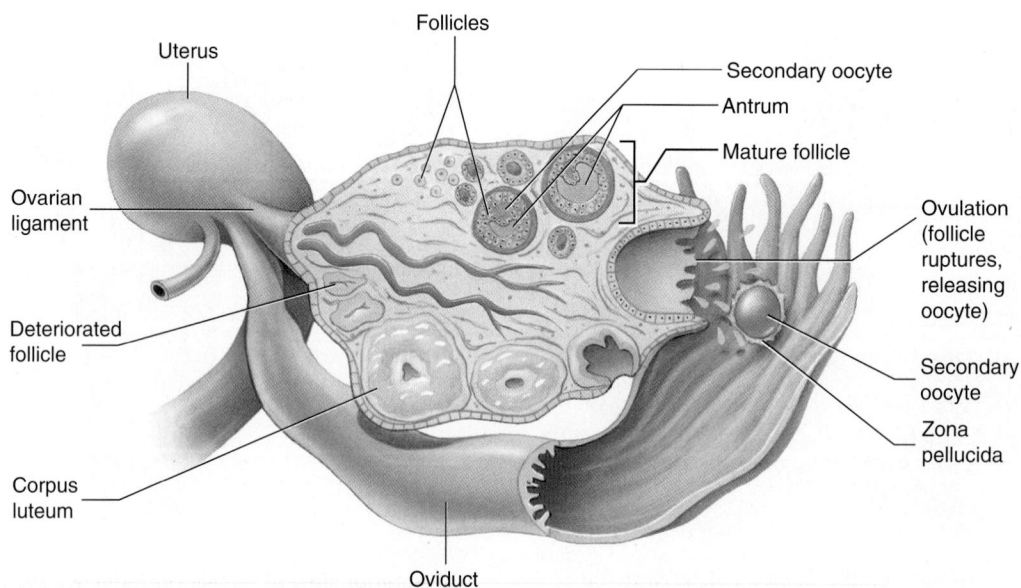

Uterus
Follicles
Secondary oocyte
Antrum
Mature follicle
Ovarian ligament
Ovulation (follicle ruptures, releasing oocyte)
Deteriorated follicle
Secondary oocyte
Zona pellucida
Corpus luteum
Oviduct

(b) Follicles in the ovary. This drawing is a composite; all these stages of development would not be present at the same time.

Figure 50-10 Development of a follicle in the ovary
© Cengage Learning

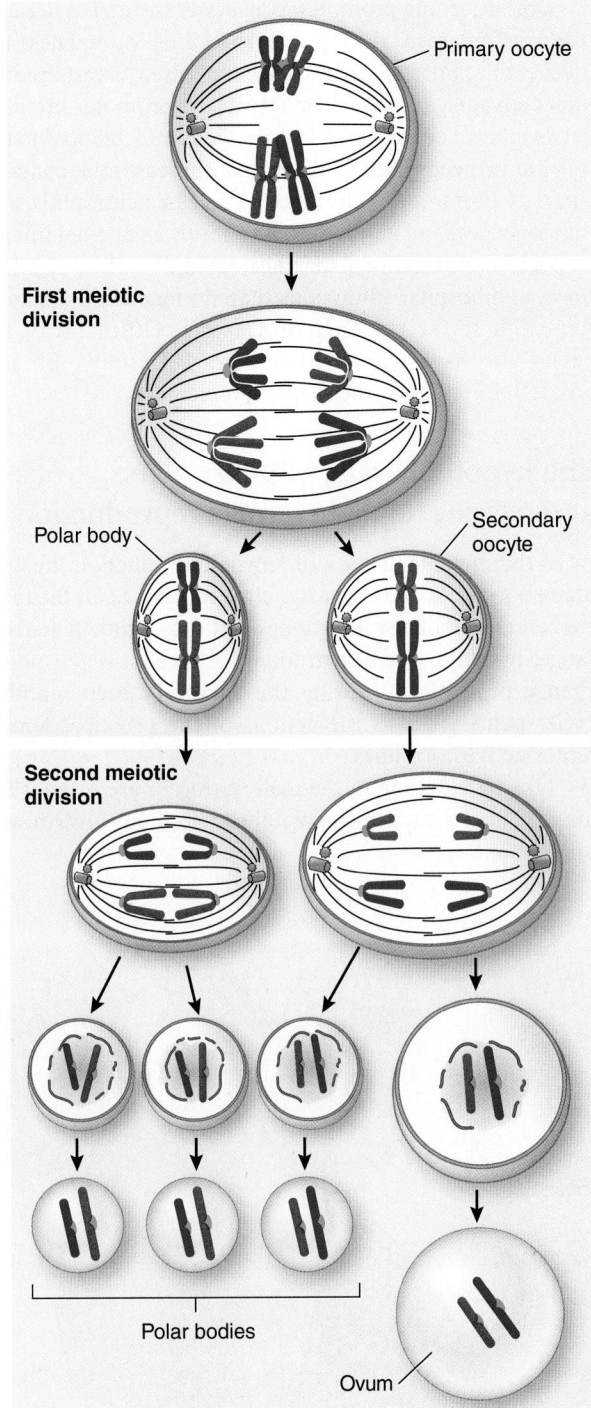

Figure 50-11 Oogenesis

Before birth, oogonia (*not shown*) divide many times by mitosis. Some oogonia differentiate to become primary oocytes that undergo meiosis. Only one functional ovum is produced from each primary oocyte. The other cells produced are polar bodies that degenerate. The first polar body sometimes divides as shown but often just deteriorates. The second meiotic division is completed after fertilization. Note in this example that four chromosomes are present in the primary oocyte (2*n*) and that meiosis produces the haploid (*n*) number (two chromosomes) in the polar bodies, secondary oocyte, and mature ovum.
© Cengage Learning

The uterus incubates the embryo

The oviducts open into the upper corners of the pear-shaped **uterus** (see Fig. 50-9b). About the size of a fist, the uterus (or womb) occupies a central position in the pelvic cavity. It has thick walls of smooth muscle and an epithelial lining, the **endometrium,** which thickens each month in preparation for possible pregnancy. About 10% of women (more than ten million women in the United States alone) are affected by *endometriosis* during their reproductive years. In this painful disorder, fragments of the endometrium migrate to other areas, such as the oviducts or ovaries. Like pelvic inflammatory disease, endometriosis causes scarring that can lead to infertility.

If a secondary oocyte is fertilized, the tiny embryo enters the uterus and implants in the endometrium. As it grows and develops, it is sustained by nutrients and oxygen delivered by surrounding maternal blood vessels. If fertilization does not occur during the monthly cycle, the endometrium sloughs off and is discharged in the process known as **menstruation.**

The lower portion of the uterus, called the **cervix,** extends slightly into the vagina. Cervical cancer is the third most common type of cancer for females in the world. Almost all cases of cervical cancer are linked to infection with human papillomavirus (HPV). (Vaccines against HPV are described later in this chapter.) Detection of cervical cancer is usually possible by the routine Papanicolaou test (Pap smear), in which a few cells are scraped from the cervix during a regular gynecological examination and studied microscopically. When cervical cancer is detected at an early stage, there is a good chance that the patient can be cured. (Interestingly, researchers have reported that in the future, Pap smears may also be used to detect mutations that put women at high risk for developing ovarian cancer or uterine cancer.)

The vagina receives sperm

The **vagina** is an elastic, muscular tube that extends from the uterus to the exterior of the body. The vaginal opening is located in the *perineal region* between the opening of the urethra (anteriorly) and the opening of the anus (posteriorly). The vagina serves as a receptacle for sperm during sexual intercourse and as part of the birth canal (see Fig. 50-9).

The vulva are external genital structures

The female external genitalia, collectively known as the **vulva,** include several structures. Liplike folds, the *labia minora,* surround the vaginal and urethral openings (**FIG. 50-12**). The area enclosed by the labia minora is the *vestibule* of the vagina. Vestibular glands secrete a lubricating mucus into the vestibule. The *hymen* is a thin ring of tissue that forms a border around the entrance to the vagina.

Anteriorly, the labia minora merge to form the prepuce of the **clitoris,** a small erectile structure comparable to the male glans penis. Like the penis, the clitoris contains erectile tissue that becomes engorged with blood during sexual excitement. Rich in nerve endings, the clitoris is highly sensitive to touch,

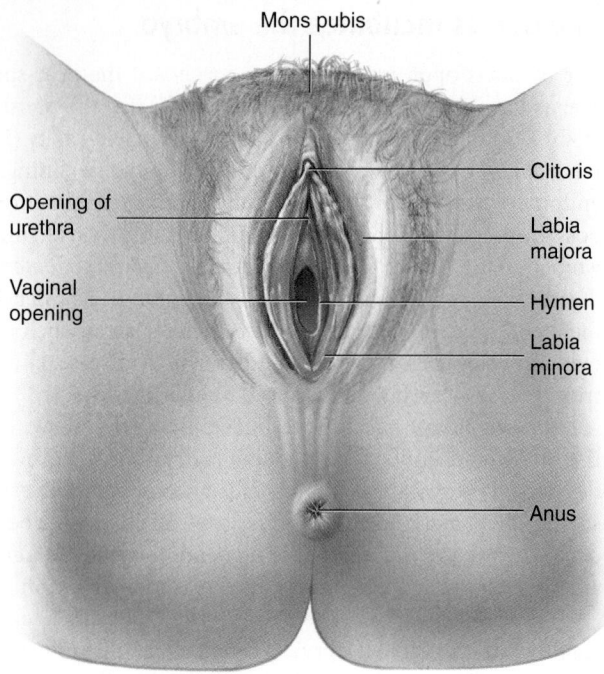

Figure 50-12 External female genital structures
Collectively, the structures shown (excluding the anus) are referred to as the vulva.
© Cengage Learning

Labels on figure: Mons pubis, Clitoris, Opening of urethra, Labia majora, Vaginal opening, Hymen, Labia minora, Anus

pressure, and temperature, and serves as the main center of sexual sensation in the female.

External to the delicate labia minora are the thicker *labia majora*. The *mons pubis* is the mound of fatty tissue just above the clitoris at the junction of the thighs and torso. At puberty the mons pubis and labia majora become covered by coarse pubic hair.

The breasts function in lactation

Each breast consists of 15 to 20 lobes of glandular tissue. The amount of adipose tissue around these lobes determines the size of the breasts and accounts for their softness. Gland cells are arranged in grapelike clusters called **alveoli** (**FIG. 50-13**). Ducts from each cluster join to form a single duct from each lobe, producing 15 to 20 tiny openings on the surface of each nipple. The breasts are the most common site of cancer in women, except for the skin (see *Inquiring About: Breast Cancer*).

Lactation is the production of milk for nourishing the young. During pregnancy, high concentrations of the female reproductive hormones, estrogen and progesterone, stimulate the breasts to increase in size. For the first few days after childbirth, the mammary glands produce a fluid called *colostrum*, which contains proteins and antibodies, but little fat. When the baby suckles, the hormone **prolactin,** secreted by the anterior pituitary gland, stimulates milk production. Nursing also stimulates the posterior pituitary gland to release **oxytocin.** This hormone stimulates ejection of milk from the alveoli into the ducts.

Breast-feeding promotes recovery of the uterus because oxytocin released during breast-feeding stimulates the uterus to contract to nonpregnant size. Breast-feeding offers advantages to the baby as well. It promotes bonding between mother and child, and provides milk tailored to the nutritional needs of the human infant. Breast milk contains a variety of immune cells (B cells, T cells, neutrophils, and macrophages) and antibodies. As a result, breast-fed infants have a lower incidence of diarrhea, ear and respiratory infections, and hospital admissions than do formula-fed babies. According to the U.S. Surgeon General's Office, breast-fed babies also have a lower lifetime risk of asthma, obesity, type 2 diabetes, and other health problems.

The hypothalamus, pituitary gland, and ovaries regulate female reproduction

As in the male, regulation of female reproduction involves many hormones and other signaling molecules. In the male concentration of sex hormones is kept within a narrow range. In contrast, concentrations of female sex hormones change dramatically during the course of each monthly cycle. **TABLE 50-2** lists the actions of the principal female reproductive hormones.

Like testosterone in the male, *estrogens* are responsible for growth of the sex organs at puberty, for body growth, and

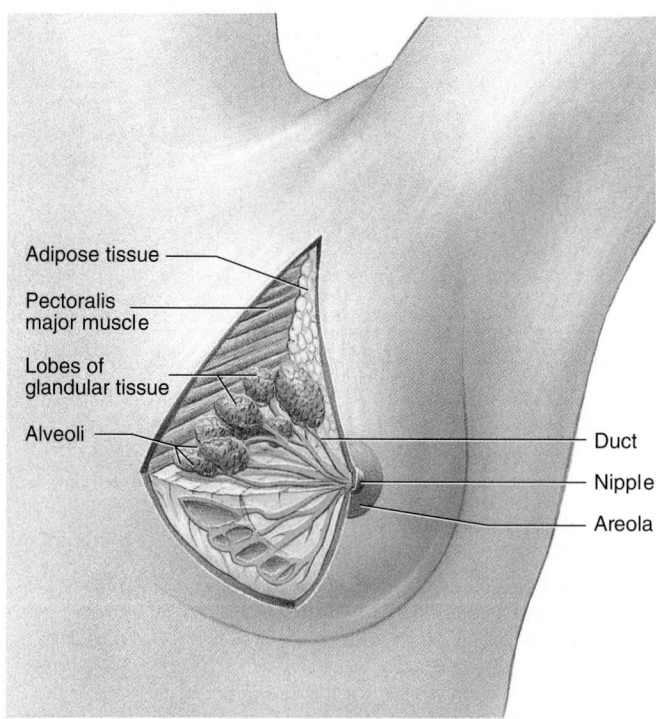

Figure 50-13 Structure of the mature female breast
The breast contains lobes of glandular tissue. This tissue consists of alveoli, clusters of gland cells.
© Cengage Learning

Labels on figure: Adipose tissue, Pectoralis major muscle, Lobes of glandular tissue, Alveoli, Duct, Nipple, Areola

How common is breast cancer? According to the American Cancer Society, breast cancer is the second most common type of cancer among women in the United States (exceeded only by skin cancer.) More than 200,000 new cases of invasive breast cancer are diagnosed each year, and about 40,000 women die from breast cancer every year. Breast cancer is the second leading cause of cancer deaths (exceeded only by lung cancer). Risk factors for breast cancer include being overweight (especially after menopause), alcohol use (more than two drinks per day), and lack of exercise. Some studies show that cigarette smoking is also a risk factor.

Women with a family history of breast cancer are at higher risk than those without such family history. An estimated 10% of breast cancers are familial, and about half of these patients have mutations in a tumor suppressor gene, either *BRCA1* or *BRCA2* (see Fig. 17-21). When the normal protein product of the *BRCA1* gene is phosphorylated by a specific protein kinase, it interacts with the normal protein product of the *BRCA2* gene and certain other compounds to repair DNA damage. Women with a family history of breast (or ovarian) cancer should seek genetic counseling and consider genetic testing. For those who test positive, there are now drugs available (such as tamoxifen) that decrease the risk of cancer. Some women choose preventive surgery.

About 50% of breast cancers begin in the upper, outer quadrant of the breast. As a malignant tumor grows, it may adhere to the deep tissue of the chest wall. Sometimes it extends to the skin, causing dimpling. Eventually, the cancer spreads to the lymphatic system. About two-thirds of breast cancers have metastasized (spread) to the lymph nodes by the time they are first diagnosed. When diagnosis and treatment begin early, about 86% of patients survive for 5 years, and 65% survive for 20 years or longer. Untreated patients have a 5-year survival rate of only 20%.

Common methods of treating breast cancer include mastectomy (surgical removal of the breast), chemotherapy, and radiation treatment. Lumpectomy (surgical removal of only the affected portion of the breast) in conjunction with radiation treatment appears to be as effective as mastectomy in some cases. *Chemotherapy* (chemo) is treatment with drugs that kill cancer cells. Chemotherapy is often administered after surgery to destroy any cancer cells that may have broken away from the main tumor and escaped surgical removal. Even in the early stages of cancer, *metastasis* can occur: cancer cells may move away from the primary tumor and establish new tumors in other parts of the body.

Targeted therapy involves the use of newer drugs that specifically target certain changes in cancer cells. Targeted therapy is often used along with chemotherapy. For example, in about one in five individuals with breast cancer, HER2, a protein that promotes growth, is present in abnormally large amounts on the surface of cancer cells. Cancer cells with HER2 spread more aggressively than other cancer cells. Monoclonal antibodies, such as Herceptin, have been developed that slow the growth of these cells. These drugs may also stimulate the immune system to more effectively destroy the cancer.

To grow, tumors must stimulate growth of new blood vessels. A monoclonal antibody has been developed that targets the growth of new blood vessels that supply the tumors. New drugs are being developed that target specific proteins that promote tumor growth.

About two-thirds of breast cancers have estrogen or progesterone receptors. Growth of these cancers is enhanced by circulating estrogens and progesterone. Removing the ovaries in patients with these tumors relieves the symptoms and may cause remission of the disease for months or even years. Drugs (e.g., tamoxifen and aromatase inhibitors) have been developed that antagonize the action of estrogen receptors.

According to the American Cancer Society, when breast cancer is confined to the breast, the 5-year survival rate is almost 100%. Because early detection of breast cancer greatly increases the chances of cure and survival, self-examination and screening are important. The American Cancer Society recommends that women in their 20s and 30s should have a clinical breast exam by a health professional about every 3 years. After age 40, women should have an annual breast exam by a health professional. Women age 40 and older should have a screening mammogram annually. *Mammography*, a soft-tissue radiological study of the breast, is helpful in detecting very small lesions that might not be identified by routine examination. In mammography lesions show on an X-ray plate as areas of increased density (see figure).

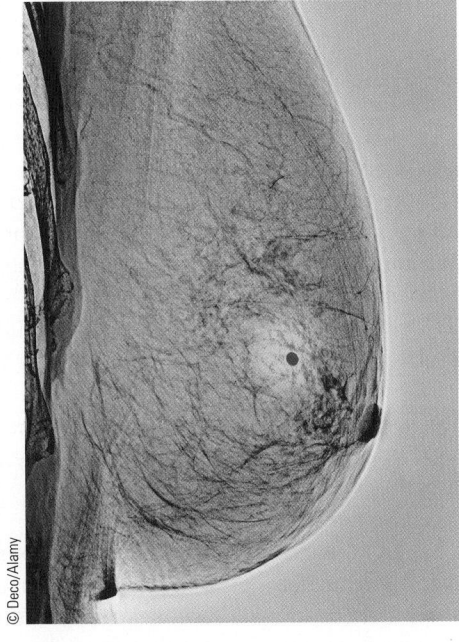

© Deco/Alamy

Mammogram showing area of breast cancer. Cancer is seen as a dense mass (*blue*) in this colorized mammogram. The healthy fibrous tissue (*lavender*) supports the glandular structures.

---

for the development of secondary sexual characteristics. In the female these characteristics include development of the breasts, broadening of the pelvis, and the development and distribution of muscle and fat responsible for the female body shape. In females *puberty* typically begins between ages 10 and 12 and continues until ages 14 to 16.

Hormones of the hypothalamus, anterior pituitary, and ovaries regulate the **menstrual cycle,** the monthly sequence of events that prepares the body for possible pregnancy. (The term *menstrual cycle* is sometimes used more narrowly to refer to the changes that occur in the uterus; we use the term here to include both the ovarian and uterine cycles.) The menstrual cycle runs its course every month from puberty until *menopause,* which

occurs at about age 50. Although wide variations exist, a typical menstrual cycle is 28 days long (**FIG. 50-14**).

The first two weeks of the menstrual cycle make up the *preovulatory phase.* Ovulation occurs on about the 14th day of the cycle. The third and fourth weeks of the menstrual cycle are the *postovulatory phase.* As you read the following description of the menstrual cycle, follow the steps illustrated in **FIGURE 50-15**.

The preovulatory phase spans the first two weeks of the cycle

The first day of the menstrual cycle is marked by the onset of *menstruation,* the monthly discharge through the vagina of blood and tissue from the endometrium. Accordingly,

TABLE 50-2 Principal Female Reproductive Hormones

ENDOCRINE GLAND AND HORMONES	PRINCIPAL TARGET TISSUE	PRINCIPAL ACTIONS
HYPOTHALAMUS		
Gonadotropin-releasing hormone (GnRH)	Anterior pituitary	Stimulates release of FSH and LH
ANTERIOR PITUITARY		
Follicle-stimulating hormone (FSH)	Ovary	Stimulates development of follicles and secretion of estrogen
Luteinizing hormone (LH)	Ovary	Stimulates ovulation and development of corpus luteum
Prolactin	Breast	Stimulates milk production (after breast has been prepared by estrogen and progesterone)
POSTERIOR PITUITARY		
Oxytocin	Uterus	Stimulates contraction and stimulates prostaglandin release
	Mammary glands	Stimulates ejection of milk into ducts
OVARIES		
Estrogens (estradiol)	General	Stimulates growth of sex organs at puberty and development of secondary sex characteristics
	Reproductive structures	Induces maturation; stimulates monthly preparation of endometrium for pregnancy; makes cervical mucus thinner and more alkaline
Progesterone	Uterus	Completes preparation of endometrium for pregnancy
Inhibin	Anterior pituitary	Inhibits secretion of FSH

© Cengage Learning

the first five days or so of the preovulatory phase are known as the *menstrual phase.* During the first part of the preovulatory phase, the hypothalamus releases *gonadotropin-releasing hormone (GnRH).* This hormone stimulates the anterior pituitary to release gonadotropic hormones: *follicle-stimulating hormone (FSH)* and *luteinizing hormone (LH).* GnRH, FSH, and LH are the same hormones that help regulate reproduction in the male.

The preovulatory phase is also called the *follicular phase* because during this period, FSH stimulates a few follicles to begin to develop in the ovary. FSH also stimulates the granulosa cells of the follicle to multiply and produce estrogen. Some estrogen diffuses into the blood, but estrogen also has an *autocrine* action on the granulosa cells that produce it and a *paracrine* effect on nearby granulosa cells (see Chapter 49 for a discussion of autocrine and paracrine regulation). The amount of estrogen secreted by the granulosa cells is enhanced by the action of LH on the theca cells. LH stimulates the theca cells to proliferate and produce androgens. The androgens diffuse into the granulosa cells, where they are converted to estrogen. Estrogen stimulates growth of the endometrium, which begins to thicken and develop new blood vessels and glands.

Typically, after the first week of the menstrual cycle, only one follicle continues to develop. Its granulosa cells become sensitive to LH as well as to FSH. This dominant follicle now secretes enough estrogen to cause a rise in the concentration of estrogen in the blood. *Although still at relatively low concentration,* estrogen inhibits secretion of FSH from the pituitary. Estrogen also acts directly on certain neurons in the hypothalamus, which inhibit secretion of GnRH. In addition, the granulosa cells secrete *inhibin,* a hormone that mainly inhibits FSH secretion. As a result of these negative feedback signals, FSH concentration decreases.

As its concentration in the blood peaks during the late preovulatory phase, estrogen signals a different set of neurons in the hypothalamus to stimulate GnRH secretion. The GnRH stimulates the anterior pituitary to secrete LH, which increases estrogen production by the follicle. This is a positive feedback mechanism. *The high estrogen concentration stimulates a surge in LH secretion.*

The surge of LH secreted at the middle of the menstrual cycle stimulates the final maturation of the follicle. The LH surge also stimulates *ovulation,* the ejection of the secondary oocyte from the ovary. The granulosa cells of the follicle decrease their estrogen output, resulting in a temporary decrease in estrogen concentration.

The corpus luteum develops during the postovulatory phase The postovulatory phase, also called the *luteal phase,* begins after ovulation. LH stimulates development of the corpus luteum, which secretes a large amount of progesterone and estrogen as well as inhibin. These hormones stimulate the uterus to continue its preparation for pregnancy. Progesterone stimulates tiny glands in the endometrium to secrete a fluid rich in nutrients.

During the postovulatory phase, the high concentration of progesterone in the blood, along with estrogen, inhibits secretion of GnRH, FSH, and LH. Progesterone acts on the hypothalamus and the anterior pituitary. Inhibin acts on the anterior pituitary to further inhibit FSH secretion. As a result of these negative feedback systems, FSH and LH concentrations are low during the postovulatory phase, and no new follicles develop.

If the secondary oocyte is not fertilized, the corpus luteum begins to degenerate after about eight days. Although the mechanism responsible for corpus luteum degeneration is not completely understood, the decrease in LH is an important factor. When the corpus luteum stops secreting progesterone and estrogen, the concentrations of these hormones in the blood fall markedly. As a result, small arteries in the

Gonadotropic hormones and ovarian hormones regulate the monthly sequence of events that take place within the ovary and uterus. The preovulatory phase begins with menstruation; the postovulatory phase begins after ovulation.

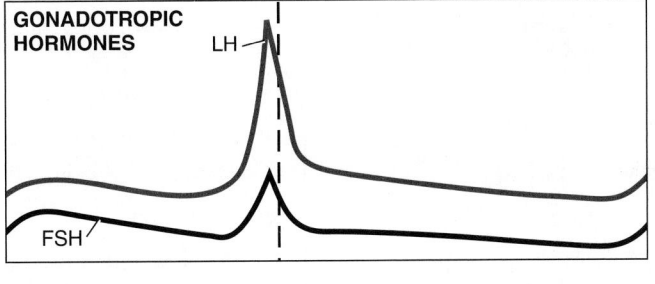

(a) Concentrations of pituitary gonadotropic hormones. Note that the concentrations of both FSH and LH peak just prior to ovulation (*dotted line*).

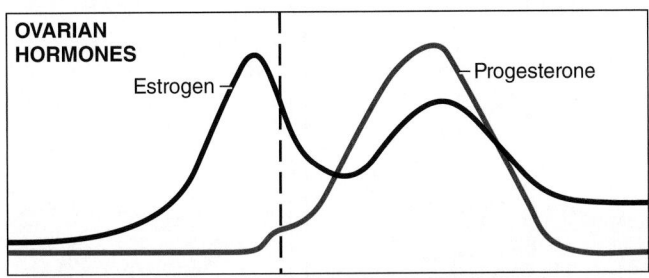

(b) Concentrations of ovarian hormones. Estrogen concentration peaks during the late preovulatory phase. Progesterone, secreted mainly by the corpus luteum, reaches its peak concentration during the postovulatory phase.

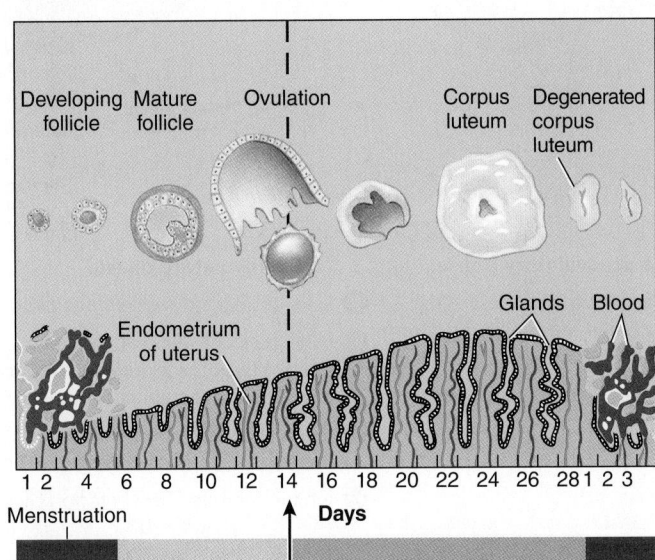

(c) Ovarian and uterine cycles. Hormone concentrations correlate with changes that take place in the ovaries and uterus. If fertilization occurs, the corpus luteum continues to secrete estrogen and progesterone, and menstruation does not occur.

Figure 50-14 *Animation* **Endocrine regulation of the menstrual cycle**

When fertilization does not occur, the menstrual cycle repeats about every 28 days. Note the changes in hormone concentrations that regulate the menstrual cycle. (FSH, follicle-stimulating hormone; LH, luteinizing hormone)

© Cengage Learning

PREDICT What happens when a woman's ovaries produce a very low amount of estrogen?

endometrium constrict, reducing the oxygen supply. Menstruation, which marks the beginning of a new cycle, begins as cells die and damaged blood vessels rupture and bleed. The concentrations of estrogen and progesterone are now too low to inhibit the anterior pituitary, and secretion of FSH and LH increases once again.

Menstrual cycles stop at menopause

At about age 50, a woman enters **menopause**, a period when ova are no longer produced and the woman becomes infertile. Apparently, a change in the hypothalamus triggers menopause. Although gonadotropic hormones are produced, the ovaries

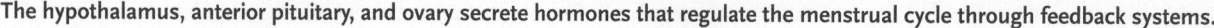

The hypothalamus, anterior pituitary, and ovary secrete hormones that regulate the menstrual cycle through feedback systems.

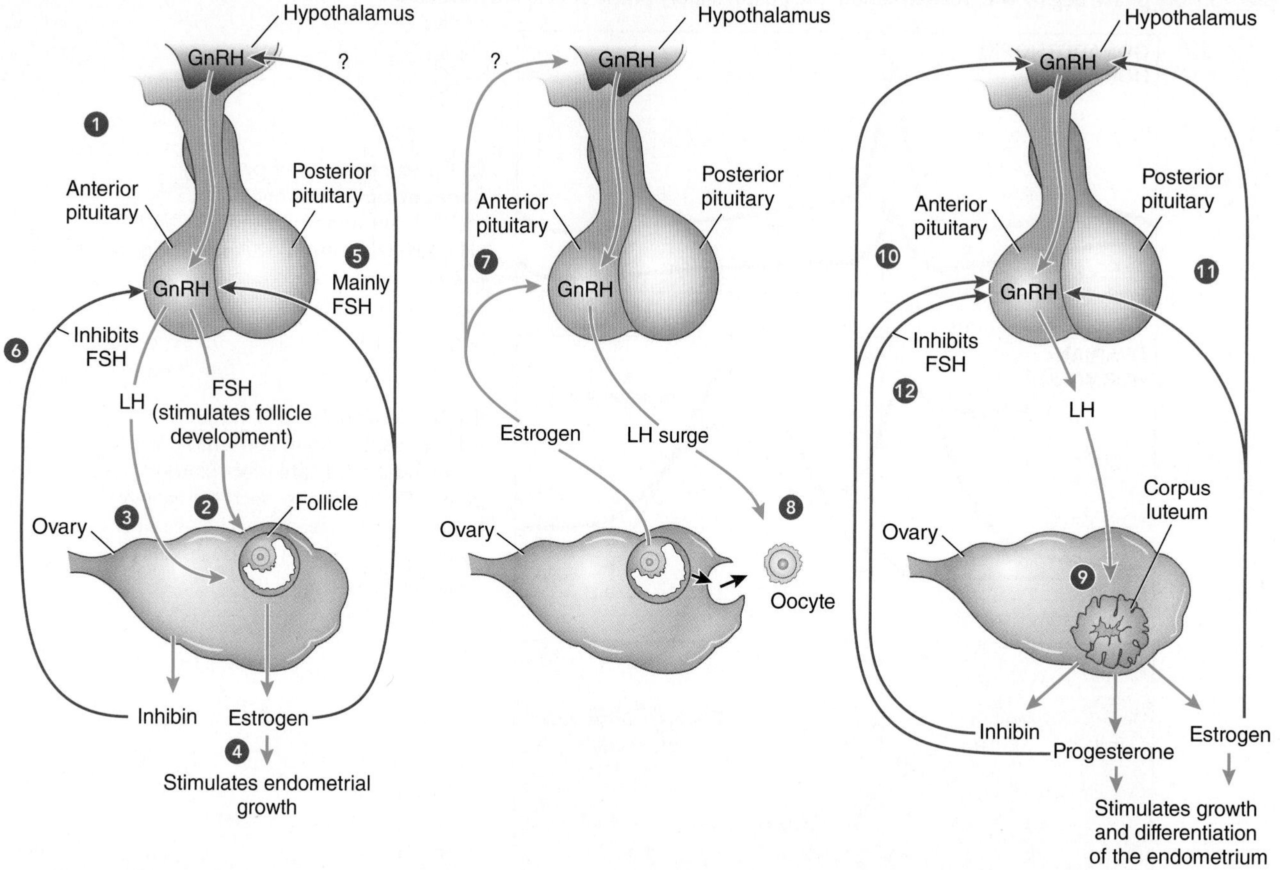

(a) Preovulatory phase.

1 Hypothalamus releases GnRH.

2 FSH stimulates follicle development.

3 LH indirectly stimulates estrogen production.

4 Estrogen stimulates thickening of endometrium.

5 Estrogen has a negative feedback effect on the pituitary and hypothalamus.

6 Inhibin inhibits FSH secretion by pituitary.

(b) Late preovulatory phase.

7 High concentration of estrogen has a positive feedback effect on the pituitary and hypothalamus.

8 Surge of LH stimulates ovulation.

(c) Postovulatory phase.

9 LH stimulates development of corpus luteum.

10 High concentration of progesterone strongly inhibits hypothalamus and anterior pituitary. FSH and LH secretion decreases.

11 Estrogen has a negative feedback effect on the hypothalamus and anterior pituitary.

12 Inhibin inhibits FSH secretion by pituitary.

Figure 50-15 *Animation* Feedback mechanisms in endocrine regulation of female reproduction

The secretion and interaction of hormones changes during the phases of the menstrual cycle. *Green arrows* indicate stimulation; *red arrows* indicate inhibition. (GnRH, gonadotropin-releasing hormone; FSH, follicle-stimulating hormone; LH, luteinizing hormone)

© Cengage Learning

PREDICT Imagine that the anterior pituitary no longer responded to inhibin. How would that affect the ovary?

become less responsive to them, and the oocytes in the ovaries begin to degenerate. The ovaries secrete less estrogen and progesterone than previously, and the menstrual cycle becomes irregular and eventually halts.

Women in menopause sometimes experience a sensation of heat ("hot flashes"), probably because of less stable regulation of blood flow through blood vessels in the skin. Periodic increases of warm blood through these blood vessels may cause the

sensation of a hot flash. Other physiological changes associated with menopause include night sweats, mood swings, and weight gain. Although its use is controversial, hormone replacement therapy with estrogen and progesterone relieves symptoms of menopause for many women.

In contrast to humans, females of other species maintain their ability to reproduce throughout the lifespan. They do not experience menopause. Researchers have suggested that the infertile postmenopausal period allows older women to help their daughters with grandchildren, increasing their survival rate. This practice may help ensure that the older woman's genes will be transmitted to future generations.

Most mammals have estrous cycles

Only humans and some other primates have menstrual cycles. Most other mammals have *estrous cycles.* Although their reproductive organs and hormones are generally similar, there are some important differences. In animals with estrous cycles, the uterus reabsorbs the thickened lining of the endometrium if conception does not occur. As we have just learned, in animals with a menstrual cycle, the endometrium is expelled during menstruation.

Female animals with estrous cycles are sexually receptive ("in heat") only during the *estrus* phase of the cycle. The frequency and length of the estrus phase vary. In dogs the estrus phase typically occurs twice a year and lasts 4 to 13 days. Estrus occurs every 4 to 5 days in rodents. In cats cycles can occur about every 3 weeks, and the estrus phase averages 7 days. The resting phase of the cycle is prolonged in the fall and early winter. (Why might that be advantageous?) Animals with estrous cycles typically display physiological and behavioral changes that signal sexual readiness to potential mates.

CHECKPOINT 50.3

- *How is oogenesis different from spermatogenesis?*
- PREDICT *What would happen if ovulation occurred but no corpus luteum developed?*
- VISUALIZE *Draw a diagram illustrating hormone interactions during the preovulatory phase. What are specific actions of FSH? of estrogens?*

50.4 FERTILIZATION, PREGNANCY, AND BIRTH

LEARNING OBJECTIVE

8 Describe the process of human fertilization and summarize the actions of hormones that regulate pregnancy and birth.

Hormones play key roles in preparing the body for fertilization and pregnancy. These signaling molecules are also critical in initiating the birth process.

Fertilization is the fusion of sperm and egg

Fertilization and establishing pregnancy together are referred to as *conception.* After ejaculation into the female reproductive tract, sperm typically remain alive and retain their ability to fertilize an ovum for about 48 hours. However, sperm can survive up to 5 days in the female reproductive tract. The ovum remains fertile for 12 to 24 hours after ovulation. Therefore, in a very regular 28-day menstrual cycle, sexual intercourse during the middle of the cycle is most likely to result in fertilization. However, many women do not have regular menstrual cycles, and many factors can cause an irregular cycle even in women who are generally regular.

When conditions in the vagina and cervix are favorable, sperm begin to arrive at the site of fertilization in the upper oviduct within 30 minutes after ejaculation. At the time of ovulation, when estrogen concentration is high, the cervical mucus has a thin consistency that permits passage of sperm from the vagina into the uterus. During the rest of the menstrual cycle, the cervical mucus is too thick and sticky for sperm to penetrate.

Once sperm enter the uterus, contractions of the muscular uterine wall help transport them. When they reach the oviduct, contractions of smooth muscle in the wall of the oviduct help move sperm toward the egg. The uterine and oviduct contractions are induced by the high estrogen concentration present just before ovulation. Prostaglandins in the semen contribute to this muscle contraction. The sperm's own motility is important, especially in approaching and fertilizing the secondary oocyte. Also, the secondary oocyte is thought to release a compound that attracts the sperm.

When a sperm encounters an egg, openings develop in the sperm acrosome, exposing enzymes that digest a path through the zona pellucida surrounding the secondary oocyte. As soon as one sperm enters the secondary oocyte, changes occur that prevent the entrance of other sperm. As the fertilizing sperm enters, it usually loses its flagellum (FIG. 50-16). Sperm entry stimulates the secondary oocyte to complete its second meiotic division. The head of the haploid sperm then swells to form the *male pronucleus* and fuses with the *female pronucleus* to form the diploid nucleus of the zygote. The process of fertilization is described in more detail in Chapter 51 (also see *Inquiring About: Novel Origins*).

If only one sperm is needed to fertilize a secondary oocyte, why are more than a million sperm ejaculated? First, many die as a result of unfavorable pH or phagocytosis by white blood cells and macrophages in the female tract. Only a few hundred succeed in making their way up the correct oviduct and reaching the vicinity of the secondary oocyte. Second, the enzymes released by the acrosomes of many sperm are needed to break down the barriers that surround the secondary oocyte.

When the secondary oocyte is fertilized, development begins while the embryo is still in the oviduct. After three or four days, the progesterone concentration is sufficiently high to induce changes that permit the oviduct to move the embryo to the uterus. The embryo is moved along by muscle contractions and by the cilia lining the oviduct. By the time it enters the uterus, cleavage has occurred and the embryo consists of a ball of about 32 cells. After floating free in the uterus for another three to four days, the embryo (which has developed into a *blastocyst*) begins to implant in the thick endometrium. Implantation occurs on about the seventh day after fertilization (FIG. 50-17 on page 1088). (See Chapter 51 for a discussion of development.)

What are some benefits of reproductive technology? About six million couples of reproductive age in the United States are affected by infertility, the inability to achieve conception after using no contraception for at least one year. More than three million infertile couples consult health care professionals each year. Some are helped with conventional treatment, such as with hormone therapy that regulates ovulation or with fertility drugs.

About 30% of infertility cases involve both male and female factors. Male infertility is often attributed to low sperm count. Sperm can be concentrated; alternatively, sperm from a donor can be used to fertilize an egg. Although the sperm donor usually remains anonymous to the prospective birth parents, his genetic qualifications are screened by physicians.

Among the common causes of female infertility are failure to ovulate, production of infertile eggs (common in older women), and oviduct scarring (often caused by pelvic inflammatory disease). Women with blocked oviducts can usually ovulate and incubate an embryo normally but need clinical assistance in getting the young embryo to the uterus.

Many couples are helped with **artificial insemination,** in which a catheter is used to inject sperm directly into the cervix or uterus. Transfer into the uterus is called *intrauterine insemination* (IUI). More than 600,000 IUI procedures are performed annually, with a success rate of about 10%.

More than 40,000 couples need more sophisticated clinical help. They turn to high-tech, *assisted reproductive techniques* (ART) that have been developed through reproduction and human embryo research. Currently, these techniques are expensive, and the success rate is low, less than one-third. According to the Centers for Disease Control and Prevention (CDC), more than 1% of all infants born in the United States are conceived using ART, including in vitro fertilization, gamete intrafallopian transfer, and zygote intrafallopian transfer.

Using **in vitro fertilization (IVF),** a woman takes a fertility drug that induces ovulation of several eggs. The eggs are retrieved by needle aspiration through the vagina. These eggs are placed in a dish with sperm, and fertilization takes place in vitro (outside the patient's body). Embryos are cultured for three to five days and then screened for chromosome or gene abnormalities. Healthy ones are transferred into the uterus through the vagina. IVF was first used in England in 1978 to help a couple who had tried unsuccessfully for several years to have a child. Since that time, thousands of babies have been conceived and born in this way. The success rate of in vitro fertilization is more than 30%; women younger than 35 years have the highest rate of pregnancy, and the rate declines with age.

In *gamete intrafallopian transfer (GIFT)*, a laparoscope (a fiber-optic instrument) is used to guide the transfer of eggs and sperm into a woman's oviduct through a small incision in her abdomen. More than 4000 of these expensive procedures are performed each year, with a success rate of about 28%.

In *zygote intrafallopian transfer (ZIFT)*, the woman's own ova may be used. However, if she does not produce fertile eggs, they can be contributed by a donor (oocyte donation). The ova are fertilized in the laboratory with the sperm of the male partner. A laparoscope is used to guide the transfer of the resulting zygotes into the oviduct. The fertilized eggs move through the oviduct into the uterus.

Another novel procedure is **host mothering.** An embryo is removed from its natural mother and implanted into a female substitute. The foster mother can support the developing embryo either until birth or temporarily until it is implanted again into the original mother or another host. This technique has proved useful to animal breeders. For example, embryos from prize sheep can be temporarily implanted into rabbits for easy shipping by air and then implanted into a host ewe, perhaps one of inferior quality. Host mothering allows an animal with superior genetic traits to produce more offspring than would be naturally possible. This procedure is also used to increase the populations of certain endangered species (see photograph).

Technology is available to freeze the gametes or embryos of many species, including humans, and then transplant them into their donors or into host mothers. Freezing eggs may be an effective strategy for women with cancer who undergo chemotherapy or for women not yet ready to become parents but who want to preserve young eggs with lower risk for chromosome abnormality. These eggs can be implanted at a later time.

Courtesy of Betsy Dresser

Newborn bongo with its surrogate mother, an eland. As a young embryo, the bongo was transplanted into the eland's uterus, where it implanted and developed. Bongos are a rare and elusive species that inhabits dense forests in Africa. The larger and more common elands inhabit open areas.

Hormones are necessary to maintain pregnancy

Estrogen and progesterone are necessary to maintain conditions required for pregnancy. Estrogen stimulates development of the uterine wall, including the muscle needed to expel the fetus during delivery. Progesterone inhibits uterine contractions so that the fetus is not expelled too soon. Because these hormones also inhibit FSH and LH, new follicles do not develop, and the menstrual cycle stops during pregnancy.

Membranes that develop around the embryo secrete **human chorionic gonadotropin (hCG),** a peptide hormone that signals

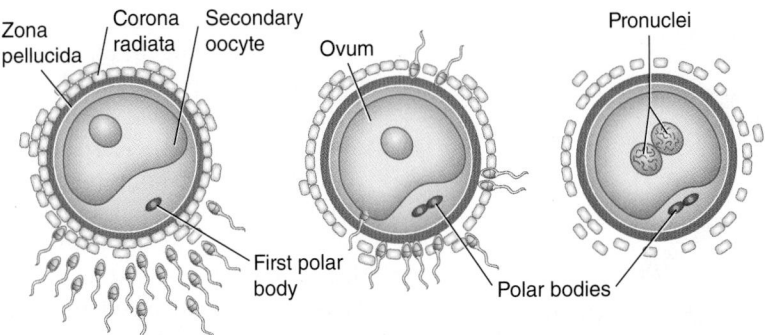

(a) Sperm release an enzyme that helps disperse the layer of follicle cells (corona radiata) surrounding the secondary oocyte.

(b) After a sperm cell enters, the secondary oocyte completes its second meiotic division, producing an ovum and a polar body.

(c) Pronuclei of sperm and ovum fuse, producing a zygote with the diploid number of chromosomes.

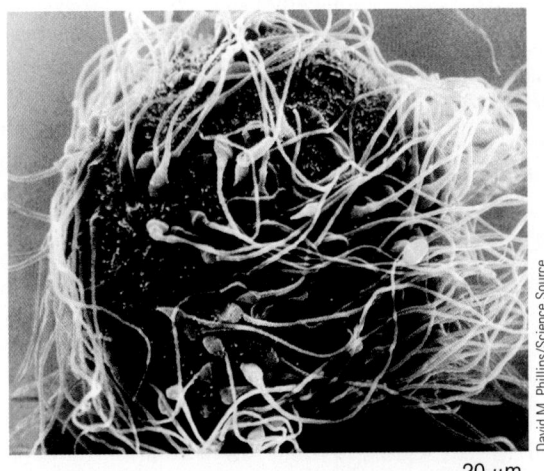

David M. Phillips/Science Source

20 μm

(d) Color-enhanced SEM showing sperm (*yellow*) clustered on the surface of a human ovum (*blue*). Only one sperm penetrates the thick surface of the ovum and fertilizes the ovum.

Figure 50-16 *Animation* **Fertilization**
© Cengage Learning

the mother's corpus luteum to continue to function. (The presence of hCG in urine or blood is used as an early pregnancy test.) Concentrations of estrogen and progesterone remain high throughout pregnancy. During the first two months, the corpus luteum secretes almost all the estrogen and progesterone necessary to maintain the pregnancy. During this time, membranes surrounding the embryo, together with uterine tissue, form the **placenta,** the organ of exchange between the mother and developing embryo. By the tenth week of pregnancy, hCG secretion decreases, and the corpus luteum begins to deteriorate. The placenta takes over and secretes large amounts of estrogen and progesterone.

The birth process depends on a positive feedback system

A normal human pregnancy is about 38 weeks long, counting from the day of fertilization, or about 40 weeks from the first day of the last menstrual cycle. The mechanisms that terminate pregnancy and initiate the birth process depend on several hormones. High levels of estrogen secreted by the placenta greatly increase the number of receptors for oxytocin in the uterine wall. As a result, the uterus becomes about one hundred times more responsive to oxytocin, the hormone that stimulates uterine contractions. (Other hormones are involved, but discussion of their roles is beyond the scope of this chapter.)

At the end of pregnancy, the stretching of the uterine muscle by the growing fetus combined with the effects of increased estrogen concentration and oxytocin produce strong uterine contractions. A long series of involuntary contractions of the uterus are experienced as *labor*. Real labor begins when uterine contractions occur in a frequent, regular pattern and cause the cervix to change in preparation for delivery.

Labor can be divided into three stages. During the first stage, which typically lasts 8 to 10 hours, the contractions of the uterus move the fetus toward the cervix, causing the cervix to *dilate* (open) to a maximum diameter of 10 cm (4 in.). The cervix also becomes *effaced*; that is, it thins out so that the fetal head can pass through. During the first stage of labor, the amnion (the membrane that forms a fluid-filled sac around the embryo/fetus) usually ruptures and releases about a liter of amniotic fluid, which flows out through the vagina.

A positive feedback cycle operates during the birth process. As the baby's head pushes against the cervix, a reflex action causes the uterus to contract. The contraction forces the head against the cervix again, resulting in another contraction, and the positive feedback cycle repeats again and again until the baby descends through the cervix.

During the second stage, which normally lasts from 20 minutes to an hour, the fetus passes through the cervix and vagina and is born, or "delivered" (**FIG. 50-18**). With each uterine contraction, the woman bears down so that the fetus is expelled by the combined forces of uterine contractions and contractions of abdominal wall muscles.

At birth, the baby is still connected to the placenta by the umbilical cord. Contractions of the uterus squeeze much of the fetal blood from the placenta into the infant. The cord is tied and cut, separating the child from the mother. (The stump of the cord gradually shrivels until nothing remains but the scar, the *navel.*)

During the third stage of labor, which may last 5 to 30 minutes after the birth, the placenta and the fetal membranes are loosened from the lining of the uterus by another series of contractions and are expelled. At this stage they are collectively called the *afterbirth* because these tissues are expelled from the vagina after the baby has been delivered.

During labor an obstetrician may administer oxytocin to increase the contractions of the uterus or may assist delivery with

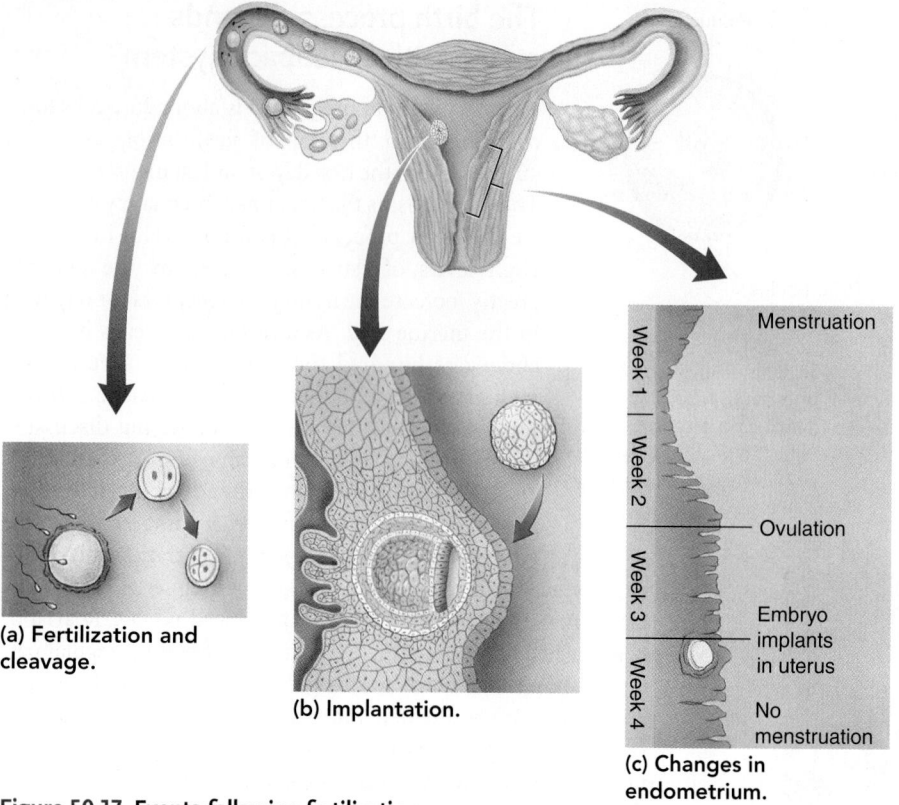

(a) Fertilization and cleavage.

(b) Implantation.

Menstruation

Week 1 | Week 2 | Week 3 | Week 4

Ovulation

Embryo implants in uterus

No menstruation

(c) Changes in endometrium.

Figure 50-17 Events following fertilization

The ovarian and uterine cycles are interrupted when pregnancy occurs. The corpus luteum does not degenerate, and menstruation does not take place. Instead, the wall of the uterus thickens even more, permitting the embryo to implant and develop within it. (Cleavage is the early series of mitotic divisions that converts the zygote to a multicellular embryo.)

© Cengage Learning

special forceps or a vacuum device. In some women the opening between the pelvic bones is too small to permit the passage of the baby vaginally. In this situation, the obstetrician performs a *cesarean section* (C-section), a surgical procedure in which the baby is delivered through an incision made in the abdominal and uterine walls. A C-section may also be performed if the baby's position prevents normal delivery or if the baby shows signs of distress.

CHECKPOINT 50.4

- *What is the function of hCG?*
- **PREDICT** *What would happen if the concentration of estrogen and progesterone decreased sharply during early pregnancy? Explain why.*
- **CONNECT** *Which hormones most affect the birth process? What is the role of each?*

50.5 HUMAN SEXUAL RESPONSE

LEARNING OBJECTIVE

9 Describe the physiological changes that take place during human sexual response.

During *sexual intercourse,* also called *coitus* or *copulation,* the male deposits semen into the upper end of the vagina. The complex structures of the male and female reproductive systems, and the physiological, endocrine, and psychological processes associated with sexual activity, are adaptations that promote fertilization of the secondary oocyte and thus perpetuation of the species.

Sexual stimulation results in two basic physiological responses: (1) vasocongestion, the concentration of blood in reproductive structures as well as in certain other areas of the body; and (2) increased muscle tension. Sexual response includes four phases: sexual excitement, plateau, orgasm, and resolution. The *desire* to have sexual activity may be motivated by fantasies or thoughts about sex. This anticipation can lead to (physical) sexual excitement and a sense of sexual pleasure.

Physiologically, the *sexual excitement phase* involves *vasocongestion,* engorging a tissue with blood, and increased muscle tension. Penile erection is the first male response to sexual excitement. Vasocongestion makes the penis erect so that it can enter the vagina and function in coitus. In the female vasocongestion occurs in the vagina, clitoris, labia, and breasts. The vaginal epithelium secretes a sticky lubricant. Vaginal lubrication is the female's first response to effective sexual stimulation. During the excitement phase, the vagina lengthens and expands in preparation for receiving the penis.

If erotic stimulation continues, sexual excitement heightens to the *plateau phase.* Vasocongestion and muscle tension increase markedly. In both sexes heart rate, blood pressure, and respiration rate increase. Sexual intercourse is usually initiated during the plateau phase. The penis creates friction as it is moved inward and outward in the vagina, in actions referred to as pelvic thrusts. Physical sensations resulting from this friction and the psychological experience of emotional intimacy lead to **orgasm,** the intense physical pleasure that is the climax of sexual excitement. In the female stimulation of the clitoris heightens the sexual excitement that leads to orgasm.

Although it lasts only a few seconds, orgasm is the phase of maximum sexual tension and its release. Musculoskeletal contractions occur, especially in the reproductive structures. Heart rate and respiration more than double, and blood pressure increases markedly, just before and during orgasm. In both sexes orgasm is marked by rhythmic contractions of the muscles of the pelvic floor and reproductive structures. These muscular contractions continue at about 0.8-second intervals for several seconds. After the first few contractions, their intensity decreases, and they become less regular and less frequent.

In the male orgasm is marked by the *ejaculation* of semen from the penis. Semen is typically ejaculated in about seven spurts that accompany the rhythmic contractions. No fluid

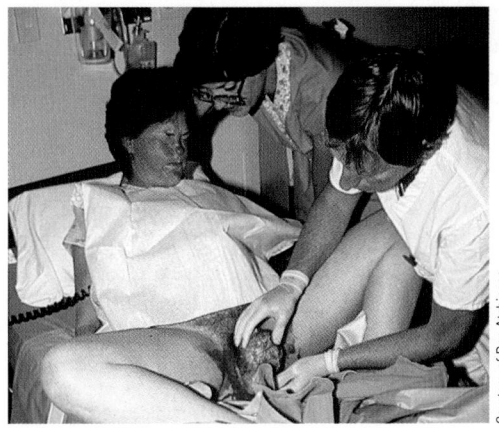

(a) The mother bears down hard with her abdominal muscles, helping push the baby out. When the head fully appears, the physician or midwife gently grasps it and guides the baby's entrance into the outside world.

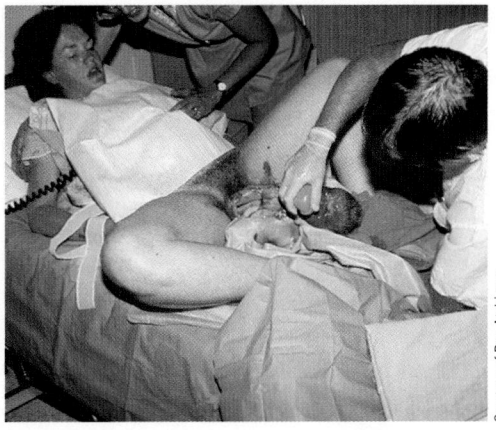

(b) Once the head has emerged, the rest of the body usually follows readily. The physician gently aspirates the mouth and pharynx to clear the upper airway of amniotic fluid, mucus, or blood. At this time the newborn takes its first breath.

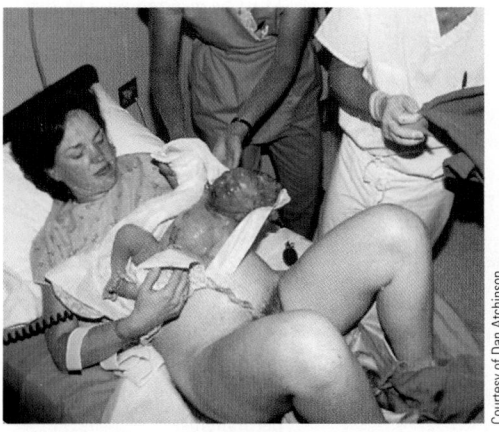

(c) The baby, still attached to the placenta by its umbilical cord, is presented to its mother.

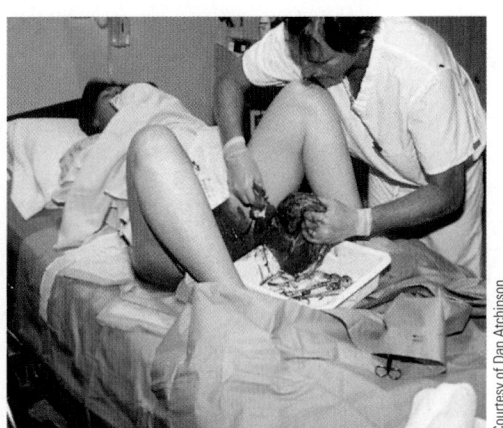

(d) During the third stage of labor, the placenta is delivered.

Figure 50-18 The birth process
Uterine contractions during labor expel the fetus from the uterus. In about 95% of human births, the baby descends through the cervix and vagina in the head-down position.

ejaculation accompanies orgasm in the female. Orgasm is followed by the *resolution phase,* a state of well-being during which the body is restored to its unstimulated state.

CHECKPOINT 50.5

- *What physiological changes take place during the sexual response cycle?*

50.6 BIRTH CONTROL METHODS AND ABORTION

LEARNING OBJECTIVE

10 Compare the modes of action, effectiveness, advantages, and disadvantages of the methods of birth control discussed, including sterilization and emergency contraception; distinguish between spontaneous and induced abortions.

Worldwide, there are about 140 million births annually. Population experts estimate that more than one-fourth of them are unplanned. An estimated 40 to 50 million induced abortions are performed annually. Although most couples worldwide might agree that it is best to have babies by choice rather than by chance, many do not have contraceptives available to them. (About one-third of married women in developing countries have an unmet need for contraception.)

In the United States, nearly half of all pregnancies are unintended. Teen pregnancy is a particularly serious problem. Many teens lack the knowledge and means of protecting themselves from unwanted pregnancy. Not surprisingly, about 80% of teen pregnancies are unplanned. Nearly one million teenagers become pregnant each year in the United States, and about 500,000 give birth. About 30% of teen pregnancies end in abortion.

Many birth control methods are available

When a fertile, heterosexually active woman uses no form of birth control, her chances of becoming pregnant during the course of a year are about 85%. Any method for deliberately separating sexual intercourse from reproduction is **contraception** (literally, "against conception").

Since ancient times, humans have searched for effective methods of preventing fertilization and pregnancy. Of course, fertilization can be avoided by abstaining from sexual intercourse, particularly around the time of ovulation. This cyclical temporary abstinence is called the *rhythm method* or *natural family planning.* Because the time of ovulation is difficult to determine, the rhythm method has a high failure rate (almost 25%).

Scientists continue to develop a variety of contraceptives with a high percentage of reliability. Some of the more common methods of birth control are described in the following paragraphs and in **TABLE 50-3**. Oral contraceptives, intrauterine devices (IUDs), and female sterilization account for more than two-thirds of all contraception practiced worldwide (**FIG. 50-19**).

TABLE 50-3 | Selected Contraceptive Methods

METHOD	FAILURE RATE*	MODE OF ACTION	ADVANTAGES	DISADVANTAGES
Oral contraceptives (combined estrogen and progestin pill)	0.3; 9	Inhibit ovulation	Highly effective; regulate menstrual cycle	Requires women to remember to take daily pill; should not be used by women over age 35 who smoke or have other risk factors
Injectable contraceptives (Depo-Provera)	0.2; 6	Inhibit ovulation	Effective; long-lasting	Irregular menstrual bleeding; fertility may not return for 6–12 months after contraceptive is discontinued
Intrauterine device (IUD)	0.6; 0.8 (for example, Copper T 380)	Prevents fertilization	Provides continuous protection; highly effective for several years	Cramps; increased menstrual flow; increased risk of pelvic inflammatory disease and infertility; not recommended for women who have not given birth
Spermicides: foams, jellies, creams	18–20; 24–28	Chemically kill sperm	No known side effects; can be used with a condom or diaphragm to improve efficacy	Messy; must be applied before intercourse
Contraceptive diaphragm (with jelly)	6; 12	Diaphragm mechanically blocks entrance to cervix; jelly is spermicidal	No side effects	Must be inserted prior to intercourse and diaphragm left in place for several hours afterward
Condom	2; 18	Mechanically prevents sperm from entering vagina	No side effects; some protection against STIs, including HIV	Slightly decreased sensation for male; could break
Rhythm† (fertility awareness methods)	7; 24	Abstinence during fertile period	No known side effects	Not very reliable
Withdrawal (coitus interruptus)	4; 22	Male withdraws penis from vagina prior to ejaculation	No side effects	Not reliable; sperm in fluid secreted before ejaculation may be sufficient for conception
Sterilization				
Tubal sterilization	0.5	Prevents oocyte from leaving oviduct	Most reliable method	Requires surgery; considered permanent
Vasectomy	0.1	Prevents sperm from leaving vas deferens	Most reliable method	Requires surgery; considered permanent
Chance (no contraception)	About 85			

*Lower figure is the failure rate of the method when it is used consistently and used exactly as prescribed; higher figure is failure rate of the method during "typical use"; that is, it includes couples who use the method inconsistently or incorrectly. Based on number of failures per 100 women who use method per year in the United States.

†There are several variations of rhythm method. For those who use the calendar method alone, the failure rate is about 35. However, if body temperature is taken daily and careful records are kept (temperature rises after ovulation), the failure rate can be reduced. When women use some method to determine time of ovulation and have intercourse *only* more than 48 hours *after* ovulation, the failure rate can be reduced to about 7.

SOURCE: Guttmacher Institute.

Most hormonal contraceptives prevent ovulation

Hormonal contraceptives include oral contraceptives, contraceptive patches, rings, implants, and injectable contraceptives. More than 80 million women worldwide (more than 11 million in the United States alone) use oral contraceptives. When used as prescribed, oral contraceptives are about 99.7% effective in preventing pregnancy. They are also used to regulate menstrual cycles and help with symptoms associated with menstruation. The most common preparations are combinations of progestin (synthetic progesterone) and synthetic estrogen. (Natural hormones are destroyed by the liver almost immediately, but synthetic ones are chemically modified during production so that they can be absorbed effectively and metabolized slowly.) In a typical regimen, a woman takes one pill each day for about three weeks. Then, for one week she takes a sugar pill that allows menstruation to occur as a result of the withdrawal of the hormones. One type of oral contraceptive works on a 91-day regimen, reducing menstruation to

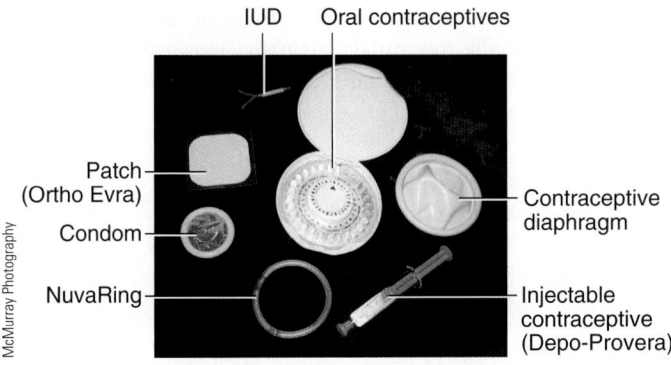

Figure 50-19 Some commonly used contraceptives

The only common male method of contraception is the condom, but researchers are developing other methods.

four times per year. Other new oral contraceptives are being marketed as methods to manipulate menstrual cycles so that menstruation is shorter.

Oral contraceptives prevent ovulation. When postovulatory levels of ovarian hormones are maintained in the blood, the body is tricked into responding as though conception has occurred. The pituitary gland is inhibited and does not produce the surge of LH that stimulates ovulation.

Studies suggest that women older than age 35 who smoke or have other risk factors, such as untreated hypertension, should not take oral contraceptives. Women in this category who take oral contraceptives have an increased risk of death from stroke and myocardial infarction. Low-dose oral contraceptive pills appear safe for nonsmokers up to the time of menopause. Oral contraceptives are linked to deaths in about 3 per 100,000 users per year, which compares favorably with the death rate of about 9 per 100,000 pregnancies.

Trends in contraception are geared toward convenience and improved compliance. The birth control patch (Ortho Evra) delivers estrogen and progestin through the skin. Another approach is a thin, flexible plastic ring (NuvaRing) that women can flatten like a rubber band and insert into the vagina once a month. The ring releases progestin and estrogen in the amounts present in low-dose birth control pills. These types of contraception are convenient because they do not require a daily regimen, while maintaining a regular, predictable menstrual cycle.

Injectable progestin (Depo-Provera) is injected intramuscularly every three months. It prevents ovulation by suppressing anterior pituitary function; it also thickens the cervical mucus (which makes it more difficult for the sperm to reach the egg). Implanon and Nexplanon are implant systems. Each consists of a single plastic progestin rod (the size of a cardboard matchstick). The progestin is released over a three-year period. Many consider these implants the most effective birth control methods available.

Intrauterine devices are widely used

The *intrauterine device (IUD)* is the most widely used method of reversible contraception in the world, used by an estimated 90 million women. In the United States, however, only about 2 million women use IUDs. The IUD is a small device that is inserted into the uterus by a medical professional. The safety of IUDs in current use has been demonstrated, and they are about 99% effective. Disadvantages of the IUD include uterine cramping and bleeding in a small percentage of women.

One IUD presently being used in the United States, the Copper T 380 (ParaGard), can be left in place for ten years. Another IUD (Mirena), also a small, plastic T-shaped device, releases a very small amount of progestin onto the inner wall of the uterus. This IUD can be left in place for up to five years and can improve the symptoms of a woman's menstrual periods. The Skyla IUD (which contains a synthetic progesterone) can be left in place for three years.

The IUD apparently prevents fertilization in a number of ways. Some IUDs cause the cervical mucus to thicken, which impedes passage of sperm into the uterus. The copper in an IUD apparently also reduces the ability of sperm to fertilize the egg. In addition, white blood cells mobilized in response to the foreign body (IUD) in the uterus may produce substances toxic to sperm.

Barrier methods of contraception include the diaphragm and condom

The *contraceptive diaphragm* mechanically blocks the passage of sperm from the vagina into the cervix. It is covered with spermicidal jelly or cream and inserted just prior to sexual intercourse.

The *condom* is also a mechanical method of birth control. The only contraceptive device currently sold for men, the condom provides a barrier that contains the semen so that sperm cannot enter the female tract. The latex condom provides some protection against infection with HIV and other sexually transmitted infections (STIs). The female condom is a strong, soft, polyurethane pouch, or sheath, inserted in the vagina before sexual intercourse. When used correctly, it provides protection against both pregnancy and sexually transmitted infections.

Emergency contraception is available

Emergency contraception is after-the-fact contraception for rape victims and others who have had unprotected sexual intercourse. The use of emergency contraception has significantly reduced the number of abortions. If physicians actively advocated for and more women used emergency contraception, many more unwanted pregnancies that occur each year (as a result of contraceptive failure, unprotected sex, or sexual assault) and thousands of abortions could be prevented in the United States alone. In 2013, it became legal to sell emergency contraception to people of all ages without a prescription.

The most common types of emergency contraception are progestin pills, known as Plan B, or combination progestin-and-estrogen pills. Emergency contraception pills are sometimes called morning-after pills, but they can actually be taken up to five days after unprotected intercourse. Emergency contraception pills prevent fertilization or ovulation. They decrease the probability of pregnancy by about 89%.

A less common type of emergency contraception is insertion of a copper IUD within five days of unprotected intercourse. This method is about 99% effective in preventing pregnancy.

Sterilization renders an individual incapable of producing offspring

Sterilization is the only method of contraception not affected by failure to use a contraceptive method correctly. Worldwide, female sterilization is the most common contraceptive method and accounts for one-third of all contraceptive use.

Male sterilization is performed by vasectomy An estimated one million *vasectomies* are performed each year in the United States. After using a local anesthetic, the physician makes a small incision on each side of the scrotum. Then each vas deferens is cut and its ends sealed so that they cannot grow back together (FIG. 50-20a). Because testosterone secretion is not affected, vasectomy does not affect masculinity. Sperm continue to be produced, although at a much slower rate, and are destroyed by macrophages in the testes. No change in amount of semen ejaculated is noticed because sperm account for very little of the semen volume.

By surgically reuniting the ends of the vasa deferentia, surgeons can successfully reverse sterilization in about 50% of males. Some sterilized men eventually develop antibodies against their own sperm and remain sterile even after their vasectomies have been surgically reversed. Therefore, fertility rates are low (about 30%) for men who undergo vasectomy reversal after ten or more years.

An alternative to vasectomy reversal is the storage of frozen sperm in sperm banks. Such banks have been established throughout the United States. If the male should decide to father another child after he has been sterilized, he "withdraws" his sperm to artificially inseminate his mate. However, sperm do not always survive freezing, and even if they remain viable, there may be an increased risk of genetic defects.

Tubal sterilization blocks the oviducts In the United States, **tubal sterilization** is the most common means of birth control. As with the male, hormone balance and sexual performance of the female are not affected by sterilization. A surgeon performs this procedure either by laparoscopy or by open incision. Tubal sterilization is often performed at the time of a C-section. During female sterilization, the tubes are cauterized, clipped, or tied off (FIG. 50-20b).

More recently, tubal sterilization can be accomplished without an incision. Essure is a nonsurgical, permanent sterilization procedure in which small, flexible inserts are placed in the uterine tubes. Small coils are inserted through the vagina, so no incision is needed. The Essure apparatus causes tissue to grow around the insert. After about three months, this tissue growth blocks sperm from reaching an egg.

Future contraceptives may control regulatory peptides

Ideal contraceptives have not yet been developed. Among the issues that must be considered in developing contraceptives are safety, effectiveness, cost, convenience, and ease of use. Risks of cancer, birth defects in case the method fails and the woman becomes pregnant, permanent infertility, and side effects such as menstrual abnormalities must be minimized.

Some researchers predict that future contraceptive methods will control regulatory peptides and the genes that code for them. For example, the genes that code for certain hormone signals could be interrupted. Another approach being studied is development of contraceptive vaccines that would stimulate the immune system to produce antibodies against hCG or some other protein essential to reproduction. Some investigators are exploring methods for interfering with sperm motility or with fertilization.

It is likely that current methods of contraception will be replaced by more sophisticated molecular methods within the next several years.

Abortions can be spontaneous or induced

Abortion is the termination of pregnancy that results in the death of the embryo or fetus. *Spontaneous abortions* (commonly referred to as miscarriages) occur without intervention. Embryos that are spontaneously aborted are frequently abnormal. *Induced abortions* are performed deliberately either for therapeutic reasons or as

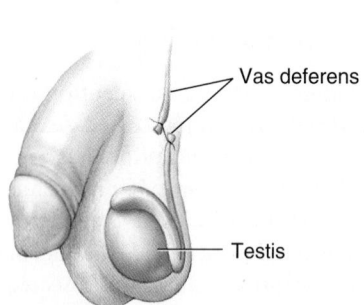

(a) Vasectomy. The vas deferens (sperm duct) on each side is cut and cauterized.

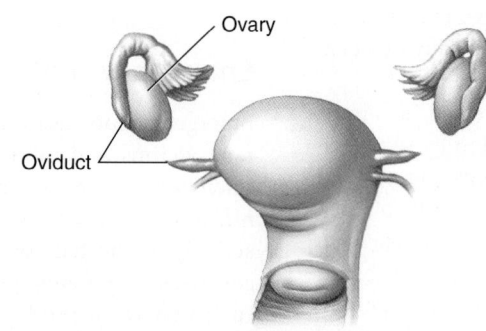

(b) Tubal sterilization. Each oviduct is cut and cauterized so that oocyte and sperm can no longer meet.

Figure 50-20 Sterilization
© Cengage Learning

a means of birth control. *Therapeutic abortions* are performed when the mother's life is in danger or when there is reason to suspect that the embryo is grossly abnormal.

Abortions are sometimes used as a means of birth control. More than 40 million induced abortions are performed each year (more than 1 million in the United States). Many first-trimester abortions (those performed during the first three months of pregnancy) are performed by administering drugs that interrupt the pregnancy and induce expulsion of the embryo. The drug *RU-486* (mifepristone) binds competitively with progesterone receptors in the uterus but does not activate them. Because the normal actions of progesterone are blocked, the endometrium breaks down, and uterine contractions occur. These conditions prevent implantation of an embryo. Prostaglandins may then be administered to induce the uterine contractions that expel the embryo. Misoprostol (Cytotec) is a synthetic prostaglandin also used to terminate pregnancies. During a surgical abortion, the cervix is dilated, and a suction aspirator is inserted into the uterus. The embryo and other products of conception are evacuated.

After the first trimester, abortions are performed mainly because of fetal abnormalities or maternal illness. The method most commonly used is dilation and evacuation. The cervix is dilated, a forceps-type instrument is used to remove the fetus, and suction is used to aspirate the remaining contents of the uterus. Drugs such as prostaglandins are also used to induce labor in second-trimester abortions.

When abortions are performed by skilled medical personnel, they are relatively safe. The mortality rate in the United States is less than 1 per 100,000. In countries where abortion is illegal (and unsafe), the death rate is as high as 450 per 100,000. These statistics can be contrasted with the death rate from pregnancy and childbirth of about 9 per 100,000. The World Health Organization estimates that about 13% of all maternal deaths result from poorly performed, illegal abortions.

CHECKPOINT 50.6

- *How do hormone methods of contraception work?*
- PREDICT *The use of which types of contraception would have the greatest effect on the number of unwanted pregnancies?*
- *What is a spontaneous abortion? a therapeutic abortion?*

50.7 SEXUALLY TRANSMITTED INFECTIONS

LEARNING OBJECTIVE

11 Identify common sexually transmitted infections and describe their symptoms, effects, and treatments.

Sexually transmitted infections (STIs), also called *sexually transmitted diseases (STDs)*, are, next to the common cold, the most prevalent communicable diseases in the world. The World Health Organization has estimated that about 500 million new curable sexually transmitted infections occur each year. More than 62 million people are infected each year with *gonorrhea,* and more than 12 million are infected with *syphilis.* Globally, more than 33 million people are living with HIV, and more than 30 million have died of AIDS since 1981 (discussed in Chapter 45).

In the United States, more than 110 million people are currently living with an STI, and more than 20 million people (including 3 million teens) are diagnosed with new STI infections each year. One in three sexually active young people will acquire an STI by age 24. These young people account for 50% of all new STIs. Some common STIs are described in **TABLE 50-4**. (See Chapter 45 for a discussion of HIV.)

Human papillomavirus (HPV) is one of the two most common STIs in the United States. An estimated 20 million people in the United States are currently infected with HPV. In most infected individuals, the immune system destroys or suppresses the virus before it causes serious health problems. However, HPV can cause abnormal Pap smears and is the main cause of cervical cancer. Some strains of HPV cause genital warts. Vaccines (Gardasil and Cervarix) prevent infection by several harmful HPV strains. Health professionals anticipate that use of these vaccines will lead to the decline of genital warts and cervical cancer. For the vaccine to be most effective, individuals must be vaccinated before they become sexually active. These vaccines are recommended for females ages 11 through 26. The vaccine Gardasil protects boys and men against genital warts and is recommended for males ages 11 through 26 years.

Chlamydia is one of the two most common STIs in the United States. Most individuals infected with chlamydia do not know it because this disease causes no symptoms in about 50% of infected men and 75% of infected women. Infected men may notice a discharge from the penis, and infected women may have a vaginal discharge. In women the infection can spread and cause *pelvic inflammatory disease (PID).* This serious infection leads to infertility in about 20% of infected women.

One out of four women and one out of five men are infected with **genital herpes,** which is caused by a virus. Herpes can be transmitted by genital contact or by oral sex. Most infected individuals do not know they have the infection because it causes minimal symptoms, or no symptoms. When outbreaks occur, symptoms may include blisters around the genital area.

CHECKPOINT 50.7

- *What are the two most common STIs in the United States?*
- CONNECT *What is PID? What causes it?*

TABLE 50-4 | Some Common Sexually Transmitted Infections

DISEASE AND CAUSATIVE ORGANISM	COURSE OF DISEASE	TREATMENT
Human papillomavirus (HPV; about 30 strains of HPV can infect reproductive organs)	Causes abnormal Pap smears and cervical cancer; some strains cause genital warts	Immune modifiers; interferon; vaccines available
Chlamydia (*Chlamydia trachomatis,* a bacterium)	Discharge and burning with urination, or asymptomatic; men ages 15–30 with multiple sex partners are most at risk; in women can cause pelvic inflammatory disease (PID), infection of reproductive organs and pelvic cavity that may lead to sterility (>15% of cases)	Antibiotics
Genital herpes (herpes simplex type 2 virus)	Symptoms may be absent or minimal; when they occur, symptoms include painful blisters in genital area that develop into ulcers; symptoms recur periodically; threat to fetus or newborn infant; can be contracted via oral sex	No effective cure; some drugs may shorten outbreaks or reduce severity; medication can be taken to prevent outbreak
Acquired immunodeficiency syndrome (AIDS)* (caused by human immunodeficiency virus, HIV)	First few weeks: no symptoms or influenza-like symptoms (fever, headache, rash, sore throat; later, other symptoms may include swollen lymph glands, fever, night sweats, weight loss; depressed immunity may lead to pneumonia, rare forms of cancer, and other diseases	No effective cure; a variety of drugs reduce symptoms and prolong life
Trichomoniasis (a protozoon)	Itching, discharge, soreness; can be contracted from dirty toilet seats and towels; may be asymptomatic in males	Metronidazole (an antibiotic)
"Yeast" infections (genital candidiasis; *Candida albicans*)	Irritation, soreness, discharge; especially common in females; rare in males; can be acquired nonsexually	Antifungal drugs
Gonorrhea (*Neisseria gonorrhoeae,* a gonococcus bacterium)	Bacterial toxin may produce redness and swelling at infection site; symptoms in males are painful urination and discharge of pus from penis; in about 60% of infected women, no symptoms initially; can spread to epididymis (in males) or oviducts and ovaries (in females), causing sterility; can cause widespread infection; damage to heart valves, meninges, and joints	Antibiotics
Syphilis (*Treponema pallidum,* a spirochete bacterium)	Bacteria enter body through defect in skin; primary chancre (small, painless ulcer) at site of initial infection; highly infectious at this stage; secondary stage, widespread rash and influenza-like symptoms; scaly lesions may occur that are highly infectious; latent stage that follows can last 20 years; eventually, lesions called gummae may form that damage liver, brain, bone, or spleen	Penicillin

*AIDS was discussed in Chapter 45.

© Cengage Learning

SUMMARY: FOCUS ON LEARNING OBJECTIVES

50.1 Asexual and Sexual Reproduction *(page 1069)*

1 Compare the benefits of asexual and sexual reproduction, and describe each mode of reproduction, giving specific examples.

- In **asexual reproduction** a single parent endows its offspring with a set of genes identical to its own. Asexual reproduction is energy efficient. It is most successful in stable environments.

- In **budding** a part of the parent's body grows and separates from the rest of the body. In **fragmentation** the parent's body may break into several pieces, and each piece can develop into a new animal. In **parthenogenesis** an unfertilized egg develops into an adult.

- In **sexual reproduction** offspring are produced by the fusion of two types of gametes: **ovum** and **sperm.** When a sperm and egg fuse, a fertilized egg, or **zygote,** is produced. Sexual reproduction promotes genetic variety and is especially adaptive in unstable, changing environments.

- In **external fertilization** mating partners typically release eggs and sperm into the water simultaneously. In **internal fertilization** the male delivers sperm into the female's body.

- In **hermaphroditism** a single individual produces both eggs and sperm.

50.2 Human Reproduction: The Male *(page 1071)*

2 Relate the structure of each organ of the human male reproductive system to its function.

- The human male reproductive system includes the testes, which produce sperm and testosterone; a series of conducting ducts; accessory glands; and the penis.

- The **testes,** housed in the **scrotum,** contain the **seminiferous tubules,** where **spermatogenesis** (sperm production) takes place. The **interstitial cells** in the testes secrete testosterone.

- Sperm complete their maturation and are stored in the **epididymis** and **vas deferens.** During **ejaculation,** sperm pass from the vas deferens to the **ejaculatory duct** and then into the **urethra,** which passes through the penis.

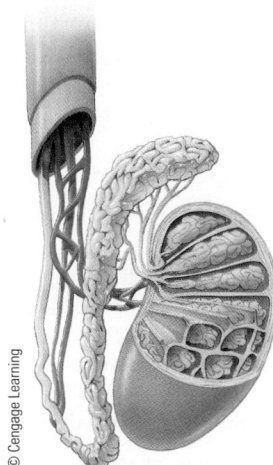

© Cengage Learning

- The **penis** consists of three columns of *erectile tissue:* two *cavernous bodies* and one *spongy body*, which surrounds the urethra. When its erectile tissue becomes engorged with blood, the penis becomes erect and functions as the copulatory organ.

3 Trace the passage of sperm cells through the human male reproductive system from their origin in the seminiferous tubules to their expulsion from the body in the semen; include a description of spermatogenesis.

- Spermatogenesis takes place in the seminiferous tubules of the testes. **Spermatogonia** divide by mitosis; some differentiate and become **primary spermatocytes,** which undergo **meiosis.** The first meiotic division produces two **secondary spermatocytes.** In the second meiotic division, each secondary spermatocyte produces two **spermatids.** Each spermatid differentiates to form a mature sperm.

- The head of a sperm consists of the nucleus and a cap, or *acrosome,* containing enzymes that help the sperm penetrate the egg.

- Sperm pass in sequence through the seminiferous tubules of the testis, epididymis, vas deferens, ejaculatory duct, and urethra.

- Each ejaculate of semen contains about 200 million sperm suspended in the secretions of the **seminal vesicles** and **prostate gland.** The *bulbourethral glands* release a mucous secretion.

4 Describe endocrine regulation of reproduction in the human male.

- **Testosterone** establishes and maintains male *primary sex characteristics*, growth of reproductive organs, and spermatogenesis. Testosterone also stimulates development of **secondary sex characteristics** at puberty: growth of facial and body hair, muscle development, and deepened voice.

- Endocrine regulation of male reproduction involves the hypothalamus, pituitary gland, and testes. The hypothalamus secretes **gonadotropin-releasing hormone (GnRH),** which stimulates the anterior pituitary gland to secrete the gonadotropic hormones: **follicle-stimulating hormone (FSH)** and **luteinizing hormone (LH).**

- FSH, LH, and testosterone directly or indirectly stimulate sperm production. LH stimulates the interstitial cells of the testes to produce testosterone. FSH stimulates the Sertoli cells to produce (1) **androgen-binding protein (ABP),** which binds to testosterone and concentrates it; and (2) **inhibin,** a hormone that inhibits FSH secretion.

50.3 Human Reproduction: The Female *(page 1076)*

5 Relate the structure of each organ of the human female reproductive system to its function.

- The **ovaries** produce gametes and the steroid hormones **estrogens** and **progesterone.** Fertilization takes place in the **oviducts** (uterine tubes).

- The **uterus** serves as an incubator for the developing embryo. The epithelial lining of the uterus, the **endometrium,** thickens each month in preparation for possible pregnancy. The lower part of the uterus, the **cervix,** extends into the vagina.

- The **vagina** receives the penis during sexual intercourse and is the lower part of the birth canal. The **vulva** includes the *labia majora, labia minora, vestibule* of the vagina, **clitoris,** and *mons pubis.*

- The breasts function in **lactation,** production of milk for the young. Each breast consists of 15 to 20 lobes of glandular tissue. The hormone **prolactin** stimulates milk production; **oxytocin** stimulates ejection of milk from the alveoli into the ducts, making it available to the nursing infant.

6 Trace the development of a human egg and trace its passage through the female reproductive system until it is fertilized.

- **Oogenesis** takes place in the ovaries. **Oogonia** differentiate into **primary oocytes.** A primary oocyte and the **granulosa cells** surrounding it make up a **follicle.**

- As the follicle grows, connective tissue cells surrounding the granulosa cells form a layer of **theca cells.** As the follicle matures, the primary oocyte undergoes the first meiotic division, giving rise to a **secondary oocyte** and a **polar body.**

- During **ovulation,** the secondary oocyte is ejected from the ovary and enters one of the paired oviducts, where it may be fertilized. The part of the follicle remaining in the ovary develops into a **corpus luteum,** a temporary endocrine gland.

7 Describe the endocrine regulation of reproduction in the human female and identify key events of the menstrual cycle, such as ovulation and menstruation.

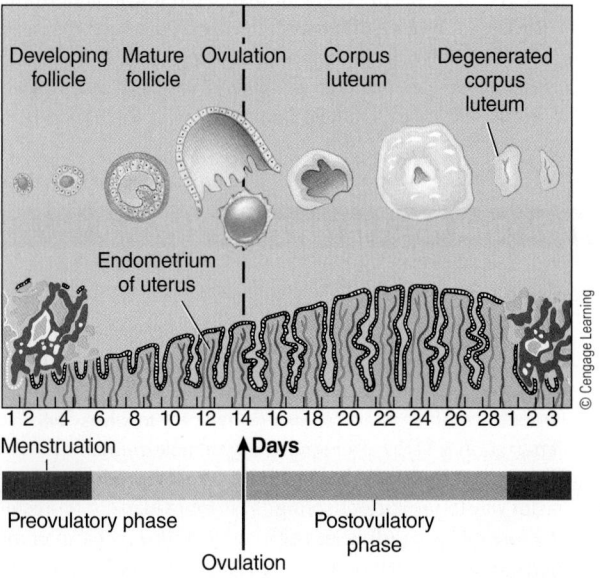

© Cengage Learning

- The hypothalamus, pituitary gland, and ovaries secrete hormones that regulate female reproduction. The first day of the **menstrual cycle** is marked by the beginning of menstrual bleeding. Ovulation occurs at about day 14 in a typical 28-day menstrual cycle.

- During the *preovulatory phase*, gonadotropin-releasing hormone (GnRH) from the hypothalamus stimulates the anterior lobe of the pituitary to secrete follicle-stimulating hormone (FSH) and luteinizing hormone (LH). FSH stimulates follicle development and stimulates the granulosa cells to produce estrogen. LH stimulates theca cells to multiply and produce androgens, which are converted to estrogen.

- Estrogen is responsible for primary and secondary female sex characteristics and stimulates development of the endometrium.

- After the first week, only one follicle continues to develop. Although at relatively low concentration, estrogen inhibits FSH secretion by negative feedback. Granulosa cells produce inhibin, which also inhibits FSH secretion.

- During the late preovulatory phase, estrogen concentration peaks. Through a positive feedback mechanism, estrogen signals the anterior pituitary to secrete LH. This hormone stimulates final maturation of the follicle and stimulates ovulation.

- During the *postovulatory phase*, LH promotes development of the corpus luteum. The corpus luteum secretes progesterone and estrogen, which stimulate final preparation of the uterus for pregnancy. During the postovulatory phase, progesterone, along with estrogen, inhibits secretion of GnRH, FSH, and LH.

- If fertilization does not occur, the corpus luteum degenerates, concentrations of estrogen and progesterone in the blood fall, and menstruation occurs.

50.4 Fertilization, Pregnancy, and Birth *(page 1085)*

8 Describe the process of human fertilization and summarize the actions of hormones that regulate pregnancy and birth.

- Human **fertilization** is the fusion of secondary oocyte and sperm to form a zygote. The processes of fertilization and establishing pregnancy together are called *conception*.

- If the secondary oocyte is fertilized, development begins and the embryo implants in the uterus. Membranes that develop around the embryo secrete **human chorionic gonadotropin (hCG)**, a hormone that maintains the corpus luteum. During the first two to three months of pregnancy, the corpus luteum secretes the large amounts of estrogen and progesterone needed to maintain pregnancy. After three months, the

placenta, the organ of exchange between mother and embryo, assumes this function.

- Several hormones, including estrogen, oxytocin, and prostaglandins, regulate the birth process. Labor can be divided into three stages; the baby is delivered during the second stage.

50.5 Human Sexual Response *(page 1088)*

9 Describe the physiological changes that take place during human sexual response.

- *Vasocongestion,* engorgement of a tissue with blood, and increased muscle tension, are physiological responses to sexual stimulation. The phases of sexual response include *sexual excitement, plateau,* **orgasm,** and *resolution.*

50.6 Birth Control Methods and Abortion *(page 1089)*

10 Compare the modes of action, effectiveness, advantages, and disadvantages of the methods of birth control discussed, including sterilization and emergency contraception; distinguish between spontaneous and induced abortions.

- Effective methods of **contraception** include *hormonal contraceptives,* such as oral contraceptives and injectable progestin; *intrauterine devices (IUDs); condoms; contraceptive diaphragms;* and **sterilization** (*vasectomy* in the male and *tubal sterilization* in the female); see Table 50-3. Emergency contraception can be used to prevent unwanted pregnancy after rape or unprotected intercourse.

- *Spontaneous* **abortions** (miscarriages) occur without intervention. *Induced abortions* include *therapeutic abortions,* performed when the mother's health is in danger or when the embryo is thought to be grossly abnormal. Abortions are also induced as a means of birth control.

50.7 Sexually Transmitted Infections *(page 1093)*

11 Identify common sexually transmitted infections and describe their symptoms, effects, and treatments.

- Among the common types of *sexually transmitted infections (STIs)* are *chlamydia* (which causes *pelvic inflammatory disease*), *human papillomavirus (HPV), genital herpes, gonorrhea, syphilis,* and **human immunodeficiency virus (HIV)** (see Table 50-4).

TEST YOUR UNDERSTANDING

Know and Comprehend

1. Hermaphroditism (a) is a form of asexual reproduction (b) occurs when an unfertilized egg develops into an adult animal (c) is a form of sexual reproduction in which an animal produces both eggs and sperm (d) typically involves self-fertilization (e) typically requires only a male animal

2. In contrast to asexual reproduction, sexual reproduction (a) is a faster way to produce offspring (b) is more efficient (c) results in more offspring (d) is less efficient at removing harmful mutations from a population (e) increases genetic variety among members of a population

3. The seminiferous tubules (a) are the site of spermatogenesis (b) produce most of the seminal fluid (c) empty directly into the vas deferens (d) are located within the cavernous body (e) receive fluid from the bulbourethral glands

4. Which of the following cells is haploid? (a) primary oocyte (b) secondary oocyte (c) oogonium (d) cells of the corpus luteum (e) follicle cell

5. The corpus luteum (a) is surrounded by ova (b) degenerates if fertilization occurs (c) develops in the preovulatory phase (d) is maintained by prostaglandins (e) serves as a temporary endocrine gland

6. After ovulation, the secondary oocyte enters the (a) oviduct (b) corpus luteum (c) cervix (d) ovary (e) vagina

7. Hormonal contraceptives (a) prevent ovulation (b) decrease estrogen concentration in the blood (c) are not as effective as barrier methods of contraception (d) decrease risk of STIs (e) a and b only

8. Pelvic inflammatory disease is commonly caused by (a) syphilis (b) gonorrhea (c) genital herpes (d) HIV (e) chlamydia

Apply and Analyze

9. Arrange the following stages into the correct sequence.
   1. spermatid 2. spermatogonium 3. sperm 4. secondary spermatocyte 5. primary spermatocyte
   (a) 2, 1, 3, 4, 5 (b) 5, 1, 3, 4, 2 (c) 1, 3, 4, 2, 5 (d) 2, 5, 4, 1, 3
   (e) 1, 4, 3, 2, 5

10. Which sequence best describes the passage of sperm?
    1. urethra 2. vas deferens 3. epididymis 4. ejaculatory duct
    5. seminiferous tubules
    (a) 3, 1, 2, 4, 5 (b) 1, 3, 2, 4, 5 (c) 5, 3, 2, 4, 1 (d) 1, 3, 4, 2, 5
    (e) 3, 1, 4, 2, 5

11. **VISUALIZE** Label the diagram.

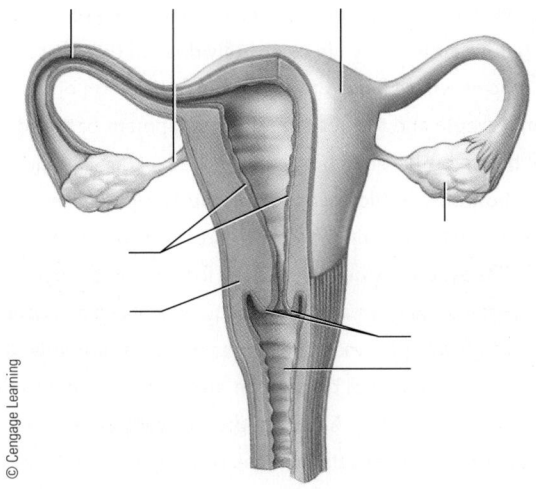

© Cengage Learning

12. **INTERPRET DATA** Examine Figure 50-14. Which hormones are present in largest concentration just before ovulation? During which part of the menstrual cycle is the endometrium most prepared to receive an embryo? Which hormones stimulate thickening of the endometrium?

Evaluate and Synthesize

13. **EVOLUTION LINK** Contrast the biological advantages of hermaphroditism that involves cross-fertilization with hermaphroditism that involves self-fertilization.

14. **EVOLUTION LINK** Asexual reproduction is more common in habitats with few parasites, and sexual reproduction is more common in habitats with larger numbers of parasites. Develop a hypothesis to explain why.

15. **SCIENCE, TECHNOLOGY, AND SOCIETY** Why do you think that a variety of effective methods of male contraception have not yet been developed? If you were a researcher, what are some approaches you might take to development of a method of male contraception?

16. **SCIENCE, TECHNOLOGY, AND SOCIETY** The incidence of abortion has declined more in countries where abortions are legal (and safe) than in countries in which they are illegal (and unsafe). Can you suggest an explanation?

To access course materials, such as Aplia and other companion resources, please visit **www.cengagebrain.com.**

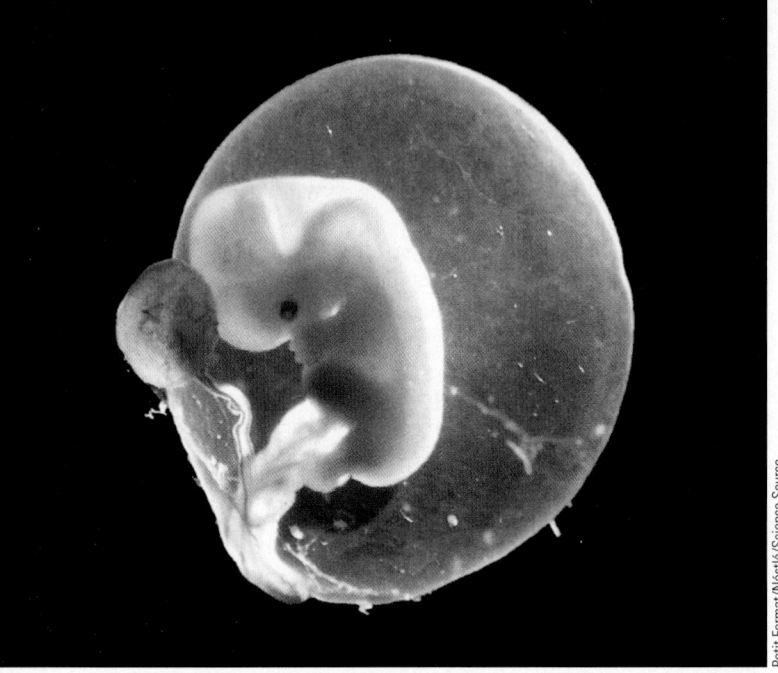

Petit Format/Nestlé/Science Source

Human embryo at 5.5 weeks of development. The embryo is 1 cm (0.4 in.) long. Note the limb buds and the eyes. The balloonlike structure, connected to the embryo by a stalk, is the yolk sac.

KEY CONCEPTS

51.1 The development of form requires not only cell division and growth but also cell determination and cell differentiation as well as pattern formation and morphogenesis.

51.2 Fertilization includes contact and recognition between egg and sperm, regulated sperm entry, activation of the fertilized egg, and fusion of egg and sperm pronuclei.

51.3 Cleavage, a series of rapid cell divisions without growth, provides cellular building blocks for development.

51.4 Gastrulation establishes ectoderm, mesoderm, and endoderm; each gives rise to specific types of tissues.

51.5 One of the earliest events in organogenesis (organ development) is neurulation, the origin of the central nervous system.

51.6 Extraembryonic membranes (amnion, chorion, allantois, and yolk sac) have evolved in terrestrial vertebrates as adaptations to reproduction on land.

51.7 Human development follows the pattern for placental mammals.

D evelopment includes all the changes that take place during the entire life of an individual. In this chapter, however, we focus mainly on the fertilization of the egg to form a zygote and the subsequent development of the young animal before birth or hatching.

Just how does a microscopic, unicellular zygote give rise to the bones, muscles, brain, and other structures of a complex animal? The zygote divides by mitosis, forming an **embryo** that subsequently undergoes an orderly sequence of cell divisions. In animals growth (i.e., increase in mass) occurs primarily by an increase in the number of cells, but it also occurs by an increase in cell size, as in fat cells. Although the very early embryo usually does not grow, later cell divisions typically contribute to growth. However, by itself cell division, which is governed by a genetic program that interacts with environmental signals (see Chapter 10), would produce only a formless heap of similar cells. We therefore begin this chapter with a discussion of the fundamental processes introduced in Chapter 17 that contribute to the development of form: cell determination and cell differentiation; and pattern formation and morphogenesis.

The photograph shows a human embryo enclosed in its fluid-filled amniotic cavity and attached to its yolk sac. In this chapter you have an opportunity to compare and contrast the sequences of developmental events in several different animals in addition to humans. Researchers have chosen these model organisms because they have certain desirable characteristics. For example, the embryos of certain types of animals, such as echinoderms and amphibians, can be obtained in large quantities, which facilitates studies in which biochemical changes are correlated with developmental landmarks. The embryos of echinoderms are particularly easy to observe because they are transparent. The embryos of echinoderms, amphibians, and birds are easily grown in the laboratory. As you read, note the intimate interrelationships among the sequential developmental processes as well as the fundamental similarities among the early developmental events in the different animal groups featured. Many decades of painstaking investigations on these similarities became the foundations of the science of evolutionary developmental biology, popularly known as **Evo Devo,** introduced in Chapters 17 and 30.

51.1 DEVELOPMENT OF FORM

LEARNING OBJECTIVES

1 Analyze the relationship between cell determination and cell differentiation and between pattern formation and morphogenesis.
2 Relate differential gene expression to nuclear equivalence.

The development of the form of an animal is a consequence of a balanced combination of several fundamental processes that go beyond cell division and growth: cell determination and cell differentiation, and pattern formation and morphogenesis.

As embryonic development proceeds, cell division gives rise to increasing numbers of cells, which serve as the building blocks of development. At various times, certain cells become biochemically and structurally specialized to carry out specific functions through a process known as **cell differentiation.** In our discussion of the genetic control of development in Chapter 17, you learned how cell differentiation occurs through **cell determination,** a series of molecular events in which the activities of certain genes are altered in ways that cause a cell to progressively commit to a particular differentiation pathway. This process proceeds even though there may not be any immediate changes in the cell's morphology. You also evaluated evidence supporting the principle of **nuclear equivalence,** which states that in most cases neither cell determination nor cell differentiation entails a loss of genetic information from the cell nucleus. That is, the nuclei of virtually all differentiated cells of an animal contain the same genetic information present in the zygote, but each cell type *expresses* a different subset of that information. In fact, nuclear equivalence is the principle that underlies the cloning of organisms.

Cell differentiation is therefore an expression of changes in the activity of specific genes, which in turn is influenced by a variety of factors inside and outside the cell. This **differential gene expression** is responsible for variations in chemistry, behavior, and structure among cells. Through this process, an embryo can develop into an organism of more than 200 types of cells, each specialized to perform specific functions. However, not all cells differentiate. Some, known as **stem cells,** remain in a relatively undifferentiated state and retain the ability to give rise to various cell types.

Cell differentiation by itself does not explain development. The differentiated cells must become progressively organized, shaping the intricate pattern of tissues and organs that characterizes a multicellular animal. This development of form, known as **morphogenesis,** proceeds through the process of pattern formation. **Pattern formation** is a series of steps requiring signaling between cells, changes in the shapes of certain cells, precise cell migrations, interactions with the extracellular matrix, and even **apoptosis** (programmed cell death; see Chapter 4) of some cells.

CHECKPOINT 51.1

- *Which typically comes first, cell differentiation or cell determination?*
- CONNECT *What are some kinds of events that contribute to pattern formation? How does pattern formation lead to morphogenesis?*

51.2 FERTILIZATION

LEARNING OBJECTIVES

3 Describe the four processes involved in fertilization.
4 Describe fertilization in echinoderms and point out some ways in which mammalian fertilization differs.

In Chapter 50 you studied **spermatogenesis** and **oogenesis,** the processes by which meiosis leads to the formation of haploid cells, which differentiate as sperm and eggs, respectively. In **fertilization** a usually flagellated, motile sperm fuses with a much larger, immotile egg cell to produce a **zygote,** or fertilized egg.

Fertilization has two important genetic consequences: restoration of the diploid chromosome number and, in mammals and many other animals, determination of the sex of the offspring (see Chapter 11). Fertilization also has profound physiological effects because it activates the egg, thus initiating reactions that permit development.

Fertilization involves four events, some of which may occur simultaneously: (1) the sperm contacts the egg and recognition occurs; (2) the sperm or sperm nucleus enters the egg; (3) the egg becomes activated, and certain developmental changes begin; and (4) the sperm and egg nuclei fuse. These events do not necessarily follow the same temporal sequence in all animals. Unless otherwise stated, our discussion applies to sea urchins and other echinoderms such as sea stars, which have been studied intensively because they produce large numbers of gametes and because fertilization is external.

The first step in fertilization involves contact and recognition

Although eggs are immotile, they are active participants in fertilization. An egg is surrounded by a plasma membrane and by one or more external coverings that are important in fertilization. For example, a mammalian egg is enclosed by a thick, noncellular **zona pellucida,** which is surrounded by a layer of granulosa cells derived from the follicle in which the egg developed.

The egg coverings not only facilitate fertilization by sperm of the same species but also bar interspecific fertilization, a particularly important function in species that practice external fertilization. External to the plasma membrane of a sea urchin egg are two acellular layers that interact with sperm: a very thin **vitelline envelope** and, outside it, a thick glycoprotein layer called the **jelly coat.** Sea urchin sperm become motile when released into sea water, and their motility increases when they reach the vicinity of a sea urchin egg. Motility improves the probability that sperm will encounter the egg, but motility alone may not be sufficient to ensure actual contact with the egg. In sea urchins as well as certain fishes and some cnidarians, the egg or one of its coverings secretes a chemotactic substance that attracts sperm from the same species. However, for many species, researchers have not found a specific chemical that attracts the sperm to the egg.

When a sea urchin sperm contacts the jelly coat, the sperm undergoes an **acrosome reaction,** in which the membranes surrounding the acrosome (the cap at the head of the sperm; see Fig. 50-6) fuse and pores in the membrane enlarge. Calcium ions from the sea water move into the acrosome, which swells and begins to disorganize. The acrosome then releases proteolytic enzymes that digest a path through the jelly coat to the vitelline envelope of the egg. If the sperm and egg are of the same species, *bindin,* a species-specific binding protein located on the acrosome, adheres to species-specific bindin receptors located on the egg's vitelline envelope.

Before a mammalian sperm participates in fertilization, it first undergoes **capacitation,** a maturation process in the female reproductive tract. During capacitation, which in humans can take several hours, sperm become increasingly motile and capable of undergoing an acrosome reaction when they encounter an egg. Evidence indicates that mammalian fertilization requires interaction between sperm and species-specific glycoproteins in the egg's zona pellucida.

Sperm entry is regulated

In sea urchins, once contact occurs between acrosome and vitelline envelope, enzymes dissolve a bit of the vitelline envelope in the area of the sperm head. The plasma membrane of the egg is covered with microvilli, several of which elongate to surround the head of the sperm. As they do so, the plasma membranes of sperm and egg fuse, and a *fertilization cone* is formed that contracts to draw the sperm into the egg (**FIG. 51-1**).

Fertilization of the egg by more than one sperm, a condition known as **polyspermy,** results in an offspring with extra sets of chromosomes, which is usually lethal. Two reactions, known as the fast and slow blocks to polyspermy, prevent this occurrence. In the *fast block to polyspermy,* the egg plasma membrane depolarizes, which prevents its fusion with additional sperm. An unfertilized egg is polarized; that is, the cytoplasm is negatively charged relative to the outside. However, within a few seconds after sperm fusion, ion channels in the plasma membrane open, permitting positively charged calcium ions to diffuse across the membrane and depolarize the egg.

The *slow block to polyspermy,* the **cortical reaction,** requires up to several minutes to complete, but it is a complete block. Binding of the sperm to receptors on the vitelline envelope activates one or more *signal transduction* pathways (see Chapter 6) in the egg. These events cause calcium ions stored in the egg endoplasmic reticulum to be released into the cytosol. This rise in the level of intracellular calcium causes

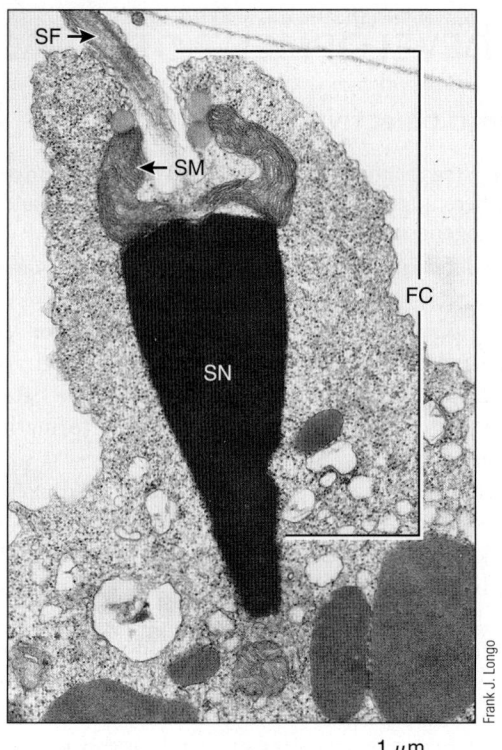

1 μm

Figure 51-1 Fertilization

In this TEM, a fertilization cone (FC) forms as a sperm enters a sea urchin egg. (SN, sperm nucleus; SM, sperm mitochondrion; SF, sperm flagellum).

thousands of cortical granules, membrane-enclosed vesicles in the egg cortex (region close to the plasma membrane), to release enzymes, various proteins, and other substances by exocytosis into the space between the plasma membrane and the vitelline envelope (**FIG. 51-2**). Some of the enzymes dissolve the protein linking the two membranes; thus, the space

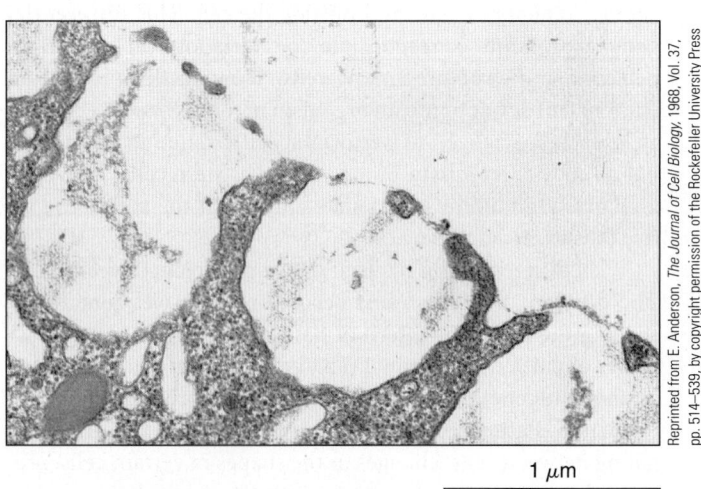

1 μm

Figure 51-2 The cortical reaction

Three cortical granules are undergoing exocytosis in this TEM of the plasma membrane of a sea urchin egg shortly after fertilization. This release of the contents of the cortical granules initiates the slow block to polyspermy.

Reprinted from E. Anderson, *The Journal of Cell Biology,* 1968, Vol. 37, pp. 514–539, by copyright permission of the Rockefeller University Press

enlarges as additional substances released by the cortical granules raise the osmotic pressure, which causes an influx of water from the surroundings. The vitelline envelope then becomes elevated away from the plasma membrane and forms the fertilization envelope, a hardened covering that prevents entry of additional sperm.

In mammals a fertilization envelope does not form. Instead, the enzymes released during exocytosis of the cortical granules alter the sperm receptors on the egg's zona pellucida so that no additional sperm bind to them.

Not all species have these types of blocks to polyspermy. In some species, such as salamanders, several sperm usually enter the egg, but only one sperm nucleus fuses with the egg nucleus; the rest degenerate.

Fertilization activates the egg

Release of calcium ions into the egg cytoplasm does more than stimulate the cortical reaction; it also triggers the *activation program,* a series of metabolic changes within the egg. Aerobic respiration increases; certain maternal enzymes and other proteins become active; and within a few minutes after sperm entry, a burst of protein synthesis occurs. In addition, the egg nucleus is stimulated to complete meiosis. (Recall from Chapter 50 that in most animal species, including mammals, at the time of fertilization an egg is actually a secondary oocyte, arrested early in the second meiotic division.)

In some species an egg can be artificially activated without sperm penetration by swabbing it with blood and pricking the plasma membrane with a needle, by calcium injection, or by certain other treatments. These haploid eggs go through some developmental stages *parthenogenetically*—that is, without fertilization (see Chapter 50)—but they cannot complete development. Special mechanisms of egg activation have evolved in species that are naturally parthenogenetic.

Sperm and egg pronuclei fuse, restoring the diploid state

After the sperm nucleus enters the egg, researchers think that it is guided toward the egg nucleus by a system of microtubules that forms within the egg. The sperm nucleus swells and forms the *male pronucleus.* The nucleus formed during completion of meiosis in the egg becomes the *female pronucleus.* (Recall from Figure 50-11 that the other nucleus formed during meiosis II in the egg becomes the second polar body.) The haploid male and female pronuclei then fuse to form the diploid nucleus of the zygote, and DNA synthesis occurs in preparation for the first cell division.

CHECKPOINT 51.2

- *How do the mechanisms of fertilization ensure control of both quality (fertilization by a sperm of the same species) and quantity (fertilization by only one sperm)?*
- CONNECT *What is the function of the cortical reaction in echinoderms? in mammals?*

51.3 CLEAVAGE

LEARNING OBJECTIVES

5 Trace the generalized pattern of early development of the embryo from zygote through early cleavage and formation of the morula and blastula.

6 Contrast early development, including cleavage, in the echinoderm (or in amphioxus), the amphibian, and the bird, paying particular attention to the importance of the amount and distribution of yolk.

Despite its simple appearance, the zygote is **totipotent;** that is, it gives rise to all the cell types of the new individual. Because the egg is very large compared with the sperm, the bulk of the zygote cytoplasm and organelles comes from the egg. However, the sperm and egg usually contribute equal numbers of chromosomes.

Shortly after fertilization, the zygote undergoes **cleavage,** a series of rapid mitotic divisions with no period of growth during each cell cycle. For this reason, although the cell number increases, the embryo does not increase in size. The zygote initially divides to form a 2-celled embryo. Then each of these cells undergoes mitosis and divides, bringing the number of cells to 4. Repeated divisions increase the number of cells, called **blastomeres,** that make up the embryo. At about the 32-cell stage, the embryo is a solid ball of blastomeres called a **morula.** Eventually, anywhere from 64 to several hundred blastomeres form the **blastula,** which is usually a hollow ball with a fluid-filled cavity, the **blastocoel.**

The pattern of cleavage is affected by yolk

Many animal eggs contain **yolk,** a mixture of proteins, phospholipids, and fats that serves as food for the developing embryo. The amount and distribution of yolk vary among different animal groups, depending on the needs of the embryo. Mammalian eggs have very little yolk because the embryo obtains maternal nutritional support throughout most of its development, whereas the eggs of birds and reptiles must contain sufficient yolk to sustain the embryo until hatching. Echinoderm eggs typically need only enough yolk to nourish the embryo until it becomes a tiny larva capable of obtaining its own food.

Most invertebrates and simple chordates have **isolecithal eggs** with relatively small amounts of yolk uniformly distributed through the cytoplasm. Isolecithal eggs divide completely in a process called **holoblastic cleavage.** Cleavage of these eggs is radial or spiral. Radial cleavage is characteristic of *deuterostomes,* such as chordates and echinoderms; spiral cleavage is common in the embryos of *protostomes,* such as annelids and mollusks (see Fig. 30-5).

In **radial cleavage** the first division is vertical and splits the egg into two equal cells. The second cleavage division, also vertical, is at right angles to the first division and separates the two cells into four equal cells. The third division is horizontal, at right angles to the other two, and separates the four cells into eight cells: four above and four below the third line of cleavage. This pattern of radial cleavage occurs in echinoderms and in the cephalochordate amphioxus (**FIGS. 51-3** and **51-4**).

Cleavage in sea stars is radial and holoblastic.

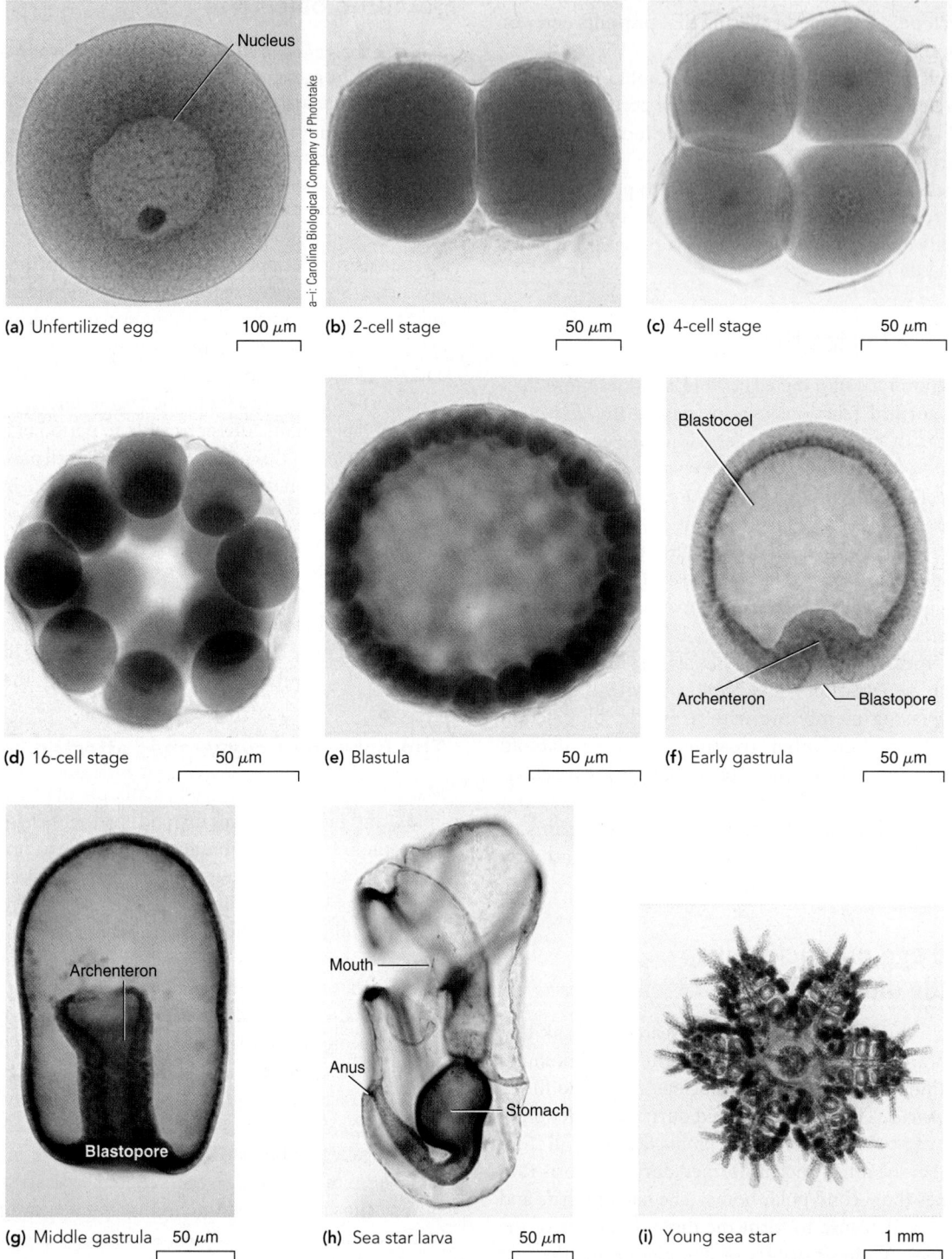

a–i: Carolina Biological Company of Phototake

(a) Unfertilized egg — Nucleus — 100 μm

(b) 2-cell stage — 50 μm

(c) 4-cell stage — 50 μm

(d) 16-cell stage — 50 μm

(e) Blastula — 50 μm

(f) Early gastrula — Blastocoel, Archenteron, Blastopore — 50 μm

(g) Middle gastrula — Archenteron, Blastopore — 50 μm

(h) Sea star larva — Mouth, Anus, Stomach — 50 μm

(i) Young sea star — 1 mm

Figure 51-3 **LMs showing sea star development**

(a) The isolecithal egg has a small amount of uniformly distributed yolk. **(b–e)** The cleavage pattern is radial and holoblastic (the entire egg becomes partitioned into cells). **(f, g)** The three germ layers form during gastrulation. The blastopore is the opening into the developing gut cavity, the archenteron. The rudiments of organs are evident in the sea star larva **(h)** and the young sea star **(i)**. All views are side views with the animal pole at the top except **(c)** and **(i)**, which are top views. Note that the sea star larva is bilaterally symmetrical, but differential growth produces a radially symmetrical young sea star.

VISUALIZE Draw a simple sketch illustrating a horizontal section through the midplane of the middle gastrula in part **(g).** Label the archenteron and blastocoel. Is the blastopore included in your drawing?

Cleavage in amphioxus is radial and holoblastic.

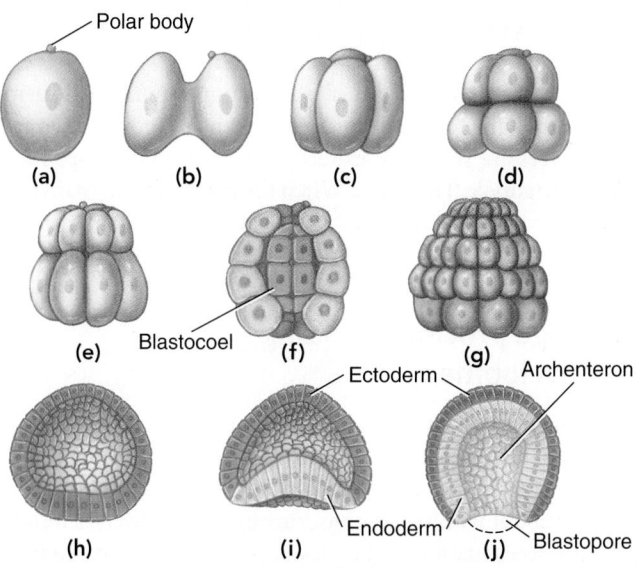

Figure 51-4 Cleavage and gastrulation in amphioxus

As in the sea star, cleavage is holoblastic and radial. The embryos are shown from the side. **(a)** Mature egg with polar body. **(b–e)** The 2-, 4-, 8-, and 16-cell stages. **(f)** Embryo cut open to show the blastocoel. **(g)** Blastula. **(h)** Blastula cut open. **(i)** Early gastrula showing beginning of invagination at vegetal pole. **(j)** Late gastrula. Invagination is completed, and the blastopore has formed.

VISUALIZE Draw a simple sketch illustrating a horizontal section through the midplane of part **(j)**. Label the archenteron, endoderm, and ectoderm. Is the blastopore included in your drawing?

© Cengage Learning

Cleavage in annelids is spiral and holoblastic.

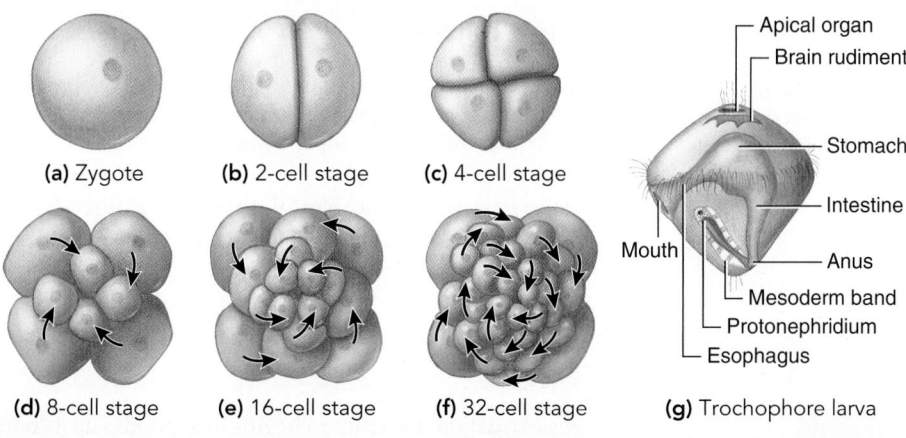

Figure 51-5 Spiral cleavage in an annelid embryo

(a–f) Top views of the animal pole. The successive cleavage divisions occur in a spiral pattern as illustrated. **(g)** A typical trochophore larva. The upper half of the trochophore develops into the extreme anterior end of the adult worm; all the rest of the adult body develops from the lower half.

VISUALIZE Draw a simple sketch illustrating how the embryo in part **(d)** (8-cell stage) would look if cleavage were radial rather than spiral.

© Cengage Learning

In **spiral cleavage** the plane of cytokinesis after the first two divisions is diagonal to the polar axis (**FIG. 51-5**). The result is a spiral arrangement of cells, with each cell located above and between the two underlying cells. This pattern is typical of annelids and mollusks.

Many vertebrate eggs are **telolecithal,** meaning that they have large amounts of yolk concentrated at one end of the cell, known as the **vegetal pole.** The opposite, more metabolically active pole is the **animal pole.** The eggs of amphibians are moderately telolecithal (mesolecithal). Although cleavage is radial and holoblastic, the divisions in the vegetal hemisphere are slowed by the presence of the inert yolk. As a result, the blastula consists of many small cells in the animal hemisphere and fewer but larger cells in the vegetal hemisphere (**FIG. 51-6**). The blastocoel is displaced toward the animal pole.

The telolecithal eggs of reptiles and birds have very large amounts of yolk at the vegetal pole and only a small amount of cytoplasm concentrated at the animal pole. The yolk of such eggs never cleaves. Cell division is restricted to the **blastodisc,** the small disc of cytoplasm at the animal pole (**FIG. 51-7**); this type of cleavage is called **meroblastic cleavage.** In birds and some reptiles, the blastomeres form two layers separated by the blastocoel cavity: an upper *epiblast* and a lower, thin layer of flat cells, the *hypoblast.*

Cleavage may distribute developmental determinants

The pattern of cleavage in a particular species depends on both yolk and other factors. Recall from Chapter 17 that some animals have relatively rigid developmental patterns, known as **mosaic development.** This type of development is largely a consequence of the unequal distribution of important materials in the cytoplasm of the zygote. Because the zygote cytoplasm is not homogeneous, cytoplasmic developmental determinants portioned out to each new cell during cleavage may be different. Such differences help determine the course of development. At the other extreme are mammals, which have zygotes with very homogeneous cytoplasm. They exhibit highly **regulative development,** in which individual cells produced by the cleavage divisions are equivalent, thus allowing the embryo to develop as a self-regulating whole. Developmental patterns of most animals fall somewhere on a continuum between these two extremes.

In some species the distribution of developmental determinants in the unfertilized egg is maintained in the zygote. In other species sperm penetration initiates rearrangement of the cytoplasm. For example, fertilization in the

Cleavage in the moderately telolecithal frog egg is radial and holoblastic but is retarded by yolk in the vegetal hemisphere.

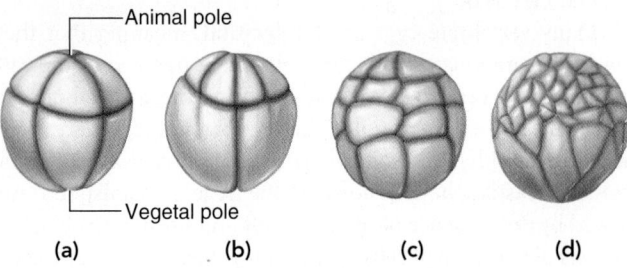

Animal pole

Vegetal pole

(a)   (b)   (c)   (d)

Figure 51-6 Cleavage pattern in a frog egg

(a–d) Although cleavage is holoblastic, the large amount of yolk concentrated in the vegetal hemisphere slows cleavage. As a result, fewer cells develop at the vegetal hemisphere than at the animal hemisphere. These embryos are shown from the side.

VISUALIZE Draw a simple sketch illustrating how the embryo in part **(a)** would look if the frog egg were isolecithal.

© Cengage Learning

amphibian egg causes a shift of some of the cortical cytoplasm. These cytoplasmic movements can be easily followed because the egg cortex contains dark pigment granules. As a result of this rearrangement, a crescent-shaped region of underlying lighter colored (gray) cytoplasm becomes evident directly opposite the point on the cell where the sperm penetrated the egg. This **gray crescent** region is thought to contain growth factors and other developmental determinants. The first cleavage bisects the gray

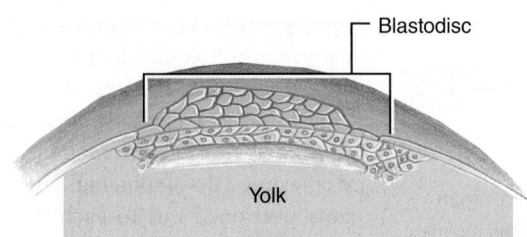

Blastodisc

Yolk

(a) Early blastodisc formation. The blastodisc is a small disc of cytoplasm on the upper surface of the egg yolk. This cutaway view shows cells on the blastodisc surface as well as in the interior.

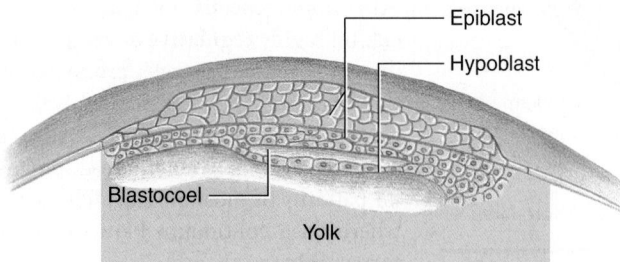

Epiblast

Hypoblast

Blastocoel

Yolk

(b) The blastodisc splits into two tissue layers, an upper epiblast and a lower hypoblast, separated by the blastocoel.

Figure 51-7 Cleavage in a bird embryo

Cleavage in bird embryos is restricted to the blastodisc (meroblastic cleavage).
© Cengage Learning

crescent, distributing half to each of the first two blastomeres; in this way the position of the gray crescent establishes the future right and left halves of the embryo. As cleavage continues, the gray crescent material becomes partitioned into certain blastomeres. Those that contain parts of the gray crescent eventually develop into the dorsal region of the embryo.

Experiments have confirmed the importance of determinants in the gray crescent to development. If the first two frog blastomeres are separated experimentally, each develops into a complete tadpole (**FIG. 51-8a**). When the plane of the first division is altered so that the gray crescent is completely absent from one of the cells, that cell does not develop normally (**FIG. 51-8b**).

Cleavage provides building blocks for development

Cleavage partitions the zygote into many small cells that serve as the basic building blocks for subsequent development. At the end of cleavage, the small cells that make up the blastula begin to move about with relative ease, arranging themselves into the patterns necessary for further development. Surface proteins are important in helping cells "recognize" one another and therefore in determining which ones adhere to form tissues.

CHECKPOINT 51.3

- **VISUALIZE** *Draw simple sketches illustrating the difference between radial cleavage and spiral cleavage.*
- *What type of cleavage is characteristic of telolecithal eggs of reptiles and birds?*
- **CONNECT** *How do the amount and distribution of cytoplasm in the fertilized egg influence early development?*

51.4 GASTRULATION

LEARNING OBJECTIVE

7 Identify the significance of gastrulation in the developmental process and compare gastrulation in the echinoderm (or in amphioxus), the amphibian, and the bird.

The process by which the blastula becomes a three-layered embryo, or **gastrula,** is called **gastrulation.** Thus, early development proceeds through the following stages:

zygote ⟶ early cleavage stages ⟶ morula ⟶ blastula ⟶ gastrula

During gastrulation, the embryo begins to approximate its body plan as cells arrange themselves into three distinct **germ layers,** or embryonic tissue layers: the outermost layer, the **ectoderm;** the innermost layer, the **endoderm;** and the **mesoderm,** which develops between them. Additional cell divisions take place during gastrulation, and the germ layers become established through a combination of processes. Many cells lose their old

Is the development of a frog egg influenced by cytoplasmic developmental determinants?

HYPOTHESIS: The gray crescent, which is normally bisected by the first cleavage division and thus establishes right and left halves of the embryo, is the location of crucial developmental determinants.

EXPERIMENT: The two blastomeres resulting from the first cleavage division are separated by the researcher and allowed to develop independently.

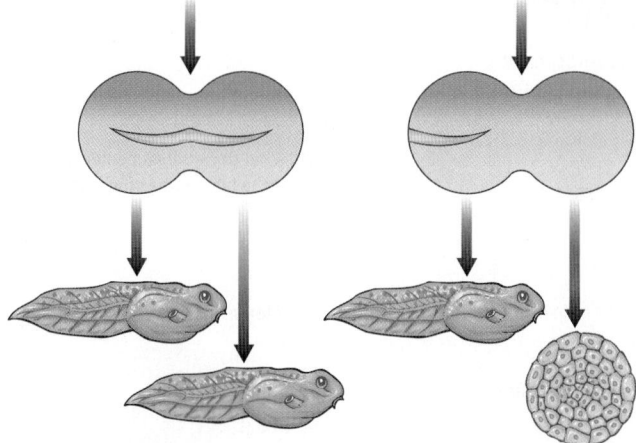

First cleavage (control)   First cleavage (experimental)

(a) The first cleavage is allowed to occur normally, and each separated blastomere includes half of the gray crescent.

(b) The plane of cleavage is altered by the experimenter such that only one blastomere contains the gray crescent.

Each of the separated blastomeres contained half of the gray crescent, and each developed into a normal tadpole.

The separated blastomere that received the entire gray crescent developed normally, but the blastomere lacking the gray crescent gave rise to a disorganized ball of cells.

RESULTS AND CONCLUSION: Because only blastomeres containing all or part of the gray crescent developed into tadpoles, the gray crescent marks the location of essential developmental determinants in the fertilized egg.

SOURCE: Spemann, H. (1938) *Embryonic Development and Induction*. Yale University Press, New Haven, CT.

Figure 51-8 *Animation* **Cytoplasmic determinants in frog development**

PREDICT Consider Gurdon's experiments demonstrating nuclear totipotency in frogs illustrated in Figure 17-3. What do you think would be the result if the same experiments were done, but substituting a blastomere lacking a gray crescent for an unfertilized egg?

© Cengage Learning

cell-to-cell contacts and establish new ones through cell recognition and adhesion processes involving interactions among the integrins and other plasma membrane proteins and the extracellular matrix (see Chapters 4 and 5). Many cells undergo cytoskeletal changes, particularly alterations in the distribution of actin microfilaments; these changes in the internal architecture of the cells allow them to change shape and/or undergo specific, directional, amoeboid movements. As a result of these movements, many cells ultimately take up new positions in the interior of the embryo.

Each of the germ layers develops into specific parts of the embryo. As you study the illustrations, keep in mind that a conventional color code accepted by developmental biologists is used to depict the germ layers: endoderm, yellow; mesoderm, red or pink; and ectoderm, blue.

The amount of yolk affects the pattern of gastrulation

The simple type of gastrulation that occurs in echinoderms and in amphioxus is illustrated in Figures 51-3f and g and 51-4i and j, respectively. Gastrulation begins when a group of cells at the vegetal pole undergoes a series of changes in shape that cause that part of the blastula wall to first flatten and then bend inward (invaginate). The invaginated wall eventually meets the opposite wall, obliterating the blastocoel.

You can roughly demonstrate this type of gastrulation by pushing inward on the wall of a partly deflated rubber ball until it rests against the opposite wall. In a similar way, the embryo is converted into a double-walled, cup-shaped structure. The new internal wall lines the **archenteron,** the newly formed cavity of the developing gut. The opening of the archenteron to the exterior, the **blastopore,** is the site of the future anus in deuterostomes.

This simple type of gastrulation cannot occur in the amphibian embryo because the large yolk-laden cells in the vegetal half of the blastula obstruct any inward movement at the vegetal pole. Instead, cells from the animal pole move down the embryo surface; when they reach the region derived from the gray crescent, they move into the interior. This inward movement is accomplished as the cells change shape, first becoming flask- or bottle-shaped (so that most of their mass is actually below the surface) and then sinking into the interior as they lose their remaining connections with other cells on the surface. This spot on the embryo's surface, referred to as the *dorsal lip of the blastopore,* is marked by a dimple, shaped like a C lying on its side (**FIG. 51-9**). As the process continues, the blastopore becomes ring-shaped as cells lateral, and then ventral, to its dorsal lip become involved in similar movements. The yolk-filled cells fill the space enclosed by the lips of the blastopore, forming the yolk plug.

The archenteron forms and is lined on all sides by cells that have moved in from the surface. At first, the archenteron is a narrow slit, but it gradually expands at the anterior end of the embryo. As a result, the blastocoel progressively shrinks and eventually disappears.

Gastrulation forms the three germ layers: endoderm, ectoderm, and mesoderm.

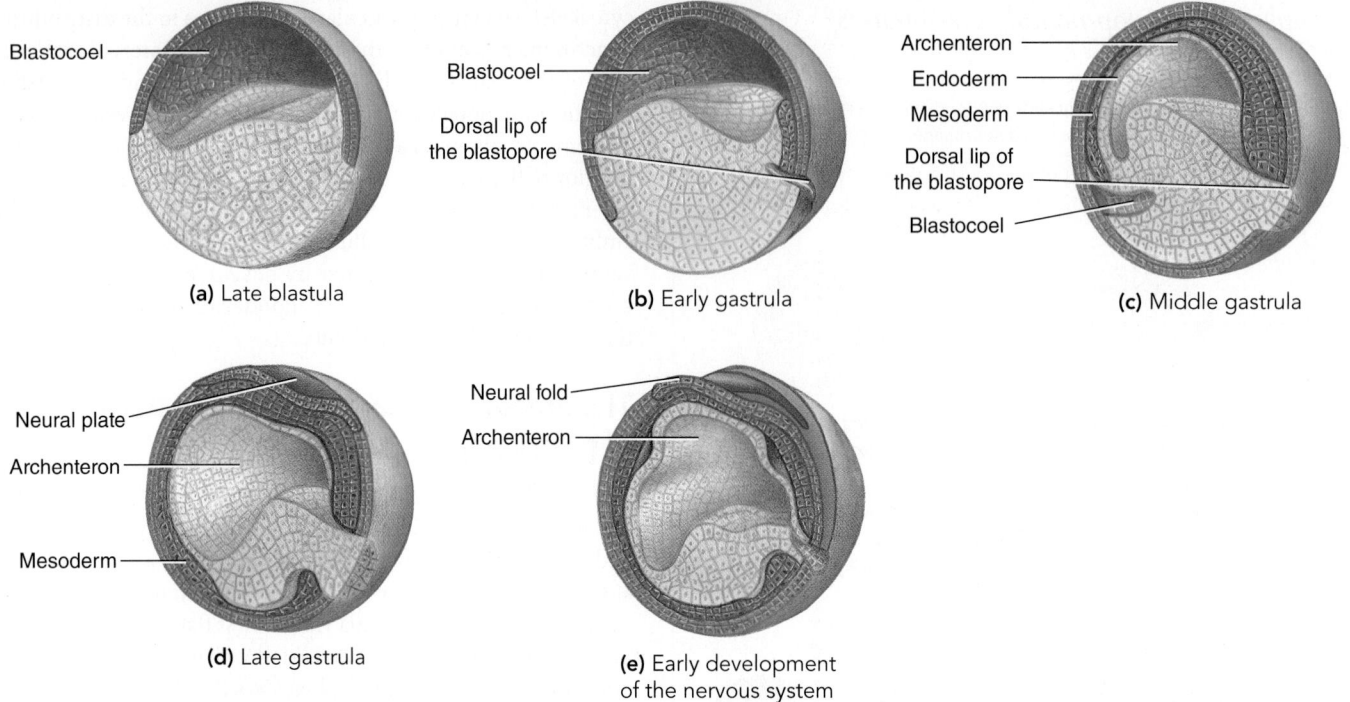

(a) Late blastula

(b) Early gastrula

(c) Middle gastrula

(d) Late gastrula

(e) Early development of the nervous system

Figure 51-9 *Animation* **Gastrulation in a frog embryo**
The embryos in the diagrams are cut in half longitudinally to show the formation of the germ layers.
© Cengage Learning

CONNECT What does the archenteron become in the tadpole? the blastopore? the blastocoel?

Although the details differ somewhat, gastrulation in birds is basically similar to amphibian gastrulation. Cells that make up the upper layer (the epiblast) migrate toward the midline to form a thickened region known as the **primitive streak,** which elongates and narrows as it develops. At its center is a narrow furrow, the **primitive groove.** The primitive streak is a dynamic structure. Its cells constantly change as they migrate in from the epiblast, sink inward at the primitive groove, and then move out laterally and anteriorly in the interior (**FIG. 51-10**). The primitive groove is the functional equivalent of the blastopore of the echinoderm, amphioxus, and amphibian embryos. However, a bird embryo contains no cavity homologous to the archenteron.

At the anterior end of the primitive streak, cells destined to form the **notochord** (supporting rod of mesodermal cartilage-like cells that serves as the flexible skeletal axis in all chordate embryos) concentrate in a thickened knot known as **Hensen's node.** These cells sink into the interior and then move anteriorly just beneath the epiblast, forming a narrow extension from the node. Cells that will form other types of mesoderm move laterally and anteriorly from the primitive streak, between the epiblast (which becomes ectoderm) and the hypoblast. Other cells that form the endoderm displace the hypoblast cells and cause them to move laterally. These displaced cells will form part of the extraembryonic membranes, discussed in a later section.

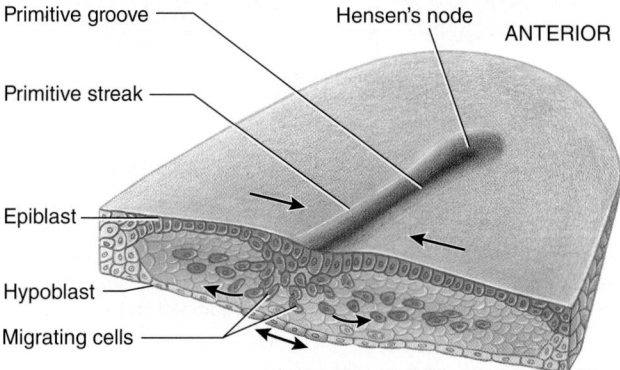

Figure 51-10 *Animation* **Gastrulation in birds**
A three-layered embryo forms as cells move toward the primitive streak, move inward at the primitive groove, and migrate laterally and anteriorly in the interior. The *red cells* will develop as mesoderm. Cells shown just above the double-headed arrow have migrated into the hypoblast; they are moving laterally and will eventually become endoderm.
© Cengage Learning

CHECKPOINT 51.4

- **CONNECT** *What is the archenteron? Is an archenteron formed in sea stars? in amphibians? in birds?*
- **CONNECT** *How do the amount and distribution of yolk influence gastrulation?*

51.5 ORGANOGENESIS

LEARNING OBJECTIVES

8 Define *organogenesis* and summarize the fate of each of the germ layers.

9 Trace the early development of the vertebrate nervous system.

Gastrulation leads to **organogenesis,** or organ formation, in which the cells of the three germ layers continue the processes of pattern formation that lead to the formation of specific structures. The ectoderm eventually forms the outer layer of the skin and gives rise to the nervous system and sense organs. Tissues that eventually line the digestive tract, and organs that develop as outgrowths of the digestive tract (including the liver, pancreas, and lungs), are all of endodermal origin. Skeletal tissue, muscle, and the circulatory, excretory, and reproductive systems all are derived from mesoderm (TABLE 51-1; see Fig. 17-1).

The notochord, brain, and spinal cord are among the first organs to develop in the early vertebrate embryo (FIG. 51-11; see also Figs. 51-9d and e). First, the notochord, which is mesodermal tissue, grows forward along the length of the embryo as a cylindrical rod of cells. The notochord is eventually replaced by the vertebral column, although notochord remnants will remain in the cartilage discs between the vertebrae.

The developing notochord has yet another crucial function. Numerous experiments in which researchers have transplanted portions of the notochord mesoderm to other locations in the embryo show that the notochord causes the overlying ectoderm to thicken and differentiate to form the precursor of the central nervous system, the **neural plate.** Such phenomena, in which certain cells stimulate or otherwise influence the differentiation of their neighbors, are examples of **induction** (see Chapter 17).

Repeated testing supports the hypothesis that induction of the neural plate cells does not require direct cell-to-cell contact with the developing notochord cells. Researchers therefore think that this induction involves the diffusion of some type of signal molecule, although they have not yet identified the chemical signal itself. The induction of the neural plate by the notochord is a good example of the importance of a cell's position in relation to other cells. That position is often critical in determining the fate of a cell because it determines the cell's exposure to substances released from other cells.

The cells of the neural plate undergo changes in shape that cause the central cells of the neural plate to move downward and form a depression called the *neural groove;* the cells flanking the

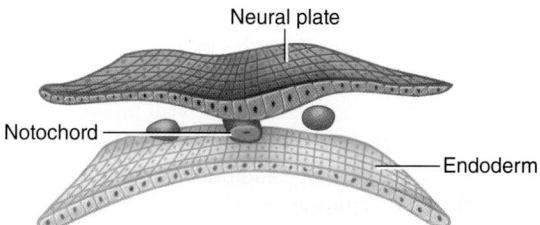

(a) Approximately 19 days. The neural plate has begun to indent and form a shallow groove that will be flanked by neural folds.

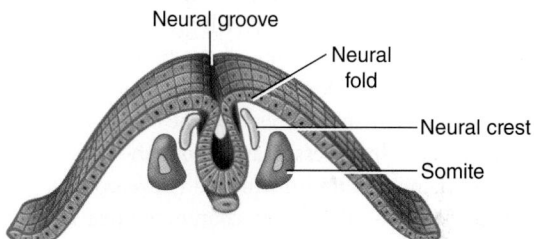

(b) Approximately 20 days. The neural folds approach one another. The neural crest cells, derived from ectoderm, will migrate in the embryo and give rise to sensory neurons; the somites are blocks of mesoderm that become the vertebrae and other segmented body parts. (The endoderm is not shown.)

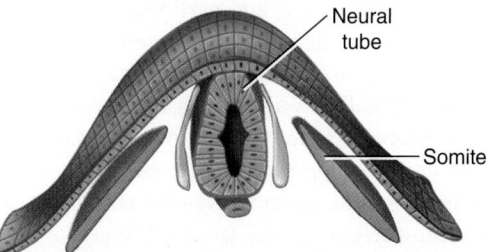

(c) Approximately 26 days. The neural folds have joined to form the hollow neural tube, which gives rise to the brain at the anterior end of the embryo and the spinal cord posteriorly. (The endoderm is not shown.)

Figure 51-11 *Animation* **Development of the human nervous system**

© Cengage Learning

TABLE 51-1	Fate of the Germ Layers Formed at Gastrulation
ECTODERM	
Nervous system and sense organs	
Outer layer of skin (epidermis) and its associated structures (nails, hair, etc.)	
Pituitary gland	
MESODERM	
Notochord	
Skeleton (bone and cartilage)	
Muscles	
Circulatory system	
Excretory system	
Reproductive system	
Inner layer of skin (dermis)	
Outer layers of digestive tube and of structures that develop from it, such as part of the respiratory system	
ENDODERM	
Lining of digestive tube and of structures that develop from it, such as lining of the respiratory system	

© Cengage Learning

groove on each side form *neural folds.* Continued changes in cell shape bring the folds closer together until they meet and fuse to form the **neural tube.** In this process, the hollow neural tube comes to lie beneath the surface. The ectoderm overlying it will form the outer layer of skin. The anterior portion of the neural tube grows and differentiates into the brain; the rest of the tube develops into the spinal cord. The brain and the spinal cord remain hollow, even in the adult; the ventricles of the brain and central canal of the spinal cord are persistent reminders of their embryonic origins.

Various motor nerves grow out of the developing brain and spinal cord, but sensory nerves have a separate origin. When the neural folds fuse to form the neural tube, bits of nervous tissue known as the *neural crest* arise from ectoderm on each side of the tube. **Neural crest cells** migrate downward from their original position and form the dorsal root ganglia of the spinal nerves and the postganglionic sympathetic neurons. From sensory cells in the dorsal root ganglia, dendrites grow out to the sense organs, and axons grow into the spinal cord. Other neural crest cells migrate to various locations in the embryo. They give rise to parts of certain sense organs, nearly all pigment-forming cells in the body, and parts of the head. Some head derivatives of the neural crest are the cartilage and bone of the skull, face, and jaw; connective tissue in the eye and face dermis; sensory nerves and other parts of the peripheral nervous system; and ganglia of the autonomic nervous system.

As the nervous system develops, other organs also begin to take shape. Blocks of mesoderm known as somites form on either side of the neural tube and will give rise to the vertebrae, muscles, and other components of the body axis. In addition to the skeleton and muscles, mesodermal structures include the kidneys, reproductive structures, and the circulatory organs. In fact, the heart and blood vessels are among the first structures to form, and they must function while still developing.

The development of the four-chambered heart of birds and mammals is a good example of organogenesis (**FIG. 51-12;** see Chapter 44). The heart originates through the fusion of paired blood vessels. Venous blood enters the single atrium; passes into the single ventricle, which is located above the atrium; and is then pumped into the embryo. During subsequent development, the atrium undergoes a process of torsion that brings it to a position above the ventricle, and partitions form that divide the atrium and the ventricle into right and left chambers.

The digestive tract is first formed as a separate foregut and hindgut as the body wall grows and folds; this process cuts them off as two simple tubes. As the embryo grows, these tubes, which are lined with endoderm, grow and become greatly elongated. The liver, pancreas, and trachea originate as hollow, tubular outgrowths from the gut. As the trachea grows downward, it gives rise to the paired lung buds, which develop into lungs. The most anterior part of the foregut becomes the pharynx. A series of small outpocketings of the pharynx, the *pharyngeal pouches,* bud out laterally. These pouches meet the *branchial grooves,* a corresponding set of inpocketings from the overlying ectoderm. The arches of tissue formed between the grooves are called *branchial arches.* They contain the rudimentary skeletal, neural, and vascular elements of the face, jaws, and neck.

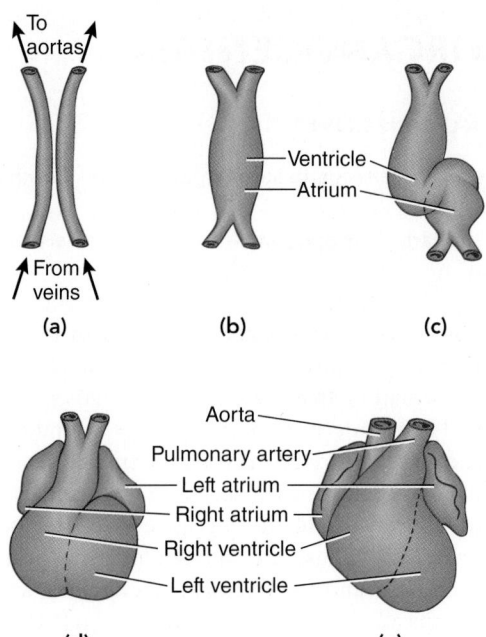

Figure 51-12 Formation of the heart in birds and mammals

These ventral views of successive stages show that the heart forms from the fusion of two blood vessels and that at first the end that receives blood from the veins is at the bottom. The developing heart twists to carry the atrium to a position above the ventricle, and the chambers then divide to form the four-chambered heart.

© Cengage Learning

In fishes and some amphibians, the pharyngeal pouches and branchial grooves meet and form a continuous passage from the pharynx to the outside; these gill slits function as respiratory organs. In terrestrial vertebrates each branchial groove remains separated from the corresponding pharyngeal pouch by a thin membrane of tissue, and these structures give rise to organs more appropriate for life on land. For example, the middle-ear cavity and its connection to the pharynx, the eustachian tube, derive from a pharyngeal pouch.

CHECKPOINT 51.5

- **CONNECT** *What adult structures develop from each germ layer?*
- **CONNECT** *How do cell division, growth, cell differentiation, and morphogenesis interact in the formation of the neural tube?*
- *What is the developmental significance of the neural crest cells?*

51.6 EXTRAEMBRYONIC MEMBRANES

LEARNING OBJECTIVE

10 Give the origins and functions of the chorion, amnion, allantois, and yolk sac.

In terrestrial vertebrates the three germ layers also give rise to four **extraembryonic membranes:** the chorion, amnion, allantois, and yolk sac (**FIG. 51-13**). Although they develop from the germ layers, these membranes are not part of the embryo itself and are discarded at hatching or birth. The extraembryonic membranes are adaptations to the challenges of embryonic development on land. They protect the embryo, prevent it from drying out, and help in obtaining food and oxygen and eliminating wastes.

The outermost membrane, the **chorion,** encloses the entire embryo as well as the other membranes. Lying underneath the eggshell in birds and reptiles, it functions as the major organ of gas exchange. As you will see in the next section, its functions have been extended in humans and most other mammals.

Like the chorion, the **amnion** encloses the entire embryo. The amniotic cavity, the space between the embryo and the amnion, becomes filled with amniotic fluid secreted by the membrane. (Recall from Chapter 32 that terrestrial vertebrates are known as *amniotes* because their embryos develop within this pool of fluid.) The amniotic fluid prevents the embryo from drying out and permits it a certain freedom of motion. The fluid also serves as a protective cushion that absorbs shocks and prevents the amniotic membrane from sticking to the embryo. In current medical practice, a sample of amniotic fluid can be obtained through a procedure known as *amniocentesis* and

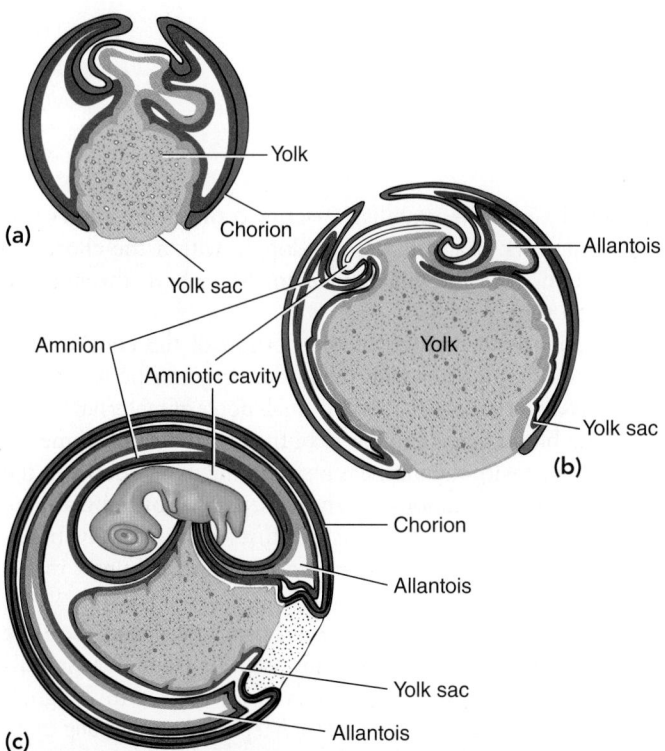

(a)

Yolk

Chorion

Yolk sac

Amnion

Amniotic cavity

(b)

Allantois

Yolk

Yolk sac

Chorion

Allantois

Yolk sac

Allantois

(c)

Figure 51-13 Extraembryonic membranes

The formation of the extraembryonic membranes of the chick is illustrated at **(a)** four days, **(b)** five days, and **(c)** nine days of development. Each of the membranes develops from a combination of two germ layers. The chorion and amnion form from lateral folds of the ectoderm and mesoderm that extend over the embryo and fuse. The allantois and the yolk sac develop from endoderm and mesoderm.

© Cengage Learning

analyzed for biochemical or chromosomal abnormalities that indicate a possible birth defect (see Fig. 16-13).

The **allantois** is an outgrowth of the developing digestive tract. In reptiles and birds, it stores nitrogenous wastes. In humans the allantois is small and nonfunctional except that its blood vessels contribute to the formation of umbilical vessels joining the embryo to the placenta. In vertebrates with yolk-rich eggs, the **yolk sac** encloses the yolk, slowly digests it, and makes it available to the embryo. A yolk sac, connected to the embryo by a yolk stalk, develops even in mammalian embryos that have little or no yolk. Its walls serve as temporary centers for the formation of blood cells.

CHECKPOINT 51.6

- What are the roles of the chorion and amnion in terrestrial vertebrate embryos?
- **CONNECT** What are the functions of the allantois and yolk sac in birds? Do they have comparable roles in mammals? Explain your answer.

51.7 HUMAN DEVELOPMENT

LEARNING OBJECTIVES

11 Describe the general course of early human development, including fertilization, the fates of the trophoblast and inner cell mass, implantation, and the role of the placenta.

12 Contrast postnatal and prenatal life, describing several changes that occur at or shortly after birth that allow the neonate to live independently.

The human *gestation period,* the duration of pregnancy, averages 266 days (38 weeks, or about 9 months) from the time of fertilization to the birth of the baby (**TABLE 51-2**). We will use this system of timing when discussing the events of early development. Because the time of fertilization is not easily marked, obstetricians usually time the pregnancy by counting from the date of onset of the woman's last menstrual period; by this calculation, an average pregnancy is 280 days (40 weeks).

Fertilization occurs in the oviduct, and within 24 hours the human zygote has divided to become a 2-celled embryo (**FIG. 51-14**). Cleavage continues as the embryo is propelled along the oviduct by ciliary action.

When the embryo enters the uterus on about the fifth day of development, the *zona pellucida* (its surrounding coat) is dissolved. During the next few days, the embryo floats free in the uterine cavity and is nourished by a nutritive fluid secreted by the glands of the uterus. Its cells arrange themselves, forming a blastula, which in mammals is called a **blastocyst.** The outer layer of cells, the **trophoblast,** eventually forms the chorion and amnion that surround the embryo. A little cluster of cells, the **inner cell mass,** projects into the cavity of the blastocyst. The inner cell mass gives rise to the embryo proper.

On about the seventh day of development, the embryo begins the process of **implantation,** in which it embeds in the

TABLE 51-2 Some Important Developmental Events in the Human Embryo

TIME FROM FERTILIZATION	EVENT
24 hours	Embryo reaches 2-cell stage
3 days	Morula reaches uterus
7 days	Blastocyst begins to implant
2.5 weeks	Notochord and neural plate are formed; tissue that will give rise to heart is differentiating; blood cells are forming in yolk sac and chorion
3.5 weeks	Neural tube forming; primordial eye and ear visible; pharyngeal pouches forming; liver bud differentiating; respiratory system and thyroid gland just beginning to develop; heart tubes fuse, bend, and begin to beat; blood vessels are laid down
4 weeks	Limb buds appear; three primary divisions of brain forming
2 months	Muscles differentiating; embryo capable of movement; gonad distinguishable as testis or ovary; bones begin to ossify; cerebral cortex differentiating; principal blood vessels assume final positions
3 months	Sex can be determined by external inspection; notochord degenerates; lymph glands develop
4 months	Face begins to look human; lobes of cerebrum differentiate; eyes, ears, and nose look more "normal"
Third trimester	Downy hair covers the fetus, then later is shed; neuron myelination begins; tremendous growth of body
266 days	Birth

© Cengage Learning

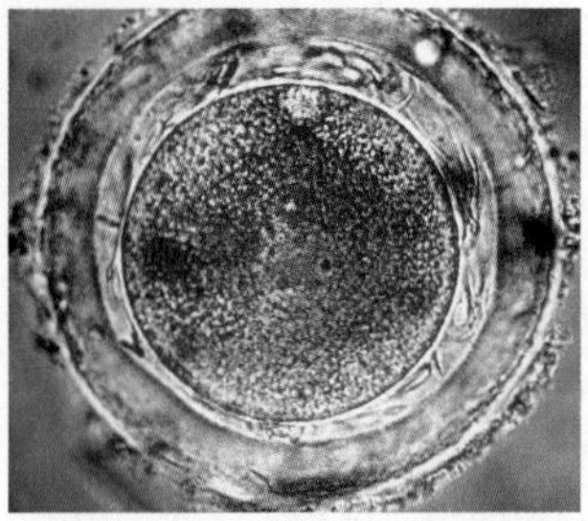

(a) A fertilized egg prior to cell division

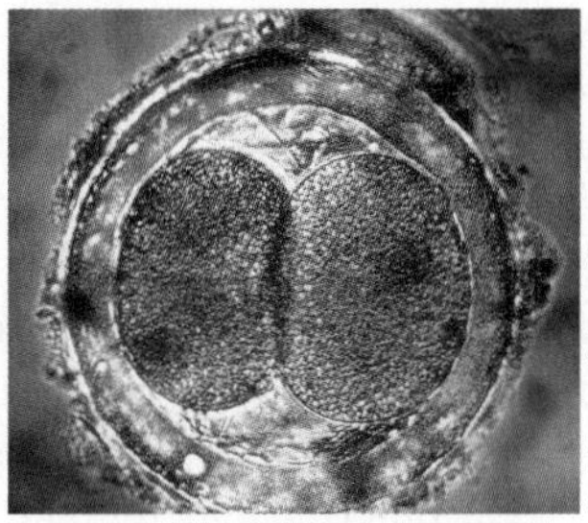

(b) 2-cell stage

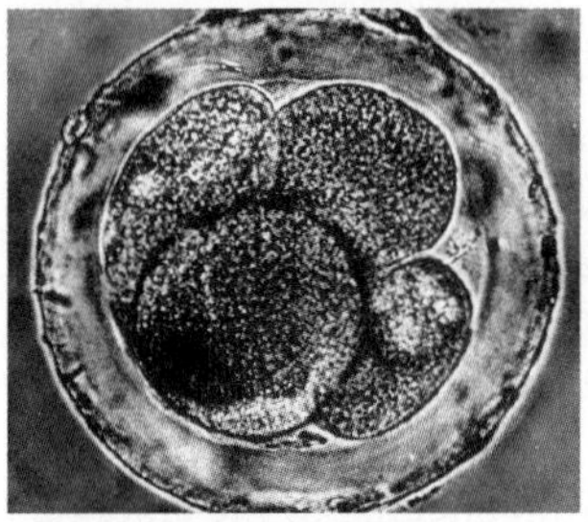

50 μm

50 μm

(c) Cleavage divisions continue

50 μm

Figure 51-14 Cleavage in a human embryo

endometrium of the uterus (**FIG. 51-15**). The trophoblast cells in contact with the endometrium secrete enzymes that erode an area just large enough to accommodate the tiny embryo. As the embryo slowly works its way into the underlying connective and vascular tissues, the opening in the endometrium repairs itself. All further development of the embryo takes place *within* the endometrium.

The placenta is an organ of exchange

In placental mammals the **placenta** is the organ of exchange between mother and embryo (see Fig. 51-15). The placenta provides nutrients and oxygen for the fetus and removes wastes, which the mother then excretes. In addition, the placenta is an endocrine organ that secretes estrogens and progesterone to maintain pregnancy. The placenta develops from both the chorion of the embryo and the uterine tissue of the mother. In early development the chorion grows rapidly, invading the endometrium and forming fingerlike projections known as *chorionic villi*. The chorionic villus sampling technique allows prenatal detection of certain genetic disorders (see Fig. 16-14). The villi become vascularized (infiltrated with blood vessels) as the embryonic circulation develops.

As the human embryo grows, the **umbilical cord** develops and connects the embryo to the placenta (see Fig. 51-15e). The umbilical cord contains the two umbilical arteries and the umbilical vein. The umbilical arteries connect the embryo to a vast network of capillaries developing within the chorionic villi. Blood from the villi returns to the embryo through the umbilical vein.

The placenta consists of the portion of the chorion that develops villi, together with the underlying uterine tissue that contains maternal capillaries and small pools of maternal blood. The fetal blood in the capillaries of the chorionic villi comes in close contact with the mother's blood in the tissues between the villi. The two circulations are always separated by a membrane through which substances may diffuse or be actively transported. Maternal and fetal blood do not normally mix in the placenta or any other place.

The placenta produces several hormones. From the time the embryo first begins to implant itself, its trophoblastic cells release **human chorionic gonadotropin (hCG),** which signals the corpus luteum (see Chapter 50) that pregnancy has begun. In response, the corpus luteum increases in size and releases large amounts of progesterone and estrogens. These hormones stimulate continued development of the endometrium and placenta.

Without hCG, the corpus luteum would degenerate and the embryo would abort and be flushed out with the menstrual flow. The woman would probably not even know she had been pregnant. If the corpus luteum is removed before about the 11th week after fertilization, the embryo spontaneously aborts.

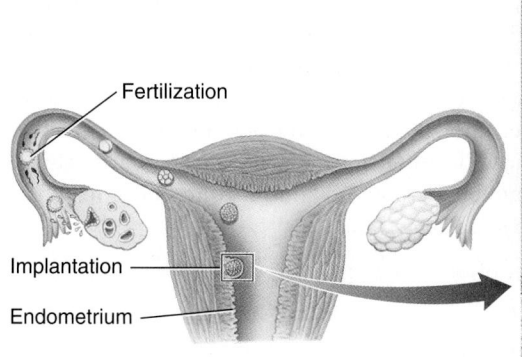

Fertilization

Implantation

Endometrium

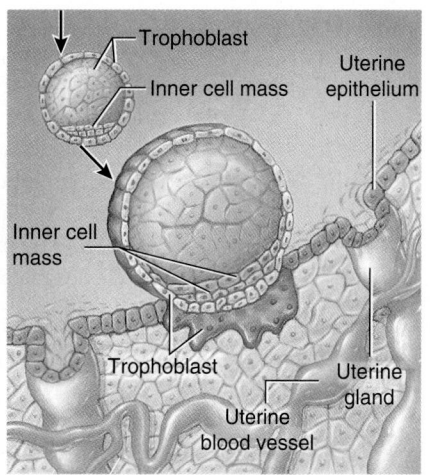

Trophoblast

Inner cell mass

Uterine epithelium

Inner cell mass

Trophoblast

Uterine blood vessel

Uterine gland

(a) 7 days. About 7 days after fertilization, the blastocyst drifts to an appropriate site along the uterine wall and begins to implant. The cells of the trophoblast proliferate and invade the endometrium.

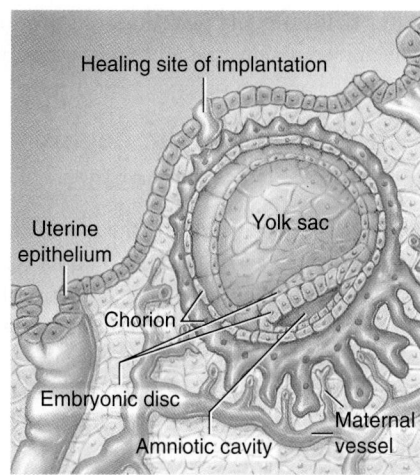

Healing site of implantation

Uterine epithelium

Yolk sac

Chorion

Embryonic disc

Amniotic cavity

Maternal vessel

(b) 10 days. About 10 days after fertilization, the chorion has formed from the trophoblast.

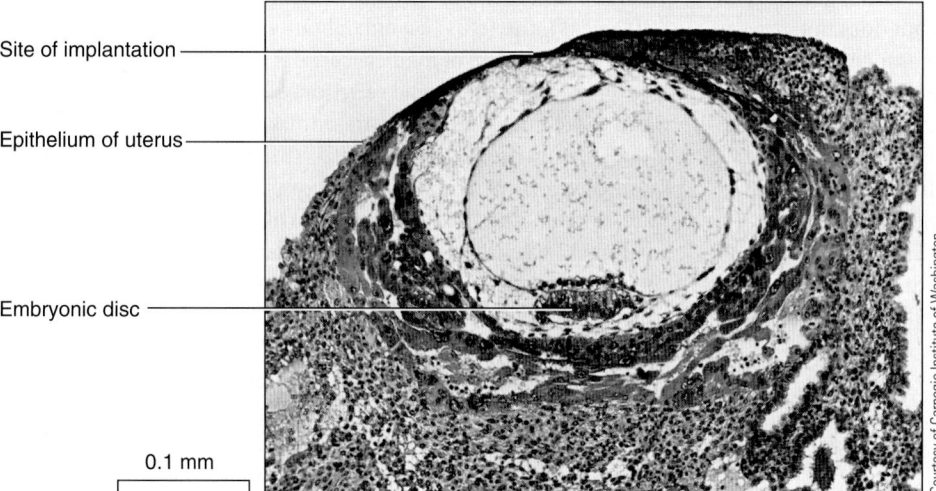

Site of implantation

Epithelium of uterus

Embryonic disc

0.1 mm

Courtesy of Carnegie Institute of Washington

(c) 12 days. This LM shows an implanted blastocyst at about 12 days after fertilization.

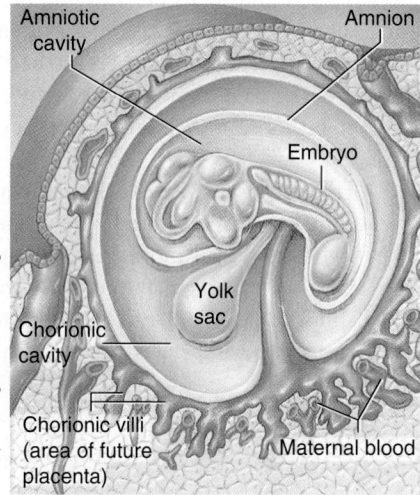

Amniotic cavity

Amnion

Embryo

Yolk sac

Chorionic cavity

Chorionic villi (area of future placenta)

Maternal blood

(d) 25 days. After 25 days, the maternal blood vessels provide the embryo with oxygen and nutrients. Note the specialized region of the chorion that will become the placenta.

(e) 45 days. At about 45 days, the embryo and its membranes together are about the size of a Ping-Pong ball, and the mother still may be unaware of her pregnancy. The amnion filled with amniotic fluid surrounds and cushions the embryo. The yolk sac has been incorporated into the umbilical cord. Blood circulation has been established through the umbilical cord to the placenta.

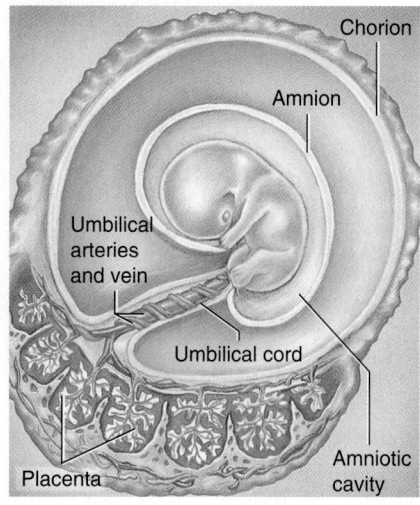

Chorion

Amnion

Umbilical arteries and vein

Umbilical cord

Placenta

Amniotic cavity

Figure 51-15 *Animation* **Implantation and early development in the uterus**
© Cengage Learning

After that time, the placenta itself produces enough progesterone and estrogens to maintain pregnancy.

Organ development begins during the first trimester

Gastrulation occurs during the second and third weeks of development. Then the notochord begins to form and induces formation of the neural plate. The neural tube develops, and the forebrain, midbrain, and hindbrain are evident by the fifth week of development. A week or so later the forebrain begins to grow outward, forming the rudiments of the cerebral hemispheres.

The heart begins to develop and after 3.5 weeks begins to beat spontaneously because the embryo is growing large enough to require a functional circulatory system. Pharyngeal pouches, branchial grooves, and branchial arches form in the region of the developing pharynx. In the floor of the pharynx, a tube of cells grows downward to form the primordial trachea, which gives rise to the lung buds. The digestive system also gives rise to outgrowths that develop into the liver, gallbladder, and pancreas. A thin tail becomes evident but does not grow as rapidly as the rest of the body and so becomes inconspicuous by the end of the second month. Near the end of the fourth week, the limb buds begin to differentiate; they eventually give rise to arms and legs.

All the organs continue to develop during the second month (see chapter opening photograph). Muscles develop, and the embryo becomes capable of movement. The brain begins to send impulses that regulate the functions of some organs, and a few simple reflexes are evident. After the first two months of development, the embryo is referred to as a **fetus** (FIG. 51-16).

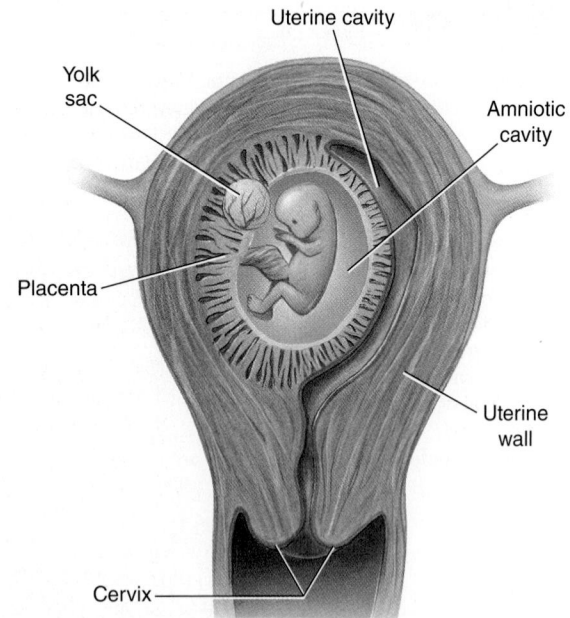

Figure 51-16 Human fetus at 10 weeks

Note the position of the fetus within the uterine wall.

© Cengage Learning

Labels: Uterine cavity, Yolk sac, Amniotic cavity, Placenta, Uterine wall, Cervix

By the end of the *first trimester* (the first three months of development), the fetus is about 56 mm (2.2 in.) long and weighs about 14 g (0.5 oz). Although small, it looks human. The external genital structures have differentiated, indicating the sex of the fetus. Ears and eyes approach their final positions. Some of the skeleton becomes distinct, and the developing vertebral column has replaced the notochord. The fetus performs breathing movements, pumping amniotic fluid into and out of its lungs, and even makes sucking motions.

Development continues during the second and third trimesters

During the second trimester (months four through six), the fetal heart can be heard with a stethoscope. The fetus moves freely within the amniotic cavity, and during the fifth month the mother usually becomes aware of relatively weak fetal movements ("quickening"). The fetus grows rapidly during the final trimester (months seven through nine), and final differentiation of tissues and organs occurs. If born at 24 weeks (out of 40 weeks), the fetus has only about a 50% chance of surviving, even with the best of medical care. Its brain is not yet sufficiently developed to sustain vital functions such as rhythmic breathing, and the kidneys and lungs are immature.

During the seventh month, the cerebrum grows rapidly and develops convolutions. The grasping and sucking reflexes are evident, and the fetus may suck its thumb. Any infant born before 37 weeks (out of 40) is considered premature. According to statistics made available by the March of Dimes, premature birth is an increasing problem in the United States, its incidence having risen to one in nine births in 2012, a 36% increase since 1987. Each year, well over half a million premature babies are born in the United States and more than 15 million worldwide. Babies born at 28 or fewer weeks are at greatest risk of death and severe defects. Babies born closer to 37 weeks have a greater chance of survival with competent medical care but may still face multiple long-term health problems.

At birth the average full-term baby weighs about 3000 g (6.6 lb) and measures about 52 cm (20 in.) in total length. A baby that weighs less than 2500 g (5.5 lb) is considered low birth weight and is at risk even if not premature.

More than one mechanism can lead to a multiple birth

Occasionally, the cells of the 2-cell embryo separate, and each cell develops into a complete organism. Alternatively, sometimes the inner cell mass subdivides, forming two groups of cells, each of which develops independently. Because these cells have identical sets of genes, the individuals formed are exactly alike and are called **monozygotic,** or identical, twins. Very rarely, the two inner cell masses do not completely separate and so give rise to *conjoined twins,* who are physically attached and usually share one or more body parts.

Dizygotic twins, also called fraternal twins, develop when two eggs are ovulated and each is fertilized by a different sperm.

Each zygote has its own distinctive genetic endowment, so the individuals produced are not identical. They may not even be of the same sex. Similarly, triplets (and other multiple births) may be either identical or fraternal.

Before fertility-inducing agents became available in the United States, twins were born once in about 80 births (about 30% of twins are monozygotic), triplets once in 80^2 (or 1 in 6400), and quadruplets once in 80^3 (or 1 in 512,000). However, widespread use of drugs to improve fertility has been linked to an increase in multiple births. The March of Dimes reports that between 1980 and 2004 the rate of twin births increased by about 70% and the rate of triplet births increased fivefold. These rates may now be stabilizing; the rate of twin births does not appear to be increasing, and the rate of triplet births has declined slightly.

A family history of twinning increases the probability of having dizygotic twins. However, giving birth to monozygotic twins is probably not influenced by heredity, age of the mother, or other known factors. An estimated two-thirds of multiple pregnancies end in the birth of a single baby; the other embryo(s) may be absorbed within the first ten weeks of pregnancy, or spontaneously aborted. Ultrasound imaging techniques are used to produce a type of image known as a **sonogram,** which can provide valuable data regarding the presence of multiple embryos. This information is medically important because multiple births are associated with higher infant mortality, which is largely a consequence of an increased risk of prematurity and low birth weight.

Environmental factors affect the embryo

We all know that the growth and development of babies are influenced by the food they eat, the air they breathe, the disease organisms that infect them, and the chemicals or drugs to which they are exposed. **Prenatal** development is also affected by these environmental influences. Life before birth is even more sensitive to environmental changes than it is for the fully formed baby. Although there is no direct mixing of maternal and fetal blood, diffusion and various other mechanisms allow many substances—nutrients, drugs, pathogens, and gases—to travel across the placenta.

TABLE 51-3 describes some environmental influences on development. Some of them are **teratogens,** drugs or other substances that interfere with morphogenesis and so cause malformations. Many factors, such as smoking, alcohol use, and poor nutrition, contribute to low birth weight, a condition responsible for many infant deaths.

About one out of every 33 newborns (more than 120,000 babies per year) in the United States has a defect of clinical significance. Such birth defects account for more than 20% of deaths among newborns. Birth defects may be caused by genetic and/or environmental factors. Genetic factors were discussed in Chapter 17. In this section we examine some environmental conditions that affect the well-being of the embryo.

Timing is important. Each developing structure has a critical period during which it is most susceptible to unfavorable conditions. Generally, this critical period occurs early in the structure's development, when interference with cell movements or divisions may prevent formation of normal shape or size, resulting in permanent malformation. Because most structures form during the first three months of embryonic life, the embryo is most susceptible to environmental factors during this early period. During part of this time, the woman may not even realize that she is pregnant and so may not take special precautions to minimize potential dangers.

Physicians diagnose some defects while the embryo is in the uterus. In some cases, treatment is possible before birth. Amniocentesis and chorionic villus sampling, discussed in Chapter 16, are techniques used to detect certain defects. Ultrasound imaging techniques mentioned previously help diagnose defects and determine the position of the fetus. Ultrasound has been used for many years, and unlike imaging techniques that use radiation, no risk of harm to the fetus has been traced to its use. New methods currently under development or refinement include special types of magnetic resonance imaging, or MRI, and high-resolution, three-dimensional ultrasound imaging (**FIG. 51-17**).

The neonate must adapt to its new environment

Important changes take place after birth. During prenatal life the fetus received both food and oxygen from the mother through the placenta. Now the newborn's own digestive and respiratory systems must function. Correlated with these changes are several major changes in the circulatory system.

Normally, the **neonate** (newborn infant) begins to breathe within a few seconds of birth and cries within half a minute. If anesthetics have been given to the mother, however, the fetus may also be anesthetized, and its breathing and other activities may be depressed. Some infants may not begin breathing until several minutes have passed, which is one reason many women request childbirth methods that minimize the use of medication.

Researchers think that the neonate's first breath is initiated by the accumulation of carbon dioxide in the blood after the umbilical cord is cut. Carbon dioxide stimulates the respiratory centers in the medulla. The resulting expansion of the lungs enlarges its blood vessels (which in the uterus were partially collapsed). Blood from the right ventricle flows in increasing amounts through these larger pulmonary vessels. (During fetal life, blood bypasses the lungs in two ways: by flowing through an opening, the *foramen ovale,* which shunts blood from the right atrium to the left atrium; and by flowing through an arterial duct connecting the pulmonary artery and aorta. Both of these routes close off after birth.)

Aging is not a uniform process

We have examined briefly the development of the embryo and fetus, the birth process, and the adjustments required of the neonate. The human life cycle then proceeds through the stages of

TABLE 51-3	Environmental Influences on the Embryo	
FACTOR	**EXAMPLE AND EFFECT**	**COMMENT**
NUTRITION	Severe protein malnutrition doubles number of defects, fewer brain cells are produced, and learning ability may be permanently affected; deficiency of folic acid (a vitamin) linked to CNS defects such as spina bifida (open spine), low birth weight.	Growth rate mainly determined by rate of net protein synthesis by embryo's cells.
MEDICATIONS	Many medications, even aspirin, affect development of the fetus.	Herbal drugs and dietary supplements are unregulated.
Excessive vitamins	Vitamin D essential, but excessive amounts may result in a form of intellectual disability; an excess of vitamins A and K may also be harmful.	Vitamin supplements are normally prescribed for pregnant women, but only the recommended dosage should be taken.
Thalidomide	Thalidomide, marketed in Europe as a mild sedative, was responsible for serious malformations in about 10,000 babies born in the late 1950s in 20 countries; principal defect was phocomelia, a condition in which babies are born with extremely short limbs, often with no fingers or toes.	This teratogenic drug interferes with the development of blood vessels and nerves; most hazardous when taken during fourth to fifth weeks, when limbs are developing; thalidomide has been approved for the treatment of leprosy and certain cancers (e.g., myeloma), and clinical trials are being conducted to test its usefulness in the treatment of other cancers and some of the complications of AIDS.*
Accutane (isotretinoin)	Accutane, a synthetic derivative of vitamin A, is used for the treatment of cystic acne. A woman who takes Accutane during pregnancy has a one in five chance of having a child with serious malformations of the brain (causing intellectual disability), head and face, thymus gland (interfering with immune function), or heart.	Although Accutane is marketed with severe restrictions to prevent pregnant women from taking this teratogenic drug, several children are born each year with birth defects attributable to Accutane exposure.
PATHOGENS		
Rubella	Rubella (German measles) virus crosses placenta and infects embryo; interferes with normal metabolism and cell movements; causes syndrome that involves blinding cataracts, deafness, heart malformations, and intellectual disability; risk is greatest (about 50%) when rubella is contracted during the first month of pregnancy; risk declines with each succeeding month.	Rubella epidemic in the United States in 1963–1965 resulted in about 20,000 fetal deaths and 30,000 infants born with serious defects; immunization is available but must be administered several months before conception.
HIV*	HIV can be transmitted from mother to baby before birth.	See discussion of AIDS in Chapter 45.
Syphilis	Syphilis is transmitted to fetus in about 40% of infected women; fetus may die or be born with defects and congenital syphilis.	Pregnant women are routinely tested for syphilis during prenatal examinations; most cases can be safely treated with antibiotics.
IONIZING RADIATION	When a pregnant woman is subjected to X-rays or other forms of radiation, infant has a higher risk of birth defects and leukemia.	Radiation was one of the earliest causes of birth defects to be recognized.
RECREATIONAL AND/OR ABUSED SUBSTANCES		
Alcohol	When a woman drinks heavily during pregnancy, the baby may be born with fetal alcohol syndrome (FAS), which includes certain physical deformities and intellectual disability; low birth weight and structural abnormalities have been associated with as little as two drinks a day; some cases of hyperactivity and learning disabilities have occurred.	FAS is the leading cause of preventable intellectual disability in the United States.
Cigarette smoking	Cigarette smoking reduces the amount of oxygen available to the fetus because some of the maternal hemoglobin is combined with carbon monoxide; may slow growth and can cause subtle forms of damage.	Mothers who smoke cigarettes deliver babies with lower-than-average birth weights and have a higher incidence of spontaneous abortions, stillbirths, and neonatal deaths, especially from SIDS;* studies also indicate a possible relationship between maternal smoking and slower intellectual development in offspring.
Cocaine	Constricts fetal arteries, resulting in retarded development and low birth weight; in severe cases there may be learning and behavioral problems, heart defects, and other medical problems.	Cocaine users frequently abuse alcohol as well, so it is difficult to separate the effects.
Heroin	High rates of mortality and prematurity; low birth weight.	Infants who survive are born addicted and must be treated for weeks or months; risk of SIDS is increased ten-fold.

*AIDS, acquired immunodeficiency syndrome; HIV, human immunodeficiency virus; SIDS, sudden infant death syndrome

© Cengage Learning

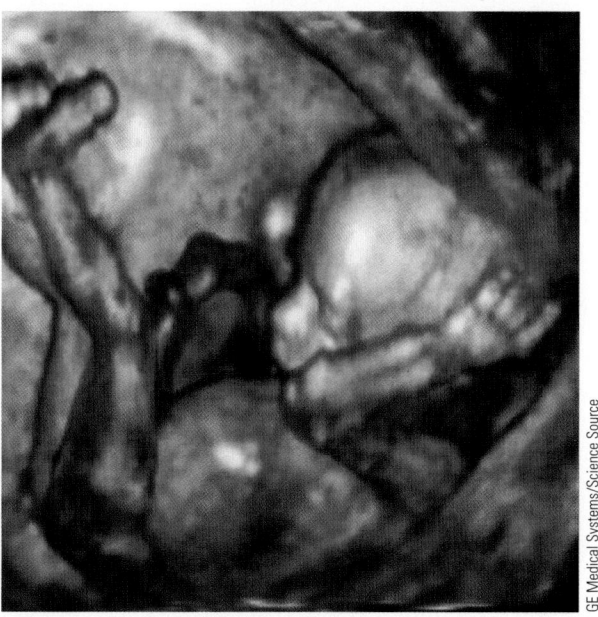

Figure 51-17 Three-dimensional ultrasound image of a human fetus at about 16 weeks gestation
Note the enhanced soft tissue detail.

infant, child, adolescent, young adult, middle-aged adult, and elderly adult.

Development encompasses any biological change within an organism over time, including the changes commonly called **aging.** Changes during the aging process decrease function in the older individual. The declining capacities of the various systems in the human body, although most apparent in the elderly, may begin much earlier in life. Relatively little is known about the aging process itself; it is now an active field of investigation, and researchers are gaining considerable insight through genetic studies on model organisms such as the nematode worm, *Caenorhabditis elegans;* the fruit fly, *Drosophila melanogaster;* and the laboratory mouse, *Mus musculus* (see Chapter 17).

CHECKPOINT 51.7

- *How does a human blastocyst form and implant?*
- *How does the placenta form, and what is its function?*
- *During which trimester does the cerebrum develop convolutions? When do the limb buds begin to differentiate?*
- CONNECT *What differences between fetal life and the life of a neonate require adaptations immediately after birth?*

SUMMARY: FOCUS ON LEARNING OBJECTIVES

51.1 Development of Form *(page 1099)*

1 Analyze the relationship between cell determination and cell differentiation and between pattern formation and morphogenesis.

- **Cell determination** is a series of molecular events that lead to **cell differentiation,** in which a cell becomes specialized and carries out specific functions. Relatively undifferentiated cells are known as **stem cells.**
- **Morphogenesis,** the development of form, occurs through **pattern formation,** a series of events in which cells may communicate by signaling, migrate, undergo changes in shape, and even undergo **apoptosis** (programmed cell death).

2 Relate differential gene expression to nuclear equivalence.

- Usually no genetic changes occur in cell determination and cell differentiation (the principle of **nuclear equivalence**). Instead, various types of differentiated cells express different subsets of their genes (**differential gene expression**).

51.2 Fertilization *(page 1099)*

3 Describe the four processes involved in fertilization.

- Contact and recognition occur between noncellular egg coverings and sperm.
- Sperm entry is regulated to prevent interspecific fertilization and **polyspermy,** which is fertilization of the egg by more than one sperm.
- Fertilization activates the egg, triggering the events of early development.
- Sperm and egg pronuclei fuse and initiate DNA synthesis.

4 Describe fertilization in echinoderms and point out some ways in which mammalian fertilization differs.

- The coverings of echinoderm eggs are the **vitelline envelope** and the **jelly coat;** a **zona pellucida** encloses the mammalian egg.
- On contact, a sperm undergoes an **acrosome reaction,** which facilitates penetration of the egg coverings; in mammals the acrosome reaction is preceded by **capacitation,** a maturation process that results in the ability of a sperm to fertilize an egg.
- Sea urchin fertilization is followed by a fast block to polyspermy (depolarization of the plasma membrane) and a slow block to polyspermy (the **cortical reaction**); changes in the zona pellucida prevent polyspermy in mammals.

51.3 Cleavage *(page 1101)*

5 Trace the generalized pattern of early development of the embryo from zygote through early cleavage and formation of the morula and blastula.

- The zygote undergoes **cleavage,** a series of rapid cell divisions without a growth phase. The main effect of cleavage is to partition the zygote into many small **blastomeres.**
- Cleavage forms a solid ball of cells called the **morula** and then usually a hollow ball of cells called the **blastula.**

6 Contrast early development, including cleavage, in the echinoderm (or in amphioxus), the amphibian, and the bird, paying particular attention to the importance of the amount and distribution of yolk.

- The **isolecithal eggs** of most invertebrates and simple chordates have evenly distributed yolk. They undergo **holoblastic cleavage,** which involves division of the entire egg.

- In the moderately **telolecithal eggs** of amphibians, a concentration of yolk at the **vegetal pole** slows cleavage so that only a few large cells form there compared with a large number of smaller cells at the **animal pole.**
- The highly telolecithal eggs of reptiles and birds, with a large concentration of yolk at one end, undergo **meroblastic cleavage,** which is restricted to the **blastodisc.**
- Animals whose zygotes have relatively homogeneous cytoplasm exhibit **regulative development,** in which the embryo develops as a self-regulating whole. The relatively rigid developmental patterns of some animals (**mosaic development**) is a consequence of the unequal distribution of cytoplasmic components.
- The **gray crescent** of an amphibian zygote determines the body axis of the embryo.

51.4 Gastrulation *(page 1104)*

7 Identify the significance of gastrulation in the developmental process and compare gastrulation in the echinoderm (or in amphioxus), the amphibian, and the bird.

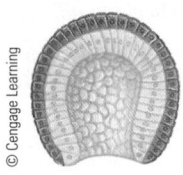

- In **gastrulation** the basic body plan is laid down as three germ layers form: the outer **ectoderm,** the inner **endoderm,** and the **mesoderm** between them.
- The **archenteron,** the forerunner of the digestive tube, forms in some groups; its opening to the exterior is the **blastopore.**
- In gastrulation in the sea star and in amphioxus, cells from the blastula wall invaginate and eventually meet the opposite wall, forming the archenteron.

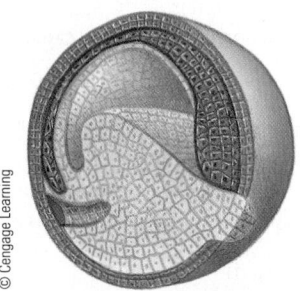

- In the amphibian invagination at the vegetal pole is obstructed by large, yolk-laden cells; instead, cells from the animal pole move down over the yolk-rich cells and invaginate, forming the *dorsal lip of the blastopore.*
- In the bird invagination occurs at the primitive streak, and no archenteron forms.

51.5 Organogenesis *(page 1107)*

8 Define *organogenesis* and summarize the fate of each of the germ layers.
- **Organogenesis** is the process of organ formation.
- Ectoderm becomes the nervous system, sense organs, and outer layer of skin (epidermis).

- Mesoderm becomes the notochord, skeleton, muscles, circulatory system, and inner layer of skin (dermis).
- Endoderm becomes the lining of the digestive tube.

9 Trace the early development of the vertebrate nervous system.

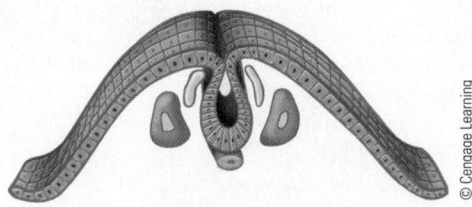

- The developing **notochord** is responsible for **induction,** which causes the overlying ectoderm to differentiate as the **neural plate,** the precursor of the central nervous system.
- The cells of the neural plate undergo changes in shape to form the hollow **neural tube,** from which the brain and spinal cord develop.

51.6 Extraembryonic Membranes *(page 1108)*

10 Give the origins and functions of the chorion, amnion, allantois, and yolk sac.
- The **chorion** and **amnion** derive from ectoderm and mesoderm; the **allantois** and **yolk sac** derive from endoderm and mesoderm.
- The chorion is used in gas exchange. The amnion is a fluid-filled sac that surrounds the embryo and keeps it moist; it also acts as a shock absorber. The allantois stores nitrogenous wastes. The yolk sac makes food available to the embryo.

51.7 Human Development *(page 1109)*

11 Describe the general course of early human development, including fertilization, the fates of the trophoblast and inner cell mass, implantation, and the role of the placenta.
- Fertilization occurs in the oviduct.
- Cleavage takes place as the embryo is moved down the oviduct.
- In the uterus the embryo develops into a **blastocyst** consisting of an outer **trophoblast,** which gives rise to the chorion and amnion, and an **inner cell mass,** which becomes the embryo proper. The blastocyst undergoes **implantation** in the endometrium.
- The **umbilical cord** connects the embryo to the **placenta,** the organ of exchange between the maternal and fetal circulation. The placenta derives from the embryonic chorion and maternal tissue.

12 Contrast postnatal and prenatal life, describing several changes that occur at or shortly after birth that allow the neonate to live independently.
- Human **prenatal** development requires 266 days from the time of fertilization; organogenesis begins during the *first trimester.* After the first two months of development, the embryo is referred to as a **fetus.** Growth and refinement of the organs continue in the second and third trimesters.
- The **neonate** (newborn) must undergo rapid adaptations, especially changes in the respiratory and digestive systems.

Know and Comprehend

1. The mechanism that leads directly to morphogenesis is (a) differential gene expression (b) pattern formation (c) nuclear equivalence (d) cell differentiation (e) cell determination

2. The main function of the acrosome reaction is to (a) activate the egg (b) improve sperm motility (c) prevent interspecific fertilization (d) facilitate penetration of the egg coverings by the sperm (e) cause fusion of the sperm and egg pronuclei

3. The fast block to polyspermy in sea urchins (a) is a depolarization of the egg plasma membrane (b) requires exocytosis of the cortical granules (c) includes the elevation of the fertilization envelope (d) involves the hardening of the jelly coat (e) is a complete block

4. The cleavage divisions of a sea urchin embryo (a) occur in a spiral pattern (b) do not include DNA synthesis (c) do not include cytokinesis (d) are holoblastic (e) do not occur at the vegetal pole

5. Meroblastic cleavage is typical of embryos formed from ___ eggs. (a) moderately telolecithal (b) highly telolecithal (c) isolecithal (d) b and c (e) a, b, and c

6. The primitive groove of the bird embryo is the functional equivalent of the ___ in the amphibian embryo. (a) yolk plug (b) archenteron (c) blastocoel (d) gray crescent (e) blastopore

7. Which of the following are mismatched? (a) endoderm; lining of the digestive tube (b) ectoderm; circulatory system (c) mesoderm; notochord (d) mesoderm; reproductive system (e) ectoderm; sense organs

8. Which of the following has three germ layers? (a) morula (b) gastrula (c) blastula (d) blastocyst (e) trophoblast

9. An unidentified substance (or substances) released from the developing notochord causes the overlying ectoderm to form the neural plate. This phenomenon is known as (a) activation (b) determination (c) induction (d) implantation (e) mosaic development

10. Which of the following consists of both fetal and maternal tissues? (a) umbilical cord (b) placenta (c) amnion (d) allantois (e) yolk sac

11. The embryo proper of a mammal develops from the (a) trophoblast (b) umbilical cord (c) inner cell mass (d) entire blastocyst (e) yolk sac

Apply and Analyze

12. Place the following events of sea urchin fertilization in the proper sequence.
    1. fusion of egg and sperm pronuclei 2. DNA synthesis
    3. increased protein synthesis 4. release of calcium ions into the egg cytoplasm
    (a) 4, 3, 1, 2 (b) 3, 2, 4, 1 (c) 2, 3, 1, 4 (d) 1, 2, 3, 4 (e) 4, 1, 2, 3

Evaluate and Synthesize

13. For almost 200 years, scientists debated whether an egg or sperm cell contains a completely formed, miniature human (preformation) or whether structures develop gradually from a formless zygote (epigenesis). Relate these views to current concepts of development.

14. Not all teratogenic medications are banned by the U.S. Food and Drug Administration. Why?

15. **EVOLUTION LINK** Consider the chordate characteristics described in Chapter 32 (notochord, dorsal hollow nerve cord, postanal tail, and pharynx with slits, a deuterostome characteristic). How are they evident in a human embryo? Do they have the same functions in humans as they do in a less specialized chordate, such as amphioxus? Why have they persisted in humans?

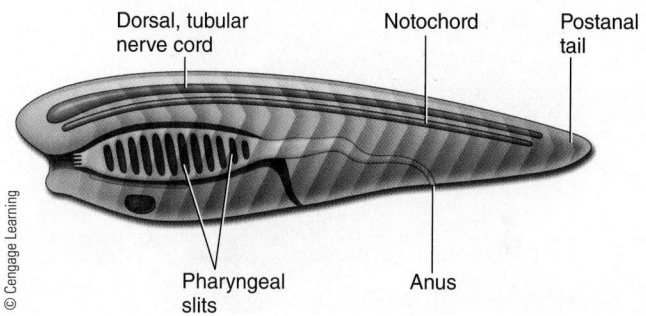

16. **EVOLUTION LINK** What is the adaptive value of developing a placenta?

17. **SCIENCE, TECHNOLOGY, AND SOCIETY** As a public health official, what policies might you promote to reduce the incidence of prematurity and/or low birth weight? of malformations at birth?

aplia To access course materials, such as Aplia and other companion resources, please visit **www.cengagebrain.com**.

E. S. Ross

The sand wasp (*Philanthus triangulum*) digging a burrow. How does the wasp "know" how to carry out the complex set of behaviors that ensure her reproductive success?

S uppose that your professor gave you a hypodermic syringe full of poison and told you to find a particular type of insect, one that you had never seen before and that was armed with active defenses. You then had to inject the ganglia of your prey's nervous system (about which you had been taught nothing) with just enough poison to paralyze but not kill it. You would have difficulty accomplishing these tasks, but a solitary wasp no larger than the first joint of your thumb does it all with elegance and surgical precision, without instruction.

Philanthus, the sand wasp (also called beewolf), captures a bee, stings it, and places the paralyzed insect in a burrow excavated in the sand (see photograph). She then lays an egg on her prey. The larva that hatches from that egg devours the bee. *Philanthus* brings provisions to her hidden burrow until autumn, when the larva becomes a hibernating pupa. Her offspring will repeat this behavior, precisely executing each step without ever having seen it done.

An animal's **behavior** is *what it does* and *how it does it*, usually in response to stimuli in its environment. A dog may wag its tail, a bird may sing, a butterfly may release a volatile sex attractant. Behavior is just as diverse as biological structure and just as characteristic of a given species as its anatomy or physiology. Like its morphology and physiology, an animal's behavior is the product of natural selection on phenotypes and indirectly on the genotypes that code for those phenotypes. Thus, an animal's repertoire of behavior is a set of adaptations that equip it for survival in a particular environment.

Whether biologists study behavior in an animal's natural environment or in the laboratory, they must consider that what an animal does cannot be isolated from the way in which it lives. **Behavioral ecology** is the study of behavior in natural environments from an evolutionary perspective. For more than two decades, behavioral ecology has been the main approach of biologists who study animal behavior. Before this approach emerged, the study of animal behavior was called *ethology*, and this term is sometimes still used to refer to the overall study of animal behavior.

In this chapter we consider how an animal's behavior contributes to its reproductive success and to the survival of its species. We discuss types of learning and then focus on the *hows* and *whys* of some key types of behavior, including migration, foraging, and social behavior. We also explore sexual selection and helping behavior.

KEY CONCEPTS

52.1 Proximate causes of behavior are the immediate causes that explain *how* an animal carries out a specific behavior. Ultimate causes of behavior are the evolutionary explanations for *why* a certain behavior occurs. If its benefits are greater than its costs, a behavior is adaptive.

52.2 The capacity for behavior is inherited, but behavior is modified in response to environmental experience; the changed behavior is learned behavior.

52.3 Biological clocks regulate biological rhythms. Migration is an example of a behavior that involves interactions among biological rhythms, physiology, and the environment.

52.4 Optimal foraging is the most efficient way for an animal to obtain food.

52.5 Social behavior requires communication among animals in the group. Dominance hierarchies and defense of a territory benefit many social groups.

52.6 Sexual selection is a type of natural selection in which individuals with reproductive advantages are selected over others of the same sex and species.

52.7 Altruism is a type of cooperative behavior in which one animal behaves in a way that appears to benefit others rather than itself. Kin selection is a form of natural selection that increases inclusive fitness through the successful reproduction of close relatives.

52.8 Some bird and mammal species transmit culture, learned behavior common to a population that one generation can teach to the next.

52.1 BEHAVIOR AND ADAPTATION

LEARNING OBJECTIVES

1 Distinguish between proximate and ultimate causes of behavior and apply the concepts of ultimate cause and cost–benefit analysis to decide whether a particular behavior is adaptive.

2 Describe the interactions of heredity, environment, and maturation in animal behavior.

In considering the reproductive behavior of *Philanthus*, we might wonder *how* she accomplishes her task and *why* she behaves as she does. Early investigators of animal behavior focused on *how* questions. These questions address **proximate causes,** immediate causes such as the genetic, developmental, and physiological processes that permit the animal to carry out the particular behavior.

Today, biologists also ask *why* questions that address the **ultimate causes** of behavior. These questions, which have evolutionary explanations, ask *why* a particular behavior has evolved. Ultimate considerations address costs and benefits of behavior patterns. When studying ultimate causes, we may ask what the adaptive value of a particular behavior might be. An understanding of behavior requires consideration of both proximate and ultimate causes.

Behaviors have benefits and costs

Behavioral ecologists use **cost–benefit analysis** to understand specific behaviors. A behavior may help an animal obtain food or water, protect itself, reproduce, or acquire and maintain territory in which to live. The benefits typically contribute to **direct fitness,** which is an individual's reproductive success, measured by the number of viable offspring it produces. Reproduction is, of course, the key to evolutionary success.

Behaviors also involve costs. For example, while a parent is off hunting for food for its offspring, the youngsters may be killed by predators. Behaviors also have energy costs and costs of missed opportunities to carry out other behaviors. When an animal is out foraging for food, it may be missing opportunities to mate. If the benefits are greater than the costs, a behavior is adaptive.

The ultimate cause of a behavior is to increase the probability that the genes of the individual animal will be passed to future generations. Certain responses may lead to the death of the individual while increasing the chance that copies of its genes will survive through the enhanced production or survival of its offspring or other relatives.

Genes interact with environment

Early biologists debated nature versus nurture, that is, the relative importance of genes compared with environmental experience. They defined **innate behavior** (inborn behavior, popularly referred to as *instinct*) as genetically programmed and **learned behavior** as behavior that has been modified in response to environmental experience. More recently, behavioral ecologists have recognized that no true dichotomy exists. All behavior has a genetic basis. Even the *capacity* for learned behavior is inherited. However, behavior is modified by the environment in which an animal lives; behavior is a product of the interaction between genetic capacity and environmental influences. Thus, behavior begins with an inherited framework that experience can modify.

We can think of a range of behaviors from the more rigidly genetically programmed types through those that, although they have a genetic component, are extensively developed through experience. The sand wasp *Philanthus*, discussed in the chapter introduction, efficiently carries out a sequence of behaviors that depends mainly on genetic programming. How to dig the burrow, how to cover it, how to kill the bees—these behaviors are genetically determined. Yet some of her behavior is learned. There is no way that her ability to locate the burrow could be genetically programmed. Because a burrow can be dug only in a suitable spot, its location must be learned *after* it is dug. When *Philanthus* covers her burrow with sand, she takes precise bearings of its location by circling the area a few times before flying off again to hunt.

Dutch ethologist Niko Tinbergen studied this behavior of *Philanthus*. He surrounded the wasp's burrow with a circle of pine cones as potential landmarks (**FIG. 52-1**). Before the wasp returned with another bee, Tinbergen rearranged or removed the pine cones. Without them, the wasp could not find her burrow. When Tinbergen moved the pine cones to an area where there was no burrow, the female wasp responded as though the burrow were there. When the investigator completely removed them, the female seemed very confused. Only when Tinbergen restored the cones to their original location could the wasp find her burrow.

When Tinbergen substituted a ring of stones for the cones, the wasp responded as though the burrow were in the center of the stones. Thus, *Philanthus* responded to the *arrangement* of the cones rather than to the cones themselves. Tinbergen's findings demonstrated that for *Philanthus*, learning the landmarks surrounding the burrow was critical for burrow locating.

Studies of fruit fly courtship and mating have provided interesting examples of interaction between genes and behavior. A ritual consisting of a complex sequence of steps, almost like a dance, must occur before mating takes place. This courtship ritual involves an exchange of visual, auditory, tactile, and chemical signals between the male and female. J. B. Hall of Brandeis University and his colleagues identified more than a dozen genes controlling these actions, suggesting that courtship behavior is largely inherited and preprogrammed. Nevertheless, the fruit fly has the capacity to learn from experience, a capacity that, of course, is also inherited.

The interaction between genes and environment has been studied in many vertebrates. Several species of the lovebird (*Agapornis*) differ not only in appearance but in behavior. One species uses its bill to transport small pieces of bark for building a nest. Another species tucks nest-building materials under its rump feathers. Hybrid birds attempt to tuck material under their feathers and then try to carry it in their bills, repeating

How does Philanthus *locate her burrow when she returns to provision it with food?*

HYPOTHESIS: *Philanthus* uses visual cues to locate her burrow.

EXPERIMENT:

❶ While *Philanthus* was in the burrow, Tinbergen placed a circle of pine cones around its entrance.

❷ While the wasp was out foraging, Tinbergen moved the cones to a new position (away from the burrow entrance).

❸ In a second experiment, Tinbergen left the pine cones around the burrow but rearranged them in a triangle shape in the wasp's absence. Nearby, he made a circle of stones.

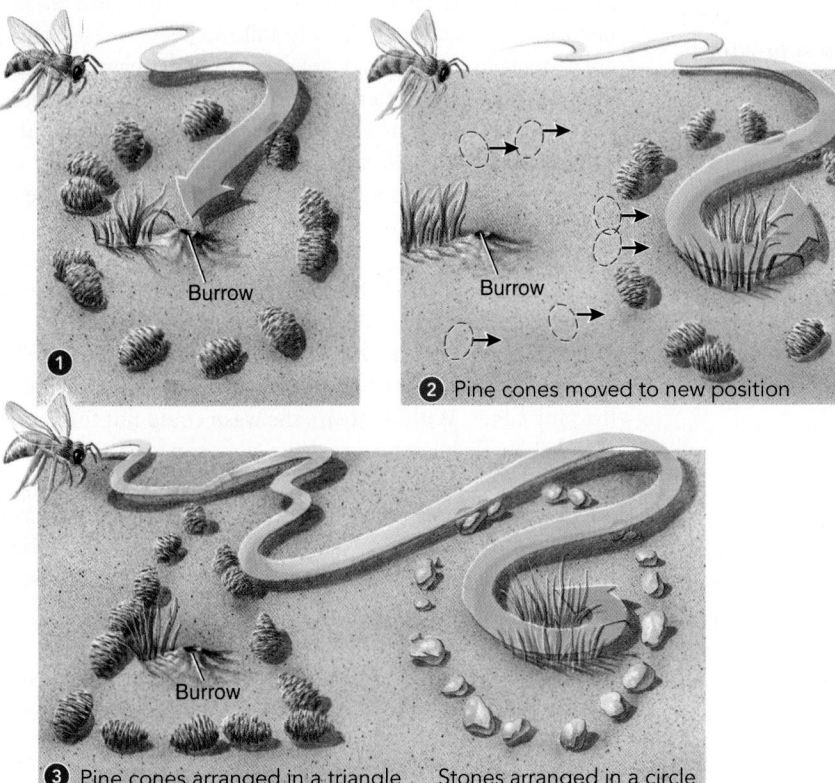

❶ Burrow

❷ Pine cones moved to new position

❸ Pine cones arranged in a triangle   Stones arranged in a circle

Burrow

RESULTS AND CONCLUSION:

❶ When *Philanthus* emerged from her burrow, she responded to the circle of pine cones by circling the site, an orientation flight. When she returned with prey, she flew to the middle of the circle of pine cones to locate the small, concealed entrance to her burrow.

❷ When Tinbergen relocated the circle of pine cones away from the burrow, *Philanthus* flew to the middle of the new circle because she had learned the position of the burrow in relation to the cones.

❸ When Tinbergen rearranged the cones in the shape of a triangle and made a circle of stones nearby, *Philanthus* flew to the middle of the circle of stones.

Philanthus used visual cues (a circle of cones) to learn the location of her burrow. Later she looked for the circle of cones to locate her burrow. She responded to the *arrangement* of cones rather than to the cones themselves. Learning landmarks (visual cues) is critical for locating the burrow.

SOURCE: Tinbergen, N. *The Animal in Its World*, Vol. 1. Harvard University Press, Cambridge, MA, 1972.

Figure 52-1 Niko Tinbergen's experiments determining how the wasp *Philanthus* locates her burrow

PREDICT Imagine that when *Philanthus* first emerged from her nest, the investigator scared her away before she could orient herself to any visual cues. Would she be able to find her burrow?

© Cengage Learning

the pattern several times. Eventually, most of the birds carry the material in their bills, but it takes them up to three years to perfect this behavior, and most continue to make futile attempts to tuck material into their feathers. Thus, the method of transporting materials is inherited but somewhat flexible. (In nature hybrids would likely experience dramatically reduced fitness because of the long delay in breeding onset.)

Behavior depends on physiological readiness

Although behavior involves all body systems, it is influenced mainly by the nervous and endocrine systems. The capacity for behavior depends on the genetic characteristics that govern the development and functions of these systems. Before an animal can exhibit any pattern of behavior, it must be physiologically ready to produce the behavior. For example, breeding behavior does not ordinarily occur among birds or most mammals unless sufficient concentrations of sex hormones are present in their blood. A human baby cannot walk until its neurons and muscles are sufficiently developed. These states of physiological readiness are themselves produced by a continuous interaction with the environment. The level of sex hormones in a bird's blood may be determined by seasonal variations in day length. The baby's muscles develop with experience as well as age.

Several factors influence the development of song in male white-crowned sparrows (generally only male songbirds sing a complex song). These birds show considerable regional variation in their song. During early development, young sparrows normally hear adult males sing the distinctive song of their population. Days 10 to 52 after hatching are a critical period for learning the song. When he is several months old, a young male sparrow "practices" the song over several weeks until he eventually sings in the local "dialect."

In laboratory experiments birds kept in isolation and deprived of the acoustic experience of hearing the song of mature males eventually sing a poorly developed but recognizable white-crowned sparrow song. When a young white-crowned sparrow is permitted to interact socially with a strawberry finch (which belongs to a different genus of birds), it learns the song of the finch rather than its own species-specific song. This learning occurs even if the white-crowned sparrow can hear the song of other sparrows but does not interact with them socially. From these experiments, investigators have concluded that the white-crowned sparrow is hatched equipped with a rough genetic pattern of its song, but social and acoustic stimuli are important in developing the ability to sing its specific song.

Many behavior patterns depend on motor programs

Many behaviors that we think of as automatic depend on coordinated sequences of muscle actions called **motor programs.** Some motor programs—for example, walking in newborn gazelles—seem mainly innate. Others, such as walking in human infants, have a greater learned component.

A classic example of a motor program in vertebrates is egg rolling in the European graylag goose (**FIG. 52-2**). Once activated by a simple sensory stimulus, egg-rolling behavior continues to completion regardless of sensory feedback. There is little flexibility. Behavioral ecologists call this behavior a **behavioral pattern.** Certain behavioral patterns can be elicited by a **sign stimulus,** or *releaser,* a simple signal that triggers a specific behavioral response. A wooden egg is a sign stimulus that elicits egg-rolling behavior in the graylag goose.

In a set of classic experiments performed in 1951, Tinbergen showed that the red stripe on the ventral surface (belly) of a male stickleback fish is a sign stimulus. The ventral surface of the male stickleback is normally silver but turns red at the beginning of the breeding season. The male stickleback selects a territory, builds a nest, and defends his territory against rival males. After a female lays her eggs, she leaves; the male then guards the eggs and, later, the newly hatched fish.

Tinbergen showed that the red stripe triggers aggressive behavior by a male when his territory is being invaded. Tinbergen found that crude models painted with a red belly were more likely to be attacked than more realistic models lacking the red belly (**FIG. 52-3**). Aggressive behavior is triggered by the red belly rather than by recognition based on a combination of features.

CHECKPOINT 52.1

- *In what ways are the behaviors of* Philanthus, *the sand wasp, adaptive?*
- **CONNECT** *Give an example showing that behavior capacity is inherited and is modified by learning.*
- **CONNECT** *How does physiological readiness affect innate behavior? How does it affect learned behavior?*

52.2 LEARNING: CHANGING BEHAVIOR AS A RESULT OF EXPERIENCE

LEARNING OBJECTIVE

3   Discuss the adaptive significance of habituation, imprinting, classical conditioning, operant conditioning, and cognition.

Based on our discussion thus far, it should be clear that much inherited behavior can be modified by experience. **Learning** involves persistent changes in behavior that result from experience. The capacity to learn new responses to new situations is adaptive, enabling animals to survive as their environment changes.

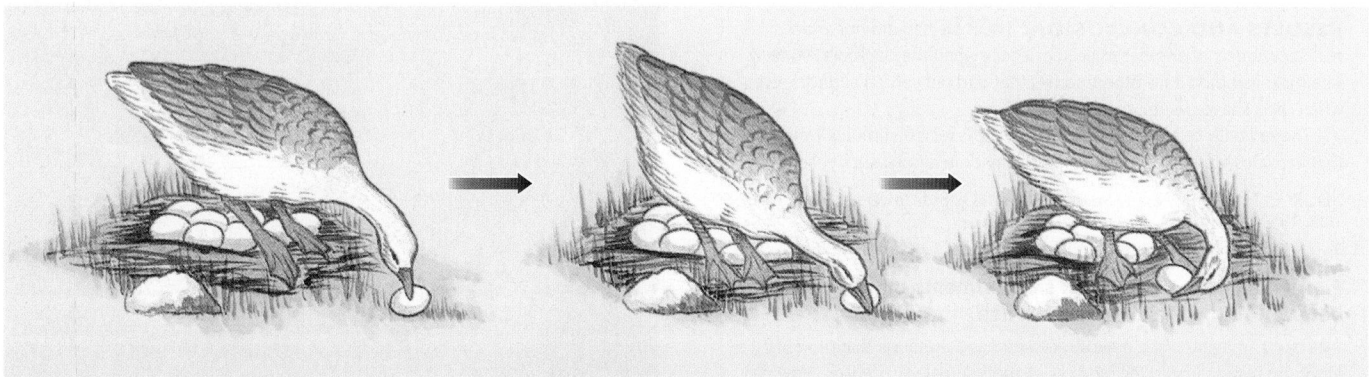

Figure 52-2 Egg-rolling behavior in the European graylag goose (*Anser anser*)

In this rigid behavioral pattern, the goose reaches out by extending her neck and uses her bill to pull the egg back into the nest. If the investigator quickly removes the egg while the goose is in the process of reaching for it or pulling it back, the goose continues the behavior to completion, as though pulling the now absent egg back to the nest.

© Cengage Learning

What triggers aggressive behavior in the male stickleback?

HYPOTHESIS: The red ventral surface (belly) of a male stickleback is a sign stimulus that triggers aggressive behavior in other male sticklebacks.

EXPERIMENT: Niko Tinbergen presented male sticklebacks with two series of models.

❶ The models in the neutral series were the neutral color of the male in the nonbreeding season.

❷ The models in the red series were red on the ventral surface.

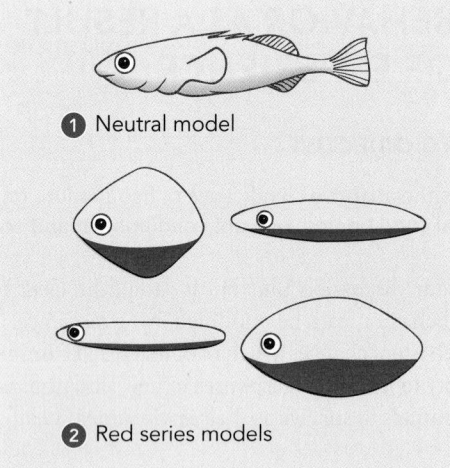

❶ Neutral model

❷ Red series models

RESULTS AND CONCLUSION: The male fish did not attack realistic models of another male stickleback when they lacked a red belly. The male fish did attack other models that had red ventral surfaces, even when their shapes were unrealistic.

The red belly of another male stickleback is a sign stimulus (releaser) that activates stereotyped aggressive behavior in a male stickleback.

SOURCE: Tinbergen, N. *The Study of Instinct.* Oxford University Press, New York, 1951.

Figure 52-3 Niko Tinbergen's experiments on aggressive behavior in the male stickleback fish

PREDICT Imagine that a male stickleback has a mutation that prevents him from responding to the red belly of another male stickleback. How might that affect his direct fitness?

© Cengage Learning

Learning abilities are biased; information most important to survival appears most easily learned. It may have taken a rat a dozen trials to learn the artificial task of pushing a lever to get a

reward. However, the same rat learns from *one* experience to avoid a food that has made it ill. People who poison rats to get rid of them can readily appreciate the adaptive value of this learning ability. Such quick learning in response to an unpleasant experience forms the basis of **warning coloration,** or **aposematic coloration,** which is found in many poisonous insects and brilliantly colored, but distasteful, bird eggs. Once made ill by such a meal, predators quickly learn to avoid them. In the following sections, we consider several types of learning, including habituation, imprinting, classical conditioning, operant conditioning, and cognition.

An animal habituates to irrelevant stimuli

Habituation is a type of learning in which an animal learns to ignore a repeated, irrelevant stimulus, that is, one that neither rewards nor punishes. Pigeons gathered in a city park learn by repeated harmless encounters that humans are not dangerous to them and behave accordingly. This behavior benefits them. A pigeon intolerant of people might waste energy by flying away each time a human approached and might not get enough to eat. Many African animals habituate to humans on photo safari and to the vans that transport them (**FIG. 52-4**). Urban humans habituate to the noise of traffic. In fact, many urban dwellers report that they do not sleep well when they visit a quiet rural area.

Imprinting occurs during an early critical period

Anyone who has watched a mother goose with her goslings must have wondered how she can "keep track" of such a horde of almost identical little creatures tumbling about in the grass, let alone distinguish them from those of another goose (**FIG. 52-5**). Although she can recognize her offspring to some extent,

McMurray Photography

Figure 52-4 Habituation

After repeated safe encounters with vans transporting humans on photo safari, many animals, including giraffes, zebras, antelope, and lions in the Serengeti, learn to ignore them. Elephants typically ignore the vans unless the driver provokes them by moving too close. In that event an elephant may challenge and even charge the van.

Figure 52-5 Imprinting
Parent–offspring bonds generally form during an early critical period. The goslings have imprinted on their mother, an Egyptian goose (*Alopochen aegyptiacus*). Photographed in Ndutu, Tanzania, Africa.

Doug Cheeseman/Getty Images

basically *they* have the responsibility of keeping track of *her*. The survival of a gosling requires that it quickly learn to discriminate its mother (care provider) from other geese.

Imprinting, a type of social learning based on early experience, has been studied in some mammals as well as birds. It occurs during a *critical period,* usually within a few hours or days after birth (or hatching). Konrad Lorenz, an Austrian physician and early ethologist, discovered that a newly hatched bird imprints on the first moving object it sees, even a human or an inanimate object such as a colored sphere or light. Although the process of imprinting is genetically determined, the bird *learns* to respond to a particular animal or object.

Among many types of birds, especially ducks and geese, older embryos can exchange calls with their nest mates and parents through the porous eggshell. When they hatch, at least one parent is normally on hand, emitting the characteristic sounds with which the hatchlings are already familiar. If the parent moves, the chicks follow. This movement and the sounds produce imprinting. During a brief critical period after hatching, the chicks learn the appearance of the parent.

Imprinting in some mammals depends on scent. Baby shrews, for example, become imprinted on the odor of their mother (or any female nursing them). In many species the mother also learns to distinguish her offspring during a critical period. In some species of hoofed mammals, such as sheep, the mother will accept her offspring for only a few hours after its birth; if the mother and her offspring are kept apart past that time, the young are rejected. Normally, the mother learns to distinguish her own offspring from those of others by olfactory cues.

In classical conditioning a reflex becomes associated with a new stimulus

In **classical conditioning** an association forms between some normal body function and a new stimulus. If you have observed dog or cat behavior, you know that the sound of a

Can an animal learn to respond in a certain way to a new (previously neutral) stimulus by associating it with a familiar stimulus?

HYPOTHESIS: A dog can be conditioned to salivate at the sound of a bell.

EXPERIMENT: Ivan Pavlov presented food (the unconditioned stimulus) to a dog and observed that the dog salivated (unconditioned response). He then rang a bell (conditioned stimulus) just before he gave food to the dog. Pavlov repeated this pairing several times. Then he sounded the bell in the absence of food.

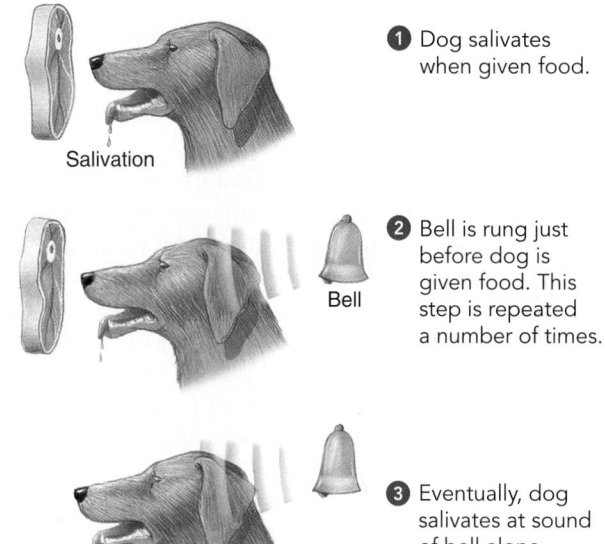

Salivation

❶ Dog salivates when given food.

Bell

❷ Bell is rung just before dog is given food. This step is repeated a number of times.

❸ Eventually, dog salivates at sound of bell alone.

RESULTS AND CONCLUSION: The dog became conditioned to salivate at the sound of the bell alone. When a bell (conditioned stimulus) is paired with food (an unconditioned stimulus), an animal learns to associate them. The animal then salivates to the sound of the bell alone (conditioned response).

SOURCE: Pavlov, I. P. *Conditioned Reflexes: An Investigation of the Physiological Activity of the Cerebral Cortex* (translated by G. V. Anrep). Oxford University Press, London, 1927.

Figure 52-6 Classical conditioning: Pavlov's dog
PREDICT A young boy was bitten by a dog. He was startled, scared, and hurt, and he cried. A few weeks later, he sees another dog. How do you think he is likely to respond?
© Cengage Learning

can opener at dinnertime can captivate a pet's attention! Early in the 20th century, Ivan Pavlov, a Russian physiologist, discovered that if he rang a bell just before he fed a dog, the dog learned to associate the sound of the bell with food. Eventually, the dog salivated even when the bell was rung in the absence of food (**FIG. 52-6**).

Pavlov called the physiologically meaningful stimulus (food, in this case) the *unconditioned stimulus*. The normally irrelevant stimulus (the bell) that became a substitute for it was the *conditioned stimulus*. Salivating to the food was an *unconditioned response,* whereas salivating to the bell was a *conditioned response.* Because a dog does not normally salivate at the sound of a bell, the association was clearly learned. It could also be forgotten. If the bell no longer signaled food, the dog eventually stopped responding to it. Pavlov called this latter process **extinction.** Classical conditioning can be adaptive. For example, if an animal is threatened by a predator in a particular area, it associates the place with danger and avoids that area in the future. In this case the animal learns to *avoid* a negative stimulus rather than to expect a reward.

In operant conditioning spontaneous behavior is reinforced

In **operant conditioning** the animal must do something to gain a reward (positive reinforcement) or avoid punishment. Operant conditioning has been studied in many animals, including flatworms, insects, spiders, birds, and mammals. In a typical laboratory experiment, a rat is placed in a cage containing a movable bar. When random actions of the rat result in pressing the bar, a pellet of food rolls down a chute to the rat. Thus, the rat is positively reinforced for pressing the bar. Eventually, the rat learns the association and presses the bar to obtain food.

In negative reinforcement removal of a stimulus increases the probability that a behavior will occur. For example, a rat may be subjected to an unpleasant stimulus, such as a low-level electric shock. When the animal presses a bar, this negative reinforcer is removed.

Many variations of these techniques have been developed. A pigeon may be trained to peck at a lighted circle to obtain food, a chimpanzee may learn to perform some task to get tokens that can be exchanged for food, and children may learn to sit quietly in their seats at school to obtain the teacher's praise.

Operant conditioning plays a role in the development of some behaviors that appear genetically programmed. An example is the feeding behavior of gull chicks. Herring gull chicks peck the beaks of the parents, which stimulates the parents to regurgitate partially digested food for them. The chicks are attracted by two stimuli: the general appearance of the parent's beak with its elongated shape and distinctive red spot, and its downward movement as the parent lowers its head. Like the rat's chance pressing of the bar, the chicks' initial exploratory pecking behavior is sufficiently functional for them to get their first meal, but they waste a lot of energy in pecking. Some pecks are off target and are therefore not rewarded. However, pecking behavior becomes more efficient with practice. Thus, a behavior that might appear to be entirely genetic is improved by learning (**FIG. 52-7**).

Figure 52-7 Operant conditioning

When the gull chick pecks its parent's beak, the chick is positively reinforced with food. After many tries, the chick learns to be more accurate in begging for food from its parent.

Animal cognition is controversial

Cognition is the process of gaining and using knowledge; it includes thinking, processing information, learning, reasoning, and awareness of thoughts, perceptions, and self. Cognition allows us to carry out high-level mental functions. For example, we can make difficult decisions; solve problems, including mathematical problems; plan; and develop and comprehend languages. How closely animal cognition approximates human thought processes has been hotly debated and is the focus of a great deal of research.

Cognition allows some animals to *problem solve*. For example, a chimpanzee presented with a new problem is likely to have the insight to make new associations from tasks it has learned previously. The chimp can adapt prior knowledge and experience to solve the current problem (**FIG. 52-8**). Primates are especially skilled at cognition and problem solving, but some other mammals and a few birds, notably some crows (e.g., ravens), also appear to problem solve to some degree.

Play may be practice behavior

Perhaps you have watched a kitten pouncing on a dead leaf or practicing a carnivore neck bite or a hind-claw disemboweling stroke on a littermate without causing injury. Many animals, especially young birds and mammals, appear to practice

Figure 52-8 Solving a problem

Confronted with the problem of cracking a nut, a female bonobo (*Pan paniscus*; also known as the dwarf chimpanzee) uses a rock as a tool.

adult patterns of behavior while they play (**FIG. 52-9**). They may improve their ability to escape, hunt, kill prey, or perform sexual behavior. Play may be an example of operant conditioning in action. Hypotheses for the ultimate causes of play behavior include exercising, learning to coordinate movements, and learning social skills. Of course, there could be other unknown explanations.

CHECKPOINT 52.2

- **CONNECT** *How is imprinting adaptive?*
- **CONNECT** *How is operant conditioning adaptive?*

Figure 52-9 Cheetah cubs playing

Play may serve as a means of practicing behavior that will be useful later in life, such as when hunting, fighting for territory, or competing for mates. These cheetah cubs (*Acinonyx jubatus*) were photographed in Kenya, Masai Mara National Reserve.

52.3 BEHAVIORAL RESPONSES TO ENVIRONMENTAL STIMULI

LEARNING OBJECTIVES

4 Give examples of biological rhythms and describe some of the physiological mechanisms responsible for them.
5 Analyze costs and benefits of migrations and distinguish between directional orientation and navigation.

Think of the many variations that take place in our natural surroundings: cycles of day and night, changes in length of day with changing seasons, lunar cycles, high and low tides, changes in weather, and variations in amount and distribution of food. Biologically successful animals are physiologically and behaviorally adapted to the continuous changes that take place in their environment.

Biological rhythms regulate many behaviors

The biochemical, physiological, and behavioral responses that animals make to the periodic changes in the environment are referred to as **biological rhythms.** Biologists have identified many types of biological rhythms, including daily, monthly, and annual rhythms. Among many animals, physiological cycles such as body-temperature fluctuations and hormone secretion are rhythmic. Human body temperature, for example, follows a typical daily curve, with the lowest temperature early in the morning and the highest in the late afternoon. (Women also have a monthly temperature rhythm, with the highest body temperature during the last half of the menstrual cycle.) Biological rhythms regulate many behaviors, including activity, sleep, feeding, drinking, reproduction, and migration.

The behavior of many animals, like the activities of many plants, is organized around **circadian rhythms** (circadian means "approximately one day"), which are daily, 24-hour cycles of activity. **Diurnal animals,** such as honeybees and pigeons, are most active during the day. Most bats, moths, and cats are **nocturnal animals** and are most active during the hours of darkness. **Crepuscular animals,** like many mosquitoes and fiddler crabs, are busiest at dawn, dusk, or both. Generally, there are ecological reasons for these patterns. If an animal's food is most plentiful in the early morning, for example, its cycle of activity must be regulated so that it becomes active shortly before dawn.

Some biological rhythms of animals reflect the *lunar cycle* (moon). The most striking rhythms are those in marine organisms that are attuned to changes in tides and phases of the moon. For instance, a combination of tidal, lunar, and annual rhythms governs the reproductive behavior of the grunion, a small fish that lives off the Pacific coast of North America. Grunions swarm from April through June on those three or four nights when the highest tides of the year occur. At precisely the high point of the tide, the fish squirm onto the beach and deposit eggs and sperm in the sand. They return to the sea in the next wave. By the time the next tide reaches that portion of the beach 15 days later, the young fish have hatched in the damp sand and are ready to enter the sea. This synchronization may help protect fish eggs from aquatic predators.

An animal's metabolic processes and behavior are typically synchronized with the cyclic changes in its external environment. Its behavior anticipates these regular changes. The little fiddler crabs of marine beaches often emerge from their burrows at low tide to engage in social activities such as territorial disputes. To avoid being washed away, they must return to their burrows before the tide returns. How do the crabs "know" that high tide is about to occur? One might guess that the crabs recognize clues present in the seashore. However, when the crabs are isolated in the laboratory away from any known stimulus that could relate to time and tide, their characteristic behavioral rhythms persist.

Many biological rhythms are regulated by *internal* timing mechanisms that serve as **biological clocks.** As illustrated by the fiddler crabs, these timing mechanisms do not simply respond to environmental cues but are capable of sustaining biological rhythms independently. Such internal clocks have been identified in almost every eukaryote as well as in some prokaryotes. Molecular biologists have demonstrated that genes control biological clocks. In *Drosophila* seven genes produce clock proteins that seem to interact in feedback loops in many cell types both inside and outside the nervous system. Researchers have identified similar genes and proteins in many animal groups, including mammals. The principal clock is located in specific areas of the central nervous system.

In mammals the master clock is located in the **suprachiasmatic nucleus (SCN)** in the hypothalamus (see Chapter 42). This SCN clock generates approximately 24-hour cycles even without input from the environment. However, the SCN-generated cycles are normally adjusted every day based on visual input received from the retina. Signals from the retina reflecting changes in light intensity adjust internal circadian rhythms to light–dark cycles in the environment. The SCN sends rhythmic signals, in the form of neuropeptides, to the **pineal gland,** an endocrine gland located in the brain. In response, the pineal gland secretes *melatonin,* a hormone that promotes sleep in humans. Clock genes in the SCN are turned on and off by the very proteins they encode, setting up feedback loops that have a 24-hour cycle.

In addition to the master clock, most cells have timing mechanisms that use many of the same clock proteins. These peripheral clocks may function independently of the SCN.

Environmental signals trigger physiological responses that lead to migration

Ruby-throated hummingbirds cross the vast distance of the Gulf of Mexico twice each year. The sooty tern travels across the entire South Atlantic from Africa to reach its tiny island breeding grounds south of Florida. Birds, butterflies, fishes, sea turtles, wildebeest, zebras, and whales are among the many animals that travel long distances. Behavioral ecologists define **migration** as a periodic long-distance travel from one location to another. Many migrations involve astonishing feats of endurance and navigation.

Why do animals migrate? Ultimate causes of migration apparently involve the advantages of moving from an area that seasonally becomes less hospitable to a region more likely to support reproduction or survival. Seasonal changes include shifts in climate, availability of food resources, and safe nesting sites. For example, as winter approaches, many birds migrate to a warmer region.

Proximate causes of migration include signals from the environment that trigger physiological responses leading to migration. In migratory birds, for example, the pineal gland senses changes in day length and then releases hormones that cause restless behavior. The birds show an increased readiness to fly and to fly for longer periods.

Migration has costs and benefits An interesting example of migration is the annual journey of millions of monarch butterflies (*Danaus plexippus*) from Canada and the continental United States to Mexico, a journey of 2520 km (about 1520 mi) for some. Their dramatic annual migration appears related to the availability of milkweed plants on which females lay their eggs. On hatching, the larvae (caterpillars) feed on the leaves of the milkweed plant. In cold regions these plants die in late autumn and grow again with the warmer weather of spring.

The wintering destinations of monarchs also may be determined by temperature and humidity. Possible benefits include the opportunity to winter in an area that offers an abundant food supply, nonfreezing temperatures, and moist air. Thus, the investment made by migrating monarch butterflies in their long journey increases their chances of surviving the winter.

The benefits of migration are not without costs in time, energy, and even survival. Many weeks may be spent each year on energy-demanding journeys. Some animals may become lost or die along the way from fatigue or predation. When in unfamiliar areas, migrating individuals are often at greater risk from predators.

In recent years human activities have interfered with migrations of many kinds of animals. For example, after millions of years of biological success, survival of sea turtles is threatened by fishing and shrimping nets in which they become entangled. As a result, thousands of migrating turtles drown each year. Sea turtles also die as a result of contamination from oil spills, entanglement in discarded fishing line, and ingestion of floating plastic bags that they mistake for jellyfish.

Animals find their way using directional orientation and navigation How do migrating animals find their way? *Directional orientation* refers to travel in a specific direction. To travel in a straight line toward a destination requires a sense of direction, or **compass sense.** Many animals use the sun to orient themselves. Because the sun appears to move across the sky each day, an animal must have a sense of time. Biological clocks seem to sense time and regulate circadian rhythms.

Navigation is more complex than traveling in a specific direction. Navigation requires both compass sense and **map sense,** which is an "awareness" of location. Navigation involves

Figure 52-10 Navigation by light and magnetic field
To study turtle navigation, researcher Kenneth Lohmann harnessed leatherback turtles (*Dermochelys coriacea*), such as this hatchling, and wired them to a computer that recorded their swimming direction.

the use of cues to change direction when necessary to reach a specific destination. When navigating, an animal must integrate information about distance and time, as well as direction.

DNA tests confirm that loggerhead sea turtles that hatch on Florida beaches along the Atlantic swim hundreds of miles across the ocean to the Mediterranean Sea, an area rich in food. Several years later, those that survive to become adults mate, and the females navigate back, often to the same beach, to lay their eggs. Their journey requires both compass sense and map sense.

Marine biologist Kenneth Lohmann of the University of North Carolina fitted hatchling turtles with harnesses connected to a swivel arm in the center of a large tank (FIG. 52-10). A computer connected to the swivel arm recorded the turtles' swimming movements. By manipulating light and magnetic fields, Lohmann demonstrated that both these environmental cues are important in turtle migration. The turtles swam toward the east until Lohmann reversed the magnetic field. The turtles then reversed their direction to swim toward the new "magnetic east," which was now actually west. Subsequent research suggests that young turtles use wave direction to help set magnetic direction preference.

Some biologists suspect that turtles also use their sense of smell to guide them, particularly to guide the females to the very same beach to lay their eggs. Similarly, adult salmon use the unique odors of different streams to help find their way back to the same stream from which they hatched.

Birds and some other animals that navigate by day rely on the position of the sun (our local star); those that travel at night use the stars to guide them. When birds see the star patterns of the night sky, they fly in the appropriate migration direction for their species, even when they have had no opportunity to learn this feat from other birds.

Studies by Stephen Emlen of Cornell University showed that young indigo buntings learn the constellations using the position of the North Star as their reference point. (Other stars in the Northern Hemisphere appear to rotate about the North Star.) When Emlen rearranged the constellations in a planetarium sky, birds in the planetarium learned the altered patterns of stars and later attempted to fly outside in a direction consistent with the

artificial pattern. These and other studies suggested that birds have a genetic ability to learn constellation patterns and use them to orient during migration.

Investigators have long observed that some species of birds navigate even when the sky is overcast and they cannot see the stars. They hypothesized that these birds use Earth's magnetic field to navigate. Studies of garden warblers indicate that as these birds make their way from central Europe to Africa each winter, they navigate both by the stars and by Earth's magnetic field. The warblers use stars to determine the general direction of travel and then use magnetic information to refine and correct their course. In addition to birds and sea turtles, honeybees, some fishes, and some other animals are sensitive to Earth's magnetic field and use it as a guide.

CHECKPOINT 52.3

- CONNECT *Why is it adaptive for some species to be diurnal but for others to be nocturnal or crepuscular?*
- *What is the difference between directional orientation and navigation?*

52.4 FORAGING BEHAVIOR

LEARNING OBJECTIVE

6   Define *optimal foraging behavior* and explain how it is adaptive.

Feeding behavior, or **foraging,** involves locating and selecting food as well as gathering and capturing food. Some behavioral ecologists study the costs and benefits of searching for and selecting certain types of food as well as the mechanisms used to locate prey. For example, many camouflage strategies have evolved that make potential prey difficult to detect. As predators experience repeated success in locating a particular prey species, they are thought to develop a *search image,* a constellation of cues that help them identify hidden prey.

Why do grizzly bears spend hours digging arctic ground squirrels out of burrows while ignoring larger prey such as caribou? It is more energy efficient to dig for squirrels because the bears' efforts most probably will be rewarded, whereas caribou are more likely to escape, leaving the bears hungry. This behavior is an example of **optimal foraging,** the most efficient way for an animal to obtain food. According to *optimal foraging theory,* when animals forage, they maximize their energy intake per unit of foraging time. When animals maximize energy obtained per unit of foraging time, they may maximize their reproductive success. Many factors, such as avoiding predators while foraging, must be considered in determining efficient or optimal strategies.

In habitats where the most-preferred food items are abundant and an animal does not have to travel far to obtain them, animals can afford to be selective. In contrast, the optimal strategy in poor habitats, where it takes longer to find the best food items, is to select less-preferred items. Animals may learn to forage efficiently through operant conditioning, that is, by randomly trying various strategies and selecting the one associated with the most rewards (and the fewest hunger pangs at the end of the day).

Foraging is also affected by the forager's risk of predation. British biologist Guy Cowlishaw studied foraging behavior in a population of baboons in Namibia. He found that the baboons spent more time foraging in a habitat in which food was relatively scarce than in an area with more-abundant food that had a high risk of predation by lions and leopards.

Lions live, and often forage, in social units called *prides*. A pride typically consists of a group of related adult females, their cubs, and unrelated males. Biologist Craig Packer of the University of Minnesota and his research team studied lion behavior in Serengeti National Park, Tanzania. Packer radio-collared at least one female in each of 21 prides and tracked them for several years. During the season when prey was abundant, lions hunted wildebeest, gazelle, and zebra that migrated into the area. When these herds migrated out of the area during the dry season, these prey become scarce, and lions fed mainly on warthog and Cape buffalo.

Packer reported that when prey was abundant, the size of the foraging group had little effect on daily food intake. Whether a lion hunted individually or in small (two to four females) or large (five to seven) groups, food was captured by individuals or groups of any size. During the season when prey was scarce, however, lions were more successful if they hunted alone or in large groups (**FIG. 52-11**). Nevertheless, Packer's data from radio-collared lions indicated that females typically foraged in as large a group as they could, even if the group consisted of three or four females, an approach that decreased foraging efficiency. Thus, selection pressure seems to be stronger to protect cubs and defend territory against larger prides than to forage maximally (alone).

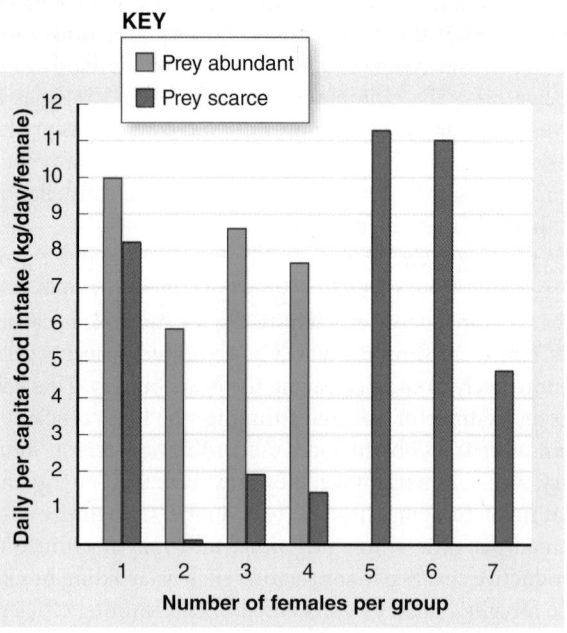

KEY
- Prey abundant
- Prey scarce

Figure 52-11 Optimal foraging and group size in lions

During times of prey scarcity, optimal foraging is related to group size: lions hunt most successfully alone or in large groups of five to seven. Observations of group size suggest that the benefits of optimal defense often outweigh the benefits of optimal foraging. (Based on data from Packer, C., D. Scheel, and A. E. Pusey. "Why Lions Form Groups: Food Is Not Enough." *American Naturalist*, Vol. 136, July 1990.)

- *What is optimal foraging?*
- **CONNECT** *How is optimal foraging adaptive?*

52.5 COSTS AND BENEFITS OF SOCIAL BEHAVIOR

LEARNING OBJECTIVES

7 Analyze social behavior in terms of costs and benefits, and compare different types of social groups, including societies of social insects.

8 Describe common modes of animal communication, including signaling by pheromones.

9 Compare the costs and benefits of dominance hierarchies and of territoriality.

The mere presence of more than one individual does not mean that a behavior is social. Many factors of the physical environment bring animals together in *aggregations*, but whatever interaction they experience may be circumstantial. A light shining in the dark is a stimulus that draws large numbers of moths, and the high humidity under a log attracts wood lice. Although these aggregations may have adaptive value, they are not truly social because the animals are not communicating with one another.

We can define **social behavior** as the interaction of two or more animals, usually of the same species. In the same way that natural selection led to the evolution of behaviors in response to environmental factors, natural selection has resulted in the evolution of social behavior. In this chapter we examine several familiar types of social behavior, including communication, living together, dominance behaviors, territoriality, mating rituals, parenting, and helping behavior.

An organized **society** is an actively cooperating group of individuals belonging to the same species and often closely related. Many insects, such as tent caterpillars, which spin communal nests, cooperate socially. A school of fish, a flock of birds, a pack of wolves, and a hive of bees are other examples of societies. Some societies are loosely organized, whereas others are elaborately structured.

Many animals benefit from living in groups. Schools of fishes tend to confuse predators, so individuals within the school may be less vulnerable to predators than a solitary fish would be (see Fig. 53-1b). Zebra stripes appear to be a visual antipredator adaptation. When viewed from a distance, the stripes tend to visually break up the form of the animal so that individuals cannot be distinguished. A herd of zebras confuses predators, whereas a solitary animal is easy prey for lions, cheetahs, or spotted hyenas (**FIG. 52-12**).

Social foraging is an adaptive strategy used routinely by many animal species. A pack of wolves has greater success in hunting than an individual wolf would have. Among some birds of prey, such as certain hawks, group hunting locates prey more quickly than hunting alone.

Figure 52-12 Social behavior in zebras

The social unit consists of a fairly stable group of females with young and a dominant stallion. The stripes of a group of zebras confuse predators.

(a) A male spring peeper frog (*Hyla crucifer*) calling to locate a mate.

Social behavior has benefits in addition to defense against predators and in some cases more successful foraging. Living together provides opportunities for interfering with reproduction of competitors, such as destroying their eggs. Social living also provides the opportunity to receive help from others in the group. Thus, for many types of animals, social behavior increases direct fitness. However, social behavior also has certain costs. Living together means increased competition for food and habitats, increased risk of attracting predators, and increased risk of transmitting disease. In addition, time and energy must often be invested in gaining and maintaining social status.

Communication is necessary for social behavior

One animal can influence the behavior of another only if the two of them can exchange mutually recognizable signals (**FIG. 52-13**). *Communication* is most evident when one animal performs an act that changes the behavior of another. Animals may communicate to hold a group together, warn of danger, signal social status, indicate willingness to accept or provide care, identify members of the same species, or indicate sexual maturity or readiness to mate. Communication may also be important in finding food, as in the elaborate dances of honeybees. Animals use electrical, tactile, visual, auditory, and chemical signals to transmit information to one another.

Certain fishes (gymnotids) emit electric pulses that generate electric fields in the water. These electric pulses can be used for navigation and communication, including territorial threat, in a fashion similar to bird vocalization. As Harvard sociobiologist Edward O. Wilson said, "The fish, in effect, sing electrical songs."

Examples of animal touch include grooming and sharing food. Social insects, such as ants, groom and stroke one another with their antennae and mouthparts. Many types of bird chicks stimulate their parents to feed them by pecking at their beaks.

(b) Communicating with language. Chimpanzee Tatu (*top*) is signing "food" to Washoe (*below*), the first chimp to learn American Sign Language from a human. Washoe then taught other chimps how to sign.

Figure 52-13 Animal communication

Visual signals can be transmitted quickly Visual signals are fast, communicate a great deal of information, and indicate the location and position of the animal sending the signal. An animal's own body, such as its coloration, may serve as the visual message. Numerous species communicate with visual displays that serve as social signals. For example, many birds present visual displays along with their songs.

Auditory signals can be conveyed in the dark Most animals cannot communicate effectively with visual signals when it is dark. However, animals can hear at night as well as during the day. Auditory signals can also be transmitted over longer distances than visual signals. Sound can move in all directions at once, and an auditory signal can be rapidly terminated. Most animals that communicate with auditory signals use **calls,** short, simple sounds. Birds and mammals use more complex auditory signals. In many bird species, males announce their presence and willingness to interact socially by singing. Some birds respond to a particular sound by matching it.

Among nonhuman mammals, only bottlenose dolphins are known to match sounds in communicating. These dolphins respond to the whistle of a member of its own species by imitating and emitting the same sound. (Researchers suggest that vocal matching was important in the evolution of human language.)

Orcas (formerly known as killer whales) communicate by sounds and songs. Members of a given pod (social group) have an average of 12 different calls, which vary in pitch and duration and appear to reflect their "emotional" state.

Some animals communicate by scent Dogs and wolves mark territory by frequent urination. Antelopes, deer, and cats rub facial gland secretions on conspicuous objects in their vicinity and urinate on the ground. **Pheromones** are chemical signals secreted into the environment that convey information between members of a species. These small, volatile molecules provide a simple, widespread means of communication. Animals use pheromones to communicate danger, ownership of territory, and availability for mating. For example, female moths release pheromones that attract males. Ants mark their trails with pheromones.

Most pheromones elicit a very specific, immediate, but transitory type of behavior. Others trigger hormonal activities that result in slow but longer-lasting responses. Still other pheromones may elicit both transitory and longer-lasting responses. An advantage to pheromone communication is that relatively little energy is needed to synthesize the simple, but distinctive, organic compounds involved. Members of the same species have receptors that fit the molecular configuration of the pheromone; other species usually ignore it or do not detect it at all.

Pheromones are effective in the dark, can pass around obstacles, and last for several hours or longer. A major disadvantage of pheromone communication is that pheromones are transmitted slowly. Also, their information content is limited. However, some animals secrete several pheromones with different meanings.

Pheromones are important in attracting the opposite sex and in sex recognition in many species. Many social insects use these chemical signals to regulate reproduction. Humans have taken advantage of some sex-attractant pheromones to help control pests such as gypsy moths. They lure the males to traps baited with synthetic versions of female gypsy moth pheromones.

In honeybees certain fatty acids are mixed with hydrocarbons from wax glands, and this mixture is transferred onto worker bees when they touch the comb. This mixture serves as a pheromone that identifies all the bees that belong to a particular hive. Should a bee from another hive attempt to enter, guard bees sting and may even kill the foreigner.

In vertebrates pheromones affect sexual cycles and reproductive behavior, including choice of a mate. Pheromones may also play a role in defending territory. Among some mammals, an ovulating female (one physiologically ready to mate) releases pheromones as part of her vaginal secretion. When males detect these chemical odors, their sexual interest increases. When the odor of a male mouse is introduced among a group of females, the reproductive cycles of the female mice synchronize. In some species of mice, the odor of a strange male, a sign of high population density, causes a newly impregnated female to abort.

The extent to which humans respond to pheromones is the focus of research. One interesting finding suggests that an unconsciously perceived body odor can synchronize the menstrual cycles of women who associate closely (e.g., college roommates or prison cell mates). Human steroids used in commercial fragrances such as perfumes may act as subtle signals that modulate mood and behavior.

As discussed in Chapter 43, mammals detect pheromones with specialized chemoreceptor cells that make up the **vomeronasal organ** in the epithelium of the nose. The vomeronasal sensory neurons signal the amygdala and hypothalamus, brain structures that regulate emotional responses and certain endocrine processes. When neurons of the vomeronasal system are damaged in virgin mice, they do not mate. Biologists have identified about 100 genes that code for pheromone receptors in the mouse and rat. These receptors initiate **signal transduction** pathways that involve G proteins.

Dominance hierarchies establish social status

In the spring female paper wasps awaken from hibernation and begin to build a nest together. During the early course of construction, a series of squabbles among the females takes place in which the combatants bite one another's bodies or legs. Finally, one of the wasps emerges as dominant. After that, she is rarely challenged. This queen wasp spends more and more time tending the nest and less and less time out foraging for herself. She takes the food she needs from the others as they return.

The queen then begins to take an interest in raising a family, her family. Because she is almost always in the nest, she can prevent other wasps from laying eggs in the brood cells by rushing at them, jaws agape. Because of her supreme dominance, the queen can bite any other wasp without serious risk of retaliation.

The other wasps of the nest are further organized into a **dominance hierarchy,** a ranking of social status in which each wasp has more status than the wasps that are lower in rank. Wasps lower in the hierarchy are subordinate to those above them, as follows:

queen $\longrightarrow$ wasp A $\longrightarrow$ wasp B $\longrightarrow$ ... wasp M

Once a dominance hierarchy is established, little or no time is wasted in fighting. When challenged, subordinate wasps communicate with submissive poses that inhibit the queen's aggressive behavior. Consequently, few or no colony members are lost through wounds sustained in fighting one another. This social organization ensures greater reproductive success for the colony.

In many species males and females have separate dominance systems. For example, both female and male chimpanzees establish dominance hierarchies. Some females apparently become dominant through aggressive behavior, whereas others achieve high rank by virtue of their mother's status. Dominant females are more successful reproductively than those with lower social status. In many monogamous animals, especially birds, the female acquires the dominance status of her mate by virtue of their relationship and the male's willingness to defend her.

Like many fishes and some invertebrates, certain coral reef fishes (labrids) are capable of sex reversal. The largest, most dominant individual of a group is always male, and the remaining fish within his territory are all female. If the male dies or is removed, the most dominant female becomes the new male. Should any harm come to him, the next-ranking female undergoes sex reversal, takes charge, and protects the group territory. Other fishes exhibit the reverse behavior, in which the most dominant fish is always female. In such species size is less important for aggressive defense than for maximal egg production.

In establishing dominance males may expend energy in posturing, roaring, leaping about, or sometimes fighting fiercely. These behaviors appear to be a test of male quality. The male with the greatest endurance will likely gain dominance and will have the greatest opportunity to mate and perpetuate his genes. In many animals social dominance is a function of aggressiveness (**FIG. 52-14**). Tremendous energy is often needed for the fights engaged in by some birds and many mammals, including baboons, rams, and elephant seals. Establishing dominance is often strenuous and dangerous.

Sex hormones, particularly the male reproductive hormone testosterone, increase aggressiveness. The female hormone estrogen sometimes reduces dominance. Among chickens, the rooster is the most dominant. If a hen receives testosterone injections, her place in the dominance hierarchy shifts upward. When male rhesus monkeys are dominant, their testosterone levels are much higher than when they have been defeated. Not only can testosterone increase dominance, but dominance may even increase testosterone production. It is not always easy to determine cause and effect. (Studies suggest that the effects of testosterone on human mood and behavior may be different from those in other vertebrates.)

Social status affects many organs and systems, including brain structure. For example, dominant male rats have more new neurons in the hippocampus than subordinate rats or controls.

Many animals defend a territory

Most animals have a **home range,** a geographic area they seldom leave (**FIG. 52-15**). Because the animal has the opportunity to

Paul Brough/Science Source

Figure 52-14 Communicating dominance

This "sacred" baboon (*Papio hamadryas*) bares his teeth and screams in an unmistakable show of aggression, a signal that allows him to establish and maintain dominance. In many social groups, animals send signals that help them establish and maintain dominance.

become familiar with everything in that range, it has an advantage over its competitors, predators, and prey in negotiating the terrain and finding food. Some animals exhibit **territoriality.** They defend a *territory,* a portion of the home range, often against other individuals of their species and sometimes against individuals of other species. Many species are territorial for only part of the year, often during the breeding season, but others are territorial throughout the year. Territoriality has been positively correlated with the availability of needed resources that occur in small areas that can be defended.

Territoriality is easily studied in birds. Typically, the male chooses a territory at the beginning of the breeding season. This behavior results from high sex hormone concentrations in his blood. The males of adjacent territories fight until territorial boundaries become established. Generally, the dominance of a male is directly associated with how close he is to the center of his territory. When close to home, he acts like a lion, but when invading some other bird's territory, he may behave more like a lamb. Sometimes males respect a neutral line, an area between their territories in which neither is dominant.

Birds use songs to announce their territory and songs often substitute for fighting. Songs also announce to eligible females that a propertied male resides in the territory. Typically, male birds take up a conspicuous station, sing, and sometimes display striking patterns of coloration or aerial acrobatics to their neighbors, their rivals, and sometimes their mates.

The costs of territoriality include the time and energy expended in staking out and defending a territory and the risks involved in fighting for it. Benefits often include exclusive rights to food within the territory and greater reproductive success. The males of a lion pride father all the cubs born within that pride. Their ability to defend their territory from invasion

Figure 52-15 Home range

The Cape buffalo (*Syncerus caffer*) lives in herds of several hundred animals. Each herd, such as this one photographed in the Serengeti, has a fairly constant home range. Buffalo protect herd members, especially calves, from potential predators.

McMurray Photography

by strange males ensures their reproductive success. Among many species, animals that fail to establish territories fail to reproduce. Territoriality also tends to reduce conflict among members of the same species and ensures efficient use of environmental resources by encouraging individuals to spread throughout a habitat.

Usually, territorial behavior is related to the specific lifestyle of the animal and to whatever aspect of its environment is most critical to its reproductive success. For instance, seabirds may range over hundreds of square kilometers of open water but exhibit territorial behavior only at crowded nesting sites on an island. The nesting sites are their scarcest resource and the one for which competition is keenest.

Some insect societies are highly organized

Social insects form elaborate societies. Characteristics of a highly organized society include cooperation and division of labor among animals of different sexes, age groups, or castes. An intricate system of communication reinforces the organization of the society. The members tend to remain together and to resist attempts by outsiders to enter the group.

By cooperation and division of labor, some insects construct elaborate nests and raise young by mass-production methods. Bees, ants, wasps, and termites form the most organized insect societies. (The first three of these all belong, not coincidentally, to the same order, Hymenoptera.) Insect societies are held together by a complex system of sign stimuli that are keyed to social interaction, and their behaviors tend to be quite rigid.

The social organization of honeybees has been studied more extensively than that of any other social insect. Instructions for the honeybee society are inherited and preprogrammed, and the size and structure of the bee's nervous system permit only a limited range of behavioral variation. These insects are not automatons, however. Within its inherited limits, the complex bee society can respond with some flexibility to food and other stimuli in the environment.

A honeybee society generally consists of a single adult queen, up to 80,000 worker bees (all female), and at certain times, a few males called drones that fertilize newly developed queens. The queen, the only female in the hive capable of reproduction, deposits about 1000 fertilized eggs per day in the wax cells of a comb.

The composition of a bee society is generally controlled by a pheromone secreted by the queen. It inhibits the workers from raising a new queen and prevents development of ovaries in the workers (FIG. 52-16). If the queen dies or if the colony becomes so large that the inhibiting effect of the pheromone dissipates, the workers begin to feed some larvae special food that promotes their development into new queens.

The queen bee stores sperm cells from previous matings in a seminal receptacle. If she releases sperm to fertilize an egg as it is laid, the resulting offspring is female; otherwise, it is male. Thus, males develop by parthenogenesis from unfertilized eggs and are haploid. Because a drone is haploid, each of his sperm cells has *all* his chromosomes; that is, meiosis does not occur during sperm production. The queen bee stores this sperm throughout her lifetime and uses it to produce worker bees.

The worker bees of a hive are more closely related to one another than would be sisters born of a diploid father. Indeed, they have up to three-fourths of their genes in common. (Assuming the queen bred with a single drone, they share half of the queen's chromosomes and all the drone's chromosomes.) As a consequence, they are more closely related to one another than they would be to their own offspring, if they could have any. (A worker bee's female offspring would have only half of its genes in common with its worker mother.) New queens are also their sisters. Worker bees are therefore more likely to pass on copies of their genes to the next generation by raising these individuals than they would if they were to produce their own offspring.

Division of labor among worker bees is mostly determined by age. The youngest worker bees serve as nurses that nourish larval bees. After about a week as nurse bees, workers begin to produce wax and build and maintain the wax cells. Older workers are foragers, bringing home nectar and pollen. Most worker bees die at the ripe old age of 42 days.

Bees have the most sophisticated mode of communication known among nonmammals. They perform a stereotyped series of body movements called a *dance*. The dance re-enacts, in miniature, the bee's flight to the food. When a honeybee scout locates a rich source of nectar, it apparently observes the angles among the food, hive, and the sun. The forager bee then communicates the direction and distance of the food source relative to the hive (FIG. 52-17). If the food supply is within about 52 m (about 55 yd), the scout performs a *round dance,* which generally excites the other bees and prompts them to fly short distances in all directions from the hive until they find the nectar. If the food is distant, however, the scout performs a *waggle dance,* which follows a figure-eight pattern.

In the 1940s, German zoologist Karl von Frisch pioneered studies in bee communication. He found that the waggle dance conveys information about both distance and direction. During the straight run part of the figure eight, the number and frequency of the waggles indicate the distance. The orientation

Treat Davidson/Science Source

Figure 52-16 Maintaining a complex social structure
Numerous worker honeybees surround their queen (*center*). Workers constantly lick secretions from the queen bee. These secretions, transmitted throughout the hive, suppress the activity of the workers' ovaries.

of the movements indicates the direction of the food source in reference to the position of the sun. For example, if the bee dances straight up, the nectar is located directly toward the sun. If the bee dances 40 degrees to the left of the vertical surface of the comb, the nectar is located 40 degrees to the left of a line between the hive and the sun.

The forager bee typically performs the dance in the dark hive on the vertical surface of the comb. How, then, does its audience "see" the dance? Investigators have shown that the bee produces sounds as it dances. The auditory signals apparently communicate the position of the dancing bee.

(a) Round dance. If the scout performs the round dance, it indicates the food source is close to the hive (within about 52 m).

(b) Waggle dance. If the scout performs the waggle dance, it indicates the food source is more distant from the hive. Distance is further indicated by how many abdominal waggles are performed in the straight part of the dance.

CHECKPOINT 52.5

- *Compare the costs and benefits of living in groups.*
- *What are some advantages of pheromones? some limitations?*
- *What costs and benefits accrue to dominant individuals in a dominance hierarchy? to subordinate individuals?*
- *What are some costs and benefits of territoriality?*
- CONNECT *How does the dance of bees compare with human language?*

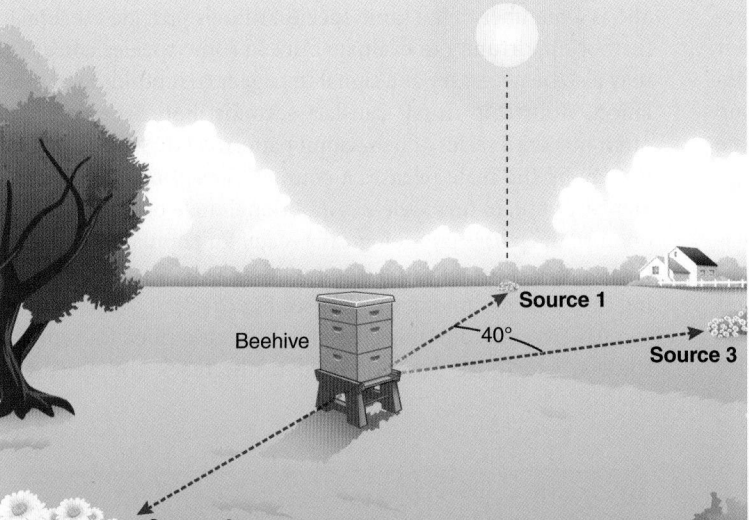

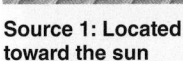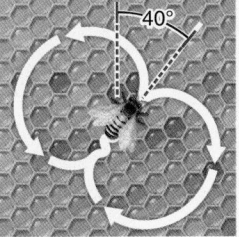

Source 1: Located toward the sun

Source 2: Located opposite the sun

Source 3: Located 40 degrees to the right of the sun

(c) Direction coded in waggle dance. If the scout waggles upward, it indicates the food source is toward the sun. If the scout waggles downward, it indicates the food source is in the opposite direction of the sun. If the scout waggles 40 degrees to the right, it indicates the food source is at a 40° angle to the right of the sun.

Figure 52-17 *Animation* **The honeybee dance**
© Cengage Learning

52.6 SEXUAL SELECTION

LEARNING OBJECTIVE

10 Define *sexual selection* and describe different types of mating systems and approaches to parental care.

Sexual selection is a type of natural selection for successful mating. Animals vary in their ability to compete for mates. Individuals with some inherited advantage, such as males with large size, have a greater chance to mate and pass on their genes to the next generation. The beneficial trait becomes more common in the population over time. Thus, sexual selection results in the reproductive advantage that some individuals have over others of the same species and sex.

For a male, reproductive success depends on how many females he can impregnate. For a female, reproductive success depends on how many eggs she can produce during her reproductive lifetime, on the quality of the sperm that fertilize them, and on the survival of her offspring to reproductive age. Recall that an animal's reproductive success is a measure of its *direct fitness*. We can distinguish between *intrasexual selection* and *intersexual selection*.

Animals of the same sex compete for mates

In **intrasexual selection** individuals of the same sex actively compete for mates. Typically, in such populations abundant males compete for a limited number of receptive females. Competition may take the form of physical combat. Among many species—for example, baboons and elephant seals—the largest male has the advantage.

Often, males establish dominance in more subtle ways than direct competition. Many males use elaborate ornamental displays to establish dominance. For example, the male deer with the largest antlers may discourage his competitors. The male bird with the most dramatic plumage and brightest colors may have a psychological advantage over competing males.

Several studies confirm that males ranking higher in a dominance hierarchy mate more frequently than males that rank lower. However, exceptions demonstrate the complexity of reproductive behavior among some species. For example, among baboons with a definite dominance hierarchy, males lower in the hierarchy copulated with females as frequently as did males of higher status. Investigators observed, though, that dominant males copulated more frequently with females that were in *estrus,* their fertile period.

Animals select quality mates

Females frequently have the opportunity to select a sexual partner from among several males. In **intersexual selection** females select their mates on the basis of some physical trait or some resource offered by the winning suitors. The physical trait typically indicates genetic quality or good health. In many species the success of a male in dominance encounters with other males indicates his quality to the female, and she allows the victorious male to court her. Some lower-ranking males develop alternative strategies for attracting females. For instance, a male may win a female's interest by protecting her baby, even though it is not his own.

Females of some species select their mates based on ornamental displays. A female fish may select the most brightly colored male. Female birds often select males with the showiest plumage and brightest colors, and female deer prefer males that display elaborate antlers. Expression of ornamental traits may

give the female important information about the male's physical condition and ability to fight. The size of the antlers of red deer, for example, may indicate combat effectiveness as well as proper nutrition and good health. Female lions are attracted to males with thick, dark manes, an indicator of good nutrition and plenty of testosterone (which regulates growth of hair and melanin production). Thus, in some species ornaments signal good health and good genes.

Among crickets and many other insect species, a courting male offers a gift of food to a prospective mate. Studies show that the larger or higher quality the food offering is, the better the chances the male will be accepted.

Among some species of insects, birds, and bats, males gather in a small display area called a **lek,** where they compete for females. When a receptive female appears, males may excitedly display themselves and compete for her attention. Among some species, the female selects a male based on his location in the lek rather than on his appearance. The dominant male may occupy a central position and be chosen by most of the females. The female selects a male, mates with him, and then leaves the lek. The male remains to woo other females.

Courtship rituals ensure that the male is indeed a male and is a member of the same species. Rituals provide the female further opportunity to evaluate him. In some species courtship may also be necessary as a signal to trigger nest building or ovulation. Courtship rituals can last seconds, hours, or days and often involve a series of behavioral patterns (**FIG. 52-18**). The first display by the male releases a counter behavior by the female. Her response, in turn, releases additional male behavior, and so on until the pair is physiologically ready for copulation. Specific cues enable courtship rituals to function as reproductive isolating mechanisms among species (see Fig. 20-3).

An extreme courtship ritual has been described for redback spiders, which are closely related to the black widow spider.

(a) The male great frigate bird (*Frigata minor*) inflates his red throat sac in display as part of a courtship ritual. Photographed on Christmas Island in the Pacific.

(b) Egrets (*Egretta rufescens*) performing a mating dance.

Figure 52-18 Courtship rituals
Elaborate courtship rituals help animals determine that a potential mate is of the same species and gives each animal a chance to evaluate the other as a potential mate.

During copulation, the small male spider positions himself above his larger mate's jaws. During 65% of matings, the finale is that the female eats her suitor. The apparent explanation for this behavior is that the risk-taking male is able to copulate for a longer period and thus fertilize more eggs than males that are not cannibalized. Also, the female is more likely to reject additional mates.

Sexual selection favors polygynous mating systems

In most species males make little parental investment in their offspring apart from providing sperm. Males ensure reproductive success by impregnating many females, increasing the probability that their genes will be propagated in multiple offspring. Thus, sexual selection often favors **polygyny,** a mating system in which males fertilize the eggs of many females during a breeding season.

In the mating system known as **polyandry,** one female mates with several males. Benefits may include receiving gifts from several males or enlisting several males' help in caring for the young. Data collected at Jane Goodall's research center at Gombe National Park in Tanzania indicate that female chimpanzees practice polyandry. Females mate with males of their own group, but when they are most fertile, many females slip off to neighboring communities and copulate with less familiar males. Perhaps these sexual rendezvous protect against inbreeding.

Investigators also hypothesize that mating with many males provides insurance against infanticide. Males do not aggress against infants of mothers with which they have copulated. Studies indicate that females with multiple mates are more fertile and produce more offspring than those without multiple mates.

Polyandry and polygyny sometimes occur in the same species. After mating, a female giant water bug attaches a clutch of eggs to her mate's back (FIG. 52-19a). He cares for the eggs until they hatch. She then may mate with a different male and glue the new clutch of eggs to his back. However, if the male has more space for eggs, he may mate with another female.

Apparently, in most species it is not certain which male fathered the offspring. However, raising some other male's offspring is a genetic disadvantage, so males of some species may compromise mate-chasing behavior in favor of **mate guarding.** The male guards his partner after copulation to ensure that she does not copulate with another male.

Mate-guarding behavior is likely to occur when the female is receptive and has eggs that might be fertilized by another male. For example, dominant male African elephants guard a female only during the phase of estrus when she is most likely to be fertile. Before that time, or later in estrus after mating has already occurred, younger male elephants of lesser social status can copulate with the female. A high cost of mate guarding is the loss of opportunity for a dominant male to mate with other females.

Sexual selection favors males that inseminate many females and produce many offspring. Perhaps for this reason, **monogamy,** a mating system in which a male mates with only one female during a breeding season, is not common. Less than 10% of mammals are monogamous.

Researchers long thought that monogamy was common among birds because many species form **pair bonds,** which are stable relationships that ensure cooperative behavior in mating and the rearing of the young. However, genetic evidence shows that some offspring are fathered by males other than the one caring for them. For example, DNA tests show that among eastern bluebirds, 15% to 20% of the chicks are fathered by other males. The extra-pair male contributes new genes. The female then produces offspring with greater genetic variability, increasing the chance that at least some of her offspring will survive. Some researchers distinguish between *sexual monogamy* and *social monogamy,* in which animals form a pair bond.

Monogamy does occur in some species, typically when males are needed to protect and feed the young. For example, the California mouse is sexually as well as socially monogamous. The offspring need their parents' body heat to survive, and the parents take turns keeping them warm.

Some animals care for their young

Parental investment is the contribution that each parent makes in producing and rearing offspring. The benefit of parental investment is the greater probability that each offspring will survive. Despite this, most animals do not invest time and energy in caring for their offspring because the costs of parenting are high. For example, an animal busy feeding and protecting offspring might produce fewer additional offspring. There are also risks in protecting the young from predators.

Natural selection has favored parental care in species in which the female produces few young or reproduces only once during a breeding season. Parental care is an important part of successful reproduction in some invertebrates, including some cnidarians (jellyfish), rotifers, mollusks, and arthropods (crustaceans, insects, spiders, and scorpions), and is a common practice among vertebrates (FIG. 52-19b–d).

Females of many vertebrate species produce relatively few, large eggs. Because of the time and energy invested in producing eggs and carrying the developing embryo, the female has more to lose than the male if the young do not develop. Thus, females are more likely than males to brood eggs and young, and usually females invest more in parental care. Parental care is especially skewed toward the female in mammals because females provide milk to nourish their young. Investing time and effort in care of the young is usually less advantageous to a male than to a female (assuming that the female can handle the job by herself) because time spent in parenting is time lost from inseminating other females.

In some situations a male benefits by helping rear his own young or even those of a genetic relative. Receptive females may be scarce, breeding territories may be difficult to establish or guard, and gathering sufficient food may require more effort than one parent can provide. In some habitats the young need protection against predators or cannibalistic males of the same species. Among wolves and many other carnivores, for example, males defend the young and the food supply.

(a) Giant water bug (*Belostoma* sp.) male carrying eggs and a few newly hatched offspring on his back.

(b) A female crocodile in South Africa transports hatchlings in her mouth from their nest site to Lake St. Lucia.

(c) Adult robins invest energy in feeding their young.

(d) Female cheetah (*Acinonyx jubatus*) stands watch as her cubs eat the Thompson's gazelle she has hunted and killed for them.

Figure 52-19 Caring for the young
Parental investment in offspring increases the probability that the young will survive.

Among many species of fish, the male cares for the young. Apparently, parenting costs to the males are less than they would be for the females. A male fish must have a territory to attract a female. While guarding his territory, the male guards the fertilized eggs as well. Having eggs also appears to make the male more attractive to potential mates. In contrast, when a female is caring for her eggs or young, she is not able to feed optimally. As a result, her body size is smaller, and her fertility is reduced.

Behavioral ecologist Bryan D. Neff of the University of Western Ontario showed that male bluegill sunfish (*Lepomis macrochirus*) adjust their level of caregiving according to their degree of certainty that the offspring are indeed their own. Adult males, called parentals, defend a nest, attract females, and care for the eggs and newly hatched offspring. Some males, called *sneaker males* (or cuckolders), sneak into nests and fertilize eggs

during spawning. Other sneaker males have developed in a way that gives them characteristics of females. The parental male is tricked into regarding the sneaker male as another female, allowing it access to the eggs (**FIG. 52-20**).

Neff performed a series of experiments in which he placed sneaker males in transparent containers so that they could not fertilize eggs but the parentals could see them. In response, parentals reduced their level of protecting the eggs (**FIG. 52-21**). However, after the eggs hatched, the parentals increased their level of care. Apparently, olfactory cues indicated that these young fish were indeed the offspring of the parental caring for them. (They smelled right.)

In a second experiment, Neff exchanged about one-third of the eggs between two nests. Parentals continued to protect the eggs, but after the eggs hatched, parentals decreased their level of care. This investigator showed that parentals responded

Figure 52-20 Deception in mating
A large parental (adult male) bluegill sunfish (*Lepomis macrochirus*) courts a female (*center*). Another male, impersonating a female, swims close to the female (*right*). The parental is not aware that the sneaker is a rival male. In about 20% of spawnings, a sneaker fertilizes some of the eggs and then swims away.

Courtesy Dr. Bryan D. Neff

to olfactory cues. At least some of their offspring did not smell right. (The odor may be produced by the offspring's urine.) Neff suggested that an animal learns its own odor (or appearance) and forms a template of what its relatives should smell (or look) like. If they don't match up closely enough, they may be rejected.

CHECKPOINT 52.6

- *What is sexual selection? What are some factors that influence mate choice?*
- *Under what conditions does sexual selection favor polygynous (or polyandrous) mating systems? Why is monogamy uncommon?*
- *What are some benefits of courtship rituals?*
- **CONNECT** *Why do more females than males invest in caring for their offspring? Explain in terms of costs, benefits, and fitness.*

52.7 HELPING BEHAVIOR

LEARNING OBJECTIVE

11 Relate the concepts of inclusive fitness and kin selection to altruism.

Among wild turkeys and many other species, subordinate males help dominant males attract a mate, but the subordinate males do not reproduce. If an animal's principal evolutionary mandate is to perpetuate its genes, we must wonder why some animals spend time and energy in helping others.

Altruistic behavior can be explained by inclusive fitness

In a type of cooperative behavior known as **altruistic behavior,** one individual behaves in a way that seems to benefit others rather than itself, with no potential payoff. In fact, the helper experiences a fitness cost, whereas the recipient of the altruistic behavior benefits. How could true altruistic behavior be adaptive if the animal's response decreases its reproductive success relative to individuals that do not exhibit the altruistic behavior? In 1964, British evolutionary biologist William D. Hamilton offered a plausible explanation. He suggested that evolution does not distinguish between genes transmitted directly from parent to offspring and those transmitted indirectly through close relatives.

According to Hamilton, natural selection favors animals that help a relative because the relative's offspring carries some of the helper's alleles. **Inclusive fitness** is the sum of direct fitness (which can be measured by the number of alleles an animal perpetuates in its offspring) plus indirect fitness (the alleles it helps perpetuate in the offspring of its kin).

We can measure the potential genetic gain from helping relatives by the *coefficient of relatedness,* the probability that two individuals inherit the same uncommon allele from a recent common ancestor. The coefficient of relatedness for two animals that are not related is 0, whereas between a parent and its offspring this coefficient is 0.5, indicating that half of their genes are the same. The coefficient of relatedness for siblings is also 0.5; for first cousins, it is 0.125. The higher the coefficient of relatedness, the more likely an animal will help a relative.

Hamilton developed a mathematical model to show how an allele that promotes altruistic behavior can increase in frequency in a population. According to *Hamilton's rule,* an altruistic act is adaptive if (1) its indirect fitness benefits are high for the animals that profit, (2) the recipients are close relatives of the altruist, and (3) the direct fitness cost to the altruistic animal is low.

Behavioral ecologists have made DNA fingerprints of lions in the Serengeti National Park and in the Ngorongoro Crater (both in Tanzania). Their findings confirm that lion brothers that cooperate in prides have a greater probability of perpetuating their genes than those that go off on their own. This finding is true even if the lion perpetuates his genes only by proxy, through his brother. **Kin selection,** also called *indirect selection,*

Does a male bluegill sunfish adjust his level of caregiving if he perceives that he is not the genetic parent of the eggs or offspring that he is guarding?

HYPOTHESIS: Behavioral ecologist Bryan D. Neff hypothesized that male bluegill sunfish adjust the level of their parental care according to their level of perceived paternity.

EXPERIMENTS: Neff manipulated visual and olfactory cues.

In **Experiment 1,** Neff manipulated sneaker males. He placed two transparent containers with sneaker males near the nests of 34 parentals (adult males) and left them there during spawning. Empty containers were placed around 20 control nests. After the fish spawned, a potential egg predator (a pumpkinseed sunfish in a clear plastic container) was presented, and the parental's defensive behavior was observed and quantified for both experimental and control groups.

In **Experiment 2,** the investigator removed one-third of the eggs from the nests of 20 parentals and replaced them with eggs from other nests. Sham egg swaps were carried out on eggs of 15 control parentals. Parental defensive behavior was observed before and after the eggs hatched.

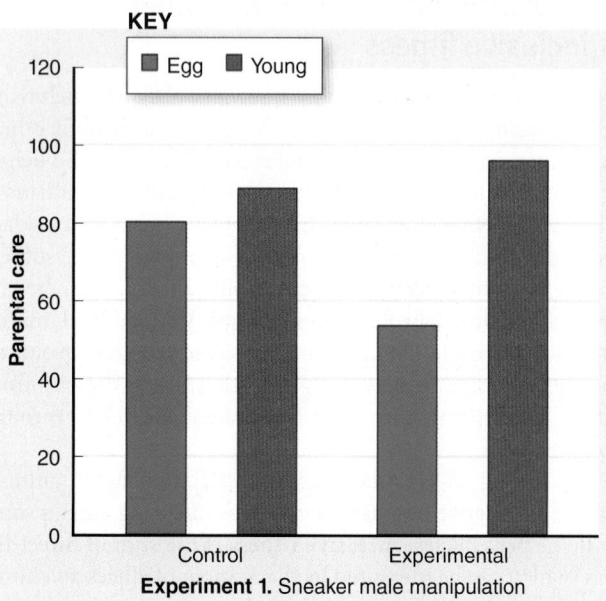

Experiment 1. Sneaker male manipulation

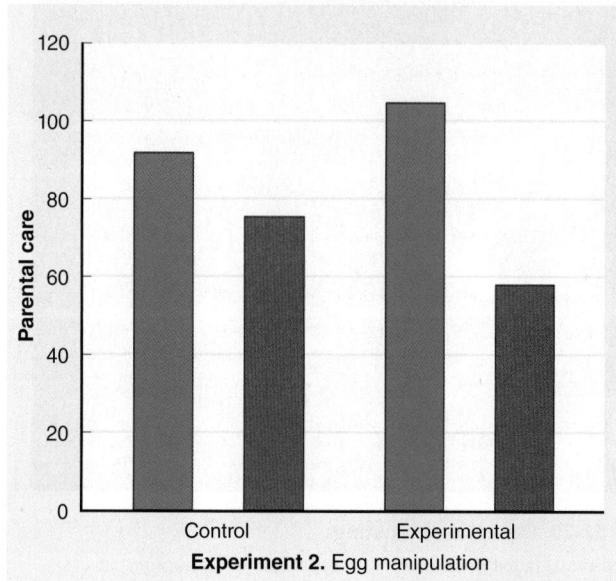

Experiment 2. Egg manipulation

RESULTS:

Experiment 1: As indicated on the y-axis, which measures level of parental care, parentals reduced their level of guarding the eggs. Eight of the males in the experimental group abandoned their nests, and egg defense was significantly lower in this group compared with that in the control group. However, after eggs hatched, there was little difference in parental care of the young between the two groups.

Experiment 2: During the egg phase, there was little difference in level of parental care between experimental and control groups. However, after eggs hatched, the experimental parentals significantly decreased their level of guarding the nest.

CONCLUSION: Male bluegill sunfish adjust their level of parental care according to their level of perceived paternity.

In **Experiment 1,** parentals provided less care when they perceived that the eggs may have been fertilized by sneaker males. After the eggs hatched, olfactory cues indicated that the offspring were indeed their own, and their level of care increased.

In **Experiment 2,** parentals cared for the eggs even though some had been swapped. However, after they hatched, olfactory cues from the offspring indicated that they were not the parental's own offspring. The level of parental care decreased significantly.

SOURCE: Neff, B. D. *Nature,* Vol. 422, pp. 716–719, April 17, 2003.

Figure 52-21 Decisions about parental care

PREDICT Imagine the investigator had temporarily blinded the male sunfish in Experiment 1. How might the results have been different?

is a form of natural selection that increases inclusive fitness through the breeding success of close relatives.

The bee society provides an interesting example of kin selection. Recall that worker bees do not reproduce, but they do nourish larval bees. The worker bees have up to three-fourths of their genes in common with the larval bees. For this reason, worker bees are more likely to perpetuate their genes by raising larval bees than they would by producing their own offspring.

Among some birds (such as Florida jays), nonreproducing individuals aid in the rearing of the young. Nests tended by these additional helpers as well as parents produce more young than do nests with the same number of eggs overseen only by parents. The nonreproducing helpers are close relatives that increase their own biological success by ensuring the successful, although indirect, perpetuation of their genes. Helpers may be prevented from producing their own offspring by limiting factors such as a shortage of mates or territories.

Helping behavior may have alternative explanations

Some behavioral ecologists are questioning the role of kin selection in helping behavior. One of the time-honored examples of kin selection has been sentinel behavior, keeping watch for predators and warning other members of the group when danger threatens. Until recently, biologists thought that sentinels were at greater risk for predation than other members of that group, yet selflessly engaged in risky behavior that benefited their relatives. This explanation appears valid for some species, including prairie dogs and ground squirrels. However, some studies suggest an opposing hypothesis for guarding behavior in other species.

U.S. biologist Peter A. Bednekoff developed a model to explain how guarding could result from selfish behavior. Bednekoff observed that for some species sentinels are not at higher risk than others in their group. In fact, the opposite may be true, and sentinels may actually have an advantage. By detecting predators first, the sentinels can escape more rapidly than others in the group.

Studies of sentinel behavior in the African mongoose, commonly known as the meerkat, support Bednekoff's model (**FIG. 52-22**). During 2000 hours of researcher observation, no sentinel was killed. In fact, investigators concluded that guarding is a selfish behavior in meerkat populations because sentinels were able to escape quickly into their nearby burrows. This research is an important reminder that it is risky for biologists to generalize the meanings of complex social behavior from one species to another. Many explanations are possible. Helpers often derive direct benefits in safety and eventual reproductive success. For example, among some species younger animals that help are permitted to remain in the social group and derive the benefits of group protection. These helpers are also positioned to take over territories and mates when older members of the group die.

Some animals help nonrelatives

Occasionally, an animal helps a nonrelative in a fight, grooms it, or even shares food. Evolutionary biologist Robert L. Trivers of Rutgers University explained this behavior in 1971 with his hypothesis of **reciprocal altruism.** According to this hypothesis, one animal helps a nonrelative with no immediate benefit, but at some later time the animal that was helped repays the debt. In this way, the original helper experiences a net fitness benefit. A vampire bat that has just fed regurgitates blood for hungry bats that share the roost. If a bat does not share blood with a neighbor that has fed it in the past, the "cheater" will probably not be helped in the future.

Martin Harvey/Getty Images

Figure 52-22 Sentinel behavior in the meerkat (*Suricata suricatta*), also known as the African mongoose

Studies of sentinel behavior in meerkats suggest that meerkat sentinels are not being altruistic and their behavior is not about kin selection. In fact, guarding may actually be a selfish behavior, because when danger threatens, sentinels can quickly escape into the burrow.

CHECKPOINT 52.7

- **CONNECT** *How does kin selection explain the evolution of altruistic behavior?*
- **PREDICT** *How might an animal benefit by helping a nonrelative?*

52.8 CULTURE IN VERTEBRATE SOCIETIES

LEARNING OBJECTIVE

12 Contrast a vertebrate society with a society of social insects, give examples of transmission of culture, and briefly describe sociobiology.

Vertebrate societies usually have nothing comparable to the physically and behaviorally specialized castes of honeybees or ants. An exception is the naked mole rat of southern Africa, a rodent that has a social structure closely resembling that of the social insects. Although most vertebrate societies seem simpler than insect societies, they are also more flexible. Vertebrate societies share a great range and plasticity of potential behaviors and can effectively modify behavior to meet environmental challenges.

Some vertebrates transmit culture

The behavioral plasticity of vertebrates makes possible the transmission of culture in some bird and mammal species. **Culture** is behavior common to a population, learned from other members of the group, and transmitted from one generation to another. Culture is not inherited. It is maintained by **social learning,** such as through observing others and imitating them or by teaching.

About 17 behaviors that may be considered cultural have been described among cetaceans (whales, dolphins, porpoises). For example, female orcas teach their offspring to hunt seals according to the custom of their particular group. Orcas also have local dialects that have been documented for at least six generations, suggesting that they are passed on from parent to offspring. Separate dialects are maintained even when various populations of orcas interact socially.

Biologists have reported that humpback whales learn new strategies for feeding by observing one another. Bottlenose dolphins are well known for their ability to learn by imitation. Behavioral ecologists have documented foraging practices that appear to be cultural, that is, transmitted by social learning.

Teaching is an efficient form of social learning. If adult animals can teach younger ones, critical information can be rapidly transferred from one generation to the next. University of Cambridge researchers Alex Thornton and Katherine McAuliffe reported that helper meerkats teach pups skills for capturing prey. Meerkats feed on a variety of animals, many of which are difficult to subdue. Some, such as scorpions, are potentially dangerous. Young meerkat pups cannot find and capture their own prey. Instead, they emit begging calls that induce helpers to feed them. Helpers monitor and nudge the pups to handle the prey efficiently. As they grow older, the pups' begging calls change. In response, helpers modify their teaching, giving the pups the opportunity to practice a greater range of skills. The helpers gradually introduce the pups to live prey.

Studies of chimpanzee populations demonstrate that some groups develop local customs and teach them to their offspring. Using data provided by researchers at seven chimpanzee research sites, animal behaviorists identified 39 chimpanzee behaviors that they considered cultural variations. These behaviors included various ways of using tools, courtship rituals, and grooming techniques. Each of these local customs was learned from other chimpanzees in the population, and customs varied among different populations.

For example, chimpanzees on the western side of the Sassandra-N'Zo River use stone hammers to crack coula nuts. Researchers have photographed mother chimpanzees teaching this behavior to their offspring. Just a few miles away on the eastern side of the river, chimpanzees do not crack nuts, even though they are available.

When a chimp migrates and joins a new population, it can transmit knowledge learned in its previous culture. In this way, cracking nuts and other customs spread from one population to another. An example of cultural transmission has been observed in a chimp community at Gombe. (Anthropologist Jane Goodall has studied the chimps there since 1960.) An adult female chimp that joined the Gombe chimp community in the early 1990s had apparently learned to use tools to fish out carpenter ants in her former community. Some of the young chimps of her new community observed this behavior and learned how to insert sticks into holes to fish out ants.

As researchers continue to study cultural variation among chimpanzee populations, learning ever more about our closest relatives, it is important to remember that the capacity to learn such behaviors is the product of natural selection. The behaviors that are maintained are adaptive for these animals in their environment. Studying culture in nonhuman primates may provide insights into the evolutionary origins of human culture. Human society is based to a great extent on the symbolic transmission of culture through spoken and written language.

Sociobiology explains human social behavior in terms of adaptation

Sociobiology focuses on the evolution of social behavior through natural selection. In his landmark book, *Sociobiology: The New Synthesis,* published in 1975, Edward O. Wilson combined principles of population genetics, evolution, and animal behavior to present a comprehensive view of the evolution of social behavior. Like Darwin and many other biologists of the past, Wilson suggested that human behavior can be studied in evolutionary terms. Many of the concepts discussed in this chapter, such as paternal investment in care of the young, are based on contributions made by Wilson and other sociobiologists. Wilson's work greatly influenced the development of behavioral ecology, evolutionary psychology, and evolutionary anthropology.

Sociobiology has been controversial, at least in part because of its application to human behavior and the possible ethical implications of such application. Some critics have complained that according to sociobiology human behavior may not be flexible enough to permit substantial improvements in the quality of our social lives. Sociobiologists, however, agree with their critics that human behavior *is* flexible. At least some of the debate focuses on the *degree* to which human behavior is genetically determined and the extent to which it can be modified.

As sociobiologists acknowledge, people can, through culture, change their way of life far more profoundly in a few years than could a hive of bees or a troop of baboons in hundreds of generations of genetic evolution. This capacity to make changes is indeed genetically determined, and it is a great gift. How we use it and what we accomplish with it are not a gift but, rather, a responsibility on which our own well-being and the survival of other species depend.

CHECKPOINT 52.8

- **CONNECT** *How are vertebrate societies different from those of social insects?*
- *What are some examples of culture?*
- *According to sociobiologists, what determines social behavior?*

SUMMARY: FOCUS ON LEARNING OBJECTIVES

52.1 Behavior and Adaptation *(page 1119)*

1 Distinguish between proximate and ultimate causes of behavior and apply the concepts of ultimate cause and cost–benefit analysis to decide whether a particular behavior is adaptive.

- **Behavior** is what an animal does and how it does it, usually in response to stimuli in its environment. **Proximate causes** of behavior are immediate causes, such as genetic, developmental, and physiological processes that permit the animal to carry out a specific behavior. Proximate causes answer *how* questions. **Ultimate causes** are the evolutionary explanations for *why* a certain behavior occurs.

- We can use **cost–benefit analysis** to determine whether a behavior is adaptive. Benefits contribute to **direct fitness,** the animal's reproductive success measured by the number of viable offspring it produces. If benefits outweigh costs, the behavior is adaptive.

2 Describe the interactions of heredity, environment, and maturation in animal behavior.

- Behavior results from the interaction of genes (**innate behavior**) and environmental factors. The capacity for behavior is inherited, but behavior is modified in response to environmental experience.

- An organism must be mature—physiologically ready to produce a given behavior—before it can perform that pattern of behavior. Walking and many other behaviors that we view as automatic are **motor programs,** coordinated sequences of muscle actions.

- A **behavioral pattern** is an automatic behavior that, once activated by a simple sensory stimulus, continues to completion regardless of sensory feedback. A behavioral pattern can be triggered by a specific unlearned **sign stimulus,** or *releaser.*

52.2 Learning: Changing Behavior as a Result of Experience *(page 1121)*

3 Discuss the adaptive significance of habituation, imprinting, classical conditioning, operant conditioning, and cognition.

- **Learning** is a change of behavior that results from experience. **Habituation** is a type of learning in which an animal learns to ignore a repeated, irrelevant stimulus so that it can focus on finding food and carrying out other life activities.

- **Imprinting** establishes a parent–offspring bond during a *critical period* early in development, ensuring that the offspring recognizes its mother.

- In **classical conditioning** an association is formed between some normal body function and a new stimulus. This type of learning allows an animal to make an association between two stimuli.

- In **operant conditioning** an animal learns a behavior so as to receive positive reinforcement or to avoid punishment. This type of learning is important in many natural situations; for example, young herring gulls perfect their pecking behavior so as to obtain food.

- **Cognition** is the process of gaining and using knowledge; it includes knowing, thinking, processing information, learning, reasoning, and awareness of thoughts, perceptions, and self. Cognition allows some animals to adapt past experiences to solve new problems that may involve different stimuli.

52.3 Behavioral Responses to Environmental Stimuli *(page 1125)*

4 Give examples of biological rhythms and describe some of the physiological mechanisms responsible for them.

- It is adaptive for an organism's metabolic processes and behavior to be synchronized with cyclical changes in the environment. In many species physiological processes and activities follow **circadian rhythms,** which are daily cycles of activity.

- **Diurnal animals** are most active during the day; **nocturnal animals,** at night; and **crepuscular animals,** at dawn or dusk. Some biological rhythms—for example, the changes in tides due to changes in the phase of the moon—reflect the *lunar cycle.*

- Many biological rhythms are regulated by internal timing mechanisms that serve as **biological clocks.** In mammals the master clock is located in the **suprachiasmatic nucleus (SCN)** in the hypothalamus.

5 Analyze costs and benefits of migrations and distinguish between directional orientation and navigation.

- **Migration** is periodic long-distance travel from one location to another. Proximate causes of migration include physiological responses that are triggered by environmental changes. Ultimate causes of migration include the benefit of moving away from an area that seasonally becomes too cold, dry, or depleted of food to a more hospitable area. Costs of migration include time, energy, and greater risk of predation.

- *Directional orientation* is travel in a specific direction and requires **compass sense,** a sense of direction. Many migrating animals use the sun to orient themselves.

- **Navigation** requires both compass sense and **map sense,** an awareness of location. Birds that navigate at night use the stars as guides. Birds and many other animals also use Earth's magnetic field to navigate.

52.4 Foraging Behavior (page 1127)

6 Define *optimal foraging behavior* and explain how it is adaptive.

- **Optimal foraging,** the most efficient strategy for an animal to get food, can enhance reproductive success.

52.5 Costs and Benefits of Social Behavior (page 1128)

7 Analyze social behavior in terms of costs and benefits, and compare different types of social groups, including societies of social insects.

- **Social behavior** is the interaction of two or more animals, usually of the same species. An organized **society** is an actively cooperating group of individuals belonging to the same species and often closely related. In some species social organization ensures greater reproductive success than do noncooperative arrangements.

- Many animal societies are characterized by a means of communication, cooperation, division of labor, and a tendency to stay together. Benefits include cooperative foraging or hunting and defense from predators. Costs include increased competition for food and habitats and increased risks of attracting predators and transmitting disease.

8 Describe common modes of animal communication, including signaling by pheromones.

- Animal *communication* involves the exchange of mutually recognizable signals, which can be electrical, tactile, visual, auditory, or chemical. **Pheromones** are chemical signals that convey information between members of a species.

9 Compare the costs and benefits of dominance hierarchies and of territoriality.

- A **dominance hierarchy** is a ranking of status within a group in which more dominant members are accorded benefits (such as food or mates) by subordinates. Establishing dominance often, but not always, involves aggressive behavior. Social ranking reduces costly physical fights.

- Animals often inhabit a **home range,** a geographic area that they seldom leave but do not necessarily defend. A defended area within a home range is called a *territory,* and the defensive behavior is **territoriality.** Costs of territoriality include time and energy expended in staking out and defending a territory and risks in fighting for it. Benefits include rights to food in the territory and reduction in conflict among members of a population.

52.6 Sexual Selection (page 1133)

10 Define *sexual selection* and describe different types of mating systems and approaches to parental care.

- **Sexual selection,** a type of natural selection, occurs when individuals vary in their ability to compete for mates. Individuals with reproductive advantages are selected over others of the same sex and species. In **intrasexual selection** animals of the same sex actively compete for mates.

- In **intersexual selection** females may choose their mates based on dominance, gifts, ornaments, or courtship displays. Males of some species gather in a **lek,** a small display area where they compete for females. **Courtship rituals** ensure that the male is a member of the same species and permit the female to assess the quality of the male.

- Sexual selection often favors **polygyny,** a mating system in which a male mates with many females. In **polyandry** a female

mates with several males. **Monogamy,** mating with a single partner during a breeding season, is less common. Among many species, males engage in **mate guarding,** especially when the female is most fertile, to prevent other males from fertilizing her eggs.

- A **pair bond** is a stable relationship between a male and a female that may involve cooperative behavior in mating and in rearing the young. **Parental investment** in care of eggs and offspring increases the probability that offspring will survive. A high investment in parenting is typically less advantageous to the male than to the female.

McMurray Photography

52.7 Helping Behavior (page 1137)

11 Relate the concepts of inclusive fitness and kin selection to altruism.

- **Altruism** is a type of cooperative behavior in which one individual appears to behave in a way that benefits others rather than itself. **Inclusive fitness** is the sum of an individual's direct fitness and indirect fitness (number of offspring of kin). This concept suggests that natural selection favors animals that help a relative because it is an indirect way to perpetuate some of the helper's own alleles.

- According to *Hamilton's rule,* an altruistic act is adaptive if its indirect fitness benefits are high for the animals that are helped, if the recipients are close relatives of the altruist, and if the direct fitness cost to the altruist is low.

- **Kin selection** is a type of natural selection that increases inclusive fitness through successful reproduction of close relatives. Some types of cooperative behavior may increase the direct fitness of the helper.

- In a type of cooperative behavior known as **reciprocal altruism,** the helper does not immediately benefit but is helped later by the animal it helped.

52.8 Culture in Vertebrate Societies (page 1139)

12 Contrast a vertebrate society with a society of social insects, give examples of transmission of culture, and briefly describe sociobiology.

- Compared with insect societies, vertebrate societies are more flexible and the role of the individual less narrowly defined. Some bird and mammal species develop **culture,** learned behavior common to a population that is transmitted from one generation to the next. Symbolic transmission of culture is important in human societies.

- **Sociobiology** focuses on the evolution of social behavior through natural selection.

Know and Comprehend

1. The responses of an organism to signals from its environment are its (a) behavior (b) culture (c) ultimate behavior (d) releasers (e) motor programs

2. A behavioral pattern is elicited by (a) a motor program (b) a sign stimulus (c) an imprint (d) a, b, and c (e) b and c

3. A form of learning in which an animal forms a strong attachment to a moving object (usually its parent) within a few hours of birth is (a) classical conditioning (b) operant conditioning (c) imprinting (d) insight learning (e) parental investment

4. Both compass sense and map sense are necessary for (a) navigation (b) migration (c) directional orientation (d) both navigation and migration (e) the action of biological clocks

5. In optimal foraging (a) animals always hunt in social groups (b) an animal obtains food in the most efficient way (c) animals can rarely afford to be selective (d) groups of five to seven animals are most successful (e) animals are typically most successful when they hunt alone

6. Chemical signals that convey information between members of a species are (a) pheromones (b) hormones (c) neurotransmitters (d) leks (e) neuropeptides

7. The benefits of territoriality include (a) right to defend a home range (b) higher potential for monogamy (c) energy investment in staking out and defending the area (d) increased reproductive success (e) pair bonding

8. Sexual selection (a) occurs mainly among animals that practice polyandry (b) occurs when animals are very similar in their ability to compete for mates (c) results in animals that have lower direct fitness (d) occurs mainly among animals that practice polygyny (e) is a form of natural selection

9. In most species mate choice is least likely influenced by (a) ornamental displays such as antlers (b) fidelity of the male (c) gifts (d) courtship behavior (e) dominance

10. Behavior that appears to have no payoff—that is, an individual appears to act to benefit others rather than itself—is known as (a) mutualism (b) helping behavior (c) reciprocal altruism (d) inclusive fitness (e) altruism

11. Kin selection (a) increases inclusive fitness through the breeding success of close relatives (b) is a way of perpetuating genes of nonrelatives (c) accounts for some forms of migration (d) typically involves mate guarding (e) involves ornamental displays and use of a lek

12. Culture (a) is common among invertebrate groups (b) does not vary within the same species (c) is behavior learned from other members of the group and shared by members of a population (d) has a strong genetic component in primates (e) is a shared derived character of humans

Apply and Analyze

13. Behavioral ecologists have demonstrated that young tiger salamanders can become cannibals and devour other salamanders. Investigators observed that the cannibal salamanders preferred eating unrelated salamanders over their own cousins and preferred eating cousins over their siblings. Explain this behavior based on what you have learned in this chapter.

Evaluate and Synthesize

14. **INTERPRET DATA** Look at the two graphs in Figure 52-21. In which experiment did the parentals in the experimental group guard the eggs more closely? In which experiment did the experimental parentals guard the young more closely? Account for these differences.

15. **INTERPRET DATA** Look at the diagram (Fig. 52-11) and identify the groups most likely to be hungry at the end of the day's hunt during times of prey scarcity. Develop a hypothesis to explain why these groups hunt as they do.

16. How is the society of a social insect different from human society? What are some similarities between the transmission of information by heredity and by culture? What are some differences?

17. **EVOLUTION LINK** What might be the adaptive value of sea turtle migration?

18. **SCIENCE, TECHNOLOGY, AND SOCIETY** Consider how the adaptive value of sea turtle migration has changed if, as a result of human activities, migration now puts sea turtles at greater risk than if they restricted their habitat to a single location. Discuss the possible evolutionary mechanisms by which the behavior of these species may (or may not) adapt to these environmental pressures. What conservation efforts should we take to increase the probability of successful migration?

To access course materials, such as Aplia and other companion resources, please visit **www.cengagebrain.com.**

Introduction to Ecology:
Population Ecology

A population of Mexican poppies. Mexican poppies (*Eschsolzia mexicana*), which thrive on gravelly desert slopes, bloom in the desert after the winter rains.

KEY CONCEPTS

53.1 A population can be described in terms of its density, dispersion, birth and death rates, growth rate, survivorship, and age structure.

53.2 Changes in population size are caused by natality, mortality, immigration, and emigration.

53.3 Population size may be influenced by density-dependent factors and density-independent factors.

53.4 Life history traits of a population are adaptations that affect the ability of individuals to survive and reproduce.

53.5 A metapopulation consists of two or more local populations with dispersal occurring among them.

53.6 Human population structure differs among countries in ways that are primarily related to differences in level of development.

The science of **ecology** is the study of how living organisms and the physical environment interact in an immense and complicated web of relationships. Biologists call the interactions among organisms **biotic factors** and those between organisms and their nonliving, physical environment **abiotic factors.** Abiotic factors include precipitation, temperature, pH, wind, and chemical nutrients. Ecologists formulate hypotheses to explain such phenomena as the distribution and abundance of life, the ecological roles of specific species, the interactions among species in communities, and the importance of ecosystems in maintaining the health of the biosphere. They then test these hypotheses.

The focus of ecology can be local or global, specific or generalized, depending on what questions the scientist is asking and trying to answer. Ecology is the broadest field in biology, with explicit links to evolution and every other biological discipline. It includes studies on transfer of information among organisms and analyses of the transfer of energy for life. Its universality encompasses subjects that are not traditionally part of biology. Earth science, geology, chemistry, oceanography, climatology, and meteorology are extremely important to ecology, especially when ecologists examine the abiotic environment of planet Earth. Because humans are part of Earth's web of life, all our activities, including economics and politics, have profound ecological implications. *Environmental science*, a scientific discipline with ties to ecology, focuses on how humans interact with the environment.

As you learned in Chapter 1, most ecologists are interested in the levels of biological organization including and above the level of the individual organism: population, community, ecosystem, landscape, and biosphere. Each level has its own characteristic composition, structure, and functioning.

An individual belongs to a **population,** a group consisting of members of the same species that live together in a prescribed area at the same time. The boundaries of the area are defined by the ecologist performing a particular study. A population ecologist might study a population of microorganisms, animals, or plants, like the Mexican poppies in the photograph, to see how individuals within it live and interact with one another, with other species in their community, and with their physical environment.

In this chapter we begin our study of ecological principles by focusing on the study of populations as functioning systems and end with a discussion of the human population. Subsequent chapters examine the interactions among different populations within communities (Chapter 54), the dynamic exchanges between communities and their physical environments (Chapter 55), the characteristics of Earth's major biological ecosystems (Chapter 56), and biological diversity and conservation biology (Chapter 57).

53.1 FEATURES OF POPULATIONS

LEARNING OBJECTIVE

1 Define *population density* and *dispersion,* and describe the main types of population dispersion.

Populations exhibit characteristics distinctive from those of the individuals of which they are composed. Some features discussed in this chapter that characterize populations are population density, population dispersion, birth and death rates, growth rates, survivorship, and age structure.

Although communities consist of all the populations of all the different species that live together within an area, populations have properties that communities lack. Populations, for example, share a common gene pool (see Chapter 19). Consequently, natural selection can cause changes in allele frequencies in populations. As a result, allele frequency changes resulting from natural selection occur in populations. Natural selection therefore acts directly to produce adaptive changes in populations and only indirectly affects the community level.

Population ecology considers both the number of individuals of a particular species that are found in an area and the dynamics of the population. **Population dynamics** is the study of changes in populations: how and why those numbers increase or decrease over time. Population ecologists try to determine the processes common to all populations. They study how a population interacts with its environment, such as how individuals in a population compete for food or other resources and how predation, disease, and other environmental pressures affect the population. Population growth, whether of bacteria, maples, or giraffes, cannot increase indefinitely because of such environmental pressures.

Additional aspects of populations that interest biologists are their reproductive success or failure (extinction), their evolution, their genetics, and the way they affect the normal functioning of communities and ecosystems. Biologists in applied disciplines, such as forestry, agronomy (crop science), and wildlife management, must understand population ecology to manage populations of economic importance, such as forests, field crops, game animals, or fishes. Understanding the population dynamics of endangered and threatened species plays a key role in efforts to prevent their slide to extinction. Knowledge of population ecology helps in efforts to prevent the increase of pest populations to levels that cause significant economic or health effects.

Density and dispersion are important features of populations

The concept of population size is meaningful only when the boundaries of that population are defined. Consider, for example, the difference between 1000 mice in 100 hectares (250 acres) and 1000 mice in 1 hectare (2.5 acres). Often a population is too large to study in its entirety. Researchers examine such a population by sampling a part of it and then expressing the population in terms of density. Examples include the number of dandelions per square meter of lawn, the number of water fleas per liter of pond water, and the number of cabbage aphids per square centimeter of cabbage leaf. **Population density,** then, is the number of individuals of a species per unit of area or volume at a given time.

Different environments vary in the population density of any species they can support. This density may also vary in a single habitat from season to season or year to year. For example, red grouse are ground-dwelling game birds whose populations are managed for hunting. Consider two red grouse populations in the treeless moors of northwestern Scotland, at locations only 2.5 km (1.5 mi) apart. At one location the population density remained stationary during a three-year period, but at the other site it almost doubled in the first two years and then declined to its initial density in the third year. The reason was likely a difference in habitat. Researchers had experimentally burned the area where the population density increased initially and then decreased. Young heather shoots (*Calluna vulgaris*) produced after the burn provided nutritious food for the red grouse. So, population density may be determined in large part by biotic or abiotic factors in the environment that are external to the individuals in the population.

The individuals in a population often exhibit characteristic patterns of **dispersion,** or spacing, relative to one another. Individuals may be spaced in a random, clumped, or uniform dispersion. **Random dispersion** occurs when individuals in a population are spaced throughout an area in a manner that is unrelated to the presence of others (FIG. 53-1a). Of the three major types of dispersion, random dispersion is least common and hardest to observe in nature, leading some ecologists to question its existence. Trees of the same species, for example, sometimes appear to be distributed randomly in a tropical rain forest. However, an international team of 13 ecologists studied six tropical forest plots that were 25 to 52 hectares (62 to 130 acres) in area and reported that most of the 1000 tree species observed were clumped and not randomly dispersed. (Ecologists determine both clumped and uniform dispersion by statistically testing for differences from an assumed random distribution.) Random dispersion may occur infrequently because important environmental factors affecting dispersion usually do not occur at random. Flour beetle larvae in a container of flour are randomly dispersed, but their environment (flour) is unusually homogeneous.

Perhaps the most common spacing is **clumped dispersion,** also called **aggregated distribution** or **patchiness,** which occurs when individuals are concentrated in specific parts of the habitat. Clumped dispersion often results from the patchy

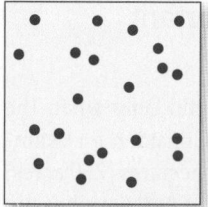

(a) Random dispersion, illustrated for comparison, rarely, if ever, occurs in nature.

Andrew G. Wood/Science Source

(b) Clumped dispersion is evident in the schooling behavior of certain fish species. Shown are bluestripe snappers (*Lutjanus kasmira*), photographed in Hawaii. This introduced fish, which grows to 30 cm (12 in.), may be displacing native fish species in Hawaiian waters.

Robert Hernandez/Science Source

(c) Uniform dispersion is characteristic of these nesting Cape gannets (*Morus capensis*) on the coast of South Africa. The birds space their nests more or less evenly.

Figure 53-1 Dispersion of individuals within a population
© Cengage Learning

distribution of resources in the environment. It also occurs among animals because of the presence of family groups and pairs, and among plants because of limited seed dispersal or asexual reproduction. An entire grove of aspen trees, for example, may originate asexually from a single plant. Clumped dispersion may sometimes be advantageous because

social animals derive many benefits from their association. Many fish species, for example, associate in dense schools for at least part of their life cycle, possibly because schooling may reduce the risk of predation for any particular individual (**FIG. 53-1b**). The many pairs of eyes of schooling fish tend to detect predators more effectively than a single pair of eyes of a single fish. When threatened, schooling fish clump together more closely, making it difficult for a predator to single out an individual.

Uniform dispersion occurs when individuals are more evenly spaced than would be expected from a random occupation of a given habitat. A nesting colony of seabirds, in which the birds are nesting in a relatively homogeneous environment and place their nests at a more or less equal distance from one another, is an example of uniform dispersion (**FIG. 53-1c**). What might this spacing pattern tell us? In this case, uniform dispersion may occur as a result of nesting territoriality. Aggressive interactions among the nesting birds as they peck at one another from their nests cause each pair to place its nest just beyond the reach of nearby nesting birds. Uniform dispersion also occurs when competition among individuals is severe, when plant roots or leaves that have been shed produce toxic substances that inhibit the growth of nearby plants, or when animals establish feeding or mating territories.

Some populations have different spacing patterns at different ages. Competition for sunlight among same-aged sand pine in a Florida scrub community resulted in a change over time from either random or clumped dispersion when the plants were young to uniform dispersion when the plants were old. Sand pine is a fire-adapted plant with cones that do not release their seeds until they have been exposed to high temperatures (45°C to 50°C or higher). As a result of seed dispersal and soil conditions following a fire, the seedlings grow back in dense stands that exhibit random or slightly clumped dispersion. Over time, however, many of the more crowded trees tend to die from shading or competition, resulting in uniform dispersion of the surviving trees (**TABLE 53-1**).

CHECKPOINT 53.1

- *What is the difference between population density and dispersion?*

- **CONNECT** *What are some biological advantages of a clumped dispersion? What are some disadvantages?*

TABLE 53-1	Dispersion in a Sand Pine Population in Florida	
TREE TRUNKS EXAMINED	DENSITY (per m²)	DISPERSION
All (alive and dead)	0.16	Random
Alive only	0.08	Uniform

Source: Adapted from Laessle, A.M. "Spacing and Competition in Natural Stands of Sand Pine." *Ecology*, Vol. 46, pp. 65–72, 1965. Data were collected 51 years after a fire.

53.2 CHANGES IN POPULATION SIZE

LEARNING OBJECTIVES

2 Explain the four factors (natality, mortality, immigration, and emigration) that produce changes in population size and solve simple problems involving these changes.

3 Define *intrinsic rate of increase* and *carrying capacity,* and explain the differences between J-shaped and S-shaped growth curves.

One goal of science is to discover common patterns among separate observations. As mentioned previously, population ecologists wish to understand general processes that are shared by many different populations, so they develop mathematical models based on equations that describe the dynamics of a single population. Population models are not perfect representations of a population, but models help illuminate complex processes. Moreover, mathematical modeling enhances the scientific process by providing a framework with which experimental population studies can be compared. We can test a model and see how it fits or does not fit with existing data. Data that are inconsistent with the model are particularly useful because they demand that we ask how the natural system differs from the mathematical model that we developed to explain it. As more knowledge accumulates from observations and experiments, the model is refined and made more precise.

Population size, whether of sunflowers, elephants, or humans, changes over time. On a global scale, this change is ultimately caused by two factors, expressed on a per capita (i.e., per individual) basis: **natality,** the average per capita birth rate; and **mortality,** the average per capita death rate. In humans the birth rate is usually expressed as the number of births per 1000 people per year and the death rate as the number of deaths per 1000 people per year.

To determine the rate of change in population size, we must also take into account the time interval involved, that is, the change in time. To express change in equations, we employ the Greek letter delta (Δ). In equation (1), ΔN is the change in the number of individuals in the population, Δt the change in time, N the number of individuals in the existing population, b the natality, and d the mortality.

$$(1)\ \Delta N/\Delta t = N(b - d)$$

The **growth rate (r),** or rate of change (increase or decrease) of a population on a per capita basis, is the birth rate minus the death rate:

$$(2)\ r = b - d$$

As an example, consider a hypothetical human population of 10,000 in which there are 200 births per year (i.e., by convention, 20 births per 1000 people) and 100 deaths per year (10 deaths per 1000 people):

$$r = 20/1000 - 10/1000 = 0.02 - 0.01 = 0.01, \text{ or 1\% per year}$$

A modification of equation (1) tells us the rate at which the population is growing at a particular instant in time, that is, its instantaneous growth rate (dN/dt). (The symbols dN and dt are the mathematical differentials of N and t, respectively; they are not products, nor should the d in dN or dt be confused with the death rate, d.) Using differential calculus, this growth rate can be expressed as follows:

$$(3)\ dN/dt = rN$$

where N is the number of individuals in the existing population, t the time, and r the per capita growth rate.

Because $r = b - d$, if individuals in the population are born faster than they die, r is a positive value and population size increases. If individuals in the population die faster than they are born, r is a negative value and population size decreases. If r is equal to zero, births and deaths match, and population size is stationary despite continued reproduction and death.

Dispersal affects the growth rate in some populations

In addition to birth and death rates, **dispersal,** which is movement of individuals among populations, must be considered when examining changes in populations on a *local* scale. There are two types of dispersal: immigration and emigration. **Immigration** occurs when individuals enter a population and thus increase its size. **Emigration** occurs when individuals leave a population and thus decrease its size. The growth rate of a local population must take into account birth rate (b), death rate (d), immigration rate (i), and emigration rate (e) on a per capita basis. The per capita growth rate equals the birth rate minus the death rate, plus the immigration rate minus the emigration rate:

$$(4)\ r = (b - d) + (i - e)$$

For example, the growth rate of a human population of 10,000 that has 200 births (by convention, 20 per 1000), 100 deaths (10 per 1000), 10 immigrants (1 per 1000), and 100 emigrants (10 per 1000) in a given year would be calculated as follows:

$$r = (20/1000 - 10/1000) + (1/1000 - 10/1000)$$
$$= 0.001, \text{ or 0.1\% per year}$$

Each population has a characteristic intrinsic rate of increase

The maximum rate at which a population of a given species could increase under ideal conditions, when resources are abundant and its population density is low, is known as its **intrinsic rate of increase (r_{max}).** Different species have different intrinsic rates of increase. A particular species' intrinsic rate of increase is influenced by several factors. They include the age at which reproduction begins, the fraction of the **lifespan** (duration of the individual's life) during which the individual is capable of reproducing, the number of reproductive periods per lifetime, and the number of offspring the individual is capable of producing during each period of reproduction. These factors, which we discuss in greater detail later in the chapter, determine whether a particular species has a large or small intrinsic rate of increase.

Generally, large species such as blue whales and elephants have the smallest intrinsic rates of increase, whereas microorganisms have the greatest intrinsic rates of increase. Under ideal conditions (an environment with unlimited resources), certain bacteria can reproduce by binary fission every 20 minutes. At this rate of growth, a single bacterium would increase to a population of more than one billion in just ten hours!

If we plot the population size versus time under optimal conditions, the graph has a J shape that is characteristic of **exponential population growth,** which is the accelerating population growth rate that occurs when optimal conditions allow a constant per capita growth rate (**FIG. 53-2**). When a population grows exponentially, it grows faster as it gets larger.

Regardless of which species we are considering, whenever a population is growing at its intrinsic rate of increase, population size plotted versus time gives a curve of the same shape. The only variable is time. It may take longer for an elephant population than for a bacterial population to reach a certain size (because elephants do not reproduce as rapidly as bacteria), but both populations will always increase exponentially as long as their per capita growth rates remain constant.

No population can increase exponentially indefinitely

Certain populations may grow exponentially for brief periods. Exponential growth has been experimentally demonstrated in certain insects and in cultures of bacteria or protists (by continually supplying nutrients and removing waste products). However, organisms cannot reproduce indefinitely at their intrinsic

rate of increase because the environment sets limits. These limits include such unfavorable environmental conditions as the limited availability of food, water, shelter, and other essential resources (resulting in increased competition) as well as limits imposed by disease and predation.

In the earlier example, bacteria in nature would never be able to reproduce unchecked for an indefinite period because they would run out of food and living space, and poisonous wastes would accumulate in their vicinity. With crowding, bacteria would also become more susceptible to parasites (high population densities facilitate the spread of infectious organisms such as viruses among individuals) and predators (high population densities increase the likelihood of a predator catching an individual). As the environment deteriorated, their birth rate (b) would decline and their death rate (d) would increase. Conditions might worsen to a point where d would exceed b, and the population would decrease. The number of individuals in a population, then, is controlled by the ability of the environment to support it. As the number of individuals in a population (N) increases, environmental limits act to control population growth.

Over longer periods, the rate of population growth may decrease to nearly zero. This leveling out occurs at or near the limits of the environment to support the population. The **carrying capacity (K)** represents the largest population that can be maintained for an indefinite period by a particular environment, assuming that there are no changes in that environment. In nature the carrying capacity is dynamic and changes in response to environmental changes. An extended drought, for example, could decrease the amount of vegetation growing in an area; this change, in turn, would lower the carrying capacity for deer and other herbivores in that environment.

When a population regulated by environmental limits is plotted over longer periods, the curve has a characteristic S shape (**FIG. 53-3**). The curve shows the population's initial exponential increase (note the curve's J shape at the start, when environmental limits are few), followed by a leveling out as the carrying capacity of the environment is approached. The S-shaped growth curve, also called **logistic population growth,** can be modeled by a modified growth equation called a *logistic equation.* The logistic model of population growth was developed to explain population growth in continually breeding populations. Similar models exist for populations that have specific breeding seasons.

The logistic model describes a population increasing from a small number of individuals to a larger number of individuals that are ultimately limited by the environment. The logistic equation takes into account the carrying capacity of the environment:

$$(5)\ dN/dt = rN[(K - N)/K]$$

Note that part of the equation is the same as equation (3). The added element, $[(K - N)/K]$, reflects a decline in growth as a population size approaches its carrying capacity. When the number of organisms (N) is small, the rate of population growth is unchecked by the environment because the expression

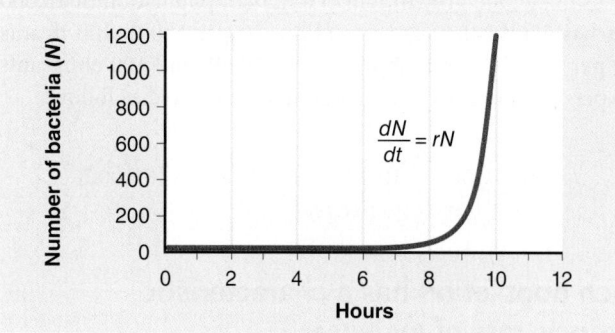

Figure 53-2 *Animation* **Exponential population growth**

When bacteria divide every 20 minutes, their numbers (expressed in millions) increase exponentially. The curve of exponential population growth has a characteristic J shape. The ideal conditions under which bacteria or other organisms reproduce exponentially rarely occur in nature, and when they do, they are of short duration.

© Cengage Learning

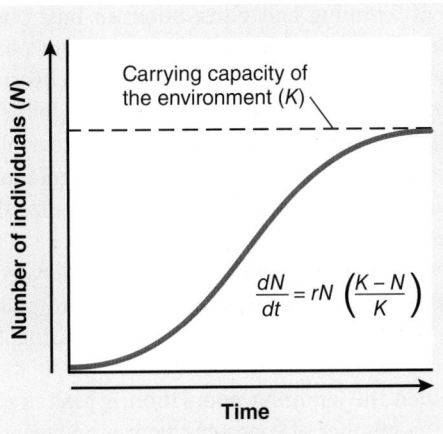

Figure 53-3 *Animation* **Carrying capacity and logistic population growth**

In many laboratory studies, exponential population growth slows as the carrying capacity (*K*) of the environment is approached. The logistic model of population growth, when plotted, has a characteristic S-shaped curve.
© Cengage Learning

$[(K - N)/K]$ has a value of almost 1. As the population (*N*) begins to approach the carrying capacity (*K*), however, the growth rate declines because the value of $[(K - N)/K]$ approaches zero.

Although the S curve is an oversimplification of how most populations change over time, it does appear to fit some populations that have been studied in the laboratory as well as a few that have been studied in nature. For example, Georgyi F. Gause, a Russian ecologist who conducted experiments during the 1930s, grew a population of a single species, *Paramecium caudatum,* in a test tube. He supplied a limited amount of food (bacteria) daily and replenished the growth medium occasionally to eliminate the accumulation of metabolic wastes. Under these conditions, the population of *P. caudatum* increased exponentially at first; then its growth rate declined to zero, and the population size leveled off (see Fig. 54-5, middle graph).

A population rarely stabilizes at *K* (carrying capacity), but it may temporarily rise higher than *K*. It will then drop back to, or below, the carrying capacity. Sometimes a population that overshoots *K* will experience a *population crash,* an abrupt decline from high to low population density. Such an abrupt change is commonly observed in bacterial cultures, zooplankton, and other populations whose resources have been exhausted.

The carrying capacity for reindeer, which live in cold northern habitats, is determined largely by the availability of winter forage. In 1910, humans introduced a small herd of 26 reindeer onto one of the Pribilof Islands of Alaska. The herd's population increased exponentially for about 25 years until there were approximately 2000 reindeer, many more than the island could support, particularly in winter. The reindeer overgrazed the vegetation until the plant life was almost wiped out. Then, in slightly longer than a decade, as reindeer died from starvation, the number of reindeer plunged to 8, one-third the size of the original introduced population. Recovery of subarctic and arctic vegetation after overgrazing by reindeer can take 15 to 20 years, during which time the carrying capacity for reindeer is greatly reduced.

○ **CHECKPOINT 53.2**

● *What effect does each of the following have on population size: natality, mortality, immigration, and emigration?*

● **CONNECT** *How does a J-shaped population growth curve differ from an S-shaped curve in terms of intrinsic rate of increase and carrying capacity?*

● **VISUALIZE** *Sketch simple graphs representing the long-term growth of two populations of bacteria cultured in test tubes, one in which the nutrient medium is replenished and the other in which it is not replenished.*

53.3 FACTORS INFLUENCING POPULATION SIZE

○ **LEARNING OBJECTIVE**

4 Contrast the influences of density-dependent and density-independent factors on population size and give examples of each.

Certain natural mechanisms influence population size. Factors that affect population size fall into two categories: density-dependent factors and density-independent factors. These two sets of factors vary in importance from one species to another and, in most cases, probably interact simultaneously to determine the size of a population.

Density-dependent factors regulate population size

Sometimes the influence of an environmental factor on the individuals in a population varies with the density or crowding of that population. If a change in population density alters how an environmental factor affects that population, the environmental factor is said to be a **density-dependent factor.**

As population density increases, density-dependent factors tend to slow population growth by causing an increase in death rate and/or a decrease in birth rate. The effect of these density-dependent factors on population growth increases as the population density increases; that is, density-dependent factors affect a larger proportion, not just a larger number, of the population. Density-dependent factors can also affect population growth when population density declines by decreasing the death rate and/or increasing the birth rate. Thus, density-dependent factors tend to regulate a population at a relatively constant size that is near the carrying capacity of the environment. (Keep in mind, however, that the carrying capacity of the environment frequently changes.) Density-dependent factors are an excellent example of a **negative feedback system** (FIG. 53-4).

Predation, disease, and competition are examples of density-dependent factors. As the density of a population increases, predators are more likely to find an individual of a given prey species than before. When population density is high, the members of a population encounter one another

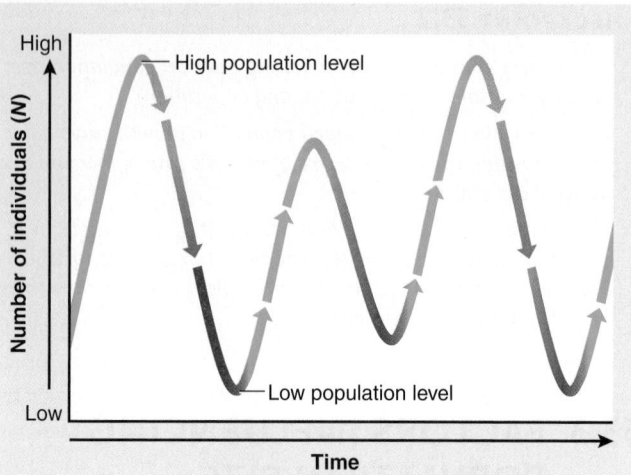

KEY

→ Density-dependent factors are increasingly severe: Population peaks and begins to decline.

→ Density-dependent factors are increasingly relaxed: Population bottoms out and begins to increase.

Figure 53-4 Density-dependent factors and negative feedback

When the number of individuals in a population increases, density-dependent factors cause a decline in the population. When the number of individuals in a population decreases, a relaxation of density-dependent factors allows the population to increase.

© Cengage Learning

more frequently than previously, and the chance of their transmitting parasites and infectious disease organisms increases. As population density increases, so does competition for resources such as living space, food, cover, water, minerals, and sunlight; eventually, the point may be reached at which many members of a population fail to obtain the minimum amount of whatever resource is in shortest supply. At higher population densities, density-dependent factors raise the death rate and/or lower the birth rate, inhibiting further population growth. The opposite effect occurs when the density of a population decreases. Predators are less likely to encounter individual prey, parasites and infectious diseases are less likely to be transmitted from one host to another, and competition among members of the population for resources such as living space and food declines.

Density-dependent factors may explain what makes certain populations fluctuate cyclically over time Lemmings are small, stumpy-tailed rodents that are found in colder regions of the Northern Hemisphere (FIG. 53-5). They are herbivores that feed on sedges and grasses in the arctic tundra. It has long been known that lemming populations have a three- to four-year cyclical oscillation that is often described as "boom or bust." That is, the population increases dramatically and then crashes; the population peaks may be 100 times as high as the low points in the population cycle. Many other populations, such as snowshoe hares and red grouse, also exhibit cyclic fluctuations.

What is the driving force behind these fluctuations? Several hypotheses have been proposed to explain the cyclical

periodicity of lemming and other boom-or-bust populations, and many involve density-dependent factors. One possibility is that as a prey population becomes more dense, it overwhelms its food supply; as a result, the population declines. In the lemming example, researchers have studied the shape of several lemming population curves during their oscillations; the data suggest that lemming populations crash because they overgraze the plants, not because predators eat them.

Another explanation is that the population density of predators, such as long-tailed jaegers (birds related to gulls that eat lemmings), increases in response to the increasing density of prey. Few jaegers breed when the lemming population is low. However, when the lemming population is high, most jaegers breed, and the number of eggs per clutch is greater than usual. As more predators consume the abundant prey, the prey population declines. Later, with fewer prey in an area, the population of predators declines (some disperse out of the area, and fewer offspring are produced).

In a long-term study of collared lemmings in Greenland, researchers at the University of Helsinki, Finland, observed four-year cycles of lemming population density that were not affected by availability of food or living space. They developed a model in which the fluctuations in the lemming population are most accurately predicted by those of another species, the stoat. The population of the stoat, a member of the weasel family that preys almost exclusively on lemmings, peaks a year after a peak in the lemming population. On the other hand, populations of three other lemming predators—snowy owls, arctic foxes, and long-tailed skuas (birds related to gulls)—respond more immediately to changes in the lemming population and thus tend to stabilize the population cycle established by the stoats.

Climate change may also be affecting lemming population dynamics. Researchers from the University of Oslo, Norway,

Tom McHugh/Science Source

Figure 53-5 Lemming

The brown lemming (*Lemmus trimucronatus*) lives in the arctic tundra. Although lemming populations have been studied for decades, much about the cyclic nature of lemming population oscillations and their effects on the rest of the tundra ecosystem is still not well understood. This species ranges from Alaska eastward to the Hudson Bay.

and a group of international collaborators have observed a correlation between climate and lemming cycles. They found that lemming populations increase in the years in which they can survive by eating moss in the space between the ground and the snow pack. Populations are reduced in warmer years, when slight melting and refreezing of the snow eliminates this space.

Parasites may also interact with their hosts to cause regular cyclic fluctuations. Detailed studies of red grouse have shown that even managed populations in wildlife preserves may have significant cyclic oscillations. Reproduction in red grouse is related to the density of parasitic nematodes (roundworms) living in adult intestines. Fewer birds breed successfully when adults are infected with worms; thus, a high density of worms leads to a population crash. Hypothesizing that red grouse populations fluctuate in response to the parasites, ecologists at the University of Stirling in Scotland successfully reduced or eliminated population fluctuations in several red grouse populations. They did so by catching and orally treating the birds with a chemical that causes worms to be ejected from their bodies.

Competition is an important density-dependent factor

Competition is an interaction among two or more individuals that attempt to use the same essential resource, such as food, water, sunlight, or living space, that is in limited supply. The use of the resource by one of the individuals reduces the availability of that resource for other individuals. Competition occurs both within a given population (**intraspecific competition**) and among populations of different species (**interspecific competition**). We consider the effects of intraspecific competition here; interspecific competition is discussed in Chapter 54.

Individuals of the same species compete for a resource in limited supply by interference competition or by exploitation competition. In **interference competition,** also called **contest competition,** certain dominant individuals obtain an adequate supply of the limited resource at the expense of other individuals in the population; that is, the dominant individuals actively interfere with other individuals' access to resources. In **exploitation competition,** also called **scramble competition,** all the individuals in a population "share" the limited resource more or less equally so that at high population densities none of them obtains an adequate amount. The populations of species in which exploitation competition operates often oscillate over time, and there is always a risk that the population size will drop to zero. In contrast, those species in which interference competition operates experience a relatively small drop in population size, caused by the death of individuals that are unable to compete successfully.

Intraspecific competition among red grouse involves interference competition. When red grouse populations are small, the birds are less aggressive, and most young birds establish a feeding territory (an area defended against other members of the same species). However, when the population is large, establishing a territory is difficult because there are more birds than there are territories and the birds are therefore much more

aggressive. Those birds without territories often die from predation or starvation. Thus, birds with territories use a larger share of the limited resource (the territory with its associated food and cover), whereas birds without territories cannot compete successfully.

The moose population on Isle Royale, Michigan, the largest island in Lake Superior, provides a vivid example of exploitation competition that is similar to that of the reindeer population on the Pribilof Islands (discussed earlier). Isle Royale differs from most islands in that large mammals can walk to it when the lake freezes over in winter. The minimum distance to be walked is 24 km (15 mi), however, so this movement has happened infrequently. Around 1900, a small herd of moose wandered across the ice of frozen Lake Superior and reached the island for the first time. By 1934, the moose population on the island had increased to about 3000 and had consumed almost all the edible vegetation. In the absence of this food resource, there was massive starvation in 1934. More than 60 years later, in 1996, a similar die-off claimed 80% of the moose after they had again increased to a high density. Thus, exploitation competition for scarce resources can result in dramatic population oscillations.

The effects of density-dependent factors are difficult to assess in nature

Most studies of density dependence have been conducted in laboratory settings where all density-dependent (and density-independent) factors except one are controlled experimentally. Populations in natural settings, however, are exposed to a complex set of variables that continually change. As a result, in natural communities it is difficult to evaluate the relative effects of different density-dependent factors.

Ecologists from the University of California, Davis noted that few spiders occur on tropical islands inhabited by lizards, whereas more spiders and more species of spiders are found on lizard-free islands. Deciding to study these observations experimentally, David Spiller and Thomas Schoener staked out plots of vegetation (mainly seagrape shrubs) and enclosed some of them with lizard-proof screens (**FIG. 53-6**). Some of the plots were emptied of all lizards; each control enclosure had approximately nine lizards. Nine web-building spider species were observed in the enclosures. Spiders were counted approximately 30 times from 1989 to 1994. During the 4.5 years of observations reported here, spider population densities were higher in the lizard-free enclosures than in enclosures with lizards. Moreover, the enclosures without lizards had more species of spiders. Therefore, we might conclude that lizards control spider populations.

Even this relatively simple experiment, though, may be explained by a combination of two density-dependent factors: predation (lizards eat spiders) and interspecific competition (lizards compete with spiders for insect prey; i.e., both spiders and lizards eat insects). In this experiment the effects of the two density-dependent factors in determining spider population size cannot be evaluated separately. Additional lines of evidence support the actions of both competition and predation at these sites.

What is the influence of density-dependent factors on population size in a natural habitat?

HYPOTHESIS: Lizards reduce the size of spider populations.

EXPERIMENT: Researchers conducted a field experiment in which they constructed enclosed plots (see photograph); spiders and insects could pass freely into and out of the enclosures, but the lizards could not. Some plots included lizards; the other plots were lizard-free. The spiders were periodically counted over a period of 4.5 years. An earlier experiment indicated that the enclosures themselves have no effect on the number of web-building spiders.

KEY

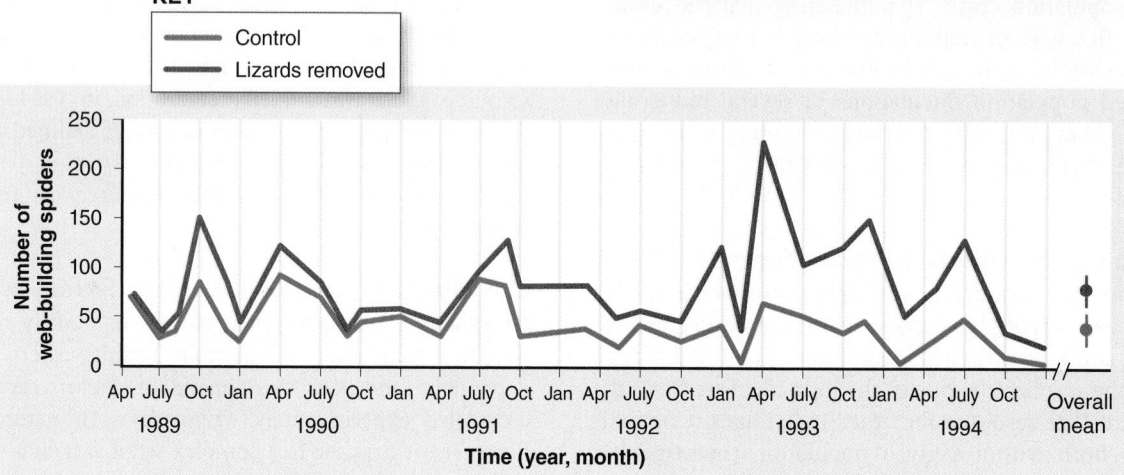

RESULTS AND CONCLUSION: The graph of the mean number of web-building spiders at each count per enclosure with lizards (*control, blue*) and per enclosure with lizards removed (*red*) demonstrates that the number of web-building spiders was consistently higher in the absence of lizards. Lizards may control spider populations by preying on them, by competing with them for insect prey, or by a combination of these factors.

SOURCE: Graph from Spiller, D.A., and T.W. Schoener. "Lizards Reduce Spider Species Richness by Excluding Rare Species." *Ecology*, Vol. 79, No. 2, 1998. Copyright © 1998 Ecological Society of America. Reprinted with permission.

Figure 53-6 Interaction of density-dependent factors

PREDICT Note that the spider populations fluctuated during the experiment, although the lizard-free populations were consistently larger. Formulate a hypothesis relating to whether such fluctuations are due to biotic factors, abiotic factors, or a combination, and devise a way to test it.

Density-independent factors are generally abiotic

Any environmental factor that affects the size of a population but is not influenced by changes in population density is called a **density-independent factor.** Such factors are typically abiotic. Random weather events that reduce population size serve as density-independent factors. They often affect population density in unpredictable ways. A killing frost, severe blizzard, or hurricane, for example, may cause extreme and irregular reductions in a vulnerable population, regardless of its size, and thus may be considered largely density-independent.

Consider a density-independent factor that influences mosquito populations in arctic environments. These insects produce several generations per summer and achieve high population densities by the end of the season. A shortage of food does not seem to be a limiting factor for mosquitoes, nor is there any shortage of ponds in which to breed. What puts a stop to the skyrocketing mosquito population is winter. Not a single adult mosquito survives winter, and the entire population must grow afresh the next summer from the few eggs and hibernating larvae that survive. Thus, severe winter weather is a density-independent factor that affects arctic mosquito populations.

Density-independent and density-dependent factors are often inter-related. Social animals, for example, often resist dangerous weather conditions by collective behavior, as in the case of sheep huddling together in a snowstorm. In this case it appears that the greater the population density of the sheep, the better their ability to resist the environmental stress of a density-independent event (such as a snowstorm).

CHECKPOINT 53.3

- *What are three examples of density-dependent factors that affect population growth?*
- *What are three density-independent factors?*

53.4 LIFE HISTORY TRAITS

LEARNING OBJECTIVES

5 Contrast semelparous and iteroparous reproduction.
6 Distinguish among species exhibiting an *r* strategy, those with a *K* strategy, and those that do not easily fit either category.
7 Describe Type I, Type II, and Type III survivorship curves and explain how life tables and survivorship curves indicate mortality and survival.

Each species is uniquely suited to its lifestyle. Many years pass before a young magnolia tree flowers and produces seeds, whereas an annual plant grows from seed, flowers, and dies within a single season. A mating pair of black-browed albatrosses produces a single chick every year, but a mating pair of gray-headed albatrosses produces a single chick biennially (every other year).

Species that expend their energy in a single, immense reproductive effort are said to be **semelparous.** Most insects and invertebrates, many plants, and some species of fish exhibit semelparity. Pacific salmon, for example, hatch in fresh water and swim to the ocean, where they live until they mature. Adult salmon swim from the ocean back into the same rivers or streams in which they hatched to spawn (reproduce). After they spawn, the salmon die.

Agaves are semelparous plants that are common in arid tropical and semitropical areas. The thick, fleshy leaves of the agave plant are crowded into a rosette at the base of the stem. Commonly called the century plant because it was mistakenly thought to flower only once in a century, agaves can flower after they are ten years old or so, after which the entire plant dies (FIG. 53-7).

Many species are **iteroparous** and exhibit repeated reproductive cycles—that is, reproduction during several breeding seasons—throughout their lifetimes. Iteroparity is common in most vertebrates as well as in perennial herbaceous plants, shrubs, and trees. The timing of reproduction, earlier or later in life, is a crucial aspect of iteroparity and involves trade-offs. On the one hand, reproducing earlier in life may mean a reduced likelihood of survival (because an individual is expending energy toward reproduction instead of its own growth), which reduces the potential for later reproduction. On the other hand, reproducing later in life means that the individual has less time for additional reproductive events.

R. Gustafson/Visuals Unlimited

Figure 53-7 Semelparity
Agaves flower once and then die. The agave is a succulent with sword-shaped leaves arranged as a rosette around a short stem. Shown is *Agave shawii*, whose leaves grow to 61 cm (2 ft). Note the floral stalk, which can grow more than 3 m (10 ft) tall.

Ecologists try to understand the adaptive consequences of various **life history traits,** such as semelparity and iteroparity. Adaptations such as reproductive rate, age at maturity, and **fecundity** (potential capacity to produce offspring), all of which are a part of a species' life history traits, influence an organism's survival and reproduction. The ability of an individual to reproduce successfully, thereby making a genetic contribution to future generations of a population, is called its **fitness** (recall the discussion of fitness in Chapter 19).

Although many different life histories exist, some ecologists recognize two extremes: *r*-selected species and *K*-selected species. As you read the following descriptions of *r* selection and *K* selection, keep in mind that these concepts, although useful, oversimplify most life histories. Species tend to possess a combination of *r*-selected and *K*-selected traits as well as traits that cannot be classified as either *r*-selected or *K*-selected. In addition, some populations within a species may exhibit the characteristics of *r* selection, whereas other populations in different environments may assume *K*-selected traits.

Populations described by the concept of *r* selection have traits that contribute to a high population growth rate. Recall that *r* designates the per capita growth rate. Because such organisms have a high *r*, biologists call them *r* strategists or *r*-selected species. Small body size, early maturity, short lifespan, large broods, and little or no parental care are typical of many *r* strategists, which are usually opportunists found in variable, temporary, or unpredictable environments where the probability of long-term survival is low. Some of the best examples of *r* strategists are insects, such as mosquitoes, and plants such as the poppies shown in the chapter-opening photograph. Following a rainy period, these desert annuals rapidly grow from seed, flower and set seed, and then die.

In populations described by the concept of *K* selection, traits maximize the chance of surviving in an environment where the number of individuals is near the carrying capacity (*K*) of the environment. These organisms, called *K* strategists or *K*-selected species, do not produce large numbers of offspring. They characteristically have long lifespans with slow development, late reproduction, large body size, and a low reproductive rate. *K* strategists tend to be found in relatively constant or stable environments, where they have a high competitive ability. Redwood trees as well as most tropical rainforest trees are classified as *K* strategists. Animals that are *K* strategists typically invest in parental care of their young. Tawny owls (*Strix aluco*), for example, are *K* strategists that pair-bond for life, with both members of a pair living and hunting in adjacent, well-defined territories. Their reproduction is regulated in accordance with the resources, especially the food supply, present in their territories. In an average year, 30% of the birds do not breed at all. If food supplies are more limited than initially indicated, many of those that do breed fail to incubate their eggs. Rarely do the owls lay the maximum number of eggs that they are physiologically capable of laying, and breeding is often delayed until late in the season, when the rodent populations on which they depend have become large. Thus, the behavior of tawny owls ensures better reproductive success of the individual and leads to a stable population at or near the carrying capacity of the environment. Starvation, an indication that the tawny owl population has exceeded the carrying capacity, rarely occurs.

Life tables and survivorship curves indicate mortality and survival

A **life table** can be constructed to show the mortality and survival data of a population or **cohort,** a group of individuals of the same age, at different times during their lifespan. Insurance companies were the first to use life tables, using them to calculate the relationship between a client's age and

the likelihood of the client surviving to pay enough insurance premiums to cover the cost of the policy. Ecologists construct such tables for animals and plants based on data that rely on a variety of population sampling methods and age determination techniques.

TABLE 53-2 shows a life table for a cohort of 530 gray squirrels. The first two columns show the units of age (years) and the number of individuals in the cohort that were alive at the beginning of each age interval (the actual data collected in the field by the ecologist). The values for the third column (the proportion alive at the beginning of each age interval) are calculated by dividing each number in column 2 by 530, the number of squirrels in the original cohort. The values in the fourth column (the proportion dying during each age interval) are calculated using the values in the third column and subtracting the number of survivors at the beginning of the next interval from those alive at the beginning of the current interval. For example, the proportion dying during interval 0–1 years is $1.000 - 0.253 = 0.747$. The far right column, the death rate for each age interval, is calculated by dividing the proportion dying during the age interval (column 4) by the proportion alive at the beginning of the age interval (column 3). For example, the death rate for the age interval 1–2 years is $0.147 \div 0.253 = 0.581$.

Survivorship is the probability that a given individual in a population or cohort will survive to a particular age. Plotting the logarithm (base 10) of the number of surviving individuals against age, from birth to the maximum age reached by any individual, produces a **survivorship curve. FIGURE 53-8** shows the three main survivorship curves that ecologists recognize.

In *Type I survivorship,* as exemplified by bison and humans, the young and those at reproductive age have a high probability of surviving. The probability of survival decreases more rapidly with increasing age; mortality is concentrated later in life. **FIGURE 53-9** shows a survivorship curve for a natural population of Drummond phlox, an annual native to

TABLE 53-2	Life Table for a Cohort of 530 Gray Squirrels (*Sciurus carolinensis*)			
AGE INTERVAL (YEARS)	NUMBER ALIVE AT BEGINNING OF AGE INTERVAL	PROPORTION ALIVE AT BEGINNING OF AGE INTERVAL	PROPORTION DYING DURING AGE INTERVAL	DEATH RATE FOR AGE INTERVAL
0–1	530	1.000	0.747	0.747
1–2	134	0.253	0.147	0.581
2–3	56	0.106	0.032	0.302
3–4	39	0.074	0.031	0.418
4–5	23	0.043	0.021	0.488
5–6	12	0.022	0.013	0.591
6–7	5	0.009	0.006	0.666
7–8	2	0.003	0.003	1.000
8–9	0	0.000	0.000	—

Source: Adapted from Smith, R.L., and T.M. Smith. *Elements of Ecology,* 4th ed., Table 13.1, p. 150. Benjamin/Cummings Science Publishing, San Francisco, 1998.

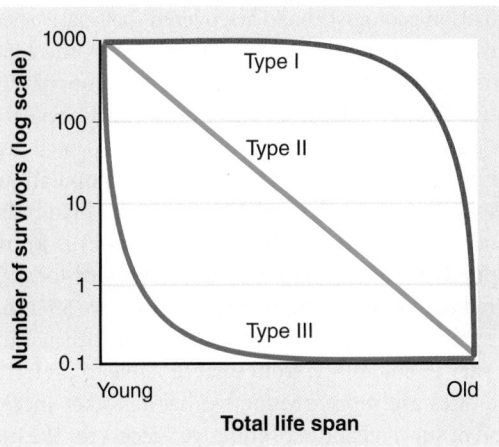

Figure 53-8 *Animation* **Survivorship curves**

These curves represent the ideal survivorships of species in which mortality is greatest in old age (Type I), spread evenly across all age groups (Type II), and greatest among the young (Type III). The survivorship of most organisms can be compared to these curves.

© Cengage Learning

East Texas that became widely distributed in the southeastern United States after it escaped from cultivation. Because most Drummond phlox seedlings survive to reproduce, after germination the plant exhibits a Type I survivorship that is typical of annuals.

In *Type III survivorship* the probability of mortality is greatest early in life, and those individuals that avoid early death subsequently have a high probability of survival, that is, the probability of survival increases with increasing age. Type III survivorship is characteristic of oysters; young oysters have three free-swimming larval stages before settling down and secreting a shell. These larvae are vulnerable to predation, and few survive to adulthood.

In *Type II survivorship,* which is intermediate between Types I and III, the probability of survival does not change with age. The probability of death is equally likely across all age groups, resulting in a linear decline in survivorship. This constancy probably results from essentially random events that cause death with little age bias. Although this relationship between age and survivorship is rare, some lizards have Type II survivorship.

The three survivorship curves are generalizations, and few populations exactly fit one of the three. Some species have one type of survivorship curve early in life and another type as adults. Herring gulls, for example, start out with a Type III survivorship curve but develop a Type II curve as adults (**FIG. 53-10**). The survivorship curve shown in this figure is characteristic of birds in general. Note that most death occurs almost immediately after hatching, despite the protection and care given to the chicks by the parent bird. Herring gull chicks die from predation or attack by other herring gulls, inclement weather, infectious diseases, or starvation following death of the parent. Once the chicks become independent, their survivorship increases dramatically, and death occurs at about the same rate throughout their remaining lives. Few or no herring gulls die from the degenerative diseases of "old age" that cause death in most humans.

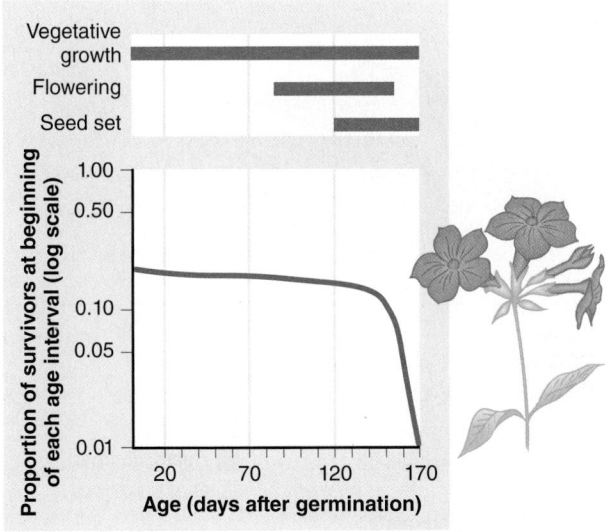

Figure 53-9 Survivorship curve for a Drummond phlox population

Drummond phlox has a Type I survivorship after germination of the seeds. *Bars above the graph* indicate the various stages in the Drummond phlox life history. Data were collected from Nixon, Texas, in 1974 and 1975. Survivorship on the y-axis begins at 0.296 instead of 1.00 because the study took into account death during the seed dormancy period prior to germination (*not shown*). (From Leverich, W.J., and D.A. Levin. "Age-Specific Survivorship and Reproduction in *Phlox drummondii." American Naturalist,* Vol. 113, No. 6, p. 1148, 1979. Copyright © 1979 University of Chicago Press. Reprinted with permission.)

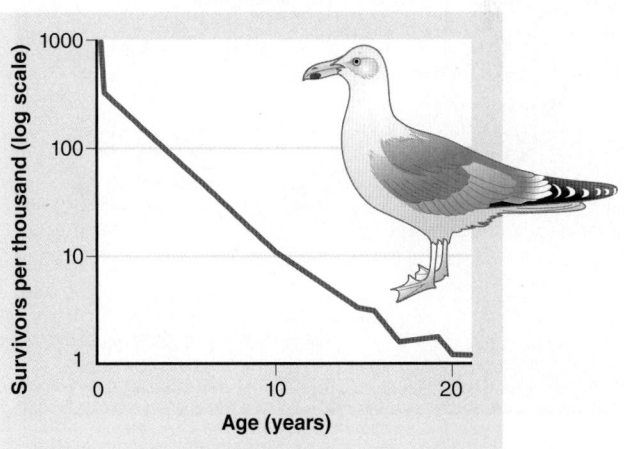

Figure 53-10 Survivorship curve for a herring gull population

Herring gulls (*Larus argentatus*) have Type III survivorship as chicks and Type II survivorship as adults. Data were collected from Kent Island, Maine, during a five-year period in the 1930s; baby gulls were banded to establish identity. The very slight increase just prior to 20 years of age is due to sampling error. (Paynter, R.A., Jr. "A New Attempt to Construct Lifetables for Kent Island Herring Gulls." *Bulletin of the Museum of Comparative Zoology,* Vol. 133, No. 11, pp. 489–528, 1966.)

- *What are the advantages of semelparity? of iteroparity? Are there disadvantages?*
- CONNECT *Why is parental care of young a common characteristic of K strategists?*
- *Do all survivorship curves neatly fit the Type I, II, or III models? Explain.*

53.5 METAPOPULATIONS

LEARNING OBJECTIVE

8 Define *metapopulation* and distinguish between source habitats and sink habitats.

The natural environment is a heterogeneous **landscape** consisting of interacting ecosystems that provide a variety of habitat patches. Landscapes, which are typically several to many square kilometers in area, cover larger land areas than individual ecosystems. Consider a forest, for example. The forest landscape is a mosaic of different elevations, temperatures, levels of precipitation, soil moisture, soil types, and other properties. Because each species has its own habitat requirements, this heterogeneity in physical properties is reflected in the different organisms that occupy the various patches in the landscape (**FIG. 53-11**). Some species occur in very narrow habitat ranges, whereas others have wider habitat distributions.

Population ecologists have discovered that many species are not distributed as one large population across the landscape. Instead, many species exist as a series of local populations distributed in distinct habitat patches. Each local population has its own characteristic demographic features, such as birth, death, emigration, and immigration rates. A population that is divided into several local populations among which individuals occasionally disperse (emigrate and immigrate) is known as a **metapopulation.** For example, note the various local populations of red oak on the mountain slope in Figure 53-11b.

The spatial distribution of a species occurs because different habitats vary in suitability, from unacceptable to preferred. The preferred sites are more productive habitats that increase the likelihood of survival and reproductive success for the individuals living there. Good habitats, called **source habitats,** are areas where local reproductive success is greater than local mortality. *Source populations* generally have greater population densities than populations at less suitable sites, and surplus individuals in the source habitat disperse and find another habitat in which to settle and reproduce.

Individuals living in lower quality habitats may suffer death or, if they survive, poor reproductive success. Lower quality habitats, called **sink habitats,** are areas where local reproductive success is less than local mortality. Without immigration from other areas, a *sink population* declines until extinction occurs. If a local population becomes extinct, individuals from a source habitat may recolonize the vacant habitat at a later time. Source and sink habitats, then, are linked to one another by dispersal (**FIG. 53-12**).

(a) This early spring view from Newfound Gap Rd., Great Smoky Mountains National Park, gives some idea of the heterogeneity of the landscape. During the summer, when all the vegetation is a deep green, the landscape appears homogeneous.

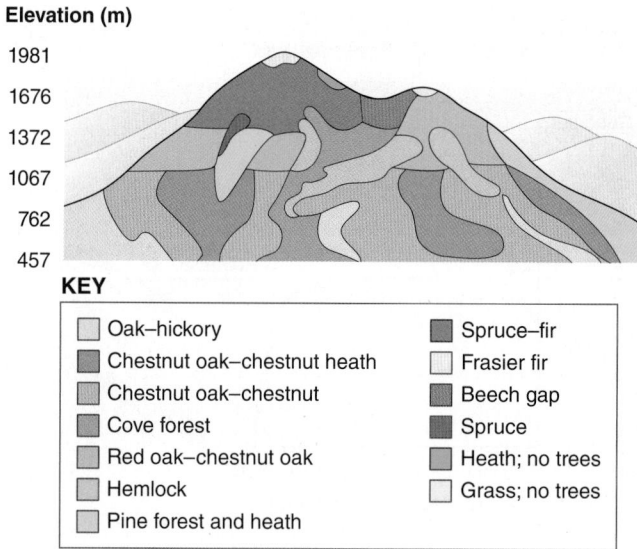

Elevation (m)

1981
1676
1372
1067
762
457

KEY

☐ Oak–hickory	■ Spruce–fir
■ Chestnut oak–chestnut heath	☐ Frasier fir
☐ Chestnut oak–chestnut	■ Beech gap
☐ Cove forest	■ Spruce
☐ Red oak–chestnut oak	■ Heath; no trees
☐ Hemlock	☐ Grass; no trees
☐ Pine forest and heath	

(b) An evaluation of the distribution of vegetation on a typical west-facing slope in the Great Smoky Mountains National Park reveals that the landscape consists of patches. Chestnut oak is a species of oak (*Quercus prinus*); chestnut heath is an area within chestnut oak forest where the trees are widely scattered and the slopes underneath are covered by a thick growth of laurel (*Kalmia*) shrubs; cove forest is a mixed stand of deciduous trees.

Figure 53-11 The mosaic nature of landscapes

(Adapted from Whittaker, R.H. "Vegetation of the Great Smoky Mountains." *Ecological Monographs*, Vol. 26, 1956.)

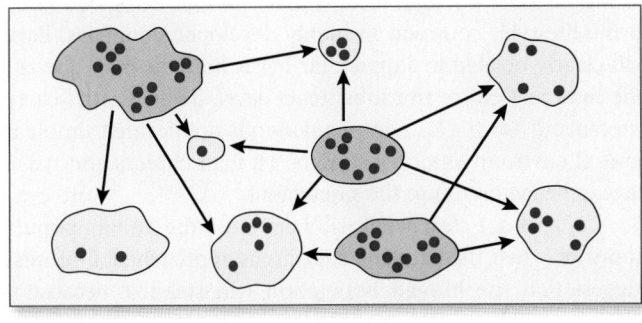

KEY

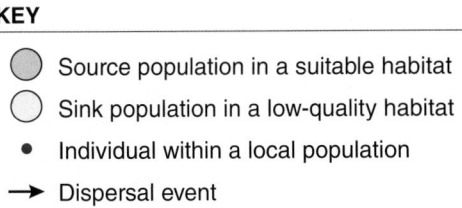

- ⬤ Source population in a suitable habitat
- ◯ Sink population in a low-quality habitat
- • Individual within a local population
- → Dispersal event

Figure 53-12 Source and sink populations in a hypothetical metapopulation

The local populations in the habitat patches shown here collectively make up the metapopulation. Source habitats provide individuals that emigrate and colonize sink habitats. Studied over time, the metapopulation is shown to exist as a shifting pattern of occupied and vacant habitat patches.
© Cengage Learning

Metapopulations are becoming more common as humans alter the landscape by fragmenting existing habitats to accommodate homes and factories, agricultural fields, and logging. As a result, the concept of metapopulations, particularly as it relates to endangered and threatened species, has become an important area of study in conservation biology.

CHECKPOINT 53.5

- **CONNECT** *What roles do source habitats and sink habitats play in the dynamic nature of a metapopulation?*

53.6 HUMAN POPULATIONS

LEARNING OBJECTIVES

9 Summarize the history of human population growth.

10 Explain how highly developed and developing countries differ in population characteristics such as infant mortality rate, total fertility rate, replacement-level fertility, and age structure.

11 Distinguish between people overpopulation and consumption overpopulation.

Now that we have examined some of the basic concepts of population ecology, we can apply those concepts to the human population. Examine **FIGURE 53-13**, which shows the world increase in the human population since the development of agriculture approximately 10,000 years ago. Now look back at Figure 53-2 and compare the two curves. The characteristic J curve of exponential population growth shown in Figure 53-13 reflects the decreasing

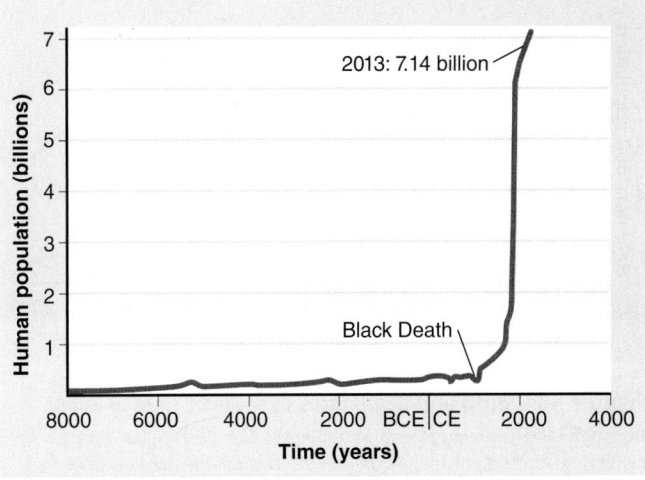

Figure 53-13 Human population growth

During the last 1000 years, the human population has been increasing nearly exponentially. Population experts predict that the population will level out during the 21st century, forming the S curve observed in other species. (Black Death refers to a devastating disease, probably bubonic plague, that decimated Europe and Asia in the 14th century.)
© Cengage Learning

amount of time it has taken to add each additional billion people to our numbers. It took thousands of years for the human population to reach 1 billion, a milestone that took place around 1800. It took 130 years to reach 2 billion (in 1930), 30 years to reach 3 billion (in 1960), 15 years to reach 4 billion (in 1975), 12 years to reach 5 billion (in 1987), and 12 years to reach 6 billion (in 1999). The world population reached 7.14 billion in 2013, and the United Nations projects that it will rise to more than 8 billion by 2025.

Thomas Malthus (1766–1834), a British clergyman and economist, was one of the first to recognize that the human population cannot continue to increase indefinitely (see Chapter 18). He pointed out that human population growth is not always desirable (a view contrary to the beliefs of his day and to those of many people even today) and that the human population is capable of increasing faster than the food supply. He maintained that the inevitable consequences of population growth are famine, disease, and war.

The world population is currently increasing by about 84 million people per year. This rise is not caused by an increase in the birth rate (*b*). In fact, the world birth rate has actually declined during the past 200 years. The increase in population is due instead to a dramatic decrease in the death rate (*d*). This decrease in mortality has occurred primarily because greater food production, better medical care, and improved sanitation practices have increased the life expectancies of a great majority of the global population. For example, from 1920 to 2000, the death rate in Mexico fell from approximately 40 per 1000 individuals to 4 per 1000, whereas the birth rate dropped from approximately 40 per 1000 individuals to 24 per 1000 (**FIG. 53-14**). As of 2013 Mexico's birth rate had continued to fall, to about 19 per 1000, and its death rate remained at 4 per 1000.

The human population has reached a turning point. Although our numbers continue to increase, the world per capita growth rate (*r*) has declined over the past several years, from a peak of

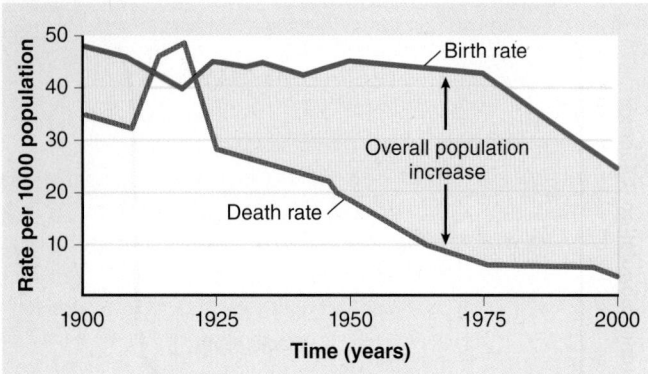

Figure 53-14 Birth and death rates in Mexico, 1900 to 2000
Both birth and death rates declined during the 20th century, but because the death rate declined much more than the birth rate, Mexico has experienced a high growth rate. (The high death rate prior to 1920 was caused by the Mexican Revolution; Population Reference Bureau.)
© Cengage Learning

2.2% per year in the mid-1960s to 1.2% per year in 2013. Population experts at the United Nations and the World Bank have projected that the growth rate may continue to slowly decrease until zero population growth is attained. Thus, exponential growth of the human population may end, and if it does, the S curve will replace the J curve. Demographers project that **zero population growth,** the point at which the birth rate equals the death rate ($r = 0$), will occur toward the end of the 21st century.

The latest (2013) projections available through the Population Reference Bureau, based on data from the United Nations, forecast that the human population will exceed 9.7 billion in the year 2050. This forecast is a "medium" projection; such population projections are "what-if" exercises. Given certain assumptions about future tendencies in natality, mortality, and dispersal, population experts can calculate an area's population for a given number of years into the future. Population projections indicate the changes that may occur, but they must be interpreted with care because they vary depending on what assumptions have been made. Small differences in fertility as well as death rates produce large differences in population forecasts.

The main unknown factor in any population growth scenario is Earth's carrying capacity. According to Joel Cohen of the Earth Institute at Columbia University, most published estimates of how many people Earth can support range from 4 billion to 16 billion. These estimates vary so widely because of the assumptions that are made about standard of living, resource consumption, technological innovations, and waste generation. If we want

all people to have a high level of material well-being equivalent to the lifestyles common in highly developed countries, Earth will clearly be able to support far fewer humans than if everyone lives just above the subsistence level. Thus, Earth's carrying capacity for the human population is not decided simply by natural environmental constraints. Human choices and values have to be factored into the assessment.

It is also not clear what will happen to the human population if or when the carrying capacity is approached. Optimists suggest that the human population will stabilize because of a decrease in the birth rate. Some experts take a more pessimistic view and predict that the widespread degradation of our environment caused by our ever-expanding numbers will make Earth uninhabitable for humans and other species. These experts contend that a massive wave of human suffering and death will occur. Some experts think that the human population has already exceeded the carrying capacity of the environment, a potentially dangerous situation that threatens our long-term survival as a species.

Not all countries have the same growth rate

Although world population figures illustrate overall trends, they do not describe other important aspects of the human population story, such as population differences from country to country. Human **demographics,** the science that deals with human population statistics such as size, density, and distribution, provides information on the populations of various countries. As you probably know, not all parts of the world have the same rates of population increase. Countries can be classified into two groups, highly developed and developing, based on their rates of population growth, degrees of industrialization, and relative prosperity (**TABLE 53-3**).

Highly developed countries, such as the United States, Canada, France, Germany, Sweden, Australia, and Japan, have low rates of population growth and are highly industrialized relative to the rest of the world. Highly developed countries have the lowest birth rates in the world. Indeed, some highly developed countries such as Germany have birth rates just below that needed to sustain the population and are thus

TABLE 53-3	Comparison of 2013 Population Data in Selected Developed and Developing Countries		
	DEVELOPED	DEVELOPING	
	(HIGHLY DEVELOPED) UNITED STATES	(MODERATELY DEVELOPED) BRAZIL	(LESS DEVELOPED) ETHIOPIA
Fertility rate	1.9	1.8	4.8
Doubling time	140 years	78 years	27 years
Infant mortality rate	5.9 per 1000	21 per 1000	52 per 1000
Life expectancy at birth	79 years	74 years	62 years
GNI PPP per capita (U.S. $)	$50,610	$11,720	$1040
Married women 15–49 using modern contraception	73%	77%	27%

Source: Population Reference Bureau.

declining slightly in numbers. Highly developed countries also have low **infant mortality rates** (the number of infant deaths per 1000 live births). The infant mortality rate of the United States in 2013, for example, was 5.9, compared with a world infant mortality rate of 40.

Highly developed countries also have longer life expectancies (79 years at birth in the United States versus 70 years worldwide) and higher average GNI PPP per capita ($50,610 in the United States versus $11,690 worldwide) than other countries. GNI PPP per capita is gross national income (GNI) in purchasing power parity (PPP) divided by midyear population. It indicates the amount of goods and services an average citizen of that particular region or country could buy in the United States.

Developing countries fall into two subcategories: moderately developed and less developed. Mexico, Turkey, Thailand, and most countries of South America are examples of *moderately developed countries.* Their birth rates and infant mortality rates are generally higher than those of highly developed countries, but the rates are declining. Moderately developed countries have a medium level of industrialization, and their average GNI PPP per capita is lower than those of highly developed countries.

Bangladesh, Niger, Ethiopia, Laos, and Cambodia are examples of *less developed countries.* These countries have the highest birth rates, the highest infant mortality rates, the lowest life expectancies, and the lowest average GNI PPP per capita in the world.

One way to represent the population growth of a country is to determine the **doubling time,** the amount of time it would take for its population to double in size assuming that its current growth rate did not change. A simplified formula for doubling time (t_d) is $t_d = 70 \div r$. (The actual formula involves calculus and is beyond the scope of this text. This simplified formula actually has many practical applications; e.g., you can use it to estimate how long it will take to double your money in a savings account at a specific rate of compound interest.)

A look at a country's doubling time identifies it as a highly, moderately, or less developed country: the shorter the doubling time, the less developed the country. At rates of growth in 2013, the approximate doubling time is 30 years for Ethiopia, 35 years for Laos, 58 years for Turkey, 175 years for Thailand, 140 years for the United States, and 175 years for France.

It is also instructive to examine **replacement-level fertility,** the number of children a couple must produce to "replace" themselves. Replacement-level fertility is usually given as 2.1 children in highly developed countries and 2.7 children in developing countries. The number is always greater than 2.0 because some children die before they reach reproductive age. Higher infant mortality rates are the main reason that replacement levels in developing countries are greater than in highly developed countries. The **total fertility rate**—the average number of children born to a woman during her lifetime—is 2.5 worldwide, which is well above replacement levels. However, it has fallen below replacement levels, to 1.6, in the highly developed countries. The total fertility

rate in the United States is 1.9, which is higher than in other industrialized countries.

The population in many developing countries is beginning to approach stabilization. The fertility rate must decline for the population to stabilize (**TABLE 53-4**; note the general decline in total fertility rate from the 1960s to 2013 in selected developing countries). The total fertility rate in developing countries has decreased from an average of 6.1 children per woman in 1970 to 3.0 in 2013 (excluding China), or 2.6 if China is included. Fertility is highly variable among developing countries, and although the fertility rates in most of these countries have declined, it should be remembered that most still exceed replacement-level fertility. Consequently, the populations in these countries are still increasing. Also, even when fertility rates equal replacement-level fertility, population growth will still continue for some time. To understand why, we now examine the age structure of various countries.

The age structure of a country helps predict future population growth

To predict the future growth of a population, it is important to know its **age structure,** which is the number and proportion of people at each age in a population. The number of males and number of females at each age, from birth to death, are represented in an **age structure diagram.**

The overall shape of an age structure diagram indicates whether the population is increasing, stationary, or shrinking (**FIG. 53-15**). The age structure diagram for less developed countries is shaped like a pyramid. Because the largest percentage of the population is in the pre-reproductive age group (i.e., 0 to 14 years of age), the probability of future population growth is great. A strong **population growth momentum** exists because when all these children mature they will become the parents of the next generation, and this group of parents will be larger than the previous group. Thus, even if the fertility rates in

TABLE 53-4	Fertility Changes in Selected Developing Countries	
	TOTAL FERTILITY RATE	
COUNTRY	1960–1965	2013
Bangladesh	6.7	2.3
Brazil	6.2	1.8
China	5.9	1.5
Egypt	7.1	3.0
Guatemala	6.9	3.9
India	5.8	2.4
Kenya	8.1	4.5
Mexico	6.8	2.2
Nepal	5.9	2.6
Nigeria	6.9	6.0
Thailand	6.4	1.6

Source: Population Reference Bureau.

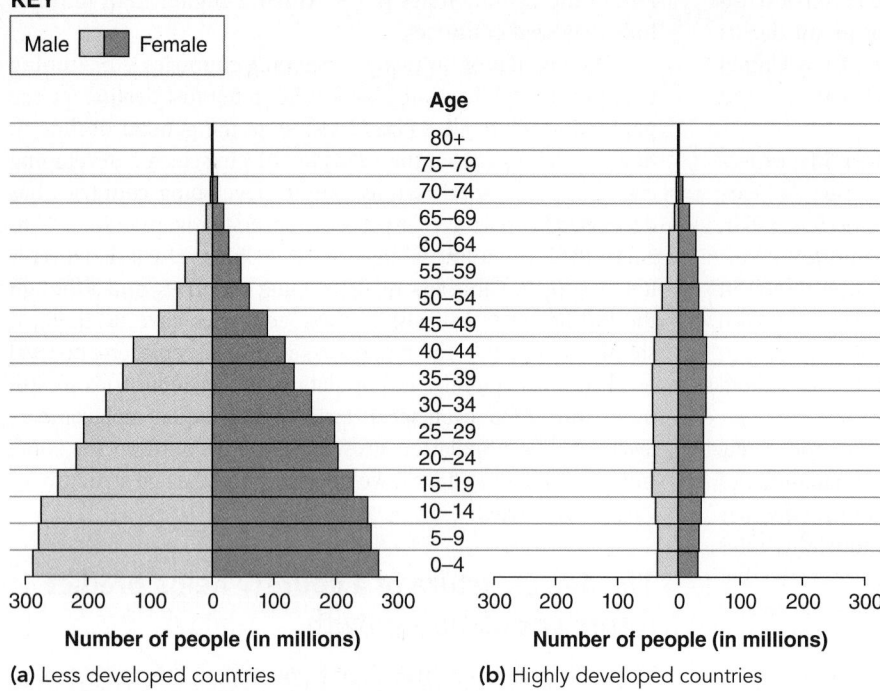

Age

80+
75–79
70–74
65–69
60–64
55–59
50–54
45–49
40–44
35–39
30–34
25–29
20–24
15–19
10–14
5–9
0–4

300 200 100   0   100 200 300

Number of people (in millions)

(a) Less developed countries

300 200 100   0   100 200 300

Number of people (in millions)

(b) Highly developed countries

Figure 53-15 *Animation* **Age structure diagrams**

These age structure diagrams for **(a)** less developed countries and **(b)** highly developed countries indicate that less developed regions have a greater percentage of young people than do highly developed countries. As a result, less developed countries are projected to have greater population growth than are highly developed countries. (Adapted from Population Reference Bureau using data from United Nations, World Population Prospects: The 2002 Revision, 2003.)

these countries decline to replacement level (i.e., couples have smaller families than their parents did), the population will continue to grow for some time. Population growth momentum can have either a positive value (i.e., the population will grow) or a negative value (i.e., the population will decline). However, it is usually discussed in a positive context, to explain how the future growth of a population is affected by its present age distribution.

In contrast, the more tapered bases of the age structure diagrams of highly developed countries with slowly growing, stable, or declining populations indicate that a smaller proportion of the population will become the parents of the next generation. The age structure diagram of a stable population, one that is neither growing nor shrinking, demonstrates that the number of people at pre-reproductive and reproductive ages are approximately the same. Also, a larger percentage of the population is older—that is, post-reproductive—than in a rapidly increasing population. Many countries in Europe have stable populations. In a population that is shrinking in size, the pre-reproductive age group is smaller than either the reproductive or post-reproductive group. Germany, Russia, and Bulgaria are examples of countries with slowly shrinking populations.

Worldwide, 26% of the human population is younger than age 15. When these people enter their reproductive years, they have the potential to cause a large increase in the growth rate. Even if the birth rate does not increase, the growth rate will increase simply because more people are reproducing.

Most of the world population increase since 1950 has taken place in developing countries as a result of the younger age structure and the higher-than-replacement-level fertility rates of their populations. In 1950, 66.8% of the world's population was in the developing countries in Africa, Asia (minus Japan), and Latin America. Between 1950 and 2013, the world's population more than doubled in size, but most of that growth occurred in developing countries. As a reflection of this growth, in 2013 the people in developing countries had increased to more than 82% of the world's population. Most of the population increase that will occur during the 21st century will also take place in developing countries, largely the result of their young age structures. These countries, most of which are poor, are least able to support such growth.

Environmental degradation is related to population growth and resource consumption

The relationships among population growth, use of natural resources, and environmental degradation are complex, but we can make two useful generalizations.

First, although the amount of resources essential to an individual's survival may be small, a rapidly increasing population tends to overwhelm and deplete a country's soils, forests, and other natural resources. Thus, **people overpopulation** occurs when the environment is worsening from too many people, even if those people consume few resources per person. People overpopulation is the current problem in many developing nations.

Second, in affluent highly developed nations, individual resource demands are large, far above requirements for survival. **Consumption overpopulation** occurs when people in more affluent nations exhaust resources and degrade the global environment through excessive consumption and "throwaway" lifestyles.

The effects of human populations on the environment are explored in more detail in Chapter 57.

CHECKPOINT 53.6

- *What is the reason for the dramatic increase in the world population over the last 200 years?*
- CONNECT *Why is replacement-level fertility greater in developing countries than in highly developed countries? Explain your answer.*
- *How can a single child born in the United States have a greater effect on the environment and natural resources than a dozen children born in Kenya?*

53.1 Features of Populations *(page 1145)*

1 Define *population density* and *dispersion,* and describe the main types of population dispersion.

- **Population density** is the number of individuals of a species per unit of area or volume at a given time.

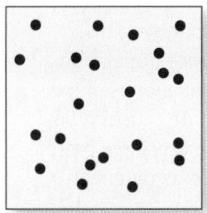 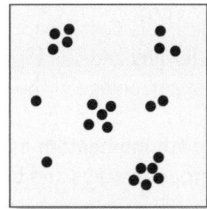 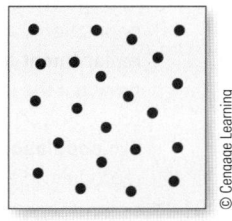

Random dispersion   **Clumped dispersion**   **Uniform dispersion**

- Population **dispersion** (spacing) may be **random dispersion** (unpredictably spaced), **clumped dispersion** (clustered in specific parts of the habitat), or **uniform dispersion** (evenly spaced).

53.2 Changes in Population Size *(page 1147)*

2 Explain the four factors (natality, mortality, immigration, and emigration) that produce changes in population size and solve simple problems involving these changes.

- Population size is affected by the average per capita birth rate (*b*), average per capita death rate (*d*), and two measures of **dispersal:** average per capita **immigration** rate (*i*) and average per capita **emigration** rate (*e*).
- The **growth rate (r)** is the rate of change (increase or decrease) of a population on a per capita basis. On a global scale (when dispersal is not a factor), $r = b - d$. Populations increase in size as long as the average per capita birth rate (**natality**) is greater than the average per capita death rate (**mortality**).
- For a local population (where dispersal is a factor), $r = (b - d) + (i - e)$.

3 Define *intrinsic rate of increase* and *carrying capacity,* and explain the differences between J-shaped and S-shaped growth curves.

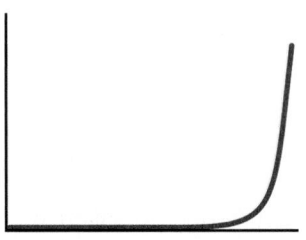 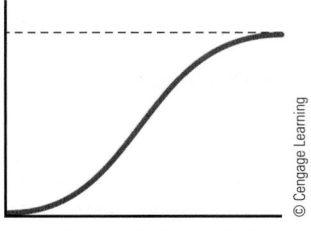

Exponential population growth   **Logistic population growth**

- **Intrinsic rate of increase (r_{max})** is the maximum rate at which a species or population could increase in number under ideal conditions.
- Although certain populations exhibit an accelerated pattern of growth known as **exponential population growth** for limited periods of time (the J-shaped curve), eventually the growth rate decreases to around zero or becomes negative.
- Population size is modified by limits set by the environment. The **carrying capacity (K)** of the environment is the largest population that can be maintained for an indefinite time by

a particular environment. **Logistic population growth,** when plotted, shows a characteristic S-shaped curve. Seldom do natural populations follow the logistic growth curve very closely.

53.3 Factors Influencing Population Size *(page 1149)*

4 Contrast the influences of density-dependent and density-independent factors on population size and give examples of each.

- **Density-dependent factors** regulate population growth by affecting a larger proportion of the population as population density rises. Predation, disease, and competition are examples.
- **Density-independent factors** limit population growth but are not influenced by changes in population density. Hurricanes and blizzards are examples.

53.4 Life History Traits *(page 1153)*

5 Contrast semelparous and iteroparous reproduction.

- **Semelparous** species expend their energy in a single, immense reproductive effort. **Iteroparous** species exhibit repeated reproductive cycles throughout their lifetimes.

6 Distinguish among species exhibiting an *r* strategy, those with a *K* strategy, and those that do not easily fit either category.

- Although many different combinations of life history traits exist, some ecologists recognize two extremes: *r* strategy and *K* strategy.
- An *r* strategy emphasizes a high growth rate. Organisms characterized as **r strategists** often have small body sizes, high reproductive rates, and short **lifespans,** and they typically inhabit variable environments.
- A *K* strategy maintains a population near the carrying capacity of the environment. Organisms characterized as **K strategists** often have large body sizes, low reproductive rates, and long lifespans, and they typically inhabit stable environments.
- The two strategies oversimplify most life histories. Many species combine *r*-selected and *K*-selected traits as well as traits that cannot be classified as either *r*-selected or *K*-selected.

7 Describe Type I, Type II, and Type III survivorship curves and explain how life tables and survivorship curves indicate mortality and survival.

- A **life table** shows the mortality and survival data of a population or **cohort**—a group of individuals of the same age—at different times during the population's lifespan.
- **Survivorship** is the probability that a given individual in a population or cohort will survive to a particular age. There are three general **survivorship curves:** *Type I survivorship,* in which mortality is greatest in old age; *Type II survivorship,* in which mortality is spread evenly across all age groups; and *Type III survivorship,* in which mortality is greatest among the young.

53.5 Metapopulations *(page 1156)*

8 Define *metapopulation* and distinguish between source habitats and sink habitats.

- Many species exist as a **metapopulation,** a set of local populations among which individuals are distributed in distinct habitat patches across a **landscape,** which is a large area of terrain (several to many square kilometers) composed of interacting ecosystems.

- Within a metapopulation, individuals occasionally disperse from one local habitat to another by emigration and immigration. **Source habitats** are preferred sites where local reproductive success is greater than local mortality. Surplus individuals disperse from source habitats. **Sink habitats** are lower quality habitats where individuals may suffer death or, if they survive, poor reproductive success. If extinction of a local *sink population* occurs, individuals from a *source population* may recolonize the vacant habitat at a later time.

53.6 Human Populations *(page 1157)*

9 Summarize the history of human population growth.

- The world population increases by more than 84 million people per year and reached 7.14 billion in 2013.
- Although our numbers continue to increase, the per capita growth rate (r) has declined over the past several years, from a peak in 1965 of about 2% per year to a 2013 growth rate of 1.2% per year.
- Scientists who study human **demographics** (human population statistics) project that the world population will become stationary ($r = 0$, or **zero population growth**) by the end of the 21st century.

10 Explain how highly developed and developing countries differ in population characteristics such as infant mortality rate, total fertility rate, replacement-level fertility, and age structure.

- *Highly developed countries* have the lowest birth rates, lowest infant mortality rates, lowest total fertility rates, longest life expectancies, and highest GNI PPP per capita (a measure of the amount of goods and services the average citizen could purchase in the United States) of all countries. *Developing countries* have the highest birth rates, highest infant mortality rates, highest total fertility rates, shortest life expectancies, and lowest GNI PPP per capita.
- The **age structure** of a population greatly influences population dynamics. It is possible for a country to have **replacement-level fertility** and still experience population growth if the largest percentage of the population is in the pre-reproductive years. A young age structure causes a positive **population growth momentum** as the very large pre-reproductive age group matures and becomes parents.

11 Distinguish between people overpopulation and consumption overpopulation.

- Developing countries tend to have **people overpopulation,** in which population increase degrades the environment even though each individual uses few resources.
- Highly developed countries tend to have **consumption overpopulation,** in which each individual in a slow-growing or stationary population consumes a large share of resources, which results in environmental degradation.

TEST YOUR UNDERSTANDING

Know and Comprehend

1. Population _____ is the number of individuals of a species per unit of habitat area or volume at a given time. (a) dispersion (b) density (c) survivorship (d) age structure (e) demographics
2. The per capita growth rate of a population where dispersal is not a factor is expressed as (a) $i + e$ (b) $b - d$ (c) dN/dt (d) $rN(K - N)$ (e) $(K - N) \div K$
3. The maximum rate at which a population could increase under ideal conditions is known as its (a) total fertility rate (b) survivorship (c) intrinsic rate of increase (d) doubling time (e) age structure
4. When r is a positive number, the population size is (a) stable (b) increasing (c) decreasing (d) either increasing or decreasing, depending on interference competition (e) either increasing or stable, depending on whether the species is semelparous
5. In a graph of population size versus time, a J-shaped curve is characteristic of (a) exponential population growth (b) logistic population growth (c) zero population growth (d) replacement-level fertility (e) population growth momentum
6. The largest population that can be maintained by a particular environment for an indefinite period is known as a (a) semelparous population (b) population undergoing exponential growth (c) metapopulation (d) population's carrying capacity (e) source population
7. Giant bamboos live many years without reproducing, then send up a huge flowering stalk and die shortly thereafter. Giant bamboo is therefore an example of (a) iteroparity (b) a source population (c) a metapopulation (d) an r strategist (e) semelparity
8. Predation, disease, and competition are examples of _____ factors. (a) density-dependent (b) density-independent (c) survivorship (d) dispersal (e) semelparous

9. _____ competition occurs within a population, and _____ competition occurs among populations of different species. (a) Interspecific; intraspecific (b) Intraspecific; interspecific (c) Type I survivorship; Type II survivorship (d) Interference; exploitation (e) Exploitation; interference
10. A highly developed country has a (a) long doubling time (b) low infant mortality rate (c) high GNI PPP per capita (d) a and b (e) a, b, and c
11. The continued growth of a population with a young age structure, even after its fertility rate has declined, is known as (a) population doubling (b) iteroparity (c) population growth momentum (d) r selection (e) density dependence

Apply and Analyze

12. Which of the following patterns of cars parked along a street is an example of uniform dispersion? (a) five cars parked next to one another in the middle, leaving two empty spaces at one end and three empty spaces at the other end (b) five cars parked in this pattern: car, empty space, car, empty space, and so on (c) five cars parked in no discernible pattern, sometimes having empty spaces on each side and sometimes parked next to another car
13. A female elephant bears a single offspring every two to four years. Based on this information, which survivorship curve do you think is representative of elephants? Explain your answer.

Evaluate and Synthesize

14. Explain why the population size of a species that competes by interference competition is often near the carrying capacity, whereas the population size of a species that competes by exploitation competition is often greater than or below the carrying capacity.

15. **EVOLUTION LINK** In developing his scientific theory of evolution by natural selection, Charles Darwin considered four main observations: heritable variation among individuals in a population, overproduction of offspring, limits on population growth, and differential reproductive success. Compare how these observations may apply to populations that tend to fit the concepts of *r* selection and *K* selection.

16. **INTERPRET DATA** In Bolivia, 35% of the population is younger than age 15, and 5% is older than 65. In Austria, 14% of the population is younger than 15, and 18% is older than 65. Which country will have the highest growth rate over the next two decades? Explain your answer.

17. **INTERPRET DATA** The 2013 population of the Netherlands was 16.8 million, and its land area is 15,768 square miles. The 2013 population of the United States was 316.2 million, and its land area is 3,717,796 square miles. Which country has the greater population density?

18. **INTERPRET DATA** The population of India in 2013 was 1,276.5 million, and its growth rate was 1.5% per year. Calculate the 2014 population of India.

19. **INTERPRET DATA** The world population in 2013 was 7.14 billion, and its annual growth rate was 1.2%. If the birth rate was 20 per 1000 people in the year 2013, what was the death rate, expressed as number per 1000 people?

20. **INTERPRET DATA** Consider the age structure diagrams for counties (a) and (b). Which diagram is consistent with negative growth momentum? Why?

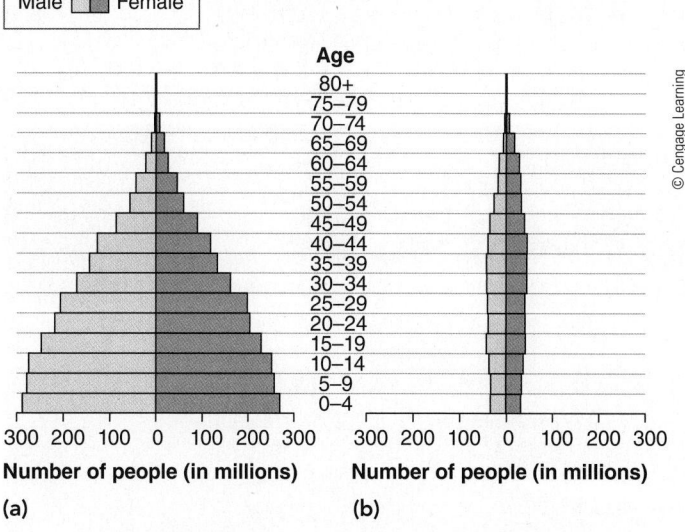

21. **SCIENCE, TECHNOLOGY, AND SOCIETY** In what ways has technology contributed to consumption overpopulation? Can you propose some applications of technology that might help alleviate its effects?

aplia   To access course materials, such as Aplia and other companion resources, please visit **www.cengagebrain.com.**

54 | Community Ecology

Mangrove root community. Spreading mangrove roots at the water and land interface provide shelter for many species of animals including fish, invertebrates, and amphibians.

Luis Javier Sandoval/Getty Images

I n Chapter 53 we examined the dynamics of populations. In the natural world, most populations are part of a **community,** which consists of an association of populations of different species that live and interact in the same place at the same time. The definition for *community* is deliberately broad because it refers to ecological categories that vary greatly in size, lack precise boundaries, and are rarely completely isolated.

Smaller communities nest within larger communities. A mangrove forest is a community, but so are the interlacing roots that support each tree (see photograph). Mangrove roots, anchored in sediment made up of sand and peat, are breeding grounds and nurseries for crabs, shrimp, and many commercially important fish. Blue crabs, lizards, and turtles feed on insects, algae, and dead animal material. Small mammals scurry among the roots, and manatees and alligators find shelter from the tropical sun. Mangrove root surfaces are home to invertebrates, algae, bacteria, archaea, and fungi, which provide nutrients for other organisms. Mangroves also filter excess nutrients and some toxic pollutants from the water.

Organisms exist in an abiotic (nonliving) environment that is as essential to their lives as their interactions with one another. Minerals, air, water, and sunlight are just as much a part of a bat's environment, for example, as the flowers it pollinates and from which it takes nectar and insects. A biological community and its abiotic environment together compose an **ecosystem.** Like communities, ecosystems are broad entities that refer to ecological units of various sizes.

We now present some of the challenges in trying to find common patterns and processes that govern communities. The living community is emphasized in this chapter. The abiotic components of ecosystems, including energy flow and trophic structure (feeding relationships), nutrient cycling, and climate, are considered in Chapter 55.

54.1 COMMUNITY STRUCTURE AND FUNCTIONING

LEARNING OBJECTIVES

1 Define *ecological niche* and distinguish between an organism's fundamental niche and its realized niche.

2 Define *competition* and distinguish between interspecific and intraspecific competition.

3 Summarize the concepts of the competitive exclusion principle, resource partitioning, and character displacement.

4 Define *predation* and describe the effects of natural selection on predator–prey relationships.

5 Distinguish among mutualism, commensalism, and parasitism, and give examples of each.

Communities exhibit characteristic properties that populations lack. These properties, known collectively as *community structure* and *community functioning,* include the number and types of species present, the relative abundance of each species, the interactions among different species, community resilience to disturbances, energy and nutrient flow throughout the community, and productivity. **Community ecology** is the description and analysis of patterns and processes within the community. Finding common patterns and processes in a wide variety of communities—for example, a pond community, a pine forest community, and a sagebrush desert community—helps ecologists understand community structure and functioning.

Communities are exceedingly difficult to study because a large number of individuals of many different species interact with one another and are interdependent in a variety of ways. Species compete with one another for food, water, living space, and other resources. (Used in this context, a *resource* is anything from the environment that meets needs of a particular species.) Some organisms kill and eat other organisms or cause disease.

Some species form intimate associations with one another, whereas other species seem only distantly connected. Certain species interact in positive ways in a process known as **facilitation,** which modifies and enhances the local environment for other species. For example, alpine plants in harsh mountain environments grow faster and larger and reproduce more successfully when certain other plants are growing nearby (**FIG. 54-1**).

(a) *Laretia acaulis* plants increase the survival rate of certain nearby plants at 2800 m (9100 ft) in the Andes Mountains of Chile. The effect is less pronounced at higher elevations.

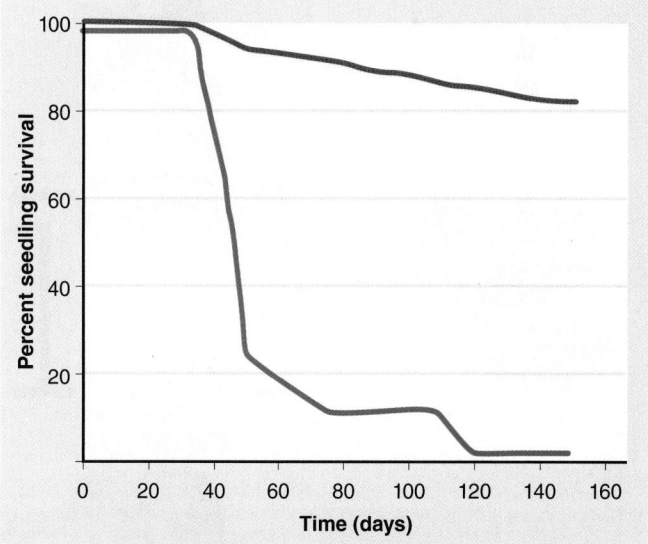

(b) Field mouse-ear chickweed (*Cerastium arvense*) plants were planted and grown within (*red*) and outside (*blue*) clumps of *Laretia acaulis.*

(c) Alpine barley (*Hordeum comosum*) plants were planted and grown within (*red*) and outside (*blue*) clumps of *Laretia acaulis.*

Figure 54-1 Facilitation in alpine communities

(**b** and **c,** Adapted from Cavieres, L.A., et al. "Positive Interactions between Alpine Plant Species and the Nurse Cushion Plants *Laretia acaulis* Do Not Increase with Elevation in the Andes of Central Chile." 2006, *New Phytologist,* 169:59–69.)

Also, each organism plays one of three main roles in community life: producer, consumer, or decomposer. Unraveling the many positive and negative, direct and indirect interactions of organisms living together as a community is one of the goals of community ecologists.

Community interactions are complex and often not readily apparent

During the late 1990s, an intricate relationship emerged among acorn production, white-footed mice, deer, gypsy moth population growth, and the potential occurrence of Lyme disease in humans (**FIG. 54-2**). Large-scale experiments conducted in oak forests of the northeastern United States linked bumper acorn crops, which occur every three to four years, to booming mouse populations (the mice eat the acorns). Because the mice also eat gypsy moth pupae, high acorn conditions also lead to low populations of gypsy moths. This outcome helps the oaks because gypsy moths cause serious defoliation. However, abundant acorns also attract tick-bearing deer to oak forests. The ticks' hungry offspring feed on the mice, which often carry the Lyme disease–causing bacterium. The bacterium infects the maturing ticks and, in turn, spreads to humans who are bitten by affected ticks. In 2006, scientists concluded that it is possible to predict which years pose the greatest potential threat of Lyme disease to humans based on when oaks are most productive. They determined that the risk of Lyme disease is greater two years following a bumper acorn crop than in other years.

Within a community, no species exists independently of other species. As the preceding example shows, the populations of a community interact with and affect one another in complex ways that are not always obvious. Three main types of interactions occur among species in a community: competition, predation, and symbiosis. Before we address these interactions, however, we need to examine the way of life of a given species in its community.

The niche is a species' ecological role in the community

Every species is thought to have its own ecological role within the structure and function of a community; we call this role its **ecological niche.** Although the concept of ecological niche has been in use in ecology since early in the 20th century, Yale ecologist G. E. Hutchinson first described in 1957 the multidimensional nature of the niche that is accepted today. An ecological niche takes into account all biotic and abiotic aspects of the species' existence, that is, all the physical, chemical, and biological factors that the species needs to survive, remain healthy, and reproduce. A niche includes the local environment in which a species lives, its **habitat.** A niche also encompasses what a species eats, what eats it, what organisms it competes with, and how it interacts with and is influenced by the abiotic components of its environment, such as light, temperature, and moisture. The niche thus represents the totality of adaptations by a species to its environment, its use of resources, and the lifestyle to which it is suited. Although

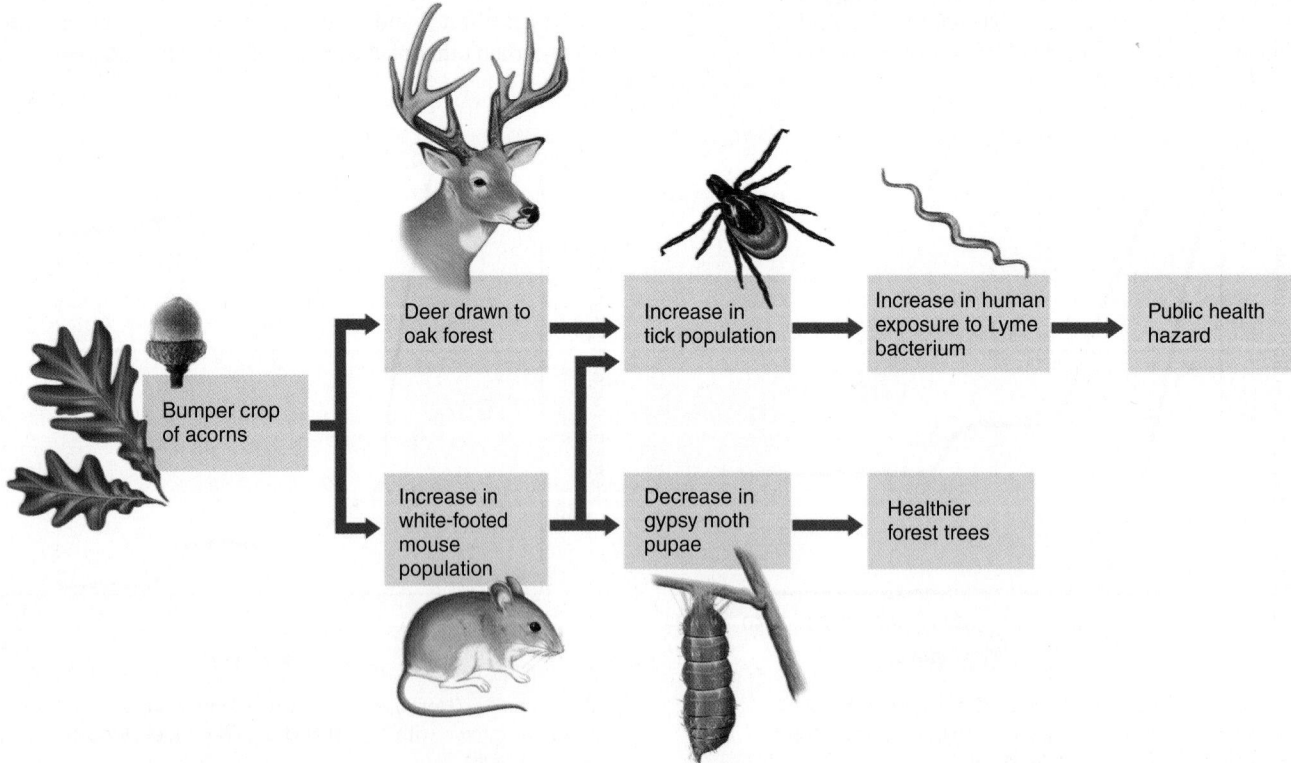

Figure 54-2 Connections to the size of an acorn crop

When there is a bumper crop of acorns, more mice survive and breed in winter, and more deer are attracted to oak forests. Both mice and deer are hosts of ticks that may carry the Lyme disease bacterium to humans. Abundant mice also reduce gypsy moth populations, thereby improving the health of forest trees.

© Cengage Learning

a complete description of an organism's ecological niche involves many dimensions and is difficult to define precisely, ecologists usually confine their studies to one or a few niche variables, such as feeding behaviors or ability to tolerate temperature extremes.

The ecological niche of a species is far broader hypothetically than in actuality. A species is usually capable of using much more of its environment's resources or of living in a wider assortment of habitats than it actually does. The potential ecological niche of a species is its **fundamental niche,** but various factors, such as competition with other species, may exclude it from part of this fundamental niche. Thus, the lifestyle that a species actually pursues and the resources it actually uses make up its **realized niche.**

An example may help make this distinction clear. The green anole, a lizard native to Florida and other southeastern states, perches on trees, shrubs, walls, or fences during the day, waiting for insect and spider prey (FIG. 54-3a). In the past, these little lizards were widespread in Florida. Several decades ago, however, a related species, the brown anole introduced from Cuba into southern Florida, spread northward up the Florida peninsula and quickly became common (FIG. 54-3b). Suddenly, the green anoles became rare, apparently driven out of their habitat by competition from the slightly larger brown anoles. Careful

investigation, however, disclosed that green anoles were still around. They were now confined largely to the vegetation in wetlands and to the foliated crowns of trees, where they were less easily seen.

The habitat portion of the green anole's fundamental niche includes the trunks and crowns of trees, exterior house walls, and many other locations. Once the brown anoles became established in the green anole habitat, they drove the green anoles from all but wetlands and tree crowns; competition between the two species shrank the green anoles' realized niche (FIGS. 54-3c and d). Because communities consist of numerous species, many of which compete to some extent, the interactions among them produce each species' realized niche.

Limiting resources restrict the ecological niche of a species A species' structural, physiological, and behavioral adaptations determine its tolerance for environmental extremes. If any feature of an environment lies outside the bounds of its tolerance, the species cannot live there. Just as you would not expect to find a cactus living in a pond, you would not expect water lilies in a desert.

The environmental factors that actually determine a species' ecological niche are difficult to identify. For this reason, the concept of ecological niche is largely abstract, although some of its dimensions can be experimentally determined. Any environmental resource that, because it is scarce or unfavorable, tends to restrict the ecological niche of a species is called a **limiting resource.**

Most of the limiting resources that have been studied are simple variables, such as the soil's mineral content, temperature extremes, and precipitation amounts. Such investigations have disclosed that any factor either exceeding the tolerance of a species or present in quantities smaller than the required minimum limits the presence of that species in a community. By their interaction, such factors help define the ecological niche for a species.

Limiting resources may affect only part of an organism's life cycle. For example, although adult blue crabs live in fresh or slightly brackish water, they do not become permanently established in such areas because their larvae (immature forms) require salt water. Similarly, the ring-necked pheasant, a popular game bird native to Eurasia, has been widely introduced in North America but has not become

(a) Green anole lizard with dragonfly prey. The green anole (*Anolis carolinensis*) is the only anole species native to North America. Males are about 12.5 cm (5 in.) long; females are slightly smaller. (Photographed on Jefferson Island, Louisiana.)

Danita Delimont/Getty Images

(b) The brown anole (*A. sagrei*), which is 15.2 cm (6 in.) long, was introduced to Florida. It is native to Cuba and the Bahamas.

Robert Clay/Visuals Unlimited, Inc.

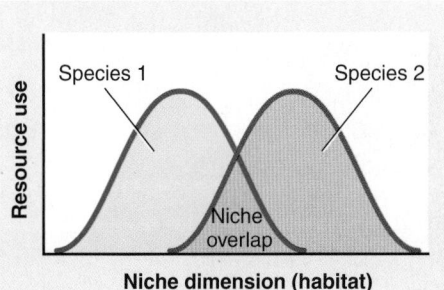

(c) Positions of the two species along a single niche dimension (in this case, habitat). Species 1 represents the green anole, and Species 2 represents the brown anole. The fundamental niches of the two lizards overlap.

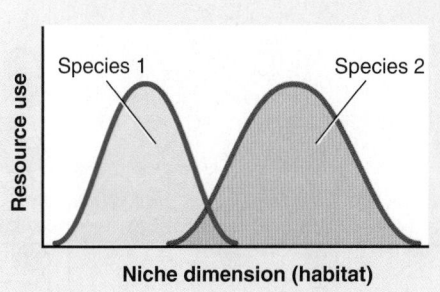

(d) The brown anole outcompetes the green anole in the area where their niches overlap, restricting the niche of the green anole.

Figure 54-3 *Animation* **Effect of competition on an organism's realized niche**
© Cengage Learning

established in the southern United States. The adult birds do well there, but the eggs do not develop properly in the warmer southern temperatures.

Biotic and abiotic factors influence a species' ecological niche In the 1960s, U.S. ecologist Joseph H. Connell investigated biotic and abiotic factors that affect the distribution of two barnacle species in the rocky *intertidal zone* along the coast of Scotland. The intertidal zone is a challenging environment, and organisms in the intertidal zone must tolerate exposure to the drying air during low tides. Barnacles are sessile crustaceans whose bodies are covered by a shell of calcium carbonate (see Fig. 31-21a). When the shell is open, feathery appendages extend to filter food from the water.

Along the coast of Scotland, more adults of one barnacle species, *Semibalanus balanoides* (formerly called *Balanus balanoides*), are attached on lower rocks in the intertidal zone than are adults of the other species, *Chthamalus stellatus* (**FIG. 54-4**). The distributions of the two species do not overlap, although immature larvae of both species live together in the intertidal zone. Connell manipulated the two populations to determine what factors were affecting their distribution. When he removed *Chthamalus* from the upper rocks, *Semibalanus* barnacles did not expand into the vacant area. Connell's experiments showed that *Chthamalus* is more resistant than *Semibalanus* to desiccation

when the tide retreats. However, when Connell removed *Semibalanus* from the lower rocks, *Chthamalus* expanded into the lower parts of the intertidal zone. The two species compete for space, and *Semibalanus*, which is larger and grows faster, outcompetes the smaller *Chthamalus* barnacles.

Competition among barnacle species for the limiting resource of living space was one of the processes that Connell's research demonstrated. We now examine other aspects of competition that various ecologists have revealed in both laboratory and field studies.

Competition is intraspecific or interspecific

Competition occurs when two or more individuals attempt to use the same essential resource, such as food, water, shelter, living space, or sunlight. Because resources are often in limited supply in the environment, their use by one individual decreases the amount available to others (**TABLE 54-1**). If a tree in a dense forest grows taller than surrounding trees, for example, it absorbs more of the incoming sunlight. Less sunlight is therefore available for nearby trees that are shaded by the taller tree. Competition occurs among individuals within a population (**intraspecific competition**) or between different species (**interspecific competition**). (Intraspecific competition was discussed in Chapter 53.)

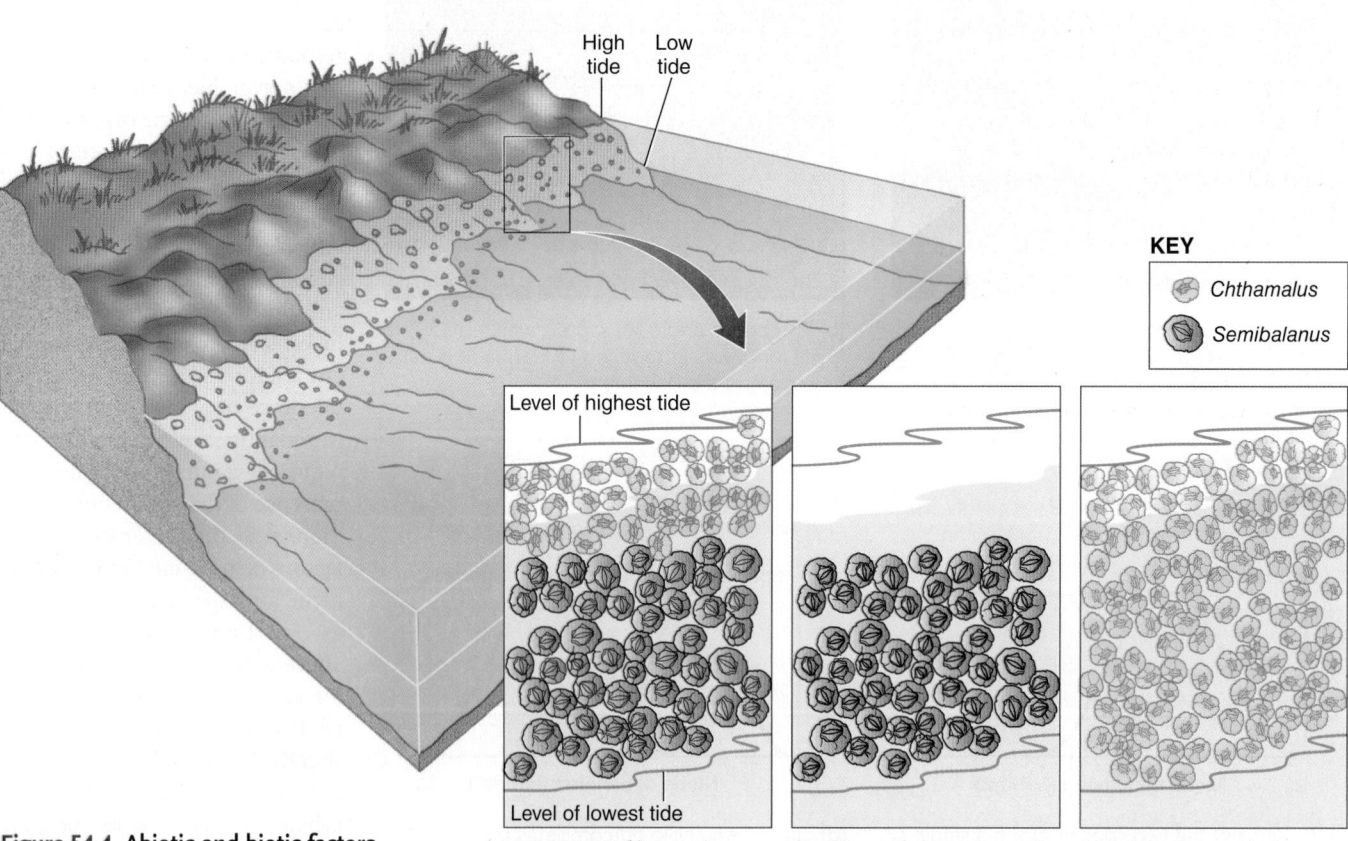

Figure 54-4 Abiotic and biotic factors that affect barnacle distribution

(After Connell, J.H. "The Influence of Interspecific Competition and Other Factors on the Distribution of the Barnacle *Chthamalus stellatus*." *Ecology*, Vol. 42, 1961.)

(a) Species of barnacles belonging to two genera, *Chthamalus* and *Semibalanus*, grow in the intertidal zone of a rocky shore in Scotland.

(b) When *Chthamalus* individuals were experimentally removed, *Semibalanus* individuals did not expand into their section of the rock.

(c) When *Semibalanus* individuals were experimentally removed, *Chthamalus* individuals spread into the empty area.

TABLE 54-1   Ecological Interactions among Species

INTERACTION	EFFECT ON SPECIES 1	EFFECT ON SPECIES 2
Competition between species 1 and species 2	Harmful	Harmful
Predation of species 2 (prey) by species 1 (predator)	Beneficial	Harmful
Symbiosis		
Mutualism of species 1 and species 2	Beneficial	Beneficial
Commensalism of species 1 with species 2	Beneficial	No effect
Parasitism by species 1 (parasite) on species 2 (host)	Beneficial	Harmful

© Cengage Learning

Ecologists traditionally assumed that competition is the most important determinant of both the number of species found in a community and the size of each population. Today ecologists recognize that competition is only one of many interacting biotic and abiotic factors that affect community structure. Furthermore, competition is not always a straightforward, direct interaction. Several kinds of flowering plants, for example, live in a young pine forest and presumably compete with the conifers for such resources as soil moisture and soil minerals. Their relationship, however, is more complex than simple competition. The flowers produce nectar that is consumed by some insect species that also prey on needle-eating insects, thereby reducing the number of insects feeding on pines. It is therefore difficult to assess the overall effect of flowering plants on pines. If the flowering plants were removed from the community, would the pines grow faster because they were no longer competing for necessary resources? Or would the increased presence of needle-eating insects (caused by fewer predatory insects) inhibit pine growth?

Short-term experiments in which one competing plant species is removed from a forest community have often demonstrated an improved growth for the remaining species. However, very few studies have tested the long-term effects on forest species of removing a single competing species. These long-term effects may be subtle, indirect, and difficult to ascertain; they may lessen or negate the short-term effects of competition for resources.

Competition between species with overlapping niches may lead to competitive exclusion When two species are similar, as are the green and brown anoles or the two species of barnacles, their fundamental niches may overlap. However, based on experimental and modeling work, many ecologists think that no two species indefinitely occupy the same niche in the same community. According to the **competitive exclusion principle,** it is hypothesized that one species excludes another from its niche as a result of interspecific competition. Although it is possible for species to compete for some necessary resource without being total competitors, two species with absolutely identical ecological niches cannot coexist. Coexistence occurs, however, if the overlap between the two niches is reduced. In the lizard example, direct competition between the two species

was reduced as the brown anole excluded the green anole from most of its former habitat.

The initial evidence that interspecific competition contributes to a species' realized niche came from a series of laboratory experiments by Russian biologist Georgyi F. Gause in the 1930s. In one study Gause grew populations of two species of protozoa, *Paramecium aurelia* and the larger *P. caudatum,* in controlled conditions (**FIG. 54-5**). When grown in separate test tubes—that is, in the absence of the second species—the population of each species quickly increased to a level imposed by the resources and remained there for some time thereafter. When grown together, however, only *P. aurelia* thrived, whereas *P. caudatum* dwindled and eventually died out. Under different sets of culture conditions, *P. caudatum* prevailed over *P. aurelia.* Gause interpreted these results to mean that although one set of conditions favored one species, a different set favored the other. Nonetheless, because both species were so similar, in time one or the other would eventually triumph. Similar experiments with competing species of fruit flies, mice, beetles, and annual plants have supported Gause's results: one species thrives, and the other eventually dies out.

Thus, competition has adverse effects on species that use a limited resource and may result in competitive exclusion of one or more species. It therefore follows that over time natural selection should favor individuals of each species that avoid or reduce competition for environmental resources. Reduced competition among coexisting species as a result of each species' niche differing from the others in one or more ways is called **resource partitioning.** Resource partitioning is well-documented in animals; studies in tropical forests of Central America and South America demonstrate little overlap in the diets of fruit-eating birds, primates, and bats that coexist in the same habitat. Although fruits are the primary food for several hundred bird, primate, and bat species, the wide variety of fruits available has allowed fruit eaters to specialize, which reduces competition.

Resource partitioning may also include timing of feeding, location of feeding, nest sites, and other aspects of a species' ecological niche. Princeton ecologist Robert MacArthur's study of five North American warbler species is a classic example of resource partitioning (**FIG. 54-6**). Although initially the warblers' niches seemed nearly identical, MacArthur found that individuals of each species spend most of their feeding time in different portions of the spruces and other conifer trees they frequent. They also move in different directions through the canopy, consume different combinations of insects, and nest at slightly different times.

Difference in root depth is an example of resource partitioning in plants. For example, three common annuals found in certain abandoned fields are smartweed, Indian mallow, and bristly foxtail. Smartweed roots extend deep into the soil, Indian mallow roots grow to a medium depth, and bristly foxtail roots are shallow. This difference reduces competition for the same soil resources—water and minerals—by allowing the plants to exploit different portions of the resource.

Character displacement is an adaptive consequence of interspecific competition Sometimes populations of two similar species occur together in some locations and separately

Can competitive exclusion be demonstrated in the laboratory under controlled conditions?

HYPOTHESIS: When two species of paramecia (ciliated protozoa) are grown together in a mixed culture, one species outcompetes the other.

EXPERIMENT: Gause grew two species of paramecia, both separately and together.

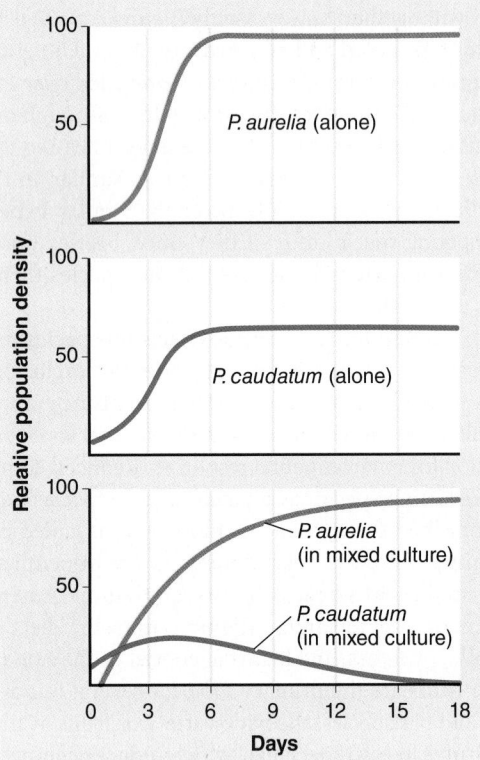

RESULTS AND CONCLUSION: The *top* and *middle* graphs show how each species of *Paramecium* flourishes when grown alone. The *bottom graph* shows how they grow together, in competition with each other. In a mixed culture, *P. aurelia* outcompetes *P. caudatum*, resulting in competitive exclusion.

SOURCE: Adapted from Gause, G.F. *The Struggle for Existence.* Williams and Wilkins, Baltimore, 1934.

Figure 54-5 *Animation* G.F. Gause's classic experiment on interspecific competition

PREDICT What would happen to the two competing cultures shown in the bottom graph if a medium containing a nutrient that could be metabolized only by *P. caudatum* replaced the existing growth medium at day 6?

in others. Where their geographic distributions overlap, the two species tend to differ more in their structural, ecological, and behavioral characteristics than they do where each occurs in separate geographic areas. Such divergence in traits in two similar species living in the same geographic area is known as **character displacement.** Biologists think that character

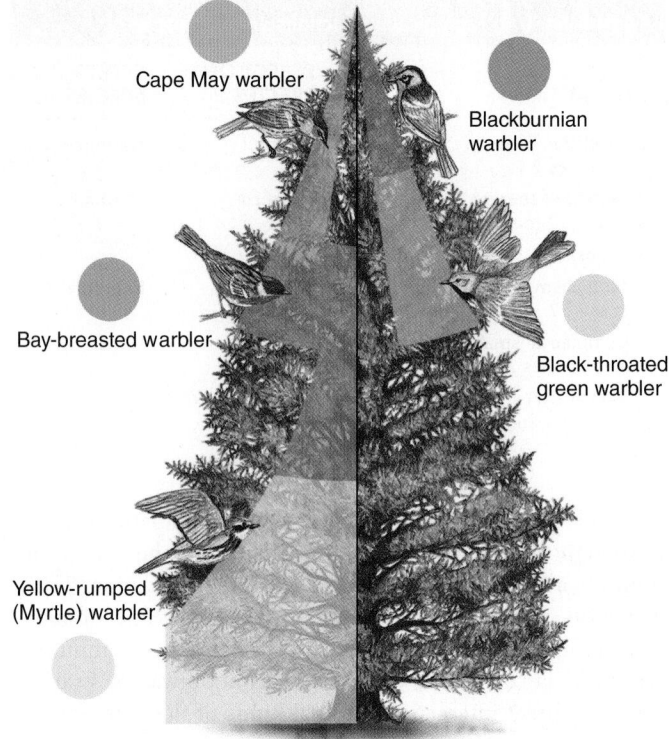

Figure 54-6 Resource partitioning

Each warbler species spends at least half its foraging time in its designated area of a spruce tree, thereby reducing the competition among warbler species. (After MacArthur, R.H. "Population Ecology of Some Warblers of Northeastern Coniferous Forests." *Ecology*, Vol. 39, 1958.)

displacement reduces competition between two species because their differences give them somewhat different ecological niches in the same environment.

There are several well-documented examples of character displacement between two closely related species. The flowers of two *Solanum* species in Mexico are quite similar in areas where either one or the other occurs. However, where their distributions overlap, the two species differ significantly in flower size and are pollinated by different kinds of bees. In other words, character displacement reduces interspecific competition, in this case for the same animal pollinator.

The bill sizes of Darwin's finches provide another example of character displacement (**FIG. 54-7**). On large islands in the Galápagos where the medium ground finch (*Geospiza fortis*) and the small ground finch (*G. fuliginosa*) occur together, their bill depths are distinctive. *Geospiza fuliginosa* has a smaller bill depth that enables it to crack small seeds, whereas *G. fortis* has a larger bill depth that enables it to crack medium-sized seeds. However, *G. fortis* and *G. fuliginosa* also live on separate islands. Where the two species live separately, bill depths are about the same intermediate size, perhaps because there is no competition from the other species.

Although these examples of the coexistence of similar species are explained in terms of character displacement, the character displacement hypothesis has been demonstrated in nature in only a few instances. In 2006, Peter and Rosemary

Can character displacement be demonstrated in the wild?

HYPOTHESIS: Two species of finches are more different where they occur together than where they occur separately.

EXPERIMENT: Two species of Darwin's finches in the Galápagos were observed on islands where they occur separately and on an island where they exist together. Bill depth **(a)**, which varies considerably in the two species, was measured in the different environments.

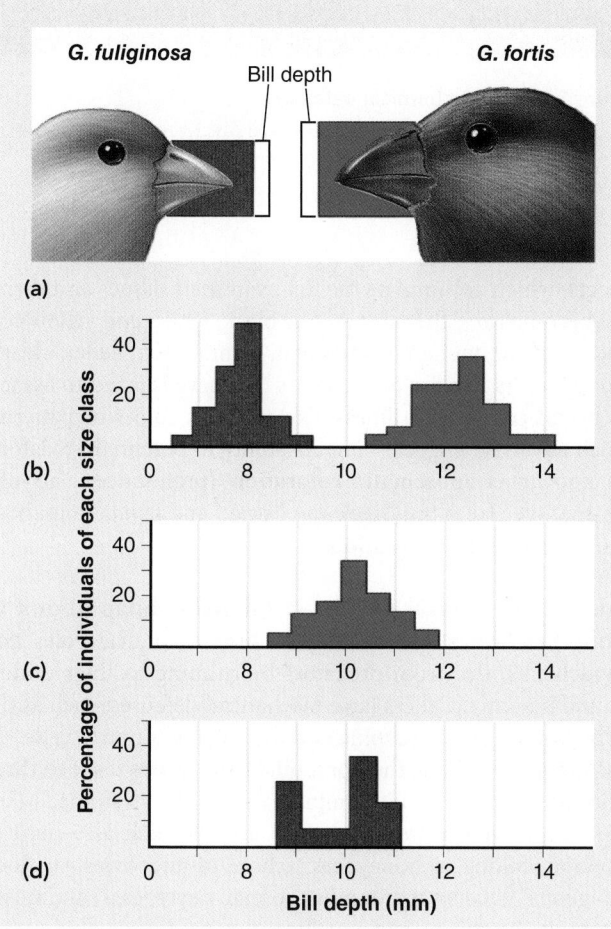

RESULTS AND CONCLUSION: (b) Where the two species are found on the same island (Santa Cruz), *G. fuliginosa* (red) has a smaller average bill depth than *G. fortis* (blue). Where they occur on separate islands (**c,** on Daphne Major, and **d,** on Los Hermanos), the average bill depths of each are similar. Thus, they illustrate character displacement.

SOURCE: After Lack, D. *Darwin's Finches*. Cambridge University Press, Cambridge, 1947.

Figure 54-7 Character displacement

PREDICT If *G. fuliginosa* were removed from the shared island, would you expect the beaks of the *G. fortis* population to change over time? In what way?

Grant, who have studied Darwin's finches in the Galápagos Islands since the 1970s, reported in the journal *Science* another example of character displacement in Darwin's finches. The medium ground finch (*G. fortis*) originally lived on one of the small Galápagos islands by itself. Different *G. fortis* individuals had varying beak sizes that allowed them to eat small, medium, or large seeds. However, in recent years the large ground finch (*G. magnirostris*) colonized the island and began to compete with the medium ground finch for the larger seeds. A severe drought reduced the seed supply, putting additional pressure on the medium finch population. In response, the medium ground finch's beak quickly diverged, becoming significantly smaller in length, depth, and width on average than it had been. The altered population of medium ground finches now consumes small seeds almost exclusively; their small beaks cannot crack the larger seeds preferred by the large ground finch.

Natural selection shapes the bodies and behaviors of both predator and prey

Predation is the consumption of one species, the *prey*, by another, the *predator* (see Table 54-1). Predation includes animals eating other animals as well as animals eating plants (*herbivory*). Predation has resulted in an evolutionary "arms race," with the evolution of predator strategies (more efficient ways to catch prey) and prey strategies (better ways to escape the predator). A predator that is more efficient at catching prey exerts a strong selective force on its prey. Over time, adaptations that reduce the probability of being captured may evolve in the prey species. In turn, these adaptations exert a strong selective force on the predator. This type of interdependent evolution of two interacting species is known as **coevolution** (see discussion of coevolution in Chapter 37).

We now consider several adaptations related to predator–prey interactions, including predator strategies (pursuit and ambush) and prey strategies (plant defenses and animal defenses). Keep in mind as you read these descriptions that such strategies are not "chosen" by the respective predators or prey. New traits arise randomly in a population as a result of genetic changes. Some new traits are beneficial, some are harmful, and others have no effect. Beneficial strategies, or traits, persist in a population because such characteristics make the individuals that have them well suited to thrive and reproduce. In contrast, characteristics that make the individuals that have them poorly suited to their environment tend to disappear in a population.

Pursuit and ambush are two predator strategies A brown pelican sights its prey—a fish—while in flight. Less than 2 seconds after diving into the water at a speed as great as 72 km/h (45 mph), it has its catch. Orcas, formerly known as killer whales, hunt in packs and often herd salmon or tuna into a cove so that they are easier to catch. Any trait that increases hunting efficiency, such as the speed of brown pelicans or the intelligence of orcas, favors predators that pursue their prey. Because these carnivores must process information quickly

Figure 54-8 Ambush

A yellow crab spider (*Misumena vatia*) blends into its surroundings, waiting for an unwary insect to visit the flower. An effective predator strategy, ambush relies on surprising the prey.

Figure 54-9 Plant chemical defenses

Toxic chemicals protect the common milkweed (*Asclepias syriaca*). Its leaves are poisonous to most herbivores except monarch caterpillars (*Danaus plexippus*) and a few other insects. Monarch caterpillars have bright aposematic coloration.

during the pursuit of prey, their brains are generally larger, relative to body size, than those of the prey they pursue.

Ambush is another effective way to catch prey. The yellow crab spider, for example, is the same color as the white or yellow flowers on which it hides (**FIG. 54-8**). This camouflage keeps unwary insects that visit the flower for nectar from noticing the spider until it is too late. It also fools birds that prey on the crab spider.

Predators that *attract* prey are particularly effective at ambushing. For example, a diverse group of deep-sea fishes called anglerfish possess rodlike bioluminescent lures close to their mouths to attract prey (see Fig. 32-14c).

Chemical protection is an effective plant defense against herbivores

Plants cannot escape predators by fleeing, but they have several physical and chemical adaptations that protect them from being eaten. The presence of spines, thorns, tough leathery leaves, or even thick wax on leaves discourages foraging herbivores from grazing. Other plants produce an array of protective chemicals that are unpalatable or even toxic to herbivores. The active ingredients in such plants as marijuana and tobacco affect hormone activity and nerve, muscle, liver, and kidney functions and may discourage foraging by herbivores. Interestingly, many of the chemical defenses in plants are useful to humans. India's neem tree, for example, contains chemicals effective against more than 100 species of herbivorous insects, mites, and nematodes. Nicotine from tobacco, pyrethrum from chrysanthemum, and rotenone from the derris plant are other examples of chemicals extracted and used as insecticides.

Milkweeds are an excellent example of the biochemical coevolution between plants and herbivores (**FIG. 54-9**). Milkweeds produce alkaloids and cardiac glycosides, chemicals that are poisonous to all animals except a small group of insects. The ability to either tolerate or metabolize the milkweed toxins has evolved in these insects. They eat milkweeds and avoid competition from other herbivorous insects because few others tolerate milkweed toxins. Predators also learn to avoid these

insects, which accumulate the toxins in their tissues and therefore become toxic themselves. The black, white, and yellow coloration of the monarch caterpillar, a milkweed feeder, clearly announces its toxicity to predators that have learned to associate bright colors with illness. Conspicuous colors or patterns, which advertise a species' unpalatability to potential predators, are known as **aposematic coloration** (pronounced "ap′-uh-suh-mat′-ik"; from the Greek *apo*, "away," and *semat*, "a mark or sign"), or **warning coloration.**

Animal prey possess various defensive adaptations to avoid predators

Many animals, such as prairie voles and woodchucks, flee from predators by running to their underground burrows. Others have mechanical defenses, such as the barbed quills of a porcupine and the shell of a pond turtle. To discourage predators, the porcupine fish inflates itself to three times its normal size by pumping water into its stomach (see Fig. 32-14a). Some animals live in groups, such as a herd of antelope, colony of honeybees, school of anchovies, or flock of pigeons. Because a group has so many eyes, ears, and noses watching, listening, and smelling for predators, this social behavior decreases the likelihood of a predator catching any one of them unaware.

Chemical defenses are also common among animal prey. The South American poison arrow frog (*Dendrobates*) stores poison in its skin. (These frogs obtain the toxins from ants and other insects in their diet.) The frog's bright aposematic coloration prompts avoidance by experienced predators (see Fig. 32-18a). Snakes and other animals that have tried once to eat a poisonous frog do not repeat their mistake! Other examples of aposematic coloration occur in the striped skunk, which sprays acrid chemicals from its anal glands, and the bombardier beetle, which spews harsh chemicals at potential predators (see Fig. 7-8).

Some animals have **cryptic coloration,** colors or markings that help them hide from predators by blending into their physical surroundings. Certain caterpillars resemble twigs so closely that

Figure 54-10 Cryptic coloration

The mossy leaf-tailed gecko (*Uroplatus sikorae*) hunts for insects by night and sleeps pressed against a tree branch by day. It is virtually invisible when it sleeps (look closely to see its head and front foot). Photographed in a rain forest in southern Madagascar.

you would never guess that they were animals unless they moved. Pipefish are slender green fish that are almost perfectly camouflaged in green eelgrass. Leaf-tailed geckos in southern Madagascar resemble dead leaves or mossy bark, depending on the species (FIG. 54-10). Such cryptic coloration has been preserved and accentuated by means of natural selection. Predators are less likely to capture animals with cryptic coloration than other animals, Such

animals are therefore more likely to live to maturity and produce offspring that also carry the genes for cryptic coloration.

Sometimes a defenseless species (a *mimic*) is protected from predation by its resemblance to a species that is dangerous in some way (a *model*). Such a strategy is known as **Batesian mimicry.** Many examples of this phenomenon exist. For example, a harmless scarlet king snake looks so much like a venomous coral snake that predators may avoid it (FIG. 54-11). Interestingly, the range of the scarlet king snake extends far beyond that of the coral snake. In an area where coral snakes do not occur, having the coloration of the model confers no special advantage to the king snake and may be harmful, triggering natural selection. This seems to be the case: scientists have reported that scarlet king snake populations located far from coral snakes have undergone natural selection and look less like their model.

In **Müllerian mimicry** different species (*co-models*), all of which are poisonous, harmful, or distasteful, resemble one another. Although their harmfulness protects them as individual species, their similarity in appearance works as an added advantage because potential predators more easily learn a single common aposematic coloration. Scientists hypothesize that viceroy and monarch butterflies are an example of Müllerian mimicry (see *Inquiring About: Mimicry in Butterflies*).

Symbiosis involves a close association between species

Symbiosis is any intimate relationship or association between members of two or more species. Usually, symbiosis involves one species living on or in another species. The partners of a symbiotic relationship, called **symbionts,** may benefit from, be unaffected by, or be harmed by the relationship (see Table 54-1). Examples of symbiosis occur across all the domains and kingdoms of life. Most of the thousands, or perhaps even millions, of symbiotic associations are products of coevolution. Symbiosis takes three forms: mutualism, commensalism, and parasitism.

(a) Scarlet king snake.

(b) Eastern coral snake.

Figure 54-11 Batesian mimicry

In this example **(a)** the scarlet king snake (*Lampropeltis triangulum elapsoides*) is the mimic, and **(b)** the eastern coral snake (*Micrurus fulvius fulvius*) is the model. Note that the red and yellow warning colors touch on the coral snake but do not touch on the harmless mimic.

Why do some butterfly species resemble one another? The monarch butterfly (*Danaus plexippus*), for example, is an attractive insect found throughout much of North America (see figure, *left side*). As a caterpillar, it feeds exclusively on milkweed leaves. The milky white liquid produced by the milkweed plant contains poisons that the insect tolerates but that remain in its tissues for life. When a young bird encounters and tries to eat its first monarch butterfly, the bird becomes sick and vomits. Thereafter, the bird avoids eating the distinctively marked insect.

Many people confuse the viceroy butterfly (*Limenitis archippus;* see figure, *right side*) with the monarch. The viceroy, which is found throughout most of North America, is approximately the same size, and the color and markings of its wings are almost identical to those of the monarch. As caterpillars, viceroys eat cottonwood, aspen, and willow leaves, which apparently do not contain poisonous substances.

During the past century, it was thought that the viceroy butterfly was a tasty food for birds but that its close resemblance to monarchs gave it some protection against being eaten. In other words, birds that had learned to associate the distinctive markings and coloration of the monarch butterfly with its bad taste tended to avoid viceroys because they were similarly marked. The viceroy butterfly

was therefore considered a classic example of Batesian mimicry.

In 1991, ecologists David Ritland and Lincoln Brower of the University of Florida reported the results of an experiment that tested the long-held notion that birds like the taste of viceroys but avoid eating them because of their resemblance to monarchs. They removed the wings of different kinds of butterflies—monarchs, viceroys, and several tasty species—and fed the seemingly identical wingless bodies to red-winged blackbirds. The results were surprising: monarchs and viceroys were equally distasteful to the birds.

As a result of this work, ecologists are re-evaluating the evolutionary significance of different types of mimicry. Rather than an example of Batesian mimicry, monarchs and viceroys may be an example of Müllerian mimicry, in which two or more different species that are distasteful or poisonous have come to resemble each other during the course of evolution. This likeness provides an adaptive advantage because predators learn quickly to avoid all butterflies with the coloration and markings of monarchs and viceroys. As a result, fewer butterflies of either species die, and more individuals survive to reproduce.

Thomas C. Emmel

Müllerian mimicry. Evidence suggests that monarch (*left*) and viceroy (*right*) butterflies are an example of Müllerian mimicry, in which two or more poisonous, harmful, or distasteful organisms resemble each other.

Benefits are shared in mutualism **Mutualism** is a symbiotic relationship in which both partners benefit. Mutualism is either obligate (essential for the survival of both species) or facultative (either partner can live alone under certain conditions).

The association between *nitrogen-fixing bacteria* of the genus *Rhizobium* and legumes (plants such as peas, beans, and clover) is an example of mutualism (see Fig. 55-9). Nitrogen-fixing bacteria, which live inside nodules on the roots of legumes, supply the plants with most of the nitrogen they require to manufacture such nitrogen-containing compounds as chlorophylls, proteins, and nucleic acids. The legumes supply sugars and other energy-rich organic molecules to their bacterial symbionts.

Another example of mutualism is the association between reef-building animals and dinoflagellates called **zooxanthellae.** These symbiotic protists live inside cells of the coral polyp (the coral forms a vacuole around the algal cell), where they photosynthesize and provide the animal with carbon and nitrogen compounds as well as oxygen (**FIG. 54-12**). Calcium carbonate skeletons form around the coral bodies much faster when zooxanthellae are present; their faster growth helps the corals dominate their location on a reef. The corals, in turn, supply the zooxanthellae with waste products

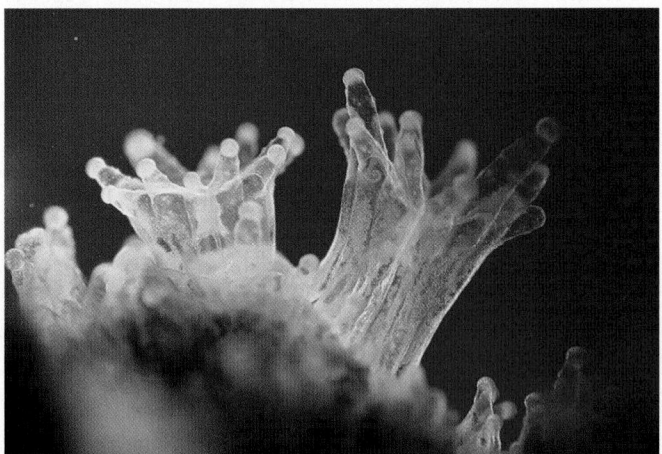

Oxford Scientific/Getty Images

Figure 54-12 Mutualism

The greenish brown specks in these polyps of stony coral (*Pocillopora*) are zooxanthellae, algae that live symbiotically within the coral's translucent cells and supply the coral with carbon and nitrogen compounds. In return, the coral provides the zooxanthellae with nitrogen in the form of ammonia.

such as ammonia, which the algae use to make nitrogen compounds for both partners.

Mycorrhizae are mutualistic associations between fungi and the roots of plants. The association is common; biologists think that nearly 90% of all plant species have mycorrhizae. The fungus, which grows around and into the root as well as into the surrounding soil, absorbs essential minerals, especially phosphorus, from the soil and provides them to the plant. In return, the plant provides the fungus with organic molecules produced by photosynthesis. Plants grow more vigorously in the presence of mycorrhizal fungi (see Figs. 29-19 and 36-11), and they better tolerate environmental stressors such as drought and high soil temperatures. Indeed, some plants cannot maintain themselves under natural conditions if the fungi with which they normally form mycorrhizae are not present.

Humans and bacteria also share mutualistic relationships. Recall in Chapter 25 that we discussed the intestinal component of the human microbiome, which receives shelter and nutrients in exchange for breaking down complex carbohydrates and vitamin production. In addition, human intestinal microbiota ferment unused materials, prevent colonization of pathogenic bacteria, and strengthen the immune system. Recent research also suggests that intestinal microbiota are involved in endocrine and neurological functions and may play a major role in diseases such as diabetes and obesity.

Commensalism is taking without harming Commensalism is a type of symbiosis in which one species benefits and the other one is neither harmed nor helped. One example of commensalism is the relationship between social insects and scavengers, such as mites, beetles, or millipedes, which live in the social insects' nests. Certain types of silverfish, for example, move along in permanent association with marching columns of army ants and share the abundant food caught in the ant raids. The army ants derive no apparent benefit or harm from the silverfish.

Another example of commensalism is the relationship between a host tree and its epiphytes, which are smaller plants, such as orchids, ferns, or Spanish moss, attached to the host's branches (FIG. 54-13). The epiphyte anchors itself to the tree but does not obtain nutrients or water directly from it. Living on the tree enables it to obtain adequate light, water (as rainfall dripping down the branches), and required minerals (washed out of the tree's leaves by rainfall). Thus, the epiphyte benefits from the association, whereas the tree is apparently unaffected. (Epiphytes harm their host, however, if they are present in a large enough number to block sunlight from the host's leaves; in this instance, the relationship is no longer commensalism.)

Parasitism is taking at another's expense Parasitism is a symbiotic relationship in which one member, the **parasite,** benefits, whereas the other, the **host,** is adversely affected. The parasite obtains nourishment from its host. A parasite rarely kills its host directly but may weaken it, rendering it more vulnerable to predators, competitors, or abiotic stressors. When a parasite causes disease and sometimes the death of a host, it is called a **pathogen.**

Ticks and other parasites that live outside the host's body are called **ectoparasites.** Figure 32-9 shows a lamprey, an ectoparasite on fish. Parasites such as tapeworms that live within the host are called **endoparasites.** Parasitism is a successful lifestyle; by one estimate, more than two-thirds of all species are parasites, and nearly 1000 species of parasites (including fungi, bacteria, and

Figure 54-13 Commensalism

Spanish moss (*Tillandsia usneoides*) is a gray-colored epiphyte that hangs suspended from larger plants in the southeastern United States. Spanish moss is not a moss but a flowering plant in the bromeliad, or pineapple, family. It often grows 6 m (20 ft) or longer, but it is not a parasite and rarely harms the host tree. Spanish moss provides shelter to small animals such as bats, spiders, and snakes.

Pete Oxford/Getty Images

viruses) live in or on the human species alone! Examples of human parasites include *Entamoeba histolytica,* an amoeba that causes amoebic dysentery; *Plasmodium,* an apicomplexan that causes malaria; a variety of parasitic worms, such as blood flukes, tapeworms, pinworms, and hookworms (see Figs. 31-9 and 31-11); and *Pseudomonas aeruginosa,* an opportunistic bacterium responsible for many urinary tract and respiratory system infections.

Since the 1980s, wild and domestic honeybees in the United States have been dying off. According to the U.S. Department of Agriculture, many beekeepers have lost one-third to one-half of their bees each winter from a condition called colony collapse disorder. Habitat loss, severe weather, pesticide use, and higher pathogen load, as well as the mite *Varroa destructor* and smaller tracheal mites, have contributed to the problem. Although research continues, many scientists believe that a combination of stressors may be the major reason for the continued honeybee decline (FIG. 54-14).

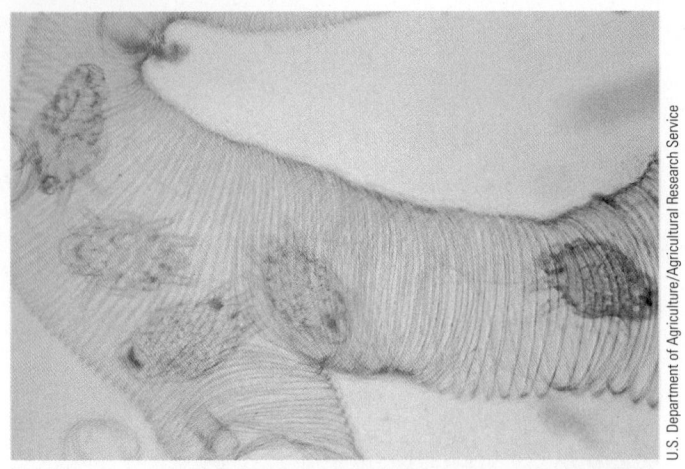

100 μm

Figure 54-14 Parasitism

Microscopic tracheal mites (*Acarapis woodi*) are endoparasites that live in the tracheal tubes of honeybees, clogging their airways so that they cannot breathe efficiently. Tracheal mites also suck the bees' circulatory fluid, weakening and eventually killing them. Entomologists think that the larger varroa mites (*not shown*), which are ectoparasites that also feed on the circulatory fluid, are more devastating to honeybee populations than are tracheal mites. Mites may also transmit viruses to their honeybee hosts.

CHECKPOINT 54.1

- **CONNECT** *How are acorns, gypsy moths, and Lyme disease related?*
- *Why is an organism's realized niche usually narrower, or more restricted, than its fundamental niche?*
- **CONNECT** *Which principle of community ecology is illustrated by the following example: Two closely related species of small fish occupy the upper two feet of water when they occur in separate ponds, but if they are in the same pond, species a is more often found in the uppermost foot, and species b in the foot below.*
- *Name the three kinds of symbiosis and give an example of each.*

54.2 STRENGTH AND DIRECTION OF COMMUNITY INTERACTIONS

LEARNING OBJECTIVES

6   Distinguish between keystone species and dominant species.
7   Distinguish between bottom-up and top-down processes in ecosystem regulation.

There is a great deal of variation in the strength and direction of interactions among species within ecological communities. Some species, known as *keystone species,* are present in a community in low abundance and/or biomass but have strong effects on their communities. Other species, known as *dominant species,* exert strong effects on their communities because they are present in great abundance and/or biomass.

Two types of control affect ecological communities, bottom-up processes and top-down processes. In *bottom-up processes* nutrient availability or food organisms whose population is directly affected by nutrient availability affect the abundance of other community organisms. In *top-down processes* predators impact the abundance of other populations in the community. We now examine these aspects of interactions among species within ecological communities.

Other species of a community depend on or are greatly affected by keystone species

Certain species, called **keystone species,** are crucial in determining the nature of the entire community, that is, its species composition and ecosystem functioning. Keystone species are usually not the most abundant species in the community. Although present in relatively small numbers, the individuals of a keystone species profoundly influence the entire community because they often impact the amount of available food, water, or some other resource. Thus, the effect of keystone species is greatly disproportionate to their abundance. Identifying and protecting keystone species are crucial goals of conservation biology (see Chapter 57). If a keystone species disappears from a community, many other species in that community may become more common or rarer, or may even disappear.

The term *keystone* species was coined by ecologist Robert T. Paine in 1969, based on his experimental studies along the Pacific coast in the state of Washington (**FIG. 54-15**). Paine removed a predatory sea star, *Pisaster ochraceus,* from a rocky intertidal community that included barnacles, mussels, limpets, and chitons, all of which the sea star preyed on. He observed the changes in community structure that resulted and compared these changes to a nearby control area in which the sea stars were left intact. After the sea stars were removed, about half of the 15 species disappeared from the test area, crowded out by certain mussel and barnacle populations, which increased rapidly when freed from predation by the sea stars. Paine noted that when a keystone predator such as the sea star is removed from a community, the species diversity of that community changes dramatically.

One problem with the concept of keystone species is that it is often difficult to measure all the direct and indirect effects of a keystone species on an ecosystem. Consequently, most evidence for the existence of keystone species is based on indirect observations rather than on experimental manipulations. For example, consider the fig tree. Because fig trees produce a continuous crop of fruits, they may be keystone species in tropical rain forests of Central America and South America. Fruit-eating monkeys, birds, bats, and other fruit-eating vertebrates of the forest do not normally consume large quantities of figs in their diets.

During that time of the year when other fruits are less plentiful, however, fig trees become important in sustaining fruit-eating vertebrates. It is therefore assumed that should the fig trees disappear, most of the fruit-eating vertebrates would also disappear. In turn, should the fruit eaters disappear, the

Figure 54-15 Rocky intertidal community at low tide

Pisaster sea stars and chitons cling to the rock during low tide. *Pisaster* sea stars range in color from purple to brown to orange. Photographed in the summer in the Pacific Northwest.

spatial distribution of other fruit-bearing plants would become more limited because the fruit eaters help disperse their seeds. Thus, protecting fig trees in such tropical rainforest ecosystems probably increases the likelihood that monkeys, birds, bats, and many other tree species will survive. The question is whether this anecdotal evidence of fig trees as keystone species is strong enough for policy makers to grant special protection to fig trees.

Like the *Pisaster* sea stars, many keystone species are top predators; for example, consider the gray wolf. Where wolves were hunted to extinction, populations of elk, deer, and other larger herbivores increased exponentially. As these herbivores overgrazed the vegetation, many plant species that could not tolerate such grazing pressure disappeared. Smaller animals such as rodents, rabbits, and insects declined in number because the plants they depended on for food were now less abundant. The number of foxes, hawks, owls, and badgers that prey on these small animals decreased, as did the number of ravens, eagles, and other scavengers that eat wolf-kill. Thus, the disappearance of the wolf resulted in communities with considerably less biological diversity.

The reintroduction of wolves to Yellowstone National Park in 1995 has given ecologists a unique opportunity to study the impact of a keystone species. The wolf's return has already caused substantial changes for other residents in the park. The top predator's effects have ranged from altering relationships among predator and prey species to transforming vegetation profiles. Coyotes are potential prey for wolves, and wolf packs have decimated some coyote populations. A reduction in coyotes has allowed populations of the coyotes' prey, such as ground squirrels and chipmunks, to increase. Scavengers such as ravens, bald eagles, and grizzly bears have benefited from dining on scraps from wolf-kills.

Biologists believe prey populations of elk in Yellowstone may now be in balance with the wolf population. Researchers have also recorded increases in the numbers of bear and cougar, which also feed on elk, resulting in a more balanced, healthier ecosystem.

Dominant species influence a community as a result of their greater size or abundance

In contrast to keystone species, which have a large impact out of proportion to their abundance, **dominant species** greatly affect the community because they are very common. Trees, the dominant species of forests, change the local environment. Trees provide shade, which changes both the light and moisture availability on the forest floor. Trees provide numerous habitats and *microhabitats* (such as a hole in a tree trunk or a rotting log) for other species. Forest trees also provide food for many organisms and therefore play a large role in energy flow through the forest ecosystem. Similarly, cordgrass (*Spartina*) is the dominant species in salt marshes, prairie grass in grasslands, and kelp in kelp beds. Animals are also dominant species. Corals, for example, are dominant species in coral reefs, and cattle are dominant species in overgrazed rangelands. Typically, a community has one or a few dominant species, and most other species are relatively rare.

Ecosystem regulation occurs from the bottom up and top down

One question that ecologists have recently considered is which regulatory process—bottom-up or top-down—is more significant in the regulation of various ecosystems. Both energy flow (see Fig. 1-8) and cycles of matter are involved in bottom-up and top-down processes. **Bottom-up processes** are based on **food webs,** the interconnected series of organisms through which energy flows in an ecosystem. Food webs always have plants or other producers at the first (lowest) trophic level (**FIG. 54-16a**). (A *trophic level* is an organism's position in a food web; e.g., producers are the first trophic level, and plant-eating herbivores are the second level.) In a sense, the biogeochemical cycles that regenerate necessary nutrients such as nitrates and phosphates for producers to assimilate are located "under" the first trophic level. Thus, bottom-up

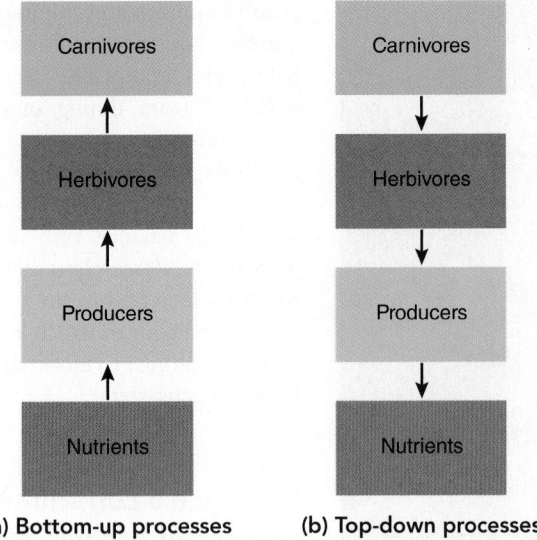

Figure 54-16 Bottom-up and top-down processes
© Cengage Learning

(a) Bottom-up processes   (b) Top-down processes

processes regulate ecosystem function by nutrient cycling and by availability of other resources. If bottom-up processes dominate an ecosystem, the availability of resources such as water or soil minerals controls the number of producers, which controls the number of herbivores, which controls the number of carnivores. (Recall from Chapter 53 that *limiting resources* tend to restrict the ecological niches of organisms, thereby affecting population size.)

Bottom-up processes apparently predominate in certain aquatic ecosystems in which nitrogen or phosphorus is limiting. An experiment in which phosphorus was added to a phosphorus-deficient river (the Kaparuk River in Alaska) resulted in an increase in algae, followed over time by increased populations of aquatic insects, other invertebrates, and fishes.

In contrast, **top-down processes** regulate ecosystem function by trophic interactions, particularly from the highest trophic level (**FIG. 54-16b**). Ecosystem regulation by top-down processes occurs because carnivores eat herbivores, which eat producers, which impacts levels of nutrients. If top-down processes dominate an ecosystem, the effects of an increase in the population of top predators cascade down the food web through the herbivores and producers. In fact, top-down processes are also known as a *trophic cascade.*

A change in the feeding preferences of orcas, off the coast of Alaska, provides an excellent example of a trophic cascade. A few decades ago, when orcas began preying on sea otters, the sea otter population sharply declined. As sea otters have declined, the number of sea urchins, which sea otters eat, has increased. Sea urchins eat kelp, the producers at the base of the food web, and the increase in the number of sea urchins has caused a decline in kelp populations. Kelp forests help reduce levels of carbon dioxide, a greenhouse gas that contributes to climate warming (discussed in Chapter 57).

Top-down regulation appears to predominate in ecosystems with few trophic levels and low species richness. Such ecosystems may have only one or a few species of dominant herbivores, but those species have a strong impact on the producer populations. An excellent example of top-down regulation—reindeer introduced to the Pribilof Islands of Alaska overgrazed the vegetation until the plants were almost wiped out—was discussed in Chapter 53.

It may be that top-down and bottom-up regulatory processes are not mutually exclusive. In a 13-year study of a thorn-scrub community in north-central Chile, ecologists demonstrated that top-down regulation predominates in certain small desert mammals and plant species. However, during periodic El Niño events (discussed later in the chapter), the increase in precipitation resulted in bottom-up increases in producers and consumers. In this ecosystem it appears that both top-down and bottom-up regulatory processes are important in trophic dynamics over an extended period.

CHECKPOINT 54.2

- *Both dominant and keystone species exert strong effects on the character of a community, but in different ways. How do dominant species and keystone species differ?*

- PREDICT *Biologists think that the reintroduction of wolves to Yellowstone National Park will ultimately result in a more varied and lush plant composition. How could that occur? Is it an example of a bottom-up or top-down process? Why?*

54.3 COMMUNITY BIODIVERSITY

LEARNING OBJECTIVES

8 Summarize the main determinants of species richness in a community and relate species richness to community stability.

9 State the results of the analysis of the distance effect in South Pacific bird species.

Species diversity, species richness, and species evenness vary greatly from one community to another and are influenced by many biotic and abiotic factors. **Species diversity** is a measure of both the number of species within a community (*species richness*) and the relative importance of each species, based on its abundance, productivity, or size.

Species richness, the number of species in a community, is determined by counting the species of interest. Tropical rain forests and coral reefs are examples of communities with extremely high species richness. In contrast, geographically isolated islands and mountaintops exhibit low species richness. **Species evenness** tells us about the relative abundance of one species compared to other species (**FIG. 54-17**).

Ecologists have developed various mathematical expressions, such as the *Shannon index,* to represent species diversity quantitatively. These *diversity indices* enable ecologists to compare species diversity in different communities. Conservation biologists use diversity indices as part of a comprehensive approach to saving biodiversity.

(a) Low species evenness. In this community the abundance of broomsedge is high relative to the abundance of other wildflowers (burdock, yarrow, and oxeye daisy). Therefore, the community has low species evenness.

(b) High species evenness. In this community each species has the same abundance, and the community has high species evenness.

Figure 54-17 Species evenness

Shown are two hypothetical communities with different species evenness but the same species richness (i.e., each has four species of wildflowers).
© Cengage Learning

A community in which the vegetation is structurally complex generally provides animals with more kinds of food and hiding places than a community with a lower structural complexity.

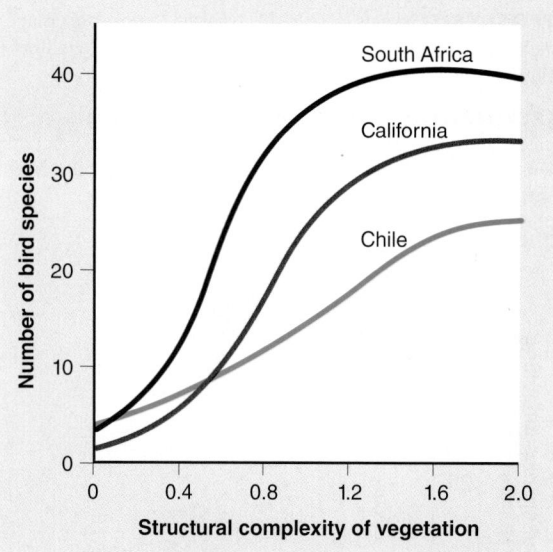

Figure 54-18 Effect of community complexity on species richness

Data were compiled in comparable chaparral habitats (shrubby and woody areas) in Chile, South Africa, and California. The structural complexity of vegetation (*x*-axis) is a numerically assigned gradient of habitats, based on height and density of vegetation, from low complexity (very dry scrub) to high complexity (woodland). (After Cody, M.L., and J.M. Diamond, eds. *Ecology and Evolution of Communities.* Harvard University Press, Cambridge, MA, 1975.)

PREDICT What would a species richness graph look like if the data were compiled from a tundra habitat?

Ecologists seek to explain why some communities have more species than others

What determines the number of species in a community? No single conclusive answer exists, but several explanations appear plausible. They include the structural complexity of habitats, geographic isolation, habitat stress, closeness to the margins of adjacent communities, dominance of one species over others, and geologic history. Although these and other environmental factors have positive or negative effects on species richness, there are exceptions and variations in every explanation. Some explanations vary at different spatial scales: an explanation that seems to work at a large geographic scale (such as a continent) may not work at a smaller local scale (such as a meadow).

In many habitats species richness is related to the structural complexity of habitats. In terrestrial environments the types of plants growing in an area typically determine structural complexity. A structurally complex community, such as a forest, offers a greater variety of potential ecological niches than does a simple community, such as an arid desert or semiarid grassland

(**FIG. 54-18**). An already complex habitat, such as a coral reef, may become even more complex if species potentially capable of filling vacant ecological niches evolve or migrate into the community because these species create "opportunities" for additional species. Thus, it appears that species richness is self-perpetuating to some degree.

Species richness is inversely related to the geographic isolation of a community. Isolated island communities are generally much less diverse than are communities in similar environments found on continents. This difference is due partly to the *distance effect,* the difficulty encountered by many species in reaching and successfully colonizing the island (**FIG. 54-19**). Also, sometimes species become locally extinct as a result of random events. In isolated habitats such as islands or mountaintops, locally extinct species are not readily replaced. Isolated areas are usually small and have fewer potential ecological niches.

Generally, species richness is inversely related to the environmental stress of a habitat. Only those species capable of tolerating extreme conditions live in an environmentally stressed community. Thus, the species richness of a highly polluted

Does the geographic isolation of a community affect its species richness?

HYPOTHESIS: Islands close to the mainland (or to a large island) have a higher species richness than the islands that are farther from the mainland.

EXPERIMENT: The number of bird species was cataloged on South Pacific islands that are various distances from New Guinea, a source of colonizing species for these smaller islands. Each point on the graph represents a different island.

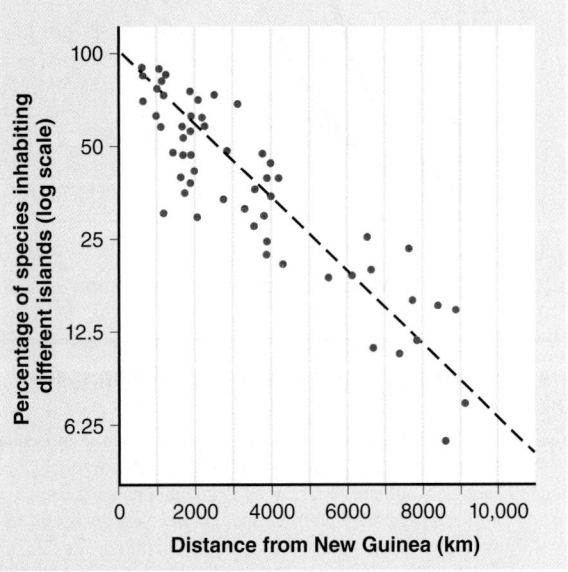

RESULTS AND CONCLUSION: The percentage of South Pacific bird species found on each island in the South Pacific is related to its distance from New Guinea. (New Guinea has 100% of the bird species living in that region.) Thus, species richness declines as the distance from New Guinea increases.

SOURCE: After Diamond, J.M. "Biogeographic Kinetics: Estimation of Relaxation Times for Avifaunas of Southwest Pacific Islands." *Proceedings of the National Academy of Sciences*, Vol. 69, 1972.

Figure 54-19 The distance effect

PREDICT Approximately what percentage of species have colonized islands 5000 km from New Guinea? 8000 km from New Guinea? Based on your answers, explain the relationship between species richness and geographic isolation.

Ecuador, and Peru occupy only 2% of Earth's land, they contain a remarkable 46,000 native plant species. The continental United States and Canada, with a significantly larger land area, host a total of 19,000 native plant species. Ecuador alone contains more than 1600 native species of birds, twice as many as the United States and Canada combined.

Species richness is usually greater at the margins of distinct communities than in their centers. The reason is that an **ecotone,** a transitional zone where two or more communities meet, contains all or most of the ecological niches of the adjacent communities as well as some that are unique to the ecotone (see Fig. 56-24). This change in species composition produced at ecotones is known as the **edge effect.**

Species richness is reduced when any one species enjoys a position of dominance within a community; a dominant species may appropriate a disproportionate share of available resources, thus crowding out, or outcompeting, other species. Ecologist James H. Brown of the University of New Mexico has addressed species composition and richness in experiments conducted since 1977 in the Chihuahuan Desert of southeastern Arizona. In one experiment the removal of three dominant species, all kangaroo rats, from several plots resulted in an increased diversity of other rodent species. This increase was ascribed both to lowered competition for food and to an altered habitat because the abundance of grass species increased dramatically after the removal of the kangaroo rats.

Species richness is greatly affected by geologic history. Many scientists think that tropical rain forests are old, stable communities that have undergone relatively few widespread disturbances through Earth's history. (In ecology a **disturbance** is any event in time that disrupts community or population structure.) During this time, myriad species evolved in tropical rain forests. In contrast, glaciers have repeatedly altered temperate and arctic regions during Earth's history. An area recently vacated by glaciers will have low species richness because few species have had a chance to enter it and become established. The idea that older, more stable habitats have greater species richness than habitats subjected to frequent, widespread disturbances is known as the *time hypothesis.*

Species richness may promote community stability

Traditionally, most ecologists assumed that community stability—the ability of a community to withstand disturbances—is a consequence of community complexity. That is, ecologists hypothesized that a community with considerable species richness is more stable than a community with less species richness. According to this view, the greater the species richness, the less critically important any single species should be. With many possible interactions within the community, it appears unlikely that any single disturbance could affect enough components of the system to make a significant difference in its functioning.

Supporting evidence for this hypothesis is found in destructive outbreaks of pests being more common in cultivated fields, which have a low species richness, than in natural

---

stream is low compared with that of a nearby pristine stream. Similarly, the species richness of high-latitude (farther from the equator) communities exposed to harsh climates is lower than that of lower latitude (closer to the equator) communities with milder climates (**FIG. 54-20**). This observation, known as the *species richness–energy hypothesis,* suggests that different latitudes affect species richness because of variations in solar energy. Greater energy may permit more species to coexist in a given region. Although the equatorial countries of Colombia,

Species richness increases along a polar ⟶ equatorial gradient. This correlation has been observed for many organisms, from plants to primates, in both terrestrial and marine environments.

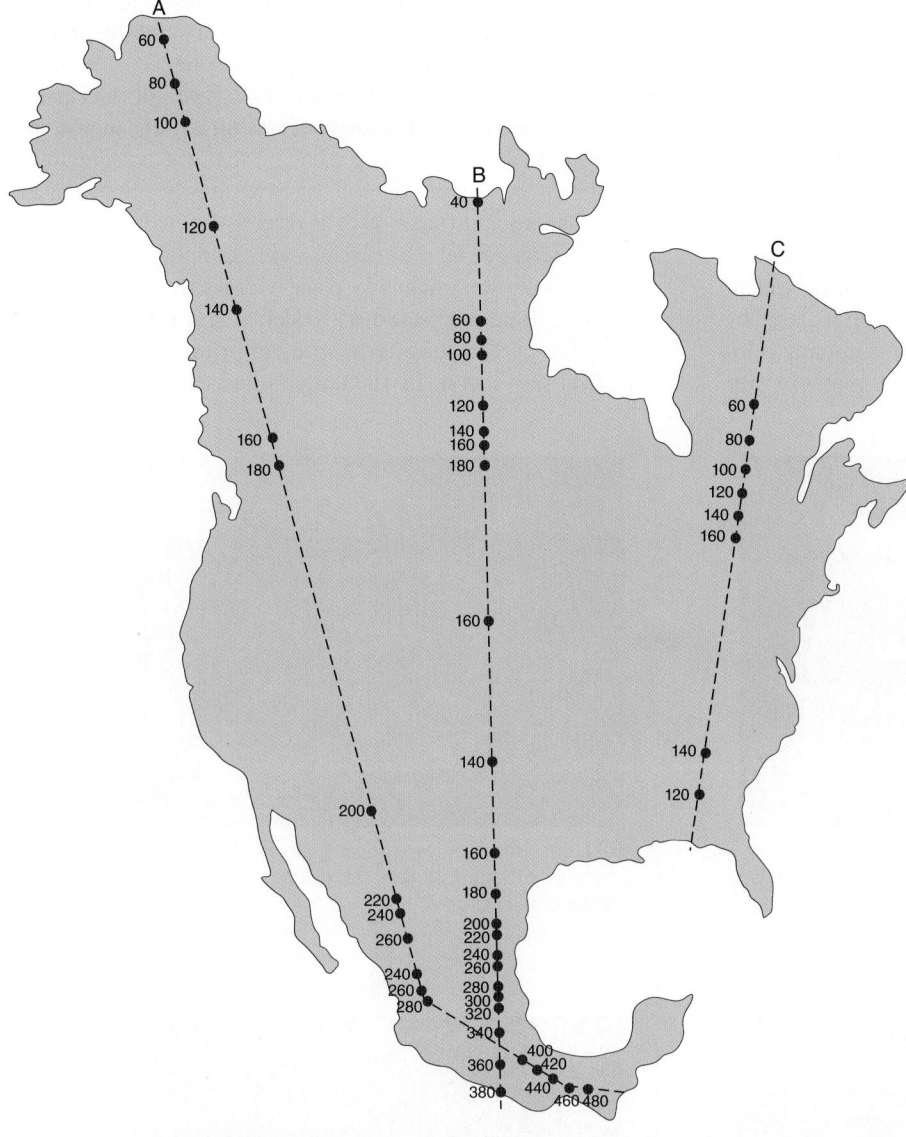

Figure 54-20 Effect of latitude on species richness of breeding birds in North America

The species richness for three north–south transects (A, B, and C) are shown. Note that the overall number of breeding bird species is greater at lower latitudes toward the equator than at higher latitudes. However, this pattern is strongly modified by other factors, such as precipitation and surface features (e.g., mountains). (After Cook, R.E. "Variation in Species Density of North American Birds." *Systematic Zoology*, Vol. 18, 1969.)

PREDICT How might an increase in temperature over time (climate change) affect species richness in Alaska?

strengthened the link between species richness and community stability. In their initial study, they established and monitored 207 plots of various grassland species in Minnesota for seven years. During the study period, Minnesota's worst drought in 50 years occurred (1987–1988). The ecologists found that those plots with the greatest number of plant species lost less ground cover, as measured by dry weight, and recovered faster than species-poor plots. Later studies by Tilman and his colleagues supported these conclusions and showed a similar effect of species richness on community stability during non-drought years. Similar work by almost three dozen ecologists at eight grassland sites in Europe also supports the link between species richness and community stability.

Some scientists do not agree with the conclusions of Tilman and other research groups that the presence of more species confers greater community stability. These critics argue that it is difficult to separate species number from other factors that could affect productivity. They suggest that it would be better to start with established ecosystems and study what happens to their productivity when plants are removed.

Another observation that adds a layer of complexity to the species richness–community stability debate is that populations of individual species within a species-rich community often vary significantly from year to year. It may seem paradoxical that variation within populations of individual species relates to the stability of the entire community. When you consider all the interactions among the organisms in a community, however, it is obvious that some species benefit at the expense of others. If one species declines in a given year, other species that compete with it may flourish. Thus, if an ecosystem contains more species, it is likely that at least some will be resistant to any given disturbance.

communities with a greater species richness. As another example, the almost complete loss of the American chestnut tree to the chestnut blight fungus had little ecological effect on the moderately diverse Appalachian woodlands of which it was formerly a part.

Ongoing studies by David Tilman of the University of Minnesota and John Downing of the University of Iowa have

CHECKPOINT 54.3

- **CONNECT** *How is the species richness of a community related to geographic isolation?*
- **CONNECT** *How is the structural complexity of habitats relevant to species richness?*
- **CONNECT** *How is the species richness of a community related to the environmental stress of a habitat?*

54.4 COMMUNITY DEVELOPMENT

LEARNING OBJECTIVES

10 Define *succession* and distinguish between primary and secondary succession.
11 Describe the intermediate disturbance hypothesis.
12 Discuss the two traditional views of the nature of communities: Clements's organismic model and Gleason's individualistic model.

A community does not spring into existence full-blown; rather, it develops gradually through a series of stages, each dominated by different organisms. The process of community development over time, which involves species in one stage being replaced by different species, is called **succession.** An area is initially colonized by certain early successional species that give way over time to others, and then, in turn, give way much later to late-successional species.

Succession is usually described in terms of the changes in the species composition of an area's vegetation, although each successional stage also has its own characteristic kinds of animals and other species. The time involved in succession is on the order of tens, hundreds, or thousands of years, not the millions of years involved in the evolutionary time scale.

Ecologists distinguish between two types of succession: primary and secondary. **Primary succession** is the change in species composition over time in a habitat that was not previously inhabited by organisms. No soil exists when primary succession begins. Bare rock surfaces, such as recently formed volcanic lava and rock scraped clean by glaciers, are examples of sites where primary succession might take place (**FIG. 54-21**).

The Indonesian island of Krakatoa has provided scientists with a perfect long-term study of primary succession in a tropical rain forest. In 1883, a volcanic eruption destroyed

(a) After the glacier's retreat, lichens initially colonize the gravel moraine. Mosses and shallow rooted herbs such as mountain avens (*Dryas*) follow.

(b) Dense mats of low-growing *Dryas* plants dominate the landscape, stabilizing the thin layer of developing soil.

(c) Alders replace the *Dryas* community. Both alders (*Alnus*) and *Dryas* have nitrogen-fixing bacteria in root nodules that improve the nitrogen content of the soil.

(d) As the condition of the soil improves, Sitka spruce (*Picea sichensis*) replaces the alder community. After several hundred years, the forest consists of Sitka spruce and western hemlock (*Tsuga heterophylla*).

Figure 54-21 *Animation* **Primary succession following retreating glaciers**

Ecologists have documented primary succession as glaciers have retreated at Glacier Bay, Alaska.

virtually all life on the island. Ecologists have surveyed the ecosystem during the years since the devastation to document the return of life-forms. In the 1990s, ecologists found that the progress of primary succession was extremely slow, in part because of Krakatoa's isolation (recall the distance effect discussed earlier in the chapter). Many species are limited in their ability to disperse over water. Krakatoa's forest, for example, may have only one-tenth the tree species richness of undisturbed tropical rain forest of nearby islands. The lack of plant diversity has, in turn, limited the number of colonizing animal species. In a forested area of Krakatoa where zoologists would expect more than 100 butterfly species, for example, there are only 2.

Secondary succession is the change in species composition that takes place after some disturbance removes the existing vegetation; soil is already present at these sites. Abandoned agricultural fields or open areas produced by forest fires are common examples of sites where secondary succession occurs. During the summer of 1988, wildfires burned approximately one-third of Yellowstone National Park. This natural disaster provided a valuable chance for ecologists to study secondary succession in areas that had been forests. After the conflagration, gray ash covered the forest floor, and most of the trees, although standing, were charred and dead. Secondary succession in Yellowstone has occurred rapidly since 1988. Less than a year later, in the spring of 1989, trout lily and other herbs sprouted and covered much of the ground. By 1998, a young forest of knee-high to shoulder-high lodgepole pines dominated the area. Douglas fir seedlings also began appearing in 1998. Ecologists continue to monitor the changes in Yellowstone as secondary succession unfolds.

Disturbance influences succession and species richness

Early studies suggested that succession inevitably progressed to a stable and persistent community, known as a *climax community,* which was determined solely by climate. Periodic disturbances, such as fires or floods, were not thought to exert much influence on climax communities. If the climax community were disturbed in any way, it would return to a self-sustaining, stable equilibrium in time.

This traditional view of stability has fallen out of favor. The apparent end-point stability of species composition in a "climax" forest is probably the result of how long trees live relative to the human lifespan. It is now recognized that forest communities never reach a state of permanent equilibrium but instead exist in a state of continual disturbance. The species composition and relative abundance of each species vary in a mature community over a range of environmental gradients, although the community retains a relatively uniform appearance overall.

Because all communities are exposed to periodic disturbances, both natural and human-induced, ecologists have long tried to understand the effects of disturbance on species richness. A significant advance was the development of the **intermediate disturbance hypothesis** by Joseph H. Connell.

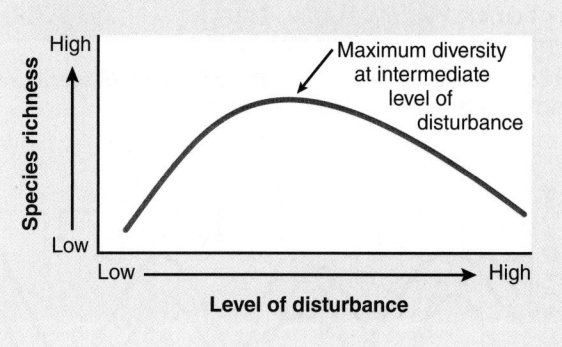

Figure 54-22 Intermediate disturbance hypothesis
Species richness is greatest at an intermediate level of disturbance. (After Connell, J.H. "Diversity in Tropical Rain Forests and Coral Reefs." *Science,* Vol. 199, 1978.)

When he examined species richness in tropical rain forests and coral reefs, he proposed that species richness is greatest at moderate levels of disturbance (**FIG. 54-22**). At a moderate level of disturbance, the community is a mosaic of habitat patches at different stages of succession; a range of sites exists, from those that were recently disturbed to those that have not been disturbed for many years. When disturbances are frequent or intense, only those species best adapted to earlier stages of succession persist, whereas low levels of disturbance allow late-successional species to dominate to such a degree that other species disappear. For example, when periodic wildfires are suppressed in forests, reducing disturbance to a low level, some of the "typical" forest herbs decline in number or even disappear.

One of the difficulties with the intermediate disturbance hypothesis is defining precisely what constitutes an "intermediate" level of disturbance. Despite this problem, the intermediate disturbance hypothesis has important ramifications for conservation biology because it tells us that we cannot maintain a particular community simply by creating a reserve around it. Both natural and human-induced disturbances will cause changes in species composition, and humans may have to intentionally intervene to maintain the species richness of the original community. What kind of and how much human intervention is necessary to maintain species richness is controversial (we discuss this issue in Chapter 57).

Ecologists continue to study community structure

One of the major issues in community ecology, from the early 1900s to the present, is the nature of communities. Are communities highly organized systems of predictable species, or are they abstractions produced by the minds of ecologists?

U.S. botanist Frederick E. Clements (1874–1945) was struck by the worldwide uniformity of large tracts of vegetation, such as tropical rain forests in South America, Africa, and Southeast Asia. He also noted that even though the species

Most communities are individualistic associations of species rather than distinct units that act like "superorganisms."

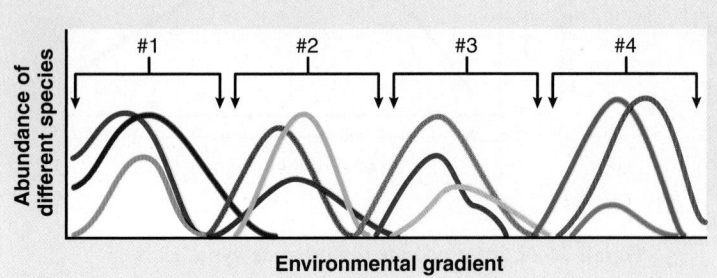

(a) Organismic model. According to this hypothesis, communities are organized as distinct units. Each of the four communities shown (*numbered brackets*) consists of an assemblage of distinct species (*each colored curve represents a single species*). The arrows indicate ecotones, regions of transition along community boundaries.

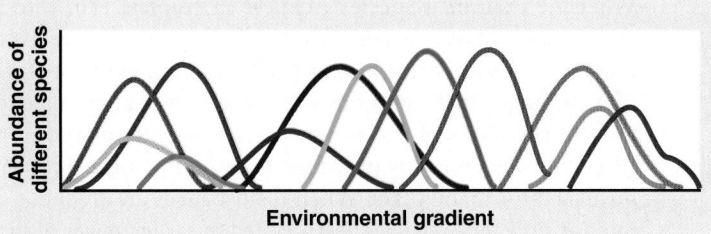

(b) Individualistic model. This model predicts a more random assemblage of species along a gradient of environmental conditions.

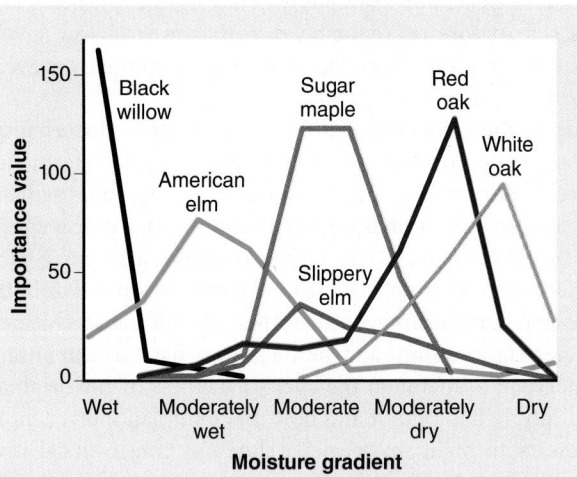

(c) Tree species in Wisconsin forests, which are distributed along a moisture gradient, more closely resemble the individualistic model. The "importance value" (*y-axis*) of each species in a given location combines three aspects (population density, frequency, and size).

Figure 54-23 Nature of communities

(**a** and **b:** © Cengage Learning; **c:** Adapted from Curtis, J.T. *The Vegetation of Wisconsin.* University of Wisconsin Press, Madison, 1959.)

VISUALIZE Create a graph with the same *x*-axis and *y*-axis titles as Figure 54-23c and show what the curves for a palm tree and a cactus might look like.

composition of a community in a particular habitat may be different from that of a community in a habitat with a similar climate elsewhere in the world, overall the components of the two communities are usually similar. He viewed communities as something like "superorganisms" whose member species cooperated with one another in a manner that resembled the cooperation of the parts of an individual organism's body. Clements's view was that a community went through certain stages of development, like the embryonic stages of an organism, and eventually reached an adult state; the developmental process was succession, and the adult state was the climax community. This cooperative view of the community, called the **organismic model,** stresses the interaction of the members, which tend to cluster in tightly knit groups within discrete community boundaries (**FIG. 54-23a**).

Opponents of the organismic model, particularly U.S. ecologist Henry A. Gleason (1882–1975), held that biological interactions are less important in the production of communities than are environmental gradients (such as climate and soil) or even chance. Indeed, the concept of a community is questionable. It may be a classification category with no reality, reflecting little more than the tendency of organisms with similar environmental requirements to live in similar places. This school of thought, called the **individualistic model,** emphasizes species individuality, with each species having its own particular abiotic living requirements. It holds that communities are therefore not interdependent associations of organisms. Rather, each species is independently distributed across a continuum of areas that meets its own individual requirements (**FIG. 54-23b**).

Debates such as this one over the nature of communities are an integral part of the scientific process because they fuel discussion and research that lead to a better understanding of broad scientific principles. Studies testing the organismic and individualistic hypotheses of communities do not support Clements's interactive concept of communities as discrete units. Instead, most studies favor the individualistic model. As shown in **FIGURE 54-23c,** tree species in Wisconsin forests are distributed in a gradient from wet to dry environments. Also, studies of plant and animal movements during the past 14,000 years support the individualistic model because it appears that individual species, not entire communities, became redistributed in response to changes in climate.

CHECKPOINT 54.4

- *How does primary succession differ from secondary succession?*
- *What is Connell's intermediate disturbance hypothesis?*
- *How do the organismic and individualistic models of the nature of communities differ?*

54.1 Community Structure and Functioning (page 1165)

1 Define *ecological niche* and distinguish between an organism's fundamental niche and its realized niche.

- The distinctive lifestyle and role of an organism in a community is its **ecological niche.** An organism's ecological niche takes into account all abiotic and biotic aspects of the organism's existence. An organism's **habitat** (where it lives) is one of the parameters used to describe the niche.

- Organisms potentially exploit more resources and play a broader role in the life of their community than they actually do. The potential ecological niche for an organism is its **fundamental niche,** whereas the niche it actually occupies is its **realized niche.**

2 Define *competition* and distinguish between interspecific and intraspecific competition.

- **Competition** occurs when two or more individuals attempt to use the same essential resource, such as food, water, shelter, living space, or sunlight.

- Competition occurs among individuals within a population (**intraspecific competition**) or between different species (**interspecific competition**).

3 Summarize the concepts of the competitive exclusion principle, resource partitioning, and character displacement.

- According to the **competitive exclusion principle,** two species cannot occupy the same niche in the same community for an indefinite period; one species is excluded by another as a result of competition for a limiting resource.

- The evolution of differences in resource use, known as **resource partitioning,** reduces competition between similar species.

- Competition among some species is reduced by **character displacement,** in which their structural, ecological, and behavioral characteristics diverge where their ranges overlap.

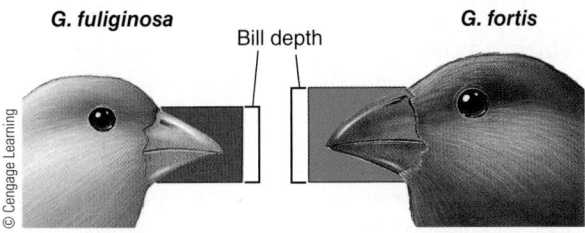

4 Define *predation* and describe the effects of natural selection on predator–prey relationships.

- **Predation** is the consumption of one species (the *prey*) by another (the *predator*). During **coevolution** between predator and prey, more efficient ways to catch prey evolve in the predator species, and better ways to escape the predator evolve in the prey species.

5 Distinguish among mutualism, commensalism, and parasitism, and give examples of each.

- **Symbiosis** is any intimate or long-term association between two or more species. The three types of symbiosis are mutualism, commensalism, and parasitism.

- In **mutualism** both partners benefit. Three examples of mutualism are *nitrogen-fixing bacteria* and legumes, **zooxanthellae** and corals, and **mycorrhizae** (fungi and plant roots).

- In **commensalism** one organism benefits and the other is unaffected. Two examples of commensalism are silverfish and army ants, and epiphytes and larger plants.

- In **parasitism** one organism (the **parasite**) benefits and the other (the **host**) is harmed. One example of parasitism is mites that grow in or on honeybees. Some parasites are **pathogens** that cause disease.

54.2 Strength and Direction of Community Interactions (page 1176)

6 Distinguish between keystone species and dominant species.

- **Keystone species** are present in relatively small numbers but are crucial in determining the species composition and ecosystem functioning of the entire community.

- In contrast to keystone species, which have an effect that is out of proportion to their abundance, **dominant species** greatly affect the community of which they are a part because they are very common.

7 Distinguish between bottom-up and top-down processes in ecosystem regulation.

- If **bottom-up processes** dominate an ecosystem, the availability of resources such as minerals controls the number of producers (i.e., the lowest trophic level), which controls the number of herbivores, which controls the number of carnivores.

- **Top-down processes** regulate ecosystems from the highest trophic level, by consumers eating producers. If top-down processes dominate an ecosystem, an increase in the number of top predators cascades down the food web through the herbivores and producers.

54.3 Community Biodiversity (page 1178)

8 Summarize the main determinants of species richness in a community and relate species richness to community stability.

- Community complexity is expressed in terms of **species richness,** the number of species within a community, and **species diversity,** a measure of the relative importance of each species within a community based on abundance, productivity, or size.

- Species richness is often great in a habitat that has structural complexity; in a community that is not isolated (the distance effect) or severely stressed; in a community where more energy is available (the species richness–energy hypothesis); in **ecotones** (transition zones between communities); and in communities with long histories without major **disturbances,** events that disrupt community or population structure.

- Several studies suggest that species richness may promote community stability.

9 State the results of the analysis of the distance effect in South Pacific bird species.

- The number of bird species found on each island in the South Pacific is inversely related to its geographic isolation, that is, to its distance from New Guinea. Species richness declines as the distance from New Guinea increases.

54.4 Community Development *(page 1182)*

10 Define *succession* and distinguish between primary and secondary succession.

- **Succession** is the orderly replacement of one community by another. **Primary succession** occurs in an area that has not previously been inhabited (such as bare rock). **Secondary succession** begins in an area where there was a pre-existing community and well-formed soil (such as abandoned farmland).

11 Describe the intermediate disturbance hypothesis.

- Disturbance affects succession and species richness. According to the **intermediate disturbance hypothesis,** species richness is greatest at moderate levels of disturbance, which create a mosaic of habitat patches at different stages of succession.

12 Discuss the two traditional views of the nature of communities: Clements's organismic model and Gleason's individualistic model.

- The **organismic model** views a community as a "super-organism" that goes through certain stages of development (succession) toward adulthood (climax). According to this view, biological interactions are primarily responsible for species composition, and organisms are highly interdependent.

- Most ecologists support the **individualistic model,** which challenges the concept of a highly interdependent community. According to this model, abiotic environmental factors are the primary determinants of species composition in a community, and organisms are somewhat independent of one another.

TEST YOUR UNDERSTANDING

Know and Comprehend

1. A symbiotic association in which organisms are beneficial to one another is known as (a) predation (b) interspecific competition (c) intraspecific competition (d) commensalism (e) mutualism

2. A species' _____ is the totality of its adaptations, its use of resources, and its lifestyle. (a) habitat (b) ecotone (c) ecological niche (d) competitive exclusion (e) coevolution

3. Primary succession occurs on (a) bare rock (b) newly cooled lava (c) abandoned farmland (d) a and b (e) a, b, and c

4. The tendency for two similar species to differ from each other more markedly in areas where they occur together is known as (a) Müllerian mimicry (b) Batesian mimicry (c) resource partitioning (d) competitive exclusion (e) character displacement

5. Competition with other species helps determine an organism's (a) ecotone (b) fundamental niche (c) realized niche (d) limiting resource (e) ecosystem

6. "Complete competitors cannot coexist" is a statement of the principle of (a) primary succession (b) limiting resources (c) Müllerian mimicry (d) competitive exclusion (e) character displacement

7. Based on current evidence, monarch and viceroy butterflies are probably an example of (a) Batesian mimicry (b) character displacement (c) resource partitioning (d) Müllerian mimicry (e) cryptic coloration

8. The _____ signifies that species richness is greater where two communities meet than at the center of either community. (a) edge effect (b) fundamental niche (c) character displacement (d) realized niche (e) limiting resource

9. An unpalatable species demonstrates its threat to potential predators by displaying (a) character displacement (b) limiting resources (c) cryptic coloration (d) aposematic coloration (e) competitive exclusion

10. A limiting resource does all the following *except* that it (a) tends to restrict the ecological niche of a species (b) is in short supply relative to a species' need for it (c) limits the presence of a species in a given community (d) results in an intermediate disturbance (e) may be limiting for only part of an organism's life cycle

Apply and Analyze

11. An ecologist studying several forest-dwelling, insect-eating bird species does not find any evidence of interspecific competition. The most likely explanation is (a) lack of a keystone species (b) low species richness (c) pronounced intraspecific competition (d) coevolution of predator–prey strategies (e) resource partitioning

12. Support for the individualistic model of community structure includes (a) the decline of honeybees because of two species of parasitic mites (b) the identification of fig trees as a keystone species in tropical forests (c) the competitive exclusion of one *Paramecium* species by another (d) the distribution of trees along a moisture gradient in Wisconsin forests (e) the effects of the removal of a dominant rodent species from an Arizona desert

Evaluate and Synthesize

13. In your opinion, are humans a dominant species or a keystone species? Explain your answer.

14. Many plants that produce nodules for nitrogen-fixing bacteria are common on disturbed sites. Explain how these plants might simultaneously compete with and facilitate other plant species.

15. **EVOLUTION LINK** The rough-skinned newt, which lives in western North America, stores a poison in its skin and is avoided by predators. However, several populations of garter snakes have undergone one or a few mutations that enable them to tolerate the toxin, and these snakes eat the newts with no ill effects. How has natural selection affected this predator–prey relationship? Based on what you have learned about evolutionary arms races, predict what may happen to the newts and the poison-resistant garter snakes over time.

16. **EVOLUTION LINK** Competition is an important part of Darwin's scientific theory of evolution by natural selection, and the evolution of features that reduce competition increases a population's overall fitness. Relate this idea to character displacement and resource partitioning in Darwin's finches.

17. **INTERPRET DATA** Examine the top and middle graphs in Figure 54-5. Are these examples of exponential or logistic population growth? Where is *K* in each graph? (You may need to refer to Chapter 53 to answer these questions.)

18. **SCIENCE, TECHNOLOGY, AND SOCIETY** Describe the ecological niche of humans. How have science and technology changed our realized niche during the past 1000 years?

☺ **aplia** To access course materials, such as Aplia and other companion resources, please visit **www.cengagebrain.com.**

Ecosystems and the Biosphere | 55

Planet Earth has often been compared to a vast spaceship, inhabited by diverse communities of organisms. As Earth orbits around the sun, these organisms use the sun's energy to produce oxygen, transfer energy, and recycle water and minerals (inorganic nutrients) with great efficiency. Yet none of these ecological processes would be possible without the abiotic (nonliving) environment of spaceship Earth. As the sun warms the planet, it powers the hydrologic cycle (causes precipitation), drives ocean currents and atmospheric circulation patterns, and produces much of the climate to which organisms have adapted. The sun also supplies the energy that almost all organisms use to carry on life processes.

Individual communities and their abiotic environments are **ecosystems,** which are the basic units of ecology. An ecosystem encompasses all the interactions among organisms living together in a particular place and among those organisms and their abiotic environment. **Ecosystem ecology** is a subfield of ecology that studies energy flow and the cycling of chemicals among the interacting biotic and abiotic parts of an ecosystem.

Ecosystem interactions are complex because each organism responds not only to other organisms but to conditions in the atmosphere, soil, and water. In turn, organisms exert an effect on the abiotic environment, as when a beaver dam creates a pond in a formerly forested area (see photograph). The pond is formed as the beaver builds an island lodge that will be safe from predators. However, the beaver dam also regulates the flow of water in the stream or river: it holds back water during rainy periods and releases a controlled amount of water throughout the year, even during periods of drought.

Like communities, ecosystems vary in size, lack precise boundaries, and are nested within larger ecosystems. Earth's largest ecosystem is the biosphere, which consists of all Earth's communities and their interactions and connections with the planet's abiotic environment: its water, soil, rock, and atmosphere.

Beaver pond. American beavers (*Castor canadensis*) constructed this dam out of branches and mud, converting a creek into a lake. Photographed on Lake Silver Horn, Alaska.

KEY CONCEPTS

55.1 Studying the energy content of the different trophic levels provides insight into how energy flows through the biotic and abiotic components of ecosystems.

55.2 Carbon, nitrogen, water, and other materials cycle through both biotic and abiotic parts of ecosystems.

55.3 The abiotic environment—including solar radiation, the atmosphere, the ocean, climate, and fire—helps shape the biotic portion of ecosystems.

55.4 Ecosystem ecologists focus on chemical, physical, and biological processes of ecosystems to learn how ecosystems function.

55.1 ENERGY FLOW THROUGH ECOSYSTEMS

LEARNING OBJECTIVES

1 Summarize the concept of energy flow through a food web.
2 Explain typical pyramids of numbers, biomass, and energy.
3 Distinguish between gross primary productivity and net primary productivity.

The passage of energy in a one-way direction through an ecosystem is known as **energy flow.** Energy enters an ecosystem as radiant energy (sunlight), a tiny portion (less than 1%) of which *producers* trap and use during photosynthesis. The energy, now in chemical form, is stored in the bonds of organic (carbon-containing) molecules such as glucose. When cellular respiration breaks these molecules apart, energy becomes available (in the form of ATP) to do work, such as repairing tissues, producing body heat, moving about, or reproducing. As the work is accomplished, energy escapes the organisms and dissipates into the environment as heat. Ultimately, this heat energy radiates into space. Thus, once an organism has used energy, the energy is unavailable for reuse (**FIG. 55-1**). (See also the discussion of the second law of thermodynamics in Chapter 7.)

In an ecosystem energy flow occurs in **food chains,** in which energy from food passes from one organism to the next in a sequence. **Primary producers,** also called *autotrophs,* form the beginning of the food chain by capturing the sun's energy through photosynthesis. Producers, by incorporating the chemicals they manufacture into their own *biomass* (living material), become potential food resources for other organisms. Plants are the most significant producers on land, whereas algae and cyanobacteria are important producers in aquatic environments. (Alternative producers are photoheterotrophs and chemoautotrophs discussed in Chapters 9 and 25.) All other organisms in a community are **consumers,** also called *heterotrophs,* which extract energy from organic molecules produced by other organisms.

Food chains are divided into **trophic levels** (from the Greek *tropho,* which means "nourishment"). Producers occupy the first trophic level. **Herbivores,** which occupy the second trophic level, are *primary consumers* that eat plants. They obtain the chemical energy of the producers' molecules and the building materials used to construct their own tissues. Herbivores are, in turn, consumed by **carnivores,** *secondary consumers* that reap the energy stored in the herbivores' molecules. *Tertiary consumers* are carnivores that eat secondary consumers. **Omnivores** are consumers that eat a variety of organisms, both plant and animal. Omnivores and consumers occupy third or higher trophic levels depending on what they eat.

KEY POINT

Ecologists gain insights into how ecosystems function by examining energy flow and the energy content of each trophic level.

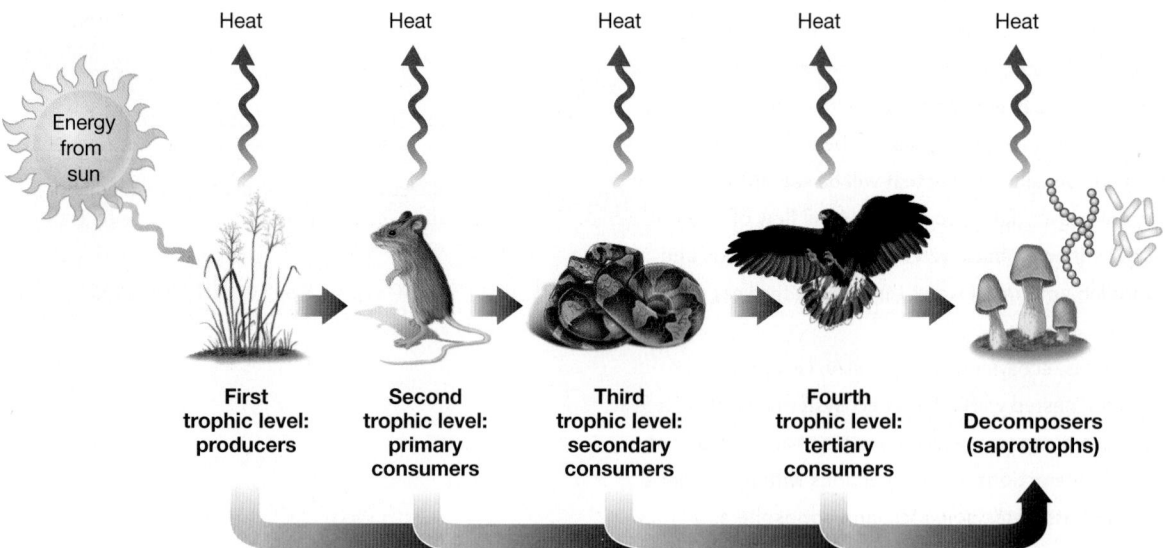

Figure 55-1 *Animation* **Energy flow through ecosystems**
Energy enters ecosystems from an external source (the sun) and exits as heat loss.
PREDICT Which organism(s) would have the longest survival rate if the sun ceased to generate energy?
© Cengage Learning

Some consumers, called **detritus feeders,** or **detritivores,** eat **detritus,** which is dead organic matter that includes animal carcasses, leaf litter, wood, and feces. Detritivores and microbial decomposers break down dead organisms and waste products. **Decomposers,** also called **saprotrophs,** are a subset of detritus feeders. Decomposers include microbial heterotrophs and fungi. These organisms supply themselves with energy by externally breaking down organic molecules in remains (carcasses and body wastes) and ingesting the inorganic products. They typically release simple inorganic molecules, such as carbon dioxide and mineral salts, which may be reused by producers. Most bacteria and fungi are important decomposers of all members of the food chain. (See Chapters 25 and 29.)

Simple food chains such as the one described rarely occur in nature because few organisms eat just one kind, or are eaten by just one other kind, of organism. More typically, the flow of energy and materials through ecosystems takes place in accordance with a range of food choices for each organism. In an ecosystem of average complexity, hundreds of alternative pathways are possible. Thus, a **food web,** which is a complex of interconnected food chains, is a more realistic model of the flow of energy and materials through ecosystems (FIG. 55-2).

Because food chains and food webs are descriptions of "who eats whom," they indicate the negative effects predators have on their prey. For example, consider a simple food chain: grass ⟶ field mouse ⟶ owl. The owl, which kills and eats mice, obviously exerts a negative effect on the mouse population; in like manner, field mice, which eat grass seeds, reduce the grass population.

A trophic level in a food web also influences other trophic levels to which it is not directly linked. Producers and top carnivores do not usually exert direct effects on one another, yet each indirectly affects the other. In our example, the owls help the grasses by keeping the population of seed-eating mice under control. Likewise, the grasses benefit owls by supporting a population of mice on which the owl population feeds. These indirect interactions may be as important in food-web dynamics as direct predator–prey interactions.

The most important thing to remember about energy flow in ecosystems is that it is linear, or a one-way system. That is, energy moves along a food web from one trophic level to the next trophic level. Once an organism has used energy, however, it is lost as heat and is unavailable to any other organism in the ecosystem.

Figure 55-2 *Animation* **A food web at the edge of an Eastern deciduous forest**
This diagram illustrates only a few of the thousands of species and links that might be part of a food web at the edge of a deciduous forest.
PREDICT What could happen to this food web if drought conditions occurred?
© Cengage Learning

Ecological pyramids illustrate how ecosystems work

Ecologists sometimes compare trophic levels by determining the number of organisms, the biomass, or the relative energy found at each level. This information is presented graphically as **ecological pyramids.** The base of each ecological pyramid represents the producers, the next level is the primary consumers (herbivores), the level above that is the secondary consumers (carnivores), and then tertiary consumers (carnivores), and so on. The relative area of each bar of the pyramid is proportional to what is being demonstrated.

A **pyramid of numbers** shows the number of organisms at each trophic level in a given ecosystem, with a larger area illustrating greater numbers for that section of the pyramid. In most pyramids of numbers, fewer organisms occupy each successive trophic level. Thus, in African grasslands the number of herbivores, such as zebras and wildebeests, is greater than the number of carnivores, such as lions. Inverted pyramids of numbers, in which higher trophic levels have more organisms than lower trophic levels, are often observed among decomposers, parasites, and herbivorous insects. One tree provides food for thousands of leaf-eating insects, for example. Pyramids of numbers are of limited usefulness because they do not indicate the biomass of the organisms at each level, nor do they indicate the amount of energy transferred from one level to another.

A **pyramid of biomass** illustrates the total biomass at each successive trophic level. **Biomass** is a quantitative estimate of the total mass, or amount, of living material; it indicates the amount of fixed energy at a particular time. Biomass units of measure vary: biomass may be represented as total volume, dry weight, or live weight. Typically, these pyramids illustrate a progressive reduction of biomass in succeeding trophic levels (**FIG. 55-3a**). Assuming an average biomass reduction of about 90% for each trophic level, 10,000 kg of grass should support 1000 kg of grasshoppers, which in turn support 100 kg of frogs. (The 90% reduction in biomass is an approximation; actual biomass reduction from one trophic level to the next varies widely.) If we use this logic, the biomass of frog eaters (such as snakes) could weigh,

at most, only about 10 kg. From this brief exercise, you see that although carnivores do not eat producers, a large producer biomass is required to support carnivores in a food web.

Occasionally, we find an inverted pyramid of biomass in which the primary consumers outweigh the producers (**FIG. 55-3b**). In these instances, herbivores such as fishes and zooplankton (protozoa, tiny crustaceans, and immature stages of many aquatic animals) consume large numbers of producers, which are usually unicellular algae that are short-lived and reproduce quickly. Thus, although at any point in time relatively few algae are present, the rate of biomass production of the primary consumers is much less than that of the producers.

A **pyramid of energy** indicates the energy content, often expressed as kilocalories (or kilojoules), per square meter per year of the biomass of each trophic level. A common method ecologists use to measure energy content is to burn a sample of tissue in a calorimeter; the heat released during combustion is measured to determine the energy content of the organic material in the sample. Energy pyramids always have large bases and get progressively smaller through succeeding trophic levels to show that most energy dissipates into the environment when there is a transition from one trophic level to the next. Less energy reaches each successive trophic level from the level beneath it because those organisms at the lower level use some of the energy to perform work, and some of it is lost as heat (**FIG. 55-4**). (Remember that no biological process is 100% efficient.) The second law of thermodynamics explains why there are few trophic levels: energy pyramids are short because of the dramatic reduction in energy content that occurs at each successive trophic level.

Ecosystems vary in productivity

The **gross primary productivity (GPP)** of an ecosystem is the rate at which energy is captured during photosynthesis.[1] Thus,

---

[1] Gross and net primary productivities are referred to as primary because plants and other producers occupy the first position in food webs.

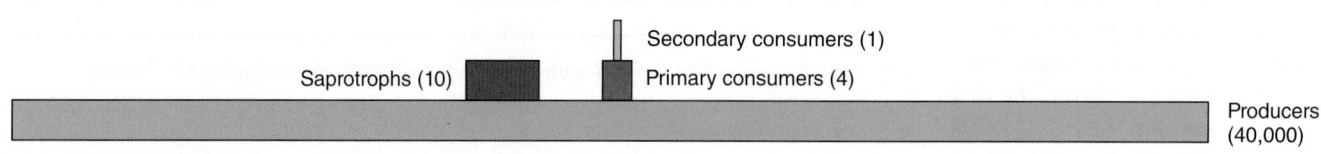

Saprotrophs (10)   Secondary consumers (1)
Primary consumers (4)
Producers (40,000)

(a) A pyramid of biomass for a tropical forest in Panama.

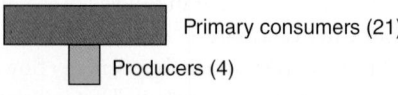

Primary consumers (21)

Producers (4)

(b) An inverted biomass pyramid, such as that for plankton in the English Channel, occurs when a highly productive lower trophic level experiences high rates of turnover. Plankton are free-floating, mainly microscopic algae and animals.

Figure 55-3 Pyramids of biomass

These pyramids are based on the biomass at each trophic level and generally have a pyramid shape with a large base and progressively smaller areas for each succeeding trophic level. Biomass values are in grams of dry weight per square meter. (**a, b,** Adapted from Odum, E. P. *Fundamentals of Ecology,* 3rd ed., W. B. Saunders Company, Philadelphia, 1971, and based on studies by F. B. Golley and G. I. Child **[a]** and H. W. Harvey **[b]**.)

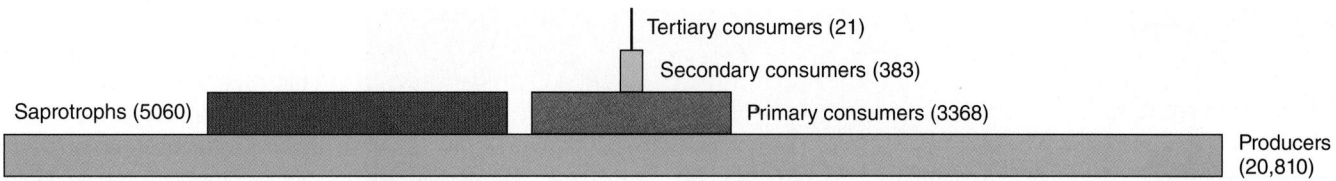

Figure 55-4 *Animation* **Pyramid of energy**

A pyramid of energy for Silver Springs, Florida, represents energy flow, the functional basis of ecosystem structure. Energy values are in kilocalories per square meter per year. Note the substantial loss of usable energy from one trophic level to the next. The Silver Springs ecosystem is complex, but tape grass (producers), spiral-shelled snails (primary consumers), young river turtles (secondary consumers), gar (fish; tertiary consumers), and bacteria and fungi (saprotrophs) are representative organisms. When they are young, river turtles are carnivores and consume snails, aquatic insects, and worms; as adults, river turtles are herbivores. (Based on Odum, H. T. "Trophic Structure and Productivity of Silver Springs, Florida." *Ecological Monographs*, Vol. 27, 1957.)

GPP is the total amount of photosynthetic energy captured in a given period. Of course, plants and other producers must respire to provide energy for their life processes, and cellular respiration acts as a drain on photosynthetic output. Energy that remains in plant tissues after cellular respiration has occurred is called **net primary productivity (NPP).** That is, NPP is the amount of biomass (the energy stored in plant tissues) found in excess of that broken down by a plant's cellular respiration for normal daily activities. NPP represents the rate at which this organic matter is actually incorporated into plant tissues to produce growth.

$$\begin{matrix} \text{net} \\ \text{primary} \\ \text{productivity} \end{matrix} = \begin{matrix} \text{gross} \\ \text{primary} \\ \text{productivity} \end{matrix} - \begin{matrix} \text{plant} \\ \text{respiration} \end{matrix}$$

(plant growth per unit area per unit time)   (total photosynthesis per unit area per unit time)   (per unit area per unit time)

Only the energy represented by net primary productivity is available for consumers, and of this energy only a portion is actually used by them. Both GPP and NPP are expressed as energy per unit area per unit time (e.g., kilojoules of energy fixed by photosynthesis per square meter per year) or as dry weight (e.g., grams of carbon incorporated into tissue per square meter per year).

Ecologists use different methods to measure net primary productivity, such as through calorimetry (previously described) and change in dry biomass per unit time. Scientists also determine NPP by measuring carbon dioxide uptake and oxygen uptake over time. Another method to assess NPP in plants is through recording the change in chlorophyll *a* concentration, which measures daily carbon dioxide fixation.

Herbivores and other consumers eventually consume all of a plant's net primary production. What happens to this energy? Consider the transfer of net primary production from a plant to a deer that eats the plant. Much of the energy stored in the plant material that the deer consumes—about 25%—is not digested and is lost in its feces. (This energy is not lost from the ecosystem because detritivores and decomposers will make use of it; it is lost from the deer's point of view, however.) Perhaps 55% of the energy that the deer takes in as food is released during cellular respiration and used to do work such as muscle contraction and to maintain and repair the deer's body. The remaining energy—less than 20%—is available to produce new biomass, that is, new tissues. This net energy available for biomass production by consumer organisms is called **secondary productivity.** An ecosystem's secondary productivity is based on its primary productivity.

Many factors may interact to determine primary productivity. Some plants are more efficient than others in fixing carbon. Environmental factors are also important. They include the availability of solar energy, minerals, and water; other climate factors; the degree of maturity of the community; and the severity of human modification of the environment.

These factors are difficult to assess, particularly on a large scale. The summer of 2003, which was extremely hot and dry in Europe, provided ecologists with a chance to measure primary productivity changes in response to these unusual conditions. Based on their measurements, they estimate that GPP throughout Europe was reduced by 30% as a result of the heat and drought.

Ecosystems differ strikingly in their primary productivities (**FIG. 55-5** and **TABLE 55-1**). On land, tropical rain forests have the highest productivity, probably as a result of the abundant rainfall, warm temperatures, and intense sunlight. As you might expect, tundra, because of its short, cool growing season, and deserts, because of their lack of precipitation, are the least productive terrestrial ecosystems. In ecosystems with comparable annual temperatures (e.g., temperate deciduous forest, temperate grassland, and temperate desert), water availability affects NPP. Availability of essential minerals such as nitrogen and phosphorus also affects NPP.

Wetlands (swamps and marshes) connect terrestrial and aquatic environments and are extremely productive. The most productive aquatic ecosystems are algal beds, coral reefs, and estuaries. The lack of available minerals in the sunlit region of the open ocean makes this area extremely unproductive, equivalent to an aquatic desert. Earth's major aquatic and terrestrial ecosystems are discussed in Chapter 56.

As primary productivity increases, species richness declines Ecologists have observed that an ecosystem's species richness declines with increasing productivity. For example, the resource-poor depths of the Atlantic Ocean's abyssal plain have

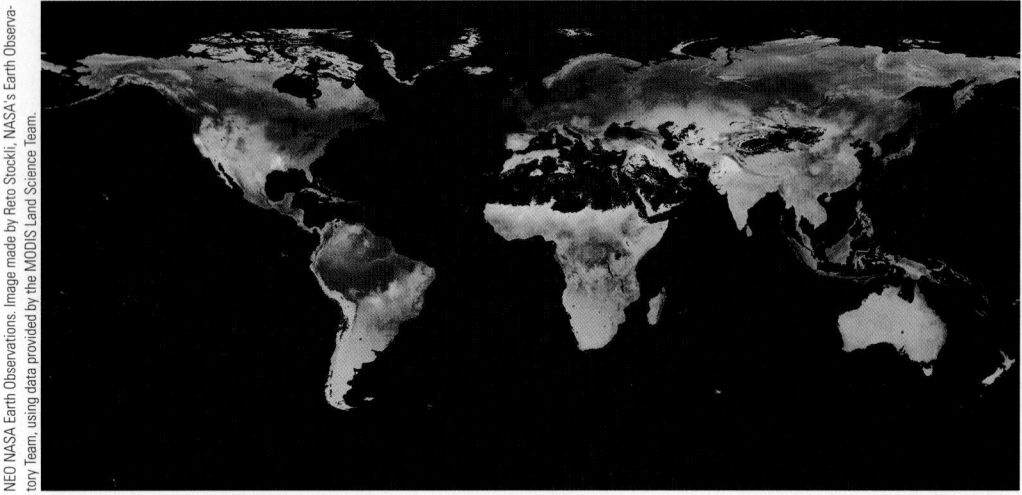

Figure 55-5 View of Earth's net primary productivity

The colors on this NASA satellite image indicate how quickly plants took carbon in for every square meter of land during August, 2013. Values range from −1.0 gram of carbon per square meter per day (*tan*) to 6.5 grams per square meter per day (*dark green*). A negative value indicates that decomposition or respiration was greater than carbon production; more carbon was released to the atmosphere than plants captured. The least productive areas are deserts. The most productive regions are tropical rain forests. Images like this allow scientists to assess changes in productivity over time.

phosphorus inputs from fossil fuels, fertilizers, and livestock. This continual enrichment may make Earth's ecosystems increasingly productive, a shift that some ecologists think could cost the world a substantial loss of species richness. (Other factors that affect species richness were discussed in Chapter 54.)

Humans consume an increasingly greater percentage of global primary productivity

Humans consume far more of Earth's resources than do any of the other millions of animal species. Peter Vitousek and co-workers at Stanford University calculated in 1986 how much of the global NPP is appropriated for the human economy. When both direct and indirect human effects are accounted for, Vitousek estimated that humans use 32% to 40% of land-based annual NPP. Since 1986, scientists have done additional research on global ecology, resulting in improved data sets. In 2001, Stuart Rojstaczer and co-workers at Duke University used satellite-based data to determine a conservative estimate of land-based annual NPP appropriation by humans at 32%.

The take-home message from Vitousek's and Rojstaczer's research is simple. Essentially, human use of global productivity is competing with other species' needs for energy. Our use of so much of the world's productivity may contribute to the loss, through extinction or genetic impoverishment, of many species that have unique roles in maintaining functional ecosystems. Clearly, at these levels of consumption and exploitation of Earth's resources, human population growth threatens the planet's ability to support all its occupants.

more species richness than productive shallow waters near the coasts. Ecologists are designing experiments to help explain the pattern, which has been documented with rodents in Israel, birds in South America, and large mammals in Africa. Mathematical ecosystem models suggest that a less productive environment has a *patchy distribution of resources* that reduces competition and allows a greater variety of organisms to coexist.

The bad news for global biodiversity is that humans are constantly enriching the environment, such as with nitrogen and

TABLE 55-1	Net Primary Productivity (NPP) for Selected Ecosystems
ECOSYSTEM	**AVERAGE NPP (g dry matter/m^2/year)**
Algal beds and reefs	2500
Tropical rain forest	2200
Swamp and marsh	2000
Estuaries	1500
Temperate evergreen forest	1300
Temperate deciduous forest	1200
Savanna	900
Boreal (northern) forest	800
Woodland and shrubland	700
Agricultural land	650
Temperate grassland	600
Upwelling zones in ocean	500
Lake and stream	250
Arctic and alpine tundra	140
Open ocean	125
Desert and semidesert scrub	90
Extreme desert (rock, sand, ice)	3

Source: Based on Whittaker, R. H. *Communities and Ecosystems*, 2nd ed. Macmillan, New York, 1975.

Some toxins persist in the environment

You have seen how energy flows through food chains in ecosystems. Before leaving the discussion of food chains, let us consider how certain toxins, including some pesticides, radioactive isotopes, heavy metals such as mercury, and industrial chemicals such as polychlorinated biphenyls (PCBs), enter and pass through food chains. The effects of the pesticide DDT on some bird species first drew attention to the problem. Falcons, pelicans, bald eagles, ospreys, and many other birds are sensitive to traces of DDT in their tissues. A substantial body of scientific evidence indicates that one effect of DDT on these birds is that their eggs have extremely thin, fragile shells that usually break during incubation, causing the chicks' deaths. In 1962, U.S. biologist Rachel Carson published *Silent Spring*, which heightened public awareness about the dangers of DDT and other pesticides. After 1972, the year DDT was banned in the United States, the reproductive success of many birds gradually improved.

Does DDT exhibit biological magnification as it moves through a food chain?

HYPOTHESIS: DDT, a persistent insecticide, increases in concentration at each level of a food chain.

EXPERIMENT: Biologists sampled the concentration of DDT in various organisms of a Long Island salt marsh.

TROPHIC LEVEL	AMOUNT OF DDT IN TISSUE
Tertiary consumer	Ring-billed gull (75.5 ppm)
Secondary consumer	Atlantic needlefish (2.07 ppm)
Secondary consumer	American eel (0.28 ppm)
Primary consumer	Shrimp (0.16 ppm)
Producers, primary consumers	Plankton (0.04 ppm)

RESULTS AND CONCLUSION: The level of DDT increased in the tissues of various organisms as DDT moved through the food chain from producers to consumers. Ring-billed gulls at the top of the food chain had approximately 1 million times as much DDT in their tissues as the concentration of DDT in the water (0.00005 ppm). (Plankton consisted of a mixture of phytoplankton and zooplankton.)

SOURCE: Based on data from Woodwell, G. M., C. F. Worster Jr., and P. A. Isaacson. "DDT Residues in an East Coast Estuary: A Case of Biological Concentration of a Persistent Insecticide." *Science*, Vol. 156, May 12, 1967.

Figure 55-6 Biological magnification of DDT (expressed as parts per million) in a Long Island salt marsh

PREDICT How might DDT be passed on by the ring-billed gull, and what might be the consequences?

The effect of DDT on birds is the result of three characteristics of DDT (and other toxins that cause problems in food webs): its persistence, bioaccumulation, and biological magnification. Some toxins are extremely stable and may take many years to break down into less toxic forms. The **persistence** of synthetic pesticides and industrial chemicals is a result of their novel chemical structures. These toxins accumulate in the environment because ways to degrade them have not evolved in natural decomposers such as bacteria.

When an organism does not metabolize (break down) or excrete a persistent toxin, the toxin simply gets stored, usually in fatty tissues. Over time, the organism may accumulate high concentrations of the toxin. The buildup of such a toxin in an organism's body is known as **bioaccumulation.**

Organisms at higher trophic levels in food webs tend to store greater concentrations of bio-accumulated toxins in their bodies than do those at lower levels. The increase in concentration as the toxin passes through successive levels of the food web is known as **biological magnification.**

As an example of the concentrating characteristic of persistent toxins, consider a food chain studied in a Long Island salt marsh that was sprayed with DDT over a period of years for mosquito control (**FIG. 55-6**). Although this example involved a bird at the top of the food chain, it is important to recognize that

all top carnivores, from fishes to humans, are at risk from biological magnification of persistent toxins. Because of this risk, currently approved pesticides have been tested to ensure that they do not persist and accumulate in the environment.

CHECKPOINT 55.1

- **VISUALIZE** *Draw a diagram tracing energy flow through a food web such as one found in a deciduous forest.*
- **CONNECT** *What are trophic levels, and how are they related to ecological pyramids?*
- *How do gross primary productivity (GPP) and net primary productivity (NPP) differ?*

55.2 CYCLES OF MATTER IN ECOSYSTEMS

LEARNING OBJECTIVE

4 Describe the main steps in each of these biogeochemical cycles: the carbon, nitrogen, phosphorus, and hydrologic cycles.

Matter moves in numerous cycles from one part of an ecosystem to another; that is, it moves from one organism to another (in food chains) and from living organisms to the abiotic environment and back again. We call these cycles of matter **biogeochemical cycles** because they involve biological, geologic, and chemical interactions. For all practical purposes, matter cannot escape from Earth's boundaries. The materials organisms use cannot be "lost," although this matter can end up in locations outside the reach of organisms for a long period. Usually, materials are reused and often recycled both within and among ecosystems.

We discuss four different biogeochemical cycles of matter—carbon, nitrogen, phosphorus, and water—as representative of all biogeochemical cycles. These four cycles are particularly important to organisms because they involve materials used to make the chemical components of cells.

Carbon dioxide is the pivotal molecule in the carbon cycle

Proteins, nucleic acids, lipids, carbohydrates, and other molecules essential to life contain carbon. Carbon is present in the atmosphere as the gas carbon dioxide (CO_2), which makes up approximately 0.04% of the atmosphere. It is also present in the ocean and fresh water as dissolved carbon dioxide, that is, carbonate (CO_3^{2-}) and bicarbonate (HCO_3^-); other forms of dissolved inorganic carbon; and dissolved organic carbon from decay processes. Carbon is also present in rocks such as limestone ($CaCO_3$). The global movement of carbon between the abiotic environment, including the atmosphere and ocean, and organisms is known as the **carbon cycle** (FIG. 55-7).

During photosynthesis, plants, algae, and cyanobacteria remove carbon dioxide from the air and *fix,* or incorporate, it into organic compounds such as glucose. Plants use much of the glucose to make cellulose, starch, amino acids, nucleic acids, and other compounds.

Many of these compounds are used as fuel for cellular respiration by the producer that made them, by a consumer that eats the producer, or by a decomposer that breaks down the remains of the producer or consumer. The process of cellular respiration returns carbon dioxide to the atmosphere. A similar carbon cycle occurs in aquatic ecosystems between aquatic organisms and dissolved carbon dioxide in the water (not shown in Figure 55-7).

Some archaea fix carbon using atmospheric CO_2. Other archaea, the methanogens, are decomposers and are the major source of methane in the atmosphere. The methanogens significantly contribute to global climate change (see discussion of methanogens in Chapter 25).

Sometimes the carbon in biological molecules is not recycled back to the abiotic environment for a long time and remains in reservoirs, or sinks, indefinitely. Carbon stored in the wood

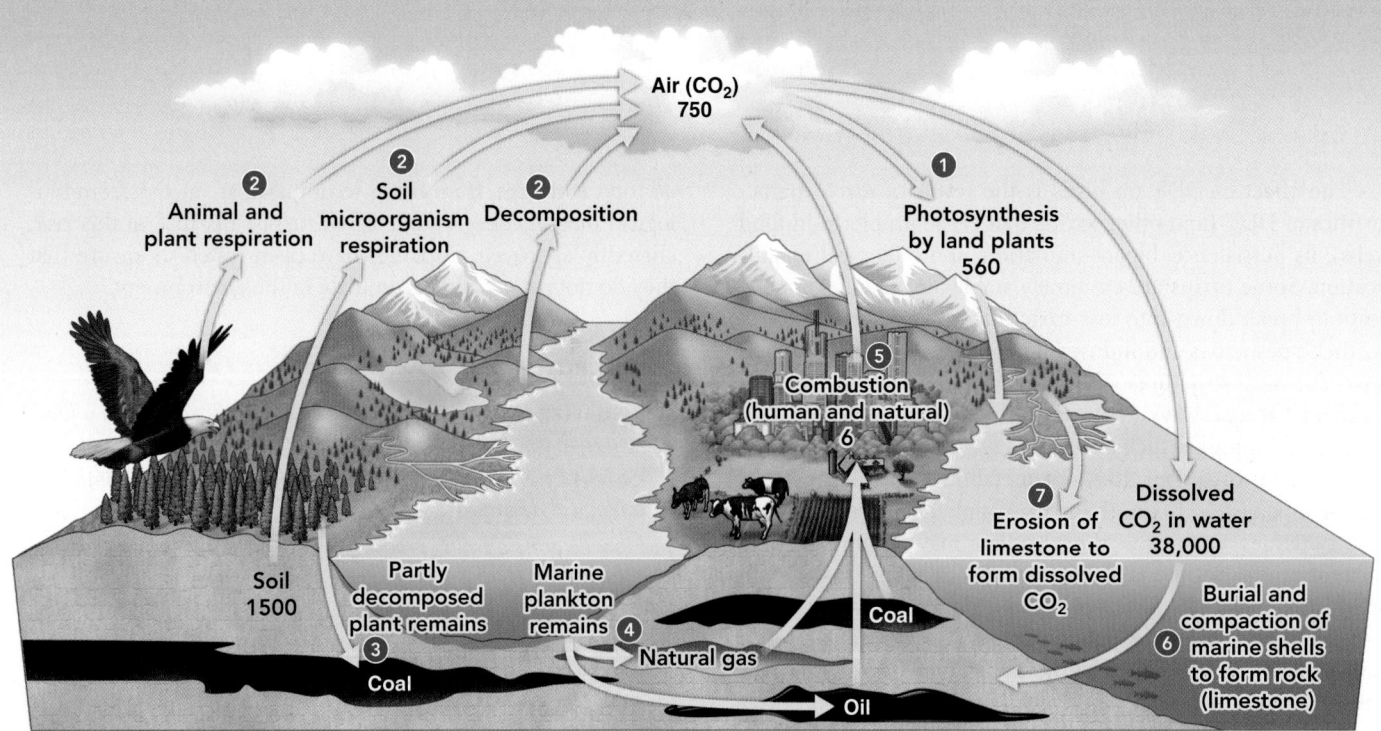

Figure 55-7 *Animation* **A simplified diagram of the carbon cycle**

All but a tiny fraction of Earth's estimated 10^{23} g of carbon is buried in sedimentary rocks and fossil fuel deposits. The values shown for some of the active pools in the global carbon budget are expressed as 10^{15} g of carbon. For example, the soil contains an estimated 1500×10^{15} g of carbon. (Values from Schlesinger, W. H. *Biogeochemistry: An Analysis of Global Change,* 2nd ed., Academic Press, San Diego, 1997, and several other sources.)

© Cengage Learning

of trees may stay for several hundred years or even longer. Also, millions of years ago vast coal beds formed from the bodies of ancient trees that were buried and subjected to anaerobic conditions before they had fully decayed. Similarly, the oils of unicellular marine organisms probably gave rise to the underground deposits of oil and natural gas that accumulated in the geologic past. Coal, oil, and natural gas, called **fossil fuels** because they formed from the remains of ancient organisms, are vast deposits of carbon compounds, the end products of photosynthesis that occurred millions of years ago.

The process of burning, or combustion, may return the carbon in coal, oil, natural gas, and wood to the atmosphere. In combustion organic molecules are rapidly oxidized (combined with oxygen) and converted to carbon dioxide and water with an accompanying release of light and heat.

An even greater amount of carbon that is stored for millions of years is incorporated into the shells of marine organisms. When these organisms die, their shells sink to the ocean floor, and sediments cover them, forming seabed deposits thousands of meters thick. The deposits are eventually cemented together to form limestone, a sedimentary rock. Earth's crust is dynamically active, and over millions of years, sedimentary rock on the bottom of the seafloor may lift to form land surfaces. When the process of geologic uplift exposes limestone, chemical and physical weathering processes slowly erode it away. This returns carbon to the water and atmosphere, where it is available to participate in the carbon cycle once again.

Human activities have disturbed the global carbon budget Before the Industrial Revolution, around 1750, the global carbon cycle was in a steady state. Enormous amounts of carbon moved to and from the atmosphere, ocean, and terrestrial ecosystems, but these movements within the global carbon cycle just about canceled out one another.

Since 1750, our industrial society has required a lot of energy, and we have burned increasing amounts of fossil fuels—coal, oil, and natural gas—to obtain this energy. This trend, along with a greater combustion of wood as a fuel and the burning of large sections of tropical forest, has released CO_2 into the atmosphere at a rate greater than the natural carbon cycle can handle.

Earth's ocean absorbs much of this excess CO_2 from the atmosphere. In the ocean some dissolved CO_2 is converted to carbonic acid (H_2CO_3), which is acidifying surface ocean waters. The pH of modern surface waters is about 0.1 pH unit lower than it was in preindustrial times, and models predict up to 1.4 pH units of further acidification during the next 300 years. *Ocean acidification* harms marine organisms, particularly those that produce skeletons and shells of calcium carbonate ($CaCO_3$), which dissolve in the presence of acid.

The level of atmospheric CO_2 increased dramatically beginning in the last half of the 20th century (see Fig. 57-16), and this rise of CO_2 has initiated human-induced changes in global climate. Global climate change will result in a rise in sea level, changes in precipitation patterns, death of forests, extinction of organisms, and problems for agriculture. It will force the displacement of thousands or even millions of people, particularly from coastal areas. (Chapter 57 contains a more thorough discussion of increasing atmospheric CO_2 and the potential effects of global climate change.)

Bacteria and archaea are essential to the nitrogen cycle

Nitrogen is crucial for all organisms because it is an essential part of proteins, nucleic acids, and chlorophyll. Because Earth's atmosphere is about 78% nitrogen gas (N_2), it would appear that there could be no possible shortage of nitrogen for organisms. However, molecular nitrogen is so stable that it does not readily combine with other elements. Therefore, the N_2 molecule must be broken apart before the nitrogen atoms combine with other elements to form proteins, nucleic acids, and chlorophyll. Chemical reactions that break up N_2 and combine nitrogen with other elements require a great deal of energy.

The **nitrogen cycle,** in which nitrogen cycles between the abiotic environment and organisms, has five steps: nitrogen fixation, nitrification, assimilation, ammonification, and denitrification (**FIG. 55-8**). Certain bacteria and archaea are exclusively involved in all these steps except assimilation.

The first step in the nitrogen cycle, biological **nitrogen fixation,** involves conversion of gaseous nitrogen (N_2) to ammonia (NH_3). This process fixes nitrogen into a form that organisms can use. Combustion, volcanic action, lightning discharges, and industrial processes also fix nitrogen as nitrate (NO_3^-). Certain archaea and nitrogen-fixing bacteria, including cyanobacteria and certain other free-living and symbiotic bacteria, carry on biological nitrogen fixation in soil and aquatic environments. These nitrogen-fixing prokaryotes employ an enzyme called **nitrogenase** to break up molecular nitrogen and combine the resulting nitrogen atoms with hydrogen.

Because nitrogenase functions only in the absence of oxygen, the prokaryotes that fix nitrogen insulate the enzyme from oxygen in some way. Some nitrogen-fixing prokaryotes live beneath layers of oxygen-excluding slime on the roots of several plant species. Other important nitrogen-fixing bacteria, in the genus *Rhizobium,* live in oxygen-excluding swellings, or **nodules,** on the roots of legumes such as beans and peas and some woody plants (**FIG. 55-9**; also see Table 25-3).

In aquatic environments bacteria and archaea perform most nitrogen fixation. Filamentous cyanobacteria have special oxygen-excluding cells called **heterocysts** that function to fix nitrogen (see Table 25-3). Some water ferns have cavities in which cyanobacteria live, in a manner comparable to the way *Rhizobium* lives in root nodules of legumes. Other cyanobacteria fix nitrogen in symbiotic association with cycads and other terrestrial plants or as the photosynthetic partner of certain lichens.

The second step of the nitrogen cycle is **nitrification** (see Fig. 55-8), the conversion of ammonia (NH_3) or ammonium (NH_4^+), formed when water reacts with ammonia, to nitrate (NO_3^-). Soil bacteria are responsible for the two-phase process of nitrification, which furnishes these bacteria, called *nitrifying bacteria,* with energy.

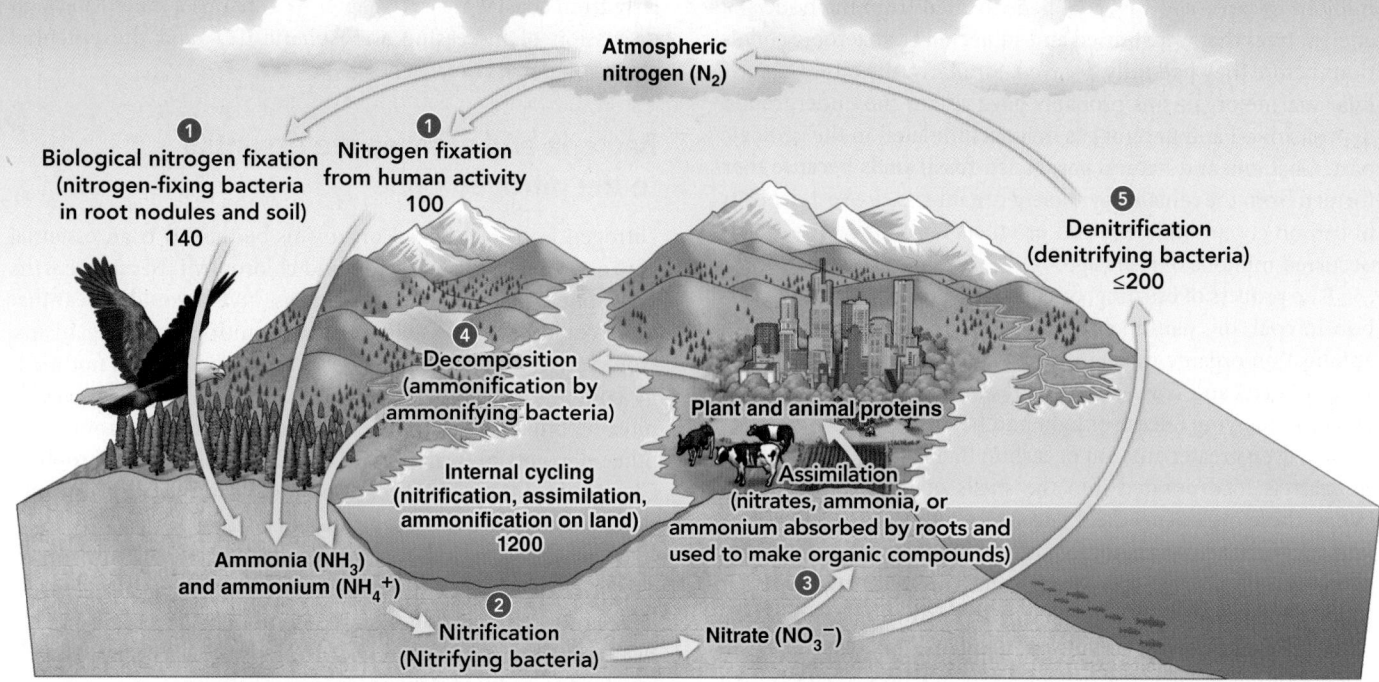

Figure 55-8 *Animation* **A simplified diagram of the nitrogen cycle**

The largest pool of nitrogen, estimated at 3.9×10^{21} g, is in the atmosphere. The values shown for selected nitrogen fluxes in the global nitrogen budget are expressed as 10^{12} g of nitrogen per year and represent terrestrial values. For example, each year humans fix an estimated 100×10^{12} g of nitrogen. (Values from Schlesinger, W. H. *Biogeochemistry: An Analysis of Global Change,* 2nd ed., Academic Press, San Diego, 1997, and several other sources.)

© Cengage Learning

In the third step, **assimilation,** roots absorb ammonia (NH_3), ammonium (NH_4^+), or nitrate (NO_3^-) formed by nitrogen fixation and nitrification, and incorporate the nitrogen into proteins, nucleic acids, and chlorophyll. When animals consume plant tissues, they assimilate nitrogen by taking in plant nitrogen compounds and converting them to animal nitrogen compounds.

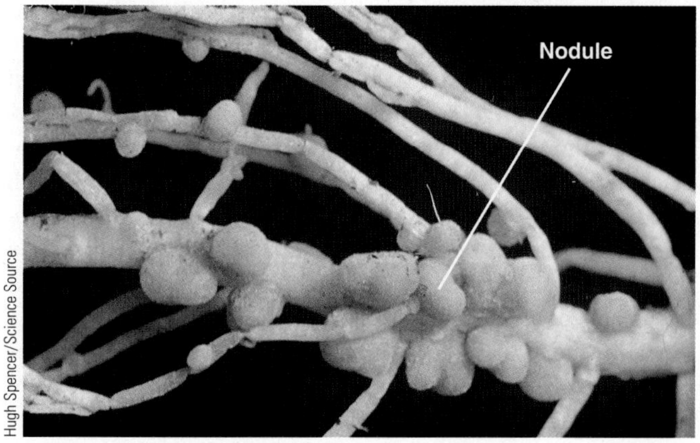

The fourth step, **ammonification,** is the conversion of organic nitrogen compounds into ammonia (NH_3) and ammonium ions (NH_4^+). Ammonification begins when organisms produce nitrogen-containing wastes such as urea in urine and uric acid in the wastes of birds (see Fig. 48-1). As these substances, along with the nitrogen compounds in dead organisms, decompose, nitrogen is released into the abiotic environment as ammonia (NH_3). The bacteria that perform ammonification in both the soil and aquatic environments are called *ammonifying bacteria.* Most available nitrogen in the soil derives from the recycling of organic nitrogen by ammonification.

The fifth step of the nitrogen cycle is **denitrification,** the reduction of nitrate (NO_3^-) to gaseous nitrogen (N_2). Denitrifying prokaryotes reverse the action of nitrogen-fixing and nitrifying prokaryotes by returning nitrogen to the atmosphere as nitrogen gas. Denitrifying prokaryotes are anaerobic and therefore live and grow best where there is little or no free oxygen. For example, they are found deep in the soil near the water table, an environment that is nearly oxygen-free.

Human activities have changed the global nitrogen budget Human activities have disturbed the balance of the global nitrogen cycle. During the 20th century, humans more than doubled the amount of fixed nitrogen (nitrogen that has been chemically combined with hydrogen, oxygen, or carbon) entering the global nitrogen cycle. The excess nitrogen is seriously altering many terrestrial and aquatic ecosystems.

Figure 55-9 Root nodules and nitrogen fixation

Root nodules on the roots of a pea plant provide an oxygen-free environment for nitrogen-fixing *Rhizobium* bacteria that live in them.

Large quantities of nitrogen fertilizer, for commercial farming operations and for residential use, are produced from nitrogen gas for agriculture. The increasing use of fertilizer has resulted in higher crop yields, but there are negative environmental effects from human-produced nitrogen. Nitrogen fertilizer is extremely mobile and is easily transferred from the land to rivers to estuaries to the ocean. Thus, the overuse of commercial fertilizer on the land causes water-quality problems that may help explain long-term declines in many coastal fisheries. The amount of nitrate or ammonium in most aquatic ecosystems is in limited supply and limits the growth of algae. Rain washes fertilizer into rivers and lakes, where it stimulates the growth of algae, some of which are toxic. As these algae die, their decomposition by bacteria robs the water of dissolved oxygen, which in turn causes other aquatic organisms, including many fishes, to suffocate.

Nitrates from fertilizer also leach (dissolve and wash down) through the soil and contaminate groundwater. Many people who live in rural areas drink groundwater. Groundwater contaminated by nitrates is dangerous, particularly for infants and small children.

Another human activity that affects the nitrogen cycle is the combustion of fossil fuels. When fossil fuels are burned, the nitrogen locked in organic compounds in the fuel is chemically altered and transferred to the atmosphere. In addition, the high temperature of combustion converts some atmospheric nitrogen to **nitrogen oxides.** Automobile exhaust is one of the main sources of nitrogen oxides. Nitrogen oxides are a necessary ingredient in the production of **photochemical smog,** a mixture of several air pollutants that injure plant tissues, irritate eyes, and cause respiratory problems in humans.

Nitrogen oxides also react with water in the atmosphere to form nitric acid (HNO_3) and nitrous acid (HNO_2). When these and other acids leave the atmosphere as precipitation (rain, sleet, snow, or hail), they decrease the pH of surface waters (lakes and streams) and soils. **Acid precipitation** includes rain, snow, fog, or dust that is unusually acidic (pH below 5.6). Acid precipitation has been linked to declining animal populations in aquatic ecosystems. On land, acid precipitation alters soil chemistry: certain essential minerals, such as calcium and potassium, wash out of the soil and are therefore unavailable for plants. Nitrous oxide (N_2O), one of the nitrogen oxides, retains heat in the atmosphere (like CO_2) and so promotes global climate change.

Nitrous oxide also contributes to the depletion of ozone in the stratosphere. (See Chapter 2 for a discussion of acids and pH, and Chapter 57 for a discussion of global climate change and stratospheric ozone depletion.)

The phosphorus cycle lacks a gaseous component

Phosphorus does not exist in a gaseous state and therefore does not enter the atmosphere. In the **phosphorus cycle,** phosphorus cycles from the land to sediments in the ocean and back to the land (**FIG. 55-10**).

As water runs over rocks containing phosphorus, it gradually erodes the surface and carries off inorganic phosphate (PO_4^{3-}). The erosion of phosphorus rocks releases phosphate into the soil, where it is taken up by roots in the form of inorganic phosphates. Once in cells, phosphates are incorporated into a variety of biological molecules, including nucleic acids, ATP, and the phospholipids that make up cell membranes. Animals obtain most of their required phosphorus from the food they eat, although in some places drinking water may contain a substantial amount of inorganic phosphate. Phosphate released by decomposers becomes part of the pool of inorganic phosphate in the soil that plants reuse. Thus, like carbon and nitrogen, phosphorus moves through the food web as one organism consumes another.

Phosphorus cycles through aquatic ecosystems in much the same way as through terrestrial ecosystems. Dissolved phosphate enters aquatic ecosystems through absorption by algae and aquatic plants, which zooplankton and larger organisms consume. In turn, a variety of fishes and mollusks eat the zooplankton. Ultimately, decomposers break down wastes and dead organisms to release inorganic phosphate into the water, making it available for use again by aquatic producers.

Phosphate can be lost for varying time periods from biological cycles. Streams and rivers carry some phosphate to the ocean, where it is deposited on the seafloor and remains for millions of years. The geologic process of uplift may someday expose these reservoirs of seafloor sediments as new land surfaces, from which phosphate will be once again eroded. Phosphate deposits are also mined for agricultural use in phosphate fertilizers.

Humans affect the natural cycling of phosphorus In natural terrestrial communities, very little phosphorus is lost from the cycle, but few communities today are in a natural state, that is, unaltered in some way by humans. Land-denuding practices, such as the clear-cutting of timber, and erosion of agricultural and residential lands accelerate phosphorus loss from the soil into waterways. Excess phosphorus enriches the water, causing algal blooms that rob the water of dissolved oxygen. For practical purposes, phosphorus that washes from the land into the ocean is permanently lost from the terrestrial phosphorus cycle (and from further human use) because it remains in the ocean for millions of years.

Ecologists are also concerned that we are mining phosphorus faster than is sustainable. At its current rate of use, U.S. deposits of phosphorus will probably be depleted during the 21st century. There are currently no substitutes for phosphorus or synthetic methods to produce it.

Water moves among the ocean, land, and atmosphere in the hydrologic cycle

Life would be impossible without water, which makes up a substantial part of the mass of most organisms. All species, prokaryotes and eukaryotes, use water as a medium for chemical

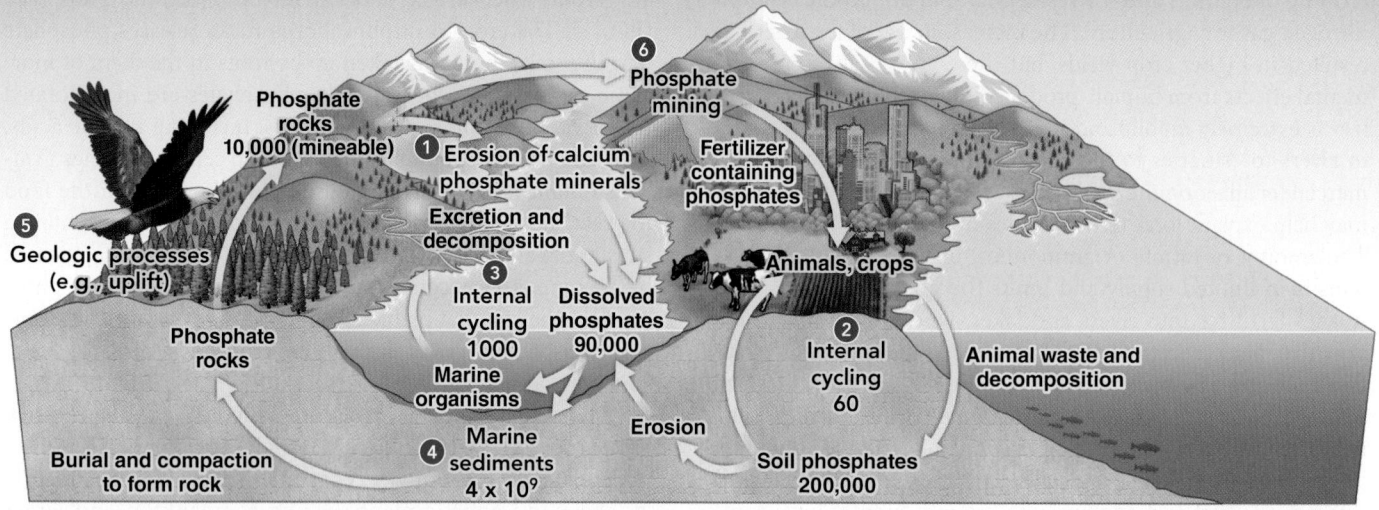

Figure 55-10 *Animation* **A simplified diagram of the phosphorus cycle**

Some values of the global phosphorus budget are given, in units of 10^{12} g phosphorus per year. For example, each year an estimated 60×10^{12} g of phosphorus cycles from the soil to terrestrial organisms and back to the soil. (Values from Schlesinger, W. H. *Biogeochemistry: An Analysis of Global Change*, 2nd ed., Academic Press, San Diego, 1997, and several other sources.)

© Cengage Learning

reactions as well as for the transport of materials within and among cells. (Recall from Chapter 2 that water has many unique properties that help shape the continents, moderate climate, and allow organisms to survive.)

In the **hydrologic cycle,** water continuously circulates from the ocean to the atmosphere to the land and back to the ocean (**FIG. 55-11**). Water moves from the atmosphere to the land and ocean in the form of precipitation. Water that evaporates from the ocean surface and from soil, streams, rivers, and lakes eventually condenses and forms clouds in the atmosphere. In addition, **transpiration,** the loss of water vapor from land plants, adds a considerable amount of water vapor to the atmosphere. Roughly 97% of the water a plant absorbs from the soil is transported to the leaves, where it is lost by transpiration.

Water may evaporate from land and re-enter the atmosphere directly. Alternatively, it may flow in rivers and streams to coastal **estuaries,** where fresh water meets the ocean. The movement of surface water from land to ocean is called *runoff,* and the area of land drained by runoff is called a *watershed.* Water also percolates (seeps) downward in the soil to become *groundwater,* where it is trapped and held for a time. The underground caverns and porous layers of rock in which groundwater is stored are called *aquifers.* Groundwater may reside in the ground for hundreds to many thousands of years, but eventually it supplies water to the soil, streams and rivers, plants, and the ocean. The human removal of more groundwater than precipitation or melting snow recharges, called *aquifer depletion,* eliminates groundwater as a water resource.

Regardless of its physical form (solid, liquid, or vapor) or location, every molecule of water eventually moves through the hydrologic cycle. Tremendous amounts of water cycle annually between Earth and its atmosphere. The volume of water entering the atmosphere from the ocean each year is estimated at about 425,000 km³. Approximately 90% of this water re-enters the ocean directly as precipitation over water; the remainder falls on land. As is true of the other cycles, water (in the form of glaciers, polar ice caps, and certain groundwater) can be lost from the cycle for thousands of years.

CHECKPOINT 55.2

- **CONNECT** *What are the roles of the following processes in the carbon cycle: photosynthesis, cellular respiration, combustion, and erosion?*

- **VISUALIZE** *Draw a diagram showing the five steps in the nitrogen cycle and explain what happens in each step.*

- *How does the phosphorus cycle proceed without a gaseous component?*

55.3 ABIOTIC FACTORS IN ECOSYSTEMS

LEARNING OBJECTIVES

5   Summarize the effects of solar energy on Earth's temperatures.
6   Discuss the roles of solar energy and the Coriolis effect in the generation of global air and water flow patterns.
7   Give two causes of regional precipitation differences.
8   Discuss the effects of fire on certain ecosystems.

You have seen how ecosystems depend on the abiotic environment to supply energy and essential materials (in biogeochemical cycles). Other abiotic factors such as solar radiation, the

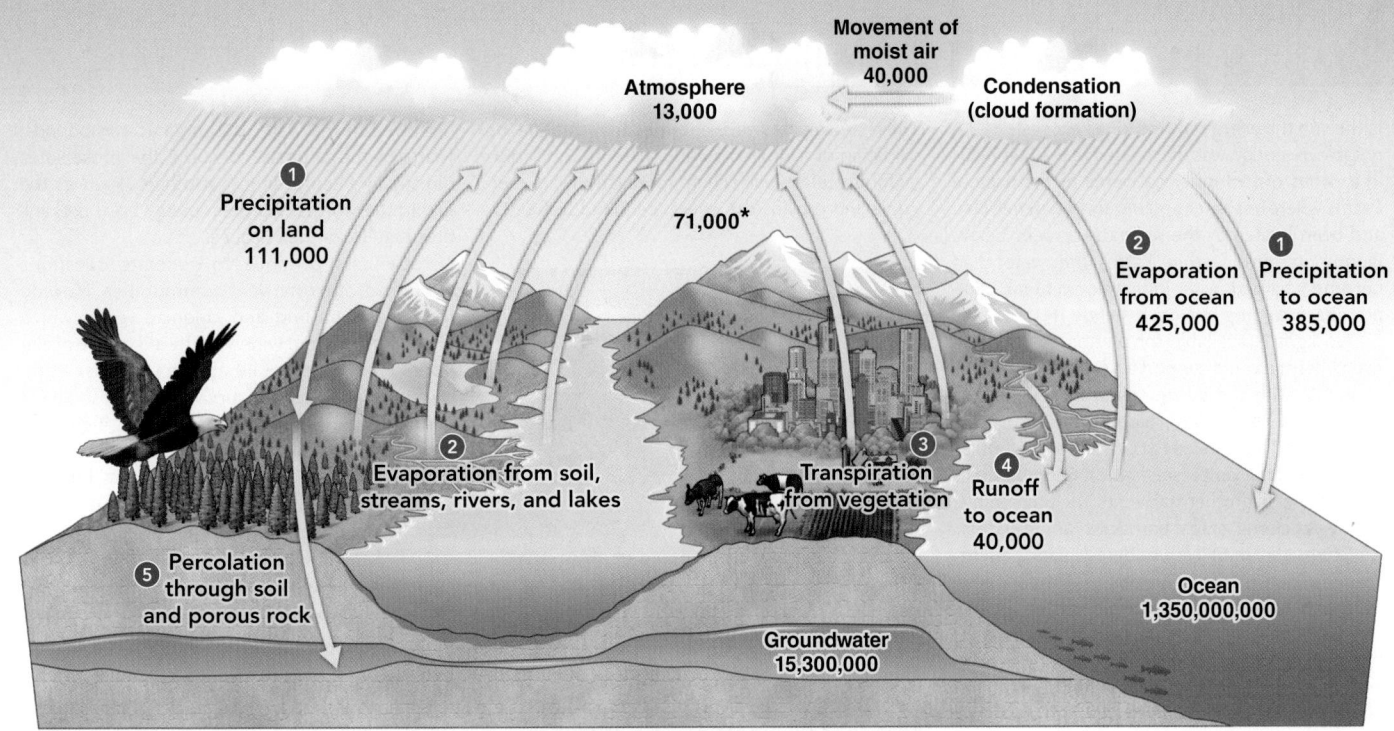

Figure 55-11 *Animation* **A simplified diagram of the hydrologic cycle**

The global water budget values shown for pools are expressed as cubic kilometers; values for fluxes (movements associated with arrows) are in cubic kilometers per year. The *starred value* (71,000 km³/yr) is the sum of both transpiration and evaporation from soil, streams, rivers, and lakes. (Values from Schlesinger, W. H. *Biogeochemistry: An Analysis of Global Change*, 2nd ed., Academic Press, San Diego, 1997, and several other sources.)

© Cengage Learning

atmosphere, the ocean, climate, and fire also affect ecosystems. For a given abiotic factor, each organism living in an ecosystem has an optimal range in which it survives and reproduces. Water and temperature are probably the two abiotic factors that most affect organisms in ecosystems.

The sun warms Earth

The sun makes life on Earth possible. Without the sun's energy, the temperature on planet Earth would approach absolute zero (0 K or −273°C), and all water would be frozen, even in the ocean. The sun powers the hydrologic cycle, carbon cycle, and other biogeochemical cycles and is the primary determinant of climate. Photosynthetic organisms capture the sun's energy and use it to make organic compounds that almost all forms of life require. Most of our fuels, such as wood, oil, coal, and natural gas, represent solar energy captured by photosynthetic organisms. Without the sun, almost all life on Earth would cease (see *Inquiring About: Life without the Sun* for an interesting exception).

The sun's energy, which is the product of a massive nuclear fusion reaction, is emitted into space in the form of electromagnetic radiation, especially ultraviolet, visible, and infrared radiation. About one-billionth of the total energy that the sun releases strikes the atmosphere, and of this tiny trickle of energy, a minute part operates the biosphere. On average, clouds and surfaces—especially snow, ice, and the ocean—immediately reflect away

30% of the solar radiation that falls on Earth (**FIG. 55-12**). Earth's surface and atmosphere absorb the remaining 70%, which runs the water cycle, drives winds and ocean currents, powers photosynthesis, and warms the planet. Ultimately, the continual radiation of long-wave infrared (heat) energy returns all this energy to space. If heat gains did not exactly balance losses, Earth would heat up or cool down.

Temperature changes with latitude The most significant local variations in Earth's temperature are produced because the sun's energy does not uniformly reach all places. Our planet's roughly spherical shape and the tilted angle of its axis produce significant variation in the exposure of the surface to sunlight. The sun's rays strike almost vertically near the equator, concentrating the energy and producing warmer temperatures. Near the poles the sun's rays strike more obliquely and, as a result, are spread over a larger surface area. Also, rays of light entering the atmosphere obliquely near the poles must pass through a deeper envelope of air than those entering near the equator. This angle causes more of the sun's energy to be scattered and reflected back into space, which further lowers temperatures near the poles. Thus, the solar energy that reaches polar regions is less concentrated and produces lower temperatures than elsewhere.

Temperature changes with season Earth's inclination on its axis (23.5 degrees from a line drawn perpendicular to the orbital plane) primarily determines the seasons.

Is the sun the energy source for all ecosystems? A notable exception was discovered in the late 1970s in a series of hydrothermal vents in the eastern Pacific where seawater apparently had penetrated and been heated by the radioactive rocks below. During its time within Earth, the water had become charged with inorganic mineral compounds, including hydrogen sulfide (H_2S).

No light is available for photosynthesis, but hydrothermal vents support a rich ecosystem that contrasts with the surrounding "desert" of the deep-ocean floor. Giant, blood-red tube worms almost 3 m (10 ft) in length cluster in great numbers around the vents (see figure). Other animals around the hydrothermal vents include unique species of clams, crabs, barnacles, and mussels.

Scientists initially wondered what energy source sustains the organisms in this dark environment. Most deep-sea communities depend on the organic matter that drifts down from the surface waters; in other words, they rely on energy ultimately derived from photosynthesis. Hydrothermal vent communities, however, are too densely clustered and too productive to be dependent on chance encounters with organic material from surface waters.

Instead, chemoautotrophic prokaryotes occupy the base of the food web in these aquatic oases.

These prokaryotes have enzymes that catalyze the oxidation of hydrogen sulfide, yielding water plus sulfur or sulfate. Such chemical reactions are exergonic and provide the energy required to fix CO_2

dissolved in the water into organic compounds. Many of the animals consume the prokaryotes directly by filter-feeding, but others, such as the giant tube worms, get their energy from prokaryotes that live in their tissues.

Scientists continue to generate questions about hydrothermal vent communities. How do the organisms find and colonize vents, which are ephemeral and widely scattered on the ocean floor? How have the inhabitants of these communities adapted to survive the harsh living conditions, including high pressure, high temperatures, and toxic chemicals? As vent community research continues, scientists hope to discover answers to these and other questions.

Hydrothermal vent ecosystem. Chemoautotrophic prokaryotes living in the tissues of these tube worms (*Riftia pachyptila*) extract energy from hydrogen sulfide to manufacture organic compounds. These worms lack digestive systems and depend on the organic compounds the prokaryotes provide, along with materials filtered from the surrounding water. Also visible in the photograph are some filter-feeding clams (*yellow*) and a crab (*white*).

Less than one-billionth of the sun's total energy reaches Earth's outer atmosphere.

30% reflected back into space immediately

47% absorbed by the atmosphere

23% runs the hydrologic cycle

Less than 1% drives the winds and ocean currents

All solar energy is ultimately re-radiated to space as heat

0.02% captured by photosynthesis

Figure 55-12 *Animation* The fate of solar radiation that reaches Earth
Most of the energy released by the sun never reaches Earth. The solar energy that does reach Earth warms the planet's surface, drives the hydrologic and other biogeochemical cycles, produces the climate, and powers almost all life through the process of photosynthesis.
© Cengage Learning

During half of the year (March 21 to September 22), the Northern Hemisphere tilts toward the sun, concentrating the sunlight and making the days longer (**FIG. 55-13**). During the other half of the year (September 22 to March 21), the Northern Hemisphere tilts away from the sun, giving it a lower concentration of sunlight and shorter days. The orientation of the Southern Hemisphere is just the opposite at these times. Summer in the Northern Hemisphere corresponds to winter in the Southern Hemisphere.

The atmosphere contains several gases essential to organisms

The atmosphere is an invisible layer of gases that envelops Earth. Oxygen (21%) and nitrogen (78%) are the predominant gases in the atmosphere, accounting for about 99% of dry air; other gases, including argon, carbon dioxide, neon, and

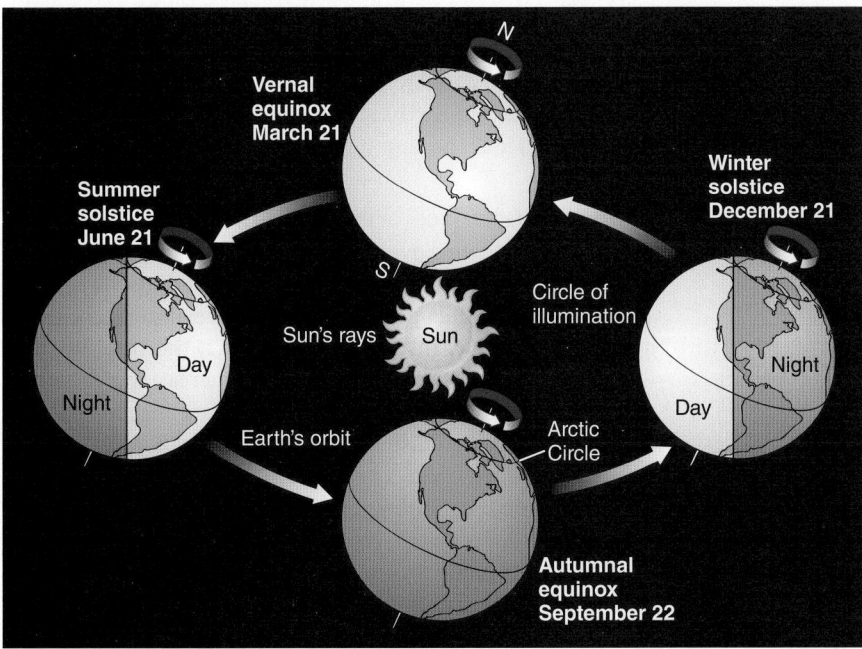

Figure 55-13 Seasonal changes in temperature

Earth's inclination on its axis remains the same as Earth travels around the sun. Thus, the sun's rays hit the Northern Hemisphere obliquely during winter months and more directly during summer months. In the Southern Hemisphere, the sun's rays are oblique during the winter, which corresponds to the Northern Hemisphere's summer. At the equator, the sun's rays are approximately vertical on March 21 and September 22.

© Cengage Learning

helium, make up the remaining 1%. In addition, water vapor and trace amounts of various air pollutants, such as methane, ozone, dust particles, pollen, microorganisms, and chlorofluorocarbons (CFCs), are present. Atmospheric oxygen is essential to plants, animals, and other organisms that respire aerobically; and plants and other photosynthetic organisms also require carbon dioxide.

The atmosphere performs several essential ecological functions. It protects Earth's surface from most of the sun's ultraviolet radiation and X-rays as well as from lethal amounts of cosmic rays from space. Without this atmospheric shielding, life as we know it would cease. Although the atmosphere protects Earth from high-energy radiation, visible light and some infrared radiation can penetrate it, and they warm the surface and the lower atmosphere. This interaction between the atmosphere and solar energy is responsible for weather and climate.

Organisms depend on the atmosphere, but they also help maintain and, in certain instances, modify its composition. For example, atmospheric oxygen increased to its present level as a result of billions of years of photosynthesis. Today, an approximate balance between oxygen-producing photosynthesis and oxygen-using aerobic respiration helps maintain the level of atmospheric oxygen.

The sun drives global atmospheric circulation In large measure, differences in temperature that are due to variations in the amount of solar energy at different locations on Earth drive the circulation of the atmosphere. The warm surface near the equator heats the air with which it comes into contact, causing this air to expand and rise. As the warm air rises, it flows away from the equator, cools, and sinks again (**FIG. 55-14**). Much of it recirculates to the same areas it left, but the remainder splits and flows in two directions, toward the poles. The air chills enough to sink to the surface at about 30 degrees north and south latitudes. Similar upward movements of warm air and its subsequent flow toward the poles occur at higher latitudes, farther from the equator. At the poles, the cold polar air sinks and flows toward the lower latitudes, generally beneath the warm air that simultaneously flows toward the poles. The constant motion of air transfers heat from the equator toward the poles, and as the air returns, it cools the land over which it passes. This continuous turnover moderates temperatures over Earth's surface.

The atmosphere exhibits complex horizontal movements In addition to global circulation patterns, the atmosphere exhibits complex horizontal movements called **winds.** The nature of wind, with its turbulent gusts, eddies, and lulls, is difficult to understand or predict. It results in part from differences in

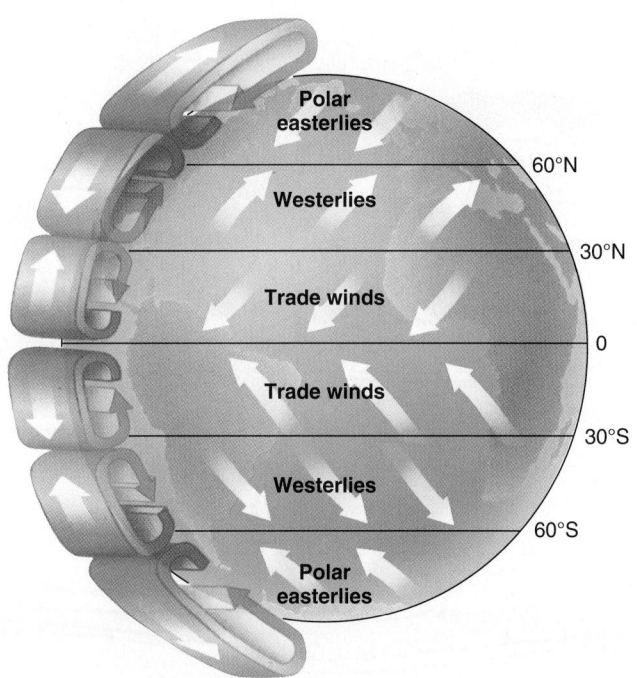

Figure 55-14 *Animation* **Atmospheric circulation**

The greatest solar energy input occurs at the equator and heats air most strongly in that area. The air rises and travels poleward (*left*) but is cooled in the process, so much of it descends again around 30 degrees latitude in both hemispheres. At higher latitudes, the patterns of air movement are more complex.

© Cengage Learning

atmospheric temperature and pressure changes, Earth's rotation, and uneven heating of the oceans and continents.

The gases that constitute the atmosphere have weight and exert a pressure that is, at sea level, about 1013 millibars (14.7 lb/in.²). Air pressure is variable, however, and changes with altitude, temperature, and humidity. Winds tend to blow from areas of high atmospheric pressure to areas of low pressure; the greater the difference and proximity between the high and low pressure areas, the stronger the wind.

Earth's rotation influences the direction that wind blows. Because Earth rotates from west to east, wind swerves to the right in the Northern Hemisphere and to the left in the Southern Hemisphere. This tendency of moving air to be deflected from its path by Earth's rotation is known as the **Coriolis effect.**

The global ocean covers most of Earth's surface

The global ocean is a huge body of salt water that surrounds the continents and covers almost three-fourths of Earth's surface.

It is a single, continuous body of water, but geographers divide it into four sections separated by the continents: the Pacific, Atlantic, Indian, and Arctic Oceans. The Pacific Ocean, which covers one-third of Earth's surface and contains more than half of Earth's water, is the largest by far.

Winds drive surface ocean currents The persistent prevailing winds blowing over the ocean produce mass movements of surface ocean water known as **ocean currents** (FIG. 55-15). The prevailing winds generate circular ocean currents called *gyres*. For example, in the North Atlantic, the tropical trade winds tend to blow toward the west, whereas the westerlies in the midlatitudes blow toward the east (see Fig. 55-14). This movement helps establish a clockwise gyre in the North Atlantic. Thus, surface ocean currents and winds tend to move in the same direction, although there are many variations on this general rule.

The Coriolis effect is partly responsible for the paths that surface ocean currents travel. Earth's rotation from west to east causes surface ocean currents to swerve to the right in the

---

KEY POINT

The driving force of wind produces Earth's major surface ocean currents.

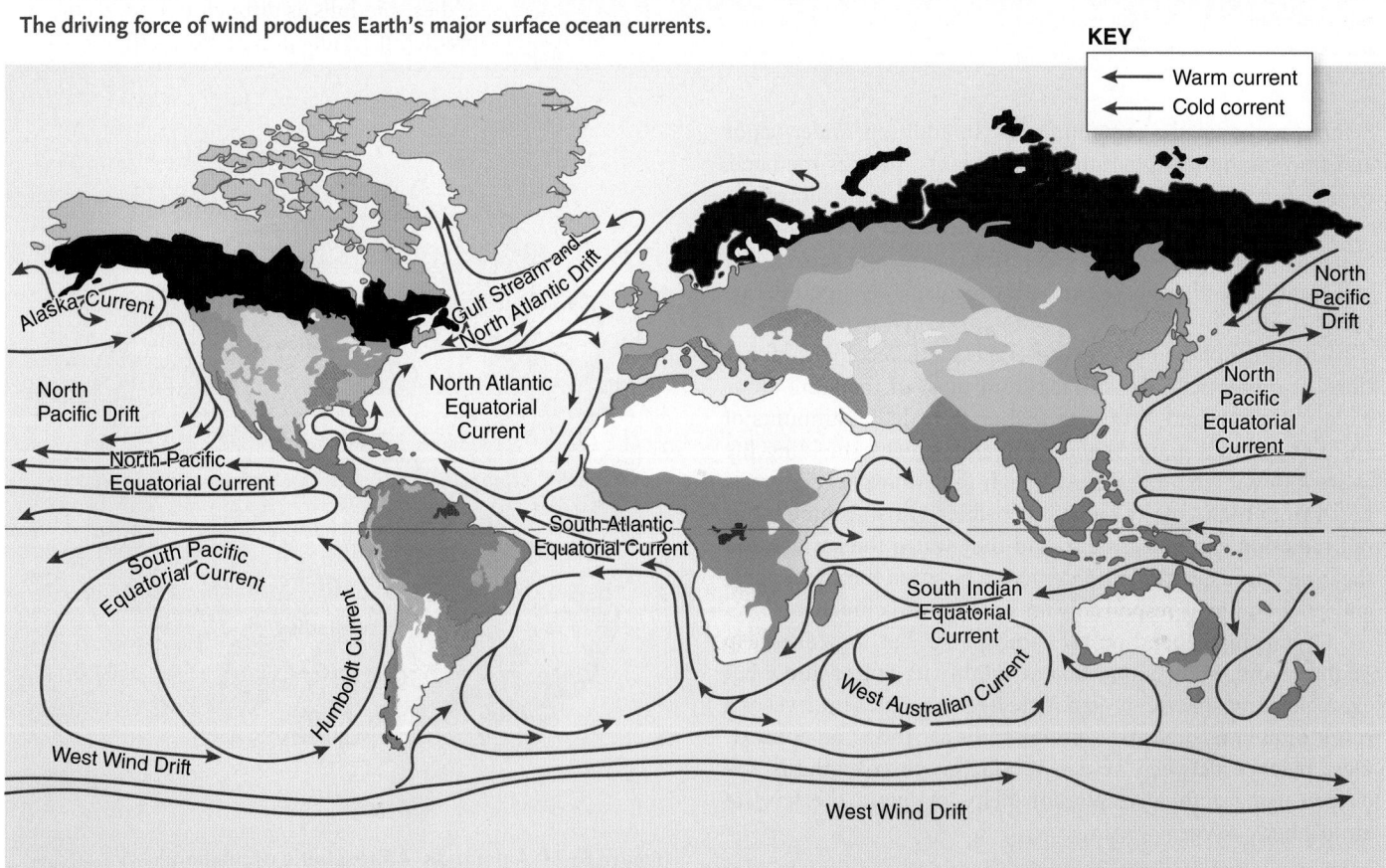

Figure 55-15 Major surface ocean currents

Each current has characteristic temperature and flow patterns.

© Cengage Learning

PREDICT How could currents be affected by global climate change?

Northern Hemisphere, producing a clockwise gyre of water currents. In the Southern Hemisphere, ocean currents swerve to the left, producing a counterclockwise gyre.

The ocean interacts with the atmosphere The ocean and the atmosphere are strongly linked. Wind from the atmosphere affects the ocean currents, and heat from the ocean affects atmospheric circulation. One of the best examples of the interaction between ocean and atmosphere is the **El Niño–Southern Oscillation (ENSO)** event. ENSO is a periodic warming of surface waters of the tropical eastern Pacific that alters both oceanic and atmospheric circulation patterns and results in unusual weather in areas far from the tropical Pacific. Normally, westward-blowing trade winds restrict the warmest waters to the western Pacific (near Australia). Every three to seven years, however, the trade winds weaken, and the warm water mass expands eastward to South America, raising surface temperatures in the eastern Pacific. Ocean currents, which normally flow westward in this area, slow down, stop altogether, or even reverse and go eastward. The phenomenon is called El Niño, which is Spanish for "the (Christ) child," because the warming usually reaches the fishing grounds off Peru just before Christmas. Most ENSOs last from one to two years. La Niña (Spanish for "the girl"), the opposite phase, is the cold phase, of the ENSO cycle and occurs when cold ocean waters return.

An ENSO event changes biological productivity in parts of the ocean. The warmer sea-surface temperatures and accompanying changes in ocean circulation patterns off the west coast of South America prevent colder, nutrient-laden deeper waters from **upwelling** (coming to the surface) (**FIG. 55-16**). The lack of nutrients in the water results in a severe decrease in the populations of anchovies and many other marine fishes. Other species, such as shrimp and scallops, thrive during an ENSO event. Along the Pacific coast of North America, ENSO shifts the distribution of tropical fishes northward and even affects the salmon run in Alaska.

Climate profoundly affects organisms

Climate is the average weather conditions, plus extremes (records), that occur in a given place over a period of years. The two most important factors that determine an area's climate are temperature (both average temperature and temperature extremes) and precipitation (both average precipitation and seasonal distribution). Other climate factors include wind, humidity, fog, cloud cover, and lightning-caused wildfires. Unlike weather, which changes rapidly, climate generally changes slowly, over hundreds or thousands of years.

Day-to-day variations, day-to-night variations, and seasonal variations are also important dimensions of climate that affect organisms. Latitude, elevation, topography, vegetation, distance from the ocean or other large bodies of water, and location on a continent or other landmass all influence temperature, precipitation, and other aspects of climate.

Earth has many different climates, and because they are relatively constant for many years, organisms have adapted to them. The wide variety of organisms on Earth evolved in part

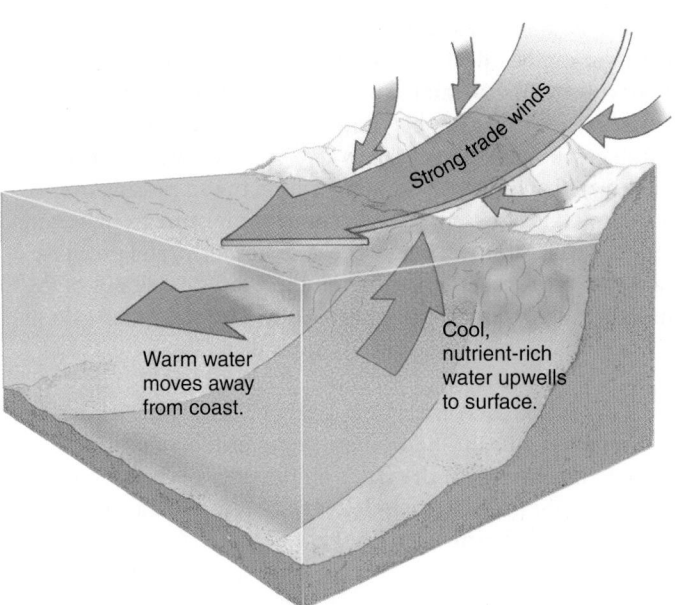

Figure 55-16 *Animation* **Upwelling**
Coastal upwelling, where deeper waters come to the surface, occurs in the Pacific Ocean along the South American coast. Upwelling provides nutrients for microscopic algae, which in turn support a complex food web. Coastal upwelling weakens considerably during years with ENSO events, temporarily reducing fish populations.
© Cengage Learning

because of the many different climates, ranging from cold, snow-covered, polar climates to hot, tropical climates where it rains almost every day.

Air and water movements and surface features affect precipitation patterns Precipitation varies from one location to another and has a profound effect on the distribution and kinds of organisms present. One of the driest places on Earth is in the Atacama Desert in Chile, where the average annual rainfall is 0.05 cm (0.02 in.). In contrast, Mount Waialeale in Hawaii, Earth's wettest spot, receives an average annual precipitation of 1200 cm (472 in.).

Differences in precipitation depend on several factors. The heavy-rainfall areas of the tropics result mainly from the uplifting of moisture-laden air. High surface-water temperatures (recall the enormous amount of solar energy striking the equator) cause the evaporation of vast quantities of water from tropical parts of the ocean. Prevailing winds blow the resulting moist, warm air over landmasses. Land surfaces warmed by the sun heat the air and cause moist air to rise. As it rises, the air cools, and its moisture-holding ability decreases. (Cool air holds less water vapor than warm air.) When air reaches its saturation point, it cannot hold any additional water vapor; clouds form, and water is released as precipitation. The air eventually returns to the surface on both sides of the equator near the Tropics of Cancer and Capricorn (latitudes 23.5 degrees north and south, respectively). By then, most of the moisture has precipitated so that dry air returns to the equator. This dry air makes little biological difference over the ocean, but its lack of moisture produces some of the great subtropical deserts, such as the Sahara.

Air is also dried during long journeys over landmasses. Near the windward (side from which the prevailing wind blows) coasts of continents, rainfall may be heavy. However, in the temperate zones—the areas between the tropical and polar zones—continental interiors are usually dry because they are far from the ocean that replenishes water vapor in the air passing over it.

Mountains force air to rise, removing moisture from humid air. As it gains altitude, the air cools, clouds form, and precipitation typically occurs, primarily on the windward slopes of the mountains. As the air mass moves down on the other side of the mountain, it is warmed, and clouds then evaporate, thereby lessening the chance of precipitation of any remaining moisture. This situation exists on the west coast of North America, where precipitation falls on the western slopes of mountains that are close to the coast. The dry lands on the sides of the mountains away from the prevailing wind (in this case, east of the mountain range) are called **rain shadows** (FIG. 55-17).

Microclimates are local variations in climate Differences in elevation, in the steepness and direction of slopes, and therefore in exposure to sunlight and prevailing winds may produce local variations in climate known as **microclimates,** which are sometimes quite different from their overall surroundings. Patches of sun and shade on a forest floor, for example, produce a variety of microclimates for plants, animals, and microorganisms living there. The microclimate of an organism's habitat is of primary importance because it is the climate that an organism actually experiences and must cope with. (Keep in mind, however, that microclimates are largely affected by the regional climates in which they are located.)

Sometimes organisms modify their own microclimate. For example, trees modify the local climate within a forest so that in summer the temperature is usually lower, and the relative humidity greater, than outside the forest. The temperature and humidity beneath the litter of the forest floor differ still more; in the summer the microclimate of this area is considerably cooler and moister than the surrounding forest. As another example, many desert-dwelling animals burrow to avoid surface climate conditions that would kill them in minutes. The cooler daytime microclimate in their burrows permits them to survive until night, when the surface cools off and they come out to forage or hunt.

Fires are a common disturbance in some ecosystems

Wildfires, which are fires started by lightning, are an important ecological force in many geographic areas. Those areas most prone to wildfires have wet seasons followed by dry seasons. Vegetation that grows and accumulates during the wet season dries out enough during the dry season to burn easily. When lightning hits vegetation or ground litter, it ignites the dry organic material, and a fire spreads through the area.

Fires have several effects on organisms. First, combustion frees the minerals that were locked in dry organic matter. The ashes remaining after a fire are rich in potassium, phosphorus, calcium, and other minerals essential for plant growth. With the arrival of precipitation, vegetation flourishes following a fire. Second, fire removes plant cover and exposes the soil. This change stimulates the germination and establishment of seeds requiring bare soil as well as encourages the growth of shade-intolerant plants. Third, fire causes increased soil erosion because it removes plant cover, leaving the soil more vulnerable to wind and water.

African savanna, California chaparral, North American grasslands, and ponderosa pine forests of the western United States are some fire-adapted ecosystems (see Chapter 56). Fire helps maintain grasses as the dominant vegetation in grasslands by removing fire-sensitive hardwood trees.

Humans try to prevent fires, and sometimes this effort has disastrous consequences. When fire is excluded from a fire-adapted ecosystem, deadwood and other plant litter accumulate. As a result, when a fire does occur, it can be very destructive. The sometimes deadly wildfires in Colorado and other western states and provinces are blamed in part on decades of suppressing fires in the region. Prevention of fire also converts grassland to woody vegetation and facilitates the invasion of fire-sensitive trees into fire-adapted forests.

Controlled burning is a tool of ecological management in which the undergrowth and plant litter are deliberately burned under controlled conditions before they have accumulated to dangerous levels (FIG. 55-18). Controlled burns are also used to suppress fire-sensitive trees, thereby maintaining the natural fire-adapted ecosystem. In a 2010 study of fire reintroduction in Sierra Nevada forests reported in *Ecosphere,* Karen Webster and Charles Halpern explored two decades of changes in plant diversity and abundance following reintroduction and repeated use of fire in previously unmanaged forests. Results from their research suggest that repeated controlled burning can gradually enhance the diversity and abundance of understory species and may enhance the dispersal of those species impacted by fire exclusion. More research needs to be conducted on when controlled burning is appropriate or how often it should be carried out. Moreover, there are major practical, political, and educational issues that must be addressed in implementing controlled burns.

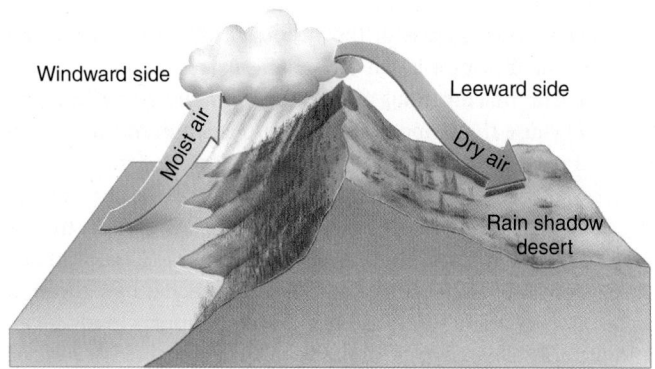

Figure 55-17 *Animation* **Rain shadow**

A rain shadow is the arid or semiarid land that occurs on the leeward side of a mountain. Such a rain shadow occurs east of the Cascade Range in Washington State. The western side of the range receives more than 500 cm of precipitation annually, whereas the eastern side receives 40 to 50 cm.

© Cengage Learning

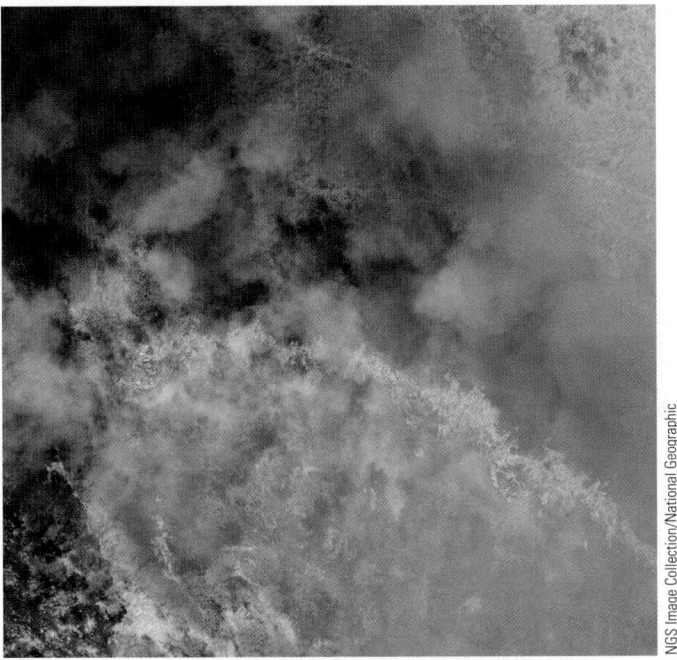

Figure 55-18 Controlled burn as a tool of ecological management
This aerial view of a controlled burn shows the leading edge of the fire, or fire line.

CHECKPOINT 55.3

- *What basic forces determine the circulation of the atmosphere?*
- *What basic forces produce the main ocean currents?*
- **CONNECT** *What are some of the factors that produce regional differences in precipitation?*

55.4 STUDYING ECOSYSTEM PROCESSES

LEARNING OBJECTIVE

9 Briefly describe some of the long-term ecological research conducted at Hubbard Brook Experimental Forest.

Ecologists conduct detailed ecosystem studies in laboratory simulations and in the field to measure such processes as energy flow, the cycling of nutrients, and the effects of natural and human-induced disturbances (e.g., air pollution, tree harvesting, and land-use changes). Some ecosystem studies, such as those performed at the Hubbard Brook Experimental Forest (HBEF), a 3100-hectare (7750-acre) reserve in the White Mountain National Forest in New Hampshire, are long term. Beginning in the late 1950s and continuing to the present, HBEF has been the site of numerous studies that address the hydrology (e.g., precipitation, surface runoff, and groundwater flow), biology, geology, and chemistry of forests and associated aquatic ecosystems. The National Science Foundation (NSF) has designated HBEF as one of its 24 long-term ecological research sites.

Many of the experiments at HBEF are based on field observations. For example, salamander populations were originally surveyed in forest communities in 1970 and have been resurveyed in recent years. Other studies involve manipulative experiments. In 1978, scientists added dilute sulfuric acid to a small stream in HBEF to study the chemical and biological effects of acidification. This experiment was of practical value because acid precipitation, a form of air pollution, has acidified numerous lakes and streams in industrialized countries.

Several researchers have studied the effects on HBEF stream ecosystems of **deforestation,** the clearance of large expanses of forest for agriculture or other uses. When a forest is removed, the total amount of water and minerals that flow into streams increases drastically. A concrete dam called a *catchment* can be constructed across a stream so that ecologists can measure the flow of water and chemical components out of the ecosystem. Catchments help scientists to measure the quantity, timing, and quality of water flowing from a forested watershed. Typically, outflow is measured in two separate ecosystems: one serves as a control, and one is experimentally manipulated. These studies demonstrate that deforestation causes soil erosion and leaching of essential minerals that result in decreased soil fertility. The summer temperatures in streams running through deforested areas are higher than in shady streams running through uncut forests. Many stream organisms do not fare well in deforested areas, in part because they are adapted to cooler temperatures.

Detailed studies such as those at HBEF provide ecologists with insights into how ecological processes function in individual ecosystems. Ecologists compare these data with similar information from other ecosystem studies to develop generalized insights into how ecosystems are structured and how they function. Long-term ecosystem experiments enable ecologists to evaluate and predict the effects of environmental change, including human-induced change.

Ecosystem experiments also contribute to our practical knowledge about how to maintain water quality, wildlife habitat, and productive forests. **Ecosystem management,** a conservation approach that emphasizes restoring and maintaining the quality of an entire ecosystem rather than the conservation of individual species, makes use of such knowledge.

Ecologists are also studying the causes of food-web flips and ecosystem collapse to recognize tipping points. A food-web flip occurs when there is a major persistent change in the sequences of organisms eating and being eaten by one another. As reported in *Science,* a team from University of Wisconsin–Madison led by Stephen Carpenter examined 30 years of research on Peter Lake, Wisconsin, and developed mathematical models to recognize early signs of tipping points. The researchers identified signs of change in an endangered food web 15 months before it flipped. Continued research in this area may enhance our ability to recognize warning signs to protect ecosystems from natural and human effects before irreversible or costly damage occurs.

CHECKPOINT 55.4

- *What are some of the environmental effects observed in the deforestation study at Hubbard Brook Experimental Forest?*

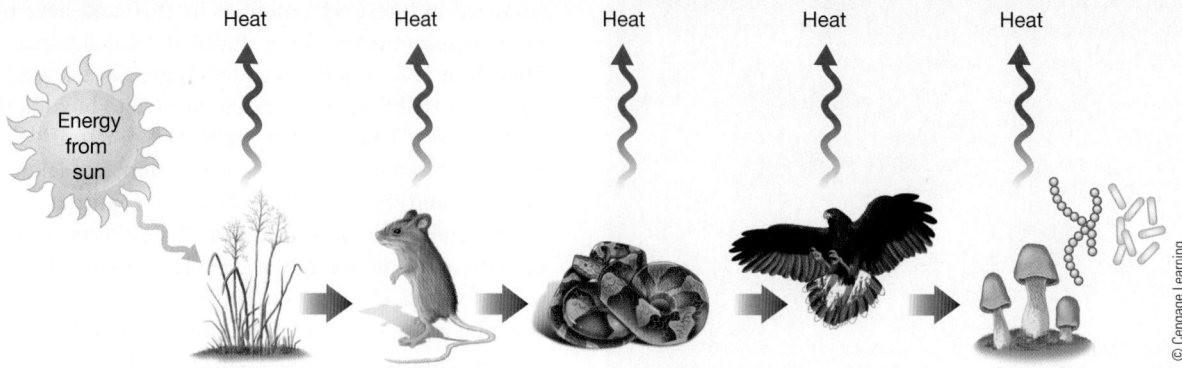

55.1 Energy Flow Through Ecosystems *(page 1188)*

1 Summarize the concept of energy flow through a food web.

- **Energy flow** through an ecosystem is linear, from the sun to producer to consumer to decomposer. Much of this energy is converted to heat as it moves from one organism to another, so organisms occupying the next **trophic level** cannot use it.
- Trophic relationships may be expressed as **food webs,** which show the many alternative pathways that energy may take among the producers, consumers, and decomposers of an ecosystem.

2 Explain typical pyramids of numbers, biomass, and energy.

- **Ecological pyramids** typically express the progressive reduction in numbers of organisms, biomass, and energy found in successive trophic levels. A **pyramid of numbers** shows the number of organisms at each trophic level in a given ecosystem. A **pyramid of biomass** shows the total biomass at each successive trophic level. A **pyramid of energy** indicates the energy content of the biomass of each trophic level.

3 Distinguish between gross primary productivity and net primary productivity.

- **Gross primary productivity (GPP)** of an ecosystem is the rate at which photosynthesis captures energy. **Net primary productivity (NPP)** is the energy that remains (as biomass) after plants and other producers carry out cellular respiration.

55.2 Cycles of Matter in Ecosystems *(page 1193)*

4 Describe the main steps in each of these biogeochemical cycles: the carbon, nitrogen, phosphorus, and hydrologic cycles.

- Carbon dioxide is the important gas of the **carbon cycle.** Carbon enters plants, algae, and cyanobacteria as CO_2, which photosynthesis incorporates into organic molecules. Cellular respiration, combustion, and erosion of limestone return CO_2 to the water and atmosphere, where it is again available to producers.
- The **nitrogen cycle** has five steps. **Nitrogen fixation** is the conversion of nitrogen gas to ammonia. **Nitrification** is the conversion of ammonia or ammonium to nitrate. **Assimilation** is the conversion of nitrates, ammonia, or ammonium to proteins and other nitrogen-containing compounds by plants; conversion of plant proteins into animal proteins is also assimilation. **Ammonification** is the conversion of organic nitrogen to ammonia and ammonium ions. **Denitrification** is the conversion of nitrate to nitrogen gas.

- The **phosphorus cycle** has no biologically important gaseous compounds. Phosphorus erodes from rock as inorganic phosphate, which the roots of plants absorb from the soil. Animals obtain the phosphorus they need from their diets. Decomposers release inorganic phosphate into the environment. When phosphorus washes into the ocean and is subsequently deposited in seabeds, it is lost from biological cycles for millions of years.
- The **hydrologic cycle** involves an exchange of water between the land, ocean, atmosphere, and organisms. Water enters the atmosphere by evaporation and **transpiration** and leaves the atmosphere as precipitation. On land, water filters through the ground or runs off to lakes, rivers, and the ocean. Aquifers are underground caverns and porous layers of rock in which groundwater is stored.

55.3 Abiotic Factors in Ecosystems *(page 1198)*

5 Summarize the effects of solar energy on Earth's temperatures.

- Of the solar energy that reaches Earth, 30% is immediately reflected away; the atmosphere and surface absorb the remaining 70%. Ultimately, all absorbed solar energy is reradiated into space as infrared (heat) radiation.
- A combination of Earth's roughly spherical shape and the tilted angle of its axis concentrates solar energy at the equator and dilutes it at the poles; the tropics are hotter and less variable in climate than are temperate and polar areas.

6 Discuss the roles of solar energy and the Coriolis effect in the generation of global air and water flow patterns.

- Visible light and some infrared radiation warm the surface and the lower part of the atmosphere. Atmospheric heat transferred from the equator to the poles produces movement of warm air toward the poles and of cool air toward the equator.
- **Winds** result in part from differences in atmospheric pressure and from the **Coriolis effect,** the tendency of moving air or water, because of Earth's rotation, to be deflected to the right in the Northern Hemisphere and to the left in the Southern Hemisphere.
- Surface **ocean currents** result in part from prevailing winds and the Coriolis effect.

7 Give two causes of regional precipitation differences.

- Latitude, elevation, topography, vegetation, distance from the ocean or other large bodies of water, and location on a continent or other landmass influence precipitation.

- Precipitation is greatest where warm air passes over the ocean, absorbs moisture, and then cools, such as in areas where mountains force humid air upward. Deserts develop in the **rain shadows** of mountain ranges or in continental interiors.

8 Discuss the effects of fire on certain ecosystems.

- Fire frees the minerals locked in dry organic matter, removes plant cover and exposes the soil, and increases soil erosion. Many ecosystems, such as savanna, chaparral, grasslands, and certain forests, contain fire-adapted organisms.

55.4 Studying Ecosystem Processes (page 1205)

9 Briefly describe some of the long-term ecological research conducted at Hubbard Brook Experimental Forest.

- Hubbard Brook Experimental Forest (HBEF) in New Hampshire is the site of numerous studies that address the hydrology (precipitation, surface runoff, and groundwater flow), biology (effects of deforestation and changes in salamander populations), geology, and chemistry (acid precipitation) of forests and associated aquatic ecosystems.

TEST YOUR UNDERSTANDING

Know and Comprehend

1. The movement of matter is _____ in ecosystems, and the movement of energy is _____. (a) linear; linear (b) linear; cyclic (c) cyclic; cyclic (d) cyclic; linear (e) cyclic; linear or cyclic

2. A complex of interconnected food chains in an ecosystem is called (a) an ecosystem (b) a pyramid of numbers (c) a pyramid of biomass (d) a biosphere (e) a food web

3. The quantitative estimate of the total amount of living material is called (a) biomass (b) energy flow (c) gross primary productivity (d) plant respiration (e) net primary productivity

4. Which of the following equations shows the relationship between gross primary productivity (GPP) and net primary productivity (NPP)? (a) GPP = NPP − photosynthesis (b) NPP = GPP − photosynthesis (c) GPP = NPP − plant respiration (d) NPP = GPP − plant respiration (e) NPP = GPP − animal respiration

5. Which of the following processes increase(s) the amount of atmospheric carbon in the carbon cycle? (a) photosynthesis (b) cellular respiration (c) combustion (d) a and c (e) b and c

6. In the nitrogen cycle, gaseous nitrogen is converted to ammonia during (a) nitrogen fixation (b) nitrification (c) assimilation (d) ammonification (e) denitrification

7. The conversion of ammonia to nitrate, known as _____, is a two-step process performed by soil bacteria. (a) nitrogen fixation (b) nitrification (c) assimilation (d) ammonification (e) denitrification

8. Which biogeochemical cycle does not have a gaseous component but cycles from the land to sediments in the ocean and back to the land? (a) carbon cycle (b) nitrogen cycle (c) phosphorus cycle (d) hydrologic cycle (e) neither a nor c has a gaseous component

9. Which of the following processes is *not* directly involved in the hydrologic cycle? (a) transpiration (b) evaporation (c) precipitation (d) nitrification (e) condensation

10. The periodic warming of surface waters of the tropical eastern Pacific that alters both oceanic and atmospheric circulation patterns is known as (a) upwelling (b) prevailing wind (c) ocean current (d) El Niño–Southern Oscillation (e) Coriolis effect

11. A mountain range may produce a downwind arid (a) upwelling (b) rain shadow (c) ocean current (d) microclimate (e) ecological pyramid

Apply and Analyze

12. **VISUALIZE** Draw the simplest stable ecosystem you can imagine.
13. **PREDICT** How might a food web change if all decomposers were eliminated from it?

14. Why is the cycling of matter essential to the long-term continuance of life?
15. What would happen to the nitrogen cycle if all bacteria were absent? Explain your answer.

Evaluate and Synthesize

16. Would the microclimate of an ant be the same as that of an elephant living in the same area? Why or why not?
17. **INTERPRET DATA** Examine Figure 55-3b (shown here). Explain why this pyramid of biomass is inverted. In other words, how can 4 g of producers support 21 g of primary consumers?

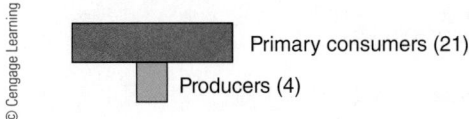

Primary consumers (21)

Producers (4)

© Cengage Learning

18. **INTERPRET DATA** Scientists have compiled databases of large forest wildfires in the western United States and compared them to climate and land-surface data. Examine the graph showing wildfire frequency compared to average spring–summer temperature. Do you see a correlation? If so, describe it. Based on these data, do you think that climate warming is causing more wildfires? Explain your answer.

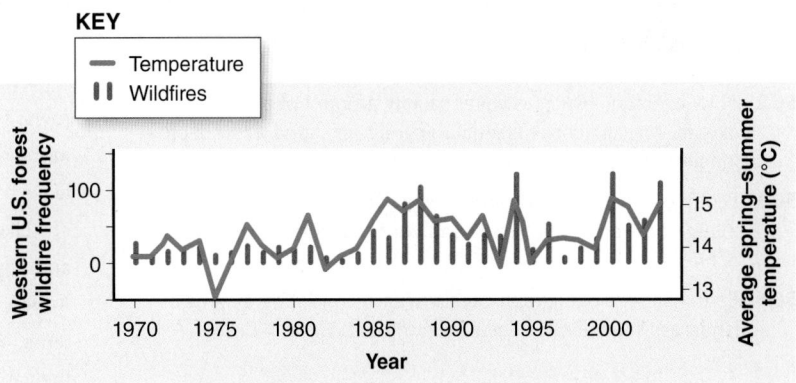

Source: Adapted from Westerling, A. L., H. G. Hidalgo, D. R. Cayan, and T. W. Swetnam. "Warming and Earlier Spring Increase Western U.S. Forest Wildfire Activity." *Science*, Vol. 313, Aug. 18, 2006.

19. **SCIENCE, TECHNOLOGY, AND SOCIETY** How do humans alter the nitrogen cycle, and what can scientists, engineers, and the public do to mitigate the damage?

aplia To access course materials, such as Aplia and other companion resources, please visit **www.cengagebrain.com**.

56 | Ecology and the Geography of Life

Black-tailed prairie dog. Prairie dogs (*Cynomys ludovicianus*) never wander far from their burrows, which they use to escape from predators.

© Ed Endicott/Alamy

KEY CONCEPTS

56.1 Climate, particularly temperature and precipitation, affects the distribution of Earth's major biomes, such as tropical rain forests and tundra.

56.2 Abiotic factors—such as water salinity, amount of dissolved oxygen, availability of essential minerals, light, and water depth—influence the distribution of organisms in aquatic ecosystems.

56.3 Ecotones—areas of transition where two communities meet and intergrade—provide diverse conditions that encourage species richness.

56.4 Earth has six biogeographic realms, each consisting of a major landmass separated by deep water, mountains, or a desert.

arth has many different environments. *Natural selection* affects an organism's ability to survive and reproduce in a given environment. In natural selection both **abiotic** (nonliving) and **biotic** (living) factors eliminate the least-fit individuals in a population. Over time, succeeding generations of organisms that live in each biome or major aquatic ecosystem become better adapted to local environmental conditions.

Black-tailed prairie dogs are superbly adapted to their environment. Their teeth and digestive tracts are modified to eat and easily digest the seeds and leaves of grasses that grow in great profusion on the Great Plains of western North America.

Prairie dogs live in large colonies of about 500 individuals. The eyes of every individual in the colony watch for potential danger, and when they see it, prairie dogs call out to warn the rest of the colony (see photograph). When danger approaches, each prairie dog dives into its underground home. Each burrow has at least two openings and consists of an elaborate network of long tunnels with several chambers: a nursery for the young, a sleeping chamber, a toilet chamber, and a listening chamber (close to an entrance). Piles of excavated soil surround the burrow entrances and help prevent flooding during rainstorms.

To survive winter, black-tailed prairie dogs sleep in their burrows. Their metabolism slows, and they subsist on the stored fat in their bodies. They do not truly hibernate, however, and may leave their burrows to look for food when the weather warms.

Like prairie dogs, each species has structural, behavioral, and physiological adaptations for its own particular environment. As you examine Earth's major terrestrial and aquatic ecosystems, including the species characteristic of each, think about the variety of adaptations that natural selection has produced in organisms in response to their particular environments.

56.1 BIOMES

LEARNING OBJECTIVES

1 Define *biome* and briefly describe the nine major terrestrial biomes, giving attention to the climate, soil, and characteristic plants and animals of each.

2 Describe at least one human effect on each of the biomes discussed.

A **biome** is a large, relatively distinct terrestrial region that has similar climate, soil, plants, and animals regardless of where it occurs. Because it covers such a large geographic area, a biome encompasses many interacting landscapes. Recall from Chapter 53 that a **landscape** is a large land area (several to many square kilometers) composed of interacting ecosystems.

Biomes largely correspond to major climate zones, with temperature and precipitation being most important (**FIG. 56-1**). Near the poles, temperature is generally the overriding climate factor, whereas in tropical and temperate regions, precipitation becomes more significant than temperature. Other abiotic factors to which biomes are sensitive include temperature extremes, rapid temperature changes, floods, droughts, strong winds, and fires (see section on fires in Chapter 55).

We discuss nine major biomes in this chapter: tundra, boreal forest, temperate rain forest, temperate deciduous forest, temperate grassland, chaparral, desert, savanna, and tropical rain forest. Although we discuss each biome as a distinct entity, biomes intergrade into one another at their boundaries.

Tundra is the cold, boggy plains of the far north

Tundra (also called **arctic tundra**) occurs in extreme northern latitudes wherever snow melts seasonally (**FIG. 56-2**). In the Southern Hemisphere, the **Antarctic tundra,** located on the Antarctic Peninsula and nearby islands, is home to animals such as penguins and elephant seals. Plant life in this warmer, slightly moister region of Antarctica consists of many mosses and lichens and some liverworts and fungi, but only two species of vascular plants. A third similar ecosystem, located in the higher elevations of mountains, above the tree line, is called **alpine tundra** (see *Inquiring About: The Distribution of Vegetation on Mountains*).

Arctic tundra has long, harsh winters and extremely short summers. Although the growing season, with its warmer temperatures, is as short as 50 days, the days are long. Above the Arctic Circle, the sun does not set at all for many days in midsummer, although the amount of light at midnight is only one-tenth that at noon. There is little precipitation (10 to 25 cm, or 4 to 10 in., per year) over much of the tundra, and most of it falls during summer months.

Tundra soils tend to be geologically young (with the exception of Antarctic tundra) because most were formed only after

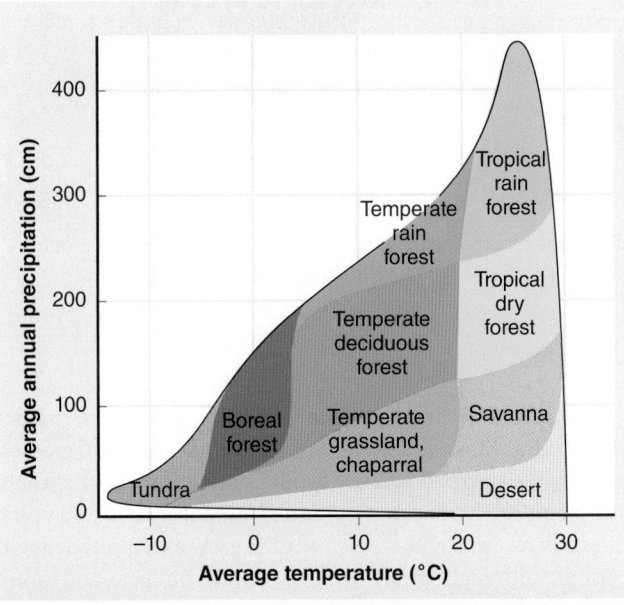

KEY POINT

The distribution of the world's biomes is largely the result of climate patterns, which determine an area's water (measured as precipitation) and energy (measured as temperature).

Figure 56-1 Using precipitation and temperature to identify biomes

Factors such as soil type, fire, and seasonality of climate affect whether temperate grassland or chaparral develops. (Adapted from Whittaker, R. H. *Communities and Ecosystems,* 2nd ed. Macmillan, New York, 1975.)

PREDICT As global climate continues to change, how might Earth's biomes be affected?

Figure 56-2 Arctic tundra

Because of the short growing season and permafrost in arctic tundra, only small, hardy plants grow in this northernmost biome that encircles the Arctic Ocean. Photographed during autumn in Alaska.

Does the type of vegetation change at different elevations on a mountain? Hiking up a mountain is similar to traveling toward the North Pole with respect to the major life zones encountered (see figure). This elevation–latitude similarity occurs because the temperature drops as one climbs a mountain, just as it does when one travels north; the temperature drops about 6°C (11°F) with each 1000-m increase in elevation. The types of species growing on the mountain change as the temperature changes.

Deciduous trees, which shed their leaves every autumn, may cover the base of a mountain in Colorado, for example. At higher elevations, where the climate is colder and more severe, a coniferous *subalpine forest* resembling boreal forest grows. Spruces and firs are the dominant trees here.

Higher still, the forest thins, and the trees become smaller, gnarled, and shrublike. These twisted, shrublike trees, called *krummholz* (a German word meaning "crooked wood"), are found at their elevational limit (the *tree line*). The exact elevation at which the tree line occurs depends on the latitude and distance from the ocean. In the Rocky Mountains, between 35° and 50° north latitude, the tree line drops 100 m with each 1° latitude northward.

Above the tree line, where the climate is quite cold, a kind of tundra occurs, with vegetation composed of grasses, sedges, and small tufted plants, most of which are hardy perennials. Some alpine plants (e.g., buttercups) are lowland species that have adapted to the alpine environment, whereas other plants (e.g., mountain douglasia) live exclusively in the mountains. This tundra is called *alpine tundra* to distinguish it from arctic tundra. At the top of the mountain, a permanent ice cap or snowcap might be found, similar to the nearly lifeless polar land areas.

Important environmental differences exist between high elevations and high latitudes that affect the types of organisms found in each place. Alpine tundra typically lacks permafrost and receives more precipitation than does arctic tundra. High elevations of temperate mountains do not have the great extremes of day length that are associated with the changing seasons in high-latitude biomes. The intensity of solar radiation is greater at high elevations than at high latitudes. At high elevations, the sun's rays pass through less atmosphere, which results in greater exposure to ultraviolet radiation (less is filtered out by the atmosphere) than occurs at high latitudes.

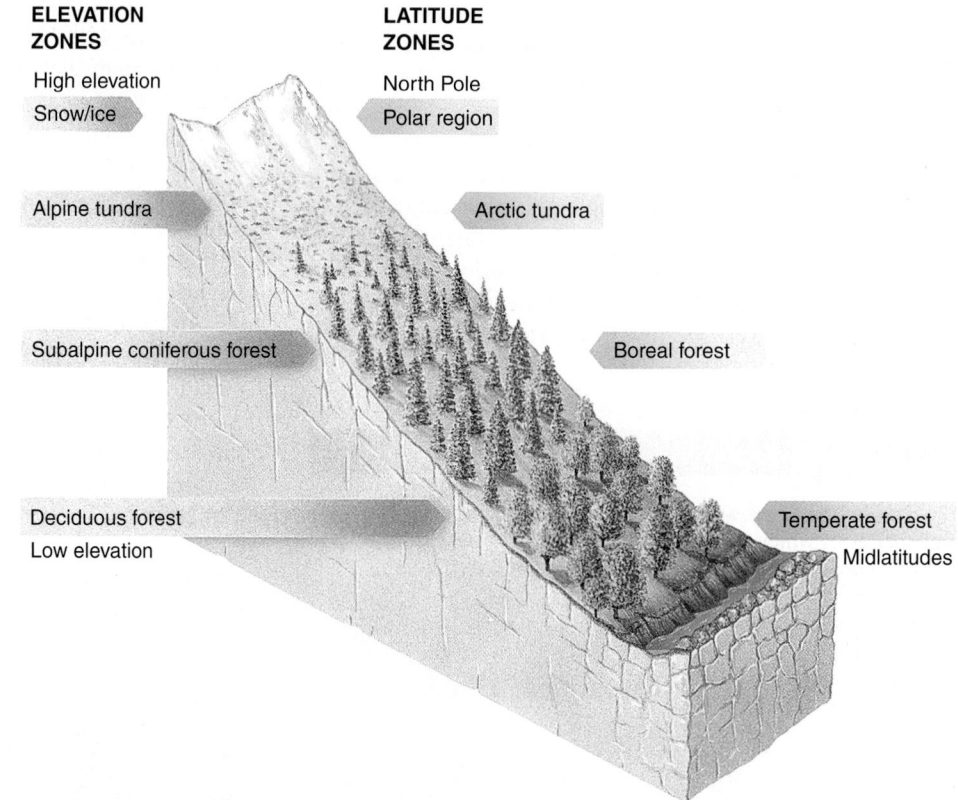

Comparison of elevation and latitude zones. The cooler temperatures at higher elevations of a mountain produce a series of ecosystems similar to those encountered when going toward the North Pole.
© Cengage Learning

the last Ice Age.[1] These soils are usually nutrient poor and have little organic litter (dead leaves and stems, animal droppings, and remains of organisms) in the uppermost layer of soil. Although the soil surface melts during the summer, tundra has a layer of permanently frozen ground called **permafrost** that varies in depth and thickness. Because permafrost interferes with drainage, the thawed upper zone of soil is usually waterlogged during the summer. Limited precipitation, combined with low temperatures, flat topography (surface features), and permafrost, produces a landscape of broad, shallow lakes, sluggish streams, and bogs.

Low species richness and low primary productivity characterize tundra. Mosses, lichens (such as reindeer moss), grasses, and grasslike sedges dominate tundra vegetation; most of these short plants are herbaceous perennials that live 20 to 100 years. No readily recognizable trees or shrubs grow except in sheltered locations, although dwarf willows, dwarf birches, and other dwarf trees are common.

Year-round animal life of the tundra includes voles, weasels, arctic foxes, gray wolves, snowshoe hares, ptarmigan, snowy owls, musk oxen, and lemmings (see Figure 53-5 and the discussion of lemming population cycles in Chapter 53). In the summer caribou migrate north to the tundra to graze on sedges, grasses, and dwarf willow. Dozens of bird species also migrate north in summer to nest and feed on abundant insects.

---

[1] Glacier ice, which occupied about 30% of Earth's land during the last Ice Age, began retreating about 17,000 years ago. Today, glacier ice occupies about 10% of the land surface.

Mosquitoes, blackflies, and deerflies survive the winter as eggs or pupae and occur in great numbers during summer weeks.

Tundra regenerates quite slowly after it has been disturbed. Even casual use by hikers causes damage. Long-lasting injury, likely to persist for hundreds of years, has been done to large portions of the arctic tundra as a result of oil exploration and military use.

Boreal forest is the evergreen forest of the north

Just south of the tundra lies the **boreal forest,** also known as *coniferous forest* or **taiga.** (Conifers are cone-bearing evergreens.) The boreal forest, which stretches across both North America and Eurasia, is the world's largest biome, covering approximately 11% of Earth's land (**FIG. 56-3**). A biome comparable to the boreal forest is not found in the Southern Hemisphere because it has no land at the corresponding latitudes. Winters are extremely cold and severe, although not as harsh as in the tundra. Boreal forest receives little precipitation, perhaps 50 cm (20 in.) per year, and its soil is typically acidic, low in minerals (inorganic nutrients), and has a deep layer of partly decomposed conifer needles at the surface. Boreal forest contains numerous ponds and lakes in water-filled depressions dug by grinding ice sheets during the last Ice Age.

White and black spruces, balsam fir, eastern larch, and other conifers dominate the boreal forest, but **deciduous** trees such as aspen or birch, which shed their leaves in autumn, form striking stands. Conifers have many drought-resistant adaptations, such as needlelike leaves with a minimal surface area to reduce water loss (see Fig. 34-9). Such an adaptation enables conifers to withstand the "drought" of the northern winter months, when roots do not absorb water because the ground is frozen. Natural selection also favors conifers in the boreal forest because, being evergreen, they resume photosynthesis as soon as warmer temperatures return.

Animal life of the boreal forest includes some larger species, such as caribou (which migrate from the tundra to the boreal forest for winter), wolves, bears, and moose. However, most animal life is medium-sized to small and includes rodents, rabbits, and fur-bearing predators such as lynx, sable, and mink. Most species of birds are seasonally abundant but migrate to warmer climates for winter. Insects are numerous, but there are few amphibians and reptiles except in the southern boreal forest.

Most of the boreal forest is not suitable for agriculture because of its short growing season and mineral-poor soil. The boreal forest, which is harvested primarily by clear-cutting, is currently the primary source of the world's industrial wood and wood fiber.

Temperate rain forest has cool weather, dense fog, and high precipitation

Coniferous **temperate rain forest** grows on the northwestern coast of North America. Similar vegetation exists in southeastern Australia and in southwestern South America. Annual precipitation in this biome is high, from 200 to 380 cm (80 to 150 in.); condensation of water from dense coastal fog augments the annual precipitation. The proximity of temperate rain forest to the coastline moderates the temperature so that seasonal fluctuation is narrow; winters are mild, and summers are cool. Temperate rain forest has a relatively nutrient-poor soil, although its organic content may be high. Cool temperatures slow the activity of bacterial and fungal decomposers. Thus, needles and large fallen branches and trunks accumulate on the ground as litter that takes many years to decay and release inorganic minerals to the soil.

The dominant vegetation type in the North American temperate rain forest is large evergreen trees, such as western hemlock, Douglas fir, Sitka spruce, and western red cedar. Temperate rain forest is rich in epiphytic vegetation, which consists of smaller plants that grow nonparasitically on the trunks and branches of large trees (**FIG. 56-4**). Epiphytes in this biome are mainly mosses, lichens, and ferns, all of which also carpet the ground. Squirrels, wood rats, mule deer, elk, numerous bird species (such as jays, nuthatches, and chickadees), several species of reptiles (such as painted turtles and western terrestrial garter snakes), and amphibians (such as Pacific giant salamanders and Pacific treefrogs) are common temperate rainforest animals.

Temperate rain forest, one of the richest wood producers in the world, supplies us with lumber and pulpwood. It is also one of the most complex ecosystems in terms of species richness. Care must be taken to avoid overharvesting original old-growth forest because such an ecosystem takes hundreds of years to develop. When the logging industry harvests old-growth forest, it typically replants the area with a monoculture (a single species) of trees that it harvests in 40- to 100-year cycles. Thus, the old-growth forest ecosystem, once harvested, never has a chance to redevelop. A small fraction of the original, old-growth temperate rain forest in Washington, Oregon, and northern California remains untouched. These stable forest ecosystems provide biological habitat for many organisms, including 40 endangered and threatened species.

Beth Davidow/Visuals Unlimited

Figure 56-3 Boreal forest
Boreal forest is coniferous forest that occurs in cold regions of the Northern Hemisphere adjacent to the tundra. Photographed in Yukon, Canada.

Figure 56-4 Temperate rain forest
Large amounts of precipitation characterize temperate rain forest. Note the epiphytes hanging from the branches of coniferous trees. Photographed in Olympic National Park in Washington State.

Figure 56-5 Temperate deciduous forest
The broad-leaf trees that dominate the temperate deciduous forest shed their leaves before winter. Photographed during autumn at the Durance River in France.

Temperate deciduous forest has a canopy of broad-leaf trees

Seasonality (hot summers and cold winters) is characteristic of **temperate deciduous forest,** which occurs in temperate areas where precipitation ranges from about 75 to 126 cm (30 to 50 in.) annually. Typically, the soil of a temperate deciduous forest consists of both a topsoil rich in organic material and a deep, clay-rich lower layer. As organic materials decay, mineral ions are released. If roots of living trees do not absorb these ions, they leach into the clay, where they may be retained.

Broad-leaf hardwood trees, such as oak, hickory, maple, and beech, which lose their foliage annually dominate temperate deciduous forests of the northeastern and Mid-Atlantic United States (FIG. 56-5). The trees of the temperate deciduous forest form a dense canopy that overlies saplings and shrubs.

Temperate deciduous forests originally contained a variety of large mammals such as mountain lions, wolves, bison, and other species now regionally extinct, plus deer, bears, and many small mammals and birds (such as wild turkeys, blue jays, and scarlet tanagers). Both reptiles (such as box turtles and rat snakes) and amphibians (such as spotted salamanders and wood frogs) abounded, together with a denser and more varied insect life than exists today.

In Europe and North America, logging and land clearing for farms, tree plantations, and cities have removed much of the original temperate deciduous forest. Where it has regenerated, temperate deciduous forest is often in a seminatural state, highly modified by humans for recreation, livestock foraging, timber harvest, and other uses. Although these returning forests do not have the biological diversity of virgin stands, many forest organisms have successfully become re-established. For example, in the eastern United States, white-tailed deer populations were decimated by hunting by 1900. As large parts of forests were modified for agricultural use and large predators (wolves and panthers) were eliminated, the white-tailed deer population rebounded. In fact, many regions consider the animals an agricultural or suburban nuisance and have had variable success in controlling the deer population.

Worldwide, temperate deciduous forest was among the first biomes to be converted to agricultural use. In Europe and Asia, many soils that originally supported temperate deciduous forest have been cultivated by traditional agricultural methods for thousands of years without a substantial loss in fertility. During the 20th century, however, intensive agricultural practices were adopted; along with overgrazing and deforestation, these methods have contributed to the degradation of some agricultural lands.

Temperate grasslands occur in areas of moderate precipitation

Summers are hot, winters are cold, fires help shape the landscape, and rainfall is often uncertain in **temperate grasslands.** Annual precipitation averages 25 to 75 cm (10 to 30 in.). In grasslands with less precipitation, minerals tend to accumulate in a marked layer just below the topsoil. These minerals tend to leach out of the soil in areas with more precipitation. Grassland soil contains considerable organic material because surface parts of many grasses die off each winter and contribute to the organic content of the soil (the roots and rhizomes survive underground). Many grasses are sod formers: their roots and rhizomes form a thick, continuous underground mat.

Moist temperate grasslands, also known as *tallgrass prairies,* occur in the United States in Iowa, western Minnesota, eastern Nebraska, and parts of other midwestern states and across Canada's prairie provinces. Although few trees grow except

Figure 56-6 Temperate grassland

The Nature Conservancy owns this tallgrass prairie preserve in Oklahoma. Like other moist temperate grasslands, it is mostly treeless but contains a profusion of grasses and other herbaceous flowering plants. As bison graze on the plants, they affect community structure and diversity.

Harvey Payne

near rivers and streams, grasses, some as tall as 2 m (6.5 ft), grow in great profusion in the deep, rich soil (FIG. 56-6). Before most of this area was converted to arable land, it was covered with herds of grazing animals, particularly bison. The principal predators were wolves, although in sparser, drier areas, coyotes took their place. Smaller fauna included prairie dogs and their predators (foxes, black-footed ferrets, and birds of prey such as prairie falcons), western meadowlarks, bobolinks, reptiles (such as gopher snakes and short-horned lizards), and great numbers of insects.

Shortgrass prairies, in which the dominant grasses are less than 0.5 m (1.6 ft) tall, are temperate grasslands that receive less precipitation than the moister grasslands just described but more precipitation than deserts. In the United States, shortgrass prairies occur in the eastern half of Montana, the western half of South Dakota, and parts of other midwestern states; they also appear in western Alberta in Canada. The plants grow in less abundance than in the moister grasslands, and occasionally some bare soil is exposed.

The North American grassland, particularly the tallgrass prairie, was so well suited to agriculture that little of it remains. More than 90% has vanished under the plow, and the remainder is so fragmented that almost nowhere can we see even an approximation of what European settlers saw when they settled in the Midwest. Today, the tallgrass prairie is considered North America's rarest biome.

In recent decades, woody shrubs and small trees such as juniper have invaded many of the world's grasslands in North and South America, Africa, and Australia. These plants displace native grasses, reducing the amount of food available for grazing animals. Fires in such altered ecosystems are more catastrophic, and the deep roots of these invasive plants tap into water, altering the area's hydrology. Scientists are studying why this change is occurring. Many think that climate change, increased droughts, overgrazing, and fire suppression all play a role.

Chaparral is a thicket of evergreen shrubs and small trees

Some hilly temperate environments have mild winters with abundant rainfall combined with extremely dry summers. Such Mediterranean climates, as they are called, occur not only in the area around the Mediterranean Sea, but also in California, Western Australia, portions of Chile, and South Africa. In southern California this environment is called **chaparral.** This vegetation type is also known as maquis in the Mediterranean region, mallee scrub in Australia, matorral in Chile, and Cape scrub in Africa. Chaparral soil is thin and infertile. Frequent fires occur naturally in this environment, particularly in late summer and autumn.

Chaparral vegetation looks strikingly similar in different areas of the world, even though the individual species are quite different. A dense growth of evergreen shrubs, often of drought-resistant pine or scrub oak trees, dominates chaparral (FIG. 56-7). During the rainy winter season the landscape may be lush and green, but during the hot, dry summer, the plants lie dormant. Trees and shrubs often have hard, small, leathery leaves that resist water loss. Many plants are also fire-adapted and grow best in the months following a fire. Such growth is possible because fire releases minerals that were tied up in the plants that burned. With the new availability of essential minerals, plants sprout vigorously during winter rains. Mule deer, wood rats, brush rabbits, skinks and other lizards, and many species of birds (such as Anna's hummingbird, scrub jay, and bushtit) are common animals of the chaparral.

Fires, which occur at irregular intervals in California chaparral vegetation, often consume expensive homes built on the hilly chaparral landscape. Unfortunately, efforts to control naturally occurring fires sometimes backfire. Denser, thicker vegetation tends to accumulate when periodic fires are prevented; then, when a fire does occur, it is much more severe. Removing the chaparral vegetation, whose roots hold the soil in place, also causes problems; witness the mud slides that sometimes occur during winter rains in these areas.

Figure 56-7 Chaparral

Chaparral, which consists primarily of drought-resistant evergreen shrubs and small trees, develops where hot, dry summers alternate with mild, rainy winters. Photographed in northwest Arizona.

Deserts are arid ecosystems

Deserts are dry areas found in temperate (*cold deserts*) and subtropical or tropical regions (*warm deserts*). North America has four distinct deserts. The Great Basin Desert in Nevada, Utah, and neighboring states is a cold desert dominated by sagebrush. The Mojave Desert in Nevada and California is a warm desert known for its Joshua trees (a type of yucca with an erect, woody stem). The Chihuahuan Desert, home of century plants (agaves; see Fig. 53-7), is a warm desert found in Texas, New Mexico, and Mexico. The warm Sonoran Desert, with its many species of cacti, is found in Arizona, California, and Mexico (FIG. 56-8).

The low water-vapor content of the desert atmosphere leads to daily temperature extremes of heat and cold, so a major change in temperature occurs in each 24-hour period. Deserts vary greatly depending on the amount of precipitation they receive, which is generally less than 25 cm (10 in.) per year. A few deserts are so dry that virtually no plant life occurs in them. As a result of sparse vegetation, desert soil is low in organic material but often high in mineral content, particularly the salts NaCl, $CaCO_3$, and $CaSO_4$.

Desert vegetation includes both perennials (cacti, yuccas, Joshua trees, and sagebrushes) and, after a rain, flowering annuals. Desert plants tend to have reduced leaves or no leaves, an adaptation that conserves water. In cacti such as the giant saguaro, for example, the stem carries out photosynthesis and also expands accordion-style to store water; the leaves are modified into spines, which discourage herbivores. Other desert plants shed their leaves for most of the year, growing only during the brief moist season.

Desert animals tend to be small. During the heat of the day, they remain under cover or return to shelter periodically, whereas at night they come out to forage or hunt. In addition to desert-adapted insects, there are many specialized desert reptiles (such as desert iguanas, desert tortoises, and rattlesnakes) and a few desert-adapted amphibians

Figure 56-8 Desert

Summer rainfall characterizes the warmer deserts of North America, such as the Sonoran Desert shown here. The Sonoran Desert contains many species of cacti, including the large, treelike saguaro (*Carnegiea gigantea*), which grows 15 to 18 m (50 to 60 ft) tall. Photographed in Arizona.

(such as western spadefoot toads). Mammals include such rodents as the American kangaroo rat, which does not need to drink water but subsists solely on the water content of its food (primarily seeds and insects). American deserts are also home to jackrabbits, and kangaroos live in Australian deserts. Carnivores such as the African fennec fox and some birds of prey, especially owls, live on rodents and rabbits. During the driest months of the year, many desert insects, amphibians, reptiles, and mammals tunnel underground, where they remain inactive; this period of dormancy is known as *estivation*.

Humans have altered North American deserts in several ways. Off-road vehicles damage desert vegetation, which sometimes takes years to recover. When the top layer of desert soil is disturbed, erosion occurs more readily than when the soil is undamaged, and less vegetation grows to support native animals. Another problem is that certain cacti and desert tortoises have become rare as a result of poaching. Houses, factories, and farms built in desert areas require vast quantities of water, which must be imported from distant areas. Irrigation of desert soils often causes them to become salty and unfit for crops or native vegetation. Increased groundwater consumption by many desert cities has caused groundwater levels to drop. Aquifer depletion in U.S. deserts is particularly critical in central and southern Arizona and southwestern New Mexico and is an increasing concern in southern California, Nevada, and Utah.

Savanna is a tropical grassland with scattered trees

The **savanna** biome is a tropical grassland with widely scattered clumps of low trees (FIG. 56-9). Savanna is found in areas of relatively low or seasonal rainfall with prolonged dry periods. Temperatures in savannas vary little throughout the year, and precipitation, not temperature (as in temperate grasslands), regulates seasons. Annual precipitation is 85 to 150 cm (34 to 60 in.). Savanna soil is low in essential minerals, in part because it is strongly leached. Savanna soil is often rich in aluminum, which resists leaching, and in places the aluminum reaches levels that are toxic to many plants. Although the African savanna is best known, savanna also occurs in South America and northern Australia.

Wide expanses of grasses interrupted by occasional trees characterize savanna. Trees such as *Acacia* bristle with thorns that provide protection against herbivores. Both trees and grasses have fire-adapted features, such as extensive underground root systems, that enable them to survive seasonal droughts as well as periodic fires that sweep through the savanna.

The world's greatest assemblage of hoofed mammals occurs in the African savanna. Here live great herds of herbivores, including wildebeests, antelopes, giraffes, zebras, and elephants. Large predators, such as lions and hyenas, kill and scavenge the herds. In areas of seasonally varying rainfall, the herds and their predators may migrate annually.

Savanna is rapidly being fragmented and converted to rangeland for cattle and other domesticated animals, which are replacing the big herds of wild animals. The problem is particularly acute in Africa, which has the most rapidly growing human population of any continent. In some places severe overgrazing by domestic animals has contributed to the conversion of marginal savanna into desert, a process known as **desertification.** In desertification the reduced grass cover caused by overgrazing allows wind and water to erode the soil; erosion removes the topsoil and decreases the soil's ability to support crops or livestock. Desertification has also become a serious problem in South America, Mexico, and the Caribbean due to overgrazing, farming, and drought. (Desertification is not restricted to savanna. Temperate grasslands and tropical dry forests can also be degraded to desert.)

There are two basic types of tropical forests

There are many kinds of tropical forests, but ecologists generally classify them as one of two types: tropical dry forests or tropical rain forests. **Tropical dry forests** occur in regions with a wet season and a dry season (usually two to three months each year). Annual precipitation is 150 to 200 cm (60 to 80 in.). During the dry season, many tropical trees shed their leaves and remain dormant, much as temperate trees do during the winter. India, Brazil, Thailand, and Mexico are some of the countries

Carlyn Iverson

Figure 56-9 Savanna

In this photograph of African savanna in Tanzania, the smaller animals in the foreground are Thomson's gazelles (*Gazella thomsonii*), and the larger ones near the trees are wildebeests (*Connochaetes taurinus*).

that have tropical dry forests. Tropical dry forests intergrade with savanna on their dry edges and with tropical rain forests on their wet edges. Logging and overgrazing by domestic animals have fragmented and degraded many tropical dry forests.

The annual precipitation of **tropical rain forests** is 200 to 450 cm (80 to 180 in.). Much of this precipitation, which occurs almost daily, comes from locally recycled water that enters the atmosphere by transpiration from the forest's own trees. Tropical rain forests are often located in areas with ancient, highly weathered, mineral-poor soil. Little organic matter accumulates in such soils. Because temperatures are high and soil moisture is abundant year-round, decay organisms and detritus-feeding ants and termites decompose organic litter quite rapidly. Vast networks of roots and mycorrhizae quickly absorb minerals from decomposing materials. Thus, minerals of tropical rain forests are tied up in the vegetation rather than in the soil.

Tropical rain forests are found in Central and South America, Africa, and Southeast Asia. A tropical rain forest is very productive despite the scarcity of minerals in the soil. Its plants, stimulated by abundant solar energy and precipitation, capture considerable energy by photosynthesis. Of all the biomes, the tropical rain forest is unrivaled in species richness.

Most trees of tropical rain forests are evergreen flowering plants. A fully developed rain forest has several distinct stories of vegetation (**FIG. 56-10**). The topmost story, called the *emergent layer,* consists of the crowns of the oldest, tallest trees, some 40 m (130 ft) or more in height; these trees are exposed to direct sunlight and are subject to the warmest temperatures, lowest humidities, and strongest winds. The next story, the *canopy,* reaches a height of 30 to 35 m (100 to 115 ft) and lets in little sunlight for the support of the sparse *understory,* shrub layer, and ground layer, all of which consist of smaller plants specialized for life in the shade as well as seedlings of taller trees. Vegetation of tropical rain forests is generally not dense at ground level except near stream banks or where a fallen tree has opened the canopy.

Tropical rainforest trees support extensive epiphytic communities of smaller plants such as orchids and bromeliads. Although epiphytes grow in crotches of branches, on bark, or even on the leaves of their hosts, most use their host trees only for physical support, not for nourishment.

Figure 56-10 Tropical rain forest

Tropical rainforest vegetation is stratified. Except at river banks, tropical rain forest has a closed canopy that admits little light to the forest floor. The animals highlighted in this figure occupy a variety of ecological niches in the rain forest. (Adapted from Miller, G. T. and Spoolman, S. E. *Living in the Environment,* 17th ed. Cengage/Brooks Cole, Belmont, CA, 2012, p. 162, Figure 7-15.)

Because little light penetrates to the understory, many plants living there are adapted to climb already-established host trees. Lianas (woody tropical vines), some as thick as a human thigh, twist up through the branches of tropical rainforest trees. Once in the canopy, lianas grow from the upper branches of one tree to another, connecting the tops of the trees and providing a walkway for many of the canopy's residents.

Not counting bacteria and other soil-dwelling organisms, about 90% of tropical rainforest organisms live in the middle and upper canopies. Rainforest animals include the most abundant and varied insect, reptile, and amphibian fauna on Earth. Birds, too, are diverse, with some specialized to consume fruits (such as parrots), some to consume nectar (such as hummingbirds and sunbirds), and others to consume insects. Most rainforest mammals, such as sloths and monkeys, live only in the trees and never climb down to the ground. Some large ground-dwelling mammals, including elephants, are also found in tropical rain forests.

Unless strong conservation measures are initiated soon, human population growth and agricultural and industrial expansion in tropical countries may spell the end of tropical rain forests by the early 22nd century. Many rainforest species may become extinct before they are even identified and scientifically described. (Tropical rainforest destruction is discussed in detail in Chapter 57.)

To review, examine **FIGURE 56-11**, which shows the geographic distribution of the world's biomes.

CHECKPOINT 56.1

- **CONNECT** *What climate and soil factors produce the major biomes?*

- **CONNECT** *What representative organisms are found in each of these forest biomes: boreal forest, temperate deciduous forest, temperate rain forest, and tropical rain forest?*

- **CONNECT** *In which biome do you live? Does it match the description given in this text? If not, explain the discrepancy.*

- **CONNECT** *How does tundra compare to desert? How does temperate grassland compare to savanna?*

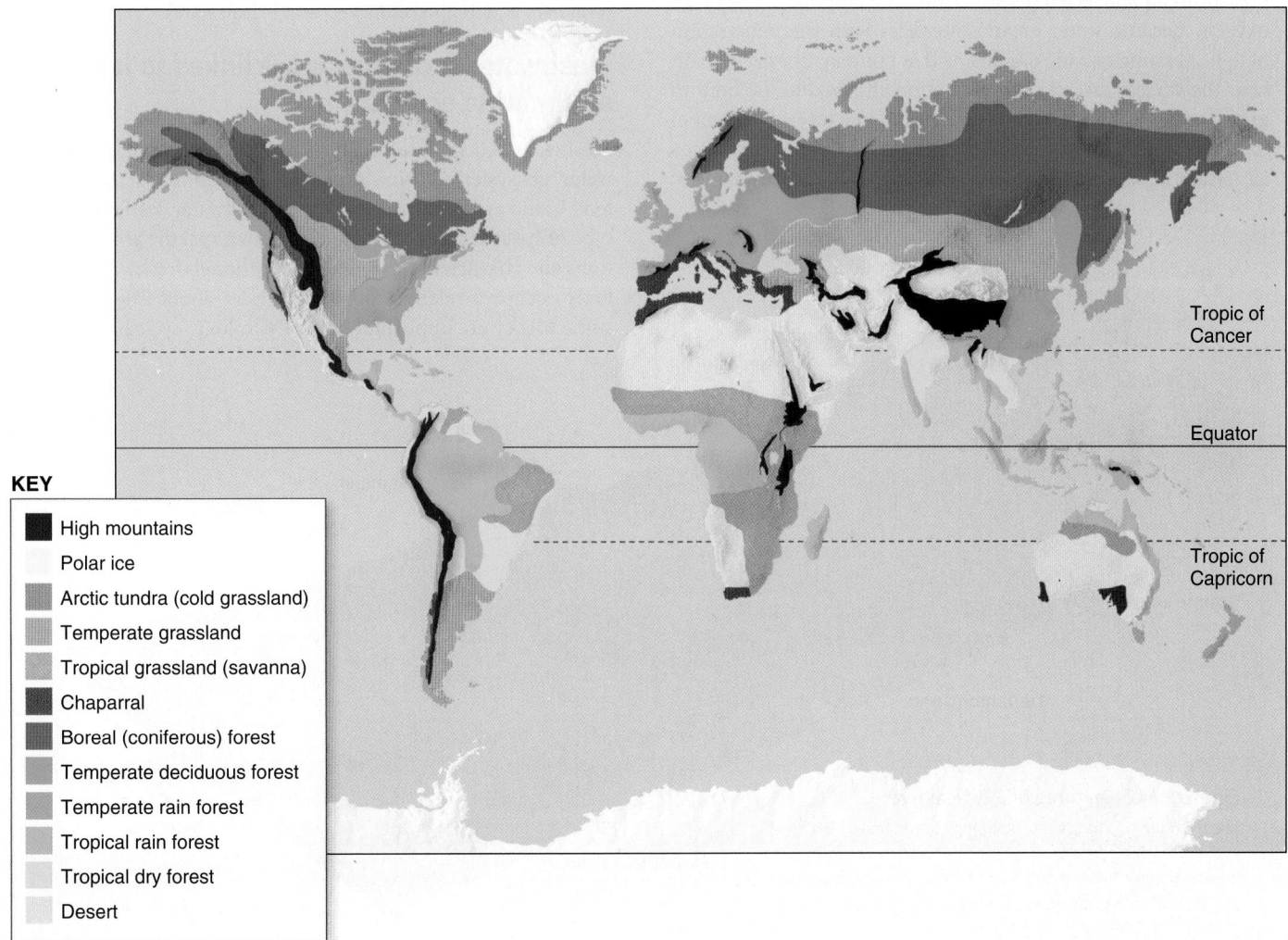

KEY

- High mountains
- Polar ice
- Arctic tundra (cold grassland)
- Temperate grassland
- Tropical grassland (savanna)
- Chaparral
- Boreal (coniferous) forest
- Temperate deciduous forest
- Temperate rain forest
- Tropical rain forest
- Tropical dry forest
- Desert

Figure 56-11 The world's major biomes

This simplified diagram shows sharp boundaries between biomes. Biomes actually intergrade at their boundaries, sometimes over large areas. Note that mountains are keyed separately because they have variable vegetation. (Adapted from Miller, G. T. and Spoolman, S. E. *Living in the Environment,* 17th ed. Cengage/Brooks Cole, Belmont, CA, 2012, p. 153, Figure 7-7.)

56.2 AQUATIC ECOSYSTEMS

Aquatic "biomes" do not exist in the sense that aquatic ecologists do not distinguish aquatic ecosystems based on the dominant form of vegetation. Aquatic ecosystems are classified primarily on abiotic factors, such as salinity, that help determine an aquatic life zone's boundaries. **Salinity,** the concentration of dissolved salts in a body of water, affects the kinds of organisms present in aquatic ecosystems, as does the amount of dissolved oxygen. Because water greatly interferes with the penetration of light, floating aquatic organisms that photosynthesize remain near the water's surface, and vegetation attached to the bottom grows only in shallow water. In addition, low levels of essential minerals often limit the number and distribution of organisms in certain aquatic environments. Other abiotic determinants of species composition in aquatic ecosystems include water depth, temperature, pH, and presence or absence of waves and currents.

Aquatic ecosystems contain three main ecological categories of organisms: free-floating plankton, strongly swimming nekton, and bottom-dwelling benthos. **Plankton** are usually small or microscopic organisms that are relatively feeble swimmers. For the most part, they are carried about at the mercy of currents and waves. They are unable to swim far horizontally, but some species are capable of large vertical migrations and are found at different depths of water at different times of the day or at different seasons. Plankton are generally subdivided into two major categories: phytoplankton and zooplankton. **Phytoplankton** (photosynthetic cyanobacteria and free-floating algae) are producers that form the base of most aquatic food webs. **Zooplankton** are nonphotosynthetic organisms that include protozoa, tiny crustaceans, and the larval stages of many animals. **Nekton** are larger, actively swimming organisms such as fishes, turtles, and whales. **Benthos** are bottom-dwelling organisms that fix themselves to one spot (sponges, oysters, and barnacles), burrow into the sand (many worms and echinoderms), or walk or swim about on the bottom (crayfish, aquatic insect larvae, and brittle stars).

Freshwater ecosystems are linked to land and marine ecosystems

Freshwater ecosystems include streams and rivers (flowing-water ecosystems), ponds and lakes (standing-water ecosystems), and marshes and swamps (freshwater wetlands). Each type of freshwater ecosystem has its own specific abiotic conditions and characteristic organisms. Although freshwater ecosystems occupy a relatively small portion—about 2%—of Earth's surface, they are important in the hydrologic cycle because they

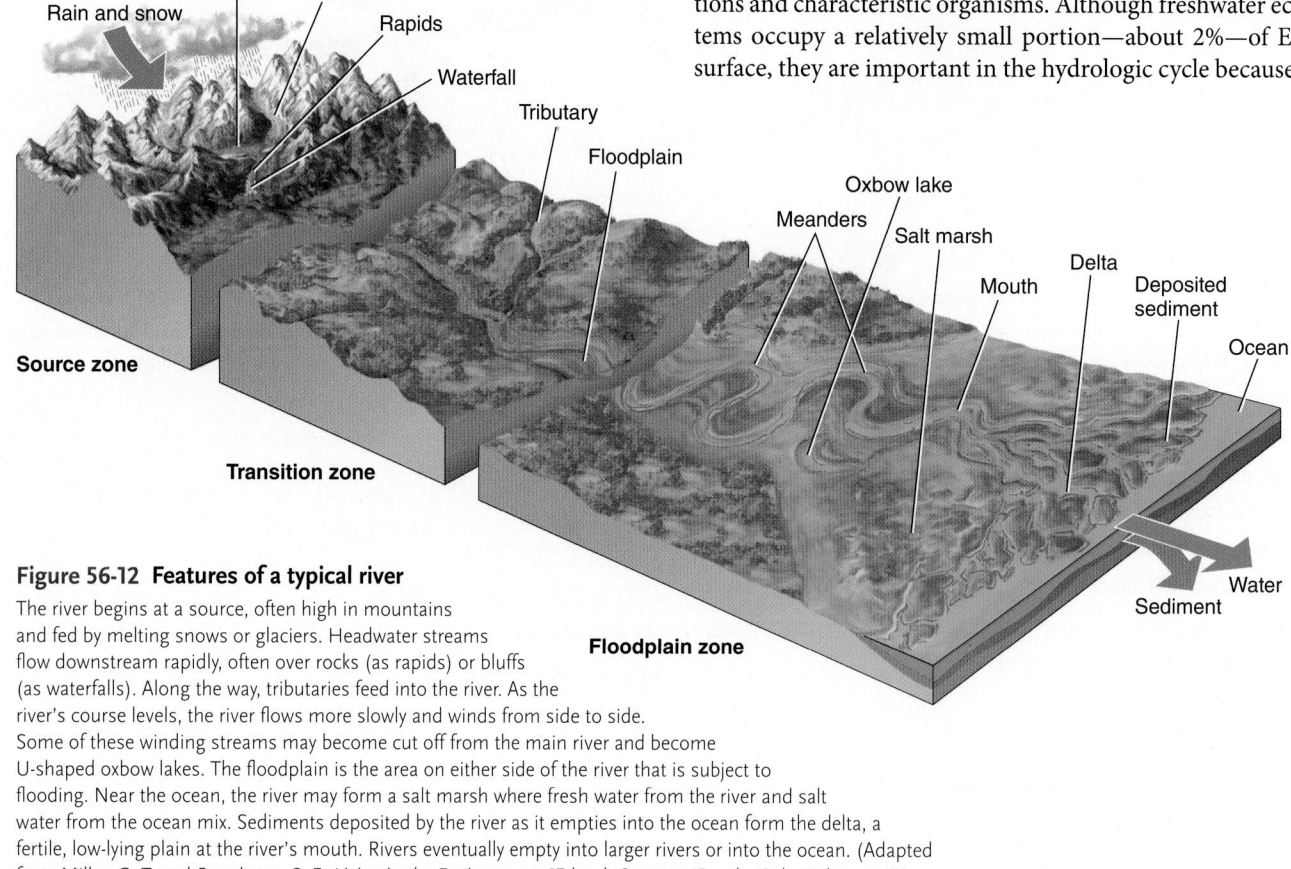

Figure 56-12 Features of a typical river

The river begins at a source, often high in mountains and fed by melting snows or glaciers. Headwater streams flow downstream rapidly, often over rocks (as rapids) or bluffs (as waterfalls). Along the way, tributaries feed into the river. As the river's course levels, the river flows more slowly and winds from side to side. Some of these winding streams may become cut off from the main river and become U-shaped oxbow lakes. The floodplain is the area on either side of the river that is subject to flooding. Near the ocean, the river may form a salt marsh where fresh water from the river and salt water from the ocean mix. Sediments deposited by the river as it empties into the ocean form the delta, a fertile, low-lying plain at the river's mouth. Rivers eventually empty into larger rivers or into the ocean. (Adapted from Miller, G. T. and Spoolman, S. E. *Living in the Environment,* 17th ed. Cengage/Brooks Cole, Belmont, CA, 2012, p. 183, Figure 8-18.)

assist in recycling precipitation that flows as surface runoff to the ocean (see discussion of hydrologic cycle in Chapter 55). Freshwater habitats also provide homes for many species.

Streams and rivers are flowing-water ecosystems Many different conditions exist along the length of a stream or river (**FIG. 56-12**). The nature of a **flowing-water ecosystem** changes greatly from its source (where it begins) to its mouth (where it empties into another body of water). Headwater streams (small streams that are the sources of a river) are usually shallow, clear, cold, swiftly flowing, and highly oxygenated. In contrast, rivers downstream from the headwaters are wider and deeper, cloudy (i.e., they contain suspended particulates), not as cold, slower flowing, and less oxygenated than headwater streams. Surrounding forest may shade certain parts of the stream or river, whereas other parts may be exposed to direct sunlight. Along parts of a stream or river, groundwater wells up through sediments on the bottom. This local input of water moderates the water temperature so that summer temperatures are cooler and winter temperatures are warmer than in adjacent parts of the flowing-water ecosystem.

The kinds of organisms in flowing-water ecosystems vary greatly from one stream to another, depending primarily on the strength of the current. In streams with fast currents, inhabitants have adaptations such as suckers to attach themselves to rocks so that they are not swept away. The larvae of blackflies, for example, attach themselves with suction discs located on the ends of their abdomens. Some stream inhabitants, such as immature water-penny beetles, have flattened bodies that enable them to slip under or between rocks. The water-penny beetle larva gets its common name from its flattened, nearly circular shape. Alternatively, inhabitants such as the brown trout are streamlined and muscular enough to swim in the current.

Streams and rivers depend on land for much of their energy. In headwater streams up to 99% of the energy input comes from detritus (dead organic material such as leaves) carried from the land into streams and rivers by wind or surface runoff. Downstream, rivers contain more producers and therefore depend slightly less on detritus as a source of energy than do the headwaters.

Human activities have several adverse effects on rivers and streams, including water pollution and the impacts of dams built to contain the water of rivers or streams. Pollution, such as sewage and fertilizer runoff, alters the physical environment of a flowing-water ecosystem and changes the biotic component downstream from the pollution source. Dams change the nature of flowing-water ecosystems, both upstream and downstream from the dam location. A dam causes water to back up, resulting in flooding of large areas of land and formation of a reservoir, which destroys terrestrial habitat. Below the dam, the once-powerful river is reduced to a relative trickle, so the flowing-water ecosystem is altered.

Ponds and lakes are standing-water ecosystems Zonation characterizes **standing-water ecosystems.** A large lake has four basic zones: the littoral, limnetic, profundal, and benthic zones (**FIG. 56-13**). Smaller lakes and ponds typically lack a profundal zone.

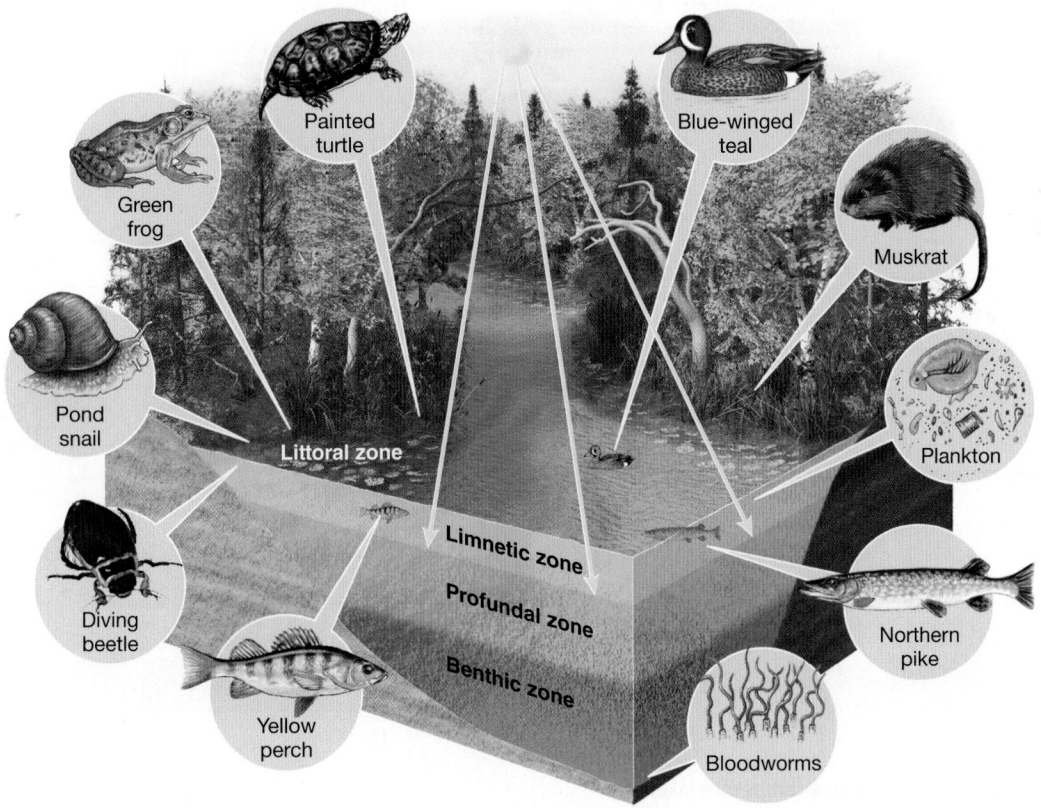

Figure 56-13 *Animation* **Zonation in a large, deep, temperate lake**

A lake is a standing-water ecosystem surrounded by land. A large lake has littoral, limnetic, profundal, and benthic zones. (Adapted from Miller, G. T. and Spoolman, S. E. *Living in the Environment*, 17th ed. Cengage/Brooks Cole, Belmont, CA, 2012, p. 182, Figure 8-16.)

The **littoral zone** is a shallow-water area along the shore of a lake or pond. It includes rooted, emergent vegetation, such as cattails and bur-reeds (flowering wetland plants), plus several deeper-dwelling aquatic plants and algae. The littoral zone is the most productive zone of the lake. Photosynthesis is greatest in the littoral zone, in part because light is abundant and because the littoral zone receives nutrient inputs from surrounding land that stimulate the growth of plants and algae. In addition, ponds and lakes—like streams and rivers—depend on detritus carried from the land for much of their energy. Animals of the littoral zone include frogs and their tadpoles, turtles, worms, crayfish and other crustaceans, insect larvae, and many fishes such as perch, carp, and bass. Surface dwellers, such as water striders and whirligig beetles, are found in the quieter areas.

The **limnetic zone** is the open water beyond the littoral zone, that is, away from the shore; it extends down as far as sunlight penetrates to permit photosynthesis. The main organisms of the limnetic zone are microscopic phytoplankton and zooplankton. Larger fishes also spend some of their time in the limnetic zone, although they may visit the littoral zone to feed and reproduce. Because of the depth of this zone, less vegetation grows in the limnetic zone than in the littoral zone.

Beneath the limnetic zone of a large lake is the **profundal zone.** Because light does not penetrate effectively to this depth, plants and algae do not live in this zone. Food drifts into the profundal zone from the littoral and limnetic zones. Bacteria and archaea decompose dead plants and animals that reach the profundal zone, thus liberating minerals. These minerals are not effectively recycled because no photosynthetic organisms are present to absorb them and incorporate them into the food web. As a result, the profundal zone tends to be both mineral rich and anaerobic (oxygen deficient), with few organisms other than anaerobic bacteria and archaea occupying it.

The **benthic zone** is the bottom layer of the lake. Decomposers, detritus feeders, and some fish live here. These organisms depend mainly on organic matter that drifts down from the upper zones.

Thermal stratification in temperate lakes The marked layering of large temperate lakes caused by light penetration is accentuated by **thermal stratification,** in which the temperature changes sharply with depth (FIG. 56-14a). Thermal stratification occurs because the summer sunlight penetrates and warms surface water, making it less dense. (Recall from Chapter 2 that the density of water is greatest at 4°C; both above and below this temperature, water is less dense.) In summer, cool (and therefore *more* dense) water remains at the lake bottom and is separated from warm (and therefore *less* dense) water above by an abrupt temperature transition called the **thermocline.** Seasonal distribution of temperature and oxygen (more oxygen dissolves in water at cooler temperatures) affects the distribution of fish in the lake.

In temperate lakes falling temperatures in autumn cause a mixing of the lake waters called the **fall turnover** (FIG. 56-14b). As surface water cools, its density increases, and it sinks and displaces the less dense, warmer, mineral-rich water beneath. Warmer water then rises to the surface where it, in turn, cools and sinks. This process of cooling and sinking continues until the lake reaches a uniform temperature throughout.

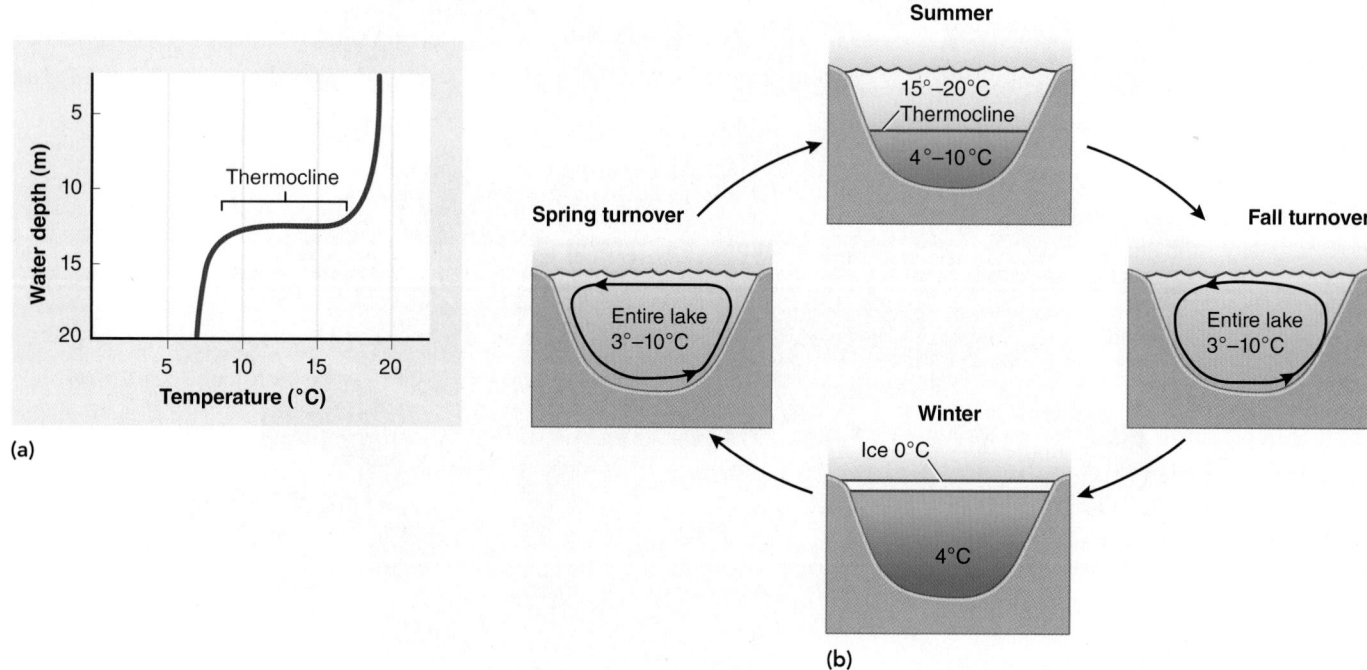

Figure 56-14 Thermal stratification in a temperate lake

(a) Temperature varies at different depths during the summer. There is an abrupt temperature transition, the thermocline. **(b)** During fall and spring turnovers, a mixing of upper and lower layers of water brings oxygen to the oxygen-depleted depths of the lake and minerals to the mineral-deficient surface waters.

© Cengage Learning

In winter, surface water cools to below 4°C, its temperature of greatest density. Ice, which forms at 0°C, is less dense than cold water. Thus, ice forms on the surface, and the water on the lake bottom is warmer than the ice on the surface. In spring, a **spring turnover** occurs as ice melts and surface water reaches 4°C. Surface water again sinks to the bottom, and bottom water returns to the surface. As summer arrives, thermal stratification occurs once again.

The mixing of deeper, nutrient-rich water with surface, nutrient-poor water during the fall and spring turnovers brings essential minerals to the surface and oxygenated water to the bottom. The sudden presence of large amounts of essential minerals in surface waters encourages the development of large algal and cyanobacterial populations, which may form temporary **blooms** in the fall and spring.

Increased nutrients and algal growth The presence of high levels of plant and algal nutrients such as nitrogen and phosphorus causes *enrichment,* the fertilization of a body of water. Excess amounts of these nutrients enter waterways from sewage and from fertilizer runoff from lawns and fields. The water in an enriched pond or lake is cloudy because of the vast numbers of algae and cyanobacteria that the nutrients support.

The species composition is different in enriched and unenriched lakes. For example, an unenriched lake in the northeastern United States may contain pike, sturgeon, and whitefish in the deeper, colder part of the lake where there is a higher concentration of dissolved oxygen. In contrast, the deeper, colder levels of water in enriched lakes are depleted of dissolved oxygen because of the greater amount of decomposition on the lake floor. Fishes such as pike, sturgeon, and whitefish die out, and fishes such as catfish and carp, which tolerate lower concentrations of dissolved oxygen, replace them.

Enrichment is reversible and has declined in North America since the 1970s because legislation limits the phosphate content of detergents and because better sewage treatment plants have been constructed. Agriculture is the leading source of water-quality impairment of U.S. surface waters today. Major causes of enrichment problems in waterways include fertilizer runoff, animal wastes, and sewage plant residues. States and communities continue to improve storm water management practices, and many have passed laws to ban phosphate fertilizers.

Freshwater wetlands are transitional between aquatic and terrestrial ecosystems **Freshwater wetlands,** which are usually covered by shallow water for at least part of the year, have characteristic soils and water-tolerant vegetation. They include marshes, dominated by grasslike plants, and swamps, in which woody trees or shrubs dominate (FIG. 56-15). Freshwater wetlands also include hardwood bottomland forests (lowlands along streams and rivers that are periodically flooded), prairie potholes (small, shallow ponds that formed when glacial ice melted at the end of the last Ice Age), and peat moss bogs (peat-accumulating wetlands where sphagnum moss dominates).

Wetlands are valued as a wildlife habitat for migratory waterfowl and many other bird species, beavers, otters, muskrats, and game fishes. Wetlands are holding areas for excess water when rivers flood their banks. The floodwater stored in wetlands then drains slowly back into the rivers, providing a steady flow of water throughout the year. Wetlands also serve as groundwater recharging areas. One of their most important roles is to trap and hold pollutants in the flooded soil, thereby cleansing and purifying the water. Such important environmental functions as these are known as **ecosystem services.**

Wetlands were once considered wastelands, areas to be filled in or drained so that farms, housing developments, and industrial plants could be built on them. Wetlands are also breeding places for mosquitoes and therefore were viewed as a menace to public health. The crucial ecosystem services that wetlands provide are widely recognized today, and wetlands have some legal protection. Despite this recognition of their value, agriculture, pollution, engineering (dams), and urban and suburban development continue to threaten wetlands.

Estuaries occur where fresh water and salt water meet

Where the ocean meets the land, there may be one of several kinds of ecosystems: a rocky shore, a sandy beach, an intertidal mudflat, or a tidal estuary. An **estuary** is a coastal body of water, partly surrounded by land, with access to the open ocean and a large supply of fresh water from rivers. Water levels in an estuary rise and fall with the tides, and salinity fluctuates with tidal cycles, the time of year, and precipitation. Salinity also changes gradually within the estuary, from fresh water at the river entrance to salty ocean water at the mouth of the estuary. Because estuaries undergo marked daily, seasonal, and annual variations in temperature, salinity, and other physical properties, estuarine organisms have a wide tolerance to such changes.

Figure 56-15 Freshwater swamp

Trees, such as bald cypress (*shown*), dominate freshwater swamps. In this wetland, photographed in northeastern Texas, a floating carpet of tiny aquatic plants covers the water's surface.

Gregory J. Dimijian/Science Source

Figure 56-16 Salt marsh
Cordgrass (*Spartina alterniflora*) is the dominant vegetation in this salt marsh in Georgia.

and nurseries for commercially important fish and shellfish species, such as blue crabs, shrimp, mullet, and spotted sea trout. Mangrove branches are nesting sites for many species of birds, such as pelicans, herons, egrets, and roseate spoonbills. Mangroves are under assault from coastal development, including aquaculture facilities, and unsustainable logging. Some countries, such as the Philippines, Bangladesh, and Guinea-Bissau, have cut down more than two-thirds of their mangrove forests. In the United States, commercial developers, home builders, and some residents continue to sidestep conservation laws by destroying mangroves to provide waterfront properties with unobstructed views.

Estuaries are among the most fertile ecosystems in the world, often having much greater productivities than the adjacent ocean or freshwater river (see Table 55-1). This high productivity is the result of four factors: (1) the action of tides promotes a rapid circulation of nutrients and helps remove waste products; (2) minerals, transported from land into streams and rivers emptying into the estuary, provide nutrients; (3) the high level of light that penetrates the shallow water supports photosynthesis; and (4) the presence of many plants provides an extensive photosynthetic carpet and also mechanically traps detritus, forming the basis of detritus food webs. Most commercially important fishes and shellfish spend their larval stages in estuaries among the protective tangle of decaying stems.

Temperate estuaries usually contain **salt marshes,** shallow wetlands in which salt-tolerant grasses dominate (**FIG. 56-16**). Uninformed people have often considered salt marshes as worthless, empty stretches of land. As a result, people have used them as dumps, severely polluting them, or filled them with dredged bottom material to form artificial land for residential and industrial development. A large part of the estuarine environment is lost in this way, along with many of its ecosystem services, such as biological habitats, sediment and pollution trapping, groundwater supply, and storm buffering (salt marshes absorb much of the energy of a storm surge and therefore prevent flood damage elsewhere).

Mangrove forests, the tropical equivalent of salt marshes, cover perhaps 70% of tropical and subtropical coastal mudflats where tides and waves fluctuate (**FIG. 56-17**; Chapter 54 introduction). Like salt marshes, mangrove forests provide valuable ecosystem services. Mangrove roots stabilize the sediments, preventing coastal erosion and providing a barrier against the ocean during storms. Their interlacing roots are breeding grounds

Marine ecosystems dominate Earth's surface

Although lakes and the ocean are comparable in many ways, they have many differences. Depths of even the deepest lakes

Figure 56-17 Mangroves
Red mangroves (*Rhizophora mangle*) have stiltlike roots that support the tree. Many animals live in the complex root systems of mangrove forests. Photographed at low tide along the coast of Florida in Biscayne National Park.

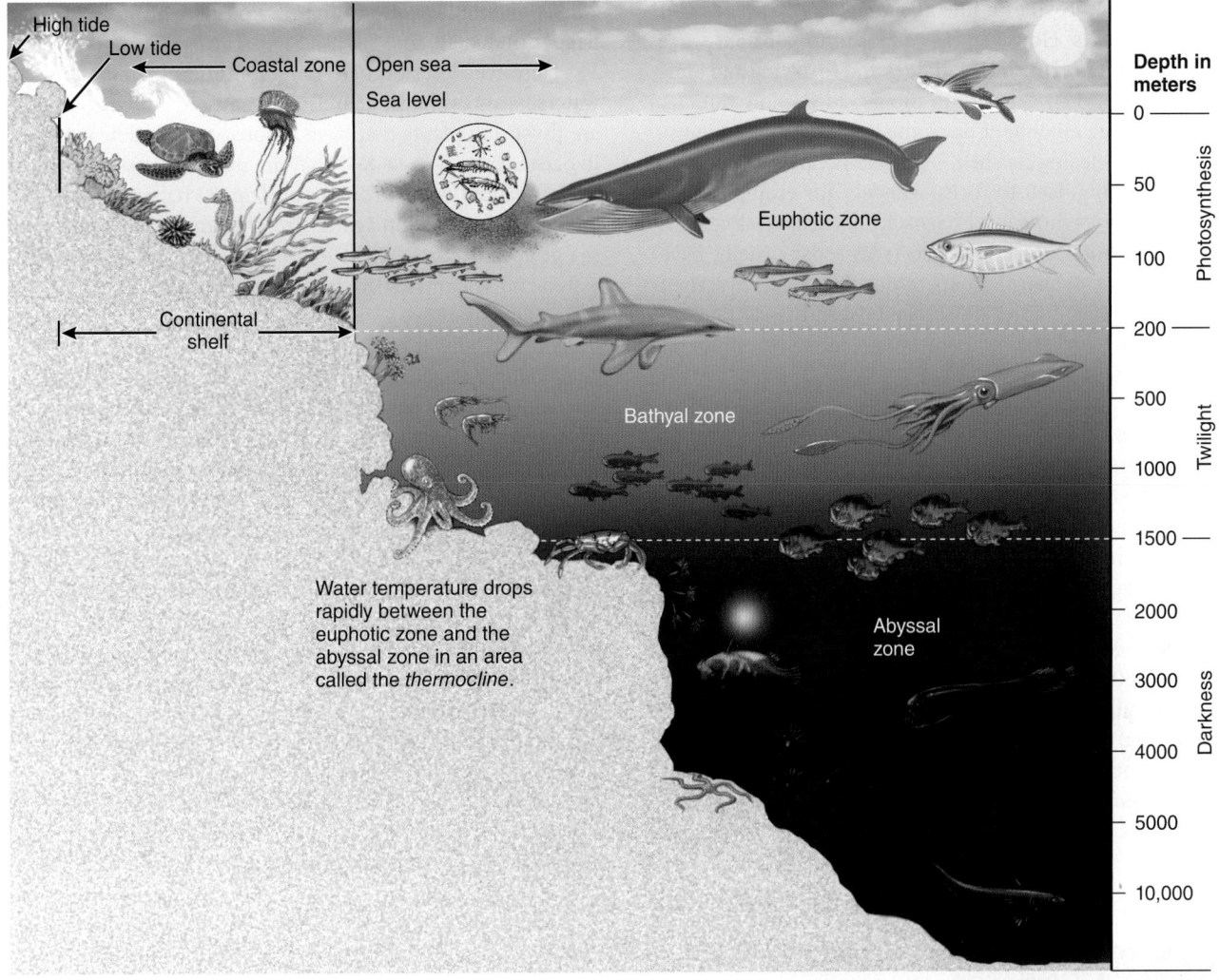

Figure 56-18 *Animation* **Vertical zones in an ocean**

The euphotic zone, the brightly lit upper region of an ocean, is inhabited by plankton and plankton-eating nekton. The bathyal zone, which receives little sunlight, does not support phytoplankton. Zooplankton and some invertebrates and fish live in the bathyal zone. The dark, cold abyssal zone is inhabited by some sponges, clams, oysters, squid, worms, and fish. Many organisms that inhabit the abyssal zone are filter feeders that depend on organic material that drifts down from the upper layers.

Note that the slopes of the ocean floor are not as steep as shown; they are exaggerated to save space. (Adapted from Miller, G. T. and Spoolman, S. E. *Living in the Environment*, 17th ed. Cengage/Brooks Cole, Belmont, CA, 2012, p. 173, Figure 8-6.)

do not approach those of the ocean, which has areas that extend more than 6 km (3.6 mi) below the sunlit surface. Tides and currents profoundly influence the ocean. Gravitational pulls of both the sun and the moon produce two tides a day throughout most areas of the ocean, but the height of those tides varies with season, local topography, and phases of the moon (full moons and new moons cause the highest tides).

The immense, complex marine environment is subdivided into several zones: the intertidal zone, the benthic (ocean floor) environment, and the pelagic (ocean water) environment (**FIG. 56-18**). The pelagic environment is, in turn, divided into two provinces: the neritic province and the oceanic province.

The intertidal zone is transitional between land and ocean The **intertidal zone** is the shoreline area between low tide and high tide. Although high levels of light and nutrients, together with an abundance of oxygen, make the intertidal zone a biologically productive environment, it is also a stressful one. If an intertidal beach is sandy, inhabitants must contend with a constantly shifting environment that threatens to engulf them and gives them scant protection against wave action. Consequently, most sand-dwelling organisms, such as mole crabs, are continual and active burrowers. Because they follow the tides up and down the beach, most do not have any notable adaptations to survive drying or exposure.

A rocky shore provides a fine anchorage for seaweeds and invertebrate animals. However, it is exposed to constant wave action when immersed during high tides and to drying and temperature changes when exposed to the air during low tides. A typical rocky-shore inhabitant has some way of sealing in moisture, perhaps by closing its shell, if it has one, plus a powerful means of anchoring itself to rocks. Mussels, for example, have horny, threadlike anchors, and barnacles have special cement glands. Rocky-shore intertidal algae (seaweeds) usually have thick, gummy polysaccharide coats, which dry out slowly when exposed, and flexible bodies not easily broken by wave action (FIG. 56-19). Some rocky-shore community inhabitants hide in burrows or crevices at low tide.

Seagrass beds, kelp forests, and coral reefs are part of the benthic environment The **benthic environment** is the ocean floor. It is divided into zones based on distance from land, light availability, and depth. The benthic environment consists of sediments (mostly sand and mud) in which many marine animals, such as worms and clams, burrow. Archaea and bacteria are common in marine sediments, and living archaea and bacteria have been found in deeply buried ocean sediments at least 800 m (2600 ft) below the ocean floor at several different sites in the Pacific Ocean.

The **abyssal zone** is that part of the benthic environment that extends from a depth of 4000 to 6000 m (2.5 to 3.7 mi). (In Chapter 55 *Inquiring About: Life without the Sun* describes some of the unusual organisms in hydrothermal vents in the abyssal

zone.) The **hadal zone** is that part of the benthic environment deeper than 6000 m.

Here we describe benthic communities in shallow ocean waters: seagrass beds, kelp forests, and coral reefs. **Sea grasses** are flowering plants that have adapted to complete submersion in ocean water (FIG. 56-20). They are not true grasses. Sea grasses live in shallow water, to depths of 10 m (33 ft), where they receive enough light to photosynthesize efficiently. Extensive beds of sea grasses occur in quiet temperate, subtropical, and tropical waters; no sea grasses live in polar waters.

Sea grasses have a high primary productivity and are therefore ecologically important in shallow marine areas. Their roots and rhizomes stabilize the sediments, reducing surface erosion.

Sea grasses provide food and habitat for many marine organisms. In temperate waters ducks and geese eat sea grasses; in tropical waters manatees, green turtles, parrot fish, sturgeon fish, and sea urchins eat them. These herbivores consume only about 5% of the sea grasses. The remaining 95% eventually enter the detritus food web when the sea grasses die and bacteria decompose them. In turn, a variety of animals such as mud shrimp, lug worms, and mullet consume the bacteria. Seagrass beds have been severely damaged and depleted by water pollution and by boat propellers. Fortunately, many coastal communities and environmental organizations are replanting seagrass beds and educating boaters about the importance of healthy seagrass ecosystems.

Kelps, which may reach lengths of 60 m (200 ft), are the largest brown algae (see Fig. 26-11b). Kelps are common in cooler temperate marine waters of both the Northern and Southern Hemispheres. They are especially abundant in relatively shallow waters (depths of about 25 m, or 82 ft) along rocky coastlines. Kelps are photosynthetic and are therefore the primary food producers for the kelp forest ecosystem. Kelp forests also provide habitats for many marine animals. Tube worms, sponges, sea cucumbers, clams, crabs, fishes (such as tuna), and mammals

Figure 56-19 Seaweeds in a rocky intertidal zone
Sea palms (*Postelsia*), which are 50 to 75 cm (20 to 30 in.) tall, are common on the rocky Pacific coast from Vancouver Island to California. The bases of these brown algae are firmly attached to the rocky substrate, enabling them to withstand heavy surf action. Photographed at low tide.

Figure 56-20 Seagrass bed
Turtle grasses (*Thalassia*) have numerous invertebrates and algal epiphytes attached to their leaves. These shallow underwater meadows of marine flowering plants are ecologically important for shelter and food for many organisms. Photographed in the Red Sea, Egypt.

(such as sea otters) find refuge in the algal blades. Some animals eat the blades, but kelps are consumed mainly in the detritus food web. Bacteria that decompose dead kelp provide food for sponges, tunicates, worms, clams, and snails. Kelp beds support a diversity of life that almost rivals that found in coral reefs.

Coral reefs, which are built from accumulated layers of calcium carbonate ($CaCO_3$), are found in warm (temperature usually greater than 21°C), shallow sea water. The living portions of coral reefs grow in shallow waters where light penetrates. Many coral reefs are composed principally of red coralline algae that require light for photosynthesis. Coral animals also require light for the large number of symbiotic dinoflagellates, known as **zooxanthellae,** which live and photosynthesize in their tissues (see Fig. 54-12). Although species of coral without zooxanthellae exist, only those with zooxanthellae build reefs. In addition to obtaining nitrogen compounds from the zooxanthellae living inside them, coral animals capture food at night, using their stinging tentacles to paralyze small animals that drift nearby.

Coral reefs grow slowly as coral organisms build on the calcareous remains of countless organisms before them. The warm, shallow waters in which coral reefs are found are often poor in nutrients. Other factors favor high productivity, however, including the presence of symbiotic zooxanthellae, warm temperatures, and plenty of sunlight.

Coral reef ecosystems are the most diverse of all marine environments and contain hundreds or even thousands of species of fishes and invertebrates, such as giant clams, sea urchins, sea stars, sponges, brittle stars, sea fans, and shrimp (**FIG. 56-21**). The Great Barrier Reef, along the northeastern coast of Australia, occupies only 0.1% of the ocean's surface, but 8% of the world's fish species live there. The multitude of relationships and interactions that occur at coral reefs is comparable only to those in tropical rain forests among terrestrial ecosystems. As in the rain forest, competition is intense, particularly for light and space to grow.

In addition to providing habitat for a wide variety of marine organisms, coral reefs are ecologically important because they protect coastlines from shoreline erosion. They also provide humans with seafood, pharmaceuticals, and income from tourism and recreation. Although coral formations are important ecosystems, they are rapidly being degraded and destroyed. According to the United Nations Environment Program, 27% of the world's coral reefs are at high risk. Coral reefs of Southeast Asia, which contain the most species of all coral reefs, are the most threatened of any region.

In some areas, silt washing downstream from clear-cut inland forests has smothered reefs under a layer of sediment. Some scientists hypothesize that high salinity resulting from the diversion of fresh water to supply the growing human population is killing Florida reefs. Overfishing, pollution from sewage discharge and agricultural runoff, oil spills, boat groundings, fishing with dynamite or cyanide, hurricane damage, disease, coral bleaching, land reclamation, tourism, and the mining of corals for building material are also taking a heavy toll. (Coral reefs are also discussed in Chapter 31.)

The neritic province consists of shallow waters close to shore

The **neritic province** is the open ocean that overlies the continental shelves, that is, the ocean floor from the shoreline to a depth of 200 m (650 ft). Organisms that live in the neritic province are floaters or swimmers (**FIG. 56-22**). The upper reaches of the neritic province make up the **euphotic zone,** which extends from the surface to a depth of approximately 100 m (325 ft). Enough light penetrates the euphotic zone to support photosynthesis.

Large numbers of phytoplankton, particularly diatoms in cooler waters and dinoflagellates in warmer waters, produce food by photosynthesis and are thus the base of food webs. Zooplankton (including tiny crustaceans; jellyfish; comb jellies; protists such as foraminiferans; and larvae of barnacles, sea urchins, worms, and crabs) feed on phytoplankton.

Plankton-eating nekton, such as herring, sardines, squid, manta rays, and baleen whales, consume zooplankton. These, in turn, become prey for carnivorous nekton such as sharks, tuna, dolphins, and toothed whales. Nekton are thought to be mostly confined to the shallower neritic waters (less than 60 m, or 195 ft, deep) because that is where their food is. However,

Figure 56-21 Coral reef organisms

A panoramic view of a coral reef in the Indian Ocean off the coast of the Maldives shows a few of the many animals that live on and around coral reefs. The fish in the foreground are blue-green damselfish (*Chromis viridis*).

Reinhard Dirscherl/Getty Images

Figure 56-22 Fish in the open ocean

Rough triggerfish (*Canthidermis maculatus*) are tropical fish usually found in the open ocean to depths of 92 m (300 ft). Sometimes they spend time on deep rocky slopes. Photographed in the Pacific Ocean near Hawaii.

not much is known about the behavior and migration patterns of marine nekton. Many fishes appear to be wide-ranging. For example, an individual fish tagged on one side of an ocean may be recaptured on the other side a few months later. Whether the fish traveled alone or in a school is not known at this time.

The oceanic province makes up most of the ocean The average depth of the world's ocean is 4000 m (2.4 mi). The **oceanic province** is that part of the open ocean that covers the deep-ocean basin, that is, the ocean floor at depths more than 200 m (650 ft). It is the largest marine environment and contains about 75% of the ocean's water. Cold temperatures, high hydrostatic pressure, and absence of sunlight characterize the oceanic province. These environmental conditions are uniform throughout the year.

Most organisms of the oceanic province depend on **marine snow,** organic debris that drifts down into the unlit **aphotic region** from the upper, lighted regions. Organisms of this little-known realm are filter-feeders, scavengers, or predators. Many are invertebrates, some of which attain great sizes. The giant squid, for example, measures up to 18 m (59 ft) in length, including its

tentacles. Fishes of the oceanic province are strikingly adapted to darkness and food scarcity. For example, the gulper eel's huge jaws enable it to swallow large prey (**FIG. 56-23**). (An organism that encounters food infrequently needs to eat as much as possible when it has the chance.) The lizardfish has strong pelvic fins that stabilize it in deep underwater currents. The many, long, folding teeth in its jaw and on its tongue maintain a tight hold on its prey. Many animals of the oceanic province have illuminated organs that enable them to see one another to mate or to capture food. Adapted to drifting or slow swimming, these fishes often have reduced bone and muscle mass.

Human activities are harming the ocean Because the ocean is so vast, many individuals have difficulty realizing how seriously human activities negatively impact and harm it. Development of resorts, cities, industries, transportation, and agriculture along coasts alters or destroys many coastal ecosystems, including mangrove forests, salt marshes, seagrass beds, and coral reefs. Coastal and marine ecosystems receive pollution from land, from rivers emptying into the ocean, and from

Figure 56-23 Gulper eel from the oceanic province

The gulper eel (*Saccopharynx lavenbergi*) uses its "trap-door" jaws to swallow prey as large as itself. The tail, of which only a small portion is shown, makes up most of the length of a gulper eel's body. Gulper eels grow to 1.8 m (6 ft). Shown is a live specimen, photographed in an aquarium aboard ship after being captured at a depth of 1500 m (4921 ft) off southern California.

atmospheric contaminants that enter the ocean via precipitation. Disease-causing viruses and bacteria from human sewage contaminate shellfish and other seafood and sometimes threaten public health. Millions of tons of trash, including plastic, fishing nets, and packaging materials, end up in coastal and marine ecosystems; some of this trash entangles and kills marine organisms.

Less visible ocean contaminants, including fertilizers, pesticides, heavy metals, pharmaceuticals, and synthetic chemicals from agriculture and industry, continue to be detected in increasing amounts. Coastal dead zones, caused by runoff of fertilizers and burning fossil fuels, have been reported in more than 400 locations. These dead zones have severely depleted dissolved oxygen levels in bottom waters.

Offshore mining and oil drilling pollute the neritic province with oil and other contaminants. Millions of ships dump oily ballast and other wastes overboard in the neritic and oceanic provinces. Fishing is highly mechanized, and new technologies can remove every single fish in a targeted area of the ocean. Scallop dredges and shrimp trawls are dragged across the benthic environment, destroying entire communities with a single swipe. Most countries recognize the importance of the ocean to life on this planet, but few have the resources or programs to protect and manage the ocean effectively.

CHECKPOINT 56.2

- **PREDICT** *Imagine that the ocean plankton suddenly disappeared. What effect would that have on the nekton and benthos?*

- **CONNECT** *What environmental factors are most important in determining the adaptations of organisms that live in aquatic environments?*

- *How do you distinguish between freshwater wetlands and estuaries? between flowing-water and standing-water ecosystems?*

- **CONNECT** *Compare and contrast the four main marine environments.*

- **CONNECT** *Which aquatic ecosystem is often compared to tropical rain forests? Why?*

56.3 ECOTONES

LEARNING OBJECTIVE

7 Define *ecotone* and describe some of its features.

We have discussed the various terrestrial biomes and aquatic ecosystems as if they were distinct and separate entities, but within landscapes, ecosystems intergrade with one another at their boundaries. The transition zone where two communities or biomes meet and intergrade is called an **ecotone** (see Chapter 54). Ecotones range in size from quite small, such as the area where an agricultural field meets a woodland or where a stream flows through a forest, to continental in scope. For example, at the

border between tundra and boreal forest, an extensive ecotone exists that consists of tundra vegetation interspersed with small, scattered conifers. Such ecotones provide habitat diversity. In fact, often a greater variety and density of organisms populate ecotones than live in either adjacent ecosystem (**FIG. 56-24**).

Ecologists who study ecotones look for adaptations that enable organisms to survive there. They also examine the relationship between species richness and ecotones and how

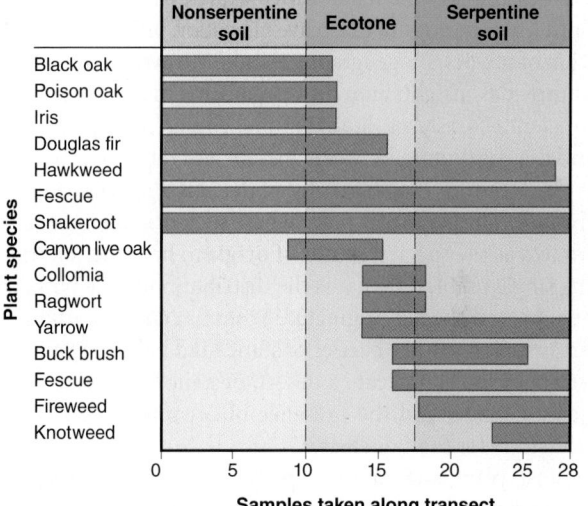

KEY EXPERIMENT

Does species richness vary among ecotones and their adjoining communities?

HYPOTHESIS: Ecotones have greater species richness than the communities they connect.

EXPERIMENT: Plant species were sampled for two communities in southwestern Oregon and for the ecotone between them. The communities, one with nonserpentine soil and one with serpentine soil, are defined largely by soil conditions. Nonserpentine soils (samples 1 to 10) are "normal" soils. Serpentine soils (samples 18 to 28) contain high levels of elements such as chromium, nickel, and magnesium that are toxic to many plants.

RESULTS AND CONCLUSION: The various plant species found in the two communities (*yellow* and *blue*) and in the ecotone between them (*green*) are shown in the graph. The ecotone had a greater richness than either adjoining community.

SOURCE: Modified from White, C. D. "Vegetation-Soil Chemistry Correlations in Serpentine Ecosystems." Ph.D. dissertation, University of Oregon, Eugene, 1971. Reprinted with permission of Dr. Charles D. White.

Figure 56-24 Ecotones and species richness

Note that the two fescues in the figure are different species.

CONNECT Why were fireweed and knotweed samples not found in the ecotone area?

ecotones change over time. Long-term, or longitudinal, studies of ecotones have revealed they are far from static. The ecotone boundary between desert and semiarid grassland in southern New Mexico, for example, has moved since the 1950s as the desert ecosystem has continued its expansion into the grassland.

CHECKPOINT 56.3

- *What habitat feature is characteristic of an ecotone?*

56.4 BIOGEOGRAPHY

LEARNING OBJECTIVE

8 Define *biogeography* and briefly describe Wallace's biogeographic realms.

The study of the geographic distribution of plants and animals is called **biogeography** (see Chapter 18). Biogeographers search for patterns in geographic distribution and try to explain how such patterns arose, including where populations originated, how they spread, and when. Biogeographers recognize that geologic and climate changes such as mountain building, continental drift (encompassed in the theory of *plate tectonics,* which describes the continents as moving plates floating on Earth's mantle), and periods of extensive glaciation influence the distribution of species. Biogeography is linked to evolutionary history and provides insights into how organisms may have interacted in ancient ecosystems. Studying biogeography helps us understand the continuum between ancient and modern ecosystems.

One tenet of biogeography is that each species originated only once. The particular place where such origination occurred is known as the species' **center of origin.** The center of origin is not a single point; rather, it is the distribution of the population when the new species originated. From its center of origin, each species spreads until a barrier of some kind halts it. Examples of barriers include an ocean, a desert, or a mountain range; unfavorable climate; and the presence of organisms that compete successfully for food or shelter.

Most plant and animal species have characteristic geographic distributions. The **range** of a particular species is that portion of Earth in which it is found. The range of some species may be a relatively small area. For example, wombats, the marsupial equivalent of groundhogs, are found only in drier parts of southeastern Australia and nearby islands. Such localized, native species are said to be **endemic;** that is, they are not found anywhere else in the world. In contrast, some species have a nearly worldwide distribution and occur on more than one continent or throughout much of the ocean. Such species are said to be **cosmopolitan.**

One of the early observations of biogeographers was that the ranges of different species do not include everywhere that they *could* survive. Central Africa has elephants, gorillas, chimpanzees, lions, antelopes, umbrella trees, and guapiruvu trees, whereas areas in South America with a similar climate have none of them. These animals and plants originated in Africa after continental drift had already separated the supercontinent Pangaea into several landmasses. The organisms could not expand their range into South America because the Atlantic Ocean was an impassable barrier. Likewise, the ocean was a barrier to South American monkeys, sloths, tapirs, balsa trees, and snakewood trees, none of which are found in Africa.

Land areas are divided into six biogeographic realms

As the various continents were explored and their organisms studied, biologists observed that the world could be divided into major blocks of vegetation, such as forests, grasslands, and deserts, and that these vegetation types corresponded to specific climates. The relationship between animal distribution, geography, and climate was not deduced until 1876. At that time, Alfred Russel Wallace, who independently and concurrently discovered the same scientific theory of evolution by natural selection as Charles Darwin, divided Earth's land areas into six major biogeographic realms: the Palearctic, Nearctic, Neotropical, Ethiopian, Oriental, and Australian (FIG. 56-25). A major barrier separates each of the six biogeographic realms from the others and helps maintain each region's biological distinctiveness.

Many biologists quickly embraced Wallace's classification system, which is still considered valid today. However, human activities, such as the intentional and unintentional introduction of foreign species, are contributing to a homogenization of the biogeographic realms (see Chapter 57).

Refer to Figure 56-25 as we briefly consider some of the characteristic animals in each realm. The *Nearctic* and *Palearctic realms* are more closely related than the other regions, especially in their northern parts, where they share many animals such as wolves, hares, and caribou. This similarity may be due to a land bridge that has periodically connected Siberia and Alaska. This bridge was present until late in the Pleistocene epoch, about 10,000 years ago. Animals adapted to cold environments likely dispersed between Asia and North America along this bridge.

The *Neotropical realm* was almost completely isolated from the Nearctic realm and other landmasses for most of the past 70 million years. During this time, many marsupial species evolved. The isthmus of Panama, which formed a dryland connection about 3 million years ago, linked North and South America and provided a route for dispersal. Only three species—the opossum, armadillo, and porcupine—are descendants of animals that survived the northward dispersal from South America, but many species, such as the tapir and llama, are descendants of animals that survived the southward dispersal from North America. Competition from these species caused many of South America's marsupial species to go extinct.

The Sahara Desert separates the *Ethiopian realm* from other landmasses. The Ethiopian realm contains the most varied vertebrates of all six realms. Some overlap exists between the Ethiopian realm and *Oriental realm* because a land bridge with a moist climate linked Africa to Asia during the Miocene and Pliocene epochs. The Oriental realm has the fewest endemic species of all the tropical realms.

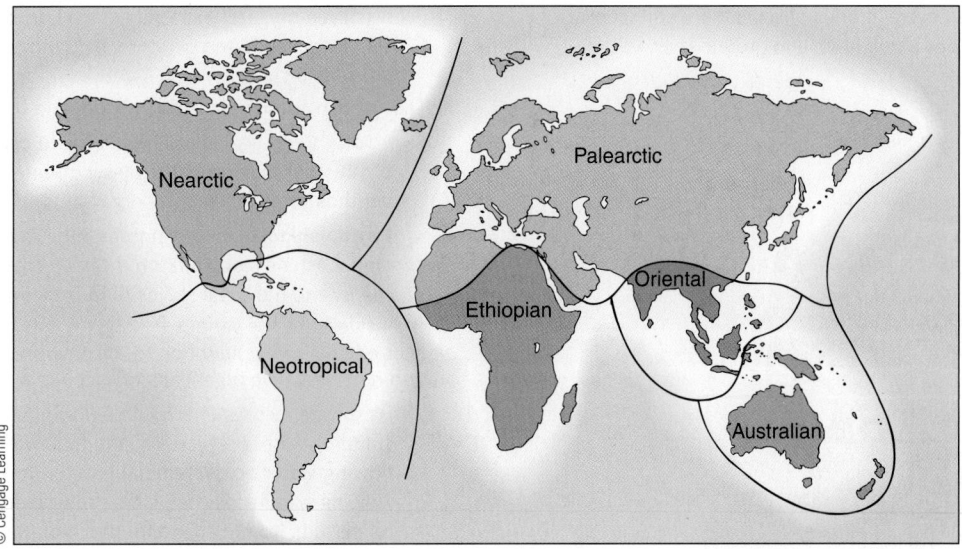

Figure 56-25 Wallace's biogeographic realms

Certain unique species characterize each of the six biogeographic realms. The boundary between the Oriental and Australian realms, for example, is a deep-water gap. Wallace noted that hornbills, Sumatran tigers, orangutans, and tree shrews live west of the boundary, whereas tree kangaroos, cockatoos, birds of paradise, and cuscus live east of the boundary. Many of the Australian species are endemic; that is, they evolved in Australia and are found nowhere else. Wallace's studies of biogeography helped him develop his scientific theory of evolution by natural selection.

The *Australian realm* has not had a land connection with other regions for more than 85 million years. It has no native placental mammals, and marsupials and monotremes, including the duck-billed platypus and the spiny anteater, dominate it. Adaptive radiation of the marsupials during their long period of isolation led to species with ecological niches similar to those of placental mammals of other realms (see Fig. 32-28).

CHECKPOINT 56.4

- *What is biogeography, and what do biogeographers do?*
- *Which biogeographic realm has been separated from the other biogeographic realms for the longest period? What animals characterize this biogeographic realm?*

SUMMARY: FOCUS ON LEARNING OBJECTIVES

56.1 Biomes *(page 1209)*

1 Define *biome* and briefly describe the nine major terrestrial biomes, giving attention to the climate, soil, and characteristic plants and animals of each.

- A **biome** is a large, relatively distinct terrestrial region with characteristic climate, soil, plants, and animals.
- A frozen layer of subsoil (**permafrost**) and low-growing vegetation that is adapted to extreme cold and a short growing season characterize **tundra** (also called **arctic tundra**), the northernmost biome. **Alpine tundra,** at higher elevations above the tree line, and **Antarctic tundra** of the Southern Hemisphere are also included in the tundra biome classification.
- Conifers—which are adapted to cold winters, a short growing season, and acidic, mineral-poor soil—dominate the **boreal forest,** or **taiga.**
- Large conifers dominate **temperate rain forest,** which receives high precipitation.

- **Temperate deciduous forest** occurs where precipitation is relatively high and soils are rich in organic matter. Broad-leaf trees that lose their leaves seasonally dominate temperate deciduous forest.
- **Temperate grassland** typically has a deep, mineral-rich soil and has moderate but uncertain precipitation.
- Thickets of small-leaf evergreen shrubs and trees and a climate of wet, mild winters and dry summers characterize **chaparral.**
- **Desert,** found in both temperate (cold deserts) and subtropical or tropical regions (warm deserts) with low levels of precipitation, is inhabited by organisms with specialized water-conserving adaptations.
- Tropical grassland, called **savanna,** has widely scattered trees interspersed with grassy areas. Savanna occurs in tropical areas with low or seasonal rainfall.
- Mineral-poor soil and high rainfall that is evenly distributed throughout the year characterize **tropical rain forest.** Tropical rain forest has high species richness and high productivity.

A rain forest has several stories of vegetation: emergent layer, canopy, understory, shrub layer, and ground layer.

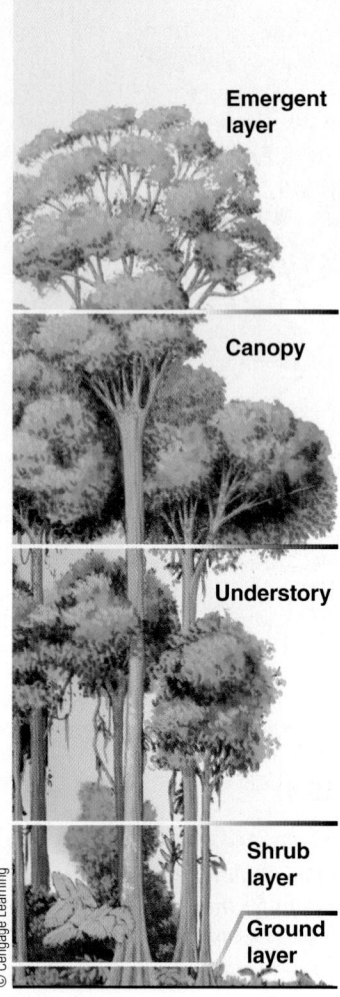

Emergent layer

Canopy

Understory

Shrub layer

Ground layer

© Cengage Learning

2 Describe at least one human effect on each of the biomes discussed.

- Oil exploration and military exercises result in long-lasting damage to tundra. Clear-cut logging destroys boreal and temperate rain forests. Temperate deciduous forests are removed for logging, farms and tree plantations, and land development.

- Human population growth and its accompanying agricultural and industrial expansion threaten most of the world's tropical rain forests.

- Farmland has replaced most temperate grasslands; savannas are increasingly converted to rangeland for cattle. Development of hilly chaparral results in mud slides and costly fires. Land development in deserts reduces wildlife habitat.

56.2 Aquatic Ecosystems *(page 1218)*

3 Explain the important environmental factors that affect aquatic ecosystems.

- In aquatic ecosystems important environmental factors include **salinity** (concentration of dissolved salts), amount of

dissolved oxygen, availability of light, levels of essential minerals, water depth, temperature, pH, and presence or absence of waves and currents.

4 Distinguish among plankton, nekton, and benthos.

- Aquatic life is ecologically divided into **plankton** (free-floating organisms), **nekton** (strongly swimming organisms), and **benthos** (bottom-dwelling organisms).

- **Phytoplankton** includes photosynthetic algae and cyanobacteria; **zooplankton** includes protozoa, tiny crustaceans, and the larval stages of many animals.

5 Briefly describe the various freshwater, estuarine, and marine ecosystems, giving attention to the environmental characteristics and representative organisms of each.

- Freshwater ecosystems include flowing-water ecosystems, standing-water ecosystems, and freshwater wetlands. In **flowing-water ecosystems,** the water flows in a current. Flowing-water ecosystems have few phytoplankton and depend on detritus from the land for much of their energy.

- Large **standing-water ecosystems** (freshwater lakes) are divided into zones on the basis of water depth. The marginal **littoral zone** contains emergent vegetation and algae. The **limnetic zone,** open water away from the shore that extends as far down as sunlight penetrates, contains phytoplankton, zooplankton, and larger fishes. The deep, dark **profundal zone** holds little life other than bacterial decomposers.

- **Freshwater wetlands** are transitional between fresh water and terrestrial ecosystems. They are usually covered at least part of the year by shallow water and have characteristic soils and vegetation. Freshwater wetlands perform many valuable **ecosystem services.**

- An **estuary** is a coastal body of water, partly surrounded by land, with access to both the ocean and a large supply of fresh water from rivers. Salinity fluctuates with tidal cycles, the time of year, and precipitation. Temperate estuaries usually contain **salt marshes,** whereas **mangrove forests** dominate tropical coastlines.

- Four important marine environments are the intertidal zone, the benthic environment, the neritic province, and the oceanic province. The **intertidal zone** is the shoreline area between low tide and high tide. Organisms of the intertidal zone have adaptations to resist wave action and being exposed to air during low tide.

- The **benthic environment** is the ocean floor. **Sea grasses, kelps,** and **coral reefs** are important benthic communities in shallow ocean waters.

- The **neritic province** is open ocean from the shoreline to a depth of 200 m. Organisms that live in the neritic province are all floaters or swimmers. The **euphotic zone** is the upper part of the neritic province, where enough light penetrates to support photosynthesis. Phytoplankton are the base of the food web in the euphotic zone.

- The **oceanic province** is that part of the open ocean that is deeper than 200 m. Its uniform environment is one of darkness, cold temperature, and high pressure. Animal inhabitants of the oceanic province are either predators or scavengers that subsist on **marine snow,** detritus that drifts down from other areas of the ocean.

6 Describe at least one human effect on each of the aquatic ecosystems discussed.

- Water pollution and dams adversely affect flowing-water ecosystems. Increased nutrients, supplied by human activities, stimulate algal growth, resulting in enriched ponds and lakes.
- Agriculture, pollution, and land development threaten wetlands and estuaries. Pollution, coastal development, offshore mining and oil drilling, and overfishing threaten marine ecosystems.

56.3 Ecotones (page 1227)

7 Define *ecotone* and describe some of its features.

- An **ecotone** is the transition zone where two communities or biomes meet and intergrade. Ecotones provide habitat diversity and are often populated by a greater variety of organisms than lives in either adjacent ecosystem.

56.4 Biogeography (page 1228)

8 Define *biogeography* and briefly describe Wallace's biogeographic realms.

- **Biogeography** is the study of the geographic distribution of plants and animals, including where populations came from, how they got there, and when. Each species originated only once, at its **center of origin.** From its center of origin, each species spreads until a physical, environmental, or biological barrier halts it. The **range** of a particular species is that portion of Earth in which it is found.
- Alfred Russel Wallace divided Earth's land areas into six major biogeographic realms: the Palearctic, Nearctic, Neotropical, Ethiopian, Oriental, and Australian. Each realm is biologically distinct because a mountain range, desert, deep water, or other barrier separates it from the others.

TEST YOUR UNDERSTANDING

Know and Comprehend

1. The northernmost biome, known as _____, typically has little precipitation, a short growing season, and permafrost. (a) chaparral (b) boreal forest (c) tundra (d) northern deciduous forest (e) savanna

2. Forests of the northeastern and middle Atlantic United States, which have broad-leaf hardwood trees that lose their foliage annually, are called (a) temperate deciduous forests (b) tropical dry forests (c) boreal forests (d) temperate rain forests (e) tropical rain forests

3. The deepest, richest soil in the world occurs in (a) temperate rain forest (b) tropical rain forest (c) savanna (d) temperate grassland (e) chaparral

4. This biome, with its thicket of evergreen shrubs and small trees, is found in areas with Mediterranean climates. (a) temperate rain forest (b) tropical rain forest (c) savanna (d) temperate grassland (e) chaparral

5. This biome is a tropical grassland interspersed with widely spaced trees. (a) temperate rain forest (b) tropical rain forest (c) savanna (d) temperate grassland (e) chaparral

6. This biome has the greatest species richness. (a) temperate rain forest (b) tropical rain forest (c) savanna (d) temperate grassland (e) chaparral

7. Organisms in aquatic environments fall into three categories: free-floating _____, strongly swimming _____, and bottom-dwelling _____. (a) nekton; benthos; plankton (b) nekton; plankton; benthos (c) plankton; benthos; nekton (d) plankton; nekton; benthos (e) benthos; nekton; plankton

8. Emergent vegetation grows in the _____ zone of freshwater lakes. (a) littoral (b) limnetic (c) profundal (d) neritic (e) intertidal

9. The _____ is open ocean from the shoreline to a depth of 200 m. (a) benthic environment (b) intertidal zone (c) neritic province (d) oceanic province (e) aphotic region

10. The transition zone where two ecosystems or biomes meet and intergrade is called (a) a biosphere (b) an aphotic region (c) a thermocline (d) a biogeographic realm (e) an ecotone

11. Which biogeographic realm has been separated from the other landmasses for more than 85 million years? (a) Ethiopian (b) Palearctic (c) Nearctic (d) Oriental (e) Australian

Apply and Analyze

12. **INTERPRET DATA** Develop a hypothesis to explain why animals adapted to the desert are usually small. How would you test your hypothesis?

13. **INTERPRET DATA** Examine Figure 56-1. What is the lowest average annual precipitation characteristic of tropical rain forests? the highest? What is the range of average temperature in tropical rain forests?

14. **INTERPRET DATA** Examine Figure 56-24. How many of the sampled species are found in the nonserpentine soil? in the serpentine soil? in the ecotone? What generalization about ecotones do these data support?

Evaluate and Synthesize

15. **PREDICT** What would happen to the organisms in a river with a fast current if a dam were built? Would there be any differences in habitat if the dam were upstream or downstream of the organisms in question? Explain your answers.

16. **EVOLUTION LINK** When a black-tailed prairie dog or other small animal dies, other prairie dogs bury it. Develop a hypothesis to explain how this behavior may be adaptive. How would you test this hypothesis?

17. **SCIENCE, TECHNOLOGY, AND SOCIETY** Imagine that you are a graduate student working with university scientists on the effects on sea turtles of the 2010 oil spill in the Gulf of Mexico. What technologies might you employ to collect data for your research? Knowing that the turtles' life cycle includes laying eggs on beaches, how would you involve local communities in your research?

aplia To access course materials, such as Aplia and other companion resources, please visit **www.cengagebrain.com.**

57 Biological Diversity and Conservation Biology

Cross River gorilla, *Gorilla gorilla diehli*. Will this critically endangered species survive?

KEY CONCEPTS

57.1 Never before in Earth's history has biodiversity been threatened with mass extinction in such a compressed time period by one species: *Homo sapiens*.

57.2 Conservation biology is a multidisciplinary science that studies human effects on biodiversity and devises practical solutions to address these effects.

57.3 Globally, removal of large expanses of forest for timber, agriculture, and other uses is contributing to the loss of biological diversity and valuable ecosystem services.

57.4 Scientists continue to provide evidence of global climate change, which is adversely affecting biodiversity, human health, and agriculture.

The Cross River gorilla, *Gorilla gorilla diehli*, was discovered early in the 20th century (see photograph). It was eventually identified as a fourth subspecies of gorilla and second member of the western lowland gorilla family. This rarest of all the great apes lives in small groups in highland forests of Nigeria and western Cameroon. Cross River gorillas, listed as critically endangered (estimated population 250 to 300), are rapidly facing extinction due to habitat loss caused by human destruction. Local people have cut down, or destroyed, much of the forests for lumber, cattle grazing, and agriculture. The gorillas are hunted and poached for bush meat.

The World Wildlife Federation (WWF) reports that these gorillas also face the risk of inbreeding and loss of genetic diversity due to their small population size and the low rate of genetic exchange between the different subpopulations. The WWF has partnered with the Wildlife Conservation Society and the governments of Nigeria and Cameroon to protect the fragile Cross River gorilla groups by establishing a sanctuary, enforcing laws, and supporting government management of the Campo Ma'an National Park in Cameroon and the surrounding buffer zone. Both Cameroon and Nigeria are working with local people to educate about sustainable use of forest resources and to promote alternative sources of income, such as tourism. Plans for the survival of the Cross River gorillas will focus on developing protected corridors of forest habitat that will allow safe movement of gorillas between different groups to increase genetic diversity.

Humans are very closely related to gorillas, both in our evolutionary development and genetically. In 2012, researchers reported in *Nature* that when the western lowland gorilla genome was sequenced, they found that 15% of the human genome was more closely related to gorillas than to chimpanzees. Scientists also confirmed that the gorilla is about 98% genetically identical to us. Further, we share some rapidly evolving genes with gorillas, which may facilitate medical research of such ailments as heart disease and dementia. The study increases our understanding of human evolution and, according to the researchers, "connects us to a time when our existence was more tenuous, and in doing so, highlights the importance of protecting and conserving these remarkable species."

The Cross River gorilla decline underscores how humans have negatively affected biological diversity. For example, according to the International Union for the Conservation of Nature (IUCN), which monitors the status of Earth's biological diversity, as many as one-fourth of the world's mammals are at risk of extinction. Our recognition of this serious situation and our efforts to address the damage we have caused may well be what ultimately ensures our own human survival.

In this chapter we focus on declining biological diversity and conservation biology. We discuss deforestation and climate change, two major environmental issues that negatively affect biological diversity.

57.1 THE BIODIVERSITY CRISIS

LEARNING OBJECTIVES

1 Identify various levels of biodiversity: genetic diversity, species richness, and ecosystem diversity.
2 Distinguish among threatened species, endangered species, and extinct species.
3 Discuss at least four causes of declining biological diversity and identify the most important causes.

The human species (*Homo sapiens*) has been present on Earth for about 195,000 years (see Chapter 22), which is a brief span of time compared with the age of our planet, some 4.6 billion years. Despite our relatively short tenure on Earth, our biological effect on other species is unparalleled. Our numbers have increased dramatically—the human population reached 7.14 billion in 2013—and we have expanded our biological range, moving into almost every habitat on Earth.

Wherever we have lived, we have altered the environment and shaped it to meet our needs. In only a few generations, we have transformed the face of Earth, placed a great strain on Earth's resources and resilience, and profoundly affected other species. As a result of these changes, we must all be concerned about **environmental sustainability** (or simply, *sustainability*), the ability to meet humanity's current needs without compromising the ability of future generations to meet their needs. The effect of humans on the environment merits special study in biology, not merely because we ourselves are humans but because our impact on the rest of the biosphere is so extensive.

Extinction, the death of a species, occurs when the last individual member of a species dies. (We introduced the important concept of extinction in Chapter 20.) Although extinction is a natural biological process, human activities have greatly accelerated it (**FIG. 57-1**). According to the 2005 Millennium Ecosystem Assessment report, extinction is happening at about 1000 times more rapidly as the normal background rate of extinction. The burgeoning human population has forced us to spread to almost all areas of Earth. Whenever humans invade an area, the habitats of many plants and animals are disrupted or destroyed, which can contribute to their extinction.

Biological diversity, also called **biodiversity,** is the variation among organisms (**FIG. 57-2**). Biological diversity includes much more than simply **species richness,** the number of species of archaea, bacteria, protists, plants, fungi, and animals. Biological diversity occurs at all levels of ecological organization, from populations to ecosystems. It takes into account **genetic diversity,** the genetic variety within a species, both among individuals within a given population and among geographically separate populations. (An individual species may have hundreds of genetically distinct populations.) Biological diversity also includes **ecosystem diversity,** the variety of ecosystems found on Earth: the forests, prairies, deserts, lakes, coastal estuaries, coral reefs, and other ecosystems.

Biologists must consider all three levels of biological diversity—species richness, genetic diversity, and ecosystem diversity—as they address the human impact on biodiversity. For example, the disappearance of populations (i.e., a decline in

Passenger pigeon   Great auk   Dodo   Golden toad   Aepyornis

Figure 57-1 Selected animal extinctions

These animals became extinct largely as a result of human activities such as destruction of their habitats and hunting. (Adapted from Miller, G. T., and S. E. Spoolman. *Living in the Environment,* 16th ed. Brooks/Cole, Belmont, CA, 2009.)

Genetic diversity affects a population's ability to thrive, which in turn affects species richness within an ecosystem. Ultimately, species richness helps determine ecosystem diversity across a large area.

(a) Genetic diversity in an American oystercatcher population.

(b) Species richness in an East Coast sandy shore ecosystem.

(c) Ecosystem diversity across an entire region (ocean and shore).

Figure 57-2 Levels of biodiversity

The single unduplicated chromosome below each oystercatcher represents the genetic variation among individuals within a population. Conservation at all three ecological levels must occur to protect biological diversity.

PREDICT What might happen to this oystercatcher population if another population of oystercatchers migrates to the area and competes for food and space?

© Cengage Learning

genetic diversity) indicates an increased risk that a species will become extinct (a decline in species richness). Biologists perform detailed analyses to quantify how large a population must be to ensure a given species' long-term survival. The smallest population that has a high chance of enduring into the future is known as the **minimum viable population (MVP).**

Biological diversity is currently decreasing at an unprecedented rate (see *Inquiring About: Declining Amphibian Populations*). For example, the IUCN Red List of Threatened Plants, which is based on 20 years of data collection and analysis around the world, lists about 34,000 species of plants currently threatened with extinction. Because Red Lists are not yet available for many tropical countries, it is difficult to know the true scale of the biodiversity crisis.

We may have entered the greatest period of mass extinction in Earth's history, but the current situation differs from previous periods of mass extinction in several respects. First, its cause is directly attributable to human activities. Second, it is occurring in a tremendously compressed period (just a few decades as opposed to hundreds of thousands of years), much faster than rates of speciation (or replacement). Perhaps even more sobering, larger numbers of plant species are becoming extinct today than in previous mass extinctions. Because plants are the base of terrestrial food webs, extinction of animals that depend on plants is not far behind. It is crucial that we determine how the loss of biodiversity affects the stability and functioning of ecosystems, which make up our life-support system.

The legal definition of an **endangered species,** as stipulated by the U.S. Endangered Species Act, is a species in imminent danger of extinction throughout all or a significant part of its range. (The area in which a particular species is found is its **range.**) Unless humans intervene, an endangered species will probably become extinct.

When extinction is less imminent but the population of a particular species is quite small, the species is classified as threatened. The legal definition of a **threatened species** is a species likely to become endangered in the foreseeable future, throughout all or a significant part of its range.

Endangered and threatened species represent a decline in biological diversity because as their numbers decrease, their genetic diversity is severely diminished. Endangered and threatened species are at greater risk of extinction than species with greater genetic variability because long-term survival and evolution depend on genetic diversity (see the section on genetic drift in Chapter 19).

Endangered species have certain characteristics in common

Many threatened and endangered species share certain characteristics that may make them more vulnerable to extinction. Some of these characteristics include having an extremely small (localized) range, requiring a large territory, living on islands, having a low reproductive success, needing specialized breeding areas, and possessing specialized feeding habits.

What is happening to our amphibian populations? Since the 1970s, many of the world's frog populations have dwindled or disappeared. According to the IUCN Global Amphibian Assessment, as many as 168 amphibian species may have disappeared since 1980 and are presumed extinct. About one-third of all amphibian species are in decline, and more than 40% are at risk of becoming extinct by 2060 if they are not taken into captivity.

The declines are not limited to areas with obvious habitat destruction (such as drainage of wetlands where frogs live), degradation from pollutants, or overharvesting. Some remote, pristine locations also show dramatic declines in amphibians. Researchers have reported evidence that pollutants, infectious diseases, and climate change are all factors in these declines.

Agricultural chemicals are implicated in amphibian declines in California's Sierra Nevadas. Frog populations on the eastern slopes are relatively healthy, but about eight species are declining on the western slopes, where prevailing winds carry residues of 15 different pesticides from the Central Valley's agricultural region. Agricultural chemicals are also implicated in amphibian declines on the eastern shore of Maryland and in Ontario, Canada. Researchers at the University of South Florida found that atrazine (an herbicide) combined with phosphate fertilizer increased trematode

(a parasitic flatworm) infection and suppressed the immune system, making it difficult for amphibians to fight parasitic infections.

Infectious diseases are strongly implicated in some of the declines. In 1998, a chytrid (fungus) was first connected to amphibian declines. Climate change may be exacerbating chytrid-induced amphibian deaths, and this factor continues to be investigated. At certain altitudes and temperatures, the chytrid has infected and killed 85% of the amphibians.

The discovery of amphibians with deformities adds another layer of complexity to the amphibian crisis (see photograph). Frogs with extra legs, extra toes, eyes located on the shoulder or back, deformed jaws, missing legs, missing toes, or missing eyes usually die before they reproduce. Predators easily catch frogs with extra or missing legs. Deformed amphibians have been reported on four continents and in almost all states in the United States.

Biologists have investigated many possible causes of amphibian deformities during development, and no single factor explains the deformities found in all locations. Several pesticides affect normal development in frog embryos. Also, infecting tadpoles with a trematode causes the adults that develop from the tadpoles to exhibit limb deformities. Multiple environmental stressors, such as habitat loss, disease, and pollution, may interact synergistically with one another to cause deformities.

Frans Lanting/Corbis

Frog deformity. Pollution and parasites are implicated in developmental abnormalities in amphibians, such as this Pacific tree frog.

Many endangered species have a small natural range, which makes them particularly prone to extinction if their habitat is altered. The Tiburon mariposa lily, for example, is found nowhere in nature except on a single hilltop near San Francisco. Wildfire or development of that hilltop would almost certainly cause the extinction of this species.

Species that need extremely large territories may be threatened with extinction when all or part of their territory is modified by human activity. The California condor, for example, is a scavenger bird that lives off carrion and requires hundreds of square kilometers of undisturbed territory to find adequate food. The condor is slowly recovering from the brink of extinction. From 1987 to 1992, it existed only in zoos and was not found in nature. A program to reintroduce zoo-bred California condors into the wild began in 1992. Currently, there are more than 350 condors, with more than half of them living in the wild in California, Arizona, Utah, and Baja California. The condor story is not an unqualified success, however; the condors have yet to become a self-sustaining population that replaces its numbers without captive breeding.

Many **endemic** island species are endangered. (*Endemic* means that they are local species not found anywhere else in the world.) These species often have small populations that cannot be replaced by immigration if their numbers decline. Because

they evolved in isolation from competitors, predators, and disease organisms, island species have few defenses when such organisms are introduced, usually by humans. It is not surprising that of the 171 bird species that have become extinct in the past few centuries, 155 of them lived on islands.

For a species to survive, its members must be present within their range in large enough numbers for males and females to mate. The minimum viable population that ensures reproductive success varies from one type of organism to another. For all species, when the population size falls below the MVP, the population goes into a decline and becomes susceptible to extinction.

Endangered species often have low reproductive rates. The female blue whale produces a single calf every other year, whereas no more than 6% of swamp pinks, an endangered species of small flowering plant, produce flowers in a given year (**FIG. 57-3**). Some endangered species breed only in specialized areas; for example, the green sea turtle lays its eggs on just a few beaches.

Highly specialized feeding habits endanger a species. In nature, the giant panda eats only bamboo. Periodically, all the bamboo plants in a given area flower and die together; when that occurs, panda populations face starvation. Like many other endangered species, giant pandas are also endangered

Figure 57-3 The swamp pink, an endangered species
The swamp pink (*Helonias bullata*) lives in boggy areas of the eastern United States. Photographed in Killens Pond State Park, Delaware.

KEY POINT

Because direct and indirect causes of decreased biodiversity interact in complex ways, declining biodiversity is addressed from a systems biology perspective rather than from a more-specific perspective.

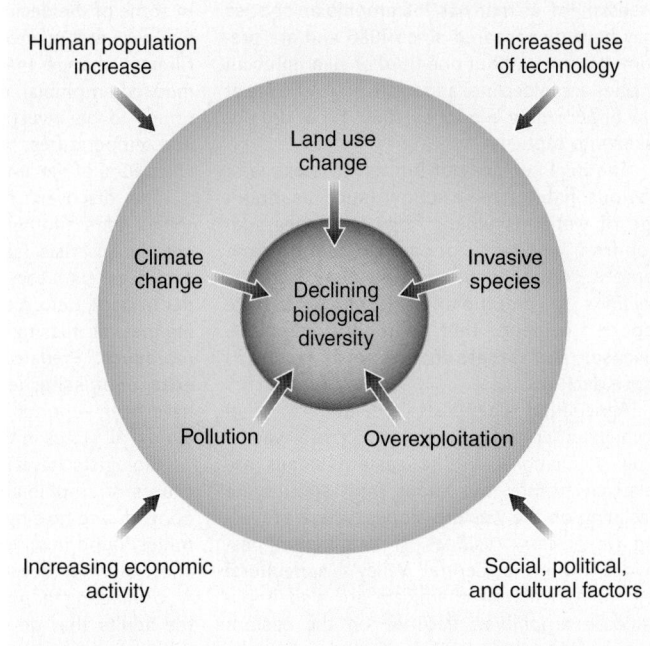

Figure 57-4 *Animation* **Causes of declining biological diversity**
Indirect causes (*blue*) of decreasing biodiversity interact with and amplify the effects of one another and of direct causes (*green*). Note that each of these causes can be traced to human activities.

CONNECT Give an example of how an indirect effect could positively change a direct effect.

© Cengage Learning

because their habitat has been fragmented, and there are few intact habitats where they can survive. China's 1600 wild giant pandas live in 24 isolated habitats that represent a small fraction of their historic range.

Human activities contribute to declining biological diversity

Species become endangered and extinct for a variety of reasons, including the destruction or modification of habitats (i.e., land use change) and the production of pollution, including the greenhouse gases that cause climate change. Humans also upset the delicate balance of organisms in a given area by introducing invasive species, which compete with native organisms. Overexploitation through overfishing, hunting, and poaching is also a factor. **FIGURE 57-4** shows the interactions between these direct causes of declining biological diversity and indirect human factors such as human population increase.

To save species, we must protect their habitats Most species facing extinction today are endangered by destruction, fragmentation, or degradation of natural habitats. Building roads, parking lots, bridges, and buildings; clearing forests to grow crops or graze domestic animals; and logging forests for timber all take their toll on natural habitats. Draining marshes converts aquatic habitats to terrestrial ones, whereas building dams and canals floods terrestrial habitats (**FIG. 57-5**).

Because most organisms require a particular type of environment, habitat destruction reduces their biological range and ability to survive.

Humans often leave small, isolated patches of natural landscape that roads, fences, fields, and buildings completely surround. In ecological terms *island* refers not only to any landmass surrounded by water but also to any isolated habitat surrounded by an expanse of unsuitable territory. Accordingly, a small patch of forest surrounded by agricultural and suburban lands is considered an island.

Habitat fragmentation, the breakup of large areas of habitat into small, isolated segments (i.e., islands), is a major threat to the long-term survival of many populations and species (**FIG. 57-6**). Species from the surrounding "developed" landscape may intrude into the isolated habitat (the *edge effect,* discussed in Chapter 54), whereas rare and wide-ranging species that require a large patch of undisturbed habitat may disappear altogether. Habitat fragmentation encourages the spread of *invasive species,* disease organisms, and other "weedy" species and makes it difficult for organisms to migrate. Generally, habitat fragments support only

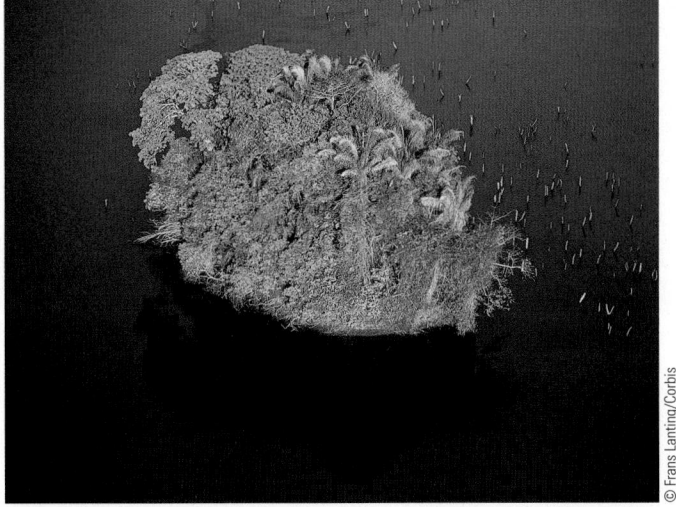

Figure 57-5 Habitat destruction

This tiny island lies in the Panama Canal. It was once a hilltop in a forest that was flooded when the canal was built.

a fraction of the species found in the original, unaltered environment. However, much remains to be learned about precisely how habitat fragmentation affects specific populations, species, and ecosystems.

Human activities that produce acid precipitation and other forms of pollution indirectly modify habitats left undisturbed and in their natural state. Acid precipitation, also called acid rain or industrial precipitation, has contributed to the decline of large stands of forest trees and to the biological death of many freshwater lakes in, for example, the Adirondack Mountains and Nova Scotia. Acid rain is especially problematic in much of Eastern Europe and around industrial regions of China. Other types of pollutants, such as chemicals from industry and agriculture, organic pollutants from sewage, acid wastes seeping from mines, thermal pollution from the heated wastewater of industrial plants, and radioactive contamination, also adversely affect organisms.

To save native species, we must control intrusions of invasive species **Biotic pollution,** the introduction of a foreign species into an area where it is not native, often upsets the balance among the organisms living in that area and interferes with the ecosystem's normal functioning. Unlike other forms of pollution, which may be cleaned up, biotic pollution is usually permanent. The foreign species may prey on native species or compete with them for food or habitat. If the foreign species causes economic or environmental harm, it is known as an **invasive species.** Generally, a foreign competitor or predator harms local organisms more than do native competitors or predators. Most invasive species lack natural agents (such as parasites, predators, and competitors) that would otherwise control them. Also, without a shared evolutionary history, most native species typically are less equipped to cope with invasive species. Although foreign species sometimes spread into new areas on their own, humans are

KEY EXPERIMENT

How does habitat fragmentation affect predation of juvenile bay scallops in a marine environment?

HYPOTHESIS: Predation of juvenile bay scallops is greater in patchy areas of sea grass than in undisturbed (continuous) seagrass meadows.

EXPERIMENT: Predation rates of juvenile bay scallops were measured over a four-week period in three seagrass environments: continuous, patchy, and very patchy.

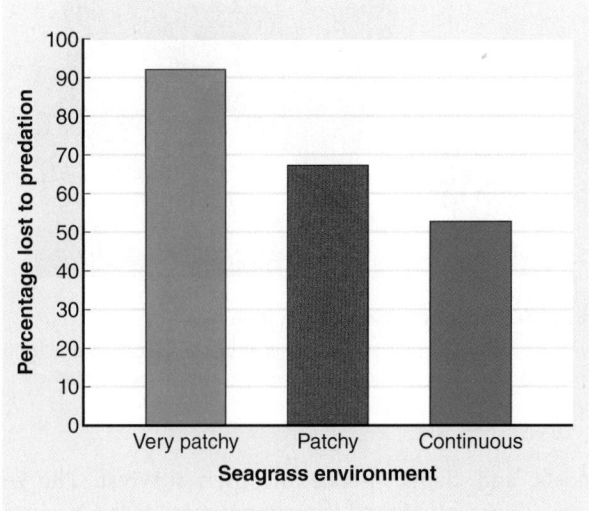

RESULTS AND CONCLUSION: The lowest rate of predation (52%) occurred in the continuous seagrass meadow, and the highest predation rate (92%) occurred in the very patchy environment. The patchy (fragmented) areas provided the predators with easier access to the young scallops than did the continuous area. It was also more difficult for the predators to intrude from surrounding areas into the continuous seagrass meadow.

SOURCE: Irlandi, E. A., W. G. Ambrose, and B. A. Orando. "Landscape Ecology and the Marine Environment: How Spatial Configuration of Seagrass Habitat Influences Growth and Survival of the Bay Scallop." *Oikos*, Vol. 72, 1995.

Figure 57-6 Effects of habitat fragmentation in a shallow marine environment

PREDICT What is likely to happen to bay scallop populations if seagrass environments continue to deteriorate and become more fragmented?

usually responsible for such introductions, either knowingly or accidentally.

One of North America's greatest biological threats is the zebra mussel, a native of the Caspian Sea. It was probably introduced by a foreign ship that flushed ballast water into the Great Lakes in 1985 or 1986. Since then, the tiny freshwater mussel, which clusters in extraordinary densities, has massed on hulls of boats, piers, buoys, water intake systems, and, most damaging of all, on native clam and mussel shells. The zebra mussel's large appetite for algae, phytoplankton, and zooplankton is also cutting into the food supply of native fishes,

Figure 57-7 Zebra mussels clog a pipe
Thumbnail-sized zebra mussels (*Dreissena polymorpha*) have caused billions of dollars in damage in addition to displacing native clams and mussels.

mussels, and clams, threatening their survival. The zebra mussel is currently found throughout most of the Mississippi River and its tributary rivers. According to the U.S. Coast Guard, the United States spends about $5 billion each year to control the spread of the zebra mussel and to repair damage such as clogged pipes (**FIG. 57-7**).

Islands are particularly susceptible to the introduction of invasive species. In Hawaii the introduction of sheep has imperiled both the mamane tree (because the sheep eat it) and a species of honeycreeper, an endemic bird that relies on the tree for food. Hawaii's plants evolved in the absence of herbivorous mammals and therefore have no defenses against introduced sheep, pigs, goats, and deer.

Other human activities affect biodiversity directly or indirectly Sometimes species become endangered or extinct as a result of deliberate efforts to eradicate or control their numbers, often because they prey on game animals or livestock. In the past, ranchers, hunters, and government agents decimated populations of large predators such as the wolf, mountain lion, and grizzly bear. Some animals are killed because their lifestyles cause problems for humans. The Carolina parakeet, a beautiful green, red, and yellow bird endemic to the southeastern United States, was extinct by 1920, exterminated by farmers because it ate fruit from their trees.

Unregulated hunting, or overhunting, has caused the extinction of certain species in the past but is now strictly controlled in most countries. The passenger pigeon was one of the most common birds in North America in the early 1800s, but a century of overhunting resulted in its extinction in the early 1900s.

Illegal commercial hunting, or *poaching,* endangers larger animals such as the tiger, cheetah, and snow leopard, whose beautiful furs are quite valuable. Rhinoceroses are slaughtered for their horns (used for ceremonial dagger handles in the Middle East and for purported medicinal purposes in Asian medicine). Elephants are poached for their ivory tusks. Bears are killed for their gallbladders (which Asian doctors use to treat ailments ranging from indigestion to hemorrhoids). Bushmeat—meat from wild animals, including rare primates, elephants, anteaters, and great apes—is sold to urban restaurants. Although laws protect these animals, demand for their products on the black market has promoted illegal hunting, particularly in impoverished countries where a sale of contraband products can support a family for months.

Shark finning, removing a shark's fin and throwing the animal back in the ocean to die, has caused a third of the world's shark species to become endangered, according to the International Union for the Conservation of Nature (IUCN). The fins are sold to China and other Asian countries for shark fin soup and medicinal purposes. One million sharks were reportedly harvested in 2012. Sharks play an important role in maintaining ecosystems and many are keystone species. Many countries have passed laws banning any trading in shark fins, or restricting harvest of depleted species, but laws have been difficult to enforce and are easy to circumvent. Recently, phylogenetic DNA testing has been developed to identify illegal catches of banned species using samples of confiscated fins and shark soup.

Commercial harvest is the collection of live organisms from nature. Most commercially harvested organisms end up in zoos, aquaria, biomedical research labs, circuses, and pet stores. For example, several million birds are commercially harvested each year for the pet trade, but many die in transit, and many more die from improper treatment in their owners' homes. At least 40 parrot species are now threatened or endangered, in part because of commercial harvest. Although it is illegal to capture endangered animals from the wild, a thriving black market exists, mainly because collectors in the United States, Europe, and Japan pay large sums for rare tropical birds (**FIG. 57-8**).

Commercial harvest also threatens plants. Many unique or rare plants have been so extensively collected from the wild that they are now classified as endangered. Included are certain carnivorous plants, wildflower bulbs, cacti, and orchids. In contrast, carefully monitored and regulated commercial use of animal and plant resources creates an economic incentive to ensure that these resources do not disappear.

CHECKPOINT 57.1

- *What are the three levels of biological diversity?*
- **PREDICT** *Which organism is more likely to become extinct: an endangered species or a threatened species? Explain your answer.*
- **CONNECT** *How does habitat fragmentation contribute to declining biological diversity?*
- **CONNECT** *How do invasive species contribute to the biodiversity crisis?*

Figure 57-8 Illegal commercial harvesting

These hyacinth macaws (*Anodorhynchus hyacinthus*) were seized in French Guiana in South America as part of the illegal animal trade there. The hyacinth macaw population has been seriously reduced in South America.

57.2 CONSERVATION BIOLOGY

LEARNING OBJECTIVES

4   Define *conservation biology* and compare in situ and ex situ conservation measures.

5   Describe the benefits and shortcomings of the U.S. Endangered Species Act and the Convention on International Trade in Endangered Species of Wild Flora and Fauna.

Conservation biology is the scientific study of how humans impact organisms and of the development of strategies to protect biological diversity. Conservation biologists develop models, design experiments, and perform fieldwork to address a wide range of questions. For example, what are the processes that influence a decline in biological diversity? How do we protect and restore populations of endangered species? If we are to preserve entire ecosystems and landscapes, which ones are the most important to save?

Conservation biologists have determined that a single large area of habitat capable of supporting several populations is more effective at safeguarding an endangered species than several habitat fragments, each capable of supporting a single population. Also, a large area of habitat typically supports greater species richness than several habitat fragments.

Conservation is more successful when habitat areas for a given species are in close proximity rather than far apart. If an area of habitat is isolated from other areas, individuals may not effectively disperse from one habitat to another. Because the presence of humans adversely affects many species, habitat areas that lack roads or are inaccessible to humans are better than human-accessible areas.

According to conservation biologists, it is more effective and, ultimately, more economical to preserve intact ecosystems in which many species live than to try to preserve individual species. Conservation biologists generally consider it a higher priority to preserve areas with greater biological diversity.

Conservation biology includes two problem-solving approaches that save organisms from extinction: in situ and ex situ conservation. **In situ conservation,** which includes the establishment of parks and reserves, concentrates on preserving biological diversity in nature. A high priority of in situ conservation is identifying and protecting sites that harbor a great deal of diversity.

With increasing demands on land, in situ conservation cannot preserve all types of biological diversity. Sometimes only ex situ conservation can save a species. **Ex situ conservation** conserves individual species in human-controlled settings. Breeding captive species in zoos and storing seeds of genetically diverse plant crops are examples of ex situ conservation.

In situ conservation is the best way to preserve biological diversity

Protecting animal and plant habitats—that is, conserving and managing ecosystems as a whole—is the single best way to protect biological diversity. Many nations have set aside areas for wildlife habitats. Such natural ecosystems offer the best strategy for the long-term protection and preservation of biological diversity. Currently, more than 160,000 national parks, marine sanctuaries, wildlife refuges, forests, and other areas are protected throughout the world. Nearly 15% of the world's land area is protected, but only 1.7% of ocean area is set aside.

Unfortunately, many protected areas have multiple uses that sometimes conflict with the goal of preserving species. National parks provide for recreational needs, for example, whereas national forests are used for logging, grazing, and mineral extraction. The mineral rights to many wildlife refuges are privately owned, and oil, gas, and other mineral development has occurred on some wildlife refuges.

Protected areas are not always an effective strategy to preserve biological diversity, particularly in developing countries where biological diversity is greatest, because there is little money or expertise to manage them. Another shortcoming of the world's protected areas is that many are in lightly populated mountain areas, tundra, and the driest deserts, places that often have spectacular scenery but relatively few kinds of species. In reality, such remote areas are often designated reserves because they are unsuitable for commercial development.

In contrast, ecosystems in which biological diversity is greatest often receive little attention. Protected areas are urgently needed in tropical rain forests, the tropical grasslands and savannas of Brazil and Australia, and dry forests that are widely scattered around the world. Desert organisms are underprotected in northern Africa and Argentina, and the species of many islands and temperate river basins also need protection.

Many of these unprotected areas are part of what some biologists have identified as the world's 25 **biodiversity hotspots** (FIG. 57-9). The hotspots collectively make up 1.4% of Earth's land but contain as many as 44% of all vascular plant species, 29% of the world's endemic bird species, 27% of endemic mammal species, 38% of endemic reptile species, and 53% of endemic amphibian species. More than 20% of the world's human population lives in the hotspots. Fifteen of the 25 hotspots are tropical, and 9 are mostly or solely islands. The 25 hotspots are somewhat arbitrary because additional hotspots are recognized when different criteria are used.

Not all conservation biologists support focusing limited conservation resources on protecting biodiversity hotspots because doing so ignores the many **ecosystem services** provided by entire ecosystems that are rapidly vanishing due to habitat destruction. For example, the vast boreal forests in Canada and Russia are critically important to the proper functioning of the global carbon and nitrogen cycles. Conservation organizations do not recognize them as hotspots, however, and these forests are being logged at unprecedented rates. Landscape ecology takes a larger view of ecosystem conservation.

Landscape ecology considers ecosystem types on a regional scale The subdiscipline of ecology that studies the connections in a heterogeneous **landscape** consisting of multiple interacting ecosystems is known as **landscape ecology**. Increasingly, biologists are focusing efforts on preserving biodiversity in landscapes. What, though, is the minimum ecosystem/landscape size needed to preserve species numbers and distributions?

One long-term study addressing this question in tropical rain forests is the Biological Dynamics of Forest Fragment Project (formerly the Minimum Critical Size of Ecosystems Project), which is studying a series of Amazonian rainforest fragments varying in size from 1 to 100 hectares (One hectare is equal to 2.471 acres.) (FIG. 57-10). Preliminary data from this study, which began in 1979, indicate that smaller forest fragments do not maintain their ecological integrity. For example, large trees

---

KEY POINT

Biodiversity hotspots are areas of the world that are critically important because they contain a disproportionate number of the world's endemic species.

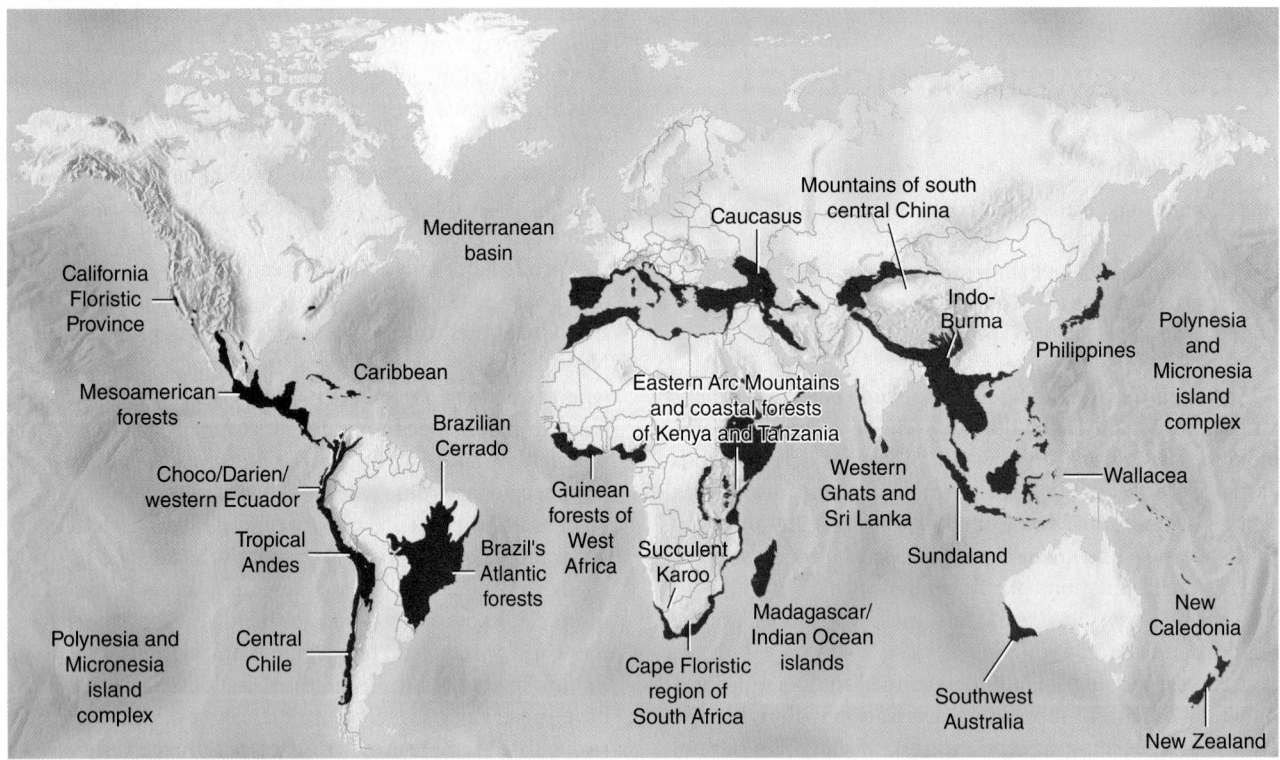

Figure 57-9 *Animation* **Biodiversity hotspots**

Rich in endemic species, these hotspots are under pressure from the number of humans living in them. (Adapted from Miller, G. T. and S. E. Spoolman. *Living in the Environment*, 17th ed., 2012, p. 243, Figure 10-27. Brooks Cole/Cengage Learning. Data from Center for Applied Biodiversity Science at Conservation International.)

PREDICT How could a large human population in a hotspot area be turned into an advantage?

Figure 57-10 Biological Dynamics of Forest Fragment Project

Shown are 1-hectare and 10-hectare plots of a long-term (about three decades) study under way in Brazil on the effects of fragmentation on Amazonian rain forest. Plots with an area of 100 hectares are also under study, along with identically sized sections of intact forest, which are controls.

near forest edges often die or are damaged from exposure to wind, desiccation (from lateral exposure of the forest fragment to sunlight), and invasion by parasitic woody vines. In addition, biologists have documented that various species adapted to forest interiors do not prosper in the smaller fragments and eventually die out, whereas species adapted to forest edges invade the smaller fragments and thrive.

Unfortunately, human colonization and other activities are threatening the Biological Dynamics of Forest Fragment Project itself. Recently, a 1,100-km (683.1 mi) road was paved nearby, expanding agriculture, logging, and hunting opportunities. Should human expansion spread into the research site, the potential benefits that might be derived from understanding the effects of forest fragmentation will be lost.

To remedy widespread habitat fragmentation and its associated loss of biodiversity, conservation biologists have proposed linking isolated fragments with **habitat corridors,** strips of habitat connecting isolated habitat patches. Habitat corridors allow wildlife to move about so they can feed, mate, and recolonize habitats after local extinctions take place. Research has also shown that habitat fragments linked by habitat corridors retain more native plant species than do isolated fragments.

Habitat corridors range from a local scale (such as an overpass that allows wildlife to cross a road safely) to a landscape scale that connects separate reserves. The minimum corridor width for landscape-scale corridors varies depending on the size of home ranges for various species. For example, to link two reserves and protect wolves in Minnesota, the corridor must have a minimum width of 12 km (7.2 mi), whereas the minimum corridor width for bobcats in South Carolina must be 2.5 km (1.5 mi).

The United Nations Educational, Scientific, and Cultural Organization (UNESCO) Program on Man and the Biosphere (MAB) has established **biosphere reserves** to protect biodiversity. Each reserve consists of three zones: core, buffer, and transition. The core zone contains a protected ecosystem and allows for nondestructive research and education. The surrounding buffer zone is used for environmental education, ecotourism, recreation, and research. The transition zone consists of local farms, fisheries, and towns where stakeholders collaborate to oversee and sustainably develop the biosphere's resources. Maintaining biodiversity is a major principle of environmental sustainability.

Restoring damaged or destroyed habitats is the goal of restoration ecology Increasingly, countries are reclaiming disturbed lands and converting them into areas with high biological diversity. **Restoration ecology,** in which the principles of ecology are used to return a degraded environment to one that is more functional and sustainable, is an important part of in situ conservation. Costa Rica is an important success story. By the 1980s, ranchers had cleared so much forest to provide land for grazing and farmers had cleared so much land for growing coffee trees that only about 20% of Costa Rica's tropical forests remained. Then, this small, but ecologically important, country made a commitment to restore and protect its rich biodiversity. Economic and other incentives were offered for restoring and protecting its lands. Once again, forests cover more than 50% of its land area and Costa Rica has become a global leader in environmental sustainability (**FIG. 57-11**).

One of the oldest and most extensive ecological restoration projects in the United States was undertaken in the early 1930s by the University of Wisconsin–Madison Arboretum. The university ecologists have developed several distinct natural communities on damaged agricultural land (**FIG. 57-12**). These communities include a tallgrass prairie, a xeric (dry) prairie, savannas, and several types of pine and maple forests native to Wisconsin. More than 300 species have been restored.

Restoration of disturbed lands not only creates biological habitats but also has additional benefits, such as the regeneration of soil that agriculture or mining damaged. Restoration ecology is an important aspect of conservation biology.

Ex situ conservation attempts to save species on the brink of extinction

Zoos, aquaria, and botanical gardens are valuable because they provide the public with informal learning and education opportunities to understand the importance of biodiversity and conservation. These facilities work to save certain endangered species from extinction. Eggs are collected from nature, or the remaining few animals are captured and bred in zoos and other research facilities.

Artificial insemination and host mothering can be used to increase the number of offspring. In **artificial insemination** sperm can be collected from a suitable male of a rare species and used to impregnate a selected female, which may be located in another zoo, or even another country. In **host mothering** a female of a rare species is treated with fertility drugs, which cause her to produce multiple eggs. Eggs are collected, fertilized with sperm, and surgically implanted into females of a related but less rare species, which later give birth to offspring of the rare species (see *Inquiring About: Novel Origins,* in Chapter 50). Plans are under way to clone endangered species, such as the giant panda, that do not reproduce well in captivity. Hormone patches are being developed to stimulate reproduction in endangered birds; the patch is attached under the female bird's wing.

A few spectacular successes have occurred in captive-breeding programs, in which large enough numbers of a species have been produced to re-establish small populations in the wild. Conservation efforts, for example, let the bald eagle make a remarkable comeback in the contiguous states, and in 2007, the U.S. Fish and Wildlife Service (FWS) removed the bald eagle from the endangered list.

Attempting to save a species on the brink of extinction is usually expensive, and only a small proportion of endangered species can be saved. Moreover, zoos, aquaria, and botanical gardens do not have the space to try to save all endangered species. Conservation biologists must therefore set priorities on which species to attempt to save. Zoos have traditionally focused on large, charismatic animals, such as pandas, bald eagles, and whooping cranes, because the public is interested in them. However, such conservation efforts ignore millions of ecologically important species. Clearly, controlling unchecked human development so that species do not become endangered in the first place is a more effective way to protect and maintain natural habitat.

The Endangered Species Act provides some legal protection for species and habitats

In 1973, the *Endangered Species Act (ESA)* was passed in the United States, authorizing the FWS to protect endangered and threatened species in the United States and abroad. Most other countries now have similar legislation. The FWS conducts a detailed study of a species to determine if it should be listed as endangered or threatened (TABLE 57-1). The ESA provides legal protection to listed species, reducing their danger of extinction. For example, the act makes it illegal to sell or buy any product made from an endangered or threatened species.

The ESA requires officials of the FWS to select critical habitats and design a recovery plan for each species listed. The recovery plan includes an estimate of the current population size, an

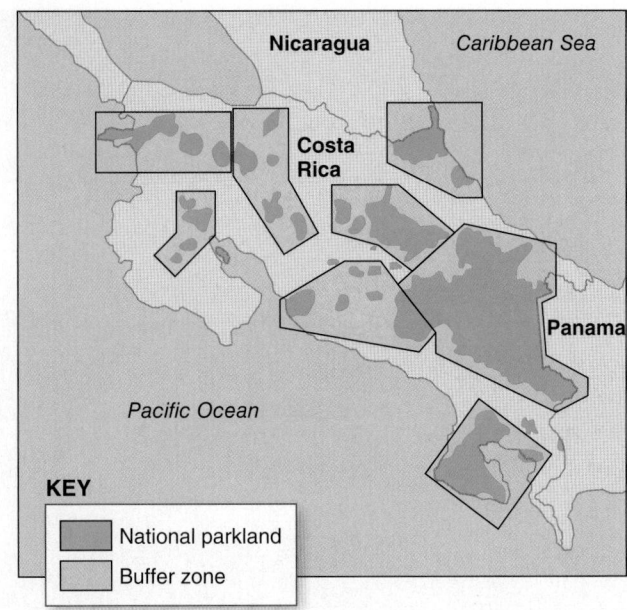

Figure 57-11 Restoration ecology and biodiversity conservation in Costa Rica

Costa Rica has won international awards for restoring its tropical forests and protecting its rich biodiversity. Its protected natural parklands (*green*) are surrounded by buffer zones (*yellow*), which are used for *sustainable* agriculture, cattle grazing, forestry, logging, fishing, and ecotourism. (Adapted from Miller, G. T. and S. E. Spoolman. *Living in the Environment,* 17th ed., 2012, p. 241, Figure 10-26. Brooks Cole/Cengage Learning.)

analysis of what factors contributed to its endangerment, and a list of activities that may help the species recover.

The ESA is considered one of the strongest pieces of environmental legislation in the United States, in part because species are designated as endangered or threatened entirely on biological grounds. Economic considerations cannot influence the designation of endangered or threatened species. Biologists generally agree that as a result of passage of the ESA in 1973, fewer species became extinct than would have had the law not been passed.

The ESA is also a very controversial piece of environmental legislation. For example, the ESA does not provide compensation for private property owners who suffer financial losses because they cannot develop their land if a threatened or endangered species lives there. The ESA has also interfered with some federally funded development projects.

It currently takes about 15 years for a candidate species with declining numbers to be listed as threatened or endangered. This backlog is due to both the high cost of evaluating each species and the limited funds allocated to the FWS for these studies. Meanwhile, more than 80 species on the ESA's "candidate list" have become extinct since 1973, without ever being classified as threatened or endangered.

(a) The restoration of the prairie by the University of Wisconsin–Madison Arboretum was at an early stage in 1935. The men are digging holes to plant prairie grass sod.

(b) The prairie as it looks today. This picture was taken at about the same location as the 1935 photograph.

Figure 57-12 Restoring damaged lands

TABLE 57-1	Organisms Listed as Endangered or Threatened in the United States, 2012	
TYPE OF ORGANISM	NUMBER OF ENDANGERED SPECIES	NUMBER OF THREATENED SPECIES
Mammals	69	16
Birds	78	15
Reptiles	14	22
Amphibians	16	11
Fishes	80	51
Snails	33	13
Clams	72	12
Crustaceans	20	3
Insects	55	10
Spiders	12	0
Flowering plants	668	150
Conifers and cycads	2	1
Ferns, other plants	28	2
Corals	0	2
Lichens	2	0

Source: U.S. Fish and Wildlife Service.

The ESA is geared more to saving a few popular or unique endangered species than to saving the much larger number of less popular species that perform valuable ecosystem services. About one-third of the annual funding for the ESA is used to help just ten species. Less glamorous plants, fungi, bacteria and archaea, and insects, however, play central roles in ecosystems and contribute. Prokaryotes and fungi, for example, provide the critically important ecosystem service of breaking down dead organic materials into simple substances (CO_2, water, and minerals) that are subsequently recycled to plants and other autotrophs.

Conservation biologists and numerous conservation organizations support a stronger ESA to manage whole ecosystems and maintain complete biological diversity rather than attempt to save endangered species as isolated entities. This approach offers collective protection to many declining species rather than to a few specific species.

International agreements provide some protection for species and habitats

The *Convention on International Trade in Endangered Species of Wild Flora and Fauna (CITES)* is an international agreement that went into effect in 1975. Originally drawn up to protect endangered animals and plants considered valuable in the highly lucrative international wildlife trade, CITES bans hunting, capturing, and selling of endangered or threatened species and regulates trade of organisms listed as potentially threatened. Unfortunately, enforcement of this treaty varies from country to country, and even where enforcement exists, the penalties are minimal. As a result, illegal trade in rare, commercially valuable species continues.

The goals of CITES often stir up controversy over such issues as who actually owns the world's wildlife and whether global conservation concerns take precedence over competing local interests. These conflicts often highlight socioeconomic differences between wealthy consumers of products protected by CITES and poor people who trade the endangered organisms.

Another international treaty, the *Convention on Biological Diversity,* requires that each signatory nation inventory its own biodiversity and develop a **national conservation strategy,** a detailed plan for managing and preserving the biological diversity of that specific country.

CHECKPOINT 57.2

- **PREDICT** *Imagine that conservation biology did not exist. What would be the effect on biodiversity?*
- **CONNECT** *Are landscape biology and restoration ecology examples of in situ or ex situ conservation?*
- *Which type of conservation measure, in situ or ex situ, helps the greatest number of species? Why?*
- **CONNECT** *Why is the U.S. Endangered Species Act so controversial?*

57.3 DEFORESTATION

LEARNING OBJECTIVES

6 Discuss the ecosystem services of forests and describe the consequences of deforestation.

7 State at least three reasons forests (tropical rain forests and boreal forests) are disappearing today.

The most serious problem facing the world's forests and their biological diversity is **deforestation,** the temporary or permanent clearance of large expanses of forest for agriculture or other uses (**FIG. 57-13**). According to the UN Food and Agriculture Organization, between 2000 and 2010 forests declined by about 5.2 million hectares each year (an area equal to the size of Costa Rica).

When forests are destroyed, their valuable ecosystem services are no longer available to the environment or to people who depend on them. Deforestation increases soil erosion and thus decreases soil fertility. Soil erosion causes increased sedimentation of waterways, which harms downstream aquatic ecosystems by reducing light penetration, covering aquatic organisms, and filling in waterways. Uncontrolled soil erosion, particularly on steep deforested slopes, causes mudflows that

Figure 57-13 *Animation* **Deforestation**
Aerial view of clear-cut areas in the Gifford Pinchot National Forest in southwestern Washington State. The lines are roads built at taxpayer expense to haul away logs.

endanger human lives and property and reduces production of hydroelectric power as silt builds up behind dams. In drier areas deforestation can lead to the formation of deserts.

Deforestation contributes to the loss of biological diversity. Many species have limited ranges within a forest, particularly in the tropics, making these species especially vulnerable to habitat destruction or modification. Migratory species, such as birds and butterflies, also suffer from tropical deforestation.

By trapping and absorbing precipitation, forests on hillsides and mountains help protect nearby lowlands from floods. When a forest is cut down, the watershed cannot absorb and hold water as well, and the total amount of surface runoff flowing into rivers and streams increases. This increased runoff not only causes soil erosion, but also puts lowland areas at extreme risk of flooding.

Deforestation may affect regional and global climate changes. Transpiring trees release substantial amounts of moisture into the air. This moisture falls back to the surface in the hydrologic cycle. When a large forest is removed, rainfall may decline, and droughts may become common in that region. Studies suggest that the local climate has become drier in parts of Brazil where tracts of the rain forest have been burned. Temperatures rise slightly in a deforested area because there is less evaporative cooling from the trees.

Deforestation increases global temperature by releasing carbon stored in the trees into the atmosphere as carbon dioxide, which enables the air to retain heat. The carbon in forests is released immediately if the trees are burned or more slowly when unburned parts decay. If trees are harvested and logs are removed, roughly one-half of the forest carbon remains as dead materials (branches, twigs, roots, and leaves) that decompose, releasing carbon dioxide. When an old-growth forest is harvested, it may take 200 years for the replacement forest to accumulate the amount of carbon that was stored in the original forest.

Why are tropical rain forests continuing to disappear?

Most of the remaining undisturbed tropical rain forests, in the Amazon and Congo River basins of South America and Africa, continue to be cleared and burned at a rate unprecedented in human history. Tropical rain forests are also being rapidly destroyed in southern Asia, Indonesia, Central America, and the Philippines.

Several studies show a strong statistical correlation between population growth and deforestation. More people need more food and fuel, so they clear forests for agricultural expansion and burn wood for personal use. However, tropical deforestation cannot be attributed simply to population pressures. The main causes of deforestation vary from place to place, and a variety of economic, social, and governmental factors interact to cause deforestation. Government policies sometimes provide incentives that favor the removal of forests. For example, in the late 1950s the Brazilian government constructed the Belem–Brasilia Highway, which cut through the Amazon Basin and opened the Amazonian frontier for settlement (FIG. 57-14). Sometimes economic conditions encourage deforestation. The farmer who converts more forest to pasture can maintain a larger herd of cattle, which is a good hedge against inflation.

If we keep in mind that tropical deforestation is a complex problem, three agents are probably the most immediate causes of deforestation in tropical rain forests: subsistence agriculture, commercial logging, and cattle ranching. Other reasons for the destruction of tropical forests include the development of hydroelectric power, which inundates large areas of forest; mining, particularly when ore smelters burn charcoal produced from rainforest trees; and plantation-style agriculture of crops such as sugarcane, bananas, and palm oil.

Subsistence agriculture, in which a family produces enough food to feed itself, accounts for nearly 50% of tropical deforestation. In many developing countries where tropical rain forests are located, many people do not own the land on which they live and work and must clear the forest to grow food. Land reform in Brazil, Madagascar, Mexico, the Philippines, Thailand, and many other countries is slowly easing some of the pressure of subsistence farmers on tropical forests as deforestation is decoupled from economic growth.

Subsistence farmers often follow loggers' access roads, cut down the trees, burn the area, and plant crops immediately. This method is known as **slash-and-burn agriculture.** Yields from the first crop are often quite high because the nutrients that were in the trees are now available in the soil. In a few years, however, soil productivity declines, and the farmer must move to a new part of the forest and repeat the process. Cattle ranchers often claim the abandoned land because it can still support livestock.

Slash-and-burn agriculture carried out on a small scale, with periods of 20 to 100 years between cycles, is sustainable. The forest regrows rapidly after a few years of farming. However, when millions of people try to obtain a living in this way, the land is not allowed to lie uncultivated long enough to recover. Globally, at least 180 million subsistence farmers obtain a living from slash-and-burn agriculture.

About 14% of tropical deforestation is the result of commercial logging, and vast tracts of tropical rain forests, particularly in Southeast Asia, are harvested for export abroad. Some tropical countries still allow commercial logging to proceed much faster than is sustainable because it supplies them with much-needed revenues. In the final analysis, uncontrolled tropical deforestation does not contribute to economic development; rather, it reduces or destroys the value of an important natural resource.

Approximately 10% of tropical rainforest destruction is carried out to provide open rangeland for cattle. Cattle ranching is particularly important in Central America. Much of the beef raised on these ranches, which foreign companies often own, is exported to restaurant chains in North America and Europe. After the forests are cleared, cattle graze on the land for as long as 20 years, after which time the soil fertility is depleted. When that occurs, shrubby plants, or *scrub savanna*, take over the range.

Why are boreal forests disappearing?

Tropical rain forests are not the only forests at risk from deforestation. Extensive logging of certain boreal forests began in the late 1980s and continues today. Coniferous evergreen

Figure 57-14 Human settlements in Brazil's tropical rain forest
This satellite photograph shows numerous smaller roads extending perpendicularly from the main roads. As people settle along the roads, they clear out more and more forest (*dark green*) for their croplands and pastures (*tans and pinks*).

NRSC/Science Source

trees such as spruce, fir, cedar, and hemlock dominate these northern forests of Alaska, Canada, Scandinavia, and northern Russia.

Boreal forests, harvested primarily by clear-cut logging, are currently the primary source of the world's industrial wood and wood fiber. More than 1 million hectares (2.5 million acres) of Canadian forests are logged annually, and most of Canada's forests are under logging tenures. (*Tenures* are agreements between provinces and companies that give companies the right to cut timber.) Canada is the world's biggest timber exporter, and most of its forest products are exported to China and the United States. In 2010, Canada reached an agreement between loggers and environmentalists to protect a huge part of the Canadian boreal forest, 1.4 billion acres of wilderness containing mostly never-harvested woodland. Extensive tracts of boreal forests in Russia are also harvested, although exact estimates are unavailable. Alaska's boreal forests are also at risk because the U.S. government may increase logging on public lands there in the near future.

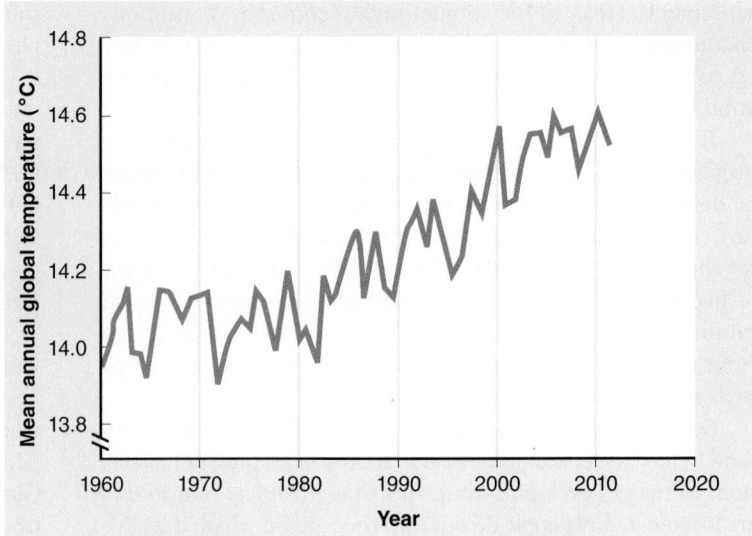

Figure 57-15 Mean annual global temperature, 1960 to 2011

Data are presented as surface temperatures (°C) for 1960, 1965, and every year thereafter. The measurements, which naturally fluctuate, clearly show the warming trend of the last several decades. (Data from Surface Air Temperature Analysis, Goddard Institute for Space Studies, NASA.)

CHECKPOINT 57.3

- *What are three ecosystem services that forests provide?*
- *What are two reasons for deforestation in tropical rain forests? What is the main reason for deforestation of boreal forests?*

57.4 CLIMATE CHANGE

LEARNING OBJECTIVES

8 Name at least three greenhouse gases and explain how greenhouse gases contribute to climate change.

9 Describe how climate change may affect sea level, precipitation patterns, organisms (including humans), and food production.

Earth's average temperature is based on daily measurements from several thousand land-based meteorological stations around the world as well as data from weather balloons, orbiting satellites, transoceanic ships, and hundreds of sea-surface buoys with temperature sensors. *Data indicate that the ten warmest years since the mid-1800s have occurred between 1998 and 2012.* The 1990s was the warmest decade of the 20th century, and the early 2000s continued the warming trend (FIG. 57-15). Earth's rapidly changing climate imposes stress on many living organisms and is already having a negative effect on biological diversity.

Scientists worldwide have studied **climate change** for several decades. As the evidence has accumulated, a strong consensus was reached that the 21st century will experience significant rapid climate change and that this change has been caused mainly by human activities. Governments around the world organized the United Nations' Intergovernmental Panel on Climate Change (IPCC) in 1988 to review all published literature about global climate change, especially recent research, and provide an update of knowledge on the scientific, technical, and socioeconomic aspects of climate change. Observed and projected effects of climate change in each IPCC report have been progressively worse. The 2013 IPCC Fifth Assessment Report concluded that human-produced air pollutants caused most of the climate change and increased ocean water temperatures observed in the prior 55 years. The IPCC report projects global surface temperature increase is likely to exceed 1.5 to 2.0°C (2.7 to 3.6°F) by the year 2100. Additionally, the global ocean will continue to warm, which will affect ocean circulation, and will undergo further acidification.

Sea levels will rise more rapidly during the 21st century (compared with observations from 1970–2010) as glaciers and ice sheet shrinking will continue to rapidly increase. Thus, Earth will likely become warmer during the 21st century than it has been for several million years.

Greenhouse gases cause climate change

The World Meteorological Organization recently reported a 29% increase in the warming effect of greenhouse gases between 1990 and 2010; carbon dioxide accounted for 80% of this increase. Carbon dioxide (CO_2), methane (CH_4), and nitrous oxide (N_2O) are the three most prevalent and long-lived greenhouse gases, excluding water vapor. Surface ozone (O_3) and hydrochlorofluorocarbons (HCFCs) also continue to accumulate in the atmosphere as a result of human activities (TABLE 57-2). The concentration of atmospheric CO_2 has increased 39% since the start of the industrial era in 1750, recorded at 400 parts per million in 2013 (FIG. 57-16). Burning carbon-containing fossil fuels—coal, oil, and natural gas—accounts for about three-fourths of human-made carbon dioxide emissions to the atmosphere. Land conversion, such as when forests are logged or burned, also releases carbon dioxide. Trees normally remove carbon dioxide from the atmosphere during photosynthesis, but tree removal prevents this process.

TABLE 57-2	Changes in Selected Atmospheric Greenhouse Gases, Preindustrial Times to Present	
GAS	ESTIMATED PREINDUSTRIAL CONCENTRATION	PRESENT CONCENTRATION
Carbon dioxide	288 ppm*	400 ppm
Methane	848 ppb†	1866 ppb
Nitrous oxide	285 ppb	324 ppb
Tropospheric ozone	25 ppb	34 ppb
CFC-12	0 ppt‡	530 ppt
CFC-11	0 ppt	237 ppt

Note: The preindustrial value is for the 17th and 18th centuries.

*ppm = parts per million

†ppb = parts per billion

‡ppt = parts per trillion

Source: Carbon Dioxide Information Analysis Center, Environmental Sciences Division, Oak Ridge National Laboratory.

Climate change occurs because greenhouse gases absorb infrared radiation (heat) in the atmosphere. This absorption slows the natural heat flow into space, warming the lower atmosphere. Some of the heat from the lower atmosphere is transferred to the ocean and raises its temperature as well. This atmospheric retention of heat is a natural phenomenon that has made Earth habitable for its millions of species.

Because carbon dioxide and other gases trap the sun's radiation somewhat like glass does in a greenhouse, the natural trapping of heat in the atmosphere is called the **greenhouse effect,** and the gases that absorb infrared radiation are known as **greenhouse gases.** This natural heating of the atmosphere prevents Earth from becoming a frozen planet. However, the

additional warming produced when increased levels of gases produced by human activities absorb additional infrared radiation, called the **enhanced greenhouse effect,** has potentially catastrophic implications as atmospheric and ocean temperatures continue to rise (FIG. 57-17).

Although current rates of fossil fuel combustion and deforestation are high, causing the carbon dioxide level in the atmosphere to increase markedly, scientists think that the warming trend is slower than the increasing level of carbon dioxide might indicate. The reason is that water requires more heat to raise its temperature than gases in the atmosphere do. (Recall the high specific heat of water discussed in Chapter 2.) As a result, the ocean takes longer to warm than the atmosphere. Most climate scientists think that warming will be more pronounced in the second half of the 21st century than in the first half.

Methane gas, the second most prevalent greenhouse gas, has increased 158% since 1750, mostly due to cattle grazing, rice planting, fossil fuel exploitation, and landfills. Human activities currently account for 60% of methane emissions, whereas nearly 40% of emissions are naturally occurring. The natural methane emissions are also growing and are expected to increase still more as northern permafrost thaws and tropical wetlands emissions increase.

Nitrous oxide (N_2O) is now the third most important greenhouse gas and is found at levels 20% higher than in 1750. Its concentration is increasing rapidly due to use of nitrogen containing fertilizers and manure. The effect of nitrous oxide over the past century is nearly 300 times greater than equal emissions of carbon dioxide and is contributing to the destruction of the stratospheric ozone layer, which protects us from harmful solar ultraviolet rays. (See *Inquiring About: Stratospheric Ozone Depletion from Industrial Chemicals.*)

What are the probable effects of climate change?

We now consider some of the probable effects of climate change, including changes in sea level; changes in precipitation patterns; effects on biological diversity, including humans; and effects on agriculture. These changes will persist for centuries because many greenhouse gases remain in the atmosphere for hundreds of years. Furthermore, even after greenhouse gas concentrations have stabilized, scientists think that Earth's mean surface temperature will continue to rise because the ocean adjusts to climate change on a delayed time scale.

With climate change, the global average sea level is rising As Earth's overall temperature increases, major thawing of glaciers and the polar ice caps is taking place. In addition to sea-level rise caused by the retreat of glaciers and thawing of polar ice, the sea level will rise due to thermal expansion of the warming ocean. Water, like other substances, expands as it warms.

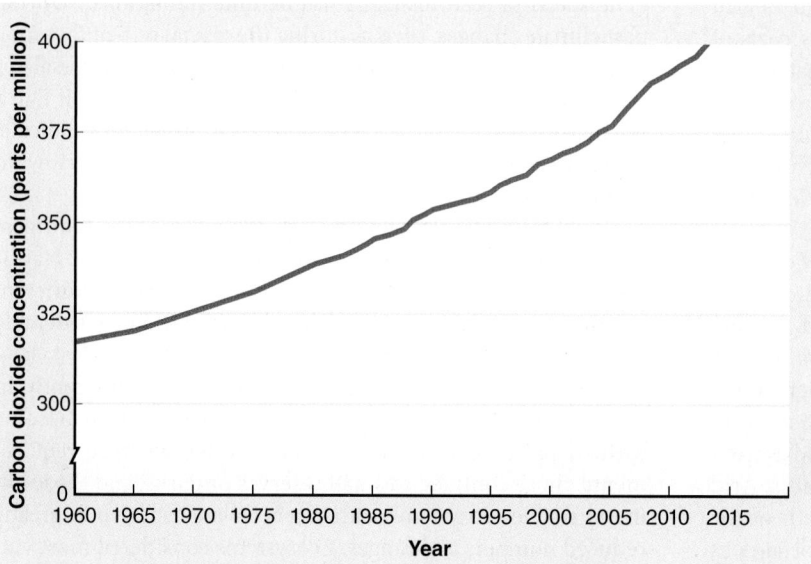

Figure 57-16 Carbon dioxide in the atmosphere, 1960 to 2013

Note the steady increase in the concentration of atmospheric carbon dioxide. Measurements are taken at the Mauna Loa Observatory, Hawaii, far from urban areas where factories, power plants, and motor vehicles emit carbon dioxide. (Data from Scripps Institution of Oceanography, University of California, La Jolla, California.)

Greenhouse gases accumulating in the atmosphere are causing an enhanced greenhouse effect, resulting in climate change. Human activity is largely responsible for these changes.

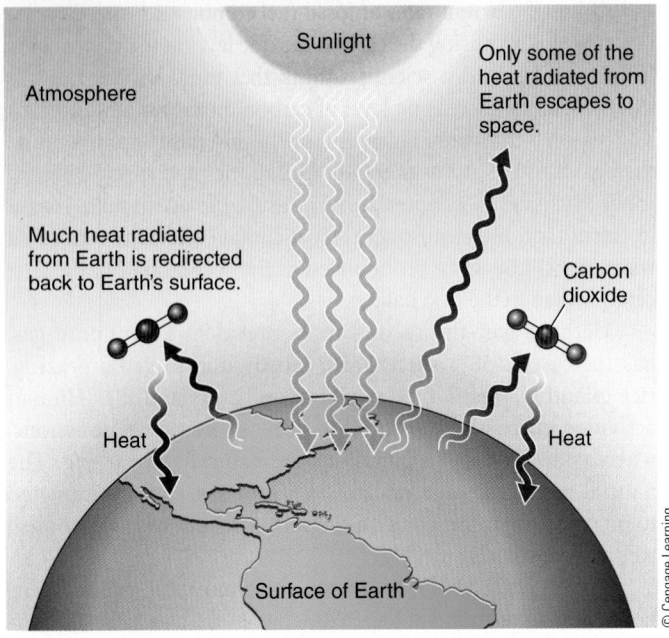

Figure 57-17 *Animation* **Enhanced greenhouse effect**

The buildup of carbon dioxide and other greenhouse gases in the atmosphere absorbs some of the outgoing infrared (heat) radiation and redirects it back to Earth's surface. As a result, the atmosphere, land, and ocean are warming.

CONNECT How can we slow the accumulation of greenhouse gases in the atmosphere?

The IPCC estimates that sea level will rise 18 to 59 cm (0.6 to 1.9 ft) by 2100 and notes that it could be even higher. Such an increase will flood low-lying coastal areas, such as parts of southern Louisiana and south Florida. Coastal areas that are not inundated will likely suffer erosion and other damage from more frequent and more intense weather events such as hurricanes. Countries particularly at risk include Bangladesh, Egypt, Vietnam, Mozambique, and many island nations such as the Maldives.

With climate change, precipitation patterns will change Computer simulations of weather changes from climate warming indicate that precipitation patterns will be altered, causing some areas such as midlatitude continental interiors to have more frequent droughts. At the same time, heavier snowstorms and rainstorms may cause more frequent flooding in other areas. Changes in precipitation patterns could impact the availability and quality of fresh water in many places. Arid or semiarid areas, such as the Sahel region just south of the Sahara Desert, may have the most serious water shortages as the climate warms. Closer to home, water experts predict more water shortages in the western United States because warmer winter temperatures will cause more precipitation to fall as rain rather than snow; melting snow currently provides 70% of stream flows in that region during summer months.

The frequency and intensity of storms over warm surface waters may also increase. A computer model developed by National Oceanic and Atmospheric Administration (NOAA) scientists predicts how climate change may affect hurricanes. When the model was run with a sea-surface temperature 2.2°C warmer than today, more intense hurricanes resulted. (The question of whether hurricanes will occur *more frequently* in a warmer climate continues to be studied.) Increases in storm frequency and intensity are expected because as the atmosphere warms, more water evaporates, which in turn releases more energy into the atmosphere. (Recall the discussion of water's heat of vaporization in Chapter 2.) This energy generates unusually powerful storms.

With climate change, the ranges of organisms are changing Dozens of studies report on the effects of climate change on organisms, from flowering times in plants to breeding times in birds. For example, researchers determined that populations of zooplankton in the California Current have declined 80% since 1951, apparently because the current has warmed slightly. (The California Current flows from Oregon southward along the California coast.) The decline in zooplankton has affected the entire ecosystem's food web, and populations of seabirds and plankton-eating fishes have also declined.

Warmer temperatures in Antarctica—since 1960, the average annual temperature on the Antarctic Peninsula has increased 2.6°C (5°F)—have contributed to reproductive failure in Adélie penguins. The birds normally lay their eggs in snow-free rocky outcrops, but the warmer temperatures have caused increased snowfall (recall that warmer air holds more moisture), which melts when the birds incubate the eggs. The melted snow forms cold pools of slush that kill the developing chick embryos.

Biologists generally agree that climate change will have an especially severe effect on plants, which cannot migrate as quickly as animals when environmental conditions change. (The speed of seed dispersal has definite limitations.) During past climate changes, such as during the glacial retreat that took place some 12,000 years ago, the upper limit of dispersal for tree species was probably 200 km (124 mi) per century. If Earth warms as much during the 21st century as projections indicate, the ranges for some temperate tree species may shift northward as much as 480 km (300 mi).

The U.S. Department of Agriculture reports hardiness zones, based on the average low temperature in an area. Hardiness zones, which indicate which plants will do well at different locations around the country, have already shifted considerably over the past twenty-five years.

Each species reacts to changes in temperature differently. In response to climate change, some species will probably become extinct, particularly those with narrow temperature requirements, those confined to small reserves or parks, and those living in fragile ecosystems. Other species may survive in greatly reduced numbers and ranges. Ecosystems considered most vulnerable to species loss in the short term are polar seas, coral reefs and atolls, prairie wetlands, coastal wetlands, tundra, boreal forests, tropical forests, and mountains, particularly alpine tundra.

In response to climate change, some species may disperse into new environments or adapt to the changing conditions in

The **stratosphere,** which encircles planet Earth some 10 to 45 km (6 to 28 mi) above the surface, contains a layer of **ozone (O_3).** The ozone layer shields Earth's surface from about 95% of the harmful ultraviolet (UVA and UVB) radiation from the sun (see Figure 1). Ozone is a form of oxygen that is a human-made pollutant in the lower atmosphere, but is a naturally produced, essential part of the stratosphere. Ozone in the lower atmosphere is converted back to oxygen in a few days and so does not replenish the ozone depleted in the stratosphere.

A slight thinning in the ozone layer over Antarctica forms naturally for a few months each year. In 1985, however, scientists observed more thinning than usual. This increased ozone thinning, which begins each September, is commonly referred to as the "ozone hole" (see Figure 2). During the 1990s, the ozone-thinned area continued to grow, and by 2000 it had reached the record size of 28.3 million km^2 (11.3 million mi^2), larger than the North American continent.

Certain chemicals, such as chlorine- and bromine-containing substances, catalyze ozone destruction, destroying stratospheric ozone. The primary chemicals responsible for ozone loss in the stratosphere are a group of chlorine compounds called *chlorofluorocarbons (CFCs),* which were used as propellants in aerosol cans, coolants in air conditioners and refrigerators, and solvents and cleaners for the electronics industry. Additional compounds that also attack ozone include halocarbons (used in many fire extinguishers), methyl bromide (a pesticide), methyl chloroform (an industrial solvent), and carbon tetrachloride (used in many industrial processes, including the manufacture of pesticides and dyes).

After release into the troposphere (lowest part of Earth's atmosphere), CFCs and similar compounds slowly drift up to the stratosphere. There, ultraviolet radiation breaks them down, releasing chlorine. (Similarly, methyl bromide releases bromine.)

The thinning in the ozone layer over Antarctica occurs annually between September and November (spring in the Southern Hemisphere). At this time, two important conditions occur: sunlight returns to the polar region, and the *circumpolar vortex,* a mass of cold air that circulates around the southern polar region and isolates it from the warmer air in the rest of the planet, is well developed.

The cold air causes polar stratospheric clouds to form; these clouds contain ice crystals to which chlorine and bromine adhere, making them available to destroy ozone. The sunlight promotes the chemical reaction in which chlorine or bromine breaks ozone molecules apart, converting them into O_2 molecules. The chemical reaction in which ozone is destroyed does not alter the chlorine or bromine, and thus a single chlorine or bromine atom breaks down many thousands of ozone molecules. The chlorine and bromine remain in the stratosphere for many years. When the circumpolar vortex breaks up each year, the ozone-depleted air spreads northward, diluting ozone levels in the stratosphere over New Zealand, Australia, South America, and South Africa.

As the ozone layer is depleted, more ultraviolet radiation reaches Earth's surface. Increased levels of UV radiation disrupt ecosystems. For example, the productivity of Antarctic phytoplankton, the microscopic drifting algae that are the base of the Antarctic food web, has declined

from increased exposure to UVB. Biologists have also documented direct damage to natural populations of Antarctic fish. High levels of UV radiation may also damage crops and forests. Excessive exposure to UV radiation has been linked to human health problems, such as cataracts, severe sunburns, skin cancer, and a weakened immune system.

If ozone were to disappear from the stratosphere, Earth would become unlivable for most forms of life. The world's nations have addressed this serious environmental problem. In 1987, representatives from many countries signed the **Montreal Protocol,** an agreement that limited production of CFCs. Industrial companies that manufacture CFCs quickly developed substitutes.

International cooperation has continued to be a major force in repairing the ozone layer. Production of CFCs, carbon tetrachloride, and methyl chloroform has been completely phased out in the United States and other countries. Methyl bromide will be phased out by 2015 in most countries. Hydrochlorofluorocarbons will be phased out in 2020 in high-income countries and in 2030 in lower-income countries. Satellite measurements taken in 1997 provided the first evidence that the levels of ozone-depleting chemicals were starting to decline in the stratosphere. International agreements regarding ozone have now been signed by all independent countries in the world (about 196). However, reversing environmental damage is often a very slow process. Stratospheric ozone levels are not expected to return to 1980 levels until 2068, and 1950 levels are not expected until 2108!

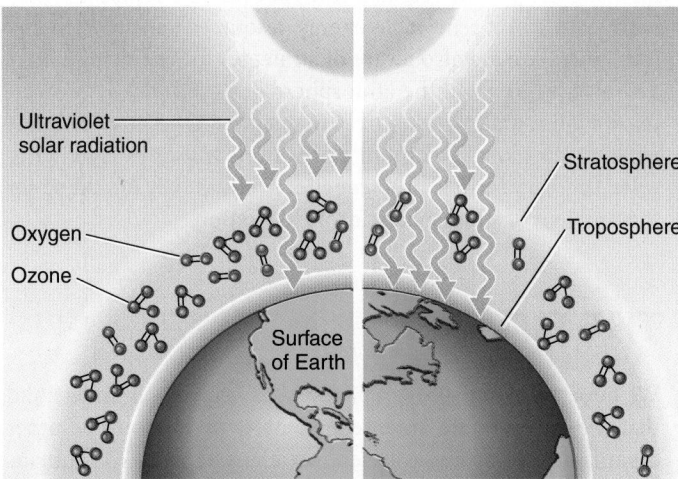

Figure 1 Ultraviolet radiation and the ozone layer. Stratospheric ozone absorbs 99% of incoming UV radiation, effectively shielding Earth's surface (*left*). When stratospheric ozone is reduced (*right*), more high-energy UV radiation penetrates the atmosphere to the surface, where it harms organisms. Stratospheric ozone thinning is accelerated by certain human-produced chemical compounds containing chlorine or bromine.
© Cengage Learning

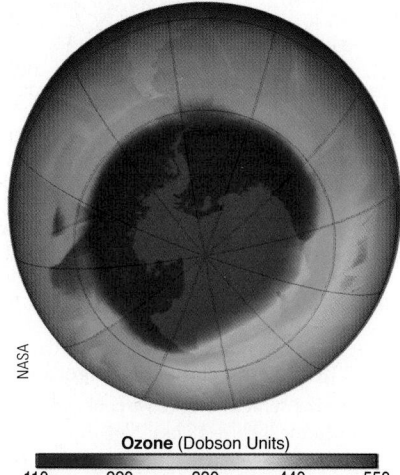

Ozone (Dobson Units)
110   220   330   440   550

Figure 2 Ozone thinning. A computer-generated image of part of the Southern Hemisphere, taken in 2009, showing ozone thinning (*purple* and *blue areas*). The ozone-thinned area is moved about by air currents. (Dobson units measure stratospheric ozone. They are named after British scientist Gordon Dobson, who studied ozone levels over Antarctica during the late 1950s.)

their present habitats. Climate change may not affect certain species, whereas other species may emerge as winners, with greatly expanded numbers and ranges. Those considered most likely to prosper include weeds, pests, and disease-carrying organisms, all of which are generalists that are already common in many different environments.

Climate change will have a more pronounced effect on human health in developing countries Data linking climate change and human health problems continue to accumulate. Since 1950, the United States has experienced an increased frequency of extreme heat-stress events, which are extremely hot, humid days during summer months. Heat-related deaths among elderly and other vulnerable people have been increasing now for many years worldwide.

Climate change has also affected human health indirectly. Mosquitoes and other disease carriers have expanded their range into the newly warm areas and are spreading malaria, dengue fever, yellow fever, Rift Valley fever, and viral encephalitis. As many as 50 million to 80 million additional cases of malaria are predicted to occur annually in tropical, subtropical, and temperate areas.

High-income countries are less vulnerable to such disease outbreaks because of better housing (which keeps mosquitoes outside), medical care, pest control, and public health measures such as water treatment plants. Dengue fever, however, has been increasing in Texas and the Florida Keys. The Centers for Disease Control recently reported a strain of dengue fever unique to the Florida Keys, indicating that the disease has been evolving in the region for many years. In 2012, the United States reported a 70% increase in cases of dengue fever. Mexico has recognized thousands of new cases of dengue fever, especially in the Yucatan and surrounding areas where tourism is prevalent.

Climate change may increase problems for agriculture Numerous studies show that the rising sea level will inundate river deltas, which are some of the world's best agricultural lands. The Nile River (Egypt), Mississippi River (United States), and Yangtze River (China) are examples of river deltas that have been studied. Certain agricultural pests and disease-causing organisms will probably proliferate. As mentioned earlier, climate change will likely increase the frequency and duration of droughts and, in some areas, crop-damaging floods.

On a regional scale, current climate-change models forecast that agricultural productivity will increase in some areas and decline in others. Models have suggested that Canada and Russia will increase their agricultural productivity in a warmer climate, whereas tropical and subtropical regions, where many of the world's poorest people live, will decline in agricultural productivity. Central America and Southeast Asia may experience some of the greatest declines in agricultural productivity.

We know how to prevent emission of greenhouse gases, for example, by reducing use of fossil fuels and replacing them with low-carbon renewable energy resources and solar energy (**FIG. 57-18**). We are also aware of strategies for removing excess carbon dioxide from the atmosphere, for example, by protecting wetlands and planting trees that store CO_2. The question is whether the global community of nations will work together to solve this very serious problem effectively, as they are doing for the problem of stratospheric ozone depletion.

Prevention

Cut fossil fuel use (especially coal)

Shift from coal to natural gas

Put a price on greenhouse gas emissions

Improve energy efficiency

Shift to renewable energy resources

Transfer energy efficiency and renewable energy technologies to developing countries

Reduce deforestation

Use more sustainable agriculture and forestry

Reduce poverty

Slow population growth

Cleanup

Remove CO_2 from smokestack and vehicle emissions

Store (sequester) CO_2 by planting trees

Sequester CO_2 in soil by using no-till cultivation and taking cropland out of production

Sequester CO_2 deep underground (with no leaks allowed)

Sequester CO_2 in the deep ocean (with no leaks allowed)

Repair leaky natural gas pipelines and facilities

Use animal feeds that reduce CH_4 emissions from cows (belching)

Figure 57-18 Slowing climate change: solutions
This figure shows some strategies we could use to slow atmospheric warming. Reducing the use of fossil fuel, especially coal, is a major prevention strategy. Shifting to renewable energy resources, including solar energy, is another critical action for decreasing greenhouse gas emissions. Because trees and wetlands remove carbon dioxide from the atmosphere, protecting forests and wetlands and planting trees will reduce carbon dioxide in the atmosphere. (Adapted from Miller, G. T. and S. E. Spoolman. *Living in the Environment*, 17th ed., 2012, p. 513, Figure 19-16. Brooks Cole/Cengage Learning.)

THE FUTURE?

We hope that your study of biology has helped you understand that we humans are interdependent with millions of other organisms: our fellow travelers on Planet Earth. We humans have impacted the biosphere more than any other species, and we have seriously disrupted its fragile web of interconnected ecosystems. We are not immune to the environmental damage that we have produced. In fact, we will share the fate of other species on the planet.

We humans differ from other organisms, however, in our capacity to reflect on the consequences of our actions and to alter our behavior accordingly. Humans, both individually and collectively, can bring about change. A global commitment to sustainable living is the key to ensuring the biosphere's survival. The fate of future generations and of the biosphere depends on our decisions and our actions.

- *What is the enhanced greenhouse effect? What causes the enhanced greenhouse effect?*
- **PREDICT** *What are some of the significant problems that climate change may cause during the 21st century?*

SUMMARY: FOCUS ON LEARNING OBJECTIVES

57.1 The Biodiversity Crisis *(page 1233)*

1 Identify various levels of biodiversity: genetic diversity, species richness, and ecosystem diversity.

- **Genetic diversity** is the genetic variety within a species, both within a given population and among geographically separate populations. **Species richness** is the number of species of archaea, bacteria, protists, plants, fungi, and animals. **Ecosystem diversity** is the variety of Earth's ecosystems, such as forests, prairies, deserts, lakes, coastal estuaries, and coral reefs.

2 Distinguish among threatened species, endangered species, and extinct species.

- A species' **extinction** occurs when its last individual member dies. A species whose severely reduced numbers throughout all or a significant part of its **range** put it in imminent danger of extinction is classified as an **endangered species.** When extinction is less imminent but the population is quite small, a species is classified as a **threatened species.**

Golden toad

© Cengage Learning

3 Discuss at least four causes of declining biological diversity and identify the most important causes.

- Human activities that reduce biological diversity include habitat loss and **habitat fragmentation,** pollution, introduction of **invasive species,** pest and predator control, illegal commercial hunting, and **commercial harvest.** Of these, habitat loss and fragmentation are the most significant.

57.2 Conservation Biology *(page 1239)*

4 Define *conservation biology* and compare in situ and ex situ conservation measures.

- **Conservation biology** is the study of how humans affect organisms and of the development of ways to protect biological diversity.

- Efforts to preserve biological diversity in the wild, known as **in situ conservation,** are urgently needed in the world's **biodiversity hotspots.** Increasingly, biologists are focusing efforts on preserving biodiversity in entire ecosystems and **landscapes,** which consist of multiple interacting ecosystems.

- **Ex situ conservation** involves conserving individual species in human-controlled settings. Breeding captive species in zoos and storing seeds of genetically diverse plant crops are examples.

5 Describe the benefits and shortcomings of the U.S. Endangered Species Act and the Convention on International Trade in Endangered Species of Wild Flora and Fauna.

- The Endangered Species Act (ESA) authorizes the U.S. Fish and Wildlife Service to protect endangered and threatened species, both in the United States and abroad. Conservationists would like to strengthen the ESA to manage whole ecosystems rather than endangered species and individual entities.

- At the international level, the Convention on International Trade in Endangered Species of Wild Flora and Fauna (CITES) protects endangered animals and plants considered valuable in the highly lucrative international wildlife trade. Enforcement varies from country to country; where enforcement exists, penalties are not severe, so illegal trade in rare species continues.

57.3 Deforestation *(page 1244)*

6 Discuss the ecosystem services of forests and describe the consequences of deforestation.

- Forests provide many **ecosystem services,** including wildlife habitat, protection of watersheds, prevention of soil erosion, moderation of climate, and protection from flooding.

- **Deforestation** is the temporary or permanent clearance of forests for agriculture or other uses. Deforestation increases soil erosion and decreases soil fertility; contributes to loss of biological diversity; adversely affects watersheds; and may affect regional and global climate changes.

7 State at least three reasons why forests (tropical rain forests and boreal forests) are disappearing today.

- Forests are destroyed to provide subsistence farmers with agricultural land, to produce timber, to provide open rangeland for cattle, and to supply fuel wood. **Subsistence agriculture,** in which a family produces enough food to feed itself, accounts for much of tropical rainforest deforestation. Extensive logging of boreal forests in Alaska, Canada, and Russia is the world's primary source of industrial wood and wood fiber.

- In **slash-and-burn agriculture,** trees are cut down and burned so that crops can be grown in the soil. Yields from the first crop are high because the nutrients that were in the trees are now available in the soil. When soil productivity declines, the farmer moves to a new part of the forest and repeats the process.

57.4 Climate Change *(page 1246)*

8 Name at least three major greenhouse gases and explain how greenhouse gases contribute to climate change.

- **Greenhouse gases**—carbon dioxide, methane, nitrous oxide, surface ozone, and chlorofluorocarbons—cause the **greenhouse effect,** in which the atmosphere retains heat and warms Earth's surface. Increased levels of human-produced greenhouse gases in the atmosphere are causing concerns about an **enhanced greenhouse effect,** which is additional warming produced by increased levels of gases that absorb infrared radiation.

9 Describe how climate change may affect sea level, precipitation patterns, organisms (including humans), and food production.

- During the 21st century, **climate change** is causing a rise in sea level. Precipitation patterns are changing, resulting in more frequent droughts in some areas and more frequent flooding in other areas.

- Biologists think that climate change will cause some species to go extinct, some to be unaffected, and others to expand their numbers and ranges. Data linking climate change and human health problems (particularly in developing countries) are accumulating.

- Problems for agriculture include increased flooding, increased droughts, and declining agricultural productivity in tropical and subtropical areas.

TEST YOUR UNDERSTANDING

Know and Comprehend

1. Which of the following statements about extinction is *not* correct? (a) extinction is the permanent loss of a species (b) extinction is a natural biological process (c) once a species is extinct, it never reappears (d) human activities have little impact on extinctions (e) thousands of plant and animal species are currently threatened with extinction

2. An endangered species (a) is severely reduced in number (b) is in imminent danger of becoming extinct throughout all or a significant part of its range (c) usually does not have reduced genetic variability (d) is not in danger of extinction in the foreseeable future (e) a and b

3. The most important reason for declining biological diversity is (a) air pollution (b) introduction of foreign (invasive) species (c) habitat destruction and fragmentation (d) illegal commercial hunting (e) commercial harvesting

4. Habitat corridors (a) surround a given habitat (b) cut through a continuous habitat, producing an edge effect (c) vary in width depending on the species they are designed to protect (d) are an important strategy of ex situ conservation (e) have been widely adopted by restoration ecologists

5. In situ conservation (a) includes breeding captive species in zoos (b) includes seed storage of genetically diverse crops (c) concentrates on preserving biological diversity in the wild (d) focuses exclusively on large, charismatic animals (e) a and b

6. Restoration ecology (a) is the study of how humans impact organisms (b) returns a degraded environment as close as possible to its former state (c) is an example of ex situ conservation (d) has been used to successfully reverse the decline in amphibian populations (e) is an important provision of the Endangered Species Act

7. About 60% of tropical rainforest deforestation is the result of (a) commercial logging (b) cattle ranching (c) hydroelectric dams (d) mining (e) subsistence agriculture

8. Climate change occurs because (a) carbon dioxide and other greenhouse gases react chemically to produce excess heat (b) Earth has too many greenhouses and other glassed buildings (c) volcanic eruptions produce large quantities of sulfur and other greenhouse gases (d) carbon dioxide and other greenhouse gases trap infrared radiation in the atmosphere (e) carbon dioxide and other greenhouse gases allow excess heat to pass out of the atmosphere

9. **CONNECT** What gas is a human-made pollutant in the lower (surface) atmosphere but a natural and beneficial gas in the stratosphere? (a) CO_2 (b) CH_4 (c) O_3 (d) CFCs (e) N_2O

10. Where is stratospheric ozone depletion most pronounced? (a) over Antarctica (b) over the equator (c) over South America (d) over North America and Europe (e) over Alaska and Siberia

Apply and Analyze

11. Why might captive-breeding programs that reintroduce species into natural environments fail?

12. **CONNECT** Conservation biologists often say that their discipline is less about biology than it is about economics and human decision making. What do you think that they mean?

13. A more descriptive name for *Homo sapiens* is *Homo dangerous*. Explain this specific epithet, given what you have learned in this chapter.

Evaluate and Synthesize

14. **EVOLUTION LINK** Because new species will eventually evolve to replace those that humans are driving to extinction, why is declining biological diversity such a threat?

15. **EVOLUTION LINK** Biologists have wondered how introduced species that would probably have limited genetic variation (due to the founder effect) survive and adapt so successfully that they become invasive. Part of the answer may be that invasive species are the result of multiple introductions instead of a single one. Explain how multiple introductions from a species' native

area to an introduced area could increase that species' invasion success.

16. **EVOLUTION LINK** Conservation biologists have altered the evolution of salmon populations in captive-breeding programs. Wild female salmon tend to produce fewer but larger eggs because the large eggs contain more nutrients for the offspring than smaller eggs, giving each individual a greater chance to survive. After just a few generations, however, captive-bred females now lay greater numbers of small eggs. Suggest a possible adaptive advantage for many small eggs in the captive-bred environment. What would you predict regarding the reproductive success of captive-bred females released in the wild?

17. **INTERPRET DATA** Study the graph, which shows the combined effects of various factors on biological diversity in 12 different terrestrial and aquatic ecosystems. Which factor is most important overall? Why do you think that climate change and increasing atmospheric CO_2 are represented as separate factors? What is nitrogen pollution? (Adapted from Sala, O. E., et al. *Science*, Vol. 287, 2000.)

18. **SCIENCE, TECHNOLOGY, AND SOCIETY** If you were given the task of developing a policy for the United States to deal with global climate change during the next 50 years, what would you propose? Explain your answer by describing how science, technology, and society would play a role in your policy.

To access course materials, such as Aplia and other companion resources, please visit **www.cengagebrain.com.**

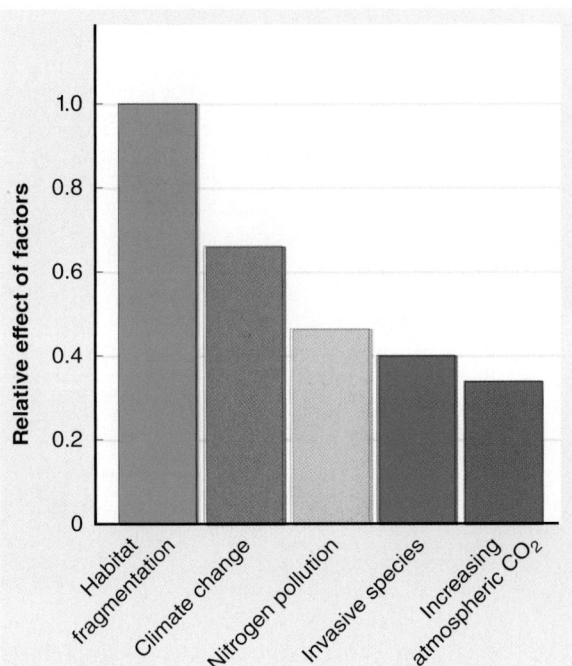

Appendix A
Periodic Table of the Elements

1	S
H	
Hydrogen	
1.0079	

Atomic number — 92
Symbol — **U** (S)
Uranium
Atomic mass — 238.0289

State:
- S Solid
- L Liquid
- G Gas
- X Not found in nature

- Main Group metals
- Transition metals, lanthanide series, actinide series
- Metalloids
- Nonmetals, noble gases

1A

3 S	4 S
Li	**Be**
Lithium	Beryllium
6.941	9.0122

2A

11 S	12 S
Na	**Mg**
Sodium	Magnesium
22.9898	24.3050

3B

19 S	20 S	21 S	22 S	23 S	24 S	25 S	26 S	27 S	28 S	29 S	30 S
K	**Ca**	**Sc**	**Ti**	**V**	**Cr**	**Mn**	**Fe**	**Co**	**Ni**	**Cu**	**Zn**
Potassium	Calcium	Scandium	Titanium	Vanadium	Chromium	Manganese	Iron	Cobalt	Nickel	Copper	Zinc
39.0983	40.078	44.9559	47.88	50.9415	51.9961	54.9380	55.847	58.9332	58.693	63.546	65.39

37 S	38 S	39 S	40 S	41 S	42 S	43 X	44 S	45 S	46 S	47 S	48 S
Rb	**Sr**	**Y**	**Zr**	**Nb**	**Mo**	**Tc**	**Ru**	**Rh**	**Pd**	**Ag**	**Cd**
Rubidium	Strontium	Yttrium	Zirconium	Niobium	Molybdenum	Technetium	Ruthenium	Rhodium	Palladium	Silver	Cadmium
85.4678	87.62	88.9059	91.224	92.9064	95.94	(98)	101.07	102.9055	106.42	107.8682	112.411

55 S	56 S	57 S	72 S	73 S	74 S	75 S	76 S	77 S	78 S	79 S	80 L
Cs	**Ba**	**La**	**Hf**	**Ta**	**W**	**Re**	**Os**	**Ir**	**Pt**	**Au**	**Hg**
Cesium	Barium	Lanthanum	Hafnium	Tantalum	Tungsten	Rhenium	Osmium	Iridium	Platinum	Gold	Mercury
132.9054	137.327	138.9055	178.49	180.9479	183.85	186.207	190.2	192.22	195.08	196.9665	200.59

87 S	88 S	89 S	104 X	105 X	106 X	107 X	108 X	109 X	110 X	111 X	112 X
Fr	**Ra**	**Ac**	**Rf**	**Db**	**Sg**	**Bh**	**Hs**	**Mt**	—	—	—
Francium	Radium	Actinium	Rutherfordium	Dubnium	Seaborgium	Bohrium	Hassium	Meitnerium			
(223)	227.0254	227.0278	(261)	(262)	(263)	(262)	(265)	(266)	(269)	(272)	(277)

8A

2 G
He
Helium
4.0026

3A 4A 5A 6A 7A

5 S	6 S	7 G	8 G	9 G	10 G
B	**C**	**N**	**O**	**F**	**Ne**
Boron	Carbon	Nitrogen	Oxygen	Fluorine	Neon
10.811	12.011	14.0067	15.9994	18.9984	20.1797

13 S	14 S	15 S	16 S	17 G	18 G
Al	**Si**	**P**	**S**	**Cl**	**Ar**
Aluminum	Silicon	Phosphorus	Sulfur	Chlorine	Argon
26.9815	28.0855	30.9738	32.066	35.4527	39.948

1B 2B

31 S	32 S	33 S	34 S	35 L	36 G
Ga	**Ge**	**As**	**Se**	**Br**	**Kr**
Gallium	Germanium	Arsenic	Selenium	Bromine	Krypton
69.723	72.61	74.9216	78.96	79.904	83.80

49 S	50 S	51 S	52 S	53 S	54 G
In	**Sn**	**Sb**	**Te**	**I**	**Xe**
Indium	Tin	Antimony	Tellurium	Iodine	Xenon
114.82	118.710	121.757	127.60	126.9045	131.29

81 S	82 S	83 S	84 S	85 S	86 G
Tl	**Pb**	**Bi**	**Po**	**At**	**Rn**
Thallium	Lead	Bismuth	Polonium	Astatine	Radon
204.3833	207.2	208.9804	(209)	(210)	(222)

58 S	59 S	60 S	61 X	62 S	63 S	64 S	65 S	66 S	67 S	68 S	69 S	70 S	71 S
Ce	**Pr**	**Nd**	**Pm**	**Sm**	**Eu**	**Gd**	**Tb**	**Dy**	**Ho**	**Er**	**Tm**	**Yb**	**Lu**
Cerium	Praseodymium	Neodymium	Promethium	Samarium	Europium	Gadolinium	Terbium	Dysprosium	Holmium	Erbium	Thulium	Ytterbium	Lutetium
140.115	140.9076	144.24	(145)	150.36	151.965	157.25	158.9253	162.50	164.9303	167.26	168.9342	173.04	174.967

90 S	91 S	92 S	93 X	94 S	95 X	96 X	97 X	98 X	99 X	100 X	101 X	102 X	103 X
Th	**Pa**	**U**	**Np**	**Pu**	**Am**	**Cm**	**Bk**	**Cf**	**Es**	**Fm**	**Md**	**No**	**Lr**
Thorium	Protactinium	Uranium	Neptunium	Plutonium	Americium	Curium	Berkelium	Californium	Einsteinium	Fermium	Mendelevium	Nobelium	Lawrencium
232.0381	231.0359	238.0289	237.0482	(244)	(243)	(247)	(247)	(251)	(252)	(257)	(258)	(259)	(260)

Appendix B
Classification of Organisms

The system of cataloging organisms used in this book is described in Chapter 1 and in Chapters 23 through 32. In this tenth edition of *Biology*, we use the three-domain classification. The three domains are Bacteria, Archaea, and Eukarya (eukaryotes). We also discuss the five "supergroups" of eukaryotes based on molecular data. The kingdoms Plantae, Fungi, and Animalia are assigned to domain Eukarya. The protists are members of domain Eukarya, but are no longer considered a kingdom. We assign them to several supergroups.

In this classification overview, we have included select groups. (We have also omitted many groups, especially extinct ones.) We have omitted viruses from this survey because they are not considered living organisms and are not assigned to any of the three domains.

PROKARYOTES

Domains Bacteria and Archaea are made up of prokaryotic organisms. They are distinguished from eukaryotic organisms by their smaller ribosomes and absence of membranous organelles, including the absence of a discrete nucleus surrounded by a nuclear envelope. Prokaryotes reproduce mainly asexually by binary fission. When present, flagella are simple and solid; they do not have the 9 + 2 microtubule structure typical of eukaryotes.

DOMAIN BACTERIA, KINGDOM BACTERIA

Very large, diverse group of prokaryotic organisms. Typically unicellular, but some form colonies or filaments. Mainly chemoheterotrophic, but some groups are photoautotrophic or photoheterotrophic, and some are chemoautotrophic. Bacteria are nonmotile or move by rotating flagella. Typically have peptidoglycan in their cell walls. Estimated 100,000 to 200,000 species. Bacterial nomenclature and taxonomic practices are controversial and changing. See Table 25-3.

Proteobacteria   Large, diverse group of gram-negative bacteria. Five subgroups are designated alpha, beta, gamma, delta, and epsilon. The alpha proteobacteria group includes *Rhizobium* and rickettsias. Beta proteobacteria include the bacterium that causes gonorrhea. Gamma proteobacteria include the enterobacteria, the group to which the intestinal bacterium *Escherichia coli* belongs, and the purple sulfur bacteria. The delta proteobacteria group includes the myxobacteria, and the epsilon bacteria include *Helicobacter,* which can cause peptic ulcers.

Gram-positive bacteria   Diverse group; includes actinomycetes, lactic acid bacteria, mycobacteria, streptococci, staphylococci, clostridia. Thick cell wall of peptidoglycan; many produce spores.

Mycoplasmas   Lack cell walls. Extremely small bacteria bounded by plasma membrane. May have evolved from gram-positive bacteria.

Cyanobacteria   Gram negative, photosynthetic.

Chlamydias   Gram negative; lack peptidoglycan in their cell walls. Energy parasites dependent on host for ATP.

Spirochetes   Gram negative; spiral-shaped bacteria with flexible cell walls.

DOMAIN ARCHAEA, KINGDOM ARCHAEA

Prokaryotes with unique cell membrane structure and cell walls lacking peptidoglycan. Also distinguished by their ribosomal RNA, lipid structure, and specific enzymes. About 225 named species. Based on molecular data, there are four main clades: **Crenarchaeota, Euryarchaeota, Korarchaeota,** and **Nanoarchaeota.** Archaea are common to most environments, including the human digestive tract, but many are found in extreme environments such as hot springs, sea vents, dry and salty seashores, boiling mud, and near ash-ejecting volcanoes. Three main types of archaea (not clades) based on their metabolism and ecology are *methanogens, extreme halophiles,* and *extreme thermophiles.* See Fig. 25-10.

Methanogens   Anaerobes that produce methane gas from simple carbon compounds.

Extreme halophiles   Inhabit saturated salt solutions.

Extreme thermophiles   Grow at 70°C or higher; some thrive above boiling point. Antarctic archaea live in very cold (1.8°C) environments.

DOMAIN EUKARYA, PROTISTS

Primarily unicellular or simple multicellular eukaryotic organisms that do not form tissues and that exhibit relatively little division of labor. Most modes of nutrition occur in this group. Life cycles may include both sexually and asexually reproducing phases and may be extremely complex, especially in parasitic forms. Locomotion is by cilia, flagella, amoeboid movement, or by other means; flagella and cilia have 9 + 2 structure. Many biologists recognize five supergroups: Excavates, Chromalveolates, Rhizarians, Archaeplastids, and Unikonts.

Excavates Unicellular protists that have deep, or excavated, oral (feeding) groove. Have atypical, greatly modified mitochondria or lack them. Many are endosymbionts. Currently include diplomonads, parabasilids, euglenoids, and trypanosomes.

Diplomonads Excavates with one or two nuclei, no functional mitochondria, no Golgi complex, and up to eight flagella.

Parabasilids Anaerobic, flagellated excavates that often live in animals. Include trichonymphs and trichomonads.

Euglenoids and trypanosomes Unicellular flagellates with crystalline rod in flagella; some free-living and some pathogenic. Many are heterotrophic, but some (about one-third) have plastids and are photosynthetic. At least 900 species.

Chromalveolates Diverse protists that may have originated as a result of secondary endosymbiosis in which an ancestral cell engulfed a red alga. Include alveolates (dinoflagellates, ciliates, and apicomplexans) and stramenopiles (water molds, diatoms, golden algae, and brown algae).

Alveolates Chromalveolates with alveoli, flattened vesicles located inside plasma membrane.

Dinoflagellates Unicellular (some colonial), photosynthetic, biflagellate. Cell walls composed of overlapping cell plates; contain cellulose. Contain chlorophylls *a* and *c* and carotenoids, including fucoxanthin. About 2000 to 4000 species.

Ciliates Unicellular protists that move by means of cilia. Reproduction is asexual by binary fission or sexual by conjugation. About 7200 species.

Apicomplexans Parasitic unicellular protists that lack specific structures for locomotion. At some stage in life cycle, they develop spores (small infective agents). Some pathogenic. About 3900 species.

Stramenopiles Diverse group; most have motile cells with two flagella, one with tiny, hairlike projections off shaft; no flagella in some. Include water molds, diatoms, golden algae, and brown algae.

Water molds Consist of branched, coenocytic mycelia. Cellulose and/or chitin in cell walls. Produce biflagellate asexual spores. Sexual stage involves production of oospores. Some parasitic. About 700 species.

Diatoms Unicellular (some colonial), photosynthetic. Most nonmotile, but some move by gliding. Cell walls composed of silica rather than cellulose. Contain chlorophylls *a* and *c* and carotenoids, including fucoxanthin. At least 100,000 species estimated.

Golden algae Unicellular (some colonial), photosynthetic, biflagellate (some lack flagella). Cells covered by tiny scales of either silica or calcium carbonate. Contain chlorophylls *a* and *c* and carotenoids, including fucoxanthin. About 1000 species.

Brown algae Multicellular, often quite large (kelps). Photosynthetic; contain chlorophylls *a* and *c* and carotenoids, including fucoxanthin. Biflagellate reproductive cells. About 1500 species.

Rhizarians Amoeboid cells that often have tests (shells). Include forams and actinopods.

Foraminiferans (forams) Unicellular protists that produce calcareous tests (shells) with pores through which cytoplasmic projections extend, forming sticky net to entangle prey.

Actinopods Unicellular protists that produce axopods (long, filamentous cytoplasmic projections) that protrude through pores in their siliceous shells.

Archaeplastids Photosynthetic organisms with chloroplasts bounded by outer and inner membranes. This monophyletic group includes Plantae (land plants) and protists (red algae and green algae).

Red algae Most multicellular (some unicellular), mainly marine. Some (coralline algae) have bodies impregnated with calcium carbonate. No motile cells. Photosynthetic; contain chlorophyll a, carotenoids, phycocyanin, and phycoerythrin. About 5000 species.

Green algae Unicellular, colonial, and multicellular forms. Some motile and flagellate. Photosynthetic; contain chlorophylls a and b and carotenoids. About 17,000 species.

Unikonts Cells that have a single flagellum or are amoebas with no flagella. Have a triple-gene fusion that is lacking in other eukaryotes. Includes organisms currently classified in two kingdoms, Animalia and Fungi, and certain protists (amoebozoa and opisthokonts).

Amoebozoa Amoeboid protists that lack tests and move by means of lobose pseudopodia. Include amoebas, plasmodial slime molds, and cellular slime molds.

Amoebas Free-living and parasitic unicellular protists whose movement and capture of food are associated with pseudopodia.

Plasmodial slime molds Spend part of life cycle as thin, streaming, multinucleate plasmodium that creeps along on decaying leaves or wood. Flagellate or amoeboid reproductive cells; form spores in sporangia. About 700 species.

Cellular slime molds Vegetative (nonreproductive) form is unicellular; move by pseudopods. Amoeba-like cells aggregate to form multicellular pseudoplasmodium that eventually develops into fruiting body that bears spores. About 50 species.

Opisthokonts Opisthokonts (Greek opistho, "rear," and kontos, "pole") have a single posterior flagellum in flagellate cells. Includes animals, fungi, and choanoflagellates.

Choanoflagellates Have a single flagellum surrounded at its base by a collar of microvilli. Probably related to the common ancestor of animals.

DOMAIN EUKARYA, KINGDOM PLANTAE

Multicellular eukaryotic organisms with differentiated tissues and organs. Cell walls contain cellulose. Cells frequently contain large vacuoles; photosynthetic pigments in plastids. Photosynthetic pigments are chlorophylls *a* and *b* and carotenoids. Nonmotile. Reproduce both asexually and sexually, with alternation of gametophyte (*n*) and sporophyte (*2n*) generations. Some biologists classify plants as Archaeplastids (along with red algae and green algae).

Phylum Bryophyta *Mosses*   Nonvascular plants that lack xylem and phloem. Marked alternation of generations with dominant gametophyte generation. Motile sperm. Gametophytes generally form dense green mat consisting of individual plants. At least 9900 species.

Phylum Hepatophyta *Liverworts*   Nonvascular plants that lack xylem and phloem. Marked alternation of generations with dominant gametophyte generation. Motile sperm. Gametophytes of certain species have flat, liverlike thallus; other species more mosslike in appearance. About 6000 species.

Phylum Anthocerophyta *Hornworts*   Nonvascular plants that lack xylem and phloem. Marked alternation of generations with dominant gametophyte generation. Motile sperm. Gametophyte is small, flat, green thallus with scalloped edges. Spores produced on erect, hornlike stalk. About 100 species.

Phylum Pteridophyta   Vascular plants with dominant sporophyte generation. Reproduce by spores. Motile sperm.

> *Ferns*   Generally homosporous. Gametophyte is free-living and photosynthetic. About 11,000 species.
>
> *Whisk ferns*   Homosporous. Sporophyte stem branches dichotomously; lacks true roots and leaves. Gametophyte is subterranean and nonphotosynthetic and forms mycorrhizal relationship with fungus. About 12 species.
>
> *Horsetails*   Homosporous. Sporophyte has hollow, jointed stems and reduced, scalelike leaves. Gametophyte is tiny photosynthetic plant. About 15 species.

Phylum Lycopodiophyta *Club mosses*   Sporophyte plants are vascular with branching rhizomes and upright stems that bear microphylls. Although modern representatives are small, some extinct species were treelike. Some homosporous; others heterosporous. Motile sperm. About 1200 species.

Phylum Coniferophyta *Conifers*   Heterosporous vascular plants with woody tissues (trees and shrubs) and needle-shaped or scalelike leaves. Most are evergreen. Seeds usually borne naked on surface of cone scales. Nutritive tissue in seed is haploid female gametophyte tissue. Nonmotile sperm. About 630 species.

Phylum Cycadophyta *Cycads*   Heterosporous, vascular, dioecious plants that are small and shrubby or larger and palmlike. Produce naked seeds in conspicuous cones. Flagellate sperm. About 140 species.

Phylum Ginkgophyta *Ginkgo*   Broadleaf deciduous trees that bear naked seeds directly on branches. Dioecious. Contain vascular tissues. Flagellate sperm. Ginkgo tree is only living representative. One species.

Phylum Gnetophyta *Gnetophytes*   Woody shrubs, vines, or small trees that bear naked seeds in cones. Contain vascular tissues. Possess many features similar to flowering plants. About 70 species.

Phylum Anthophyta *Flowering plants or angiosperms*   Largest, most successful group of plants. Heterosporous; dominant sporophytes with extremely reduced gametophytes. Contain vascular tissues. Bear flowers, fruits, and seeds (enclosed in fruit; seeds contain endosperm as nutritive tissue). Double fertilization. More than 300,000 species.

DOMAIN EUKARYA, KINGDOM FUNGI

Eukaryotic, mainly multicellular organisms with cell walls containing chitin. Heterotrophs that secrete digestive enzymes onto food source and then absorb predigested food. Most decomposers, but some parasites. Body form typically a mycelium. Cells usually haploid or dikaryotic, with brief diploid period following fertilization. Reproduce by means of spores, which may be produced sexually or asexually. No flagellate stages except in chytrids. Classified as opisthokonts (along with choanoflagellates and animals) because flagellate cells, where present, have a single posterior flagellum (Greek *opistho,* "rear," and *kontos,* "pole").

Phylum Chytridiomycota *Chytridiomycetes or chytrids*   Parasites and decomposers found mainly in fresh water. Motile cells (gametes and zoospores) contain single, posterior flagellum. Reproduce both sexually and asexually. About 1000 species.

Phylum Zygomycota *Zygomycetes (molds)*   Important decomposers; some are insect parasites. Produce sexual resting spores called *zygospores* and nonmotile, haploid, asexual spores in sporangium. Hyphae are coenocytic. Many are heterothallic (two mating types). About 1100 species.

Phylum Glomeromycota *Glomeromycetes*   Symbionts that form intracellular mycorrhizal associations within roots of most trees and herbaceous plants. Reproduce asexually with large, multinucleate spores called *blastospores.* About 200 species.

Phylum Ascomycota *Ascomycetes or sac fungi (yeasts, powdery mildews, molds, morels, truffles)*   Important symbionts; 98% of lichen-forming fungi are ascomycetes; some form mycorrhizae. Sexual reproduction: form ascospores in sacs called *asci.* Asexual reproduction: produce spores called *conidia,* which pinch off from conidiophores. Hyphae usually have perforated septa. Dikaryotic stage. About 32,000 species.

Phylum Basidiomycota *Basidiomycetes or club fungi (mushrooms, bracket fungi, puffballs)*   Many form mycorrhizae with tree roots. Sexual reproduction: form basidiospores on basidium. Asexual reproduction uncommon. Heterothallic. Hyphae usually have perforated septa. Dikaryotic stage. More than 30,000 species.

DOMAIN EUKARYA, KINGDOM ANIMALIA

Eukaryotic, multicellular heterotrophs with differentiated cells. In most animals, cells are organized to form tissues, tissues form organs, and tissues and organs form specialized organ systems that carry on specific functions. Most have a well-developed nervous system and respond adaptively to changes in their environment. Most are capable of locomotion during some time in their life cycle. Most diploid and reproduce sexually; flagellate haploid sperm unites with large, nonmotile, haploid egg, forming diploid zygote that undergoes cleavage. Classified as opisthokonts (along with choanoflagellates and fungi) because flagellate cells, when present, have a single posterior flagellum (Greek *opistho*, "rear," and *kontos*, "pole").

Phylum Porifera *Sponges*   Mainly marine; solitary or colonial. Body bears many pores through which water circulates. Food is filtered from water by collar cells (choanocytes). Asexual reproduction by budding; external sexual reproduction in which sperm are released and swim to internal egg. Larva is motile. About 10,000 species.

Phylum Cnidaria *Hydras, jellyfish, sea anemones, corals*   Marine, with a few freshwater species; solitary or colonial; polyp and medusa forms. Radial symmetry. Tentacles surrounding mouth. Stinging cells (cnidocytes) contain stinging structures called *nematocysts*. Planula larva. About 10,000 species.

Phylum Ctenophora *Comb jellies*   Marine; free-swimming. Biradial symmetry. Two tentacles and eight longitudinal rows of cilia resembling combs; animal moves by means of these bands of cilia. About 150 species.

Protostomes: Lophotrochozoa

Most are true coelomates (characterized by a body cavity completely lined with mesoderm). Spiral, determinate cleavage; mouth typically develops from blastopore. There are two branches of protostomes: Lophotrochozoa and Ecdysozoa. The Lophotrochozoa include the platyhelminths, nemerteans (ribbon worms), mollusks, annelids, the lophophorate phyla, and the rotifers. The name *Lophotrochozoa* comes from two characteristics of some animals in this clade: the lophophore, a ciliated ring of tentacles surrounding the mouth in three small groups of animals, and the trochophore larva that characterizes its two major groups: the mollusks and annelids.

Phylum Platyhelminthes *Flatworms*   Acoelomate (no body cavity); region between body wall and internal organs filled with tissue. Planarians are free-living; flukes and tapeworms are parasitic. Body dorsoventrally flattened; cephalization; three tissue layers. Simple nervous system with ganglia in head region. Excretory organs are protonephridia with flame cells. About 20,000 species.

Phylum Nemertea *Proboscis worms (also called ribbon worms)*   Long, dorsoventrally flattened body with complex proboscis used for defense and for capturing prey. Functionally acoelomate but have small true coelom in proboscis. Definite organ systems. Complete digestive tract. Circulatory system with blood. About 1200 species.

Phylum Mollusca *Snails, clams, squids, octopods*   Unsegmented, soft-bodied true coelomate animals usually covered by dorsal shell. Have ventral, muscular foot. Most organs located above foot in visceral mass. Shell-secreting mantle covers visceral mass and forms mantle cavity, which contains gills. Trochophore and/or veliger larva. More than 80,000 species.

Phylum Annelida *Segmented worms: polychaetes, earthworms, leeches*   True coelomates; both body wall and internal organs are segmented. Body segments separated by septa. Some have nonjointed appendages. Setae used in locomotion. Closed circulatory system; metanephridia; specialized regions of digestive tract. Trochophore larva. About 15,000 species.

Phylum Brachiopoda *Lamp shells*   One of lophophorate phyla. Marine; body enclosed between two shells. About 325 species.

Phylum Phoronida   One of lophophorate phyla. Tube-dwelling marine worms. About 20 species.

Phylum Bryozoa   One of lophophorate phyla. Mainly marine; sessile colonies produced by asexual budding. About 4500 species.

Phylum Rotifera *Wheel animals*   Aquatic, microscopic. Anterior end has ciliated crown that looks like wheel when cilia beat. Posterior end tapers to foot. Characterized by pseudocoelom (body cavity not completely lined with mesoderm). Constant number of cells in adult. About 2000 species.

Protostomes: Ecdysozoa

Animals in this group of protostomes characterized by ecdysis (molting).

Phylum Nematoda *Roundworms: ascaris, hookworms, pinworms*   Slender, elongated, cylindrical worms; covered with cuticle. Characterized by pseudocoelom (body cavity not completely lined with mesoderm). Free-living and parasitic forms. More than 25,000 species.

Phylum Arthropoda *Arachnids (spiders, mites, ticks), crustaceans (lobsters, crabs, shrimp), insects, centipedes, millipedes*   Segmented animals with paired, jointed appendages and hard exoskeleton made of chitin. Open circulatory system with dorsal heart. Hemocoel occupies most of body cavity, and coelom is reduced. More than 1 million species.

Deuterostomes

True coelomates with radial, indeterminate cleavage. Blastopore develops into anus, and mouth forms from second opening. At some time in their life cycle most deuterostomes (except echinoderms) develop pharyngeal slits, openings that connect the pharynx with the outside environment.

Phylum Echinodermata *Sea stars, sea urchins, sand dollars, sea cucumbers*   Marine. Pentaradial symmetry as adults; bilateral symmetry as larvae. Endoskeleton of small, calcareous plates. Water vascular system; tube feet for locomotion. About 7000 species.

Phylum Hemichordata *Acorn worms*   Marine with ring of cilia around mouth. Anterior muscular proboscis is connected by collar region to long, wormlike body. Larval form resembles echinoderm larva. About 100 species.

Phylum Chordata *Subphylum Urochordata (tunicates), subphylum Cephalochordata (lancelets), subphylum Vertebrata (fishes, amphibians, reptiles, including birds, mammals)*   Notochord; dorsal, tubular nerve cord; postanal tail; and endostyle. About 60,000 species.

Appendix C
Understanding Biological Terms

Your task of mastering new terms will be greatly simplified if you learn to dissect each new word. Many terms can be divided into a prefix, the part of the word that precedes the main root; the word root itself; and often a suffix, a word ending that may add to or modify the meaning of the root. As you progress in your study of biology, you will learn to recognize the more common prefixes, word roots, and suffixes. Such recognition will help you analyze new terms so that you can more readily determine their meaning and will help you remember them.

Prefixes

a-, ab- from, away, apart (*abduct,* move away from the midline of the body)

a-, an-, un- less, lack, not (*asymmetrical,* not symmetrical)

ad- (also **af-, ag-, an-, ap-**) to, toward (*adduct,* move toward the midline of the body)

allo- different (*allometric growth,* different rates of growth for different parts of the body during development)

ambi- both sides (*ambidextrous,* able to use either hand)

andro- a man (*androecium,* the male portion of a flower)

anis- unequal (*anisogamy,* sexual reproduction in which the gametes are of unequal sizes)

ante- forward, before (*anteflexion,* bending forward)

anti- against (*antibody,* proteins that have the capacity to react against foreign substances in the body)

auto- self (*autotroph,* organism that manufactures its own food)

bi- two (*biennial,* a plant that takes two years to complete its life cycle)

bio- life (*biology,* the study of life)

circum-, circ- around (*circumcision,* a cutting around)

co-, con- with, together (*congenital,* existing with or before birth)

contra- against (*contraception,* against conception)

cyt- cell (*cytology,* the study of cells)

di- two (*disaccharide,* a compound made of two sugar molecules chemically combined)

dis- apart (*dissect,* cut apart)

ecto- outside (*ectoderm,* outer layer of cells)

end-, endo- within, inner (*endoplasmic reticulum,* a network of membranes found within the cytoplasm)

epi- on, upon (*epidermis,* upon the dermis)

ex-, e-, ef- out from, out of (*extension,* a straightening out)

extra- outside, beyond (*extraembryonic membrane,* a membrane that encircles and protects the embryo)

gravi- heavy (*gravitropism,* growth of a plant in response to gravity)

hemi- half (*cerebral hemisphere,* lateral half of the cerebrum)

hetero- other, different (*heterozygous,* having unlike members of a gene pair)

homeo- unchanging, steady (*homeostasis,* reaching a steady state)

homo-, hom- same (*homologous,* corresponding in structure; *homozygous,* having identical members of a gene pair)

hyper- excessive, above normal (*hypersecretion,* excessive secretion)

hypo- under, below, deficient (*hypotonic,* a solution whose osmotic pressure is less than that of a solution with which it is compared)

in-, im- not (*incomplete flower,* a flower that does not have one or more of the four main parts)

inter- between, among (*interstitial,* situated between parts)

intra- within (*intracellular,* within the cell)

iso- equal, like (*isotonic,* equal osmotic pressure)

macro- large (*macronucleus,* a large, polyploid nucleus found in ciliates)

mal- bad, abnormal (*malnutrition,* poor nutrition)

mega- large, great (*megakaryocyte,* giant cell of bone marrow)

meso- middle (*mesoderm,* middle tissue layer of the animal embryo)

meta- after, beyond (*metaphase,* the stage of mitosis after prophase)

micro- small (*microscope,* instrument for viewing small objects)

mono- one (*monocot,* a group of flowering plants with one cotyledon, or seed leaf, in the seed)

oligo- small, few, scant (*oligotrophic lake,* a lake deficient in nutrients and organisms)

oo- egg (*oocyte,* cell that gives rise to an egg cell)

paedo- a child (*paedomorphosis,* the preservation of a juvenile characteristic in an adult)

para- near, beside, beyond (*paracentral,* near the center)

peri- around (*pericardial membrane,* membrane that surrounds the heart)

photo- light (*phototropism,* growth of a plant in response to the direction of light)

poly- many, much, multiple, complex (*polysaccharide,* a carbohydrate composed of many simple sugars)

post- after, behind (*postnatal,* after birth)

pre- before (*prenatal,* before birth)

pseudo- false (*pseudopod,* a temporary protrusion of a cell, i.e., "false foot")

retro- backward (*retroperitoneal,* located behind the peritoneum)

semi- half (*semilunar,* half-moon)

sub- under (*subcutaneous tissue,* tissue immediately under the skin)

super-, supra- above (*suprarenal,* above the kidney)

sym- with, together (*sympatric speciation,* evolution of a new species within the same geographic region as the parent species)

syn- with, together (*syndrome,* a group of symptoms that occur together and characterize a disease)

trans- across, beyond (*transport,* carry across)

Suffixes

-able, -ible able (*viable,* able to live)

-ad used in anatomy to form adverbs of direction (*cephalad,* toward the head)

-asis, -asia, -esis condition or state of (*euthanasia,* state of "good death")

-cide kill, destroy (*biocide,* substance that kills living things)

-emia condition of blood (*anemia,* a blood condition in which there is a lack of red blood cells)

-gen something produced or generated or something that produces or generates (*pathogen,* an organism that produces disease)

-gram record, write (*electrocardiogram,* a record of the electrical activity of the heart)

-graph record, write (*electrocardiograph*, an instrument for recording the electrical activity of the heart)

-ic adjective-forming suffix that means *of* or *pertaining to* (*ophthalmic*, of or pertaining to the eye)

-itis inflammation of (*appendicitis*, inflammation of the appendix)

-logy study or science of (*cytology*, study of cells)

-oid like, in the form of (*thyroid*, in the form of a shield, referring to the shape of the thyroid gland)

-oma tumor (*carcinoma*, a malignant tumor)

-osis indicates disease (*psychosis*, a mental disease)

-pathy disease (*dermopathy*, disease of the skin)

-phyll leaf (*mesophyll*, the middle tissue of the leaf)

-scope instrument for viewing or observing (*microscope*, instrument for viewing small objects)

Some Common Word Roots

abscis cut off (*abscission*, the falling off of leaves or other plant parts)

angi, angio vessel (*angiosperm*, a plant that produces seeds enclosed within a fruit or "vessel")

apic tip, apex (*apical meristem*, area of cell division located at the tips of plant stems and roots)

arthr joint (*arthropods*, invertebrate animals with jointed legs and segmented bodies)

aux grow, enlarge (*auxin*, a plant hormone involved in growth and development)

blast a formative cell, germ layer (*osteoblast*, cell that gives rise to bone cells)

brachi arm (*brachial artery*, blood vessel that supplies the arm)

bry grow, swell (*embryo*, an organism in the early stages of development)

cardi heart (*cardiac*, pertaining to the heart)

carot carrot (*carotene*, a yellow, orange, or red pigment in plants)

cephal head (*cephalad*, toward the head)

cerebr brain (*cerebral*, pertaining to the brain)

cervic, cervix neck (*cervical*, pertaining to the neck)

chlor green (*chlorophyll*, a green pigment found in plants)

chondr cartilage (*chondrocyte*, a cartilage cell)

chrom color (*chromosome*, deeply staining body in nucleus)

cili small hair (*cilium*, a short, fine cytoplasmic hair projecting from the surface of a cell)

coleo a sheath (*coleoptile*, a protective sheath that encircles the stem in grass seedlings)

conjug joined (*conjugation*, a sexual phenomenon in certain protists)

cran skull (*cranial*, pertaining to the skull)

decid falling off (*deciduous*, a plant that sheds its leaves at the end of the growing season)

dehis split (*dehiscent fruit*, a fruit that splits open at maturity)

derm skin (*dermatology*, study of the skin)

ecol dwelling, house (*ecology*, the study of organisms in relation to their environment, i.e., "their house")

enter intestine (*enterobacteria*, a group of bacteria that includes species that inhabit the intestines of humans and other animals)

evol to unroll (*evolution*, descent with modification, gradual directional change)

fil a thread (*filament*, the thin stalk of the stamen in flowers)

gamet a wife or husband (*gametangium*, the part of a plant, protist, or fungus that produces reproductive cells)

gastr stomach (*gastrointestinal tract*, the digestive tract)

glyc, glyco sweet, sugar (*glycogen*, storage form of glucose)

gon seed (*gonad*, an organ that produces gametes)

gutt a drop (*guttation*, loss of water as liquid "drops" from plants)

gymn naked (*gymnosperm*, a plant that produces seeds that are not enclosed with a fruit, i.e., "naked")

hem blood (*hemoglobin*, the pigment of red blood cells)

hepat liver (*hepatic*, of or pertaining to the liver)

hist tissue (*histology*, study of tissues)

hydr water (*hydrolysis*, a breakdown reaction involving water)

leuk white (*leukocyte*, white blood cell)

menin membrane (*meninges*, the three membranes that envelop the brain and spinal cord)

morph form (*morphogenesis*, development of body form)

my, myo muscle (*myocardium*, muscle layer of the heart)

myc a fungus (*mycelium*, the vegetative body of a fungus)

nephr kidney (*nephron*, microscopic unit of the kidney)

neur, nerv nerve (*neuromuscular*, involving both the nerves and muscles)

occiput back part of the head (*occipital*, back region of the head)

ost bone (*osteology*, study of bones)

path disease (*pathologist*, one who studies disease processes)

ped, pod foot (*bipedal*, walking on two feet)

pell skin (*pellicle*, a flexible covering over the body of certain protists)

phag eat (*phagocytosis*, process by which certain cells ingest particles and foreign matter)

phil love (*hydrophilic*, a substance that attracts, i.e., "loves," water)

phloe bark of a tree (*phloem*, food-conducting tissue in plants that corresponds to bark in woody plants)

phyt plant (*xerophyte*, a plant adapted to xeric, or dry, conditions)

plankt wandering (*plankton*, microscopic aquatic protists that float or drift passively)

rhiz root (*rhizome*, a horizontal, underground stem that superficially resembles a root)

scler hard (*sclerenchyma*, cells that provide strength and support in the plant body)

sipho a tube (*siphonous*, a type of tubular body form found in certain algae)

som body (*chromosome*, deeply staining body in the nucleus)

sor heap (*sorus*, a cluster or "heap" of sporangia in a fern)

spor seed (*spore*, a reproductive cell that gives rise to individual offspring in plants, protists, and fungi)

stom a mouth (*stoma*, a small pore, i.e., "mouth," in the epidermis of plants)

thigm a touch (*thigmotropism*, plant growth in response to touch)

thromb clot (*thrombus*, a clot within a blood vessel)

troph nourishment (*heterotroph*, an organism that must depend on other organisms for its nourishment)

tropi turn (*thigmotropism*, growth of a plant in response to contact with a solid object, such as a tendril "turning" or wrapping around a wire fence)

visc pertaining to an internal organ or body cavity (*viscera*, internal organs)

xanth yellow (*xanthophyll*, a yellowish pigment found in plants)

xyl wood (*xylem*, water-conducting tissue in plant, the "wood" of woody plants)

zoo an animal (*zoology*, the science of animals)

Appendix D
Abbreviations

The biological sciences use a great many abbreviations and with good reason. Many technical terms in biology and biological chemistry are both long and difficult to pronounce. When confronted with something like NADPH or EPSP, it can be difficult for beginners to understand the reference. Here, for your ready reference, are some of the common abbreviations used in biology.

A adenine
ABA abscisic acid
ABC transporters ATP-binding cassette transporters
ABP androgen-binding protein
ACTH adrenocorticotropic hormone
AD Alzheimer's disease
ADA adenosine deaminase
ADH antidiuretic hormone
ADP adenosine diphosphate
AIDS acquired immunodeficiency syndrome
AMP adenosine monophosphate
amu atomic mass unit (dalton)
ANP atrial natriuretic peptide
APC anaphase-promoting complex *or* antigen-presenting cell
AS Angelman syndrome
ATP adenosine triphosphate
AV node or **valve** atrioventricular node or valve (of heart)
B lymphocyte or **B cell** lymphocyte responsible for antibody-mediated immunity
BAC bacterial artificial chromosome
BH brain hormone (of insects)
BMI body mass index
BMR basal metabolic rate
BR brassinosteroid
bya billion years ago
C cytosine
C$_3$ three-carbon pathway for carbon fixation (Calvin cycle)
C$_4$ four-carbon pathway for carbon fixation
cal calories
CAM crassulacean acid metabolism
cAMP cyclic adenosine monophosphate
CAP catabolite activator protein
CCK cholecystokinin
CD4 T cell T helper cell (T$_H$); T cell with a surface marker designated CD4
CD8 T cell T cell with a surface marker designated CD8; includes T cytotoxic cells (T$_C$)
Cdk cyclin-dependent protein kinase
cDNA complementary deoxyribonucleic acid
CFCs chlorofluorocarbons
CFTR cystic fibrosis transmembrane conductance regulator

cGMP cyclic guanosine monophosphate
CITES Convention on International Trade in Endangered Species of Wild Flora and Fauna
CNS central nervous system
CNVs copy number variations
CO cardiac output
CoA coenzyme A
COPD chronic obstructive pulmonary disease
CP creatine phosphate
CPR cardiopulmonary resuscitation
CREB cyclic AMP response element binding protein
CRF corticotropin releasing factor
CSF cerebrospinal fluid
CVA cardiovascular accident
CVS cardiovascular system
CVS chorionic villus sampling
DAG diacylglycerol
DNA deoxyribonucleic acid
DOC dissolved organic carbon
E_A activation energy (of an enzyme)
ECG electrocardiogram
ECM extracellular matrix
EEG electroencephalogram
EKG electrocardiogram
EM electron microscope or micrograph
ENSO El Niño–Southern Oscillation
EPSP excitatory postsynaptic potential (of a neuron)
ER endoplasmic reticulum
ERK extracellular signal-regulated kinases
ES cells embryonic stem cells
EST expressed sequence tag
F$_1$ first filial generation
F$_2$ second filial generation
Fab portion the part of an antibody that binds to an antigen
Factor VIII blood-clotting factor (absent in certain hemophiliacs)
FAD/FADH$_2$ flavin adenine dinucleotide (oxidized and reduced forms, respectively)
FAP fixed action pattern
F$_c$ portion the part of an antibody that interacts with cells of the immune system
FISH fluorescent in situ hybridization
FSH follicle-stimulating hormone
G free energy
G guanine
G protein cell-signaling molecule that requires GTP
G$_1$ phase first gap phase (of the cell cycle)
G$_2$ phase second gap phase (of the cell cycle)
G3P glyceraldehyde-3-phosphate

GA gibberellin

GAA Global Amphibian Assessment (by World Conservation Union)

GABA gamma-aminobutyric acid

GH growth hormone (somatotropin)

GHIH growth hormone–inhibiting hormone

GHRH growth hormone–releasing hormone

G_i A G protein that inhibits adenylyl cyclase

GINA Genetic Information Nondiscrimination Act

GIP glucose-dependent insulinotropic peptide

GnRH gonadotropin-releasing hormone

GPP gross primary productivity

G_s A G protein that stimulates adenylyl cyclase

GTP guanosine triphosphate

GWA genome-wide association

H enthalpy

Hb hemoglobin

HBEF Hubbard Brook Experimental Forest

HBO$_2$ oxyhemoglobin

HCFCs hydrochlorofluorocarbons

hCG human chorionic gonadotropin

HD Huntington disease

HDL high-density lipoprotein

HFCs hydrofluorocarbons

HGH human growth hormone

HIV human immunodeficiency virus

HLA human leukocyte antigen

HPV human papillomavirus

HUGO Human Genome Organization

IAA indole acetic acid (natural auxin)

Ig immunoglobulin, as in IgA, IgG, etc.

IGF insulin-like growth factor

IP$_3$ inositol trisphosphate

IPCC United Nations Intergovernmental Panel on Climate Change

iPSCs induced pluripotent stem cells

IPSP inhibitory postsynaptic potential (of a neuron)

IUCN World Conservation Union

IUD intrauterine device

J joules

JH juvenile hormone (of insects)

kb kilobase

kcal kilocalories

kJ kilojoules

LDH lactate dehydrogenase enzyme

LDL low-density lipoprotein

LH luteinizing hormone

LM light microscope or micrograph

lncRNA long noncoding RNA

LSD lysergic acid diethylamide

LTP long-term potentiation

LUCA last universal common ancestor

MAO monoamine oxidase

MAP mitogen-activated protein

MAPs microtubule-associated proteins

MH molting hormone (ecdysone)

MHC major histocompatibility complex

MI myocardial infarction

miRNA microribonucleic acid

miRNA microRNA

MPF mitosis-promoting factor

MRI magnetic resonance imaging

mRNA messenger RNA

MSAFP maternal serum *a*-fetoprotein

MSH melanocyte-stimulating hormone

mtDNA mitochondrial DNA

MTOC microtubule organizing center

MVP minimum viable population

mya million years ago

9 + 2 structure cilium or flagellum (of a eukaryote)

9 × 3 structure centriole or basal body (of a eukaryote)

n, 2n the chromosome number of a gamete and of a zygote, respectively

NAD$^+$/NADH nicotinamide adenine dinucleotide (oxidized and reduced forms, respectively)

NADP$^+$/NADPH nicotinamide adenine dinucleotide phosphate (oxidized and reduced forms, respectively)

NAG *N*-acetyl glucosamine

NK cell natural killer cell

NLS nuclear localization signal

NMDA *N*-methyl-D aspartate (an artificial ligand)

NO nitric oxide

NPP net primary productivity

NPY neuropeptide Y

NSF National Science Foundation

P generation parental generation

P53 a tumor suppressor gene

P680 reaction center of photosystem II

P700 reaction center of photosystem I

PABA para-aminobenzoic acid

PAMPs pathogen-associated molecular patterns

PCR polymerase chain reaction

PEP phosphoenolpyruvate

Pfr phytochrome (form that absorbs far-red light)

PGA phosphoglycerate

PGD preimplantation genetic diagnosis

PID pelvic inflammatory disease

PIF3 phytochrome-interacting factor 3

piRNA piwi-associated RNA

PKU phenylketonuria

PNS peripheral nervous system

Pr phytochrome (form that absorbs red light)

pre-mRNA precursor messenger RNA (in eukaryotes)

PRR pattern recognition receptor

PTH parathyroid hormone

PWS Prader-Willi syndrome

r growth rate of a population

RAO recent African origin

RAS reticular activating system

RBC red blood cell (erythrocyte)

REM sleep rapid-eye-movement sleep
RFLP restriction fragment length polymorphism
r_{max} intrinsic rate of increase (of a population)
RNA ribonucleic acid
RNAi RNA interference
rRNA ribosomal RNA
rubisco ribulose bisphosphate carboxylase/oxygenase
RuBP ribulose bisphosphate
S entropy
S phase DNA synthetic phase (of the cell cycle)
SA node sinoatrial node (of heart)
SAR systemic acquired resistance (in plants)
SCID severe combined immunodeficiency
SEM scanning electron microscope or micrograph
siRNA small interfering RNA
snoRNA small nucleolar RNA
SNPs single nucleotide polymorphisms ("snips")
snRNA small nuclear RNA
snRNPs small nuclear ribonucleoprotein complexes ("snurps")
SRP signal-recognition particle
SRP RNA signal recognition particle ribonucleic acid
SSB protein single-strand binding protein
ssp subspecies

STD sexually transmitted disease
STI sexually transmitted infection
STR short tandem repeat
T thymine
T_3 triiodothyronine
T_4 thyroxine
T lymphocyte or **T cell** lymphocyte responsible for cell-mediated immunity
T_C **lymphocyte** T cytotoxic cell (CD8 T cell)
T_H **lymphocyte** T helper cell (CD4 T cell)
TATA box base sequence in eukaryotic promoter
TCA cycle tricarboxylic acid cycle (citric acid cycle, Krebs cycle)
TCR T-cell receptor
TEM transmission electron microscope or micrograph
Tm tubular transport maximum
TNF tumor necrosis factor
TRH thyroid-releasing hormone
tRNA transfer RNA
TSH thyroid-stimulating hormone
U uracil
UV light ultraviolet light
WBC white blood cell (leukocyte)

Appendix E
Answers

CHAPTER 1

CHECKPOINT

1.1 • Information transmission, energy transfer, and evolution are considered basic to life because organisms must be able to send and receive information within their own bodies and with other organisms and their environment; organisms must be able to transfer energy from one form to another to perform life functions; populations of organisms must change over time, or evolve, to survive as their environment changes. • Consumers are dependent on producers for food and energy. Producers and consumers are dependent on decomposers for recycling nutrients. Organisms are interdependent for maintaining the needed balance of gases in the atmosphere. **1.2** • Living organisms differ from a rock because they are characterized by cellular organization, growth and development, self-regulated metabolism, the ability to respond to stimuli, reproduction, and adaptation to environmental change. • If an organism's homeostatic mechanisms failed, the organism could not maintain metabolic processes and it would die. **1.3** • The levels of organization within an organism are the chemical level (atoms and molecules), cells, tissues, organs, and organ systems. • Ecosystems include all of the other categories; because they are the most complex, they interact with the most other biological systems. **1.4** • The function of DNA is to transmit genetic information from one generation to the next through information coded in sequences of nucleotides. • A nervous system transmits information through electrical impulses and chemical compounds, called neurotransmitters. **1.5** • A forest ecosystem would include producers (flowering plants, bushes, trees), primary consumers (grasshoppers, rabbits, worms), secondary consumers (birds, coyote, snakes), and decomposers (bacteria, fungi, archaea). • Consumers depend on producers to provide energy captured from the sun by photosynthesis and stored as carbohydrates. Consumers depend on decomposers to break down complex compounds, releasing nutrients and elements for reuse by producers and consumers. **1.6** • *Python* is the genus name for the African rock python. • The tree showing the major forms of life may be modified as scientists obtain additional data about evolutionary relationships. • The sharp claws and teeth of tigers may have been an evolutionary advantage (improved predation and defensive ability) that began as a random mutation and was selected for in the tiger's environment and then passed on to future generations, becoming an important adaptation in the tiger population. **1.7** • A good hypothesis is reasonably consistent with well-established facts, generates predictions that can be tested, yields test results that are repeatable by independent researchers, and is falsifiable, or can be proved false through testing. • A controlled experiment consists of an experimental group and a control group, which are treated identically except for a variable treatment of the experimental group. • Systems biology depends on information generated by the reductionist approach to yield basic information about components of biological systems.

FIGURE QUESTIONS

FIG. 1-9 Smog and discharge from factories, utilities, and automobiles can reduce sunshine over broad areas, causing decreased photosynthetic activity and death in plants and algae, which results in less food, or available energy, for animals. Gases released into the atmosphere can cause acidified precipitation, which damages leaves of producers and increases carbon dioxide levels, which further increases climate change. **FIG. 1-10** The mushroom would first be classified into a domain (Eukarya), then kingdom (Fungi). Physical and molecular information will help to determine phylum, class, order, family, genus, and species of the mushroom. **FIG. 1-11** The protists do not share enough common characteristics to be classified as a kingdom and have been placed in five "supergroups" based on their commonalities to better reflect how they are evolutionally related. **FIG. 1-16** The cell may be able to carry out life processes under the direction of the donor nucleus. **FIG. 1-17** Results from the repeated experiment might be similar but may not be identical to the result from the previous experiment.

TEST YOUR UNDERSTANDING

1. c 2. e 3. c 4. d 5. b 6. c 7. e 8. d 9. See Fig. 1-11. 10. Failure of the homeostatic mechanism could cause the organism to die. Further explanation relies on a logical description and discussion of the homeostatic mechanism selected for example. 11. A good hypothesis is reasonably consistent with well-established facts, generates predictions that can be tested, yields test results that are repeatable by independent researchers, and is falsifiable, or can be proved false through testing. Examples here will vary. 12. Any answer is acceptable if logically presented and defended. 13. Reductionists study the simplest components of biological processes to understand the whole, while systems biology researchers take the large amounts of reductionist-derived data and do computer analyses to understand the big picture of how biological systems function. Systems biology depends on reductionism because it relies on information generated by the reductionist approach to yield basic information about components and large data sets. The systems approach is more likely to consider emergent properties. 14. The second graph shows the number of chimpanzees who successfully employed learned tool use and retained the knowledge, or remembered, after two months. As time passes, a learned skill may be forgotten. 15. Evolution depends on transfer of information because organisms must be able to recognize changes in their environment to be able to respond and for the population to adapt. Transfer of information depends on evolution because chemical, electrical, and behavioral components must change, or adapt, as the environment changes in order for life to continue. 16. Any response that is defended by logical argument is acceptable. 17. Any subjective response that is defended by logical argument is acceptable.

CHAPTER 2

CHECKPOINT

2.1 • All atoms of an element have the same atomic number, corresponding to the number of protons in the nucleus. Different isotopes of the same element have different atomic masses, depending on their number of neutrons. • A radioisotope is an unstable isotope of an element that can decay, releasing a radioactive particle, such as a β particle. Radioisotopes are used in research to label biological molecules used to trace biochemical pathways, follow the transport of molecules in cells and tissues, and determine the sequence of genetic information in DNA. • Electrons occupying different orbitals of the same electron shell are at the same principal energy level (i.e., have similar energies). **2.2** • A radioisotope can substitute for a nonradioactive atom of the same element because both isotopes have the same chemical characteristics. • Structural formulas contain the most information because they illustrate the structural arrangement of atoms in the compound, as well as their types and numbers. • One gram of hydrogen atoms would contain one mole, or 6.02×10^{23} particles of hydrogen atoms. Two grams of hydrogen molecules (one mole of H_2) would also contain 6.02×10^{23} particles. **2.3** • A chemical compound consists of atoms of two or more different elements combined in a fixed ratio. Not all compounds are made up of molecules. For example, sodium chloride is not a molecule because the sodium and chloride atoms are not joined by covalent bonds. • An atom or molecule can become an anion by gaining one or more electrons; a molecule can also become an anion by losing one or more protons. An atom or molecule can become a cation by losing one or more electrons; a molecule can also become a cation by gaining one or more protons. • Ionic bonds are formed by an attraction between the positive charge of a cation and the negative charge of an anion. Covalent bonds are formed between atoms that share electrons. • Weak forces, such as hydrogen bonds and van der Waals interactions, can produce strong forces when large numbers of these bonds occur over short distances between regions of molecules. **2.4** • In redox reactions energy is transferred in the form of electrons. **2.5** • Water molecules are formed by polar covalent bonds between oxygen and two hydrogen atoms. The region around the oxygen atom is slightly negative and the regions around the hydrogen atoms are slightly positive, allowing hydrogen bonds to form between the oxygen atom of one water molecule and a hydrogen atom of another water molecule. • Hydrogen bonding is responsible for water's high specific heat, which affects the freezing and boiling points of water. Hydrogen bonding also affects the spacing of water molecules in ice, making ice less dense and allowing ice to float on liquid water. The high water content of organisms helps them maintain relatively constant internal temperatures. The high heat of vaporization of water also allows organisms to dissipate heat through evaporative cooling. • Weak forces such as hydrogen bonds have collective strength when many work together. **2.6** • A solution with a $[H^+]$ of 10^{-2} has a pH of 2.0. Its hydroxide ion concentration is 10^{-12}. The solution is strongly acidic. Its $[H^+]$ is one-tenth that of a solution with a pH of 1.0. • Adding a reactant or removing a product would shift the system to the right, resulting in the formation of more product. Adding a product or removing a reactant would shift the system to the left, resulting in the formation of more reactant. • Buffers prevent excessive acidity or alkalinity from occurring in organisms, helping to maintain cellular and extracellular environments within a narrow range of pH. Strong acids or bases cannot work as buffers because there are few undissociated molecules present relative to

the dissociated components of the system. This prevents a dynamic equilibrium from forming, which can allow associated or dissociated components to be produced. • Electrolytes, such as acids, bases, and salts, dissociate to form anions and cations in solution, allowing the solution to conduct electric current.

FIGURE QUESTIONS

FIG. 2-1 Examine the Bohr model for oxygen (atomic number 8) in Figure 2-1. A Bohr model for fluorine would have an additional electron in its valence shell. **FIG. 2-5** An N_2 molecule has 2 electrons in each inner shell and 5 electrons in each valence shell. There are 3 pairs of electrons bonded between the two atoms.

TEST YOUR UNDERSTANDING

1. c 2. e 3. a 4. b 5. c 6. d 7. a 8. a 9. d 10. e 11. e 12. e 13. Refer to Figure 2-1 to construct your diagram. Element A would be expected to donate electrons. Element B would most likely share electrons. Element C would be expected to accept electrons. 14. HCl is the reactant. H^+ and Cl^- are the products. The expression indicates that this is an irreversible reaction. HCl is a strong acid and could not be used as a buffer because there would be very little undissociated HCl remaining after the reaction. 15. You should not immerse your hand. The solution has a pH of 12, which is strongly basic. 16. If hydrogen bonds were stronger, the freezing point of water would be higher, its boiling point would be higher, and it would have a higher specific heat and heat of vaporization. If hydrogen bonds were weaker, these properties would be lower than for existing water. 17. Water is the most fundamental indicator because its unique properties make it unlikely that life could evolve in its absence.

CHAPTER 3

CHECKPOINT

3.1 • Carbon–carbon bonds can produce a wide variety of 3-D molecular shapes because the four possible covalent bonds do not form in a single plane. Carbon–carbon bonds can consist of single, double, or triple bonds that are strong and not easily broken. These bonds can form unbranched or branched chains as well as rings that can be joined, forming complex 3-D structures. • See Figure 3-3 for examples of different types of isomers. Isomers with the same molecular formula have different physical and chemical properties. Cells can distinguish between isomers and usually only one will be biologically active. • See Table 3-1 for the structures and properties of functional groups. • Nonpolar groups lack

distinct charged regions and do not interact readily with water. Polar, acidic, and basic groups are hydrophilic. They have charged regions that associate with polar water molecules. • See Figure 3-5 for an illustration of how water molecules participate in condensation and hydrolysis reactions. **3.2** • Hydrogen bonding between polysaccharides occurs between the hydroxyl groups of the sugar units. Your drawing should show that these bonds occur within a single molecule of the storage polysaccharide, resulting in a helical structure (see Fig. 3-9c). The hydrogen bonds in structural polysaccharides occur between different molecules, forming bundles of fibers (see Fig. 3-10a). **3.3** • Saturated fatty acids contain flexible, unbranched hydrocarbon chains. Adjacent saturated fatty acids can align with each other, and associate by van der Waals interactions along the length of the chains. This makes them more solid at a given temperature by limiting the motions of the chains. Unsaturated fatty acids contain double bonds that produce bends in the chains, preventing them from aligning closely with adjacent chains, reducing the number of van der Waals interactions. *Cis* fatty acids are more liquid than saturated fatty acids at a given temperature. *Trans* fatty acids contain double bonds that do not produce a bend in the chain. These mimic the properties of saturated fatty acids and are more solid at a given temperature than *cis* fatty acids. • Phospholipids are amphipathic molecules. These molecules can form a bilayer in which their hydrophobic fatty acid chains associate in the center of the two layers and the hydrophilic head groups interact with water on the outer surfaces of the bilayer. Triacylglycerols and diacylglycerols are more hydrophobic molecules that tend to form spherical oil droplets in water. This minimizes their interactions with surrounding water molecules and maximizes interactions between their hydrophobic hydrocarbon chains. **3.4** • See Figures 3-16 and 3-17 for representative amino acid structures and the properties of their side chains (*R* groups). The amino and carboxyl groups participate in peptide bond formation between amino acid subunits. If they are exposed (not part of a peptide bond), the amino group can assume positive charge and the carboxyl group can have a negative charge, depending on the pH of the medium. The *R* group can consist of any of a number of different functional groups that confer different shapes and properties to the amino acid. Nonpolar *R* groups are found on hydrophobic amino acids; polar *R* groups are found on more hydrophilic amino acids. Acidic and basic amino acids have negatively and positively charged side chains, respectively. • The primary structure of a polypeptide influences its secondary and tertiary structures through the differing structural and chemical properties of the amino acids in the polypeptide chain. Interactions between amino acids within the chain, such as hydrogen bonding, ionic bonding, hydrophobic interactions, and disulfide bonding can result in the formation of complex folds and structural domains within the polypeptide. **3.5** • See Figs. 3-24a and 3-25. Pyrimidine subunits in RNA consist of cytosine and uracil. The 5-carbon sugar in RNA is ribose. To convert this sugar to deoxyribose found in DNA, the 3′ hydroxyl group of the ribose would be removed, and if the pyrimidine base were uracil, a methyl group would have to be added to convert it to thymine. **3.6** • Refer to the following pairs of figures to compare differences between the indicated molecules: pentose and hexose sugars—Figs. 3-6b and c; disaccharides and sterols—Figs. 3-8 and 3-15; amino acids and monosaccharides—Figs. 3-17 and 3-6; phospholipid and triacylglycerol—Figs. 3-13a and 3-12b; protein and polysaccharide—Figs. 3-18, 3-19, and 3-21, as well as 3-9b and 3-10b; nucleic acid and protein—Figs. 3-18, 3-19, and 3-21 and 3-25.

FIGURE QUESTIONS

FIG. 3-13 Unlike bilayer-forming phospholipids, triacylglycerols lack the hydrophilic head groups needed to associate with water. They would not be able to form organized bilayer structures. **FIG. 3-20** Secondary structure is based on a regular repeating sequence of atoms in the backbone. Insertion of an additional carbon atom between the α-carbon and the peptide bond carbon in some amino acids would disrupt secondary structure by altering the distances between the atoms involved in hydrogen bond formation. **FIG. 3-21** Leucine and valine would associate through hydrophobic interactions. Alanine (a nonpolar amino acid) would not interact with serine (a polar amino acid). Lysine and arginine are positively charged, basic amino acids and would repel each other. Glutamic acid has a negative charge and would form an ionic interaction with positively charged lysine.

TEST YOUR UNDERSTANDING

1. d 2. b 3. e 4. c 5. c 6. a 7. a 8. b 9. c 10. e 11. b 12. b and c or c and d 13. Sulfur would not be able to form hydrogen or ionic bonds with positively charged atoms found in proteins and other biological molecules. 14. Hydrogen bonds and van der Waals interactions can produce strong bonding when many of these bonds and interactions occur within a region of a biological molecule. Examples include α-helical regions and β-sheet regions of proteins and van der Waals interactions between fatty acid chains of membrane lipids. 15. All forms of life share many common components of biological molecules, including the nucleotides of nucleic acids, amino acids of proteins, fatty acids of membrane and storage lipids, and monosaccharides found in complex carbohydrates. These suggest that the use of these building blocks to form many different types of biological structures evolved early in the history of life. 16. Although there are very large combinations of amino acid sequences that can be generated for polypeptides, evolutionary processes select only the small number of sequences that have structures and functions that provide a selective advantage to the organism. 17. The existence of only one of two different kinds of enantiomers for amino acids suggests that the first cells to evolve had L-amino acids and that this feature has been conserved in all their descendants.

CHAPTER 4

CHECKPOINT

4.1 • Because cells, the basic units of life, only come from pre-existing cells, it follows that their lineage extends to very ancient times. • Membrane barriers are required to maintain the appropriate internal chemical environments for cellular functions. • DNA is used for information storage in all cells. • ATP is one convenient form of chemical energy used by all cells. • Small surface areas relative to large internal volumes prevent/restrict the movement of materials into, out of, and throughout the cells. **4.2** • Magnification that exceeds the limits of resolution (the equivalent of sharp focus) cannot improve the observation of fine structural details. • Each method provides different information about the structure or function of a cellular component. For example, microscopy can reveal the location of a protein within the cell and its association with cellular structures. Immunoprecipitation provides information about the quantity of the protein and other proteins that bind to it. Biochemical analyses provide structural and functional information about the protein. **4.3** • DNA, cell membranes, protein synthesis. • Plants have cell walls, chloroplasts, vacuoles, and plasmodesmata. Animal cells have centrioles. • Membrane-enclosed organelles provide cellular compartments with unique chemical environments for specialized functions. **4.4** • DNA is highly organized as chromatin (with RNA and protein); ribosomal RNA is synthesized in the nucleoli; and the nuclear lamina helps organize nuclear contents. • The nuclear envelope and nuclear pores provide selective entry and exit for large macromolecules and macromolecular complexes, such as RNAs, ribosomes, nuclear proteins, etc. • Large molecules, such as large proteins, RNA molecules, and macromolecular complexes are bound to specialized proteins that selectively import their "cargo" into or out of the nucleus through the nuclear pores. Small molecules such as ions and small polypeptides can freely diffuse through the nuclear pores. **4.5** • The rough ER typically consists of flattened membranous sheets that are studded with ribosomes on the cytosolic membrane surfaces. A primary function of the rough ER involves the synthesis of membrane proteins and proteins that are translocated through the ER membrane into the ER lumen. The smooth ER is typically made up of tubular membranous structures. Some regions of the smooth ER carry out membrane and storage lipid biosynthesis; others are involved in numerous functions including detoxification reactions, calcium storage, and various protein modifications. • The *cis* face of the Golgi complex receives proteins from membrane vesicles derived from the ER and initiates modification of those proteins. The proteins are sequentially modified as they move through the Golgi complex. At the *trans* face of the Golgi complex, proteins are sorted into specialized transport vesicles that are targeted to the plasma membrane or other organelles in the cell such as lysosomes. • See Figure 4-15 for steps in the manufacture and secretion of cellular proteins. • Chloroplasts and mitochondria are organelles that contain multiple membrane compartments involved in the conversion of energy from one form to another. A major function of mitochondria is the breakdown (oxidation) of carbon compounds to form ATP. A primary function of chloroplasts is the capture of energy from light that is then stored in reduced carbon compounds. (See Fig. 4-19.) **4.6** • Cytoskeletal structures maintain cell shape, anchor intracellular structures and membranes, and are involved in the movement of cells and intracellular organelles. The endomembrane system is involved in many different cellular functions, including protein biosynthesis and modification, membrane formation, and membrane trafficking. • Microtubules and microfilaments are fiber-like structures that can be rapidly assembled and disassembled. Both are involved in different types of cellular and intracellular movements. Microtubules are hollow cylinders that can serve as intracellular tracks for organellar movement using molecular motors. Microtubules are also components of cilia and flagella used for cellular locomotion. Microfilaments can generate force and intracellular movement by rapidly assembling and disassembling. Combined with motor proteins such as myosins, microfilaments can also generate force and intracellular movement. • Microtubules are bundled, cylinder-like elements of cilia and flagella; they provide movement when bent by motor proteins such as dynein. **4.7** • The glycocalyx surrounds most cells as a cell coat that extends from the plasma membrane. It is formed by polysaccharides. • Fibronectins are extracellular matrix (ECM) glycoproteins that bind to integrins, which are plasma membrane receptor proteins.

FIGURE QUESTIONS

FIG. 4-11 Ribosomes are assembled in the nucleus and move from the inside of the nucleus to the cytosol through the nuclear pores. **FIG. 4-12** The protein fragment containing the hypothetical NLS sequence would bind to the importin attached to the beads. **FIG. 4-15** The vesicles would be unable to dock with the plasma membrane and release their contents. **FIG. 4-19** Cells would break down the available glucose, but would not be able to produce glucose in the chloroplasts. **FIG. 4-22** The cell would not be able to maintain its appropriate shape or move. **FIG. 4-26** Cilia would not be able to bend if the linking proteins were inflexible or absent.

TEST YOUR UNDERSTANDING

1. d 2. d 3. d 4. e 5. d 6. b 7. a 8. c 9. See Figs. 4-20 and 4-21. 10. None of the measurements agree. Observer 2 is most likely correct if the cells are prokaryotic (0.002 mm = 2 μm); observer 3 is most likely correct if the cells are eukaryotic (2 × 10^4 nm = 20 μm). 11. Consider the following in formulating your answer: Membranous organelles provide the cell with compartments with unique environments. Fibrous cytoskeletal components provide for structure and movement. 12. Mutant a is probably defective in the formation of ER-derived transport vesicles or in movement of vesicles to the Golgi; mutant b probably has a defect in the formation of secretory vesicles on the *trans* face of the Golgi complex. 13. In formulating your answer, consider features common to all cells, including plasma membranes, ribosomes, DNA, and ATP. 14. Consider the origins and applications of different methods used to study the structures and functions of cells.

CHAPTER 5

CHECKPOINT

5.1 • Phospholipids and sterols are the primary determinants of membrane lipid bilayer fluidity. • See Figs. 5-4 and 5-6. • Sugars are added to proteins in the lumens of the ER and Golgi complexes. **5.2** • Specific molecules move across bilayers through transmembrane proteins that form highly selective channels or pumps. • Some membrane proteins respond to signals. Extracellular parts of transmembrane proteins can bind or link to proteins on the surface of neighboring cells. **5.3** • Water, as well as very small, nonpolar molecules. • Channels function as highly selective pores. Transporters bind specific ions or molecules and change shape to move

bound substrates across membranes. • Aquaporins are specific channel proteins that move water across the bilayer. **5.4** • Isotonic conditions result in no change in plant or animal cells. Under hypertonic conditions plant cells undergo plasmolysis due to water loss; they become turgid under hypotonic conditions. Hypertonic conditions cause animal cells to shrink; under hypotonic conditions they swell and may possibly burst. • The energy source for diffusion and facilitated diffusion is a chemical concentration gradient. • The net movement of particles in a concentration gradient is from a region of high concentration to low concentration. Simple diffusion and facilitated diffusion do not differ in the direction of net particle movement. **5.5** • Active transport requires a source of metabolic energy, such as ATP. • In cotransport, the transport of a substance down its concentration gradient provides the energy to drive the transport of the other substance against its concentration gradient. **5.6** • Both exocytosis and endocytosis move materials across the plasma membrane by vesicular transport mechanisms. • Phagocytosis moves large solid particles across the membrane; pinocytosis moves dissolved molecules. • The steps for the sequence of events for receptor mediated endocytosis are outlined in Figure 5-22. **5.7** • Both desmosomes and tight junctions form strong attachments between neighboring cells. Desmosomes form strong sheets of cells; tight junctions prevent passage of materials through intercellular spaces. • Both plasmodesmata and gap junctions allow the movement of water, ions, and small molecules between the cytosols of neighboring cells. Gap junctions form protein-based channels between neighboring cells; plasmodesmata form continuous membrane-lined channels between the plasma membranes of neighboring cells.

FIGURE QUESTIONS

FIG. 5-5 If the proteins labeled with the green fluorescent antibody were linked to cytoplasmic microtubules, they would not immediately become distributed over the entire hybrid cell. **FIG. 5-16** If equal numbers of Na$^+$ and K$^+$ ions were moved in each pump cycle, chemical gradients of the two different ions would be generated, but there would be no electrical gradient produced. **FIG. 5-18** The energy source for glucose transport is the cotransport of Na$^+$ ions down their concentration gradient. If the Na$^+$/K$^+$ pump were inhibited, the Na$^+$ gradient would disappear, blocking the ability of the symporter to transport glucose. **FIG. 5-22** Humans who produce an LDL-receptor protein with a missing membrane-spanning region exhibit very high levels of cholesterol in the blood because their cells are unable to bind and remove it.

TEST YOUR UNDERSTANDING

1. c 2. e 3. a 4. b 5. d 6. e 7. b 8. e 9. e 10. The stimulated cells have ten times as many GLUT4 proteins in their plasma membranes as unstimulated cells. 11. Consider that plant cells are held in relatively fixed positions by their cell walls, and also that plant cell walls may act as barriers to many types of molecules. 12. Consider that the properties of cell membranes include their ability to form compartments that maintain unique chemical environments. Can you think of other ways this could have been accomplished in the origin of life? 13. Early evolution of transport proteins would have greatly improved the ability of cells to maintain high internal concentrations of essential molecules and allowed cells to interact with their external molecular environments in a controlled way. These functions would have needed to be provided by some other mechanism if transport proteins had evolved later.

CHAPTER 6

CHECKPOINT

6.1 • The sequence of events involved in cell signaling are: *signal transmission* (synthesis and release of signal molecules), *signal reception* (binding of signals to target cell receptors), *signal transduction* (conversion and amplification of extracellular signals to intracellular signals), and *response* (conversion of the final signal into a cellular response that alters some cellular process). • Signal transduction results in the conversion of an extracellular signal into an intracellular form. Multiple intracellular signals are typically produced in the process, activating and/or deactivating multiple genes and proteins. **6.2** • Neurotransmitters are molecules that transmit signals across synapses between adjacent nerve cells. • Animal hormones are typically secreted by cells into the interstitial fluid. Some may bind to neighboring target cells, whereas others enter the blood to be transported to distant targets. • Endocrine and paracrine regulation are similar in that both involve the production of hormones secreted by ductless glands into the interstitial fluid. Paracrine hormones act locally, diffusing directly to their target cells. Endocrine regulation typically involves the transport of hormones through the blood to other parts of the body. **6.3** • Receptor proteins have binding sites that recognize specific signal molecules. • Receptor up-regulation occurs in response to low concentrations of signaling molecules, increasing the sensitivity of that cell to that particular signal. Receptor down-regulation occurs when concentrations of signaling molecules remain high over extended times, decreasing the sensitivity of the down-regulated cell to that specific signal • Cell-surface receptors contain an external *signal-binding domain*, a *membrane-spanning domain*, and an *internal cytoplasmic domain*. Each type of cell-surface receptor contains a unique signal-binding domain that binds a specific type of signal molecule. • Intracellular receptors bind signal molecules that can diffuse into the cell through the plasma membrane. The activated intracellular receptors are typically transcription factors that regulate the expression of specific genes. **6.4** • The binding of a signal molecule to a cell surface receptor alters the shape of the receptor protein, resulting in conformational changes in its intracellular domain. These changes trigger the activation of intracellular enzymes that produce or activate intracellular signaling molecules. For example, a signaling molecule binds to a receptor that activates a G protein; adenylyl cyclase is activated and catalyzes the conversion of many ATPs to cyclic AMP; cyclic AMP activates many protein kinases, which phosphorylate many proteins, leading to one or more responses. • Second messenger molecules are produced intracellularly in large quantities in response to an extracellular signal. • Scaffold proteins tightly organize the proteins involved in a signal transduction pathway so that they can interact efficiently. Loss of a scaffold protein could result in significant slowing of the signaling pathway, reducing the effectiveness of the response. **6.5** • Cells respond to signals by opening or closing ion channels, altering enzyme activities, or by changing patterns of gene expression • Refer to Figure 6-14 as an example of signal amplification. • The failure to terminate signal transmission can lead to maintenance of the signal, resulting in continuation of a now-inappropriate response such as a failure to close certain ion channels. **6.6** • The observation that choanoflagellates and animals have similar protein kinases suggests that cell signaling became important at very early stages of animal evolution.

FIGURE QUESTIONS

FIG. 6-2 If a cell were exposed to a drug that blocked the activity of signaling molecule A, signal molecule B and other signaling molecules downstream of B would not be activated. There would be no changes in membrane permeability, metabolic processes, and gene expression controlled by that signaling pathway. **FIG. 6-12** If the Ras protein is continuously activated, any of the downstream components of the signaling pathway (Raf, Mek, or ERK) might prove to be good candidates for a drug target aimed to prevent cancerous growth. All of these components act as steps in the pathway to activation of cellular growth.

FIG. 6-13 If the altered LH receptor were activated in the absence of LH, its signaling pathway would be continuously activated, resulting in the constant activation of oocyte maturation and ovulation.

TEST YOUR UNDERSTANDING

1. a 2. e 3. d 4. c 5. a 6. a 7. c 8. d 9. c 10. Binding of a hormone to a G protein–linked receptor results in the formation of many second messengers along with the activation of a number of different protein kinases. Each of these types of signals can result in the activation (or deactivation) of different sets of genes and enzymes, leading to the blocking of DNA synthesis, directional growth toward a mating partner, and other physiological events associated with mating in yeast. 11. The identification of hundreds of protein kinase genes in humans indicates that these enzymes play major roles in the regulation of many different physiological functions in human cells. Since each kinase activates or deactivates a specific set of proteins in response to specific cellular signals, the large number of different kinases indicates that many different types of responses are controlled by these enzymes. 12. d 13. b 14. You might hypothesize that plants and animals have common ancestors. Because they share many of the basic molecular components involved in membrane and organellar structure, growth, metabolism, and cell division, a number of these molecules might have been selected as molecular signals. Similarities in plant and animal signaling include protein kinases and G protein receptors that employ GTP or trigger the synthesis of modified forms of ATP (cAMP) as signals. For the differences in signaling, consider the different structures and functions of plant and animal cells. Plants have cell walls and cells communicate through plasmodesmata, whereas animals can communicate via the extracellular fluid through receptors on membrane surfaces. Plants also respond to different types of stimuli, including light, which could lead to the evolution of receptors and different types of light-sensitive signal molecules. 15. The occurrence of G protein–linked receptors and signal transduction pathways found in plants and animals as well as fungi and algae suggests that these signal pathways may have evolved very early in a common ancestor of all these organisms. This suggests that these molecules and pathways existed in ancient organisms early in the evolutionary process. 16. Tyrosine kinase signaling proteins trigger different types of signal transduction pathways in different cell types, depending on which types of signal transduction proteins are present. Drugs that inhibit tyrosine kinases in cancer cells may also inhibit tyrosine kinases that may be required for certain essential cellular functions in normal cells.

CHAPTER 7

CHECKPOINT

7.1 • See Fig. 7-1. Kinetic energy is used to stretch the spring. Work is performed. The stretched spring has stored potential energy. Potential energy is converted to kinetic energy as the spring is released. **7.2** • First law of thermodynamics: energy cannot be created or destroyed, but can be converted from one form of energy to another. Second law of thermodynamics: when energy is converted from one form to another, some of the usable energy is converted to heat and dissipated into the surroundings. This results in the amount of usable energy in the universe available to do work decreasing with time. • Organisms succeed in the struggle with the second law because they are open systems. They can capture energy from their surroundings to do work. Although this allows living systems to increase in order, the total entropy of the universe (organisms plus their surroundings) always increases over time. **7.3** • In a reaction in which enthalpy decreases and entropy increases, ΔG has a negative value. The reaction is exergonic. • Reaction 2 can do work. **7.4** • (1) X + ATP $\longrightarrow$ X–P + ADP; (2) X–P $\longrightarrow$ Y + P_i; If the ratio of ATP to ADP were to become 1:1, ATP-requiring reactions would not work effectively because the ATP/ADP system would be near equilibrium. **7.5** • Reduced compounds have more energy available than oxidized forms. They contain electrons that can be transferred to oxidizing agents, transferring energy in the process. **7.6** • An enzyme reduces the energy of activation of a chemical reaction, speeding up the reaction. • An active site of an enzyme is a region involved in the conversion of a substrate to its product. An allosteric site is a binding site on the enzyme for a molecule that regulates the enzyme's activity. • Temperature and pH affect an enzyme's structure, and thus its activity, by promoting the formation or breaking of hydrogen bonds and number of electrical charges within the protein. • Allosteric inhibition is noncompetitive. The allosteric regulatory molecule binds at a site other than the active site of the enzyme.

FIGURE QUESTION

FIG. 7-10 The reaction would not proceed if ΔG had a positive value. A reaction cannot proceed spontaneously if the products will have more energy than the reactants.

TEST YOUR UNDERSTANDING

1. e 2. d 3. b 4. c 5. b 6. c 7. b 8. b 9. b 10. e 11. a 12. The use of ATP/ADP by all organisms suggests that their role in energy metabolism evolved very early in the history of life among

ancestors that are common for all living organisms. 13. The statement misinterprets the second law of thermodynamics. Entropy always increases only in a closed system such as the entire universe. Organisms are open systems that can extract energy from their surroundings, which can be used to build greater complexity. 14. The reaction is exergonic. The free energy of the products is less than the free energy of the reactants. 15. Reaction 2 is capable of performing work. 16. If the inhibition is competitive, increasing the succinate concentration should reverse the inhibition of the enzyme.

CHAPTER 8

CHECKPOINT

8.1 • Glucose loses electrons, becoming oxidized. • Oxygen serves as an electron acceptor in most cells. **8.2** • ATPs available from one molecule of glucose: (1) glycolysis produces a net yield of 2 ATPs from substrate-level phosphorylation in the cytosol; (2) acetyl CoA formation in the mitochondria does not produce ATP; (3) the citric acid cycle produces a net yield of 2 ATPs from substrate-level phosphorylation in the mitochondria; (4) oxidative phosphorylation produces a net yield of 32–34 ATPs produced from NADH and $FADH_2$ in the mitochondria through electron transport and chemiosmosis. • The electron transport chain functions in the transport of H^+ ions from the mitochondrial matrix across the inner mitochondrial membrane, establishing a proton gradient. Energy stored across the mitochondrial inner membrane in the form of the proton gradient powers ATP synthesis as protons flow back across the inner membrane through the ATP synthase complex. The ATP synthase complex catalyzes the formation of ATP using the energy from proton movement through the complex. • NAD^+ and FAD accept electrons derived from the breakdown of glucose and its products, forming NADH and $FADH_2$. These high-energy electrons are then passed to membrane proteins in the electron transport chain, powering the transport of H^+ ions across the inner membrane to establish the proton gradient used to synthesize ATP. Oxygen is the terminal electron acceptor from the electron transport chain. In the process it is reduced to H_2O. • If most of the ADP in the cell were converted to ATP, oxidative phosphorylation would slow because of a lack of ADP to be converted to ATP. **8.3** • Humans obtain energy from a low carbohydrate diet by converting lipids and other biological molecules in their food into molecular intermediates of glycolysis or the citric acid cycle. • Amino acids from food supplies must be deaminated before the carbon chain can be

metabolized through glycolysis or the citric acid cycle. • Refer to Figure 8-13. Show in your diagram that fatty acids are broken down to form acetyl coenzyme A before being metabolized through aerobic respiration in the citric acid cycle. **8.4** • An equivalent number of hydrogen atoms removed from glucose during glycolysis are transferred to NAD^+ to form NADH. When oxygen is present, the NADH is recycled to NAD^+ through electron transport and oxidative phosphorylation. An equivalent number of these hydrogen atoms is used in the formation of H_2O when the electrons are passed to molecular oxygen during that process. In the absence of oxygen, NAD^+ is recycled from NADH by passing the electrons (and an equivalent number of hydrogen atoms) to organic molecules, forming ethanol and carbon dioxide or lactate. • The ATP yield from fermentation is much lower than that from aerobic respiration because the energy released from the glucose in the form of reduced NADH and $FADH_2$ cannot be used to form large amounts of ATP through electron transport and oxidative phosphorylation. • Chemiosmosis is not involved in fermentation, but anaerobic respiration does involve an electron transport chain (with a terminal electron acceptor other than oxygen) and chemiosmosis.

FIGURE QUESTIONS

FIG. 8-3 In glycolysis, 2 molecules of G3P will yield 4 ATPs, whereas glucose will consume 2 molecules of ATP in the formation of G3P, yielding a net gain of only 2 ATPs. Two molecules of G3P will yield 4 molecules of ATP and 2 molecules of NADH in glycolysis; no energy is captured from pyruvate in glycolysis because it is the end product in the pathway. **FIG. 8-8** If complex III were missing, complex I and complex II would not be able to pass on their electrons down the chain to complex IV and molecular oxygen. Complex I and complex II would remain in a reduced state and would not be able to recycle NADH and $FADH_2$ to the NAD^+ and FAD needed to accept new electrons generated from the citric acid cycle. **FIG. 8-9** If the cells were placed in a basic (low H^+ concentration) environment, a proton gradient would not exist and ATP synthesis would not take place. **FIG. 8-11** If ATP synthases were removed, electron transport would still create a proton gradient across the membrane, but ATP synthesis would not take place. **FIG. 8-14** Anaerobic respiration, because in that process high-energy electrons are transferred from NADH to an electron transport chain that synthesizes ATP by chemiosmosis.

TEST YOUR UNDERSTANDING

1. d 2. a 3. a 4. b 5. c 6. d 7. d 8. e 9. b 10. c 11. d 12. The proton gradient represents a state of low entropy because the proton gradient is highly organized with high H^+ concentrations

in the intermembrane space and low H^+ in the matrix. 13. The first phase of glycolysis involves the coupling of the exergonic reactions of ATP hydrolysis to the endergonic phosphorylation of glucose to form fructose 1,6 bisphosphate. The second phase of glycolysis involves the coupling of the exergonic reactions of the hydrolysis of phosphate and the removal of electrons from G3P to the endergonic phosphorylation of ADP to form ATP and the reduction of NAD^+ to form NADH. 14. If the inner mitochondrial membrane were readily permeable to H^+ ions, the coupling of electron transport to ATP synthesis could not occur because a proton gradient could not be formed across the membrane. 15. Refer to Figure 8-11 in constructing your diagram. 16. When you lose weight, carbon-containing compounds (especially fatty acids) become oxidized, yielding carbon dioxide and water. 17. The universal occurrence of the identical reactions of glycolysis in all organisms indicates that glycolysis evolved at early stages in the origin of life, in cells that became the ancestors of all modern organisms. 18. The need for photosynthetic organisms to continuously replenish oxygen today indicates that aerobic respiration could not have evolved without the previous evolution of oxygen-releasing photosynthetic processes.

CHAPTER 9

CHECKPOINT

9.1 • Red light has the longer wavelength. • Violet light has the higher energy per photon. **9.2** • Refer to Figure 9-4b. The line should be colored green. One side of the line should be labeled the thylakoid lumen. The other side should be labeled the chloroplast stroma. • The matching of the action spectrum of photosynthesis with combined spectra of chlorophylls *a* and *b* indicates that these molecules are the primary mechanism for capturing light energy for photosynthesis. Photosynthesis would be less efficient if their absorption spectra coincided exactly because fewer wavelengths of light in the red and blue regions would be captured, losing those sources of energy. • Fluorescence results from release of a low-energy photon from a molecule that has captured a photon with higher energy. That low-energy photon cannot be used to drive photosynthesis. **9.3** • Molecular oxygen is more oxidized than the oxygen in a water molecule. • Carbon fixation reactions are dependent on ATP and NADPH molecules produced during the light-dependent reactions of photosynthesis. **9.4** • Molecular oxygen is released as a waste product through the splitting of water as electrons are transferred to photosystem II. In aerobic respiration molecular oxygen is the terminal electron acceptor for the mitochondrial electron transport chain, forming water in the process. • Photophosphorylation of ADP to form ATP involves the transport of electrons energized in photosystem II. Their energy is used to form a proton gradient across the thylakoid membrane. ATP is formed when protons diffuse back, through the ATP synthase complexes in the thylakoid membrane. • Cyclic electron transport through photosystem I cannot drive photosynthesis. While ATP can be generated through this process, it does not produce NADPH required for photosynthesis. **9.5** • Phase 1 of the Calvin cycle (CO_2 uptake phase) involves the fixation of carbon when CO_2 reacts with RuBP to form PGA. In phase 2 (carbon reduction phase) PGA is converted to G3P using energy from ATP and NADPH formed during the light-dependant reactions. G3Ps are converted to carbohydrate or remain in the cycle in phase 3 (RuBP regeneration phase, which also requires ATP energy). • A decrease in entropy occurs during CO_2 uptake due to the fixation of the CO_2 into a carbon skeleton by combining it with RuBP. This reaction is exergonic because RuBP is a high-energy molecule. Its formation required energy from ATP and NADPH earlier in the cycle. • Photorespiration requires oxygen and produces CO_2 and H_2O; it differs from aerobic cellular respiration in that ATP is not produced. • C_3, C_4, and CAM plants all have rubisco, which is used to fix CO_2 through the reaction with RuBP at the beginning of the Calvin cycle. Only C_4 and CAM plants employ PEP carboxylase as a way of initially capturing CO_2 prior to the Calvin cycle. **9.6** • Green plants obtain energy from light. They obtain carbon through the fixation of CO_2. Humans obtain energy and carbon from reduced carbon compounds derived from food. **9.7** • Root cells obtain energy and organic molecules from sugars transported through plant vascular tissues from photosynthetic tissues. • Molecular oxygen in the atmosphere is produced by photosynthetic organisms.

FIGURE QUESTIONS

FIG. 9-7 We would expect the bacteria to be distributed in the area corresponding to the yellow/green part of the spectrum if their photosynthetic pigments had a maximal absorption at those wavelengths. **FIG. 9-11** If photosystem I were missing, there would be no terminal electron acceptor to receive the electrons derived from water in photosystem II, so all photosynthesis would stop. **FIG. 9-13** If an inhibitor prevented protons from being pumped across the thylakoid membrane, ATP could not be synthesized using energy captured by photosystem I. As long as $NADP^+$ is available, NADPH could be produced by

electron transport through photosystems II and I. **FIG. 9-14** If the 2:10 ratio of G3P molecules leaving the Calvin cycle were to become a 4:8 ratio, the cycle would run down because the pool of G3P would become depleted with each turn of the cycle, leaving insufficient G3P molecules to continue the formation of RuBP.

TEST YOUR UNDERSTANDING

1. a 2. d 3. a 4. e 5. c 6. a 7. c 8. b 9. Refer to Figure 9-13. The interior compartment should be labeled thylakoid lumen. The exterior should be chloroplast stroma. The proton gradient should show high levels in the lumen and low in the stroma. ATP is synthesized in the stroma. 10. Compare your sketch to Figure 8-11. For chloroplasts, protons are pumped into the innermost compartment (the thylakoid lumen); in mitochondria protons are pumped out of the innermost compartment (the matrix). The direction doesn't matter as long as the ATP synthase complex is oriented correctly with respect to the proton gradient. 11. Not all autotrophs use energy from light. Chemoautotrophs capture energy from reduced inorganic compounds such as hydrogen sulfide, nitrite, or ammonia. 12. All actively metabolizing plant cells require energy (in the form of ATP, NADH, and other reduced compounds) for growth, development, and maintenance that is generated through mitochondrial functions. 13. High-energy electrons in glucose became energized during the light-dependant reactions of photosynthesis. 14. Photoautotrophs could sustain life on Earth without the presence of chemoheterotrophs. Chemoheterotrophs, however, could not survive without sources of energy and carbon skeletons provided by photoautotrophs. 15. The common occurrence of ATP synthase complexes among bacteria, chloroplasts, and mitochondria suggests that these enzyme systems must have evolved early in the history of life. 16. The absorption spectrum indicates that these pigments absorb purple and blue light. Their color would be a mixture of reflected green, yellow, orange, and red light that is not absorbed by the pigments. 17. Some strategies for increasing the food supply might involve engineering food crops to more efficiently carry out photosynthesis and carbon fixation. One strategy might be to produce a more efficient form of the carbon-fixing enzyme rubisco. Another might be to increase the productivity of traditional C_3 food crops for growth in arid and hot climates by incorporating C_4 and CAM enzyme systems into the plant photosynthetic tissues.

CHAPTER 10

CHECKPOINT

10.1 • Informational units on chromosomes are called genes. Genes consist of sequences within DNA molecules that encode information needed to carry out one or more specific cell functions. • DNA found within a nucleus is very long compared to the diameter of the nucleus. In order to address the large differences between chromosome length and nuclear dimensions, DNA molecules are packaged in a highly organized way in chromosomes. **10.2** • The stages of the cell cycle consist of G_1, S, G_2, and M phases. DNA is replicated during the S phase. • The stages in mitosis are: *prophase*—chromosomes shorten and condense; *prometaphase*—chromosomes begin to move to cell's midplane; *metaphase*—duplicated chromosomes align on the metaphase plate; *anaphase*—chromatids (each now a chromosome) separate and move toward opposite poles of the cell; and *telophase*—two nuclei are formed and cytokinesis usually occurs, giving rise to two cells. **10.3** • Cell-cycle checkpoints temporarily block key events in the cell cycle from being activated, ensuring that the previous phase of the cell cycle has been completed before the cell cycle can proceed. • Two types of molecular controls that control the onset of different stages of the cell cycle are (1) the association of Cdk with cyclin to form M-Cdk, which triggers the M phase, and (2) the degradation of M-Cdk, allowing daughter cells formed by mitosis to enter G_1. **10.4** • Homologous chromosome pairs are present in diploid cells. Only one member of a homologous chromosome pair is present in a haploid cell. • Meiosis is a process of cell division resulting in the formation of haploid cells. Mitosis is a process of cell division that produces daughter cells that are genetically identical to the original cell. • Haploid cells can divide by mitosis. They cannot undergo meiotic cell division because they do not contain pairs of homologous chromosomes. **10.5** • In animals, somatic cells are produced by mitosis, and gametes by meiosis. In plants, cells of the 2n sporophyte and n gametophyte generations are produced by mitosis. Plant spores (n) are produced by sporophyte cells by meiosis; plant gametes are produced by gametophyte cells by mitosis. • Many protists remain haploid throughout most of their life cycle. Two haploid gametes can fuse through fertilization, however, to form a diploid zygote, which subsequently undergoes meiosis to restore the haploid chromosome number.

FIGURE QUESTIONS

FIG. 10-4 Histone H1 is involved in the packing of 10 nm nucleosomes, forming 30 nm packed nucleosome fibers. **FIG. 10-8** If cohesins failed to separate, two daughter chromosomes might move to the same pole, resulting in one cell that has two copies of a given chromosome and a sister cell that has no copies of that same chromosome. If the chromatids dissociated prematurely in late prophase, the separated chromosomes would not arrange properly on the metaphase plate and might not separate equally into the two daughter cells. **FIG. 10-10** If spindle microtubules disassembled at their polar ends, the distances between the bleached areas and the poles would be decreased. **FIG. 10-13** Refer to Figure 10-9. If the cell were arrested at the metaphase–anaphase checkpoint, chromosomes would stay aligned at the metaphase plate of the cell. **FIG. 10-14** If cyclin were not degraded in step 4, cells formed by mitosis would not enter into the G_1 phase.

TEST YOUR UNDERSTANDING

1. c 2. d 3. e 4. d 5. c 6. c 7. b 8. b 9. Refer to Figures 10-6 and 10-7. 10. The DNA content of a cell doubles as the cells progress through the S phase of interphase. The number of chromosomes does not change. At the end of the S phase, each chromosome consists of two identical DNA molecules as chromatids. 11. For a cell with 5 chromosomes:

	Number of Duplicated Chromosomes	Number of Unduplicated Chromosomes	Number of Kinetochores
Prophase (number per cell)	5	0	10
Metaphase (number per cell)	5	0	10
Telophase (number per nucleus)	0	5	5

The cell is haploid because an odd number of chromosomes indicates that there are no pairs of homologous chromosomes. 12. For a diploid cell with 14 chromosomes:

	Number of Duplicated Chromosomes	Number of Unduplicated Chromosomes	Number of Tetrads
Beginning of prophase I (number per cell)	14	0	0
End of prophase I (number per cell)	14	0	7
End of telophase II (number per nucleus)	0	7	0

13. Honeybee drones are produced by an asexual mechanism with no fusion of gametes. 14. Seeds develop from a flower pollinated by the same plant through sexual reproduction resulting from the fusion of two haploid gametes. 15. The chromosomal mechanisms of mitosis are similar in organisms as diverse as seaweeds and mammals, indicating an evolutionary relationship between these groups of organisms. 16. During unfavorable conditions, sexual reproduction involving meiotic recombination and fusion of gametes would ensure a more genetically diverse generation than the identical progeny produced by asexual reproduction. Greater genetic diversity within a generation increases the chance of survival of at least some offspring as new environmental conditions emerge.

CHAPTER 11

CHECKPOINT

11.1 • A single diploid individual can have a maximum of two different alleles for a genetic locus. • Segregation cannot be demonstrated between crosses of either two homozygous dominant individuals or two homozygous recessive individuals. All progeny in those crosses will either show a homozygous dominant or homozygous recessive phenotype. • A monohybrid cross involves only one locus and cannot be used to demonstrate independent assortment. To observe independent assortment, you need individuals heterozygous for two different loci. **11.2** • The F_1 individuals crossed would be $YyRr \times YyRr$. To produce yellow round seeds a plant must be $Y\_R\_$ (probability in this cross = $\frac{9}{16}$). • The product rule and the sum rule were used to obtain the result. **11.3** • If two loci are unlinked, the ratio of genotypes to phenotypes in a two-point testcross is 1:1:1:1. • Unlinked genes reside on different chromosomes. Crossing-over occurs only between homologous chromosomes. The mechanism of recombination for unlinked genes is independent assortment. • The map distance between the two loci is 5 map units. • The Y chromosome determines male sex in humans and most other mammals. • Dosage compensation is required in mammals because females have roughly twice as many genes on their two X chromosomes as males, who have only one X chromosome. In females, one X chromosome is inactivated in most cells to balance the number of active X chromosome genes between males and females. One of the two X chromosomes is active in some cells, and the other X is active in other cells, leading to variegated expression (on a cell by cell basis) of many X-linked genes. **11.4** • For ABO blood types, the A allele and the B allele are codominant, and are both fully expressed in heterozygotes. For traits with incomplete dominance, both alleles are expressed, but the phenotype is a mixture of the two traits, as in the pink flower color of heterozygous four o'clocks. • A trait with multiple alleles means that more than two different alleles for a locus can occur in a population. Only two alleles, however, can occur within an individual. Polygenic inheritance refers to a phenotypic trait that is produced by the additive action of alleles of more than one locus in an individual. • Pleiotropy is the ability of a gene to have multiple effects on the phenotype. Epistasis refers to the interaction between alleles of two different loci. • The *norm of reaction* refers to the range of phenotypic possibilities that can develop from a single genotype under different environmental conditions. In the case of the hydrangeas, the range may include a variety of pink, purple, and blue hues, but may not extend to other colors such as yellow.

FIGURE QUESTIONS

FIG. 11-3 All F_2 generation short plants are homozygous recessive. You would be able to discover the genotype of an F_2 generation tall plant by crossing it with an F_2 short plant. If the tall plant is heterozygous, you will obtain a 1:1 ratio of tall and short progeny. If it is homozygous dominant, all progeny will be tall (but heterozygous). **FIG. 11-4** The haploid cells would be spores if meiosis were occurring in a plant and gametes if it were occurring in an animal. **FIG. 11-6** You would expect the same results if the genotypes in the P generation were reversed. **FIG. 11-7** If only one black individual were produced from the test cross, the genotype of the black parent could be either homozygous or heterozygous. If a single brown individual were produced, the genotype of the black parent would be heterozygous. **FIG. 11-8** The results would be the same. Segregation and independent assortment do not act differently depending on parental genotype. **FIG. 11-9** The diagram would be drawn the same (refer to the answer for Figure 11-8). **FIG. 11-11** You would expect a similar excess of parental class progeny and a deficiency of recombinant class progeny, but the classes would be reversed. The parental classes would be *Bbvv* and *bbVv*. The recombinant classes would be *BbVv* and *bbvv*. **FIG. 11-12** Your diagram would be similar to that shown in Figure 11-12. The red homologue would have the genotype *bV* on each chromatid and the blue homologue would have the genotype *Bv*. Following the crossover, the recombinant chromatid of the red homologue would have a blue tip with *B* and the recombinant chromatid of the blue homologue would have a red tip with *b*. **FIG. 11-16** The calico cat would be female. If the cat were black or orange, it could be either a male, or a female homozygous for black or orange. **FIG. 11-17** A pink-flowered (heterozygous) plant crossed with a white-flowered (homozygous) plant would yield a 1:1 ratio of pink- and white-flowered plants. Crossing a pink plant with a red (homozygous) plant would yield a ratio of 1:1 red and pink progeny. **FIG. 11-18** Two *PpRr* chickens would produce offspring in the ratio of 9 walnut, 3 pea, 3 rose, and 1 single-combed. **FIG. 11-19** Two black labs could produce a yellow puppy if the parents were heterozygous (*BbEe*), producing a puppy with a *B_ee* or *bbee* genotype. **FIG. 11-20** Possible genotypes from the class in the middle include all combinations that include 3 "dark" alleles and 3 "light" alleles (*AaBbCc, AABbcc, AAbbCc, aaBBCc, aaBbCC, AaBBcc,* or *AabbCC*).

TEST YOUR UNDERSTANDING

1. d 2. c 3. c 4. e 5. c 6. b 7. The short-winged genotype is recessive. The parents were both heterozygous for wing length. 8. Hair length: S = short hair, s = long hair. Coat color: B = black, b = brown/tan. *BBss* × *bbSS* ⟶ *BbSs* F$_1$ progeny (black, short hair). If two F$_1$s are mated, probability of long hair is ¼, and the probability of brown/tan is ¼. ¼ × ¼ = $\frac{1}{16}$. The probability of producing a long-haired, brown/tan cat is $\frac{1}{16}$. 9. Yes, Richard could be the child of the parents if both were heterozygous ($I^A i \times I^B i$). 10. Rooster: *PpRr*; Hen A: *PpRR*; Hen B: *Pprr*; Hen C: *PPRR*. 11. The cross is a two-point test cross; the loci are linked. Parental classes are *Aabb* and *aaBb*; recombinant classes are *AaBb* and *aabb*. The percentage of recombination between the two loci is 4.6%. The two loci are 4.6 map units apart. 12. Gene *A*

is in the middle if *B* and *C* are 10 map units apart. Gene *C* is in the middle if *B* and *C* are 2 map units apart. 13. (a) Refer to Figure 11-4 to construct a diagram of gene segregation. (b) Refer to Figure 11-9 to diagram alleles that undergo independent assortment. (c) Refer to Figure 11-12 to construct a diagram of two linked loci that undergo recombination. 14. You cannot usually discriminate between the phenotypes of heterozygous and homozygous individuals for genes that exhibit simple Mendelian dominant and recessive inheritance. You may be able to predict the phenotype of an individual if you know the genotype, but keep in mind that the phenotype may be influenced by the environment or by other genes (epistasis). 15. Linked genes recombine through crossing-over during meiosis. Unlinked genes are on nonhomologous chromosomes and recombine through independent assortment during meiosis. 16. The inherited variation among individuals that Darwin observed is caused by the different combinations of alleles of one or more loci found in each individual. 17. Sixty-four individuals were involved in the hypothetical cross. Of those, 1.6% (¹⁄₆₄) had the lightest skin possible and 1.6% had the darkest skin possible.

CHAPTER 12

CHECKPOINT

12.1 • Griffith's work showed that a genetic trait (virulence) could be transferred from one bacterial strain to another. This allowed Avery and his colleagues to determine that the trait was transferred from the DNA that was extracted from the donor strain. • Hershey and Chase knew that proteins of bacteriophages could be specifically labeled with radioactive sulfur and that bacteriophage DNA could be specifically labeled with radioactive phosphorus, allowing them to determine which type of molecule entered a bacterial cell upon infection. **12.2** • Adenine, thymine, guanine, and cytosine nucleotides make up a single strand of DNA. They are linked together through their 5′ phosphate groups that are covalently bonded to the 3′ carbon of the deoxyribose subunit of the next base in the polynucleotide chain. • Watson and Crick determined that double-stranded DNA consisted of two single strands of DNA running in opposite directions to form a helix, with the nitrogenous bases stacked on the inside and their phosphodiester chains on the outside; the bases in the interior of the helix are hydrogen bonded to their complementary bases (A:T and G:C) on the opposite strand. • Chargaff's rules showed that in DNA molecules extracted from different organisms, the amount of adenine was equal to the amount of thymine and the amount of guanine was equal to the amount of cytosine. This corresponded to the base pairing rules for

double-stranded DNA in Watson and Crick's model. **12.3** • Meselson and Stahl were able to separate and identify DNA strands from different generations of bacteria by labeling them with either ^{14}N or ^{15}N as they were being replicated. By showing that ^{15}N-labeled DNA progressed from a dense to intermediate and lighter densities in subsequent generations of growth on ^{14}N-labeled medium, they demonstrated that DNA was replicated by a semiconservative mechanism. • The two strands of a DNA molecule run in opposite directions. Because DNA polymerase can only add bases to a growing 3′ end, one strand is replicated in a continuous manner, and the opposite strand is replicated discontinuously. • Eukaryotic cells require telomerase because their DNA is linear, resulting in shortened ends with each round of DNA replication. Bacterial DNA is circular and is replicated completely with each round of replication.

FIGURE QUESTIONS

FIG. 12-1 Heat-killed S cells treated by methods 1 or 2 would still transform the rough cells to virulence. Method 3 would prevent the S cells from transforming R cells into S cells. **FIG. 12-3** The Hershey–Chase experiments reinforced Avery, McCleod, and McCarthy's findings that genes are composed of nucleic acids by showing that only bacteriophage DNA is required to form new bacteriophage particles composed of both DNA and protein. **FIG. 12-9** G:C base pairs are linked by three double bonds; A:T base pairs are linked by two double bonds. It would require more energy to separate the strands (to break the bonds) of G:C-rich DNA than A:T-rich DNA. **FIG. 12-11** Refer to Figures 12-10b and c. Conservative replication would produce a light and heavy band after the first round of replication (no intermediate band). Dispersive replication would produce a smear of bands in the centrifuge after one and two replication cycles.

TEST YOUR UNDERSTANDING

1. c 2. d 3. d 4. e 5. b 6. d 7. b 8. e 9. d 10. b 11. a 12. b 13. Cancer cells are locked in a constant state of growth and cell division. They require active telomerase to maintain continuous cell division without losing essential chromosomal material. 14. Refer to Figures 12-15 and 12-16 to construct your diagram. 15. In drafting your response, consider the properties the genetic material must have, including the ability to store information, the ability to be copied accurately from one generation to the next, and the ability to undergo stable mutations that persist in following generations. Then explain how these properties are found in the structure of DNA. 16. If the radioactive-labeled proteins were found inside the cells, Hershey and Chase would have to argue that the genetic information needed to replicate and assemble bacteriophages was encoded by proteins. 17. DNA is

the universal molecule of inheritance in all cells. This supports evolutionary theory providing a common mechanism for inheritance and the ability to undergo stable changes in inherited traits, providing a source of variable traits for natural selection to act upon.

CHAPTER 13

CHECKPOINT

13.1 • The one gene, one enzyme hypothesis argued that each gene encodes the information to produce a single enzyme. • Garrod showed that a single recessive gene was associated with a missing enzyme; Beadle and Tatum showed that there was a one-to-one correspondence of genes with enzymes; Pauling demonstrated that a mutation in a single gene altered the structure of a single polypeptide chain, refining the definition of a gene (at that time) to "one gene, one polypeptide chain." **13.2** • DNA (transcription) $\longrightarrow$ RNA (translation) $\longrightarrow$ polypeptide. • DNA and RNA are both polynucleotides consisting of strands of nucleotides linked together by phosphodiester bonds. They differ in that DNA molecules are typically double-stranded molecules with deoxyribose as the sugar. RNA molecules are typically single-stranded molecules with ribose as the sugar component. **13.3** • DNA and RNA polymerases are enzymes that assemble polynucleotide chains by adding nucleotides to a growing 3′ end. They form phosphodiester linkages from nucleoside triphosphate subunits. DNA polymerases can only polymerize nucleotides from an existing 3′ end of DNA molecule that is hydrogen-bonded to complementary strand; RNA polymerases do not require a 3′ end of a primer to copy a template DNA strand. • The transcribed RNA would read 5′ AUG ACG UAU UAC UAA 3′. The nontemplate DNA strand would read 5′ ATG ACG TAT TAC TAA 3′. • Eukaryotic mRNAs have a 3′ $\longrightarrow$ 5′ 7-methylguanosine cap at the 5′ end of the molecule and a poly-A tail at the 3′ end of the message. **13.4** • Ribosomes are composed of two subunits. The small subunit contains a single RNA molecule and over 20 different proteins; the large subunit contains two RNA molecules and over 30 proteins. The ribosomal RNAs have catalytic functions and do not encode information about the amino acid sequence of proteins. • Initiation of protein synthesis involves binding of the small ribosomal subunit to the start codon region of the mRNA. The initiator tRNA binds to the start codon, followed by the assembly of the complete ribosome. Elongation is a cyclic process in which amino acids are added, one by one, linked by peptide bonds to the growing polypeptide chain. Elongation proceeds from the 5′ end to the 3′ end of the mRNA and from the amino end to the carboxyl terminal end of the polypeptide chain. Termination occurs when the ribosome encounters a stop codon in the mRNA. A release factor binds to the ribosomal A site, releasing the polypeptide chain and disassociating the ribosomal subunits and translation complex. • The anticodons are 3′ UAC UGC AUA UUG AAA, and the corresponding amino acid sequence is NH$_2$–Met–Thr–Tyr–Asn–Phe–COOH. **13.5** • The main types of mutations are base-pair substitution mutations (either silent, missense, or nonsense mutations), frameshift mutations, or mutations involving mobile genetic elements. • Silent mutations do not alter the amino acid composition of the protein; missense mutations substitute one amino acid for another in the polypeptide; a nonsense mutation inserts a stop codon, resulting in a shortened polypeptide. Frameshift mutations alter the reading frame, usually resulting in the insertion of a stop codon a short distance from the mutation. **13.6** • RNA interference occurs when small RNA molecules interfere with the expression of genes or their transcripts. • A gene is a unit of inheritance. In molecular terms it is a DNA sequence that contains information used to produce an RNA or protein product. • Retroviruses use reverse transcriptase to synthesize a DNA strand that is complementary to the viral RNA. After the double-stranded DNA provirus is completed, it is integrated into the host cell's DNA.

FIGURE QUESTIONS

FIG. 13-2 Citrulline or arginine would support the growth of a strain missing both enzyme 1 and enzyme 2. **FIG. 13-4** mRNA strand: 5′ AUG ACU UGC GAA UGU UUC 3′; template strand: 3′ TAC TGA ACG CTT ACA AAG 5′ **FIG. 13-9** RNA polymerase will require a helicase activity to unwind the DNA, exposing the template strand; a polymerase activity to form the phosphodiester bonds of the mRNA; and topoisomerase activities to reduce strain on the DNA as the double helix is uncoiled. Each of these activities may involve the actions of two or more proteins. **FIG. 13-12** A mutation in the polyadenylation signal of an mRNA would prevent it from being polyadenylated. This would probably prevent the mRNA from being exported from the nucleus. The absence of a poly-A tail would also affect the stability of the mRNA, as well as prevent it from being properly translated.

TEST YOUR UNDERSTANDING

1. e 2. a 3. b 4. d 5. b 6. a 7. e 8. c 9. a 10. d 11. b 12. e 13. Bacterial and eukaryotic mRNAs are both transcribed by RNA polymerase II enzymes. Transcription is started at initiation sequences that are downstream of the promoter site. Both types of mRNAs are transcribed in a 5′ $\longrightarrow$ 3′ direction as the polymerase travels down the template strand in a 3′ $\longrightarrow$ 5′ direction. Unlike bacterial mRNAs, the 5′ ends of eukaryotic mRNAs are quickly capped after the initiation of transcription. Intron sequences in the eukaryotic pre-mRNA molecule are removed by spliceosomes and the remaining exons are spliced together. After the poly-A site (which is located 3′ to the mRNA coding sequences) is transcribed, the pre-mRNA is cut and poly-A sequences are added to the 3′ end of the transcript. Transcription of eukaryotic mRNA requires more energy because many of the nucleotides incorporated into the pre-mRNA are removed by intron splicing and are not required for protein synthesis. The formation of the poly-A tail also requires the use of more nucleotides to form the functional mRNA than that found on a corresponding bacterial mRNA molecule. The eukaryotic mRNA offers more opportunities to control gene expression because regulation could occur at the levels of transcription, capping, intron splicing, polyadenylation, and mRNA export from the nucleus. 14. The genetic code was deciphered by adding artificial mRNAs with specific nucleotide sequences to an artificial protein synthesis system. By analyzing the amino acid composition of the resulting polypeptides, researchers could connect specific nucleotide sequences to the genetic code. 15. Transposons would eventually lose the ability to replicate and jump between genes as they acquired mutations that would disrupt their transposable element functions. 16. Refer to Figure 13-23. RNA $\longrightarrow$ (reverse transcriptase) $\longrightarrow$ DNA. Your diagram should show that reverse transcriptase polymerizes the DNA strand in a 5′ $\longrightarrow$ 3′ direction as it reads the RNA strand from a 3′ $\longrightarrow$ 5′ direction. 17. If the irradiated fungus grows on all three media, then it does not have a mutation in a gene affecting its nutritional requirements. If it does not grow on any of the three media, its mutation is unknown, but it does not affect vitamin or amino acid biosynthesis. 18. Prokaryotes may have lost their introns early in evolution because the synthesis of intron-free mRNA may have conferred a selective advantage on the ancestral organisms, by supporting more rapid, energy-efficient growth.

CHAPTER 14

CHECKPOINT

14.1 • Regulation of transcription is the most efficient means of gene regulation in bacteria. • Prokaryotic gene regulation acts primarily through the control of transcription. Eukaryotic gene regulation can act at many different levels of gene expression such as control of transcription, and posttranscriptional mechanisms that include regulation of mRNA processing, nuclear export and stability, and controls of translation, protein activity, and protein stability. **14.2** • Promoter—RNA polymerase binding site; CAP binding site—binds catabolite activator protein when cells have insufficient supply of glucose; operator sequence—binding site for *lac* repressor protein; structural genes—encode enzymes and proteins that function in lactose metabolism. • The *trp* operon encodes a promoter, operator (repressor binding site), and structural genes as does the *lac* operon. • The *trp* operon is a repressible operon because it is usually turned "on" and is only repressed when the cells have sufficient levels of tryptophan; the *lac* operon is inducible because it is only activated when cells have an insufficient supply of glucose and are exposed to the sugar lactose. • Glucose is the preferred carbon source; cells sense glucose levels, and when they are low, cAMP is produced, activating the CAP protein, which binds near the lactose promoter to speed up transcription of the *lac* operon. **14.3** • Bacterial genes are primarily regulated through mechanisms that control the rate of transcription. Eukaryotic genes are regulated at many different levels of gene expression, including transcriptional, posttranscriptional, and posttranslational controls. • Certain eukaryotic genes (e.g., ribosomal RNA genes, egg storage protein genes) that must produce high levels of their products in cells are present in multiple copies to provide high levels of RNA needed to produce high levels of their encoded product. • Chromosome structure affects the expression of eukaryotic genes by controlling its packing. Genes are inactivated in tightly packed and coiled chromosomes (e.g., heterochromatin). Active genes are associated with more loosely packed euchromatin. Chromosome packing is controlled through chemical modifications of histones and the methylation of DNA. • Alternate splicing can produce different proteins by removing or leaving intact some of the pre-mRNA molecule during the splicing process. Refer to Figure 14-15 in constructing your diagram. • See Table 13-1 and Fig. 14-14.

FIGURE QUESTIONS

FIG. 14-2 If the *lac* repressor gene were deleted, the *lac* operon would be activated in the presence or absence of lactose. **FIG. 14-4** If the *trp* operator region were mutated or deleted, the operon would remain active in cells grown with high intracellular levels of tryptophan. **FIG. 14-5** If adenylyl cyclase were activated in the absence of glucose, the *lac* operon would be activated to high levels in cells grown on lactose, with no glucose.

TEST YOUR UNDERSTANDING

1. a 2. c 3. c 4. b 5. c 6. e 7. a 8. c 9. a 10. c 11. c 12. (a) constitutive; (b) inducible; (c) repressible. 13. *Mutant a:* mutation lies within the *trp* operator; *mutant b:* mutation lies within the *trp* promoter; *mutant c:* mutation lies within the *trp* repressor coding sequence. 14. The data suggest that the deletion removed an enhancer sequence that interacts with the gene. 15. The repressor gene needs to be

constitutively transcribed and not under the control of the *trp* operon. If it were under the control of the *trp* promoter, it would not be transcribed under repressed conditions. 16. Multiple levels of control probably evolved in eukaryotes due to cell differentiation and the long lifetimes of cells. A eukaryotic cell might encounter many different changes to physiological, metabolic, and nutritional states during its lifetime, requiring different regulatory mechanisms to respond to those changes. 17. Gene regulation is essential in all cells to ensure efficient growth and adaptability to environmental changes. Loss of an essential regulatory element would most likely result in the loss of selective advantage or death.

CHAPTER 15

CHECKPOINT

15.1 • 3′ CCTAG↓G. • Both types of libraries contain fragments of DNA inserted into plasmids that are maintained in a large population of cells (in many cases, *E. coli*). Genomic libraries contain DNA fragments corresponding to parts of chromosomal DNA. Gene sequences in eukaryotic genomic libraries may contain introns. cDNA libraries contain DNA fragments that are copies of mRNA molecules—those DNA fragments do not contain introns. • DNA probes are used to identify specific DNA or RNA fragments in Southern blots and Northern blots by hybridization to base sequences that are complementary to the probe molecules. • If the maximum insert size were 1×10^4 base pairs, the minimum number of plasmids (or colonies of bacteria each containing a single type of plasmid) that would contain the entire genome would be 3×10^5. If the insert size were 350 kb, 8571 plasmids (or colonies) would be required. Actually, many more colonies would be required to isolate a complete genome because some genomic DNA would be inserted multiple times into a plasmid by chance, whereas others might be missed if only the minimum number were used. • PCR amplification allows one to specifically amplify the desired fragment of DNA to be inserted into a plasmid, whereas gene cloning would require sorting through many different clones to find the required gene from a library. **15.2** • For Northern blot analysis, equal amounts of RNAs isolated from the two cell lines are separated according to size by gel electrophoresis. The separated RNAs are transferred to a filter and the filter is exposed to a radioactive probe complementary to the gene of interest. The amount of radioactivity that binds to each of the RNA samples is compared to determine the relative levels of expression between the two cell types. For RT–PCR analysis, the cDNAs of the gene of interest from each cell type are amplified by PCR

using fluorescent-labeled primers. The relative levels of fluorescence are compared between the two cell types. For a DNA chip, mRNAs from each cell type are copied as fluorescent-labeled cDNAs using different colors for each cell type. The two cDNA populations are incubated on a DNA chip containing dots of DNA fragments from each gene in the genome. The level of gene expression in the cells is determined by comparing the relative fluorescence intensity of the two colors found in each dot on the chip. • Identify your gene by comparing its sequence to data in the GenBank. Using your derived protein sequence data, search for similar proteins (or parts of your protein) in organisms in which their function is known. **15.3** • The gene might be involved in one of a number of different developmental functions relating to touch, including the development of touch receptor cells, or of nerve cells used to signal touch, or of muscle cells that might respond to the touch signals. **15.4** • The significant medical advantages of recombinant human insulin include the lack of allergic reactions produced by insulin produced by animal sources. • STRs are stretches of short DNA sequences repeated multiple times within the human genome. The number of repeated sequences in each region is highly variable between individuals. By measuring the number of repeated sequences in multiple STR regions, investigators can identify an individual with a very high degree of probability. • Mice have many genes in common with humans. By using targeted gene modifications and mutagenesis of genes in mice that are also found in humans, investigators can gain insights into their functions and develop potential treatments for genetic disorders involving those genes. **15.5** • Environmental risks include the potential for genetically modified organisms to pass on engineered foreign genes to wild relatives or different species, causing unknown environmental problems, such as the production of "super weeds" from herbicide-resistant crops, or the harmful killing of useful insects from crops engineered to produce pesticides.

FIGURE QUESTIONS

FIG. 15-9 Lane 1: homozygous mutant; lane 2: homozygous wild type; lane 3: heterozygote. **FIG. 15-13** Suspect 2's DNA matches the crime scene DNA.

TEST YOUR UNDERSTANDING

1. a 2. b 3. e 4. e 5. a 6. b 7. a 8. a 9. c 10. b 11. d 12. Many eukaryotic genes contain introns, which must be removed from pre-mRNAs. Bacterial cells do not have the spliceosomes required to remove introns and splice together exons to form functional mRNAs. Expressing eukaryotic genes in transgenic plants or animals allows functional mRNAs to be formed through processing by the host organism's splicing apparatus. 13. Knowledge of bacterial genetics led to the discovery of the enzymes,

gene regulation mechanisms, and vector plasmids that were used (and are still used) to develop recombinant DNA technology. We would not have been able to develop genetic engineering methods employed today without the information derived from the years of basic research into bacterial genetics. 14. Suspect 2's DNA matches the evidence. 15. The universality of genetic systems is strong support for the view that all organisms evolved from common ancestors. 16. DNA technologies used today have provided many advances in fields of medicine, pharmaceutics, agriculture, and criminal justice.

CHAPTER 16

CHECKPOINT

16.1 • Human karyotypes provide information about the chromosome composition of the individual. It can identify the existence of chromosome abnormalities such as extra or missing chromosomes, translocated parts of chromosomes, deletions, and inversions. • Pedigree analysis can assist in determining the pattern of inheritance of a specific trait; it complements karyotype analysis, which can identify chromosome abnormalities, and molecular genetic methods, which can identify specific types of mutations in affected genes. • Comparison of an individual's gene sequences with other sequences stored in gene databases can assist in identifying the specific mutations that are involved in a genetic disease. Comparative genome database information can also shed light on the specific function of an affected gene if the function of that gene is known in another organism. • Mice and humans share many genes that have common structures and functions. Comparing a human gene to a similar mouse gene with a known function can give insights into the function of its human counterpart. If the mouse gene's function is not known, genetic manipulation of the mouse gene (by mutation and observing the effect on the mouse's phenotype) can provide insights into function of the homologous human gene. **16.2** • Most cases of Down syndrome are produced by nondisjunction in meiosis associated with an extra chromosome 21. Figure 16-4 illustrates how disjunction works with cases involving sex chromosomes. Draw a similar diagram illustrating how meiotic nondisjunction would produce a gamete with two chromosome 21s. • Cri du chat syndrome involves a deletion in chromosome 5. • Fragile X syndrome involves a weak region near the tip of an X chromosome, associated with many repeats of a CGG sequence that disrupts a gene involved in nerve cell function. • Genomic imprinting is a form of epigenetic inheritance (see Chapter 14), usually due to methylation

of specific regions of DNA within a chromosome. **16.3** • Phenylketonuria and Tay-Sachs diseases are autosomal recessive genetic diseases. • Huntington's disease is inherited as an autosomal dominant trait. • Hemophilia A is inherited as an X-linked recessive disease. **16.4** • Gene therapy is a strategy for treating genetic disease by introducing a normal, therapeutic allele into the individual, usually by employing a viral vector to move the allele into target cells. A potential concern is that the viral vectors might be toxic or trigger severe reactions by the patient's immune system. **16.5** • The advantages of amniocentesis, chorionic villus sampling, and preimplantation genetic diagnoses are that genetic diseases in the fetus can be detected during a pregnancy. Disadvantages are that none of these tests are foolproof. With amniocentesis, the results of the test are not obtained until well into the second trimester. Most of the conditions it detects are incurable, making it more difficult to safely terminate the pregnancy. Chorionic villus sampling results are obtained earlier, reducing those potential risks. Preimplantation diagnosis is not as accurate as the first two methods but is used to detect potential diseases in embryos prior to implantation. It is used for patients who have possible risk factors, allowing physicians to screen healthy embryos for implantation. • The goals of genetic screening for newborns are to detect potential genetic diseases that can be addressed by early treatment of the disease. The goals of genetic screening in adults are to determine potential risk factors of having children with inherited diseases. **16.6** • Many different types of genetic disorders occur in all human populations. It would be incorrect to assume that certain individuals or populations carry most of the abnormal alleles. • The incidence of autosomal recessive diseases depends on the frequency of the abnormal allele in the population. Each parent must contribute a recessive allele to produce offspring with the disease. Parents who are closely related have a higher probability of each having the same abnormal allele, resulting in a higher risk of offspring who will develop that disease. • Health and life insurance companies base their insurance rates on the probability that an individual will be healthy or live to a certain age. If the companies knew that a person were harboring genes that might cause poor health or death at an early age, they would be in a position to refuse to insure that individual.

FIGURE QUESTION

FIG. 16-4 An XY sperm would give rise to Klinefelter syndrome, a 0 sperm (no sex chromosome) would give rise to Turner syndrome, and a YY sperm would give rise to an XYY karyotype.

TEST YOUR UNDERSTANDING

1. e 2. c 3. e 4. a 5. d 6. c 7. e 8. a 9. d 10. e 11. a 12. (a) Test the potential parents for the Tay-Sachs allele. If both carry the recessive allele, perform preimplantation diagnosis for in vitro fertilization, or chorionic villi sampling during early pregnancy. (b) There is low risk of genetic disease if the young man is unaffected. (c) There is no significant risk of inheriting X-linked hemophilia A given that the young woman's male parent is unaffected. (d) Offer the young man testing for the Huntington's gene, particularly if he intends to have children. He should receive counseling regarding the implications for his own future, should he test positive. (e) There is a duty to respect the patient's wishes. Some would argue that there is a moral duty to advise the patient that she should inform the sons of her condition if they intend to have children before the mother's diagnosis is revealed to them. 13. Genes contribute to a large part of the physical makeup of the individual. Many types of disorders and behaviors, however, are complex and may have multiple genetic and environmental components. 14. Chromosome deletions usually result in the removal of many genes, whereas a frameshift mutation typically results in the shortening of some part of a single gene's product. If a small deletion were to occur within a specific gene, then the expression of that gene might be affected in a similar way to a frameshift mutation within the same gene. 15. It is likely that the approximately 500 identical DNA segments shared by mice and humans contain genetic information that is essential to both species and has been conserved in evolution. 16. About one-half of the individuals carrying a Huntington's disease allele will have developed symptoms by the age of 48; approximately three-fourths will have symptoms by the age of 58.

CHAPTER 17

CHECKPOINT

17.1 • The principle of nuclear equivalence is demonstrated by the Steward, Gurdon, and Wilmut experiments that show that a differentiated cell nucleus can be reprogrammed to support the development of an organism to adulthood. This demonstrates that the genes in a differentiated cell nucleus are equivalent to the genes found in embryonic cells. • Wilmut's team recognized that the embryonic nucleus was arrested in metaphase II of meiosis, whereas the donor cell for the adult nucleus is usually in the middle of the S phase of the cell cycle. By withholding nutrients, the group was able to force the donor cells into the G_0 state, synchronizing the cell cycles of the donor nucleus and the egg cytoplasm. • Induced pluripotent stem cells are produced from differentiated cells by reactivating genes that are "locked" in an inactive state during development through epigenetic changes involving DNA methylation, histone modifications, and the production of inhibitory microRNAs. The ability to activate genes that can convert differentiated cells to pluripotent cells supports the principle of nuclear equivalence because it demonstrates that these genes have not been lost or irreparably altered during the differentiation process. **17.2** • *Drosophila* development involves an anterior–posterior body plan that is found in many invertebrates and vertebrates. *C. elegans* has a rigid developmental plan and transparent cells that allow investigators to follow the lineage of every somatic cell. Both *Drosophila* and *C. elegans* have short life cycles and are genetically well characterized, facilitating the isolation and analysis of developmental mutations. Many genes affecting the development of multicellular organisms were first discovered with *Drosophila*. *Arabidopsis* is a small, fast-growing plant with a small genome used to study development in plants. *M. musculus* is a model organism for studying development in mammals, including humans. It has a short lifespan compared to other mammals, it is easily raised in captivity, and methods have been developed for the production and analysis of strains with developmental mutations. • Transcription factors control the activity of genes. They influence development by activating genes required for the differentiation of specific cells. • Induction involves developmental interactions between cells in which specific cells signal other cells to organize into specialized structures. • Apoptosis refers to programmed cell death. For example, during development, certain cells in budding appendages may be programmed to die, forming fingers. **17.3** • Oncogenes are typically growth-activating genes that produce continuous, abnormal growth when the control of their expression is altered. Tumor suppressor genes normally prevent cells from undergoing inappropriate growth and cell division. In cancer cells tumor suppressor genes are inactivated. • When reprogrammed cells interact with older surrounding cells, they may develop as tumors. The use of highly expressed transcription factors that are coded for by oncogenes may contribute to this effect if their expression remains at high levels in the reprogrammed cells.

FIGURE QUESTIONS

FIG. 17-1 Vertebrate organs can contain cells derived from more than one lineage. For example, the cortex of the adrenal gland is derived from mesoderm and the medulla is derived from ectoderm. **FIG. 17-2** Only some of the undifferentiated cells from the root discs produced plantlets, suggesting that not all differentiated plant cells are totipotent. **FIG. 17-3** The egg nucleus would be haploid and it is not clear that it would play a role in the development of

the tadpole. If the nucleus in the unfertilized egg had not been killed, one could argue that the development of the adult frog may have proceeded from the activation of genes in the egg nucleus, rather than those of the differentiated intestinal cell nucleus. It would not be possible to clearly interpret the results from that experiment. **FIG. 17-4** The egg nucleus was haploid; the nucleus from the cultured mammary cell was diploid. **FIG. 17-13** *Hox* genes that are active in the mouse, but not in *Drosophila*, are genes 3, 10, 11, 12, and 13. **FIG. 17-15** Germ line cells in *C. elegans* do not give rise to other types of cells in the individual. They give rise to gametes. Zygotes produced from the fusion of gametes, however, give rise to all of the different cell types in a new individual.

TEST YOUR UNDERSTANDING

1. c 2. d 3. b 4. c 5. d 6. c 7. b 8. d 9. c 10. a 11. b 12. e 13. Almost all developmental processes are controlled through mechanisms that regulate some aspect of gene expression. Specific sets of genes are activated or inactivated during development. Many developmental genes code for transcription factors. 14. Each model organism provides specific characteristics suitable for the efficient analysis of different types of developmental processes. Most have well-characterized genetic systems and short life cycles, allowing for the isolaton of many different types of mutant strains affecting development. 15. A number of genes responsible for the production and maintenance of stem cells are proto-oncogenes that code for transcription factors. Inappropriate expression of these genes causes them to become cancer-driving oncogenes. 16. Many genes involved in the development of *Drosophila* and *C. elegans* also control development in mammals. The common chromosomal organization and function of *Hox* genes in almost all animals with an anterior–posterior axis is a strong argument that modern organisms evolved from a common ancestor. 17. Evo Devo argues that developmental genes (such as the *Hox* genes) that are common in both structure and function represent "molecular fossils" that are important drivers of evolutionary change. The occurrence of mutations in these genes can lead to dramatic morphological changes. This could explain how a new species with a different body plan could rapidly evolve. 18. The resulting flowers will consist of sepals and carpels.

CHAPTER 18

CHECKPOINT

18.1 • Microevolution is the accumulation of small genetic changes in a population; macroevolution involves greater genetic changes,

such as those leading to an ancestral population giving rise to a new species. • Evolution involves aggregate genetic changes that occur among organisms in a population over generations, not the changes that may occur in a particular individual during its life. **18.2** • Aristotle visualized organisms changing over time. His ideas differed from modern evolutionary theory because he did not propose any type of natural process to explain these changes and he envisioned changes as progression or a "scale of nature" in which organisms became more perfect. • Darwin proposed natural selection as the mechanism for evolution; Lamarck's explanation invoked a type of "vital force" that drove organisms to change to become more adapted to their environment during their lifetimes and to pass these adaptations to their descendants. **18.3** • Darwin invoked the well-recognized power of artificial selection in the breeding of new varieties of plants and animals as the basis for his explanation of evolution by natural selection. • Variations that cannot be transmitted to offspring are a dead end and cannot play a role in evolution. • Although Darwin recognized that evolution required natural selection acting on heritable variation in populations, he did not know the mechanism of inheritance. The modern synthesis couples Mendel's explanation of inheritance with an understanding of the genetics of populations. **18.4** • Scientists usually date fossils through comparisons of the sedimentary rock layers in which they are found. • *Mesosaurus* lived prior to the separation of southern Africa and South America by continental drift. • Homologous features are evidence that two groups share a common ancestor. Homoplastic features, which are superficially similar because they are used in similar ways, are evidence of convergent evolution. • In addition to common embryonic morphology among vertebrates, genetic studies (as discussed in Chapter 17) provide strong evidence that evolution builds on pre-existing developmental patterns. • Researchers moved populations of guppies from a high-predation habitat (in which there is natural selection against larger guppies) to a low-predation habitat (in which such selection was absent). Eighteen generations later, they were able to demonstrate heritable changes in the relocated populations such that the guppies matured at a larger size compared to their ancestral population.

FIGURE QUESTIONS

FIG. 18-11 Northwest coast of Europe. **FIG. 18-12** Australia. **FIG. 18-19** Hippopotamus. **FIG. 18-21** You might predict that larger predators introduced into the low-predation habitat would convert it to a high-predation habitat in which there would be natural selection against larger guppies. The expected result might be heritable changes in the population, causing the guppies of future generations to mature at a smaller size on average. Can you think of alternative predictions?

TEST YOUR UNDERSTANDING

1. d 2. b 3. c 4. c 5. b 6. d 7. e 8. b 9. b 10. b 11. Changes that affect developmental patterns. 12. Comparative anatomy; molecular comparisons (including proteins as well as the virtual universality of the genetic code); common developmental patterns and their genetic basis. 13. Darwin meant that a species that possesses heritable variation enabling it to adapt to new challenges is most likely to survive (and reproduce). 14. (a) Natural selection can only work in the present; it cannot anticipate the future. (b) Evolution results from natural selection acting on heritable variation within a population. Although individuals do not themselves evolve, the genetic makeup of the population over time will be influenced by the contributions of individuals possessing heritable characteristics that cause them to be more likely to survive and produce offspring that exhibit those same characteristics. (c) Every organism is a descendant, but not every individual organism becomes an ancestor. Every individual alive today exists because every single one of its direct ancestors in the entire history of life on planet Earth survived long enough to reproduce. (d) Replacement of a single nucleotide can result in the replacement of one amino acid by another, with the potential of producing an altered protein. The new protein may be nonfunctional, or it may possess novel functions. Single nucleotide changes affecting RNAs, particularly those with regulatory functions, may have even greater evolutionary significance. (e) Genetic changes in a population over time constitute microevolution. 15. The information that "limbless" salamanders occupy similar habitats (shallow water) is a clue that that limblessness may have been selected as an adaptation to this particular environment. Think about the kinds of additional information that would help you answer this question more fully. For example: Are there any limbed salamanders that occupy the same habitats? If so, do they have special adaptations? Do the limbless salamanders have any close relatives (perhaps as identified by molecular analysis) that have limbs? If so, what types of habitats do they occupy? 16. Tarsier. 17. Those individuals in a pest population that are genetically resistant to a particular pesticide are more likely to survive exposure and pass on their genes for resistance to their offspring. If exposure to the pesticide continues, there will also be selection for any additional new genes for resistance that may arise through mutations. Therefore, over time the population will consist of a higher proportion of resistant individuals, rendering the pesticide much less effective.

CHAPTER 19

CHECKPOINT

19.1 • Populations. • No. • Frequency of $T = 0.6$. **19.2** • $T = 0.6$; $t = 0.4$; $TT = 0.36$; $Tt = 0.48$. • $A = 0.8$; $a = 0.2$; $AA = 0.64$; $Aa = 0.32$; $aa = 0.04$. • No. The genotype frequencies clearly do not match Hardy–Weinberg expectations. **19.3** • Natural selection. • Mutation is the source of new genetic variation in a population. • Mutation and gene flow are associated with an increase in variation within a population. Populations diverge genetically due to such forces as nonrandom mating, genetic drift, and differences in natural selection. • Stabilizing selection favoring a heterozygous genotype as well as disruptive selection against a heterozygous genotype can change genotype frequencies without changing allele frequencies. Alleles will be maintained in the favored genotypes. **19.4** • There is strong selection against individuals with sickle cell anemia (homozygous recessive). In regions where malaria is a problem, the sickle cell allele is maintained in the population because it confers resistance to malaria on heterozygous persons (relative to homozygous individuals with normal hemoglobin). • Frequency-dependent selection tends to maintain genetic variation within a population over time. • Researchers can test the genetic basis for clinal variation by conducting studies similar to those of Clausen, Keck, and Hiesey (see Fig. 19-9).

FIGURE QUESTIONS

FIG. 19-2 40 mice. **FIG. 19-4** The relationship between wing color and antenna length is unknown, so selection for antenna length would be impossible to predict. **FIG. 19-8** $a = 0.7$; $A = 0.3$.

TEST YOUR UNDERSTANDING

1. b 2. b 3. b 4. b 5. b 6. c 7. a 8. e 9. b 10. c 11. d 12. Compare your graphs to those in Figure 19-4. 13. Mutations are the source of all new genetic variation. 14. Natural selection acts on individuals, each of which has a particular phenotype. That phenotype is determined partly by the genotype, so selection on the genotype is indirect. 15. Compared to the United States, the populations of Finland and Iceland are genetically much more homogeneous and occupy more similar environments. This relative lack of variability makes it much easier for researchers to analyze their data and draw meaningful conclusions. 16. The phrase "survival of the fittest" has become a poor way to think about evolution because it is widely misinterpreted by nonscientists. Evolutionary biologists think about fitness in terms of reproductive success in response to selective forces. 17. Amish population: 4.9 out of 1000; general population: 1.0 out of 1 million. 18. If

due to genetic factors, the differences in average weights observed in the natural populations would be similar in the aquarium-raised populations. 19. New information on genetic polymorphism in humans will be useful in understanding the genetic basis of various diseases, with the expectation that new therapeutic approaches may follow. However, this information will be challenging, particularly to individuals, because risk estimates may not be accompanied by useful recommendations and may even create needless worry.

CHAPTER 20

CHECKPOINT

20.1 • The members of a biological species interbreed in nature to produce fertile offspring, but do not interbreed with other species. • The biological species concept does not work well for extinct species, asexually reproducing species, or species in captivity. • A population that meets all the criteria of the biological species concept shares a common gene pool. To the extent that various morphological characteristics are genetically determined, the population is likely to meet the definition of a morphological species. **20.2** • Wood frogs and leopard frogs breed at different times (temporal isolation). They are also subject to behavioral isolation because males use different vocalizations to attract mates. • Temporal isolation involves breeding at different times; behavioral isolation, also known as sexual isolation, is based on differences in reproductive behaviors. • Mechanical isolation is based on differences in reproductive structures. In gametic isolation, incompatibilities between eggs and sperm prevent fertilization. • Hybrid sterility. **20.3** • Mountains, glaciers, rivers, canyons, land barriers, deserts, etc. • Individuals produced by allopolyploidy (hybridization followed by polyploidy) are reproductively isolated. They can form a new species because they can only reproduce with themselves (if self-fertilizing) or with each other, and not with individuals of the two parent species. • Pupfishes: allopatric speciation among isolated springs; cichlids: sympatric speciation within individual lakes. **20.4** • Phyletic gradualism and punctuated equilibrium are not mutually exclusive. Periods of phyletic gradualism, stasis, and rapid evolutionary change can all occur in the course of the evolution of a particular group. **20.5** • The large-scale phenotypic changes that occur when pre-existing structures become modified as evolutionary novelties are associated with the evolution of new species and higher taxa (macroevolution). • The modification of a pre-existing structure (a preadaptation) can occur through relatively minor changes in genetic regulation, thereby quickening the pace of macroevolution. • Paedomorphosis allows a species to occupy a particular habitat for the entire lifespan, without competition from adults of related species. • Mass extinction leaves many ecological niches available to the surviving species. Adaptive radiation may occur as these species evolve to utilize these now unoccupied habitats.

FIGURE QUESTIONS

FIG. 20-9 The $2n = 10$ hybrids would produce $n = 5$ gametes. It is unlikely that it could produce fertile offspring crossed with either species A ($n = 3$ gametes) or species B ($n = 2$ gametes). **FIG. 20-13** Your hypothesis would be supported if the F_1 hybrid females were found to choose red and blue males with equal frequency in well-lit aquaria. **FIG. 20-15** Long-term environmental stability, such as that observed in tropical rain forests, is usually associated with phyletic gradualism. **FIG. 20-18** Future adaptive radiation in the Hawaiian honeycreepers (if it occurs at all) will likely be restricted to human-created habitats because traditional natural habitats are disappearing.

TEST YOUR UNDERSTANDING

1. a 2. a 3. c 4. d 5. b 6. c 7. b 8. e 9. c 10. d 11. Compare your diagrams with Figure 20-9. The F_1 hybrid ($2n = 9$) is unlikely to be fertile. However, if the chromosome number is doubled ($2n = 18$), it is likely that the hybrid could produce $n = 9$ gametes. It would be able to breed with other such hybrids, but not with either of the two parental species. 12. If the original isolated population is small, it could become extinct. However, if it survives it is more likely to diverge genetically from the parent population due to genetic drift and the founder effect. 13. In one sense hawthorn maggot flies and apple maggot flies can be considered an example of speciation that originated through positive assortative mating because they are reproductively isolated by their mate choice behaviors. 14. The human-caused mass extinction that is occurring today may be on a scale that is unprecedented in geological history and many of the potential effects on the future of the human population are unpredictable. 15. April 1 is the only date (of those given) in which both populations are within their breeding season.

CHAPTER 21

CHECKPOINT

21.1 • The origin of life required a reducing atmosphere (i.e., the virtual absence of free molecular oxygen) because of oxygen's reactive and destructive properties. Chemical sources of important elements (particularly carbon, hydrogen, oxygen, and nitrogen) were needed along with energy to power chemical reactions. Time was the fourth requirement, although not as much time as one might think because there is evidence that life arose early in the history of the planet. • According to the prebiotic soup hypothesis, life arose on or close to Earth's surface; the iron–sulfur world hypothesis envisions life originating from molecules produced in hydrothermal vents on the ocean floor. Both hypotheses seek to explain how life could have arisen as the chemical reactions of simple molecules led to greater complexity. **21.2** • Major steps that probably occurred in the origin of cells were the evolution of simple metabolism with an energy source and occurring within a boundary, as well as the evolution of molecular reproduction. • Molecular oxygen (produced by photosynthetic cyanobacteria) increased slowly in the atmosphere. Eventually oxygen rose to levels that poisoned anaerobic organisms but also provided conditions for the evolution of aerobes. • Compare your sketch to Figure 21-8. **21.3** • Prokaryotic cells → eukaryotic cells → multicellular organisms • Fishes → amphibians → reptiles → mammals • Ferns → gymnosperms → flowering plants.

FIGURE QUESTIONS

FIG. 21-2 There would have been no source of carbon for organic molecules if methane had been omitted in the experiment. **FIG. 21-5** Without mutations the selective process would have had nothing to act upon. **FIG. 21-7** The chemical reactions that destroy ozone cause this protective layer to be less effective in screening out ultraviolet radiation, which is harmful to most organisms.

TEST YOUR UNDERSTANDING

1. a 2. b 3. d 4. c 5. e 6. c 7. a 8. a 9. b 10. b 11. c 12. c 13. There was little or no molecular oxygen for aerobic metabolism in bacteria prior to the evolution of oxygen-producing photosynthetic bacteria. 14. Keep in mind that life requires metabolism as well as self-replication. 15. A successful experiment such as the one described would be a demonstration that the origin of life could have occurred through chemical evolution but would not constitute proof. 16. Molecular oxygen (produced by photosynthesis) is the basis for aerobic metabolism. In contrast to anaerobic metabolism, aerobic metabolism makes large amounts of energy available to organisms (see Chapter 8), enabling them to grow to much larger sizes. 17. One hypothesis could be that the molecular fossils (in the form of lipids) were not actually produced by cyanobacteria and therefore do not provide an accurate date for their early presence. An opposing hypothesis might be that cyanobacteria evolved at the early date (2.7 bya), but conditions for formation of structural fossils did not exist or that such fossils are yet to be found. Similarly, one might hypothesize that over 700 million years were required for their numbers to reach a level at which structural fossils would have been reliably produced.

CHAPTER 22

CHECKPOINT

22.1 • Opposable thumb (and possibly big toe); highly flexible digits with fleshy pads at the ends and nails instead of claws. • Eyes positioned at the front of the head. **22.2** • Suborder Prosimii; suborder Tarsiiformes; suborder Anthropoidea. • Monkeys possess tails, which hominoids (apes and humans) lack. **22.3** • The foramen magnum of an ape is situated in the rear of the skull; that of the human is located in the skull base. Unlike modern humans, apes have pronounced supraorbital ridges. The ape jaw is relatively rectangular compared to the more U-shaped jaw of humans. • The human skeleton exhibits multiple adaptations for bipedal posture, including complex curvature of the spine (in contrast to the simply curved ape spine), a pelvis that is shorter and broader than that of an ape, and alignment of all the toes (unlike the opposable big toe of many apes). • Members of genus *Homo* are distinguished from the australopithecines by larger brains, smaller premolars and molars, and evidence of tool use. • *H. erectus* had a larger brain than *H. habilis,* was taller, and made more sophisticated tools. • Compared to *H. sapiens, H. neanderthalensis* had larger brains and sturdier builds. They also possessed heavy supraorbital ridges, which are lacking in *H. sapiens.* **22.4** • Early human societies consisted of nomadic hunters and gatherers. Agriculture began with the cultivation of plants around 10,000 years ago, which was usually followed by the domestication of animals. Agriculture freed much of the population from the need to produce food, fostering the cultural evolution that led to the industrial society in which we live today. Not all human societies have gone through these transitions in this way.

FIGURE QUESTIONS

FIG. 22-1 Gorilla. **FIG. 22-2** Gorillas. **FIG. 22-9** Consider the following in placing *Homo floresiensis* and the Denisovans in the chart: The researchers who have been most involved in studies of *H. floresiensis* fossils think that it was an evolutionary offshoot of *H. erectus* that existed from about 38,000–12,000 years ago. Based on fossil dating and molecular evidence, the Denisiovans are currently thought to represent a sister group of *H. neanderthalensis,* existing as recently as 50,000 years ago (and possibly as early as 80,000 years ago).

1. c 2. a 3. a 4. b 5. e 6. a 7. b 8. b 9. d 10. c 11. c 12. e 13. Scientists have found Neanderthal bones in Europe showing signs of cannibalism and dating from a time when no modern humans were present in that region. You might look for caves with evidence of occupation by *H. sapiens* and Neanderthals at close to the same time. Evidence of violent injuries to the Neanderthal remains, but not the *H. sapiens* remains, would possibly implicate modern humans in the Neanderthals' demise. The evidence would be stronger if the injuries could be shown to differ from the types of injuries Neanderthals have been implicated in inflicting on their own species. 14. Climate change since the last Ice Age has altered the range of reindeer. Humans are an extremely adaptable species, able to live in a wide variety of habitats. 15. Chimpanzees and humans did not exist at the time their lineages are thought to have diverged about 4 million years ago; therefore, their as yet undiscovered most recent common ancestor was a hominoid that was neither a chimpanzee nor a human. 16. As you form your opinion, consider the concept of fitness (both direct fitness and inclusive fitness) as understood by evolutionary biologists today. Do these ideas correspond to popular notions of "survival of the fittest"? Does natural selection apply to humans? Do you think artificial selection plays any role? Do you think that your ideas regarding future human evolution will be influenced by new information regarding the human genome? 17. Your cladogram should show *Ardipithecus* giving rise to *Australopithecus,* which in turn gave rise to *Homo.* You would not include the genus *Paranthropus* because there is no evidence that it was ancestral to *Homo.* 18. (a) African; (b) first migrants out of Africa; (c) Eurasian; (d) Amerindian (native Americans).

CHAPTER 23

CHECKPOINT

23.1 • *Amanita phalloides.* • The categories range from species to domain; as you move up the hierarchy, each taxon is more inclusive than the one below it. • Domain, kingdom, phylum, class, order, family, genus, species. **23.2** • Archaea (archaeons) belong to Domain Archaea; bacteria belong to Domain Bacteria; protists, plants, fungi, and animals belong to Domain Eukarya; white willow belongs to Domain Eukarya, Kingdom Plantae; *Escherichia coli* belongs to Domain Bacteria, Kingdom Bacteria; tapeworms belong to Domain Eukarya, Kingdom Animalia; black bread mold belongs to Domain Eukarya, Kingdom Fungi. • Nodes in a cladogram represent the splitting of two or more new groups from a

common ancestor; each node represents the most recent common ancestor of each clade, represented by branches; and the branches diverging from a node share a set of characteristics. **23.3** • Shared ancestral characters are characteristics that were present in an ancestral group and remain present in all groups descended from that ancestor; shared derived characters were present in a *recent* common ancestor and are present in its descendants; homology refers to traits organisms share as a result of their shared ancestry. • Shared ancestral characters are present in all groups descended from a particular ancestor. For example, all mammals have hair, so hair does not help us distinguish between carnivores and marsupials (e.g., kangaroos). However, shared derived characters of marsupials separate them from carnivores. For example, marsupial young are born at an immature stage and complete their development in the mother's pouch. • Molecular biology is helping systematists establish evolutionary relationships. The more similar the macromolecules of various groups of organisms, the more closely related the species are considered to be. • A monophyletic taxon includes an ancestral species and all of its descendants. **23.4** • Systematists use shared derived characters to determine monophyletic groups. • Outgroup analysis is a research method used for estimating which features are shared derived characters in a group of organisms. • In applying the principle of parsimony, systematists select the simplest explanation to interpret the data. **23.5** • Modern systematics helps researchers understand the relationships and origins of viruses and other pathogens; this information helps health professionals understand the source and spread of infections.

FIGURE QUESTIONS

FIG. 23-2 The Archaea are more closely related to the Eukarya; they share a common ancestor as indicated by node B. **FIG. 23-6** The coyote. **FIG. 23-7** Taxa 4, 5, and 6.

TEST YOUR UNDERSTANDING

1. a 2. e 3. c 4. a 5. c 6. c 7. e 8. a 9. d 10. Shared derived characters; molecular data. 11. Domains have been determined based on molecular data and cladistics analyses; historically, animals were assigned to kingdoms less rigorously; for example, systematists have determined that the protists are not a monophyletic group and now assign them to several "supergroups." 12. Similar because they share a common ancestor. 13. Suggests archaea are more closely related to eukaryotes than to bacteria; such data suggests that archaea and eukaryotes share a common ancestor. 14. Figure 23-7a without branch 4 has similar branching; mosses would be represented by branch 1; horsetails and ferns by branches 2 and 3; and gymnosperms and angiosperms by branches

5 and 6. 15. Paraphyletic; monophyletic; polyphyletic.

CHAPTER 24

CHECKPOINT

24.1 • Viruses are not cellular and cannot reproduce without a host; they lack the enzymes and other proteins needed to carry on cellular respiration and other metabolic activities; most viruses lack the components necessary to synthesize proteins. • A virus consists of a core of nucleic acid surrounded by a capsid made of protein subunits. Many viruses are surrounded by an envelope external to the capsid. **24.2** • Viruses can be classified based on their home range (the type of host species the virus can infect), on the type of nucleic acid (DNA or RNA) the virus contains, and whether its nucleic acid is single stranded or double stranded. They can also be classified based on size, shape, and presence of an envelope. **24.3** • Steps in a lytic cycle include attachment, penetration, replication and synthesis, assembly, and release. See Fig. 24-3. • In a lysogenic cycle, a temperate virus integrates into the host DNA. The host is not immediately destroyed. **24.4** • Viruses can enter plants through damaged cells. Viruses can be transmitted from one plant generation to the next through infected seeds or by asexual reproduction. • Some viruses enter animal cells by fusing with the cell's plasma membrane; the entire virus, including the capsid, is released into the animal cell. Other viruses enter by endocytosis. • See Fig. 24-7. **24.5** • Viruses could have evolved from free-living complex cellular ancestors. Later, they became parasitic and through time, lost some of the genes needed for protein synthesis and other cellular activities. • Polydnaviruses do not have genes for making the proteins needed to replicate and produce new viruses. Genes needed for viral replication are found in the wasp genome; the viruses can replicate only in the wasp ovary cells. The wasp injects polydnaviruses along with her eggs into certain caterpillars. These viruses express toxins that interfere with the caterpillar's immune defenses and development. This action allows the wasp eggs to hatch and develop inside the caterpillar. The young wasps feed on the caterpillar. **24.6** • Both viroids and prions are subviral agents; they are smaller and simpler than viruses. Viroids consist of a very short, circular, single strand of naked RNA; they do not have a protective protein coat or proteins to assist in duplication. Viroids infect plants. In contrast, prions are misfolded proteins. They arise spontaneously, mostly as a result of mutation. Prions apparently can aggregate and accumulate in the brain and in certain other tissues where they cause serious damage.

FIGURE QUESTIONS

FIG. 24-3 No, because the tail has proteins that recognize and bind to the host cell. Also, the capsid would not surround phage DNA, leaving it vulnerable to being destroyed. **FIG. 24-4** By integrating into the bacterial genome, the virus is reproduced in all of the cell's daughter cells for, potentially, many generations. In this way, the virus population is expanded and disseminated. **FIG. 24-6** The cytoplasm contains the materials and organelles necessary to synthesize proteins needed by the new viruses. Also glycoproteins must be transported to the cell membrane in order to envelope new viruses before release. **FIG. 24-7** The infected cells replicate, producing daughter cells with integrated proviral DNA. These daughter cells continue to transcribe the proviral DNA and they bud progeny virions. **FIG. 24-8** The caterpillar population would increase; the wasp population would decrease; the virus could become extinct because it would no longer have wasps in which to reproduce.

TEST YOUR UNDERSTANDING

1. e 2. a 3. b 4. a 5. c 6. e 7. b 8. b 9. c 10. c 11. Release of viruses from host cell; see Fig. 24-3a, step 5. 12. Viral genomes are single stranded and can be DNA or RNA. 13. Neither has a protective coat; viroids have no proteins; prions have no nucleic acid 14. Polydnaviruses are mutualistic partners with certain wasps. Their relationship is so complex that it is likely they coevolved. Researchers have concluded that the polydnavirus evolved from a nudivirus that infected wasps millions of years ago. In time, the virus genome became incorporated into the wasp genome. Proteins needed for viral replication are now part of the wasp's DNA, and the virus can replicate only in the wasp's ovaries. 15. The regressive hypothesis is supported by discovery of giant viruses. Viruses may have evolved from complex, cellular, free-living ancestors. Later, they became parasitic and through time, lost some of the genes needed for protein synthesis and other cellular activities. 16. Influenza viruses mutate frequently, necessitating ongoing development of new vaccines. The latest advances in nucleic acid technology must be applied to meeting this challenge.

CHAPTER 25

CHECKPOINT

25.1 • Bacteria and archaea differ genetically, in their chemical composition, and anatomically. Unlike archaea, bacteria contain peptidoglycan in their cell walls and can form endospores. Prokaryotes are exclusively single-celled organisms that, in contrast to eukaryotic cells, do not have membrane-enclosed organelles; most have a cell wall that surrounds the plasma membrane, and most have a single, circular DNA molecule instead of linear DNA. • The cell walls of gram-positive bacteria are very thick and consist primarily of peptidoglycan, while the cell walls of gram-negative bacteria have two layers: a thin peptidoglycan layer and a thick outer membrane containing polysaccharides bonded to lipids. Distinguishing between gram-positive and gram-negative bacteria is important for treating certain diseases. For example, penicillin interferes with peptidoglycan synthesis and weakens cell walls of gram-positive bacteria. • Bacterial and eukaryotic flagella are different in structure, mechanism of propulsion, and protein composition. **25.2** • Prokaryotes reproduce asexually by binary fission, budding, or fragmentation. Gene transfer among prokaryotes takes place by three different mechanisms: transformation, transduction, and conjugation. • Transduction contributes to the rapid evolution of bacterial populations through horizontal gene transfer. The transferred bacterial chromosome becomes a recombination of its own original DNA and the DNA from another bacterium, which provides new DNA allowing diversification and adaptation. • The steps that take place during conjugation can be found in Figure 25-8. **25.3** • Chemoheterotrophs obtain energy from organic molecules. • Facultative anaerobes use oxygen for cellular respiration if it is available, but can metabolize anaerobically when necessary. Obligate anaerobes carry out anaerobic respiration using terminal electron acceptors other than oxygen. Some obligate anaerobes are killed by even low concentrations of oxygen. Facultative anaerobes differ from aerobes because they do not require oxygen to carry out cellular respiration. • Prokaryotes obtain nitrogen needed to produce amino acids and nucleic acids by nitrogen fixation in which atmospheric nitrogen is reduced to ammonia. **25.4** • The four phyla of archaea were once thought to inhabit only extreme environments, such as hydrothermal vents and springs, extreme salt water, and oxygen-free environments. Scientists now think archaea may be found everywhere other organisms live. Methanogens produce methane from simple carbon compounds and inhabit oxygen-free environments in sewage, swamps, and the digestive tracts of humans and other animals. Methanogens are important in recycling components of organic products of organisms that inhabit swamps and contribute more than 80% of greenhouse gases to Earth's atmosphere. • See Table 25-3. **25.5** • Bacteria form symbiotic relationships with other organisms. The three forms of symbiosis are mutualism, commensalism, and parasitism. • Biofilms are common along surfaces in aquatic environments and are also found on teeth, contact lenses, pipelines, catheters, and surgical implants. Biofilms consist of many species of bacteria, archaea, fungi, and protists. • Chemoheterotrophs break down dead organisms and waste to their organic components, which recycles nitrogen, oxygen, carbon, phosphorus, and other minerals. Many prokaryotes are important parts of biogeochemical cycles. Cyanobacteria fix carbon dioxide and generate oxygen. • Bioremediation is the process of using microorganisms (and sometimes other organisms) to detoxify or remove oil, gasoline, and other pollutants or toxic chemicals from the environment. The microorganisms break down certain toxins, leaving behind harmless metabolic byproducts. **25.6** • Each of Koch's postulates is important because all four are part of a system, which demonstrate that a specific pathogen causes specific disease symptoms. • Exotoxins are strong poisons produced by bacteria that are either secreted from the cell or leak out when the bacterial cell is destroyed. Endotoxins are components of the cell walls of most gram-negative bacteria, which affect the host only when they are released when the bacteria dies. Exotoxins cause specific symptoms and are destroyed by heating; endotoxins affect the entire body by binding to macrophages of the immune system and are not destroyed by heat.

FIGURE QUESTIONS

FIG. 25-2 A nucleus might be an unnecessary organelle that could interfere with or slow replication, causing prokaryotes to lose the advantage of rapid replication. **FIG. 25-8** The F plasmid contains the genetic information for making a sex pilus. Each time an F plasmid is transferred to another cell, that recipient cell can then produce a sex pilus; genetic recombination increases.

TEST YOUR UNDERSTANDING

1. d 2. b 3. b 4. a 5. a 6. c 7. e 8. d 9. e 10. a 11. a 12. See Fig. 25-2. 13. Characteristics of archaea are no peptidoglycan in their cell walls, plasma membranes with branched-chain hydrocarbons (synthesized from isoprene units) bonded to glycerol by ether linkages, and flagella structurally different from bacteria. 14. Consider the ecological and industrial roles prokaryotes have in order to answer this question. 15. Antibiotics impose selective pressure on bacteria by killing susceptible bacteria and leaving those that are resistant to pass on their genes for antibiotic resistance to future generations of bacteria. 16. The public would need to be educated about how to properly take prescribed antibiotics so that they are not contributing to the problem of antibiotic resistance. Society would need to fund research and support the development of new antibiotics.

CHAPTER 26

CHECKPOINT

26.1 • Protists may be photoautotrophic or heterotrophic (obtaining nutrition by absorption or by ingestion). • Protists that are not free living may be involved in symbiotic relationships with other organism (ranging from mutualism, to commensalism, to parasitism). **26.2** • The structure of chloroplasts in different groups of eukaryotes (some with two external membranes and others with three) is evidence that chloroplasts arose through several endosymbiotic events. These events began with primary endosymbiosis of ancient cyanobacteria and continued with subsequent secondary endosymbiosis in some evolutionary lines. • The various groups informally called protists differ widely in their ultrastructural features and molecular composition. These differences are strong evidence that, as a group, the protists do not include all the descendants of a common eukaryote ancestor; hence, they are a paraphyletic group. **26.3** • The mitochondria of excavates are atypical. • Diplomonads include *Giardia intestinalis,* which causes severe diarrhea. *Trichomonas vaginalis,* which causes a sexually transmitted disease in humans, is a parabasilid. The trypanosomes include *Trypanosoma brucei,* which causes African sleeping sickness. **26.4** • The main groups of chromalveolates are the alveolates (characterized by alveoli, flattened vesicles inside their plasma membranes) and stramenopiles (most of which have motile cells with two flagella, one of which bears tiny hairlike projections). • Like dinoflagellates, apicomplexans have alveoli. There is evidence that the chloroplasts of photosynthetic dinoflagellates were derived from a red alga through secondary endosymbiosis; the apicomplexans carry a nonphotosynthetic chloroplast remnant also thought to have originated from a red alga. • A water mold, *Phytophthora infestans,* causes light blight of potatoes and was responsible for the 19th-century Irish potato famine. • Diatoms are major producers in the ocean. Brown algae are also producers and provide habitat for many marine organisms. **26.5** • The term *rhizarian* ("root") refers to the threadlike cytoplasmic projections extended by forams and actinopods. **26.6** • Red algae and green algae share similarities with land plants at the molecular level, and, like land plants, have chloroplasts surrounded by only two external membranes. **26.7** • The cytoplasmic projections of the amoebozoa are rounded and wide (i.e., lobose) as opposed to those of the rhizarians, which are thin and threadlike. • The choanoflagellates are the unikonts most structurally and genetically similar to animals.

FIGURE QUESTIONS

FIG. 26-2 Two. **FIG. 26-3** The common ancestor of the brown algae and the water molds is most likely a stramenopile that evolved before the water mold clade branched off. The likely common ancestor of animals and fungi would be an opisthokont that evolved before the fungal clade branched off. **FIG. 26-17** Unikont.

TEST YOUR UNDERSTANDING

1. c 2. a 3. c 4. b 5. d 6. d 7. b 8. a 9. d 10. e 11. c 12. Compare your sketches to Figures 26-5b (*Euglena*) and 26-8b (*Paramecium*). 13. The protists include many, but not all, of the descendants of the ancestral eukaryote at the base of the cladogram (i.e., land plants, fungi, and animals are not protists). 14. Actin (*red*) is found in the microvilli in the choanoflagellate collar, tubulin (*green*) in the flagellum and as part of the cytoskeleton, and DNA (*blue*) in the nucleus. 15. The available evidence supports the hypothesis that the chloroplast remnant in the cells of the malaria parasite *Plasmodium* evolved by serial endosymbiosis from a red algal chloroplast. Although nonphotosynthetic, it is essential to the parasite's survival and therefore a potential target for drug therapy. Unfortunately, the use of chemotherapies against malaria, and insecticides against the *Anopheles* mosquito, are selecting for the evolution of resistant strains of both the parasite and the vector.

CHAPTER 27

CHECKPOINT

27.1 • Challenges for terrestrial plants include preventing desiccation, obtaining CO_2 for photosynthesis without excessive water loss, providing a means for sperm and egg to meet, and protection of the embryo. • The presence of a waxy cuticle helps prevent desiccation of the plant body, and gas exchange is facilitated by stomata. Various mechanisms have evolved for sperm and egg to meet and embryos are protected within a multicellular female gametangium. • Many lines of molecular evidence link the land plants more closely to the charophytes than to other groups of Archaeplastids. • Compare your diagram to Figure 27-2. **27.2** • The haploid gametophyte generation in mosses includes the following: spores, protonema (each develops from a spore), archegonia (produce egg cells by mitosis), and antheridia (produce sperm by mitosis) • Mosses, liverworts, and hornworts lack vascular tissue and share similar gametophyte dominant life cycles. They most obviously differ in their body forms: mosses (leafy gametophytes), liverworts (thalloid or leafy gametophytes), and hornworts (thalloid gametophytes that give rise to sporophytes that exhibit indeterminate

growth). **27.3** • Unlike algae and bryophytes, ferns have dominant sporophytes that possess vascular tissue, as well as stems, leaves (megaphylls), and roots. • Microphylls are small, with a single vascular strand; megaphylls are typically larger and each possesses more than one vascular strand. • The diploid sporophyte generation of ferns includes the following: zygote, roots, rhizome, frond, sorus, and sporangium. • Whisk ferns and horsetails are classified with ferns based on characteristics such as similarities in DNA and sperm structure. • The heterosporous life cycle is characterized by the formation of two kinds of spores and hence two kind of gametophytes: megaspores (give rise to female gametophytes) and microspores (give rise to male gametophytes).

FIGURE QUESTIONS

FIG. 27-2 Both diploid (sporophyte generation) and haploid (gametophyte generation) cells undergo mitosis in the plant life cycle. **FIG. 27-5** The cladogram indicates that all land plants have multicellular embryos (unlike their green algal most recent common ancestor). **FIG. 27-7** Compare your sketch to Figure 27-7, step 3. The tissue of the archegonium belongs to the haploid gametophyte generation and the developing embryo belongs to the new sporophyte generation. **FIG. 27-11** You would not expect *Physcomitrella* to possess genes required to produce functional xylem and phloem because it lacks these tissues. However, evolution builds on pre-existing genetic variation, so it is possible that gene sequences that were important in the evolution of xylem and phloem may be found in the *Physcomitrella* genome. **FIG. 27-13** Ferns. **FIG. 27-16** The fern life cycle is sporophyte dominant, unlike the gametophyte dominant life cycle of mosses. **FIG. 27-19** The heterosporous life cycle, like the basic plant life cycle, consists of a multicellular haploid gametophyte generation and a multicellular diploid sporophyte generation.

TEST YOUR UNDERSTANDING

1. d 2. b 3. d 4. d 5. a 6. b 7. c 8. a 9. c 10. d 11. c 12. c 13. Compare your diagram to Figure 27-19. 14. (a) Plants that are dependent on water for fertilization are limited to aquatic environments or those in which sperm can be transported in a film of water. (b) Unlike homospory, heterospory can lead to a life cycle in which seeds can be formed. Seeds are advantageous for life on land because they provide both protection and nutrition for new sporophyte plants. 15. Nonvascular bryophytes. 16. Based on current evidence, you would probably place the rhyniophytes in the fern clade because they possessed xylem, although their relationship to other ferns remains unclear. Their line would not extend to the tips of the rest of the cladogram because they are all extinct, and it is not clear if any extant group descended directly from them.

CHAPTER 28

CHECKPOINT

28.1 • An ovule is a megasporangium surrounded by integuments. • Unlike the naked seeds of gymnosperms, those of angiosperms are surrounded by an ovary wall. **28.2** • The sporophyte is the dominant generation in the pine. Gymnosperms are commonly wind pollinated, although some, such as cycads, may be pollinated by insects. • Gymnosperms possess seeds, lack free-living gametophytes, and, unlike most seedless plants, are heterosporous. • The four groups of gymnosperms are the conifers, cycads, ginkgoes, and gnetophytes. • Unlike ginkgoes and gnetophytes, cycads produce seeds in cones. Both cycads and ginkgoes have flagellated motile sperm cells (unlike gnetophytes). Ginkgoes have unique fan-shaped leaves and produce exposed seeds; gnetophytes also produce exposed seeds, and unlike all other gymnosperms, have vessel elements in their xylem. **28.3** • The vascular tissue of angiosperms has more efficient components than that of gymnosperms; it includes vessel elements in the xylem and sieve tube elements in the phloem. • The flowering plant life cycle is characterized by a unique double fertilization process, which provides endosperm for the nutrition of the new sporophyte. Angiosperm seeds are enclosed in a fruit, which protects the seeds and promotes their dispersal. • Monocots have embryos with one cotyledon; eudicot embryos have two cotyledons. Refer to Table 28-2 to compare additional distinguishing characteristics. • Fertilization in gymnosperms yields a diploid zygote; double fertilization in angiosperms yields a diploid zygote, plus a triploid cell that develops into the endosperm. **28.4** • Unlike the seed ferns, progymnosperms reproduced by spores (not seeds). • *Archaefructus,* the oldest known fossil angiosperm, had carpels that enclosed ovules. • Based on molecular data, monocots are considered core angiosperms (a sister clade to the eudicots).

FIGURE QUESTIONS

FIG. 28-2 Gnetophytes. **FIG. 28-4** A pine pollen grain is a haploid immature microgametophyte (not a gamete), which will mature to produce sperm (male gametes) by mitosis. **FIG. 28-13** Microspores and megaspores are haploid cells that divide by mitosis to give rise to multicellular microgametophytes and megagametophytes, respectively. Megagametophytes produce eggs (female gametes) and microgametophytes produce sperm (male gametes) by mitosis. **FIG. 28-18** The basal angiosperm *Amborella* lacks vessel elements, which are present in water lilies and star anise. **FIG. 28-19** In gymnosperms the haploid female gametophyte forms the nutritive tissue in the seed.

TEST YOUR UNDERSTANDING

1. e 2. b 3. d 4. a 5. e 6. d 7. b 8. d 9. a 10. c 11. a 12. c 13. Eudicot angiosperm: five petals, netted leaf veination. 14. Compare your sketch to Figures 28-4 and 28-13. Embryos (new sporophytes) and seed coats (parent sporophyte tissue) are diploid. The gymnosperm nutritive tissue is the haploid female gametophyte; that of the monocot is triploid endosperm. 15. Both seedless and seed plants have the same basic life cycle, with an alteration of multicellular haploid gametophyte and multicellular diploid sporophyte generations. Most seedless plants are homosporous; all seed plants are heterosporous. 16. Bisexual flowers increase the odds that an individual will be able to reproduce sexually because the individual has the potential to become a male parent, a female parent, or both. 17. Algae, mosses, and ferns all require at least a thin film of water for sperm transport in sexual reproduction. Both gymnosperms and angiosperms enclose their gametes and embryos in structures that provide some measure of protection from desiccation, an important adaptation to life on land. 18. Given the information currently available in the fossil record, the progymnosperm clade would branch off from the cladogram after the ferns because they had gymnosperm-like vascular tissue; it would branch off before the evolution of seeds, which they lacked.

CHAPTER 29

CHECKPOINT

29.1 • Fungi are distinguished from other simple eukaryotes by nutrition by absorption from the environment and by cell walls of chitin. • Yeasts are unicellular; molds form filamentous mycelia composed of hyphae. **29.2** • A diploid cell contains two sets of chromosomes in a single nucleus. A dikaryotic cell contains two kinds of nuclei, each with a single set of chromosomes. • Compare your sketch to Figure 29-4. **29.3** • Chytrids, like other opisthokonts but unlike all other fungi, produce flagellate cells at some stage of their life cycle. • Zygomycetes: produce sexual zygospores. Glomeromycetes: sexual spores are large, multinucleate blastospores. Ascomycetes: asci produce sexual ascospores. Basidiomycetes: basidia produce sexual basidiospores. • Compare your diagrams to Figures 29-13 and 29-17. • An ascocarp is a fruiting body that produces ascospores (sexual spores) in asci. A basidiocarp is a fruiting body that produces basidiospores (sexual spores) on basidia. **29.4** • Fungal decomposers recycle important nutrients back into the environment and are particularly important in breaking down certain materials such as cellulose and lignin. • Plants provide organic nutrients to mycorrhizal fungi, which in turn make soil nutrients available by decomposing organic matter; through their larger surface area, they increase the ability of plant roots to absorb water and minerals. • The photoautotroph is the likely host because it can typically grow well when separated from the fungal partner. The fungal partner does not grow well on its own, so it is the likely parasite. **29.5** • Studies of fungi such as *Saccharomyces, Aspergillus,* and *Neurospora* as model organisms have yielded important insights in biology and medicine. Some fungi produce valuable medications, while others are responsible for causing serious diseases. • Some fungi benefit plants by participating in mycorrhizal associations; by recycling nutrients in the environment, making them available to plants; by degrading toxins in the environment; and by controlling insects and other pests. Other fungi cause serious plant diseases.

FIGURE QUESTIONS

FIG. 29-4 Beadle and Tatum irradiated asexual spores of *Neurospora* to induce mutations, and then selected the mutant strains with differing nutritional requirements. **FIG. 29-7** Animals produce gametes by meiosis. **FIG. 29-9** Haploid gametangia function as gametes in the zygomycete life cycle. **FIG. 29-13** Conidia and ascospores are haploid; the ascocarp is dikaryotic; and the zygote is diploid. **FIG. 29-17** The basidiospore and primary mycelium are haploid; secondary mycelium and basidiocarp are dikaryotic; and the zygote is diploid. **FIG. 29-19** The potential loss of mycorrhizal fungi (negatively affected by acid precipitation) will reduce the ability of their plant partners to obtain water and essential mineral nutrients. The problem can be alleviated by controls on industries responsible for releasing acid-causing pollutants into the atmosphere.

TEST YOUR UNDERSTANDING

1. a 2. b 3. e 4. d 5. d 6. a 7. e 8. c 9. c 10. c 11. b 12. e 13. (a) Dikaryotic. (b) Dikaryotic, preparing for sexual reproduction. (c) Compare your sketches to steps 4 and 5 of Figure 29-17. 14. Fungal mycelia form an extensive, but not easily seen, network of microscopic hyphae in the soil. 15. (a) Fungi are classified as opisthokonts based on molecular data and the presence of platelike cristae in their mitochondria. The flagella of chytrids (the only fungi that have a life cycle that includes flagellate cells) are single and posterior, as in flagellate opisthokonts. (b) Microsporidia are classified as fungi based on gene sequences. (c) Ascomycetes and basidiomycetes are sister clades, thought to have diverged after the evolution of the dikaryotic stage in the life cycle. 16. Fungi, like humans, are eukaryotes. Consequently, there are fewer differences between humans and fungi that can be exploited in the development of safe and

effective treatment options for fungal diseases. 17. *Pro:* The genomes of the fungi that are model organisms provide important insights into the genetic basis of a vast array of processes in other organisms, including humans. Knowledge of the genomes of economically important fungi (both beneficial and harmful) can be exploited in making these fungi more useful or to provide ways to combat them. *Con:* The scientific resources required can be costly and expenditures must be balanced with other important societal and scientific goals (but keep in mind that the latest advances in genome sequencing technology and bioinformatics are reducing, not increasing, costs).

CHAPTER 30

CHECKPOINT

30.1 • Like other animals, sponges are multicellular, eukaryotic heterotrophs; they have an extracellular matrix containing collagen; sponges exchange gases, circulate materials, and dispose of waste; they are capable of locomotion and can reproduce sexually. **30.2** • Advantages of the marine environment for animals are the buoyancy of sea water, which provides support; the large volume of the ocean, which keeps the water temperature relatively stable; and the similar osmotic concentration of animal body fluids and sea water, which helps maintain fluid and salt homeostasis. • Animal adaptations to the terrestrial environment are a body covering that minimizes desiccation, internal respiratory organs that reduce fluid loss; internal fertilization; protective egg, shell, or internal embryo development; skeletal and muscular support; and behavioral and physiological adaptations for maintaining body temperature. **30.3** • The Cambrian radiation was the rapid appearance of a great variety of body plans evidenced by fossils dating to the Early and Middle Cambrian time periods. • According to the current hypothesis, most major groups of animals evolved several hundred million years before they appeared in the fossil record, suggesting that the Cambrian radiation was a rapid evolution of new animal body plans among clades that already existed. • The discovery of *Hox* genes helped biologists understand that similarities in molecular development among different animal groups suggest that they had a common ancestor because all the *Hox* gene groups had evolved by the beginning of the Cambrian period. Further study has shown that regulation by *Hox* gene groups has been linked with the development of wings or legs. **30.4** • Animals are classified based on radial symmetry, in which the animal can be divided into two mirror images by multiple planes through the central axis, or bilateral symmetry,

in which an animal can be divided into two mirror images by only one plane through the central axis. • Some differences between protostomes and deuterostomes are in the patterns of cleavage. For example, in early development, early cell division consists of spiral cleavage in protostomes and radial cleavage in deuterostomes. Also, the early four-cell embryo of protostomes exhibits a pattern of determinate cleavage, while the first embryonic cells of deuterostomes undergo indeterminate cleavage. Further, in protostomes the blastula develops into a mouth, while in the deuterostomes the blastula develops into an anus, and the mouth, a second opening, forms later. • Refer to Figure 30-7 for a cladogram illustrating the evolutionary relationships among and major differences between the three main clades of bilateral animals.

FIGURE QUESTION

FIG. 30-6 Nematodes, tardigrades, onychophorans, and arthropods are the animal groups assigned to the Ecdysozoa. The shared derived character present in the Ecdysozoa clade is molting.

TEST YOUR UNDERSTANDING

1. d 2. e 3. b 4. b 5. d 6. c 7. a 8. b 9. c 10. See Fig. 30-6b. 11. You could decide that the organism is an animal if the following criteria are observed: multicellular, eukaryotic, heterotrophic, extracellular matrix containing collagen, locomotion, and sexual reproduction. 12. Large molecular data sets, physical and molecular examination of fossils and existent animals, morphology, and patterns of development were some of the types of data that biologists used to determine these phylogenetic relationships. 13. Any hypothesis that is logically defended is acceptable. 14. The branch would occur just below the node where Lophotrochozoa and Ecdysozoa branch, at the spot labeled protostome pattern.

CHAPTER 31

CHECKPOINT

31.1 • Choanocytes resemble choanoflagellates (protists) in form and function. Many systematists think that sponges and other animals share a common ancestor with choanoflagellates. Choanocytes create water current in the sponge, bringing food and oxygen in and carrying out carbon dioxide and wastes. Choanocytes also phagocytize food particles. • Cnidarians are more highly organized than sponges; they have two-tissue cell layers, a nerve net with simple sense organs, a hydrostatic skeleton for body support and movement,

and cnidocytes, with stinging organelles for defense and predation. • See Fig. 31-5. • Ctenophores are like cnidarians in that both have a similar body plan, two cell layers, and a nerve net. Ctenophores are different from cnidarians because they lack nematocysts; their digestive system has a mouth for intake and anal pores for waste removal, while cnidarians have only one opening. **31.2** • Cephalization is an advantage because forward-moving organisms with concentration of sense organs in a head area are better able to search for food, shelter, mates, and be alerted to predators and dangerous conditions. Advantages of having a coelom include room for more complex organ systems with organs attached to the body wall and space for gonads to develop. • Nemerteans are classified as lophotrochozoans because they have a rhynchocoel, which is a remnant of a coelomate ancestor; recent molecular evidence and cladistic analysis place them in the lophotrochozoan clade. • Gastropods have a large, flat foot for locomotion, a mantle that functions as a lung (for land snails), and torsion, which protects the head. They have a slow-moving lifestyle and retreat to their shell when threatened. Cephalopods have very complex eyes; a body containing a siphon, enabling rapid swimming for predation and evasion; and defense mechanisms such as chromatophores and an ink sac. They live a fast-moving predator lifestyle, and many exhibit highly adaptable vertebrate-like behavior. • Flatworms have a simple nervous system and are acoelomate. Mollusks have a muscular foot and a mantle. Annelids have segmented bodies in which each segment has its own muscles, segmented organs, and most members have setae (with the exception of leeches). **31.3** • Nematodes differ from flatworms because they molt, have a thick cuticle, and a pseudocoelom. • Arthropod characteristics for success are segmentation for specialized body functions; an exoskeleton to provide protection from predators and desiccation; paired, jointed appendages, which are modified for diverse functions; and a more developed nervous system, which may include hearing organs, antennae, and compound eyes. • Crustaceans have three body divisions, while chelicerates have two; crustaceans have biramous appendages, while chelicerates are uniramous; crustaceans have antennae, while chelicerates do not; crustaceans develop through larval stages, while most chelicerates exhibit direct development. • Adaptations that have made insects very successful include the specialized insect body plan, ability of many insects to fly, great reproductive capacity, and offensive and defensive adaptations (cryptic coloration, toxins, pheromones).

FIGURE QUESTION

FIG. 31-32 All four stages are diploid; the adult produces haploid gametes.

1. a 2. c 3. b 4. d 5. a 6. e 7. b 8. d 9. b 10. e 11. d 12. See Fig. 31-4. 13. Having a coelom resulted in advantages such as highly developed, complex organ systems. Many internal organs are suspended from folds of tissue lining the coelom; these organs, for example, the digestive tube, can move independently of the outer body wall. In some animals, fluid in the coelom helps transport food, oxygen, and wastes; cells bathed by the coelomic fluid can exchange materials such as food, oxygen, and wastes with it. The coelom also provides space for production and storage of gametes. 14. Cephalization was an advantage because centralized sense organs efficiently promoted greater nervous and sensory adaptations (offensive and defensive), but it was a disadvantage to have the majority of sensory organs in the head, which could be easily harmed and destroyed. The arthropod exoskeleton was an evolutionary advantage that provided strong armor and muscle support but was a disadvantage when the animal molted, leaving it vulnerable while waiting for its new exoskeleton to harden. Specialized segmentation was advantageous in the development of legs and wings for locomotion but was a disadvantage if that area became damaged where other segments could no longer substitute for the function. 15. The animal would be classified in the lophotrochozoan clade due to segmented body, bristles, and no observable head. Based on Figure 30-6a, the animal would likely be classified as an annelid, between the unsegmented mollusks and lophophorates, which do have an obvious mouth end. 16. Biologists have recently recognized that animals with very similar body cavities may have great evolutionary and genetic differences based on molecular phylogenetic analyses. 17. Some animals, such as parasites, became adapted to environments in which they no longer used some of their organs, or even organ systems. For example, the tapeworm lives inside the digestive system of its host and absorbs food through its body wall. Through the course of evolution these structures were not selected for, eventually resulting in their loss. Tapeworms no longer have a mouth or digestive system. 18. Insects that undergo complete metamorphosis may be more successful than other insects because the larval form and the adult form are very different and occupy different habitats. They do not compete with each other for food and other resources. 19. Coral reef health is an indicator of the quality and temperature of our oceans. We humans are as dependent on healthy, pollution-free, oceans as are all living organisms on this planet we share. Damage to coral reefs from human-caused water temperature increase, higher acidity from carbon dioxide release, reduced water clarity from contaminants, introduction of toxic pollutants, and physical destruction of reefs are all problems society must address in partnership with scientists to develop successful technologies for remediation.

CHAPTER 32

CHECKPOINT

32.1 • Three shared derived characteristics of deuterostomes are radial cleavage, indeterminate cleavage in which blastopore develops into anus and mouth develops from second opening of embryo, and pharyngeal slits. **32.2** • Three derived characters of echinoderms are adult pentaradial symmetry, allowing them to respond to all directions of their environment; a water vascular system for feeding, gas exchange, and hydrostatic support for locomotion; and a calcium carbonate endoskeleton covered by a ciliated epidermis with spines, which protects them from predation. • Sea stars are different from crinoids because their mouths are on the underside of the disc, not the upper side, and most sea stars are active predators and scavengers, while crinoids move little or are sessile and obtain food by removing microorganisms in suspension. Sea stars are different from sea urchins because they have movable arms, while sea urchins have flattened, fused plates and no arms but do have movable, long, sharp spines that aid locomotion and provide protection. **32.3** • Four shared derived characters of chordates are a notochord; the dorsal, tubular nerve cord; the postanal tail; and an endostyle or thyroid gland. • The chordate nerve chord differs from most other animals because it is dorsal, hollow, and single rather than ventral, solid, and double. • Human jaws are related to early chordate pharyngeal slits because the anterior pharyngeal arches of the slits evolved into jaws during early vertebrate evolution. **32.4** • A tunicate larva has the four chordate characters (notochord; dorsal, tubular nerve chord; postanal tail; and endostyle), while a tunicate adult loses the notochord, postanal tail, and most of the nervous system. • If the organism lacks paired fins, eyes, and jaws, it would be identified as a lancelet instead of a fish. **32.5** • The four shared derived characters of vertebrates are vertebral column, cranium, neural crest cells, and pronounced cephalization with 10 to 12 pairs of cranial nerves and developed sensory organs. • The main classes of fishes are Myxini, Petromyzontida, Chondrichthyes, Actinopterygii, Actinistia, and Dipnoi. The main groups of tetrapods are amphibians, mammals, and reptiles. **32.6** • Like ostracoderms, lampreys and hagfishes do not have paired fins or jaws. • Unlike most other fishes, hagfishes do not have jaws, paired fins, or vertebrae. **32.7** • Placoderms were the first fish to have jaws and paired fins, characteristics that were passed on to modern fishes. Ray-finned fishes underwent adaptive radiation, aided by additional clusters of *Hox* genes, which increased amount of genetic material for evolution and promoted the diversification of modern bony fishes. Lungfishes retained lungs from their sarcopterygian ancestors and evolved fleshy fins from which limbs of tetrapods developed. *Tiktaalik* was a fish that had limblike fins and tetrapod characteristics (movable neck and ribs that supported lungs); it was a transitional form between fishes and tetrapods. • Amphibians undergo metamorphosis in which they lose gills, gill slits, and tail, and grow limbs. Simple lungs and moist skin allow for respiration/gas exchange on land. Mucus-producing glands develop to help maintain skin moisture and make the animal slippery to evade land predators. Many adult amphibians also have glands that secrete poison for protection. **32.8** • Adaptations that allow amniotes to be terrestrial are the amniotic egg, internal fertilization, a body covering that retards water loss, and physiological mechanisms to retain water. • Birds and mammals are endotherms. Advantages of endothermy are high metabolic rate allowing for a high level of activity and decreased vulnerability to temperature fluctuations. • Birds and reptiles are now correctly classified as a single clade based on fossil records and molecular analyses; they evolved from feathered, saurischian dinosaurs, bipedal theropods, which share characteristics such as feet with three digits, thin-walled, hollow bones, and a wishbone. Modern birds also retain reptilian scales on their feet and legs. • Protherian mammals, or monotremes, lay eggs and secrete milk for young (do not have nipples). Metatherian mammals, or marsupials, give birth to undeveloped young that move to a pouch on the mother to continue development. They obtain milk from a nipple in the pouch. Eutherian mammals are attached to a placenta before birth and are born much more developed. Babies obtain milk from nipples on the mother's chest or abdomen.

FIGURE QUESTIONS

FIG. 32-4 The vertebrate nerve cord differentiates into the brain and spinal cord. **FIG. 32-7** The pronounced cephalization that was part of the evolution of vertebrates from jawless fish to tetrapods allowed for a larger, more complex brain. As the brain and sensory organs further evolved, various regulatory and cognitive systems could become more complex; increased brain complexity resulted in greater success in competing with other animals for food and mates. **FIG. 32-20** Birds are more closely related to the extinct branches of reptiles because they share the same archosaur ancestor. Mammals share a distant relationship with all members of the diapsids through the amniotic ancestor at the root of the cladogram.

1. b 2. a 3. d 4. e 5. e 6. c 7. c 8. See Fig. 32-25. 9. Sea urchins are not cnidarians because they have the characteristics of deuterostomes and echinoderms. Radial symmetry may have evolved as an adaptation to the sessile lifestyle of early echinoderms. 10. Birds share the same archosaur ancestor based on fossil records and molecular analyses; they are most closely related to saurischian dinosaurs (now extinct) and share characteristics of the dinosaurs such as bipedal locomotion, feet with three digits, thin-walled, hollow bones, a wishbone, and scales on their legs. 11. Echinoderms and chordates are both members of the deuterostomes and share derived characters at some stage of life such as radial cleavage, indeterminate cleavage in which blastopore develops into anus and mouth develops from second opening of embryo, and pharyngeal slits. 12. Dorsal tubular nerve cord would identify the animal as a chordate. Cranium would place the animal in vertebrates. Skin with scales and two atria/two ventricles would indicate the animal is closely related to a crocodilian, bird, or saurischian dinosaur. The animal discovery would be placed somewhere between crocodilians and birds on the cladogram. 13. Sea turtles can be protected by scientists sharing research about the animals and their needs with the public; educators can inform society about sea turtle conservation and lead activities to improve habitat, reduce poaching, and pass conservation laws; technology can be developed to support more complex research, rehabilitate habitat, control poaching, and improve health care for turtles.

CHAPTER 33

CHECKPOINT

33.1 • Roots absorb water and minerals; shoots bear the leaves, which absorb CO_2 and serve as the main organs of photosynthesis. • The *ground tissue system* carries out photosynthesis, and provides for storage and support. The *vascular tissue system* conducts water, minerals, and carbohydrates. The *dermal tissue system* provides a covering for the plant body. • Parenchyma tissue contains thin-walled parenchyma cells that carry out such functions as photosynthesis, storage, and secretion. Collenchyma tissue consists of collenchyma cells, which provide flexible support by means of their thickened primary cell walls. The sclerenchyma cells (sclerids and fibers) in sclerenchyma tissue have thickened secondary cell walls that provide strong support. • Xylem is the tissue responsible for conduction of water and minerals by means of tracheids (with pits) and vessel elements (with perforations in end walls as well as pits). Sieve tube elements in phloem are responsible for conducting dissolved sugar, which has been moved into them by companion cells. • The epidermis,

the outer covering of the plant body, functions in protection and also includes guard cells in leaves that regulate stomata as well as trichomes, which carry out varied functions. The underlying periderm, like the epidermis, has protective functions. It also contains meristematic and storage cells. **33.2** • Plant meristems provide new cells for growth. Animal growth does not involve meristems. • Secondary growth is an increase in girth; primary growth is an increase in length. • Apical meristems are responsible for primary growth; lateral meristems provide for secondary growth. **33.3** • Cell expansion typically precedes cell differentiation. • Changes in gene activity (differential gene expression) are responsible for pattern formation. • Cells alter the expression of their genes response to positional information received from their environment. These changes bring about pattern formation, which eventually leads to morphogenesis. • Many developmental mutants of *Arabidopsis,* a small plant with a fast life cycle, are available for study. Information gained from these investigations can then be applied to research on other plants.

FIGURE QUESTIONS

FIG. 33-1 Air: CO_2; soil: water. **FIG. 33-2** For transport to take place, the vascular tissue must be continuous throughout the plant. The plant would die because there would be no transport between the leaves and the roots.

TEST YOUR UNDERSTANDING

1. a 2. c 3. d 4. b 5. c 6. c 7. a 8. b 9. b 10. e 11. a 12. Compare your labels to Figure 33-2a. 13. Compare your sketch to Figure 33-11. 14. Compare your sketch to Figure 33-12. Note that your arrow should be perpendicular to the cellulose microfibrils. 15. The initials will still be 4 ft above the ground. The increase in the height of the tree occurred at multiple apical meristems. 16. Unlike dead cells (schlerenchyma), living cells (bone) require maintenance in the form of nutrition, gas exchange, and waste removal. On the other hand, dead cells cannot be repaired or replaced. 17. Vessel elements are more efficient water conductors than are tracheids. It is thought that possession of both tracheids and vessel elements has contributed to the adaptive radiation of angiosperms into a wide variety of environments. 18. The relationship of the genome to various plant functions can be studied more easily in *Arabidopsis* than in most other plants. The pace of research is enhanced when these insights are applied to investigations on commercially important plants.

CHAPTER 34

CHECKPOINT

34.1 • Leaves are covered with a waxy cuticle and may also have trichomes that are specialized to reduce water loss. Gas exchange is

limited to stomata, tiny pores whose opening and closing can be controlled. • Mesophyll, the leaf's photosynthetic ground tissue, typically consists of palisade mesophyll under the upper epidermis, and spongy mesophyll below. • A vascular bundle contains xylem (upper) and phloem (below). • Leaf mesophyll cells receive water and dissolved minerals from the xylem; the carbohydrate products of photosynthesis are transported away by the phloem. CO_2 enters the leaf through stomata, which also provide for the exit of excess O_2 (i.e., that not used for respiration by leaf cells). **34.2** • Blue light begins the process of stomatal opening by causing the activation of proton pumps in the plasma membranes of guard cells. It also triggers synthesis of malic acid and hydrolysis of starch. • In stomatal opening, proton pumps (activated by blue light) actively transport H^+ out of the guard cells. This resulting electrochemical gradient is responsible for the opening of voltage-activated ion channels through which K^+ and Cl^- enter the guard cells by facilitated diffusion; water then enters by osmosis. The resulting turgor pressure in the guard cells causes them to change shape, resulting in stomatal opening. The stomata are maintained in an open position during the day due to maintenance of osmotic pressure by sucrose produced by photosynthesis. When sucrose concentration declines at the end of the day, the guard cells lose turgor as they lose water by osmosis and the stomata close. **34.3** • The leaf cuticle and stomata limit water loss by transpiration. • Sunlight, higher temperatures, lower relative humidity, and higher wind speeds all increase the rate of transpiration. • Transpiration benefits plants by making water and minerals available throughout the plant. It is harmful when it results in excess water loss. • In guttation, liquid water is forced out of vein endings in the leaves of small plants as a result of water entering the roots when soil moisture is high and transpiration is negligible. Transpiration is the evaporation of water from aerial parts of the plant. **34.4** • Loss of leaves in autumn is a mechanism to reduce water loss during winter when cold temperatures (and perhaps freezing) make water less available from the soil. • Leaf abscission occurs after enzymes have dissolved the middle lamella between the cells in the abscission zone. • The abscission zone is an area where a leaf petiole attaches to a stem. It is composed mainly of thin-walled parenchyma cells. It also includes cork cells, which will remain with the stem as a protective layer after the leaf has abscissed. **34.5** • Spines, protection from herbivores; tendrils; aid in climbing; bud scales; protection of meristematic tissue. • Bulbs; storage of food; succulent leaves; storage of water. • Certain leaves of carnivorous plants are specialized to capture and digest small animals such as insects.

FIGURE QUESTIONS

FIG. 34-3 Control of developmental processes is not limited to genes that code for transcription factors. Other examples include genes that code for components of signaling pathways, regulatory RNAs, and a great many more. **FIG. 34-10** During photosynthesis during the day, guard cells produce sucrose, which is osmotically active and helps maintain their turgor pressure, keeping the stomata open.

TEST YOUR UNDERSTANDING

1. c 2. d 3. d 4. e 5. d 6. a 7. e 8. c 9. a 10. c 11. a 12. You would expect a higher concentration of guard cells in the lower epidermis. 13. In the stem, an arrangement with the xylem toward the center and the phloem toward the periphery would account for the leaf arrangement (xylem toward the upper epidermis and the phloem toward the lower epidermis). 14. Compare your simple drawing to Figure 34-4. Mesophyll cells and guard cells should contain chloroplasts. 15. A few large leaves present a large area for photosynthesis, but may effectively shade leaves below them and also present a large area for potential water loss. This might not be a great disadvantage in a humid climate. Many small leaves may have less area for photosynthesis, but they will shade each other less. Collectively they can be oriented in many directions to capture light and may lose much less water. 16. Leaf abscission helps prevent water loss during the winter when photosynthesis is negligible and less water is available from the soil. The smaller surface area, thick cuticle, and sunken stomata of conifer leaves (needles) are adaptations that allow conifers to survive winter without leaf abscission. 17. Agricultural plant yields depend on photosynthesis, which in turn depends on availability of CO_2 (through open stomata). Water loss (also through open stomata) reduces plant yields.

CHAPTER 35

CHECKPOINT

35.1 • In a stem cross section, the vascular tissue of a primary eudicot stem is arranged in a circle; vascular bundles of a monocot stem are scattered. • Vascular cambium gives rise to secondary xylem (to the inside) and secondary phloem (to the outside). Cork cambium produces periderm (outer bark). • When secondary growth occurs, the primary tissues become nonfunctional (e.g., primary xylem) or become destroyed (e.g., primary phloem). • Growth rings are a consequence of the differences between springwood (with large-diameter tracheids and vessel elements formed when water is plentiful) and summerwood (with smaller water-conducting cells, formed when water is less available). • Terminal buds form at the tips of stems; axillary buds form in the axils of

plant leaves. **35.2** • Water moves from a region of higher water potential (generally less negative) to a region of lower water potential (generally more negative). • It has been demonstrated that tension–cohesion is strong enough to pull water as high as 130 m. The tallest known tree is 115.7 m high. **35.3** • Xylem transport occurs upward from roots to leaves. Phloem transport is from a source to a sink (upward or downward). • In the pressure–flow model, companion cells load sugar into sieve tube elements at the source and water follows by osmosis. This increases the hydrostatic pressure inside the sieve tube, pushing the contents toward the sink. The hydrostatic pressure is lower at the sink because companion cells unload sugar from the sieve tube elements, causing water to leave by osmosis.

FIGURE QUESTIONS

FIG. 35-3 The irregularly arranged (scattered) vascular bundles of a monocot stem would not provide for regular positioning of a vascular cambium within the stem. **FIG. 35-7** The branch is two years old (two sets of bud scale scars). There are two twigs with terminal buds at the end. **FIG. 35-11** Water would not be able to move from the soil to the root (and might even move out of the root) if the water potential of the soil were more negative than that of the root. **FIG. 35-12** The direction of translocation is from a source (in this case a root where starch is broken down) to a sink (in this case a developing leaf cell). Therefore the direction of transport would be upward in the stem. **FIG. 35-13** Outward movement of fluid would not occur because, unlike phloem, the xylem of a large plant is not under positive pressure.

TEST YOUR UNDERSTANDING

1. d 2. e 3. b 4. b 5. a 6. b 7. e 8. c 9. b 10. a 11. b 12. Compare your drawing to Figure 35-4. The daughter cell formed to the outside of the meristematic cell should be labeled secondary phloem. 13. Periderm or outer bark (which is usually most easily peeled away) consists of remnants of primary phloem, cortex, and epidermis. 14. A tree 150 m tall would exceed the calculated maximum height for xylem transport by tension–cohesion (130 m). 15. One hypothesis would be that tropical vines must be strong and long lived to obtain adequate sunlight by growing to the tops of tall rainforest trees. 16. Water will flow from the cell with higher (less negative) water potential (i.e., −1.5 MPa) to the cell with lower (more negative) water potential (i.e., −1.8 MPa).

CHAPTER 36

CHECKPOINT

36.1 • A fibrous root system is adapted to obtain water from a larger area of the soil surface; a taproot system has access to water

deep underground. • A primary eudicot root has a central stele containing xylem and phloem. The center of the stele of a typical monocot root is occupied by pith surrounded by xylem and phloem arranged in alternating bundles. • Water and dissolved minerals move through the apoplast by way of interconnected cell walls. Water and minerals must cross at least one plasma membrane to enter the symplast, and then move from cell to cell through plasmodesmata. • The Casparian strip prevents materials from entering the center of the root without passing through the plasma membrane of an endodermal cell. • As in the stem, the root vascular cambium produces secondary xylem toward the inside and secondary phloem toward the outside. Cork cambium also contributes to the increase in girth as it produces cork cells and cork parenchyma cells. • Prop roots provide support; buttress roots provide support and help distribute shallow roots around a tree; pneumatophores may assist in gas exchange by submerged roots. **36.2** • Mycorrhizae and root nodules are symbiotic associations that benefit plants. Mycorrhizal fungi aid plant roots in water and mineral uptake and rhizobial bacteria fix nitrogen in root nodules. **36.3** • Minerals supply essential inorganic nutrients; organic matter provides many nutrients and holds water; air supplies oxygen for aerobic respiration and nitrogen that can be used by nitrogen-fixing bacteria; soil water makes water available to plants and dissolves nutrients and gases. • In cation exchange, H^+ pumped into the soil by plant roots releases various mineral cations from negatively charged clay particles, making them available for absorption. • In weathering processes, rock is broken down by acids produced by organisms (particularly carbonic acid formed from carbon dioxide) and by alternating cycles of freezing and thawing. • Macronutrients are required in relatively large quantities (greater than 0.05% dry weight), whereas trace amounts of micronutrients are required. • Human mismanagement can damage soil through mineral depletion, soil erosion, and salt accumulation.

FIGURE QUESTIONS

FIG. 36-4 Many kinds of minerals are carried along dissolved in the water traveling through the nonselective apoplast. To enter the symplast, all minerals must pass through at least one plasma membrane. The plasma membranes therefore act as "gatekeepers" that promote entry of minerals the plant requires, and typically exclude others. **FIG. 36-5** Water enters the endodermal cells by osmosis, which does not require a direct expenditure of energy. Energy is supplied indirectly by the energy-requiring transport of minerals, which creates relatively hypertonic conditions (i.e., lowers the effective water concentration) inside the endodermal cells. **FIG. 36-9** Stem vascular cambium, like root vascular cambium, produces secondary xylem inward and secondary phloem outward.

1. c 2. b 3. b 4. a 5. c 6. c 7. d 8. a 9. d 10. e 11. Compare your sketch to Figure 36-9. 12. The roots of woody plants, like their stems, produce annual rings. You should take a core sample of the root (as you would for a tree trunk) close to the entrance of the mineshaft and count the rings. 13. Without a microscope, look for the presence of root hairs. With a microscope, examine the internal structure in cross section and determine if a stele (characteristic of roots) is present. 14. In roots with secondary growth, all absorption occurs close to the tips, where only primary growth occurs and where the root hairs are found. 15. A shallow root system enables a cactus to capture more water from the infrequent rains that occur in the desert before the water runs off. 16. Understanding development of root hairs could lead to ways of modifying agricultural crops to improve water and mineral absorption. Mycorrhizal associations benefit many plants, and it is possible these benefits could be extended to additional types of agricultural crops. 17. The roots of the common reeds (*Phragmites australis*) absorbed almost all of the selenium. The plants should be uprooted and disposed of in a way that does not release selenium into the environment.

CHAPTER 37

CHECKPOINT

37.1 • Sepals, the outermost flower whorl, are typically more leaflike than the petals, which form the second whorl and are typically colored to attract pollinators. • Stamens and carpels are modified for reproduction. Stamens produce pollen and carpels produce ovules. • The female gametophyte of a flowering plant, produced within an ovule, is an embryo sac containing a haploid egg (which is fertilized to become the zygote) and two haploid polar nuclei (which are fertilized to become the triploid endosperm). • The immature male gametophytes of flowering plants are pollen grains, produced in anthers. **37.2** • Bee. • Wind. • Nectar or pollen are important rewards for many pollinators. Flowers have evolved to have colors that are visible to particular pollinators. Specific plant odors attract specific pollinators. **37.3** • Pollination is delivery of pollen to a flower stigma. Fertilization occurs when a pollen tube delivers sperm to fuse with the egg (producing an embryo) and with the two polar nuclei (producing the endosperm). • A eudicot embryo proceeds through the following stages: proembryo, globular stage embryo, heart stage embryo, torpedo stage embryo, and maturing embryo with two cotyledons. • A fruit is a mature ripened ovary. • Simple fruit develops from a single ovary with a single carpel or several fused carpels; aggregate fruit develops from a flower that contains several unfused carpels; multiple fruit develops from ovaries of many flowers; accessory fruit develops from other tissues in addition to the ovary. • Animal-dispersed fruits and seeds tend to be edible, or they have barbs or other structures that attach them to the bodies of animals. **37.4** • All plant seeds require water for germination; the young plant would not be able to grow in the absence of water. Many seeds require a period of low temperature before they germinate; this prevents them from germinating in the fall and being unable to survive the winter. Some very small seeds require light for germination, ensuring that they will germinate near the soil surface. • The stem of a eudicot seedling protects its delicate tip by forming a hook as it emerges from the soil. The stem tip of a monocot seedling is enclosed by a protective coleoptile. **37.5** • Rhizomes are horizontal underground stems that produce new aerial shoots. Tubers are storage organs that develop from rhizomes and have axillary buds that can develop into new plants. • A bulb is composed mainly of fleshy leaves attached to short underground stem. The underground stem of a corm is enlarged relative to the papery leaves. • Apomixis is the production of seeds enclosed in fruits without sexual reproduction. It has the advantages of the protections and dispersal methods usually provided by seeds and fruits. **37.6** • As in all organisms, asexual reproduction in plants results in offspring that are like their parents. Conversely, asexual reproduction results in genetically varied offspring. • Genetic variety in the offspring is considered adaptive because it typically results in at least some offspring that are better suited for survival and subsequent reproduction than their parents. However, it is costly because it also results in some less well-adapted offspring.

FIGURE QUESTIONS

FIG. 37-3 Haploid: megaspore, microspore, egg, polar nucleus, generative cell, sperm cell. Diploid: megasporocyte, microsporocyte. **FIG. 37-5** Some additional traits that might affect pollination in monkeyflowers would be scent (important to bumblebees but less so to hummingbirds) and rewards (nectar important to both, pollen more important to bumblebees).

TEST YOUR UNDERSTANDING

1. c 2. d 3. c 4. a 5. c 6. b 7. b 8. d 9. c 10. d 11. a 12. Compare your sketch with Figure 37.1. 13. Compare your sketches with Figures 37-1, 37-11, and 37-12. 14. The offspring of asexual reproduction might be expected to develop close to the parent plant because they are well suited to the same environmental conditions. You could test this hypothesis by growing the offspring in various locations, including close to the parent. Do you think there could be competition between the parent and offspring? 15. (a) Likely asexual. (b) Likely sexual. (c) Asexual (because all the offspring would be adapted to the parent's climate range) or could be sexual (because some of the offspring might be able to live in a somewhat different climate). 16. Plants produce seeds with elaiosomes to attract ants, which have coevolved to bury the seeds as they make use of the food reward. 17. Because almost all agricultural plants are flowering plants, an improved understanding of the control of their flowering can yield important benefits in agriculture. Similarly, almost every new crop begins with the sowing of seeds, so an understanding of their germination processes can be equally valuable.

CHAPTER 38

CHECKPOINT

38.1 • Phototropins absorb wavelengths other than yellow (which they reflect); they respond to blue light, which is one of the wavelengths they absorb. • Phototropism: directional growth in response to the direction of light; gravitropism: directional growth in response to gravity (either positive or negative); thigmotropism: growth response to a mechanical stimulus. **38.2** • The signal provided by auxin activates a pathway that results in cell walls of target cells becoming acidified. These acidified cell walls are more plastic, enabling them to expand due to the force of the cell's turgor pressure. This cell expansion is a form of growth without cell division. • Auxin is transported from cell to cell in a plant in a unidirectional manner (polar transport). In phototropism, auxin is transported from the lighted side of the stem to the shaded side. Expansion of the cells on the shaded side causes the stem to bend toward the light. • (1) Seed germination: gibberellins, cytokinins, ethylene, brassinosteroids; (2) stem elongation: auxins, gibberellins; (3) fruit ripening: ethylene; (4) leaf abscission: ethylene; (5) seed dormancy: abscisic acid. **38.3** • Phytochrome is the main photoreceptor for photoperiodism and for shade avoidance. • One or more signal transduction pathways are activated when phytochrome undergoes changes in shape in response to light. Some of these pathways activate specific transcription factors that control gene expression. • Cryptochrome plays a role in resetting the biological clock responsible for maintaining circadian rhythms. **38.4** • The production of necrotic lesions is associated with the hypersensitive response. These areas of dead tissue limit or slow the spread of an infection.

FIGURE QUESTIONS

FIG. 38-5 One possible experiment would be to expose the uncovered coleoptiles to directional light for varying times, then cover their tips, and record bending responses. **FIG. 38-6** Coleoptile

tips previously exposed to light could be placed on agar blocks, which would then be exposed to heat at varying temperatures and for varying durations of time. The agar blocks could then be placed on decapitated coleoptiles and the bending responses recorded.

TEST YOUR UNDERSTANDING

1. c 2. a 3. b 4. c 5. e 6. e 7. d 8. b 9. e 10. b 11. 1 $\longrightarrow$ 4 $\longrightarrow$ 3 $\longrightarrow$ 5 $\longrightarrow$ 2 12. (a) Plant will not flower; night length does not exceed 14 hours. (b) Plant will flower; night length exceeds 14 hours. (c) Plant will not flower; night length does not exceed *uninterrupted* 14 hours. 13. The plant will not flower at the equator (night length does not exceed 14 hours). 14. Negatively geotropic shoots grow up, into to the air where CO_2 and light are available to support photosynthesis; positive gravitropic roots grow into the soil where they can absorb water and minerals. 15. A plant with a phytochrome gene mutation that affects flowering time would be at an evolutionary disadvantage for many reasons. For example, it might flower at time when pollen or pollinators are not available. Its seeds might not have time to mature during the growing season or might mature too early and germinate during an unfavorable season. 16. Flowering is 100% when the day length is 10 hours and the night length is 14 hours; flowering is 0% when the day length is 16 hours and the night length is 8 hours. According to the graph, the critical night length for this plant is about 11 hours. 17. There is a growing body of evidence that plant-signaling molecules are involved in a wide range of processes affecting plant growth and development, all of which are of significant agricultural importance.

CHAPTER 39

CHECKPOINT

39.1 • Epithelial tissue forms a continuous sheet of cells covering a body surface or lining a body cavity. Its functions are to protect, absorb, secrete, and sense. Connective tissue consists of few scattered cells separated by intercellular substance, composed of fibers scattered throughout a matrix. Connective tissue joins other tissues of the body, supports the body and its organs, and protects underlying organs. • The air sacs of the lungs are lined with simple squamous epithelium. The cells are flat and arranged in a single layer to permit diffusion and passage of materials where little protection is required. • Skeletal muscle is striated with each elongated muscle fiber containing several nuclei. Skeletal muscles contract voluntarily; they move body parts. For example, they lift and lower arms and legs on demand. Cardiac muscle is striated, but the elongated fibers are branched and fused with only one or two nuclei. Cardiac muscle

contracts involuntarily for the heart to beat; it is a complex network that pumps blood continuously. Smooth muscle is not striated; it is made up of elongated spindle-like fibers, each with one central nucleus. Smooth muscle contracts involuntarily to move body organs; it contracts in the walls of the digestive tract to apply pressure to move food onward; it constricts arteries causing blood pressure to increase. • The respiratory system supplies oxygen to the blood and excretes carbon dioxide. The urinary system removes metabolic waste and excess materials from the blood, produces urine, and helps regulate blood chemistry. The endocrine system works with the nervous system to regulate metabolic activities and other body functions. **39.2** • In a negative feedback system, a change in a steady state triggers a response that counteracts, or reverses, the change. A sensor detects a change, a deviation from the normal condition, or set point. The sensor signals an integrator, or control center. Based on the input of the sensor, the integrator activates homeostatic mechanisms that restore the steady state. The response counteracts the inappropriate change, which restores the steady state. Unlike a negative feedback system, a positive feedback system does not reverse the change; a change in a steady state sets off a response that intensifies the changing condition. **39.3** • A cost of ectothermy is that temperature conditions, daily and seasonal, can limit an animal's activity. A benefit for ectotherms is that they survive on less food and convert more energy from food to growth and reproduction because they do not need to maintain a high metabolic rate. To adjust body temperature, ectotherms may move in or out of the sun to raise or lower temperature; they may migrate as seasons change or hibernate. • A cost of endothermy is that animals must expend considerable energy to maintain body temperature, even when inactive. Benefits of endothermy are a rapid response to internal and external stimuli, and a greater level of activity in cold weather. To help maintain body temperature, endotherms have structural adaptations (feathers, hair, insulating fat), behavioral adaptations, and several physiological mechanisms (shivering, sweating, panting, bringing heat to skin surface) mediated by nervous and endocrine systems. • See Fig. 39-9.

FIGURE QUESTIONS

FIG. 39-5 Glucose concentration in your blood would increase. Your body would respond by negative feedback, releasing insulin into your blood to counteract the glucose until the steady state, or normal condition, is restored. **FIG. 39-7** If we could convert to a negative feedback system, the loss of blood and decrease in blood pressure would initiate factors that would slow the loss of blood by clotting at the wound site and constricting blood vessels.

TEST YOUR UNDERSTANDING

1. a 2. e 3. d 4. b 5. c 6. d 7. See Tables 39-1 and 39-2. 8. Cells $\longrightarrow$ Tissues $\longrightarrow$ Organs $\longrightarrow$ Organ Systems. 9. Loss of epithelium would allow pathogens to enter the body, causing infection; loss of body fluids would lead to dehydration; skin damage could impact sensation and cause extreme pain. If the skin damage is not repaired and systems are not returned to normal state, the patient could go into shock and may die. 10. Joint connections, body, and organ support would deteriorate, and the body would collapse. 11. Rapid breathing is typically shallow. Less oxygen would be inhaled and less carbon dioxide would be exhaled. As a result carbon dioxide concentration would further increase, disrupting homeostasis even more. 12. Most animals are ectotherms; they do not use metabolic energy to maintain a constant body temperature. Compared to endotherms, ectotherms can survive on less food and they expend more energy on growth and reproduction. 13. A health care worker has an obligation to recommend any and all tests that may benefit the patient. A genetic test can provide information necessary for making medical decisions, especially if your family has a history of cancer. Knowing that you may be predisposed toward certain types of cancer can help you and your caregivers make educated decisions to improve your chances for survival.

CHAPTER 40

CHECKPOINT

40.1 • Invertebrates and vertebrates both have epithelial coverings that protect the body. The epithelial coverings may be specialized for gas exchange, secretion, or temperature regulation; skin may contain sensory receptors that obtain information from the surrounding environment. • Birds have feathers and mammals have hair, which provide insulation and help maintain body temperature; amphibians have skin that allows gas exchange and poison secretion; mammals have claws and nails as well as sweat glands, oil glands, and sensory receptors. • Keratin gives skin mechanical strength and reduces water loss. • Melanin is a pigment that contributes to skin color and has a role in reducing ultraviolet radiation damage. **40.2** • The septa separating segments in the annelid worm allow the worm to move with more flexibility, as each segment is independent. • Advantages of the exoskeleton are to protect, transmit force, and reduce water loss; arthropods have modified parts of their exoskeleton that function as tools or weapons. The exoskeleton is a disadvantage for arthropods because their body continues to grow and they must molt, or shed their exoskeleton, which makes them weak and vulnerable to predation, while a new covering grows and hardens. • Axial skeleton main bones are skull, sternum, rib cage, and vertebrae. Appendicular skeleton main bones are clavicle, scapula, humerus, radius, ulna, carpals, metacarpals, phalanges, pelvic girdle, femur, patella, fibula, tibia, tarsals, and metatarsals. • Osteoblasts build up bone by secreting the protein collagen that forms strong bone fibers; hydroxyapatite forms around the fibers and hardens into bone matrix. Osteoclasts break down (resorb) bone by secreting hydrogen ions that dissolve crystals and enzymes that digest collagen. Osteoblasts and osteoclasts work synergistically to shape bones. **40.3** • See Figs. 40-9 and 40-10. Actin is the thin filament; myosin is the thick filament. • For a muscle fiber to contract: (1) Acetycholine, released by a motor neuron, binds with receptors on the muscle fiber, causing depolarization and generation of an action potential. (2) The action potential spreads through the T tubules, releasing calcium ions from the sarcoplasmic reticulum. (3) Calcium ions bind to troponin in the actin filaments, causing the troponin to change shape. The troponin pushes tropomyosin away from the binding sites on the actin filaments. (4) ATP (bound to myosin) is split, putting the myosin head in a high-energy state. Energized myosin heads attach to the exposed binding sites on the actin filaments, forming cross bridges that link the myosin and actin filaments. (5) After myosin attaches to the actin filament, P_i is released, triggering the power stroke. (6) During the power stroke, myosin heads pull the actin filaments toward the center of the sarcomere. ADP is released during the power stroke. (7) The myosin head binds a new ATP, which lets the myosin head detach from the actin. As long as the calcium ion concentration remains elevated, the new ATP is split and the sequence repeats. • ATP is the immediate source of energy for muscle contraction. The hydrolysis of ATP provides the energy to "cock" the myosin. Creatine phosphate is an intermediate energy storage compound for muscle tissue. Glycogen is the fuel stored in muscle fibers. • Skeletal muscle is striated, with each elongated muscle fiber contracting voluntarily to move body parts; arms and legs, for example, lift and lower on demand. Cardiac muscle is striated with elongated fibers that are branched and fused. Cardiac muscle contracts involuntarily for the heart to beat; it is a complex tissue that pumps blood continuously. Smooth muscle is not striated; it is made up of elongated spindle-like fibers. Smooth muscle contracts involuntarily to move body organs; it contracts in the walls of the digestive tract to move food onward; it constricts arteries causing blood pressure to increase.

FIGURE QUESTIONS

FIG. 40-10 The greater the volume of myofibrils in a muscle fiber, the greater the force that the muscle can generate. **FIG. 40-11** If the body stopped producing acetylcholine, action potentials could not be generated. The person would quickly die because the muscles necessary for breathing could not contract.

TEST YOUR UNDERSTANDING

1. c 2. c 3. e 4. b 5. e 6. c 7. The high rate of metabolism in insect flight muscles is adaptive because a great amount of energy must be generated to support flight. This energy is generated by the numerous mitochondria in these muscles. Flight increases the opportunities to exploit additional habitats. 8. See Fig. 40-9. 9. The resulting curve would be similar to the curve in Figure 40-13b, except there would be a third hump added to the graph slightly higher than the second twitch hump. 10. A skeletal muscle shifts function—agonist or antagonist—depending on what action is needed. If each muscle could only perform one function, twice as many muscles would be needed. 11. Muscles are attached to the bones by tendons; the muscles pull on the bones, which causes the bones to act as levers. As muscles weaken, or are not exercised, the bones also lose strength and weaken. 12. Benefits of the arthropod exoskeleton are protection, reduction of water loss, and modified parts that function as tools or weapons; a disadvantage of the exoskeleton is that as their bodies continue to grow, they must molt, making them weak and vulnerable to predation while a new covering grows and hardens. Advantages of the vertebrate skeleton are structural support for organs and locomotion, continued growth and change as the body ages, protection, transmission of muscle forces, and calcium storage for maintaining homeostatic levels of calcium in the blood. Disadvantages of the vertebrate skeleton are degenerative bone and joint diseases, and lack of external protection for many soft tissues. The insect skeleton is adapted to its terrestrial lifestyle: a light, but strong, chitin covering that resists water protects tissues; wings for flight, modified from thoracic exoskeleton; high degree of leverage for powerful muscle use. 13. Human muscle contraction and cellular movement in single-celled organisms both involve actin and myosin. 14. As science has elucidated how joints function, technology has used this knowledge to develop devices and materials to replace destroyed and deteriorated joints. Because humans live longer, joint replacement surgery has become common in developed countries to maintain peoples' active lifestyles. The return of wounded war veterans has driven new demands for high-tech, sophisticated replacement joints. Governments and corporations are investing in needed medical research and development in bioengineering and robot science.

CHAPTER 41

CHECKPOINT

41.1 • The sequence of processes that would allow you to escape the shark are: (1) reception by a sensory receptor; (2) transmission by afferent neuron; (3) integration by interneurons in CNS; (4) transmission by efferent neuron; (5) action by effectors: you yell for help and swim away from the shark. • After neural signals are integrated in the CNS, a neural message is transmitted from the CNS by efferent neurons to effectors. The actions of the effectors are the response to the stimulus. **41.2** • See Fig. 41-2. • A neuron produces and transmits an action potential; it is a cell of the nervous system, made of a cell body containing nucleus and organelles, many branched dendrites extending out from the cell body, and a single long axon. A nerve is a bundle of several hundred axons wrapped in connective tissue. • Astrocytes provide physical support and nutrients for neurons, help regulate the composition of the extracellular fluid in the CNS, communicate with other neurons, and induce synapse formation. Microglia are specialized macrophages that mediate responses to injury and disease by moving to the affected area and removing bacteria and cellular debris. They also release signaling molecules that mediate inflammation and help regulate the number of neurons developing in the brain. **41.3** • Resting potential is mainly due to K^+ easily flowing out of ion channels. Sodium–potassium pumps move Na^+ out and K^+ into the neuron. For some neurons, resting potential is maintained at about -70 mV. • When the threshold level is reached, Na^+ channels open and Na^+ enters the neuron, causing depolarization. The action potential transmits a signal. After signal transmission, the neuron repolarizes; K^+ channels are opened, and K^+ diffuses out of neuron; the neuron then returns to its resting state. • An action potential is an all-or-none response; it is either initiated and continues (self-propagating) or there is no response. A graded potential is a local response that varies in magnitude depending on the strength of the stimulus and fades out within a few millimeters of its origin. • The benefits of saltatory conduction are that it is more rapid and requires less energy than continuous conduction. **41.4** • Biogenic amines have important roles in regulating mood. Dopamine also has a role in motor function. GABA functions as a widespread inhibitory neurotransmitter. • See Fig. 41-11. • EPSPs are produced by a change in membrane potential that brings a neuron closer to the firing level. IPSPs are produced by a change in membrane potential that moves a neuron further from the firing level. **41.5** • Neural integration is the process of summing, or integrating, incoming signals. • Temporal summation occurs when repeated stimuli cause new EPSPs to develop before previous

EPSPs have decayed. Spatial summation occurs when several closely spaced synaptic terminals release neurotransmitters simultaneously, stimulating the postsynaptic neuron at several different places. • In vertebrates most neural integration takes place in the CNS (the brain or spinal cord). **41.6** • See Fig. 41-13.

FIGURE QUESTIONS

FIG. 41-8 Because charybdotoxin blocks K^+ channels, it would not let the neuron repolarize. **FIG. 41-10** In multiple sclerosis, an autoimmune response destroys and scars myelin along nerves, causing gaps in the myelin covering; this lack of myelin slows or stops the transmission of neural impulses; it can also lead to abnormal nerve signaling ("cross-talk" between neurons). **FIG. 41-11** The neurotransmitter would not be able to bind; the ion channels would not open or close, so no action potential would be generated.

TEST YOUR UNDERSTANDING

1. c 2. d 3. a 4. c 5. d 6. b 7. See Fig. 41-11c. 8. See Fig. 41-11c. The antidepressant/spheres could block reuptake of the neurotransmitter into the presynaptic neuron. 9. (a) Na^+ ions entering the postsynaptic neuron cause a graded depolarization, or EPSPs. (b) K^+ diffusing out of a postsynaptic neuron cause IPSPs. (c) Cl^- entering a postsynaptic neuron would produce IPSPs. 10. The acetylcholine affects the skeletal and cardiac muscle differently because of the different chemical composition of the two muscles. 11. A person could make future plans, change lifestyle, and obtain help earlier if he/she had a higher risk for AD and knew about the potential risk. Psychological care would need to partner with medical care for the future AD patient. 12. Diseases related to these neurotransmitters could be studied in actual cells as they are produced, and cells could be engineered to provide neurotransmitters for therapeutic reasons. 13. Convergent circuits (signals from many neurons sent to a few) and divergent circuits (signals from one neuron sent to many others) allow for greater neural flexibility and control of stimuli/responses to the environment.

CHAPTER 42

CHECKPOINT

42.1 • The cerebral ganglia of some arthropods differ from the ganglia of flatworms in that the arthropod ganglia have specific functional regions. These areas are specialized for integrating information transmitted to the ganglia from sense organs. • Trends in the evolution of nervous systems include: the number of nerve cells increased; nerve cell bodies became concentrated to form a brain and ganglia; neurons formed nerves and nerve cords; neurons became specialized to perform specific functions; the number of interneurons and complex synaptic contacts increased. • With cephalization, sense organs became concentrated at the front end of the body; this allows the animal to find food and detect enemies and mates more effectively. With the brain located close by, responses can be rapid. **42.2** • See Fig. 42-4. • The CNS consists of the brain where most decisions are made and the spinal cord, which integrates some information itself, and also transmits information from afferent neurons to the brain and from the brain to efferent neurons. The PNS consists of the sensory receptors, the afferent nerves that transmit signals from the sensory receptors to the CNS, and the efferent neurons that transmit signals from the CNS to the muscles and glands. **42.3** • The vertebrate embryonic forebrain subdivides to form the telencephalon, which gives rise to the cerebrum and olfactory bulbs, and to the diencephalon, which gives rise to the thalamus and hypothalamus. The midbrain gives rise to the superior colliculi and the inferior colliculi. The hindbrain gives rise to the cerebellum, pons, and medulla. The cerebrum is the center of intellect, memory, and language, and controls sensory and motor functions. The thalamus is a relay center for motor and sensory messages. The hypothalamus is a major coordinating center for regulating autonomic and somatic responses. It controls body temperature and regulates appetite and water balance. The hypothalamus is important in emotional and sexual responses, and is the link between the nervous and endocrine systems; it also produces certain hormones. The superior colliculi in the midbrain are centers for visual reflexes, and the inferior colliculi are centers for auditory reflexes. The cerebellum regulates motor coordination, muscle tone, posture, and equilibrium. It is important in planning and executing voluntary movements. The pons connects various parts of the brain with one another, and contains respiratory and sleep centers. The medulla contains centers that regulate respiration, heartbeat, and blood pressure. Other reflex centers in the medulla regulate activities such as swallowing, coughing, and vomiting. • The midbrain is the most prominent part of the amphibian brain, and it serves as the main association area. In mammals, the cerebrum is the largest, most complex part of the brain; it serves as the main association area. The cerebellum is also much larger in the mammalian brain than in the amphibian brain. **42.4** • The CNS is encased in bone and covered by three layers of connective tissue, the meninges (dura mater, arachnoid, pia mater). The CNS is also protected by the shock-absorbing cerebrospinal fluid. • The spinal cord transmits information to and from the brain and controls many reflex actions. • See Figs. 42-10a and 42-11a. • The amygdala, part of the limbic system, filters incoming information and evaluates threat. When it perceives danger, the amygdala signals other parts of the brain. Following a traumatic experience, the amygdala may become hypersensitive to possible danger and may over-respond to harmless "triggers." • Learning depends on synaptic plasticity, the ability of synaptic connections to change in response to experience. LTP changes the strength of a synaptic connection. NMDA receptors respond to the neurotransmitter glutamate by opening Ca^{2+} channels. Calcium ions act as second messengers that initiate long-term changes ultimately responsible for LTP. Gene expression and protein synthesis take place during the process of establishing long-term memory. (See Fig. 42-16.) **42.5** • The somatic division of the PNS innervates skeletal muscles and helps maintain posture and balance. The autonomic division of the PNS helps maintain homeostasis in the internal environment; its effectors are smooth muscle, cardiac muscle, and glands. • In general, the sympathetic system stimulates organs to mobilize energy, e.g., it speeds the heart; the parasympathetic system has the opposite effect. Typically, it stimulates organs to conserve and restore energy, e.g., it slows the heart. Sympathetic preganglionic neurons secrete acetylcholine and postganglionic neurons release norepinephrine. Both preganglionic and postganglionic parasympathetic neurons secrete acetylcholine. **42.6** • Alcohol is a CNS depressant; antipsychotic drugs block dopamine receptors and have a calming effect; many SSRI antidepressants elevate mood by blocking serotonin reuptake; amphetamines stimulate release and block reuptake of dopamine and norepinephrine; opiates inhibit dopamine reuptake; and MDMA is a CNS stimulant. (See Table 42-4.)

FIGURE QUESTIONS

FIG. 42-6 During future evolution, the cerebellum and other motor areas of the brain may become smaller because physical activity may continue to decline. Scientists may interfere with evolution by using genetic engineering and technology to produce brains that function more effectively into old age (that is, without cognitive decline); and they may also develop brains with larger, more complex frontal lobes. **FIG. 42-16** Calcium ions activate pathways that lead to LTP. For example, Ca^{2+} activate a pathway leading to insertion of more AMPA receptors in the postsynaptic membrane; more AMPA receptors increase sensitivity to glutamate, resulting in more EPSPs. The EPSPs strengthen the synapse, maintaining the LTP.

TEST YOUR UNDERSTANDING

1. b 2. c 3. d 4. e 5. b 6. a 7. a 8. See Fig. 42-11a. 9. During the evolution of the vertebrate brain, there has been a trend toward greater complexity, particularly of the cerebrum and cerebellum. 10. Parents can stimulate their child's intellectual development by exposing the child to a wide variety of experiences from a very

early age and by giving the child the opportunity to manipulate objects, including electronic devices such as computers. Providing a calm, safe environment enhances learning. 11. The limbic system would quickly assess the danger and send signals to other parts of the brain that would initiate protective action (e.g., running away). 12. The presence of CREB in diverse animals suggests a common ancestor (and that common pathways developed for learning and memory). 13. Because genes are activated during information processing (a critical measure of intelligence), and because genes are inherited, intelligence depends, at least in part, on inheritance of genes that code for information processing.

CHAPTER 43

CHECKPOINT

43.1 • A receptor potential is a change in membrane potential; it is a graded response in which the magnitude of change depends on the strength of the stimulus. In contrast, an action potential is generated only when a neuron becomes sufficiently depolarized to reach its threshold level. In a sensory neuron, receptor potentials depolarize the neuron to threshold level, generating an action potential that transmits the signal to the CNS. • As you continue to hike, your perception of the odor will quickly decrease as sensory adaptation occurs. (The frequency of action potentials decreases, and decreases in neurotransmitter release may occur at the synapses.) • Sensory information is initially integrated in the receptor where receptor potentials produced by various stimuli are integrated by summation. However, integration occurs mainly in the spinal cord and at various locations in the brain, which selects, interprets, and organizes sensations and then converts them to perceptions of stimuli. • Refer to Table 43-1: Classification of Receptors by Type of Energy They Transduce. **43.2** • Mosquitos have thermoreceptors, which can locate an endothermic host. Snakes use thermoreceptors to locate prey or to detect predators. Mammals have thermoreceptors, which detect outside environmental temperature changes. Mammals also have thermoreceptors in the hypothalamus to detect internal changes in temperature and receive and integrate information from receptors on the body surface. **43.3** • Electroreceptors sense differences in electrical potential to detect electric fields for locating prey and for navigation. Electroreceptors also aid in communication, as males and females have different frequencies of electric discharge. • Some animals use electromagnetic reception to facilitate orientation and navigation. **43.4** • Mechanical nociceptors respond to strong tactile stimuli such as cutting, crushing, or pinching; thermal nociceptors respond to temperature extremes (above 45°C or below 5°C); polymodal nociceptors respond to a variety of damaging stimuli, including certain chemicals. • Substance P activates interneurons that transmit the pain message to the opposite side of the spinal cord. Endorphins bind with opiate receptors on the terminals of sensory neurons in the spinal cord to inhibit release of substance P, stopping transmission of the pain signal. **43.5** • Mechanoreceptors help to maintain balance and body position as you walk. Tactile receptors aid orientation and the contact you have with the floor. Proprioceptors help you perceive movement of your legs, arms, and head as you move, and they help maintain balance. • The vestibular apparatus helps maintain equilibrium. The three semicircular canals provide information about turning movements, called angular acceleration. • The sequence of events involved in hearing are as follows: sound waves enter external auditory canal; tympanic membrane vibrates; malleus, incus, and stapes amplify vibrations; oval window vibrates; vibrations are conducted through fluid; basilar membrane vibrates; hair cells in organ of Corti are stimulated; cochlear nerve transmits impulses to the brain. **43.6** • Olfaction and gustation are closely related functions; they occur through chemoreceptors, which allow animals to detect substances in food, air, and water. Information from both systems is sent to the limbic system in the brain. **43.7** • The insect compound eye is structurally and functionally different from the vertebrate eye. The insect eye has many faceted units called ommatidia, each of which consists of a cornea and lens unit made up of photoreceptors called retinular cells. The insect eye can form only coarse mosaic images, and it compensates by superbly detecting movement. Because an insect's eye is sensitive to wavelengths of light in the range from red to ultraviolet (UV), it can see UV light well. The vertebrate eye has an adjustable lens, which can be focused for different distances, an iris, which regulates the amount of light entering the eye, and a retina made up of rod and cone photoreceptor cells. The vertebrate eye can adjust focus to see sharp images. The vertebrate's eye is sensitive to wavelengths ranging from red to violet. • When light strikes rhodopsin, light energy is transduced. Refer to Figure 43-22 for description of the signal transduction pathway. • Light passes through the cornea and then through aqueous fluid, through the lens and through the vitreous body. An image forms on photoreceptor cells in the retina, signaling bipolar cells and then ganglion cells. The optic nerve transmits signals to the thalamus and visual areas of cerebral cortex. • All visual images constructed by the retina are integrated and interpreted in the brain.

FIGURE QUESTIONS

FIG. 43-9 The utricle and saccule would not be able to sense changes in head position and would have no function in a weightless environment. **FIG. 43-10** When you reverse your spinning direction, the endolymph flows in the opposite direction. If you then stop spinning, the position of the stereocilia would return to normal, and action potentials signaling circular rotation would stop. **FIG. 43-11** If the basilar membrane were damaged at its higher, stiff end, the individual would no longer be able to hear high frequency sounds. If the basilar membrane were damaged at its wide, flexible end, the individual would no longer be able to hear lower frequency sounds. (High frequency hearing is greatly impacted by excessive exposure to high frequency and high impact [loud] sounds common to our environment [loud music, cell phones], which damage and destroy cilia.) **FIG. 43-20** Cones, which are responsible for color vision, respond best in daylight. We depend on rods to detect shape and movement in dim light.

TEST YOUR UNDERSTANDING

1. c 2. d 3. a 4. d 5. b 6. c 7. See Fig. 43-18. 8. Although there are only a few types of taste receptors, each different substance sensed by taste receptors is a unique combination of molecules with its own signature. 9. We distinguish quality and intensity of sensory information based on the destination of the sensory neuron (that is, to which structure is the information delivered?); total number of neurons transmitting the signal; total number of action potentials transmitted by the neuron; and frequency of the action potentials. 10. The same proteins in the photoreceptors of many diverse organisms indicate that opsins evolved early and the genes coding for them are highly conserved. 11. The rapid evolution of image-forming eyes and their presence in most animals was an advantage, or edge, for survival. Animals that could see better could detect prey or avoid capture and passed this genetic advantage to their offspring. Most animals without image-forming eyes were unable to compete and many became extinct. 12. Any answer is acceptable if a logical argument is presented.

CHAPTER 44

CHECKPOINT

44.1 • Gases are exchanged by diffusion in hydras and flatworms; earthworms have a closed circulatory system, and gas exchange takes place through the thin walls of capillaries; in insects, gas is exchanged through a system of tubes that are separate from the circulatory system; frogs have a respiratory system with lungs for gas exchange. • In an open circulatory system, a heart pumps hemolymph into vessels that have open ends. The hemolymph bathes the cells directly, and then re-enters the circulatory system through openings in the heart (in arthropods) or through open-ended blood vessels (in mollusks). In closed circulatory systems, blood flows through a continuous circuit of blood vessels. The walls of capillaries are thin enough to permit diffusion of gases, nutrients, and wastes between blood in the vessels and the interstitial fluid that bathes the cells. • The vertebrate circulatory system does the following: transports nutrients from the digestive system and from storage depots to each cell; transports oxygen from respiratory structures to the cells; transports metabolic wastes from each cell to organs that excrete them; transports hormones from endocrine glands to target tissues; helps maintain fluid balance; helps distribute metabolic heat within the body; helps maintain appropriate pH. **44.2** • Alpha globulins include certain hormones and proteins that transport hormones and other compounds, prothrombin, and HDL. Beta globulins include other lipoproteins that transport fats and cholesterol and proteins that transport certain vitamins and minerals. The gamma globulin fraction of plasma contains antibodies. Globulins and albumins help regulate distribution of fluid between plasma and interstitial fluid. Plasma proteins help maintain the pH of the blood within a narrow range. • Red blood cells transport oxygen; neutrophils, the main phagocytic cells in the blood, ingest bacteria. • A shortage of fibrinogen in the blood would prevent clotting. **44.3** • Arteries transport blood away from the heart; veins transport blood toward the heart; and capillaries are the exchange vessels. • If arterioles could not constrict, the body could not regulate blood pressure; also, the body would not be able to regulate the amount of blood going to each tissue or organ. **44.4** • During the evolution of the vertebrate cardiovascular system, structural changes evolved that allow separation of oxygen-rich from oxygen-poor blood; most nonavian reptiles, all avian reptiles (birds), and all mammals have a double circuit of blood flow. In avian reptiles and mammals, the heart has four chambers: two atria and two ventricles. **44.5** • During cardiac conduction, the SA node initiates contraction → action potential spread through atrial muscle fibers (atria contract) → action potential delayed briefly in AV node → action potential spreads into AV bundle → right and left bundle branches → Purkinje fibers → conduct impulses to muscle fibers of both ventricles → ventricles contract. • When the mitral valve does not close completely, some blood leaks backward into the atrium when the ventricle contracts; less blood flows to the tissues of the body. As a result, the heart pumps harder. This condition can lead to congestive heart failure. • Cardiac output = stroke volume (100 ml per stroke) × heart rate (100 contractions/minute) = 10,000 ml/min (10 L/min). **44.6** • Arteriole constriction increases blood pressure. • Blood pressure in arteries is higher than blood pressure in veins. **44.7** • Blood flowing from the carotid artery to the superior vena cava passes through: carotid artery → capillaries in brain → jugular vein → superior vena cava. Blood

flowing from the renal vein to the renal artery passes through: renal vein → inferior vena cava → right atrium → right ventricle → pulmonary arteries → pulmonary capillaries (in lungs) → pulmonary vein → left atrium → left ventricle → aorta → renal artery. • Recall that most veins conduct blood to larger veins or to the superior or inferior vena cava. The hepatic portal system is an exception. This vein transports nutrients from the intestine to the liver. In the liver, the hepatic portal vein branches into a network of hepatic sinuses from which nutrients leave the circulation and enter liver cells. The liver stores nutrients that are not immediately needed by other cells. The hepatic sinuses then merge to form hepatic veins. The sequence of blood flow: hepatic portal vein → hepatic sinuses (in liver) → hepatic veins. **44.8** • See Figs. 44-19 and 44-20. • The lymphatic system collects excess interstitial fluid and returns it to the blood. **44.9** • Joe is at high risk for cardiovascular disease. • During development of atherosclerotic plaque, endothelial cells respond to oxidized LDLs by releasing compounds that attract monocytes. These white blood cells enter the tissues and trigger an inflammatory response. Cells of the immune system migrate to the area and produce substances that maintain chronic inflammation in the inner layer of the arterial wall. Some macrophages become foam cells, which stimulate the development of fatty streaks. These streaks may be the initial lesion in atherosclerosis.

FIGURE QUESTIONS

FIG. 44-6 Lymph vessels have very small dead ends that infiltrate most tissues in the body. They are one-way vessels that pick up fluid and deliver it into the blood circulation through two small ducts in the shoulder region. **FIG. 44-8** The separation of oxygen-rich blood from oxygen-poor blood in avian reptiles and mammals enables their cardiovascular systems to deliver more oxygen to their tissues. As a result, these animals can maintain a higher metabolic rate than other animals and a constant high body temperature even in cold surroundings. **FIG. 44-11** When the SA node increases its rate of firing, the cardiac cycle speeds up accordingly. **FIG. 44-16** The left ventricle needs a thicker wall than the right ventricle because the left ventricle must contract strongly enough to pump blood at a higher pressure through the higher resistance and longer arterial system. If the right ventricle contracted that powerfully, the lung tissue would be damaged.

TEST YOUR UNDERSTANDING

1. a 2. d. 3. e 4. d 5. e 6. e 7. d 8. d 9. See Fig. 44-6. The thick walls of arteries are adapted to the high pressure they must withstand. Compared to arterial walls, the walls of veins are thin. Capillary walls are only one cell thick, allowing nutrients, oxygen, and other substances to pass through them. 10. The heart rate is carefully

regulated by sympathetic and parasympathetic nerves, which are controlled by a cardiac center in the brain stem. 11. The four-chambered heart allows efficient oxygenation of body tissues, which in turn allows these animals to be endothermic. 12. Fish have a single circuit of blood vessels. Blood leaving the heart passes through the gills, where it is oxygenated. 13. Blood vessels with atherosclerotic plaque deliver less oxygen to the tissues, including the heart itself. In ischemic heart disease the heart muscle is not receiving enough oxygen, particularly during exercise or emotional stress. The remaining two parts of this question ask your opinion. As you reflect on the questions, consider the human misery caused by cardiovascular disease, as well as the monetary cost to society.

CHAPTER 45

CHECKPOINT

45.1 • Adaptive immunity consists of specific defenses that target specific macromolecules. In adaptive immunity the body typically produces antibodies that recognize and bind to specific antigens. Also unlike innate immunity, adaptive immunity is characterized by immunological memory. • Like invertebrates, vertebrates depend on innate immunity for their initial defense against invading pathogens. Unlike invertebrates, vertebrates have highly developed adaptive immunity. **45.2** • The function of pattern recognition receptors is to recognize nonself molecules; these receptors recognize pathogen-associated molecular patterns (PAMPs). • If you had no NK cells, you would be much more vulnerable to virus infections and developing tumors. Macrophages are the main phagocytes in the body. If you had no macrophages, both your innate and adaptive immune systems would be severely impaired. • Cytokines are important signaling molecules with a wide variety of functions. For example, they enhance the inflammatory response, stimulate NK cell activity, and activate lymphocytes. Complement proteins destroy viruses and bacteria; they attract white blood cells to the site of infection; they enhance the inflammatory response • See Fig. 45-3. Main steps in the inflammatory response would include activation of macrophages and mast cells, vasodilation, increased capillary permeability, and increased phagocytosis. **45.3** • Cell-mediated immunity and antibody-mediated immunity are two types of adaptive immunity. Lymphocytes are the key cells in both. T cells (and APCs) are the main cells in cell-mediated immunity. T cells attack foreign cells, cancer cells, and body cells infected by invading pathogens. B cells are responsible for antibody-mediated immunity. B cells mature into plasma cells, which produce specific antibodies. • Antigen-presenting cells (APCs) activate T helper cells. Macrophages, dendritic cells, and B cells function as APCs. • A kidney transplant from a first cousin has a higher probability of success than a transplant from an unknown donor because first cousins are more likely to have more HLA alleles in common. (First cousins have 12.5% of their genes in common.) **45.4** • APCs activate T_H cells (and also activate T_C cells). Activated T cells divide by mitosis to form a clone of activated cells. T_H cells release cytokines that activate T_C cells (and macrophages). T_C cells release proteins that destroy infected cells. **45.5** • See highlighted sequence of events on page 967. • See Fig. 45-8 or 45-9. Antigen–antibody complexes: inactivate pathogens or their toxins; stimulate phagocytes to ingest pathogens; IgG and IgM activate the complement system. • Immunological memory depends on memory B cells and memory T cells; after an infection, these cells remain strategically stationed in the body and can respond rapidly to antigens they have encountered before. • Passive immunity is borrowed immunity; memory cells are not produced and its effects are temporary. **45.6** • Some types of cancer cells can block T_C cells. Other cancer cells decrease their expression of class I MHC molecules, preventing T_C cells from recognizing them. Still others do not produce co-stimulatory molecules needed to activate T_C cells. • HIV enters the body through the mucosa that lines passageways into the body. A protein on the outer envelope of the virus attaches to CD4 on the surface of T_H cells. HIV enters the T_H cell and is reverse-transcribed to produce DNA. Its DNA is then incorporated into the host genome. The virus reproduces itself, causing death of the host cell. Innate immune responses include production of antimicrobial peptides and action by NK cells, dendritic cells, and the complement system. Adaptive immune responses include activation and mobilization of T_C cells. • An autoimmune disease is a failure in self-tolerance. Antibodies and T cells attack the body's own tissues. Viral or bacterial infection often precedes the onset of an autoimmune disease. Molecular mimicry has evolved in some pathogens; these pathogens trick the body by producing molecules that look like self-molecules. Immune cells may not discriminate and may attack the body's similar molecules. Overactivation of cells of the immune system can result in inflammation and autoimmune disease. • See Fig. 45-17. • When tissue from a donor is transplanted to the body of another person, several of the HLA antigens are likely to be different. The recipient's T_H cells and T_C cells recognize the graft as nonself (i.e., foreign) and launch an immune response known as graft rejection. T_C cells, the complement system, and cytokines secreted by T_H cells attack the transplanted tissue and can destroy it within a week.

FIGURE QUESTIONS

FIG. 45-3 Several signs of infection would be evident due to the inflammatory response. The foot would be swollen, red, and warm to the touch at and near the site of the puncture. You would experience pain from your foot and might have a low-grade fever. Pus might be seeping from the wound. **FIG. 45-5** The malignant cells would crowd out other cells in the bone marrow, severely slowing the production of antigen-presenting cells and lymphocytes. **FIG. 45-6** Macrophages, dendritic cells, and B cells function as antigen-presenting cells (APCs). **FIG. 45-8** No. Immunity would be depressed if T_C cells collected in the lymph tissues as plasma cells do. T_C cells are the cellular infantry; they must spread out strategically to discover and destroy incoming pathogens. In contrast, plasma cells remain in the bone marrow and release antibodies into the circulation.

TEST YOUR UNDERSTANDING

1. b 2. c 3. e 4. c 5. a 6. e 7. c 8. b 9. a 10. d 11. See Fig. 45-7, step 3. This diagram shows a T_H cell activating a B cell, which then divides, producing two B cells. 12. Specificity helps make the immune system very efficient. T cells and antibodies are produced that are highly specific for destroying the particular pathogens that invade the body. Diversity gives the immune system the capability of protecting the body from almost any new pathogen that invades (see Fig. 45-11). Immunological memory allows the immune system to "remember" each type of pathogen against which it has already launched an immune response. When it encounters the same pathogen again, the immune system launches a rapid attack. 13. Macrophages are important in both innate and adaptive immune responses. Reduction in macrophages would critically decrease the body's ability to launch an immune response. A significant decrease in macrophages would probably have a more serious negative impact than a decrease in memory B cells because macrophages are first responders. The body could compensate for a decrease in memory B cells by launching a primary immune response. 14. In kindergarten, John has been immunized so is unlikely to develop measles again. Jack will likely develop measles, and his body will develop immunity to the measles virus. Five years later when Judy (who has the measles virus) sneezes on both of them, it is improbable that either John or Jack will develop measles because both of them have immunological memory of the virus. 15. These facts suggest that the genes that encode the enzymes necessary for antibody diversity, an important aspect of adaptive immunity, are very old and that echinoderms and jawed vertebrates share a common ancestor. 16. In forming your opinion, you might consider how we could use technology to better inform the millions of people living in poverty in low-income countries. Also, how could we more rapidly apply new technology to development of new vaccines and treatments?

CHAPTER 46

CHECKPOINT

46.1 • Air has a higher concentration of oxygen; oxygen diffuses faster in air; animals that breathe air expend significantly less energy to obtain oxygen than animals in an underwater habitat. **46.2** • Flatworms can depend on diffusion for gas exchange, but the larger tadpole body requires a respiratory system for gas exchange. • An earthworm exchanges gasses through the body surface. A grasshopper exchanges gases through a system of tracheal tubes, or tracheae. Fish exchange gases through internal or external gills. Birds exchange gases with lungs. • The countercurrent exchange system increases the efficiency of gas exchange because much more oxygen is able to diffuse into the blood when the blood flows through the capillaries in the opposite direction compared with the water. **46.3** • In mammals air enters through the nostrils, then nasal cavities, pharynx, larynx, trachea, bronchi, then bronchioles, and lastly the alveoli. • Gas exchange takes place in the alveoli. Oxygen concentration is lower in the alveoli than in atmospheric air because cells of the body use oxygen during cellular respiration; carbon dioxide concentration is higher in the alveoli than in atmospheric air because carbon dioxide is produced during cellular respiration. • Increased carbon dioxide in the blood lowers the pH (blood becomes more acidic). When blood is more acidic, hemoglobin can unload oxygen more readily. • Breathing would increase as more oxygen is required for cellular respiration. You would also breathe harder to increase ventilation to dispose of excess carbon dioxide. Chemoreceptors in the walls of the aorta and carotid arteries sense lower blood pH, which stimulates chemoreceptors to up-regulate breathing. As the lungs remove carbon dioxide, the blood pH returns to normal and homeostasis is restored. **46.4** • The ciliated mucous lining of the airway traps inhaled particles to remove pollutants from the respiratory system. • When the respiratory defense system is overwhelmed by pollutants, diseases such as chronic bronchitis, pulmonary emphysema, and lung cancer can occur.

FIGURE QUESTIONS

FIG. 46-2 The highly efficient tracheal system of insects delivers oxygen to all parts of the body. **FIG. 46-3** The countercurrent flow functions in the same way for carbon dioxide removal; carbon dioxide is transferred from blood in the capillaries to the surrounding water at the same time oxygen is transferred from the water to the blood in capillaries. **FIG. 46-7** The capillary walls are only one cell thick; millions of pulmonary capillaries act as an interface to readily exchange gases in the alveoli. **FIG. 46-11** If the carbonic anhydrase enzyme were inhibited, carbon dioxide would build up in the red blood cells; net diffusion of carbon dioxide from the body cells into the red blood cells would slow and then cease.

TEST YOUR UNDERSTANDING

1. a 2. a 3. b 4. c 5. d 6. e 7. c 8. e 9. c 10. d 11. See Fig. 46-6. 12. Having millions of alveoli immensely increases the surface area for gas exchange. 13. Strenuous muscle activity would increase acidity of the blood and hemoglobin would unload oxygen more readily. The oxygen–hemoglobin curve would shift to the right. 14. If a terrestrial animal had gills, the gills would not function properly on land; water is required to support the gill structure. The animal with gills would not be able to breathe on land and would die. 15. It would be advantageous for fish that live at the land–water interface to have gills and lungs. The lungs of modern fish serve as swim bladders to control buoyancy. 16. Whales and dolphins were originally land mammals with lungs; when they returned to the sea, they continued to require air to breathe. 17. Further increasing tobacco taxes could encourage users to seek help for their expensive addiction; the collected taxes could pay for addiction treatment research. Also, reducing and gradually eliminating places to smoke may continue to be an effective method to curtail smoking.

CHAPTER 47

CHECKPOINT

47.1 • Carnivorous mammals have canine teeth for stabbing prey; their pointed incisors and canines are adaptations for ripping flesh. • Flatworms have a gastrovascular cavity with only one opening; digestion is completed intracellularly within food vacuoles. Earthworms have a complete digestive tract with a mouth and anus; some regions of the digestive tract are specialized to perform specific food-processing functions. • Various regions are specialized to perform specific functions, and different functions can be carried on simultaneously—for example, at the same time food is being digested in the stomach, nutrients can be absorbed in the small intestine and wastes can be eliminated from the large intestine. **47.2** • Mouth—mechanical and enzymatic digestion begins; pharynx—food swallowed; esophagus—moves food to the stomach; stomach—mechanical and enzymatic digestion; small intestine—digestion and absorption of digested nutrients; large intestine—absorbs water and sodium from the chyme, and bacteria there produce vitamin K and certain B vitamins; anus—opening for

elimination of wastes. • Rugae increase the surface area in the stomach; villi increase the surface area in the small intestine. • Trypsin digests proteins and polypeptides; bile salts mechanically digest fats; gastrin stimulates gastric motility and stimulates gastric glands to secrete pepsinogen and HCl; and lacteals, the central lymph vessel in each villus, absorb chylomicrons (fat droplets). • See Fig. 47-4. **47.3** • Carbohydrates, fats, proteins, vitamins, minerals. • Glucose provides energy; essential amino acids are required to synthesize proteins. • LDLs deliver cholesterol to the cells; they are the main source of the cholesterol that builds up in the walls of arteries, causing cardiovascular disease. **47.4** • When energy input = energy output, body weight remains constant; when energy input is less than energy output, body weight decreases; when energy input is greater than energy output, body weight increases. • BMI is an index of weight in relation to height. • Melanocortins suppress appetite, resulting in decreased food intake and weight loss; NPY stimulates appetite; ghrelin stimulates appetite by activating neurons in the hypothalamus to produce NPY. Leptin signals the hypothalamus about the status of energy stores in adipose tissue. It slows the release of NPY and stimulates release of melanocortins.

FIGURE QUESTIONS

FIG. 47-4 The esophagus would need to be connected to the duodenum. **FIG. 47-9** Lymph tissue in the wall of the gastrointestinal tract contains macrophages, T cells, and other cells of the immune system. These cells are closely positioned to destroy pathogens that enter the body in contaminated food or water. **FIG. 47-10** If gluten remains part of the diet, the mucosa of the small intestine becomes damaged and the villi cannot absorb enough nutrients. This can lead to malnutrition.

TEST YOUR UNDERSTANDING

1. c 2. d 3. d 4. e 5. b 6. d 7. a 8. b 9. c 10. c 11. See Fig. 47-4. 12. You could include an experimental and a control group with about 20 mice in each group. Both groups would be fed the same diet *except* the diet of the experimental group would lack pyridoxine. You could weigh each mouse each day and also note specific health factors, including condition of skin, GI tract function, and presence of convulsions. 13. In their inactive form, proteolytic enzymes do not digest the lining of the GI tract. 14. You could examine the animal's tooth structure, and look for other clues, e.g., sharp claws, size of jaws. 15. Identifying and understanding the actions of the many hormones and other signaling molecules that regulate energy metabolism might lead to the development of new compounds that could inhibit appetite-stimulating compounds or stimulate appetite-suppressing compounds in the body. A cure for obesity would lead to a healthier society and also save billions of dollars each year in health care costs.

CHAPTER 48

CHECKPOINT

48.1 • Osmoregulation maintains salt and water concentrations within homeostatic limits. • Excretory systems help maintain homeostasis through osmoregulation and excretion of metabolic wastes. **48.2** • Some aquatic animals excrete ammonia; insects, certain reptiles, and birds excrete uric acid; and urea is the principal nitrogenous waste excreted by amphibians and mammals. • Excretion of nitrogenous wastes as uric acid conserves water in terrestrial animals. Uric acid and urea are less toxic than ammonia. The body can store concentrated urea without causing damage to tissues. **48.3** • Nephridia consist of simple or branching tubes that open to the outside of the body through pores. The protonephridia of flatworms and nemerteans have blind ends that consist of flame cells with cilia. The cilia propel fluid through the tubules. The metanephridia of annelids and mollusks are open at both ends. As fluid moves through the tubule of a nephridium, water and certain other substances (e.g., glucose) are reabsorbed by capillaries, leaving the waste to be excreted. The Malpighian tubules of insects excrete wastes more selectively than nephridia. Malpighian tubules are slender extensions of the gut wall; they have blind ends that extend into the hemocoel. Cells of the tubule wall actively transport uric acid, potassium ions, and some other substances from the hemolymph into the tubule. Malpighian tubules empty into the gut. Water, some salts, and other solutes are reabsorbed into the hemolymph by a specialized epithelium in the rectum. **48.4** • Marine fishes lose water osmotically and take in salt. To compensate they drink and retain sea water. Specialized cells in their gills excrete salt. Marine cartilaginous fishes retain a high urea concentration, which makes their body fluids hypertonic to sea water, so there is a net inflow of sea water. Freshwater fishes are hypertonic to the fresh water around them, and water continuously moves into the body. Their kidneys excrete large amounts of dilute urine. These fishes also lose salts by diffusion through their gills into the surrounding water. Special gill cells actively transport salt back into the body. • Marine birds ingest sea water and take in a lot of salt in their food. Glands in their heads excrete salt in response to osmotic stress. Marine mammals ingest sea water along with their food. Their kidneys produce concentrated urine, saltier than sea water. Terrestrial birds and mammals have efficient kidneys that conserve water. Birds conserve water by reabsorbing water from the intestine and by excreting uric acid. Mammalian kidneys produce hypertonic urine. **48.5** • The liver produces urea; kidneys produce urine; urinary bladder stores urine; urethra conducts urine out of body. • Filtration is the job of the glomerulus; reabsorption is the job of the renal tubules; secretion takes place in the renal tubules and collecting ducts. • Bowman's capsule → proximal convoluted tubule → loop of Henle → distal convoluted tubule → collecting duct → ureter → urinary bladder. See also Figs. 48-8 and 48-10. • The main actions of the renin–angiotensin–aldosterone pathway are to increase sodium reabsorption, which raises blood pressure. Angiotensin II also raises blood pressure directly by constricting blood vessels, and it stimulates the posterior pituitary to release ADH. In contrast, ANP decreases blood pressure by increasing sodium excretion, dilating afferent arterioles, and inhibiting aldosterone secretion.

FIGURE QUESTIONS

FIG. 48-5 If a shark were swimming in fresh water, its kidneys would need to work harder in order to produce a large volume of dilute urine to compensate for the increase in water entering the shark through osmosis. FIG. 48-10 Mild dehydration would affect kidney function by reducing the volume of the blood plasma. More water would need to be reabsorbed, and a low volume of more concentrated urine would be produced. FIG. 48-14 ADH increases water reabsorption, which increases blood volume. By this mechanism, ADH raises arterial blood pressure.

TEST YOUR UNDERSTANDING

1. c 2. d 3. e 4. d 5. a 6. a 7. c 8. d 9. e 10. b 11. e 12. See Fig. 48-10. 13. In someone with diabetes mellitus, the concentration of glucose in the blood exceeds its tubular transport maximum. The excess glucose that cannot be reabsorbed is excreted in the urine. Because the urine is more hypertonic, more water enters it by osmosis. The urine volume increases. 14. Dehydration is a familiar example of a human osmoregulatory challenge. Dehydration results in a lower blood volume and resulting lower blood pressure. These events stimulate increased secretion of ADH and aldosterone. Decreased blood pressure also stimulates renin release, which leads to increased angiotensin II. These hormones increase blood pressure. 15. In planaria, when the salinity of the surrounding water decreases, the number of protonephridia increase. When its environment is less salty, more water enters the planarian's body by osmosis. The protonephridia expel the excess water. 16. The kangaroo rat must conserve water by producing a very low volume of urine. 17. Mammalian kidneys are more efficient than bird kidneys; they can produce more concentrated urine. There is a correlation between the length of the loop of Henle and the ability to concentrate urine. 18. Answers will vary based on your opinion.

CHAPTER 49

CHECKPOINT

49.1 • Target cells are those that are influenced by a particular hormone and have specific binding receptors. • The nervous and endocrine systems work together to maintain homeostasis. Neurons signal gland cells, including endocrine cells, and the nervous system helps regulate many endocrine responses. Some signal molecules can function as either a neurotransmitter or a hormone, depending on their source. • When calcium concentration in the blood decreases below the homeostatic level, a negative feedback system is engaged; the parathyroid glands release more parathyroid hormone, returning the calcium concentration to the homeostatic level. • The four major chemical groups of hormones are fatty acid derivatives, which include prostaglandins; steroid hormones such as cortisol; amino acid derivatives such as epinephrine; and peptide hormones such as glucagon. **49.2** • See Figs. 49-3a and b. • In autocrine signaling a hormone acts on the cells that produce it, while in paracrine signaling a hormone diffuses through interstitial fluid and acts on nearby target cells. • Prostaglandins modify cyclic AMP levels and interact with other hormones to regulate various metabolic activities. **49.3** • Receptors convert an extracellular hormone signal into an intracellular signal that affects some cell process. The hormone does not enter the cell but serves as the first messenger and relays information to a second messenger. The second messenger signals effector molecules that carry out the action. • Steroid hormones are relatively small, lipid-soluble molecules that easily pass through the plasma membrane of the target cell. Peptide hormones, not lipid soluble, bind to specific cell-surface receptors in the plasma membrane. • Many protein hormones bind to G protein–linked receptors. The receptor activates a second messenger such as cyclic AMP, which then catalyzes the conversion of ATP to cAMP. This second messenger relays the signal, activating a protein, such as protein kinase, which leads to a response or change in the cell. **49.4** • Four actions of hormones in invertebrates are helping to regulate metabolism, growth and development, regeneration, and molting. • Juvenile hormone suppresses metamorphosis at each larval molt. With each successive molt, the level of juvenile hormone decreases. **49.5** • The hypothalamus is the link between the nervous and endocrine systems because it regulates endocrine activity and connects both systems anatomically. The anterior pituitary regulates growth and other endocrine glands. The posterior pituitary secretes oxytocin and antidiuretic hormone produced by the hypothalamus.

• See Fig. 49-10. • The beta cells in the pancreas would secrete insulin, lowering blood glucose concentration. Insulin decreases blood glucose concentration, while glucagon increases it; both are regulated by negative feedback systems. • The adrenal glands would be triggered to release large amounts of epinephrine and norepinephrine so you could think quicker, fight harder, and run faster to escape the reptilian jaws. • The adrenal medulla is regulated by the hypothalamus and sympathetic nervous system, while the adrenal cortex is regulated by negative feedback involving the hypothalamus and adrenocorticotropic hormone.

FIGURE QUESTIONS

FIG. 49-4 Because there would be fewer receptors that could become activated, specific protein synthesis would be slowed. FIG. 49-5 Less cAMP (the second messenger) would be produced, so fewer proteins would be activated, and the effect of the hormone would be decreased. FIG. 49-13 Less glucagon would be secreted and the glucose concentration in the blood would decrease below normal. The person would experience hypoglycemia.

TEST YOUR UNDERSTANDING

1. c 2. d 3. c 4. e 5. a 6. b 7. e 8. a 9. d 10. See Fig. 49-13. 11. Cells require glucose for cellular respiration. There are many challenges to glucose homeostasis, and multiple hormones work together to regulate carbohydrate metabolism. For example, when the body is under stress, cortisol promotes conversion of other nutrients to glucose. 12. The brain does not store glucose; however, neurons require a steady supply of this fuel molecule. When neurons in the brain are deprived of glucose, the brain is not able to function. The individual may pass out, or if the situation is prolonged, could become comatose or die. 13. Receptors only bind with a specific hormone. An advantage of having complexity in hormone action is the precision that is achieved among and between different body systems and regulatory functions. 14. The receptor for aldosterone may have been a receptor for some ancient, no longer present, similar hormone before aldosterone evolved. 15. Humans have evolved over tens of thousands of years (or longer) to store fat and consume a great deal of food when it is available. This physiological characteristic cannot be rapidly changed, but the dilemma can be addressed psychologically and socially. Scientists and doctors can partner with educators to facilitate public understanding about healthy food consumption, physical exercise, and our human proclivity to stuff ourselves with fatty, sugary foods. More technologies can be developed to stimulate pleasurable physical exercise and to retool the food-processing industry without sacrificing taste.

CHAPTER 50

CHECKPOINT

50.1 • In budding (e.g., in cnidarians) a small part of the parent's body separates from the rest and develops into a new individual. In fragmentation, the body of the parent breaks into several pieces; each piece regenerates the missing parts and develops into a new animal. In parthenogenesis, an unfertilized egg develops into an adult animal. • Compared with sexual reproduction, asexual reproduction is the fastest and most efficient way to reproduce. For example, each cell can produce two new cells. The disadvantage of asexual reproduction is that the offspring has all of the same genes as the single parent (except for mutations); as a result, there is a lack of variability in the population. **50.2** • The testes produce sperm and the male hormone testosterone. • Seminiferous tubules ⟶ epididymis ⟶ vas deferens ⟶ ejaculatory duct ⟶ urethra ⟶ release from body (see Fig. 50-3). • Testosterone stimulates development of the male's primary and secondary sex characteristics, maintains the secondary sex characteristics, and stimulates spermatogenesis. • See Fig. 50-8. **50.3** • In oogenesis, a primary oocyte gives rise to only one ovum; compare to a primary spermatocyte giving rise to four sperm cells during spermatogenesis. • If the secondary oocyte were fertilized, it could not develop without the hormones produced by a corpus luteum. • See Fig. 50-15a. FSH stimulates follicle development. Estrogens stimulate the endometrium to thicken; estrogens are necessary for the development of the sex organs at puberty, and for the development and maintenance of female secondary sex characteristics. **50.4** • If a woman becomes pregnant, hCG signals the corpus luteum to continue to function. • Without sufficient estrogen and progesterone, the embryo would be spontaneously aborted. Estrogen is necessary for the development and maintenance of the uterine wall, and progesterone inhibits uterine contractions that might expel the embryo. • High levels of estrogen secreted by the placenta greatly increase the number of receptors for oxytocin in the uterine wall. As a result, the uterus becomes more responsive to oxytocin, the hormone that stimulates uterine contractions. **50.5** • Muscle tension increases and certain tissues (including tissues in the penis, vagina, and clitoris) become engorged with blood (vasocongestion); heart rate and respiration increase. **50.6** • Hormone methods of contraception inhibit ovulation. • Sterilization would have the greatest impact on the number of unwanted pregnancies; other methods with low failure rates are injectable or implanted hormone contraceptives and IUDs. • A spontaneous abortion, or miscarriage, occurs naturally; a therapeutic abortion is performed when the mother's life is in danger or when there is evidence that the embryo will be grossly abnormal.

50.7 • The two most common STIs in the United States are human papillomavirus (HPV) and *Chlamydia*. • Pelvic inflammatory disease (PID), a serious infection that can lead to infertility, can be caused by chlamydia.

FIGURE QUESTIONS

FIG. 50-8 The testes could not produce testosterone, which would stop sperm production, lower his sex drive, and affect his secondary sex characteristics. **FIG. 50-14** Ovulation does not take place and symptoms of menopause may occur. **FIG. 50-15** FSH would continue to be released and new follicles would develop.

TEST YOUR UNDERSTANDING

1. c 2. e 3. a 4. b 5. e 6. a 7. a 8. e 9. d 10. c 11. See Fig. 50-9b. 12. LH and estrogen; mid to late postovulatory phase; estrogen and progesterone. 13. Cross-fertilization increases variability in the offspring, increasing their chance of surviving in a changing environment. 14. Some animals may have mutations that help them resist parasites. Sexual reproduction allows such beneficial mutations to spread through the population. 15. Sociological factors—historically women have been held responsible for contraception; biological factors—there are more opportunities for intervention in the female. Development of a method of male contraception might focus on reduction of sperm production in the seminiferous tubules, or on methods that would decrease sperm motility or viability. 16. Countries where abortions are legal and safe are most likely higher income countries in which people have easier access to contraception (and can also afford contraception).

CHAPTER 51

CHECKPOINT

51.1 • Cell determination generally precedes cell differentiation. • Pattern formation is the process that leads to morphogenesis through the organization of differentiated cells. The steps in pattern formation include cell-to-cell signaling, cell migrations, changes in cell shape, and apoptosis. **51.2** • Species-specific recognition molecules ensure that sperm and egg are of the same species. Events occurring at the time of fertilization ensure that more than one sperm cannot enter the egg (i.e., prevent polyspermy). • Although the details differ, the cortical reaction serves as a block to polyspermy in both echinoderms and mammals. **51.3** • Compare your sketches to Figure 30-5. • Meroblastic cleavage (cleavage restricted to the blastodisc). • Large amounts of yolk in the egg cytoplasm tend to slow down cleavage in the embryo. As a consequence, the vegetal pole cells of embryos developing from moderately telolecithal eggs tend to be fewer and larger than the animal pole cells. The yolk of embryos developing from extremely telolecithal eggs does not cleave at all (meroblastic cleavage in reptiles and birds). **51.4** • The archenteron is the embryonic forerunner of the digestive tract. It is formed in sea stars and amphibians, but not in birds. • Large amounts of yolk prevent the simple type of gastrulation that occurs in echinoderms and amphioxus. Instead, the establishment of the three embryonic cell layers is accomplished as cells migrate from the cell surface and move into the interior of the embryo. **51.5** • Ectoderm: outer skin layer, nervous system, and sense organs; mesoderm: notochord, skeleton, muscles, circulatory system, reproductive system; endoderm: lining of digestive tube. Refer to Table 51-1 for a more complete list. • Neural plate cells differentiate in response to inductive signals from the underlying notochord tissue. These cells undergo a process of morphogenesis in which their shapes change, converting the neural plate into the neural tube, which eventually sinks below the surface. The cells of the neural tube subsequently differentiate into specialized cells of the central nervous system. • Neural crest cells are extremely versatile cells that give rise to parts of spinal nerves; they also migrate to various locations in the embryo, giving rise to many structures, including parts of the peripheral nervous system, various sense organs, parts of the head, and pigment-forming cells. **51.6** • Chorion: gas exchange and contributes to placenta in mammals; amnion: protection from desiccation and shocks. • In birds the allantois stores wastes and the yolk sac stores nutrients. In mammals the allantois contributes to the blood vessels in the umbilical cord and blood cells form in the walls of the yolk sac. **51.7** • The human embryo undergoes cleavage while being propelled down the oviduct, developing into a blastocyst (a blastula) after it enters the uterus. Trophoblast cells (the outer cell layer of the blastocyst) secrete enzymes that implant the blastocyst in the uterine wall. • The placenta develops from embryonic tissue (the chorion, which develops from trophoblast cells) and maternal uterine tissue, and serves as the organ of exchange between mother and embryo. • The cerebrum develops lobes at four months; limb buds appear at four weeks. • Rapid changes in the circulatory system are required as the embryo begins to breathe air. Most notably the blood that bypassed the lungs during fetal life becomes redirected to the lungs. In addition, the neonate's digestive system must begin to function because it is no longer supplied by nutrients from the mother.

FIGURE QUESTIONS

FIG. 51-3 Compare your sketch to Figure 51-3g. The innermost cavity is the archenteron, surrounded by the blastocoel. Your sketch should not include the blastopore.

FIG. 51-4 Compare your sketch to Figure 51-4j. It should not include the blastopore. **FIG. 51-5** Compare your sketch to Figures 51-4d and 30-5b. **FIG. 51-6** Compare your sketch to Figure 51-3c. **FIG. 51-8** It is unlikely that development would proceed beyond a few divisions because a blastomere lacking a gray crescent would lack cytoplasmic determinants required to support normal development. **FIG. 51-9** Archenteron: gut; blastopore: anus; blastocoel: nothing.

TEST YOUR UNDERSTANDING

1. b 2. d 3. a 4. d 5. b 6. e 7. b 8. b 9. c 10. b 11. c 12. a 13. The modern view is that eggs and sperm contain genetic instructions that interact with developmental determinants in the egg cytoplasm to direct development. 14. Some teratogenic medications such as thalidomide are of significant medical value, but their use must be carefully regulated to prevent birth defects. 15. A human embryo possesses a notochord, which does not support the adult body as in amphioxus, but is responsible for the induction of the overlying cells to form the neural plate, the forerunner of the dorsal hollow nerve cord, which then develops into the brain and spinal cord. A postanal tail develops, but does not persist (although it is evident in many of our primate relatives). Pharyngeal pouches, pharyngeal grooves, and branchial arches develop in the region of the pharynx. Slits do not form between the pouches and grooves, and they are not used for filter feeding (as in amphioxus) or for respiration (as in fishes). However, they and the branchial arches of supporting tissue give rise to many structures in the head, jaws, and neck (see Section 51.5). These structures have persisted because evolution is conservative and commonly builds pre-existing structures. 16. The placenta allows the embryo to develop within the body of the mother, which affords much more protection than an egg. 17. Consult Table 51-3 and the March of Dimes website for information on environmental influences to be avoided during (and even before) pregnancy. Your challenge as a public health official will be to devise programs that will effectively improve and extend maternal education and care.

CHAPTER 52

CHECKPOINT

52.1 • *Philanthus* performs a complex series of behaviors as she provides for her larva. Her behaviors increase the probability that her offspring will survive. The number of viable offspring an animal produces is a measure of its direct fitness. Therefore, the behaviors of *Philanthus* are adaptive. • Sample answer: A child inherits the capacity to run. However, individuals who want to be Olympic athletes must train for many years to improve their basic capacity to run. These athletes must learn how to modify and improve their muscle structure, form, and speed. • Innate behavior occurs when the animal is physiologically ready. Physiological readiness also affects the capacity for learned behavior. For example, a baby cannot learn to walk or play the piano until her body is physiologically ready. **52.2** • Imprinting is adaptive because the young learn to stay close to their mother, increasing their chance of survival. • Operant conditioning is adaptive because animals learn how to obtain food and other necessities; they also learn how to avoid dangerous or unpleasant stimuli. **52.3** • It is adaptive for some species to be diurnal while others are nocturnal or crepuscular because it decreases competition for food and other resources. • Directional orientation refers to travel in a specific direction. In navigation, an animal must integrate information about distance and time, as well as direction. **52.4** • Optimal foraging is the most efficient way for an animal to obtain food. • Optimal foraging is adaptive because the animals maximize their energy intake per unit of foraging time. This approach may maximize their reproductive success. **52.5** • The costs of living in groups include increased competition for food and habitats, increased risk of attracting predators, increased risk of transmitting disease, and time and energy used to gain and maintain social status. The benefits include defense against predators, help in foraging, and opportunities to receive help from others in the group. • Advantages of pheromones include that they are simple, widespread means of communication, and very little energy is required for their synthesis. They are effective in the dark and can be used to communicate danger, ownership of a territory, and availability for mating. Disadvantages include that they are transmitted slowly and the information that can be conveyed is limited. • The benefits of being dominant in a dominance hierarchy often include greater opportunity to mate and the right to eat first. Limitations include the need to maintain the position, which may require fighting. Subordinate individuals may be protected by the dominant animal, but may not have as much opportunity to mate or eat. • Benefits of territoriality often include a higher probability of mating and exclusive rights to food within the territory. Costs of territoriality include the time and energy expended in staking out and defending a territory and the risks involved in fighting for it. • Although it is the most sophisticated communication known among nonmammals, the dance of bees is highly stereotyped compared with human language. **52.6** • Sexual selection is a type of natural selection for successful mating. Factors that may influence mate choice include rank in a dominance hierarchy, ornamental displays, physical attributes that indicate good health or strength, and gifts presented to the female. • Polygyny is favored when the male is not needed to feed and protect the young; polyandry is favored when more than one male may help in caring for the young and when the males give gifts to the female. Monogamy is uncommon because it reduces direct fitness. • Benefits of courtship rituals include ensuring that the male is really a male and that he is a member of the same species. Rituals also provide opportunity for a female to evaluate a male. • Females who invest in parenting have more to gain than males because females of many vertebrate species produce relatively few, large eggs. The time and energy invested in producing eggs and carrying the developing embryo make it important for the female to continue to protect her investment. In addition, female mammals provide milk to nourish their young. The benefit is viable offspring and increased direct fitness. For males of most species, the costs of investing in parental care are outweighed by the benefits of mating with multiple females, thus increasing their direct fitness. **52.7** • Kin selection increases inclusive fitness through the breeding success of close relatives. Natural selection favors animals that help a relative (altruistic behavior) because the relative's offspring carries some of the helper's alleles. • An animal may benefit from helping a nonrelative because according to the hypothesis of reciprocal altruism, at some later time the animal will be paid back by the animal it helped. When paid back, the animal that is helped may benefit in terms of safety, obtaining food, or in direct fitness. **52.8** • Vertebrate societies are simpler than insect societies, but they are more flexible. • Examples of culture include female orcas teaching their offspring to hunt seals according to the custom of their particular group and chimps teaching others in their group how to use rocks to crack nuts. • According to sociobiologists, social behavior is determined by natural selection.

FIGURE QUESTIONS

FIG. 52-1 *Philanthus* would be unable to find her nest because she would be prevented from learning the necessary visual clues/landmarks. **FIG. 52-3** The impaired male stickleback would be less likely to father offspring so his direct fitness would decrease. **FIG. 52-6** The boy would likely be scared and would cry because he has become conditioned to associate dogs with pain. **FIG. 52-21** The males would not have reduced their level of guarding the eggs.

TEST YOUR UNDERSTANDING

1. a 2. b 3. c 4. d 5. b 6. a 7. d 8. e 9. b 10. e 11. a 12. c 13. The young salamanders described preferred eating more distant relatives, which increased the probability that their own relatives would survive to reproduce; this is an example of inclusive fitness. 14. In experiment 1, the parentals guarded the young

more closely than the eggs. After the young hatched, the parentals perceived that (despite the sneaker males lurking nearby during the egg stage) they were indeed the parents. The increase in care was based on olfactory cues. In experiment 2, the parentals guarded the eggs more closely than the young. After the young hatched, olfactory cues from the offspring indicated lack of paternity. 15. During times of prey scarcity, the groups with three and four females were most likely to be hungry at the end of the day. Perhaps when there are more animals in the group, they can venture farther in search of prey or take down larger prey. 16. The society of a social insect is more rigid and less flexible than human society. Information about behavior is largely transmitted genetically; like culture, genetic information is passed from parent to offspring. The young share the behavior of the parents. An important difference is that the insect society remains rigid and unlikely to change, whereas the human society is more flexible and more likely to change. 17. Sea turtle migration may be adaptive because it reduces competition between sea turtles of various ages. Young turtles are more likely to find food when they do not need to compete with older turtles. Migration may also be adaptive in that turtles return to the beach where their ancestors hatched. Those ancestors survived the conditions on that beach, suggesting that the new generation of offspring will also be able to survive. 18. It is unlikely that sea turtles could adapt to environmental pressures created by humans over a short period of time. Base your opinion on what you have learned in your biology course.

CHAPTER 53

CHECKPOINT

53.1 • Population density is the number of individuals per unit of area or volume. Population dispersion is spacing of individuals relative to one another. • A clumped dispersion can be advantageous to social animals, but it can be disadvantageous if there is competition within the population. **53.2** • Natality (the birth rate) and immigration contribute to population growth; mortality (the death rate) and emigration contribute to population decline. • A J-shaped population growth curve (exponential growth) describes a population at its intrinsic rate of increase (r_{max}). An S-shaped growth curve (logistic growth) describes the decline in a population's growth rate as it approaches its carrying capacity (K), the largest population that can be maintained indefinitely. • Compare your graphs to Figure 53-2 (bacteria could grow exponentially for a time if the medium is to provide nutrients and waste removal) and Figure 53-3 (bacteria

would exhibit logistic growth as they approach their carrying capacity). **53.3** • Three density-dependent factors that influence population growth are predation, disease, and competition. • Density-independent factors include abiotic events such as extreme weather, fires, and floods. **53.4** • In semelparity an organism puts all of its resources into a single reproductive event. These resources tend to improve the chances of success, but all is lost if the effort fails. Iteroparity provides many chances for reproduction; however, the organism's reproductive resources may become depleted by repeated reproductive cycles. • Parental care improves survival of the young when the population is at or near its carrying capacity. • Survivorship curves do not neatly fit the three models: Type I, Type II, and Type III. For example, a species' survivorship curve can be different early in life compared to later in life. **53.5** • In a metapopulation, excess individuals tend to move from good or source habitats, where population growth is greatest, to less favorable sink habitats where population success is poorer. **53.6** • The dramatic increase in the world population over the last 200 years is attributable to a dramatic decrease in the death rate. • Replacement-level fertility is higher in developing countries because they have higher infant mortality rates than do developed countries. • Unlike a child born in a developing country, a single child born in the United States is a major contributor to consumption overpopulation.

FIGURE QUESTION

FIG. 53-6 As is often the case in science, you will need more information to develop your hypothesis. You will want gather information on such abiotic factors as changes in climate and specific weather events. You will also need information on biotic factors such as organisms other than spiders and lizards (including microorganisms) that may be living within the enclosures and affecting the spider populations.

TEST YOUR UNDERSTANDING

1. b 2. b 3. c 4. b 5. a 6. d 7. e 8. a 9. b 10. e 11. c 12. b 13. The pattern of repeated reproduction exhibited by female elephants is characteristic of Type I survivorship, in which individuals of reproductive age have a high probability of surviving. 14. Interference competition, in which a few individuals obtain adequate resources and others do not, is more sustainable than exploitation competition, which can lead to population booms (exceeding the carrying capacity) and crashes. 15. A population that can be described as mainly characterized by overproduction of offspring and is kept in check by limits on population growth. A K-selected population is more stable; reproductive success is achieved with production of fewer offspring who are more likely to survive and reproduce. 16. Bolivia, with a much higher

percentage of its population entering reproductive age, has greater population growth momentum than Austria, and can be expected to have a higher growth rate over the next two decades. 17. With about 1065 people per square mile, the population density of the Netherlands is much greater than that of the United States (about 85 per square mile). 18. 1295.6 million. 19. 8 per 1000. 20. The diagram depicting population (b), with a very small proportion of its population at reproductive age or prereproductive age is consistent with negative population growth momentum. 21. A few examples of technologies that contribute to consumption overpopulation are those that typically involve the exploitation of fossil fuels (electricity, planes, trains, motor vehicles), rare elements (electronics), and those that produce pollution as a byproduct. New technologies to address some of these problems are being developed all the time. Which ones do you think might be most promising?

CHAPTER 54

CHECKPOINT

54.1 • Abundant acorn crops attract mice to oak forests. Mice eat the acorns, so more mice survive and breed in winter. Gypsy moths cause serious damage to oak trees, but mice also eat gypsy moth pupae, reducing their numbers. Both deer and mice are hosts to ticks. The mice often carry the bacteria that cause Lyme disease. The bacteria infect the ticks and are spread to humans when they are bitten by infected ticks. • The realized niche is usually narrower because factors, such as competition with other species, may exclude an organism from a part of its fundamental niche. • The principle of community ecology illustrated by the example is resource partitioning. • The three kinds of symbiosis are mutualism, commensalism, and parasitism. Refer to Section "Symbiosis involves a close association between species" for examples of each. **54.2** • Dominant species are present in great abundance in a community, while keystone species are present in low abundance. • Reintroduced wolves would reduce the herbivore population (such as deer and elk) which, in turn, would allow more and varied plant species to grow. It is an example of a top-down process because the predators (wolves) impact the abundance of herbivores and producers in the community. **54.3** • Species richness is reduced by geographic isolation due to the distance effect, where species are unable to colonize islands due to distance; species richness can also be reduced by random events in mountain or island habitats, which cause species extinction. Locally extinct species are not readily replaced in these isolated habitats. • The structural complexity of habitats such as forests and coral reefs can enhance species richness because they provide numerous diverse ecological niches. New species may migrate to these areas; and over time, new species may evolve there. • Species richness is inversely related to the environmental stress of a habitat because only species that can tolerate the particular type of environmental stress survive, while species that are unable to adapt to the environmental stress become extinct. **54.4** • Primary succession is the change in species composition over time in a habitat without soil that was not previously inhabited by organisms; secondary succession is the change in species composition that takes place after some disturbance removes the existing vegetation from a habitat containing soil. • Connell's intermediate disturbance hypothesis proposes that species richness is greatest at moderate levels of disturbance where the community is a mosaic of habitats at different stages of succession. • Organismic models of the nature of communities propose that member species cooperate with one another in a manner similar to the cooperation of the parts of an individual organism's body. The community passes through stages of development, like the embryonic stages of an organism, eventually reaching an adult state, or climax community. Individualistic models of the nature of communities propose species individuality, with each species having its own particular abiotic living requirements; communities are not interdependent associations of organisms, and each species is independently distributed across a continuum of areas that meets its own individual requirements.

FIGURE QUESTIONS

FIG. 54-5 *P. caudatum* would show increase in population density similar to the second graph, while *P. aurelia* would decrease in population density. **FIG. 54-7** If the *G. fuliginosa* population were removed from the shared island, the smaller seeds it ate would become more available to *G. fortis*. Over time, a smaller bill depth might evolve in the *G. fortis* population, adapting these finches to the wider seed selection. **FIG. 54-18** A species richness graph compiled from a tundra habitat would likely have fewer bird species because the community would have a lower structural complexity of vegetation. **FIG. 54-19** About 60 percent of species have colonized islands at 2000 km and about 20 percent at 6000 km have colonized. The greater the geographic isolation, the lower the species richness. **FIG. 54-20** A gradual increase in temperature could result in an increase in species richness as more bird species move into a warmer, more favorable habitat. **FIG. 54-23** Palm tree: The curve on the graph would begin at a high number on the *y*-axis at the wet end of the *x*-axis; the slope would decrease down to the dry end of

the *x*-axis. Cactus: The curve for the cactus would begin at a low number on the *y*-axis at the wet end of the *x*-axis; the slope would increase as the line approaches the dry side of the *x*-axis.

TEST YOUR UNDERSTANDING

1. e 2. c 3. d 4. e 5. c 6. d 7. d 8. a 9. d 10. d 11. e 12. d 13. Humans are a keystone species as we determine the species composition and functioning of our ecosystem, for better or to our detriment. 14. Plants containing nitrogen-fixing nodules can provide nitrogen in the surrounding soil for other plants to grow but may grow better and faster, excluding other plant species. 15. Natural selection has favored the garter snake and made it a predator of the newt, a role reversal for their predator–prey relationship. Over time, the newt may develop (through natural selection) a new defense to escape the garter snake, again changing the predator–prey balance. 16. Competition among finch populations drives natural selection. Character displacement and resource partitioning both reduce interspecific competition, increasing the overall fitness of the different species. 17. The top and middle graph show logistic population growth. The *K,* or carrying capacity, for the two *Paramecium* growth curves occurs at a point where there is no further increase in population and is indicated by a flat line. 18. The human ecological niche constantly changes and expands as we learn about our surroundings and adapt to the challenges of survival. Science and technology provide us with the means to understand and thrive in our ecological niche. Failure of society to promote an understanding of science and misuse of technology could cause our ecological niche to shrink and endanger our survival. (Answers will vary as to how science and technology have changed our ecological niche over the past 1000 years.)

CHAPTER 55

CHECKPOINT

55.1 • See Fig. 55-2. • A trophic level is each sequential level of matter and energy in a food web, from producers to primary, to secondary, to tertiary consumers; each organism is assigned to a trophic level based on its primary source of nourishment. Ecological pyramids illustrate how food webs work and where members of each trophic level fit. • Gross primary productivity is a measurement of the rate at which energy accumulates in an ecosystem during photosynthesis, while net primary productivity is the energy that remains after plant cellular respiration has occurred. Gross primary productivity minus plant cellular respiration is net primary productivity. **55.2** • During

photosynthesis, CO_2 in the air is fixed by plants and some prokaryotes; cellular respiration, combustion, and erosion return CO_2 to the atmosphere. • See Fig. 55-8. • Phosphorus cycles from land and terrestrial organisms to ocean and marine organisms and back to land and terrestrial organisms. **55.3** • The circulation of the atmosphere is primarily driven by the energy from the sun; different temperatures at different locations cause warm air to rise and cooler air to sink, creating movement. Earth's rotation and ocean currents also play a role in atmospheric circulation. • Ocean currents are produced by persistent prevailing winds blowing over the ocean surface; the Coriolis effect causes currents to swerve right in the Northern Hemisphere and left in the Southern Hemisphere. • Factors that produce regional differences in precipitation are wind movement over air and water; landmass features such as mountains, valleys, and deserts; ocean water temperatures; currents; and upwelling. **55.4** • The deforestation study observed that when a forest is removed, the total amount of water and minerals that flow into streams increases drastically. Also deforestation causes soil erosion and leaching of essential minerals, resulting in decreased soil fertility.

FIGURE QUESTIONS

FIG. 55-1 Once the sun ceased to provide light for plant growth, all the organisms would starve. The decomposers would be left to consume the remains before eventually dying themselves. **FIG. 55-2** Plants would produce less and gradually die, leaving many insects and small animals without enough food or shelter. Larger animals would leave as smaller animals disappeared or flew away. Decomposers would consume remains. What was left of the dry, weakened forest would be more susceptible to fire. **FIG. 55-6** The gulls might be consumed by larger birds, which could eventually die from the high bioaccumulation of DDT. When the gulls die, decomposers could be poisoned by consuming the remains. The DDT would not disappear from the salt marsh ecosystem. **FIG. 55-15** Increases in land temperature could cause wind speeds to increase, which would produce stronger currents, causing greater upwelling and dead zones. Water temperature increase would alter density of water near the shore, causing fresh water to flow into oceans and change surface current patterns.

TEST YOUR UNDERSTANDING

1. d 2. e 3. a 4. d 5. e 6. a 7. b 8. c 9. d 10. d 11. b 12. See Fig. 55-1. 13. If all decomposers were eliminated from a food web, organic material from waste and remains would not be broken down to simple inorganic compounds. All other life would eventually cease to exist as inorganic compounds became depleted. 14. The cycling of matter is essential because life requires

accessible building materials for long-term continuance; materials must be reused and recycled within and among ecosystems. 15. The nitrogen cycle would collapse if bacteria were absent because the five steps bacteria perform (nitrogen fixation, nitrification, assimilation, ammonification, and denitrification) are all essential to the cycle. 16. The microclimate for the ant and for the elephant would be different. The ant inhabits a microclimate close to the ground, in moist air and under shade from trees. The elephant's microclimate is a drier atmosphere, surrounded by sunlight and wind. 17. The pyramid is inverted because the plankton (microscopic algae and animals) are highly productive and experience high rates of turnover (are short lived and reproduce rapidly). 18. The graph indicates that wildfires increase as temperatures increase. Climate warming may be causing more wildfires because the second half of the graph (1985–2005) indicates higher temperatures and more fires; more data for a longer period of time may support this observation. 19. Humans alter the nitrogen cycle through overuse of nitrogen fertilizer, raw sewage dumping, and automobile/power plant/factory emissions. Scientists can partner with educators to explain the problem to the public while working with engineers to develop methods to mitigate the current damage and prevent future problems. The public can, in turn, encourage legislators to enact laws and set money aside for reducing nitrogen pollution and for education.

CHAPTER 56

CHECKPOINT

56.1 • Tundra: long, cold winters and short summers, little precipitation; geologically young soil with permanent layer of permafrost. Boreal forest: cold winters and short growing season; acidic, mineral-poor soil. Temperate rainforest: cool weather, dense fog, high precipitation; nutrient-poor soil, which may have high organic content. Temperate deciduous forest: hot summers and cold winters, high precipitation; soils rich in organic matter at surface and deeper clay layer. Temperate grasslands: hot summers and cold winters, moderate to unpredictable precipitation; soils rich in organic matter; roots/rhizomes. Chaparral: mild, wet winters and hot, dry summers; thin and infertile soil. Deserts: cold or warm temperatures, very low precipitation, and soil low in organic material, but high in mineral content. Savanna: tropical temperatures, low rainfall with dry periods; soil low in essential minerals, may be high in aluminum. Tropical dry forests: wet season and dry season, high precipitation; soil with some organic matter but low in essential minerals. Tropical rain

forest: tropical temperatures, high year-round precipitation; old, highly weathered, mineral-poor soil with low organic content. • Boreal forest: caribou, wolves, bears, moose, rodents, lynx, seasonal birds, and few reptiles and amphibians. Temperate deciduous forest: mountain lions, wolves, bison, deer, many small animals, insects, birds, reptiles, and amphibians. Temperate rain forest: mule deer, elk, small rodents, insects, many birds, some reptiles and amphibians. Tropical rain forest: greatest number and diversity of insects, bird, reptiles, and amphibians; sloths, monkeys, elephants. • Answers will vary. • Tundra is generally cold and dry, while deserts may be cold or hot as well as dry. Animals and plants in both tundra and desert are adapted to extreme temperatures and limited water. Temperate grassland has cold winters and more rainfall, while savanna is warm year round and has low rainfall and dry periods. Animals and plants are abundant; drought, fire, and overgrazing are significant in both biomes. **56.2** • If ocean plankton disappeared, plankton-eating nekton would lose their food source and die off. Carnivorous nekton (sharks, whales) would lose much of their food source and die. Benthos would, in turn, lose most of their food sources that float down from the surface, and their population would be decimated as well. The ocean would rapidly lose most living organisms. • Salinity, dissolved oxygen, light availability, water depth, essential minerals, temperature, pH, and presence/absence of waves and currents are environmental factors that determine the adaptations of organisms that live in aquatic environments. • Freshwater wetlands are covered part of the year by shallow water, while estuaries are coastal areas near rivers where fresh water mixes with tidal salt water. In flowing-water ecosystems there is a flowing current; in standing-water ecosystems the top layer, due to light penetration, contains phytoplankton, zooplankton, and fishes, while the lower, dark zone contains mostly prokaryotic decomposers. • The intertidal zone, shoreline area between low tide and high tide, has organisms with adaptations that resist wave action and exposure to air during low tide. The benthic environment, the ocean floor, has sea grasses, kelps, and coral reefs in shallow ocean waters. The neritic province is open ocean from shoreline to 200 m. Organisms are all floaters or swimmers. The euphotic zone, the upper part of the neritic province, has phytoplankton that are the base of the food web. The oceanic province is the part of the open ocean deeper than 200 m. It is dark, cold, and has a high pressure. Animals are either predators or scavengers that subsist on marine snow. • Coral reefs are often compared to tropical rain forests because of their immense diversity of life and high productivity. **56.3** • An ecotone is the transition area where two communities, or biomes, meet and

intergrade. Ecotones occur where agricultural land meets forest, tundra borders a boreal forest, or a river touches land. **56.4** • Biogeography is the study of the geographic distribution of plants and animals. Biogeographers look for patterns in geographic distribution and try to explain how such patterns arose, including where populations originated, how they spread, and when. • The Australian realm has been separated from the other biogeographic realms for the longest period; animals living in the Australian realm include marsupials and monotremes, such as the kangaroo, platypus, and spiny anteater.

FIGURE QUESTIONS

FIG. 56-1 As climate changes, plants, animals, and other organisms will disappear if they are unable to adapt. They may be replaced or outcompeted by species that can adapt more rapidly to these climate changes. **FIG. 56-24** Fireweed and knotweed have adapted to the serpentine soil and would have too much competition to colonize the ecotone area.

TEST YOUR UNDERSTANDING

1. c 2. a 3. d 4. e 5. c 6. b 7. d 8. a 9. c 10. e 11. e 12. Desert animals may be small because they have limited food and water resources. To test this hypothesis, a desert animal experimental group could be provided with abundant amounts of food and water for numerous generations. 13. The lowest average annual precipitation characteristic of tropical rain forests is 225 cm. The highest average annual precipitation is 450 cm. The range of average temperature in tropical rain forests is 20°C–28°C. 14. Eight sampled species are found in the nonserpentine soil. Ten sampled species are found in the serpentine soil. Thirteen are found in the ecotone. The ecotone supports more plant species than serpentine or nonserpentine areas alone. 15. Organisms adapted to a fast current would be displaced by organisms that thrive in still water. Upstream organisms' habitats would probably have still water and flooding close to the dam, while downstream organisms' habitats would now have still water and less water. 16. Prairie dogs may bury the dead animal to keep predators from smelling it as a food source and approaching the hole. Researchers could prevent prairie dogs in an experimental group from burying dead animals; the dead animals could be protected in a cage that would allow odors to escape. Prairie dogs in a control group would be permitted to bury dead animals. 17. Technologies employed for data collection could be coastal and ocean buoys, satellite pictures (visual and IR), and research vessels. Local communities could receive instruction from science educators to count turtle eggs and record breeding data to assist with the research.

CHAPTER 57

CHECKPOINT

57.1 • The three levels of biological diversity are genetic diversity, species richness, and ecosystem diversity. • An endangered species would be at greater immediate risk of extinction because there are fewer surviving members of the species throughout its range compared with a threatened species. • Habitat fragmentation contributes to declining biological diversity because it encourages disease organisms, as well as invasive species that compete with native species. • Invasive species contribute to the biodiversity crisis because they cause more damage to native species than do native competitors or predators. Native species are not usually evolutionarily equipped to deal with invasive species. **57.2** • If conservation biology did not exist, many species would not be protected and could become extinct; biodiversity would plummet. • Landscape biology and restoration ecology are examples of in situ conservation. • In situ conservation helps the greatest number of species because it keeps the ecosystems intact and maintains biodiversity on a large scale. • The U.S. Endangered Species Act is controversial because it pits landowners against conservationists, in most cases without developing effective means of fostering education and cooperation between the groups. It is also a very slow, political, litigious process to list species. **57.3** • Three ecosystem services that forests provide are wildlife habitat, protection of watersheds, and prevention of soil erosion and flooding. • Deforestation in tropical rain forests is caused by subsistence agriculture and slash-and-burn agriculture. The main reason for deforestation of boreal forests is clear-cut logging. **57.4** • The enhanced greenhouse effect involves increased levels of greenhouse gases that absorb infrared (heat) radiation, combined with additional warming of the atmosphere. The enhanced greenhouse effect is caused by human actions, such as industrialization without mitigation of pollutants. • Climate change may cause oceans to rise, ice caps and glaciers to melt and disappear, increased water and atmospheric temperatures, reduction of species and diversity, higher frequency and more intense storm systems and wildfires, and increased disease.

FIGURE QUESTIONS

FIG. 57-2 The oystercatcher population would experience changes in genetic diversity if it interbreeds with the newcomer population or could be reduced or eliminated if competition became extreme. **FIG. 57-4** Effectively addressing social, political, and cultural factors, for example by education, could result in greater societal commitment to addressing and changing the direct effects of declining biodiversity, such as pollution and climate change. **FIG. 57-6** Bay scallop populations will decrease, and the species may no longer be able to survive in that habitat. **FIG. 57-9** A large human population could insulate and protect the endemic species through education, research, and well-planned land use solutions. **FIG. 57-17** Accumulation of greenhouse gases can be slowed by humans changing their lifestyles, such as reducing carbon footprint, purchasing low emission energy and automobiles, and maximizing recycling opportunities.

TEST YOUR UNDERSTANDING

1. d 2. e 3. c 4. c 5. c 6. b 7. e 8. d 9. c 10. a 11. Captive breeding may fail because reintroduced species may no longer be able to compete in their original habitats. 12. Human decisions, good and bad, ultimately affect the outcome of survival of organisms and protection of our planet. 13. Modern humans are the most dangerous species, to ourselves and other inhabitants of Earth. 14. Any loss of species is permanent, and every genetic code and genome could potentially be of future benefit for medicine, research, or other economic interests. 15. Multiple introductions, each under slightly different circumstances (weather, predators, nutrient level), could give some of the invasives the edge to thrive. 16. The production of many smaller eggs could be due to more protected surroundings and nourishment not afforded to eggs in the wild, which would need to be bigger and heartier. Captive-bred females might have lower survival rates than wild females. 17. Habitat fragmentation is most important. Climate change and increasing atmospheric CO_2 are represented as separate factors because they have multiple causes. Nitrogen pollution relates to overuse of fertilizer and nitrogen-containing emissions that cause air pollution. 18. Education and dialogue between the public, scientists, and developers of technology in all countries would be essential to developing and implementing a plan for all residents of Earth that would be sustainable for developed (high income) and developing (lower income) countries.

Glossary

ABC transporters ATP-binding cassette transporters; use ATP energy to transport certain ions, sugars, and polypeptides across cell membranes.

abiotic factors Elements of the nonliving, physical environment that affect a particular organism. Compare with *biotic factors.*

abortion Termination of pregnancy, resulting in the death of the embryo or fetus; abortions can occur spontaneously or can be induced.

abscisic acid (ABA) (ab-sis′-ik) A plant hormone involved in dormancy and responses to stress.

abscission (ab-sizh′-en) The normal (usually seasonal) fall of leaves or other plant parts, such as fruits or flowers.

abscission zone The area at the base of the petiole where the leaf will break away from the stem. Also known as *abscission layer.*

absorption (ab-sorp′-shun) (1) The movement of nutrients and other substances through the wall of the digestive tract and into the blood or lymph. (2) The process by which chlorophyll takes up light for photosynthesis.

absorption spectrum A graph of the amount of light at specific wavelengths that is absorbed as light passes through a substance. Each type of molecule has a characteristic absorption spectrum. Compare with *action spectrum.*

abyssal zone Part of benthic environment that extends from a depth of 4000 to 6000 m.

acanthodians (ak-an-tho′-de-uns) An extinct group of armored jawed fishes; possessed paired spines and pectoral and pelvic fins.

accessory fruit A fruit consisting primarily of tissue other than ovary tissue, e.g., apple, pear. Compare with *aggregate, simple,* and *multiple fruits.*

acclimatization Adjustment to seasonal changes.

acetyl coenzyme A (acetyl CoA) (as′-uh-teel) A key intermediate compound in metabolism; consists of a two-carbon acetyl group covalently bonded to coenzyme A.

acetylcholine (ah″-see-til-koh′-leen) A common neurotransmitter released by cholinergic neurons, including motor neurons.

acetyl group A two-carbon group derived from acetic acid (acetate).

achene (a-keen′) A simple, dry fruit with one seed in which the fruit wall is separate from the seed coat, e.g., sunflower fruit.

acid A substance that is a hydrogen ion (proton) donor; acids unite with bases to form salts. Compare with *base.*

acidic solution A solution in which the concentration of hydrogen ions [H^+] exceeds the concentration of hydroxide ions [OH^-]. An acidic solution has a pH less than 7. Compare with *basic solution* and *neutral solution.*

acid precipitation Precipitation that is acidic as a result of both sulfur and nitrogen oxides forming acids when they react with water in the atmosphere.

acoelomate (a-seel′-oh-mate) An animal lacking a body cavity (coelom). Compare with *coelomate* and *pseudocoelomate.*

acquired immune responses See *adaptive immune responses.*

acquired immunodeficiency syndrome (AIDS) A serious, potentially fatal disease caused by the human immunodeficiency virus (HIV).

acromegaly (ak″-roh-meg′-ah-lee) A condition characterized by overgrowth of the extremities of the skeleton, fingers, toes, jaws, and nose. It may be produced by excessive secretion of growth hormone by the anterior pituitary gland.

acrosome reaction (ak′-roh-sohm) A series of events in which the acrosome, a caplike structure covering the head of a sperm cell, releases proteolytic (protein-digesting) enzymes and undergoes other changes that permit the sperm to penetrate the outer covering of the egg.

actin (ak′-tin) The protein of which microfilaments consist. Actin, together with the protein myosin, is responsible for muscle contraction.

actin filaments Thin filaments consisting mainly of the protein actin; actin and myosin filaments make up the myofibrils of muscle fibers.

actinopods (ak-tin′-o-podz) Protists characterized by axopods that protrude through pores in their shells. See *rhizarians.*

action potential An electrical signal resulting from depolarization of the plasma membrane in a neuron or muscle cell. Compare with *resting potential.*

action spectrum A graph of the effectiveness of light at specific wavelengths in promoting a light-requiring reaction. Compare with *absorption spectrum.*

activation energy (E_A) The kinetic energy required to initiate a chemical reaction.

activator protein A positive regulatory protein that stimulates transcription when bound to DNA. Compare with *repressor protein.*

active immunity Immunity that develops as a result of exposure to antigens; it can occur naturally after recovery from a disease or can be artificially induced by immunization with a vaccine. Compare with *passive immunity.*

active site A specific region of an enzyme (generally near the surface) that accepts one or more substrates and catalyzes a chemical reaction. Compare with *allosteric site.*

active transport Transport of a substance across a membrane that does not rely on the potential energy of a concentration gradient for the substance being transported and therefore requires an additional energy source (often ATP); includes carrier-mediated active transport, endocytosis, and exocytosis. Compare with *diffusion* and *facilitated diffusion.*

adaptation (1) An evolutionary modification that improves an organism's chances of survival and reproductive success. (2) A decline in the response of a receptor subjected to repeated or prolonged stimulation.

adaptive immune responses Defense mechanisms that target specific macromolecules associated with a pathogen. Includes cell-mediated immunity and antibody-mediated immunity. Also known as *acquired immune responses* or *specific immune responses.* Compare with *innate immune responses.*

adaptive immunity See *adaptive immune responses.*

adaptive radiation The evolution of a large number of related species from an unspecialized ancestral organism.

adaptive zone A new ecological opportunity that was not exploited by an ancestral organism; used by evolutionary biologists to explain the ecological paths along which different taxa evolve.

addiction Physical dependence on a drug, generally based on physiological changes that take place in response to the drug; when the drug is withheld, the addict may suffer characteristic withdrawal symptoms.

Addison's disease Caused by insufficient secretion of aldosterone and cortisol by the adrenal cortex and characterized by progressive anemia, low blood pressure, and inability to respond to stress.

adenine (ad′-eh-neen) A nitrogenous purine base that is a component of nucleic acids and ATP.

adenosine diphosphate (ADP) See *adenosine triphosphate (ATP).*

adenosine triphosphate (ATP) (a-den′-oh-seen) An organic compound containing adenine, ribose, and three phosphate groups; hydrolysis of the terminal phosphate yields adenosine diphosphate (ADP); of prime importance for energy transfers in cells.

adenylyl cyclase Enzyme responsible for catalyzing the conversion of ATP to cyclic AMP (cAMP).

adhering junction A type of anchoring junction between cells; connects epithelial cells.

adhesion The property of sticking to some other substance. Compare with *cohesion.*

adipose tissue (ad′-i-pohs) Tissue in which fat is stored.

adrenal cortex (ah-dree′-nul kor′-teks) The outer region of each adrenal gland; secretes steroid hormones, including mineralocorticoids and glucocorticoids.

adrenal glands (ah-dree′-nul) Paired endocrine glands, one located just superior to each kidney; secrete hormones that help regulate metabolism and help the body cope with stress.

adrenal medulla (ah-dree′-nul meh-dull′-uh) The inner region of each adrenal gland; secretes epinephrine and norepinephrine.

adrenergic neuron (ad-ren-er′-jik) A neuron that releases norepinephrine or epinephrine as a neurotransmitter. Compare with *cholinergic neuron.*

adrenocorticotropic hormone (ACTH) Hormone secreted by anterior pituitary that regulates glucocorticoid and aldosterone secretion.

adult stem cells Cells derived from body tissues that can differentiate into some specific cell types, e.g., blood cells, neuronal cells, or muscle cells, but not all types of cells found in an organism. See *pluripotent stem cells* and *totipotent stem cells.*

adventitious (ad″-ven-tish′-us) Of plant organs, such as roots or buds, that arise in an unusual position on a plant.

aerobe Organism that grows or metabolizes only in the presence of molecular oxygen. Compare with *anaerobe.*

aerobic (air-oh′-bik) Growing or metabolizing only in the presence of molecular oxygen. Compare with *anaerobic.*

aerobic cellular respiration See *respiration.*

aerobic respiration See *respiration.*

afferent (af′fer-ent) Leading toward some point of reference. Compare with *efferent.*

afferent arteriole In the mammalian kidney, an arteriole that conducts blood into the capillaries that make up a glomerulus; afferent arterioles branch from the renal artery. Compare with *efferent arteriole.*

afferent neurons Neurons that transmit action potentials from sensory receptors to the brain or spinal cord. Compare with *efferent neurons.*

age structure The number and proportion of people at each age in a population. Age structure diagrams represent the number of males and females at each age, from birth to death, in the population.

age structure diagram See *age structure.*

aggregated distribution See *clumped dispersion.*

aggregate fruit A fruit that develops from a single flower with many separate carpels, e.g., raspberry. Compare with *simple, accessory,* and *multiple fruits.*

aging Progressive changes in development in an adult organism.

agnathans (ag-na′-thanz) Jawless fishes; term is mainly used to refer to the earliest vertebrates, which are now extinct.

AIDS See *acquired immunodeficiency syndrome.*

albinism (al′-bih-niz-em) A hereditary inability to form melanin pigment, resulting in light coloration.

albumin (al-bew′-min) A class of protein found in most animal tissues; a fraction of plasma proteins.

aldehyde An organic molecule containing a carbonyl group bonded to at least one hydrogen atom. Compare with *ketone.*

aldosterone (al-dos′-tur-ohn) A steroid hormone produced by the vertebrate adrenal cortex; stimulates sodium reabsorption. See *mineralocorticoids.*

algae (al′-gee) (sing., *alga*) An informal group of unicellular, or simple multicellular, photosynthetic protists that are important producers in aquatic ecosystems.

allantois (a-lan′-toe-iss) An extraembryonic membrane of reptiles, birds, and mammals that stores the embryo's nitrogenous wastes; most of the allantois is detached at hatching or birth.

allele frequency The proportion of a specific allele in the population. Compare with *genotype frequency* and *phenotype frequency.*

alleles (al-leelz′) Genes governing variation of the same character that occupy corresponding positions (loci) on homologous chromosomes; alternative forms of a gene.

allelopathy (uh-leel′-uh-path″-ee) An adaptation in which toxic substances secreted by roots or shed leaves inhibit the establishment of competing plants nearby.

allergen A substance that stimulates an allergic reaction.

allergy A hypersensitivity to some substance in the environment, manifested as hay fever, skin rash, asthma, food allergies, etc.

allometric growth Variation in the relative rates of growth for different parts of the body during development.

allopatric speciation (al-oh-pa′-trik) Speciation that occurs when one population becomes geographically separated from the rest of the species and subsequently evolves. Compare with *sympatric speciation.*

allopolyploid (al″-oh-pol′-ee-ployd) A polyploid whose chromosomes are derived from two species. Compare with *autopolyploid.*

all-or-none response The principle that no variation exists in the strength of a single action potential; only a stimulus strong enough to depolarize the membrane to its critical threshold level results in transmission of an action potential.

allosteric regulators Substances that affect protein function by binding to allosteric sites.

allosteric site (al-oh-steer′-ik) A site on an enzyme other than the active site, to which a specific substance binds, thereby changing the shape and activity of the enzyme. Compare with *active site.*

alpha (α) carbon Asymmetrical carbon in an amino acid to which an amino group, a carboxyl group, a side chain (*R* group), and a hydrogen are covalently bonded.

alpha cells Pancreatic cells that secrete glucagon.

alpha (α) helix A regular, coiled type of secondary structure of a polypeptide chain, maintained by hydrogen bonds. Compare with *beta (β)-pleated sheet.*

alpine tundra An ecosystem located in the higher elevations of mountains, above the tree line and below the snow line. Compare with *tundra.*

alternation of generations A type of life cycle characteristic of plants and a few algae and fungi in which they spend part of their life in a multicellular *n* gametophyte stage and part in a multicellular 2*n* sporophyte stage.

alternative splicing A genetic mechanism in which exons are spliced in different ways to produce different mRNAs from a single gene.

altruism A type of cooperative behavior in which an individual appears to behave in a way that benefits another rather than itself.

altruistic behavior Behavior in which one individual helps another, seemingly at its own risk or expense.

alveolates Protists that have alveoli, flattened vesicles located just inside the plasma membrane; include the dinoflagellates, apicomplexans, and ciliates. See *chromalveolates;* compare with *stramenopiles.*

alveolus (al-vee′-o-lus) (pl., *alveoli*) (1) An air sac of the lung through which gas exchange with the blood takes place. (2) Saclike unit of some glands, e.g., mammary glands. (3) One of several flattened vesicles located just inside the plasma membrane in alveolate protists.

Alzheimer's disease (AD) A progressive, degenerative brain disorder characterized by amyloid plaques and neurofibrillary tangles.

amacrine cell A lateral interneuron in the retina of the eye; receives signals from bipolar cells and sends signals back to them or to ganglion cells.

amino acid (uh-mee′-no) An organic compound containing an amino group (—NH₂) and a carboxyl group (—COOH); may be joined by peptide bonds to form a polypeptide chain.

amino acid derivatives Simplest hormones, such as epinephrine and melatonin, that are derived from amino acids.

aminoacyl–tRNA (uh-mee″-no-ace′-seel) Molecule consisting of an amino acid covalently linked to a transfer RNA.

aminoacyl–tRNA synthetase One of a family of enzymes, each responsible for covalently linking an amino acid to its specific transfer RNA.

amino group A weakly basic functional group; abbreviated —NH₂.

ammonification (uh-moe″-nuh-fah-kay′-shun) The conversion of nitrogen-containing organic compounds to ammonia (NH₃) by certain soil bacteria (ammonifying bacteria); part of the nitrogen cycle.

amniocentesis (am″-nee-oh-sen-tee′-sis) Sampling of the amniotic fluid surrounding a fetus to obtain information about its development and genetic makeup. Compare with *chorionic villus sampling*.

amnion (am′-nee-on) In terrestrial vertebrates, an extraembryonic membrane that forms a fluid-filled sac for the protection of the developing embryo.

amniotes Terrestrial vertebrates: reptiles, birds, and mammals; animals whose embryos are enclosed by an amnion.

amniotic egg Egg that contains an amnion and other membranes that surround and protect the developing embryo and keep it moist; the evolution of the amniotic egg was one of the adaptations that allowed animals to become completely terrestrial.

amoeba (a-mee′-ba) (pl., *amoebas*) A unicellular protist that moves by means of pseudopodia.

amoebozoa Members of the unikont clade characterized by lobose pseudopodia at some time in the life cycle; include amoebas, plasmodial slime molds, and cellular slime molds. Compare with *opisthokonts*.

amphibians Members of vertebrate class that includes salamanders, frogs, and caecilians.

amphipathic lipid See *amphipathic molecule*.

amphipathic molecule (am″-fih-pa′-thik) A molecule containing both hydrophobic and hydrophilic regions.

ampulla Any small, saclike extension, e.g., the expanded structure at the end of each semicircular canal of the ear.

amygdala Part of the limbic system; filters incoming information and interprets it in the context of emotional needs and survival.

amylase (am′-uh-laze) Starch-digesting enzyme, e.g., human salivary amylase or pancreatic amylase.

amyloid plaques Extracellular deposits in the brain of the peptide amyloid-beta surrounded by degenerating dendrites and axons.

amyloplasts See *leukoplasts*.

anabolic hormone Hormone that promotes tissue growth.

anabolic steroids Synthetic androgens that increase muscle mass, physical strength, endurance, and aggressiveness but cause serious side effects; these drugs are often abused.

anabolism (an-ab′-oh-lizm) The aspect of metabolism in which simpler substances are combined to form more complex substances, resulting in the storage of energy, the production of new cell materials, and growth. Compare with *catabolism*.

anaerobe Organism that can grow or metabolize in the absence of molecular oxygen. See *facultative anaerobe* and *obligate anaerobe*. Compare with *aerobe*.

anaerobic (an″-air-oh′-bik) Growing or metabolizing in the absence of molecular oxygen. Compare with *aerobic*.

anaerobic respiration See *respiration*.

anaphase (an′uh-faze) Stage of mitosis in which the chromosomes move to opposite poles of the cell; anaphase occurs after metaphase and before telophase.

anaphase I See *meiosis* and *meiosis I*.

anaphase II See *meiosis* and *meiosis II*.

anaphase-promoting complex (APC) Enzyme complex responsible for initiating the transition from metaphase to anaphase.

anaphylaxis (an″-uh-fih-lak′-sis) An acute allergic reaction following sensitization to a foreign substance or other substance.

ancestral characters See *shared ancestral characters*.

androgen (an′-dro-jen) Any substance that has masculinizing properties, such as a sex hormone. See *testosterone*.

androgen-binding protein (ABP) A protein produced by Sertoli cells in the testes; binds and concentrates testosterone.

anemia (uh-nee′-mee-uh) A deficiency of hemoglobin or red blood cells.

aneuploidy (an′-you-ploy-dee) Any chromosomal aberration in which there are either extra or missing copies of certain chromosomes.

angiosperms (an′-jee-oh-spermz″) The traditional name for flowering plants, a very large (more than 300,000 species), diverse phylum of plants that form flowers for sexual reproduction and produce seeds enclosed in fruits; include monocots and eudicots.

angiotensin I (an-jee-o-ten′-sin) A polypeptide produced by the action of renin on the plasma protein angiotensinogen.

angiotensin II A peptide hormone formed by the action of angiotensin-converting enzyme on angiotensin I; stimulates aldosterone secretion by the adrenal cortex.

animal pole The non-yolky, metabolically active pole of a vertebrate or echinoderm egg. Compare with *vegetal pole*.

anion (an′-eye-on) A particle with one or more units of negative charge, such as a chloride ion (Cl^-) or hydroxide ion (OH^-). Compare with *cation*.

anisogamy (an″-eye-sog′-uh-me) Sexual reproduction involving motile gametes of similar form but dissimilar size. Compare with *isogamy* and *oogamy*.

annelids (an′-eh-lids) Segmented worms with bilateral symmetry and a tubular body that may be partitioned into ringlike segments.

annual plant A plant that completes its entire life cycle in one year or less. Compare with *perennial plant* and *biennial plant*.

Antarctic tundra Treeless biome of Antarctica Peninsula and nearby islands characterized by extremely cold temperatures and containing lichens, mosses, and few vascular plants.

antenna complex The arrangement of chlorophyll, accessory pigments, and pigment-binding proteins into light-gathering units in the thylakoid membranes of photoautotrophic eukaryotes. See *reaction center* and *photosystem*.

antennae (sing., *antenna*) Sensory structures characteristic of some arthropod groups.

anterior Toward the head end of a bilaterally symmetrical animal. Compare with *posterior*.

anterior pituitary The anterior, glandular lobe of the pituitary gland; functions as a classical endocrine gland; secretes growth hormone, prolactin, melanocyte-stimulating hormone, and several tropic hormones. Compare with *posterior pituitary*.

anther (an′-thur) The part of the stamen in flowers that produces microspores and, ultimately, pollen grains.

antheridium (an″-thur-id′-ee-im) (pl., *antheridia*) In plants, the multicellular male gametangium (sex organ) that produces sperm cells. Compare with *archegonium*.

anthocyanins Class of red water-soluble plant pigments.

anthropoid (an′-thra-poid) A member of a suborder of primates that includes monkeys, apes, and humans.

antibody (an′-tee-bod″-ee) A specific protein (immunoglobulin) that recognizes and binds to specific antigens; produced by plasma cells.

antibody-mediated immunity A type of specific immune response in which B cells differentiate into plasma cells and produce antibodies that bind with foreign antigens, leading to the destruction of pathogens. Compare with *cell-mediated immunity*.

anticodon (an″-tee-koh′-don) A sequence of three nucleotides in transfer RNA that is complementary to, and combines with, the three-nucleotide codon.

antidiuretic hormone (ADH) (an″-ty-dy-uh-ret′-ik) A hormone secreted by the posterior lobe of the pituitary that controls the rate of water reabsorption by the kidney.

antigen (an′-tih-jen) Any molecule, usually a protein or large carbohydrate, that is specifically recognized as foreign by cells of the immune system.

antigen–antibody complex The combination of antigen and antibody molecules.

antigen-presenting cell (APC) A cell that displays foreign antigens as well as its own surface proteins. Dendritic cells, macrophages, and B cells are APCs.

antimicrobial peptides Soluble molecules that destroy pathogens.

anti-oncogene See *tumor suppressor gene*.

antioxidants Certain enzymes (e.g., catalase and peroxidase), vitamins, and other substances that destroy free radicals and other reactive molecules. Compare with *oxidants*.

antiparallel Said of a double-stranded nucleic acid in which the 5′ to 3′ direction of the sugar–phosphate backbone of one strand is reversed in the other strand.

antiporter Membrane carrier protein that transports two types of substances in opposite directions. Compare with *uniporter* and *symporter*.

anus (ay′-nus) The distal end and outlet of the digestive tract.

aorta (ay-or′-tah) The largest and main systemic artery of the vertebrate body; arises from the left ventricle and branches to distribute blood to all parts of the body except the lungs.

aphotic region (ay-fote′-ik) The lower layer of the ocean (deeper than 100 m or so) where light does not penetrate.

apical dominance (ape′-ih-kl) The inhibition of axillary (lateral) buds by a shoot tip.

apical meristem (mehr′-ih-stem) An area of dividing tissue, located at the tip of a shoot or root, that gives rise to primary tissues; apical meristems cause an increase in the length of the plant body. Compare with *lateral meristems.*

apicomplexans A group of parasitic protists that lack structures for locomotion and that produce sporozoites as infective agents; malaria is caused by an apicomplexan.

apoenzyme (ap″-oh-en′-zime) Protein portion of an enzyme; requires the presence of a specific coenzyme to become a complete functional enzyme.

apomixis (ap″-uh-mix′-us) A type of reproduction in which fruits and seeds are formed asexually.

apoplast A continuum consisting of the interconnected, porous plant cell walls, along which water moves freely. Compare with *symplast.*

apoptosis (ap-uh-toe′-sis) Programmed cell death; apoptosis is a normal part of an organism's development and maintenance. Compare with *necrosis.*

aposematic coloration The conspicuous coloring of a poisonous or distasteful organism that enables potential predators to easily see and recognize it. Also called *warning coloration.* Compare with *cryptic coloration.*

aquaporin One of a family of transport proteins located in the plasma membrane that facilitate the rapid movement of water molecules into or out of cells.

aquifer Underground caverns and porous layers of rock in which groundwater is stored.

aquifer depletion Human removal of more groundwater than precipitation or melting snow can replace.

arachnids (ah-rack′-nidz) Eight-legged arthropods, such as spiders, scorpions, ticks, and mites.

arachnoid The middle of the three meningeal layers that cover and protect the brain and spinal cord; see *pia mater* and *dura mater.*

arboreal Living in trees.

arbuscules Tree-like branched structures inside root cells produced by hyphae of endomycorrhizal fungi known as glomeromycetes.

Archaea (ar′-key-ah) One of the two prokaryotic domains. The absence of peptidoglycan in their cell walls sets them apart from the bacteria. Compare with *bacteria.*

Archaean eon The period of Earth's history from the formation of the crust approximately 4 billion years ago, until 2.5 billion years ago; life originated during the Archaean.

archaeplastids A monophyletic supergroup of eukaryotes with chloroplasts bounded by outer and inner membranes; include red algae, green algae, and land plants.

archegonium (ar′-ke-go′-nee-um) (pl., *archegonia*) In plants, the multicellular female gametangium (sex organ) that contains an egg. Compare with *antheridium.*

archenteron (ark-en′-ter-on) The central cavity of the gastrula stage of embryonic development that is lined with endoderm; primitive digestive system.

archosaurs (ar′-kuh-sors) Lineage of diapsid vertebrates that includes the extinct dinosaurs and flying reptiles, as well as the extant crocodiles (order Crocodilia) and the birds.

arctic tundra See *tundra.*

Ardipithecus ramidus*.* See *australopithecines.*

arterial pulse See *pulse, arterial.*

arteriole (ar-teer′-ee-ole) A very small artery. Vasoconstriction and vasodilation of arterioles help regulate blood pressure.

artery A thick-walled blood vessel that carries blood away from a heart chamber and toward the body organs. Compare with *vein.*

arthropod (ar′-throh-pod) An invertebrate that belongs to phylum Arthropoda; characterized by a hard exoskeleton; a segmented body; and paired, jointed appendages.

artificial insemination The impregnation of a female by artificially introducing sperm from a male.

artificial selection The selection by humans of traits that are desirable in plants or animals and breeding only those individuals that have the desired traits. Compare with *natural selection.*

ascocarp (ass′-koh-karp) The fruiting body of an ascomycete.

ascomycete (ass″-koh-my′-seat) Member of a phylum of fungi characterized by the production of nonmotile asexual conidia and sexual ascospores.

ascospore (ass′-koh-spor) One of a set of sexual spores, usually eight, contained in a special spore case (an ascus) of an ascomycete.

ascus (ass′-kus) A saclike spore case in ascomycetes that contains sexual spores called *ascospores.*

asexual reproduction Reproduction in which there is no fusion of gametes and in which the genetic makeup of parent and of offspring is usually identical. Compare with *sexual reproduction.*

assimilation (of nitrogen) The conversion of inorganic nitrogen (nitrate, NO_3^-, or ammonia, NH_3) to the organic molecules of living things; part of the nitrogen cycle.

association areas Areas of the brain that link sensory and motor areas; responsible for thought, learning, memory, language abilities, judgment, and personality.

association neuron See *interneuron.*

assortative mating Sexual reproduction in which individuals pair nonrandomly, i.e., select mates on the basis of phenotype; selection may be positive (mates with the same phenotype) or less commonly negative (mates with opposite phenotypes).

asters Clusters of microtubules radiating out from the poles in dividing cells that have centrioles.

astrocyte A type of glial cell; some are phagocytic; others regulate the composition of the extracellular fluid in the central nervous system.

atherosclerosis (ath″-ur-oh-skle-row′-sis) A progressive disease in which lipid deposits accumulate in the inner lining of arteries, leading eventually to impaired circulation and heart disease.

atom The smallest quantity of an element that retains the chemical properties of that element.

atomic mass The total number of protons and neutrons in an atom; expressed in atomic mass units or daltons.

atomic mass unit (amu) The approximate mass of a proton or neutron; also called a *dalton.*

atomic number The number of protons in the atomic nucleus of an atom, which uniquely identifies the element to which the atom corresponds.

ATP See *adenosine triphosphate.*

ATP synthase Large enzyme complex that catalyzes the formation of ATP from ADP and inorganic phosphate by chemiosmosis; located in the inner mitochondrial membrane, the thylakoid membrane of chloroplasts, and the plasma membrane of prokaryotes.

atrial natriuretic peptide (ANP) A hormone released by the atrium of the heart; helps regulate sodium excretion and lowers blood pressure.

atrioventricular (AV) node (ay″-tree-oh-ven-trik′-you-lur) Mass of specialized cardiac tissue that receives an impulse from the sinoatrial node (pacemaker) and conducts it to the ventricles.

atrioventricular (AV) valve (of the heart) A valve between each atrium and its ventricle that prevents backflow of blood. The right AV valve is the tricuspid valve; the left AV valve is the mitral valve.

atrium (of the heart) (ay′-tree-um) A heart chamber that receives blood from the veins.

australopithecines Early hominids that lived between about 6 mya and 1 mya, based on fossil evidence. Include several species in three genera: *Ardipithecus, Australopithecus,* and *Paranthropus.*

Australopithecus afarensis See *australopithecines.*

Australopithecus africanus See *australopithecines.*

Australopithecus anamensis See *australopithecines.*

Australopithecus sediba See *australopithecines.*

autocrine regulation A type of regulation in which a signaling molecule (e.g., a hormone) is secreted into interstitial fluid and then acts on the cells that produce it. Compare with *paracrine regulation.*

autoimmune disease (aw″-toh-ih-mune′) A disease in which the body produces antibodies against its own cells or tissues. Also called *autoimmunity.*

autonomic division (of the PNS) (aw-tuh-nom′-ik) The portion of the peripheral nervous system that controls the visceral functions of the body, e.g., regulates smooth muscle, cardiac muscle, and glands. Its divisions are the sympathetic and parasympathetic systems. Compare with *somatic division* of the PNS.

autopolyploid A polyploid whose chromosomes are derived from a single species. Compare with *allopolyploid*.

autoradiography Method for detecting radioactive decay; radiation causes the appearance of dark silver grains in special X-ray film.

autosome (aw′-toh-sohm) A chromosome other than the sex (X and Y) chromosomes.

autotroph (aw′-toh-trof) An organism that synthesizes complex organic compounds from simple inorganic raw materials; also called *producer* or *primary producer*. Compare with *heterotroph*. See *chemoautotroph* and *photoautotroph*.

auxin (awk′-sin) A plant hormone involved in various aspects of growth and development, such as stem elongation, apical dominance, and root formation on cuttings, e.g., indole acetic acid (IAA).

avirulence Properties that render an infectious agent nonlethal, i.e., unable to cause disease in its host. Compare with *virulence*.

Avogadro's number The number of units (6.02 × 10^{23}) present in one mole of any substance.

axillary bud A bud in the axil of a leaf; also called *lateral bud*. Compare with *terminal bud*.

axon (aks′-on) The long extension of the neuron that transmits nerve impulses away from the cell body. Compare with *dendrite*.

axopods (aks′-o-podz) Long, filamentous, cytoplasmic projections characteristic of actinopods.

B cell (B lymphocyte) The type of white blood cell responsible for antibody-mediated immunity. When stimulated, B cells differentiate to become plasma cells that produce antibodies. Compare with *T cell*.

bacillus (bah-sill′-us) (pl., *bacilli*) A rod-shaped bacterium. Compare with *coccus, spirillum, vibrio*, and *spirochete*.

background extinction The continuous, low-level extinction of species that has occurred throughout much of the history of life. Compare with *mass extinction*.

bacteria (bak-teer′-ee-ah) Prokaryotic organisms that have peptidoglycan in their cell walls; most are decomposers, but some are parasites and others are autotrophs. Bacteria is the name of one of the two prokaryotic domains. Compare with *archaea*.

bacterial artificial chromosome (BAC) Genetically engineered segments of chromosomal DNA with the ability to carry large segments of foreign DNA.

bacteriophage (bak-teer′-ee-oh-fayj) A virus that infects a bacterium (literally, "bacteria eater"). Also called *phage*.

balanced polymorphism (pol″-ee-mor′-fizm) The presence in a population of two or more genetic variants that are maintained in a stable frequency over several generations.

bark The outermost covering over woody stems and roots; consists of all plant tissues located outside the vascular cambium.

baroreceptors (bare″-oh-ree-sep′-torz) Receptors within certain blood vessels that are stimulated by changes in blood pressure.

Barr body A condensed and inactivated X chromosome appearing as a distinctive dense spot in the interphase nucleus of certain cells of female mammals.

basal angiosperms (bay′-sl) Three clades of angiosperms that are thought to be ancestral to all other flowering plants. Compare with *core angiosperms*.

basal body Structure involved in the organization and anchorage of a cilium or flagellum. Structurally similar to a centriole; each is in the form of a cylinder composed of nine triplets of microtubules (9 × 3 structure).

basal metabolic rate (BMR) The amount of energy expended by the body at resting conditions, when no food is being digested and no voluntary muscular work is being performed.

base (1) A substance that is a hydrogen ion (proton) acceptor; bases unite with acids to form salts. Compare with *acid*. (2) A nitrogenous base in a nucleotide or nucleic acid. See *purines* and *pyrimidines*.

basement membrane The thin, noncell layer of an epithelial membrane that attaches to the underlying tissues; consists of tiny fibers and polysaccharides produced by the epithelial cells.

base-pair substitution mutation A change in one base pair in DNA. See *missense mutation* and *nonsense mutation*.

basic solution A solution in which the concentration of hydroxide ions [OH⁻] exceeds the concentration of hydrogen ions [H⁺]. A basic solution has pH greater than 7. Compare with *acidic solution* and *neutral solution*.

basidiocarp (ba-sid′-ee-o-karp) The fruiting body of a basidiomycete, e.g., a mushroom.

basidiomycete (ba-sid″-ee-o-my′-seat) Member of a phylum of fungi characterized by the production of sexual basidiospores.

basidiospore (ba-sid′-ee-o-spor) One of a set of sexual spores, usually four, borne on a basidium of a basidiomycete.

basidium (ba-sid′-ee-um) The clublike spore-producing organ of basidiomycetes that bears sexual spores called *basidiospores*.

basilar membrane The multicellular tissue in the inner ear that separates the cochlear duct from the tympanic canal; the sensory cells of the organ of Corti rest on this membrane.

Batesian mimicry (bate′-see-un mim′-ih-kree) The resemblance of a harmless or palatable species to one that is dangerous, unpalatable, or poisonous so that predators are more likely to avoid them. Compare with *Müllerian mimicry*.

B-cell receptors The receptors on B cells that recognize specific antigens.

behavior What an animal does and how the animal does it, usually in response to stimuli in the environment.

behavioral ecology The scientific study of behavior in natural environments from the evolutionary perspective.

behavioral isolation A prezygotic reproductive isolating mechanism in which reproduction between similar species is prevented because each group exhibits its own characteristic courtship behavior; also called *sexual isolation*.

behavior pattern A fixed behavior that, once activated, continues to completion regardless of sensory feedback.

benthic environment Ocean floor divided into zones based on distance from land, light availability, and depth.

benthos (ben′-thos) Bottom-dwelling sea organisms that fix themselves to one spot, burrow into the sediment, or simply walk about on the ocean floor.

berry A simple, fleshy fruit in which the fruit wall is soft throughout, e.g., tomato, banana, grape.

beta cells Pancreatic cells that secrete insulin.

beta (β) oxidation Process by which fatty acids are converted to acetyl CoA before entry into the citric acid cycle.

beta (β)-pleated sheet A regular, folded, sheetlike type of protein secondary structure, resulting from hydrogen bonding between two different polypeptide chains or two regions of the same polypeptide chain. Compare with *alpha (α) helix*.

biennial plant (by-en′-ee-ul) A plant that takes two years to complete its life cycle. Compare with *annual plant* and *perennial plant*.

bikonts One of two main clades of all eukaryotes; had a common ancestor with two flagella. Compare with *unikonts*.

bilateral symmetry A body shape with right and left halves that are approximately mirror images of each other. Compare with *radial symmetry*.

bile The fluid secreted by the liver; emulsifies fats.

binary fission (by′-nare-ee fish′-un) Equal division of a prokaryotic cell into two; a type of asexual reproduction.

binomial system of nomenclature (by-nome′-ee-ul) System of naming a species by the combination of the genus name and a specific epithet.

bioaccumulation The buildup of a persistent toxic substance, such as certain pesticides, in an organism's body.

biodiversity See *biological diversity*.

biodiversity hotspot Biogeographic region that is a significant reservoir of endemic biodiversity and is threatened with continued habitat destruction.

biofilm An irregular layer of microorganisms embedded in the slime they secrete and concentrated at a solid or liquid surface.

biogenic amines A class of neurotransmitters that includes norepinephrine, serotonin, and dopamine.

biogeochemical cycle (bye″-o-jee″-o-kem′-ee-kl) Process by which matter cycles from the living world to the nonliving, physical environment and back again, e.g., the carbon cycle, the nitrogen cycle, and the phosphorus cycle.

biogeography The study of the past and present geographic distributions of organisms.

bioinformatics The storage, retrieval, and comparison of biological information, particularly DNA or protein sequences within a given species and among different species.

biological clocks Mechanisms by which activities of organisms are adapted to regularly recurring changes in the environment. See *circadian rhythm.*

biological diversity The variety of living organisms considered at three levels: genetic diversity, species diversity, and ecosystem diversity. Also called *biodiversity.*

biological magnification The increased concentration of toxic chemicals, such as PCBs, heavy metals, and certain pesticides, in the tissues of organisms at higher trophic levels in food webs.

biological species concept See *species.*

biomass (bye'-o-mas) A quantitative estimate of the total mass, or amount, of living material in a particular ecosystem.

biome (by'-ohm) A large, relatively distinct terrestrial region characterized by a similar climate, soil, plants, and animals, regardless of where it occurs on Earth.

bioremediation A method to clean up a hazardous waste site that uses microorganisms to break down toxic pollutants, or plants to selectively accumulate toxins.

biosphere All Earth's communities of living organisms and their physical environments.

biosphere reserves Zones designed and set aside to protect biodiversity, promote sustainable development, and foster research and conservation education.

bioterrorism The intentional use of biological agents, such as microorganisms or toxins derived from living organisms, to cause death or disease.

biotic factors Elements of the living world that affect a particular organism, i.e., its relationships with other organisms. Compare with *abiotic factors.*

biotic pollution Introduction of a foreign species into an area where it is not native.

biotic potential See *intrinsic rate of increase.*

bipedal Walking on two feet. Compare with *quadripedal.*

bipolar cell A type of neuron in the retina of the eye; receives input from the photoreceptors (rods and cones) and synapses on ganglion cells.

biramous appendages Appendages with two jointed branches extending from their base. Compare with *uniramous appendages.*

blade (1) The thin, expanded part of a leaf. (2) The flat, leaflike structure of certain multicellular algae.

blastocoel (blas'-toh-seel) The fluid-filled cavity of a blastula.

blastocyst The mammalian blastula. See *blastula.*

blastodisc A small disc of cytoplasm at the animal pole of a reptile or bird egg; cleavage is restricted to the blastodisc (meroblastic cleavage).

blastomere A cell of an early embryo.

blastopore (blas'-toh-pore) The primitive opening into the body cavity of an early embryo that may become the mouth (in protostomes) or anus (in deuterostomes) of the adult organism.

blastula (blas'-tew-lah) In animal development, a hollow ball of cells produced by cleavage of a fertilized ovum. In mammalian development, known as a *blastocyst.*

blood A fluid, circulating connective tissue that transports nutrients and other materials through the bodies of many types of animals.

blood pressure The force exerted by blood against the inner walls of the blood vessels.

bloom The sporadic occurrence of huge numbers of algae in freshwater and marine ecosystems.

body mass index (BMI) An index of weight in relation to height; calculated by dividing the square of the weight (square kilograms) by height (meters).

Bohr effect Increased oxyhemoglobin dissociation due to lowered pH; occurs as carbon dioxide concentration increases.

bolting The production of a tall flower stalk by a plant that grows vegetatively as a rosette (growth habit with a short stem and a circular cluster of leaves).

bond energy The energy required to break a particular chemical bond.

bone Principal vertebrate skeletal tissue; a type of connective tissue that consists of cells (osteocytes) embedded in a hard matrix of collagen fibers and minerals, including calcium and phosphate.

bone marrow The soft, cellular tissue that fills the internal spaces of bones; red bone marrow in certain bones produces blood cells.

boreal forest (bor'-ee-uhl) The northern coniferous forest biome found primarily in Canada, northern Europe, and Siberia; also called *taiga.*

bottleneck A sudden decrease in a population size caused by adverse environmental factors; may result in genetic drift; also called *genetic bottleneck* or *population bottleneck.*

bottom-up processes Control of ecosystem function by nutrient cycles and other parts of the abiotic environment. Compare with *top-down processes.*

Bowman's capsule A double-walled sac of cells that surrounds the glomerulus of each nephron.

brachiopods (bray'-kee-oh-podz) The phylum of solitary marine invertebrates having a pair of shells and, internally, a pair of coiled arms with ciliated tentacles; one of the lophophorate phyla.

bracts Modified leaves associated with flower clusters; may be showy and petal-like.

brain A concentration of nervous tissue that controls neural function; in vertebrates, the anterior, enlarged portion of the central nervous system.

brain stem The part of the vertebrate brain that includes the medulla, pons, and midbrain.

branchial Pertaining to the gills or gill region.

brassinosteroid (BR) One of a group of steroids that function as plant hormones and are involved in several aspects of growth and development.

Broca's area Located near motor areas in left frontal lobe; controls our ability to speak; important in language processing and speech comprehension.

bronchiole (bronk'-ee-ole) Air duct in the lung that branches from a bronchus; divides to form air sacs (alveoli).

bronchus (bronk'-us) (pl., *bronchi*) One branch of the trachea and its immediate branches within the lung.

brown alga One of a group of predominantly marine algae that are multicellular and contain the pigments chlorophyll *a* and *c*, and carotenoids, including fucoxanthin.

bryophytes (bry'-oh-fites) Nonvascular plants including mosses, liverworts, and hornworts.

bryozoans Animals belonging to phylum Bryozoa, one of the three lophophorate phyla; form sessile colonies by asexual budding.

bud An undeveloped shoot that develops into flowers, stems, or leaves. Buds are enclosed in bud scales; see *axillary bud* and *terminal bud.*

budding Asexual reproduction in which a small part of the parent's body separates from the rest and develops into a new individual; characteristic of yeasts and certain other organisms.

bud scale A modified leaf that covers and protects a dormant bud.

bud scale scar Scar on a twig left when a bud scale abscises from the terminal bud.

buffer A substance in a solution that tends to lessen the change in hydrogen ion concentration (pH) that otherwise would be produced by adding an acid or base.

bulb A globose, fleshy, underground bud that consists of a short stem with fleshy leaves, e.g., onion.

bundle scar Marks on a leaf scar left when vascular bundles of the petiole break during leaf abscission.

bundle sheath Group of tightly packed cells that form a sheath around the veins of a leaf.

bundle sheath extension Group of support cells that extend from the bundle sheath of a leaf vein toward the upper and/or lower epidermis.

buttress root A bracelike root at the base of certain trees that provides upright support.

C_3 pathway See *C_3 plant.*

C_3 plant Plant that carries out carbon fixation solely by the Calvin cycle. Compare with *C_4 plant* and *CAM plant.*

C_4 pathway See *C_4 plant.*

C_4 plant Plant that fixes carbon initially by a pathway in which the reaction of CO_2 with phosphoenolpyruvate is catalyzed by PEP carboxylase in leaf mesophyll cells; the products are transferred to the bundle sheath cells, where the Calvin cycle takes place. Compare with *C₃ plant* and *CAM plant*.

cadherins Transmembrane proteins that are components of adhering junctions between animal cells.

calcitonin (kal-sih-toh′-nin) A hormone secreted by the thyroid gland that rapidly lowers the calcium content in the blood.

calls Short, simple sounds animals use to communicate.

callus (kal′-us) Undifferentiated tissue formed on an explant (excised tissue or organ) in plant tissue culture.

calmodulin A calcium-binding protein; when bound, it alters the activity of certain enzymes or transport proteins.

calorie The amount of heat energy required to raise the temperature of 1 g of water 1°C; equivalent to 4.184 joules. Compare with *kilocalorie*.

Calvin cycle Cyclic series of reactions in the chloroplast stroma in photosynthesis; fixes carbon dioxide and produces carbohydrate. See *C₃ plant*.

calyx (kay′-liks) The collective term for the sepals of a flower.

cambium See *lateral meristems*.

Cambrian explosion See *Paleozoic era*.

Cambrian period See *Paleozoic era*.

Cambrian radiation See *Paleozoic era*.

cAMP See *cyclic AMP*.

CAM plant Plant that carries out crassulacean acid metabolism; carbon is initially fixed into organic acids at night in the reaction of CO_2 and phosphoenolpyruvate, catalyzed by PEP carboxylase; during the day the acids break down to yield CO_2, which enters the Calvin cycle. Compare with *C₃ plant* and *C₄ plant*.

cGMP See *cyclic GMP*.

cancer cells See *malignant cells*.

cancer driver gene A gene that, when its expression is altered, either by a particular mutation or by certain epigenetic changes, gives a cell a selective growth advantage. See *oncogene*, *proto-oncogene*, and *tumor suppressor gene*.

canines The long pointed teeth of a mammal adapted for piercing prey and tearing food. Compare with *incisors*, *premolars*, and *molars*.

CAP See *catabolite activator protein*.

capacitation Maturation process that enables mammalian sperm to participate in fertilization; takes place in the female reproductive tract.

capillaries (kap′-i-lare-eez) Microscopic blood vessels in the tissues that permit exchange of materials between cells and blood.

capillary action The ability of water to move in small-diameter tubes as a consequence of its cohesive and adhesive properties.

capping See *mRNA cap*.

capsid Protein coat surrounding the nucleic acid of a virus.

capsule (1) The portion of the moss sporophyte that contains spores. (2) A simple, dry, dehiscent fruit that develops from two or more fused carpels and opens along many sutures or pores to release seeds. (3) A gelatinous coat that surrounds some bacteria.

carbohydrate Compound containing carbon, hydrogen, and oxygen, in the approximate ratio of C:2H:O, e.g., sugars, starch, and cellulose.

carbon cycle The worldwide circulation of carbon from the abiotic environment into living things and back into the abiotic environment.

carbon fixation reactions Reduction reactions of photosynthesis in which carbon from carbon dioxide becomes incorporated into organic molecules, leading to the production of carbohydrate. See *Calvin cycle*.

Carboniferous period See *Paleozoic era*.

carbonyl group A polar functional group consisting of a carbon attached to an oxygen by a double bond; found in aldehydes and ketones.

carboxyl group A weakly acidic functional group; abbreviated —COOH.

carcinogen (kar-sin′-oh-jen) An agent that causes cancer or accelerates its development.

cardiac cycle One complete heartbeat.

cardiac muscle Involuntary, striated type of muscle found in the vertebrate heart. Compare with *smooth muscle* and *skeletal muscle*.

cardiac output The volume of blood pumped by the left ventricle into the aorta in 1 minute.

cardiovascular disease Disease of the heart or blood vessels; the leading cause of death in most industrial societies.

cardiovascular system The closed circulatory system of vertebrates that transports oxygen, nutrients, hormones, and other substances, and protects the body against disease; consists of heart and blood vessels.

carnivore (kar′-ni-vor) An animal that mainly feeds on other animals; a mammal belonging to order Carnivora, e.g., wolves or seals.

carotenoids (ka-rot′-n-oidz) A group of yellow to orange plant pigments synthesized from isoprene subunits; include carotenes and xanthophylls.

carpel (kar′-pul) The female reproductive unit of a flower; carpels bear ovules. Compare with *pistil*.

carrier-mediated active transport Transport across a membrane of a substance from a region of low concentration to a region of high concentration; requires both a transport protein with a binding site for the specific substance and an energy source (often ATP).

carrier-mediated transport Any form of transport across a membrane that uses a membrane-bound transport protein with a binding site for a specific substance; includes both facilitated diffusion and carrier-mediated active transport.

carrying capacity (K) The largest population that a particular habitat can support and sustain for an indefinite period, assuming there are no changes in the environment.

cartilage A flexible skeletal tissue of vertebrates; a type of connective tissue.

caryopsis (pl., *caryopses*) See *grain*.

Casparian strip (kas-pare′-ee-un) A band of waterproof material around the radial and transverse walls of endodermal root cells.

caspase Any of a group of proteolytic enzymes that are active in the early stages of apoptosis.

catabolism The aspect of metabolism in which complex substances are broken down to form simpler substances; catabolic reactions are particularly important in releasing chemical energy stored by the cell. Compare with *anabolism*.

catabolite activator protein (CAP) A positively acting regulator that becomes active when bound to cAMP; active CAP stimulates transcription of the *lac* operon and other operons that code for enzymes used in catabolic pathways.

catalyst (kat′-ah-list) A substance that increases the speed at which a chemical reaction occurs without being used up in the reaction. Enzymes are biological catalysts.

catecholamine (cat″-eh-kole′-ah-meen) A class of compounds including dopamine, epinephrine, and norepinephrine; these compounds serve as neurotransmitters and hormones.

cation A particle with one or more units of positive charge, such as a hydrogen ion (H^+) or calcium ion (Ca^{2+}). Compare with *anion*.

cation exchange Process in which root cells secrete protons (H^+), which replace cations adhering to soil particles, thereby releasing them for absorption by the roots.

cecum (see′-kum) The blind pouch formed at the junction of the small and large intestine.

CD4 T cell See *T helper cell*.

CD8 T cell See *T cytotoxic cell*.

cDNA library A collection of recombinant plasmids that contain complementary DNA (cDNA) copies of mRNA templates. The cDNA, which lacks introns, is synthesized by reverse transcriptase. Compare with *genomic DNA library*.

cell The basic structural and functional unit of life, which consists of living material enclosed by a membrane.

cell body The large portion of a neuron that contains most of the cytoplasm, the nucleus, and most of the other organelles.

cell cycle Cyclic series of events in the life of a dividing eukaryotic cell; consists of mitosis, cytokinesis, and the stages of interphase.

cell-cycle checkpoints See *cell-cycle control system*.

cell-cycle control system Regulatory molecules that control key events (cell-cycle checkpoints) in the cell cycle; common to all eukaryotes.

cell determination See *determination*.

cell differentiation See *differentiation*.

cell fractionation The technique used to separate the components of cells by subjecting them to centrifugal force. See *differential centrifugation* and *density gradient centrifugation.*

cell-mediated immunity A type of specific immune response carried out by T cells. Compare with *antibody-mediated immunity.*

cell plate The structure that forms during cytokinesis in plants, separating the two daughter cells produced by mitosis.

cell signaling Mechanisms of communication between cells. Cells signal one another with secreted signaling molecules, or a signaling molecule on one cell combines with a receptor on another cell. See *signal transduction.*

cell theory The scientific theory that the cell is the basic unit of life, of which all living things are composed, and that all cells are derived from pre-existing cells.

cellular respiration See *respiration.*

cellular slime mold A group of funguslike amoebozoan protists whose feeding stage consists of unicellular, amoeboid organisms that aggregate to form a pseudoplasmodium during reproduction; compare with *plasmodial slime mold.*

cellulose (sel′-yoo-lohs) A structural polysaccharide consisting of beta glucose subunits; the main constituent of plant primary cell walls.

cell wall The structure outside the plasma membrane of certain cells; may contain cellulose (plant cells), chitin (most fungal cells), peptidoglycan and/or lipopolysaccharide (most bacterial cells), or other material.

Cenozoic era Span of geologic time extending from 66 mya to the present; marked by diversification of mammals, birds, insects, and flowering plants. Divided into Paleogene (66–23 mya), Neogene (23–2.6 mya), and Quaternary (2.6 mya to present) periods.

center of origin The geographic area where a given species originated.

central nervous system (CNS) In vertebrates, the brain and spinal cord. Compare with *peripheral nervous system.*

centrifuge A device used to separate cells or their components by subjecting them to centrifugal force.

centriole (sen′-tree-ohl) One of a pair of small, cylindrical organelles lying at right angles to each other near the nucleus in the cytoplasm of animal cells and certain protist and plant cells; each centriole is in the form of a cylinder composed of nine triplets of microtubules (9 × 3 structure).

centromere (sen′-tro-meer) A specialized constricted region of a chromatid; contains the kinetochore. In cells at prophase and metaphase, sister chromatids are joined in the vicinity of their centromeres.

centrosome (sen′-tro-sowm) An organelle in animal cells that is the main microtubule-organizing center; typically contains a pair of centrioles and is important in cell division.

cephalization The evolution of a head; the concentration of nervous tissue and sense organs at the front end of the animal.

cephalochordates Members of the chordate subphylum that includes the lancelets.

cerebellum (ser-eh-bel′-um) A convoluted subdivision of the vertebrate brain concerned with coordination of muscular movements, muscle tone, and balance.

cerebral cortex (ser-ee′-brul kor′-teks) The outer layer of the cerebrum composed of gray matter and consisting mainly of nerve cell bodies.

cerebrospinal fluid (CSF) The fluid that bathes the central nervous system of vertebrates.

cerebrum (ser-ee′-brum) A large, convoluted subdivision of the vertebrate brain; in humans, it functions as the center for learning, voluntary movement, and interpretation of sensation.

cervix The lower region of the uterus; extends slightly into the vagina.

chaparral (shap″-uh-ral′) A biome with a Mediterranean climate (mild, moist winters and hot, dry summers). Chaparral vegetation is characterized by drought-resistant, small-leaved evergreen shrubs and small trees.

chaperones See *molecular chaperones.*

character An attribute or characteristic of an organism.

character displacement The tendency for two similar species to diverge (become more different) in areas where their ranges overlap; reduces interspecific competition.

Chargaff's rules A relationship in DNA molecules based on nucleotide composition data; the number of adenines equals the number of thymines, and the number of guanines equals the number of cytosines.

charophytes Group of green algae closely related to land plants.

chelicerae (keh-lis′-er-ee) The first pair of appendages in certain arthropods; clawlike appendages located immediately anterior to the mouth and used to manipulate food into the mouth.

chemical bond A force of attraction between atoms in a compound. See *covalent bond, hydrogen bond,* and *ionic bond.*

chemical compound Two or more elements combined in a fixed ratio.

chemical evolution The origin of life from nonliving matter.

chemical formula A representation of the composition of a compound; the elements are indicated by chemical symbols with subscripts to indicate their ratios. See *molecular formula, structural formula,* and *simplest formula.*

chemical symbol The abbreviation for an element; usually the first letter (or first and second letters) of the English or Latin name.

chemical synapse A junction between two neurons in which an action potential in the presynaptic neuron triggers release of a neurotransmitter that crosses the gap and binds to a receptor on the postsynaptic neuron; specific gated ion channels open or close which changes the permeability of the postsynaptic neuron. When threshold level is reached, the postsynaptic neuron transmits an action potential. Compare with *electrical synapse.*

chemiosmosis Process by which phosphorylation of ADP to form ATP is coupled to the transfer of electrons down an electron transport chain; the electron transport chain powers proton pumps that produce a proton gradient across the membrane; ATP is formed as protons diffuse through transmembrane channels in ATP synthase.

chemoautotroph (kee″-moh-aw′-toh-trof) Organism that obtains energy from inorganic compounds and synthesizes organic compounds from inorganic raw materials; includes some bacteria and many archaea. Compare with *photoautotroph, photoheterotroph,* and *chemoheterotroph.*

chemoheterotroph (kee″-moh-het′-ur-oh-trof) Organism that uses organic compounds as a source of energy and carbon; includes animals, fungi, many bacteria, and a few archaea. Compare with *photoautotroph, photoheterotroph,* and *chemoautotroph.*

chemoreceptor (kee″-moh-ree-sep′-tor) A sensory receptor that responds to chemical stimuli.

chemotaxis Movement of a cell or organism in response to certain chemicals.

chemotroph (kee′-moh-trof) Organism that uses organic compounds or inorganic substances, such as iron, nitrate, ammonia, or sulfur, as sources of energy. Compare with *phototroph.* See *chemoautotroph* and *chemoheterotroph.*

chiasma (ky-az′-muh) (pl., *chiasmata*) An X-shaped site in a tetrad usually marking the location where homologous (nonsister) chromatids previously crossed over.

chimera (ky-meer′-uh) An organism consisting of two or more kinds of genetically dissimilar cells.

chitin (ky′-tin) A nitrogen-containing structural polysaccharide that forms the exoskeleton of insects and the cell walls of many fungi.

chlorophyll (klor′-oh-fil) A group of light-trapping green pigments found in most photosynthetic organisms.

chlorophyll *a* See *chlorophyll.*

chlorophyll *b* See *chlorophyll.*

chlorophyll-binding proteins About 15 different proteins associated with chlorophyll molecules in the thylakoid membrane.

chloroplasts (klor′-oh-plazts) Membranous organelles that are the sites of photosynthesis in eukaryotes; occur in some plant and algal cells.

choanocyte (ko-an′-uh-site) A collared, flagellate cell that traps and phagocytizes food particles; found in sponges; also called a *collar cell.*

choanoflagellates Collared flagellate protists; opisthokonts closely related to animals.

cholinergic neuron (kohl″-in-air′-jik) A neuron that releases acetylcholine as a neurotransmitter. Compare with *adrenergic neuron.*

chondrichthyes (kon-drik′-thees) The class of cartilaginous fishes that includes the sharks, rays, and skates.

chondrocytes Cartilage cells.

chordates (kor′-dates) Deuterostome animals that, at some time in their lives, have a cartilaginous, dorsal skeletal structure called a *notochord;* a dorsal, tubular nerve cord; *pharyngeal slits;* a postanal tail; and an *endostyle* (or its derivative, a thyroid gland).

chorion (kor-′ee-on) An extraembryonic membrane in reptiles, birds, and mammals that forms an outer cover around the embryo and in mammals contributes to the formation of the placenta.

chorionic villus sampling (CVS) (kor″ee-on′ik) Study of extraembryonic cells that are genetically identical to the cells of an embryo, making it possible to assess its genetic makeup. Compare with *amniocentesis.*

choroid layer A layer of cells filled with black pigment that absorbs light and prevents reflected light from blurring the image that falls on the retina; the layer of the eyeball outside the retina.

chromalveolates A supergroup composed of diverse protists with few shared characters; most are photosynthetic, and heterotrophic chromalveolates, such as the water molds and ciliates, probably descended from autotrophic ancestors. Divided into two main groups; see *alveolates* and *stramenopiles.*

chromatid (kroh′-mah-tid) One of the two identical halves of a duplicated chromosome; the two chromatids that make up a chromosome are referred to as *sister chromatids.*

chromatin (kro′-mah-tin) The complex of DNA and protein that makes up eukaryotic chromosomes.

chromoplasts Pigment-containing plastids; found mainly in flowers and fruits.

chromosome theory of inheritance A basic principle in biology that states that inheritance can be explained by assuming that genes are linearly arranged in specific locations along the chromosomes.

chromosomes Structures in the cell nucleus that consist of chromatin and contain the genes. The chromosomes become visible under the microscope as distinct structures during cell division.

chylomicrons (kie-low-my′-kronz) Protein-covered fat droplets produced in the intestinal cells; they enter the lymphatic system and are transported to the blood.

chytrid See *chytridiomycete.*

chytridiomycete (ki-trid″-ee-o-my′-seat) A member of a phylum of fungi characterized by the production of flagellate cells at some stage in their life history. Also called *chytrid.*

ciliate (sil′-e-ate) A unicellular protist covered by many short cilia.

cilium (sil′-ee-um) (pl., *cilia*) One of many short, hairlike structures that project from the surface of some eukaryotic cells and are used for locomotion or movement of materials across the cell surface.

circadian rhythm (sir-kay′-dee-un) An internal rhythm that approximates the 24-hour day. See *biological clocks.*

circulatory system The body system that functions in internal transport and protects the body from disease.

cisternae (sing., *cisterna*) Stacks of flattened membranous sacs that make up the Golgi complex.

citrate (citric acid) A six-carbon organic acid.

citric acid cycle Series of chemical reactions in aerobic cellular respiration in which acetyl coenzyme A is completely degraded to carbon dioxide and water with the release of metabolic energy that is used to produce ATP; also known as the *Krebs cycle* and the *tricarboxylic acid (TCA) cycle.*

clade A group of organisms containing a common ancestor and all its descendants; a monophyletic group.

cladistics An approach to classification based on recency of common ancestry rather than degree of structural similarity. Also called *phylogenetic systematics.* Compare with *phenetics* and *evolutionary systematics.*

cladogram A branching diagram that illustrates taxonomic relationships based on the principles of cladistics.

class A taxonomic category made up of related orders.

classical conditioning A type of learning in which an association is formed between some normal response to a stimulus and a new stimulus, after which the new stimulus elicits the response.

cleavage Series of mitotic cell divisions, without growth, that converts the zygote to a multicellular blastula.

cleavage furrow A constricted region of the cytoplasm that forms and progressively deepens during cytokinesis of animal cells, thereby separating the two daughter cells.

climate Average weather conditions, plus extremes (records), that occur in a given place over a period of years.

climate change Change in Earth's climate patterns; today mostly caused by increased levels of atmospheric carbon dioxide produced by use of fossil fuels.

cline Gradual change in phenotype and genotype frequencies among contiguous populations that is the result of an environmental gradient.

clitoris (klit′-o-ris) A small, erectile structure at the anterior part of the vulva in female mammals; homologous to the male penis.

cloaca (klow-a′-ka) An exit chamber in some animals that receives digestive wastes and urine; may also serve as an exit for gametes.

clonal expansion The increase in number of T cells or B cells specific for an antigen; occurs when specific T cells or B cells are activated by the antigen.

clonal selection Lymphocyte activation in which a specific antigen causes activation, cell division, and differentiation only in cells that express receptors with which the antigen binds.

clone (1) A population of cells descended by mitotic division from a single ancestral cell. (2) A population of genetically identical organisms asexually propagated from a single individual. Also see *DNA cloning.*

cloning The process of forming a clone.

closed circulatory system A type of circulatory system in which the blood flows through a continuous circuit of blood vessels; characteristic of annelids, cephalopods, and vertebrates. Compare with *open circulatory system.*

closed system An entity that does not exchange energy with its surroundings. Compare with *open system.*

club mosses A phylum of seedless vascular plants with a life cycle similar to that of ferns.

clumped dispersion The spatial distribution pattern of a population in which individuals are more concentrated in specific parts of the habitat. Also called *aggregated distribution* and *patchiness.* Compare with *random dispersion* and *uniform dispersion.*

cnidarians (ni-dah′-ree-anz) Phylum of animals that have stinging cells called *cnidocytes,* two tissue layers, and radial symmetry; include hydras and jellyfish.

cnidocytes Stinging cells characteristic of cnidarians.

coated pit A depression in the plasma membrane, the cytosolic side of which is coated with the protein clathrin; important in receptor-mediated endocytosis.

coccus (kok′-us) (pl., *cocci*) A bacterium with a spherical shape. Compare with *bacillus, spirillum, vibrio,* and *spirochete.*

cochlea (koke′-lee-ah) The structure of the inner ear of mammals that contains the auditory receptors (organ of Corti).

codominance (koh″-dom′-in-ants) Condition in which two alleles of a locus are expressed in a heterozygote.

codon (koh′-don) A triplet of mRNA nucleotides. The 64 possible codons collectively constitute a universal genetic code in which each codon specifies an amino acid in a polypeptide, or a signal to either start or terminate polypeptide synthesis.

coelacanths A genus of lobe-finned fishes that has survived to the present day.

coelom (see′-lum) The main body cavity of most animals; a true coelom is lined with mesoderm. Compare with *pseudocoelom.*

coelomate (seel′-oh-mate) Animal that has a true coelom. Compare with *acoelomate* and *pseudocoelomate.*

coenocyte (see'-no-site) An organism consisting of a multinucleate cell, i.e., the nuclei are not separated from one another by septa.

coenzyme (koh-en'-zime) An organic cofactor for an enzyme; generally participates in the reaction by transferring some component, such as electrons or part of a substrate molecule.

coenzyme A (CoA) Organic cofactor responsible for transferring groups derived from organic acids.

coevolution The reciprocal adaptation of two or more species that occurs as a result of their close interactions over a long period.

cofactor A nonprotein substance needed by an enzyme for normal activity; some cofactors are inorganic (usually metal ions); others are organic (coenzymes).

cognition The mental processes involved in gaining and using knowledge; cognition is learning and knowing, including awareness, thinking, processing information, reasoning, and awareness of thoughts, perceptions, and self.

cohesin Ring-shaped protein complex responsible for linking sister chromatids during prophase.

cohesion The property of sticking together. Compare with *adhesion*.

cohort A group of individuals of the same age.

colchicine A drug that blocks the division of eukaryotic cells by binding to tubulin subunits, which make up the microtubules, the major component of the mitotic spindle.

coleoptile (kol-ee-op'-tile) A protective sheath that encloses the young stem in certain monocots.

collagens (kol'-ah-gen) Proteins found in the collagen fibers of connective tissues.

collar cell See *choanocyte.*

collecting duct A tube in the kidney that receives filtrate from several nephrons and conducts it to the renal pelvis.

collenchyma (kol-en'-kih-mah) Living cells with moderately but unevenly thickened primary cell walls; collenchyma cells help support the herbaceous plant body.

colon The region of the large intestine that extends from the cecum to the rectum.

colony An association of loosely connected cells or individuals of the same species.

commensalism (kuh-men'-sul-izm) A type of symbiosis in which one organism benefits and the other one is neither harmed nor helped. Compare with *mutualism* and *parasitism.*

commercial harvest The collection of commercially important organisms from the wild. Examples include the commercial harvest of parrots (for the pet trade) and cacti (for houseplants).

community An association of populations of different species living together in a defined habitat with some degree of interdependence. Compare with *ecosystem.*

community ecology The description and analysis of patterns and processes within the community.

compact bone Dense, hard bone tissue found mainly near the surfaces of a bone.

companion cell A cell in the phloem of flowering plants that governs loading and unloading sugar into the sieve tube element for translocation.

compass sense The sense of direction an animal requires to travel in a straight line toward a destination.

competition The interaction among two or more individuals that attempt to use the same essential resource, such as food, water, sunlight, or living space. See *interspecific* and *intraspecific competition.* See *interference* and *exploitation competition.*

competitive exclusion principle The concept that no two species with identical living requirements can occupy the same ecological niche indefinitely.

competitive inhibition Binding of a substance (the competitive inhibitor) to the active site of an enzyme, thus lowering the rate of the reaction catalyzed by the enzyme. Compare with *noncompetitive inhibition.*

competitive inhibitor See *competitive inhibition.*

complement A group of proteins in blood and other body fluids that are activated by an antigen–antibody complex and then destroy pathogens.

complementary base pairing The pairing of bases by hydrogen bond formation in double-stranded nucleic acids according to the base pairing rules: adenine pairs with thymine (in DNA) or uracil (in RNA), and guanine pairs with cytosine.

complementary DNA (cDNA) DNA synthesized by reverse transcriptase, using RNA as a template.

complete flower A flower that has all four parts: sepals, petals, stamens, and carpels. Compare with *incomplete flower.*

complete metamorphosis In insects, the transition during the life cycle from egg to larva, to pupa, and finally, to adult. Larvae have different forms and lifestyles than adults. Compare with *incomplete metamorphosis.*

compound eye An eye, such as that of an insect, consisting of many light-sensitive units called *ommatidia.*

concentration gradient A difference in the concentration of a substance from one point to another, as for example, across a cell membrane.

condensation reaction A reaction in which two monomers are combined covalently through the removal of the equivalent of a water molecule; also called *condensation synthesis.* Compare with *hydrolysis reaction.*

condensin Collective name for protein complex that binds DNA and forms it into coiled loops as part of the chromosome compaction process required for cell division.

cone (1) In botany, a reproductive structure in many gymnosperms that produces either microspores or megaspores. (2) In zoology, one of the conical photoreceptive cells of the retina that is particularly sensitive to bright light and, by distinguishing light of various wavelengths, mediates color vision. Compare with *rod.*

conidiophore (kah-nid'-e-o-for″) A specialized hypha that bears conidia.

conidium (kah-nid'-e-um) (pl., *conidia*) An asexual spore that is usually formed at the tip of a specialized hypha called a *conidiophore.*

conifer (kon'-ih-fur) Any of a large phylum of gymnosperms that are woody trees and shrubs with needlelike, mostly evergreen leaves and with seeds in cones.

conjugation (kon″-jew-gay'-shun) (1) A sexual process in ciliate protists that involves exchange of haploid nuclei with another cell. (2) A mechanism for DNA exchange in bacteria that involves cell-to-cell contact.

connective tissue Animal tissue consisting mostly of an intercellular substance (fibers scattered through a matrix) in which the cells are embedded, e.g., bone.

conodonts Extinct, simple fishlike chordates with large eyes and tooth-like hooks; may have been early vertebrates.

conservation biology A multidisciplinary science that focuses on the study of how humans impact organisms and on the development of ways to protect biological diversity.

constitutive gene A gene that is constantly transcribed.

consumer See *heterotroph.*

consumption overpopulation A situation in which each individual in a human population consumes too large a share of resources; results in pollution, environmental degradation, and resource depletion. Compare with *people overpopulation.*

contest competition See *interference competition.*

continental drift The scientific theory that continents were once joined and later split and drifted apart.

contraception Any method used to intentionally prevent pregnancy.

contractile root (kun-trak'-til) A specialized type of root that contracts and pulls a bulb or corm deeper into the soil.

contractile vacuole A membrane-enclosed organelle found in certain freshwater protists, such as *Paramecium;* appears to have an osmoregulatory function.

control group In a scientific experiment, a group in which the experimental variable is kept constant. The control group, which is as closely matched to the experimental group as possible, provides a standard of comparison used to verify the results of the experiment. Compare with *experimental group.*

controlled burning Tool of ecological management in which the undergrowth and plant litter are deliberately burned under controlled conditions before they have accumulated to dangerous levels.

controlled mating A mating in which the genotypes of the parents are known.

convergence The functional arrangement of neurons in which a neuron receives signals from two or more presynaptic neurons. Compare with *divergence.*

convergent circuit (kun-vur'-jent) A neural pathway in which a postsynaptic neuron is controlled by signals coming from two or more presynaptic neurons. Compare with *divergent circuit.*

convergent evolution (kun-vur'-jent) The independent evolution of structural or functional similarity in two or more distantly related species, usually as a result of adaptations to similar environments.

convolutions The numerous folds of the cerebral cortex.

copy number variations (CNVs) Genetic variation among individuals in a population as measured by segments of DNA that have been gained or lost, compared to a reference genome.

coral bleaching The stress-induced loss of the symbiotic algae that inhabit coral cells.

coral reef Reef found in warm, shallow seas; built by coral which has solidified into layers of calcium carbonate ($CaCO_3$).

core angiosperms The clade to which most angiosperm species belong. Core angiosperms are divided into three subclades: magnoliids, monocots, and eudicots. Compare with *basal angiosperms.*

corepressor Substance that binds to a repressor protein, converting it to its active form, which is capable of preventing transcription.

Coriolis effect (kor"-e-o'-lis) The tendency of moving air or water to be deflected from its path to the right in the Northern Hemisphere and to the left in the Southern Hemisphere. Caused by the direction of Earth's rotation.

cork cambium (kam'-bee-um) A lateral meristem that produces cork cells and cork parenchyma; cork cambium and the tissues it produces make up the *periderm* (outer bark) of a woody plant. Compare with *vascular cambium.*

cork cell A cell in the bark that is produced outwardly by the cork cambium; cork cells are dead at maturity and function for protection and reduction of water loss.

cork parenchyma (par-en'-kih-mah) One or more layers of parenchyma cells produced inwardly by the cork cambium.

corm A short, thickened underground stem specialized for food storage and asexual reproduction, e.g., crocus, gladiolus.

cornea (kor'-nee-ah) The transparent covering of an eye.

corolla (kor-ohl'-ah) Collectively, the petals of a flower.

coronary circulation The system of blood vessels that delivers blood to the heart muscle and returns blood from the heart muscle to the right atrium.

corpus callosum (kah-loh'-sum) In mammals, a large bundle of nerve fibers interconnecting the two cerebral hemispheres.

corpus luteum (loo'-tee"-um) The temporary endocrine tissue in the ovary that develops from the ruptured follicle after ovulation; secretes progesterone and estrogen.

cortex (kor'-teks) (1) The outer part of an organ, such as the cortex of the kidney. Compare with *medulla.* (2) The tissue between the epidermis and vascular tissue in the stems and roots of many herbaceous plants.

cortical reaction Process occurring after fertilization that prevents additional sperm from entering the egg; also known as the "slow block to polyspermy."

corticotropin-releasing factor (CRF) Secreted by the hypothalamus in response to stress; stimulates the anterior pituitary to secrete adrenocorticotropic hormone (ACTH).

cortisol A steroid hormone, secreted by the adrenal cortex, that helps the body adjust to long-term stress; stimulates conversion of other nutrients to glucose in the liver, resulting in increased blood glucose concentration.

cosmopolitan species Species that have a nearly worldwide distribution and occur on more than one continent or throughout much of the ocean. Compare with *endemic species.*

cost–benefit analysis An analysis of the costs versus the benefits of a particular behavior; the costs and benefits are typically assessed in terms of direct fitness, an individual's reproductive success.

cotransport The active transport of a substance from a region of low concentration to a region of high concentration by coupling its transport to the transport of a substance down its concentration gradient.

cotyledon (kot"-uh-lee'-dun) The seed leaf of a plant embryo, which may contain food stored for germination.

cotylosaurs The first reptiles; also known as *stem reptiles.*

countercurrent exchange system A biological mechanism that enables maximum exchange between two fluids. The two fluids must be flowing in opposite directions and have a concentration gradient between them.

coupled reactions A set of reactions in which an exergonic reaction provides the free energy required to drive an endergonic reaction; energy coupling generally occurs through a common intermediate.

courtship rituals Behaviors intended to attract or maintain the attention of a potential mate. A courtship ritual may be a signal that triggers nest building, ovulation, or mating.

covalent bond The chemical bond involving shared pairs of electrons; may be single, double, or triple (with one, two, or three shared pairs of electrons, respectively). Compare with *ionic bond* and *hydrogen bond.*

covalent compound A compound in which atoms are held together by covalent bonds; covalent

compounds consist of molecules. Compare with *ionic compound.*

cranial nerves The 10 to 12 pairs of nerves in vertebrates that emerge directly from the brain; they transmit sensory information to the CNS: some cranial nerves transmit motor information from the CNS to effectors.

cranium The bony framework that protects the brain in vertebrates.

crassulacean acid metabolism (CAM) pathway See *CAM plant.*

creatine phosphate An energy-storing compound in muscle cells.

crepuscular animals Animals most active during dusk, dawn, or both, e.g., fiddler crabs and mosquitoes. Compare with *diurnal animals* and *nocturnal animals.*

Cretaceous period See *Mesozoic era.*

cretinism (kree'-tin-izm) A chronic condition caused by lack of thyroid secretion during fetal development and early childhood; results in physical and intellectual disability if untreated.

cri du chat syndrome A human genetic disease caused by losing part of the short arm of chromosome 5 and characterized by intellectual disability, a cry that sounds like a kitten mewing, and death in infancy or childhood.

cristae (kris'-tee) (sing., *crista*) Shelflike or fingerlike inward projections of the inner membrane of a mitochondrion.

cross bridges The connections between myosin and actin filaments in muscle fibers; formed by the binding of myosin heads to active sites on actin filaments.

crossing-over A process in which genetic material (DNA) is exchanged between paired, homologous chromosomes.

crustaceans Members of a subphylum that includes lobsters, crabs, shrimp, barnacles, and their relatives.

cryptic coloration Colors or markings that help some organisms hide from predators by blending into their physical surroundings. Compare with *aposematic coloration.*

cryptochrome A proteinaceous pigment that strongly absorbs blue light; implicated in resetting the biological clock in plants, fruit flies, and mice.

ctenophores (ten'-oh-forz) Phylum of marine animals (comb jellies) whose bodies consist of two layers of cells enclosing a gelatinous mass. The outer surface is covered with comblike rows of cilia, by which the animal moves.

culture Behavior common to a population, that is learned from members of the population, and transmitted from one generation to the next.

cupula In certain mechanoreceptors, the mass of gelatinous material that encloses the tips of stereocilia; secreted by the hair cells.

Cushing's syndrome Hormone disorder caused by overexposure to glucocorticoids; characterized by fat deposited around trunk, facial edema, and high blood glucose level, which causes *adrenal diabetes*. Long-term exposure may lead to permanent diabetes mellitus.

cuticle (kew'-tih-kl) (1) A noncell covering over the epidermis of the aerial parts of plants that reduces water loss; composed mainly of *cutin*, a waxy substance. (2) The outer covering of some animals, such as roundworms.

cyanobacteria (sy-an"-oh-bak-teer'-ee-uh) Prokaryotic photosynthetic microorganisms that possess chlorophyll and produce oxygen during photosynthesis.

cycad (sih'-kad) Any of a phylum of gymnosperms that live mainly in tropical and semitropical regions and have stout stems (to 20 m in height) and fernlike leaves.

cyclic adenosine monophosphate See *cyclic AMP*.

cyclic AMP (cAMP) A form of adenosine monophosphate in which the phosphate is part of a ring-shaped structure; acts as a regulatory molecule and second messenger in organisms ranging from bacteria to humans.

cyclic electron transport In photosynthesis, the cyclic flow of electrons through photosystem I; ATP is formed by chemiosmosis, but no photolysis of water occurs, and O_2 and NADPH are not produced. Compare with *noncyclic electron transport*.

cyclic GMP (cGMP) A form of guanosine monophosphate in which the phosphate is part of a ring-shaped structure; involved in certain cell signaling processes.

cyclic guanosine monophosphate See *cyclic GMP*.

cyclin–Cdk complex See *cyclins*.

cyclin-dependent kinases (Cdks) Protein kinases involved in controlling the cell cycle.

cyclins Regulatory proteins whose levels oscillate during the cell cycle; cyclins associate with cyclin-dependent kinases to form cyclin–Cdk complexes.

cystic fibrosis A genetic disease with an autosomal recessive inheritance pattern; characterized by secretion of abnormally thick mucus, particularly in the respiratory and digestive systems.

cytochromes (sy'-toh-krohmz) Iron-containing heme proteins of an electron transport system.

cytokines Signaling proteins that regulate interactions between cells in the immune system. Important groups include interferons, interleukins, tumor necrosis factors, and chemokines.

cytokinesis (sy"-toh-kih-nee'-sis) Stage of cell division in which the cytoplasm divides to form two daughter cells.

cytokinin (sy"-toh-ky'-nin) A plant hormone involved in various aspects of plant growth and development, such as cell division and delay of senescence.

cytoplasm The plasma membrane and cell contents with the exception of the nucleus.

cytosine A nitrogenous pyrimidine base that is a component of nucleic acids.

cytoskeleton The dynamic internal network of protein fibers that includes microfilaments, intermediate filaments, and microtubules.

cytosol The fluid component of the cytoplasm in which the organelles are suspended.

cytotoxic T cell See *T cytotoxic cell*.

dalton See *atomic mass unit (amu)*.

day-neutral plant A plant whose flowering is not controlled by variations in day length that occur with changing seasons. Compare with *long-day*, *short-day*, and *intermediate-day plants*.

deamination (dee-am-ih-nay'-shun) The removal of an amino group ($—NH_2$) from an amino acid or other organic compound.

decapods Lobsters, crabs, shrimp, crayfish, and other species assigned to order Decapoda, the largest order of crustaceans.

decarboxylation A reaction in which a molecule of CO_2 is removed from a carboxyl group of an organic acid.

deciduous A term describing a plant that sheds leaves or other structures at regular intervals, e.g., during autumn. Compare with *evergreen*.

declarative memory Factual knowledge of people, places, or objects; requires conscious recall of the information.

decomposers Microbial heterotrophs that break down dead organic material and use the decomposition products as a source of energy. Also called *saprotrophs* or *saprobes*.

deductive reasoning The reasoning that operates from generalities to specifics and can make relationships among data more apparent. Compare with *inductive reasoning*. See *hypothetico-deductive approach*.

deforestation The temporary or permanent removal of forest for agriculture or other uses.

dehydrogenation (dee-hy"-dro-jen-ay'-shun) A form of oxidation in which hydrogen atoms are removed from a molecule.

deletion (1) A chromosome abnormality in which part of a chromosome is missing, e.g., cri du chat syndrome. (2) The loss of one or more base pairs from DNA, which can result in a frameshift mutation.

demographics The science that deals with human population statistics, such as size, density, and distribution.

denature (dee-nay'-ture) To alter the physical properties and three-dimensional structure of a protein, nucleic acid, or other macromolecule by treating it with excess heat, strong acids, or strong bases.

dendrites (den'-dritez) Branches of a neuron that receive and conduct nerve impulses toward the cell body. Compare with *axon*.

dendritic cells A set of immune cells present in many tissues that capture antigens and present them to T cells.

dendrochronology (den"-dro-kruh-naal'-uh-gee) A method of dating that uses the annual rings of trees.

denitrification (dee-nie"-tra-fuh-kay'-shun) The conversion of nitrate (NO_3^-) to nitrogen gas (N_2) by certain bacteria (denitrifying bacteria) in the soil; part of the nitrogen cycle.

dense connective tissue A type of tissue that may be irregular, as in the dermis of the skin, or regular, as in tendons.

density-dependent factor An environmental factor whose effects on a population change as population density changes; tends to retard population growth as population density increases and enhance population growth as population density decreases. Compare with *density-independent factor*.

density gradient centrifugation Procedure in which cell components are placed in a layer on top of a density gradient, usually a sucrose solution and water. Cell structures migrate during centrifugation, forming a band at the position in the gradient where their own density equals that of the sucrose solution.

density-independent factor An environmental factor that affects the size of a population but is not influenced by changes in population density. Compare with *density-dependent factor*.

deoxyribonucleic acid (DNA) Double-stranded nucleic acid; contains genetic information coded in specific sequences of its constituent nucleotides.

deoxyribose Pentose sugar lacking a hydroxyl (—OH) group on carbon-2'; a constituent of DNA.

depolarization (dee-pol"-ar-ih-zay'-shun) A decrease in the charge difference across a plasma membrane; may result in an action potential in a neuron or muscle cell.

deposit feeder Animal that consumes nutrients by ingesting soil or sediments.

derived characters See *shared derived characters*.

dermal tissue system The tissue that forms the outer covering over a plant; the epidermis or periderm.

dermis (dur'-mis) The layer of dense connective tissue beneath the epidermis in the skin of vertebrates.

desert A temperate or tropical biome in which lack of precipitation limits plant growth.

desertification The degradation of once-fertile land into nonproductive desert; caused partly by soil erosion, deforestation, and overgrazing by domestic animals.

desmosomes (dez'-moh-sohmz) Buttonlike plaques, present on two opposing cell surfaces, that hold the cells together by means of protein filaments that span the intercellular space.

determinate cleavage Type of cleavage, characteristic of protostomes, in which the fate of each cell is determined when the cell is produced. Compare with *indeterminate cleavage*.

determinate growth Growth of limited duration, as for example, in flowers and leaves. Compare with *indeterminate growth.*

determination The developmental process by which one or more cells become progressively committed to a particular fate. Determination is a series of molecular events usually leading to differentiation. Also called *cell determination.*

detritivore (duh-try´-tuh-vore) An organism, such as an earthworm or crab, that consumes fragments of freshly dead or decomposing organisms; also called *detritus feeder.*

detritus (duh-try´-tus) Organic debris from decomposing organisms.

detritus feeder See *detritivore.*

deuteromycetes (doo˝-ter-o-my´-seats) An artificial grouping of fungi characterized by the absence of sexual reproduction but usually having other traits similar to ascomycetes; also called *imperfect fungi.*

deuterostomes (doo´-ter-oh-stomes) Animals that belong to the Deuterostomia, one of the major animal clades; include the echinoderms and chordates. Compare with *protostomes.*

development All the progressive changes that take place throughout the life of an organism.

developmental genetics Study of how genes and gene expression affect the differentiation of cells and the development of an organism.

Devonian period See *Paleozoic era.*

diabetes mellitus (mel´-i-tus) The most common endocrine disorder. In type 1 diabetes, there is a marked decrease in the number of beta cells in the pancreas, resulting in insulin deficiency. In the more common type 2 diabetes, insulin receptors on target cells do not bind with insulin (insulin resistance).

diacylglycerol (DAG) (di˝-as-il-glis´-er-ol) A lipid consisting of glycerol combined chemically with two fatty acids; also called *diglyceride.* Compare with *monoacylglycerol* and *triacylglycerol.*

dialysis The diffusion of certain solutes across a selectively permeable membrane.

diaphragm In mammals, the muscular floor of the chest cavity; contracts during inhalation, expanding the chest cavity.

diapsids (di-ap´-sids) Members of a clade of amniotes in which the skull has two pairs of temporal openings; includes all extant reptilian groups (including birds) as well as most extinct reptiles. Compare with *synapsids.*

diastole (di-ass´-toh-lee) Phase of the cardiac cycle in which the heart is relaxed. Compare with *systole.*

diatom (die´-eh-tom˝) A usually unicellular stramenopile alga that is covered by an ornate, siliceous shell consisting of two overlapping halves; an important component of plankton in both marine and fresh waters.

dichotomous branching (di-kaut´-uh-mus) In botany, a type of branching in which one part always divides into two more or less equal parts.

diencephalon See *forebrain.*

differential centrifugation Separation of cell particles according to their mass, size, or density. In differential centrifugation, the supernatant is spun at successively higher revolutions per minute.

differential gene expression The expression of different subsets of genes at different times and in different cells during development.

differentiated cell A specialized cell; carries out unique activities, expresses a specific set of proteins, and usually has a recognizable appearance.

differentiation (dif˝-ah-ren-she-ay´-shun) Development toward a more mature state; a process changing a young, relatively unspecialized cell to a more specialized cell. Also called *cell differentiation.*

diffusion The net movement of particles (atoms, molecules, or ions) from a region of higher concentration of that type of particle to a region of lower concentration (i.e., down a concentration gradient), resulting from random motion; also called *simple diffusion.* Compare with *facilitated diffusion* and *active transport.*

digestion The breakdown of food to small molecules.

diglyceride See *diacylglycerol.*

dihybrid cross (dy-hy´-brid) A genetic cross that takes into account the behavior of alleles of two loci. Compare with *monohybrid cross.*

dikaryotic (dy-kare-ee-ot´-ik) Condition of having two nuclei per cell (i.e., *n* + *n*), characteristic of certain fungal hyphae. Compare with *monokaryotic.*

dimer An association of two monomers (e.g., a disaccharide or a dipeptide).

dinoflagellate (dy˝-noh-flaj´-eh-late) A unicellular, biflagellate, typically marine protist that is an important component of plankton; usually photosynthetic.

dioecious (dy-ee´-shus) Having male and female reproductive structures on separate plants; compare with *monoecious.*

dipeptide See *peptide.*

diploblastic (dip-lo-blas´-tik) Animal body plan in which there are only two embryonic tissue layers, the ectoderm and endoderm. Compare with *triploblastic.*

diploid (dip´-loyd) The condition of having two sets of chromosomes per nucleus. Compare with *haploid* and *polyploid.*

diplomonads Small, mostly parasitic excavates with one or two nuclei, no functional mitochondria, and one to four flagella.

directed evolution See *in vitro evolution.*

direct fitness An individual's reproductive success, measured by the number of viable offspring it produces. Compare with *inclusive fitness.*

directional selection The gradual replacement of one phenotype with another because of environmental change that favors phenotypes at one of the extremes of the normal distribution. Compare with *stabilizing selection* and *disruptive selection.*

disaccharide (dy-sak´-ah-ride) A sugar produced by covalently linking two monosaccharides (e.g., maltose or sucrose).

disomy The normal condition in which both members of a chromosome pair are present in a diploid cell or organism. Compare with *monosomy* and *trisomy.*

dispersal The movement of individuals among populations. See *immigration* and *emigration.*

dispersion The pattern of distribution in space of the individuals of a population relative to their neighbors; see *clumped dispersion, random dispersion,* and *uniform dispersion.*

disruptive selection A special type of directional selection in which changes in the environment favor two or more variant phenotypes at the expense of the mean. Compare with *stabilizing selection* and *directional selection.*

distal Remote; farther from the point of reference. Compare with *proximal.*

distal convoluted tubule The part of the renal tubule that extends from the loop of Henle to the collecting duct. Compare with *proximal convoluted tubule.*

disturbance In ecology, any event that disrupts community or population structure.

diurnal animals Animals most active during the day, e.g., pigeons. Compare with *crepuscular animals* and *nocturnal animals.*

divergence The functional arrangement of neurons in which a neuron sends signals to two or more postsynaptic neurons. Compare with *convergence.*

divergent circuit A neural pathway in which a presynaptic neuron stimulates many postsynaptic neurons. Compare with *convergent circuit.*

diving reflex A group of physiological mechanisms, such as decrease in metabolic rate, that are activated when a mammal dives to its limit.

dizygotic twins Twins that arise from the separate fertilization of two eggs; commonly known as *fraternal twins.* Compare with *monozygotic twins.*

DNA See *deoxyribonucleic acid.*

DNA chip See *DNA microarray.*

DNA cloning The process of selectively amplifying DNA sequences so their structure and function can be studied.

DNA fingerprinting The analysis of DNA fragments extracted from an individual; the resulting pattern is unique to that individual.

DNA ligase Enzyme that catalyzes the joining of the 5′ and 3′ ends of two DNA fragments; essential in DNA replication and used in recombinant DNA technology.

DNA methylation A process in which gene inactivation is perpetuated by enzymes that add methyl groups to DNA.

DNA microarray A diagnostic test involving thousands of DNA molecules placed on a glass slide or chip.

DNA polymerases Family of enzymes that catalyze the synthesis of DNA from a DNA template by adding nucleotides to a growing 3′ end.

DNA primase Enzyme that begins the process of DNA replication by synthesizing a short RNA segment complementary to the DNA template strand; this RNA strand is subsequently degraded and replaced with DNA

DNA probe A labeled single-stranded DNA fragment used to detect a gene of interest in a hybridization experiment.

DNA provirus Double-stranded DNA molecule that is an intermediate in the life cycle of an RNA tumor virus (retrovirus).

DNA replication The process by which DNA is duplicated; ordinarily a semiconservative process in which a double helix gives rise to two double helices, each with an "old" strand and a newly synthesized strand.

DNA sequencing Procedure by which the sequence of nucleotides in DNA is determined.

domain (1) A structural and functional region of a protein. (2) The broadest taxonomic category; each domain includes one or more kingdoms.

dominance hierarchy A linear "pecking order" into which animals in a population may organize according to status; regulates aggressive behavior within the population.

dominant allele (al-leel′) An allele that is always expressed when it is present, regardless of whether it is homozygous or heterozygous. Compare with *recessive allele*.

dominant species In a community, a species that as a result of its large biomass or abundance exerts a major influence on the distribution of populations of other species.

dopamine A neurotransmitter of the biogenic amine group; important in motor function.

dormancy A temporary period of arrested growth in plants or plant parts such as spores, seeds, bulbs, and buds.

dorsal (dor′-sl) Toward the uppermost surface or back of an animal. Compare with *ventral*.

dosage compensation Genetic mechanism by which the expression of X-linked genes is made equivalent in XX females and XY males; in mammals this is accomplished by rendering all but one X chromosome inactive.

double fertilization A process in the flowering plant life cycle in which there are two fertilizations; one fertilization results in formation of a zygote, whereas the second results in formation of endosperm.

double helix The structure of DNA, which consists of two antiparallel polynucleotide chains twisted around each other.

doubling time The amount of time it takes for a population to double in size, assuming that its current rate of increase does not change. Doubling time can be estimated by dividing 70 by the rate of increase (percent increase per unit of time).

Down syndrome An inherited condition in which individuals have abnormalities of the face, eyelids, tongue, and other parts of the body and exhibit physical and intellectual disability; usually results from trisomy of chromosome 21.

drupe (droop) A simple, fleshy fruit in which the inner wall of the fruit is a hard stone, e.g., peach, cherry.

duodenum (doo″-o-dee′-num) The portion of the small intestine into which the contents of the stomach first enter.

duplication An abnormality in which a set of chromosomes contains more than one copy of a particular chromosomal segment; the translocation form of Down syndrome is an example.

dura mater The tough, outer meningeal layer that covers and protects the brain and spinal cord. Also see *arachnoid* and *pia mater*.

dynamic equilibrium The condition of a chemical reaction when the rate of change in one direction is exactly the same as the rate of change in the opposite direction, i.e., the concentrations of the reactants and products are not changing, and the difference in free energy between reactants and products is zero.

dynein See *microtubule-associated proteins (MAPs)*.

ecdysis Molting; shedding outer skin; common process in insects, crustaceans, and snakes.

ecdysone (ek′-dih-sone) See *molting hormone*.

Ecdysozoa A branch of the protostomes that includes animals that molt, such as the rotifers, nematodes, and arthropods.

echinoderms (eh-kine′-oh-derms) Phylum of spiny-skinned marine deuterostome invertebrates characterized by a water vascular system and tube feet; include sea stars, sea urchins, and sea cucumbers.

echolocation Determination of the position of objects by detecting echoes of high-pitched sounds emitted by an animal; a type of sensory system used by bats and dolphins.

ecological niche See *niche*.

ecological pyramid A graphical representation of the relative energy value at each trophic level. See *pyramid of biomass* and *pyramid of energy*.

ecological succession See *succession*.

ecology (ee-kol′-uh-jee) A discipline of biology that studies the interrelations among living things and their environments.

ecosystem (ee′-koh-sis-tem) The interacting system that encompasses a community and its nonliving, physical environment. Compare with *community*.

ecosystem diversity Variety of ecosystems found on Earth such as forests, prairies, deserts, lakes, coastal estuaries, and coral reefs.

ecosystem ecology A subfield of ecology that studies energy flow and the cycling of chemicals among the interacting biotic and abiotic parts of an ecosystem.

ecosystem management A conservation focus that emphasizes restoring and maintaining ecosystem quality rather than the conservation of individual species.

ecosystem services Important environmental services, such as clean air to breathe, clean water to drink, and fertile soil in which to grow crops, that ecosystems provide.

ecotone The transition zone where two communities meet and intergrade.

ectoderm (ek′-toh-derm) The outer germ layer of the early embryo; gives rise to the skin and nervous system. Compare with *mesoderm* and *endoderm*.

ectomycorrhizal fungi See *mycorrhizae*.

ectoparasite A tick or other parasite that lives outside its host's body. Compare with *endoparasite*.

ectotherm An animal whose temperature fluctuates with that of the environment; may use behavioral adaptations to regulate temperature; sometimes referred to as *cold-blooded*. Compare with *endotherm*.

edge effect The ecological phenomenon in which ecotones between adjacent communities often contain a greater number of species or greater population densities of certain species than either adjacent community.

Ediacaran period (ee-dee-ack′-uh-ran″) The last (most recent) period of the Proterozoic eon, from 635 million to 541 million years ago; named for early animal fossils found in the Ediacara Hills in South Australia.

effector (1) A muscle or gland that contracts or secretes in direct response to nerve impulses. (2) In homeostasis, an organ or process that helps restore a steady state.

efferent (ef′-fur-ent) Leading away from some point of reference. Compare with *afferent*.

efferent arteriole In the mammalian kidney, an arteriole that conducts blood away from a glomerulus. Compare with *afferent arteriole*.

efferent neurons Neurons that transmit action potentials from the brain or spinal cord to muscles or glands. Compare with *afferent neurons*.

ejaculation (ee-jak″-yoo-lay′-shun) A sudden expulsion, as in the ejection of semen from the penis.

ejaculatory duct A short duct that passes through the prostate gland and opens into the urethra; receives semen from the vas deferens.

electrical synapse A junction between two neurons in which an action potential in the presynaptic neuron spreads directly to the postsynaptic neuron by way of gap junctions. Compare with *chemical synapse*.

electrochemical gradient A difference in charge and chemical concentration existing between two regions.

electrolyte A substance that dissociates into ions when dissolved in water; the resulting solution can conduct an electric current.

electron A particle with one unit of negative charge and negligible mass, located outside the atomic nucleus. Compare with *neutron* and *proton*.

electron configuration The arrangement of electrons around the atom. In a Bohr model, the electron configuration is depicted as a series of concentric circles.

electronegativity A measure of an atom's attraction for electrons.

electron microscope A microscope capable of producing high-resolution, highly magnified images through the use of an electron beam (rather than light). Transmission electron microscopes (TEMs) produce images of thin sections; scanning electron microscopes (SEMs) produce images of surfaces.

electron shell Group of orbitals of electrons with similar energies.

electron transport system A series of redox reactions during which hydrogens or their electrons are passed along an electron transport chain from one acceptor molecule to another, with the release of energy.

electron transport chain See *electron transport system*.

electrophoresis, gel See *gel electrophoresis*.

electroreceptor A receptor that responds to electrical stimuli.

element A substance that cannot be changed to a simpler substance by a normal chemical reaction.

elimination Ejection of undigested food from the body. Compare with *excretion*.

elongation (in protein synthesis) Cyclic process by which amino acids are added one by one to a growing polypeptide chain. See *initiation* and *termination*.

El Niño–Southern Oscillation (ENSO) (el nee′-nyo) A recurring climatic phenomenon that involves a surge of warm water in the Pacific Ocean and unusual weather patterns elsewhere in the world.

embryo (em′-bree-oh) (1) A young organism before it emerges from the egg, seed, or body of its mother. (2) Developing human until the end of the second month, after which it is referred to as a fetus. (3) In plants, the young sporophyte produced following fertilization and subsequent development of the zygote.

embryonic stem cell (ES cell) A pluripotent stem cell derived from an early stage embryo.

embryo sac The female gametophyte generation in flowering plants.

embryo transfer See *host mothering*.

emergent properties Characteristics of an object, process, or behavior that could not be predicted from its component parts; emergent properties can be identified at each level as we move up the hierarchy of biological organization.

emerging diseases Diseases new to the human population; often appear suddenly. Compare with *re-emerging diseases*.

emigration The movement of individuals out of a population. Compare with *immigration*.

enantiomers (en-an′-tee-oh-merz) Two isomeric chemical compounds that are mirror images.

Encode Project (ENCyclopedia of DNA Elements) An international collaborative project to determine all the functional elements within the human genome. Initial results published in 2012 showed that at least 80% of non-protein-coding human DNA has biochemical functions.

3′ end End of a nucleic acid strand that has a 3′ carbon of deoxyribose (DNA) or ribose (RNA) attached to a hydroxyl group. Compare with *5′ end*.

5′ end End of a nucleic acid strand that has a 5′ carbon of deoxyribose (DNA) or ribose (RNA) attached to a phosphate group. Compare with *3′ end*.

endangered species A species whose numbers are so severely reduced that it is in imminent danger of extinction throughout all or part of its range. Compare with *threatened species*.

endemic species Localized, native species that are not found anywhere else in the world. Compare with *cosmopolitan species*.

endergonic reaction (end′-er-gon″-ik) A nonspontaneous reaction; a reaction requiring a net input of free energy. Compare with *exergonic reaction*.

endocrine gland (en′-doh-crin) A gland that secretes hormones directly into the blood or tissue fluid instead of into ducts. Compare with *exocrine gland*.

endocrine system The body system that helps regulate metabolic activities; consists of ductless glands and tissues that secrete hormones.

endocrinology Study of the endocrine system and endocrine activity, including the production and actions of chemical messengers produced by organs, tissues, and cells.

endocytosis (en″-doh-sy-toh′-sis) The active transport of substances into the cell by the formation of invaginated regions of the plasma membrane that pinch off and become cytoplasmic vesicles. Includes *phagocytosis, pinocytosis,* and *receptor-mediated endocytosis*. Compare with *exocytosis*.

endoderm (en′-doh-derm) The inner germ layer of the early embryo; becomes the lining of the digestive tract and the structures that develop from the digestive tract—liver, lungs, and pancreas. Compare with *ectoderm* and *mesoderm*.

endodermis (en″-doh-der′-mis) The innermost layer of the plant root cortex. Endodermal cells have a waterproof Casparian strip around their radial and transverse walls that ensures that water and minerals enter the xylem only by passing through the endoderm cells.

endolymph (en′-doh-limf) The fluid of the membranous labyrinth and cochlear duct of the ear.

endomembrane system The group of membranous structures in eukaryotic cells that interact through direct connections by vesicles; includes the endoplasmic reticulum, outer membrane of the nuclear envelope, Golgi complex, lysosomes, and the plasma membrane; also called *internal membrane system*.

endometrium (en″-doh-mee′-tree-um) The uterine lining.

endomycorrhizal fungi See *mycorrhizae*.

endoparasite A parasite such as a tapeworm that lives within the host. Compare with *ectoparasite*.

endoplasmic reticulum (ER) (en′-doh-plaz″-mik reh-tik′-yoo-lum) An interconnected network of internal membranes in eukaryotic cells enclosing a compartment, the ER lumen. Rough ER has ribosomes attached to the cytosolic surface; smooth ER, a site of lipid biosynthesis, lacks ribosomes.

endorphins (en-dor′-finz) Neuropeptides that block pain signals; released by certain neurons in the CNS.

endoskeleton (en″-doh-skel′-eh-ton) Bony and/or cartilaginous structures within the body that provide support. Compare with *exoskeleton*.

endosperm (en′-doh-sperm) The 3*n* nutritive tissue that is formed at some point in the development of all angiosperm seeds.

endospore A resting cell formed by certain bacteria; highly resistant to heat, radiation, and disinfectants.

endostyle A shared derived character of chordates; a groove in the floor of the pharynx that secretes mucus and traps food particles in sea water passing through the pharynx. In vertebrates, the thyroid gland is derived from the endostyle.

endosymbiont (en″-doe-sim′-bee-ont) An organism that lives inside the body or a cell of another kind of organism. Endosymbionts may benefit their host (mutualism) or harm their host (parasitism).

endosymbiosis Type of symbiosis in which one organism lives inside another. See *endosymbiont*.

endothelium (en-doh-theel′-ee-um) The tissue that lines the cavities of the heart, blood vessels, and lymph vessels.

endotherm (en′-doh-therm) An animal that uses metabolic energy to maintain a constant body temperature despite variations in environmental temperature; e.g., birds and mammals. Compare with *ectotherm*.

endotoxin A poisonous substance in the cell walls of gram-negative bacteria. Compare with *exotoxin*.

end product inhibition See *feedback inhibition*.

energy The capacity to do work; expressed in kilojoules or kilocalories.

energy flow Passage of energy in a one-way direction through an ecosystem.

energy of activation See *activation energy*.

enhanced greenhouse effect See *greenhouse effect*.

enhancers Regulatory DNA sequences that enhance gene transcription; can be located long distances away from the actual coding regions of a gene.

enkephalins (en-kef′-ah-linz) Neuropeptides that block pain signals; released by certain neurons in the CNS.

enterocoely (en′-ter-oh-seely) The process by which the coelom forms as a cavity within mesoderm produced by outpocketings of the primitive gut (archenteron); characteristic of many deuterostomes. Compare with *schizocoely*.

enthalpy The total potential energy of a system; sometimes referred to as the "heat content of the system."

entropy (en′-trop-ee) Disorderliness; a quantitative measure of the amount of the random, disordered energy that is unavailable to do work.

environmental resistance Unfavorable environmental conditions, such as crowding, that prevent organisms from reproducing indefinitely at their intrinsic rate of increase.

environmental sustainability The ability to meet humanity's current needs without compromising the ability of future generations to meet their needs.

enzyme (en′-zime) An organic catalyst (usually a protein) that accelerates a specific chemical reaction by lowering the activation energy required for that reaction.

enzyme-linked receptors Transmembrane proteins with a hormone-binding site outside the cell and an enzyme site inside the cell.

enzyme–substrate complex The temporary association between enzyme and substrate that forms during the course of a catalyzed reaction; also called *ES complex*.

eon The largest division of the geologic time scale; eons are divided into eras.

eosinophil (ee-oh-sin′-oh-fil) A type of white blood cell whose cytoplasmic granules absorb acidic stains; functions in parasitic infestations and allergic reactions.

ependymal cells Ciliated cells that line internal cavities of CNS; help produce and circulate cerebrospinal fluid.

epidermis (ep-ih-dur′-mis) (1) An outer layer of cells that covers the body of plants and functions primarily for protection. (2) The outer layer of vertebrate skin.

epididymis (ep-ih-did′-ih-mis) (pl., *epididymides*) A coiled tube that receives sperm from the testis and conveys it to the vas deferens.

epigenetic inheritance Inheritance that involves changes in how a gene is expressed without any change in that gene's nucleotide sequence.

epiglottis A thin, flexible structure that guards the entrance to the larynx, preventing food from entering the airway during swallowing.

epinephrine (ep-ih-nef′-rin) Hormone produced by the adrenal medulla; stimulates the sympathetic nervous system.

epiphyte Plant that grows attached to another plant.

epistasis (ep-ih-sta′-sis) Condition in which certain alleles of one locus alter the expression of alleles of a different locus.

epithelial tissue (ep-ih-theel′-ee-al) The type of animal tissue that covers body surfaces, lines body cavities, and forms glands; also called *epithelium*.

epoch The smallest unit of geologic time; a subdivision of a period.

equilibrium See *dynamic equilibrium, genetic equilibrium,* and *punctuated equilibrium*.

equilibrium potential The membrane potential at which the flow of a particular type of ion inward (due to the electrical gradient) equals the flow of that ion outward (because of the concentration gradient).

era An interval of geologic time that is a subdivision of an eon; eras are divided into periods.

erythroblastosis fetalis (eh-rith″-row-blas-toe′-sis fi-tal′-is) Serious condition in which Rh⁺ red blood cells (which bear antigen D) of a fetus are destroyed by maternal anti-D antibodies.

erythrocyte (eh-rith′-row-site) A vertebrate red blood cell; contains hemoglobin, which transports oxygen. Compare with *leukocyte*.

erythropoietin (eh-rith″-row-poy′-ih-tin) A peptide hormone secreted mainly by kidney cells; stimulates red blood cell production.

ES complex See *enzyme–substrate complex*.

esophagus (e-sof′-ah-gus) The part of the digestive tract that conducts food from the pharynx to the stomach.

essential amino acid An amino acid that must be provided in the diet because the body cannot make it or cannot make it in sufficient quantities to meet nutritional needs.

essential nutrient A nutrient that must be provided in the diet because the body cannot make it or cannot make it in sufficient quantities to meet nutritional needs, e.g., essential amino acids and essential fatty acids.

ester linkage Covalent linkage formed by the reaction of a carboxyl group and a hydroxyl group, with the removal of the equivalent of a water molecule; the linkage includes an oxygen atom bonded to a carbonyl group.

estivation A state of torpor caused by lack of food or water during periods of high temperature. Compare with *hibernation*.

estrogens (es′-troh-jenz) Female sex hormones produced by the ovary; promote the development and maintenance of female reproductive structures and of secondary sex characteristics.

estuary (es′-choo-wear-ee) A coastal body of water that connects to an ocean, in which fresh water from the land mixes with salt water.

ethology (ee-thol′-oh-jee) The study of animal behavior under natural conditions from the point of view of adaptation.

ethyl alcohol A two-carbon alcohol.

ethylene (eth′-ih-leen) A gaseous plant hormone involved in various aspects of plant growth and development, such as leaf abscission and fruit ripening.

euchromatin (yoo-croh′-mah-tin) A loosely coiled chromatin that is generally capable of transcription. Compare with *heterochromatin*.

eudicot (yoo-dy′-kot) One of the two clades of flowering plants; eudicot seeds contain two cotyledons, or seed leaves. Compare with *monocot*.

euglenoids (yoo-glee′-noidz) A group of mostly freshwater unicellular excavates that move by means of an anterior flagellum and are usually photosynthetic.

Eukarya (yoo″-kar′-ee-ah) The domain that includes all eukaryotes: protists, fungi, plants, and animals.

eukaryote (yoo″-kar′-ee-ote) An organism whose cells have nuclei and other membrane-enclosed organelles. Compare with *prokaryote*.

eukaryotic cell See *eukaryote*.

Eumetazoa (yu″-met-uh-zo′-ah) A clade of animals with true tissues; almost all animals except sponges are eumetazoans.

euphotic zone The upper reaches of the ocean, in which enough light penetrates to support photosynthesis.

eustachian tube (yoo-stay′-shee-un) The auditory tube passing between the middle-ear cavity and the pharynx in vertebrates; permits the equalization of pressure on the tympanic membrane.

eutherians A clade of mammals characterized by a well-developed placenta; the young are more developed at birth than are marsupials.

eutrophic lake A lake enriched with nutrients such as nitrate and phosphate and consequently overgrown with plants or algae.

evergreen A plant that sheds leaves over a long period, so some leaves are always present. Compare with *deciduous*.

Evo Devo The study of the evolution of the genetic control of development.

evolution Any cumulative genetic changes in a population from generation to generation. Evolution leads to differences in populations and explains the origin of all the organisms that exist today or have ever existed.

evolutionary species concept See *phylogenetic species concept*.

evolutionary systematics An approach to classification that considers both evolutionary relationships and the extent of divergence that has occurred since a group branched from an ancestral group. Compare with *cladistics* and *phenetics*.

excavates Flagellated unicellular protists, many of which have a deep (excavated) oral groove; include diplomonds, parabasilids, and euglenoids.

excitatory postsynaptic potential (EPSP) A change in membrane potential that brings a neuron closer to the firing level. Compare with *inhibitory postsynaptic potential (IPSP)*.

excretion (ek-skree′-shun) The discharge from the body of a waste product of metabolism (not to be confused with the elimination of undigested food materials). Compare with *elimination*.

excretory system The body system in animals that functions in osmoregulation and in the discharge of metabolic wastes.

exergonic reaction (ex′-er-gon″-ik) A reaction characterized by a release of free energy. Also called *spontaneous reaction*. Compare with *endergonic reaction*.

exocrine gland (ex′-oh-crin) A gland that excretes its products through a duct that opens onto a free surface, such as the skin (e.g., sweat glands). Compare with *endocrine gland*.

exocytosis (ex″-oh-sy-toh′-sis) The active transport of materials out of the cell by fusion of cytoplasmic vesicles with the plasma membrane. Compare with *endocytosis*.

exon (1) A protein-coding region of a eukaryotic gene. (2) The mRNA transcribed from such a region. Compare with *intron*.

exoskeleton (ex″-oh-skel′-eh-ton) An external skeleton, such as the shell of mollusks or outer covering of arthropods; provides protection and sites of attachment for muscles. Compare with *endoskeleton*.

exotoxin A poisonous substance released by certain bacteria. Compare with *endotoxin*.

experimental group In a scientific experiment, a group in which the experimental variable is manipulated. Compare with *control group*.

exploitation competition Intraspecific competition in which all the individuals in a population "share" the limited resource equally so that at high population densities none of them obtains an adequate amount. Also called *scramble competition*. Compare with *interference competition*.

exponential population growth The accelerating population growth that occurs when optimal conditions allow a constant per capita growth rate. Compare with *logistic population growth*.

ex situ conservation Conservation efforts that involve conserving individual species in human-controlled settings, such as zoos. Compare with *in situ conservation*.

external fertilization The process in which gametes meet outside the body, typically when females release eggs and males release sperm into the surrounding water; occurs in most aquatic invertebrates, bony fishes, and amphibians. Compare with *internal fertilization*.

exteroceptor (ex′-tur-oh-sep″-tor) One of the sense organs that receives sensory stimuli from the outside world, such as the eyes or touch receptors. Compare with *interoceptor*.

extinction The elimination of a species; occurs when the last individual member of a species dies; see *background extinction* and *mass extinction*.

extracellular matrix (ECM) A network of proteins and carbohydrates that surrounds many animal cells.

extraembryonic membranes Multicellular membranous structures that develop from the germ layers of a terrestrial vertebrate embryo but are not part of the embryo itself. See *chorion, amnion, allantois,* and *yolk sac*.

extreme halophiles Certain types of archaea, and few types of bacteria and protists, that inhabit environments with very high salt concentrations.

extreme thermophiles Certain types of archaea that require a very high temperature or very low temperature for growth.

F_1 generation (first filial generation) The first generation of hybrid offspring resulting from a cross between parents from two different true-breeding lines.

F_2 generation (second filial generation) The offspring of the F_1 generation.

facilitated diffusion The passive transport of ions or molecules by a specific carrier protein in a membrane. As in simple diffusion, net transport is down a concentration gradient, and no additional energy has to be supplied. Compare with *diffusion* and *active transport*.

facilitation In ecology, a situation in which one species has a positive effect on other species in the community, for example, by enhancing the local environment.

facultative anaerobe An organism capable of carrying out aerobic cellular respiration but able to switch to fermentation when oxygen is unavailable, e.g., yeast. Compare with *obligate anaerobe*.

FAD/FADH$_2$ Oxidized and reduced forms, respectively, of flavin adenine dinucleotide, a coenzyme that transfers electrons (as hydrogen) in metabolism, including cellular respiration.

fallopian tube See *oviduct*.

fall turnover Mixing of temperate lake waters in autumn, caused by falling temperatures, in which surface water sinks to the bottom and bottom water rises to the surface. Compare with *spring turnover*.

family A taxonomic category made up of related genera.

fast-glycolytic fibers White muscle fibers that generate a lot of power for a brief period; they contract rapidly, fatigue quickly, and obtain most of their ATP from glycolysis. Compare with *fast-oxidative fibers* and *slow-oxidative fibers*.

fast-oxidative fibers Muscle fibers specialized for rapid response; they contract quickly, have an intermediate rate of fatigue, and obtain most of their ATP from aerobic respiration. Compare with *slow-oxidative fibers* and *fast-glycolytic fibers*.

fatty acid A lipid that is an organic acid containing a long hydrocarbon chain, with no double bonds (saturated fatty acid), one double bond (mono-unsaturated fatty acid), or two or more double bonds (polyunsaturated fatty acid); components of triacylglycerols and phospholipids, as well as monoacylglycerols and diacylglycerols.

fatty acid derivatives Prostaglandins, juvenile hormones of insects, and other compounds derived from fatty acids.

fecundity The potential capacity of an individual to produce offspring.

feedback inhibition A type of enzyme regulation in which the accumulation of the product of a reaction inhibits an earlier reaction in the sequence; also known as *end product inhibition*.

fermentation An anaerobic process by which ATP is produced by a series of redox reactions in which organic compounds serve both as electron donors and terminal electron acceptors.

fern One of a phylum of seedless vascular plants that reproduce by spores produced in sporangia; ferns undergo an alternation of generations between the dominant sporophyte and the gametophyte (prothallus).

fertilization The fusion of two *n* gametes; results in the formation of a 2*n* zygote. Compare with *double fertilization*.

fetus The unborn human offspring from the third month of pregnancy to birth.

fiber (1) In plants, a type of sclerenchyma cell; fibers are long, tapered cells with thick walls. Compare with *sclereid*. (2) In animals, an elongated cell such as a muscle or nerve cell. (3) In animals, the microscopic, threadlike protein and carbohydrate complexes scattered through the matrix of connective tissues.

fibrin An insoluble protein formed from the plasma protein fibrinogen during blood clotting.

fibroblasts Connective tissue cells that produce the fibers and the protein and carbohydrate complexes of the matrix of connective tissues.

fibronectins Glycoproteins of the extracellular matrix that bind to integrins (receptor proteins in the plasma membrane).

fibrous root system A root system consisting of several adventitious roots of approximately equal size that arise from the base of the stem. Compare with *taproot system*.

Fick's law of diffusion A physical law governing rates of gas exchange in animal respiratory systems; states that the rate of diffusion of a substance across a membrane is directly proportional to the surface area and to the difference in pressure between the two sides.

fibrinogen A soluble plasma protein that is converted to the insoluble protein fibrin during blood clotting.

filament (1) In flowering plants, the thin stalk of a stamen; the filament bears an anther at its tip. (2) In muscles, a myofilament; when myofilaments slide past one another, muscle contraction occurs.

fimbriae Hairlike structures that project from the cell surface of some prokaryotes; help bacteria to adhere to one another and to attach to the surfaces of cells they infect.

first law of thermodynamics The law of conservation of energy, which states that the total energy of any closed system (any object plus its surroundings, i.e., the universe) remains constant. Compare with *second law of thermodynamics*.

first messenger A signaling molecule that binds to a receptor on the cell surface, initiating a signal transduction process that leads to the formation of an intracellular *second messenger.*

fissure A groove or furrow, for example the grooves between the convolutions of the cerebral cortex.

fitness See *direct fitness.*

fixed action pattern (FAP) An innate behavior triggered by a sign stimulus.

flagellate A unicellular nonphotosynthetic protist that has one or more long, whiplike flagella.

flagellum (flah-jel′-um) (pl., *flagella*) A long, whiplike structure extending from certain cells and used in locomotion. (1) Eukaryote flagella consist of two central, single microtubules surrounded by nine double microtubules (9 + 2 structure), all covered by a plasma membrane. (2) Prokaryote flagella are filaments rotated by special structures located in the plasma membrane and cell wall.

flame cells Collecting cells that have cilia; part of the osmoregulatory system of flatworms.

flavin adenine dinucleotide See *FAD/FADH₂.*

flowering plants See *angiosperms.*

flowing-water ecosystem A river or stream ecosystem.

fluid mosaic model The currently accepted model of the plasma membrane and other cell membranes, in which protein molecules "float" in a fluid phospholipid bilayer.

fluorescence The emission of light of a longer wavelength (lower energy) than the light originally absorbed.

fluorescent in situ hybridization (FISH) A technique to detect specific DNA segments by hybridization directly to chromosomes; visualized microscopically by using a fluorescent dye.

follicle (fol′-i-kl) (1) A simple, dry, dehiscent fruit that develops from a single carpel and splits open at maturity along one suture to liberate the seeds. (2) A small sac of cells in the mammalian ovary that contains a maturing egg. (3) The pocket in the skin from which a hair grows.

follicle-stimulating hormone (FSH) A gonadotropic hormone secreted by the anterior lobe of the pituitary gland; stimulates follicle development in the ovaries of females and sperm production in the testes of males.

food chain The series of organisms through which energy flows in an ecosystem. Each organism in the series eats or decomposes the preceding organism in the chain. See *food web.*

food web A complex interconnection of all the food chains in an ecosystem.

foraging Feeding behavior; involves locating and selecting food, as well as capturing and gathering it.

foram See *formaminiferan.*

foramen magnum The opening in the vertebrate skull through which the spinal cord passes.

foraminiferan (for″-am-in-if′-er-an) A marine protist that produces a shell, or test, that encloses

an amoeboid body. Also called *foram.* See *rhizarians.*

forebrain In the early embryo, one of the three divisions of the developing vertebrate brain; subdivides to form the telencephalon, which gives rise to the cerebrum, and the diencephalon, which gives rise to the thalamus and hypothalamus. Compare with *midbrain* and *hindbrain.*

forest decline A gradual deterioration (and often death) of many trees in a forest; can be caused by a combination of factors, such as acid precipitation, toxic heavy metals, and surface-level ozone.

fossil Parts or traces of an ancient organism usually preserved in rock.

fossil fuel Combustible deposits in Earth's crust that are composed of the remnants of prehistoric organisms that existed millions of years ago, e.g., oil, natural gas, and coal.

founder cell A cell from which a particular cell lineage is derived.

founder effect Genetic drift that results from a small population colonizing a new area.

fovea (foe′-vee-ah) The area of sharpest vision in the retina; cone cells are concentrated here.

fragile site A weak point at a specific location on a chromosome where part of a chromatid appears attached to the rest of the chromosome by a thin thread of DNA.

fragile X syndrome A human genetic disorder caused by a fragile site that occurs near the tip on the X chromosome; effects range from mild learning disabilities to severe intellectual disability and hyperactivity.

fragmentation A method of asexual reproduction in some bacteria, archaea, animals, or plants; a cell or an organism separates into several pieces and each piece develops into a new organism.

frameshift mutation A mutation that results when one or two nucleotide pairs are inserted into or deleted from the DNA. The change causes the mRNA transcribed from the mutated DNA to have an altered reading frame such that all codons downstream from the mutation are changed.

fraternal twins See *dizygotic twins.*

free energy The maximum amount of energy available to do work under the conditions of a biochemical reaction.

free radicals Toxic, highly reactive compounds with unpaired electrons that bond with other compounds in the cell and interfere with normal function.

frequency-dependent selection Selection in which the relative fitness of different genotypes is related to how frequently they occur in the population.

freshwater wetlands Land that is transitional between freshwater and terrestrial ecosystems and is covered with water for at least part of the year, e.g., marshes and swamps.

frontal lobes Include the region of the cerebral cortex anterior to the central sulcus; contain important association and motor areas.

fruit In flowering plants, a mature, ripened ovary. Fruits contain seeds and usually provide seed protection and dispersal.

fruiting body A multicellular structure that contains the sexual spores of certain fungi; refers to the ascocarp of an ascomycete and the basidiocarp of a basidiomycete.

fucoxanthin (few″-koh-zan′-thin) The brown carotenoid pigment found in brown algae, golden algae, diatoms, and dinoflagellates.

functional genomics The study of the roles of genes in cells.

functional group A group of atoms that confers distinctive properties on an organic molecule (or region of a molecule) to which it is attached, e.g., hydroxyl, carbonyl, carboxyl, amino, phosphate, and sulfhydryl groups.

fundamental niche The potential ecological niche that an organism could occupy if there were no competition from other species. Compare with *realized niche.*

fungus (pl., *fungi*) A heterotrophic eukaryote belonging to the opisthokont clade, with chitinous cell walls and a body usually in the form of a mycelium of branched, threadlike hyphae. Most fungi are decomposers; some are parasitic.

G protein One of a group of proteins that bind GTP and are involved in the transfer of signals across the plasma membrane.

G protein–linked receptors Cell surface receptors that activate a G protein in response to binding by a signaling molecule.

G₁ phase The first gap phase within the interphase stage of the cell cycle; G₁ occurs before DNA synthesis (S phase) begins. Compare with *S phase* and *G₂ phase.*

G₂ phase Second gap phase within the interphase stage of the cell cycle; G₂ occurs after DNA synthesis (S phase) and before mitosis. Compare with *S phase* and *G₁ phase.*

gallbladder A small sac that stores bile.

gametangium (gam″-uh-tan′-gee-um) Special multicellular or unicellular structure of plants, protists, and fungi in which gametes are formed.

gamete (gam′-eet) A sex cell; in plants and animals, an egg or sperm. In sexual reproduction, the union of gametes results in the formation of a zygote. The chromosome number of a gamete is designated *n.*

gametic isolation (gam-ee′-tik) A prezygotic reproductive isolating mechanism in which sexual reproduction between two closely related species cannot occur because of chemical differences in the gametes.

gametogenesis The process of gamete formation. See *spermatogenesis* and *oogenesis.*

gametophyte generation (gam-ee′-toh-fite) The *n,* gamete-producing stage in the life cycle of a plant. Compare with *sporophyte generation.*

gamma-aminobutyric acid (GABA) A neurotransmitter that has an inhibitory effect.

ganglion (gang'-glee-on) (pl., *ganglia*) A mass of neuron cell bodies; in vertebrates, refers to aggregations of cell bodies in the peripheral nervous system. Compare with *nucleus* (definition 3).

ganglion cell A type of neuron in the retina of the eye; receives input from bipolar cells.

gap junction Structure consisting of specialized regions of the plasma membrane of two adjacent cells; contains numerous pores that allow the passage of certain small molecules and ions between them.

gastric glands Glands in the wall of the stomach; contain cells that secrete hydrochloric acid, intrinsic factor, and pepsinogen, the precursor of the enzyme pepsin.

gastrin (gas'-trin) A hormone released by the stomach mucosa; stimulates the gastric glands to secrete pepsinogen.

gastrovascular cavity A central digestive cavity with a single opening that functions as both mouth and anus; characteristic of cnidarians and flatworms.

gastrula (gas'-troo-lah) A three-layered embryo formed by the process of gastrulation.

gastrulation (gas-troo-lay'-shun) Process in embryonic development during which the three germ layers (ectoderm, mesoderm, and endoderm) form.

gel electrophoresis Procedure by which proteins or nucleic acids are separated on the basis of size and charge as they migrate through a gel in an electric field.

gemmae (sing., *gemma*) Tiny balls of tissue (borne in *gemmae cups*) used for asexual reproduction by liverworts.

gene A segment of DNA that serves as a unit of hereditary information; includes a transcribable DNA sequence (plus associated sequences regulating its transcription) that yields a protein or RNA product with a specific function.

gene amplification The developmental process in which certain cells produce multiple copies of a gene by selective replication, thus allowing for increased synthesis of the gene product. Compare with *nuclear equivalence* and *genomic rearrangement*.

gene flow The movement of alleles between local populations due to the migration of individuals; can have significant evolutionary consequences.

gene locus See *locus*.

gene pool All the alleles of all the genes present in a freely interbreeding population.

gene therapy Any of a variety of methods designed to correct a disease or alleviate its symptoms through the introduction of normal, therapeutic genes into the affected person's cells.

genetic bottleneck See *bottleneck*.

genetic code See *codon*.

genetic counseling Medical and genetic information provided to couples who are concerned about the risk of abnormality in their children.

genetic diversity Genetic variety within a species both among individuals within a given population and among geographically separate populations.

genetic drift A random change in allele frequency in a small breeding population.

genetic engineering Manipulation of genes, often through recombinant DNA technology. Also called *molecular modification.*

genetic equilibrium The condition of a population that is not undergoing evolutionary change, i.e., in which allele and genotype frequencies do not change from one generation to the next. See *Hardy–Weinberg principle.*

genetic polymorphism (pol″-ee-mor′-fizm) The presence in a population of two or more alleles for a given gene locus.

genetic recombination See *recombination, genetic.*

genetics The science of heredity; includes genetic similarities and genetic variation between parents and offspring or among individuals of a population.

genetic screening A systematic search through a population for individuals with a genotype or karyotype that might cause a serious genetic disease in them or their offspring.

genetic variation Inherited differences among individuals.

genome (jee′-nome) Originally, all the genetic material in a cell or individual organism. The term is used in more than one way depending on context, e.g., an organism's haploid genome is all the DNA contained in one haploid set of its chromosomes, and its mitochondrial genome is all the DNA in a mitochondrion. See *human genome.*

genome-wide association study (GWAS) A comparison of differences in the entire genomes of a population of individuals, e.g., single nucleotide polymorphisms (SNPs), DNA insertions, deletions, or copy number variations.

genomic DNA library A collection of recombinant plasmids in which all the DNA in the genome is represented. Compare with *cDNA library.*

genomic imprinting See *imprinting* (definition 1).

genomic rearrangement A physical change in the structure of one or more genes that occurs during the development of an organism and leads to an alteration in gene expression; compare with *nuclear equivalence* and *gene amplification.*

genomics The emerging field of biology that studies the entire DNA sequence of an organism's genome to identify all the genes, determine their RNA or protein products, and ascertain how the genes are regulated.

genotype (jeen′-oh-type) The genetic makeup of, or combination of alleles in, an individual. Compare with *phenotype.*

genotype frequency The proportion of a particular genotype in the population. Compare with *allele frequency* and *phenotype frequency.*

genus (jee′-nus) A taxonomic category made up of related species.

geographic variation Genetic differences that may exist among different populations within the same species. See *cline* for an example of one type of geographic variation.

geometric isomer One of two or more chemical compounds having the same arrangement of covalent bonds but differing in the spatial arrangement of their atoms or groups of atoms.

germination Resumption of growth of an embryo or spore; occurs when a seed or spore sprouts.

germ layers In animals, three embryonic tissue layers: endoderm, mesoderm, and ectoderm.

germ line cell In animals, a cell that is part of the line of cells that will ultimately undergo meiosis to form gametes. Compare with *somatic cell.*

germplasm Any plant or animal material that may be used in breeding; includes seeds, plants, and plant tissues of traditional crop varieties and the sperm and eggs of traditional livestock breeds.

gibberellin (jib″-ur-el′-lin) A plant hormone involved in many aspects of plant growth and development, such as stem elongation, flowering, and seed germination.

gigantism Abnormally tall growth caused when anterior pituitary secretes excessive amounts of human growth hormone during childhood.

gills (1) Moist, thin structures that extend outward from the body surface onto a respiratory medium (water or air); surfaces for gas exchange, mainly in aquatic animals. (2) The spore-bearing, platelike structures under the caps of mushrooms.

ginkgo (ging′-ko) A member of an ancient gymnosperm group that consists of a single living representative (*Ginkgo biloba*), a hardy, deciduous tree with broad, fan-shaped leaves and naked, fleshy seeds (on female trees).

gland See *endocrine gland* and *exocrine gland.*

glial cells (glee′-ul) In nervous tissue, cells that support and nourish neurons; they also communicate with neurons and have several other functions; also see *astrocyte, oligodendrocyte, ependymal cells,* and *microglia.*

globulin (glob′-yoo-lin) One of a class of proteins in blood plasma, some of which (gamma globulins) function as antibodies.

glomeromycetes A group of endomycorrhizal fungi that form arbuscular mycorrhizae with the roots of many plants; belong to phylum Glomeromycota.

glomerular filtration The process by which plasma is forced out of the capillaries and into Bowman's capsule as blood flows through the glomerular capillaries at high pressure; it is the first step in the production of urine.

glomerulus (glom-air′-yoo-lus) The cluster of capillaries at the proximal end of a nephron; the glomerulus is surrounded by Bowman's capsule.

glucagon (gloo′-kah-gahn) A hormone secreted by the pancreas that stimulates glycogen breakdown, thereby increasing the concentration of glucose in the blood. Compare with *insulin.*

glucocorticoids Steroid hormones, particularly *cortisol*, produced by the adrenal cortex that are involved in carbohydrate, protein, and fat metabolism. Glucocorticoids have anti-inflammatory properties.

glucose A hexose aldehyde sugar that is central to many metabolic processes.

glutamate An amino acid that functions as the major excitatory neurotransmitter in the vertebrate brain.

glyceraldehyde-3-phosphate (G3P) Phosphorylated three-carbon compound that is an important intermediate in glycolysis and in the Calvin cycle.

glycerol A three-carbon alcohol with a hydroxyl group on each carbon; a component of triacylglycerols and phospholipids, as well as monoacylglycerols and diacylglycerols.

glycine An amino acid that is the main inhibitory neurotransmitter in the spinal cord.

glycocalyx (gly″-koh-kay′-lix) A coating on the outside of an animal cell, formed by the polysaccharide portions of glycoproteins and glycolipids associated with the plasma membrane.

glycogen (gly′-koh-jen) The principal storage polysaccharide in animal cells; formed from glucose and stored primarily in the liver and, to a lesser extent, in muscle cells.

glycolipid A lipid with covalently attached carbohydrates.

glycolysis (gly-kol′-ih-sis) The first stage of cellular respiration as well as fermentation; literally the "splitting of sugar." The metabolic conversion of glucose into pyruvate, accompanied by the production of ATP.

glycoprotein (gly′-koh-pro-teen) A protein with covalently attached carbohydrates.

glycosidic linkage Covalent linkage joining two sugars; includes an oxygen atom bonded to a carbon of each sugar.

glyoxysomes (gly-ox′-ih-sohmz) Membrane-enclosed structures in cells of certain plant seeds; contain a large array of enzymes that convert stored fat to sugar.

gnathostomes (nath′-o-stomes) Vertebrates with jaws.

gnetophyte (nee′-toe-fite) One of a small phylum of unusual gymnosperms that have some features similar to those of flowering plants.

goblet cells Unicellular glands that secrete mucus.

goiter (goy′-ter) An enlargement of the thyroid gland.

golden alga A member of a group of algae, most of which are biflagellate, are unicellular, and contain pigments, including chlorophylls *a* and *c*, and carotenoids, including fucoxanthin.

Golgi complex (goal′-jee) Organelle composed of stacks of flattened, membranous sacs. Mainly responsible for modifying, packaging, and sorting proteins that will be secreted or targeted to other organelles of the internal membrane system or to the plasma membrane; also called *Golgi body* or *Golgi apparatus*.

Golgi tendon organ (goal′-jee) A proprioceptor that responds to tension in contracting muscles and in the tendons that attach muscle to bone.

gonad (goh′-nad) A gamete-producing gland; an ovary or a testis.

gonadotropic hormones (go-nad-oh-troh′-pic) Hormones produced by the anterior pituitary gland that stimulate the testes and ovaries; include follicle-stimulating hormone (FSH) and luteinizing hormone (LH).

gonadotropin-releasing hormone (GnRH) A hormone secreted by the hypothalamus that stimulates the anterior pituitary to secrete the gonadotropic hormones: follicle-stimulating hormone (FSH) and luteinizing hormone (LH).

graded potential A local change in electrical potential that varies in magnitude depending on the strength of the applied stimulus.

gradualism See *phyletic gradualism*.

graft rejection An immune response directed against a transplanted tissue or organ.

grain A simple, dry, one-seeded fruit in which the fruit wall is fused to the seed coat, e.g., corn and wheat kernels. Also called *caryopsis*.

gram-negative bacteria Bacteria with cell walls consisting of a thin peptidoglycan layer and a thick outer membrane. Compare with *gram-positive bacteria*.

gram-positive bacteria Bacteria with cell walls that are very thick and consist mainly of peptidoglycan. Compare with *gram-negative bacteria*.

granulosa cells In mammals, cells that surround the developing oocyte and are part of the follicle; produce estrogens and inhibin.

granum (pl., *grana*) A stack of thylakoids within a chloroplast.

Graves' disease Common form of hyperthyroidism, an autoimmune disease, characterized by extreme weight loss and emotional irritability.

gravitropism (grav″-ih-troh′-pizm) Growth of a plant in response to gravity.

gray crescent The grayish area of cytoplasm that marks the region where gastrulation begins in an amphibian embryo.

gray matter Nervous tissue in the brain and spinal cord that contains cell bodies, dendrites, and unmyelinated axons. Compare with *white matter*.

green alga A member of a diverse group of archaeplastid algae that contain the same pigments as plants (chlorophylls *a* and *b* and carotenoids).

greenhouse effect The natural global warming of Earth's atmosphere caused by the presence of carbon dioxide and other gases that trap the sun's radiation. The additional warming produced when increased levels of greenhouse gases absorb infrared radiation is known as the *enhanced greenhouse effect*.

greenhouse gases Trace gases in the atmosphere that allow the sun's energy to penetrate to Earth's surface but do not allow as much of it to escape as heat.

gross primary productivity (GPP) The rate at which energy accumulates (is assimilated) in an ecosystem during photosynthesis. Compare with *net primary productivity (NPP)*.

ground state The lowest energy state of an atom.

ground tissue system All tissues in the plant body other than the dermal tissue system and vascular tissue system; consists of parenchyma, collenchyma, and sclerenchyma.

groundwater Water stored underground in soil and crevasses, or an aquifer.

growth factors A group of more than 50 extracellular peptides that signal certain cells to grow and divide.

growth hormone (GH) A hormone secreted by the anterior lobe of the pituitary gland; stimulates growth of body tissues; also called *somatotropin*.

growth hormone–inhibiting hormone (GHIH) Hormone secreted by hypothalamus that inhibits growth hormone (GH) secretion.

growth hormone–releasing hormone (GHRH) Hormone secreted by hypothalamus that permits growth hormone (GH) secretion.

growth rate (*r*) The rate of change of a population's size on a per capita basis.

guanine (gwan′-een) A nitrogenous purine base that is a component of nucleic acids and GTP.

guanosine diphosphate (GDP) See *guanosine triphosphate*.

guanosine triphosphate (GTP) An energy transfer molecule similar to ATP that releases free energy with the hydrolysis of its terminal phosphate group, yielding guanosine diphosphate (GDP).

guard cell One of a pair of epidermal cells that adjust their shape to form a stomatal pore for gas exchange.

guttation (gut-tay′-shun) The appearance of water droplets on leaves, forced out through leaf pores by root pressure.

gymnosperm (jim′-noh-sperm) Any of a group of seed plants in which the seeds are not enclosed in an ovary; gymnosperms frequently bear their seeds in cones. Includes four phyla: conifers, cycads, ginkgoes, and gnetophytes.

habitat The natural environment or place where an organism, population, or species lives.

habitat corridors Strips of land connecting isolated habitat patches that allow wildlife to pass through freely.

habitat fragmentation The division of habitats that formerly occupied large, unbroken areas into smaller pieces by roads, fields, cities, and other human land-transforming activities.

habitat isolation A prezygotic reproductive isolating mechanism in which reproduction between similar species is prevented because they live and breed in different habitats.

habituation (hab-it″-yoo-ay′-shun) A type of learning in which an animal becomes accustomed to a repeated, irrelevant stimulus and no longer responds to it.

hadal zone Part of benthic environment of ocean deeper than 6000 m.

hagfishes Jawless marine fishes with a notochord for axial support; lack vertebrae and paired appendages.

hair cells Vertebrate mechanoreceptors that detect movement.

half-life The period of time required for a radioisotope to change into a different material.

haploid (hap'-loyd) The condition of having one set of chromosomes per nucleus. Compare with *diploid* and *polyploid*.

"hard-wiring" Refers to how neurons signal one another, how they connect, and how they carry out basic functions such as regulating heart rate, blood pressure, and sleep–wake cycles.

Hardy–Weinberg principle The mathematical prediction that allele frequencies do not change from generation to generation in a large population in the absence of microevolutionary processes (mutation, genetic drift, gene flow, natural selection).

haustorium (hah-stor'-ee-um) (pl., *haustoria*) In parasitic fungi, a specialized hypha that penetrates a host cell and obtains nourishment from the cytoplasm.

Haversian canals (ha-vur'-zee-un) Channels extending through the matrix of bone; contain blood vessels and nerves.

heat The total amount of kinetic energy in a sample of a substance.

heat energy The thermal energy that flows from an object with a higher temperature to an object with a lower temperature.

heat of vaporization The amount of heat energy that must be supplied to change one gram of a substance from the liquid phase to the vapor phase.

helicases Enzymes that unwind the two strands of a DNA double helix.

helper T cell See *T helper cell*.

hemichordates A phylum of sedentary, wormlike deuterostomes.

hemizygous (hem″-ih-zy'-gus) Possessing only one allele for a particular locus; a human male is hemizygous for all X-linked genes. Compare with *homozygous* and *heterozygous*.

hemocoel Blood cavity characteristic of animals with an open circulatory system.

hemocyanin A hemolymph pigment that transports oxygen in some mollusks and arthropods.

hemoglobin (hee'-moh-gloh″-bin) The red, iron-containing protein pigment in blood that transports oxygen and carbon dioxide and aids in regulation of pH.

hemolymph (hee'-moh-limf) The fluid that bathes the tissues in animals with an open circulatory system, e.g., arthropods and most mollusks.

hemophilia (hee″-moh-feel'-ee-ah) A hereditary disease in which blood does not clot properly; the forms known as *hemophilia A* (deficient for clotting factor VIII) and *hemophilia B* (deficient for clotting factor IX) exhibit an X-linked, recessive inheritance pattern.

Hensen's node See *primitive streak*.

hepatic (heh-pat'-ik) Pertaining to the liver.

hepatic portal system The portion of the circulatory system that carries blood from the intestine through the liver.

herbivore (erb'-uh-vore) An animal that feeds on plants or algae. Also called *primary consumer*.

heredity The transmission of genetic information from parent to offspring.

hermaphrodite (her-maf'-roh-dite) An organism that has both male and female sex organs; hermaphroditism is the form of sexual reproduction in which a single individual produces both eggs and sperm.

hermaphroditism See *hermaphrodite*.

heterochromatin (het″-ur-oh-kroh'-mah-tin) Highly coiled and compacted chromatin in an inactive state. Compare with *euchromatin*.

heterocyst (het'-ur-oh-sist″) An oxygen-excluding cell of cyanobacteria that is the site of nitrogen fixation.

heterogametic A term describing an individual that produces two classes of gametes with respect to their sex chromosome constitutions. Human males (XY) are heterogametic, producing X and Y sperm. Compare with *homogametic*.

heterospory (het″-ur-os'-pur-ee) Production of two types of *n* spores, microspores (male) and megaspores (female). Compare with *homospory*.

heterothallic (het″-ur-oh-thal'-ik) Pertaining to certain algae and fungi that have two mating types; only by combining a plus strain and a minus strain can sexual reproduction occur. Compare with *homothallic*.

heterotroph (het'-ur-oh-trof) An organism that cannot synthesize its own food from inorganic raw materials and therefore must obtain body-building materials from other organisms. Also called *consumer*. Compare with *autotroph*. See *chemoheterotroph* and *photoheterotroph*.

heterozygote advantage A phenomenon in which the heterozygous condition confers some special advantage on an individual that either homozygous condition does not (i.e., *Aa* has a higher degree of fitness than does *AA* or *aa*).

heterozygous (het-ur″-oh-zye'-gus) Having a pair of unlike alleles for a particular locus. Compare with *homozygous*.

hexose A monosaccharide containing six carbon atoms.

hibernation Long-term torpor in response to winter cold and scarcity of food. Compare with *estivation*.

high-density lipoprotein (HDL) See *lipoprotein*.

hindbrain In the early embryo, one of the three divisions of the developing vertebrate brain; subdivides to form the metencephalon, which gives rise to the cerebellum and pons, and the myelencephalon, which gives rise to the medulla. Compare with *forebrain* and *midbrain*.

hippocampus Part of the limbic system; important in forming and retrieving declarative memories.

histamine (his'-tah-meen) Substance released from mast cells that causes inflammation.

histones (his'-tohnz) Small, positively charged (basic) proteins in the cell nucleus that bind to the negatively charged DNA. See *nucleosomes*.

holdfast The basal structure for attachment to solid surfaces found in multicellular algae.

holoblastic cleavage A cleavage pattern in which the entire embryo cleaves; characteristic of eggs with little or moderate yolk (isolecithal or moderately telolecithal), e.g., the eggs of echinoderms, amphioxus, and mammals. Compare with *meroblastic cleavage*.

homeobox A short (180-nucleotide) DNA sequence that characterizes many homeotic genes as well as some other genes that play a role in development.

homeodomain A functional region of certain transcription factors; consists of approximately 60 amino acids specified by a homeobox DNA sequence and includes a recognition alpha helix, which binds to specific DNA sequences and affects their transcription.

homeostasis (home″-ee-oh-stay'-sis) The balanced internal environment of the body; the automatic tendency of an organism to maintain such a steady state.

homeostatic mechanisms (home″-ee-oh-stat'-ik) The regulatory mechanisms that maintain homeostasis.

homeotic gene (home″-ee-ah'-tik) A gene that controls the formation of specific structures during development. Such genes were originally identified through insect mutants in which one body part is substituted for another.

home range A geographic area that an individual animal seldom or never leaves. Compare with *range*.

hominid See *hominin*.

hominin (hah'-min-in) Any of a group of extinct and living humans. Also called *hominid*.

hominoid (hah'-min-oid) The apes and hominins (hominids).

Homo antecessor Early humans living in about 1.2 mya to 800,000 years ago; some scientists believe that they should be classified with *Homo heidelbergensis*.

Homo erectus Species that originated in Africa about 1.7 mya and then spread to Europe and Asia; may have persisted until 200,000 years ago or later; may be an evolutionary dead end.

Homo ergaster African species that lived from about 2.0 to 1.4 mya; may be a direct ancestor of later humans.

Homo habilis Oldest known species in genus *Homo;* lived in Africa from approximately 2.5 to 1.6 mya.

Homo heidelbergensis Early humans living from about 600,000 to 300,000 years ago; some scientists believe that fossils classified as *Homo antecessor* should be included in *Homo heidelbergensis*.

Homo neanderthalensis Human species living in Europe and western Asia from about 250,000 to 28,000 years ago; may have interbred with modern humans (*Homo sapiens*).

Homo sapiens Modern human species.

homogametic Term describing an individual that produces gametes with identical sex chromosome constitutions. Human females (XX) are homogametic, producing all X eggs. Compare with *heterogametic*.

homologous chromosomes (hom-ol′-ah-gus) Chromosomes that are similar in morphology and genetic constitution. In humans there are 23 pairs of homologous chromosomes; one member of each pair is inherited from the mother, and the other from the father.

homologous features See *homology*.

homologous traits See *homology*.

homology Similarity in different species that results from their derivation from a common ancestor. The features that exhibit such similarity are called *homologous features* or *homologous traits*. Compare with *homoplasy*.

homoplastic features See *homoplasy*.

homoplastic traits See *homoplasy*.

homoplasy Similarity in the characters in different species that is due to convergent evolution, not common descent. Characters that exhibit such similarity are called *homoplastic features* or *homoplastic traits*. Compare with *homology*.

homospory (hoh″-mos′-pur-ee) Production of one type of *n* spore that gives rise to a bisexual gametophyte. Compare with *heterospory*.

homothallic (hoh″-moh-thal′-ik) Pertaining to certain algae and fungi that are self-fertile. Compare with *heterothallic*.

homozygous (hoh″-moh-zy′-gous) Having a pair of identical alleles for a particular locus. Compare with *heterozygous*.

horizontal cell A lateral interneuron in the retina of the eye; receives signals from the photoreceptor cells and sends signals to bipolar cells and amacrine cells.

horizontal gene transfer The transfer of genetic material from one genome to another across species and even domains. Also called *lateral gene transfer*. Compare with *vertical gene transfer*.

hormone A chemical messenger, often produced in one region of the body of a multicellular organism and transported to another region where it signals cells to alter some aspect of growth, development, or metabolism.

hornwort A phylum of spore-producing, nonvascular, thallose plants with a life cycle similar to that of mosses.

horsetails Fern relatives with hollow, jointed stems, to which tiny leaves (reduced megaphylls) are attached in whorls.

host Organism that is adversely affected by a parasite member in a symbiotic relationship.

host mothering The introduction of an embryo from one species into the uterus of another species, where it implants and develops; the host mother subsequently gives birth and may raise the offspring as her own.

host range The specific types of host species a specific infectious agent can infect.

Hox genes Clusters of homeobox-containing genes that specify the anterior–posterior axis of various animals during development.

human chorionic gonadotropin (hCG) A hormone secreted by cells surrounding the early embryo; signals the mother's corpus luteum to continue to function.

human genetics The science of inherited variation in humans.

human genome The totality of genetic information in human cells; includes the DNA content of both the nucleus and mitochondria. See *genome*.

Human Genome Project Initiative completed in 2003 to determine the sequence of bases for a human haploid reference genome of more than 3 billion nucleotide base pairs.

human immunodeficiency virus (HIV) The retrovirus that causes AIDS (acquired immunodeficiency syndrome).

human leukocyte antigen complex (HLA) The MHC in humans. See *major histocompatibility complex*.

humus (hew′-mus) Organic matter in various stages of decomposition in the soil; gives soil a dark brown or black color.

Huntington's disease A genetic disease that has an autosomal dominant inheritance pattern and causes mental and physical deterioration.

hybrid The offspring of two genetically dissimilar parents.

hybrid breakdown A postzygotic reproductive isolating mechanism in which, although an interspecific hybrid is fertile and produces a second (F_2) generation, the F_2 has defects that prevent it from successfully reproducing.

hybrid inviability A postzygotic reproductive isolating mechanism in which the embryonic development of an interspecific hybrid is aborted.

hybridization (1) Interbreeding between members of two different taxa. (2) Interbreeding between genetically dissimilar parents. (3) In molecular biology, complementary base pairing between nucleic acid (DNA or RNA) strands from different sources.

hybrid sterility A postzygotic reproductive isolating mechanism in which an interspecific hybrid cannot reproduce successfully.

hybrid vigor The genetic superiority of an F_1 hybrid over either parent, caused by the presence of heterozygosity for a number of different loci.

hybrid zone An area of overlap between two closely related populations, subspecies, or species in which interbreeding occurs.

hydration Process of association of a substance with the partial positive and/or negative charges of water molecules.

hydrocarbon An organic compound composed solely of hydrogen and carbon atoms.

hydrogen bond A weak attractive force existing between a hydrogen atom with a partial positive charge and an electronegative atom (usually oxygen or nitrogen) with a partial negative charge. Compare with *covalent bond* and *ionic bond*.

hydrologic cycle The water cycle, which includes evaporation, precipitation, and flow to the ocean; supplies terrestrial organisms with a continual supply of fresh water.

hydrolysis reaction Reaction in which a covalent bond between two subunits is broken through the addition of the equivalent of a water molecule; a hydrogen atom is added to one subunit and a hydroxyl group to the other. Compare with *condensation reaction*.

hydrophilic Interacting readily with water; having a greater affinity for water molecules than they have for each other. Compare with *hydrophobic*.

hydrophobic Not readily interacting with water; having less affinity for water molecules than they have for each other. Compare with *hydrophilic*.

hydrophobic interactions The tendency of hydrophobic substances to cluster together due to strong cohesive interactions among surrounding water molecules.

hydroponics (hy″-dra-paun′-iks) Growing plants in an aerated solution of dissolved inorganic minerals, i.e., without soil.

hydrostatic skeleton A type of skeleton found in some invertebrates in which contracting muscles push against a tube of fluid.

hydrothermal vents Fissures in the deep ocean floor that release hot water and inorganic compounds.

hydroxide ion An anion (negatively charged particle) consisting of oxygen and hydrogen; usually written OH⁻.

hydroxyl group (hy-drok′-sil) Polar functional group; abbreviated —OH.

hyperpolarize To change the membrane potential so that the inside of the cell becomes more negative than its resting potential.

hyperpolarized The condition of the neuron membrane when the inside of the cell becomes more negative than its resting potential.

hypersecretion Abnormally increased output of a secreted material. Compare with *hyposecretion*.

hypertension Sustained, above normal, mean arterial blood pressure.

hypertonic A term referring to a solution having an osmotic pressure (or solute concentration) greater than that of the solution with which it is compared. Compare with *hypotonic* and *isotonic*.

hypha (hy'-fah) (pl., *hyphae*) One of the threadlike filaments composing the mycelium of a water mold or fungus.

hypocotyl (hy'-poh-kah"-tl) The part of the axis of a plant embryo or seedling below the point of attachment of the cotyledons.

hyposecretion Abnormally reduced output of a secreted material. Compare with *hypersecretion.*

hypothalamus (hy-poh-thal'-uh-mus) Part of the vertebrate brain; in mammals it regulates the pituitary gland, the autonomic system, emotional responses, body temperature, water balance, and appetite; located below the thalamus.

hypothesis A testable statement about the nature of an observation or relationship. Compare with *scientific theory.*

hypothetico-deductive approach Emphasizes the use of deductive reasoning to test hypotheses. Compare with *hypothetico-inductive approach.* See *deductive reasoning.*

hypothetico-inductive approach Emphasizes the use of inductive reasoning to discover new general principles. Compare with *hypothetico-deductive approach.* See *inductive reasoning.*

hypotonic A term referring to a solution having an osmotic pressure (or solute concentration) less than that of the solution with which it is compared. Compare with *hypertonic* and *isotonic.*

hypotrichs A group of dorsoventrally flattened ciliates that exhibit an unusual creeping–darting locomotion.

identical twins See *monozygotic twins.*

illuviation The deposition of material leached from the upper layers of soil into the lower layers.

imaginal discs Paired structures in an insect larva that develop into specific adult structures during complete metamorphosis.

imago (ih-may'-go) The adult form of an insect.

imbibition (im"-bi-bish'-en) The absorption of water by a seed prior to germination.

immigration The movement of individuals into a population. Compare with *emigration.*

immune response Process of recognizing foreign macromolecules and mounting a response aimed at eliminating them. See *innate immune responses* and *adaptive immune responses; primary immune responses* and *secondary immune responses.*

immune system The system of molecules and cells that protect the body against pathogens and other foreign agents. In vertebrates, includes lymphatic vessels and other lymph structures.

immunity The ability to recognize and destroy foreign or dangerous macromolecules.

immunodeficiency disease A condition that increases susceptibility to infection; the condition can be inherited or acquired. HIV is the major cause of acquired immunodeficiency in adults.

immunofluorescence Method used to localize molecules in cells or quantify specific molecules in biochemical assays by observing or measuring light emitted from a fluorescent dye joined to an antibody that binds to the specific molecule under study.

immunoglobulin (im-yoon"-oh-glob'-yoo-lin) See *antibody.*

immunological memory The ability of the immune system to remember antigen after contact with them so that it can respond more rapidly if it encounters the same antigens again. Memory B cells and memory T cells are responsible for immunological memory.

imperfect flower A flower that lacks either stamens or carpels. Compare with *perfect flower.*

imperfect fungi See *deuteromycetes.*

implantation The embedding of a developing embryo in the inner lining (endometrium) of the uterus.

implicit memory The unconscious memory for perceptual and motor skills, e.g., riding a bicycle.

imprinting (1) The expression of a gene based on its parental origin; also called *genomic imprinting* or *parental imprinting.* (2) A type of learning by which a young bird or mammal forms a strong social attachment to an individual (usually a parent) or object within a few hours after hatching or birth.

inborn error of metabolism A metabolic disorder caused by the mutation of a gene that codes for an enzyme needed for a biochemical pathway.

inbreeding depression The phenomenon in which inbred offspring of genetically similar individuals have lower fitness (e.g., decline in fertility and high juvenile mortality) than do noninbred individuals.

inbreeding The mating of genetically similar individuals. Homozygosity increases with each successive generation of inbreeding. Compare with *outbreeding.*

incisors Chisel-shaped teeth of a mammal; adapted for biting and cutting food. Compare with *canines, premolars,* and *molars.*

inclusive fitness The total of an individual's direct and indirect fitness; includes the genes contributed directly to offspring and those contributed indirectly by kin selection. Compare with *direct fitness.* See *kin selection.*

incomplete dominance A condition in which neither member of a pair of contrasting alleles is completely expressed when the other is present.

incomplete flower A flower that lacks one or more of the four parts: sepals, petals, stamens, and/or carpels. Compare with *complete flower.*

incomplete metamorphosis In insects, a life cycle during which the larva resembles the adult in many ways, but lacks functional wings and reproductive structures. Compare with *complete metamorphosis.*

independent assortment, principle of The genetic principle, first noted by Gregor Mendel, that states that the alleles of unlinked loci are randomly distributed to gametes.

indeterminate cleavage Type of cleavage characteristic of deuterostomes; during early cleavage each cell has the potential of developing into a complete embryo. Compare with *determinate cleavage.*

indeterminate growth Unrestricted growth, as for example, in stems and roots. Compare with *determinate growth.*

index fossils Fossils restricted to a narrow period of geologic time and found in the same sedimentary layers in different geographic areas.

individualistic model Independent view of community, described by Gleason, in which biological interactions are less important in the production of communities than are environmental gradients (such as climate and soil) or even chance. Compare with *organismic model.*

indoleacetic acid (IAA) See *auxin.*

induced fit Conformational change in the active site of an enzyme that occurs when it binds to its substrate.

induced pluripotent stem cell (iPSC) A pluripotent stem cell derived by "reprogramming" or modifying the gene expression patterns of an adult somatic cell.

inducer A molecule that binds to a repressor protein, converting it to its inactive form, which is unable to prevent transcription.

inducible operon An operon that is normally inactive because a repressor molecule is attached to its operator; transcription is activated when an inducer binds to the repressor, making it incapable of binding to the operator. Compare with *repressible operon.*

induction The process by which the differentiation of a cell or group of cells is influenced by interactions with neighboring cells.

inductive reasoning The reasoning that uses specific examples to draw a general conclusion or discover a general principle. Compare with *deductive reasoning.* See *hypothetico-inductive approach.*

infant mortality rate The number of infant deaths per 1000 live births. (A child is an infant during his or her first two years of life.)

inferior vena cava In humans, a very large vein that delivers blood from the lower part of the body to the right atrium. Compare with *superior vena cava.*

inflammation The response of body tissues to injury or infection; the body recruits white blood cells and plasma proteins from the blood to the location of an infection resulting in increased dilation of blood vessels and increased phagocytosis.

inflammatory response See *inflammation.*

inflorescence A cluster of flowers on a common floral stalk.

ingestion The process of taking food (or other material) into the body.

ingroup See *outgroup.*

inhibin A hormone that inhibits FSH secretion; produced by Sertoli cells in the testes and by granulosa cells in the ovaries.

inhibiting hormone A hormone secreted by the hypothalamus that inhibits secretion of a specific hormone by the anterior lobe of the pituitary gland.

inhibitory postsynaptic potential (IPSP) A change in membrane potential that takes a neuron farther from the firing level. Compare with *excitatory postsynaptic potential (EPSP)*.

initiation (of protein synthesis) The first steps of protein synthesis, in which the large and small ribosomal subunits and other components of the translation machinery bind to the 5′ end of mRNA. See *elongation* and *termination*.

initiation codon See *start codon*.

innate behavior Behavior that is inherited and typical of the species; also called *instinct*.

innate immune responses Mechanisms such as physical barriers (e.g., the skin) and phagocytosis that provide immediate and general protection against pathogens. Also called *nonspecific immune responses* or *nonspecific immunity*. Compare with *adaptive immune responses*.

innate immunity See *innate immune responses*.

inner cell mass The cluster of cells in the early mammalian embryo that gives rise to the embryo proper.

inorganic compound A simple substance that does not contain a carbon backbone. Compare with *organic compound*.

inositol trisphosphate (IP₃) A second messenger that increases intracellular calcium concentration and activates enzymes.

insight learning A complex learning process in which an animal adapts past experience to solve a new problem that may involve different stimuli.

in situ conservation Conservation efforts that concentrate on preserving biological diversity in the wild. Compare with *ex situ conservation*.

instinct See *innate behavior*.

insulin (in′-suh-lin) A hormone secreted by the pancreas that lowers blood glucose concentration. Compare with *glucagon*.

insulin resistance See *diabetes mellitus*.

insulin shock A condition in which the blood glucose concentration is so low that the individual may appear intoxicated or may become unconscious and even die; caused by the injection of too much insulin or by certain metabolic malfunctions.

insulin-like growth factors (IGFs) Proteins that mediate responses to growth hormone.

insulin-like growth factor 2 (IGF2) See *insulin-like growth factors*.

integral membrane protein A protein that is tightly associated with the lipid bilayer of a biological membrane; a transmembrane integral protein spans the bilayer. Compare with *peripheral membrane protein*.

integration The process of summing (adding and subtracting) incoming neural signals.

integrins Receptor proteins that bind to specific proteins in the extracellular matrix and to membrane proteins on adjacent cells; transmit signals into the cell from the extracellular matrix.

integumentary system (in-teg″-yoo-men′-tar-ee) The body's covering, including the skin and its nails, glands, hair, and other associated structures.

integuments The outer cell layers that surround the megasporangium of an ovule; develop into the seed coat.

intercellular substance In connective tissues, the combination of matrix and fibers in which the cells are embedded.

interference competition Intraspecific competition in which certain dominant individuals obtain an adequate supply of the limited resource at the expense of other individuals in the population. Also called *contest competition*. Compare with *exploitation competition*.

interferons (in″-tur-feer′-onz) Cytokines produced by animal cells when challenged by a virus; prevent viral reproduction and enable cells to resist a variety of viruses.

integrator In homeostasis, a control center.

intercalated discs (in-ter′-kuh-lay″-ted) The junctions between cells in cardiac muscle; contain gap junctions.

interkinesis The stage between meiosis I and meiosis II. Interkinesis is usually brief; the chromosomes may decondense, reverting at least partially to an interphase-like state, but DNA synthesis and chromosome duplication do not occur.

interleukins A diverse group of cytokines produced mainly by macrophages and lymphocytes.

intermediate disturbance hypothesis In community ecology, the idea that species richness is greatest at moderate levels of disturbance, which create a mosaic of habitat patches at different stages of succession.

intermediate filaments Cytoplasmic fibers that are part of the cytoskeletal network and are intermediate in size between microtubules and microfilaments.

intermediate-day plant A plant that flowers when it is exposed to days and nights of intermediate length but does not flower when the day length is too long or too short. Compare with *long-day, short-day,* and *day-neutral plants*.

internal fertilization The process in which the male delivers sperm into the body of the female (or close to the entrance of the reproductive tract). Compare with *external fertilization*.

internal membrane system See *endomembrane system*.

interneuron (in″-tur-noor′-on) A nerve cell that carries impulses from one nerve cell to another and is not directly associated with either an effector or a sensory receptor. Most interneurons are *association neurons*.

internode The region on a stem between two successive nodes. Compare with *node*.

intersexual selection The process by which females select their mates on the basis of some physical trait that indicates genetic quality or good health; sometimes selection depends on some

resource offered by the males. Compare with *intrasexual selection*.

interoceptor (in′-tur-oh-sep″-tor) A sense organ within a body organ that transmits information regarding chemical composition, pH, osmotic pressure, or temperature. Compare with *exteroceptor*.

interphase The stage of the cell cycle between successive mitotic divisions; its subdivisions are the G_1 (first gap), S (DNA synthesis), and G_2 (second gap) phases.

interspecific competition The interaction between members of different species that vie for the same resource in an ecosystem (e.g., food or living space). Compare with *intraspecific competition*.

interstitial cells (of testis) The cells between the seminiferous tubules that secrete testosterone.

interstitial fluid The fluid that bathes the tissues of the body; also called *tissue fluid*.

intertidal zone The marine shoreline area between the high-tide mark and the low-tide mark.

intrasexual selection The process by which individuals of the same sex actively compete for mates. Typically numerous males compete for a limited number of receptive females. Compare with *intersexual selection*.

intraspecific competition The interaction between members of the same species that vie for the same resource in an ecosystem (e.g., food or living space). Compare with *interspecific competition*.

intrinsic rate of increase (r_{max}) The theoretical maximum rate of increase in population size occurring under optimal environmental conditions. Also called *biotic potential*.

intron A non-protein-coding region of a eukaryotic gene and also of the pre-mRNA transcribed from such a region. Introns do not appear in mRNA. Compare with *exon*.

invasive species A foreign species that, when introduced into an area where it is not native, upsets the balance among the organisms living there and causes economic or environmental harm.

inversion A chromosome abnormality in which the breakage and rejoining of chromosome parts results in a chromosome segment that is oriented in the opposite (reverse) direction.

invertebrate An animal without a backbone (vertebral column); invertebrates account for about 95% of animal species.

in vitro Occurring outside a living organism (literally "in glass"). Compare with *in vivo*.

in vitro evolution Test tube experiments that demonstrate that RNA molecules in the RNA world could have catalyzed the many different chemical reactions needed for life. Also called *directed evolution*.

in vitro fertilization The fertilization of eggs in the laboratory prior to implantation in the uterus for development.

in vivo Occurring in a living organism. Compare with *in vitro*.

ion An atom or group of atoms bearing one or more units of electric charge, either positive (cation) or negative (anion).

ion channel–linked receptors Cell surface receptors that open or close specific ion channels in response to binding by a signaling molecule (ligand). Also called *ligand-gated channels*.

ion channels Channels for the passage of ions through a membrane; formed by specific membrane proteins.

ionic bond The chemical attraction between a cation and an anion. Compare with *covalent bond* and *hydrogen bond*.

ionic compound A substance consisting of cations and anions, which are attracted by their opposite charges; ionic compounds do not consist of molecules. Compare with *covalent compound*.

ionization The dissociation of a substance to yield ions, e.g., the ionization of water yields H$^+$ and OH$^-$.

iris The pigmented portion of the vertebrate eye.

iron–sulfur world hypothesis The hypothesis that simple organic molecules that are the precursors of life originated at hydrothermal vents in the deep-ocean floor. Compare with *prebiotic soup hypothesis*.

irreversible inhibitor A substance that permanently inactivates an enzyme. Compare with *reversible inhibitor*.

islets of Langerhans (eye′-lets of lahng′-er-hanz) The endocrine portion of the pancreas that secretes glucagon and insulin, hormones that regulate the concentration of glucose in the blood.

isogamy (eye-sog′-uh-me) Sexual reproduction involving motile gametes of similar form and size. Compare with *anisogamy* and *oogamy*.

isolecithal egg An egg containing a relatively small amount of uniformly distributed yolk. Compare with *telolecithal egg*.

isomer (eye′-soh-mer) One of two or more chemical compounds having the same chemical formula but different structural formulas, e.g., structural and geometrical isomers and enantiomers.

isoprene units Five-carbon hydrocarbon monomers that make up certain lipids such as carotenoids and steroids.

isotonic (eye″-soh-ton′-ik) A term applied to solutions that have identical concentrations of solute molecules and hence the same osmotic pressure. Compare with *hypertonic* and *hypotonic*.

isotope (eye′-suh-tope) An alternative form of an element with a different number of neutrons but the same number of protons and electrons. See *radioisotopes*.

iteroparous Having repeated reproductive cycles throughout a lifetime. Compare with *semelparous*.

jasmonic acid One of a group of lipid-derived plant hormones that affect several processes, such as pollen development, root growth, fruit ripening, and senescence; also involved in defense against insect pests and disease-causing organisms.

jelly coat One of the acellular coverings of the eggs of certain animals, such as echinoderms.

joint The junction between two or more bones of the skeleton.

joint receptors Proprioceptors that detect movement in ligaments.

joule A unit of energy, equivalent to 0.239 calorie.

Jurassic period See *Mesozoic era*.

juvenile hormone (JH) An arthropod hormone that preserves juvenile structure during a molt. Without it, metamorphosis toward the adult form takes place.

juxtaglomerular apparatus (juks″-tah-glo-mer′-yoo-lar) A structure in the kidney that secretes renin in response to a decrease in blood pressure.

K selection A reproductive strategy recognized by some ecologists in which a species typically has a large body size, slow development, and long lifespan and does not devote a large proportion of its metabolic energy to the production of offspring. Compare with *r selection*.

K strategist See *K selection*.

karyogamy (kar-e-og′-uh-me) The fusion of two haploid nuclei; follows fusion (plasmogamy) of cells from two sexually compatible mating types.

karyotype (kare′-ee-oh-type) The chromosomal composition of an individual.

kelp Large brown alga that grows in shallow ocean areas and forms underwater forests.

keratin (kare′-ah-tin) A horny, water-insoluble protein found in the epidermis of vertebrates and in nails, feathers, hair, and horns.

ketone An organic molecule containing a carbonyl group bonded to two carbon atoms. Compare with *aldehyde*.

keystone species A species whose presence in an ecosystem largely determines the species composition and functioning of that ecosystem.

kidney The paired vertebrate organ important in excretion of metabolic wastes and in osmoregulation.

killer T cell See *T cytotoxic cell*.

kilobase (kb) 1000 bases or base pairs of a nucleic acid.

kilocalorie The amount of heat required to raise the temperature of 1 kg of water 1°C; also called *Calorie*, which is equivalent to 1000 calories.

kilojoule 1000 joules. See *joule*.

kinases Enzymes that catalyze the transfer of phosphate groups from ATP to acceptor molecules. See *protein kinases*. Compare with *phosphatases*.

kinesin See *microtubule-associated proteins (MAPs)*.

kinetic energy Energy of motion. Compare with *potential energy*.

kinetochore (kin-eh′-toh-kore) The portion of the chromosome centromere to which the mitotic spindle fibers attach.

kinetoplastid A single mitochondrion with an organized deposit of DNA; characteristic of trypanosomes.

kinocilium The single, long cilium that projects from the surface of a vertebrate hair cell.

kingdom A broad taxonomic category made up of related phyla; many biologists currently assign living organisms to five kingdoms and several "supergroups."

kin selection A type of natural selection that favors altruistic behavior toward relatives (kin), thereby ensuring that although the chances of an individual's survival are lessened, some of its genes will survive through successful reproduction of close relatives; increases inclusive fitness.

Klinefelter syndrome Inherited condition in which the affected individual is a sterile male with an XXY karyotype.

knockout mice Mouse strain in which both alleles of a gene are functionally disabled in one or more tissues by genetic engineering methods.

Koch's postulates A set of guidelines used to demonstrate that a specific pathogen causes specific disease symptoms.

Krebs cycle See *citric acid cycle*.

krummholz The gnarled, shrublike growth habit found in trees at high elevations, near their upper limit of distribution.

K-selected species See *K selection*.

labyrinth The system of interconnecting canals of the inner ear of vertebrates.

labyrinthodonts The first successful group of tetrapods.

lactate (lactic acid) A three-carbon organic acid.

lactation (lak-tay′-shun) The production or release of milk from the breast.

lacteal (lak′-tee-al) One of the many lymphatic vessels in the intestinal villi that absorb fat.

lagging strand A strand of DNA that is synthesized as a series of short segments, called *Okazaki fragments*, which are then covalently joined by DNA ligase. Compare with *leading strand*.

lamins Polypeptides attached to the inner surface of the nuclear envelope that provide a type of skeletal framework.

lampreys Jawless freshwater and marine fishes with skeletons of cartilage, complete cranium, and rudimentary vertebrae.

landscape A large land area (several to many square kilometers) composed of interacting ecosystems.

landscape ecology The subdiscipline in ecology that studies the connections in a heterogeneous landscape.

large intestine The portion of the digestive tract of humans (and other vertebrates) consisting of the cecum, colon, rectum, and anus.

larva (pl., *larvae*) An immature form in the life history of some animals; may be unlike the parent.

larynx (lare'-inks) The organ at the upper end of the trachea that contains the vocal cords.

lateral line organ In fishes, a sensory organ that detects vibrations caused by waves and other movement in the water.

lateral meristems Areas of localized cell division on the side of a plant that give rise to secondary tissues. Lateral meristems, including the vascular cambium and the cork cambium, cause an increase in the girth of the plant body. Compare with *apical meristem.*

leaching The process by which dissolved materials are washed away or carried with water down through the various layers of the soil.

leader sequence Noncoding sequence of nucleotides in mRNA that is transcribed from the region that precedes (is upstream to) the coding region.

leading strand Strand of DNA that is synthesized continuously. Compare with *lagging strand.*

leaf scar Marks on a stem left when the petiole of a leaf breaks during abscission.

learning The process by which knowledge or skills are acquired as a result of experience; a change in the behavior of an animal that results from experience.

legume (leg'-yoom) (1) A simple, dry fruit that develops from a single carpel and splits open at maturity along two sutures to release seeds. (2) Any member of the pea family, e.g., pea, bean, peanut, alfalfa.

lek A small territory in which males compete for females.

lens The oval, transparent structure located behind the iris of the vertebrate eye; bends incoming light rays and brings them to a focus on the retina.

lenticels (len'-tih-sels) Porous swellings of cork cells in the stems of woody plants; facilitate the exchange of gases.

lepidosaurs (lep'-ih-do-sorz) A lineage of diapsid vertebrates that includes snakes, lizards, and tuataras.

leptin A hormone produced by adipose tissue that signals brain centers about the status of energy stores.

leukocytes (loo'-koh-sites) White blood cells; colorless amoeboid cells that defend the body against disease-causing organisms.

leukoplasts Colorless plastids; include amyloplasts, which are used for starch storage in cells of roots and tubers.

lichen (ly'-ken) A compound organism consisting of a symbiotic fungus and an alga or cyanobacterium.

life history traits Significant features of a species' life cycle, particularly traits that influence survival and reproduction.

lifespan The maximum duration of life for an individual of a species.

life table A table showing mortality and survival data by age of a population or cohort.

ligament (lig'-uh-ment) A connective tissue cable or strap that connects bones to each other or holds other organs in place.

ligand A molecule that binds to a specific site in a receptor or other protein.

ligand-gated channels See *ion channel–linked receptors.*

light-dependent reactions Reactions of photosynthesis in which light energy absorbed by chlorophyll is used to synthesize ATP and usually NADPH. Include *cyclic electron transport* and *noncyclic electron transport.*

light microscope (LM) Microscope in which light is refracted (bent) by glass lenses to produce a magnified image.

lignin (lig'-nin) A substance found in many plant cell walls that confers rigidity and strength, particularly in woody tissues.

limbic system In vertebrates, an action system of the brain. In humans, plays a role in emotional responses, motivation, sexual behavior, autonomic responses, and biological rhythms.

limiting resource An environmental resource that because it is scarce or unfavorable tends to restrict the ecological niche of an organism.

limnetic zone (lim-net'-ik) The open water away from the shore of a lake or pond extending down as far as sunlight penetrates. Compare with *littoral zone* and *profundal zone.*

linkage The tendency for a group of genes located on the same chromosome to be inherited together in successive generations.

linked genes See *linkage.*

lipase (lip'-ase) A fat-digesting enzyme.

lipid Any of a group of organic compounds that are insoluble in water but soluble in nonpolar solvents; lipids serve as energy storage and are important components of cell membranes.

lipoprotein (lip-oh-proh'-teen) A large molecular complex consisting of lipids and protein; transports lipids in the blood. High-density lipoproteins (HDLs) transport cholesterol to the liver; low-density lipoproteins (LDLs) deliver cholesterol to many cells of the body.

littoral zone (lit'-or-ul) The region of shallow water along the shore of a lake or pond. Compare with *limnetic zone* and *profundal zone.*

liver A large, complex organ that secretes bile, helps maintain homeostasis by removing or adding nutrients to the blood, and performs many other metabolic functions.

liverworts A phylum of spore-producing, nonvascular, thallose or leafy plants with a life cycle similar to that of mosses.

local hormones See *local regulators.*

local regulators Prostaglandins (a group of local hormones), growth factors, cytokines, and other soluble molecules that act on nearby cells by paracrine regulation or act on the cells that produce them (autocrine regulation).

locus (pl., *loci*) The place on the chromosome at which the gene for a given trait occurs, i.e., a segment of the chromosomal DNA containing information that controls some feature of the organism; also called *gene locus.*

logistic population growth Population growth that initially occurs at a constant rate of increase over time (i.e., exponential) but then levels out as the carrying capacity of the environment is approached. Compare with *exponential population growth.*

long-day plant A plant that flowers in response to shortening nights; also called *short-night plant.* Compare with *short-day, intermediate-day,* and *day-neutral plants.*

long-night plant See *short-day plant.*

long noncoding RNAs (lncRNAs) Molecules longer than 200 bases responsible for controlling genes by regulating chromatin structure.

long-term memory Memory that stores information for long periods of time; the hippocampus temporarily holds new information and helps place our experiences into categories so they can be consolidated and transferred to the cerebral cortex where they are stored along with similar memories. Compare with *short-term memory.*

long-term potentiation (LTP) Long-lasting increase in the strength of synaptic connections that occurs in response to a series of high-frequency electrical stimuli. Compare with *long-term synaptic depression (LTD).*

long-term synaptic depression (LTD) Long-lasting decrease in the strength of synaptic connections that occurs in response to low-frequency stimulation of neurons. Compare with *long-term potentiation (LTP).*

loop of Henle (hen'-lee) The U-shaped loop of a mammalian kidney tubule that extends down into the renal medulla.

loose connective tissue A type of connective tissue that is widely distributed in the body; consists of fibers strewn through a semifluid matrix.

lophophorates (lof-ah-for'ates) Members of three related invertebrate protostome phyla, characterized by a lophophore, a ciliated ring of tentacles that surrounds the mouth.

Lophotrochozoa A branch of the protostomes that includes the flatworms, nemerteans (proboscis worms), mollusks, annelids, and the lophophorate phyla.

low-density lipoprotein (LDL) See *lipoprotein.*

LUCA Acronym referring to the hypothesized **l**ast **u**niversal **c**ommon **a**ncestor of all living things on Earth today.

lumen (loo'-men) (1) The space enclosed by a membrane, such as the lumen of the endoplasmic reticulum or the thylakoid lumen. (2) The cavity or channel within a tube or tubular organ, such as a blood vessel or the digestive tract. (3) The space left within a plant cell after the cell's living material dies, as in tracheids.

lung An internal respiratory organ that functions in gas exchange; enables an animal to breathe air.

luteinizing hormone (LH) (loot′-eh-ny-zing) Gonadotropic hormone secreted by the anterior pituitary; stimulates ovulation and maintains the corpus luteum in the ovaries of females; stimulates testosterone production in the testes of males.

lymph (limf) The colorless fluid within the lymphatic vessels that is derived from blood plasma; contains white blood cells; ultimately lymph is returned to the blood.

lymphatic system A subsystem of the cardiovascular system; returns excess interstitial fluid (lymph) to the circulation; defends the body against disease organisms.

lymph node A mass of lymph tissue surrounded by a connective tissue capsule; manufactures lymphocytes and filters lymph.

lymphocyte (lim′-foh-site) White blood cell with nongranular cytoplasm that governs immune responses. See *B cell* and *T cell*.

lysis (ly′-sis) The process of disintegration of a cell or some other structure.

lysogenic conversion The change in properties of bacteria that results from the presence of a prophage.

lysogenic cycle (ly-so jen′-ik) A type of phage reproductive cycle in which the viral genome usually becomes integrated into the host DNA as a prophage; it is replicated along with the host DNA and does not kill the host cell. Compare with *lytic cycle.*

lysosomes (ly′-soh-sohmz) Intracellular organelles present in many animal cells; contain a variety of hydrolytic enzymes.

lysozyme An enzyme found in many tissues and in tears and other body fluids; attacks the cell wall of many gram-positive bacteria.

lytic cycle (lit′-ik) A type of phage reproductive cycle in which the release of phages from the bacterial cell results in rapid cell lysis (destruction). Compare with *lysogenic cycle.*

M phase That part of the cell cycle in which mitosis and cytokinesis occur.

macroevolution Large-scale evolutionary events over long time spans. Macroevolution results in phenotypic changes in populations that are significant enough to warrant their placement in taxonomic groups at the species level and higher. Compare with *microevolution.*

macromolecule A very large organic molecule, such as a protein or nucleic acid.

macronucleus A large nucleus found, along with one or several micronuclei, in ciliates. The macronucleus regulates metabolism and growth. Compare with *micronucleus.*

macronutrient An essential element required in fairly large amounts for normal growth. Compare with *micronutrient.*

macrophage (mak′-roh-faje) A large phagocytic cell capable of ingesting and digesting bacteria

and cell debris. Macrophages are also antigen-presenting cells.

magnification The ratio of the size of an image as seen with a microscope to its actual size.

magnoliid One of the clades of flowering plants; magnoliids are core angiosperms that were traditionally classified as "dicots," but molecular evidence indicates they are neither eudicots or monocots.

major histocompatibility complex (MHC) A large cluster of loci that encode the self-antigens present on the surface of most cells; these self-antigens are slightly different in each individual. The term refers to both the loci and to the self-antigens. Also see *human leukocyte antigen complex.*

malignant cells Cancer cells; tumor cells that are able to invade tissue and metastasize.

malignant transformation See *transformation.*

malnutrition Poor nutritional status; results from dietary intake that is either below or above required needs.

Malpighian tubules (mal-pig′-ee-an) The excretory organs of many arthropods.

mammals The class of vertebrates characterized by hair, mammary glands, a diaphragm, and differentiation of teeth.

mammary glands Exocrine glands that secrete milk for the young; a derived character of mammals.

mandible (man′-dih-bl) (1) The lower jaw of vertebrates. (2) Jawlike, external mouthparts of insects.

mangrove forest A tidal wetland dominated by mangrove trees in which the salinity fluctuates between that of sea water and fresh water.

mantle In the mollusk, a fold of tissue that covers the visceral mass and that usually produces a shell.

map unit A unit of a genetic map of a chromosome; 1% recombination between two loci equals one map unit.

marine snow The organic debris (plankton, dead organisms, fecal material, etc.) that "rains" into the dark area of the oceanic province from the lighted region above; the primary food of most organisms that live in the ocean's depths.

marsupials (mar-soo′-pee-ulz) A subclass of mammals, characterized by the presence of an abdominal pouch in which the young, which are born in a very undeveloped condition, are carried for some time after birth.

mass extinction The extinction of numerous species during a relatively short period of geologic time. Compare with *background extinction.*

mast cell A type of cell found in connective tissue; contains histamine and is important in an inflammatory response and in allergic reactions.

mate guarding The male guards his partner after copulation to ensure that she does not copulate with another male.

maternal effect genes Genes of the mother that are transcribed during oogenesis and subsequently

affect the development of the embryo. Compare with *zygotic genes.*

matrix (may′-triks) (1) In cell biology, the interior of the compartment enclosed by the inner mitochondrial membrane. (2) In zoology, nonliving material secreted by and surrounding connective tissue cells; contains a network of microscopic fibers.

matter Anything that has mass and takes up space.

maxillae Appendages used for manipulating food; characteristic of crustaceans.

mechanical isolation A prezygotic reproductive isolating mechanism in which fusion of the gametes of two species is prevented by morphological or anatomical differences.

mechanoreceptor (meh-kan′-oh-ree-sep″-tor) A sensory cell or organ that perceives mechanical stimuli, e.g., touch, pressure, gravity, stretching, or movement.

medulla (meh-dul′-uh) (1) The inner part of an organ, such as the medulla of the kidney. Compare with *cortex.* (2) The most posterior part of the vertebrate brain, lying next to the spinal cord.

medusa A jellyfish-like animal; a free-swimming, umbrella-shaped stage in the life cycle of certain cnidarians. Compare with *polyp.*

megaphyll (meg′-uh-fil) Type of leaf found in horsetails, ferns, gymnosperms, and angiosperms; contains multiple vascular strands (i.e., complex venation). Compare with *microphyll.*

megasporangium (pl., *megasporangia*) In a heterosporous plant, structure in which megaspores are produced. Megasporangia of seed plants are *ovules* that develop into seeds following fertilization. Compare with *microsporangium.*

megaspore (meg′-uh-spor) The *n* spore in heterosporous plants that gives rise to a female gametophyte. Compare with *microspore.*

megasporocyte Cell within a megasporangium that undergoes meiosis to produce four *n* megaspores; also called *megaspore mother cell.* Compare with *microsporocyte.*

meiosis (my-oh′-sis) Process in which a 2*n* cell undergoes two successive nuclear divisions (meiosis I and meiosis II), potentially producing four *n* nuclei; leads to the formation of gametes in animals and spores in plants.

meiosis I The first meiotic division; consists of prophase I, metaphase I, anaphase I, and telophase I; see *meiosis.*

meiosis II The second meiotic division; follows interkinesis and consists of prophase II, metaphase II, anaphase II, and telophase II; see *meiosis.*

melanin A dark pigment present in many animals; contributes to the color of the skin.

melanocortins A group of peptides that appear to decrease appetite in response to increased fat stores.

melanocyte-stimulating hormones (MSH) See *melanocortins.*

melatonin (mel-ah-toh'-nin) A hormone secreted by the pineal gland that plays a role in setting circadian rhythms.

membrane potential A difference in electric charge between the two sides of a membrane.

memory The process of encoding, storing, and retrieving information or learned skills.

memory B cells B cells that continue to produce antibodies after the immune system overcomes an infection.

memory cells B or T cells (lymphocytes) that permit rapid mobilization of immune response on second or subsequent exposure to a particular antigen. See *memory B cells* and *memory T cells.*

memory T cells Cytotoxic or helper T cells that, after participating in an immune response, enter an inactive state until activated by exposure to the same type of antigen.

meninges (meh-nin'-jeez) (sing., *meninx*) The three membranes that protect the brain and spinal cord: the dura mater, arachnoid, and pia mater.

menopause The period (usually occurring between 45 and 55 years of age) in women when the recurring menstrual cycle ceases.

menstrual cycle (men'-stroo-ul) In the human female, the monthly sequence of events that prepares the body for pregnancy.

menstruation (men-stroo-ay'-shun) The monthly discharge of blood and degenerated uterine lining in the human female; marks the beginning of each menstrual cycle.

meristem (mer'-ih-stem) A localized area of mitotic cell division in the plant body. See *apical meristem* and *lateral meristems.*

meroblastic cleavage Cleavage pattern observed in the telolecithal eggs of reptiles and birds, in which cleavage is restricted to a small disc of cytoplasm at the animal pole. Compare with *holoblastic cleavage.*

mesencephalon See *midbrain.*

mesenchyme (mes'-en-kime) A loose, often jelly-like connective tissue containing undifferentiated cells; found in the embryos of vertebrates and the adults of some invertebrates.

mesoderm (mez'-oh-derm) The middle germ layer of the early embryo; gives rise to connective tissue, muscle, bone, blood vessels, kidneys, and many other structures. Compare with *ectoderm* and *endoderm.*

mesoglea (mez-o-glee'-ah) Jellylike layer between the epidermis and gastrodermis in cnidarians and ctenophores.

mesophyll (mez'-oh-fil) Photosynthetic tissue in the interior of a leaf; sometimes differentiated into palisade mesophyll and spongy mesophyll.

Mesozoic era Span of geologic time commonly known as the Age of Reptiles; divided into Triassic (252–201 mya), Jurassic (201–145 mya), and Cretaceous (145–66 mya) periods; ended with the mass extinction of the dinosaurs and many other species.

messenger RNA (mRNA) RNA that specifies the amino acid sequence of a protein; transcribed from DNA.

metabolic pathway A series of chemical reactions in which the product of one reaction becomes the substrate of the next reaction.

metabolic rate Energy use by an organism per unit time. See *basal metabolic rate (BMR).*

metabolic syndrome Condition that often precedes type 2 diabetes. Symptoms include obesity, large waist circumference, elevated blood glucose concentration, high triglyceride and low HDL ("good" cholesterol) concentrations, and high blood pressure.

metabolism The sum of all the chemical processes that occur within a cell or organism; the transformations by which energy and matter are made available for use by the organism. See *anabolism* and *catabolism.*

metabolism first hypothesis An explanation of the origin of life, in which life began as a self-sustaining, organized system consisting of chemical reactions between simple molecules enclosed within a boundary.

metamorphosis (met"-ah-mor'-fuh-sis) Transition from one developmental stage to another, such as from a larva to an adult. See also *complete metamorphosis* and *incomplete metamorphosis.*

metanephridia (sing., *metanephridium*) The excretory organs of annelids and mollusks; each consists of a tubule open at both ends; at one end a ciliated funnel opens into the coelom, and the other end opens to the outside of the body.

metaphase (met'-ah-faze) The stage of mitosis in which the chromosomes line up on the midplane of the cell (metaphase plate). Occurs after prometaphase and before anaphase.

metaphase I See *meiosis* and *meiosis I.*

metaphase II See *meiosis* and *meiosis II.*

metaphase plate See *metaphase.*

metapopulation A population that is divided into several local populations among which individuals occasionally disperse.

metarterioles Small blood vessels that directly link arterioles with veinules.

metastasis (met-tas'-tuh-sis) The spreading of cancer cells from one organ or part of the body to another.

metencephalon Part of the vertebrate embryonic brain; gives rise to the cerebellum and pons.

methanogens (meth-an'-o-jens) Archaeons that inhabit oxygen-free environments, are obligate anaerobes, and produce methane, an important greenhouse gas.

methyl group A nonpolar functional group; abbreviated —CH_3.

microbiome A community of microorganisms, including their genomes and all their interactions.

microbiota A community of microorganisms that inhabits a particular habitat or site.

microclimate Local variations in climate produced by differences in elevation, in the steepness and direction of slopes, and in exposure to prevailing winds.

microevolution Small-scale evolutionary change caused by changes in allele or genotype frequencies that occur within a population over a few generations. Compare with *macroevolution.*

microfilaments Thin fibers consisting of actin protein subunits; form part of the cytoskeleton.

microfossils Ancient traces (fossils) of microscopic life.

microglia Phagocytic glial cells found in the CNS.

micronucleus One or more smaller nuclei found, along with the macronucleus, in ciliates. The micronucleus is involved in sexual reproduction. Compare with *macronucleus.*

micronutrient An essential element that is required in trace amounts for normal growth. Compare with *macronutrient.*

microphyll (mi'-kro-fil) Type of leaf found in club mosses; contains one vascular strand (i.e., simple venation). Compare with *megaphyll.*

microRNA (miRNA) Single-stranded RNA molecules about 22 nucleotides in length that control the expression of genes at the posttranscriptional level by binding mRNA molecules with base sequences complementary to the miRNA, blocking translation of the mRNA or causing it to be degraded. Also involved in control of transcription of some genes. See also *RNA interference (RNAi).*

microsphere A protobiont produced by adding water to abiotically formed polypeptides.

microsporangium (pl., *microsporangia*) In a heterosporous plant, structure in which microspores are produced; compare with *megasporangium.*

microspore (mi'-kro-spor) The *n* spore in heterosporous plants that gives rise to a male gametophyte. Compare with *megaspore.*

microsporocyte Cell within a microsporangium that undergoes meiosis to produce four *n* microspores; also called *microspore mother cell.* Compare with *megasporocyte.*

microsporidia Small, unicellular, fungal parasites that infect eukaryotic cells; classified with the zygomycetes.

microtubule-associated proteins (MAPs) Structural proteins that help regulate microtubule assembly and cross-link microtubules to other cytoskeletal polymers; and motors, such as kinesin and dynein, that use ATP to produce movement.

microtubule-organizing center (MTOC) The region of the cell from which microtubules are anchored and possibly assembled. The MTOCs of many organisms (including animals, but not flowering plants or most gymnosperms) contain a pair of centrioles.

microtubules (my-kroh-too'-bewls) Hollow, cylindrical fibers consisting of tubulin protein subunits; major components of the cytoskeleton and found in mitotic spindles, cilia, flagella, centrioles, and basal bodies.

microvilli (sing., *microvillus*) Minute projections of the plasma membrane that increase the surface area of the cell; found mainly in cells concerned with absorption or secretion, such as those lining the intestine or the kidney tubules.

midbrain In vertebrate embryos, one of the three divisions of the developing brain. Also called *mesencephalon*. Compare with *forebrain* and *hindbrain*.

middle lamella The layer composed of pectin polysaccharides that serves to cement together the primary cell walls of adjacent plant cells.

midvein The main, or central, vein of a leaf.

migration (1) The periodic or seasonal movement of an organism (individual or population) from one place to another, usually over a long distance. See *dispersal*. (2) In evolutionary biology, a movement of individuals that results in a transfer of alleles from one population to another. See *gene flow*.

mineralocorticoids (min″-ur-al-oh-kor′-tih-koidz) Hormones produced by the adrenal cortex that regulate mineral metabolism and, indirectly, fluid balance. The principal mineralocorticoid is aldosterone.

minerals Inorganic nutrients ingested as salts dissolved in food and water.

minimum viable population (MVP) The smallest population size at which a species has a high chance of sustaining its numbers and surviving into the future.

mismatch repair A DNA repair mechanism in which special enzymes recognize the incorrectly paired nucleotides and remove them. DNA polymerases then fill in the missing nucleotides.

missense mutation A type of base-pair substitution mutation that causes one amino acid to be substituted for another in the resulting protein product. Compare with *nonsense mutation*.

mitochondria (my″-toh-kon′-dree-ah) (sing., *mitochondrion*) Intracellular organelles that are the sites of oxidative phosphorylation in eukaryotes; include an outer membrane and an inner membrane.

mitochondrial DNA (mtDNA) DNA present in mitochondria that is transmitted maternally, from mothers to their offspring. Mitochondrial DNA mutates more rapidly than nuclear DNA.

mitosis (my-toh′-sis) The division of the cell nucleus resulting in two daughter nuclei, each with the same number of chromosomes as the parent nucleus; mitosis consists of prophase, prometaphase, metaphase, anaphase, and telophase. Cytokinesis usually overlaps the telophase stage.

mitotic spindle Structure consisting mainly of microtubules that provides the framework for chromosome movement during cell division.

mitral valve See *atrioventricular valve*.

mobile genetic element See *transposon*.

model organism A species chosen for biological studies because it has characteristics that allow for the efficient analysis of biological processes. Most model organisms are small, have short generation times, and are easy to grow and study under controlled conditions.

modern synthesis A comprehensive, unified explanation of evolution based on combining previous scientific theories, especially of Mendelian genetics with Darwin's scientific theory of evolution by natural selection.

molars Posterior teeth of a mammal; flattened for crushing and grinding; compare with *incisors, canines,* and *premolars*.

mole The atomic mass of an element or the molecular mass of a compound, expressed in grams; one mole of any substance has 6.02×10^{23} units (Avogadro's number).

molecular anthropology The branch of science that compares genetic material from individuals of regional human populations to help unravel the origin and migrations of modern humans.

molecular chaperones Proteins that help other proteins fold properly. Although chaperones do not dictate the folding pattern, they make the process more efficient.

molecular clock analysis A comparison of the DNA nucleotide sequences of related organisms to estimate when they diverged from one another during the course of evolution.

molecular formula The type of chemical formula that gives the actual numbers of each type of atom in a molecule. Compare with *simplest formula* and *structural formula*.

molecular mass The sum of the atomic masses of the atoms that make up a single molecule of a compound; expressed in atomic mass units (amu) or daltons.

molecular systematics The science that focuses on molecular structure to clarify evolutionary relationships.

molecule The smallest particle of a covalently bonded element or compound; two or more atoms joined by covalent bonds.

mollusks A phylum of coelomate protostome animals characterized by a soft body, visceral mass, mantle, and foot.

molting The shedding and replacement of an outer covering such as an exoskeleton.

molting hormone A steroid hormone that stimulates growth and molting in insects. Also called *ecdysone*.

monoacylglycerol (mon″-o-as-il-glis′-er-ol) Lipid consisting of glycerol combined chemically with a single fatty acid. Also called *monoglyceride*. Compare with *diacylglycerol* and *triacylglycerol*.

monoclonal antibodies Identical antibody molecules produced by cells cloned from a single cell.

monocot (mon′-oh-kot) One of two classes of flowering plants; monocot seeds contain a single cotyledon, or seed leaf. Compare with *eudicot*.

monocyte (mon′-oh-site) A type of white blood cell; a large, phagocytic, nongranular leukocyte that enters the tissues and differentiates into a macrophage.

monoecious (mon-ee′-shus) Having male and female reproductive parts in separate flowers or cones on the same plant; compare with *dioecious*.

monogamy A mating system in which a male animal mates with a single female during a breeding season.

monoglyceride See *monoacylglycerol*.

monohybrid cross A genetic cross that takes into account the behavior of alleles of a single locus. Compare with *dihybrid cross*.

monokaryotic (mon″-o-kare-ee-ot′-ik) The condition of having a single *n* nucleus per cell, characteristic of certain fungal hyphae. Compare with *dikaryotic*.

monomer (mon′-oh-mer) A molecule that can link with other similar molecules; two monomers join to form a dimer, whereas many form a polymer. Monomers are small (e.g., sugars or amino acids) or large (e.g., tubulin or actin proteins).

monophyletic group (mon″-oh-fye-let′-ik) A group of organisms that includes a recent common ancestor and all its descendants; a clade. Compare with *polyphyletic group* and *paraphyletic group*.

monosaccharide (mon-oh-sak′-ah-ride) A sugar that cannot be degraded by hydrolysis to a simpler sugar (e.g., glucose or fructose).

monosomy A type of aneuploidy in which an individual lacks one member of a pair of chromosomes; designated $2n - 1$.

monotremes (mon′-oh-treemz) Egg-laying mammals such as the duck-billed platypus of Australia.

monounsaturated fatty acid See *fatty acid*.

monozygotic twins Genetically identical twins that arise from the division of a single fertilized egg; commonly known as *identical twins*. Compare with *dizygotic twins*.

Montreal Protocol International agreement designed to protect the ozone layer by gradually eliminating the production of ozone-depleting substances.

morphogen Any chemical agent thought to govern the processes of cell differentiation and pattern formation that lead to morphogenesis.

morphogenesis (mor-foh-jen′-eh-sis) The development of the form and structures of an organism and its parts; proceeds through a series of steps known as *pattern formation*.

mortality The rate at which individuals die; the average per capita death rate.

morula (mor′-yoo-lah) An early embryo consisting of a solid ball of cells.

mosaic development A rigid developmental pattern in which the fates of cells become restricted early in development. Compare with *regulative development*.

mosses A phylum of spore-producing nonvascular plants with an alternation of generations in which the dominant *n* gametophyte alternates with a 2*n* sporophyte that remains attached to the gametophyte.

motor neuron An efferent neuron that transmits impulses away from the central nervous system to skeletal muscle.

motor program A coordinated sequence of muscle actions responsible for many behaviors we think of as automatic.

motor unit Functional unit consisting of a single motor neuron and the muscle fibers it innervates.

mRNA cap An unusual nucleotide, 7-methylguanylate, that is added to the 5′ end of a eukaryotic messenger RNA. Capping enables eukaryotic ribosomes to bind to mRNA.

mucosa (mew-koh′-suh) See *mucous membrane.*

mucous membrane A type of epithelial membrane that lines a body cavity that opens to the outside of the body, e.g., the digestive and respiratory tracts; also called *mucosa.*

mucus (mew′-cus) A sticky secretion composed of covalently linked protein and carbohydrate; serves to lubricate body parts and trap particles of dirt and other contaminants. (The adjectival form is spelled *mucous.*)

Müllerian mimicry (mul-ler′-ee-un mim′-ih-kree) The resemblance of dangerous, unpalatable, or poisonous species to one another so that potential predators recognize them more easily. Compare with *Batesian mimicry.*

multiple alleles (al-leelz′) Three or more alleles of a single locus (in a population), such as the alleles governing the ABO series of blood types.

multiple fruit A fruit that develops from many ovaries of many separate flowers, e.g., pineapple. Compare with *simple, aggregate,* and *accessory fruits.*

muscle (1) A tissue specialized for contraction. (2) An organ that produces movement by contraction.

muscle fibers Muscle cells; elongated cells that contain myofibrils, the contractile units.

muscle spindles Proprioceptors that detect muscle movement.

muscle tone The continuous, partial contraction of individual muscles and muscle groups.

mutagen (mew′-tah-jen) Any agent capable of entering the cell and producing mutations.

mutation Any change in DNA; may include a change in the nucleotide base pairs of a gene, a rearrangement of genes within the chromosomes so that their interactions produce different effects, or a change in the chromosomes themselves.

mutualism In ecology, a symbiotic relationship in which both partners benefit from the association. Compare with *parasitism* and *commensalism.*

mycelium (my-seel′-ee-um) (pl., *mycelia*) The vegetative body of most fungi and certain protists (water molds); consists of a branched network of hyphae.

mycology The study of fungi.

mycorrhizae (my″-kor-rye′-zee) Mutualistic associations of fungi and plant roots that aid in the plant's absorption of essential minerals from the soil. Hyphae of ectomycorrhizal fungi coat plant roots; those of endomycorrhizal fungi penetrate plant cell walls (see *arbuscules*).

mycotoxins Poisonous chemical compounds produced by fungi, e.g., aflatoxins that harm the liver and are known carcinogens.

myelencephalon Part of the vertebrate embryonic brain; gives rise to the medulla.

myelin sheath (my′-eh-lin) The white, fatty material that forms a sheath around the axons of certain nerve cells, which are then called *myelinated fibers.*

myocardial infarction (MI) Heart attack; serious consequence occurring when the heart muscle receives insufficient oxygen.

myofibrils (my-oh-fy′-brilz) Tiny threadlike structures in the cytoplasm of striated and cardiac muscle that are composed of myosin filaments and actin filaments; these filaments are responsible for muscle contraction; see *myosin filaments* and *actin filaments.*

myofilament See *filament.*

myoglobin (my′-oh-glo″-bin) A hemoglobin-like, oxygen-transferring protein found in muscle.

myosin (my′-oh-sin) A protein that together with actin is responsible for muscle contraction.

myosin filaments Thick filaments consisting mainly of the protein myosin; actin and myosin filaments make up the myofibrils of muscle fibers.

myxedema Hypothyroid condition in which there is almost no thyroid function, characterized by slowing down of physical and mental activity.

n The chromosome number of a gamete. The chromosome number of a zygote is $2n$. If an organism is not polyploid, the n gametes are haploid and the $2n$ zygotes are diploid.

NAD⁺/NADH Oxidized and reduced forms, respectively, of nicotinamide adenine dinucleotide, a coenzyme that transfers electrons (as hydrogen), particularly in catabolic pathways, including cellular respiration.

NADP⁺/NADPH Oxidized and reduced forms, respectively, of nicotinamide adenine dinucleotide phosphate, a coenzyme that acts as an electron (hydrogen) transfer agent, particularly in anabolic pathways, including photosynthesis.

nanoplankton Extremely minute (< 10 μm in length) algae that are major producers in the ocean because of their great abundance; part of phytoplankton.

natality The rate at which individuals produce offspring; the average per capita birth rate.

national conservation strategy Detailed plan for managing and preserving the biological diversity of a specific country.

natural killer cell (NK cell) A large, granular lymphocyte that functions in both nonspecific and specific immune responses; releases cytokines and proteolytic enzymes that target tumor cells and cells infected with viruses and other pathogens.

natural selection The mechanism of evolution proposed by Charles Darwin; the tendency of organisms that have favorable adaptations to their environment to survive and become the parents of the next generation. Evolution occurs when natural selection results in changes in allele frequencies in a population. Compare with *artificial selection.*

navigation Involves the use of cues to change direction when necessary in order to reach a specific destination; requires a sense of direction, called compass sense, and an awareness of location, called map sense.

Neandertals See *Homo neanderthalensis.*

necrosis Uncontrolled cell death that causes inflammation and damages other cells. Compare with *apoptosis.*

nectary (nek′-ter-ee) In plants, a gland or other structure that secretes nectar.

negative control A genetic control system in which a regulatory protein binds to DNA, turning off transcription of a gene. Compare with *positive control.*

negative feedback system A regulatory system in which a change in some steady state triggers a response that counteracts, or reverses, the change, restoring homeostasis, e.g., how mammals maintain body temperature. Compare with *positive feedback system.*

nekton (nek′-ton) Free-swimming aquatic organisms such as fish and turtles. Compare with *plankton.*

nematocyst (nem-at′-oh-sist) A stinging structure found within cnidocytes (stinging cells) in cnidarians; used for anchorage, defense, and capturing prey.

nematodes The phylum of animals commonly known as *roundworms.*

nemerteans The phylum of animals commonly known as *ribbon worms;* each has a proboscis (tubular feeding organ) for capturing prey.

Neogene period See *Cenozoic era.*

neonate Newborn individual.

neoplasm See *tumor.*

nephridial organ (neh-frid′-ee-al) The excretory organ of many invertebrates; consists of simple or branching tubes that usually open to the outside of the body through pores; also called *nephridium.*

nephron (nef′-ron) The functional, microscopic unit of the vertebrate kidney.

neritic province (ner-ih′-tik) Ocean water that extends from the shoreline to where the bottom reaches a depth of 200 m. Compare with *oceanic province.*

nerve A bundle of axons (or dendrites) wrapped in connective tissue that conveys impulses between the central nervous system and some other part of the body.

nerve cord A bundle of nerves that extends from a central ganglion or brain and gives rise to smaller nerves. In vertebrates the nerve cord is dorsal and tubular.

nerve net The network of neurons found in cnidarians and ctenophores.

nervous system In most invertebrates, consists of ganglia, sensory cells or sense organs, and networks of neurons. In vertebrates, consists of the brain, spinal cord, sense organs, and nerves; the body's main regulatory system.

nervous tissue A type of animal tissue specialized for transmitting electrical and chemical signals.

net primary productivity (NPP) The energy that remains in an ecosystem (as biomass) after cellular respiration has occurred; net primary productivity equals gross primary productivity minus respiration. Compare with *gross primary productivity (GPP)*.

neural circuit A functional group of neurons, typically consisting of one or more afferent neurons, one or more interneurons, and one or more efferent neurons.

neural crest cells (noor'-ul) A group of embryonic cells found only in vertebrates; develop along the neural tube and migrate to various parts of the embryo where they give rise to (or influence the development of) nerves, head muscles, cranium, jaws, and other structures.

neural plasticity The ability of the nervous system to change in response to experience.

neural plate See *neural tube*.

neural signaling The transmission of information by neurons (nerve cells).

neural transmission The conduction of a neural impulse, or action potential, along a neuron or from one neuron to another.

neural tube The hollow, longitudinal structure in the early vertebrate embryo that gives rise to the brain and spinal cord. The neural tube forms from the neural plate, a flattened, thickened region of the ectoderm that rolls up and sinks below the surface.

neuroendocrine cells Neurons that produce neurohormones.

neuroendocrine signaling Neuroendocrine cells produce neurohormones that are transported down axons and released into the interstitial fluid.

neurofibrillary tangles Abnormal fibrous deposits in the cytoplasm of neurons in the brains of individuals with Alzheimer's disease; interfere with neural signaling.

neurogenesis The production of new neurons.

neurohormones Hormones produced by neuroendocrine cells; transported down axons and released into interstitial fluid; common in invertebrates; in vertebrates, the hypothalamus produces neurohormones.

neuron (noor'-on) A nerve cell; a conducting cell of the nervous system that typically consists of a cell body, dendrites, and an axon.

neuropeptide One of a group of peptides produced in neural tissue that function as signaling molecules; many are neurotransmitters.

neuropeptide Y (NPY) A signaling molecule produced by the hypothalamus that increases appetite and slows metabolism; helps restore energy homeostasis when leptin levels and food intake are low.

neurotransmitters Chemical signals used by neurons to transmit impulses across a synapse.

neutral solution A solution of pH 7; there are equal concentrations of hydrogen ions [H$^+$] and hydroxide ions [OH$^-$]. Compare with *acidic solution* and *basic solution*.

neutral variation Variation that does not appear to confer any selective advantage or disadvantage to the organism.

neutron (noo'-tron) An electrically neutral particle with a mass of 1 atomic mass unit (amu) found in the atomic nucleus. Compare with *proton* and *electron*.

neutrophil (new'-truh-fil) A type of granular leukocyte important in immune responses; a type of phagocyte that engulfs and destroys bacteria and foreign matter.

niche (nich) The totality of an organism's adaptations, its use of resources, and the lifestyle to which it is fitted in its community; how an organism uses materials in its environment as well as how it interacts with other organisms; also called *ecological niche*. See *fundamental niche* and *realized niche*.

nicotinamide adenine dinucleotide See *NAD$^+$/ NADH*.

nicotinamide adenine dinucleotide phosphate See *NADP$^+$/NADPH*.

nitric oxide (NO) A gaseous signaling molecule; a neurotransmitter.

nitrification (nie''-tra-fuh-kay'-shun) The conversion of ammonia (NH$_3$) to nitrate (NO$_3^-$) by certain bacteria (nitrifying bacteria) in the soil; part of the nitrogen cycle.

nitrogenase (nie-traa'-jen-ase) The enzyme responsible for nitrogen fixation under anaerobic conditions.

nitrogen cycle The worldwide circulation of nitrogen from the abiotic environment into living things and back into the abiotic environment.

nitrogen fixation The conversion of atmospheric nitrogen (N$_2$) to ammonia (NH$_3$) by certain bacteria; part of the nitrogen cycle.

nitrogen oxides Binary compounds of oxygen and nitrogen, produced by combustion, that are a necessary ingredient in photochemical smog.

nociceptors (no'-sih-sep-torz) Pain receptors; free endings of certain sensory neurons whose stimulation is perceived as pain.

nocturnal animals Animals that are most active at night. Compare with *crepuscular animals* and *diurnal animals*.

node The area on a stem where each leaf is attached. Compare with *internode*.

nodes of Ranvier (ron'-ve-a) Gaps in the myelin sheath of myelinated neurons where an action potential can be generated. In saltatory conduction, the action potential appears to "leap" along from one node of Ranvier to the next.

nodules Swellings on the roots of plants, such as legumes, in which symbiotic nitrogen-fixing bacteria (*rhizobia*) live.

noncompetitive inhibition The binding of a substance (the noncompetitive inhibitor) to a site other than the active site of an enzyme, thereby lowering the rate at which the enzyme catalyzes a reaction. Compare with *competitive inhibition*.

noncompetitive inhibitor See *noncompetitive inhibition*.

noncyclic electron transport In photosynthesis, the linear flow of electrons, produced by photolysis of water, through photosystems II and I; results in the formation of ATP (by chemiosmosis), NADPH, and O$_2$. Compare with *cyclic electron transport*.

nondisjunction Abnormal separation of sister chromatids or of homologous chromosomes caused by their failure to disjoin (move apart) properly during mitosis or meiosis.

nonpolar covalent bond Chemical bond formed by the equal sharing of electrons between atoms of approximately equal electronegativity. Compare with *polar covalent bond*.

nonpolar molecule Molecule that does not have a positively charged end and a negatively charged end; nonpolar molecules are generally insoluble in water. Compare with *polar molecule*.

nonsense mutation A base-pair substitution mutation that results in an amino acid–specifying codon being changed to a termination (stop) codon; when the abnormal mRNA is translated, the resulting protein is usually shortened and nonfunctional. Compare with *missense mutation*.

nonspecific immune responses See *innate immune responses*.

nonspecific immunity See *innate immune responses*.

norepinephrine (nor-ep-ih-nef'-rin) A neurotransmitter that is also a hormone secreted by the adrenal medulla.

norm of reaction The range of phenotypic possibilities that can develop from a single genotype under different environmental conditions.

Northern blot A technique in which RNA fragments, previously separated by gel electrophoresis, are transferred to a nitrocellulose membrane and detected by autoradiography or chemical luminescence. Compare with *Southern blot* and *Western blot*.

notochord (no'-toe-kord) The flexible, longitudinal rod in the anterior–posterior axis that serves as an internal skeleton in the embryos of all chordates and in the adults of some.

nuclear area Region of a prokaryotic cell that contains DNA; not enclosed by a membrane. Also called *nucleoid*.

nuclear envelope The double membrane system that encloses the cell nucleus of eukaryotes.

nuclear equivalence The concept that the nuclei of all differentiated cells of an adult organism are genetically identical to one another and to the nucleus of the zygote from which they were derived. Compare with *genomic rearrangement* and *gene amplification*.

nuclear pores Structures in the nuclear envelope that allow passage of certain materials between the cell nucleus and the cytoplasm.

nucleoid See *nuclear area.*

nucleolus (new-klee′-oh-lus) (pl., *nucleoli*) Specialized structure in the cell nucleus formed from regions of several chromosomes; site of assembly of the ribosomal subunits.

nucleoplasm The contents of the cell nucleus.

nucleoside triphosphate Molecule consisting of a nitrogenous base, a pentose sugar, and three phosphate groups, e.g., adenosine triphosphate (ATP).

nucleosomes (new′-klee-oh-sohmz) Repeating units of chromatin structure, each consisting of a length of DNA wound around a complex of eight histone molecules. Adjacent nucleosomes are connected by a DNA linker region associated with another histone protein.

nucleotide (noo′-klee-oh-tide) A molecule consisting of one or more phosphate groups, a five-carbon sugar (ribose or deoxyribose), and a nitrogenous base (purine or pyrimidine).

nucleotide excision repair A DNA repair mechanism commonly used to repair a damaged segment of DNA caused by the sun′s ultraviolet radiation or by harmful chemicals.

nucleus (new′-klee-us) (pl., *nuclei*) (1) The central region of an atom that contains the protons and neutrons. (2) A cell organelle in eukaryotes that contains the DNA and serves as the control center of the cell. (3) A mass of nerve cell bodies in the central nervous system. Compare with *ganglion.*

nut A simple, dry fruit that contains a single seed and is surrounded by a hard fruit wall.

nutrients The chemical substances in food that are used as components for synthesizing needed materials and/or as energy sources.

nutrition The process of taking in and using food (nutrients).

obesity Excess accumulation of body fat; a person is considered obese if the body mass index (BMI) is 30 or higher.

obligate anaerobe An organism that grows only in the absence of oxygen. Compare with *facultative anaerobe.*

occipital lobes Posterior areas of the mammalian cerebrum; interpret visual stimuli from the retina of the eye.

ocean currents Mass movements of surface ocean water produced by winds blowing over the ocean.

oceanic province That part of the open ocean that overlies an ocean bottom deeper than 200 m. Compare with *neritic province.*

Okazaki fragment One of many short segments of DNA, each 100 to 1000 nucleotides long, that must be joined by DNA ligase to form the lagging strand in DNA replication.

olfactory bulbs Olfactory relay centers where axons from the olfactory nerves transmit information to neurons that signal higher centers.

olfactory epithelium Tissue containing odor-sensing neurons.

olfactory nerve The first cranial nerve; consists of the axons of the olfactory receptor cells and extends to the olfactory bulb in the brain.

oligodendrocyte A type of glial cell that forms myelin sheaths around neurons in the CNS.

omega-3 fatty acids Essential fatty acids found in salmon and other fatty fishes, walnuts, and certain vegetable oils (e.g., canola, soybean); studies suggest that omega-3 fatty acids help lower triglycerides and blood pressure, and have other health benefits.

ommatidium (om″-ah-tid′-ee-um) (pl., *ommatidia*) One of the light-detecting units of a compound eye, consisting of a lens and a crystalline cone that focus light onto photoreceptors called *retinular cells.*

omnivore (om′-nih-vore) An animal that eats a variety of plant and animal materials.

oncogene (on′-koh-jeen) An abnormally functioning gene implicated in causing cancer. Compare with *proto-oncogene* and *tumor suppressor gene.*

1000 Genomes Project An international collaborative initiative completed in 2012 to provide a detailed catalogue of human genetic variation by determining the genomic DNA sequences of over 1000 anonymous individuals from a number of different ethnic groups. Compare with *Human Genome Project.*

oocytes (oh′-oh-sites) Meiotic cells that give rise to egg cells (ova).

oogamy (oh-og′-uh-me) The fertilization of a large, nonmotile female gamete by a small, motile male gamete. Compare with *isogamy* and *anisogamy.*

oogenesis (oh″-oh-jen′-eh-sis) Production of female gametes (eggs) by meiosis. Compare with *spermatogenesis.*

oogonia Undifferentiated cells in the ovaries that give rise to primary oocytes.

oospore A thick-walled, resistant spore formed from a zygote during sexual reproduction in water molds.

open circulatory system A type of circulatory system in which the blood bathes the tissues directly; characteristic of arthropods and many mollusks. Compare with *closed circulatory system.*

open system An entity that exchanges energy with its surroundings. Compare with *closed system.*

operant conditioning A type of learning in which an animal is rewarded or punished for performing a behavior it discovers by chance.

operator site One of the control regions of an operon; the DNA segment to which a repressor binds, thereby inhibiting the transcription of the adjacent structural genes of the operon.

operculum In bony fishes, a protective flap of the body wall that covers the gills.

operon (op′-er-on) In prokaryotes, a group of structural genes that are coordinately controlled and transcribed as a single message, plus their adjacent regulatory elements.

opisthokonts Members of the unikont clade in which motile cells possess a single posterior flagellum; include fungi, choanoflagellates, and animals. Compare with *amoebozoa.*

opposable thumb The arrangement of the fingers so that they are positioned opposite the thumb, enabling the organism to grasp objects.

optic chiasm The X-shaped structure formed by the crossing of the optic nerves in the floor of the hypothalamus.

optic nerve The second cranial nerve; transmits visual information from the retina to the brain; composed of the axons of ganglion cells.

optimal foraging The process of obtaining food in a manner that maximizes benefits and/or minimizes costs.

orbital Region in which electrons occur in an atom or molecule.

orbital hybridization A rearrangement of the orbitals in the valence shell that may occur when an atom forms covalent bonds with other atoms.

order A taxonomic category made up of related families.

Ordovician period See *Paleozoic era.*

organ A specialized structure, such as the heart or liver, or a flower, made up of tissues and adapted to perform a specific function or group of functions.

organelle One of the specialized structures within the cell, such as the mitochondria, Golgi complex, ribosomes, or contractile vacuole; many organelles are membrane-enclosed.

organic compound A compound consisting of a backbone made up of carbon atoms. Compare with *inorganic compound.*

organism Any living system consisting of one or more cells.

organismic model Cooperative view of community, described by Clements, in which a community goes through certain stages of development (succession), like the embryonic stages of an organism, and eventually reaches an adult state (climax community). Compare with *individualistic model.*

organismic respiration See *respiration.*

organ of Corti The structure within the inner ear of vertebrates that contains receptor cells that sense sound vibrations.

organogenesis The process of organ formation.

organ system An organized group of tissues and organs that work together to perform a specialized set of functions, e.g., the digestive system or circulatory system.

orgasm (or′-gazm) The climax of sexual excitement.

origin of replication A specific site on the DNA where replication begins.

orthologous genes Homologous genes found in different species because they were inherited from a common ancestor.

osculum (os′kyuh-lum) In a sponge, the main opening between the spongocoel and the external environment.

osmoconformer An animal in which the salt concentration of body fluids varies along with changes in the surrounding sea water so that it stays in osmotic equilibrium with its surroundings. Compare with *osmoregulator*.

osmolarity The number of osmoles of solute per liter of solution. See *osmole*.

osmole The number of solute particles (molecules or ions) produced when a mole of solute dissolves. One mole of glucose is one osmol; one mole of sodium chloride (NaCl), which dissolves to give two types of particles, is two osmols.

osmoregulation (oz″-moh-reg-yoo-lay′-shun) The active regulation of the osmotic pressure of body fluids so that they do not become excessively dilute or excessively concentrated.

osmoregulator An animal that maintains an optimal salt concentration in its body fluids despite changes in salinity of its surroundings. Compare with *osmoconformer*.

osmosis (oz-moh′-sis) The net movement of water (the principal solvent in biological systems) by diffusion through a selectively permeable membrane from a region of higher concentration of water (a hypotonic solution) to a region of lower concentration of water (a hypertonic solution).

osmotic pressure The pressure that must be exerted on the hypertonic side of a selectively permeable membrane to prevent diffusion of water (by osmosis) from the side containing pure water.

osteichthyes (os″-tee-ick′-thees) Historically, the vertebrate class of bony fishes. Biologists now divide bony fishes into three classes: Actinopterygii, the ray-finned fishes; Actinistia, the lobe-finned fishes; and Dipnoi, the lungfishes.

osteoblast (os′-tee-oh-blast) A type of bone cell that secretes the protein matrix of bone. Also see *osteocyte*.

osteoclast (os′-tee-oh-clast) Large, multinucleate cell that helps sculpt and remodel bones by dissolving and removing part of the bony substance.

osteocyte (os′-tee-oh-site) A mature bone cell; an osteoblast that has become embedded within the bone matrix and occupies a lacuna.

osteon (os′-tee-on) The spindle-shaped unit of bone composed of concentric layers of osteocytes organized around a central Haversian canal containing blood vessels and nerves.

otoliths (oh′-toe-liths) Small calcium carbonate crystals in the saccule and utricle of the inner ear; sense gravity and are important in static equilibrium.

outbreeding The mating of individuals of unrelated strains, also called *outcrossing*. Compare with *inbreeding*.

outcrossing See *outbreeding*.

outgroup In cladistics, a taxon that represents an approximation of the ancestral condition; the outgroup is related to the *ingroup* (the members of the group under study) but separated from the ingroup lineage before they diversified.

ovary (oh′-var-ee) (1) In animals, one of the paired female gonads responsible for producing eggs and sex hormones. (2) In flowering plants, the base of the carpel that contains ovules; ovaries develop into fruits after fertilization.

oviduct (oh′-vih-dukt) The tube that carries ova from the ovary to the uterus, cloaca, or body exterior. Also called *fallopian tube* or *uterine tube*.

oviparous (oh-vip′-ur-us) Bearing young in the egg stage of development; egg laying. Compare with *viviparous* and *ovoviviparous*.

ovoviviparous (oh″-voh-vih-vip′-ur-us) A type of development in which the young hatch from eggs incubated inside the mother's body. Compare with *viviparous* and *oviparous*.

ovulation (ov-u-lay′-shun) The release of an egg from the ovary.

ovule (ov′-yool) The structure in the plant ovary that develops into the seed following fertilization; also called *megasporangium*.

ovum (pl., *ova*) Female gamete of an animal.

oxaloacetate Four-carbon compound; important intermediate in the citric acid cycle and in the C_4 and CAM pathways of carbon fixation in photosynthesis.

oxidants Highly reactive molecules such as free radicals, peroxides, and superoxides that are produced during normal cell processes that require oxygen; can damage DNA and other molecules by snatching electrons. Compare with *antioxidants*.

oxidation The loss of one or more electrons (or hydrogen atoms) by an atom, ion, or molecule. Compare with *reduction*.

oxidative phosphorylation (fos″-for-ih-lay′-shun) The production of ATP using energy derived from the transfer of electrons in the electron transport system of mitochondria; occurs by chemiosmosis.

oxygen-carrying capacity The maximum amount of oxygen transported by hemoglobin.

oxygen debt The oxygen necessary to metabolize the lactic acid produced during strenuous exercise.

oxygen–hemoglobin dissociation curve A curve depicting the percentage saturation of hemoglobin with oxygen, as a function of certain variables such as oxygen concentration, carbon dioxide concentration, or pH.

oxyhemoglobin Hemoglobin that has combined with oxygen.

oxytocin (ok″see-tow′-sin) Hormone secreted by the hypothalamus and released by the posterior lobe of the pituitary gland; stimulates contraction of the pregnant uterus and the ducts of mammary glands.

ozone (O_3) A blue gas with a distinctive odor that is a human-made pollutant near Earth's surface (in the troposphere) but a natural and essential component of the stratosphere.

P generation (parental generation) Members of two different true-breeding lines that are crossed to produce the F_1 generation.

P680 Chlorophyll *a* molecules that serve as the reaction center of photosystem II, transferring photoexcited electrons to a primary acceptor; named by their absorption peak at 680 nm.

P700 Chlorophyll *a* molecules that serve as the reaction center of photosystem I, transferring photoexcited electrons to a primary acceptor; named by their absorption peak at 700 nm.

pacemaker (of the heart) See *sinoatrial (SA) node*.

Pacinian corpuscle (pah-sin′-ee-an kor′-pus-el) A receptor located in the dermis of the skin that responds to pressure.

paedomorphosis Retention of juvenile or larval features in a sexually mature animal.

pair bond A stable relationship between animals of opposite sex that ensures cooperative behavior in mating and rearing the young.

paleoanthropology (pay″-lee-o-an-thro-pol′-uh-gee) The study of human evolution.

Paleogene period See *Cenozoic era*.

Paleozoic era That part of geologic time extending from roughly 541 million to 252 million years ago; divided into six periods: Cambrian (appearance of many new animal groups, the *Cambrian radiation* or *Cambrian explosion*); Ordovician (diversification of sea life); Silurian (appearance of jawed fishes, terrestrial plants, and air-breathing animals); Devonian (appearance of sharks and bony fishes); Carboniferous (great swamp forests formed coal deposits); Permian (ended with the greatest mass extinction of all time).

palindromic Reading the same forward and backward; DNA sequences are palindromic when the base sequence of one strand reads the same as its complement when both are read in the 5′ to 3′ direction.

palisade mesophyll (mez′-oh-fil) The vertically stacked, columnar mesophyll cells near the upper epidermis in certain leaves. Compare with *spongy mesophyll*.

pancreas (pan′-kree-us) Large gland located in the vertebrate abdominal cavity. The pancreas produces pancreatic juice containing digestive enzymes; also serves as an endocrine gland, secreting the hormones insulin and glucagon.

parabasilids Anaerobic, flagellated excavates that often live in animals; examples include trichonymphs and trichomonads.

parabronchi (sing., *parabronchus*) Thin-walled ducts in the lungs of birds; gases are exchanged across their walls.

paracrine regulation A type of regulation in which a signal molecule (e.g., certain hormones) diffuses through interstitial fluid and acts on nearby target cells. Compare with *autocrine regulation*.

paraphyletic group A group of organisms made up of a common ancestor and some, but not all, of its descendants. Compare with *monophyletic group* and *polyphyletic group*.

parapodia (par″-uh-poh′-dee-ah) (sing., *parapodium*) Paired, thickly bristled paddlelike appendages extending laterally from each segment of polychaete worms.

parasite A heterotrophic organism that obtains nourishment from the living tissue of another organism (the host).

parasitism (par′-uh-si-tiz″-m) A symbiotic relationship in which one member (the parasite) benefits and the other (the host) is adversely affected. Compare with *commensalism* and *mutualism*.

parasympathetic nervous system A division of the autonomic nervous system concerned with the control of the internal organs; functions to conserve or restore energy. Compare with *sympathetic nervous system*.

parathyroid glands Small, pea-sized glands closely adjacent to the thyroid gland; they secrete parathyroid hormone, which regulates calcium and phosphate metabolism.

parathyroid hormone (PTH) A hormone secreted by the parathyroid glands; regulates calcium and phosphate metabolism.

parenchyma (par-en′-kih-mah) Highly variable living plant cells that have thin primary walls; function in photosynthesis, the storage of nutrients, and/or secretion.

parental imprinting See *imprinting* (definition 1).

parental investment The contribution that each parent makes in producing and rearing offspring; increases the probability that their offspring will survive.

parietal lobes The region of the cerebrum between the frontal lobes and the occipital lobes; contains the general sensory association areas.

parsimony The principle based on the experience that the simplest explanation is most probably the correct one.

parthenogenesis (par″-theh-noh-jen′-eh-sis) The development of an unfertilized egg into an adult organism; common among honeybees, wasps, and certain other arthropods.

partial pressure (of a gas) The pressure exerted by a gas in a mixture, which is the same pressure it would exert if alone. For example, the partial pressure of atmospheric oxygen (P_{O_2}) is 160 mm Hg at sea level.

passive immunity Temporary immunity that depends on the presence of immunoglobulins produced by another organism. Compare with *active immunity*.

passive ion channel A channel in the plasma membrane that permits the passage of specific ions such as Na^+, K^+, or Cl^-.

patch clamp technique A method that allows researchers to study the ion channels of a tiny patch of membrane by tightly sealing a micropipette to the patch and measuring the flow of ions through the channels.

patchiness See *clumped dispersion*.

pathogen (path′-oh-gen) An organism, usually a microorganism, capable of producing disease.

pathogen-associated molecular patterns (PAMPs) Distinctive molecules on bacteria and other pathogens that are not found in animals; pattern recognition receptors bind them.

pattern formation See *morphogenesis*.

pattern recognition receptors Receptors on certain types of animal cells that bind pathogen-associated molecular patterns (PAMPs); binding activates innate immune responses.

pedigree A chart constructed to show an inheritance pattern within a family through multiple generations.

peduncle The stalk of a flower or inflorescence.

pellicle A flexible outer covering consisting of protein; characteristic of certain protists, e.g., ciliates and euglenoids.

penis The male sexual organ of copulation in reptiles, mammals, and a few birds.

pentose A sugar molecule containing five carbons.

people overpopulation A situation in which there are too many people in a given geographic area; results in pollution, environmental degradation, and resource depletion. Compare with *consumption overpopulation*.

PEP carboxylase Enzyme with an extremely high affinity for CO_2; responsible for initial fixation of CO_2 in C_4 plants and CAM plants.

pepsin (pep′-sin) An enzyme produced in the stomach that initiates digestion of protein.

pepsinogen The precursor of pepsin; secreted by chief cells in the gastric glands of the stomach.

peptide (pep′-tide) A compound consisting of a chain of amino acid groups linked by peptide bonds. A dipeptide consists of two amino acids, a polypeptide of many.

peptide bond A distinctive covalent carbon-to-nitrogen bond that links amino acids in peptides and proteins.

peptide hormones Water-soluble hormones composed of proteins, such as oxytocin and antidiuretic hormone (ADH).

peptidoglycan (pep″-tid-oh-gly′-kan) A modified protein or peptide having an attached carbohydrate; component of the bacterial cell wall.

peptidyl transferase The ribosomal enzyme (a ribozyme) that catalyzes the formation of a peptide bond.

perception Conscious awareness of our internal and external environments as a result of processing sensory input and comparing present sensory experience with memories of past experience.

perennial plant (purr-en′-ee-ul) A woody or herbaceous plant that grows year after year, i.e., lives more than two years. Compare with *annual plant* and *biennial plant*.

perfect flower A flower that has both stamens and carpels. Compare with *imperfect flower*.

pericentriolar material Fibrils surrounding the centrioles in the microtubule-organizing centers in cells of animals and other organisms having centrioles.

pericycle (pehr′-eh-sy″-kl) A layer of meristematic cells typically found between the endodermis and phloem in roots.

periderm (pehr′-ih-durm) The outer bark of woody stems and roots; composed of cork cells, cork cambium, and cork parenchyma, along with traces of primary tissues.

period An interval of geologic time that is a subdivision of an era; each period is divided into epochs.

periodic table A chart of the elements arranged in order by atomic number.

peripheral membrane protein A protein associated with one of the surfaces of a biological membrane. Compare with *integral membrane protein*.

peripheral nervous system (PNS) In vertebrates, the nerves and receptors that lie outside the central nervous system. Compare with *central nervous system (CNS)*.

peristalsis (pehr″-ih-stal′-sis) Rhythmic waves of muscular contraction and relaxation in the walls of hollow tubular organs, such as the ureter or parts of the digestive tract, that serve to move the contents through the tube.

peritubular capillaries In the mammalian kidney, a capillary network that surrounds each renal tubule; receives blood from the efferent arterioles.

permafrost Permanently frozen subsoil characteristic of frigid areas such as the tundra.

Permian period See *Paleozoic era*.

peroxisomes (pehr-ox′-ih-sohmz) In eukaryotic cells, membrane-enclosed organelles containing enzymes that produce or degrade hydrogen peroxide.

persistence A characteristic of certain chemicals that are extremely stable and may take many years to be broken down into simpler forms by natural processes.

petal One of the parts of the flower attached inside the whorl of sepals; petals are usually colored.

petiole (pet′-ee-ohl) The part of a leaf that attaches to a stem.

Pfr See *phytochrome*.

pH The negative logarithm of the hydrogen ion concentration of a solution (expressed as moles per liter). Neutral pH is 7, values less than 7 are acidic, and those greater than 7 are basic.

phage See *bacteriophage*.

phagocytosis (fag″-oh-sy-toh″-sis) Literally, "cell eating"; a type of endocytosis by which certain cells engulf food particles, microorganisms, foreign matter, or other cells. Compare with *pinocytosis* and *receptor-mediated endocytosis*.

pharmacogenetics A field of gene-based medicine in which drugs are personalized to match a patient's genetic makeup.

pharyngeal slits (fair-in′-jel) Openings that lead from the pharyngeal cavity to the outside; evolved as part of a filter-feeding system in chordates and later became modified for other functions, including gill slits in many aquatic vertebrates.

pharynx (fair′-inks) Part of the digestive tract. In complex vertebrates, it is bounded anteriorly by the mouth and nasal cavities and posteriorly by the esophagus and larynx; the throat region in humans.

phenetics (feh-neh′-tiks) An approach to classification based on measurable similarities in phenotypic characters without consideration of homology or other evolutionary relationships. Compare with *cladistics* and *evolutionary systematics.*

phenotype (fee′-noh-type) The physical or chemical expression of an organism's genes. Compare with *genotype.*

phenotype frequency The proportion of a particular phenotype in the population. Compare with *allele frequency* and *genotype frequency.*

phenylketonuria (PKU) (fee″-nl-kee″-toh-noor′-ee-ah) An inherited disease in which there is a deficiency of the enzyme that normally converts phenylalanine to tyrosine; results in intellectual disability if untreated.

pheromone (fer′-oh-mone) A substance secreted by an organism to the external environment that influences the development or behavior of other members of the same species.

phloem (flo′-em) The vascular tissue that conducts dissolved sugar and other organic compounds in plants.

phosphatases Enzymes that catalyze the removal of phosphate groups from proteins and other molecules. Compare with *kinases* and *protein kinases.*

phosphate group A weakly acidic functional group that can release one or two hydrogen ions.

phosphodiesterase Enzyme that breaks a phosphodiester bond; responsible for inactivating cyclic AMP by converting it to AMP.

phosphodiester linkage Covalent linkage between two nucleotides in a strand of DNA or RNA; includes a phosphate group bonded to the sugars of two adjacent nucleotides.

phosphoenolpyruvate (PEP) Three-carbon phosphorylated compound that is an important intermediate in glycolysis and is a reactant in the initial carbon fixation step in C_4 and CAM photosynthesis.

phosphoglycerate (PGA) Phosphorylated three-carbon compound that is an important metabolic intermediate.

phospholipids (fos″-foh-lip′-idz) Lipids in which two fatty acids and a phosphorus-containing group are attached to glycerol; major components of cell membranes.

phosphorus cycle The worldwide circulation of phosphorus from the abiotic environment into living things and back into the abiotic environment.

phosphorylation (fos″-for-ih-lay′-shun) The introduction of a phosphate group into an organic molecule. See *kinases.*

photoautotroph An organism that obtains energy from light and synthesizes organic compounds from inorganic raw materials; includes plants,

algae, and some bacteria. Compare with *photoheterotroph, chemoautotroph,* and *chemoheterotroph.*

photochemical smog Mixture of several air pollutants that injure plant tissues, irritate eyes, and cause respiratory problems in humans.

photoheterotroph An organism that can carry out photosynthesis to obtain energy but cannot fix carbon dioxide and therefore requires organic compounds as a carbon source; includes some bacteria and archaea. Compare with *photoautotroph, chemoautotroph,* and *chemoheterotroph.*

photolysis (foh-tol′-uh-sis) The photochemical splitting of water in the light-dependent reactions of photosynthesis; a specific enzyme is needed to catalyze this reaction.

photon (foh′-ton) A particle of electromagnetic radiation; one quantum of radiant energy.

photoperiodism (foh″-teh-peer′-ee-o-dizm) The physiological response (such as flowering) of plants to variations in the length of daylight and darkness.

photophosphorylation (foh″-toh-fos-for-ih-lay′-shun) The production of ATP in photosynthesis.

photoreceptor (foh″-toh-ree-sep′-tor) (1) A sense organ specialized to detect light. (2) A pigment that absorbs light before triggering a physiological response.

photorespiration (foh″-toh-res-pur-ay′-shun) The process that reduces the efficiency of photosynthesis in C_3 plants during hot spells in summer; consumes oxygen and produces carbon dioxide through the degradation of Calvin cycle intermediates.

photosynthesis The biological process that captures light energy and transforms it into the chemical energy of organic molecules (e.g., carbohydrates), which are manufactured from carbon dioxide and water.

photosystem One of two photosynthetic units responsible for capturing light energy and transferring excited electrons; photosystem I strongly absorbs light of about 700 nm, whereas photosystem II strongly absorbs light of about 680 nm. See *antenna complex* and *reaction center.*

phototaxis Movement of a cell or organism in response to light.

phototroph (foh′-toh-trof) Organism that uses light as a source of energy. Compare with *chemotroph.* See *photoautotroph* and *photoheterotroph.*

phototropins (foh″-toh-troh′-pinz) Family of yellow pigments responsible for such blue light responses as the phototropic response and stomatal opening.

phototropism (foh″-toh-troh′-pizm) The growth of a plant in response to the direction of light.

phycocyanin (fy″-koh-sy-ah′-nin) A blue pigment found in cyanobacteria and red algae.

phycoerythrin (fy″-koh-ee-rih′-thrin) A red pigment found in cyanobacteria and red algae.

phyletic gradualism The idea that evolution occurs by a slow, steady accumulation of genetic changes over time; also called *gradualism.* Compare with *punctuated equilibrium.*

phylogenetic species concept An alternative to the biological species concept in which for a population to be declared a separate species, it must have undergone evolution long enough for statistically significant differences to emerge. Also called *evolutionary species concept.* Compare with *species.*

phylogenetic systematics See *cladistics.*

phylogenetic tree A branching diagram that shows lines of descent among a group of related species.

phylogeny (fy-loj′-en-ee) The complete evolutionary history of a group of organisms.

phylum (fy′-lum) (pl., *phyla*) A taxonomic grouping of related, similar classes; a category beneath the kingdom and above the class.

phytoalexins Antimicrobial compounds produced by plants that limit the spread of pathogens such as fungi.

phytochemicals Non-nutrient compounds found in plants that are biologically active in the body and may promote health.

phytochrome (fy′-toh-krome) A blue-green, proteinaceous pigment involved in a wide variety of physiological responses to light; occurs in two interconvertible forms, Pr (red-absorbing phytochrome) and Pfr (far-red-absorbing phytochrome).

phytoplankton (fy″-toh-plank′-tun) Microscopic floating algae and cyanobacteria that are the base of most aquatic food webs. Compare with *zooplankton.* See *plankton* and *nanoplankton.*

pia mater (pee′-a may′-ter) The inner membrane covering the brain and spinal cord; the innermost of the meninges; also see *dura mater* and *arachnoid.*

pigment A substance that selectively absorbs light of specific wavelengths.

pili (pie′-lie) (sing., *pilus*) Hairlike structures on the surface of many bacteria; function in conjugation or attachment. A sex pilus forms a cytoplasmic bridge with another bacterium and transfers DNA via this bridge.

pineal gland (pie-nee′-al) Endocrine gland located in the brain.

pinocytosis (pin″-oh-sy-toh′-sis) Cell drinking; a type of endocytosis by which cells engulf and absorb droplets of liquids. Compare with *phagocytosis* and *receptor-mediated endocytosis.*

pioneer The first organism to colonize an area and begin the first stage of succession.

pistil The female reproductive organ of a flower; consists of either a single carpel or two or more fused carpels. See *carpel.*

pith The innermost tissue in the stems and roots of many herbaceous plants; primarily a storage tissue.

pituitary gland (pi-too′-ih-tehr″-ee) An endocrine gland located below the hypothalamus; secretes several hormones that influence a wide range of physiological processes.

piwi-associated RNA (piRNA) Small RNAs (26–31 bases) involved in gene silencing of *retrotransposons* and regulating gene activity in animal germ line cells.

placenta (plah-sen′-tah) The partly fetal and partly maternal organ whereby materials are exchanged between fetus and mother in the uterus of placental mammals.

placoderms (plak′-oh-durmz) A group of extinct jawed fishes.

placoid scales (pla′-koid) The toothlike scales found in sharks and other cartilaginous fishes.

plankton Free-floating, mainly microscopic aquatic organisms found in the upper layers of the water; consisting of phytoplankton and zooplankton. Compare with *nekton*.

plants Members of the archaeplastid supergroup, distinguished by development of sporophytes from multicellular embryos enclosed in maternal tissues; also called *land plants*.

planula larva (plan′-yoo-lah) A ciliated larval form found in cnidarians.

plasma The fluid portion of blood in which red blood cells, white blood cells, and platelets are suspended.

plasma cell Cell that secretes antibodies; a differentiated B lymphocyte (B cell).

plasma membrane The selectively permeable surface membrane that encloses the cell contents and through which all materials entering or leaving the cell must pass.

plasma proteins Proteins such as albumins, globulins, and fibrinogen that circulate in the blood plasma.

plasmid (plaz′-mid) Small, circular, double-stranded DNA molecule that carries genes separate from those in the main DNA of a cell.

plasmodesmata (sing., *plasmodesma*) Cytoplasmic channels connecting adjacent plant cells and allowing for the movement of molecules and ions between cells.

plasmodial slime mold (plaz-moh′-dee-uhl) A funguslike amoebozoan protist whose feeding stage consists of a plasmodium; compare with *cellular slime mold*.

plasmodium (plaz-moh′-dee-um) A multinucleate mass of living matter that moves and feeds in an amoeboid fashion.

plasmogamy (1) Fusion of the cytoplasm of two cells without fusion of nuclei. (2) A stage in the asexual reproduction of some fungi; hyphae of two compatible mating types come together, and their cytoplasm fuses.

plasmolysis (plaz-mol′-ih-sis) The shrinkage of cytoplasm and the pulling away of the plasma membrane from the cell wall when a plant cell (or other walled cell) loses water, usually in a hypertonic environment.

plastids (plas′-tidz) A family of membrane-enclosed organelles occurring in photosynthetic eukaryotic cells; include chloroplasts, chromoplasts, and amyloplasts and other leukoplasts.

platelets (playt′-lets) Cell fragments in vertebrate blood that function in clotting; also called *thrombocytes*.

platyhelminths The phylum of acoelomate animals commonly known as *flatworms*.

pleiotropy The ability of a single gene to have multiple effects.

pleural membrane (ploor′-ul) The membrane that lines the thoracic cavity and envelops each lung.

ploidy The number of chromosome sets in a nucleus or cell. See *haploid, diploid,* and *polyploid*.

plumule (ploom′-yool) The embryonic shoot apex, or terminal bud, located above the point of attachment of the cotyledon(s).

pluripotent (ploor-i-poh′-tent) A term describing a stem cell that can divide to give rise to many, but not all, types of cells in an organism. Compare with *totipotent*.

pneumatophore (noo-mat′-uh-for″) Roots that extend up out of the water in swampy areas and are thought to provide aeration between the atmosphere and submerged roots.

polar body A small *n* cell produced during oogenesis in female animals that does not develop into a functional ovum.

polar covalent bond Chemical bond formed by the sharing of electrons between atoms that differ in electronegativity; the end of the bond near the more electronegative atom has a partial negative charge, and the other end has a partial positive charge. Compare with *nonpolar covalent bond*.

polar molecule Molecule that has one end with a partial positive charge and the other with a partial negative charge; polar molecules are generally soluble in water. Compare with *nonpolar molecule*.

polar nucleus In flowering plants, one of two *n* nuclei in the embryo sac that fuse with a sperm during double fertilization to form the 3*n* endosperm.

pollen grain The immature male gametophyte of seed plants (gymnosperms and angiosperms) that produces sperm capable of fertilization.

pollen tube In gymnosperms and flowering plants, a tube or extension that forms after germination of the pollen grain and through which male gametes (sperm cells) pass into the ovule.

pollination (pol″-uh-nay′-shen) In seed plants, the transfer of pollen from the male to the female part of the plant.

polyadenylation (pol″-ee-a-den-uh-lay′-shun) That part of eukaryotic mRNA processing in which multiple adenine-containing nucleotides (a poly-A tail) are added to the 3′ end of the molecule.

polyandry A mating system in which a female mates with several males during a breeding season. Compare with *polygyny*.

poly-A tail See *polyadenylation*.

polydnaviruses Virus particles that consist of multiple circles of dsDNA surrounded by capsid proteins and an envelope; they lack DNA coding for proteins needed for viral replication, but this DNA is provided by ovary cells of the wasps they inhabit.

polygenic inheritance (pol″-ee-jen′-ik) Inheritance in which several independently assorting or loosely linked nonallelic genes modify the intensity of a trait or contribute to the phenotype in additive fashion.

polygyny A mating system in which a male animal mates with many females during a breeding season. Compare with *polyandry*.

polymer (pol′-ih-mer) A molecule built up from repeating subunits of the same general type (monomers); examples include proteins, nucleic acids, or polysaccharides.

polymerase chain reaction (PCR) A method by which a targeted DNA fragment is amplified in vitro to produce millions of copies.

polymorphism (pol″-ee-mor′-fizm) (1) The existence of two or more phenotypically different individuals within a population. (2) The presence of detectable variation in the genomes of different individuals in a population.

polyp (pol′-ip) A hydralike animal; the sessile stage of the life cycle of certain cnidarians. Compare with *medusa*.

polypeptide See *peptide*.

polyphyletic group (pol″-ee-fye-let′-ik) A group made up of organisms that evolved from two or more different ancestors. Compare with *monophyletic group* and *paraphyletic group*.

polyploid (pol′-ee-ployd) The condition of having more than two sets of chromosomes per nucleus. Compare with *diploid* and *haploid*.

polyribosome A complex consisting of a number of ribosomes attached to an mRNA during translation; also known as a *polysome*.

polysaccharide (pol-ee-sak′-ah-ride) A carbohydrate consisting of many monosaccharide subunits (e.g., starch, glycogen, and cellulose).

polysome See *polyribosome*.

polyspermy The fertilization of an egg by more than one sperm.

polytene A term describing a giant chromosome consisting of many parallel DNA double helices. Polytene chromosomes are typically found in cells of the salivary glands and some other tissues of certain insects, such as the fruit fly, *Drosophila*.

polyunsaturated fatty acid See *fatty acid*.

pons (ponz) The white bulge that is the part of the brain stem between the medulla and the midbrain; connects various parts of the brain.

population A group of organisms of the same species that live in a defined geographic area at the same time.

population bottleneck See *bottleneck*.

population crash An abrupt decline in the size of a population.

population density The number of individuals of a species per unit of area or volume at a given time.

population dynamics The study of changes in populations, such as how and why population numbers change over time.

population ecology That branch of biology that deals with the numbers of a particular species that are found in an area and how and why those numbers change (or remain fixed) over time.

population genetics The study of genetic variability in a population and of the forces that act on it.

population growth momentum The continued growth of a population after fertility rates have declined, as a result of a population's young age structure.

positional information The exposure of cells to different concentrations of signaling molecules that specify where the cell is located relative to the body's axes; this information affects cell differentiation and tissue formation.

positive control A genetic control system in which a regulatory protein binds to DNA, turning on transcription of a gene. Compare with *negative control.*

positive feedback system A system in which a change in some steady state triggers a response that intensifies the changing condition. Compare with *negative feedback system.*

posterior Toward the tail end of a bilaterally symmetrical animal. Compare with *anterior.*

posterior pituitary The posterior, neuroendocrine lobe of the pituitary gland; secretes oxytocin and antidiuretic hormone (ADH). Compare with *anterior pituitary.*

postsynaptic neuron A neuron that transmits an impulse away from a synapse. Compare with *presynaptic neuron.*

posttranslational control Regulation of gene expression that controls activity or functions of proteins, usually by phosphorylation or other chemical modifications of the protein. Compare with *transcriptional-level control, translational-level control,* and *posttranscriptional control.*

posttranscriptional control Regulation of gene expression that occurs after the transcription of an mRNA, usually referring to pre-mRNA processing, mRNA export, or mRNA stability. Compare with *transcriptional-level control, translational-level control,* and *posttranslational control.*

postzygotic barrier One of several reproductive isolating mechanisms that prevent gene flow between species after fertilization has taken place, e.g., *hybrid inviability, hybrid sterility,* and *hybrid breakdown.* Compare with *prezygotic barrier.*

potential energy Stored energy; energy that can do work as a consequence of its position or state. Compare with *kinetic energy.*

potentiation A form of synaptic enhancement (increase in neurotransmitter release) that can last for several minutes; occurs when a presynaptic neuron continues to transmit action potentials at a high rate for a minute or longer.

Pr See *phytochrome.*

preadaptation A novel evolutionary change in a pre-existing biological structure that enables it to have a different function; feathers, which evolved from reptilian scales, represent a preadaptation for flight.

prebiotic soup hypothesis The hypothesis that simple organic molecules that are the precursors of life originated and accumulated at Earth's surface, in shallow seas or on rock or clay surfaces. Compare with *iron–sulfur world hypothesis.*

predation Relationship in which one organism (the predator, a secondary or higher level consumer) devours another organism (the prey).

predator See *predation.*

prefrontal cortex Region of the frontal lobes anterior to the motor areas; important in intellect, memory, judgment, language, emotion; and in interpreting incoming sensory information.

prehensile tail Tail adapted for grasping.

preimplantion genetic diagnosis (PGD) Technique by which single cells of embryos conceived by in vitro fertilization are tested for certain genetic diseases prior to the implantation of the embryo in the uterus.

premolars In mammals, teeth positioned between the canines and molars; adapted for crushing and grinding; compare with *incisors, canines,* and *molars.*

pre-mRNA RNA precursor to mRNA in eukaryotes; contains both introns and exons.

prenatal Pertaining to the time before birth.

preprophase band In plant cells, a dense array of microtubules just inside of the plasma membrane that appears just prior to mitosis and determines the plane in which the cell will divide.

pressure–flow model The mechanism by which dissolved sugar is thought to be transported in phloem; caused by a pressure gradient between the source (where sugar is loaded into the phloem) and the sink (where sugar is removed from phloem).

presynaptic neuron A neuron that transmits an impulse to a synapse. Compare with *postsynaptic neuron.*

prezygotic barrier One of several reproductive isolating mechanisms that interfere with fertilization between male and female gametes of different species, e.g., *temporal isolation, habitat isolation, behavioral isolation, mechanical isolation,* and *gametic isolation.* Compare with *postzygotic barrier.*

primary cilium Single nonmotile cilium on the cell surface of animal cells, especially those of vertebrates; binds specific molecules and serves as a "cellular antenna" in many signaling pathways.

primary consumer An animal that eats producers, e.g., plants or algae.

primary growth An increase in the length of a plant that occurs at the tips of the shoots and roots due to the activity of apical meristems. Compare with *secondary growth.*

primary immune response The response of the immune system to first exposure to an antigen. Compare with *secondary immune response.*

primary mycelium A mycelium in which the cells are monokaryotic and haploid; a mycelium that grows from either an ascospore or a basidiospore. Compare with *secondary mycelium.*

primary oocytes Diploid cells that develop from oogonia; undergo a first meiotic division, giving rise to haploid secondary oocytes.

primary producer See *autotroph.*

primary productivity The amount of light energy converted to organic compounds by autotrophs in an ecosystem over a given period. Compare with *secondary productivity.* See *gross primary productivity (GPP)* and *net primary productivity (NPP).*

primary spermatocytes Diploid cells that develop from spermatogonia; undergo a first meiotic division, giving rise to haploid secondary spermatocytes.

primary structure (of a protein) The complete sequence of amino acids in a polypeptide chain, beginning at the amino end and ending at the carboxyl end. Compare with *secondary, tertiary,* and *quaternary protein structure.*

primary succession An ecological succession that occurs on land that has not previously been inhabited by plants; no soil is present initially. See *succession.* Compare with *secondary succession.*

primates Mammals that share such traits as flexible hands and feet with five digits; a strong social organization; and front-facing eyes; includes lemurs, tarsiers, monkeys, apes, and humans.

primer See *RNA primer.*

primitive groove See *primitive streak.*

primitive streak Dynamic, constantly changing structure that forms at the midline of the blastodisc in birds, mammals, and some other vertebrates. It is active in gastrulation as cells migrate to a narrow furrow at its center, the primitive groove, and sink into the interior of the embryo. The anterior end of the primitive streak is Hensen's node.

primosome A complex of proteins responsible for synthesizing the RNA primers required in DNA synthesis.

principle In science, a statement of a rule that explains how something works. A scientific principle has withstood repeated testing.

prion (pri′-on) An infectious agent that consists only of protein.

producer See *autotroph.*

product Substance formed by a chemical reaction. Compare with *reactant.*

product rule The rule for combining the probabilities of independent events by multiplying their individual probabilities. Compare with *sum rule.*

profundal zone (pro-fun′-dl) The deepest zone of a large lake, located below the level of penetration by sunlight. Compare with *littoral zone* and *limnetic zone.*

progesterone (pro-jes′-ter-own) A steroid hormone secreted by the ovary (mainly by the corpus luteum) and placenta; stimulates the uterus (to prepare the endometrium for implantation) and breasts (for milk secretion).

progymnosperm (pro-jim′-noh-sperm) An extinct group of plants that may have been the ancestors of gymnosperms.

prokaryote (pro-kar′-ee-ote) A cell that lacks a nucleus and other membrane-enclosed organelles; includes the bacteria and archaea (kingdoms Bacteria and Archaea). Compare with *eukaryote*.

prokaryotic cell See *prokaryote*.

prolactin Hormone secreted by pituitary gland that stimulates cells of the mammary glands to produce milk in nursing mother. Helps maintain water and electrolyte balance in fishes and amphibians. Inhibits metamorphosis in amphibians and stimulates molting in reptiles. Has role in immune function and in formation of new blood vessels.

prometaphase Stage of mitosis during which spindle microtubules attach to kinetochores of chromosomes, which begin to move toward the cell's midplane; occurs after prophase and before metaphase.

promoter The nucleotide sequence in DNA to which RNA polymerase attaches to begin transcription.

prop root An adventitious root that arises from the stem and provides additional support for a plant such as corn.

prophage (pro′-faj) Bacteriophage nucleic acid inserted into the bacterial DNA.

prophase The first stage of mitosis. During prophase the chromosomes become visible as distinct structures, the nuclear envelope breaks down, and a spindle forms.

prophase I See *meiosis* and *meiosis I*.

prophase II See *meiosis* and *meiosis II*.

proplastids Organelles that are plastid precursors; may mature into various specialized plastids, including chloroplasts, chromoplasts, or leukoplasts.

proprioceptors (pro″-pree-oh-sep′-torz) Receptors in muscles, tendons, and joints that respond to changes in movement, tension, and position; enable an animal to perceive the position of its body.

prostaglandins (pros″-tah-glan′-dinz) A group of local regulators derived from fatty acids; synthesized by most cells of the body and produce a wide variety of effects; sometimes called *local hormones*.

prostate gland A gland in male animals that produces an alkaline secretion that is part of the semen.

protease An enzyme that breaks down proteins by cleaving peptide bonds.

proteasome A large multiprotein structure that recognizes and degrades protein molecules tagged with ubiquitin into short, nonfunctional peptide fragments.

protein A large, complex organic compound composed of covalently linked amino acid subunits; contains carbon, hydrogen, oxygen, nitrogen, and sulfur.

protein domain See *domain* (definition 1).

protein kinases Enzymes that activate or inactivate other proteins by phosphorylating (adding phosphate groups to) them.

proteomics The study of all the proteins expressed by a cell at a given time.

Proterozoic eon The period of Earth's history that began approximately 2.5 billion years ago and ended 541 million years ago; marked by the accumulation of oxygen and the appearance of the first multicellular eukaryotic life-forms.

prothallus (pro-thal′-us) (pl., *prothalli*) The free-living, *n* gametophyte in ferns and other seedless vascular plants.

protists (pro′-tists) Eukaryotic organisms that may be unicellular, colonial, or simple multicellular; for example, algae, amoebas, ciliates, and slime molds.

protobionts (pro″-toh-by′-ontz) Assemblages of organic polymers that spontaneously form under certain conditions. Protobionts may have been involved in chemical evolution.

proton A particle present in the nuclei of all atoms that has one unit of positive charge and a mass of 1 atomic mass unit (amu). Compare with *electron* and *neutron*.

protonema (pro″-toh-nee′-mah) (pl., *protonemata*) In mosses, a filament of *n* cells that grows from a spore and develops into leafy moss gametophytes.

protonephridia (pro″-toh-nef-rid′-ee-ah) (sing., *protonephridium*) The flame-cell excretory organs of flatworms and some other simple invertebrates.

proto-oncogene A gene that normally promotes cell division in response to the presence of certain growth factors; when mutated, it may become an oncogene, possibly leading to the formation of a cancer cell. Compare with *oncogene*.

protostomes (pro′-toh-stomes) Animals that belong to the Protostomia, one of the major animal clades. Include the annelids, arthropods, and mollusks. Compare with *deuterostomes*.

protozoa (proh″-toh-zoh′-a) (sing., *protozoon*) An informal group of unicellular, animal-like protists, including amoebas, foraminiferans, actinopods, ciliates, flagellates, and apicomplexans. (The adjectival form is *protozoan*.)

provirus (pro-vy′-rus) A part of a virus, consisting of nucleic acid only, that was inserted into a host genome. See *DNA provirus*.

proximal Closer to the point of reference. Compare with *distal*.

proximal convoluted tubule The part of the renal tubule that extends from Bowman's capsule to the loop of Henle. Compare with *distal convoluted tubule*.

proximate causes (of behavior) The immediate causes of behavior, such as genetic, developmental, and physiological processes that permit the animal to carry out a specific behavior. Compare with *ultimate causes of behavior*.

pseudocoelom (sue″-doh-see′-lom) A body cavity between the mesoderm and endoderm; derived from the blastocoel. Compare with *coelom*.

pseudocoelomate (sue″-doh-seel′-oh-mate) An animal having a pseudocoelom. Compare with *coelomate* and *acoelomate*.

pseudoplasmodium (sue″-doe-plaz-moh′-dee-um) In cellular slime molds, an aggregation of amoeboid cells that forms a spore-producing fruiting body during reproduction.

pseudopodium (sue″-doe-poe′-dee-um) (pl., *pseudopodia*) A temporary extension of an amoeboid cell that is used for feeding and locomotion.

pterosaurs (ter′-uh-sawrs) Flying reptiles of the Mesozoic era.

puberty The period of sexual maturation during which secondary sex characteristics begin to develop and the individual becomes capable of reproducing.

puff In a polytene chromosome, a decondensed region that is a site of intense RNA synthesis.

pulmonary circulation The part of the circulatory system that delivers blood to and from the lungs for oxygenation. Compare with *systemic circulation*.

pulse, arterial The alternate expansion and recoil of an artery.

punctuated equilibrium The idea that evolution proceeds with periods of little or no genetic change, followed by very active phases, so that major adaptations or clusters of adaptations appear suddenly in the fossil record. Compare with *phyletic gradualism*.

Punnett square The grid structure, first developed by Reginald Punnett, that allows direct calculation of the probabilities of occurrence of all possible offspring of a genetic cross.

pupa (pew′-pah) (pl., *pupae*) A stage in the development of an insect, between the larva and the imago (adult); a form that neither moves nor feeds and may be in a cocoon.

purines (pure′-eenz) Nitrogenous bases with carbon and nitrogen atoms in two attached rings, e.g., adenine and guanine; components of nucleic acids, ATP, GTP, NAD$^+$, and certain other biologically active substances. Compare with *pyrimidines*.

pyramid of biomass Ecological pyramid that illustrates the total biomass, as, for example, the total dry weight, of all organisms at each trophic level in an ecosystem. Compare with *pyramid of energy* and *pyramid of numbers*.

pyramid of energy Ecological pyramid that shows the energy flow through each trophic level of an ecosystem. Compare with *pyramid of biomass* and *pyramid of numbers*.

pyramid of numbers Ecological pyramid that shows the number of organisms at each trophic level in a given ecosystem. Compare with *pyramid of biomass* and *pyramid of energy*.

pyrimidines (pyr-im′-ih-deenz) Nitrogenous bases, each composed of a single ring of carbon and nitrogen atoms, e.g., thymine, cytosine, and uracil; components of nucleic acids. Compare with *purines*.

pyruvate (pyruvic acid) A three-carbon compound; the end product of glycolysis.

quadrupedal (kwad′-roo-ped″-ul) Walking on all fours. Compare with *bipedal*.

quantitative trait A trait that shows continuous variation in a population (e.g., human height) and typically has a polygenic inheritance pattern.

Quaternary period See *Cenozoic era.*

quaternary structure (of a protein) The overall conformation of a protein produced by the interaction of two or more polypeptide chains. Compare with *primary, secondary,* and *tertiary protein structure.*

***r* selection** A reproductive strategy recognized by some ecologists, in which a species typically has a small body size, rapid development, and short lifespan and devotes a large proportion of its metabolic energy to the production of offspring. Compare with *K selection.*

***r* strategist** See *r selection.*

radial cleavage The pattern of blastomere production in which the cells are located directly above or below one another; characteristic of early deuterostome embryos. Compare with *spiral cleavage.*

radial symmetry A body plan in which any section through the mouth and down the length of the body divides the body into similar halves. Jellyfish and other cnidarians have radial symmetry. Compare with *bilateral symmetry.*

radicle (rad′-ih-kl) The embryonic root of a seed plant.

radioactive decay The process in which a radioactive element emits radiation, and as a result, its nucleus changes into the nucleus of a different element.

radioisotopes Unstable isotopes that spontaneously emit radiation; also called *radioactive isotopes.*

radiolarians Those actinopods that secrete elaborate shells of silica (glass).

radula (rad′-yoo-lah) A rasplike structure in the digestive tract of chitons, snails, squids, and certain other mollusks.

rain shadow An area that has very little precipitation, found on the downwind side of a mountain range. Deserts often occur in rain shadows.

random dispersion The spatial distribution pattern of a population in which the presence of one individual has no effect on the distribution of other individuals. Compare with *clumped dispersion* and *uniform dispersion.*

range The area where a particular species occurs. Compare with *home range.*

Ras proteins A group of small G proteins, named by their discovery in **ra**t **s**arcoma cancer cells; important in many signaling pathways.

ratite (rat′-it) A flightless bird, such as an ostrich or kiwi.

ray A chain of parenchyma cells (one to many cells thick) that functions for lateral transport in stems and roots of woody plants.

ray-finned fishes A class (Actinopterygii) of modern bony fishes; contains about 95% of living fish species.

reabsorption The selective removal of certain substances from the glomerular filtrate by the renal tubules and collecting ducts of the kidney, and their return into the blood.

reactant Substance that participates in a chemical reaction. Compare with *product.*

reaction center The portion of a photosystem that includes chlorophyll *a* molecules capable of transferring electrons to a primary electron acceptor, which is the first of several electron acceptors in a series. See *antenna complex* and *photosystem.*

realized niche The lifestyle that an organism actually pursues, including the resources that it actually uses. An organism's realized niche is narrower than its fundamental niche because of interspecific competition. Compare with *fundamental niche.*

Recent Africa Origin model (RAO) See *(recent) out-of-Africa model.*

(recent) out-of-Africa model (also called *Recent Africa Origin* or *RAO*) Hypothesis that modern humans (*Homo sapiens*) originated in Africa and subsequently migrated to other parts of the world.

receptacle The end of a flower stalk where the flower parts (sepals, petals, stamens, and carpels) are attached.

reception Process of detecting a stimulus.

receptor (1) In cell biology, a molecule on the surface of a cell, or inside a cell, that serves as a recognition or binding site for signaling molecules such as hormones, antibodies, or neurotransmitters. (2) A sensory receptor. See *sensory receptor.*

receptor down-regulation The process by which some hormone receptors decrease in number, thereby suppressing the sensitivity of target cells to a hormone. Compare with *receptor up-regulation.*

receptor-mediated endocytosis A type of endocytosis in which extracellular molecules become bound to specific receptors on the cell surface and then enter the cytoplasm enclosed in vesicles. Compare with *phagocytosis* and *pinocytosis.*

receptor potential The change in membrane potential by a sensory receptor in response to a stimulus.

receptor up-regulation The process by which some hormone receptors increase in number, thereby increasing the sensitivity of the target cells to a hormone. Compare with *receptor down-regulation.*

recessive allele (al-leel′) An allele that is not expressed in the heterozygous state. Compare with *dominant allele.*

reciprocal altruism A type of altruistic behavior in which an animal helps nonrelatives if they are likely to repay the debt in the future.

recombinant DNA Any DNA molecule made by combining genes from different organisms.

recombination, genetic The appearance of new allele combinations. Recombination in eukaryotes generally results from meiotic events, either crossing-over or shuffling of chromosomes.

rectum The last region of the large intestine.

red alga A member of a diverse group of archaeplastid algae that contain the pigments chlorophyll *a*, carotenoids, phycocyanin, and phycoerythrin.

red blood cell (RBC) See *erythrocyte.*

redox reaction (ree′-dox) The chemical reaction in which one or more electrons are transferred from one substance (the substance that becomes oxidized) to another (the substance that becomes reduced). See *oxidation* and *reduction.*

red tide A red or brown coloration of ocean water caused by a population explosion, or bloom, of dinoflagellates.

reduction The gain of one or more electrons (or hydrogen atoms) by an atom, ion, or molecule. Compare with *oxidation.*

reductionism Learning about a structure or process by studying its simplest components.

re-emerging diseases Diseases that have been almost eradicated and then suddenly recur, causing an epidemic. Compare with *emerging diseases.*

reflex action An automatic, involuntary response to a given stimulus that generally functions to restore homeostasis.

refractory period The brief period that elapses after the response of a neuron or muscle fiber, during which it cannot respond to another stimulus.

regulative development The very plastic developmental pattern in which each individual cell of an early embryo retains totipotency. Compare with *mosaic development.*

regulatory gene Gene that turns the transcription of other genes on or off.

regulatory T cells (T$_{regs}$) A population of T cells that help regulate immune responses.

reinforcement The increase in reproductive isolation that often occurs over time in a hybrid zone because natural selection strengthens and increases the number of prezygotic barriers between the two species.

releasing hormone A hormone secreted by the hypothalamus that stimulates secretion of a specific hormone by the anterior lobe of the pituitary gland.

renal (ree′-nl) Pertaining to the kidney.

renal artery An artery that branches from the aorta and delivers blood to the kidneys.

renal cortex The outer region of the mammalian kidney; surrounds the renal medulla. Compare with *renal medulla.*

renal medulla The inner region of the mammalian kidney. Compare with *renal cortex.*

renal pelvis The funnel-shaped chamber of the kidney that receives urine from the collecting ducts; urine then moves into the ureters.

renin (reh′-nin) An enzyme released by the kidney in response to a decrease in blood pressure; activates a pathway leading to production of angiotensin II, a hormone that increases aldosterone release; aldosterone increases blood pressure.

replacement-level fertility The number of children a couple must produce to "replace" themselves. The average number is greater than two, because some children die before reaching reproductive age.

replication See *DNA replication.*

replication fork Y-shaped structure produced during the semiconservative replication of DNA.

repolarization The process of returning membrane potential to its resting level.

repressible operon An operon that is normally active, but can be controlled by a repressor protein, which becomes active when it binds to a corepressor; the active repressor binds to the operator, making the operon transcriptionally inactive. Compare with *inducible operon.*

repressor protein A negative regulatory protein that inhibits transcription when bound to DNA; some repressors require a corepressor to be active; some other repressors become inactive when bound to an inducer molecule. Compare with *activator protein.*

reproduction The process by which new individuals are produced. See *asexual reproduction* and *sexual reproduction.*

reproductive isolating mechanisms The reproductive barriers that prevent a species from interbreeding with another species; as a result, each species' gene pool is isolated from those of other species. See *prezygotic barrier* and *postzygotic barrier.*

reptiles Vertebrates characterized by dry skin with horny scales and adaptations for terrestrial reproduction; include turtles, snakes, and alligators; reptiles are a paraphyletic group unless birds are included.

residual volume The volume of air that remains in the lungs at the end of a normal exhalation.

resin A viscous organic material that certain plants produce and secrete into specialized ducts; may play a role in deterring disease organisms or plant-eating insects.

resolution See *resolving power.*

resolving power The ability of a microscope to show fine detail, defined as the minimum distance between two points at which they are seen as separate images; also called *resolution.*

resource partitioning The reduction of competition for environmental resources such as food that occurs among coexisting species as a result of each species' niche differing from the others in one or more ways.

respiration (1) Cellular respiration is the process by which cells generate ATP through a series of redox reactions. In aerobic cellular respiration the terminal electron acceptor is molecular oxygen; in anaerobic cellular respiration the terminal acceptor is an inorganic molecule other than oxygen. (2) Organismic respiration is the process of gas exchange between a complex animal and its environment, generally through a specialized respiratory surface, such as a lung or gill.

respiratory centers Centers in the medulla and pons that regulate breathing.

respiratory system The body system that carries on gas exchange; in vertebrates, consists of the lungs and airways.

response The action of effectors in response to a stimulus.

resting potential The membrane potential (difference in electric charge between the two sides of the plasma membrane) of a neuron in which no action potential is occurring. The typical resting potential is about −70 millivolts. Compare with *action potential.*

restoration ecology The scientific field that uses the principles of ecology to help return a degraded environment as closely as possible to its former undisturbed state.

restriction enzyme One of a class of enzymes that cleave DNA at specific base sequences; produced by bacteria to degrade foreign DNA; used in recombinant DNA technology.

restriction fragment length polymorphism (RFLP) A difference, or variation, in the genomic DNA of a population caused by the presence or absence of a restriction enzyme site.

restriction map A physical map of DNA in which sites cut by specific restriction enzymes serve as landmarks.

reticular activating system (RAS) (reh-tik′-yoo-lur) A diffuse network of neurons in the brain stem; responsible for maintaining consciousness.

retina (ret′-ih-nah) The innermost of the three layers (retina, choroid layer, and sclera) of the eyeball, which is continuous with the optic nerve and contains the light-sensitive rod and cone cells.

retinal Visual pigment derived from vitamin A; present in eyes of insects, mollusks, and vertebrates.

retinular cells See *ommatidium.*

retrotransposon A chromosomal DNA segment that can move to a different site within the genome by forming an RNA intermediate that is converted to a DNA fragment using reverse transcriptase; also called a *mobile genetic element.*

retrovirus (ret′-roh-vy″-rus) An RNA virus that uses reverse transcriptase to produce a DNA intermediate, known as a *DNA provirus,* in the host cell. See *DNA provirus.*

reverse transcriptase An enzyme produced by retroviruses that catalyzes the production of DNA using RNA as a template.

reversible inhibitor A substance that forms weak bonds with an enzyme, temporarily interfering with its function; a reversible inhibitor is either competitive or noncompetitive. Compare with *irreversible inhibitor.*

Rh factors Red blood cell antigens, known as *D antigens,* first identified in *Rhesus* monkeys. People who have these antigens are Rh⁺; people lacking them are Rh⁻. See *erythroblastosis fetalis.*

rhizarians A diverse supergroup of amoeboid cells that often have hard outer shells, called tests, through which cytoplasmic projections extend; include forams, actinopods, and certain shell-less amoebas.

rhizobia See *nodules.*

rhizome (ry′-zome) A horizontal underground stem that bears leaves and buds and often serves as a storage organ and a means of asexual reproduction, e.g., iris.

rhodopsin (rho-dop′-sin) Visual purple; a light-sensitive pigment found in the rod cells of the vertebrate eye; a similar molecule is employed by certain bacteria in the capture of light energy to make ATP.

ribonucleic acid (RNA) A family of single-stranded nucleic acids that function mainly in protein synthesis

ribose The five-carbon sugar present in RNA and in important nucleoside triphosphates such as ATP.

ribosomal RNA (rRNA) See *ribosomes.*

ribosomes (ry′-boh-sohmz) Organelles that are part of the protein synthesis machinery of both prokaryotic and eukaryotic cells; consist of a larger and smaller subunit, each composed of ribosomal RNA (rRNA) and ribosomal proteins.

ribozyme (ry′-boh-zime) A molecule of RNA that has catalytic properties.

ribulose bisphosphate (RuBP) A five-carbon phosphorylated compound with a high energy potential that reacts with carbon dioxide in the initial step of the Calvin cycle.

ribulose bisphosphate carboxylase/oxygenase See *rubisco.*

RNA interference (RNAi) Phenomenon in which certain small RNA molecules interfere with the expression of genes or their RNA transcripts; RNA interference involves small interfering RNAs (siRNAs), microRNAs (miRNAs), piwi-associated RNAs (piRNAs), and other kinds of short RNA molecules.

RNA polymerase An enzyme that catalyzes the synthesis of RNA from a DNA template.

RNA primer The sequence of about five RNA nucleotides that are synthesized during DNA replication to provide a 3′ end to which DNA polymerase adds nucleotides. The RNA primer is later degraded and replaced with DNA.

RNA world A model that proposes that during the evolution of cells, RNA was the first informational molecule to evolve, followed at a later time by proteins and DNA.

rod One of the rod-shaped, light-sensitive cells of the retina that are particularly sensitive to dim light and mediate black-and-white vision. Compare with *cone.*

root cap A covering of cells over the root tip that protects the delicate meristematic tissue directly behind it.

root graft The process of roots from two different plants growing together and becoming permanently attached to each other.

root hair An extension, or outgrowth, of a root epidermal cell. Root hairs increase the absorptive capacity of roots.

root pressure The pressure in xylem sap that occurs as a result of the active absorption of mineral ions followed by the osmotic uptake of water into roots from the soil.

root system The underground portion of a plant that anchors it in the soil and absorbs water and dissolved minerals. Compare with *shoot system.*

rough ER See *endoplasmic reticulum.*

r-selected species See *r selection.*

rubisco The common name of ribulose bisphosphate carboxylase/oxygenase, the enzyme that catalyzes the fixation of carbon dioxide in the Calvin cycle.

rugae (roo'-jee) Folds, such as those in the lining of the stomach.

runner See *stolon.*

runoff Movement of excess surface water over land toward bodies of water.

S phase Stage in interphase of the cell cycle during which DNA and other chromosomal constituents are synthesized. Compare with G_1 *phase* and G_2 *phase.*

saccule The structure within the vestibule of the inner vertebrate ear that along with the utricle houses the receptors of static equilibrium.

salicylic acid A signaling molecule that helps plants defend against insect pests and pathogens such as viruses by helping activate systemic acquired resistance.

salinity The concentration of dissolved salts (e.g., sodium chloride) in a body of water.

salivary amylase An enzyme in saliva that hydrolyzes starch to the disaccharide maltose.

salivary glands Accessory digestive glands found in vertebrates and some invertebrates; in humans there are three pairs.

salt An ionic compound consisting of an anion other than a hydroxide ion and a cation other than a hydrogen ion. A salt is formed by the reaction between an acid and a base.

saltatory conduction The transmission of a neural impulse along a myelinated neuron; ion activity at one node depolarizes the next node along the axon.

salt marsh A wetland dominated by grasses in which the salinity fluctuates between that of sea water and fresh water; salt marshes are usually located in estuaries.

saprobe See *decomposer.*

saprotroph (sap'-roh-trof) See *decomposer.*

sarcolemma (sar"-koh-lem'-mah) The muscle cell plasma membrane.

sarcomere (sar'-koh-meer) A segment of a striated muscle cell located between adjacent Z lines that serves as a unit of contraction.

sarcoplasmic reticulum The system of vesicles in a muscle cell that surrounds the myofibrils and releases calcium in muscle contraction; a modified endoplasmic reticulum.

satellites Subviral agents that require a helper virus to reproduce.

saturated fatty acid See *fatty acid.*

savanna (suh-van'-uh) A tropical grassland containing scattered trees; found in areas of low rainfall or seasonal rainfall with prolonged dry periods.

scaffolding proteins Nonhistone proteins that help maintain the structure of a chromosome.

scaffold proteins Proteins that organize groups of intracellular signaling molecules into signaling complexes.

scanning electron microscope (SEM) See electron microscope.

schizocoely (skiz'-oh-seely) The process of coelom formation in which the mesoderm splits into two layers, forming a cavity between them; characteristic of protostomes. Compare with *enterocoely.*

Schwann cells Supporting cells found in nervous tissue outside the central nervous system; produce the myelin sheath around peripheral neurons.

scientific method The process that scientists use to investigate the natural world; includes observing, recognizing a problem or stating a critical question, developing a hypothesis, making a prediction that can be tested, making further observations, performing experiments, interpreting results, and drawing conclusions that support or falsify the hypothesis.

scientific theory In science, a widely accepted explanation supported by a large body of observations and experiments. A scientific theory relates facts that appear unrelated; it predicts new facts and suggests new relationships. Compare with *hypothesis.*

sclera (skler'-ah) The outer coat of the eyeball; a tough, opaque sheet of connective tissue that protects the inner structures and helps maintain the rigidity of the eyeball.

sclereid (skler'-id) In plants, a sclerenchyma cell that is variable in shape but typically not long and tapered. Compare with *fiber.*

sclerenchyma (skler-en'-kim-uh) Cells that provide strength and support in the plant body, are often dead at maturity, and have extremely thick walls; includes fibers and sclereids.

scramble competition See *exploitation competition.*

scrotum (skroh'-tum) The external sac of skin found in most male mammals that contains the testes and their accessory organs.

sea grasses Flowering plants that have adapted to complete submersion in ocean water.

secondary consumer An animal that eats herbivores (primary consumers).

secondary growth An increase in the girth of a plant due to the activity of the vascular cambium and cork cambium; secondary growth results in the production of secondary tissues, i.e., wood and bark. Compare with *primary growth.*

secondary immune response The rapid production of antibodies induced by a second exposure to an antigen several days, weeks, or even months after the initial exposure. Compare with *primary immune response.*

secondary mycelium A dikaryotic mycelium formed by the fusion of two primary hyphae. Compare with *primary mycelium.*

secondary oocytes Haploid cells that undergo a second meiotic division, giving rise to haploid ootids.

secondary productivity The amount of food molecules converted to biomass by consumers in an ecosystem over a given period. Compare with *primary productivity.*

secondary sex characteristics Changes that develop during puberty; in males, include growth of facial and body hair, muscle development, and the increase in vocal cord length and thickness that causes the voice to deepen; in females, include development of the breasts, broadening of the pelvis, and the development and distribution of muscle and fat responsible for the female body shape.

secondary spermatocytes Haploid cells that undergo a second meiotic division, giving rise to haploid spermatids.

secondary structure (of a protein) A regular geometric shape produced by hydrogen bonding between the atoms of the uniform polypeptide backbone; includes the alpha helix and the beta-pleated sheet. Compare with *primary, tertiary,* and *quaternary protein structure.*

secondary succession An ecological succession that takes place after some disturbance destroys the existing vegetation; soil is already present. See *succession.* Compare with *primary succession.*

second law of thermodynamics The physical law stating that the total amount of entropy in the universe continually increases. Compare with *first law of thermodynamics.*

second messenger A substance, e.g., cyclic AMP or calcium ions, that relays a message from a signaling molecule (the *first messenger*) bound to a cell-surface receptor; leads to some change in the cell.

secretory vesicles Small cytoplasmic vesicles that move substances from an internal membrane system to the plasma membrane.

seed A plant reproductive body consisting of a young, multicellular plant and nutritive tissue (food reserves), enclosed by a seed coat.

seed coat The outer protective covering of a seed.

seed fern An extinct group of seed-bearing woody plants with fernlike leaves; seed ferns probably descended from progymnosperms and gave rise to cycads and possibly ginkgoes.

segmentation A body plan in which the body is divided into a series of compartments (segments).

segmentation genes In *Drosophila,* genes transcribed in the embryo that are responsible for generating a repeating pattern of body segments within the embryo and adult fly.

segregation, principle of The genetic principle, first noted by Gregor Mendel, that states that two alleles of a locus become separated into different gametes.

selectively permeable membrane A membrane that allows some substances to cross it more easily than others. Biological membranes are generally permeable to water but restrict the passage of many solutes.

self-incompatibility A genetic condition in which the pollen cannot fertilize the same flower or flowers on the same plant.

semelparous Having a single reproductive effort in a lifetime. Compare with *iteroparous*.

semen The fluid consisting of sperm suspended in various glandular secretions that is ejaculated from the penis during orgasm.

semicircular canals The passages in the vertebrate inner ear containing structures that control the sense of equilibrium (balance).

semiconservative replication See *DNA replication*.

semilunar valves Valves between the ventricles of the heart and the arteries that carry blood away from the heart; aortic and pulmonary valves.

seminal vesicles (1) In mammals, glandular sacs that secrete a component of semen (seminal fluid). (2) In some invertebrates, structures that store sperm.

seminiferous tubules (sem-ih-nif´-er-ous) Coiled tubules in the testes in which spermatogenesis takes place in male vertebrates.

senescence (se-nes´-cents) The aging process.

sense organ Sensory receptors along with other types of cells make up sense organs, e.g., eyes, ears, taste buds.

sensor In homeostasis, a receptor that detects a change, i.e., a deviation from the homeostatic state.

sensory adaptation The condition in which the response to a stimulus is reduced even though the stimulus continues at the same intensity.

sensory neuron A neuron that transmits an impulse from a receptor to the central nervous system.

sensory receptor The ending of an afferent neuron that is specialized to detect specific energy stimuli in its environment.

sepal (see´-pul) One of the outermost parts of a flower, usually leaflike in appearance, that protect the flower as a bud.

septum (pl., *septa*) A cross wall or partition. (1) In fungi, the walls that divide a hypha into cells. (2) In annelids, the partitions that separate the segments. (3) In birds and mammals, the wall that separates the two ventricles (and the two atria) of the heart.

sequencing See *DNA sequencing*.

serial endosymbiosis The hypothesis that certain organelles such as mitochondria and chloroplasts originated as symbiotic prokaryotes that lived inside other, free-living prokaryotic cells.

serotonin A neurotransmitter of the biogenic amine group; important in mood and helps regulate the sleep–wake cycle.

serous membrane (sir´-us) An epithelial membrane that lines a body cavity that does not open to the outside of the body.

Sertoli cells (sur-tole´-ee) Supporting cells of the tubules of the testis.

serum Plasma minus fibrinogen and other clotting agents.

sessile (ses´-sile) Permanently attached to one location, e.g., coral animals.

setae (sing., *seta*) Bristlelike structures that aid in annelid locomotion.

set point A normal condition maintained by homeostatic mechanisms.

sex chromosome Chromosome that plays a role in sex determination.

sex pili See *pili*.

sex-influenced trait A genetic trait that is expressed differently in males and females.

sex-linked gene A gene carried on a sex chromosome. In mammals almost all sex-linked genes are borne on the X chromosome, i.e., are X-linked.

sexual dimorphism Marked phenotypic differences between the two sexes of the same species.

sexual isolation See *behavioral isolation*.

sexual reproduction A type of reproduction in which two gametes (usually, but not necessarily, contributed by two different parents) fuse to form a zygote. Compare with *asexual reproduction*.

sexual selection A type of natural selection that occurs when individuals of a species vary in their ability to compete for mates; individuals with reproductive advantages are selected over others of the same sex.

shade avoidance The tendency of plants that are adapted to high light intensities to grow taller when they are closely surrounded by other plants.

shared ancestral characters Traits that were present in an ancestral species that have remained essentially unchanged; suggest a distant common ancestor. Also called *plesiomorphic characters*. Compare with *shared derived characters*.

shared derived characters Homologous traits found in two or more taxa that are present in their most recent common ancestor but not in earlier common ancestors. Also called *synapomorphic characters*. Compare with *shared ancestral characters*.

shoot system The aboveground portion of a plant, such as the stem and leaves. Compare with *root system*.

short-day plant A plant that flowers in response to lengthening nights; also called *long-night plant*. Compare with *long-day*, *intermediate-day*, and *day-neutral plants*.

short-night plant See *long-day plant*.

short tandem repeats (STRs) Molecular markers that are short sequences of repetitive DNA; because STRs vary in length from one individual to another, they are useful in identifying individuals with a high degree of certainty.

short-term memory Memory that stores only about seven chunks of information and allows us to recall that information only for seconds or for a few minutes. Compare with *long-term memory*.

sickle cell anemia An inherited form of anemia in which there is an abnormality in the hemoglobin beta chains; the inheritance pattern is autosomal recessive.

sieve tube Structure consisting of sieve tube elements joined end to end.

sieve tube elements Cells that conduct dissolved sugar in the phloem of flowering plants.

signal amplification The process by which a few signaling molecules can elicit major responses in the cell; the strength of each signaling molecule is magnified.

signaling molecule A molecule such as a hormone, local regulator, or neurotransmitter that transmits information when it binds to a receptor on the cell surface or within the cell.

signal-recognition particle (SRP) A protein–RNA complex that directs the ribosome–mRNA–polypeptide complex to the surface of the endoplasmic reticulum.

signal-recognition particle RNA See *signal-recognition particle (SRP)*.

signal transduction A process in which a cell converts and amplifies an extracellular signal into an intracellular signal that affects some function in the cell. Also see *cell signaling*.

sign stimulus Any stimulus that elicits a fixed action pattern in an animal.

silencer A regulatory element that associates with a eukaryotic gene and can decrease its transcription.

Silurian period See *Paleozoic era*.

simple diffusion See *diffusion*.

simple fruit A fruit that develops from a single ovary. Compare with *aggregate, accessory,* and *multiple fruits*.

simplest formula A type of chemical formula that gives the smallest whole-number ratio of the component atoms. Compare with *molecular formula* and *structural formula*.

single nucleotide polymorphisms (SNPs) Genetic variation among individuals in a population as measured by alleles of specific loci that differ by as little as a single nucleotide. The abbreviation is usually pronounced "snips."

single-strand binding proteins (SSBs) Proteins involved in DNA replication that bind to single DNA strands and prevent the double helix from re-forming until the strands are copied.

sink habitat A lower-quality habitat in which local reproductive success is less than local mortality. Compare with *source habitat*.

sinoatrial (SA) node The mass of specialized cardiac muscle in which the impulse triggering the heartbeat originates; the pacemaker of the heart.

sister chromatids See *chromatid*.

sister taxa Groups of organisms that share a more recent common ancestor with one another than either taxon does with any other group shown on a cladogram.

skeletal muscle The voluntary striated muscle of vertebrates, so called because it usually is directly or indirectly attached to some part of the skeleton. Compare with *cardiac muscle* and *smooth muscle*.

slash-and-burn agriculture A type of agriculture in which tropical rain forest is cut down, allowed to dry, and burned. The crops that are planted immediately afterward thrive because the ashes provide nutrients; in a few years, however, the soil is depleted and the land must be abandoned.

slime layer A protective layer surrounding the cell wall of many prokaryotic species; more loosely attached to the cell wall than a capsule. See *capsule*, definition 3.

slow block to polyspermy See *cortical reaction*.

slow-oxidative fibers Muscle fibers specialized for endurance activities; they contract slowly, fatigue slowly, and obtain most of their ATP from aerobic respiration. Compare with *fast-oxidative fibers* and *fast-glycolytic fibers*.

small interfering RNAs (siRNAs) Double-stranded RNA molecules about 23 nucleotides in length that silence genes at the posttranscriptional level by selectively cleaving mRNA molecules with base sequences complementary to the siRNA. See also *microRNA (miRNA)*, *RNA interference (RNAi)*, and *piwi-associated RNA (piRNA)*.

small intestine Portion of the vertebrate digestive tract that extends from the stomach to the large intestine.

small nuclear ribonucleoprotein complexes (snRNPs) Aggregations of protein and small nuclear RNA (snRNA) that associate to form a spliceosome that binds to pre-mRNA in eukaryotes and catalyzes the excision of introns and splicing of exons. The abbreviation is usually pronounced "snurps."

small nuclear RNA (snRNA) See *small ribonucleo-protein complexes (snRNPs)*.

small nucleolar RNA (snoRNA) RNA molecules involved in the processing of pre-ribosomal RNA during the formation of ribosome subunits in the nucleolus.

smooth ER See *endoplasmic reticulum*.

smooth muscle Involuntary muscle tissue that lacks transverse striations; found mainly in sheets surrounding hollow organs, such as the intestine. Compare with *cardiac muscle* and *skeletal muscle*.

SNP See *single nucleotide polymorphism*.

social behavior Interaction of two or more animals, usually of the same species.

social learning Learning through observing others and imitating them.

society A group of individuals belonging to the same species and often closely related; an organized society is an actively cooperating group.

sociobiology The branch of biology that focuses on the evolution of social behavior through natural selection.

sodium–potassium pump ATP-binding cassette transporter in the plasma membrane of all animal cells; uses ATP energy to transport Na^+ out of the cell and K^+ into the cell.

soil erosion The wearing away or removal of soil from the land; although soil erosion occurs naturally from precipitation and runoff, human activities (such as clearing the land) accelerate it.

solute A dissolved substance. Compare with *solvent*.

solvent Substance capable of dissolving other substances. Compare with *solute*.

somatic cell In animals, a cell of the body not involved in formation of gametes. Compare with *germ line cell*.

somatic division (of the PNS) That part of the vertebrate peripheral nervous system that keeps the body in adjustment with the external environment; includes sensory receptors on the body surface and within the muscles, and the nerves that link them with the central nervous system. Compare with *autonomic division* of the PNS.

somatomedins See *insulin-like growth factors*.

somatotropin See *growth hormone*.

somites A series of paired blocks of mesoderm that develop on each side of the notochord in cephalochordates and vertebrates. Somites define the segmentation of the embryo, and in vertebrates, they give rise to the vertebrae, ribs, and certain skeletal muscles.

sonogram See *ultrasound imaging*.

soredium (sor-id′e-um) (pl., *soredia*) In lichens, a type of asexual reproductive structure that consists of a cluster of algal cells surrounded by fungal hyphae.

sorus (soh′rus) (pl., *sori*) In ferns, a cluster of spore-producing sporangia.

source habitat A good habitat in which local reproductive success is greater than local mortality. Surplus individuals in a source habitat may disperse to other habitats. Compare with *sink habitat*.

Southern blot A technique in which DNA fragments, previously separated by gel electrophoresis, are transferred to a nitrocellulose or nylon membrane and detected by autoradiography or chemical luminescence. Compare with *Northern blot* and *Western blot*.

spatial summation The summing of several postsynaptic potentials resulting when several presynaptic neurons release neurotransmitter simultaneously. Compare with *temporal summation*.

speciation Evolution of a new species.

species According to the biological species concept, one or more populations whose members are capable of interbreeding in nature to produce fertile offspring and do not interbreed with members of other species. Compare with *phylogenetic species concept*.

species diversity A measure of the relative importance of each species within a community; represents a combination of species richness and species evenness.

species evenness Relative abundance of one species compared to other species in a community.

species richness The number of species in a community.

specific epithet The second part of the name of a species; designates a specific species belonging to that genus.

specific heat The amount of heat energy that must be supplied to raise the temperature of 1 g of a substance 1°C.

specific immune responses See *adaptive immune responses*.

specific immunity See *adaptive immune responses*.

sperm The motile male gamete of animals and some plants and protists; also called a *spermatozoan*.

spermatid (spur′-ma-tid) An immature sperm cell.

spermatocyte (spur-mah′-toh-site) A meiotic cell that gives rise to spermatids and ultimately to mature sperm cells.

spermatogenesis (spur″-mah-toh-jen′-eh-sis) The production of male gametes (sperm) by meiosis and subsequent cell differentiation. Compare with *oogenesis*.

spermatogonia Undifferentiated cells in the testes that give rise to primary spermatocytes.

spermatozoan (spur-mah-toh-zoh′-un) See *sperm*.

sphincter (sfink′-tur) A group of circularly arranged muscle fibers, the contractions of which close an opening, e.g., the pyloric sphincter at the exit of the stomach.

spicules (spik′-yuls) Slender skeletal spikes made of calcium carbonate or silica; found in sponges and some other animals.

spinal cord In vertebrates, the dorsal, tubular nerve cord.

spinal nerves In vertebrates, the nerves that emerge from the spinal cord; there are 31 pairs in humans.

spindle See *mitotic spindle*.

spine A leaf that is modified for protection, such as a cactus spine.

spiracle (speer′-ih-kl) An opening for gas exchange, such as the opening of a trachea on the body surface of an insect.

spiral cleavage A distinctive spiral pattern of blastomere production in an early protostome embryo. Compare with *radial cleavage*.

spirillum (pl., *spirilla*) A long, rigid, helical bacterium. Compare with *spirochete*, *vibrio*, *bacillus*, and *coccus*.

spirochete A long, flexible, helical bacterium. Compare with *spirillum*, *vibrio*, *bacillus*, and *coccus*.

spleen An abdominal organ located just below the diaphragm that removes worn-out blood cells and bacteria from the blood and plays a role in immunity.

spliceosome A large ribonucleoprotein particle that catalyzes the reactions that remove introns from pre-mRNA.

spongocoel (spon′-jo-seel) The central cavity of a sponge.

spongy bone Consists of a network of thin strands of bone; provides mechanical strength.

spongy mesophyll (mez′-oh-fil) The loosely arranged mesophyll cells near the lower epidermis in certain leaves. Compare with *palisade mesophyll*.

spontaneous reaction See *exergonic reaction*.

sporangium (spor-an′-jee-um) (pl., *sporangia*) A spore case, found in plants, certain protists, and fungi.

spore A reproductive cell that gives rise to individual offspring in plants, fungi, and certain algae and protozoa.

sporophyll (spor′-oh-fil) A leaflike structure that bears spores.

sporophyte generation (spor′-oh-fite) The 2*n*, spore-producing stage in the life cycle of a plant. Compare with *gametophyte generation*.

sporozoite The infective sporelike state in apicomplexans.

spring turnover Mixing of temperate lake waters in spring, caused by rising temperatures, in which surface water sinks to the bottom and bottom water rises to the surface. Compare with *fall turnover*.

stabilizing selection Natural selection that acts against extreme phenotypes and favors intermediate variants; associated with a population well adapted to its environment. Compare with *directional selection* and *disruptive selection*.

stamen (stay′-men) The male part of a flower; consists of a filament and anther.

standing-water ecosystem A lake or pond ecosystem.

starch A polysaccharide composed of alpha glucose subunits; made by plants for energy storage.

start codon The codon AUG, which signals the beginning of translation of messenger RNA. Compare with *stop codon*.

stasis Long periods in the fossil record in which there is little or no evolutionary change.

statocyst (stat′-oh-sist) An invertebrate sense organ containing one or more granules (statoliths); senses gravity and motion.

statoliths (stat′-uh-liths) Granules of loose sand or calcium carbonate found in statocysts.

stele The cylinder in the center of roots and stems that contains the vascular tissue; also called *vascular cylinder*.

stem cell A relatively undifferentiated cell capable of repeated cell division. At each division at least one of the daughter cells usually remains a stem cell, whereas the other may differentiate as a specific cell type. Compare with *embryonic stem cell (ES cell)* and *induced pluripotent stem cell (iPSC)*.

stereocilia Hairlike projections of hair cells; microvilli that contain actin filaments.

sterilization A procedure that renders an individual incapable of producing offspring; the most common surgical procedures are vasectomy in the male and tubal ligation in the female.

steroid hormones Hormones, such as insect molting hormone, cortisol, and male/female sex hormones, that are derived from cholesterol.

steroids (steer′-oids) Complex molecules containing carbon atoms arranged in four attached rings, three of which contain six carbon atoms each and the fourth of which contains five, e.g., cholesterol and certain hormones, including the male and female sex hormones of vertebrates.

stigma The portion of the carpel where pollen grains land during pollination (and before fertilization).

stipe A short stalk or stemlike structure that is a part of the body of certain multicellular algae.

stipule (stip′-yule) One of a pair of scalelike or leaflike structures found at the base of certain leaves.

stolon (stow′-lon) An aboveground, horizontal stem with long internodes; stolons often form buds that develop into separate plants, e.g., strawberry; also called a *runner*.

stomach Muscular region of the vertebrate digestive tract, extending from the esophagus to the small intestine.

stomata (sing., *stoma*) Small pores located in the epidermis of plants that provide for gas exchange for photosynthesis; each stoma is flanked by two guard cells, which are responsible for its opening and closing.

stop codon Any of the three codons in mRNA that do not code for an amino acid (UAA, UAG, or UGA) but signal the termination of translation. Compare with *start codon*.

stramenopiles Protists that have motile cells with two flagella, one of which has tiny hairlike projections off the shaft; include water molds, diatoms, golden algae, and brown algae. See *chromalveolates*; compare with *alveolates*.

stratosphere The layer of the atmosphere between the troposphere and the mesosphere. It contains a thin ozone layer that protects life by filtering out much of the sun's ultraviolet radiation.

stratum basale (strat′-um bah-say′-lee) The deepest sublayer of the human epidermis, consisting of cells that continuously divide. Compare with *stratum corneum*.

stratum corneum The most superficial sublayer of the human epidermis. Compare with *stratum basale*.

strobilus (stroh′-bil-us) (pl., *strobili*) In certain plants, a conelike structure that bears spore-producing sporangia.

stroke volume The volume of blood pumped by one ventricle during one contraction.

stroma A fluid space of the chloroplast, enclosed by the chloroplast inner membrane and surrounding the thylakoids; site of the reactions of the Calvin cycle.

stromatolite (stroh-mat′-oh-lite) A columnlike rock that consists of many minute layers of prokaryotic cells, usually cyanobacteria.

structural formula A type of chemical formula that shows the spatial arrangement of the atoms in a molecule. Compare with *simplest formula* and *molecular formula*.

structural isomer One of two or more chemical compounds having the same chemical formula but differing in the covalent arrangement of their atoms, e.g., glucose and fructose.

style The neck connecting the stigma to the ovary of a carpel.

subsidiary cell In plants, a structurally distinct epidermal cell associated with a guard cell.

subsistence agriculture Farming practice in which crops are grown only to meet the needs of the farm family.

substance P A peptide neurotransmitter released by certain sensory neurons in pain pathways; signals the CNS regarding pain; also stimulates other structures, including smooth muscle in the digestive tract.

substrate A substance on which an enzyme acts; a reactant in an enzymatically catalyzed reaction.

substrate-level phosphorylation The formation of ATP by the transfer of a phosphate to ADP from a phosphorylated intermediate.

subviral agents Infective agents that are smaller and simpler than viruses; include satellites, viroids, and prions.

succession The sequence of changes in the species composition of a community over time. See *primary succession* and *secondary succession*.

sucker A shoot that develops adventitiously from a root; a type of asexual reproduction.

sulcus (sul′-kus) (pl., *sulci*) A groove, trench, or depression, especially one occurring on the surface of the brain, separating the convolutions.

sulfhydryl group Functional group abbreviated —SH; found in organic compounds called thiols.

summation The process of adding together excitatory postsynaptic potentials (EPSPs).

sum rule The rule for combining the probabilities of mutually exclusive events by adding their individual probabilities. Compare with *product rule*.

superior vena cava In humans, a very large vein that delivers blood from the upper part of the body to the right atrium. Compare with *inferior vena cava*.

suppressor T cell T lymphocyte that suppresses the immune response.

suprachiasmate nucleus Main biological clock; located in the hypothalamus. Important in setting circadian rhythms.

supraorbital ridge (soop″-rah-or′-bit-ul) The prominent bony ridge above the eye socket; ape skulls have prominent supraorbital ridges.

surface tension The attraction that the molecules at the surface of a liquid may have for one another.

survivorship The probability that a given individual in a population or cohort will survive to a particular age; usually presented as a survivorship curve.

survivorship curve A graph of the number of surviving individuals of a cohort, from birth to the maximum age attained by any individual.

suspensor (suh-spen′-sur) In plant embryo development, a multicellular structure that anchors the embryo and aids in nutrient absorption from the endosperm.

sustainability See *environmental sustainability.*

swim bladder The hydrostatic organ in bony fishes that permits the fish to hover at a given depth.

symbionts The partners in a symbiotic relationship.

symbiosis (sim-bee-oh′-sis) An intimate relationship between two or more organisms of different species. See *commensalism, mutualism,* and *parasitism.*

sympathetic nervous system A division of the autonomic nervous system; its general effect is to mobilize energy, especially during stress situations; prepares the body for fight-or-flight response. Compare with *parasympathetic nervous system.*

sympatric speciation (sim-pa′-trik) The evolution of a new species within the same geographic region as the parental species. Compare with *allopatric speciation.*

symplast A continuum consisting of the cytoplasm of many plant cells, connected from one cell to the next by plasmodesmata. Compare with *apoplast.*

symporter Membrane carrier protein that transports two types of substances in one direction. Compare with *uniporter* and *antiporter.*

synapomorphic characters See *shared derived characters.*

synapse (sin′-aps) The junction between two neurons or between a neuron and an effector (muscle or gland).

synapsids (sin-ap′-sids) Members of a clade of amniotes in which the skull has one pair of temporal openings; include the extinct therapsid reptiles and mammals. Compare with *diapsids.*

synapsis (sin-ap′-sis) The process of physical association of homologous chromosomes during prophase I of meiosis.

synaptic enhancement An increase in neurotransmitter release thought to occur as a result of calcium ion accumulation inside the presynaptic neuron.

synaptic plasticity The ability of synapses to change in response to certain types of stimuli. Synaptic changes occur during learning and memory storage.

synaptonemal complex The structure, visible with the electron microscope, produced when homologous chromosomes undergo synapsis.

synthesis phase See *S phase.*

systematics The scientific study of the diversity of organisms and their evolutionary relationships. Taxonomy is an aspect of systematics. See *taxonomy.*

systemic acquired resistance (SAR) A defensive response in infected plants that helps fight infection and promote wound healing.

systemic anaphylaxis A rapid, widespread allergic reaction that can lead to death.

systemic circulation The part of the circulatory system that delivers blood to and from the tissues and organs of the body. Compare with *pulmonary circulation.*

systems biology A field of biology that synthesizes knowledge of many small parts to understand the whole. Also referred to as *integrative biology* or *integrative systems biology.*

systole (sis′-tuh-lee) The phase of the cardiac cycle when the heart is contracting. Compare with *diastole.*

T cell (T lymphocyte) The type of white blood cell responsible for a wide variety of immune functions, particularly cell-mediated immunity. T cells are processed in the thymus. Compare with *B cell.*

T cytotoxic cells (T_C) T cells that recognize and destroy cells infected with viruses or other intracellular pathogens; also destroy cancer cells and other pathogenic cells. Also known as *CD8 T cells* and *killer T cells.*

T helper cells (T_H) T cells that activate B cells (B lymphocytes) and stimulate T cytotoxic cell production. Also known as *CD4 T cells.*

T tubules Transverse tubules; system of inward extensions of the muscle fiber plasma membrane.

taiga (tie′-gah) See *boreal forest.*

taproot system A root system consisting of a prominent main root with smaller lateral roots branching off it; a taproot develops directly from the embryonic radicle. Compare with *fibrous root system.*

target cell or tissue A cell or tissue with receptors that bind a hormone.

TATA box A component of a eukaryotic promoter region; consists of a sequence of bases located about 30 base pairs upstream from the transcription initiation site.

taxon A formal taxonomic group at any level, e.g., phylum or genus.

taxonomy (tax-on′-ah-mee) The science of naming, describing, and classifying organisms; see *systematics.*

Tay-Sachs disease A serious genetic disease in which abnormal lipid metabolism in the brain causes mental deterioration in affected infants and young children; inheritance pattern is autosomal recessive.

T-cell receptors (TCR) The receptors on T cells that recognize specific antigens.

tectorial membrane (tek-tor′-ee-ul) The roof membrane of the organ of Corti in the cochlea of the ear.

telencephalon See *forebrain.*

telolecithal egg An egg with a large amount of yolk, concentrated at the vegetal pole. Compare with *isolecithal egg.*

telomerase A special DNA replication enzyme that can lengthen telomeric DNA by adding repetitive nucleotide sequences to the ends of eukaryotic chromosomes; typically present in cells that divide an unlimited number of times.

telomeres The protective end caps of chromosomes that consist of short, simple, noncoding DNA sequences that repeat many times.

telophase (teel′-oh-faze or tel′-oh-faze) The last stage of mitosis and of meiosis I and II when, having reached the poles, chromosomes become decondensed, and a nuclear envelope forms around each group.

telophase I See *meiosis* and *meiosis I.*

telophase II See *meiosis* and *meiosis II.*

temperate deciduous forest A forest biome that occurs in temperate areas where annual precipitation ranges from about 75 cm to 125 cm.

temperate grassland A grassland characterized by hot summers, cold winters, and less rainfall than is found in a temperate deciduous forest biome.

temperate phage See *temperate virus.*

temperate rain forest A coniferous biome characterized by cool weather, dense fog, and high precipitation, e.g., the north Pacific coast of North America.

temperate virus A virus that integrates into the host DNA as a prophage.

temperature The average kinetic energy of the particles in a sample of a substance.

template A pattern or guide; e.g., one strand of DNA functions as a template for the synthesis of a complementary DNA or RNA strand.

temporal isolation A prezygotic reproductive isolating mechanism in which genetic exchange is prevented between similar species because they reproduce at different times of the day, season, or year.

temporal lobes Region of the cerebrum that contains the auditory areas.

temporal summation The summing of several postsynaptic potentials produced when a single presynaptic neuron fires multiple times in rapid succession. Compare with *spatial summation.*

tendon A connective tissue structure that joins a muscle to another muscle, or a muscle to a bone. Tendons transmit the force generated by a muscle.

tendril A leaf or stem that is modified for holding or attaching onto objects.

tension–cohesion model The mechanism by which water and dissolved inorganic minerals are thought to be transported in xylem; water is pulled upward under tension because of transpiration while maintaining an unbroken column in xylem because of cohesion; also called *transpiration–cohesion model.*

teratogen Any agent capable of interfering with normal morphogenesis in an embryo, thereby causing malformations; examples include radiation, certain chemicals, and certain infectious agents.

terminal bud A bud at the tip of a stem. Compare with *axillary bud.*

termination (of protein synthesis) The final stage of protein synthesis, which occurs when a termination (stop) codon is reached, causing the completed polypeptide chain to be released from the ribosome. See *initiation* and *elongation.*

termination codon See *stop codon.*

territoriality Behavior pattern in which one organism (usually a male) stakes out a territory of its own and defends it against intrusion by other members of the same species and sex.

tertiary consumer See *carnivore.*

tertiary structure (of a protein) (tur′-she-air″-ee) The overall three-dimensional shape of a polypeptide that is determined by interactions involving the amino acid side chains. Compare with *primary, secondary,* and *quaternary protein structure.*

test A shell.

test cross The genetic cross in which either an F_1 individual, or an individual of unknown genotype, is mated to a homozygous recessive individual.

testis (tes′-tis) (pl., *testes*) The male gonad that produces sperm and the male hormone testosterone; in humans and certain other mammals; the testes are located in the scrotum.

testosterone (tes-tos′-ter-own) The principal male sex hormone (androgen); a steroid hormone produced by the interstitial cells of the testes; stimulates spermatogenesis and is responsible for primary and secondary sex characteristics in the male.

tetrad The chromosome complex formed by the synapsis of a pair of homologous chromosomes (i.e., four chromatids) during meiotic prophase I.

tetrapods (tet′-rah-podz) Four-limbed vertebrates: the amphibians, reptiles, birds, and mammals.

thalamus (thal′-uh-mus) The part of the vertebrate brain that serves as a main relay center, transmitting information between the spinal cord and the cerebrum.

thallus (thal′-us) (pl., *thalli*) The simple body of an alga, fungus, or nonvascular plant that lacks root, stems, or leaves, e.g., a liverwort thallus or a lichen thallus.

theca cells The layer of connective tissue cells that surrounds the granulosa cells in an ovarian follicle; stimulated by luteinizing hormone (LH) to produce androgens, which are converted to estrogen in the granulosa cells.

theory See *scientific theory.*

therapods Bipedal, saurischian dinosaurs that were predators; included *Tyrannosaurus.*

therapsids (ther-ap′-sidz) A group of mammal-like reptiles of the Permian period; gave rise to the mammals.

thermal stratification The marked layering (separation into warm and cold layers) of temperate lakes during the summer. See *thermocline.*

thermocline (thur′-moh-kline) A marked and abrupt temperature transition in temperate lakes between warm surface water and cold deeper water. See *thermal stratification.*

thermodynamics Principles governing energy transfer (often expressed in terms of heat transfer). See *first law of thermodynamics* and *second law of thermodynamics.*

thermoreceptor A sensory receptor that responds to heat.

thermoregulation The process of maintaining body temperature within certain limits despite changes in the surrounding temperature.

thigmomorphogenesis (thig″-moh-mor-foh-jen′-uh-sis) An alteration of plant growth in response to mechanical stimuli, such as wind, rain, hail, and contact with passing animals.

thigmotropism (thig″-moh-troh′-pizm) Plant growth in response to contact with a solid object, such as the twining of plant tendrils.

threatened species A species in which the population is small enough for it to be at risk of becoming extinct throughout all or part of its range but not so small that it is in imminent danger of extinction. Compare with *endangered species.*

threshold level The potential that a neuron or other excitable cell must reach for an action potential to be initiated.

thrombocytes Small, nucleated cells that function in blood clotting; found in most vertebrates other than mammals. Also see *platelets.*

thylakoid lumen See *thylakoids.*

thylakoids (thy′-lah-koidz) An interconnected system of flattened, saclike, membranous structures inside the chloroplast; the thylakoid membranes contain chlorophyll and enclose an internal space, the thylakoid lumen.

thymine (thy′-meen) A nitrogenous pyrimidine base found in DNA.

thymosin Hormone produced by the thymus gland that plays a role in immune responses.

thymus gland (thy′-mus) An endocrine gland that functions as part of the lymphatic system; processes T cells; important in cell-mediated immunity.

thyroid gland An endocrine gland that lies anterior to the trachea and releases hormones that regulate the rate of metabolism.

thyroid hormones Hormones, including thyroxin (T_4) and triiodothyronine (T_3), secreted by the thyroid gland; stimulate rate of metabolism.

thyroid-releasing hormone (TRH) Hormone secreted by the hypothalamus to increase thyroid hormone secretion when mammals are exposed to extreme cold. Compare with *thyroid-stimulating hormone (TSH).*

thyroid-stimulating hormone (TSH) Hormone secreted by the anterior pituitary gland to regulate thyroid hormone secretion. Compare with *thyroid-releasing hormone (TRH).*

thyroxine (T_4) See *thyroid hormones.*

tidal volume The volume of air moved into and out of the lungs with each normal resting breath.

tight junctions Specialized structures that form between some animal cells, producing a tight seal that prevents materials from passing through the spaces between the cells.

tissue A group of closely associated, similar cells that work together to carry out specific functions.

tissue culture The growth of tissue or cells in a synthetic growth medium under sterile conditions.

tissue engineering A developing technology that is striving to grow human tissues and organs (for transplantation) in cell cultures.

tissue fluid See *interstitial fluid.*

tolerance A decreased response to a drug over time.

Toll-like receptors A type of pattern recognition receptor. Cell-surface receptors on several cell types that recognize pathogen-associated molecular patterns (certain common features of classes of pathogens). See also *pathogen-associated molecular patterns (PAMPs).*

tonoplast The membrane surrounding a vacuole.

top-down processes Control of ecosystem function by trophic interactions, particularly from the highest trophic level. Compare with *bottom-up processes.*

topoisomerases (toe-poe-eye-sahm′-er-ases) Enzymes that relieve twists and kinks in a DNA molecule by breaking and rejoining the strands.

torpor An energy-conserving state of low metabolic rate and inactivity. See *estivation* and *hibernation.*

torsion The twisting of the visceral mass characteristic of gastropod mollusks.

total fertility rate The average number of children born to a woman during her lifetime.

totipotent (toh-ti-poh′-tent) A term describing a cell or nucleus that contains the complete set of genetic instructions required to direct the normal development of an entire organism. Compare with *pluripotent.*

trace element An element required by an organism in very small amounts.

trachea (tray′-kee-uh) (pl., *tracheae*) (1) Principal thoracic air duct of terrestrial vertebrates; windpipe. (2) One of the microscopic air ducts (or tracheal tubes) branching throughout the body of most terrestrial arthropods and some terrestrial mollusks.

tracheal tubes See *trachea.*

tracheid (tray′-kee-id) A type of water-conducting and supporting cell in the xylem of vascular plants.

tract A bundle of nerve fibers within the central nervous system.

trait A heritable difference.

transcription The synthesis of RNA from a DNA template.

transcriptional-level control Regulation of gene expression by controlling RNA synthesis. Compare with *translational-level control, posttranslational control,* and *posttranscriptional control.*

transcription factors DNA-binding proteins that regulate transcription in eukaryotes; include positively acting activators and negatively acting repressors.

transduction (1) The transfer of a genetic fragment from one cell to another, e.g., from one bacterium to another, by a virus. (2) In the nervous system, the conversion of energy of a stimulus to electrical signals.

transfer RNA (tRNA) RNA molecules that bind to specific amino acids and serve as adapter molecules in protein synthesis. The tRNA anticodons bind to complementary mRNA codons.

transformation (1) The incorporation of genetic material into a cell, thereby changing its phenotype. (2) The conversion of a normal cell to a cancer cell (called a malignant transformation).

transgenic organism A plant or animal that has foreign DNA incorporated into its genome.

translation The conversion of information provided by mRNA into a specific sequence of amino acids in a polypeptide chain; process also requires transfer RNA and ribosomes.

translational control See *translational-level control.*

translational-level control Regulation of gene expression by controlling the translation of mRNA into protein. Compare with *transcriptional-level control, posttranscriptional control,* and *posttranslational control.*

translocation (1) The movement of organic materials (dissolved food) in the phloem of a plant. (2) Chromosome abnormality in which part of one chromosome has become attached to another. (3) Part of the elongation cycle of protein synthesis in which a transfer RNA attached to the growing polypeptide chain is transferred from the A site to the P site.

transmembrane protein An integral membrane protein that spans the lipid bilayer.

transmission electron microscope (TEM) See *electron microscope.*

transpiration The loss of water vapor from the aerial surfaces of a plant (i.e., leaves and stems).

transpiration–cohesion model See *tension–cohesion model.*

transport vesicles Small cytoplasmic vesicles that move substances from one membrane system to another.

transposon (tranz-poze'-on) A DNA segment that is capable of moving from one chromosome to another or to different sites within the same chromosome; also called a *mobile genetic element.*

transverse tubules See *T tubules.*

tree line Zone at high altitudes or high latitudes beyond which trees are unable to grow.

triacylglycerol (try-ace"-il-glis'-er-ol) The main storage lipid of organisms, consisting of a glycerol combined chemically with three fatty acids; also called *triglyceride.* Compare with *monoacylglycerol* and *diacylglycerol.*

Triassic period See *Mesozoic era.*

tricarboxylic acid (TCA) cycle See *citric acid cycle.*

trichome (try'-kohm) A hair or other appendage growing out from the epidermis of a plant.

tricuspid valve See *atrioventricular valve.*

triglyceride See *triacylglycerol.*

triiodothyronine (T₃) See *thyroid hormones.*

triose A sugar molecule containing three carbons.

triple-gene fusion The fusion of three separate genes into a single unit early in the course of eukaryote evolution; provides evidence of a bifurcation; characteristic of unikonts.

triplet A sequence of three nucleotides that serves as the basic unit of genetic information.

triplet code The sequences of three nucleotides that compose the codons, the units of genetic information in mRNA that specify the order of amino acids in a polypeptide chain.

triplobastic (trip-lo-blas'-tik) An animal body plan in which there are three embryonic tissue layers, the ectoderm, mesoderm, and endoderm. Compare with *diploblastic.*

trisomy (try'-sohm-ee) A type of aneuploidy in which an individual has two copies of each chromosome except for one, which has three copies; designated $2n + 1$. Compare with *monosomy* and *disomy.*

trisomy 21 See *Down syndrome.*

trochophore larva (troh'-koh-for) A larval form found in mollusks and many polychaetes.

trophic level (troh'-fik) Each sequential step of matter and energy in a food web, from producers to primary, secondary, or tertiary consumers; each organism is assigned to a trophic level based on its primary source of nourishment.

trophoblast (troh'-foh-blast) The outer cell layer of a late blastocyst, which in placental mammals gives rise to the chorion and to the fetal contribution to the placenta.

tropical dry forest A tropical forest where enough precipitation falls to support trees but not enough to support the lush vegetation of a tropical rain forest; often occurs in areas with pronounced rainy and dry seasons.

tropical rain forest A lush, species-rich forest biome that occurs in tropical areas where the climate is very moist throughout the year. Tropical rain forests are also characterized by old, infertile soils.

tropic hormone (trow'-pic) A hormone that regulates the secretion of another hormone.

tropism (troh'-pizm) In plants, a directional growth response that is elicited by an environmental stimulus.

tropomyosin (troh-poh-my'-oh-sin) A muscle protein involved in regulation of contraction.

troponin (tro-po'-nun) A regulatory protein in the actin filaments of muscle fibers.

true-breeding Refers to a genetic strain of an organism in which all individuals are homozygous at the loci under consideration.

trypsin (trip'-sin) A pancreatic enzyme that digests polypeptides to short peptides.

tube feet Structures characteristic of echinoderms; function in locomotion and feeding.

tuber A thickened end of a rhizome that is fleshy and enlarged for food storage, e.g., white potato.

tubular reabsorption The process by which epithelial cells lining the renal tubules selectively reabsorb about 99% of the glomerular filtrate and return it to the blood; an essential step in adjusting the chemical composition of the blood.

tubular secretion The selective transfer of substances from the blood in the peritubular capillaries into the renal tubule.

tubulin See *microtubules.*

tubular transport maximum (Tm) The maximum rate at which a substance is reabsorbed from the renal tubules of the kidney.

tumor A mass of tissue that grows in an uncontrolled manner; a neoplasm.

tumor necrosis factors (TNFs) Cytokines that kill tumor cells and stimulate immune cells to initiate an inflammatory response.

tumor suppressor gene A gene (also known as an *anti-oncogene*) whose normal role is to block cell division in response to certain growth-inhibiting factors; when mutated, may contribute to the formation of a cancer cell. Compare with *oncogene.*

tundra (tun'-dra) A treeless biome between the boreal forest in the south and the polar ice cap in the north that consists of boggy plains covered by lichens and small plants. Also called *arctic tundra.* Compare with *alpine tundra.*

tunicates Chordates belonging to subphylum Urochordata; sea squirts.

turgor pressure (tur'-gor) Hydrostatic pressure that develops within a walled cell and presses outward against the plasma membrane.

Turner syndrome An inherited condition in which only one sex chromosome (an X chromosome) is present in cells; karyotype is designated X0; affected individuals are sterile females.

two-point test cross A genetic cross used to test for linkage; individuals heterozygous at two loci are crossed with individuals who are homozygous recessive at those loci.

tyrosine kinase An enzyme that phosphorylates the tyrosine part of proteins.

tyrosine kinase receptor A plasma membrane receptor that phosphorylates the tyrosine part of proteins; when a ligand binds to the receptor, the conformation of the receptor changes and it may phosphorylate itself as well as other molecules; important in immune function and serves as a receptor for insulin.

ubiquitin A small eukaryotic regulatory polypeptide that can be covalently bonded to proteins, targeting the protein for destruction and recycling. Also used to regulate protein activities.

ultimate causes (of behavior) Evolutionary explanations for why a certain behavior occurs. Compare with *proximate causes of behavior.*

ultrasound imaging A technique in which high-frequency sound waves (ultrasound) are used to provide an image (sonogram) of an internal structure.

ultrastructure The fine detail of a cell, generally only observable by use of an electron microscope.

umbilical cord In placental mammals, the organ that connects the embryo to the placenta.

uniform dispersion The spatial distribution pattern of a population in which individuals are regularly spaced. Compare with *random dispersion* and *clumped dispersion.*

unikonts One of two main clades of all eukaryotes; had a common ancestor with a single posterior flagellum. See *opisthokonts* and *amoebozoa.* Compare with *bikonts.*

uniporter Membrane carrier protein that transports one type of substance in one direction. Compare with *symporter* and *antiporter.*

uniramous appendages Unbranched appendages. Compare with *biramous appendages.*

unsaturated fatty acid See *fatty acid.*

upwelling An upward movement of water that brings nutrients from the ocean depths to the surface. Where upwelling occurs, the ocean is very productive.

uracil (yur′-ah-sil) A nitrogenous pyrimidine base found in RNA.

urea (yur-ee′-ah) The principal nitrogenous excretory product of mammals; one of the water-soluble end products of protein metabolism.

ureter (yur′-ih-tur) One of the paired tubular structures that conducts urine from the kidney to the bladder.

urethra (yoo-ree′-thruh) The tube that conducts urine from the bladder to the outside of the body.

uric acid (yoor′-ik) The principal nitrogenous excretory product of insects, birds, and reptiles; a relatively insoluble end product of protein metabolism; also occurs in mammals as an end product of purine metabolism.

urinary bladder An organ that receives urine from the ureters and temporarily stores it.

urinary system The body system in vertebrates that consists of the kidneys, urinary bladder, and associated ducts.

urochordates A subphylum of chordates; includes the tunicates.

uterine tube (yoo′-tur-in) See *oviduct.*

uterus (yoo′-tur-us) The hollow, muscular organ of the female reproductive tract in which the fetus undergoes development.

utricle The structure within the vestibule of the vertebrate inner ear that, along with the saccule, houses the receptors of static equilibrium.

vaccine (vak-seen′) A commercially produced, weakened or killed antigen associated with a particular disease that stimulates the body to make antibodies.

vacuole (vak′-yoo-ole) A fluid-filled, membrane-enclosed sac found within the cytoplasm; may function in storage, digestion, or water elimination.

vagina The elastic, muscular tube, extending from the cervix to its external opening, that receives the penis during sexual intercourse and serves as the birth canal.

valence electrons The electrons in the outer electron shell, known as the *valence shell,* of an atom; in the formation of a chemical bond, an atom can accept electrons into its valence shell or donate or share valence electrons.

van der Waals interactions Weak attractive forces between atoms; caused by interactions among fluctuating charges.

vascular cambium A lateral meristem that produces secondary xylem (wood) and secondary phloem (inner bark). Compare with *cork cambium.*

vascular cylinder See *stele.*

vascular tissue system The tissues specialized for translocation of materials throughout the plant body, i.e., the xylem and phloem.

vas deferens (vas def′-ur-enz) (pl., *vasa deferentia*) One of the paired sperm ducts that connects the epididymis of the testis to the ejaculatory duct.

vasoconstriction Narrowing of the diameter of blood vessels.

vasodilation Expansion of the diameter of blood vessels.

vector (1) Any carrier or means of transfer. (2) Agent, e.g., a plasmid or virus, that transfers genetic information. (3) Agent that transfers a parasite from one host to another.

vegetal pole The yolky pole of a vertebrate or echinoderm egg. Compare with *animal pole.*

vein (1) A blood vessel that carries blood from the tissues toward a chamber of the heart (compare with *artery*). (2) A strand of vascular tissue that is part of the network of conducting tissue in a leaf; also called a *vascular bundle.*

veliger larva The larval stage of many marine gastropods (snails) and bivalves (e.g., clams); often is a second larval stage that develops after the trochophore larva.

ventilation The process of actively moving air or water over a respiratory surface.

ventral Toward the lowermost surface or belly of an animal. Compare with *dorsal.*

ventricle (1) One of four interconnected chambers in the brain through which cerebrospinal fluid flows. (2) One of the chambers of the heart that receives blood from an atrium and pumps blood into arteries.

vertebrae (vert′teh-bray) (sing., *vertebra*) The bones that make up the spine.

vertebral column The spine, the rigid, bony structure in the midline of the back; composed of vertebrae.

vertebrates A subphylum of chordates that includes fishes, amphibians, reptiles, birds, and mammals; possess a bony vertebral column.

vertical gene transfer The transfer of genetic material from parent to offspring. Compare with *horizontal gene transfer.*

vesicle (ves′-ih-kl) Any small sac, especially a small, spherical, membrane-enclosed compartment, within the cytoplasm.

vessel In plants, a stack of vessel elements.

vessel element A type of water-conducting cell in the xylem of vascular plants; a stack of vessel elements is a vessel.

vestibular apparatus Collectively, the saccule, utricle, and semicircular canals of the inner ear.

vestigial (ves-tij′-ee-ul) Rudimentary; an evolutionary remnant of a formerly functional structure.

vestigial structure See *vestigial.*

vibrio A spirillum (spiral-shaped bacterium) that is shaped like a comma. Compare with *spirillum, spirochete, bacillus,* and *coccus.*

villus (pl., *villi*) A multicellular, minute, elongated projection from the surface of an epithelial membrane, e.g., villi of the mucosa of the small intestine.

virion A complete virus particle that is outside a cell.

viroid (vy′-roid) A tiny, naked, infectious particle consisting only of nucleic acid.

virulence Properties that render an infectious agent pathogenic (and often lethal) to its host. Compare with *avirulence.*

virus A tiny pathogen consisting of a core of nucleic acid usually encased in protein and capable of infecting living cells; a virus is characterized by total dependence on a living host.

viscera (vis′-ur-uh) The internal body organs, especially those located in the abdominal or thoracic cavities.

visceral mass The concentration of body organs (viscera) located above the foot in mollusks.

vital capacity The maximum volume of air a person exhales after filling the lungs to the maximum extent.

vitamin A complex organic molecule required in very small amounts for normal metabolic functioning.

vitelline envelope An acellular covering of the eggs of certain animals (e.g., echinoderms), located just outside the plasma membrane.

viviparous (vih-vip′-er-us) Bearing living young that develop within the body of the mother. Compare with *oviparous* and *ovoviviparous.*

voltage-activated ion channels Ion channels in the plasma membrane of neurons that are regulated by changes in voltage. Also called *voltage-gated channels*.

vomeronasal organ In mammals, an organ in the epithelium of the nose, made up of specialized chemoreceptor cells that detect pheromones.

vulva The external genital structures of female mammals.

warning coloration See *aposematic coloration*.

water mold A funguslike stramenopile protist with a body consisting of a coenocytic mycelium that reproduces asexually by forming motile zoospores and sexually by forming oospores.

water potential Free energy of water; the water potential of pure water is zero and that of solutions is a negative value. Differences in water potential are used to predict the direction of water movement (always from a region of less negative water potential to a region of more negative water potential).

watershed Area of land where all the water that drains off of it (runoff) goes into the same body of water.

water vascular system Unique hydraulic system of echinoderms; functions in locomotion and feeding.

wavelength The distance from one wave peak to the next; the energy of electromagnetic radiation is inversely proportional to its wavelength.

weathering processes Chemical or physical processes that help form soil from rock; during weathering processes, the rock is gradually broken into smaller and smaller pieces.

Wernicke's area Located in the left temporal lobe; center for language and comprehension.

Western blot A technique in which proteins, previously separated by gel electrophoresis, are transferred to paper. A specific labeled antibody is generally used to mark the location of a particular protein. Compare with *Southern blot* and *Northern blot*.

whisk ferns Fern relatives lacking true roots and leaves; characterized by dichotomous branching.

white blood cells (WBC) See *leukocytes*.

white matter Nervous tissue in the brain and spinal cord that contains myelinated axons. Compare with *gray matter*.

wild type The phenotypically normal (naturally occurring) form of a gene or organism.

winds Complex horizontal atmospheric movements caused in part by differences in atmospheric temperature and pressure changes, Earth's rotation, and uneven heating of the oceans and continents.

wobble hypothesis. Describes the ability of some tRNA anticodons to associate with more than one mRNA codon; in these cases the 5′ base of the anticodon is capable of forming hydrogen bonds with more than one kind of base in the 3′ position of the codon.

work Any change in the state or motion of matter.

X chromosome One of the two sex chromosomes of mammals and some other organisms; normal human females are XX, normal males are XY. Compare with *Y chromosome*.

X-linked gene A gene carried on an X chromosome.

X-ray diffraction A technique for determining the spatial arrangement of the components of a crystal.

xylem (zy′-lem) The vascular tissue that conducts water and dissolved minerals in plants.

XYY karyotype Chromosome constitution that causes affected individuals (who are fertile males) to be unusually tall, with severe acne.

Y chromosome One of the two sex chromosomes of mammals and some other organisms; the Y chromosome determines male gender in humans; normal females are XX, normal males are XY. Compare with *X chromosome*.

yeast A unicellular fungus (ascomycete) that reproduces asexually by budding or fission, and sexually by ascospores.

yolk In many animal eggs, a mixture of nutrients (mostly proteins and lipids) that nourish the developing embryo.

yolk sac One of the extraembryonic membranes; a pouchlike outgrowth of the digestive tract of embryos of certain vertebrates (e.g., birds) that grows around the yolk and digests it. Embryonic blood cells are formed in the mammalian yolk sac, which lacks yolk.

zero population growth Point at which the birth rate equals the death rate. A population with zero population growth does not change in size.

zona pellucida (pel-loo′-sih-duh) The thick, transparent covering that surrounds the plasma membrane of a mammalian ovum.

zoonotic disease An animal disease that crosses the species barrier and infects humans.

zooplankton (zoh″-oh-plank′-tun) The nonphotosynthetic organisms present in plankton, e.g., protozoa, tiny crustaceans, and the larval stages of many animals. See *plankton*. Compare with *phytoplankton*.

zoosporangium Structure in which zoospores form.

zoospore (zoh′-oh-spore) A flagellated motile spore produced asexually by chytrids, certain algae, water molds, and other protists.

zooxanthellae (zoh″-oh-zan-thel′-ee) (sing., *zooxanthella*) Endosymbiotic, photosynthetic dinoflagellates found in certain marine invertebrates; their mutualistic relationship with corals enhances the corals' reef-building ability.

zygomycetes (zy″-gah-my′-seats) Fungi characterized by the production of nonmotile asexual spores and sexual zygospores.

zygosporangium (zy″-gah-spor-an′-gee-um) A thick-walled sporangium containing a zygospore.

zygospore (zy′-gah-spor) A sexual spore produced by a zygomycete.

zygote The 2*n* cell that results from the union of *n* gametes in sexual reproduction. Species that are not polyploid have haploid gametes and diploid zygotes.

zygotic genes Genes that are transcribed after fertilization, either in the zygote or in the embryo. Compare with *maternal effect genes*.

Index

The letter i designates illustrations; t designates tables; **bold** designates defined or introduced term.

barbiturates, 867, 898
bark, **716**
  outer, 742 (*See also* periderm)
  willow, 812
barnacles, 662
baroreceptors, **947**
Barr body, **244, 343**
basal angiosperms, 593–**594,** 594i
basal body, **99**
basal cells, 784–785
basal deuterostomes, 671
basal metabolic rate (BMR), **1021**
base, in cladogram, 489
Basel Institute for Immunology, 970
basement membrane, **816**
base pairing
  complementary, 258–259
    hydrogen bonding and, 257,
    258i, 258–259
base-pair substitution, **288,** 289i
bases, **39**–41
  dissociation of, 39
  mixed with acids, 41
basic solution, **40**
basidia (basdium), **608,** 611i
basidiocarp, **608**–609. *See also*
  mushrooms
basidiomycetes, 600, **608**
  characteristics of, 608
  enzymes in, 615
  fairy rings produced by, 610, 613i
  fruiting bodies of, 608, 610i
  life cycle of, 608, 612i
  reproduction of, 608–611
Basidiomycota, 602, 608
basidiospores, **608**
basilar membrane, **915**
basket stars, as echinoderms,
  672–674
basophils, **934**–935
basswood (*Tilia americana*), 742,
  745i
Batesian mimicry, **1173**
B cells, 963–965
  activation of, 966i
  memory, 965, 967
  survival gene in, 972
Beadle, George, 271–273, 272i
Beadle–Tatum experiments,
  271–273, 272i
Beagle voyage, 383i, 383–384
The Beak of the Finch: A Story of
  Evolution in Our Time
  (Weiner), 398
bearded fireworm (*Hermodice*
  carunculata), 635i
bears, 1238
Bednekoff, Peter A., 1139
bee hummingbird, 695

bee's purple, 782
behavior, **1118**–1141
  altruistic, 1137
  cost–benefit analysis for
    understanding, 1119
  dance of bees, 1132–1135, 1133i
  emotional aspects of, 890–891
  feeding, 1127 (*See also* foraging)
  innate, 1119
  learned, 1119
  proximate causes of, 1119
  sentinel, 1139
  social, 1128–1141
  ultimate causes of, 1119
behavioral ecology, **1118**
behavioral isolation, **420,** 420i, 422t
behavioral pattern, **1121**
Beijerinck, Martinus, 495
Belem–Brasilia Highway, 1244
beneficial bacteria, management of,
  957
Bennett, Joe Claude, 971
Benson, Andrew, 196
benthic environment, 1224
benthic zone, **1220**
benthos, **1218**
benzodiazepines, 867
Bergey's Manual of Systematic
  Bacteriology, 521
berry, **786**
Berthold, A. A., 1044
beta blockers, 944
beta cells, **1059**
beta-chains (ß-chains), 63
beta-endorphin, 867
beta-glucose (ß-glucose), 50, 51,
  51i, 53, 53i
beta-oxidation (ß-oxidation), 180
beta-particle (ß-particle), 28
beverage, fungus and, 615–616
Bézier, Annie, 506, 507
bias, 18
bicarbonate, 1194
biennial plants, **704**
biflagellates, unicellular, 546–547.
  See also golden algae
Big Bang theory, 21, 25
bikonts, **550**
bilateral acoelomate, 643–646. *See*
  also flatworms
bilateral animals
  clades of, 629–630
  mesoderm of, 628–629
bilateral symmetry, **626,** 627i, **643**
Bilateria, 629
bile, 1011, **1013**
bile salts, 1014
bills, 695
binary fission, **214,** 511i, **516**–517

binomial system of nomenclature,
  11, 475
bioaccumulation, **1193**
biochemical coevolution, 1172
biodiversity, **474, 1233**–1251
  deforestation, contribution to
    loss of, 1244
  in ecosystems, 1239–1240
  extinction and, 474
  hotspots, **1240**
  importance of, 474
  levels of, 1233–1234
  See also biological diversity
biodiversity hotspots, **1240**
biofeedback, 896–897
  system, 828
biofilms, **129**
  archaea, formation of, 523
  bacteria, formation of, 523
  dental plaque as, 523, 523i
  Escherichia coli, formation of,
    529–530
biogenic amines, **867**
biogeochemical cycles, 522, **1194,**
  1199
biogeographic realm, 1228–1229,
  1229i
biogeography, **391**–392, **1228**
bioinformatics, **322**
biological clocks, **1126**
biological diversity. *See* biodiversity
Biological Dynamics of Forest
  Fragment Project,
  1240–1241
biological growth, **3**
biological magnification, **1193**
biological membranes, 104–128
  active transport and, 118–121
  cell junctions and, 123, 125–126
  closed vesicles and, 108
  endocytosis and, 121–123
  exocytosis and, 121
  fluid mosaic model and, 106, 107i
  functions of, 111–112, 112i
  passive transport and, 113–118
  permeability and, 112–113
  phospholipids and, 105–106,
    106i
  polarized, 119
  polar molecules and, 112–113
  proteins and, 108–111, 109i
  structure of, 104–111, 107i
  as two-dimensional fluids, 106,
    107–108, 108i
biological molecules. *See* organic
  compounds
biological movement, 4i
biological organization, levels of,
  6–7, 7i

biological processes, ions and, 34i
biological rhythms, **1125**
biological species concept, **418,**
  418i
biology, **1**
  characteristics of life, 3–6
  conservation, 1239
  develomental, evolutionary
    patterns and, 396–397,
    398i
  forensic, 526
  molecular, 526
  reductionist approach to, 21
  science, process of, 15–22
  systems approach to, 21
  themes of, 2
biomass, **1190**
biome, **1209**–1217, 1209i, 1217i
  aquatic, 1218
  types, 1209–1217
  See also specific types of
biomolecules, 526
bioremediation, **382, 526,** 526i
biosonar, 905. *See also* echolocation
biosphere, **6,** 7i, 522
  reserves, 1241
  zones in, 1241
biosphere reserves, **1241**
bioterrorism, **501**
biotic factors, **1144, 1208**
bipedal, **463**
bipolar cells, **924,** 925
  depolarization of, 925
  glutamate in, hyperpolarization
    of, 925
  hyperpolarization of, 924
biradial symmetry, of ctenophores,
  629
biramous appendages, **658**
birds
  as amniotes, 693–696
  blood circulation in, 940
  calls of, 696
  classifying, 486, 487i
  colors of, 696
  digestive system of, 695–696
  as dinosaurs, 693–694
  electroreceptors in, 910
  evolution of, 694, 694i
  flight, adaptations to, 694–696
  lungs of, 695, 988–990, 990i
  magnetic fields of, 910
  migration of, 696
  modern, 694, 695i
  nervous system of, 696
  origin of flight in, 453
  reproduction of, 624
  respiratory system of, 988–990
  septum of, 940

gene transfer, 516, 957
  horizontal, 480–481, 516–517
  vertical, 480, 516
genital herpes, 501, 502t, 1093,
  1094, 1094t
genital warts, 1093
genomes, 316–318, 317i, 318i, 625
genome-wide association study
  (GWAS), **327, 339**
genomic DNA library, 316, 317i
genomic imprinting, **304, 345**–347,
  346i
genomics, **327**–328
genotype, **228**
genotype frequency, **403**
genus (genera), **475**
genus (in taxonomy), **11,** 12i
genus name, 475
geographic range, 482
geographic variation, **414,** 414i
geometric isomers, 46i, **46**–47
Geospiza fortis, 1170–1171
Geospiza fuliginosa, 1170
Geospiza magnirostris, 1171
geranium, secondary growth of, 740
germination, **790**
  of date seed, 786
  temperature, affect on, 791
germ layers, 628, 629, **1104**–1105
germ line cells, **224, 359**
gestation period, 1109
Getz, Wayne, 1070–1071
ghrelin, **1022**
giant panda, 148i
giant sequois (*Sequoiadendron
  giganteum*), 580, 739i
Giardia intestinalis, 538
Gibberella fujikuroi, 803
gibberellin (GA), **803**–804
Gibbs, J. W., 151
Gibbs free energy, 151
gigantism, **1056**
gill filaments, 988
gills, **608, 988**
  of amphibians, 988
  of annelids, 988
  of chordates, 988
  of crustaceans, 988
  dermal, 988
  of fish, 988
  functions of, 986–988
Gilman, Alfred G., 137
Ginkgo biloba, **584**
  charatistics of, 584
  dioecious, 584
  fossils of, 584, 591
  as herbal remedy, 584
Ginkgophyta, 580, 584
gizzard, 653, 696

glaciers, 1198
glands, **816, 817,** 817i
  adrenal, 944, 1062
  bulbourethral, 1074
  endocrine, 680, 817, 1045, 1052
  epithelium, formation of, 816
  exocrine, 817, 1045
  gastric, 1010, 1045
  oil, 837
  parathyroid, 1045, 1059
  pineal, 887, 1064, 1126
  pituitary, 1053
  prostate, 1074
  prothoracic, 1052
  salivary, 1009
  sweat, 1045
  thymus, 964
  thyroid, 687, 1056, 1058–1059
glans, 1074
Gleason, Henry A., 1184
glial cells, **823**–824, 854, **857**–859,
  858t, 858i
  neural function, critical role in,
    857–859
  types of, 857–859
global climate change, 1195
  coral reefs, threats to, 641
  deforestation, adverse impact on,
    1244
global temperature, deforestation
  and, 1244
globin, 995
globular embryo, 785
globulins, **933**
  alpha, 933
  beta, 933
  gamma, 933
glomeromycetes, **605**
  characteristics of, 605
  endomycorrhizal connections
    formed by, 613
  plant roots, symbiotic
    relationship with, 605–607
Glomeromycota, 602, 605
glomerular filtrate, 1034–1035,
  1038
glomerular filtration, 1035
glomeruli, 1030–1031, 1034
glomerulus, **1034**
glucagon, 1046, 1060
glucocorticoid, **1063**
gluconeogenesis, 1060
glucose, 49–50, 50i, 919
  decreased use of, 1062
  liver, transportation to, 1015
  mobilization of, 1060
glucose transporter 1 (GLUT 1),
  116–117, 118i
glutamate, **867, 910**

inhibitory, 924
glutamic acid, 60i, 65
glutamine, 60i
glyceraldehyde, 49, 50i
glyceraldehyde-3-phosphate (G3P),
  196, **198**
glycerol, **54,** 55i, 1014
glycine, 58, 61i, 62i, **867**
glycocalyx, 100, **101**
glycogen, **53, 849,** 1014
glycolipids, **54**
glycolysis, 167, 167i, **168**
  ATP and, 168
  detailed look at, 170i–171i
  NADH and, 168
  overview of, 169i
  pyruvates and, 168–169
  summary of, 167t
glycoproteins, **54,** 90, 91i, **109,** 110,
  111i, 960
  envelope, 497
glycosidic linkage, **51**
glycylalanine, 62i
glyoxysomes, 92–93
gnathostomes, **682**
Gnetophyta, 580, 584
gnetophytes, **584**
Gnetum, 584–585
Gnetum gnemon, 589
goblet cells, **817**
goiter, **1058**–1059
golden algae, 546i, **546**–547
  as chromalveolates, 546–547
  unicellular, 546
Golgi, Camillo, 89
Golgi complex, 82i, 83i, 87, 88t,
  89–90, 90i, 91i
  glycoprotein in, 110, 111i
Glogi tendon organs, **912**
Gombe National Park, 1135,
  1140
gonadotropin-releasing hormone
  (GnRH), **1075**
gonads, 1077
G_1 phase, **208**
gonorrhea, 1093, 1094t
Goodall, Jane, 1135, 1140
Gorgonzola, 616
Gould, Elizabeth, 856
G protein–linked receptors, 134i,
  135, 137
G proteins, **870, 1050**
  evolution of cell signaling and,
    144–145
  phospholipids as second
    messengers and, 138–139,
    140i
  Ras pathways and, 142, 143i
  signal amplification and, 142

signal response and, 141–142
  termination of cell signals and,
    143–144
graded potential, **861, 906**
gradients, 78
  auxin, 802
  concentration, 860–861
  electrical, 859–861
  electrochemical, 730
graft rejection, **981**
grains, **786**
Gram, Christian, 513
gram-negative bacteria, **513,** 522,
  960
  lipopolysaccharide in, 959
gram-positive bacteria, **513,** 522,
  526
  pathogen-associated molecular
    patterns in, 957
Gram staining procedure, 513
grana, **95,** 95i, **187**
Grand Prismatic Spring, 159i
Grant, Peter, 385–386, 1170–1171
Grant, Rosemary, 385–386,
  1170–1171
granulosa cells, **1077,** 1082
grasses, sea, 1224
grasshoppers, 664, 664i
grasslands
  semiarid, 1228
  temperate, 1212–1213
Graves' disease, **1058**
gravitropic response, 799
  of *Arabidopsis thaliana,* 799
  of auxin, 799
gravitropism, **798**
gray crescent, **1104**
gray matter, **882**
Great Barrier Reef, 1225
Great Basin Desert, 1214
great condors, 695
Great Lakes, 1237
Great Salt Lake, 522
green algae, **549,** 549i, 557, 559,
  567, 607
  as archaeplastids, 549
  ecosystem, importance of, 549
  reproduction of, 549
green cones, 925
green fluorescent protein (GFP), 76
greenhouse effect, **1247**
greenhouse gases, **1247**
  emission of, 1246, 1250
  warming effect of, 1246
Greider, Carol, 266
Griffith, Frederick, 252, 255t
grizzly bear, 165i, 1177
gross national income (GNI), 1159

microtubules (*continued*)
  flagella and, 97–99
  9 × 3 structures of, 97, 99
  9 + 2 arrangement of, 98, 98i
  organization of, 96i, 97
  polarity and, 96
  structure and function of, 89t
microvilli, 73, **1011**, 1012i
midbrain, **880–881**
  of amphibians, 880–881
  of fish, 880–881
middle ear, 916
middle-ear bone, 696
middle lamella, **101**, 101i, **733**
midvein, **724**
Miescher, Friedrich, 255t
migration, **1126**
  of birds, 696
mildew, powdery, 619
milkweeds, 1172
Millennium Ecosystem Assessment, 1233
Miller, Stanley, 440, 440i
millimoles (mmol), 31
milliosmole, 1027
millipedes (Diplopoda), 659
millivolts (mV), 859, 862
Mimivirus, 496, 507
Mimulus cardinalis, 782
Mimulus lewisii, 782
mineralocorticoid, **1063**
minerals, **1020**, 1020t
minimal viable population (MVP), **1234**
Minimum Critical Size of Ecosystems Project, 1240–1241
Mirena, 1091
mismatch repair, **265**
misoprostol (Cytotec), 1093
missense mutations, **288**, 289i
Mississippi River, 1238, 1250
Mitchell, Peter, 174, 175
mites, 660–661, 1175
mitochondria, 82i, 83i, 87i, 89t, **93**, 1072
  ATP made through aerobic respiration, 93–94, 94i
  endosymbiosis, origin of, 535
  as energy-converting organelles, 93
mitochondrial DNA, 535
mitosis, **208**, 605
mitotic spindle, **97**, **209**
mitral valve, **940**
M line, 844
mobile genetic elements, **288**, 290, 290i
model organisms, **365**, 366i

moderately developed countries, 1159
modern synthesis, **386**, **435**
Mojave Desert, 1214
molars, **1009**
molds, 598
  black bread (*Rhizopus stolonifer*), 597, 603–604
  coenocytic, 543
  heterothallic, 603–604
  plasmodial, feeding stage of, 551, 552i
  slime, 602
  toxins of, exposure of, 618
  water, 543–544, 544i, 602
mole (mol), **30–31**
molecular anthropology, **469**
molecular biology, 25, 526
molecular chaperones, **64**, **87**, **286**, **302**
molecular clocks, **484**
molecular formula, **30**
molecular mimicry, 978
molecular switches, 136i, 137
molecular systematics, **484**, **625**
molecules, **6**, 7i, **30**, **34**
  interactions between light and, 186, 187i
  organic, on primitive Earth, 439–441
  shape of, chemical bonds and, 32–33
mollusks (Mollusca), **647**
  bivalves as, 650
  blood flow patterns in, 648
  body plan of, 647, 648i
  cephalopod, 877
  cephalopods as, 650–651
  characteristics of, 647
  chitons as, 649
  classes of, 647, 649t
  closed circulatory system of, 932
  coelom in, 931
  digestive system of, 647–648
  electroreceptors in, 910
  evolution of, 647
  eyes in, 920–921
  ganglia, 877
  gastropods as, 649–650
  gills of, 988
  groups of, 647
  immune response of, 957
  larval stages of, 649
  as lophotrochozoa, 643
  mantle of, 647–649
  muscular foot of, 647–649
  nervous system of, 877
  open circulatory system of, 648, 931, 932i

radula of, 647–648
  rhodopsins in, 920
  visceral mass of, 647–649
  See also cephalopods
molting, **656**, **839**
  Ecdysozoa, characteristic of, 839
monarch butterflies (*Danaus plexippus*), 1126
monoacylglycerols (monoglycerides), **54**, 1014
monoclonal antibodies, **971**
monocots, 586
  coleoptile of, 791
  core angiosperms, 594
  cotyledons in, 785
  early growth of, 791
  examples of, 724
  parallel venation in, 724
  roots of, 761
Monocotyledones, **586**
monoculture, 1211
monocytes, **935**, 959
Monod, Jacques, 297, 298
monogamy, **1135**
  sexual, 1135
  social, 1135
Monogenea, 643, 645
monohybrid cross, **231**–233, 232i, 233i
monokaryotic, **601**
monomers, **47–49**, 96
monophyletic groups, 485–**486**, 535, **567**, **622**
Monopteros, mutation of, 720, 720i
monosaccharides, **49**–50, 50i
monosaturated fatty acids, 55i, 55–56
monosodium glutamate (MSG), 919
monosomy, **340**
monotremes, **697**
monounsaturated fatty acids, **55**
monozygotic, **1112**
mons pubis, 1080
Montreal Protocol, **1249**
mood-altering drugs, 898
morels, 616
Morgan, Thomas Hunt, 236
morning-after pills, 1092
morphine, 867
morphogen, 369
morphogenesis, **359**, 719, **1099**
mortality, **1147**
morula, **1101**
mosaic development, 372, **1103**
mosaic picture, 921
mosquitoes, 666, 1250
  thermoreception, use of, 909

moss, 560, **562–565**
  alternation of generations in, 562
  as bryophytes, 562–565, 563i
  characteristics of, 562
  club, 560, 565, 569, 572
  commercial use of, 563–564
  environment, important role in, 563
  fertilization of, 562–563
  gametophyte stage of, 758
  leaves, differentiated into, 562–565
  life cycle of, 562, 564i
  peat, 563–564, 564i
  reindeer, 565, 613–614
  Spanish, 565
  stems, differentiated into, 562–565
moss animals, 654
mother-of-pearl, 650
moths
  adaptations of, 664–665
  fertilization of, 664
  pheromones in, 920
  Polyphemus, metamorphosis of, 664, 665i
motile cells, 548–549
motility, **1007**
motor areas, 883
motor neurons, **847**, **855**
motor programs, **1121**
motor unit, **850**, 850i
Mount Waialeale, 1203
Mousterian tools, 468, 469i
movements
  fetal, 1112
  sleep, 811
M phase, 208
MRSA (methicillin-resistant *Staphylococcus aureus*), 836
Mucana holtonii, 782
mucin, 959
mucosa, **1008**
mucous membrane, **817**
mud puppies, 687
Müllerian mimicry, **1173**
multicellular organisms, 3, 3i
multiple alleles, 245–**246**, 245t
multiple fruit, **786–787**
multiple sclerosis, **859**, 978
  immune reponses to, 956
mumps, 495, 501
Münch, Ernst, 751
muscle fibers, 822, 843, 942
  skeletal, 849, 849t, 860
muscles, **822**, 845i
  from actin, 847
  of animals, 842–851
  atrophies of, 850

oxygen (O)
atmospheric, 1201
Bohr model of, 27i
covalent bonds formed by, 32
diffusion, 932
function of, 26t
oxygen-carrying capacity, **995**
oxygen content, 995
oxygen debt, **849**
oxygen–hemoglobin dissociation
curve, 995–996, 995i
oxyhemoglobin (HbO$_2$), **995**
oxytocin, 1046, **1053**, **1080**,
1087–1088
oysters, habitat of, 650
ozone (O$_3$), **445**, 445i, 1201, **1249**

P700, 192
pacemaker, 851, 940, **941**. *See also*
sinoatrial nodes (SA)
Pacific hagfish (*Eptatretus stoutii*),
682i
Pacific Ocean, 1202
Pacinian corpuscles, 911i, **912**
Packer, Craig, 1128
Paedocypris progenetica, 685
paedomorphosis, **431–432**, 432i,
687–**688**
Paine, Robert T., 1176
pair bonds, **1135**
pair-rule genes, 368, 369i, 369t
palaeognaths, 694
Palearctic biogeographic realm,
1228, 1229i
paleoanthropology, **457**
paleocortex, 882
Paleogene period, 448t, **453**
Paleozoic era, **447**, 448t, 449–450,
560
palindromic sequences, **314**–315
palisade mesophyll, **727**
palmitic acid, 55, 55i
pancreas, **1013**, 1013i
pancreatic lipase, 1013
Pancrustacea, 664
pandoraviruses, 496
Pangaea, 391–392, 392i
Pan troglodytes. See chimpanzee
(*Pan troglodytes*)
Papanicolaou test (Pap smear),
1079, 1093
paper birch, cork cambium in, 744
papillae, 919
para-aminobenzoic acid (PABA),
162, 162i
parabasilids, **538–539**, 539i
in animals, 538–539
as excavates, 538–539
parabronchi, **990**

paracrine agents, 960
paracrine effect, 1082
paracrine regulation, **131**–132, 132i
paradigm shift, 21–22
Paramecium, 115, 542
Paramecium aurelia, 1169
Paramecium caudatum, 1149, 1169
paraphyletic group, 485–486, 535.
See also protists
parapodia (parapodium), **652,** 653
parasite, 522, 1175
of animals, spore-forming,
541–542
blood-sucking, 653 (*See also*
leeches)
free-living, 539 (*See also*
euglenoids; trypanosomes)
host, **523**
obligate intracellular, 496
spore-forming, 541–542 (*See also*
apicomplexans)
parasitic flatworms, 644
parasitism, **522, 534,** 1175
parasympathetic nerves, 944
parasympathetic system, **896**
parathyroid glands, **1059**
parathyroid hormone (PTH), **1059**
paravertebral sympathetic ganglion
chains, 896
Parazoa, 636
parenchyma, **758**
cork, 714, 716
xylem, 710
parenchyma cells, 705–706, 710i, 758
parental (P) generation, **228,** 229i
parental imprinting, 345–347, 346i
parental investment, **1135**
parental types, 238
parietal cells, 1010
parietal lobes, **883**
Parkinson's disease, 886
parsimony, **490,** 625
parthenogenesis, 523, **1069,** 1101
passive ion channels, **860**
passive transport, 112i
biological membranes and,
113–118
diffusion and, 113–118, 114i
Pasteur, Louis, 527, 974
patch clamp technique, **730, 861,**
862i
patchiness, **1145**–1146
pathogen-associated molecular
patterns (PAMPs), **957,**
959–960
pathogens, **500, 511, 956,** 1175
adaptations, success related to,
527–529
of animals, 534

bacteria as, 511, 522, 527, 528t, 956
functions of, 956
phylogeny, clues to origin and
spread of, 482
types of, 956
pathways, 857, 882
mesolimbic dopamine, 891, 898
reflex, 883, 896
renin–angiotensin–aldosterone,
947, 1039
signal transduction, 925
visual, 924
pattern formation, **359, 719, 1099**
pattern recognition receptors
(PRRs), **957**
patterns
alpha rhythm, 888
beta rhythm, 888
pathogen-associated molecular
(PAMPs), **957,** 959–960
Pauling, Linus, 273
Pavlov, Ivan, 1123
pea plant studies, 227–228,
227–228i, 229i
peat, 564, 564i
pectins, 101, 101i, 710
pectoral girdle, 840
pedicellariae, 671
pedigree, **338**
pedipalps, 660
peduncle, 586
pellet, 77, 78
pellicle, **539**
pelvic girdle, 840
pelvic inflammatory disease (PID),
1078, 1093
penicillin, 615
Penicillium, 16, 616
Penicillium notatum, 617
Penicillium roquefortii, 616
penis, **1074,** 1074i
components of, 1074
erection of, 1088
pentaradial symmetry, 671
pentoses, 49, 50i
PEP carboxylase, **199,** 200i
pepsin, **1010**
pepsinogen, **1010**
peptide bonds, 58–**59,** 62i, 62–63
peptide hormones, **1046**
peptides, 1013
antimicrobial, 957–958
atrial natriuretic (ANP), 947
cytokines as, 960
glucose-dependent insulinotropic
(GIP), 1014
peptide YY, 1022
peptidoglycan, **513,** 957
peptidyl transferase, **285**

perception, sensory, 908
perennial
deciduous, 705
evergreen, 705
plants, **704**
perfect flower, **587**
perforations, 710
pericardial cavity, 940
pericardium, 940
pericentriolar material, **209**
pericycle, **760**
periderm, 714, **742–744, 762**
cork cambium, production of,
742–744
dermal tissue system, 714
perilymph, 915
perineal region, 1079
periodic table, **26,** 26t, 27i
periods, **447**
periosteum, 840
peripheral membrane proteins, **109,**
110i
peripheral nervous system (PNS),
855, 878, 879i
peristalsis, **1007,** 1010i
peritoneal cavity, 1009
peritoneum, 1009
peritonitis, 1009
peritubular capillaries, **1034,** 1035
permafrost, **1210**
permeability
biological membranes and,
112–113
of capillaries, 961
Permian period, 448t, **449**–450
pernicious anemia, 934
peroxidases, 934
peroxisomes, 82i, 83i, 86, 89t,
92–**93,** 93i
pesticides, 1192
DDT, 1192–1193
persistence of, **1193**
synthetic, 1193
petals, 586, **588,** 590, 777
petiole, **724**
Petromyzontida, 681
pH, **40**–41
acidity measured by, 40–41, 40t
of blood, 933
buffers, 41
enzymes and, 159i, 159–160
values of common solutions, 40,
40i, 40t
phages, **253, 498.** *See also*
bacteriophages (phages)
phagocytes, 957
bacteria, counteracting, 959
innate immune reponses of, 959
phagocytic cells, 859, 934

Ras proteins, **142**
Ras signaling pathway, 142, 143i
rate of increase, intrinsic (r_{max}), 1147
ratites, **694**, 695
rattlesnakes, thermoreception and, 909
ray-finned fish (Actinopterygii), **685**
rays
  endoskeletons of, 839–840
  magnetic fields of, 910
reactants, **31**
reaction center, **191**
reactions
  acrosome, 1100
  alarm, 1063
  allergic, 979–980
realized niche, **1167**
receptacle, **586, 777**
reception, **855**
reception, signaling, **130**, 131i, 132–135
  regulation of, 133
  types of receptors, 133–135, 134i
receptive field, 925
receptor cells
  light-sensitive, 920
  taste, 918i, 919
receptor down-regulation, **133, 1049**
receptor-mediated endocytosis, **122**, 124i
receptors, **130**
  acetylcholine, 869
  beta-adrenergic, 944
  in *Drosophila*, 957, 959
  electromagnetic, 908, 910
  enzyme-linked, 799–800, 1050, 1051
  excitatory, 871
  glutamate, 893
  G protein–linked, 1050
  gravity, 912–913
  inhibitory, 871
  N-methyl-D-aspartate (NMDA), 893
  olfactory, 919
  "orphan" nuclear, 1044
  pain, 908, 910
  pattern recognition (PRRs), 957–958, 959
  sound, 915 (*See also* auditory receptors)
  synthesis of, 1049
  tactile, 911
  T-cell (TCR), 965
  TIR1, 799–800
  Toll-like, 957, 959

touch, 911
tyrosine kinase, 1051
receptors, sensory, 855, 905, **906**
  electrical signals, transduction of, 906
  in fish, 913
  impulses from, 908
  stimulation of, 908
  transduction of, 908
receptor types, 133–135, 134i
  in cell, 133i, 135
  enzyme-linked receptors, 134i, 135
  G protein–linked receptors, 134i, 135
  ion channel-linked receptors, 133, 134i, 135
receptor up-regulation, **133, 1049**
recessive, **229**
recessive alleles, **230**
reciprocal altruism, **1139**
recombinant DNA, **313**
recombinant types, 238
recombination, **234**
rectum, **1015**
red algae, 548i, **548–549**
  as archaeplastids, 548–549
  environment, adaptations to, 481
  multicellular, 548
  reproduction of, 548
  unicellular, 548
red blood cells (RBCs), **822, 933**
  life cycle of, 933
red bone marrow, 933, 934
red color, of blood, 933
red cones, 925
redox reactions, 35–**36**, 155–156, **166**
red tides, **541**
reduced ferns, 570–571. *See also* whisk ferns
reduced state, 156
reducing agent, 36
reduction, **36**
reduction hypothesis, 506
reductionism, **6**, 21
redwoods (*Sequoia sempervirens*), transpiration in, 750
re-emerging diseases, **503**
reflex
  action, **882**
  diving, 999
  withdrawal, 882–883
refractory period, **864**
  absolute, 864–865
  relative, 865
regressive hypothesis, 506
regulative development, **373, 1103**
regulators, **828**
regulatory proteins, 306, 306i

regulatory T cells, **965**
reinforcement, 429
release, **499**
releaser, 1121
releasing hormones, **1053**
remodeling of bone, 841
renal arteries, 948, **1034**
renal cortex, **1032**
renal medulla, **1032**
renal papilla, 1032
renal pelvis, **1032**
renal pyramids, 1032
renal tubule, **1034**
renal vein, **1035**
renin, **947, 1034, 1039**
repeatable experiments, 20
replacement-level fertility, **1159**
replication, 83, 259–267, 499
replication fork, 261
repolarization, 862–864, **863**
repressible operons, **299**
repressor protein, **297**
reproduction
  of animals, 623
  of aquatic animals on land, 624
  of ascomycetes, 607–608
  ascospores, formation of, 607–608
  asexual, 545, 1069
  of basidiomycetes, 608–611
  basidiospores, formation of, 608–611
  of birds, 624
  of brown algae, 545
  of diatoms, 545
  of earthworms, 624
  sexual, 607–608, 608–611, 610i, 1069
reproductive isolation, **419–421**, 422t
  behavioral isolation, 420, 420i, 422t
  gametic isolation, 420, 422t
  habitat isolation, 419–420, 422t
  mechanical isolation, 420, 420i, 422t
  prezygotic barriers, 419i, 419–421
  temporal isolation, 419, 419i, 422t
reproductive system, **824**, 827i, 1071–1085, 1071i, 1077i
  of algae, 558, 558i
reptiles, **688**, 692i
  as amniotes, 689–690
  clades of, 690–691
  classifying, 486, 487i
  desert, 1214, 1215
  electroreceptors in, 910
  evolution of, 688
  extinction of, 697

flying, 689, 693 (*See also* pterosaurs)
  lizards as, 691–693
  lungs of, 988
  reproduction of, 624
  snakes as, 691–693
  terrestrial adaptations of, 689–690
research
  on bacteria, 522
  on cardiovascular system, 930
reserves, in biosphere, 1241
residual volume, **992**
resin, conifer production and, **580**
resistance
  to drugs, 529–530
  peripheral, 945
  vascular, 945
resolution, 74
resolution phase, 1088–1089
resolving power, **74**
resources
  limiting, 772, 1167–1168, 1178
  partitioning, **1169**
  patchy distribution of, 1192
respiration, **985**
  aerobic, 985
  cellular, 985
  cutaneous, 651
  organismic, 985
respiratory acidosis, 996
respiratory centers, **996**
respiratory system, **824**, 827i, 991–999, 991i
  of birds, 988–990, 990i
response, 130, **855**
  all-or-none, 865
  conditioned, 1124
  unconditioned, 1124
restoration ecology, 1241
restriction enzymes, 314i, **314–315**, 499
restriction fragment length polymorphism (RFLP), **321**, 323i
reticular activating system (RAS), **887**
reticular fibers, 817, 820t
retina, **921–927**, 924i, 925i
  amacrine, 923
  bipolar, 923
  cells in, 923–924
  ganglion, 923–924
  horizontal, 923
  neurons in, 926
  photoreceptors in, 923
  visual images, construction of, 926
retinal, **57**, 923, **924**
  isomers of, 924

retinular cells, **921**

retrograde amnesia, 893

retrograde messenger, 867

retrograde transport, 97

retrotransposons, **290**

retrovirus, **292, 330,** 502t, **503,** 505i

reuptake of dopamine, 871

reuptake process, 871

reversal, **483**

reverse transcriptase, **290, 318, 503**

reverse transcription, 323, 324–325

reversible inhibition, 161

reward centers, 891

Reznick, D. N., 397–398, 399i

R factors, 529

rhabdome, 921

rheumatoid arthritis, 956

Rh factor, **978**

Rh incompatibility, **979,** 979i

rhinoceroses, 1238

rhizarians, 547i, **547–548**

   actinopods as, 548

   characteristics of, 547

   foraminiferans as, 547

rhizobia, **766**

rhizobial bacteria, 765, 766

Rhizobium, 1174, 1195

rhizoids, 603

rhizome, 569, **791–792**

rhodopsins, **133, 920,** 923–924

rhynchocoel, 647

Rhynia gwynne-vaughanii, 575, 575i

Rhynia major, 575–576

Rhyniophyta, 575

rhythm method, 1089

rhythms

   biological, 1125–1126

   circadian, 731, 1125

rib cage, 840

ribonuclease, 1013

ribonucleic acid (RNA), **66, 273.** *See also* RNA (ribonucleic acid)

ribose, 49, 50i, 66, **273**

ribosomal RNA (rRNA), 11, 13, **83,** 84, **274,** 484, 535

ribosomes, **80,** 275, 283, 283i

   of archaea, 484

   of bacteria, 484

   in Golgi complex, 91i

   in prokaryotic cells *vs.* eukaryotic cells, 80, 81, 81i, 82i

   proteins manufactured by, 84, 85

   protein synthesis and, 83

   on rough ER, 87

   structure and function of, 88t

ribozymes, **66, 280, 442**

ribulose bisphosphate (RuBP), **196,** 198

ribulose bisphosphate carboxylase/oxygenase, 196

rice, disease of, 803

rickettsia, 506, 666

Rift Valley fever, 1250

rigor mortis, 847–849

ring of life, 481

rings

   annual, 745

   fairy, 610, 613i

RNA (ribonucleic acid), 66, 67i, **262,** 273, 292t, 1013

   DNA transcribed to form, 274, 274i

   nucleotide structure of, 273i

   single-stranded (ssRNA), 500

   synthesis of, 503

   translated to form polypeptide, 274–275, 275i

   of viroids, 507

   in viruses, 496

   in vitro evolution of molecules, 441–443, 443i

RNA interference (RNAi), **291,** 308, 327–**328**

RNA polymerases, **277**

RNA primer, 262

RNA tumor virus, 292, 292i

RNA world, **442**

roan, 245

Rockefeller University, 856

Rodbell, Martin, 137

rod cells, 924

   depolarization of, 924

   hyperpolarization of, 925

rodents, desert, 1215

rods, **923**

Rojstaczer, Stuart, 1192

root cap, **715, 757**

root cortex, 758

root hairs, in *Arabidopsis thaliana,* 757–758

root-hair zone, 757

roots, **480**

   adventitious, 757

   aerial "breathing," 762

   associations with, 763–767

   buttress, 762

   caps of, **715,** 757–756

   carbon dioxide (CO_2), production of, 767

   carbonic acid (H_2CO_3), production of, 767

   in cladogram, 480, 489

   cork cambium in, 762

   cortex of, 760

   of desert-dwelling tamarisk, 756

   embryonic of, 785

   endodermis of, 760

   epidermis, 758, 760

   eudicots, 758–761

   function of, 757–763

   fungus, 605

   hairs of, **714,** 757–756

   herbaceous, epidermis of, 740

   interactions with, 763–767

   lateral, 760

   of monocots, 758–761

   mycorrhizae of, 767–768

   nitrogen in, 766–767

   periderm, 758

   pressure, **751**

   prop, **762**

   rhizobial bacteria of, 766–767

   secondary growth of, 761–762

   for specialized functions, 762

   structure of, 757–763

   tissue of, 760

   vascular tissue system, 758–761

   woody plants, 761–762

root system, **705**

root tip, 715, 715i

   onion, 718i

Roquefort, 616

Ross, Ronald, 541

rotifers, 655i

   cells in, 655

   characteristics of, 655

   cilia of, crown of, 655

   as lophotrochozoa, 643

   nervous system of, 655

   as pseudocoelomates, 629

rough ER, 87, 87i, 91i

round dance, 1132

round window, 916

roundworms, 657i

   characteristics of, 656

   circulatory system of, 931

   as ecdysozoa, 656

   ecological importance of, 656

   species of, 656

   See also nematodes (Nematoda)

r-selected species, 1153–**1154**

r selection, **1154**

r strategists, **1154**

RU-486 (mifepristone), 1093

rubella (German measles), 495, 501

rubeola (measles), 501

rubisco, **196, 331,** 810

RuBP regeneration, in Calvin cycle, 198

rugae, **1010**

ruminants, stomachs of, 1005

runners, **793**

runoff, 1198

Rutgers University, 1139

Saccharomyces, 615–616. *See also* ascomycetes

Saccharomyces cerevisiae, 616

saccule, **914,** 915i

sacrum, 840

sagebrushes, 1214

sagittal plane, 627

Sahara Desert, 1228, 1248

salamanders, 687–688

salicylic acid, **812**

salinity, **1218**

salinization, 773

saliva, 1009

salivary amylase, **1009**

salivary glands, **1009**

Salix alba, 475

Salmonella typhi, 960

salt (NaCl), 34, 39–**41**

saltatory conduction, **866,** 866i

salt marshes, **1222**

sampling error, 20

sand, 768

sand dollars, as echinoderms, 674

saprotrophs, **1189**

sapwood, 746i

   xylem in, 745

sarcolemma, **844**

sarcomeres, **844**

sarcoplasm, 844

sarcoplasmic reticulum, **844**

sarcopterygians, 685–686

   tetrapods, evolution from, 685–687

Sassandra-N'Zo River, 1140

satellites, **507**

saturated fat, 1018

saturated fatty acids, 55i, **55–56**

saurischian dinosaurs, 451–452, 452i, 689

savanna, **1215**

   animals in, 1215

   characteristics of, 1215

   fragmentation of, 1215

   mammals in, 1215

   precipitation in, 1215

   rangeland, conversion to, 1215

   scrub, 1245

scaffold proteins, **140,** 142i

scaffolding proteins, **206**

scales, placoid, 683

scanning electron microscope (SEM), **76,** 77

scapulas, 840

scarification, 791

scars, **744**

   bud scale, 744

   bundle, 744

   leaf, 744

scavengers, 1177

Scientific Measurement

Examples

kilo	1000	a kilogram is 1000 grams
centi	0.01	a centimeter is 0.01 meter
milli	0.001	a milliliter is 0.001 liter
micro (μ)	one-millionth	a micrometer is 10^{-6} (one-millionth) of a meter
nano (n)	one-billionth	a nanogram is 10^{-9} (one-billionth) of a gram
pico (p)	one-trillionth	a picogram is 10^{-12} (one-trillionth) of a gram

The relationship between mass and volume of water (at 20°C):
$1\,g = 1\,cm^3 = 1\,mL$

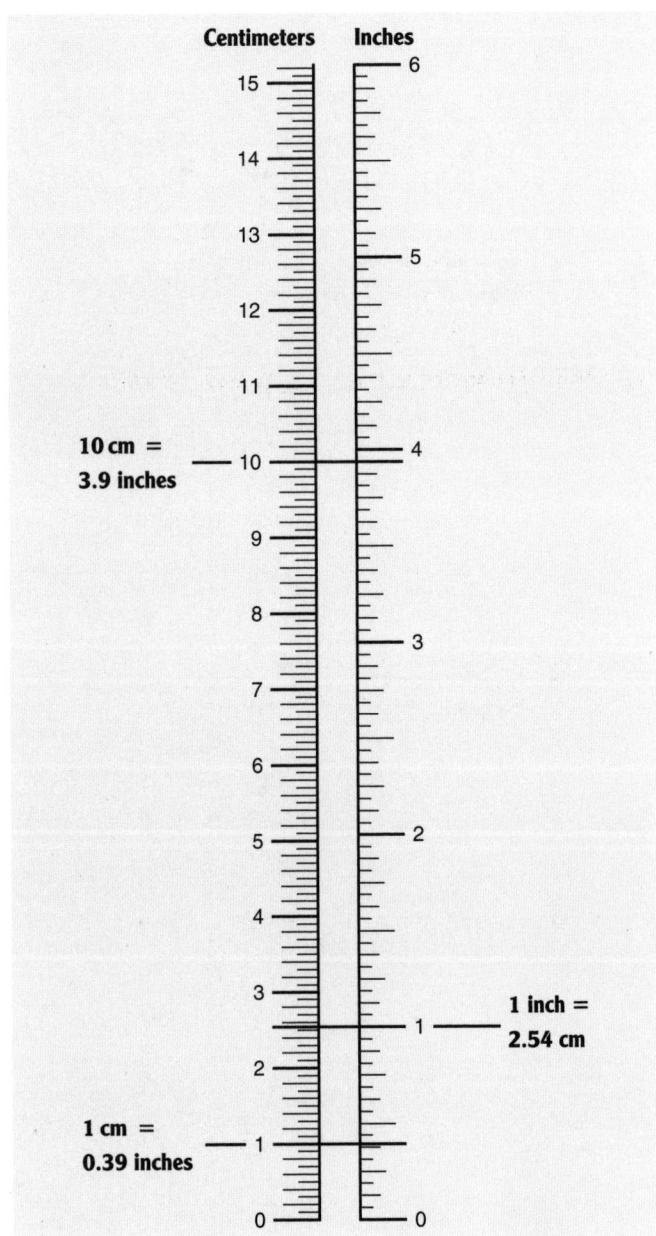

Centimeters / Inches

10 cm = 3.9 inches

1 inch = 2.54 cm

1 cm = 0.39 inches

Some Common Units of Length

Unit	Abbreviation	Equivalent
meter	m	approximately 39 in.
centimeter	cm	10^{-2} m
millimeter	mm	10^{-3} m
micrometer	μm	10^{-6} m
nanometer	nm	10^{-9} m

Length Conversions

1 in. = 2.5 cm	1 mm = 0.039 in.
1 ft = 30.48 cm	1 cm = 0.39 in.
1 yd = 0.9 m	1 m = 39 in.
1 mi = 1.6 km	1 m = 1.094 yd
	1 km = 0.6 mi

To convert	Multiply by	To obtain
inches	2.54	centimeters
feet	30.48	centimeters
centimeters	0.39	inches
millimeters	0.039	inches

Standard Metric Units

		Abbreviation
Standard unit of mass	gram	g
Standard unit of length	meter	m
Standard unit of volume	liter	L